DIRECTORY OF RESEARCH GRANTS

37th Edition • 2014

Schoolhouse Partners LLC
West Lafayette, Indiana

Why 'Partners'? Partners work together as a team, with each person contributing different skills and expressing his or her individual interests and opinions to the unity and efficiency of the group in order to achieve common goals. Our aim is to make worthwhile resources available to you so that you and your institution or organization can and will achieve those goals.

Schoolhouse Partners, 1281 Win Hentschel Blvd, West Lafayette, IN 47906
www.schoolhousepartners.net
Published 1975, 37th edition 2014

Printed and bound in the United States of America

ISBN 978-0-9860357-6-0
0-9860357-6-9

Table of Contents

Introduction iv
How to Use This Directory v
Government and Organization Acronyms viii
Grant Programs 1
Subject Index 759
Program Type Index 947
Geographic Index 1027

Introduction

For over three decades the GRANTS Database and its print complements, including the annual Directory of Research Grants, have provided the research community with current, accurate information regarding funding for research-related programs and projects, scholarships, fellowships, conferences, and internships. To meet the increased information needs of professionals, the GRANTS Database, researched and edited by members of the Schoolhouse Partners editorial staff, has grown significantly. This 37th edition of the annual *Directory of Research Grants* offers factual and concise descriptions of more than 4.900 research funding programs. With this new edition both experienced and novice grantseekers will be better able to target only the most applicable programs for their research needs.

Directory of Research Grants 2014

The *Directory of Research Grants 2014* features listings of programs that offer non-repayable research funding for projects in medicine, the physical and social sciences, the arts and humanities, education, community and economic development, and children and youth programs. Listings in the main section of the *Directory* contain annotations describing each program's focuses and goals, program requirements explaining eligibility, funding amounts, deadlines, *Catalog of Federal Domestic Assistance* program number (for U.S. government programs only), sponsor name and address, contact information, and the sponsor's web address. Grantseekers with access to the Internet can use the addresses to locate further information about the organizations and their application procedures. Internet addresses are also provided, when available, within listings. Some of the grant programs listed in the main section have geographic restrictions for applicants.

A number of people have devoted a great deal of patience, hard work and dedication to keeping this project and the database an informative service for the user. It is especially important to mention the GrantSelect team – Louis S. Schafer, Ed.S., and Anita Schafer – for their diligent research, editorial work, development and assignment of index terms, and production – as well as all database editors, who contribute to the grants editorial effort on a daily basis.

Indexes. The Subject Index of this *Directory* lists all program titles—with accession numbers—under their applicable subject terms. Other indexes follow, including the Grants by Program Type Index, which lists 38 program categories, such as Basic Research, Fellowships, Travel Grants, etc., with the grants that fall within their scope; and the Geographic Index, which lists programs that have state, regional, or international focus. See How to Use This Directory on pages v-vii for sample index entries.

Using the *Directory* for Grantseeking

By using the *Directory of Research Grants*, grantseekers can match the needs of their particular programs with those sponsors offering funding in the researchers' area of interest. The information listed here is meant to eliminate the costs incurred by both grantseekers and grantmakers when inappropriate proposals are submitted for a sponsor's funding program. However, because the GRANTS Database is updated daily basis, with program listings continually added, deleted, and revised, grantseekers using this *Directory* may also search the GRANTS database online through *GrantSelect* (www.grantselect.com).

All new and revised information within the *Directory* has been taken from (1) sponsors' updates of previously published program statements, (2) questionnaires sent to sponsors whose programs were not listed in previous editions, or (3) other materials published by the sponsor and furnished to Schoolhouse Partners. Updated information for U.S. government programs includes new and revised program information published in the *Federal Register*; the latest edition of the *Catalog of Federal Domestic Assistance*; the *NIH Guide*, published weekly by the National Institutes of Health; and the *NSF E-Bulletin*, a monthly publication of the National Science Foundation. Included in this edition are identifying document numbers from the NIH and NSF publications. Located at the ends of the program descriptions in certain entries, the numbers indicated the ongoing NIH program number (PA) or the request for applications number (RFA). For programs of the National Science Foundation, the *NSF Bulletin* number appears. This information will help users identify the programs when seeking additional information from program staff.

While Schoolhouse Partners has made every effort to verify that all information is both accurate and current within the confines of format and scope, the publisher does not assume and hereby disclaims any liability to any party for loss or damages caused by errors or omissions in this *Directory*, whether such errors or omissions result from accident, negligence, or any other cause. Anyone having questions regarding the content, format, or any other aspect of the *Directory of Research Grants 20123* GrantSelect, or the GRANTS Database should contact the Chief Editors at Schoolhouse Partners via email–ChiefEditor@schoolhousepartners.net.

How to Use This Directory

The *Directory of Research Grants 2014* is designed to allow the user quick and easy access to information regarding funding programs in a researcher's specific area of interest. This *Directory* is composed of a main section, Grant Programs, which lists grant programs in alphabetical order, and three indexes: the Subject Index, the Grants by Program Type Index, and the Geographic Index.

GRANT PROGRAMS

Each listing in this section consists of the following elements: an annotation describing each program's focus and goals, requirements explaining eligibility, funding amounts, application and renewal dates, the *Catalog of Federal Domestic Assistance* program number (for U.S. government programs only), sponsor information, contact information, and Internet address.

GRANT TITLE — **Paul G. Allen Family Foundation Grants** **4298** — ACCESSION NUMBER

The single foundation, created through the consolidation of Allen's six previous foundations (The Allen Foundation for the Arts, The Paul G. Allen Charitable Foundation, The Paul G. Allen Foundation for Medical Research, The Paul G. Allen Forest Protection Foundation, The Allen Foundation for Music, and The Paul G. Allen Virtual Education Foundation), will continue to focus on the Allen family's philanthropic interests in the areas of arts and culture, youth engagement, community development and social change, and scientific and technological innovation. The Arts and Culture Program fosters creativity and promotes critical thinking by helping strong arts organizations become sustainable and supporting projects that feature innovative and diverse artistic forms. — GRANT DESCRIPTION The Youth Engagement Program improves the way young people learn by supporting organizations that use innovative teaching strategies and provide opportunities for children to address issues relevant to their lives. The Community Development and Social Change Program promotes individual and community development by supporting initiatives and organizations that provide access to resources and opportunities. The Scientific and Technological Innovation Program advances promising scientific and technology research that has the potential to enhance understanding and stewardship of the world in which we live. Organizations may only receive one grant per year. Organizations must not have any delinquent final reports due to any of the Paul G. Allen Foundations for previous grants. Grantseekers are encouraged to apply through the online application process, where basic organizational and project information will be requested. Guidelines are available online.

REQUIREMENTS — *Requirements* 501(c)3 tax-exempt organizations, status from the Internal Revenue government entities, and IRS-recognized tribes are eligible. Eligible organizations must be located in, or serving populations of, the Pacific Northwest, which includes Alaska, Idaho, Montana, Oregon, and Washington.

RESTRICTIONS — *Restrictions* In general, the foundation will not consider requests for general fund drives, annual appeals, or federated campaigns; special events or sponsorships; direct grants, scholarships, or loans for the benefit of specific individuals; projects of organizations whose policies or practices discriminate on the basis of race, ethnic origin, sex, creed, or sexual orientation; contributions to sectarian or religious organizations whose principle activity is for the benefit of their own members or adherents; loans or debt retirement; projects that will benefit the students of a single school; general operating support for ongoing activities; or projects not aligned with the foundation's specified program areas. 509(a) private foundations are ineligible.

APPLICATION/ DUE DATE — *Date(s) Application Is Due* Mar 31; Sep 30.

Contact Grants Administrator, (206) 342-2030; fax: (206) 342-3030; email: info@pgafamilyfoundation.org — CONTACT

INTERNET ADDRESS — *Internet* http://www.pgafamilyfoundation.org

Sponsor Paul G. Allen Family Foundation — SPONSOR INFORMATION
505 Fifth Ave S, Ste 900
Seattle, WA 98104

SUBJECT INDEX

The most effective way to access specific funding programs is through the Subject Index. This index lists the subject terms with applicable grants program titles – and their accession numbers – alphabetically under each term. Terms were assigned to target the specific area of research designated in the description of each program. Cross-references are used to link subjects and assist the user in finding specific grant information.

Following are general guidelines that can make your search of this index more successful. First, check under the specific topic of interest rather than a more general term. For instance, if you are interested in chemical engineering, look under "Chemical Engineering" rather than "Engineering." Items indexed under "Engineering" indicate funding in broad areas of engineering.

Use general headings when you want grants covering broader areas or if you can't find a specific topic. For example, many grants list funding for humanities research, health programs, or science and technology. To find these grants use such headings as "Humanities," "Medical Programs," "Science," or "Technology." For additional grant information on more specific humanities research opportunities, such as in American History or Cultural Anthropology, also check under the topics "United States History" and "Anthropology, Cultural."

Many of the grant programs provide funding for research-related scholarships, faculty fellowships, dissertations, undergraduate education, conferences, or internships. If grant funds are designated for specific disciplines, you will find the items under the specific subject. Scholarships and fellowships are also listed under the terms "Native American Education," "African Americans (Student Support)," "Hispanic Education," "Minority Education," and "Women's Education."

Grants concerning study of a particular country are listed under the name of the country. Grants concerning the history, literature, art and language of a country are listed under the name of the country also, e.g., "Chinese Art" and "Chinese Language/Literature."

SUBJECT TERM —— **Education**

A.L. Mailman Family Foundation Grants, 5
AAAS Science & Technology Policy Fellowships - Health, Education & Human Services, 22
AARP Andrus Foundation Grants, 104
Abbott Laboratories Fund Grants, 125
ACIE Host University Edmund S. Muskie/Freedom Support Act Graduate Fellowships, 156
ACT Awards, 189
Akonadi Foundation Anti-Racism Grants, 351 —— **PROGRAM TITLE**
Akron Community Foundation Grants, 352
Albert and Margaret Alkek Foundation Grants, 370
Albuquerque Community Foundation Grants, 379
Alcoa Foundation Grants, 380
Alcon Foundation Grants Program, 382

GEOGRAPHIC INDEX

This index lists programs that have state, regional, or international geographic focus. The Geographic Index is arranged by state, followed by Canadian programs, then by international programs by country, and lists grant program titles and their corresponding accession numbers.

COUNTRY —— **United States**

Alabama —— **STATE**

3M Fndn Grants, 2
Alabama Humanities Fndn Grants Program, 368
Arkema Inc. Fndn Science Teachers Program, 705
CDC Injury Control Research Centers Grants, 1198
PROGRAM TITLE —— DOE Experimental Program to Stimulate Competitive Research (EPSCoR), 1653
Hill Crest Fndn Grants, 2293
Linn-Henley Charitable Trust Grants, 2777
NOAA Community-Based Restoration Program (CRP) Grants, 3843
Southern Company's Longleaf Pine Reforestation Fund, 4726

PROGRAM TYPE INDEX

This index is broken into 44 categories according to the type of program funded:

- Adult Basic Education
- Adult/Family Literacy Training
- Awards/Prizes
- Basic Research
- Building Construction and/or Renovation
- Capital Campaigns
- Centers: Research/Demonstration/Service
- Citizenship Instruction
- Community Development
- Consulting/Visiting Personnel
- Cultural Outreach
- Curriculum Development/Teacher Training
- Demonstration Grants
- Development (Institutional/Departmental)
- Dissertation/Thesis Research Support
- Educational Programs
- Emergency Programs
- Endowments
- Environmental Programs
- Exchange Programs
- Exhibitions, Collections, Performances, Video/Film Production
- Faculty/Professional Development
- Fellowships
- General Operating Support
- Graduate Assistantships
- Grants to Individuals
- International Exchange Programs
- International Grants
- Job Training/Adult Vocational Programs
- Land Acquisition
- Matching/Challenge Funds
- Materials/Equipment Acquisition (Computers, Books, Tapes, etc.)
- Preservation/Restoration
- Professorships
- Publishing/Editing/Translating
- Religious Programs
- Scholarships
- Seed Grants
- Service Delivery Programs
- Symposia, Conferences, Workshops, Seminars
- Technical Assistance
- Training Programs/Internships
- Travel Grants
- Vocational Education

Government and Organization Acronyms

AAAAI	American Academy of Allergy Asthma and Immunology
AAAS	American Association for the Advancement of Science
AACAP	American Academy of Child and Adolescent Psychiatry
AACN	American Association of Critical Care Nurses
AACR	American Association of Cancer Research
AAFCS	American Association of Family and Consumer Sciences
AAF	American Architectural Foundation
AAFP	American Academy of Family Physicians Foundation
AAFPRS	American Academy of Facial Plastic and Reconstructive Surgery
AAP	American Academy of Pediatrics
AAR	American Academy in Rome
AAS	American Antiquarian Society
AASL	American Association of School Libraries
AAUW	American Association of University Women
ABA	American Bar Association
ACC	Asian Cultural Council
ACF	Administration on Children, Youth and Families
ACLS	American Council of Learned Societies
ACM	Association for Computing Machinery
ACE	American Council on Education
ACMP	Amateur Chamber Music Players
ACS	American Cancer Society
ADA	American Diabetes Association
ADHF	American Digestive Health Foundation
AF	Arthritis Foundation
AFAR	American Federation for Aging Research
AFOSR	Air Force Office of Scientific Research
AFUD	American Foundation for Urologic Disease
AFUW	Australian Federation of University Women
AGS	American Geriatrics Society
AHA	American Heart Association
AHAF	American Health Assistance Foundation
AFHMR	Alberta Heritage Foundation for Medical Research
AHRQ	Agency for Healthcare Research and Quality
AICR	American Institute for Cancer Research
AIIS	American Institute for Indian Studies
AJA	American Jewish Archives
AJL	American Jewish Libraries
ALA	American Library Association
ALISE	Association for Library and Information Science Education
AMNH	American Museum of Natural History
AMS	American Musicological Society
ANL	Argonne National Library
ANS	American Numismatic Society
AOA	American Osteopathic Association
AOCS	American Oil Chemists' Society
APA	American Psychological Association
APAP	Association of Performing Arts Presenters
APEAL	Asian Pacific Partners for Empowerment and Leadership
APS	Arizona Public Service
APSA	American Political Science Association
ARIT	American Research Institute in Turkey
ARO	Army Research Office
ASA	American Statistical Association
ASCSA	American School of Classical Studies at Athens
ASECS	American Society for Eighteenth-Century Studies
ASF	American-Scandinavian Foundation
ASHA	American Speech-Language-Hearing Association
ASHRAE	American Society of Heating, Refrigerating, and Air Conditioning Engineers
ASME	American Society of Mechanical Engineers
ASNS	American Society for Nutritional Sciences
ASPRS	American Society of Photogrammetry and Remote Sensing
ASTA	American String Teachers Association
ATA	Alberta Teachers Association
AWHONN	Association of Women's Health, Obstetric, and Neonatal Nurses
AWU	Associated Western Universities
AWWA	American Water Works Association
BA	British Academy
BBF	Barbara Bush Foundation
BCBS	Blue Cross Blue Shield
BCBSM	Blue Cross Blue Shield of Michigan
BCBSNC	Blue Cross Blue Shield of North Carolina
BWF	Burroughs Wellcome Fund
CBIE	Canadian Bureau for International Education
CCF	Catholic Community Foundation
CCF	Common Council Foundation
CCFF	Canadian Cystic Fibrosis Foundation
CCFF	Christopher Columbus Fellowship Foundation
CDC	Centers for Disease Control and Prevention
CDECD	Connecticut Department of Economic and Community Development
CDI	Children's Discovery Institute
CEC	Council for Exceptional Children
CEF	Chemical Educational Foundation
CES	Council for European Studies
CF	The Commonwealth Fund
CFF	Cystic Fibrosis Foundation
CFFVR	Community Foundation for the Fox Valley Region
CFKF	Classic for Kids Foundation

CFNCR Community Foundation for the National Capital Region
CFPC College of Family Physicians of Canada
CFUW Canadian Federation of University Women
CHCF California Health Care Foundation
CHEA Canadian Home Economics Association
CICF Central Indiana Community Foundation
CIES Council for International Exchange of Scholars
CIUS Canadian Institute of Ukrainian Studies
CLA Canadian Lung Association
CLF Canadian Liver Foundation
CMS Centers for Medicare and Medicaid Services
CNCS Corporation for National and Community Service
CRI Cancer Research Institute
CTCNet Community Technology Centers Network
DAAD Deutscher Akademische Austauschdienst (German Academic Exchange Service)
DHHS Department of Health and Human Services
DOA Department of Agriculture
DOC Department of Commerce
DOD Department of Defense
DOE Department of Energy
DOI Department of the Interior
DOJ Department of Justice
DOL Department of Labor
DOS Department of State
DOT Department of Transportation
EFA Epilepsy Foundation of America
EIF Entertainment Industry Foundation
EPA Environmental Protection Agency
ESF European Science Foundation
ETS Educational Testing Service
FCAR Formation de Chercheurs et L'Aide a la Recherche
FCD Foundation for Child Development
FDA Food and Drug Administration
FIC Fogarty International Center
GAAC German American Academic Council
GCA Garden Club of America
GEF Green Education Foundation
GNOF Greater New Orleans Foundation
HAF Humboldt Area Foundation
HBF Herb Block Foundation
HHS Health and Human Services
HHMI Howard Hughes Medical Institute
HRSA Health Resources and Services Administration
HUD Department of Housing and Urban Development
ICC Indiana Campus Compact
IIE Institute of International Education
IRA International Reading Association
IRC International Rescue Committee
IREX International Research and Exchanges Board
IUCP Indiana University Center on Philanthropy
IYI Indiana Youth Institute
JDF Juvenile Diabetes Foundation International
JMO John M. Olin Foundation
JSPS Japan Society for the Promotion of Science
KFC Kidney Foundation of Canada
LISC Local Initiatives Support Corporation
LSA Leukemia Society of America
MFRI Military Family Research Institute
MHRC Manitoba Health Research Council
MLA Medical Library Association
MLB Major League Baseball
MMA Metropolitan Museum of Art
MMS Massachusetts Medical Society
MSSC Multiple Sclerosis Society of Canada
NAA Newspaper Association of America
NAACP National Association for the Advancement of Colored People
NAGC National Association for Gifted Children
NAPNAP National Association of Pediatric Nurse Associates and Practitioners
NARSAD National Alliance for Research on Schizophrenia and Depression
NASA National Aeronautics and Space Administration
NASE National Association for the Self-Employed
NASM National Air and Space Museum
NATO North Atlantic Treaty Organization
NCCAM National Center for Complementary and Alternative Medicine
NCFL National Center for Family Literacy
NCI National Cancer Institute
NCIC National Cancer Institute of Canada
NCRR National Center for Research Resources
NCSS National Council for the Social Studies
NEA National Education Association
NEH National Endowment for the Humanities
NEI National Eye Institute
NFID National Foundation for Infectious Diseases
NFL National Football League
NFWF National Fish and Wildlife Foundation
NGA National Gardening Association
NHGRI National Human Genome Research Institute
NHLBI National Heart, Lung and Blood Institute
NHSCA New Hampshire State Council on the Arts
NIA National Institute on Aging
NIAF National Italian American Foundation
NIAAA National Institute on Alcohol Abuse and Alcoholism

NIAF	National Italian American Foundation
NIAID	National Institute of Allergy and Infectious Diseases
NIAMS	National Institute of Arthritis and Musculoskeletal Skin Diseases
NICHD	National Institute of Child Health and Human Development
NIDA	National Institute on Drug Abuse
NIDCD	National Institute on Deafness and Other Communication Disorders
NIDCR	National Institute of Dental and Craniofacial Research
NIDDK	National Institute of Diabetes, and Digestive and Kidney Diseases
NIDRR	National Institute on Disability and Rehabilitation Research
NIEHS	National Institute of Environmental Health Sciences
NIGMS	National Institute of General Medical Sciences
NIH	National Institutes of Health
NIJ	National Institute of Justice
NIMH	National Institute of Mental Health
NINDS	National Institute of Neurological Disorders and Strokes
NINR	National Institute of Nursing Research
NIOSH	National Institute for Occupational Safety and Health
NIST	National Institute of Standards and Technology
NJSCA	New Jersey State Council on the Arts
NKF	National Kidney Foundation
NL	Newberry Library
NLM	National Library of Medicine
NMF	National Medical Fellowships, Inc.
NMSS	National Multiple Sclerosis Society
NNEDVF	National Network to End Domestic Violence Fund
NOAA	National Oceanic and Atmospheric Administration
NRA	National Rifle Association
NRC	National Research Council
NSERC	Natural Sciences and Engineering Research Council of Canada
NSF	National Science Foundation
NSTA	National Science Teachers Association
NYCH	New York Council for the Humanities
NYCT	New York Community Trust
NYFA	New York Foundation for the Arts
NYSCA	New York State Council on the Arts
OAH	Organization of American Historians
ODKF	Outrigger Duke Kahanamoku Foundation
OJJDP	Office of Juvenile Justice and Delinquency Prevention
ONF	Oncology Nursing Foundation
ONR	Office of Naval Research
OREF	Orthopaedic Research and Education Foundation
ORISE	Oak Ridge Institute for Science and Education
OSF	Open Society Foundation
PAS	Percussive Arts Society
PCA	Pennsylvania Council on the Arts
PDF	Peace Development Fund
PDF	Parkinson's Disease Foundation
PhRMA	Pharmaceutical Research and Manufacturers of American Foundation
PHSC	The Photographic Historical Society of Canada
PSEG	Public Service Enterprise Group
RCF	Richland County Foundation
RCPSC	Royal College of Physicians and Surgeons of Canada
RSC	Royal Society of Canada
RWJF	Robert Wood Johnson Foundation
SAMHSA	Substance Abuse and Mental Health Services Administration
SLA	Special Libraries Association
SME	Society of Manufacturing Engineers
SOCFOC	Sisters of Charity Foundation of Cleveland
SORP	Society of Biological Psychiatry
SSHRC	Social Sciences and Humanities Research Council of Canada
SSRC	Social Science Research Council
STTI	Sigma Theta Tau International
SVP	Social Venture Partners
SWE	Society of Women Engineers
TAC	Tennessee Arts Commission
TOMF	Tucson Osteopathic Medical Foundation
TRCF	Three Rivers Community Fund
TSYSF	Teemu Selanne Youth Sports Foundation
UPS	United Parcel Service
USHMM	United States Holocaust Memorial Museum Research Institute
USIA	United States Information Agency
USAID	United States Agency for International Development
USDA	United States Department of Agriculture
USFA	United States Fencing Association
USGA	United States Golf Association
USIP	United States Institute of Peace
USTA	United States Tennis Association
UUA	Unitarian Universalist Association
WAWH	Western Association of Woman Historians
WHO	Women Helping Others

Grant Programs

1 in 9: Long Island Breast Cancer Action Coalition Grants 1

The Coalition's mission is to promote awareness of the breast cancer epidemic through education, outreach, advocacy, and direct support of research which is being done to find the causes of and cures for breast cancer and other related cancers. The Coalition's major goals are to: raise awareness of the epidemic of breast cancer and keep the disease in the forefront; obtain funding for research; investigate the link between pesticides and breast cancer since Long Island is a farm area; increase detection of breast cancer; and find the causes, a cure and how to prevent breast cancer.
Geographic Focus: New Jersey, New York, Pennsylvania
Contact: Geri Barish, President; (516) 374-3190; fax (516) 569-1894; info@1in9.org
Internet: http://www.1in9.org/
Sponsor: Long Island Breast Cancer Action Coalition
86 East Rockaway Road
Hewlett, NY 11557

1st Touch Foundation Grants 2

Established by Derek L. Lee in 2006 in an effort to help his daughter, Jada, giving is centered around the testing and researching of Leber's Congenital Amaurosis (LCA), an extremely rare disease which causes severe vision loss and blindness. The Foundation's primary long-range mission is to help researchers find a cure and eradicate this disease. It will also continue to provide grants for higher education, and help children and youth organizations within its local communities. There are no specific guidelines, application forms, or deadlines with which to adhere, so applicants should contact the Foundation office directly.
Geographic Focus: All States
Contact: Mark Martella, (415) 421-0535 or (818) 501-4421; Mmartella@1sttouch.org
Internet: http://1sttouch.org/
Sponsor: 1st Touch Foundation
5098 Foothills BOulevard, Suite 3, #492
Roseville, CA 95747

3M Canada National Teaching Fellowships 3

In 1986, the Society for Teaching and Learning in Higher Education (STLHE) and 3M Canada joined to reward exceptional contributions to teaching and learning at Canadian universities. Nominations of faculty members teaching at Canadian universities for one-year teaching fellowships are invited. The program embodies the highest ideals of teaching excellence and scholarship with a commitment to enhance the educational experience of every learner. Up to ten awards, presented annually at the STLHE conference in June, recognize exemplary contributions to educational and teaching excellence in Canadian universities. The award includes lifetime membership in the Society. 3M Fellowship winners will attend an all expense paid 3-day retreat at the Fairmont Le Chateau Montebello, to celebrate and share exceptional achievements in teaching and provide an opportunity for outstanding teachers to share experiences and ideas.
Requirements: Awards are open to all individuals currently teaching at a Canadian university, regardless of discipline or level of appointment.
Geographic Focus: Canada
Date(s) Application is Due: Aug 31
Contact: Sylvia Avery; (905) 525-9140, ext. 20130; riselays@mcmaster.ca
Ron Marken, Program Coordinator; (306) 934-0792; ron.marken@usask.ca
Internet: http://www.mcmaster.ca/3Mteachingfellowships/index.html
Sponsor: Society for Teaching and Learning in Higher Education (STLHE)
1280 Main Street West
Hamilton, ON L8S 4L6 Canada

3M Company Higher Education Grants 4

Parallel to its Foundation, the 3M Company makes charitable contributions to higher education organizations directly. Support is given on a national basis. Most giving is initiated through a Request for Proposal process that allows the company to focus our giving and maximize results. The Company invests primarily in technical, engineering and business schools, and efforts that further teaching and learning. Its grants make a difference for students and their success through support of new curriculum, scholarship and retention programs. The overall goal is to help students prepare for future opportunities in an ever-changing world. Areas of interest for Higher Education include: to advance teaching and learning in science, engineering and business; to increase participation by underrepresented people in these fields; and to establish a link with K-12 education through professional development for teachers and college readiness for students. Types of support include: building and renovation; capital campaigns; donated equipment; donated land; donated products; employee volunteer services; general operating support; in-kind gifts; internship funds; program development; seed money; technical assistance; and use of facilities.
Restrictions: No support is offered for: political, fraternal, social, veterans, or military organizations; propaganda or lobbying organizations; religious organizations not of direct benefit to the entire community; animal-related organizations; or disease-specific organizations. No grants are given to individuals, or for electronic media promotion or sponsorships, athletic events, non-3M equipment, endowments, emergency needs, conferences, seminars, workshops, symposia, fund raising or testimonial events, travel, or film or video production; no cause-related marketing.
Geographic Focus: Alabama, Alaska, Arkansas, California, Connecticut, Georgia, Hawaii, Illinois, Indiana, Iowa, Kentucky, Massachusetts, Michigan, Minnesota, Missouri, Nebraska, New Jersey, New York, Ohio, South Carolina, South Dakota, Texas, Utah, Wisconsin
Samples: South Dakota School of Mines & Technology, Rapid City, South Dakota, $200,000-New Biochemistry Course and Laboratory (2011); University of Illinois Foundation, Urbana, Illinois, $50,000-Research Experience for Undergraduates (2011); Coe College, Cedar Rapids, Iowa, $50-College/ University Match Grant (2011).
Contact: Cynthia F. Kleven, Secretary; (651) 733-0144 or (651) 736-8146; fax (651) 737-3061; cfkleven@mmm.com
Internet: http://solutions.3m.com/wps/portal/3M/en_U.S./CommunityAffairs/CommunityGiving/U.S./HigherEduc/
Sponsor: 3M Company Contributions
3M Center, Building 225-01-S-23
Saint Paul, MN 55144-1000

3M Foundation Community Giving Grants 5

The foundation supports organizations involved with arts and culture, K-12 education, higher education, the environment, and health and human services. Special emphasis is directed toward programs designed to help prepare individuals and families for success. Fields of interest are: arts; arts education; business education; disaster relief, preparedness, and services; economics; elementary and secondary education; employment training; engineering; environmental causes; family services; federated giving programs; health care; higher education; human services; mathematics; minorities; science programs; youth development; and youth, services. Types of support include: building construction and renovation, capital campaigns, curriculum development, employee matching gifts, general operating support, in-kind gifts, program development, and scholarship funds. The foundation utilizes an invitational Request For Proposal (RFP) process for organizations located in Minneapolis and St. Paul, Minnesota, and Austin, Texas. Application forms are not required.
Requirements: 501(c)3 tax-exempt organizations in all 3M communities are eligible.
Restrictions: The 3M Foundation does not accept unsolicited proposals in St. Paul/Minneapolis, Minnesota, and Austin, Texas. No support for religious organizations, conduit agencies, political groups, fraternal organizations, social groups, or veteran organizations. No funding for hospitals, K-12 schools, military organizations, animal-related organizations, or disease-specific organizations. No grants to individuals, or for endowments, emergency operating support, advocacy and lobbying efforts, fundraising events and associated advertising, travel, publications, start-up needs, non -equipment, debt reduction, conferences, athletic events, or film or video production; no loans or investments.
Geographic Focus: Alabama, Alaska, Arkansas, California, Connecticut, Georgia, Hawaii, Illinois, Indiana, Iowa, Kentucky, Massachusetts, Michigan, Minnesota, Missouri, Nebraska, New York, Ohio, South Carolina, South Dakota, Texas, Utah, Wisconsin
Samples: MacPhail Center for Music, Minneapolis, Minnesota, $7,000-Learning with Music (2011); University of Wisconsin Foundation, Madison, Wisconsin, $90,000-Continuing the 3M Science & Technology (2011); Indiana University Foundation, Bloomington, Indiana, $2,250-College/University Match Grant (2011).
Contact: Cynthia F. Kleven, Secretary; (651) 733-0144 or (651) 736-8146; fax (651) 737-3061; cfkleven@mmm.com
Internet: http://solutions.3m.com/wps/portal/3M/en_U.S./3M-Sustainability/Global/Stakeholders/Giving/
Sponsor: 3M Foundation
3M Center, Building 225-01-S-23
Saint Paul, MN 55144-1000

25th Anniversary Foundation Grants 6

Established by Arista Records, which was founded by Clive Jay Davis in the late 1970s, the 25th Anniversary Foundation was named to honor a quarter-century of company number one hits. The Foundation's primary field of interest is supporting AIDS research at nationally recognized organizations. There are no particular application formats or deadlines with which to adhere, and potential applicants should first approach the Foundation in writing, outlining their overall need.
Geographic Focus: All States
Amount of Grant: 10,000 USD
Contact: Robert J. Sorrentino; (212) 782-1137 or (212) 782-1000; (212)782-1010
Sponsor: 25th Anniversary Foundation (formerly Arista Records Foundation)
1745 Broadway
New York, NY 10019

1976 Foundation Grants 7

Established in Pennsylvania in 1976, the 1976 Foundation gives primarily to organizations that support services for the blind, pediatric medicine, health care, and research. The average grant ranges up to $10,000, though larger grants will be considered. Though there are no specific guidelines or application forms, the annual deadline is June 30. Applicants should contact the office in writing, explaining the project and amount needed.
Geographic Focus: Pennsylvania

Date(s) Application is Due: Jun 30
Amount of Grant: Up to 10,000 USD
Samples: Children's Hospital of Philadelphia, Philadelphia, Pennsylvania, $10,000; Purnell School 51, Pottersville, New Jersey, $10,000.
Contact: Nathaniel Peter Hamilton, President; (610) 254-9401; fax (610) 254-9404
Sponsor: 1976 Foundation
200 Eagle Road, Suite 308
Wayne, PA 19087-3115

A-T Children's Project Grants 8
The A-T Children's Project strives to assist respected scientists in developing a clearer understanding of ataxia-telangiectasia. The Project is determined to find a timely cure, or life-improving treatments, for this serious disease. Grant awards are made through a careful and detailed selection process. The members of the Scientific Advisory Board examine each proposal and make their independent recommendations to its Board of Directors which then votes on each proposed project. Proposals from junior investigators, from scientists in related disciplines, and from individuals with innovative new ideas for A-T research are particularly encouraged, as are laboratories and teams working together from industry as well as teaching universities. The Project provides competitive grant awards for basic and translational research grants related to A-T. One- and two-year projects are funded up to a maximum total direct cost of $75,000 per year. Grants of $75,000 per year, however, are rare; grants in the $25,000 to 50,000 per year range are much more common. Budgets for up to $150,000 for a two-year project are acceptable.
Requirements: Proposals from junior investigators, from scientists in related disciplines, and from individuals with innovative new ideas for A-T research are particularly encouraged, as are laboratories and teams working together from industry as well as teaching universities. A Letter of Intent is not required. However, prior to submission of a full-length proposal, applicants seeking Scientific Advisory Board input may submit a Letter of Intent directly to the A-TCP Science Coordinator. This letter (not to exceed two pages) should include a brief abstract describing the proposed research, specific aims and an estimated budget. Applicants must submit an electronic copy of their Proposal in either MSWord or PDF formats to grants@atcp.org.
Restrictions: The sponsor does not pay for administrative overhead and indirect costs. The sponsor does not pay for institutional construction or renovation; purchase of major capital equipment other than directly needed for proposed experiments; office equipment or furniture; travel (except as required to perform the project); tuition fees; journal subscriptions; dues or memberships; and printing or publishing costs.
Geographic Focus: All States
Date(s) Application is Due: Mar 1; Sep 1
Amount of Grant: Up to 150,000 USD
Contact: Cynthia Rothblum-Oviatt, PhD, Science Coordinator; (703) 765-1223 or (954) 481-6611; fax (954) 725-1153; cynthia@atcp.org or grants@atcp.org
Internet: http://www.communityatcp.org/page.aspx?pid=3538
Sponsor: A-T Children's Project
5300 W. Hillsboro Boulevard, Suite 105
Coconut Creek, FL 33073

A-T Children's Project Post Doctoral Fellowships 9
The A-T Children's Project strives to assist respected scientists in developing a clearer understanding of ataxia-telangiectasia. Post doctorals with one year experience or less post degrees are eligible and must be nominated for this award by their principal investigator (PI). Any interested PI who would like to nominate a new post doc for the A-T Post Doctoral Fellowship Award must send a Letter of Intent to the A-T Children's Project. This letter (not to exceed two pages) should include a brief abstract describing the proposed research, specific aims and an estimated budget. Fellowship award applications are subject to the same conditions and guidelines as the Project's regular investigator initiated grants. However, the level of funding will be in the range of $30,000 to $40,000 per year for two years.
Restrictions: The sponsor does not pay for administrative overhead and indirect costs. The sponsor does not pay for institutional construction or renovation; purchase of major capital equipment other than directly needed for proposed experiments; office equipment or furniture; travel (except as required to perform the project); tuition fees; journal subscriptions; dues or memberships; and printing or publishing costs.
Geographic Focus: All States
Date(s) Application is Due: Mar 1; Sep 1
Amount of Grant: 30,000-40,000 USD
Contact: Cynthia Rothblum-Oviatt, Science Coordinator; (703) 765-1223 or (954) 481-6611; fax (954) 725-1153; cynthia@atcp.org or grants@atcp.org
Internet: http://www.communityatcp.org/page.aspx?pid=3539
Sponsor: A-T Children's Project
5300 W. Hillsboro Boulevard, Suite 105
Coconut Creek, FL 33073

A-T Medical Research Foundation Grants 10
The Foundation is a national 501(c) tax-exempt organizations offering a comprehensive program of information and support services for patients and their families on ataxia-telangiectasia, a lethal genetic disease that attacks children, causing progressive loss of muscle control, immune system problems, and a strikingly high rate of cancer, especially leukemia and lymphoma. The Foundation also offers patient and professional information and education materials, sponsors meetings and scientific workshops, funds research, and provides referrals to chapters and support groups. Seed grants and research grants are available.
Geographic Focus: All States
Amount of Grant: 35,000-255,000 USD
Contact: Pamela Smith, President; (818) 906-2861; fax (818) 906-2870; becca4435@aol.com or atmrf@aol.com
Sponsor: Ataxia Telangiectasia Medical Research Foundation
16224 Elisa Place
Encino, CA 91436-3320

A.L. Mailman Family Foundation Grants 11
The foundation awards grants on the state, regional, and national levels with a focus on childcare, especially on issues of quality, caregiver training, teacher/student ratios, and connecting quality childcare to later achievements. The foundation also supports efforts to stimulate moral growth and the development of social responsibility and for research in and refinement of developmental, individualized education of young children. Types of support include curriculum development, matching/challenge grants, program development, publication, research grants, seed grants, and technical assistance. Prospective applicants should write a letter describing the proposed project. The letter should include a project summary, contact information, fit with Foundation objectives, and an estimated budget and timeframe. The Foundation board meets in April and October to authorize grants. Proposals should be sent to the foundation by June 15 for fall review and by January 15 for spring review. Letters of inquiry should be submitted by May 1 for fall proposals and by December 1 for spring proposals.
Restrictions: Grants do not support direct service organizations, childcare centers and schools, locally focused organizations, individuals, capital campaigns and endowments, scholarships, general operating expenses or deficit reduction, or organizations or projects outside of the United States.
Geographic Focus: All States
Date(s) Application is Due: Jan 15; May 1; Jun 15; Dec 1
Amount of Grant: 5,000-50,000 USD
Contact: Luba H. Lynch; (914) 683-8089; fax (914) 686-5519; info@mailman.org
Internet: http://www.mailman.org/guidelines/index.html
Sponsor: A.L. Mailman Family Foundation
707 Westchester Avenue
White Plains, NY 10604

A.O. Smith Community Grants 12
The foundation supports nonprofit organizations in communities where A.O. Smith Corporation has facilities. The foundation supports elementary and secondary school projects that focus on the quality of educational programs and curriculum development; civic, cultural, and social welfare of communities; and medical research and improved local health services. Types of support include continuing support, annual campaigns, building/renovation, scholarship funds, and employee matching gifts. Proposals should describe the project's benefits and constituency, budget information including other sources of funding, and how results will be reported. There is no deadline for applications, although proposals to be considered for the following year's foundation budget must be received by October 30.
Requirements: Nonprofit organizations in company areas may apply.
Restrictions: Grants do not support political organizations or organizations whose chief purpose is to influence legislation.
Geographic Focus: All States
Date(s) Application is Due: Oct 30
Amount of Grant: 500-20,000 USD
Contact: Edward J. O'Connor, (414) 359-4000; fax (414) 359-4064
Internet: http://www.aosmith.com/About/Detail.aspx?id=132
Sponsor: A.O. Smith Foundation
P.O. Box 245008
Milwaukee, WI 53224-9510

A. V. Stout Fund Research Grants 13
The fund is dedicated to supporting research grants for New England wildlife and its habitat as well as improving management techniques through innovative ideas and programs. The research, ideas and programs supported by the fund will give resource administrators and policy makers better understanding in managing the political and environmental issues involved. The fund is open to qualified students, interns, resource managers and organizations. Grants average approximately $1,000 and usually will not exceed $3,000. In certain cases, as determined by the selection committee, grant amounts could exceed $3,000.
Requirements: All students, individuals and organizations whose projects are within the New England area and who are interested in the fund's objectives are eligible. All applications must be completed on the application form (see website) and attachments will not be accepted. Applicants are encouraged to seek other sources of funding to share project costs.
Geographic Focus: Connecticut
Amount of Grant: Up to 3,000 USD
Contact: Karen Outlaw, Executive Director; (212) 362-4831
Internet: http://www.norcrossws.org/Grants/Andystout.htm
Sponsor: Norcross Wildlife Foundation
250 W 88th Street, #806
New York, NY 10024

AAAAI ARTrust Mini Grants for Allied Health 14
Grants are awarded to AAAAI members for research in allergy and immunology. Awards and deadlines vary according to availability. Applicants should refer to the website for the most current information and application process.
Requirements: Applicants must be members of the AAAAI.

Geographic Focus: All States
Contact: Jerome Schultz; (414) 272-6071; fax (414) 276-6070; jschultz@aaaai.org
Internet: http://www.aaaai.org/professional-education-and-training/grants-awards.aspx
Sponsor: American Academy of Allergy, Asthma, and Immunology
555 East Wells Street, Suite 1100
Milwaukee, WI 53202-3823

AAAAI ARTrust Mini Grants for Clinical Research 15

The ARTrust invites AAAAI members to apply for mini-grants for for projects to advance the knowledge and/or treatment of allergy, asthma, and immunology for: small clinical or research projects; new educational initiatives for professional and public audiences; and other research including equipment needs, supplemental funding, and FIT projects. Funding is available for up to $15,000 for one year.
Requirements: Applicants must be U.S. citizens or permanent residents and AAAAI members in good standing. Those with a membership status of Fellow, Member, or In-training must be ABAI certified or ABAI eligible. Applicants with a membership status of Allied Health may apply. Applicants must submit a one page demographic/CV, budget, and a 1-2 page proposal with the following as section headers: hypothesis (if applicable); rational; methods; and expected outcome. Proposals are submitted online and must not exceed two pages.
Restrictions: Funding is not available for indirect costs, or the salary and benefits for the project's principal investigator. Individuals currently receiving an ARTrust grant are not eligible.
Geographic Focus: All States
Date(s) Application is Due: Nov 12
Amount of Grant: Up to 15,000 USD
Contact: Jerome Schultz; (414) 272-6071; fax (414) 276-6070; jschultz@aaaai.org
Internet: http://www.aaaai.org/Aaaai/media/MediaLibrary/PDF%20Documents/Education%20and%20Training/Awards/2013-ARTrust-Mini-Grants.pdf
Sponsor: American Academy of Allergy, Asthma, and Immunology
555 East Wells Street, Suite 1100
Milwaukee, WI 53202-3823

AAAAI ARTrust Mini Grants in Faculty Development 16

The Faculty Development Grant offers an individual three years of financial support to concentrate his/her research efforts in either basic or clinical allergy and immunology. The objectives of this funding are to recognize and support young researchers and promote the specialty of allergy/immunology. The recipient is expected to participate in a seminar at the AAAAI annual meeting, where awardees present their research.
Requirements: The award is open to researchers of junior faculty status—either instructor, assistant professor, or the equivalent grade. Funding is directed toward MDs, but in unusual circumstances, PhDs will be considered. Applicants must be U.S. or Canadian citizens or permanent residents who are AAAAI members (or have applied for membership by September 1).
Restrictions: Past recipients are ineligible.
Geographic Focus: All States
Date(s) Application is Due: Nov 15
Amount of Grant: 150,000 USD
Contact: Jerome Schultz; (414) 272-6071; fax (414) 276-6070; jschultz@aaaai.org
Internet: http://www.aaaai.org/global/ARTrust/award-recipients.aspx
Sponsor: American Academy of Allergy, Asthma, and Immunology
555 East Wells Street, Suite 1100
Milwaukee, WI 53202-3823

AAAAI Distinguished Clinician Award 17

The Distinguished Clinician Award is given to a person or persons whose clinically relevant activities have made a significant constructive impact on the field of allergy and clinical immunology. The awardee(s) should have a primary commitment to patient care of clinical investigation focused on diagnosis, treatment and prevention of asthma and allergic disease. The individual(s) should have a record of substantial clinical research as evidenced by papers published in referenced journals (such as JACI); evidence of scholarship by presentations on the programs of national or international meetings, published books, editorship of supplements to referenced journals or equivalent activity. Especially desirable, but not mandatory, the individual should show dedicated service to the field of allergy and clinical immunology, e.g. service to the American Board of Allergy/Immunology, a voluntary foundation or government agency, or public education, outreach, or community activities in the field of allergy\immunology. Alternatively, the recipient(s) could have worked actively in the AAAAI or similar professional society, serving on major committees, participating in workshops, and/or serving as an officer in the AAAAI.
Requirements: Nomination forms are available at the AAAAI website. The nomination letter, along with a copy of the nominee's CV and two letters of recommendation, should be mailed to the Awards Subcommittee at the AAAAI corporate office.
Geographic Focus: All States
Date(s) Application is Due: Jan 14
Contact: Dr. Stuart L; (414) 272-6071; fax (414) 276-6070
Internet: http://www.aaaai.org/professional-education-and-training/grants—-awards/past-aaaail-honorary-award-recipients.aspx#dca
Sponsor: American Academy of Allergy, Asthma, and Immunology
555 East Wells Street, Suite 1100
Milwaukee, WI 53202-3823

AAAAI Distinguished Layperson Award 18

The Distinguished Layperson Award is granted once a year to the citizen who has made the most substantial contribution to the field of allergy. The AAAAI recognizes that many scientists and physicians who are not clinical allergists have made contributions to the understanding and practice of the specialty. Non-medical citizens have worked in various ways to promote the field. Many are members of voluntary health organizations dedicated to public education and fundraising activities to support research. The nomination instructions are available at the AAAAI website.
Requirements: The recipient should be a non-medical person who has made a substantial contribution in service or substance to education, research, or to the practice. The contribution need not be directly related to programs sponsored by the AAAAI. To be eligible, a candidate for the award should still be living at the time the award is made. At least one member of the AAAAI should have first-hand knowledge of the candidate's contribution.
Restrictions: Employees of physicians' or health institutions shall not be eligible for the award unless their contribution has been made outside of their duties of employment.
Geographic Focus: All States
Date(s) Application is Due: Jan 14
Contact: Dr. Stuart L. Abramson, Chair; (414) 272-6071; fax (414) 276-3349
Internet: http://www.aaaai.org/professional-education-and-training/grants—-awards/past-aaaail-honorary-award-recipients.aspx#dla
Sponsor: American Academy of Allergy, Asthma, and Immunology
555 East Wells Street, Suite 1100
Milwaukee, WI 53202-3823

AAAAI Distinguished Scientist Award 19

The Distinguished Scientist Award is presented to recognize scientific contributions to the field of allergy/immunology that have advanced research in the specialty, in addition to leadership contributions to the specialty. Nomination forms are available at the AAAAI website.
Requirements: Nominations should include a letter of nomination, with a copy of the nominee's CV and two letters of recommendation.
Geographic Focus: All States
Date(s) Application is Due: Jan 14
Contact: Dr. Stuart L. Abramson; (414) 272-6071; fax (414) 276-3349
Internet: http://www.aaaai.org/professional-education-and-training/grants—-awards/past-aaaail-honorary-award-recipients.aspx#dsca
Sponsor: American Academy of Allergy, Asthma, and Immunology
555 East Wells Street, Suite 1100
Milwaukee, WI 53202-3823

AAAAI Distinguished Service Award 20

The Distinguished Service Award recognizes long and distinguished service to the AAAAI rather than primarily for scientific achievement. Past-Presidents are eligible for this award. The number of Distinguished Service Awards should be limited to one per year.
Requirements: Nomination forms are available at the AAAAI website. Nomination letters must include the nominee's CV and two letters of recommendation.
Geographic Focus: All States
Date(s) Application is Due: Jan 14
Contact: Dr. Stuart L. Abramson; (414) 272-6071; fax (414) 276-3349
Internet: http://www.aaaai.org/professional-education-and-training/grants—-awards/past-aaaail-honorary-award-recipients.aspx#dsa
Sponsor: American Academy of Allergy, Asthma, and Immunology
555 East Wells Street, Suite 1100
Milwaukee, WI 53202-3823

AAAAI Fellows-in-Training Abstract Award 21

Fellows-in-Training are eligible for the Abstract Award. Accepted abstracts are presented at the Annual Meeting. Authors of accepted abstracts will receive a reduced registration rate for the National Conference. Additional information is available at the website, including instructions for online abstract submission.
Requirements: Abstracts must be submitted using the online abstract submission process. Submission by members and non-members are welcome, and participation is open to health professionals in any field. However, some sections require sponsors for papers whose authors do not include a member of the respective section. The text of each abstract is limited to 450 words. Images and tables are permitted. Images will not count toward the word limit, but any words or numbers within an included table will be counted accordingly. All authors will be required to complete conflict of interest disclosures. Authors whose abstracts are accepted for presentation will be expected to attend the meeting and make their presentation.
Restrictions: An abstract may be submitted to one section or council only.
Geographic Focus: All States
Date(s) Application is Due: Sep 12
Contact: Mariana Duran; (414) 272-6071; fax (414) 276-6070; mduran@aaaai.org
Internet: http://www.aaaai.org/professional-education-and-training/fellows-in-training.aspx
Sponsor: American Academy of Allergy, Asthma, and Immunology
555 East Wells Street, Suite 1100
Milwaukee, WI 53202-3823

AAAAI Fellows-in-Training Grants 22

Fellowship status in the AAAAI recognizes the professional achievements in a member's A/I specialty and provides several benefits available only to Fellows. Benefits include: voting, proposing motions, and holding office on the Board of Directors; recognition as a leader in the specialty's premier A/I society; first opportunity to register for seminars,

workshops, and courses; complimentary subscriptions to professional journals and magazines; and networking opportunities with other A/I professionals around the world. Applications are accepted year round and reviewed in March, June, and November of each year. Applicants are notified of their status within four weeks after the meeting. The application is available at the corporate website or http://www.aaaai.org/Aaaai/media/MediaLibrary/PDF%20Documents/Member%20Applications/Fellowship-Application.pdf. No application fee is required for current members.
Requirements: All applications must be accompanied by a letter of intent and CV from the applicant. Applicants must also be sponsored by one Fellow of the AAAAI. Applicants should meet the following criteria for consideration: full membership with AAAAI for at least three years; certification by a subspeciality board, the American Board of Allergy and Immunology, or its foreign equivalent; and continuing training and experience in the field of A/I.
Geographic Focus: All States, All Countries
Contact: Mariana Duran, Fellows-in-Training Contact; (414) 272-6071; fax (414) 276-6070; mduran@aaaai.org or membership@aaaai.org
Internet: http://www.aaaai.org/professional-education-and-training/fellows-in-training.aspx
Sponsor: American Academy of Allergy, Asthma, and Immunology
555 East Wells Street, Suite 1100
Milwaukee, WI 53202-3823

AAAAI Fellows-in-Training Travel Grants 23
The AAAAI will provide funding to assist Fellows-in-Training (FITs) and non-faculty PhD Post Doc FITS in Allergy/Immunology (A/I) to attend AAAAI's Annual Meeting. Grant awards vary according to the level of the applicant. First year in-training FITs without an accepted abstract for presentation will receive up to $800. Third or fourth year in-training FITs without an accepted abstract will receive up to $650. In-Training or Post Doc FITs who are first author and present their accepted abstract will receive up to $1,100. Candidates will apply through the online application process at the AAAAI website.
Requirements: First, second, third, or fourth year FITS enrolled in an A/I program approved by the ACGME or by the Royal College of Physicians and Surgeons of Canada may apply. Accepted abstracts are not required (although funding will typically be higher for accepted abstracts). Post-Doctoral Research FITs may apply if their training has taken place in the U.S. or Canada, and involves A/I related research. Abstracts must be submitted and accepted for presentation at the annual meeting. Fellows-in-Training in a medical or surgical specialty other A/I may apply if they have submitted an abstract and it has been accepted for presentation.
Restrictions: Only AAAAI members may apply. A maximum of three FIT awards can be received during the fellowship training years.
Geographic Focus: All States
Date(s) Application is Due: Nov 2
Amount of Grant: 650-1,100 USD
Contact: Mariana Duran, Fellows-in Training Contact; (414) 272-6071; fax (414) 276-6070; mduran@aaaai.org
Internet: http://www.aaaai.org/professional-education-and-training/fellows-in-training.aspx
Sponsor: American Academy of Allergy, Asthma, and Immunology
555 East Wells Street, Suite 1100
Milwaukee, WI 53202-3823

AAA Foundation for Traffic Safety Grants 24
The foundation sponsors research that not only identifies critical traffic safety problems, but also searches for underlying causes and possible solutions, as well as identifying early trends and offering solutions before the general public is aware of the problem. The foundation continues research into ways to help the growing numbers of teen drivers and older drivers evaluate and improve their driving performance. A sampling of recent research includes drunk driving, seeking additional solutions; seated for safety; headlight glare countermeasures; supplemental transportation programs for seniors; distracted driving, phase I; and longer combination vehicle safety study. Prior to submitting unsolicited proposals, applicants may email a one-page preproposal to gauge the foundation's general interest in the proposed topic.
Requirements: Research institutions and nonprofits with 501(c)3 status are eligible.
Restrictions: The foundation does not fund research to develop new devices. No grants are made for community action initiatives or other purely local traffic safety programs.
Geographic Focus: All States
Samples: Guidebook for Observing Occupant Restraint System Use; Older Driver Involvement in Injury Crashes; Drunk Driving—Seeking Additional Solutions; Children Seated for Safety
Contact: Fairley Mahlum, Communications Director; (202) 638-5944; fax (202) 638-5943; info@aaafoundation.org
Internet: http://www.aaafoundation.org/about_us/index.cfm
Sponsor: AAA Foundation for Traffic Safety
607 14th Street NW, Suite 201
Washington, DC 20005

AAA Minority Dissertation Fellowship Program 25
The association invites minority doctoral candidates in anthropology to apply for a nonrenewable dissertation writing fellowship. Students of any subfield or specialty in anthropology will receive equal consideration. Dissertation topics in all areas of the discipline are welcome. Doctoral students who require financial assistance to complete the write-up phase of the dissertation are urged to apply. The dissertation proposal must be approved by the applicant's committee prior to application. Application and guidelines are available online.
Requirements: An applicant must be a U.S. citizen; a member of an historically underrepresented ethnic minority group; enrolled in a full-time academic program leading to a doctoral degree in anthropology at the time of application; admitted to degree candidacy before the dissertation fellowship is awarded; and a member of the American Anthropological Association.
Geographic Focus: All States
Date(s) Application is Due: Feb 15
Amount of Grant: 10,000 USD
Contact: Kathleen Terry-Sharp; (703) 528-1902, ext. 3010; ksharp@aaanet.org
Internet: http://www.aaanet.org/profdev/awards/
Sponsor: American Anthropological Association
2200 Wilson Boulevard, Suite 600
Arlington, VA 22201-3357

AAAS Award for Public Engagement with Science 26
AAAS invites nominations for this annual award, which honors working scientists and engineers who make outstanding contributions to the popularization of science. It is intended to encourage technical professionals to communicate with the public by popularizing their own work. Such efforts can include publications or broadcasts, lectures, museum presentations or exhibit designs, or any other outreach project. The Award shall be given annually to scientists or engineers who, while working in their fields, have also contributed substantially to public understanding of science and technology. A monetary prize of $5,000, a commemorative plaque, complimentary registration, and reimbursement for reasonable travel and hotel expenses to attend the AAAS Annual Meeting to receive the prize are given to the recipient. Nominators are encouraged to identify candidates whose contributions reach broad audiences that include women, minorities, disabled persons, and senior citizens.
Requirements: Eligible individuals include scientists and engineers (individual or a small group) from all disciplines (including social sciences and medicine) engaged in research, teaching, and related activities that have contributed substantially to the public's understanding of science or technology. Nominations should include name, position, institution, professional address and phone, and home address and phone of the candidate; name, position, institution, and professional address and phone of the nominator; a summary of the action or actions that form the basis for the nomination (about 250 words); a longer statement (no more than three pages) providing additional details of the actions for which the candidate is nominated, including one paragraph that describes the nominee's impact on the general public; the candidate's vita (three-page maximum); two letters of support; and at least one representative sample that illustrates the nominee's contribution. Books, videotapes, brochures, magazine articles, or other materials are welcome.
Restrictions: Only materials produced for general audiences, as opposed to professional or trade audiences, will be considered. Employees of the AAAS are not eligible. All materials become the property of AAAS.
Geographic Focus: All States
Date(s) Application is Due: Sep 7
Amount of Grant: 5,000 USD
Contact: Dione Rossiter; (202) 326-6445; fax (202) 371-9849; spasco@aaas.org
Internet: http://www.aaas.org/aboutaaas/awards/public/
Sponsor: American Association for the Advancement of Science
1200 New York Avenue NW
Washington, DC 20005-3920

AAAS Award for Science Diplomacy 27
AAAS, in collaboration with its affiliated organizations, seeks to recognize an individual or a limited number of individuals working together in the scientific or engineering community for making an outstanding contribution to furthering international cooperation in science and engineering. Nominations for the award may be originated by anyone. CAIP administers the screening and selection process for the awarding of the AAAS Award for Science Diplomacy. The ISC award is presented each February at the AAAS Annual Meeting. A monetary prize of $5,000, a commemorative plaque, complimentary registration, and reimbursement for reasonable travel and hotel expenses to attend the AAAS annual meeting are given to the recipient.
Requirements: Nominations should be typed and should include the following information: the nominator's name, address, and phone number; the nominee's name and title, institutional affiliation, and address; a summary of the actions that form the basis for the nomination (about 250 words); a longer statement (no more than three pages) providing additional details of the actions for which the candidate is nominated; at least two letters of support with addresses and phone numbers; and the candidate's vita (no more than three pages) containing professional positions held. Any documentation—books, articles, etc.—that illuminates the significance of the nominee's achievements may also be submitted.
Restrictions: The award is open to all regardless of nationality or citizenship. Nominees must be living at the time of their nomination. Any individual or small group in the scientific and engineering community that has contributed substantially to the understanding or development of science or engineering across national boundaries is eligible for this award.
Geographic Focus: All States, All Countries
Date(s) Application is Due: Sep 1
Amount of Grant: 5,000 USD
Contact: Linda Stroud, CAIP Coordinator; (202) 326-6659 or (202) 326-6650; fax (202) 289-4958; lstroud@aaas.org
Internet: http://www.aaas.org/aboutaaas/awards/int/
Sponsor: American Association for the Advancement of Science
1200 New York Avenue NW
Washington, DC 20005-3920

AAAS Award for Scientific Freedom and Responsibility 28

The award is presented to honor scientists and engineers whose exemplary actions have served to foster scientific freedom and responsibility and recognizes scientists and engineers who have: acted to protect the public's health, safety, or welfare; or focused public attention on important potential impacts of science and technology on society by their responsible participation in public policy debates; or established important new precedents in carrying out the social responsibilities or in defending the professional freedom of scientists and engineers. Nominations should include nominator's name, address, and phone number; and the name and address of the nominee; a summary of the action or actions that form the basis for the nomination (about 250 words); a longer statement (no more than three pages) providing additional details of the actions for which the candidate is nominated; at least two letters of support with addresses and phone numbers; and the candidate's vita (no more than three pages). Any documentation—books, articles, etc.—that illuminates the significance of the nominee's achievements may also be submitted.
Requirements: The award is open to all regardless of nationality or citizenship. Eligible nominees are scientists and engineers who have acted to protect the public's health, safety, or welfare; or to focus public attention on potentially serious impacts of science and technology on society by their responsible participation in public policy debates; or to establish important new precedents in social responsibility or in defending the professional freedom of scientists and engineers.
Restrictions: All materials become the property of AAAS.
Geographic Focus: All States, All Countries
Date(s) Application is Due: Sep 1
Amount of Grant: 5,000 USD
Contact: Deborah Runkle, (202) 326-6794; fax (202) 289-4950; drunkle@aaas.org
Internet: http://www.aaas.org/aboutaaas/awards/freedom/
Sponsor: American Association for the Advancement of Science
1200 New York Avenue NW
Washington, DC 20005-3920

AAAS Early Career Award for Public Engagement with Science 29

The AAAS Early Career Award for Public Engagement with Science, established in 2010, recognizes early-career scientists and engineers who demonstrate excellence in their contribution to public engagement with science activities. A monetary prize of $5,000, a commemorative plaque, complimentary registration to the AAAS Annual Meeting, and reimbursement for reasonable hotel and travel expenses to attend the AAAS Annual Meeting to receive the prize are given to the recipient. For the purposes of this award, public engagement activities are defined as the individual's active participation in efforts to engage with the public on science- and technology-related issues and promote meaningful dialogue between science and society, as highlighted in this video. The award will be given at the AAAS Annual Meeting.
Requirements: Nominee must be an early-career scientist or engineer in academia, government or industry actively conducting research in any scientific discipline (including social sciences and medicine). Groups or institutions will not be considered for this award. AAAS employees are ineligible. One scientist or engineer will be chosen to receive the award on an annual basis. Nominee will have demonstrated excellence in his/her contribution to public engagement with science activities, with a focus on interactive dialogue between the individual and a non-scientific, public audience(s).
Geographic Focus: All States
Date(s) Application is Due: Oct 15
Amount of Grant: 5,000 USD
Contact: Linda Cendes, (202) 326-6656; fax (202) 289-4950; lcendes@aaas.org
Internet: http://www.aaas.org/aboutaaas/awards/public_engagement/
Sponsor: American Association for the Advancement of Science
1200 New York Avenue NW
Washington, DC 20005-3920

AAAS Entry Point Internships 30

Entry Point is a program of the American Association for the Advancement of Science (AAAS), offering outstanding internship opportunities for students with disabilities in science, engineering, mathematics, computer science, and some fields of business. To meet the challenge of the competitive global economy in the new millennium, private industry and government research agencies must expand the pool of technical talent. The American Association for the Advancement of Science (AAAS) has developed unique partnerships with IBM, NASA, Merck, NOAA, Google, Lockheed Martin, CVS, NAVAIR, Pfizer, Infosys, and university science laboratories to meet their human resources needs. Working with its partners, AAAS identifies and screens undergraduate and graduate students with disabilities who are pursuing degrees in science, engineering, mathematics, computer science, and some fields of business, and places them in paid summer internships.
Requirements: Students must: be full-time undergraduate or graduate student, with a disability, majoring in a science or engineering field; maintain a B average or better; and be a U.S. citizen or have a right-to-work permit.
Geographic Focus: All States
Contact: Laureen Summers, Program Coordinator; (202) 326-6649; fax (202) 371-9849; lsummers@aaas.org or entrypoint@aaas.org
Internet: http://ehrweb.aaas.org/entrypoint/
Sponsor: American Association for the Advancement of Science
1200 New York Avenue NW
Washington, DC 20005-3920

AAAS Eppendorf and Science Prize for Neurobiology 31

The Eppendorf and Science Prize for Neurobiology acknowledges the increasingly active and important role of neurobiology in advancing our understanding of the functioning of the brain and the nervous system — a quest that seems destined for dramatic expansion in the coming decades. This international prize, established in 2002, encourages the work of promising young neurobiologists by providing support in the early stages of their careers. It is awarded annually for the most outstanding neurobiological research by a young scientist, as described in a 1,000-word essay based on research performed during the past three years. The winner is awarded money and publication of his or her essay in 'Science'. The essay and those of up to three finalists are also published on 'Science Online'. The award is announced and presented at a ceremony at the annual meeting of the Society for Neuroscience. Eppendorf provides financial support to help enable the grand prize winner to attend the meeting.
Requirements: Entrants must be a neurobiologist with an advanced degree and not older than 35 years. The research described in the entrant's essay must be based on the methods of molecular and cell biology. The entrant must have performed or directed the work described in the essay. The research must have been performed during the previous three years.
Restrictions: Employees of Eppendorf AG, its subsidiaries, Science and AAAS, and their relatives are not eligible for the prize.
Geographic Focus: All States
Date(s) Application is Due: Jun 15
Amount of Grant: 25,000 USD
Contact: Monica M. Bradford, Executive Editor; (202) 326-6550; fax (202) 289-7562; mmadrid@aaas.org or eppendorfscienceprize@aaas.org
Internet: http://www.sciencemag.org/site/feature/data/prizes/eppendorf/
Sponsor: American Association for the Advancement of Science
1200 New York Avenue NW
Washington, DC 20005-3920

AAAS GE and Science Prize for Young Life Scientists 32

The GE and Science Prize for Young Life Scientists has been established to provide support to scientists at the beginning of their careers, because Science/AAAS and GE Healthcare believe that such support is critical for continued scientific progress. The prize will recognize outstanding graduate students in molecular biology from all regions of the world. This international prize will be awarded for the outstanding thesis in the general area of molecular biology as described in a 1000-word essay. The winning essay will be published in Science; essays of the regional award winners will appear in the online version of Science. For judging purposes, the essays will be grouped according to the geographic location of the degree-granting institution: North America, Europe, Japan, and all other countries. Initial screening of the submissions will be done by regional judges. Essays will be judged on the quality of the research and the entrant's ability to articulate the contribution of the research to the field of molecular biology. The top five essays from each geographic region will be forwarded to a panel of judges. All regional winners will compete for the grand prize of U.S.$25,000. The regional winners who do not receive the grand prize will be awarded U.S.$5,000. Winners will be announced in Science and the prize will be awarded in a location to be announced. The grand prize essay will be published in 'Science', and essays of the regional winners will be published on the online version of 'Science'.
Requirements: Entrants must have been awarded their Ph.D. between January, 1 and, December 31, of the previous year. Candidates for M.D./Ph.D. degrees are eligible to compete for the prize in either the year the Ph.D. is awarded or the year the final degree is awarded. The research described in the entrant's thesis must be in the field of molecular biology as described above. The prize will recognize only work that was performed while the entrant was a graduate student. The prize will be awarded without regard to sex, race, or nationality.
Restrictions: Employees of GE, Science and AAAS, and their relatives are not eligible for the prize.
Geographic Focus: All States
Date(s) Application is Due: Aug 1
Amount of Grant: 25,000 USD
Contact: Sylvia Kihara, Prize Coordinator; (202) 326-6507; fax (202) 289-7562; skihara@aaas.org or gescienceprize@aaas.org
Internet: http://www.sciencemag.org/site/feature/data/prizes/ge/index.xhtml
Sponsor: American Association for the Advancement of Science
1200 New York Avenue NW
Washington, DC 20005-3920

AAAS Kavli Science Journalism Awards 33

The awards, which are cosponsored by AAAS and Johnson & Johnson Pharmaceutical Research & Development, recognize outstanding reporting for a general audience and honor individuals (rather than institutions, publishers or employers) for their coverage of the sciences, engineering and mathematics. The winning journalists have helped to foster the public's understanding and appreciation of science. Independent screening and judging committees select the winning journalists and their entries based on scientific accuracy, initiative, originality, clarity of interpretation and value in fostering a better understanding of science by the public. One award in each of six categories is given annually: newspapers with daily circulation more than 100,000, newspapers with circulation of less than 100,000, general circulation magazines, radio, and television, and online. Additional International category-Children's Science News (Provide English translations of foreign-language entries whenever possible).
Requirements: Persons other than the author may submit entries in accordance with the rules of the competition. The contest is open to journalists who have published or broadcast in media outlets accessible to the general public. Except for the Children's

Science News award, which is open to international media outlets, the submitted materials must have been published, broadcast or posted online within the United States by a U.S.-based news organization. Print articles must be available to the public by subscription or newsstand sales.
Restrictions: Items must be: exclusively concerning health or medical treatment; published originally in AAAS publicaitons or produced by AAAS; and by employees of AAAS or Johnson and Johnson Pharmaceutical Research Development. Winners from the previous year are not eligible. Individuals winning three times are no longer eligible.
Geographic Focus: All States
Date(s) Application is Due: Aug 1
Amount of Grant: 3,000 USD
Contact: Awards Coordinator; (202) 326-6434; fax (202) 789-0455; media@aaas.org
Internet: http://www.aaas.org/aboutaaas/awards/sja/
Sponsor: American Association for the Advancement of Science
1200 New York Avenue NW
Washington, DC 20005-3920

AAAS Martin and Rose Wachtel Cancer Research Award 34

AAAS and Science Translational Medicine invite applications for the inaugural AAAS Martin and Rose Wachtel Cancer Research Award. This annual award, funded by an endowment established through a generous bequest from Martin L. Wachtel, will honor an early-career investigator who has performed outstanding work in the field of cancer research. The award winner will be invited to deliver a public lecture on his or her research and will receive a cash award of $25,000. The award winner's Essay will be published as a Perspective in Science Translational Medicine. The deadline for application is February 15.
Requirements: Entrants must be a researcher in the field of cancer, with an advanced degree (Ph.D., M.D., D.V.M.) received within the last 10 years. The research must have been performed during the previous 10 years. The entrant must have performed or personally directed the work described in the Essay.
Restrictions: Employees of AAAS and their relatives are not eligible for the prize.
Geographic Focus: All States
Amount of Grant: 25,000 USD
Contact: Stephen Nelson, (202) 326-6617 or (202) 326-6700; fax (202) 289-4950; snelson@aaas.org
Internet: http://www.aaas.org/aboutaaas/awards/wachtel/
Sponsor: American Association for the Advancement of Science
1200 New York Avenue NW
Washington, DC 20005-3920

AAAS Mentor Awards 35

The two categories of the AAAS Mentor Awards (Lifetime Mentor Award and Mentor Award) both honor individuals who during their careers demonstrate extraordinary leadership to increase the participation of underrepresented groups in science and engineering fields and careers. These groups include: women of all racial or ethnic groups; African American, Native American, and Hispanic men; and people with disabilities. Both awards recognize an individual who has mentored and guided significant numbers of students from underrepresented groups to the completion of doctoral studies or who has impacted the climate of a department, college, or institution to significantly increase the diversity of students pursuing and completing doctoral studies. Nominations should include name, position, institution, professional address and phone, and home address and phone of the candidate; name, position, institution, and professional address and phone of the nominator; a summary of the actions that form the basis for the nomination (about 250 words); a letter of nomination that enumerates the ways in which the person reflects the purpose of this award; the candidate's vita (three-page maximum); the total number of students the candidate mentored at the bachelor's or master's level who went on to the doctoral level at other institutions, and the total number of underrepresented students the candidate mentored at the doctoral level; a maximum of five supporting letters from students and three supporting letters from colleagues representative of the different spheres in which the candidate has demonstrated effort, results, and commitment.
Requirements: The award is open to all regardless of nationality or citizenship. Nominees must be living at the time of their nomination. Commitment and extraordinary effort may be demonstrated by the number and diversity of students mentored; assisting students to present and publish their work, to find financial aid, and to provide career guidance; providing psychological support, encouragement, and essential strategies for life in the scholarly community; and continued interest in the individual's professional advancement. The Lifetime Mentor award winner will have served in the role of mentor for 25 or more years. The Mentor Award winner will have served in the role of mentor for less than 25 years.
Geographic Focus: All States, All Countries
Date(s) Application is Due: Jul 31
Contact: Yolanda George; (202) 326-6670; fax (202) 371-9849; ygeorge@aaas.org
Internet: http://www.aaas.org/aboutaaas/awards/mentor/
Sponsor: American Association for the Advancement of Science
1200 New York Avenue NW
Washington, DC 20005-3920

AAAS Newcomb Cleveland Prize 36

The Association's oldest award, the AAAS Newcomb Cleveland Prize, supported by The Fodor Family Trust, was established in 1923 with funds donated by Newcomb Cleveland of New York City and was originally called the AAAS Thousand Dollar Prize. It is now known as the AAAS Newcomb Cleveland Prize, and its value has been raised to $25,000. The winner also receives a commemorative plaque, complimentary registration and reimbursement for reasonable travel and hotel expenses to attend the AAAS Annual Meeting. The prize is awarded to the author or authors of an outstanding paper published in the Research Articles or Reports sections of Science. Each annual contest starts with the first issue of June and ends with the last issue of the following May.
Requirements: An eligible paper is one that includes original research data, theory, or synthesis; is a fundamental contribution to basic knowledge or is a technical achievement of far-reaching consequence; and is a first-time publication of the author's own work. Reference to pertinent earlier work by the author may be included to give perspective. Readers of 'Science' are invited to nominate papers appearing in the Articles, Research Articles, or Reports sections. Nominations must be typed and provide the following information: title of paper, author's name, date of issue in which it was published, page number, and a brief statement of justification for nomination.
Restrictions: Self-nominations will not be accepted.
Geographic Focus: All States
Date(s) Application is Due: Jun 30
Amount of Grant: 25,000 USD
Contact: Alison Crawford, Executive Assistant to the Editor-in-Chief; (202) 326-6560; fax (202) 289-7562; acrawfor@aaas.org
Internet: http://www.aaas.org/aboutaaas/awards/newcomb/index.shtml
Sponsor: American Association for the Advancement of Science
1200 New York Avenue NW
Washington, DC 20005-3920

AAAS Philip Hauge Abelson Award 37

The prize was established by the AAAS Board of Directors and is presented at the AAAS Annual Meeting. It is awarded annually to either a public servant, in recognition of sustained exceptional contributions to advancing science, or to a scientist whose career has been distinguished both for scientific achievement and for other notable services to the scientific community. The annual nomination deadline is September 1.
Requirements: The award is open to all regardless of nationality or citizenship. Nominees must be living at the time of their nomination. Any individual or small group in the scientific and engineering community that has contributed substantially to the understanding or development of science or engineering across national boundaries is eligible for this award.
Restrictions: Each nomination must be seconded by at least two other AAAS members. The winner will be selected by a five-member judging panel.
Geographic Focus: All States
Date(s) Application is Due: Sep 1
Amount of Grant: 5,000 USD
Contact: Stephen Nelson; (202) 326-6600; fax (202) 289-4950; snelsond@aaas.org
Internet: http://www.aaas.org/aboutaaas/awards/abelson/index.shtml
Sponsor: American Association for the Advancement of Science
1200 New York Avenue NW
Washington, DC 20005-3920

AAAS Roger Revelle Fellowship in Global Stewardship 38

The Fellowship is designed to provide a unique opportunity for an accomplished scientist to address global stewardship issues by applying his or her broad, multidisciplinary background toward solutions to these important societal problems. The Revelle Fellow will engage with professionals in the public policy arena to strengthen his or her understanding of the intricacies of environmental policymaking and make practical contributions to the more effective use of scientific and technical knowledge in federal decision-making. The fellowship provides the unique opportunity to address sustainability challenges across multiple disciplines. The focus will be on human interaction with ecosystems, which includes, but is not limited to, ecosystem, population, sustainable development, food, oceans, global climate change, and related environmental concerns.
Requirements: Applications are invited from candidates in any physical, biological, or social science; any field of engineering; or any relevant interdisciplinary field. All applicants must have a PhD or an equivalent doctoral-level degree and at least three years of post-degree professional experience by the application deadline. Individuals with a master's degree in engineering and six years of post-degree, professional experience may apply. All applicants must be U.S. citizens.
Restrictions: Federal employees are not eligible.
Geographic Focus: All States
Date(s) Application is Due: Dec 5
Amount of Grant: 64,000 USD
Contact: Cynthia Robinson, Director; (202) 326-6700; fax (202) 289-4950; crobinso@aaas.org or fellowships@aaas.org
Internet: http://fellowships.aaas.org/02_Areas/02_index.shtml#Revelle
Sponsor: American Association for the Advancement of Science
1200 New York Avenue NW
Washington, DC 20005-3920

AAAS Science and Technology Policy Fellowships: Congressional 39

Fellows spend one year working as special legislative assistants on the staffs of members of Congress or congressional committees, beginning in September. The program includes an orientation on congressional and executive branch operations, and a year-long seminar program on issues involving science and public policy. The fellowships are designed to provide a unique public policy learning experience, to demonstrate the value of science-government interaction, and to bring technical backgrounds and external perspectives to the decision-making process in government. AAAS will select and sponsor two Congressional Fellows. Annual stipends are approximately $70,000, with allowances for health insurance, relocation, and professional travel.

Requirements: Prospective Fellows must demonstrate exceptional competence in some area of science or engineering; have a good scientific and technical background; be cognizant of and demonstrate sensitivity toward political and social issues; and, perhaps most importantly, have a strong interest and some experience in applying personal knowledge toward the solution of societal problems. Applications are invited from individuals in any physical, biological, or social science, any field of engineering, or any relevant interdisciplinary field. All applicants must have a Ph.D. or an equivalent doctoral-level degree by the application deadline (December 10). Individuals with a master's degree in engineering and at least three years of post-degree professional experience may apply. Persons from underrepresented minority groups and persons with disabilities are encouraged to apply.
Restrictions: Applicants must be U.S. citizens. Federal employees are not eligible. Will accept only online application submissions. Applications with fewer than three recommendation letters will not be considered.
Geographic Focus: All States
Date(s) Application is Due: Dec 5
Amount of Grant: 70,000 USD
Contact: Tamisha Marsh, Senior Program Coordinator; (202) 326-6700; fax (202) 289-4950; tmarsh@aaas.org or fellowships@aaas.org
Internet: http://fellowships.aaas.org/02_Areas/02_Congressional.shtml
Sponsor: American Association for the Advancement of Science
1200 New York Avenue NW
Washington, DC 20005-3920

AAAS Science and Technology Policy Fellowships: Diplomacy, Security, and Development 40

Selected fellows apply their scientific and technical expertise into policy development and program planning, implementation and evaluation in the areas of: foreign policy and international trade, treaty engagement, and multilateral cooperation; disaster preparedness and response; infrastructure, environmental, cyber and health security, terrorism and warfare prevention, and nonproliferation; and international aid, capacity building, and development assistance. Placement opportunities are available at Fogerty International Center, National Institutes of Health; Foreign Agriculture Service, U.S. Department of Agriculture; U.S. Agency for International Development; or U.S. Department of State.
Requirements: Applicants must have a PhD or equivalent doctoral level degree at the time of application. Persons with a master's degree in engineering and at least three years of postdegree professional experience will also be considered. All applicants must be U.S. citizens. Prospective fellows must demonstrate exceptional competence in some area of science and engineering (multidisciplinary expertise is especially valued); be cognizant of the ways in which science and technology affect a broad range of international development and foreign policy issues; communicate and work effectively with decision makers and others outside of the scientific and engineering communities; exhibit a willingness and flexibility to tackle problems in a number of nonscientific areas; demonstrate sensitivity toward political, economic, and social issues; and have some experience and/or strong interest in applying knowledge toward the solution of problems in the area of foreign affairs or international development. Applications are invited from individuals in any physical, biological, or social science; any field of engineering; or any relevant interdisciplinary field. Some program areas and agencies seek additional qualifications. See the website for more specific guidelines.
Restrictions: Federal employees are not eligible.
Geographic Focus: All States
Date(s) Application is Due: Dec 5
Amount of Grant: 55,000-85,000 USD
Contact: Rick Kempinski, Project Director; (202) 326-6700; fax (202) 289-4950; rkempinski@aaas.org or fellowships@aaas.org
Internet: http://fellowships.aaas.org/02_Areas/02_index.shtml#DSD
Sponsor: American Association for the Advancement of Science
1200 New York Avenue NW
Washington, DC 20005-3920

AAAS Science and Technology Policy Fellowships: Energy, Environment, and Agriculture 41

Fellows chosen for this program engage in projects, programs, policies and outreach initiatives to protect environmental and human health, tackle energy challenges and opportunities, and to safeguard our air, water, land, and natural resources. Placement opportunities are available with the national Oceanic Atmospheric Association, National Science Foundation, U.S. Department of Agriculture, USDA Forest Service, U.S. Department of Energy, and U.S. Environmental Protection Agency. Fellowships are awarded to highly qualified individuals interested in learning about the science-policy interface while applying their scientific and technical knowledge and analytical skills to the federal policy realm. Stipends range from $55,000 to more than $85,000.
Requirements: Prospective fellows must have a PhD or equivalent doctoral level degree. Persons with a master's degree in engineering and at least three years of post-degree professional experience may apply. Candidates must demonstrate exceptional competence in some area of science or engineering, and an interest in applying their expertise to the economic and technical assessment of problems relating to human health, agriculture, and the environment. Applications are invited from individuals in any physical, biological, or social science; any field of engineering; or any relevant interdisciplinary field. Persons with a DVM, MD, or a PhD in the natural sciences or economics are especially encouraged to apply. Applicants must be U.S. citizens.
Restrictions: Federal employees are ineligible.
Geographic Focus: All States
Date(s) Application is Due: Dec 5
Amount of Grant: 55,000-85,000 USD
Contact: Kira E. Mock, Program Manager; (202) 326-6700; fax (202) 289-4950; kmock@aaas.org or fellowships@aaas.org
Internet: http://fellowships.aaas.org/02_Areas/02_index.shtml#EEA
Sponsor: American Association for the Advancement of Science
1200 New York Avenue NW
Washington, DC 20005-3920

AAAS Science and Technology Policy Fellowships: Global Health and Development 42

The AAAS Science and Technology Policy Fellowships have partnered with the Bill and Melinda Gates Foundation to create the Global Health & Development (GHD) Fellowships. This new fellowship opportunity places Fellows at the Gates Foundation in Seattle, Washington, for one year, with the potential to renew for a second year. The GHD Fellowships provide opportunities for Fellows to learn from and contribute to the international development initiatives of the Foundation, by helping to identify and support strategies and programs in which science and technology can be harnessed to advance development in designated countries and/or regions.
Requirements: This fellowship opportunity is open to current and alumni AAAS Science & Technology Policy Fellows. All applicants must have a doctoral-level degree (PhD, MD, DVM, DSc, PharmD, and other terminal degrees), in any physical, biological, health/medical or social science, any field of engineering, or any relevant interdisciplinary field (individuals with a master's degree in engineering and at least three years of post-degree professional experience also may apply). Note: All requirements for the degree must be completed by the application deadline. A prospective fellow must demonstrate exceptional competence in their specialty appropriate to their career stage, and have the strong endorsement of three references. Show an understanding of the opportunities for science and engineering to support a broad range of non-scientific issues, and display a commitment to apply their scientific or technical expertise to serve society; Exhibit awareness and sensitivity to the political, economic and social issues that influence policy; Be articulate communicators, both verbally and in writing, to decision-makers and non-scientific audiences, and have the ability to work effectively with individuals and groups outside the scientific community; Demonstrate integrity and good judgment, and the flexibility and willingness to address policy issues outside their scientific realm.
Restrictions: Federal employees are ineligible. All applicants must be U.S. citizens (dual citizenship from the United States and another country is acceptable). Some program areas and host agencies seek additional qualifications. Holding a Title 42 position is considered federal employment and therefore renders an applicant ineligible for a AAAS Science and Technology Policy Fellowship. Individuals working in postdoctoral positions at government labs or agencies, for private contractors, and in state government posts are not federal employees and therefore are eligible for the AAAS Fellowships.
Geographic Focus: All States
Date(s) Application is Due: Dec 5
Amount of Grant: 70,000-90,000 USD
Contact: Sage Russell, Associate Director; (202) 326-6700; fax (202) 289-4950; srussell@aaas.org or fellowships@aaas.org
Internet: http://fellowships.aaas.org/02_Areas/02_index.shtml#GHD
Sponsor: American Association for the Advancement of Science
1200 New York Avenue NW
Washington, DC 20005-3920

AAAS Science and Technology Policy Fellowships: Health, Education and Human Services 43

Fellows chosen for this program focus their experience to support improved programs, policies, planning, and risk analysis for initiatives in: preventive and community health, disease identification and response, and medical research; individual, family and community social services, systems and support; food, processing and distribution safety; and science education, research, and innovation. Placement opportunities are available with the National Institutes of Health, National Science Foundation, USDA Food Safety Inspection Services and U.S. Department of Health & Human Services. Fellowships are awarded to highly qualified individuals interested in learning about the science-policy interface while applying their scientific and technical knowledge and analytical skills to the federal policy realm.
Requirements: Prospective fellows must have a PhD or equivalent doctoral level degree. Persons with a master's degree in engineering and at least three years of post-degree professional experience may apply. Candidates must demonstrate exceptional competence in some area of science or engineering, and an interest in applying their expertise to the economic and technical assessment of problems relating to human health, agriculture, and the environment. Applications are invited from individuals in any physical, biological, or social science; any field of engineering; or any relevant interdisciplinary field. Persons with a DVM, MD, or a PhD in the natural sciences or economics are especially encouraged to apply. Applicants must be U.S. citizens.
Restrictions: Federal employees are not eligible.
Geographic Focus: All States
Date(s) Application is Due: Dec 5
Amount of Grant: 55,000-85,000 USD
Contact: Robert Harper, Program Manager; (202) 326-6700; fax (202) 289-4950; rharper@aaas.org or fellowships@aaas.org
Internet: http://fellowships.aaas.org/02_Areas/02_index.shtml#HEHS
Sponsor: American Association for the Advancement of Science
1200 New York Avenue NW
Washington, DC 20005-3920

AAAS Science and Technology Policy Overseas Fellowships at USAID Missions 44

Re-launched in 2011, the AAAS Overseas Fellowships offer Fellows the experience of living and working overseas at a U.S. Embassy or U.S. Agency for International Development (USAID) Mission , as well as the opportunity to expand both the service and professional development opportunities of the S&T Policy Fellowship experience. AAAS supports Fellows to pursue international assignments via the Overseas Fellowships. AAAS Overseas Fellowships are currently available only at USAID Missions. Overseas Fellowships offer one- to two-year international assignments. The Fellow's primary responsibility will be to help the Mission identify key opportunities to develop strategies and programs that contribute to development in science and technology and areas where S&T can be harnessed to advance development in the host country and/or region. Anticipated placement opportunities will vary from year to year.
Requirements: This fellowship opportunity is limited to current and alumni AAAS Science & Technology Policy Fellows. Fellows may have served in any program area.
Restrictions: Applicants must be U.S. citizens. Federal employees are not eligible. Will accept only online application submissions. Applications with fewer than three recommendation letters will not be considered.
Geographic Focus: All States
Amount of Grant: 55,000-85,000 USD
Contact: Chitra Kalyandurg, Program Manager; (202) 326-6700; fax (202) 289-4950; ckalyand@aaas.org or fellowships@aaas.org
Internet: http://fellowships.aaas.org/02_Areas/02_index.shtml#overseas
Sponsor: American Association for the Advancement of Science
1200 New York Avenue NW
Washington, DC 20005-3920

AAAS Science Prize for Inquiry-Based Instruction 45

Science and engineering education are being redefined in ways that encourage all students to actively experience "science as inquiry" and "engineering as design under constraint." And, for the first time, it is possible for advanced high school and undergraduate students to work with some of the same data and tools as practicing scientists and engineers. How can science and engineering education capitalize on these new and unprecedented opportunities? One approach is through the Science Prize for Inquiry-Based Instruction, a prize established to encourage innovation and excellence in education by recognizing outstanding, inquiry-based science and design-based engineering education modules. This type of education is a form of active learning wherein the instructor provides a question, or a challenge, and a general set of procedures that can be used to answer it. The students then produce an explanation or answer that is based on the evidence that they collect from appropriate resource materials or experimental processes that are, at least in part, of the students own proposal. The contest will recognize any inquiry-based or design based module associated with either an advanced high school level or an introductory-level college course in science or engineering. Only one entry per academic department will be accepted.
Requirements: Winning applicants must provide complete instructions on how to implement their module. The resource must contain no copyright restrictions. And, to make the module easily portable, it should require only standard resources (supplies, equipment, or specialized expertise) that are normally available for such courses.
Geographic Focus: All States
Date(s) Application is Due: Apr 30
Contact: Dr. Melissa McCartney, (202) 326-6700; fax (202) 289-4950; mmccartn@aaas.org
Internet: http://www.sciencemag.org/site/feature/data/prizes/inquiry/
Sponsor: American Association for the Advancement of Science
1200 New York Avenue NW
Washington, DC 20005-3920

AABB Dale A. Smith Memorial Award 46

The AABB Dale A. Smith Memorial Award, created in 2002 and funded by Fenwal, Inc., honors Dale A. Smith, a long-time Baxter Healthcare executive who was responsible for establishing the Fenwal Division of Baxter. This award will be presented to an individual or institution to recognize groundbreaking work performed in the application of technology to the practice of transfusion medicine or cellular therapies. The recipient will receive a $7,500 honorarium. The Award is presented annually.
Requirements: Award recipients are not required to be members of AABB. A nomination form must be completed for each nomination submission.
Geographic Focus: All States
Amount of Grant: 7,500 USD
Contact: Awards Coordinator; (301) 907-6977; fax (301) 907-6895; aabb@aabb.org
Internet: http://www.aabb.org/development/awardsscholarships/awards/Pages/descriptions.aspx#smith
Sponsor: American Association of Blood Banks
8101 Glenbrook Road
Bethesda, MD 20814-2749

AABB Emily Cooley Memorial Award and Lectureship 47

The AABB Emily Cooley Memorial Award and Lectureship began as a lectureship in 1963, and was designated as a Memorial Award in 1983. The person who receives this award shall have demonstrated teaching ability and have made a major contribution to the field of transfusion medicine or cellular therapies. Acceptance of the award requires attendance at the Annual Meeting and the presentation of a lecture. Recipient will receive a $5,000 honorarium. The Award and Lectureship is presented annually and the recipient is chosen one year in advance by a Joint Committee composed of leaders from the Cellular Therapies Section coordinating committee and Transfusion Medicine Section coordinating committee.
Geographic Focus: All States
Amount of Grant: 5,000 USD
Contact: Awards Coordinator; (301) 907-6977; fax (301) 907-6895; aabb@aabb.org
Internet: http://www.aabb.org/development/awardsscholarships/awards/Pages/descriptions.aspx#cooley
Sponsor: American Association of Blood Banks
8101 Glenbrook Road
Bethesda, MD 20814-2749

AAB Berlin Prize Fellowships 48

The American Academy in Berlin was established in 1994 with the primary goal of fostering greater understanding and dialogue between the people of the United States and the people of Germany. The Academy offers residential fellowships at its Hans Arnhold Center to American scholars, writers, policymakers, and artists who wish to engage in independent study in Berlin. Around two dozen of these various Berlin Prize Fellowships are conferred annually. The academy welcomes younger as well as established scholars, artists, and professionals. Fellowships have been awarded in a variety of disciplines, including history, political science, economics, German studies, art history, musicology, anthropology, law, and linguistics, as well as to writers and poets, public policy experts, journalists, composers, and artists. Fellows are in residence for one academic semester or occasionally for a full academic year (Bosch Fellows in Public Policy usually stay for six to eight weeks). Fellowship benefits include round-trip airfare, housing at the Academy, partial board, and a stipend of $5,000 per month. Furnished apartments at the Hans Arnhold Center are available for individuals and couples; limited accommodations are also available for families with children. Fellows are encouraged to work in association with a Berlin institution such as a museum, library, archive, university, government agency, studio, or media organization. Current guidelines and applications are made available at the Academy's website in mid-June. Application deadlines may vary from year to year. Candidates should verify current deadline dates at the Academy's website.
Requirements: All candidates should be permanently based in the United States and must be either U.S. citizens or permanent residents. They must have completed their doctorate or equivalent professional degree. Writers must have published at least one book at the time of application. Candidates need not work on German topics, but their project descriptions should explain how a residency in Berlin will contribute to further professional development.
Restrictions: The Academy does not accept project proposals in mathematics and the hard sciences. The Academy does not accept applications from visual artists. Instead a small number of candidates are contacted individually by members of the Academy's Visual Arts Jury. Only one fellowship in music composition will be awarded per selection cycle. Applications in music composition are due in the Academy's New York office on October 1. Music composition applicants should use the Application for Music Composition, not the general application for fellowships. Music composition applicants may not apply online.
Geographic Focus: All States
Date(s) Application is Due: Sep 28; Oct 1
Amount of Grant: 7,500-60,000 USD
Contact: Michael Hoffman and Gary Smith, Managing Directors; (212) 588-1755; fax (212) 588-1758; js@americanacademy.de
Internet: http://www.americanacademy.de/home/fellows/applications/
Sponsor: American Academy in Berlin
14 E 60th Street, Suite 604
New York, NY 10022

AABB Hemphill-Jordan Leadership Award 49

The AABB Hemphill-Jordan Leadership Award, renamed in 2005 after Bernice Hemphill, W. Quinn Jordan, and Joel Solomon, honors leaders from the transfusion medicine and cellular therapy community. The award recognizes an individual who made significant contributions in the areas of administration, quality programs, law and/or government affairs. The individual shall have demonstrated leadership qualities and a consistent willingness to lend his/her expertise to his/her peers. It may recognize one particular act or an accumulation of years of contributions. Acceptance of the award requires attendance at the Annual Meeting and the presentation of a lecture. The recipient will receive a $5,000 honorarium. The Award is presented annually and the recipient is chosen one year in advance.
Geographic Focus: All States
Amount of Grant: 5,000 USD
Contact: Awards Coordinator; (301) 907-6977; fax (301) 907-6895; aabb@aabb.org
Internet: http://www.aabb.org/development/awardsscholarships/awards/Pages/descriptions.aspx#hemphill
Sponsor: American Association of Blood Banks
8101 Glenbrook Road
Bethesda, MD 20814-2749

AABB John Elliott Memorial Award 50

The AABB John Elliott Memorial Award, established in 1956, recognizes an individual who has given outstanding service to AABB by demonstrating a willingness to lend his/her expertise to the association through work on committees, the AABB Board of Directors and other areas. The recipient will receive a $1,000 honorarium. The Award is presented in odd years only, and the recipient is selected by a Joint Committee composed of leaders from the Cellular Therapies Section coordinating committee and Transfusion Medicine Section coordinating committee.

Requirements: Award recipients are not required to be members of AABB. A nomination form must be completed for each nomination submission.
Geographic Focus: All States
Amount of Grant: 1,000 USD
Contact: Awards Coordinator; (301) 907-6977; fax (301) 907-6895; aabb@aabb.org
Internet: http://www.aabb.org/development/awardsscholarships/awards/Pages/descriptions.aspx#elliott
Sponsor: American Association of Blood Banks
8101 Glenbrook Road
Bethesda, MD 20814-2749

AABB Karl Landsteiner Memorial Award and Lectureship 51
Initiated in 1954 to honor Karl Landsteiner, MD, whose lifetime research laid the foundation for modern blood transfusion therapy, the AABB Karl Landsteiner Memorial Award and Lectureship recognizes a scientist whose original research resulted in an important contribution to the body of scientific knowledge. The scientist who receives the award shall have an international reputation in transfusion medicine or cellular therapies. The recipient must agree to lecture at the AABB Annual Meeting. The recipient will receive a $7,500 honorarium. The Award and Lectureship will be awarded as distinguished candidates are identified. The recipient is chosen one year in advance by AABB's Board of Directors.
Requirements: Award recipients are not required to be members of AABB. A nomination form must be completed for each nomination submission.
Geographic Focus: All States
Amount of Grant: 7,500 USD
Contact: Awards Coordinator; (301) 907-6977; fax (301) 907-6895; aabb@aabb.org
Internet: http://www.aabb.org/development/awardsscholarships/awards/Pages/descriptions.aspx#landsteiner
Sponsor: American Association of Blood Banks
8101 Glenbrook Road
Bethesda, MD 20814-2749

AABB Sally Frank Memorial Award and Lectureship 52
The Sally Frank Memorial Award and Lectureship was established in 1982 in memory of Ms. Sally Frank and her dedication to red cell serology and education. This award recognizes an individual who is, or has been, a medical technologist involved with these fields and has demonstrated quality research, teaching and/or service abilities in the technical aspects of immunohematology. Recipient will receive a $1,000 honorarium. The Award and Lectureship is presented annually and the recipient is chosen one year in advance by AABB's Transfusion Medicine Section Coordinating Committee with Formal Approval by AABB's Board of Directors.
Requirements: Award recipients are not required to be members of AABB. A nomination form must be completed for each nomination submission.
Geographic Focus: All States
Amount of Grant: 1,000 USD
Contact: Awards Coordinator; (301) 907-6977; fax (301) 907-6895; aabb@aabb.org
Internet: http://www.aabb.org/development/awardsscholarships/awards/Pages/descriptions.aspx#frank
Sponsor: American Association of Blood Banks
8101 Glenbrook Road
Bethesda, MD 20814-2749

AABB Tibor Greenwalt Memorial Award and Lectureship 53
The AABB Tibor Greenwalt Memorial Award and Lectureship honors Tibor Greenwalt, MD, who was the first registrant at the first AABB Annual Meeting and founding editor of Transfusion. The award recognizes an individual who made major scientific or clinical contributions to hematology, transfusion medicine or cellular therapies, and succinctly communicated these advances. Acceptance of this award requires attendance at the AABB Annual Meeting and the presentation of a lecture on a current topic of interest. Recipient will receive a $1,000 honorarium. The Award and Lectureship is presented annually and the recipient is chosen one year in advance by a Joint Committee Composed of Leaders from the Cellular Therapies Section Coordinating Committee and Transfusion Medicine Section Coordinating Committee.
Requirements: Award recipients are not required to be members of AABB. A nomination form must be completed for each nomination submission.
Geographic Focus: All States
Amount of Grant: 1,000 USD
Contact: Awards Coordinator; (301) 907-6977; fax (301) 907-6895; aabb@aabb.org
Internet: http://www.aabb.org/development/awardsscholarships/awards/Pages/descriptions.aspx#greenwalt
Sponsor: American Association of Blood Banks
8101 Glenbrook Road
Bethesda, MD 20814-2749

AACAP-NIDA K12 Career Development Awards 54
The grant provides up to five years of salary support and mentored addiction research training for qualified child and adolescent psychiatrists (CAPs) who intend to establish careers as independent investigators in mental health and addiction research. Letters of Intent are required and due on or before January 5. Full applications are due March 30, with new grants awarded on June 1. Applicants are strongly encouraged to contact program staff regarding the program and candidate qualifications before submitting Letters of Intent.
Requirements: Eligible applicants must be a child and adolescent psychiatrist with a commitment and goal of becoming an independent investigator in addiction-related research focused on children and adolescents. Applicants must also be U.S. citizens, non-citizen nationals, and/or individuals with permanent residence status. Women and minority candidates are especially encouraged to apply.
Geographic Focus: All States
Date(s) Application is Due: Jan 5
Amount of Grant: Up to 140,000 USD
Contact: Stacia Fleisher, (202) 966-7300; fax (202) 966-2891; sfleisher@aacap.org or research@aacap.org
Internet: http://www.aacap.org/cs/root/research_and_training_awards/aacap_nida12
Sponsor: American Academy of Child and Adolescent Psychiatry
3615 Wisconsin Avenue, NW
Washington, DC 20016-3007

AACAP Beatrix A. Hamburg Award for the Best New Research Poster by a Child and Adolescent Psychiatry Resident 55
This award recognizes the author of the best new research poster presented at the AACAP Annual Meeting. The award provides a $1,000 honorarium. The recipient will receive a plaque at the Young Leader Awards Ceremony during the AACAP Annual Meeting. There is no application process for this award. The AACAP Program Committee will review all new research poster submissions by child and adolescent psychiatry residents to determine the recipient of this award. The recipient will be notified through an acceptance letter. Non-recipients will not be notified.
Requirements: The author must be a child and adolescent psychiatry resident at an accredited institution and AACAP member. Applicants must: be a child and adolescent psychiatry resident; attend an accredited institution; be an AACAP member; and attend and present a new research poster at the AACAP Annual Meeting.
Geographic Focus: All States
Date(s) Application is Due: Jun 14
Contact: Stacie Hall; (202) 966-7300, ext. 113; fax (202) 966-2891; shall@aacap.org
Internet: http://www.aacap.org/cs/root/research_and_training_awards/aacap_beatrix_a_hamburg_award_for_the_best_new_research_poster_by_a_child_and_adolescent_psychiatry_resident_at_the_aacap_annual_meeting
Sponsor: American Academy of Child and Adolescent Psychiatry
3615 Wisconsin Avenue, NW
Washington, DC 20016-3007

AACAP Educational Outreach Program for Child and Adolescent Psychiatry Residents 56
The Educational Outreach Program (EOP) funding provides the opportunity for up to 50 child and adolescent psychiatry residents to receive a formal overview to the field of child and adolescent psychiatry, establish child and adolescent psychiatrists as mentors and experience the AACAP Annual Meeting. Participants will be exposed to the breadth and depth of the field of child and adolescent psychiatry, including research opportunities, access to mentors, and various networking opportunities. Participation in this program provides up to $1,000 for travel expenses, which includes airfare, hotel, and meals (maximum $75/day).
Requirements: Applicants must: be child and adolescent psychiatry residents at the time of the AACAP Annual Meeting; be currently enrolled in a residency program in the United States; be residents in their first or second year of child fellowship training are eligible (Triple Boarders in their fourth or fifth year of training in their triple board programs are eligible); either be members of the AACAP or have a membership application pending at the time of application; and attend all AACAP Annual Meeting events specified by AACAP.
Geographic Focus: All States
Date(s) Application is Due: Jul 12
Contact: Ashley Partner, (202) 966-7300, ext. 117; apartners@aacap.org
Internet: http://www.aacap.org/cs/residents/eop-capfellows
Sponsor: American Academy of Child and Adolescent Psychiatry
3615 Wisconsin Avenue, NW
Washington, DC 20016-3007

AACAP Educational Outreach Program for General Psychiatry Residents 57
The Educational Outreach Program (EOP) for General Psychiatry residents provides the opportunity for up to 20 general psychiatry residents to receive a formal overview to the field of child and adolescent psychiatry, establish child and adolescent psychiatrists as mentors and experience the AACAP Annual Meeting. Participants will be exposed to the breadth and depth of the field of child and adolescent psychiatry, including research opportunities, access to mentors, and various networking opportunities. Participation in this program provides up to $1,500 for travel expenses to the AACAP Annual Meeting.
Requirements: Applicants must: be general psychiatry residents at the time of the AACAP Annual Meeting; be currently enrolled in a residency program in the United States; be residents in their first, second or third year of general psychiatry training (Triple Boarders in their first, second or third year of training in their triple board programs are eligible); either be members of the AACAP or have a membership application pending at the time of application; and attend all AACAP Annual Meeting events specified by AACAP.
Geographic Focus: All States
Date(s) Application is Due: Jul 12
Contact: Ashley Partner, (202) 966-7300, ext. 117 or (202) 587-9663; fax (202) 966-2891; apartner@aacap.org or training@aacap.org
Internet: http://www.aacap.org/cs/residents/eop-generalresidents
Sponsor: American Academy of Child and Adolescent Psychiatry
3615 Wisconsin Avenue, NW
Washington, DC 20016-3007

AACAP Educational Outreach Program for Residents in Alcohol Research 58

The Educational Outreach Program (EOP) for Residents in Alcohol Research provides the opportunity for up to 10 child and adolescent psychiatry and general psychiatry residents to pursue advanced knowledge in adolescent alcohol research. The program also allows residents to expand their knowledge of adolescent alcohol-related issues by attending the AACAP Annual Meeting. EOP recipients will meet with an assigned alcohol research mentor and with a representative from NIAAA at the AACAP Annual Meeting to encourage mentorship connections and career development. Participants will be exposed to the field of child and adolescent psychiatry, including research opportunities, access to mentors, and various networking opportunities. Participation in this program provides up to $1,000 for travel expenses to the AACAP Annual Meeting.
Requirements: Applicants must be child and adolescent psychiatry or general psychiatry residents at the time of the AACAP Annual Meeting. Participants must: be currently enrolled in a child and adolescent psychiatry, general psychiatry, or triple board residency program in the United States; either be members of the AACAP or have a membership application pending at the time of application; and attend all Annual Meeting events specified by AACAP.
Geographic Focus: All States
Date(s) Application is Due: Jul 12
Amount of Grant: Up to 1,000 USD
Contact: Ashley Partner, (202) 966-7300, ext. 117 or (202) 587-9663; fax (202) 966-2891; apartner@aacap.org or training@aacap.org
Internet: http://www.aacap.org/cs/award_opportunities/details/aacap_educational_outreach_program_for_residents_in_alcohol_research
Sponsor: American Academy of Child and Adolescent Psychiatry
3615 Wisconsin Avenue, NW
Washington, DC 20016-3007

AACAP Elaine Schlosser Lewis Award for Research on Attention-Deficit Disorder 59

This annual award acknowledges outstanding leadership and continuous contributions in the field of research by giving $5,000 for the best paper published in the Journal on attention-deficit disorder written by a child and adolescent psychiatrist. The award winner will be recognized at a Distinguished Awards Luncheon and make an Honors Presentation about his or her work during the AACAP Annual Meeting.
Requirements: Nominations must be accompanied by a CV for the individual nominated.
Geographic Focus: All States
Date(s) Application is Due: May 7
Amount of Grant: 5,000 USD
Contact: Alyssa Sommer; (202) 966-7300, ext. 157; asommer@aacap.org
Internet: http://www.aacap.org/cs/root/research_and_training_awards/aacap_journal_awards
Sponsor: American Academy of Child and Adolescent Psychiatry
3615 Wisconsin Avenue, NW
Washington, DC 20016-3007

AACAP George Tarjan Award for Contributions in Developmental Disabilities 60

The AACAP George Tarjan Award for Contributions in Developmental Disabilities recognizes a child and adolescent psychiatrist and Academy member who has made significant contributions in a lifetime career or single seminal work to the understanding or care of those with mental retardation and developmental disabilities. These contributions must have national and/or international stature and clearly demonstrate lasting effects. The contributions may be in areas of teaching, research, program development, direct clinical service, advocacy or administrative commitment. A cash prize of up to $1,000 will be awarded. The award winner will be recognized at a Distinguished Awards Luncheon and make an Honors Presentation about his or her work during the AACAP Annual Meeting.
Requirements: Nominations must be accompanied by a CV for the individual nominated.
Geographic Focus: All States
Date(s) Application is Due: Apr 30
Amount of Grant: 1,000 USD
Contact: Bryan King, Chair; (202) 966-7300; fax (202) 966-2891; bking@aacap.org
Internet: http://www.aacap.org/cs/distinguishedmembers
Sponsor: American Academy of Child and Adolescent Psychiatry
3615 Wisconsin Avenue, NW
Washington, DC 20016-3007

AACAP Irving Philips Award for Prevention 61

The AACAP Irving Philips Award for Prevention recognizes a child and adolescent psychiatrist and Academy member who has made significant contributions in a lifetime career or single seminal work to the prevention of mental illness in children and adolescents. These contributions must have national and/or international stature and clearly demonstrate lasting effects. The contributions may be in the areas of teaching, research, program development, direct clinical service, advocacy or administrative commitment. The award pays $2,500 to the winner and a $2,000 donation to a prevention program or center of the awardee's choice. The award winner will be recognized at a Distinguished Awards Luncheon and make an Honors Presentation about his or her work during the AACAP Annual Meeting.
Requirements: Nominations must be accompanied by a CV for the individual nominated.
Geographic Focus: All States
Date(s) Application is Due: Apr 30
Contact: James Hudziak; (202) 966-7300; fax (202) 966-2891; jhudziak@aacap.org
Internet: http://www.aacap.org/cs/distinguishedmembers
Sponsor: American Academy of Child and Adolescent Psychiatry
3615 Wisconsin Avenue, NW
Washington, DC 20016-3007

AACAP Jeanne Spurlock Lecture and Award on Diversity and Culture 62

The Jeanne Spurlock Lecture and Award on Diversity and Culture reflects the spirit of the nomination for outstanding contributions to diversity and culture and in some way encourages individuals from diverse cultural backgrounds to become child and adolescent psychiatrists. Nominees should be individuals who have made contributions in the areas of social awareness including: civil rights; spirituality and/or religion; social welfare; public information; scientific research; education and mentoring; and the arts (literature, theatre, music, painting, sculpture or photography). The award includes an honorarium of $2,500, with the award winner being recognized at a Distinguished Awards Luncheon where he or she will make an honors award presentation about his or her work during the AACAP Annual Meeting.
Geographic Focus: All States
Date(s) Application is Due: Apr 30
Amount of Grant: 2,500 USD
Contact: Adriano Boccanelli, Clinical Practice Manager; (202) 966-7300, ext. 133; fax (202) 966-9518; aboccanelli@aacap.org or clinical@aacap.org
Internet: http://www.aacap.org/cs/distinguishedmembers
Sponsor: American Academy of Child and Adolescent Psychiatry
3615 Wisconsin Avenue, NW
Washington, DC 20016-3007

AACAP Jeanne Spurlock Minority Medical Student Clinical Fellowship in Child and Adolescent Psychiatry 63

Up to fourteen (14) fellowships are available each year for minority medical students to explore a career in child and adolescent psychiatry, gain valuable work experience, and meet leaders in the child and adolescent psychiatry field. The fellowship opportunity provides up to $3,500 for 12 weeks of clinical training under a child and adolescent psychiatrist mentor. Research assignments may include responsibility for part of the observation or evaluation, developing specific aspects of the research mechanisms, conducting interviews or tests, use of rating scales, and psychological or cognitive testing of subjects. Contact AACAP for applications.
Requirements: African American, Asian American, Native American, Alaskan Native, Mexican American, Hispanic, and Pacific Islander students in accredited U.S. medical schools are eligible.
Geographic Focus: All States
Date(s) Application is Due: Mar 1
Amount of Grant: 3,500 USD
Contact: Gabe Robbins, (202) 966-7300, ext. 117; fax (202) 966-2891; grobbins@aacap.org
Internet: http://www.aacap.org/cs/students/opportunities/SpurlockClinical
Sponsor: American Academy of Child and Adolescent Psychiatry
3615 Wisconsin Avenue, NW
Washington, DC 20016-3007

AACAP Jeanne Spurlock Research Fellowship in Substance Abuse and Addiction for Minority Medical Students 64

The Jeanne Spurlock Research Fellowship in Substance Abuse and Addiction for Minority Medical Students offers a unique opportunity for minority medical students to explore a research career in substance abuse in relation to child and adolescent psychiatry, gain valuable work experience, and meet leaders in the child and adolescent psychiatry field. The fellowship opportunity provides up to $4,000 for 12 weeks of summer research under a child and adolescent psychiatrist researcher/mentor. The research training plan must provide for significant contact between the student and the mentor and for exposure to state-of-the-art drug abuse and addiction research. The plan should include program planning discussions, instruction in research planning and implementation, regular meetings with the mentor, laboratory director, and the research group, and assigned readings. Research assignments may include responsibility for part of the observation or evaluation, developing specific aspects of the research mechanisms, conducting interviews or tests, use of rating scales, and psychological or cognitive testing of subjects.
Requirements: All fellowship participants must attend the AACAP Annual Meeting. Applications are considered from African-American, Native American, Alaskan Native, Mexican American, Hispanic, Asian, and Pacific Islander students in accredited U.S. medical schools.
Geographic Focus: All States
Date(s) Application is Due: Feb 15
Amount of Grant: Up to 4,000 USD
Contact: Ashley Partner, (202) 966-7300, ext. 117; fax (202) 966-2891; apartner@aacap.org or training@aacap.org
Internet: http://www.aacap.org/cs/students/opportunities/SpurlockResearch
Sponsor: American Academy of Child and Adolescent Psychiatry
3615 Wisconsin Avenue, NW
Washington, DC 20016-3007

AACAP Junior Investigator Awards 65

The AACAP Junior Investigator Award offers two awards of up to $30,000 a year for two years for child and adolescent psychiatry junior faculty (assistant professor level or equivalent). The program is intended to facilitate innovative research. The research may be basic or clinical in nature but must be relevant to our understanding, treatment and prevention of child and adolescent mental health disorders. The award also includes the cost of attending the AACAP Annual Meeting for five days.
Requirements: Recipients are required to submit a poster or oral presentation on his or her research for the AACAP's Annual Meeting. Applicants must: be board eligible or certified in child and adolescent psychiatry; have a doctoral level degree and be in a faculty or independent research position; and have an on-site mentor who has had

experience in the type of research that is being proposed that will normally include work with children and adolescents. Candidates must either be AACAP members or have a membership application pending (not paid by the award).
Restrictions: Applicants who have served as Principal Investigator on an NIH R01 grant are not eligible.
Geographic Focus: All States
Date(s) Application is Due: Mar 1
Amount of Grant: Up to 30,000 USD
Contact: Alyssa Sommer, Research Coordinator; (202) 966-7300, ext. 157; fax (202) 364-5925; asommer@aacap.org or research@accap.org
Internet: http://www.aacap.org/cs/root/research_and_training_awards/aacap_junior_investigator_awards
Sponsor: American Academy of Child and Adolescent Psychiatry
3615 Wisconsin Avenue, NW
Washington, DC 20016-3007

AACAP Klingenstein Third Generation Foundation Award for Research in Depression or Suicide 66
This annual award acknowledges outstanding leadership and continuous contributions in the field of research by giving $5,000 for the best paper on suicide and/or depression published in the Journal during the past year. The award winner will be recognized at a Distinguished Awards Luncheon and make an Honors Presentation about his or her work during the AACAP Annual Meeting.
Requirements: Nominations must be accompanied by a CV for the individual nominated.
Geographic Focus: All States
Date(s) Application is Due: May 7
Amount of Grant: 5,000 USD
Contact: Gabe Robbins, (202) 966-7300, ext. 117; fax (202) 966-2891; grobbins@aacap.org
Internet: http://www.aacap.org/cs/root/research_and_training_awards/aacap_journal_awards
Sponsor: American Academy of Child and Adolescent Psychiatry
3615 Wisconsin Avenue, NW
Washington, DC 20016-3007

AACAP Life Members Mentorship Grants for Medical Students 67
The Life Members Mentorship Grants for Medical Students (MGM) provides the opportunity for seven medical students to attend the AACAP Annual Meeting and receive an introduction into the field of child and adolescent psychiatry through the AACAP Mentorship Program. MGM recipients also participate in programs sponsored by the Life Members, which is a group of the oldest and most distinguished members of AACAP, all having been members for at least 30 years. Many of those in this group served as AACAP leadership and also pioneered many of the significant discoveries and developments in the field of child and adolescent psychiatry. Partnered with the Mentorship Program, this program provides participants with networking opportunities, exposure to varying specialties, and interaction with Life Members. Participation in this program provides up to $1,000 for travel expenses to the AACAP Annual Meeting, which includes airfare, hotel, and meals (maximum $75/day).
Requirements: Applicants must be enrolled in a medical school in the United States at the time of the AACAP Annual Meeting. Recipients are required to: share the AACAP recruitment video with fellow medical students in their program within the six months following the Annual Meeting; and write a follow-up report on the experience that they would be encouraged to submit to their program's listserve and/or website and share a copy of this with AACAP. Participants must: be in good standing at their medical school; and attend all AACAP Annual Meeting events specified by AACAP.
Geographic Focus: All States
Date(s) Application is Due: Jul 12
Amount of Grant: Up to 1,000 USD
Contact: Ashley Partner, (202) 966-7300, ext. 117 or (202) 587-9663; fax (202) 966-2891; apartner@aacap.org or training@aacap.org
Internet: http://www.aacap.org/cs/students/opportunities/life_members_mentorship_grants_for_medical_students
Sponsor: American Academy of Child and Adolescent Psychiatry
3615 Wisconsin Avenue, NW
Washington, DC 20016-3007

AACAP Mary Crosby Congressional Fellowships 68
The Mary Crosby Congressional Fellowships program is designed to educate policy makers and Congressional staff about child and adolescent psychiatry, and to foster awareness of children's mental health issues. This experience will allow the Fellow to develop a keen understanding of how public policy affects patient care, education, and health insurance issues. The AACAP Mary Crosby Congressional Fellow will be placed with a Congressional office or Committee where they will gain invaluable experience as they assist in the development of legislative and public policy initiatives. The award includes an $85,000 stipend, up to $3,000 in relocation or moving expenses, and health insurance for the length of the fellowship program.
Requirements: Successful applicants must: be an AACAP member; be a PGY-4 psychiatry resident or beyond; have completed training in or be scheduled to complete training in one of the child and adolescent psychiatry programs (CAP training program, Combined General Psychiatry/CAP Program, or Triple Board Program); have a demonstrated interest in public policy issues; exhibit an interest in applying medical and scientific knowledge to the resolution of public policy issues; and be a U.S. citizen or permanent resident.
Geographic Focus: All States
Amount of Grant: 85,000-88,000 USD
Contact: Ashley Partner, (202) 966-7300, ext. 117; fax (202) 966-2891; apartner@aacap.org
Internet: http://www.aacap.org/cs/root/legislative_action/2008_congressional_fellowship_program
Sponsor: American Academy of Child and Adolescent Psychiatry
3615 Wisconsin Avenue, NW
Washington, DC 20016-3007

AACAP Norbert and Charlotte Rieger Award for Scientific Achievement 69
This award is given annually for the most significant article published in the Journal of the American Academy of Child and Adolescent Psychiatry (JAACAP) in the past year (July-May). The article must be written by a psychiatrist specializing in child and/or adolescent psychiatry and cannot be a review.
Requirements: The only submissions accepted are papers from child and adolescent psychiatrists that have been published in the Journal during the past year.
Geographic Focus: All States
Date(s) Application is Due: May 7
Amount of Grant: Up to 4,500 USD
Samples: Gail Bernstein, M.D., University of Minnesota, Minneapolis, MN—School-Based Interventions for Anxious Children.
Contact: Gabe Robbins, (202) 966-7300, ext. 117; fax (202) 966-2891; grobbins@aacap.org
Internet: http://www.aacap.org/cs/root/research_and_training_awards/aacap_journal_awards
Sponsor: American Academy of Child and Adolescent Psychiatry
3615 Wisconsin Avenue, NW
Washington, DC 20016-3007

AACAP Pilot Research Award for Attention-Deficit Disorder 70
The AACAP Pilot Research Award for Attention Disorders, supported by The Elaine Schlosser Lewis Fund, offers $15,000 for child and adolescent psychiatry residents and junior faculty who have an interest in beginning a career in child and adolescent mental health research. By providing one award to a child and adolescent psychiatry junior faculty member or resident for pilot research on attention disorders, the AACAP supports a young investigator at a critical stage, encouraging a future career in child and adolescent psychiatry research. The recipient has the opportunity to submit a poster presentation on his or her research for AACAP's Annual Meeting, and present at the Elaine Schlosser Lewis Luncheon.
Requirements: Candidates must be board eligible, certified in child and adolescent psychiatry, or enrolled in a child psychiatry residency or fellowship program. Candidates must also have a faculty appointment in an accredited medical school or be in a fully accredited child and adolescent psychiatry clinical research or training program.
Restrictions: At the time of application, candidates may not have more than two years experience following graduation from residency/fellowship training. Candidates must not have any previous significant, individual research funding in the field of child and adolescent mental health. These include the following: NIMH/NIH Funding (Small Grants, T Award or R-01) or similar foundation or industry research funding.
Geographic Focus: All States
Date(s) Application is Due: Apr 29
Amount of Grant: 15,000 USD
Contact: Ashley Partner, Research Coordinator; (202) 966-7300, ext. 117; fax (202) 364-5925; apartner@aacap.org
Internet: http://www.aacap.org/cs/root/research_and_training_awards/pilot_research_award_for_attention_disorder_for_a_junior_faculty_or_child_psychiatry_resident_supported_by_the_elaine_schlosser_lewis_fund
Sponsor: American Academy of Child and Adolescent Psychiatry
3615 Wisconsin Avenue, NW
Washington, DC 20016-3007

AACAP Pilot Research Award for Learning Disabilities, Supported by the Elaine Schlosser Lewis Fund 71
The Award for Learning Disabilities and Psychiatric Disorders is available for a junior faculty or child psychiatry resident, and is supported by the Elaine Schlosser Lewis Fund. The award encourages work with a child and adolescent psychiatric investigator with expertise in his or her particular area of interest. Work must be completed within one year of receipt of the award. Please note that travel to both the Annual Meeting and the Research Update Luncheon is not included in the award and is the responsibility of the recipient.
Requirements: Candidates must be board eligible, certified in child and adolescent psychiatry, or enrolled in a child psychiatry residency or fellowship program. Candidates must also have a faculty appointment in an accredited medical school or be in a fully accredited child and adolescent psychiatry clinical research or training program.
Restrictions: Candidates must not have any previous significant, individual research funding in the field of child and adolescent mental health. These include the following: NIMH/NIH Funding (Small Grants, T Award or R-01) or similar foundation or industry research funding. The recipient must be an AACAP member at the time of application and agree to present his or her research.
Geographic Focus: All States
Date(s) Application is Due: May 7
Amount of Grant: 15,000 USD
Contact: Alyssa Sommer; (202) 966-7300, ext. 157; asommer@aacap.org
Internet: http://www.aacap.org/cs/root/research_and_training_awards/pilot_research_award_for_learning_disabilities_for_a_junior_faculty_or_child_psychiatry_resident_supported_by_the_elaine_schlosser_lewis_fund
Sponsor: American Academy of Child and Adolescent Psychiatry
3615 Wisconsin Avenue, NW
Washington, DC 20016-3007

AACAP Pilot Research Awards, Supported by Eli Lilly and Company 72

Eight awards are available each year to junior faculty and child psychiatry fellows for pilot research with child and adolescent psychiatric researchers. Award recipients will be matched with child and adolescent psychiatric investigators working in their particular area of interest to act as consultants or mentors during the course of the award. Work must be completed within one year of receipt of the award.

Requirements: Candidates must be board eligible, certified in child and adolescent psychiatry, or enrolled in a child psychiatry residency or fellowship program. Candidates must have a faculty appointment in an accredited medical school or be in a fully accredited child and adolescent psychiatry clinical research or training program.

Restrictions: At the time of application, candidates may not have more than two years experience following graduation from residency/fellowship training. Candidates must not have any previous significant, individual research funding in the field of child and adolescent mental health. These include the following: NIMH/NIH Funding (Small Grants, T Award or R-01) or similar foundation or industry research funding.

Geographic Focus: All States
Date(s) Application is Due: May 7
Amount of Grant: 15,000 USD
Contact: Alyssa Sommer; (202) 966-7300, ext. 157; asommer@aacap.org
Internet: http://www.aacap.org/cs/root/research_and_training_awards/aacap_pilot_research_award_for_junior_faculty_and_child_psychiatry_fellows_supported_by_lilly_usa_llc
Sponsor: American Academy of Child and Adolescent Psychiatry
3615 Wisconsin Avenue, NW
Washington, DC 20016-3007

AACAP Quest for the Test Bipolar Disorder Pilot Research Award 73

The AACAP Quest for the Test Bipolar Disorder Pilot Research Award, supported by The Ryan Licht Sang Bipolar Foundation, offers $15,000 for child and adolescent psychiatry residents and junior faculty who have an interest in beginning a career in child and adolescent mental health research. Proposed research projects should address questions concerning the pathology or diagnosis of children and adolescents with bipolar disorder. The award also includes the cost of attending the AACAP Annual Meeting for five days.

Requirements: The recipient is required to submit a poster presentation on his or her research for the AACAP's Annual Meeting. Candidates must be board eligible, certified in child and adolescent psychiatry, or enrolled in a child psychiatry residency or fellowship program. Candidates must also have a faculty appointment in an accredited medical school or be in a fully accredited child and adolescent psychiatry clinical research or training program.

Restrictions: At the time of application, candidates may not have more than two years experience following graduation from residency/fellowship training. Candidates must not have any previous significant, individual research funding in the field of child and adolescent mental health. These include the following: NIH Funding (K or R level grants) or similar foundation or industry research funding.

Geographic Focus: All States
Date(s) Application is Due: May 7
Amount of Grant: 15,000 USD
Contact: Alyssa Sommer; (202) 966-7300, ext. 157; asommer@aacap.org
Internet: http://www.aacap.org/cs/root/research_and_training_awards/aacap_quest_for_the_test_bipolar_disorder_pilot_research_award
Sponsor: American Academy of Child and Adolescent Psychiatry
3615 Wisconsin Avenue, NW
Washington, DC 20016-3007

AACAP Rieger Psychodynamic Psychotherapy Award 74

The AACAP Rieger Psychodynamic Psychotherapy Award recognizes the best published or unpublished paper, written by an AACAP member that addresses the use of psychodynamic psychotherapy in clinical practice and fosters development, teaching, and practice of psychodynamic psychotherapy within child and adolescent psychiatry. Papers that express a novel hypothesis, raise questions about existing theory, or integrate new neuroscience and developmental psychotherapy research with psychodynamic principles may be nominated. Unpublished, new papers and papers published within the last three years may be submitted by their authors. Published papers may be nominated by any member of the AACAP. Authors may be senior or junior faculty members or residents. Delivery of the winning paper will be made at the AACAP Annual Meeting Honors Presentation.

Geographic Focus: All States
Date(s) Application is Due: Apr 30
Amount of Grant: 4,500 USD
Contact: Tim Dugan, Co-Chair; (202) 966-7300; fax (202) 966-2891; tdugan@aacap.org or clinical@aacap.org
Internet: http://www.aacap.org/cs/awards/riegerpsychotherapy
Sponsor: American Academy of Child and Adolescent Psychiatry
3615 Wisconsin Avenue, NW
Washington, DC 20016-3007

AACAP Rieger Service Program Award for Excellence 75

The AACAP Rieger Service Program Award for Excellence recognizes innovative programs that address prevention, diagnosis, or treatment of mental illnesses in children and adolescents, and serve as model programs to the community. This award of $4,500 is shared among the awardee and his or her service program. The award winner will be recognized at a Distinguished Awards Luncheon and make an Honors Presentation about his or her work during the AACAP Annual Meeting.

Requirements: Nomination letters must be accompanied by a CV and any support materials for the individual or organization nominated.

Geographic Focus: All States
Date(s) Application is Due: Apr 30
Amount of Grant: 4,500 USD
Contact: Kaye McGinty, Co-Chair; (202) 966-7300; fax (202) 966-2891; kmcginty@aacap.org or clinical@aacap.org
Mark Chenven, Co-Chair; (202) 966-7300; fax (202) 966-2891; mchenven@aacap.org or clinical@aacap.org
Internet: http://www.aacap.org/cs/distinguishedmembers
Sponsor: American Academy of Child and Adolescent Psychiatry
3615 Wisconsin Avenue, NW
Washington, DC 20016-3007

AACAP Robert Cancro Academic Leadership Award 76

The AACAP Cancro Academic Leadership Award recognizes a currently serving General Psychiatry Training Director, Medical School Dean, CEO of a Training Institution, Chair of a Department of Pediatrics or Chair of a Department of Psychiatry for his or her contributions to the promotion of child and adolescent psychiatry. Named in honor of Robert Cancro, M.D., Chairman at New York University, this award offers a $2,000 honorarium to the awardee. The award is presented in even-numbered years. Nominations for the award may be made by Child and Adolescent Training Directors or Division Directors and must include a CV for the individual nominated. The recipient of this award will receive a plaque and be recognized at the AACAP Annual Meeting in San Francisco, CA. The award recipient will be honored at the Distinguished Awards Luncheon, Residency Program Directors Luncheon, and will provide an Honors Presentation on his or her work during the Annual Meeting.

Requirements: Nominations must include a CV for the individual nominated

Geographic Focus: All States
Date(s) Application is Due: Apr 30
Amount of Grant: 2,000 USD
Contact: Jeffrey Hunt, Co-Chair; (202) 966-7300; fax (202) 966-2891; jhunt@aacap.org or training@aacap.org
Howard Liu, Co-Chair; (202) 966-7300; fax (202) 966-2891; hliu@aacap.org or clinical@aacap.org
Internet: http://www.aacap.org/cs/distinguishedmembers
Sponsor: American Academy of Child and Adolescent Psychiatry
3615 Wisconsin Avenue, NW
Washington, DC 20016-3007

AACAP Robinson-Cunningham Award 77

The Award is given for the best manuscript written by a child and adolescent psychiatrist during residency training. The paper must involve children, adolescents, or their families and be published in a professional, peer-reviewed journal within 3-5 years of graduation from a residency training program. The recipient will receive a $100 honorarium and a plaque at the Young Leaders Awards Ceremony during the AACAP Annual Meeting.

Geographic Focus: All States
Date(s) Application is Due: May 7
Amount of Grant: 100 USD
Contact: Gabe Robbins, (202) 966-7300, ext. 117; fax (202) 966-2891; grobbins@aacap.org
Internet: http://www.aacap.org/cs/root/research_and_training_awards/the_robinson cunningham_award
Sponsor: American Academy of Child and Adolescent Psychiatry
3615 Wisconsin Avenue, NW
Washington, DC 20016-3007

AACAP Sidney Berman Award for the School-Based Study and Intervention for Learning Disorders and Mental Ilness 78

The AACAP Sidney Berman Award for the School-Based Study and Treatment of Learning Disorders and Mental Illness recognizes an individual or program that has shown outstanding achievement in the school-based study or delivery of intervention for learning disorders and mental illness. A cash prize of $4,500 will be awarded. The award winner will be recognized at a Distinguished Awards Luncheon and make an Honors Presentation about his or her work during the AACAP Annual Meeting.

Requirements: Nomination letters must be accompanied by a CV for the individual nominated or program information.

Geographic Focus: All States
Date(s) Application is Due: Apr 30
Amount of Grant: 4,500 USD
Contact: Sheryl Kataoka, Co-Chair; (202) 966-7300; fax (202) 966-2891; skataoka@aacap.org or clinical@aacap.org
Internet: http://www.aacap.org/cs/distinguishedmembers
Sponsor: American Academy of Child and Adolescent Psychiatry
3615 Wisconsin Avenue, NW
Washington, DC 20016-3007

AACAP Simon Wile Leadership in Consultation Award 79

The Simon Wile Leadership in Consultation Award recognizes innovative programs that address prevention, diagnosis, or treatment of mental illnesses in children and adolescents, and serve as model programs to the community. This award of $4,500 is shared among the awardee and his or her service program. Nomination letters must be accompanied by a CV and any support materials for the individual or organization nominated. The recipient will be recognized at a Distinguished Awards Luncheon and make an Honors Presentation about his or her work during the AACAP Annual Meeting.

Requirements: Practicing child and adolescent psychiatrists are eligible for nomination.

Geographic Focus: All States
Date(s) Application is Due: Apr 30
Amount of Grant: 500 USD
Contact: Kaye McGinty, Co-Chair; (202) 966-7300; fax (202) 966-2891; kmcginty@aacap.org or clinical@aacap.org
Internet: http://www.aacap.org/cs/distinguishedmembers
Sponsor: American Academy of Child and Adolescent Psychiatry
3615 Wisconsin Avenue, NW
Washington, DC 20016-3007

AACAP Summer Medical Student Fellowships 80
AACAP Summer Medical Student Fellowships offer a chance for medical students to explore a career in child and adolescent psychiatry, gain valuable work experience, and meet leaders in the child and adolescent psychiatry field. The fellowship opportunity provides up to $3,500 for 12 weeks of clinical or research training under a child and adolescent psychiatrist mentor. Fellowship stipends will be distributed in two installments. Upon receipt of the fellowship, $2,500 will be sent to the student's mentor to be disbursed at the onset of the summer fellowship. Upon meeting all program requirements, the last installment will be sent directly to the recipient and will be pro-rated according to the total amount of time spent completing the fellowship. Fellowships lasting the full 12 weeks will receive the maximum amount. Participants are required to attend the AACAP Annual Meeting. The application deadline for this fellowship is February 15.
Requirements: Applicants must be students enrolled in accredited U.S. medical schools.
Geographic Focus: All States
Date(s) Application is Due: Feb 15
Amount of Grant: 3,500 USD
Contact: Ashley Partner, (202) 966-7300, ext. 117; fax (202) 966-2891; apartner@aacap.org or training@aacap.org
Internet: http://www.aacap.org/cs/students/summerfellowship
Sponsor: American Academy of Child and Adolescent Psychiatry
3615 Wisconsin Avenue, NW
Washington, DC 20016-3007

AACAP Systems of Care Special Program Scholarships 81
The American Academy of Child and Adolescent Psychiatry (AACAP) is pleased to announce the opportunity for child and adolescent psychiatry residents to apply for the 2012 Systems of Care Special Program Scholarship. The Special Program is a day-long event that explores the psychological, psychosocial and psycho-educational needs of children and adolescents as manifest in their experiences in the schools. Policy and best practice considerations will be reviewed as regards health promotion, screening, assessment and treatment intervention for students and student groups, with the goal of advancing the knowledge, skills and attitudes of child psychiatrists and other mental health providers so that they can better function as service providers and systems consultants for youth and their school systems. The scholarship includes $750 for travel expenses to the AACAP's 59th Annual Meeting. This includes airfare, hotel, and meals, maximum $75 per day.
Geographic Focus: All States
Date(s) Application is Due: Jul 12
Amount of Grant: 750 USD
Contact: Adriano Boccanelli, Clinical Practice Assistant; (202) 966-7300, ext. 133 or (202) 587-9671; fax (202) 966-2891; aboccanelli@aacap.org or clinical@aacap.org
Internet: http://www.aacap.org/cs/root/medical_students_and_residents/residents/2012_systems_of_care_special_program_scholarship
Sponsor: American Academy of Child and Adolescent Psychiatry
3615 Wisconsin Avenue, NW
Washington, DC 20016-3007

AACC International Alsberg-French-Schoch Memorial Lectureship Awards 82
The Alsberg-French-Schoch Memorial Lectureship Award, sponsored by the Corn Refiners Association, is the organization's highest award given every two to four years in recognition of scientists who have made significant and superior contributions to fundamental starch science. The award is given every two to four years and includes a $2,000 honorarium and a plaque, as well as the opportunity to present a lecture on some phase of starch science before an annual meeting of the Association. To be nominated for an award, contact information for both the person suggesting the nominee and the individual being nominated, as well as a summary of the nominee's career, accomplishments, and contributions should be forwarded directly to AACC International headquarters. Nominations should address the activities most relevant to the award, with a clear statement of the impact the nominee has had in his/her specific area of expertise. Nomination forms are available at the AACC website.
Requirements: Nominee must be an AACC International member. All members are strongly encouraged to submit nominations for these awards at any time. All nominations are considered for three years.
Geographic Focus: All States, All Countries
Date(s) Application is Due: Jan 1
Amount of Grant: 2,000 USD
Contact: Linda Schmitt, AACC International Award Chairperson; (651) 994-3828 or (651) 454-7250; fax (651) 454-0766; lschmitt@scisoc.org or aacc@scisoc.org
Internet: http://www.aaccnet.org/membership/awards/Pages/AlsbergFrenchSchochMemorialLectureship.aspx
Sponsor: American Association of Cereal Chemists International
3340 Pilot Knob Road
St. Paul, MN 55121

AACC International Best Student Paper and Best Student Poster Award 83
The Protein Division selects the best protein chemistry papers presented by a student at the AACC International Annual Meeting each year. The Protein Division Best Student Paper Awards consists of an engraved plaque, an honorarium, and recognition at the annual meeting. The Best Student Poster Award consists of an engraved plaque, an honorarium, and recognition at the annual meeting.
Geographic Focus: All States, All Countries
Date(s) Application is Due: Jan 1
Contact: Bin X. Fu; (800) 853-6705 or (204) 983-2770; fax (204) 983-2751
Internet: http://www.aaccnet.org/membership/divisions/Pages/Protein.aspx
Sponsor: American Association of Cereal Chemists International
3340 Pilot Knob Road
St. Paul, MN 55121

AACC International Bruce Wasserman Young Investigator Award 84
The purpose of the Bruce Wasserman Young Investigator Award is to recognize young scientists who have made outstanding contributions to the field of cereal biotechnology. The research recognized by this award should be relevant to the broad aims of the AACC International. The work can be either basic (e.g., improved understanding of structure-function relationships) or applied (e.g., new approaches for cereal production, breeding or utilization). For the purposes of this award, 'Cereal Biotechnology' is broadly defined, and encompasses any significant body of research using plants, microbes, genes, proteins or other biomolecules. Contributions in the disciplines of genetics, molecular biology, biochemistry, microbiology and fermentation engineering are all included. The research will be evaluated by its impact as measured as either enhanced knowledge of cereal functionality or improved cereal production or utilization. Achievements will be considered in relation to the age and experience of the nominee. AACC International membership is not required.
Requirements: Nominees must be no older than 40 by July 1, but nominations of younger scientists are particularly encouraged.
Geographic Focus: All States, All Countries
Date(s) Application is Due: Jan 15
Amount of Grant: 1,000 USD
Contact: Tandace A. Bell, Committee Chair; (816) 891-0459 or (651) 454-7250; fax (651) 454-0766; tandace.a.bell@usda.gov
Internet: http://www.aaccnet.org/membership/divisions/Pages/Biotechnology.aspx
Sponsor: American Association of Cereal Chemists International
3340 Pilot Knob Road
St. Paul, MN 55121

AACC International Carl Wilhelm Brabender Award 85
The Carl Wilhelm Brabender Award was established to honor the distinguished contributions to the application of rheology in milling and baking. Every three years, the Award provides a plaque and a travel grant to enable the recipient to visit scientists and laboratories in other parts of the world to increase the Awardee's knowledge and experience in the field. The only qualification is to be professionally active in rheology, specifically those engaged in the milling and baking industry. The Awardee shall present an address at the AACC International Annual Meeting or at one of the three annual meetings of the AG at which the Award is given. The nomination form is available at the AACC website.
Requirements: All members are strongly encouraged to submit nominations for these Awards at any time. All Awards approved by the Board of Directors will be presented at the annual meeting. All nominations for AACCI Awards are considered for three years if the nominee meets the Award criteria.
Geographic Focus: All States, All Countries
Date(s) Application is Due: Jan 1
Contact: Linda Schmitt, AACC International Award Chairperson; (651) 994-3828 or (651) 454-7250; fax (651) 454-0766; lschmitt@scisoc.org or aacc@scisoc.org
Internet: http://www.aaccnet.org/membership/awards/Pages/Brabender.aspx
Sponsor: American Association of Cereal Chemists International
3340 Pilot Knob Road
St. Paul, MN 55121

AACC International Cecil F. Pinney Travel Awards 86
The Cecil F. Pinney Travel Awards are presented to individuals establishing a career in grain science and technology. The candidate can apply for the travel award or can be nominated. The application process consists of an application form, nomination letter (or letter of context if self nominating), and the candidate's resume. Achievements and participation in AACCI activities will be considered in relation to the age and experience of the candidate. Up to two $750 awards are available. In addition, a one-year membership to the division of their choice will be awarded to the recipient(s) and the registration fee to the annual meeting will be waived, as well as admission to a division function at the annual meeting. The application is available at the AACCI website, and must also include a cover sheet, resume, and supporting data and descriptions of previous AACCI involvement, along with activities in grain science and technology.
Requirements: Candidates must be AACCI members interested in some aspect of grain science and technology. This award is for new professionals who have been in their career less than eight years as of March 1 of the year of application. Preference will be given to nominees or applicants with a demonstrated need for the support to attend the AACCI Annual Meeting. Previous Cecil F. Pinney Travel Award recipients are eligible to apply for future awards, but they are limited to receiving this award three times within the eight-year time frame.
Geographic Focus: All States, All Countries
Date(s) Application is Due: Mar 1
Amount of Grant: 750 USD

Contact: Linda Schmitt, Pinney Award Coordinatorr; (651) 994-3828 or (651) 454-7250; fax (651) 454-0766; lschmitt@scisoc.org or aacc@scisoc.org
Internet: http://www.aaccnet.org/membership/foundation/awards/Pages/PinneyTravelAward.aspx
Sponsor: American Association of Cereal Chemists International
3340 Pilot Knob Road
St. Paul, MN 55121

AACC International Edith A. Christensen Award 87

In recognition of the central role that analytical methodology and approved methods have played in the history of AACC International and their future importance, the Edith A. Christensen Award for outstanding contributions in analytical methodology was created to recognize scientific contributions in analytical methodology, and service to the society through approved methods leadership. The award consists of a suitably inscribed plaque and a $1,500 honorarium.
Requirements: The recipient must have contributed to the development of new analytical technologies, and/or the application of new analytical technologies to cereal grains and products, and/or the application of existing analytical technologies to solving detection and measurement problems in the field of cereal science. Further, the recipient must have demonstrated involvement and leadership in approved methods activity within AACC International. Nomination must be made by letter of nomination detailing the nominee's accomplishments in analytical methodology and the approved methods process, and two additional letters of support for the nominee. Awardees will have the option to present a lecture during an annual meeting of the Association when the Award is presented.
Geographic Focus: All States, All Countries
Date(s) Application is Due: Jan 1
Amount of Grant: 1,500 USD
Contact: Linda Schmitt, AACC International Award Chairperson; (651) 994-3828 or (651) 454-7250; fax (651) 454-0766; lschmitt@scisoc.org or aacc@scisoc.org
Internet: http://www.aaccnet.org/membership/awards/Pages/EdithAChristensen.aspx
Sponsor: American Association of Cereal Chemists International
3340 Pilot Knob Road
St. Paul, MN 55121

AACC International Engineering and Processing Division's Stanley Watson Award 88

The Stanley Watson Award recognizes outstanding AACC International members who, through the application of engineering principles, have made significant contributions in the area of cereal/grain processing. The Award is presented every two years, and includes a $1,000 honorarium, engraved plaque, and a lecture presentation at the division luncheon. Nominations are accepted through June 1.
Requirements: Nominations must include: name of nominee; position; complete address and contact information; area of research; and a nominating statement detailing the nominee's significant contributions in the area of cereal/grain processing via application of engineering principles.
Geographic Focus: All States, All Countries
Date(s) Application is Due: Jun 1
Amount of Grant: 1,000 USD
Contact: Heather Maskus, AACC Division Chair; (204) 984-3139 or (204) 983-5344; fax (204) 983-2642; hmaskus@cigi.ca
Internet: http://www.aaccnet.org/divisions/divisionsdetail.cfm?CODE=ENG
Sponsor: American Association of Cereal Chemists International
3340 Pilot Knob Road
St. Paul, MN 55121

AACC International Excellence in Teaching Award 89

The Excellence in Teaching Award is to be presented to an AACC International member and current teacher who has made significant contributions through teaching in the broad field of Cereal Science and Technology. The Award consists of an honorarium of $1,500 and a suitably inscribed plaque. To be nominated for an Award, contact information for both the person suggesting the nominee and the individual being nominated, as well as a summary of the nominee's career, accomplishments, and contributions should be forwarded directly to AACC International headquarters. Nominations should address the activities most relevant to the Award, with a clear statement of the impact the nominee has had in his/her specific area of expertise. Nomination forms are available at the AACC website.
Requirements: Nominee must be an AACC International member. All members are strongly encouraged to submit nominations for these awards at any time.
Geographic Focus: All States, All Countries
Date(s) Application is Due: Jan 1
Amount of Grant: 1,500 USD
Contact: Linda Schmitt, AACC International Award Chairperson; (651) 994-3828 or (651) 454-7250; fax (651) 454-0766; lschmitt@scisoc.org or aacc@scisoc.org
Internet: http://www.aaccnet.org/membership/awards/Pages/ExcellenceinTeaching.aspx
Sponsor: American Association of Cereal Chemists International
3340 Pilot Knob Road
St. Paul, MN 55121

AACC International George W. Scott Blair Lecture Award 90

The Rheology Division of AACC International sponsors several awards to scientists and students working in the area of rheology, in order to recognize their accomplishments, subsidize their efforts and publicize the field of rheology in general. The purpose of the George W. Scott Blair Lecture Award is to recognize scientists who have demonstrated outstanding ability in research in the area of rheology and texture as related to cereal based products. The Award consists of an engraved plaque, $500 honorarium and travel expenses (if necessary) to attend the AACC International Annual Meeting. This Award is presented in even years.
Geographic Focus: All States, All Countries
Amount of Grant: 500 USD
Contact: Hulya Dogan, Chair; (785) 532-2628; fax (785) 532-7010; dogan@ksu.edu
Internet: http://www.aaccnet.org/membership/divisions/Pages/Rheology.aspx
Sponsor: American Association of Cereal Chemists International
3340 Pilot Knob Road
St. Paul, MN 55121

AACC International Graduate Fellowships 91

Graduate fellowships of between $2,000 and $3,000 will be available, supported by endowment funds of the AACCI Foundation, and from annual contributions by companies and AACCI divisions. The purpose of the fellowship program is to encourage graduate research in grain-based food science and technology. Graduate awards are paid directly to the winners in September after AACCI receives verification of registration for the fall semester or quarter from the appropriate department head. If the student is already receiving an assistantship or fellowship, they will still receive the amount of the AAEE Fellowship. The application and additional guidelines are available at the AAEE website.
Requirements: Applicants must be a current AACCI Student Division member. The candidate must be enrolled in graduate studies by the time the Fellowship becomes effective, or be a current graduate student pursuing a course of study leading to an M.S. or a Ph.D. degree. The educational institution in which the recipient is enrolled must be conducting fundamental investigations for the advancement of cereal science and technology, including oilseeds. The recipient must be enrolled in an academic schedule that meets the minimum requirements of the university involved, for full-time graduate study.
Restrictions: Research in such disciplines as genetics, horticulture, nutrition, microbiology, biochemistry, engineering, or chemistry is only eligible if it is directly related to the student's primary research program dealing with a grain-based food science and technology related areas.
Geographic Focus: All States, All Countries
Date(s) Application is Due: Mar 5
Amount of Grant: 2,000-3,000 USD
Contact: Linda Schmitt, AACC International Award Chairperson; (651) 994-3828 or (651) 454-7250; fax (651) 454-0766; lschmitt@scisoc.org or aacc@scisoc.org
Internet: http://www.aaccnet.org/membership/foundation/awards/Pages/GraduateProgram.aspx
Sponsor: American Association of Cereal Chemists International
3340 Pilot Knob Road
St. Paul, MN 55121

AACC International Milling & Baking Division's John C. Halverson Memorial Lectureship 92

The John C. Halverson Memorial Lectureship is an award presented by the Milling and Baking Division to honor an individual for their contributions to the technology of milling and baking. The award consists of an honorarium and suitably inscribed certificate. The recipient is given the opportunity to present a technical paper detailing their findings and accomplishments.
Geographic Focus: All States, All Countries
Contact: Barbara B. Heidolph, Chair; (314) 983-7500; fax (314) 983-7636
Internet: http://www.aaccnet.org/membership/divisions/Pages/MillingBaking.aspx
Sponsor: American Association of Cereal Chemists International
3340 Pilot Knob Road
St. Paul, MN 55121

AACC International Milling & Baking Division's M. Rella Dwyer Graduate Fellowship 93

The M. Rella Dwyer Graduate Fellowship is offered annually to a deserving student preparing a career in milling and baking. Applications for AACCI Scholarships and Fellowships are solicited from appropriate University department heads and other sources by AACCI. Candidates may review the AACC website for additional information.
Geographic Focus: All States, All Countries
Amount of Grant: 3,000 USD
Contact: Barbara B. Heidolph, Chair; (314) 983-7500; fax (314) 983-7636
Internet: http://www.aaccnet.org/membership/divisions/Pages/MillingBaking.aspx
Sponsor: American Association of Cereal Chemists International
3340 Pilot Knob Road
St. Paul, MN 55121

AACC International Phil Williams Applied Research Award 94

The AACC International Phil Williams Applied Research Award is presented for a significant body of distinguished contributions to the application of science in the cereals area. Of primary interest is the development of novel machinery, processes, products, patents, tests, methodologies, or procedures taking advantage of the wealth of accumulated knowledge from basic research as available in the literature together with basic/applied research performed by the applicant in developing the product/patent. The methodologies, applications, products and processes must have a clear and sustained record of adoption and/or commercialization and must have contributed positively to the development of science and knowledge of relevance to the cereal/food industry. The Award will consist of a $2,000 honorarium and a suitably inscribed medal in the form of a plaque that will be given to an individual scientist or a team of scientists. Recipients of the Award are also accorded the status of AACC International Fellow for their contributions

leading to this Award. Awardees will have the opportunity to present a lecture during an annual meeting of the Association.
Requirements: Nominee must be an AACC International member. All members are strongly encouraged to submit nominations for these awards at any time.
Geographic Focus: All States, All Countries
Date(s) Application is Due: Jan 1
Amount of Grant: 2,000 USD
Contact: Linda Schmitt, AACC International Award Chairperson; (651) 994-3828 or (651) 454-7250; fax (651) 454-0766; lschmitt@scisoc.org or aacc@scisoc.org
Internet: http://www.aaccnet.org/membership/awards/Pages/PhilWilliamsApplied Research.aspx
Sponsor: American Association of Cereal Chemists International
3340 Pilot Knob Road
St. Paul, MN 55121

AACC International Student Division's Best Research Paper Awards 95
The objectives of the Student Division's Best Research Paper Award competition are to challenge students to demonstrate superior presentation skills, encourage students to present their research work at AACC International, highlight the best research conducted and presented by students, and offer an opportunity for students to interact with the AACC International community at an early stage in their career. The competition will be judged in two stages. The Best Student Research Paper Competition jury will review the nomination forms, abstracts, and initial presentations. Based on the judge's scores, six students will be chosen to participate in the competition. Winners of the competition will be announced at the annual meeting where the competition took place. Each of the six finalists will receive a $650 travel grant for attending the the annual meeting. First place awards include $1,250 cash award, $650 travel grant, and the opportunity to publish their paper in Cereal Chemistry, Cereal Foods World, or the Cereal Science Knowledge Database. Second place awards include $750 cash award, $650 travel grant, and the opportunity to publish their paper in Cereal Chemistry, Cereal Foods World, or the Cereal Science Knowledge Database. Third place award includes $500 cash award, $650 travel grant, and the opportunity to publish their paper in Cereal Chemistry, Cereal Foods World, or the Cereal Science Knowledge Database. Academic departments may submit one nominee each for the competition. More than one student may be nominated from a university as long as the nomination comes from different departments. It is up to the department to determine which student will be chosen for the competition. Submitters need not be AACCI members. Additional information and guidelines are available at the AACC website.
Requirements: Applicants must be a full-time student at the the time of submission date. Applicants must provide proof of student status by way of department chair signature on the nomination form. The student presenting the paper must be an author of the presentation, must have made a major contribution to the research, must present it, and must take primary responsibility for its content.
Geographic Focus: All States, All Countries
Date(s) Application is Due: Feb 28
Amount of Grant: 650-1,250 USD
Contact: Cindy Scheller; (651) 454-7250; fax (651) 454-0766; cscheller@scisoc.org
Internet: http://www.aaccnet.org/membership/divisions/Pages/AACCIBestStudent ResearchPaperCompetition.aspx
Sponsor: American Association of Cereal Chemists International
3340 Pilot Knob Road
St. Paul, MN 55121

AACC International Student Division's Product Development Awards 96
The AACC International Student Division Product Development Competition (PDC) is organized by students and sponsored by food companies who are members of AACC International. The PDC provides an opportunity for students to apply the knowledge and skills acquired in school to create new food products professionally and intelligently in the real business world. The competition is designed to encourage innovative thinking of students in the cereal science area, expose students to a professional experience, and promote networking with food industry representatives. Students can work individually or in a team to develop a novel food product containing at least one major cereal/legume ingredient. A total of three prizes will be awarded: first place ($2,550); second place ($1,700); and third place ($850). All finalists receive $300 per team for travel expenses. The entry form and additional guidelines are available at the AACC website.
Requirements: All participating team members must be student members of AACC International. Each participant must be enrolled at a four-year accredited college or a graduate student during the spring semester.
Geographic Focus: All States, All Countries
Date(s) Application is Due: Jun 8
Amount of Grant: 850-2,550 USD
Contact: Tom Jondiko; (651) 454-7250; fax (651) 454-0766; jondiko@yahoo.com
Internet: http://www.aaccnet.org/membership/divisions/Pages/AACCIProduct DevelopmentCompetition.aspx
Sponsor: American Association of Cereal Chemists International
3340 Pilot Knob Road
St. Paul, MN 55121

AACC International Student Travel Grant 97
AACC International's Protein Division awards a student travel grant of $250 to a student attending the annual meeting. Students presenting a poster or paper in the area of protein chemistry are eligible to apply.
Geographic Focus: All States, All Countries
Date(s) Application is Due: Jan 1
Amount of Grant: 250 USD
Contact: Bin X. Fu; (800) 853-6705 or (204) 983-2770; fax (204) 983-2751
Internet: http://www.aaccnet.org/membership/divisions/Pages/Protein.aspx
Sponsor: American Association of Cereal Chemists International
3340 Pilot Knob Road
St. Paul, MN 55121

AACC International Thomas Burr Osborne Medal 98
The Thomas Burr Osborne Medal was established in 1926, and is awarded to an individual whose research in the field of cereal chemistry has contributed significantly to the progress of the science. The award consists of an honorarium of $2,000 and a suitably inscribed medal in the form of a plaque. The recipient shall present an address at the AACC International Annual Meeting at which the medal is presented. Recipients of the Medal are also accorded the status of AACC International Fellow.
Requirements: AACCI Membership is required.
Geographic Focus: All States, All Countries
Date(s) Application is Due: Jan 1
Amount of Grant: 2,000 USD
Contact: Linda Schmitt, AACC International Award Chairperson; (651) 994-3828 or (651) 454-7250; fax (651) 454-0766; lschmitt@scisoc.org or aacc@scisoc.org
Internet: http://www.aaccnet.org/membership/awards/Pages/ThomasBurrOsborne Medal.aspx
Sponsor: American Association of Cereal Chemists International
3340 Pilot Knob Road
St. Paul, MN 55121

AACC International Walter Bushuk Graduate Research Award 99
The AACC International Walter Bushuk Graduate Research Award in Cereal Protein Chemistry for Ph.D. students is presented to an individual for outstanding contributions in basic and/or applied research in cereal protein chemistry. This Award recognizes research in relevant to the broad aims of AACC International. The research can be either fundamental/basic (such as improved understanding of structure-function relationships) or applied (such as development of new products, processes, techniques). The Awardee will receive an honorarium of $750 and a plaque, and will be required to present a lecture at the AACC International annual meeting of the year in which the award is given. Nomination forms are available at the AACC website, and are due January 1.
Geographic Focus: All States, All Countries
Date(s) Application is Due: Jan 1
Amount of Grant: 750 USD
Contact: Bin X. Fu; (800) 853-6705 or (204) 983-2770; fax (204) 983-2751
Internet: http://www.aaccnet.org/membership/divisions/Pages/Protein.aspx
Sponsor: American Association of Cereal Chemists International
3340 Pilot Knob Road
St. Paul, MN 55121

AACC International Young Scientist Research Award 100
The Award is presented to an individual for outstanding contributions in basic and applied research to cereal science with the expectation that contributions will continue. This Award recognizes research relevant to the broad aims and interests of AACC International. The research can be either basic (such as improved understanding of structure-function relationships) or applied (such as development of new products, processes, or techniques) and nominees may be from government, academia or industry. Evidence of accomplishments can include awards, publications, invitations to lecture, patents and commercialized foods, processes and/or testing procedures. The research will be evaluated based on impact on advancing cereal science knowledge or improved production or utilization. Achievements will be considered in relation to the age and experience of the nominee. The recipient will receive a $1,000 honorarium, a plaque, and will be encouraged to present a lecture at the AACC International meeting of the year in which the award is given.
Requirements: Nominees must be no older than 40 years by June 1 of the year the award is sought, but nominations of younger scientists are particularly encouraged. AACC International membership is required.
Geographic Focus: All States, All Countries
Date(s) Application is Due: Jan 1
Contact: Linda Schmitt, AACC International Award Chairperson; (651) 994-3828 or (651) 454-7250; fax (651) 454-0766; lschmitt@scisoc.org or aacc@scisoc.org
Internet: http://www.aaccnet.org/membership/awards/Pages/YoungScientistResearch.aspx
Sponsor: American Association of Cereal Chemists International
3340 Pilot Knob Road
St. Paul, MN 55121

AACC Texture Technologies Quality Research Award 101
Established in 2012, the Texture Technologies Quality Research Award is awarded annually to two authors who write and present high quality research which broadly involves instrumental texture analysis. The objective of the award is to increase awareness among professors and students of the importance of publishing impacting research in sufficient high quality and detail that fellow researchers can replicate and build upon the work. One award will be presented to an author whose paper was published in Cereal Chemistry or Cereal Foods World the previous year. The second award will be presented to an oral or poster during the upcoming AACCI Annual Meeting.
Requirements: Award recipients must be either Professional or Student AACCI members. Submitted papers and posters will be judged by the Texture Award Committee.

Geographic Focus: All States, All Countries
Date(s) Application is Due: Jan 1
Contact: Linda Schmitt, AACC International awards Chairperson; (651) 994-3828 or (651) 454-7250; fax (651) 454-0766; lschmitt@scisoc.org or aacc@scisoc.org
Internet: http://www.aaccnet.org/membership/awards/Pages/TextureTechnologiesQualityResearchAward.aspx
Sponsor: American Association of Cereal Chemists International
3340 Pilot Knob Road
St. Paul, MN 55121

AACN-Edwards Lifesciences Nurse-Driven Clinical Practice Outcomes Grants 102
Co-sponsored by Edwards Lifesciences and the American Association of Critical-Care Nurses, each grant supports a nurse experienced in research who is conducting a clearly articulated research study that relates to use of protocol-based care driven by nurses. A commitment is required to produce results that will be available and disseminated in a short timeframe. Proposals submitted by an interdisciplinary team or an experienced researcher mentoring a novice will also be considered. Funds may be used for original research or replication of existing research. Especially desirable are proposals that focus on any of the following at any point in the continuum of acute or critical care: implementation of nurse-driven protocols using technology, including minimally invasive devices, with evaluation of patient outcomes and nurses' role in implementing the protocols; or development and clinical evaluation of nurse-driven protocols. Two awards are available each year. Each proposal can be funded up to of $5,000.
Requirements: Principal Investigators must be nurses holding current AACN membership and must remain AACN members throughout the life of the grant funding.
Restrictions: Funds may not be used to pay salaries or institutional overhead or augment funding from formal grants. Principal investigators who have received funding from AACN are ineligible to receive additional funding from AACN during the lifetime of the original award.
Geographic Focus: All States
Date(s) Application is Due: Oct 1
Amount of Grant: Up to 5,000 USD
Contact: Pamela Shellner, Clinical Practice Specialist; (949) 362-2050, ext. 321 or (800) 394-5995, ext. 321; fax (949) 362-2020; research@aacn.org or info@aacn.org
Internet: http://www.aacn.org/WD/Practice/Content/grant-Edwards.pcms?pid=1@menu=practice
Sponsor: American Association of Critical-Care Nurses
101 Columbia
Aliso Viejo, CA 92656-4109

AACN-Philips Medical Systems Clinical Outcomes Grants 103
Co-sponsored by Philips Medical Systems and the American Association of Critical-Care Nurses, each grant supports experienced nurses in conducting clearly articulated research studies. The research must be relevant to clinical nursing practice in acute or critical care. Proposals submitted by an interdisciplinary team or an experienced researcher mentoring a novice will also be considered. Funds may be used for original research or replication of existing research. Qualified proposals should seek to achieve improved outcomes and/or system efficiencies in care of acute or critically ill individuals of any age. Especially desirable are proposals for hospital-based inquiry that focuses on any of the following: intervention strategies addressing technology integration into patient care; use of computerized medical record systems in assessing patient outcomes and managing care; improving specific patient care outcomes (e.g., clinical, safety, financial); use of simulation in clinical education of nurses; and one or more of the 2006 National Patient Safety Goals issued by the Joint Commission on Accreditation of Healthcare Organizations.
Requirements: Principal Investigators must be nurses holding current AACN membership and must remain AACN members throughout the life of the grant funding.
Restrictions: Funds may not be used to pay salaries or institutional overhead or augment funding from formal grants. Principal investigators who have received funding from AACN are ineligible to receive additional funding from AACN during the lifetime of the original award.
Geographic Focus: All States
Date(s) Application is Due: Oct 1
Amount of Grant: Up to 10,000 USD
Contact: Pamela Shellner, Clinical Practice Specialist; (949) 362-2050, ext. 321 or (800) 394-5995, ext. 321; fax (949) 362-2020; research@aacn.org or info@aacn.org
Internet: http://www.aacn.org/WD/Practice/Content/grant-Philips.pcms?pid=1@menu=practice
Sponsor: American Association of Critical-Care Nurses
101 Columbia
Aliso Viejo, CA 92656-4109

AACN-Sigma Theta Tau Critical Care Grant 104
The grant, cosponsored by AACN and Sigma Theta Tau International, supports critical care nursing research and is awarded annually for a study relevant to critical care nursing practice. The proposed study may be used to meet requirements of an academic degree.
Requirements: Principal investigators must be a member of AACN and/or of Sigma Theta Tau International. The PI must have a minimum Masters Degree.
Restrictions: Funds may not be used to pay salaries or institutional overhead or augment funding from formal grants. Principal investigators who have received funding from AACN are ineligible to receive additional funding from AACN during the lifetime of the original award.
Geographic Focus: All States
Date(s) Application is Due: Oct 1
Amount of Grant: Up to 10,000 USD
Contact: Pamela Shellner, Clinical Practice Specialist; (949) 362-2050, ext. 321 or (800) 394-5995, ext. 321; fax (949) 362-2020; research@aacn.org or info@aacn.org
Internet: http://www.aacn.org/WD/Practice/Content/grant-STT.pcms?pid=1@menu=practice
Sponsor: American Association of Critical-Care Nurses
101 Columbia
Aliso Viejo, CA 92656-4109

AACN Clinical Inquiry Fund Grants 105
The fund provides small awards to qualified individuals carrying out clinical research projects that directly benefit patients and/or families. Interdisciplinary projects are encouraged. Funds may be awarded for new projects, projects in progress, and projects required for an academic degree as long as all other project criteria are met. The funds may be used to cover direct project expenses—i.e., printed materials, small equipment, and supplies including computer software.
Requirements: Principal Investigators must be nurses holding current AACN membership and must remain AACN members throughout the life of the grant funding.
Restrictions: Funds may not be used to pay salaries or institutional overhead or augment funding from formal grants.
Geographic Focus: All States
Date(s) Application is Due: Oct 1
Amount of Grant: 500 USD
Contact: Pamela Shellner, Clinical Practice Specialist; (949) 362-2050, ext. 321 or (800) 394-5995, ext. 321; fax (949) 362-2020; research@aacn.org or info@aacn.org
Internet: http://www.aacn.org/WD/Practice/Docs/Grant_Table09-10.pdf
Sponsor: American Association of Critical-Care Nurses
101 Columbia
Aliso Viejo, CA 92656-4109

AACN Clinical Practice Grants 106
This $6,000 grant supports research focused on one or more AACN Research priorities. Research conducted in fulfillment of an academic degree is acceptable. Funds may be awarded for new projects, projects in progress and projects required for an academic degree as long as all other project criteria are met. Eligible projects may include research utilization studies, CQI projects or outcomes evaluation projects. Interdisciplinary and collaborative projects are encouraged and may involve interdisciplinary teams, multiple nursing units, home health, subacute and transitional care, other institutions or community agencies. Funds may be used to cover direct project expenses, such as printed materials, small equipment or supplies including computer software. Six awards are available each year. One award is funded up to $6,000 each year.
Requirements: Principal Investigators must be nurses holding current AACN membership and must remain AACN members throughout the life of the grant funding.
Restrictions: Funds may not be used to pay salaries or institutional overhead or augment funding from formal grants. Principal investigators who have received funding from AACN are ineligible to receive additional funding from AACN during the lifetime of the original award.
Geographic Focus: All States
Date(s) Application is Due: Oct 1
Amount of Grant: Up to 6,000 USD
Contact: Pamela Shellner, Clinical Practice Specialist; (949) 362-2050, ext. 321 or (800) 394-5995, ext. 321; fax (949) 362-2020; research@aacn.org or info@aacn.org
Internet: http://www.aacn.org/WD/Practice/Content/grant-CP.pcms?pid=1@menu=practice
Sponsor: American Association of Critical-Care Nurses
101 Columbia
Aliso Viejo, CA 92656-4109

AACN Critical Care Grants 107
AACN's research agenda promotes the creation of cultures of inquiry, broad sharing and evidence-based practice. This $15,000 grant supports research focused on one or more of AACN Research priorities. Research priority areas include: effective and appropriate use of technology to achieve optimal patient assessment, management, and/or outcomes; creating a healing, humane environment; processes and systems that foster the optimal contribution of critical care nurses; effective approaches to symptom management; and prevention and management of complication.
Requirements: Principal Investigators must be nurses holding current AACN membership and must remain AACN members throughout the life of the grant funding.
Restrictions: The proposed research may not be used to meet the requirements of an academic degree. Funds may not be used to pay salaries or institutional overhead or augment funding from formal grants. Principal investigators who have received funding from AACN are ineligible to receive additional funding from AACN during the lifetime of the original award.
Geographic Focus: All States
Date(s) Application is Due: Oct 1
Contact: Pamela Shellner, Clinical Practice Specialist; (949) 362-2050, ext. 321 or (800) 394-5995, ext. 321; fax (949) 362-2020; research@aacn.org or info@aacn.org
Internet: http://www.aacn.org/WD/Practice/Content/grant-CC.pcms?pid=1@menu=practice
Sponsor: American Association of Critical-Care Nurses
101 Columbia
Aliso Viejo, CA 92656-4109

AACN End of Life/Palliative Care Small Projects Grants 108

The end of life/palliative care grant provides an award to a qualified individual(s) carrying out a project focusing on end of life and/or palliative care outcomes in acute and/or critical care settings. Eligible projects may focus on any age group (neonatal to elderly) in the acute or critical care arena. They may include patient education programs, staff development programs, competency-based educational programs, CQI/outcomes evaluation projects, or small clinical research studies. Funds may be awarded for new projects, projects in progress and projects required for an academic degree as long as all other project criteria are met. Collaborative projects are encouraged and may involve interdisciplinary teams, multiple nursing units, home health, subacute and transitional care, other institutions or community agencies. Funds may be used to cover direct project expenses, such as printed materials, small equipment, supplies including computer software. Two awards are available each year. Each proposal can be funded for up to $500.
Requirements: Principal Investigators must be nurses holding current AACN membership and must remain AACN members throughout the life of the grant funding.
Restrictions: Funds may not be used to pay for salaries, speaker honorarium or costs, institutional overhead nor augment funding from formal large grants.
Geographic Focus: All States
Date(s) Application is Due: Oct 1
Amount of Grant: Up to 500 USD
Contact: Pamela Shellner, Clinical Practice Specialist; (949) 362-2050, ext. 321 or (800) 394-5995, ext. 321; fax (949) 362-2020; research@aacn.org or info@aacn.org
Internet: http://www.aacn.org/WD/Practice/Content/grant-EOL.pcms?pid=1@menu=Practice
Sponsor: American Association of Critical-Care Nurses
101 Columbia
Aliso Viejo, CA 92656-4109

AACN Evidence-Based Clinical Practice Grants 109

The Evidence-Based Clinical Practice Grant provides awards to grants that stimulate the use of patient-focused data and/or previously generated research findings to develop, implement and evaluate changes critical and acute care nursing practice. Funds may be awarded for new projects, projects in progress and projects required for an academic degree as long as all other project criteria are met. Eligible projects may include research utilization studies, CQI projects or outcomes evaluation projects. Interdisciplinary and collaborative projects are encouraged and may involve interdisciplinary teams, multiple nursing units, home health, subacute and transitional care, other institutions or community agencies. Funds may be used to cover direct project expenses, such as printed materials, small equipment or supplies including computer software. Six awards are available each year. Each proposal can be funded for up to $1,000.
Requirements: Principal Investigators must be nurses holding current AACN membership and must remain AACN members throughout the life of the grant funding.
Restrictions: Funds may not be used to pay salaries or institutional overhead or augment funding from formal grants. Principal investigators who have received funding from AACN are ineligible to receive additional funding from AACN during the lifetime of the original award.
Geographic Focus: All States
Date(s) Application is Due: Oct 1
Contact: Pamela Shellner, Clinical Practice Specialist; (949) 362-2050, ext. 321 or (800) 394-5995, ext. 321; fax (949) 362-2020; research@aacn.org or info@aacn.org
Internet: http://www.aacn.org/WD/Practice/Content/grant-EBP.pcms?pid=1@menu=Practice
Sponsor: American Association of Critical-Care Nurses
101 Columbia
Aliso Viejo, CA 92656-4109

AACN Mentorship Grant 110

This award, cosponsored by Mallinckrodt Inc. and AACN, facilitates critical care nursing practice research between a novice and an experienced researcher. The novice researcher is a beginning researcher with limited or no research experience in the area of the proposed investigation. The novice researcher is also a registered nurse with current AACN membership. The grant may be used to fund research for an academic degree. The mentor must show strong evidence of research expertise in the proposed area of research to be pursued by the novice investigator.
Requirements: Principal Investigators must be nurses holding current AACN membership and must remain AACN members throughout the life of the grant funding.
Restrictions: The mentor may not be designated mentor in two consecutive years and may not be conducting research as part of an academic degree.
Geographic Focus: All States
Date(s) Application is Due: Oct 1
Contact: Pamela Shellner, Clinical Practice Specialist; (949) 362-2050, ext. 321 or (800) 394-5995, ext. 321; fax (949) 362-2020; research@aacn.org or info@aacn.org
Internet: http://www.aacn.org/WD/Practice/Content/grant-Mentor.pcms?pid=1@menu=practice
Sponsor: American Association of Critical-Care Nurses
101 Columbia
Aliso Viejo, CA 92656-4109

AACN Physio-Control Small Projects Grants 111

The program provides awards to qualified individuals carrying out projects focusing on aspects of acute myocardial infarction, resuscitation, or sudden cardiac death. Such projects may include the use of defibrillation, synchronized cardioversion, noninvasive pacing, or interpretive 12-lead electrocardiogram. Eligible projects may include patient education programs, staff development programs, competency-based educational programs, continuous quality improvement projects, outcomes evaluation projects, or small clinical research studies. Funds may be awarded for new projects, projects in progress, and projects required for an academic degree as long as all other project criteria are met. Collaborative projects are encouraged and may involve interdisciplinary teams, multiple nursing units, home health, subacute and transitional care, other institutions, or community agencies. Funds may be used to cover direct project expenses, such as printed materials, small equipment, or supplies including computer software.
Requirements: Principal Investigators must be nurses holding current AACN membership and must remain AACN members throughout the life of the grant funding.
Restrictions: Funds may not be used to pay for salaries, speaker honorarium or costs, institutional overhead nor augment funding from formal large grants.
Geographic Focus: All States
Date(s) Application is Due: Oct 1
Amount of Grant: Up to 1,500 USD
Contact: Pamela Shellner, Clinical Practice Specialist; (949) 362-2050, ext. 321 or (800) 394-5995, ext. 321; fax (949) 362-2020; research@aacn.org or info@aacn.org
Internet: http://www.aacn.org/WD/Practice/Docs/Grant_Table09-10.pdf
Sponsor: American Association of Critical-Care Nurses
101 Columbia
Aliso Viejo, CA 92656-4109

AACR-American Cancer Society Award for Research Excellence in Cancer Epidemiology and Prevention 112

The AACR and the American Cancer Society established this Award in 1992 to honor outstanding research accomplishments in the fields of cancer epidemiology, biomarkers, and prevention. Nominations may be made via letter from any scientist, whether an AACR member or nonmember, who is now or has been affiliated with any institution involved in cancer research, cancer medicine, or cancer-related biomedical science. The winner will receive an honorarium of $5,000 and give a 50-minute lecture during the AACR Annual Meeting.
Requirements: Candidacy is open to all cancer researchers who are affiliated with any institution involved in cancer research, cancer medicine, or cancer-related biomedical science anywhere in the world. Such institutions include those in academia, industry, or government.
Restrictions: Institutions or organizations are not eligible for the Award. Candidates may not nominate themselves.
Geographic Focus: All States
Date(s) Application is Due: Oct 15
Contact: Monique P. Eversley, Program Associate; (267) 646-0576 or (215) 440-9300, ext. 1400; fax (267) 646-0576; monique.eversley@aacr.org
Internet: http://www.aacr.org/home/scientists/scientific-achievement-awards/acs-award.aspx
Sponsor: American Association for Cancer Research
615 Chestnut Street, 17th Floor
Philadelphia, PA 19106-4404

AACR-Colorectal Cancer Coalition Fellows Grant 113

The Fellowship program supports innovative research by a meritorious young investigator by presenting the Fellow with research funds to pursue an independent line of investigation within the context of his/her current placement. This is a one-year grant of $30,000. Research projects are restricted to translational or clinical cancer research that has an ultimate goal of developing or improving therapeutic interventions for patients with metastatic colorectal cancer. One grant will be awarded this cycle.
Requirements: Candidates must be, at the start of the grant term, in their third, fourth, or fifth year of their fellowship status.
Geographic Focus: All States
Date(s) Application is Due: Dec 18
Amount of Grant: 30,000 USD
Contact: Hanna Hopfinger; (267) 646-0665; hanna.hopfinger@aacr.org
Internet: http://www.aacr.org/home/scientists/research-funding—fellowships/fellows-grants.aspx
Sponsor: American Association for Cancer Research
615 Chestnut Street, 17th Floor
Philadelphia, PA 19106-4404

AACR-FNAB Fellows Grant for Translational Pancreatic Cancer Research 114

The Fellows Grants support innovative research by a meritorious young investigator by presenting the Fellow with research funds to pursue an independent line of investigation within the context of his/her current placement. This is a one-year grant of $35,000. Research projects are restricted to translational pancreatic cancer research that includes the use of human tissue and has implications for therapeutic application and individualized medicine.
Requirements: Candidates must be, at the start of the grant term, in their third, fourth, or fifth year of their fellowship status.
Geographic Focus: All States
Date(s) Application is Due: Dec 18
Contact: Julia Laurence, Program Assistant; (215) 440-9300, ext. 102 or (267) 646-0655; fax (215) 440-9372; julia.laurence@aacr.org or laurence@aacr.org
Internet: http://www.aacr.org/home/scientists/research-funding—fellowships/fellows-grants.aspx
Sponsor: American Association for Cancer Research
615 Chestnut Street, 17th Floor
Philadelphia, PA 19106-4404

AACR-FNAB Foundation Career Development Award for Translational Cancer Research 115

Career development awards are two-year awards that support research by scientists who are engaged in meritorious cancer research at academic institutions. These awards provide important transitional support for direct research expenses as researchers move from the ranks of early career scientists to faculty status. The Awards are open to junior faculty who completed postdoctoral studies or clinical fellowships between specified dates and who are at an academic or medical institution. Research projects in this area should have direct applicability to translational cancer and be focused on any individualized therapeutic area. Projects should have implications for individualized cancer treatment and make use of human biopsies or samples, such as needle biopsies or circulating cancer cells. One grant will be presented annually. The submission deadline is December 3, with the decision date March 31.
Requirements: Candidates must have acquired a doctoral degree in a related field and may not currently be a candidate for a further doctoral or professional degree. If a candidate has obtained an equivalent degree at a foreign institution, information on the nature of the degree must be provided at the time of application.
Restrictions: Awards are restricted to investigators or institutions within the U.S. Employees of a national government and employees of private industry are ineligible.
Geographic Focus: All States
Date(s) Application is Due: Dec 3
Amount of Grant: 50,000 USD
Samples: Tara A. Young, M.D., Ph.D., University of California, Los Angeles, Los Angeles, CA, $50,000-Project: High Resolution Cytogenetic Study of Archival Metastatic Choroidal Melanoma (2008-2010).
Contact: Julia Laurence, Program Assistant; (215) 440-9300, ext. 102 or (267) 646-0655; fax (215) 440-9372; julia.laurence@aacr.org or awards@aacr.org
Internet: http://www.aacr.org/home/scientists/research-funding—fellowships/research-funding/research-funding—fellowships/career-development-award-recipients.aspx
Sponsor: American Association for Cancer Research
615 Chestnut Street, 17th Floor
Philadelphia, PA 19106-4404

AACR-Genentech BioOncology Career Development Award for Cancer Research on the HER Family Pathway 116

Career development awards are two-year awards that support research by scientists who are engaged in meritorious cancer research at academic institutions. These awards provide important transitional support for direct research expenses as researchers move from the ranks of early career scientists to faculty status. The Awards are open to junior faculty who completed postdoctoral studies or clinical fellowships between specified dates and who are at an academic or medical institution. Research projects in this area should have direct applicability to basic, translational, or clinical cancer research focused on the HER Family Pathway. One grant will be presented annually. The submission deadline is December 3, with the decision date March 31.
Requirements: Candidates must be affiliated with an institution in the U.S. and must hold an M.D. or combined M.D./Ph.D. If a candidate has obtained an equivalent degree at a foreign institution, information on the nature of the degree must be provided at the time of application.
Restrictions: Employees of a national government and private industry are ineligible.
Geographic Focus: All States
Date(s) Application is Due: Dec 3
Amount of Grant: 50,000 USD
Contact: Julia Laurence, Program Assistant; (215) 440-9300, ext. 102 or (267) 646-0655; fax (215) 440-9372; julia.laurence@aacr.org or awards@aacr.org
Internet: http://www.aacr.org/
Sponsor: American Association for Cancer Research
615 Chestnut Street, 17th Floor
Philadelphia, PA 19106-4404

AACR-Genentech BioOncology Fellowship for Cancer Research in Angiogenesis 117

AACR Research Fellowships are open to Postdoctoral Fellows and Clinical Research Fellows at an academic facility, teaching hospital, or research institution who will be in the 1st, 2nd, or 3rd year of their postdoctoral training at the start of the fellowship term on July 1. This is a two-year term with a grant of $35,000 per year. One Fellowship will be awarded.
Requirements: The recipients of the Fellowships must attend the AACR Annual Meeting to accept the award.
Restrictions: Research projects are restricted to cancer research with direct applicability to angiogenesis.
Geographic Focus: All States
Date(s) Application is Due: Dec 3
Amount of Grant: 35,000 USD
Contact: Hanna Hopfinger, Program Assistant; (267) 646-0665 or (215) 440-9300; fax (215) 440-9372; hanna.hopfinger@aacr.org or awards@aacr.org
Internet: http://www.aacr.org/
Sponsor: American Association for Cancer Research
615 Chestnut Street, 17th Floor
Philadelphia, PA 19106-4404

AACR-GlaxoSmithKline Clinical Cancer Research Scholar Awards 118

These awards provide support of $4,000 which may be used by the recipient to support attendance at the Annual Meeting and any other AACR meeting within a two-year period. Up to $2,000 per meeting may be used, and reimbursement checks are provided after the meeting upon submission of proper documentation of travel, housing, registration, and/or subsistence expenses. Award recipients are still required to pay the Annual Meeting registration fee and make their own travel and housing arrangements. Selection is made by the Program Committee based upon the novelty, quality, and significance of the abstract submitted. Award recipients will receive notification separate from abstract acceptance and scheduling information.
Requirements: Graduate students, medical students and residents, clinical fellows or equivalent, and postdoctoral fellows (AACR members and nonmembers) who are presenters of a meritorious abstract in clinical cancer research are eligible.
Restrictions: Employees or subcontractors of private industry are not eligible.
Geographic Focus: All States
Date(s) Application is Due: Nov 28
Amount of Grant: 2,000-4,000 USD
Contact: Mona E. Shater, Staff Assistant; (215) 440-9300 or (267) 646-0654; fax (215) 440-9372; mona.shater@aacr.org
Internet: http://www.aacr.org/home/scientists/meetings—workshops/travel-grants/aacr-glaxosmithkline-clinical-cancer-research-scholar-awards.aspx
Sponsor: American Association for Cancer Research
615 Chestnut Street, 17th Floor
Philadelphia, PA 19106-4404

AACR-Minorities in Cancer Research Jane Cooke Wright Lectureship Awards 119

The AACR-Minorities in Cancer Research Jane Cooke Wright Lectureship was first presented in 2006. The Lectureship is intended to give recognition to an outstanding scientist who has made meritorious contributions to the field of cancer research and who has, through leadership or by example, furthered the advancement of minority investigators in cancer research. The recipient will present the AACR-Minorities in Cancer Research Jane Cooke Wright Lectureship during the AACR Annual Meeting, receive an honorarium and commemorative plaque, and receive support for the winner and a guest to attend the Annual Meeting.
Requirements: Candidacy is open to all cancer researchers who are affiliated with any institution involved in cancer research, cancer medicine, or cancer-related biomedical science anywhere in the world. Such institutions include those in academia, industry, or government. Nominations may be made by any scientist, whether an AACR member or nonmember, who is now or has been affiliated with any institution involved in cancer research, cancer medicine, or cancer-related biomedical science. Candidates may not nominate themselves. Nominations must be submitted online at https://proposalcentral.altum.com, no later than 4:00 p.m. United States Eastern Time on October 15th. Paper nominations will not be accepted.
Restrictions: Institutions or organizations are not eligible for the Lectureship.
Geographic Focus: All States
Date(s) Application is Due: Oct 15
Contact: Monique P. Eversley, Staff Associate; (267) 646-0576 or (215) 440-9300, ext. 1400; fax (215) 440-9372; monique.eversley@aacr.org or awards@aacr.org
Internet: http://www.aacr.org/home/scientists/scientific-achievement-awards/micr-wright-lectureship.aspx
Sponsor: American Association for Cancer Research
615 Chestnut Street, 17th Floor
Philadelphia, PA 19106-4404

AACR-National Brain Tumor Foundation Fellows Grant 120

The AACR-National Brain Tumor Society Fellowship, in memory of Bonnie Brooks, is open to Postdoctoral Fellows and Clinical Research Fellows at an academic facility, teaching hospital, or research institution. This fellowship provides a one-year grant of $40,000 to support the salary and benefits of Fellows engaged in Glioblastoma multiforme (GBM) research. A partial amount of funds may be designated for research support. It is anticipated that two fellowships will be awarded. The AACR requires applicants to submit both an online and a paper application. The online application must be submitted by 12:00 noon (United States Eastern Time) on August 10, the application must be submitted using the proposalCENTRAL website at https://proposalcentral.altum.com. The paper application with original signatures and all supporting documents must be postmarked and sent no later then August 12.
Requirements: Candidates must be, at the start of the grant term, in their third, fourth, or fifth year of their fellowship status.
Restrictions: Investigators holding the rank of Instructor, Adjunct Professor, Assistant Professor, Research Assistant Professor, the equivalent or higher are not eligible.
Geographic Focus: All States
Date(s) Application is Due: Aug 10; Aug 12
Amount of Grant: 40,000 USD
Contact: Hanna Hopfinger; (267) 646-0665; hanna.hopfinger@aacr.org
Internet: http://www.aacr.org/home/scientists/research-funding—fellowships/post doctoral-fellowships.aspx
Sponsor: American Association for Cancer Research
615 Chestnut Street, 17th Floor
Philadelphia, PA 19106-4404

AACR-NCI International Investigator Opportunity Grants 121

Recognizing that there is no substitute for the collegial interaction and scholarly discussion that is a cornerstone of scientific advancement, these grants provide significant financial support for meeting registration, housing, travel, and incidentals to attend the annual American Association for Cancer Research Meeting. Grants are available for ten cancer researchers from countries where such opportunities are limited.
Requirements: Applicants must be nationals or permanent residents of, and must reside and conduct cancer research within, low and middle-income countries as defined by

the World Bank. Applicants should also hold an doctorate or equivalent degree and be actively engaged in a program of cancer research with a record of cancer-focused publications in peer-reviewed scientific journals.
Restrictions: Funds may not be used for entertainment expenses (such as for tours or in-room movies), meals for other individuals, souvenirs, and other non-meeting expenses.
Geographic Focus: All States
Date(s) Application is Due: Sep 28
Samples: Hakan Akbulut, M.D., Professor, Medical Oncology, Ankara University, Ankara, Turkey; Soo H. Teo, D.Phil., Chief Executive, Cancer Research Initiatives Foundation, Subang Jaya, Malaysia; Jozsef Tovari, Ph.D., Lab Head, Department of Tumor Progression, National Institute of Oncology, Budapest, Hungary.
Contact: Susan Waskey, Program Assistant; (215) 440-9300 or (866) 423-3965; fax (215) 440-9313; susan.waskey@aacr.org
Internet: http://www.aacr.org/home/scientists/meetings—workshops/travel-grants.aspx
Sponsor: American Association for Cancer Research
615 Chestnut Street, 17th Floor
Philadelphia, PA 19106-4404

AACR-Pancreatic Cancer Action Network Career Development Award for Pancreatic Cancer Research 122
The Pancreatic Cancer Action Network-AACR Career Development Award represents a joint effort to encourage and support early career scientist who are in the first four years of a faculty appointment to conduct pancreatic cancer research and establish successful career paths in this field. The research proposed for funding may be basic, translational, clinical or epidemiological in nature and must have direct applicability and relevance to pancreatic cancer. The Award provides a two year grant of $200,000 ($100,000 per year) for direct research expenses, which may include salary and benefits of the grant recipient, postdoctoral or clinical research fellows, and/or research assistants, research/laboratory supplies and equipment. It is anticipated that three Career Development Awards will be funded.
Requirements: Candidates must have acquired a doctoral degree in a related field and may not currently be a candidate for a further doctoral or professional degree. If a candidate has obtained an equivalent degree at a foreign institution, information on the nature of the degree must be provided at the time of application.
Restrictions: Awards are restricted to investigators or institutions within the U.S. Employees of a national government and employees of private industry are ineligible.
Geographic Focus: All States
Date(s) Application is Due: Oct 28
Amount of Grant: 200,000 USD
Contact: Julia Laurence, Program Assistant; (215) 440-9300, ext. 102 or (267) 646-0655; fax (215) 440-9372; julia.laurence@aacr.org or awards@aacr.org
Internet: http://www.aacr.org/home/scientists/research-funding—fellowships/career-development-awards.aspx
Sponsor: American Association for Cancer Research
615 Chestnut Street, 17th Floor
Philadelphia, PA 19106-4404

AACR-Pancreatic Cancer Action Network Innovative Grants 123
The AACR-PanCAN Pilot Grants represent a joint effort of the AACR and the Pancreatic Cancer Action Network to support innovative research focusing on the early detection or treatment of pancreatic cancer. Research projects may have direct application to pancreatic cancer or may demonstrate relevance to pancreatic cancer in order to encourage investigators with experience in other areas of research to apply their ideas to this challenging field. Proposals will be accepted for innovative research projects that have a direct application or demonstrate relevance to the early detection or treatment of pancreatic cancer; projects may be in any discipline of basic, translational, clinical, or epidemiological cancer research. Proposals that will develop preliminary data necessary to prepare and submit a competitive research grant application to a major federal funding agency will also be accepted. Special emphasis will be placed on research that is not duplicative of other efforts and has the potential for national application. Researchers who are affiliated with any institution involved in cancer research, cancer medicine, or cancer-related biomedical science within the United States may apply. There are no residency status restrictions. Each grantee will be awarded a total of $200,000, paid over the two-year term.
Requirements: Applicants must have acquired a doctoral degree in a related field. If an applicant has obtained an equivalent degree at a foreign institution, information on the nature of the degree must be provided at the time of application.
Restrictions: Employees or subcontractors of a national government or the for-profit private industry are not eligible to apply. Members of the Scientific Review Committee are not eligible for these grants. Current or past recipients of a PanCAN Pilot Grant are not eligible.
Geographic Focus: All States
Date(s) Application is Due: Oct 5; Nov 30
Amount of Grant: 200,000 USD
Contact: Julia Laurence, Program Assistant; (215) 440-9300, ext. 102 or (267) 646-0655; fax (215) 440-9372; julia.laurence@aacr.org or laurence@aacr.org
Internet: http://www.aacr.org/home/scientists/research-funding—fellowships/pancreatic-cancer-action-network-aacr-innovative-grants.aspx
Sponsor: American Association for Cancer Research
615 Chestnut Street, 17th Floor
Philadelphia, PA 19106-4404

AACR-Prevent Cancer Foundation Award for Excellence in Cancer Prevention Research 124
he Award for Excellence in Cancer Prevention Research will be given to a scientist residing in any country in the world for his or her seminal contributions to the field of cancer prevention. Such investigations must have been conducted in basic, translational, clinical, epidemiological, or behavioral science in cancer prevention research. Further, these studies must have had not only a major impact on the field, but must also have stimulated new directions in this important area. The recipient of the Award will receive a $5,000 honorarium and will present a lecture at the Annual AACR International Conference. Nominations for the Award must be submitted electronically no later than 4:00 p.m. United States Eastern Time on Monday, July 31.
Requirements: All cancer researchers who are affiliated with any institution involved in cancer research, cancer medicine, or cancer-related biomedical science anywhere in the world. Such institutions include those in academia, industry, or government may be nominated for the Award. Candidates must currently maintain an active research program, have a record of recent publications, and be able to present the Award lecture at the Conference.
Restrictions: Institutions or organizations are not eligible for the Award. Paper nominations will not be accepted.
Geographic Focus: All States
Date(s) Application is Due: Jul 31
Contact: Monique P. Eversley, Staff Associate; (267) 646-0576 or (215) 440-9300, ext. 1400; fax (215) 440-9372; monique.eversley@aacr.org or awards@aacr.org
Internet: http://www.aacr.org/home/scientists/scientific-achievement-awards/prevent-cancer-award.aspx
Sponsor: American Association for Cancer Research
615 Chestnut Street, 17th Floor
Philadelphia, PA 19106-4404

AACR-WICR Scholar Awards 125
AACR-WICR Scholar Awards are awarded to ten members of Women in Cancer Research who are scientists-in-training and presenters of meritorious scientific papers at AACR Annual Meetings.
Requirements: Eligible candidates for these awards must fit the following criteria, candidates must be: members of Women in Cancer Research (WICR); full time scientists-in-training who are graduate students, medical students, residents, clinical fellows or equivalent, or postdoctoral fellows; travelling from within the United States or abroad; first authors on abstracts submitted for consideration for presentation at the AACR Annual Meeting.
Geographic Focus: All States
Date(s) Application is Due: Dec 1
Contact: Lauren Medvetz; (267) 646-0689; lauren.medvets@aacr.org
Internet: http://www.aacr.org/home/scientists/meetings—workshops/travel-grants/wicr-scholar-awards.aspx
Sponsor: American Association for Cancer Research
615 Chestnut Street, 17th Floor
Philadelphia, PA 19106-4404

AACR-Women in Cancer Research Charlotte Friend Memorial Lectureship Awards 126
The AACR-Women in Cancer Research Charlotte Friend Memorial Lectureship was established in 1998 in honor of renowned virologist and discoverer of the Friend virus, Dr. Charlotte Friend, for her pioneering research on viruses, cell differentiation, and cancer. The lecture is intended to give recognition to an outstanding female or male scientist who has made meritorious contributions to the field of cancer research and who has, through leadership or by example, furthered the advancement of women in science. The winner of the Lectureship will present a 50-minute lecture during the AACR Annual Meeting, receive an honorarium and commemorative plaque, and receive support for the winner and a guest to attend the Annual Meeting.
Requirements: Candidacy is open to all cancer researchers who are affiliated with any institution involved in cancer research, cancer medicine, or cancer-related biomedical science anywhere in the world. Such institutions include those in academia, industry, or government.
Restrictions: Nominations may be made by any scientist, whether an AACR member or nonmember, who is now or has been affiliated with any institution involved in cancer research, cancer medicine, or cancer-related biomedical science. Candidates may not nominate themselves. Nominations must be submitted online at https://proposalcentral.altum.com, no later than 4:00 p.m. United States Eastern Time on October 15th. Paper nominations will not be accepted. Institutions or organizations are not eligible for the Lectureship.
Geographic Focus: All States
Date(s) Application is Due: Oct 15
Contact: Lauren Medvetz; (267) 646-0689; lauren.medvets@aacr.org
Internet: http://www.aacr.org/home/scientists/scientific-achievement-awards/wicr-friend-lectureship.aspx
Sponsor: American Association for Cancer Research
615 Chestnut Street, 17th Floor
Philadelphia, PA 19106-4404

AACR Award for Lifetime Achievement in Cancer Research 127
The AACR Award for Lifetime Achievement in Cancer Research was established and first given in 2004 to honor an individual who has made significant fundamental contributions to cancer research, either through a single scientific discovery or a body of work. These contributions, whether they have been in research, leadership, or mentorship, must have had a lasting impact on the cancer field and must have demonstrated a lifetime

commitment to progress against cancer. Nominations may be made by any scientist, whether an AACR member or nonmember, who is now or has been affiliated with any institution involved in cancer research, cancer medicine, or cancer-related biomedical science. Nominations for the Award must be submitted electronically to the AACR no later than 4:00 p.m. United States Eastern Time on November 16.
Requirements: Nominations may be made on behalf of individuals who are living at the time of the nomination. Candidates need not be members of the AACR. Candidates need not be currently engaged in cancer research.
Restrictions: Institutions or organizations are not eligible for the Award. Paper nominations will not be accepted.
Geographic Focus: All States
Date(s) Application is Due: Nov 16
Contact: Monique P. Eversley, Staff Associate; (267) 646-0576 or (215) 440-9300, ext. 1400; fax (215) 440-9372; monique.eversley@aacr.org or awards@aacr.org
Internet: http://www.aacr.org/home/scientists/scientific-achievement-awards/lifetime-achievement-award.aspx
Sponsor: American Association for Cancer Research
615 Chestnut Street, 17th Floor
Philadelphia, PA 19106-4404

AACR Award for Outstanding Achievement in Cancer Research **128**
This award is given in recognition of a young investigator on the basis of meritorious achievement in cancer research. The winner receives an honorarium of $5,000, presents a 50-minute lecture, and is given full support for the winner and a guest to attend the AACR Annual Meeting.
Requirements: Candidacy is open to all cancer researchers who are affiliated with any institution involved in cancer research, cancer medicine, or cancer-related biomedical science anywhere in the world. Such institutions include those in academia, industry, or government. In accordance with the wishes of the donor, the recipient must be no more than 40 years of age by the time the award is received.
Restrictions: Institutions or organizations are not eligible for the Award.
Geographic Focus: All States
Date(s) Application is Due: Oct 15
Amount of Grant: 5,000 USD
Contact: Monique P. Eversley, Program Associate; (267) 646-0576 or (215) 440-9300, ext. 1400; fax (267) 646-0576; monique.eversley@aacr.org
Internet: http://www.aacr.org/home/scientists/scientific-achievement-awards/outstanding-achievement-award.aspx
Sponsor: American Association for Cancer Research
615 Chestnut Street, 17th Floor
Philadelphia, PA 19106-4404

AACR Award for Outstanding Achievement in Chemistry in Cancer Research **129**
The Award will be given for outstanding, novel, and significant chemistry research, which has led to important contributions to the fields of basic cancer research; translational cancer research; cancer diagnosis; the prevention of cancer; or the treatment of patients with cancer. Such research may include, but is not limited to, drug discovery and design; structural biology; proteomics, metabolomics and biological mass spectrometry; chemical aspects of carcinogenesis; imaging agents and radiotherapeutics; and chemical biology. The winner of the award will give a 50-minute lecture during the AACR Annual Meeting, will receive a commemorative plaque and a $10,000 honorarium, and receive support for the winner and a guest to attend the Annual Meeting. Nominations for the Award must be submitted electronically no later than 4:00 p.m. United States Eastern Time on October 15.
Requirements: Candidacy is open to all researchers who are affiliated with any institution involved in cancer research, cancer medicine, or cancer-related biomedical science anywhere in the world. Such institutions include those in academia, industry, or government.
Restrictions: Institutions or organizations are not eligible for the Award. Paper nominations will not be accepted.
Geographic Focus: All States
Date(s) Application is Due: Oct 15
Amount of Grant: 10,000 USD
Contact: Lauren Medvetz; (267) 646-0689; lauren.medvets@aacr.org
Internet: http://www.aacr.org/home/scientists/scientific-achievement-awards/chemistry-in-cancer-research-award.aspx
Sponsor: American Association for Cancer Research
615 Chestnut Street, 17th Floor
Philadelphia, PA 19106-4404

AACR Basic Cancer Research Fellowships **130**
The AACR Basic Cancer Research Fellowships consist of two Fellowships, each one-year grants. The AACR Anna D, Barker Fellowship in Basic Cancer Research, funds $40,000, and the AACR-Astellas U.S.A Foundation Fellowships in Basic Cancer Research provides funding in the amount of $30,000. These grants are for salary and benefits to the Fellows. A partial amount of funds may be designated for direct research support. The recipients must attend the AACR Annual Meeting to accept the grants. The AACR requires applicants to submit both an online and a paper application. The online application must be completed by 12:00 noon (United States Eastern Time) on December 10, using the proposalCENTRAL website at http://propasalcentral.altum.com. The paper application with original signatures and all supporting documents must be postmarked and sent no later then December 15.
Requirements: Applications are open to Postdoctoral Fellows and Clinical Research Fellows at an academic facility, teaching hospital, or research institution who will be in the 1st, 2nd, or 3rd year of their postdoctoral training at the start of the fellowship term.
Restrictions: Academic faculty holding the rank of assistant professor or higher, graduate or medical students, government employees, and employees of private industry are not eligible.
Geographic Focus: All States
Date(s) Application is Due: Dec 10; Dec 15
Amount of Grant: 30,000-40,000 USD
Contact: Hanna Hopfinger, Grants Program Assistant; (267) 646-0665; fax (215) 440-9372; hanna.hopfinger@aacr.org
Internet: http://www.aacr.org/home/scientists/research-funding—fellowships/postdoctoral-fellowships.aspx
Sponsor: American Association for Cancer Research
615 Chestnut Street, 17th Floor
Philadelphia, PA 19106-4404

AACR Brigid G. Leventhal Scholar in Cancer Research Awards **131**
The purpose of this award program is to enhance the education and training of early career scientists by providing financial support for their participation in AACR Annual Meetings and Special Conferences. The AACR-WICR Brigid G. Leventhal Scholar Award Committee makes selection for these competitive awards after careful consideration of the candidate's application and accompanying materials.
Requirements: Candidates must be: Women in Cancer Research members; full time scientists in training who are graduate students, medical students, residents, clinical fellows or equivalent, or postdoctoral fellows; and first authors on abstracts submitted for consideration for presentation at the AACR Annual Meeting or Special Conference which the applicant wants to attend.
Geographic Focus: All States
Date(s) Application is Due: Sep 10
Contact: Mona E. Shater, Staff Assistant; (215) 440-9300 or (267) 646-0654; fax (215) 440-9372; mona.shater@aacr.org or wicr@aacr.org
Internet: http://www.aacr.org/home/membership-/association-groups/women-in-cancer-research.aspx
Sponsor: American Association for Cancer Research
615 Chestnut Street, 17th Floor
Philadelphia, PA 19106-4404

AACR Career Development Awards for Pediatric Cancer Research **132**
Two Career Development Awards will be presented, the AACR-Aflac Incorporated Career Development Award for Pediatric Cancer Research and the AACR Centennial Career Development Award for Childhood Cancer Research. These Awards are open to junior faculty who are in their first full-year, faculty appointment. The Awards provide two-year grants of $100,000 ($50,000 per year) for direct research expenses and/or salary support. Research projects must have direct applicability to pediatric cancer.
Requirements: Candidates must have acquired a doctoral degree and have completed postdoctoral studies or clinical fellowships on or after July 2 (two years prior to the start of the grant term). Candidates must be in their first full-time, faculty appointment and hold the title of Instructor, Research Assistant Professor, Assistant Professor, or an equivalent full-time faculty appointment at the start of the grant term (July 1).
Restrictions: Non AACR members must submit a satisfactory application for Active Membership with the grant application by December 10th. Employees or subcontractors of a government or a for-profit private industry are not eligible.
Geographic Focus: All States
Date(s) Application is Due: Dec 18
Amount of Grant: 100,000 USD
Contact: Julia Laurence, Program Assistant; (215) 440-9300, ext. 102 or (267) 646-0655; fax (215) 440-9372; julia.laurence@aacr.org or awards@aacr.org
Internet: http://www.aacr.org/home/scientists/research-funding—fellowships/career-development-awards.aspx
Sponsor: American Association for Cancer Research
615 Chestnut Street, 17th Floor
Philadelphia, PA 19106-4404

AACR Centennial Postdoctoral Fellowships in Cancer Research **133**
AACR Centennial Postdoctoral Research Fellowships in Cancer Research will provide $60,000 per year for up to three years to provide salary support and research funding for clinical and postdoctoral fellows in the first, second, or third year of their fellowship status. In addition to financial flexibility these fellowships also provide career flexibility. Fellowship funds may be transferred to move with the grantee if the grantee's fellowship is completed during the grant term-a valuable commodity for a newly-independent investigator. Under this award mechanism, the postdoctoral trainee is considered the Lead Investigator and should write the proposal independently with appropriate direction from the mentor. Proposals may be submitted in any area of cancer research whether laboratory-based, translational, clinical, epidemiological, behavioral, or other. Applications must be completed online using the proposalCENTRAL website, with one paper copy submitted to the AACR office. Application instructions and program guidelines are available online.
Requirements: Applicants must: be in the 1st, 2nd, or 3rd year of their postdoctoral training at the start of the fellowship term; be members of the AACR or must submit an application for membership prior to the fellowship application deadline; have acquired a doctoral degree in a biological sciences-related field; and hold the title of Postdoctoral Fellow or Clinical Research Fellow at an academic facility, teaching hospital, or research institution.
Restrictions: Employees, trainees, or subcontractors of a national government or private industry are not eligible.
Geographic Focus: All States
Date(s) Application is Due: Feb 1

Amount of Grant: 60,000-180,000 USD
Contact: Hanna Hopfinger, Program Assistant; (267) 646-0665 or (215) 440-9300; fax (215) 440-9372; hanna.hopfinger@aacr.org or awards@aacr.org
Internet: http://www.aacr.org/home/scientists/research-funding—fellowships/research-funding.aspx
Sponsor: American Association for Cancer Research
615 Chestnut Street, 17th Floor
Philadelphia, PA 19106-4404

AACR Centennial Pre-doctoral Fellowships in Cancer Research 134
The AACR Centennial Pre-doctoral Research Fellowships will provide full-time graduate students a rare opportunity to invest in their future cancer research careers. The grant terms are flexible, allowing applicants to tailor their grant funds to fit their individual research funding and salary needs. Grantees may also transfer their funds to a postdoctoral fellowship position, should they advance in their careers during the grant term. Under this award mechanism, the pre-doctoral trainee is considered the Lead Investigator and should write the proposal independently with appropriate direction from the mentor. Proposals may be submitted in any area of cancer research whether laboratory-based, translational, clinical, epidemiological, behavioral, or other. A grant of $30,000 per year over a maximum period of three years will be awarded for salary support and research funding.
Requirements: Applicants must: be full-time graduate students who are pursuing a doctoral degree and engaged in a program of cancer research; be members of the AACR or must submit an application for membership prior to the fellowship application deadline; have completed at least one year of research training at the graduate level or have equivalent research experience; and be engaged in basic, clinical, translational or epidemiological cancer research. At the time of award acceptance the awardee must have advanced to degree candidacy.
Restrictions: Summer student positions and undergraduate summer studentships do not qualify. Employees, trainees, or subcontractors of a national government or private industry are not eligible.
Geographic Focus: All States
Date(s) Application is Due: Feb 1
Amount of Grant: 30,000-90,000 USD
Contact: Hanna Hopfinger, Program Assistant; (267) 646-0665 or (215) 440-9300; fax (215) 440-9372; hanna.hopfinger@aacr.org or awards@aacr.org
Internet: http://www.aacr.org/home/scientists/research-funding—fellowships/research-funding.aspx
Sponsor: American Association for Cancer Research
615 Chestnut Street, 17th Floor
Philadelphia, PA 19106-4404

AACR Clinical and Translational Research Fellowships 135
The AACR Clinical and Translational Research Fellowships consist of three different Fellowships: AACR-Astellas U.S.A Foundation Fellowship in Clinical/Translational Cancer Research, which is a one-year grant of $30,000; AACR-AstraZeneca Fellowship for Translational Lung Cancer Research, which is a two-year grant of $90,000; and, AACR-Bristol-Myers Squibb Oncology Fellowship in Clinical Cancer Research, which is a one-year grant of $40,000. These fellowships support the salary and benefits of the Fellow. A partial amount of funds may be designated for direct research support. The AACR requires applicants to submit both an online and a paper application. The online application deadline is December 10th, the application must be submitted using the proposalCENTRAL website at https://proposalcentral.altum.com. The paper application with original signatures and all supporting documents must be postmarked and sent no later then December 15.
Requirements: Applications are open to Postdoctoral Fellows and Clinical Research Fellows at an academic facility, teaching hospital, or research institution who will be in the 1st, 2nd, or 3rd year of their postdoctoral training at the start of the fellowship term.
Restrictions: Academic faculty holding the rank of assistant professor or higher, graduate or medical students, government employees, and employees of private industry are not eligible.
Geographic Focus: All States
Date(s) Application is Due: Dec 10; Dec 15
Amount of Grant: 30,000-90,000 USD
Contact: Hanna Hopfinger, Program Assistant; (267) 646-0665 or (215) 440-9300; fax (215) 440-9372; hanna.hopfinger@aacr.org or awards@aacr.org
Internet: http://www.aacr.org/home/scientists/research-funding—fellowships/postdoctoral-fellowships.aspx
Sponsor: American Association for Cancer Research
615 Chestnut Street, 17th Floor
Philadelphia, PA 19106-4404

AACR Fellows Grants 136
The Fellowship program supports innovative research by a meritorious young investigator by presenting the Fellow with research funds to pursue an independent line of investigation within the context of his/her current Fellowship placement. By allowing a Fellow to acquire the equipment and supplies needed to pursue a new direction in his/her research program, the Fellows Grant assists the Fellow in developing preliminary data to support a future project or investigating a new technique that otherwise would not be possible in the absence of this funding. The Selection Committee favors proposals that provided evidence of an applicant's research initiative and creativity, and the support of a scientific mentor. The Committee also weighed the potential of the candidate to make meaningful contributions to the field, and promise of success as a cancer researcher in future years.
Requirements: Candidates must be, at the start of the grant term, in their third, fourth, or fifth year of their fellowship status.
Geographic Focus: All States
Date(s) Application is Due: Dec 18
Amount of Grant: 30,000-40,000 USD
Contact: Julia Laurence, Program Assistant; (215) 440-9300, ext. 102 or (267) 646-0655; fax (215) 440-9372; julia.laurence@aacr.org or awards@aacr.org
Internet: http://www.aacr.org/home/scientists/research-funding—fellowships/fellows-grants.aspx
Sponsor: American Association for Cancer Research
615 Chestnut Street, 17th Floor
Philadelphia, PA 19106-4404

AACR Gertrude Elion Cancer Research Award 137
This prestigious award provides a one-year research grant to an assistant professor at an academic or nonprofit research institute worldwide for salary and benefits, laboratory supplies, and limited domestic travel to support research in cancer etiology, diagnosis, treatment, or prevention (basic, translational, or clinical cancer research). The AACR requires applicants to submit both an online and a paper application. The online application must be submitted by 12:00 noon (United States Eastern Daylight Time) on February 27, using the proposalCENTRAL website at https://proposalcentral.altum.com. The paper application with original signatures and all supporting documents must be postmarked and sent no later then March 3.
Requirements: Candidates must be nominated by an AACR member and submit a detailed application. Candidates must have completed postdoctoral studies or clinical fellowships no later than July 1 of the award year and ordinarily not more than five years earlier. Candidates must be tenure-tracked scientists at the level of assistant professor at an academic institution anywhere in the world.
Restrictions: Tenured faculty in academia, employees of a national government, and employees of private industry are not eligible.
Geographic Focus: All States
Date(s) Application is Due: Feb 27; Mar 3
Amount of Grant: 50,000 USD
Contact: Julia Laurence, Program Assistant; (215) 440-9300, ext. 102 or (267) 646-0655; fax (215) 440-9372; julia.laurence@aacr.org or laurence@aacr.org
Internet: http://www.aacr.org/home/scientists/research-funding—fellowships/elion-award.aspx
Sponsor: American Association for Cancer Research
615 Chestnut Street, 17th Floor
Philadelphia, PA 19106-4404

AACR Henry Shepard Bladder Cancer Research Grants 138
These two-year Grants will provide up to $250,000 in total support for innovative cancer research projects designed to accelerate the discovery, development, and application of new agents to treat bladder cancer and/or for pre-clinical research with direct therapeutic intent. Laboratory-based projects must present plans with clinical collaborators indicating how the work will be translated into the clinic. Similarly, clinical studies must show how the work was derived from basic preclinical work and how results will be channeled back to laboratory-based collaborators. Four grants will be awarded. The submission deadline is July 25, at 12:00 noon, U.S. Eastern Time.
Requirements: Independent investigators who are affiliated with any institution involved in cancer research, cancer medicine, or cancer-related biomedical science anywhere in the world may apply. There are no geographic, national, or residency status restrictions. Applicants must have acquired a doctoral degree in a related field.
Restrictions: Pre- and Postdoctoral fellows are not eligible. Employees or subcontractors of a national government or the for-profit private industry are not eligible to serve as the Principal Investigator for the purpose of these Grants. However, collaborations with such individuals are encouraged. Neither the members of the Scientific Review Committee nor the members of their individual laboratories are eligible for these Grants.
Geographic Focus: All States
Date(s) Application is Due: Jul 25
Amount of Grant: Up to 250,000 USD
Contact: Hanna Hopfinger, Program Assistant; (267) 646-0665 or (215) 440-9300; fax (215) 440-9372; hanna.hopfinger@aacr.org or awards@aacr.org
Internet: http://www.aacr.org/home/scientists/research-funding—training-grants/research-funding/research-funding—fellowships/bladder-cancer-research-grants.aspx
Sponsor: American Association for Cancer Research
615 Chestnut Street, 17th Floor
Philadelphia, PA 19106-4404

AACR Joseph H. Burchenal Memorial Award 139
This Award was established in 1996 to recognize outstanding achievements in clinical cancer research. It is named for the late Dr. Joseph H. Burchenal, Honorary Member and Past President of the AACR, and a major figure in clinical cancer research and chemotherapy. The winner of the Award for Outstanding Achievement in Clinical Cancer Research will give a 50-minute lecture during the AACR Annual Meeting, and receive an honorarium and support for attendance at the meeting.
Requirements: Candidacy is open to all cancer researchers who are affiliated with any institution involved in cancer research, cancer medicine, or cancer-related biomedical science anywhere in the world. Such institutions include those in academia, industry, or government.
Restrictions: Institutions or organizations are not eligible for the Award. Candidates may not nominate themselves.
Geographic Focus: All States

Date(s) Application is Due: Oct 15
Samples: Joseph R. Bertino, M.D., Interim Director & Chief Scientific Officer, The Cancer Institute of New Jersey.
Contact: Monique P. Eversley, Program Associate; (267) 646-0576 or (215) 440-9300, ext. 1400; fax (267) 646-0576; monique.eversley@aacr.org
Internet: http://www.aacr.org/home/scientists/scientific-achievement-awards/burchenal-award.aspx
Sponsor: American Association for Cancer Research
615 Chestnut Street, 17th Floor
Philadelphia, PA 19106-4404

AACR Judah Folkman Career Development Award for Anti-Angiogenesis Research 140

The AACR Judah Folkman Career Development Award for Angiogenesis Research is open to junior faculty who completed postdoctoral studies or clinical fellowships no more then three years prior to the start of the grant term (Dec 1); who are in their first full-time, faculty appointment and hold the title of Instructor, Research Assistant Professor, Assistant Professor, or an equivalent full-time faculty appointment; and who are at an academic, medical, or research Institution. The Career Development Award provides a two-year grant of $100,000 ($50,000 per year) for direct research expenses and/or salary support. One Award will be presented.
Requirements: Applicants must have acquired a medical degree or hold a combined M.D./Ph.D and may not currently be a candidate for a further doctoral or professional degree. The Award is open to AACR members. Non-members must submit a satisfactory application for Active Membership and additional Curriculum Vitae with the grant application.
Restrictions: Employees, trainees, or subcontractors of a national government or private industry are not eligible. Exceptions may apply if an applicant holds a full-time position at a Veterans' Hospital or national laboratory in the U.S. Current recipients of any AACR grant are also not eligible.
Geographic Focus: All States
Date(s) Application is Due: Aug 10
Amount of Grant: 100,000 USD
Contact: Julia Laurence, Program Assistant; (215) 440-9300, ext. 102 or (267) 646-0655; fax (215) 440-9372; julia.laurence@aacr.org or awards@aacr.org
Internet: http://www.aacr.org/home/scientists/research-funding—fellowships/career-development-awards.aspx
Sponsor: American Association for Cancer Research
615 Chestnut Street, 17th Floor
Philadelphia, PA 19106-4404

AACR Kirk A. Landon and Dorothy P. Landon Foundation Prizes 141

These two major international awards recognize outstanding scientists who have made seminal basic and translational cancer research discoveries at the cutting edge of scientific novelty and significance, which have accelerated progress against cancer and have implications for future discoveries and contributions to cancer research. The Kirk A. Landon-AACR Prize for Basic Cancer Research recognizes significant, fundamental contributions to laboratory research. The Dorothy P. Landon-AACR Prize for Translational Cancer Research recognizes extraordinary achievement in translational cancer research—the interface between basic research and its application to the clinic in the areas of diagnosis, treatment, or the prevention of cancer. Both prizes bring heightened public attention to landmark achievements in the continuing effort to prevent and cure cancer through the presentation of dynamic lectures during the AACR annual meeting; and promote and reward continued, productive cancer research.
Requirements: The prizes are open to all cancer researchers who are affiliated with any institution involved in cancer research, cancer medicine, or cancer-related biomedical science anywhere in the world. Such institutions include those in academia, industry, or government. Candidates must be active researchers and have a record of recent publications.
Restrictions: Institutions or organizations are ineligible.
Geographic Focus: All States
Date(s) Application is Due: Aug 25
Amount of Grant: 100,000 USD
Samples: Arnold J. Levine, Ph.D., Professor, The Cancer Institute of New Jersey—Kirk A. Landon-AACR Prize for Basic Cancer Research; John Mendelsohn, M.D., President, UT M.D. Anderson Cancer Center, Houston, Texas—Dorothy P. Landon-AACR Prize for Translational Cancer Research.
Contact: Monique P. Eversley, Staff Associate; (267) 646-0576 or (215) 440-9300, ext. 1400; fax (215) 440-9372; monique.eversley@aacr.org or awards@aacr.org
Internet: http://www.aacr.org/home/scientists/scientific-achievement-awards/landon-prizes.aspx
Sponsor: American Association for Cancer Research
615 Chestnut Street, 17th Floor
Philadelphia, PA 19106-4404

AACR Margaret Foti Award for Leadership and Extraordinary Achievements in Cancer Research 142

This Award will recognize a true champion of cancer research, who embodies the sustained commitment of Margaret Foti to the prevention and cure of cancer. The Award will be given to an individual whose leadership and extraordinary achievements in cancer research have made a major impact in the field. Such achievements may include contributions to the acceleration of progress in cancer research; raising national or international awareness of cancer research; or in other ways demonstrating a sustained commitment to cancer research. Nominations for the Award must be submitted electronically no later than 4:00 p.m. U.S. Eastern Time on November 16. The recipient will receive an honorarium, a commemorative plaque, and support for the winner to attend the AACR Annual Meeting.
Requirements: Candidacy is open to all individuals who are affiliated with any organization whose mission supports cancer research or any institution involved in cancer research, cancer medicine, or cancer-related biomedical science anywhere in the world. Such institutions include those in academia, industry, or government.
Restrictions: Institutions or organizations are not eligible for the Award. Paper nominations will not be accepted.
Geographic Focus: All States
Date(s) Application is Due: Nov 16
Contact: Monique P. Eversley, Program Associate; (267) 646-0576 or (215) 440-9300, ext. 1400; fax (215) 440-9372; monique.eversley@aacr.org or awards@aacr.org
Internet: http://www.aacr.org/home/scientists/scientific-achievement-awards/margaret-foti-award.aspx
Sponsor: American Association for Cancer Research
615 Chestnut Street, 17th Floor
Philadelphia, PA 19106-4404

AACR Minority-Serving Institution Faculty Scholar in Cancer Research Awards 143

The AACR is pleased to announce the availability of Scholar Awards in Cancer Research for full-time faculty members of Minority-Serving Institutions [Historically Black Colleges and Universities (HBCUs), Hispanic Serving Institutions (HSIs), and Tribal Colleges and Universities and other post-secondary institutions as defined by the U.S. Department of Education]. The purposes of this Award program is to increase the scientific knowledge base of faculty members at Minority-Serving Institutions, and to encourage them and their students to pursue careers in cancer research. AACR Minority-Serving Institution Faculty Scholar in Cancer Research Awards are presented by the American Association for Cancer Research to scientists at the level of Assistant Professor or above at a Minority-Serving Institution, who are engaged in meritorious basic, clinical, or translational cancer research.
Requirements: Candidates must have completed doctoral studies or clinical fellowships and hold full-time faculty status at an institution designated as a Minority-Serving Institution. Candidates must have acquired doctoral degrees in fields relevant to cancer research. Candidates must be citizens or permanent residents of the United States or Canada.
Geographic Focus: All States
Date(s) Application is Due: Jul 20
Samples: Joseph Agyin, Ph.D., UT Health Science Center, San Antonio, Texas; Wellington Ayensu, M.D., Jackson State University, Jackson, Mississippi; Rebecca Hartley, Ph.D., University of New Mexico, Albuquerque, New Mexico.
Contact: Mona E. Shater, Staff Assistant; (215) 440-9300 or (267) 646-0654; fax (215) 440-9372; mona.shater@aacr.org or micr@aacr.org
Internet: http://www.aacr.org/home/scientists/meetings—workshops/special-conferences/metabolism-and-cancer/financial-support-for-attendance.aspx
Sponsor: American Association for Cancer Research
615 Chestnut Street, 17th Floor
Philadelphia, PA 19106-4404

AACR Minority Scholar in Cancer Research Awards 144

The AACR administers this award program that is supported by a grant from the Comprehensive Minority Biomedical Branch of the National Cancer Institute to provide funds for participation of meritorious minority scientists in the AACR Special Conferences Series (U.S. domestic conferences only). These awards are intended to enhance the education and training of minority researchers and to increase the visibility and recognition of minorities involved in cancer research.
Requirements: Candidates must be full-time graduate students, medical students, residents, clinical or postdoctoral fellows, or junior faculty members who are either engaged in cancer research or who have the training and potential to make contributions to this field. Candidates must also be citizens or permanent residents of the United States or Canada.
Restrictions: This program applies only to racial/ethnic minority groups identified by the National Cancer Institute as being traditionally underrepresented in cancer and biomedical research. These groups include African Americans/Blacks, Alaskan Natives, Hispanic Americans, Native Americans, and Native Pacific Islanders. Only citizens of the United States or Canada or scientists who are permanent residents of these countries may receive one of these awards.
Geographic Focus: All States
Date(s) Application is Due: Jul 20
Samples: Irene Alvarez, B.S., University of Arizona, Tucson, Arizona; Cristina M. Contreras, B.S., UT Southwestern Medical Center, Dallas, Texas; Vivian A. Galacia, B.S., University of Southern California, Los Angeles, California.
Contact: Mona E. Shater, Staff Assistant; (215) 440-9300 or (267) 646-0654; fax (215) 440-9372; mona.shater@aacr.org or micr@aacr.org
Internet: http://www.aacr.org/home/scientists/meetings—workshops/special-conferences/metabolism-and-cancer/financial-support-for-attendance.aspx
Sponsor: American Association for Cancer Research
615 Chestnut Street, 17th Floor
Philadelphia, PA 19106-4404

AACR Outstanding Investigator Award for Breast Cancer Research 145

The AACR Outstanding Investigator Award for Breast Cancer Research, funded by Susan G. Komen for the Cure, will recognize an investigator of no more than 50 years of age whose novel and significant work has had or may have a far-reaching impact on the etiology, detection, diagnosis, treatment, or prevention of breast cancer. Such work may involve any discipline across the continuum of biomedical research, including basic,

translational, clinical, and epidemiological studies. The recipient of the Award will receive a $10,000 honorarium and present a 25-minute lecture at the Annual San Antonio Breast Cancer Symposium. Nominations may be made by any scientist, whether an AACR member or nonmember, who is now or has been affiliated with any institution involved in cancer research, cancer medicine, or cancer-related biomedical science. Candidates may not nominate themselves. Nominations must be submitted online at https://proposalcentral.altum.com, no later than 4:00 p.m. United States Eastern Time on May 15th.
Requirements: All cancer researchers who are affiliated with any institution involved in cancer research, cancer medicine, or cancer-related biomedical science anywhere in the world may be nominated. Such institutions include those in academia, industry, or government. Candidates must be no more than 50 years of age at the time the Award is received.
Restrictions: Institutions or organizations are not eligible for the Award. Paper nominations will not be accepted.
Geographic Focus: All States
Date(s) Application is Due: May 15
Amount of Grant: 10,000 USD
Contact: Monique P. Eversley, Program Associate; (267) 646-0576 or (215) 440-9300, ext. 1400; fax (215) 440-9372; monique.eversley@aacr.org or awards@aacr.org
Internet: http://www.aacr.org/home/scientists/scientific-achievement-awards/outstanding-investigator-award.aspx
Sponsor: American Association for Cancer Research
615 Chestnut Street, 17th Floor
Philadelphia, PA 19106-4404

AACR Scholar-in-Training Awards **146**
Scholar-in-Training Awards provide financial support for early-career scientists to attend the AACR Annual Meeting. Since its inception in 1986, the program has provided over 3,200 grants and has received support from more than 30 cancer research foundations, corporations, individuals, and other organizations dedicated to the fight against cancer. A stipend (from $400-$2,000 depending on the geographic location of the recipient or the type of award) will be presented to recipients on-site. Recipients may pick up their a check and welcome package at the Associate Member Resource and Career Center.
Requirements: Eligible candidates are graduate students, medical students and residents, clinical fellows or equivalent, and postdoctoral fellows. Eligible candidates must be the presenter of a proffered paper. Eligible candidates may be traveling within the U.S. or from abroad. Some awards are specifically designated for those traveling from Asia, Europe, and countries with emerging economies.
Restrictions: Employees or subcontractors of private industry are not eligible.
Geographic Focus: All States
Date(s) Application is Due: Jul 20
Amount of Grant: 400-2,000 USD
Contact: Mona E. Shater, Staff Assistant; (215) 440-9300 or (267) 646-0654; fax (215) 440-9372; mona.shater@aacr.org
Internet: http://www.aacr.org/home/scientists/meetings—workshops/special-conferences/metabolism-and-cancer/financial-support-for-attendance.aspx
Sponsor: American Association for Cancer Research
615 Chestnut Street, 17th Floor
Philadelphia, PA 19106-4404

AACR Scholar-in-Training Awards-Special Conferences **147**
The AACR offers Scholar-in-Training Awards to enhance the education and training of early career scientists by providing financial support for their attendance at AACR Special Conferences. These awards are provided to offset a portion of the registration, travel, and subsistence expenses incurred in attending these conferences. A stipend (typically $400-$1,400) will be presented to award recipients at the conference.
Requirements: All graduate and medical students, postdoctoral fellows, and physicians-in-training who submit abstracts may be considered for a Scholar-in-Training Award.
Restrictions: Employees or subcontractors of private industry are not eligible.
Geographic Focus: All States
Amount of Grant: 400-1,400 USD
Contact: Mona E. Shater, Staff Assistant; (215) 440-9300 or (267) 646-0654; fax (215) 440-9372; mona.shater@aacr.org or programs@aacr.org
Internet: http://www.aacr.org/home/scientists/meetings—workshops/travel-grants/scholar-in-training-awards—-special-conferences.aspx
Sponsor: American Association for Cancer Research
615 Chestnut Street, 17th Floor
Philadelphia, PA 19106-4404

AACR Team Science Award **148**
The Award has been established by the AACR and Eli Lilly and Company to acknowledge and catalyze the growing importance of interdisciplinary teams to the understanding of cancer and/or the translation of research discoveries into clinical cancer applications. In addition, through the presentation of this Award, the AACR and Eli Lilly seek to effect change within the traditional cancer research culture by recognizing those institutions that value and foster interdisciplinary team science. These institutions will have demonstrated their support of a team science environment by creating mechanisms to enhance the required infrastructure, such as through pilot funding, technology transfer offices, shared resources, etc., and by presenting awards, honors, appointments, and promotions to those who participate in interdisciplinary teams. Nominations for the 2009 Award will be accepted until November 16.
Requirements: For the purpose of this Award, a team is defined as a group of individuals representing interdisciplinary expertise, each of whom have made substantive and quantifiable contributions to the research being recognized. Team members may be working within the same institution or at several institutions. Candidacy is open to all cancer researchers who are affiliated with any institution involved in cancer research, cancer medicine, or cancer-related biomedical science anywhere in the world. Such institutions include those in academia, industry, or government.
Geographic Focus: All States
Date(s) Application is Due: Nov 16
Amount of Grant: 50,000 USD
Contact: Monique P. Eversley, Program Associate; (267) 646-0576 or (215) 440-9300, ext. 1400; fax (215) 440-9372; monique.eversley@aacr.org or awards@aacr.org
Internet: http://www.aacr.org/home/scientists/scientific-achievement-awards/team-science-award.aspx
Sponsor: American Association of Cancer Research
615 Chestnut Street, 17th Floor
Philadelphia, PA 19106-4404

AACR Thomas J. Bardos Science Education Awards **149**
AACR-Thomas J. Bardos Science Education Awards for Undergraduate Students are available to full-time third-year undergraduate students majoring in science. The purpose of these awards is to inspire young science students to enter the field of cancer research. The award consists of a waiver of registration fees for two annual AACR meetings and $1500/year for research and travel expenses. The award is for a two-year period.
Requirements: Candidates must be full-time, third year undergraduate students majoring in science. Applications from students who are not yet committed to cancer research are welcome.
Geographic Focus: All States
Date(s) Application is Due: Dec 7
Amount of Grant: 1,500 USD
Contact: Mona E. Shater, Staff Assistant; (215) 440-9300 or (267) 646-0654; fax (215) 440-9412; mona.shater@aacr.org or scienceeducation@aacr.org
Internet: http://www.aacr.org/home/scientists/meetings—workshops/travel-grants/aacr-thomas-j-bardos-science-education-awards-for-undergraduate-students-.aspx
Sponsor: American Association for Cancer Research
615 Chestnut Street, 17th Floor
Philadelphia, PA 19106-4404

AAES Engineering Journalism Award **150**
The purpose of this American Association of Engineering Societies (AAES) award is to recognize outstanding reporting of an event or issue that furthers public understanding of engineering and is given in one of these three categories: daily newspapers, general circulation print media, and broadcast radio or television. Nominations must be articles published or broadcast in English between January 1 and December 31. Nominations are welcome from reporters, editors, publishers, and members of the engineering community. Print entries must be easily reproducible; electronic entries must include both a VHS or audio cassette and a script. The nomination cycle runs from October 1 to December 31.
Geographic Focus: All States
Date(s) Application is Due: Dec 31
Amount of Grant: Up to 5,000 USD
Samples: Daniel Polin; Kenneth Mandel; Jon Palfreman.
Contact: William Koffel, Executive Director; (202) 296-2237, ext. 201; fax (202) 296-1151; wkoffel@aaes.org or info@aaes.org
Internet: http://www.aaes.org/communications/awards.asp
Sponsor: American Association of Engineering Societies
1620 I Street NW, Suite 210
Washington, DC 20006-4028

AAES National Engineering Award **151**
Established in 1979, the National Engineering Award is presented on behalf of the engineering profession to those engineers whose leadership and accomplishments have particularly benefited humanity. The Award is presented for inspirational leadership and tireless devotion to the improvement of engineering education and to the advancement of the engineering profession, as well as to the development of sound public policies as an engineer-statesman. The award may be jointly presented to two or more engineers if a team effort is worthy of recognition.
Geographic Focus: All States
Date(s) Application is Due: Dec 31
Samples: Harold (Bud) E. Nelson; E. Walter LeFevre; Daniel R. Benigni.
Contact: William Koffel, Executive Director; (202) 296-2237, ext. 201; fax (202) 296-1151; wkoffel@aaes.org or info@aaes.org
Internet: http://www.aaes.org/communications/awards.asp
Sponsor: American Association of Engineering Societies
1620 I Street NW, Suite 210
Washington, DC 20006-4028

AAES Norm Augustine Award **152**
Established in 1998, the Norm Augustine Award for Outstanding Achievement in Engineering Communications is presented annually to an engineering who has demonstrated the capacity for communicating the excitement and wonder of engineering. The Award is named a for one of the titans of the engineering profession-Norm Augustine-whose leadership within the engineering community and service to society has established him as one our nation's greatest engineers. The Award is to be conferred on those rare individuals who can speak with passion about engineering-its promise as well as its responsibility-so that the public may have a better understanding of engineering and a better appreciation for how engineers improve our quality of life.

Geographic Focus: All States
Date(s) Application is Due: Dec 31
Samples: Celeste Baine; Bernard Amadei; William Gene Corely, Ph.D.
Contact: William Koffel, Executive Director; (202) 296-2237, ext. 201; fax (202) 296-1151; wkoffel@aaes.org or info@aaes.org
Internet: http://www.aaes.org/communications/awards.asp
Sponsor: American Association of Engineering Societies
1620 I Street NW, Suite 210
Washington, DC 20006-4028

AAFA Investigator Research Grants 153
AAFA is dedicated to finding the causes, new treatments and cures for asthma and allergic diseases. AAFA offers a program that sponsors seed grants for investigators wishing to explore new areas of scientific merit related to asthma and allergic diseases. Support has enabled 80% of the researchers who have received the AAFA research grant to win additional research funding from other sources totaling more than $18 million. The Asthma and Allergy of Foundation of America is the only patient organization in the United States that funds research grants for both asthma and allergies. At this time, the application process for grants is through an NIH referral system.
Geographic Focus: All States
Samples: Stephen C. Dreskin, MD, PhD, University of Colorado, Boulder, Colorado; Anthony A. Horner, MD, University of California, San Diego, CA—Analysis of airway; David Lewis, MD, Stanford University School of Medicine—Mechanisms of viral induced asthma.
Contact: Mary Brasler, Director; (202) 466-7643, ext. 238; mbrasler@aafa.org
Internet: http://www.aafa.org/display.cfm?id=6&sub=64
Sponsor: Asthma and Allergy Foundation of America
8201 Corporate Drive, Suite 100
Landover, MD 20785

AAFCS International Graduate Fellowships 154
The Association encourages graduate study in family and consumer sciences and its subspecialties, such as textiles and clothing, nutrition, institutional management/food service systems administration, consumer studies, Cooperative Extension, and family and consumer sciences communication, education, and administration. International graduate fellowships are awarded to individuals who have exhibited the potential to make contributions to the family and consumer sciences profession.
Requirements: Applicants must be citizens or permanent residents of the United States. Applicant must show clearly defined plans for full-time graduate or undergraduate study during the time for which the fellowship is awarded. Each request for appropriate application form must be accompanied by an application fee of $40.
Geographic Focus: All States
Contact: Fellowships Committee; (703) 706-4600; fax (703) 706-4663; staff@aafcs.org
Internet: http://www.aafcs.org/programs/scholarships.htm
Sponsor: American Association of Family and Consumer Sciences
400 N Columbus Street, Suite 202
Alexandria, VA 22314

AAFCS National Graduate Fellowships 155
AAFCS annually awards national graduate fellowships to support graduate study in areas such as family and consumer sciences, nutrition, textiles, and home economics. The fellowships are awarded to individuals who have exhibited the potential for making contributions to these professions. It is recommended that applicants first complete at least one year of professional family and consumer science experience by the time of application.
Requirements: Applicants must be citizens or permanent residents of the United States. Applicant must show clearly defined plans for full-time graduate or undergraduate study during the time for which the fellowship is awarded. Each request for appropriate application form must be accompanied by an application fee of $40.
Geographic Focus: All States
Date(s) Application is Due: Jan 7
Amount of Grant: Up to 5,000 USD
Contact: Fellowships Committee; (703) 706-4600; fax (703) 706-4663; staff@aafcs.org
Internet: http://www.aafcs.org/programs/fellowshipseven.html
Sponsor: American Association of Family and Consumer Sciences
400 N Columbus Street, Suite 202
Alexandria, VA 22314

AAFP Foundation Joint Grants 156
The foundation supports clinical research projects within family practice that ultimately will result in improved patient care. This program includes a research grant stimulation program that is designed to stimulate research in family practice by supporting small and first-time research projects. Longer research proposals must be prepared in accordance with standard guidelines for grant proposals.
Requirements: Eligible to apply are individual family physicians, residents in family practice, and medical students planning to pursue the specialty of family practice; educational and health care institutions or organizations that will use the grant exclusively for programs or projects directly involved with family practice or family medicine; departments of family practice; family practice residency programs; and family practice organizations or associations. Research stimulation proposals may be reviewed at any time during the year depending on availability of funds.
Restrictions: Proposals seeking support for the cost of instituting programs or support for such activities as videotape production, designing curriculum, or implementing a project, etc., will be denied.
Geographic Focus: All States
Amount of Grant: Up to 30,000 USD
Contact: Susie Morantz, Senior Program Manager; (913) 906-6000, ext. 4470 or (800) 274-2237; fax (913) 906-6095; smorantz@aafp.org
Internet: http://www.aafpfoundation.org/x270.xml
Sponsor: American Academy of Family Physicians Foundation
11400 Tomahawk Creek Parkway, Suite 440
Leawood, KS 66211-2672

AAFP Foundation Pfizer Teacher Development Awards 157
The purpose of the awards is to honor community-based physicians who give up time from their practice to teach family medicine students and/or residents part time. Eligible candidates may serve as preceptors or as volunteer teachers at other sites, including family medicine teaching centers. Guidelines are available online.
Requirements: Applicant must have graduated from an ACGME-approved family practice residency program within the last seven years; have entered family medicine; be a member of the American Academy of Family Physicians; have entered or plan to enter part-time family medicine teaching (not less than four hours per month and no more than 32 hours per month—averaged over a year); and teach voluntarily or receive no more than $18,000 compensation for the educational time devoted to residents and/or students.
Geographic Focus: All States
Amount of Grant: 2,000 USD
Contact: Sondra Goodman, Grants Program Manager; (913) 906-6000, ext. 4457 or (800) 274-2237; fax (913) 906-6095; sgoodman@aafp.org
Internet: http://www.aafpfoundation.org/x269.xml
Sponsor: American Academy of Family Physicians Foundation
11400 Tomahawk Creek Parkway, Suite 440
Leawood, KS 66211-2672

AAFP Foundation Wyeth Immunization Awards 158
This program offers family medicine residency programs the opportunity to receive recognition for identifying and developing creative solutions to overcoming barriers to childhood immunizations, thus increasing immunization rates, and to share the best practices that have benefited the communities in which they live. Applications are accepted annually for one of the following award track categories: Best Practices Award—overall achievement with systems already in place to overcome immunization barriers and achieve high rates within a certain time frame; Most Improved Award—overcame barriers and other challenges to greatly enhance immunization rates; and Implement New System Grant—residency programs seeking to implement a new system to help increase immunization rates in under-served children. Grant winners will receive: one of 10 monetary awards (four at $5,000 and six at $10,000) determined by rank of application as scored by review panel; an education scholarship for $1,000 to send a resident to the annual AAFP National Conference of Family Medicine Residents and Medical Students; an additional travel scholarship to send a resident to the AAFP Scientific Assembly (top six); and a plaque of recognition.
Requirements: Family medicine residency programs that have increased immunization rates within a defined time frame and can demonstrate the best practices implemented to achieve the higher compliance are eligible.
Geographic Focus: All States
Date(s) Application is Due: Apr 21
Amount of Grant: 1,000-10,000 USD
Contact: Sondra Goodman, Grants Program Manager; (913) 906-6000, ext. 4457 or (800) 274-2237; fax (913) 906-6095; sgoodman@aafp.org
Internet: http://www.aafpfoundation.org/x655.xml
Sponsor: American Academy of Family Physicians Foundation
11400 Tomahawk Creek Parkway, Suite 440
Leawood, KS 66211-2672

AAFP Research Stimulation Grants 159
This Program provides relatively small grants to stimulate research in family practice in the areas of preliminary efforts leading to a larger research project, pilot projects, and data collection for a larger research project. The applicant must include a discussion of how the proposed project is anticipated to lead to a larger project. Grant applications will be accepted twice a year: by March 1 and September 1.
Requirements: Eligible to apply are individual family physicians, family practice/medicine organizations or associations, residents in family practice, departments of family medicine, and educational and health care institutions directly involved in family practice/medicine.
Geographic Focus: All States
Date(s) Application is Due: Mar 1; Sep 1
Amount of Grant: 7,500 USD
Contact: Susie Morantz, Senior Program Manager; (913) 906-6000, ext. 4470 or (800) 274-2237; fax (913) 906-6095; smorantz@aafp.org
Internet: http://www.aafpfoundation.org/x446.xml
Sponsor: American Academy of Family Physicians Foundation
11400 Tomahawk Creek Parkway, Suite 440
Leawood, KS 66211-2672

AAFPRS Ben Shuster Memorial Award 160
This award is given annually for the best paper based on clinical work or research in the field of facial plastic surgery by a resident or fellow in training. The paper must have been delivered at a national meeting (or its equivalent) within the preceding two years prior to the February 1 date of submission. Studies prepared during the first year after completion of residency training will be considered provided that the research was

conducted during the author's residency training program or fellowship. A certificate and an award of $1,000 are presented.
Requirements: The competition is open to U.S. or Canadian residents or fellows in otolaryngology who are members of AAFPRS. Each entrant must be the sole or senior author and an AAFPRS member.
Geographic Focus: All States
Date(s) Application is Due: Feb 1
Amount of Grant: 1,000 USD
Contact: Awards Coordinator; (703) 299-9291; fax (703) 299-8898; info@aafprs.org
Internet: http://www.aafprs.org/physician/awards_grants/awards.html
Sponsor: American Academy of Facial, Plastic, and Reconstructive Surgery
310 S Henry Street
Alexandria, VA 22314

AAFPRS Bernstein Grant 161
The purpose of the grant is to encourage original research projects that will advance facial plastic and reconstructive surgery. The grant may be used as seed money for research projects. Applicants should submit an application, a detailed proposal, the investigator's curriculum vita, and letters from the appropriate institutional review committees certifying institutional conformity to the U.S. government guidelines for human and animal experiments. The recipient should complete his or her work within three (3) years with interim reports due each year. All research grant monies awarded are intended to fund direct costs related to the research as described in the grant proposal.
Requirements: Applicants should be U.S. or Canadian AAFPRS fellow members. The primary criteria are that the research be original and have direct application to facial plastic and reconstructive surgery.
Geographic Focus: All States
Date(s) Application is Due: Jan 15
Amount of Grant: 25,000 USD
Contact: Grant Coordinator; (703) 299-9291; fax (703) 299-8898; info@aafprs.org
Internet: http://www.aafprs.org/physician/awards_grants/bernstein_grant.html
Sponsor: American Academy of Facial, Plastic, and Reconstructive Surgery
310 S Henry Street
Alexandria, VA 22314

AAFPRS Fellowships 162
The Educational and Research Foundation for the American Academy of Facial Plastic and Reconstructive Surgery fellowship program provides postgraduate training in facial plastic surgery. The objectives of the fellowship program are to: provide an outstanding academic opportunity for the acquisition of specialized knowledge and skills in facial plastic surgery; develop trained specialists who will contribute to the ongoing development of facial plastic and reconstructive surgery; foster development of facial plastic and reconstructive surgery educators, especially in residency programs; and encourage the development of new skills and knowledge in facial plastic and reconstructive surgery through basic research and clinical trials. Eligible physicians may apply for 40 positions available each year.
Requirements: Applicants to the Fellowship Program must be physicians who are in or have completed an otolaryngology or plastic surgery residency program accredited by the Accreditation Council for Graduate Medical Education (ACGME) or Royal College of Physicians and Surgeons of Canada (RCPSC) or board-certified in otolaryngology-head and neck surgery or plastic surgery. Applicants must be members of the AAFPRS before submitting an application.
Restrictions: Applicants should not be full-time faculty members holding the rank of assistant professor or higher at the institution where the fellowship will take place.
Geographic Focus: All States
Date(s) Application is Due: Feb 1
Contact: Fellowship Coordinator; (703) 299-9291; fax (703) 299-8898; info@aafprs.org or fsanders@aafprs.org
Internet: http://www.aafprs.org/physician/benefits/ben_fellow.html
Sponsor: American Academy of Facial, Plastic, and Reconstructive Surgery
310 S Henry Street
Alexandria, VA 22314

AAFPRS Investigator Development Grant 163
The purpose of the Investigator Development Grant is to support the work of a young faculty member in facial plastic surgery conducting significant clinical or laboratory research and involved in the training of resident surgeons in research. One $15,000 grant may be awarded each year upon receipt of the signed agreement. Recipients should complete their work within three (3) years with interim reports due each year. An extension is possible upon written request to the Research Committee. The original proposal and all other requirements should be submitted to C.O.R.E. no later than January 15 or the next business day thereafter.
Requirements: An applicant must be an AAFPRS member at any level involved in the training of resident surgeons.
Geographic Focus: All States
Date(s) Application is Due: Jan 15
Amount of Grant: Up to 15,000 USD
Contact: Grant Coordinator; (703) 299-9291; fax (703) 299-8898; info@aafprs.org
Internet: http://www.aafprs.org/physician/awards_grants/investigator_grant.html
Sponsor: American Academy of Facial, Plastic, and Reconstructive Surgery
310 S Henry Street
Alexandria, VA 22314

AAFPRS Ira J. Tresley Research Award 164
This annual award is given to a U.S. or Canadian academy member for the best paper based on any research study in the field of facial plastic surgery. The competition is open to any U.S. or Canadian physician who is an AAFPRS member and who has been board certified for at least three years. The paper must have been presented at a national meeting or its equivalent within the preceding two years prior to the February 1 date of submission. A certificate and an award of $1,000 are presented.
Geographic Focus: All States
Date(s) Application is Due: Feb 1
Amount of Grant: 1,000 USD
Contact: Awards Coordinator; (703) 299-9291; fax (703) 299-8898; info@aafprs.org
Internet: http://www.aafprs.org/physician/awards_grants/awards.html
Sponsor: American Academy of Facial, Plastic, and Reconstructive Surgery
310 S Henry Street
Alexandria, VA 22314

AAFPRS John Orlando Roe Award 165
This award, named for John Orlando Roe, the surgeon who accomplished the first rhinoplasty in 1887, includes a certificate and an award of $1,000 to be presented each year to an AAFPRS Fellow in an AAFPRS Foundation Fellowship Program who submits the best clinical research paper written during the current fellowship year.
Geographic Focus: All States
Date(s) Application is Due: Feb 1
Amount of Grant: 1,000 USD
Contact: Awards Coordinator; (703) 299-9291; fax (703) 299-8898; info@aafprs.org
Internet: http://www.aafprs.org/physician/awards_grants/awards.html
Sponsor: American Academy of Facial, Plastic, and Reconstructive Surgery
310 S Henry Street
Alexandria, VA 22314

AAFPRS Leslie Bernstein Resident Research Grants 166
Two resident research grants, tenable in the United States or Canada, may be awarded each year to stimulate resident research in facial plastic surgery on projects that are well conceived and scientifically valid. Residents are encouraged to enter early in their training so that their applications may be revised and resubmitted if not accepted the first time. This award may be integrated with other funding to complete a project. Interested residents should write for entry guidelines.
Requirements: U.S. and Canadian residents at any level who are AAFPRS members are eligible to apply, even if the research work will be done during their fellowship year. Grant recipients must complete their work in one (1) to two (2) years with interim reports due each year.
Geographic Focus: All States
Date(s) Application is Due: Jan 15; Dec 15
Amount of Grant: 5,000 USD
Contact: Grant Coordinator; (703) 299-9291; fax (703) 299-8898; info@aafprs.org
Internet: http://www.aafprs.org/physician/awards_grants/grants.html
Sponsor: American Academy of Facial, Plastic, and Reconstructive Surgery
310 S Henry Street
Alexandria, VA 22314

AAFPRS Residency Travel Awards 167
Two Residency Travel Awards may be given each year for two outstanding papers in facial plastic and reconstructive surgery primarily authored by a resident or medical student in training. The paper must be submitted by February 1 for consideration, which will be presented at the Annual Fall Meeting. Entries should conform to the guidelines of the Archives of Facial Plastic Surgery, the official journal of the AAFPRS. Papers submitted will be judged anonymously. Only the cover page should contain the applicant's name or institutional identification.
Geographic Focus: All States
Date(s) Application is Due: Feb 1
Contact: Awards Coordinator; (703) 299-9291; fax (703) 299-8898; info@aafprs.org
Internet: http://www.aafprs.org/physician/awards_grants/awards.html
Sponsor: American Academy of Facial, Plastic, and Reconstructive Surgery
310 S Henry Street
Alexandria, VA 22314

AAFPRS Sir Harold Delf Gillies Award 168
This award is named for Sir Harold Delf Gillies, a British otolaryngologist who in September 1917 described the tubed pedicle flap. Dr. Gillies frequently visited the U.S. and lectured widely to surgeons of various specialties and was given the title Father of Plastic Surgery. A certificate and an award of $1,000 will be presented each year to an AAFPRS Fellow in an AAFPRS Foundation Fellowship Program who submits the best basic science research paper written during the current fellowship year.
Geographic Focus: All States
Date(s) Application is Due: Feb 1
Amount of Grant: 1,000 USD
Contact: Awards Coordinator; (703) 299-9291; fax (703) 299-8898; info@aafprs.org
Internet: http://www.aafprs.org/physician/awards_grants/awards.html
Sponsor: American Academy of Facial, Plastic, and Reconstructive Surgery
310 S Henry Street
Alexandria, VA 22314

AAF Richard Morris Hunt Architecture Fellowship 169

The Richard Morris Hunt Fellowship, c0-sponsored by the American Architectural Foundation and the French Heritage Society as part of a commitment to stewardship of American and French heritage, is awarded to architects pursuing a career in historic preservation. The fellowship permits young architects from France and the United States to observe and practice the latest preservation technologies and techniques in each other's countries. Inaugurated in 1990 as part of the organizations' commitment to foster stewardship of architectural heritage, the recipients represent the best their countries have to offer in the field of historic preservation. The six-month fellowship alternates each year between a French and an American architect and carries a stipend of $25,000.

Geographic Focus: All States, France
Amount of Grant: 25,000 USD
Contact: Liz Blazevich, Program Director; (202) 787-1001; fax (202) 787-1002; eblazevich@archfoundation.org or info@archfoundation.org
Internet: http://www.archfoundation.org/2011/11/richard-morris-hunt-fellowship-overview/
Sponsor: American Architectural Foundation
2101 L Street NW, Suite 670
Washington, DC 20037

AAG Anne U. White Fund Grants 170

The Fund's income is to be used to encourage field research conducted by a member of the Association jointly with her or his partner. The Fund will serve the purposes Anne White held dear if it enables people, regardless of any formal training in geography, to engage in useful field studies and to have the joy of working along side their partners. In the past, two awards have been made each year of approximately $1,500 each. Digital submissions are required. All applicants will be notified of the status of their submission within approximately 3 months after the deadline. Awardees may anticipate receiving their grants by late Spring.

Geographic Focus: All States
Date(s) Application is Due: Dec 31
Amount of Grant: 1,500 USD
Contact: Ehsan Khater, Coordinator; (202) 234-1450; fax (202) 234-2744; ekhater@aag.org or grantsawards@aag.org
Internet: http://www.aag.org/Grantsawards/Annewhitefund.html
Sponsor: Association of American Geographers
1710 Sixteenth Street NW
Washington, DC 20009-3198

AAG Dissertation Research Grants 171

The following funds provide financial assistance to PhD candidates preparing dissertations in geography. The Robert D. Hodgson Fund gives preference to those topics that best reflect Dr. Hodgson's belief that the pursuit and understanding of geographic knowledge may lead to international cooperation. The Paul Vouras Fund gives preference to minority applicants for support of dissertation work. The Otis Paul Starkey Fund gives preference to regional study or significant problem areas in the United States or its possessions. Funds are to be used for direct research expenses and may not be used to cover overhead costs. Applicants may obtain application forms from the association.

Requirements: Any person is eligible to apply for a grant who has been a member of AAG for at least a year at the time of submitting the application, does not have a PhD degree at the time of the award, and has completed all PhD requirements except for the dissertation research.

Geographic Focus: All States
Date(s) Application is Due: Dec 31
Amount of Grant: Up to 500 USD
Contact: Ehsan Khater, Coordinator; (202) 234-1450; fax (202) 234-2744; ekhater@aag.org or grantsawards@aag.org
Internet: http://www.aag.org/Grantsawards/Dissertationresearch.html
Sponsor: Association of American Geographers
1710 Sixteenth Street NW
Washington, DC 20009-3198

AAG E. Willard and Ruby S. Miller Award in Geography 172

This annual award recognizes members of the Association who have made truly outstanding contributions to the geographic field due to their special competence in teaching or research. Funding for the award comes from the estate of Ruby S. Miller. More than one award may be awarded each year. Each award includes $1,000 and a commemorative plaque. Electronic submission is encouraged.

Requirements: Members from all fields of geography are eligible for the award. Nominations (including self-nominations) are welcome.

Geographic Focus: All States
Date(s) Application is Due: Dec 31
Amount of Grant: 1,000 USD
Contact: Ehsan Khater, Coordinator; (202) 234-1450; fax (202) 234-2744; ekhater@aag.org or grantsawards@aag.org
Internet: http://www.aag.org/Grantsawards/miller.html
Sponsor: Association of American Geographers
1710 Sixteenth Street NW
Washington, DC 20009-3198

AAG International Geographic Information Fund Graduate Research Grants 173

The Association of American Geographers will award one or more small grants of up to $500 to support graduate research proposed by a student in any area of spatial analysis or geographic information science or systems. The award(s) will be presented at the AAG Annual Meeting. The program is designed to support college and university student career development in the academic areas of applied spatial data analysis or geographic information systems (GIS). Digital submissions are encouraged.

Requirements: AAG IGIF Graduate Research Awards are available to full-time students who are currently registered in an undergraduate or graduate degree program providing either a degree or explicit specialization in some area of applied spatial data analysis or GIS study at a duly accredited and recognized college, university or other educational institution located within the United States.

Geographic Focus: All States
Date(s) Application is Due: Dec 31
Amount of Grant: 500 USD
Contact: Ehsan Khater, Coordinator; (202) 234-1450; fax (202) 234-2744; ekhater@aag.org or grantsawards@aag.org
Internet: http://www.aag.org/Grantsawards/IGIF.cfm
Sponsor: Association of American Geographers
1710 Sixteenth Street NW
Washington, DC 20009-3198

AAG International Geographic Information Fund Student Travel Grants 174

The Association of American Geographers will award one or more small grants of approximately $500 for the support of student travel to national and international symposia or specialized meetings sponsored by recognized professional organizations supporting activities associated with specific aspects of geographic data analysis and handling and geographic information systems. Support of student travel means reimbursement for the cost of transportation, accommodations, meals, and any other cost or expense reasonable related to travel for attendance at such meetings. The program is designed to support college and university student career development in the academic areas of applied spatial data analysis or geographic information systems (GIS).

Requirements: Grants are available to full-time students who are currently registered in an undergraduate or graduate degree program providing either a degree or explicit specialization in some area of applied spatial data analysis or GIS study at a duly accredited and recognized college, university or other educational institution located within the United States.

Geographic Focus: All States
Date(s) Application is Due: Dec 31
Amount of Grant: 500 USD
Contact: Ehsan Khater, Coordinator; (202) 234-1450; fax (202) 234-2744; ekhater@aag.org or grantsawards@aag.org
Internet: http://www.aag.org/Grantsawards/IGIF.cfm
Sponsor: Association of American Geographers
1710 Sixteenth Street NW
Washington, DC 20009-3198

AAG John Brinkerhoff Jackson Prize 175

The association offers this award for the best nonfiction book about the human geography of the United States. The prize will go to the writer who can best interpret geographical research in language that lay readers can understand. Publishers are invited to submit entries for the Jackson Prize competition, and should forward one copy of the published book to each of the four committee members listed at the web site. The prize, which carries an award of $1,000 to the author, is administered by the Association of American Geographers.

Requirements: The prize is restricted to books written by geographers, with preference given to those by U.S. citizens or permanent residents.

Restrictions: Textbooks, dissertations, and scholarly articles will not be considered.

Geographic Focus: All States
Date(s) Application is Due: Dec 31
Amount of Grant: 1,000 USD
Contact: Michael P. Conzen, Chair; (312) 702-8308; m-conzen@uchicago.edu
Internet: http://www.aag.org/Grantsawards/jackson_prize.cfm
Sponsor: Association of American Geographers
1710 Sixteenth Street NW
Washington, DC 20009-3198

AAG Mel Marcus Fund Grants for Physical Geography 176

The objective of The Mel Marcus Fund for Physical Geography is to carry on this tradition of excellence and humanity in field work. Grants from the Mel Marcus Fund for Physical Geography will foster personally formative participation by students in field-based physical geography research in challenging outdoor environments. Marcus Fund grants of up to $2000 will be awarded to faculty applicants to support travel and logistical costs of including students in field research in physical geography. Research sites may include international as well as national locations. Marcus grant funds may be used to cover travel to the research site, as well as expenses related to the logistical support of the student, such as food, fuel and lodging. Digital submissions are encouraged.

Restrictions: Student stipends are not eligible expenses under this grant.

Geographic Focus: All States
Date(s) Application is Due: Dec 31
Amount of Grant: Up to 2,000 USD
Contact: Ehsan Khater, Coordinator; (202) 234-1450; fax (202) 234-2744; ekhater@aag.org or grantsawards@aag.org
Internet: http://www.aag.org/Grantsawards/marcus.cfm
Sponsor: Association of American Geographers
1710 Sixteenth Street NW
Washington, DC 20009-3198

AAG Research Grants 177

Grants are given to support research and field work done by AAG members. Preference will be given to projects that may generate substantial support from private or federal granting agencies; may be pertinent to questions affecting important gaps in the spectrum of geographic knowledge and research; or will open up new areas, topics, or techniques for geographic research. Applicants should obtain forms from AAG.

Requirements: Any person is eligible to apply for a grant who has been a member of the association for at least two years at the time of submitting an application.

Restrictions: This grant will not be made to support doctoral dissertations.

Geographic Focus: All States

Date(s) Application is Due: Dec 31

Amount of Grant: Up to 1,000 USD

Contact: Ehsan Khater, Coordinator; (202) 234-1450; fax (202) 234-2744; ekhater@aag.org or grantsawards@aag.org

Internet: http://www.aag.org/Grantsawards/Generalresearch.html

Sponsor: Association of American Geographers
1710 Sixteenth Street NW
Washington, DC 20009-3198

AAHN Grant for Historical Research 178

The American Association for the History of Nursing is offering a research grant of $3,000 for new researchers. (Indirect costs of 8% are also available). It is expected that the research and new materials produced by the award winner will help ensure the growth of scholarly work focused on the history of nursing. The deadline for submission of applications is April 1. The AAHN Panel will make its decision about the award by May 15 and the recipient will be notified by June 1 of each year. Funding will start July 1 of the grant year and last for one year. A no-cost extension may be granted on request.

Requirements: Applicants must be members of AAHN and hold the doctorate. They may be faculty members or independent researchers.

Geographic Focus: All States

Date(s) Application is Due: Apr 1

Amount of Grant: 3,000 USD

Contact: David Stumph; (303) 422-2685; fax (303) 422-8894; aahn@aahn.org

Internet: http://www.aahn.org/grants.html

Sponsor: American Association of the History of Nursing
10200 W 44th Avenue, Suite 304
Wheat Ridge, CO 80033

AAHN Pre-Doctoral Research Grant 179

This grant is designed to encourage and support graduate training and historical research at the Masters and Doctoral levels. The grant will be in the amount of $2,000. Proposals will focus on a significant question in the history of nursing. Selection criteria include the scholarly merit of the proposal, consideration of the student's preparation for this study, the advisor's qualifications for guiding the study and the project's potential for contributing to scholarship in the field of nursing history.

Requirements: The student must be: enrolled in an accredited masters program or doctoral program; and a member of AAHN. The research advisor must be doctorally prepared with scholarly activity in the field of nursing history and prior experience in guidance of research training.

Geographic Focus: All States

Date(s) Application is Due: Apr 1

Contact: David Stumph; (303) 422-2685; fax (303) 422-8894; aahn@aahn.org

Internet: http://www.aahn.org/grants.html

Sponsor: American Association of the History of Nursing
10200 W 44th Avenue, Suite 304
Wheat Ridge, CO 80033

AAHPM Hospice and Palliative Medicine Fellowships 180

The program offers training programs to prepare physicians for careers in academic or community-based palliative medicine. Fellowships provide clinical experience and supervision for palliative medicine in hospitals, homes, hospices, chronic care institutions, and outpatient settings, and include consultational and longitudinal care exposure, as well as training in bereavement care. Some programs also provide opportunities for mentored research or teaching. Most programs begin in July, and clinical training generally lasts one year, allowing for board eligibility upon successful completion. For a complete list of U.S. palliative medicine fellowship training programs, members should visit the web site.

Requirements: Trainees must have already completed residency training in a primary specialty, such as internal or family medicine, or have similar clinical training and experience.

Geographic Focus: All States

Contact: Fellowships Director; (847) 375-4712; fax (847) 375-6475; info@aahpm.org

Internet: http://www.aahpm.org/fellowship/index.html

Sponsor: American Academy of Hospice and Palliative Medicine
4700 W Lake Avenue
Glenview, IL 60025-1485

AAJA Fellowships 181

The fellowships are designed to enable members of the association to attend short-term professional training and skills development programs. These fellowships can be used to defray tuition, travel, food, lodging, and other related costs. News organizations of the recipients must grant them a leave of absence to attend the program. Fellowships must be used within 12 months of the award date. Applications are accepted on a case-by-case basis year-round. Grants will be awarded contingent upon proof of registration at, or acceptance to, the training program.

Requirements: Applicants must be AAJA members with at least three years of professional journalism experience.

Geographic Focus: All States

Amount of Grant: Up to 1,000 USD

Contact: Albert Lee; (415) 346-2051, ext. 107; fax (415) 346-6343; albertl@aaja.org

Internet: http://www.aaja.org/programs/professional/fellowships

Sponsor: Asian American Journalists Association
1182 Market Street, Suite 320
San Francisco, CA 94102

AALAS Foundation Grants 182

The foundation educational grants have a twofold purpose: to attract talented people to the field of laboratory animal science and to develop new training materials that promote the continuing education of AALAS members. The foundation awards grants for student clerkships to enhance the training of future laboratory animal personnel. The head of the training and/or research program sponsoring the clerkship should submit the application, not the student who will receive the education. Applications are judged on the strength of the training program, not the credentials of the students. Three aspects are considered—clinical experience, teaching experience, and research experience—in evaluating student training programs. The foundation also supports the development of training materials that address the continuing educational needs of AALAS members. After completion, AALAS will have unrestricted use of these materials and will make them available to the members. Interested persons should write for the brochure containing information required to apply for either type of grant.

Requirements: Persons currently in the field of laboratory animal science and those persons who are considering the field as a career are eligible. Applications for student clerkship programs must be submitted by the program sponsor.

Geographic Focus: All States

Date(s) Application is Due: Apr 15; Sep 15

Amount of Grant: Up to 5,000 USD

Contact: John McCutchen, (901) 754-8620; fax (901) 753-0046; john.mccutchen@aalas.org

Internet: http://aalasfoundation.org/about_grants_and_funding.html

Sponsor: American Association for Laboratory Animal Science
9190 Crestwyn Hills Drive
Memphis, TN 38125

AAMC Award for Distinguished Research 183

The Award for Distinguished Research in the Biomedical Sciences honors outstanding biomedical research related to health and disease. The research recognized should have contributed to the substance of medicine. Nominations may be made by anyone on the faculty or staff of a medical school or teaching hospital or by a member of an academic society.

Requirements: The nominee must be an individual who serves on the faculty of an AAMC member medical school or teaching hospital.

Geographic Focus: All States

Date(s) Application is Due: May 1

Contact: Sandra D. Gordon, (202) 828-0472; fax (202) 828-1125; sgordon@aamc.org

Internet: http://www.aamc.org/about/awards/research.htm

Sponsor: Association of American Medical Colleges
2450 N Street NW
Washington, DC 20037-1127

AANS Neurosurgery Research and Education Foundation/Spine Section Young Clinician Investigator Award 184

Established by the Neurosurgery Research and Education Foundation (NREF) of the American Association of Neurological Surgeons (AANS), the Joint Section on Disorders of the Spine and Peripheral Nerves Award grants support to young faculty who are pursuing careers in the area of spinal cord, vertebral column or peripheral nerve disorders. The purpose of the award is to fund pilot studies that provide preliminary data used to strengthen applications for more permanent funding from other sources. The one-year award totals $40,000.

Requirements: Applicants must be physicians or neurosurgeons who are full-time faculty in North American teaching institutions and in the early years of their careers.

Restrictions: Applicant may not accept another award for the same project during the same time period. No more than one award per year will be awarded to the same institution.

Geographic Focus: All States

Date(s) Application is Due: Oct 31

Contact: Michele S. Gregory, Director of Development; (847) 378-0540 or (847) 378-0500; fax (847) 378-0600; msg@aans.org or info@aans.org

Internet: http://www.aans.org/research/fellowship/nref_y.asp

Sponsor: Research Foundation of the American Association of Neurological Surgeons
5550 Meadowbrook Drive
Rolling Meadows, IL 60008-3852

AANS Neurosurgery Research and Education Foundation Young Clinician Investigator Award 185

This one-year, renewable award is available to North American neurosurgeons who are full-time faculty in teaching institutions and are in the early years of their careers. The purpose of the program is to fund pilot studies that could provide preliminary data to be used for strengthening applications for more permanent funding from other sources. Applications related to any field of neurosurgery are encouraged. The NREF encourages both patient-oriented clinical research and basic science research projects. The one-year award totals $40,000.

Restrictions: Applicant may not accept another award for the same project during the same time period. No more than one award per year will be awarded to the same institution.

Geographic Focus: All States

Date(s) Application is Due: Oct 31
Amount of Grant: 40,000 USD
Contact: Michele S. Gregory, Director of Development; (847) 378-0540 or (847) 378-0500; fax (847) 378-0600; msg@aans.org or info@aans.org
Internet: http://www.aans.org/research/fellowship/nref_y.asp
Sponsor: Research Foundation of the American Association of Neurological Surgeons
5550 Meadowbrook Drive
Rolling Meadows, IL 60008-3852

AAO-HNSF CORE Research Grants **186**
CORE Research Grants are available in three different categories: Research Project Grants—an independent or assisted project using existing skills and resources, and is hypothesis-based, with specific aims and timetable; Research Training Grants—a research project that is designed to help you refine existing research skill set or acquire new skills, and requires mentor and evidence of training process; and Career Development Grants—a formal, mentored period of research combined with coursework, designed to help researcher become independent researcher. Applicants should complete a Letter of Intent by December 15th, using the online application system, and should then submit the completed application before 12 o'clock midnight EST on January 15.
Geographic Focus: All States
Date(s) Application is Due: Jan 15; Dec 15
Contact: Stephanie L. Jones; (703) 519-1586 or (703) 836-4444; sljones@entnet.org
Internet: http://www.entnet.org/EducationAndResearch/coreGrants.cfm
Sponsor: American Academy of Otolaryngology-Head and Neck Surgery Foundation
1650 Diagonal Road
Alexandria, VA 22314-2857

AAO-HNSF Health Services Research Grants **187**
The purpose of the Health Services Research Grant is to foster research that will improve the effectiveness and appropriateness of medical practice. Projects supported under this program will develop and disseminate scientific information on the effects of otolaryngology services and procedures on patients' survival, health status, functional capacity, and quality of life. The award is intended to promote increased participation by otolaryngologists in the rapidly expanding area of health services research. Proposed projects may be related to any area of otolaryngology-head and neck surgery, but must have direct or potential clinical significance for patients seen by otolaryngologist-head and neck surgeons. Grants range up to $10,000 maximum, are for one year, and are not renewable. Projects must be completed within two years of the award date; no-cost extensions are available upon written request. Letters of intent must be received by December 15, and completed applications must be submitted by January 15.
Requirements: Any otolaryngologist in the United States or Canada is eligible to apply for the Health Services Research Grant. Applicants may be independent practitioners, residents or fellows in an approved training program, or practitioners affiliated with academic or similar institutions. All applicants must be members in good standing of AAO-HNSF; Associate and Corresponding Members are not eligible to apply as Principal Investigator, but may participate actively in the proposed project.
Geographic Focus: All States
Date(s) Application is Due: Jan 15; Dec 15
Amount of Grant: Up to 10,000 USD
Contact: Stephanie L. Jones; (703) 519-1586 or (703) 836-4444; sljones@entnet.org
Internet: http://www.entnet.org/EducationAndResearch/coreGrants.cfm
Sponsor: American Academy of Otolaryngology-Head and Neck Surgery Foundation
1650 Diagonal Road
Alexandria, VA 22314-2857

AAO-HNSF Maureen Hannley Research Training Awards **188**
The purpose of the Maureen Hannley Research Training Award program is to foster the acquisition of contemporary basic or clinical research skills among new full-time academic otolaryngologist-head and neck surgeons. The award is intended as a preliminary step in clinical investigator career development and is expected to facilitate the recipient's preparation of a more comprehensive individualized research training plan suitable for submission to one of the National Institutes of Health Clinical Investigator Career Development Award (CIA) programs (K series). Grants range up to $15,000 maximum. The award is one year, and is non-renewable. Projects must be completed within two years of the award date; no-cost extensions are available upon written request. Letters of intent must be received by December 15, and completed applications must be submitted by January 15.
Requirements: Applicants in the United States or Canada and members in good standing of the AAO-HNSF are welcome. Applicants must have demonstrated potential for excellence in research and teaching and a serious commitment to an academic research career in otolaryngology-head and neck surgery. Priority will be given to senior residents, fellows or faculty who have completed residencies or fellowships within two years of the application receipt date. All candidates must be sponsored by the Chairperson of his/her Division or Department and by an official representative of the institution which would administer the Award and in whose name the application is formally submitted.
Geographic Focus: All States
Date(s) Application is Due: Jan 15; Dec 15
Amount of Grant: Up to 15,000 USD
Contact: Stephanie L. Jones; (703) 519-1586 or (703) 836-4444; sljones@entnet.org
Internet: http://www.entnet.org/EducationAndResearch/coreGrants.cfm
Sponsor: American Academy of Otolaryngology-Head and Neck Surgery Foundation
1650 Diagonal Road
Alexandria, VA 22314-2857

AAO-HNSF Percy Memorial Research Award **189**
The Percy Memorial Research Award is an annual grant-in-aid of a worthy research project proposed in any area within the scope of otolaryngology-head and neck surgery. The Award was established in memory of A. Edward Percy, Jr., M.D. and his parents. Projects must have direct or potential clinical significance for patients seen by otolaryngologists-head and neck surgeons. They must be designed so as to yield useful information within the period of award, but priority will be given to projects which are also innovative and promise to develop into new long range research programs which will attract funding from other sources. Grants range up to $10,000 maximum, are for one year, and are not renewable. Letters of intent must be received by December 15, and completed applications must be submitted by January 15.
Requirements: Applicants in the United States or Canada and members in good standing of AAO-HNSF are welcome to apply. Applicants should be experienced independent otolaryngologist investigators affiliated with academic or similar institutions eligible to apply for and administer Federal research awards. Fellows with substantial research experience are eligible to apply, as are individuals who have already competed successfully for independent research grant support from a private or Federal funding agency.
Restrictions: Funds may not be requested to pay any portions of the salaries of the principal investigator or of any support personnel with strictly secretarial or clerical responsibilities. Funds may also not be used for the purchase of any item of equipment costing more than $500.
Geographic Focus: All States
Date(s) Application is Due: Jan 15; Dec 15
Amount of Grant: Up to 25,000 USD
Contact: Stephanie L. Jones; (703) 519-1586 or (703) 836-4444; sljones@entnet.org
Internet: http://www.entnet.org/EducationAndResearch/coreGrants.cfm
Sponsor: American Academy of Otolaryngology-Head and Neck Surgery Foundation
1650 Diagonal Road
Alexandria, VA 22314-2857

AAO-HNSF Rande H. Lazar Health Services Research Grant **190**
The purpose of the Rande H. Lazar Health Services Research Grant is to support the gathering of socioeconomic data for otolaryngology. The award is intended to promote increased participation by otolaryngologists in the rapidly expanding area of health services research. Proposed projects may be related to any area of otolaryngology-head and neck surgery, but must have direct or potential clinical significance for patients seen by otolaryngologist-head and neck surgeons. Grants range up to $10,000 maximum, are for one year, and are not renewable. Letters of intent must be received by December 15, and completed applications must be submitted by January 15.
Requirements: Any otolaryngologist in the United States or Canada is eligible to apply for the Rande H. Lazar Health Services Research Grant. Applicants may be independent practitioners, residents or fellows in an approved training program, or practitioners affiliated with academic or similar institutions. All applicants must be members in good standing of AAO-HNSF; Associate and Corresponding Members are not eligible to apply as Principal Investigator, but may participate actively in the proposed project.
Geographic Focus: All States
Date(s) Application is Due: Jan 15; Dec 15
Amount of Grant: Up to 10,000 USD
Contact: Stephanie L. Jones; (703) 519-1586 or (703) 836-4444; sljones@entnet.org
Internet: http://www.entnet.org/EducationAndResearch/coreGrants.cfm
Sponsor: American Academy of Otolaryngology-Head and Neck Surgery Foundation
1650 Diagonal Road
Alexandria, VA 22314-2857

AAO-HNSF Resident Research Awards **191**
The purpose of this grant is to stimulate original resident research in otolaryngology projects that are well-conceived and scientifically valid, with the potential to advance otolaryngology. Proposed projects may be related to any area of otolaryngology-head and neck surgery, and should be designed in collaboration with a preceptor investigator and approved by the candidate's department chairperson and institution. Grants range up to $10,000 maximum. Projects must be completed within one year of the award date; no-cost extensions are available upon written request. Letters of intent must be received by December 15, and completed applications must be submitted by January 15. Up to eight awards are available annually.
Requirements: The award is open to any AAO-HNS member in good standing who is a resident in an accredited otolaryngology-head and neck surgery training program in the U.S. and Canada.
Geographic Focus: All States
Date(s) Application is Due: Jan 15; Dec 15
Amount of Grant: Up to 10,000 USD
Contact: Stephanie L. Jones; (703) 519-1586 or (703) 836-4444; sljones@entnet.org
Internet: http://www.entnet.org/EducationAndResearch/coreGrants.cfm
Sponsor: American Academy of Otolaryngology-Head and Neck Surgery Foundation
1650 Diagonal Road
Alexandria, VA 22314-2857

AAO-HNS Medical Student Research Prize **192**
The Medical Student Research Prize is given annually for a manuscript submitted to the Editor of Otolaryngology-Head and Neck Surgery for review and consideration for publication in the journal. An award winning manuscript will be automatically accepted as an oral or poster presentation at the Translational and Basic Research Presentations Program (previously called the Research Forum). First prize is $1,000, and the award is available in both clinical and basic science categories through the generosity of GYRU.S. ACMI ENT Division.

Requirements: All manuscripts submitted for this competition must have a medical student as senior author and must describe work that has not been placed under active consideration for publication other than in Otolaryngology-Head and Neck Surgery.
Geographic Focus: All States
Date(s) Application is Due: Mar 9
Amount of Grant: 1,000 USD
Contact: Stephanie L. Jones; (703) 519-1586 or (703) 836-4444; sljones@entnet.org
Internet: http://www.entnet.org/educationandresearch/grantsandfellowships.cfm
Sponsor: American Academy of Otolaryngology-Head and Neck Surgery Foundation
1650 Diagonal Road
Alexandria, VA 22314-2857

AAO-HNS Resident Research Prizes 193
Resident Research Prizes are given annually for manuscripts submitted to the Editor of Otolaryngology-Head and Neck Surgery for review and consideration for publication in the journal. All award winning manuscripts will be automatically accepted as oral or poster presentations at the Translational and Basic Research Presentations Program (previously called the Research Forum). First prize is $2,000, second prize is $1,500, and third prize is $1,000. All awards are available in both clinical and basic science categories through the generosity of GYRU.S. ACMI ENT Division.
Requirements: All manuscripts submitted for this competition must have a resident as senior author and must describe work that has not been placed under active consideration for publication other than in Otolaryngology-Head and Neck Surgery.
Geographic Focus: All States
Date(s) Application is Due: Mar 9
Amount of Grant: 1,000-2,000 USD
Contact: Stephanie L. Jones; (703) 519-1586 or (703) 836-4444; sljones@entnet.org
Internet: http://www.entnet.org/educationandresearch/grantsandfellowships.cfm
Sponsor: American Academy of Otolaryngology-Head and Neck Surgery Foundation
1650 Diagonal Road
Alexandria, VA 22314-2857

AAOA Foundation/AAO-HNSF Combined Research Grants 194
The purpose of this award is to support a collaborative AAOA Foundation research project by fostering basic or clinical research related to otolaryngic allergy, rhinology, or related immunology by otolaryngologists-head and neck surgeons. Grants range up to $10,000 maximum. Projects must be completed within two years of the award date; no-cost extensions are available upon written request. Letters of intent must be received by December 15, and completed applications must be submitted by January 15.
Requirements: Applicants must have demonstrated potential for excellence in research. There are no restrictions on career stage, education, or country of residence. AAOA members are preferred but not required. All candidates must be sponsored by the Chairperson of his/her Division or Department and by an official representative of the institution that would administer the Award and in whose name the application is formally submitted.
Geographic Focus: All States
Date(s) Application is Due: Jan 15; Dec 15
Amount of Grant: Up to 10,000 USD
Contact: Stephanie L. Jones; (703) 519-1586 or (703) 836-4444; sljones@entnet.org
Internet: http://www.entnet.org/EducationAndResearch/coreGrants.cfm
Sponsor: American Academy of Otolaryngic Allergy Foundation
1990 M. Street NW, Suite 680
Washington, DC 20036

AAOHN Foundation Experienced Researcher Grants 195
The foundation funds research focused on improving the health and safety of the nation's workforce. Research grants are awarded to support research in the field of occupational and environmental health by an experienced occupational and environmental health registered nurse. Proposals may be submitted that are in the early stages of development (prior to data collection). Two $10,000 grants are awarded annually.
Requirements: Experienced occupational and environmental health registered nurse researchers are eligible to apply.
Restrictions: Completed research projects will not be accepted for consideration.
Geographic Focus: All States
Date(s) Application is Due: Dec 1
Amount of Grant: 10,000 USD
Contact: Ann Cox; (770) 455-7757; fax (770) 455-7271; ann@aaohn.org
Internet: http://www.aaohn.org/foundation/grants/index.cfm
Sponsor: American Association of Occupational Health Nurses Foundation
2920 Brandywine Road, Suite 100
Atlanta, GA 30341

AAOHN Foundation New Investigator Researcher Grants 196
The foundation funds research, scholarship, and leadership development grants focused on improving the health and safety of the nation's workforce. Grants are awarded to encourage research in the field of occupational and environmental health by a new or novice registered nurse principal investigator. Proposals may be submitted that are in the early stages of development (prior to data collection).
Restrictions: Completed research projects are ineligible.
Geographic Focus: All States
Date(s) Application is Due: Dec 1
Amount of Grant: 3,000 USD
Contact: Ann Cox; (770) 455-7757; fax (770) 455-7271; ann@aaohn.org
Internet: http://www.aaohn.org/foundation/grants/index.cfm
Sponsor: American Association of Occupational Health Nurses Foundation
2920 Brandywine Road, Suite 100
Atlanta, GA 30341

AAP International Travel Grants 197
The AAP has set aside a minimum of twelve (12) five-hundred dollar ($500.00) grants to be awarded to categorical pediatric or combined-training pediatric residents who wish to complete a clinical pediatric elective in the developing world during residency. The selection committee is composed of members from the AAP Section on International Child Health and the Section on Residents. After the selection committee reviews the applications the recipients will be notified mid-May and mid-November.
Requirements: Must be a categorical pediatric or combined-training pediatric resident at a U.S. or Canadian university.
Geographic Focus: All States
Date(s) Application is Due: Mar 15; Sep 15
Amount of Grant: 500 USD
Samples: Gunjan Kamdar, New York Presbyterian; Bill Finn, University of Rochester; Raina Paul, Boston Combined Residency Program.
Contact: Kimberley VandenBrook, Grants Administrator; (800) 433-9016, ext. 7134; kvandenbrook@core.com
Internet: http://www.aap.org/sections/ypn/r/funding_awards/international_travel.html
Sponsor: American Academy of Pediatrics
141 Northwest Point Boulevard
Elk Grove Village, IL 60007-1098

AAP Nutrition Award 198
One award is given annually for outstanding achievements in research relating to nutrition of infants and children. The award is made to an individual or for one project. There is no age restriction for the award; however, it is hoped that younger persons will be considered. The award is made possible by a grant from the Infant Formula Council. In addition to the honorarium, the award includes round-trip airfare and two days' lodging for the recipient and a guest to attend the annual meeting and receive the award.
Requirements: The competition is open to residents of the United States or Canada whose research has been completed and publicly reported. Separate letters should be written for each individual nominated; a letter should contain a description of the nominee's achievements, stating clearly the basis for the recommendation, including references to the literature that describes her/his work. Nominee's bibliography is to be submitted with the nominating letter together with copies of available reprints. Letters supporting the nomination are to be screened by the nominator and forwarded with the nomination.
Restrictions: An individual may not submit more than five letters supporting the nomination. Current members of the AAP Committee on Nutrition are not eligible for this award.
Geographic Focus: All States
Date(s) Application is Due: Mar 1
Amount of Grant: 3,000 USD
Samples: Virginia A Stallings; Dennis M Bier; William Dietz; William W. Hay, Jr..
Contact: Debra Burrows; (847) 434-4000; fax (847) 434-8000; research@aap.org
Internet: http://www.aap.org/visit/nutrannouncemts.htm
Sponsor: American Academy of Pediatrics
141 Northwest Point Boulevard
Elk Grove Village, IL 60007-1098

AAP Resident Research Grants 199
The Program is designed to give pediatric residents with limited research experience an opportunity to initiate and complete research projects related to their professional interests. Projects may be related to the full spectrum of child health research, such as behavioral sciences, biomedical sciences, epidemiology, health services, perinatal/neonatal health, prevention, public health, quality improvement, quality measurement, or basic laboratory-based science. Research projects can be conducted for a maximum of 2 years and should be completed during the residency program.
Requirements: Applicants must be legal residents of the United States or Canada. Although awards are primarily intended for support of first- and second-year pediatric residents, individuals who have secured a position for a third or fourth year of residency beginning in July may also apply.
Geographic Focus: All States
Date(s) Application is Due: Feb 27
Amount of Grant: Up to 2,000 USD
Contact: Jeannine Hess; (800) 433-9016, ext. 7876; fax (847) 434-8000; jhess@aap.org
Internet: http://www.aap.org/sections/ypn/r/funding_awards/research_grants.html
Sponsor: American Academy of Pediatrics
141 Northwest Point Boulevard
Elk Grove Village, IL 60007-1098

Aaron & Cecile Goldman Family Foundation Grants 200
The Aaron and Cecile Goldman Family Foundation, established in 1962 in the District of Columbia, offers its primary support to Jewish and local agencies, including education and human services, and to local children's programs and camps. Fields of interest include: arts and humanities; children and youth services; education; human services; Jewish agencies and synagogues; Jewish federated giving programs; recreational programs; and camps. Types of support include building and renovation, capital campaigns, emergency funding, endowments, equipment purchase, fellowships, matching/challenge grants, support for professions, program development, and scholarships. Applicants should forward a letter of proposal to the Foundation office, whose board members meet in the spring and fall of each year.

Requirements: One copy of the proposal letter is required. 501(c)3 organizations supporting the residents of Washington, DC, and its surrounding area are eligible to apply.
Restrictions: No grants are offered to individuals.
Geographic Focus: District of Columbia
Amount of Grant: Up to 20,000 USD
Samples: The Institute for Jewish Leadership and Values, Rockville, Maryland, $18,000; Ethiopian Community Services and Development Council, Washington, DC, $15,000; Save a Childs Heart Foundation, Potomac, Maryland, $10,000.
Contact: Jennifer Margolius Fisher, (202) 332-6600; fax (202) 332-1800; jennifer@themargoliusfirm.com
Sponsor: Aaron and Cecile Goldman Family Foundation
3000 Connecticut Avenue NW, Suite 100
Washington, DC 20008-2509

AAR Postdoctoral Fellowships 201
The Academy offers 11-month postdoctoral fellowships in the following areas: architecture, design, historic preservation and conservation, landscape architecture, literature, musical composition, visual arts, ancient studies, medieval studies, Renaissance and early modern studies, and modern Italian studies. Applications may be downloaded from the Web site or obtained by contacting the office.
Requirements: All applicants for postdoctoral fellowships must have received the PhD at the time of application. U.S. citizens and those foreign nationals who have lived in the U.S. for the three years immediately preceding the application deadline are eligible to apply for the NEH postdoctoral fellowships and are not required to submit the application fee. Applicants for academy postdoctoral awards must be assistant professors, associate professors appointed within the prior two years, or independent scholars who have received the PhD within the prior seven years. Applicants for pre- and postdoctoral Rome Prize fellowships in the humanities must include the following materials in support of their applications: Six copies of the completed and signed application form; six copies of a current curriculum vitae, which includes languages read or spoken and level of fluency; six copies of a project proposal (on 8.5 x 11 paper), not to exceed four pages, describing the nature of the work you plan to undertake in Rome and identifying the special resources in Italy that will be important to your work (please include bibliographical notes for any sources cited, [not counted towards the four-page proposal statement limit]); and six copies of a supporting paper in the proposed field of study, not to exceed 20 pages, excluding bibliography and footnotes.
Geographic Focus: All States
Date(s) Application is Due: Nov 1
Amount of Grant: Up to 21,000 USD
Contact: Shawn Miller; (212) 751-7200; fax (212) 751-7220; info@aarome.org
Internet: http://www.aarome.org/rome_prize/index.htm
Sponsor: American Academy in Rome
7 East 60th Street
New York, NY 10022-1001

AAR Rome Prize Fellowships 202
The American Academy in Rome is a leading American overseas center for independent studies and advanced research in the fine arts and humanities. It began as a collaborative effort in 1893 at the World's Columbian Exhibition when a small group, including architects Charles Follen McKim and Daniel Burham, painters John LaFarge and Francis Millet, and sculptors Augustus Saint-Gaudens and Daniel Chester French resolved to create a center to study art amid the classical tradition of ancient Rome. Today the academy awards up to thirty Rome Prize Fellowships to emerging artists and scholars in the early or middle stages of their careers. Fellows are chosen from the disciplines of architecture, design, historic preservation and conservation, landscape architecture, literature, musical composition, visual arts, ancient studies, medieval studies, renaissance and early modern studies, and modern Italian studies. Prize recipients are invited to Rome for six months, eleven months, or two years to immerse themselves in the Academy community. Each Rome Prize winner is provided with a stipend, meals, a bedroom with private bath, and a study or studio. Recipients with a spouse receive a double room. Recipients with children under eighteen live in partially subsidized apartments nearby. Stipends for six-month and eleven-month fellowships are $15,000 and $27,000 respectively. Winners of two-year fellowships receive $27,000 annually. Eleven-month fellowships generally begin in early-September and end at the beginning of August. Winners of six-month fellowships may choose to begin in early September and end in early March or begin in early February and end in early August. The deadline for applications is November 1 but applications will be accepted through November 15 for a higher application fee. Application fees are based upon date of application receipt and number of disciplines for which an applicant applies. The deadline for receipt of letters of recommendation (when required) is November 30. The Academy makes application forms available on their website in early September of the current funding cycle.
Requirements: Applicants for the Rome Prize fellowships must be U.S. citizens at the time of application, with the exception of those competing for postdoctoral fellowships sponsored by the National Endowment for the Humanities. U.S. citizens and those foreign nationals who have lived in the United States for the three years immediately preceding the application deadline may apply for the NEH postdoctoral fellowships. Applicants applying solely for NEH post-doctoral fellowships are not required to submit application fees.
Restrictions: Graduate students in the humanities may apply only for pre-doctoral fellowships. Previous winners of the Rome Prize are not eligible to reapply. Undergraduate students are not eligible for Rome Prize fellowships. The Rome Prize for Literature Fellowship is awarded only by nomination through the American Academy of Arts and Letters. Rome Prize recipients may not hold full-time jobs while at the Academy.
Geographic Focus: All States
Date(s) Application is Due: Nov 1; Nov 15
Amount of Grant: 15,000-54,000 USD
Contact: Shawn Miller; (212) 751-7200; fax (212) 751-7220; info@aarome.org
Internet: http://www.aarome.org/apply/rome-prize/procedure-requirements
Sponsor: American Academy in Rome
7 East 60th Street
New York, NY 10022-1001

AASF Lydia Donaldson Tutt-Jones Memorial Research Grant 203
The African American Success Foundation (AASF) is a 501(c)3 organization with a mission to promote the positive image of African Americans by funding research and collecting and disseminating information that sheds light on the many successes being attained by African Americans. Holder of a bachelor of science degree in nursing and a masters of science degree in human resources development, Ms. Tutt-Jones (for whom this AASF research grant is named) was committed to excellence and achievement in her life. She dedicated many years of her professional career traveling around the country to recruit highly talented and motivated teachers to work in the Fort Lauderdale public-school system. The Tutt-Jones grant provides financial support to students and professionals who conduct research activities to study African-American success particularly in the area of education. The research can focus upon early childhood education or student performance in the elementary, middle, or high-school years as well as the attitudes and behaviors that help individuals finish college and graduate school. The research can also focus upon parental behaviors and home and community variables that contribute to student success and promote excellence in educational performance. The foundation's interest is to increase the body of knowledge about African-American students who are excelling in school to aid in the replication of that success for others. The duration of the Tutt-Jones grant is for one year; former applicants may re-apply. No special application form, recommended format, or length of material is required. Interested individuals should submit a letter of interest, curriculum vita, a description of the proposed research project (including a timeline and a letter of recommendation from their faculty mentor, if students, or from their department chairperson, agency head, or officer of their professional association, if professionals). An original and five copies of the application materials should be submitted, along with a stamped, self-addressed postcard that will be used to acknowledge receipt of the application. Deadlines may vary. All proposals must be received by 5 pm on the deadline date. Detailed guidelines are available on the website.
Requirements: Graduate students and professionals who reside in the United States are eligible to apply. Recipients must prepare a publishable caliber paper upon the completion of the project and provide a copy to the foundation. Student applicants must be recommended by a faculty mentor who agrees to oversee the project and submission of a publishable caliber paper.
Geographic Focus: All States
Date(s) Application is Due: Jun 14
Amount of Grant: 5,000 USD
Contact: Cynthia Wilson, Ph.D., Chairperson of the Grant Selection Committee; (954) 792-1117; fax (954) 792-9191; info@blacksuccessfoundation.org
Internet: http://blacksuccessfoundation.org/lcdtj%20research%20grant.htm
Sponsor: African American Success Foundation
7027 West Broward Boudlevard, #313
Fort Lauderdale, FL 33317

AAS Kate B. and Hall J. Peterson Fellowships 204
The society will award a number of fellowships for one to three months to scholars for research and writing-- including doctoral dissertations-- in any field of American history and culture through 1876. All awards will be made not only on the basis of the applicant's scholarly qualifications and the general interest of the project, but also on the appropriateness of the inquiry to the society's holdings.
Requirements: The fellowships are open to individuals, including foreign nationals, and students working on dissertations.
Restrictions: All recipients are expected to be in regular and continuous residence at the society's library during the period of the grant.
Geographic Focus: All States
Date(s) Application is Due: Jan 15
Amount of Grant: 1,700-5,100 USD
Contact: Caroline Sloat, Director of Scholarly Programs; (508) 471-2130 or (508) 755-5221; fax (508) 753-3311; csloat@mwa.org or academicfellowships@mwa.org
Internet: http://www.americanantiquarian.org/peterson.htm
Sponsor: American Antiquarian Society
185 Salisbury Street
Worcester, MA 01609-1634

AAS Legacy Fellowships 205
The Legacy Fellowship is awarded for research on any topic supported by the collections. It is funded by the gifts of former fellows and research associates. This fellowship is awarded to an individual engaged in scholarly research and writing—including doctoral dissertations—in any field of American history and culture through 1876. The Fellow is selected on the basis of the applicant's scholarly qualifications, the scholarly significance or importance of the project, and the appropriateness of the proposed study to the Society's collections.
Requirements: The fellowships are open to individuals and students working on dissertations.
Restrictions: All recipients are expected to be in regular and continuous residence at the society's library during the period of the grant.
Geographic Focus: All States
Date(s) Application is Due: Jan 15
Amount of Grant: 1,700-20,400 USD

Contact: Caroline Sloat, Director of Scholarly Programs; (508) 471-2130 or (508) 755-5221; fax (508) 753-3311; csloat@mwa.org or academicfellowships@mwa.org
Internet: http://www.americanantiquarian.org/legacy.htm
Sponsor: American Antiquarian Society
185 Salisbury Street
Worcester, MA 01609-1634

AAS Long-Term Visiting Academic Research Fellowships **206**
Society offers long-term visiting academic research fellowships tenable for four to twelve months each year. All awards are for a period of residence to use the AAS library's resources for research and writing. Fellowships are awarded not only on the basis of the applicant's scholarly qualifications and the general interest of his/her project, but also on the appropriateness of the inquiry to the society's holdings—early U.S. history and culture through 1876. Recipients are expected to be in regular residence at the society's library during the period of the grant. NEH fellows must devote full time to their study and may not accept teaching assignments nor undertake any other major activities during the tenure of the award. Recipients may not hold any other major fellowships, except sabbaticals or other grants from their own institutions.
Requirements: These Fellowships are available for scholars beyond the doctorate.
Restrictions: The fellowships may not be awarded to degree candidates or for study leading to advanced degrees, nor may they be granted to foreign nationals unless they have been residents of the United States for at least three years preceding their award.
Geographic Focus: All States
Date(s) Application is Due: Jan 15; Oct 3; Oct 15
Amount of Grant: 35,000-40,000 USD
Contact: Caroline Sloat, Director of Scholarly Programs; (508) 471-2130 or (508) 755-5221; fax (508) 753-3311; csloat@mwa.org
Internet: http://www.americanantiquarian.org/longterm.htm
Sponsor: American Antiquarian Society
185 Salisbury Street
Worcester, MA 01609-1634

AASL Research Grant **207**
The AASL Research Grants are given to up to two school librarians, library educators, library information science, or education professors to conduct innovative research aimed at measuring and evaluating the impact of school library programs on learning and education. The study should have the potential to serve as a model for future school library research, and researchers should furnish documentation of the results of their work. Up to two grants of $2,500 each are given every year. Recipients of the Research Grant will be invited to present the results of their projects at a concurrent session during the AASL National Conference. Recipients may also be invited to publish a summary of their findings in School Library Research and/or Knowledge Quest. Winners are announced in the spring.
Requirements: Applications are evaluated on the following criteria: potential to measure and/or evaluate the impact of school library programs on learning and education; originality of the research project and methodology; potential for replication; demonstrated ability of the applicant(s) to undertake and successfully complete the project; evidence that sufficient time and resources have been allocated to the effort; support and commitment (financial and otherwise) by institutions and organizations to the project; clarity and completeness of the proposal; personal resume of individual applicant attached; and personal membership in ALA/AASL at the time of application by the applicant.
Geographic Focus: All States
Date(s) Application is Due: Feb 1
Amount of Grant: Up to 2,500 USD
Contact: Melissa Jacobsen, Manager, Professional Development; (312) 280-4381 or (800) 545-2433, ext. 4381; fax (312) 280-5276; mjacobsen@ala.org
Internet: http://www.ala.org/awardsgrants/research-grant-0
Sponsor: American Association of School Librarians
50 East Huron Street
Chicago, IL 60611-2795

AAS Short-Term Visiting Academic Research Fellowships **208**
The society awards short-term fellowships for one to three months. Fellowships are awarded not only on the basis of the applicant's scholarly qualifications and the general interest of his/her project, but also on the appropriateness of the inquiry to the society's holdings—early U.S. history and culture through 1876. Recipients are expected to be in regular residence at the society's library during the period of the grant. NEH fellows must devote full time to their study and may not accept teaching assignments nor undertake any other major activities during the tenure of the award. Recipients may not hold any other major fellowships, except sabbaticals or other grants from their own institutions.
Requirements: Short-term fellowships are available for scholars holding the PhD and for doctoral candidates engaged in dissertation research.
Restrictions: The fellowships may not be awarded to degree candidates or for study leading to advanced degrees, nor may they be granted to foreign nationals unless they have been residents of the United States for at least three years preceding their award.
Geographic Focus: All States
Date(s) Application is Due: Jan 15
Amount of Grant: 1,000-3,000 USD
Contact: Caroline Sloat, Director of Scholarly Programs; (508) 471-2130 or (508) 755-5221; fax (508) 753-3311; csloat@mwa.org
Internet: http://www.americanantiquarian.org/acafellowship.htm
Sponsor: American Antiquarian Society
185 Salisbury Street
Worcester, MA 01609-1634

AAUW American Dissertation Fellowships **209**
Dissertation Fellowships are available to women who will complete their dissertation writing between July 1 and June 30 of the upcoming year. Degree conferral must be between April 1 and September 15. To qualify, applicants must have completed all course work, passed all required preliminary examinations, and received approval for their research proposal or plan by November 15. Open to applicants in all fields of study. Scholars engaged in science, technology, engineering and math and also researching gender issues are especially encouraged to apply.
Requirements: Applicants must be U.S. citizens or permanent residents by the application postmark deadline. When comparing proposals of equal merit, special consideration will be given to women holding junior academic appointments who are seeking research leave, women who have held the doctorate for at least three years, and women whose educational careers have been interrupted.
Restrictions: Candidates may apply for only one of the awards described. Former recipients of these awards are not eligible to apply for additional American Fellowships or publication grants. Students holding any fellowship for writing a dissertation in the year prior to the AAUW Educational Foundation fellowship year are not eligible.
Geographic Focus: All States
Date(s) Application is Due: Nov 15
Amount of Grant: 20,000 USD
Contact: Fellowship Contact; (319) 337-1716; aauw@act.org
Internet: http://www.aauw.org/learn/fellowships_grants/american.cfm
Sponsor: American Association of University Women
1111 16th Street NW
Washington, DC 20036

AAUW American Postdoctoral Research Leave Fellowships **210**
Postdoctoral Research Leave Fellowships offer support for women in tenure-track faculty positions in support of their earning tenure and further promotions by enabling them to spend a year pursuing independent research. Candidates must have earned a doctoral degree by November 15. Postdoctoral fellowships are available in the arts and humanities, social sciences, and natural sciences.
Requirements: Applicants must be U.S. citizens or permanent residents by the application deadline. Candidates are evaluated on the basis of scholarly excellence; the quality and originality of project design; and active commitment to helping women and girls through service in their communities, professions, or fields of research.
Restrictions: Candidates may apply for only one of the awards described. Former recipients of these awards are not eligible to apply for additional American Fellowships or publication grants.
Geographic Focus: All States
Date(s) Application is Due: Nov 15
Amount of Grant: 30,000 USD
Contact: Fellowship Contact; (319) 337-1716, ext. 60; aauw@act.org
Internet: http://www.aauw.org/learn/fellowships_grants/american.cfm
Sponsor: American Association of University Women
1111 16th Street NW
Washington, DC 20036

AAUW American Summer/Short-Term Research Publication Grants **211**
The Grants fund women college and university faculty and independent researchers to prepare research data for publication. Applicants may be tenure track, part-time, or temporary faculty or new or established scholars and researchers at universities. Time must be available for eight consecutive weeks of final writing, editing, and responding to issues raised in critical reviews.
Requirements: Applicants must be U.S. citizens or permanent residents by the application postmark deadline. Applicants must have received their doctorates by the application deadline.
Restrictions: Funds are not available for research. Candidates may apply for only one of the awards described. Former recipients of these awards are not eligible to apply for additional American Fellowships or publication grants.
Geographic Focus: All States
Date(s) Application is Due: Nov 15
Amount of Grant: 6,000 USD
Contact: Fellowship Contact; (319) 337-1716; aauw@act.org
Internet: http://www.aauw.org/learn/fellowships_grants/american.cfm
Sponsor: American Association of University Women
1111 16th Street NW
Washington, DC 20036

AAUW Career Development Grants **212**
The Career Development Grants support women currently holding a bachelor's degree who are preparing to advance their careers, change careers, or re-enter the work force. Special consideration is given to AAUW members, women of color, and women pursuing their first advanced degree or credentials in nontraditional fields. Funds are available for tuition, fees, books, supplies, local transportation, and dependent care. Grants provide support for course work beyond a bachelor's degree, including a master's degree, second bachelor's degree, or specialized training in technical or professional fields. Funds are available for distance learning. Course work must be taken at an accredited two- or four-year college or university, or at a technical school that is fully licensed or accredited by an agency recognized by the U.S. Department of Education.
Requirements: Applicants must be women who are U.S. citizens or permanent residents of the U.S. who hold at least a baccalaureate degree and who have completed their most recent degree at least five years before the July 1 effective date of the award. Preference is

given to AAUW members. Applicants must have applied or been accepted at a two- or four-year accredited college or university; technical schools must be licensed or accredited by an agency recognized by the U.S. Department of Education.
Restrictions: Academic grants are not available for doctoral level work. Institute grants may not be used for internships, research, or independent study.
Geographic Focus: All States
Contact: Customer Service; (319) 337-1716, ext. 60; aauw@act.org
Internet: http://www.aauw.org/learn/fellows_directory/cd.cfm
Sponsor: American Association of University Women
1111 16th Street NW
Washington, DC 20036

AAUW Community Action Grants 213
Community Action grants provide funds to individuals, AAUW branches and AAUW state organizations as well as local community-based nonprofit organizations for innovative programs or non-degree research projects that promote education and equality for women and girls. One-year grants from $2,000 to $7,000 provide funding for community-based projects. Two-year grants from $5,000 to $10,000 provide start-up funds for new projects that address the particular needs of the community and develop girls' sense of efficacy through leadership or advocacy opportunities. Topic areas are unrestricted, but should include a clearly defined activity that promotes education and equality for women and girls. Special consideration is given to projects focused on K-12 and community college girls' and women's achievements in science, technology, engineering, or math.
Requirements: Applicants must be women who are U.S. citizens or permanent residents. Grant projects must have direct public impact, be nonpartisan, and take place within the U.S. or its territories.
Geographic Focus: All States
Amount of Grant: 2,000-10,000 USD
Contact: Customer Service; (319) 337-1716, ext. 60; aauw@act.org
Internet: http://www.aauw.org/learn/fellowships_grants/community_action.cfm
Sponsor: American Association of University Women
1111 16th Street NW
Washington, DC 20036

AAUW International Doctoral Fellowships 214
International Fellowships are awarded for full-time study or research in the United States to women who are not U.S. citizens or permanent residents. Both graduate and postgraduate studies at accredited U.S. institutions are supported. Funding may be renewable for a second year. The application and additional information are available at the website.
Requirements: Candidates must have an academic degree equipment to the U.S. bachelor's degree; a specific plan of study or research that will advance the applicant's professional competence; the intention to return home to pursue a professional career; devote full time to graduate work; and must have satisfactory English proficiency. Applicants whose native language is not English must submit up-to-date scores on one of four language proficiency examinations. Preference will be given to applicants who have specific positions to return to in their own countries.
Geographic Focus: All Countries
Date(s) Application is Due: Dec 1
Amount of Grant: 20,000 USD
Contact: Fellowship Contact; (319) 337-1716; aauw@act.org
Internet: http://www.aauw.org/learn/fellowships_grants/international.cfm
Sponsor: American Association of University Women
1111 16th Street NW
Washington, DC 20036

AAUW International Master's/First Professional Degree Fellowships 215
International Fellowships are awarded for full-time study or research in the United States to women who are not U.S. citizens or permanent residents. Both graduate and postgraduate studies at accredited U.S. institutions are supported. Funding may be renewable for a second year. The application and additional information are available at the website.
Requirements: Candidates must have an academic degree equivalent to the U.S. bachelor's degree; a specific plan of study or research that will advance the applicant's professional competence; the intention to return home to pursue a professional career; must devote full time to graduate work; and must have satisfactory English proficiency. Applicants whose native language is not English must submit up-to-date scores on one of four language proficiency examinations. Preference will be given to applicants who have specific positions to return to in their own countries.
Restrictions: Previous and current recipients of AAUW fellowships are not eligible to apply.
Geographic Focus: All Countries
Date(s) Application is Due: Dec 1
Amount of Grant: 18,000 USD
Contact: Fellowship Contact; (319) 337-1716; aauw@act.org
Internet: http://www.aauw.org/learn/fellowships_grants/international.cfm
Sponsor: American Association of University Women
1111 16th Street NW
Washington, DC 20036

AAUW International Postdoctoral Fellowships 216
International Fellowships are awarded for full-time study or research in the United States to women who are not U.S. citizens or permanent residents. Funding may be renewable for a second year. The application and additional information are available at the website.
Requirements: Candidates must have an academic degree equivalent to the U.S. bachelor's degree; a specific plan of study or research that will advance the applicant's professional competence; the intention to return home to pursue a professional career; devote full time to the work pursued with the fellowship; and must have satisfactory English proficiency. Applicants whose native language is not English must submit up-to-date scores on one of four language proficiency examinations. Preference will be given to applicants who have specific positions to return to in their own countries.
Restrictions: Previous and current recipients of AAUW fellowships are not eligible to apply.
Geographic Focus: All Countries
Date(s) Application is Due: Dec 1
Amount of Grant: 30,000 USD
Contact: Fellowship Contact; (319) 337-1716, ext. 60; aauw@act.org
Internet: http://www.aauw.org/learn/fellowships_grants/international.cfm
Sponsor: American Association of University Women
1111 16th Street NW
Washington, DC 20036

AAUW Selected Professions Fellowships 217
Selected Professions Fellowships are awarded to women who intend to pursue a full-time course of study at accredited U.S. institutions during the fellowship year in one of the designated degree programs where women's participation traditionally has been low (architecture, computer/information sciences, engineering, and mathematics/statistics).
Requirements: Applicants must be U.S. citizens or permanent residents of the U.S.
Restrictions: Fellowships in business administration, law, and medicine are restricted to women of color, who have been underrepresented in these fields.
Geographic Focus: All States
Date(s) Application is Due: Jan 10
Amount of Grant: 5,000-18,000 USD
Contact: Customer Service; (319) 337-1716, ext. 60; aauw@act.org
Internet: http://www.aauw.org/learn/fellowships_grants/selected.cfm
Sponsor: American Association of University Women
1111 16th Street NW
Washington, DC 20036

Abbott-ASM Lifetime Achievement Award 218
The award, sponsored by Abbott Laboratories, is given to honor a distinguished scientist for a lifetime of outstanding contributions in fundamental biomedical research in any of the microbiological sciences. The award consists of a cash prize, framed certificate, medallion, and national or international travel expenses incidental to receiving the award at the ASM general meeting. Guidelines are available online. All nomination materials must be submitted electronically.
Requirements: Mature scientists, both active and retired, from all areas of microbiology are eligible.
Restrictions: Self-nominations are not be accepted.
Geographic Focus: All States
Date(s) Application is Due: Oct 1
Amount of Grant: Up to 20,000 USD
Samples: Norman R. Pace, $20,000—award recipient; R. John Collier, $20,000—award recipient; Jonathan Beckwith, $20,000—award recipient.
Contact: Awards Committee; (202) 942-9226; fax (202) 942-9353; awards@asmusa.org
Internet: http://www.asm.org/Academy/index.asp?bid=2587
Sponsor: American Society for Microbiology
1752 N Street, N.W.
Washington, DC 20036-2904

Abbott Fund Global AIDS Care Grants 219
Abbott has been a significant contributor to the fight against HIV/AIDS for more than two decades. Since 2000, Abbott and the Abbott Fund have invested $225 million in grants and product donations targeted to resource poor countries most impacted by HIV/AIDS. The focus of these efforts includes expanding access to care, testing and treatment; strengthening HIV/AIDS health care systems; preventing mother-to-child transmission; and supporting children and families affected by HIV/AIDS. In addition, Abbott and the Abbott Fund have helped pioneer innovative model programs to combat the disease. For example, we helped build the Baylor International Pediatric AIDS Initiative's first pediatric outpatient clinic in Romania in 2001 that has served an average of 600 patients per year. The clinic has reduced pediatric HIV mortality rates by more than 90 percent. Today Baylor has replicated the Romania model in additional clinics throughout Africa and now serves nearly 60,000 children and young people with HIV – including those at two clinics built by and supported by the Abbott Fund in Malawi and Tanzania.
Requirements: Grants are made to tax-exempt organizations supporting global HIV/AIDS programs.
Restrictions: Contributions will not be made to individuals; for-profit entities; purely social organizations; political parties or candidates; sectarian religious organizations; advertising; symposia, conferences, and meetings; ticket purchases; memberships; business-related purposes; volunteer efforts of non-Abbott employees; or marketing sponsorships.
Geographic Focus: Arizona, California, Illinois, Kansas, Massachusetts, Michigan, New Jersey, New York, North Carolina, Ohio, Puerto Rico, Texas, Utah, Virginia
Amount of Grant: 10,000-100,000 USD
Contact: Cindy Schwab; (847) 937-7075; cindy.schwab@abbott.com
Internet: http://www.abbottfund.org/project/20/30/Helping-Children-and-Young-People-Living-with-HIV-AIDS
Sponsor: Abbott Fund
100 Abbott Park Road, Department 379, Building 6D
Abbott Park, IL 60064-3500

Abbott Fund Science Education Grants 220

The world urgently needs people who are well-trained in science and technology, and the Abbott Fund is committed to doing its part to address this challenge. Serving as a catalyst by stimulating community investment and engagement, Abbott's investment in science education: engages and inspires students, families and teachers in scientific exploration in out-of-school informal settings; encourages young people to be more proficient in science and attracts more scientists to the field; and builds strong partnerships that are systemic, replicable and sustainable for multiple years and multiple locations. Complete and submit the Abbott Fund online grant application, posted when available. It should include your organization's Federal Tax ID. At times when the Abbott Fund is accepting unsolicited grant applications, it will acknowledge receipt of an online application via email. The Fund will notify an applicant of its decision on a funding request within six to eight weeks.
Requirements: Grants are made to tax-exempt organizations supporting company operating areas in Arizona, California, Illinois, Kansas, Massachusetts, Michigan, New Jersey, New York, North Carolina, Ohio, Puerto Rico, Texas, and Virginia, and Utah. It also supports the communities of: Abingdon, England; Brockville, Canada; Campoverde, Italy; Clonmel, Ireland; Cootehill, Ireland; Delkenheim, Germany; Kanata, Canada; Katsuyama, Japan; Ludwigshafen, Germany; Queenborough, England; Rio de Janeiro, Brazil; Sligo, Ireland; and Zwolle, the Netherlands.
Restrictions: Contributions will not be made to individuals; for-profit entities; purely social organizations; political parties or candidates; sectarian religious organizations; advertising; symposia, conferences, and meetings; ticket purchases; memberships; business-related purposes; volunteer efforts of non-Abbott employees; or marketing sponsorships.
Geographic Focus: Arizona, California, Illinois, Kansas, Massachusetts, Michigan, New Jersey, New York, North Carolina, Ohio, Puerto Rico, Texas, Utah, Virginia, Brazil, Canada, Germany, Great Britain, Ireland, Italy, Japan, Netherlands
Amount of Grant: 20,000-100,000 USD
Contact: Cindy Schwab; (847) 937-7075; cindy.schwab@abbott.com
Internet: http://www.abbottfund.org/tags/science/1
Sponsor: Abbott Fund
100 Abbott Park Road, Department 379, Building 6D
Abbott Park, IL 60064-3500

Abdus Salam ICTP-ELETTRA Users Grants 221

The ICTP-ELETTRA Users Program is offering access to the synchrotron radiation facility ELETTRA in Trieste in the years 2008-2013 to scientists who are citizens of developing countries and work in those countries. A minimum of 1500 hours of beamtime can be made available within this program for measurement and applications at any of the existing ELETTRA beamlines. The program offers a limited number of grants to cover travel and living expenses of individuals and small groups who will use beamtime shift at ELETTRA. The number of scientists per group who receive support depends on the number of allocated shifts and available funds: up to 2 scientists for experiments which received less than 12 shifts; and up to 3 scientists for experiments which received 12 shifts or more. In order to allow experiment preparation and/or sample disposal, support can be offered towards stays up to 3 days before the beginning date of the experiment and 1 day after its completion. In exceptional cases, training on a specific topic for longer periods can be arranged if it is meant to improve the expertise of relevance to a project. In that case a quota is deducted from the total of annual hours.
Geographic Focus: All Countries
Date(s) Application is Due: Mar 15; Sep 15
Contact: Patrizia Passarella; (+39) 040 2240 389; itlabs@ictp.trieste.it
Internet: http://www.elettra.trieste.it/userarea/ictp-elettra-users-programme.html
Sponsor: Abdus Salam International Centre for Theoretical Physics
Strada Costiera 11
Trieste, I-34014 Italy

Abdus Salam ICTP-OEA Affiliated Centre Grants 222

An ICTP Affiliated Centre (ICAC) is an institute or university department of physics or mathematics that agrees to carry out a specific long-term research project on a definite subject with well-defined purposes. An ICAC is expected to have a regional character and to be strongly supported by the local authorities and the hosting Institution. The proposer is expected to submit a scientific project to the ICTP, in collaboration with the leading scientists of that department, and possibly, of other departments in the country. Many physicists participate in the ICAC activities. As a result of such collaboration, an ICAC scientific advisory board is normally formed. The scientific project must consist of: a general program for activities planned over a period of three years; and a detailed project and itemized budget for the first year. Funding will be determined by scope of the project.
Requirements: An institute or university department of physics or mathematics in a developing country that agrees to carry out a specific long-term research project on a definite subject with well defined purposes may apply.
Geographic Focus: All Countries
Contact: Patrizia Passarella, Program Assistant; (+39) 040 2240 323 or (+39) 040 2240 322; fax (+39) 040 2240 443; itlabs@ictp.trieste.it or oea@ictp.it
Internet: http://oea.ictp.it/
Sponsor: Abdus Salam International Centre for Theoretical Physics
Strada Costiera 11
Trieste, I-34014 Italy

Abdus Salam ICTP-OEA Projects Grants 223

The OEA promotes initiatives aimed at "training for research" of young scientists. A Project offers scientific advice and financial support towards the establishment of new Ph.D. courses in physics or mathematics especially at universities in the less developed countries. A Project may also provide support towards fellowships for young scientists from less developed countries to work in institutions in more advanced developing countries in their region. These are often organized on the basis of cooperation agreements between the ICTP and those institutions. In general a Project is meant to undertake actions which are likely to develop into a long term research program. Thus a Project could also be the preliminary phase leading eventually to an Affiliated Center or to the formation of research groups.
Requirements: Developing country university physics or mathematics departments, especially those in less developed countries may apply.
Geographic Focus: All Countries
Contact: Patrizia Passarella, Program Assistant; (+39) 040 2240 323 or (+39) 040 2240 322; fax (+39) 040 2240 433; itlabs@ictp.trieste.it or oea@ictp.it
Internet: http://oea.ictp.it/oea/programmes/projects/
Sponsor: Abdus Salam International Centre for Theoretical Physics
Strada Costiera 11
Trieste, I-34014 Italy

Abdus Salam ICTP-OEA Scientific Meetings Grants 224

The Abdus Salam ICTP Office of External Activities (O.E.A.) encourages the organization of international and regional scientific meetings in developing countries by offering financial assistance to the organizers of conferences, workshops, and schools held in O.E.A. supported countries. Grants are offered for meetings in the fields of physics and mathematics. A similar program sponsored by the Third World Academy of Sciences offers support for meetings in the following fields of natural sciences: agricultural, biological, chemical, engineering, geological and medical sciences.
Requirements: Institutions in developing countries that want to organize international and regional scientific meetings in developing countries should apply. The application should include a scientific program and list of invited speakers with the topics indicated in detail together with title of lectures. Importance will be given to applications with most of the key speakers already confirmed.
Geographic Focus: All Countries
Contact: Eleonora Crotta, (+39) 040 2240 323 or (+39) 040 2240 322; fax (+39) 040 2240 443; crotta@ictp.it or oea@ictp.it
Internet: http://oea.ictp.it/oea/programmes/sm/
Sponsor: Abdus Salam International Centre for Theoretical Physics
Strada Costiera 11
Trieste, I-34014 Italy

Abdus Salam ICTP-OEA Visiting Scholars/Consultants Grants 225

The Visiting Scholar/Consultant Program provides assistance to Institutes or research groups in a developing country requiring expert advice to enhance existing activities or to initiate a new research program. The objective of the program is to establish a permanent link between an outside expert in a particular field of physics or mathematics and the applying Institute on a long-term basis. The Visiting Scientist is an established researcher whose visits will provide a clear benefit to the research group of the applicant in particular and to the host institution in general. To ensure a long-term scientific collaboration the visiting scientist will be expected to deliver a series of topical lectures/seminars and to interact closely with the hosting scientists. The visiting scientist should be the same throughout the project. Only rarely may different scientists carry out the same project, provided they are from the same scientific group as the visiting scientist. It is important to note that the grant is awarded to the proposer and not to the visiting scientist. Round trip travel costs of the visiting scientist are allowed. The amount provided depends on the application, but usually does not exceed Euro 2,500 for each visit.
Requirements: The visiting scientist should make a commitment for two or three visits of about one month each over a period of three years. The Assignments will automatically terminate three years after they are granted.
Geographic Focus: All Countries
Date(s) Application is Due: Sep 30
Amount of Grant: Up to 7,500 EUR
Contact: Patrizia Passarella, Program Assistant; (+39) 040 2240 323 or (+39) 040 2240 322; fax (+39) 040 2240 443; itlabs@ictp.trieste.it or oea@ictp.it
Internet: https://oea.ictp.it/oea/programmes/vs/
Sponsor: Abdus Salam International Centre for Theoretical Physics
Strada Costiera 11
Trieste, I-34014 Italy

Abdus Salam ICTP/IAEA Sandwich Training Education Grants 226

The Sandwich Training Educational Program (STEP) aims at offering fellowship opportunities to Ph.D. students from developing countries in the fields of physics and mathematics. The goal of the STEP is to strengthen the scientific capability of young scientists and researchers from developing countries to better contribute and serve the scientific, technical, and economic development of their home countries. The STEP offers fellowships to candidates from the International Atomic Energy Agency (IAEA) member states in the following fields: atomic and nuclear physics; nuclear, isotope and laser techniques applied to human health, agriculture, biotechnology and biophotonics, and water resources and environmental studies; synchrotron radiation and applications; and medical radiation physics. Fellowships are also provided by the ICTP, within its scientific and technical competence, to candidates from developing countries in the fields listed below: condensed matter and statistical physics; high energy physics; mathematics; earth system physics; and applied physics, including micro- and nano-electronics, fluid dynamics, and dense plasma source physics. Applications are expected from students in developing countries who are enrolled in a PhD course at their home university. PhD students who are supervised by an ICTP Associate are encouraged to apply. The program aims at an equitable participation of fellows, and therefore especially welcomes

applications from woman scientists. The STEP fellowship includes a stipend of Euro 1,400 per month during each stay at the host institute, plus travel costs and medical insurance. Additional information is available at the ICTP website.
Geographic Focus: All Countries
Date(s) Application is Due: Feb 15; Aug 15
Contact: Patrizia Passarella, Program Assistant; (+39) 040 2240 389; fax (+39) 040 2240 388; itlabs@ictp.trieste.it or calligar@ictp.it
Internet: http://portal.ictp.it/step/
Sponsor: Abdus Salam International Centre for Theoretical Physics
Strada Costiera 11
Trieste, I-34014 Italy

Abdus Salam ICTP Federation Arrangement Scheme Grants 227
Federation Arrangements are agreements signed by the ICTP and a scientific institute (Federated Institute) in developing countries. This enables the Institute to send junior scientists-up to the age of 40-to the ICTP for a total of 60 to 150 days, during the validity of the contract (i.e. three years), on a cost-sharing basis. While living expenses are always covered by the Center, travel expenses are either covered by the Federated Institute or shared with the Center. The Federation Arrangement Scheme gives the opportunity to junior scientists from developing countries to participate in the Center's scientific programs. Their participation is proposed by the Federated Institute and approved by the ICTP. Presently there are five different standard types of Federation Arrangements depending upon geographical areas. While the junior scientists' living expenses are always covered by the Centre, travel expenses are either covered by the Federated Institute or shared with the Center.
Requirements: Scientific institutes from developing countries (to send junior scientists to ICTP) are eligible to apply.
Geographic Focus: All Countries
Contact: Patrizia Passarella, Program Assistant; (+39) 040 2240 389; fax (+39) 040 2240 388; itlabs@ictp.trieste.it or diploma@ictp.it
Internet: http://assoc.ictp.it/federation-scheme/description.html
Sponsor: Abdus Salam International Centre for Theoretical Physics
Strada Costiera 11
Trieste, I-34014 Italy

Abdus Salam ICTP Group Associates Grants 228
The ICTP Associateship Scheme was created at the very inception of the Centre. It represents one of the means through which the vocation of the ICTP for the promotion and the development of scientific knowledge in Third World Countries has concretely been applied to reality. In the context of the fundamental goal of the ICTP of promoting research in developing countries it has become apparent that, in many countries, due in part to historical reasons combined with budgetary restrictions, research is hampered not only by isolation but even more by the atomization of research interests. For this reason, research collaboration groups rather than individual awards were established. Group Association awards last six years and are composed of a certain number of Associate awards, according to age and therefore, its members are either Junior, Regular or Senior Associates. Each Group will avail of a number of Associateship privileges depending on the number of components of the Group such as to allow each member at least one visit at ICTP during the allocation period. Depending on composition of group, each member receives entitlements as either Junior, Regular or Senior Associates.
Requirements: Research Groups must be operating in developing countries and have people working and living in the same or in nearby institutes. Each member of the group must fill in an individual online application form, and specify that they are applying for a group association.
Geographic Focus: All Countries
Date(s) Application is Due: Sep 30
Contact: Patrizia Passarella, Program Assistant; (+39) 040 2240 389; fax (+39) 040 2240 388; itlabs@ictp.trieste.it or diploma@ictp.it
Internet: http://assoc.ictp.it/associateship-scheme/group.html
Sponsor: Abdus Salam International Centre for Theoretical Physics
Strada Costiera 11
Trieste, I-34014 Italy

Abdus Salam ICTP Joint Laurea Magistralis Grants 229
There are a limited number of scholarships available to students from developing countries selected to follow the training track in Condensed Matter Physics. Each scholarship consists of: 800 Euro a month, travel grant, insurance and tuition fees. ICTP cooperates with the University of Trieste's Faculty of Sciences in providing physics education to graduate students from developing countries. This international study programme for the "Laurea Magistralis Interateneo" (Universities of Trieste and Udine) in physics, includes curricula in: microphysics and condensed matter physics; theoretical physics; astrophysics and cosmology; and earth and environmental physics. The Programme covers two academic years and is designed to provide young promising students in physics from developing countries with graduate training suitable for pursuing teaching and research in physics. Courses are held in English and each are followed by an examination. A course on Italian language will be organized during the first year. After having passed all exams, participants prepare and defend a dissertation. Upon successful completion of the entire study plan, the students will obtain the "Laurea Magistralis" from the University of Trieste. The participants may also be required to take part in ongoing activities of ICTP.
Requirements: The Program is open to young, qualified graduates from all countries that are members of the United Nations, UNESCO or IAEA. The minimum qualification for applicants is a degree equivalent to an M.Sc. (or an exceptionally good B.Sc.) in physics. The selection of candidates will be based on their university record and academic recommendations. Fluency in both spoken and written English is required.
Geographic Focus: All Countries
Date(s) Application is Due: May 31
Amount of Grant: 9,600 EUR
Contact: Patrizia Passarella, Program Assistant; (+39) 040 2240 389; fax (+39) 040 2240 388; itlabs@ictp.trieste.it or diploma@ictp.it
Internet: http://www.ictp.it/programmes/joint-laurea-description.aspx
Sponsor: Abdus Salam International Centre for Theoretical Physics
Strada Costiera 11
Trieste, I-34014 Italy

Abdus Salam ICTP Joint PhD Programme in Environmental Fluid Mechanics Grants 230
The Doctorate School in Environmental and Industrial Fluid Mechanics trains students in the field of Fluid Dynamics with special emphasis on basic fluid dynamics, mathematical methods, the physics of large scale flows and application to industrial problems. Due to the interdisciplinarity of the program, students have different backgrounds in the fields of engineering, physics and mathematics, and acquire skills useful both for the academic environment and for a career in high-tech companies. The PhD school is based at the University of Trieste in collaboration with the following institutes: International Center for Theoretical Physics; Istituto Nazionale di Oceanografia e Geofisica Sperimentale; Agenzia Regionale Protezione Ambientale-Friuli Venezia Giulia; and Istituto di Scienze Marine.
Geographic Focus: All Countries
Date(s) Application is Due: Oct 31
Contact: Patrizia Passarella, Program Assistant; (+39) 040 2240 389; fax (+39) 040 2240 388; itlabs@ictp.trieste.it or diploma@ictp.it
Internet: http://phdfluidmechanics.appspot.com/
Sponsor: Abdus Salam International Centre for Theoretical Physics
Strada Costiera 11
Trieste, I-34014 Italy

Abdus Salam ICTP Junior Associates Grants 231
The ICTP Associateship Scheme was created at the very inception of the Centre. It represents one of the means through which the vocation of the ICTP for the promotion and the development of scientific knowledge in Third World Countries has concretely been applied to reality. Junior Associates are intended for promising young scientists up to the age of 35. The award lasts for six years and entitles them to spend from 35 to 70 days at the Center, three times during the validity of the award, with fare and living allowance covered by the Associateship. After the three visits are completed, should the Associate Member still have a certain number of days remaining from his 210 day allocation, then he/she is entitled to visit the ICTP with living expenses covered.
Requirements: Scientists up to age 35 from and working in developing countries are eligible.
Geographic Focus: All Countries
Date(s) Application is Due: Sep 30
Contact: Patrizia Passarella, Program Assistant; (+39) 040 2240 389; fax (+39) 040 2240 388; itlabs@ictp.trieste.it or diploma@ictp.it
Internet: http://assoc.ictp.it/associateship-scheme/junior.html
Sponsor: Abdus Salam International Centre for Theoretical Physics
Strada Costiera 11
Trieste, I-34014 Italy

Abdus Salam ICTP Mathematical Research Fellowships 232
The Abdus Salam International Centre for Theoretical Physics (ICTP) invites applications for a small number of fellowships for outstanding mathematicians from developing countries. There are two types: Visiting Fellowships are open to mathematicians of any age and fund visits of between 1 and 12 months to ICTP during the next year; Post-Doctoral Fellowships are intended for younger mathematicians who have obtained a PhD in mathematics after January 1. Fellowships have a 12 month duration with a possibility to be renewed for a further 12 months. Preference is given to candidates who will benefit most from the time spent at ICTP, in the sense of pursuing their own research, using the ICTP facilities, interacting with other mathematicians, and ultimately turning the fellowship into a positive opportunity for their home institution or country as well as for themselves. Candidates should apply using the online application.
Geographic Focus: All Countries
Date(s) Application is Due: Jan 13
Contact: Patrizia Passarella, Program Assistant; (+39) 040 2240 201; fax (+39) 040 2240 490; math@ictp.it
Internet: http://www.ictp.it/research/math/visiting-fellowships.aspx
Sponsor: Abdus Salam International Centre for Theoretical Physics
Strada Costiera 11
Trieste, I-34014 Italy

Abdus Salam ICTP Networks Grants 233
A Network is a system of research groups in an entire region, or among different regions, that pursues a common scientific project over an extended period. The network program represents an efficient approach to overcoming the problem of isolation and counteracting the "brain-drain" of scientists in the Third World. Scientific network projects receive support in the following ways: as institutes or individuals with common scientific interests who agree upon a well defined joint project that emphasizes collaboration and the sharing of expertise and facilities. The proposer of the network is expected to act as coordinator and his/her Institution as headquarters. A network may organize seminars or workshops, arrange for visiting scientists and maintain joint links with the rest of the

international scientific community. The ICTP contributes towards the travel expenses of scientists who belong to the network to visit other member Institutes. Grants can be offered to young scientists to work for extended periods at institutes that are part of the network. The fields of study are physics or mathematics.
Requirements: A network project proposal should include: a detailed scientific and financial plan for one year; the list of member institutions/individuals; a description of the facilities of the member institutions that will be involved in the network; a justification for the proposed network; and the commitment of the member institutions to contribute substantially to the expenses of the project, in particular, to offer hospitality and local services. Academic institutions, nonprofit organizations, and Ph.D./M.D./ other professionals may apply.
Geographic Focus: All Countries
Contact: Patrizia Passarella, Program Assistant; (+39) 040 2240 323 or (+39) 040 2240 322; fax (+39) 040 2240 433; itlabs@ictp.trieste.it or oea@ictp.it
Internet: http://oea.ictp.it/programmes/net/index_html
Sponsor: Abdus Salam International Centre for Theoretical Physics
Strada Costiera 11
Trieste, I-34014 Italy

Abdus Salam ICTP Postdoctoral Positions in Theoretical High Energy Physics 234
Up to four new postdoctoral positions will be available in the Abdus Salam ICTP High Energy, Cosmology and Astroparticle Physics (HECAP) section for the upcoming academic years. The award amount is unspecified. Appointments will be made initially for two years, with the possibility of an extension for a third year. Members of the section are currently working on: string theory, including D-branes, gauge/string theory duality, phenomenological string theory/M theory; particle phenomenology including grand unification, supersymmetry, extra dimensions, LHC/collider physics; cosmology including large scale structures, physics of the early universe, present acceleration; topological field theories; and neutrino physics and astrophysics.
Geographic Focus: All Countries
Date(s) Application is Due: Dec 1
Contact: Patrizia Passarella, Program Assistant; (+39) 040 2240 323 or (+39) 040 2240 322; fax (+39) 040 2240 443; itlabs@ictp.trieste.it or rosanna@ictp.it
Internet: http://www.ictp.it/about-ictp/media-centre/news/postdoctoral-positions-in-theoretical-high-energy-physics.aspx
Sponsor: Abdus Salam International Centre for Theoretical Physics
Strada Costiera 11
Trieste, I-34014 Italy

Abdus Salam ICTP Postgraduate Diploma Grants 235
Talented young science students in developing countries are sometimes limited in achieving their full potential by the absence of advanced training at a postgraduate level, and up to standard international criteria. The late Professor Abdus Salam, Nobel Laureate and former Director of ICTP instituted an intense 12-month program in the main research fields of the center, with the aim of filling this gap. Among the projects on capacity building that ICTP has undertaken over time is the so-called Postgraduate Diploma Programme. The Programme consists of a rigorous, one-year long, pre-doctoral course work, with a small part devoted to independent projects. The Program initially had three streams: High Energy Physics, Condensed Matter Physics and Mathematics. A fourth stream has recently been added, Earth System Physics. The goal in these programs is to take good students from the least developed countries generally – and for the Physics Program students from only sub-Saharan Africa – and educate them so they can compete favorably for graduate studies in any center of learning in the world.
Geographic Focus: All Countries
Date(s) Application is Due: Dec 31
Contact: Patrizia Passarella, Program Assistant; (+39) 040 2240 389; fax (+39) 040 2240 388; itlabs@ictp.trieste.it or diploma@ictp.it
Internet: http://diploma.ictp.it/
Sponsor: Abdus Salam International Centre for Theoretical Physics
Strada Costiera 11
Trieste, I-34014 Italy

Abdus Salam ICTP Ramanujan Prize for Young Mathematicians 236
ICTP has created the Ramanujan Prize for young mathematicians from developing countries. The Prize is awarded annually to a researcher from a developing country less than 45 years of age on 31 December of the year of the award, who has conducted outstanding research in a developing country. Researchers working in any branch of the mathematical sciences are eligible. The Prize carries a $15,000 cash award and travel and subsistence allowance to visit ICTP for a meeting where the winner will be required to deliver a lecture. The Selection Committee will take into account not only the scientific quality of the research, but also the background of the candidate and the environment in which the work was carried out. The Committee will in particular favor candidates who have overcome adversity to achieve distinction in mathematics. The Prize is usually awarded to one person, but may be shared equally among recipients who have contributed to the same body of work.
Requirements: Nominations should include a CV and a list of publications, as well as a letter of recommendation. Additional supporting letters are encouraged.
Restrictions: Self nominations are not eligible.
Geographic Focus: All Countries
Date(s) Application is Due: Feb 1
Amount of Grant: 15,000 USD
Contact: Patrizia Passarella, Program Assistant; (+39) 040 2240 323 or (+39) 040 2240 322; fax (+39) 040 2240 443; director@ictp.it
Internet: http://www.ictp.it/about-ictp/prizes-awards/the-ramanujan-prize.aspx
Sponsor: Abdus Salam International Centre for Theoretical Physics
Strada Costiera 11
Trieste, I-34014 Italy

Abdus Salam ICTP Regular Associates Grants 237
The ICTP Associateship Scheme represents one of the means through which the promotion and the development of scientific knowledge in Third World countries has concretely been applied to reality. Regular Associates between the age of 36-45 are entitled to visit for 35 to 70 days, three times in a period of six years. For three visits the Centre provides their fare and a living allowance. After the three visits are completed, should the Associate Member still have a certain number of days remaining from his 210 day allocation, then he/she is entitled to visit the ICTP with living expenses covered.
Requirements: Scientists up to age 35 from and working in developing countries are eligible.
Geographic Focus: All Countries
Date(s) Application is Due: Sep 30
Contact: Patrizia Passarella, Program Assistant; (+39) 040 2240 389; fax (+39) 040 2240 388; itlabs@ictp.trieste.it or diploma@ictp.it
Internet: http://assoc.ictp.it/associateship-scheme/regular.html
Sponsor: Abdus Salam International Centre for Theoretical Physics
Strada Costiera 11
Trieste, I-34014 Italy

Abdus Salam ICTP Schools, Conferences, Workshops and Extended Research Workshops Grants 238
Each year, the ICTP organizes training and research activities in all areas of physica; sciences including mathematics. The subjects cover theory, simulations, and experiments. Proposals may be areas where there is in-house interest or research at ICTP, as follows: high energy, cosmology and astroparticle physics; condensed matter physics including statistical physics; fundamental and applied mathematics; earth system physics including geophysics, ocean cynamics, weather and climate modeling, upper atmosphere; and applied physics including fluid dynamics, plasma physics, nuclear data and applications, laser physics, energy systems, medical physics, ecological and environmental economics, machine learning, computer algorithms, soft matter, biophysics and quantitative aspects of life sciences and radio propagation. Proposals on web technologies and distance education, information and communication technologies, open source for science and development are also welcome. Subjects are not restricted to theory and the activities are selected on the basis of scientific novelty, the impact on the international community with special emphasis on bringing together scientists from South and North, basic training for younger scientists, and hands-on training for computer-intensive subjects. Where appropriate, proposals that combine pedagogical material with advanced research results, or those that spur new results in an emerging or multidisciplinary field, or those that resolve an outstanding or controversial issue will be preferred.
Requirements: Proposals are welcome at any time. Decisions concerning the program for next year are taken around March or April. Since the decision process involves several stages, early submission of proposals with a lead time of a few months is strongly encouraged. Proposals arriving after February 15 will as a rule not be processed for inclusion in the next year program, and might be considered for subsequent years only if there is room available.
Geographic Focus: All Countries
Date(s) Application is Due: Feb 15
Contact: Patrizia Passarella, Program Assistant; (+39) 040 2240 389; fax (+39) 040 2240 388; itlabs@ictp.trieste.it or diploma@ictp.it
Internet: http://www.ictp.it/call-for-proposals.aspx
Sponsor: Abdus Salam International Centre for Theoretical Physics
Strada Costiera 11
Trieste, I-34014 Italy

Abdus Salam ICTP Senior Associates Grants 239
The ICTP Associateship Scheme represents one of the means through which the vocation of the ICTP for the promotion and the development of scientific knowledge in Third World countries has concretely been applied to reality. All candidates over the age of 45 will automatically be considered for a Senior Associateship award. However, a Senior Associateship may be granted to candidates below the age of 45 who opt only to apply for a Senior Associateship award by stating this very clearly on their application form. The upper age limit for Senior Associates is 62 at the time of award. During the six years, Senior Associate Members may apply to visit the Center as often and for as long as they wish, until the allocation is exhausted, although the maximum foreseen duration of any visit is 70 days. Payment is Euros 8,000, which must be used to cover travel and living expenses during the 6-year appointment.
Geographic Focus: All Countries
Date(s) Application is Due: Sep 30
Amount of Grant: 8,000 EUR
Contact: Patrizia Passarella, Program Assistant; (+39) 040 2240 389; fax (+39) 040 2240 388; itlabs@ictp.trieste.it or diploma@ictp.it
Daniela Valentini, Senior Assistant; (+39) 040 2240 551; fax (+39) 040 2240 511; valentin@ictp.it
Internet: http://assoc.ictp.it/associateship-scheme/senior.html
Sponsor: Abdus Salam International Centre for Theoretical Physics
Strada Costiera 11
Trieste, I-34014 Italy

Abdus Salam ICTP Training and Research in Italian Laboratories Grants 240

The ICTP Programme gives scientists from Developing Countries in Central and Eastern Europe the opportunity to spend periods ranging from a few months to one year at Italian research laboratories of universities, governmental and private institutions, which operate in different branches of Physics. This includes academic studies as well as practical applications and industrial projects. Follow up visits and collaboration programs are also envisioned. Agreements of collaboration have been established with Italian research institutions, i.e. CNR (National Research Council), INFN (National Institute for Nuclear Physics), ENEA (Italian Commission for New Technologies, Energy and Environment) and INFM (National Institute for the Physics of Matter), APAT (Agency for Environmental Protection and Technical Services), Sincrotrone Trieste (Elettra Synchrotron Light Source), National Institute of Oceanography and Experimental Geophysics (OGS) and others. They offer hospitality and a special financial contribution. The fields covered, which reflect activities held at the ICTP, are presently: physics of condensed matter-solid state physics and materials science, computational methods, atomic and molecular physics, laser spectroscopy, semiconductor physics and technology; physics and energy-plasma physics and fusion, non-conventional energy sources, applied nuclear physics; physics and technology-optical physics and lasers, communication physics and technology, optical fibers, microprocessors and informatics, synchrotron radiation applications; earth and environmental sciences-geophysics and seismology, oceanography, physics of the atmosphere and aeronomy, meteorology and climatology, soil physics, environmental control, mathematical ecology, modelling of environmental systems; physics of the living state-biophysics, medical physics, neurophysics; and miscellaneous-instrumentation for nuclear and subnuclear physics, space physics, topics at the interface with chemistry, engineering, or biology. The minimum amount of these grants will be Euro 1,300 per month. Limited funds for travel expenses will be available in case these are not covered by other sources. Health insurance coverage will be provided for visitors and their accompanying family members. A monthly family allowance for those having dependent family members will be granted. The average length of a visit is 6 to 12 months. Additional information, including the application form, is available at the ICTP website.

Requirements: Less developed countries in Central and Eastern Europe may apply.
Geographic Focus: All Countries
Amount of Grant: Up to 15,600 EUR
Contact: Patrizia Passarella, Program Assistant; (+39) 040 2240 389; fax (+39) 040 2240 388; itlabs@ictp.trieste.it or diploma@ictp.it
Internet: http://users.ictp.it/www_users/ItaLab/index.html
Sponsor: Abdus Salam International Centre for Theoretical Physics
Strada Costiera 11
Trieste, I-34014 Italy

Abdus Salam ITCP Postgraduate Diploma Programme in General Physics 241

In each Postgraduate Diploma Programme, scholarships and travel grants will be awarded to those 10 students selected from developing countries. There are no course fees. The scholarships cover living expenses for the entire year of study (around 800 Euro per month). A limited number of qualified candidates who meet the same selection criteria as others, may attend the courses at their own cost. They must supply written evidence from the funding agency, or banking institution that funds are available for their entire year's stay, and are equal to those offered by the Centre.

Requirements: B.Sc. (but a M.Sc. is preferred) candidates in Physics or Mathematics are preferred. Applicants under 26 are also preferred from less developed countries, particularly sub-Sahara Africa.
Geographic Focus: All States, Angola, Benin, Botswana, Burkina Faso, Burundi, Cameroon, Cape Verde, Central African Republic, Chad, Comoros, Congo, Congo, Democratic Republic of, Cote d' Ivoire (Ivory Coast), Djibouti, Equatorial Guinea, Eritrea, Ethiopia, Gabon, Gambia, Ghana, Guinea, Guinea-Bissau, Kenya, Lesotho, Liberia, Madagascar, Malawi, Mali, Mauritania, Mozambique, Namibia, Niger, Nigeria, Reunion, Rwanda, Sao Tome & Principe, Senegal, Seychelles, Sierra Leone, Somalia, South Africa, Sudan, Swaziland, Tanzania, Togo, Uganda, Western Sahara, Zambia, Zimbabwe
Date(s) Application is Due: Dec 31
Amount of Grant: Up to 9,600 EUR
Contact: Patrizia Passarella, Program Assistant; (+39) 040 2240 389; fax (+39) 040 2240 388; itlabs@ictp.trieste.it or diploma@ictp.it
Internet: http://www.ictp.it/programmes.aspx
Sponsor: Abdus Salam International Centre for Theoretical Physics
Strada Costiera 11
Trieste, I-34014 Italy

Abdus Salam Joint ICTP/SISSA PhD Programme in Physics and Mathematics 242

The Abdus Salam International Centre for Theoretical Physics (ICTP) and the International School for Advanced Studies (SISSA) offer the opportunity for carrying out Ph.D. studies in the following fields: Applied Mathematics; Mathematical Analysis; Geometry; Mathematical Physics; Astroparticle Physics; Astrophysics; Elementary Particles; Physics and Chemistry of Biological Systems; Statistical Physics; and Theory and Numerical Simulation of the Condensed Matter. Admitted candidates will be awarded a fellowship of the amount of Euro 13,638.47 gross per year (approx. Euro 1030,00 net per month) for the entire duration of the course (3 years extendable to a fourth, subject to availability of funds and approval by ICTP/SISSA Governing bodies). A contribution towards living expenses of the amount of Euro 100,00 gross per month will be awarded to those who will subscribe a registered rent contract. A contribution towards the expenses to be enrolled in the health insurance system up to Euro 198.77 per year will be granted.

Requirements: Applicants interested in participating in this selection must be possess an "ICTP Diploma" or be enrolled in the last year of the "ICTP diploma Program". All the admitted candidates must hold the "ICTP Diploma" by the date of the beginning of the courses (3 October for Applied Mathematics, Astroparticle Physics, Geometry, Mathematical Analysis, Mathematical Physics and Theory of Elementary Particles, 7 November for all other courses). Admitted candidates will have to pass a qualifying examination within the first year of study.
Geographic Focus: All Countries
Date(s) Application is Due: Mar 31
Amount of Grant: 14,838-15,037 EUR
Contact: Patrizia Passarella, Program Assistant; (+39) 040 2240 389; fax (+39) 040 2240 388; itlabs@ictp.trieste.it or diploma@ictp.it
Internet: http://www.ictp.it/programmes/degree-programmes.aspx
Sponsor: Abdus Salam International Centre for Theoretical Physics
Strada Costiera 11
Trieste, I-34014 Italy

Abington Foundation Grants 243

The Abington Foundation awards grants to Ohio nonprofit organizations serving Cuyahoga County in its areas of interest, including pre-primary and higher education, geriatric healthcare and nursing, the promotion or sustenance of individual and family economic independence, and cultural activities. Priority is given to funding requests for specific programs or projects that represent critical periods in a child's development from birth to age 5, and to early adolescence, ages 10-15. Requests for endowment or general operating support are discouraged. Foundation staff is available for consultation during the proposal preparation process.

Requirements: Ohio nonprofit organizations serving Cuyahoga County are eligible.
Geographic Focus: Ohio
Date(s) Application is Due: May 1; Sep 1; Dec 1
Amount of Grant: 5,000-50,000 USD
Samples: Cleveland Play House, Cleveland, Ohio, $50,000-to build the Allen Theatre Complex, a three theatre center for performing arts and arts education, in partnership with Playhouse Square and Cleveland State University (2011); North Coast Health Ministry, Cleveland, Ohio, $33,000-to hire a chronic care nurse case manager (2011); St. Martin de Porres High School, Cleveland, Ohio, $23,000-for summer programming and tutoring for incoming freshmen (2011).
Contact: Janet E. Narten; (216) 621-2901 or (216) 621-2632; fax (216) 621-8198
Internet: http://www.fmscleveland.com/abington
Sponsor: Abington Foundation
1422 Euclid Avenue, Suite 627
Cleveland, OH 44115-1952

ACAAI Foundation Research Grants 244

The Foundation develops, promotes and funds clinical research and educational programs related to allergy, asthma & immunology. The Foundation is dedicated to: strengthening, supporting and funding allergy/immunology training programs; heightening awareness of the critical role of the allergist in cost-effective treatment of asthma and other allergic and immunologic diseases; and supporting asthma camp programs to teach children and their parents how to live with asthma. Programs include clinical fellowship stipends, fellow-in-training research grants, young faculty support awards, scholar's return awards, and support for asthma camps.

Geographic Focus: All States
Amount of Grant: Up to 250,000 USD
Contact: Program Administrator; (847) 427-1200; fax (847) 427-1294; mail@acaai.org
Internet: http://www.acaai.org/Member/ACAAI_Foundation/ACAAI_Foundation.htm
Sponsor: American College of Allergy, Asthma and Immunology Foundation
85 West Algonquin Road, Suite 550
Arlington Heights, IL 60005

AcademyHealth Alice S. Hersh New Investigator Awards 245

The Alice S. Hersh New Investigator Award recognizes the contribution of new scholars to the field of health services research. Eligible nominees are those whose first research appointments were no earlier than January 1, 2002. Post-doctoral positions will be considered research appointments unless these years provided training necessary for entry into the field. In addition, the eligibility period may be extended for a nominee who has taken a formal leave of absence from their employer. The awards will be presented at the Annual Research Meeting in June. All awardees receive complimentary registration and lodging to attend the meeting and a $2,500 prize from the Alice S. Hersh Memorial Fund.

Requirements: Nominators must be members of AcademyHealth although nominees need not be. Nominees from academic and non-academic institutions are encouraged. Nominations must include one letter of nomination (no more than two pages) that reviews the nominee's contribution to the field and specifies the date of the nominee's first research appointment, and must include curriculum vitae.
Geographic Focus: All States
Date(s) Application is Due: Mar 2
Amount of Grant: 2,500 USD
Contact: Jennifer Muldoon, Director; (202) 292-6770; fax (202) 292-6870; jennifer.muldoon@academyhealth.org
Internet: http://www.academyhealth.org/content.cfm?ItemNumber=954@navItemNumber=2342
Sponsor: AcademyHealth
1150 17th Street NW, Suite 600
Washington, DC 20036

AcademyHealth HSR Impact Awards 246

The program requests nominations of health services research that has made a positive impact on health policy and/or practice. The winning impacts will be disseminated widely as part of AcademyHealth's ongoing efforts to promote the field of health services research and communicate its value for health care decision-making. Nominated research may be published or unpublished, a single study or a body of work, the work of an individual or a team. The time frame for when the research was conducted is open, but the impact should have occurred recently. Nominators and nominees are not required to be AcademyHealth members. Self-nominations are accepted. Award winners are announced at the National Health Policy Conference, February 6 and 7, and the winners receive complimentary registration, travel, and lodging to the conference.
Requirements: Each nomination must include a summary describing the research used, the method of translation, and the impact on policy and/or practice (no more than 500 words); a letter of recommendation from the user of this research discussing the impact the research has had; contact information for the lead researcher; and contact information for the nominator (may be the lead researcher).
Geographic Focus: All States
Date(s) Application is Due: Jul 29
Amount of Grant: 2,000 USD
Contact: Jennifer Muldoon, Director; (202) 292-6770; fax (202) 292-6870; jennifer.muldoon@academyhealth.org
Internet: http://www.academyhealth.org/awards/hsrimpacts.htm
Sponsor: AcademyHealth
1150 17th Street NW, Suite 600
Washington, DC 20036

AcademyHealth Nemours Child Health Services Research Awards 247

The Nemours Child Health Services Research Award recognizes the scientific work of emerging scholars in the field of child health services research, particularly research on quality improvement of pediatric health services. Recipients must be within seven years of entry into the field of child health services research as of December 31, 2007. Year of entry will be judged by the nominee's first publication in child health services research. The winner will receive $1,000 in recognition of his/her contribution to child health services research, and the award will be presented at the Child Health Services Research Meeting, held in in Washington, D.C., in conjunction with AcademyHealth's Annual Research Meeting in June.
Requirements: Nominations should include: Letter of nomination that identifies when the candidate first entered the field (i.e., title and year of first publication in child health services research); Curriculum vitae that includes a detailed list of accomplishments that merit this recognition. Self-nominations will be accepted.
Geographic Focus: All States
Date(s) Application is Due: Apr 15
Amount of Grant: 1,000 USD
Contact: Jennifer Muldoon, Director; (202) 292-6770; fax (202) 292-6870; jennifer.muldoon@academyhealth.org
Internet: http://www.academyhealth.org/content.cfm?ItemNumber=982&navItemNumber=2342
Sponsor: AcademyHealth
1150 17th Street NW, Suite 600
Washington, DC 20036

AcademyHealth PHSR Research Article of the Year Awards 248

Public Health Systems Research (PHSR) is a field of inquiry examining the organization, financing, performance, and impact of health systems defined as the constellation of governmental and non-governmental actors that influence population health, including health care providers, insurers, purchasers, public health agencies, community-based organizations and entities that operate outside the traditional sphere of health care. The Article of the Year Award recognizes the best scientific work that the field of PHSR has produced and published in the previous year. The awardee will receive complimentary registration for both the PHSR Interest Group Annual Meeting and AcademyHealth's Annual Research Meeting, travel assistance, and a $1,000 cash award. Nominated articles must have been published in a peer-reviewed journal between January 1 and December 31 of the previous year, using either qualitative or quantitative techniques, articles must present, analyze, and comment on new data or synthesize and analyze data that have already been collected. The selection will be based on the article's contribution to the understanding of public health systems; provision of new insights to the field of PHSR; and potential to advance the field and/or challenge current thinking. Nominators must be members of AcademyHealth, although nominees need not be. Self-nominations are acceptable.
Requirements: Nominations must: be received at AcademyHealth by April 15; include the name and contact information for both the nominator and nominee and a copy of the article(s). Nominations may be submitted by email or postal mail. email to: PHSR@academyhealth.org. Mail to: AcademyHealth, Attention: PHSR Article of the Year Committee, 1150 17th Street, NW, Suite 600, Washington, DC 20036.
Geographic Focus: All States
Date(s) Application is Due: Apr 15
Amount of Grant: 1,000 USD
Contact: Jennifer Muldoon, Director; (202) 292-6770; fax (202) 292-6870; jennifer.muldoon@academyhealth.org
Internet: http://www.academyhealth.org/Communities/content.cfm?ItemNumber=2536
Sponsor: AcademyHealth
1150 17th Street NW, Suite 600
Washington, DC 20036

Academy of American Franciscan History Dissertation Fellowships 249

The Academy of American Franciscan History is accepting applications for four dissertation fellowships, each worth $10,000. As many as two of these fellowships will be awarded for projects dealing with some aspect of the history of the Franciscan Family in Latin America, including the United States Borderlands, Mexico, Central and South America. Up to another two fellowships will be awarded to support projects dealing with some aspect of the history of the Franciscan Family in the rest of the United States and Canada. Projects may deal with any aspect of the history of the Franciscan Family, including any of the branches of the Family, male, female, tertiary, Capuchin, etc. The deadline for applications is February 1 each year.
Requirements: The recipient must be engaged in full-time research during the period of the fellowship. The applicant must be a doctoral candidate at a University in the Americas, and the bulk of the research should be conducted in the Americas.
Geographic Focus: All States, Argentina, Belize, Bolivia, Brazil, Canada, Chile, Colombia, Costa Rica, Ecuador, El Salvador, Falkland Islands (Malvinas), French Guiana, Guatemala, Guyana, Honduras, Mexico, Nicaragua, Panama, Paraguay, Peru, Suriname, Uruguay, Venezuela
Date(s) Application is Due: Feb 1
Amount of Grant: 10,000 USD
Contact: Dr. Jeffrey M. Burns; (510) 548-1755; acadafh@aol.com or acadafh@fst.edu
Internet: http://aafh.org/Scholarships.html
Sponsor: Academy of American Franciscan History
1712 Euclid Avenue
Berkley, CA 94709-1208

ACC American Artists and Museum Professionals in Asia Fellowships 250

American Artists and Museum Professionals in Asia is a grant program established at the ACC in 2007 through the support of a three-year donation from The Henry Luce Foundation. The program provides individual fellowships to American artists and American museum professionals for research, study, and creative and collaborative projects in those countries of Asia extending from Burma eastward through Korea and Indonesia.
Requirements: Americans seeking aid to undertake activities in Asia are eligible to apply. Completed applications must be submitted at least six months prior to the planned date of project implementation. Applicants should send a brief description of the activity for which assistance is sought. If the proposed activity falls within the council's guidelines, additional application materials will be provided.
Restrictions: The Council is unable to consider proposals for the following: personal exhibitions; individual performance tours; undergraduate study; or activities conducted by individuals in their home countries.
Geographic Focus: All States
Date(s) Application is Due: Feb 1
Contact: Program Director; (212) 843-0403; fax (212) 843-0343; acc@accny.org
Internet: http://www.asianculturalcouncil.org/programs.html
Sponsor: Asian Cultural Council
6 West 48th Street, 12th Floor
New York, NY 10036-1802

ACC Asian Art and Religion Fellowships 251

This Fellowship program focusing on the relationship between the artistic and religious traditions of Asia enables American scholars, specialists, and artists to conduct research and undertake projects in Asia involving the interdisciplinary analysis of religion and the arts. The Council awards up to five research fellowships or travel grants each year through this program. Asian Art and Religion Fellowships have supported research in India on the iconography of Hindu deities, in Indonesia on the relationship between tantric religious thought and masked performance, and in Thailand on Buddhist architecture of northern Thailand.
Requirements: Asian individuals in the visual and performing arts seeking grant assistance to conduct research, study, receive specialized training, undertake observation tours, or pursue creative activity in the United States are eligible to apply for fellowship support from the Council. Americans seeking aid to undertake activities in Asia are also eligible to apply. Completed applications must be submitted at least six months prior to the planned date of project implementation. Applicants should send a brief description of the activity for which assistance is sought. If the proposed activity falls within the council's guidelines, additional application materials will be provided.
Restrictions: The Council is unable to consider proposals for the following: personal exhibitions; individual performance tours; undergraduate study; or activities conducted by individuals in their home countries.
Geographic Focus: All States
Date(s) Application is Due: Feb 1
Contact: Program Director; (212) 843-0403; fax (212) 843-0343; acc@accny.org
Internet: http://www.asianculturalcouncil.org/programs.html
Sponsor: Asian Cultural Council
6 West 48th Street, 12th Floor
New York, NY 10036-1802

ACC Fellowships and Projects 252

Many of the Asian Cultural Council's fellowships are awarded with unrestricted grant funds, primarily to support artists and specialists from Asia pursuing research, training, and creative programs in the United States. On average each year, unrestricted funds provide support for sixteen individual fellowships and seventeen project grants.
Requirements: Asian individuals in the visual and performing arts seeking grant assistance to conduct research, study, receive specialized training, undertake observation tours, or pursue creative activity in the United States are eligible to apply for project or fellowship

support from the Council. Americans seeking aid to undertake activities in Asia are also eligible to apply. Completed applications must be submitted at least six months prior to the planned date of project implementation. Applicants should send a brief description of the activity for which assistance is sought. If the proposed activity falls within the council's guidelines, additional application materials will be provided.
Restrictions: The Council is unable to consider proposals for the following: personal exhibitions; individual performance tours; undergraduate study; or activities conducted by individuals in their home countries.
Geographic Focus: All States
Date(s) Application is Due: Feb 1
Contact: Program Director; (212) 843-0403; fax (212) 843-0343; acc@accny.org
Internet: http://www.asianculturalcouncil.org/programs.html
Sponsor: Asian Cultural Council
6 West 48th Street, 12th Floor
New York, NY 10036-1802

ACC Ford Foundation Fellowships 253

This fellowship program provides support for training, travel, and research in the United States by individuals from Asia engaged in the study, documentation, and preservation of Asian traditional arts. Funded by an endowment grant from the Ford Foundation, this program offers awards in archaeology, art history, conservation, dance ethnology, ethnomusicology, museology, and other disciplines involving traditional Asian culture. Fellowships are generally awarded for periods ranging from three to 12 months. Completed applications must be submitted at least six months prior to the planned date of project implementation. Applicants should send a short description of the activity for which assistance is being requested. If the proposed activity falls within the council's guidelines, application materials will be provided by ACC.
Requirements: Asian individuals in the visual and performing arts seeking grant assistance to conduct research, study, receive specialized training, undertake observation tours, or pursue creative activity in the United States are eligible to apply for project or fellowship support from the Council. Americans seeking aid to undertake activities in Asia are also eligible to apply. Completed applications must be submitted at least six months prior to the planned date of project implementation. Applicants should send a brief description of the activity for which assistance is sought. If the proposed activity falls within the council's guidelines, additional application materials will be provided.
Restrictions: The Council is unable to consider proposals for the following: personal exhibitions; individual performance tours; undergraduate study; or activities conducted by individuals in their home countries.
Geographic Focus: All States
Date(s) Application is Due: Feb 1
Contact: Program Director; (212) 843-0403; fax (212) 843-0343; acc@accny.org
Internet: http://www.asianculturalcouncil.org/programs.html#4
Sponsor: Asian Cultural Council
6 West 48th Street, 12th Floor
New York, NY 10036-1802

ACC Hong Kong Arts Program Fellowships 254

The Program, made possible by donations from a broad group of donors in Hong Kong, provides fellowships to especially promising artists, students, and scholars from Hong Kong and other parts of China for research, study, and creative work in the United States. Limited support is also available for projects sponsored by Hong Kong institutions that involve Hong Kong-United States or Hong Kong-Asia cultural exchange. The ACC is able to award grant support to approximately ten individual and institutional applicants each year. Applicants should send a short description of the activity for which assistance is being requested. If the proposed activity falls within the council's guidelines, further application materials will be provided.
Requirements: Asian individuals in the visual and performing arts seeking grant assistance to conduct research, study, receive specialized training, undertake observation tours, or pursue creative activity in the United States are eligible to apply for project or fellowship support from the Council. Americans seeking aid to undertake activities in Asia are also eligible to apply. Completed applications must be submitted at least six months prior to the planned date of project implementation. Applicants should send a brief description of the activity for which assistance is sought. If the proposed activity falls within the council's guidelines, additional application materials will be provided.
Restrictions: The Council is unable to consider proposals for the following: personal exhibitions; individual performance tours; undergraduate study; or activities conducted by individuals in their home countries.
Geographic Focus: All States, China
Date(s) Application is Due: Feb 1
Contact: Program Director; (212) 843-0403; fax (212) 843-0343; acc@accny.org
Internet: http://www.asianculturalcouncil.org/programs.html#5
Sponsor: Asian Cultural Council
6 West 48th Street, 12th Floor
New York, NY 10036-1802

ACC Humanities Fellowship Program 255

The program assists American scholars, doctoral students, and specialists in the humanities to undertake research, training, and study in Asia in the following fields: archaeology; conservation; museology; and the theory, history, and criticism of architecture, art, dance, film, music, photography, and theater. Fellowship grants range in duration from one to nine months. The program also supports American and Asian scholars participating in international conferences, exhibitions, visiting professorships, and similar projects. Funding includes international travel, maintenance, per diem, and related research expenses. Completed applications must be submitted at least six months prior to the planned date of implementation. Applicants should send a short description of the activity for which assistance is being sought. If the proposed activity falls within the council's guidelines, application materials will be provided.
Requirements: American scholars, doctoral students, and specialists in the humanities are eligible to apply.
Restrictions: The Council is unable to consider proposals for the following: personal exhibitions; individual performance tours; undergraduate study; or activities conducted by individuals in their home countries.
Geographic Focus: All States
Date(s) Application is Due: Feb 1
Contact: Program Director; (212) 843-0403; fax (212) 843-0343; acc@accny.org
Internet: http://www.asianculturalcouncil.org/programs.html#6
Sponsor: Asian Cultural Council
6 West 48th Street, 12th Floor
New York, NY 10036-1802

ACC Japan-U.S. Arts Program Fellowships 256

The Program provides grant fellowships to individuals and institutions in Japan and the United States for exchange activities that encourage the study and understanding of Japanese art and culture. The individual fellowships enable Japanese artists, scholars, and specialists to visit the United States for research, observation, and creative work and allow their American counterparts to visit Japan for similar purposes. The council also provides assistance for performances, exhibitions, and other projects of unusual importance for the development of Japanese-American cultural exchange. The program is administered with the assistance of the ACC's office in Tokyo. Annually, the program supports approximately thirteen individuals and seven institutional projects. Applicants should send a short description of the activity for which assistance is being requested. If the proposed activity falls within the council's guidelines, application materials will be provided by the ACC.
Requirements: Asian individuals in the visual and performing arts seeking grant assistance to conduct research, study, receive specialized training, undertake observation tours, or pursue creative activity in the United States are eligible to apply for project or fellowship support from the Council. Americans seeking aid to undertake activities in Asia are also eligible to apply. Completed applications must be submitted at least six months prior to the planned date of project implementation. Applicants should send a brief description of the activity for which assistance is sought. If the proposed activity falls within the council's guidelines, additional application materials will be provided.
Restrictions: The Council is unable to consider proposals for the following: personal exhibitions; individual performance tours; undergraduate study; or activities conducted by individuals in their home countries.
Geographic Focus: All States
Date(s) Application is Due: Feb 1
Contact: Program Director; (212) 843-0403; fax (212) 843-0343; acc@accny.org
Internet: http://www.asianculturalcouncil.org/programs.html#7
Sponsor: Asian Cultural Council
6 West 48th Street, 12th Floor
New York, NY 10036-1802

ACC John D. Rockefeller III Award 257

This annual award is presented to an individual from Asia or the United States who has made an especially significant contribution to the understanding, practice, or study of the visual or performing arts of Asia. The award will enable recipients to pursue work in some aspect of the arts of Asia through international travel and research. Applications are accepted anytime.
Requirements: Individuals from Asia and the United States who are active in any field of the visual or performing arts of Asia, whether affiliated with an institution or working independently, are eligible for award consideration. Candidates are nominated by artists, scholars, and others professionally involved in Asian art and culture.
Geographic Focus: All States
Amount of Grant: 30,000 USD
Contact: Program Director; (212) 843-0403; fax (212) 843-0343; acc@accny.org
Internet: http://www.asianculturalcouncil.org/programs.html#12
Sponsor: Asian Cultural Council
6 West 48th Street, 12th Floor
New York, NY 10036-1802

ACC Mekong Region Fellowship 258

The Mekong Region Fellowship program is funded by the Rockefeller Foundation. The program assists individual artists and specialists from Burma, Cambodia, Laos, Thailand, Vietnam, and China's Yunnan Province in undertaking research, study, and creative projects in the United States and Asia. The program also provides funds for special projects that encourage regional collaboration and partnership.
Requirements: Asian individuals in the visual and performing arts seeking grant assistance to conduct research, study, receive specialized training, undertake observation tours, or pursue creative activity in the United States are eligible to apply for project or fellowship support from the Council. Americans seeking aid to undertake activities in Asia are also eligible to apply. Completed applications must be submitted at least six months prior to the planned date of project implementation. Applicants should send a brief description of the activity for which assistance is sought. If the proposed activity falls within the council's guidelines, additional application materials will be provided.
Restrictions: The Council is unable to consider proposals for the following: personal exhibitions; individual performance tours; undergraduate study; or activities conducted by individuals in their home countries.

Geographic Focus: All States
Date(s) Application is Due: Feb 1
Contact: Program Director; (212) 843-0403; fax (212) 843-0343; acc@accny.org
Internet: http://www.asianculturalcouncil.org/programs.html#8
Sponsor: Asian Cultural Council
6 West 48th Street, 12th Floor
New York, NY 10036-1802

ACCP Frontiers Career Development Research Awards 259
Supports previously unmet or under-served areas of Pharmacy-Based Health Services Research, Clinical Research, or Translational Research. Through its awards, the Institute is especially interested in supporting: research that helps to pursue one or more of the priority areas outlined in the Research Agenda of the American College of Clinical Pharmacy (see below); the career development of clinical pharmacy investigators; health services research that assesses the impact of clinical pharmacy services on the use, costs, quality, accessibility, delivery, organization, financing, and outcomes of health care, including the development of payment models for clinical pharmacists' services; and clinical and translational research, including areas that extend beyond those funded by other ACCP awards and grants.
Requirements: Any ACCP Full Member, Associate Member, or Affiliate Member is eligible to apply.
Restrictions: Proposals submitted concurrently with an application for membership will not be considered.
Geographic Focus: All States
Date(s) Application is Due: Jun 1
Amount of Grant: 10,000-30,000 USD
Contact: Cathy Englund; (913) 492-3311; fax (913) 492-0088; cenglund@accp.com
Internet: http://www.accp.com/frontiers/research.php#resfel
Sponsor: American College of Clinical Pharmacy
13000 W 87th Street Parkway
Lenexa, KS 66215-4530

ACCP Hematology/Oncology PRN Minisabbatical Program 260
The Minisabbatical Program is an opportunity for members of the Hematology/Oncology PRN exclusively to gain or expand their skills in practice or research under the guidance of experts in oncology pharmacotherapy. The major purposes are: to learn new clinical skills or approaches to patient care that will enhance the applicant's practice in oncology pharmacotherapy (e.g., patient management, clinic development, payment for services); and to learn new research skills that will enhance or expand the applicant's research program in oncology pharmacotherapy (e.g., laboratory techniques, pharmacokinetic modeling).
Requirements: The applicant for the program must be a current member of the Hematology/Oncology PRN.
Restrictions: The following individuals will not be considered as candidates: current residents or fellows; members of the Minisabbatical Awards Committee; or persons in receipt of a minisabbatical award in the past 5 years.
Geographic Focus: All States
Date(s) Application is Due: Jun 2
Contact: Cathy Englund; (913) 492-3311; fax (913) 492-0088; cenglund@accp.com
Internet: http://www.accp.com/frontiers/rihmonmini.php
Sponsor: American College of Clinical Pharmacy
13000 W 87th Street Parkway
Lenexa, KS 66215-4530

ACC Philippines Fellowship Program 261
The Program was formally inaugurated in 2000 in association with the ACC Philippines Foundation, a foundation established in Manila to collaborate with the ACC in raising funds and making grants for Philippines-U.S. cultural exchange. Through the ACC Philippines Foundation a range of donors in both countries are contributing to the Program, which emphasizes support for artists, scholars, and specialists from the Philippines undertaking research, study, and creative work in the United States. Some grants are also made to Americans pursuing similar activities in the Philippines and to institutions engaged in Philippines-U.S. or Philippines-Asia exchange projects. Applicants should send a short description of the activity for which assistance is being requested. If the proposed activity falls within the council's guidelines, application materials will be provided by the ACC.
Requirements: Asian individuals in the visual and performing arts seeking grant assistance to conduct research, study, receive specialized training, undertake observation tours, or pursue creative activity in the United States are eligible to apply for project or fellowship support from the Council. Americans seeking aid to undertake activities in Asia are also eligible to apply. Completed applications must be submitted at least six months prior to the planned date of project implementation. Applicants should send a brief description of the activity for which assistance is sought. If the proposed activity falls within the council's guidelines, additional application materials will be provided.
Restrictions: The Council is unable to consider proposals for the following: personal exhibitions; individual performance tours; undergraduate study; or activities conducted by individuals in their home countries.
Geographic Focus: All States, Philippines
Date(s) Application is Due: Feb 1
Contact: Program Director; (212) 843-0403; fax (212) 843-0343; acc@accny.org
Internet: http://www.asianculturalcouncil.org/programs.html#9
Sponsor: Asian Cultural Council
6 West 48th Street, 12th Floor
New York, NY 10036-1802

ACCP Infectious Diseases PRN Minisabbatical Program 262
The minisabbatical program is an opportunity for members of the ID PRN exclusively to gain or expand their skills in practice or research under the guidance of experts in infectious diseases pharmacotherapy. The major purposes are: to learn new clinical skills or approaches to patient care that will enhance the applicant's practice in infectious diseases (e.g., patient evaluation techniques, practice management, antimicrobial surveillance, etc.); and to learn new research skills that will enhance or expand the applicant's research program in infectious diseases (e.g., laboratory techniques, pharmacokinetic and pharmacodynamic modeling, epidemiology, ID outcomes research).
Requirements: The applicant for the ID PRN Minisabbatical Program must be a current member and maintain membership with the ID PRN for the duration of project.
Restrictions: The following individuals will not be considered as candidates: current students, residents or fellows at time of application; members of the Minisabbatical Awards Committee; if currently funded by the NIH or have other multi-year peer reviewed funding; or persons in receipt of a minisabbatical award in the past 5 years.
Geographic Focus: All States
Date(s) Application is Due: Jun 2
Amount of Grant: 1,000-3,000 USD
Contact: Cathy Englund, Executive Secretary; (913) 492-3311; fax (913) 492-0088; cenglund@accp.com
Internet: http://www.accp.com/frontiers/riinfdmini.php
Sponsor: American College of Clinical Pharmacy
13000 W 87th Street Parkway
Lenexa, KS 66215-4530

ACCP Infectious Diseases Research Fellowships 263
The purpose of ACCP fellowships is to support the development of Clinical Scientists through postgraduate fellowship experiences, or through the support of graduate students who have at least completed their qualifying examinations. Research activities must relate directly to the stated areas of emphasis. Fellowships support a research training experience and not a clinical pharmacy residency program. A pharmacy fellowship is a directed, highly individualized postgraduate program designed to prepare the participant to become an independent researcher. Fellowships exist primarily to develop competency in the scientific research process. The ACCP Research Institute believes that clinical pharmacy fellowships should be at least two years in length, and will give strong preference to funding applications that describe a multi-year training program. This grant will provide the fellow's stipend for either the first, second, or third year. Preceptors will have the option to receive the fellowship's entire funding over one year, or half funding for each of two years.
Requirements: Preceptor: An ACCP Full Member at the time application is made. Fellowship applications submitted concurrently with an application for Full Membership will not be considered. Fellow: An ACCP Member, Associate Member, or applicant for member. If a graduate student, (a) preference will be given to supporting students who have received or are concurrently enrolled in a Pharm.D. degree program; and (b) the Fellow must at least have completed his/her qualifying examinations.
Restrictions: Ineligible applicants include: ACCP Board of Regent members; ACCP Research Institute Board of Trustee members; Grants and Fellowship Selection Committee members; preceptor of an ACCP Research Institute fellowship within the previous year.
Geographic Focus: All States
Date(s) Application is Due: Mar 23
Amount of Grant: 40,500 USD
Contact: Cathy Englund, Executive Secretary; (913) 492-3311; fax (913) 492-0088; cenglund@accp.com
Internet: http://www.accp.com/frontiers/research.php#resfel
Sponsor: American College of Clinical Pharmacy
13000 W 87th Street Parkway
Lenexa, KS 66215-4530

ACCP Oncology Research Fellowships 264
This fellowship, supported by a grant from sanofi-aventis, supports the development of Clinical Scientists in oncology through postgraduate fellowship experiences, or through the support of graduate students who have at least completed their qualifying examinations. Research activities must related to the stated area of emphasis. This grant provides the fellow's stipend of $38,000, travel support of $1,000, and personnel fringe benefits of $1,500.
Requirements: Preceptor: An ACCP Full Member at the time application is made. Fellowship applications submitted concurrently with an application for Full Membership will not be considered. Fellow: An ACCP Member, Associate Member, or applicant for member. If a graduate student, (a) preference will be given to supporting students who have received or are concurrently enrolled in a Pharm.D. degree program; and (b) the Fellow must at least have completed his/her qualifying examinations.
Restrictions: Ineligible applicants include: ACCP Board of Regent members; ACCP Research Institute Board of Trustee members; Grants and Fellowship Selection Committee members; preceptor of an ACCP Research Institute fellowship within the previous year.
Geographic Focus: All States
Date(s) Application is Due: Mar 3
Amount of Grant: 40,500 USD
Contact: Cathy Englund; (913) 492-3311; fax (913) 492-0088; cenglund@accp.com
Internet: http://www.accp.com/frontiers/research.php#resfel
Sponsor: American College of Clinical Pharmacy
13000 W 87th Street Parkway
Lenexa, KS 66215-4530

ACCP Pain & Palliative Care PRN Minisabbatical Program 265

The Minisabbatical Program is an opportunity for members of the Pain & Palliative Care PRN exclusively to gain or expand their skills in practice or research under the guidance of experts in pain & palliative care pharmacotherapy. The Pain & Palliative Care minisabbatical program is made possible by an unrestricted educational grant from Amgen. The major purposes are: to learn new clinical skills or approaches to patient care that will enhance the applicant's practice in pain and palliative care pharmacotherapy; and to learn new research skills that will enhance or expand the applicant's research program in pain and palliative care pharmacotherapy.
Requirements: The applicant for the minisabbatical program must be a current member of the Pain & Palliative Care PRN.
Restrictions: The following individuals will not be considered as candidates: current residents or fellows; members of the Minisabbatical Awards Committee; or persons in receipt of a minisabbatical award in the past 5 years.
Geographic Focus: All States
Date(s) Application is Due: Jun 1
Amount of Grant: 750-2,750 USD
Contact: Cathy Englund; (913) 492-3311; fax (913) 492-0088; cenglund@accp.com
Internet: http://www.accp.com/frontiers/rinephmini.php
Sponsor: American College of Clinical Pharmacy
13000 W 87th Street Parkway
Lenexa, KS 66215-4530

ACCP Pharmacotherapy Investigator Development Research Awards 266

The purpose of these awards is to provide funding for research projects that will: help pursue one or more of the priority areas outlined in the Research Agenda of the American College of Clinical Pharmacy; contribute to the development of the principal investigator's research career; promote the safe, effective, and cost-effective use of medications; and advance the practice of clinical pharmacy. This includes, but is not limited to, investigations aimed at generating preliminary data or seed money for pilot projects. The award may serve as the total support for a project or supplement an existing research effort as long as a specific portion of the research is identified as being made possible by this award, and provided that the investigator states specifically how the balance will be funded and provides evidence of its guaranteed availability. The overall goal of the awards is to assist developing researchers in improving their competitiveness for extramural funding and to support their evolving research careers. In this context, projects that involve the use of animals or in vitro models will be considered as long as the applicant adequately justifies the request.
Requirements: The principal investigator of the award application must: be an ACCP member; be ten or fewer years since completion of his/her formal training or first academic appointment; not be currently funded by the NIH or have other multi year peer reviewed funding; or not be a student, resident, or fellow.
Restrictions: The award must not duplicate funding for another research project.
Geographic Focus: All States
Date(s) Application is Due: Jul 16
Amount of Grant: 20,000 USD
Contact: Cathy Englund; (913) 492-3311; fax (913) 492-0088; cenglund@accp.com
Internet: http://www.accp.com/frontiers/research.php#resfel
Sponsor: American College of Clinical Pharmacy
13000 W 87th Street Parkway
Lenexa, KS 66215-4530

ACC Residency Program in Asia 267

The program in Asia assists individual American artists, scholars, and professionals undertaking research, teaching, and creative residencies at cultural and educational institutions in East and Southeast Asia. Projects supported in the Program demonstrate close collaboration in design and execution between the visiting American specialist and the host organization and produce tangible results such as publication, course development, or the creation of new artistic work. The program helps to foster the growth of Asian cultural studies in the U.S. and of American cultural studies in Asia, as well as to stimulate long-term relationships in the arts and humanities between American individuals and institutions and their colleagues in Asia. Eligible fields for consideration include archaeology, architecture (design, history, and theory), art history, conservation, crafts, dance, film, museology, music, painting, photography, printmaking, sculpture, theater, and video.
Requirements: Applications may be submitted by the participating American specialist or by the host institution in Asia.
Restrictions: The Council is unable to consider proposals for the following: personal exhibitions; individual performance tours; undergraduate study; or activities conducted by individuals in their home countries.
Geographic Focus: All States
Date(s) Application is Due: Jan 15; Aug 1
Contact: Program Director; (212) 812-4300; fax (212) 812-4299; acc@accny.org
Internet: http://www.asianculturalcouncil.org/programs.html#1
Sponsor: Asian Cultural Council
6 West 48th Street, 12th Floor
New York, NY 10036-1802

ACC Starr Foundation Fellowship Program 268

The contemporary visual arts of Asia are the focus of this fellowship program, established with an endowment grant from the Starr Foundation. Assistance is provided to artists and arts specialists from Asia for creative activity, research, training, and observation in the United States in the fields of art criticism, crafts, design, film, museology, painting, photography, printmaking, and sculpture. Assistance is provided for periods ranging from one month to one year. Approximately seventeen Fellowship grants are given annually supporting artists, curators, and critics from Asia traveling to the United States. Completed applications must be submitted at least six months prior to the planned date of project implementation.
Requirements: Asian individuals in the visual and performing arts seeking grant assistance to conduct research, study, receive specialized training, undertake observation tours, or pursue creative activity in the United States are eligible to apply for project or fellowship support from the Council. Americans seeking aid to undertake activities in Asia are also eligible to apply. Completed applications must be submitted at least six months prior to the planned date of project implementation. Applicants should send a brief description of the activity for which assistance is sought. If the proposed activity falls within the council's guidelines, additional application materials will be provided.
Restrictions: The Council is unable to consider proposals for the following: personal exhibitions; individual performance tours; undergraduate study; or activities conducted by individuals in their home countries.
Geographic Focus: All States
Date(s) Application is Due: Feb 1
Contact: Program Director; (212) 843-0403; fax (212) 843-0343; acc@accny.org
Internet: http://www.asianculturalcouncil.org/programs.html#10
Sponsor: Asian Cultural Council
6 West 48th Street, 12th Floor
New York, NY 10036-1802

ACC Taiwan Fellowship Program 269

The program makes grants to artists, scholars, and specialists from Taiwan for research, study, and creative work in the United States. The program also supports American and Asian arts specialists visiting Taiwan to participate in cultural exchange projects organized by local arts institutions. Application procedures and grantee programming are administered with the assistance of the ACC-SAACF office in Taipei. The program supports up to eight grant recipients each year. Applicants should send a short description of the activity for which assistance is being sought. If the proposed activity falls within the council's guidelines, application materials will be provided.
Requirements: Asian individuals in the visual and performing arts seeking grant assistance to conduct research, study, receive specialized training, undertake observation tours, or pursue creative activity in the United States are eligible to apply for project or fellowship support from the Council. Americans seeking aid to undertake activities in Asia are also eligible to apply. Completed applications must be submitted at least six months prior to the planned date of project implementation. Applicants should send a brief description of the activity for which assistance is sought. If the proposed activity falls within the council's guidelines, additional application materials will be provided.
Restrictions: The Council is unable to consider proposals for the following: personal exhibitions; individual performance tours; undergraduate study; or activities conducted by individuals in their home countries.
Geographic Focus: All States
Date(s) Application is Due: Feb 1
Contact: Program Director; (212) 843-0403; fax (212) 843-0343; acc@accny.org
Internet: http://www.asianculturalcouncil.org/programs.html#11
Sponsor: Asian Cultural Council
6 West 48th Street, 12th Floor
New York, NY 10036-1802

ACF Adoption Opportunities Project Grants 270

The Adoption Opportunities program provides discretionary funds for projects designed to eliminate barriers to adoption and help find permanent families for children who would benefit from adoption, particularly children with special needs. Some of the major program areas include: the development and implementation of a national adoption information exchange system; increasing services in support of the placement in adoptive families of minority children who are in foster care and have the goal of adoption, with a special emphasis on the recruitment of minority families; increasing post-legal adoption services for families who have adopted children with special needs; and, supporting the placement of children in kinship care arrangements, pre-adoptive, or adoptive homes.
Requirements: Eligible entities include States, local government entities, federally recognized Indian Tribes and tribal organizations, faith-based and community organizations, colleges and universities, public or private non-profit licensed child welfare or adoption agencies, and adoption exchanges.
Geographic Focus: All States
Amount of Grant: 300,000 USD
Contact: Geneva Ware-Rice, (202) 205-8354
Internet: http://www.acf.hhs.gov/programs/fbci/progs/fbci_adoption.html
Sponsor: Administration on Children, Youth and Families
P.O. Box 13505
Silver Spring, MD 20911-3505

ACGT Investigators Grants 271

The Investigators Award in Clinical Translation of Gene Therapy for Cancer distributes funds over 3-5 years, inclusive of a maximum of 10% indirect costs. Funds may be used at the recipient's discretion for salary, technical assistance, supplies, animals or capital equipment, but may not support staff not directly related to the project, e.g. secretaries or administrative assistants. Purchase of equipment is not allowed in the final year of the grant. ACGT is accepting grant applications to produce and release-testing of the clinical trial agents under cGMP, conduct the necessary pre-clinical pharmacological and toxicological studies in appropriate animal models, and/or conducting the clinical translational trials in patients in support of an Investigative New Drug application to the FDA.

Requirements: The candidate must hold an MD, PhD, or equivalent degree and be a tenure-track or tenured faculty. The investigator must be conducting original research as an independent faculty member. ACGT has no citizenship restrictions, and research supported by the award must be conducted at medical schools and research centers only in the United States. While the unambiguous demonstration of preclinical efficacy in cancer treatment by gene therapy is a pre-requisite, entering into the clinical trial during the funding period is also a requirement.
Geographic Focus: All States
Amount of Grant: Up to 1,000,000 USD
Contact: Grace Pedersen, Foundation Administrator; (203) 358-8000, ext. 495; aneslage@acgtfoundation.org
Internet: http://www.acgtfoundation.org/grants.html
Sponsor: Alliance for Cancer Gene Therapy
96 Cummings Point Road
Stamford, CT 06902

ACGT Young Investigator Grants 272
The alliance funds research aimed at furthering the development of gene therapy approaches to the treatment of cancer. The overall objectives of this grant are to advance gene therapy into the causes, treatment, and prevention of all types of cancer by promoting development of novel and innovative studies by young investigators. The emphasis of this initiative is to promote basic, and preclinical research approaches utilizing cells and genes as medicine. The six main areas of research ACGT will support are: tumor-specific replicating viruses and bacteria, anti-angiogenesis, immune-mediated gene therapy and cancer vaccines, oncogenes/suppressor oncogenes/apoptosis, tumor targeting and vector development, and other cancer gene therapy research. The three-year grant may be used at the recipient's discretion for salary, technical assistance, supplies, animals or capital equipment, but may not support staff not directly related to the project, e.g. secretaries or administrative assistants. Purchase of equipment is not allowed in the third year of the grant. Continued support is contingent upon submission and approval of a noncompetitive renewal application each year. Guideline and applications are available online.
Requirements: Candidates must hold an MD, MPH, PhD, or equivalent degree and be a tenure-track assistant professor within five years of their initial appointment to this rank, at the time of award activation. The investigator must be conducting original research as an independent faculty member. ACGT has no citizenship restrictions, and research supported by the award must be conducted at medical schools and research centers only in the United States.
Geographic Focus: All States
Date(s) Application is Due: Sep 21
Amount of Grant: Up to 500,000 USD
Contact: Grace Pedersen; (203) 358-8000, ext. 495; aneslage@acgtfoundation.org
Internet: http://www.acgtfoundation.org/grants.html
Sponsor: Alliance for Cancer Gene Therapy
96 Cummings Point Road
Stamford, CT 06902

A Charitable Foundation Grants 273
Founded in 1997 in Nevada and funded by the Tactical Investment Management Corporation, A Charitable Foundation offers support primarily in California, Hawaii, and New York, as well as Brazil. Its major fields of interest include: educational research; the environment; food banks; human services; and international development. Types of support include annual campaigns, building and renovation projects, emergency assistance, basic research, and grants to individuals. Applications should include a detailed description of the project, purpose of the research (if applicable), and a detailed budget. There are no annual deadlines, and grants typically range from $2000 to $35,000.
Geographic Focus: California, Hawaii, New York, Brazil
Amount of Grant: 250-30,000 USD
Samples: Self-Realization Fellowship, Los Angeles, California, $35,000-support of a fellowship program (2012); Wailua United Church of Christm Wailua, Hawaii, $18,000-for public charity work (2012); Hawaii Food Bank, Honolulu, Hawaii, $10,000-public charities (2012).
Contact: David S. Druz, President; (702) 248-8184; ddruz@aol.com
Sponsor: A Charitable Foundation
2657 Windmill Parkway, Suite 220
Henderson, NV 89014

Achelis Foundation Grants 274
Elisabeth Achelis was born in Brooklyn Heights in 1880. She used her inheritance from her father Fritz Achelis, who was President of the American Hard Rubber Company, to establish the Achelis Foundation in 1940 to aid and contribute to charitable, benevolent, educational and religious uses and purposes for the moral, ethical, physical, mental and intellectual well-being and progress of mankind; to aid and contribute to methods for the peaceful settlement of international differences; to aid and contribute to the furtherance of the objects and purposes of any charitable, benevolent, educational or religious institution or agency; and to establish and maintain charitable, benevolent and educational institutions and agencies. The Achelis Foundation shares trustees, staff, office space, and even a website with the Bodman Foundation which has a similar mission and geographic area of concentration (both foundations give in New York City, while the Bodman Foundation also gives in New Jersey). Funding is concentrated in six program areas: Arts and Culture, Education, Employment, Health, Public Policy, and Youth and Families.
Requirements: 501(c)3 organizations based in New York City that fall within the foundation's areas of interest are welcome to submit an inquiry or proposal letter by regular mail (initial inquiries by email or fax are not accepted, nor are CDs, DVDs, computer discs, or video tapes). An initial inquiry to the foundation should include only the following items: a proposal letter that briefly summarizes the history of the project, need, objectives, time period, key staff, project budget, and evaluation plan; the applicant's latest annual report and complete set of audited financial statements; and the applicant's IRS 501(c)3 tax-exemption letter. Applications may be submitted at any time during the year. Each request is reviewed by staff and will usually receive a written response within thirty days. Those requests deemed consistent with the interests and resources of the foundation will be evaluated further and more information will be requested. Foundation staff may request a site visit, conference call, or meeting. All grants are reviewed and approved by the Trustees at one of their three board meetings in May, September, or December.
Restrictions: The foundation generally does not make grants for the following purposes or program areas: nonprofit organizations outside of New York; annual appeals, dinner functions, and fundraising events; endowments and capital campaigns; loans and deficit financing; direct grants to individuals; individual day-care and after-school programs; housing; organizations or projects based outside the U.S; films or video projects; small art, dance, music, and theater groups; individual K-12 schools (except charter schools); national health and mental health organizations; and government agencies or nonprofit organizations significantly funded or reimbursed by government agencies. Limited resources prevent the foundations from funding the same organization on an ongoing annual basis.
Geographic Focus: New York
Samples: American Ballet Theatre, New York, New York, $25,000-general operating support (2012); Achievement First, Brooklyn, New York, $50,000-general operating support (2012); Academy for Teachers, New York, New York, $15,000-general operating support (2012).
Contact: John B. Krieger, Executive Director; (212) 644-0322; fax (212) 759-6510; main@achelis-bodman-fnds.org
Internet: http://www.achelis-bodman-fnds.org/guidelines.html
Sponsor: Achelis Foundation
767 Third Avenue, 4th Floor
New York, NY 10017-2023

ACIE Title VIII Combined Research and Language Training Fellowships 275
Funded by the U.S. Department of State's Program for Research and Training on Eastern Europe and the Independent States of the Former Soviet Union (Title VIII), the American Councils Combined Research and Language Training (CRLT) Program serves graduate students and scholars who, in addition to support for research in the independent states of the former Soviet Union, require supplemental language instruction. The synergistic nature of the research and language training aspects of the Combined Research and Language Training Program promises both more productive research and efficient language acquisition. The total value of Title VIII CRLT fellowships, administered by American Councils, ranges from $5,000 to $25,000. Typical awards include: international airfare from the scholar's home city to his/her host city overseas; academic affiliation at a leading local university; visa(s) arranged by American Councils in direct collaboration with academic host institutions in order to facilitate archive access and guarantee timely visa registration; a housing stipend; a monthly living stipend; health insurance of up to $100,000 per accident or illness; and ongoing logistical support from American Councils offices throughout the region, including in-country orientation and 24-hour emergency aid. The application deadline is October 1.
Requirements: Applicants must conduct research and language study for a minimum of three months in the field to a maximum of nine months in the field. Participants must have attained at least an intermediate level of language proficiency to apply. Applicants must be U.S. citizens. The competition for funding is open and merit-based. All applications will receive consideration without regard to any factor such as race, color religion, sex, sexual orientation, national origin, martial status, political affiliation, or disability.
Geographic Focus: All States
Contact: Liliya Kholod, Program Officer, Teacher Programs; (202) 833-7522; fax (202) 833-7523; lkholod@americancouncils.org
Internet: http://researchfellowships.americancouncils.org/crlt
Sponsor: American Councils for International Education
1828 L Street, NW, Suite 1200
Washington, DC 20036

ACIE Title VIII Research Scholar Fellowships 276
Funded by the U.S. Department of State's Program for Research and Training on Eastern Europe and the Independent States of the Former Soviet Union (Title VIII), the American Councils Research Scholar Program provides full support for graduate students, faculty, and independent scholars seeking to conduct in-country, independent research for three consecutive months to nine consecutive months in Central Asia, Moldova, Russia, the South Caucasus, Southeast Europe, and Ukraine. The total value of Title VIII Research Scholar fellowships, administered by American Councils, ranges from $5,000 to $25,000. Typical awards include: international airfare from the scholar's home city to his/her host city overseas; academic affiliation at a leading local university; visa(s) arranged by American Councils in direct collaboration with academic host institutions in order to facilitate archive access and guarantee timely visa registration; a housing stipend; a monthly living stipend; health insurance of up to $100,000 per accident or illness; and ongoing logistical support from American Councils offices throughout the region, including in-country orientation and 24-hour emergency aid. The application deadline is October 1.
Requirements: Scholars in the social sciences and humanities are eligible to apply for the program. While a wide-range of topics receive support each year, all funded research must contribute to a body of knowledge enabling the U.S. to better understand the region and formulate effective policies within it. All applicants should clearly describe the policy-relevance of their work, be it in anthropology, history, international relations, political science, or some other field.

Geographic Focus: All States
Date(s) Application is Due: Oct 1
Amount of Grant: 5,000-25,000 USD
Contact: Liliya Kholod; (202) 833-7522; fax (202) 833-7523; lkholod@americancouncils.org
Internet: http://researchfellowships.americancouncils.org/researchscholar
Sponsor: American Councils for International Education
1828 L Street, NW, Suite 1200
Washington, DC 20036

ACLS/New York Public Library Fellowships 277

The American Council of Learned Societies and The New York Public Library offer a collaborative program to support up to five residential fellowships per year at the Library's Dorothy and Lewis B. Cullman Center for Scholars and Writers. The Center provides opportunities for up to five Fellows to explore the rich, diverse collections in the Library's Stephen A. Schwarzman Building. The Center also serves as a forum for the exchange of ideas among Fellows, invited guests, the wider academic and cultural communities, and the interested public. It provides individual office space and common areas in the Library building. Fellows are required to be in residence from September through May, and to participate in Center activities such as lunches, panel discussions, public conversations, symposiums, and interviews. Specific guidelines, the online ACLS application, and link to the Cullman Center application (where candidates must also apply) is available at the website.
Requirements: Applicants must be U.S. citizens or permanent residents by the application date. They must also have earned a Ph.D. at least two years before the application deadline. An applicant's most recent supported research leave must have been two years before the application date.
Geographic Focus: All States
Date(s) Application is Due: Sep 30
Amount of Grant: 65,000 USD
Contact: Nicole Stahlmann, Director of Fellowship Programs; (212) 697-1505, ext. 134; fax (212) 949-8058; nstahlmann@acls.org
Internet: http://www.acls.org/programs/acls/
Sponsor: American Council of Learned Societies
633 Third Avenue
New York, NY 10017-6795

ACLS African Humanities Program for Dissertation Completion 278

The African Humanities Program (AHP) seeks to revitalize the humanities in Ghana, Nigeria, South Africa, Tanzania, and Uganda through fellowship competitions and meetings associated with them. Approximately 40 fellowships are awarded annually. Applicants for Dissertation-Completion Fellowships should be in the final year of completing the dissertation at a university in Ghana, Nigeria, Tanzania, or Uganda. Projects must be in the humanities and must be carried out in sub-Saharan Africa. An initiative to publish the best manuscripts produced under the terms of AHP fellowships is now being established. All fellows will be encouraged to submit their completed manuscripts to the African Humanities Series, which will be a cooperative venture between AHP and a major African university press. AHP fellowships may not be used for travel outside the continent. Specific guidelines, the application with instructions, a recommendation letter form, and a list of residential sites is available at the website.
Restrictions: Dissertation Completion Fellowships are not available in South Africa.
Geographic Focus: Ghana, Nigeria, Tanzania, Uganda
Date(s) Application is Due: Nov 1
Amount of Grant: 11,000 USD
Contact: Nicole Stahlmann, Director of Fellowship Programs; (212) 697-1505; nstahlmann@acls.org
Internet: http://www.acls.org/grants/Default.aspx?id=3210
Sponsor: American Council of Learned Societies
633 Third Avenue
New York, NY 10017-6795

ACLS African Humanities Program in Ghana, Nigeria, South Africa, Tanzania, and Uganda 279

The African Humanities Program (AHP) seeks to revitalize the humanities in Ghana, Nigeria, South Africa, Tanzania, and Uganda through fellowship competitions and meetings associated with them. Fellowship awards to promising African scholars are the centerpiece of the African Humanities Program. Approximately 40 such fellowships will be awarded annually. An initiative to publish the best manuscripts produced under the terms of AHP fellowships is now being established. All fellows will be encouraged to submit their completed manuscripts to the African Humanities Series, which will be a cooperative venture between AHP and a major African university press. The application form with instructions, a list of proposed residential sites, and a recommendation letter form are posted on the website.
Requirements: Proposed projects must be in the humanities, defined by the study of history, language, and culture, and by qualitative approaches in research. The list of humanities disciplines includes anthropology, archaeology, studies of the fine and performing arts, history, linguistics, literature studies, studies of religion, and philosophy. A project in social sciences such as economics, sociology, or political science, as well as in law or international relations, is not eligible unless it is clearly humanistic in content and focus. Applicants must propose a written scholarly product such as a book manuscript. Research and writing under terms of the fellowship award must be undertaken in sub-Saharan Africa. Applicants must be nationals of a country in sub-Saharan Africa, residing and working in Ghana, Nigeria, South Africa, Tanzania, or Uganda. At the time of the submission, applicants must have defended the dissertation and must be no more than five years past the date of receiving the Ph.D. degree. Applicants may propose either a combination of research and writing, or writing only. They must be able to devote full time to the AHP project throughout the fellowship period.
Restrictions: Fellowships may not be used for travel outside the continent.
Geographic Focus: Ghana, Nigeria, South Africa, Tanzania, Uganda
Date(s) Application is Due: Nov 1
Amount of Grant: 18,000 USD
Contact: Nicole Stahlmann, Director of Fellowship Programs; (212) 697-1505, ext. 134; fax (212) 949-8058; nstahlmann@acls.org
Sponsor: American Council of Learned Societies
633 Third Avenue
New York, NY 10017-6795

ACLS American Research in the Humanities in China Fellowships 280

The American Council of Learned Societies American Research in the Humanities in China Fellowships are open to scholars in the humanities and related social services. Applicants must submit a carefully formulated research proposal that reflects an understanding of the present Chinese academic and research environment. The proposal should include a persuasive statement of the need to conduct the research in China. Stipends are calculated based upon the awardee's academic rank and the length of stay in China (from four to twelve months). Stipends for four months of research cannot exceed $25,000. The maximum award is $50,000. Specific guidelines, the online application, and a sample application are available at the website.
Requirements: An applicant must be a citizen or permanent resident of the United States who has lived in the U.S. continuously for at least three years when applying. Applicants must also hold a Ph.D. in the humanities or related social sciences. Those submitting a joint proposal must apply individually. Support is offered to specialists in all fields of the humanities and humanities-related social sciences, and is not limited to China scholars.
Restrictions: The award may be reduced if the candidate also receives full or partial salary during the fellowship period. There is no financial support for dependents.
Geographic Focus: All States
Date(s) Application is Due: Sep 28
Amount of Grant: Up to 50,000 USD
Contact: Nicole Stahlmann, Director of Fellowship Programs; (212) 697-1505, ext. 134; fax (212) 949-8058; nstahlmann@acls.org
Internet: http://www.acls.org/grants/Default.aspx?id=524
Sponsor: American Council of Learned Societies
633 Third Avenue
New York, NY 10017-6795

ACLS Charles A. Ryskamp Research Fellowships 281

The American Council of Learned Societies Ryskamp Fellowships support advanced assistant professors and non-tenured associate professors in the humanities and related social sciences who have well designed plans for new research, and whose scholarly contributions have advanced their fields. Appropriate fields of specialization include but are not limited to: anthropology, archeology, art history, economics, geography, history, languages and literatures, law, linguistics, musicology, philosophy, political science, psychology, religion, and sociology. Proposals in the social science fields listed are eligible only if they employ predominantly humanistic approaches (e.g., economic history, law and literature, political philosophy). Proposals in interdisciplinary and cross-disciplinary studies are welcome, as are proposals focused on any geographic region or on any cultural or linguistic group. The ultimate goal of the project should be a major piece of scholarly work by the applicant that will take the form of a monograph or other equally substantial form of scholarship. The fellowships support an academic year of research (nine months), plus an additional summer's research (two months) if justified. Recipients have three years to use the funds awarded them. Completed applications must be submitted through the ACLS Online Fellowship Application system. A sample application and instructions for writing ACLS proposals is also located at the website.
Requirements: The program is open to tenure-track assistant professors and non-tenured associate professors who by September 25 will have successfully completed their institution's last reappointment review before tenure review (if the institution does not have multi-year contracts, the applicant should have passed three annual reappointment reviews), and whose tenure review will not be complete before February 1. Applicants must hold the Ph.D. or equivalent, and be employed at academic institutions in the United States for the duration of the fellowship.
Restrictions: ACLS does not fund creative work (e.g., novels or films), textbooks, straightforward translation, or pedagogical projects.
Geographic Focus: All States
Date(s) Application is Due: Sep 28
Amount of Grant: 66,500 USD
Contact: Nicole Stahlmann, Director of Fellowship Programs; (212) 697-1505, ext. 134; fax (212) 949-8058; nstahlmann@acls.org
Internet: http://www.acls.org/programs/ryskamp/
Sponsor: American Council of Learned Societies
633 Third Avenue
New York, NY 10017-6795

ACLS Collaborative Research Fellowships 282

The American Council of Learned Societies (ACLS) Collaborative Research Fellowships support collaborative research in the humanities and related social sciences. Collaborations need not be interdisciplinary or inter-institutional, but must involve at least two scholars. Applicants at the same institution must demonstrate why local funding is insufficient to support the project. It is hoped that projects of successful

applicants will help demonstrate the range and value of both collaborative research and inquiry in the humanities, and model how such collaboration may be carried out successfully. Collaborations that involve the participation of assistant and associate faculty members, or that of scholars at different kinds of institutions, are particularly encouraged. Award amounts will range from $60,000 to $140,000 in total, depending on the nature and duration of the collaboration, the kinds of expenses projected to carry out the research, and the number of collaborators. Salary replacement stipends are based on academic rank: up to $35,000 for Assistant Professor and career equivalent; up to $40,000 for Associate Professor and career equivalent; and up to $60,000 for full Professor and career equivalent, as well as up to $20,000 in collaboration funds (which may be used for travel, materials, or research assistance). Applications must be filled out through the online system. Specific guidelines, a sample budget and application, and advice on how to write ACLS proposals are located at the website.
Requirements: A collaborative project consists of at least two scholars who are each seeking salary-replacement stipends for six to twelve continuous months of supported research leave to pursue full-time collaborative research during the fellowship tenure. The Project Coordinator must have an appointment at a U.S.-based institution of higher education. Other project members may be at institutions outside the United States or may be independent scholars. All project collaborators must hold a Ph.D. degree or its equivalent in publications and professional experience at the time of application.
Restrictions: ACLS does not fund creative work (e.g., novels or films), textbooks, straightforward translation, or pedagogical projects. The Fellowship does not fund the following: large research clusters that do not produce any collaborative-authored publications; collaborative projects that result in an anthology or edited volume of secondary scholarship; projects that are not primarily focused on research; projects whose primary aim is to transform existing research results into digital format; and projects whose primary emphasis is on organization of events (workshops, lectures, exhibitions).
Geographic Focus: All States
Date(s) Application is Due: Sep 28
Amount of Grant: Up to 140,000 USD
Contact: Nicole Stahlmann, Director of Fellowship Programs; (212) 697-1505, ext. 134; fax (212) 949-8058; nstahlmann@acls.org
Internet: http://www.acls.org/programs/collaborative/
Sponsor: American Council of Learned Societies
633 Third Avenue
New York, NY 10017-6795

ACLS Comparative Perspectives on Chinese Culture and Society Grants **283**

The Comparative Perspectives on Chinese Culture and Society Grants support projects that bridge disciplinary or geographic boundaries, engage new kinds of information, develop fresh approaches to traditional materials and issues, or bring innovative perspectives to the study of Chinese culture and society. The program aims to provide opportunities for interchange among scholars who may not otherwise have chances to work with one another. The grants support planning meetings, workshops, and/or conferences leading to publication of scholarly volumes. Proposals are expected to be empirically grounded, theoretically informed, and methodologically explicit. The program will support collaborative work of three types: formal research conferences intended to produce significant new research published in a conference volume; workshops or seminars, designed to less formally facilitate new research on newly available or inadequately researched problems, data, or texts; and planning meetings to organizers of the above-described types of projects. Grants up to $6,000 will be offered for one-day meetings to plan conferences or workshops, or for less structured explorations, e.g., brainstorming sessions. Grants of $10,000 to $15,000 will be offered for workshops designed to promote discussion and the exchange of ideas on newly available or inadequately researched data or texts in a collegial, seminar-like setting. Grants up to $25,000 will be offered for formal research conferences intended to produce significant new research that will be published in a conference volume. There are no application forms, but the submitted materials must include the following: the online application information sheet; a description of the project and its purposes; a budget for the proposed event; a short bibliography of relevant sources; a two page CV of the principal organizer; a list of those invited with their CVs; and, for conferences only, an appendix containing abstracts for each paper to be presented at the conference.
Requirements: The principal organizer must be affiliated with a university or research institution and must hold a Ph.D. Proposals must include at least one scholar from Taiwan as a participant. There are no restrictions to citizenship of participants or location of the project; however, it is expected the scholars from academic institutions in Taiwan will participate in conferences, workshops, and planning meetings.
Restrictions: The program will not support regularly scheduled meetings or conventions. Funds awarded in these competitions cannot be used to cover the following expenses: direct research expenses, such as those of field work, obtaining research materials, or analysis of data; salaries or released time for organizers or participants; honoraria or speaker fees; purchase of equipment; or institutional overhead (although direct administrative costs are allowed).
Geographic Focus: All States
Date(s) Application is Due: Sep 28
Amount of Grant: 6,000-25,000 USD
Contact: Nicole Stahlmann, Director of Fellowship Programs; (212) 697-1505, ext. 134; fax (212) 949-8058; nstahlmann@acls.org
Internet: http://www.acls.org/grants/Default.aspx?id=528
Sponsor: American Council of Learned Societies
633 Third Avenue
New York, NY 10017-6795

ACLS Digital Innovation Fellowships **284**

The American Council of Learned Societies (ACLS) Digital Innovation Fellowships supports digitally based research projects in all disciplines of the humanities and humanities-related social sciences. It is hoped that projects of successful applicants will help advance digital humanistic scholarship by broadening understanding of its nature and exemplifying the robust infrastructure necessary for creating such works. The Fellowships are intended to support an academic year (nine months) dedicated to work on a major scholarly project that takes a digital form. Up to $60,000 is available for stipends, and up to $25,00 is available for project costs. Projects may: address a consequential scholarly question through new research methods, new ways of representing the knowledge produced by research, or both; create new digital research resources; increase the scholarly utility of existing digital resources by developing new means of aggregating, navigating, searching, or analyzing those resources; propose to analyze and reflect upon the new forms of knowledge creation and representation made possible by the digital transformation of scholarship. The Fellowships are intended as salary replacement and may be held concurrently with other fellowships and grants and any sabbatical pay up to an amount equal to the candidate's current academic year salary. All applications must include the endorsement of a senior administrator of the applicant's institution. This endorsement should include discussion of how the institution's existing cyber infrastructure complements and supports the technologies to be developed for the specified project. Completed applications must be submitted through the ACLS online application system. Specific uidelines, FAQs, a sample application, and advice on how to write an ACLS proposal are available at the website.
Requirements: Fellowships are open to scholars in all fields of the humanities and the humanistic social sciences. Applicants must have a Ph.D. degree prior to the application deadline. (An established scholar who can demonstrate the equivalent of the Ph.D. in publications and professional experience may also qualify.) United States citizenship or permanent resident status is required as of application deadline.
Restrictions: ACLS does not support creative works (e.g., novels or films), textbooks, straightforward translations, or purely pedagogical projects.
Geographic Focus: All States
Date(s) Application is Due: Oct 2
Amount of Grant: Up to 85,000 USD
Contact: Nicole Stahlmann; (212) 697-1505, ext. 134; nstahlmann@acls.org
Internet: http://www.acls.org/programs/digital/
Sponsor: American Council of Learned Societies
633 Third Avenue
New York, NY 10017-6795

ACLS Dissertation Fellowships in East European Studies **285**

The American Council of Learned Societies offers support for writing dissertations in East European Studies in all disciplines of the humanities and the social sciences. Dissertation Fellowships in East European Studies are offered in these disciplines relating to Albania, Bosnia and Herzegovina, Bulgaria, Croatia, Czech Republic, Estonia, Hungary, Kosovo, Latvia, Lithuania, Macedonia, Montenegro, Poland, Romania, Serbia, Slovakia, and Slovenia. Applicants may propose comparative work considering more than one country of Eastern Europe or relating East European societies or those of other parts of the world. Funding is offered for two types of support: research fellowships for use in Eastern Europe to conduct fieldwork or archival investigation; and writing fellowships for use outside of Eastern Europe, after all research is complete, to write the dissertation. Fellowships will be granted on the basis of the scholarly potential of the applicant, the quality and scholarly importance of the proposed work, and its importance to the development of scholarship on Eastern Europe. ACLS selection committees consider language competence essential to research. Therefore, applicants will be asked to describe their command of the language(s) required for their proposed projects. As a condition of the award, the applicant's home university will be required (consistent with its policies and regulations) to provide or to waive normal academic year tuition payments or to provide alternative cost-sharing support. Specific guidelines for the application packet and the online application is available on the website.
Requirements: Applicants must be U.S. citizens who have completed all requirements for their Ph.D. except their dissertation. Applications for research fellowships should state the questions or hypotheses driving research, the methods to be used for gathering relevant evidence, and preliminary versions of the dissertation's main argument. Applications for writing fellowships should state what materials have been collected, how research questions may have been answered or modified, and the direction that analysis will take once writing has begun. Projects must meet all of the following *Requirements:* pertain to at least one of the East European countries posted; humanities or social sciences related and investigate aspects of East European societies, histories, and cultures for how they illuminate contemporary issues or the necessary background for understanding such issues. (This excludes projects in the natural sciences, mathematics, archaeology, music theory, pure linguistics, and philosophy. However, proposals in such disciplines as the history or sociology of science, music, etc. and sociolinguistics may be eligible, if they address the impact of a given topic on East European societies.); aim to produce a written scholarly product written in English; and state a plan of work that can be accomplished during the fellowship period.
Restrictions: Applicants may apply for one-year research and writing fellowships in sequence, but may not apply for a second year of funding in either category.
Geographic Focus: All States
Date(s) Application is Due: Nov 9
Contact: Nicole Stahlmann; (212) 697-1505, ext. 134; nstahlmann@acls.org
Internet: http://www.acls.org/grants/Default.aspx?id=532
Sponsor: American Council of Learned Societies
633 Third Avenue
New York, NY 10017-6795

ACLS Early Career Postdoctoral Fellowships in East European Studies **286**

The Early Career Postdoctoral Fellowships in East European Studies fund postdoctoral research and training in the social sciences and humanities relating to Albania, Bosnia and Herzegovina, Bulgaria, Croatia, Czech Republic, Estonia, Hungary, Kosovo, Latvia, Lithuania, Macedonia, Montenegro, Poland, Romania, Serbia, Slovakia, and Slovenia. Scholars may apply for Fellowships funding full-time research and writing for periods of six to 12 consecutive months of the following year. The Fellowships are intended primarily as salary replacement to provide free time for research. The funds may be used to supplement sabbatical salaries or awards from other sources, provided they intensify or extend the contemplated research. Five to seven Fellowships are awarded each year. In awarding these grants, primary considerations are the scholarly merit of the proposal, its importance to the development of the field, and the scholarly potential and accomplishments of the applicant. Further, ACLS selection committees consider language competence essential to research. Therefore, applicants will be asked to describe their command of the language(s) required for their proposed projects. All proposals should be for scholarly work, the product of which is to be disseminated in English. Specific guidelines, the online application, and a sample application are available at the website.

Requirements: An applicant must: be a citizen or permanent resident of the United States as of the application deadline date; hold a Ph.D. degree prior to the application deadline (however, an established scholar who can demonstrate the equivalent of the Ph.D. in publications and professional experience may also qualify); and be at an early career stage. It is essential that all projects: state a plan of work that can be accomplished during the fellowship year; pertain to at least one of the East European countries posted; investigate aspects of East European societies, histories, and cultures with a special regard for how they illuminate contemporary issues or the necessary background for understanding such issues. This excludes projects in the natural sciences, mathematics, archaeology, music theory, pure linguistics, and philosophy. However, proposals in such disciplines as the history or sociology of science, music, etc. and sociolinguistics may be eligible if they address the impact of a given topic on East European societies. Projects must aim to produce a written scholarly product such as a book manuscript to be written in English.

Restrictions: These fellowships are to be used for work outside East Europe, although short visits to the area may be proposed as part of a coherent program primarily based elsewhere. Tenured faculty are not eligible.

Geographic Focus: Albania, Bosnia & Herzegovina, Bulgaria, Croatia, Czech Republic, Estonia, Hungary, Kosovo, Latvia, Lithuania, Macedonia, Montenegro, Poland, Romania, Serbia, Slovakia, Slovenia

Date(s) Application is Due: Nov 9

Amount of Grant: Up to 25,000 USD

Contact: Nicole Stahlmann; (212) 697-1505, ext. 134; nstahlmann@acls.org

Internet: http://www.acls.org/grants/Default.aspx?id=534

Sponsor: American Council of Learned Societies
633 Third Avenue
New York, NY 10017-6795

ACLS Fellowships **287**

The American Council of Learned Societies Fellowships fund research in all disciplines of the humanities and humanities-related social sciences. Appropriate fields of specialization include but are not limited to: anthropology, archeology, art history, economics, geography, history, languages and literatures, law, linguistics, musicology, philosophy, political science, psychology, religion, and sociology. Proposals in the social science fields listed are eligible only if they employ predominantly humanistic approaches (e.g., economic history, law and literature, political philosophy). Proposals in interdisciplinary and cross-disciplinary studies are welcome, as are proposals focused on any geographic region or on any cultural or linguistic group. The Fellowships are intended as salary replacement and may be held concurrently with other fellowships and grants and any sabbatical pay to reach that goal. Specific guidelines, the online application, a sample application, a suggestions for writing ACLS proposals are located at the website.

Requirements: Applicants must have a Ph.D. degree completed two years prior to the application deadline. The applicant must have U.S. citizenship or permanent resident status as of the application deadline. The most recent supported research leave must have concluded two years prior to July 1. (Supported research leave is defined as the equivalent of one semester or more of time free from teaching or other employment to pursue scholarly research or writing supported by sabbatical pay or other institutional funding, fellowships and grants, or a combination of these.) This definition will apply to independent scholars as well as those with institutional affiliations.

Restrictions: ACLS does not fund creative work (e.g., novels or films), textbooks, straightforward translation, or pedagogical projects.

Geographic Focus: All States

Date(s) Application is Due: Sep 28

Amount of Grant: Up to 60,000 USD

Contact: Nicole Stahlmann; (212) 697-1505, ext. 134; nstahlmann@acls.org

Internet: http://www.acls.org/programs/acls/

Sponsor: American Council of Learned Societies
633 Third Avenue
New York, NY 10017-6795

ACLS Frederick Burkhardt Residential Fellowships for Recently Tenured Scholars **288**

The American Council of Learned Societies (ACLS) Burkhardt Fellowships support long-term, unusually ambitious projects in the humanities and related social sciences. The ultimate goal of the project should be a major piece of scholarly work by the applicant. Burkhardt Fellowships are intended to support an academic year (nine months) of residence at any one of the national residential research centers (listed on the website) participating in the program. Such an environment, beyond providing free time, encourages exchanges across disciplinary lines that can be especially helpful to expanding the significance of projects in the humanities and related social sciences. Appropriate fields of specialization include but are not limited to: anthropology, archaeology, art history, economics, geography, history, languages and literature, law, linguistics, musicology, philosophy, political science, psychology, religion, and sociology. Proposals in the social science fields listed above are eligible only if they employ predominantly humanistic approaches (e.g., economic history, law and literature, political philosophy). Proposals in interdisciplinary and cross-disciplinary studies are welcome, as are proposals focused on any geographic region or on any cultural or linguistic group. Applications must be submitted online. Specific guidelines also are available at the website.

Requirements: The Burkhardt Fellowship Program is open to recently tenured humanists scholars who will have begun their first tenured contracts by the application deadline but began their first tenured contracts no earlier than the fall 2002 semester or quarter. An applicant must be employed in a tenured position at a degree-granting academic institution in the U.S., remaining so for the duration of the fellowship. American citizenship or permanent residency is not required, and previous supported research leaves do not affect eligibility for a Burkhardt Fellowship.

Restrictions: ACLS does not fund creative work (e.g., novels or films), textbooks, straightforward translation, or pedagogical projects.

Geographic Focus: All States, All Countries

Date(s) Application is Due: Sep 28

Contact: Nicole Stahlmann; (212) 697-1505, ext. 134; nstahlmann@acls.org

Internet: http://www.acls.org/grants/Default.aspx?id=480

Sponsor: American Council of Learned Societies
633 Third Avenue
New York, NY 10017-6795

ACLS International and Area Studies Fellowships **289**

In order to encourage humanistic research in area studies, special funding by the National Endowment for the Humanities and the ACLS has been set aside for up to ten ACLS/SSRC/NEH International and Area Studies Fellowships. Scholars pursuing research and writing on the societies and cultures of Asia, Africa, the Middle East, Latin America and the Caribbean, Eastern Europe, and the former Soviet Union are eligible for these special fellowships. Specific guidelines, the online application, a sample application, and suggestions for writing ACLS proposals are located at the website.

Requirements: Applicants must be hold a Ph.D., and be U.S. citizens or permanent residents who have lived in the U.S. at least three years prior to application. These fellows must also submit a final report to the National Endowment for the Humanities and the American Council of Learned Societies.

Geographic Focus: All States

Date(s) Application is Due: Sep 28

Contact: Nicole Stahlmann; (212) 697-1505, ext. 134; nstahlmann@acls.org

Internet: http://www.acls.org/programs/acls/

Sponsor: American Council of Learned Societies
633 Third Avenue
New York, NY 10017-6795

ACLS New Faculty Fellows **290**

The American Council of Learned Societies New Faculty Fellowships allow recent PhDs in the humanities to take up two-year positions at U.S. universities where their particular research and teaching expertise augments departmental offerings. Up to 40 New Faculty Fellow awards will be made for appointments that will begin in the fall of the academic year. New Faculty Fellows will receive an annual stipend of $50,000 plus a research/travel allowance of $5,000, health insurance, and a $1,500 one-time moving allowance. The teaching load is three courses in the undergraduate curriculum over two semesters of the academic year. Additionally, mentors at the receiving institutions help integrate Fellows into their scholarly communities. Specific guidelines and a list of participating institutions is available at the website.

Requirements: Applications are accepted by nomination only from participating institutions (see the extensive list at the Fellows website). Candidates must hold a PhD in a humanities discipline or humanistic social science, which includes history, anthropology, and such areas as political theory, historical sociology, and economic history. Applicants must have completed their PhD by the time they fill out the application. There is no citizenship requirement.

Restrictions: PhDs who have already secured tenure-track positions are not eligible. Candidates must be willing to relocate to take up offered positions. ACLS does not accept applications from candidates whose work is primarily creative, such as music composition or performance.

Geographic Focus: All States

Date(s) Application is Due: Oct 7

Amount of Grant: 111,500 USD

Contact: Nicole Stahlmann; (212) 697-1505, ext. 134; nstahlmann@acls.org

Internet: http://www.acls.org/programs/newfaculty/

Sponsor: American Council of Learned Societies
633 Third Avenue
New York, NY 10017-6795

ACM / AAAI Allen Newell Award **291**

The award is presented to an individual selected for career contributions that have breadth within computer science, or that bridge computer science and other disciplines. This endowed award is accompanied by a prize of $10,000, and is supported by the American Association for Artificial Intelligence, and by individual contributions.

Requirements: Each nomination should consist of name, address, and phone number of person making the suggestion; name and address of candidate; statement of 200 to 500 words explaining why the candidate deserves the award; and if possible, names and addresses or telephone numbers of others who agree with the recommendation.
Geographic Focus: All States
Date(s) Application is Due: Oct 31
Amount of Grant: 10,000 USD
Samples: Karen Sparck Jones—award recipent, $10,000; Jack Minker—award recipent, $10,000; Richard Gabriel—award recipent, $10,000;
Contact: John Seely Brown, (650) 324-3090; jsb@johnseelybrown.com
Internet: http://awards.acm.org/homepage.cfm?awd=150
Sponsor: Association for Computing Machinery
2 Penn Plaza, Suite 701
New York, NY 10121-0701

ACM A.M. Turing Award 292

This award is the association's most prestigious technical award and is given to an individual selected for contributions of a technical nature made to the computing community. The contributions should be of lasting importance to the computer field. Nomination forms should be obtained from ACM headquarters.
Requirements: Each nomination should consist of name, address, and phone number of person making the suggestion; name and address of candidate; statement of 200 to 500 words explaining why the candidate deserves the award; and if possible, names and addresses or telephone numbers of others who agree with the recommendation.
Geographic Focus: All States
Amount of Grant: 100,000 USD
Samples: Francis E. Allen—award recipient, $100,000; Vinton Cerf—award recipient, $100,000; Robert Kahn—award recipient, $100,000.
Contact: Rosemary McGuinness; (212) 626-0561; mcguinness@acm.org
Internet: http://www.acm.org/awards/taward.html
Sponsor: Association for Computing Machinery
2 Penn Plaza, Suite 701
New York, NY 10121-0701

ACM Distinguished Service Award 293

Awarded on the basis of value and degree of services to the computing community. The contribution should not be limited to service to the Association, but should include activities in other computer organizations and should emphasize contributions to the computing community at large.
Requirements: Each nomination should consist of name, address, and phone number of person making the suggestion; name and address of candidate; statement of 200 to 500 words explaining why the candidate deserves the award; and if possible, names and addresses or telephone numbers of others who agree with the recommendation. Supporting letters from such persons are useful.
Geographic Focus: All States
Samples: Susan L. Graham—award recipient; Mary Jane Irwin—award recipient.
Contact: Global Office; (212) 342-6626
Internet: http://awards.acm.org/homepage.cfm?awd=144
Sponsor: Association for Computing Machinery
2 Penn Plaza, Suite 701
New York, NY 10121-0701

ACM Doctoral Dissertation Award 294

This award is given to the individual who has been judged to have prepared the most outstanding doctoral dissertation on computer science and engineering during the previous year. Dissertations will be reviewed for technical depth and significance of the research contribution, potential impact on theory and practice, and quality of presentation. A committee of five individuals serving staggered five-year terms performs an initial screening to generate a short list, followed by an in-depth evaluation to determine the winning dissertation. Nomination forms are available from ACM headquarters. The winning dissertation is published by Springer.
Requirements: Each nominated dissertation must have been accepted by the department within a 12-month period prior to August 31. Exceptional dissertations completed in August, but too late for submission last year will be considered. Only English language versions will be accepted.
Restrictions: Nominations are limited to one per university or college, from any country, unless more than 10 Ph.D.s are granted in one year, in which case two may be nominated.
Geographic Focus: All States
Date(s) Application is Due: Aug 31
Amount of Grant: 5,000 USD
Samples: Yi-Ren Ng—award recipient, $5000; Ben Liblit—award recipient, $5000 Boaz Barak—award recipient, $5000; AnHai Doan—award recipient, $5000.
Contact: Priya Narasimhan, (412) 268-8801; fax (412) 268-1194; Priya@cs.cmu.edu
Internet: http://awards.acm.org/homepage.cfm?awd=146
Sponsor: Association for Computing Machinery
2 Penn Plaza, Suite 701
New York, NY 10121-0701

ACM Eckert-Mauchly Award 295

This award is given for contributions to computer and digital systems architecture where the field of computer architecture is considered at present to encompass the combined hardware-software design and analysis of computing and of digital systems. The award is administered jointly by ACM and the IEEE Computer Society.
Requirements: Nomination should consist of name, address, and phone number of person making the suggestion; name and address of candidate; a statement of 200-500 words explaining why the candidate deserves the award; and, if possible, the names and addresses or telephone numbers of others who agree with the recommendation.
Geographic Focus: All States
Date(s) Application is Due: Oct 31
Amount of Grant: 5,000 USD
Samples: James H. Pomerene—award recipient, $5000; Robert P. Colwell—award recipient, $5000; Frederick Brooks, Jr.—award recipient, $5000; Joseph Fisher—award recipient, $5000.
Contact: Alan Berenbaum, (631) 435-6480; aberenbaum@acm.org
Internet: http://awards.acm.org/homepage.cfm?awd=148
Sponsor: Association for Computing Machinery
2 Penn Plaza, Suite 701
New York, NY 10121-0701

ACM Eugene L. Lawler Award 296

This award is to recognize an individual or a group who have made a significant contribution through the use of computing technology. It will be given once every two years, assuming that there are worthy recipients. The award is intentionally defined broadly. The professional credentials of the recipient(s) are not important. The recipient(s) need never to have earned a degree or published a paper, or even be considered to be a computer professional. The emphasis of the Award Committee will be on the significance of the contribution itself, within the prescribed areas of technology for humanitarian contributions in the field of computing. Some examples of the types of contributions that this award is created to recognize are: application of computer technology to aid the disabled; making an educational contribution using computers or Computer Science in inner city schools; creative research concerning intellectual property issues; expansion of educational opportunities in Computer Science for women and underrepresented minorities; application of computers or computing techniques to problems of developing countries.
Requirements: Each nomination should consist of name, address, and phone number of person making the suggestion; name and address of candidate; statement of 200 to 500 words explaining why the candidate deserves the award; and if possible, names and addresses or telephone numbers of others who agree with the recommendation.
Geographic Focus: All States
Date(s) Application is Due: Oct 31
Amount of Grant: 5,000 USD
Samples: Albrecht Ehrensperger, Solomon Mbuguah, and Ernest Siva—award recipent, $5000; Patrick Ball—award recipent, $5000.
Contact: Nina Bhatti, (650) 857-3390; fax (650) 852-3791; nina.bhatti@hpl.hp.com
Internet: http://awards.acm.org/homepage.cfm?awd=141
Sponsor: Association for Computing Machinery
2 Penn Plaza, Suite 701
New York, NY 10121-0701

ACM Gordon Bell Prize 297

The prize is awarded each year to recognize outstanding achievement in high-performance computing. The purpose of the award is to track the progress over time of parallel computing, with particular emphasis on rewarding innovation in applying high-performance computing to applications in science. Prizes are awarded for peak performance as well as special achievements in scalability, time-to-solution on important science and engineering problems, and low price/performance.
Requirements: Nomination should consist of name, address, and phone number of person making the suggestion; name and address of candidate; a statement of 200-500 words explaining why the candidate deserves the award; and, if possible, the names and addresses or telephone numbers of others who agree with the recommendation.
Geographic Focus: All States
Date(s) Application is Due: Apr 6
Amount of Grant: 10,000 USD
Contact: David H. Bailey, (510) 495-2773; dhbailey@lbl.gov
Internet: http://awards.acm.org/homepage.cfm?awd=160
Sponsor: Association for Computing Machinery
2 Penn Plaza, Suite 701
New York, NY 10121-0701

ACM Grace Murray Hopper Award 298

This award is given to the outstanding young computer professional of the year, selected on the basis of a single recent major technical or service contribution. This candidate must have been 35 years of age or less at the time the qualifying contribution was made. Nominating forms should be obtained from ACM headquarters. This award is provided by Google.
Requirements: Each nomination submitted should consist of name, address, and phone number of person making the nomination; name and address of candidate; a statement of 200-500 words explaining why the candidate deserves the award; and if possible, the names and addresses or telephone numbers of others who agree with the recommendation.
Geographic Focus: All States
Date(s) Application is Due: Oct 31
Amount of Grant: 15,000 USD
Contact: Gabriel Silberman, (212) 415-6908; Gabby.Silberman@ca.com
Internet: http://awards.acm.org/homepage.cfm?awd=145
Sponsor: Association for Computing Machinery
2 Penn Plaza, Suite 701
New York, NY 10121-0701

ACM Karl V. Karlstrom Outstanding Educator Award 299

This award is presented annually to an outstanding educator who is appointed to a recognized educational baccalaureate institution, recognized for advancing new teaching methodologies or effecting new curriculum development or expansion in computer science and engineering, or making a significant contribution to the educational mission of the ACM. Those who have been teaching for 10 years or less will be given special consideration. The prize is supplied by Prentice-Hall Publishing Company.
Requirements: Each nomination should consist of name, address, and phone number of person making the suggestion; name and address of candidate; statement of 200 to 500 words explaining why the candidate deserves the award; and if possible, names and addresses or telephone numbers of others who agree with the recommendation.
Geographic Focus: All States
Date(s) Application is Due: Oct 31
Amount of Grant: 5,000 USD
Samples: Stuart J. Russell—award winner, $5000,; Sartaj Sahni—award winner, $5000,; John Gorgone—award winner, $5000,.
Contact: Jeffrey Ullman, (650) 494-8016; ullman@gmail.com
Internet: http://awards.acm.org/homepage.cfm?awd=147
Sponsor: Association for Computing Machinery
2 Penn Plaza, Suite 701
New York, NY 10121-0701

ACM Outstanding Contribution Award 300

This award may be given to up to three individuals per year, for entirely different activities, and they are selected based on the value and degree of service to the Association for Computing Machinery.
Requirements: Each nomination should consist of name, address, and phone number of person making the suggestion; name and address of candidate; statement of 200 to 500 words explaining why the candidate deserves the award; and if possible, names and addresses or telephone numbers of others who agree with the recommendation.
Geographic Focus: All States
Date(s) Application is Due: Oct 31
Samples: David S. Wise—award recipent; Don Gotterbarn—award recipent; Richard T. Snodgrass—award recipent.
Contact: Ronald Boisvert, (301) 975-3812; fax (301) 975-3553; Boisvert@acm.org
Internet: http://awards.acm.org/homepage.cfm?awd=143
Sponsor: Association for Computing Machinery
2 Penn Plaza, Suite 701
New York, NY 10121-0701

ACM Paris Kanellakis Theory and Practice Award 301

The award honors specific theoretical accomplishments that have had a significant and demonstrable effect on the practice of computing. This award is endowed by contributions from the Kanellakis family, with additional financial support provided by ACM's Special Interest Groups on Algorithms and Computational Theory (SIGACT), Design Automaton (SIGDA), Management of Data (SIGMOD), and Programming Languages (SIGPLAN), the ACM SIG Projects Fund, and individual contributions.
Requirements: Each nomination should consist of name, address, and phone number of person making the suggestion; name and address of candidate; statement of 200 to 500 words explaining why the candidate deserves the award; and if possible, names and addresses or telephone numbers of others who agree with the recommendation.
Geographic Focus: All States
Date(s) Application is Due: Oct 31
Amount of Grant: 5,000 USD
Samples: Robert Brayton—award recipent, $5000; Yoav Freund, Columbia University, and Robert Schapire, Princeton University—award recipients, $5000; Gary Miller, Michael Rabin, Robert Solovay, Volker Strassen—award recipents, $5000; Peter Franaszek—award recipients, $5000.
Contact: Anna Karlin, (206) 543-9344; karlin@cs.washington.edu
Internet: http://awards.acm.org/homepage.cfm?awd=147
Sponsor: Association for Computing Machinery
2 Penn Plaza, Suite 701
New York, NY 10121-0701

ACM Software System Award 302

Provided by IBM, this award goes to an institution or individual(s) recognized for developing a software system that has had a lasting influence, reflected in contributions to concepts, in commercial acceptance, or both.
Requirements: Each nomination should consist of name, address, and phone number of person making the suggestion; name and address of candidate; statement of 200 to 500 words explaining why the candidate deserves the award; and if possible, names and addresses or telephone numbers of others who agree with the recommendation.
Geographic Focus: All States
Date(s) Application is Due: Oct 31
Samples: Bertrand Meyer—award recipient, $10,000; Robert S. Boyer, Matt Kaufmann, and J. Strother Moore—award recipients, $10,000; Raghuram Bindignavle, Simon Lam, Shaowen Su, and Thomas Y. C. Woo—award recipients, $10,000; Stuart Feldman—award recipient, $10,000; James Gosling—award recipient, $10,000.
Contact: Frank Tompa, (519) 888-4567; fwtompa@uwaterloo.ca
Internet: http://awards.acm.org/homepage.cfm?awd=149
Sponsor: Association for Computing Machinery
2 Penn Plaza, Suite 701
New York, NY 10121-0701

ACRL Academic/Research Librarian of the Year Award 303

This award is presented by the Association of College and Research Libraries. Funded by the Baker and Taylor Company, it is intended to recognize an individual member of the library profession who is making an outstanding national or international contribution to academic or research librarianship and library development. Individuals nominated should have demonstrated achievements in such areas as service to the organized profession through ACRL and related organizations, significant and influential research on academic or research library service, publication of a body of scholarly and/or theoretical writing contributing to academic or research library development, or planning and implementing a library program of such exemplary quality that it has served as a model for others.
Requirements: Individuals may nominate themselves or others.
Geographic Focus: All States, All Countries
Date(s) Application is Due: Dec 7
Amount of Grant: 5,000 USD
Contact: Casey Kinson, Program Coordinator; (312) 280-2511 or (312) 280-2523; fax (312) 280-2520; ckinson@ala.org
Internet: http://www.ala.org/acrl/awards/achievementawards/acrllibrarian
Sponsor: Association of College and Research Libraries
50 East Huron Street
Chicago, IL 60611-2795

ACRL Instruction Section Innovation Award 304

This annual award recognizes a project that demonstrates creative, innovative, or unique approaches to information literacy instruction or programming. Projects nominated for the award should demonstrate recognized creativity, quality, and innovation within the context of national trends in information literacy instruction or programming. A certificate and a $3,000 award, donated by ProQuest, is given. Recipients(s) are recognized during the Instruction Section Program at the ALA Annual Conference. The award is administered by the Instruction Section of ACRL.
Requirements: Academic librarians or academic project teams that include an academic librarian are eligible to receive the award. Recipients must have implemented their project in an academic or research library or through the aegis of a professional library organization no more than two years prior to the nomination submission deadline.
Geographic Focus: All States
Date(s) Application is Due: Dec 7
Amount of Grant: 3,000 USD
Contact: Robin Kear, IS Innovation Award Co-Chair; (412) 648-7728; rlk25@pitt.edu
Erin L. Ellis, IS Innovation Award Co-Chair; (785) 864-5589; eellis@ku.edu
Internet: http://www.ala.org/acrl/awards/achievementawards/innovationaward
Sponsor: Association of College and Research Libraries
50 East Huron Street
Chicago, IL 60611-2795

ACS Cancer Control Career Development Awards for Primary Care Physicians 305

ACS annually awards three three-year career development awards to physicians specializing in primary care (e.g., family practice, internal medicine, pediatrics, and obstetrics and gynecology). These awards are intended to encourage and assist in the development of promising candidates who will pursue academic careers in primary care specialties. It is anticipated that physicians trained under these awards will improve cancer control through involvement in primary care practice, education, and research activities related to cancer control. Awards are made for three years with progressive stipends.
Requirements: Candidates must be U.S. citizens or permanent residents; must hold an MD, DO, or equivalent degree; and must have completed residency requirements of the appropriate primary care specialty board. Candidate must hold academic rank from instructor to assistant professor and must not be more than 10 years out of training.
Restrictions: Applicants may not have academic rank above that of assistant professor and must not be tenured or be the section head (or equivalent) in his/her discipline. Additionally, the applicant may not be training as a fellow at the time of the award.
Geographic Focus: All States
Date(s) Application is Due: Oct 15
Amount of Grant: Up to 100,000 USD
Contact: Extramural Grants Department; (404) 329-7558 or (800) 875-2562; fax (404) 321-4669; grants@cancer.org
Internet: http://www.cancer.org/docroot/RES/content/RES_5_2x_Cancer_Control_Career_Development_Awards_for_Primary_Care_Physicians.asp?sitearea=RES
Sponsor: American Cancer Society
3100 Wyman Park Drive, Suite W400
Baltimore, MD 21211

ACS Clinical Research Professor Grants 306

The American Cancer Society offers a limited number of grants to established investigators in mid-career who have made seminal contributions that have changed the direction of clinical, psychosocial, behavioral, health policy or epidemiologic cancer research. Furthermore, it is expected that these investigators will continue to provide leadership in their research area. Up to two awards are made annually for a five-year term that can be renewed once. The award of up to $80,000 per year (direct costs only) may be used for salary or research project support. Interested individuals should submit their curriculum vitae with a complete bibliography and a letter of intent that briefly describes their seminal contributions to research. This information must be submitted via the Research Professor letter of intent on proposalCENTRAL at: https://proposalcentral.altum.com. The Letter of Intent for the Clinical Research Professor Award must be submitted no later than August 1. Upon receipt, the ACS will provide appropriate guidance either by telephone

or email. Clinical Research Professor candidates whose Letter of Intent has been approved must submit their application for the award before the October 15 deadline.
Geographic Focus: All States
Date(s) Application is Due: Aug 1; Oct 15
Amount of Grant: Up to 80,000 USD
Contact: John J. Stevens, Vice President; (404) 329-7550; john.stevens@cancer.org
Internet: http://www.cancer.org/docroot/RES/content/RES_5_2x_Clinical_Research_Professorships.asp?sitearea=RES
Sponsor: American Cancer Society
3100 Wyman Park Drive, Suite W400
Baltimore, MD 21211

ACS Doctoral Scholarships in Cancer Nursing 307
Doctoral degree scholarships in cancer nursing are awarded to graduate students pursuing doctoral study in the following cancer nursing fields: research, education, administration, or clinical practice. Awards are made for up to four years.
Requirements: Applicants must be currently enrolled in or applying to a doctoral degree program and have current licensure to practice as a registered nurse. American citizens or permanent residents are eligible.
Geographic Focus: All States
Date(s) Application is Due: Oct 15
Amount of Grant: 15,000 USD
Contact: Grants Department; (404) 329-7558; fax (404) 321-4669; grants@cancer.org
Internet: http://www.cancer.org/docroot/RES/content/RES_5_2x_Doctoral_Degree_Scholarships_in_Cancer_Nursing.asp?sitearea=RES
Sponsor: American Cancer Society
3100 Wyman Park Drive, Suite W400
Baltimore, MD 21211

ACS Doctoral Training Grants in Oncology Social Work 308
This grant is awarded to qualifying doctoral students at schools of social work that train individuals to conduct research related to the psychosocial needs of persons with cancer and their families. Initial 2-year grant providing a stipend of $20,000 per year with possibility of a 2-year competitive renewal.
Requirements: The following eligibility requirements must be met: the applicant must have master's degree in social work; at least one year of social work experience in a health care setting (oncology/cancer control experience preferred, but not required); a demonstrated commitment to a career in oncology social work as evidenced by recent experience, education, and/or research; the applicant must be currently enrolled in or applying to a doctoral degree program in social work; if an applicant is not yet enrolled in a Ph.D. program, the applicant must meet requirements for doctoral study and must have been accepted by the institution to which s/he has applied at the time of funding; the applicant must project a full-time program of study that integrates oncology social work and provides evidence of faculty support for the program of study; to qualify for funding for the dissertation phase of the doctoral program, the applicant must conduct a dissertation relevant to Oncology Social Work. Successful applicants must also agree to attend the American Cancer Society activities at the annual Society for Social Work Research meeting. (Recipients at the dissertation phase will be expected to present their research at this meeting.)
Geographic Focus: All States
Date(s) Application is Due: Oct 15
Amount of Grant: 20,000 USD
Contact: Virginia Krawiec, MPA Director; (404) 329-7612; fax (404) 321-4669; ginger.krawiec@cancer.org
Internet: http://www.cancer.org/docroot/RES/content/RES_5_2x_Post-Masters_Training_Grants_in_Clinical_Oncology_Social_Work.asp?sitearea=RES
Sponsor: American Cancer Society
3100 Wyman Park Drive, Suite W400
Baltimore, MD 21211

ACSE J. Waldo Smith Hydraulic Fellowship 309
The J. Waldo Smith Hydraulic Fellowship is restricted to research in the field of experimental hydraulics as distinguished from research concentrating on purely theoretical hydraulics. Emphasis is to be placed on practical experiments designed and executed for the purpose of advancing knowledge with respect to the laws of hydraulic flow, rather than to the type of research that proceeds on the theory of mathematical analysis based on assumptions of unknown validity. The purpose of the research is to test the assumptions that are currently made, and also to develop a better understanding of fluid flow. The fellowship is offered every third year, and runs for one full academic year. Administration is in part through the institution that invites cooperation through its engineering faculty. Application packets are available through the student chapter advisor or from ASCE.
Requirements: Fellowships are awarded to graduate students, who are members of the society, who give promise of best fulfilling the ideals of the fellowship. There is a brief application form. Each application shall include a statement in general terms of the purposes for which the funds are expected to be used.
Geographic Focus: All States
Date(s) Application is Due: Feb 9
Amount of Grant: Up to 5,000 USD
Contact: John Casazza; (703) 295-6155; fax (703) 295-6132; jcasazza@asce.org
Internet: http://www.asce.org/Student-Organizations/Scholarships-and-Fellowships/J—Waldo-Smith-Hydraulic-Fellowship/
Sponsor: American Society of Civil Engineers
1801 Alexander Bell Drive
Reston, VA 20191-4400

ACS Institutional Research Grants 310
The IRG program provides seed money for the initiation of promising new projects by independent junior investigators so they may obtain preliminary results that will enable them to compete successfully for national research grants.
Requirements: Only full-time, tenure-track JHU faculty members at the ranks of Instructor or Assistant Professor or equivalent who are within the first six years of their first independent research or faculty appointment may apply. Relevance to the cancer problem is important and must be well documented. Within this research scope, a wide breadth of research approaches will be considered; these include basic laboratory studies and clinical research.
Restrictions: Senior investigators, postdoctoral fellows and junior investigators who have competitive national research grants or who have received prior support form the IRG are not eligible.
Geographic Focus: All States
Date(s) Application is Due: Apr 1
Amount of Grant: 90,000 USD
Contact: Grace Bigelow; (410) 516-5256; fax (410) 516-7775; graceb@jhu.edu
Internet: http://www.cancer.org/docroot/RES/content/RES_5_2x_Institutional_Research_Grants.asp?sitearea=RES
Sponsor: American Cancer Society
3100 Wyman Park Drive, Suite W400
Baltimore, MD 21211

ACS Master's Training Grants in Clinical Oncology Social Work 311
These grants are awarded annually to qualifying institutions that train clinical oncology social workers to provide cancer patients and their families with psychosocial services. Grants are available to second-year students in a master's program and post-master's social workers within five years of graduation. The master's training must introduce social workers to the special needs of cancer patients and their families. Contact the society for application forms.
Requirements: Applications will be considered from institutions that identify clinical oncology social work training activities, cancer programs with defined psychosocial support services, and relationships with schools accredited by the Council on Social Work Education.
Geographic Focus: All States
Date(s) Application is Due: Oct 15
Amount of Grant: Up to 12,000 USD
Contact: Grants Department; (404) 329-7558; fax (404) 321-4669; grants@cancer.org
Internet: http://www.cancer.org/docroot/RES/content/RES_5_2x_Masters_Training_Grants_in_Clinical_Oncology_Social_Work.asp?sitearea=RES
Sponsor: American Cancer Society
3100 Wyman Park Drive, Suite W400
Baltimore, MD 21211

ACSM Carl V. Gisolfi Memorial Fund Grant 312
Carl V. Gisolfi, Ph.D., FACSM, was one of the most influential leaders of ACSM in its 50+ year history. With Gatorade, ACSM created the Gisolfi Memorial Fund to honor his many contributions to ACSM and the exercise science field. This grant is designated to encourage research in thermoregulation, exercise, and hydration. One award will be dispersed to ACSM doctoral students in the amount of $5,000 each for a one-year period.
Requirements: Applicants for student research grants must have graduate student status during the term of the grant to be considered for funding. Applicants must be current members of ACSM at the time of submitting an application to be funded. Grants are open to all ACSM members, including international members.
Geographic Focus: All States, All Countries
Date(s) Application is Due: Jan 17
Contact: Jane Senior, (317) 637-9200, ext. 143; fax (317) 634-7817; jsenior@acsm.org
Internet: http://acsm.org/find-continuing-education/awards-grants/research-grants/2011/08/15/carl-v.-gisolfi-memorial-fund
Sponsor: American College of Sports Medicine
401 West Michigan Street
Indianapolis, IN 46202-3233

ACSM Coca-Cola Company Doctoral Student Grant on Behavior Research 313
For research addressing behavioral strategies and techniques to promote the adoption and maintenance of physical activity/exercise. Studies that focus on innovative strategies to improve individual-level adoption and/or maintenance of physical activity will be given first priority. The proposed study can focus on youth, adolescents, adults, or older adults 55 and over. One award of $5,000 is available.
Requirements: Applicants for student research grants must have graduate student status during the term of the grant to be considered for funding. Applicants must be current members of ACSM at the time of submitting an application to be funded. Grants are open to all ACSM members, including international members. However, the NASA initiative is open to U.S. residents only.
Geographic Focus: All States, All Countries
Date(s) Application is Due: Jan 17
Contact: Jane Senior, (317) 637-9200, ext. 143; fax (317) 634-7817; jsenior@acsm.org
Internet: http://acsm.org/find-continuing-education/awards-grants/research-grants/2011/08/16/the-coca-cola-company-doctoral-student-grant-on-behavior-research
Sponsor: American College of Sports Medicine
401 West Michigan Street
Indianapolis, IN 46202-3233

ACS Mentored Research Scholar Grant in Applied and Clinical Research 314

This funding opportunity supports mentored research by full-time faculty, typically within the first four years of their appointment, with the goal of becoming independent investigators in clinical, cancer control and prevention, epidemiologic, psychosocial, behavioral, health services and health policy research. Awards are made up to five years and for up to $135,000 per year (direct costs) plus 8% allowable indirect costs. A maximum of $10,000 per year for the mentor(s) (regardless of the number of mentors) is included in the $135,000. These grants are not renewable.
Geographic Focus: All States
Date(s) Application is Due: Apr 1; Oct 15
Amount of Grant: 135,000 USD
Contact: Grants Department; (404) 329-7558; fax (404) 417-5974; grants@cancer.org
Internet: http://www.cancer.org/docroot/RES/content/RES_5_2x_Mentored_Research_Scholar_Grant_in_Applied_and_Clinical_Research.asp?sitearea=RES
Sponsor: American Cancer Society
3100 Wyman Park Drive, Suite W400
Baltimore, MD 21211

ACSM Foundation Clinical Sports Medicine Endowment Grants 315

ACSM's Clinical Sports Medicine Endowment will fund an annual award of $5,000 for research that is directly related to clinical sports medicine. The purpose of the award is to stimulate clinical research in sports medicine and research in clinical sports medicine. Clinical research in sports medicine must be derived from active care of athletes, such as prevention strategies, treatment protocols, or treatment outcomes. Research in clinical sports medicine must be directly related to issues of active care of athletes, such as pathogenesis, pathophysiology, biomechanics, or environmental issues. The proposals must demonstrate the study's direct relationship to one of these purposes. The active patient care must be provided to physically active recreational, competitive or elite athletes. It is anticipated that the completed research will be presented at a highlighted session at the ACSM Annual Meeting.
Requirements: Grants are open to all ACSM members, including international members. This clinical research award is open to MDs, DOs, PTs, ATCs, and other medical professionals (ACSM members only) involved in the conduct of patient-based clinical research. Applicants for student research grants must have graduate student status during the term of the grant to be considered for funding. Applicants must be current members of ACSM at the time of submitting an application to be funded.
Geographic Focus: All States, All Countries
Date(s) Application is Due: Jan 17
Amount of Grant: 5,000 USD
Contact: Jane Senior, (317) 637-9200, ext. 143; fax (317) 634-7817; jsenior@acsm.org
Internet: http://acsm.org/find-continuing-education/awards-grants/research-grants/2011/08/18/clinical-sports-medicine-endowment
Sponsor: American College of Sports Medicine
401 West Michigan Street
Indianapolis, IN 46202-3233

ACSM Foundation Doctoral Student Research Grants 316

ACSM Foundation Research Grants for doctoral students awards up to $5,000 for a one-year period. The awards are to be used for experimental subjects, supplies, and small equipment needs.
Requirements: Doctoral students enrolled in full-time programs are eligible to apply. Applicants for student research grants must have graduate student status during the term of the grant to be considered for funding. Applicants must be current members of ACSM at the time of submitting an application to be funded. Grants are open to all ACSM members, including international members. However, the NASA initiative is open to U.S. residents only.
Geographic Focus: All States, All Countries
Date(s) Application is Due: Jan 17
Amount of Grant: 5,000 USD
Contact: Jane Senior, (317) 637-9200, ext. 143; fax (317) 634-7817; jsenior@acsm.org
Internet: http://acsm.org/find-continuing-education/awards-grants/research-grants/2011/08/18/acsm-foundation-doctoral-student-research-grant
Sponsor: American College of Sports Medicine
401 West Michigan Street
Indianapolis, IN 46202-3233

ACSM Foundation Research Grants 317

The research endowment has been made possible by charitable funds and individual donations derived from the annual campaign, specifically targeted for research. The intent of the research endowment is to use a portion of the interest derived from these funds to support basic and applied research in exercise science. The primary goal is to fund mechanistic, hypothesis driven, basic and applied research. A $10,000 grant is available and is applicable to all relevant fields of exercise science.
Requirements: Funding is primarily targeted for new or junior investigators, within seven years of attaining a terminal degree (e.g., Ph.D., Ed.D.). It is the intent of this grant to provide seed money support after which further funding would be sought from other sources. Only one application per person is allowed. Applicants for student research grants must have graduate student status during the term of the grant to be considered for funding. Applicants must be current members of ACSM at the time of submitting an application to be funded. Grants are open to all ACSM members, including international members. However, the NASA initiative is open to U.S. residents only.
Geographic Focus: All States, All Countries
Date(s) Application is Due: Jan 17
Amount of Grant: 10,000 USD
Contact: Jane Senior, (317) 637-9200, ext. 143; fax (317) 634-7817; jsenior@acsm.org
Internet: http://acsm.org/find-continuing-education/awards-grants/research-grants/2011/08/18/research-endowment
Sponsor: American College of Sports Medicine
401 West Michigan Street
Indianapolis, IN 46202-3233

ACSM International Student Awards 318

ACSM international awards provide funding for professionals and students to participate in clinical and research exchange opportunities. The International Student Award honors students from countries outside of the United States and helps fund travel expenses for those traveling to the ACSM Annual Meeting to present their scholarly work. Awards will be determined by the quality of research to be presented at the ACSM Annual Meeting. Award recipients will receive $1,000 USD to assist with travel costs for attending ACSM's Annual Meeting.
Requirements: Applicants may not be residents of the United States. Applicants must be a student at the time of the ACSM Annual Meeting and must not hold a completed doctoral degree of any type (Ph.D., Ed.D., M.D., DPH, etc.) at the time of application. Applicants must be ACSM members in good standing and must have been accepted to present research at the ACSM Annual Meeting. Applicants should be the first author of the accepted abstract and should plan to personally present the research.
Restrictions: Previous International Student Award winners are not eligible to apply.
Geographic Focus: All Countries
Date(s) Application is Due: Feb 1
Amount of Grant: 1,000 USD
Contact: Jane Senior, (317) 637-9200, ext. 143; fax (317) 634-7817; jsenior@acsm.org
Internet: http://acsm.org/find-continuing-education/awards-grants/international-awards/2011/08/16/international-student-award
Sponsor: American College of Sports Medicine
401 West Michigan Street
Indianapolis, IN 46202-3233

ACSM NASA Space Physiology Research Grants 319

Through the generosity of the National Aeronautics and Space Administration (NASA), $10,000 is available for research grants in the area of exercise, weightlessness, and musculoskeletal physiology. Grants can range up to $5,000 and are available for a one-year period.
Requirements: Doctoral students (U.S. residents only) enrolled in full-time programs are eligible to apply. Applicants for student research grants must have graduate student status during the term of the grant to be considered for funding. Applicants must be current members of ACSM at the time of submitting an application to be funded. Grants are open to all ACSM members, including international members. However, the NASA initiative is open to U.S. residents only.
Geographic Focus: All States
Date(s) Application is Due: Jan 17
Amount of Grant: Up to 5,000 USD
Contact: Jane Senior or Michael F. Dell, (317) 637-9200, ext. 143; fax (317) 634-7817; jsenior@acsm.org or mdell@acsm.org
Internet: http://acsm.org/find-continuing-education/awards-grants/research-grants/2011/08/16/nasa-space-physiology-research-grants
Sponsor: American College of Sports Medicine
401 West Michigan Street
Indianapolis, IN 46202-3233

ACSM Oded Bar-Or International Scholar Awards 320

The Oded Bar-Or International Scholar Award allows professionals to gain technical expertise and/or scientific knowledge through an international exchange program. Award recipients from the United States and Canada are required to travel to an institution outside of the United States or Canada. Similarly, award recipients from countries other than Canada and the United States are required to travel to an institution within the United States or Canada. Membership in ACSM is not a requirement. Awards will be determined by the likelihood that the activities will open avenues for professional growth, the intrinsic quality of the proposed research activity and evidence the project can be completed. Award recipients will receive round-trip air transportation to the host institution, a lodging and living stipend, medical health insurance for two months, complimentary registration and lodging for the ACSM Annual Meeting.
Requirements: The award is open to all ACSM members, including international members. Applicants must: be professionals in a discipline recognized by ACSM (minimum of a bachelor's degree is required; plan to engage in a professional activity that will advance the professional standing and reputation of scientists (Though many proposals will be for research-based activities, requests for other types of support will be considered; choose a host scientist who is a member of ACSM (ACSM will not match applicants with potential host scientists).
Geographic Focus: All States, Canada
Date(s) Application is Due: Feb 1
Contact: Jane Senior, (317) 637-9200, ext. 143; fax (317) 634-7817; jsenior@acsm.org
Internet: http://acsm.org/find-continuing-education/awards-grants/research-grants/2011/08/16/oded-bar-or-international-scholar-award
Sponsor: American College of Sports Medicine
401 West Michigan Street
Indianapolis, IN 46202-3233

ACSM Paffenbarger-Blair Fund for Epidemiological Research on Physical Activity Grants 321

Dr. Ralph S. Paffenbarger, Jr., a co-recipient of the first Olympic Prize in Exercise Science and Sports Medicine, made an extraordinarily generous gift to the American College of Sports Medicine by donating his share of the award to the College's capital campaign with the monies added to the College's epidemiology endowment. The interest on this contribution makes this grant possible. The intent of this award is to encourage researchers early in their career to become involved with physical activity epidemiology. The applications may focus on observational studies of physical activity and health outcomes, or on randomized controlled trials that are clearly focused on physical activity and important public health issues. Applicants are expected to apply within two (2) years of receiving a postgraduate degree or completion of clinical training. The award will be in the amount of $10,000 for a one year period.

Requirements: Applicants for student research grants must have graduate student status during the term of the grant to be considered for funding. Applicants must be current members of ACSM at the time of submitting an application to be funded. Grants are open to all ACSM members, including international members. However, the NASA initiative is open to U.S. residents only.

Geographic Focus: All States, All Countries

Date(s) Application is Due: Jan 17

Amount of Grant: 10,000 USD

Contact: Jane Senior, (317) 637-9200, ext. 143; fax (317) 634-7817; jsenior@acsm.org

Internet: http://acsm.org/find-continuing-education/awards-grants/research-grants/2011/08/18/paffenbarger-blair-fund-for-epidemiological-research-on-physical-activity

Sponsor: American College of Sports Medicine
401 West Michigan Street
Indianapolis, IN 46202-3233

ACSM Raymond and Rosalee Weiss Research Endowment Grant 322

Wishing to advance research into the health benefits-physical, mental and emotional-of physical activity and sports, the Weiss Research Endowment has been established for ACSM students studying in this field. One project will be funded for applied rather than for basic research, with the intent of applying the results to programs involving physical activity and sports. The psychological and emotional benefits of physical activity are of particular importance to the benefactors of this endowment, Drs. Raymond and Rosalee Weiss, and proposals addressing those issues will be given priority. One reward of $1,500 is available.

Requirements: Applicants for student research grants must have graduate student status during the term of the grant to be considered for funding. Applicants must be current members of ACSM at the time of submitting an application to be funded. Grants are open to all ACSM members, including international members. However, the NASA initiative is open to U.S. residents only.

Geographic Focus: All States, All Countries

Date(s) Application is Due: Jan 17

Amount of Grant: 1,500 USD

Contact: Jane Senior, (317) 637-9200, ext. 143; fax (317) 634-7817; jsenior@acsm.org

Internet: http://acsm.org/find-continuing-education/awards-grants/research-grants/2011/08/16/raymond-and-rosalee-weiss-research-endowment

Sponsor: American College of Sports Medicine
401 West Michigan Street
Indianapolis, IN 46202-3233

ACS Physician Training Awards in Preventive Medicine 323

The Physician Training Award in Preventive Medicine (PTAPM) is intended to encourage and assist the development of promising individuals who will pursue careers in preventive medicine. This program is designed to create a cadre of preventive medicine specialists who are expert in cancer prevention and control, and with the potential to become leaders in research, education, and intervention in this area. Through the Physician Training Award in Preventive Medicine, the Society seeks to support physicians in accredited residency programs that will lead to eligibility for certification in preventive medicine; such programs must provide cancer prevention and control research and practice opportunities. Awards are for four years in the total amount of $300,000, based on an average of $50,000 per resident training year. These grants are renewable.

Requirements: Requirements for the institution are available on the ACS website. Eligibility requirements for the residents are: nominated residents must be citizens or non-citizen nationals of the United States or its possessions and territories, or must have been lawfully admitted to the United States for permanent residence at the time of application; must have an MD, DO, or equivalent degree; must have completed the clinical year of a residency program in preventive medicine, or have at least one year of postgraduate clinical training; must be accepted by or applying to the sponsoring residency program. It is preferred that the candidates not have completed more than half of the required academic work at the beginning of the award period. Residents that have completed the MPH degree will be considered on a case-by-case basis, and must propose course work in cancer prevention and control.

Geographic Focus: All States

Date(s) Application is Due: Apr 1

Amount of Grant: Up to 300,000 USD

Contact: Grants Department; (404) 329-7558; fax (404) 417-5974; grants@cancer.org

Internet: http://www.cancer.org/docroot/RES/content/RES_5_2x_Physician_Training_Awards_in_Preventive_Medicine.asp?sitearea=RES

Sponsor: American Cancer Society
3100 Wyman Park Drive, Suite W400
Baltimore, MD 21211

ACS Pilot and Exploratory Projects in Palliative Care of Cancer Patients and Their Families Grants 324

This request for application (RFA) provides funding for investigators performing pilot and exploratory research studies whose purpose is to test interventions, develop research methodologies, and explore novel areas of research in palliative care of cancer patients and their families. A condition of funding is a clearly defined plan as to how the investigator will use the results of the project to develop larger, extramurally funded research projects. This RFA is limited to applications that focus on palliative care research projects for seriously ill cancer patients and their families in three specific areas: exploring the relationship of pain and other distressing symptoms on quality and quantity of life, independence, function, and disability and developing interventions directed at their treatment in patients with advanced and chronic illnesses; studying methods of improving communication between adults living with serious illness, their families and their health care providers; evaluating models and systems of care for patients living with advanced illness and their families. As a condition of accepting the award, each recipient of this grant, will agree to: attend the required meetings of the NPCRC in Fall during the award period; present results of the funded research at this required NPCRC meeting; prepare annual progress reports for each year of funding and a final report at the conclusion of the award period; list the American Cancer Society as funding this study on all publications and presentations.

Requirements: Applications may be submitted by not-for-profit institutions located within the United States, its territories and the Commonwealth of Puerto Rico. Applicants must: be United States citizens, non-citizen nationals or permanent residents of the United States; hold a doctorate degree (MD, PhD, or equivalent) and have a full-time faculty position or equivalent at a college, university, medical school, or other fiscally responsible not-for-profit organization within the United States. Independent investigators at all stages of their career are eligible to apply. Applicants must submit their full application electronically and in paper format per guidelines on cancer.org no later than 5pm (Eastern) on October 15th.

Restrictions: Awards may not exceed $60,000 per year (direct costs) plus 20% indirect costs. Salary support for the Principal Investigator may not exceed 20% of the direct costs. Awards may not exceed a period of one to two years duration.

Geographic Focus: All States

Date(s) Application is Due: Oct 15

Amount of Grant: Up to 60,000 USD

Contact: Ronit Elk; (404) 417-5957; fax (404) 321-4669; ronit.elk@cancer.org

Internet: http://www.cancer.org/docroot/RES/content/RES_5_2x_RFA__Pilot_and_Exploratory_Projects_in_Palliative_Care_of_Cancer_Patients_and_Their_Families.asp?sitearea=RES

Sponsor: American Cancer Society
3100 Wyman Park Drive, Suite W400
Baltimore, MD 21211

ACS Postdoctoral Fellowships 325

Fellowships are designed to enable a new investigator to qualify for an independent career in cancer research. Postdoctoral fellowships may be made for one, two, or three years. Stipend payments are made directly to the individual, or the institution if requested, at the beginning of each month. Travel funds are paid to the individual. Institutional allowances are paid annually at the start of the grant and on the anniversary date thereafter. Awards are made for one to three years with progressive stipends.

Requirements: Applicants must be citizens or permanent residents of the United States (the latter must provide notarized evidence of their legal resident alien status) and shall have been awarded a doctoral degree prior to the activation date of the grant. Application must be endorsed by applicant's mentor and the head of the department in which the training will be received. A plan of training must be formulated and agreed upon by the mentor and the applicant and described in detail in the application.

Restrictions: Awards will not be made to applicants who have completed five or more years of postdoctoral training prior to the start date of the fellowship.

Geographic Focus: All States

Date(s) Application is Due: Apr 1; Oct 15

Amount of Grant: 142,000 USD

Contact: Grants Department; (404) 329-7558; fax (404) 321-4669; grants@cancer.org

Internet: http://www.cancer.org/docroot/RES/content/RES_5_2x_Postdoctoral_Fellowships.asp?sitearea=RES

Sponsor: American Cancer Society
3100 Wyman Park Drive, Suite W400
Baltimore, MD 21211

ACS Research Professor Grants 326

The American Cancer Society offers a limited number of grants to investigators in mid-career who have made seminal contributions that have changed the direction of cancer research. Furthermore, it is expected that these investigators will continue to provide leadership in their research area. Up to two awards are made annually for a five-year term that can be renewed once. The award of up to $80,000 per year (direct costs only) may be used for salary or research project support. Interested individuals should submit their curriculum vitae with a complete bibliography and a letter of intent that briefly describes their seminal contributions to research. This information must be submitted no later then February 1, via the Research Professor letter of intent on proposalCENTRAL at: https://proposalcentral.altum.com. Upon receipt, ACS will provide appropriate guidance either by telephone or email.

Geographic Focus: All States

Date(s) Application is Due: Feb 1; Apr 1

Amount of Grant: Up to 80,000 USD

Contact: John J. Stevens, Vice President; (404) 329-7550; john.stevens@cancer.org
Internet: http://www.cancer.org/docroot/RES/content/RES_5_2x_Research_Professorships.asp?sitearea=RES
Sponsor: American Cancer Society
3100 Wyman Park Drive, Suite W400
Baltimore, MD 21211

ACS Research Scholar Grants for Health Services and Health Policy Research 327

The grants support research projects centered on health services and health policy initiated by investigators at any stage in their careers. The initial award includes 20 percent for indirect costs for up to four years, and may be renewed once for up to four years.
Geographic Focus: All States
Date(s) Application is Due: Apr 1; Oct 15
Amount of Grant: Up to 200,000 USD
Contact: Grants Department; (404) 329-7558; fax (404) 321-4669; grants@cancer.org
Internet: http://www.cancer.org/docroot/RES/content/RES_5_2x_Research_Scholar_Grants_For_Health_Services_and_Health_Policy_and_Outcomes_Research.asp?sitearea=RES
Sponsor: American Cancer Society
3100 Wyman Park Drive, Suite W400
Baltimore, MD 21211

ACS Research Scholar Grants in Basic, Preclinical, Clinical and Epidemiology Research 328

This grant supports investigator-initiated research projects in basic, preclinical, clinical and epidemiologic research. Awards are for up to four years and for up to $200,000 per year (direct costs), plus 20% allowable indirect costs.
Geographic Focus: All States
Date(s) Application is Due: Apr 1; Oct 15
Amount of Grant: 200,000 USD
Contact: Grants Department; (404) 329-7558; fax (404) 417-5974; grants@cancer.org
Internet: http://www.cancer.org/docroot/RES/content/RES_5_2x_Research_Scholar_Grants_for_Beginning_Investigators.asp?sitearea=RES
Sponsor: American Cancer Society
3100 Wyman Park Drive, Suite W400
Baltimore, MD 21211

ACS Research Scholar Grants in Psychosocial and Behavioral and Cance Control Research 329 r

The grants support research projects focusing on the psychosocial and behavioral aspects of cancer by independent investigators at any stage in their careers. Applications are encouraged in which an individual at an early career stage is coprincipal investigator with an established researcher. The initial award supports four years of research and may be renewed once for four additional years.
Requirements: Individuals must not have held an independent position for more than six years at the time the application is submitted.
Geographic Focus: All States
Date(s) Application is Due: Apr 1; Oct 15
Amount of Grant: Up to 200,000 USD
Contact: Grants Department; (404) 329-7558; fax (404) 417-5974; grants@cancer.org
Internet: http://www.cancer.org/docroot/RES/content/RES_5_2x_Research_Scholar_Grants_in_Psychological_and_Behavioral_Research_for_Beginning_and_Senior_Investigators.asp?sitearea=RES
Sponsor: American Cancer Society
3100 Wyman Park Drive, Suite W400
Baltimore, MD 21211

ACT Awards 330

The program is designed to benefit programs that aim to help middle and high school students reach their career and educational goals. The awards target programs that work with students whose circumstances and environments could keep them from continuing their education or career training. Winners receive awards in the form of ACT programs, services, consultation and resources. Awards are provided based on the benefit a program is likely to offer its defined at-risk population; overall program design; how readily it could be replicated elsewhere; and how much support a program has.
Requirements: Proposed projects may be programs or studies undertaken with at-risk students in grades eight through 12 or in two- and four-year colleges and universities.
Restrictions: Awards do not include salaries or stipends for award recipients or other personnel, travel expenses, conference fees, tuition costs, or any similar types of direct expenses.
Geographic Focus: All States
Date(s) Application is Due: May 15
Contact: Director of Policy Research; (319) 337-1293; noethr@act.org
Internet: http://www.act.org
Sponsor: ACT, Inc.
500 ACT Drive, P.O. Box 168
Iowa City, IA 52243-0168

Acton Institute for the Study of Religion and Liberty Calihan Academic Fellowships 331

The Calihan Academic Fellowships provide scholarships and research grants to future scholars and religious leaders whose academic work shows outstanding potential. Applicants studying theology, philosophy, economics, or related fields must demonstrate the potential to advance understanding in the relationship between theology and the principles of the free and virtuous society. Such principles include recognition of human dignity, the importance of the rule of law, limited government, religious liberty, and freedom in economic life. Awards are open to seminarians or graduate students in theology, philosophy, religion, economics, or in related fields.
Requirements: Strong academic performance is essential. Candidates must: have a demonstrated interest in the themes of the Acton Institute; and display the potential to advance the themes of the Acton Institute.
Geographic Focus: All States
Date(s) Application is Due: Jul 15; Oct 31
Amount of Grant: Up to 3,000 USD
Contact: Kara Eagle; (616) 454-3080; fax (616) 454-9454; keagle@acton.org
Internet: http://www.acton.org/programs/students/calihan_academic_scholarship.php
Sponsor: Acton Institute for the Study of Religion and Liberty
161 Ottawa NW, Suite 301
Grand Rapids, MI 49503

Acton Institute for the Study of Religion and Liberty Novak Award 332

Named after distinguished American theologian Michael Novak, this $10,000 award rewards new outstanding scholarly research into the relationship between religion, economic freedom, and the free and virtuous society. This award recognizes those scholars early in their academic career who demonstrate outstanding intellectual merit in advancing the understanding of theology's connection to human dignity, the importance of the rule of law, limited government, religious liberty, and freedom in economic life. The nomination deadline is November 1, and the application deadline is December 1.
Requirements: Professors, university faculty members, and other scholars may nominate qualified individuals for the Novak Award by completing the online nomination form.
Geographic Focus: All States
Date(s) Application is Due: Nov 30
Amount of Grant: 10,000 USD
Samples: Professor Carlos Hoevel—for A Rosminian Vision for the Post-Crisis Global Economy, $10,000; Dr. Andrea Schneider—for Centesimus Annus: Law, Human Rights, and the Free Society, $10,000; Dr. Jan Klos—for Spiritualizing the World: Secularism as a Religion-Challenges for Faith and Liberty in Europe, $10,000.
Contact: Kara Eagle; (616) 454-3080; fax (616) 454-9454; keagle@acton.org
Internet: http://www.acton.org/programs/students/novak.php
Sponsor: Acton Institute for the Study of Religion and Liberty
161 Ottawa NW, Suite 301
Grand Rapids, MI 49503

Acuity Charitable Foundation Grants 333

The Acuity Charitable Foundation was established in 2003 with a major donation from the Acuity Mutual Insurance Company. The foundation primarily supports Sheboygan, Wisconsin, organizations involved with arts and culture, elementary and secondary education, engineering, health care, cancer, and youth development. Types of support include general operating support, program development, and sponsorships. Though formal application forms are not required, proposals should be submitted using the organization's letterhead. There are no deadlines, and applicants should submit the entire proposal, which should include: need for the project, group being served, evaluation process, and necessary funding.
Requirements: 501(c)3 organizations based in or serving residents of Sheboygan, Wisconsin, can apply.
Geographic Focus: Wisconsin
Samples: Sheboygan Arts Foundation, Sheboygan, Wisconsin, $65,000; Junior Achievement of Wisconsin, Milwaukee, Wisconsin, $35,000.
Contact: Lynn Yunger, (920) 458-9131
Sponsor: Acuity Charitable Foundation
2800 S. Taylor Drive, P.O. Box 58
Sheboygan, WI 53082-0058

ACVIM Foundation Clinical Investigation Grants 334

The Foundation improves the health and well-being of animals by funding humane studies and communicating information about this work to the veterinary community and the general public. It supports leading veterinary scientists as they take on a range of critical animal health issues within the specialties of small and large animal internal medicine, cardiology, neurology, and oncology. This work, in turn, holds promise for better preventive care, better nutrition, and more sensitive screening and improved treatment for such devastating diseases as cancer, epilepsy, cardiomyopathy, and kidney failure. The Foundation's main goal is to support clinical investigations that lead to new diagnostic, treatment, and prevention techniques. Resident applicants may request up to $12,500 per year, while diplomate applicants may request up to $15,000. One $20,000 grant will be awarded to a collaborative/multi-center proposal, with preference toward private practice/academia partnerships. Funding for multi-year projects up to 3 years will be considered, subject to annual non-competitive renewal based on availability of funds and demonstration of adequate progress on the project. The foundation will consider funding for multi-year projects up to three years.
Geographic Focus: All States
Date(s) Application is Due: Sep 15
Amount of Grant: Up to 20,000 USD
Samples: Dr. Jill Beech, University of Pennsylvania School of Veterinary Medicine—Evaluation of Endogenous Alpha MSH Concentrations at Different Seasons and in Response to TRH in Normal Horses and Those with Pituitary Hyperplasia; Dr. Stephanie Kottler, University of Missouri-Columbia College of Veterinary Medicine—Prevalence of Staphylococcus aureus and MRSA carriage in three populations.

Contact: Angela E. Frimberger, Executive Director; (303) 231-9933 or (800) 245-9081; info@acvimfoundation.org
Internet: http://www.acvimfoundation.org/grants/info.html
Sponsor: American College of Veterinary Internal Medicine
997 Wadsworth Boulevard, Suite A
Lakewood, CO 80214

ADA Foundation Dentsply International Research Fellowships 335
The American Dental Association Foundation's Dentsply International Research Fellowships may be applied for by finalists in the Annual Dental Student Research Competition conducted by Dentsply International. Fellowship recipients receive a ten-week research opportunity, a $4,000 stipend, and room and board expenses in order to work in the Foundation's Paffenbarger Research Center located in Gaithersburg, Maryland. Applications are available at the dental school research advisor and/or dean's office.
Geographic Focus: All States
Amount of Grant: 4,000 USD
Contact: Dr. Clifton M. Carey, Director of Administration; (301) 975-6806; fax (301) 926-9143; clif.carey@nist.gov
Internet: http://www.ada.org/applyforassistance.aspx#research
Sponsor: American Dental Association Foundation
211 East Chicago Avenue
Chicago, IL 60611

ADA Foundation George C. Paffenbarger Student Research Awards 336
The ADA Foundation George C. Paffenbarger Student Research Awards are offered in collaboration with the Academy of Operative Dentistry. Awards honors George C. Paffenbarger, an internationally recognized authority on dental materials and a pioneer in the development of specifications and standards for testing materials. The Foundation's Paffenbarger Research Center is named in his honor. Awards are open to dental students at all levels who wish to undertake novel research relevant to contemporary operative dentistry. The budget for the proposed research may not exceed $6,000. The awardee will also receive up to $1,000 to attend and present the research findings at the annual meeting. Applications must be submitted electronically to the Research Committee Chair.
Requirements: The application should take the form of a protocol outlining the background, aims and hypothesis of the proposed research, the methodology to be employed, and the anticipated work on the clinical practice of operative dentistry. Application information, including deadline and submission format, can be found at the Foundation's website.
Geographic Focus: All States
Date(s) Application is Due: Jan 13
Contact: Cristina Garcia, Grants Program Coordinator; (312) 440-2763; fax (312) 440-3526; garciac@ada.org or adaf@ada.org
Internet: http://www.ada.org/applyforassistance.aspx#tarrson
Sponsor: American Dental Association Foundation
211 East Chicago Avenue
Chicago, IL 60611

ADA Foundation Scientist in Training Fellowship 337
The American Dental Association Foundation's Scientist in Training Fellowships are conducted at the Foundation's Paffenbarger Research Center (PRC) in Gaithersburg, Maryland. The program includes a full-time fellow working in conjunction with the PRC's scientific research staff. Fellowships have been made possible by the Great-West Life and Annuity Insurance Company. Individuals interested in learning more about the program, application process or deadlines are invited to contact Paffenbarger Research Center.
Geographic Focus: All States
Contact: Dr. Clifton M. Carey, Director of Administration; (301) 975-6806; fax (301) 926-9143; clif.carey@nist.gov or prc@ada.org
Internet: http://www.ada.org/applyforassistance.aspx#grants
Sponsor: American Dental Association Foundation
211 East Chicago Avenue
Chicago, IL 60611

ADA Foundation Summer Scholars Fellowships 338
The ADA Foundation's Paffenbarger Research Center annually appoints two young investigators to its summer scholars program. The program brings industrial and dental researchers together in an environment outside the dental school. This program is made possible by Dentsply International. Individuals interested in learning more about the program, including the application process and deadlines, are invited to contact the Program Coordinator at the Paffenbarger Research Center.
Geographic Focus: All States
Contact: Gretchen Duppins, Summer Scholars Program Coordinator; (301) 975-6806; fax (301) 926-9143; gretchen@dupins.nist.gov
Internet: http://www.ada.org/applyforassistance.aspx
Sponsor: American Dental Association Foundation
211 East Chicago Avenue
Chicago, IL 60611

Adam Richter Charitable Trust Grants 339
The Trust, which is administered by the Bank of America in Dallas, Texas, was established in California in 1994. Offering grants on a national basis, its primary fields of interest include: the arts; animal welfare; Christian agencies and churches; the environment; health organizations; higher education; hospitals; and human services. There are no deadlines, though a formal application is required. Interested applicants should begin by sending a letter of inquiry to the Trust office.
Restrictions: Funding is not available for grants to individuals.
Geographic Focus: California
Amount of Grant: 3,000-20,000 USD
Contact: Michael Schlebach, Specialty Asset Manager; (866) 461-7281
Sponsor: Adam Richter Charitable Trust
c/o Bank of America, N.A.
Dallas, TX 75201-3115

Adaptec Foundation Grants 340
The corporate contributions program is primarily directed, although not exclusively, to organizations and programs serving the communities where its employees work and live, with special emphasis in the Silicon Valley area. Funding priorities include education and research—programs that lower the drop-out rate of youth in accredited schools and colleges, address career development and preparation, job training and placement, educational scholarship assistance, and investment in higher education (primarily in engineering-related university programs); health and human services—research cures for devastating diseases, emergency shelter and subsistence for the homeless, treatment and assistance for abused spouses and children, drug and alcohol rehabilitation, and supporting organizations that strengthen youths and families; helping people enjoy life more fully through music, dance, art museums, opera, theater, and performing arts groups; and $25-$500 employee matching gifts. The Corporate Contributions Committee meets quarterly to consider requests.
Restrictions: In general, Adaptec does not make contributions to any organization that does not have 501(c)3 or comparable status; individuals, except through the Adaptec Scholarship program; churches, synagogues, and other religious groups. (Requests are considered from religious-sponsored and other groups whose activities do not support any specific religious doctrine or ethnic/cultural group.); political or fraternal organizations; underwrite or assist with the development of films, television, or video/radio productions, whether for commercial, independent, or nonprofit ventures, unless such activity is in support of the computer I/O software/hardware industry and is directly related to the business of the company; tickets for raffles, contests, or other fundraising activities (except requests from major fundraising organizations that meet the defined objectives of the contribution program, such as United Way, the Silicon Valley Charity Ball and the San Jose Symphony); organizations whose primary objective is to assist, benefit, and address animal welfare or environmental issues, such as the preservation of endangered species and plants or conservation efforts to preserve the quality of wildlife, water, soil, or air; reduce debts, fund general operating expenses or capital improvement projects, or retroactively fund activities or programs that are already completed; and school sports leagues and events, and other non-educational related activities such as band, debate teams, and goodwill trips.
Geographic Focus: All States
Amount of Grant: 250-2,500 USD
Contact: Contribution Program Manager; (408) 945-8600; fax (408) 262-2533
Internet: http://cms.adaptec.com/ko-KR/company/about/_corporate/adaptec_corpcontributions.htm?nc=/ko-KR/company/about/_corporate/adaptec_corpcontributions.htm
Sponsor: Adaptec Foundation
691 South Milpitas Boulevard, MS-15
Milpitas, CA 95035-5473

AEG Foundation Legget Fund Grants 341
The Fund was established by the Foundation (then the Engineering Geology Foundation) in 1993 to support publications and public outreach in engineering geology and environmental geology that serve as information resources for the professional practitioner, students, faculty, and the public. The fund also supports public education about the interactions between the works of mankind and the geologic environment. Public outreach activities that the fund can support include K-12 teacher resources, publications to inform the public about the interactions between the works of humankind and the geologic environment, and signage or explanatory materials at public locations of interest for their engineering geology or environmental geology significance. The grants are intended to support technical and production editing of eligible publications, production costs, page charges, and conversion of print media to electronic form. Public education project grants from the Legget Fund are made to support technical and production editing, and production costs, of eligible works.
Restrictions: The Fund does not support research costs of individuals or organizations leading to publication or production, nor does it support general and administrative (overhead) costs, technical or scientific consulting fees, or equipment purchase or maintenance costs of organizations or individuals receiving grants from it.
Geographic Focus: All States
Amount of Grant: 1,000-50,000 USD
Contact: Becky Roland; (303) 757-2926; fax (303) 757-2969; broland@aegweb.org
Internet: http://www.aegfoundation.org/funds/legget
Sponsor: Association of Engineering Geologists Foundation
3773 Cherry Creek N Drive, Suite 575
Denver, CO 80209

AERA Dissertation Grants 342
AERA invites education-related dissertation proposals using NCES, NSF, and other federal databases. Dissertation Grants are available for advanced doctoral students and are intended to support the student while writing the doctoral dissertation. Applications are encouraged from a variety of disciplines, such as but not limited to, education, sociology, economics, psychology, demography, statistics, and psychometrics. Awards for Dissertation Grants are up to $20,000 for 1-year projects. In addition to the dissertation grant award, grantees will participate in a 2-day conference in Washington, DC. The

conference will provide unique professional development experiences for grantees, including highly qualified speakers on topics of education policy and career development, presentations of dissertation research by former grantees, and interaction with the Governing Board and federal agency staff. This conference is specifically for AERA grantees, and travel expenses will be paid by AERA.
Requirements: Applicants may be U.S. citizens or U.S. permanent residents enrolled in a doctoral program. Non-U.S. citizens enrolled in a doctoral program at a U.S. institution are also eligible to apply. Applicants should be advanced doctoral students at the dissertation writing stage. Underrepresented racial and ethnic minority researchers are strongly encouraged to apply. Grantees will present their research at a poster session during the AERA Annual Meeting and will participate in a one-day capstone workshop. Grantees must include travel funds (up to $1,000) in their grant budget to attend the AERA Annual Meeting held in Spring.
Restrictions: Grants are not renewable. In accordance with AERA's agreement with the funding agency, institutions may not charge indirect costs on these awards.
Geographic Focus: All States
Date(s) Application is Due: Jan 14
Amount of Grant: Up to 20,000 USD
Contact: Kevin Dieterle, Program Manager; (202) 238-3200; fax (202) 238-3250; grantsprogram@aera.net or kdieterle@aera.net
Internet: http://www.aera.net/ProfessionalOpportunitiesFunding/FundingOpportunities/AERAGrantsProgram/DissertationGrants/tabid/12812/Default.aspx
Sponsor: American Educational Research Association
1430 K Street, NW, Suite 1200
Washington, DC 20005

AERA Minority Fellowships 343

In 1991, the Council of the American Educational Research Association (AERA) established the AERA Minority Dissertation Fellowship in Education Research to provide support for doctoral dissertation research. The purposes of the program are to advance education research by outstanding minority graduate students and to improve the quality and diversity of university faculties. This program offers doctoral fellowships to enhance the competitiveness of outstanding minority scholars for academic appointments at major research universities. It supports fellows conducting education research and provides mentoring and guidance toward the completion of their doctoral studies.
Requirements: This program is open to U.S. citizens and permanent residents. Fellows will be required to provide proof of advancement to candidacy at the beginning of the grant period. In the letter of recommendation from the applicant's advisor, the dissertation advisor will attest to the proposal constituting research for the doctoral degree. Applicants must work full-time on their dissertations and course requirements. This program is targeted for members of racial and ethnic groups historically underrepresented in higher education (e.g., African Americans, Alaskan Natives, American Indians, Asian Americans, Hispanics or Latinos, and Native Hawaiian or Pacific Islanders). Each fellowship award is for one year, beginning July 1 or later, and is nonrenewable. Fellowships are awarded for doctoral dissertation research conducted under faculty sponsorship in any accredited university in the United States.
Geographic Focus: All States
Date(s) Application is Due: Nov 15
Contact: George Wimberly; (202) 238-3200; fellowships@aera.net
Internet: http://www.aera.net/ProfessionalOpportunitiesFunding/FundingOpportunities/AERAMinorityFellowshipProgram/tabid/10243/Default.aspx
Sponsor: American Educational Research Association
1430 K Street, NW, Suite 1200
Washington, DC 20005

AERA Research Grants 344

The AERA Grants Program, with support from the National Science Foundation (NSF), announces calls for proposals each year for its Research Grants competition. The program seeks to stimulate research on U.S. education issues using data from the large-scale, national and international data sets supported by the National Center for Education Statistics (NCES), NSF, and other federal agencies, and to increase the number of education researchers using these data sets. The program supports research projects that are quantitative in nature, include the analysis of existing data from NCES, NSF or other federal agencies, and have U.S. education policy relevance. AERA invites education-related research proposals using NCES, NSF, and other federal databases. Research Grants are available for faculty at institutions of higher education, postdoctoral researchers, and other doctoral-level scholars. Applications are encouraged from a variety of disciplines, such as but not limited to, education, sociology, economics, psychology, demography, statistics, and psychometrics. Awards for Research Grants are up to $20,000 for 1-year projects, or up to $35,000 for 2-year projects.
Requirements: Applicants may be U.S. citizens or U.S. permanent residents. Non-U.S. citizens working at a U.S. institution are also eligible to apply. Applicants must have received the doctoral degree by the start date of the grant. Underrepresented racial and ethnic minority researchers are strongly encouraged to apply. Applicants are encouraged to submit proposals that: develop or benefit from new quantitative measures or methodological approaches for addressing education issues; include interdisciplinary teams with subject matter expertise, especially when studying science, technology, engineering and mathematics (STEM) learning; analyze TIMSS, PISA, or other international data resources; include the integration and analysis of more than one data set. Proposals for Research Grants will be reviewed twice a year, once in the fall and once in the winter.
Restrictions: In accordance with AERA's agreement with the funding agencies, institutions may not charge indirect costs or overhead on these awards.
Geographic Focus: All States
Date(s) Application is Due: Jan 24; Oct 1
Amount of Grant: Up to 35,000 USD
Contact: Kevin Dieterle, Program Manager; (202) 238-3200; fax (202) 238-3250; kdieterle@aera.net or grantsprogram@aera.net
Internet: http://www.aera.net/ProfessionalOpportunitiesFunding/FundingOpportunities/AERAGrantsProgram/ResearchGrants/tabid/12813/Default.aspx
Sponsor: American Educational Research Association
1430 K Street, NW, Suite 1200
Washington, DC 20005

AES-Grass Young Investigator Travel Awards 345

To recognize outstanding young investigators conducting research in basic or clinical neuroscience related to epilepsy, the Grass Foundation and the American Epilepsy Society (AES) have partnered to present the Young Investigator Travel Awards for up to eight candidates to help support travel costs to present their research at the annual AES meeting. The AES will announce the awardees and their talk/poster titles in an award ceremony during the annual program. Following the initial review of application abstracts, as part of the application process, a recommendation may be required from one of the applicant's trainers/mentors. This will be solicited by the AES review committee should the application be deemed competitive for the award. This recommendation should include the applicant's research focus, role in the abstract, promise as a future investigator, and any other pertinent information.
Requirements: Candidates will be investigators early in their careers, including graduate students, postdoctoral fellows, residents, clinical fellows or junior faculty (M.D., M.D./Ph.D., Ph.D., or the equivalent), who are conducting basic or clinical research relating to epilepsy. Candidates must also be no more than five years out of postgraduate training. Applicants should also be first authors on an abstract submitted to the AES. Applicants can self-nominate, or nominations from second parties will be accepted. Applicants submit an abstract through the AES online submission program as first author, then indicate the Junior Investigator Travel Award box. Candidates must be present at the AES Annual Meeting.
Geographic Focus: All States
Date(s) Application is Due: Jun 15
Amount of Grant: 1,000 USD
Contact: Cheryl-Ann Tubby, Assistant Executive Director; (860) 586-7505, ext. 542; fax (860) 586-7550; ctubby@aesnet.org
Internet: http://www.aesnet.org/research/aes-sponsored-grant/grass-travel-award/
Sponsor: American Epilepsy Society
342 North Main Street
West Hartford, CT 06117-2507

AES Robert S. Morison Fellowship 346

The American Epilepsy Society, in partnership with the Grass Foundation, offers the Robert S. Morison Fellowship for the training of academic clinician scientists in epilepsy. This two-year post-doctoral Fellowship will be awarded to a promising young investigator possessing an M.D. degree who intends to continue training in basic science in an epilepsy research laboratory. The Fellowship was created to honor the contributions of Dr. Morison, one of the founding Trustees of The Grass Foundation. The Fellowship funds $40,000 per year for two years of salary with an additional $10,000 (maximum) per year to be used for institutional fringe benefits plus $1,000 for the Fellow to travel to the AES Annual Meeting to present his/her results. The award is announced and presented during the AES Annual Meeting.
Requirements: Applicants must apply at the Epilepsy Foundation website for either a Post-doctoral Fellowship or a Research and Training Fellowship for Clinicians. Candidates should note in the application their desire to be considered specifically for the Morison Fellowship. Applicants for this Fellowship can at the same time be considered for a Fellowship in the category under which he/she applied.
Geographic Focus: All States, All Countries
Amount of Grant: Up to 50,000 USD
Contact: Cheryl-Ann Tubby, Assistant Executive Director; (860) 586-7505, ext. 542; fax (860) 586-7550; ctubby@aesnet.org
Internet: http://www.aesnet.org/go/research/aes-sponsored-grant-program/grass-foundation/aes-robert-s-morison-fellowship
Sponsor: American Epilepsy Society
342 North Main Street
West Hartford, CT 06117-2507

AES Susan S. Spencer Clinical Research Epilepsy Training Fellowship 347

The Spencer Fellowship is a two-year fellowship to support clinical research training in the field of epilepsy. The Fellowship is supported by the American Academy of Neurology Foundation (AAN), the American Epilepsy Society, and the Epilepsy Foundation. Each Fellowship will consist of $55,000 per year for two years, plus $10,000 per year for tuition to support formal education in clinical research methodology at the applicant's institution or elsewhere. Supplementation of the stipend with other grants or by the fellowship institution is permissible, but Fellows may not accept other Fellowships, similar awards, or have another source of support for more than 50% of their research salary. Only direct costs will be funded by this award. Detailed information about the application and its process is available at the AES website. All materials must be submitted online.
Requirements: For the purpose of this Fellowship, clinical research is defined as "patient-oriented research conducted with human subjects, or translational research specifically designed to develop treatments or enhance diagnosis of neurological disease." These areas of research include epidemiologic or behavioral studies, clinical trials, studies of disease mechanisms, the development of new technologies, and health services and outcomes research." Disease related studies not directly involving humans or human tissue also are

encouraged if the primary goal is the development of therapies, diagnostic tests, or other tools to prevent or mitigate neurological diseases. Applicants must be AAN members interested in an academic career in clinical research who have completed a residency or a post-doctoral Fellowship (for a Ph.D.) within the past five years. Those early in their clinical research careers will be given priority.
Geographic Focus: All States
Date(s) Application is Due: Oct 1
Amount of Grant: 65,000 USD
Contact: Terry Heinz, Grants Administrator; (651) 695-2746; theinz@aan.com
Internet: http://www.aesnet.org/go/research/aes-sponsored-grant-program
Sponsor: American Epilepsy Society
342 North Main Street
West Hartford, CT 06117-2507

Aetna Foundation Integrated Health Care Grants 348

The Aetna Foundation is advancing an integrated health care agenda that focuses on care coordination and builds on a foundation of strong primary care. The Aetna Foundation seeks to support projects that promote evidence-based models of care coordination that can lead to high-quality, patient-centered health care services, improve health outcomes and lower costs. The Foundation's goal is to demonstrate the key components, best practices and benefits of care coordination that is centered on strong primary care. Examples of grants supported include projects or studies that: develop standards and metrics of care coordination in ambulatory care settings; evaluate models of care coordination that enhance providers' communication with each other and with their patients and lead to improved patient outcomes and experiences with their care; evaluate methods of care coordination for engaging patients as partners in their care, particularly in the management of chronic conditions; identify best practices to align financial and other incentives for achieving well-coordinated care; and identify and assess models of care coordination that reduce the cost of care while improving patient health outcomes.
Requirements: Grants will be made only to nonprofit organizations with evidence of IRS 501(c)3 designation or de facto tax-exempt status.
Restrictions: The Foundation generally does not fund: endowment or capital costs, including construction, renovation, or equipment; direct delivery of reimbursable health care services; basic biomedical research; grants or scholarships to individuals; work for which results and impact cannot be measured; advertising; golf tournaments; advocacy, political causes or events; sacramental or theological functions of religious organizations; operational expenses; or existing deficits.
Geographic Focus: All States
Amount of Grant: 25,000-100,000 USD
Contact: Melenie O. Magnotta, Grants Manager; (860) 273-1012 or (860) 273-0123; fax (860) 273-7764; aetnafoundation@aetna.com
Internet: http://www.aetna-foundation.org/foundation/aetna-foundation-programs/integrated-healthcare/index.html
Sponsor: Aetna Foundation
151 Farmington Avenue
Hartford, CT 06156-3180

Aetna Foundation Minority Scholars Grants 349

Diversifying the next generation of health care professionals and researchers is an important part of the Aetna Foundation's efforts to support high-quality health care for all population groups. Founded in 2010, the AcademyHealth/Aetna Foundation Minority Scholars Program provides professional development, mentoring and networking opportunities for graduate-level students, post-doctoral fellows and other researchers to attract more men and women from underrepresented groups to the field of racial and ethnic disparities research in health outcomes and access to health care.
Restrictions: The Foundation generally does not fund: endowment or capital costs, including construction, renovation, or equipment; direct delivery of reimbursable health care services; basic biomedical research; grants or scholarships to individuals; work for which results and impact cannot be measured; advertising; golf tournaments; advocacy, political causes or events; sacramental or theological functions of religious organizations; operational expenses; or existing deficits.
Geographic Focus: All States
Contact: Melenie O. Magnotta, Grants Manager; (860) 273-1012 or (860) 273-0123; fax (860) 273-7764; aetnafoundation@aetna.com
Internet: http://www.aetna-foundation.org/foundation/aetna-foundation-programs/scholars/index.html
Sponsor: Aetna Foundation
151 Farmington Avenue
Hartford, CT 06156-3180

Aetna Foundation National Medical Fellowship in Healthcare Leadership 350

Diversifying the next generation of health care professionals and researchers is an important part of the Aetna Foundation's efforts to support high-quality health care for all population groups. In 2011, in partnership with National Medical Fellowships (NMF), the Aetna Foundation established the Aetna Foundation/NMF Healthcare Leadership program to provide scholarships to second- and third-year medical students from underrepresented minority groups. Recipients commit to practice medicine in medically underserved communities and are distinguished by their community service and leadership potential.
Restrictions: The Foundation generally does not fund: endowment or capital costs, including construction, renovation, or equipment; direct delivery of reimbursable health care services; basic biomedical research; grants or scholarships to individuals; work for which results and impact cannot be measured; advertising; golf tournaments; advocacy, political causes or events; sacramental or theological functions of religious organizations; operational expenses; or existing deficits.
Geographic Focus: All States
Contact: Melenie O. Magnotta, Grants Manager; (860) 273-1012 or (860) 273-0123; fax (860) 273-7764; aetnafoundation@aetna.com
Internet: http://www.aetna-foundation.org/foundation/aetna-foundation-programs/scholars/index.html
Sponsor: Aetna Foundation
151 Farmington Avenue
Hartford, CT 06156-3180

AFAR CART Fund Grants 351

The goal of the CART Fund is to encourage exploratory and developmental AD research projects within the United States by providing support for the early and conceptual plans of those projects that may not yet be supported by extensive preliminary data but have the potential to substantially advance biomedical research. This proposal should be distinct from those projects designed to increase knowledge in a well established area unless it is intended to extend previous discoveries toward new directions or applications. Applications may encompass a project period of up to two years with a combined budget for direct cost up to $250,000. No indirect costs are allowed. Domestic public and private institutions are eligible, such as universities, colleges, hospitals and laboratories. This is for NEW projects only. Up to two $250,000 awards will be available this year. Dr. John Trojanowski will chair a scientific review group that will triage the Letters-of-Intent and select a maximum of fifteen deemed to have the highest merit. Those selected will be invited to submit a subsequent standard grant application from which the final recommendation will be made by the review group. The final selection will be made by the CART Fund selection committee. Notifications of the finalists (maximum of 15) will be mailed by January 7th. Investigators whose proposals are accepted for further consideration will be sent a full application form which must be completed and returned by February 16th.
Requirements: The CART Fund is inviting interested applicants from within the United States to submit a Letter-of-Intent that includes sufficient detail to communicate the importance of your study as well as information on its feasibility. Submit email questions and Letters-of-Intent to Dr. James B. Puryear, Chairman, CART Grants Program: jimpuryear@comcast.net.
Restrictions: Applications will be deemed ineligible from for-profit organizations as well as those already supported by regular or program grants.
Geographic Focus: All States
Date(s) Application is Due: Feb 16
Amount of Grant: 250,000 USD
Samples: Sanjay W. Pimplikar, Ph.D.-The Cleveland Clinic, OH; Todd E. Golde, M.D., Ph.D.-Mayo Clinic, Jacksonville, FL; Claudio Soto, Ph.D.-University of Texas Medical Branch, TX;
Contact: Grants Manager; (212) 703-9977 or (888) 582-2327; fax (212) 997-0330; grants@afar.org or info@afar.org
Internet: http://www.afar.org/CART.html
Sponsor: American Federation for Aging Research
55 West 39th Street, 16th Floor
New York, NY 10018

AFAR Medical Student Training in Aging Research Program 352

The Medical Student Training in Aging Research Program (MSTAR) provides medical students, early in their training, with an enriching experience in aging related research and geriatrics, under the mentorship of top experts in the field. This program introduces students to research and academic experiences that they might not otherwise have during medical school. Students participate in an eight to twelve week structured research, clinical, and didactic program in geriatrics, appropriate to their level of training and interests. Students may train at a National Training Center supported by the National Institute on Aging or, for a limited number of medical schools, at their own institution. Refer to the AFAR website for a complete listing of the participating institutions.
Requirements: Any allopathic or osteopathic medical student in good standing, who will have successfully completed one year of medical school at a U.S. institution by June, may apply. Applicants must be citizens or non-citizen nationals of the United States, or must have been lawfully admitted for permanent residence (i.e., in possession of a currently valid Alien Registration Receipt Card I-551, or some other legal verification of such status.)
Restrictions: Individuals on temporary or student visas and individuals holding PhD, MD, DVM, or equivalent doctoral degrees in the health sciences are not eligible.
Geographic Focus: All States
Date(s) Application is Due: Feb 6
Contact: Grants Manager; (212) 703-9977 or (888) 582-2327; fax (212) 997-0330; grants@afar.org or info@afar.org
Internet: http://www.afar.org/medstu.html
Sponsor: American Federation for Aging Research
55 West 39th Street, 16th Floor
New York, NY 10018

AFAR Paul Beeson Career Development Awards in Aging Research for the Island of Ireland 353

The Paul Beeson Career Development Awards In Aging Research Program offers faculty development awards to outstanding junior physician faculty committed to academic careers in aging-related research, teaching, and practice. The goals of the program are: to encourage and assist the development of future leaders in the field of aging by supporting faculty members early in their careers to gain additional research training as needed and to establish independent programs in aging research; to deepen the commitment of research

institutions to academic research in aging and to translating research outcomes to geriatric medicine by involving mentor and recipient in establishing and advancing the recipient's career in aging research; to expand medical research on aging broadly defined as including the biology of aging, maintenance of health and independence in old age, diseases and disabilities of old age and issues in their clinical management, and systems of care for the elderly. To maximize the educational/training opportunities of this fellowship program, Beeson Ireland Scholars are encouraged to spend three to six months abroad with a research team in the U.S. in the second or the third year of the award. The goals of this rotation are varied and can be for purposes of training in techniques, data collection, data analysis, etc. to enhance the Scholars' transition to independence. This is unique to the Beeson Ireland Program. To date, the program has provided awards to 149 very promising junior faculty at institutions in the United States as well as 3 junior faculty on the Island of Ireland. The program was established in 1994 by The Atlantic Philanthropies, The John A. Hartford Foundation, The Commonwealth Fund and The Starr Foundation, and administered by the American Federation for Aging Research (AFAR.) In 2004, The Atlantic Philanthropies, The John A. Hartford Foundation, The Starr Foundation and AFAR entered a partnership with The National Institute on Aging (NIA.) Until 2006, the program soley funded Scholars at U.S.-based institutions but beginning in 2007, with support from The Atlantic Philanthropies, Scholars are also funded in Ireland (The Republic of Ireland and Northern Ireland.) The scholar will receive a grant for three years. The salary will be at specialty registrar/clinical lecturer level to protect a minimum of 75% of the scholar's time for research, with the remainder available annually for research support, including an institutional overhead on the research support component.
Requirements: Nominations for the Paul Beeson Career Development Awards In Aging Research Program are to be made by the deans of medical schools (or equivalent) in Ireland (the Republic of Ireland and Northern Ireland.) Institutions may submit as many applications as they wish. To be eligible for nomination, a candidate must: be a U.S. physician or an Irish citizen physician of the Republic of Ireland, a UK citizen of Northern Ireland, an EU citizen living and working in the Island of Ireland or a non-EU citizen resident in the island of Ireland with a valid work permit; commit at least 75% of his/her full-time professional effort to the goals of this award; be a medical graduate who has recently completed or is about to complete a doctoral degree (PhD/MD) and be undertaking or have recently completed higher clinical training; have at least one research publication in a high-impact journal. If the candidate is not the first author, he/she must provide a short description of his/her contribution to the research and the writing of the paper; this should be substantial. For each scholar, a senior faculty member at the scholar's institution must be selected to serve as a mentor to help guide the scholar's research and career planning and provide access to organizations, programs, and colleagues helpful to the scholar's efforts. More than one mentor may be selected. Letters of endorsement including specific information on institutional support for the scholar should be provided by the dean, the relevant department chairperson (or equivalent-may be one person), and the mentor. In addition, three letters of reference should be provided by other faculty members and/or senior professionals with whom the scholar has worked and who are well acquainted with his/her capabilities, potential, accomplishments, and commitment. All candidates must submit applications endorsed by the Dean of School of Medicine (or equivalent).
Geographic Focus: All States
Date(s) Application is Due: Jan 22
Contact: Grants Manager; (212) 703-9977 or (888) 582-2327; fax (212) 997-0330; beeson@afar.org
Internet: http://afar.org/BeesonIreland.html
Sponsor: American Federation for Aging Research
55 West 39th Street, 16th Floor
New York, NY 10018

AFAR Research Grants **354**

The major goal of this program is to assist in the development of the careers of junior investigators committed to pursuing careers in the field of aging research. Of particular interest are research projects concerned with understanding the basic mechanisms of aging; projects investigating age-related diseases, especially if approached from the point of view of how basic aging processes may lead to these outcomes; and projects concerning mechanisms underlying common geriatric functional disorders, as long as these include connections to fundamental problems in the biology of aging. Examples of promising areas of research include: aging and immune function; genetic control of longevity; neurobiology and neuropathology of aging; invertebrate or vertebrate animal models; cardiovascular aging; aging and cellular stress resistance; metabolic and endocrine changes; age-related changes in cell proliferation; caloric restriction and aging; DNA repair and control of gene expression; biology of the menopause; and aging and apoptosis. Application and instruction sheet are available online.
Requirements: The applicant must be an independent investigator with assigned independent space and must be within the first four years of a junior faculty appointment (instructor, assistant professor or equivalent) by July 1st.
Restrictions: The program does not provide support for postdoctoral fellows in the laboratory of a senior investigator; investigators who have already received major independent funding for research on aging, such as an R01 grant or a grant of equal to or greater than $100,000 from another private funding source; senior faculty, i.e., at the rank of associate professor level or higher; and projects that deal strictly with clinical problems, such as the diagnosis and treatment of disease, health outcomes, or the social context of aging. Former AFAR grant recipients are not eligible to reapply. Applicants for the Glenn/AFAR Breakthroughs in Gerontology (BIG) program cannot also submit an application for this research grant.
Geographic Focus: All States
Date(s) Application is Due: Dec 16
Amount of Grant: 75,000 USD
Contact: Grants Manager; (212) 703-9977 or (888) 582-2327; fax (212) 997-0330; grants@afar.org or info@afar.org
Internet: http://afar.org/afar99.html
Sponsor: American Federation for Aging Research
55 West 39th Street, 16th Floor
New York, NY 10018

Affymetrix Corporate Contributions Grants **355**

Affymetrix makes charitable donations to nonprofit organizations through its corporate philanthropy program. The company focuses its giving in three main areas: education—programs that support science and math education, focusing on K-12 students and their teachers; ethics—organizations that help foster an ongoing public dialog about genetic-related ethics; and cancer research and advocacy—nonprofit organizations working in the areas of disease research and advocacy, with a specific emphasis on cancer. Requests are reviewed four times per year. The average grant size is approximately $2500.
Requirements: 501(c)3 nonprofit organizations are eligible.
Restrictions: The corporation does not provide funding for advertising journals or booklets; fundraising events such as telethons, walkathons, and races; specific performances or concerts; sporting events; endowment campaigns; film, video, television, or radio projects; grants to individuals; political causes or candidates; or organizations that practice discrimination or limit membership on the basis of race, creed, gender, age, sexual orientation, or national origin.
Geographic Focus: All States
Contact: Contributions Manager, outreach@affymetrix.com
Internet: http://www.affymetrix.com/corporate/outreach/corporate.affx
Sponsor: Affymetrix
3420 Central Expressway
Santa Clara, CA 95051

AFP Research Grants **356**

The council awards annual research grants to individuals who wish to undertake research projects that enrich our knowledge of philanthropic fundraising and improve its practice. Preference will be given to research focused on Canada and the United States, regardless of where the research is based. Guidelines are available online.
Geographic Focus: All States
Date(s) Application is Due: Mar 1; Sep 1
Amount of Grant: Up to 10,000 USD
Contact: Administrator; (703) 519-8469; fax (703) 684-0540; cwilliams@afpnet.org
Internet: http://www.afpnet.org/tier3_cd.cfm?folder_id=930&content_item_id=1531
Sponsor: Association of Fundraising Professionals
1101 King Street, Suite 700
Alexandria, VA 22314

African American Heritage Grants **357**

Grants are awarded to assist organizations in the preservation and promotion of historic African American properties and sites in Indiana. Awards are made on a four-to-one matching basis, funding 80% of the total project cost up to $2,500, whichever is less.
Requirements: Civic groups, schools, libraries, historical societies, and other nonprofit agencies are eligible to apply for grants for organizational assistance, studies assisting in or leading to the preservation of a historic African American place, and programs promoting the preservation, interpretation, and/or visitation of a historic African American place. Contact the regional community preservation specialist that serves your community (see website for list of regional offices or contact the state headquarters office) for guidelines and forms.
Restrictions: Properties must be located in Indiana.
Geographic Focus: Indiana
Contact: Carla Jones, Receptionist, State Headquarters; (317) 639-4534; fax (317) 639-6734; info@historiclandmarks.org
Internet: http://www.historiclandmarks.org/help/grants.html
Sponsor: Historic Landmarks Foundation of Indiana
340 W Michigan Street
Indianapolis, IN 46202

AFRL/NRC Resident Research Associateship (RRA) Program **358**

The council administers postdoctoral and senior research awards through its associateship programs office, part of the Policy and Global Affairs Division. The programs are sponsored by 30 federal laboratories and NASA Research Centers at over 100 locations in the United States and overseas. Awards are made to doctoral level scientists and engineers who can apply their special knowledge and research talents to research areas that are of interest to them and to the host laboratories and centers. Each awardee works in collaboration with a research adviser, who is a staff member of the federal laboratory. In addition to traditional postdoctoral and senior awards, the associateship programs offer summer faculty awards, combined teaching and research awards, research management awards, opportunities in the NASA Astrobiology Institute, and international opportunities including a joint program with the Alexander von Humboldt Foundation.
Requirements: Postdoctoral research associateships are awarded to U.S. citizens and permanent residents who have held doctorates for less than five years at the time of application. Senior research associateships are awarded to individuals who have held doctorates for more than five years, have significant research experience, and are recognized internationally as experts in their specialized fields as evidenced by numerous publications in reviewed journals, invited presentations, authorship of books or book chapters, and professional society awards of international stature. U.S. citizenship is not a requirement for senior research associateships.

Geographic Focus: All States
Date(s) Application is Due: Feb 1; May 1; Aug 1; Nov 1
Amount of Grant: 36,000-65,000 USD
Contact: Programs Coordinator; (202) 334-2760; fax (202) 334-2759; rap@nas.edu
Internet: http://www4.nationalacademies.org/pga/rap.nsf
Sponsor: National Research Council
550 Fifth Street NW
Washington, DC 20001

AFUW Diamond Jubilee Bursary 359
The program offers a bursary to assist men and women to undertake and complete higher degrees at Australian universities and to encourage advanced scholarship and original research by university men and women. The bursary may be used for the purchase or hire of equipment, field trip or research expenses, thesis publication costs, fees, dependent care expenses incurred because of study commitments, or short-term assistance with living expenses. The Bursary must be used within twelve months of the date of the award. There is no restriction on field of study or citizenship. The bursary may be held concurrently with awards offered by organizations other than the AFUW-SA Inc. Trust Fund. Selection of winners is based primarily on academic merit, but also on financial need, the importance of the purpose for which the bursary will be used to the progress or completion of the degree, community activities and other interests.
Requirements: Applications are invited from men and women who are enrolled in MA or PhD programs at Australian universities and who are Australian or overseas citizens.
Restrictions: Applicants may not be in full-time paid employment or on fully paid study leave during the tenure of the bursary.
Geographic Focus: All States, Australia
Date(s) Application is Due: Mar 2
Contact: Gaynor Reeves, Fellowships Convener; (02) 4952 3174; fax (02) 4952 3070; gay.reeves@bigpond.com or gaynor.reeves@gmail.com
Internet: http://www.afuwsa-bursaries.com.au/
Sponsor: Australian Federation of University Women-South Australia Trust Fund
GP.O. Box 634
Adelaide, SA 5001 Australia

AFUW Doreen McCarthy, Barbara Crase, Cathy Candler & Brenda Nettle Bursaries 360
The program offers a bursary to assist women to undertake and complete higher degrees at Australian universities and to encourage advanced scholarship and original research by university women. The bursary may be used for the purchase or hire of equipment, field trip or research expenses, thesis publication costs, fees, dependent care expenses incurred because of study commitments, or short-term assistance with living expenses. The Bursary must be used within twelve months of the date of the award. Selection of winners is based primarily on academic merit, but also on financial need, the importance of the purpose for which the bursary will be used to the progress or completion of the degree, community activities and other interests.
Requirements: Applications are invited from women who are enrolled in MA or PhD programs at Australian universities and who are Australian or overseas citizens.
Restrictions: Applicants may not be in full-time paid employment or on fully paid study leave during the tenure of the bursary.
Geographic Focus: All States, Australia
Date(s) Application is Due: Mar 2
Contact: Gaynor Reeves, Fellowships Convener; (02) 4952 3174; fax (02) 4952 3070; gay.reeves@bigpond.com or gaynor.reeves@gmail.com
Internet: http://www.afuwsa-bursaries.com.au/
Sponsor: Australian Federation of University Women-South Australia Trust Fund
GP.O. Box 634
Adelaide, SA 5001 Australia

AFUW Fellowships 361
Regular AFUW Fellowships are offered annually through special donations from AFUW-Qld. The Fellowships are available to women who are Australian citizens or permanent residents and enrolled for a PhD degree by research in any Australian university. Funding is to be used for a specific project or purpose for PhD research in any field. Awards are made both by AFUW Inc and by the State and Territory Associations that comprise the Federation.
Requirements: Applications must be made electronically on the form supplied, either as a Word document or in rich text format.
Geographic Focus: All States, Australia
Date(s) Application is Due: May 31
Contact: Gaynor Reeves, Fellowships Convener; (02) 4952 3174; fax (02) 4952 3070; gay.reeves@bigpond.com or gaynor.reeves@gmail.com
Internet: http://www.afuw.org.au/ScholarshipsFS.htm
Sponsor: Australian Federation of University Women-South Australia Trust Fund
53 Verulam Road
Lambton, NSW 2299 Australia

AFUW Georgina Sweet Fellowships 362
Georgina Sweet Fellowships are available to women postgraduates enrolled in or accepted by an Australian university for advanced post-first degree study or research in any field. Members of the Australian Federation of University Women (AFUW Inc.) should forward their application direct to the Fellowships Officer. Members of other International Federation of University Women (IFUW) Federations or Associations must forward their applications through their National Federation or Association.
Requirements: Applicants must be a financial member of a National Federation or Association of the International Federation of University Women or an independent member of IFUW.
Geographic Focus: All States, Australia
Date(s) Application is Due: May 31
Contact: Gaynor Reeves, Fellowships Convener; (02) 4952 3174; fax (02) 4952 3070; gay.reeves@bigpond.com or gaynor.reeves@gmail.com
Internet: http://www.afuw.org.au/ScholarshipsFS.htm
Sponsor: Australian Federation of University Women-South Australia Trust Fund
53 Verulam Road
Lambton, NSW 2299 Australia

AFUW Heather Latz Bursary 363
The AFUW Heather Latz Bursary is intended to assist women to undertake and complete higher degrees at Australian universities and to encourage advanced scholarship and original research by university women. The bursary may be used for the purchase or hire of equipment, field trip or research expenses, thesis publication costs, fees, dependent care expenses incurred because of study commitments, or short-term assistance with living expenses. The Bursary must be used within twelve months of the date of the award. There is no restriction on field of study or citizenship. The bursary may be held concurrently with awards offered by organizations other than the AFUW-SA Inc. Trust Fund. Selection of winners is based primarily on academic merit, but also on financial need, the importance of the purpose for which the bursary will be used to the progress or completion of the degree, community activities and other interests.
Requirements: Applications are invited from women who are enrolled in MA or PhD programs at Australian universities and who are Australian or overseas citizens.
Restrictions: Applicants may not be in full-time paid employment or on fully paid study leave during the tenure of the bursary.
Geographic Focus: All States, Australia
Date(s) Application is Due: Mar 2
Contact: Gaynor Reeves, Fellowships Convener; (02) 4952 3174; fax (02) 4952 3070; gay.reeves@bigpond.com or gaynor.reeves@gmail.com
Internet: http://www.afuwsa-bursaries.com.au/
Sponsor: Australian Federation of University Women-South Australia Trust Fund
GP.O. Box 634
Adelaide, SA 5001 Australia

AFUW Padnendadlu Graduate Bursary 364
The aim of this bursary is to assist Aboriginal and/or Torres Strait Islander Australian women towards the completion of either a Graduate Diploma or Graduate Certificate at a South Australian university. A Bursary may be used for the purchase or hire of equipment, field trip or research expenses, thesis publication costs, fees incurred because of study commitments, conference attendance or short-term assistance with living expenses. The Bursary must be used within twelve months of the date of the award. The amount of the bursary will not exceed $2,000 but may be less depending on the requirements of the successful applicant.
Requirements: This application form should be used by Aboriginal and/or Torres Strait Islander Australian women undertaking a postgraduate degree at a South Australian university either via research or coursework.
Restrictions: Applicants may not be in full-time paid employment or on fully paid study leave during the tenure of the bursary.
Geographic Focus: All States, Australia
Date(s) Application is Due: Mar 2
Contact: Gaynor Reeves, Fellowships Convener; (02) 4952 3174; fax (02) 4952 3070; gay.reeves@bigpond.com or gaynor.reeves@gmail.com
Internet: http://www.afuwsa-bursaries.com.au/
Sponsor: Australian Federation of University Women-South Australia Trust Fund
GP.O. Box 634
Adelaide, SA 5001 Australia

AFUW Padnendadlu Postgraduate Bursary 365
The aim of the Padnendadlu Postgraduate Bursary is to assist Aboriginal and/or Torres Strait Islander Australian women to undertake and complete higher degrees at South Australian universities and to encourage advanced scholarship and original research. A Bursary may be used for the purchase or hire of equipment, field trip or research expenses, thesis publication costs, fees incurred because of study commitments, conference attendance or short-term assistance with living expenses. The Bursary must be used within twelve months of the date of the award.
Requirements: This application form should be used by Aboriginal and/or Torres Strait Islander Australian women undertaking a postgraduate degree at a South Australian university either via research or coursework.
Restrictions: Applicants may not be in full-time paid employment or on fully paid study leave during the tenure of the bursary.
Geographic Focus: All States, Australia
Date(s) Application is Due: Mar 2
Contact: Gaynor Reeves, Fellowships Convener; (02) 4952 3174; fax (02) 4952 3070; gay.reeves@bigpond.com or gaynor.reeves@gmail.com
Internet: http://www.afuwsa-bursaries.com.au/
Sponsor: Australian Federation of University Women-South Australia Trust Fund
GP.O. Box 634
Adelaide, SA 5001 Australia

AFUW Padnendadlu Undergraduate Bursary 366

The bursaries are envisaged as a short-term aid to assist with the completion of the academic year. Examples of purposes for which the Bursary may be used are equipment or book costs, fees incurred because of study commitments, short-term assistance with living expenses or any other purpose that will assist with the completion of the academic qualification for which the applicant is enrolled. The Bursary may be held concurrently with awards offered by organizations other than The AFUW-SA Inc. Trust Fund.
Requirements: This Bursary is open to Aboriginal and/or Torres Strait Islander Australian women undergraduates at any of the South Australian universities. Applicants must be undertaking subjects for the final year of their Bachelor degree, or undertaking an Honors year.
Restrictions: Women in full-time paid employment or on fully-paid study leave during the tenure of the Bursary are not eligible to apply, nor are women who won this bursary in the previous year.
Geographic Focus: All States, Australia
Date(s) Application is Due: Mar 2
Contact: Gaynor Reeves, Fellowships Convener; (02) 4952 3174; fax (02) 4952 3070; gay.reeves@bigpond.com or gaynor.reeves@gmail.com
Internet: http://www.afuwsa-bursaries.com.au/
Sponsor: Australian Federation of University Women-South Australia Trust Fund
GP.O. Box 634
Adelaide, SA 5001 Australia

AFUW South Australia Jean Gilmore, Thenie Baddams & Daphne Elliott Bursaries 367

The program offers a bursary to assist women to undertake and complete higher degrees at Australian universities and to encourage advanced scholarship and original research by university women. The bursary may be used for the purchase or hire of equipment, field trip or research expenses, thesis publication costs, fees, dependent care expenses incurred because of study commitments, or short-term assistance with living expenses. The bursary must be used within 12 months of the date of the award. Bursaries may be shared by two successful applicants.
Requirements: Applications are invited from women who are enrolled in MA or PhD programs at Australian universities and who are Australian or overseas citizens.
Restrictions: Applicants may not be in full-time paid employment or on fully paid study leave during the tenure of the bursary.
Geographic Focus: All States, Australia
Date(s) Application is Due: Mar 2
Contact: Gaynor Reeves, Fellowships Convener; (02) 4952 3174; fax (02) 4952 3070; gay.reeves@bigpond.com or gaynor.reeves@gmail.com
Internet: http://www.afuwsa-bursaries.com.au/
Sponsor: Australian Federation of University Women-South Australia Trust Fund
GP.O. Box 634
Adelaide, SA 5001 Australia

AFUW Winifred E. Preedy Postgraduate Bursary 368

The program offers a bursary to assist women to undertake and complete higher degrees in dentistry or allied fields at Australian universities. The bursary may be used for the purchase or hire of equipment, field trip or research expenses, thesis publication costs, fees, dependent care expenses incurred because of study commitments, or short-term assistance with living expenses. The bursary must be used within 12 months of the date of the award. Bursaries may be shared by two successful applicants.
Requirements: Applications are invited from women who are enrolled in MA or PhD programs in dentistry or a related field at Australian universities and have completed at least one year of a postgraduate degree.
Restrictions: Applicants may not be in full-time paid employment or on fully paid study leave during the tenure of the bursary, nor be previous winners of the Winifred E. Preedy Bursary.
Geographic Focus: All States, Australia
Date(s) Application is Due: Mar 2
Contact: Gaynor Reeves, Fellowships Convener; (02) 4952 3174; fax (02) 4952 3070; gay.reeves@bigpond.com or gaynor.reeves@gmail.com
Internet: http://www.afuwsa-bursaries.com.au/
Sponsor: Australian Federation of University Women-South Australia Trust Fund
GP.O. Box 634
Adelaide, SA 5001 Australia

AGHE Graduate Scholarships and Fellowships 369

The association awards graduate scholarships and fellowships for research on aging. Gerontology and aging studies faculty are invited to nominate qualified students. This program is funded by the AARP Andrus Foundation. Faculty nomination and student application forms are available online. Annual deadline dates may vary; contact program staff for exact dates.
Requirements: Gerontology and aging studies faculty are invited to nominate qualified students for these awards.
Geographic Focus: All States
Amount of Grant: 2,000-15,000 USD
Contact: M. Angela Baker; (202) 289-9806, ext. 125; abaker@aghe.org
Internet: http://www.aghe.org/677858
Sponsor: Association for Gerontology in Higher Education
1220 L Street, NW, Suite 901
Washington, DC 20005-4015

AHAF Alzheimer's Disease Research Grants 370

AHAF funds outstanding scientists and physicians in neurobiology, physiology, pathology, molecular and developmental biology, chemistry, pharmacology, epidemiology, and surgery who are conducting research to better understand and/or treat Alzheimer's disease. Junior or senior investigators will be considered. Grants are awarded for up to two years and are renewable. Grant applications are reviewed by a scientific review committee on a competitive peer-review system.
Requirements: Grants are awarded to universities, medical centers, and independent research institutions. AHAF funds grants for research at nonprofit organizations only. The principal investigator must hold the academic rank of assistant professor (or equivalent) or higher.
Restrictions: Funding is not provided for overhead costs, construction, or building expenses.
Geographic Focus: All States
Date(s) Application is Due: Oct 19
Amount of Grant: Up to 300,000 USD
Contact: Kara Hurst, Grants Coordinator; (800) 437-2423 or (301) 948-3244; fax (301) 258-9454; khurst@ahaf.org
Internet: http://www.ahaf.org/alzdis/research/grants.htm
Sponsor: American Health Assistance Foundation
22512 Gateway Center Drive
Clarksburg, MD 20871

AHAF Macular Degeneration Research Grants 371

The foundation awards research grants to advance study of macular degeneration. Applications are evaluated based on the scientific merit of the proposal, the feasibility of the proposed research, the potential of the research to lead to better understanding and treatment of eye diseases, and the demonstrated ability of the investigator to complete the research. Application materials are available on the Web site.
Geographic Focus: All States
Date(s) Application is Due: Jul 10; Oct 29
Amount of Grant: Up to 50,000 USD
Contact: Kara Hurst, Grants Coordinator; (800) 437-2423 or (301) 948-3244; fax (301) 258-9454; khurst@ahaf.org
Internet: http://www.ahaf.org/macular/research/grants.htm
Sponsor: American Health Assistance Foundation
22512 Gateway Center Drive
Clarksburg, MD 20871

AHAF National Glaucoma Research Grants 372

AHAF funds outstanding scientists and physicians with expertise in cell and molecular biology, physiology, biochemistry, endocrinology, and pharmacology. Grants are awarded on the basis of scientific merit of the proposal, the relevance of the research, and the potential impact of the proposed study on better understanding and/or treatment of glaucoma. Applications are reviewed by a scientific review committee on a competitive peer-review system. AHAF is interested in receiving focused research grant applications from investigators at all stages of their careers. AHAF is particularly interested in new investigators with little or no previous grant support and established investigators with new ideas or directions for their research.
Requirements: AHAF grants are awarded to universities, medical centers, and independent research institutions. AHAF provides grants for research at nonprofit organizations only.
Restrictions: Grants are not made to individuals. Funding is not provided for overhead costs, construction, or building expenses.
Geographic Focus: All States
Date(s) Application is Due: Oct 27
Amount of Grant: Up to 100,000 USD
Contact: Kara Hurst, Grants Coordinator; (800) 437-2423 or (301) 948-3244; fax (301) 258-9454; khurst@ahaf.org
Internet: http://www.ahaf.org/glaucoma/research/glresrch.htm
Sponsor: American Health Assistance Foundation
22512 Gateway Center Drive
Clarksburg, MD 20871

AHEPA Educational Foundation Scholarships 373

The AHEPA Educational Foundation's scholarship programs promote, encourage, induce, and advance education at the college, university, and graduate school level. A number of named undergraduate and graduate scholarships are available. Application and guidelines are available online. Annual deadline dates may vary; contact program staff for exact dates.
Geographic Focus: All States
Date(s) Application is Due: Jun 1
Amount of Grant: 3,000-5,000 USD
Contact: Headquarters; (202) 232-6300; fax (202) 232-2140; ahepa@ahepa.org
Internet: http://www.ahepa.org/educ_foundation/index.html
Sponsor: American Hellenic Educational Progressive Association
1909 Q Street NW, Suite 500
Washington, DC 20009

AHIMA Dissertation Assistance Grants 374

The Dissertation Assistance Award program supports research undertaken as part of an academic program to qualify for a doctorate in areas relevant to health information management (HIM). Each recipient shall be limited to one funded grant per year. Ordinarily, the scope of the proposal shall be such that it can be completed within 18

months from the date of funding. There are no minimum or maximum grant request expectations, but the range of grants historically has been between $5,000 and $10,000. Submissions that address one or more of the AHIMA research priorities will receive priority consideration for funding.
Requirements: To qualify for an award under this program the student (principal investigator) must be enrolled in an accredited doctoral degree program in an area related to HIM (computer science, business management, education, public health, and so forth) and must be an active, associate, or student member of AHIMA. All requirements for the doctoral degree, other than the dissertation, must be completed by the award date.
Geographic Focus: All States
Date(s) Application is Due: Oct 31
Amount of Grant: 5,000-10,000 USD
Contact: Carol Nielsen, Director; (312) 233-1175; carol.nielsen@ahimafoundation.org
Internet: http://www.ahimafoundation.org/Scholarships/dissertation.aspx
Sponsor: Foundation of Research and Education of the American Health Information Management Association
233 N Michigan Avenue, 21st Floor
Chicago, IL 60601

AHIMA Faculty Development Stipends 375
FORE Faculty Development Stipends are intended to encourage excellence in effective teaching and leadership in areas related to e-HIM. Program goals are to: expedite curriculum innovation and development in e-HIM content areas; research and develop new methods of teaching, evaluation, and assessment in the HIM learning process; support faculty in maintaining the highest standards of knowledge and practice in e-HIM; integrate new technologies effectively into the HIM learning process; and strengthen the capacity of HIM academic programs to address emerging issues in design and implementation of the electronic health record at all levels of health care delivery, in support of an electronic health information infrastructure.
Requirements: Application should be for activity planned during the four (4) month period following the expected notification date.
Geographic Focus: All States
Date(s) Application is Due: Mar 14; Jul 11; Oct 3
Contact: Susan H. Fenton, Director of Research; (312) 233-1532 or (312) 233-1100; fax (210) 479-1043; susan.fenton@ahima.org or fore@ahima.org
Internet: http://www.ahimafoundation.org/Scholarships/Faculty.aspx
Sponsor: Foundation of Research and Education of the American Health Information Management Association
233 N Michigan Avenue, 21st Floor
Chicago, IL 60601

AHIMA Grant-In-Aid Research Grants 376
The FORE Grant-in-Aid program is directed toward supporting the development of HIM professionals as leaders in defining and validating the unique body of knowledge encompassed by HIM. The results of these studies not only provide information for the HIM professional to apply in meeting current and future challenges, but also support policy initiatives and the redefinition of the roles of HIM practitioners. Ordinarily, the scope of the proposal shall be such that it can be completed within 18 months from the date of funding. There are no minimum or maximum grant request expectations, but the range of grants historically has been between $15,000 and $40,000. Submissions that address one or more of the AHIMA research priorities will receive priority consideration for funding.
Requirements: The primary or secondary investigator must be an active, associate, or student member of AHIMA. Each recipient shall be limited to one funded grant per year.
Geographic Focus: All States
Date(s) Application is Due: Sep 19
Amount of Grant: 15,000-40,000 USD
Contact: Susan H. Fenton, Director of Research; (312) 233-1532 or (312) 233-1100; fax (210) 479-1043; susan.fenton@ahima.org or fore@ahima.org
Internet: http://www.ahima.org/fore/research/grantinaid.asp
Sponsor: Foundation of Research and Education of the American Health Information Management Association
233 N Michigan Avenue, 21st Floor
Chicago, IL 60601

Ahmanson and Getty Postdoctoral Fellowships 377
Theme-based, residential fellowships are awarded for periods of two consecutive academic quarters for participation in the interdisciplinary, cross-cultural programs of the Center for 17th- & 18th-Century Studies/William Andrews Clark Memorial Library. The theme for each academic year is announced the preceding fall. An organized research unit of the University of California, the center provides a forum for the discussion of central issues in the study of the 17th and 18th centuries, facilitates research and publication, and encourages the creation of programs that advance understanding of this time period. The William Andrews Clark Memorial Library, which is administered by the center, is known for its collections on 17th- and 18th-century Britain, Oscar Wilde and the 1890s, the history of printing, and certain aspects of Western Americana. Application materials are available upon request.
Requirements: Scholars who have received a Ph.D. in the last six years and are engaged in research pertaining to the announced theme are eligible to apply. Fellows are expected to make a substantive contribution to the Center's workshops and seminars. Awards are for one full academic year in residence at the Clark Library.
Geographic Focus: All States
Date(s) Application is Due: Feb 1
Amount of Grant: 35,000 USD
Contact: Coordinator; (310) 206-8552; fax (310) 206-8577; c1718cs@humnet.ucla.edu
Internet: http://www.humnet.ucla.edu/humnet/c1718cs/Postd.htm#AhmGet
Sponsor: Center for 17th- and 18th-Century Studies
310 Royce Hall
Los Angeles, CA 90095-1404

Ahmed Zewail Award in Ultrafast Science and Technology 378
The Award is intended to recognize outstanding and creative contributions to fundamental discoveries or inventions in ultrafast science & technology in areas of physics, chemistry, biology, or related fields. Areas of interest include physics, chemistry, biology, or related fields without regard to age or nationality. The recipient will have conducted original and insightful research that has had a significant impact on the field of ultrafast science and technology. The award will consist of $5,000 and a certificate.
Requirements: Any individual, except a member of the award committee, may submit one nomination or seconding letter for the award in any given year. Nominating documents consist of a letter of not more than 1000 words containing an evaluation of the nominee's accomplishments and a specific identification of the work to be recognized, a biographical sketch including date of birth, and a list of publications and patents authored by the nominee.
Restrictions: Self-nominations are not accepted.
Geographic Focus: All States
Date(s) Application is Due: Nov 1
Amount of Grant: 5,000 USD
Contact: Felicia Dixon, Awards Administrator; (800) 227-5558 or (202) 872-4408; fax (202) 776-8008; f_dixon@acs.org or awards@acs.org
Internet: http://portal.acs.org/portal/acs/corg/content?_nfpb=true&_pageLabel=PP_ARTICLEMAIN&node_id=1319&content_id=CTP_004488&use_sec=true&sec_url_var=region1
Sponsor: American Chemical Society
1155 Sixteenth Street, NW
Washington, DC 20036-4801

Ahn Family Foundation Grants 379
The Ahn Family Foundation, established by Sangwoo and Laura Ahn in 1997, has as its primary purposes to support aging, religion, human and social services, art, culture, education, and the environment. Giving is primarily limited to Connecticut, Massachusetts, and New York, although the Foundation has also supported projects in Maryland and Washington. There are no particular guidelines or deadlines with which to adhere, and applicants should begin by contacting the Foundation in writing. This letter of inquiry should outline the purpose for the request, the mission of the applicant organization, and the amount requested.
Geographic Focus: Connecticut, Massachusetts, New York
Amount of Grant: 1,000-20,000 USD
Contact: Alison D. Ahn, Treasurer; fax (203) 869-4875
Sponsor: Ahn Family Foundation
901 Hillsboro Mile
Hillsboro Beach, FL 33062-2801

AHNS/AAO-HNSF Surgeon Scientist Combined Award 380
The Grant is open to surgeons beginning a clinician-scientist career track to support research in the pathogenesis, pathophysiology, diagnosis, prevention, or treatment of head and neck neoplastic disease. This grant is only available during odd numbered years. Awards are two year, nonrenewable, with a possible $70,000 maximum total costs ($35,000 per year). One award available per year. Letters of intent must be received by December 15, and completed applications must be submitted by January 15.
Requirements: Applicants must be members or candidate members of the American Academy of Otolaryngology-Head and Neck Surgery and/or the American Head and Neck Society.
Geographic Focus: All States
Date(s) Application is Due: Jan 15; Dec 15
Amount of Grant: Up to 70,000 USD
Contact: Stephanie L. Jones, (703) 519-1586 or (703) 836-4444; sljones@entnet.org
Internet: http://www.headandneckcancer.org/research/grants.php
Sponsor: American Head and Neck Society
11300 W Olympic Boulevard, Suite 600
Los Angeles, CA 90064

AHNS/AAO-HNSF Young Investigator Combined Award 381
The purpose of this award is to support a collaborative AHNS/AAO-HNSF research project by fostering the development of contemporary basic or clinical research skills focused on neoplastic disease of the head and neck among new full-time academic surgeons. The award is intended as a preliminary step in clinical investigator career development and is expected to facilitate the recipient's preparation of a more comprehensive individualized research plan suitable for submission to the National Institutes of Health or comparable funding agency. This is a two year, non-renewable award, offering $40,000 maximum ($20,000 per year). One award is available annually. Letters of intent must be received by December 15, and completed applications must be submitted by January 15.
Requirements: Applicants must be physicians with demonstrated potential for excellence in research and teaching and serious commitment to an academic research career in head and neck surgery. Applicants must be members or candidate members of the American Academy of Otolaryngology-Head and Neck Surgery and/or the American Head and Neck Society. Priority will be given to fellows or junior faculty who have completed residencies or fellowships within four years of the application receipt date. All candidates must be sponsored by the Chair of his/her Division or Department and by an official

representative of the institution which would administer the Award and in whose name the application is formally submitted.
Geographic Focus: All States
Date(s) Application is Due: Jan 15; Dec 15
Amount of Grant: Up to 40,000 USD
Contact: Stephanie L. Jones; (703) 519-1586 or (703) 836-4444; sljones@entnet.org
Internet: http://www.headandneckcancer.org/research/grants.php
Sponsor: American Head and Neck Society
11300 W Olympic Boulevard, Suite 600
Los Angeles, CA 90064

AHNS Alando J. Ballantyne Resident Research Pilot Grant 382
The purpose of this award is to support basic, translational, or clinical research projects in head and neck oncology. Clinical or translational research studies are strongly encouraged and should be specifically related to the prevention, diagnosis, treatment, outcomes, or pathophysiology of head and neck neoplastic disease. Research supported by this award should be specifically directed toward the pathogenesis, pathophysiology, diagnosis, prevention, or treatment of head and neck neoplastic disease, and may be either basic or clinical/translational in approach. While not specifically required, proposals which aim to introduce new knowledge and methodology from other disciplines to research in head and neck disease, or which demonstrate collaborative effort with members of other related disciplines are encouraged. Grants range up to $10,000 maximum, are for one year, and are not renewable. Letters of intent must be received by December 15, and completed applications must be submitted by January 15.
Requirements: This grant is open to resident in U.S. or Canadian training programs. Previous AHNS or AAO-HNS Foundation research grant recipients are eligible to compete for this grant. However, candidates who have successfully obtained funding from a private or federal funding agency for the same research are ineligible. Candidates who have applied for support of the same research from other funding sources, and who are notified of an award from both another agency and from AHNS must choose only one of the awards.
Geographic Focus: All States
Date(s) Application is Due: Jan 15; Dec 15
Amount of Grant: Up to 10,000 USD
Contact: Stephanie L. Jones, Assistant Director; (703) 519-1586 or (703) 836-4444; sljones@entnet.org
Internet: http://www.headandneckcancer.org/research/grants.php
Sponsor: American Head and Neck Society
11300 W Olympic Boulevard, Suite 600
Los Angeles, CA 90064

AHNS Pilot Grant 383
The purpose of this award is to support basic, translational, or clinical research projects in head and neck oncology. Clinical or translational research studies are strongly encouraged and should be specifically related to the prevention, diagnosis, treatment, outcomes, or pathophysiology of head and neck neoplastic disease. Research supported by this award should be specifically directed toward the pathogenesis, pathophysiology, diagnosis, prevention, or treatment of head and neck neoplastic disease, and may be either basic or clinical/translational in approach. While not specifically required, proposals which aim to introduce new knowledge and methodology from other disciplines to research in head and neck disease, or which demonstrate collaborative effort with members of other related disciplines are encouraged. Grants range up to $10,000 maximum, are for one year, and are not renewable. Letters of intent must be received by December 15, and completed applications must be submitted by January 15.
Requirements: Candidates for this award should reside in the U.S. or Canada, be medical students, residents, Ph.D.s or faculty members at the rank of associate professor or below.
Geographic Focus: All States
Date(s) Application is Due: Jan 15; Dec 15
Amount of Grant: Up to 10,000 USD
Contact: Stephanie L. Jones; (703) 519-1586 or (703) 836-4444; sljones@entnet.org
Internet: http://www.headandneckcancer.org/research/grants.php
Sponsor: American Head and Neck Society
11300 W Olympic Boulevard, Suite 600
Los Angeles, CA 90064

AHRF Eugene L. Derlacki, M.D. Research Grants 384
The Eugene L. Derlacki, M.D. Grant is awarded for excellence in the field of hearing research. This grant provides $25,000 per year for two years, for a total award of $50,000. It is dedicated to hearing research and is intended for more significant research projects requiring funding that extends beyond the normal AHRF grant of $25,000. Researchers are invited to apply for this grant using the Research Grant Application Guidelines. According to the rules of the grant, the recipient will receive the first year funding of $25,000. After the first year, a progress report must be submitted to the Research Committee. If the report is approved, the second year funding of $25,000 will then be awarded.
Geographic Focus: All States
Date(s) Application is Due: Aug 3
Amount of Grant: 25,000 USD
Contact: Kristen Madhuizen; (312) 726-9670; fax (312) 726-9695; kristen.madhuizen@american-hearing.org or ahrf@american-hearing.org
Internet: http://www.american-hearing.org/research/derlacki_grant.html
Sponsor: American Hearing Research Foundation
8 South Michigan Avenue, Suite 814
Chicago, IL 60603-4539

AHRF Georgia Birtman Grant 385
The American Hearing Research Foundation (AHRF), together with the Northwestern Memorial Foundation (NMF) give the Georgia Birtman Grant, named in honor of long-time supporter of the AHRF. The one-year $75,000 grant ($25,000 each from the Birtman Fund, the AHRF and Northwestern Memorial Foundation) supports the advancement of research and education in otology and neurotology. The grant is awarded to an exceptional researcher in audiology, otology or neurotology who will work in a lab at Northwestern University. The research topic involves some aspect of the diagnosis, treatment, and rehabilitation of hearing and balance disorders related to the inner ear. The research has the potential to generate clinical care innovations, facilitate translational clinical studies, and develop creative educational programs.
Geographic Focus: All States
Date(s) Application is Due: Aug 3
Amount of Grant: Up to 75,000 USD
Contact: Kristen Madhuizen; (312) 726-9670; fax (312) 726-9695; kristen.madhuizen@american-hearing.org or ahrf@american-hearing.org
Internet: http://www.american-hearing.org/research/birtman_grant.html
Sponsor: American Hearing Research Foundation
8 South Michigan Avenue, Suite 814
Chicago, IL 60603-4539

AHRF Regular Research Grants 386
The American Hearing Research Foundation funds five to ten $20,000 research grants each year. Research Grants should relate to the hearing or balance functions of the ear. Both basic and clinical studies may be proposed. Priority is given to providing start-up funds for new projects. Applications are reviewed by a Research Committee and awards begin in January. Applications are due no later that noon on August 3 of the previous year.
Geographic Focus: All States
Date(s) Application is Due: Aug 3
Amount of Grant: Up to 20,000 USD
Contact: Kristen Madhuizen; (312) 726-9670; fax (312) 726-9695; kristen.madhuizen@american-hearing.org or ahrf@american-hearing.org
Internet: http://www.american-hearing.org/research/grant_guidelines.html
Sponsor: American Hearing Research Foundation
8 South Michigan Avenue, Suite 814
Chicago, IL 60603-4539

AHRF Wiley H. Harrison Memorial Research Award 387
The purpose of this award is to support clinical research projects in otology or neurotology designed to increase understanding of hearing disorders. Research supported by this award should be specifically directed toward the clinical identification, diagnosis, prevention, or treatment of diseases, disorders, or conditions of the ear. While not specifically required, proposals which aim to introduce new knowledge and methodology from other disciplines to research in otology or neurotology, or which demonstrate collaborative effort with members of other related disciplines are encouraged. This is a one year, non-renewable grant of $25,000 maximum, with only one award available annually. Completed applications must be submitted by August 3.
Requirements: Candidates for this award should be physicians (M.D.) at the resident, fellow, or junior faculty stage, or PhD scientists. Previous AHRF or AAO-HNS Foundation research grant recipients are eligible to compete for this grant.
Geographic Focus: All States
Date(s) Application is Due: Aug 3
Amount of Grant: Up to 25,000 USD
Contact: Kristen Madhuizen; (312) 726-9670; fax (312) 726-9695; kristen.madhuizen@american-hearing.org or ahrf@american-hearing.org
Internet: http://www.american-hearing.org/research/harrisongrant.html
Sponsor: American Hearing Research Foundation
8 South Michigan Avenue, Suite 814
Chicago, IL 60603-4539

AHRQ Independent Scientist Award 388
The award is a special salary-only grant designed to provide protected time for newly independent scientists who currently have non-research obligations such as heavy teaching loads, clinical work, committee assignments, service, and administrative duties that prevent them from having a period of intensive research focus. The award is targeted to persons with doctoral degrees who have completed their research training, have independent peer-reviewed research support, and who need a period of protected research time in order to foster their research career development. New applications are due February 12, June 12, and October 12, while renewal applications are due March 12, July 12, and November 12. Up to $90,000 per year of base salary plus fringe benefits is available. No other research development support funds are provided.
Requirements: The following organizations and institutions are eligible to apply: public/state controlled institutions of higher education; private institutions of higher education; nonprofits with 501(c)3 IRS status; nonprofits without 501(c)3 IRS status; small businesses; for-profit organizations; State governments; U.S. territories or possessions; Indian/Native American tribal governments (Federally recognized and other than Federally recognized); Indian/Native American tribally designated organizations; Hispanic-serving institutions; historically Black colleges and universities (HBCUs); tribally controlled colleges and universities (TCCUs); Alaska Native and Native Hawaiian serving institutions; regional organizations; and faith-based or community based organizations. The candidate must have a doctoral degree and peer-reviewed, independent research support at the time the award is made. The candidate must spend a minimum of 75 percent effort conducting research during the period of the award.

Restrictions: The award is not intended for investigators who already have full time to perform research, or have substantial publication records or considerable research support indicating that they are well established in their fields. Foreign institutions are not eligible to apply.
Geographic Focus: All States
Date(s) Application is Due: Feb 12; Mar 12; Jun 12; Jul 12; Oct 12; Nov 12
Contact: Kay Anderson, (301) 427-1555; Kay.Anderson@ahrq.hhs.gov
Internet: http://grants.nih.gov/grants/guide/pa-files/PAR-07-444.html
Sponsor: Agency for Healthcare Research and Quality
540 Gaither Road
Rockville, MD 20850

AHRQ Individual Awards for Postdoctoral Fellows Ruth L. Kirschstein National Research Service Awards (NRSA) 389

The purpose of the postdoctoral fellowship award is to provide support to promising postdoctoral applicants who have the potential to become productive and successful independent research investigators. The proposed postdoctoral training must offer an opportunity to enhance the applicant's understanding of health services research and must be responsive to AHRQ's mission, which is to improve the quality, safety, efficiency, and effectiveness of health care for all Americans. The research sponsored and conducted by AHRQ develops and presents scientific evidence regarding all aspects of health care. It addresses issues of organization, delivery, financing, utilization, patient and provider behavior, outcomes, effectiveness and cost. It evaluates both clinical services and the system in which these services are provided. These scientific results improve the evidence base to enable better decisions about health care, including such areas as disease prevention, appropriate use of medical technologies, improving diagnosis and treatment utilizing comparative effectiveness research, and reducing racial and ethnic disparities. In addition, AHRQ is interested in the application of health information technology (health IT), as well as reducing medical errors and improving patient safety.
Requirements: The following organizations and institutions are eligible to apply: public or non-profit private institutions, such as a university, college, or a faith-based or community-based organization; units of local or State government; eligible agencies of the Federal government; Indian/Native American tribal governments (Federally recognized); Indian/Native American tribal governments (other than Federally recognized); and Indian/Native American tribally designated organizations.
Restrictions: Awards for fellowship training may only be made to domestic institutions.
Geographic Focus: All States
Date(s) Application is Due: Apr 8; Aug 8; Dec 8
Contact: Shelley M. Benjamin, (301) 427-1528; Shelley.Benjamin@ahrq.hhs.gov
Internet: http://grants.nih.gov/grants/guide/pa-files/PA-09-229.html
Sponsor: Agency for Healthcare Research and Quality
540 Gaither Road
Rockville, MD 20850

AIA Latrobe Prize 390

Named in honor of one of America's first professional architects, the Latrobe Prize supports path-finding research designed to advance the art and science of architectural practice. The prize winner will receive $100,000 to support a two-year program of research selected by jury review for its promise to advance professional knowledge in architecture.
Requirements: Criteria for the evaluation of proposals include: (a) relevance of the proposed research program to the general goals and objectives of the College of Fellows, the Latrobe Prize program, and the current theme; (b) breadth, depth, and innovativeness of the research program; (c) projected benefits of research outcomes to practice; (d) benefits of research outcomes to professional and public constituencies; (e) qualifications, expertise, prior achievements, maturity, and performance record of the applicant(s); (f) demonstrated capacity to administer a sustained research program, including the formulation and management of the budget; (g) quality and content of supporting documentation; and, (h) quality and content of letters of reference. See the website for detailed listing of required proposal components.
Geographic Focus: All States
Date(s) Application is Due: Dec 1
Contact: Pauline Porter; (202) 626-7521; fax (202) 626-7527; pporter@aia.org
Internet: http://www.aia.org/latrobe_prize/index.html
Sponsor: American Institute of Architects
1735 New York Avenue NW
Washington, DC 20006-5292

AIAR/Council of American Overseas Research Centers Fellowships for Advanced Multi-Country Research 391

The fellowships support scholars pursuing research on broad questions of multicountry significance in the fields of humanities, social sciences, and related natural sciences in countries in the Near and Middle East and South Asia. For information and application, contact CAORC, (202) 842-8636; email: caorc@caorc.si.edu; web: www.caorc.org.
Requirements: Doctoral candidates and established scholars with U.S. citizenship, as individuals or as teams, are eligible.
Geographic Focus: All States
Date(s) Application is Due: Oct 1
Contact: Grants Administrator; (202) 633-1599
Internet: http://www.caorc.org/fellowships/multi/index.html, fellowships@caorc.org, 202-633-1599.
Sponsor: W.F. Albright Institute of Archaeological Research Jerusalem
P.O. Box 37012
Washington, DC 20013-7012

AIAR Andrew W. Mellon Foundation Fellowships 392

Fellowships are open to those in ancient Near Eastern studies, including the fields of archaeology, anthropology, art history, Bible, epigraphy, historical geography, history, language, literature, philology, and religion or related disciplines from Pre-history, through the early Islamic period. The program is open to Bulgarian, Czech, Estonian, Hungarian, Latvian, Lithuanian, Polish, Romanian, and Slovak scholars. Guidelines are available online.
Requirements: Fellowships are available to Bulgarian, Czech, Estonian, Hungarian, Latvian, Lithuanian, Polish, Romanian, and Slovak scholars. Candidates should not be permanently resident outside the nine countries concerned, and should have obtained a doctorate by the time the fellowship is awarded.
Geographic Focus: All States, Bulgaria, Czech Republic, Estonia, Poland, Romania, Slovakia
Date(s) Application is Due: Apr 2
Amount of Grant: 34,500 USD
Contact: Joan Branham, (401) 865-1789; fax (401) 865-1036; jbranham@providence.edu
Internet: http://www.aiar.org/deadlines.html
Sponsor: W.F. Albright Institute of Archaeological Research Jerusalem
Providence College
Providence, RI 02918

AIAR Annual Professorships 393

Open to postdoctoral scholars in Near Eastern archaeology, geography, history, and Biblical studies. Residence at the institute is required. Period of appointment is 10 months. The professorship period should be continuous, without frequent trips outside the country. The professorship carries a stipend plus room and half-board for appointee and spouse. Annual deadline dates may vary; contact program staff for exact dates.
Requirements: Postdoctoral scholars may apply. U.S. citizens are eligible for the entire award. Non-U.S. citizens may apply but are only eligible for nongovernmental funds. Residence at the institute is required.
Geographic Focus: All States
Date(s) Application is Due: Oct 1
Amount of Grant: 30,000 USD
Contact: Dr. Joan Branham, Department of Art and Art History; (401) 865-1789; fax (401) 865-1036; jbranham@providence.edu
Internet: http://www.aiar.org/deadlines.html
Sponsor: W.F. Albright Institute of Archaeological Research Jerusalem
Providence College
Providence, RI 02918

AIAR Ernest S. Frerichs Fellow and Program Coordinator Grant 394

The program supports predoctoral students and postdoctoral scholars specializing in near Eastern archaeology, geography, history, and biblical studies. The recipient is expected to assist the Albright's director in planning and implementing the Ernest S. Frerichs Program for Albright Fellows. Residence at the institute for the 10-month research period is required. The research period should be continuous, without frequent trips outside the country. The award covers stipend, room, and one-half board at the institute. Annual deadline dates may vary; contact program staff for exact dates.
Geographic Focus: All States
Date(s) Application is Due: Oct 1
Amount of Grant: 19,000 USD
Contact: Dr. Joan Branham, Department of Art and Art History; (401) 865-1789; fax (401) 865-1036; jbranham@providence.edu
Internet: http://www.aiar.org/deadlines.html
Sponsor: W.F. Albright Institute of Archaeological Research Jerusalem
Providence College
Providence, RI 02918

AIAR George A. Barton Fellowships 395

This fellowship is open to seminarians, predoctoral students, and recent PhD recipients specializing in Near Eastern archaeology, geography, history, and biblical studies. Research period is for five months. Awards consist of a stipend plus room and half-board at the institute. The research period should be continuous, without frequent trips outside the country.
Geographic Focus: All States
Date(s) Application is Due: Oct 1
Amount of Grant: 7,000 USD
Contact: Dr. Joan Branham, Department of Art and Art History; (401) 865-1789; fax (401) 865-1036; jbranham@providence.edu
Internet: http://www.aiar.org/deadlines.html
Sponsor: W.F. Albright Institute of Archaeological Research Jerusalem
Providence College
Providence, RI 02918

AIChE Alpha Chi Sigma Award 396

The AIChE Alpha Chi Sigma Award for Chemical Engineering Research are sponsored by The Alpha Chi Sigma Educational Foundation. The award recognizes an individual's outstanding accomplishments in fundamental or applied chemical engineering research. The award recognizes research carried out during the past ten years. This research will normally have been carried out in North America. When appropriate, the winner will be invited to present his or her work at a symposium arranged in his or her honor. When this is not practical, the recipient will be invited to present a paper at an AIChE meeting.
Requirements: The recipient does not have to be a member of AIChE or the Alpha Chi Sigma Fraternity.

Geographic Focus: All States
Date(s) Application is Due: Feb 15
Amount of Grant: 5,000 USD
Contact: Gordon Ellis; (646) 495-1348; fax (646) 495-1503; gorde@aiche.org
Internet: http://www.aiche.org/About/Awards/AlphaChiSigmaAward.aspx
Sponsor: American Institute of Chemical Engineers
3 Park Avenue
New York, NY 10016

AIChE Andrew Chase Forest Bioproducts Division Award in Chemical Engineering 397

The AIChE Andrew Chase Forest Bioproducts Division Award in Chemical Engineering is sponsored by Forest Bioproducts Division. The award recognizes an individual's outstanding chemical engineering contribution in the forest products and related industries. The award is presented at a Division event held at the AIChE Annual Meeting
Requirements: Selection criteria include: significant discoveries, research, or development, successfully implemented on a commercial scale; outstanding chemical engineering contributions in the field of design, operation or production management which led to significant technological improvements; distinguished service as a chemical engineering educator, with an emphasis on application of chemical engineering principles to forest products industry technologies; outstanding service to the division; nominees must have presented a paper at one of the division's symposia describing his or her original contributions in the application of chemical engineering in the forest products or related industries.
Geographic Focus: All States
Date(s) Application is Due: Jun 1
Contact: Gordon Ellis; (646) 495-1348; fax (646) 495-1503; gorde@aiche.org
Internet: http://www.aiche.org/About/Awards/Divisions/ForestProductsDivision.aspx
Sponsor: American Institute of Chemical Engineers
3 Park Avenue
New York, NY 10016

AIChE CAST Outstanding Young Researcher Award 398

This Award will be given to an individual under the age of 40 to recognize outstanding contributions to the chemical engineering computing and systems technology literature. The individual must be age 39 or less on December 31 of the award year. An individual age 40 or over will be eligible for this award if, on December 31 of the award year, 12 years or less have elapsed since this individual received the Ph. D. degree.
Geographic Focus: All States
Date(s) Application is Due: Apr 15
Amount of Grant: 3,000 USD
Contact: Gordon Ellis; (646) 495-1348; fax (646) 495-1503; gorde@aiche.org
Internet: http://www.aiche.org/About/Awards/Divisions/CASTOutstandingYoungResearcher.aspx
Sponsor: American Institute of Chemical Engineers
3 Park Avenue
New York, NY 10016

AIChE Charles M.A. Stine Award 399

The Charles M. A. Stine Award is sponsored by E.I. duPont de Nemours & Company. It recognizes an individual's outstanding contribution to the scientific, technological, educational or service areas of materials engineering and science.
Requirements: Nominees must be members of AIChE.
Geographic Focus: All States
Date(s) Application is Due: Feb 15
Amount of Grant: 1,500 USD
Contact: Gordon Ellis; (646) 495-1348; fax (646) 495-1503; gorde@aiche.org
Internet: http://www.aiche.org/About/Awards/Divisions/CharlesMAStineAward.aspx
Sponsor: American Institute of Chemical Engineers
3 Park Avenue
New York, NY 10016

AIChE Clarence (Larry) G. Gerhold Award 400

The AIChE Clarence (Larry) G. Gerhold Award is sponsored by UOP. The award recognizes an individual's outstanding contribution in research, development, or in the application of chemical separations technology. The award is presented at the Separations Division Dinner held during the AIChE Annual Meeting, where the awardee is invited to give an address.
Requirements: Selection criteria include: a sustained record (at least 15 years) of contributions to separations technology. Such activity may include demonstrated leadership in research, teaching, or engineering; a unique and valuable contribution to separations technology which has had a major impact on one of the separation technologies. The contribution should represent a turning point in a particular technology; acontinuing and extensive record of service to the Separations Division.
Restrictions: Nominees must be members of AIChE.
Geographic Focus: All States
Date(s) Application is Due: Jun 1
Amount of Grant: 3,000 USD
Contact: Gordon Ellis; (646) 495-1348; fax (646) 495-1503; gorde@aiche.org
Internet: http://www.aiche.org/About/Awards/Divisions/ClarenceGGerholdAward.aspx
Sponsor: American Institute of Chemical Engineers
3 Park Avenue
New York, NY 10016

AIChE Computational Molecular Science and Engineering Forum Graduate Student Award 401

The award recognizes significant contributions to research in computational molecular science and engineering by graduate students. The award consists of a plaque and a $500 honorarium.
Requirements: The nominee must be a graduate student at the time of the poster presentation, and the faculty nominator must be a member of CoMSEF. Nominations should consist of a nominating letter from the student's research advisor and the curriculum vitae of the nominee.
Geographic Focus: All States
Date(s) Application is Due: Sep 1
Amount of Grant: 500 USD
Contact: Gordon Ellis; (646) 495-1348; fax (646) 495-1503; gorde@aiche.org
Internet: http://www.aiche.org/About/Awards/Divisions/CoMSEFGradAward.aspx
Sponsor: American Institute of Chemical Engineers
3 Park Avenue
New York, NY 10016

AIChE Computing in Chemical Engineering Award 402

The Computing in Chemical Engineering Award is sponsored by The Dow Chemical Company & Mitsubishi Chemical Company. The Computing in Chemical Engineering Award recognizes outstanding contributions in the application of computing and systems technology to chemical engineering. The award consists of a plaque and $3,000.
Requirements: The awardee is invited to deliver an address at the CAST Division dinner at the AIChE Annual Meeting.
Geographic Focus: All States
Date(s) Application is Due: Apr 15
Amount of Grant: 3,000 USD
Contact: Gordon Ellis; (646) 495-1348; fax (646) 495-1503; gorde@aiche.org
Internet: http://www.aiche.org/About/Awards/DivisionsComputingChemicalEngineeringAward.aspx
Sponsor: American Institute of Chemical Engineers
3 Park Avenue
New York, NY 10016

AIChE Computing Practice Award 403

The award, which is sponsored by Aspen Technology, Inc., and ExxonMobil Chemical Company, recognizes outstanding contributions in the application of chemical engineering to computing and systems technology.
Geographic Focus: All States
Date(s) Application is Due: Apr 15
Amount of Grant: 3,000 USD
Contact: Gordon Ellis; (646) 495-1348; fax (646) 495-1503; gorde@aiche.org
Internet: http://www.aiche.org/awards/awarddtl.asp?AwardID=29
Sponsor: American Institute of Chemical Engineers
3 Park Avenue
New York, NY 10016

AIChE CRE Travel Awards 404

The Catalysis and Reaction Engineering Division will make up to 10 travel awards to assist graduate students with their travel expenses so that they can attend the AIChE Annual meeting and present the results of their research. The awards will consist of $400, plus a ticket to the CRE Division Dinner, where the recipients will be officially recognized.
Requirements: To be eligible, the student must be a member of the AIChE and present a paper at the Annual Meeting in one of the CRE Division sessions.
Geographic Focus: All States
Date(s) Application is Due: Jul 31
Amount of Grant: 400 USD
Contact: Gordon Ellis; (646) 495-1348; fax (646) 495-1503; gorde@aiche.org
Internet: http://www.aiche.org/About/Awards/Divisions/CRETravelAwards.aspx
Sponsor: American Institute of Chemical Engineers
3 Park Avenue
New York, NY 10016

AIChE David Himmelblau Award for Innovations in Computer-Based Chem Engineering Education 405

The David Himmelblau Award for Innovations in Computer-Based Chemical Engineering Education is sponsored by CAChE Corporation. The award recognizes an individual or group making new and novel contributions to computer aids for chemical engineering education. Educational innovators working in industry or in a company that develops computer-based educational aids are also eligible for the award. Specifically, the nomination and the citation for the award must refer to a significant contribution to computer-based chemical engineering education within the past decade. The award is presented at the CAST Division dinner at the AIChE Annual Meeting. The award winner will give an invited talk in the CAChE-sponsored session at the AIChE Annual Meeting.
Geographic Focus: All States
Date(s) Application is Due: Apr 15
Amount of Grant: 1,000 USD
Contact: Gordon Ellis; (646) 495-1348; fax (646) 495-1503; gorde@aiche.org
Internet: http://www.aiche.org/About/Awards/DivisionsCASTHimmelblauAward.aspx
Sponsor: American Institute of Chemical Engineers
3 Park Avenue
New York, NY 10016

AIChE Donald Q. Kern Award 406

The AIChE Donald Q. Kern Award is sponsored by Heat Transfer Research, Inc. The award will be presented at the AIChE Spring National Meeting or the Summer Heat Transfer Conference. The award is given in honor of Donald Q. Kern, pioneer in process heat transfer, the Division recognizes an individual's expertise in a given field of heat transfer or energy conversion.

Requirements: Selection criteria include: significant contributions to applied heat transfer or conversion or in the translation of research results into useful technological applications; the recipient is required to prepare a written review of a topic selected by the recipient. The review is usually published in an AIChE publication.
Geographic Focus: All States
Date(s) Application is Due: Mar 16
Amount of Grant: 1,500 USD
Contact: Gordon Ellis; (646) 495-1348; fax (646) 495-1503; gorde@aiche.org
Internet: http://www.aiche.org/About/Awards/Divisions/DonaldQKernAward.aspx
Sponsor: American Institute of Chemical Engineers
3 Park Avenue
New York, NY 10016

AIChE Environmental Division Graduate Student Paper Award 407

The Environmental Division Graduate Student Paper Award recognizes outstanding graduate student contributions to environmental protection through chemical engineering. The award is based on the annual graduate student paper competition.

Requirements: Selection criteria include: the graduate student must be a member of AIChE at the time of submission; the work presented must be carried out while the graduate student is enrolled at a university with an accredited chemical engineering program; the graduate student must be the primary author,; the paper must describe original research or design; papers consisting of literature reviews are not permitted; the paper must be suitable for publication in a refereed journal; the paper must represent a contribution to environmental protection through the application of chemical engineering. Nomination for the award should include: A single nomination letter detailing the student's strengths and accomplishments and a single paper contributing to environmental science and engineering fundamentals or applications. Papers will be judged on quality, technical content, writing and organization. Nominations may be submitted by any member of AIChE.
Geographic Focus: All States
Amount of Grant: 200-500 USD
Contact: Gordon Ellis; (646) 495-1348; fax (646) 495-1503; gorde@aiche.org
Internet: http://www.aiche.org/About/Awards/Divisions/EnvironmentalDivision GraduateStudentPaper.aspx
Sponsor: American Institute of Chemical Engineers
3 Park Avenue
New York, NY 10016

AIChE Environmental Division Undergraduate Student Paper Award 408

The Environmental Division Undergraduate Student Paper Award is awarded to a full-time undergraduate student who prepares the best original paper based on the results of research or an investigation related to the environment.

Requirements: The nominee must be a full-time undergraduate in good standing at a college or university with an accredited program in chemical engineering, and must be a member of an AIChE Student Chapter; the work must be performed during the student's undergraduate enrollment, and the paper must be submitted prior to or within six months of graduation; the student must be the sole author of the paper.
Geographic Focus: All States
Amount of Grant: 100-300 USD
Contact: Gordon Ellis; (646) 495-1348; fax (646) 495-1503; gorde@aiche.org
Internet: http://www.aiche.org/About/Awards/Divisions/EnvironmentalDivision UndergraduateStudentPaper.aspx
Sponsor: American Institute of Chemical Engineers
3 Park Avenue
New York, NY 10016

AIChE Excellence in Process Development Research Award 409

The AIChE Excellence in Process Development Research Award is sponsored by Pfizer. The award recognizes individuals who have made significant technical contributions to the advancement of process development within research, teaching, or regulatory activities. Accomplishments must be disseminated by means of well-documented materials. Emphasis will be placed on accomplishments and advances made within the last ten years although the award can also be given to someone for an outstanding career. The awardee is invited to deliver an address at the Process Development Division Dinner at the AIChE Annual Meeting
Geographic Focus: All States
Date(s) Application is Due: Jun 15
Amount of Grant: 1,000 USD
Contact: Gordon Ellis; (646) 495-1348; fax (646) 495-1503; gorde@aiche.org
Internet: http://www.aiche.org/About/Awards/Divisions/ExcellenceProcess DevelopmentResearchAward.aspx
Sponsor: American Institute of Chemical Engineers
3 Park Avenue
New York, NY 10016

AIChE Food, Pharmaceutical and Bioengineering Division Awards in Chemical Engineering 410

The AIChE Food, Pharmaceutical and Bioengineering Division Awards in Chemical Engineering is sponsored by Merck. The award recognizes an individual's outstanding chemical engineering contribution in the food, pharmaceutical and/or bioengineering industry. These contributions may have been made in industry, government, or academic areas, or with other organizations. The award is presented at the Food, Pharmaceutical & Bioengineering Division plenary session of the AIChE Annual Meeting.
Geographic Focus: All States
Date(s) Application is Due: Feb 15
Amount of Grant: 4,000 USD
Contact: Gordon Ellis; (646) 495-1348; fax (646) 495-1503; gorde@aiche.org
Internet: http://www.aiche.org/About/Awards/Divisions/FoodPharmaceuticalBio engineeringDivisionAwardChemicalEngineering.aspx
Sponsor: American Institute of Chemical Engineers
3 Park Avenue
New York, NY 10016

AIChE Food, Pharmaceutical and Bioengineering Division Distinguished Service Award in Chemical Engineering 411

The purpose of this award is to recognize a division member for his or her dedication and service to the profession of food engineering, pharmaceutical engineering, and/or bioengineering in general, and to Division 15 in particular. The award consists of a plaque and a monetary portion, the amount of which is determined by the executive committee. The award is to be given to the individual in a public session at the Annual Meeting of AIChE.
Geographic Focus: All States
Date(s) Application is Due: Apr 15
Contact: Gordon Ellis; (646) 495-1348; fax (646) 495-1503; gorde@aiche.org
Internet: http://www.aiche.org/About/Awards/Divisions/FoodPharmaceuticalBio engineeringDistinguishedService.aspx
Sponsor: American Institute of Chemical Engineers
3 Park Avenue
New York, NY 10016

AIChE Institute Award for Excellence in Industrial Gases Technology 412

The AIChE Institute Award for Excellence in Industrial Gases Technology is sponsored by Praxair, Inc. The award recognizes an individual's sustained excellence in contributing to the advancement of technology in the production, distribution and application of industrial gases. The award is presented at the Honors Ceremony of the AIChE Annual Meeting and consists of a plaque and $3,000, plus a $500 travel allowance.

Requirements: The recipient will have a record of sustained contributions that have advanced the frontier of industrial gases technology. These contributions may be characterized by a sustained record of important fundamental research, innovation, technological development or the novel application of technology, either fostering or leading to important commercial results. The recipient does not have to be a member of AIChE.
Geographic Focus: All States
Date(s) Application is Due: Feb 15
Amount of Grant: 3,000 USD
Contact: Gordon Ellis; (646) 495-1348; fax (646) 495-1503; gorde@aiche.org
Internet: http://www.aiche.org/About/Awards/InstituteAwardforExcellencein IndustrialGasesTechnology.aspx
Sponsor: American Institute of Chemical Engineers
3 Park Avenue
New York, NY 10016

AIChE John A. Tallmadge Award for Contributions to Coating Technology 413

The AIChE John A. Tallmadge Award for Contributions to Coating Technology is sponsored by Eastman Kodak Company; International Society of Coating Science and Technology, and AIChE. The award recognizes an individual's significant contributions to the understanding or improvement of the technology of the coating of continuous webs. The award is presented bi-annually, in even numbered years.

Requirements: Nominations must follow the form posted on the ISCST website.
Geographic Focus: All States
Date(s) Application is Due: Jul 1
Amount of Grant: 1,000 USD
Contact: Gordon Ellis; (646) 495-1348; fax (646) 495-1503; gorde@aiche.org
Internet: http://www.aiche.org/About/Awards/Committee/JohnATallmadgeAward.aspx
Sponsor: American Institute of Chemical Engineers
3 Park Avenue
New York, NY 10016

AIChE Kazutoshi Fujimura Award for Lifetime Achievement 414

The Kazutoshi Fujimura Award for Lifetime Achievement is sponsored by Hyperion Catalysis International and Bioveris, Cambridge, Massachusetts, U.S.A. This award recognizes the work of an individual in helping to develop and propagate new technology internationally. The individual should have played a prominent role for many years in identifying, developing and introducing new chemical engineering technologies to the international market place. The award is not intended to honor a single event but instead to recognize the lifetime achievement of an individual in international chemical engineering development. The Fujimura Award is overseen by AIChE's Management Division. A selection committee consisting of representatives of the three international chemical engineering confederations and Dr. Norman Li, the first recipient, plus Management Division representatives and AIChE staff will make the final selection.

Geographic Focus: All States
Date(s) Application is Due: Jan 31
Amount of Grant: 10,000 USD
Contact: Gordon Ellis; (646) 495-1348; fax (646) 495-1503; gorde@aiche.org
Internet: http://www.aiche.org/About/Awards/Divisions/KazutoshiFujimuraAward.aspx
Sponsor: American Institute of Chemical Engineers
3 Park Avenue
New York, NY 10016

AIChE Lawrence K. Cecil Award in Environmental Chemical Engineering **415**
The Lawrence K. Cecil Award in Environmental Chemical Engineering is sponsored by B.P. America, Inc. The award recognizes an individual's outstanding chemical engineering contribution and achievement in the preservation or improvement of the environment.
Requirements: The recipient must be a member of AIChE, have 15 years of chemical engineering experience in the environmental field, and demonstrate leadership in research, teaching, engineering, or regulatory activities in either the public or private sector.
Geographic Focus: All States
Date(s) Application is Due: Feb 28
Amount of Grant: 2,500 USD
Contact: Gordon Ellis; (646) 495-1348; fax (646) 495-1503; gorde@aiche.org
Internet: http://www.aiche.org/About/Awards/Divisions/LawrenceKCecilAward EnvironmentalChemicalEngineering.aspx
Sponsor: American Institute of Chemical Engineers
3 Park Avenue
New York, NY 10016

AIChE Lectureship Award in Fluidization **416**
The AIChE Lectureship Award in Fluidization is sponsored by Particulate Solid Research, Inc. The award recognizes an individual's outstanding scientific/technical research contributions with impact in the field of fluidization and fluid-particle flow systems. The award is presented at a Forum sponsored event during the AIChE Annual Meeting.
Requirements: Selection criteria include: an outstanding contribution advancing fluidization or fluid-particle flow systems; the awardee is required to deliver a keynote paper at the Fundamentals of Fluidization and Fluid-Particle Systems session of Area 3b during the AIChE Annual Meeting. The awardee is also required to submit a written manuscript; membership in the Particle Technology Forum or AIChE is not required.
Geographic Focus: All States
Date(s) Application is Due: Jun 15
Amount of Grant: 1,000 USD
Contact: Gordon Ellis; (646) 495-1348; fax (646) 495-1503; gorde@aiche.org
Internet: http://www.aiche.org/LectureshipAwardFluidization.aspx
Sponsor: American Institute of Chemical Engineers
3 Park Avenue
New York, NY 10016

AIChE Nanoscale Science and Engineering Forum Young Investigator Award **417**
This award recognizes outstanding interdisciplinary research in nanoscience and nanotechnology by engineers or scientists in the early stages of their professional careers (within 10 years of completion of highest degree). The awardee is invited to deliver an address during the NSEF Plenary Lectures at the AIChE Annual Meeting.
Geographic Focus: All States
Date(s) Application is Due: May 31
Amount of Grant: 750-1,750 USD
Contact: Gordon Ellis; (646) 495-1348; fax (646) 495-1503; gorde@aiche.org
Internet: http://www.aiche.org/About/Awards/Divisions/NSEFYoungInvestigator.aspx
Sponsor: American Institute of Chemical Engineers
3 Park Avenue
New York, NY 10016

AIChE North American Mixing Forum Award **418**
The AIChE North American Mixing Forum Award is sponsored by The Procter & Gamble Company. It is Awarded for Excellence and Sustained Contributions to Mixing Research and Practice. The award is presented at the award lecture and a dinner ceremony during the AIChE Annual Meeting.
Geographic Focus: All States
Date(s) Application is Due: May 1
Amount of Grant: 1,500 USD
Contact: Gordon Ellis; (646) 495-1348; fax (646) 495-1503; gorde@aiche.org
Internet: http://www.aiche.org/About/Awards/Divisions/NorthAmericanMixing ForumAward.aspx
Sponsor: American Institute of Chemical Engineers
3 Park Avenue
New York, NY 10016

AIChe North American Mixing Forum Start-up Grant **419**
The AIChE North American Mixing Forum Start-Up Grant is sponsored by North American Mixing Forum. The purpose of the grant is to support a young, non-tenured professor seeking to do research in the field of mixing or mixing related processes.
Geographic Focus: All States
Date(s) Application is Due: May 1
Contact: Gordon Ellis; (646) 495-1348; fax (646) 495-1503; gorde@aiche.org
Internet: http://www.aiche.org/About/Awards/Divisions/NorthAmericanMixing ForumGrant.aspx
Sponsor: American Institute of Chemical Engineers
3 Park Avenue
New York, NY 10016

AIChE North American Mixing Forum Student Award **420**
The AIChE North American Mixing Forum Student Award is sponsored by Rohm & Haas and AIChE's North American Mixing Forum. The NAMF Student Award was established to encourage, recognize and reward students for quality research in the area of mixing. Any graduate or undergraduate student doing research in the field of Fluid Mixing at an accredited University in North America is eligible. The winner will be notified during the conference and will be presented with a plaque and an award of $500 at the NAMF Award Banquet.
Requirements: An applicant must submit a paper in electronic form, describing his or her work in mixing to the NAMF Student Awards Committee Chair. See website for the contact information.
Geographic Focus: All States
Date(s) Application is Due: Mar 15
Amount of Grant: 500 USD
Contact: Gordon Ellis; (646) 495-1348; fax (646) 495-1503; gorde@aiche.org
Internet: http://www.aiche.org/About/Awards/Divisions/NorthAmericanMixing ForumStudent.aspx
Sponsor: American Institute of Chemical Engineers
3 Park Avenue
New York, NY 10016

AIChE Nuclear Engineering Division Outstanding Student Paper & Presentation Awards **421**
The AIChE Nuclear Engineering Division Outstanding Student Paper & Presentation Awards are sponsored by Westinghouse Electric Company. The awards were established to encourage, recognize and reward chemical engineering students for quality research in the nuclear science and technology. Any graduate or undergraduate student doing research in this field is eligible for these awards.
Requirements: All eligible applicants must submit a paper describing their work in nuclear science and technology to the NED Outstanding Student Awards Committee Chair. Finalist(s) are selected and invited to present the paper(s) at the Annual AIChE meeting. The finalist(s) are selected no later than 4 weeks before the conference. NED covers both the travel expenses (up to $700) and the registration fee for the finalist(s). The winner for the Student Paper Award will be selected by the Award Committee Members before the conference. See website for detailed description of the application process.
Geographic Focus: All States
Date(s) Application is Due: May 19; May 19; Jun 31; Jun 31; Sep 18; Sep 18
Amount of Grant: 700 USD
Contact: Gordon Ellis; (646) 495-1348; fax (646) 495-1503; gorde@aiche.org
Internet: http://www.aiche.org/About/Awards/Divisions/NuclearEngineeringDivi sionStudentAward.aspx
Sponsor: American Institute of Chemical Engineers
3 Park Avenue
New York, NY 10016

AIChE Process Development Practice Award **422**
The AIChE Process Development Practice Award is sponsored by Zeton Inc. The award recognizes individuals with outstanding contributions in the practice or application of chemical engineering to process development. This award hopes to focus on applicants from industrial settings where documentation of contributions is not as readily provided.
Requirements: Awardees will be selected based on their contributions to the discovery and application of innovative solutions to technological problems, and /or commercialization of new products and processes. The award may be given to someone for an outstanding career.
Restrictions: Candidate must be an AIChE member.
Geographic Focus: All States
Date(s) Application is Due: Dec 15
Amount of Grant: 1,000 USD
Contact: Gordon Ellis; (646) 495-1348; fax (646) 495-1503; gorde@aiche.org
Internet: http://www.aiche.org/About/Awards/Divisions/ProcessDevelopmentPrac ticeAward.aspx
Sponsor: American Institute of Chemical Engineers
3 Park Avenue
New York, NY 10016

AIChE Professional Progress Award for Outstanding Progress in Chemical Engineering **423**
This award recognizes outstanding progress in the field of chemical engineering.
Requirements: The awardee will have made a significant contribution to the science of chemical engineering through one of the following means: a theoretical discovery or development of a new principle in the chemical engineering field; development of a new process or product in the chemical engineering field; an invention or development of new equipment in the chemical engineering field; distinguished service rendered to the field or profession of chemical engineering.
Restrictions: The recipient must be less than 45 years of age at the end of the calendar year in which the award is presented.
Geographic Focus: All States
Date(s) Application is Due: Feb 15
Amount of Grant: 4,500 USD
Contact: Gordon Ellis; (646) 495-1348; fax (646) 495-1503; gorde@aiche.org

Internet: http://www.aiche.org/awards/awarddtl.asp?AwardID=60
Sponsor: American Institute of Chemical Engineers
3 Park Avenue
New York, NY 10016

AIChE Research Excellence In Sustainable Engineering Award 424
The AIChE Research Excellence In Sustainable Engineering Award is annually presented for basic or applied research results relative to the sustainability of products, processes, or the environment. The award is to recognize the one who has made significant technical contributions to the advancement of sustainable engineering in research, teaching, and development activities. Emphasis should be placed on accomplishments and advances made within the last five years, although the award can also be made to someone who has had an outstanding career. The awardee will receive $1,000 and a plaque.
Geographic Focus: All States
Date(s) Application is Due: Aug 1
Amount of Grant: 1,000 USD
Contact: Gordon Ellis; (646) 495-1348; fax (646) 495-1503; gorde@aiche.org
Internet: http://www.aiche.org/About/Awards/Divisions/SEFExcellenceAward.aspx
Sponsor: American Institute of Chemical Engineers
3 Park Avenue
New York, NY 10016

AIChE Separations Division Graduate Student Research Award 425
AIChE Separations Division Graduate Student Research Award recognizes six outstanding graduate students in any of the Separations Division's areas: adsorption and ion exchange; crystallization and evaporation; distillation and absorption; extraction; fluid-particle separations; membrane-based separations; bioseparations and general separations. Awards will be made at the next AIChE Annual Meeting.
Requirements: For consideration for the award, the following should be submitted: a single paper contributing to separations fundamentals or applications; a single nomination letter detailing the student's strengths and accomplishments; the student's CV. To be eligible for this year's awards, the nominees must currently be graduate students or have been graduate students since the last AIChE Annual meeting last November. The selection criteria include: the paper must report on research, investigation or design, and must be part of the student's work for a graduate degree; he paper may be co-authored by others, but the student nominee must have been the primary author; the paper should be of a quality acceptable for publication in journals such as the AIChE Journal and Chemical Engineering Science; papers will be judged on technical content, quality of writing and organization; the person nominating a student must be a member of AIChE.
Geographic Focus: All States
Date(s) Application is Due: Jun 1
Amount of Grant: 300 USD
Contact: Gordon Ellis; (646) 495-1348; fax (646) 495-1503; gorde@aiche.org
Internet: http://www.aiche.org/Students/Awards/SeparationsDivisionGraduateStudentResearchAward.aspx
Sponsor: American Institute of Chemical Engineers
3 Park Avenue
New York, NY 10016

AIChE Thomas Baron Award in Fluid-Particle Systems 426
The AIChE Thomas Baron Award in Fluid-Particle Systems is sponsored by Shell Global Solutions, Inc. The award recognizes an individual's outstanding scientific/technical accomplishment which has made a significant impact in the field of fluid-particle systems or in a related field with potential for cross-fertilization. The award is presented at a Forum sponsored event at the AIChE Annual Meeting.
Requirements: Selection criteria include: an outstanding contribution advancing fluid-particle systems, or a related field; the awardee is invited to deliver a Plenary Lecture at an AIChE Annual Meeting session. The awardee is also required to submit a written manuscript.
Geographic Focus: All States
Date(s) Application is Due: Jun 15
Amount of Grant: 1,000 USD
Contact: Gordon Ellis; (646) 495-1348; fax (646) 495-1503; gorde@aiche.org
Internet: http://www.aiche.org/About/Awards/Divisions/ThomasBaronAwardFluidParticleSystems.aspx
Sponsor: American Institute of Chemical Engineers
3 Park Avenue
New York, NY 10016

AIChE Transport and Energy Processes Division Award 427
The AIChE Transport and Energy Processes Division Award is sponsored by Fauske & Associates. The award recognizes an individual's outstanding chemical engineering contribution and achievement in heat transfer or energy conversion. The recipient is invited to speak at a Transport Energy Processes Division dinner, held at the AIChE Annual Meeting.
Requirements: Selection criteria include: distinguished service of a chemical engineering nature, in heat transfer or energy conversion as a professional engineer, educator, or AIChE leader; outstanding contributions of a chemical engineering nature, in the field of design, construction, operation, or management of heat transfer or energy conversion facilities or enterprises; research or development of new processes or equipment relating to chemical engineering applications in heat transfer or energy conversion.
Geographic Focus: All States
Amount of Grant: 500 USD
Contact: Gordon Ellis; (646) 495-1348; fax (646) 495-1503; gorde@aiche.org
Internet: http://www.aiche.org/About/Awards/Divisions/TransportEnergyProcessesDivisionAward.aspx
Sponsor: American Institute of Chemical Engineers
3 Park Avenue
New York, NY 10016

AIChE W. David Smith, Jr. Graduate Student Paper Award 428
This Award will be given to an individual for outstanding published work in chemical engineering computing and systems technology. The publication shall consist of a single paper based on work done by the individual while pursuing graduate or undergraduate studies.
Geographic Focus: All States
Date(s) Application is Due: Apr 15
Amount of Grant: 1,500 USD
Contact: Gordon Ellis; (646) 495-1348; fax (646) 495-1503; gorde@aiche.org
Internet: http://www.aiche.org/About/Awards/Divisions/WDavidSmithGraduateStudentPaperAward.aspx
Sponsor: American Institute of Chemical Engineers
3 Park Avenue
New York, NY 10016

AIChE Warren K. Lewis Award for Contributions to Chemical Engineering Education 429
Sponsored by ExxonMobil Research and Engineering Company and administered by AIChE, this award is given to recognize distinguished and continuing contributions to chemical engineering education. Recipients need not be members of the institute.
Geographic Focus: All States
Date(s) Application is Due: Feb 15
Amount of Grant: 5,000-5,500 USD
Contact: Gordon Ellis; (646) 495-1348; fax (646) 495-1503; gorde@aiche.org
Internet: http://www.aiche.org/About/Awards/WarrenKLewisAward.aspx
Sponsor: American Institute of Chemical Engineers
3 Park Avenue
New York, NY 10016

AIChE William H. Walker Award for Excellence in Contributions to Chemical Engineering Literature 430
AIChE William H. Walker Award for Excellence in Contributions to Chemical Engineering Literature is sponsored by John Wiley and Sons. The award is presented to a member of AIChE who has made an outstanding contribution to chemical engineering literature. The contribution may consist of a review, a history of the development of a process, a theoretical contribution, a research report, or other material of interest and importance to the chemical engineering profession.
Requirements: The recipient must be the author or co-author of an outstanding work in chemical engineering.
Geographic Focus: All States
Date(s) Application is Due: Feb 15
Amount of Grant: 5,500 USD
Contact: Gordon Ellis; (646) 495-1348; fax (646) 495-1503; gorde@aiche.org
Internet: http://www.aiche.org/About/Awards/WilliamHWalkerAward.aspx
Sponsor: American Institute of Chemical Engineers
3 Park Avenue
New York, NY 10016

Aid to Scholarly Publications (ASPP) 431
The ASPP is designed to assist the publication of works of advanced scholarship which make an important contribution to knowledge, but which are unlikely to be self-supporting. Manuscripts in the various disciplines in the humanities and social sciences are considered for the ASPP process if they meet ASPP eligibility criteria. An award from the ASPP is designed to help offset costs of marketing and promoting a scholarly work. All funds are paid directly to a book's publisher. Also, five subventions per year will be devoted to translations. Successful candidates applying for a translation subvention will also be eligible for a grant of up to $4,000 to help offset the costs of translating the work.
Requirements: With few exceptions, ASPP-funded authors must be Canadian citizens or landed immigrants. In general, only book-length manuscripts (at least 100 typescript pages) written by Canadian citizens or landed immigrants / permanent residents, and to be published in Canada by approved Canadian scholarly presses, are eligible for consideration. Authors wishing to work with overseas presses should consult page 3 of these Guidelines, item 3, Submitting Party. Authors seeking ASPP support are strongly advised not to sign a contract with a non-Canadian publisher before obtaining ASPP approval of that publisher. The only modern languages in which submissions will be considered are English and French. Classical-language works may also be admissible. Guidelines and registrations forms are available at the sponsor's website.
Restrictions: The ASPP does not award research grants. The ASPP does not provide grants for the publication of: text books; technical reports; original works of poetry, fiction and drama; scholarly journals and articles; or, conference proceedings. Not all publishers are eligible to work with the ASPP (please see List of Eligible Publishers available at the website).
Geographic Focus: All States, Canada
Contact: Secretariat; (613) 238-6112, ext. 350; fax (613) 238-6114; secaspp@fedcan.ca
Internet: http://www.fedcan.ca/english/aspp/about/general/
Sponsor: Canadian Federation for the Humanities and Social Sciences
151 Slater Street, Suite 415
Ottawa, ON K1P 5H3 Canada

AIHP Sonnedecker Visiting Scholar Grants 432

The Sonnedecker Visiting Scholar Program offers assistance for short-term historical research related to the history of pharmacy, including the history of drugs, at the University of Wisconsin-Madison. The program provides assistance for travel, maintaining temporary residence in Madison, and meeting research expenses associated with utilizing the collection. A brochure is available on request that describes the pharmaco-historical collections, which have been developed in Madison during more than a century by the University of Wisconsin-Madison, the State Historical Society of Wisconsin, and the American Institute of the History of Pharmacy. At least $1,000 becomes available annually to defray part of the expenses of a recipient, for whatever period of residence is appropriate. Grants are made throughout the year on the basis of the merit of previous historical work and on the appropriateness of historical resources on the University of Wisconsin campus to the research proposed.

Requirements: Eligible applicants are: Historians, Pharmacists, and other scholars working in the Pharmacy field.
Geographic Focus: All States
Amount of Grant: 1,000 USD
Contact: Gregory Higby, Director; (608) 262-5378; grants@aihp.org
Internet: http://cms.pharmacy.wisc.edu/aihp/programs/sonnedecker
Sponsor: American Institute of the History of Pharmacy
777 Highland Avenue
Madison, WI 53705-2222

AIHP Thesis Support Grants 433

The Program offers a grant-in-aid totaling $2,000 or more annually to a graduate student to reinforce historical investigations of some aspect of pharmacy, whether ancient or modern, to pay research expenses not normally met by the university granting the degree. Any thesis project devoted to the history of pharmacy, history of drugs, or other humanistic study utilizing a pharmaco-historical approach, is eligible if based in an institution of higher learning of the U.S.A. The maximum grant in this program of the Institute will be $2,500.

Requirements: Any graduate student in good standing at an institution of the United States may apply, regardless of the department through which the Doctor of Philosophy degree will be granted. The graduate student need not be an American citizen; nor does the research topic need to be in the field of American history.
Restrictions: Examples of ineligible expenses would be: living expenses of the applicant at the home university; routine typing to produce research notes or the thesis manuscript; routine illustrations for the manuscript; and publication of research results. Indirect expenses, such as overhead and other institution-related costs, may not be included in a grant application.
Geographic Focus: All States
Date(s) Application is Due: Feb 1
Amount of Grant: Up to 2,500 USD
Contact: Gregory Higby, Director; (608) 262-5378; grants@aihp.org
Internet: http://cms.pharmacy.wisc.edu/aihp/programs/thesisgrant
Sponsor: American Institute of the History of Pharmacy
777 Highland Avenue
Madison, WI 53705-2222

AIHS/Mitacs Health Pilot Partnership Internship Grants 434

The AIHS/Mitacs Health Pilot Partnership is a partnership between AIHS, Mitacs and not-for-profit organizations. This is a pilot program created to address the existing gap and lack of funding available to engage highly qualified graduate students in the public and not-for-profit sector, specifically in health-related research. The health pilot partnership will fund up to three internships for graduate and postgraduate health research and innovation trainees in Alberta. The AIHS/Mitacs Health Pilot Partnership provides a grant of $15,000 for a four-month internship administered by Mitacs-Accelerate and a partner university. The Partnership will meet the following objectives: connect not-for-profit organizations with peer-reviewed research expertise, ensuring high quality research that meets the organization's needs; apply knowledge by interns to real-world issues to create solutions to challenging problems; and provide research labs with a unique opportunity to train graduate and postgraduate trainees in new and applied skills that can be brought back to the lab. Interns may work in the areas of health economics; remote health monitoring; health policy; e-health; chronic disease management; or knowledge transfer/translation. A minimum $10,000 stipend will be provided to the intern, and the remaining $5,000 will be available for other project-related expenses. A longer eight-month double internship can be developed if there is strong justification to accommodate larger projects. Interns will spend approximately 50% of their time with the partnered not-for-profit organization, with the remaining time spent at the university, advancing the research under the guidance of their faculty supervisor.

Requirements: An applicant must be officially accepted into, or currently enrolled in, graduate or postgraduate studies at the University of Alberta or the University of Calgary; able to participate in a four-month single internship or an eight-month double internship with 50% of the intern's time being spent outside of the lab; and have a defined research project that they will undertake with the partnered not-for-profit organization.
Geographic Focus: All States, Canada
Date(s) Application is Due: May 15
Amount of Grant: 15,000 USD
Contact: Dr. Ryan Perry; (780) 423-5727; ryan.perry@albertainnovates.ca
Internet: http://www.aihealthsolutions.ca/grants/training-and-early-career-development-programs/mitacs/
Sponsor: Alberta Innovates-Health Solutions
10104-103 Avenue, Suite 1500
Edmonton, AB T5J 4A7 Canada

AIHS Alberta/Pfizer Translational Research Grants 435

The Alberta/Pfizer Translational Research Fund Opportunity is a partnership between Pfizer Canada Inc. (Pfizer), Alberta Innovates-Health Solutions (AIHS), Alberta's Ministry of Enterprise and Advanced Education, and Western Economic Diversification Canada. This partnership will provide opportunities to focus on the development and commercialization of innovations in health. The Alberta/Pfizer Translational Research Fund Opportunity will support innovative translational research projects in areas of unmet health or health system need that have a strong likelihood for technology transfer and commercialization within the next two to five years. Grants will support the development and commercialization of innovations in priority areas identified by Pfizer and the province of Alberta. Those areas include the following: neuroscience/Alzheimer's Disease/ neurodegeneration/ neurological diseases; pain and sensory disorders; immunology and autoimmunity; inflammation and remodeling; diabetes and cardiovascular or other chronic disease issues as they relate to the consequences of diabetes; and orphan and genetic diseases. Preference will be given to research projects focused on the development, use and/or application of innovative technological platforms. These platforms include, but are not limited to: metabolomics and other 'omics' platforms that facilitate personalized medicine strategies; drug discovery; information and communication technologies; nanotechnology; and health-systems-based research approaches in the creation of new knowledge and its potential use. Translational research grants will provide up to $200,000 for a period of up to 18 months. Eligible expenses include salaries of research personnel, purchase and/or leasing of equipment, and research material required for the project. Payments will be based on agreed upon milestones and deliverables, and the submission of periodic reports outlining progress.

Requirements: Organizations will submit a letter of intent and full application. Additional guidelines are available at the AIHS website.
Geographic Focus: All States, Canada
Amount of Grant: Up to 200,000 USD
Contact: Tara McCarthy, Grant Coordinator, Programs; (780) 423-5727; fax (780) 429-3509; tara.mccarthy@albertainnovates.ca
Internet: http://www.aihealthsolutions.ca/grants/industry-partnered-translational-fund/pfizer/
Sponsor: Alberta Innovates-Health Solutions
10104-103 Avenue, Suite 1500
Edmonton, AB T5J 4A7 Canada

AIHS Clinical Fellowships 436

The AIHS Clinical Fellowships program is designed for highly qualified individuals who hold an MD or DDS degree, and who anticipate undertaking a career in health related or clinical research in Alberta. The awards consist of a stipend and a research allowance. A Clinical Fellowship award is normally tenable for a maximum of three years. However, if the trainee is registered in a graduate program, the term may be extended.

Requirements: Normally, an award will be held within the Province of Alberta; however, candidates who are Canadian citizens or permanent residents with records of outstanding performance in postgraduate training may seek research training elsewhere, if sponsored by an Alberta faculty with an expressed interest in future recruitment of the candidate. Candidates must hold an MD or DDS degree and have received a significant portion of their postgraduate training in Alberta.
Restrictions: Clinical fellows are expected to commit to a minimum of two years of full time research.
Geographic Focus: Canada
Date(s) Application is Due: Mar 1; Oct 1
Amount of Grant: 20,000-50,000 USD
Contact: Pamela Valentine, Vice President of Research and Innovation; (780) 423-5727, ext. 230; fax (780) 429-3509; pamela.valentine@albertainnovates.
Internet: http://www.aihealthsolutions.ca/grants/Clin-fellow.php
Sponsor: Alberta Innovates-Health Solutions
10104-103 Avenue, Suite 1500
Edmonton, AB T5J 4A7 Canada

AIHS Collaborative Research and Innovation Grants-Collaborative Program 437

The AIHS Collaborative Research and Innovation Opportunities (CRIO) aims to bring together experts in different disciplines, fields, and areas to tackle health research problems in areas of strategic priority that would benefit from an interdisciplinary approach. The objectives of the CRIO are to: catalyze and support collaborative, interdisciplinary, multi-sectoral and/or multi-institutional research with a focus on achieving solutions that address complex health problems or issues; support areas of research that are aligned with the thematic priority areas defined in Alberta's Health Research and Innovation Strategy; engage end users in the process to enhance impact on the health of Albertans and/or the health care system; provide opportunities for interdisciplinary research training and mentorship; and encourage the use of the AIHS collaborative opportunities to connect Alberta to national and international initiatives and leverage additional opportunities. A maximum of up to $500,000 per year for five years may be awarded. The Collaborative Program involves three to five principal collaborators who may already have an informal arrangement or shared funding to support interdisciplinary work, or have specific plans to develop such arrangements. The Collaborative Program may be considered for a one time renewal, based on agreed-upon deliverables set up in the management plan for years 1 through 5.

Requirements: The application process involves registration, a letter of intent, and full application. Each CRIO must involve Collaborative Lead(s) who are established individual(s) with demonstrated leadership. Collaborative Leads can lead more than one CRIO Project, Program and/or Team simultaneously. There are no limits on the number of applications a Collaborative Lead can submit. Collaborative members may be

defined as needed to address the research objectives, and may include individuals from outside the Alberta area. Interdisciplinary, multi-institutional activity is encouraged. The focus is on a complex program of health research and innovation. A full list of eligibility requirements is available at the AIHS website.
Geographic Focus: All States, Canada
Date(s) Application is Due: Jan 18
Amount of Grant: Up to 500,000 USD
Contact: Kathy Morrison, Grants Coordinator, Programs; (780) 423-5727; fax (780) 429-3509; kathy.morrison@albertainnovates.ca
Internet: http://www.aihealthsolutions.ca/grants/crio/#project
Sponsor: Alberta Innovates-Health Solutions
10104-103 Avenue, Suite 1500
Edmonton, AB T5J 4A7 Canada

AIHS Collaborative Research and Innovation Grants-Collaborative Project 438
The AIHS Collaborative Research and Innovation Opportunities (CRIO) aims to bring together experts in different disciplines, fields, and areas to tackle health research problems in areas of strategic priority that would benefit from an interdisciplinary approach. The objectives of the CRIO are to: catalyze and support collaborative, interdisciplinary, multi-sectoral and/or multi-institutional research with a focus on achieving solutions that address complex health problems or issues; support areas of research that are aligned with the thematic priority areas defined in Alberta's Health Research and Innovation Strategy; engage end users in the process to enhance impact on the health of Albertans and/or the health care system; provide opportunities for interdisciplinary research training and mentorship; and encourage the use of the AIHS collaborative opportunities to connect Alberta to national and international initiatives and leverage additional opportunities. The CRIO Collaborative Project allows approximately three collaborators to come together to complete a defined project. There is no expectation that the collaboration will continue beyond the duration of this non-renewable grant. The opportunity also provides a framework for collaborators who choose to evolve their project into a more complex Collaborative Program or Collaborative Team. The application process involves registration, a letter of intent, and full application. Up to $250,000 per year may be funded for three years.
Requirements: The Collaborative Lead must be an individual with an established career and proven leadership skills and experience. Collaborative members will have an extensive record of success, be creative and original in their approach to research and its translation, and have experience working in teams. Members may also be located outside the Alberta area. Co-funding from philanthropic, national or international organizations is encouraged and should be declared as part of the submission. A full list of eligibility requirements is available at the AIHS website.
Geographic Focus: Canada
Date(s) Application is Due: Aug 31
Amount of Grant: Up to 250,000 USD
Contact: Kathy Morrison, Grants Coordinator, Programs; (780) 423-5727; fax (780) 429-3509; kathy.morrison@albertainnovates.ca
Internet: http://www.aihealthsolutions.ca/grants/crio/
Sponsor: Alberta Innovates-Health Solutions
10104-103 Avenue, Suite 1500
Edmonton, AB T5J 4A7 Canada

AIHS Collaborative Research and Innovation Grants-Collaborative Team 439
The AIHS Collaborative Research and Innovation Opportunities (CRIO) aims to bring together experts in different disciplines, fields, and areas to tackle health research problems in areas of strategic priority that would benefit from an interdisciplinary approach. The objectives of the CRIO are to: catalyze and support collaborative, interdisciplinary, multi-sectoral and/or multi-institutional research with a focus on achieving solutions that address complex health problems or issues; support areas of research that are aligned with the thematic priority areas defined in Alberta's Health Research and Innovation Strategy; engage end users in the process to enhance impact on the health of Albertans and/or the health care system; provide opportunities for interdisciplinary research training and mentorship; and encourage the use of the AIHS collaborative opportunities to connect Alberta to national and international initiatives and leverage additional opportunities. The Collaborative Team fosters and support a large interdisciplinary collaborative team or network with national or international stature. The complexity of the health issues of interest will require the involvement of many stakeholder groups. Up to $1,000,000 per year for five years may be funded.
Requirements: The application process involves registration, a letter of intent, and a full application. Each CRIO must involve a Collaborative Lead() who are established individuals with demonstrated leadership. Collaborative members may be defined as needed to address the research objectives and may include individuals from outside Alberta's knowledge or end users. Interdisciplinary, multi-institutional activity is encouraged. A full list of eligibility requirements is available at the website.
Geographic Focus: All States, All Countries
Date(s) Application is Due: Jan 18
Amount of Grant: Up to 1,000,000 USD
Contact: Tara McCarthy, Grants Coordinator, Programs; (780) 423-5727; fax (780) 429-3509; tara.mccarthy@albertainnovates.ca
Internet: http://www.aihealthsolutions.ca/grants/crio/#project
Sponsor: Alberta Innovates-Health Solutions
10104-103 Avenue, Suite 1500
Edmonton, AB T5J 4A7 Canada

AIHS Fast-Track Fellowships 440
AIHS will accept, at any time, applications from Alberta-based institutions for full-time Fellowship awards for outstanding candidates currently training outside of Alberta. The fast-track program is intended to assist these institutions in recruiting highly qualified trainees to the province. Up to a maximum of 10 awards per calendar year will be awarded through the fast-track review process in each category. Applications to the program will be reviewed on a first come-first serve basis. Fast-track awards are tenable for up to 12 months.
Requirements: Applications must include a letter from the Dean/designate of the faculty indicating that the proposal for a fast-track award has been reviewed and stating why a fast-track application should be considered for the candidate in question. Please refer to the sections on Full-Time Fellowships or Full-Time Studentships for further details regarding eligibility for training awards.
Geographic Focus: Canada
Amount of Grant: 20,000-50,000 USD
Contact: Pamela Valentine, Vice President of Research and Innovation; (780) 423-5727, ext. 230; fax (780) 429-3509; pamela.valentine@albertainnovates.ca or grants.health@albertainnovates.ca
Internet: http://www.aihealthsolutions.ca/grants/fasttrack.php
Sponsor: Alberta Innovates-Health Solutions
10104-103 Avenue, Suite 1500
Edmonton, AB T5J 4A7 Canada

AIHS ForeFront Executive-in-Residence Award 441
ForeFront supports career, industry and innovation development. Alberta's universities are integral to the innovation system and the creation of the medical and health industries through research, technology transfer, and spin-off company creation. The Executive-in-Residence Award enables the technology transfer offices to recruit senior-level management who have relevant industry experience and appropriate networks to assist university spin-off companies with the commercialization of technologies. The goal is to advance technologies and build better companies that will both improve health care and diversify and contribute to the economy.
Geographic Focus: All States, Canada
Contact: Carla Weyland, Manager, Training and Development Programs; (780) 423-5727; fax (780) 429-3509; carla.weyland@albertainnovates.ca
Internet: http://www.aihealthsolutions.ca/forefront/EIR.php
Sponsor: Alberta Innovates-Health Solutions
10104-103 Avenue, Suite 1500
Edmonton, AB T5J 4A7 Canada

AIHS ForeFront Internships 442
The AIHS Internship Program addresses the need for highly trained staff and management of the Alberta-based health, medical products and biotechnology industries. This program is intended to support technology commercialization training and experience for individuals with an appropriate background in science and/or business. Although the work plan should identify the specific area of focus, the intern should receive a broad exposure to all of the following areas: product development; intellectual property strategy; regulatory requirements; market research and strategy; and business planning and development. Interns are awarded a $5,000 training allowance, which may be used for expenses related to educational workshops/seminars/courses necessary for relevant commercialization training, plus a $45,000 to $50,000 stipend. The application and additional guidelines are available at the AIHS website.
Requirements: The applicant must hold, or be in the final year of a degree in business, management, or science. The applicants should have some relevant experience in the business or private sector. Applicants must be sponsored by an Alberta organization actively engaged in the commercialization of medical or health related technology. The organization must have the experience and resources to provide an appropriate training environment, and designate one person within the organization who will be the primary supervisor for the intern. The organization must agree to provide direct and continuous supervision.
Geographic Focus: All States, Canada
Amount of Grant: 45,000-50,000 USD
Contact: Carla Weyland, ForeFront Officer; (780) 423-5727; fax (780) 429-3509; carla.weyland@albertainnovates.ca
Internet: http://www.aihealthsolutions.ca/forefront/internship.php
Sponsor: Alberta Innovates-Health Solutions
10104-103 Avenue, Suite 1500
Edmonton, AB T5J 4A7 Canada

AIHS ForeFront MBA Studentship Award 443
This award will enable students with a background in the medical/health/life sciences to apply their knowledge and pursue a career in management in the medical/health industry by supporting their education and hands-on training in the University of Alberta Master of Business Administration program, specializing in Technology Commercialization. The Tech Com MBA builds and delivers a tool kit that emphasizes the importance of capital, management practices, infrastructure, information, and networks. The goal of the Tech Com MBA is to position graduates for a leadership role in today's leading edge, high technology industries. The main objectives are to: encourage Albertans with a background in science to explore careers that will give them the opportunity to apply their knowledge to commercializing medical discoveries that will lead to improved health; and build capacity of knowledgeable, highly-skilled technology managers to meet the needs of the Alberta medical/health industry. The MBA Studentship Award includes a contribution towards tuition fees and a stipend of $2,000 per month for the duration of the student's full-time MBA studies. The MBA Program may be completed in two years on a full time basis. Students have a maximum of six years to complete the course

requirements. Any extensions to the two-year full-time term must be approved by AIHS. Additional guidelines and terms of the award are available at the AIHS website.
Requirements: Students must hold a graduate degree in medical/health/life sciences (graduated within the last two years) and meet the minimum eligibility criteria for admission to the MBA program.
Geographic Focus: All States, Canada
Contact: Carla Weyland, (780) 423-5727; carla.weyland@albertainnovates.ca
Internet: http://www.aihealthsolutions.ca/forefront/mba.php
Sponsor: Alberta Innovates-Health Solutions
10104-103 Avenue, Suite 1500
Edmonton, AB T5J 4A7 Canada

AIHS ForeFront MBT Studentship Awards 444

The award will enable students to pursue a career in the medical, health, and biotechnology industries by supporting their education and hands-on training in the Master of Biomedical Technology (MBT) Program at the University of Calgary. The MBT Program takes a multi-disciplinary approach and exposes students to a broad range of courses and cutting-edge technologies that will ensure the students acquire practical competencies and extensive knowledge of concepts in biomedical and bioinformatics disciplines, including the application of this knowledge through commercialization of medical and health-related innovations. The main objectives are to: encourage individuals with a background in science to explore careers that will give them the opportunity to apply their knowledge to commercializing medical discoveries that will lead to improved health; and build capacity of knowledgeable, highly-skilled technology managers to meet the needs of the Alberta medical/health industry. The MBT Studentship Award includes a contribution towards tuition fees and a stipend for the practicum (up to a maximum of $3,000). The MBT Program may be completed in one year on a full time basis. Any extensions to the one-year full-time term must be approved by AIHS. Application information is available at the AIHS website.
Requirements: Candidates must hold at least an undergraduate degree in an area of relevance to the objectives of Alberta Innovates – Health Solutions (AIHS).
Geographic Focus: All States, Canada
Contact: Kathy Morrison; (780) 423-5727; kathy.morrison@albertainnovates.ca
Internet: http://www.aihealthsolutions.ca/forefront/mbt.php
Sponsor: Alberta Innovates-Health Solutions
10104-103 Avenue, Suite 1500
Edmonton, AB T5J 4A7 Canada

AIHS Full-Time Fellowships 445

The AIHS Full-Time Fellowships program is designed to enable highly qualified doctoral graduates to prepare for careers in medical or health research as independent investigators. A fellowship award will provide to the host institution funding for one year's stipend, its associated benefits, and a research allowance. Fellows may engage in teaching activities related to their research discipline a maximum of 20 percent of their time. The maximum term is three years
Requirements: Candidates must have a Ph.D., M.D., D.D.S., D.V.M. or D.Pharm. degree. Normally, support will not be provided beyond 6 years after receipt of the Ph.D. degree, or beyond 8 years after receipt of the M.D., D.D.S., D.V.M. or D.Pharm. degrees.
Geographic Focus: Canada
Date(s) Application is Due: Mar 1; Oct 1
Amount of Grant: 35,000-50,000 USD
Contact: Pamela Valentine; (780) 423-5727, ext. 230; fax (780) 429-3509; pamela.valentine@albertainnovates.ca or grants.health@albertainnovates.ca
Internet: http://www.aihealthsolutions.ca/grants/FT-fellow.php
Sponsor: Alberta Innovates-Health Solutions
10104-103 Avenue, Suite 1500
Edmonton, AB T5J 4A7 Canada

AIHS Full-Time Health Research Studentships 446

The Health Research Studentship program enables academically superior students to undertake full-time training in health research. The award supports training in: research on the organization and delivery of health care; technology assessment; community health; health promotion; disease prevention; and related disciplines. The Health Research Studentship consists of a stipend and a research allowance. The term of the award is a maximum of five years (three years at the graduate level).
Requirements: Candidates must have been accepted into, or be currently studying in, a full-time, thesis-based, graduate program at an Alberta-based university in a health-related discipline leading to a Master's or doctoral degree. Normally, support will not be provided beyond six years of enrollment in graduate school. Candidates who have interrupted their training for parenting or other reasons and who have consequently exceeded this time limit, may apply for studentship support. In such cases, however, the candidate is advised to clearly explain the nature of their particular circumstances at the time of application.
Restrictions: The award is not available to students registered in a course-based program.
Geographic Focus: Canada
Date(s) Application is Due: Mar 1; Oct 1
Amount of Grant: 20,000 USD
Contact: Pamela Valentine; (780) 423-5727, ext. 230; fax (780) 429-3509; pamela.valentine@albertainnovates.ca or grants.health@albertainnovates.ca
Internet: http://www.aihealthsolutions.ca/grants/FT-student.php#HRS
Sponsor: Alberta Innovates-Health Solutions
10104-103 Avenue, Suite 1500
Edmonton, AB T5J 4A7 Canada

AIHS Full-Time M.D./Ph.D. Studentships 447

The AIHS Full-Time M.D./Ph.D. Studentships are intended to provide an opportunity for exceptional candidates, who wish to pursue careers as Clinical Investigators, and study simultaneously for the M.D. and the Ph.D. degrees. Support is complementary to the formal M.D./Ph.D. programs at the University of Alberta and the University of Calgary. Awards are tenable only at an Alberta-based university. The term of the award is a maximum of six years.
Requirements: Applicants are limited to those students that formally enter the MD/PhD program no later than the start of their second year of PhD studies at either the University of Alberta or the University of Calgary. AIHS's offer of support is contingent on the granting of complementary stipend support from the Faculty of not less than 15% of the value of the AIHS award, for the duration of the award. Furthermore, it is expected that the universities will undertake to provide administrative support to the offices of the coordinators of the M.D./Ph.D. programs.
Geographic Focus: Canada
Date(s) Application is Due: Mar 1; Oct 1
Amount of Grant: 20,000 USD
Contact: Pamela Valentine; (780) 423-5727, ext. 230; fax (780) 429-3509; pamela.valentine@albertainnovates.ca or grants.health@albertainnovates.ca
Internet: http://www.aihealthsolutions.ca/grants/FT-student.php#MPS
Sponsor: Alberta Innovates-Health Solutions
10104-103 Avenue, Suite 1500
Edmonton, AB T5J 4A7 Canada

AIHS Full-Time Studentships 448

AIHS Full-Time Studentships enable academically superior students to undertake full-time research training in the basic biomedical sciences or in clinical research. The award consists of a stipend and a research allowance. Approved uses of the research allowance include: the purchase of scientific materials, supplies and expendables; the purchase of minor equipment; computer software programs; costs for the use of libraries, or computers; costs associated with the publication of research results; travel expenses to attend scientific meetings; and purchase of books, periodicals and journals. The term of the award is a maximum of five years (three years at the graduate level).
Requirements: Candidates must have been accepted into, or be currently engaged in, a full-time, thesis-based, graduate program at an Alberta-based university in a health-related discipline leading to a Master's or doctoral degree.
Restrictions: This award is not available to students registered in a course-based program. Normally, support will not be provided beyond 6 years of enrollment in graduate school. Candidates who have interrupted their training for parenting or other reasons and who have consequently exceeded this time limit, may apply for studentship support. In such cases, however, the candidate is advised to clearly explain the nature of their particular circumstances at the time of application.
Geographic Focus: Canada
Date(s) Application is Due: Mar 1; Oct 1
Amount of Grant: 20,000 USD
Contact: Pamela Valentine; (780) 423-5727, ext. 230; fax (780) 429-3509; pamela.valentine@albertainnovates.ca or grants.health@albertainnovates.ca
Internet: http://www.aihealthsolutions.ca/grants/FT-student.php#fts_description
Sponsor: Alberta Innovates-Health Solutions
10104-103 Avenue, Suite 1500
Edmonton, AB T5J 4A7 Canada

AIHS Heritage Youth Researcher Summer Science Program 449

The Heritage Youth Researcher Summer Program (HYRS) is an intensive six-week summer science program for high school student in Alberta. The program is designed to give exemplary students from around the province first-hand biomedical and health research experience at a university campus, and introduce them to the many educational and career options available in this field. Students work on health research projects in a variety of work environments of researchers who are on faculty at the University of Alberta, the University of Calgary, and the University of Lethbridge. Schedules vary slightly at each University, but are generally from July 4 through August 15. Students will be supervised by a faculty member or by a member of their research team. Research projects could be in cardiology, cell biology, genetics, biomedical engineering, population health, epidemiology, or other health research areas. The application and additional guidelines are available at the AIHS website.
Requirements: Applicants require at least an 85% standing in each of math 20-1 or math 20-2, biology 20, and one other grade 11 science class, which is the minimum requirement to work in a university laboratory. Applicants must have two reference forms completed by two of their science teachers, or one math and one science teacher. They must also submit a reference form completed by an adult who is not a relative and is from outside their school and who has known the student for at least one year. Additional guidelines are available at the website.
Restrictions: Students must be willing to commit to a full six weeks. Any time taken off will disqualify the student for the program.
Geographic Focus: All States, Canada
Date(s) Application is Due: Mar 23
Contact: Danica Wolkow, HYRS Program Provincial Coordinator; (780) 423-5727, ext 237; fax (780) 429-3509; hyrs@albertainnovates.ca
Internet: http://www.aihealthsolutions.ca/HYRS/
Sponsor: Alberta Innovates-Health Solutions
10104-103 Avenue, Suite 1500
Edmonton, AB T5J 4A7 Canada

AIHS Interdisciplinary Team Grants 450

The AIHS Interdisciplinary Team Grants Program provides opportunities for high-quality, internationally recognized teams of investigators to complete research initiatives with defined health outcomes. Funds available in this competition are to support collaborative, interdisciplinary and multi-institutional teams that address important research questions, health problems or issues in defined areas of research that are aligned with strategic research priorities of the AIHS and Alberta. The Program will provide up to $1 million per year per team for up to five years. The area of research to be pursued must be relevant to one of a number of areas of special interest identified by AIHS in collaboration with other stakeholders in Alberta. Those areas are: maternal, fetal and child health; mental health, mental illness and addictions; health system sustainability; modern lifestyles and health; health issues in rural and remote environments; health, genes and the environment; health and injury; modern techniques and technologies (including IT) and health; health and behavior: disease prevention; health and infectious diseases; and food and health. The LOI and full application must clearly identify under which of these theme areas the proposed program of research falls and how the proposed research is relevant to the theme. More than one theme can be identified as appropriate. Applicants are asked to make special reference as to how their program of research will be relevant to health issues facing vulnerable populations.
Requirements: Each eligible Interdisciplinary Team Grant application will include: A Team Leader—the Team Leader must be an established researcher with proven leadership skills and experience who will act as research program director and who will assume administrative responsibility for the grant. It is expected that the Team Leader will devote a significant and appropriate portion of their time to these tasks. The Team Leader will have their primary academic appointment at an Alberta-based university; at least two additional independent investigators with established research track records. Teams with a nucleus of experienced investigators are encouraged to include promising new investigators as part of their group; team members who collectively have an extensive record of success, are creative and original in their approach to research and its translation, and who have experience working in research teams. The specific contribution of each team member and end-user partner, where applicable, must be described; representation from more than one research discipline and from more than one "research pillar" (i.e. biomedical; clinical science; health systems and services; and the social, cultural and other factors that affect the health of populations.); and representation from more than one Alberta-based university. Multi-institutional collaboration is strongly encouraged in this Program. In addition, the members of the Team may pursue other research in addition to their commitment to the Interdisciplinary Team Grant. However, each individual investigator must contribute sufficient time to the Team Grant research program to ensure the achievement of its research and translational objectives. Collaborators from outside of Alberta who make a substantial intellectual contribution to the research program may be listed as team members. Specific and justified requests that AIHS funding be used for work performed outside of Alberta will be considered. However, it is expected that the great majority of the funds will remain in Alberta.
Geographic Focus: All States, Canada
Amount of Grant: Up to 1,000,000 USD
Contact: Pamela Valentine, Vice President of Research and Innovation; (780) 423-5727; fax (780) 429-3509; pamela.valentine@albertainnovates.ca
Internet: http://www.aihealthsolutions.ca/grants/team_guidelines.php
Sponsor: Alberta Innovates-Health Solutions
10104-103 Avenue, Suite 1500
Edmonton, AB T5J 4A7 Canada

AIHS Knowledge Exchange-Visiting Professorships 451

Visiting Professorships are awarded to particularly distinguished scientists to contribute to the education and training of students and fellows, and promote the progress of medical or health research in Alberta. Visiting Professorships may be awarded in order to assist the Alberta-based universities in planning or initiating new research directions, or for the purpose of evaluating AIHS-funded programs. The award includes an honorarium and per diem, plus travel costs. The length of the award is expected to be no longer than seven days.
Requirements: Applications should be submitted directly to AIHS at least three months prior to the proposed visit.
Geographic Focus: All States, Canada
Contact: Donna Angus, Director, Knowledge Transfer; (780) 423-5727, ext 208; fax (780) 429-3509; donna.angus@albertainnovates.ca
Internet: http://www.aihealthsolutions.ca/grants/knowledge.php
Sponsor: Alberta Innovates-Health Solutions
10104-103 Avenue, Suite 1500
Edmonton, AB T5J 4A7 Canada

AIHS Knowledge Exchange-Visiting Scientists 452

The Visiting Scientist program is intended to bring productive scientists with expertise not currently available in provincial institutions to Alberta. This expertise may consist of special knowledge, new approaches to research, and the management of new technology. The program is also intended to allow Alberta-based institutions to sponsor Alberta scientists to visit major medical or health research centers, and acquire knowledge abut new research concepts or technology. The visit must result in important contributions to medical or health related science in Alberta, and an increase in the visitor's research productivity. The sponsoring institution must assure the return of the Alberta scientist to their regular position at the completion of the award. Visiting Scientists will receive travel assistance and a research allowance to cover some of the costs of materials, supplies, and expenses critical for their participation in the program. The length of the visit ranges from one month to 12 months.
Requirements: The sponsoring institution must identify the candidate's special expertise, as well as the expected long-term research benefits of the visit. This applies to both visiting scientists to and visiting scientists from Alberta. Applications demonstrating that arrangements have been made for the visitor to meet with scientists in other Alberta universities or institutions will receive preference.
Geographic Focus: All States, Canada
Contact: Donna Angus, Director, Knowledge Transfer; (780) 423-5727, ext 208; fax (780) 429-3509; donna.angus@albertainnovates.ca
Internet: http://www.aihealthsolutions.ca/grants/knowledge.php
Sponsor: Alberta Innovates-Health Solutions
10104-103 Avenue, Suite 1500
Edmonton, AB T5J 4A7 Canada

AIHS Media Summer Fellowship 453

The AIHS Media Fellowship offers University of Alberta and University of Calgary students with strong biomedical science backgrounds the opportunity to spend 12 weeks during the summer (May through August) working with CBC Radio. Media Fellows will work as reporters, researchers and productions assistants with CBC Radio in either Edmonton or Calgary. Through the Media Fellowship, students will have an opportunity to impact the discussion of biomedical science in popular media and will be contributing to the translation of knowledge from the scientists to the media and the general public. The Media Fellowship will strengthen the relationship between scientists and the media; provide young scientists, at a critical stage in their careers, the opportunity to observe and participate in how events and ideas become news; improve the Fellow's skills in communicating complex technical subjects to the public; increase the Fellow's understanding of editorial decision-making and how news is communicated to the public; enhance the coverage of science and technology by the media; and improve public understanding and appreciation of science and medical research. The Media Fellowship provides a stipend to successful applicants, to be determined at the time of acceptance.
Requirements: An applicant must be an undergraduate- or graduate-level university student at either the University of Alberta or the University of Calgary with a strong background in biomedical science; continuing university studies in the fall in any discipline; demonstrate excellent writing skills; and work well independently as well as part of a team.
Geographic Focus: All States, Canada
Date(s) Application is Due: Mar 23
Contact: Dwayne Brunner, Interim Media Relations Manager; (780) 423-5727, 224; fax (780) 429-3509; dwayne.brunner@albertainnovates.ca
Internet: http://www.aihealthsolutions.ca/grants/training-and-early-career-development-programs/media-fellowship/
Sponsor: Alberta Innovates-Health Solutions
10104-103 Avenue, Suite 1500
Edmonton, AB T5J 4A7 Canada

AIHS Part-Time Fellowships 454

The AIHS Part-Time Fellowships enable highly qualified doctoral graduates to prepare for careers in medical or health research as independent investigators doing part-time research training during the regular academic year while enrolled in a full-time professional degree program in Alberta. Part-time awards will consist of a stipend only and will be pro-rated to the amount of time the candidate is prepared to commit to research. The full-time fellowship rate will be used as the basis for this calculation. The maximum term for the award is three years.
Requirements: Candidates must hold a PhD and must be engaged in a professional degree program (usually the MD program).
Restrictions: Part-time fellowship awards are tenable only at an Alberta-based university.
Geographic Focus: Canada
Date(s) Application is Due: Mar 1; Oct 1
Amount of Grant: 5,000-30,000 USD
Contact: Pamela Valentine; (780) 423-5727, ext. 230; fax (780) 429-3509; pamela.valentine@albertainnovates.ca or grants.health@albertainnovates.ca
Internet: http://www.aihealthsolutions.ca/grants/PT-fellow.php
Sponsor: Alberta Innovates-Health Solutions
10104-103 Avenue, Suite 1500
Edmonton, AB T5J 4A7 Canada

AIHS Part-Time Studentships 455

The Alberta Innovates-Health Solutions (AIHS) Part-Time Studentships enable students enrolled in a full-time professional degree program (usually the M.D. program) in Alberta to engage in part-time research training during the regular academic year. The award consists of a stipend only. Part-time students may elect to engage in full-time research during the summer; these students should contact AIHS to request that full-time summer trainee status be implemented.
Requirements: Candidates must normally have been accepted into, or be currently engaged in, a full-time graduate program at an Alberta-based university in a health-related discipline leading to a Master's or doctoral degree. It is expected that the amount of time the candidate will commit to research will not be less than the equivalent of one day per week, taken either as one full day or two half days.
Restrictions: The award is tenable at an Alberta-based university only.
Geographic Focus: Canada
Date(s) Application is Due: Mar 1; Oct 1
Contact: Pamela Valentine; (780) 423-5727, ext. 230; fax (780) 429-3509; pamela.valentine@albertainnovates.ca or grants.health@albertainnovates.ca
Internet: http://www.aihealthsolutions.ca/grants/PT-student.php
Sponsor: Alberta Innovates-Health Solutions
10104-103 Avenue, Suite 1500
Edmonton, AB T5J 4A7 Canada

AIHS Proposals for Special Initiatives 456

The AIHS welcomes proposals for innovative initiatives that accelerate the achievement of its overall objectives. Institutions recruiting internationally-recognized researchers to lead new initiatives should consult AIHS, if current programs are inappropriate. AIHS will also consider proposals for pilot projects in patient or population-based research. These projects must have the potential to develop information, and the investigators must have the expertise to attract grants from other agencies. The review process and financial support for special initiatives may be shared by AIHS and other agencies.
Restrictions: Applications that are normally eligible for regular operating grants from other agencies will not be considered under this program.
Geographic Focus: All States, Canada
Contact: Jacques Magnan; (780) 423-5727; jmagnan@albertainnovates.ca
Internet: http://www.aihealthsolutions.ca/grants/initiatives.php
Sponsor: Alberta Innovates-Health Solutions
10104-103 Avenue, Suite 1500
Edmonton, AB T5J 4A7 Canada

AIHS Research Prize 457

The AIHS Research Prize is intended to maintain and improve AIHS Personnel Awards as a potent vehicle for the recruitment and retention of highly qualified, internationally competitive investigators at Alberta institutions. The Prize will be made to every eligible investigator supported by AIHS (Population Health Investigator, Clinical Investigator, Scholar, Senior Scholar, Scientist). The Research Prize is not determined by institutional rank and salary scale. It is provided in recognition of the individual investigator's research accomplishments, and of his/her success in obtaining an AIHS Independent Investigator award. The Research Prize is a personal prize to the recipient. No application is required.
Requirements: To be eligible, the investigator must be currently supported by an unconditional AIHS investigator award. Investigators receiving AIHS terminal support or those on leaves of absence are not eligible to receive the Research Prize.
Geographic Focus: All States, Canada
Amount of Grant: 10,000-20,000 USD
Contact: Pamela Valentine, Vice President of Research and Innovation; (780) 423-5727; fax (780) 429-3509; pamela.valentine@albertainnovates.ca
Internet: http://www.aihealthsolutions.ca/grants/resprize.php
Sponsor: Alberta Innovates-Health Solutions
10104-103 Avenue, Suite 1500
Edmonton, AB T5J 4A7 Canada

AIHS Summer Studentships 458

The AIHS Summer Studentships offer motivated students with exceptional academic records an opportunity to participate in medical or health research in Alberta during the summer months. The award is meant to encourage students to consider pursuing formal training and a career in health research. The award is tenable for a minimum of two months, and a maximum of four months during the period of May to August. The award consists of a stipend of $1300 per month. Applications are available at the AIHS website.
Requirements: Candidates must meet one of the following criteria to be eligible to apply: registered in an Alberta-based undergraduate degree program in a medical or health-related field; registered in an undergraduate degree program outside of Alberta, and desiring to engage in research during the summer at an Alberta institution; registered in an M.D. program, and who may also hold an undergraduate or graduate degree; exceptional high school students with records of participation in the health care system, and a clear interest in pursuing a health research career; or currently in the last term of their undergraduate degrees and who have applied either to medical school or to a graduate program that would start in the coming fall semester.
Restrictions: A faculty supervisor possessing both a record of productive health-oriented research and sufficient resources to ensure the satisfactory conduct of the research must sponsor the candidate. The supervisor may sponsor no more than two summer students per year, and each student must have a separate project. These awards are not renewable, but students may reapply in subsequent years.
Geographic Focus: All States, Canada
Contact: Pamela Valentine, Vice President of Research and Innovation; (780) 423-5727; fax (780) 429-3509; pamela.valentine@albertainnovates.ca
Internet: http://www.aihealthsolutions.ca/grants/Sum-student.php
Sponsor: Alberta Innovates-Health Solutions
10104-103 Avenue, Suite 1500
Edmonton, AB T5J 4A7 Canada

AIHS Sustainability Grants 459

The Sustainability Funding Opportunity provides bridge funding for Alberta health researchers who have submitted renewal applications that, although highly rated, were not funded in the CIHR Open Operating Grants or NSERC Discovery Grant competitions. The objective of these bridge grants is to ensure the maintenance of excellent research programs without loss of momentum, staff, or trainees until the Principal Investigators have the opportunity to resubmit their research proposals. A webinar discussing the grants is available at the AIHS website.
Requirements: The grant recipient must be a full-time university-based independent investigator. The recipient must also have applied to a Tri-Council agency for a peer-reviewed, individual research grant renewal in the CIHR Open Operating Grant or NSERC Discovery Grant competitions within the prior year, and was deemed excellent but was not funded. The applicant must have at least one more opportunity for reapplication of grant renewal with the external agency. The initial renewal application to the Tri-Council agency's operating grant competition was internally peer-reviewed according to processes defined at the institution prior to submission to the Tri-Council agency. Recipients must provide specific plans for a future application to the Vice President Research of their university.
Restrictions: The Sustainability Funding Program will not supplement existing grants or cover shortfalls in approved grant support. Each institution will work through their Office of the Vice President Research for the review of applications and selection of health researchers to be offered support under the Sustainability Funding Opportunity according to the defined eligibility criteria. Applications will be evaluated by the university receiving a Sustainability Funding allocation, not by AIHS.
Geographic Focus: All States, Canada
Amount of Grant: Up to 50,000 USD
Contact: Pamela Valentine, Vice President of Research and Innovation; (780) 423-5727, ext 230; fax (780) 429-3509; pamela.valentine@albertainnovates.ca
Internet: http://www.aihealthsolutions.ca/grants/sustainability.php
Sponsor: Alberta Innovates-Health Solutions
10104-103 Avenue, Suite 1500
Edmonton, AB T5J 4A7 Canada

AINSE Honours Scholarships 460

The Australian Institute of Nuclear Science and Engineering will award ten Honours scholarships annually. Eligible students will be enrolled in an Honours or Masters program, and their university supervisor will hold a AINSE Research Award. AINSE Honour Scholars receive a $5,000 stipend.
Geographic Focus: Australia
Date(s) Application is Due: Apr 15
Amount of Grant: 5,000 AUD
Contact: Frank Bruhn, Managing Director; +61 02 9717 3388 or +61 02 9717 3376; fax +61 02 9717 9268; frankb@ainse.edu.au or ainse@ansto.gov.au
Internet: http://www.ainse.edu.au/grad_students2/honours_scholarships
Sponsor: Australian Institute of Nuclear Science and Engineering
Locked Mail Bag 2001
Kirrawee, NSW 2232 Australia

AINSE International Conference Travel Scholarships 461

Travel scholarships will be awarded to students and post doctoral fellows from AINSE member universities who wish to present their AINSE supported research at an international meeting which is being conducted outside Australia. Successful applicants will be awarded up to $900. More support will be given when you are able to indicate an intention to publish the results being presented in a prestigious journal.
Requirements: Applicants should provide their personal details, name, university, supervisor, etc. as well as the following information: full name of the conference; location of the conference; dates of the conference; are you presenting an oral or poster? If so, provide evidence of the status of your presentation oral/poster, refereed/not refereed, accepted/not accepted; details of (or plans for) publication of the results; note that the committee looks more favorably on applications with an intention to publish in high impact journals; the AINSE Award number(s) which relate to the research being presented; abstract of your presentation; and, your budget including any additional support you have secured for the conference. Applications should be sent by email to ainse@ansto.gov.au.
Geographic Focus: Australia
Amount of Grant: Up to 900 AUD
Contact: Frank Bruhn, Managing Director; +61 2 9717 3388 or +61 02 9717 3376; fax +61 02 9717 9268; frankb@ainse.edu.au or ainse@ansto.gov.au
Internet: http://www.ainse.edu.au/grad_students2/travel_scholarships
Sponsor: Australian Institute of Nuclear Science and Engineering
Locked Mail Bag 2001
Kirrawee, NSW 2232 Australia

AINSE John Ferris Memorial Award 462

Dr. John Ferris was a cherished identity at ANSTO. He graduated from the University of Adelaide in South Australia with a B.Sc (Hons) and from the University of Tasmania in 1986 with a Ph.D in Limnology. John joined ANSTO in 1991, following post-doctoral work with the Australian Antarctic Division and the Centre for Water Research at the University of Western Australia. He brought considerable expertise in ecosystem studies involving algae which was to be the basis of both his research contributions within ANSTO's Institute for Environmental Research and more recently in environmental management in safety and radiation services. John became involved in bio monitoring and environmental management at ANSTO and was responsible for ANSTO's Annual Environmental Report for many years. He was a significant contributor to ANSTO's ISO 14001 certification. In more recent times, John's research interests focused on radiation effects on non-human biota and his work contributed to international thinking in this area. A John Ferris Memorial Award has been established to honour John's life and work. The award is designed assist in supporting the best post-graduate student whose application is in environmental sciences.
Geographic Focus: Australia
Date(s) Application is Due: Apr 15
Contact: Frank Bruhn, Managing Director; +61 2 9717 3388 or +61 02 9717 3376; fax +61 02 9717 9268; frankb@ainse.edu.au or ainse@ansto.gov.au
Internet: http://www.ainse.edu.au/grad_students2/john_ferris_memorial_award
Sponsor: Australian Institute of Nuclear Science and Engineering
Locked Mail Bag 2001
Kirrawee, NSW 2232 Australia

AINSE Postgraduate Research Awards 463

Awards are offered by the Australian Institute of Nuclear Science and Engineering (AINSE) for suitably qualified persons wishing to undertake studies in AINSE's field of interest for a higher degree at a member university of AINSE. Applications for PGRAs will also be considered, as in past years, where the research aligns with the broad spectrum of research foci supported by AINSE. A full list of over 100 major and minor facilities facilities is available on request. Research Supplements ($7,500 pa) are the principal form of award and are offered to scholars who are (or will be) in receipt of an Australian Postgraduate Award (APA) or equivalent scholarship. The award also provides $10,000 pa for costs involved in using the facilities and services at Lucas Heights. Certain travel and accommodation costs to enable students to work at Lucas Heights are also provided.
Requirements: Candidates for an AINSE Postgraduate Research Supplement must be nominated by the Australian or New Zealand university where PhD enrollment is held or proposed for tenure commencing before 30 June of the year of application. The university must be a member of AINSE. Guidelines and applications can be found at the website.
Geographic Focus: Australia
Date(s) Application is Due: Apr 15
Amount of Grant: 7,500-17,500 AUD
Contact: Frank Bruhn, Managing Director; +61 02 9717 3388 or +61 02 9717 3376; fax +61 02 9717 9268; frankb@ainse.edu.au or ainse@ansto.gov.au
Internet: http://www.ainse.edu.au/grad_students2/postgraduate_awards
Sponsor: Australian Institute of Nuclear Science and Engineering
Locked Mail Bag 2001
Kirrawee, NSW 2232 Australia

AINSE Research Fellowships 464

AINSE Research Fellowship are offered at academic level B for outstanding and enthusiastic scientists with three to eight years of postdoctoral experience to undertake research in areas of science and technology requiring use of the research facilities at the Australian Nuclear Science and Technology Organisation (ANSTO). Fellows will be expected to build a research group in one of the AINSE member universities and be attached to ANSTO for experiments on the new, state of the art facilities on the OPAL research reactor or other facilities at ANSTO.
Requirements: AINSE is highlighting three research directions in calling for applications for research fellowships: neutron scattering; radiopharmaceuticals; and, high resolution climate records using nuclear techniques. Applications will also be considered which fall outside these topics which but still require use of ANSTO facilities and which align with the joint research interests of ANSTO and AINSE member universities.
Geographic Focus: Australia
Date(s) Application is Due: May 31
Contact: Frank Bruhn, Managing Director; +61 2 9717 3388 or +61 02 9717 3376; fax +61 02 9717 9268; frankb@ainse.edu.au or ainse@ansto.gov.au
Internet: http://www.ainse.edu.au/academic_researchers2
Sponsor: Australian Institute of Nuclear Science and Engineering
Locked Mail Bag 2001
Kirrawee, NSW 2232 Australia

AINSE Research Fellowships with Institute for Science and International Security Support 465

AINSE Research Fellowship are offered at academic level B for outstanding and enthusiastic scientists with 3-8 years of postdoctoral experience to undertake research in areas of science and technology requiring use of the research facilities at the Australian Nuclear Science and Technology Organisation (ANSTO). Fellows will be expected to build a research group in one of the AINSE member universities and be attached to ANSTO for experiments on the new, state of the art facilities on the OPAL research reactor or other facilities at ANSTO. Since the demise of the Access to Major Research Facilities Program (AMRFP), there is no mechanism for Australian researchers to gain funds to travel to the United Kingdom. AINSE has received a number of requests from researchers needing funding for international travel. Following discussion, AINSE will now support travel for researchers to attend ISIS, on the same basis that the AMRF program allocated funding (one trip per individual per 6 months). ANSTO research scientists will also receive the same level of support as university researchers. Maximum grant available is $5,000 for any one visit.
Requirements: AINSE is highlighting three research directions in calling for applications for research fellowships: neutron scattering; radiopharmaceuticals; and, high resolution climate records using nuclear techniques. Applications will also be considered which fall outside these topics which but still require use of ANSTO facilities and which align with the joint research interests of ANSTO and AINSE member universities.
Geographic Focus: All States
Date(s) Application is Due: May 31
Contact: Frank Bruhn, Managing Director; +61 02 9717 3388 or +61 02 9717 3376; fax +61 02 9717 9268; frankb@ainse.edu.au or ainse@ansto.gov.au
Internet: http://www.ainse.edu.au/academic_researchers2/isis_support
Sponsor: Australian Institute of Nuclear Science and Engineering
Locked Mail Bag 2001
Kirrawee, NSW 2232 Australia

AIR/IES-NCES Postdoctoral Policy Fellowships in Washington 466

The program is a project funded by the National Center for Education Statistics (NCES). Postdoctoral Fellow projects will be responsible for undertaking analysis that results in improvements to the quality, comparability, and usefulness of the Integrated Postsecondary Education Data System (IPEDS). Fellows will be introduced to the Washington higher education policy community through meetings, seminars and conversations. Fellowships provide $5,000 monthly for support while the Fellow is in residence in Washington, for a period of from nine to twelve months. An additional $5,000 will be provided for expenses for the life of the fellowship. It is expected that Fellows will be advanced graduate students completing dissertations, postdoctoral students, or institutional research professionals. The award amount is set with the expectation that the Fellowship will provide support for an individual working in residence at NCES for a period of 9 to 12 months. Proposals with other residency formats may be acceptable, and individuals may propose a project of shorter duration.
Requirements: Applications may be submitted at any time. Institutional researchers, graduate students, postdoctoral students, faculty, and others may submit applications. Fellows must be U.S. citizens or permanent residents, and will be employees of AIR during the period of the Fellowship. Before formal submission, the application should be discussed with the AIR staff. Detailed guidelines and application instructions are available at the website. Applications must be submitted electronically, as a Microsoft Word or PDF email attachment. Individuals unable to submit in this manner should contact the Project Manager for an alternative method.
Restrictions: Financial assistance for Fellowships is subject to certain statutory and other general requirements, including compliance with Circular A-133 Audit requirements and various federal statutes: Title VI of the Civil Rights Act of 1964, Title IX of the Education Amendments of 1972, Section 504 of the Rehabilitation Act of 1973, the Age Discrimination Act of 1975, and other laws and regulations prohibiting discrimination; prohibition of misconduct in science and engineering; Drug-Free Workplace requirements; restrictions on lobbying; patent and copyright requirements; cost-sharing; and the use of U.S.-flag carriers for international travel. Intellectual property developed under the Postdoctoral Policy Fellow is in the public domain.
Geographic Focus: All States
Amount of Grant: 50,000-60,000 USD
Contact: Bethany Sumrow, Project Manager; (850) 385-4155, ext. 105; fax (850) 385-5180; bsumrow@airweb2.org
Internet: http://www.airweb.org/?page=1349
Sponsor: Association for Institutional Research
1435 E Piedmont Drive, Suite 211
Tallahassee, FL 32308

AIR/NCES Fellowships for Graduate Study 467

The Association for Institutional Research (AIR), as part of its contract to improve the quality of institutional research in U.S. postsecondary education systems and hence, the Integrated Postsecondary Education Data System (IPEDS), offers a graduate fellowship program funded by the National Center for Education Statistics (NCES). The fellowships are for graduate study advancing or leading to careers in institutional research, thereby increasing the national level of expertise of institutional research officers and the data that they produce. The program awards individual fellowships for full- and part-time study for up to $10,000 annually for full-time study, $3,000 annually for part-time study. Award funds may be used for tuition, fees and other educational expenses while enrolled in a graduate program.
Requirements: Candidates must demonstrate knowledge and commitment to IR through work experience (current or prior) or previous coursework. Currently employed IR professionals are urged to apply. Fellowships may be used at any regionally accredited U.S. institution. The applicant's proposal must make the case that the course of study is appropriate for preparing for a career in institutional research, by showing that the program addresses the core competencies of institutional research. Three levels of credit-bearing graduate work will be considered eligible for the awards: certificate, masters and doctorate.
Restrictions: There is a three year maximum for the fellowship.
Geographic Focus: All States
Date(s) Application is Due: Mar 14
Amount of Grant: 9,000-30,000 USD
Contact: Bethany Sumrow, Project Manager; (850) 385-4155, ext. 105; fax (850) 385-5180; bsumrow@airweb2.org
Internet: http://www.airweb.org/?page=1381
Sponsor: Association for Institutional Research
1435 E Piedmont Drive, Suite 211
Tallahassee, FL 32308

AIR/NRC Summer Institute on First-Year Assessment Undergraduate Student Fellowships 468

The National Conference on First-Year Assessment, formerly the Summer Institute on First-Year Assessment, is designed to maximize opportunities for participant learning and networking and structured to create a cumulative learning experience. The Conference's purpose is to provide a comprehensive introduction to the theory and practice of first-year assessment. Exploring both quantitative and qualitative assessment practices, conference concurrent sessions allow participants to gain a breadth of knowledge while workshops allow an in-depth examination of major issues. The National Resource Center for The First-Year Experience and Students in Transition, in collaboration with the Association for Institutional Research (AIR), is offering up to five fellowships that will cover the registration fee for the National Conference on First-Year Assessment. The fellowships pay the participant's conference registration fee, but no travel or lodging expenses.
Requirements: The Annual Conference on The First-Year Experience Fellowship Program is open to all undergraduate students who are enrolled for the current academic year. All students who are interested in first-year programs, mentoring first-year students, and/or any other aspect of first-year student life and leadership are encouraged to apply. The purpose and aim of this fellowship program is to advance the leadership skills and knowledge base of undergraduate students so that they may in turn use what they learn at the conference to enhance and encourage first-year student development on their respective campuses.
Restrictions: Only one applicant from a single institution is eligible to receive a Fellowship.
Geographic Focus: All States

Date(s) Application is Due: Oct 24
Amount of Grant: 270-580 USD
Contact: Tricia Kennedy, (803) 777-3984; KENNEDTL@mailbox.sc.edu
Internet: http://www.airweb.org/?page=1189
Sponsor: Association for Institutional Research
1435 E Piedmont Drive, Suite 211
Tallahassee, FL 32308

AIR National Summer Data Policy Institute Fellowships **469**
The Association for Institutional Research (AIR), with support from the National Center for Educational Statistics (NCES) and the National Science Foundation (NSF), has developed a program, Improving Institutional Research in Postsecondary Educational Institutions. The goals of the program are to provide professional development opportunities to doctoral students, institutional researchers, educators and administrators, and to foster the use of the federal databases for institutional research in postsecondary education. The program has four main components: a National Data Policy Institute held each summer in the Washington, D.C. area to study the national databases of NSF and NCES; dissertation fellowships for doctoral students; research grants for practitioners and faculty; and a focused grant program supported by the National Postsecondary Education Cooperative. Dissemination of research findings is an integral part of this program. Fellowships are available to cover roundtrip transportation from the Fellow's home to Washington, D.C., hotel accommodations, a fixed per diem reimbursement for meals and incidental expenses for the duration of the Institute, and a one-year AIR membership. Acceptance to the Institute is made on the basis of a short proposal.
Requirements: Applicants must have a basic knowledge of statistical methods and be experienced in the use of software packages (i.e., SPSS, SAS). Applicants should also have an interest in using national databases for studies in institutional research. The National Data Policy Institute is open to institutional researchers, faculty, graduate students, and educators affiliated with a U.S. postsecondary institution or governance agency. A limited number of Fellowships are available to similar personnel in non-U.S. institutions. Unaffiliated individuals should contact AIR before preparing a proposal for submission. Prospective Institute Fellows are strongly encouraged to study the databases of NSF and NCES before submitting a proposal. Proposals must be submitted via the online proposal submission tool(http://airweb.org/DataPolicySubmission.html). Guidelines and detailed instructions are available at the website.
Restrictions: Proposals may not be considered for review by AIR for the following reasons: (1) inappropriate for AIR funding; (2) failure to meet AIR requirements for proposal content, format, etc; or (3) failure to meet announced proposal deadline requirements.
Geographic Focus: All States
Date(s) Application is Due: Jan 0
Contact: Bethany Sumrow; (850) 385-4155, ext. 105; bsumrow@airweb2.org
Internet: http://www.airweb.org/?page=1084
Sponsor: Association for Institutional Research
1435 E Piedmont Drive, Suite 211
Tallahassee, FL 32308

AIR Research Grants/Dissertation Fellowships **470**
The Association for Institutional Research operates a grant program to support research on a wide range of issues of critical importance to U.S. higher education. The funding for these grants comes from the National Center for Education Statistics (NCES), the National Science Foundation (NSF) and the National Postsecondary Education Cooperative (NPEC). The program supports research of national relevance for the improvement of higher education theory and practice. Proposals that use one or more of the national datasets of NCES and NSF or that address the current NPEC focus topic will be considered. Grant funds may be used to pay for salary support, benefits (if allowable by the institution), graduate assistant salaries, travel to conferences, dissemination of results (printing, etc.), materials and supplies needed for the project, consultant services, and computer software.
Requirements: Doctoral students enrolled at a U.S. postsecondary institution are eligible for dissertation fellowships of up to $20,000 for one year to support dissertation research under the guidance of a faculty dissertation advisor. Fellowships are based on a submitted budget and paid in two installments (May 1st and January 3rd). Faculty and practitioners are eligible for research grants of up to $40,000 for one year of independent research. All grant recipients must be affiliated with a U.S. postsecondary institution or relevant nonprofit higher education organization without regard to citizenship. Unaffiliated individuals may be eligible to submit a proposal, but should contact the AIR Grant Staff before submitting a proposal. Research grants are issued to the applicant's sponsored research unit to be dispersed according to an approved budget. To qualify for funding, the proposal submission must meet one or both of the following criteria: (1) Use data from one or more of the national NCES and/or NSF datasets (Appendix A); and/or (2) Address the current NPEC focus topic (Appendix B). Proposals may only be submitted electronically at http://airweb.org/grants/. The online submission site will be available in late October. The guidelines, online submission, and review process have been completely revised and streamlined, so please review them carefully.
Restrictions: Independent research grants are not available to students. Grant recipients may not purchase computer hardware, but may purchase software necessary for the project.
Geographic Focus: All States
Date(s) Application is Due: Jan 16
Contact: Randy Swing; (850) 385-4155; fax (850) 385-5180; rswing@airweb2.org
Internet: http://www.airweb.org/?page=40
Sponsor: Association for Institutional Research
1435 E Piedmont Drive, Suite 211
Tallahassee, FL 32308

AJA Ethel Marcus Memorial Fellowship in American Jewish Studies **471**
The Marcus Center's Fellowship Program was founded with the intent of creating a forum where students and scholars of the American Jewish experience could gather together to research, discuss, and study their chosen topics. Under the auspices of this unique program scholars come to Cincinnati to conduct in-depth research at the American Jewish Archives and to take part in the academic community of the Hebrew Union College-Jewish Institute of Religion. The program provides fellows with an opportunity not only to pursue their own research, but also to interact and exchange ideas with research peers as well as with the faculty and students of HUC-JIR.
Requirements: This fellowship is available to ABDs for one month of research or writing at the American Jewish Archives. Applicant must provide an up-to-date curriculum vita, a research proposal, and two faculty recommendations (including one from the dissertation supervisor). Typically, Marcus Center fellowships will be awarded to postdoctoral candidates, PhD candidates who are completing dissertations, and senior or independent scholars.
Geographic Focus: All States
Date(s) Application is Due: Mar 18
Contact: Kevin Proffitt, Program Contact; (513) 221-7444, ext. 304; fax (513) 221-7812; kproffitt@huc.edu or aja@huc.edu
Internet: http://www.americanjewisharchives.org/aja/programs/index.html
Sponsor: American Jewish Archives
3101 Clifton Avenue
Cincinnati, OH 45220-2408

AJA Lowenstein-Wiener Summer Fellowship Awards in American Jewish Studies **472**
Fellowships are available to ABDs or postdoctoral candidates for one month of research or writing at the American Jewish Archives during the stipend year. Applicant must provide an up-to-date curriculum vita, a research proposal, and evidence of published research, where possible. ABDs must provide three faculty recommendations (including one from the dissertation supervisor), and postdoctoral candidates must provide two recommendations from academic colleagues.
Requirements: Typically, Marcus Center fellowships will be awarded to postdoctoral candidates, PhD candidates who are completing dissertations, and senior or independent scholars.
Geographic Focus: All States
Date(s) Application is Due: Mar 18
Contact: Kevin Proffitt; (513) 221-7444, ext. 304; fax (513) 221-7812; kproffitt@huc.edu
Internet: http://www.americanjewisharchives.org/aja/programs/index.html
Sponsor: American Jewish Archives
3101 Clifton Avenue
Cincinnati, OH 45220-2408

AJA Marguerite R. Jacobs Memorial Postdoctoral Award in American Jewish Studies **473**
This award is made to postdoctoral candidates for one month of active research or writing at the American Jewish Archives. Applicant must provide an up-to-date curriculum vita; a research proposal; evidence of published research, where possible; and two recommendations from academic colleagues.
Requirements: Typically, Marcus Center fellowships will be awarded to postdoctoral candidates, PhD candidates who are completing dissertations, and senior or independent scholars.
Geographic Focus: All States
Date(s) Application is Due: Mar 18
Amount of Grant: 2,000 USD
Contact: Kevin Proffitt; (513) 221-7444, ext. 304; fax (513) 221-7812; AJA@cn.huc.edu
Internet: http://www.americanjewisharchives.org/aja/programs/index.html
Sponsor: American Jewish Archives
3101 Clifton Avenue
Cincinnati, OH 45220-2408

AJA Rabbi Frederic A. Doppelt Memorial Fellowship in American Jewish Studies **474**
This award is available to ABDs for one month of research or writing at the American Jewish Archives during the stipend year. Applicant must provide an up-to-date curriculum vita, a research proposal, and two faculty recommendations (including one from the dissertation supervisor).
Requirements: Typically, Marcus Center fellowships will be awarded to postdoctoral candidates, PhD candidates who are completing dissertations, and senior or independent scholars.
Geographic Focus: All States
Date(s) Application is Due: Mar 18
Amount of Grant: 1,000 USD
Contact: Kevin Proffitt; (513) 221-7444, ext. 304; fax (513) 221-7812; AJA@cn.huc.edu
Internet: http://www.americanjewisharchives.org/aja/programs/index.html
Sponsor: American Jewish Archives
3101 Clifton Avenue
Cincinnati, OH 45220-2408

AJA Starkoff Fellowship in American Jewish Studies **475**
This fellowship is available to ABDs for one month of research or writing at the American Jewish Archives. Applicant must provide an up-to-date curriculum vita, a research proposal, and two faculty recommendations.
Requirements: Typically, Marcus Center fellowships will be awarded to postdoctoral candidates, PhD candidates who are completing dissertations, and senior or independent scholars.
Geographic Focus: All States
Date(s) Application is Due: Mar 18
Amount of Grant: 1,000 USD

Contact: Kevin Proffitt, Director of the Fellowship Program, The Jacob Rader Marcus Center; (513) 221-7444, ext. 304; fax (513) 221-7812; AJA@cn.huc.edu
Internet: http://www.americanjewisharchives.org/aja/programs/index.html
Sponsor: American Jewish Archives
3101 Clifton Avenue
Cincinnati, OH 45220-2408

AJCC Cancer Staging Grants 476
AJCC seeks proposals to develop and evaluate improved staging algorithms for specific cancer sites and types. Successful proposals will present new approaches and utilize data sets available to the applicant to test and validate current or revised staging algorithms. Applicants may use elements currently used for AJCC/UICC TNM staging, additional or modified anatomic and non-anatomic factors to define prognosis, and incorporate validated factors that identify response to treatment. Proposals that address and validate the use of non-anatomic factors to improve the utility of staging and define the response to therapy are encouraged. The AJCC will provide funding to a number of groups for a variety of disease sites to promote innovative strategies for enhancing the clinical utility of staging while maintaining compatibility with current staging data collection systems. The listed application deadline is for letters of intent. Guidelines are available online.
Requirements: Eligible institutions include for-profit and nonprofit organizations; public or private institutions, such as universities, colleges, laboratories, and hospitals; units of state and local government; eligible agencies of the federal government; and domestic or foreign institutions/organizations.
Geographic Focus: All States
Date(s) Application is Due: Dec 15
Contact: Valerie Vesich, (312) 202-5420; fax (312) 202-5009; AJCC@facs.org
Internet: http://www.cancerstaging.org/initiatives/rfp.html#
Sponsor: American Joint Committee on Cancer
633 N Saint Clair Street
Chicago, IL 60611

Akonadi Foundation Anti-Racism Grants 477
The foundation's mission is to work with others to eliminate racism, with a particular focus on structural and institutional racism. Grants have supported programmatic approaches including research, policy work, advocacy, litigation, organizing, media, arts, diversity training, education, and other tools in their anti-racism work. The foundation awards general operating grants and project grants. Grants are made to organizations in the San Francisco Bay area and to national organizations with national reach. Letters of interest will be accepted year round. Full proposals should only be submitted upon request.
Geographic Focus: California
Date(s) Application is Due: Feb 18
Contact: Grants Administrator; (510) 663-3867; info@akonadi.org
Internet: http://www.akonadi.org/application_guidelines.html
Sponsor: Akonadi Foundation
469 9th Street, Suite 210
Oakland, CA 94607

ALA ACRL Doctoral Dissertation Fellowship 478
This award fosters research in academic librarianship by encouraging and assisting doctoral students in the field with their dissertation research.
Requirements: The recipient of the fellowship must meet the following qualifications: be an active doctoral student enrolled in an accredited degree-granting institution; be engaged in researching a topic related to academic librarianship; have completed all coursework; have had a dissertation proposal accepted by the institution; a recipient of the fellowship may not receive it a second time; applicant need not be an ACRL member
Geographic Focus: All States
Date(s) Application is Due: Dec 5
Contact: Megan Griffin, (800) 545-2433, ext. 2514 or (312) 280-2514; mgriffin@ala.org
Internet: http://www.ala.org/ala/mgrps/divs/acrl/awards/doctoraldissertation.cfm
Sponsor: ALA ACRL Doctoral Dissertation Fellowship
50 E Huron Street
Chicago, IL 60611

ALA Annual Diversity Research Grant 479
Continuing the American Library Association's commitment to diversity, the ALA Office for Diversity began sponsorship of a Diversity Research Grant program in 2002 to address critical gaps in the knowledge of diversity issues within library and information science. The Diversity Research Grant consists of a one-time $2000 annual award for original research and a $500 travel grant to attend and present at ALA Annual Conference. Three grants are awarded each year.
Requirements: Persons submitting a proposal must be a current member of the American Library Association. You will be required to supply your membership ID number with your proposal. Only proposals demonstrating relevance to the year's designated research topics will be fully considered. See website for complete proposal requirements.
Geographic Focus: All States
Date(s) Application is Due: Apr 30
Amount of Grant: 2,500 USD
Contact: Tracie D. Hall, Director; (800) 545-2433; thall@ala.org.
Internet: http://www.ala.org/Template.cfm?Section=awards&template=/ContentManagement/ContentDisplay.cfm&ContentID=132643
Sponsor: ALA Office for Diversity
50 E Huron Street
Chicago, IL 60611

Alabama Humanities Foundation Mini Grants 480
Designed for flexibility, Mini Grants support a wide variety of projects from workshops to radio programs and reading/discussion series. Up to $2,000 in outright and $2,000 in matching funds is available. Preliminary application can be completed anytime, with final proposals due the first working day of the following months: February, May, August, and November.
Requirements: Applications must be submitted three months prior to start date of the program. Grants are made to nonprofit public or private organizations. Funds are to be used in state only.
Restrictions: Grants are not made to individuals.
Geographic Focus: Alabama
Amount of Grant: Up to 2,000 USD
Contact: Susan Perry, Grants Director; (205) 558-3980 or (205) 558-3989; fax (205) 558-3981; sperry@ahf.net
Internet: http://www.ahf.net/programs/grantsProgram.html
Sponsor: Alabama Humanities Foundation
1100 Ireland Way, Suite 101
Birmingham, AL 35205

Alabama Humanities Foundation Planning/Consultant Grants 481
Available for planning Major Grants only, these Planning/Consultant Grants provide funds for humanities scholar honoraria and travel, as well as other planning expense. These grants are awarded in support of organizations that do not have extensive grant writing experience or expertise in planning public humanities projects. Application deadlines are the first working day of February, May, August, and November.
Requirements: Grants are made to nonprofit public or private organizations. Funds are to be used in state only.
Restrictions: Grants are not made to individuals. Colleges and universities are ineligible.
Geographic Focus: Alabama
Amount of Grant: Up to 1,000 USD
Contact: Susan Perry; (205) 558-3980 or (205) 558-3989; sperry@ahf.net
Internet: http://www.ahf.net/programs/grantsProgram.html
Sponsor: Alabama Humanities Foundation
1100 Ireland Way, Suite 101
Birmingham, AL 35205

Alabama Library Association CU.S. Research Grants 482
Sponsored by the Research Committee of the College, University, and Special Libraries (CU.S.) Division, Alabama Library Association, research supported through this program must relate to college, university, or special libraries. The Committee looks for the following: originality, validity, relevance to the profession, and applicant's need for funding support. Grant funds may be used for expenses such as duplicating, mailing, travel, office supplies, consulting services, and computer services. The deadline for submitting a grant application is normally December 1.
Requirements: Competition for the grant is open to CU.S. members employed in Alabama college, university, and special libraries, and to CU.S. members attending an ALA-accredited library school.
Geographic Focus: Alabama
Date(s) Application is Due: Dec 1
Amount of Grant: 250 USD
Contact: Sonja McAbee; (256) 782-5757; fax (256) 782-5872; smcabee@jsu.edu
Internet: http://alla.associationsonline.com/membership_grants.cfm
Sponsor: Alabama Library Association
Houston Cole Library, 700 Pelham Road N
Jacksonville, AL 36265

ALA Carroll Preston Baber Research Grant 483
The Carroll Preston Baber Research Grant is given annually to one or more librarians or library educators who will conduct innovative research that could lead to an improvement in services to any specified group(s) of people. The project should aim to answer a question that is of vital importance to the library community and the researchers should plan to provide documentation of the results of their work. Up to $3,000 is available. Half of the amount will be paid within one month of the Annual Conference. The remaining half will be provided approximately six months later after a satisfactory progress report. The application cover sheet and information about additional materials required are available at the website.
Requirements: Any ALA member may apply. The jury would welcome projects that involve both a practicing librarian and a researcher. Applications must be clearly defined, with the potential to address a currently important issue in library service that is national in scope. Applications must also demonstrate the ability to undertake and successfully complete the project, with a reasonable dissemination plan.
Restrictions: Institutional overhead is not an acceptable budget item within the parameters of the Baber Grant, nor should overhead be listed as institutional support. If travel for dissemination of results is in the budget, it must be less than 10% of the total.
Geographic Focus: All States
Date(s) Application is Due: Jan 10
Amount of Grant: Up to 3,000 USD
Contact: Rudolph Rose; (800) 545-2433; fax (312) 280-4392; nrose@ala.org
Internet: http://www.ala.org/awardsgrants/awards/55/apply
Sponsor: American Library Association
50 East Huron Street
Chicago, IL 60611-2795

ALA Donald G. Davis Article Award 484
The Donald G. Davis Article Award is presented by the Library History Round Table of the American Library Association every second year to recognize the best article written in English in the field of United States and Canadian library history including the history of libraries, librarianship, and book culture. Entries are judged on quality of scholarship, clarity of style, depth of research, and ability to place research findings in a broad social, cultural, and political context. One award will be given every second year unless the jury does not find a suitable candidate for that biennial period. Any member of the Library History Round Table may nominate one or more articles by sending a recommendation to the Chair of the Davis Award Committee. Also, the editor of the LHRT Newsletter Library History Bibliography will submit a list of candidates. The winner will be announced in a press release on or about June 1st of the award year. Certificates honoring the author(s) and journal will be presented at a Library History Round Table awards ceremony during the American Library Association Annual Conference in the year of the award.
Requirements: Entries for each biennial award must have been published between January 1 and December 31 of the two years preceding the award year.
Restrictions: Papers that have won the Justin Winsor or Jesse Shera Awards are not eligible for consideration.
Geographic Focus: All States
Date(s) Application is Due: Jan 15
Contact: Norman Rose; (312) 280-4283; nrose@ala.org or ors@ala.org
Internet: http://www.ala.org/awardsgrants/awards/138/apply
Sponsor: American Library Association
50 East Huron Street
Chicago, IL 60611-2795

ALA Emerald Research Grant Awards 485
The Emerald Research Grant is awarded to individuals seeking support to conduct research in business librarianship. Two awards of $5,000 each are given each year. The awards (check and citation) will be presented at the Reference and User Services Association (RU.S.A) Awards Ceremony at the annual conferences each year. Recipients will be required to attend the awards ceremony. Recipients may be asked to present their findings at a Business Reference and Services Section (BRASS) event within two years of receiving the award (at the discretion of the BRASS Executive Committee). Recipients will also be required to acknowledge the Emerald Research Grant when publishing or presenting their research.
Requirements: Proposals will be accepted from both individual researchers and those working collaboratively. Applicants must be members of ALA. Candidates must submit a detailed proposal outlining their proposed research project; methodology, scope and timetable; how this project fits into the existing literature; and projected outcomes, including a statement outlining how this research will benefit the library profession.
Geographic Focus: All States
Date(s) Application is Due: Dec 15
Amount of Grant: 5,000 USD
Contact: Greg Fleming, Committee Chair; (773) 834-8987; gdf@uchicago.edu
Internet: http://www.ala.org/awardsgrants/awards/132/apply
Sponsor: American Library Association
50 East Huron Street
Chicago, IL 60611-2795

ALA Excellence in Academic Libraries Award 486
The Excellence in Academic Libraries Award recognizes academic librarians and staff who work together to develop academic libraries that are outstanding in furthering the educational missions of their insitutions. The program will support three annual awards of $3,000 each by type of academic library (community, college, and university). Awards are judged on the following criteria: how academic librarians and staff work together to develop an academic library that is outstanding in furthering the educational mission of its parent institution; creativity and innovation in meeting the needs of their academic community; leadership in developing and implementing exemplary programs that other libraries can emulate; and substantial and productive relationships with classroom faculty and students. Awards will be made by an ACRL representative at a presentation at each institution receiving the award. Further recognition of award recipients will be made at the ACRL President's Program at the annual ALA conference. Funding will cover the outright awards and provide $3,000 for travel to presentation ceremonies. Applications are kept in the award pool for three years. Applications may be updated each year, as long as they are received by the award deadline. Application forms are available at the website.
Requirements: Academic and research libraries may apply. The award winners will agree to take responsibility for organizing and funding a recognition ceremony on campus to receive the award.
Geographic Focus: All States
Date(s) Application is Due: Dec 2
Contact: Megan Griffin; (800) 545-2433, ext. 2514; fax (312) 440-9374; mgriffin@ala.org
Internet: http://www.ala.org/awardsgrants/awards/234/apply
Sponsor: American Library Association
50 East Huron Street
Chicago, IL 60611-2795

ALA Frederick G. Kilgour Award for Research in Library and Information Technology 487
The Frederick G. Kilgour Award for Research in Library and Information Technology is sponsored by OCLC Online Computer Center, Inc. and the Library and Information Technology Association (LITA), a division of the American Library Association. The purpose of this award is to bring attention to research relevant to the development of information technologies, especially work which shows promise of having a positive and substantive impact on any aspect of the publication, storage, retrieval and dissemination of information, or the processes by which information and data is manipulated and managed. The award recipient receives $2,000 cash and an expense paid trip to the ALA conference (airfare and two nights lodging).
Requirements: Nominations will be accepted from any member of the American Library Association. Nominating letters should briefly address: how the research is relevant to libraries; is creative in its design or methodology; builds on existing research or enhances potential for future exploration; and/or solves an important current problem in the provision of information resources. In addition, a complete curriculum vita and a copy of several seminal publications by the nominee must be included. Nominees may be asked to submit additional information to the Kilgour Committee. Preference will be given to completed research over works in progress.
Restrictions: Currently-serving officers and elected officials of LITA, members of Kilgour Committee and OCLC employees and their immediate family members are not eligible.
Geographic Focus: All States
Date(s) Application is Due: Dec 31
Amount of Grant: 2,000 USD
Contact: Melissa Prentice; (312) 280-4268; fax (312) 280-3257; mprentice@ala.org
Internet: http://www.ala.org/awardsgrants/awards/227/apply
Sponsor: American Library Association
50 East Huron Street
Chicago, IL 60611-2795

ALA Gale Cengage History Research and Innovation Award 488
The Gale Cengage History Research and Innovation Award consists of up to $2,500 and a citation recognizing an MLS degreed librarian to further research relating to history and history librarianship. The project will be evaluated with the following criteria: value of the project to the history librarianship community; quality, depth, comprehensiveness of the literature reviewed; feasibility of the project, with a combination of the appropriateness of the research method/research design and the reasonableness of the timeline; the applicant's qualifications; and the raasonableness of proposed expenditures. The nomination form is available at the website.
Requirements: Librarians with an ALA accredited degree are encouraged to apply. Historical research must consist of an emphasis in an area reflected by the History Section's subject-oriented committees: American history; genealogy, local history, instruction and research services, and historical materials. Candidates must be members of the RU.S.A History section by the time the award is presented.
Geographic Focus: All States
Date(s) Application is Due: Dec 15
Amount of Grant: Up to 2,500 USD
Contact: Janice Schultz, Coordinator, Gale Cengage History Research and Innovation Award; (816) 252-7228; jschultz@mymcpl.org
Internet: http://www.ala.org/awardsgrants/awards/31876/apply
Sponsor: American Association of School Librarians
50 East Huron Street
Chicago, IL 60611-2795

ALA Ingenta Research Awards 489
The Ingenta Research Award is given annually by the Library Research Round Table of the American Library Association to support research projects about acquisition, use, and preservation of digital information. The grant consists of up to $6,000 for research and up to $1,000 for travel to a national or international conference to present the results of the research. Expenditures must directly support research; the award does not cover indirect costs or overhead. Half of the research amount will be paid within one month of the selection of the awardee; the remaining half will be provided approximately six months later upon the receipt of a satisfactory progress report as determined by the Ingenta Award Jury Chair and the ALA staff liaison to the Ingenta Award Jury.
Requirements: Applications are welcome from practicing librarians, faculty and students at schools of library and information science, and independent scholars. Open to non-U.S. researchers. International projects not coordinated through a U.S. institution may be required to have a U.S. account for funds deposits. Please contact the award liasion for additional information.
Geographic Focus: All States
Date(s) Application is Due: Jan 31
Amount of Grant: 7,000 USD
Contact: Denise M. Davis; (800) 545-2433, ext. 4273; dmdavis@ala.org
Internet: http://www.ala.org/ala/aboutala/offices/ors/orsawards/ingentaresearchaward/ingentaresaward.cfm
Sponsor: American Library Association
50 East Huron Street
Chicago, IL 60611-2795

ALA Innovation Award 490
This annual award recognizes a project that demonstrates creative, innovative, or unique approaches within the context of national trends to information literacy instruction or programming. The winner receives $3,000 and a certificate.
Requirements: Academic librarians or academic project teams that include an academic librarian are eligible to receive the award. Recipients must have implemented their project in an academic or research library, or through the sponsorship of a professional library organization no more than two years prior to the nomination submission deadline. Electronic submissions are required. Nominations must include a description of the innovative project, a letter supporting the project, and sufficient supporting documentation for the committee to understand its purpose, content, impact, and innovative aspects.

Geographic Focus: All States
Date(s) Application is Due: Dec 3
Amount of Grant: 3,000 USD
Contact: Casey Kinson; (800) 545-2433, ext. 2511; fax (312) 440-9374; ckinson@ala.org
Internet: http://www.ala.org/awardsgrants/awards/244/apply
Sponsor: American Library Association
50 East Huron Street
Chicago, IL 60611-2795

ALA Jan Merrill-Oldham Professional Development Grant 491
The Merrill-Oldham Award was established by the Preservation and Reformatting Section (PARS) of the Association for Library Collections & Technical Services (ALCTS) to honor the career and influence of Jan Merrill-Oldham, distinguished leader, author, and mentor in the field of library and archives preservation. This annual award consists of a $1,250 grant donated by the Library Binding Institute, and a citation. The grant is applicable toward airfare, lodging, and registration fees related to the ALA Annual Conference attendance. Application instructions are available at the website.
Requirements: Applicants must have no more than five years of experience in the field of library and archives preservation, and no previous attendance to an ALA Annual Conference. They should currently work as a librarian or para-professional within a library or archives preservation department, or have preservation responsibilities within their institution, or be currently enrolled in a preservation-related graduate program. They must be willing to participate in designated conference events, and submit a summary of their conference experience to the ALCTS Newsletter Online no later than 30 days after the conference.
Restrictions: Previous winners are not eligible.
Geographic Focus: All States
Date(s) Application is Due: Dec 1
Amount of Grant: 1,250 USD
Contact: Charles Wilt; (312) 280-5030; fax (312) 280-5033; cwilt@ala.org
Internet: http://www.ala.org/awardsgrants/awards/31536/apply
Sponsor: American Library Association
50 East Huron Street
Chicago, IL 60611-2795

ALA Margaret T. Lane/Virginia F. Saunders Memorial Research Award 492
The award is given annually to an author or shared among collaborative authors of an outstanding research article in which government documents, either published or archival in nature, form a substantial part of the documented research. Preference may be given to articles published in library literature and those that appeal to a broader audience. A plaque and $2,000 is awarded to the recipient(s). The nomination form is available at the website. Awards are announced following the ALA Midwinter meeting and presented at the Government Documents Round Table (GODORT) reception at the ALA Annual Conference.
Requirements: The article must be published, either in print or online, no more than 18 months prior to the due date for nominations. The award is not restricted to articles in library journals.
Restrictions: Only articles are eligible for the award. Book chapters, books, and dissertations are not eligible.
Geographic Focus: All States
Date(s) Application is Due: Dec 3
Amount of Grant: 2,000 USD
Contact: Andrea Sevetson; (541) 992-5461; asevetson@hotmail.com
Internet: http://www.ala.org/awardsgrants/awards/31624/apply
Sponsor: American Library Association
50 East Huron Street
Chicago, IL 60611-2795

ALA Marta Lange/CQ Press Award 493
This annual award, supported by Congressional Quarterly, recognizes an academic or law librarian who, through research, other creative activity, or service to the profession, makes distinguished contributions to bibliography and information service in law or political science. Nominees should have achieved distinction in one or more of the following areas: planning and implementing a model bibliography/information services program in a law or political science library; history of contributions to the field through research, publications, and other activities displaying active participation in the advancement of law/political science librarianship; service to the profession through ACRL or related regional and national organizations; and promotion or development of an education program for law and political science librarianship that has served as a model for other courses and programs.
Requirements: There are no restrictions on who may apply. Individuals may nominate themselves or others. Electronic submissions are required. Nominators should submit a letter of nomination including the name, address, and phone number of both the nominating party and the nominee, as well as a narrative supporting the nomination, and a current vita.
Geographic Focus: All States
Date(s) Application is Due: Dec 2
Amount of Grant: 1,000 USD
Contact: Casey Kinson; (312) 280-2511 or (312) 280-2523; ckinson@ala.org
Internet: http://www.ala.org/awardsgrants/awards/73/apply
Sponsor: American Library Association
50 East Huron Street
Chicago, IL 60611-2795

ALA NewsBank/Readex Catharine J. Reynolds Award 494
The NewsBank/Readex Catharine J. Reynolds Award provides funding for research in the field of documents librarianship or in a related area that would benefit the individual's performance as a documents librarian or make a contribution to the field. This award is named for Catharine J. Reynolds, former Head of Government Publications at the University of Colorado. It is supported by an annual contribution from the NewsBank/Readex Corporation. Up to three grants a year are awarded. Recipients are selected on the basis of proven or potential ability, promise of future usefulness and permanence in the profession, financial need, and benefit of the project to the profession as well as the individual. The application is available at the website. Award recipient(s) will be announced as soon as all recipients are informed following the ALA Midwinter meeting. Awards are presented at the Government Documents Round Table (GODORT) reception at the ALA Annual Conference.
Requirements: Any documents librarian may apply. The awards support research in all areas of government information, U.S., foreign, international, state, regional, or local. The grant can be used to finance research costs, such as computer time or travel.
Restrictions: The Awards Committee may designate the amount of each award at its discretion.
Geographic Focus: All States
Date(s) Application is Due: Dec 3
Amount of Grant: 2,000 USD
Contact: Andrea Sevetson; (541) 992-5461; asevetson@hotmail.com
Internet: http://www.ala.org/awardsgrants/awards/207/apply
Sponsor: American Library Association
50 East Huron Street
Chicago, IL 60611-2795

ALA Phyllis Dain Library History Dissertation Award 495
The award, named in honor of a library historian widely known as a supportive advisor and mentor as well as a rigorous scholar and thinker, recognizes outstanding dissertations in the general area of library history. An award of $500 is given for a work that embodies original research on a significant topic relating to the history of books, libraries, librarianship or information science. The award is given in the odd numbered years. Dissertations completed and accepted during the preceding two academic years are eligible.
Requirements: Entries are judged on: clear definition of the research questions and/or hypotheses; use of appropriate source materials; depth of research; superior quality of writing; ability to place the subject within its broader historical context; and significance of the conclusions. Four copies of the dissertation and a letter of support from the doctoral advisor or from another faculty member at the degree-granting institution are required.
Geographic Focus: All States
Date(s) Application is Due: Jan 14
Amount of Grant: 500 USD
Contact: Rudolph Rose, Staff Liaison; (800) 545-2433, ext. 4283; fax (312) 440-9374; nrose@ala.org or ala@ala.org
Internet: http://www.ala.org/awardsgrants/awards/208/apply
Sponsor: American Library Association
50 East Huron Street
Chicago, IL 60611-2795

ALA ProQuest Documents to the People Award 496
The ProQuest Documents to the People Award is a tribute to an individual, library, institution, or other non-commercial group that has most effectively encouraged the use of government documents in support of library service. A framed plaque with the individual's name and achievement plus a $3,000 monetary prize to be assigned to a project of the recipient's choice is given annually. The nomination form is available at the website.
Restrictions: The recipient is responsible for any travel expenses to attend the ALA Annual Conference.
Geographic Focus: All States
Date(s) Application is Due: Dec 3
Amount of Grant: 3,000 USD
Contact: Andrea Sevetson; (541) 992-5461; asevetson@hotmail.com
Internet: http://www.ala.org/awardsgrants/awards/435/apply
Sponsor: American Library Association
50 East Huron Street
Chicago, IL 60611-2795

ALA Samuel Lazerow Fellowship For Research in Collections and Technical Services in Academic and Research Libraries 497
This award fosters advances in acquisitions or technical services by providing fellowships to librarians for travel or writing in those fields. Proposals will be judged with an emphasis on potential significance to acquisitions or technical services, originality and creativity, clarity and completeness of the proposal, and evidence of an interest in scholarship (previous publication record). Fellowship recipients are required to submit reports of the results of their research to ACRL for possible publication in C&RL News.
Requirements: Brief proposal (five pages or less) should include a description of research, travel, or writing project; project schedule; an estimate of expenses; and an up-to-date curriculum vitae.
Restrictions: Research projects in collection development or the compilation of bibliographies will not be supported.
Geographic Focus: All States
Date(s) Application is Due: Dec 4
Samples: Sara Marcus $1,000-for her research project on the change of terms and terminology over several editions of the Sears List of Subject Headings; Ping Situ

and Shuyong Jiang $1,000-for their research project on vendor-provided records and the experience of a research library in outsourcing cataloging service for its Chinese language materials;
Contact: Megan Griffin, (800) 545-2433, ext. 2514 or (312) 280-2514; mgriffin@ala.org
Internet: http://www.ala.org/ala/mgrps/divs/acrl/awards/samuellazerow.cfm
Sponsor: American Library Association
50 East Huron Street
Chicago, IL 60611-2795

Alaska Conservation Foundation Watchable Wildlife Conservation Trust Grants 498
The Watchable Wildlife Conservation Trust is a donor-advised fund developed in cooperation with the Alaska Department of Fish and Game. The purpose of the Trust is to fund those projects that emphasize non-consumptive use of wildlife, expand wildlife conservation efforts, and broaden public support for conserving Alaska's wildlife resources. Areas of interest include: interpretation (e.g., signage, viewing guides, media programming); education (e.g., curriculum development or publication development and distribution); facilities construction; applied ecosystem research; and projects involving a state wildlife sanctuary. The Trust makes grants once per calendar year. The annual deadline is October 1. Prospective grantees should contact ACF to determine if the project fits within the grant making program and when to submit their proposal. The Trust generally funds projects in the $1,000 to $4,000 range.
Requirements: As a 501(c)3 public foundation, ACF accepts requests for funding from incorporated, tax-exempt organizations, non-incorporated organizations, and individuals.
Restrictions: Ineligible requests include: projects outside the state of Alaska; general operating support; litigation or lobbying; endowments; and scholarships.
Geographic Focus: Alaska
Date(s) Application is Due: Oct 1
Amount of Grant: 1,000-4,000 USD
Contact: Polly Carr, Program Officer; (907) 276-1917; fax (907) 274-4145; pcarr@alaskaconservation.org or grants@akcf.org
Internet: http://alaskaconservation.org/_pages/grants_amp_awards/types_of_grants/watchable_wildlife.php
Sponsor: Alaska Conservation Foundation
441 West Fifth Avenue, Suite 402
Anchorage, AK 99501-2340

ALA STS Innovation in Science and Technology Librarianship Award 499
The purpose of the award is to recognize creative, innovative approaches to solving problems or improving products and services in science and technology librarianship. The award recognizes and promotes leadership and initiative that demonstrate impact on the academic and research community. Projects nominated for the award should inspire information professionals to overcome obstacles in the profession, and should demonstrate creativity, quality and innovation within the context of national trends in science and technology librarianship. The award is administered by the ACRL Science and Technology Section and is presented at the ALA Annual Conference each year.
Requirements: Any library-related project is eligible to be nominated for the award. Nominations must include a description of the innovative project, and sufficient supporting documentation for the committee to understand the purpose, content, impact, and innovative aspects of the program. Electronic submissions are required.
Restrictions: Projects from the award sponsors are ineligible for consideration. Projects of the STS Chair/Vice-Chair while those persons are in office are not eligible to apply.
Geographic Focus: All States
Date(s) Application is Due: Dec 2
Amount of Grant: 3,000 USD
Contact: Casey Kinson, Program Coordinator; (312) 280-2511 or (312) 280-2523; fax (312) 280-2520; ckinson@ala.org
Internet: http://www.ala.org/awardsgrants/awards/411/apply
Sponsor: Association of College and Research Libraries
50 East Huron Street
Chicago, IL 60611-2795

ALA Ulrich's Serials Librarianship Award 500
This award, donated by the CSA/Ulrich Company, is given for contributions to serials librarianship in areas of professional association participation, library education, serials literature, research, or development of tools leading to better understanding of the field of serials. This annual award consists of $1,500 and a citation. The winner is notified after the Midwinter meeting.
Requirements: Applicants must be members of the Association for Library Collections and Technical Services (ALCTS). They must show a record of outstanding contributions to serials and continuing resources librarianship over the course of a professional career. Nominations should include a resume and letters of recommendation.
Restrictions: Employees of the CSA/Ulrich Company are not eligible to apply.
Geographic Focus: All States
Date(s) Application is Due: Dec 1
Amount of Grant: 1,500 USD
Contact: Charles Wilt; (312) 280-5030; fax (312) 280-5033; cwilt@ala.org
Internet: http://www.ala.org/awardsgrants/awards/216/apply
Sponsor: American Library Association
50 East Huron Street
Chicago, IL 60611-2795

ALA WESS De Gruyter European Librarianship Study Grant 501
The De Gruyter Study Grant supports research in European studies with an emphasis on librarianship, the book trade, resource documentation, and similar information-science related topics. The grant funds travel to and from Europe, in addition to transportation, room, and board in Europe for up to thirty consecutive days. The purpose of the grant is to support research on the acquisition, organization, or use of library resources from or relating to Europe. Proposals are reviewed with the following in mind: the work to be accomplished; the need for and value of the proposal research; the research methodology; whether the work can be accomplished within the time frame proposed; and whether the applicant's qualifications are sufficient to carry out the research. Additional application information is available at the website.
Requirements: Each applicant must be a member of ACRL and employed as a librarian or information professional in a university, college, community college, or research library in the year prior to application for the award. The application must include a proposal; tentative travel itinerary of up to thirty days; travel budget; and current curriculum vitae. Applications and supporting materials must be emailed.
Geographic Focus: All States
Date(s) Application is Due: Jan 13
Amount of Grant: 3,000 USD
Contact: Casey Kinson; (312) 280-2511 or (312) 280-2523; ckinson@ala.org
Internet: http://www.ala.org/awardsgrants/awards/426/apply
Sponsor: American Library Association
50 East Huron Street
Chicago, IL 60611-2795

Albany Medical Center Prize in Medicine and Biomedical Research 502
The prize recognizes a physician or biomedical scientist (or group of physicians or scientists) who has made extraordinary and sustained leadership contributions to improving healthcare and patient care; or who has successfully pursued innovative biomedical research with demonstrated translational benefits applied to improved patient care. Each year's prize winner will have demonstrated significant outcomes that offer medical value of national or international importance. Prize winner activities will include but not be limited to disease and injury management, clinical research, and basic science investigations of diseases and injuries, leading to new discoveries and improved clinical outcomes. Nomination guidelines are available online.
Requirements: Any physician or scientist or group whose work has led to significant advances in the fields of health care and scientific research with demonstrated translational benefits applied to improved patient care may be nominated. Those honored will be practitioners and/or scientists whose accomplishments and outcomes have been demonstrated in the past quarter century, with preference to demonstrated accomplishments in the past decade.
Geographic Focus: All States
Date(s) Application is Due: Jan 8
Amount of Grant: 500,000 USD
Contact: Teri A. Cerveny; (518) 262-8043; AMCprize@mail.amc.edu
Internet: http://www.amc.edu/Academic/AlbanyPrize/prize_criteria.html
Sponsor: Albany Medical Center
628 Madison Avenue, Center Building, 1st Floor
Albany, NY 12208

Alberta Law Foundation Grants 503
The objectives of the foundation are to conduct research into and recommend reform of law and the administration of justice; establish, maintain, and operate law libraries; contribute to the legal education and knowledge of the people of Alberta and provide programs and facilities for those purposes; provide assistance to native people's legal programs, student legal aid programs, and programs of like nature; and contribute to the costs incurred by the Legal Aid Society of Alberta to administer a plan to provide legal aid. To be considered for funding, programs or projects must fall within these objectives. Operating grants and project grants are awarded. The application process begins with a discussing the program or project idea with the executive director. This will be followed by an exchange of drafts during the development of the application.
Restrictions: Grants will not be made to an individual or for the support of a commercial venture. Funds are not available for bursaries, fellowships, sabbatical leave support, endowments, building funds, etc.
Geographic Focus: All States, Canada
Contact: David Aucoin, Executive Director; (403) 264-4701; fax (403) 294-9238; contact@albertalawfoundation.org
Internet: http://www.albertalawfoundation.org/Apply/general.html
Sponsor: Alberta Law Foundation
407 8th Avenue SW, Suite 300
Calgary, AB T2P 1E5 Canada

Albert and Margaret Alkek Foundation Grants 504
The foundation awards grants to Texas nonprofit organizations to support charitable, religious, scientific (primarily medical), literary, cultural and educational organizations and programs. Preference will be given to research and education-related projects that will pay lasting dividends in terms of new discoveries and improved quality of life. One application per 12-month period will be considered. There are no application deadlines or forms. Applicants should submit a one- to two-page letter of inquiry that includes a brief description of the organization and the project for which funds are being considered, and the amount of funding need in total as well as the amount being requested. Inquiries should be sent via U.S. postal mail; fax or email applications are not accepted.
Requirements: Texas nonprofit organizations are eligible.

Restrictions: The foundation does not make grants to individuals or loans of any type. The foundation does not make direct scholarships to students. All scholarship programs are administered through educational institutions. The foundation prefers not to fund: organizations that in turn make grants to others; grants intended to influence legislation or to support candidates for political office; fund-raising events such as luncheons, dinners, galas, advertising in programs, or other similar activities; charities operated by service clubs; memorials for individuals; student organizations; or purchase of uniforms, equipment, or trips for school-related organizations or sports teams.
Geographic Focus: Texas
Contact: Grants Administrator; (713) 951-0019; fax (713) 951-0043; info@alkek.org
Internet: http://www.alkek.org/grantguidelines.htm
Sponsor: Albert and Margaret Alkek Foundation
1221 McKinney, Suite 4525
Houston, TX 77010-2023

Albert and Mary Lasker Foundation Awards **505**
The major purpose of the awards is to honor the individual(s) who have made significant contributions in basic or clinical research in diseases that are the main cause of death and disability. Four categories of awards are made. The Albert Lasker Basic Medical Research Award honors the scientist or scientists who have made fundamental investigations that open new areas of biomedical science. The Lasker-DeBakey Clinical Medical Research Award honors the scientist(s) whose contributions, directly or indirectly, have led to the improvement of the clinical management or treatment of patients and to the alleviation or elimination or one of the major medical causes of disability or death. The Lasker-Koshland Special Achievement Award in Medical Science honors a scientist whose contributions to research are of unique magnitude and immeasurable influence on the course of science, health, or medicine, and whose professional career has engendered extreme respect within the biomedical community. And the Lasker~Bloomberg Public Service Award honors men and women who have helped make possible the federal legislation and funding that supports research, and who have created public communication, public health, and advocacy programs of major importance. Nominations must be received by the listed application deadline.
Geographic Focus: All States
Date(s) Application is Due: Feb 1
Amount of Grant: 25,000-50,000 USD
Contact: David Keegan; (212) 286-0222; dkeegan@laskerfoundation.org
Internet: http://www.laskerfoundation.org/awards/index.htm
Sponsor: Albert and Mary Lasker Foundation
110 East 42nd Street, Suite 1300
New York, NY 10017

Albert and Mary Lasker Foundation Clinical Research Scholars **506**
The Lasker Foundation and the National Institutes of Health have joined together in an innovative partnership to nurture the next generation of clinical researchers. The Lasker Clinical Research Scholars Program is an NIH-funded initiative providing up to 11 years of support for medical doctors to bridge the widening gap between cutting-edge research and improved patient care. Under the program, NIH Institutes will specify clinical research topics that merit support and accept applications each year from early-career investigators in those areas. Applications will be reviewed for scientific merit by a peer review group of scientists, convened by the NIH Center for Scientific Review, and make recommendations to the NIH scientific and clinical directors who will annually select outstanding scholars for placement on the NIH tenure-track. The NIH expects to appoint at least three Lasker Scholars in the first year of the program, and to ultimately reach a steady-state of 20-30 active awardees. The Lasker Foundation will assist with providing mentors for the scholars, fund travel costs for NIH lecturers and the attendance by scholars at the Lasker Awards program. The Foundation will also support annual meetings for presentations and discussions on topics related to the program. Scholars' work will receive regular evaluation at NIH. After five to seven years, each scholar will be reviewed again by a group of distinguished clinical investigators. By mutual agreement between the NIH and the candidate, a decision will be made as to whether the scholar will stay on as a tenured senior investigator at the NIH, or whether the candidate will be eligible for outside research grant support worth $500,000 a year plus overhead for four years.
Geographic Focus: All States
Contact: David Keegan; (212) 286-0222; dkeegan@laskerfoundation.org
Internet: http://www.laskerfoundation.org/programs/index.htm
Sponsor: Albert and Mary Lasker Foundation
110 East 42nd Street, Suite 1300
New York, NY 10017

Albert M. Greenfield Foundation Dissertation Fellowship **507**
This fellowship supports dissertation research in residence at the Library Company on any subject relevant to its collections. The term of the fellowship is from September to May, with a stipend of $20,000. The award may be divided between two applicants, each of whom would spend a semester in residence. To apply, an applicant should send five copies each of a brief resume, a two- to four-page description of the proposed research, two letters of reference, and a writing sample of about 25 pages.
Geographic Focus: All States
Date(s) Application is Due: Mar 1
Amount of Grant: 20,000 USD
Contact: James Green, (215) 546-3181; fax (215) 546-5167; jgreen@librarycompany.org
Internet: http://www.librarycompany.org/fellowships/greenfield.htm
Sponsor: Albert M. Greenfield Foundation
1314 Locust Street
Philadelphia, PA 19107

Albuquerque Community Foundation Grants **508**
The foundation seeks to improve the quality of life in the greater Albuquerque, New Mexico, area by providing support for projects and organizations that serve the community in arts and culture, education, environmental and historic preservation, children and youth, and health and human services. Through its grant program, the foundation supports projects that are innovative, meet the needs of underserved segments of the community, encourage matching funds or additional gifts, promote cooperation among agencies, empower the disadvantaged and disabled, and enhance the effectiveness of local charitable organizations. Types of support include continuing support, general operating support, program development, publication, seed grants, scholarships funds and scholarships to individuals, and technical assistance.
Requirements: IRS 501(c)3 organizations based in Albuquerque, NM, are eligible. Proposals are reviewed on the basis of the following priorities: impact, innovation, leverage, management, and nonduplication.
Restrictions: Grants are generally not made to or for individuals, political or religious purposes, debt retirement, payment of interest or taxes, annual campaigns, endowments, emergency funding, to influence legislation or elections, scholarships, awards, or to private foundations and other grantmaking organizations.
Geographic Focus: New Mexico
Date(s) Application is Due: Apr 16; Aug 15
Amount of Grant: Up to 10,000 USD
Contact: R. Randall Royster; (505) 883-6240; rroyster@albuquerquefoundation.org
Internet: http://www.albuquerquefoundation.org/grants/grant-home.htm
Sponsor: Albuquerque Community Foundation
P.O. Box 36960
Albuquerque, NM 87176-6960

Alcatel-Lucent Technologies Foundation Grants **509**
The Alcatel-Lucent Foundation is the philanthropic arm of Alcatel-Lucent and it leads the company's charitable activities. With a focus on volunteerism, the Foundation's mission is to support the commitment of Alcatel-Lucent to social responsibility by serving and enhancing the communities where its employees and customers live and work. To accomplish its mission, the Foundation manages grants and employee volunteerism on a global level. It receives its income from the corporation-Alcatel-Lucent-whose name it bears. However, legally the Foundation is an independent, charitable, non-profit and private entity and is governed by its own board of trustees that is separate from the corporate board of directors. Global Foundation grants are dedicated to the main focus areas of the Foundation and are managed by the Foundation.
Requirements: Giving is on an international basis.
Geographic Focus: All States, All Countries
Amount of Grant: 500-5,000 USD
Contact: Bishalakhi Ghosh, Executive Director; +91-99-58418547 or +91-22-66798700; fax +91-22-26598542; bishalakhi.ghosh@alcatel-lucent.com
Internet: http://www.alcatel-lucent.com/wps/portal/foundation
Sponsor: Alcatel-Lucent Technologies Foundation
600 Mountain Avenue, Room 6F4
Murray Hill, NJ 07974-2008

Alces Foundation Grants **510**
The Alces Foundation, established in Massachusetts and managed by Welch and Forbes, is focused on the support of multi-purpose art centers, elementary and secondary education, higher education, the environment, performing arts, theater, and interfaith religious issues in Massachusetts. There are no specific application formats or deadlines with which to adhere, and potential applicants should forward a letter of inquiry to the office.
Geographic Focus: Massachusetts
Contact: Jay Emmons, Executive Vice President; (617) 523-1635; fax (617) 742-6243; info@welchforbes.com
Sponsor: Alces Foundation
45 School Street, Old City Hall, 5th Floor
Boston, MA 0002108-3204

Alcoa Foundation Grants **511**
General priorities of the foundation include safe and healthy children and families—ensuring that children and their families have the tools, the knowledge and the services to remain healthy and safe at home, in the community and in the workplace; conservation and sustainability—educating young leaders on conservation issues, protecting forests, promoting sound public policy research, and understanding the linkages between business and the environment; skills today for tomorrow—providing individuals with critical skills and services to be economically connected, workplace-ready, and productive in a changing economy; business and community partnerships—strengthening the nonprofit sector and developing meaningful partnerships among nonprofits, the private sector, and local government; and global education in science, engineering, technology and business—broadening student participation in areas central to Alcoa to prepare a diverse cross-section of our communities for a global workplace. Types of support include capital grants, building funds, challenge grants, matching gifts, general support, research grants, scholarships, and seed money. Initial contact should be a letter of inquiry.
Requirements: The foundation awards grants to nonprofit public charities in communities where Alcoa has a presence. Local Alcoans work within their communities to evaluate organizations and make recommendations for funding to Alcoa Foundation. Nonprofit organizations that serve localized communities should find the Alcoa facility nearest to them and write a one-page letter describing their mission, nature of request, connection to the areas of excellence and offering contact information. If interested, the Alcoa location contact will notify the requesting organization and invite them to submit

more information. Areas of operation include western Pennsylvania; Davenport, Iowa; Evansville, Indiana; Massena, New York; New Jersey; Cleveland, Ohio; Knoxville, Tennessee; and Rockdale, Texas.
Restrictions: The foundation does not make gifts to local projects other than those near Alcoa plant or office locations; endowment funds, deficit reduction, or operating reserves; hospital capital campaign programs unless the hospital presents a comprehensive area analysis that justifies, on a regional rather than an individual institutional basis, the need for the capital improvement; individuals, except for the scholarship program for children of Alcoa employees; tickets and other promotional activities; trips, tours, or student exchange programs; or documentaries and videos.
Geographic Focus: All States
Amount of Grant: 1,000-50,000 USD
Contact: Meg McDonald; (412) 553-2348; fax (412) 553-4498; alcoa.foundation@alcoa.com
Internet: http://www.alcoa.com/global/en/community/foundation.asp
Sponsor: Alcoa Foundation
201 Isabella Street
Pittsburgh, PA 15212-5858

Alcohol Misuse and Alcoholism Research Grants 512
The foundation's research interests include factors influencing transitions in drinking patterns and behavior, effects of moderate use of alcohol on health and well-being, mechanisms underlying the behavioral and biomedical effects of alcohol, biobehavioral/interdisciplinary research on the etiology of alcohol misuse. Applications may be obtained from the Web site or by contacting the office. Grantees are usually notified within two weeks following the advisory council meetings, which are held in April and November.
Requirements: Applications may be submitted by public or private nonprofit organizations such as universities, colleges, hospitals, research institutes and organizations, and governmental research agencies and laboratories in the United States and Canada.
Restrictions: Non-research activities such as education projects, public awareness efforts and treatment or referral services are not eligible for support. The Foundation also does not support the training of pre- and post-doctoral fellows, undergraduates, graduate students, medical students, interns or residents. It does not fund thesis or dissertation research. The Foundation does not encourage applications on treatment of the complications of advanced alcoholism. However, research involving treatment intended to elucidate the pathogenesis of alcohol-related problems will be considered.
Geographic Focus: All States
Date(s) Application is Due: Feb 1; Sep 1
Amount of Grant: Up to 100,000 USD
Contact: Grants Administrator; (410) 821-7066; fax (410) 821-7065; info@abmrf.org
Internet: http://www.abmrf.org/grants.htm
Sponsor: Alcoholic Beverage Medical Research Foundation (ABMRF)
1122 Kenilworth Drive, Suite 407
Baltimore, MD 21204

Alcon Foundation Grants 513
The foundation supports organizations in the fields of health care, leadership programs, research, education and community responsibility. Programs that advance the education and skill levels of eye care professionals are given special consideration. General operating grants to organizations and institutions improving education and research in the areas of specialization of Alcon Laboratories—ophthalmology and vision care. Grants are also awarded to community activities that benefit company employees. Applications may be submitted at any time.
Requirements: Grants are not made for building programs.
Restrictions: Non 501(c)3organizations, individuals and scholarship programs, religious, veterans or fraternal organizations, political causes, capital campaigns, matching gifts, trips, tournaments and tours, and endowments are not supported.
Geographic Focus: All States
Amount of Grant: 100-50,000 USD
Contact: Mary Dulle, Chair; (817) 293-0450; Mary.Dulle@Alconlabs.com
Internet: http://www.alconlabs.com/corporate-responsibility/alcon-foundation.asp
Sponsor: Alcon Foundation
6201 S Freeway
Fort Worth, TX 76134

Alexander and Margaret Stewart Trust Grants 514
The trust awards grants in the greater Washington, DC, area for cancer treatment, especially equipment used in diagnosis and treatment; and caring for children who are physically ill, mentally ill, or disabled. Grants also support research, education, and prevention of common childhood diseases, including negative societal behavioral patterns that impact children. Proposals for projects aiding the economically deprived receive preference. The trust awards start-up funding. Applications may be submitted at any time; requests received by the listed deadline date are reviewed by the end of the year.
Requirements: Nonprofits in the greater District of Columbia area are eligible.
Restrictions: Requests for endowments, buildings, or capital campaigns are denied.
Geographic Focus: District of Columbia
Date(s) Application is Due: Sep 15
Amount of Grant: 50,000-150,000 USD
Contact: William J. Bierbower, (202) 785-9892; wbierbower@stewart-trust.org
Internet: http://www.stewart-trust.org/guidelines.htm
Sponsor: Alexander and Margaret Stewart Trust
888 17th Street NW, Brawner Building, Suite 610
Washington, DC 20006-3321

Alexander von Humboldt Foundation Anneliese Maier Award 515
Nominations for the Anneliese Maier Award can be made for researchers from abroad in the humanities and social sciences. Researcher achievements should be internationally recognized, along with research collaboration with specialist colleagues in Germany. A sustainable contribution is expected by these researchers toward the further internationalization of the humanities and social sciences in Germany. The Award is also aimed at researchers who are not yet so advanced in their scientific careers, but who are already known in the shaping of the humanities and social sciences in Germany. Award winners are expected to spend up to five years cooperating on a long-term research project with the nominator and/or specialist colleagues at a research institution in Germany. Up to eight Maier Research Awards are given annually, and are valued at 250,000 EUR each. Additional information is available at the website.
Requirements: Nominations may be submitted by established academics in Germany. The nominees must be recognized as established leaders in their subject. Researchers who are not yet advanced, but internationally established may also be recognized. One international expert and one German expert review by collaborative partners and/or academics are necessary.
Restrictions: Direct applications are not accepted.
Geographic Focus: All States, All Countries
Date(s) Application is Due: Apr 30
Amount of Grant: 250,000 EUR
Contact: Gerhard Lauer; (+49) 0228-833-0; fax (+49) 0228-833-199; info@avh.de
Internet: http://www.humboldt-foundation.de/web/anneliese-maier-award.html
Sponsor: Alexander von Humboldt Foundation
Jean-Paul-Street 12
Bonn, D-53173 Germany

Alexander von Humboldt Foundation CAPES-Humboldt Fellowships for Experienced Researchers 516
The CAPES-Humboldt Fellowships for Experienced Researchers enable highly qualified scientists from Brazil to spend extended periods of research (6-18 months) in Germany. Research projects are carried out in cooperation with academic hosts at research institutions in Germany. Candidates are expected to have their own, clearly defined research profile. The Fellowship is worth 3,100 EUR per month and includes the following benefits: an allowance for research costs; language fellowship for the Fellow and marital partner at selected language institutes in Germany immediately prior to the actual period of research; family and child allowances; a Europe allowance for a research stay at a research institute if necessary to ensure the success of the research project; and an extensive alumni sponsorship once the successful research stay has come to an end. A candidate's academic eligibility is assessed on the following criteria: academic record and performance; quality of key publications; quality of research plan; and the academic's future potential. Additional information is available at the website.
Requirements: Candidates should be working at the level of assistant professor, junior research group leader, or be able to document independent research work over a number of years. Scientists and scholars from all disciplines may apply. Applicants must hold a doctorate from Brazil or abroad with an official revalidation in Brazil completed less than twelve years prior to the application date. They must also hold Brazilian nationality or permanent residence visa in Brazil, with proven teaching or research activity in Brazil. Natural scientists and engineers must have a good knowledge of German or English. Physicians and researchers from humanities and social sciences must have a good knowledge of German if is necessary to carry out the project (otherwise a good knowledge of English is required). Eligibility requirements also include confirmation that research facilities are available, with a mentoring agreement and detailed expert's report by the academic host.
Restrictions: Research already funded in other programs cannot be financed again within the CAPES Fellowships. Potential applicants who have been in Germany for more than six months at the time of application are not eligible.
Geographic Focus: Brazil
Date(s) Application is Due: Jan 31; Aug 31
Amount of Grant: 37,200 EUR
Contact: Denise de Menezes Neddermeyer, Directorate of International Relations; (61) 2022-6900; fax (61) 2022-6902; dri@capes.gov.br
Internet: http://www.humboldt-foundation.de/web/capes-humboldt-fellowship.html
Sponsor: Alexander von Humboldt Foundation
Jean-Paul-Street 12
Bonn, D-53173 Germany

Alexander von Humboldt Foundation CAPES-Humboldt Fellowships for Postdoctoral Researchers 517
The CAPES-Humboldt Fellowships for postdoctoral researchers enable highly qualified scientists from Brazil to spend extended periods of research (6-24 months) in Germany. The Fellowship is worth 2,600 EUR per month. Additional benefits include an allowance for research costs, a language stipend for the Fellow and marital partner at selected language institutes in Germany, family and child allowances, and extensive alumni sponsorship. Scientists and scholars from all disciplines may apply, and there are no quotas for fields of disciplines. Candidate selection is based on academic record and performance; quality of key publications; quality of research plan; and the applicant's future potential and career prospects. The application documents and additional information are available at the website.
Requirements: Postdoctoral researchers must be scientists or scholars from Brazil at the beginning of their career with above average qualifications, who have earned a doctoral degree within the last four years. Candidates choose their own research projects and their host in Germany, and prepare their own research plan. Details of the research project and time schedule must be agreed upon with the prospective host in advance. They

must also have an academic publication file, with proven teaching or research activity in Brazil, along with Brazilian nationality or permanent residence visa in the country. Applicants must confirm that research facilities are available. A mentoring agreement with a detailed expert's report and proposed timetable should be discussed within the application. Natural scientists and engineers must have a good knowledge of German or English; physicians and researchers in the humanities and social sciences must have a good knowledge of German if it is necessary to carry out the project successfully; otherwise a good knowledge of English is required.
Restrictions: Short term visits for study and training purposes or for attending conferences are not eligible. Potential applicants who have been in Germany for more than six months are not eligible to apply. Research already funded in other programs cannot be financed again with the CAPES Fellowship.
Geographic Focus: Brazil
Date(s) Application is Due: Jan 31; Aug 31
Contact: Denise de Menezes Neddermeyer, Directorate of International Relations; (61) 2022-6900; fax (61) 2022-6902; dri@capes.gov.br or humboldt@capes.gov.br
Internet: http://www.humboldt-foundation.de/web/capes-humboldt-fellowship.html
Sponsor: Alexander von Humboldt Foundation
Jean-Paul-Street 12
Bonn, D-53173 Germany

Alexander von Humboldt Foundation Feodor-Lynen Research Fellowships for Experienced Researchers 518

The Feodor Lynen Research Fellowships for experienced researchers allows a researcher to carry out a long-term project (6 to 18 months) selected in cooperation with an academic host at a research institution in Germany. Candidates choose their own research projects and their host abroad and prepare their own research plan. The fellowship is flexible and can be divided into as many as three stays within three years. The Foundation tries to involve the host in financing the research fellowship. Scientists and scholars from all disciplines may apply to the Foundation directly at any time. The provisions of the fellowship comprise a basic monthly sum and a monthly foreign allowance which may differ according to location and marital status. Marital and child allowances may be paid in addition. Additional benefits may include: return travel expenses; a reintegration allowance; a return fellowship; and an alumni sponsorship. Applicants can apply if he/she has been living in Germany continuously for at least five years at the time of application and Germany is their principal country of residence and work. The Foundation awards up to 150 Fellowships for postdoctoral researchers and experienced researchers annually. There is no closing date for submitting applications.
Requirements: Applicants must furnish proof of work published in recognized academic journals; submit a research plan agreed with the foreign host, including confirmation that research facilities can be made available; and possess a good knowledge of the host country's language or at least a very good knowledge of English, provided they can carry out their research project in that language. The host must be an academic working abroad who has already been sponsored by the Humboldt Foundation.
Geographic Focus: All States, Germany
Contact: Program Contact; (+49) 0228-833-0; fax (+49) 0228-833-199; info@avh.de
Internet: http://www.humboldt-foundation.de/en/programme/stip_deu/flf_e.htm
Sponsor: Alexander von Humboldt Foundation
Jean-Paul-Street 12
Bonn, D-53173 Germany

Alexander von Humboldt Foundation Feodor-Lynen Research Fellowships for Postdoctoral Researchers 519

Feodor Lynen Research Fellowships for postdoctoral researchers are the instrument with which the Alexander von Humboldt Foundation enables highly-qualified scientists and scholars from Germany who are just embarking on their academic careers and who completed their doctorates less than four years ago1 to spend extended periods of research (6-24 months) abroad. Applicants can apply if he/she has been living in Germany continuously for at least five years at the time of application and Germany is the candidate's principal country of residence and work. Scientists and scholars from all disciplines may apply for any target country abroad. Candidates choose their own research projects and their host abroad and prepare their own research plan. The provisions of the fellowship comprise a basic monthly sum and a monthly foreign allowance which may differ according to location and marital status. Marital and child allowances may be paid in addition. Additional benefits may include: return travel expenses; reintegration allowance; a return fellowship; and alumni sponsorship. There is no closing date for submitting applications.
Requirements: Applicants must be less than 38 years old; have successfully completed their doctorate with very good or good ratings; furnish proof of work published in recognized academic journals; submit a research plan agreed with the foreign host, including confirmation that research facilities can be made available; and possess a good knowledge of the host country's language or at least a very good knowledge of English, provided they can carry out their research project in that language. The academic host must be an academic working abroad who has already been sponsored by the Humboldt Foundation. Details of the research project and the time schedule must be agreed upon with the prospective host in advance.
Restrictions: Short-term visits for study and training purposes or for attending conferences are not eligible for sponsorship.
Geographic Focus: All States, Germany
Contact: Program Contact; (+49) 0228-833-0; fax (+49) 0228-833-199; info@avh.de
Internet: http://www.humboldt-foundation.de/web/2983.html
Sponsor: Alexander von Humboldt Foundation
Jean-Paul-Street 12
Bonn, D-53173 Germany

Alexander von Humboldt Foundation Feodor Lynen Research Fellowships for Experienced Researchers 520

Each year up to 150 Feodor Lynen Research Fellowships are awarded to German researchers. The Fellowship allows experienced researchers to carry out a research project from six to 18 months with an academic host they have selected at an international research institute. The Fellowship is flexible and can be divided up into as many as three stays within three years. Scientists and scholars from all disciplines may apply to the Alexander von Humboldt Foundation directly at any time. The application and additional information about the stipend and benefits are available at the website.
Requirements: Applicants must meet the following *Requirements:* hold a doctorate or comparable degree that was completed less than twelve years prior to the Fellowship application date; a research profile documented by academic publications; confirmation that research facilities are available and a mentoring agreement and detailed expert's report from an academic host; three recommendation letters; and a good knowledge of the language spoken at the chosen research facility.
Restrictions: Short-term study visits, participation in congresses, and training courses are not eligible. Applications to continue a research stay already underway can only be considered in exceptional cases. The applicant may not have spent an extended period in the host country prior to application. Potential applicants who have been in the host country for more than six months at the time of application are not eligible to apply.
Geographic Focus: Germany
Contact: Anne Bohnenkamp-Renken, Selection Committee Chairperson; (+49) 0228-833-0; fax (+49) 0228-833-199; info@avh.de
Internet: http://www.humboldt-foundation.de/web/feodor-lynen-fellowship-experienced.html
Sponsor: Alexander von Humboldt Foundation
Jean-Paul-Street 12
Bonn, D-53173 Germany

Alexander von Humboldt Foundation Fraunhofer-Bessel Research Awards 521

The Fraunhofer-Gesellschaft for the Promotion of Applied Research and the Alexander von Humboldt Foundation grant up to three research awards annually to internationally renowned academics from all non-European countries in recognition of their outstanding academic performance in applied research to date. Scientists and scholars from all non-European countries, who have completed their doctorates less than twelve years ago, who are internationally recognized for their achievements in the field of applied research, and who in future are expected to continue producing cutting-edge achievements which will have a seminal influence on their discipline beyond their immediate field of work, are eligible to be nominated for an Award. Winners are invited to spend a period of up to one year cooperating on a long-term research project with specialist colleagues at one of the Fraunhofer Institutes in Germany. The stay may be divided up into blocks. The Foundation and the Fraunhofer Society for the Advancement of Applied Research jointly grant up to three Awards annually. The award is valued at 45,000 EUR.
Requirements: Academics of all disciplines can be nominated for the Fraunhofer-Bessel Research Award, provided that they have received their doctoral degree within the past 18 years. The Foundation particularly encourages the nomination of qualified female academics. At the time of nomination, the candidate must have been predominantly living and working abroad for at least five years. Nominations may be submitted by the heads of institutes or senior academics at institutes belonging to the Fraunhofer Society. Direct applications are not accepted.
Restrictions: The Fraunhofer-Bessel Research Award can only be granted once to any one academic. Researchers who have already received an award for their academic achievement from the Humboldt Foundation are not eligible to be nominated. Furthermore, the nominee should not have recently won any awards or fellowships in Germany for his or her academic achievements.
Geographic Focus: All States, Germany
Amount of Grant: 45,000 EUR
Contact: Dr. Anke Hellwig, Department of International Business Development; (+49) 0891-205-4710 or (+49) 0228-833-0; fax (+49) 0228-833-199; anke.hellwig@zv.fraunhofer.de or info@avh.de
Internet: http://www.humboldt-foundation.de/web/5174.html
Sponsor: Alexander von Humboldt Foundation
Jean-Paul-Street 12
Bonn, D-53173 Germany

Alexander von Humboldt Foundation Friedrich Wilhelm Bessel Research Awards 522

The Alexander von Humboldt Foundation awards annually about 25 Friedrich Wilhelm Bessel Research Awards, funded by the Federal Ministry for Education and Research, to internationally renowned scientists and scholars from abroad in recognition of their outstanding accomplishments in research to date and their exceptional promise for the future. Scientists and scholars of all disciplines or research areas, regardless of their nationality, may be nominated for an Award, provided they have received their doctoral degree within the past 12 years. The Foundation particularly encourages the nomination of female scholars. In addition, awardees are invited to conduct an original research project of their own design in close collaboration with an appropriate colleague in Germany over a period of six to twelve months, which may be divided into segments. The award is valued at 45,000 Euro, and nominations may be submitted by regular mail at any time..
Requirements: A candidate must be nominated by distinguished scientists/scholars employed by any university or research institution in Germany. Direct applications are not accepted. Scientists and scholars who are nominated must have attained international reputation through research in their chosen field; they must also show promise that their further research will remain on the cutting edge and have impact beyond their area of specialization.
Geographic Focus: All States, All Countries, Germany
Amount of Grant: 45,000 EUR

Contact: Program Contact; (+49) 0228-833-0; fax (+49) 0228-833-199; info@avh.de
Internet: http://www.humboldt-foundation.de/web/bessel-award.html
Sponsor: Alexander von Humboldt Foundation
Jean-Paul-Street 12
Bonn, D-53173 Germany

Alexander von Humboldt Foundation Georg Forster Fellowships for Experienced Researchers 523

In providing Georg Forster Research Fellowships for experienced researchers, the Alexander von Humboldt Foundation enables highly-qualified scientists and scholars from abroad, who completed their doctorates less than twelve years ago to spend extended periods of research (6-18 months; may be divided up into a maximum of three blocks) in Germany. Candidates are expected to have their own, clearly defined research profile. This means they should usually be working at least at the level of Assistant Professor or Junior Research Group Leader or be able to document independent research work over a number of years. The fellowship is worth 3,150 EUR per month. This includes a mobility lump sum and a contribution towards health and liability insurance. Additional benefits may include: return travel expenses; language fellowship; family allowances; additional extension; a subsidy towards research costs; a Europe allowance; and an extensive alumni sponsorship. There is no closing date for submitting applications. Applications are processed as part of an ongoing procedure.
Requirements: Scientists and scholars from all disciplines from developing countries, emerging economies and transition states may apply. Applicants must have completed their doctorates less than twelve years ago. Scholars in the humanities or social sciences and physicians must have a good knowledge of German if it is necessary to carry out the project successfully; otherwise a good knowledge of English; scientists and engineers must have a good knowledge of German or English.
Restrictions: Short-term study tours, participation in conferences, or educational visits cannot be funded. Applications for extension of research stays already commenced in Germany cannot be considered.
Geographic Focus: All Countries
Amount of Grant: Up to 37,800 EUR
Contact: Program Contact; (+49) 0228-833-0; fax (+49) 0228-833-199; info@avh.de
Internet: http://www.humboldt-foundation.de/web/georg-forster-fellowship-experienced.html
Sponsor: Alexander von Humboldt Foundation
Jean-Paul-Street 12
Bonn, D-53173 Germany

Alexander von Humboldt Foundation Georg Forster Fellowships for Postdoctoral Researchers 524

Fellowships for postdoctoral researchers are the instrument with which the Foundation enables highly-qualified scientists and scholars who are just embarking on their academic careers and who completed their doctorates less than four years ago to spend extended periods of research (6-24 months) in Germany. The research proposal should deal with issues of major relevance to the future development of the candidate's country of origin and, in this context, particularly suited to transferring knowledge and methods to developing and threshold countries. Research projects are carried out in cooperation with academic hosts at research institutions in Germany. Candidates choose their own research projects and their host in Germany and prepare their own research plan. The fellowship is worth 2,650 euro per month. This includes a mobility lump sum and a contribution towards health and liability insurance. There is no closing date for submitting applications. Applications are processed as part of an ongoing procedure.
Requirements: Scientists and scholars from all disciplines from developing and threshold countries (excluding People's Republic of China, India and Turkey) may apply.
Geographic Focus: All States
Amount of Grant: 31,800 EUR
Contact: Program Contact; (+49) 0228-833-0; fax (+49) 0228-833-199; info@avh.de
Internet: http://www.humboldt-foundation.de/web/georg-forster-research-fellowship-postdoc.html
Sponsor: Alexander von Humboldt Foundation
Jean-Paul-Street 12
Bonn, D-53173 Germany

Alexander von Humboldt Foundation Georg Forster Research Award 525

The Georg Forster Research Award is granted in recognition of a researcher's entire achievements to academics of all disciplines whose fundamental discoveries, new theories, or insights have had a significant impact on their own discipline and beyond, and who are expected to continue developing research-based solutions to the specific challenges facing transition and developing countries. Up to four awards are offered every year at 60,000 EUR each. Award winners are invited to conduct a research project of their own choosing in Germany in close collaboration with a specialist colleague. The project duration of about six to twelve months may be divided into segments. In addition, further expenditures (such as travel or language courses) are covered in the context of the invitation to stay in Germany for research purposes. To support the collaboration, the Alexander von Humboldt Foundation may grant additional funding of up to 25,000 EUR, particularly for participating in scientific conferences, additional material resources, such as specialist literature and scientific equipment for the nominee's own institute, as well as for involving junior researchers. An established academic employed by a university or research institution in Germany must take the initiative to nominate a candidate for the Georg Forster Research Award. Nominations may also be initiated by award winners of the Humboldt Foundation working abroad, provided that the nomination is made jointly with a colleague working in Germany. Nomination information is posted on the website.
Requirements: Nominees must be nationals of a developing or transition country (a full list of qualifying countries is available at the website). At the time of nomination, they must have had their main residence and place of work in one of these countries for at least five years. They must be recognized internationally, and continue developing research based solutions for the challenges facing transition and developing countries.
Restrictions: Self nomination is not acceptable.
Geographic Focus: All States, All Countries
Date(s) Application is Due: May 15
Amount of Grant: Up to 85,000 EUR
Contact: Bernhard Fleischer, Selection Committee Chairperson; (+49) 0228-833-0; fax (+49) 0228-833-199; info@avh.de
Internet: http://www.humboldt-foundation.de/web/georg-forster-award.html
Sponsor: Alexander von Humboldt Foundation
Jean-Paul-Street 12
Bonn, D-53173 Germany

Alexander von Humboldt Foundation German Chancellor Fellowships for Prospective Leaders from the People's Republic of China 526

The Alexander von Humboldt Foundation's German Chancellor Fellowships are for university graduates from the People's Republic of China with an interest in international issues and demonstrated leadership potential. Up to ten Fellowships are given annually. Fellows are mentored by hosts in Germany chosen by the Fellows on the basis of their expertise in the respective areas. The program is targeted at accomplished young professionals who are likely to become decision-makers, thought leaders, and influential voice in their respective fields. During their time in Germany, Fellows conduct independent projects at their host institutions. Projects should be relevant to modern societies, have a long term and visible impact, and help to advance Fellow's careers and professional development. It should also involve original exploration of a topic or issue or research in the Fellow's field of interest. Through their experience in Germany, Fellows will acquire greater knowledge of their fields, gain new international experience, and strengthen their intercultural competence. In addition to a stipend, Fellows also receive extensive benefits which are discussed in detail in the guidelines at the website. The application is also available at the website.
Requirements: Applicants may apply from a broad range of areas such as politics and public policy, law, media, business, the non-governmental sector, and the arts. Candidates will spend one year in Germany where they will network with other prospective leaders from abroad to explore new solutions to global issues. Candidates must meet the following *Requirements:* citizenship from the People's Republic of China; a bachelor's or comparable academic degree completed less than twelve years prior to the Fellowship's starting date; demonstrated leadership potential; either English or German language skills; a project plan detailing the project's subject matter, goals, and a time plan; a detailed host's statement, including a confirmation of support; and two to three expert recommendations from those qualified to comment on the candidate's professional, personal, and/or academic background.
Geographic Focus: China
Date(s) Application is Due: Sep 15
Amount of Grant: 25,800-33,000 EUR
Contact: Jianhua Zhu, Selection Committee Chairperson; (+49) 0228-833-0; fax (+49) 0228-833-212; chancellor-chn.select@avh.de or buka@avh.de
Internet: http://www.humboldt-foundation.de/web/4074.html
Sponsor: Alexander von Humboldt Foundation
Jean-Paul-Street 12
Bonn, D-53173 Germany

Alexander von Humboldt Foundation German Chancellor Fellowships for Prospective Leaders from the Russian Federation 527

The Alexander von Humboldt Foundation's German Chancellor Fellowships are for university graduates from the Russian Federation with an interest in international issues and demonstrated leadership potential. Up to ten Fellowships are given annually. Fellows are mentored by hosts in Germany, whom the Fellows have chosen on the basis of their expertise in their respective areas. The program is targeted at accomplished young professionals who are likely to become decision-makers, thought leaders, and influential voices in their respective fields. During their time in Germany, Federal Chancellor Fellows conduct independent projects at their host institutions. Projects should be of relevance to modern societies, have a long-term and visible impact, and help to advance Fellows' careers and professional development. It should also involve original exploration of a topic or issue, or research in the Fellow's respective field of interest. Through their experience in Germany, Fellows will acquire greater knowledge of their fields, gain new international experience, and strengthen their intercultural competence. In addition to a stipend, Fellows also receive extensive benefits which are discussed in detail in the guidelines at the website. The application is also available at the website.
Requirements: Applicants may apply from a broad range of areas such as politics and public policy, law, media, business, the non-governmental sector, and the arts. Candidates will spend one year in Germany where they will network with other prospective leaders from abroad to explore new solutions to global issues. Candidates must meet the following *Requirements:* citizenship from the Russian Federation; a bachelor's or comparable academic degree, completed less than twelve years prior to the starting date of the Fellowship; demonstrated leadership potential; either English or German language skills; a project plan detailing the project's subject matter, goals and a time plan; a detailed host's statement, including a confirmation of support; and two or three expert recommendations from those qualified to comment on the candidate's professional, personal, and/or academic background.
Geographic Focus: Russian Federation

Date(s) Application is Due: Sep 15
Amount of Grant: 25,800-33,000 EUR
Contact: Leonid Zhmud, Selection Committee Chairperson; (+49) 0228-833-0; fax (+49) 0228-833-212; info@avh.de
Internet: http://www.humboldt-foundation.de/web/4074.html
Sponsor: Alexander von Humboldt Foundation
Jean-Paul-Street 12
Bonn, D-53173 Germany

Alexander von Humboldt Foundation Hezekiah Wardwell Fellowships 528

Up to ten Hezekiah Wardwell Fellowships are offered annually to young, highly-gifted Spanish musicians and musicologists for further training or advanced studies, primarily in Germany. The Fellowship is aimed at candidates from Spain in transition to their professional career. Fellows are supervised by recognized music teachers at colleges of music or conservatories, or by specialist scholars at universities. The Foundation expects candidates to establish contact with supervisors prior to submitting applications, and to obtain binding clarification for acceptance and academic supervision. Fellowships offer EUR 800 per month. In addition, the Foundation pays a single lump sum on arrival as well as travel expenses to Fellows traveling directly from Spain. Initially, applications can be made for a Fellowship lasting two academic semesters or ten months. Additional information is available at the website.
Requirements: Applicants should have a good knowledge of German. The Fellowship period must begin no later than three years from the end of the applicant's graduation. Preference will be given to candidates who have previously had no opportunity to continue their music studies in Germany.
Geographic Focus: Spain
Date(s) Application is Due: Jan 10
Amount of Grant: 8,250-16,250 EUR
Contact: Paulo Sergio Guimares Alvares, Fellowship Committee Chairperson; (+49) 0228-833-0; fax (+49) 0228-833-199; info@avh.de
Internet: http://www.humboldt-foundation.de/web/wardwell-es.html
Sponsor: Alexander von Humboldt Foundation
Jean-Paul-Street 12
Bonn, D-53173 Germany

Alexander von Humboldt Foundation International Climate Protection Fellowships 529

The International Climate Protection Fellowships enable prospective leaders to conduct a research-related project of their own choice during a one-year stay in Germany. Up to 15 Fellowships are awarded. Any prospective leader from a non-European or developing country working the climate protection and resource conservation in academia, business, or administration may apply. A complete list of countries is available at the website (Africa, East Asia and the Pacific, Eurasia, Near East, South and Central Asia, Latin America, and the Caribbean). The Fellowship will enable the recipients to conduct a research-related project of their own choice with hosts in Germany that they have chosen. The program includes a preparatory intensive language course in Germany, a several week introductory seminar, a several day training course, and a final meeting in Berlin. These activities will enable Fellows to gain additional insights into academic, social, cultural, economic and political life in Germany. Applicants who have been trained in the natural and engineering sciences, and candidates who have been engaged in legal, economic and societal issues relating to climate change are encouraged to apply. The application and additional information, including specific information about the Fellowship stipend and funding allowances, are available at the website.
Requirements: Applicants must provide a clearly visible leadership potential either by experience in a first leadership position or be able to provide appropriate references. They must also have completed their first university degree less than 12 years prior to the start of the Fellowship with outstanding results. They must also have extensive professional experience in a leadership role (at least 48 months at the time of application) or hold a further academic or professional qualification. Fellows are expected to have gained initial practical experience through involvement in projects related to climate protection and resource conservation.
Geographic Focus: All Countries
Date(s) Application is Due: Dec 1
Contact: Francois Claude Buscot, Selection Committe Chairperson; (+49) 0228-833-0; fax (+49) 0228-833-199; info@avh.de
Internet: http://www.humboldt-foundation.de/web/icf.html
Sponsor: Alexander von Humboldt Foundation
Jean-Paul-Street 12
Bonn, D-53173 Germany

Alexander von Humboldt Foundation JSPS Research Fellowships 530

The Alexander von Humboldt Foundation JSPS Research Fellowships assist promising, highly-qualified, young foreign researchers from Germany who hold a doctoral degree. These Fellowships allow researchers to pursue collaborative research under the leadership of Japanese host researchers in Japanese universities and research institutes. This allows them to advance their own research while promoting scientific advancement in Japan and the counterpart countries through close collaboration in scientific activities. JSPS offers two funding lines for a stay in Japan. JSPS Postdoctoral Fellowships (short-term) provide support for doctoral candidates and post-doctoral researchers of all disciplines to spend 1-12 months working on research in Japan. For sponsorship periods exceeding six months, the Humboldt Foundation preselects post-doctoral candidates on behalf of JSPS. Up to 20 JSPS Postdoctoral Fellowships are granted annually. Research stays lasting between 12 and 24 months are funded in the framework of JSPS Postdoctoral Fellowships (standard). Pre-selection of candidates for this funding line is conducted exclusively by the Alexander von Humboldt Foundation. Up to 16 JSPS Postdoctoral Fellowships (standard) are granted annually. Fellows receive 362,000 EUR per month, plus a 200,000 EUR relocation allowance. The application and additional detailed guidelines are available at the website.
Requirements: All fields of the humanities, social sciences, and natural sciences may apply. Each candidate for the Fellowship must also meet the following requirements to be eligible: possess either the nationality, citizenship or permanent residency of the nominating country; hold a doctorate degree which must have been received within six years prior to April 1 of the Fellowship year; and have arranged in advance a research plan with his/her Japanese host.
Restrictions: Short-term study visits, participation in congresses and training courses are not eligible for funding. Medical doctors without a Ph.D. are not eligible. Persons of Japanese nationality are not eligible to apply, nor are those of dual nationality if one is Japanese.
Geographic Focus: Germany
Amount of Grant: 362,000 EUR
Contact: Matthias Kleiner; (+49) 0228-833-0; fax (+49) 0228-833-199; info@avh.de
Internet: http://www.humboldt-foundation.de/web/jsps-fellowship-postdoc.html
Sponsor: Alexander von Humboldt Foundation
Jean-Paul-Street 12
Bonn, D-53173 Germany

Alexander von Humboldt Foundation Konrad Adenauer Research Award 531

The Konrad Adenauer Research Award is made annually to one Canadian scholar in the humanities or social sciences. The aim of the Award is to promote academic relations between Canada and Germany. The Award is given to a highly qualified Canadian scholar whose research work in the humanities or social sciences has brought international recognition, and who belongs to a group of leading scholars in the respective area of specialization. The Awardee will carry out research work at German research institutes for up to one year. The winner is included in the Foundation's sponsorship program, and will be invited to take part in all functions arranged by the Foundation. The amount of the Award includes travel for Awardee and family members. Additional information is available at the website.
Requirements: Candidates are nominated by their universities for this award. Nomination should include a letter from the candidate's institution regarding the candidate's academic qualifications; a brief statement of the candidate's research proposal; three references who can provide information on the candidate's academic qualifications; and the names of German scholars who would provide guidance to the scholar during his/her stay in Germany.
Restrictions: Self-nominations will not be accepted. Applications will not be considered if the candidate holds or has held another award from the Humboldt Foundation within the previous three years.
Geographic Focus: Canada
Date(s) Application is Due: Jan 31
Amount of Grant: Up to 50,000 EUR
Contact: Barbara Konig; (+49) 0228-833-0; fax (+49) 0228-833-199; info@avh.de
Internet: http://www.humboldt-foundation.de/web/31764.html
Sponsor: Alexander von Humboldt Foundation
Jean-Paul-Street 12
Bonn, D-53173 Germany

Alexander von Humboldt Foundation Max Planck Research Award 532

The Max Planck Award is conferred upon exceptionally highly-qualified international scientists. The Award is given to one international researcher and one researcher working in Germany every other year. The researchers are expected to have already achieved international recognition and to continue to produce outstanding academic results in international collaboration. On an annually-alternating basis, the call for nominations addresses areas within the natural and engineering sciences, the life sciences, and the humanities. Recipients of the Award will be granted funding of up to 750,000 EUR over a period of three to a maximum of five years in order to pursue research of their own choosing. Nomination information is available at the website.
Geographic Focus: All States, All Countries
Date(s) Application is Due: Jan 31
Amount of Grant: Up to 750,000 EUR
Contact: Michaela Kreilos, Selection Department Coordinator; (+49) 0228-833-0; fax (+49) 0228-833-212; michaela.kreilos@avh.de
Internet: http://www.humboldt-foundation.de/web/6674.html
Sponsor: Alexander von Humboldt Foundation
Jean-Paul-Street 12
Bonn, D-53173 Germany

Alexander von Humboldt Foundation NSC Taiwan Fellowships 533

The NSC Taiwan Fellowships assist promising, highly-qualified, young foreign researchers who have held a doctoral degree for less than twelve years. This program allows researchers to pursue collaborative research at Taiwan universities and research institutes. This activity allows them to advance their own research while promoting scientific advancement in Taiwan and the counterpart countries through close collaboration in scientific activities. All fields of humanities, social sciences, and natural sciences are eligible. Up to two Fellowships are granted each year. The application and additional guidelines are available at the website.
Requirements: Applicants must be a post-doctoral researcher from Germany. They must also have completed their doctorate within the last twelve years. Candidates must have a good knowledge of Chinese or English, and have academic publications. They must also have confirmation that research facilities are available, with a detailed expert's report from an academic host from a Taiwan institute.

Geographic Focus: Germany
Contact: Coordinator; (+49) 0228-833-0; fax (+49) 0228-833-199; info@avh.de
Internet: http://www.humboldt-foundation.de/web/3626.html
Sponsor: Alexander von Humboldt Foundation
Jean-Paul-Street 12
Bonn, D-53173 Germany

Alexander von Humboldt Foundation Philipp Franz von Siebold Award for Japanese Researchers 534

The Philipp Franz von Siebold Award is given annually to a Japanese academic for outstanding services to enhancing mutual understanding of culture and society in Germany and Japan. The award winner will be invited to spend up to a year researching in Germany. The prize is valued at 50,000 EUR. Nominations are accepted from presidents of Japanese universities, former Siebold award winners, the heads of the Goethe Institutes in Japan, and the German Institute for Japanese Studies, as well as the German Ambassador in Tokyo and the German Consul General in Osaka. Humboldt Fellows and award winners may make nominations via the Vice Chancellor of a Japanese university. A decision on the award is made by a special committee under the chairmanship of the German Ambassador in Japan.
Requirements: The scientist or scholar should be under 50 years of age, and have a working knowledge of German.
Restrictions: Self nominations are not accepted.
Geographic Focus: Japan
Contact: Barbara Konig; (+49) 0228-833-0; fax (+49) 0228-833-199
DAAD Tokyo Office; (03) 3582-5962; fax (03) 3582-5554; daad-tokyo@daadjp.com
Internet: http://www.humboldt-foundation.de/web/31771.html
Sponsor: Alexander von Humboldt Foundation
Jean-Paul-Street 12
Bonn, D-53173 Germany

Alexander von Humboldt Foundation Professorships 535

Academics of all disciplines from abroad, who are internationally recognized as leaders in their field and who are expected to contribute to enhancing Germany's sustained international competitiveness as a research location in consequence of the award, are eligible to be nominated. The Professorship, which is being financed by the Federal Ministry of Education and Research through the International Research Fund for Germany, enables award winners to carry out long-term and ground-breaking research at universities and research institutions in Germany. The prize money, totaling 5 million EUR for academics in experimental disciplines and 3.5 million EUR for researchers in theoretical disciplines, is being made available for a period of five years.
Requirements: Nominations may be made by German universities; non-university research institutions may also submit nominations jointly with a German university. The nominee must be: recognized internationally as an outstandingly qualified academic; be established as an academic abroad; and be nominated by an institution entitled to make nominations.
Geographic Focus: All States, All Countries, Germany
Date(s) Application is Due: Apr 15; Oct 15
Contact: Contact; (+49) 0228-833-0; fax (+49) 0228-833-199; avh-professur@avh.de
Internet: http://www.humboldt-foundation.de/en/programme/preise/ahp.htm
Sponsor: Alexander von Humboldt Foundation
Jean-Paul-Street 12
Bonn, D-53173 Germany

Alexander von Humboldt Foundation Reimar Lust Award 536

The Award recognizes the contribution of outstanding humanities scholars and social scientists from abroad to promoting cultural and academic relations between Germany and their own countries. Up to two Awards are given each year. Nominations for the Award may be made on behalf of highly respected academics who, as multipliers in and through the field of academic studies, have made an exceptional contribution to the enduring promotion of bilateral relations between Germany and their own countries. These services include the transfer of traditions of and approaches to academic studies from Germany to other countries and from other countries to Germany, translation work, or the promotion of institutional cooperation. In order to strengthen and extend bilateral cooperative relations to their partners and other specialist colleagues in Germany on a sustainable basis, the award winners are invited to spend up to a year in Germany, which may be divided up into blocks. The Award is valued at 50,000 EUR. Additional information is available at the website.
Restrictions: Academics who have already been granted an award by the Alexander von Humboldt Foundation are not eligible to be nominated for the Award.
Geographic Focus: All States, All Countries, Germany
Date(s) Application is Due: Oct 31
Contact: Gerhard Lauer; (+49) 0228-833-0; fax (+49) 0228-833-199; info@avh.de
Internet: http://www.humboldt-foundation.de/web/luest-award.html
Sponsor: Alexander von Humboldt Foundation
Jean-Paul-Street 12
Bonn, D-53173 Germany

Alexander von Humboldt Foundation Research Awards 537

Each year the Alexander von Humboldt Research Awards honor up to 100 internationally renowned scientists and scholars from abroad. These awards are based primarily on the scientist's/scholar's entire academic record. Scientists and scholars who are nominated for an Award must have contributed fundamental discoveries, new theories, or insights that significantly impact their own research area and also reach beyond their specialized field of research. Nominees should be active researchers who are likely to achieve the same amount high level of achievement in the future. A candidate must be nominated by distinguished scientists/scholars employed by any university or research institution in Germany. The Award amounts to 60,000 EUR. Additional information about the nomination process is available at the website.
Requirements: Scientists and scholars of all disciplines or research areas, regardless of their nationality, may be nominated.
Restrictions: Direct applications are not accepted.
Geographic Focus: All States, All Countries
Amount of Grant: 60,000 EUR
Contact: Barbara Konig; (+49) 0228-833-0; fax (+49) 0228-833-199; info@avh.de
Internet: http://www.humboldt-foundation.de/web/6446.html
Sponsor: Alexander von Humboldt Foundation
Jean-Paul-Street 12
Bonn, D-53173 Germany

Alexander von Humboldt Foundation Research Awards for Top Researchers in Germany 538

Research awards for Top Researchers in Germany are granted by partner organizations of the von Humboldt Foundation. Foreign researchers who wish to nominate a colleague in Germany in their field for a research award must do so through the relevant programs of the respective partner organization in their country. The respective partner organizations are independent from the Humboldt Foundation and follow their own procedures concerning the selection and final decision on the awards. As a rule, the award sum is conferred in conjunction with an invitation to spend up to one year working on research in the host country, which may be divided up into several blocks. Additional information about the agreements for each country is available at the website.
Requirements: The nominator should contact the partner organization posted on the website, since nomination procedures and regulations vary from country to country.
Geographic Focus: Germany
Contact: Barbara Konig; (+49) 0228-833-0; fax (+49) 0228-833-199; info@avh.de
Internet: http://www.humboldt-foundation.de/web/award-researchers-in-germany.html
Sponsor: Alexander von Humboldt Foundation
Jean-Paul-Street 12
Bonn, D-53173 Germany

Alexander von Humboldt Foundation Research Fellowships for Postdoctoral Researchers 539

The Research Fellowships for Postdoctoral Researchers are the instrument with which the Foundation enables highly-qualified scientists and scholars from abroad who are just embarking on their academic careers and who completed their doctorates less than four years ago to spend extended periods of research (6-24 months) in Germany. Scientists and scholars from all disciplines and countries may apply. Projects are carried out in cooperation with academic hosts at research institutions in Germany. Candidates choose their own research projects and their host in Germany and prepare their own research plan. Details of the research project and the time schedule must be agreed upon with the prospective host in advance. Candidates are selected solely on the basis of their academic record. There are no quotas for individual disciplines or countries. The Fellowship is worth 2,600 EUR per month. This includes a mobility lump sum and a contribution towards health and liability insurance. The application and additional guidelines are available at the website. Applications may be submitted at any time.
Requirements: Application requirements include the following: a doctorate or comparable academic degree (Ph.D., C.Sc. or equivalent), completed less than twelve years prior to the date of application; the candidate's own research profile documented by a comprehensive list of academic publications reviewed according to international standards and printed in journals and/or by publishing houses; and a good knowledge of German or English if it is necessary to carry out the project successfully.
Restrictions: Short-term visits for study and training purposes or for attending conferences are not eligible for sponsorship.
Geographic Focus: All States, All Countries
Amount of Grant: 31,200 EUR
Contact: Matthias Kleiner; (+49) 0228-833-0; fax (+49) 0228-833-212; info@avh.de
Internet: http://www.humboldt-foundation.de/web/humboldt-fellowship-postdoc.html
Sponsor: Alexander von Humboldt Foundation
Jean-Paul-Street 12
Bonn, D-53173 Germany

Alexander von Humboldt Foundation Sofja Kovalevskaja Awards 540

Up to eight Sofja Kovalevskaja Awards are granted in recognition of the outstanding academic achievements of exceptionally promising junior researchers from abroad. It enables them to establish their own groups of junior researchers at research establishments in Germany. Scientists and scholars from abroad, whose research records to date have already qualified them to be recognized as top-level junior researchers, and who are expected to continue producing outstanding results as recipients of the Award, are eligible to apply. The award winners are able to concentrate on high-level, innovative research work of their own choice in Germany without restraint and strengthen the internationalization of research in Germany. The Award allows the winners to finance their own working group at a university or non-university research establishment of their own choice in Germany. It also covers their living expenses. The application and additional information is available at the website.
Requirements: Application is open to scientists and scholars from all countries and disciplines who have completed their doctorates with distinction within the past six years, and have published work in prestigious international journals or publishing houses.

Foreign academics staying in Germany are eligible to apply if they have not been working in an academic capacity in Germany for more than two years at the time of application. German academics are eligible to apply if they have been working in an academic capacity abroad for at least five years at the time of application.
Geographic Focus: All States, All Countries
Date(s) Application is Due: Jul 31
Amount of Grant: Up to 1,650,000 EUR
Contact: Barbara Konig; (+49) 0228-833-0; fax (+49) 0228-833-199; info@avh.de
Internet: http://www.humboldt-foundation.de/web/7360.html
Sponsor: Alexander von Humboldt Foundation
Jean-Paul-Street 12
Bonn, D 53173 Germany

Alex C. Walker Educational and Charitable Foundation Grants 541
The foundation awards grants to investigate the causes of economic imbalances; investigate the effect of the monetary system in fostering a sustainable economy; investigate causes tending to destroy or impair the free-market system; explore and develop free-market solutions; and disseminate information on these issues. The foundation funds projects dealing with the research and development of innovative ideas, education of market principles, the application of ideas developed through seed grants and research, and ecological economics with a free-market orientation. Projects with the following qualities are favored: innovation, impact, longevity, collaboration, solution oriented, trackability, objectivity, and sustainability. Grant requests must be submitted on a formal application form through this website. Guidelines are available online.
Requirements: U.S. 501(c)3 tax-exempt organizations are eligible.
Restrictions: Ideological or political activities and general educational programs are not funded. Grants do not support endowment, building funds, or individuals.
Geographic Focus: All States
Date(s) Application is Due: Apr 1; Oct 1
Amount of Grant: 5,000-89,500 USD
Contact: Grants Administrator, c/o Barret P. Walker; (404) 378-2752
Internet: http://walker-foundation.org/page.aspx?s=5534.0.69.5316
Sponsor: Alex C. Walker Educational and Charitable Foundation
1729 Coventry Place
Decatur, GA 30030

Alex Stern Family Foundation Grants 542
The foundation awards grants to North Dakota and Minnesota nonprofits in its areas of interest, including arts and culture, child welfare, the elderly, alcohol abuse, community funds, family and social services, education, minorities, hospices, and cancer research. Types of support include general operating support, continuing support, annual campaigns, building construction/renovation, equipment acquisition, emergency funds, program development, scholarship funds, research, and matching funds. Applications are reviewed in June and November.
Requirements: Moorhead, MN, and Fargo, ND, nonprofit organizations are eligible.
Restrictions: Grants are not awarded to individuals or for endowments.
Geographic Focus: Minnesota, North Dakota
Date(s) Application is Due: Mar 31; Aug 31
Contact: Donald Scott, Executive Director; (701) 237-0170
Sponsor: Alex Suitern Family Foundation
609 1/2 1st Avenue N, Suite 205
Fargo, ND 58102

Alfred P. Sloan Foundation Barcode of Life Grants 543
The goal of the Barcode of Life Grant program is to speed the building and use of a library of short DNA sequences for identifying species of animals and plants reliably and cheaply. The Barcode of Life (BoL) refers to a short DNA sequence, from a uniform locality on the genome, used for identifying species. A mere 600 or so base pairs present in every animal cell can prove a hair or a scale comes from a homo sapiens or a particular species of fish. Grant requests can be made at any time for support of activities related to Foundation program areas and interests. The Foundation is generally limited to supporting tax-exempt organizations. The Foundation has no deadlines or standard forms. The Foundation accepts proposals sent by email. A brief letter of inquiry, rather than a fully developed proposal, is an advisable first step for an applicant, conserving his or her time and allowing for a preliminary response regarding the possibility of support.
Requirements: Concise, well-organized proposals are preferred. In no case should the body of the proposal exceed 20 double-spaced pages.
Restrictions: The Foundation's activities do not normally extend to religion, the creative or performing arts, elementary or secondary education, medical research or health care, the humanities or to activities outside the United States. Grants are not made for endowments or for buildings or equipment.
Geographic Focus: All States
Contact: Jesse H. Ausubel; (212) 649-1650 or (212) 649-1649; ausubel@sloan.org
Internet: http://www.sloan.org/major-program-areas/basic-research/barcode-of-life/
Sponsor: Alfred P. Sloan Foundation
630 Fifth Avenue, Suite 2550
New York, NY 10111-0242

Alfred P. Sloan Foundation Behavioral Economics and Household Finance Grants 544
This program supports research on the structure, behavior, and performance of the U.S. economy with the goal of providing objective and nonpartisan insights that can inform and strengthen critical decisions facing leaders, policymakers, and the public. Projects in the Behavioral Economics and Household Finance sub-program study individuals and households, specifically with regard to the quality of their economic decision-making. Appropriate research topics include the annuity paradox; the energy efficiency paradox; insurance markets; risk-taking, savings, and personal bankruptcy; cognitive biases; public understanding of economics and markets for financial advice. Successful grant applications to the Foundation's program in Economic Institutions, Behavior and Performance have several features in common. Typically, they are: empirical and hypothesis-driven; policy-relevant, but neither policy research nor advocacy; motivated by non-ideological questions rather than preconceived answers; engaged with fundamental puzzles but using fresh approaches; unbiased, statistically significant, and replicable; careful about baselines, controls, and econometrics; savvy about markets, institutions, transaction costs, regulatory arbitrage, etc; contributors to research infrastructure, datasets, or resources for general use; generators of highly-cited results in high-quality journals; and ultimately concerned with the quality of life in the United States. The Foundation encourages projects by multidisciplinary teams where appropriate.
Requirements: Concise, well-organized proposals are preferred. In no case should the body of the proposal exceed 20 double-spaced pages.
Restrictions: The Foundation does not make grants: to individuals; to for-profit institutions (except in rare instances); in religion, medical research or the humanities; to projects aimed at pre-college students; to projects in the creative or performing arts, except when those projects are related to educating the public about science, technology or economics; for endowments, buildings or equipment, or fundraising drives, including fundraising dinners; or to political campaigns, to support political activities or to lobby for or against particular pieces of legislation.
Geographic Focus: All States
Contact: Daniel L. Goroff, Program Director; (212) 649-1676 or (212) 649-1649; fax (212) 757-5117; goroff@sloan.org
Internet: http://www.sloan.org/major-program-areas/economic-performance-and-quality-of-life/economic-institutions-behavior-and-performance/
Sponsor: Alfred P. Sloan Foundation
630 Fifth Avenue, Suite 2550
New York, NY 10111-0242

Alfred P. Sloan Foundation Civic Initiatives Grants 545
The goal of the Program is to make a contribution to the Foundation's home area, New York City. There are two directions to the Program: to respond to special opportunities in New York City; and to fund high-leverage projects in New York City that are related to other parts of our program. Interested readers should refer to the descriptions of other program areas in the Foundation website. Grant requests can be made at any time for support of activities related to Foundation program areas and interests. The Foundation is generally limited to supporting tax-exempt organizations.
Requirements: Concise, well-organized proposals are preferred. In no case should the body of the proposal exceed 20 double-spaced pages.
Restrictions: The Foundation's activities do not normally extend to religion, the creative or performing arts, elementary or secondary education, medical research or health care, the humanities or to activities outside the United States. Grants are not made for endowments or for buildings or equipment.
Geographic Focus: New York
Contact: Paula J. Olsiewski; (212) 649-1658 or (212) 649-1649; olsiewski@sloan.org
Internet: http://www.sloan.org/major-program-areas/select-national-issues/civic-initiatives/
Sponsor: Alfred P. Sloan Foundation
630 Fifth Avenue, Suite 2550
New York, NY 10111-0242

Alfred P. Sloan Foundation Data and Computational Research Grants 546
From the natural sciences to the social sciences to the humanities to the arts, the availability of more data and cheaper computing is transforming research. As costs for sensors, sequencing, and other forms of data collection decline, researchers can generate data at greater and greater scale, relying on parallel increases in computational power to make sense of it all and allowing the investigation of phenomena too large or complex for conventional observation. Grants in this sub-program aim to help researchers develop tools, establish norms, and build the institutional and social infrastructure needed to take full advantage of these important developments in data-driven, computation-intensive research. Emphasis is placed on projects that encourage access to and sharing of scholarly data, that promote the development of standards and taxonomies necessary for the interoperability of datasets, that enable the replication of computational research, and that investigate models of how researchers might deal with the increasingly central role played by data management and curation.
Requirements: Concise, well-organized proposals are preferred. In no case should the body of the proposal exceed 20 double-spaced pages.
Restrictions: The Foundation's activities do not normally extend to religion, the creative or performing arts, elementary or secondary education, medical research or health care, the humanities or to activities outside the United States. Grants are not made for endowments or for buildings or equipment.
Geographic Focus: All States
Contact: Joshua M. Greenberg, Program Director; (212) 649-1633 or (212) 649-1649; fax (212) 757-5117; greenberg@sloan.org
Internet: http://www.sloan.org/major-program-areas/digital-information-technology/data-and-computational-research/
Sponsor: Alfred P. Sloan Foundation
630 Fifth Avenue, Suite 2550
New York, NY 10111-0242

Alfred P. Sloan Foundation Deep Carbon Observatory Grants 547

The relatively new Deep Carbon Observatory program aims to revolutionaize our understanding of the carbon deep in the Earth, including its connections to the origins of life and to the origins, distribution, and abundance of fossile fuels, through a multidisciplinary international network that develops and applies new instruments, takes observations, and performs analyses. A three-year 2009 grant to the Carnegie Institution of Washington supported the initial phase of the program, which was focused on developing instruments to meet the severe technical challenges associated with probing the high-pressure, high-temperature rocesses in Earth's deep interior and on building an organizational infrastructure to set strategic priorities, engage a network of researchers, and secure funding commitments from institutional partners.

Requirements: Concise, well-organized proposals are preferred. In no case should the body of the proposal exceed 20 double-spaced pages.

Restrictions: The Foundation's activities do not normally extend to religion, the creative or performing arts, elementary or secondary education, medical research or health care, the humanities or to activities outside the United States. Grants are not made for endowments or for buildings or equipment.

Geographic Focus: All States

Contact: Gail Pesyna; (212) 649-1655 or (212) 649-1649; pesyna@sloan.org

Internet: http://www.sloan.org/major-program-areas/basic-research/deep-carbon-observatory/

Sponsor: Alfred P. Sloan Foundation
630 Fifth Avenue, Suite 2550
New York, NY 10111-0242

Alfred P. Sloan Foundation Economic Analysis of Science and Technology Grants 548

This program supports research on the structure, behavior, and performance of the U.S. economy with the goal of providing objective and nonpartisan insights that can inform and strengthen critical decisions facing leaders, policymakers, and the public. Projects in the Economic Analysis of Science and Technology sub-program study universities and groundbreaking industries, specifically regarding human capital development and applications of information technology. Appropriate research topics include labor markets for scientists and engineers; high-skilled immigration; patterns of scientific publication, collaboration, and intellectual property protection; the economics of digitization; and the international distribution of returns on high-tech investments. Successful grant applications to the Foundation's program in Economic Institutions, Behavior and Performance have several features in common. Typically, they are: empirical and hypothesis-driven; policy-relevant, but neither policy research nor advocacy; motivated by non-ideological questions rather than preconceived answers; engaged with fundamental puzzles but using fresh approaches; unbiased, statistically significant, and replicable; careful about baselines, controls, and econometrics; savvy about markets, institutions, transaction costs, regulatory arbitrage, etc; contributors to research infrastructure, datasets, or resources for general use; generators of highly-cited results in high-quality journals; and ultimately concerned with the quality of life in the United States. The Foundation encourages projects by multidisciplinary teams where appropriate.

Requirements: Concise, well-organized proposals are preferred. In no case should the body of the proposal exceed 20 double-spaced pages.

Restrictions: The Foundation does not make grants: to individuals; to for-profit institutions (except in rare instances); in religion, medical research or the humanities; to projects aimed at pre-college students; to projects in the creative or performing arts, except when those projects are related to educating the public about science, technology or economics; for endowments, buildings or equipment, or fundraising drives, including fundraising dinners; or to political campaigns, to support political activities or to lobby for or against particular pieces of legislation.

Geographic Focus: All States

Contact: Daniel L. Goroff; (212) 649-1676 or (212) 649-1649; goroff@sloan.org

Internet: http://www.sloan.org/major-program-areas/economic-performance-and-quality-of-life/economic-institutions-behavior-and-performance/

Sponsor: Alfred P. Sloan Foundation
630 Fifth Avenue, Suite 2550
New York, NY 10111-0242

Alfred P. Sloan Foundation Economic Implications of the Great Recession Grants 549

This program supports research on the structure, behavior, and performance of the U.S. economy with the goal of providing objective and nonpartisan insights that can inform and strengthen critical decisions facing leaders, policymakers, and the public. Projects in Economic Implications of the Great Recession sub-program study markets and governments, specifically with regard to lessons we can draw from the recent financial crisis and Great Recession. Appropriate research topics include systemic stability; international regulatory coordination; risk measurement, capital requirements, and credit ratings; labor market recovery rates and liquidity; and dataset and model development concerning labor trends. Successful grant applications to the Foundation's program in Economic Institutions, Behavior and Performance have several features in common. Typically, they are: empirical and hypothesis-driven; policy-relevant, but neither policy research nor advocacy; motivated by non-ideological questions rather than preconceived answers; engaged with fundamental puzzles but using fresh approaches; unbiased, statistically significant, and replicable; careful about baselines, controls, and econometrics; savvy about markets, institutions, transaction costs, regulatory arbitrage, etc; contributors to research infrastructure, datasets, or resources for general use; generators of highly-cited results in high-quality journals; and ultimately concerned with the quality of life in the United States. The Foundation encourages projects by multidisciplinary teams where appropriate.

Requirements: Concise, well-organized proposals are preferred. In no case should the body of the proposal exceed 20 double-spaced pages.

Restrictions: The Foundation does not make grants: to individuals; to for-profit institutions (except in rare instances); in religion, medical research or the humanities; to projects aimed at pre-college students; to projects in the creative or performing arts, except when those projects are related to educating the public about science, technology or economics; for endowments, buildings or equipment, or fundraising drives, including fundraising dinners; or to political campaigns, to support political activities or to lobby for or against particular pieces of legislation.

Geographic Focus: All States

Contact: Daniel L. Goroff; (212) 649-1676 or (212) 649-1649; goroff@sloan.org

Internet: http://www.sloan.org/major-program-areas/economic-performance-and-quality-of-life/economic-institutions-behavior-and-performance/

Sponsor: Alfred P. Sloan Foundation
630 Fifth Avenue, Suite 2550
New York, NY 10111-0242

Alfred P. Sloan Foundation Empirical Economic Research Enablers Grants 550

This program supports research on the structure, behavior, and performance of the U.S. economy with the goal of providing objective and nonpartisan insights that can inform and strengthen critical decisions facing leaders, policymakers, and the public. Projects in the Empirical Economic Research Enablers sub-program study economic researchers, specifically with regard to their needs, opportunities, incentives, and professional practices. Appropriate research topics include legal entity identifiers; data citation standards; identification and tracking systems for scholars; federal statistics; smart disclosure platforms for obfuscated markets; data and metadata management protocols; the replicability of empirical research; and the economics of knowledge contribution and distribution. Successful grant applications to the Foundation's program in Economic Institutions, Behavior and Performance have several features in common. Typically, they are: empirical and hypothesis-driven; policy-relevant, but neither policy research nor advocacy; motivated by non-ideological questions rather than preconceived answers; engaged with fundamental puzzles but using fresh approaches; unbiased, statistically significant, and replicable; careful about baselines, controls, and econometrics; savvy about markets, institutions, transaction costs, regulatory arbitrage, etc; contributors to research infrastructure, datasets, or resources for general use; generators of highly-cited results in high-quality journals; and ultimately concerned with the quality of life in the United States. The Foundation encourages projects by multidisciplinary teams where appropriate.

Requirements: Concise, well-organized proposals are preferred. In no case should the body of the proposal exceed 20 double-spaced pages.

Restrictions: The Foundation does not make grants: to individuals; to for-profit institutions (except in rare instances); in religion, medical research or the humanities; to projects aimed at pre-college students; to projects in the creative or performing arts, except when those projects are related to educating the public about science, technology or economics; for endowments, buildings or equipment, or fundraising drives, including fundraising dinners; or to political campaigns, to support political activities or to lobby for or against particular pieces of legislation.

Geographic Focus: All States

Contact: Daniel L. Goroff; (212) 649-1676 or (212) 649-1649; goroff@sloan.org

Internet: http://www.sloan.org/major-program-areas/economic-performance-and-quality-of-life/economic-institutions-behavior-and-performance/

Sponsor: Alfred P. Sloan Foundation
630 Fifth Avenue, Suite 2550
New York, NY 10111-0242

Alfred P. Sloan Foundation Encyclopedia of Life Grants 551

The goal of the Encylopedia of Life (EOL) program is to build a reliable online encyclopedia with a Web page for every species. The EOL should serve both as a handy field guide for millions of users and as a macroscope to discover patterns in ecology and evolution otherwise too large or complex to perceive. Grant requests can be made at any time for support of activities related to Foundation program areas and interests. The Foundation is generally limited to supporting tax-exempt organizations. The Foundation has no deadlines or standard forms. The Foundation accepts proposals sent by email. A brief letter of inquiry, rather than a fully developed proposal, is an advisable first step for an applicant, conserving his or her time and allowing for a preliminary response regarding the possibility of support.

Requirements: Concise, well-organized proposals are preferred. In no case should the body of the proposal exceed 20 double-spaced pages.

Restrictions: The Foundation's activities do not normally extend to religion, the creative or performing arts, elementary or secondary education, medical research or health care, the humanities or to activities outside the United States. Grants are not made for endowments or for buildings or equipment.

Geographic Focus: All States

Contact: Jesse H. Ausubel; (212) 649-1650 or (212) 649-1649; ausubel@sloan.org

Internet: http://www.sloan.org/major-program-areas/basic-research/encyclopedia-of-life/

Sponsor: Alfred P. Sloan Foundation
630 Fifth Avenue, Suite 2550
New York, NY 10111-0242

Alfred P. Sloan Foundation Energy and Environment Grants 552

Grantmaking in this small interdisciplinary program looks for unique opportunities to expand our understanding of the economic, technological, organizational, regulatory, national security, and environmental consequences of energy production and consumption. Past grantmaking in this program has led to the publication of the influential MIT reports, The Future of Nuclear Power, and The Future of Coal and the Foundation is currently supporting a similar project which focuses on economic,

technological, and institutional issues associated with the use of solar energy. Other recent Foundation grants in this area support the exploration of strategies for the safe, responsible expansion of nuclear power around the world, a project to examine the feasibility of extending the life of existing nuclear plants, a project to disseminate safety guidelines for the responsible use of radioactive materials at academic, industrial and medical institutions, a project on potential improvements to the measurement of carbon sequestered in forests, and two projects on natural gas: one examining the economics of state-of-the-art natural gas extraction based on detailed analyses of the geological and economic attributes of five important U.S. shale gas deposits, and one examining environmental and regulatory issues surrounding shale gas exploration and extraction. Due to the significant funding available from both public and private sources for energy and environmental research, the Foundation is very selective in the grants it makes in this area, supporting only projects for which non-partisan funding is not readily available, projects related to other Foundation programs or priorities, or projects where Foundation support could be leveraged to significantly raise the chances of the project's success.
Requirements: Concise, well-organized proposals are preferred. In no case should the body of the proposal exceed 20 double-spaced pages.
Restrictions: The Foundation's activities do not normally extend to religion, the creative or performing arts, elementary or secondary education, medical research or health care, the humanities or to activities outside the United States. Grants are not made for endowments or for buildings or equipment.
Geographic Focus: All States
Contact: Gail Pesyna; (212) 649-1655 or (212) 649-1649; pesyna@sloan.org
Internet: http://www.sloan.org/major-program-areas/select-national-issues/energy-and-environment/?L=iqohhobrq
Sponsor: Alfred P. Sloan Foundation
630 Fifth Avenue, Suite 2550
New York, NY 10111-0242

Alfred P. Sloan Foundation History of Science and Technology Grants 553
The goal of this program is to preserve the raw material of history by supporting archival projects now centered on Charles Darwin, Thomas A. Edison, and Kurt G'del, and via new projects based on the World Wide Web. Two specific areas are of interest: the archival program; and recent history of science and engineering on the web. Grant requests can be made at any time for support of activities related to Foundation program areas and interests. The Foundation is generally limited to supporting tax-exempt organizations. The Foundation has no deadlines or standard forms. The Foundation accepts proposals sent by email. A brief letter of inquiry, rather than a fully developed proposal, is an advisable first step for an applicant, conserving his or her time and allowing for a preliminary response regarding the possibility of support.
Requirements: Concise, well-organized proposals are preferred. In no case should the body of the proposal exceed 20 double-spaced pages.
Restrictions: The Foundation's activities do not normally extend to religion, the creative or performing arts, elementary or secondary education, medical research or health care, the humanities or to activities outside the United States. Grants are not made for endowments or for buildings or equipment.
Geographic Focus: All States
Contact: Doron Weber; (212) 649-1652 or (212) 649-1649; weber@sloan.org
Internet: http://www.sloan.org/major-program-areas/select-national-issues/history-of-science-and-technology/
Sponsor: Alfred P. Sloan Foundation
630 Fifth Avenue, Suite 2550
New York, NY 10111-0242

Alfred P. Sloan Foundation Microbiology of the Built Environment Grants 554
Formerly known as the Indoor Environment Grant, the goal of the Alfred P. Sloan Foundation Microbiology of the Built Environment program is to understand the human indoor environment at the microbial level. The Foundation believes that this understanding may ultimately help to make this artificial environment more hospitable to human life or resistant to biological attacks. The Foundation plan has three parts. First, the Foundation wants to establish a basic understanding of the indoor microbial environment. Second, it wants to help develop tools for understanding and probing the indoor environment. And lastly, it wants to apply the knowledge and tools to various human indoor environments, such as hospitals. Grant requests can be made at any time for support of activities related to Foundation program areas and interests. The Foundation is generally limited to supporting tax-exempt organizations. The Foundation has no deadlines or standard forms. The Foundation accepts proposals sent by email. A brief letter of inquiry, rather than a fully developed proposal, is an advisable first step for an applicant, conserving his or her time and allowing for a preliminary response regarding the possibility of support.
Requirements: Concise, well-organized proposals are preferred. In no case should the body of the proposal exceed 20 double-spaced pages.
Restrictions: The Foundation's activities do not normally extend to religion, the creative or performing arts, elementary or secondary education, medical research or health care, the humanities or to activities outside the United States. Grants are not made for endowments or for buildings or equipment.
Geographic Focus: All States, Canada
Contact: Paula J. Olsiewski, Program Director; (212) 649-1658 or (212) 649-1649; fax (212) 757-5117; olsiewski@sloan.org
Internet: http://www.sloan.org/major-program-areas/basic-research/mobe/
Sponsor: Alfred P. Sloan Foundation
630 Fifth Avenue, Suite 2550
New York, NY 10111-0242

Alfred P. Sloan Foundation Public Understanding of Science and Technology Book Grants 555
The Alfred P. Sloan Foundation supports a wide range of books aimed at public understanding of science and technology. Popular categories include books that elucidate the scientific basis of issues that are often confusing or controversial or unnecessarily obscure. The Foundation also supports books that profile scientific and technological figures from varying angles, but with an emphasis on the human story. Books about the relevance of technology to daily life and about the relationship between women and technology also find occasional support. Inventing America: A History of the United States, a popular college textbook supported by the Foundation, explores the central role of science, technology and business in the nation's development. The Foundation's Book program aims to reach a wide, lay audience.
Requirements: Concise, well-organized proposals are preferred. In no case should the body of the proposal exceed 20 double-spaced pages.
Restrictions: The Foundation's activities do not normally extend to religion, the creative or performing arts, elementary or secondary education, medical research or health care, the humanities or to activities outside the United States. Grants are not made for endowments or for buildings or equipment.
Geographic Focus: All States
Contact: Doron Weber; (212) 649-1652 or (212) 649-1649; weber@sloan.org
Internet: http://www.sloan.org/major-program-areas/public-understanding-of-science-and-technology/books/
Sponsor: Alfred P. Sloan Foundation
630 Fifth Avenue, Suite 2550
New York, NY 10111-0242

Alfred P. Sloan Foundation Public Understanding of Science and Technology Other Efforts 556
The Alfred P. Sloan Foundation occasionally sponsors innovative efforts to reach a broad, cross-cultural audience or to target specific segments of the public. These efforts may take the form of, but are not limited to, Internet projects, conferences, multimedia events, performances, and science and arts festivals. The Alfred P. Sloan Foundation makes grants year-round. Interested grantseekers with a relevant project idea should email a one page letter of inquiry to Program Director.
Requirements: Concise, well-organized proposals are preferred. In no case should the body of the proposal exceed 20 double-spaced pages.
Restrictions: The Foundation's activities do not normally extend to religion, the creative or performing arts, elementary or secondary education, medical research or health care, the humanities or to activities outside the United States. Grants are not made for endowments or for buildings or equipment. Telephone inquiries are not accepted.
Geographic Focus: All States
Contact: Doron Weber; (212) 649-1652 or (212) 649-1649; weber@sloan.org
Internet: http://www.sloan.org/major-program-areas/public-understanding-of-science-and-technology/other-efforts/?L=iqohhobrq
Sponsor: Alfred P. Sloan Foundation
630 Fifth Avenue, Suite 2550
New York, NY 10111-0242

Alfred P. Sloan Foundation Public Understanding of Science and Technology Public Television Grants 557
The Foundation is a major supporter of public television documentaries, docudramas and dramas about science, technology and the lives of the men and women involved in scientific and technological pursuit. Foundation-supported shows highlight the role of engineering and technology in society or broaden our view of the nation's history and of the central role of science and technology in the country's narrative. Some programs celebrate scientific and technological breakthroughs and milestones, while others reveal just how little we know, and how far we still have to go. The Foundation has a longtime interest in the underappreciated role of women and minorities in science and technology. The Foundation also supports television programs based on projects it has sponsored in other media, such as books and plays. The Foundation also continues to support commercial television by developing various film projects with cable and broadcast networks.
Requirements: Concise, well-organized proposals are preferred. In no case should the body of the proposal exceed 20 double-spaced pages. Grant requests can be made at any time for support of activities related to Foundation program areas and interests. The Foundation is generally limited to supporting tax-exempt organizations. The Foundation has no deadlines or standard forms. The Foundation accepts proposals sent by email. A brief letter of inquiry, rather than a fully developed proposal, is an advisable first step for an applicant, conserving his or her time and allowing for a preliminary response regarding the possibility of support. Interested grant seekers with a relevant project idea should email a one page letter of inquiry to Program Director Doron Weber.
Restrictions: The Foundation's activities do not normally extend to religion, the creative or performing arts, elementary or secondary education, medical research or health care, the humanities or to activities outside the United States. Grants are not made for endowments or for buildings or equipment. Telephone inquiries are not accepted.
Geographic Focus: All States
Contact: Doron Weber; (212) 649-1652 or (212) 649-1649; weber@sloan.org
Internet: http://www.sloan.org/program/23
Sponsor: Alfred P. Sloan Foundation
630 Fifth Avenue, Suite 2550
New York, NY 10111-0242

Alfred P. Sloan Foundation Public Understanding of Science and Technology Radio Grants 558

The Alfred P. Sloan Foundation supports a range of radio programs that deal with science and technology and seeks to increase both the quantity and the quality of science and technology coverage. Grant requests can be made at any time for support of activities related to Foundation program areas and interests. The Foundation is generally limited to supporting tax-exempt organizations. The Foundation has no deadlines or standard forms. The Foundation accepts proposals sent by email. To begin the application process, send a one page letter of inquiry to program director Doron Weber.
Requirements: Concise, well-organized proposals are preferred. In no case should the body of the proposal exceed 20 double-spaced pages.
Restrictions: The Foundation's activities do not normally extend to religion, the creative or performing arts, elementary or secondary education, medical research or health care, the humanities or to activities outside the United States. Grants are not made for endowments or for buildings or equipment. Telephone inquiries are not accepted.
Geographic Focus: All States
Contact: Doron Weber; (212) 649-1652 or (212) 649-1649; weber@sloan.org
Internet: http://www.sloan.org/major-program-areas/public-understanding-of-science-and-technology/radio/
Sponsor: Alfred P. Sloan Foundation
630 Fifth Avenue, Suite 2550
New York, NY 10111-0242

Alfred P. Sloan Foundation Public Understanding of Science and Technology Theater Grants 559

In this program, the Foundation makes grants to encourage playwrights and theater companies to write and produce new plays about scientists, engineers and mathematicians that will break down the barrier between the two cultures. The Foundation's major partners include: the Ensemble Studio Theatre; the Manhattan Theatre Club and; Playwrights Horizons. The Foundation has been a leading force in commissioning and producing new science and technology plays. In this role, it has helped to stimulate numerous playwrights to tackle these subjects while also creating a more hospitable environment among the theater-going public for such fare, heralding an important cultural shift.
Geographic Focus: All States
Contact: Doron Weber; (212) 649-1652 or (212) 649-1649; weber@sloan.org
Internet: http://www.sloan.org/program/24
Sponsor: Alfred P. Sloan Foundation
630 Fifth Avenue, Suite 2550
New York, NY 10111-0242

Alfred P. Sloan Foundation Research Fellowships 560

Alfred P. Sloan Foundation Research Fellowships stimulate fundamental research by young scholars of outstanding promise at a time in their careers when their creative abilities are especially high and when government or other support is difficult to obtain. Fellows are free to pursue whatever lines of inquiry that are of the most compelling interest to them. Funds are awarded directly to the fellow's institution and may be used by the fellow for such purposes as equipment, technical assistance, professional travel, trainee support, or activities directly related to the fellow's research. The foundation welcomes nominations of all candidates who meet the traditional high standards of this program and strongly encourages the participation of women and members of underrepresented minority groups. Awards are made in the fields of physics, chemistry, neuroscience, economics, pure mathematics, applied mathematics, computer science, and computational and evolutionary molecular biology. Fellowships are awarded for a two-year period, with possible extension for another two years. Each year 118 fellows are selected from the nominations received. Nominations are due September 15 for awards to begin the following September. Nomination forms are available upon request.
Requirements: Candidates must hold a PhD or equivalent in chemistry, physics, mathematics, computer science, economics, neuroscience or computational and evolutionary molecular biology, or in a related interdisciplinary field, and must also be a faculty member at a U.S. or Canadian university. Candidates must be no more than six years from the completion of the most recent PhD or equivalent as of the year of their nomination.
Restrictions: Direct applications are not accepted. Candidates must be nominated by department heads or other senior scholars.
Geographic Focus: All States, Canada
Date(s) Application is Due: Sep 15
Amount of Grant: 45,000 USD
Contact: Daniel L. Goroff, Program Director; (212) 649-1676 or (212) 649-1649; fax (212) 757-5117; goroff@sloan.org
Internet: http://www.sloan.org/sloan-research-fellowships/?L=gaajyejruhmak
Sponsor: Alfred P. Sloan Foundation
630 Fifth Avenue, Suite 2550
New York, NY 10111-0242

Alfred P. Sloan Foundation Scholarly Communication Grants 561

The shift to digitally-mediated forms of scholarship has been characterized by a substantial growth in channels for and diversity of scholarly work. We see this in the flourishing of content in preprint servers and rapid-publication channels like arXiv, PLoS ONE, and the Social Science Research Network alongside unconventional forms of scholarly communication like research blogs and personal websites, all of which enable scholars to put their work out for broad access. Grants in this sub-program aim to ease this transition by supporting the development of new models of filtering and curating online scholarly materials and by engaging the emerging community of stakeholders and practitioners tackling similar issues in widely divergent disciplinary contexts. Grantseekers with general inquiries or with a relevant project idea regarding the Data and Computational Research or Scholarly Communications sub-programs should send a one-page letter of inquiry to Program Director.
Requirements: Concise, well-organized proposals are preferred. In no case should the body of the proposal exceed 20 double-spaced pages.
Restrictions: The Foundation does not make grants: to individuals; to for-profit institutions (except in rare instances); in religion, medical research or the humanities; to projects aimed at pre-college students; to projects in the creative or performing arts, except when those projects are related to educating the public about science, technology or economics; for endowments, buildings or equipment, or fundraising drives, including fundraising dinners; or to political campaigns, to support political activities or to lobby for or against particular pieces of legislation.
Geographic Focus: All States
Contact: Dr. Gail Pesyna, Fellowship Administrator; (212) 649-1633 or (212) 649-1649; fax (212) 757-5117; greenberg@sloan.org
Internet: http://www.sloan.org/major-program-areas/digital-information-technology/scholarly-communication/
Sponsor: Alfred P. Sloan Foundation
630 Fifth Avenue, Suite 2550
New York, NY 10111-0242

Alfred P. Sloan Foundation Science of Learning STEM Grants 562

Grantmaking in the Science of Learning STEM program aims to improve the quality of higher education in STEM fields through the support of original, high-quality research on the factors affecting undergraduate and graduate student learning and retention in STEM fields. Grants primarily support consortia of colleges, universities, and other educational institutions with plans to develop and to study the impact and effectiveness of new approaches to STEM pedagogy, especially in "gateway" courses, with an explicit commitment to institutionalize successful initiatives. Successful proposals are expected to be hypothesis-driven, sensitive to the heterogeneity of STEM disciplines, attentive to differences in student motivations to choose STEM majors and persist in STEM careers, and concerned with the dissemination and portability of results to other institutions.
Geographic Focus: All States
Contact: Elizabeth Boylan, Program Director; (212) 649-1634 or (212) 649-1649; fax (212) 757-5117; boylan@sloan.org
Internet: http://www.sloan.org/major-program-areas/stem-higher-education/the-science-of-learning-stem/
Sponsor: Alfred P. Sloan Foundation
630 Fifth Avenue, Suite 2550
New York, NY 10111-0242

Alfred P. Sloan Foundation Selected National Issues Grants 563

The Alfred P. Sloan Foundation recognizes that there are select opportunities outside of science, education and economics in which it can create an important benefit to society. Its National Issues program area looks for unique opportunities where Foundation funds promise to advance a significant national interest. Grants in the Select National Issues program are funding work to increase America's biosecurity and investigate how recent advances in information technology affect the spread of knowledge and the structure of scientific endeavor. The Foundation has no deadlines or standard forms. The Foundation accepts proposals sent by email. A brief letter of inquiry, rather than a fully developed proposal, is an advisable first step for an applicant, conserving his or her time and allowing for a preliminary response regarding the possibility of support.
Requirements: Concise, well-organized proposals are preferred. In no case should the body of the proposal exceed 20 double-spaced pages.
Restrictions: The Foundation does not make grants: to individuals; to for-profit institutions (except in rare instances); in religion, medical research or the humanities; to projects aimed at pre-college students; to projects in the creative or performing arts, except when those projects are related to educating the public about science, technology or economics; for endowments, buildings or equipment, or fundraising drives, including fundraising dinners; or to political campaigns, to support political activities or to lobby for or against particular pieces of legislation.
Geographic Focus: All States
Contact: Paula J. Olsiewski, Program Director; (212) 649-1658 or (212) 649-1649; fax (212) 757-5117; olsiewski@sloan.org
Internet: http://www.sloan.org/about-the-foundation/staff-directory/show-staff/show/people/paula-j-olsiewski/
Sponsor: Alfred P. Sloan Foundation
630 Fifth Avenue, Suite 2550
New York, NY 10111-0242

Alfred P. Sloan Foundation Sloan Digital Sky Survey Grants 564

The Sloan Digital Sky Survey (SDSS) program is expanding our understanding of the evolution and structure of the universe by providing the largest uniform, detailed archive of objects in the skies that has ever existed. In cooperation with the Astrophysical Research Consortium, the Foundation has helped build and operate a specially designed telescope to observe and archive galaxies, quasars and other cosmological phenomena. Data collected through the SDSS is providing an increasingly detailed picture of the universe, including data bearing on the curvature of the universe itself, on the existence of dark energy, and on the features of the Milky Way. Grant requests can be made at any time for support of activities related to Foundation program areas and interests. The Foundation is generally limited to supporting tax-exempt organizations. The Foundation has no deadlines or standard forms. The Foundation accepts proposals sent by email. A

brief letter of inquiry, rather than a fully developed proposal, is an advisable first step for an applicant, conserving his or her time and allowing for a preliminary response regarding the possibility of support.
Requirements: Concise, well-organized proposals are preferred. In no case should the body of the proposal exceed 20 double-spaced pages.
Restrictions: The Foundation's activities do not normally extend to religion, the creative or performing arts, elementary or secondary education, medical research or health care, the humanities or to activities outside the United States. Grants are not made for endowments or for buildings or equipment.
Geographic Focus: All States
Contact: Gail Pesyna; (212) 649-1655 or (212) 649-1649; pesyna@sloan.org
Internet: http://www.sloan.org/major-program-areas/basic-research/sloan-digital-sky-survey/
Sponsor: Alfred P. Sloan Foundation
630 Fifth Avenue, Suite 2550
New York, NY 10111-0242

Alfred P. Sloan Foundation Synthetic Biology Grants **565**
The goal of Sloan's Synthetic Biology initiative is to identify the risks associated with research in and applications of synthetic biology and to assess the ethical, regulatory, and public policy implications of these risks. Grantmaking aims to educate scientists, policy makers, journalists and the public about synthetic biology, improve biosecurity and biosafety within the field, lay groundwork to address issues in regulation and governance, and help develop a cadre of scholars and practitioners to evaluate the ethical, social, and public policy consequences of synthetic biology research. Recent grantmaking in this program has focused on informing key audiences about synthetic biology. A grant to The Hastings Center aims to engage the ethical community to identify and articulate ethical issues associated with synthetic biology research and provide a basis for informed policy discussion. A Sloan-funded project at the J. Craig Venter Institute is educating the scientific community about societal concerns regarding synthetic biology and educating the policy and journalism communities about the science underlying synthetic biology research. A grant to the Woodrow Wilson International Center aims to identify risks associated with synthetic biology, evaluate the adequacy of existing regulatory mechanisms, and educate policy makers and the public through events and through its website
Requirements: Concise, well-organized proposals are preferred. In no case should the body of the proposal exceed 20 double-spaced pages.
Restrictions: The Foundation's activities do not normally extend to religion, the creative or performing arts, elementary or secondary education, medical research or health care, the humanities or to activities outside the United States. Grants are not made for endowments or for buildings or equipment.
Geographic Focus: All States
Contact: Paula J. Olsiewski; (212) 649-1658 or (212) 649-1649; olsiewski@sloan.org
Internet: http://www.sloan.org/major-program-areas/basic-research/synthetic-biology/
Sponsor: Alfred P. Sloan Foundation
630 Fifth Avenue, Suite 2550
New York, NY 10111-0242

Alfred P. Sloan Foundation Working Longer Grants **566**
Americans are working later in life, delaying retirement for a variety of reasons. While some continue working by choice, others needto remain in the workforce for financial reasons. Regardless of their motivation, the result is that people in the United States are working beyond what conventionally is thought of as retirement age. While most studies related to this issue have focused on how individually based factors such as health status, pension plans, and financial incentives influence the decision to retire, relatively little is known about employment patterns, obstacles to employment, or the ensuing economic impact. The Foundation's Working Longer program began grant making in 2010 to expand and deepen our understanding of aging Americans' work patterns. The goal is to understand; employer practices by industry and sector; obstacles to continued employment of older Americans; and the economic consequences for both individuals and for the federal budget.
Requirements: Concise, well-organized proposals are preferred. In no case should the body of the proposal exceed 20 double-spaced pages.
Restrictions: The Foundation does not make grants: to individuals; to for-profit institutions (except in rare instances); in religion, medical research or the humanities; to projects aimed at pre-college students; to projects in the creative or performing arts, except when those projects are related to educating the public about science, technology or economics; for endowments, buildings or equipment, or fundraising drives, including fundraising dinners; or to political campaigns, to support political activities or to lobby for or against particular pieces of legislation.
Geographic Focus: All States
Contact: Kathleen Christensen; (212) 649-1628; christensen@sloan.org
Internet: http://www.sloan.org/major-program-areas/economic-performance-and-quality-of-life/working-longer/
Sponsor: Alfred P. Sloan Foundation
630 Fifth Avenue, Suite 2550
New York, NY 10111-0242

Alfred W. Bressler Prize in Vision Science **567**
The Prize recognizes a professional in the field of vision science whose leadership, research and service have resulted in important advancements in the treatment of eye disease or rehabilitation of persons with vision loss. A panel of distinguished vision science professionals will select the winner who receives a prize of $40,000.
Requirements: The application process is open to established professionals in the field of vision science whose contributions have advanced vision care, the treatment of eye disease, or the rehabilitation of persons with visual disabilities or blindness and whose further work is expected to contribute significantly. Candidates from the United States and countries around the world are eligible for the award.
Geographic Focus: All States
Date(s) Application is Due: Dec 31
Amount of Grant: 40,000 USD
Contact: Program Administrator; (212) 769-7801; bressler@jgb.org
Internet: http://www.jgb.org/programs_bressler.asp
Sponsor: Jewish Guild for the Blind
15 W 65th Street
New York, NY 10023

Alice Fisher Society Fellowships **568**
The Barbara Bates Center for the Study of the History of Nursing offers a fellowship of $2,500 to support two weeks in residence at the Center and ongoing collaboration with Center historians. Selection of Alice Fisher Society scholars will be based on evidence of interest in and aptitude for historical research related to nursing. It is expected that the research and new materials produced by Alice Fisher Society scholars will help ensure the growth of scholarly work focused on the history of nursing. Fisher scholars will participate in Center activities and will present their research at a Center seminar.
Requirements: The scholarships are open to individuals with master's and doctoral level preparation. The application should be sent via email to either Patricia D'Antonio, PhD, RN, FAAN (dantonio@nursing.upenn.edu) or Barbra Mann Wall, PhD, RN (wallbm@nursing.upenn.edu).
Geographic Focus: All States
Date(s) Application is Due: Dec 31
Amount of Grant: 2,500 USD
Contact: Dr. Karen Buhler-Wilkerson, Director; (215) 898-4725; fax (215) 573-2168; karenwil@nursing.upenn.edu
Internet: http://www.nursing.upenn.edu/history/Pages/Alice_Fisher_Fellowship.aspx
Sponsor: University of Pennsylvania
School of Nursing, 418 Curie Boulevard
Philadelphia, PA 19104-6020

Alice Tweed Tuohy Foundation Grants **569**
The foundation promotes organizations that promote young people; that provide outstanding opportunities for performance, growth, and creativity; that nurture personal integrity and ambition; and that reward high achievement. The foundation assists organizations offering services to children whose choices might otherwise be unfairly restricted by need; supported are activities both academic and extracurricular that challenge young people while encouraging the growth of responsibility and personal integrity. Organizations dedicated to improving the quality of life by meeting the vital needs of the community are also supported. Types of support include building construction/ renovation, scholarship funds, and matching funds. Financing priority is accorded those organizations with the least in-house capacity to raise capital, assisting these groups to surmount critical monetary obstacles and continue productive service to the community. Proposals may be submitted annually between July 1 and September 15.
Requirements: Applications are considered only from Santa Barbara, CA public, tax-exempt organizations. Priority consideration is given to applications from organizations serving young people, education, health and medicine, community affairs, and the arts.
Restrictions: Excluded from consideration are applications for the benefit of specific individuals, organizations outside the Santa Barbara area, organizations in overpopulated nonprofit areas, national campaigns, fund-raising normally carried out by the organization, operating expenses, or budgetary support.
Geographic Focus: California
Date(s) Application is Due: Sep 15
Amount of Grant: 750-98,000 USD
Contact: Program Contact; (805) 962-6430; fax (805) 962-7135; atuohyfdn@aol.com
Sponsor: Alice Tweed Tuohy Foundation
205 E Carrillo Street, Room 219, P.O. Box 1328
Santa Barbara, CA 93102-1328

Allen Foundation Educational Nutrition Grants **570**
The foundation supports projects that benefit human nutrition in the areas of education, training, and research. Priorities include training programs for children and young adults to improve their health and development; training programs for educators and demonstrators concerned with good nutritional practices; programs for the education and training of mothers during pregnancy and after the birth of their children, so that good nutritional habits can be formed at an early age; and programs that aid in the dissemination of information regarding healthful nutrition practices. Applications are available by fax or mail. There are no application deadlines; proposals are reviewed throughout the year.
Requirements: 501(c)3 tax-exempt organizations nationwide may apply. In certain circumstances, the foundation will consider requests from the following: hospitals or medical clinics; social, religious, fraternal, or community organizations; private foundations; and K-12 public, parochial, or private schools.
Geographic Focus: All States
Date(s) Application is Due: Dec 31
Amount of Grant: 5,000-250,000 USD
Contact: Dr. Dale Baum, Secretary; (989) 832-5678 or (979) 695-1132; fax (989) 832-8842; dbaum@allenfoundation.org or Lucille@allenfoundation.org
Internet: http://www.allenfoundation.org/commoninfo/aboutus.asp
Sponsor: Allen Foundation
P.O. Box 1606
Midland, MI 48641-1606

Allen Lane Foundation Grants 571

The foundation operates in the United Kingdom and Ireland in the field of social welfare and awards grants to institutions at the local, regional, and national levels. Areas of interest include projects in Scotland, Northern Ireland, Wales, and regions outside the London area; groups supporting refugees and asylum-seekers; advisory and information services; and the coordination of small groups. The broad areas of work which are priorities for the foundation include the provision of advice, information, and advocacy; community development; employment and training; mediation, conflict resolution, and alternatives to violence; research and education aimed at changing public attitudes or policy; and social welfare.
Requirements: Grants may be made for project costs or revenue costs. The foundation no longer has closing dates.
Restrictions: The foundation only very rarely makes grants to national organizations with an income of more than $500,000 per annum or to local organizations with an income of more than about $150,000. Grants are not made to individuals.
Geographic Focus: All States, United Kingdom
Amount of Grant: 500-15,000 GBP
Contact: Gill Aconley; 01-904-613-223; fax 01-904-613-133; info@allenlane.org.uk
Internet: http://www.allenlane.org.uk
Sponsor: Allen Lane Foundation
90 The Mount
York, YO24 1AR England

All Saints Educational Trust Grants 572

The main purposes of the trust is to help increase the number of qualified teachers, improve the skills and qualifications of teachers, encourage research that would assist teachers in their work, and support specifically the teaching of Religious Studies and Home Economics and related areas. There is no deadline date for the preliminary questionnaire form but there is a deadline for the full application form. Annual deadline dates may vary; contact program staff for exact dates. The Trustees normally only give financial assistance when the course of study is undertaken at recognized educational establishments within the United Kingdom.
Requirements: Applicants must be over 18 years of age when they begin their studies.
Restrictions: Applications are not accepted from people outside the U.K. except residents of Commonwealth nations. The trust cannot support general or core funds of any organization, public appeals, school buildings, equipment or supplies, the establishment of courses or departments in universities and colleges, and general Bursary funds of other organizations.
Geographic Focus: All States, United Kingdom
Contact: Dr. K. G. Riglin; 44 (0) 20 7283 4485; fax 44 (0) 20 7621 9758; aset@aset.org.uk
Internet: http://www.aset.org.uk
Sponsor: All Saints Educational Trust
St. Katharine Cree Church, 86 Leadenhall Street
London, EC3A 3DH England

Allstate Foundation Safe and Vital Communities Grants 573

Established in 1952, The Allstate Foundation is an independent, charitable organization made possible by subsidiaries of The Allstate Corporation. The Foundation is dedicated to fostering safe and vital communities where people live, work and raise families. We strive to foster communities that are economically strong, crime-free, and give residents a sense of belonging and commitment. Our Safe and Vital Communities funding priorities are: teen safe driving-helping to save young lives and instill a lifetime of safe driving attitudes and behaviors; catastrophe response-rebuilding lives after a natural disaster strikes; and neighborhood revitalization-nurturing safe, strong, and healthy communities. Proposals are accepted throughout the year, though the regional grant proposal deadline is May 31. The average grant amount ranges from $5,000 to $20,000 for regional grants.
Requirements: The Allstate Foundation makes grants to nonprofit, tax-exempt organizations under Section 501(c)3 of the Internal Revenue Code, or be a municipal, state or federal government entity.
Restrictions: The Foundation does not support the following: individuals; fundraising events, sponsorships; capital and endowment campaigns; equipment purchase unless part of a community outreach program; athletic events; memorial grants; athletic teams, bands, and choirs; organizations that advocate religious beliefs or restrict participation on the basis of religion; groups or organizations that will re-grant the Foundation's gift to other organizations or individuals; scouting groups; private secondary schools; requests to support travel; grant requests for production of audio, film, or video; multiyear pledge requests; or non-domestic (international) causes.
Geographic Focus: All States
Date(s) Application is Due: May 31
Amount of Grant: 5,000-20,000 USD
Contact: Vicky Dinges, Vice President of Corporate Social Responsibility; (847) 402-5600 or (847) 402-7893; fax (847) 326-7517; allfound@allstate.com
Internet: http://www.allstatefoundation.org/grant-focus-areas
Sponsor: Allstate Foundation
2775 Sanders Road, Suite F4
Northbrook, IL 60062-6127

Allyn Foundation Grants 574

The mission of the Allyn Foundation is to improve the quality of life in Central New York. To accomplish this, the foundation focuses its grant making in the following areas: health facilities and services: promote the delivery of quality health care in our communities; higher education in the Onondaga and Cayuga counties: expand access and success in education beyond high school, particularly for students of low income or other underrepresented backgrounds; basic human services: work to ensure that all people have basic daily needs of food, clothing, and shelter and enable the means to provide for themselves; family planning medical and educational services: promote education in family planning and improved access to service; quality of life for youth: improve quality of life for children and families through organizations that are focused on at risk youth and families.
Requirements: Nonprofit organizations in Onondaga and Cayuga Counties, NY, may submit grant applications.
Restrictions: Grants are not awarded for religious purposes, endowment funds, loans, or to individuals.
Geographic Focus: New York
Samples: Auburn Memorial Hospital, $350,000; Frank Hiscock Legal Aid Society, $18,000; Child Care Council of the Finger Lakes, $4,000.
Contact: Margaret O'Connell; (315) 685-5059; info@allynfoundation.org
Internet: http://www.allynfoundation.org/apply.html
Sponsor: Allyn Foundation
14 West Genesee Street, P.O. Box 22
Skaneateles, NY 13152

Alpha Kappa Alpha Educational Advancement Foundation Fellowships 575

Educational Advancement Foundation fellowships are awarded during even numbered years. These awards help to improve the quality of life of others by funding research and projects with practical applications. Foundation fellowships have been awarded to individuals for a wide range of purposes, from assistance in finding non-violent solutions to human disputes to conducting research on hypertension and the impact of nuclear testing on marine life. The Helen Cromer Cooper/Rosa Parks Fellowship is earmarked to support individuals and projects involved in finding non-violent solutions to human problems. The Dr. Dorri Phipps Fellowship is for students pursuing degrees in medicine or research pertaining to lupus. Applications, along with supporting documents, must be postmarked by January 15 of the award year.
Geographic Focus: All States
Date(s) Application is Due: Jan 15
Contact: Andrea Kerr, Program/Scholarship Coordinator; (800) 653-6528 or (773) 947-0026; fax (773) 947-0277; akerr@akaeaf.net or akaeaf@akaeaf.net
Internet: http://www.akaeaf.org/programsandinitiatives/
Sponsor: Alpha Kappa Alpha Educational Advancement Foundation
5656 South Stony Island Avenue
Chicago, IL 60637

Alpha Omega Foundation Grants 576

The Foundation was established to advance the dental profession through research, grants, and scholarships. Fields of interest include: dental care, dental education, and medical research. Grants are given to individuals, program development, and higher education institutions. The Foundation was responsible for dispersing almost $250,000 this past year for projects in Israel and worldwide. The application must be typed and submitted via email by September 1 each year. Notification will be completed by the following December.
Geographic Focus: All States, All Countries
Date(s) Application is Due: Sep 1
Amount of Grant: 1,000-50,000 USD
Contact: Heidi Weber, Association Director; (301) 738-6400; fax (301) 738-6403; hweber@ao.org or foundation@ao.org
Internet: http://www.ao.org/index.php?option=com_content&task=view&id=80&Itemid=147
Sponsor: Alpha Omega Foundation
50 W Edmonston Drive, Suite 303
Rockville, MD 20852-1274

Alpha Research Foundation Grants 577

The primary purpose of the Foundation is to encourage and support biomedical research for the welfare of the general public. There are no specific deadlines or application formats with which to adhere. Applicants should contact the foundation prior to submission, in order to discuss their proposed project. If approved, applicants should then submit an outline of the proposed research along with a proposed budget.
Geographic Focus: Maryland
Amount of Grant: Up to 2,000 USD
Contact: Barry Rueben Fierst, (301) 762-8872; fax (301) 762-8874; bfierst@aol.com
Sponsor: Alpha Research Foundation
7118 Glenbrook Road
Bethesda, MD 20814-1238

Alpha Sigma Nu Book Award Competition 578

The book awards operate on a three-year cycle. The category for 2007 is The Professional Studies, and includes the following disciplines: Architecture, Business and Administration, Communication, Education, Engineering, Foreign Service, Law, and Social Work. Four prizes will be awarded. The category for 2008, will be The Humanities with one award each for Theology, Philosophy/Ethics, Literature/Fine Arts, and History. In 2009, the category will be The Sciences with one award for Social Sciences, Natural Sciences, Mathematics/Computer Science, and Health Sciences. Information and application forms are available from academic vice presidents and deans of the Jesuit colleges and universities, or from the Association of Jesuit Colleges and Universities.
Requirements: Candidates must be full- or part-time staff or faculty or anyone who has emeritus status at a Jesuit college or university in the United States.
Geographic Focus: All States
Date(s) Application is Due: Mar 1
Amount of Grant: 1,000 USD

Contact: Program Contact; (202) 862-9893; agarner@ajcunet.edu
Internet: http://www.marquette.edu/dept/ASN/book_award_competition.html
Sponsor: Alpha Sigma Nu
One Dupont Cir, Suite 405
Washington, DC 20036

ALSAM Foundation Grants 579

The foundation awards grants in its areas of interest, including agriculture, Christian agencies and churches, higher education, health care and medical research, human services, minorities, and the economically disadvantaged. Types of support include building construction/renovation, general operating costs, and scholarships. The board meets in January and October. Contact the office for application forms.
Requirements: Higher education institutions, nonprofit organizations, religious organizations, and research institutions are eligible.
Restrictions: No grants to individuals.
Geographic Focus: All States
Amount of Grant: 5,000-50,000 USD
Contact: Ron Cutshall, Chair; (801) 266-4950
Sponsor: ALSAM Foundation
6190 Moffat Farm Lane
Salt Lake City, UT 84121

Alston Bannerman Sabbatical Fellowships 580

The Center for Social Inclusion (CSI) recognizes that working for social change can lead to pressures that result in a loss of creative and critical leadership. Because of this, Sabbatical Fellows receive a $25,000 award to take three months off for reflection and renewal. Sabbaticals strengthen organizational planning and leadership development, encourage innovation and collaboration, promote the value of reflection, and foster connections between organizers and organizations. By reinvigorating experienced grassroots organizers and encouraging younger ones to see organizing as a long-term career, the Fellowship helps build the necessary infrastructure for systemic and sustainable change. Fellows are selected through an annual application process. The Selection Committee is comprised of leaders from around the country working in or for communities of color. Fellows are required to attend pre- and post-sabbatical retreats; begin their sabbaticals within one year of the awarding of the Fellowship; and report on their sabbaticals within four months of their completion to describe what they did, what they learned that will improve their leadership, and what impact had on their work. The application is available at the CSI website.
Requirements: A Fellowship applicant must be a person of color; have more than ten years of community organizing experience; be committed to social change work in communities of color; and live in the United States, Puerto Rico, Guam, American Samoa, or the U.S. Virgin Islands. Beyond the basic eligibility, the Alston Bannerman Program seeks applicants whose work attacks root causes of inequity by organizing those affected to take collective strategic action; challenges the systems that perpetrate injustice and effects institutional and structural change; builds community capacity for democratic participation and develops grassroots leadership; acknowledges the cultural values of the community; creates accountable participatory structures in which community members have decision-making power; and contributes to building a movement for social change by making connections between issues, developing alliances with other constituencies, and collaborating with other organizations.
Restrictions: Awards are not made to individuals who exclusively provide services, such as substance abuse counseling, after school programs, HIV-AIDS outreach or shelter for the homeless; or who advocate on behalf of a community without directly involving the members of that community in asserting their own interests and choosing their own leadership.
Geographic Focus: All States, Guam, Puerto Rico, U.S. Virgin Islands, American Samoa
Amount of Grant: 25,000 USD
Contact: Madeleine Adamson; (212) 248-2785; fax (212) 248-6409; info@thecsi.org
Internet: http://www.centerforsocialinclusion.org/leadership/alston-bannerman-sabbatical-fellows/
Sponsor: Center for Social Inclusion
150 Broadway, Suite 303
New York, NY 10038

Alternatives Research and Development Foundation Alternatives in Education Grants 581

The mission of the Alternatives Research and Development Foundation is to fund and promote the development, validation and adoption of non-animal methods in biomedical research, product testing and education. ARDF has announced a special grant initiative seeking proposals to replace traditional use of animals for education and training purposes. Up to $10,000 in funding is available to support individual projects. the deadline for Letters of Intent is January 8, while invited full proposals are due March 12.
Requirements: Grants from the Foundation are made only to individuals affiliated with a non-profit, tax-exempt, institution, organization, or foreign equivalent.
Geographic Focus: All States
Date(s) Application is Due: Jan 8; Mar 12
Amount of Grant: Up to 10,000 USD
Contact: Sue A. Leary, President and Executive Director; (215) 887-8076; fax (215) 887-0771; grants@ardf-online.org or info@ardf-online.org
Internet: http://www.ardf-online.org/
Sponsor: Alternatives Research and Development Foundation
801 Old York Road, Suite 316
Jenkintown, PA 19046

Alternatives Research and Development Foundation Grants 582

The foundation awards grants to research centers, educational institutions, and other nonprofit organizations exploring non-animal testing. Grants will be awarded for research projects that use human, rather than nonhuman, vertebrae tissue; do not use intact, vertebrae animals; and can be completed within one year. Multiyear projects are considered on a case-by-case basis. The foundation often provides partial grants to initiate projects, with continuation funding available at a later date. The foundation will not consider proposals if program staff uses animals acquired from a shelter, or if individuals are employed who use such animals in their personal research programs. Winners are announced on July 15. Telephone inquiries are discouraged.
Requirements: Individuals attending or employed by U.S. universities and research institutions may apply. Applications from non-U.S. institutions or investigators may be considered.
Restrictions: The foundation does not make grants for in vitro projects that use nonhuman animal serum, indirect costs, purchase of personal computers, salary supplements, fringe benefits, travel expenses, or publication costs for the principal investigators. Phone calls to the foundation are discouraged. Guidelines and application are available online.
Geographic Focus: All States
Date(s) Application is Due: Apr 30
Amount of Grant: Up to 40,000 USD
Contact: Sue A. Leary, President and Executive Director; (215) 887-8076; fax (215) 887-0771; grants@ardf-online.org or info@ardf-online.org
Internet: http://www.ardf-online.org
Sponsor: Alternatives Research and Development Foundation
801 Old York Road, Suite 316
Jenkintown, PA 19046

Alton Ochsner Award 583

The Award is presented to one or more clinical or basic science investigators, without regard to age, race, gender, or nationality, for outstanding and exemplary original scientific investigations that relate tobacco consumption and health. This scientific work may be clinical, fundamental, epidemiological or preventive in scope. The prime criterion for award selection is its scientific impact on this major health threat. The $15,000 award is presented at the Annual Convocation of the American College of Chest Physicians.
Requirements: All nominations, whatever the category of scientific inquiry, must be supported by letters and copies of peer-reviewed scientific publications.
Geographic Focus: All States
Date(s) Application is Due: Mar 31
Amount of Grant: 15,000 USD
Contact: Edward D. Frohlich; (504) 842-3000; fax (504) 842-3258
Internet: http://www.ochsner.org/homepage.cfm
Sponsor: Ochsner Clinic Foundation
1514 Jefferson Highway
New Orleans, LA 70121

Alvin and Lucy Owsley Foundation Grants 584

The foundation gives grants in to organizations in the state of Texas in various interest areas including (but not limited to) arts, community development, education.
Requirements: There are no specific deadlines, but the board meets in March, June, September and December. Submit your proposal before the board meets. The application process is a letter no more than 2 pages in length with no attachments.
Geographic Focus: Texas
Amount of Grant: 500-100,000 USD
Contact: Alvin Owsley, Trustee; (713) 229-1272
Sponsor: Alvin and Lucy Owsley Foundation
65 Brair Hollow Lane
Houston, TX 77027

ALVRE Casselberry Award 585

The Casselberry Award has been established to encourage the advancement of the art and science of Laryngology and Rhinology. The award is given for outstanding manuscripts or accomplishments in Laryngology and Rhinology and consists of a $1,000 award and a certificate from the Association. Competition for this award will be limited to those persons whose abstracts are selected in consideration for inclusion in the Annual Scientific Program.
Geographic Focus: All States
Date(s) Application is Due: Nov 30
Amount of Grant: 1,000 USD
Contact: Maxine Cunningham, Administrator; (615) 322-6326; fax (615) 322-9102; Maxine@alahns.org or ala-hns@comcast.net
Sponsor: American Laryngological Association / American Laryngological Voice and Research Education Foundation
1215 21st Avenue South, 7302 MCE South
Nashville, TN 37232-8783

ALVRE Grant 586

The purpose of this award is to support basic, translational, or clinical research projects in laryngology, voice, outcomes, and related subjects. Research supported by this award should be specifically directed toward the pathogenesis, pathophysiology, diagnosis, prevention, or treatment of diseases, disorders, or conditions of the larynx and may be either basic or clinical/translational in approach. While not specifically required, proposals which aim to introduce new knowledge and methodology from other disciplines to research in laryngology or neurolaryngology, or which demonstrate collaborative effort

with members of other related disciplines are encouraged. Projects must be designed so as to yield useful information within the period of award, but priority will be given to projects that are also innovative with promise to develop into new long-range or expanded research programs capable of attracting funding from other sources. A single, one-year, non-renewable award of $25,000 maximum is available annually. The foundation will consider requests to cover travel expenses up to $1,000 for the principal investigator to present his/her results at the ALA annual meeting.
Requirements: Candidates for this award should be otolaryngologists in the U.S. or Canada who have completed their training at an ACGME accredited program in otolaryngology—head and neck surgery. The prinicipal investigator should be a physician faculty member of a recognized department, division, or section of otolaryngology-head and neck surgery.
Geographic Focus: All States
Amount of Grant: Up to 25,000 USD
Contact: Maxine Cunningham, Administrator; (615) 322-6326; fax (615) 322-9102; Maxine@alahns.org or ala-hns@comcast.net
Internet: http://www.entnet.org/EducationAndResearch/coreGrants.cfm
Sponsor: American Laryngological Association / American Laryngological Voice and Research Education Foundation
1215 21st Avenue South, 7302 MCE South
Nashville, TN 37232-8783

ALVRE Seymour R. Cohen Award 587
The Seymour R. Cohen Award was established with a bequest by the Cohen family during Dr. Cohen's presidency in 1989. In order to qualify for the Award, the candidate must perform basic science research in the area of pediatric laryngology and/or pediatric neurolaryngology and be a citizen of the United States or Canada. The material must be approved and accepted for presentation at the Annual Meeting. Should multiple candidates collaborate in the research, all authors must comply with the citizenship requirements. Dr. Cohen wishes this Award only to be used for the funding of research in pediatric laryngology and/or pediatric neurolaryngology. The Award is to be presented bi-annually.
Requirements: The scope of the Award is limited to citizens of the United States of America and Canada.
Geographic Focus: All States
Contact: Maxine Cunningham, Administrator; (615) 322-6326; fax (615) 322-9102; Maxine@alahns.org or ala-hns@comcast.net
Internet: http://www.alahns.org/i4a/pages/index.cfm?pageid=3332
Sponsor: American Laryngological Association / American Laryngological Voice and Research Education Foundation
1215 21st Avenue South, 7302 MCE South
Nashville, TN 37232-8783

ALZA Corporate Contributions Grants 588
The corporation awards grants to eligible California nonprofit organizations in four areas of interest: education—K-12 after school programs, professional development for teachers, and health care workers, nonprofit educational outreach programs; health and human services—projects that particularly access to healthcare for underserved and underrepresented members of our community; arts & culture—arts programs that have an education component or art related to health and healing are of particular interest; and environment—projects that focus on sustainable solutions, stewardship, & education.
Requirements: Nonprofits in ALZA's California operating areas are eligible.
Restrictions: Grants will not be made to individuals, or to organizations whose activities or policies include sectarian or denominational religious activities, political campaigns, or organizations that discriminate on the basis of religion, race, nationality, sexual preference, or gender.
Geographic Focus: California
Contact: Ellen Rose, Director; (650) 564-5000; fax (650) 564-7070
Internet: http://www.alza.com/alza/community
Sponsor: ALZA Corporation
1900 Charleston Road, P.O. Box 7210
Mountain View, CA 94043

Alzheimer's Association Conference Grants 589
The Alzheimer's Association has a long history of supporting scientific conferences that advance research on Alzheimer's disease. One of the principal goals of the Association from its inception has been to increase public awareness and to facilitate the exchange of information through the scientific and clinical communities. The support of conferences, workshops and meetings has been a key vehicle in achieving this goal. The objectives for conference support are to: facilitate and speed the exchange of information relevant to Alzheimer's disease research; convene experts to address emerging issues in Alzheimer's disease research; offer opportunities for new investigators and graduate students to participate in scientific meetings; facilitate the creation of networks among investigators in related areas; and increase visibility of the research interests and programs of the Alzheimer's Association. At this time, support requests must be limited to no more than $10,000 per conference. Most awarded conference support requests have been in the range of $2000 to $5000.
Requirements: Requests for conference support may be submitted at any time. It is recommended that requests be submitted at least three months before the conference.
Geographic Focus: All States
Date(s) Application is Due: Jan 4; Mar 6; May 1; Jun 7; Aug 7; Oct 2; Nov 1
Contact: Angela Worlds, (312) 335-5807; fax (312) 335-4034; angela.worlds@alz.org
Internet: http://www.alz.org/research/alzheimers_grants/types_of_grants.asp
Sponsor: Alzheimer's Association
225 North Michigan Avenue, Suite 1700
Chicago, IL 60601-7633

Alzheimer's Association Development of New Cognitive and Functional Instruments Grants 590
The Alzheimer's Association is launching a new initiative to stimulate the scholarly investigation and development of cognitive or functional evaluation instruments that can capture the earliest changes in the disease, are sensitive to change over time, and/or could be used in clinical trials. This RFA is designed to enable pilot research or proof-of-principle studies that can provide preliminary data for subsequent inquiry. Allowable costs under this award include: purchase and care of laboratory animals; small pieces of laboratory equipment and laboratory supplies; computer equipment if used strictly for data collection; travel (up to $1,000 per year); and salary for the principal investigator, scientific (including post-doctoral fellows) and technical staff (including laboratory technicians and administrative support staff whose work is directly related to the funded project). The Association anticipates funding up to 2 approaches addressing needs in cognitive and functional instrument development. Each award is limited to $400,000 (direct and indirect costs) for two to three years. Requests in any given year may not exceed $200,000 (direct and indirect costs). Indirect costs are capped at 10 percent (rent for laboratory or office space is expected to be covered by indirect costs paid to the institution). Letters of Intent (LOIs) must be received by December 20, with completed applications due by February 7.
Requirements: Researchers with full-time faculty appointments are encouraged to apply.
Restrictions: Applications from post-doctoral candidates will not be accepted. Costs not allowed under this award include: tuition; computer hardware or software for investigators; rent for laboratory or office spaces; or construction or renovation costs.
Geographic Focus: All States
Date(s) Application is Due: Feb 7
Amount of Grant: Up to 400,000 USD
Contact: Nico Stanculescu, Program Administrator; (312) 335-5747 or (312) 335-5862; fax (312) 335-4034; nico.stanculescu@alz.org or grantsapp@alz.org
Internet: http://www.alz.org/research/alzheimers_grants/types_of_grants.asp
Sponsor: Alzheimer's Association
225 North Michigan Avenue, Suite 1700
Chicago, IL 60601-7633

Alzheimer's Association Everyday Technologies for Alzheimer Care Grants 591
Everyday Technologies for Alzheimer Care (ETAC) is a cooperative research funding initiative sponsored by the Alzheimer's Association and Intel Corporation. ETAC seeks proposals on personalized diagnostics, preventive tools and interventions for adults coping with the spectrum of cognitive aging and neurodegenerative disease, particularly Alzheimer's disease. ETAC is designed to support exploratory multidisciplinary research that would not typically be funded by national health and science granting foundations. Minor iterations in testing plans or populations will not be considered for funding. Collaboration between social science/medical/public health and computer science/engineering researchers is valued. Mobile computing, high bandwidth sensing, robotics, imaging, face recognition, natural language processing, statistical modeling and a host of other technology advances allow unprecedented opportunities to study disease progression and therapeutic strategies in the context of everyday life. ETAC supports research that integrates such emerging technology capabilities with leading directions in behavioral science and biomedical research. Grants that merely create Internet-based versions of existing services or paper tools will not be considered. Submissions must be original ideas, not continuations of previously funded ETAC projects. Please see links provided below for examples of studies that have been funded by ETAC. Letters of Intent (LOIs) must be received by December 20, with completed applications due by February 7. The Association anticipates funding two awards under this program.
Requirements: Researchers with full-time faculty appointments are encouraged to apply.
Restrictions: The Alzheimer's Association will not accept new research grant applications from currently funded Alzheimer's disease investigators who are delinquent in submitting interim/final scientific or interim/final financial reports on active grants. ETAC applications from post-doctoral candidates will not be accepted. Each total award is limited to $200,000 (direct and indirect costs) for up to three years. Requests in any given year may not exceed $90,000 (direct and indirect costs). Indirect costs are capped at 10 percent (rent for laboratory/office space is expected to be covered by indirect costs paid to the institution).
Geographic Focus: All States
Date(s) Application is Due: Feb 7
Amount of Grant: Up to 200,000 USD
Contact: Nico Stanculescu, Program Administrator; (312) 335-5747 or (312) 335-5862; fax (312) 335-4034; nico.stanculescu@alz.org or grantsapp@alz.org
Internet: http://www.alz.org/research/alzheimers_grants/types_of_grants.asp
Sponsor: Alzheimer's Association
225 North Michigan Avenue, Suite 1700
Chicago, IL 60601-7633

Alzheimer's Association Investigator-Initiated Research Grants 592
The program is structured to provide one to three years of sustained project support for independent, ongoing research. Proposals are solicited for basic, clinical, and social/behavioral research relevant to degenerative brain diseases such as Alzheimer's disease. Allowable costs for this award include the purchase and care of laboratory animals; small pieces of laboratory equipment and laboratory supplies; and salary for the principal investigator, scientific (including postdoctoral fellows) and technical staff (including laboratory technicians and modest secretarial support). It is required that most of the funds awarded under this program be used for direct research support. Allowable costs under this award include: purchase and care of laboratory animals; small pieces of laboratory equipment and laboratory supplies; computer software if used strictly for data collection; salary for the principal investigator, scientific (including post-doctoral

fellows) and technical staff (including laboratory technicians and administrative support related directly to the funded project); and travel to scientific and professional meetings, not to exceed $1,000 per year. Letters of Intent (LOIs) must be received by December 20, with completed applications due by February 7.
Requirements: Public, private, domestic and foreign research laboratories, medical centers and hospitals, and universities are eligible to apply. Investigators from all stages of their research career development are encouraged to apply.
Restrictions: Your budget must not exceed the maximum amount of the award, $240,000 or $100,000 per year. Costs not allowed under this award include: tuition; computer hardware or standard software (e.g., Microsoft Office) for investigators; rent for laboratory or office space; or construction or renovation costs.
Geographic Focus: All States
Date(s) Application is Due: Feb 7
Amount of Grant: 100,000-240,000 USD
Contact: Nico Stanculescu, Program Administrator; (312) 335-5747 or (312) 335-5862; fax (312) 335-4034; nico.stanculescu@alz.org or grantsapp@alz.org
Internet: http://www.alz.org/research/alzheimers_grants/types_of_grants.asp
Sponsor: Alzheimer's Association
225 North Michigan Avenue, Suite 1700
Chicago, IL 60601-7633

Alzheimer's Association Mentored New Investigator Research Grants to Promote Diversity 593
The Mentored New Investigator Research Grant to Promote Diversity (MNIGD) is a three-year award intended to be a research-based and mentoring investment in the process of closing the health disparities gap between diverse and non-diverse investigator populations. The Alzheimer's Association feels strongly that the mentoring and involvement of diverse researchers in independently funded Alzheimer's research is a pressing need. The MNIRGD is intended to enhance the capacity of diverse and non-diverse scientists to conduct basic, clinical and social/behavioral research. Each MNIRGD award is limited to $170,000. A total of $150,000 will be awarded for costs related to the proposed research for up to three years (direct and indirect costs). Allowable costs under this award include: purchase and care of laboratory animals; small pieces of laboratory equipment and laboratory supplies; computer equipment if used strictly for data collection; travel (up to $1,000 per year); and salary for the principal investigator, scientific (including post-doctoral fellows) and technical staff (including laboratory technicians and administrative support related directly to the funded project) Letters of Intent (LOIs) must be received by December 20, with completed applications due by February 7.
Requirements: Eligibility for this grant competition is restricted to investigators who have less than 10 years of research experience after receipt of their terminal degree.
Restrictions: Each MNIRGD award is limited to $170,000. A total of $150,000 will be awarded for costs related to the proposed research for up to three years (direct and indirect costs). Requests in any given year may not exceed $60,000 (direct and indirect costs). Indirect costs are capped at 10 percent (rent for laboratory and/or office space is expected to be covered by indirect costs paid to the institution). Costs not allowed under this award include: tuition; computer hardware or software for investigators; rent for laboratory/office space; or construction or renovation costs.
Geographic Focus: All States
Date(s) Application is Due: Feb 7
Amount of Grant: Up to 170,000 USD
Contact: Nico Stanculescu, Program Administrator; (312) 335-5747 or (312) 335-5862; fax (312) 335-4034; nico.stanculescu@alz.org or grantsapp@alz.org
Internet: http://www.alz.org/research/alzheimers_grants/types_of_grants.asp
Sponsor: Alzheimer's Association
225 North Michigan Avenue, Suite 1700
Chicago, IL 60601-7633

Alzheimer's Association Neuronal Hyper Excitability and Seizures in Alzheimer's Disease Grants 594
The Alzheimer's Association is launching a new initiative to stimulate the development of new pharmacological strategies to prevent or treat seizures and abnormal neural network activity in Alzheimer's disease (AD). The Association's Request for Applications (RFAs) is aimed at the identification, screening and development of therapeutic strategies to reduce seizures and other types of abnormal neural network activity and at the evaluation of drug safety and efficacy at the preclinical and clinical levels. The RFA is designed to enable preliminary pilot research or proof-of-principle studies that can provide data for further research support by other funding agencies. Allowable costs under this award include: purchase and care of laboratory animals; small pieces of laboratory equipment and laboratory supplies; computer equipment if used strictly for data collection; travel (up to $1,000 per year); and salary for the principal investigator, scientific (including post-doctoral fellows) and technical staff (including laboratory technicians and administrative support staff whose work is directly related to the funded project). The Association anticipates funding up to two (2) Neuronal Hyper Excitability and Seizures in AD awards. Each award is limited to $400,000 (direct and indirect costs) for two to three years. Requests in any given year may not exceed $200,000 (direct and indirect costs). Indirect costs are capped at 10 percent (rent for laboratory/office space is expected to be covered by indirect costs paid to the institution). Letters of Intent (LOIs) must be received by December 20, with completed applications due by February 7.
Requirements: Researchers with full-time faculty appointments are encouraged to apply.
Restrictions: Applications from post-doctoral candidates will not be accepted. Costs not allowed under this award include: tuition; computer hardware or software for investigators; rent for laboratory/office spaces; or construction or renovation costs.
Geographic Focus: All States
Date(s) Application is Due: Feb 7
Amount of Grant: Up to 400,000 USD
Contact: Nico Stanculescu, Program Administrator; (312) 335-5747 or (312) 335-5862; fax (312) 335-4034; nico.stanculescu@alz.org or grantsapp@alz.org
Internet: http://www.alz.org/research/alzheimers_grants/types_of_grants.asp
Sponsor: Alzheimer's Association
225 North Michigan Avenue, Suite 1700
Chicago, IL 60601-7633

Alzheimer's Association New Investigator Research Grants 595
The purpose of this program is to provide new investigators with funding that will allow them to develop preliminary or pilot data, to test procedures, and develop hypotheses which will then underpin the preparation of research grant applications to NIH, NSF, and other funding agencies and groups, including the Alzheimer's Association. All applications submitted to the program must focus on a question or questions in interventions for Alzheimer's disease to be considered responsive to the program announcement. Thirty awards will be made under this program. Letters of Intent (LOIs) must be received by December 20, with completed applications due by February 7.
Requirements: Public, private, domestic and foreign research laboratories, medical centers and hospitals, and universities are eligible to apply. Eligibility is restricted to investigators who have less than 10 years of research experience, including postdoctoral fellowships or residencies, after receipt of the doctoral degree. Applications from graduate and doctoral students for research projects, which will be used for the thesis or dissertation, will be accepted.
Restrictions: The Association anticipates funding 45 awards under this competition. Each total award is limited to $100,000 (direct and indirect costs) for up to two years. Requests in any given year may not exceed $60,000 (direct and indirect costs). Indirect costs are capped at 10 percent (rent for laboratory/office space is expected to be covered by indirect costs paid to the institution).
Geographic Focus: All States
Date(s) Application is Due: Feb 7
Amount of Grant: Up to 100,000 USD
Contact: Nico Stanculescu, Program Administrator; (312) 335-5747 or (312) 335-5862; fax (312) 335-4034; nico.stanculescu@alz.org or grantsapp@alz.org
Internet: http://www.alz.org/research/alzheimers_grants/types_of_grants.asp
Sponsor: Alzheimer's Association
225 North Michigan Avenue, Suite 1700
Chicago, IL 60601-7633

Alzheimer's Association New Investigator Research Grants to Promote Diversity 596
The New Investigator Research Grant to Promote Diversity (NIRGD) in Alzheimer's research is a two-year award to investigators who are currently underrepresented at academic institutions in Alzheimer's or related dementias research. The objective of this award is to increase the number of highly trained investigators from diverse backgrounds whose basic, clinical and social/behavioral research interests are grounded in the advanced methods and experimental approaches needed to solve problems related to Alzheimer's and related dementias in general and in health disparities populations. Each NIRGD award is limited to $100,000 (direct and indirect costs) for up to two years. Requests in any given year may not exceed $60,000 (direct and indirect costs). Letters of Intent (LOIs) must be received by December 20, with completed applications due by February 7.
Requirements: Eligibility to apply for this grant competition is restricted to investigators who have less than 10 years of research experience after receipt of their terminal degree.
Restrictions: Costs not allowed under this award include: tuition; computer hardware or software for investigators; rent for laboratory/office space; or construction or renovation costs. Your budget must not exceed the maximum amount of the award, $100,000 or $60,000 per year.
Geographic Focus: All States
Date(s) Application is Due: Feb 7
Amount of Grant: Up to 100,000 USD
Contact: Nico Stanculescu, Program Administrator; (312) 335-5747 or (312) 335-5862; fax (312) 335-4034; nico.stanculescu@alz.org or grantsapp@alz.org
Internet: http://www.alz.org/research/alzheimers_grants/types_of_grants.asp
Sponsor: Alzheimer's Association
225 North Michigan Avenue, Suite 1700
Chicago, IL 60601-7633

Alzheimer's Association U.S.-U.K. Young Investigator Exchange Fellowships 597
The U.S.-U.K. Young Investigator Exchange Fellowship provides a three-year grant to fund quality scientific research into the causes, diagnosis and treatment of Alzheimer's disease. The fellowship aims to address important research questions as well as to help and encourage promising scientists as they establish their careers within Alzheimer's research internationally. It is also hoped that by supporting meaningful scientific collaboration between scientists in the United Kingdom and the United States, there will be mutual benefit to the research output of both countries. In addition to an exchange of ideas, an important aim is to promote the learning of new experimental techniques and methodologies. The purpose of this fellowship is to provide new investigators with funding that will allow them to develop preliminary or pilot data, to test procedures and to develop hypotheses on an international level. The intent is to support international early-career development that will lay the groundwork for future research grant applications to other international funding agencies and groups. Allowable costs under this award include: purchase and care of laboratory animals; small pieces of laboratory equipment and laboratory supplies; computer equipment if used strictly for data collection; travel (up to $1,000 per year); and salary for the principal investigator, scientific (including post-doctoral fellows) and technical staff (including laboratory technicians and administrative

support staff whose work is directly related to the funded project). The Alzheimer's Association and the Alzheimer's Reearch United Kingdom anticipate funding up to four awards (two originating from each country) under this competition. Each total award is limited to $300,000 USD. A total of $260,000 USD will be awarded for costs related to the proposed research for up to three years (direct and indirect costs). Requests in any given year may not exceed $90,000 USD (direct and indirect). Indirect costs are capped at 10 percent (rent for laboratory/office space is expected to be covered by indirect costs paid to the institution). The PI must commit to a 75 percent effort toward the proposed project over the funding period. Letters of Intent (LOIs) must be received by December 20, with completed applications due by February 7.
Requirements: Researchers with full-time faculty appointments are encouraged to apply.
Restrictions: Applications from post-doctoral candidates will not be accepted. Costs not allowed under this award include: tuition; computer hardware or software for investigators; rent for laboratory/office spaces; or construction or renovation costs.
Geographic Focus: All States, United Kingdom
Date(s) Application is Due: Feb 7
Amount of Grant: Up to 300,000 USD
Contact: Nico Stanculescu, Program Director; (312) 335-5747 or (312) 335-5862; fax (312) 335-4034; nico.stanculescu@alz.org or grantsapp@alz.org
Internet: http://www.alz.org/research/alzheimers_grants/types_of_grants.asp
Sponsor: Alzheimer's Association
225 North Michigan Avenue, Suite 1700
Chicago, IL 60601-7633

Alzheimer's Association Zenith Fellows Awards 598

The Zenith Fellows award was initiated in 1991 to provide a vehicle for research support for donors with a substantial personal commitment to the advancement of Alzheimer's disease research. The objective of the Zenith Fellows Awards competition is to provide major support for investigators who have: contributed significantly to the field of Alzheimer's disease research or made significant contributions to other areas of science and are now beginning to focus more directly on problems related to Alzheimer's disease; and are likely to make substantial contributions in the future. We anticipate funding up to four awards under this competition. Each award is limited to $450,000 (direct and indirect costs) for three years. Requests in any given year may not exceed $250,000 (direct and indirect costs). Indirect costs are capped at 10 percent (rent for laboratory/office space is expected to be covered by indirect costs paid to the institution).
Requirements: Only established independent investigators are eligible as evidenced by: academic appointment; major, peer-reviewed, external multi-year grant support on which the applicant is the principal investigator (PI); independent laboratory operation; and quality and independence of publication record.
Restrictions: Previous recipients of Zenith Awards, Alzheimer's Disease Center Directors (P50 and P30), Medical and Scientific Advisory Council, and members of the National Board of the Alzheimer's Association are not eligible to apply.
Geographic Focus: All States
Date(s) Application is Due: Feb 7
Amount of Grant: Up to 450,000 USD
Contact: Nico Stanculescu, Program Administrator; (312) 335-5747 or (312) 335-5862; fax (312) 335-4034; nico.stanculescu@alz.org or grantsapp@alz.org
Internet: http://www.alz.org/research/alzheimers_grants/types_of_grants.asp
Sponsor: Alzheimer's Association
225 North Michigan Avenue, Suite 1700
Chicago, IL 60601-7633

AMA-MSS Research Poster Award 599

The MSS/RFS Joint Research Poster Symposium is held annually at the MSS Interim Meeting (November). Research is presented in eight categories (biochemistry/cell biology, cancer biology, cardiology/vascular biology, clinical/epidemiological/health care, immunology/microbiology, neurobiology/neuroscience, radiology/imaging, surgery). The overall winner receives a free trip to the MSS Annual Meeting in Chicago.
Geographic Focus: All States
Date(s) Application is Due: Sep 1
Contact: Rebecca Gierhahn, Director; (800) 262-3211, ext. 4753; rebecca.gierhahn@ama-assn.org or mss@ama-assn.org
Internet: http://www.ama-assn.org/ama/pub/about-ama/our-people/member-groups-sections/medical-student-section/opportunities/grants-awards-scholarships.shtml
Sponsor: American Medical Association
515 N State Street
Chicago, IL 60654

AMA-WPC Joan F. Giambalvo Memorial Scholarships 600

The American Medical Association (AMA) Foundation in association with the AMA Women Physicians Congress (WPC) has established the Joan F. Giambalvo Memorial Scholarship Fund with the goal of advancing the progress of women in the medical profession and strengthening the ability of the AMA to identify and address the needs of women physicians and medical students. Proposals for the Joan F. Giambalvo Memorial Scholarship Fund will be accepted between Nov. 1 and Feb. 15. The AMA-WPC Joan F. Giambalvo Memorial Scholarship seeks innovative research proposals focusing on professional work/practice issues that affect women physicians, including, but not limited to: part-time working strategies; leadership training protocols; gender-based physician practice patterns; physician satisfaction or burnout; retention incentives; practice re-entry issues. Proposals for projects with concurrent/complementary funding and/or plans to use this grant as seed money for larger studies to follow will be received favorably, as will proposals for independent new projects. This award is for a maximum of $10,000. A budget submitted with each applicant's proposal should reflect the anticipated use of funds for the amount requested. Proposals will be evaluated and awardee(s) selected based on a variety of factors, including, but not limited to: the innovation, quality and/or feasibility of the idea; project/research methodology; potential to produce action, change or more comprehensive studies; and career goals of the applicant.
Requirements: Applicants/awardees shall be: female or male; working alone or collaboratively on the specific research project; physicians, medical students, other health professionals, or individuals working or doing graduate work in an applicable profession such as public health, sociology, psychology, etc.
Restrictions: Members of the AMA-WPC Governing Council, AMA Foundation Board of Directors, AMA Board of Trustees and Joan F. Giambalvo Memorial Scholarship Fund selection committee, and AMA and AMA Foundation staff are not eligible for this award.
Geographic Focus: All States
Date(s) Application is Due: Feb 15
Amount of Grant: Up to 10,000 USD
Contact: Alice Reed, Scholarship Contact; (312) 464-5523; alice.reed@ama-assn.org
Internet: http://www.ama-assn.org/ama/pub/about-ama/our-people/member-groups-sections/women-physicians-congress/about-wpc/joan-f-giambalvo-memorial-scholarship.shtml
Sponsor: American Medical Association
515 N State Street
Chicago, IL 60654

AMA Foundation Seed Grants for Research 601

The AMA Foundation established the Seed Grant Research Program to encourage medical students, physician residents and fellows to enter the research field. The program provides $2,500-$5,000 grants to help them conduct small basic science, applied, or clinical research projects. These funds will round out new project budgets, rather than sustain current initiatives. One-year grants will be awarded in the following research categories: Cardiovascular/pulmonary diseases; HIV/AIDS; Leukemia; Neoplastic diseases; Secondhand smoke. Grants in Cardiovascular/Pulmonary Diseases, HIV/AIDS, Leukemia, and Neoplastic Diseases will be $2,500. Grants in the Secondhand smoke category will be $5,000.
Requirements: Applicants must be a medical student, physician resident or fellow of an accredited U.S. medical school or institution; they must also be either a U.S. citizen or a permanent resident of the U.S.. Projects must be applicant-conceived, rather than ongoing research of their mentor or Principal Investigator.
Restrictions: Seed grant funds cannot be used for salary or stipend, indirect/administrative costs, to hire a consultant or contractor, and solely for travel expenses. Seed grants will not be awarded to any applicant who has previously received an AMA Foundation seed grant in the research category in which they are applying.
Geographic Focus: All States
Date(s) Application is Due: Dec 15
Amount of Grant: 2,500-5,000 USD
Contact: Supervisor; (312) 464-4200; fax (312) 464-5973; seedgrants@ama-assn.org
Internet: http://www.ama-assn.org/ama/pub/category/7785.html
Sponsor: American Medical Association Foundation
515 N State Street
Chicago, IL 60654

American-Italian Cancer Foundation Fellowships 602

The mission of the program is to further the advancement of cancer research by sponsoring the research and training of promising young scientists in Italy and in the United States. A major activity is the award of fellowships to Italian and American scientists so that they might spend a period of research and advanced training in preclinical or clinical research centers in the United States or Italy. The fellowships are intended to augment the training of motivated scientists who have the potential to conduct path-breaking original research upon the return to their respective countries. The one-year fellowships are renewable for an additional year.
Requirements: Individuals in the United States and Italy who have been awarded the degrees of PhD, MD, DSc, or DVM may apply in conjunction with a host institution. Candidates must have received their degree no more than three years prior to the date of application.
Geographic Focus: All States
Date(s) Application is Due: Mar 1
Amount of Grant: 25,000-30,000 USD
Contact: Franca Gaudio, (212) 628-9090; fax (212) 517-6089; fgaudio@aicfonline.org
Internet: http://www.aicfonline.org
Sponsor: American-Italian Cancer Foundation
112 E 71st Street, Suite 2B
New York, NY 10021

American-Scandinavian Foundation Awards for Advanced Study or Research in the United States 603

ASF offers a limited number of awards, usually at the graduate level, for Scandinavians wishing to undertake research or studies in the United States. The program also is designed to foster understanding and further cooperation between the nations. Candidates are recommended by cooperating Scandinavian organizations such as the Denmark-America Foundation, the League of Finnish-American Societies, the Icelandic-American Society, the Norway-America Association, and the Sweden-America Foundation.
Requirements: Applicants must be citizens of a Scandinavian country (Denmark, Finland, Iceland, Norway, or Sweden).
Geographic Focus: All States, Scandinavia, Sweden

Contact: Ellen McKey, Director of Fellowships and Grants; (212) 879-9779; fax (212) 249-3444; grants@amscan.org
Internet: http://www.amscan.org/fellowship.html
Sponsor: American-Scandinavian Foundation
58 Park Avenue
New York, NY 10016

American-Scandinavian Foundation Fellowship for Study in Scandinavia 604

The American-Scandinavian Foundation (ASF) offers fellowships (up to $23,000) to individuals to pursue research or study in one or more Scandinavian country for up to one year. Awards are made in all fields.
Requirements: Applicants must: have a well-defined research or study project that makes a stay in Scandinavia essential; be United States citizens or permanent residents; have completed their undergraduate education by the start of their project in Scandinavia. Team projects are eligible, but each member must apply as an individual, submitting a separate, fully-documented application. The ASF considers it desirable that all candidates have at least some ability in the language of the host country, even if it is not essential for the execution of the research plan.
Restrictions: Application may be made for either an ASF fellowship or a grant, not for both. Funds will not be provided for research assistants, dependent support, loan obligations, publication costs, equipment purchase, institutional overhead charges, study at English-language institutions, beginning studies of any subject matter, conference attendance, foregone salary, or supplementation of substantial sabbatical support.
Geographic Focus: All States
Date(s) Application is Due: Nov 2
Amount of Grant: Up to 23,000 USD
Contact: Ellen McKey; (212) 879-9779; fax (212) 249-3444; grants@amscan.org
Internet: http://www.amscan.org/study_scandinavia_details.html
Sponsor: American-Scandinavian Foundation
58 Park Avenue
New York, NY 10016

American-Scandinavian Foundation Grant for Study in Scandinavia 605

ASF encourages advanced study and research in Scandinavian countries. Grants are considered especially suitable for postgraduate scholars, professionals, and candidates in the arts to carry out research or study visits of one to three months. Outstanding proposals from all sources are invited. ASF desires that all candidates have at least some ability in the language of the country in which they plan to study.
Requirements: Applicants must have: a well-defined research or study project that makes a stay in Scandinavia essential; be United States citizens or permanent residents; completed their undergraduate education by the start of their project in Scandinavia. Team projects are eligible, but each member must apply as an individual, submitting a separate, fully documented application. The ASF considers it desirable that all candidates have at least some ability in the language of the host country, even if it is not essential for the execution of the research plan.
Restrictions: Application may be made for either an ASF grant or a fellowship, not for both.
Geographic Focus: All States
Date(s) Application is Due: Nov 2
Amount of Grant: 5,000 USD
Contact: Ellen McKey, Director of Fellowships and Grants; (212) 879-9779; fax (212) 249-3444; grants@amscan.org
Internet: http://www.amscan.org/study_scandinavia_details.html
Sponsor: American-Scandinavian Foundation
58 Park Avenue
New York, NY 10016

American-Scandinavian Foundation Short-Term Training Grants 606

Grants enable U.S. citizens and permanent residents who are at least 20 years of age to live, work, and train in Scandinavia on a temporary basis. Work assignments of eight to 12 weeks or longer (from spring through fall), are available to U.S. students in certain fields, principally engineering, chemistry, and agriculture. The emphasis of the program is the cultural and educational experience rather than financial gain.
Requirements: An applicant should be a full-time student majoring in the field in which training is sought; have at least three years of undergraduate studies completed; and have some previous related work experience. Knowledge of a Scandinavian language is not necessary.
Restrictions: Applications are not accepted in the medically related professions involving patient care.
Geographic Focus: All States
Contact: Ellen McKey; (212) 879-9779; fax (212) 249-3444; grants@amscan.org
Internet: http://www.amscan.org/training.html
Sponsor: American-Scandinavian Foundation
58 Park Avenue
New York, NY 10016

American Academy of Advertising Distinguished Service Award 607

The American Academy of Advertising (AAA), founded in the late 1950's, is an organization of advertising scholars and professionals. The academy fosters research that is relevant to the advertising field and provides a forum for the exchange of ideas among its academic and professional members. The Distinguished Service Award is given by AAA to individuals who have rendered distinguished service to the organization and/or to advertising education. Nominees should be champions of advertising education, active in providing opportunities for the professional development of students and faculty, and should unselfishly perform the service for which they are nominated. Nominations and supporting documentation should be submitted via email to the Chair of the Awards Committee on or before the submission deadline. Deadlines may vary from year to year; Academy members who wish to nominate are encouraged to visit AAA's website to confirm current deadline dates and to obtain complete guidelines for submission of nominations.
Requirements: A nominee need not be a member of the AAA to be considered for the Distinguished Service Award; however, only AAA members may submit a nomination.
Geographic Focus: All States, All Countries
Date(s) Application is Due: Nov 15
Contact: Dr. Jef I. Richards; (512) 471-1101; fax (512) 471-7018; jef@mail.utexas.edu
Internet: http://www.aaasite.org/Award_DistinguishedService.html
Sponsor: American Academy of Advertising
24710 Shaker Boulevard
Beachwood, OH 44122

American Academy of Advertising Doctoral Dissertation Competition Awards 608

The American Academy of Advertising (AAA), founded in the late 1950's, is an organization of advertising scholars and professionals. The academy fosters research that is relevant to the advertising field and provides a forum for the exchange of ideas among its academic and professional members. Members' research findings and scholarly contributions are disseminated through the academy's annual conference proceedings and journals (the Journal of Advertising, the Journal of Current Issues and Research in Advertising, and the Journal of Interactive Advertising). The purpose of the academy's annual Doctoral Dissertation Competition is to promote doctoral research in advertising. Awards are based on a competitive review of dissertation proposals; any topic in advertising may be addressed. Each award is in the range of $1,000 to $2,500. In addition to the standard awards, the competition gives the Dunn Award for outstanding proposals in the area of international advertising. Recipients receive half their award at the time of selection and half when their dissertation has been successfully defended. All submissions must reach the Chair of the Research Committee electronically by 5 p.m. on the deadline date. Complete guidelines for submission and format of proposals are on the academy's website. Application deadlines may vary from year to year and should be verified at the academy's website.
Requirements: Only members of the American Academy of Advertising working on their dissertation at the time of proposal submission are eligible for these awards. Award recipients must maintain their membership until they complete their project. Applicants must be currently enrolled in a graduate program.
Restrictions: Submissions based on completed or near completed dissertations are not eligible; submissions should be in the proposal stage. Recipients have three years from the time of the award to complete their dissertation.
Geographic Focus: All States, All Countries
Date(s) Application is Due: Nov 2
Amount of Grant: 1,000-2,500 USD
Contact: Dr. Eric Haley; (865) 974-5102; haley@utk.edu
Internet: http://www.aaasite.org/Doctoral_Dissertation_Competition.html
Sponsor: American Academy of Advertising
24710 Shaker Boulevard
Beachwood, OH 44122

American Academy of Advertising Ivan Preston Outstanding Contribution to Research on Advertising Award 609

The American Academy of Advertising (AAA), founded in the late 1950's, is an organization of advertising scholars and professionals. AAA fosters research that is relevant to the advertising field and provides a forum for the exchange of ideas among its academic and professional members. AAA's Ivan Preston Outstanding Contribution to Research on Advertising Award is designed to honor an individual who has made an outstanding contribution to the discipline of advertising through a systematic and sustained program of published research. The award carries a $1,000 prize. Nominations must be submitted electronically to the AAA Research Committee Chair by the deadline date; nominations should include a letter of nomination, at least two supporting letters from leading scholars, the nominee's curriculum vitae, and copies of the nominee's five most influential works (in the case of books, a representative chapter my be submitted). The winner will be notified prior to the AAA annual conference and is required to attend the conference to receive the award. Deadlines may vary from year to year; nominators are encouraged to check the AAA website for current deadline dates and complete guidelines for submitting a nomination.
Requirements: To be eligible for the award, a nominee must satisfy the following *Requirements:* have received a doctorate more than 20 years ago; have been promoted to rank of Full Professor prior to nomination; and have been an active member of AAA for the last five consecutive years. Active membership is defined as having participated in AAA annual conference activities as a reviewer, a presenter, a track chair, a session chair, a discussant, and/or a special-topics-session participant. Additionally active membership should include service in at least one, but preferably two of the following ways: as an AAA committee member or chair; as an AAA officer; as a Webmaster; as a newsletter editor; or as a Journal of Advertising reviewer or Review Board member.
Restrictions: The Ivan Preston award will be given only to active and contributing AAA members for exemplary research on advertising and is not necessarily awarded every year.
Geographic Focus: All States, All Countries
Date(s) Application is Due: Nov 2
Contact: Dr. Eric Haley; (865) 974-5102; haley@utk.edu
Internet: http://www.aaasite.org/IPAward_Research.html
Sponsor: American Academy of Advertising
24710 Shaker Boulevard
Beachwood, OH 44122

American Academy of Advertising Kim Rotzoll Award for Advertising Ethics and Social Responsibility 610

The American Academy of Advertising (AAA), founded in the late 1950's, is an organization of advertising scholars and professionals. The academy fosters research that is relevant to the advertising field and provides a forum for the exchange of ideas among its academic and professional members. The Kim Rotzoll Award was established by AAA in 2004 and recognizes individuals or organizations who have demonstrated an outstanding commitment to advertising ethics and social responsibility. Nominations and supporting documentation should be submitted via email to the Chair of the Awards Committee on or before the submission deadline. Deadlines may vary from year to year; Academy members who wish to nominate are encouraged to visit AAA's website to confirm current deadline dates and to obtain complete guidelines for submission of nominations.
Requirements: Nominees need not be members of AAA, but only AAA members may submit a nomination. Nominees must evidence lifetime achievement in the fields of advertising ethics or social responsibility. For academics, such evidence may be in the form of scholarly publications; for advertising practitioners or organizations, the results of specific programs or activities that have had a positive impact on society may serve as evidence.
Restrictions: The Rotzoll Award is not to be considered an annual award and will be given only to individuals or organizations who have demonstrated an outstanding commitment to advertising ethics and social responsibility.
Geographic Focus: All States, All Countries
Date(s) Application is Due: Nov 15
Contact: Dr. Jef I. Richards; (512) 471-1101; fax (512) 471-7018; jef@mail.utexas.edu
Internet: http://www.aaasite.org/Award_KimRotzoll.html
Sponsor: American Academy of Advertising
24710 Shaker Boulevard
Beachwood, OH 44122

American Academy of Advertising Research Fellowships 611

The American Academy of Advertising (AAA), founded in the late 1950's, is an organization of advertising scholars and professionals. The academy fosters research that is relevant to the advertising field and provides a forum for the exchange of ideas among its academic and professional members. The academy's Research Fellowship Competition promotes the continued scholarship of professors and advertising professionals who have completed their education and are doing research in advertising. Awards are based on a competitive review of research proposals; any topic that is appropriate for potential publication in the academy's Journal of Advertising may be addressed. Each award is typically in the range of $1,000 to $3,000. Fellows receive half their award at time of selection and half when the first completed paper is submitted to the Journal of Advertising. Complete guidelines for submission and format of proposals are at the academy's website. All submissions must reach the Chair of the Research Committee electronically by 5 p.m. on the deadline date. Winners will be announced at the academy's annual conference and communicated to the membership through the AAA newsletter. Application deadlines may vary from year to year and should be verified at the academy's website.
Requirements: Applicants must comply with the following *Requirements:* be current members of the academy; maintain membership until they complete the project; grant the Journal of Advertising first right of refusal on any papers resulting from the supported research; and acknowledge the AAA Fellowship in all publications resulting from the project. Winners are asked to publicize the award on their campuses and in their communities as widely as possible. Researchers who previously have received an AAA Research Fellowship are eligible for the award upon completion or termination of the previous award. Researchers who have submitted proposals that have not been awarded support in the past may resubmit proposals for similar projects.
Restrictions: While members may submit as many single or joint proposals as desired within a given year, authors cannot receive funding for more than one proposal; should more than one proposal containing an author's name receive high rankings from the reviewers, only the highest ranking proposal will be funded. Winners have three years from the time of the award to complete their project and are not eligible for additional AAA research fellowships while the award is active. Doctoral students are not eligible for these fellowships. Joint research with doctoral students is likewise excluded unless the research will not be part of the student's dissertation.
Geographic Focus: All States, All Countries
Date(s) Application is Due: Sep 14
Amount of Grant: 1,000-3,000 USD
Contact: Dr. Eric Haley; (865) 974-5102; haley@utk.edu
Internet: http://aaasite.org/Award_ResearchFellowship.html
Sponsor: American Academy of Advertising
24710 Shaker Boulevard
Beachwood, OH 44122

American Academy of Arts and Sciences Hellman Fellowships 612

As part of the Initiative for Science, Engineering, and Technology, the American Academy has established the the Hellman Fellowship in Science and Technology Policy for an early-career professional with training in science or engineering who is interested in transitioning to a career in public policy and administration. While in residence, the Hellman Fellow will work with senior scientists and policy experts on critical national and international policy issues related to science, engineering, and technology. The focus of the work will be on one or more of the ongoing projects under the Academy's Initiative for Science, Engineering, and Technology to which the Hellman Fellow will contribute substantively. The mission of the Initiative is to examine, in broad terms, how the world of science and technology is evolving, how to help the public understand these changes, and how society can better adapt. The Initiative brings together scientists and public policy experts in a neutral setting, outside of the constraints of the political process.
Requirements: Applicants must have a Ph.D. in an area of science or engineering and have some experience or a demonstrated interest in an area related to science and technology policy. Masters degrees may be considered in the fields of engineering and computer science. Strong writing and organizational skills are desired. Candidates must be U.S. citizens, permanent residents, or current employees of an academic or professional organization in the United States.
Geographic Focus: All States
Date(s) Application is Due: Jan 17
Contact: Hillary Dobel, Program Coordinator; (617) 576-5024; fax (617) 576-5050; hdobel@amacad.org or vsp@amacad.org
Internet: http://www.amacad.org/hellman.aspx
Sponsor: American Academy of Arts and Sciences
136 Irving Street
Cambridge, MA 02143

American Academy of Arts and Sciences Visiting Scholars 613

The Academy's Visiting Scholars Program provides residential fellowships for junior faculty members and postdoctoral scholars in the humanities and social sciences. The fellowship program offers scholars a year for research and writing free from teaching and administrative duties, a collaborative work environment, and the opportunity to interact with Academy members. It also creates a national network for these scholars, assisting them in their research and professional development. The Academy seeks proposals in the humanities and social sciences relating to American history, culture, and public policy from the founding period to the present. The academy also welcomes proposals that explore the impact of scientific and technological advances over the past two centuries on American institutions, humanities and culture in America, American foreign policy, and global security. Proposals should take into account the academy's emphasis on interdisciplinary work, as well as its interest in broadening public understanding of important intellectual trends and contemporary policy choices. In addition to pursuing individual projects, scholars participate in activities of the academy; collaboration with academy fellows is encouraged. Visiting scholars remain in residence at the academy throughout the academic year (September-May), with some fellowships eligible for a one-year renewal. Faculty remain in residence for one academic year, although one-semester fellowships may be considered. Scholars receive office space, computer services, library privileges, and assistance in locating housing. Health benefits can be arranged. Applicants are responsible for contacting three references and ensuring that their letters are mailed to the academy by the postmark deadline. Faculty can receive up to $65,000 (not to exceed one-half of current salary). Postdoctoral students receive an annual stipend of up to $45,000. The Academy provides office space, computer services, library privileges, and information on locating housing.
Requirements: Preference is given to untenured junior faculty but the program is also open to qualified postdocs. Candidates must be U.S. citizens, permanent residents, or current employees of an academic or professional organization in the United States. The Ph.D., J.D, or equivalent professional training (e.g., public policy) should have been completed within the last 10 years (although exceptional circumstances will be taken into consideration). Graduate student applicants must complete all degree requirements by August 1.
Restrictions: Research trips, interviews, attendance at scholarly meetings or speaking engagements must be limited to no more than twenty days during the fellowship term, and scholars must not accept teaching appointments or other major commitments during their fellowship. Applications must be submitted electronically.
Geographic Focus: All States
Date(s) Application is Due: Oct 22
Amount of Grant: 45,000-65,000 USD
Contact: Hilary Dobel, Program Coordinator; (617) 576-5024; fax (617) 576-5050; hdobel@amacad.org or vsp@amacad.org
Internet: https://www.amacad.org/content/about/about.aspx?d=363&t=4&s=0
Sponsor: American Academy of Arts and Sciences
136 Irving Street
Cambridge, MA 02143

American Academy of Nursing Building Academic Geriatric Nursing Capacity Scholarships 614

The Foundation's overall goal is to increase the nation's capacity to provide effective and affordable care to its rapidly increasing older population. Specifically, the Foundation seeks to enhance the training of physicians, nurses, social workers and other health professionals who care for older adults, and promote innovations in the integration and delivery of services. The goal of the Scholarship program is to increase academic geriatric nursing capacity in the United States. BAGNC focuses on the development of academic leadership in gerontological nursing through strong mentorship in the components of academic geriatric nursing (research, teaching and community service); leadership development, a national network of scholars and academic geriatric nurses; and exposure to a wide range of experts in gerontology and geriatrics. Scholars in collaboration with their identified mentor will design and implement a tailored professional development plan designed to support development of new competencies and enhanced effectiveness as an academic leader.
Requirements: Registered nurses who are U.S. citizens or U.S. permanent residents and who hold a degree(s) in nursing are eligible. Predoctoral applicants must: be registered nurses; hold degree(s) in nursing; be United States citizens or permanent U.S. residents; plan an academic and research career; and demonstrate potential for long-term contributions to geriatric nursing.
Geographic Focus: All States
Date(s) Application is Due: Jan 15
Amount of Grant: Up to 50,000 USD
Contact: Patricia Archbold, (202) 777-1172; fax (202) 777-0107; parchbold@aannet.org

Internet: http://www.aannet.org/i4a/pages/Index.cfm?pageID=3295
Sponsor: American Academy of Nursing
888 17th Street NW, Suite 800
Washington, DC 20006

American Academy of Nursing Claire M. Fagin Fellowships 615

The Fellowship supports two years of full time advanced research and leadership training for doctorally prepared faculty committed to careers in academic geriatric nursing by providing $120,000 for the 2-year fellowship ($60,000 per annum). Program focuses on the development of academic leadership in gerontological nursing through such activities as: research; focused study; networking among scholars, mentors and colleagues in other fields; demonstration of growth in ability to transform self and organizations by moving outside of traditional modes of success; completion and write-up of a significant research project; and by success in achieving funding from other sources. Selected fellows, in collaboration with their mentor, will design and implement an individual professional development plan that will support them in developing new competencies and enhanced effectiveness as an academic leader and researcher. Award programs must begin between July 1st and September 1st of the award year. The program is committed to advancing well-qualified applicants from under-represented minority groups to improve the nation's ability to provide culturally competent care to its increasingly diverse aging population.
Requirements: Applicants must: be doctorally-prepared registered nurses; hold degree(s) in nursing; be United States citizens or permanent U.S. residents; be doctorally-prepared registered nurses,; have the potential to develop into independent investigators; and demonstrate potential for long-term contributions to geriatric nursing. Applications will be accepted from doctoral students who will complete their doctoral program prior to the award. Faculty members in accredited Schools of Nursing who hold the rank of assistant professor or associate professor may apply for fellowships.
Geographic Focus: All States
Date(s) Application is Due: Jan 13
Amount of Grant: 120,000 USD
Contact: Patricia Archbold; (202) 777-1172; fax (202) 777-0107; parchbold@aannet.org
Internet: http://www.geriatricnursing.org/applications/cmf-fellowship.asp
Sponsor: American Academy of Nursing
888 17th Street NW, Suite 800
Washington, DC 20006

American Academy of Nursing Mayday Fund Grants 616

The Foundation's overall goal is to increase the nation's capacity to provide effective and affordable care to its rapidly increasing older population. Specifically, the Foundation seeks to enhance the training of physicians, nurses, social workers and other health professionals who care for older adults, and promote innovations in the integration and delivery of services. This program is aimed at candidates whose research includes the study of pain in the elderly. Award programs must begin between July 1st and September 1st.
Requirements: Predoctoral applicants must: be registered nurses; hold degree(s) in nursing; be United States citizens or permanent U.S. residents; plan an academic and research career; and demonstrate potential for long-term contributions to geriatric nursing.
Geographic Focus: All States
Date(s) Application is Due: Jan 9
Contact: Patricia Archbold; (202) 777-1172; fax (202) 777-0107; parchbold@aannet.org
Internet: http://www.geriatricnursing.org/applications/applications.asp
Sponsor: American Academy of Nursing
888 17th Street NW, Suite 800
Washington, DC 20006

American Academy of Religion Collaborative Research Grants 617

These grants are intended to stimulate cooperative research among scholars in different institutions, with a focus on a clearly identified research project. They may also be used for interdisciplinary work with scholars outside the field of religion, especially when such work shows promise of continuing beyond the year funded. Grants can provide funds for networking and communication. In addition, grants may be used to support small research conferences. Conference proposals will be considered only if they are designed primarily to advance research. Conferences presenting papers that report on previous research will not be considered. Award notification letters will be sent by the end of October. There are no application forms; guidelines are available online.
Requirements: Applicants must be current AAR members who have been in good standing for the previous three years. In the case of proposals involving scholars from other disciplines, not all participants need to hold AAR membership.
Restrictions: Applicants will not be considered who have received an AAR research grant in the previous five years.
Geographic Focus: All States
Date(s) Application is Due: Aug 1
Contact: Jessica Davenport, (404) 727-4707; jdavenport@aarweb.org
Internet: http://www.aarweb.org/Programs/Grants/Research/collaborative.asp
Sponsor: American Academy of Religion
825 Houston Mill Road, Suite 300
Atlanta, GA 30329-4205

American Academy of Religion Individual Research Grants 618

The grants provide support to academy members for important aspects of research, such as travel to archives and libraries, research assistance, field work, and release times. Proposals are judged on the project's contribution to scholarship in religious studies and on the significance of that contribution for advancing the understanding of religion or for advancing discussion between religion and other humanistic and social science disciplines. Scholars in small institutions, departments, or programs without research support and unaffiliated scholars are encouraged to apply.
Requirements: Applicants: must be current AAR members who have been in good standing for the previous three years; will not be considered if they have received an AAR Research Award in the previous five years; must have completed the doctorate.
Restrictions: Funds are not provided for dissertation research, nor for travel to attend the AAR annual meeting.
Geographic Focus: All States
Date(s) Application is Due: Aug 1
Amount of Grant: 500-5,000 USD
Contact: Jessica Davenport, (404) 727-4707; jdavenport@aarweb.org
Internet: http://www.aarweb.org/Programs/Grants/Research/individual.asp
Sponsor: American Academy of Religion
825 Houston Mill Road, Suite 300
Atlanta, GA 30329-4205

American Australian Association Australia To U.S.A Fellowships 619

The fellowships are available for Australian scholars doing advanced research or study in the fields of business, science, technology, medicine or engineering in the United States. In return, the fellows are expected to add to Australia's intellectual capital as well as contribute to the country's overall social and economic well-being upon their return. The Fellowships will support part of the costs of one year of research or study in the United States. Applicants must submit a complete budget. The Association's budget form lists acceptable expenses.
Requirements: Applicants must: be Australian citizens doing research or study at the graduate level. Applicants may already be in the U.S; conduct a well-defined research project or study in one of the following fields—business, science, technology, medicine or engineering; demonstrate why travel to the United States is important to their research or study; demonstrate how their research or study will benefit Australia and intend to return to work in Australia; show proof of the arrangements made at a United States university or institution where the research or study will be conducted for the academic year beginning in September of that year; are expected to devote full time to their research or study; and be fluent in English.
Geographic Focus: All States, Australia
Date(s) Application is Due: Mar 15
Contact: Diane Sinclair, Director; (212) 338-6860; diane.sinclair@aaanyc.org
Internet: http://www.americanaustralian.org/Educational/australia-usa.php
Sponsor: American Australian Association
599 Lexington Avenue, 18th Floor
New York, NY 10022

American Australian Association U.S.A to Australia Fellowships 620

Fellowships are available for Americans who will benefit from doing advanced research or study in the fields of life sciences, engineering, medicine, or mining; there will be particular interest in the fields of oceanography/marine sciences and stem cell research. In return, the fellows are expected to add to the United States and Australia's intellectual capital as well as contribute to those countries' overall social and economic well-being upon their return.
Requirements: Applicants must: be American citizens or permanent residents of the U.S.A doing research or study at the graduate level (applicants may already be in Australia); conduct a well-defined research project or study in one of the following fields: life sciences, engineering, medicine or mining. There will be particular interest in the fields of oceanography/marine sciences and stem cell research; demonstrate how their research or study will benefit the United States and Australia and that they intend to return to work in the United States; demonstrate why travel to Australia is important to their research or study; show proof of the arrangements made at an Australian university or institution where the research or study will be conducted for the next academic year; devote full time to their research or study; and be fluent in English.
Geographic Focus: All States
Date(s) Application is Due: Oct 31
Contact: Diane Sinclair, Director; (212) 338-6860; diane.sinclair@aaanyc.org
Internet: http://www.americanaustralian.org/Educational/usa-australia.php
Sponsor: American Australian Association
599 Lexington Avenue, 18th Floor
New York, NY 10022

American Automatic Control Council Control Practice Award 621

The Control Engineering Practice Award is given to one individual or one team to be selected from those nominated for significant contribution to the advancement of control practice. The primary criterion for selection will be for the application and implementation of innovative control concepts, methodology, and technology, for the planning, design, manufacture, and operation of control systems. Achievement and usefulness will be evidenced by the benefit to society and by the degree of acceptance by those who use control as a tool. The work on which the nomination is based must have been performed while the nominated individual or at least one member of the team was a resident of the U.S.A. The award consists of a certificate and an honorarium. In the event that the winner is a team, each member of the team will receive a certificate and the honorarium will be divided equally among the team members. Nomination forms as well as additional information may be obtained from the AACC secretariat.
Requirements: Nomination packages should consist of a letter of nomination, three to five reference letters, current resume, current list of publications and patents, any other supporting material, and a nomination form. The nomination letter should clearly identify the primary reason that the nominee should receive the award, as well as four ancillary reasons. All materials should be submitted in a single package.

Geographic Focus: All States
Date(s) Application is Due: Dec 15
Contact: B. Wayne Bequette, AACC Secretariat; (518) 276-6683; fax (518) 276-4030
Internet: http://a2c2.org/awards/control-engineering-practice-award
Sponsor: American Automatic Control Council
2145 Sheridan Road
Evanston, IL 60208-3118

American Automatic Control Council Donald P. Eckman Award 622

The Donald P. Eckman Award is given for outstanding accomplishments by a young engineer in the field of automatic control. Nominees must be younger than 35 years on January 1 of the year of the award and have made contributions to the field while residing in the United States. Contributions may be technical or scientific publications, theses, patents, inventions, or combinations or the above in the field of automatic control. The award consists of a certificate and honorarium, which are presented at the annual conference. Nomination forms as well as additional information may be obtained from the AACC secretary.
Requirements: Nomination packages should consist of letter of nomination, three to five letters of reference, current resume, current list of publications and patents, any other supporting material, and nomination form. The nomination letter should clearly identify the primary reason that the nominee should receive the award, as well as up to four ancillary reasons. Supporting documentation must include a full endorsement by at least one responsible supervisor.
Geographic Focus: All States
Contact: B. Wayne Bequette, AACC Secretariat; (518) 276-6683; fax (518) 276-4030
Internet: http://a2c2.org/awards/donald-p-eckman-award
Sponsor: American Automatic Control Council
2145 Sheridan Road
Evanston, IL 60208-3118

American Automatic Control Council Hugo Schuck Award 623

The Hugo Schuck Award is given to recognize the best two papers presented at the previous ACC. One award is for a paper emphasizing contributions to theory and the other emphasizing significant or innovative applications. The papers must have been presented by the awardee or a coauthor. Criteria for selection include the quality of the written and oral presentation, the technical contribution, timeliness, and practicality. Additional information may be obtained from the AACC secretariat.
Geographic Focus: All States
Date(s) Application is Due: Dec 15
Contact: B. Wayne Bequette, AACC Secretariat; (518) 276-6683; fax (518) 276-4030
Internet: http://a2c2.org/awards/o-hugo-schuck-best-paper-award
Sponsor: American Automatic Control Council
2145 Sheridan Road
Evanston, IL 60208-3118

American Automatic Control Council John R. Ragazzini Education Award 624

The John R. Ragazzini Education Award is given to recognize outstanding contributions to automatic control education in any form. These contributions can be from any source and in any media, i.e., electronic, publications, courses, etc. Nomination forms as well as additional information may be obtained from the AACC secretariat.
Requirements: Nomination packages should consist of a letter of nomination, three to five reference letters, current resume, current list of publications and patents, any other supporting material, and a nomination form. The nomination letter should clearly identify the primary reason that the nominee should receive the award, as well as four ancillary reasons. All materials should be submitted in a single package.
Geographic Focus: All States
Date(s) Application is Due: Dec 15
Contact: B. Wayne Bequette, AACC Secretariat; (518) 276-6683; fax (518) 276-4030
Internet: http://a2c2.org/awards/john-r-ragazzini-education-award
Sponsor: American Automatic Control Council
2145 Sheridan Road
Evanston, IL 60208-3118

American Automatic Control Council Richard E. Bellman Control Heritage Award 625

The Richard E. Bellman Control Heritage Award recognizes distinguished career contributions to the theory or applications of automatic control. It is the highest recognition of professional achievement for U.S. control systems engineers and scientists. The nominee must have spent a significant part of his/her career in the United States. The award consists of a certificate and honorarium, which are presented at the council's annual conference. The awardee is expected to make a short acceptance speech at the AACC Awards Ceremonies. Contact the AACC secretary for the nomination form and further information.
Requirements: Nomination packages should consist of a letter of nomination, three to five reference letters, current resume, current list of publications and patents, any other supporting material, and a nomination form. The nomination letter should clearly identify the primary reason that the nominee should receive the award, as well as four ancillary reasons. All materials should be submitted in a single package.
Geographic Focus: All States
Contact: B. Wayne Bequette, AACC Secretariat; (518) 276-6683; fax (518) 276-4030
Internet: http://a2c2.org/awards/richard-e-bellman-control-heritage-award
Sponsor: American Automatic Control Council
2145 Sheridan Road
Evanston, IL 60208-3118

American Chemical Society AGFD Advancement of the Application of Agricultural and Food Chemistry Award 626

The purpose of the American Chemical Society Division of Agricultural and Food Chemistry (AGFD) Advancement of the Application of Agricultural and Food Chemistry Award is to recognize and encourage outstanding contributions to pure or applied agricultural and food chemistry. The Award consists of $3,000, a plaque, and meeting travel expenses, and is co-sponsored by the AGFD and International Flavors and Fragrances. The annual deadline is November 1.
Requirements: Applicants must have made an outstanding application of chemistry or chemical technology to the solution of agricultural or food problems of importance to the nourishment and health of mankind or outstanding contributions to the advancement of pure or applied agricultural and food chemistry, special consideration given to independence of thought and the originality shown, or to the importance of the work when applied to public welfare or the needs of humanity, must be actively engaged in the line of work, granted without regard to sex, creed, or nationality.
Geographic Focus: All States
Date(s) Application is Due: Nov 1
Amount of Grant: 3,000-4,500 USD
Contact: Neil Da Costa, AGFD Chair; (732) 335-2110; neil.dacosta@iff.com
Internet: http://webapps.acs.org/findawards/detail.jsp?ContentId=WPCP_012439
Sponsor: American Chemical Society
1155 Sixteenth Street, NW
Washington, DC 20036-4801

American Chemical Society AGFD Donald A. Withycombe Graduate Fellowship Award in Food Chemistry 627

The purpose of the Division of Agricultural and Food Chemistry (AGFD) Donald A. Withycombe Graduate Fellowship Award in Food Chemistry is to support and encourage graduate students who display outstanding potential early in their program. One prize will be awarded comprising of a cash award of $1000. Travel expenses will be paid up to $1,000. The annual deadline for applications is October 31.
Requirements: Applicants must have outstanding and beginning graduate grade point averages (GPA) and show promise of an excellent graduate research career.
Geographic Focus: All States
Date(s) Application is Due: Oct 31
Amount of Grant: 1,000-2,000 USD
Contact: Neil Da Costa, AGFD Chair; (732) 335-2110; neil.dacosta@iff.com
Internet: http://webapps.acs.org/findawards/detail.jsp?ContentId=WPCP_012447
Sponsor: American Chemical Society
1155 Sixteenth Street, NW
Washington, DC 20036-4801

American Chemical Society AGFD Teranishi Graduate Fellowship Award 628

The purpose of the Division of Agricultural and Food Chemistry (AGFD) Teranishi Graduate Fellowship Award is to recognize an AGFD member who has made outstanding scientific contributions to the field of agricultural and food chemistry. The award includes $2,500 and an appropriate plaque. The annual deadline for applications is February 1.
Requirements: Only AGFD member may apply.
Geographic Focus: All States
Date(s) Application is Due: Feb 1
Amount of Grant: 2,000 USD
Contact: Neil Da Costa, AGFD Chair; (732) 335-2110; neil.dacosta@iff.com
Internet: http://webapps.acs.org/findawards/detail.jsp?ContentId=WPCP_012454
Sponsor: American Chemical Society
1155 Sixteenth Street, NW
Washington, DC 20036-4801

American Chemical Society AGFD Undergraduate Research Symposium Travel Grants 629

American Chemical Society AGFD Undergraduate Research Symposium Travel Grants are offered to those students chosen to participate in the symposium, and each will receive travel expenses up to $1,000 to attend and present their research at the Spring meeting of the ACS. The AGFD Advisory Board will choose the students who will participate in the symposium. Two prizes will be awarded. First prize will receive a cash award of $750. Second place will receive a cash award of $250. The annual deadline for applications is October 31.
Geographic Focus: All States
Date(s) Application is Due: Oct 31
Contact: Neil Da Costa, AGFD Chair; (732) 335-2110; neil.dacosta@iff.com
Internet: http://agfd.sites.acs.org/awards.htm
Sponsor: American Chemical Society
1155 Sixteenth Street, NW
Washington, DC 20036-4801

American Chemical Society AGFD Young Scientist Award 630

The primary purpose of the American Chemical Society AGFD Young Scientist Award is to recognize outstanding research in agricultural and food chemistry by a young scientist. The winner receives a $1000 honorarium, an engraved plaque and paid expenses (up to $1500) to the ACS meeting to present a paper and to receive the award.
Requirements: Each applicant must be an AGFD member and have earned last degree within eight years of October 1 of the presentation year.
Geographic Focus: All States
Amount of Grant: 1,000 USD
Contact: Neil Da Costa, Chair; (732) 335-2110; neil.dacosta@iff.com

Internet: http://webapps.acs.org/findawards/detail.jsp?ContentId=WPCP_012462
Sponsor: American Chemical Society
1155 Sixteenth Street, NW
Washington, DC 20036-4801

American Chemical Society AGRO Award for Innovation in Chemistry of Agriculture 631
The American Chemical Society AGRO Award for Innovation in Chemistry of Agriculture is given to an active researcher working in North America for a chemical innovation that significantly enhances agricultural or veterinary pest management and productivity. The awardee will be asked to give an award address at the National ACS meeting. Nominations must be received by the committee chair by September 30 of each year. Balloting will be conducted in November/December and results will be announced shortly thereafter.
Geographic Focus: All States
Date(s) Application is Due: Sep 30
Contact: Dr. James Seiber, Awards Committee Chair; (530) 752-1465 or (530) 752-1141; jnseiber@ucdavis.edu
Internet: http://www.agrodiv.org/awards/
Sponsor: American Chemical Society
1155 Sixteenth Street, NW
Washington, DC 20036-4801

American Chemical Society AGRO Education Awards 632
The AGRO Division has established an endowment fund in collaboration with Bayer CropScience that will be used to promote an understanding of the role of chemistry in agriculture. The purpose of the American Chemical Society AGRO Education Awards is to promote an understanding of the role of chemistry in agriculture as embraced in the following areas related to pest management chemistry: synthesis; metabolism; regulatory; biotechnology; delivery; risk assessment; resistance; residues; mode of action; and fate and/or behavior. To apply, a graduate student should submit the application materials to be received no later than March 1.
Requirements: An applicant must be a graduate student.
Geographic Focus: All States
Date(s) Application is Due: Mar 1
Contact: Dr. James Seiber; (530) 752-1465 or (530) 752-1141; jnseiber@ucdavis.edu
Internet: http://webapps.acs.org/findawards/detail.jsp?ContentId=WPCP_012468
Sponsor: American Chemical Society
1155 Sixteenth Street, NW
Washington, DC 20036-4801

American Chemical Society AGRO Fellow Awards 633
The American Chemical Society Division of Agrochemicals has established the Fellow Awards to recognize its members whose dedicated and enthusiastic service has kept the Division moving forward. Nomination criteria shall be continued and substantial contributions of time, talents, and service to the AGRO Division of ACS, and to the agrochemical science over a period of at least six years. Nominations include a letter, noting the contributions to the Division, and a current curriculum vitae. Deadline for submitting nominations is May 31 of each year.
Geographic Focus: All States
Date(s) Application is Due: May 31
Contact: Dr. James Seiber; (530) 752-1465 or (530) 752-1141; jnseiber@ucdavis.edu
Internet: http://www.agrodiv.org/fellow-awards/
Sponsor: American Chemical Society
1155 Sixteenth Street, NW
Washington, DC 20036-4801

American Chemical Society AGRO International Award for Research in Agrochemicals 634
The ACS International Award for Research in Agrochemicals is given to a scientist who has made outstanding contributions to the field of agrochemicals at the international level. Their vision and sustained contributions will have opened new horizons for other investigators in their field and beyond. Nominations are due by December 31.
Requirements: The award is given to an individual without regard to nationality for outstanding contributions to research in agrochemicals.
Geographic Focus: All States
Date(s) Application is Due: Dec 31
Contact: Dr. James Seiber; (530) 752-1465 or (530) 752-1141; jnseiber@ucdavis.edu
Internet: http://www.agrodiv.org/awards/
Sponsor: American Chemical Society
1155 Sixteenth Street, NW
Washington, DC 20036-4801

American Chemical Society AGRO Sterling B. Hendricks Memorial Lectureship 635
The USDA-Agricultural Research Service (ARS) is seeking nominations for the Sterling B. Hendricks Memorial Lectureship Award. This Lectureship was established in 1981 by ARS to honor the memory of Sterling B. Hendricks and to recognize scientists who have made outstanding contributions to the chemical science of agriculture. Dr. Hendricks contributed to many diverse scientific disciplines, including soil science, mineralogy, agronomy, plant physiology, geology, and chemistry. He is most frequently remembered for discovering phytochrome, the light-activated molecule that regulates many plant processes. The lecture should address a scientific topic, trend, or policy issue related to agriculture. The lectureship is presented at an AGFD symposium in even-numbered years, and in an AGRO symposium in odd-numbered years. The award includes an honorarium of $2000, a bronze medallion, and expenses to attend the meeting. The annual deadline for nominations is November 5.
Requirements: Nominees will be outstanding senior scientists in industry, universities, consulting, or government positions.
Restrictions: Current ARS employees are not eligible.
Geographic Focus: All States
Date(s) Application is Due: Nov 5
Amount of Grant: 2,000-3,500 USD
Contact: Dr. James Seiber; (530) 752-1465 or (530) 752-1141; jnseiber@ucdavis.edu
Kim Kaplan, Lecture Coordinator; (301) 504-1637
Internet: http://webapps.acs.org/findawards/detail.jsp?ContentId=WPCP_012471
Sponsor: American Chemical Society
1155 Sixteenth Street, NW
Washington, DC 20036-4801

American Chemical Society AGRO Young Scientist Research Awards 636
AGRO, a Division of the American Chemical Society, brings together a worldwide community of scientists and stakeholders to advance knowledge and promote innovative solutions for the protection of agricultural productivity, public health and the environment. The purpose of the American Chemical Society AGRO Young Scientist Research Awards is to recognize outstanding agrochemical research by pre- or postdoctoral students.
Requirements: Eligibility includes only pre- and postdoctoral students whose work has been conducted primarily at U.S. institutions.
Geographic Focus: All States
Contact: Dr. James Seiber; (530) 752-1465 or (530) 752-1141; jnseiber@ucdavis.edu
Internet: http://webapps.acs.org/findawards/detail.jsp?ContentId=WPCP_012474
Sponsor: American Chemical Society
1155 Sixteenth Street, NW
Washington, DC 20036-4801

American Chemical Society Alfred Bader Award in BioInorganic or BioOrganic Chemistry 637
This Alfred Bader Award in BioInorganic or BioOrganic Chemistry is given annually to recognize outstanding contributions to bioorganic or bioinorganic chemistry. The award will be granted for outstanding research accomplishments without regard to age or nationality. The award is intended to recognize significant accomplishments that are at the interface between biology and organic or inorganic chemistry. Special consideration will be given to applications of the fundamental principles and experimental methodology of chemistry to areas of biological significance. The award consists of $5,000 and a certificate. Up to $2,500 for travel expenses to the meeting at which the award will be presented will be reimbursed.
Requirements: Any individual, except a member of the awards committee, may submit one nomination or seconding letter for the award in any given year. The nominating documents consist of a letter of not more than 1000 words containing an evaluation of the nominee's accomplishments and a specific identification of the work to be recognized, a biographical sketch including date of birth, and a list of publications and patents authored by the nominee.
Restrictions: Self-nominations are not accepted.
Geographic Focus: All States
Date(s) Application is Due: Nov 1
Amount of Grant: 5,000-7,500 USD
Contact: Administrator; (800) 227-5558 or (202) 872-4408; awards@acs.org
Internet: http://webapps.acs.org/findawards/detail.jsp?ContentId=CTP_004491
Sponsor: American Chemical Society
1155 Sixteenth Street, NW
Washington, DC 20036-4801

American Chemical Society Alfred Burger Award in Medicinal Chemistry 638
This American Chemical Society Alfred Burger Award in Medicinal Chemistry, supported by GlaxoSmithKline in 1978, is awarded biennially in even-numbered years to recognize outstanding contributions to research in medicinal chemistry. The award is granted without regard to age or nationality, and the recipient presents an award address at the spring meeting of the ACS Division of Medicinal Chemistry. Applications are accepted in odd-numbered years. The award consists of $3,000 and a certificate. Up to $2,500 for travel expenses to the meeting at which the award will be presented will be reimbursed. The award is presented biennially in even-numbered years, and the recipient will present an award address at the spring meeting of the ACS Division of Medicinal Chemistry.
Requirements: Any individual, except a member of the award committee, may submit one nomination or seconding letter for the award in any given year. The nominating documents consist of a letter of not more than 1000 words containing an evaluation of the nominee's accomplishments and a specific identification of the work to be recognized, a biographical sketch including date of birth, and a list of publications and patents authored by the nominee.
Restrictions: Self-nominations are not accepted.
Geographic Focus: All States
Date(s) Application is Due: Nov 1
Amount of Grant: 3,000-5,500 USD
Contact: Administrator; (800) 227-5558 or (202) 872-4408; awards@acs.org
Internet: http://webapps.acs.org/findawards/detail.jsp?ContentId=CTP_004491
Sponsor: American Chemical Society
1155 Sixteenth Street, NW
Washington, DC 20036-4801

American Chemical Society ANYL Arthur F. Findeis Award for Achievements by a Young Analytical Scientist 639

The purpose of the American Chemical Society ANYL Arthur F. Findeis Award for Achievements by a Young Analytical Scientist is to recognize and encourage outstanding contributions to the fields of analytical chemistry by a young analytical scientist. Evidence shall be presented for one or more of the following outstanding accomplishments: conceptualization and development of unique instrumentation that has had an enabling impact upon analytical chemistry and has substantively advanced the field; development of novel and important analytical methods or methodologies that have found significant beneficial applications in the chemical sciences; elucidation of fundamental events or processes involved in or important to analytical chemistry; authorship of books, patents, and/or research papers that have had an influential role in the development of analytical chemistry; or other significant contributions to the furtherance of analytical chemical sciences role in the use of chemical instrumentation. The Award consists of a plaque and $2,500.
Requirements: The awardee must have earned his or her highest degree within ten years of January 1 of the year of the award. Both the nationality of the young analytical scientist and the arena (e.g., academic, industrial, national laboratory) in which the contributions of the young analytical scientist have been made are unrestricted.
Geographic Focus: All States
Date(s) Application is Due: Nov 1
Amount of Grant: 2,500 USD
Contact: Miquela Sena, (505) 820-0443; fax (505) 989-1073; office@analyticalsciences.org
Internet: http://www.analyticalsciences.org/awards.php#Findeis
Sponsor: American Chemical Society
2019 Galisteo Street, Building I-1
Santa Fe, NM 87505

American Chemical Society ANYL Award for Distinguished Service in the Advancement of Analytical Chemistry 640

The purpose of the American Chemical Society ANYL Award for Distinguished Service in the Advancement of Analytical Chemistry is to honor an individual who through professional service in activities such as teaching, writing, research, and administration has substantially and uniquely enhanced the field of analytical chemistry. Nominations should present evidence for one or more of the following outstanding accomplishments: enhance the positive perception of analytical chemistry in the public eye; foster the development of analytical chemistry research in academic institutions, government laboratories, or in private industries; develop and implement programs that benefit the analytical community (these can be but are not limited to efforts within the Division of Analytical Chemistry or within the American Chemical Society; advance and promote the careers of analytical chemists in any area of employment; and play a central role in improving the way analytical chemistry is practiced role in the use of chemical instrumentation. The Award consists of a plaque and $2,500.
Geographic Focus: All States
Date(s) Application is Due: Nov 1
Amount of Grant: 2,500 USD
Contact: Miquela Sena, (505) 820-0443; fax (505) 989-1073; office@analyticalsciences.org
Internet: http://www.analyticalsciences.org/awards.php#Service
Sponsor: American Chemical Society
2019 Galisteo Street, Building I-1
Santa Fe, NM 87505

American Chemical Society ANYL Award in Chemical Instrumentation 641

The purpose of the American Chemical Society ANYL Award in Chemical Instrumentation is to recognize and encourage achievement in the origination or improvement of analytical methods of instrumentation. Advancing the field of chemical instrumentation through achievement of one or more of the following: conceptualization and development of unique instrumentation that has made a significant impact on the field; demonstration of innovative use of instrumentation in chemical measurement; stimulation of other researchers to use instrumentation in chemical measurement; and authorship of research papers or books that have had an influential role in the use of chemical instrumentation. This Award consists of a plaque and $2,500.
Requirements: Candidates must be an ANYL member and must have made original and eminent contributions to chemical instrumentation.
Geographic Focus: All States
Date(s) Application is Due: Nov 1
Amount of Grant: 2,500 USD
Contact: Miquela Sena, (505) 820-0443; fax (505) 989-1073; office@analyticalsciences.org
Internet: http://www.analyticalsciences.org/awards.php#Instrumentation
Sponsor: American Chemical Society
2019 Galisteo Street, Building I-1
Santa Fe, NM 87505

American Chemical Society ANYL Award in Electrochemistry 642

The purpose of the American Chemical Society ANYL Award in Electrochemistry is to recognize an individual who through scholarly activity has definitely and uniquely advanced the field of electrochemistry. Advancing the field of electrochemical analysis in one or more of the following: conceptualization and development of unique instrumentation that has made a significant impact on the field; development of novel and important instrumentation; elucidation of fundamental events or processes important to the field; and authorship of important research papers and/or books that have had an influential role in the development of the field of chemical instrumentation. This Award consists of a plaque and $2,500.
Geographic Focus: All States
Date(s) Application is Due: Nov 1
Amount of Grant: 2,500 USD
Contact: Miquela Sena, (505) 820-0443; fax (505) 989-1073; office@analyticalsciences.org
Internet: http://www.analyticalsciences.org/awards.php#Electrochemistry
Sponsor: American Chemical Society
2019 Galisteo Street, Building I-1
Santa Fe, NM 87505

American Chemical Society ANYL Award in Spectrochemical Analysis 643

The purpose of the American Chemical Society ANYL Award in Spectrochemical Analysis is to recognize an individual who through scholarly activity has definitely and uniquely advanced the field of spectrochemical analysis and optical spectrometry. Advancing the fields of spectrochemical analysis and optical spectrometry in one or more of the following: conceptualization and development of unique instrumentation that has made a significant impact on the field; development of novel and important instrumentation; elucidation of fundamental events or processes important to the field; and authorship of important research papers and/or books that have had an influential role in the development of the field of chemical instrumentation. This Award consists of a plaque and $2,500.
Geographic Focus: All States
Date(s) Application is Due: Nov 1
Amount of Grant: 2,500 USD
Contact: Miquela Sena, (505) 820-0443; fax (505) 989-1073; office@analyticalsciences.org
Internet: http://www.analyticalsciences.org/awards.php#Spectrochemical
Sponsor: American Chemical Society
2019 Galisteo Street, Building I-1
Santa Fe, NM 87505

American Chemical Society ANYL I.M. Kolthoff Enrichment Awards for Undergraduate Students 644

The purpose American Chemical Society ANYL I.M. Kolthoff Enrichment Awards are given annually to deserving undergraduates who have done undergraduate research in the field of analytical chemistry, broadly defined, in celebration of the life and accomplishments of the late Professor Izaak Maurits Kolthoff, viewed by many as the father of modern analytical chemistry. The Kolthoff awards provide travel funding up to $750 for undergraduate students to present a poster on their undergraduate research this year
Requirements: Only U.S. citizens and permanent residents are eligible, and preference will be given to those who have not made a previous presentation at a national scientific meeting.
Restrictions: Prior recipients are not eligible.
Geographic Focus: All States
Date(s) Application is Due: Nov 1
Amount of Grant: 750 USD
Contact: Miquela Sena, (505) 820-0443; fax (505) 989-1073; office@analyticalsciences.org
Internet: http://www.analyticalsciences.org/awards.php?action=kolthoff
Sponsor: American Chemical Society
2019 Galisteo Street, Building I-1
Santa Fe, NM 87505

American Chemical Society ANYL J. Calvin Giddings Award for Excellence in Education 645

The purpose of the American Chemical Society ANYL J. Calvin Giddings Award for Excellence in Education is to recognize the enhancement of the personal and professional development of students in the study of analytical chemistry. Enhancing the personal and professional development of students in the study of analytical chemistry in one or more of the following ways: authorship of an influential textbook for an analytical chemistry course; design and implementation of a successful new approach to teaching analytical chemistry; stimulation through teaching or research mentorship a significant number of students to become analytical chemists; development and publication of innovative experiments; design of improved equipment for teaching labs; and publication of widely quoted articles on teaching analytical chemistry. This Award consists of a plaque and $2,500.
Geographic Focus: All States
Date(s) Application is Due: Nov 1
Contact: Miquela Sena, (505) 820-0443; fax (505) 989-1073; office@analyticalsciences.org
Internet: http://www.analyticalsciences.org/awards.php#Teaching
Sponsor: American Chemical Society
2019 Galisteo Street, Building I-1
Santa Fe, NM 87505

American Chemical Society ANYL Pfizer Graduate Travel Awards in Analytical Chemistry 646

The purpose of the American Chemical Society ANYL Pfizer Graduate Travel Awards in Analytical Chemistry is to provide funding for graduate students to travel to an ACS national meeting and to present a poster. The award maximum is $750.
Requirements: Only U.S. citizens and permanent residents are eligible, and preference will be given to those who have not made a previous presentation at a national scientific meeting.
Geographic Focus: All States
Date(s) Application is Due: Nov 1
Amount of Grant: 750 USD
Contact: Miquela Sena, (505) 820-0443; fax (505) 989-1073; office@analyticalsciences.org
Internet: http://webapps.acs.org/findawards/detail.jsp?ContentId=WPCP_012500
Sponsor: American Chemical Society
2019 Galisteo Street, Building I-1
Santa Fe, NM 87505

American Chemical Society Arthur C. Cope Award 647

The American Chemical Society Arthur C. Cope Award is given to recognize outstanding achievement in the field of organic chemistry, the significance of which has become apparent within the five years preceding the year in which the award will be considered. The award consists of $25,000, a medallion with a presentation box, and a certificate. Up to $2,500 for travel expenses to the spring national meeting at which the award will be presented and to the fall national meeting to deliver an award address will be reimbursed. In addition to the award to the recipient, an unrestricted grant in aid for $150,000 for research in organic chemistry, under the direction of the recipient, will be made to any university or nonprofit institution selected by the recipient. The recipient may choose to assign the Arthur C. Cope Fund Grant to an institution for use by others for research or education in organic chemistry.
Requirements: Any individual, except a member of the award committee, may submit one nomination or seconding letter for the award in any given year. The nominating documents consist of a letter of not more than 1000 words containing an evaluation of the nominee's accomplishments and a specific identification of the work to be recognized, a biographical sketch including date of birth, and a list of publications and patents authored by the nominee.
Restrictions: Self-nominations are not accepted.
Geographic Focus: All States
Date(s) Application is Due: Nov 1
Amount of Grant: 175,000-177,500 USD
Contact: Administrator; (800) 227-5558 or (202) 872-4408; awards@acs.org
Internet: http://webapps.acs.org/findawards/detail.jsp?ContentId=CTP_004495
Sponsor: American Chemical Society
1155 Sixteenth Street, NW
Washington, DC 20036-4801

American Chemical Society Arthur C. Cope Scholar Awards 648

The American Chemical Society Arthur C. Cope Scholar Awards are given to recognize and encourage excellence in organic chemistry. Ten (10) Arthur C. Cope Scholars will be named annually: four between the ages of thirty-six and forty-nine, four age fifty or older, and two thirty-five years old and younger all inclusive before November 1 of the year in which the award is presented. The award consists of $5,000, a certificate and an unrestricted research grant in the amount of $40,000 to be assigned by the recipient to any university or nonprofit institution. The recipient is required to deliver a lecture at the annual Arthur C. Cope Symposium; traveling expenses incidental to participation in the symposium will also be paid.
Requirements: Any individual qualified to evaluate the nominee's accomplishments may submit one nomination for the award in any year; nomination should include a letter containing an evaluation of the nominee's contributions to organic chemistry with identification of the work to be recognized including literature and/or patent references; a biographical sketch of the nominee including date of birth and record of employment; and a list of publications and patents authored by the nominee.
Restrictions: Self-nominations are not accepted. An individual may not receive the award a second time. Recipients of the Arthur C. Cope Award are ineligible to be named Arthur C. Cope Scholars.
Geographic Focus: All States
Date(s) Application is Due: Nov 1
Contact: Administrator; (800) 227-5558 or (202) 872-4408; awards@acs.org
Internet: http://webapps.acs.org/findawards/detail.jsp?ContentId=CTP_004496
Sponsor: American Chemical Society
1155 Sixteenth Street, NW
Washington, DC 20036-4801

American Chemical Society Award for Achievement in Research for the Teaching and Learning of Chemistry 649

The award honors qualitative, quantitative, experimental research or case study research that investigates some aspect of the teaching and learning of chemistry. The research contributions may include the testing of research hypotheses, basing experiments on existing theories, incorporating what is known in the literature into the research, using an experimental design or recognized data-collection procedure, analyzing quantitative or qualitative data, and drawing valid conclusions from the observations and data. Experimental research should include one or more recognized techniques and be designed as control-group designs with random assignments of subjects or the use of in-tact class sections; factorial designs, bi-variate or multivariate correlation studies, etc. The research must show independence of thought and originality. The award consists of $5,000 and a certificate. Up to $2,500 for travel expenses to the meeting at which the award will be presented will be reimbursed. This is an international award.
Requirements: Any individual, except a member of the award committee, may submit one nomination or seconding letter for the award in any given year. The nominating documents consist of a letter of not more than 1000 words containing an evaluation of the nominee's accomplishments and a specific identification of the work to be recognized, a biographical sketch including date of birth, and a list of publications and patents authored by the nominee.
Restrictions: Self-nominations are not accepted.
Geographic Focus: All States
Date(s) Application is Due: Nov 1
Amount of Grant: 5,000-7,500 USD
Contact: Administrator; (800) 227-5558 or (202) 872-4408; awards@acs.org
Internet: http://webapps.acs.org/findawards/detail.jsp?ContentId=CTP_004484
Sponsor: American Chemical Society
1155 Sixteenth Street, NW
Washington, DC 20036-4801

American Chemical Society Award for Affordable Green Chemistry 650

The purposes of the Award are: to recognize outstanding scientific discoveries that lay the foundation for environmentally friendly products or manufacturing processes at a cost comparable to or less than that of current technologies, or discoveries that deliver new applications with compelling cost/benefit profiles; and to identify and recognize discovery of new eco-friendly chemistries with the potential to enable products or manufacturing processes that are less expensive than existing alternatives. Nominees may include: individuals; research teams (maximum 3 individuals); representatives from an organization (company, university or national lab); or professors or other academic-based teams whose work has been commercialized. The award will consist of $5,000 and a certificate. Up to $2,500 for travel expenses to the meeting at which the award will be presented will be reimbursed.
Requirements: Nominee eligibility is neither contingent on the timeframe in which the technology was developed, nor on demonstrated commercial success. Discoveries must have significant commercial potential, however, to be considered for award. Any individual, except a member of the award committee, may submit one nomination or seconding letter for the award in any given year. The nominating documents consist of a letter of not more than 1000 words containing an evaluation of the nominee's accomplishments and a specific identification of the work to be recognized, a biographical sketch including date of birth, and a list of publications and patents authored by the nominee.
Restrictions: Self-nominations are not accepted.
Geographic Focus: All States
Date(s) Application is Due: Nov 1
Amount of Grant: 5,000-7,500 USD
Contact: Administrator; (800) 227-5558 or (202) 872-4408; awards@acs.org
Internet: http://webapps.acs.org/findawards/detail.jsp?ContentId=CTP_004487
Sponsor: American Chemical Society
1155 Sixteenth Street, NW
Washington, DC 20036-4801

American Chemical Society Award for Computers in Chemical or Pharmaceutical Research 651

The purpose of the Award is to recognize and encourage the use of computers in the advancement of the chemical and biological sciences. The award consists of $5,000 and a certificate. Up to $1,000 for travel expenses to the meeting at which the award will be presented will be reimbursed. The annual deadline is November 1.
Requirements: Any individual, except a member of the award committee, may submit one nomination or seconding letter for the award in any given year. The nominating documents consist of a letter of not more than 1000 words containing an evaluation of the nominee's accomplishments and a specific identification of the work to be recognized, a biographical sketch including date of birth, and a list of publications and patents authored by the nominee.
Restrictions: Self-nominations are not accepted.
Geographic Focus: All States
Date(s) Application is Due: Nov 1
Amount of Grant: 5,000-6,000 USD
Contact: Administrator; (800) 227-5558 or (202) 872-4408; awards@acs.org
Internet: http://webapps.acs.org/findawards/detail.jsp?ContentId=CTP_004503
Sponsor: American Chemical Society
1155 Sixteenth Street, NW
Washington, DC 20036-4801

American Chemical Society Award for Creative Advances in Environmental Science and Technology 652

The purpose of the Award is to encourage creativity in research and technology or methods of analysis to provide a scientific basis for informed environmental control decision-making processes, or to provide practical technologies that will reduce health risk factors. The award consists of $5,000 and a certificate. Up to $2,500 for travel expenses to the meeting at which the award will be presented will be reimbursed.
Requirements: Any individual, except a member of the award committee, may submit one nomination or seconding letter for the award in any given year. The nominating documents consist of a letter of not more than 1000 words containing an evaluation of the nominee's accomplishments and a specific identification of the work to be recognized, a biographical sketch including date of birth, and a list of publications and patents authored by the nominee.
Restrictions: The award will be granted without regard to age or nationality. Self-nominations are not accepted.
Geographic Focus: All States
Date(s) Application is Due: Nov 1
Amount of Grant: 5,000-7,500 USD
Contact: Administrator; (800) 227-5558 or (202) 872-4408; awards@acs.org
Internet: http://webapps.acs.org/findawards/detail.jsp?ContentId=CTP_004504
Sponsor: American Chemical Society
1155 Sixteenth Street, NW
Washington, DC 20036-4801

American Chemical Society Award for Creative Invention 653

The purpose of the Award is to recognize a single inventor for the successful application of research in chemistry and/or chemical engineering that contributes to the material prosperity and happiness of people. The award consists of $5,000 and a certificate. Up to $1,000 for travel expenses to the meeting at which the award will be presented will be reimbursed.
Requirements: An inventor/nominee must be a resident of the United States or Canada. A patent must have been granted for the work to be recognized, with the nominee listed as an inventor, and it will have been developed during the 17 years ending January 1 of

the award year. A copy of the patent must be submitted with the nominating documents. Any individual, except a member of the award committee, may submit one nomination or seconding letter for the award in any given year. The nominating documents consist of a letter of not more than 1000 words containing an evaluation of the nominee's accomplishments and a specific identification of the work to be recognized, a biographical sketch including date of birth, and a list of publications and patents authored by the nominee.
Restrictions: Self-nominations are not accepted.
Geographic Focus: All States
Date(s) Application is Due: Nov 1
Amount of Grant: 5,000-6,000 USD
Contact: Administrator; (800) 227-5558 or (202) 872-4408; awards@acs.org
Internet: http://webapps.acs.org/findawards/detail.jsp?ContentId=CTP_004506
Sponsor: American Chemical Society
1155 Sixteenth Street, NW
Washington, DC 20036-4801

American Chemical Society Award for Creative Research and Applications of Iodine Chemistry 654
The purpose of the Award is to support, promote, and motivate global research of iodine chemistry and develop its use and knowledge through applications. Applications may include but are not limited to its uses in medicine, catalysis, food, and photography. The award consists of $10,000 and a certificate. Up to $1,000 for travel expenses to the meeting at which the award will be presented will also be reimbursed. The award is presented biennially in odd-numbered years.
Requirements: A nominee must have performed outstanding and creative research related to iodine chemistry or its applications. Any individual, except a member of the award committee, may submit one nomination or seconding letter for the award in any given year. The nominating documents consist of a letter of not more than 1000 words containing an evaluation of the nominee's accomplishments and a specific identification of the work to be recognized, a biographical sketch including date of birth, and a list of publications and patents authored by the nominee.
Restrictions: Self-nominations are not accepted.
Geographic Focus: All States
Date(s) Application is Due: Nov 1
Amount of Grant: 10,000-11,000 USD
Contact: Administrator; (800) 227-5558 or (202) 872-4408; awards@acs.org
Internet: http://webapps.acs.org/findawards/detail.jsp?ContentId=CTP_004507
Sponsor: American Chemical Society
1155 Sixteenth Street, NW
Washington, DC 20036-4801

American Chemical Society Award for Creative Work in Fluorine Chemistry 655
The purpose of the Award is to recognize outstanding contributions to the advancement of the chemistry of fluorine. The award consists of $5,000 and a certificate. Up to $2,500 for travel expenses to the meeting at which the award will be presented will be reimbursed. The award is presented in odd-numbered years at the Biennial Winter Fluorine Conference. The award is presented in even-numbered years during an ACS national meeting at which the ACS Division of Fluorine Chemistry meets.
Requirements: A nominee must have made an outstanding contribution or contributions to the field of fluorine chemistry. Any individual, except a member of the award committee, may submit one nomination or seconding letter for each award in any given year. The nominating documents consist of a letter of not more than 1000 words containing an evaluation of the nominee's accomplishments and a specific identification of the work to be recognized, a biographical sketch including date of birth, and a list of publications and patents authored by the nominee.
Restrictions: Self-nominations are not accepted.
Geographic Focus: All States
Date(s) Application is Due: Nov 1
Amount of Grant: 5,000-7,500 USD
Contact: Administrator; (800) 227-5558 or (202) 872-4408; awards@acs.org
Internet: http://webapps.acs.org/findawards/detail.jsp?ContentId=CTP_004505
Sponsor: American Chemical Society
1155 Sixteenth Street, NW
Washington, DC 20036-4801

American Chemical Society Award for Creative Work in Synthetic Organic Chemistry 656
The purpose of the Award is to recognize and encourage creative work in synthetic organic chemistry. The award consists of $5,000 and a certificate. Up to $1,000 for travel expenses to the meeting at which the award will be presented will be reimbursed.
Requirements: A nominee must have accomplished outstanding creative work in synthetic organic chemistry that has been published. Any individual, except a member of the award committee, may submit one nomination or seconding letter for the award in any given year. The nominating documents consist of a letter of not more than 1000 words containing an evaluation of the nominee's accomplishments and a specific identification of the work to be recognized, a biographical sketch including date of birth, and a list of publications and patents authored by the nominee.
Restrictions: Self-nominations are not accepted.
Geographic Focus: All States
Date(s) Application is Due: Nov 1
Amount of Grant: 5,000-6,000 USD
Contact: Administrator; (800) 227-5558 or (202) 872-4408; awards@acs.org
Internet: http://webapps.acs.org/findawards/detail.jsp?ContentId=CTP_004508
Sponsor: American Chemical Society
1155 Sixteenth Street, NW
Washington, DC 20036-4801

American Chemical Society Award for Distinguished Service in the Advancement of Inorganic Chemistry 657
The purpose of the Award is to recognize individuals who advanced inorganic chemistry by significant service in addition to performance of outstanding research. Activities recognized by the award may include such fields as teaching, writing, research, and administration. The award consists of $5,000 and a certificate. Up to $1,000 for travel expenses to the meeting at which the award will be presented will be reimbursed. The annual deadline is November 1.
Requirements: A nominee must have demonstrated extensive contributions to the advancement of inorganic chemistry. Any individual, except a member of the award committee, may submit one nomination or seconding letter for the award in any given year. The nominating documents consist of a letter of not more than 1000 words containing an evaluation of the nominee's accomplishments and a specific identification of the work to be recognized, a biographical sketch including date of birth, and a list of publications and patents authored by the nominee.
Restrictions: Self-nominations are not accepted.
Geographic Focus: All States
Date(s) Application is Due: Nov 1
Amount of Grant: 5,000-6,000 USD
Contact: Administrator; (800) 227-5558 or (202) 872-4408; awards@acs.org
Internet: http://webapps.acs.org/findawards/detail.jsp?ContentId=CTP_004486
Sponsor: American Chemical Society
1155 Sixteenth Street, NW
Washington, DC 20036-4801

American Chemical Society Award for Encouraging Disadvantaged Students into Careers in the Chemical Sciences 658
The purpose of this Award is to recognize significant accomplishments by individuals in stimulating students, underrepresented in the profession, to elect careers in the chemical sciences and engineering. The award consists of $5,000 and a certificate. A grant of $10,000 will be made to an academic institution, designated by the recipient, to strengthen its activities in meeting the objectives of the award. Up to $1,500 for travel expenses to the meeting at which the award will be presented will be reimbursed. The annual deadline for nominations is November 1.
Requirements: Nominees for the award may come from any professional setting: academia, industry, government, or other independent facility. The award is intended to recognize significant accomplishments in the United States by individuals in stimulating students, especially those currently underrepresented in the profession, to elect careers in the chemical sciences and engineering, and in generating a broader appreciation of chemistry as the central science.
Restrictions: Self-nominations are not accepted.
Geographic Focus: All States
Date(s) Application is Due: Nov 1
Amount of Grant: Up to 16,500 USD
Contact: Administrator; (800) 227-5558 or (202) 872-4408; awards@acs.org
Internet: http://webapps.acs.org/findawards/detail.jsp?ContentId=CTP_004509
Sponsor: American Chemical Society
1155 Sixteenth Street, NW
Washington, DC 20036-4801

American Chemical Society Award for Encouraging Women into Careers in the Chemical Sciences 659
The purpose of the Award is to recognize individuals who have significantly stimulated or fostered the interest of women in chemistry, thereby promoting their professional development as chemists or chemical engineers, and/or increasing their appreciation of chemistry as the central science. Nominees for the award may come from any professional setting: academia, industry, government, or other independent facility. The award is intended to recognize significant accomplishments by individuals in stimulating women to elect careers in the chemical sciences and engineering, and in generating a broader appreciation of chemistry as the central science. The award consists of $5,000 and a certificate. A grant of $10,000 will be made to an academic institution, designated by the recipient, to strengthen its activities in meeting the objectives of the award. Up to $1,500 for travel expenses to the meeting at which the award will be presented will be reimbursed. The annual deadline is November 1.
Requirements: Any individual, except a member of the award committee, may submit one nomination or seconding letter for the award in any given year. Nominating documents consist of a letter of not more than 1000 words containing an evaluation of the nominee's accomplishments and a specific identification of the work to be recognized, a biographical sketch including date of birth, and a list of publications and patents authored by the nominee.
Restrictions: Self-nominations are not accepted.
Geographic Focus: All States
Date(s) Application is Due: Nov 1
Amount of Grant: 15,000-16,500 USD
Contact: Administrator; (800) 227-5558 or (202) 872-4408; awards@acs.org
Internet: http://webapps.acs.org/findawards/detail.jsp?ContentId=CTP_004517
Sponsor: American Chemical Society
1155 Sixteenth Street, NW
Washington, DC 20036-4801

American Chemical Society Award for Research at an Undergraduate Institution 660
The purpose of the Award is to recognize the importance of research with undergraduates. The award will honor a chemistry faculty member whose research in an undergraduate setting has achieved wide recognition and contributed significantly to chemistry and to the professional development of undergraduate students. This award recognizes fundamental research that constitutes advances in science as evidenced by refereed publications with undergraduate coauthors in leading scientific research journals, external research grant support, and the subsequent professional development of students who have participated in the research program. The award will be given for significant work over a long period of time rather than for a specific, limited project. The award consists of $5,000 and a certificate. Up to $2,500 for travel expenses to the meeting at which the award will be presented will be reimbursed. The program will also provide a grant of $5,000 directly to the recipient's institution.
Requirements: The nominee's department may offer work leading to the master's degree but cannot have a doctoral program. A nominee must be a tenured faculty member of a predominantly undergraduate institution. Any individual, except a member of the award committee, may submit one nomination or seconding letter for the award in any given year. The nominating documents consist of a letter of not more than 1000 words containing an evaluation of the nominee's accomplishments and a specific identification of the work to be recognized, a biographical sketch including date of birth, and a list of publications and patents authored by the nominee.
Restrictions: Self-nominations are not accepted.
Geographic Focus: All States
Date(s) Application is Due: Nov 1
Amount of Grant: 10,000-12,500 USD
Contact: Administrator; (800) 227-5558 or (202) 872-4408; awards@acs.org
Internet: http://webapps.acs.org/findawards/detail.jsp?ContentId=CTP_004549
Sponsor: American Chemical Society
1155 Sixteenth Street, NW
Washington, DC 20036-4801

American Chemical Society Award for Team Innovation 661
This award was established to highlight the value and importance of technical teams and teamwork to the chemical and allied industries by recognizing a multidisciplinary team for successfully moving an innovative idea to a product now in commercial use. For each team member, the award consists of $3,000 and a certificate. Up to $1,000 for travel expenses to the meeting at which the award will be presented will be reimbursed to each team member. A certificate will also be provided to the employers of the team.
Requirements: The team shall be multidisciplinary in nature and consist of not fewer than two nor more than five members. The team's work leading to this award must have been carried out primarily in the United States and the technical accomplishments of the team must be documented in the technical literature as a publication(s) or a patent(s). The output of the team must also demonstrate innovation, commercialization of a product or process, be of a special value to society, and of a nature that could only be achieved by professionals working together effectively. The rate of commercialization of the team's output will also be considered in the selection of the recipients of this award.
Restrictions: Application or self-nomination is not acceptable.
Geographic Focus: All States
Date(s) Application is Due: Nov 1
Amount of Grant: 3,000-6,000 USD
Contact: Administrator; (800) 227-5558 or (202) 872-4408; awards@acs.org
Internet: http://webapps.acs.org/findawards/detail.jsp?ContentId=CTP_004554
Sponsor: American Chemical Society
1155 Sixteenth Street, NW
Washington, DC 20036-4801

American Chemical Society Award in Analytical Chemistry 662
This award, sponsored by Battelle Memorial Institute and administered by the American Chemical Society, is given annually to recognize and encourage outstanding contributions to the science of analytical chemistry, pure or applied, carried out in the United States or Canada. Special consideration will be given to the independence of thought and the originality shown, or to the importance of the work when applied to public welfare, economics, or the needs and desires of humanity. The award consists of $5,000 and a certificate. Up to $2,500 for travel expenses to the meeting at which the award will be presented will be reimbursed.
Requirements: A nominee must be a resident of the United States or Canada and must have made an outstanding contribution to analytical chemistry. Any individual, except a member of the award committee, may submit one nomination or seconding letter for the award in any given year. The nominating documents consist of a letter of not more than 1000 words containing an evaluation of the nominee's accomplishments and a specific identification of the work to be recognized, a biographical sketch including date of birth, and a list of publications and patents authored by the nominee.
Restrictions: Self-nominations are not accepted.
Geographic Focus: All States
Date(s) Application is Due: Nov 1
Amount of Grant: 5,000-7,500 USD
Contact: Administrator; (800) 227-5558 or (202) 872-4408; awards@acs.org
Internet: http://webapps.acs.org/findawards/detail.jsp?ContentId=CTP_004492
Sponsor: American Chemical Society
1155 Sixteenth Street, NW
Washington, DC 20036-4801

American Chemical Society Award in Applied Polymer Science 663
This award, sponsored by Eastman Chemical Company and administered by the American Chemical Society, is given annually to recognize and encourage outstanding achievements in the science or technology of plastics, coatings, polymer composites, adhesives, and related fields. The recipient will be selected primarily on the basis of scientific contributions made to the specified fields during the 10-year period preceding date of selection. To avoid repeating specific areas of technology, preference will be given whenever recognized by the granting of this award in the two preceding years. The award consists of $5,000 and a certificate. Up to $2,500 for travel expenses to the meeting at which the award will be presented will be reimbursed.
Requirements: Any individual, except a member of the award committee, may submit one nomination or seconding letter for the award in any given year. The nominating documents consist of a letter of not more than 1000 words containing an evaluation of the nominee's accomplishments and a specific identification of the work to be recognized, a biographical sketch including date of birth, and a list of publications and patents authored by the nominee.
Restrictions: Self-nominations are not accepted.
Geographic Focus: All States
Date(s) Application is Due: Nov 1
Amount of Grant: 5,000-7,500 USD
Contact: Administrator; (800) 227-5558 or (202) 872-4408; awards@acs.org
Internet: http://webapps.acs.org/findawards/detail.jsp?ContentId=CTP_004493
Sponsor: American Chemical Society
1155 Sixteenth Street, NW
Washington, DC 20036-4801

American Chemical Society Award in Chromatography 664
The purpose of this Award is to recognize outstanding contributions to the fields of chromatography. The award consists of $5,000 and a certificate. Up to $2,500 for travel expenses to the meeting at which the award will be presented will be reimbursed. The annual deadline is November 1.
Requirements: A nominee must have made an outstanding contribution to the field with particular consideration given to developments of new methods. Any individual, except a member of the award committee, may submit one nomination or seconding letter for the award in any given year. The nominating documents consist of a letter of not more than 1000 words containing an evaluation of the nominee's accomplishments and a specific identification of the work to be recognized, a biographical sketch including date of birth, and a list of publications and patents authored by the nominee.
Restrictions: Self-nominations are not accepted.
Geographic Focus: All States
Date(s) Application is Due: Nov 1
Amount of Grant: 5,000-7,500 USD
Contact: Administrator; (800) 227-5558 or (202) 872-4408; awards@acs.org
Internet: http://webapps.acs.org/findawards/detail.jsp?ContentId=CTP_004500
Sponsor: American Chemical Society
1155 Sixteenth Street, NW
Washington, DC 20036-4801

American Chemical Society Award in Colloid and Surface Chemistry 665
The award, sponsored by the Procter and Gamble Company, is given annually to recognize and encourage outstanding scientific contributions to colloid or surface chemistry in the United States or Canada. Recognition will also be given to originality and independence of thought, and to the technological impact of the nominee's contribution. In odd-numbered years, awards are made for contributions in surface chemistry. In even-numbered years, awards are made for advances in colloid chemistry. The award consists of $5,000 and a certificate. Up to $2,500 for travel expenses to the meeting at which the award will be presented will be reimbursed.
Requirements: The nominee must be a resident of the United States or Canada and must have made outstanding scientific contributions to colloid or surface chemistry. Any individual, except a member of the award committee, may submit one nomination or seconding letter for the award in any given year. The nominating documents consist of a letter of not more than 1000 words containing an evaluation of the nominee's accomplishments and a specific identification of the work to be recognized, a biographical sketch including date of birth, and a list of publications and patents authored by the nominee.
Restrictions: Self-nominations are not accepted.
Geographic Focus: All States
Date(s) Application is Due: Nov 1
Contact: Administrator; (800) 227-5558 or (202) 872-4408; awards@acs.org
Internet: http://webapps.acs.org/findawards/detail.jsp?ContentId=CTP_004502
Sponsor: American Chemical Society
1155 Sixteenth Street, NW
Washington, DC 20036-4801

American Chemical Society Award in Industrial Chemistry 666
The award, sponsored by ACS Division of Business Development and Management, is given annually to recognize outstanding contributions to industrial chemistry resulting in the commercialization of an economically significant new product or process. Any field of chemical, chemical engineering, or biochemical research is appropriate if it is of general interest and reflects the concerns of modern society. Any chemical research, whether industrial, governmental, or academic, is eligible, provided the work was done in North America and yielded significant commercial results for a period of more than one year. The award consists of $5,000 and a certificate. Up to $2,500 for travel expenses to the meeting at which the award will be presented will be reimbursed.

Requirements: Any individual, except a member of the award committee, may submit one nomination or seconding letter for the award in any given year. The nominating documents consist of a letter of not more than 1000 words containing an evaluation of the nominee's accomplishments and a specific identification of the work to be recognized, a biographical sketch including date of birth, and a list of publications and patents authored by the nominee.
Restrictions: Self-nominations are not accepted.
Geographic Focus: All States
Date(s) Application is Due: Nov 1
Amount of Grant: 5,000-7,500 USD
Contact: Administrator; (800) 227-5558 or (202) 872-4408; awards@acs.org
Internet: http://webapps.acs.org/findawards/detail.jsp?ContentId=CTP_004531
Sponsor: American Chemical Society
1155 Sixteenth Street, NW
Washington, DC 20036-4801

American Chemical Society Award in Inorganic Chemistry **667**
This award, sponsored by the Aldrich Chemical Company and administered by the American Chemical Society, is given annually to recognize and encourage fundamental research in the field of inorganic chemistry. A nominee must have accomplished outstanding research in the preparation, properties, reactions, or structure of inorganic substances. Special consideration shall be given to the independence of thought and originality shown. The award consists of $5,000 and a certificate. Up to $2,500 for travel expenses to the meeting at which the award will be presented will be reimbursed.
Requirements: Any individual, except a member of the award committee, may submit one nomination or seconding letter for the award in any given year. The nominating documents consist of a letter of not more than 1000 words containing an evaluation of the nominee's accomplishments and a specific identification of the work to be recognized, a biographical sketch including date of birth, and a list of publications and patents authored by the nominee.
Restrictions: Self-nominations are not accepted.
Geographic Focus: All States
Date(s) Application is Due: Nov 1
Amount of Grant: 5,000-7,500 USD
Contact: Administrator; (800) 227-5558 or (202) 872-4408; awards@acs.org
Internet: http://webapps.acs.org/findawards/detail.jsp?ContentId=CTP_004532
Sponsor: American Chemical Society
1155 Sixteenth Street, NW
Washington, DC 20036-4801

American Chemical Society Award in Organometallic Chemistry **668**
The Award, sponsored by the Dow Chemical Company Foundation and administered by the American Chemical Society, is given annually to recognize a recent advancement that is having a major impact on research in organometallic chemistry. A nominee must have shown outstanding research in the preparation, reactions, properties, or structure of organometallic substances. Special consideration will be given to demonstrated creativity and independence of thought. The award consists of $5,000 and a certificate. Up to $1,000 for travel expenses to the meeting at which the award will be presented will be reimbursed.
Requirements: Preference will be given to U.S. citizens. Any individual, except a member of the award committee, may submit one nomination or seconding letter for the award in any given year. The nominating documents consist of a letter of not more than 1000 words containing an evaluation of the nominee's accomplishments and a specific identification of the work to be recognized, a biographical sketch including date of birth, and a list of publications and patents authored by the nominee.
Restrictions: Self-nominations are not accepted.
Geographic Focus: All States
Date(s) Application is Due: Nov 1
Amount of Grant: 5,000-6,000 USD
Contact: Administrator; (800) 227-5558 or (202) 872-4408; awards@acs.org
Internet: http://webapps.acs.org/findawards/detail.jsp?ContentId=CTP_004542
Sponsor: American Chemical Society
1155 Sixteenth Street, NW
Washington, DC 20036-4801

American Chemical Society Award in Polymer Chemistry **669**
The Award, supported by the ExxonMobil Chemical Company, is intended to recognize outstanding contributions to polymer chemistry. This award is specifically intended to acknowledge the impact in basic science of a researcher's body of work in the broad field of polymer chemistry on our global needs for advanced polymeric systems and materials. The award consists of $5,000 and a certificate. Up to $1,000 for travel expenses to the meeting at which the award will be presented will be reimbursed. The annual deadline is November 1.
Requirements: The nominees for this award are expected to have demonstrated pioneering research accomplishments in synthetic methods of unique polymeric materials and/or achieved profound discoveries in polymer physical chemistry leading to recognized advancements in our understanding of complex macromolecular behavior.
Restrictions: Self-nominations are not accepted.
Geographic Focus: All States
Date(s) Application is Due: Nov 1
Contact: Administrator; (800) 227-5558 or (202) 872-4408; awards@acs.org
Internet: http://webapps.acs.org/findawards/detail.jsp?ContentId=CTP_004544
Sponsor: American Chemical Society
1155 Sixteenth Street, NW
Washington, DC 20036-4801

American Chemical Society Award in Pure Chemistry **670**
This award, sponsored by Alpha Chi Sigma Fraternity and the Alpha Chi Sigma Educational Award Foundation, is given annually to recognize and encourage fundamental research in pure chemistry carried out in North America by young men and women. Special consideration is given to independence of thought and the originality shown in the research. The award consists of $5,000 and a certificate. Up to $1,000 for travel expenses to the meeting at which the award will be presented will be reimbursed. The annual deadline is November 1.
Requirements: A nominee must have been born after April 30, 1978 and must have accomplished research of unusual merit for an individual on the threshold of her or his career. Any individual, except a member of the award committee, may submit one nomination or seconding letter for the award in any given year. The nominating documents consist of a letter of not more than 1000 words containing an evaluation of the nominee's accomplishments and a specific identification of the work to be recognized, a biographical sketch including date of birth, and a list of publications and patents authored by the nominee.
Restrictions: Self-nominations are not accepted.
Geographic Focus: All States
Date(s) Application is Due: Nov 1
Amount of Grant: 5,000-6,000 USD
Contact: Administrator; (800) 227-5558 or (202) 872-4408; awards@acs.org
Internet: http://webapps.acs.org/findawards/detail.jsp?ContentId=CTP_004546
Sponsor: American Chemical Society
1155 Sixteenth Street, NW
Washington, DC 20036-4801

American Chemical Society Award in Separations Science and Technology **671**
This award is given annually to recognize outstanding accomplishments in fundamental or applied research directed to separations science and technology. The award shall be granted to an individual without regard to age or nationality. The scope of the award is to be as broad as possible, covering all fields where separations science and technology is practiced, including but not limited to biology, chemistry, engineering, geology, and medicine. The award consists of $5,000 and a certificate. Up to $2,500 for travel expenses to the meeting at which the award will be presented will be reimbursed. The recipient will deliver a lecture at the annual Division of Industrial and Engineering Chemistry Separations Science and Technology Symposium.
Requirements: Any individual, except a member of the award committee, may submit one nomination or seconding letter for the award in any given year. The nominating documents consist of a letter of not more than 1000 words containing an evaluation of the nominee's accomplishments and a specific identification of the work to be recognized, a biographical sketch including date of birth, and a list of publications and patents authored by the nominee.
Restrictions: Self-nominations are not accepted.
Geographic Focus: All States
Date(s) Application is Due: Nov 1
Amount of Grant: 5,000-7,500 USD
Contact: Administrator; (800) 227-5558 or (202) 872-4408; awards@acs.org
Internet: http://webapps.acs.org/findawards/detail.jsp?ContentId=CTP_004552
Sponsor: American Chemical Society
1155 Sixteenth Street, NW
Washington, DC 20036-4801

American Chemical Society Award in the Chemistry of Materials **672**
This award, supported by E.I. du Pont de Nemours and Company, is given annually to recognize and encourage creative work in the chemistry of materials. A nominee must have made outstanding contributions to the chemistry of materials. Particular emphasis will be placed on research relating to materials of actual or potential technological importance, where a fundamental understanding of the chemistry associated with materials preparation, processing, or use is critical. The award consists of $5,000 and a certificate. Up to $2,500 for travel expenses to the meeting at which the award will be presented will be reimbursed. The annual deadline is November 1.
Requirements: Any individual, except a member of the award committee, may submit one nomination or seconding letter for the award in any given year. The nominating documents consist of a letter of not more than 1000 words containing an evaluation of the nominee's accomplishments and a specific identification of the work to be recognized, a biographical sketch including date of birth, and a list of publications and patents authored by the nominee.
Restrictions: Self-nominations are not accepted.
Geographic Focus: All States
Date(s) Application is Due: Nov 1
Amount of Grant: 5,000 USD
Contact: Administrator; (800) 227-5558 or (202) 872-4408; awards@acs.org
Internet: http://webapps.acs.org/findawards/detail.jsp?ContentId=CTP_004499
Sponsor: American Chemical Society
1155 Sixteenth Street, NW
Washington, DC 20036-4801

American Chemical Society Award in Theoretical Chemistry **673**
The purpose of this annual award, which is sponsored by Dell Incorporated, is to recognize innovative research in theoretical chemistry that either advances theoretical methodology or contributes to new discoveries about chemical systems. Emphasis in the selection process will be on work characterized by depth, originality, and scientific significance. The award is given without regard to age or nationality, and consists of

$5,000 and a certificate, with an additional maximum of $2,500 in travel expenses. The annual deadline is November 1.
Requirements: Any individual, except a member of the award committee, may submit one nomination or seconding letter for each award in any given year. The letter of nomination may not exceed 1000 words and should contain an evaluation of the nominee's accomplishments and a specific identification of the work to be recognized.
Restrictions: Self-nominations are not accepted.
Geographic Focus: All States
Date(s) Application is Due: Nov 1
Amount of Grant: 5,000-7,500 USD
Contact: Administrator; (800) 227-5558 or (202) 872-4408; awards@acs.org
Internet: http://webapps.acs.org/findawards/detail.jsp?ContentId=CTP_004555
Sponsor: American Chemical Society
1155 Sixteenth Street, NW
Washington, DC 20036-4801

American Chemical Society Chemical Technology Partnership Mini Grants 674
The American Chemical Society (ACS) offers mini grants to help collaboration activities among industry, academia, and the community that are essential to the education and professional development of chemical technicians. The program aims to: (1) Raise community and industry awareness of the changing needs of professional technicians; (2) Highlight opportunities for industry, academia, professional societies, and the community to collaborate on meeting those needs; (3) Increase involvement of current and future technicians in the American Chemical Society.
Requirements: Eligible programs or activities must meet the following conditions: (1) Program/activity must support technician education or career development; (2) Representatives from at least two different sectors of the chemical enterprise (academia, industry, workforce organizations, professional societies, etc.) must be involved; (3) Program/activity must take place during the year of application.
Geographic Focus: All States
Date(s) Application is Due: Feb 20
Contact: Blake J. Aronson, (800) 227-5558 or (202) 872-4600; b_aronson@acs.org
Internet: http://portal.acs.org/portal/acs/corg/content?_nfpb=true&_pageLabel=PP_TRANSITIONMAIN&node_id=1772&use_sec=false&sec_url_var=region1
Sponsor: American Chemical Society
1155 Sixteenth Street, NW
Washington, DC 20036-4801

American Chemical Society Claude S. Hudson Awards 675
This award, supported by the National Starch and Chemical Company, is given biennially in odd-numbered years to recognize outstanding contributions to carbohydrate chemistry, whether in education, research, or application. The award is granted without regard to age or nationality. Applications are accepted in even-numbered years. The award consists of $5,000 and a certificate. Up to $1,000 for travel expenses to the meeting at which the award will be presented will be reimbursed.
Requirements: Any individual, except a member of the award committee, may submit one nomination or seconding letter for the award in any given year. Nominating documents consist of a letter of not more than 1000 words containing an evaluation of the nominee's accomplishments and a specific identification of the work to be recognized, a biographical sketch including date of birth, and a list of publications and patents authored by the nominee.
Restrictions: Self-nominations are not accepted.
Geographic Focus: All States
Date(s) Application is Due: Nov 1
Amount of Grant: 6,000 USD
Contact: Felicia Dixon, Awards Administrator; (202) 872-4408 or (202) 872-4408; fax (202) 776-8008; f_dixon@acs.org or awards@acs.org
Internet: http://portal.acs.org/portal/acs/corg/content?_nfpb=true&_pageLabel=PP_ARTICLEMAIN&node_id=1319&content_id=CTP_004501&use_sec=true&sec_url_var=region1
Sponsor: American Chemical Society
1155 Sixteenth Street, NW
Washington, DC 20036-4801

American Chemical Society Corporate Associates Seed Grants 676
The American Chemical Society recognizes activities at the community level and provides grants to support programs that advance the public's understanding of chemistry. Awards are generally $1,000 up to a maximum of $10,000 per calendar year. CA Seed Grants should not be considered as continuous support.
Requirements: Proposals will be considered in the following areas: (a) Education in the field of chemical sciences; (b) Education of the public regarding the chemical industry; (c) Enhancement of professionalism or safety in chemistry. Specific proposal guidelines and application instructions are available at the sponsor's website.
Geographic Focus: All States
Date(s) Application is Due: Feb 15; Jul 1
Amount of Grant: Up to 10,000 USD
Contact: Nelufar Mohajeri; (202) 872-4441; n_mohajeri@acs.org
Internet: http://portal.acs.org/portal/acs/corg/content?_nfpb=true&_pageLabel=PP_TRANSITIONMAIN&node_id=1454&use_sec=false&sec_url_var=region1
Sponsor: American Chemical Society
1155 Sixteenth Street, NW
Washington, DC 20036-4801

American Chemical Society GCI Pharmaceutical Roundtable Research Grants 677
The Roundtable is a partnership with the ACS GCI (Green Chemistry Institute) and pharmaceutical corporations united by a shared commitment to integrate the principles of green chemistry and engineering into the business of drug discovery and production. The program was created to support research in the key green chemistry research areas. (Ten chemical transformations and two process related operations were identified and subsequently published-download the list from the sponsor's website). The primary purpose of this grant is to publish research to make information publicly available. One grant is planned to be awarded. The total award is limited to $150,000 for a grant period of 12 to 24 months.
Requirements: Proposals will only be accepted from public and private institutions of higher education. The grant is not limited to institutions in the United States. Proposals must be submitted through the appropriate institutional office for external funding. Proposed research must focus on the key targeted areas identified by the Roundtable. Proposals must clearly describe how the proposed research will provide improvements over existing technology that are consistent with the goals of Green Chemistry including improved atom economy, use of less hazardous reagents, more energy efficient processes, elimination of by-products, reduced waste generation, improved environmental, health, and safety and life cycle impacts, catalytic versus stoichiometric methods, and development of inherently safer transformations. Illustrations of the relevance of the proposed technology to synthetic and process related problems within the pharmaceutical industry are encouraged. Complete guidelines can be downloaded at the sponsor's website.
Restrictions: Applicants may have only one research grant with the ACS GCI Pharmaceutical Roundtable at a time. Current research grant holders may not apply for a new grant until their active grant or no-cost time extension officially expires.
Geographic Focus: All States
Date(s) Application is Due: Sep 14
Amount of Grant: Up to 150,000 USD
Contact: Gayle Peterman, Finance and Grants Manager; (202) 872-6092 or (202) 872-4481; fax (202) 872-6319; g_peterman@acs.org or gcipr@acs.org
Internet: http://portal.acs.org/portal/acs/corg/content?_nfpb=true&_pageLabel=PP_TRANSITIONMAIN&node_id=1456&use_sec=false&sec_url_var=region1
Sponsor: American Chemical Society
1155 Sixteenth Street, NW
Washington, DC 20036-4801

American Chemical Society Green Chemistry Chapter Award for Student Affiliates 678
The Education Division of the American Chemical Society (ACS) sponsors the award and the ACS Green Chemistry Institute assists with evaluating the green chemistry activities. The award's purpose is to recognize Student Affiliates Chapters of the ACS that engage in at least three green chemistry activities during the academic year. The Student Affiliates of the American Chemical Society is the official ACS organization for undergraduate chemical science majors. Each chapter receiving the green chemistry chapter award receives a plaque during the Student Affiliates Awards Ceremony, held annually during the Spring ACS National Meeting.
Geographic Focus: All States
Contact: Ron Siatkowski; (202) 872-6091; fax (202) 872-6319; r_siatkowski@acs.org
Internet: http://webapps.acs.org/findawards/detail.jsp?ContentId=WPCP_012552
Sponsor: American Chemical Society
1155 Sixteenth Street, NW
Washington, DC 20036-4801

American Chemical Society Green Chemistry Grants 679
The American Chemical Society Green Chemistry Institute sponsors a green chemistry grant program by promoting funding, increasing opportunities, and developing information on the benefits of green chemistry. Its efforts span the spectrum from theory and basic research through application and commercialization of science and technology. The program also sponsors a number of research fellowships and scientific exchanges through its grants programs.
Geographic Focus: All States
Contact: Gayle Peterman, Finance and Grants Manager; (202) 872-6092 or (202) 872-4481; fax (202) 872-6319; g_peterman@acs.org
Internet: http://portal.acs.org/portal/acs/corg/content?_nfpb=true&_pageLabel=PP_TRANSITIONMAIN&node_id=1456&use_sec=false&sec_url_var=region1&__uuid=c14e3223-89d3-4263-8072-51ccffe52007
Sponsor: American Chemical Society
1155 Sixteenth Street, NW
Washington, DC 20036-4801

American Chemical Society Herman Frasch Foundation Grants 680
The foundation grants are intended for research in the field of agricultural chemistry. Supported research is primarily chemical or biochemical in nature. Because such research is often interdisciplinary, a joint effort by two scientists with expertise in different complementary fields may achieve the desired results; joint projects have been endorsed by the Frasch Committee. Projects which focus on nutrition, agronomy, genetics, or entomology are not usually favored. Proposals must provide opportunities for training students at the graduate, undergraduate, and/or postdoctoral levels. Grants will be made in the amount of $50,000 per year or $250,000 for the maximum five year period. Proposals are considered only once in five years and grants are allocated at five-year intervals.
Requirements: Research must be conducted in the field of Agricultural Chemistry. The results obtained must be of potential practical benefit to the agricultural development of the United States. Grantee must be a non-profit institution incorporated and existing within the United States.

Restrictions: Funds cannot be used for tuition, overhead, administrative expense, large items of equipment, salary expenses or benefits for the principal investigator.
Geographic Focus: All States
Amount of Grant: 50,000 USD
Contact: World Wide Grants Administrator; (202) 872-4600; fraschinfo@acs.org
Internet: http://portal.acs.org/portal/acs/corg/content?_nfpb=true&_pageLabel=PP_TRANSITIONMAIN&node_id=1290&use_sec=false&sec_url_var=region1
Sponsor: American Chemical Society
1155 Sixteenth Street, NW
Washington, DC 20036-4801

American Chemical Society Irving Langmuir Award in Chemical Physics 681
This award, supported by the General Electric Global Research, is presented in even-numbered years to recognize and encourage outstanding interdisciplinary research in chemistry and physics. A nominee must have made an outstanding contribution to chemical physics or physical chemistry within the 10 years preceding the year in which the award is made. The award will be granted without restriction, except that the recipient must be a resident of the United States and the monetary prize must be used in the United States or its possessions. Applications are accepted in odd-numbered years.
Requirements: Any individual, except a member of the award committee, may submit one nomination or seconding letter for the award in any given year. Nominating documents consist of a letter of not more than 1000 words containing an evaluation of the nominee's accomplishments and a specific identification of the work to be recognized, a biographical sketch including date of birth, and a list of publications and patents authored by the nominee.
Restrictions: Self-nominations are not accepted.
Geographic Focus: All States
Date(s) Application is Due: Nov 1
Amount of Grant: 10,000 USD
Contact: Felicia Dixon, Awards Administrator; (800) 227-5558 or (202) 872-4408; fax (202) 776-8008; f_dixon@acs.org or awards@acs.org
Internet: http://portal.acs.org/portal/acs/corg/content?_nfpb=true&_pageLabel=PP_ARTICLEMAIN&node_id=1319&content_id=CTP_004534&use_sec=true&sec_url_var=region1
Sponsor: American Chemical Society
1155 Sixteenth Street, NW
Washington, DC 20036-4801

American Chemical Society Joseph Breen Memorial Fellowship in Green Chemistry 682
The American Chemical Society (ACS) established the Joseph Breen Memorial Fund through the ACS International Endowment Fund in 2000. This fund commemorates Joe Breen's commitment and accomplishments for the advancement of green chemistry. The purpose of the Joseph Breen Memorial Fellowship in Green Chemistry is to sponsor a young international green chemistry scholar to participate in an international green chemistry technical meeting, conference or training program. The award is a travel reimbursement up to a limit decided by the award judges. The amount of the award is based on the estimated cost of travel, accommodation, and fees associated with the event.
Geographic Focus: All States
Contact: Ron Siatkowski, Director, Office of Research Grants; (202) 872-6091; fax (202) 872-6319; r_siatkowski@acs.org
Internet: http://webapps.acs.org/findawards/detail.jsp?ContentId=WPCP_012550
Sponsor: American Chemical Society
1155 Sixteenth Street, NW
Washington, DC 20036-4801

American Chemical Society Kenneth G. Hancock Memorial Student Award in Green Chemistry 683
The ACS Division of Environmental Chemistry and the National Institute of Standards and Technology (NIST) sponsor the Kenneth G. Hancock Memorial Student Award in Green Chemistry. The award is administered by the ACS Green Chemistry Institute, and its purpose is to recognize and honor outstanding student contributions to furthering the goals of green chemistry through research and/or studies. The award provides national recognition for outstanding student contributions to furthering the goals of green chemistry (i.e., the research, development, and implementation of fundamental and innovative chemical technologies that incorporate the principles of green chemistry into chemical design, manufacture, and use, and that have the potential to be utilized in achieving national pollution prevention goals). The recipient of the award receives $1000. The award is presented in conjunction with the annual ceremony for the Presidential Green Chemistry Challenge Awards, held in Washington, D.C. in June.
Requirements: The award is open to undergraduate and graduate students, regardless of citizenship or country of study. Students who completed their degree prior to May 1 of the year before the application year are not eligible for the award.
Geographic Focus: All States
Date(s) Application is Due: Feb 1
Amount of Grant: 1,000 USD
Contact: Ron Siatkowski; (202) 872-6091; fax (202) 872-6319; r_siatkowski@acs.org
Internet: http://webapps.acs.org/findawards/detail.jsp?ContentId=WPCP_012551
Sponsor: American Chemical Society
1155 Sixteenth Street, NW
Washington, DC 20036-4801

American Chemical Society Local Section YCC Starter Grants 684
The Younger Chemists Committee (YCC) funds starter grants to help Local Section Younger Chemists Committees (LSYCCs) sponsor their first big event or launch a new program. Recipients receive a cash award to start a LSYCC; reports and share the program ideas with other LSYCCs; and opportunities to network with scientists in their area while engaging in fun activities with other chemists.
Requirements: Applications are reviewed year round. Creativity is encouraged.
Geographic Focus: All States
Amount of Grant: 250 USD
Contact: Alvin Collins, Ph.D; (800) 227-5558, ext. 8724 or (202) 872-8724; fax (202) 776-8003; a_collins@acs.org
Internet: http://portal.acs.org/portal/acs/corg/content?_nfpb=true&_pageLabel=PP_ARTICLEMAIN&node_id=120&content_id=WPCP_006959&use_sec=true&sec_url_var=region1
Sponsor: American Chemical Society
1155 Sixteenth Street, NW
Washington, DC 20036-4801

American Chemical Society Mentoring Grants 685
The American Chemical Society Committee on Minority Affairs offers minigrants of up to $500 to local sections to support their implementing programs that encourage the participation of underrepresented minority scientists.
Requirements: To find the local section in your area, go to http://membership.acs.org/y/ycc/lsycc/lsycc.htm.
Geographic Focus: All States
Amount of Grant: 500 USD
Contact: Brenda Thompson, (800) 227-5558, ext. 4065 or (202) 872-4065; fax (202) 776-8003; b_thompson@acs.org
Internet: http://portal.acs.org/portal/acs/corg/content?_nfpb=true&_pageLabel=PP_ARTICLEMAIN&node_id=120&content_id=WPCP_006959&use_sec=true&sec_url_var=region1
Sponsor: American Chemical Society
1155 Sixteenth Street, NW
Washington, DC 20036-4801

American Chemical Society Overcoming Challenges Award 686
The Women Chemists Committee of the American Chemical Society established this award to recognize a woman undergraduate from a two-year or four-year institution for her efforts in overcoming hardship to achieve success in chemistry. One award is offered annually. The awardee receives a monetary prize of $250 and up to $1,000 for travel expenses to attend the fall ACS national meeting. The awardee is presented with a plaque and recognized at the WCC Luncheon at the fall meeting. The current application deadline is April 1.
Requirements: To be considered a candidate, students must meet the following criteria: applicant must be a female undergraduate currently enrolled in a two-year chemistry-related program or pursuing a major or minor in a four-year chemistry program at a school that does not grant doctorates in chemistry; applicant must have completed one semester of college-level chemistry; applicant must demonstrate that she has overcome hardships (economic, personal, or academic) in pursuit of her education; and applicants are reviewed on: improvement, initiative, successes and grades from the previous two semesters (not cumulative grade point average).
Geographic Focus: All States
Date(s) Application is Due: Apr 1
Amount of Grant: 250-1,000 USD
Contact: Ron Siatkowski, Director, Office of Research Grants; (202) 872-6091; fax (202) 872-6319; r_siatkowski@acs.org
Internet: http://webapps.acs.org/findawards/detail.jsp?ContentId=WPCP_010370
Sponsor: American Chemical Society
1155 Sixteenth Street, NW
Washington, DC 20036-4801

American Chemical Society PCD Award in Experimental Physical Chemistry 687
The purpose of the Physical Chemistry Division's Award in Experimental Physical Chemistry is to recognize outstanding contributions in experimental physical chemistry. The recipient will receive a $2,000 honorarium, a plaque, and will be invited to present his/her work at the fall ACS meeting that immediately follows the announcement of the award. At that meeting, he/she will present an invited talk and receive the award at the annual reception sponsored by the PHYS Division and the Journal of Physical Chemistry. The Division will pay up to $1,500 to cover the travel, lodging, and meal expenses for the award winner while at that ACS meeting. The annual deadline is November 1.
Requirements: Eligibility is restricted to Physical Chemistry Division members who, at the time of the nomination, have not yet won a national award from a scientific society that is based on the nominee's scientific accomplishments. Members of the National Academy of Science are also ineligible, but fellowship in a professional society is not considered a national award in this context. Current serving members of the PHYS Division Executive Committee or Executive Committees of any of the Subdivisions of the PHYS Division are ineligible for nomination for this award until after their term of service.
Geographic Focus: All States
Date(s) Application is Due: Nov 1
Contact: Administrator; (202) 872-4575 or (202) 872-4408; awards@acs.org
Internet: http://webapps.acs.org/findawards/detail.jsp?ContentId=CNBP_030731
Sponsor: American Chemical Society
1155 Sixteenth Street, NW
Washington, DC 20036-4801

American Chemical Society PCD Award in Theoretical Chemistry 688
Sponsored by the Physical Chemistry Division, Telluride School on Theoretical Chemistry, the Award in Theoretical Chemistry's purpose is to recognize outstanding contributions in theoretical chemistry. The recipient will receive a $2,000 honorarium, a plaque, and will be invited to the next Telluride School on Theoretical Chemistry (TSTC), which are held every other summer. At that meeting, he/she will present a plenary lecture and receive the award. The TSTC will pay up to $1,500 to cover travel, lodging, and meal expenses for the Award winner while at the TSTC. The recipient will also be invited to present his/her work at the fall ACS meeting that immediately follows the announcement of the award. At that meeting, he/she will also be recognized at the annual reception sponsored by the PHYS Division and the Journal of Physical Chemistry. The Division will pay up to $1,500 to cover the travel, lodging, and meal expenses for the award winner while at that ACS meeting. The annual deadline is November 1.
Requirements: Eligibility is restricted to Physical Chemistry Division members who, at the time of the nomination, have not yet won a national award from a scientific society that is based on the nominee's scientific accomplishments. Members of the National Academy of Science are also ineligible, but fellowship in a professional society is not considered a national award in this context. Currently serving members of the PHYS Division Executive Committee or Executive Committees of any of the Subdivisions of the PHYS Division are ineligible for nomination for this award until after their term of service.
Geographic Focus: All States
Date(s) Application is Due: Nov 1
Amount of Grant: 2,000-3,500 USD
Contact: Administrator; (202) 872-4575 or (202) 872-4408; awards@acs.org
Internet: http://webapps.acs.org/findawards/detail.jsp?ContentId=CNBP_030732
Sponsor: American Chemical Society
1155 Sixteenth Street, NW
Washington, DC 20036-4801

American Chemical Society PCD Early Career Awards 689
Established in 2012, the purpose of the Physical Chemistry Division's Early Career Awards is to recognize outstanding contributions in experimental and theoretical physical chemistry by young investigators. Two awards will be given each year – one each in theoretical and experimental physical chemistry – to honor achievements by investigators in the early stages of their careers. Each recipient will receive a plaque and will be invited to present a lecture at the divisional award symposium held at the fall ACS meeting that immediately follows the announcement of the award. Award winners will be introduced at the annual reception sponsored by the PHYS Division and the Journal of Physical Chemistry. The Division will pay up to $1,500 to cover the travel, lodging, and meal expenses for the award winner while at that ACS meeting. Winners will also be invited to the Sunday evening dinner following the PHYS Executive Committee meeting. The annual deadline is November 1.
Requirements: Eligibility is restricted to Physical Chemistry Division members who, at the time of the nomination, are in the first ten years of their careers as independent researchers.
Geographic Focus: All States
Date(s) Application is Due: Nov 1
Amount of Grant: Up to 1,500 USD
Contact: Administrator; (202) 872-4575 or (202) 872-4408; awards@acs.org
Internet: http://webapps.acs.org/findawards/detail.jsp?ContentId=CNBP_030736
Sponsor: American Chemical Society
1155 Sixteenth Street, NW
Washington, DC 20036-4801

American Chemical Society Petroleum Research Fund Doctoral New Investigator Grants 690
The goals of the American Chemical Society Petroleum Research Fund are: the support of fundamental research in the petroleum and energy fields; and development of the next generation of engineers and scientists through support of advanced scientific education. The emphasis of the DNI grants is focused on providing start-up funding for scientists and engineers who are within the first three years of their first academic appointment, those who have limited or no preliminary results for a research project they wish to pursue, with the intention of using the preliminary results obtained to seek continuation funding from other agencies. An estimated 90 awards will be given each year. Proposals are considered at three advisory board meetings per year (early fall, mid-winter, and late spring), which are coordinated with the ACS Grants and Awards board meetings. The amount of grants will be a maximum of $100,000 over two years, and the average number of awards is 90 annually.
Requirements: Regularly-appointed faculty members at U.S. academic institutions are eligible to apply. Research or adjunct faculty holding non-tenured, long-term scientific appointments also may be eligible. Current eligibility criteria are: the non-profit institution submitting the DNI proposal must certify that the individual listed as a principal investigator on the cover page qualifies as a principal investigator under the institution's policies; each principal investigator must be eligible to serve as the formal, official supervisor of the students for whom he/she is seeking support; the terms of appointment of the principal investigator must promise reasonable continuity of service. The appointment should continue at least through the period of funding requested in the proposal.
Restrictions: Although a PI may send the same proposal to more than one agency, PRF will not support a project having overlap, or partial overlap, with research funded by another agency. New investigators may submit only one research proposal for consideration per meeting and may not hold more than one active PRF research grant at a time. New investigators may have only three (3) DNI grant applications considered in their career. Thereafter, the new investigator may apply to the New Directions (ND) Grant Program. A principal investigator with an active PRF research grant, including a grant on time extension, may not submit an application for a new grant. No overhead costs may be charged, which includes secretarial and/or administrative salaries. Funds may not be used to support laboratory technicians, contractors, consultants, or visiting faculty.
Geographic Focus: All States
Date(s) Application is Due: Mar 16
Amount of Grant: 100,000 USD
Contact: Ron Siatkowski, Director, Office of Research Grants; (202) 872-6091; fax (202) 872-6319; r_siatkowski@acs.org
Internet: http://portal.acs.org/portal/acs/corg/content?_nfpb=true&_pageLabel=PP_SUPERARTICLE&node_id=1264&use_sec=false&sec_url_var=region1&__uuid=95d2e453-63db-4752-b1bc-abbd6a0ff8df
Sponsor: American Chemical Society
1155 Sixteenth Street, NW
Washington, DC 20036-4801

American Chemical Society Petroleum Research Fund New Directions Grants 691
The New Directions (ND) grants program provides funds to scientists and engineers with limited—or even no—preliminary results for a research project they wish to pursue, who intend to use the PRF-driven preliminary results to seek continuation funding from other agencies. ND grants are to be used to illustrate proof of concept/feasibility. Accordingly, they are to be viewed as seed money for new research ventures. The amount of grants will be a maximum of $100,000 over two years, and the average number of awards is 100 annually.
Requirements: The lead principal investigator must demonstrate that this is an entirely new research direction from what the lead principal investigator has done to date. Proposals that are considered to not be a new direction of research will be denied without external review, or may be rejected. Regularly-appointed faculty members at academic institutions in countries where ACS PRF can administer grants are eligible. In addition, certain other long-term scientific appointments may be eligible. Current eligibility criteria are: (a) The non-profit institution submitting the ND proposal must certify that each individual listed as a principal investigator on the cover page qualifies as a principal investigator under the institution's Principal Investigator policies. (b) In view of the long-standing policy of The Petroleum Research Fund to give priority to support of students (undergraduate, graduate, or postdoctoral), each principal investigator (lead and co-PI) must be eligible to serve as the formal, official supervisor of the students for whom he/she is seeking support. (c) The terms of appointment of the principal investigator (and if appropriate the co-PI) must promise reasonable continuity of service through the life of the grant. Proposals are due by 5:00 pm of the published deadline date.
Restrictions: Principal investigators (lead or co-) may submit only one research proposal per meeting. Principal investigators may not hold more than one active PRF research grant at a time. Principal investigators with an active PRF research grant, including a grant on time extension, may not submit an application for a new grant. Principal investigators may have only one ND grant application considered in a 12-month period. Therefore, a principal investigator who has an ND proposal denied may not submit another ND proposal until the Advisory Board meeting one year later. No overhead costs may be charged, which includes secretarial and/or administrative salaries. Funds may not be used to support laboratory technicians, contractors, consultants, or visiting faculty.
Geographic Focus: All States
Date(s) Application is Due: Mar 16
Contact: Ron Siatkowski, Director, Office of Research Grants; (202) 872-6091; fax (202) 872-6319; r_siatkowski@acs.org
Internet: http://portal.acs.org/portal/acs/corg/content?_nfpb=true&_pageLabel=PP_SUPERARTICLE&node_id=1262&use_sec=false&sec_url_var=region1&__uuid=64a4c12d-a548-4e81-a299-89007bd0aeb3
Sponsor: American Chemical Society
1155 Sixteenth Street, NW
Washington, DC 20036-4801

American Chemical Society Petroleum Research Scientific Education Grants 692
The program assists with the travel expenses of foreign (defined as non-U.S. and non-Canadian) scientists who participate in major symposia in this country (or in Canada). The symposium topic must be fundamental research in the petroleum field to be eligible for support. The PRF Advisory Board makes recommendations on Type SE proposals during meetings three times per year: September, January, and May. Funding for foreign travel assistance is limited to $1,500 per foreign speaker with a maximum of $4,500 per symposium (three or more visiting speakers). Download the application packet from the sponsor's website.
Requirements: Researchers at colleges and universities are eligible. Type SE proposals for ACS symposia should be written on behalf of an ACS division, rather than the principal investigator's institutional affiliation; proposals for support of non-ACS symposia should be written on behalf of the institution organizing the symposium. In order to be eligible, researchers must meet the following *Requirements:* the symposium topic must be 'fundamental research in the petroleum field'; each eligible speaker must be from a foreign (non-U.S. and non-Canadian) non-profit institution, and must give a petroleum-relevant or alternate-energy presentation; the sponsor of the symposium must be a non-profit scientific or educational institution; the meeting or symposium must be accessible and open; symposia whose attendance is by invitation only, or where facilities sharply limit attendance have not been favored for support by the PRF Advisory Board; and the results of the meeting or symposium must be publishable in peer-reviewed literature.
Restrictions: The program does not support travel of American scientists to similar meetings in foreign countries, on the rationale that more American scientists derive benefit from having foreign speakers travel to North America, and that support helps American organizations to enhance the value of their programs and, in some instances, to meet their obligations as hosts for international meetings.

Geographic Focus: All States
Amount of Grant: 1,500-4,500 USD
Contact: Ron Siatkowski; (202) 872-6091; fax (202) 872-6319; r_siatkowski@acs.org
Internet: http://portal.acs.org/portal/acs/corg/content?_nfpb=true&_pageLabel=PP_SUPERARTICLE&node_id=1265&use_sec=false&sec_url_var=region1&__uuid=7e98ac5c-1769-493b-ba5b-086690c3c85d
Sponsor: American Chemical Society
1155 Sixteenth Street, NW
Washington, DC 20036-4801

American Chemical Society Petroleum Research Fund Undergraduate Faculty Sabbatical Grants 693

The Undergraduate Faculty Sabbatical grant program is designed to provide year-long, full-time, research sabbaticals for qualified faculty performing fundamental research in the petroleum field during a sabbatical. This is a matching grants program that can provide up to $45,000 in faculty salary and qualified benefits, and up to $5,000 in relocation and research expenses for a faculty member's sabbatical year. The maximum individual award is $50,000. Applications are considered each year at the September PRF Advisory Board meeting.
Requirements: An eligible sabbatical must be at least one academic year (i.e., 9 months or more) and must be carried out in its entirety at a location other than the applicant's home institution. Sabbaticals may be taken anywhere in the world. Academic, national lab, and industrial sabbaticals are eligible, as are sabbaticals that involve more than one institution. Faculty with tenured appointments in departments that do not offer a doctoral degree and faculty at two- and four-year institutions are eligible to apply. Current ACS PRF grant recipients, as well as those with pending PRF proposals are also eligible. Complete eligibility requirements and application packet can be downloaded from the sponsor's website.
Restrictions: Fundamental research is required. ACS PRF cannot support applied research or methods development. Note also that the proposed research need not be directly connected to petroleum, but rather may provide a basis for subsequent research directly connected with the petroleum field. No overhead or indirect costs may be charged to the grant.
Geographic Focus: All States
Amount of Grant: Up to 50,000 USD
Contact: Ron Siatkowski; (202) 872-6091; fax (202) 872-6319; r_siatkowski@acs.org
Internet: http://portal.acs.org/portal/acs/corg/content?_nfpb=true&_pageLabel=PP_SUPERARTICLE&node_id=1266&use_sec=false&sec_url_var=region1&__uuid=b9dc3d16-4e42-41c6-b235-86d242b3a346
Sponsor: American Chemical Society
1155 Sixteenth Street, NW
Washington, DC 20036-4801

American Chemical Society Petroleum Research Fund Undergraduate New Investigator Grants 694

The goals of the American Chemical Society Petroleum Research Fund are: to support fundamental research in the petroleum field; and to develop the next generation of engineers and scientists through support of advanced scientific education. The Undergraduate New Investigator Grants program is intended to initiate the research program of new scientists and engineers who are faculty members at undergraduate research institutions and to provide financial incentives for students at those institutions to become involved in research activities leading to employment or continued study in graduate school. These grants provide funds for scientists and engineers who are beginning their independent careers in academia and have limited or no preliminary results for a research project they wish to pursue. The UNI grants are to be used to illustrate proof of principle, i.e., feasibility, and accordingly, are to be viewed as seed money for generating preliminary results that can be used to apply for continuation funding from other agencies. The amount of grants will be a maximum of $50,000 over two years, and the average number of awards is 45 annually.
Requirements: Eligibility for a UNI grant requires that a PI is in a department without a doctoral program, and that the students receiving stipends for the work to be done are undergraduates (M.S.-level students can also be supported IF one or more undergraduates are also supported from this grant). Accordingly, the research being proposed need not be high risk but it should be of publishable quality. The research opportunities afforded must be of the highest caliber, and provide a compelling educational experience for the student. To be eligible as a principal investigator for a UNI grant, the applicant must be a member of the faculty of a college or university within the United States; be within the first three years of their first academic appointment as a regular faculty member; have completed all requirements for the Ph.D. (however, an application may be submitted before the degree has been awarded); and be appointed at the rank of Assistant Professor or the equivalent. In addition, applicants must meet the following three criteria: (a) The non-profit institution submitting the UNI proposal must certify that the person listed as a principal investigator on the cover page qualifies as a principal investigator under the institution's policies. (b) In view of the long-standing policy of The Petroleum Research Fund to give priority to support of students, the principal investigator must be eligible to serve as the formal, official supervisor of the students for whom he/she is seeking support. (c) The term of appointment of the principal investigator must promise reasonable continuity of service. The appointment should continue at least through the period of funding requested in the UNI proposal.
Restrictions: A principal investigator may submit only one research proposal per meeting. Only one UNI grant may be held during an investigator's career.
Geographic Focus: All States
Date(s) Application is Due: Mar 16
Amount of Grant: 50,000 USD
Contact: Ron Siatkowski; (202) 872-6091; fax (202) 872-6319; r_siatkowski@acs.org
Internet: http://portal.acs.org/portal/acs/corg/content?_nfpb=true&_pageLabel=PP_SUPERARTICLE&node_id=1796&use_sec=false&sec_url_var=region1&__uuid=30190564-11ef-4b06-a76b-71cdf4d4c22c
Sponsor: American Chemical Society
1155 Sixteenth Street, NW
Washington, DC 20036-4801

American Chemical Society Petroleum Research Fund Undergraduate Research Grants 695

The goals of the American Chemical Society Petroleum Research Fund are: to support fundamental research in the petroleum field; and to develop the next generation of engineers and scientists through support of advanced scientific education. The Undergraduate Research (UR) grants program supports the research programs of established scientists and engineers at non-doctoral departments and provides financial support for students at those institutions to become involved in advanced research activities, in preparation for continued study in graduate school or employment. Demonstration of productivity is important, but a UR grant may also be used for a project with limited or no preliminary results in a new research area the PI wishes to pursue, with the intention of using the preliminary results obtained to seek continuation funding from other agencies. The amount of each grant over a three-year period is $65,000, and the estimated number of awards is 45 annually.
Requirements: Eligibility for a UR grant requires that a PI is in a department without a doctoral program, and that undergraduates are involved in the project. Investigators from Master's degree-granting departments are eligible, and support can be provided for M.S. level research, but undergraduates must be included in the project. Accordingly, the research being proposed need not be comparable in volume and scope of research in doctoral settings, but it should be of publishable quality, and the research opportunities afforded the students must be of the highest caliber.
Restrictions: A principal investigator may have only one research proposal active or under consideration per meeting. This limitation extends to those who have co-PI status on an application. A PI must close an active grant, after the full grant period has been completed, before submitting another proposal.
Geographic Focus: All States
Date(s) Application is Due: Mar 16
Amount of Grant: Up to 65,000 USD
Contact: Ron Siatkowski; (202) 872-6091; fax (202) 872-6319; r_siatkowski@acs.org
Internet: http://portal.acs.org/portal/acs/corg/content?_nfpb=true&_pageLabel=PP_SUPERARTICLE&node_id=1263&use_sec=false&sec_url_var=region1&__uuid=b499d85c-9121-440b-bbec-7a0976442e2e
Sponsor: American Chemical Society
1155 Sixteenth Street, NW
Washington, DC 20036-4801

American Chemical Society Presidential Green Chemistry Challenge Awards 696

The Office of Pollution Prevention and Toxics of the United States Environmental Protection Agency sponsors and administers these awards and the American Chemical Society (ACS Green Chemistry Institute) convenes the panel of technical experts to evaluate the nominations. The Award's purpose is to recognize outstanding chemical technologies that incorporate the principles of green chemistry into chemical design, manufacture, and use. The program invites nominations that describe the science and innovation of a green chemistry technology, its human health and environmental benefits, and its applicability and impact. The five award categories are: Greener Synthetic Pathways; Greener Reaction Conditions; Designing Greener Chemicals; Academic; and Small Business. The recipient of the award is recognized at an awards ceremony in Washington, D.C., typically held during June. A crystal sculpture will be presented to the primary sponsor(s) of the winning green chemistry technology in each of the five award categories. Certificates will be presented to individuals (as identified by the primary sponsor) who contributed to the research, development, or implementation of the chemistry.
Requirements: Nominations are accepted from individuals, groups, non-profit and for-profit organizations, academia, small business, and industry. The nominated green chemistry technology must have reached a significant milestone within the past five years (i.e., it must been researched, demonstrated, implemented, applied, patented, etc.). It must also have a significant component within the United States.
Geographic Focus: All States
Date(s) Application is Due: Dec 31
Contact: Ron Siatkowski; (202) 872-6091; fax (202) 872-6319; r_siatkowski@acs.org
Internet: http://webapps.acs.org/findawards/detail.jsp?ContentId=WPCP_012553
Sponsor: American Chemical Society
1155 Sixteenth Street, NW
Washington, DC 20036-4801

American Chemical Society Travel Grants for Chemists with Disabilities 697

The Committee on Chemists with Disabilities can assist students and postdoctoral researchers with disabilities with Travel Grant funding to defray some or all of the additional travel and lodging expenses to attend scientific meetings. Recipients receive: up to $2,500 to cover additional travel costs often incurred when attending technical meetings/conferences by people with a disability; and, opportunities to build lasting professional relationships through networking.
Geographic Focus: All States
Amount of Grant: Up to 2,500 USD
Contact: Tony Scurry, (800) 227-5558, ext. 6262 or (202) 872-6262; w_scurry@acs.org

Internet: http://portal.acs.org/portal/acs/corg/content?_nfpb=true&_pageLabel=PP_ARTICLEMAIN&node_id=120&content_id=CTP_005836&use_sec=true&sec_url_var=region1
Sponsor: American Chemical Society
1155 Sixteenth Street, NW
Washington, DC 20036-4801

American Chemical Society WCC Rising Star Award 698

The ACS Women Chemists Committee (WCC) serves the membership of the American Chemical Society with its mission to be leaders in attracting, retaining, developing, promoting, and advocating for women in the chemical sciences. Established in 2011, the WCC Rising Star Award recognizes exceptional early to mid-career women chemists across all areas of chemistry on a national level. To be given annually, this award is also intended to help promote retention of women in science. The Award includes the following: an opportunity to present contributions at a WCC-sponsored symposium at the spring ACS national meeting; a $1,000 stipend to cover spring national meeting travel expenses; and visibility and networking opportunities at the spring national meeting—awardees are recognized at the WCC Just Cocktails reception and at the WCC Luncheon, where awardees are seated at the head tables along with ACS governance, WCC members, and fellow award winners. The deadline for receipt of nominations is June 15.
Requirements: The award is open to all female ACS members in chemistry and chemical engineering working in academic, industrial, government, non-profit or other employment sectors. Appropriate candidates will typically be no more than 15 years from receipt of their terminal scientific degree and have demonstrated outstanding promise for contributions to their respective fields. Applicants can either be self-nominated or can be nominated by another individual for this award.
Restrictions: The following are not eligible: those who are already widely acknowledged and recognized in their area of chemistry or chemical engineering and are at the pinnacle of their careers; ACS national award winners; those who have received a prior award under this program; or members of the Women Chemists Committee.
Geographic Focus: All States
Date(s) Application is Due: Jun 15
Contact: Ron Siatkowski; (202) 872-6091; fax (202) 872-6319; r_siatkowski@acs.org
Internet: http://webapps.acs.org/findawards/detail.jsp?ContentId=CNBP_029245
Sponsor: American Chemical Society
1155 Sixteenth Street, NW
Washington, DC 20036-4801

American College of Bankruptcy Grants 699

The college awards grants for existing bankruptcy or consumer-debtor programs, programs in development, and research projects to improve the delivery of pro bono bankruptcy services. Grants also support organizations for broad educational purposes. There are no application guidelines. Guidelines and application are available online.
Geographic Focus: All States
Date(s) Application is Due: Mar 1
Amount of Grant: 1,000-10,000 USD
Contact: Shari Bedker, Executive Director; (703) 934-6154; fax (703) 802-0207; sbedker@amercol.org or college@amercol.org
Internet: http://www.amercol.org/acb_foundation.cfm
Sponsor: American College of Bankruptcy
PMB 626A, 11350 Random Hills Road, Suite 800
Fairfax, VA 22030-6044

American College of Rheumatology Fellows-in-Training Travel Scholarships 700

The American College of Rheumatology Fellows-in-Training (FIT) Travel Scholarship is designed to provide funding to assist Fellows- In-Training at U.S. and Canadian programs in rheumatology attend the Annual Meeting. Preference will be given to those applicants who are currently a first or second year adult or first, second or third year pediatric fellows in a training program approved by the Accreditation Council of Graduate Medical Education or the Royal College of Physicians and Surgeons of Canada. Applicants in a third year of an adult or fourth year of a pediatric fellowship will be given consideration after the registration deadline, space permitting on a first come first serve basis.
Requirements: Applicant must be first or second year adult or first, second or third year pediatric fellows in a training program approved by the Accreditation Council of Graduate Medical Education or the Royal College of Physicians and Surgeons of Canada. Applicants must be in an accredited fellowship position at the time of the Annual Meeting.
Restrictions: Individuals with a faculty appointment will not be considered.
Geographic Focus: All States
Date(s) Application is Due: Jul 24
Contact: Coordinator; (404) 633-3777; fax (404) 633-1870; fittravel@rheumatology.org
Internet: http://www.rheumatology.org/fellows/fitscholarship.asp
Sponsor: American College of Rheumatology
1800 Century Place, Suite 250
Atlanta, GA 30345-4300

American College of Surgeons and The Triological Society Clinical Scientist Development Awards 701

The Triological Society (TRIO) and the American College of Surgeons (ACS) announce a competitive research career development program that will provide supplemental funding to otolaryngologist-head and neck surgeons who receive NIH Mentored Clinical Scientist Development Awards (K08) or Mentored Patient-Oriented Research Development Awards (K23). The TRIO and ACS are offering these awards as a means to facilitate the research career development of otolaryngologists-head and neck surgeons, with the expectation that awardees will have sufficient pilot data to submit a competitive R01 proposal prior to the conclusion of the K awards. Awards will use the K08 or K23 mechanisms. Planning, direction, and execution of the program will be the responsibility of the candidate and her/his mentor on behalf of the applicant institution. The project period will be for a period of up to five years with a minimum of three years. Awards are not renewable. TRIO and ACS will provide salary and fringe benefits for the NIH K08/K23 recipient. A combined total of $80,000 per year in direct costs will be awarded to the institution of the awardee.
Requirements: Applicants must be MD otolaryngologist-head and neck surgeons and members of the American College of Surgeons. Applicants must meet all requirements as set forth in the K08 and K23 application criteria and must be recipients of a new K08 or K23 award or have an existing K08/K23 award with a minimum of three years remaining in the funding period as of June.
Restrictions: Pre-existing applications and awards are not eligible for consideration.
Geographic Focus: All States
Date(s) Application is Due: Apr 15
Amount of Grant: 80,000 USD
Contact: Kate Early; (312) 202-5281; fax (312) 202-5021; kearly@facs.org
Internet: http://www.facs.org/memberservices/acstriol.html
Sponsor: American College of Surgeons
633 N Saint Clair Street
Chicago, IL 60611-3211

American College of Surgeons Co-Sponsored K08/K23 NIH Supplement Awards 702

The American College of Surgeons announces a program that will provide supplemental funding to up to five individuals who receive a Mentored Clinical Scientist Development Award (K08/K23). It is directed at surgeon-scientists working in the early stages of their research careers. The award requires co-sponsorship with an approved surgical society of a three, four or five year period of supervised research experience that may integrate didactic studies with laboratory or clinically based research. This award program will offer a means to facilitate the career development of individuals pursuing careers in surgical research by enhancing salary support over and above that offered by the K08/K23 mechanism. The application deadline is June 12, with funding to begin July 1.
Requirements: Awardees must be members in good standing of both the College and a co-sponsoring surgical society. Participating surgical societies include American Association of Plastic Surgeons (AAPS), American Head and Neck Society (AHNS), American Society of Transplant Surgeons (ASTS), American Vascular Association (AVA), Society of Gynecologic Oncologists (SGO), Society of University Surgeons (SU. S.), and Thoracic Surgery Foundation for Research and Education (TSFRE).
Geographic Focus: All States
Date(s) Application is Due: Jun 12
Contact: Kate Early; (312) 202-5281; fax (312) 202-5021; kearly@facs.org
Internet: http://www.facs.org/memberservices/acs-nih.html
Sponsor: American College of Surgeons
633 N Saint Clair Street
Chicago, IL 60611-3211

American College of Surgeons Faculty Career Development Award for Neurological Surgeons 703

The American College of Surgeons (ACS) and the Neurosurgery Research and Education Foundation of the American Association of Neurological Surgeons (NREF-AANS) are offering a two-year faculty career development award to neurological surgeons. The award is to support the establishment of a new and independent research program in an area of neurological surgery. The award is $40,000 per year for each of two years, to support the research, and is not renewable thereafter.
Requirements: The award is open to surgeons who are members or candidate members in good standing of both the American College of Surgeons (ACS) and the American Association of Neurological Surgeons (AANS); and have completed specialty training within the preceding five years and have received a full-time faculty appointment at a medical school accredited by the Liaison Committee on Medical Education in the United States or by the Committee for Accreditation of Canadian Medical Schools in Canada.
Restrictions: Applicants may not be current recipients of major research grants.
Geographic Focus: All States
Date(s) Application is Due: Dec 1
Amount of Grant: 40,000 USD
Contact: Michele S. Gregory, Director of Development; (847) 378-0540 or (847) 378-0500; fax (847) 378-0600; msg@aans.org or info@aans.org
Internet: http://www.facs.org/memberservices/acsnrefaans07.pdf
Sponsor: American College of Surgeons
633 N Saint Clair Street
Chicago, IL 60611-3211

American College of Surgeons Faculty Research Fellowships 704

The American College of Surgeons is offering two-year faculty research fellowships to surgeons entering academic careers in surgery or a surgical specialty. The fellowship is to assist a surgeon in the establishment of a new and independent research program. Applicants should have demonstrated their potential to work as independent investigators. The fellowship award is $40,000 per year for each of two years, to support the research.
Requirements: The fellowship is restricted to surgeons who have recently (usually within three years) completed formal surgical education and entered the field of full-time academic surgery and received a faculty appointment in a department of surgery, or one of the surgical specialties, at an approved medical school in the United States or Canada. Preference will be given to applicants who directly enter academic surgery following

residency or fellowships. Approval of the application is required from the administration (dean or fiscal officer) and the head of the department under whom the recipient will be studying during the fellowship.
Geographic Focus: All States
Date(s) Application is Due: Nov 2
Contact: Kate Early, (312) 202-5281; fax (312) 202-5021; kearly@facs.org
Internet: http://www.facs.org/memberservices/acsfaculty.html
Sponsor: American College of Surgeons
633 N Saint Clair Street
Chicago, IL 60611-3211

American College of Surgeons Resident Research Scholarships 705
ACS scholarships will be awarded each year to encourage residents to pursue careers in academic surgery. Priority will be given to residents beginning full-time investigative activities, with no clinical responsibilities, for the two-year period of the scholarship. Study outside the United States or Canada is permissible. The scholarship is for the personal use of the recipient and is not to diminish or replace the usual or expected compensation; award is made directly to the scholar and not to the institution. Renewal for the second year is contingent upon acceptable progress and study protocol for the second year.
Requirements: The applicant must have completed two postdoctoral years in an accredited surgical training program in the United States or Canada at the time the scholarship is awarded and shall not complete formal residency training before the end of the scholarship. Approval of the application is required from the administration (dean or fiscal officer) and the head of the department under whom the recipient will be studying.
Restrictions: Only in exceptional circumstances will more than one ACS scholarship be granted in a single year to applicants from the same institution.
Geographic Focus: All States
Date(s) Application is Due: Sep 1
Contact: Administrator; (800) 621-4111 or (312) 202-5000; fax (312) 202-5001
Internet: http://www.facs.org/memberservices/acsresident.html
Sponsor: American College of Surgeons
633 N Saint Clair Street
Chicago, IL 60611-3211

American College of Surgeons Traveling Fellowship to Germany 706
The International Relations Committee of the American College of Surgeons announces the availability of the ACS Traveling Fellowship to Germany. The purpose of this fellowship is to encourage international exchange of surgical scientific information. The ACS Traveling Fellow will visit Germany, and a German Traveling Fellow will visit North America.
Requirements: The scholarship is available to a Fellow of the American College of Surgeons in any of the surgical specialties who meets the following *Requirements:* has a major interest and accomplishment in clinical and basic science related to surgery; holds a current full-time academic appointment in Canada or the United States; is under 45 years of age on the date the application is filed; and is enthusiastic and personable and possesses good communication skills. The Fellow is required to spend a minimum of two weeks in Germany.
Geographic Focus: All States
Date(s) Application is Due: Apr 1
Contact: Kate Early, (312) 202-5281; fax (312) 202-5021; kearly@facs.org
Internet: http://www.facs.org/memberservices/acsgermany.html
Sponsor: American College of Surgeons
633 N Saint Clair Street
Chicago, IL 60611-3211

American College of Surgeons Traveling Fellowship to Japan 707
The International Relations Committee of the American College of Surgeons announces the availability of the ACS Traveling Fellowship to Japan. The purpose of this fellowship is to encourage international exchange of surgical scientific information. The ACS Traveling Fellow will visit Japan, and a Japanese Traveling Fellow will visit North America. The College will provide the sum of $7,500 to the successful applicant, who will also be exempted from registration fees for the annual meeting of the Japan Surgical Society.
Requirements: The scholarship is available to a Fellow of the American College of Surgeons in any of the surgical specialties who meets the following *Requirements:* has a major interest and accomplishment in clinical and basic science related to surgery; holds a current full-time academic appointment in Canada or the United States; is under 45 years of age on the date the application is filed; and is enthusiastic and personable and possesses good communication skills. The Fellow is required to spend a minimum of two weeks in Japan.
Geographic Focus: All States
Date(s) Application is Due: Jun 1
Contact: Kate Early, (312) 202-5281; fax (312) 202-5021; kearly@facs.org
Internet: http://www.facs.org/memberservices/acsjapan.html
Sponsor: American College of Surgeons
633 N Saint Clair Street
Chicago, IL 60611-3211

American College of Surgeons Wound Care Management Award 708
The American College of Surgeons is offering a two-year faculty research award through KCI U.S.A, to a general surgeon engaged in a research project addressing wound care management. The purpose of this fellowship will be to acquire knowledge leading to new clinical applications or projects that will provide the medical community with a better understanding of the use of advanced wound healing therapies. The fellowship award is $85,000 per year, with the possibility of an extension for a second year if satisfactory progress is made. Specifically, translational projects such as new methods to mechanically stimulate wound healing, design of clinically applicable skin substitutes, methods to enhance specific, difficult-to-heal wounds related to general or extremity trauma, various soft-tissue injuries, quality limb salvage, or methods to reduce surgical intervention to heal a variety of wounds, are being sought. This award may be used by the recipient for support of his/her research or academic enrichment in any fashion that the recipient deems maximally supportive of his/her investigations. This may include faculty salary replacement, support for clinical research personnel, supplies directly related to the research activity, consumable equipment, and research related travel costs up to $2,000 per year.
Requirements: The fellowship is open to Fellows or Associate Fellows of the College who have: (1) completed the chief residency year or accredited fellowship training within the preceding 10 years; and (2) received a fulltime faculty appointment in a department of surgery or a surgical specialty at a medical school accredited by the Liaison Committee on Medical Education in the United States or by the Committee for Accreditation of Canadian Medical Schools in Canada. Preference will be given to applicants who directly enter academic surgery following residency or fellowship.
Geographic Focus: All States
Date(s) Application is Due: May 1
Amount of Grant: 85,000 USD
Contact: Administrator; (800) 621-4111 or (312) 202-5000; fax (312) 202-5001
Internet: http://www.facs.org/memberservices/woundcareposter2008.pdf
Sponsor: American College of Surgeons
633 N Saint Clair Street
Chicago, IL 60611-3211

American Family Insurance Corporate Giving Grants 709
The corporation awards grants to nonprofits in its areas of interest, including education and youth, arts and culture, health, and human services.
Requirements: Nonprofit organizations and K-12 and higher education institutions located in the companies area of interest.
Restrictions: Individuals or teams and organizations outside the 18-state operating territory will not be supported.
Geographic Focus: All States
Amount of Grant: Up to 2,000,000 USD
Contact: Community Relations Manager; (608) 249-2111; info@amfam.com
Internet: http://www.amfam.com/company/commitment_contribution.asp
Sponsor: American Family Insurance Corporation
6000 American Parkway
Madison, WI 53783

American Foodservice Charitable Trust Grants 710
American Foodservice Charitable Trust provides funding in the following areas of interest: Christian agencies & churches; community/economic development; higher education; hospitals (general) and; youth development. The types of support available are: general/operating support; program development; scholarship funds. Giving is primarily available in the areas of: King of Prussia, Pennsylvania; Thomasville, Georgia; and Fort Worth, Texas.
Requirements: Qualifying IRA 501(c)3 nonprofit organizations are eligible to apply. There's no application form nor is there a deadline date to adhere to when submitting a proposal to the Foundation.
Restrictions: No grants to individuals.
Geographic Focus: Georgia, Pennsylvania, Texas
Amount of Grant: 200-8,000 USD
Contact: Richard S. Downs, Trustee; (610) 933-9792
Sponsor: American Foodservice Charitable Trust
860 First Avenue, Suite 9A
King of Prussia, PA 19406-1404

American Historical Association Albert B. Corey Prize in Canadian-American Relations 711
The prize is awarded biennially in even-numbered years to an American or Canadian resident for the best book on the history of Canadian-American relations or on the history of both countries. The prize is awarded jointly by the Canadian Historical Association and the American Historical Association. Books bearing an imprint during the previous two years prior to the prize year are eligible.
Requirements: Books bearing an imprint of the previous two years are eligible for the prize.
Geographic Focus: All States
Date(s) Application is Due: May 15
Contact: Julia Brookins, Special Projects Coordinator; (202) 544-2422, ext. 119; fax (202) 544-8307; jbrookins@theaha.org or aha@theaha.org
Internet: http://www.historians.org/prizes/index.cfm?PrizeAbbrev=COREY
Sponsor: American Historical Association
400 A Street SE
Washington, DC 20003

American Historical Association Albert J. Beveridge Award 712
The Award is given annually for the best book in English on the history of the United States, Latin America, or Canada from 1492 to the present. Books that employ new methodological or conceptual tools or that constitute a significant reinterpretation of an important historical problem are given preference in the awarding of this prize. Biographies, monographs, and works of synthesis or interpretation are eligible.
Requirements: U.S. and Canadian citizens and permanent residents are eligible. Books published after May 1 of the previous year and before April 30 of the award year are eligible.

Restrictions: Translations, anthologies, and collections of documents are not eligible.
Geographic Focus: All States
Date(s) Application is Due: May 15
Contact: Julia Brookins, Special Projects Coordinator; (202) 544-2422, ext. 119; fax (202) 544-8307; jbrookins@theaha.org or aha@theaha.org
Internet: http://www.historians.org/prizes/index.cfm?PrizeAbbrev=Beveridge
Sponsor: American Historical Association
400 A Street SE
Washington, DC 20003

American Historical Association Albert J. Beveridge Grants for Research in the History of the Western Hemisphere 713

The Association announces the availability of modest grants to support research in the history of the Western hemisphere (United States, Canada, and Latin America). The funds for this program come from the earnings of the Albert J. Beveridge Memorial Fund. The grants are intended to further research in progress and may be used for travel to a library or archive, for microfilms, photographs, or photocopying-a list of purposes that is meant to be merely illustrative, not exhaustive (other expenses, such as child care, can be included). Of particular interest are those applicants with specific research needs, such as the completion of a project or completion of a discrete segment thereof. The annual deadline is February 15.
Requirements: Preference will be given to Ph.D. candidates and junior scholars.
Restrictions: Only members of the Association are eligible. Within a five-year period, no individual is eligible to receive more than a combined total of $1,000 from the Beveridge, Kraus, and Littleton-Griswold grant programs.
Geographic Focus: All States
Date(s) Application is Due: Feb 15
Amount of Grant: Up to 1,000 USD
Contact: Julia Brookins, Special Projects Coordinator; (202) 544-2422, ext. 119; fax (202) 544-8307; jbrookins@theaha.org or aha@theaha.org
Internet: http://www.historians.org/prizes/BeveridgeGrantInfo.htm
Sponsor: American Historical Association
400 A Street SE
Washington, DC 20003

American Historical Association Award for Scholarly Distinction 714

In 1984 the AHA Council established the American Historical Association Award for Scholarly Distinction to honor senior historians in the United States. According to the selection criteria of this Award, recipients must be senior historians of the highest distinction who have spent the bulk of their professional careers in the United States. Generally, they must also be of emeritus rank, if from academic life, or equivalent standing otherwise. Under normal circumstances the award is not intended to go to former presidents of the Association; rather, the intent is to honor persons not otherwise recognized by the profession to an extent commensurate with their contributions.
Geographic Focus: All States
Date(s) Application is Due: Apr 30
Contact: Sharon K. Tune, Assistant Director; (202) 544-2422, ext. 101; fax (202) 544-8307; stune@theaha.org or aha@theaha.org
Internet: http://www.historians.org/prizes/ScholarlyDistinction.htm
Sponsor: American Historical Association
400 A Street SE
Washington, DC 20003

American Historical Association Bernadotte E. Schmitt Research Grants in European, African, or Asian History 715

Modest grants support research in the history of Europe, Africa, and Asia. The funds for this program come from the earnings of a bequest from Bernadotte E. Schmitt, president of the Association in 1960. The grants are intended to further research in progress and may be used for: travel to a library or archive; microfilms, photographs, or photocopying; coding and keypunching; and other approved research-related activities.
Requirements: Preference will be given to those with specific research needs, such as the completion of a project or completion of a discrete segment thereof. Preference will be given to Ph.D. candidates and junior scholars.
Restrictions: Only members of the Association are eligible to apply.
Geographic Focus: All States
Date(s) Application is Due: Feb 15
Amount of Grant: Up to 1,000 USD
Contact: Julia Brookins, Special Projects Coordinator; (202) 544-2422, ext. 119; fax (202) 544-8307; jbrookins@theaha.org or aha@theaha.org
Internet: http://www.historians.org/prizes/SchmittGrantInfo.htm
Sponsor: American Historical Association
400 A Street SE
Washington, DC 20003

American Historical Association Clarence H. Haring Prize 716

The Clarence H. Haring Prize is awarded every five years to a Latin American who, in the opinion of the committee, has published the most outstanding book on Latin American history during the preceding five years ending June 1 of the award year. There is no language limitation on works submitted. There is no language limitation on works submitted.
Requirements: Books published between May 1, 2011, and April 30, 2016, will be eligible for the prize in 2016.
Geographic Focus: All States
Date(s) Application is Due: May 16
Contact: Julia Brookins, Special Projects Coordinator; (202) 544-2422, ext. 119; fax (202) 544-8307; jbrookins@theaha.org or aha@theaha.org
Internet: http://www.historians.org/prizes/index.cfm?PrizeAbbrev=HARING
Sponsor: American Historical Association
400 A Street SE
Washington, DC 20003

American Historical Association Fellowship in Aerospace History 717

The American Historical Association will fund one Fellow to undertake a research project related to aerospace history. It will provide a Fellow with an opportunity to engage in significant and sustained advanced research in all aspects of the history of aerospace from the earliest human interest in flight to the present, including cultural and intellectual history, economic history, history of law and public policy, and the history of science, engineering, and management. The Fellowship term is for a period of at least six months, but not more than one year. The Fellow will be expected to devote the term entirely to the proposed research project. The Fellow will have (and be encouraged to take advantage of) the opportunity to use the documentary resources of the National Aeronautics and Space Administration, and may also spend the Fellowship in residence at the NASA headquarters or one of the NASA centers. The stipend is $20,000 for a 6 to 9 month fellowship. This amount is adjustable to the length of the fellowship term. Applications and letters of recommendation must be postmarked by April 1.
Requirements: Applicants must possess a doctorate degree in history or in a closely related field, or be enrolled as a student (having completed all coursework) in a doctoral degree-granting program. The applicant must complete an application form and offer a specific and detailed research proposal that will be the basis of the Fellow's research during the term. At the term's conclusion, the Fellow will be expected to write a report, and to present a paper or a public lecture on the Fellowship experience.
Restrictions: Funds may not be used to support tuition or fees. A Fellow may not hold other major fellowships or grants during the fellowship term, except sabbatical and supplemental grants from their own institutions, and small grants from other sources for specific research expenses. Sources of anticipated support must be listed in the application form.
Geographic Focus: All States
Date(s) Application is Due: Apr 1
Amount of Grant: Up to 20,000 USD
Contact: Julia Brookins, Special Projects Coordinator; (202) 544-2422, ext. 119; fax (202) 544-8307; jbrookins@theaha.org or aha@theaha.org
Internet: http://www.historians.org/prizes/NASA.htm
Sponsor: American Historical Association
400 A Street SE
Washington, DC 20003

American Historical Association Folger Shakespeare Library Fellowship 718

The American Historical Association/Folger Shakespeare Library Fellowship is sponsored jointly by the AHA and the Folger Shakespeare Library. It is awarded for research on seventeenth- and eighteenth-century western European history; the fellow will be awarded a one-month fellowship to be taken at the Folger Shakespeare Library. Apply directly to the Folger by March 1.
Requirements: Applicants must hold the PhD at the time of application and must be a member in good standing of the AHA.
Geographic Focus: All States
Date(s) Application is Due: Mar 1
Contact: Julia Brookins, Special Projects Coordinator; (202) 544-2422, ext. 119; fax (202) 544-8307; jbrookins@theaha.org or aha@theaha.org
Internet: http://www.historians.org/prizes/Folger.cfm
Sponsor: American Historical Association
400 A Street SE
Washington, DC 20003

American Historical Association George L. Mosse Prize 719

The American Historical Association awards the George L. Mosse Prize annually for an outstanding major work of extraordinary scholarly distinction, creativity, and originality in the intellectual and cultural history of Europe since the Renaissance. This prize was established with funds donated by former students, colleagues, and friends of Dr. Mosse. Only books of a high scholarly distinction should be submitted. Research accuracy, originality, and literary merit are important selection factors.
Requirements: Books published after May 1 of the previous year and before April 30 of the award year are eligible.
Geographic Focus: All States
Date(s) Application is Due: May 15
Contact: Julia Brookins, Special Projects Coordinator; (202) 544-2422, ext. 119; fax (202) 544-8307; jbrookins@theaha.org or aha@theaha.org
Internet: http://www.historians.org/prizes/index.cfm?PrizeAbbrev=MOSSE
Sponsor: American Historical Association
400 A Street SE
Washington, DC 20003

American Historical Association George Louis Beer Prize 720

The American Historical Association offers the George Louis Beer Prize in recognition of outstanding historical writing in European international history since 1895. This prize was established in accordance with the terms of a bequest by George Louis Beer (d. 1920), historian of the British colonial system before 1765, to be awarded annually for the best work on any phase of European international history since the year 1895 that is submitted by a scholar who is a United States citizen or permanent resident. The phrase "European international history since the year 1895" may be understood to mean any study of

international history since the year 1895 with a significant European dimension. Only books of a high scholarly historical nature should be submitted. Research accuracy, originality, and literary merit are important factors. Entries must be postmarked by or on May 15.
Requirements: Books published after May 1 of the previous year and before April 30 of the award year are eligible. Only books of a high scholarly historical nature should be submitted.
Geographic Focus: All States
Date(s) Application is Due: May 15
Contact: Julia Brookins, Special Projects Coordinator; (202) 544-2422, ext. 119; fax (202) 544-8307; jbrookins@theaha.org or aha@theaha.org
Internet: http://www.historians.org/prizes/index.cfm?PrizeAbbrev=BEER
Sponsor: American Historical Association
400 A Street SE
Washington, DC 20003

American Historical Association Helen and Howard R. Marraro Prizes 721
The Helen and Howard R. Marraro Prizes are given annually by the American Historical Association, the American Catholic Historical Association, and the Society for Italian Historical Studies for the best works on any epoch of Italian cultural history or on Italian-American relations written by American and Canadian resident citizens. The associations appoint a joint selection committee. One copy of each work, together with a brief curriculum vita and bibliography of the author, must be submitted to designated committee members. The American Historical Association will send instructions upon request.
Requirements: Books published after May 1 of the previous year and before April 30 of the award year are eligible. Entries must first have been published in English by a historian whose usual residence is North America.
Geographic Focus: All States
Date(s) Application is Due: May 15
Contact: Julia Brookins, Special Projects Coordinator; (202) 544-2422, ext. 119; fax (202) 544-8307; jbrookins@theaha.org or aha@theaha.org
Internet: http://www.historians.org/prizes/index.cfm?PrizeAbbrev=MARRARO
Sponsor: American Historical Association
400 A Street SE
Washington, DC 20003

American Historical Association Herbert Baxter Adams Prize 722
The American Historical Association offers the Herbert Baxter Adams Prize annually for a distinguished book by an American author in the field of European history. The prize was established in memory of the first secretary of the Association, Herbert Adams of Johns Hopkins University, who was also one of the Association's founders. Together with the Leo Gershoy Award, the Adams Prize is the most important distinction bestowed by the profession in the field of European history and thus has considerable prestige.
Requirements: The author must be a citizen or permanent resident of the United States or Canada. Since the prize is designed especially to encourage scholars who have not yet obtained an established reputation, the entry must be the author's first substantial book.
Restrictions: Textbooks in the strict sense of the word are not eligible, but a work of wide scope which interprets a major period or area would certainly qualify. Pamphlets, anthologies, edited works, and other small-scale efforts will not qualify.
Geographic Focus: All States
Date(s) Application is Due: May 15
Contact: Julia Brookins, Special Projects Coordinator; (202) 544-2422, ext. 119; fax (202) 544-8307; jbrookins@theaha.org or aha@theaha.org
Internet: http://www.historians.org/prizes/index.cfm?PrizeAbbrev=ADAMS
Sponsor: American Historical Association
400 A Street SE
Washington, DC 20003

American Historical Association Herbert Feis Award 723
Established in 1984, the Herbert Feis Award is offered annually to recognize the scholarly interests of historians outside academe and the importance of the work of independent scholars in the United States. The terms of the award define both contribution and public history broadly. Contributions could include work as the administrator of a public history group or agency (such as a historical society, a historic site, or a community history project) or as the creator or producer of a public history product or products (such as a museum exhibit, radio script, web site, oral history collection, or film). The contribution may be the result of years of effort in the field, but the prize might also recognize a singular contribution of major importance such as a path breaking museum exhibit. Public history is defined as work primarily directed at non-academic, non-school-based audiences. Those audiences could be very broad (e.g., television viewers) or highly specialized (e.g., policymakers). Although the audience should be primarily outside of academia, the recipient of the award could be employed at a university. Nominations must be postmarked by or on August 1.
Requirements: Works published or issued in-house between May 1 of the previous year and April 30 of the upcoming year are eligible.
Geographic Focus: All States
Date(s) Application is Due: Aug 1
Contact: Julia Brookins, Special Projects Coordinator; (202) 544-2422, ext. 119; fax (202) 544-8307; jbrookins@theaha.org or aha@theaha.org
Internet: http://www.historians.org/prizes/index.cfm?PrizeAbbrev=FEIS
Sponsor: American Historical Association
400 A Street SE
Washington, DC 20003

American Historical Association J. Franklin Jameson Award In Outstanding Editorial Achievement 724
Commencing in 1980, this prize is awarded biennially in odd-numbered years for outstanding achievement in the editing of historical sources. The prize is named in honor of J. Franklin Jameson, a founder of the Association and longtime managing editor of the American Historical Review.
Requirements: To be eligible for consideration, works must be of a scholarly, historical nature; review/journal editing is ineligible. Books published after May 1 of the previous year and before April 30 of the award year are eligible.
Geographic Focus: All States
Date(s) Application is Due: May 15
Contact: Julia Brookins, Special Projects Coordinator; (202) 544-2422, ext. 119; fax (202) 544-8307; jbrookins@theaha.org or aha@theaha.org
Internet: http://www.historians.org/prizes/index.cfm?PrizeAbbrev=Jameson
Sponsor: American Historical Association
400 A Street SE
Washington, DC 20003

American Historical Association J. Franklin Jameson Fellowship 725
The Franklin Jameson Fellowship in American History is offered annually by the Library of Congress and the American Historical Association to support significant scholarly research for one semester in the collections of the Library of Congress by scholars at an early stage in their careers in history. The fellowship is named in honor of J. Franklin Jameson, a founder of the Association, longtime managing editor of the American Historical Review, formerly Chief of the Manuscript Division of the Library of Congress, and the first incumbent of the Library's Chair of American History. It is designed to assist scholars early in their careers. The applicant's project in American history must be one for which the general and special collections of the Library of Congress offer unique research support. Applicants should include a statement substantiating this relationship. The fellowship will be awarded for 2 to 3 months, as the Jameson Fellow desires, to spend in full-time residence at the Library of Congress. Working space will be provided by the Library of Congress, but the Jameson Fellow is required to spend at least two months in residence. The total stipend may vary from year to year. Currently the amount will be $5,000. This will not be prorated according to the proposed length of tenure of the fellowship.
Requirements: At the time of application, applicants must hold the Ph.D. degree or equivalent; must have received this degree within the past seven years, and must not have published or had accepted for publication a book-length historical work.
Restrictions: The fellowship will not be awarded for completion of a doctoral dissertation.
Geographic Focus: All States
Date(s) Application is Due: Mar 15
Amount of Grant: 5,000 USD
Contact: Julia Brookins, Special Projects Coordinator; (202) 544-2422, ext. 119; fax (202) 544-8307; jbrookins@theaha.org or aha@theaha.org
Internet: http://www.historians.org/prizes/Jameson_fellowship.htm
Sponsor: American Historical Association
400 A Street SE
Washington, DC 20003

American Historical Association J. Russell Major Prize 726
The American Historical Association awards the J. Russell Major Prize annually for the best work in English on any aspect of French history. The prize was established in memory of J. Russell Major, the distinguished scholar of French history who died on December 12, 1998 at the age of 77. Major served on the history faculty at Emory University from 1949 until his retirement in 1990, and wrote 10 books, including Representative Government in Early Modern France and From Renaissance Monarchy to Absolute Monarchy: French Kings, Nobles and Estates.
Requirements: Books published after May 1 of the previous year and before April 30 of the award year are eligible.
Geographic Focus: All States
Date(s) Application is Due: May 15
Contact: Julia Brookins, Special Projects Coordinator; (202) 544-2422, ext. 119; fax (202) 544-8307; jbrookins@theaha.org or aha@theaha.org
Internet: http://www.historians.org/prizes/index.cfm?PrizeAbbrev=MAJOR
Sponsor: American Historical Association
400 A Street SE
Washington, DC 20003

American Historical Association James A. Rawley Prize in Atlantic History 727
The American Historical Association announced the establishment of the James A. Rawley Prize in Atlantic History in recognition of outstanding historical writing that explores aspects of integration of Atlantic worlds before the 20th century. The prize was established in accordance with the terms of a gift from James A. Rawley, Carl Adolph Happold Professor of History Emeritus at the University of Nebraska at Lincoln. Only books of high scholarly and literary merit will be considered. Research accuracy and originality also will be important factors in the evaluation of the books.
Requirements: Books published after May 1 of the previous year and before April 30 of the award year are eligible. Only books of a high scholarly historical nature should be submitted.
Geographic Focus: All States
Date(s) Application is Due: May 15
Contact: Julia Brookins, Special Projects Coordinator; (202) 544-2422, ext. 119; fax (202) 544-8307; jbrookins@theaha.org or aha@theaha.org

Internet: http://www.historians.org/prizes/index.cfm?PrizeAbbrev=Rawley
Sponsor: American Historical Association
400 A Street SE
Washington, DC 20003

American Historical Association James Harvey Robinson Prize 728

The James Harvey Robinson Prize was established by the AHA Council in 1978 and is awarded biennially for the teaching aid which has made the most outstanding contribution to the teaching and learning of history in any field for public or educational purposes. Teaching aid encompasses textbooks, source and reference materials, audiovisuals, computer-assisted instruction, and public history or museum materials. Monographs and revisions will not be considered. The winner will receive a one-year membership in the Association. The work should: have the potential to influence history education; demonstrate recent and good historical scholarship; and be well written and attractively presented.
Requirements: Monographs and revisions will not be considered. Books published after May 1 two years prior and before April 30 of the award year are eligible.
Restrictions: Monographs and revisions will not be considered.
Geographic Focus: All States
Date(s) Application is Due: May 15
Contact: Julia Brookins, Special Projects Coordinator; (202) 544-2422, ext. 119; fax (202) 544-8307; jbrookins@theaha.org or aha@theaha.org
Internet: http://www.historians.org/prizes/index.cfm?PrizeAbbrev=ROBINSON
Sponsor: American Historical Association
400 A Street SE
Washington, DC 20003

American Historical Association James Henry Breasted Prize 729

Established in 1985, this prize, named in honor of James Henry Breasted, a pioneer in ancient Egyptian and Near Eastern history and president of the Association in 1928, is offered for the best book in English in any field of history prior to CE 1000 and rotates annually among the following geographical areas: Near East and Egypt; Far East and South Asia; Africa, North America, and Latin America; and Europe. Only works of high scholarly and literary merit will be considered. One copy of each entry must be submitted to designated committee members.
Requirements: Books published after May 1 of the previous year and before April 30 of the award year are eligible.
Geographic Focus: All States
Date(s) Application is Due: May 15
Contact: Julia Brookins, Special Projects Coordinator; (202) 544-2422, ext. 119; fax (202) 544-8307; jbrookins@theaha.org or aha@theaha.org
Internet: http://www.historians.org/prizes/index.cfm?PrizeAbbrev=BREASTED
Sponsor: American Historical Association
400 A Street SE
Washington, DC 20003

American Historical Association Joan Kelly Memorial Prize in Women's History 730

Established in 1984, this prize, named in memory of Joan Kelly (1928-82), is awarded annually for the book in women's history and/or feminist theory that best reflects the high intellectual and scholarly ideals exemplified by the life and work of Joan Kelly. The prize was established by the Coordinating Committee on Women in the Historical Profession and the Conference Group on Women's History (now the Coordinating Council for Women in History), and is administered by the American Historical Association.
Requirements: Books in any chronological period, any geographical location, or in an area of feminist theory that incorporates a historical perspective are eligible for consideration. Books should demonstrate originality of research, creativity of insight, graceful stylistic presentation, analytical skills, and a recognition of the important role of sex and gender in the historical process. The inter-relationship between women and the historical process should be addressed. Books published after May 1 of the previous year and before April 30 of the award year are eligible.
Geographic Focus: All States
Date(s) Application is Due: May 15
Contact: Julia Brookins, Special Projects Coordinator; (202) 544-2422, ext. 119; fax (202) 544-8307; jbrookins@theaha.org or aha@theaha.org
Internet: http://www.historians.org/prizes/index.cfm?PrizeAbbrev=KELLY
Sponsor: American Historical Association
400 A Street SE
Washington, DC 20003

American Historical Association John F. Richards Prize in South Asian History 731

The John F. Richards Prize in South Asian History recognizes the most distinguished work of scholarship on South Asian history published in English during the previous calendar year. South Asia is defined as the geographic area included in the modern states of Afghanistan, Bangladesh, Bhutan, Burma, India, the Maldives, Nepal, Pakistan, and Sri Lanka. Eligibility will be defined quite broadly, including books on any period or field of South Asian historical studies and works which integrate South Asian history with broader global issues and movements. In making its selection, the prize committee will pay particular attention to depth of research, methodological innovation, conceptual originality and literary excellence. Works that reinterpret old themes or develop new theoretical perspectives are welcome.
Requirements: Books published after May 1 of the previous year and before April 30 of the award year are eligible. Only books of a high scholarly historical nature should be submitted.
Restrictions: Anthologies, encyclopedias and other edited volumes will not be considered.
Geographic Focus: All States
Date(s) Application is Due: May 15
Contact: Julia Brookins, Special Projects Coordinator; (202) 544-2422, ext. 119; fax (202) 544-8307; jbrookins@theaha.org or aha@theaha.org
Internet: http://www.historians.org/prizes/index.cfm?PrizeAbbrev=Richards
Sponsor: American Historical Association
400 A Street SE
Washington, DC 20003

American Historical Association John H. Dunning Prize 732

In recognition of outstanding historical writing in United States history, the American Historical Association offers the John H. Dunning Prize. Established by a bequest from Mathilde Dunning in 1927 to be awarded for an outstanding monograph on any subject relating to United States history, this prize is offered biennially in odd-numbered years. To be eligible for consideration, an entry must be of a scholarly historical nature.
Requirements: An entry must be the author's first or second book published or completed. Books published after May 1 of the previous year and before April 30 of the deadline year are eligible.
Geographic Focus: All States
Date(s) Application is Due: May 15
Contact: Julia Brookins, Special Projects Coordinator; (202) 544-2422, ext. 119; fax (202) 544-8307; jbrookins@theaha.org or aha@theaha.org
Internet: http://www.historians.org/prizes/index.cfm?PrizeAbbrev=DUNNING
Sponsor: American Historical Association
400 A Street SE
Washington, DC 20003

American Historical Association John K. Fairbank Prize in East Asian History 733

The John K. Fairbank Prize in East Asian History is awarded annually for an outstanding book on the history of China proper, Vietnam, Chinese Central Asia, Mongolia, Manchuria, Korea, or Japan substantially since the year 1800. Only books of high scholarly and literary merit will be considered. One copy of each entry must be submitted to each committee member. Application forms are available from the association.
Requirements: Books published after May 1 of the previous year and before April 30 of the award year are eligible.
Restrictions: Anthologies, edited works, and pamphlets are ineligible for the competition.
Geographic Focus: All States
Date(s) Application is Due: May 15
Contact: Julia Brookins, Special Projects Coordinator; (202) 544-2422, ext. 119; fax (202) 544-8307; jbrookins@theaha.org or aha@theaha.org
Internet: http://www.historians.org/prizes/index.cfm?PrizeAbbrev=FAIRBANK
Sponsor: American Historical Association
400 A Street SE
Washington, DC 20003

American Historical Association Leo Gershoy Award 734

In recognition of outstanding historical writing, the American Historical Association offers the Leo Gershoy Award in 17th- and 18th-century western European history. This award was established in 1975 by a gift from Mrs. Ida Gershoy in memory of her late husband, a specialist in European history. Professor Gershoy was associated with the graduate faculty of New York University for more than 25 years. The prize is awarded annually to the author of the most outstanding work published in English on any aspect of the fields of 17th- and 18th-century western European history. Only books of high scholarly and literary merit will be considered.
Requirements: Books published after May 1 of the previous year and before April 30 of the award year are eligible.
Geographic Focus: All States
Date(s) Application is Due: May 15
Contact: Julia Brookins, Special Projects Coordinator; (202) 544-2422, ext. 119; fax (202) 544-8307; jbrookins@theaha.org or aha@theaha.org
Internet: http://www.historians.org/prizes/index.cfm?PrizeAbbrev=GERSHOY
Sponsor: American Historical Association
400 A Street SE
Washington, DC 20003

American Historical Association Littleton-Griswold Prize in American Law and Society 735

Established in 1985, the Littleton-Griswold Prize is offered annually for the best book in any subject on the history of American law and society. The $1,000 prize is administered by a joint committee of the American Historical Association and the American Society for Legal History. Only books of high scholarly and literary merit will be considered. Recipients are announced at the association's annual meeting. Submission guidelines must be obtained from the association.
Requirements: Books published between May 1 of the previous year and April 30 of the award year are eligible.
Geographic Focus: All States
Date(s) Application is Due: May 15
Amount of Grant: 1,000 USD
Contact: Julia Brookins; (202) 544-2422, ext. 119; jbrookins@theaha.org
Internet: http://www.historians.org/prizes/index.cfm?PrizeAbbrev=LITTLETON%2DGRISWOLD
Sponsor: American Historical Association
400 A Street SE
Washington, DC 20003

American Historical Association Littleton-Griswold Research Grants in U.S. Legal History 736

The Littleton-Griswold Research Grants in U.S. Legal History are given annually for research in U.S. legal history and in the general field of law and society. The funds for this program come from the earnings of the Littleton-Griswold Fund. The grants are intended to further research in progress and may be used for travel to a library or archive, for microfilms, photographs, or photocopying-a list of purposes that is meant to be merely illustrative, not exhaustive (other expenses, such as child care, can be included).
Requirements: Preference will be given to those with specific research needs, such as the completion of a project or completion of a discrete segment thereof. Preference will be given to Ph.D. candidates and junior scholars.
Restrictions: Only members of the Association are eligible to apply. Specifically excluded from funding are use of research assistants except in cases of technical skills such as data entry, coding of data, and other procedures relating to the preparation of data for machine computation; help with typing except for typing of camera-ready copy for an article, dissertation, or book, which would qualify as a technical expense; and use of funds for partial salary replacement.
Geographic Focus: All States
Date(s) Application is Due: Feb 15
Amount of Grant: Up to 1,000 USD
Contact: Julia Brookins; (202) 544-2422, ext. 119; jbrookins@theaha.org
Internet: http://www.historians.org/prizes/Littleton-GriswaldGrantInfo.htm
Sponsor: American Historical Association
400 A Street SE
Washington, DC 20003

American Historical Association Martin A. Klein Prize in African History 737

The Martin A. Klein Prize in African History will recognize the most distinguished work of scholarship on African history published in English during the previous calendar year. The books must focus primarily on continental Africa (including those islands usually treated as countries of Africa). Eligibility will otherwise be defined quite broadly, to include books on any period of African history and from any disciplinary field that incorporates an historical perspective. In making its selection, the prize committee will pay particular attention to methodological innovation, conceptual originality and literary excellence. Works that reinterpret old themes or develop new theoretical perspectives are welcome. The prize is named for Martin A. Klein, who is currently professor of history at the University of Toronto. Funding for the prize was completed thanks to a substantial donation from Dr. Mougo Nyaggah of California State University at Fullerton and his wife Dr. Lynette Nyaggah. Mougo Nyaggah was Klein's first graduate student at the University of California Berkeley. Nyaggah credits the completion of his doctorate to Klein's mentoring, guidance, enthusiasm, and commitment to the research and teaching of African history. He observed that, "There are many Martins who have or will mentor and inspire many Africanist students in American universities. Those mentors will be honored by this prize for their human and scholarly contribution." Books may be submitted by authors or publishers, by sending one copy to each of the members of the committee. Anthologies, encyclopedias and other edited volumes will not be considered.
Requirements: Books published after May 1 of the previous year and before April 30 of the award year are eligible.
Geographic Focus: All States
Date(s) Application is Due: May 15
Contact: Julia Brookins; (202) 544-2422, ext. 119; jbrookins@theaha.org
Internet: http://www.historians.org/prizes/index.cfm?PrizeAbbrev=Klein
Sponsor: American Historical Association
400 A Street SE
Washington, DC 20003

American Historical Association Michael Kraus Research Grant in American Colonial History 738

The American Historical Association announces the availability of the Michael Kraus Research Grant to recognize the most deserving proposal relating to work in progress on a research project in American colonial history, with particular reference to the intercultural aspects of American and European relations. Only members of the Association are eligible. The grants are intended to further research in progress and may be used for travel to a library or archive, for microfilms, photographs, or photocopying—a list of purposes that is meant to be merely illustrative, not exhaustive.
Requirements: Applicants must be members of the association. Preference will be given to those with specific research needs such as the completion of a project or a discrete segment thereof. Preference will be given to Ph.D. candidates and junior scholars.
Restrictions: Specifically excluded from funding are use of research assistants except in cases of technical skills such as keypunching, coding of data, and other procedures relating to the preparation of data for machine computation; help with typing except for typing of camera-ready copy for an article, dissertation, or book, which would qualify as a technical expense; and use of funds for partial salary replacements. Within a five-year period, no individual is eligible to receive more than a combined total of $1,000 from the Beveridge, Kraus, and Littleton-Griswold grant programs.
Geographic Focus: All States
Date(s) Application is Due: Feb 15
Amount of Grant: Up to 800 USD
Contact: Julia Brookins; (202) 544-2422, ext. 119; jbrookins@theaha.org
Internet: http://www.historians.org/prizes/KrausGrantInfo.htm
Sponsor: American Historical Association
400 A Street SE
Washington, DC 20003

American Historical Association Morris D. Forkosch Prize 739

The American Historical Association offers the Morris D. Forkosch Prize annually in recognition of the best book in English in the field of British, British Imperial, or British Commonwealth history since 1485. Submission of books relating to the shared common law heritage of the English-speaking world are particularly encouraged in memory of the late Professor Forkosch's contributions to the field of legal studies and legal history.
Requirements: Books published after May 1 of the previous year and before April 30 of the award year are eligible.
Geographic Focus: All States
Date(s) Application is Due: May 15
Contact: Julia Brookins; (202) 544-2422, ext. 119; jbrookins@theaha.org
Internet: http://www.historians.org/prizes/index.cfm?PrizeAbbrev=FORKOSCH
Sponsor: American Historical Association
400 A Street SE
Washington, DC 20003

American Historical Association Paul Birdsall Prize in European Military and Strategic History 740

Commencing in 1986 this prize, endowed by Hans Gatzke, Yale University and named in honor of the late Paul Birdsall of Williams College (d. 1970), is offered biennially for a major work in European military and strategic history since 1870. Preference will be given to the international aspects of military history (military/diplomatic), but the political, economic, and social impact of technological developments, strategic planning, and military events on society will also qualify.
Requirements: Authors must be citizens of the United States or Canada. Preference will be given to younger academics, but older scholars and nonacademic candidates will not be excluded. Books published after May 1 of the previous year and before April 30 of the award year are eligible.
Geographic Focus: All States
Date(s) Application is Due: May 15
Contact: Julia Brookins; (202) 544-2422, ext. 119; jbrookins@theaha.org
Internet: http://www.historians.org/prizes/index.cfm?PrizeAbbrev=BIRDSALL
Sponsor: American Historical Association
400 A Street SE
Washington, DC 20003

American Historical Association Premio Del Rey 741

The American Historical Association awards the Premio Del Rey biennially for a distinguished book in English in the field of early Spanish history. It was endowed by a gift of Robert I. Burns S.J., from his Llull and Catalonia prizes and covers the medieval period in Spain's history and culture CE 500-1516. The terms of the prize include works on Hispanic history and culture, including the Islamic and Jewish communities of Medieval Spain as well as early New World topics prior to 1516. Only books of a high scholarly historical nature should be submitted. Research accuracy, originality, and literary merit are important factors.
Requirements: Books published after May 1 of the previous year and before April 30 of the award year are eligible.
Geographic Focus: All States
Date(s) Application is Due: May 15
Contact: Julia Brookins; (202) 544-2422, ext. 119; jbrookins@theaha.org
Internet: http://www.historians.org/prizes/index.cfm?PrizeAbbrev=DEL-REY
Sponsor: American Historical Association
400 A Street SE
Washington, DC 20003

American Historical Association Raymond J. Cunningham Prize 742

The Raymond J. Cunningham Prize is given for the best article published in a history department journal written by an undergraduate student. The prize was established in memory of Raymond J. Cunningham who was an associate professor of history at Fordham University. He was an authority on American Historian Herbert Baxter Adams. The deadline is May 15.
Geographic Focus: All States
Date(s) Application is Due: May 15
Contact: Julia Brookins; (202) 544-2422, ext. 119; jbrookins@theaha.org
Internet: http://www.historians.org/prizes/index.cfm?PrizeAbbrev=Cunningham
Sponsor: American Historical Association
400 A Street SE
Washington, DC 20003

American Historical Association Waldo J. Leland Prize 743

The prize is awarded every five years (next in 2016) by the American Historical Association to an American resident for the most outstanding reference tool in the field of history. The award is honorific. The term reference tool encompasses bibliographies, indexes, encyclopedias, and other scholarly apparatus.
Requirements: Books published after May 1 of the previous award year and before April 30 of the current award year are eligible.
Geographic Focus: All States
Date(s) Application is Due: May 16
Contact: Julia Brookins; (202) 544-2422, ext. 119; jbrookins@theaha.org
Internet: http://www.historians.org/prizes/index.cfm?PrizeAbbrev=LELAND
Sponsor: American Historical Association
400 A Street SE
Washington, DC 20003

American Historical Association Wesley-Logan Prize 744

The Wesley-Logan Prize in African diaspora history is jointly sponsored by the American Historical Association and the Association for the Study of African American Life & History. The prize is awarded annually for an outstanding book in African diaspora history. The AHA Committee on Minority Historians established the prize in 1992 in memory of two early pioneers in the field, Charles H. Wesley and Rayford W. Logan. The prize is offered for a book on some aspect of the history of the dispersion, settlement and adjustment, and/or return of peoples originally from Africa. Eligible for consideration are books in any chronological period and any geographical location.

Requirements: Books published between May 1 of the previous year and April 30 of the current year are eligible. Only books of high scholarly and literary merit will be considered.

Geographic Focus: All States

Date(s) Application is Due: May 15

Contact: Julia Brookins; (202) 544-2422, ext. 119; jbrookins@theaha.org

Internet: http://www.historians.org/prizes/index.cfm?PrizeAbbrev=WESLEY%2DLOGAN

Sponsor: American Historical Association
400 A Street SE
Washington, DC 20003

American Indian College Fund STEM Leadership Fellowships 745

The American Indian College Fund and the Alfred P. Sloan Foundation are funding Fellowships to provide Tribal Colleges and Universities (TCUs) with more Ph.D. professors while conducting and disseminating research relevant to TCUs and Indian Country in general by increasing and retaining the number of credentialed diverse and underrepresented faculty. Fellowship awards are for the completion of dissertation and requirements for graduation to current TCU faculty in the final stages of a terminal degree and recognized as "All But Dissertation".

Requirements: Applicants must be current faculty at a TCU accredited for higher education and a full member of the American Indian Higher Education Consortium; be actively enrolled in an accredited doctoral or other terminal degree program in the STEM discipline; have completed all course work, having advanced to doctoral candidacy and considered "All But Dissertation"; be committed to finishing the terminal degree within one academic year of the award; show commitment to American Indian education and scholarship; demonstrate persistent and respectful engagement with communities served by the TCU and/or the subjects of the research; have institutional approval for dissertation research; have an approved plan for protection of any human research subjects; have a senior faculty mentor.

Restrictions: Applicants are expected to pursue their degree full-time and complete all requirements within one academic year. Fellowships will not be renewed or extended. Applicants must agree to continue as faculty at a TCU for at least two years after completion of the terminal degree.

Geographic Focus: All States

Date(s) Application is Due: Mar 15

Amount of Grant: 36,000 USD

Contact: Dennis L. Carter, J.D.; (800) 776-3863 or (303) 430-5346; dcarder@collegefund.org

Internet: http://www.collegefund.org/students_and_alumni/content/leadership_fellowship

Sponsor: American Indian College Fund
8333 Greenwood Boulevard
Denver, CO 80221

American Indian Studies Research Institute Fellowship 746

Each year, the Institute offers a fellowship to a worthy graduate student interested in studying American Indian culture, language, or history. The fellowship is designed to attract and retain top-quality graduate students. Usually, though not always, it is awarded to an incoming graduate student. The fellowship carries with it a tuition waiver, university health insurance, and a stipend. There are no duties formally required of the student; however, the fellowship recipient is encouraged to engage in the ongoing scholarly research at the Institute.

Geographic Focus: Indiana

Contact: Program Director; (812) 855-4086; fax (812) 855-7529; aisri@indiana.edu

Internet: http://www.indiana.edu/~aisri/index.shtml

Sponsor: American Indian Studies Research Institute / Indiana University
422 North Indiana Avenue
Bloomington, IN 47408

American Jewish Historical Society Awards and Prizes 747

In cooperation with the Jewish Historical Society of New York, the society grants awards and prizes to assist researchers to carry on projects in the area of American Jewish history, including immigration, economic, social, political, and religious history. For current awards, details, and deadlines, see the AJHS website (above).

Geographic Focus: All States

Contact: David Solomon; (212) 294-6160; fax (212) 294-6161; dsolomon@ajhs.cjh.org

Internet: http://www.ajhs.org/academic/Awards.cfm

Sponsor: American Jewish Historical Society
15 W 16th Street
New York, NY 10011

American Legacy Foundation National Calls for Proposals Grants 748

The American Legacy Foundation provides grants to support innovations in tobacco control. National calls for proposals, such as the Priority Populations Initiative, are issued annually to address a variety of tobacco prevention and control issues. Comprehensive assistance and training are provided to the foundation's grantees as well as to state and local tobacco programs. Application forms, guidelines, and procedures are available online. There are no application deadlines.

Requirements: Funding is available only to state or local political subdivisions and legally constituted tax-exempt 501(c)3 organizations based in the 46 states, the District of Columbia, and five territories (American Samoa, Guam, Northern Mariana Islands, Puerto Rico, and the Virgin Islands) identified in the MSA with tobacco product manufacturers. An Indian reservation, Indian tribe, or tribal organization located within the 46 settling states or a non-governmental entity that serves such a reservation may also apply for funding.

Restrictions: The foundation will not award grants to applicants that are in current receipt of grant monies or in-kind contributions from any tobacco manufacturer, distributor, or other tobacco-related entity.

Geographic Focus: All States

Contact: Katherine Wilson, Assistant Vice President of Grants; (202) 454-5555; fax (202) 454-5599; grantsinfo@americanlegacy.org

Internet: http://www.americanlegacy.org/64.aspx

Sponsor: American Legacy Foundation
1724 Massachusetts Avenue NW
Washington, DC 20036

American Legacy Foundation Small Innovative Grants 749

The program supports projects that advance creative, promising solutions based on sound principles of tobacco control to remedy the harm caused by tobacco use in America. The program was created to seed new projects or enable an organization to pilot a new idea or approach. The proposed project must demonstrate an element of creativity, ingenuity or innovation and must distinguish itself from the large number of solid programs proposed to the foundation in each grant round.

Requirements: Funding is available only to state or local political subdivisions and legally constituted tax-exempt 501(c)3 organizations based in the 46 states, the District of Columbia, and five territories (American Samoa, Guam, Northern Mariana Islands, Puerto Rico, and the Virgin Islands) identified in the MSA with tobacco product manufacturers. An Indian reservation, Indian tribe, or tribal organization located within the 46 settling states or a nongovernmental entity that serves such a reservation may also apply for funding. Successful applications submitted under the Small Innovative Grants Program must: address one or both of Legacy's goals; demonstrate innovative or new tobacco prevention or cessation efforts; demonstrate a strong likelihood for a sustainable effort after the grant period; demonstrate that the project may be replicated; address the Healthy People 2010 risk reduction objectives related to tobacco use, and; incorporate the CDC's Best Practices for Comprehensive Tobacco Control Programs as appropriate. The foundation will give special consideration to applications addressing these current areas of interest.

Restrictions: The foundation will not award a grant to any applicant that is in current receipt of any grant monies or in-kind contribution from any tobacco manufacturer, distributor, or other tobacco-related entity. In addition, the foundation expects that a grantee will not accept any grant monies or in-kind contribution from any tobacco manufacturer, distributor, or other tobacco-related entity over the duration of the grant. Additionally, will not consider applications for: Projects focusing on youth prevention, cessation, activism, or education (up to 18 years old); Nicotine replacement therapy (NRT) and pharmaceuticals (as the sole or primary focus of the grant); Conference support (as the sole or primary focus of the grant); Media or marketing campaigns (as the sole or primary focus of the grant); Research projects EXCEPT community-based participatory research, which is allowed; Projects focusing on substances other than tobacco; Replication of an existing program; Expansion of an existing program; and Replacement funds; Grants to individuals, for religious activities, to build endowments, to support operating deficits, to retire debt, for capital purchases for building improvements, for construction, for lobbying, or for real estate purchase or development.

Geographic Focus: All States

Date(s) Application is Due: Feb 27; Aug 8

Amount of Grant: 20,000-100,000 USD

Contact: Katherine Wilson, Assistant Vice President of Grants; (202) 454-5555; fax (202) 454-5599; grantsinfo@americanlegacy.org

Internet: http://www.americanlegacy.org/2530.aspx

Sponsor: American Legacy Foundation
1724 Massachusetts Avenue NW
Washington, DC 20036

American Mathematical Society AAAS Congressional Fellowships 750

The American Mathematical Society (AMS), in conjunction with the American Association for the Advancement of Science (AAAS), sponsors a Congressional Fellow each year who spends the year working on the staff of a Member of Congress or a congressional committee, working as a special legislative assistant in legislative and policy areas requiring scientific and technical input. The program includes an orientation on congressional and executive branch operations, and a year-long seminar series on issues involving science, technology and public policy. The fellowship stipend for the Fellowship from September of next year to August of the following year is $74,872 for the fellowship period, with allowances for relocation and professional travel and a contribution towards health insurance.

Requirements: Prospective Fellows must demonstrate expertise in some area of the mathematical sciences; have a good scientific and technical background; be cognizant of and demonstrate sensitivity toward political and social issues; and, most importantly, have a strong interest and some experience in applying personal knowledge toward the solution of societal problems. Eligible applicants are individuals in the mathematical sciences with a Ph.D. or an equivalent doctoral-level degree in the mathematical sciences by the application deadline of February 15.

Geographic Focus: All States
Contact: Dr. Donald McClure, Executive Director; (401) 455-4105 or (401) 455-4000; fax (401) 331-3842; ams@ams.org
Internet: http://www.ams.org/programs/ams-fellowships/ams-aaas/ams-aaas-congressional-fellowship
Sponsor: American Mathematical Society
201 Charles Street
Providence, RI 02904-2294

American Mathematical Society AAAS Mass Media Fellowships 751
The American Mathematical Society (AMS) sponsors a Mass Media Fellow each summer through the Mass Media Science & Engineering Fellowship program organized by the American Association for the Advancement of Science (AAAS). It is a highly competitive program designed to improve public understanding of science and technology by placing advanced science, mathematics and engineering students in newsrooms nationwide. Fellows work with media professionals to improve their communication skills and increase their understanding of the editorial process by which events and ideas become news. Fellows receive a weekly stipend of $450, plus travel expenses, to work for ten weeks during the summer months as reporters, researchers, and production assistants in media organizations. They observe and participate in the process by which events and ideas become news, improve their ability to communicate about complex technical subjects in a manner understandable to the public, and increase their understanding of editorial decision making and of how information is effectively disseminated. Eligible candidates should review information on the application procedures and submit their application by the January 15 deadline.
Requirements: The program is available to college or university students (in their senior year, or in any graduate or post graduate level) in the natural, physical, health, engineering, computer or social sciences or mathematics with outstanding written and oral communication skills and a strong interest in learning about the media.
Geographic Focus: All States
Amount of Grant: 4,500 USD
Contact: Dr. Donald McClure, Executive Director; (401) 455-4105 or (401) 455-4000; fax (401) 331-3842; ams@ams.org
Internet: http://www.ams.org/programs/ams-fellowships/media-fellow/massmediafellow
Sponsor: American Mathematical Society
201 Charles Street
Providence, RI 02904-2294

American Mathematical Society Centennial Fellowships 752
The AMS Centennial Research Fellowship Program makes awards annually to outstanding mathematicians to help further their careers in research. From 1997-2001, the fellowship program was aimed at recent PhDs. Recently, the AMS Council approved changes in the rules for the fellowships. The primary selection criterion for the Centennial Fellowship is the excellence of the candidate's research. Applications will be accepted from those currently holding a tenured, tenure track, post-doctoral, or comparable (at the discretion of the selection committee) position at an institution in North America. The stipend for most recent fellowships awarded was $82,000, with an additional expense allowance of about $8,200. Applications will be accepted from October 1 through December 1.
Requirements: Preference will be given to candidates who have not had extensive fellowship support in the past. A recipient of the fellowship shall have held his or her doctoral degree for at least three years and not more than twelve years at the inception of the award.
Restrictions: Recipients may not hold the Centennial Fellowship concurrently with another major research award such as a Sloan fellowship, NSF Postdoctoral fellowship, or CAREER award. Under normal circumstances, the fellowship cannot be deferred.
Geographic Focus: All States
Date(s) Application is Due: Dec 1
Amount of Grant: 90,200 USD
Contact: Dr. Donald McClure, Executive Director; (401) 455-4105 or (401) 455-4000; fax (401) 331-3842; ams@ams.org
Internet: http://www.ams.org/programs/ams-fellowships/centennial-fellow/emp-centflyer
Sponsor: American Mathematical Society
201 Charles Street
Providence, RI 02904-2294

American Mathematical Society China Exchange Travel Grants 753
The purpose of these American Mathematical Society China Exchange Travel Grants is to support travel of mathematical scientists at institutions in the U.S. or Canada on visits to research colleagues in China, and to support travel of (generally, less senior) mathematical scientists from China to the U.S. or Canada for visits with U.S. or Canadian research colleagues at U.S. or Canadian institutions. In the case of American or Canadian travelers, preference will be given to senior mathematicians whose visit will have a positive impact on the mathematical research community in China. In the case of Chinese travelers, preference will be given to scientists in the beginning stages of their research careers. For U.S. or Canadian institutions applying, U.S.$1000 will be set aside from the total award and given to the home institution (in China) of the visitor (in the case of multiple visitors, the amount should be divided equally), in lieu of honorarium. The annual deadline is March 15 for travel that takes place the following academic year (between September and August).
Requirements: It is expected that the host institution will provide some type of additional support for the travel or subsistence of the visitor to ensure a stay of suitable length.
Restrictions: Funds from this program may not be used specifically to support conference attendance, but trips may sometimes be combined with conference attendance.
Geographic Focus: All States, Canada
Date(s) Application is Due: Mar 15
Amount of Grant: 1,000 USD
Contact: Steven Ferrucci, (401) 455-4113 or (401) 455-4000; ams@ams.org
Internet: http://www.ams.org/programs/travel-grants/china-exchange/china-exchange
Sponsor: American Mathematical Society
201 Charles Street
Providence, RI 02904-2294

American Mathematical Society Graduate Student Travel Grants 754
The AMS, with support from a private gift, accepts applications for partial support for two types of AMS meetings: Joint Mathematics Meetings-available to selected doctoral students in mathematics who are in their last year of study (that is, have not received the PhD before the application deadline, but do expect to receive the PhD degree within the twelve month period following the application deadline); and Sectional Meetings-program accepts applications from current full time doctoral students. The deadline for Joint Mathematics Meetings applications is September 26, while the deadlines for Sectional Meetings are July 18 and November 14. Support level for the January JMM is $500. Support level for sectional meetings is $250.
Requirements: The current institution of the applicant must be more than 50 miles from the site of the meeting per google maps. A representative of the university must submit a statement verifying the expected date of graduation. Each awardee's university is strongly encouraged, but not required, to provide matching support.
Geographic Focus: All States
Date(s) Application is Due: Jul 18; Sep 26; Nov 14
Amount of Grant: 250-500 USD
Contact: Steven Ferrucci, (401) 455-4113 or (401) 455-4000; ams@ams.org
Internet: http://www.ams.org/programs/travel-grants/grad-students/emp-student-JMM
Sponsor: American Mathematical Society
201 Charles Street
Providence, RI 02904-2294

American Mathematical Society International Congress of Mathematicians Travel Grants 755
The American Mathematical Society has applied to the National Science Foundation (NSF) for funds to permit partial travel support for U.S. mathematicians attending the International Congress of Mathematicians. The application period will be September 1 through November 15. This program is open to U.S. mathematicians (those who are U.S. citizens or affiliated with a U.S. institution).
Requirements: Early career mathematicians (those within six years of their doctorate), women and underrepresented minorities will be especially encouraged to apply. Early career mathematicians will be required to supply one letter from an individual who has made a commitment to serve as a mentor to the applicant during the period leading up to the International Congress, advising him or her about the appropriate sessions and activities in which to participate at the ICM.
Restrictions: Mathematicians accepting grants for partial support of the travel to ICM may not supplement them with any other NSF funds.
Geographic Focus: All States
Contact: Steven Ferrucci, (401) 455-4113 or (401) 455-4000; ams@ams.org
Internet: http://www.ams.org/programs/travel-grants/icm
Sponsor: American Mathematical Society
201 Charles Street
Providence, RI 02904-2294

American Mathematical Society Simons Travel Grants 756
Each AMS-Simons Travel Grant provides an early-career mathematician with $2,000 per year for two years to be used for research-related travel. Sixty awards will be made. Each applicant will choose a mentor with whom to interact during the two-year period of the grant. The ideal mentor is someone who is an active senior participant in the applicant's research area and or a trusted senior colleague in the applicant's department. The mentor will be expected to interact with the awardee on a regular basis throughout the two-year period of the grant. In particular, the mentor will consider the appropriateness of proposed travel with the awardee. The awardee must receive approval from his or her mentor before each trip.
Requirements: Applicants must be located in the United States (or be U.S. citizens employed outside the U.S.) and must have completed the PhD within the last four years.
Geographic Focus: All States
Amount of Grant: Up to 2,000 USD
Contact: Steven Ferrucci, (401) 455-4113 or (401) 455-4000; ams@ams.org
Internet: http://www.ams.org/programs/travel-grants/AMS-SimonsTG
Sponsor: American Mathematical Society
201 Charles Street
Providence, RI 02904-2294

American Mathematical Society Travel Grant for the Mathematical Congress of the Americas 757
The American Mathematical Society Travel Grant Program for the Mathematical Congress of the Americas is intended to provide partial travel support for up to sixty U.S.-based mathematicians to attend the inaugural Mathematical Congress of the Americas. It is expected that this travel grant program will provide travel support for both U.S.-based invited speakers (senior mathematicians) and early career mathematians; in addition, it is expected that the standard award amount will be up to $1,600 for invited speakers (senior mathematicians) and up to $1,800 for early career mathematicians. Applications will be evaluated by a panel of mathematical scientists under the terms of a proposal submitted to the National Science Foundation (NSF) by the Society.

Requirements: Support is available for U.S. mathematicians who are in one of the following categories: early career mathematicians within six years of their doctorate; MCA Invited Speakers from U.S. institutions who desire funding.
Restrictions: Mathematicians accepting grants for partial support of the travel to MCA may not supplement them with any other NSF funds.
Geographic Focus: All States
Amount of Grant: Up to 1,800 USD
Contact: Steven Ferrucci, (401) 455-4113 or (401) 455-4000; ams@ams.org
Internet: http://www.ams.org/programs/travel-grants/mca
Sponsor: American Mathematical Society
201 Charles Street
Providence, RI 02904-2294

American Montessori Society Thesis and Dissertation Awards 758
The American Montessori Society offers annual awards for graduate level work that furthers the public understanding of Montessori education. The awards are-Outstanding Doctoral Dissertation: First Place: $1,000, Second Place: $500; Outstanding Master's Thesis: First Place, $750, Second Place: $250.
Requirements: Candidates whose works have been accepted by an accredited university and meet Assessment Criteria (see website) are invited to submit a CD (plus hard copy) of their dissertation or thesis, for consideration. To ensure anonymity in reading by the committee, all identifying factors should be removed from the copies wherever possible.
Geographic Focus: All States
Date(s) Application is Due: Nov 1
Amount of Grant: 250-1,000 USD
Samples: 2009: Erin Hennigan, Assessment and Instructional Decision-Making in Montessori Early Childhood Classrooms (Master's thesis, first-place award); 2008: Linda Gatewood Massey, Pilgrims and Guides: A Phenomenological Study of Montessori Teachers Guiding and Being Guided by Children in Public Montessori Schools (Doctoral Dissertation); Michelle K. Yezbick, How Montessori Educators in the U.S. Address Culturally Responsive Education (Master's thesis, first-place award); Alison Stern, Observational Assessment of Literacy Development: The Use of Running Records in the Montessori Classroom (Master's thesis, second-place award).
Contact: Dr. Phyllis Povell; (212) 358-1250; fax (212) 358-1256; wwprof@optonline.net
Internet: http://www.amshq.org/society_thesis.htm
Sponsor: American Montessori Society
281 Park Avenue South
New York, NY 10010-6102

American Numismatic Society Doctoral Fellowships 759
The ANS administers a doctoral fellowship program, as well as a museum internship, and other grants. The society will award a doctoral fellowship to a university graduate student in the United States or Canada in the fields of the humanities or the social sciences, who will be writing a dissertation during the academic year on a topic in which the use of numismatic evidence plays a significant part. Contact the Society for details.
Requirements: Candidates must have completed the general examinations or the equivalent for the doctorate and must have attended the society's graduate seminar.
Geographic Focus: All States
Date(s) Application is Due: Mar 1
Contact: Joanne Isaac, Museum Administrator; (212) 571-4470, ext. 112; fax (212) 571-4479; isaac@numismatics.org or info@amnumsoc.org
Internet: http://www.amnumsoc.org/about/study.html
Sponsor: American Numismatic Society
75 Varick Street, 11th Floor
New York, NY 10013

American Numismatic Society Donald Groves Numismatic Research Grants 760
The fund supports research and publication in the field of early U.S. numismatics involving material dated no later than 1800. Funds are available for travel and other expenses in association with research, as well as for publication costs. Applications in letter form are welcome at any time, setting forth the proposed research project, the method of accomplishing the research, and a budget supporting the funding requested.
Geographic Focus: All States
Contact: Joanne Isaac, Museum Administrator; (212) 571-4470, ext. 112; fax (212) 571-4479; isaac@numismatics.org or info@amnumsoc.org
Internet: http://www.amnumsoc.org/about/grove.html
Sponsor: American Numismatic Society
75 Varick Street, 11th Floor
New York, NY 10013

American Numismatic Society Eric P. Newman Graduate Seminar Grants 761
The program provides graduate students and university instructors in the United States and Canada with a deeper understanding of the indispensable contributions numismatics makes to other fields of study. The program includes attendance at lectures and research on individual topics at museum headquarters, culminating in oral and written presentations. Applications are due on February 13th.
Requirements: Applicants must have completed at least one year of graduate study at a U.S. or Canadian university in classics, archaeology, history, art history, or related disciplines. Applications will also be accepted from junior college or university instructors with a degree in one of those fields.
Geographic Focus: All States
Date(s) Application is Due: Feb 13
Amount of Grant: 4,000 USD
Contact: Dr. Peter Van Alfen, Seminar Director; (212) 571-4470, ext. 153; fax (212) 571-4479; vanalfen@numismatics.org or info@amnumsoc.org
Internet: http://www.numismatics.org/about/study.html
Sponsor: American Numismatic Society
75 Varick Street, 11th Floor
New York, NY 10013

American Numismatic Society Frances M. Schwartz Fellowship 762
The fellowship supports the work and study of numismatic and museum methodology at the society. The purpose of establishing the fellowship was so that the funds would be used to educate qualified students in museum practice and to train them in numismatics, as well as to provide for curatorial assistance in the Greek, Roman, and Byzantine departments. The stipend will vary with the term of tenure (normally the academic year) but will not exceed $5,000.
Requirements: Candidates must have completed the BA or the equivalent.
Geographic Focus: All States
Date(s) Application is Due: Mar 1
Amount of Grant: Up to 5,000 USD
Contact: Joanne Isaac, Museum Administrator; (212) 571-4470, ext. 112; fax (212) 571-4479; isaac@numismatics.org or info@amnumsoc.org
Internet: http://www.numismatics.org/about/schwafell.html
Sponsor: American Numismatic Society
75 Varick Street, 11th Floor
New York, NY 10013

American Philosophical Society Daland Fellowships in Clinical Investigation 763
A limited number of fellowships are awarded annually for research in clinical medicine including the fields of internal medicine, neurology, psychiatry, pediatrics, and surgery. Patient-oriented research is emphasized. Essentially 100 percent of the fellow's time will be devoted to research. Teaching or clinical service of a limited amount is permitted. Additional salary may be granted by the institution at which the fellow is located, from another fellowship, or from a similar award during the tenure of the fellowship. The term of the fellowship is one year, with possible renewal for another year. Applications are available on the Web site, or may be obtained by written request, stating when the MD or MD/PhD degree was awarded; include a self-addressed mailing label.
Requirements: These fellowships are designed for qualified persons who have held an MD, or MD/PhD degree for less than eight years. The fellowship is generally intended to be the first post-clinical fellowship; but each case will be decided on its merits. Preference is generally given to candidates who have not more than two years of post-doctoral training and research. Applicants must expect to perform their research at an institution in the United States, under the supervision of a scientific adviser. Candidates are to be nominated by their department chairman, in a letter providing assurance that the nominee will work with the guidance of a scientific adviser of established reputation who has guaranteed adequate space, supplies, etc. for the Fellow. The adviser need not be a member of the department nominating the Fellow, nor need the activities of the Fellow be limited to the nominating department. As a general rule, no more than one fellowship will be awarded to a given institution in the same year of competition.
Restrictions: The society does not provide funds for institutional overhead.
Geographic Focus: All States
Date(s) Application is Due: Sep 1
Amount of Grant: Up to 40,000 USD
Contact: Linda Musumeci; (215) 440-3429; LMusumeci@amphilsoc.org
Internet: http://www.amphilsoc.org/grants/daland.htm
Sponsor: American Philosophical Society
104 South Fifth Street
Philadelphia, PA 19106-3387

American Philosophical Society Franklin Research Grants 764
The society awards small grants to scholars to support the cost of research leading to publication in all areas of knowledge. The program is particularly designed to help meet the costs of travel to libraries and archives for research purposes; the purchase of microfilm, photocopies, or equivalent research materials; the costs associated with fieldwork; or laboratory research expenses. Franklin grants are made for noncommercial research. They are not intended to meet the expenses of attending conferences or the costs of publication. The society is particularly interested in supporting the work of young scholars who have recently received their PhDs. Applicants who have received Franklin grants may reapply after an interval of two years.
Requirements: Applicants are expected to have a doctorate, or to have published work of doctoral character and quality. American citizens and residents of the United States may use their Franklin awards at home or abroad. Foreign nationals must use their Franklin awards for research in the United States.
Restrictions: Predoctoral graduate students are not eligible. The society does not pay overhead or indirect costs to any institution. Grants will not be made to replace salary during a leave of absence or earnings from summer teaching; pay living expenses while working at home; cover the costs of consultants or research assistants; or purchase permanent equipment such as computers, cameras, tape recorders, or laboratory apparatus.
Geographic Focus: All States
Date(s) Application is Due: Oct 1; Dec 1
Contact: Linda Musumeci; (215) 440-3429; LMusumeci@amphilsoc.org
Internet: http://www.amphilsoc.org/grants/franklin.htm
Sponsor: American Philosophical Society
104 South Fifth Street
Philadelphia, PA 19106-3387

American Philosophical Society John Hope Franklin Dissertation Fellowships 765

The fellowship is designed to support an outstanding African-American graduate student attending any PhD- granting institution in the United States, in any field of knowledge. There is no residential requirement. The stipend for this fellowship is $25,000 for a twelve-month period, plus $5,000 to support the cost of residency in Philadelphia, for a total award of $30,000. The twelve-month period is flexible.

Requirements: Candidates must have completed all course work and examinations preliminary to the doctoral dissertation and be prepared to devote full time for 12 months—with no teaching obligations—to research on their dissertation projects or the writing of their dissertations.
Geographic Focus: All States
Date(s) Application is Due: Apr 1
Amount of Grant: 30,000 USD
Contact: Linda Musumeci; (215) 440-3429; LMusumeci@amphilsoc.org
Internet: http://www.amphilsoc.org/grants/johnhopefranklin.htm
Sponsor: American Philosophical Society
104 South Fifth Street
Philadelphia, PA 19106-3387

American Philosophical Society Judson Daland Prize 766

The $20,000 Judson Daland Prize recognizes outstanding achievement in patient-oriented research. The prize will be awarded at the meeting of the American Philosophical Society in November of each year. The Society reserves the right to award a shared prize. Nominations and all three letters of support must be received by March 15.

Requirements: Nominees must have done their work in an institution in the United States. Nominees need not, however, be citizens of the United States. Nominees should be no more than 15 years beyond receipt of the M.D. degree.
Geographic Focus: All States
Date(s) Application is Due: Mar 15
Amount of Grant: 20,000 USD
Contact: Linda Musumeci; (215) 440-3429; LMusumeci@amphilsoc.org
Internet: http://www.amphilsoc.org/prizes/daland.htm
Sponsor: American Philosophical Society
104 South Fifth Street
Philadelphia, PA 19106-3387

American Philosophical Society Lewis and Clark Fund for Exploration and Field Research 767

The program encourages exploratory field studies for the collection of specimens and data and to provide the imaginative stimulus that accompanies direct observation. Applications are invited from disciplines with a large dependence on field studies, such as archeology, anthropology, biology, ecology, geography, geology, linguistics, and paleontology, but grants will not be restricted to these fields. Budgets should be limited to travel and related expenses, including personal field equipment. Application and guidelines are available online.

Requirements: Grants are available to graduate students, postdoctoral students, junior and senior scientists, and social scientists who wish to participate in field studies for their theses or for other purposes. A graduate student applicant should ask his or her academic supervisor or field trip leader to write one of the two letters of recommendation, specifying the role of the student in the field trip and the educational contribution of the trip. The competition is open to U.S. residents wishing to carry out research anywhere in the world. Foreign applicants must either be based at a U.S. institution or plan to carry out their work in the United States.
Restrictions: Undergraduates are not eligible.
Geographic Focus: All States
Date(s) Application is Due: Feb 15
Amount of Grant: Up to 5,000 USD
Contact: Linda Musumeci; (215) 440-3429; LMusumeci@amphilsoc.org
Internet: http://www.amphilsoc.org/grants/lewisandclark.htm
Sponsor: American Philosophical Society
104 South Fifth Street
Philadelphia, PA 19106-3387

American Philosophical Society Lewis and Clark Fund for Exploration and Field Research in Astrobiology 768

The Lewis and Clark Fund for Exploration and Field Research in Astrobiology is open to field studies in any area of interest to astrobiology. Astrobiology is the study of the origin, evolution, distribution, and future of life on Earth and in the universe. It encompasses research in, among others, the fields of astronomy, chemistry, evolutionary biology, field and population biology, geology, microbiology, molecular biology, oceanography, paleontology, and planetary science. Astrobiology includes investigations of the geologic and fossil record to understand the conditions of the early Earth when life arose. Its scope also includes research of contemporary locations on Earth that might be similar to early earth and to environments elsewhere in our Solar System (such as on Mars, Europa, and Titan), which may be, or have been in the past, suitable for life. Astrobiology is also about understanding the characteristics of life, which requires investigations into extreme natural environments on Earth and, eventually, elsewhere. Budgets should be limited to travel and related expenses, including personal field equipment. Amounts will depend on travel costs, but will ordinarily be in the range of several hundred dollars up to about $5,000.

Requirements: Grants will be available to graduate students and post-doctoral and junior scientists who wish to participate in field studies for their theses or for other purposes. Eligibility for applicants with doctorates is limited to those five years or fewer beyond their Ph.D. or equivalent degree, although, rarely, exceptions may be made. The competition is open to those affiliated with a U.S.-based institution, who may carry out research anywhere in the world.
Geographic Focus: All States
Date(s) Application is Due: Feb 1
Amount of Grant: Up to 5,000 USD
Contact: Linda Musumeci; (215) 440-3429; LMusumeci@amphilsoc.org
Internet: http://www.amphilsoc.org/grants/astrobiology.htm
Sponsor: American Philosophical Society
104 South Fifth Street
Philadelphia, PA 19106-3387

American Philosophical Society Library Resident Research Fellowships 769

Short-term residential fellowships for up to three months encourage research in the library's collections. Outstanding historical collections and subject areas include the papers of Benjamin Franklin; the American Revolution; 18th- and 19th-century natural history; western scientific expeditions and travel including the journals of Lewis and Clark; polar exploration; the papers of Charles Wilson Peale, including family and descendants; history of genetics, eugenics, and evolution; history of biochemistry, physiology, and biophysics; 20th-century medical research; and history of physics. Application procedures are undergoing revision, and will be entirely online. You may submit a message of interest to Libfellows AT amphilsoc DOT org and you will be notified when the new procedures are in place and ready for use. The next application deadline is March 1st.

Requirements: U.S. citizens and foreign nationals who hold the PhD or the equivalent, PhD candidates who have passed their preliminary exams, and independent scholars are eligible to apply. The fellowships are intended to encourage research in the library collections by scholars who reside beyond a 75-mile radius of Philadelphia.
Geographic Focus: All States
Date(s) Application is Due: Mar 1
Amount of Grant: 2,000-6,000 USD
Contact: Coordinator; (215) 440-3443; fax (215) 440-3423; Libfellows@amphilsoc.org
Internet: http://www.amphilsoc.org/grants/resident.htm
Sponsor: American Philosophical Society
104 South Fifth Street
Philadelphia, PA 19106-3387

American Philosophical Society Phillips Fund Grants for Native American Research 770

The Phillips Fund of the American Philosophical Society provides grants for research in Native American linguistics, ethnohistory, and the history of studies of Native Americans, in the continental United States and Canada. Grants are not made for projects in archaeology, ethnography, psycholinguistics, or for the preparation of pedagogical materials. The committee distinguishes ethnohistory from contemporary ethnography as the study of cultures and culture change through time. The grants are intended for such costs as travel, tapes, films, and consultants' fees but not for the purchase of books or permanent equipment. The average award is about $2,500; grants do not exceed $3,500.

Requirements: Applicants may be graduate students who have passed their qualifying examinations for either the master's or doctorate degrees; postdoctoral applicants are eligible.
Restrictions: Grants are not made for projects in archaeology, ethnography, or psycholinguistics, nor for the preparation of pedagogical materials.
Geographic Focus: All States
Date(s) Application is Due: Mar 1
Amount of Grant: Up to 3,500 USD
Contact: Linda Musumeci; (215) 440-3429; LMusumeci@amphilsoc.org
Internet: http://www.amphilsoc.org/grants/phillips.htm
Sponsor: American Philosophical Society
104 South Fifth Street
Philadelphia, PA 19106-3387

American Philosophical Society Sabbatical Fellowship in the Humanities and Social Sciences 771

The society supports fellowships in the humanities and the social sciences for faculty of U.S. universities and four-year colleges. The awards are intended for those, usually in mid-career, who have been granted a sabbatical year/research leave by the university or college, but for whom full financial support from the parent institution is available for only half an academic year. The fellowship helps defray expenses for the second half of an awarded sabbatical year. There is no restriction on where the fellow resides during the fellowship year, but an indication of the appropriateness of the available library resources should be given.

Requirements: Applicants should be based in the United States, but need not be U.S. citizens, and must not have had a financially supported leave during the past three years. The application requires a form for the cover sheet and three letters of support; details are available on the Web site.
Restrictions: The fellowship cannot be used to supplement external awards of similar purposes. Institutions are not eligible to apply.
Geographic Focus: All States
Date(s) Application is Due: Oct 15
Amount of Grant: 30,000-40,000 USD
Contact: Linda Musumeci; (215) 440-3429; LMusumeci@amphilsoc.org
Internet: http://www.amphilsoc.org/grants/sabbatical.htm
Sponsor: American Philosophical Society
104 South Fifth Street
Philadelphia, PA 19106-3387

American Psychiatric Association/AACDP Research Mentorship Award 772
This award honors an academic psychiatrist who has in a significant traditional or innovative manner, fostered the pursuit of student research within his/her university department. The nominee's contribution may be through direct mentorship of individual students, or by the promotion of novel research-oriented training activities within a department or residency program. The award consists of an inscribed plaque and a $1,500 honorarium, to be presented at the APA Annual Meeting.
Requirements: Candidates for the award need not be limited to senior, well-established candidates; innovativeness and dedication will be honored any academic level. Nominees must be members of APA.
Restrictions: Only one nomination from any department will be considered.
Geographic Focus: All States
Date(s) Application is Due: Aug 29
Amount of Grant: 1,500 USD
Contact: Harold Goldstein, (703) 907-8623; goharold@psych.org
Internet: http://psych.org/MainMenu/Research/ResearchTrainingandFunding/APAResearchAwards/APAAACDPAwardforResearchMentorship.aspx
Sponsor: American Psychiatric Association
1000 Wilson Boulevard, Suite 1825
Arlington, VA 22209-3901

American Psychiatric Association/AstraZeneca Young Minds in Psychiatry International Awards 773
The program recognizes and supports promising international young psychiatrists within five years of completing a psychiatric residency. Four unrestricted career development awards of $45,000 (USD) will be available. Awards will be made to two promising physicians from the U.S. with one in Bipolar Disorder research and one on research in Schizophrenia. An additional two awards will be made to promising physicians from countries outside the U.S., with one in Bipolar Disorder research and one on research in Schizophrenia. Three other awards of $30,000 (USD) in either Bipolar Disorder or Schizophrenia research will specifically focus on applicants from developing countries whose economies are classified by the World Bank as low income or lower middle income.
Requirements: The U.S. applicants must be citizens or permanent residents of the United States. Applications are evaluated based on evidence of academic promise; how the proposal will advance the applicant's career; and innovative or original concepts, approaches, or methods for developing the applicant's career.
Geographic Focus: All States
Date(s) Application is Due: Nov 30
Amount of Grant: 45,000 USD
Contact: Ernesto Guerra, (703) 907-7300; fax (703) 907-1085; eguerra@psych.org
Internet: http://psych.org/MainMenu/Research/ResearchTrainingandFunding/ResearchFellowships/APAAstraZenecaYoungMindsinPsychiatryInternationalAward.aspx
Sponsor: American Psychiatric Association
1000 Wilson Boulevard, Suite 1825
Arlington, VA 22209-3901

American Psychiatric Association/Janssen Resident Psychiatric Research Scholars 774
The program is intended to identify promising PGY-1, PGY-2, and PGY-3 psychiatric residents with the potential to become leaders in clinical and health services research in all areas of psychiatric research. Emphasis will be placed on special mentoring and career enrichment programs both at the APA Annual Meeting and throughout the year. An individual research mentor will be assigned to oversee the resident's fellowship. The mentors will be chosen among nationally known leaders in clinical and health services research. The mentor will advise and encourage the Scholar during the two-year fellowship. The scholar will receive $2,500 during the second year of the fellowship, to assist in their research career development (e.g., for use in developing a pilot research project, obtaining statistical consultation, or visiting potential research training programs). The program also provides funding for travel to the American Psychiatric Association (APA) Annual Meeting during both years of the fellowship.
Geographic Focus: All States
Date(s) Application is Due: Jan 15
Amount of Grant: 2,500 USD
Contact: Ernesto Guerra, (703) 907-7300; fax (703) 907-1085; eguerra@psych.org
Internet: http://psych.org/MainMenu/Research/ResearchTrainingandFunding/ResearchFellowships/APIREJanssenResidentPsychiatricResearchScholars.aspx
Sponsor: American Psychiatric Association
1000 Wilson Boulevard, Suite 1825
Arlington, VA 22209-3901

American Psychiatric Association/Kempf Award for Research Development in Psychobiological Psychiatry 775
This award recognizes a senior researcher who has made a significant contribution to research on the causes and treatment of schizophrenia as both a researcher and a mentor. A stipend will support the research career development of a young research psychiatrist working in a mentor-trainee relationship with the award winner on further research in this field. Submissions will be judged on the excellence of the nominee's overall contribution to the body of research in schizophrenia, including submission of the nominee's most significant paper or book; a description of the nominee's role as a mentor to younger colleagues in the field; and a description of the qualifications of the young research psychiatrist, as well as a detailed description of his or her career development plan.
Geographic Focus: All States
Date(s) Application is Due: Oct 14
Amount of Grant: 1,500-20,000 USD
Contact: Ernesto Guerra, (703) 907-7300; fax (703) 907-1085; eguerra@psych.org
Internet: http://psych.org/MainMenu/Research/ResearchTrainingandFunding/ResearchFellowships/KempfFundAwardforResearchDevelopmentinPsychobiologicalPsychiatry.aspx
Sponsor: American Psychiatric Association
1000 Wilson Boulevard, Suite 1825
Arlington, VA 22209-3901

American Psychiatric Association/Lilly Psychiatric Research Fellowship 776
This one-year fellowship is awarded to two postgraduate psychiatry trainees specifically to focus on research and personal scholarship. Minimal time will be devoted to teaching, patient care, consultation, or other duties. The fellowship is designed for a resident who demonstrates significant research potential, has not had extensive research training prior to residency, and is not already an established investigator. Each chairman of a department of psychiatry is invited to nominate one outstanding eligible resident for fellowship.
Requirements: Individuals who have received their MD or DO degree and who have completed residency training in general psychiatry or child psychiatry immediately prior to the time the fellowship commences are eligible. The individual must also be a member of APA.
Geographic Focus: All States
Date(s) Application is Due: Oct 14
Amount of Grant: 45,000 USD
Contact: Ernesto Guerra, (703) 907-7300; fax (703) 907-1085; eguerra@psych.org
Internet: http://psych.org/MainMenu/Research/ResearchTrainingandFunding/ResearchFellowships/APIRELillyPsychiatricResearchFellowship.aspx
Sponsor: American Psychiatric Association
1000 Wilson Boulevard, Suite 1825
Arlington, VA 22209-3901

American Psychiatric Association/Lilly Resident Research Awards 777
This award, supported by an unrestricted educational grant from Eli Lilly and Company provides an honorarium of $1,500 to five psychiatry residents who submit the best, original, unpublished scientific paper. In addition, a $1,000 award is given to the residency program in which the resident was in training at the time the work was completed. The award will be presented to the recipients during the APA Annual Meeting in New Orleans, LA. Each resident may submit one document which outlines the scope and results of the research conducted. The research study should be in the form of a finished paper suitable for publication. Please note that the paper must describe original scientific research; case reports and literature reviews will not be accepted. In order to ensure a fair and impartial review, the Award Board reviews each submission anonymously with no identifying information on the resident or the affiliated training program. It would therefore be helpful if the resident's name is not printed on each page of the document. The nomination packet should include the following: research paper; updated copy of Curriculum Vitae; signed statement by chair or program director. Electronic submissions should be sent to jalesbury@psych.org.
Geographic Focus: All States
Date(s) Application is Due: Dec 1
Amount of Grant: 2,500 USD
Contact: Ernesto Guerra, (703) 907-7300; fax (703) 907-1085; eguerra@psych.org
Internet: http://www.psych.org/MainMenu/Research/ResearchTrainingandFunding/ResearchTrainingAwards/APALilly-Resident-Research-Award.aspx
Sponsor: American Psychiatric Association
1000 Wilson Boulevard, Suite 1825
Arlington, VA 22209-3901

American Psychiatric Association/Merck Co. Early Academic Career Research Award 778
The Award is intended to help support the research of a junior faculty member with an interest in sleep disorders or schizophrenia. Two separate awards will be made to candidates who have completed a psychiatry residency, at least one year of a psychiatry research fellowship, and are seeking to make a commitment to a research career with the end goal of successfully transitioning to that of an independent investigator. The Award is intended to assist in this key transition period by providing one year of funding of $45,000. Salary support provided by this award will allow junior faculty in departments of psychiatry at U.S. academic institutions to devote more time to research.
Requirements: Candidates must be citizens or permanent residents of the U.S. and should also be APA members. Eligible candidates will be trained psychiatrists with an MD, MD/Ph.D. or DO degree who have completed residency training in general psychiatry or child psychiatry in the U.S. Candidates also should have completed at least one year of a psychiatry research fellowship but are not 3 or more years post fellowship completion, and do not currently hold an academic rank higher than Assistant Professor.
Restrictions: Individuals who have obtained K-awards or are considered an independent Principal Investigator on R-01 or R-21 research grants from NIH are not eligible.
Geographic Focus: All States
Date(s) Application is Due: Oct 14
Amount of Grant: 45,000 USD
Contact: Ernesto Guerra, (703) 907-7300; fax (703) 907-1085; eguerra@psych.org
Internet: http://www.psych.org/research/apire/res_careerdev/apamerck.cfm
Sponsor: American Psychiatric Association
1000 Wilson Boulevard, Suite 1825
Arlington, VA 22209-3901

American Psychiatric Association/SmithKline Beecham Young Faculty Award for Research Development in Biological Psychiatry 779

This award is designed to support research by a junior faculty member in the biology and psychopharmacology of mood disorders and/or anxiety disorders. Applicants must hold a tenure track position as an assistant professor in the psychiatry department at a school of medicine in the United States.
Requirements: APA members with a MD or DO degree who have completed residency training in general or child psychiatry are eligible for this award.
Geographic Focus: All States
Date(s) Application is Due: Oct 14
Amount of Grant: 45,000 USD
Contact: Darrel A. Regier, (202) 682-6316 or (800) 852-1390; apa@psych.org
Internet: http://www.psych.org/research/apire/training_fund/beecham.cfm
Sponsor: American Psychiatric Association
1000 Wilson Boulevard, Suite 1825
Arlington, VA 22209-3901

American Psychiatric Association/Wyeth Pharmaceuticals M.D./Ph.D. Psychiatric Research Fellowships 780

The Fellowship provides one year of funding and is designed to support two post-graduate psychiatry trainees with research experience, specifically to focus on research and personal scholarship. Minimal time will be devoted to teaching, patient care, consultation, or other duties. The protection of time for research should be assured by the department chairman.
Requirements: Individuals who have received their MD or DO and a PhD degree and who have completed residency training in general psychiatry or child psychiatry immediately prior to the time the fellowship commences are encouraged to apply. The Fellowship is designed for residents who: (a) have demonstrated significant research potential, (b) have had research training (i.e. as part of their work towards their PhD), and (c) are not already an established investigators. These individuals must also be members of the APA.
Geographic Focus: All States
Date(s) Application is Due: Oct 14
Amount of Grant: 45,000 USD
Contact: Darrel A. Regier, (202) 682-6316 or (800) 852-1390; apa@psych.org
Internet: http://www.psych.org/MainMenu/EducationCareerDevelopment/ResidentsMembersinTraining/AwardsandFellowships.aspx
Sponsor: American Psychiatric Association
1000 Wilson Boulevard, Suite 1825
Arlington, VA 22209-3901

American Psychiatric Association Award for Research in Psychiatry 781

The Award is given in recognition of a single distinguished contribution, a body of work, or a lifetime contribution that has had a major impact on the field and/or altered the practice of psychiatry. The Award is intended to cover the full spectrum of psychiatric research. The Award consists of a $5,000 prize and an honorary plaque to be presented at APA's Annual Meeting in May. The Award also includes an honorary lecture by the awardee.
Requirements: Candidates for the Award must be citizens of the United States or Canada and be nominated by a sponsor. Sponsors must be members of the American Psychiatric Association.
Restrictions: Members of the Award Committee are excluded from submitting nominations.
Geographic Focus: All States
Date(s) Application is Due: Aug 28
Amount of Grant: 5,000 USD
Contact: Harold Goldstein, (703) 907-8623; goharold@psych.org
Internet: http://www.psych.org/research/apire/training_fund/psychaward.cfm
Sponsor: American Psychiatric Association
1000 Wilson Boulevard, Suite 1825
Arlington, VA 22209-3901

American Psychiatric Association Early Career Award 782

APA's early career award recognizes the best nominated paper published during the past year by an early career psychiatrist (defined by the APA as under 40 years of age or within five years of training). Nominations can be either from the individual or a colleague in the field, such as a department chairperson, division chief, or other health services researcher. While the proposed applicant must be an APA member, the nominating individual need not be a member and may be from any discipline. The award will be presented at the Health Services Research Breakfast which is held in conjunction with APA's Institute for Psychiatric Services (IPS) meeting.
Requirements: Nomination letters should succinctly indicate the contributions that are the basis for the nomination and the nature of the relationship of the nominator and nominee. A curriculum vita of the nominee should accompany the letter, along with the nominated paper (in the case of early career award) or one to two papers of greatest significance (for senior scholar award).
Geographic Focus: All States
Date(s) Application is Due: Aug 29
Amount of Grant: 1,000 USD
Contact: Harold Goldstein, (703) 907-8623; goharold@psych.org
Internet: http://psych.org/MainMenu/Research/ResearchTrainingandFunding/APAResearchAwards/EarlyCareerandSeniorScholarHealthServicesResearchAwards.aspx
Sponsor: American Psychiatric Association
1000 Wilson Boulevard, Suite 1825
Arlington, VA 22209-3901

American Psychiatric Association Minority Fellowships Program 783

The APA offers psychiatric residents one-year fellowships in the APA organization to increase the knowledge of cultural factors influencing psychiatric diagnosis and treatment, provide opportunities for these residents to participate in the deliberations and decision-making processes, and provide role models for the fellows. Each fellow is appointed to a component of the APA's organizational structure and attends the association's annual meeting as an observer and active participant. Fellows are selected on the basis of their commitment to serving underrepresented populations, demonstrated leadership abilities, and interest in the interrelationship between mental health/illness and transcultural factors.
Requirements: Psychiatric residents who are starting their second year of psychiatric training, or third year if they are in a four-year residency program, are eligible to apply.
Geographic Focus: All States
Contact: Marilyn King; (703) 907-8653; fax (703) 907-7849; mking@psych.org
Internet: http://www.psych.org/Resources/OMNA/MFP.aspx
Sponsor: American Psychiatric Association
1000 Wilson Boulevard, Suite 1825
Arlington, VA 22209-3901

American Psychiatric Association Minority Medical Student Fellowship in HIV Psychiatry 784

The APA invites ethnic minority medical students who have an interest in psychiatric issues to apply. The program is intended to identify minority medical students who have primary interests in services related to HIV/AIDS and substance abuse and its relationship to the mental health or psychological well being of ethnic minorities.
Requirements: These programs are open to ethnic minority medical students who are U.S. citizens or permanent residents currently enrolled in a U.S. medical school.
Geographic Focus: All States
Date(s) Application is Due: Mar 31
Contact: Diane Pennessi; (703)907-8668; fax (703)907-8668; dpennessi@psych.org
Internet: http://www.psych.org/Resources/OMNA/MFP/HIV-Information-10.aspx
Sponsor: American Psychiatric Association
1000 Wilson Boulevard, Suite 1825
Arlington, VA 22209-3901

American Psychiatric Association Minority Medical Student Summer Externship in Addiction Psychiatry 785

The APA invites ethnic minority medical students who have an interest in psychiatric issues to apply. This clinical shadowing program identifies minority medical students who may have a specific interest in services related to substance abuse treatment/prevention and provide a setting where the student can work closely with a mentor who specializes in addiction psychiatry for one month.
Requirements: These programs are open to ethnic minority medical students who are U.S. citizens or permanent residents currently enrolled in a U.S. medical school.
Geographic Focus: All States
Date(s) Application is Due: Feb 28
Amount of Grant: 1,500 USD
Contact: Marilyn King, (703) 907-8653; fax (703) 907-7852; mking@psych.org
Internet: http://www.psych.org/Resources/OMNA/MFP/minoritymedicalstudentsummerexternshipinaddictionpsychiatry.aspx
Sponsor: American Psychiatric Association
1000 Wilson Boulevard, Suite 1825
Arlington, VA 22209-3901

American Psychiatric Association Minority Medical Student Summer Mentoring 786

The APA invites ethnic minority medical students who have an interest in psychiatric issues to apply. This program is intended to identify ethnic minority medical students who have an interest in psychiatric issues and expose students to a setting where they can work closely with a psychiatrist mentor for one month.
Requirements: These programs are open to ethnic minority medical students who are U.S. citizens or permanent residents currently enrolled in a U.S. medical school.
Geographic Focus: All States
Date(s) Application is Due: Feb 28
Amount of Grant: 1,500 USD
Contact: Marilyn King; (703) 907-8653; fax (703) 907-7852; mking@psych.org
Internet: http://www.psych.org/Resources/OMNA/MFP/minoritymedicalstudentsummermentoringprogram.aspx
Sponsor: American Psychiatric Association
1000 Wilson Boulevard, Suite 1825
Arlington, VA 22209-3901

American Psychiatric Association Minority Research Training in Psychiatry 787

The program is designed to increase the number of underrepresented minority men and women in the field of psychiatric research. Research training offers the opportunity to engage in scientific investigation across the full array of disciplines, from basic neuroscience, genetics, and pharmacology to the cognitive behavioral, and social sciences, clinical psychiatry, and mental health services research. Research exposure can help students and trainees develop sound skills for clinical assessment and treatment planning. The program provides funding for short and long-term training opportunities at three levels: Medical School, Residency and Post-residency. National competitions also enable qualified mini-fellows to attend research-oriented meetings of psychiatric organizations. For medical students and residents, the duration can be two months to one year. For post-residency fellows, the duration is generally two years. A third year of fellowship support

may be available if appropriate to a trainee's career development. Support from PMRTP falls into three categories: stipends, travel, and tuition and fees. Stipends are based on the trainee's years of relevant experience and the length of the research training experience.
Requirements: Preference in selection is given to underrepresented minorities such as American Indians, Asian-Americans, Blacks/African-Americans, Hispanics, Pacific Islanders, or other ethnic or racial group members found to be underrepresented in biomedical or behavioral research.
Geographic Focus: All States
Date(s) Application is Due: Apr 1; Dec 1
Amount of Grant: 20,772-51,036 USD
Contact: Ernesto Guerra, (703) 907-7300; fax (703) 907-1085; eguerra&@psych.org
Internet: http://www.psych.org/MainMenu/Research/ResearchTrainingandFunding/ResearchFellowships/ProgramforMinorityResearchTraininginPsychiatry.aspx
Sponsor: American Psychiatric Association
1000 Wilson Boulevard, Suite 1825
Arlington, VA 22209-3901

American Psychiatric Association Research Colloquium for Junior Investigators 788
The purpose of the colloquium is to provide guidance, mentorship and encouragement to young investigators in the early phases of their training. Junior investigators will have an opportunity to obtain feedback about their past, present, and future research interests from mentors who are tops in their field in a small group setting as well as general information about research career development and grantsmanship. Candidates whose research interests are similar to those listed below are also encouraged to apply. An all-day workshop for junior psychiatric investigators will focus on these three areas: childhood disorders including autism, ADHD, conduct, eating, and mood disorders; genomics, epigenetics, and proteomics; and treatment of major psychiatric disorders: from substance abuse to schizophrenia.
Requirements: Psychiatrists who are senior residents, fellows, or junior faculty, and who have an interest and potential in developing research careers in the areas of research listed above. Participants should hold a medical degree or be a member of the APA; or be eligible to become members of the APA.
Restrictions: Those with individual federal research awards are not eligible.
Geographic Focus: All States
Date(s) Application is Due: Nov 16
Amount of Grant: 1,000 USD
Contact: Ernesto Guerra, (703) 907-7300; fax (703) 907-1085; eguerra&@psych.org
Internet: http://psych.org/MainMenu/Research/ResearchTrainingandFunding/ResearchColloquiumforJuniorInvestigators.aspx
Sponsor: American Psychiatric Association
1000 Wilson Boulevard, Suite 1825
Arlington, VA 22209-3901

American Psychiatric Association Senior Scholar Health Services Research Award 789
The Award recognizes singular or sustained research accomplishments by a researcher beyond early career status which have made an important contribution to the field of mental health services research. Nominations can be either from the individual or a colleague in the field, such as a department chairperson, division chief, or other health services researcher. While the proposed applicant must be an APA member, the nominating individual need not be a member and may be from any discipline. The award will be presented at the Health Services Research Breakfast which is held in conjunction with APA's Institute for Psychiatric Services (IPS) meeting.
Requirements: Nomination letters should succinctly indicate the contributions that are the basis for the nomination and the nature of the relationship of the nominator and nominee. A curriculum vita of the nominee should accompany the letter, along with the nominated paper (in the case of early career award) or one to two papers of greatest significance (for senior scholar award).
Geographic Focus: All States
Date(s) Application is Due: Aug 29
Amount of Grant: 1,000 USD
Contact: Harold Goldstein, (703) 907-8623; goharold@psych.org
Internet: http://psych.org/MainMenu/Research/ResearchTrainingandFunding/APAResearchAwards/EarlyCareerandSeniorScholarHealthServicesResearchAwards.aspx
Sponsor: American Psychiatric Association
1000 Wilson Boulevard, Suite 1825
Arlington, VA 22209-3901

American Psychiatric Association Travel Scholarships for Minority Medical Students 790
The APA invites ethnic minority medical students who have an interest in psychiatric issues to apply. The program supports travel and related costs for approximately 10 minority medical students interested in psychiatric to attend either the APA annual meeting in May or the Institute on Psychiatric Services (IPS) meeting in October. This program is a way for medical students to witness organized psychiatry at work and to learn more about the field. Not only will students attend sessions for experts and trainees alike, but they will be assigned to a mentor who will help them maximize their annual meeting or IPS experience and discuss career plans and resident training programs.
Requirements: These programs are open to U.S. ethnic minority medical students who are U.S. citizens or permanent residents currently enrolled in a U.S. medical school. Ethnic minorities are: American-Indian, Alaska Native, Native Hawaiian, Asian-American, African-American, and Hispanic/Latino.
Geographic Focus: All States
Date(s) Application is Due: Jan 29
Contact: Marilyn King, (703) 907-8653; fax (703) 907-7852; mking@psych.org
Internet: http://www.psych.org/Resources/OMNA/MFP/MedicalStudentTravelScholarships.aspx
Sponsor: American Psychiatric Association
1000 Wilson Boulevard, Suite 1825
Arlington, VA 22209-3901

American Psychiatric Foundation Call for Proposals 791
APF support allows organizations across the country to make a difference through unique educational, informational and outreach initiatives that promote the early recognition and treatment of mental illness and encourage leadership in the field of psychiatry. The Call for Proposals program is intended to fund public education programs, as well as information and outreach initiatives that promote the early recognition and treatment of mental illness. Up to $750,000 in grant funds are available over the course of three years. Average grants are in the $50,000 range. Grant making under this program has been temporarily suspended while a new Call for Proposals is being developed. Contact the office for further updates.
Requirements: Organizations that have been in existence for at least two years and currently maintain a 501(c)3 charitable status and American Psychiatric Association District Branches and subsidiaries. Organizations need not be mental health programs.
Geographic Focus: All States
Amount of Grant: Up to 100,000 USD
Contact: Paul T. Burke, Executive Director; (703) 907-8518 or (703) 907-8512; fax (703) 907-7851; pburke@psych.org or apf@psych.org
Internet: http://www.psychfoundation.org/GrantAndAwards/Grants/CallforProposals.aspx
Sponsor: American Psychiatric Foundation
1000 Wilson Boulevard, Suite 1825
Arlington, VA 22209

American Psychiatric Foundation James H. Scully Jr., M.D., Educational Grants 792
The fund supports education and research within the American Psychiatric Association (APA), its subsidiary organizations, or affiliated organizations, including district branches, state associations, or other psychiatric organizations that are in partnership with the APA. The fund can be used to support activities that improve the quality of care for patients with mental illness, advance the prevention of mental illness, raise awareness of mental illness and its treatments, and increase access to mental health services. Requests for funding should be sent in writing to the foundation and include the following: identification and significance of need; description of program/activity to address that need; expected outcomes and evaluation process; budget and time line; and background of key project personnel. Funding decisions are made by the foundations board of directors.
Restrictions: The fund will not support public policy or lobbying activities.
Geographic Focus: All States
Contact: Paul T. Burke, Executive Director; (703) 907-8518 or (703) 907-8512; fax (703) 907-7851; pburke@psych.org or apf@psych.org
Internet: http://www.psychfoundation.org/GrantAndAwards/Grants/JamesHScullyEducationalFund.aspx
Sponsor: American Psychiatric Foundation
1000 Wilson Boulevard, Suite 1825
Arlington, VA 22209

American Psychiatric Foundation Jeanne Spurlock Congressional Fellowship 793
The Jeanne Spurlock Congressional Fellowship provides general psychiatry and child psychiatry residents an opportunity to work in a congressional office or committee on federal health policy, particularly policy related to child and minority issues. The recipient will serve a ten-month fellowship in Washington, DC. The fellow will be introduced to the structure and development of federal and congressional health policy procedures, with a focus on mental health issues affecting minorities and under-served populations, including children. Fellows traditionally help develop legislative proposals, track and analyze legislative initiatives, arrange hearings, and brief Members of Congress and their staff, and interact with their constituents. During the fellowship, recipients have opportunities to interact with health policymakers and advocacy/professional groups, including the APA.
Requirements: PGY-II and III general psychiatry residents, child psychiatry residents, or child psychiatrists who will be out of training for less than one year at the time of the fellowship may apply. Applicants must be U.S. citizens or permanent residents. The recipient will be required to submit a written summary of the Fellowship experience at the end of the fellowship and may make recommendations or suggestions for improving the fellowship.
Geographic Focus: All States
Date(s) Application is Due: Mar 14
Amount of Grant: 35,000-37,500 USD
Contact: Paul T. Burke, Executive Director; (703) 907-8518 or (703) 907-8512; fax (703) 837-5451; pburke@psych.org or apf@psych.org
Internet: http://www.psychfoundation.org/GrantAndAwards/Fellowships/SpurlockFellowship.aspx
Sponsor: American Psychiatric Foundation
1000 Wilson Boulevard, Suite 1825
Arlington, VA 22209

American Psychiatric Foundation Minority Fellowships 794
The Minority Fellowships Program (MFP) endeavors to eliminate racial and ethnic disparities in mental health and substance abuse care by providing specialized training to psychiatry residents and medical students interested in serving minority communities. The MFP is designed to: provide fellowship recipients with enriching training experiences through participation in the APA September and Annual Meetings; provide recipients

with resources to support activities that enhance culturally relevant aspects of their training program; stimulate their interest in pursuing training in areas of psychiatry where minority groups are underrepresented, such as research, child psychiatry, and addiction psychiatry; and develop leadership to improve the quality of mental health care for the following federally recognized ethnic minority groups: American Indians, Native Alaskans, Asian Americans, Native Hawaiians, Native Pacific Islanders, African Americans and Hispanics/Latinos.
Geographic Focus: All States
Contact: Paul T. Burke, Executive Director; (703) 907-8518 or (703) 907-8512; fax (703) 837-5451; pburke@psych.org or apf@psych.org
Internet: http://www.psychfoundation.org/GrantAndAwards/Fellowships/MFP.aspx
Sponsor: American Psychiatric Foundation
1000 Wilson Boulevard, Suite 1825
Arlington, VA 22209

American Rivers Anthony A. Lapham River Conservation Fellowship **795**
The Anthony A. Lapham River Conservation Fellowship at American Rivers provides an excellent professional development opportunity for talented post-graduates pursuing careers as leaders in the field of conservation advocacy. The program will develop the next generation of skilled conservation leaders who can promote practical solutions that benefit natural and human communities. Recent master's degree graduates will focus on an applied research project that will make a tangible contribution to AR's mission. The 12-month Fellowship will be supported by a team of conservation staff and members of our Scientific and Technical Advisory Committee, which includes some of the nation's foremost experts on freshwater and other conservation science and policy.
Geographic Focus: All States
Date(s) Application is Due: Feb 15
Amount of Grant: 35,000 USD
Contact: Ann C. Mills, (202) 347-7550; fax (202) 347-9240; amills@americanrivers.org or rivergrants@americanrivers.org
Internet: http://www.americanrivers.org/newsroom/press-releases/2007/new-lapham-fellowship.html
Sponsor: American Rivers
1101 14th Street NW, Suite 1400
Washington, DC 20005

American Schlafhorst Foundation Grants **796**
The foundation awards grants to eligible North Carolina nonprofit organizations in its areas of interest, including arts, children and youth, elderly, education, health care, science, and social services delivery. Types of support include building construction/renovation, equipment acquisition, general operating grants, research grants, scholarships, and seed money grants.
Requirements: 501(c)3 organizations serving the greater Charlotte area are eligible.
Restrictions: Individuals are not eligible.
Geographic Focus: North Carolina
Amount of Grant: 1,000-40,000 USD
Contact: Grants Administrator; (704) 554-0800; info@schlafhorst.com
Sponsor: American Schlafhorst Foundation
8801 South Boulevard
Charlotte, NC 28224

American Society for Enology and Viticulture Scholarships **797**
Numerous scholarships are awarded annually to undergraduate and graduate students in North America pursuing a degree in enology, viticulture, or in a curriculum emphasizing a science basic to the wine and grape industry. Candidates are required to submit transcripts of previous education, information on financial need, and a written statement of intention to pursue a career in the wine or grape industry. Previous recipients are eligible each year in open competition with all other applicants.
Requirements: Applicant must be a resident of a North American country and accepted in a full-time college or university degree program in the required fields at the junior unit level or higher. Undergraduates must have at least a 3.0 GPA; graduate students must have at least a 3.2 GPA.
Geographic Focus: All States
Date(s) Application is Due: Mar 1
Contact: Executive Director; (530) 753-3142; fax (530) 753-3318; society@asev.org
Internet: http://www.asev.org
Sponsor: American Society for Enology and Viticulture
P.O. Box 1855
Davis, CA 95617-1855

American Society for Yad Vashem Grants **798**
Established in the State of New York in 1981, the American Society for Yad Vashem gives on an international basis with emphasis on Israel. The society is the development arm of Yad Vashem, Jerusalem, which serves to remember and honor all victims of the Holocaust. Its primary fields of interest include: historical activities; genealogy; historic preservation; support of historical societies; and Jewish agencies and synagogues. There are no specified application forms or annual deadlines, and applicants should begin by forwarding a letter of application to the Society.
Geographic Focus: All States, Israel
Contact: Eli Zborowski; (212) 220-4304; ezborowski@yadvashemusa.org
Internet: http://www.yadvashemusa.org/history.html
Sponsor: American Society for Yad Vashem
500 5th Avenue, 42nd Floor
New York, NY 10110-1699

American Society of Criminology Gene Carte Student Paper Competition **799**
This competition is held to recognize outstanding scholarly work done by students in the field of criminology. To enter the competition, students must submit papers, conceptual and/or empirical, directly related to criminology. Papers must be 7500 words or less, typewritten double-spaced, on standard-size white paper using standard format for the organization of text, citations, and references. Submissions must be accompanied by a letter indicating the author(s) enrollment status and cosigned by the dean, department chair, or program director. Author name(s), department(s), and advisor(s) must appear only on the title page since papers will be evaluated by blind review. Eight copies of the paper must be provided. Papers should be submitted to the office.
Requirements: Any student currently enrolled on a full-time basis in an academic program at either the undergraduate or graduate level may apply.
Geographic Focus: All States
Date(s) Application is Due: Apr 15
Amount of Grant: 200-500 USD
Contact: Awards Committee; (614) 292-9207; fax (614) 292-6767
Internet: http://www.asc41.com/cartesp.html
Sponsor: American Society of Criminology
1314 Kinnear Road, Suite 212
Columbus, OH 43212

American Society of Naval Engineers Scholarship Program **800**
This scholarship program is intended to encourage college students to enter into the field of naval engineering and to seek advanced education in this field. Programs of study that are supported include naval architecture; marine, mechanical, civil, aeronautical, ocean, electrical, and electronic engineering; and the physical sciences. Naval engineering includes the design, construction, and repair of ships and their installed systems and equipment, as well as research, logistic support, and the management of acquisition and maintenance. Candidates may apply for support for either the last year of a full-time or co-op undergraduate program or for one year of full-time graduate study leading to a designated engineering or physical science degree in an accredited college or university of the student's choice. Application forms are available on the Web site.
Requirements: This scholarship program is open to U.S. citizens who have expressed an interest in a career in naval engineering.
Restrictions: Doctoral candidates or those who have an advanced degree are ineligible for this scholarship. A scholarship leading to an undergraduate degree will not be continued for enrollment in a graduate program; however, a student may apply for a new award for a graduate scholarship.
Geographic Focus: All States
Date(s) Application is Due: Feb 15
Amount of Grant: 2,500-3,500 USD
Contact: Dennis Pignotti, Operations Manager; (703) 836-6727; fax (703) 836-7491; dpignotti@navalengineers.org
Internet: http://www.navalengineers.org/Programs/Scholarships/sc_info.htm
Sponsor: American Society of Naval Engineers
1452 Duke Street
Alexandria, VA 22314-3458

American Society on Aging Graduate Student Research Award **801**
The award is presented annually to a graduate student for research relevant to aging and applicable to practice. The winner will be expected to attend the joint conference of the ASA and the National Council on the Aging to present the winning paper. The award comprises a certificate, complimentary registration and one-night's lodging for the conference, presentation of research findings at a highlighted session at the conference, and complimentary one-year membership in ASA. Submission guidelines and application are available online.
Requirements: Only graduate-level research will be considered. Findings must be from completed research; submission of the conceptual framework alone is not sufficient. Applicants must either be currently enrolled in a graduate degree program or must have completed their studies no more than one year before the time of submission. Applicants must be sponsored by a faculty member. Applicants and faculty sponsors need not be ASA members.
Restrictions: Papers that have been published are not eligible for submission.
Geographic Focus: All States
Date(s) Application is Due: Oct 1
Contact: Nancy Decia, Coordinator, Education & Training; (415) 974-9610; fax (415) 974-0300; awards@asaging.org
Internet: http://www.asaging.org/asav2/awards/description_grad.cfm?submenu1=grad
Sponsor: American Society on Aging
833 Market Street, Suite 511
San Francisco, CA 94103-1824

American Sociological Association Awards **802**
Each year the American Sociological Association honors outstanding scholars and scholarship in the field of Sociology. There are curently nine ASA awards presented each August, including: Distinguished Book Award; Dissertation Award; Excellence in the Reporting of Social Issues; Jessie Bernard Award; Cox-Johnson-Frazier Award; Award for the Public Understanding of Sociology; Distinguished Career Award for the Practice of Sociology; Distinguished Contributions to Teaching Award; and W.E.B. DuBois Career of Distinguished Scholarship Award. he deadline for submission of nominations is January 31st of each year unless noted otherwise in the individual award criteria.
Geographic Focus: All States
Contact: Margaret Weigers Vitullo, Director; (202) 383-9005, ext. 323; fax (202) 638-0882; spivack@asanet.org or apap@asanet.org

Internet: http://www.asanet.org/cs/awards
Sponsor: American Sociological Association
1430 K Street NW, Suite 600
Washington, DC 20005

American Sociological Association Congressional Fellowships 803

The Fellowship brings a PhD-level sociologist to Washington, DC, to work as a staff member on a congressional committee or in a congressional office. Some congressional agencies may allow fellow placements as well. This intensive six month experience reveals the intricacies of the policy making process to the sociological fellow, and shows the usefulness of sociological data and concepts to policy issues. Each applicant should have a general idea about the area of interest, some experience in client-driven work, good writing skills, and a commitment to the policy process. The stipend for the Fellowship is $20,000 for six months and $30,000 for 11 months.
Geographic Focus: All States
Date(s) Application is Due: Feb 1
Amount of Grant: 20,000-30,000 USD
Contact: Margaret Weigers Vitullo, Director; (202) 383-9005, ext. 323; fax (202) 638-0882; spivack@asanet.org or apap@asanet.org
Internet: http://www.asanet.org/cs/root/leftnav/funding/asa_congressional_fellowship
Sponsor: American Sociological Association
1430 K Street NW, Suite 600
Washington, DC 20005

American Sociological Association Fund for the Advancement of the Discipline 804

The American Sociological Association invites submissions for the Fund for the Advancement of the Discipline (FAD) awards. Supported by the American Sociological Association through a matching grant from the National Science Foundation, the goal of this project is to nurture the development of scientific knowledge by funding small, groundbreaking research initiatives and other important scientific research activities such as conferences. FAD awards provide scholars with seed money for innovative research that has the potential for challenging the discipline, stimulating new lines of research, and creating new networks of scientific collaboration. The award is intended to provide opportunities for substantive and methodological breakthroughs, broaden the dissemination of scientific knowledge, and provide leverage for acquisition of additional research funds. The amount of each award shall not exceed $7,000. Applications must be received online.
Geographic Focus: All States
Date(s) Application is Due: Jun 15; Dec 15
Amount of Grant: Up to 7,000 USD
Contact: Margaret Weigers Vitullo, Director; (202) 383-9005, ext. 323; fax (202) 638-0882; apap@asanet.org
Internet: http://www.asanet.org/cs/funding/FAD
Sponsor: American Sociological Association
1430 K Street NW, Suite 600
Washington, DC 20005

American Sociological Association Minority Fellowships 805

Through its Minority Fellowship Program (MFP), the American Sociological Association (ASA) supports the development and training of sociologists of color in mental health and drug abuse research. Funded by a training grant sponsored by the National Institute of Mental Health (NIMH) and co-funded by the National Institute of Drug Abuse (NIDA), MFP seeks to attract talented doctoral students to ensure a diverse and highly trained workforce is available to assume leadership roles in research related to the nation's mental health and drug abuse research agendas.
Requirements: American citizens and permanent visa residents including, but not limited to, persons who are African American, Latino/Hispanic (e.g., Chicano, Cuban, Puerto Rican), Native American, and Asian American (e.g., Chinese, Japanese, Korean), and Pacific Islanders (e.g., Hawaiian, Guamanian, Samoan, Filipino) are eligible. The program is open to students beginning or continuing study in graduate sociology departments. New students must qualify for acceptance at accredited institutions of higher learning and express a commitment to sociological research on mental health. Upon completion of their support, recipients are expected to engage in behavioral research or teaching, or a combination thereof, for a period equal to the period of support beyond 12 months.
Geographic Focus: All States
Date(s) Application is Due: Jan 31
Amount of Grant: 20,000-25,000 USD
Contact: Margaret Weigers Vitullo, Director; (202) 383-9005, ext. 323; fax (202) 638-0882; spivack@asanet.org or apap@asanet.org
Internet: http://www.asanet.org/cs/root/leftnav/funding/minority_fellowship_program
Sponsor: American Sociological Association
1430 K Street NW, Suite 600
Washington, DC 20005

American Sociological Association Sydney S. Spivack Program in Applied Social Research and Social Policy 806

The program is intended to encourage sociologists to undertake community action projects that bring social science knowledge, methods, and expertise to bear in addressing community-identified issues and concerns. Sociologists are expected to work in relevant community organizations, local public interest groups, or community action projects doing such activities as needs assessments, empirical research relevant to community activities or action planning, the design and implementation of evaluation studies, or analytic review of the social science literature related to a policy issue or problem. Grants cover direct costs associated with the project. Approximately four awards are made annually.
Requirements: Applications are encouraged from sociologists in academic settings, research institutions, private and nonprofit organizations, and government.
Restrictions: Advanced graduate students are eligible, but the funding cannot be used to support doctoral dissertation research.
Geographic Focus: All States
Date(s) Application is Due: Feb 1
Amount of Grant: 1,000-2,500 USD
Contact: Margaret Weigers Vitullo, Director; (202) 383-9005, ext. 323; fax (202) 638-0882; spivack@asanet.org or apap@asanet.org
Internet: http://www.asanet.org/cs/root/leftnav/funding/community_action_research_initiative
Sponsor: American Sociological Association
1430 K Street NW, Suite 600
Washington, DC 20005

American Sociological Association Teaching Enhancement Fund Grants 807

The goal of this program is to support projects that extend the quality of teaching in the United States and Canada. TEF small grants can support an individual, a program, a department, or a committee of a state/regional association. ASA will award up to two grants, each up to $2,000. Proposals limited to a maximum of five pages should describe the project and the intended audience or beneficiaries, explain how the financial support would be used, describe the expected benefits of the project including systemic impacts, and indicate how the project might have a lasting benefit.
Requirements: An individual, department, program, or committee of a state/regional association may apply. Individuals applying for the award must be members of ASA.
Geographic Focus: All States
Date(s) Application is Due: Feb 2
Amount of Grant: Up to 2,000 USD
Contact: Margaret Weigers Vitullo, Director; (202) 383-9005, ext. 323; fax (202) 638-0882; apap@asanet.org
Internet: http://www.asanet.org/cs/root/leftnav/funding/teaching_enhancement_fund
Sponsor: American Sociological Association
1430 K Street NW, Suite 600
Washington, DC 20005

American Wine Society Educational Foundation Scholarship Program 808

The foundation provides academic scholarships and research grants to graduate students based on academic excellence and genuine interest in pursuing a career in enology, viticulture, or the responsible use and health aspects of wine.
Requirements: Awards are made to North American (U.S., Canada, Mexico, and islands of Caribbean) citizens who have been accepted into a graduate program in enology, viticulture, or a wine-related area. Applicants must complete the undergraduate degree with a 2.8 overall academic average and a 3.0 in their major area of study.
Geographic Focus: All States, Canada
Date(s) Application is Due: Mar 31
Amount of Grant: 2,500 USD
Contact: Dr. Les Sperling, (610) 758-3845 or (610) 865-2401; fax (610) 758-3526; lhs0@lehigh.edu or Sperling@AmericanWineSociety.com
Internet: http://www.americanwinesociety.com/web/scholarship_eligibility.htm
Sponsor: American Wine Society Educational Foundation
1134 Prospect Avenue
Bethlehem, PA 18018-4914

amfAR Fellowships 809

An amfAR fellowship is a grant that encourages the postdoctoral (M.D., Ph.D., or equivalent) investigator with limited experience in the field to advance a career in HIV/AIDS research. Fellowships are awarded for two years and may not be renewed for additional funding. amfAR fellows and mentors must be affiliated with the same nonprofit institution. The applicant's interest in a career in HIV/AIDS will be demonstrated by previous relevant work at the postdoctoral fellow or instructor level and will be carefully evaluated. Each fellowship is funded at a total of up to $125,000: a maximum of $110,000 is allowed for personnel (salary and fringe benefits) and other research-related direct costs. It is expected that a fellow will devote the decided majority of his or her time to the approved fellowship project. Personnel costs supported by the fellowship must represent a minimum of 85% effort and be consistent with institution policy for other institution personnel of similar rank and title, regardless of source(s) of support. An additional $3,636 is provided to support attendance at amfAR-approved professional development activities, for a direct cost maximum of $113,636. The period of performance for fellowships awarded under this RFP will be for two years starting January 1.
Requirements: The fellowship applicant must be mentored by an experienced investigator who: is qualified to oversee the proposed research; has successfully supervised postdoctoral fellows; and is at the associate professor level or higher.
Restrictions: Institutional indirect costs may not exceed 10% of direct costs.
Geographic Focus: All States
Date(s) Application is Due: Jul 7
Amount of Grant: Up to 125,000 USD
Contact: Grant Administrator; (212) 806-1696 or (212) 806-1600; fax (212) 806-1601; grants@amfar.org
Internet: http://www.amfar.org/lab/grants/default.aspx?id=7455&terms=Fellowships
Sponsor: American Foundation for AIDS Research
120 Wall Street, 13th Floor
New York, NY 10005-3908

amfAR Mathilde Krim Fellowships in Basic Biomedical Research 810

amfAR, The Foundation for AIDS Research, is pleased to announce the availability of support for Mathilde Krim Fellowships in Basic Biomedical Research. Each fellowship is funded at a total of up to $125,000 (phase I): a maximum of $110,000 is allowed for personnel (salary and fringe benefits) and other research-related direct costs. It is expected that a Krim fellow will devote the decided majority of his or her time to the approved fellowship project. Personnel costs supported by the fellowship must represent a minimum of 85% effort and be consistent with institution policy for other institution personnel of similar rank and title, regardless of source(s) of support. An additional $3,636 is provided to support attendance at amfAR-approved professional development activities, for a direct cost maximum of $113,636. Contingent upon subsequent application and peer review, phase II funding for an additional consecutive year may be approved to support basic biomedical HIV/AIDS research costs (up to $50,000 total) during the first year of a tenure-track position at any U.S. or international nonprofit research institution. The period of performance for Mathilde Krim Fellowships awarded under this RFP will be from January 1 to December 31 of the following year.
Restrictions: Institutional indirect costs may not exceed 10% of direct costs.
Geographic Focus: All States
Date(s) Application is Due: Jul 7
Amount of Grant: Up to 125,000 USD
Contact: Administrator; (212) 806-1696 or (212) 806-1600; grants@amfar.org
Internet: http://www.amfar.org/lab/grants/default.aspx?id=7455&terms=Fellowships
Sponsor: American Foundation for AIDS Research
120 Wall Street, 13th Floor
New York, NY 10005-3908

amfAR Research Grants 811

Grants are awarded to support basic and clinical research projects in biomedical, humanistic, and social sciences research relevant to AIDS. In general, funds are applied to direct costs of salaries and fringe benefits for professional and technical personnel, laboratory supplies and equipment, travel, and the publication of findings. Research grants are awarded for one year without assurance of continued funding. $100,000 in direct costs plus up to 20% for indirect costs is available. Investigators are required to submit a pre-application letter of intent (LOI). Any postdoctoral investigator who is affiliated with a nonprofit institution may submit an LOI.
Requirements: Principal investigators for research grants must be faculty-level researchers affiliated with a nonprofit institution. Research grants are given to nonprofit institutions worldwide to support investigator-led projects approved by the Foundation.
Geographic Focus: All States
Amount of Grant: 100,000 USD
Samples: Columbia U (New York, NY)—to support the work of Susana Valente, for studies on microbicide development and AIDS-related research, $90,000.
Contact: Administrator; (212) 806-1696 or (212) 806-1600; grants@amfar.org
Internet: http://www.amfar.org/rfp/
Sponsor: American Foundation for AIDS Research
120 Wall Street, 13th Floor
New York, NY 10005-3908

Amgen Foundation Grants 812

Amgen seeks to: advance science education, improve quality of care and access for patients, and support resources that create sound communities where Amgen staff members live and work. Requests must be received at least 90 days in advance of the desired contribution date. Guidelines are available online.
Requirements: 501(c)3 tax-exempt organizations located in Amgen communities are eligible. Eligible grantees may include public elementary and secondary schools, as well as public colleges and universities, public libraries and public hospitals.
Restrictions: In general, Amgen does not consider requests for the following: support to individuals, fundraising or sports-related events, corporate sponsorship requests, religious organizations unless the program is secular in nature and benefits a broad range of the community, political organization or lobbying activity, labor unions, fraternal, service or veterans' organizations, private foundations, or organizations that are discriminatory.
Geographic Focus: All States
Contact: Program Contact; (805) 447-4056 or (805) 447-1000; fax (805) 447-1010
Internet: http://wwwext.amgen.com/citizenship/apply_for_grant.html
Sponsor: Amgen Foundation
1 Amgen Center Drive, MS 38-3-B
Thousand Oaks, CA 91320-1799

AMI Semiconductors Corporate Grants 813

The company makes grants to nonprofit organizations in support of the performing arts, economic development, business education, health cost containment, and social services for senior citizens. Types of support include conferences and seminars, general operating support, matching grants, multiyear grants, professorships, research, and scholarships. There are no application deadlines. Submit a letter of inquiry that includes a description of the organization and program, amount of funds requested, purpose of the request, recently audited financial statement, and proof of tax-exempt status. Contact office for grant availability.
Restrictions: The company does not support political or lobbying groups.
Geographic Focus: All States
Contact: Tamera Drake, (208) 234-6890; tamera_Drake@amis.com
Internet: http://www.amis.com/about
Sponsor: AMI Semiconductors
2300 Buckskin Road
Pocatello, ID 83201

AMNH Frank M. Chapman Memorial Fellowships 814

These fellowships are intended for postdoctoral scientists to support one year of ornithological research, both neontological and paleontological, with the possibility of renewal for a second year. These are salaried positions and have a benefits package. Fellows are expected to do the majority of their work at the museum.
Requirements: Graduate students and established ornithologists are eligible to apply for grants. Recent postdoctoral scholars and distinguished ornithologists may apply for fellowships to support a year of research at the museum or one of its field stations.
Geographic Focus: All States
Date(s) Application is Due: Nov 15
Amount of Grant: 500-2,000 USD
Contact: Grants Administrator; (212) 769-5467; fax (212) 769-5495; grants@amnh.org
Internet: http://research.amnh.org/ornithology/grants.htm
Sponsor: American Museum of Natural History
Central Park W at 79th Street
New York, NY 10024-5192

AMNH Graduate Student Fellowships 815

The program is an educational partnership with selected universities and is dedicated to the training of PhD candidates in those scientific disciplines practiced at the Museum. The university exercises educational jurisdiction over the program and awards the degree. The museum curator serves as a graduate advisor, co-major professor, or major professor. Joint programs are with Columbia University, providing students opportunities in vertebrate and invertebrate paleontology, astrophysics, earth and planetary sciences, and evolutionary biology; Cornell University in entomology; City University of New York in the Evolutionary Biology Program; Yale University in molecular biology/systematics. Fellowships cover stipend and health insurance, and awards are for one year, renewable annually for up to a maximum of four years.
Requirements: U.S. and non-U.S. citizens are eligible. Students in developing nations are particularly encouraged to apply. Applicants must have bachelors' degrees and be able to fulfill university admission requirements, which include TOEFL and Graduate Record Examinations.
Restrictions: Candidates for the master's degree are ineligible.
Geographic Focus: All States
Date(s) Application is Due: Nov 30
Contact: Grants Administrator; (212) 769-5467; fax (212) 769-5495; grants@amnh.org
Internet: http://research.amnh.org/grants/gradprog.html
Sponsor: American Museum of Natural History
Central Park W at 79th Street
New York, NY 10024-5192

AMNH Lerner-Gray Grants for Marine Research 816

The grants provide financial assistance to highly qualified persons starting careers in marine zoology. Support is limited to projects dealing with systematics, evolution, ecology, and field-oriented behavioral studies of marine animals. Research projects need not be carried out at the American Museum. Application forms and guidelines are available from the museum. Awards are announced in May.
Requirements: Eligible to apply are advanced graduate students and postdoctoral researchers at the beginning of their careers.
Geographic Focus: All States
Date(s) Application is Due: Mar 15
Contact: Grants Administrator; (212) 769-5467; fax (212) 769-5495; grants@amnh.org
Internet: http://research.amnh.org/grants/grantsprog.html
Sponsor: American Museum of Natural History
Central Park W at 79th Street
New York, NY 10024-5192

AMS/Industry Government Graduate Fellowships 817

The graduate fellowships are sponsored by major high-technology firms and government agencies and are designed to attract promising young scientists to prepare for careers in the meteorological, oceanic, and hydrologic fields. Candidates currently studying chemistry, computer sciences, engineering, environmental sciences, mathematics, and physics who intend to pursue careers in the atmospheric, oceanic, or hydrologic sciences also are encouraged to apply. Awards are based on the applicant's performance as an undergraduate student and his or her qualifications to pursue a career in the atmospheric and related oceanic and hydrologic sciences. Applications may be downloaded from the Web site.
Requirements: Students entering their first year of graduate study who wish to pursue advanced degrees in the atmospheric and related oceanic and hydrologic sciences are eligible.
Geographic Focus: All States
Date(s) Application is Due: Feb 22
Contact: Donna Sampson, Program Manager; (617) 227-2426, ext. 246; fax (617) 742-8718; dfernand@ametsoc.org
Internet: http://www.ametsoc.org/amsstudentinfo/scholfeldocs
Sponsor: American Meteorological Society
45 Beacon Street
Boston, MA 0002108-3693

AMS/Industry Minority Scholarships 818

The American Meteorological Society (AMS) Minority Scholarships help support the college educations of minority students traditionally underrepresented in the sciences, especially Hispanic, Native American, and Black/African American students, who intend to pursue careers in the atmospheric or related oceanic and hydrologic sciences. The two-year scholarships are for $3,000 for a nine-month period in the freshman year

and an additional $3,000 for a nine-month period in the sophomore year. Second year funding depends on successful completion of the first academic year.
Requirements: Minority students who will be entering their freshman year of college in the fall of the application yearare eligible to apply.
Geographic Focus: All States
Date(s) Application is Due: Feb 22
Amount of Grant: 6,000 USD
Contact: Donna Sampson, Program Manager; (617) 227-2426, ext. 246; fax (617) 742-8718; dfernand@ametsoc.org
Internet: http://www.ametsoc.org/amsstudentinfo/scholfeldocs
Sponsor: American Meteorological Society
45 Beacon Street
Boston, MA 0002108-3693

AMS Alfred Einstein Award **819**
This award is given annually to the author of the article on a musicological subject selected by a committee of scholars to be the most significant published in a periodical during the preceding calendar year. Nominations should include the name of the author, the title of the article, and the name and year of the periodical or other collection in which it was published. Additionally, the award committee will solicit the curriculum vita of each nominee.
Requirements: Eligible are Canadian and U.S. scholars in the early stages of their careers.
Geographic Focus: All States
Date(s) Application is Due: May 1
Amount of Grant: 400 USD
Contact: Robert Judd, Executive Director; (207) 798-4243 or (877) 679-7648; fax (207) 798-4254; rjudd@ams-net.org or ams@sas.upenn.edu
Internet: http://www.ams-net.org/awards/einstein.php
Sponsor: American Musicological Society
6010 College Station
Brunswick, ME 04011-8451

AMS Alvin H. Johnson 50 Dissertation Fellowships **820**
The society supports research in the various fields of music, and makes available four dissertation-year fellowships each year. Dissertation fellowships support full-time study and completion of the dissertation within the fellowship year. Preliminary application forms are sent to directors of graduate study at all North American doctorate-granting institutions. Application must be submitted online via the AMS website.
Requirements: Anyone is eligible to apply who is registered in good standing for a doctorate at a North American university and has completed all formal degree requirements except the dissertation at the time of full application.
Geographic Focus: All States
Date(s) Application is Due: Jan 15
Amount of Grant: 19,000 USD
Contact: Georgia Cowart; (207) 798-4243; a50-recomms@ams-net.org
Internet: http://www.ams-net.org/ams50.php
Sponsor: American Musicological Society
6010 College Station
Brunswick, ME 04011-8451

AMS Claude V. Palisca Award **821**
The Claude V. Palisca Award honors each year a scholarly edition or translation in the field of musicology published during the previous year in any language and in any country by a scholar who is a member of the AMS or a citizen or permanent resident of Canada or the United States, deemed by a committee of scholars to best exemplify the highest qualities of originality, interpretation, logic and clarity of thought, and communication.Three categories of musicological works are eligible for the Palisca award: translations into English of musicologically significant texts; editions of music; or editions of musicologically significant texts. The winner will receive a monetary prize and a certificate, and the finalists a certificate, conferred at the Annual Business Meeting of the Society by the chair of the committee.
Requirements: Eligible are Canadian and U.S. scholars in the early stages of their careers.
Geographic Focus: All States
Date(s) Application is Due: May 1
Contact: Robert Judd, Executive Director; (207) 798-4243 or (877) 679-7648; fax (207) 798-4254; rjudd@ams-net.org or ams@sas.upenn.edu
Internet: http://www.ams-net.org/awards/palisca.php
Sponsor: American Musicological Society
6010 College Station
Brunswick, ME 04011-8451

AMS Eileen Southern Travel Fund Grants **822**
Beginning with the 1995 Annual Meeting in New York City, the Society and the Committee on Cultural Diversity have hosted minority students at each annual conference to expose them to various aspects of the field. These elements might include introductions to representatives of graduate programs in musicology from across the country, future colleagues and mentors, listening to papers covering a broad spectrum of musicological interests, exposure to the many interest groups within the Society, and opportunities to attend a variety of concerts. The AMS also has funds available to assist minority undergraduates and terminal masters degree candidates (African-Americans, Asian-Americans, Hispanic-Americans, Native-Americans, and other traditionally under-represented groups) with travel and lodging expenses.
Geographic Focus: All States
Samples: Joy Doan, University of Michigan; Bo kyung (Blenda) Im, University of California, Los Angeles; Luci Mok, University of Toronto.
Contact: Ingrid Monson, Co-Chair; (617) 495-2791; imonson@fas.harvard.edu
Internet: http://www.ams-net.org/ccd/ccdtf-background.php
Sponsor: American Musicological Society
6010 College Station
Brunswick, ME 04011-8451

AMS Eugene K. Wolf Travel Fund for European Research Grants **823**
The Eugene K. Wolf Travel Fund is intended to encourage and assist Ph.D. candidates in all fields of musical scholarship to travel to Europe to carry out the necessary work for their dissertation on a topic in European music. The fund will award one or more travel grants each year to students attending North American universities who have completed all requirements except the dissertation for the Ph.D. in any field of musical scholarship and who need to undertake research in Europe toward the dissertation. The application should be in the form of a statement of up to 1,000 words in length describing the dissertation topic, research plan, projected itinerary, and institutions where research would occur. It should also provide a budget for travel and lodging.
Requirements: Letters of support are required from the dissertation advisor and one other scholar.
Geographic Focus: All States
Date(s) Application is Due: Mar 3
Contact: Robert Judd, Executive Director; (207) 798-4243 or (877) 679-7648; fax (207) 798-4254; rjudd@ams-net.org or ams@sas.upenn.edu
Internet: http://www.ams-net.org/awards/wolf.php
Sponsor: American Musicological Society
6010 College Station
Brunswick, ME 04011-8451

AMS Graduate Fellowships in the History of Science **824**
The goal of the graduate fellowship is to generate a dissertation topic in the history of the atmospheric, or related oceanic or hydrologic sciences, and to foster close working relations between historians and scientists. An effort will be made to place the student into a mentoring relationship with an AMS member at an appropriate institution. The fellowship is sponsored by member donations to the AMS 21st Century campaign. Fellowships cannot be deferred and must be used for the year awarded, but can be used to support research at a location away from the student's institution provided the plan is approved by the student's thesis advisor. The award carries a $15,000 stipend and will support one year of dissertation research.
Requirements: Candidates wishing to apply must be a graduate student in good standing who proposes to complete a dissertation.
Geographic Focus: All States
Date(s) Application is Due: Feb 22
Amount of Grant: 15,000 USD
Contact: Donna Sampson; (617) 227-2426, ext. 246; dfernand@ametsoc.org
Internet: http://www.ametsoc.org/amsstudentinfo/scholfeldocs/
Sponsor: American Meteorological Society
45 Beacon Street
Boston, MA 0002108-3693

AMS H. Colin Slim Award **825**
The H. Colin Slim Award honors each year a musicological article of exceptional merit, published during the previous year in any language and in any country by a scholar who is past the early stages of her or his career and who is a member of the AMS or a citizen or permanent resident of Canada or the United States. Early stages of the career is typically indicated by time from completion of the Ph.D. degree, or academic appointment at a non-tenured level, or position of the article among the initial items of the author's bibliography. The winner will receive a monetary prize and a certificate, conferred at the Annual Business Meeting and Awards Presentation of the Society by the chair of the committee.
Geographic Focus: All States
Date(s) Application is Due: May 1
Contact: Robert Judd, Executive Director; (207) 798-4243 or (877) 679-7648; fax (207) 798-4254; rjudd@ams-net.org or ams@sas.upenn.edu
Internet: http://www.ams-net.org/awards/slim.php
Sponsor: American Musicological Society
6010 College Station
Brunswick, ME 04011-8451

AMS Harold Powers World Travel Fund Grants **826**
The Harold Powers World Travel Fund is intended to encourage and assist Ph.D. candidates, post-docs, and junior faculty in all fields of musical scholarship to travel anywhere in the world to carry out the necessary work for their dissertation or other research. The Fund honors the polymathic scholar and distinguished longtime AMS member whose publications have ranged from music and language to medieval mode to Indian music to Puccini and whose interests are wider still, but always with the communicative aspects of music at their base. The application should be in the form of a statement of up to 1,000 words describing the research topic, a research plan, projected itinerary, and institutions where research would occur. The current maximum award is set at $2,000.
Requirements: Eligible applicants must currently attend or have graduated from a doctoral program in a North American university. If they seek to conduct research for their dissertation, they must have completed all other requirements for the Ph.D. (or the equivalent doctoral degree in any field of music scholarship). If they seek to conduct post-doctoral research, they should have completed the Ph.D. within the past five years.

Geographic Focus: All States
Date(s) Application is Due: Mar 3
Amount of Grant: Up to 2,000 USD
Samples: Max Katz, University of California, Santa Barbara, $2,000; Joshua Walden, Columbia University, $2,000.
Contact: Robert Judd, Executive Director; (207) 798-4243 or (877) 679-7648; fax (207) 798-4254; rjudd@ams-net.org or ams@sas.upenn.edu
Internet: http://www.ams-net.org/awards/powers.php
Sponsor: American Musicological Society
6010 College Station
Brunswick, ME 04011-8451

AMS Howard Mayer Brown Fellowships 827

Intended to increase the presence of minority scholars and teachers in musicology, the fellowship will support one year of graduate work for a member of a group historically underrepresented in the discipline. Applications are encouraged from African Americans, Native Americans, Latinos/Hispanics, and Asian Americans. Nominations may come from a faculty member of the institution at which the student is enrolled, from a member of the AMS at another institution, or directly from the student. The fellowship is awarded in alternating years. Inquiries should be addressed to the chair of the committee. Applications should be made using the online submission form.
Requirements: The fellowship will be awarded to a student who has completed at least one year of academic work at an institution with a graduate program in musicology, and who intends to complete a PhD in the field.
Geographic Focus: All States
Date(s) Application is Due: Jan 15
Amount of Grant: 17,000 USD
Contact: Martha Feldman, Committee Chair; (207) 798-4243; fax (207) 798-4254; rore@uchicago.edu or hmb-apps@ams-net.org
Internet: http://www.ams-net.org/hmb.php
Sponsor: American Musicological Society
6010 College Station
Brunswick, ME 04011-8451

AMS Janet Levy Fund for Independent Scholars 828

The purpose of the Janet Levy Fund is to support professional travel and research expenses for independent scholars who are members of the American Musicological Society. Examples of projects supported by the Levy Fund include travel to the Annual Meeting of the AMS, to a conference to read a paper or participate in an official capacity, or to archives or research libraries; research expenses; microfilms; and specialized research materials. Award amounts will range between $500 and $2,000. Proposals are evaluated by a committee of three appointed by the AMS president.
Requirements: The award is open to members of the American Musicological Society who hold the Ph.D. or similar degree (e.g., D.M.A. in Historical Performance), who are not employed as a full-time faculty member in an institution of higher learning, and who are not emeritus faculty.
Restrictions: No individual may receive a Levy Award more than once in a three-year period.
Geographic Focus: All States
Date(s) Application is Due: Jan 25; Jul 25
Amount of Grant: 500-2,000 USD
Contact: Robert Judd, Executive Director; (207) 798-4243 or (877) 679-7648; fax (207) 798-4254; rjudd@ams-net.org or ams@sas.upenn.edu
Internet: http://www.ams-net.org/awards/levy.php
Sponsor: American Musicological Society
6010 College Station
Brunswick, ME 04011-8451

AMS Lewis Lockwood Award 829

The Lewis Lockwood Award honors each year a musicological book of exceptional merit published during the previous year in any language and in any country by a scholar in the early stages of his or her career who is a member of the AMS or a citizen or permanent resident of Canada or the United States. Early stages of the career is typically indicated by time from completion of the Ph.D. degree, or academic appointment at a non-tenured level, or position of the book among the initial items of the author's bibliography. The winner will receive a monetary prize and a certificate, and the finalists a certificate, conferred at the Annual Business Meeting and Awards Presentation of the Society by the chair of the committee.
Requirements: Eligible are Canadian and U.S. scholars in the early stages of their careers.
Geographic Focus: All States
Date(s) Application is Due: May 1
Contact: Robert Judd, Executive Director; (207) 798-4243 or (877) 679-7648; fax (207) 798-4254; rjudd@ams-net.org or ams@sas.upenn.edu
Internet: http://www.ams-net.org/awards/lockwood.php
Sponsor: American Musicological Society
6010 College Station
Brunswick, ME 04011-8451

AMS M. Elizabeth C. Bartlet Fund for Research in France 830

The M. Elizabeth C. Bartlet Fund for Research in France is a memorial to M. Elizabeth C. Bartlet, one of the foremost scholars of French music from the eighteenth and nineteenth centuries. Her research was centered in Paris, where she was a commanding and beloved presence in the libraries and archives. It is to be awarded annually to one or more doctoral students at or graduates of universities in the United States and Canada to conduct doctoral or post-doctoral musicological research in France. The application should be in the form of a statement of up to 1,000 words describing the research topic, a research plan, projected itinerary, and institutions where research would occur. The current maximum award is set at $2,000.
Requirements: Eligible applicants must currently attend or have graduated from a doctoral program in a North American university. If they seek to conduct research for their dissertation, they must have completed all other requirements for the Ph.D. If they seek to conduct post-doctoral research, they should have completed the Ph.D. within the past five years. Preference will be given to applicants whose home institutions do not offer financial support for musicological research.
Geographic Focus: All States
Date(s) Application is Due: Mar 3
Amount of Grant: 2,000 USD
Contact: Robert Judd, Executive Director; (207) 798-4243 or (877) 679-7648; fax (207) 798-4254; rjudd@ams-net.org or ams@sas.upenn.edu
Internet: http://www.ams-net.org/awards/bartlet.php
Sponsor: American Musicological Society
6010 College Station
Brunswick, ME 04011-8451

AMS Noah Greenberg Award 831

The award is given annually for a distinguished contribution to the study and performance of early music, up to the end of the 17th century. The award is intended as a grant-in-aid to stimulate active cooperation between scholars and performers by recognizing and fostering outstanding contributions to historical performing practices. Both scholars and performers may apply, since the award may subsidize the publication costs of articles, monographs or editions, as well as performance, recordings, or other projects. Applicants must submit, in triplicate, a description of the project, a detailed budget, and relevant supporting materials to the committee chair.
Geographic Focus: All States
Date(s) Application is Due: Aug 15
Amount of Grant: Up to 2,000 USD
Samples: Elisabeth Le Guin—for Audience Performance Practice: A Pilot Project; Christopher Wolverton, Vox Early Music Ensemble, and Honey Meconi, University of Rochester—for Extreme Singing: Very Low Music of the Renaissance.
Contact: Robert Judd, Executive Director; (207) 798-4243 or (877) 679-7648; fax (207) 798-4254; rjudd@ams-net.org or ams@sas.upenn.edu
Internet: http://www.ams-net.org/awards/greenberg.php
Sponsor: American Musicological Society
6010 College Station
Brunswick, ME 04011-8451

AMS Otto Kinkeldey Award 832

The Otto Kinkeldey Award will honor each year a musicological book of exceptional merit published during the previous year in any language and in any country by a scholar who is past the early stages of his or her career and who is a member of the AMS or a citizen or permanent resident of Canada or the United States. Early stages of the career is typically indicated by time from completion of the Ph.D. degree, or academic appointment at a non-tenured level, or position of the book among the initial items of the author's bibliography. Nominations, including self-nominations, may be submitted at any time.
Requirements: Eligible are Canadian and U.S. scholars in the early stages of their careers.
Geographic Focus: All States
Date(s) Application is Due: May 1
Amount of Grant: 400 USD
Contact: Robert Judd, Executive Director; (207) 798-4243 or (877) 679-7648; fax (207) 798-4254; rjudd@ams-net.org or ams@sas.upenn.edu
Internet: http://www.ams-net.org/awards/kinkeldey.php
Sponsor: American Musicological Society
6010 College Station
Brunswick, ME 04011-8451

AMS Paul A. Pisk Prize 833

This prize is awarded each year to a graduate music student for a scholarly paper, to be read at the annual meeting of the society. Any paper for which the abstract has been submitted to the program committee, and the paper accepted for inclusion in the annual meeting, is eligible. Five copies of the complete text of the paper together with supporting materials to be presented must be submitted to the Philadelphia AMS office by the listed application deadline.
Requirements: The submission must be accompanied by a statement from the student's academic advisor affirming graduate-student status of the applicant as of the date of the paper's acceptance by the program committee.
Geographic Focus: All States
Date(s) Application is Due: Oct 1
Amount of Grant: 1,000 USD
Samples: Emily Abrams Ansari, University of Western Ontario/Harvard University; Jesse Rodin, Harvard University; Paul Berry, Yale University.
Contact: Robert Judd, Executive Director; (207) 798-4243 or (877) 679-7648; fax (207) 798-4254; rjudd@ams-net.org or ams@sas.upenn.edu
Internet: http://www.ams-net.org/awards/pisk.php
Sponsor: American Musicological Society
6010 College Station
Brunswick, ME 04011-8451

AMS Robert M. Stevenson Award 834
The Robert M. Stevenson Award recognizes outstanding scholarship in Iberian music. The designation Iberian music is here meant to include music composed, performed, created, collected, belonging to, or descended from the musical cultures of Spain, Portugal, and all Latin American areas in which Spanish and Portuguese are spoken. The prize will be awarded annually to a book, monograph, edition, or journal article by a member of the AMS. The publication must be written in English and must have been published during the preceding three calendar years. The award will consist of a monetary prize and a certificate.
Geographic Focus: All States
Date(s) Application is Due: May 1
Contact: Robert Judd, Executive Director; (207) 798-4243 or (877) 679-7648; fax (207) 798-4254; rjudd@ams-net.org or ams@sas.upenn.edu
Internet: http://www.ams-net.org/awards/stevenson.php
Sponsor: American Musicological Society
6010 College Station
Brunswick, ME 04011-8451

AMS Ruth A. Solie Award 835
The Ruth A. Solie Award honors each year a collection of musicological essays of exceptional merit published during the preceding calendar year in any language and in any country and edited by a scholar or scholars who are members of the AMS or citizens or permanent residents of Canada or the United States. Established to honor the editor of Musicology and Difference: Gender and Sexuality in Music Scholarship (University of California Press, 1995), a field-defining and field-changing book, the award acknowledges the value of the individual authors' contributions to the volume while recognizing the central role of the editor(s) in conceiving and shaping the whole. The winning editor(s) will receive a monetary prize and a certificate, conferred at the Annual Business Meeting and Awards Presentation of the Society by the chair of the committee.
Geographic Focus: All States
Date(s) Application is Due: May 1
Contact: Robert Judd, Executive Director; (207) 798-4243 or (877) 679-7648; fax (207) 798-4254; rjudd@ams-net.org or ams@sas.upenn.edu
Internet: http://www.ams-net.org/awards/solie.php
Sponsor: American Musicological Society
6010 College Station
Brunswick, ME 04011-8451

AMSSM Foundation Research Grants 836
The purpose of the AMSSM Foundation Research Grant Awards is to foster original scientific investigations by members of AMSSM. in this program, the Foundation welcomes research grant proposals that investigate issues within the broad discipline of sports medicine, including clinical practice, injury prevention and rehabilitation, basic science, epidemiology, and education. Grant awards are designed to provide partial support of research projects. Grantee institutions are expected to provide all necessary basic facilities and services that normally would be expected to exist in any institution qualified to undertake such research. Overhead or indirect costs will be supported to a maximum of 10% of direct costs. It is anticipated that the requested support will fall into one of three ranges. 1) $500 to $2,500 total costs annually for mailings, supplies, equipment, and technical support of small or pilot projects; 2) $2,500 to $10,000 total costs annually for a project larger in scope than the previous category and may provide partial support of research assistants or study coordinators; and 3) $10,000 to $25,000 total costs annually for a significant project that includes well documented pilot or previous studies and may provide partial salary support for key investigators. The maximum grant award is $25,000.
Requirements: The primary investigator of the grant must be an AMSSM member at the time of grant submission. Resident and fellow AMSSM members may apply as the principal investigator but must have at least one full AMSSM member listed as a co-investigator at the time of application.
Geographic Focus: All States
Date(s) Application is Due: Oct 1
Contact: Grants Administrator; (913) 327-1415; fax (913) 327-1491; office@amssm.org
Internet: http://www.newamssm.org/Research%20Grant%20Information.html
Sponsor: American Medical Society for Sports Medicine
11639 Earnshaw
Overland Park, KS 66210

AMS Undergraduate Scholarships 837
Scholarships are available to full-time students entering their final year of undergraduate study in the fall of the application year. The evaluation of applicants will be based on applicant's performance as an undergraduate student, including academic records and recommendation. Selection will be made by the AMS Executive Committee based on recommendations from the AMS Committee of Judges for Undergraduate Awards.
Requirements: Applicants must: be majoring in the atmospheric or related oceanic or hydrologic science, and/or must show clear intent to make the atmospheric or related sciences their career; be enrolled full time in an accredited U.S. institution; and have a cumulative grade-point average of at least a 3.25 on a scale of 4.0 at the time of application.
Geographic Focus: All States
Date(s) Application is Due: Feb 22
Contact: Donna Sampson; (617) 227-2426, ext. 246; dfernand@ametsoc.org
Internet: http://www.ametsoc.org/amsstudentinfo/scholfeldocs
Sponsor: American Meteorological Society
45 Beacon Street
Boston, MA 0002108-3693

AMWA Kathryn C. Bemmann New Investigator Grant on Violence Against Women 838
The grant supports a physician early in his or her career to pursue research on violence against women. The candidate should be a recent graduated physician; have a serious interest in research on violence against women; and have the ability to pursue an independent career in research. Proposals will be judged on scientific quality, research facility, and significance to the field. The grant recipient will be recognized at the annual meeting. Transportation, one-night's hotel accommodation, and complimentary meeting registration will be provided. Guidelines and application are available online.
Requirements: Physicians who have received their medical degree within the last five years are eligible.
Geographic Focus: All States
Date(s) Application is Due: Apr 30
Amount of Grant: 3,000-5,000 USD
Contact: Julie Dogil, (703) 838-0500; fax (703) 549-3864; jdogil@amwa-doc.org
Internet: http://www.amwa-doc.org/index.cfm?objectId=E22EC179-D567-0B25-5D6E0F68BB316ECB
Sponsor: American Medical Women's Association Foundation
801 N Fairfax Street, Suite 400
Alexandria, VA 22314

ANA/Grass Foundation Award in Neuroscience 839
The ANA/Grass Foundation Award was established to honor outstanding young investigators doing research in basic or clinical neuroscience. The awardee receives a $1,000 honorarium; inclusion in the annual meeting as either a poster or platform presenter; complimentary registration; a commemorative plaque; and up to $1,500 travel reimbursement.
Requirements: Eligible candidates include: physician-scientist neurology faculty members early in their careers (M.D. or M.D./Ph.D.); those who are five years or less out of postgraduate or post fellowship training; and those conducting research in neuroscience. Nominators may send a letter of recommendation and the nominee's CV and biography. Self nominations and second party nominations are accepted.
Geographic Focus: All States
Date(s) Application is Due: Feb 15
Amount of Grant: 2,500 USD
Contact: Laurie Dixon; (952) 545-6284; fax (952) 545-6073; ana@llmsi.com
Internet: http://www.aneuroa.org/i4a/pages/index.cfm?pageid=3360
Sponsor: American Neurological Association
5841 Cedar Lake Road, Suite 204
Minneapolis, MN 55416

ANA Wolfe Neuropathy Research Prize 840
The Wolfe Neuropathy Research Prize was designed to honor an outstanding investigator who has identified a new cause or treatment of axonal peripheral neuropathy. The recipient is awarded $2,000 and a plaque.
Geographic Focus: All States
Amount of Grant: 2,000 USD
Contact: Linda Scher; (952) 545-6284; fax (952) 545-6073; lindascher@llmsi.com
Internet: http://www.aneuroa.org/i4a/pages/index.cfm?pageid=3358
Sponsor: American Neurological Association
5841 Cedar Lake Road, Suite 204
Minneapolis, MN 55416

Anderson Center Residency Program 841
The Anderson Center develops, fosters, and promotes the creation of works by artists of all kinds, and provides leadership and services that help to insure a strong and healthy arts community and a greater recognition of the value of arts in society by providing residencies to artists. The Center offers short-term residencies of two weeks or one month from May-October to artists, writers, scholars, and scientists. Through a grant from the Jerome Foundation of St. Paul, the Center annually devotes the month of August to encourage the work of emerging artists from New York City and Minnesota. Each resident is provided room, board, and workspace for the length of the stay during the residency period. Each resident is asked to serve the community of Red Wing or its surrounding area, by giving a talk, lecture, reading, performance, or classroom activity. Anderson Center residencies begin on the 1st or 16th of each month and end on the 15th or last day of each month for periods of 2-4 weeks.
Restrictions: If you have been a prior resident of the Anderson Center, you must wait two years from the time of your residency to apply again.
Geographic Focus: All States
Date(s) Application is Due: Feb 1; Mar 1
Contact: Dawn Erickson, (651) 388-2009; fax (651) 388-5538; dawn@andersoncenter.org or info@andersoncenter.com
Internet: http://www.andersoncenter.org/resapp.htm
Sponsor: Anderson Center for Interdisciplinary Studies
163 Tower View Drive, P.O. Box 406
Red Wing, MN 55066

Anderson Foundation Grants 842
Established in Ohio in 1949, the Anderson Family Foundation Trust gives grants in the greater Toledo, Ohio, region, including Maumee and Columbus. Giving also to organizations located within the areas of the Anderson plants in the following states: Champaign, Illinois; Delphi, Lafayette, and Dunkirk, Indiana; and Albion, Potterville, Webberville, and White Pigeon, Michigan. The Foundation's primary fields of interest

include: agriculture; arts; children and youth services; education; the environment; government and public administration; higher education; human services; religion; secondary school education; and federated giving programs. Types of support include: annual campaigns; building and renovation programs; capital campaigns; conferences and seminars; emergency funds; general operating support; matching/challenge grants; program development; publications; research; scholarship funds; and seed money. Specific application forms are not required, and board decisions are made on the 3rd Monday of the month in March, June, September, and December. Grant amounts range up to $100,000 on occasion, though most average between $5,000 and $10,000.
Requirements: Nonprofit 501(c)3 agencies in Ohio, Illinois, Indiana, and Michigan serving areas of company operation are eligible to apply.
Restrictions: No support is available for private foundations, public high schools, or elementary schools. No grants are given to individuals, or for endowment funds, travel, or building or operating funds for churches or elementary schools.
Geographic Focus: Illinois, Indiana, Michigan, Ohio
Date(s) Application is Due: Feb 28; May 30; Aug 30; Nov 30
Contact: Fredi Heywood, Chairman; (419) 243-1706 or (419) 893-5050; fax (419) 242-5549; fredi@toledocf.org
Sponsor: Anderson Foundation
480 W. Dussel Drive, P.O. Box 119
Maumee, OH 43537-0119

Andrew Family Foundation Opportunity Knocks Grants **843**

The Opportunity Knocks program is a specific initiative created by the Foundation to focus on educational and mentoring opportunities for disadvantaged youth and families that would not otherwise be available. The Andrew Family Foundation seeks to provide expansion capital to an established organization(s) to build a new program or to enhance an existing program that addresses the issue of literacy and focuses on teaching young children in grades K-3 how to read. The program must positively affect and enhance the lives of disadvantaged youth ("program participants") and enable them to read at grade level by the end of 3rd Grade. In addition to financial support, the board members of the Andrew Family Foundation may desire to provide expertise and resources necessary to ensure the success of the program, including the pursuit of collaborative efforts to replicate ideas and to implement proven solutions to expand the program to other geographical locations. Accordingly, providing opportunities for involvement may be a plus. The program's objectives should include a plan: to develop early childhood literacy among the program participants by properly teaching them to read and develop reading skills for the future; to maximize the impact on the greatest number of program participants; and to become self-sustaining in three to five years. Preference will be given to organizations that focus on collaborating with other organizations to replicate proven solutions. The Foundation will award grant(s) of up to $250,000 over a three-year period.
Requirements: The organization must meet the following criteria for eligibility: must be a U.S. nonprofit (federally tax-exempt) organization as defined under Section 501(c)3 of the Internal Revenue Code; must focus on literacy development specifically for grades K-3; government funding must not exceed 50% of the organization's total budget; geographic focus must remain in the United States; must have two or more years of operating experience; must have an established board of directors; must be able to provide informal progress reports on a quarterly basis; and must have ability to measure results, which reflect the success of the program. Preference will be given to those organizations and/or programs with budgets of less than $2 million.
Restrictions: The Andrew Family Foundation will not consider programs that directly impact public policy or provide funding of endowments or capital grants. The financial support will be restricted to a specific purpose or project and can include operational support.
Geographic Focus: All States
Date(s) Application is Due: Sep 15
Contact: Connor Humphrey, Grants Administrator; (602) 828-8471 or (708) 460-1288; fax (602) 385-3267; Connor.Humphrey@GenSpring.com or aff@inlignwealth.com
Internet: https://online.foundationsource.com/andrew/ok.htm
Sponsor: Andrew Family Foundation
14628 John Humphrey Drive
Orland Park, IL 60462

Andrew W. Mellon/ACLS Dissertation Completion Fellowships **844**

The Mellon/ACLS Dissertation Completion Fellowships assist graduate students in the humanities and related social sciences in the last year of Ph.D. dissertation writing. Applicants must be prepared to complete their dissertations within the period of their fellowship tenure or shortly thereafter. ACLS will award 65 Fellowships in this competition for a one-year term beginning between June and September for the following academic year. The Fellowship tenure may be carried out in residence at the Fellow's home institution, abroad, or at another appropriate site for the research. The total award of up to $33,000 includes a stipend plus additional funds for university fees and research support. Specific guidelines, the online application, a sample application, and suggestions for writing proposals are available at the website.
Requirements: Applicants must: be Ph.D. candidates in a humanities or social science department in the United States. Applicants from other departments may be eligible if their project is in the humanities or related social sciences, and their principal dissertation supervisor holds an appointment in a humanities field or related social science field. (Students completing master's degrees are not eligible, even if they are the terminal degree in the field); have all requirements for the Ph.D. except the dissertation completed before beginning fellowship tenure; and be no more than six years in the degree program (awardees can hold this Fellowship no later than their seventh year).
Restrictions: These Fellowships may not be held concurrently with any other major fellowship or grant.
Geographic Focus: All States
Date(s) Application is Due: Nov 9
Amount of Grant: Up to 33,000 USD
Contact: Nicole Stahlmann, Director of Fellowship Programs; (212) 697-1505, ext. 134; fax (212) 949-8058; nstahlmann@acls.org
Internet: http://www.acls.org/programs/dcf/
Sponsor: American Council of Learned Societies
633 Third Avenue
New York, NY 10017-6795

Andrew W. Mellon Postdoctoral Fellowship in American Cultures **845**

The college invites applications for a two-year Mellon Postdoctoral Fellowship in Native American Studies, to begin in the fall. The program seeks scholars who have an interest in teaching in an undergraduate liberal arts college and who would welcome the opportunity to participate in the development of a multidisciplinary concentration in Native American Studies within the college's American Culture Program. The fellow will teach or team-teach one course in the fall semester entitled Introduction to Native American Studies, and will help facilitate a monthly faculty seminar throughout the academic year intended to extend and deepen the interests in Native American Studies of faculty from several disciplines. The fellowship includes salary plus benefits, as well as support for research, professional travel, and relocation expenses. Application guidelines are available online.
Requirements: Candidates should be relatively new PhDs with training in Native American Studies or any relevant discipline.
Geographic Focus: All States
Date(s) Application is Due: Feb 15
Contact: Ron Sharp, Dean; (845) 437-7485; fax (845) 437-7204
Internet: http://americanculture.vassar.edu/index.html?posting=39
Sponsor: Vassar College
124 Raymond Avenue, Box 746
Poughkeepsie, NY 12604-0739

Andrew W. Mellon Postdoctoral Fellowships at UCLA Humanities Consortium **846**

These fellowships are devoted to the theme Nations and Identities. The topic for the current year is The Secularization Thesis. Fellows are required to be in residence and to participate in the consortium's Mellon Seminar and Conference. Fellows also will teach, through relevant departments or programs, one course in their first year and two courses in their second, and will be expected to participate in the intellectual life of these programs. One fellow will be appointed in each of three historical periods: AD 600 to 1600, 1600 to 1800, and 1800 to present.
Requirements: Fellows must have earned their doctorates after September 1, 2001, and must have the doctorate in hand by September 1, 2004.
Geographic Focus: All States
Date(s) Application is Due: Mar 1
Contact: Mark Pokorski, (310) 206-0559; mpok@humnet.ucla.edu
Internet: http://www.humnet.ucla.edu/humnet/consortium/fships.html
Sponsor: UCLA Center for 17th- and 18th-Century Studies
405 Hilgard Avenue, 310 Royce Hall
Los Angeles, CA 90095-1404

Andy Warhol Foundation for the Visual Arts Curatorial Research Fellowships **847**

Scholarly research undertaken in the field of contemporary art is funded through Curatorial Research Fellowships. It is assumed that research will lead to a significant exhibition, though this is not a requirement. Grants to curators do not preclude separate proposals from sponsoring institutions in any given grant round.
Requirements: Institutionally-affiliated curators at any stage of their careers are eligible to apply and must have the formal support of their director.
Geographic Focus: All States
Date(s) Application is Due: Mar 1; Sep 1
Amount of Grant: 1,000-100,000 USD
Contact: Pamela Clapp, Program Director; (212) 387-7555; fax (212) 387-7560; info@warholfoundation.org
Internet: http://www.warholfoundation.org/grant/overview.html
Sponsor: Andy Warhol Foundation for the Visual Arts
65 Bleecker Street, 7th Floor
New York, NY 10012

Angels Baseball Foundation Grants **848**

Since its inception in 2004, the Angels Baseball Foundation has awarded many grants to worthy organizations throughout the community, ranging from music programs to at-risk youth shelters. Currently, it focuses on initiatives aimed to create and improve education, health care, arts and sciences, and community related youth programs, in addition to providing children the opportunity to experience the game of baseball. Giving is limited to the greater Los Angeles, California, area. Fields of interest include: the arts, athletics/sports activities, children and youth, cancer detection, cancer research, community and economic development, education (at all levels), health care, and community service programs. Applicants should begin by contacting the Foundation directly.
Requirements: Applicants must serve the residents of the greater Los Angeles area.
Geographic Focus: California
Contact: Anne Bafus, Community Relations; (714) 940-2174 or (714) 940-2244
Internet: http://losangeles.angels.mlb.com/ana/community/baseball_foundation.jsp
Sponsor: Angels Baseball Foundation
2000 E Gene Autry Way
Anaheim, CA 92806-6100

Animal Behavior Society Charles H. Turner Award 849

In 2002 the Animal Behavior Society created the Charles H. Turner Program to encourage undergraduate participation at the annual Society meetings. With generous support from the National Science Foundation, the ABS Diversity committee brings a group of about ten undergraduates to the meetings each year, covering all expenses, and providing mentors and a full program of mentoring events, including a pre-meeting workshop. Charles Turner was one of the very first African-American researchers in animal behavior. Among other things, his research showed that insects can hear and exhibit trial-and-error learning. By naming the undergraduate program after him, the ABS Diversity Committee emphasizes its goal to increase the diversity of its membership by encouraging researchers of all ages, levels, and ethnic groups to participate in the annual meetings. Applications are submitted via an online form available from the Turner-Award webpage at the ABS website. The form is generally available from January through May. Submission deadlines may vary from year to year. The Animal Behavior Society was founded in 1964 to promote the study of animal behavior. ABS members come from all over the world, but primarily from North, Central, and South America. Current members' research activities span the invertebrates and vertebrates, both in the field and in the laboratory, and include experimental psychology, behavioral ecology, neuroscience, zoology, biology, applied ethology, and human ethology as well as many other specialized areas.

Requirements: Participants must be citizens or permanent residents of the United States. Students must attend the entire meeting (generally in June), and must not have completed their undergraduate degree the prior December. Students who are members of groups traditionally under-represented in the sciences (women, ethnic and racial minorities, those with disabilities, etc.) are particularly urged to apply. Students with special needs are also encouraged to apply but should discuss their specific requirements with the organizers as part of the application process.

Geographic Focus: All States

Date(s) Application is Due: May 30

Contact: Zuleyma Tang-Martinez, Animal Behavior Society Diversity Committee Chair; (314) 516-6203; zuleyma@umsl.edu

Internet: http://animalbehaviorsociety.org/grants-and-awards/meeting-related-awards/turner-award

Sponsor: Animal Behavior Society
402 N Park Street
Bloomington, IN 47408

Animal Behavior Society Developing Nations Research Grants 850

The Animal Behavior Society supports research in the behavioral sciences through grants in two broad classes, graduate student research and developing nations research. Applicants apply via a single-application process. The goal of the Developing Nations Research Grants is to support animal behaviorists in developing countries who may have difficulty securing financial resources to fund their research. Awards are provided for research to be conducted within a 1-year period from the date of the award. The number of awards varies annually, and amounts range from U.S.$ 500 to U.S.$ 800, depending on need and funding availability. Applicants must use ABS's online submission site to submit their applications. The site generally opens in November and stops accepting submissions in late December or early January. Please note that membership dues must be paid in full and the ABS Central Office contacted to activate access to the on-line application system. The account activation cut-off date occurs several days prior to the actual submission deadline. Detailed submission guidelines are provided at the website. Submission deadlines may vary from year to year. The Animal Behavior Society was founded in 1964 to promote the study of animal behavior. ABS members come from all over the world, but primarily from North, Central, and South America. Current members' research activities span the invertebrates and vertebrates, both in the field and in the laboratory, and include experimental psychology, behavioral ecology, neuroscience, zoology, biology, applied ethology, and human ethology as well as many other specialized areas.

Requirements: Only members of the Animal Behavior Society who are residents of developing nations and who are conducting research at an institution in a developing nation are eligible to apply. Applicants may be graduate students as well as more established members of the research community. Eligible candidates may receive one award as a student and one additional award as an established researcher.

Restrictions: The following nations shall not be considered as developing nations: the United States, Canada, Israel, Japan, Australia, New Zealand, Iceland, Norway, Sweden, Finland, Denmark, Germany, the Netherlands, Belgium, Luxembourg, Ireland, Switzerland, Austria, Italy, France, the United Kingdom, Spain, and Portugal.

Geographic Focus: All States, All Countries

Date(s) Application is Due: Dec 30

Amount of Grant: 500-800 USD

Contact: Scott Forbes; s.forbes@uwinnipeg.ca

Internet: http://animalbehaviorsociety.org/grants-and-awards/developing-nations-award

Sponsor: Animal Behavior Society
402 N Park Street
Bloomington, IN 47408

Animal Behavior Society Diversity Grant 851

Each year, the Animal Behavior Society (ABS) offers a limited number of awards from the Diversity Fund to cover registration fees for minority graduate students attending the annual ABS Meeting, with the goal of broadening minority and ethnic representation. To apply for this award, applicants must submit a web-based form that includes their email address, full name, qualifying status, name and address of their school, the title of their talk for the meeting, and their preliminary abstract. The link to the form is on the society's Diversity Grant webpage. Interested applicants may check with ABS to obtain further information about deadline dates. The Animal Behavior Society was founded in 1964 to promote the study of animal behavior. ABS members come from all over the world, but primarily from North, Central, and South America. Current members' research activities span the invertebrates and vertebrates, both in the field and in the laboratory, and include experimental psychology, behavioral ecology, neuroscience, zoology, biology, applied ethology, and human ethology as well as many other specialized areas.

Requirements: Applicants must be enrolled in a graduate program at the time of application and be a member of an under-represented minority, including African Americans, Native Americans, Asian Americans, and Hispanic Americans who live in North America or are citizens of Central or South America.

Geographic Focus: All States, All Countries

Contact: Zuleyma Tang-Martinez, Animal Behavior Society Diversity Committee Chair; (314) 516-6203; zuleyma@umsl.edu

Internet: http://animalbehaviorsociety.org/grants-and-awards/meeting-related-awards

Sponsor: Animal Behavior Society
402 N Park Street
Bloomington, IN 47408

Animal Behavior Society Latin American Travel Awards 852

The Latin American Travel Awards are intended to encourage greater participation by Latin-American researchers in the Animal Behavior Society (ABS)'s annual meetings by helping to defray the costs of international travel, housing, and/or meals. Prospective applicants who have successfully submitted an abstract for their talk or poster via the ABS annual-meeting webpage should also email their submitted abstract (including the title) and curriculum vita (CV) to the second ABS member-at-large. Each document should be attached to the email message as a separate pdf file, and each labeled with the applicant's last name, initials, and the contents of each file (i.e. abstract or CV). A list of ABS Officers with their contact information is posted at the ABS website. Checks in the amount of the award will be sent directly to recipients following announcement of the results. The Animal Behavior Society was founded in 1964 to promote the study of animal behavior. ABS members come from all over the world, but primarily from North, Central, and South America. Current members' research activities span the invertebrates and vertebrates, both in the field and in the laboratory, and include experimental psychology, behavioral ecology, neuroscience, zoology, biology, applied ethology, and human ethology as well as many other specialized areas.

Requirements: Priority will be given to Ph.D. students who are attending an ABS meeting for the first time or who have not previously received a Latin American Travel Award. Award values vary according to funds available. Candidates are prioritized according to the quality of their abstracts and CVs.

Restrictions: Latin American Travel Awards are restricted to Latin-American graduate students enrolled in Latin-American institutions, who will be presenting a talk/poster at the annual ABS meeting.

Geographic Focus: All Countries

Date(s) Application is Due: Mar 16

Contact: Gabriel Francescoli, Animal Behavior Society Latin American Affairs Committee Chair; 5982 3071442; gabo@fcien.edu.uy

Internet: http://animalbehaviorsociety.org/grants-and-awards/travel-related-grants-and-awards/latin-american-travel-award

Sponsor: Animal Behavior Society
402 N Park Street
Bloomington, IN 47408

Animal Behavior Society Student Research Grants 853

The Animal Behavior Society (ABS) offers several first-time research awards to graduate-student members of the society: the George W. Barlow Award, the E.O. Wilson Conservation Award, and the Amy R. Samuels Cetacean Behavior and Conservation (CBC) Award. A brief description of each award follows: the George W. Barlow Award's purpose is to encourage excellence in graduate-student research in the field of animal behavior and will be given to one top-ranked proposal in ABS's annual Student-Research Grant cycle; the E.O. Wilson Conservation Award annually recognizes one research proposal considered meritorious for its integration of animal behavior and conservation; the CBC Award provides financial support for graduate students studying cetacean behavior and/or cetacean conservation in natural environments. Specifically, the CBC award encourages original and significant research that focuses on endangered populations or species, or on critically important problems in cetacean behavioral biology. The award is given only if there are proposals deemed worthy by the judges in any given year; up to two awards of a maximum of $2000 each will be made to successful applicants. The award must be used for field research and cannot be used for laboratory research. Applicants apply for all three grants through a single-application process, usually at the end of the year. All eligible applicants will be considered for all three grants, but may receive only one of them. Applicants must use ABS's online submission site to submit their applications. The site generally opens around the end of October and stops accepting submissions in early January. Please note that membership dues must be paid in full and the ABS Central Office contacted to activate access to the on-line application system. The account-activation cut-off date occurs several days prior to the actual submission deadline. Detailed submission guidelines are provided at the website. The Animal Behavior Society was founded in 1964 to promote the study of animal behavior. ABS members come from all over the world, but primarily from North, Central, and South America. Current members' research activities span the invertebrates and vertebrates, both in the field and in the laboratory, and include experimental psychology, behavioral ecology, neuroscience, zoology, biology, applied ethology, and human ethology as well as many other specialized areas.

Requirements: Society members who are currently enrolled in graduate programs may apply.

Restrictions: Applicants may receive only one of these grants in a lifetime.

Geographic Focus: All States, All Countries

Date(s) Application is Due: Jan 3
Amount of Grant: 500-2,000 USD
Contact: Scott Forbes, Student Research Committee Chair; s.forbes@uwinnipeg.ca
Internet: http://animalbehaviorsociety.org/grants-and-awards/student-releated-research-awards
Sponsor: Animal Behavior Society
402 N Park Street
Bloomington, IN 47408

ANL Faculty Research Leave Program 854
Argonne National Laboratory, one of DOE's major research centers, offers faculty research leave appointments (FRLA) to college and university faculty members. Faculty members spend their sabbatical leave, typically 9-12 months, at the Argonne Laboratory. The purpose of the program is to involve college/university faculty in DOE research programs conducted at Argonne, and stimulate continuing research collaboration between faculty and Argonne scientists. Participation takes the form of individual collaboration with an Argonne staff member in some part of an ongoing project of interest to the faculty participant. The research experience is augmented by seminars, independent study, and other appropriate activities. Faculty research projects are available in the basic physical and life sciences, mathematics, computer science, and in engineering, as well as in a variety of applied research programs. Interaction of faculty with students in the research programs is strongly encouraged. Typically, Argonne reimburses the university for 50 percent of salary and fringe benefits for the academic year (9 months) and full salary and fringe benefits for the summer (three months). In addition, the laboratory may negotiate reimbursement for certain travel, moving, and housing expenses. Applications are accepted at any time. Write for additional information and application forms.
Requirements: An appointee must be a full-time faculty member of an accredited U.S. college or university and must have a commitment to continue in teaching and research as a career. The applicant's objectives for the Argonne tour should be clearly specified, and these objectives must be endorsed by the department head or dean. The expression of interest in participation by an individual faculty member, along with university endorsement of such participation, should be transmitted to the Cross Division Program Leader, Division of Educational Programs. Initial expressions of interest should include a curriculum vita, a publication list, and a brief statement of the research interests of the faculty member.
Geographic Focus: All States
Contact: Carol Przyzycki; (630) 252-5448; cprzyzy@dep.anl.gov
Internet: http://www.dep.anl.gov/crossdiv/frl.htm
Sponsor: Argonne National Laboratory
9700 S Cass Avenue, DEP 223
Argonne, IL 60439

ANL Faculty Research Participation Program 855
Faculty participants in the program spend 10-12 weeks collaborating with an Argonne staff scientist or engineer on an existing project of interest to the faculty member. Argonne research falls into four broad categories: engineering research (including advanced nuclear reactors, batteries, and fuel cells); physical research (materials science, physics, chemistry, mathematics, and computer science); energy and environmental science and technology research (biology, alternate energy systems, and environmental and economic impact assessments); and technology transfer (moving the benefits of Argonne's publicly funded research to the marketplace). Appointments are made principally for the summer; however, similar arrangements can be made during the academic year. Stipends are based on academic-year salary and round-trip travel is provided. Additional information and application forms are available upon request.
Requirements: An appointee must be a U.S. citizen or permanent resident who is a full-time faculty member of an accredited U.S. college or university and must have a commitment to continue in teaching and research as a career. The applicant's objectives for the Argonne tour should be clearly specified, and these objectives must be endorsed by the department head or dean.
Geographic Focus: All States
Date(s) Application is Due: Jan 13
Contact: Carol Przyzycki; (630) 252-5448; cprzyzy@dep.anl.gov
Internet: http://www.dep.anl.gov/p_faculty/frp.htm
Sponsor: Argonne National Laboratory
9700 S Cass Avenue, DEP 223
Argonne, IL 60439

ANL Graduate Thesis Parts Research Appointments 856
Thesis Parts appointments support qualified graduate students who wish to visit Argonne for periods from a few days to a few months so that they may utilize special laboratory facilities or capabilities during the course of their thesis research. Research areas include physical and life sciences, mathematics, computer science, engineering, conservation, environment, fission and fusion energy, and other energy technologies. Application is best made through an Argonne staff member or research division appropriate to the proposed activity.
Requirements: Qualified graduate students in U.S. universities who wish to carry out some thesis research at Argonne are eligible to apply.
Geographic Focus: All States
Contact: Lisa Reed, Lab-Grad Research Appointments, Division of Educational Programs; (630) 252-3366; Lreed@dep.anl.gov
Internet: http://www.dep.anl.gov/p_graduate/thesispa.htm
Sponsor: Argonne National Laboratory
9700 S Cass Avenue, DEP 223
Argonne, IL 60439

ANL Laboratory-Graduate Research Appointments 857
Laboratory graduate research appointments are available for qualified graduate students at U.S. universities who wish to carry out their thesis research at Argonne National Laboratory under the cosponsorship of an Argonne staff member and a faculty member. Research may be conducted in the basic physical and life sciences, mathematics, computer science, and engineering, as well as in a variety of applied areas relating to conservation, environment, fission and fusion energy, and other energy technologies. Lab-grad appointments are for a one-year term with annual renewals being contingent upon satisfactory performance by the appointee. Applications may be submitted at any time, and appointments may commence any time during the year. Completed application should be submitted at least one month prior to proposed starting date. Temporary appointments are available to qualified graduate students so they may work with an Argonne staff member and become familiar with his/her work prior to application for a year-long research appointment. The student's faculty sponsor may also receive payment for limited travel expenses.
Requirements: An applicant must currently be a full-time graduate student in a recognized graduate program in a U.S. university and must be a U.S. citizen or permanent resident alien.
Geographic Focus: All States
Amount of Grant: 5,000 USD
Contact: Lisa Reed, Lab-Grad Research Appointments, Division of Educational Programs; (630) 252-3366; Lreed@dep.anl.gov
Internet: http://www.dep.anl.gov/p_graduate/labgrad.htm
Sponsor: Argonne National Laboratory
9700 S Cass Avenue, DEP 223
Argonne, IL 60439

Anna-Greta and Holger Crafoords Foundation Grants 858
The foundation operates in Sweden and the grants promote international basic research within the following subject areas: mathematics and astronomy; geosciences; biosciences, with particular emphasis on ecology; and polyarthritis. In addition to the Crafoord Prize, research grants are awarded within the relevant disciplines to individual persons or researchers in Sweden. Award of a research grant within polyarthritis takes place every third year, but a prize within polyarthritis is only awarded when a specific enquiry has demonstrated that such scientific progress has been made that a prize is warranted.
Geographic Focus: All States, Sweden
Contact: Grants Administrator; 046-38 58 80; fax 046-38 58 85; crafoord@crafoord.se
Internet: http://www.crafoord.se/index-e.htm
Sponsor: Anna-Greta and Holger Crafoords Foundation
Malmovagen 8, P.O. Box 137
Lund, 22100 Sweden

Ann and Robert H. Lurie Family Foundation Grants 859
Founded in 1986, the foundation initially had set up six categories: medical services and research; child-related medical organizations; basic services including food and shelter; education; the arts; and so-called wild things. When Ann Lurie's husband, Robert, died in 1990, she was left with a fortune worth hundreds of millions and six kids — ages 5 to 15. She has given away well over $100-million since Robert H. Lurie died at age 48, generally spurning requests from nonprofit groups with undistinguished records and concentrating her gifts on some of the world's biggest problems, including hunger, cancer, and inadequate health care. Today, the foundation supports educating and providing resources to children so they are better prepared for the future.
Requirements: There are no specific deadlines or application forms to fill out. An organization requesting funding should have finances in good order, demonstrating that it is making good use of its resources. Note: Ann Lurie has a preference for identifying her own projects, rather than responding to solicitations.
Geographic Focus: Illinois, Michigan
Contact: Ann Lurie, President; (312) 466-3750; fax (312) 466-3700
Sponsor: Ann and Robert H. Lurie Family Foundation
2 N Riverside Plaza, Suite 1500
Chicago, IL 60606-2600

Ann Arbor Area Community Foundation Grants 860
The foundation is interested in funding projects that will improve the quality of life for citizens of the Ann Arbor, Michigan, area. Eligible projects generally fall within the categories of education, culture, social service, community development, environmental awareness, health and wellness, and youth and senior citizens. Types of support include: emergency funds, program development, conferences and seminars, publication, seed grants, scholarship funds, research, and matching funds. Higher priority is given to programs that are preventive rather than remedial, increase individual access to community resources, examine and address the underlying causes of local problems, promote independence and personal achievement, attract volunteer resources and support, strengthen the private nonprofit sector, encourage collaboration with other organizations, and build the capacity of the applying organization. Organizations interested in applying are strongly encouraged to discuss their project with the program director prior to submitting an application.
Requirements: 501(c)3 nonprofit organizations in the Ann Arbor, MI, area, which is the area that falls within the boundaries of the Ann Arbor public schools district, are eligible.
Restrictions: The foundation usually does not make grants for construction projects, annual giving campaigns or capital campaigns, normal operating expenses (except for start-up purposes), religious or sectarian purposes, computer hardware equipment, individuals, advocacy or political purposes, multiyear funding, or regranting.
Geographic Focus: Michigan
Date(s) Application is Due: Feb 11; Oct 7
Contact: Phil D'Anieri; (734) 663-0401; fax (734) 663-3514; pdanieri@aaacf.org

Internet: http://www.aaacf.org/grants.asp
Sponsor: Ann Arbor Area Community Foundation
301 N Main Street, Suite 300
Ann Arbor, MI 48104

Anne Louise Barrett Fellowship 861
This fellowship is awarded in music, preferably and primarily for study or research in musical theory, composition, or the history of music. The fellowship may be used abroad or in the United States. Awards are usually made to applicants who plan full-time graduate study for the coming year. Preference will be given to applicants who have not previously held one of Wellesley College's awards. Awards will be based on merit and need. One fellowship is awarded annually; application form must be used. Annual deadline dates may vary; contact program officer for exact dates.
Requirements: The fellowship is open to graduating seniors and graduates of Wellesley College to be used for study at institutions other than Wellesley.
Geographic Focus: All States
Amount of Grant: Up to 15,000 USD
Contact: Secretary to the Committee on Graduate Fellowships; (781) 283-3525; cws-fellowships@wellesley.edu
Internet: http://www.wellesley.edu/CWS/alumnae/wellfs.html
Sponsor: Wellesley College
106 Central Street
Wellesley, MA 02181-8200

Annenberg Foundation Grants 862
The foundation provides support for projects within its grantmaking interests of education and youth development, arts and culture, civic, community and the environment, and health and human services. It encourages the development of more effective ways to share ideas and knowledge. Letters of inquiry may be submitted at all times during the year and there are no deadlines. Please review the grants database for additional types of grants given.
Requirements: 501(c)3 tax-exempt organizations are eligible.
Restrictions: Full proposals are not accepted unless requested by a Foundation representative. The foundation is not presently considering inquiries for: individuals, individual K-12 schools, for-profit organizations, political activities or attempts to influence specific legislation, individual scholarships, projects focused exclusively on research, or programs outside of its grant-making interests.
Geographic Focus: All States
Contact: Leonard Aube, Executive Director; (310) 209-4560; fax (310) 209-1631; info@annenbergfoundation.org
Internet: http://www.annenbergfoundation.org/grants
Sponsor: Annenberg Foundation
2000 Avenue of the Stars, Suite 1000
Los Angeles, CA 90067

Ann Peppers Foundation Grants 863
The foundation awards grants to eligible California nonprofit organizations in its areas of interest, including arts and cultural programs, disabled, elderly, health care, private education, and social services. Types of support include capital grants, general operating grants on a temporary basis, matching grants, scholarships, and research grants. Grants are initiated by the foundation manager. There are no application deadlines; the board meets quarterly to consider requests.
Requirements: Southern California 501(c)3 nonprofits, colleges, and universities are eligible. Preference is given to requests from Los Angeles County.
Geographic Focus: California
Amount of Grant: 2,000-50,000 USD
Contact: Jack Alexander, Secretary; (626) 449-0793
Sponsor: Ann Peppers Foundation
625 S Fair Oaks Avenue
South Pasadena, CA 91030

ANS/AAO-HNSF Herbert Silverstein Otology & Neurotology Research Award 864
The purpose of this award, which is jointly sponsored by the AAO-HNS Foundation and the American Neurotology Society, is to support a clinical or translational research project focused on diseases, disorders, or conditions of the peripheral or central auditory and/or vestibular system among new full-time academic surgeons. The award is intended as a preliminary step in clinical investigator career development and is expected to facilitate the recipient's preparation of a more comprehensive individualized research plan suitable for submission to the National Institutes of Health or comparable funding agency.
Requirements: Applicants must be physicians in the United States or Canada with demonstrated potential for excellence in research and teaching and serious commitment to an academic research career in otology or neurotology. Priority will be given to fellows or junior faculty, who have completed residencies or fellowships within four years of the application receipt date, although otolaryngology-head and neck surgery residents are eligible. All candidates must be sponsored by the Chair of his/her Division or Department and by an official representative of the institution which would administer the Award and in whose name the application is formally submitted.
Geographic Focus: All States
Amount of Grant: Up to 25,000 USD
Contact: Stephanie L. Jones; (703) 519-1586 or (703) 836-4444; sljones@entnet.org
Internet: http://www.americanneurotologysociety.com/funding.html
Sponsor: American Academy of Otolaryngology-Head and Neck Surgery Foundation
1650 Diagonal Road
Alexandria, VA 22314-2857

ANS Neurotology Fellows Award 865
The Neurotology Fellows Award was established by the American Neurotology Society in 1996 to reward scientific excellence. The award has been generously endowed through royalties received for the textbook Neurotology, Mosby Publishers, D. Brackmann and R. Jackler, editors. One or more awards of $500 each is intended to subsidize travel expenses incurred while giving a scientific presentation at the Annual Spring Meeting of the American Neurotology Society. The material presented need not have been performed during the Fellowship year; it may derive from earlier work performed during the residency or extracurricular research experience. Applicants should submit electronically a cover letter, a supporting letter from the Fellowship director to verify qualifications, and electronic submission of the abstract by October 15.
Requirements: Applicants must be full-time participants in a post-residency Neurotology Fellowship in the United States or Canada. In addition, the Fellow must be both podium presenter and first author of the paper submitted for publication.
Geographic Focus: All States
Date(s) Application is Due: Oct 15
Amount of Grant: 500 USD
Contact: Shirley Gossard, Administrator; (352) 751-0932 or (217) 414-4868; fax (352) 751-0696; segossard@aol.com
Internet: http://www.americanneurotologysociety.com/funding.html
Sponsor: American Neurotology Society
3096 Riverdale Road
The Villages, FL 32162

ANS Nicholas Torok Vestibular Award 866
The Nicholas Torok Vestibular Award, established under the aegis of the American Neurotology Society instituting a yearly lecture, is to be presented at the Annual Scientific Spring Meeting of the Society. The subject should be an innovative observation, experience or technique in the field of vestibular basic science or clinical science. The candidate shall be selected preferably from the membership of the Society or upon specific merit of any M.D. or Ph.D. A yearly award of $1,500 will be given for the selected manuscript.
Requirements: The manuscript must be presented electronically in the publication format recommended by Otology & Neurotology, the official journal of the American Neurotology Society.
Geographic Focus: All States
Date(s) Application is Due: Oct 15
Amount of Grant: 1,500 USD
Contact: Shirley Gossard, Administrator; (352) 751-0932 or (217) 414-4868; fax (352) 751-0696; segossard@aol.com
Internet: http://www.americanneurotologysociety.com/funding.html
Sponsor: American Neurotology Society
3096 Riverdale Road
The Villages, FL 32162

ANS Trainee Award 867
The American Neurotology Society offers an award for the best clinical or basic science paper in Neurotology submitted by a Resident or Fellow in training in the field of Otolaryngology-Head and Neck Surgery. A clinical study or basic research study is acceptable. The Society has established a cash award of $1,000 for the winning paper. The author will be expected to attend the Annual ANS Spring Meeting and present the paper personally.
Requirements: Residents from any approved residency program or Fellows in Neurotology in the United States or Canada are eligible to compete. The Resident or Fellow must be the primary author. He or she must provide the main inspiration for the paper and do the literature search, data collection, original and final drafts of the article, the discussion, and the conclusions.
Restrictions: The paper must be an original contribution, not previously published.
Geographic Focus: All States
Date(s) Application is Due: Oct 15
Amount of Grant: 1,000 USD
Contact: Shirley Gossard, Administrator; (352) 751-0932 or (217) 414-4868; fax (352) 751-0696; segossard@aol.com
Internet: http://www.americanneurotologysociety.com/funding.html
Sponsor: American Neurotology Society
3096 Riverdale Road
The Villages, FL 32162

Anthem Blue Cross and Blue Shield School Nurse Grants 868
The mission of the project is to promote cardiovascular health among school-age children. Anthem's goals are to assist school nurses in initiating programs that may encourage a healthy lifestyle, reduce cardiovascular risk factors, promote physical activity, and encourage healthy eating behaviors; and facilitate the development of innovative school health programs that are based on research. No awards will be made to individuals, but will be awarded to the school. To apply, the school nurse representing the school must submit a completed application form with the appropriate attachments by the listed application deadline. Guidelines, including program contacts in Kentucky and Indiana, and application are available online.
Requirements: Eligible applicants include school nurses who are licensed in the state in which they practice and who represent a public or private school (elementary, junior high, middle, or high school) in Ohio, Kentucky, or Indiana.
Restrictions: Funds cannot be used for the salary or wages of any staff or employee.
Geographic Focus: Indiana, Kentucky, Ohio
Date(s) Application is Due: Feb 1
Amount of Grant: 500-1,000 USD

Contact: Gabrielle Karpowicz, (614) 529-5068; gkarpowicz@aol.com
Internet: http://oasn.org/Awards_Grants/awards.htm
Sponsor: Ohio Association of School Nurses
6167 Heritage Point Court
Hilliard, OH 43026

Anthony R. Abraham Foundation Grants **869**

For more than 30 years, the Anthony R. Abraham Foundation mission has been to help non-profit organizations worldwide. The foundations strives to: provide programs and services to help people around the world become self-productive and give back to their communities; ensure that no child is denied medical treatment due to a lack of insurance; provide education that breaks barriers; guarantee that research into the cure of catastrophic diseases continues; help raise the quality of life. It's been a privilege for the Foundation to be able to help ease poverty, raise hospitals, build orphanages and further medical research and the Foundation looks forward to doing even more. Some of the organizations helped include: Domestic organizations -St. Jude Children's Research Hospital; Camillus House; Miami Rescue Mission; Habitat for Humanity of Greater Miami; America's Second Harvest; Miami Children's Hospital; Jackson Memorial Hospital; Florida Heart Research Institute; Big Brothers/Big Sisters; Alonzo Mourning Charities; Honey Shine Mentoring Program; Cancer Link; The Miami Lighthouse for the Blind; Overtown Youth Center. International organizations: Brothers of the Good Shepherd; School for the Blind, Lebanon; Rene Moawad Foundation; Haitian Foundation; Children's International Network; Doctors without Borders; Little Sisters of Nazareth; Maronite Order of the Holy Family.
Requirements: Non-profit 501 (c) 3 organization requesting funding should fill out a funds-request form and submit it at the Foundation's website. The Foundation will contact you, if your organization qualifies under the Foundation's guidelines and the law.
Restrictions: Grants are not made to individuals.
Geographic Focus: All States
Amount of Grant: 100-50,000 USD
Contact: Anthony R. Abraham, Chairman; (305) 665-2222
Internet: http://www.abrahamfoundation.com/about
Sponsor: Anthony R. Abraham Foundation
1320 S Dixie Highway, Suite 241
Coral Gables, FL 33146-2937

Anthropological Center for Training in Global and Environmental Change Grad Assistantships **870**

ACT is an interdisciplinary training and research center on the human dimensions of global environment change. ACT's research and training objectives focus on how particular local populations manage resources and how those activities may be monitored using remote sensing technologies and field studies. There is no formal application process for a graduate assistantship at ACT, but if an individual is applying for a graduate program at Indiana University, and feels that their skills and research interests match that of ACT, then an applicant should share his/her interest in a graduate assistantship.
Geographic Focus: Indiana
Contact: Emilio Moran, Director; (812) 855-6181; fax (812) 855-3000; moran@indiana.edu or act@indiana.edu
Internet: http://www.indiana.edu/~act/newstudents.shtml
Sponsor: Anthropological Center for Training in Global and Environmental Change
701 East Kirkwood Avenue, Student Building, Room 331
Bloomington, IN 47405

AOCS A. Richard Baldwin Award **871**

The A. Richard Baldwin Award was given to Baldwin in recognition of his lengthy and distinguished service to the AOCS. It is the Society's highest service award. The award includes a $2,000 honorarium, a plaque, and travel expenses to the Annual Meeting. Self nominations are welcomed.
Requirements: Candidates must have made outstanding and extraordinary contributions to the Society or supported the activities of the Society through officer and/or committee participation over a substantial period of time. Candidates must also be current Society members in good standing.
Geographic Focus: All States, All Countries
Date(s) Application is Due: Oct 15
Contact: Barb Semeraro; (217) 693-4804; fax (217) 693-4849; barbs@aocs.org
Internet: http://www.aocs.org/Membership/content.cfm?ItemNumber=782
Sponsor: American Oil Chemists' Society
2710 South Boulder
Urbana, IL 61802-6996

AOCS ACI/NBB Innovation Research Award **872**

Sponsored by the American Cleaning Institute (ACI) and the National Biodiesel Board (NBB), the Glycerine Innovation Research Award recognizes outstanding achievement for research into new applications for glycerine with particular emphases on commercial viability. The Award includes a $5,000 honorarium, a plaque for the recipient, and an award lecture at the Annual Meeting. Nominations for this award are accepted from members of the Society, colleagues, or directly from individual candidates. The nomination form is available at the website.
Requirements: The research of candidates must reflect particular emphasis to existing or potential commercial importance. Candidate materials must include the following: a two-page letter of nomination, describing original research work in new applications and uses of glycerine; up to two letter of support from individuals familiar with the candidate's accomplishments; and any published works relating to the research.
Geographic Focus: All States, All Countries
Date(s) Application is Due: Nov 1
Amount of Grant: 5,000 USD
Contact: Barb Semeraro; (217) 693-4804; fax (217) 693-4849; barbs@aocs.org
Internet: http://www.aocs.org/Membership/content.cfm?ItemNumber=804
Sponsor: American Oil Chemists' Society
2710 South Boulder
Urbana, IL 61802-6996

AOCS Alton E. Bailey Award **873**

The Alton E. Bailey Award recognizes outstanding research and exceptional service in the field of lipids and associated products. This prestigious award commemorates Alton E. Bailey's outstanding contributions to the field of fats and oils as a researcher, an author of several standard books in the field and as a leader in the work of the Society. The recipient receives a $750 honorarium, a plaque, and award lecture at the Annual Meeting. The application is available at the website. Self nominations are welcomed.
Requirements: Candidates must be AOCS members in good standing, with outstanding leadership to the Society. They must also have made significant contributions to the science and technology of lipids.
Geographic Focus: All States, All Countries
Date(s) Application is Due: Oct 15
Amount of Grant: 750 USD
Contact: Barb Semeraro; (217) 693-4804; fax (217) 693-4849; barbs@aocs.org
Internet: http://www.aocs.org/Membership/content.cfm?ItemNumber=781
Sponsor: American Oil Chemists' Society
2710 South Boulder
Urbana, IL 61802-6996

AOCS Analytical Division Student Award **874**

The Analytical Division Student Award is given to graduate students in the filed of lipid analytical chemistry. Graduate students at any institution of higher learning who are doing research toward a degree are eligible. The Award involves a certificate, $250 honorarium, a $500 travel allowance to present at the Annual Meeting, and an award lecture or poster presentation. Self-nominations are welcomed.
Geographic Focus: All States, All Countries
Date(s) Application is Due: Oct 15
Amount of Grant: 750 USD
Contact: Barb Semeraro; (217) 693-4804; fax (217) 693-4849; barbs@aocs.org
Internet: http://www.aocs.org/Membership/content.cfm?ItemNumber=774
Sponsor: American Oil Chemists' Society
2710 South Boulder
Urbana, IL 61802-6996

AOCS Australasian Award for Scientific Excellence in Lipid Research **875**

The AOCS Award for Scientific Excellence in Lipid Research is given to a scientist from within the Australasian region that has made a significant research contribution toward fats and oil research, either cumulative or one major advancement. The Award is offered every two years with the Section's biannual conference. The scientist receives a plaque, $1,500 AU towards expenses to attend the AOCS Annual Meeting, complimentary registration for the biannual AAOCS meeting where the award is presented, and presentation at that meeting. Candidates may nominate themselves or others. All nomination material must be emailed to the AAOCS Award Coordinator.
Requirements: Candidates must meet the following criteria to be eligible: make a significant contribution in basic or applied research, with that knowledge having a significant effect on advances in the field or hold substantial promise for its effect in the near future; and hold membership with the AAOCS during the year they are applying for the Award. Candidate materials must include the following: letter of nomination with summary of the nominee's research accomplishments; letters of support from at least two other scientists; and professional biographical data.
Geographic Focus: Australia
Date(s) Application is Due: May 17
Amount of Grant: 1,500 ARA
Contact: Amy Logan; +61 3 9731 3478; fax +61 3 9545 2175; Amy.Logan@csiro.au
Internet: http://www.aocs.org/Membership/content.cfm?ItemNumber=18316
Sponsor: American Oil Chemists' Society
2710 South Boulder
Urbana, IL 61802-6996

AOCS Biotechnology Division Student Awards **876**

The Biotechnology Division Student Awards are given to students presenting an outstanding paper in the field of biotechnology at the Annual Meeting. The Awards are given in three categories: $300 (1st place); $200 (2nd place); and $100 (3rd place).
Geographic Focus: All States, All Countries
Date(s) Application is Due: Oct 15
Amount of Grant: 600 USD
Contact: Barb Semeraro; (217) 693-4804; fax (217) 693-4849; barbs@aocs.org
Internet: http://www.aocs.org/Membership/content.cfm?ItemNumber=786
Sponsor: American Humane Association
1400 16th Street NW, Suite 360
Washington, DC 20036

AOCS Edible Applications Technology Division Outstanding Achievement Award 877
The Edible Applications Technology Division Outstanding Achievement Award recognizes a scientist, technologist, or leader making contributions to the advancement of edible oils and/or the Division. The Award includes a plaque for the recipient, $500 cash prize, and a complimentary Division dinner ticket.
Requirements: Preference is given to AOCS members involved in the Edible Applications Technology Division, Society, or related organizations over a substantial amount of time. Candidates must also meet the following criteria: significant contribution to the knowledge, technology, and history of the industry; demonstrated character, ethics and a reputation that puts the individual above his or her peers; and service as a career and technical mentor.
Geographic Focus: All States, All Countries
Date(s) Application is Due: Oct 15
Amount of Grant: 500 USD
Contact: Barb Semeraro; (217) 693-4804; fax (217) 693-4849; barbs@aocs.org
Internet: http://www.aocs.org/Membership/content.cfm?ItemNumber=18315
Sponsor: American Oil Chemists' Society
2710 South Boulder
Urbana, IL 61802-6996

AOCS Edible Applications Technology Division Student Award 878
The Edible Applications Technology Division Student Award is given to a graduate student presenting within the Edible Applications Technology program at the Annual Meeting. The Award includes a certificate for the recipient, an award lecture or poster presentation at the Annual Meeting, and $500 in travel funds to attend the Meeting.
Geographic Focus: All States, All Countries
Date(s) Application is Due: Oct 15
Amount of Grant: 500 USD
Contact: Barb Semeraro; (217) 693-4804; fax (217) 693-4849; barbs@aocs.org
Internet: http://www.aocs.org/Membership/content.cfm?ItemNumber=796
Sponsor: American Oil Chemists' Society
2710 South Boulder
Urbana, IL 61802-6996

AOCS George Schroepfer Medal 879
The George Schroepfer Medal was established to honor the memory of George J. Schroepfer, Jr., a leader in the sterol and lipid field for more than 40 years. The award aims to foster Schroeper's ideals of personal integrity, high scientific standards, perseverance, and a strong spirit of survival, tempered by charm and wit. The award recognizes major advances in the steroid or sterol field, and is presented every even numbered year. It includes a bronze medal for the recipient, an award lecture at the Annual Meeting, and an honorarium determined by the endowed fund investment return. Self nominations are welcomed.
Requirements: The candidate's work may represent a single major achievement or a series of significant accomplishments, but should be based upon sound primary experimental data. Preference is given to accomplishments in the following: biochemistry and physiology with biomedical application; and interdisciplinary research in which rigorous chemical and analytical methods were applied to elucidate the physiological roles of steroids in animals, plants, or microorganisms. Fundamental advances that are primarily chemical, pharmacological, or analytical will also be considered
Geographic Focus: All States, All Countries
Date(s) Application is Due: Oct 15
Contact: Barb Semeraro; (217) 693-4804; fax (217) 693-4849; barbs@aocs.org
Internet: http://www.aocs.org/Membership/content.cfm?ItemNumber=934
Sponsor: American Oil Chemists' Society
2710 South Boulder
Urbana, IL 61802-6996

AOCS Hans Kaunitz Award 880
The Hans Kaunitz Award is given to a student doing research in the sciences relating to fats and oils and detergent technology. The Award includes a certificate, $1,000 honorarium, $500 travel allowance to the Annual Meeting, and a speaking engagement at the section Annual Meeting.
Requirements: The applicant must be scholastically outstanding, and a graduate student attending a post-secondary school in the U.S. The applicant must be actively performing research toward a Master's or Doctorate degree. This research must be related to fats, oils, protein co-products and/or detergents, and be interested in the areas of science and technology fostered by AOCS.
Geographic Focus: All States
Date(s) Application is Due: Oct 15
Amount of Grant: 1,500 USD
Contact: Barb Semeraro; (217) 693-4804; fax (217) 693-4849; barbs@aocs.org
Internet: http://www.aocs.org/Membership/content.cfm?ItemNumber=923
Sponsor: American Oil Chemists' Society
2710 South Boulder
Urbana, IL 61802-6996

AOCS Health and Nutrition Division Student Award 881
The Health and Nutrition Division of AOCS promotes and facilitates communication and cooperation among professionals whose interests in lipid biochemistry and physiology relate to all aspects of dietary fats and health. The Division encompasses the technical areas of dietary fats and general health. The Health and Nutrition Division Student Award is given to a graduate student in the health and nutrition field. The Award includes a $500 honorarium, certificate, and award lecture or poster presentation at the Annual Meeting. Self nominations are welcomed.
Requirements: Graduate students from any institution of higher learning who are doing research toward a degree are eligible.
Geographic Focus: All States, All Countries
Date(s) Application is Due: Oct 15
Amount of Grant: 500 USD
Contact: Barb Semeraro; (217) 693-4804; fax (217) 693-4849; barbs@aocs.org
Internet: http://www.aocs.org/Membership/content.cfm?ItemNumber=2048
Sponsor: American Oil Chemists' Society
2710 South Boulder
Urbana, IL 61802-6996

AOCS Health and Nutrition Poster Competition 882
The Health and Nutrition Division promotes and facilitates communication and cooperation among professionals whose interests in lipid biochemistry and physiology relate to all aspects of dietary fats and health; encompasses the technical areas of dietary fats and general health. The Health and Nutrition Division Poster Competition was created to stimulate additional interest in the submission of quality poster student presentations at the Annual Meeting. Awards are given for best overall poster ($300), 1st place ($300), 2nd place ($200), and 3rd place ($100). All awardees also receive certificates, complimentary division dinner tickets, and award ribbons for their poster display.
Requirements: All poster abstracts submitted and selected for the Health and Nutrition Division Poster session are eligible for the competition.
Geographic Focus: All States, All Countries
Date(s) Application is Due: Oct 15
Amount of Grant: 900 USD
Contact: Barb Semeraro; (217) 693-4804; fax (217) 693-4849; barbs@aocs.org
Internet: http://www.aocs.org/Membership/content.cfm?ItemNumber=18280
Sponsor: American Oil Chemists' Society
2710 South Boulder
Urbana, IL 61802-6996

AOCS Herbert J. Dutton Award 883
The AOCS Analytical Division promotes professional interest, communication, and expert competency in the analytical chemistry field. The Herbert J. Dutton Award recognizes significant contributions to analytical chemistry. The Award includes a plaque for the recipient, $1,000 honorarium, and an award lecture at the Annual Meeting.
Requirements: Eligible candidates include those who have fulfilled one of the following criteria: significant contributions to the analysis of fats, oil, and related products (contributions reflected in current analytical methodologies); or those whose work has resulted in major advances in the understanding of processes utilized in the fats and oils industry.
Geographic Focus: All States, All Countries
Date(s) Application is Due: Oct 15
Amount of Grant: 1,000 USD
Contact: Barb Semeraro; (217) 693-4804; fax (217) 693-4849; barbs@aocs.org
Internet: http://www.aocs.org/Membership/content.cfm?ItemNumber=948
Sponsor: American Oil Chemists' Society
2710 South Boulder
Urbana, IL 61802-6996

AOCS Honored Student Award 884
The American Oil Chemists' Society Honored Student Award is given annually to a graduate student in fats and lipid chemistry who is currently attending a college or university. The award includes a certificate, travel allowance for the Annual Meeting, and award lecture at the meeting. The application is available at the website.
Requirements: Applicants must be graduate students at any institution of higher learning, in any area of science and technology fostered by the AOCS, who are doing research toward an advanced degree.
Geographic Focus: All States, All Countries
Date(s) Application is Due: Oct 15
Contact: Barb Semeraro; (217) 693-4804; fax (217) 693-4849; barbs@aocs.org
Internet: http://www.aocs.org/member/awards/awd-honstud.asp
Sponsor: American Oil Chemists' Society
2710 South Boulder
Urbana, IL 61802-6996

AOCS Industrial Oil Products Division Student Award 885
The AOCS Industrial Oil Products Division supports professionals involved in research, development, engineering, marketing, and testing of industrial products and co-products from fats and oil, including glycerine and its derivatives. The Division's Student Award is given to a graduate student for travel to the Annual Meeting to present a paper. The Award includes a certificate for the recipient, an award lecture or poster presentation at the Annual Meeting, and $500 in travel funds for that meeting.
Requirements: Graduate students presenting within the Industrial Oil Products Division technical program at the Annual Meeting are eligible.
Geographic Focus: All States, All Countries
Date(s) Application is Due: Oct 15
Amount of Grant: 500 USD
Contact: Barb Semeraro; (217) 693-4804; fax (217) 693-4849; barbs@aocs.org
Internet: http://www.aocs.org/Membership/content.cfm?ItemNumber=810
Sponsor: American Oil Chemists' Society
2710 South Boulder
Urbana, IL 61802-6996

AOCS Lipid Oxidation and Quality Division Poster Competition **886**
The Lipid Oxidation and Quality Division supports professionals in lipid oxidation with a major focus in food applications. The Lipid Oxidation and Quality Division Poster Competition is awarded to a student for travel to present a poster at the Annual Meeting. All poster abstracts submitted by students and selected for the Lipid Oxidation and Quality Poster session are considered for the competition. The Award includes a certificate, Award ribbon for poster display, complimentary division dinner ticket, and $300.
Geographic Focus: All States, All Countries
Date(s) Application is Due: Oct 15
Amount of Grant: 300 USD
Contact: Barb Semeraro; (217) 693-4804; fax (217) 693-4849; barbs@aocs.org
Internet: http://www.aocs.org/Membership/content.cfm?ItemNumber=18281
Sponsor: American Oil Chemists' Society
2710 South Boulder
Urbana, IL 61802-6996

AOCS Manuchehr Eijadi Award **887**
The Manuchehr Eijadi Award is given to a graduate student doing research toward an advanced degree. The Award includes a $1,000 in addition to a plaque and certificate.
Requirements: Applicants must be previous recipients of the AOCS Honored Student Award, with graduate research focused on chemical engineering.
Geographic Focus: All States, All Countries
Date(s) Application is Due: Oct 15
Amount of Grant: 1,000 USD
Contact: Barb Semeraro; (217) 693-4804; fax (217) 693-4849; barbs@aocs.org
Internet: http://www.aocs.org/Membership/content.cfm?ItemNumber=797
Sponsor: American Oil Chemists' Society
2710 South Boulder
Urbana, IL 61802-6996

AOCS Peter and Claire G. Kalustian Award **888**
The Peter and Claire G. Kalustian Award is given to a graduate student doing research toward an advanced degree. The Award includes $1,000 and a certificate.
Requirements: Applicants must be graduate students in any area of science and technology fostered by AOCS, who are doing research toward an advanced degree. Students must be previous AOCS Honored Students. Preference is given to students of Armenian descent.
Geographic Focus: All States, All Countries
Date(s) Application is Due: Oct 15
Amount of Grant: 1,000 USD
Contact: Barb Semeraro; (217) 693-4804; fax (217) 693-4849; barbs@aocs.org
Internet: http://www.aocs.org/Membership/content.cfm?ItemNumber=812
Sponsor: American Oil Chemists' Society
2710 South Boulder
Urbana, IL 61802-6996

AOCS Phospholipid Best Paper Award **889**
The Phospholipid Best Paper Award recognizes an outstanding published paper or monograph in the area of phospholipid research or applications for phospholipids. The Award includes a plaque for the main author, certificates for each author, funding the attend the Annual Meeting, and an award address at that meeting. Self nominations are welcomed.
Requirements: Papers eligible for nomination include those relating to phospholipids that have been published in the past year from any journal or monograph. Papers do not need to be from an AOCS publication.
Geographic Focus: All States, All Countries
Date(s) Application is Due: Oct 15
Contact: Barb Semeraro; (217) 693-4804; fax (217) 693-4849; barbs@aocs.org
Internet: http://www.aocs.org/Membership/content.cfm?ItemNumber=904
Sponsor: American Oil Chemists' Society
2710 South Boulder
Urbana, IL 61802-6996

AOCS Processing Distinguished Service Award **890**
The AOCS Processing Distinguished Service Award recognizes and honors outstanding meritorious service to the oilseed processing industry. The Award includes a certificate and travel expenses to the Annual Meeting.
Requirements: Preference is given to AOCS members involved in Division, Society, or related organizations over a substantial amount of time. Candidates must meet the following criteria: significant contributions to the knowledge, technology, and history of the industry; demonstrated character, ethics and a reputation that puts the candidate above his or her peers; and service as a career and technical mentor.
Geographic Focus: All States, All Countries
Date(s) Application is Due: Oct 15
Contact: Barb Semeraro; (217) 693-4804; fax (217) 693-4849; barbs@aocs.org
Internet: http://www.aocs.org/Membership/content.cfm?ItemNumber=922
Sponsor: American Oil Chemists' Society
2710 South Boulder
Urbana, IL 61802-6996

AOCS Processing Division Student Award **891**
The AOCS Processing Division advances the processing knowledge and managerial skills by providing a forum of technical information and networking opportunities, to promote and facilitate communication and cooperation between members, and to mentor young professionals. The Processing Division Student Award is given to a graduate student researching oilseed handling preparation and extraction, refining and processing, oil products and packaging, feed ingredients, by-products utilization, safety and health, and environmental concerns. The Award includes a certificate for the recipient, an award lecture or poster presentation at the Annual Meeting, and a $500 honorarium.
Requirements: Graduate students presenting within the Processing Division's technical program at the Annual Meeting are eligible.
Geographic Focus: All States, All Countries
Date(s) Application is Due: Oct 15
Amount of Grant: 500 USD
Contact: Barb Semeraro; (217) 693-4804; fax (217) 693-4849; barbs@aocs.org
Internet: http://www.aocs.org/Membership/content.cfm?ItemNumber=924
Sponsor: American Oil Chemists' Society
2710 South Boulder
Urbana, IL 61802-6996

AOCS Protein and Co-Products Division Student Poster Competition **892**
The Protein and Co-Products Division Student Poster Competition is awarded to a student presenting a poster at the AOCS Annual Meeting. Awards are given to a first place winner for $300, and a second place winner for $200. Each student also receives a certificate, an award ribbon for their poster display, and a complimentary division dinner ticket.
Requirements: All poster abstracts submitted by students and selected for the Protein and Co-Products Poster session are considered for the competition.
Geographic Focus: All States, All Countries
Date(s) Application is Due: Oct 15
Amount of Grant: 500 USD
Contact: Barb Semeraro; (217) 693-4804; fax (217) 693-4849; barbs@aocs.org
Internet: http://www.aocs.org/Membership/content.cfm?ItemNumber=18314
Sponsor: American Oil Chemists' Society
2710 South Boulder
Urbana, IL 61802-6996

AOCS Ralph H. Potts Memorial Fellowship **893**
The Ralph H. Potts Memorial Fellowship is awarded to a graduate student doing research in fats and oils. The award is named for Ralph Potts in recognition for his work in fats, fatty acids, and derivatives, and the personal esteem of his colleagues. The recipient receives a plaque, a travel allowanced for the AOCS Annual Meeting, and a $2,000 honorarium. The application is available at the website.
Requirements: The applicant must be a graduate student at a North American university, with research involving fatty acids or their derivatives such as long chain alcohols, amines, or other nitrogen compounds. The applicant's research should fall in the category of synthesis, processing utilization, or characterization.
Geographic Focus: All States, Puerto Rico, Anguilla, Antigua & Barbuda, Aruba, Bahamas, Barbados, Belize, Bermuda, British Virgin Islands, Canada, Cayman Islands, Costa Rica, Cuba, Dominica, Dominican Republic, El Salvador, Grenada, Guatemala, Haiti, Honduras, Jamaica, Martinique, Mexico, Montserrat, Nicaragua, Panama, Saint Kitts And Nevis, St. Lucia, St. Vincent and the Grenadines, Trinidad and Tobago, Turks and Caicos Islands, Virgin Islands (U.S.)
Date(s) Application is Due: Oct 15
Amount of Grant: 2,000 USD
Contact: Barb Semeraro; (217) 693-4804; fax (217) 693-4849; barbs@aocs.org
Internet: http://www.aocs.org/Membership/content.cfm?ItemNumber=906
Sponsor: American Oil Chemists' Society
2710 South Boulder
Urbana, IL 61802-6996

AOCS Ralph Holman Lifetime Achievement Award **894**
The AOCS Health and Nutrition Division promotes and facilitates communication and cooperation among professionals whose interests in lipid biochemistry and physiology relate to all aspects of dietary fats and health. The Ralph Holman Lifetime Achievement Award recognizes outstanding performance and meritorious contributions. The Award is named for Ralph Holman in recognition of his lifetime service to the study of essential fatty acids and their impact on health. The Award includes a framed orchid photograph, $1,000 honorarium, and an award lecture at the Annual Meeting.
Requirements: Candidates must have made significant lifetime and meritorious achievements in areas of interest to the Health and Nutrition Division of AOCS.
Geographic Focus: All States, All Countries
Date(s) Application is Due: Oct 15
Amount of Grant: 1,000 USD
Contact: Barb Semeraro; (217) 693-4804; fax (217) 693-4849; barbs@aocs.org
Internet: http://www.aocs.org/Membership/content.cfm?ItemNumber=807
Sponsor: American Oil Chemists' Society
2710 South Boulder
Urbana, IL 61802-6996

AOCS Samuel Rosen Memorial Award **895**
The Samuel Rosen Memorial Award recognizes accomplishments in surfactant chemistry. The Award is named for Samuel Rosen, who worked for more than 40 years as an industrial chemist on the formulation of printing inks. The Award includes a plaque for the recipient, a $2,000 honorarium, and an award lecture at the Annual Meeting.
Requirements: Candidates must meet the following criteria for eligibility: significant advancements, cumulative advancements, or application of the principles of surfactant chemistry; and a scientist or technician who has worked the majority of his or her career in the industry, preferable within the last three years of the award recognition.

Geographic Focus: All States, All Countries
Date(s) Application is Due: Oct 15
Amount of Grant: 2,000 USD
Contact: Barb Semeraro; (217) 693-4804; fax (217) 693-4849; barbs@aocs.org
Internet: http://www.aocs.org/Membership/content.cfm?ItemNumber=931
Sponsor: American Oil Chemists' Society
2710 South Boulder
Urbana, IL 61802-6996

AOCS Stephen S. Chang Award **896**
The Stephen S. Change Award is given annually to recognize a scientist or technologist who has made decisive accomplishments in research for the improvement or development of products related to lipids. The award includes a lecture at the Annual Meeting, a jade sculpture of a galloping horse, and an honorarium determined by the endowed fund investment return. Self nominations are welcomed.
Requirements: Candidates must have made significant and distinguished accomplishments in basic research that have been utilized by industries for the development or improvement of products related to lipids. Applicants may be recognized for either one major breakthrough or by an accumulation of publications.
Geographic Focus: All States, All Countries
Date(s) Application is Due: Oct 15
Contact: Barb Semeraro; (217) 693-4804; fax (217) 693-4849; barbs@aocs.org
Internet: http://www.aocs.org/Membership/content.cfm?ItemNumber=788
Sponsor: American Oil Chemists' Society
2710 South Boulder
Urbana, IL 61802-6996

AOCS Supelco/Nicholas Pelick-AOCS Research Award **897**
The Supelco/Nicholas Pelick-AOCS Research Award is given to a qualified scientist for outstanding original research on fats, oils, lipid chemistry, and/or biochemistry, the results of which have been published in technical papers of high quality. The recipient receives a plaque and a $10,000 honorarium. The award winner also gives a lecture at the AOCS Annual Meeting. The application is available at the website. Self nominations are welcomed.
Requirements: Candidates must have published technical papers of high quality Preference is given to individuals who are actively associated with research and have made influential discoveries in their fields.
Geographic Focus: All States, All Countries
Date(s) Application is Due: Oct 15
Amount of Grant: 10,000 USD
Contact: Barb Semeraro; (217) 693-4804; fax (217) 693-4849; barbs@aocs.org
Internet: http://www.aocs.org/Membership/content.cfm?ItemNumber=940
Sponsor: American Oil Chemists' Society
2710 South Boulder
Urbana, IL 61802-6996

AOCS Surfactants and Detergents Division Student Award **898**
The AOCS Surfactants and Detergents Division encompasses the sciences and technologies associated with the development and production of surfactants, detergents, and soaps. The Surfactants and Detergents Division Student Award is given to a graduate student to help in their travel to present a paper at the Annual Meeting. The Award includes a certificate for the recipient(s), an award lecture or poster presentation at the Annual Meeting, and $500 in travel funds. Self nominations are welcomed.
Requirements: Graduate students presenting within the Surfactants and Detergents Division technical program at the AOCS Annual Meeting are eligible.
Geographic Focus: All States, All Countries
Date(s) Application is Due: Oct 15
Amount of Grant: 500 USD
Contact: Barb Semeraro; (217) 693-4804; fax (217) 693-4849; barbs@aocs.org
Internet: http://www.aocs.org/Membership/content.cfm?ItemNumber=942
Sponsor: American Oil Chemists' Society
2710 South Boulder
Urbana, IL 61802-6996

AOCS Thomas H. Smouse Memorial Fellowship **899**
The Thomas H. Smouse Memorial Fellowship is awarded to a graduate student doing research in areas of interest to AOCS. The award includes a $10,000 stipend, plus $5,000 for travel to professional meetings and research expenditures related to the student's graduate program. Self nominations are welcomed.
Requirements: Individuals must be scholastically outstanding with interests and involvement outside the academic discipline.
Geographic Focus: All States, All Countries
Date(s) Application is Due: Oct 15
Amount of Grant: 15,000 USD
Contact: Barb Semeraro; (217) 693-4804; fax (217) 693-4849; barbs@aocs.org
Internet: http://www.aocs.org/Membership/content.cfm?ItemNumber=938
Sponsor: American Oil Chemists' Society
2710 South Boulder
Urbana, IL 61802-6996

AOCS Timothy L. Mounts Award **900**
The Edible Applications Technology Division is comprised of professionals in the processing and utilization of lipids industry with primary focus on food applications. The Timothy L. Mounts Award recognizes research relating to the science and technology of edible oils, basic or applied in nature. The Award includes a plaque for the recipient, a $500 honorarium, and an award lecture at the Annual Meeting.
Requirements: Mid-career scientists engaged in edible oil research are eligible.
Geographic Focus: All States, All Countries
Date(s) Application is Due: Oct 15
Amount of Grant: 500 USD
Contact: Barb Semeraro; (217) 693-4804; fax (217) 693-4849; barbs@aocs.org
Internet: http://www.aocs.org/Membership/content.cfm?ItemNumber=831
Sponsor: American Oil Chemists' Society
2710 South Boulder
Urbana, IL 61802-6996

AOCS Uses of Soybean Oil Award **901**
The AOCS Uses of Soybean Oil Award is sponsored by the United Soybean Board. The Award recognizes outstanding research into new industrial applications or uses for soybean oil. The Award includes a plaque for the recipient, a $3,000 honorarium, and award lecture at the Annual Meeting. Nomination information is available at the website.
Geographic Focus: All States, All Countries
Date(s) Application is Due: Oct 15
Amount of Grant: 3,000 USD
Contact: Barb Semeraro; (217) 693-4804; fax (217) 693-4849; barbs@aocs.org
Internet: http://www.aocs.org/Membership/content.cfm?ItemNumber=943
Sponsor: American Oil Chemists' Society
2710 South Boulder
Urbana, IL 61802-6996

AOCS Young Scientist Research Award **902**
The Young Scientist Research Award recognizes substantial research contributions in one of the following areas: agricultural microscopy; biotechnology; edible applications technology; health and nutrition; industrial oil products; lipid oxidation and quality; phospholipid; protein and co-products; and surfactants and detergents. The Award includes a plaque for the recipient, a $1,000 honorarium, travel and registration for the Annual Meeting, and award lecture at the Annual Meeting. Self nominations are welcomed.
Requirements: The recipient must be younger than 36 years of age by June 1 of the year of the presentation, or have received his or her highest degree within the previous ten years. The candidate must also have significant contributions to advances in basic or applied research, or hold substantial promise for its effect in the near future.
Geographic Focus: All States, All Countries
Date(s) Application is Due: Oct 15
Amount of Grant: 1,000 USD
Contact: Barb Semeraro; (217) 693-4804; fax (217) 693-4849; barbs@aocs.org
Internet: http://www.aocs.org/Membership/content.cfm?ItemNumber=944
Sponsor: American Oil Chemists' Society
2710 South Boulder
Urbana, IL 61802-6996

AOS Research Grants **903**
Research grants are awarded for experimental projects and applied and fundamental research pertaining to orchids. Support is given in biological research including taxonomy, genetics, cytogenetics, physiology, and pathology as well as in conservation and education. The society wishes to further the study of orchids in every aspect including their classification, evolution, conservation, propagation, culture, care, and development. As a first step, prospective applicants should submit a brief letter requesting an application form and instructions for requesting a grant. Grants are usually awarded for one year, possibly renewable, but not to exceed a maximum of three years. Grants received by January 1 are reviewed at the Spring Trustees Meeting and those grants received by July 1 are reviewed at the Fall Trustees Meeting
Requirements: Graduate students or other qualified research personnel associated with accredited institutions of higher learning or appropriate research institutes may apply.
Geographic Focus: All States
Date(s) Application is Due: Jan 1; Jul 1
Amount of Grant: 500-12,000 USD
Contact: Pamela Giust, (561) 404-2000; fax (561) 404-2045; pgiust@aos.org
Internet: http://www.theaos.org/aos/research/research.aspx?sec=2
Sponsor: American Orchid Society
16700 AOS Lane
Delray Beach, FL 33446-4351

APA Culture of Service in the Psychological Sciences Award **904**
The Award recognizes departments that demonstrate a commitment to service in the psychological sciences. Departments selected for this award will show a pattern of support for service from faculty at all levels, including a demonstration that service to the discipline is rewarded in faculty tenure and promotion. Successful Departments will also demonstrate that service to the profession is an integral part of training and mentoring. Service to the discipline includes such activities as departmental release time for serving on boards and committees of psychological associations; editing journals; serving on a review panel; or chairing an IRB. Other culture of service activities that a department would encourage include mentoring students and colleagues; advocating for psychological science's best interests with state and federal lawmakers; and promoting the value of psychological science in the public eye. The focus of this award is a department's faculty service to the discipline and not their scholarly achievements. Both Undergraduate and Graduate Departments of Psychology are eligible. Self-nominations are encouraged.
Geographic Focus: All States

Amount of Grant: 5,000 USD
Samples: Davidson College, Departments of Psycholog, Davidson, North Carolina; University of Minnesota, Department of Psychology, Minneapolis, Minnesota.
Contact: Suzanne Wandersman, Director for Governance Affairs; (202) 336-6000; fax (202) 336-5953; swandersman@apa.org
Internet: http://www.apa.org/science/dept_award.html
Sponsor: American Psychological Association
750 First Street NE
Washington, DC 20002-4242

APA Dissertation Research Awards **905**
Dissertation awards are granted to graduate students in psychology programs to help offset the costs associated with dissertation research. The research may be done in any area of psychology. Applicants who have already defended their dissertations are eligible to apply for these funds, as long as they have not yet received doctoral degrees as of the application deadline. APA provides dissertation research awards to approximately 50 students annually. Dissertations must be approved by the students' committees before application. The application deadline falls in mid-September of each year. Awards are announced in late December.
Requirements: APA student affiliates enrolled in a psychology graduate program are eligible to apply.
Restrictions: Departments may not nominate more than three students per year for dissertation awards.
Geographic Focus: All States
Date(s) Application is Due: Sep 15
Amount of Grant: 1,000-5,000 USD
Contact: APA Science Directorate, (800) 374-2721 or (202) 336-5500; fax (202) 336-5953; science@apa.org
Internet: http://www.apa.org/science/dissinfo.html
Sponsor: American Psychological Association
750 First Street NE
Washington, DC 20002-4242

APA Distinguished Scientific Award for Early Career Contribution to Psychology **906**
The Distinguished Scientific Award for Early Career Contribution to Psychology recognizes excellent young psychologists. Nominations of persons who received doctoral degrees during and since 1998 are being sought in these five areas: animal learning and behavior, comparative; cognition/human learning; developmental psychology; health; and psychopathology. These categories should be interpreted broadly and are not meant to be exclusive; all of psychology is of sufficient merit to be considered for awards. The Award consists of a citation and a cash prize, which are presented at the APA Annual Convention.
Geographic Focus: All States
Samples: Robert D. Gray, PhD (applied research), Department of Applied Psychology, Arizona State University; Patrik O. Vuilleumier, MD (behavioral /cognitive neuroscience), Department of Neurology, University of California, Davis; R. Chris Fraley, PhD (individual differences), Department of Psychology at University of Illinois at Urbana-Champaign; Jorn Diedrichsen, PhD (perception/motor performance), School of Psychology, Adeilad Brigantia University, Wales; Matthew D. Lieberman, PhD (social), University of California, Los Angeles.
Contact: Suzanne Wandersman, Director for Governance Affairs; (202) 336-6000; fax (202) 336-5953; swandersman@apa.org
Internet: http://www.apa.org/science/sciaward.html
Sponsor: American Psychological Association
750 First Street NE
Washington, DC 20002-4242

APA Distinguished Scientific Award for the Applications of Psychology **907**
The Distinguished Scientific Award for the Applications of Psychology honors psychologists who have made distinguished theoretical or empirical advances in psychology leading to the understanding or amelioration of important practical problems. The Award consists of a citation and a cash prize, which are presented at the APA Annual Convention.
Geographic Focus: All States
Contact: Suzanne Wandersman, Director for Governance Affairs; (202) 336-6000; fax (202) 336-5953; swandersman@apa.org
Internet: http://www.apa.org/science/sciaward.html
Sponsor: American Psychological Association
750 First Street NE
Washington, DC 20002-4242

APA Distinguished Scientific Contribution Award **908**
The Distinguished Scientific Contribution Award honors psychologists who have made distinguished theoretical or empirical contributions to basic research in psychology. The Award consists of a citation and a cash prize, which are presented at the APA Annual Convention.
Geographic Focus: All States
Samples: Marilynn B. Brewer, PhD, Department of Psychology, Ohio State University, Columbus, OH; Jean M. Mandler, PhD, Research Professor, Department of Cognitive Science, University of California, San Diego; Paul Rozin, PhD, Professor of Psychology, University of Pennsylvania.
Contact: Suzanne Wandersman, Director for Governance Affairs; (202) 336-6000; fax (202) 336-5953; swandersman@apa.org
Internet: http://www.apa.org/science/sciaward.html
Sponsor: American Psychological Association
750 First Street NE
Washington, DC 20002-4242

APA Distinguished Service to Psychological Science Award **909**
This Award recognizes individuals who have made outstanding contributions to psychological science through their commitment to a culture of service. Award recipients will receive an honorarium of $1,000. Nominees will have demonstrated their service to the discipline by aiding in association governance; serving on boards, committees and various psychological associations; editing journals; reviewing grant proposals; mentoring students and colleagues; advocating for psychological science's best interests with state and federal lawmakers; and promoting the value of psychological science in the public eye. Nominees may be involved in one service area, many of the areas, or all of the service areas noted above. An individual's service to the discipline and not a person's scholarly achievements are the focus of this award.
Geographic Focus: All States
Amount of Grant: 1,000 USD
Contact: Suzanne Wandersman, Director for Governance Affairs; (202) 336-6000; fax (202) 336-5953; swandersman@apa.org
Internet: http://www.apa.org/science/serv_award.html
Sponsor: American Psychological Association
750 First Street NE
Washington, DC 20002-4242

APA Planning Fellowships **910**
The goals of this program are both to encourage students of certain minority backgrounds to enter the planning profession and to help such students who would otherwise be unable to continue their studies in planning. The exact amount of the awards offered is contingent upon the number and qualifications of the applicants. The program is open to first- and second-year students; first-year students who receive fellowships are eligible to compete for an award the following year as well.
Requirements: Eligible to apply are U.S. or Canadian citizens who are African Americans, Hispanics, or Native Americans currently enrolled or accepted for enrollment in a PAB-accredited graduate planning program. Applicants must document their need for financial assistance.
Geographic Focus: All States
Date(s) Application is Due: Apr 30
Amount of Grant: 1,000-5,000 USD
Contact: Kriss Blank, Program Coordinator; (312) 786-6722 or (312) 431-9100; fax (312) 431-9985; kblank@planning.org or fellowship@planning.org
Internet: http://www.planning.org/institutions/scholarship.htm
Sponsor: American Planning Association
122 S Michigan Avenue, Suite 1600
Chicago, IL 60603

APA Scientific Conferences Grants **911**
The program seeks proposals for research conferences in psychology. The purpose of this program is to promote the exchange of important new contributions and approaches in scientific psychology. Conference formats include festschrifts—conferences organized as tributes to distinguished scholars; stand-alone conferences—usually two days long, involves psychologists as the primary organizers; and add-a-day conference—meetings that occur at the beginning or end of a scientific conference other than APA (e.g., Society for Ingestive Behavior or Psychonomics Society). The conference must also be supported by the host institution with direct funds, in-kind support, or a combination of the two.
Requirements: One of the primary organizers must be a member of APA. Only academic institutions accredited by a regional body may apply. Independent research institutions must provide evidence of affiliation with such an accredited institution. Joint proposals from cooperating institutions are encouraged. Conferences may be held only in the United States, its possessions, or Canada.
Restrictions: APA governance groups, APA divisions, and related entities are not eligible for funding under this program.
Geographic Focus: All States
Date(s) Application is Due: Jun 1; Dec 1
Amount of Grant: 500-20,000 USD
Contact: Deborah McCall, (202) 336-6000; fax (202) 336-5953; dmccall@apa.org
Internet: http://www.apa.org/science/psa/sep07ann.html
Sponsor: American Psychological Association
750 First Street NE
Washington, DC 20002-4242

APA Young Investigator Grants **912**
The program provides financial support to teaching, research, and health care delivery projects in general pediatrics. The number of awards is dependent on available funds and the size of grant requests of selected projects. Proposals should address the goal of the association: to encourage, promote, and facilitate improved patient care, teaching, and research in general pediatrics.
Requirements: The principal investigator must be a member of the APA or have submitted an application for membership. Preference will be given to new investigators, including those in training.
Restrictions: Multiple year funding requests will not be considered.
Geographic Focus: All States
Date(s) Application is Due: Oct 25
Amount of Grant: Up to 10,000 USD
Contact: Administrator; (703) 556-9222; fax (703) 556-8729; info@academicpeds.org
Sponsor: Ambulatory Pediatric Association
6728 Old McLean Village Drive
McLean, VA 22101

APHA Fellowship in Printing History 913

The fellowship is awarded for research in any area of the history of printing in all its forms, including all the arts and technologies relevant to printing, the book arts, and letter forms. The subject of research has no geographical or chronological limitations, and may be national or regional in scope, biographical, analytical, technical, or bibliographical in nature. Printing history-related study with a recognized printer or book artist may also be supported. The fellowship can be used to pay for travel, living, and other expenses.

Geographic Focus: All States
Date(s) Application is Due: Dec 10
Amount of Grant: Up to 2,000 USD
Samples: Lance Hidy—to support research on Boston's Society of Printers, $2000; Dr. Susanna Ashton—for work on William Stanley Braithwaite.
Contact: Stephen Crook, Executive Secretary; sgcrook@printinghistory.org
Internet: http://printinghistory.org/htm/fellowship/2005.html
Sponsor: American Printing History Association
P.O. Box 4519, Grand Central Station
New York, NY 10163-4519

APHL Emerging Infectious Diseases Fellowships 914

The Emerging Infectious Diseases (EID) Laboratory Fellowship Program, sponsored by the Association of Public Health Laboratories (APHL) and the Centers for Disease Control and Prevention (CDC), trains and prepares scientists for careers in public health laboratories and supports public health initiatives related to infectious disease research. Two types of fellowships are offered: the EID Advanced Laboratory Training Fellowship is a one-year program designed for bachelor's or master's level scientists, with emphasis on the practical application of technologies, methodologies and practices related to emerging infectious diseases; the EID Laboratory Research Fellowship is a two-year program designed for doctoral level (Ph.D., M.D. or D.V.M.) scientists to conduct high-priority research in infectious diseases. Up to 30 pre-doctoral one-year and 15 post-doctoral two-year fellowships will be awarded. Fellows are placed in local, state or federal (CDC) public health laboratories throughout the U.S.. Fellows are provided with a stipend, medical insurance, travel to the host laboratory and a professional development allowance. The fellowship program offers a wide variety of training and research experiences. Examples of projects include research in molecular genetics, pathogenesis, epidemiology and cell biology. Fellows may be trained in specific laboratory techniques such as real-time PCR, DNA sequencing and mass spectrometry. Fellows may receive specialized training and experience with a specific pathogen, or generalized training in influenza or STD surveillance, newborn screening or chemical and biological warfare agents. Other work may include diagnostic virology, bioanalytical chemistry, foodborne disease research including Pulsed Field Gel Electrophoresis (PFGE), clinical and environmental diagnostics and epidemiologic studies of disease outbreaks. Fellows also have opportunities to participate in research seminars, teleconferences, regional and national meetings and other continuing education opportunities. All fellows participate in an orientation session at CDC in Atlanta to gain a general understanding of the public health laboratory system and how it relates to infectious disease surveillance, prevention, research and control. All applicants are required to submit one original single-sided application to APHL by the application deadline. The application may be downloaded from the EID fellowship website in writeable pdf format. All required supporting documents (listed in the application) should be collected by the applicant and forwarded with the application to APHL. Deadline dates may vary from year to year. Applicants are encouraged to check the EID fellowship website or contact the Fellowship Manager to verify current the deadline date. APHL member laboratories may apply to host one or more EID Fellows. The link to the Host Laboratory Instructions and Application page is available from the EID Fellowship website.

Requirements: EID Advanced Laboratory Training Fellowship applicants must have completed a bachelor's or master's degree in microbiology, virology or a related discipline and/or completed an accredited medical technologist program prior to the start of the fellowship. EID Laboratory Research Fellowship applicants must have received a Ph.D., M.D., or D.V.M., or have completed all requirements for such a degree prior to the start of the fellowship. Applicants without a degree in a physical science will be considered if they have significant laboratory science and/or public health coursework or experience. All bachelor's and master's level candidates must have completed their degree by the time the fellowship begins (mid-September). Postdoctoral candidates may delay their start date to February 2012. Degree requirements must be completed by by this time. All appointees must attend the August Orientation session in Atlanta.
Restrictions: All applicants must be U.S. citizens.
Geographic Focus: All States
Date(s) Application is Due: Feb 11
Contact: Heather Roney, Manager, Fellowship Program; (240) 485-2778 or (240) 485-2745; fax (240) 485-2700; fellowships@aphl.org or heather.roney@aphl.org
Internet: http://www.aphl.org/mycareer/fellowships/eid/pages/default.aspx
Sponsor: Association of Public Health Laboratories
8515 Georgia Avenue, Suite 700
Silver Spring, MD 20910

Appalachian Regional Commission Health Care Grants 915

Access to comprehensive, affordable health care is vital to social and economic growth in the Appalachian Region. ARC's health projects focus on community-based efforts to encourage health-promotion and disease-prevention activities. Strategies include: using best practices in public health to develop targeted approaches to wellness and disease prevention; supporting partnerships that educate children and families about basic health risks; using telecommunications and other technology to reduce the high cost of health-care services; and encouraging the development and expansion of health professional education services within the Region. ARC health care grants have helped provide equipment for hospitals and rural clinics, training for health care professionals, and support for community-based health education activities. ARC also works with other organizations to address the high incidence of life-threatening diseases in the Region, as in its ongoing partnership with the Centers for Disease Control and Prevention in diabetes and cancer education, prevention, and treatment programs in the Region's distressed counties.

Requirements: States, and through states, public bodies and private nonprofit organizations are eligible to apply.
Restrictions: Generally, ARC grants are limited to 50% of project costs.
Geographic Focus: Alabama, Georgia, Kentucky, Maryland, Mississippi, New York, North Carolina, Ohio, Pennsylvania, South Carolina, Tennessee, Virginia, West Virginia
Contact: Jill Wilmoth; (202) 884-7668 or (202) 884-7700; jwilmoth@arc.gov
Internet: http://www.arc.gov/funding/ARCProjectGrants.asp
Sponsor: Appalachian Regional Commission
1666 Connecticut Avenue NW, Suite 700
Washington, DC 20009-1068

Applied Biosystems Grants 916

The company is committed to the communities where its employees live and work. Financial contributions as well as in-kind and product donations are awarded to nonprofit organizations. The focus is on education, health and human services, community outreach programs, environment, arts & culture, and civic. Precollege, college, graduate, and other unique programs that use technology to improve science, math, and engineering education are education programs supported. Organizations providing access to quality health care facilities and innovative solutions to chronic problems through disease and therapeutic research are supported. Outreach programs and civic groups that improve communities with innovative science, technology, and education are supported. Programs maintaining a healthy environment, arts and culture, and civic programs are also supported. Requests are accepted at any time.

Requirements: Nonprofit organizations in communities where Applied Biosystems has operations are eligible.
Geographic Focus: All States
Contact: Program Officer; (800) 327-3002 or (650) 638-5800; fax (650) 638-5998
Internet: http://www.appliedbiosystems.com/about/community.cfm
Sponsor: Applied Biosystems
850 Lincoln Centre Drive
Foster City, CA 94404

Appraisal Institute Education Trust Scholarships 917

Graduate and undergraduate scholarships are offered to students majoring in programs in land economics, real estate, real estate appraising, or allied fields. The scholarships are awarded on the basis of scholarly qualities and are intended to help finance the cost of college work leading to a degree in one of the aforementioned professional fields. The institute also awards scholarships to minority and women college students pursuing academic degrees in real estate appraisal or related fields. Application forms must be obtained from the institute.

Requirements: Applicants must be sophomores, juniors, seniors, or graduates, as well as U.S. citizens.
Geographic Focus: All States
Date(s) Application is Due: Mar 15
Amount of Grant: 2,000-3,000 USD
Contact: Olivia Carreon, Project Coordinator; (312) 335-4100; fax (312) 335-4400; ocarreon@appraisalinstitute.org
Internet: http://www.appraisalinstitute.org/education/scolarshp.asp
Sponsor: Appraisal Institute Education Trust
550 W Van Buren Street, Suite 1000
Chicago, IL 60607

APS/British Academy Fellowship for Research in London 918

In collaboration with the British Academy, the APS offers an exchange post-doctoral fellowship for up to three months of research in the archives and libraries of London. This award includes travel expenses between the United States and United Kingdom paid by the APS and a monthly subsistence of 1,350 British pounds paid by the British Academy. Candidates should use the Franklin application form, specifying that they are asking for the British Academy Fellowship, and apply by October 1; applicants not selected for the British Academy Fellowship will be considered for a Franklin Research Grant.

Requirements: Applicants are expected to have held the doctorate for at least one year. Applications may be made by residents of the United States, by American citizens on the staffs of foreign institutions, and by foreign nationals whose research can only be carried out in the United States.
Restrictions: Grants will not be made for journalistic or other writing for general readership; the preparation of textbooks, casebooks, anthologies, or other materials for classroom use by students; nor for the work of creative and performing artists. Grants are rarely made to persons who have had the doctorate less than a year and never for predoctoral study or research. The committee will seldom approve more than two grants for the same project within any five-year period.
Geographic Focus: All States
Date(s) Application is Due: Oct 1
Amount of Grant: 1,350-4,050 GBP
Contact: Linda Musumeci; (215) 440-3429; LMusumeci@amphilsoc.org
Internet: http://www.amphilsoc.org/grants
Sponsor: American Philosophical Society
104 South Fifth Street
Philadelphia, PA 19106-3387

APS/Phoenix Suns Energy and Environmental Minigrants 919

The program supports innovative, student-based educational projects that enhance learning with the objective of improving student achievement. Projects should have the ability of being replicated or adapted by other schools. Projects addressing energy, the environment, reading, mathematics or character education will be considered. Annual deadline dates may vary; contact program staff for exact dates.

Requirements: The program is open to educators who teach in public schools in the Arizona Public Service territory. Funds may be applied to equipment, materials and field trips that would not normally be provided by the school or school district.

Geographic Focus: Arizona

Date(s) Application is Due: Oct 15

Amount of Grant: 2,500 USD

Contact: Louise Moskowitz, (602) 250-2291; louise.moskowitz@aps.com

Internet: http://www.aps.com/my_community/EducationTools/LearningCenter_1.html

Sponsor: Arizona Public Service Corporation
P.O. Box 53999, MS 8010
Phoenix, AZ 85072-3999

APSAA Fellowships 920

The mission of the Fellowship Program is to encourage interest and involvement in psychoanalysis among the future leaders, researchers and educators of mental health and academia. Early-career psychiatrists, psychologists, social workers and academics are eligible for to apply for the Fellowship. The Fellowship offers the opportunity to attend the biannual meetings of APSAA, to meet analysts and Fellows from across the country, to have a Mentor, and to present their clinical work or research at the meetings. The application for the Fellowship Program is available online September 15.

Requirements: Psychiatrists, psychologists, social workers, and academics who meet the specific eligibility requirements (available online) are eligible. Applicants may be nominated by their department chairs or program directors if applicable. When not applicable, self-nominations are encouraged. Applicants must be training or working in the United States during the fellowship year. All applicants should have demonstrated leadership ability in their discipline, or have special aptitude in research, teaching, and/or clinical endeavors; and have special interest in psychodynamics, psychoanalysis, or applied psychoanalysis.

Geographic Focus: All States

Contact: James Guimaraes, Fellowship Coordinator; (212) 752-0450; fax (212) 593-0571; jguimaraes@apsa.org or info@apsa.org

Internet: http://www.apsa.org/FELLOWSHIPPROGRAM/GENERALINFORMATION/tabid/298/Default.aspx

Sponsor: American Psychoanalytic Association
309 East 49th Street
New York, NY 10007-1601

APSAA Mini-Career Grants 921

The Fund conceives of psychoanalytic research along the broadest lines, including: scholarly and empirical investigative contributions that can advance knowledge of psychoanalytic theory, practice, and links between psychoanalysis and neighboring disciplines such as developmental psychology or neuroscience. Awards of up to $15,000 annually for up to two years are intended to support the beginning career of a psychoanalytic investigator. These mini-career awards are intended, for example, to buy time for a junior faculty member or clinician just starting a practice so that they may consult with other investigators, join an investigative team, or attend year long seminars on research methodology or specific methods relevant to their research. Applicants for these awards must provide a year-long career development plan with letters from appropriate faculty and mentors describing the candidate's career development plan and how the award will facilitate the candidate's goals.

Requirements: Applicants should meet the following criteria: proposal relevant to psychoanalysis; applicant is either a new applicant to the Fund or resubmitting a previously reviewed application; applicant has a masters degree or higher; and human subject protection and consent procedures must be specified for research studies involving ongoing data collection.

Restrictions: Applicants who have received $80,000 total from the Fund are ordinarily not eligible to apply again.

Geographic Focus: All States

Date(s) Application is Due: Apr 1; Nov 1

Amount of Grant: Up to 15,000 USD

Contact: Dean K. Stein, Executive Director; (212) 752-0450; fax (212) 593-0571; deankstein@apsa.org or info@apsa.org

Internet: http://www.apsa.org/RESEARCH/FUNDFORPSYCHOANALYTICRESEARCH/GRANTS/tabid/137/Default.aspx

Sponsor: American Psychoanalytic Association
309 East 49th Street
New York, NY 10007-1601

APSAA Research Grants 922

The Fund conceives of psychoanalytic research along the broadest lines, including: scholarly and empirical investigative contributions that can advance knowledge of psychoanalytic theory, practice, and links between psychoanalysis and neighboring disciplines such as developmental psychology or neuroscience. Grants of one or two years duration at a maximum of $20,000 yearly are intended for a specific project building upon psychoanalytic principles or directly investigating the process and/or outcome of psychoanalytically informed treatments. The Board recognizes that this amount of money is not sufficient usually to cover the costs of a large research study, particularly one that may involve a large number of subjects or a longitudinal design. Hence, the Board encourages applicants to consider feasibility of their request, that is, what they can achieve for this amount of money, and also to think about this type of award as supporting the beginning stages of a research project, the gathering of pilot data, or the refinement of design and methods.

Requirements: Applicants should meet the following criteria: proposal relevant to psychoanalysis; applicant is either a new applicant to the Fund or resubmitting a previously reviewed application; applicant has a masters degree or higher; and human subject protection and consent procedures must be specified for research studies involving ongoing data collection.

Restrictions: Applicants who have received $80,000 total from the Fund are ordinarily not eligible to apply again.

Geographic Focus: All States

Date(s) Application is Due: Apr 1; Nov 1

Amount of Grant: Up to 20,000 USD

Contact: Dean K. Stein, Executive Director; (212) 752-0450; fax (212) 593-0571; deankstein@apsa.org or info@apsa.org

Internet: http://www.apsa.org/RESEARCH/FUNDFORPSYCHOANALYTICRESEARCH/GRANTS/tabid/137/Default.aspx

Sponsor: American Psychoanalytic Association
309 East 49th Street
New York, NY 10007-1601

APSAA Small Beginning Scholar Pre-Investigation Grants 923

The Fund conceives of psychoanalytic research along the broadest lines, including: scholarly and empirical investigative contributions that can advance knowledge of psychoanalytic theory, practice, and links between psychoanalysis and neighboring disciplines such as developmental psychology or neuroscience. These small grants of less than $5,000 for one year are intended to permit a beginning scholar to gather pilot data in preparation for the submission of a full grant to the Fund or to another agency. The Fund accepts revisions of a previously submitted application up to four years after the previous submission date. Up to two revised submissions may be submitted within four years of the original submission date.

Requirements: Applicants should meet the following criteria: proposal relevant to psychoanalysis; applicant is either a new applicant to the Fund or resubmitting a previously reviewed application; applicant has a masters degree or higher; and human subject protection and consent procedures must be specified for research studies involving ongoing data collection.

Restrictions: Applicants who have received $80,000 total from the Fund are ordinarily not eligible to apply again.

Geographic Focus: All States

Date(s) Application is Due: Apr 1; Nov 1

Amount of Grant: Up to 5,000 USD

Contact: Dean K. Stein, Executive Director; (212) 752-0450; fax (212) 593-0571; deankstein@apsa.org or info@apsa.org

Internet: http://www.apsa.org/RESEARCH/FUNDFORPSYCHOANALYTICRESEARCH/GRANTS/tabid/137/Default.aspx

Sponsor: American Psychoanalytic Association
309 East 49th Street
New York, NY 10007-1601

APSAA Small Beginning Scholar Visiting and Consulting Grants 924

The Fund conceives of psychoanalytic research along the broadest lines, including: scholarly and empirical investigative contributions that can advance knowledge of psychoanalytic theory, practice, and links between psychoanalysis and neighboring disciplines such as developmental psychology or neuroscience. Small grants of $3,000 are intended to permit a beginning scholar to spend time visiting and consulting with a more experienced, senior investigator who has agreed to help the junior investigator begin their investigative work. Typically these awards are made to permit the beginning scholar the funds to travel and/or to reimburse the senior investigator for travel and consultation. These grants may also be used to support travel to a research training seminar in the U.S. or abroad. The Fund accepts revisions of a previously submitted application up to four years after the previous submission date. Up to two revised submissions may be submitted within four years of the original submission date.

Requirements: Applicants should meet the following criteria: proposal relevant to psychoanalysis; applicant is either a new applicant to the Fund or resubmitting a previously reviewed application; applicant has a masters degree or higher; and human subject protection and consent procedures must be specified for research studies involving ongoing data collection.

Restrictions: Applicants who have received $80,000 total from the Fund are ordinarily not eligible to apply again.

Geographic Focus: All States

Date(s) Application is Due: Apr 1; Nov 1

Amount of Grant: Up to 3,000 USD

Contact: Dean K. Stein, Executive Director; (212) 752-0450; fax (212) 593-0571; deankstein@apsa.org or info@apsa.org

Internet: http://www.apsa.org/RESEARCH/FUNDFORPSYCHOANALYTICRESEARCH/GRANTS/tabid/137/Default.aspx

Sponsor: American Psychoanalytic Association
309 East 49th Street
New York, NY 10007-1601

APS Abraham Pais Prize for History of Physics 925

The American Physical Society (APS) through its Forum on the History of Physics (FHP) and the American Institute of Physics (AIP) through its Center for History of Physics (CHP) established the prize in 2005. Major sponsors include Richard Lounsbery Foundation, John & Elizabeth Armstrong and Virginia Trimble. The Prize is intended to recognize outstanding scholarly achievements in the history of physics. Outstanding scholarly achievements are reflected primarily in the publication of one or more scholarly books or a series of scholarly articles on any topic or chronological period in the history of physics. These can include significant editions of works of physicists that display outstanding editorial scholarship; other edited books or collections of previously published articles will be considered but given less weight. Further outstanding professional achievements such as teaching, lecturing, editing of journals, organization of conferences, and productions on the Web or in other electronic media may be considered if they have contributed significantly to fostering excellence in the history of physics. The Award usually is given to a single person but to no more than three individuals. The prize is given annually and will consist of $10,000 and a certificate citing the contributions of the recipient, plus an allowance for travel to an APS meeting to receive the award and deliver an invited lecture on the history of physics.
Geographic Focus: All States
Date(s) Application is Due: Jul 1
Amount of Grant: 10,000 USD
Samples: Gerald Holton, Harvard University, $10,000; Max Jammer, $10,000.
Contact: Jacquelyn Beamon-Kiene, Executive Office Administrator; (301)209-3269 or (301) 209-3268; fax (301) 209-0865; beamon@aps.org or honors@aps.org
Internet: http://www.aps.org/programs/honors/prizes/pais-guidelines.cfm
Sponsor: American Physical Society
One Physics Ellipse
College Park, MD 20740-3844

APSA Congressional Fellowships for Communications Scholars and Journalists 926

The purpose of this fellowship is to give early- to mid-career scholars and journalists an opportunity to learn more about Congress and the legislative process through direct participation. Ideal candidates should have an analytical interest in politics, communications, and public policy, and show promise of making a significant contribution to the public understanding of the political process. A comprehensive orientation begins in November. Office assignments as full-time legislative aides in the House of Representatives and/or Senate run from December to August. Fellows receive a stipend of $38,000, plus a small travel allowance.
Requirements: All candidates must be U.S. citizens or permanent U.S. residents and demonstrate an interest in communications and the policymaking process. Scholars must have a Ph.D completed within the last 15 years or will have defended their dissertation by November of the fellowship year. Print journalists must have a bachelor's degree and two to ten years of continuous, full-time professional experience in newspaper or magazine at the time of application (background may be in reporting or editing). Broadcast journalists must have a bachelor's degree and two to ten years of continuous, full-time professional experience in radio or television at the time of application (background may be in reporting, producing, directing or writing). Preference is given to candidates without extensive Washington experience.
Geographic Focus: All States
Date(s) Application is Due: Dec 1
Amount of Grant: 38,000 USD
Contact: Jeffrey R. Biggs, Program Director; (202) 483-2512, ext. 521; fax (202) 483-2657; jbiggs@apsanet.org or cfp@apsanet.org
Internet: http://www.apsanet.org/content_3542.cfm
Sponsor: American Political Science Association
1527 New Hampshire Avenue, NW
Washington, DC 20036-1206

APSA Congressional Fellowships for Federal Executives 927

The purpose of this fellowship is to give senior-level federal executives an opportunity to learn more about Congress and the legislative process through direct participation. A comprehensive orientation begins in November. Office assignments as full-time legislative aides in the House of Representatives and/or Senate run from December to August. Fellows continue to receive their regular salaries from their agencies during the fellowship. Agencies with substantial activity with foreign governments are additionally invited to nominate candidates to participate in the Foreign Affairs Seminar. Foreign Affairs Fellows attend an intensive eight-week course at the Johns Hopkins University School of Advanced International Studies in Washington from September to November. They then join the rest of the Fellows for the November general orientation.
Requirements: All candidates must have a minimum grade of GS-13 or equivalent at time of application, at least two years of federal service in the executive branch, and long-term career goals relevant to a congressional experience. Additionally, candidates must demonstrate: a commitment to public service; adaptability to new, diverse working environments; and an interest in the legislative process and public affairs.
Geographic Focus: All States
Date(s) Application is Due: Mar 31
Contact: Jeffrey R. Biggs, Program Director; (202) 483-2512, ext. 521; fax (202) 483-2657; jbiggs@apsanet.org or cfp@apsanet.org
Internet: http://www.apsanet.org/content_3543.cfm
Sponsor: American Political Science Association
1527 New Hampshire Avenue, NW
Washington, DC 20036-1206

APSA Congressional Fellowships for International Scholars 928

The Program is the nation's oldest and most prestigious congressional fellowship. More than fifty years later, the program remains devoted to its original objective of expanding knowledge and awareness of Congress. For nine months, select political scientists, journalists, doctors, federal executives and international scholars gain hands on understanding of the legislative process by serving on congressional staffs. Through this unique opportunity, the Association enhances public understanding of policy-making and improves the quality of scholarship, teaching and reporting on American national politics. The CFP is currently investigating the possibilities of expanding the program. Presently, however, opportunities are limited to: German Fellows—selected by the German Marshall Fund of the United States; Asian Fellows—selected by the Asia Foundation; Senior Fulbright Scholars—selected by the State Department Bureau of Education and Cultural Affairs and the Council for International Exchange of Scholars; and Australian Fellows—selected by the American Australian Association.
Geographic Focus: All States
Date(s) Application is Due: Jan 31
Contact: Jeffrey R. Biggs, Program Director; (202) 483-2512, ext. 521; fax (202) 483-2657; jbiggs@apsanet.org or cfp@apsanet.org
Internet: http://www.apsanet.org/content_3160.cfm
Sponsor: American Political Science Association
1527 New Hampshire Avenue, NW
Washington, DC 20036-1206

APSA Congressional Fellowships for Journalists 929

The purpose of this fellowship is to give early- to mid-career journalists an opportunity to learn more about Congress and the legislative process through direct participation. A comprehensive orientation begins in November. Office assignments as full-time legislative aides in the House of Representatives and/or Senate run from December to August. Fellows receive a stipend of $38,000, plus a small travel allowance.
Requirements: All candidates must be U.S. citizens or permanent U.S. residents and hold a bachelor's degree. Print journalists must have two to ten years of continuous, full-time professional experience in newspaper or magazine at the time of application (background may be in reporting or editing). Broadcast journalists must have two to ten years of continuous, full-time professional experience in radio or television at the time of application (background may be in reporting, producing, directing or writing). Preference is given to candidates with background in political reporting but without extensive Washington experience.
Geographic Focus: All States
Date(s) Application is Due: Dec 1
Amount of Grant: 38,000 USD
Contact: Jeffrey R. Biggs, Program Director; (202) 483-2512, ext. 521; fax (202) 483-2657; jbiggs@apsanet.org or cfp@apsanet.org
Internet: http://www.apsanet.org/content_5046.cfm
Sponsor: American Political Science Association
1527 New Hampshire Avenue, NW
Washington, DC 20036-1206

APSA Congressional Fellowships for Political Scientists 930

The purpose of this fellowship is to give early- to mid-career political scientists an opportunity to learn more about Congress and the legislative process through direct participation. Office assignments as full-time legislative aides in the House of Representatives and/or Senate run from December to August. Fellows are selected in a nationwide competition. Candidates must first submit their applications for a paper review before being invited before an outside selection panel. Interviews take place at the APSA offices in Washington, D.C. APSA reimburses travel to and from interviews and arranges for hotel accommodations. Applicants not currently in the United States must fund their own transportation, should they be selected for an interview. Interviews will be announced in January. All finalists will be notified in February. Further application guidelines are available online.
Requirements: Eligibility qualifications include: U.S. citizenship or permanent U.S. residency; a Ph.D completed within the last 15 years or a dissertation near completion (ABD); and a scholarly interest in Congress and the policy-making process.
Geographic Focus: All States
Date(s) Application is Due: Dec 1
Amount of Grant: 38,000 USD
Contact: Jeffrey R. Biggs, Program Director; (202) 483-2512, ext. 521; fax (202) 483-2657; jbiggs@apsanet.org or cfp@apsanet.org
Internet: http://www.apsanet.org/content_3540.cfm
Sponsor: American Political Science Association
1527 New Hampshire Avenue, NW
Washington, DC 20036-1206

APSA Congressional Health and Aging Policy Fellowships 931

Supported by the Atlantic Philanthropies and administered by Columbia University, this national program seeks to provide professions in health and aging with the experience and skills necessary to make a positive contribution to the development and implementation of health policies that affect older Americans. The program will offer two different tracks for individual placement: a residential track that includes a nine-to-12-month placement in Washington, D.C. or at a state agency (as a legislative assistant in Congress, a professional staff member in an executive agency or in a policy organization; and a non-residential track that includes a health policy project and brief placement(s) throughout the year at relevant sites. Core program components focused on career development and professional enrichment are provided for fellows in both tracks.

Requirements: The program is open to physicians, nurses and social workers at all career stages (early, mid, and late) with demonstrated commitment to health and aging issues and a desire to be involved in health policy at the federal, state or local level. Other professionals with clinical backgrounds (e.g. pharmacists, dentists, clinical psychologists) working in the field of health and aging are also eligible to apply.
Geographic Focus: All States
Contact: Jeffrey R. Biggs, Program Director; (202) 483-2512, ext. 521; fax (202) 483-2657; jbiggs@apsanet.org or cfp@apsanet.org
Internet: http://www.apsanet.org/content_50835.cfm
Sponsor: American Political Science Association
1527 New Hampshire Avenue, NW
Washington, DC 20036-1206

APSA E.E. Schattschneider Award 932

The E.E. Schattschneider prize is awarded annually for the best doctoral dissertation completed and accepted during that year or the previous year in the field of American government. This award fund was established in 1971. It carries a prize of $750. This award was set up in honor of Elmer Eric Schattschneider, a former APSA president, and widely published and respected political scientist.
Requirements: Political science departments are invited to nominate not more than one doctoral dissertation for this award. The dissertations must have been completed and accepted during the two calendar years preceding the date of the prize.
Geographic Focus: All States
Date(s) Application is Due: Jan 15
Contact: Lauren Barry; (202) 483-2512, ext. 631; lbarry@apsanet.org
Internet: http://www.apsanet.org/content_4125.cfm
Sponsor: American Political Science Association
1527 New Hampshire Avenue, NW
Washington, DC 20036-1206

APSA Edward S. Corwin Award 933

The Edward S. Corwin prize is awarded annually for the best dissertation in the field of public law. It carries a prize of $750. The Corwin award is for the best doctoral dissertation completed and accepted during that year or the previous year in the field of public law, broadly defined to include the judicial process, judicial behavior, judicial biography, courts, law, legal systems, the American constitutional system, civil liberties, or any other substantial area, or any work which deals in a significant fashion with a topic related to or having substantial impact on the American Constitution. Edward S. Corwin was a former Association president who is nationally known and widely published. He consulted with many other academics as well as politicians involved with constitutional issues, most notably when he publicly supported Franklin D. Roosevelt's Supreme Court reorganization ("court packing") plan.
Requirements: Political science departments are invited to nominate not more than one doctoral dissertation for this award. The dissertations must have been completed and accepted during the two calendar years preceding the date of the prize.
Geographic Focus: All States
Date(s) Application is Due: Jan 15
Amount of Grant: 750 USD
Contact: Lauren Barry; (202) 483-2512, ext. 631; lbarry@apsanet.org
Internet: http://www.apsanet.org/content_4122.cfm
Sponsor: American Political Science Association
1527 New Hampshire Avenue, NW
Washington, DC 20036-1206

APSA Gabriel A. Almond Award 934

The Gabriel A. Almond prize is awarded annually for the best dissertation in the field of comparative politics. It carries a prize of $750. The award was created in recognition of Gabriel Almond's contributions to the discipline, profession, and Association. He was a long time faculty member at Stanford University and former APSA President (1966). Almond's scholarly work contributed directly to the development of theory in comparative politics and brought together work on the developing areas and Western Europe that prevented splintering into an array of disparate areas studies.
Requirements: Political science departments are invited to nominate not more than one doctoral dissertation for this award. The dissertations must have been completed and accepted during the two calendar years preceding the date of the prize.
Geographic Focus: All States
Date(s) Application is Due: Jan 15
Amount of Grant: 750 USD
Contact: Lauren Barry; (202) 483-2512, ext. 631; lbarry@apsanet.org
Internet: http://www.apsanet.org/content_4119.cfm
Sponsor: American Political Science Association
1527 New Hampshire Avenue, NW
Washington, DC 20036-1206

APSA Harold D. Lasswell Award 935

Harold D. Lasswell's extensive and wide-ranging books and essays are extraordinarily rich sources of ideas, methods, and topics for the study of political behavior. Whether and how the legacy of his writings is used by contemporary political scientists and theorists is reported here by way of an investigation of references to his work appearing in mainstream political science journals (available through JSTOR) for the 17 years following the end of his academic career. We find that most references to Lasswell are superficial (perfunctory, suggestive, deferential), although a few are more substantial (critical, extending). We conclude that Lasswell's legacy is today undervalued and underused, to the discipline's detriment. The Harold D. Lasswell prize is awarded annually for the best dissertation in the field of public policy. The award is co-sponsored by the Policy Studies Organization. It carries a prize of $1,000.
Requirements: Political science departments are invited to nominate not more than one doctoral dissertation for this award. The dissertations must have been completed and accepted during the two calendar years preceding the date of the prize.
Geographic Focus: All States
Date(s) Application is Due: Jan 15
Amount of Grant: 1,000 USD
Contact: Lauren Barry; (202) 483-2512, ext. 631; lbarry@apsanet.org
Internet: http://www.apsanet.org/content_4123.cfm
Sponsor: American Political Science Association
1527 New Hampshire Avenue, NW
Washington, DC 20036-1206

APSA Helen Dwight Reid Award 936

In 1956, Helen Dwight Reid, a political scientist who taught at Bryn Mawr and the University of Buffalo, established a nonprofit foundation to support projects in education, international affairs, and the sciences. The Helen Dwight Reid prize is awarded annually for the best dissertation successfully defended during the previous two years in the field of international relations, law, and politics. The award is supported by the Helen Dwight Reid Educational Foundation. It carries a prize of $750.
Requirements: Political science departments are invited to nominate not more than one doctoral dissertation for this award. The dissertations must have been completed and accepted during the two calendar years preceding the date of the prize.
Geographic Focus: All States
Date(s) Application is Due: Jan 15
Amount of Grant: 3,500-80,000 USD
Contact: Lauren Barry; (202) 483-2512, ext. 631; lbarry@apsanet.org
Internet: http://www.apsanet.org/content_4124.cfm
Sponsor: American Political Science Association
1527 New Hampshire Avenue, NW
Washington, DC 20036-1206

APSA Leonard D. White Award 937

Leonard Dupee White (17 January 1891-23 February 1958) was a historian of the field of public administration in the United States.[1] His technique was to study administration in the context of grouped U.S. presidential periods. An important founder of the field, White worked at the University of Chicago after service in the administrations of Franklin D. Roosevelt. The Leonard D. White prize is awarded annually for the best dissertation successfully defended during the previous two years in the field of public administration. The award is supported by the University of Chicago. It carries a prize of $750.
Requirements: Political science departments are invited to nominate not more than one doctoral dissertation for this award. The dissertations must have been completed and accepted during the two calendar years preceding the date of the prize.
Geographic Focus: All States
Date(s) Application is Due: Jan 15
Amount of Grant: 750 USD
Contact: Lauren Barry; (202) 483-2512, ext. 631; lbarry@apsanet.org
Internet: http://www.apsanet.org/content_4127.cfm
Sponsor: American Political Science Association
1527 New Hampshire Avenue, NW
Washington, DC 20036-1206

APSA Leo Strauss Award 938

The Leo Strauss prize is awarded annually for the best dissertation in the field of political philosophy. It carries a prize of $750. The Fund was developed by former students of Strauss' who sought to recognize his extraordinary influence on generations of students and his contributions to the field of political philosophy. He was a major figure in the department of political science at the University of Chicago, where he taught from 1949 to 1967.
Requirements: Political science departments are invited to nominate not more than one doctoral dissertation for this award. The dissertations must have been completed and accepted during the two calendar years preceding the date of the prize.
Geographic Focus: All States
Date(s) Application is Due: Jan 15
Amount of Grant: 750 USD
Contact: Lauren Barry; (202) 483-2512, ext. 631; lbarry@apsanet.org
Internet: http://www.apsanet.org/content_4126.cfm
Sponsor: American Political Science Association
1527 New Hampshire Avenue, NW
Washington, DC 20036-1206

APSA Minority Fellowships 939

The program is designed primarily for minority students applying to enter a doctoral program in political science for the first time. The association has re-focused and increased its efforts to assist minority students in completing their doctorates by concentrating not only on the recruitment of minorities, but also on the retention of these groups within the profession. The program designates six stipend minority fellows each year. Additional applicants who do not receive funds from the association may also be recognized and recommended for admission and financial support to graduate political science programs. Fellows with stipends receive a fellowship that is disbursed in two payments-one at the end of their first graduate year and one at the end of their second-provided that they remain in good academic standing. Awards are based on students' undergraduate course work, GPA, extracurricular activities, GRE scores, and recommendations from faculty. Guidelines are available online.

Requirements: Applicants must: be members of one of the following racial/ethnic minority groups-African Americans, Latinos/as, and Native Americans (federal and state recognized tribes); demonstrate an interest in teaching and potential for research in political science; be a U.S. citizen at time of award; and demonstrate financial need.
Geographic Focus: All States
Date(s) Application is Due: Oct 2
Amount of Grant: 4,000 USD
Contact: Jeffrey R. Biggs; (202) 483-2512, ext. 521; jbiggs@apsanet.org
Internet: http://www.apsanet.org/content_3284.cfm
Sponsor: American Political Science Association
1527 New Hampshire Avenue, NW
Washington, DC 20036-1206

APS Andreas Acrivos Dissertation Award in Fluid Dynamics **940**
The program provides recognition to exceptional young scientists who have performed original doctoral thesis work of outstanding scientific quality and achievement in the area of fluid dynamics. The annual award consists of a monetary prize, a certificate citing the accomplishments of the recipient, and an allowance for travel to attend the annual meeting of the Division of Fluid Dynamics at which the award will be presented. Each nominee will be considered in not more than two consecutive cycles. Nomination guidelines are available online.
Requirements: Nominations will be accepted for any doctoral student studying at a college or university in the United States or in a education abroad program of a college or university in the United States. The work to be considered must have been accomplished as part of the requirements for a doctoral degree. Nominees must have completed their dissertations during the previous calendar year, or by March 31 of the year of the award, at a university within the United States.
Geographic Focus: All States
Date(s) Application is Due: May 1
Amount of Grant: 1,000-1,500 USD
Contact: Jacquelyn Beamon-Kiene, Executive Office Administrator; (301)209-3269 or (301) 209-3268; fax (301) 209-0865; beamon@aps.org or honors@aps.org
Internet: http://www.aps.org/programs/honors/dissertation/acrivos.cfm
Sponsor: American Physical Society
One Physics Ellipse
College Park, MD 20740-3844

APS Andrei Sakharov Prize **941**
The Sakharov Prize is named in recognition of the courageous and effective work of Andrei Sakharov on behalf of human rights, to the detriment of his own scientific career and despite the loss of his own personal freedom. The Prize is endowed by contributions from friends of Andrei Sakharov. It is intended to recognize outstanding leadership and/or achievements of scientists in upholding human rights. The Prize consists of $10,000 and an allowance for travel to the meeting of the Society at which the Prize will be presented. It is intended that the Prize be awarded every other year, at a general meeting of the American Physical Society.
Geographic Focus: All States
Date(s) Application is Due: Jul 1
Amount of Grant: 10,000 USD
Contact: Jacquelyn Beamon-Kiene, Executive Office Administrator; (301)209-3269 or (301) 209-3268; fax (301) 209-0865; beamon@aps.org or honors@aps.org
Internet: http://www.aps.org/programs/honors/prizes/sakharov.cfm
Sponsor: American Physical Society
One Physics Ellipse
College Park, MD 20740-3844

APS Aneesur Rahman Prize for Computational Physics **942**
The annual prize recognizes and encourages outstanding achievement in computational physics research. The award consists of a monetary prize of $5,000, an allowance for travel to the meeting of the society at which the prize is awarded and at which the recipient will deliver the Rahman Lecture, and a certificate citing the contributions made by the recipient. The prize ordinarily is awarded to one person, but a prize may be shared when all recipients have contributed to the same accomplishments. Nominations are active for three years. Nomination guidelines are available online.
Requirements: Nominations are open to scientists of all nationalities regardless of the geographical site at which the work was done.
Geographic Focus: All States
Date(s) Application is Due: Jul 1
Amount of Grant: 5,000 USD
Contact: Jacquelyn Beamon-Kiene, Executive Office Administrator; (301) 209-3269 or (301) 209-3268; fax (301) 209-0865; beamon@aps.org or honors@aps.org
Internet: http://www.aps.org/programs/honors/prizes/rahman.cfm
Sponsor: American Physical Society
One Physics Ellipse
College Park, MD 20740-3844

APSA Presidency Research Fund Fellowships **943**
The Presidency Research Fund Fellowship is designed to support research on the American presidency. With an interest in explaining relationships, institutions, and environment surrounding the President, the Fund will support the work of people whose scholarly research brings them to the Washington area. Whether scholars plan to conduct archival work, library research, or interviews with key officials, the fellowship will provide a place at the Centennial Center where people are supported as they conduct their work. Applications will be reviewed, and selection made by, a three-member committee appointed by the APSA president.
Geographic Focus: All States
Contact: Jeffrey R. Biggs; (202) 483-2512, ext. 521; jbiggs@apsanet.org
Internet: http://www.apsanet.org/content_6335.cfm
Sponsor: American Political Science Association
1527 New Hampshire Avenue, NW
Washington, DC 20036-1206

APSA Robert Wood Johnson Foundation Health Policy Fellowships **944**
The purpose of this fellowship is to give outstanding mid-career health professionals in academic and community-based settings an opportunity to learn more about the legislative process through direct participation. Fellows actively contribute to the formulation of national health policies and accelerate their careers as leaders in health policy. Initiated in 1973, the program is funded by The Robert Wood Johnson Foundation (RWJF)and conducted by the Institute of Medicine (IOM) of the National Academy of Sciences. The application process typically opens in September and the application deadline typically occurs in mid-November of each year. Fellows must commit 100 percent of their time to program activities during the first 12 months. Up to six grants of up to $165,000 each will be made.
Geographic Focus: All States
Date(s) Application is Due: Nov 14
Contact: Jeffrey R. Biggs; (202) 483-2512, ext. 521; jbiggs@apsanet.org
Internet: http://www.apsanet.org/content_3544.cfm
Sponsor: American Political Science Association
1527 New Hampshire Avenue, NW
Washington, DC 20036-1206

APS Arthur L. Schawlow Prize in Laser Science **945**
The prize recognizes outstanding contributions to basic research that uses lasers to advance knowledge of the fundamental physical properties of materials and their interaction with light. Some examples of relevant areas of research are: nonlinear optics, ultrafast phenomena, laser spectroscopy, squeezed states, quantum optics, multiphoton physics, laser cooling and trapping, physics of lasers, particle acceleration by lasers, and short wavelength lasers. The award consists of a $10,000 prize plus an allowance for travel to the meeting at which the prize is awarded and a certificate citing the contributions made by the recipient. Nominations are active for three years. Nomination guidelines are available online.
Requirements: Nominations are open to candidates who have made outstanding contributions to basic research using lasers.
Geographic Focus: All States
Date(s) Application is Due: Jul 1
Contact: Jacquelyn Beamon-Kiene, Executive Office Administrator; (301)209-3269 or (301) 209-3268; fax (301) 209-0865; beamon@aps.org or honors@aps.org
Internet: http://www.aps.org/programs/honors/prizes/schawlow.cfm
Sponsor: American Physical Society
One Physics Ellipse
College Park, MD 20740-3844

APSA Small Research Grants **946**
The APSA Small Research Grant Program supports research in all fields of political science. The intent of these grants is to support the research of political scientists who are not employed at PhD granting department in the field and to help further the careers of these scholars. Grant recipients in the last five years have published over 12 books, 16 refereed articles, 8 book chapters, many working papers, and a large number of conference presentations. Others are still working on book, article, and conference projects and many recipients report great benefits for their students. Students have co-authored projects, worked as research assistants, and benefited through the professors' classroom use of their research. Many recipients also were able to use the APSA grant as seed money to get additional funding.
Requirements: Applicants must be APSA members at the time of application and at the time funds are dispersed. In addition, the principal investigator and any co-author must be one of the following: a faculty member at a college or university that does not award a PhD in Political Science, Public Administration, Public Policy, International Relations, Government, or Politics and whose primary appointment is in one of these departments; or a political scientist not affiliated with an academic institution and is either: unemployed; or working in a research organization such as a think tank.
Restrictions: This grant does not provide support for dissertation research or writing and graduate students are not eligible to apply.
Geographic Focus: All States
Date(s) Application is Due: Feb 3
Contact: Sean Twombly, Director, Member Services and Development; (202) 483-2512, ext. 513; fax (202) 483-2657; twombly@apsanet.org
Internet: http://www.apsanet.org/content_9222.cfm
Sponsor: American Political Science Association
1527 New Hampshire Avenue, NW
Washington, DC 20036-1206

APSA Special Fund for the Study of Women and Politics **947**
The Special Fund for the Study of Women and Politics is designed to provide scholars conducting research on issues of women and politics with supplemental funding. Monies awarded through this fund may support resident or non-resident stays at the Centennial Center. If applying as a Visiting Scholar, a separate application file is not necessary. Interest in this assistance may be noted on the Centennial Center Visiting Scholar

Application. Non-resident scholars interested in funding should complete the Non-Resident Scholar Application Form.
Geographic Focus: All States, Florida
Contact: Jeffrey R. Biggs; (202) 483-2512, ext. 521; jbiggs@apsanet.org
Internet: http://www.apsanet.org/content_6334.cfm
Sponsor: American Political Science Association
1527 New Hampshire Avenue, NW
Washington, DC 20036-1206

APSA Travel Grants **948**
APSA provides travel grants to assist some members in attending the Annual Meeting. U.S. graduate students, international graduate students studying in the U.S., unemployed APSA members, and international scholars are encouraged to apply. Priority will be given to presenters at the Annual Meeting, first-time applicants, or those who have not received a travel grant for three years. Application forms are available upon request.
Requirements: Only applicants who are APSA members and pre-registered for the APSA Annual Meeting are eligible to receive travel grants.
Restrictions: Travel grants will be awarded as cash only; hotel rooms will no longer be included as part of the award.
Geographic Focus: All States
Date(s) Application is Due: Jun 3
Amount of Grant: Up to 700 USD
Contact: Jeffrey R. Biggs; (202) 483-2512, ext. 521; jbiggs@apsanet.org
Internet: http://www.apsanet.org/content_11228.cfm
Sponsor: American Political Science Association
1527 New Hampshire Avenue, NW
Washington, DC 20036-1206

APSA Warren E. Miller Fellowship in Electoral Politics **949**
The Warren Miller Fellowship honors the outstanding contributions of Warren E. Miller to the study of electoral politics and other areas of political science inquiry. The fellowship will support research residencies in national and comparative electoral politics. The Miller fellowships will support scholars for a period one to six months at the Centennial Center.
Geographic Focus: All States
Contact: Jeffrey R. Biggs; (202) 483-2512, ext. 521; jbiggs@apsanet.org
Internet: http://www.apsanet.org/content_6329.cfm
Sponsor: American Political Science Association
1527 New Hampshire Avenue, NW
Washington, DC 20036-1206

APSA William Anderson Award **950**
The William Anderson prize is awarded annually for the best dissertation in the general field of federalism or intergovernmental relations, state and local politics. It carries a prize of $750. The award was set up in honor of William Anderson, former APSA president, who was a leading American authority in the areas of local government, public administration, intergovernmental relations and the history of political science. He did much to shape teaching and research in these fields not only at his own university, but throughout the country.
Requirements: Political science departments are invited to nominate not more than one doctoral dissertation for this award. The dissertations must have been completed and accepted during the two calendar years preceding the date of the prize.
Geographic Focus: All States
Amount of Grant: 750 USD
Contact: Lauren Barry, Awards Coordinator; (202) 483-2512, ext. 631; fax (202) 483-2657; lbarry@apsanet.org or awards@apsanet.org
Internet: http://www.apsanet.org/content_4121.cfm
Sponsor: American Political Science Association
1527 New Hampshire Avenue, NW
Washington, DC 20036-1206

APS Dannie Heineman Prize for Mathematical Physics **951**
The annual prize recognizes outstanding publications in the field of mathematical physics. The award consists of a $10,000 prize and a certificate citing the contributions made by the recipient, plus travel expenses to attend the meeting at which the prize is bestowed. The prize may be awarded to more than one person on a shared basis when all recipients have contributed to the same accomplishments. Nominations are active for three years. Nomination guidelines are available online.
Requirements: This prize is awarded solely for valuable published contributions made in the field of mathematical physics with no restrictions placed on a candidate's citizenship or country of residence. Publication is defined as either a single paper, a series of papers, a book, or any other communication that can be considered a publication.
Geographic Focus: All States
Date(s) Application is Due: Jul 1
Amount of Grant: 10,000 USD
Contact: Jacquelyn Beamon-Kiene, Executive Office Administrator; (301) 209-3269 or (301) 209-3268; fax (301) 209-0865; beamon@aps.org or honors@aps.org
Internet: http://www.aps.org/programs/honors/prizes/heineman.cfm
Sponsor: American Physical Society
One Physics Ellipse
College Park, MD 20740-3844

APS Davisson-Germer Prize in Atomic or Surface Physics **952**
The prize recognizes and encourages outstanding work in atomic physics or surface physics. The award consists of a $5,000 prize and a certificate citing the contributions made by the recipient or recipients. This prize will normally be awarded in even-numbered years for outstanding work in atomic physics and odd-numbered years for outstanding work in surface physics. The prize ordinarily is awarded to one person, but may be shared when all recipients have contributed to the same accomplishments. Nominations are active for three cycles. Nomination guidelines are available online.
Requirements: Nominations are open to scientists of all nationalities regardless of the geographical site at which the work was done.
Geographic Focus: All States
Date(s) Application is Due: Jul 1
Amount of Grant: 5,000 USD
Contact: Jacquelyn Beamon-Kiene, Executive Office Administrator; (301)209-3269 or (301) 209-3268; fax (301) 209-0865; beamon@aps.org or honors@aps.org
Internet: http://www.aps.org/programs/honors/prizes/davisson-germer.cfm
Sponsor: American Physical Society
One Physics Ellipse
College Park, MD 20740-3844

APS Dissertation Award in Nuclear Physics **953**
The program recognizes a recent PhD in nuclear physics. The annual award consists of a $2,500 stipend and an allowance for travel to the annual Spring Meeting of the Division of Nuclear Physics of the American Physical Society, where the award will be presented. Nomination guidelines are available online.
Requirements: Nominations are open to any person who has received a PhD degree in experimental or theoretical nuclear physics from a North American university within the two-year period preceding September 1 of the current year.
Geographic Focus: All States
Date(s) Application is Due: Jul 1
Amount of Grant: 2,500 USD
Contact: Jacquelyn Beamon-Kiene, Executive Office Administrator; (301)209-3269 or (301) 209-3268; fax (301) 209-0865; beamon@aps.org or honors@aps.org
Internet: http://www.aps.org/programs/honors/dissertation/nuclear.cfm
Sponsor: American Physical Society
One Physics Ellipse
College Park, MD 20740-3844

APS Earle K. Plyler Prize for Molecular Spectroscopy **954**
The annual prize recognizes and encourages notable contributions to the field of molecular spectroscopy. The prize may be given for experimental or theoretical achievements, for a single dramatic innovation, or for a series of research contributions which, when integrated, amounts to a major contribution to the field of molecular spectroscopy. The award consists of a $10,000 prize, an allowance for travel expenses to attend the meeting at which the prize is to be presented, and a certificate citing the contributions made by the recipient. The prize ordinarily is awarded to one person, but a prize may be shared when all the recipients have contributed to the same accomplishments. Nominations are active for three years, and guidelines are available online.
Requirements: Nominations are open to scientists in North America.
Geographic Focus: All States
Date(s) Application is Due: Jul 1
Amount of Grant: 10,000 USD
Contact: Jacquelyn Beamon-Kiene, Executive Office Administrator; (301)209-3269 or (301) 209-3268; fax (301) 209-0865; beamon@aps.org or honors@aps.org
Internet: http://www.aps.org/programs/honors/prizes/plyler.cfm
Sponsor: American Physical Society
One Physics Ellipse
College Park, MD 20740-3844

APS Einstein Prize **955**
The prize recognizes outstanding accomplishments in the field of gravitational physics. The award consists of a monetary prize and a certificate citing the contributions of the recipient. It also includes an allowance for the recipient to travel to a meeting of the society to receive the award and deliver a lecture. It will be awarded biennially. The award, usually to a single individual, is for outstanding achievement in theory, experiment, or observation in the area of gravitational physics. Nominations will remain active for three years. Nomination guidelines are available online.
Requirements: Scientists worldwide are eligible for nomination.
Restrictions: Members of the Topical Group on Gravitation Executive Committee are not eligible for nomination while sitting on the committee.
Geographic Focus: All States
Date(s) Application is Due: Jul 1
Amount of Grant: 10,000 USD
Samples: Rainer Weiss, Massachusetts Institute of Technology and Ronald Drever, California Institute of Technology—prize recipients, $10,000; Bryce DeWitt—prize recipient, $10,000; Peter G. Bergmann and John Wheeler—prize recipients, $10,000.
Contact: Jacquelyn Beamon-Kiene, Executive Office Administrator; (301)209-3269 or (301) 209-3268; fax (301) 209-0865; beamon@aps.org or honors@aps.org
Internet: http://www.aps.org/programs/honors/prizes/einstein.cfm
Sponsor: American Physical Society
One Physics Ellipse
College Park, MD 20740-3844

APS Fluid Dynamics Prize 956

The annual prize recognizes and encourages outstanding achievement in fluid dynamics research. The award consists of $10,000, an allowance for travel to the meeting at which the prize is awarded, and a certificate citing the contributions made by the recipient. The prize ordinarily is awarded to one person but may be shared when all the recipients have contributed to the same achievement. Nominations are active for three years. Nomination guidelines are available online.

Geographic Focus: All States
Date(s) Application is Due: Mar 31
Amount of Grant: 10,000 USD
Contact: Jacquelyn Beamon-Kiene, Executive Office Administrator; (301)209-3269 or (301) 209-3268; fax (301) 209-0865; beamon@aps.org or honors@aps.org
Internet: http://www.aps.org/programs/honors/prizes/fluid.cfm
Sponsor: American Physical Society
One Physics Ellipse
College Park, MD 20740-3844

APS Frank Isakson Prize for Optical Effects in Solids 957

The prize recognizes outstanding optical research that leads to breakthroughs in the condensed matter sciences. The award consists of a $5,000 prize, an allowance for travel to the meeting of the society at which the prize is being presented, and a certificate citing the contributions made by the recipient. The prize is awarded biennially in even-numbered years as a memorial to Frank Isakson. The prize ordinarily is awarded to one person, but the prize may be shared among recipients when all recipients have contributed to the same accomplishment. Preference will be given to work that has been published within the past 10 years. Nominations are active for three years. Nomination guidelines are available online.

Requirements: Nominations are open to scientists of all nations regardless of the geographical site at which the work was done.
Geographic Focus: All States
Date(s) Application is Due: Jul 1
Amount of Grant: 5,000 USD
Contact: Jacquelyn Beamon-Kiene, Executive Office Administrator; (301)209-3269 or (301) 209-3268; fax (301) 209-0865; beamon@aps.org or honors@aps.org
Internet: http://www.aps.org/programs/honors/prizes/isakson.cfm
Sponsor: American Physical Society
One Physics Ellipse
College Park, MD 20740-3844

APS George E. Pake Prize 958

The annual prize recognizes and encourages outstanding work by physicists combining original research accomplishments with leadership in the management of research or development in industry. The award consists of a $5,000 prize, an allowance for travel to the meeting at which the prize is to be awarded, and a certificate recognizing the contribution of the recipient. Nominations are active for three years, and guidelines are available online.

Requirements: This prize will be awarded to one individual for outstanding achievements in physics research combined with major success as a manager of research or development in industry.
Geographic Focus: All States
Date(s) Application is Due: Jul 1
Amount of Grant: 5,000 USD
Contact: Jacquelyn Beamon-Kiene, Executive Office Administrator; (301)209-3269 or (301) 209-3268; fax (301) 209-0865; beamon@aps.org or honors@aps.org
Internet: http://www.aps.org/programs/honors/prizes/pake.cfm
Sponsor: American Physical Society
One Physics Ellipse
College Park, MD 20740-3844

APS George E. Valley, Jr., Prize 959

The prize recognizes one individual in the early stages of his or her career for an outstanding scientific contribution to physics that is deemed to have significant potential for a dramatic impact on the field. The award consists of a $20,000 award and a certificate citing the contribution made by the recipient. Nomination documents must include a statement from the nominator or from the candidate's department certifying the date of the candidate's PhD. The prize is awarded biennially. Nomination guidelines are available online.

Requirements: The nominee must have received his/her PhD no earlier than five years before April 1 of the year of the nomination deadline. Work done by a graduate student for his/her thesis is eligible for consideration if it is demonstrated that the student's contributions have been crucial to an important piece of research.
Geographic Focus: All States
Date(s) Application is Due: Jul 1
Contact: Jacquelyn Beamon-Kiene, Executive Office Administrator; (301)209-3269 or (301) 209-3268; fax (301) 209-0865; beamon@aps.org or honors@aps.org
Internet: http://www.aps.org/programs/honors/prizes/valley.cfm
Sponsor: American Physical Society
One Physics Ellipse
College Park, MD 20740-3844

APS Hans A. Bethe Prize 960

To prize recognizes outstanding work in theory, experiment, or observation in the areas of astrophysics, nuclear physics, nuclear astrophysics, or closely related fields. The prize consists of a $7,500 prize and a certificate citing the contributions made by the recipient. It is presented annually. Nominations are active for three years, and guidelines are available online.

Requirements: This award is made annually to one individual for outstanding accomplishments in the areas of astrophysics, nuclear physics, nuclear astrophysics, or closely related fields. It is open to any scientist working in these areas, worldwide.
Geographic Focus: All States
Date(s) Application is Due: Jul 1
Amount of Grant: 7,500 USD
Contact: Jacquelyn Beamon-Kiene, Executive Office Administrator; (301)209-3269 or (301) 209-3268; fax (301) 209-0865; beamon@aps.org or honors@aps.org
Internet: http://www.aps.org/programs/honors/prizes/bethe.cfm
Sponsor: American Physical Society
One Physics Ellipse
College Park, MD 20740-3844

APS Herbert P. Broida Prize 961

The prize recognizes and enhances outstanding experimental advancements in the fields of atomic and molecular spectroscopy or chemical physics. The prize consists of a $5,000 prize, an allowance for travel to the award ceremony, and a certificate citing the contributions made by the recipient. Preference will be granted to an individual whose contributions have displayed a high degree of breadth, originality, and creativity. Nominations are active for three years, and guidelines are available online.

Requirements: The prize is awarded to one individual in recognition of an outstanding contribution to the field of atomic and molecular spectroscopy or chemical physics. Emphasis will be given to work done within the five years prior to the awarding of the prize.
Geographic Focus: All States
Date(s) Application is Due: Jul 1
Amount of Grant: 5,000 USD
Contact: Jacquelyn Beamon-Kiene, Executive Office Administrator; (301)209-3269 or (301) 209-3268; fax (301) 209-0865; beamon@aps.org or honors@aps.org
Internet: http://www.aps.org/programs/honors/prizes/broida.cfm
Sponsor: American Physical Society
One Physics Ellipse
College Park, MD 20740-3844

APS I.I. Rabi Prize in Atomic, Molecular, and Optical Physics 962

The prize recognizes and encourages outstanding research in atomic, molecular, and optical physics. It will be awarded in odd-numbered years. The award consists of a $7,500 prize and a certificate citing the contributions made by the recipient. An allowance will be provided for travel expenses of the recipient to the society meeting at which the prize is presented. The prize ordinarily is awarded to one person but may be shared when all recipients have contributed to the same accomplishment. Nominations are active for three years, and guidelines are available online.

Requirements: Nominations are open to scientists of all nationalities regardless of the geographical location at which the work was done. Investigators who have held a PhD for 10 years or less are eligible.
Geographic Focus: All States
Date(s) Application is Due: Jul 1
Amount of Grant: 7,500 USD
Contact: Jacquelyn Beamon-Kiene, Executive Office Administrator; (301)209-3269 or (301) 209-3268; fax (301) 209-0865; beamon@aps.org or honors@aps.org
Internet: http://www.aps.org/programs/honors/prizes/rabi.cfm
Sponsor: American Physical Society
One Physics Ellipse
College Park, MD 20740-3844

APS Irving Langmuir Prize in Chemical Physics 963

The prize recognizes and encourages outstanding interdisciplinary research in chemistry and physics. This biennial prize consists of a $10,000 award and a certificate citing the contributions made by the recipient. In even-numbered years, the American Chemical Society selects the prize recipient and presents the prize. In odd-numbered years, the American Physical Society selects the prize recipient and presents the prize. An allowance is provided for travel expenses of the recipient to the meeting of the society at which the prize is to be bestowed. The prize is made to one person who has made an outstanding contribution in the field of chemical physics or physical chemistry within the 10 years prior to the prize. Nominations are active for three years, and guidelines are available online.

Requirements: The prize is granted without restriction, except that the recipient must be a resident of the United States at the time of selection and the prize funds must be used in the United States or its possessions.
Geographic Focus: All States
Date(s) Application is Due: Jul 1
Amount of Grant: 10,000 USD
Contact: Jacquelyn Beamon-Kiene, Executive Office Administrator; (301)209-3269 or (301) 209-3268; fax (301) 209-0865; beamon@aps.org or honors@aps.org
Internet: http://www.aps.org/programs/honors/prizes/langmuir.cfm
Sponsor: American Physical Society
One Physics Ellipse
College Park, MD 20740-3844

APS J. J. Sakurai Prize for Theoretical Particle Physics 964

The annual prize recognizes and encourages outstanding achievement in particle theory. The award consists of a $5,000 prize, an allowance for travel to the meeting of the society at which the prize is to be awarded, and a certificate citing the contributions made by the recipient. The prize may be awarded to more than one person on a shared basis. Nominations are active for three years, and guidelines are available online.

Requirements: Nominations are open to scientists of all nationalities regardless of the geographical site at which the work was done. The prize will normally be awarded for theoretical contributions made at an early stage of the recipients research career.
Geographic Focus: All States
Date(s) Application is Due: May 1
Amount of Grant: 5,000 USD
Contact: Jacquelyn Beamon-Kiene, Executive Office Administrator; (301)209-3269 or (301) 209-3268; fax (301) 209-0865; beamon@aps.org or honors@aps.org
Internet: http://www.aps.org/programs/honors/prizes/sakurai.cfm
Sponsor: American Physical Society
One Physics Ellipse
College Park, MD 20740-3844

APS James C. McGroddy Prize for New Materials **965**

The annual prize recognizes and encourages outstanding achievement in the science and application of new materials, including the discovery of new classes of materials, the observation of novel phenomena in known materials leading to both fundamentally new applications and scientific insights, as well as theoretical and experimental work contributing significantly to the understanding of such phenomena. The award consists of a $7,500 prize, plus a certificate citing the contribution of the recipient and an allowance for travel to the meeting of the society at which the award is presented. Nominations are active for three years, and guidelines are available online.
Requirements: The prize is open to scientists of all nationalities irrespective of where their work has been carried out.
Geographic Focus: All States
Date(s) Application is Due: Jul 1
Amount of Grant: 5,000 USD
Contact: Jacquelyn Beamon-Kiene, Executive Office Administrator; (301)209-3269 or (301) 209-3268; fax (301) 209-0865; beamon@aps.org or honors@aps.org
Internet: http://www.aps.org/programs/honors/prizes/mcgroddy.cfm
Sponsor: American Physical Society
One Physics Ellipse
College Park, MD 20740-3844

APS James Clerk Maxwell Prize for Plasma Physics **966**

The annual prize recognizes outstanding contributions to the field of plasma physics. The prize is made for outstanding contributions to the advancement and diffusion of the knowledge of properties of highly ionized gases of natural or laboratory origin. The award consists of a $5,000 award and a certificate citing the contributions made by the recipient. A travel allowance to attend the meeting at which the prize is to be presented is also provided. The prize ordinarily is awarded to one person, but a prize may be shared when all the recipients have contributed to the same accomplishments. Nominations are active for three years, and guidelines are available online.
Geographic Focus: All States
Date(s) Application is Due: Apr 1
Amount of Grant: 5,000 USD
Contact: Jacquelyn Beamon-Kiene, Executive Office Administrator; (301)209-3269 or (301) 209-3268; fax (301) 209-0865; beamon@aps.org or honors@aps.org
Internet: http://www.aps.org/programs/honors/prizes/maxwell.cfm
Sponsor: American Physical Society
One Physics Ellipse
College Park, MD 20740-3844

APS Julius Edgar Lilienfeld Prize **967**

The annual prize recognizes a most outstanding contribution to physics. The award consists of a $10,000 prize, a certificate citing the contributions made by the recipient, plus expenses for the three lectures by the recipient given at an APS meeting, a research university, and a predominantly undergraduate institution. Nominations are active for three years, and guidelines are available online.
Requirements: The prize is awarded for outstanding contributions to physics by a single individual who also has exceptional skills in lecturing to diverse audiences.
Geographic Focus: All States
Date(s) Application is Due: Jul 1
Amount of Grant: 10,000 USD
Contact: Jacquelyn Beamon-Kiene, Executive Office Administrator; (301)209-3269 or (301) 209-3268; fax (301) 209-0865; beamon@aps.org or honors@aps.org
Internet: http://www.aps.org/programs/honors/prizes/lilienfeld.cfm
Sponsor: American Physical Society
One Physics Ellipse
College Park, MD 20740-3844

APS Lars Onsager Prize **968**

The prize recognizes outstanding research in theoretical statistical physics including the quantum fluids. The prize is open to researchers in statistical physics covering a wide range of physical phenomena, as long as the nominee is active at the time of the award. The award consists of a $15,000 prize as well as a certificate citing the contribution made by the recipient. Nominations are active for three years, and guidelines are available online.
Requirements: Nominations are accepted from all scientists of all nations regardless of geographical location.
Geographic Focus: All States
Date(s) Application is Due: Jul 1
Amount of Grant: 15,000 USD
Samples: Christopher Pethick, NORDITA, Gordon Baym, University of Illinois and Tin-Lun Ho, Ohio State University, $15,000; A. Brooks Harris, University of Pennsylvania, $15,000; Rodney J. Baxter, Australian National University, $15,000.
Contact: Jacquelyn Beamon-Kiene, Executive Office Administrator; (301)209-3269 or (301) 209-3268; fax (301) 209-0865; beamon@aps.org or honors@aps.org
Internet: http://www.aps.org/programs/honors/prizes/onsager.cfm
Sponsor: American Physical Society
One Physics Ellipse
College Park, MD 20740-3844

APS Marshall N. Rosenbluth Outstanding Doctoral Thesis Award **969**

The program recognizes exceptional young scientists who have performed original thesis work of outstanding scientific quality and achievement in the area of plasma physics. The award consists of a $2,000 stipend, a certificate to be presented during the award ceremony at the banquet for the Division of Plasma Physics Annual Meeting, and a $500 travel allowance to the meeting. Nomination guidelines are available online.
Requirements: Nominations will be accepted for any doctoral student of a college or university in the United States or for a United States student abroad who has successfully passed his/her final thesis defense within the preceding 24 months of the current nomination deadline. The work to be considered must have been performed as part of the requirements for a doctoral degree.
Geographic Focus: All States
Date(s) Application is Due: Apr 1
Amount of Grant: 500-2,000 USD
Contact: Jacquelyn Beamon-Kiene, Executive Office Administrator; (301)209-3269 or (301) 209-3268; fax (301) 209-0865; beamon@aps.org or honors@aps.org
Internet: http://www.aps.org/programs/honors/dissertation/rosenbluth.cfm
Sponsor: American Physical Society
One Physics Ellipse
College Park, MD 20740-3844

APS Max Delbruck Prize in Biological Physics **970**

The Max Delbruck Prize in Biological Physics is intended to recognize and encourage outstanding achievement in biological physics research. The prize consists of $10,000, an allowance for travel to attend the meeting at which the prize is awarded, and a certificate citing the contributions made by the recipient or recipients. It is presented biennially in even-numbered years. Nominations are open to scientists of all nationalities regardless of the geographical site at which the work was done. The prize may be awarded to more than one investigator on a shared basis. Nominations are active for three cycles.
Geographic Focus: All States
Date(s) Application is Due: Jul 1
Amount of Grant: 10,000 USD
Contact: Jacquelyn Beamon-Kiene, Executive Office Administrator; (301)209-3269 or (301) 209-3268; fax (301) 209-0865; beamon@aps.org or honors@aps.org
Internet: http://www.aps.org/programs/honors/prizes/delbruck.cfm
Sponsor: American Physical Society
One Physics Ellipse
College Park, MD 20740-3844

APS Mitsuyoshi Tanaka Dissertation Award in Experimental Particle Physics **971**

The award provides recognition to exceptional young scientists who have performed original doctoral thesis work of outstanding scientific quality and achievement in the area of experimental particle physics. The annual award consists of a $1,500 award, a certificate citing the accomplishments of the recipient, and an allowance for travel to attend the annual meeting of the Division of Particles and Fields, at which the award will be presented. Each nominee will be considered in not more than two consecutive cycles, and guidelines are available online.
Requirements: Nominations will be accepted for any doctoral student studying at a college or university in the United States or in an education abroad program of a college or university in the United States for dissertation research carried out at a U.S. laboratory. The work to be considered must have been accomplished as part of the requirements for a doctoral degree. Nominees must pass their thesis defense 12 months or less from the nomination deadline, for the selection cycle in which the nomination is to be considered.
Geographic Focus: All States
Date(s) Application is Due: Jun 30
Amount of Grant: 1,000-1,500 USD
Contact: Jacquelyn Beamon-Kiene, Executive Office Administrator; (301)209-3269 or (301) 209-3268; fax (301) 209-0865; beamon@aps.org or honors@aps.org
Internet: http://www.aps.org/programs/honors/dissertation/tanaka.cfm
Sponsor: American Physical Society
One Physics Ellipse
College Park, MD 20740-3844

APS Nicholas Metropolis Award for Outstanding Doctoral Thesis Work in Computational Physics **972**

The purpose of the annual award is to recognize doctoral thesis research of outstanding quality and achievement in computational physics and to encourage effective written and oral presentation of research results. The award consists of a $1,500 prize and a certificate to be presented at an awards ceremony at the Division of Computational Physics annual meeting and an additional allowance to travel to the meeting. The recipient will be invited to present his or her work in an appropriate session of the meeting. An individual can be nominated only once; however, an unsuccessful candidate will be carried over for one year. Nomination guidelines are available online.

Requirements: Nominations will be accepted for any doctoral student (present or past) in any country for work performed as part of the requirements for a doctoral degree. Nominees must pass their thesis defense not more than 18 months before the nomination deadline.
Geographic Focus: All States
Date(s) Application is Due: Nov 1
Amount of Grant: 500-1,500 USD
Contact: Jorge Jose, Committee Chair; jjose@buffalo.edu
Internet: http://www.aps.org/programs/honors/dissertation/metropolis.cfm
Sponsor: American Physical Society
One Physics Ellipse
College Park, MD 20740-3844

APS Oliver E. Buckley Condensed Matter Prize 973
The annual prize recognizes and encourages outstanding theoretical or experimental contributions to condensed matter physics. The prize ordinarily is awarded to one person, but may be shared when all the recipients have contributed to the same accomplishments. The award consists of a $10,000 prize and a certificate citing the contributions made by the recipient or recipients. Nominations are active for three years, and guidelines are available online.
Geographic Focus: All States
Date(s) Application is Due: Jul 1
Amount of Grant: 10,000 USD
Contact: Jacquelyn Beamon-Kiene, Executive Office Administrator; (301)209-3269 or (301) 209-3268; fax (301) 209-0865; beamon@aps.org or honors@aps.org
Internet: http://www.aps.org/programs/honors/prizes/buckley.cfm
Sponsor: American Physical Society
One Physics Ellipse
College Park, MD 20740-3844

APS Outstanding Doctoral Thesis Research in Atomic, Molecular, or Optical Physics 974
The annual award recognizes doctoral thesis research of outstanding quality and achievement in atomic, molecular, or optical physics and encourages effective written and oral presentation of research results. The award consists of a $2,500 prize and a certificate citing the contribution made by the recipient. All finalists will receive a travel stipend. A student may be a finalist in the competition only once. Eligible non-finalists may only be renominated by submitting an entirely new package, even if it is the same as the original package. Nomination guidelines are available online.
Requirements: Doctoral students at any university in the United States or abroad who have passed their thesis defense for the PhD in the disciplines of atomic, molecular, or optical physics any time during the prior two calendar years are eligible for the award, except for those whose thesis advisors serve on the current selection committee.
Geographic Focus: All States
Date(s) Application is Due: Dec 7
Contact: Jacquelyn Beamon-Kiene, Executive Office Administrator; (301)209-3269 or (301) 209-3268; fax (301) 209-0865; beamon@aps.org or honors@aps.org
Internet: http://www.aps.org/programs/honors/dissertation/amo.cfm
Sponsor: American Physical Society
One Physics Ellipse
College Park, MD 20740-3844

APS Outstanding Doctoral Thesis Research in Beam Physics Award 975
The Award is intended to recognize doctoral thesis research of outstanding quality and achievement in beam physics and engineering. The annual award consists of $2,500 and a certificate to be presented at an awards ceremony at the Division of Physics of Beams Annual Meeting, and an additional allowance of up to $500 for travel to the meeting. The recipient will be invited to present the work as an invited talk in an appropriate session of the meeting.
Requirements: A nomination will be accepted for any doctoral student of a university worldwide for work performed as part of the requirements for a doctoral degree. Nominees must pass their thesis defense not more than 18 months before the nomination deadline.
Geographic Focus: All States
Date(s) Application is Due: Sep 30
Amount of Grant: 500-2,500 USD
Contact: Jacquelyn Beamon-Kiene, Executive Office Administrator; (301) 209-3269 or (301) 209-3268; fax (301) 209-0865; beamon@aps.org or honors@aps.org
Internet: http://www.aps.org/programs/honors/dissertation/beamphysics.cfm
Sponsor: American Physical Society
One Physics Ellipse
College Park, MD 20740-3844

APS Polymer Physics Prize 976
The annual prize recognizes outstanding accomplishment and excellence of contributions in polymer physics research. The award consists of a $10,000 prize and a certificate citing the contributions made by the recipient. The prize ordinarily is awarded to one person, but a prize may be shared when all the recipients have contributed to the same accomplishments. Nominations are active for three years, and guidelines are available online.
Requirements: Nominations are open to all scientists of all nations regardless of membership in the society or the geographical location in which the work was carried out.
Geographic Focus: All States
Date(s) Application is Due: Jul 1
Amount of Grant: 10,000 USD
Contact: Jacquelyn Beamon-Kiene, Executive Office Administrator; (301)209-3269 or (301) 209-3268; fax (301) 209-0865; beamon@aps.org or honors@aps.org
Internet: http://www.aps.org/programs/honors/prizes/polymer.cfm
Sponsor: American Physical Society
One Physics Ellipse
College Park, MD 20740-3844

APS Prize for a Faculty Member for Research in an Undergraduate Institution 977
The annual prize honors a physicist whose research in an undergraduate setting has achieved wide recognition and contributed significantly to physics and who has contributed substantially to the professional development of undergraduate physics students. The award consists of a $5,000 stipend to the prize recipient and a separate unrestricted $5,000 grant for the research to the prize recipient's institution. An additional allowance will be provided for travel expenses to the APS meeting at which the prize ceremony will take place, and a certificate citing the contributions by the recipient. Nominations are active for three years, and guidelines are available online.
Requirements: The prize will be given to a physics faculty member at a U.S. undergraduate institution. The recipient will have been recognized as contributing substantially to physics research and providing inspirational guidance and encouragement of undergraduate students participating in this research. Nominations also should include a publication list that highlights student co-authors, including whether these students are high school, undergraduate, or graduate students.
Restrictions: The nominee's department may offer a program leading to a masters degree but must not have a doctoral program in physics.
Geographic Focus: All States
Date(s) Application is Due: Jul 1
Amount of Grant: 5,000-10,000 USD
Contact: Jacquelyn Beamon-Kiene, Executive Office Administrator; (301)209-3269 or (301) 209-3268; fax (301) 209-0865; beamon@aps.org or honors@aps.org
Internet: http://www.aps.org/programs/honors/prizes/faculty-undergraduate.cfm
Sponsor: American Physical Society
One Physics Ellipse
College Park, MD 20740-3844

APS Prize for Industrial Applications of Physics 978
The purpose of the Prize is to recognize excellence in the industrial application of physics, and thereby to publicize the value of physics in industry, to encourage physics research in industry, and to enhance students' awareness of and interest in the role of physics in commercial product development. The recognized contribution may be a product, a process, or a tool enabling practical application of physics. The Prize consists of $10,000 and a certificate citing the contribution for which the Prize was awarded. In addition, an allowance is made for the expenses incurred by the recipient(s) to attend the APS meeting at which the Prize is presented. The APS Prize will be presented in odd-numbered years.
Requirements: Nominees need not be APS members, and there is no restriction with regard to geography or nationality. It is not necessary for the application to have already achieved commercial success, but it should have demonstrated potential for significant impact. Although nominees need not now be working in industry, the contribution for which they are cited must have been made while their principal employment was in industry.
Geographic Focus: All States
Date(s) Application is Due: Apr 1
Amount of Grant: 10,000 USD
Contact: Jacquelyn Beamon-Kiene, Executive Office Administrator; (301)209-3269 or (301) 209-3268; fax (301) 209-0865; beamon@aps.org or honors@aps.org
Internet: http://www.aps.org/programs/honors/prizes/industrial.cfm
Sponsor: American Physical Society
One Physics Ellipse
College Park, MD 20740-3844

APS Robert R. Wilson Prize for Achievement in the Physics of Particle Accelerators 979
The annual prize recognizes and encourages outstanding achievement in the physics of particle accelerators. The award consists of a $7,500 prize, an allowance for travel to the meeting at which the prize is awarded, and a certificate citing the contributions made by the recipient. The prize ordinarily is awarded to one person but may be shared when all recipients have contributed to the same accomplishment. Nominations are active for three years, and guidelines are available online.
Requirements: Nominations are open to scientists of all nations regardless of the geographical site at which the work was done. The prize will normally be awarded for contributions made at an early stage of the recipient's career.
Geographic Focus: All States
Date(s) Application is Due: Jul 1
Amount of Grant: 7,500 USD
Contact: Jacquelyn Beamon-Kiene, Executive Office Administrator; (301)209-3269 or (301) 209-3268; fax (301) 209-0865; beamon@aps.org or honors@aps.org
Internet: http://www.aps.org/programs/honors/prizes/wilson.cfm
Sponsor: American Physical Society
One Physics Ellipse
College Park, MD 20740-3844

APS Tom W. Bonner Prize in Nuclear Physics 980
The prize recognizes and encourages outstanding experimental research in nuclear physics, including the development of a method, technique, or device that significantly contributes in a general way to nuclear physics research. The prize consists of $7,500 prize and a certificate citing the contributions made by the recipient. It is presented annually. The prize ordinarily is awarded to one person but may be shared when all the recipients

have contributed to the same accomplishment. Nominations are active for three years, and guidelines are available online.
Requirements: Nominations are open to physicists whose work in nuclear physics is primarily experimental or a particularly outstanding piece of theoretical work. There is no time limitations on work described.
Geographic Focus: All States
Date(s) Application is Due: Jul 1
Amount of Grant: 7,500 USD
Contact: Jacquelyn Beamon-Kiene, Executive Office Administrator; (301)209-3269 or (301) 209-3268; fax (301) 209-0865; beamon@aps.org or honors@aps.org
Internet: http://www.aps.org/programs/honors/prizes/bonner.cfm
Sponsor: American Physical Society
One Physics Ellipse
College Park, MD 20740-3844

APS W.K.H. Panofsky Prize in Experimental Particle Physics 981
The annual prize recognizes and encourages outstanding achievements in experimental particle physics. The award consists of a $10,000 prize, an allowance for travel to the meeting at which the prize is to be awarded, and a certificate citing the contributions made by the recipient. The prize ordinarily is awarded to one person, but the prize may be shared when all recipients have contributed to the same accomplishment. Nominations are active for three years, and guidelines are available online.
Requirements: Nominations are open to scientists of all nations regardless of the geographical site at which the work was accomplished. The prize will normally be awarded for contributions made at an early stage of the recipient's career.
Geographic Focus: All States
Date(s) Application is Due: May 1
Amount of Grant: 10,000 USD
Contact: Jacquelyn Beamon-Kiene, Executive Office Administrator; (301)209-3269 or (301) 209-3268; fax (301) 209-0865; beamon@aps.org or honors@aps.org
Internet: http://www.aps.org/programs/honors/prizes/panofsky.cfm
Sponsor: American Physical Society
One Physics Ellipse
College Park, MD 20740-3844

APS Will Allis Prize for the Study of Ionized Gases 982
The prize was established in 1989 by contributions from American Telephone and Telegraph, General Electric, General Telephone and Electronics, International Business Machines, and Xerox Corporations in recognition of the outstanding contributions of Will Allis to the study of ionized gases. The prize ordinarily is awarded to one person, but the prize may be shared when all recipients have contributed to the same accomplishment. The prize consists of a $5,000 award and a certificate citing the contributions made by the recipient. An allowance will be provided for travel expenses of the recipient to the society's meeting, where the prize is bestowed. The prize is awarded in even-numbered years. Nominations are active for three years, and guidelines are available online.
Requirements: Nominations are open to scientists of all nations regardless of the geographical location at which the work was done.
Geographic Focus: All States
Date(s) Application is Due: Jul 1
Amount of Grant: 5,000 USD
Contact: Jacquelyn Beamon-Kiene, Executive Office Administrator; (301)209-3269 or (301) 209-3268; fax (301) 209-0865; beamon@aps.org or honors@aps.org
Internet: http://www.aps.org/programs/honors/prizes/allis.cfm
Sponsor: American Physical Society
One Physics Ellipse
College Park, MD 20740-3844

Arcadia Foundation Grants 983
The foundation awards grants to Pennsylvania nonprofit organizations to improve the quality of life. Areas of interest include hospitals and hospital building funds, health agencies and services, nursing, hospices, early childhood, adult and higher education, libraries, child development and welfare agencies, youth organizations, and social service and general welfare agencies, including care of the handicapped, aged, and hungry. Also supported are family services, environment and conservation, wildlife and animal welfare, religious organizations, historical preservation, and music organizations. Types of support include general operating support, continuing support, annual campaigns, capital campaigns, building construction/renovation, equipment acquisition, endowment funds, program development, scholarship funds, and research. Applications are accepted between September 1 and November 1.
Requirements: Eastern Pennsylvania organizations whose addresses have zip codes of 18000-19000 are eligible. Application form not required. The initial approach should be a letter or proposal—not exceeding two pages.
Restrictions: Grants are not awarded to support individuals, deficit financing, land acquisition, fellowships, demonstration projects, publications, or conferences.
Geographic Focus: Pennsylvania
Date(s) Application is Due: Nov 1
Contact: Marilyn Lee Steinbright, President; (610) 275-8460
Sponsor: Arcadia Foundation
105 E Logan Street
Norristown, PA 19401

Arca Foundation Grants 984
The Arca Foundation believes that access to knowledge, vigorous public education and citizen engagement are essential to democracy. However, there exist structures and private interests that serve to limit the transparency of our government, stifle public debate on critical issues, and foster an environment where government is not effectively serving the interests of its citizens. In order to promote greater social equity and justice at home and abroad, the foundation supports organizations and projects that work to advance transparent, accountable, and just policies. They support strategic initiatives that work to directly affect policies by: Developing and advocating for innovative ideas; Promoting transparency and access to information; Fostering greater public debate on critical issues; Educating key stakeholders; Engaging citizens in strategic organizing and advocacy that builds power and drives change.
Requirements: Domestically, in the current sociopolitical climate, the foundation is concerned about the promotion of a more equitable, accountable, and transparent economic recovery. They are considering proposals that work to advance more just policies on this and other critical issues. Internationally, the foundation has a long history of working to promote greater dialogue between the U.S. and Cuba. They are considering proposals that advance policies that further normalized U.S.-Cuban relations, as well as proposals that work to foster more just policies on a range of international issues. The Arca Foundation has two deadlines annually, on March 1 and September 1 of every year, for consideration in June and December respectively. When deadline dates fall on a weekend, the deadline is effective on the next weekday. The foundation will not respond to letters of inquiry, but will accept complete proposals submitted according to the application guidelines on regular grant deadlines.
Restrictions: The foundation does not fund organizations that provide direct social services, scholarship funds or scholarly research, capital projects or endowments, individuals, or government programs. Proposals received via fax or email will not be considered. Late proposals will not be considered, and extensions are not available for any reason.
Geographic Focus: All States
Date(s) Application is Due: Mar 1; Sep 1
Amount of Grant: 50,000 USD
Contact: Emily Casteel; (202) 822-9193; fax (202) 785-1446; grants@arcafoundation.org
Internet: http://www.arcafoundation.org/howtoapply.htm
Sponsor: Arca Foundation
1308 19th Street NW
Washington, DC 20036

Archer Daniels Midland Foundation Grants 985
The foundation prefers to fund programs that will directly impact the communities where operating units are located. Nearly a third of the funding is directed toward educational institutions, including elementary, secondary, and higher education. The advocacy category covers world affairs and foreign relations groups that deal with international trade; also supported are projects stressing free enterprise and assistance for women. Support of social services goes to minority group development, cultural activities, and hospital and youth agencies. Health service funding supports hospital and disease association programs. Conservation funding supports programs for protecting the environment and beautification as well as conservation. Religion grants support Christian, Jewish, and Roman Catholic organizations, such as churches, colleges and universities, international ministries/missions, Jewish welfare, and the Salvation Army. Applications are accepted at any time; initial contact by letter of inquiry is encouraged.
Requirements: Tax-exempt organizations are eligible. Current United Way recipients and national organizations are also eligible.
Geographic Focus: Arkansas, California, Georgia, Illinois, Indiana, Iowa, Kansas, Kentucky, Massachusetts, Michigan, Minnesota, Mississippi, Missouri, Montana, Nebraska, New Jersey, New York, North Carolina, North Dakota, Ohio, Oklahoma, Pennsylvania, Puerto Rico, South Carolina, Tennessee, Texas, Washington, Wisconsin
Amount of Grant: 10,000 USD
Contact: Brian Peterson; (217) 424-5413 or (217) 424-5200; corpaffairs@admworld.com
Internet: http://www.adm.com/en-U.S./responsibility/social_investing/Pages/default.aspx
Sponsor: Archer Daniels Midland Foundation
4666 Faries Parkway, P.O. Box 1470
Decatur, IL 62526-5630

Archibald Waynne Dingman Memorial Graduate Scholarship 986
This scholarship is for four months of study related to the petroleum industry, awarded annually to qualified graduates of any recognized university who are registered in or admissible to a program leading to a master's or doctoral degree at the University of Calgary.
Requirements: To be eligible, applicants must be graduate students that are Canadian citizens or permanent residents.
Geographic Focus: All States
Date(s) Application is Due: Feb 1
Amount of Grant: 3,000 CAD
Contact: Connie Busch; (403) 220-5690; fax (403) 289-7635; cbusch@.ucalgary.ca
Internet: http://www.grad.ucalgary.ca/funding/internal_scholarships/lev_4/dingman.htm
Sponsor: University of Calgary
2500 University Drive NW, Earth Sciences Building, Room 720
Calgary, AB T2N 1N4 Canada

Arcus Foundation Religion and Values Grants 987
The Arcus Foundation is a private grantmaking foundation that supports organizations around the world working in two areas-lesbian, gay, bisexual, and transgender (LGBT) human rights; and conservation of the world's great apes. In the former area, the Foundation supports organizations that are working to achieve social justice that is

inclusive of sexual orientation, gender identity and race. In the latter area, it supports organizations seeking to ensure respect and survival of great apes and their natural habitat. Specifically, the Religion and Values Program has the goal of achieving long-term change in cultural attitudes and religious institutions that currently stigmatize same-gender sexual orientation and gender variance. The program engages with and advances greater understanding and acceptance of sexual orientation and gender variance by mainstream religious institutions and leaders, and helps to create a positive shift in cultural attitudes and values toward sexuality in general and GLBT (gay, lesbian, bisexual and transgender) issues in particular.
Requirements: Nonprofit organizations are eligible.
Restrictions: No grants to are given to individuals, or for religious or political activities, medical research or film/video production.
Geographic Focus: All States
Amount of Grant: 1,000-50,000 USD
Contact: Lucia Leandro Gimeno, Program Assistant; (212) 488-3000; fax (212) 488-3010; lucialeandro@arcusfoundation.org
Internet: http://www.arcusfoundation.org/pages_2/home.cfm
Sponsor: Arcus Foundation
402 East Michigan Avenue
Kalamazoo, MI 49007

Arecibo Observatory Grants 988
The National Astronomy and Ionosphere Center enables research in the areas of astronomy, planetary studies, and space and atmospheric sciences by providing unique capabilities and state-of-the-art instrumentation for data collection and analysis, together with logistical support to users. Use of the Arecibo Observatory (AO) operated by NAIC is available on an equal, competitive basis to all scientists from throughout the world to pursue research in astronomy, planetary studies and, space and atmospheric sciences. Observing time is granted on the basis of the most promising research, as ascertained by peer review of proposals by external referees.
Geographic Focus: All States
Date(s) Application is Due: Feb 1; Jun 1; Oct 1
Contact: Ellen Howell; (787) 878-2612; fax (787) 878-1861; ehowell@naic.edu
Internet: http://www.naic.edu/science/proposals_set.htm
Sponsor: Arecibo Observatory
HC03 Box 53995
Arecibo, PR 00612

Arie and Ida Crown Memorial Grants 989
The program supports programs that offer opportunities to the disadvantaged, strengthens the bond of families, and improves the quality of people's lives. As a general rule, the Foundation funds organizations that serve the greater Chicago area as well as organizations that serve the broader Jewish community. Most grants are awarded to organizations within the city of Chicago. Organizations are supported in the areas of arts and culture (concentrating on educational and enrichment programs for youth), civic affairs, education, health (stressing access to services, hospice and health promotion), and human service (focusing on programs which offer assistance for children and families).
Requirements: Nonprofit organizations in Chicago and Cook County, IL, may apply for grant support.
Restrictions: Grants are not made to support individuals, conference expenses, film projects, government programs (50 percent government funded), or research projects.
Geographic Focus: Illinois
Date(s) Application is Due: Jan 31; Jul 31
Amount of Grant: 1,000-200,000 USD
Contact: Susan Crown; (312) 236-6300; fax (312) 984-1499; AICM@crown-Chicago.com
Internet: http://www.crownmemorial.org/
Sponsor: Arie and Ida Crown Memorial
222 N LaSalle Street, Suite 2000
Chicago, IL 60601

ARIT/NEH Fellowships for Research in Turkey 990
ARIT invites applications for two or three postdoctoral fellowships made possible by support from NEH. The fields of study cover all periods in the general range of the humanities, including humanistically oriented aspects of social sciences, prehistory, history, art, archaeology, literature, and linguistics, as well as interdisciplinary aspects of cultural history. The fellowships for research in Turkey can be held for four to 12 months. ARIT maintains two research institutes in Turkey: ARIT-Istanbul, with a research library focused on Byzantine, Ottoman, and modern studies of Turkey; and ARIT-Ankara, which focuses on art, archaeology, and ancient history in its library and serves Turkish and American archaeologists through its programs. Both institutes have residential facilities for fellows and provide general assistance as well as introductions to colleagues, institutions, and authorities in Turkey.
Requirements: Scholars who have completed their formal training and plan to carry out research in Turkey may apply. They may be U.S. citizens or three-year residents of the U.S.
Geographic Focus: All States
Date(s) Application is Due: Nov 1
Amount of Grant: 13,335-40,000 USD
Contact: Nancy Leinwand, (215) 898-3474; fax (215) 898-0657; leinwand@sas.upenn.edu
Internet: http://ccat.sas.upenn.edu/ARIT/NEHFellowships.htm
Sponsor: American Research Institute in Turkey
3260 South Street
Philadelphia, PA 19104-6324

ARIT Fellowships 991
Fellowships support research in ancient, medieval, or modern times, in any field of the humanities and social sciences, including prehistory, history, art, archaeology, literature, and linguistics as well as interdisciplinary aspects of cultural history for applicants who have completed their academic training. Postdoctoral and dissertation research fellowships may be held for terms of four months to one year. Grants for tenures up to one year will be considered, but preference will be given to projects of shorter duration. ARIT maintains two research institutes in Turkey. ARIT-Istanbul has a research library focused on Byzantine, Ottoman, and modern studies of Turkey. ARIT-Ankara focuses on art, archaeology, and ancient history in its library and serves Turkish and American archaeologists through its programs. Both institutes have residential facilities for fellows and provide general assistance as well as introductions to colleagues, institutions, and authorities in Turkey. Predoctoral applicants also may qualify for ARIT's Kress fellowship, and postdoctoral scholars also may qualify for ARIT's NEH fellowships. Applicants will be notified of the fellowship committee's decision by January 25.
Requirements: Applicants must be U.S. citizens or a three-year resident of the U.S., and have completed formal academic training affiliated with U.S. or Canadian institutions. Scholars and advanced graduate students engaged in research on ancient, medieval, or modern times in Turkey, in any field of the humanities and social sciences, are eligible. Student applicants must have fulfilled all preliminary requirements for the doctorate except the dissertation. Turkish law requires all foreign scholars, prior to entering the country, to obtain formal permission for any research to be carried out in Turkey.
Geographic Focus: All States
Date(s) Application is Due: Nov 1
Amount of Grant: 16,800-50,400 USD
Contact: Nancy Leinwand, (215) 898-3474; fax (215) 898-0657; leinwand@sas.upenn.edu
Internet: http://ccat.sas.upenn.edu/ARIT/FellowshipPrograms.htm
Sponsor: American Research Institute in Turkey
3260 South Street
Philadelphia, PA 19104-6324

ARIT Fellowships for Intensive Advanced Turkish Language Study 992
ARIT offers fellowships for 10 advanced students for participation in the summer program in intensive advanced Turkish language at Bogazici University in Istanbul. This intensive program offers the equivalent of one full academic year of study in Turkish at the college level. ARIT fellowship- supported courses are offered at the advanced level. Classes are conducted in Turkish, with informal and formal styles introduced and reviewed through instruction, language laboratory work, and open conversations with teaching assistants. Students meet with teaching assistants on an informal basis for additional instruction and free conversation. Participants also attend extracurricular activities including films, lectures, and cultural events both on- and off-campus. The fellowships cover round-trip airfare to Istanbul, application and tuition fees, and a maintenance stipend. Full-time students and scholars affiliated at academic institutions are eligible to apply. Guidelines and application are available online.
Requirements: Applicant must be a U.S. citizen, national, or permanent resident; be currently enrolled in an undergraduate or graduate-level academic program, or be faculty; have a minimum B average, if still a student; and perform at the high-intermediate level on a proficiency-based admissions examination.
Geographic Focus: All States
Date(s) Application is Due: Feb 1
Contact: Erika H. Gilson, Director; ehgilson@princeton.edu
Internet: http://ccat.sas.upenn.edu/ARIT/ARITSummerLanguageProgram.htm
Sponsor: American Research Institute in Turkey
3260 South Street
Philadelphia, PA 19104-6324

ARIT Kenan T. Erim Fellowship for Research at Aphrodisias 993
The institute invites applications for a fellowship to support excavation and/or research in art history and archaeology to be carried out at the site of Aphrodisias in Turkey during the summer. Applicants must submit an application, a letter of acceptance from a director of the excavations at Aphrodisias, and two letters of reference by the listed application deadline. Graduate students should supply a copy of their graduate transcript. Guidelines are available online.
Requirements: Scholars or advanced graduate students engaged in excavation at the site of Aphrodisias or research on material from that site are eligible to apply. Fields of study include the history of art and architecture from antiquity to the present, and archaeology. Graduate student applicants must have fulfilled all preliminary requirements for the doctorate except the dissertation by June, and before beginning any ARIT-sponsored research.
Geographic Focus: All States
Date(s) Application is Due: Nov 1
Amount of Grant: Up to 2,375 USD
Contact: Nancy Leinwand, (215) 898-3474; fax (215) 898-0657; leinwand@sas.upenn.edu
Internet: http://ccat.sas.upenn.edu/ARIT/FellowshipPrograms.htm
Sponsor: American Research Institute in Turkey
3260 South Street
Philadelphia, PA 19104-6324

ARIT Mellon Postdoctoral Research Fellowships for East European Scholars 994
The postdoctoral fellowships are intended to bring scholars from Central and Eastern Europe into the broader research community, specifically into Turkey. Fellowships, offered in all areas of the humanities and social sciences, have a tenure of two to three months for research to be carried out in Turkey. Preference will be given to scholars in the early stages of their careers. Fellows are expected to devote full time to their projects and to participate

in the activities of the institute. Turkish law requires all foreign scholars, prior to entering the country, to obtain formal permission for any research to be carried out. Fellowship recipients are personally responsible for obtaining their own research permission and may acquire forms and procedures through the Turkish diplomatic office within each country included in the competition. Replies for permission may take six months or more. Applicants may contact the office for advice about research permission.
Requirements: The fellowships are available to Czech, Hungarian, Polish, Slovak, Bulgarian, and Romanian scholars holding the PhD or its equivalent who are engaged in advanced research in any field of the social sciences or the humanities involving Turkey. Scholars must be permanent residents of one of the included countries.
Restrictions: Applicants may not have held a prior fellowship under this program.
Geographic Focus: All States, Czech Republic
Date(s) Application is Due: Mar 5
Amount of Grant: 23,000-34,500 USD
Contact: Nancy Leinwand, (215) 898-3474; fax (215) 898-0657; leinwand@sas.upenn.edu
Internet: http://ccat.sas.upenn.edu/ARIT/MellonFellowships.htm
Sponsor: American Research Institute in Turkey
3260 South Street
Philadelphia, PA 19104-6324

Arizona Cardinals Grants **995**
The National Football League franchise supports programs designed to improve the quality of life and enhance opportunities for children, women, and minorities in the state of Arizona. Specific areas of interest include arts and culture, civic affairs, education, health, science, and social services. The foundation is interested in expanding its giving and looks for new charities to fund. First-time applicant organizations generally will receive grants of $5000 or less.
Requirements: Applicants must be exempt under 501(c)3 of the Internal Revenue Service code.
Geographic Focus: Arizona
Date(s) Application is Due: Aug 1
Amount of Grant: 2,000-5,000 USD
Samples: AZCenters for Comprehensive Education and Life Skills; Chrysalis Shelter for Victims of Domestic Violence; Child crisis Center-East Valley; Jewish Family and Children's Services; Juvenile Diabetes Research Foundation; United Food Bank.
Contact: Pat Tankersley, (602) 379-0101; fax (480) 785-7327
Internet: http://www.azcardinals.com/community/charities.php
Sponsor: Arizona Cardinals
P.O. Box 888
Phoenix, AZ 85001-0888

Arizona Commission on the Arts Education Projects Grants **996**
The Arizona Commission on the Arts is committed to making the arts fundamental to education, particularly in programs that serve Pre-K-12 students, classroom teachers, teaching artists, arts specialists and administrators in school, after-school and summer/inter-session programs. Our goal is that applicants present a plan that creatively stimulates arts education in their school/community organizations.
Requirements: An applicant organization must be a 501(c)3 Arizona nonprofit organization or school, or a unit of government. An unincorporated Arizona organization may apply through a fiscal agent, providing that the fiscal agent is an Arizona 501(c)3 or a governmental organization. Grant applications must be submitted online.
Restrictions: No more than one Arts Learning project per individual site is funded per year. Support for staff salaries, funding for insurance or supplies are not provided through this grant.
Geographic Focus: Arizona
Date(s) Application is Due: Mar 22
Contact: Alison Marshall; (602) 771-6523; amarshall@azarts.gov
Internet: http://www.azarts.gov/index.htm
Sponsor: Arizona Commission on the Arts
417 W Roosevelt Street
Phoenix, AZ 85003

Arizona Commission on the Arts Folklorist Residencies **997**
In this residency program, folklorists guide students and teachers in the process of researching, identifying, and documenting traditional art forms in their community. Students develop a new sense of community as they learn and experience the significance of family, community, and cultural traditions directly from parents and family members, employers and workers, senior citizens, public officials, and other community members. Residency lengths are variable and are built around a combination of services: training for students and teachers in research methodology; fieldwork conducted by the folklorist with participating students and teachers; classroom workshops; lecture-demonstrations on traditional art forms; interviews with local traditional artists; and presentation of traditional artists/art forms in school or community settings. A travel, lodging, and meal subsidy is provided for artists funded for out-of-town residencies in addition to their payment for services.
Requirements: 501(c)3 tax-exempt Arizona nonprofit organizations, schools, or units of government are eligible. Schools must select an individual artist, performing company, folklorists, or interdisciplinary artists from the Arizona Artist Roster.
Geographic Focus: Arizona
Amount of Grant: 500-3,500 USD
Contact: Robert Booker; (602) 771-6501; fax (602) 256-0282; rbooker@azarts.gov
Internet: http://www.azarts.gov/localarts/index.htm#
Sponsor: Arizona Commission on the Arts
417 W Roosevelt Street
Phoenix, AZ 85003

Arizona Public Service Corporate Giving Grants **998**
Grants are awarded to support Arizona nonprofit organizations in the areas of health and human services, community development, arts and culture, education, and environment. The foundation awards support for project grants, capital building funds, research, employee matching gifts, in-kind services, conferences and seminars, and operating support. Applications are accepted on an ongoing basis.
Requirements: Arizona 501(c)3 nonprofits are eligible.
Restrictions: APS Corporate Giving does not fund individual request, charter or private schools, religious, political fraternal, legislative or lobbying efforts to organizations, travel-related or hotel expenses, private or family foundation, private non-profit organizations, salaries and/or debt reduction.
Geographic Focus: Arizona
Contact: Cindy Slick; (602) 250-4707; fax (602) 250-2113; Cindy.Slick@aps.com
Internet: http://www.aps.com/main/community/dev/default.html
Sponsor: Arizona Public Service Corporation
P.O. Box 53999, MS 8010
Phoenix, AZ 85072-3999

Arkansas Arts Council Individual Artists Fellowships **999**
The Individual Artist Fellowships are unconditional, non-matching awards made directly to individual artists in recognition of their artistic abilities. Up to nine fellowships of $4,000 each may be awarded. Awards are based on the creative excellence of works submitted for review, and are awarded to a limited number of artists of outstanding talent. Artist categories for funding vary from year to year (literary arts, performing arts, visual arts, etc.). The application and detailed guidelines for submitting documentation are available at the website.
Requirements: Applicants must be no younger than 25 and residents of Arkansas for at least one year prior to application. They must also be legal U.S. citizens or prove eligibility for full-time, permanent employment. Work submitted must be original and created within the past five years. If the creation involves collaboration, the applicant must be the primary creator and identify his or her role and the amount and type of participation contributed in creating the work. In all categories, it is best to submit documentation materials that show a concentration rather than a range of work.
Restrictions: An application may submit only one individual artist application each year. Previous individual fellowship winners are not eligible to apply. Current Arkansas Arts Council advisory board members, Department of Arkansas Heritage staff, or an immediate family members are also not eligible.
Geographic Focus: Arkansas
Date(s) Application is Due: Apr 19
Amount of Grant: 4,000 USD
Contact: Robin Muse McClea, (501) 324-9348; robinm@arkansasheritage.org
Internet: http://www.arkansasarts.org/grants/iaf.aspx
Sponsor: Arkansas Arts Council
323 Center Street, Suite 1500
Little Rock, AR 72201-2606

Armenian Students' Association Scholarships **1000**
The association awards undergraduate and graduate scholarships to applicants of Armenian ancestry. Scholarship applications must be requested (obtain form from office or the Web site—application fee required) by February 15; applications must be received March 15. Guidelines and request form are available online.
Requirements: Students must be enrolled full-time in four-year accredited colleges or universities in the United States and must be of Armenian descent. The student must be a sophomore or beyond in the current academic year.
Geographic Focus: All States
Date(s) Application is Due: Feb 15; Mar 15
Contact: Scholarship Administrator; (401) 461-6114; headasa@aol.com
Internet: http://www.asainc.org/national/scholarships.shtml
Sponsor: Armenian Students' Association of America
333 Atlantic Avenue
Warwick, RI 02888

ARM Faculty Fellowships **1001**
The fellowships provide opportunities for college and university faculty members in science and engineering to participate in research and collaborative activities at government and industrial research facilities. Sabbatical and travel awards also are available. Fellowship duration varies with the research project needs and facility approval. Award funding may be allocated for a stipend up to the faculty member's certified college/university salary, travel expenses to and from the facility, and a modest relocation allowance. Award funding may also be allocated for research participation by undergraduate or graduate students. Applications may be submitted at any time.
Requirements: The program is open to full-time faculty members at an accredited college or university. Citizenship restrictions may apply for some awards or facilities. Selection is based on applicant's professional qualifications, references, strength of the research proposal, and the expected benefit of the fellowship experience to the applicant, the applicant's home institution, and the host facility. Preference is given to faculty redirecting their research or establishing new collaborations with facility scientists.
Geographic Focus: All States
Contact: Mariann Bleazard; (801) 273-8900; fax (801) 277-5632; info@awu.org
Internet: http://www.awu.org/faculty_default.htm
Sponsor: Academia Resource Management
535 E 4500 S, Suite D120
Salt Lake City, UT 84107

ARM Graduate Fellowships **1002**
Graduate student fellowships provide master's and doctoral degree candidates in science and engineering the opportunity to conduct research toward their thesis or dissertation at one of more than 50 cooperating government and industrial facilities. Institutional affiliation and citizenship restrictions may apply for some awards or facilities. The fellowship term ranges from one to 12 months. The award includes a monthly stipend, tuition assistance, and travel allowance.Renewals are competitive and require reapplication annually. Applications may be submitted at any time. To insure optimum consideration submit by February 22 for summer, by March 20 for fall, and by October 20 for spring.
Requirements: The program is open to master's and doctoral degree candidates enrolled in an accredited graduate program. Institutional affiliation and citizenship restrictions may apply for some awards or facilities.
Geographic Focus: All States
Amount of Grant: 1,300-15,600 USD
Contact: Mariann Bleazard; (801) 273-8900; fax (801) 277 5632; info@awu.org
Internet: http://www.awu.org/appl_forms/descriptions/lg.htm
Sponsor: Academia Resource Management
535 E 4500 S, Suite D120
Salt Lake City, UT 84107

ARM Postgraduate Fellowships **1003**
The fellowships provide opportunities for post-master's and postdoctoral research associates to participate in research at national laboratories and other research facilities. Stipends are established by the host facility, and travel and relocation expenses may be available. The initial fellowship term is up to one year but may be renewed up to two additional years upon mutual agreement by the fellow and the facility and the availability of funds. Applicants must have identified a host scientist and a research project at a cooperating facility before sending their completed application to AWU. Applications may be submitted at any time but Applications may be submitted at any time, however, it is recommended that applications are submitted two to three months prior to the requested start date.
Requirements: Candidates must have been awarded the master's or doctoral degree within four years of applying. A commitment to a professional career in science or engineering research is expected. U.S. citizenship or permanent resident status is required. Exceptions to these requirements are subject to the approval of the host facility.
Geographic Focus: All States
Contact: Mariann Bleazard; (801) 273-8900; fax (801) 277-5632; info@awu.org
Internet: http://www.awu.org/appl_forms/descriptions/pd.htm
Sponsor: Academia Resource Management
535 E 4500 S, Suite D120
Salt Lake City, UT 84107

ARM Visiting Scientist Awards **1004**
AWU administers awards to Visiting Scientists who are industrial associates and collaborators in science, mathematics, engineering, and technology. They provide the opportunity to participate in and contribute to science and technology at sponsoring facilities and are designed to encourage professional development and provide special expertise to the host facility. Stipend is established by the host facility and varies by experience and discipline; usually, it is based on the applicant's current salary. Funds for relocation, supplemental travel, and professional expenses may be available. Applicants must have identified a host scientist and a project at a cooperating facility before sending their application to AWU. Applications may be submitted at any time. However, it is recommended that you submit your application two to three months prior to the requested start date.
Requirements: Professionals with continued commitment to science and engineering. U.S. citizenship or permanent resident status is required. Nonresident aliens require the approval of the host facility and an appropriate visa status.
Geographic Focus: All States
Contact: Mariann Bleazard; (801) 273-8900; fax (801) 277-5632; info@awu.org
Internet: http://www.awu.org/appl_forms/descriptions/vs.htm
Sponsor: Academia Resource Management
535 E 4500 S, Suite D120
Salt Lake City, UT 84107

ARO Atomic, Molecular, and Optical (AMO) Physics Research Contracts **1005**
Research in atomic, molecular, and optical physics will create fundamentally new capabilities for the army, as well as providing the scientific underpinnings to enhance existing technologies. Topics of interest include unconventional optics for enhanced imaging and detection; atom optics and laser cooling and trapping for ultrasensitive detectors; nonlinear atomic and molecular processes for sensor protection and optical processing; and the investigation of plasmas for toxic gas destruction, pollution reduction, materials processing, and propellant ignition. Image science, automatic target recognition, and sensor fusion issues are also addressed in the AMO program. Proposals may be submitted at any time.
Requirements: Organizations and individuals interested in submitting research proposals are strongly encouraged to make preliminary inquiries as to the general need for the type of research effort contemplated before expending extensive effort in preparing a detailed research proposal or submitting proprietary information.
Geographic Focus: All States
Contact: Dr. Richard Hammond; (919) 549-4313; richard.hammond@us.army.mil
Internet: http://www.arl.army.mil/main/main/default.cfm?Action=29&Page=203#atomic_molecular_physics
Sponsor: Army Research Office
P.O. Box 12211
Research Triangle Park, NC 27709-2211

A Rocha U.S.A Grants **1006**
A Rocha U.S.A is part of an international family of conservation organizations working to show God's love for all creation through community-based efforts focused on science and research, practical conservation, environmental education, and environmental health. The organization works toward the conservation of habitats and species worldwide through environmental education, conservation activities, and scientific research; and seeks to build a sense of community among individuals, communities, and their environments through biblical faith. Fields of interest include: community and economic development, environmental health, environmental education, research, and Jewish agencies and synagogues. Giving is both national and international in scope. There are no specific application formats or deadlines with which to adhere, and potential applicants should contact the office directly.
Geographic Focus: All States
Contact: Thomas Rowley, Executive Director; (830) 992-7940; usa@arocha.org
Internet: http://www.arocha.org/int-en/index.html
Sponsor: A Rocha U.S.A
P.O. Box 1338
Fredericksburg, TX 78624-1338

ARO Conference and Symposia Grants **1007**
ARO supports conferences and symposia in special areas of science that bring experts together to discuss recent research or education findings or to expose other researchers or advanced graduate students to new research and educational techniques. ARO encourages the convening in the United States of major international conferences and assemblies of international alliances.
Requirements: Scientific, technical, or professional organizations that qualify for tax exemption under 501(c)3 may receive conference and symposia grants. Conference support proposals should be submitted a minimum of six months prior to the date of the conference.
Restrictions: Cosponsorship of conferences and symposia with industrial concerns is not eligible. Funds provided by the ARO cannot be used to support participants from communist countries, nor can they be used for payment to any federal government employee for support, subsistence, or services in connection with the proposed conference or symposium.
Geographic Focus: All States
Contact: ARO Legal Office; (919) 549-4292; Mark.rutter@us.army.mil
Internet: http://www.aro.army.mil
Sponsor: Army Research Office
P.O. Box 12211
Research Triangle Park, NC 27709-2211

ARO Degradation, Reactivity, and Protection of Materials Research Contracts **1008**
The program in degradation, reactivity, and protection of materials investigates the metastable behavior of materials prepared by nonequilibrium surface processing approaches, including high-pressure, shock, plasma, beam, and self-assembling techniques. The principal objective is to discover new nonequilibrium structural materials and novel smart materials, which will enhance the reliability of Army systems in service. This research includes nondestructive characterization of smart and ultra-hard/superstrong materials and in-situ process monitoring. Principal emphasis is on surface or interface control during processing of these materials, characterization of their near-surface transport behavior and surface properties, and modeling or theoretical predictions of their properties. The synthesis of novel degradation-resistant, ultra-high strength nanolaminates and other refractory materials is also of interest. Experimental and theoretical approaches are sought that provide for characterization of surfaces and interfaces, modification of surface and near-surface regions, and processing and materials characterization as this relates to the structure, wear, and environmental resistance of smart materials and ultrahard/superstrong structural materials. Other phenomena including spectroscopic and chemical analysis of low level hydrogen and hydrocarbons, where these are important for understanding the properties of such new materials during processing and in-service life assessment, are also of interest. Proposals may be submitted at any time.
Requirements: Organizations and individuals interested in submitting research proposals are strongly encouraged to make preliminary inquiries as to the general need for the type of research effort contemplated before expending extensive effort in preparing a detailed research proposal or submitting proprietary information.
Geographic Focus: All States
Contact: Dr. John Prater; (919) 549-4259; John.T.Prater@us.army.mil
Internet: http://www.aro.ncren.net/research/index.htm
Sponsor: Army Research Office
P.O. Box 12211
Research Triangle Park, NC 27709-2211

ARO Discrete Mathematics and Computer Science Research Contracts **1009**
The interests in discrete mathematics are the development and analysis of solution procedures for discrete problems in computational geometry, computational algebra, logic, network flows, graph theory and combinatorics. Specific areas of emphasis include robust geometric computation, solid modeling, multiresolution methods, parallel and distributed computing and dynamic interactive visualization techniques. Other areas of interest include distributed algorithms for network flows, randomization in computing, computational algebraic geometry techniques for solution of polynomial systems, discrete methods for combinatorial optimization, symbolic methods for differential equations, mixed symbolic-numerical methods for applied problems, parallel symbolic sparse matrix methods, and algorithmic methods in symbolic mathematics arising in, for example, automated reasoning systems, mathematical logic and formal language

theory. The interests in theoretical computer science include fundamental issues in parallel computing such as advanced data structures for parallel architectures, parallel algorithms, graph theoretic methods applied to a parallel and distributed computation and models and algorithms for the control of heterogeneous concurrent computing. Also of interest is research on tools for the development of parallel algorithms and expert systems for computation and visualization of solutions to partial differential equations. Exploring fundamental techniques that optimize I/O communication is a research area of great strategic importance. Proposals are accepted at any time.
Requirements: Organizations and individuals interested in submitting research proposals are strongly encouraged to make preliminary inquiries as to the general need for the type of research effort contemplated before expending extensive effort in preparing a detailed research proposal or submitting proprietary information.
Geographic Focus: All States
Contact: Michael Coyle, Mathematics and Computer Science Division Contact; (919) 549-4256; joseph.michael.coyle@us.army.mil
Internet: http://www.aro.ncren.net
Sponsor: Army Research Office
P.O. Box 12211
Research Triangle Park, NC 27709-2211

ARO Information Processing and Fusion and Circuits Research Contracts 1010
This program sponsors research the understanding of image background, including target-competitive clutter, and of how image background compounds the difficulty of target detection and recognition. Available models for background and clutter are currently inadequate. Objective measures of clutter and modeling paradigms that enable the quantification of image properties are needed for effective comparison of scenes, evaluation of algorithm performance, validation of synthetic imagery, and strategies for data fusion. Research is needed which addresses: (i) modeling of background and clutter, (ii) definition and assessment of clutter metrics, (iii) the manner in which clutter degrades the discrimination processes, and (iv) interaction between image background and targets. Proposals may be submitted at any time.
Requirements: Organizations and individuals interested in submitting research proposals are strongly encouraged to make preliminary inquiries as to the general need for the type of research effort contemplated before expending extensive effort in preparing a detailed research proposal or submitting proprietary information.
Geographic Focus: All States
Contact: Dr. William Sander; (919) 549-4241; william.sander@us.army.mil
Internet: http://www.aro.ncren.net
Sponsor: Army Research Office
P.O. Box 12211
Research Triangle Park, NC 27709-2211

ARO Minimum Energy Low Power Electronics Research Contracts 1011
The future army communications, information processing, imaging, and other systems must be portable, functional, versatile, and highly reliable. In addition, the systems will require ultra-high speed capability for handling complex voice, data, and video multimedia signal formats. Many of the new systems will find application in wireless communications, infrared imaging, and portable diagnostic and computational equipment. As these systems become more sophisticated, the necessary electronic circuitry becomes more complex, with increased prime power requirements. Currently, and in the near future, prime power for mobile and light weight system utilization is limited. It is necessary, therefore, to develop a new generation of electronic and other systems that can operate under minimum energy constraints with very low direct current (DC) power dissipation. This research topic goes beyond the simple desirability of efficient devices and circuits. It is aimed at new concepts resulting from considering the multilevel problem; materials, devices, circuits, coding, networking, and an overarching system of systems considerations, in order to reduce overall power dissipation in RF systems. Research can be approached from the component level or the overall design level. Proposals may be submitted at any time.
Requirements: Organizations and individuals interested in submitting research proposals are strongly encouraged to make preliminary inquiries as to the general need for the type of research effort contemplated before expending extensive effort in preparing a detailed research proposal or submitting proprietary information.
Geographic Focus: All States
Contact: Dr. William Clark; (919) 549-4314; William.W.Clark@us.army.mil
Internet: http://www.aro.ncren.net
Sponsor: Army Research Office
P.O. Box 12211
Research Triangle Park, NC 27709-2211

ARO Mobile, Wireless Communications and Networks Research Contracts 1012
The mobile, wireless communications and networks research program is concerned primarily with establishing the fundamental understanding necessary to support the Army's future mobile, wireless tactical battlefield communications needs. The research in this program primarily targets the tactical battlefield at brigade and below. The Army is interested in communication systems operating in frequency bands traditionally occupied by narrowband radios high frequency (HF), very high frequency (VHF), and ultrahigh frequency (UHF) as well as systems operating in frequencies extending into the millimeter wave region. These systems must support broad-based and highly mobile communications and must perform in environments of impressive diversity, from dense foliage to dense urban obstructions, and unintentional and intentional jamming. Future Army tactical communication systems for the digital battlefield will consist of many different types of networks and must be capable of communicating on the move. These systems will be highly mobile creating highly dynamic network topologies (mobile ad-hoc networks) and routing multimedia (voice, data and video) data. Unlike commercial systems, the communications infrastructure must be mobile. In addition to the highly mobile communications, there is interest in algorithms for small, very energy-limited, stationary, unattended ground sensors. Proposals may be submitted at any time.
Requirements: Organizations and individuals interested in submitting research proposals are strongly encouraged to make preliminary inquiries as to the general need for the type of research effort contemplated before expending extensive effort in preparing a detailed research proposal or submitting proprietary information.
Geographic Focus: All States
Contact: Dr. Robert Ulman; (919) 549-4330; Robert.Ulman@us.army.mil
Internet: http://www.aro.ncren.net
Sponsor: Army Research Office
P.O. Box 12211
Research Triangle Park, NC 27709-2211

ARO Quantum Information Science Research Contracts 1013
Quantum mechanics provides the opportunity to perform highly nonclassical operations that can result in exponential speed ups in computation or ultra-secure transmittal of information. This workpackage seeks to understand, control, and exploit such nonclassical phenomena for revolutionary advances in computation and secure communication. Major areas of interest are fundamental studies, quantum computation, and quantum communication. Proposals may be submitted at any time.
Requirements: Organizations and individuals interested in submitting research proposals are strongly encouraged to make preliminary inquiries as to the general need for the type of research effort contemplated before expending extensive effort in preparing a detailed research proposal or submitting proprietary information.
Geographic Focus: All States
Contact: Dr. Henry Everitt; (919) 549-4369; henry.o.everitt@us.army.mil
Internet: http://www.aro.ncren.net/research/index.htm
Sponsor: Army Research Office
P.O. Box 12211
Research Triangle Park, NC 27709-2211

ARO Short-Term Innovation Research Program (STIR) 1014
The objectives of the STIR program are to fund innovative ideas in basic research. Proposed research may be for the continuation of or the natural outgrowth of experimental or theoretical explorations.
Requirements: Research proposals are sought from educational institutions, nonprofit organizations, and private industry. Prospective offerors of a STIR proposal are encouraged to contact the ARO program manager in their area of interest to ascertain the extent of interest in the specific research project. Research must be completed within six months of award of the agreement.
Restrictions: No capital equipment may be purchased. Travel costs must not exceed $500. Report preparation costs must not exceed $100. The assessment of indirect costs or fee is unallowable.
Geographic Focus: All States
Amount of Grant: Up to 50,000 USD
Contact: Dr. Jim Chang, Director; (919) 549-4201; jim.chang@us.army.mil
Internet: http://www.aro.army.mil/research
Sponsor: Army Research Office
P.O. Box 12211
Research Triangle Park, NC 27709-2211

ARO Software and Knowledge-Based Systems Research Contracts 1015
The program in software and knowledge-based systems (SKBS) addresses the theoretical bases for the analysis, design, development, and evolution of advanced information-based systems that enable significant improvements in state-of-the-art engineering for software, software-based applications (including modeling and simulation), machine learning, knowledge/acquisition/representation/synthesis, and knowledgebases/databases. Currently, the research supported by the program is in software prototyping, development and evolution, formal methods for software engineering, and knowledgebase/database sciences. Proposals may be submitted at any time.
Requirements: Organizations and individuals interested in submitting research proposals are strongly encouraged to make preliminary inquiries as to the general need for the type of research effort contemplated before expending extensive effort in preparing a detailed research proposal or submitting proprietary information.
Geographic Focus: All States
Contact: Dr. David Hislop, Mathematics and Computer Science Division Contact; (919) 549-4255; David.W.Hislop@us.army.mil
Internet: http://www.aro.ncren.net
Sponsor: Army Research Office
P.O. Box 12211
Research Triangle Park, NC 27709-2211

ARO Synthesis and Processing of Materials Research Contracts 1016
The program on synthesis and processing of materials focuses on the use of innovative approaches for processing high-performance structural materials reliably and at lower costs. Emphasis is placed on the design and fabrication of new materials with specific microstructure, constitution, and properties. Research interests include experimental and theoretical modeling studies to understand the influence of fundamental parameters on phase formation, microstructural evolution, and the resulting properties, in order to predict and control materials structures at all scales ranging from atomic dimensions to

macroscopic levels. Trends in this subfield include nonequilibrium materials processing (e.g., rapid solidification); powder synthesis and consolidation; novel processing of ceramics, polymers, metals and composites; welding and joining including composite materials; elastomers; fibers and fabrics; and utilization of microstructural, compositional, or other unique signatures which may provide non-destructive in situ feedback process control to enhance product reproducibility and quality. Supercritical fluid, shock-induced chemical processing and other innovative approaches for processing materials are also of interest. Proposals may be submitted at any time.
Requirements: Organizations and individuals interested in submitting research proposals are strongly encouraged to make preliminary inquiries as to the general need for the type of research effort contemplated before expending extensive effort in preparing a detailed research proposal or submitting proprietary information.
Geographic Focus: All States
Contact: Dr. William Mullins, (919) 549-4286; William.Mullins@us.army.mil
Internet: http://www.aro.army.mil/matsc/matsc.htm
Sponsor: Army Research Office
P.O. Box 12211
Research Triangle Park, NC 27709-2211

ARO Theoretical Physics and Nonlinear Phenomena Research Contracts **1017**
The theoretical physics and nonlinear phenomena program is very closely coupled to experimental science as well as to ARO's programs in mathematics, chemistry, biological chemistry, materials science, and engineering sciences. The program thus encompasses a broad base including research in electron physics, photon physics, and classical and quantum mechanical systems, and statistical physics. It includes first-principles derivations of thermomechanical strengths of alloys for armor and armor penetrators; and electronic band structure calculations of materials for electronic, magnetic, optical, and optoelectronics applications, including those that result from quantum well and multiquantum well structures for signal generation, signal processing, and propagation and detection of signals. Also of interest are many-body theoretic approaches that address the electron correlation problem in extended molecular and condensed matter systems to provide the means to predict reaction kinetics, nonequilibrium dynamics, and application to the "alloy problem. There is interest in quantum optics research to explore the role of coherent states, squeezed states, etc. which may provide new tools for improved information processing and means to control information. Statistical physics interests go beyond thermodynamics, into nonequilibrium structures and their metastability, into information theoretic formulations, and into decision algorithms to connect the underlying physics to real world applications via proper modeling, instrumentation, and data analysis. Subtopics of interest include theoretical condensed matter physics, nonlinear dynamics, and nonequilibrium dynamics. Proposals may be submitted at any time.
Requirements: Organizations and individuals interested in submitting research proposals are strongly encouraged to make preliminary inquiries as to the general need for the type of research effort contemplated before expending extensive effort in preparing a detailed research proposal or submitting proprietary information.
Geographic Focus: All States
Contact: Dr. Mikael Ciftan; (919) 549-4236; Mikael.Ciftan@us.army.mil
Internet: http://www.aro.army.mil
Sponsor: Army Research Office
P.O. Box 12211
Research Triangle Park, NC 27709-2211

ARO Young Investigator Program/Presidential Early Career Awards **1018**
The objective of the Young Investigator Program (YIP) is to attract to army research outstanding young university faculty members, to support their research and to encourage their teaching and research careers. Outstanding YIP projects may be considered for a Presidential Early Career Award for Scientists and Engineers (PECASE). PECASE awards are the highest honor bestowed by the Army on outstanding scientists and engineers beginning their independent careers. Awards are granted in all research topic areas in the army Research Office. Support for the YIP is limited to three years; PECASE support is limited to five years. Proposals may be submitted at any time.
Requirements: This program is open to U.S. citizens holding tenure-track positions at U.S. universities and colleges who have held their graduate degrees (PhD or equivalent) for fewer than five years at the time of application. ARO strongly encourages informal discussions with the cognizant ARO technical program manager before submission of a formal proposal.
Geographic Focus: All States
Contact: Kurt Preston, (919) 549-4234; Kurt.Preston@us.army.mil
Internet: http://www.aro.ncren.net
Sponsor: Army Research Office
P.O. Box 12211
Research Triangle Park, NC 27709-2211

Arronson Foundation Grants **1019**
The foundation supports nonprofit organizations in the areas of religion, including churches (Baptist, Christian, Jewish, Roman Catholic, and Salvation Army) and religious education, higher education, health care, hospices, Jewish welfare, international ministries/missions, and youth. The foundation awards grants primarily in New York, NY; Philadelphia, PA; and Israel. Types of support include endowment funds, general operating support grants, research, scholarships, and seed money grants. Applicants should submit a brief letter of inquiry and include information on the organization and its work. There are no application deadlines.
Requirements: Nonprofit organizations in Pennsylvania, with emphasis on the Philadelphia area, are eligible to apply.
Geographic Focus: Pennsylvania
Amount of Grant: 200-25,000 USD
Contact: Joseph Kohn; (215) 238-1700 or (215) 238-1968; jkohn@kohnswift.com
Sponsor: Arronson Foundation
1 S Broad Street, Suite 2100
Philadelphia, PA 19107

ARS New Investigator Award **1020**
The purpose of this award is to support basic, translational, or clinical research projects in rhinology. Proposed projects may be related to any area of rhinology. Proposed project shall be designed in collaboration with a preceptor investigator and approved by the candidate's department chairperson and institution. One award of up to $25,000 over a two-year period will be given annually. Letters of Interest should be received no later than December 15, with full proposals due on January 15.
Requirements: Any member of the American Rhinologic Society who has not received previous, significant outside funding is eligible. The applicant must have, as a mentor, an established researcher who will provide a letter of support stating the extent of involvement in the project and provide a summary of his/her research experience.
Geographic Focus: All States
Date(s) Application is Due: Jan 15; Dec 15
Amount of Grant: Up to 25,000 USD
Contact: Stephanie L. Jones; (703) 519-1586 or (703) 836-4444; sljones@entnet.org
Internet: http://www.entnet.org/EducationAndResearch/coreGrants.cfm
Sponsor: American Rhinology Society
One Prince Street
Alexandria, VA 22314-3357

ARS Resident Research Grants **1021**
The purpose of this award is to support basic, translational, or clinical research projects in rhinology. Proposed projects may be related to any area of rhinology. Proposed project shall be designed in collaboration with a preceptor investigator and approved by the candidate's department chairperson and institution. Two awards of up to $8,000 over a one-year period will be given annually. Letters of Interest should be received no later than December 15, with full proposals due on January 15.
Requirements: Any resident in training in an approved program in the U.S. or Canada is eligible to apply for the American Rhinologic Society Research Grant. Resident applicants must have as a co-investigator a supervising faculty who is a member in good standing in the American Rhinologic Society (ARS).
Geographic Focus: All States
Date(s) Application is Due: Jan 15; Dec 15
Amount of Grant: Up to 8,000 USD
Contact: Stephanie L. Jones; (703) 519-1586 or (703) 836-4444; sljones@entnet.org
Internet: http://www.entnet.org/EducationAndResearch/upload/2009-ARS-Resideint-Research-Grant.pdf
Sponsor: American Rhinology Society
One Prince Street
Alexandria, VA 22314-3357

Arthritis Foundation of Australia Grants **1022**
The foundation awards grants in the fields of medicine and health, with emphasis on arthritis research and education. Types of awards include grants-in-aid, fellowships, and scholarships. Grants are awarded in Australia, the United Kingdom, and the United States. Annual deadline dates may vary; contact program staff for exact dates.
Requirements: Awards are available to clinical, scientific, and allied health professionals who are Australian citizens or permanent residents. Scholarships, fellowships, and grants may only be held in universities, hospitals, or recognized research institutes.
Geographic Focus: All States
Date(s) Application is Due: Jun 30
Contact: Coordinator; 61 (0)2 9552 6085; info@arthritisaustralia.com.au
Internet: http://arthritisaustralia.com.au/Research%20Grants
Sponsor: Arthritis Foundation of Australia
P.O. Box 121
Sydney, NSW 2001 Australia

Arthur and Rochelle Belfer Foundation Grants **1023**
The foundation awards grants to nonprofit organizations of the Jewish faith, with a focus on New York. Grants are targeted toward programs supporting the elderly and women; education/higher education; institutions such as seminaries, synagogues, and temples; hospitals; Jewish welfare; and medical centers. Types of support include general support grants and fellowships. There are no application deadlines. Applicants should send a brief letter of inquiry describing the program.
Restrictions: Grants are not made to individuals.
Geographic Focus: New York
Amount of Grant: 1,000-100,000 USD
Samples: Joan S. Brugge, PhD, chair, Department of Cell Biology, arvard Medical School—for breast cancer studies; Dana-Farber Cancer Institute (Boston, MA)—for general support, $1 million; Anti-Defamation League of B'nai B'rith (New York, NY)—for operating support, $22,000; American Friends of Israel Museum (New York, NY)—for operating support, $5000.
Contact: Robert Belfer, President; (212) 508-6020
Sponsor: Arthur and Rochelle Belfer Foundation
767 Fifth Avenue, 46th Floor
New York, NY 10153-0002

Arthur F. and Alice E. Adams Charitable Foundation Grants 1024
Established in Florida in 1987, the Arthur F. and Alice E. Adams Charitable Foundation is managed by a board of governors with the Wells Fargo Bank acting as corporate governor. The Foundation offers support in the arts, education, and human services. Primary fields of interest include: performing arts centers, opera, and medical research. Grants range from $5,000 to $250,000, with giving throughout the United States. The annual deadline is February 15, though board members meet in both May and November. Applicants should forward a letter describing their need and budget.
Requirements: 501(c)3 organizations located in, or serving the residents of, the following areas are eligible: Miami, Florida; New York, New York; and Memphis, Tennessee.
Restrictions: No grants are awarded to individuals.
Geographic Focus: All States
Amount of Grant: 5,000-250,000 USD
Contact: Paul L Guiabo; (336) 747-8186; grantadministration@wellsfargo.com
Sponsor: Arthur F. and Alice E. Adams Charitable Foundation
1525 W. WT Harris Boulevard
Charlotte, NC 28288-5709

Arthur J. Ennis Fellowships 1025
The Arthur J. Ennis Fellowships in the Augustine and Culture: The Villanova Seminar Program were established in 1994 by the College of Liberal Arts and Sciences. Each is a one-year fellowship renewable for up to three years. These positions are available as of August 15. Each holder of the fellowship will teach three courses each semester in the College of Arts and Sciences Villanova Center for Liberal Education (VCLE).
Requirements: Candidates for these positions must have a PhD in a Humanities discipline, a strong commitment to undergraduate teaching in the humanities, and an interest in teaching the thought of Augustine and its influence. Candidates with an interest in Hebrew Scriptures and the New Testament are also encouraged to apply.
Geographic Focus: Pennsylvania
Contact: John A. Doody; (610) 519-4691 or (610) 519-7325; john.doody@villanova.edu
Internet: http://www.villanova.edu/artsci/vcle/fellowships/ennis.htm
Sponsor: Villanova University
St. Augustine Center Liberal Arts Room 104, 800 Lancaster Avenue
Villanova, PA 19085

Arthur W. Adamson Award for Distinguished Service in the Advancement of Surface Chemistry 1026
This annual award, supported by Occidental Petroleum Corporation, recognizes distinguished services in the advancement of surface chemistry. Activities recognized by the award may include such fields as teaching, writing, research, and administration. The award consists of $5,000, a medallion with a presentation box, and a certificate.
Requirements: Any individual, except a member of the award committee, may submit one nomination or seconding letter for each award in any given year. The letter of nomination should contain an evaluation of the nominee's accomplishments and a specific identification of the work to be recognized, and may not exceed 1000 words in length.
Restrictions: Self-nominations are not accepted.
Geographic Focus: All States
Date(s) Application is Due: Nov 1
Contact: Felicia Dixon, Awards Administrator; (202) 872-6283 or (202) 872-4575; fax (202) 776-8008; f_dixon@acs.org or awards@acs.org
Internet: http://portal.acs.org/portal/acs/corg/content?_nfpb=true&_pageLabel=PP_ARTICLEMAIN&node_id=1319&content_id=CTP_004494&use_sec=true&sec_url_var=region1
Sponsor: American Chemical Society
1155 Sixteenth Street, NW
Washington, DC 20036-4801

ArvinMeritor Foundation Health Grants 1027
The foundation provides grants primarily in company-operating locations in the areas of education and training, civic and health, youth organizations, and arts and culture. In the area of Health, the foundation supports programs designed to promote health research, treatment, education, and awareness; and reduce the cost of illness. Types of support include general operating budgets, building funds, continuing support, equipment, and projects. The committee foundation meets every six to eight weeks to review grant requests. Contact program staff for current guidelines.
Requirements: Nonprofit 501(c)3 organizations should submit a one- to two-page letter outlining the purpose and needs of the program, its budget, duration, goals, leadership, and amount requested.
Restrictions: Ineligibility applies to individuals; organizations that limit participation or services based on race, gender, religion, color, creed, age, or national origin; projects without ties to a community that is home to an ArvinMeritor facility; organizations that pose any conflict with the goals and mission of ArvinMeritor, its employees, communities, or products; operating expenses for United Way local agencies, except through the foundation's support of annual United Way campaigns; sponsorships of fund raising activities by individuals (i.e., walk-a-thons); requests for loans or debt retirement; religious or sectarian programs for religious purposes; labor, political, or veterans organizations; fraternal, athletic, or social clubs; or seminars, conferences, trips, and tours.
Geographic Focus: All States
Contact: Jerry Rush; (248) 435-7907; fax (248) 245-1031; jerry.rush@arvinmeritor.com
Internet: http://www.arvinmeritor.com/community/community.asp
Sponsor: ArvinMeritor Foundation
2135 W Maple Road
Troy, MI 48084

ASA & NCHS Research Fellowships 1028
The program is designed to bridge the gap between academic scholars and the federal government's health research programs. Fellowship recipients conduct research in Washington, DC, at the National Center for Health Statistics (NCHS), where they may use NCHS data and interact with NCHS staff. The 12-month fellowship may be extended one additional year under special circumstances. The stipend received is commensurate with current faculty salaries plus moving expenses and a travel allowance. Each applicant is required to submit a curriculum vita; the names and addresses of three references who may be contacted; and two copies of a detailed research proposal that includes a one-half page abstract, statement of relevant work already accomplished, statement citing the significance of the expected results to NCHS, and resource requirements.
Requirements: Applicants must be U.S. citizens.
Restrictions: U.S. government employees are ineligible.
Geographic Focus: All States
Date(s) Application is Due: Nov 30
Contact: Joyce Narine, Grants Manager; (703) 684-1221, ext. 1837; joyce@amstat.org
Internet: http://www.amstat.org/careers/fellowshipsgrants.cfm
Sponsor: American Statistical Association
732 N Washington Street
Alexandria, VA 22314-3415

ASA/BJS Statistical Methodological Research Grants 1029
The American Statistical Association (ASA) Committee on Law and Justice Statistics announces a small grant program for the analysis of Bureau of Justice Statistics (BJS) and other justice-related data. Award preference is given to research using BJS data that addresses the priority topics listed below. This program is designed to encourage the creative and appropriate use of these data to inform substantive and methodological issues. Dissertation research may qualify for these funds and young investigators are encouraged to apply. Research is to be completed within a two-year period. Awards are typically in the range of $25,000 to $30,000, although exceptional projects as high as $40,000 will be considered.
Geographic Focus: All States
Date(s) Application is Due: May 23
Contact: Joyce Narine, Grants Manager; (703) 684-1221, ext. 1837; joyce@amstat.org
Internet: http://www.amstat.org/careers/bjs.cfm
Sponsor: American Statistical Association
732 N Washington Street
Alexandria, VA 22314-3415

ASA/NSF/BLS Fellowships 1030
ASA, under a grant from the National Science Foundation and with cosponsorship by the Bureau of Labor Statistics (BLS), seeks senior researchers for this program. Fellows conduct research in residence at BLS, gaining unique opportunities for use of BLS data and interaction with bureau staff. Fellows may pursue research in any area related to BLS data or methodology. Possible areas for research include, but are not limited to, areas under the fields of economic and social measurement and research, and statistical methodology and computing. Any proposed project should have the potential of encouraging further significant, broadly based research and should require hands-on access to BLS data and/or have direct application to BLS programs. Salary is commensurate with qualifications; fringe benefits and a travel allowance are included. A program brochure is available.
Requirements: Applicants for fellowships should have recognized research records and considerable expertise in their area of proposed research. Qualified women and members of minority groups are encouraged to apply for fellowships and associateships.
Geographic Focus: All States
Date(s) Application is Due: Dec 10
Contact: Joyce Narine, Grants Manager; (703) 684-1221, ext. 1837; joyce@amstat.org
Internet: http://www.amstat.org/careers/fellowshipsgrants.cfm
Sponsor: American Statistical Association
732 N Washington Street
Alexandria, VA 22314-3415

ASA/NSF/Census Research Fellowships 1031
The ASA/NSF/Census Bureau Research Program helps bridge the gap between government and academic science by bringing researchers closer to the production of data sets relevant to their research. The program allows for senior statisticians, social scientists, computer scientists, geographers, and others to come to the U.S. Census Bureau as research fellows for 6 to 12 months to use bureau data sets and interact with bureau staff.
Requirements: Applicants should have recognized research records and considerable expertise in their areas of proposed research. They must submit detailed research proposals for competitive evaluation.
Geographic Focus: All States
Date(s) Application is Due: Dec 10
Contact: Joyce Narine, Grants Manager; (703) 684-1221, ext. 1837; joyce@amstat.org
Internet: http://www.amstat.org/careers/fellowshipsgrants.cfm
Sponsor: American Statistical Association
732 N Washington Street
Alexandria, VA 22314-3415

ASA/NSF Federal Statistics Fellowships 1032
The ASA/NSF/Federal Statistics Fellowship program is jointly supported by the National Science Foundation, U.S. Census Bureau, and the Bureau of Labor Statistics. The general objective of the program is to foster collaborative and interdisciplinary research efforts that will continue to stimulate the development and advancement of methodology and social

science research relevant to issues on which federal statistical agencies seek to provide information. The program accomplishes this by bringing academic researchers to work with statisticians and social scientists in the three federal agencies for up to one year.
Geographic Focus: All States
Date(s) Application is Due: Mar 1
Contact: Dr. Phillip S. Kott, Chief Research Statistician; (703) 877-8000, ext. 102; fax (703) 877-8042; phillip_kott@nass.usda.gov
Internet: http://www.amstat.org/careers/fellowshipsgrants.cfm
Sponsor: American Statistical Association
732 N Washington Street
Alexandria, VA 22314-3415

ASA and USDA/NASS Research Fellow and Associate Program **1033**
The purpose of this program is to supplement research in statistics with experience at NASS. The program is designed to provide the selected Fellows and Associates with research opportunities or experience in the application of statistical theory in all phases of large scale agricultural survey operations, including design, collection, quality control, forecasting, estimation, and analysis. Participants in the program will engage in research which is of mutual interest to the participant and NASS. Fellows and Associates conduct research in residence at NASS, gaining unique opportunities for use of extensive NASS data and interaction with NASS staff.
Requirements: Applicants for the fellowship program should have a Ph.D. and have established recognized research record and considerable expertise in their area of proposed research. Applicants for Associates are expected to have completed a Ph.D. in an appropriate field or to have made significant progress toward the degree (at least two years of graduate study).
Geographic Focus: All States
Contact: Joyce Narine, Grants Manager; (703) 684-1221, ext. 1837; joyce@amstat.org
Internet: http://www.nass.usda.gov/research/ASA-NASS.html
Sponsor: American Statistical Association
732 N Washington Street
Alexandria, VA 22314-3415

ASA Deming Lecturer Award **1034**
The Deming Lecturer Award was established in 1995 to honor the accomplishments of W. Edwards Deming, recognize the accomplishments of the awardee, and enhance the awareness among the statistical community of the scope and importance of Deming's contributions. The awardee will give the Deming Lecture (an invited paper) at the Joint Statistical Meetings and receive a $1,000 honorarium, an award plaque, and travel expenses.
Requirements: The individual must have either made significant contributions in fields related to those in which Deming devoted his career-including survey sampling, statistics in the transportation industry, quality management, and quality improvement-or has made significant contributions through effective promotion of statistics and statistical thinking in business and industry.
Geographic Focus: All States
Date(s) Application is Due: Dec 31
Amount of Grant: 1,000 USD
Contact: Ronald Wasserstein; (703) 684-1221, ext. 1859; ron@amstat.org
Internet: http://www.amstat.org/careers/deminglecturer award.cfm
Sponsor: American Statistical Association
732 N Washington Street
Alexandria, VA 22314-3415

ASA Jeanne E. Griffith Mentoring Award **1035**
The Jeanne E. Griffith Mentoring Award has been established to encourage mentoring of junior staff in the Federal statistical system. It is presented annually to a supervisor who is nominated by co-workers and supervisors, and chosen by the Award Selection Committee. The award is co-sponsored by the Interagency Council on Statistical Policy, the Council for Excellence in Government, the Washington Statistical Society, the Social Statistics and Government Statistics Sections of the American Statistical Association, and the Council of Professional Associations on Federal Statistics. The award will consist of a $1,000 honorarium and a citation, which will be presented at a ceremony arranged by the co-sponsors in June.
Geographic Focus: All States
Date(s) Application is Due: Mar 27
Contact: Ronald Wasserstein; (703) 684-1221, ext. 1859; ron@amstat.org
Internet: http://www.amstat.org/sections/sgovt/jegmaann09.htm
Sponsor: American Statistical Association
732 N Washington Street
Alexandria, VA 22314-3415

ASA Outstanding Statistical Application Award **1036**
The Outstanding Statistical Application Award was established in 1986 to recognize the authors of papers that demonstrate an outstanding application of statistics in any substantive field. The award consists of an engraved award and $1,000 that is divided evenly among the winners. The award is given annually if an eligible and worthy work is nominated. The Award is bestowed upon a distinguished individual or individuals based on the following criteria: the impact of the statistical application in addressing a significant problem in a substantive field; and the ingenuity and or novelty of the statistical treatment of the problem. Eligible work includes papers, monographs, reports, and other substantive evidence appearing two years prior to the presentation of the award. All nominated work must have been subject to external peer review and, preferably, to formal refereeing. Nominations are due by April 1 each year. The award is presented at the Joint Statistical Meetings in August the same year.
Requirements: Nominations should include a letter that carefully describes the paper's significance, particularly its impact on the substantive field, along with copies of the nominated work.
Geographic Focus: All States
Date(s) Application is Due: Apr 1
Amount of Grant: 1,000 USD
Contact: Ronald Wasserstein; (703) 684-1221, ext. 1859; ron@amstat.org
Internet: http://www.amstat.org/careers/outstandingstatisticalapplicationaward.cfm
Sponsor: American Statistical Association
732 N Washington Street
Alexandria, VA 22314-3415

ASA Samuel S. Wilks Award **1037**
The Samuel S. Wilks Award, one of the ASA's most prestigious, was established in 1964 to honor the memory and distinguished career of Samuel S. Wilks. The award recognizes outstanding contributions to statistics that carry on in the spirit of his work. The Wilks Memorial Award is bestowed upon a distinguished individual who has made statistical contributions to the advancement of scientific or technical knowledge, ingenious application of existing knowledge, or successful activity in the fostering of cooperative scientific efforts that have been directly involved in matters of national defense or public interest. The Award includes two cash benefits: $500 is provided by Friends of the U.S. Army to honor Wilks' contributions to Army research programs and $1,000 is provided by a fund created by Alexander M. Mood to honor Wilks' teaching skill and life-long development of outstanding statisticians. This second award was established in 2008 at $1,000, but will increase by 3% annually. Nominations are due by April 1 each year. The award is presented at the Joint Statistical Meetings in August the same year.
Geographic Focus: All States
Date(s) Application is Due: Apr 1
Contact: Ronald Wasserstein; (703) 684-1221, ext. 1859; ron@amstat.org
Internet: http://www.amstat.org/careers/samuelwilksaward.cfm
Sponsor: American Statistical Association
732 N Washington Street
Alexandria, VA 22314-3415

ASA W. J. Youden Award in Interlaboratory Testing **1038**
The W.J. Youden Award in Interlaboratory Testing was established in 1985 to recognize the authors of publications that make outstanding contributions to the design and/or analysis of interlaboratory tests or describe ingenious approaches to the planning and evaluation of data from such tests. Each recipient receives an engraved award and a cash prize that are presented at the awards ceremony during the Joint Statistical Meetings. The current amount of the cash prize is $1,000, which is divided evenly among the winners. The award is given every year if, in the opinion of the awards committee, an eligible and worthy publication is nominated. Nominations are due by April 15 each year. The award is presented at the Joint Statistical Meetings in August.
Geographic Focus: All States
Date(s) Application is Due: Apr 15
Amount of Grant: 1,000 USD
Contact: Ronald Wasserstein; (703) 684-1221, ext. 1859; ron@amstat.org
Internet: http://www.amstat.org/careers/wjyoudenaward.cfm
Sponsor: American Statistical Association
732 N Washington Street
Alexandria, VA 22314-3415

ASCE Arthur S. Tuttle Memorial Scholarships **1039**
Arthur S. Tuttle Memorial Scholarships are for tuition assistance to students during their first year of graduate studies in civil engineering. Any student in good standing who is also a national student member of the society may apply; the funds may be applied only for tuition expenses during the first year of formal graduate civil engineering education in an accredited educational institution. Financial need and educational standing will be considered in selecting recipients. The total sum for awards available will be determined annually based on earnings available from the endowment.
Requirements: Application consists of a completed application form; applicant's justification for applying; plans for continuing formal education; official transcripts, recommendation forms, and supporting statements from the student chapter faculty advisor and two other members of the engineering school faculty; and a detailed financial plan outlining both income and expenses anticipated for the award year. Applicants must be a member of the society, in any grade, and must be in good standing at the time of application and award. Membership applications may be submitted with scholarship application.
Geographic Focus: All States
Date(s) Application is Due: Feb 9
Contact: John Casazza; (703) 295-6155; fax (703) 295-6132; jcasazza@asce.org
Internet: http://www.asce.org/Student-Organizations/Scholarships-and-Fellowships/Arthur-S—Tuttle-Memorial-Scholarship/
Sponsor: American Society of Civil Engineers
1801 Alexander Bell Drive
Reston, VA 20191-4400

ASCE Freeman Fellowships **1040**
Freeman Fellowships to aid and encourage young engineers, especially in research work, are made to fund expenses for experiments, observations, and compilations to discover new and accurate data that will be useful in engineering. The fellowships also recognize underwriting fully or in part some of the loss that may be sustained in the publication of meritorious books, papers, or translations pertaining to hydraulic science

and art that might, except for some such assistance, remain mostly inaccessible. Traveling scholarships are available to members younger than 45 years of age in recognition of achievement or promise for the purpose of aiding the candidate to visit engineering works in the United States or any other part of the world where there are good prospects of obtaining information useful to engineers. The grant may be used for assisting in the translation or publication in English of papers or books in foreign languages pertaining to hydraulics. The grant is available annually. In a given year there may be more than one recipient, or no recipient if no meritorious applications are received. Stipend is based on funds available annually from the endowment.
Requirements: Application consists of a completed form and a statement in general terms of the purposes for which the funds are expected to be used. Any member of ASCE is eligible to apply.
Geographic Focus: All States
Date(s) Application is Due: Feb 9
Amount of Grant: 2,000-5,000 USD
Contact: John Casazza; (703) 295-6155; fax (703) 295-6132; jcasazza@asce.org
Internet: http://www.asce.org/Student-Organizations/Scholarships-and-Fellowships/Freeman-Fellowship/
Sponsor: American Society of Civil Engineers
1801 Alexander Bell Drive
Reston, VA 20191-4400

ASCE Jack E. Leisch Memorial National Graduate Fellowship **1041**
The Jack E. Leisch Memorial National Graduate Fellowship is a memorial to the outstanding professional accomplishments and contributions of Jack E. Leisch, M. ASCE to the fields of geometric design, traffic engineering and transportation planning. The fellowship is supported by the income from donations in Mr. Leisch's honor. Any full-time student who is enrolled in or admitted to a transportation/traffic engineering graduate degree program (master or doctoral) in Civil Engineering at a qualified university may apply.
Requirements: Applicants must be a member of the Society (Membership applications may be submitted with scholarship applications), and must be in good standing at the time of the award. Applicants must be enrolled at a qualified university (those whose undergraduate Civil Engineering program is accredited by the Accreditation Board for Engineering and Technology).
Restrictions: Previous Jack E. Leisch Scholarship recipients are not eligible to reapply.
Geographic Focus: All States
Date(s) Application is Due: Feb 9
Amount of Grant: 4,000 USD
Contact: John Casazza; (703) 295-6155; fax (703) 295-6132; jcasazza@asce.org
Andrea C. Baker, Manager; abaker@asce.org
Internet: http://www.asce.org/Content.aspx?id=25021
Sponsor: American Society of Civil Engineers
1801 Alexander Bell Drive
Reston, VA 20191-4400

ASCE O.H. Ammann Research Fellowship in Structural Engineering **1042**
Nominations are being solicited for the annual O.H. Ammann Research Fellowship in Structural Engineering, which is bestowed annually to a member for the purpose of encouraging the creation of new knowledge in the field of structural design and construction. The recipient may apply for a renewal, additional stipend to be determined annually. Applications must be postmarked by the listed application deadline date of November 1, with notification being made in May for a fellowship beginning that same year.
Requirements: Applicants must be members of the society or applicants for membership; citizens of countries other than the United States may apply if they are eligible under this rule. Each application should include completed application form, official transcripts, a description of proposed research and its objectives, a statement from the institution at which the research is to be done that the applicant and proposed research are acceptable to that institution, and a statement in general terms of the purposes for which the funds are expected to be used.
Restrictions: During the tenure of the fellowship, the recipient may not work on research projects other than that for which this fellowship has been granted, but recipients may accept other awards if the conditions of such awards are the same as those for this fellowship.
Geographic Focus: All States
Date(s) Application is Due: Nov 1
Amount of Grant: 5,000 USD
Contact: Susan Reid; (703) 295-6299; fax (703) 295-6132; sreid@asce.org
Internet: http://content.seinstitute.org/inside/ammann.html
Sponsor: American Society of Civil Engineers
1801 Alexander Bell Drive
Reston, VA 20191-4400

ASCE Trent R. Dames and William W. Moore Engineering Fellowship **1043**
The Trent R. Dames and William W. Moore Engineering Fellowship is available to practicing engineers, earth scientists, professors, or graduate students for engineering research. Preference is given to research that will contribute to the solution of conflicting engineering issues involving social, economic, environmental, or political concerns or compromises. Other selection criteria include, but are not limited to, relevant professional experience, previous research or course of study either as a practitioner or in academia, and prior contribution to the engineering profession and/or the applied earth sciences. The fellowship may be awarded to co-researchers in a single project or divided among multiple projects.
Requirements: Application must include the completed application form and attachments; description of proposed research and its objectives, including a statement from the institution at which the research is to be done that the applicant and proposed research are acceptable to the institution; and total budget. Applicants must be practicing engineers or earth scientists, professors, or graduate students.
Geographic Focus: All States
Date(s) Application is Due: Feb 9
Amount of Grant: 2,000-6,000 USD
Contact: John Casazza; (703) 295-6155; fax (703) 295-6132; jcasazza@asce.org
Internet: http://www.asce.org/Student-Organizations/Scholarships-and-Fellowships/Trent-R—Dames-and-William-W—Moore-Fellowship/
Sponsor: American Society of Civil Engineers
1801 Alexander Bell Drive
Reston, VA 20191-4400

ASCO/UICC International Cancer Technology Transfer Fellowships **1044**
The scholarships are intended primarily for researchers in the early stages of their career as well as for experienced clinical oncologists. The ICRETT objective is to facilitate rapid international transfer of cancer research and clinical technology, and it allows fellows to exchange knowledge and enhance skills in basic, clinical, behavioral and epidemiological areas of cancer research, and cancer control and prevention. Through the program fellows also acquire appropriate clinical management, diagnostic and therapeutic expertise for effective application and use in home organizations upon return. The UICC has also developed the ICRETT for teaching faculty which funds experts to teach at host institutions throughout the world for a duration of 1 week to 1 month.
Requirements: The candidate must possess the appropriate professional qualifications and experience according to the specifications of each fellowship described on the website and must currently be engaged in cancer research, clinical oncology practice, or cancer society work. To permit effective communication at the host institute, the candidate must have adequate fluency in a common language. The candidate must also be on the staff payroll of a university, research laboratory or institute, hospital, oncology unit, or voluntary cancer society (or be accredited volunteers of such societies) to where they will return at the end of a fellowship.
Restrictions: Candidates attached to commercial entities, going to a profit organization or have associations with the tobacco industry not eligible.
Geographic Focus: All States
Amount of Grant: 3,400 USD
Contact: Nancy R. Daly, Executive Director; (571) 483-1432; dalyn@asco.org
Internet: http://www.ascocancerfoundation.org/TACF/Awards/Award+Opportunities/ASCO+%26+UICC+ICRETT+Fellowship
Sponsor: Conquer Cancer Foundation
2318 Mill Road, Suite 800
Alexandria, VA 22314

ASCO Advanced Clinical Research Award in Colorectal Cancer **1045**
The Advanced Clinical Research Award (ACRA) in Colorectal Cancer is designed to fund investigators who are committed to clinical cancer research and who wish to conduct original research not currently funded. The grant totals $450,000 and is paid in 3 annual increments of $150,000 made on or about July 1 of each year of the grant term. The grant funds will be directed to the sponsoring institution and should be used towards salary support, supplies, equipment, travel, etc. necessary for the pursuit of the recipient's research project.
Requirements: Appplicants must meet the following criteria: be a physician (MD, DO, or international equivalent) who is in the fourth to ninth year of a full-time, primary faculty appointment in a clinical department at an academic medical institution at the time of grant submission; have completed productive post doctoral/post fellowship research and demonstrated the ability to undertake independent investigator-initiated clinical research; be an active member of ASCO or have submitted a membership application with the grant application; be able to commit to 75% of full-time effort in research (applies to total research, not just the proposed project) during the award period.
Geographic Focus: All States
Date(s) Application is Due: Nov 3
Contact: Nancy R. Daly, Executive Director; (571) 483-1432; dalyn@asco.org
Internet: http://www.ascocancerfoundation.org/TACF/Grants/Grant+Opportunities/Advanced+Clinical+Research+Award+in+Colorectal+Cancer
Sponsor: Conquer Cancer Foundation
2318 Mill Road, Suite 800
Alexandria, VA 22314

ASCO Advanced Clinical Research Awards in Breast Cancer **1046**
The ASCO Foundation and Genentech BioOncology invite physicians with full-time faculty appointments to apply for a clinical oncology research award in Breast Cancer. The program funds clinical investigators who are committed to clinical cancer research. Their research must have a direct patient-oriented focus including clinical trials and/or translational research involving human subjects. The Advanced Clinical Research Award is designed to provide funding beyond the Clinical Research Career Development Award (CDA). By continuing to support proven clinical researchers who are post-CDA but at a critical stage in their early career, the foundation hopes to expand the cadre of expert clinical oncology researchers who are developing promising research initiatives.
Requirements: The applicant must meet the following criteria at the time of grant award: be a physician (MD or DO) who is 5-10 years post final sub-specialty training; have a full-time faculty appointment in a clinical department at an academic medical center; have completed productive postdoctoral/post-fellowship research and demonstrated the ability to undertake independent investigator-initiated clinical research; be an active member of the American Society of Clinical Oncology (ASCO); and expect to spend 75% of time during the award period dedicated to research.

Geographic Focus: All States
Date(s) Application is Due: Oct 20
Amount of Grant: 450,000 USD
Contact: Nancy R. Daly, Executive Director; (571) 483-1432; dalyn@asco.org
Internet: http://www.ascocancerfoundation.org/TACF/Grants/Grant+Opportunities/Advanced+Clinical+Research+Award+in+Breast+Cancer
Sponsor: Conquer Cancer Foundation
2318 Mill Road, Suite 800
Alexandria, VA 22314

ASCO Advanced Clinical Research Awards in Sarcoma **1047**
The Advanced Clinical Research Award (ACRA) in Sarcoma is designed to fund investigators who are committed to clinical cancer research and who wish to conduct original research not currently funded. The award is intended to support proposals with a patient-oriented focus, including a clinical research study and/or translational research involving human subjects. The grant totals $450,000 and is paid in three annual increments of $150,000 made on or about January 1 of each year of the grant term. Preference will be given to proposals involving therapeutic vaccine or immunotherapeutic approaches for sarcoma.
Requirements: Applicants must meet the following criteria: be a physician (MD, DO, or international equivalent) who is in the fourth to ninth year of a full-time, primary faculty appointment in a clinical department at an academic medical institution at the time of grant submission; have completed productive post doctoral/post fellowship research and demonstrated the ability to undertake independent investigator-initiated clinical research; be an active member of ASCO or have submitted a membership application with the grant application; be able to commit to 75% of full-time effort in research (applies to total research, not just the proposed project) during the award period.
Restrictions: Proposals with a predominant focus on in vitro or animal studies (even if clinically relevant) are not allowed.
Geographic Focus: All States
Date(s) Application is Due: Aug 6
Amount of Grant: 450,000 USD
Contact: Nancy R. Daly, Executive Director; (571) 483-1432; dalyn@asco.org
Internet: http://www.ascocancerfoundation.org/TACF/Grants/Grant+Opportunities/Advanced+Clinical+Research+Award+in+Sarcoma
Sponsor: Conquer Cancer Foundation
2318 Mill Road, Suite 800
Alexandria, VA 22314

ASCO Long-Term International Fellowships **1048**
The ASCO Cancer Foundation Long Term International Fellowship (LIFe) provides young oncologists in developing nations the support and resources needed to advance their training by deepening their relationship with a U.S. or Canadian colleague and his or her institution. Through a one or two year fellowship the recipient will earn valuable training and experience with which they can affect change in cancer care in their home country. Recipients must return to their home institutions after completing their fellowship and are expected to disseminate the knowledge they have gained. The ASCO Cancer Foundation (TACF) will administer fundraising and grant distribution for the fellowships. Funds will be sent from TACF to the host institution in January of the award year. The host institution will then be responsible for financial management of the fellowship. Any funds not spent on the fellowship will be returned to TACF. Recipients must attend the ASCO Annual Meeting that falls during the first year of their award term where they will be honored at The ASCO Cancer Foundation Grants Brunch. Recipients will also be expected to submit their research to ASCO via an abstract or publication submission following their fellowship. Applications are due to ASCO by October 15 for the fellowship to begin anytime in the next calendar year. Applications should be completed and returned jointly by the prospective host and fellow. A decision will be distributed by December 15.
Requirements: Hosts must: be an ASCO member in good standing; be employed at a U.S. or Canadian institution with sufficient infrastructure to provide for a 1 year fellowship; independently identify fellowship applicants; agree to cover any costs in excess of ASCO's contribution; complete the host section of the LIFe Application; complete fellowship evaluations as requested by ASCO. Applicants must: be an ASCO member in good standing; reside in a country identified by the World Bank as income level low, lower middle, and upper middle; have the support of their home institution including a letter of support from their supervisor; be a medical doctor who has finished a fellowship in clinical oncology or the equivalent; have less than 10 years experience in the field of oncology; complete the applicant section of the LIFe application including submission of a CV of up to 3 pages in length; commit to return to country of origin at the completion of their fellowship.
Geographic Focus: All States
Date(s) Application is Due: Oct 15
Contact: Nancy R. Daly, Executive Director; (571) 483-1432; dalyn@asco.org
Internet: http://www.ascocancerfoundation.org/TACF/Awards/Award+Opportunities/Long-term+International+Fellowship+%28LIFe%29
Sponsor: Conquer Cancer Foundation
2318 Mill Road, Suite 800
Alexandria, VA 22314

ASCO Merit Awards **1049**
The Merit Awards are designed to further promote clinical research by young scientists and to provide fellows with an opportunity to present their research and interact with other clinical cancer investigators at ASCO scientific meetings. A select number of Merit Awards are given annually to recognize outstanding abstracts submitted for consideration for presentation at an ASCO scientific meeting. The awards are given to oncology fellows who are first authors on selected abstracts.
Requirements: The first author of an abstract must meet the following requirements to be considered for an Award: be a physician (MD or DO) in an oncology fellowship training program or a doctoral degree candidate (such as PharmD or PhD) enrolled in an approved oncology specialty training program at the time of abstract submission; work in an oncology laboratory or clinical research setting; agree to present the abstract at the ASCO scientific meeting; check the Merit Award consideration box on the online abstract submitter; provide a letter of support from the Training Program Director indicating eligibility for the award; and submit a two-page curriculum vitae.
Geographic Focus: All States
Contact: Nancy R. Daly, Executive Director; (571) 483-1432; dalyn@asco.org
Internet: http://www.ascocancerfoundation.org/TACF/Awards/Award+Opportunities/Merit+Award
Sponsor: Conquer Cancer Foundation
2318 Mill Road, Suite 800
Alexandria, VA 22314

ASCO Young Investigator Award **1050**
This grant provides funding to promising investigators to encourage and promote quality research in clinical oncology. The purpose of this award is to fund physicians during the transition from a fellowship program to a faculty appointment. Priority consideration will be given to proposals that include patient-oriented and, ultimately, clinical research. Since many awards are supported with restricted grants from outside organizations, the ASCO is particularly interested in identifying young researchers working in the following sub-specialties and emerging disciplines: breast cancer; cancer survivorship; geriatric oncology; health disparities; kidney cancer; multiple myeloma; ovarian cancer; pancreatic cancer; sarcoma; survivorship; and young adult cancer. However, The ASCO Foundation welcomes application submissions in all oncology subspecialties.
Requirements: The recipient must be a physician (MD or DO) who, at the time of grant award is in the final year of his/her final sub-specialty training program or in the first year post his/her final sub-specialty training. The sponsoring facility must be an academic medical institution. The primary mentor must be in the candidate's proposed research field, must assume responsibility, and provide guidance for the research. The applicant must either be a member of ASCO or submit a membership application with the grant application. The applicant should spend at least 60 to 75% of his or her time in research during the award period.
Geographic Focus: All States
Date(s) Application is Due: Sep 9
Contact: Nancy R. Daly, Executive Director; (571) 483-1432; dalyn@asco.org
Internet: http://www.ascocancerfoundation.org/TACF/Grants/Grant+Opportunities/Young+Investigator+Award
Sponsor: Conquer Cancer Foundation
2318 Mill Road, Suite 800
Alexandria, VA 22314

ASCSA Alison M. Frantz Fellowship **1051**
The fellowship supports research in postclassical studies at the Gennadius Library in Greece. Fields of study include late antiquity, Byzantine studies, and modern Greek studies. A limited number of research associateships are awarded for well-defined projects (deadlines April 1, September 1, and December 1). Travel grants for graduate students or postdoctoral scholars from North American institutions working on projects in archeological science in Greece also are awarded. Guidelines and application are available online.
Requirements: PhD candidates and recent PhDs from U.S. or Canadian institutions are eligible. Fellows are expected to be in residence at the school for the full term of the fellowship.
Geographic Focus: All States
Date(s) Application is Due: Jan 15
Amount of Grant: 10,000 USD
Contact: Program Contact; (609) 683-0800; fax (609) 924-0578; ascsa@ascsa.org
Internet: http://www.ascsa.edu.gr/fellowship/fellowships.htm
Sponsor: American School of Classical Studies at Athens
6-8 Charlton Street
Princeton, NJ 08540

ASCSA Fellowships **1052**
Fellowships awarded include the Heinrich Schliemann and John Williams White Fellowships in archaeology; the Thomas Day Seymour Fellowship in history and literature; two Brunilde Ridgway Fellowships in art history; and seven Fellowships unrestricted as to field—the Virginia Grace, the Michael Jameson, the Philip Lockhart, the Lucy Shoe Meritt, the Martin Ostwald, the James and Mary Ottaway Jr., and the James Rignall Wheeler fellowships. The Bert Hodge Hill fellowship is unrestricted, but with a preference for a student in art history. Fellowships are awarded on the basis of examinations and recommendations to students of classical archaeology, ancient history, and classical languages and literature.
Requirements: Applicants must be students in an American or Canadian university and will, preferably, have taken at least one year of graduate work but will not have completed the PhD before the term of fellowship starts.
Geographic Focus: All States
Date(s) Application is Due: Jan 15
Amount of Grant: 10,000 USD
Contact: Carolyn Snively; (609) 683-0800; fax (609) 924-0578; ascsa@ascsa.org
Internet: http://www.ascsa.edu.gr/fellowship/fellowships.htm#regular
Sponsor: American School of Classical Studies at Athens
6-8 Charlton Street
Princeton, NJ 08540

ASCSA Harry Bikakis Fellowship 1053
The fellowship is available to graduate students attending North American institutions or Greek graduate students whose research subject is ancient Greek law; or Greek graduate students working on a school excavation. The fellowship is awarded periodically, but not more frequently than once a year. An outline of the proposed project should be sent to Professor Stephen V. Tracy, Director of the School, c/o the office in Athens: 54 Solidias St; Athens, Greece 106 76; 011-30-210-723-6313, fax: 011-30-210-725-0584.
Geographic Focus: All States
Date(s) Application is Due: Jan 15
Amount of Grant: 1,875 USD
Contact: Administrator; (609) 683-0800; fax (609) 924-0578; asca@ascsa.org
Internet: http://www.ascsa.edu.gr/Forms/bikakis.pdf
Sponsor: American School of Classical Studies at Athens
6-8 Charlton Street
Princeton, NJ 08540

ASCSA Jacob Hirsch Fellowship 1054
ASCSA offers this annual fellowship in preclassical, classical, or postclassical archaeology to U.S. or Israeli graduate students who are at the dissertation-writing stage, or to recent recipients of the PhD who are completing the dissertation program with dissertation publication. Project requires substantial residence in Greece. The fellowship is nonrenewable. Requests for applications should be forwarded to the Committee on Admissions and Fellowships at the school.
Requirements: Students in the United States or Israel who are PhD candidates writing their dissertations in archaeology or recent PhD's completing a project are eligible to apply.
Geographic Focus: All States
Date(s) Application is Due: Jan 15
Amount of Grant: 10,000 USD
Contact: Program Supervisor; (609) 683-0800; fax (609) 924-0578; ascsa@ascsa.org
Internet: http://www.ascsa.edu.gr/fellowship/fellowships.htm
Sponsor: American School of Classical Studies at Athens
6-8 Charlton Street
Princeton, NJ 08540

ASCSA Summer Sessions Scholarships 1055
Scholarships are awarded annually for six-week summer sessions of study and research emphasizing the topography and antiquities of Greece. Graduate and advanced undergraduate students and high school and college teachers may apply. Application information is available on the Web site.
Geographic Focus: All States
Date(s) Application is Due: Jan 15
Contact: Summer Committee; (609) 683-0800; fax (609) 924-0578; ascsa@ascsa.org
Internet: http://www.ascsa.edu.gr/fellowship/fellowships.htm#SummerSession
Sponsor: American School of Classical Studies at Athens
6-8 Charlton Street
Princeton, NJ 08540

ASCSA Weiner Laboratory Faunal Studies Research Fellowship 1056
This is a one-year fellowship at the Wiener Laboratory for graduate students or postdoctoral scholars working on projects in skeletal, faunal, geoarcheological, or environmental studies in Greece. Applicants must have a well-defined project that can be undertaken within the given time in the laboratory or in collaboration with local research institutions. In addition to the proposed research, the fellow will be expected to develop and curate the lab's comparative collection, contribute to the development of the lab, assist with queries from excavators, offer a lecture on the work undertaken while at the lab, participate in one regular member school trip, and contribute to seminars on aspects of archaeological science as part of the school's annual curriculum. Applications can be downloaded from the Web site. Athens Office address is 54 Souidias Street, GR-106 76 Athens Greece; phone: 30-1-723-6313; fax: 30-1-725-0584; email: info@ascsa.edu.gr; Web site: www.ascsa.edu.gr.
Requirements: The fellowship is open to scholars with PhDs and those working on doctoral dissertations.
Geographic Focus: All States
Date(s) Application is Due: Jan 15
Amount of Grant: 15,500-25,000 USD
Contact: Director; (609) 683-0800; fax (609) 924-0578; ascsa@ascsa.org
Internet: http://www.ascsa.edu.gr/Wiener/fellowship.htm
Sponsor: American School of Classical Studies at Athens
6-8 Charlton Street
Princeton, NJ 08540

ASECS/Clark Fellowships 1057
One-month resident fellowships at the William Andrews Clark Memorial Library are available to postdoctoral scholars with projects in the Restoration or the 18th century. The library, which is administered by the Center for 17th- and 18th-Century Studies, is known for its collections on 17th- and 18th-century Britain, Oscar Wilde and the 1890s, the history of printing, and certain aspects of the American West. The application deadline is for fellowships to be held between July 1 of the current year and June 30 of the next. Applications are available on the Web site.
Requirements: Scholars who have received a PhD in the last six years and are engaged in research pertaining to the announced theme are eligible to apply.
Geographic Focus: All States
Date(s) Application is Due: Feb 1
Amount of Grant: 2,000 USD
Contact: Program Contact; (310) 206-8577
Internet: http://www.humnet.ucla.edu/humnet/c1718cs/applic3.htm
Sponsor: William Andrews Clark Memorial Library
2520 Cimarron Street
Los Angeles, CA 90018

ASECS James L. Clifford Prize 1058
The prize is awarded annually for an outstanding article appearing in a journal, festschrift, or other serial publication; the article must be an outstanding study of some aspect of 18th-century culture interesting to any 18th-century specialist, regardless of discipline, and no longer than 15,000 words. Nominations must be accompanied by eight copies of the article.
Requirements: The article may be nominated by any member of the society, by its author, or by an editor of the publishing journal. The author must be a member of the society at the time of presentation of the award.
Geographic Focus: All States
Date(s) Application is Due: Jan 1
Amount of Grant: 500 USD
Contact: Prize Committee, (336) 727-4694; fax (336) 727-4697; ASECS@wfu.edu
Internet: http://asecs.press.jhu.edu/awards.html
Sponsor: American Society for Eighteenth-Century Studies
P.O. Box 7867, Wake Forest University
Winston-Salem, NC 27109

ASEE Army Research Laboratory Postdoctoral Fellowships 1059
The program is designed to significantly increase the involvement of creative and highly trained scientists and engineers from academia and industry in scientific and technical areas of interest and relevance to the Army. Scientists and engineers at the Army Research Laboratory (ARL) help shape and execute the Army's program for meeting the challenge of developing technologies that will support Army forces in meeting future operational needs by pursuing scientific research and technological developments in such diverse fields as applied mathematics, atmospheric characterization, simulation and human modeling, digital/optical signal processing, material science and technology, multifunctional technology, combustion processes, propulsion, and flight physics. Before writing a proposal, applicants are advised to communicate directly with an ARL advisor, who can provide more specific information on the current research and available technical facilities and offer scientific support of proposal development. Applications are available on the Web site.
Requirements: Before appointment, participants must present evidence of having received the PhD, ScD, or other earned research doctoral degree recognized in U.S. academic circles as equivalent to the PhD within seven years of the date of application, or must present evidence of having completed all formal academic requirements for one of these degrees.
Geographic Focus: All States
Contact: Rick Kempinski, (202) 331-3525; r.kempinski@asee.org
Internet: http://www.asee.org/arl
Sponsor: American Society for Engineering Education
1818 N Street NW, Suite 600
Washington, DC 20036

ASEE Naval Research Laboratory Postdoctoral Fellowships 1060
The purpose of this program is to increase the involvement of highly trained U.S. scientists and engineers in disciplines to meet the evolving needs of naval technology. To this goal, the ASEE, as the agent for the Naval Research Laboratory, will award approximately 40 postdoctoral appointments per year for research at 15 participating navy laboratories. Appointments are for one year and renewable for a second. At the discretion of the laboratory, third-year appointments may be arranged if warranted. A person who has received a prior postdoctoral appointment at a navy laboratory under any program may not be eligible to participate in this postdoctoral program. Applications may be downloaded from the Web site. Applications are accepted on an ongoing basis.
Requirements: Before appointment, participants must present evidence of having received the PhD, ScD, or other earned research doctoral degree recognized in U.S. academic circles as equivalent to the PhD within seven years of the date of application, or must present evidence of having completed all formal academic requirements for one of these degrees. Participants must be U.S. citizens.
Geographic Focus: All States
Amount of Grant: 65,000 USD
Contact: Manager; (202) 331-3525; fax (202) 265-8504; r.kempinski@asee.org
Internet: http://www.asee.org/resources/fellowships/nrl/about.cfm
Sponsor: American Society for Engineering Education
1818 N Street NW, Suite 600
Washington, DC 20036

ASGE / ConMed Award for Outstanding Manuscript by a Fellow/Resident 1061
The American Society for Gastrointestinal Endoscopy and ConMed Endoscopic Technologies sponsor the annual ASGE/ConMed Outstanding Manuscript Award. The prize consists of $10,000 cash award to the Fellow or Resident in training in a gastroenterology or a surgical gastrointestinal endoscopic program whose manuscript is accepted in for publication in Gastrointestinal Endoscopy. The award is presented during Digestive Disease Week, the largest and most prestigious meeting in the world for the GI professional.
Geographic Focus: All States
Amount of Grant: 10,000 USD
Contact: Chair, ASGE Research Committee; (630) 573-0600; fax (630) 573-0691; grants@asge.org

Internet: http://www.asge.org/index.aspx?id=4892&terms=ConMed+Award
Sponsor: American Society for Gastrointestinal Endoscopy
1520 Kensington Road, Suite 202
Oak Brook, IL 60523

ASGE Don Wilson Award 1062

The Award provides Advanced Fellows or Junior Faculty with the opportunity to train outside of their home country with a premier GI endoscopist or group in order to advance their training. The award assists in underwriting reasonable and customary travel and living expenses for a period of one to three months. The award includes a $7,500 cash stipend prior to the recipient's travel. In addition, a 20% disbursement will be made to the host institution. A total of three awards will be available annually for North American and International ASGE members. Note: One training must take place in the United States.
Requirements: Applicants must be the equivalent of Junior Faculty or Advanced Fellows and be proficient in the English Language. Applicants must also: be a current ASGE Member; be a Junior Faculty or Advanced Fellow (3rd or 4th year or international equivalent); have permission from their own institution to undertake the travel; and have permission from an ASGE member at the host institution to accept the application for training.
Restrictions: Individuals applying for 1-2 year advanced fellowships do not qualify.
Geographic Focus: All States
Date(s) Application is Due: Sep 1
Amount of Grant: 7,500 USD
Contact: Chair; (630) 573-0600; fax (630) 573-0691; membership@asge.org
Internet: http://www.asge.org/nspages/research/applications/donWilson.cfm
Sponsor: American Society for Gastrointestinal Endoscopy
1520 Kensington Road, Suite 202
Oak Brook, IL 60523

ASGE Endoscopic Research Awards 1063

These research awards are offered to physicians for projects in basic and clinical endoscopic technology research, outcomes and effectiveness of endoscopy research. The primary objective is to foster research in gastrointestinal endoscopy both within and outside of academic centers. Two categories of grants may be used for up to two years of study: grants of $1-15,000 (category A) and grants of $15,001-50,000 (category B). Requests for funding seed projects that will lead to further research as well as larger requests for definitive clinical trials will be considered. Funding requests may also include: personnel expenses for research assistant and/or faculty salary support (percentage of time for study should be specified and appropriately justified); study supplies; and equipment essential for study.
Requirements: Candidate must be: an ASGE member; an MD (or have equivalent degree); and current in a gastroenterology-related and endoscopic practice in academic institutions or private practice in North America.
Restrictions: Funding will not be provided for: salary support for trainees; computer purchases (unless a unique application is proposed); standard equipment and supplies needed for appropriate patient care (for example, sclerotherapy needles); travel to meetings; or indirect costs.
Geographic Focus: All States
Date(s) Application is Due: Sep 12
Amount of Grant: Up to 50,000 USD
Contact: Chair; (630) 573-0600; fax (630) 573-0691; grants@asge.org
Internet: http://www.asge.org/index.aspx?id=4892&terms=ConMed+Award
Sponsor: American Society for Gastrointestinal Endoscopy
1520 Kensington Road, Suite 202
Oak Brook, IL 60523

ASGE Endoscopic Research Career Development Awards 1064

These awards provide the salary and/or research support necessary for the investigator to enhance his/her career development. The award must be used to acquire new skills for furthering a career in endoscopic research. Examples of such skills include advanced training in endoscopic procedures, training in outcomes research relevant to endoscopy, training in use of large databases such as CORI, and new endoscopic research techniques, including animal models.
Requirements: Applicants must be ASGE members and hold full-time faculty positions at North American (U.S., Canada, or Mexico) universities or professional institutions at the time of application. The award is intended for faculty who have demonstrated promise and have some record of accomplishment in research. Candidates must devote at least 30 percent of their effort to research related to gastrointestinal endoscopy during the period of the award.
Restrictions: The award is not available for fellows. Faculty with principle investigator current federal funding or other concurrent career development support still active at the time of this award are not eligible. Recipients of the ASGE Career Development Awards may not be granted an ASGE Research & Outcomes & Effectiveness Award during their two-year award period.
Geographic Focus: All States
Date(s) Application is Due: Dec 12
Amount of Grant: 75,000 USD
Contact: Chair; (630) 573-0600; fax (630) 573-0691; grants@asge.org
Internet: http://www.asge.org/index.aspx?id=4892&terms=ConMed+Award
Sponsor: American Society for Gastrointestinal Endoscopy
1520 Kensington Road, Suite 202
Oak Brook, IL 60523

ASGE Given Capsule Endoscopy Research Award 1065

These research awards are offered to physicians for projects specific to capsule endoscopy both within and outside of academic centers. The award must be used for projects directly relating to capsule endoscopy. Requests for funding seed projects that will lead to further research as well as requests for definitive clinical trials will be considered. Funding requests may include: personnel (research assistant and/or faculty salary support—percentage of time for study should be specified and appropriately justified); study supplies; and equipment essential for study.
Requirements: Candidate must be an ASGE member, be an MD (or have equivalent degree), and be current in a gastroenterology-related and endoscopic practice in academic institutions or private practice in North America.
Restrictions: Funding will not be provided for: salary support for trainees; computer purchases (unless a unique application is proposed); standard equipment and supplies needed for appropriate patient care (for example, sclerotherapy needles); travel to meetings; publication costs; or indirect costs.
Geographic Focus: All States
Date(s) Application is Due: Dec 7
Amount of Grant: Up to 25,000 USD
Contact: Chair; (630) 573-0600; fax (630) 573-0691; grants@asge.org
Internet: http://www.asge.org/nspages/research/applications/ceapp.cfm
Sponsor: American Society for Gastrointestinal Endoscopy
1520 Kensington Road, Suite 202
Oak Brook, IL 60523

ASGH Award for Excellence in Human Genetics Education 1066

The American Society of Human Genetics (ASGH) established the Award for Excellence in Human Genetics Education to recognize outstanding contributions to human genetics education. Nominees should have long-standing involvement in genetics education, contributions in more than one area, and contributions of substantive influence on individuals and/or organizations. Examples of the types of contributions that might qualify a nominee include: producing a set of writings that has had a major influence on human genetics education, developing a curriculum or innovative teaching program that is widely emulated, writing a book that has been adopted by many universities, developing an educational Web site, or directing a fellowship program that has consistently produced exemplary graduates. Nominees may be individuals or groups.
Requirements: Nominees must have made contributions that are recognized nationally or internationally as being of exceptional quality and great importance to human genetics education. All nominees and winners must be current ASHG members. If a group is nominated, at least one nominee in the group must be a current member. A monetary award and plaque will be presented to the recipient(s) at the Society's Annual Meeting. Nominees for the ASHG Excellence in Human Genetics Education Award may emerge from submissions from the membership, from the Awards Committee, or from the Information and Education Committee.
Geographic Focus: All States
Contact: Joann Boughman, Executive Vice-President; (866) 486-4363 or (301) 634-7300; fax (301) 634-7079; jboughman@ashg.org
Internet: http://www.ashg.org/pages/awards_overview.shtml#education
Sponsor: American Society of Human Genetics
9650 Rockville Pike
Bethesda, MD 20814-3998

ASGH C.W. Cotterman Award 1067

Each September, the editorial board of The American Journal of Human Genetics selects two articles published in the journal in the previous year that best represent outstanding scientific contributions to the field of human genetics. Two Cotterman Awards will be given annually and a monetary award of $1,000 and a plaque will be presented to the recipients for the top two papers published in the Journal during the previous year on which the first author was either a pre- or post- doctoral trainee and an ASHG member.
Geographic Focus: All States
Amount of Grant: 1,000 USD
Contact: Joann Boughman, Executive Vice-President; (866) 486-4363 or (301) 634-7300; fax (301) 634-7079; jboughman@ashg.org
Internet: http://www.ashg.org/pages/awards_overview.shtml#cotterman
Sponsor: American Society of Human Genetics
9650 Rockville Pike
Bethesda, MD 20814-3998

ASGH Charles J. Epstein Trainee Awards for Excellence in Human Genetics Research 1068

The American Society of Human Genetics (ASGH) honors excellence in research conducted by predoctoral and postdoctoral trainees (including genetic counseling trainees) through merit-based awards that recognize highly competitive abstracts submitted and presented at the ASHG Annual Meeting. These awards were renamed in 2012 to honor the late Dr. Charles Epstein, who was a past president of ASHG, former editor of AJHG, and winner of both the William Allan Award and the McKusick Leadership Award. The Program Committee and Awards Committee, in consultation with the ASHG Board of Directors, determine the number of awards, the categories for which they are given, and the prize amounts of each award (currently $750 plus complimentary registration for approximately 60 semifinalists, plus an additional $1,000 for winners). Abstract scoring is completed by the Program Committee, and the top-ranking abstracts (semifinalists) are submitted to the Awards Committee, which selects 18 finalists prior to the Annual Meeting. Six winners are chosen after the committee

members judge the finalists' presentations. The total value of ASHG's Trainee Awards is approximately $70,000 annually.
Geographic Focus: All States
Amount of Grant: 750-1,750 USD
Contact: Joann Boughman, Executive Vice-President; (866) 486-4363 or (301) 634-7300; fax (301) 634-7079; jboughman@ashg.org
Internet: http://www.ashg.org/pages/awards_overview.shtml#student
Sponsor: American Society of Human Genetics
9650 Rockville Pike
Bethesda, MD 20814-3998

ASGH McKusick Leadership Award **1069**
The prestigious McKusick Leadership Award, established by the American Society of Human Genetics (ASHG) in honor of the late Dr. Victor A. McKusick, is presented on behalf of the Society to an individual whose professional achievements have fostered and enriched the development of various human genetics disciplines. Potential recipients should exemplify the enduring leadership and vision required to ensure that the field of human genetics will flourish and successfully assimilate into the broader context of science, medicine, and health. They also may have made major contributions to awareness or understanding of human genetics by policy makers or by the general public. A plaque and $2,500 prize will be presented in honor of the awardee at the ASHG Annual Meeting.
Geographic Focus: All States
Amount of Grant: 2,500 USD
Contact: Joann Boughman, Executive Vice-President; (866) 486-4363 or (301) 634-7300; fax (301) 634-7079; jboughman@ashg.org
Internet: http://www.grantselect.com/editor/view_grant/133352
Sponsor: American Society of Human Genetics
9650 Rockville Pike
Bethesda, MD 20814-3998

ASGH William Allan Award **1070**
The William Allan Award is the top prize given by the American Society of Human Genetics (ASHG); it was established in 1961 in memory of William Allan (1881-1943), who was one of the first American physicians to conduct extensive research in human genetics. The Allan Award is presented annually to recognize substantial and far-reaching scientific contributions to human genetics, carried out over a sustained period of scientific inquiry and productivity. The recipient is presented with an engraved medal and a monetary award of $10,000 at the ASHG Annual Meeting. The Allan Award winner is also invited to present a 30-45 minute address at the ASHG Annual Meeting, and it is customary to publish a manuscript of the presentation in The American Journal of Human Genetics. This award is given on a yearly basis, but can be omitted in any given year at the discretion of the Board. Nominations will be solicited from the ASHG Awards Committee and the general membership.
Geographic Focus: All States
Amount of Grant: 1,000 USD
Contact: Joann Boughman, Executive Vice-President; (866) 486-4363 or (301) 634-7300; fax (301) 634-7079; jboughman@ashg.org
Internet: http://www.ashg.org/pages/awards_overview.shtml#allan
Sponsor: American Society of Human Genetics
9650 Rockville Pike
Bethesda, MD 20814-3998

ASGH William Curt Stern Award **1071**
The American Society of Human Genetics (ASGH) Curt Stern Award honors the memory of Curt Stern (1902-1981) as an outstanding pioneering human geneticist. This award is presented yearly for outstanding scientific achievements in human genetics that occurred in the last 10 years. The work could be a single major discovery or may be a series of contributions on a similar or related topic. A plaque and a monetary award of $2,500 are presented to the recipient(s) at the Society's Annual Meeting. Nominations for the Stern Award will be solicited from the ASHG Awards Committee and the general membership.
Geographic Focus: All States
Amount of Grant: 2,500 USD
Contact: Joann Boughman, Executive Vice-President; (866) 486-4363 or (301) 634-7300; fax 301; jboughman@ashg.org
Internet: http://www.ashg.org/pages/awards_overview.shtml#stern
Sponsor: American Society of Human Genetics
9650 Rockville Pike
Bethesda, MD 20814-3998

ASHA Advancing Academic-Research Careers (AARC) Award **1072**
This award is given to new faculty in higher education to support their academic and research endeavors in the field of communication sciences and disorders (CSD). Up to five awards of $5,000 each will be granted to support activities such as: Improving teaching skills; Mentoring graduate students; Participating in research activities; Preparing a grant application; Preparing a manuscript or publication; Presenting at a professional meeting (e.g., the ASHA Convention).
Requirements: Candidates must possess: A research doctoral degree (PhD, ScD, CScD, or equivalent, e.g., EdD), which must be conferred by August 1; A full-time faculty appointment; Primary responsibilities of teaching and conducting research; Less than six years of teaching/research experience or are below the level of associate professor; A tenure or non-tenure track appointment. A list of required documents and a downloadable application form are available at the website.
Restrictions: Postdoctoral fellows are not eligible. Only students attending a university in the U.S. are eligible to apply for the award.
Geographic Focus: All States
Date(s) Application is Due: May 15
Amount of Grant: 5,000 USD
Contact: Margaret Rogers, Chief Staff Officer for Science & Research; (800) 498-2071 or (301) 296-5706; fax (301) 296-8580; mrogers@asha.org
Internet: http://www.asha.org/students/awards.htm#aarc
Sponsor: American Speech-Language-Hearing Association
2200 Research Boulevard, #245
Rockville, MD 20850-3289

ASHA Research Mentoring-Pair Travel Award **1073**
This award is designed to foster the professional development of students, clinicians, and emerging scientists who have expressed an interest in research in communication sciences and disorders. Up to 10 travel awards of $1,000 each will be granted to eligible mentor mentee pairs (mentee= $750; mentor= $250) to help defray the cost of attending the annual research symposium. Special consideration will be given to individuals whose research or research interests are (a) relevant to the symposium topic, (b) interdisciplinary in nature, (c) translational in nature, or (d) related to issues of cultural or linguistic diversity.
Requirements: Mentees must be (a) a current student at the bachelor's, master's, AuD, or PhD level, (b) a postdoctoral fellow, (c) a junior level faculty member (less than 6 years teaching/research experience in a tenure-track position or below the level of associate professor), or (d) a clinician. Mentors must be a seasoned investigator in an academic or health care environment. Download the Mentee and Mentory Application Forms at the website.
Geographic Focus: All States
Date(s) Application is Due: Jun 22
Amount of Grant: 1,000 USD
Contact: Margaret Rogers, Chief Staff Officer for Science & Research; (800) 498-2071 or (301) 296-5706; fax (301) 296-8580; mrogers@asha.org
Internet: http://www.asha.org/students/awards.htm#2008%20Research%20Mentoring-Pair%20Travel%20Award
Sponsor: American Speech-Language-Hearing Association
2200 Research Boulevard, #245
Rockville, MD 20850-3289

ASHA Student Research Travel Award **1074**
The awards are intended to highlight the research activities of budding scientists and encourage careers in science and research. Approximately 40 student research travel awards of $500 each are available for undergraduate, Master's, and doctoral students who are first authors on a paper (all session types are acceptable: technical, poster, one hour seminar, two hour seminar, or short course). Identification of prospective awardees will be based on the recommendation of Convention co-chairs and topic coordinators. Awardees will receive a $500 travel award; have their exceptional paper highlighted in the convention program; and, be recognized at the Researcher-Academic Town Meeting and other designated events.
Requirements: Each award recipient will be required to: register for the Convention at his or her own expense; present his or her paper at the assigned time (as noted in the Convention program booklet); secure a mentor to shadow throughout the convention; and, attend selected research-related sessions during the convention. In order to apply, students must: submit an abstract summarizing their research through the normal Call for Papers process; check the appropriate box (when prompted) indicating that their paper is 'student-authored'; and, check the appropriate box (when prompted) indicating that they desire to have this abstract considered for the Student Research Travel Award.
Restrictions: Papers must be submitted online by the deadline date and must be designated as a student-authored paper.
Geographic Focus: All States
Date(s) Application is Due: Mar 31
Amount of Grant: 500 USD
Contact: Margaret Rogers, Chief Staff Officer for Science & Research; (800) 498-2071 or (301) 296-5706; fax (301) 296-8580; mrogers@asha.org
Internet: http://www.asha.org/students/awards.htm#research
Sponsor: American Speech-Language-Hearing Association
2200 Research Boulevard, #245
Rockville, MD 20850-3289

ASHA Students Preparing for Academic & Research Careers Award **1075**
The goal of SPARC is to foster students' interest in the pursuit of PhD education and careers in academia in order to fill faculty/researcher vacancies in communication sciences and disorders (CSD). Students will identify a primary faculty mentor and propose a one-year CSD Career Mentoring Plan. As many as ten students will be awarded up to $1,500 each to be used for teaching and research enhancement activities, such as travel to a research, pedagogy conference, or meeting; travel for a visit to an off-campus site that provides learning opportunities in a research lab and/or a college classroom setting; or course registration to support the mentoring plans outlined in the application.
Requirements: Eligible applicants include junior or senior undergraduates, 1st year master's students, and 1st and 2nd year entry-level clinical doctoral (e.g., AuD) students enrolled part-time or full-time in a CSD program in the United States. At the time of the award, eligible recipients' education status for the upcoming academic year must be undergraduate senior, 1st or 2nd year master's, or 1st, 2nd or 3rd year entry-level clinical doctoral (i.e., AuD) in a full-time or part-time CSD program in the United States. Master's degree and entry-level clinical doctoral students must be in a program accredited by the Council for Academic Accreditation in Audiology and Speech-

Language Pathology (CAA). Full requirements (essay question, documentation, etc.) are listed at the website, along with the downloadable application form.
Geographic Focus: All States
Date(s) Application is Due: May 15
Amount of Grant: 5,000 USD
Contact: Arlene A. Pietranton; apietranton@asha.org or academicaffairs@asha.org
Internet: http://www.asha.org/students/awards.htm#sparc
Sponsor: American Speech-Language-Hearing Association
2200 Research Boulevard, #245
Rockville, MD 20850-3289

ASHFoundation Clinical Research Grants 1076
These $50,000-$75,000 grants support scientists with a research doctorate within the discipline of communication sciences and disorders to support investigations that will advance knowledge of the efficacy of treatment and assessment practices. Project funding is available for mentored treatment research, independent treatment research, or collaborative treatment research as specified in grant guidelines.
Geographic Focus: All States
Amount of Grant: 50,000-75,000 USD
Contact: Emily Diaz, Project Assistant; fax (301) 296-8703; ediaz@asha.org
Internet: http://www.ashfoundation.org/Foundation/grants/research_grants.htm
Sponsor: American Speech-Language-Hearing Foundation
2200 Research Boulevard
Rockville, MD 20850-3289

ASHFoundation Graduate Student Scholarships 1077
The foundation invites full-time graduate students to submit applications in competition for one of seven scholarships. Full-time master's or doctoral students in communication sciences and disorders programs demonstrating outstanding academic achievement are eligible to compete for these $4,000 awards. Supported in part by Psi Iota Xi National Philanthropic Organization and the Marni Reisberg Memorial Fund. All scholarship recipients will be announced and recognized at the annual convention held in November.
Requirements: Applicants must be accepted for graduate study in a communication sciences and disorders program (master's degree candidates must be in an ASHA Educational Standards Board accredited program; this is not mandatory for doctoral degree candidates); be enrolled for full-time study (12 or more credit hours or the full-time standard); submit official university transcripts of academic coursework, credits, and grades; and be recommended by a committee of two or more past or present college faculty and colleagues (at least one supervisor) at the student's current place of employment. Annual deadline dates may vary; contact program staff for exact dates.
Restrictions: Applicants must not have received a prior scholarship from the foundation.
Geographic Focus: All States
Amount of Grant: 4,000 USD
Contact: Emily Diaz, Project Assistant; fax (301) 296-8703; ediaz@asha.org
Internet: http://www.ashfoundation.org/Foundation/grants/GraduateScholarships.htm
Sponsor: American Speech-Language-Hearing Foundation
2200 Research Boulevard
Rockville, MD 20850-3289

ASHFoundation Graduate Student Scholarships for Minority Students 1078
Racial/ethnic minority students who are U.S. citizens, who are accepted for graduate study in speech-language pathology or audiology, and who demonstrate outstanding academic achievement are eligible to compete for this scholarship. Applicants should submit a formal paper, such as a term paper, in competition for this award. The paper must be either an integrative literature review or an opinion paper on a current professional issue. All scholarship recipients will be announced and recognized at the ASHA convention.
Requirements: Candidates must be U.S. citizens and members of a racial/ethnic minority, including Native American and Alaska Native, Asian and Pacific Islander, African American, and Hispanic. Applicants must be accepted for master's or doctoral study in a speech-language pathology or audiology program for the upcoming academic year. The student must be enrolled for full-time (12 or more credit hours) study.
Restrictions: Applicants may not have received a prior scholarship from the foundation.
Geographic Focus: All States
Amount of Grant: 2,000-4,000 USD
Contact: Emily Diaz, Project Assistant; fax (301) 296-8703; ediaz@asha.org
Internet: http://www.ashfoundation.org/Foundation/grants/GraduateScholarships.htm
Sponsor: American Speech-Language-Hearing Foundation
2200 Research Boulevard
Rockville, MD 20850-3289

ASHFoundation Graduate Student with a Disability Scholarship 1079
Full-time graduate students with a disability who are enrolled in a communication sciences and disorders program and demonstrate outstanding academic achievement are eligible to compete for a $4,000 scholarship. Applicants must demonstrate superior academic achievement and submit transcripts, essay and references with official application. The Award is intended for the blind, hearing impaired, physically handicapped, or learning impaired. Master's, but not doctoral, students must be enrolled in an ASHA Council on Academic Accreditation approved program.
Requirements: Candidates must be U.S. citizens and either blind, hearing impaired, physically handicapped, or learning impaired. Applicants must be accepted for master's study in a speech-language pathology or audiology program for the upcoming academic year. The student must be enrolled for full-time (12 or more credit hours) study.
Geographic Focus: All States
Amount of Grant: 4,000 USD
Contact: Emily Diaz, Project Assistant; fax (301) 296-8703; ediaz@asha.org
Internet: http://www.ashfoundation.org/Foundation/grants/GraduateScholarships.htm
Sponsor: American Speech-Language-Hearing Foundation
2200 Research Boulevard
Rockville, MD 20850-3289

ASHFoundation New Century Scholars Program Doctoral Scholarships 1080
The Program is a special funding initiative resulting from the ASHFoundation's Dreams and Possibilities Campaign. The Award is for students accepted or currently enrolled in a research doctoral program in communications sciences and disorders. Full-time study will be given priority; part-time status will be considered. Students should be committed to a teacher-investigator career in communication sciences and disorders in the United States.
Requirements: The Award is available to U.S. citizens only. Applicants must be accepted or currently enrolled in a research doctoral program.
Geographic Focus: All States
Amount of Grant: 10,000 USD
Contact: Emily Diaz, Project Assistant; fax (301) 296-8703; ediaz@asha.org
Internet: http://www.ashfoundation.org/Foundation/grants/GraduateScholarships.htm
Sponsor: American Speech-Language-Hearing Foundation
2200 Research Boulevard
Rockville, MD 20850-3289

ASHFoundation New Century Scholars Research Grant 1081
The Program is a special funding initiative resulting from the ASHFoundation's Dreams and Possibilities Campaign. The Grant is a one-time award intended to advance the knowledge base in communication sciences and disorders. Applicants must have a PhD or equivalent research doctorate within the discipline. Priority will be given to studies that are innovative, groundbreaking, or that meet research needs not yet addressed.
Restrictions: Students enrolled in a degree program or working on dissertation research are not eligible.
Geographic Focus: All States
Amount of Grant: 10,000 USD
Contact: Emily Diaz, Project Assistant; fax (301) 296-8703; ediaz@asha.org
Internet: http://www.ashfoundation.org/Foundation/grants/research_grants.htm
Sponsor: American Speech-Language-Hearing Foundation
2200 Research Boulevard
Rockville, MD 20850-3289

ASHFoundation Research Grant for New Investigators 1082
The grants are designed to help further research activities of new investigators and should have particular clinical relevance to speech-language pathology and audiology. The awards are designed to encourage research activities of new scientists who earned their latest degree within the last five years. Proposals, while not limited in topic, are encouraged in the area of treatment research, particularly efficacy and outcome studies. Include abstract, research plan, bibliography, management plan, and budget with application.
Requirements: The individual must have received the master's or doctoral level degree in the last five years; must not have received prior funding for research, with the exception of internal university funding; and the proposal must be for research to be initiated.
Restrictions: Open to individuals with a Master's degree or doctoral degree who are not currently enrolled in a degree program.
Geographic Focus: All States
Amount of Grant: 5,000 USD
Contact: Emily Diaz, Project Assistant; fax (301) 296-8703; ediaz@asha.org
Internet: http://www.ashfoundation.org/Foundation/grants/research_grants.htm
Sponsor: American Speech-Language-Hearing Foundation
2200 Research Boulevard
Rockville, MD 20850-3289

ASHFoundation Research Grant in Speech Science 1083
ASLH, in conjunction with the Acoustical Society of America, invites new researchers to submit proposals in competition for one grant every other year. The grant is designed to further research activities of new investigators and promulgate the work of the late Dennis Klatt, a noted researcher and scientist in the area of speech communication. Priority will be given to areas reflecting Dr. Klatt's broad interests, such as proposals studying speech perception, synthesis, and acoustics, with an emphasis on an interdisciplinary research approach. The grant can be used to initiate new research or supplement an existing research project. Funds may be requested for a variety of purposes, i.e., equipment, subjects, research assistants, or research-related travel. The grant recipient will be announced at the ASHA convention.
Requirements: Individuals having received a doctoral degree within the last five years and who wish to further research activities in the areas of speech communication are eligible to compete for the grant.
Geographic Focus: All States
Amount of Grant: 5,000 USD
Contact: Emily Diaz, Project Assistant; fax (301) 296-8703; ediaz@asha.org
Internet: http://www.ashfoundation.org/Foundation/grants/research_grants.htm
Sponsor: American Speech-Language-Hearing Foundation
2200 Research Boulevard
Rockville, MD 20850-3289

ASHFoundation Student Research Grants in Audiology 1084

Each year the foundation awards one grant for support of research to be initiated in the area of clinical and/or rehabilitative audiology. One $2,000 award to a master's or doctoral student for research to be initiated in the area of audiology. Applicants must submit up to a ten-page research plan, one-page abstract, two-page management plan and budget, and letter of support. All study must be done in the United States.

Requirements: The competition open to master's or doctoral degree students enrolled in, or accepted for, graduate study in a communication sciences and disorders program in the United States. Master's degree candidates must be in a program accredited by the Council on Academic Accreditation in Audiology and Speech-Language Pathology (CAA); this is not mandatory for doctoral degree candidates. Applicants must be enrolled for full-time study for the full academic year.

Geographic Focus: All States

Amount of Grant: 2,000 USD

Contact: Emily Diaz, Project Assistant; fax (301) 296-8703; ediaz@asha.org

Internet: http://www.ashfoundation.org/Foundation/grants/research_grants.htm

Sponsor: American Speech-Language-Hearing Foundation

2200 Research Boulevard

Rockville, MD 20850-3289

ASHFoundation Student Research Grants in Early Childhood Language Development 1085

One $2,000 grant per year is awarded to a graduate or postgraduate student to support research in early childhood language development. The applicant must submit up to a ten-page research plan, one-page abstract, two-page management plan and budget, and letter of support. All study must be done in the United States. The recipient and the recipient's mentor will be announced and recognized at the ASHA Convention.

Requirements: The competition is open to graduate or postgraduate students in the area of communication sciences and disorders. Applicants must submit a proposal according to foundation guidelines, to be received by the indicated deadline date for applications.

Geographic Focus: All States

Contact: Emily Diaz, Project Assistant; fax (301) 296-8703; ediaz@asha.org

Internet: http://www.ashfoundation.org/Foundation/grants/research_grants.htm

Sponsor: American Speech-Language-Hearing Foundation

2200 Research Boulevard

Rockville, MD 20850-3289

ASHG Public Health Genetics Fellowship 1086

The objectives of this co-sponsored Fellowship are to enhance the Center for Disease Control (CDC)'s current activities and develop new approaches in integrating genetics and genomics into public health policy and programs at the national, state, and local levels. A wide range of challenges and opportunities associated with the integration of genetics and genomics are emerging for health policy makers and program staff within these health agencies. Key issues of planning, workforce training, expansion of surveillance systems, health policy development, and changes in disease prevention programs are important program activities. The Fellow is expected to work independently and in conjunction with epidemiologists, geneticists, public health specialists, and communications specialists across CDC. Activities may include: planning and development of current and future projects in genetics and genomics with CDC; providing technical assistance to CDC programs, and collaboratively with CDC programs, providing assistance to state and local health agencies on the integration of genomics into policy and practice around specific diseases; collaborating with national public health organizations and professional organizations, including the American Society of Human Genetics (ASHG), in the planning and development of national public health strategies associated with genomics and disease prevention; participating in ASHG activities related to the work at CDC to promote timely transfer of information about policy and program development; participating in writing reports about the impact of genomics and public health policy and practice; enhancing awareness about the importance of genomics in public health by providing content for the ASHG website and materials for the Mentor network; assisting with the development and delivery of training and educational activities related to genomics and public health. This career development opportunity will offer a broad experience in the emerging role of genomics in public health practice. The Fellow will learn strategies for and practical skills in integrating genomics into specific health policies and programs associated with disease prevention and health promotion. Benefits include access to and cooperation of genetics and public health professionals to enhance collaboration between the fields. The Fellow will work closely with CDC staff and state and local health agencies to assess and develop the role of genomics in public health activities. The Fellow will gain insight and expertise through the training of, and with, public health professionals as well as development and delivery of workshops and case studies appropriate for a variety of audiences. The length of the fellowship is two years for early career candidates and one year for mid-career candidates. The Fellowship starts in July. Deadlines for application may vary from year to year. Interested applicants are encouraged to check the fellowship website and contact program staff for information on how to apply for the fellowship.

Requirements: Successful applicants must have either an M.D. or Ph.D. degree or a master's degree in genetics, genetic counseling, or a related field and experience or interest in public health genetics. Applicants must qualify as early career (3 years experience) or mid-career professionals.

Restrictions: Interested applicants are encouraged to call program staff with any questions on requirements and restrictions.

Geographic Focus: All States, All Countries

Date(s) Application is Due: Apr 23

Contact: Joann Boughman, Executive Vice-President; (866) 486-4363 or (301) 634-7300; fax (301) 238-3709; jboughman@ashg.org

Internet: http://www.ashg.org/pages/education_cdc.shtml

Sponsor: American Society of Human Genetics

9650 Rockville Pike

Bethesda, MD 20814-3998

Ashland Corporate Contributions Grants 1087

The foundation awards grants to eligible Kentucky, Ohio, and West Virginia nonprofit organizations in its areas of interest, including education, arts, communities/civic, disaster relief, environment, and health and human services. Types of support include seed money grants, project grants, endowments, matching gifts, and employee volunteers. The foundation does not provide an application. Most giving is centered on programs that best address the needs of Ashland's employees, stockholders, customers, and other constituencies, and the communities in which they live. Funding requests are not solicited but will be considered. The primary focus is on education, with an emphasis on mentoring, literacy and/or diversity.

Requirements: 501(c)3 tax-exempt organizations in Boyd and Greenup Counties, KY, Lawrence County, OH, and Cabell and Wayne Counties, WV, are eligible. Charitable groups within church or religious organizations are eligible.

Restrictions: The foundation does not support individuals, capital campaigns for building or equipment, endowments, travel, film or video production, tickets, religious or political activities, or goodwill advertising.

Geographic Focus: Kentucky, Ohio, West Virginia

Contact: Program Contact; (859) 815-3333; community@ashland.com

Internet: http://www.ashland.com/commitments/contributions.asp

Sponsor: Ashland

50 East Rivercenter Boulevard, P.O. Box 391

Covington, KY 41012-0391

ASHRAE Graduate Student Grant-in-Aid Program 1088

The Grant is intended to encourage the student to continue his or her preparation for service in the HVAC&R industry. The relevance of the research proposed by the candidate is a consideration for awarding the grant. The Grant-in-Aid is made through the university solely for the support of the student in an amount not to exceed U.S. $10,000.00 per school year per student and is not renewable. Typical expenditures include living expenses, tuition, travel to ASHRAE meetings, experimental equipment, and supplies. The basis for scoring heavily emphasizes ASHRAE involvement and is intended to stimulate advisor activity in ASHRAE and associate societies.

Requirements: Qualified graduate engineering students capable of carrying out appropriate and scholarly research are eligible to apply.

Geographic Focus: All States

Date(s) Application is Due: Dec 15

Amount of Grant: Up to 10,000 USD

Contact: Manager; (404) 636-8400; fax (404) 321-5478; mvaughn@ashrae.org

Internet: http://www.ashrae.org/template/AssetDetail/assetid/23027

Sponsor: American Society of Heating, Refrigerating, and Air Conditioning Engineers

1791 Tullie Circle NE

Atlanta, GA 30329-2305

ASHRAE Willis H. Carrier Graduate Research Fellowship 1089

This fellowship is awarded for graduate research at Purdue University. The fellowship is for one year with monies dispersed by the University to the student. Completed applications must be mailed to the ASHRAE. Application forms are available on the Web site.

Requirements: Applicants should have a cumulative college grade point average of at least 3.0 (where 4.0 is the highest); current full-time enrollment with at least one full year of studies remaining; potential service to the HVAC and/or refrigeration profession; need for financial assistance; leadership ability; and recommendations from instructors.

Geographic Focus: All States

Amount of Grant: 21,000 USD

Contact: Administrator; (765) 494-6900; PurdueME@ecn.purdue.edu

Internet: http://www.ashrae.org/template/AssetDetail/assetid/23628

Sponsor: American Society of Heating, Refrigerating, and Air Conditioning Engineers

1791 Tullie Circle NE

Atlanta, GA 30329-2305

Asia Foundation Exchange Programs 1090

The foundation seeks to encourage greater understanding between Asians and Americans with the ultimate aim of contributing towards strengthened U.S.-Asia relations. Exchange programs???administered by the San Francisco-based Asian-American Exchange (AAX) unit???help shape the perceptions of the Asians and Americans who take part in academic studies, conferences, special programs, and study tours tailored to particular needs that address U.S.-Asia interests. The foundation also sponsors the following special exchange programs: the William P. Fuller Fellowship in Conflict Resolution; the Chang-Lin Tien Distinguished Fellowship Program; the L.Z. Yuan Fellowship in Media and International Affairs; India Regional Security Studies Fellowship; and the Freeman Program for Broadening American Engagement with Southeast Asia. Guidelines are available online.

Geographic Focus: All States

Contact: Administrator; (415) 982-4640; fax (415) 392-8863; info@asiafound.org

Internet: http://www.asiafoundation.org

Sponsor: Asia Foundation

465 California Street, 9th Floor, P.O. Box 193223

San Francisco, CA 94104

Asian Pacific American Institute for Congressional Studies Fellowships 1091
The fellowships are designed to provide unique opportunities to outstanding graduate students who have a commitment to the Asian Pacific Islander American communities, and who plan to pursue a career in public service. The fellows will spend nine months in Washington, D.C., in a Congressional office, in another branch of the government, or a non-profit organization. One fellow will spend nine months in the Congressional office of the Chair of the Asian Pacific American Caucus (CAPAC). These programs are made possible by the support of the following companies: Anheuser-Busch/Frank Horton Fellowship; Wal-Mart/Governor George R. Ariyoshi Fellowship; Sodexho Health and Policy Fellow; National Association of Realtors Fellow. APAICS will provide a total stipend of $20,000 to cover travel arrangements, housing and personal expenses for nine months. A separate stipend will be provided for basic health insurance coverage. In order to receive the full-stipend, the fellow will be required to participate in mandatory APAICS events, submit a report of activities, and complete an evaluation at the end of the program.
Requirements: Postgraduate and graduate applicants must meet the following *Requirements:* 1) Demonstrated interest in the political process; 2) Demonstrated commitment to public policy issues and Asian Pacific Islander American community affairs; 3) Demonstrated leadership abilities; 4) Relevant work experience; 5) Excellent oral and written communication skills; 6) Must have a graduate or bachelor's degree from an accredited educational institution. Minimum cumulative grade point average of 3.0 (on a 4.0 scale); 7) U.S. citizenship or legal permanent residency by September 1; 8) Submission of all application materials (Part I & II) by mail and email. All materials must be postmarked by the deadline.
Restrictions: Faxed documents will not be accepted.
Geographic Focus: All States
Date(s) Application is Due: May 31
Amount of Grant: 20,000 USD
Contact: Ruby G. Moy; (202) 296-9200; rubyapaics@gmail.com
Internet: http://www.apaics.org/pages/programs/fellowships.php
Sponsor: Asian Pacific American Institute for Congressional Studies
1001 Connecticut Avenue NW, Suite 530
Washington, DC 20036

ASID Foundation Dora Brahms Award 1092
One award is made on a biennial basis to educational institutions on behalf of their students in historic preservation and/or restoration studies, to encourage and support the advancement of professional activities in historic preservation and/or restoration. Institutions must illustrate how the award will best assist the student and the historic preservation and design education community.
Geographic Focus: All States
Date(s) Application is Due: Apr 30
Amount of Grant: 3,000 USD
Contact: Jack Pruitt; (202) 546-3480 or (202) 675-2365; education@asid.org
Internet: http://www.asid.org/asidfoundation/Foundation+Scholarships+and+Awards.htm
Sponsor: American Society of Interior Designers Foundation
608 Massachusetts Avenue, NE
Washington, DC 20002-6006

ASID Foundation Grants 1093
The ASID Foundation supports endeavors that capture and disseminate knowledge, encourage innovation, and benefit the health, safety and welfare of the public through interior design research, scholarships and education. Critical topics of interest include: sustainable design; technology; design for learning environments; and the needs of an aging population. Application is open to organizations and individuals.
Geographic Focus: All States
Date(s) Application is Due: Mar 1; Sep 1
Contact: Jack Pruitt; (202) 546-3480 or (202) 675-2365; education@asid.org
Internet: http://www.asidfoundation.org/
Sponsor: American Society of Interior Designers Foundation
608 Massachusetts Avenue, NE
Washington, DC 20002-6006

ASID Foundation Irene Winifred Eno Grant 1094
The Grant award in the amount of $5,000 is intended to provide financial assistance to individuals or groups engaged in the creation of an educational program or an interior design research project dedicated to health, safety and welfare. The grant will be awarded on the basis of the project description, break down of potential use of funds and the marketing plan for the use/distribution of the end product of the project.
Requirements: The grant is open to students, educators, interior design practitioners, institutions or other interior design related group.
Geographic Focus: All States
Date(s) Application is Due: Apr 30
Contact: Jack Pruitt; (202) 546-3480 or (202) 675-2365; education@asid.org
Internet: http://www.asid.org/asidfoundation/Foundation+Scholarships+and+Awards.htm
Sponsor: American Society of Interior Designers Foundation
608 Massachusetts Avenue, NE
Washington, DC 20002-6006

ASID Foundation Joel Polsky Academic Achievement Award 1095
This prize is given annually to recognize an outstanding interior design research or thesis project by an undergraduate or graduate student. Entries should address the needs of the public, designers, and students on such topics as educational research, behavioral science, business practice, design process, theory, or other technical subjects. Material will be judged on bibliography/references, breadth of material, comprehensive coverage of topic, graphic presentation and organization, and innovative subject matter. Entries will be returned upon request with a $10 postage and handling fee. Annual deadline dates may vary.
Geographic Focus: All States
Date(s) Application is Due: Apr 30
Amount of Grant: 1,000 USD
Contact: Jack Pruitt; (202) 546-3480 or (202) 675-2365; education@asid.org
Internet: http://www.asid.org/asidfoundation/Foundation+Scholarships+and+Awards.htm
Sponsor: American Society of Interior Designers Foundation
608 Massachusetts Avenue, NE
Washington, DC 20002-6006

ASID Foundation Joel Polsky Prize 1096
The Prize is given by the ASID Foundation to recognize outstanding academic contributions to the discipline of interior design through literature or visual communication. Entries should address the needs of the public, designers and students on such topics as educational research, behavioral science, business practice, design process, theory or other technical subjects. Material will be judged on innovative subject matter, comprehensive coverage of topic, organization, graphic presentation, bibliography and references.
Geographic Focus: All States
Date(s) Application is Due: Apr 30
Amount of Grant: 1,000 USD
Contact: Jack Pruitt; (202) 546-3480 or (202) 675-2365; education@asid.org
Internet: http://www.asid.org/asidfoundation/Foundation+Scholarships+and+Awards.htm
Sponsor: American Society of Interior Designers Foundation
608 Massachusetts Avenue, NE
Washington, DC 20002-6006

ASID Foundation Mabelle Wilhelmina Boldt Memorial Scholarship 1097
One scholarship will be awarded to a student who is enrolled in or has applied for admission to a graduate-level interior design program at a degree-granting institution. The scholarship will be awarded on the basis of academic/creative accomplishment, as demonstrated by school transcripts and a letter of recommendation. Preference will be given to students with a focus on design research. Annual deadline dates may vary.
Requirements: Applicants must have been practicing designers for a period of at least five years prior to returning to graduate study.
Geographic Focus: All States
Date(s) Application is Due: Apr 30
Contact: Jack Pruitt; (202) 546-3480 or (202) 675-2365; education@asid.org
Internet: http://www.asid.org/asidfoundation/Foundation+Scholarships+and+Awards.htm
Sponsor: American Society of Interior Designers Foundation
608 Massachusetts Avenue, NE
Washington, DC 20002-6006

ASM-PAHO Infectious Diseases Epidemiology and Surveillance Fellowships 1098
Funded by the Pan American Health Organization and ASM, this award offers fellowships to young scientists from Bolivia, Colombia, Costa Rica, Dominican Republic, Ecuador, El Salvador, Guatemala, Honduras, Nicaragua, Panama, Paraguay, or Peru, who have obtained their Masters, PhD, or other equivalent academic degree, and who have at least five years of laboratory experience in the area of antimicrobial resistance, to visit a host scientist in the U.S.. The ASM-PAHO Infectious Diseases Epidemiology and Surveillance Fellowship will award a maximum grant of $4,000. The award is not intended to provide travel to obtain a degree at the host institution.
Requirements: The application must be made jointly between a host institution and the visiting fellow. The Investigator must be: a member of ASM or any other national microbiological society; actively involved in research in the microbiological sciences; have obtained, or be in the process of obtaining, their masters, PhD, or other equivalent academic degree within the last five years; a national of a resource-limited Africa, Asian or Latin American country submitting the application from an institution in their home region. Investigators from developed nations, as defined by BOTH the UN high HDI group and the World Bank high income economy group, are ineligible. Investigators submitting an application from the U.S. where they are spending a training period are ineligible; sufficiently proficient in the use of the English language. The Host Scientist must be: an ASM member; actively involved in research in the microbiological sciences in the U.S; actively involved in teaching in the microbiological sciences in the U.S; interested in sustaining international collaborations Preference will be given to: applicants who can prove three years of membership in ASM or any other national microbiological society, and; applicants who have not previously had the opportunity to travel to a facility in another country. The application should demonstrate to the Review Committee: academic excellence of the applicant-Honors & Awards, CV, and Letters of Recommendation; depth of the applicant's research experience-Statement of prior Research Experience; quality and originality of work proposed during the Fellowship-Potential Collaboration and Proposed Research Plan; relevance of the work proposed to the applicant's locale-Potential Collaboration and Proposed Research Plan; excellence of the host, and the research and training environment
Geographic Focus: All States, Argentina, Bolivia, Brazil, Chile, Colombia, Ecuador, Guyana, Paraguay, Peru
Date(s) Application is Due: Apr 15; Oct 15
Contact: Supervisor; (202) 942-9368; fax (202) 942-9328; international@asmusa.org
Internet: http://www.asm.org/International/index.asp?bid=2778
Sponsor: American Society for Microbiology
1752 N Street, N.W.
Washington, DC 20036-2904

ASM-UNESCO Leadership Grant for International Educators 1099

This new program, sponsored jointly by ASM and UNESCO, has been developed to enable a select group of educators from resource-limited countries to attend the ASM Conference for Undergraduate Educators (ASMCUE) and a pre-conference workshop to provide leaders in education with the resources to build innovative teaching modules that engage students and lead to enduring understandings in microbiology. Successful applicants will receive financial support to cover airfare (economy class), room, board and registration to the pre-conference workshop and ASMCUE. Participants will be reimbursed for airfare via wire transfer upon submission of a confirmed invoice. Room, board and registration will be arranged by ASM.

Requirements: Preference will be given to ASM members. Applicants must be a citizen of, and residing in, a resource-limited country in Latin America, Asia, Africa, and Central and Eastern Europe. Resource-limited countries are classified as upper-middle income and below by the World Bank Development Group. (To find out if your country fits this criteria, visit http://tinyurl.com/7he3u.) Participants must have a minimum of eight years teaching in microbiology/biology at a university level and must also have regular and sustained access to the Internet.

Restrictions: Members of the ASM International Board, International Education Committee, International Membership Committee, International Laboratory Capacity Building Committee, and the ASM International Ambassador Network are not eligible for this program.

Geographic Focus: All States

Date(s) Application is Due: Dec 1

Contact: Michelle Godinez, Coordinator, Education Programs and Events; (202) 942-9317; fax (202) 942-9328; mgodinez@asmusa.org

Internet: http://www.asm.org/International/index.asp?bid=61306

Sponsor: American Society for Microbiology
1752 N Street, N.W.
Washington, DC 20036-2904

ASM/CDC Fellowships in Infectious Disease and Public Health Microbiology 1100

The ASM/CDC Postdoctoral Research Fellowship Program is sponsored by the American Society for Microbiology (ASM) and the Centers of Disease Control and Prevention (CDC). Founded in 1899, the American Society for Microbiology (ASM) promotes the microbiological sciences and their application for the common good. The Society accomplishes its mission by disseminating information, stimulating research, promoting education, and advancing the profession. ASM's headquarters office is located in Washington, D.C. The Centers for Disease Control and Prevention (CDC) is the agency of the United States Public Health Service (PHS) responsible for developing and applying disease control and prevention methods and improving the health of the citizens of the United States. The CDC has been a leader in efforts to prevent and control infectious diseases for nearly half a century. CDC supports surveillance, research, prevention efforts, and training in the area of infectious diseases. The goal of the two-year ASM/CDC Fellowships is to support the development of new approaches, methodologies and knowledge in infectious disease prevention and control in areas within the public health mission of the CDC. The program awards up to ten fellowships annually. The fellowships are similar to those at the postdoctoral level in universities, industry, and other government laboratories and allow bright, highly motivated, recent doctoral graduates to perform research-in-residence while headquartered at a CDC location. Participating CDC organizational units are located in: Atlanta, Georgia; Fort Collins, Colorado; and San Juan, Puerto Rico and are comprised of: the Division of Bacterial Diseases; the Division of Foodborne, Bacterial and Mycotic Diseases; the Division of Healthcare Quality Promotion; the Division of HIV/AIDS Prevention; the Division of Influenza; the Division of Viral Hepatitis; the Division of Parasitic Diseases; the Division of Tuberculosis Elimination; the Divisions of Vector-Borne Infectious Diseases; the Division of Viral and Rickettsial Diseases; the Division of Viral Diseases; and the Division of Viral Hepatitus. The fellowship appointments are for two years assuming satisfactory progress. A stipend (up to $45,243 annually), health benefits (up to $3000 annually), relocation benefits (up to $500), and professional development funds (up to $2000 annually) are provided. Fellows are classified as fellows and are not employees of either CDC or ASM but rather have the status of a visiting scientist. Applicants must use the ASM online application system to apply for the fellowship. The link to the online system is available at the program website on the top right corner of the screen. Applicants are encouraged to review the "Full Program Catalog" to obtain complete information about the program, research opportunities, and the application process. The link to the catalog is available at the program website. The ASM/CDC Postdoctoral Research Fellowship Program is sponsored by the U.S. Department of Health and Human Services.

Requirements: The program is intended for individuals who either earned their doctorate degree or completed a primary residency within three years of their proposed start date (considerations will be given to individuals with more experience if there are compelling reasons). Non-U.S. nationals whose primary language is not English must demonstrate command of the English language through the Test of English as a Foreign Language (TOEFL). These applicants may also be subject to an oral interview. Non-U.S. nationals who are offered awards must have valid visas throughout their tenure. Fellows will be responsible for managing all their own visa matters. Fellows must devote their full-time effort to the approved research project and must be in residence at the sponsoring CDC laboratory for the entire two-year period. Fellows must begin the program between October 1 and December 31 during the year in which they are accepted. All fellows must undergo security clearance to work at CDC. Support after the first year of the Program is contingent upon satisfactory progress in the proposed research. Interim and final reports and/or presentations are required. Fellows are expected to publish results of their studies in the scientific literature and to present findings at national meetings. Fellows are subject to the general regulations of the laboratory and are expected to conduct research according to the highest scientific and ethical standards and in compliance with all applicable laws, regulations, and policies regarding protection of human research subjects, humane care and use of laboratory animals, and laboratory safety. Participation in CDC seminar programs is also expected. After completion of the two-year program, fellows are required to keep ASM informed of their current position, affiliation, and continued professional development for a minimum of ten years. This arrangement will ensure proper assessment and longitudinal studies regarding the effectiveness of such training.

Restrictions: Applicants may not have a faculty position or be enrolled in a graduate degree program during the fellowship. Monies from other appointments, fellowships, private employment, or consulting or contract work are not permitted during the two-year appointment.

Geographic Focus: All States, All Countries

Date(s) Application is Due: Jan 15

Amount of Grant: Up to 100,986 USD

Contact: Tiffani Fonseca, Coordinator, Fellowship Programs; (202) 942-9295 or (202) 942-9283; fax (202) 942-9329; fellowships@asmusa.org or tfonseca@asmusa.org

Internet: http://www.asm.org/asm/index.php/education/asmcdc-program-in-infectious-disease-and-public-health-microbiology.html

Sponsor: American Society for Microbiology
1752 N Street, N.W.
Washington, DC 20036-2904

ASM Abbott Laboratories Award in Clinical and Diagnostic Immunology 1101

The ASM Abbott Laboratories Award in Clinical and Diagnostic Immunology honors a distinguished scientist in clinical or diagnostic immunology for outstanding contributions to those fields.

Requirements: The nominee must demonstrate significant contributions to the understanding of the functioning of the host immune system in human disease, clinical approaches to diseases involving the immune system or development or clinical application of immunodiagnostic procedures.

Restrictions: Self-nominations will not be accepted.

Geographic Focus: All States

Date(s) Application is Due: Oct 1

Amount of Grant: 2,000 USD

Contact: Awards Committee; (202) 942-9226; fax (202) 942-9353; awards@asmusa.org

Internet: http://www.asm.org/Academy/index.asp?bid=2301

Sponsor: American Society for Microbiology
1752 N Street, N.W.
Washington, DC 20036-2904

ASM BD Award for Research in Clinical Microbiology 1102

The ASM BD Award for Research in Clinical Microbiology honors a a distinguished scientist for research accomplishments that form the foundation for important applications in clinical microbiology.

Requirements: The nominee must be a distinguished clinical microbiologist. All nomination components must be emailed to awards@asmusa.org. Nominations must consist of the following: curriculum vitae; letter of nomination; letters of support.

Restrictions: Self-nominations will not be accepted.

Geographic Focus: All States

Date(s) Application is Due: Oct 1

Contact: Awards Committee; (202) 942-9226; fax (202) 942-9353; awards@asmusa.org

Internet: http://www.asm.org/Academy/index.asp?bid=2304

Sponsor: American Society for Microbiology
1752 N Street, N.W.
Washington, DC 20036-2904

ASM bioMerieux Sonnenwirth Award for Leadership in Clinical Microbiology 1103

Given in memory of Alexander Sonnenwirth, Ph.D., this award recognizes a distinguished microbiologist for the promotion of innovation in clinical laboratory science, dedication to ASM, and the advancement of clinical microbiology as a profession.

Requirements: Nominations will be considered without updating for three years. email (see contact info.) all nomination components. Nominations must consist of the following: curriculum vitae, including a list of nominee's publications; letter of nomination, describe the nominee's innovation in clinical laboratory science, dedication to ASM, role as a leader in the field of diagnostic microbiology, and the advancement of clinical microbiology as a profession; letters of support, two letters of support must come from persons, other than the nominator, who are familiar with the nominee's qualifications and accomplishments. No more than one of the three letters may be from the nominee's institution or the same institution.

Restrictions: Self-nominations will not be accepted.

Geographic Focus: All States

Date(s) Application is Due: Oct 1

Contact: Awards Committee; (202) 942-9226; fax (202) 942-9353; awards@asmusa.org

Internet: http://www.asm.org/Academy/index.asp?bid=36368

Sponsor: American Society for Microbiology
1752 N Street, N.W.
Washington, DC 20036-2904

ASME Allan Kraus Thermal Management Medal 1104

The ASME Allan Kraus Thermal Management Medal recognizes individuals who have demonstrated outstanding achievements in thermal management of electronic systems and their commitment to the field of thermal sciences. The award includes a bronze medal, a certificate, $1,000, and travel expenses to attend the award presentation. Nomination forms and instructions are available at the website.

Geographic Focus: All States, All Countries
Date(s) Application is Due: Mar 1
Amount of Grant: 1,000 USD
Contact: A. Bar-Cohen, Committee Chair; (301) 405-3173; abc@umd.edu
Internet: http://www.asme.org/about-asme/honors-awards/achievement-awards/allan-kraus-thermal-management-medal
Sponsor: American Society of Mechanical Engineers
3 Park Avenue
New York, NY 10016

ASME Arthur L. Williston Medal **1105**
The ASME Arthur L. Williston Medal is presented for the best paper or thesis submitted by an undergraduate or junior engineer fostering a spirit of civil service. Candidates must be student members or associate members with an ASME sponsor for the award. Recipients are awarded a bronze medal, certificate, $1,000, and travel expenses to attend the award presentation. Second place award winners receive a certificate and $500, while third place winners receive a certificate and $250.
Requirements: Applicant must be a student member or an associate member of ASME who has not graduated more than two years prior.
Geographic Focus: All States, All Countries
Date(s) Application is Due: Feb 7
Amount of Grant: 250-1,000 USD
Contact: G.A. Anderkay, Committee Chair; (212) 591-7094; mckivorf@asme.org
Internet: http://www.asme.org/honors/ms71/list.html
Sponsor: American Society of Mechanical Engineers
3 Park Avenue
New York, NY 10016

ASME Barnett-Uzgiris Product Safety Design Award **1106**
The ASME Barnett-Uzgiris Design Award recognizes individuals who have made significant contributions to the safe design of products through teaching, research, and professional accomplishments. The award includes a certificate, $1,000, and travel expenses to attend the award presentation. Nomination forms and instructions are available at the website.
Geographic Focus: All States, All Countries
Date(s) Application is Due: Dec 1
Amount of Grant: 1,000 USD
Contact: Y. Jin, Committee Chair; (213) 740-9574; yjin@usc.edu
Internet: http://www.asme.org/about-asme/honors-awards/achievement-awards/barnett-uzgiris-product-safety-award
Sponsor: American Society of Mechanical Engineers
3 Park Avenue
New York, NY 10016

ASME Ben C. Sparks Medal **1107**
The ASME Ben C. Sparks Medal is presented to a mechanical engineer for recognition of eminent service to mechanical engineering technology education through contributions toward ASME; higher education professional service; and the teaching and/or research facility roles in mechanical engineering and mechanical engineering technology. The award includes a bronze metal, a certificate, $1,000, and travel expenses to attend the award presentation. Nomination forms and instructions are available at the website.
Requirements: Candidates must have a record of outstanding performance in all of these areas: a broad range of contributions to ASME; dedicated efforts and a record of achievements in the advancement of engineering education and engineering technology education; and contributions to the positive support, recognition, or interaction of mechanical engineering technology and mechanical engineering programs, their faculties, or students. Recipients must hold current academic appointments in ABET-accredited departments where mechanical engineering and/or mechanical engineering technology is taught.
Geographic Focus: All States, All Countries
Date(s) Application is Due: Mar 10
Contact: C. Corum, Committee Chair; (765) 494-7530; corumc@purdue.edu
Internet: http://www.asme.org/about-asme/honors-awards/achievement-awards/ben-c—sparks-medal
Sponsor: American Society of Mechanical Engineers
3 Park Avenue
New York, NY 10016

ASME Bergles-Rohsenow Young Investigator Award in Heat Transfer **1108**
The ASME Bergles-Rohsenow Young Investigator Award in Heat Transfer is given to a young engineer no older than 35 years of age who has received a Ph.D. or an equivalent degree in engineering. The individual must be committed to pursuing research and making significant contributions in heat transfer. Such contributions may include analytical/numerical methods, equipment/instrumentation, or experimentation, which should lead to peer-reviewed publications. The award includes a bronze metal, a certificate, and $1,000. Nomination forms and instructions are available at the website.
Geographic Focus: All States, All Countries
Date(s) Application is Due: Oct 1
Contact: L.A. Kennedy, Committee Chair; (312) 413-7560; lkennedy@uic.edu
Internet: http://www.asme.org/about-asme/honors-awards/achievement-awards/bergles-rohsenow-young-investigator-award-in-heat
Sponsor: American Society of Mechanical Engineers
3 Park Avenue
New York, NY 10016

ASME Bernard F. Langer Nuclear Codes and Standards Award **1109**
The ASME Bernard F. Langer Award recognizes individuals who have contributed to the nuclear power plant industry through the development and promotion of ASME nuclear codes and standards or the ASME nuclear certification program. The award includes $1,000, a crystal oracle, and a certificate. Nomination instructions and forms are available at the website.
Geographic Focus: All States, All Countries
Date(s) Application is Due: Feb 1
Contact: R.R. Weidler, Committee Chair; (704) 364-2086; rrweidler@prodigy.net
Internet: http://www.asme.org/about-asme/honors-awards/achievement-awards/bernard-f—langer-nuclear-codes—-standards-award
Sponsor: American Society of Mechanical Engineers
3 Park Avenue
New York, NY 10016

ASME Blackall Machine Tool Award **1110**
The ASME Blackall Machine Tool Award is given annually for the best paper or papers clearly concerned with or related to the design or application of machine tools or dimensional measuring instruments, submitted to ASME for presentation and publication. Awardees receive $1,000 and a plaque. Papers by multiple authors are eligible. The award is made annually if warranted. Nomination forms and instructions are available at the website.
Geographic Focus: All States, All Countries
Date(s) Application is Due: Feb 1
Contact: S.M. Athavale; (313) 248-6061; sathaval@ford.com or awards@asme.org
S.N. Melkote, Committee Chair; (404) 894-8499; Shreyes.melkote@me.gatech.edu
Internet: http://www.asme.org/about-asme/honors-awards/literature-awards/blackall-machine-tool-and-gage-award
Sponsor: American Society of Mechanical Engineers
3 Park Avenue
New York, NY 10016

ASME Burt L. Newkirk Award **1111**
The ASME Newkirk Award is given to candidates who have made a notable contribution to the field of tribology in research or development as evidenced by important tribology publications. Tribology for this award is defined as pertaining to the science and technology associated with surfaces in contact and relative motion with each other. It covers all the fundamentals associated with the field of friction, lubrication, and wear. It embraces all the technological aspects of bearings, brakes, clutches, and gears. The award winner is given a certificate and $1,000.
Requirements: Nominees must be under the age of 40 and members of ASME for at least five years.
Geographic Focus: All States, All Countries
Date(s) Application is Due: Feb 1
Amount of Grant: 1,000 USD
Contact: H.P Nixon, Committee Chair; (330) 832-7581; Harvey.Nixon@att.net
Internet: http://www.asme.org/about-asme/honors-awards/achievement-awards/burt-l—newkirk-award
Sponsor: American Society of Mechanical Engineers
3 Park Avenue
New York, NY 10016

ASME Calvin W. Rice Lecture Award **1112**
The ASME Calvin W. Rice Lecture honors Calvin Rice, who served as Secretary of the Society from 1906 to 1934. It is intended to continue the effort for which Rice was known, to increase understanding among engineers of various countries. The Subcommittee looks for a lecturer and topic to fit the purpose of the lecture. It is also important that the quality and prestige of the lecture continue on a high level. Preference is given to a lecturer whose accomplishments have been achieved overseas. The award includes a certificate, a lifetime membership with ASME, and a $4,000 honorarium to offset expenses incurred to present the lecture. Nomination forms and instructions are available at the website.
Requirements: The nomination package should include the following: a letter from the principal nominator addressing how the nominee meets the criteria for the award; supporting letters from a minimum of three references that show why the nominee should be considered for the award; curriculum vitae of the nominee listing educational and professional background, awards, publications, and other information pertinent to the nomination.
Geographic Focus: All States, All Countries
Date(s) Application is Due: Feb 1
Amount of Grant: 4,000 USD
Contact: Jacinta McComie-Cates; (212) 591-7052; mccomiej@asme.org
Internet: http://www.asme.org/about-asme/honors-awards/unit-awards/calvin-w—rice-lecture-award
Sponsor: American Society of Mechanical Engineers
3 Park Avenue
New York, NY 10016

ASME Charles Russ Richards Memorial Award **1113**
The ASME Charles Russ Richards Memorial Award, jointly sponsored by Pi Tau Sigma Honorary Mechanical Engineering Fraternity and ASME, is given annually to the engineering graduate who has demonstrated outstanding achievement in mechanical engineering 20 years or more after graduation from the regular engineering course of a recognized college or university. Achievement shall be all or in part in any field

including industrial, educational, political, research, civic, and artistic. The candidate's achievements will be examined for an application of basic engineering methods or principles. Nomination forms and instructions are available at the website. The award includes a certificate, $1,000, and travel expenses to attend the award presentation.
Geographic Focus: All States, All Countries
Date(s) Application is Due: Feb 15
Amount of Grant: 1,000 USD
Contact: A. Moutsoglou, Committee Chair; (605) 688-6323; Alex.Moutsoglou@SDSTATE.EDU or awards@asme.org
Internet: http://www.asme.org/about-asme/honors-awards/achievement-awards/charles-russ-richards-memorial-award
Sponsor: American Society of Mechanical Engineers
3 Park Avenue
New York, NY 10016

ASME Daniel C. Drucker Medal 1114
The ASME Daniel C. Drucker Medal is given in recognition of distinguished contributions to the field of applied mechanics and mechanical engineering through research, teaching and service to the community over a substantial period of time. The award includes a bronze medal, a certificate, $1,000, and travel expenses to attend the award presentation. Nomination forms and instructions are available at the website.
Geographic Focus: All States, All Countries
Date(s) Application is Due: Nov 5
Amount of Grant: 1,000 USD
Contact: A.J. Rosakis, Committee Chair; (626) 395-4523; arosakis@caltech.edu
Internet: http://www.asme.org/about-asme/honors-awards/achievement-awards/daniel-c—drucker-medal
Sponsor: American Society of Mechanical Engineers
3 Park Avenue
New York, NY 10016

ASME Dixy Lee Ray Award 1115
The ASME Dixy Lee Ray Award recognizes significant achievements and contributions in the broad field of environmental protection. Achievement in the following areas will be recognized: environmental engineering, including environmental technology and related topics; environmental health, environmental sciences, environmental management and policy. The award include a bronze metal, a certificate, $1,000, and travel expenses to attend the award presentation. Nomination forms and instructions are available at the website.
Geographic Focus: All States, All Countries
Date(s) Application is Due: Feb 1
Amount of Grant: 1,000 USD
Contact: G.A. Benda, Committee Chair; (803) 345-2170; gbenda_use@hotmail.com
Internet: http://www.asme.org/about-asme/honors-awards/achievement-awards/dixy-lee-ray-award
Sponsor: American Society of Mechanical Engineers
3 Park Avenue
New York, NY 10016

ASME Edward F. Obert Award 1116
The ASME Edward F. Obert Award recognizes an outstanding paper on thermodynamics authored during the preceding two calendar years. The award shall be made for a paper presented at the Advanced Energy Systems Division's International Mechanical Engineering Congress symposium on energy systems analysis (so long as the symposium continues). The award shall be presented at the following ASME Congress. The award includes a certificate, $5,000, and travel expenses to attend the award presentation. Nomination forms and instructions are available at the website.
Restrictions: The paper nominated must have been written during the preceding two years of the presentation.
Geographic Focus: All States, All Countries
Date(s) Application is Due: Mar 1
Amount of Grant: 5,000 USD
Contact: K.V. Wong, Committee Chair; (305) 284-3314; kwong@miami.edu
Internet: http://www.asme.org/about-asme/honors-awards/literature-awards/edward-f—obert-award
Sponsor: American Society of Mechanical Engineers
3 Park Avenue
New York, NY 10016

ASME Edwin F. Church Medal 1117
The ASME Edwin F. Church Medal is given annually to an individual who has rendered eminent service in increasing the value, importance, and attractiveness of mechanical engineering education. The award includes $2500, a bronze medal, certificate, and travel expenses to attend the award presentation. Mechanical engineering is used here in its broadest sense of preparation for any aspect or level of mechanical engineering through any appropriate mechanism including universities, technical institutes, professional societies and private groups, in-house professional development programs of industrial concerns and governmental agencies, or programmed learning and self-instruction systems. The nomination instructions and form are available at the website.
Requirements: The award may be made to one recipient of any age who need not be a member of ASME.
Geographic Focus: All States, All Countries
Date(s) Application is Due: Feb 15
Amount of Grant: 2,500 USD
Contact: J.M. Oehring, Committee Chair; (281) 345-5374; J.Oehring@ssss.com or awards@asme.org
Internet: http://www.asme.org/honors/ms71/gaa/church.html
Sponsor: American Society of Mechanical Engineers
3 Park Avenue
New York, NY 10016

ASME Engineer-Historian Award 1118
The ASME History and Heritage Award Committee recognizes outstanding published work or works by an engineer dealing with the history of mechanical engineering through its Engineer-Historian Award. Preference is given to those publications dealing with the art and science of mechanical engineering. The award is intended to encourage the active interest by mechanical engineers in the history of their profession. The award with its honorarium and certificate is presented annually, if merited, at the International Mechanical Engineering Congress. Nominations for the Engineer-Historian Award are accepted year-round for consideration. Candidates may remain under consideration in following years.
Requirements: Candidates can be nominated by any ASME member, with submission accompanied by 10 photocopies of at least one published work to be distributed to Committee members. The names of these candidates as well as previously nominated candidates are considered during the regular meeting agenda. Deadline for submission of a nomination is one month prior to the Spring meeting (approximately April 1), for the following meeting's annual presentation. Criteria includes subject matter, depth of treatment, technical level of material, and the historical significance of the subject matter. Candidate must be or have been active in the practice of the art and science of mechanical engineering, including managers, teachers, and museum professionals directly concerned with mechanical engineering historical artifacts (provided they have been trained as and practiced as mechanical engineers). Multiple co-authors are permitted as long as all authors fit the above definition. Self nominations are accepted. The award is international in scope, and open to members and nonmembers.
Restrictions: History and Heritage Committee members are not eligible. Submissions must be in English or translated.
Geographic Focus: All States, All Countries
Date(s) Application is Due: Mar 1
Amount of Grant: 1,000 USD
Contact: Edward Kudlis; (212) 591-8614; kudlis@asme.org or stenosm@asme.org
Internet: http://www.asme.org/about-asme/honors-awards/unit-awards/engineer-historian-award-%281%29
Sponsor: American Society of Mechanical Engineers
3 Park Avenue
New York, NY 10016

ASME Fluids Engineering Award 1119
The ASME Fluids Engineering Award is given for outstanding contributions over a period of years to the engineering profession and especially to the field of fluids engineering through research, practice, and/or teaching. The award includes a bronze medal, a certificate, and $1,000. Nomination forms and instructions are available at the website.
Geographic Focus: All States, All Countries
Date(s) Application is Due: Aug 30
Amount of Grant: 1,000 USD
Contact: Y.T. Lee, Committee Chair; (301) 227-1328; yu.lee@navy.mil
Internet: http://www.asme.org/about-asme/honors-awards/achievement-awards/fluids-engineering-award
Sponsor: American Society of Mechanical Engineers
3 Park Avenue
New York, NY 10016

ASME Frank Kreith Energy Award 1120
The ASME Frank Kreith Energy Award honors individuals for significant contributions to a secure energy future with particular emphasis on innovations in conservation and/or renewable energy. Contributions may be through research, education, practice, or significant service to society that will lead to a sustainable energy future. The award includes a plaque, a certificate, $2,000, and travel expenses to attend the award presentation. Nomination forms and instructions are available at the website.
Geographic Focus: All States, All Countries
Date(s) Application is Due: Dec 1
Amount of Grant: 2,000 USD
Contact: B.G. Shiva Prasad; (937) 775-5040; bellur.shivaprasad@wright.edu
Internet: http://www.asme.org/about-asme/honors-awards/achievement-awards/frank-kreith-energy-award
Sponsor: American Society of Mechanical Engineers
3 Park Avenue
New York, NY 10016

ASME Frank von Flue Award 1121
The Frank von Flue Award recognizes significant contributions by an individual or a group to the promotion of lifelong learning for mechanical engineers. Lifelong learning is defined as the continued process of education and/or professional development, following a post-secondary education degree, to maintain the competence and flexibility of mechanical engineers throughout their career. Contributions may include assessment, development, distribution or promotion of materials for lifelong learning via appropriate mechanisms, such as professional development or university extension courses and programmed or self-paced learning. The award includes a bronze medal, a certificate, $1,000, and travel expenses to attend the award presentation. Nomination forms and instructions are available at the website.

Geographic Focus: All States, All Countries
Date(s) Application is Due: Feb 1
Amount of Grant: 1,000 USD
Contact: W Bryan, Committee Chair; (724) 514-2883; William.Bryan@ansys.com
Internet: http://www.asme.org/about-asme/honors-awards/achievement-awards/frank-von-flue-award
Sponsor: American Society of Mechanical Engineers
3 Park Avenue
New York, NY 10016

ASME Freeman Scholar Award 1122
The ASME Freeman Scholar Award is given every two years in even-numbered years to a person with extensive experience in fluids engineering. The applicant is expected to review a coherent topic in their specialty and suggest key future research needs. The awardee is given $7,500. The results will be presented at the winter meeting and published in the ASME Journal of Fluids Engineering. Deadline date for applications and nominations is September 1 of odd-numbered years. The application is available on the website.
Requirements: The recipient may be from industry, government, education, or private professional practice, and need not be an ASME member.
Geographic Focus: All States, All Countries
Date(s) Application is Due: Sep 1
Amount of Grant: 7,500 USD
Contact: D.E. Stock, Committee Chair; (509) 335-3223; stock@wsu.edu
Internet: http://www.asme.org/about-asme/honors-awards/literature-awards/freeman-scholar-award
Sponsor: American Society of Mechanical Engineers
3 Park Avenue
New York, NY 10016

ASME Gas Turbine Award 1123
The ASME Gas Turbine Award is given in recognition of an outstanding individual or multiple-author contribution to the literature of combustion gas turbines or gas turbines thermally combined with nuclear or steam power plants. The paper may be devoted to the design aspects of overall gas turbines or individual components and/or systems such as compressors, combustion systems, turbines, controls and accessories, bearings, regenerators, inlet air filters, silencers, etc. It may cover topics specifically related to gas turbines, such as high-temperature materials or fuel considerations including erosion and corrosion complications. It also can be devoted to application or operational aspects of gas turbines for aircraft propulsion and ground power units, or automotive, electric utility, gas pipeline pumping, locomotive, marine, oil field pumping, petrochemical, space power, steel, and similar uses. The award includes a plaque and $1,000.
Requirements: Papers published anywhere in the world are eligible; authors are not restricted by nationality, age, profession, or membership in any engineering society or other organizations.
Geographic Focus: All States, All Countries
Date(s) Application is Due: Jun 1
Amount of Grant: 1,000 USD
Contact: A.J. Strazisar; (216) 433-5881; anthony.j.strazisar@nasa.gov
Internet: http://www.asme.org/about-asme/honors-awards/literature-awards/gas-turbine-award
Sponsor: American Society of Mechanical Engineers
3 Park Avenue
New York, NY 10016

ASME George Westinghouse Medals 1124
The ASME George Westinghouse Medals are bestowed annually for eminent achievement or distinguished service in the power field of mechanical engineering. Considering power in the broad sense, the basis of the awards shall include contributions of utilization, application, design, development, research, and the organization of such activities in the power field. Any individual member, group of members, or committee may nominate candidates for the awards. Awardees receive $1,500, a vermeil medal, a certificate, a travel expenses to attend the award presentation. A silver medal, certificate, $1,000, and travel expenses are awarded to those under 45 years of age.
Requirements: Candidates are not restricted by profession nor by membership in any engineering society or organization.
Geographic Focus: All States, All Countries
Date(s) Application is Due: Feb 1
Amount of Grant: 1,000-1,500 USD
Contact: D.G. Lilley; (405) 372-6427; DGL@firedynamics.com or awards@asme.org
Internet: http://www.asme.org/about-asme/honors-awards/achievement-awards/george-westinghouse-medals
Sponsor: American Society of Mechanical Engineers
3 Park Avenue
New York, NY 10016

ASME Graduate Teaching Fellowships 1125
The ASME Graduate Teaching Fellowships are a collaboration between ASME and mechanical engineering departments to encourage outstanding doctoral candidates in mechanical engineering education (and related engineering fields), particularly women and minorities. Fellowship awards of $5,000 a year are made for a maximum of two years. Fellows are selected or renewed annually by the ASME Board on Education. The current application is posted on the website.
Requirements: Applicants must be U.S. citizens or permanent residents, Ph.D. students in mechanical engineering with an interest in a teaching career, and members or student members of ASME. Applicants must submit undergraduate grade point averages (GPA), GRE scores, a current resume or vitae, two letters of recommendation from faculty or M.S. committee, transcripts of all academic work, and a statement about a faculty career. A Master's degree or passage of qualifying exam is required. A lecture-responsibility teaching assistantship commitment from the applicant's department is also required.
Geographic Focus: All States
Date(s) Application is Due: Feb 15
Amount of Grant: 5,000 USD
Contact: Beth Lefever; (212) 591-7790; fax (212) 591-7143; lefeverb@asme.org
Internet: http://www.asme.org/about-asme/scholarship-and-loans/graduate-teaching-fellowships
Sponsor: American Society of Mechanical Engineers
3 Park Avenue
New York, NY 10016

ASME Gustus L. Larson Memorial Award 1126
The Gustus L. Larson Memorial Award, sponsored jointly by the Pi Tau Sigma Honorary Mechanical Engineering Fraternity and ASME, is given annually to the engineering graduate who has demonstrated outstanding achievement in mechanical engineering within 10 to 20 years after graduating from a regular engineering course of a recognized college or university. Achievement shall be all or in part in any field including industrial, educational, political, research, civic, and artistic. The candidate's achievements will be examined for an application of basic engineering methods or principles. Awardees are given a certificate, $1,000, and travel expenses to attend the award presentation. The nomination form with instructions is available on the website.
Geographic Focus: All States, All Countries
Date(s) Application is Due: Feb 15
Amount of Grant: 1,000 USD
Contact: A. Moutsoglou, Special Awards Committee Chair; (605) 688-6323; Alex.Moutsoglou@SDSTATE.EDU or awards@asme.org
Internet: http://www.asme.org/about-asme/honors-awards/achievement-awards/gustus-l—larson-memorial-award
Sponsor: American Society of Mechanical Engineers
3 Park Avenue
New York, NY 10016

ASME H.R. Lissner Award 1127
The ASME H.R. Lissner Award was established in honor of H.R. Lissner for his pioneering contributions in biomechanics. The award is bestowed annually for outstanding accomplishments in the area of bioengineering. These achievements may be in the form of (1) significant research contributions in bioengineering; (2) development of new methods of measuring in bioengineering; (3) design of new equipment and instrumentation in bioengineering; (4) educational impact in the training of bioengineers; and/or (5) service to the bioengineering community, in general, and to the Bioengineering Division of ASME, in particular. The award includes a bronze medal, certificate, $1,000, and travel expenses to attend the award presentation. Nomination forms and instructions are available at the website. Nomination packets will be held and considered for three years; a new nomination packet must be submitted after that time.
Requirements: Candidate must be an active member of the bioengineering division.
Geographic Focus: All States
Date(s) Application is Due: Sep 1
Contact: T.P. Andriacchi, Committee Chair; (650) 559-9170; tandriac@stanford.edu
Internet: http://www.asme.org/about-asme/honors-awards/achievement-awards/h-r—lissner-medal
Sponsor: American Society of Mechanical Engineers
3 Park Avenue
New York, NY 10016

ASME Heat Transfer Memorial Award 1128
The ASME Heat Transfer Memorial Award is bestowed on individuals who have made outstanding contributions to the field of heat transfer through teaching, research, design, or publications. Each award is based on papers in an area of heat transfer or on a paper dealing with the science or art of heat transfer. Awards are made annually in each of the three following categories: the science of heat transfer, the art of heat transfer, and the general subject of heat transfer. The awardees receive a plaque, along with a certificate and $1,000.
Requirements: There are no restrictions by age, nationality, or society membership.
Geographic Focus: All States, All Countries
Date(s) Application is Due: Oct 1
Contact: Y.Y. Bayazitoglu; (713) 348-6291; bayaz@rice.edu or awards@asme.org
Internet: http://www.asme.org/about-asme/honors-awards/achievement-awards/heat-transfer-memorial-award
Sponsor: American Society of Mechanical Engineers
3 Park Avenue
New York, NY 10016

ASME Henry Hess Award 1129
The ASME Henry Hess Award is given for an original technical paper submitted to ASME for presentation and publication. Joint authorship is permissible provided all authors meet the requirements. The paper shall be specifically recommended for the award by a review committee or a qualified individual. Nomination forms and instructions are available at the website.

Requirements: Author must not be older than 31 at the time of submission and must be an ASME student member or associate member to be eligible to apply.
Geographic Focus: All States, All Countries
Date(s) Application is Due: Mar 1
Amount of Grant: 2,500 USD
Contact: G.A. Anderkay; (212) 591-7094; mckivorf@asme.org or awards@asme.org
Internet: http://www.asme.org/about-asme/honors-awards/literature-awards/henry-hess-award
Sponsor: American Society of Mechanical Engineers
3 Park Avenue
New York, NY 10016

ASME Henry R. Worthington Medal **1130**
The ASME Henry R. Worthington Medal is given annually for eminent achievement in the field of pumping machinery, systems and concepts. Such achievement may be in the areas of research, development, design, innovation, management, education, or literature. Awardees are presented with a Bronze Medal, certificate, $5,000, and travel expenses to attend the award presentation. Any individual member, group of members, or committee may nominate candidates for the award. Nomination forms and instructions are available at the website.
Geographic Focus: All States, All Countries
Date(s) Application is Due: Feb 1
Amount of Grant: 5,000 USD
Contact: P. Collins; (713) 753-6270; ditullio@asme.org or awards@asme.org
Internet: http://www.asme.org/honors/ms71/saa/worthington.html
Sponsor: American Society of Mechanical Engineers
3 Park Avenue
New York, NY 10016

ASME Holley Medal **1131**
The ASME Holley Medal is given to an individual who has contributed a unique engineering benefit for the public good. Nominees are not limited by education, societal or organizational membership, or an official position. The achievement should be worthy of public gratitude and admiration of engineers. More than one individual may be recognized for a single achievement provided that each individual made equal or comparable contribution. Nomination forms and instructions are available at the website. Awardees are presented with a vermeil medal, certificate, and $1,000.
Requirements: Any individual member, group of members, or committee may nominate candidates for this award.
Geographic Focus: All States, All Countries
Date(s) Application is Due: Mar 1
Amount of Grant: 1,000 USD
Contact: G.A. Anderkay; (212) 591-7736; fax (212) 705-7739; mckivorf@asme.org
Internet: http://www.asme.org/about-asme/honors-awards/achievement-awards/holley-medal
Sponsor: American Society of Mechanical Engineers
3 Park Avenue
New York, NY 10016

ASME Internal Combustion Engine Award **1132**
The ASME Internal Combustion Engine Award is given in recognition of eminent achievement or distinguished contribution over a substantial period of time, which may result from research, innovation, or education in advancing the art of engineering in the field of internal combustion engines or in directing the efforts and accomplishments of those engaged in engineering practice in the design, development, application, and operation of internal combustion engines. The awardee receives a plaque and $1,000. Nomination instructions and forms are available at the website.
Geographic Focus: All States, All Countries
Date(s) Application is Due: Feb 1
Amount of Grant: 1,000 USD
Contact: N. Blythe; (814) 875-3983; neil.blythe@ge.com or awards@asme.org
Internet: http://www.asme.org/about-asme/honors-awards/achievement-awards/internal-combustion-engine-award
Sponsor: American Society of Mechanical Engineers
3 Park Avenue
New York, NY 10016

ASME J. Hall Taylor Medal **1133**
The ASME J. Hall Taylor Medal is presented for distinguished service or eminent achievement in codes and standards pertaining to the fields of piping and pressure vessels that are sponsored or undertaken by ASME. The scope shall include contributions to technical advancement and administration. The awardee receives a Bronze Medal, certificate, and $1,000. The award was established to commemorate the pioneering work of J. Hall Taylor in the field of standardization of industrial products and safety codes for their usage. Nomination instructions and forms are available at the website.
Geographic Focus: All States, All Countries
Date(s) Application is Due: Nov 15
Amount of Grant: 1,000 USD
Contact: B.W. Roberts; (423) 892-3597; blainewroberts@aol.com or awards@asme.org
Internet: http://www.asme.org/about-asme/honors-awards/achievement-awards/j—hall-taylor-medal
Sponsor: American Society of Mechanical Engineers
3 Park Avenue
New York, NY 10016

ASME J.P. Hartog Award **1134**
The ASME J.P. Hartog Award recognizes lifetime contributions to the teaching and practice of vibration engineering. The award includes a bronze medal, certificate, and $3,000, and is given in odd-numbered years. Nomination forms and instructions are available at the website.
Geographic Focus: All States, All Countries
Date(s) Application is Due: Feb 1
Amount of Grant: 3,000 USD
Contact: B.F. Feeny, Committe Chair; (517) 353-9451; deadams@ecn.purdue.edu
Internet: http://www.asme.org/about-asme/honors-awards/achievement-awards/j-p—den-hartog-award
Sponsor: American Society of Mechanical Engineers
3 Park Avenue
New York, NY 10016

ASME James Harry Potter Gold Medal **1135**
The ASME James Harry Potter Gold Medal is awarded in recognition of eminent achievement or distinguished service in the appreciation of the science of thermodynamics in mechanical engineering. The basis of the award shall include contributions involving the teaching, appreciation, or utilization of thermodynamic principles in research, development, and design in mechanical engineering. The award includes $2,000, a vermeil medal, and a certificate. Any individual member, group of members, or committee may nominate candidates for consideration of this award. Nomination applications and instructions are available on the website.
Geographic Focus: All States, All Countries
Date(s) Application is Due: Feb 1
Amount of Grant: 2,000 USD
Contact: T. Raviguruajan; (316) 978-6370; ts.ravi@wichita.edu or awards@asme.org
Internet: http://www.asme.org/about-asme/honors-awards/achievement-awards/james-harry-potter-gold-medal
Sponsor: American Society of Mechanical Engineers
3 Park Avenue
New York, NY 10016

ASME James N. Landis Medal **1136**
The ASME James N. Landis Medal is given for outstanding personal performance related to designing, constructing, or managing the operation of major steam-powered electric stations using nuclear or fossil fuels, coupled with personal leadership in humanitarian pursuits. This may include committee activity, ASME section leadership, or the broad nontechnical professional activity of the nominee's engineering society. The award is presented preferably to a member of ASME, and includes $7,500, bronze medal, certificate, and travel expenses to attend the award presentation. Nomination forms and instructions are available at the website.
Geographic Focus: All States, All Countries
Date(s) Application is Due: Feb 1
Amount of Grant: 7,500 USD
Contact: Committee Chair; (212) 591-7736; awards@asme.org
Internet: http://www.asme.org/honors/ms71/saa/landis.html
Sponsor: American Society of Mechanical Engineers
3 Park Avenue
New York, NY 10016

ASME Kate Gleason Award **1137**
The ASME Kate Gleason Award was created to recognize the contributions of distinguished female leaders in the engineering profession. Kate Gleason was the first woman to be welcomed into ASME as a full member, and helped pave the way for other women to enter and thrive in the field of engineering. The Gleason Award seeks to honor an individual female engineer who is a highly successful entrepreneur in a field of engineering or someone who had a lifetime of achievement in the engineering profession. The award includes a bronze medal, a certificate, $5,000, and travel expenses to attend the award presentation. Nomination forms and instructions are available at the website.
Geographic Focus: All States, All Countries
Date(s) Application is Due: Feb 1
Amount of Grant: 5,000 USD
Contact: Ward Winer, Committee Chair; (404) 385-2905; ward.winer@me.gatech.edu
Internet: http://www.asme.org/about-asme/honors-awards/achievement-awards/kate-gleason-award
Sponsor: American Society of Mechanical Engineers
3 Park Avenue
New York, NY 10016

ASM Eli Lilly and Company Research Award **1138**
ASM's oldest and most prestigious prize, it rewards fundamental research of unusual merit in microbiology or immunology by an individual on the threshold of his or her career.
Requirements: The nominee's work is not judged by comparison with the research of more experienced scientists, and special consideration is given to originality and independence of thought. The nominee must not have reached his or her 45th birthday by April 30 of the year the award is given. The nominee must be working in the United States or Canada and be actively engaged in the line of research for which the award is to be made.
Restrictions: Nominations must consist of the following: curriculum vitae, including a list of publications; verification of date of birth. Photocopy or scanned file of driver's license, passport or birth certificate; letter of nomination, describe the nominee's fundamental research of unusual merit in microbiology or immunology; letters of support, two letters

of support must come from persons, other than the nominator, who are familiar with the nominee's qualifications and accomplishments. No more than one of the three letters may be from the nominee's institution or the same institution.
Geographic Focus: All States
Date(s) Application is Due: Oct 1
Amount of Grant: 5,000 USD
Contact: Awards Committee; (202) 942-9226; fax (202) 942-9353; awards@asmusa.org
Internet: http://www.asm.org/academy/index.asp?bid=2281
Sponsor: American Society for Microbiology
1752 N Street, N.W.
Washington, DC 20036-2904

ASME M. Eugene Merchant Manufacturing Medal 1139
The ASME M. Eugene Merchant Manufacturing Medal of ASME/SME is awarded to an individual who has had significant influence and responsibility for improving the productivity and efficiency (either by research or by implementation of research) of the manufacturing operation(s). The award includes a vermeil medal, a certificate, and $1,500. Nomination forms and instructions are available at the website.
Geographic Focus: All States, All Countries
Date(s) Application is Due: Feb 1
Amount of Grant: 1,500 USD
Contact: K. Rajurkar, Committee Chair; (402) 472-1385; krajurkar1@unl.edu
Internet: http://www.asme.org/about-asme/honors-awards/achievement-awards/m—eugene-merchant-manufacturing-medal-of-asme-sme
Sponsor: American Society of Mechanical Engineers
3 Park Avenue
New York, NY 10016

ASME Machine Design Award 1140
The ASME Machine Design Award is made to recognize eminent achievement or distinguished service in the field of machine design, which includes application, research, development, or teaching of machine design. Recipients receive $1,000, a plaque, and a certificate. Nomination forms and instructions are available at the website.
Requirements: Any individual member, group of members, or committee may nominate candidates for this award.
Geographic Focus: All States, All Countries
Date(s) Application is Due: Feb 1
Amount of Grant: 1,000 USD
Contact: Y. Jin, Committee Chair; (213) 740-9574; yjin@usc.edu or awards@asme.org
Internet: http://www.asme.org/about-asme/honors-awards/achievement-awards/machine-design-award
Sponsor: American Society of Mechanical Engineers
3 Park Avenue
New York, NY 10016

ASME Marshall B. Peterson Award 1141
The ASME Marshall B. Peterson Award is given in recognition of early career achievements in tribology research, as demonstrated by papers published in ASME scientific journals. It is preferred, but not required, that the awardee's research emphasis be on materials aspects of tribology, such as wear, friction materials, and solid lubricants. The award includes a certificate and $2,500. Nomination forms and instructions are available at the website.
Geographic Focus: All States, All Countries
Date(s) Application is Due: Feb 1
Amount of Grant: 2,500 USD
Contact: H.P. Nixon, Committee Chair; (330) 832-7581; Harvey.Nixon@att.net
Internet: http://www.asme.org/about-asme/honors-awards/achievement-awards/marshall-b—peterson-award
Sponsor: American Society of Mechanical Engineers
3 Park Avenue
New York, NY 10016

ASME Mayo D. Hersey Award 1142
The ASME Mayo D. Hersey Award is bestowed on an individual in recognition of distinguished contributions over a substantial period of time to the advancement of the science and engineering of tribology. Distinguished contributions may result from significant original research in one or more of the many scientific disciplines related to lubrication, from excellence and creativity in lubrication engineering practice, or from sustained efforts and dissemination of information on the theory and practice of lubrication. The award includes a plaque and $2,000. Nomination forms and instructions are available at the website.
Geographic Focus: All States, All Countries
Date(s) Application is Due: Feb 1
Amount of Grant: 2,000 USD
Contact: H.P. Nixon, Committee Chair; (330) 471-2046; Harvey.Nixon@att.net
Internet: http://www.asme.org/about-asme/honors-awards/achievement-awards/mayo-d—hersey-award
Sponsor: American Society of Mechanical Engineers
3 Park Avenue
New York, NY 10016

ASME McDonald Mentoring Award 1143
The ASME McDonald Mentoring Award recognizes the outstanding mentoring of other professionals by an engineer in industry, government, education or private practice. The award includes a bronze medal, a certificate, $1,000, and travel expenses to attend the award presentation. Nomination forms and instructions are available at the website.
Requirements: Candidates for this award must have attained a record of outstanding performance in the following areas: contribute significantly to the engineering community, and serve as an effective advocate and guide, exhibiting true concern for mentees; demonstrate a portion of their mentoring or advising activities within the past five years; demonstrate the impact of practices, processes, and programs established by the nominee; current memberships with ASME, or another ICOMES member society for at least five years; possess a baccalaureate or equivalent degree in recognized field of engineering or engineering science, but not necessary in mechanical engineering.
Geographic Focus: All States, All Countries
Date(s) Application is Due: Feb 1
Amount of Grant: 1,000 USD
Contact: J. Baxter, Committee Chair; 44 019 3 276 3933; John.Baxter@bp.com
Internet: http://www.asme.org/about-asme/honors-awards/achievement-awards/mcdonald-mentoring-award
Sponsor: American Society of Mechanical Engineers
3 Park Avenue
New York, NY 10016

ASME Medal 1144
The American Society of Mechanical Engineers (ASME) Medal is the highest award that the Society can bestow and is to recognize "eminently distinguished engineering achievement." Only one ASME Medal may be awarded annually. Although some individuals have been honored by both the ASME Medal and Honorary Membership, each award has been made on the basis of different accomplishments. Awardees receive $15,000, a gold medal, certificate, and travel expenses to the attend the award presentation. Nomination forms and instructions are available at the website.
Geographic Focus: All States, All Countries
Date(s) Application is Due: Mar 1
Amount of Grant: 15,000 USD
Contact: K.A. Thole; (212) 591-7094; mckivorf@asme.org or awards@asme.org
Internet: http://www.asme.org/about-asme/honors-awards/achievement-awards/asme-medal
Sponsor: American Society of Mechanical Engineers
3 Park Avenue
New York, NY 10016

ASME Melville Medal 1145
The ASME Melville Medal is the highest ASME honor for the best current original paper, not published elsewhere, which has been presented before ASME or approved for publication by ASME during the two previous calendar years. The paper may have more than one author but one of the authors must be an ASME corporate member (fellow, member, or associate member). The paper must be specifically recommended for the medal by a review committee or qualified individual. The medal may be awarded for a paper that has been selected for some other best paper award in the society if specifically recommended by the committee responsible for that specific award. Nomination forms and instructions are available at the website.
Geographic Focus: All States, All Countries
Date(s) Application is Due: Mar 1
Amount of Grant: 2,000 USD
Contact: G.A. Anderkay; (212) 591-7094; mckivorf@asme.org or awards@asme.org
Internet: http://www.asme.org/about-asme/honors-awards/literature-awards/melville-medal
Sponsor: American Society of Mechanical Engineers
3 Park Avenue
New York, NY 10016

ASME Melvin R. Green Codes and Standards Medal 1146
The ASME Melvin R. Green Codes and Standards Medal recognizes outstanding contributions to: (1) the development, promulgation, or management of documents, objects, or devices used in ASME programs of technical codification, standardization, and conformity assessment, or (2) the acceptance of ASME Codes and Standards within the United States or internationally. The award includes a bronze medal, a certificate, and $1,500. Nomination forms and instructions are available at the website.
Requirements: The recipient may be from industry, government, education, or private professional practice, and need not be an ASME member.
Geographic Focus: All States, All Countries
Date(s) Application is Due: Jan 1
Amount of Grant: 1,500 USD
Contact: G.E. Feigel; (860) 722-5652; feigelr@asme.org or awards@asme.org
Internet: http://www.asme.org/about-asme/honors-awards/achievement-awards/melvin-r—green-codes—-standards-medal
Sponsor: American Society of Mechanical Engineers
3 Park Avenue
New York, NY 10016

ASME Milton C. Shaw Manufacturing Research Medal 1147
The ASME Milton C. Shaw Manufacturing Research Medal recognizes significant fundamental contributions to the science and technology of manufacturing processes. The award includes a medal, certificate, and $1,500. Nomination forms and instructions are available at the website.

Geographic Focus: All States, All Countries
Date(s) Application is Due: Feb 1
Amount of Grant: 1,500 USD
Contact: R. Komanduri; (405) 744-5900; ranga.komanduri@okstate.edu
Internet: http://www.asme.org/about-asme/honors-awards/achievement-awards/milton-c—shaw-manufacturing-research-medal
Sponsor: American Society of Mechanical Engineers
3 Park Avenue
New York, NY 10016

ASME Nadai Medal **1148**
The ASME Nadai Medal is awarded in recognition of significant contributions and outstanding achievements which broaden the field of materials engineering. Such achievements may be in the areas of education, research, development, and service to the field and profession. The award includes a bronze medal, a certificate, and $1,000. The nomination forms and instructions are available at the website.
Geographic Focus: All States, All Countries
Date(s) Application is Due: Feb 1
Amount of Grant: 1,000 USD
Contact: M. Zikry, Committee Chair; (919) 515-5237; zikry@eos.ncsu.edu
Internet: http://www.asme.org/about-asme/honors-awards/achievement-awards/nadai-medal
Sponsor: American Society of Mechanical Engineers
3 Park Avenue
New York, NY 10016

ASME Nancy DeLoye Fitzroy and Roland V. Fitzroy Medal **1149**
The Nancy DeLoye Fitzroy and Roland V. Fitzroy Medal recognizes pioneering contribution to the frontiers of engineering leading to a breakthrough(s) in existing technology or leading to new applications or new areas of engineering endeavor. The award include a bronze metal, a certificate, $3,000, and travel expenses to attend the award presentation. The nomination forms and instructions are available at the website.
Geographic Focus: All States, All Countries
Date(s) Application is Due: Feb 1
Amount of Grant: 3,000 USD
Contact: N.D. Fitzroy, Committee Chair; (518) 377-9704; nfitzroy@earthlink.net
Internet: http://www.asme.org/about-asme/honors-awards/achievement-awards/nancy-deloye-fitzroy-and-roland-v—fitzroy-medal
Sponsor: American Society of Mechanical Engineers
3 Park Avenue
New York, NY 10016

ASME Old Guard Early Career Award **1150**
The ASME Old Guard Early Career Award recognizes outstanding engineers who have advanced quickly in their professional career and education, shown leadership in ASME activities, and volunteered in their communities. The Old Guard is made up of ASME members who have reached the age of 65 and retired. These members continue to contribute to the Society, and their contributions are used to support the ASME Student and Early Career Engineer Development. The first place award includes a plaque, prepaid ASME membership for life, and $5,000. The second place award includes a plaque and prepaid ASME membership for life. Nomination forms and instructions are available at the website.
Geographic Focus: All States, All Countries
Date(s) Application is Due: Feb 1
Amount of Grant: 5,000 USD
Contact: D.P. Jordan, Committee Chair; (806) 794-6188; jordand@asme.org
Internet: http://www.asme.org/about-asme/honors-awards/achievement-awards/old-guard-early-career-award
Sponsor: American Society of Mechanical Engineers
3 Park Avenue
New York, NY 10016

ASME Old Guard Prizes **1151**
The Old Guard Prizes are awarded for the best four contenders at the Old Guard Oral Presentation Competition during the ASME International Mechanical Engineering Congress and Exposition (IMECE). These competitors are student members who won first prize at the district level competitions held at ASME Student Professional Development Conferences. The district level first prizes also include partial travel support to participate in the IMECE. The Old Guard is made up of ASME dues exempt members, those who have reached the age of 65 and have retired. They continue to contribute to the Society and their contributions are used to support the "Old Guard" Prizes and similar Society activities related to Student and Early Career members. Awards include the following: 1st prize: $2,000; 2nd prize: $1,500; 3rd prize: $1,000; and 4th prize: $500. Further information including specific rules, criteria, and current deadline for the competition is available at the website.
Geographic Focus: All States, All Countries
Amount of Grant: 5,000 USD
Contact: D.P. Jordan, Committee Chair; (806) 794-6188; jordand@asme.org
Internet: http://www.asme.org/about-asme/honors-awards/achievement-awards/old-guard-prizes
Sponsor: American Society of Mechanical Engineers
3 Park Avenue
New York, NY 10016

ASME Patrick J. Higgins Award **1152**
The ASME Patrick J. Higgins Award recognizes an individual who has contributed to the enhancement of standardization through the development and promotion of ASME Codes and Standards or Conformity Assessment Programs in the following areas: plumbing materials & equipment; screw threads; tools (machine, cutting and hand); fasteners, (e.g, bolts, nuts, rivets, screws, washers); chain attachments and sprockets for power transmission and conveying; metal and metal alloy wrought mill product nominal sizes; pressure and temperature instruments and accessories; classification and designation of surface qualities; gage blanks; chemical standard pumps; dimensional metrology; gas turbine procurement; overhead hoists; measurement of fluid flow in closed conduits; steel stacks; and engineering drawing and related documentation practices. The award includes a bronze metal, a certificate, and $1,000. Nomination forms and instructions are available at the website.
Geographic Focus: All States, All Countries
Date(s) Application is Due: Feb 1
Amount of Grant: 1,000 USD
Contact: D.W. Viola, Committee Chair; (909) 996-5513; dave.viola@iapmo.org
Internet: http://www.asme.org/about-asme/honors-awards/achievement-awards/patrick-j—higgins-award
Sponsor: American Society of Mechanical Engineers
3 Park Avenue
New York, NY 10016

ASME Per Bruel Gold Medal for Noise Control and Acoustics **1153**
The ASME Per Bruel Gold Medal honors Dr. Per Bruel who pioneered the development of sophisticated noise and vibration measuring and processing equipment. The Medal is given annually in recognition of eminent achievement and extraordinary merit in the field of noise control and acoustics. The achievement must include useful applications of the principles of noise control and acoustics to the art and science of mechanical engineering. Award includes a vermeil medal, certificate, and travel expenses to attend the award presentation. Nomination instructions and forms are located on the website.
Geographic Focus: All States, All Countries
Date(s) Application is Due: Feb 1
Amount of Grant: 1,000 USD
Contact: A. Akay; (412) 268-2503; akay@andrew.cmu.edu or awards@asme.org
Internet: http://www.asme.org/about-asme/honors-awards/achievement-awards/per-bruel-gold-medal-for-noise-control-and-acousti
Sponsor: American Society of Mechanical Engineers
3 Park Avenue
New York, NY 10016

ASME Performance Test Codes Medal **1154**
The ASME Performance Test Codes Medal was established to recognize the first voluntary codes and standards activity in the Society and in the United States. The codes are recognized throughout the world for their excellence, providing industry and the engineering community with the technology that promotes the philosophy of accurate and reliable performance evaluation. The Medal is awarded to an individual (or individuals in exceptional circumstances) who has made an outstanding contribution to the development and promotion of ASME performance test codes including the supplements on instruments and apparatus. The award includes a vermeil medal, a certificate, and $1,000. Nomination forms and instructions are available at the website.
Requirements: Any individual member, group of members, or committee may nominate candidates for this award.
Geographic Focus: All States, All Countries
Date(s) Application is Due: Jan 1
Amount of Grant: 1,000 USD
Contact: J.W. Milton; (713) 497-7083; miltonj1@asme.org or awards@asme.org
Internet: http://www.asme.org/about-asme/honors-awards/achievement-awards/performance-test-codes-medal
Sponsor: American Society of Mechanical Engineers
3 Park Avenue
New York, NY 10016

ASME Pi Tau Sigma Gold Medal **1155**
The ASME Pi Tau Sigma Gold Medal, sponsored jointly by the Pi Tau Sigma Honorary Mechanical Engineering Fraternity and ASME, is awarded annually to a young engineering graduate who has demonstrated outstanding achievement in mechanical engineering no more than ten years after receiving their baccalaureate degree. Achievement shall be all or in part in any field, including industrial, educational, political, research, civic, or artistic. The candidate's achievements will be examined for an application of basic engineering methods or principles. Special nomination form must be obtained from ASME. The award includes gold medal, certificate, $1,000, and travel expenses to attend the award presentation. Nomination forms and instructions are available at the website.
Geographic Focus: All States, All Countries
Date(s) Application is Due: Feb 15
Amount of Grant: 1,000 USD
Contact: A. Moutsoglou; (605) 688-6323; Alex.Moutsoglou@SDSTATE.EDU
Internet: http://www.asme.org/about-asme/honors-awards/achievement-awards/ralph-coats-roe-medal
Sponsor: American Society of Mechanical Engineers
3 Park Avenue
New York, NY 10016

ASME Prime Movers Committee Award 1156
The ASME Prime Movers Committee Award recognizes outstanding contributions to the literature of thermal electric station practice or equipment which are available through public presentation and publication. Those papers approved by the appropriate papers review committee as meeting ASME standards and available in printed form may be considered for this award. Papers need not be current and may be written by a single author or multiple authors. The award includes a certificate and $1,000. Nomination forms and instructions are available at the website.
Geographic Focus: All States, All Countries
Date(s) Application is Due: Feb 1
Amount of Grant: 1,000 USD
Contact: W.H. Kirkenir; (919) 546-6408; Bill.Kirkenir@pgnmail.com
Internet: http://www.asme.org/about-asme/honors-awards/literature-awards/prime-movers-committee-award
Sponsor: American Society of Mechanical Engineers
3 Park Avenue
New York, NY 10016

ASME R. Tom Sawyer Award 1157
The R. Tom Sawyer Award is given to an individual who has made important contributions to advance the purpose of the gas turbine industry and to the International Gas Turbine Institute over a substantial period of time. The contribution may be in any area of institute activity, but must be marked by sustained efforts. The award includes a plaque, certificate, and $1,000. Nomination forms and instructions are available at the website.
Geographic Focus: All States, All Countries
Date(s) Application is Due: Sep 4
Amount of Grant: 1,000 USD
Contact: K. Brun, Committee Chair; (210) 522-5449; klaus.brun@swri.org
Internet: http://www.asme.org/about-asme/honors-awards/achievement-awards/r—tom-sawyer-award
Sponsor: American Society of Mechanical Engineers
3 Park Avenue
New York, NY 10016

ASME Ralph Coats Roe Medal 1158
The ASME Ralph Coats Roe Medal is awarded annually to an individual selected by the society for a significant contribution to a better public understanding and appreciation of the engineer's worth to contemporary society. The successful candidate is expected to have the attributes that qualify him/her as an authoritative lecturer on his/her contribution at a general session during the winter meeting. Candidates are not restricted by profession nor by membership in any engineering society or organization. Award includes a gold medal, $12,000, a certificate, and travel expenses to attend the award presentation. Nomination forms and instructions are available at the website.
Geographic Focus: All States, All Countries
Date(s) Application is Due: Sep 15
Amount of Grant: 12,000 USD
Contact: B.I. Hyman, Committee Chair; (206) 543-9038; hymanb@asme.org
Internet: http://www.asme.org/about-asme/honors-awards/achievement-awards/ralph-coats-roe-medal
Sponsor: American Society of Mechanical Engineers
3 Park Avenue
New York, NY 10016

ASME Robert E. Koski Medal 1159
The ASME Robert E. Koski Medal recognizes individuals who have advanced the art and practice of fluid power motion and control through education and/or innovation. The honoree does not have to be a member of ASME. The Medal was established by the Fluid Power Systems and Technology Division to honor Koski's contributions to the field of Design Engineering. The award includes a vermeil medal, a certificate, $10,000, and travel expenses to attend the award presentation. Nomination forms and instructions are available at the website.
Geographic Focus: All States, All Countries
Date(s) Application is Due: Jan 15
Amount of Grant: 10,000 USD
Contact: D. Kim, Committee Chair; (309) 677-3142; deankim@bradley.edu
Internet: http://www.asme.org/about-asme/honors-awards/achievement-awards/robert-e—koski-medal
Sponsor: American Society of Mechanical Engineers
3 Park Avenue
New York, NY 10016

ASME Robert Henry Thurston Lecture Award 1160
The ASME Robert Henry Thurston Lecture, presented annually at the International Mechanical Engineering Congress, provides an outstanding leader in pure or applied science or engineering with the honor of presenting a lecture to the Society that encourages stimulating thinking on a subject of broad technical interest to engineers. In addition to the lecture, the awardee receive a plaque, $500, and travel expenses to attend the award presentation. Nomination forms and instructions are available at the website.
Geographic Focus: All States, All Countries
Date(s) Application is Due: Jan 15
Amount of Grant: 500 USD
Contact: M.F. Horstemeyer; (662) 325-5449; mfhorst@cavs.msstate.edu
Internet: http://www.asme.org/about-asme/honors-awards/achievement-awards/robert-henry-thurston-lecture-award
Sponsor: American Society of Mechanical Engineers
3 Park Avenue
New York, NY 10016

ASME Rufus T. Oldenburger Medal 1161
The ASME Rufus T. Oldenburger Medal recognizes significant contributions and outstanding achievements in the field of automatic control. Such achievements may be in the areas of education, research, development, innovation, and service to the field and profession. The award includes a bronze medal, a certificate, and $2,000. Nomination forms and instructions are available at the website.
Geographic Focus: All States, All Countries
Date(s) Application is Due: Feb 1
Amount of Grant: 2,000 USD
Contact: C. Radcliff, Committee Chair; (517) 355-5198; radcliff@msu.edu
Internet: http://www.asme.org/about-asme/honors-awards/achievement-awards/rufus-oldenburger-medal
Sponsor: American Society of Mechanical Engineers
3 Park Avenue
New York, NY 10016

ASME Ruth and Joel Spira Outstanding Design Educator Award 1162
The ASME Ruth and Joel Spira Outstanding Design Educator Award was established to recognize a person who exemplifies the best in furthering engineering design education through interacting with students and industry, and serving as a role model for other educators to emulate. The award includes a vermeil medal, a certificate, and $1,000. Nomination forms and instructions are available at the website.
Geographic Focus: All States, All Countries
Date(s) Application is Due: Feb 1
Amount of Grant: 1,000 USD
Contact: Y. Jin, Committee Chair; (213) 740-9574; yjin@usc.edu
Internet: http://www.asme.org/about-asme/honors-awards/achievement-awards/ruth-and-joel-spira-outstanding-design-educator-aw
Sponsor: American Society of Mechanical Engineers
3 Park Avenue
New York, NY 10016

ASME S.Y. Zamrik PVP Medal 1163
The ASME S.Y. Zamrik PVP Medal is given for outstanding contributions in the field of pressure vessel and piping technology including, but not limited to, research, development, teaching, and significant advancements of the state-of-the-art. The award includes a bronze medal, a certificate, and $2,000. Nomination forms and instructions are available at the website.
Geographic Focus: All States, All Countries
Date(s) Application is Due: Jun 15
Amount of Grant: 2,000 USD
Contact: D.K. Williams; (704) 847-2482; DennisKW@sharoden.com
Internet: http://www.asme.org/about-asme/honors-awards/achievement-awards/s-y—zamrik-pvp-medal
Sponsor: American Society of Mechanical Engineers
3 Park Avenue
New York, NY 10016

ASME Soichiro Honda Medal 1164
The ASME Soichiro Honda Medal recognizes an individual for an outstanding achievement or a series of significant engineering contributions in developing improvement in the field of personal transportation. Attention shall be concentrated on the brilliance of the achievement or on the overall effect of a series of contributions, not on the individual. The achievement should be of such public importance as to be worthy of the gratitude of the nation and to call forth the admiration of engineers. The award includes a gold medal, a certificate, $7,500, and travel expenses to attend the award ceremony. Nomination forms and instructions are available at the website.
Requirements: There are no restrictions as to nominee's age, nationality, society membership, degree of education, employment, or official position.
Geographic Focus: All States, All Countries
Date(s) Application is Due: Oct 1
Amount of Grant: 7,500 USD
Contact: W.R. Wade, Committee Chair; (248) 449-4549; wrwade1@gmail.com
Internet: http://www.asme.org/about-asme/honors-awards/achievement-awards/soichiro-honda-medal
Sponsor: American Society of Mechanical Engineers
3 Park Avenue
New York, NY 10016

ASME Spirit of St. Louis Medal 1165
The ASME Spirit of St. Louis Award is given for meritorious service in the advancement of aeronautics and astronautics. The award is not limited to members of the ASME or the engineering profession. The award includes a vermeil medal, a certificate, and $1,000. Nomination forms and instructions are available at the website.
Requirements: Any individual member, group of members, or committee may nominate candidates for this award.
Geographic Focus: All States, All Countries

Date(s) Application is Due: Jan 1
Amount of Grant: 1,000 USD
Contact: J. Kudva; (310) 891-2814; jkudva@nextgenaero.com or awards@asme.org
Internet: http://www.asme.org/about-asme/honors-awards/achievement-awards/spirit-of-st—louis-medal
Sponsor: American Society of Mechanical Engineers
3 Park Avenue
New York, NY 10016

ASME Student Section Advisor Award 1166
The Student Section Advisor Award recognizes the leadership and service qualities of a student who has completed at least three academic years as a Student Section Advisor prior to nomination for the award. A nomination for this award is expected to substantiate the leadership and service qualities of the nominee by reference to accomplishments in the following areas: encouraging engineering students to become Student Members as the initial step in their program of professional development; providing guidance to the Student Section on the activities and programs that will stimulate interest in the profession; and creating a professional awareness in the Student Members which will inspire them to upgrade to Member before graduation and to maintain a continuous active membership in ASME. The award includes a silver medal, certificate, $1,500, and travel expenses to attend the award presentation.
Geographic Focus: All States, All Countries
Date(s) Application is Due: Mar 1
Amount of Grant: 1,500 USD
Contact: G.A. Anderkay, Committee Chair; (212) 591-7094; mckivorf@asme.org
Internet: http://www.asme.org/about-asme/honors-awards/service-awards/student-section-advisor-award
Sponsor: American Society of Mechanical Engineers
3 Park Avenue
New York, NY 10016

ASME Thomas A. Edison Patent Award 1167
The ASME Edison Patent Award recognizes creativity of a patented device or process that has the potential of significantly enhancing some aspect of mechanical engineering. Nominations for this award are expected to demonstrate the significant impact of the nominee's invention to the practice of mechanical engineering or the significant impact that the invention had on at least one industrial/commercial sector. Such an impact might include: creating a new business or industry; substantially improving the efficiency, method of operation, or productivity of an existing machine or process; dramatically reducing the size, weight, or energy requirements of an existing machine or device; improving the quality of life for a definable segment or all of humanity; and financially impacting an existing business or industry. The nomination should focus on a single patent rather than the inventor's entire body of work. The award includes a bronze plaque, a certificate, $1,000, and travel expenses to attend the award presentation. Nomination forms and instructions are available at the website.
Geographic Focus: All States, All Countries
Date(s) Application is Due: Feb 1
Amount of Grant: 1,000 USD
Contact: A.M. Khounsary, Committee Chair; (630) 252-3384; amk@aps.anl.gov
Internet: http://www.asme.org/about-asme/honors-awards/achievement-awards/thomas-a—edison-patent-award
Sponsor: American Society of Mechanical Engineers
3 Park Avenue
New York, NY 10016

ASME Timoshenko Medal 1168
The ASME Timoshenko Medal is given in recognition of Stephen Timoshenko, a world-renowned authority in the field of applied mechanics. The award include a bronze medal, a certificate, $1,750, and travel expenses to attend the award presentation. Nomination forms and instructions are available at the website.
Geographic Focus: All States, All Countries
Date(s) Application is Due: Nov 5
Amount of Grant: 1,750 USD
Contact: A.J. Rosakis, Committee Chair; (626) 395-4523; arosakis@aero.caltech.edu
Internet: http://www.asme.org/about-asme/honors-awards/achievement-awards/timoshenko-medal
Sponsor: American Society of Mechanical Engineers
3 Park Avenue
New York, NY 10016

ASME Van C. Mow Medal 1169
The ASME Van C. Mow Medal is given to an individual who has demonstrated meritorious contributions to the field of bioengineering through research, education, professional development, leadership in the development of the profession, mentorship to young bioengineers, and service to the bioengineering community. The award includes a bronze metal, a certificate, $1,000, and travel expenses to attend the award presentation. Nomination forms and instructions are available at the website. Nomination packets are held and considered for three years. A new nomination packet must be submitted after that time.
Requirements: Candidates must have earned a Ph.D. or equivalent degree in any field of engineering, physics, medicine, or life sciences. Nominees must have earned their degree between ten and twenty years prior to June 1 of the year of the award. Nominees must also be active members of the bioengineering division of ASME.
Geographic Focus: All States, All Countries
Date(s) Application is Due: Sep 1
Amount of Grant: 1,000 USD
Contact: John Tarbell, Committee Chair; (212) 650-6841; tarbell@ccny.cuny.edu
Internet: http://www.asme.org/about-asme/honors-awards/achievement-awards/van-c—mow-medal
Sponsor: American Society of Mechanical Engineers
3 Park Avenue
New York, NY 10016

ASME Warner T. Koiter Medal 1170
The ASME Warner T. Koiter Medal was funded by the Technical University of Delft, The Netherlands, to honor Warner T. Koiter for his fundamental work in nonlinear stability of structures, his diligence in the effective application of these theories, his international leadership in mechanics, and his effectiveness as a teacher and researcher. The Koiter Medal recognizes distinguished contributions to the field of solid mechanics with special emphasis on the effective blending of theoretical and applied elements of the discipline, and on a high degree of leadership in the international solid mechanics community. The award includes a bronze metal, a certificate, $2,000, and travel expenses to attend the award presentation. Nomination forms and instructions are available at the website.
Geographic Focus: All States, All Countries
Date(s) Application is Due: Nov 5
Amount of Grant: 2,000 USD
Contact: A.J. Rosakis, Committee Chair; (626) 395-4523; rosakis@aero.caltech.edu
Internet: http://www.asme.org/about-asme/honors-awards/achievement-awards/warner-t—koiter-medal
Sponsor: American Society of Mechanical Engineers
3 Park Avenue
New York, NY 10016

ASME William T. Ennor Manufacturing Technology Award 1171
The ASME William T. Ennor Manufacturing Technology Award is presented to an individual or team of individuals for developing or contributing significantly to an innovative manufacturing technology that contributes substantial economic and/or societal benefits. The award includes a vermeil medal, a certificate, and $1,000. Nomination forms and instructions are available at the website.
Geographic Focus: All States
Date(s) Application is Due: Feb 1
Amount of Grant: 1,000 USD
Contact: S.N. Melkote; (404) 894-8499; Shreyes.melkote@me.gatech.edu
Internet: http://www.asme.org/about-asme/honors-awards/achievement-awards/william-t—ennor-manufacturing-technology-award
Sponsor: American Society of Mechanical Engineers
3 Park Avenue
New York, NY 10016

ASME Worcester Reed Warner Medal 1172
The ASME Worcester Reed Warner Medal is awarded annually for outstanding contributions to the permanent literature of engineering. Contributions may be single papers, treatises or books, or a series of papers. These contributions should be progressive ideas relating to engineering, scientific, and industrial research associated with mechanical engineering; the design and operation of mechanical and associated equipment; industrial engineering or management, organization, operation, and the concomitants of each; or other subjects closely associated with the foregoing. A certificate, vermeil medal, and $2,000 is included with the award. Nomination forms and instructions are available at the website.
Requirements: To qualify as having permanent value, any paper or treatise must be at least five years old. Recipients may be nonmembers of ASME.
Geographic Focus: All States, All Countries
Date(s) Application is Due: Mar 1
Amount of Grant: 2,000 USD
Contact: G.A. Anderkay; (212) 591-7094; mckivorf@asme.org or awards@asme.org
Internet: http://www.asme.org/about-asme/honors-awards/literature-awards/worcester-reed-warner-medal
Sponsor: American Society of Mechanical Engineers
3 Park Avenue
New York, NY 10016

ASME Y.C. Fung Young Investigator Award 1173
The ASME Y.C. Fung Young Investigator Award was established to recognize young investigators who are committed to pursuing research and have demonstrated significant potential to make substantial contributions to the field of bioengineering. Such accomplishments may take the form of design or development of new methods, equipment or instrumentation in bioengineering, and research publications in peer-reviewed journals. The award includes a bronze medal, certificate, $1,000, and travel expenses to attend the award presentation. Nomination forms and instructions are available at the website.
Requirements: Candidates must have earned a Ph.D. or equivalent degree in any field of engineering, physics, medicine or life sciences. Candidates must have received their Ph.D. degree within seven years of their award nomination and must be no more than 35 years of age on June 1st of the year in which they are nominated. Candidate must be an active member of the Bioengineering Division.
Geographic Focus: All States, All Countries
Date(s) Application is Due: Sep 1

Amount of Grant: 1,000 USD
Contact: J.D. Humphrey, Committee Chair; (203) 432-6428; jay.humphrey@yale.edu
Internet: http://www.asme.org/about-asme/honors-awards/achievement-awards/y-c—fung-young-investigator-award
Sponsor: American Society of Mechanical Engineers
3 Park Avenue
New York, NY 10016

ASM ICAAC Young Investigator Awards **1174**
The ASM ICAAC Young Investigator Awards recognize and reward early career scientists for research excellence and potential in microbiology and infectious disease.
Requirements: Nominees must be no more than three years beyond completion of postdoctoral research training in microbiology or infectious diseases at the time of the nomination deadline. Nominations must consist of the following: curriculum vitae, including a list of publications; letter of nomination, describe the nominee's research excellence and potential in microbiology and infectious diseases; letters of support,two letters of support must come from persons, other than the nominator, who are familiar with the nominee's research excellence and potential. No more than one of the three letters may be from the nominee's institution or the same institution.
Restrictions: Self-nominations will not be accepted.
Geographic Focus: All States
Date(s) Application is Due: Apr 1
Amount of Grant: 2,500 USD
Contact: Awards Committee; (202) 942-9226; fax (202) 942-9353; awards@asmusa.org
Internet: http://www.asm.org/academy/index.asp?bid=2278
Sponsor: American Society for Microbiology
1752 N Street, N.W.
Washington, DC 20036-2904

ASM International Fellowship for Latin America and the Caribbean **1175**
Funded by ASM this award offers fellowships to promising young investigators in South and Central America and the Caribbean who are within five years of obtaining, or are in the process of obtaining, their Masters, PhD, or other equivalent academic degree, and who are working in any of the microbiology disciplines, to visit a host scientist in the U.S.. The International Fellowship Program provides funding to meet the costs of a visit to one institution for between 6 weeks to 6 months. The International Fellowship Program provides funding to meet the costs of a visit to one institution for between 6 weeks to 6 months.
Requirements: The application must be made jointly between a host institution and the visiting fellow. The Investigator must be:a member of ASM or any other national microbiological society; actively involved in research in the microbiological sciences; have obtained, or be in the process of obtaining, their masters, PhD, or other equivalent academic degree within the last five years; a national of a resource-limited Africa, Asian or Latin American country submitting the application from an institution in their home region. Investigators from developed nations, as defined by BOTH the UN high HDI group and the World Bank high income economy group, are ineligible. Investigators submitting an application from the U.S. where they are spending a training period are ineligible; sufficiently proficient in the use of the English language. The Host Scientist must be: an ASM member; actively involved in research in the microbiological sciences in the U.S; actively involved in teaching in the microbiological sciences in the U.S; interested in sustaining international collaborations Preference will be given to: applicants who can prove three years of membership in ASM or any other national microbiological society, and; applicants who have not previously had the opportunity to travel to a facility in another country. The application should demonstrate to the Review Committee: academic excellence of the applicant-Honors & Awards, CV, and Letters of Recommendation; depth of the applicant's research experience-Statement of prior Research Experience; quality and originality of work proposed during the Fellowship-Potential Collaboration and Proposed Research Plan; relevance of the work proposed to the applicant's locale-Potential Collaboration and Proposed Research Plan; excellence of the host, and the research and training environment
Geographic Focus: All States, Antigua & Barbuda, Argentina, Bahamas, Barbados, Belize, Bolivia, Brazil, Chile, Colombia, Costa Rica, Cuba, Ecuador, Guyana, Paraguay, Peru
Date(s) Application is Due: Apr 15; Oct 15
Amount of Grant: 4,000 USD
Contact: Supervisor; (202) 942-9368; fax (202) 942-9328; international@asmusa.org
Internet: http://www.asm.org/International/index.asp?bid=2778
Sponsor: American Society for Microbiology
1752 N Street, N.W.
Washington, DC 20036-2904

ASM International Fellowships for Asia and Africa **1176**
Funded by ASM this award offers a maximum of one fellowship to a promising young investigator in Asia, as well as to one young investigator in Africa who is within five years of obtaining, or in the process of obtaining, his/her Masters, PhD, or other equivalent academic degree, and who is working in any of the microbiology disciplines, to visit a host scientist in the U.S.. Both International Fellowships will award a maximum grant of $5,500. The award is not intended to provide travel to obtain a degree at the host institution.
Requirements: The application must be made jointly between a host institution and the visiting fellow. The Investigator must be: a member of ASM or any other national microbiological society; actively involved in research in the microbiological sciences; have obtained, or be in the process of obtaining, their masters, PhD, or other equivalent academic degree within the last five years; a national of a resource-limited Africa, Asian or Latin American country submitting the application from an institution in their home region. Investigators from developed nations, as defined by BOTH the UN high HDI group and the World Bank high income economy group, are ineligible. Investigators submitting an application from the U.S. where they are spending a training period are ineligible; sufficiently proficient in the use of the English language. The Host Scientist must be: an ASM member; actively involved in research in the microbiological sciences in the U.S; actively involved in teaching in the microbiological sciences in the U.S; interested in sustaining international collaborations Preference will be given to: applicants who can prove three years of membership in ASM or any other national microbiological society, and; applicants who have not previously had the opportunity to travel to a facility in another country.
Geographic Focus: All States
Date(s) Application is Due: Apr 15; Oct 15
Amount of Grant: 5,500 USD
Contact: Supervisor; (202) 942-9368; fax (202) 942-9328; international@asmusa.org
Internet: http://www.asm.org/International/index.asp?bid=2778
Sponsor: American Society for Microbiology
1752 N Street, N.W.
Washington, DC 20036-2904

ASM Merck Irving S. Sigal Memorial Awards **1177**
The ASM Merck Irving S. Sigal Memorial Awards recognize and award excellence in basic research in medical microbiology and infectious diseases. The awards are presented in memory of Irving S. Sigal, who was instrumental in the early discovery of therapies to treat HIV/AIDS.
Requirements: The nominee must be no more than five years beyond completion of postdoctoral research training in microbiology or infectious diseases at the time of the nomination deadline. Nominations must consist of the following: curriculum vitae, including a list of publications; letter of nomination, describe the nominee's excellence in basic research in medical microbiology and infectious diseases; letters of support, two letters of support must come from persons, other than the nomination, who are familiar with the nominee's qualifications and accomplishments. No more than one of the three letters may be from the nominee's institution or the same institution.
Geographic Focus: All States
Date(s) Application is Due: Oct 1
Amount of Grant: 2,500 USD
Contact: Awards Committee; (202) 942-9226; fax (202) 942-9353; awards@asmusa.org
Internet: http://www.asm.org/academy/index.asp?bid=15056
Sponsor: American Society for Microbiology
1752 N Street, N.W.
Washington, DC 20036-2904

ASM Microbiology Undergraduate Research Fellowship **1178**
The goal of the Microbiology Undergraduate Research Fellowship (MURF) program is to increase the number of underrepresented undergraduate students who wish to, and have demonstrated the ability to pursue graduate careers (Ph.D. or M.D./Ph.D.) in microbiology. Students will have the opportunity to conduct full time summer research with an ASM member at a host institution, and present research results at the Annual Biomedical Research Conference for Minority Students and the ASM General Meeting. Students will: agree to participate in an undergraduate summer research program at a U.S. based institution; conduct a research project for a minimum of 10 weeks; work with a faculty mentor who is an ASM member; submit a research abstract to the Annual Biomedical Research Conference for Minority Students (ABRCMS); submit a research abstract to ASM for presentation at the ASM General Meeting. Students decide the institution, research area, and level of activity for the summer. Based on interests, independence, and ability, students will be placed at a host U.S. Institution of the student's choice to conduct basic science research. From a list provided on the application, students interested in conducting research at a host institution will select three institutions where they would like to conduct their summer research. Every effort will be made to place fellows at their first choice.
Requirements: Eligible student candidates for the fellowship must be from groups that have been determined by the applicant's institution to be underrepresented in the microbiological sciences. The ASM encourages institutions to identify individuals that have been historically underrepresented, and remain underrepresented today in the microbiological sciences nationally. These groups include African-Americans, Hispanics, Native Americans, Alaskan Natives, and Pacific Islanders. In addition, applicants must also: be U.S. citizen or permanent U.S. resident; be enrolled as full-time matriculating undergraduate students during the academic year at an accredited U.S. institution; be either freshmen with college level research experience, sophomores, juniors, or seniors who will not graduate before the completion date of the summer program; be members of an underrepresented group in microbiology; have taken introductory courses in biology, chemistry, and preferably microbiology prior to submission of the application; have strong interests in obtaining a Ph.D., or M.D./Ph.D. in the microbiological sciences, and have lab research experience.
Restrictions: Travel funds are contingent upon acceptance of an abstract for the General Meeting.
Geographic Focus: All States
Date(s) Application is Due: Feb 1
Amount of Grant: 5,850 USD
Contact: Fellowship Committee; (202) 942-9283; fax (202) 942-9329; fellowships-careerinformation@asmusa.org
Internet: http://www.asm.org/Education/index.asp?bid=4322
Sponsor: American Society for Microbiology
1752 N Street, N.W.
Washington, DC 20036-2904

ASM Millis-Colwell Postgraduate Travel Grant 1179

This new grant, negotiated between the American Society for Microbiology (ASM U.S.A) and the Australian Society for Microbiology (ASM Australia), allows for the exchange of one student member from each society to present an abstract at the annual General Meeting of the other society and to spend a week at nearby research laboratory. The grant aims to create a long lasting bond between ASM U.S.A and ASM Australia and is designed to benefit PhD students in both countries by giving them the opportunity to travel overseas to present their work and experience the best of microbiology in the partner country. ASM U.S.A will submit the successful applicant's abstract for an oral or poster presentation to ASM Australia for their Annual Scientific Meeting & Exhibition. The applicant will also have negotiated an agreement to visit the research laboratory of an Australian scientist in the period either immediately before or immediately after the ASM Australia annual meeting. The award will be conditional upon ASM Australia's acceptance of the submitted abstract.
Requirements: The successful applicant will be: a U.S. citizen or permanent resident of the U.S; a student member of the American Society for Microbiology; and enrolled in a PhD program at an American University. The award will cover roundtrip economy airfare between the U.S. and Australia, and any necessary internal flights; per diem living allowance for up to 14 days; and, registration fee for the Australian Society for Microbiology's Annual Scientific Meeting & Exhibition.
Geographic Focus: All States
Date(s) Application is Due: Jan 1
Amount of Grant: Up to 5,000 USD
Contact: Program Coordinator; (202) 942-9282; fax (202) 942-9329
Internet: http://www.asm.org/International/index.asp?bid=53729
Sponsor: American Society for Microbiology
1752 N Street, N.W.
Washington, DC 20036-2904

ASM Procter & Gamble Award in Applied and Environmental Microbiology 1180

The ASM Procter & Gamble Award in Applied and Environmental Microbiology recognizes distinguished achievement in research and development in applied (non-clinical) and environmental microbiology.
Requirements: The nominee must demonstrate outstanding accomplishment in research or development in an appropriate field and be actively engaged in research or development at the time that the award is presented. Nominations must consist of the following: curriculum vitae, including a list of publications; letter of nomination, describe the nominee's outstanding accomplishments in research or development in the appropriate field, including a list of the ten most relevant publications to the award; letters of support, two letters of support must come from persons, other than the nominator, who are familiar with the nominee's outstanding accomplishments. No more than one of the three letters may be from the nominee's institution or the same institution.
Restrictions: Self-nominations will not be accepted.
Geographic Focus: All States
Date(s) Application is Due: Oct 1
Amount of Grant: 2,000 USD
Contact: Awards Committee; (202) 942-9226; fax (202) 942-9353; awards@asmusa.org
Internet: http://www.asm.org/academy/index.asp?bid=2288
Sponsor: American Society for Microbiology
1752 N Street, N.W.
Washington, DC 20036-2904

ASM Promega Biotechnology Research Award 1181

The ASM Promega Biotechnology Research Award honors outstanding contributions to the application of biotechnology through fundamental microbiological research and development.
Requirements: A nomination can be a single exceptionally significant achievement or the aggregate of a number of exemplary achievements. Nominations must consist of the following: curriculum vitae, including a list of publications; letter of nomination, describe the nominee's outstanding accomplishments in microbiological research and development, including a list of the ten most relevant publications to the award; letters of support, two letters of support must come from persons, other than the nominator, who are familiar with the nominee's accomplishments. No more than one of the three letters may be from the nominee's institution or the same institution.
Restrictions: Neither self-nominations nor co-nominations will be accepted.
Geographic Focus: All States
Date(s) Application is Due: Oct 1
Amount of Grant: 5,000 USD
Contact: Awards Committee; (202) 942-9226; fax (202) 942-9353; awards@asmusa.org
Internet: http://www.asm.org/academy/index.asp?bid=2291
Sponsor: American Society for Microbiology
1752 N Street, N.W.
Washington, DC 20036-2904

ASM Raymond W. Sarber Awards 1182

The ASM Raymond W. Sarber Awards recognize students at the undergraduate and predoctoral levels for research excellence and potential. The awards honor Raymond W. Sarber for his contributions to the growth and advancement of ASM.
Requirements: Nominees must be a student at the undergraduate or predoctoral level, attending an accredited institution in the United States, in an academic program involving microbiology. Two awards are presented at the ASM General Meeting, one to an undergraduate and one to a graduate student. They will be judged separately.
Restrictions: Nominations must consist of the following: curriculum vitae; statement, no longer than two pages, from the student outlining academic achievements, research accomplishments, and future career goals; letter of nomination, describe the nominee's research excellence and potential; letters of support, two letters of support: one from a supervisor or mentor, and the other letter may be from an academic advisor or colleague.
Geographic Focus: All States
Date(s) Application is Due: Oct 1
Amount of Grant: 1,500 USD
Samples: Self-nominations will not be accepted.
Contact: Awards Committee; (202) 942-9226; fax (202) 942-9353; awards@asmusa.org
Internet: http://www.asm.org/academy/index.asp?bid=2294
Sponsor: American Society for Microbiology
1752 N Street, N.W.
Washington, DC 20036-2904

ASM Robert D. Watkins Graduate Research Fellowship 1183

The goal of the fellowship is to increase the number of underrepresented groups completing doctoral degrees in the microbiological sciences. The ASM Robert D. Watkins Graduate Research Fellowship is aimed at highly competitive graduate students who are enrolled in a Ph.D. program and who have completed their graduate course work in the microbiological sciences. The fellowship encourages students to continue and complete their research project in the microbiological sciences. Students will be required to submit an abstract each year to ASM for presentation at the annual ASM General Meeting and attend the ASM Kadner Institute one time during the three-year tenure of the fellowship.
Requirements: Eligible candidates must be from groups that have been determined by the applicant's institution to be underrepresented in the microbiological sciences. The ASM encourages institutions to identify individuals that have been historically underrepresented, and remain underrepresented today in the microbiological sciences nationally. These groups include African-Americans, Hispanics, Native Americans, Alaskan Natives, and Pacific Islanders. In addition, applicants must: be formally admitted to a doctoral program in the microbiological sciences in an accredited U.S. institution; have successfully completed the first year of the graduate program (first year graduate students cannot apply); have successfully completed all graduate coursework requirements for the doctoral degree by the date of activation of the fellowship; be a student member of ASM; be mentored by an ASM member; be a U.S. citizen or a permanent resident; not have funding or have funding that will expire by the start date of the fellowship(fellowship cannot run concurrently with other national fellowships from NIH, NSF, HHMI, etc.); three letters of recommendations must be submitted with your application. One letter must be from your research advisor/mentor; Official transcripts from all colleges and universities attended.
Geographic Focus: All States
Date(s) Application is Due: May 1
Amount of Grant: 63,000 USD
Contact: Education Board; (202) 942-9283; fax (202) 942-9329; fellowships-careerinformation@asmusa.org
Internet: http://www.asm.org/Education/index.asp?bid=6278
Sponsor: American Society for Microbiology
1752 N Street, N.W.
Washington, DC 20036-2904

ASM sanofi-aventis ICAAC Award 1184

ASM's premier award in antimicrobial chemotherapy research; it stimulates research and honors outstanding accomplishment in antimicrobial chemotherapy.
Requirements: The nominee must be actively engaged in research involving development of new agents, investigation of antimicrobial action or resistance to antimicrobial agents, and/or the pharmacology, toxicology or clinical use of those agents. The nominee must not have served on the ICAAC Program Committee within the past two years. Nominations must consist of the following: curriculum vitae, including a list of publications; letter of nomination, describe the nominee's outstanding accomplishment in antimicrobial chemotherapy, including a list of the ten most relevant publications to the award; letters of support, two letters of support should come from two people, other than the nominator, who are familiar with the nominee's accomplishments. No more than one of the three letters may be from the nominee's institution or the same institution.
Restrictions: Self-nominations will not be accepted.
Geographic Focus: All States
Date(s) Application is Due: Apr 1
Amount of Grant: 20,000 USD
Contact: Awards Committee; (202) 942-9226; fax (202) 942-9353; awards@asmusa.org
Internet: http://www.asm.org/academy/index.asp?bid=2268
Sponsor: American Society for Microbiology
1752 N Street, N.W.
Washington, DC 20036-2904

ASM Siemens Healthcare Diagnostics Young Investigator Award 1185

The ASM Siemens Healthcare Diagnostics Young Investigator Award honors outstanding laboratory research in clinical microbiology or antimicrobial agents and is intended to further the career development of a young clinical scientist and promote awareness of clinical microbiology as a career.
Requirements: The nominee should be conducting outstanding research in clinical microbiology, automation in clinical laboratories, development of novel antimicrobial agents, mechanisms of action of antimicrobial agents or mechanisms of resistance to antimicrobial agents. The nominee must be no more than five years beyond completion of postdoctoral research training in microbiology, infectious diseases or related disciplines at the time of the nomination deadline. Nominations must consist of the following: curriculum vitae, including a list of publications, abstracts, and manuscripts in preparation; letter of

nomination, describe the nominee's outstanding clinical research and potential; letters of support, two letters of support must come from persons, other than the nominator, who are familiar with the nominee's qualifications and accomplishments. No more than one of the three letters may be from the nominee's institution or from the same institution.
Restrictions: Self-nominations will not be accepted.
Geographic Focus: All States
Date(s) Application is Due: Oct 1
Amount of Grant: 2,000 USD
Contact: Awards Committee; (202) 942-9226; fax (202) 942-9353; awards@asmusa.org
Internet: http://www.asm.org/Academy/index.asp?bid=2307
Sponsor: American Society for Microbiology
1752 N Street, N.W.
Washington, DC 20036-2904

ASM Undergraduate Research Fellowship 1186
The ASM Undergraduate Research Fellowship (URF) is aimed at highly competitive students who wish to pursue graduate careers (Ph.D. or MD/Ph.D) in microbiology. Students will have the opportunity to conduct full time research at their home institutions with an ASM member and present research results at the ASM General Meeting the following year. Students will: conduct a research project for a minimum of 10 weeks beginning in the summer; work with faculty mentors who are ASM members and who are employed at the students home institutions, and submit a research abstract for presentation at the ASM General Meeting. The Fellowship provides: up to $4000 for student stipend; two-year ASM student membership, and up to $1000 in travel support for students to present the results of the research project at the ASM General Meeting. Travel funds are contingent upon acceptance of an abstract for the General Meeting.
Requirements: Eligible student candidates for the fellowship must: be enrolled as full-time matriculating undergraduate students during the academic year at an accredited U.S. Institution; be involved in a research project; have an ASM member at their home institutions willing to serve as a mentor, and not receive financial support for research (i.e., Council for Undergraduate Research, Minority Access to Research Careers, Sigma Xi) during the fellowship. Applicants who do not meet all eligibility requirements will not be considered.
Restrictions: The program requires a joint application from both the student and a faculty mentor. It is student's responsibility to submit a completed application.
Geographic Focus: All States
Date(s) Application is Due: Feb 1
Amount of Grant: 4,000-5,000 USD
Contact: Fellowship Committee; (202) 942-9283; fax (202) 942-9329; fellowships-careerinformation@asmusa.org
Internet: http://www.asm.org/Education/index.asp?bid=4319
Sponsor: American Society for Microbiology
1752 N Street, N.W.
Washington, DC 20036-2904

ASM William A Hinton Research Training Award 1187
The ASM William A Hinton Research Training Award honors outstanding contributions toward fostering the research training of underrepresented minorities in microbiology. It is given in memory of William A. Hinton, a physician-research scientist and one of the first African-Americans to join the ASM.
Requirements: The nominee must contribute to the research training of undergraduate students, graduate students, postdoctoral fellows or health professional students and efforts leading to the increased participation of underrepresented minorities in microbiology. Nominations must consist of the following: curriculum vitae, including a list of nominee's publications; letter of nomination, describe the nominee's training of undergraduate students, graduate students, postdoctoral fellows, or health professional students. Detail specific examples of excellence in training and mentoring of underrepresented minorities. Describe the duration, extent, and impact of the nominee's training of underrepresented minority students, graduate students, postdoctoral fellows, or health professional students; letters of support, two letters of support must come from persons, other than the nominator, who are familiar with the nominee's outstanding contributions to the training of underrepresented minorities. No more than one of the three letters may be from the nominee's institution or from the same institution.
Geographic Focus: All States
Date(s) Application is Due: Oct 1
Contact: Awards Committee; (202) 942-9226; fax (202) 942-9353; awards@asmusa.org
Internet: http://www.asm.org/Academy/index.asp?bid=2317
Sponsor: American Society for Microbiology
1752 N Street, N.W.
Washington, DC 20036-2904

ASNS Bio-Serv Award in Experimental Animal Nutrition 1188
The award is given for meritorious research in nutrition by an investigator who received the doctoral degree in the 10 years preceding the month the award is presented. The work recognized must involve the nutrition of experimental animals used as models. The award and an engraved plaque are made available by Bio-Serv Inc. Nominations should include a letter stating the basis for the nomination, a selected bibliography that supports the nomination, and the reprint or series of reprints on which the nomination is based.
Requirements: Nominations may be made by anyone, including members of the respective nominating committees and nonmembers of the institute. Nominations will be retained for two years. Candidates may be renominated. Candidates need not be members of the institute. However, the awards are usually given to professionally active nutrition scientists.
Restrictions: An individual who has received one AIN award is not eligible to receive another award unless it is for accomplishments not covered by the first award.
Geographic Focus: All States
Date(s) Application is Due: Sep 1
Amount of Grant: 1,000 USD
Contact: Secretariat; (301) 530-7051; fax (301) 571-1892; nnotes@asns.faseb.org
Internet: http://www.asns.org/invitation.html
Sponsor: American Society for Nutritional Sciences
9650 Rockville Pike, Suite L-4500
Bethesda, MD 20814

A Social Corporation Grants 1189
The Fund, established in Wisconsin in 1987, has as its primary purpose to support research in medical physics within the Madison, Wisconsin, region. Specific application guidelines are available upon request. There are no particular deadlines with which to adhere, and applicant organizations should contact the fund administrator directly for further details.
Geographic Focus: Wisconsin
Contact: Charles Lescrenier; (608) 831-1188 or (800) 426-6391; fax (608) 836-9201
Sponsor: A Social Corporation
2500 West Beitline Highway at University Avenue, P.O. Box 620327
Middleton, WI 53562-0327

ASPEN C. Richard Fleming Grant 1190
The Foundation provides annual grant support to nutrition researchers at all stages of their careers. This grant, funded by the donations of ASPEN members, is intended to assist a nutrition investigator by providing preliminary funding for promising new research in the field of nutrition and metabolic support and related areas of clinical nutrition. Applicants still in training positions may apply. Corporate employees are eligible to receive grants if the research project is not part of their normal duties and they have at least a part-time academic appointment or are associated with a professional institute that conducts nutrition research. In making choices among otherwise high-quality applicants, the reviewers will take into account the geographic distribution of grant winners; their ethnic background and gender, should the applicant choose to supply such information; their professional discipline; and other criteria as appropriate to ensure an equitable system. Funds may be used only for technician salary, equipment, supplies, animals, clinical research costs or other expenses directly related to the conduct of the proposed research.
Requirements: Applicants must submit a letter from their supervisor or department head at the institution confirming his/her commitment to the project. If the project involves human subjects, a letter pledging support in recruiting patients from the primary care provider and the institutional review board overseeing human studies is required. For junior faculty and applicants who do not hold faculty positions, three letters of reference are required.
Restrictions: This grant is not intended to support pursuit of an additional degree. Grant funds may not be used for: comparison of commercial products; indirect costs or overhead; costs of patient care; constructing or renovating facilities; furniture or office equipment; secretarial services; honoraria or membership dues; textbooks or periodicals; repair or service contract costs on institutional equipment; entertainment; travel; or salary support for the Principal Investigator.
Geographic Focus: All States
Date(s) Application is Due: Sep 8
Amount of Grant: 3,500 USD
Contact: Paula Bowen, Research Program Administrator; (301) 587-6315, ext. 132 or (301) 920-9132; fax (301) 587-2365; paulab@aspen.nutr.org
Internet: http://www.nutritioncare.org/wcontent.aspx?id=784
Sponsor: American Society for Parenteral and Enteral Nutrition Rhoads Research Foundation
8630 Fenton Street, Suite 412
Silver Spring, MD 20910

ASPEN Douglas Wilmore Grant 1191
The Foundation provides annual grant support to nutrition researchers at all stages of their careers. The grant is intended to assist a nutrition investigator by providing preliminary funding for promising new research in the field of nutrition and metabolic support and related areas of clinical nutrition. Priority consideration is given to applications that involve a multidisciplinary team or that investigate the efficacy of parenteral or enteral nutrition. Applicants still in training positions may apply for the Fleming Grant. Corporate employees are eligible to receive grants if the research project is not part of their normal duties and they have at least a part-time academic appointment or are associated with a professional institute that conducts nutrition research. In making choices among otherwise high-quality applicants, the reviewers will take into account the geographic distribution of grant winners; their ethnic background and gender, should the applicant choose to supply such information; their professional discipline; and other criteria as appropriate to ensure an equitable system. Funds may be used only for technician salary, equipment, supplies, animals, clinical research costs or other expenses directly related to the conduct of the proposed research. The grant can be renewed for a second year of funding, assuming satisfactory progress.
Requirements: Applicants must submit a letter from their supervisor or department head at the institution confirming his/her commitment to the project. If the project involves human subjects, a letter pledging support in recruiting patients from the primary care provider and the institutional review board overseeing human studies is required. For junior faculty and applicants who do not hold faculty positions, three letters of reference are required.
Restrictions: Individuals who have received other research grants in excess of $25,000 are ineligible to apply, as are those who completed training more than 5 years prior to the start of the grant period. This grant is not intended to support pursuit of an additional degree. Grant funds may not be used for: comparison of commercial products; indirect costs or overhead; costs of patient care; constructing or renovating facilities; furniture

or office equipment; secretarial services; honoraria or membership dues; textbooks or periodicals; repair or service contract costs on institutional equipment; entertainment; travel; or salary support for the Principal Investigator.
Geographic Focus: All States
Date(s) Application is Due: Sep 8
Amount of Grant: Up to 25,000 USD
Contact: Paula Bowen, Research Program Administrator; (301) 587-6315, ext. 132 or (301) 920-9132; fax (301) 587-2365; paulab@aspen.nutr.org
Internet: http://www.nutritioncare.org/wcontent.aspx?id=784
Sponsor: American Society for Parenteral and Enteral Nutrition Rhoads Research Foundation
8630 Fenton Street, Suite 412
Silver Spring, MD 20910

ASPEN Dudrick Research Scholar Award **1192**
A.S.P.E.N. members are invited to nominate individuals to be considered by the Dudrick Award Committee for the Stanley Dudrick Research Scholar Award. A.S.P.E.N. presents this annual award to recognize and support a mid-career investigator who has shown significant achievements in nutrition support, is an A.S.P.E.N. member, and demonstrates exceptional research productivity and the potential to continue to make contributions in the field of nutrition therapy. The award includes $5,000, plus recognition during the Rhoads Lecture and Awards Symposium at Clinical Nutrition Week.
Requirements: The applicant should: be an A.S.P.E.N. member; have an advanced degree (MD, PharmD, MS, PhD, or equivalent); currently be in a training program or have completed formal training in nutritional research/therapy within the past 10-15 years, e.g. mid-career; have demonstrated exceptional research productivity during and/or following formal training.
Geographic Focus: All States
Date(s) Application is Due: Oct 15
Amount of Grant: 5,000 USD
Contact: Paula Bowen, Research Program Administrator; (301) 587-6315, ext. 132 or (301) 920-9132; fax (301) 587-2365; paulab@aspen.nutr.org
Internet: http://www.nutritioncare.org/wcontent.aspx?id=878
Sponsor: American Society for Parenteral and Enteral Nutrition
8630 Fenton Street, Suite 412
Silver Spring, MD 20910

ASPEN Harry M. Vars Award **1193**
The Award is made annually for the best research presentation by an investigator at Nutrition Week. The recipient of the award is selected by the Vars Award Selection Subcommittee based upon a review of manuscripts submitted by qualified candidates. The author of the best paper presents his or her paper at the Premier Paper Session and is given a $1,000 cash prize, a travel grant and a plaque during the Research Awards Ceremony at Nutrition Week. The winning manuscript is published in the Journal of Parenteral and Enteral Nutrition (JPEN). The Award is open to all disciplines.
Geographic Focus: All States
Amount of Grant: 1,000 USD
Contact: Michelle Spangenburg, Program Director for Education; (301) 587-6315, ext. 127; fax (301) 587-2365; michelles@aspen.nutr.org
Internet: http://www.nutritioncare.org/wcontent.aspx?id=878
Sponsor: American Society for Parenteral and Enteral Nutrition
8630 Fenton Street, Suite 412
Silver Spring, MD 20910

Aspen Institute William Randolph Hearst Endowed Fellowship for Minority Students **1194**
The Aspen Institute Program on Philanthropy and Social Innovation (PSI) in Washington, DC, offers the William Randolph Hearst Endowed Fellowship three times annually. The fellowship, which is based on academic excellence and need, is open to both undergraduate and graduate students of color. The Hearst Fellow serves as an intern with PSI. Through this fellowship, PSI seeks to introduce a diverse group of students to issues and challenges affecting philanthropy, social enterprise, nonprofit organizations, and other actors in the social sector. Recipients may arrange with their colleges or universities to receive academic credit for this experience. In his or her internship, the Hearst Fellow undertakes research, writing, logistical, and administrative support for PSI's leadership initiatives, public programs, and convenings. The student must be able to intern for 12-15 weeks at the Washington, DC, office of the Aspen Institute. Fall and Spring internships will be part-time (15-20 hours per week) and Summer internships will be full-time. A fellowship grant of approximately $2,000 will be awarded to the Fall and Spring fellows and approximately $4,000 will be awarded to the Summer fellow.
Requirements: The ideal candidate for this fellowship is a highly motivated continuing graduate or undergraduate student from an underrepresented community of color. She or he should have an excellent academic record and also have the following: demonstrated interest or experience in nonprofit organizations, philanthropy, and the social sector; excellent research and writing skills; demonstrated financial need; and U.S. citizenship or U.S. permanent residency. There is an online application form for this fellowship.
Restrictions: Travel and housing costs are not covered by the fellowship grant.
Geographic Focus: All States
Date(s) Application is Due: Feb 28; Jul 18; Nov 22
Amount of Grant: 2,000-4,000 USD
Contact: Fellowship Director; (202) 736-2500; fax (202) 467-0790; lauren.stebbins@aspeninstitute.org
Internet: http://www.aspeninstitute.org/policy-work/nonprofit-philanthropy/leadership-initiatives/hearst
Sponsor: Aspen Institute
1 Dupont Circle NW, Suite 700
Washington, DC 20036

ASPEN Maurice Shils Grant **1195**
The Foundation provides annual grant support to nutrition researchers at all stages of their careers. This grant, funded by a donation from Baxter Health Care and Nestle Clinical Nutrition, is intended to assist a nutrition investigator by providing preliminary funding for promising new research in the field of nutrition and metabolic support and related areas of clinical nutrition. Persons applying for the grant must be in a post-training position and commit at least 20% of their time to research. Corporate employees are eligible to receive grants if the research project is not part of their normal duties and they have at least a part-time academic appointment or are associated with a professional institute that conducts nutrition research. In making choices among otherwise high-quality applicants, the reviewers will take into account the geographic distribution of grant winners; their ethnic background and gender, should the applicant choose to supply such information; their professional discipline; and other criteria as appropriate to ensure an equitable system. Funds may be used only for technician salary, equipment, supplies, animals, clinical research costs or other expenses directly related to the conduct of the proposed research. Grants can be renewed for a second year of funding, assuming satisfactory progress.
Requirements: Applicants must submit a letter from their supervisor or department head at the institution confirming his/her commitment to the project. If the project involves human subjects, a letter pledging support in recruiting patients from the primary care provider and the institutional review board overseeing human studies is required. For junior faculty and applicants who do not hold faculty positions, three letters of reference are required.
Restrictions: Individuals who have received other research grants in excess of $25,000 are ineligible to apply, as are those who completed training more than 5 years prior to the start of the grant period. This grant is not intended to support pursuit of an additional degree. Grant funds may not be used for: comparison of commercial products; indirect costs or overhead; costs of patient care; constructing or renovating facilities; furniture or office equipment; secretarial services; honoraria or membership dues; textbooks or periodicals; repair or service contract costs on institutional equipment; entertainment; travel; or salary support for the Principal Investigator.
Geographic Focus: All States
Date(s) Application is Due: Sep 8
Amount of Grant: 18,000 USD
Contact: Paula Bowen, Research Program Administrator; (301) 587-6315, ext. 132 or (301) 920-9132; fax (301) 587-2365; paulab@aspen.nutr.org
Internet: http://www.nutritioncare.org/wcontent.aspx?id=784
Sponsor: American Society for Parenteral and Enteral Nutrition Rhoads Research Foundation
8630 Fenton Street, Suite 412
Silver Spring, MD 20910

ASPEN Norman Yoshimura Grant **1196**
The Foundation provides annual grant support to nutrition researchers at all stages of their careers. This grant, supported by B. Braun Medical, Inc. funding, is intended to assist a nutrition investigator by providing preliminary funding for promising new research in the field of nutrition and metabolic support and related areas of clinical nutrition. Applicants still in training positions may apply. Corporate employees are eligible to receive grants if the research project is not part of their normal duties and they have at least a part-time academic appointment or are associated with a professional institute that conducts nutrition research. In making choices among otherwise high-quality applicants, the reviewers will take into account the geographic distribution of grant winners; their ethnic background and gender, should the applicant choose to supply such information; their professional discipline; and other criteria as appropriate to ensure an equitable system. Funds may be used only for technician salary, equipment, supplies, animals, clinical research costs or other expenses directly related to the conduct of the proposed research.
Requirements: Applicants must submit a letter from their supervisor or department head at the institution confirming his/her commitment to the project. If the project involves human subjects, a letter pledging support in recruiting patients from the primary care provider and the institutional review board overseeing human studies is required. For junior faculty and applicants who do not hold faculty positions, three letters of reference are required.
Restrictions: This grant is not intended to support pursuit of an additional degree. Grant funds may not be used for: comparison of commercial products; indirect costs or overhead; costs of patient care; constructing or renovating facilities; furniture or office equipment; secretarial services; honoraria or membership dues; textbooks or periodicals; repair or service contract costs on institutional equipment; entertainment; travel; or salary support for the Principal Investigator.
Geographic Focus: All States
Date(s) Application is Due: Sep 8
Amount of Grant: 15,000 USD
Contact: Paula Bowen, Research Program Administrator; (301) 587-6315, ext. 132 or (301) 920-9132; fax (301) 587-2365; paulab@aspen.nutr.org
Internet: http://www.nutritioncare.org/wcontent.aspx?id=784
Sponsor: American Society for Parenteral and Enteral Nutrition Rhoads Research Foundation
8630 Fenton Street, Suite 412
Silver Spring, MD 20910

ASPEN Nutrition Research Grants 1197

The Foundation provides annual grant support to nutrition researchers at all stages of their careers. This grant, funded by a donation from the Abbott Foundation, is intended to assist a nutrition investigator by providing preliminary funding for promising new research in the field of nutrition and metabolic support and related areas of clinical nutrition. Persons applying for the grant must be in a post-training position and commit at least 20% of their time to research. Corporate employees are eligible to receive grants if the research project is not part of their normal duties and they have at least a part-time academic appointment or are associated with a professional institute that conducts nutrition research. In making choices among otherwise high-quality applicants, the reviewers will take into account the geographic distribution of grant winners; their ethnic background and gender, should the applicant choose to supply such information; their professional discipline; and other criteria as appropriate to ensure an equitable system. Funds may be used only for technician salary, equipment, supplies, animals, clinical research costs or other expenses directly related to the conduct of the proposed research. Grants can be renewed for a second year of funding, assuming satisfactory progress.

Requirements: Applicants must submit a letter from their supervisor or department head at the institution confirming his/her commitment to the project. If the project involves human subjects, a letter pledging support in recruiting patients from the primary care provider and the institutional review board overseeing human studies is required. For junior faculty and applicants who do not hold faculty positions, three letters of reference are required.

Restrictions: Individuals who have received other research grants in excess of $25,000 are ineligible to apply, as are those who completed training more than 5 years prior to the start of the grant period. This grant is not intended to support pursuit of an additional degree. Grant funds may not be used for: comparison of commercial products; indirect costs or overhead; costs of patient care; constructing or renovating facilities; furniture or office equipment; secretarial services; honoraria or membership dues; textbooks or periodicals; repair or service contract costs on institutional equipment; entertainment; travel; or salary support for the Principal Investigator.

Geographic Focus: All States

Date(s) Application is Due: Sep 8

Amount of Grant: Up to 25,000 USD

Contact: Paula Bowen, Research Program Administrator; (301) 587-6315, ext. 132 or (301) 920-9132; fax (301) 587-2365; paulab@aspen.nutr.org

Internet: http://www.nutritioncare.org/wcontent.aspx?id=784

Sponsor: American Society for Parenteral and Enteral Nutrition Rhoads Research Foundation
8630 Fenton Street, Suite 412
Silver Spring, MD 20910

ASPEN Scientific Abstracts Awards for Papers or Posters 1198

ASPEN will consider all Scientific Abstract submissions for awards. Applicants will be notified if their abstract is a candidate. The first authors of the three highest ranked abstracts from this category (paper or poster) will receive travel grants to attend the Research Workshop, and to present their work at Clinical Nutrition Week. In addition, cash awards in the amounts of $750, $500 and $250 will be offered to the presenters of the three best papers. Awards are presented during the Rhoads Lecture and Research Awards Ceremony at Clinical Nutrition Week.

Geographic Focus: All States

Amount of Grant: 250-750 USD

Contact: Paula Bowen, Research Program Administrator; (301) 587-6315, ext. 132 or (301) 920-9132; fax (301) 587-2365; paulab@aspen.nutr.org

Internet: http://www.nutritioncare.org/Index.aspx?id=724

Sponsor: American Society for Parenteral and Enteral Nutrition
8630 Fenton Street, Suite 412
Silver Spring, MD 20910

ASPEN Scientific Abstracts Promising Investigator Awards 1199

ASPEN will consider all Scientific Abstract submissions for awards. Applicants will be notified if their abstract is a candidate. These awards are presented to the top three Scientific Abstract authors, who are also in the early stages of their careers. Winners receive $750 in travel funds to offset their travel expenses to and from Clinical Nutrition Week. Vars Award winners are eligible for these awards if they are in the early stages of their careers.

Geographic Focus: All States

Amount of Grant: 750 USD

Contact: Paula Bowen, Research Program Administrator; (301) 587-6315, ext. 132 or (301) 920-9132; fax (301) 587-2365; paulab@aspen.nutr.org

Internet: http://www.nutritioncare.org/wcontent.aspx?id=878

Sponsor: American Society for Parenteral and Enteral Nutrition
8630 Fenton Street, Suite 412
Silver Spring, MD 20910

ASPET-Astellas Awards in Translational Pharmacology 1200

The ASPET-Astellas Awards in Translational Pharmacology are intended to extend fundamental research closer to applications directed towards improving human health. The awards will be given to: recognize those individuals whose research has the potential to lead to the introduction of novel pharmacologic approaches or technologies that may offer significant advances in clinical medicine in the future; and to facilitate that translational process. Three (3) awards of $30,000 each will be made to individuals. The money may be used for supplemental research funding, travel, training, or in any way that furthers the goals of the organization. Nominations should be made electronically.

Requirements: Any ASPET member in good standing may nominate an individual for this award.

Restrictions: Self-nominations will not be accepted.

Geographic Focus: All States

Date(s) Application is Due: Sep 15

Amount of Grant: 30,000 USD

Contact: Christine Carrico, Executive Officer; (301) 634-7060; fax (301) 634-7061; ccarrico@aspet.org or info@aspet.org

Internet: http://www.aspet.org/public/awards/Astellas_awards.html

Sponsor: American Society for Pharmacology and Experimental Therapeutics
9650 Rockville Pike
Bethesda, MD 20814-3995

ASPET Bernard B. Brodie Award in Drug Metabolism 1201

This biennial award (even-numbered years), established by CIBA-GEIGY Corporation to honor the fundamental contributions of Bernard B. Brodie, is presented to recognize outstanding original research contributions in drug metabolism and disposition, particularly those having a major impact on future research in the field. Only one nominator is necessary although more are acceptable. The award consists of a $2,000 honorarium, a commemorative medal, hotel and economy airfare to the award ceremony at the annual meeting. Nominations must be received by the deadline prior to the award year.

Requirements: Nominees must be members of ASPET; nominators need not be members of ASPET.

Restrictions: Supporting research accomplishments must not have been used to win any other major award.

Geographic Focus: All States

Date(s) Application is Due: Sep 15

Amount of Grant: 2,000 USD

Contact: Christine Carrico, Executive Officer; (301) 634-7060; fax (301) 634-7061; ccarrico@aspet.org or info@aspet.org

Internet: http://www.aspet.org/public/awards/brodie_award.html

Sponsor: American Society for Pharmacology and Experimental Therapeutics
9650 Rockville Pike
Bethesda, MD 20814-3995

ASPET Division for Drug Metabolism Early Career Achievement Award 1202

The ASPET Division for Drug Metabolism Early Career Achievement Award has been established to recognize excellent original research by early career investigators in the area of drug metabolism and disposition. The award is presented biennially in odd-numbered years. The award consists of $1,000, a plaque, and complimentary registration plus travel expenses (to a maximum of $1,000) for the winner to attend the award ceremony at the annual meeting. The awardee will deliver a lecture at the annual meeting describing their relevant research accomplishments. The awardee will be invited to publish a review article on the subject matter of the award lecture in Drug Metabolism and Disposition.

Requirements: Nominees for this award must have a doctoral degree (e.g. Ph.D., M.D., Pharm. D., D.V.M.) and must be within 15 years of having received their final degree, as of December 31 of the year of the award. There are no restrictions on institutional affiliation and a candidate need not be a member of ASPET.

Geographic Focus: All States

Date(s) Application is Due: Sep 15

Amount of Grant: 1,000 USD

Contact: Christine Carrico, Executive Officer; (301) 634-7060; fax (301) 634-7061; ccarrico@aspet.org or info@aspet.org

Internet: http://www.aspet.org/public/divisions/drugmetab/early_achievement_award.htm

Sponsor: American Society for Pharmacology and Experimental Therapeutics
9650 Rockville Pike
Bethesda, MD 20814-3995

ASPET Epilepsy Research Award for Outstanding Contributions to the Pharmacology of Antiepileptic Drugs 1203

The International League Against Epilepsy (ILAE) has sponsored this biennial award of $2,000 and a Certification of Citation to be awarded by the American Society for Pharmacology and Experimental Therapeutics, for the purpose of recognizing and stimulating outstanding research leading to better clinical control of epileptic seizures. This research may include the basic screening and testing of new therapeutic agents, studies on mechanisms of action, metabolic disposition, pharmacokinetics, and clinical pharmacology studies. The recipient will be selected by the Epilepsy Award Committee appointed by the President of ASPET, with representation of ILAE.

Requirements: Nominations for the Award may be submitted by members of any recognized scientific association, domestic or foreign.

Geographic Focus: All States

Date(s) Application is Due: Sep 15

Amount of Grant: 2,000 USD

Contact: Christine Carrico, Executive Officer; (301) 634-7060; fax (301) 634-7061; ccarrico@aspet.org or info@aspet.org

Internet: http://www.aspet.org/public/awards/epilepsy_award.html

Sponsor: American Society for Pharmacology and Experimental Therapeutics
9650 Rockville Pike
Bethesda, MD 20814-3995

ASPET Goodman and Gilman Award in Drug Receptor Pharmacology 1204

This award is given biennially (even-numbered years) to recognize and stimulate outstanding research in the pharmacology of biological receptors. It is the hope that such research might provide a better understanding of the mechanism of biological processes and potentially provide the basis for the discovery of drugs useful in the treatment of

diseases. The award includes a monetary prize and travel support for winner and spouse to attend the award ceremony. Nominations must be received September 15 of the year prior to the year in which the award will be made.
Requirements: Nominations must be made by ASPET members; however, there are no restrictions as to age, sex, nationality, or institutional affiliation. Nominee need not be a member of ASPET.
Geographic Focus: All States
Date(s) Application is Due: Sep 15
Amount of Grant: 2,500 USD
Contact: Christine Carrico; (301) 634-7060; fax (301) 634-7061; ccarrico@aspet.org
Internet: http://www.aspet.org/awards/goodman_and_gilman/
Sponsor: American Society for Pharmacology and Experimental Therapeutics
9650 Rockville Pike
Bethesda, MD 20814-3995

ASPET John J. Abel Award in Pharmacology 1205
This award, sponsored by Eli Lilly and Company and administered by ASPET, is given annually to stimulate original and outstanding fundamental research in pharmacology and experimental therapeutics by young investigators. Candidates may not have passed their 39th birthday as of April 30 of the year of the award, or previously received an award from the sponsor for the same technical accomplishment.
Requirements: Candidates need not be members of the society; however, nominations must be made by ASPET members.
Geographic Focus: All States
Date(s) Application is Due: Sep 15
Amount of Grant: 2,500 USD
Contact: Christine Carrico; (301) 634-7060; fax (301) 634-7061; ccarrico@aspet.org
Internet: http://www.aspet.org/public/awards/abel_award.html
Sponsor: American Society for Pharmacology and Experimental Therapeutics
9650 Rockville Pike
Bethesda, MD 20814-3995

ASPET P. B. Dews Lifetime Achievement Award for Research in Behavioral Pharmacology 1206
ASPET's Division of Behavioral Pharmacology sponsors the P. B. Dews Award for Research in Behavioral Pharmacology to recognize outstanding lifetime achievements in research, teaching and professional service in the field of Behavioral Pharmacology and to honor Peter Dews for his seminal contributions to the development of behavioral pharmacology as a discipline. The Award consists of $750, a plaque, and partial travel expenses to the award ceremony at the ASPET annual meeting. The recipient will be invited by the Chair of the Division of Behavioral Pharmacology to deliver a special lecture on this occasion. The lecture will be published subsequently in an appropriate ASPET-sponsored publication.
Requirements: Nominations may be made by members of ASPET or of any relevant scientific society.
Restrictions: There are no restrictions on nominees for this award.
Geographic Focus: All States
Date(s) Application is Due: Sep 15
Amount of Grant: 750 USD
Contact: Christine Carrico; (301) 634-7060; fax (301) 634-7061; ccarrico@aspet.org
Internet: http://www.aspet.org/public/awards/dews_award.html
Sponsor: American Society for Pharmacology and Experimental Therapeutics
9650 Rockville Pike
Bethesda, MD 20814-3995

ASPET Paul M. Vanhoutte Award 1207
The Paul M. Vanhoutte Award in Vascular Pharmacology was established to honor Dr. Vanhoutte's lifelong scientific contributions to our better understanding and appreciation of the importance of endothelial cells and vascular smooth muscle function in health and disease and for his mentoring of countless prominent endothelial and vascular biologists and pharmacologists. This is a biennial award, consisting an honorarium of $1,000, a custom-designed crystal bowl depicting the named Lectureship, and up to $2,000 travel expenses including registration to the annual spring ASPET meeting.
Requirements: There are no restrictions on institutional affiliation, nationality, or age of the candidate, but the recipient must be a member of the ASPET.
Geographic Focus: All States
Date(s) Application is Due: Sep 15
Contact: Christine Carrico; (301) 634-7060; fax (301) 634-7061; ccarrico@aspet.org
Internet: http://www.aspet.org/public/divisions/cardiovascular/vanhoutte_award.htm
Sponsor: American Society for Pharmacology and Experimental Therapeutics
9650 Rockville Pike
Bethesda, MD 20814-3995

ASPET Summer Undergraduate Research Fellowships (SURF) 1208
The purpose of the Fellowship program is to introduce undergraduate students to pharmacology research in order to heighten interest in science as a career, with an emphasis on pharmacology graduate training to increase the number of young scientists entering the research discipline of pharmacology. A group of at least five ASPET Regular members from one institution may wish to apply for support of up to $9,000 for an undergraduate fellowship program, to include up to five students stipends. Students are to receive at least a $2,800 stipend for a minimum of ten weeks participation. It is anticipated that the request for funding will include matching funds of at least $5,000 from local resources. Awards are normally made for three (3) years.
Restrictions: No indirect costs will be provided. All funds shall be used for student stipends/wages only.
Geographic Focus: All States
Date(s) Application is Due: Mar 1; Oct 1
Amount of Grant: 2,800-9,000 USD
Contact: Christine Carrico; (301) 634-7060; fax (301) 634-7061; ccarrico@aspet.org
Internet: http://www.aspet.org/public/surf/surf.htm
Sponsor: American Society for Pharmacology and Experimental Therapeutics
9650 Rockville Pike
Bethesda, MD 20814-3995

ASPET Torald Sollmann Award in Pharmacology 1209
This award, established by Wyeth-Herst Labs, is given in odd-numbered years to commemorate the pioneer work in the United States of Dr. Torald Sollmann in the fields of pharmacological investigation and education. The award is made to a nominee for significant contributions over many years to the advancement and extension of knowledge in the field of pharmacology. The award consists of an honorarium of $3,500, a medal, hotel and economy airfare for the winner and spouse to the annual meeting. The formal presentation of this biennial award and medal will be made at the annual meeting of ASPET. The recipient will be invited by the President of the Society to deliver a lecture to the membership that may be published in an appropriate ASPET journal.
Requirements: Nominations must be made by ASPET members; however, the nominee need not be a member. There are no restrictions as to age or institutional affiliation.
Geographic Focus: All States
Date(s) Application is Due: Sep 15
Amount of Grant: 3,500 USD
Contact: Christine Carrico; (301) 634-7060; fax (301) 634-7061; ccarrico@aspet.org
Internet: http://www.aspet.org/public/awards/sollmann_award.html
Sponsor: American Society for Pharmacology and Experimental Therapeutics
9650 Rockville Pike
Bethesda, MD 20814-3995

ASP.O. Daiichi Innovative Technology Grant 1210
The Daiichi Innovative Technology Grant will fund exploratory or hypothesis-generating projects that are not well-suited to a formal grant application (eg, based on statistical analysis and sample size specification) such as: development of new surgical or diagnostic instruments; survey or quality of life measures; new use of internet technology or computer software such as CDs or DVDs; educational brochures, materials, software for patients or physicians; and other applications of innovative technology for education or research in pediatric otolaryngology. Full patent and copyright control must be retained by the applicant and the applicant's institution. If patented innovations funded by this award generate more than $5,000, the applicant may be required to return funds to the Society. One award of up to $5,000, to be used over a one-year period, will be given annually. Letters of Intent should be received no later than December 15, with full proposals due on January 15.
Requirements: Researchers (MD, PhD, DMD, DO) in disciplines who will conduct research directly relevant to innovative technology in pediatric otolaryngology are eligible to apply. Applications submitted by otolaryngologists or demonstrating collaborations with otolaryngologists are preferred. Participation of an ASP.O. member is not required, but is preferred.
Geographic Focus: All States
Date(s) Application is Due: Jan 15; Dec 15
Amount of Grant: 5,000 USD
Contact: Stephanie L. Jones; (703) 519-1586 or (703) 836-4444; sljones@entnet.org
Internet: http://www.aspo.us/information.php?info_id=12
Sponsor: American Society for Pediatric Otolaryngology
One Prince Street
Alexandria, VA 22314-3357

ASP.O. Fellowships 1211
ASP.O. exists to foster excellence in the care of children with otorhinolaryngologic disorders through education and research and thereby enhance the profession of Pediatric Otolaryngology. Its fellowship program intends to: facilitate the creation and dissemination of knowledge about the care of infants and children with ORL disorders; serve as an advocate for infants and children with ORL disorders, ASP.O. members and others with shared goals; and preserve and promote dedication to excellent and humane care for infants and children with ORL disorders. See the web site for details and all available fellowships.
Geographic Focus: All States
Contact: Stephanie L. Jones; (703) 519-1586 or (703) 836-4444; sljones@entnet.org
Internet: http://www.aspo.us/fellowships.php
Sponsor: American Society for Pediatric Otolaryngology
One Prince Street
Alexandria, VA 22314-3357

ASP.O. Research Grants 1212
The American Society of Pediatric Otolaryngology (ASP.O.) awards funds annually to support innovative research in pediatric otolaryngology. ASP.O. will consider applications from both individuals and institutions. Preference is given to proposed projects that are to be completed within one year, although exceptional proposals that have duration in excess of one year will be considered. Two awards of up to $15,000, to be used over a one-year period, will be given annually. Letters of Intent should be received no later than December 15, with full proposals due on January 15.

Requirements: Researchers (MD, PhD, DMD) in disciplines who will conduct research directly relevant to pediatric otolaryngology are eligible to apply.
Restrictions: No portion of any grant may be used for travel expenses or for principal investigator salaries.
Geographic Focus: All States
Date(s) Application is Due: Jan 15; Dec 15
Amount of Grant: 15,000 USD
Contact: Stephanie L. Jones; (703) 519-1586 or (703) 836-4444; sljones@entnet.org
Internet: http://www.aspo.us/information.php?info_id=12
Sponsor: American Society for Pediatric Otolaryngology
One Prince Street
Alexandria, VA 22314-3357

ASPRS Leica Geosystems Internship 1213
The award is an eight-week internship for graduate students in photogrammetry. The selected intern will work with LH Systems personnel at its facilities in San Diego, Denver, Heerbrugg, or elsewhere. The internship provides the award winner with an opportunity to carry out a small research project of his or her own choice or to work on an existing LH Systems project as part of a team. Application form is available on the Web site. The intern will provide ASPRS with a final report of his or her research accomplishments during the period of the internship.
Requirements: The internship is open to graduate students of photogrammetry and remote sensing who are also members of ASPRS.
Geographic Focus: All States
Date(s) Application is Due: Dec 1
Amount of Grant: 2,500 USD
Contact: Director; (301) 493-0290, ext. 101; fax (301) 493-0208; scholarships@asprs.org
Internet: http://www.asprs.org/membership/scholar.html#LH
Sponsor: American Society for Photogrammetry and Remote Sensing
5410 Grosvenor Lane, Suite 210
Bethesda, MD 20814

ASPRS Robert E. Altenhofen Memorial Scholarship 1214
This award, given in memory of Robert E. Altenhofen, who was an outstanding practitioner of photogrammetry and who made notable contributions to the mathematical aspects of the science, is to encourage and commend college students who display exceptional interest and ability in the theoretical aspects of photogrammetry.
Requirements: Application may be made by an undergraduate or graduate student in a recognized college or university in the United States or elsewhere, who is either a student member or active member of the ASPRS. The recipient is obligated to provide a final report to ASPRS of his or her scholastic accomplishments during the period for which the award is granted.
Geographic Focus: All States
Date(s) Application is Due: Dec 1
Amount of Grant: 2,000 USD
Contact: Director; (301) 493-0290, ext. 101; fax (301) 493-0208; scholarships@asprs.org
Internet: http://www.asprs.org/membership/scholar.html#altenhofen
Sponsor: American Society for Photogrammetry and Remote Sensing
5410 Grosvenor Lane, Suite 210
Bethesda, MD 20814

ASPRS Ta Liang Memorial Award 1215
This award is given annually to one graduate student in remote sensing. The award is to be used for research-related travel. Application forms are available on the Web site.
Requirements: The award is restricted to student members of the society who are currently pursuing graduate-level studies at an accredited college or university. The recipient is obligated to provide ASPRS and Ta Liang's family, a report of his or her accomplishments during the travel for which the award is granted.
Geographic Focus: All States
Date(s) Application is Due: Dec 1
Amount of Grant: 500 USD
Contact: Award Administrator; (301) 493-0290, ext. 101; fax (301) 493-0208; scholarships@asprs.org
Internet: http://www.asprs.org/membership/scholar.html#ta%20liang
Sponsor: American Society for Photogrammetry and Remote Sensing
5410 Grosvenor Lane, Suite 210
Bethesda, MD 20814

ASPRS William A. Fischer Memorial Scholarship 1216
This scholarship, created in memory of William A. Fischer, a pioneer in the use of remote sensing from space for the study of the earth, is presented annually to a worthy student from the United States or another country to facilitate graduate studies and career goals judged to address new and innovative uses of remote sensing that relate to the natural, cultural, or agricultural resources of the earth.
Requirements: The award is restricted to society members. It is intended for a student who is currently pursuing graduate-level studies or who plans to enroll in a graduate program in an accredited college or university in the United States or elsewhere. The recipient is obligated to provide ASPRS with a final report of his/her scholastic accomplishments during the period for which the award is granted.
Geographic Focus: All States
Date(s) Application is Due: Dec 1
Amount of Grant: 2,000 USD
Contact: Director; (301) 493-0290, ext. 101; fax (301) 493-0208; scholarships@asprs.org
Internet: http://www.asprs.org/membership/scholar.html#fischer
Sponsor: American Society for Photogrammetry and Remote Sensing
5410 Grosvenor Lane, Suite 210
Bethesda, MD 20814

ASR Fichter Research Grant 1217
Applications are invited from scholars involved in promising research on women and religion, gender issues, feminist perspectives on religion, and new religious movements. Dissertation research qualifies for funding. Scholars at the beginning of their careers are particularly encouraged to apply. Send four copies of the proposal, budget, and curriculum vita to the address listed.
Requirements: Applicants must be members of the Association for the Sociology of Religion at the time of submission. A membership form is available online.
Geographic Focus: All States
Date(s) Application is Due: Mar 1
Amount of Grant: Up to 12,500 USD
Contact: William Swatos, Jr. , Executive Director; (309) 932-2727; fax (309) 932-2282
Internet: http://www.sociologyofreligion.com/FICHEXLP2004.html
Sponsor: Association for the Sociology of Religion
618 SW 2nd Avenue
Galva, IL 61434-1912

ASR Robert J. McNamara Student Award 1218
The annual competition recognizes an outstanding student paper in the sociology of religion. The paper must not have appeared in print prior to receiving the award. Students who wish their papers considered for the program must submit paper abstracts to the program chair following the guidelines of all standard paper submissions. Submissions should be by one of the following methods: IBM-formatted, virus-free disk, with the text in WordPerfect, Microsoft Word, or plain ASCII; or four paper copies, typed double spaced, and single sided. Submission should be in the form of articles with a maximum length of 40 single-sided pages inclusive of all material. The title page should include an abstract of no more than 200 words. Text should not exceed 12,000 words, i.e., approximately 36 double-spaced pages of 12 point (or 10 cpi) type. Sociology of Religion has the right of first review of award-winning papers.
Requirements: Authors must be currently enrolled students who have not defended the PhD when the paper is submitted. Membership in the Association for the Sociology of Religion is required either at the time of application or previously (membership form available online).
Geographic Focus: All States
Date(s) Application is Due: Jun 1
Amount of Grant: 500 USD
Contact: Prof. Lutz Kaelber, lkaelber@zoo.uvm.edu
Internet: http://www.sociologyofreligion.com/MCNAMARA2004.html
Sponsor: Association for the Sociology of Religion
618 SW 2nd Avenue
Galva, IL 61434-1912

ASSE Liberty Mutual Safety Research Fellowship 1219
The program supports two four- to six-week summer research fellowships at the Liberty Mutual Research Center for Safety and Health, in Massachusetts. The goals of the program are to encourage research activity in the field of safety; familiarize graduate students and faculty members with current research projects for application in teaching situations; provide a forum for linking safety professionals, industry needs, and quality research programs; and lay the groundwork for graduate students and faculty members to pursue safety/health applied research projects of their choice. Fellows receive a weekly stipend to cover transportation to and from the fellow's home to Hopkinton, MA; room and board; and rental car as needed. Members of the society receive special consideration.
Requirements: Application is open to U.S. citizens or permanent residents who possess the PhD degree or are working toward a master's or PhD degree and are enrolled or teaching at an accredited U.S. college or university in a safety or safety related field. Preference will be given to applicants holding appointments or enrolled in a department with an ABET accredited program.
Geographic Focus: All States
Date(s) Application is Due: Mar 1
Contact: Mary Goranson, (847) 699-2929; fax (847) 296-3769; mgoranson@asse.org
Internet: www.asse.org/foundat.htm
Sponsor: American Society of Safety Engineers
1800 E Oakton Street
Des Plaines, IL 60018-2187

ASSE Scholarships 1220
The foundation's mission is to promote the advancement of the safety, health, and environmental professions engaged in protecting people, property, and the environment by providing the necessary resources. The program supports a variety of undergraduate and graduate scholarships. Applications are available September 1. Winners names are posted on the foundation's website on or around April 1.
Geographic Focus: All States
Date(s) Application is Due: Dec 1
Contact: Grants Administrator; (847) 699-2929; fax (847) 768-3434
Internet: http://www.asse.org/foundat.htm
Sponsor: American Society of Safety Engineers
1800 E Oakton Street
Des Plaines, IL 60018-2187

Assisi Foundation of Memphis General Grants **1221**

The foundation supports organizations in its areas of interest, including health and human services—promote the health and well-being of the Mid-South community and help the health care system respond more effectively to community needs; education and literacy—projects/programs that build organizational capacity of provider agencies, provide professional development to service providers, promote collaboration among provider agencies, and leverage resources (local, state, and federal); social justice/ethics—projects/programs that strengthen ethical values among Mid-South citizens and promote social justice leading to a better understanding of and a more effective response to economic or social threats to the community; and cultural enrichment and the arts—projects/programs that foster an appreciation of the arts in the Greater Memphis community. Religious organizations seeking funding for religious programs also are eligible. Deadlines are set to coordinate with quarterly meetings of the Board of Directors of the Foundation.

Requirements: The Foundation makes grants only to organizations that are classified as tax-exempt under Section 501(c)3 of the Internal Revenue Code and as public charities under Section 509(a) of that Code. The Foundation uses its resources for charitable endeavors that advance the well-being of people and institutions located in Shelby, Fayette, and Tipton Counties in Tennessee; Crittenden County, Arkansas; and Desoto County, Mississippi.

Restrictions: Grants are not made for individuals, national fundraising drives, projects that address the needs of only one congregation, tickets for benefits, political organizations or candidates for public office, lobbying activities, recurring budget deficits, or tournament fees and/or travel for athletic competitions.

Geographic Focus: Arkansas, Mississippi, Tennessee

Date(s) Application is Due: Feb 15; May 17; Aug 16; Nov 15

Amount of Grant: Up to 20,000 USD

Contact: Jan Young, Executive Director; (901) 684-1564; jyoung@assisifoundation.org

Internet: http://www.assisifoundation.org/generalgrants.html

Sponsor: Assisi Foundation of Memphis
515 Erin Drive
Memphis, TN 38117

Association for the Study of Higher Education Dissertation Fellowships **1222**

With support of the Lumina Foundation for Education, the association awards fellowships to support dissertation research on the broad topics of financial aid, student retention and success, and adult learners and learning. The fellowships will support up to one year of activity that will be conducted through the students' home universities and can be used to support costs of supplying data, dissemination of project results, travel, tuition, and salary for the fellows. Eight one-year fellowships will be awarded. Guidelines are available online.

Requirements: Doctoral students affiliated with any accredited doctoral program may submit a proposal. Students in doctoral programs outside the United States may apply if their study is about student financial assistance, student retention and success, and/or adult learning and learners in the United States. Recipients of the fellowship must have completed their course work and any required qualifying examinations and have their dissertation proposal accepted by their institution.

Geographic Focus: All States

Date(s) Application is Due: May 31

Amount of Grant: 12,500 USD

Contact: Dr. Dennis Brown, Executive Director; (517) 432-8805; fax (573) 884-2197; ashemsu@msu.edu

Internet: http://35.8.168.242/fellowship/aboutfellowship.htm

Sponsor: Michigan State University
424 Erickson Hall
East Lansing, MI 48824

Association for Women in Mathematics Mentoring Travel Grants **1223**

The program helps junior women develop long-term working and mentoring relationships with senior mathematicians. The relationship should help the junior mathematician establish her research program and eventually receive tenure. The applicant's research may be conducted in any field funded by the NSF Division of Mathematical Sciences. Up to seven grants will be awarded, each to fund travel, subsistence, and other required costs for an untenured woman to travel to an institute and do research with a senior mentor for one month. Specific guidelines and instructions are available on the Web site.

Requirements: Applicants holding a doctorate or equivalent experience are eligible.

Geographic Focus: All States

Date(s) Application is Due: Feb 1

Amount of Grant: 4,000 USD

Contact: Grants Administrator; (703) 934-0163; awm@awm-math.org

Internet: http://www.awm-math.org/travelgrants.html#mentor

Sponsor: University of Maryland
4114 Computer and Space Sciences Building
College Park, MD 20742-2461

ASTA Academic Scholarships **1224**

ASTA's Tourism Cares refers to its scholarships for undergraduate or graduate level students as "Academic" Scholarships. Tourism Cares is offering 59 Academic Scholarships, ranging from $1,000 from $5,000. The Academic Scholarship Program benefits undergraduate or graduate students of travel- and-tourism-related, or hospitality-related, programs of study, at accredited schools. Each undergraduate scholarship award can be used to cover expenses for tuition, books, and educational fees only. Each graduate scholarship award can be used to cover expenses for tuition, books, educational fees, and research-related expenses only. Each graduate research scholarship award can be used to cover expenses for tuition, books, educational fees, and research-related expenses only. Scholarship offerings (for Graduating High School Seniors or Undergraduate Students include: ASTA Alaska Airlines Scholarship (one at $1,000); ASTA America Express Travel Scholarship (one at $1,000); ASTA Arizona Scholarship (one at $2,000); ASTA Holland America Line Undergraduate Scholarship (one at $1,500); ASTA Northern California Chapter Richard Epping Scholarship (one at $2,000); ASTA Pacific Northwest Chapter-William Hunt Scholarship (two at $1,500); ASTA Princess Cruises Scholarship (one at $2,500); ASTA Stan and Leone Pollard Scholarship (one at $1,000). Scholarship offerings for Graduate Students or Graduate Research Students include: ASTA Arnold Rigby Scholarship (two at $2,500); ASTA Avis Budget Group Scholarship (one at $2,000); ASTA Healy Scholarship (one at $1,000); ASTA Holland America Line Graduate Research Scholarship (one at $4,000); ASTA Joseph R. Stone Scholarship (three at $2,500). Scholarship offering for an Undergraduate or Graduate Internship (at ASTA's Research Department in the Washington, DC area fall semester): ASTA David J. Hallissey Memorial Internship (one at $2,000). Scholarships for Entering, or Presently Working, Travel Professionals include: ASTA A. J. Spielman Scholarship (one at $1,000); ASTA George Reinke Scholarship (two at $1,500). Specific ASTA Chapters also have scholarship offerings of their own, though many of these have resident requirements.

Requirements: Tourism Cares scholarship awards are merit-based, meaning applicants are evaluated on: academic performance, extracurricular and volunteer activities and experiences, motivation, and leadership potential. In general, all applicants must be enrolled in school for the upcoming fall semester or term to be eligible to receive the scholarship award. Scholarship eligibility criteria are determined by each fund donor contact. Applicants may only apply for the Academic Scholarship Program through the on-line application.

Restrictions: A student may win only one scholarship award from Tourism Cares per calendar year.

Geographic Focus: All States

Amount of Grant: 1,000-4,000 USD

Contact: Danielle Belanger, Student Programs Coordinator; 781-821-5990 x 215; scholarships@tourismcares.org or danielleb@tourismcares.org

Internet: http://www.asta.org/Education/content.cfm?ItemNumber=2552&navItemNumber=614

Sponsor: American Society of Travel Agents
1101 King Street, Suite 200
Alexandria, VA 22314

ASTD Dissertation Award **1225**

The award is given to foster and disseminate research in the practice of workplace learning and performance. It is presented annually to the person who has submitted the best dissertation completed during the previous academic year (September 21 through September 20). Illustrative areas of concentration include training and development, performance analysis, career development, organizational development/learning, workplace design, and human resource planning. All materials submitted must be in English and include the following: letter of application from candidate; recommendation from committee chair, on letterhead, with the dissertation completion date; abstract of dissertation (five to 15 pages, double-spaced) including summary of the problem addressed by the study, critique of relevant literature, synopsis of the findings, and implications for practice and research.

Requirements: The candidate must be recommended and sponsored by his or her committee chair. A committee chair may nominate more than one candidate who meets the criteria.

Geographic Focus: All States

Date(s) Application is Due: Sep 20

Amount of Grant: 500 USD

Contact: Ray Rivera, (703) 683-8144; rrivera@astd.org

Internet: http://www.astd.org/astd/About_ASTD/Awards/dissertation.htm

Sponsor: American Society for Training and Development
1640 King Street, P.O. Box 1443
Alexandria, VA 22313-2043

ASTD Research Award **1226**

The award is given to encourage the publication of research with practical implications for practitioners of workplace learning and performance. It is presented annually to the author(s) of the best article published in a refereed journal during the calendar year. Illustrative areas of concentration include training and development, career development, work design, performance analysis, organizational development/learning, and human resource planning. Applications must include letter of application; proof of publication—either a galley proof or reprint of the article in its entirety or a copy of the article with its letter of acceptance; and supporting documents, including the name and issue of the refereed journal in which the article appeared or will appear and contact information for the editor of the journal (name, address, and telephone number). The award includes a cash prize, designated place on the international conference program to present the research, and a commemorative plaque(s) presented at the awards ceremony.

Requirements: The article must have appeared, or will appear, in print in a refereed journal between January 1 and December 31 of the current year. All materials submitted, including the article, must be in English.

Geographic Focus: All States

Date(s) Application is Due: Oct 31

Amount of Grant: 500 USD

Contact: Ray Rivera, ASTD Research Department; (703) 683-8144; rrivera@astd

Internet: http://www.astd.org/astd/About_ASTD/Awards/research_award.htm

Sponsor: American Society for Training and Development
1640 King Street, P.O. Box 1443
Alexandria, VA 22313-2043

Astellas Foundation Research Grants 1227

The Foundation was established with the objectives of contributing to the improvement of national health and welfare, and to the progress of therapeutic medicines. In fulfilling the objectives, the Foundation aims to conduct pioneering research in new fields of diseases and metabolisms of medicines, by elucidating the mechanism of diseases and their treatment, in particular the relationship between therapeutic medicines and in-situ metabolisms. Major activities include: subsidizing research on metabolic disorders related to the diagnosis of diseases and therapeutic medicine; publication and editing of periodicals for research achievements; supporting and production of educational movies; and other projects to achieve the objectives of the Foundation.
Geographic Focus: All States
Contact: Grants Manager; 81 3 3244 3397; fax 81 3 5201 8512
Sponsor: Astellas Foundation
2-3-11 Nihonbashihoncho, Chuo-ku
Tokyo, 103-8411 Japan

Astronomical Society of the Pacific Amateur Achievement Award 1228

The Amateur Achievement Award is given annually to recognize significant contributions to astronomy or amateur astronomy by those not employed in the field of astronomy in a professional capacity. Nominations may be submitted by any individual or group. All nominations will remain active for three years of competition. Additional support letters or nomination updates are accepted at any time.
Restrictions: Self nominations or nomination by a family member will not be accepted.
Geographic Focus: All States
Date(s) Application is Due: Jan 1
Amount of Grant: 500 USD
Contact: Awards Coordinator; (415) 337-1100; fax (415) 337-5205
Internet: http://www.astrosociety.org/membership/awards/amateur.html
Sponsor: Astronomical Society of the Pacific
390 Ashton Avenue
San Francisco, CA 94112

Astronomical Society of the Pacific Dorothea Klumpke-Roberts Award 1229

The Dorothea Klumpke-Roberts Award is annually for outstanding contributions to public understanding and appreciation of astronomy. Nominees can include a broad range of individuals involved in science, education, writing/publishing, broadcasting, astronomy popularization, the arts, or other pursuits that advance public understanding and appreciation of astronomy. Letters of nomination should highlight how the nominee or nominees have distinguished themselves in this endeavor and/or cite exceptional achievement. Nominations are solicited from anyone familiar with or knowledgeable of the achievements of the nominee or nominees. The Award recipient will be selected by the Awards Committee appointed by the Board of Directors, with a plaque and monetary award given at the annual awards banquet. If none of the nominees are judged to be sufficiently outstanding, the Award will not be given for the year in question. In February or March, the secretary of the Society will communicate with the nominee and, if the Award is accepted, it will be announced in the spring of award year. Nominations received for the current year's competition remain under consideration for the two years following. The nomination form and a list of previous recipients are available at the website.
Geographic Focus: All States
Date(s) Application is Due: Jan 1
Amount of Grant: 500 USD
Contact: Awards Coordinator; (415) 337-1100; fax (415) 337-5205
Internet: http://www.astrosociety.org/membership/awards/klumpke.html
Sponsor: Astronomical Society of the Pacific
390 Ashton Avenue
San Francisco, CA 94112

Astronomical Society of the Pacific Las Cumbres Amateur Outreach Award 1230

The Las Cumbres Amateur Outreach Award honors outstanding educational outreach by an amateur astronomer to K-12 children and the interested lay public. "Amateur"is understood to mean someone who does not receive compensation (other than expenses) for their activity and does not receive the majority of their income from a profession in astronomy. The Award consists of a recognition plaque and $500 presented at the annual meeting awards banquet. The Award recipient will be selected by the Awards Committee appointed by the Board of Directors. If, in the opinion of the Committee, none of the nominees is sufficiently outstanding, the Award will not be given for the year in question. In February or March, the secretary of the Society will communicate with the nominee and, if the Award is accepted, it will be announced in the spring of award year.
Requirements: Nominations are made by members of the astronomical community. Nomination forms are available at the ASP website.
Geographic Focus: All States
Date(s) Application is Due: Jan 1
Amount of Grant: 500 USD
Contact: Awards Coordinator; (415) 337-1100; fax (415) 337-5205
Internet: http://www.astrosociety.org/membership/awards/cumbres.html
Sponsor: Astronomical Society of the Pacific
390 Ashton Avenue
San Francisco, CA 94112

Astronomical Society of the Pacific Maria and Eric Muhlmann Award 1231

The Maria and Eric Muhlmann Award of the ASP is given annually for recent significant observational results made possible by innovative advances in astronomical instrumentation, software, or observational infrastructure. The Award will be made to the scientist(s) who acquired the results, to the builder of the instrument/software (if a different individual), or to both as appropriate. The Award recipient(s) will be selected by the Awards Committee appointed by the Board of Directors of the ASP. Once accepted, the Award will be announced in the spring. The Award includes a $500 cash award and a plaque to be presented at the Annual Meeting Awards Banquet. Nomination forms are available at the website.
Requirements: Individuals as well as small teams of individuals are eligible. The Award will be granted for significant advances in any areas of astronomy, without restriction to wavelength or space/ground-based observations. Nominations are solicited from major observatories and departments of astronomy worldwide. Nominations will be accepted from any member of the astronomical community. The nominator(s) should submit an original and seven copies of a letter (no more than two typed pages in length) giving strong support to the choice. The letter should discuss the significance of the scientific results and clearly delineate the importance of the instrumentation/software/infrastructure advancement that enabled its achievement. Appropriate journal citations to the scientific results must also be included.
Geographic Focus: All States
Amount of Grant: 500 USD
Contact: Awards Coordinator; (415) 337-1100; fax (415) 337-5205
Internet: http://www.astrosociety.org/membership/awards/muhlmann.html
Sponsor: Astronomical Society of the Pacific
390 Ashton Avenue
San Francisco, CA 94112

Astronomical Society of the Pacific Priscilla and Bart Bok Awards 1232

The Priscilla and Bart Bok Awards are given jointly by the American Astronomical Society (AAS) and the Astronomical Society of the Pacific (ASP). The main criterion for selecting the two Bok Awards is scientific merit. Observational, instrumental, theoretical and interdisciplinary projects involving physics, mathematics, computer sciences, and engineering, are eligible. The awarded funds are intended to be used by the recipients to further their education and research efforts. There is a first prize award of $1,000, and a second prize award of $500. Applicants may contact the Society for further information about the award and how to apply.
Geographic Focus: All States
Contact: Awards Coordinator; (415) 337-1100; fax (415) 337-5205
Internet: http://www.astrosociety.org/membership/awards/bok.html
Sponsor: Astronomical Society of the Pacific
390 Ashton Avenue
San Francisco, CA 94112

Astronomical Society of the Pacific Richard H. Emmons Award for Excellence in College Astronomy Teaching 1233

The Richard A. Emmons Award of the ASP is awarded annually to an individual demonstrating outstanding achievement in the teaching of college-level introductory astronomy for non-science majors. Nominees must be teachers of college-level introductory astronomy for non-science majors. The award consists of a recognition plaque and a check for $500 which is presented at the annual meeting awards banquet. Letters of nomination should highlight how the nominee has distinguished himself or herself by citing exceptional achievement. Nomination forms are available at the ASP website.
Requirements: Nominations are solicited from colleagues, students, or others familiar with or knowledgeable of the nominee's achievements. Nominations received for the current year's competition remain under consideration for the two years following.
Geographic Focus: All States
Date(s) Application is Due: Jan 1
Amount of Grant: 500 USD
Contact: Awards Coordinator; (415) 337-1100; fax (415) 337-5205
Internet: http://www.astrosociety.org/membership/awards/emmons.html
Sponsor: Astronomical Society of the Pacific
390 Ashton Avenue
San Francisco, CA 94112

Astronomical Society of the Pacific Robert J. Trumpler Award 1234

The Trumpler Award is given each year to a recent recipient of the PhD degree in North America whose research is considered unusually important to astronomy. Nominations may be made by chairpersons of the astronomy and physics departments of these institutions. A plaque is included with the monetary award. Nomination forms and instructions for additional information are available at the website.
Requirements: To be considered for the Trumpler Award, a nominee's PhD must have been awarded within two years of the time of the nomination and most or all of the thesis must have been published in a recognized journal. The nominee remains eligible for two awards cycles — the cycle for which the nomination was initially submitted and for the cycle following. The department chairman should submit an original and seven copies of a letter of nomination, supporting the selection and discussing the extent of independence exhibited by the nominee in the choice and execution of the thesis research. Eight reprints of the published articles must accompany the nomination.
Geographic Focus: All States
Date(s) Application is Due: Jan 1
Amount of Grant: 500 USD
Contact: Awards Coordinator; (415) 337-1100; fax (415) 337-5205
Internet: http://www.astrosociety.org/membership/awards/trumpler.html
Sponsor: Astronomical Society of the Pacific
390 Ashton Avenue
San Francisco, CA 94112

ASU Graduate College Completion Fellowships **1235**

The Graduate College Completion Fellowships are intended to support the degree completion of doctoral and MFA in the arts, humanities, social sciences, and education during the semester. Nominations should be made on the basis of demonstrated ability to complete research/projects of high quality and the assurance that a semester of full-time effort will allow the student to complete their dissertation/project. The Completion Fellowship provides a spring or fall stipend of $9,500. It will not include tuition and health insurance. It is expected that recipients will have completed all course work for their degree and will only need to register for one credit hour. The application is available at the graduate college website. The deadline for spring semester is October 31, while the fall deadline is March 30.
Requirements: Applicants must be officially advanced to candidacy at the time of nomination; in good academic standing with the department and the Graduate College; and finishing his or her doctoral or MFA degree by the end of the semester. Applicants must provide a copy of their doctoral candidacy letter; a letter from their advisor describing the substance of the dissertation/practical project and its significance and completion date for the dissertation/final project; and a document from the applicant with a detailed outline of the dissertation/project; a timeline for its completion; a one page dissertation summary in non-technical language; and a copy of the latest chapter completed or other substantive work as evidence of dissertation progress.
Restrictions: Each Fellows is personally responsible for individual taxes related to this award. Additional information regarding tax issues related to awards can be found at the Fellowship website or by contacting the ASU financial aid office. Recipients holding these awards will not be allowed t hold a research or teaching assistantship, or any kind of employment during the semester awarded. International awardees may request exemption to the full-course of study. MFA student needing to register for more than one credit hour must justify this at the time of the application.
Geographic Focus: All States
Date(s) Application is Due: Mar 30; Oct 31
Amount of Grant: 9,500 USD
Contact: Pat Saldivar, Fellowship Contact; (480) 965-5906; psaldivar@asu.edu
Internet: http://graduate.asu.edu/completion
Sponsor: Arizona State University Graduate College
Interdisciplinary Building, B-Wing
Tempe, AZ 85287-1003

ASU Graduate College Dean's Fellowships **1236**

Graduate College Dean's Fellowships provide up to $10,000 of support to regularly admitted first year graduate degree students who are Arizona residents or underrepresented in their discipline, and who demonstrate academic excellence. This can include under-represented minorities, such as Hispanics, African Americans and Native Americans in all disciplines and under-represented students in particular disciplines, such as Asian Americans in the Humanities and Social Sciences, women in Mathematics, Science, Engineering and some Agricultural and Life Sciences programs, men in Nursing, Public Health or Women's Studies. The award is disbursed in two equal payments each semester. Award amount will depend on the student's demonstrated financial need. Other fellowships, teaching or research assistantships, or other scholarships can be used to supplement this award. The total amount of the offer (including this fellowship), should add up to the typical offer made to comparable students. The application is available at the Fellowship website.
Requirements: Applicants must enroll for a minimum of six graduate level credit hours each semester. Students must be: newly admitted for spring, summer, or fall of the academic year; regularly admitted to a graduate degree program, with a minimum 3.0 GPA; in their first year of study; and either a resident of Arizona or underrepresented in their discipline.
Restrictions: International students are not eligible for this support. Each fellow is personally responsible for individual taxes related to this award. Additional information regarding tax issues related to awards can be found at the graduate college financial aid website.
Geographic Focus: Arizona
Amount of Grant: Up to 10,000 USD
Contact: Catherine Lebihan, Fellowship Contact; (480) 965-6916; clebihan@asu.edu
Internet: http://graduate.asu.edu/financing/fellowships/graduate-college-fellowships/deans
Sponsor: Arizona State University Graduate College
Interdisciplinary Building, B-Wing
Tempe, AZ 85287-1003

ASU Graduate College Dissertation Fellowships **1237**

Dissertation Fellowships are competitive university-wide fellowships designed to support highly meritorious doctoral students who are in the final stages of post-candidacy doctoral work. The Fellowship provides a $17,000 award plus tuition during the final year of study. At the time of the award, the student must be in good academic standing as defined by the department and the Graduate College. The academic unit nominating the student provides health insurance. Awards will be made across the following five areas: arts, humanities, and social sciences; natural sciences and mathematics; engineering; professional programs and education; and interdisciplinary research. The application is available at the fellowship website, along with a list of required documentation.
Requirements: The nominees must have demonstrated the ability to carry out original research at a high level and advanced to doctoral candidacy at time of application; be prepared to devote full time to writing the dissertation with no other employment, including university assistantships; within one year of completion of the dissertation upon receipt of the award; and maintain a minimum of six graduate level credit hours each semester. Nine graduate level credit hours is highly recommended.
Restrictions: Each Fellow is personally responsible for individual taxes related to this award. Additional information regarding tax issues related to awards can be found at the financial aid website.
Geographic Focus: All States
Date(s) Application is Due: Feb 1
Contact: Pat Saldivar, Fellowship Contact; (480) 965-5906; psaldivar@asu.edu
Internet: http://graduate.asu.edu/dissfellowship
Sponsor: Arizona State University Graduate College
Interdisciplinary Building, B-Wing
Tempe, AZ 85287-1003

ASU Graduate College Doctoral Enrichment Fellowships **1238**

Doctoral Enrichment Fellowships support new and regularly-admitted first-year doctoral Doctor of Philosophy (Ph.D.) graduate students who demonstrate academic excellence and are underrepresented in their field of study. The fellowship provides a $17,000 award plus tuition during the first year of study. In the second year, the academic unit will provide at least a 50% TA or RA position (at the department standard program rate) or similar funding assuming satisfactory academic progress. Student Health Insurance is provided by the academic unit for the first year.
Requirements: New students must be nominated by their admitting academic unit. The nominee must: be a new Ph.D. degree seeking student who will be admitted and start the degree program the next academic year; demonstrate academic excellence and contribute to diversity in the field of study and the profession; and be a U.S. citizen or assigned an IP visa.
Restrictions: Student Health Insurance (provided by the academic unit for the first year)
Geographic Focus: All States
Contact: Jennifer Cason, (480) 965-6113; fax (480) 727-6615; Jennifer.Cason@asu.edu
Internet: http://graduate.asu.edu/financing/fellowships/graduate-college-fellowships/doctoral_enrichment
Sponsor: Arizona State University Graduate College
Interdisciplinary Building, B-Wing
Tempe, AZ 85287-1003

ASU Graduate College Earl and Lenore Tripke Fellowship **1239**

One nomination of a doctoral student per school in the College of Public Programs will be accepted for consideration of the Earl A. and Lenore H. Tripke Fellowship award. This Fellowship is awarded to one student for the spring semester from one of the following schools at ASU Graduate College: Community Resources and Development; Criminology and Criminal Justice; Public Affairs; and Social Work. A detailed list of the contents of the academic nomination packet are available at the Fellowship website.
Requirements: Nominees must be regularly admitted to an ASU graduate degree program; enrolled in spring semester with a minimum of six graduate-level credit hours (or more if required); and in good academic standing, making satisfactory academic progress. International students must have been enrolled in an ASU graduate program for one academic year immediately prior to the award period.
Geographic Focus: All States
Date(s) Application is Due: Oct 1
Contact: Pat Saldivar, Fellowship Contact; (480) 965-5906; psaldivar@asu.edu
Internet: http://graduate.asu.edu/tripke
Sponsor: Arizona State University Graduate College
Interdisciplinary Building, B-Wing
Tempe, AZ 85287-1003

ASU Graduate College Herman E. Demund Memorial Fellowships **1240**

The Herman E. DeMund Memorial Fellowship awards a $2,000 fellowship to be split between fall and spring semesters. The Fellowship covers tuition for fall and spring semesters. The award will be made using the following criteria: academic accomplishments including, but not limited to GPA, test scores, publications, exhibitions, presentations; contribution to the discipline including clarity and importance of the applicant's research questions or creative endeavors, as demonstrated in the letter of nomination and vita; progress towards the degree, with typical time-to-degree in the academic program clarified in the letter of nomination; and contributions to the academic community (e.g., teaching awards or other recommendations, substantial service activities). The nomination form is available at the Fellowship website. Nomination packets must also include a detailed letter of nomination and transcripts.
Requirements: Nominees must be regularly admitted to a graduate degree program with a minimum of six graduate level credit hours to be eligible. International students must have been enrolled for one academic year immediately prior to the award period to be eligible.
Restrictions: Each fellow is personally responsible for individual taxes related to this award. Additional information regarding tax issues related to awards can be found at the university financial aid office.
Geographic Focus: All States
Contact: Pat Saldivar, Fellowship Contact; (480) 965-5906; psaldivar@asu.edu
Internet: http://graduate.asu.edu/financing/fellowships/graduate-college-fellowships/demund
Sponsor: Arizona State University Graduate College
Interdisciplinary Building, B-Wing
Tempe, AZ 85287-1003

ASU Graduate College Lattie and Elva Coor Building Great Communities Graduate Fellowship **1241**

The Lattie and Elva Coor Fellowship is a unique opportunity created to support graduate students doing community engaged scholarly work. There is no specified disciplinary focus, but the student's work must include active engagement with communities in Arizona and include research or activities directed towards strengthening neighborhoods and/or improving the lives of Arizonans. The Fellow candidate's work should contribute to sustainable associations between ASU and one or more communities throughout the state. Projects that lead to enhanced civic participation in Arizona, improved Arizona

public policy, or reducing domestic violence in Arizona are of particular interest. Successful applicants will demonstrate a clear integration of community work with scholarly work and pragmatic application of their discipline to promote enlightenment, understanding, and solutions to current community conditions. During the Fellowship, students must satisfy the following *Requirements:* at least one briefing with Office of Public Affairs staff to review project progress; a project presentation to Lattie and Elva Coor and the Office of Public Affairs staff; and at least one presentation to a community organization or other entity demonstrating the applicability of the research region's issues and concerns. The application is available at the Fellowship website.
Geographic Focus: All States
Date(s) Application is Due: Sep 28
Contact: Kenja Hassan, Fellowship Contact; (480) 965-0695; kenja.hassan@asu.edu
Internet: http://graduate.asu.edu/financing/fellowships/graduate-college-fellowships/coor
Sponsor: Arizona State University Graduate College
Interdisciplinary Building, B-Wing
Tempe, AZ 85287-1003

ASU Graduate College Martha E. Bernal Memorial Award 1242
The Martha E. Bernal Memorial Award was established to recognize the efforts of a Ph.D. student who is doing work similar to the career accomplishments of Professor Bernal, which focused on minority mental health issues, broadly defined. Dr. Bernal was one of the leading researchers on understanding minority mental and health issues, ethnic identity development and training of ethnic minority graduate students in psychology. The Department of Psychology is responsible for the selection of the award recipient. Nominations will be sought by emailing graduate students and faculty within the Psychology Department. In addition, the directors of schools in the Social Science Division of the College of Liberal Arts and Sciences, the director of the Hispanic Research Center, and the director of the Applied Linguistics Program will be contacted. Finally, information will be posted on the Graduate College's website. The selection committee will consist of the members of the Graduate Studies Committee in the Department of Psychology. The application is available at the Award website. Applicants should also include a brief statement of purpose, a current vita, and a letter of recommendation from an ASU faculty member.
Requirements: The recipient shall be a graduate student pursuing a Ph.D. with formal standing affiliated with the Department of Psychology, a school within the Social Science Division of the College of Liberal Arts and Sciences, the Hispanic Research Center, Applied Linguistics Program, or any other discipline congruent with the spirit of Martha Bernal's work and relevant to minority mental health issues, as broadly defined. The applicant should be enrolled as a full time student, making acceptable and timely progress towards the Ph.D.
Restrictions: Each Fellow is personally responsible for individual taxes related to this award. Additional information regarding tax issues related to awards can be found at the University's financial aid office. The award is not renewable, but current recipients may be nominated for the award in the future. There is no residency or financial need requirement associated with this award.
Geographic Focus: All States
Date(s) Application is Due: Sep 30
Amount of Grant: 600 USD
Contact: Kathryn Lemery-Chalfant, Fellowship Contact; klemery@asu.edu
Internet: http://graduate.asu.edu/financing/fellowships/graduate-college-fellowships/bernal
Sponsor: Arizona State University Graduate College
Interdisciplinary Building, B-Wing
Tempe, AZ 85287-1003

ASU Graduate College Reach for the Stars Fellowship 1243
Reach for the Stars Fellowship support new and regularly-admitted first-year students who demonstrate academic excellence and are underrepresented in their field of study. The Fellowship provides a $15,000 award for the first academic year plus tuition. In the second year, the academic unit will provide at least a 50% teacher assistant or research assistant position (at the department standard program rate) or similar funding assuming satisfactory academic progress. The nomination form is available at the Fellowship website.
Requirements: The nominee must be regularly admitted into a Master's degree program at the time the award is disbursed; demonstrate academic excellence and contribute to diversity in the field of study and the profession; and be a U.S. citizen or assigned an IP visa. All Fellowship recipients are required to register and participate in the one credit Interdisciplinary Research Colloquium; enroll for a minimum of nine graduate-level credit hours during fall and spring semesters during the award period; meet regularly with a faculty mentor to discuss program goals, objectives and progress toward degree completion; meet regularly with a peer mentor to discuss getting acclimated to graduate school and student life; maintain satisfactory academic progress as defined by the academic unit and the Graduate College; not be employed as a TA, RA or in any other position at ASU during the year of this award; and submit a progress report signed by the academic mentor at the end of each semester.
Restrictions: Students must be nominated by their admitting academic unit. Mandatory fees, program fee, or any other fees are not covered as part of the fellowship.
Geographic Focus: All States
Date(s) Application is Due: May 15
Amount of Grant: 15,000 USD
Contact: Jennifer Cason; (480) 965-6113; fax (480) 727-6615; Jennifer.Cason@asu.edu
Internet: http://graduate.asu.edu/financing/fellowships/graduate-college-fellowships/stars
Sponsor: Arizona State University Graduate College
Interdisciplinary Building, B-Wing
Tempe, AZ 85287-1003

ASU Graduate College Science Foundation Arizona Bisgrove Postdoctoral Scholars 1244
The Bisgrove Postdoctoral Scholars Program (sponsored by Science Foundation Arizona) is designed to attract the nation's best early career scientists and engineers who exhibit the potential for outstanding competence and creativity in their research areas, strong communication skills, a passion for communicating the importance of their research to society, and a keen interest in educational science outreach to the community. Bisgrove Scholars will receive an annual stipend of approximately $60,000, benefits and additional funding for research expenses. The Bisgrove appointment is renewable on a year-to-year basis for a maximum initial term of two years, depending on the availability of funds. The ASU Graduate College Bisgrove Leadership Academy has been launched to provide training and guidance for Bisgrove Postdoctoral Scholars in areas of collaborative interdisciplinary research, user-inspired research, laboratory management and leadership, together with related resources, that will ensure their success as they develop as scientists in a multidisciplinary setting. The major activity of the Academy will include Strategies for Success workshops and seminars led by experts in a range of fields, with participating speakers from ASU, industry and the community. Bisgrove Scholars will also be invited to participate in the Preparing Future Faculty (PFF) program. Specific details about the application package are available at the Fellowship website.
Requirements: United States citizens or permanent residents of the U.S. should have completed a Ph.D. by the time of appointment, with no prior post-doctoral experience. They should also demonstrate research training and potential to transform ideas into value for society and the interest to work at the convergence of several disciplines. Scholars should be prepared to commit time to mentor undergraduate and graduate students in related disciplines; meet regularly with a faculty member to discuss research goals and objectives; submit a progress report every 90 days, summarizing their goals and evidence of progress; and co-write at least one federal grant application with the main research faculty mentor during the second year of the award, assuming satisfactory progress.
Geographic Focus: All States
Date(s) Application is Due: Jan 14
Amount of Grant: 60,000 USD
Contact: Jennifer Cason, Director of Graduate Support Initiatives; (480) 965-8968; Jennifer.Cason@asu.edu
Internet: http://graduate.asu.edu/bisgrove
Sponsor: Arizona State University Graduate College
Interdisciplinary Building, B-Wing
Tempe, AZ 85287-1003

ASU Graduate College Science Foundation Arizona Fellowships in Science and Engineering 1245
Arizona State University's commitment to research aimed at solving specific societal challenges is being supported by graduate fellowship funding from Science Foundation Arizona. These graduate fellowships will be awarded to biologists, chemists, physicists, bioengineers, chemical engineers, electrical engineers, materials engineers, computer scientists, social scientists and individuals with related degrees interested in addressing one of the following critical needs: diagnosis and prevention of disease; sustainable energy and the environment; and information and communications technologies at the human interface. Candidates must be nominated by the degree-granting academic unit. Science Foundation Fellowships offer up to two years of competitive support. Each year of the award carries a $25,000 personal stipend plus a $5,000 supplemental allowance to the SFAz Fellow to advance the Fellow's research and/or to offset relocation expenses. Full tuition is covered. Second year support depends upon evidence of outstanding performance and progress toward the degree.
Requirements: Nominee must be admitted into a graduate program or related degrees interested in addressing one of the following critical needs: diagnosis and prevention of disease; sustainable energy and the environment; and information and communications technologies at the human interface. Detailed information about each of these needs is available at the Fellowship website.
Geographic Focus: All States
Amount of Grant: 30,000 USD
Contact: Dr. Andrew Webber; (480) 965-5906; andrew.webber@asu.edu
Internet: http://graduate.asu.edu/financing/fellowships/graduate-college-fellowships/sfaz
Sponsor: Arizona State University Graduate College
Interdisciplinary Building, B-Wing
Tempe, AZ 85287-1003

ASU Institute for Humanities Research Fellowships 1246
The purpose of the 2012-2013 Institute for Humanities Research Fellowship at Arizona State University is to encourage serious critical explorations of home as a term that resonates on cultural, emotional, intellectual, religious, philosophical, political and spiritual levels—as a place, a space, a myth, a source of identity, a promised land, a state of being, a war zone, an impossibility, and/or an inalienable right. The program invites humanities scholars from various disciplines to apply for the IHR fellowship to conduct research projects from multiple disciplinary and transdisciplinary perspectives that illuminate and enrich our understanding of how the concept of home informs and is informed by geography and history, identities and ideologies, imagined and lived experiences. Projects may explore what it means to create a home or to be without home. For example, are diasporic communities without a home, or is home more real and vital because it resides in culture, history, and memory? And while the homeless, are obviously without a physical home, what about those who reside in literal and figurative borderlands, for whom feeling at home is always contingent, always at risk? The IHR ASU Fellows program provides funding for either individual tenured or tenure-track faculty or research teams (two to three faculty) to engage in a year of research related

to the annual theme, share their research with the academic community (via lectures, a conference or symposium), and produce a strong application for a large external grant.
Geographic Focus: All States
Date(s) Application is Due: Mar 4
Amount of Grant: Up to 20,000 USD
Contact: Sally Kitch; (480) 965-3000; fax (480) 965-4300; sally.kitch@asu.edu
Internet: http://ihr.asu.edu/funding/fellows
Sponsor: Arizona State University
P.O. Box 876505
Tempe, AZ 85287-6505

ASU Institute for Humanities Research Seed Grants 1247
In fall and spring competitions each year, projects are selected for Seed Grant awards of up to $7,500 for individual researchers or up to $12,000 per team. The IHR seed grant program is designed to provide support for projects that advance the transdisciplinary, collaborative, and issue-focused mission of the Institute and are intended for sufficient development to be competitive for national grants.These grants further advance faculty research and often improve the quality of grant proposals to external funding agencies. The Institute supports those projects that best address its mission and that have strong prospects of receiving external funding. Outcomes include: conferences and symposia; invited guest scholars; publications; external grant support; and curricular impacts.
Requirements: Seed grant funds are available for junior and senior humanities faculty members and/or collaborative teams involving non-humanists. All IHR projects should incorporate humanities methodologies and involve at least one humanities scholar.
Geographic Focus: All States
Date(s) Application is Due: Oct 22
Amount of Grant: Up to 12,000 USD
Contact: Sally Kitch; (480) 965-3000; fax (480) 965-4300; sally.kitch@asu.edu
Internet: http://ihr.asu.edu/funding/seed-grants
Sponsor: Arizona State University
P.O. Box 876505
Tempe, AZ 85287-6505

ASU Institute for Humanities Research Visiting Fellowships 1248
The Visiting Fellowship program is for tenured and tenure-track scholars from other institutions of higher education in the United States and abroad to spend spring semester in residence at the Institute for Humanities Research (IHR), participating in the intellectual life of the IHR and the university community. The Fellowship provides the opportunity to conduct research, collaborate with ASU faculty, and write. The Fellowship also promotes an exchange of ideas among visitors and ASU Fellows also working on the annual theme, which this year will be the Humanities and Home. Visitors will participate in weekly fellows meetings and give public lectures and seminars on their research topics while in residence at the IHR.
Geographic Focus: All States, All Countries
Date(s) Application is Due: Feb 18
Amount of Grant: Up to 20,000 USD
Contact: Sally Kitch; (480) 965-3000; fax (480) 965-4300; sally.kitch@asu.edu
Internet: http://ihr.asu.edu/funding/fellows
Sponsor: Arizona State University
P.O. Box 876505
Tempe, AZ 85287-6505

As You Sow 1249
As You Sow was founded in 1992 and has grown into two programs that strive to increase corporate accountability. The Environmental Enforcement Program seeks to reduce and remove carcinogenic exposures by pursuing compliance with California's Safe Drinking Water and Toxic Enforcement Act. In its consideration of grant requests, As You Sow will give priority to proposals in the following areas: education of consumers about the risks of toxic exposure and their right to know what goes into consumer products; toxics education and reduction in minority communities; enforcement of local, state, or federal consumer and environment-related laws; education of youth on environmental issues; and related areas will be considered on a case-by-case basis but the proposal must generally have an environmental enforcement or education component. The grants are generally in the range of $5,000 to $10,000.
Requirements: As You Sow gives priority to proposals that will substantially benefit citizens of California through neighborhood, regional, or statewide projects. As You Sow will also give consideration to proposals from organizations outside the state if the project or organization has a California emphasis or component.
Geographic Focus: All States
Amount of Grant: 5,000-10,000 USD
Contact: Conrad MacKerron; (415) 391-3212; fax (415) 391-3245; mack@asyousow.org
Internet: http://www.asyousow.org/grantmaking/guidelines.shtml
Sponsor: As You Sow
311 California Street, Suite 510
San Francisco, CA 94104

AT&T Bell Labs Graduate Research Program 1250
The Graduate Research Program for Women (GRPW) is designed to increase the number of minorities and women in the fields of science, math, engineering and technology. Financial support is provided to outstanding women students from the beginning of their full-time graduate studies leading to a PhD. The program includes two types of financial awards, fellowships and grants, which may be renewed on a yearly basis for the normal duration of the graduate program, subject to the participant's satisfactory progress toward the doctoral degree. Employment at AT&T Bell Laboratories is offered to fellowship and grant holders for the summer preceding their entry into graduate school as well as for subsequent summers during graduate school. If a fellowship holder chooses to remain on campus for supervised university research or study during the summer, her fellowship support will be continued through the calendar year. Application forms must be obtained from AT&T Bell Laboratories. Applications and all supporting materials are due January 13.
Requirements: Women who are applying to graduate school to eventually obtain a PhD in chemical engineering, chemistry, communications science, computer science/engineering, electrical engineering, information science, materials science, mathematics, mechanical engineering, operations research, physics, or statistics are eligible. Applicants must be U.S. citizens or permanent residents and, by the time the award is made, be admitted to full-time study in a doctoral program approved by AT&T Bell Laboratories.
Geographic Focus: All States
Date(s) Application is Due: Jan 13
Amount of Grant: 25,000 USD
Contact: Fellowship Program Manager; (908) 582-4822; coopgraduate@lucent.com
Internet: http://www.lucent.com/social/blgrfp
Sponsor: AT&T Bell Laboratories
1505 Riverview Road, P.O. Box 297
St Peter, MN 56082

AT&T Industrial Ecology Faculty Fellowships 1251
The foundation awards fellowships to academic researchers in the emerging, multidisciplinary field of industrial ecology. The program is intended to stimulate interdisciplinary research and curriculum development that involve social issues, engineering, the sciences, economics, management, business, law, and public policy issues. The overarching objectives of the program are (1) to produce university faculty and students who can contribute to solving global and regional environmental problems and help shape environmentally and economically efficient strategies that have a firm scientific and engineering basis and (2) to advance the theoretical basis for the field of industrial ecology and its application in service sectors, industrial activity, and regulatory arenas. Application guidelines are available online and applications are solicited electronically.
Requirements: Individuals working independently or in collaboration at U.S. universities in the following fields are eligible: business, economics, engineering, law, management, physical sciences, public policy, and social sciences.
Geographic Focus: All States
Date(s) Application is Due: Nov 20
Amount of Grant: 25,000 USD
Contact: Michael Blazek; (908) 221-4191; mblazek@att.com
Internet: http://www.corp.att.com/ehs/ind_ecology/fellow_guidelines.html
Sponsor: AT&T Foundation
1 AT&T Way
Bedminster, NJ 07921

ATA Doctoral Fellowships in Education 1252
Each year, the Association awards two fellowships, valued at $15,000 each, to ATA members who have been accepted into, or have already embarked upon, full-time study (as specified by the institution) in a doctoral program in education at a recognized Alberta public university. The fellowship program is intended to recognize academic excellence and to help defray the financial costs of university study. In selecting fellowship recipients, the Association takes into account the applicant's academic standing, contribution to the Association and public education, and exemplary teaching practices. The Association names a selection committee, which meets in early spring to review the applications and to select potential recipients. The committee forwards its recommendations to Provincial Executive Council, which makes the final decision at its April meeting. Applications are available at the ATA website.
Requirements: Applicants must meet the following conditions: hold a permanent Alberta teaching certificate and have completed at least five years of successful teaching in Alberta; hold the highest level of membership in ATA, with membership at that level for at least five years; entering his or her first year of full time doctoral study; and intend to continue a career in education in Alberta. Applicants cannot be previous fellowship winners and must be ATA members in good standing.
Restrictions: Fellowships are paid to recipients in two equal installments, beginning in the fall. Recipients are required to file a copy of their research with the Association.
Geographic Focus: Canada
Date(s) Application is Due: Mar 1
Contact: Corinne Anderson; (780) 447-9470; corinne.anderson@ata.ab.ca
Internet: http://www.teachers.ab.ca/Professional+Development/Grants+and+Scholarships/Fellowships+and+Scholarships.htm
Sponsor: Alberta Teachers Association
11010 142nd Street NW
Edmonton, AB T5N 2R1 Canada

ATA Educational Trust Grants-Curriculum Resource or Research Project 1253
Each year, the Trust awards a number of project grants of up to $3,000 to help individuals or groups conduct research or develop practical classroom materials for Alberta teachers. The materials and resources produced through these grants are placed in the ATA library where they can be borrowed by interested teachers. Grants are payable against receipts for expenses. Applications are available at the ATA website, but must be mailed. Faxed or emailed applications are not accepted.
Requirements: Any individual or group involved in education and planning to undertake such a project during the next school year is eligible to apply. Submissions in the areas of special education, global and environmental education, languages, Francophone education, native education and early childhood education are of particular interest to the Trust.

Geographic Focus: Canada
Date(s) Application is Due: May 1
Amount of Grant: Up to 3,000 CAD
Contact: Educational Trust Contact; (800) 232-7208 or (780) 447-9494
Internet: http://www.teachers.ab.ca/For%20Members/Programs%20and%20Services/Grants%20Awards%20and%20Scholarships/Pages/The%20ATA%20Educational%20Trust.aspx
Sponsor: Alberta Teachers Association
11010 142nd Street NW
Edmonton, AB T5N 2R1 Canada

ATA Education Research Award **1254**
The ATA Education Research Award is presented annually to a faculty member or sessional lecturer at an Alberta university who has undertaken high-quality research on classroom teaching and learning. Research must be directly related to school and classroom practice; be focused on school teaching and/or learning; be current (either ongoing or completed within the last two years); be related to critical issues; have involved classroom teachers and/or students; be applicable to the Alberta context and must be of practical benefit to teachers in improving their professional practice, and be of high quality in terms of purpose, approach, methodology, originality and clarity. The researcher(s) must also be prepared to present the research to Association subgroups. The application is available at the ATA website.
Requirements: To qualify for an award, the research must meet the following criteria: directly relate to school and classroom practice; focus on school teaching and/or learning; have current value and relate to critical issues; involve classroom teachers and/or students; apply to the Alberta context; hold practical benefit to teachers in improving their professional practice, with high quality in terms of purpose, methodology and originality.
Geographic Focus: Canada
Date(s) Application is Due: May 11
Amount of Grant: 5,000 CAD
Contact: J.C. Couture, Executive Staff Officer; (780) 447-9462 or (800) 232-7208; jc.couture@ata.ab.ca
Internet: http://www.teachers.ab.ca/For%20Members/Programs%20and%20Services/Grants%20Awards%20and%20Scholarships/Pages/Educational%20Research%20Award.aspx
Sponsor: Alberta Teachers Association
11010 142nd Street NW
Edmonton, AB T5N 2R1 Canada

ATA Nadene M. Thomas Graduate Research Bursary **1255**
The Nadene M. Thomas Graduate Research Bursary is offered annually to a member of the Alberta Teachers Association who is enrolled in a graduate program in a specialty education at a recognized Canadian university. In selecting bursary recipients, the Association takes into account these factors: the applicant's academic standing; the applicant's contribution to the Association and public education; exemplary teaching practice; and the applicability of the research on health issues affecting teachers and/or teachers' working conditions. The selected recipient is granted the bursary in two equal installments, the first upon receipt of the award and the second upon receipt of the completed research. Recipients are required to file a copy of their research with the Association. The application is available at the ATA website.
Requirements: Applicants must meet the following conditions: hold a permanent Alberta teaching certificate and have completed at least five years of successful teaching in Alberta; be at the highest level of membership possible and at the highest possible level of membership, in good standing, for at least five years; intend to continue a career in the K-12 public education system in Alberta; and be registered as a graduate student in a specialty education at a recognized Canadian university. Research must focus on health issues affecting teachers and/or on teachers' working conditions.
Restrictions: The applicant must not have received an ATA fellowship or bursary in the past.
Geographic Focus: Canada
Date(s) Application is Due: Mar 1
Amount of Grant: 5,000 CAD
Contact: Corinne Anderson; (780) 447-9470; corrine.anderson@ata.ab.ca
Internet: http://www.teachers.ab.ca/For%20Members/Programs%20and%20Services/Grants%20Awards%20and%20Scholarships/Pages/Nadene%20M%20Thomas%20Graduate%20Research%20Bursary.aspx
Sponsor: Alberta Teachers Association
11010 142nd Street NW
Edmonton, AB T5N 2R1 Canada

Athenaeum of Philadelphia Senior Fellowships **1256**
The fellowship trust supports the study, recording, and preservation of early American architecture and building technology and the teaching of conservation skills in American schools of architecture. Research is not subject to geographical restrictions, although preference is given to Delaware Valley topics. Applications are reviewed by a committee of architects, architectural historians, and educators appointed by the board of directors. Applications should be submitted in the form of a single-page letter setting forth a brief statement of the project, with attached budget, schedule for completion, and professional resume. Two letters of reference should be requested by the applicant and submitted directly to the committee. There are no application forms. A clear statement of objectives is necessary, and a final report is expected. Successful applicants may be invited to give a public lecture or participate in a seminar at the Athenaeum sharing the results of the project.
Requirements: Senior fellows must hold a terminal degree and possess a distinguished record of accomplishment.
Restrictions: Grants may not be used for international travel.
Geographic Focus: All States
Date(s) Application is Due: Mar 1
Amount of Grant: Up to 5,000 USD
Contact: Eileen Magee; (215) 925-2688; fax (215) 925-3755; magee@PhilaAthenaeum.org
Internet: http://www.philaathenaeum.org/grants.html
Sponsor: Athenaeum of Philadelphia
219 S Sixth Street
Philadelphia, PA 19106-3794

ATLA Bibliography Grant **1257**
The grant is awarded annually to one or more bibliographers or indexers to aid in the development of a work that provides access to a significant body of literature within the fields of theological and/or religious studies. The scope, subject matter, length, and format are broadly conceived, with the intent of encouraging bibliographical or indexing work at all levels, especially by persons undertaking their first major project in this area. The applicant(s) should complete and submit the application form (available online) along with the two requisite letters of reference by the specified due date.
Requirements: This grant is open to applicants inside and outside of the association.
Geographic Focus: All States
Date(s) Application is Due: Jan 15
Contact: Publications Committee Office; (888) 665-2852
Internet: http://www.atla.com/pub_com/grant.html
Sponsor: American Theological Library Association
300 S Wacker Drive, Suite 2100
Chicago, IL 60606-6701

ATS Henry Luce III Fellowships **1258**
The grants, supported by the Henry Luce Foundation, are awarded to scholars of distinction for research and publication that offer an innovative and substantial contribution to theological scholarship. The foundation supports the research of junior and senior scholars whose projects offer significant and innovative contributions to theological studies; meet high scholarly standards; enhance the theological understanding of people of faith and enrich the experience of church life in North America; and develop ways for scholarship to inform contemporary culture. No more than one fellow will be selected per the following categories: bible and the church, Christianity and contemporary culture, constructive theology, history of Christianity and the church today, ministry and practice of communities of faith, plus topics that do not fall into the above categories. Applicants must specify the category in which they are requesting appointment. Fellows present their research at the annual conference, held in the fall. Annual deadline dates may vary; contact the program officer for exact dates.
Requirements: The program is open to full-time faculty members of ATS-accredited and candidate schools who have an appropriate period of leave from all teaching and institutional responsibilities. Applicants also must be eligible to receive at least one-half salary support from their institutions.
Restrictions: Fellows may not hold any other major fellowship or grant during the tenure of the award without express approval of the association.
Geographic Focus: All States
Date(s) Application is Due: Dec 1
Contact: William Myers; (412) 788-6505, ext. 252; fax (412) 788-6510; myers@ats.edu
Internet: http://www.ats.edu/LeadershipEducation/pages/LuceFellowsInTheology.aspx
Sponsor: Association of Theological Schools
10 Summit Park Drive
Pittsburgh, PA 15275-1103

ATS Lilly Theological Collaborative Research Grants **1259**
The Theological Collaborative Research Grants, supported by the Lilly Endowment, are designed to enhance the skill and capacity of faculty in ATS schools as theological researchers and scholars. It supports research efforts of faculty, particularly at junior and non-tenured levels in order to nurture the development of their scholarship; seeks to enlarge the pool of faculty actively engaged as theological researchers; works to increase knowledge about grant seeking and the craft of theological research; and nurtures the habit of research as an ongoing aspect of scholarly life. The Program will provide grants of up to $16,000 each to as many as four collaborative research projects.
Requirements: All full-time faculty members at ATS-accredited and candidate schools are eligible to apply.
Restrictions: No more than two applicants from a single institution may receive awards in any given competition.
Geographic Focus: All States
Date(s) Application is Due: Jan 5
Contact: William Myers; (412) 788-6505, ext. 252; fax (412) 788-6510; myers@ats.edu
Internet: http://www.ats.edu/LeadershipEducation/pages/LillyTheologicalResearchGrants.aspx
Sponsor: Association of Theological Schools
10 Summit Park Drive
Pittsburgh, PA 15275-1103

ATS Lilly Theological Faculty Fellowships **1260**
The Faculty Fellowships, supported by the Lilly Endowment, encourage scholarly research that contributes to theological education, informs the life of the church, develops a greater public voice for theology in society, collaborates with other academic disciplines, and offers new perspectives on Christianity in a pluralistic setting. Faculty Fellowships will support up to six faculty members with awards of $30,000 each during an institutionally approved research leave of at least one full semester.

Requirements: All full-time faculty members at ATS-accredited and candidate schools are eligible to apply. For faculty fellowships applicants must have a research leave of at least one term from all teaching and institutional responsibilities during the award year.
Restrictions: Project proposals for less than six months will not be considered for faculty fellowships. No more than two applicants from a single institution may receive awards in any given competition.
Geographic Focus: All States
Date(s) Application is Due: Jan 5
Amount of Grant: Up to 30,000 USD
Contact: William Myers; (412) 788-6505, ext. 252; fax (412) 788-6510; myers@ats.edu
Internet: http://www.ats.edu/LeadershipEducation/pages/LillyTheologicalResearchGrants.aspx
Sponsor: Association of Theological Schools
10 Summit Park Drive
Pittsburgh, PA 15275-1103

ATS Lilly Theological Research Expense Grants **1261**
The Theological Expense Grants, supported by the Lilly Endowment, are designed to enhance the skill and capacity of faculty in ATS schools as theological researchers and scholars. It supports research efforts of faculty, particularly at junior and non-tenured levels in order to nurture the development of their scholarship; seeks to enlarge the pool of faculty actively engaged as theological researchers; works to increase knowledge about grant seeking and the craft of theological research; and nurtures the habit of research as an ongoing aspect of scholarly life. The Program will give up to $5,000 each to as many as six scholars for direct research expenses.
Requirements: All full-time faculty members at ATS-accredited and candidate schools are eligible to apply.
Restrictions: No more than two applicants from a single institution may receive awards in any given competition.
Geographic Focus: All States
Date(s) Application is Due: Jan 5
Amount of Grant: Up to 5,000 USD
Contact: William Myers; (412) 788-6505, ext. 252; fax (412) 788-6510; myers@ats.edu
Internet: http://www.ats.edu/LeadershipEducation/pages/LillyTheologicalResearchGrants.aspx
Sponsor: Association of Theological Schools
10 Summit Park Drive
Pittsburgh, PA 15275-1103

ATS Lilly Theological Scholars Grants **1262**
The Theological Scholars Grants, supported by the Lilly Endowment, are designed to enhance the skill and capacity of faculty in ATS schools as theological researchers and scholars. It supports research efforts of faculty, particularly at junior and non-tenured levels in order to nurture the development of their scholarship; seeks to enlarge the pool of faculty actively engaged as theological researchers; works to increase knowledge about grant seeking and the craft of theological research; and nurtures the habit of research as an ongoing aspect of scholarly life. This program will provide as many as six grants of up to $12,000 each for research apart from a formal research leave.
Requirements: All full-time faculty members at ATS-accredited and candidate schools are eligible to apply.
Restrictions: No more than two applicants from a single institution may receive awards in any given competition.
Geographic Focus: All States
Date(s) Application is Due: Jan 5
Amount of Grant: Up to 12,000 USD
Contact: William Myers; (412) 788-6505, ext. 252; fax (412) 788-6510; myers@ats.edu
Internet: http://www.ats.edu/LeadershipEducation/pages/LillyTheologicalResearchGrants.aspx
Sponsor: Association of Theological Schools
10 Summit Park Drive
Pittsburgh, PA 15275-1103

Auburn Foundation Grants **1263**
The purpose of the Foundation is to stimulate giving and cooperative leadership among the citizens of Auburn; help improve the lives of all community residents, especially those who are most vulnerable; and enrich the cultural environment and community life. Of special interest are projects that bring together all ages and sections of the town, or that contribute to healthy, active living. Application is available online.
Requirements: Any nonprofit organization that serves residents of Auburn is invited to apply.
Restrictions: Grants will not be awarded to for-profit businesses or expenses already incurred by the applicant.
Geographic Focus: Massachusetts
Date(s) Application is Due: Apr 15
Amount of Grant: 5,000 USD
Samples: Bancroft School (Worcester, MA)—for general support, $45,000; New England Science Ctr, Rutland House (Worcester, MA)—for repairs, $20,000; Tower Hill Botanic Garden (Boylston, MA)—for general support, $30,000.
Contact: Lois Smith; (508) 755-0980, ext. 107; lsmith@greaterworcester.org
Internet: http://www.greaterworcester.org/grants/Auburn.htm
Sponsor: Auburn Foundation
370 Main Street, Suite 650
Worcester, MA 01608-1738

Auburn University President's Graduate Opportunities Program **1264**
The major purpose of the program is to recruit, retain, and support African-American students engaged in graduate study leading to a doctoral degree from Auburn University. Successful applicants will receive a fellowship in addition to a stipend provided by the department, school, or college in which recipients are enrolled. Both the fellowship and the departmental stipend are renewable for up to four years of doctoral study. Application and guidelines are available online.
Requirements: Candidates must hold an earned degree from an accredited institution, meet admission standards of the graduate school and the doctoral program in which they seek to enroll, and be recommended for the award by that academic department or unit.
Geographic Focus: All States
Date(s) Application is Due: Feb 1
Amount of Grant: 15,000 USD
Contact: Dr. Christine Curtis, Chair; (334) 844-5771; curticw@auburn.edu
Internet: http://media.cla.auburn.edu/english/gs/handbook/policies/index.cfm?include=president
Sponsor: Auburn University
2003 RBD Library, 231 Mell Street
Auburn University, AL 36849-5168

Audio Engineering Society Graduate Scholarships **1265**
Graduate students in audio engineering and related fields may apply for the scholarships, which will be awarded on the basis of demonstrated past interest and achievements in the field and on two faculty recommendations. Awards are made annually, in August, and each successful applicant may apply for a one-time renewal after the successful completion of at least one year of graduate studies. Payments for the original award or the renewal will be sent directly to the graduate school on behalf of the student's account. Completed applications including recommendations will be accepted from March 15 to the listed application deadline.
Requirements: Applicants must have acceptance or pending application for graduate studies leading to a master's or higher degree or an internationally recognized equivalent. Applications are accepted from students worldwide.
Geographic Focus: All States
Date(s) Application is Due: May 15
Amount of Grant: 3,000 USD
Contact: Scholarship Coordinator; (212) 661-8528; fax (203) 682-0477; HQ@aes.org
Internet: http://www.aes.org/education/edu_foundation.html
Sponsor: Audio Engineering Society Educational Foundation
60 E 42nd Street
New York, NY 10165

Audubon Naturalist Society Research Grants **1266**
This program provides grants for either conservation-related field work or conservation and education projects. The project should be located outside the United States. One need not be affiliated with an institution or organization to apply. There are no restrictions on nationality. Preference given to start-up projects. Applied research projects are given preference over basic research projects.
Geographic Focus: All States
Contact: Education Program Coordinator; (301) 652-9188; fax (301) 951-7179; Contact@audubonnaturalist.org
Internet: http://www.audubonnaturalist.org
Sponsor: Audubon Naturalist Society of the Central Atlantic States
8940 Jones Mill Road
Chevy Chase, MD 20815

AUPHA Corris Boyd Scholarships **1267**
The Association of University Programs in Health Administration offers the Corris Boyd Scholarship in honor of a healthcare leader who dedicated his life to diversity and excellence in leadership, especially among people of color and women. The program currently offers two $40,000 scholarships to deserving students of color entering a graduate program in healthcare management. The award will be paid in four installments, paid in August and December of the first year, then August and December of the second year. The annual deadline for applicants is May 3.
Requirements: At the time of award, applicants must meet the following eligibility
Requirements: have applied and been accepted into an AUPHA full-member masters degree program for start in the upcoming Fall semester (not before August 1); be an under-represented minority; have a minimum 3.0 GPA (out of 4.0) in undergraduate coursework; and have U.S. citizenship or permanent resident status.
Geographic Focus: All States
Date(s) Application is Due: May 3
Amount of Grant: 40,000 USD
Contact: Lydia Middleton, President and CEO; (703) 894-0940, ext. 115; fax (703) 894-0941; lmiddleton@aupha.org or aupha@aupha.org
Internet: http://www.aupha.org/i4a/pages/Index.cfm?pageID=3541
Sponsor: Association of University Programs in Health Administration
2000 14th Street North, Suite 780
Arlington, VA 22201

Australasian Institute of Judicial Administration Seed Funding Grants **1268**
The Australasian Institute of Judicial Administration (AIJA) is a research and educational institute associated with Monash University. It is funded by the Standing Council on Law and Justice (SCLJ) and also from subscription income from its membership. The principal objectives of the Institute include research into judicial administration and

the development and conduct of educational programmes for judicial officers, court administrators and members of the legal profession in relation to court administration and judicial systems. The AIJA is interested in working with academics to develop proposals for research projects. It is offering financial assistance to provide an incentive for academics to prepare ARC grant applications in areas of AIJA interest. Broad areas of research identified as relevant to judicial administration are: the proportional use of judicial resources for optimum practical effectiveness and efficiency; and public perceptions and understanding of the justice system. Included in these broad areas are: the examination of the relationship between the principles of adjudicatory independence and administrative accountability; a comparative analysis of reforms in the justice system in other jurisdictions; performance measurement, including the creation of best practice models for the introduction of benchmarking and yardstick competition models in the courts; and proportionality, namely, the proportion of the value of claims expended on legal costs and the expenditure of court resources to meet the instances and areas of greatest need and access to justice.
Geographic Focus: All States, All Countries
Contact: Gregory Reinhardt, Executive Director; +61 3 9600 1311; fax +61 3 9606 0366; Gregory.Reinhardt@monash.edu
Internet: http://www.aija.org.au/index.php/research/seed-funding
Sponsor: Australasian Institute of Judicial Adminstration
555 Lonsdale Street, Ground Floor
Melbourne, VIC 3000 Australia

Australian Academy of Science Grants **1269**
The objectives of the Academy are to promote science through a range of activities. It has defined four major program areas as their focus. The Academy focus is on recognition of outstanding contributions to science, education and public awareness, science policy, and international relations. Guidelines are available online.
Geographic Focus: All States, Australia
Contact: Executive Secretary; 61-2-6201-9400; fax 61-2-6201-9494; eb@science.org.au
Internet: http://www.science.org.au
Sponsor: Australian Academy of Science
GP.O. Box 783
Canberra, ACT 2601 Australia

Australian Postdoctoral Research Fellowships **1270**
The specific purpose of the fellowship is to strengthen Australia's national research and development capability by providing opportunities for postdoctoral level researchers to undertake research of national and international significance and to broaden their research experience.
Requirements: Applicants must have submitted their PhD thesis before commencement of the fellowship. Candidates must obtain Australian citizenship or temporary residency status at the time of commencing the fellowship.
Geographic Focus: All States, Australia
Contact: Dr. Andrew Smith, Australian Postdoctoral Fellowships Coordinator; 61 2 6284 6629; andrew.smith@arc.gov.au
Internet: http://www.arc.gov.au/grant_programs/linkage_australian.htm
Sponsor: Australian Research Council
GP.O. Box 2702
Canberra, ACT 2601 Australia

Australian Postgraduate Research Fellowships **1271**
The program supports postgraduate training with stipends available through Linkage—Projects (APAIs), ARC Centres of Excellence (APACs), and Discovery—Projects. International opportunities for postgraduate researchers are also offered through Linkage—International awards, and project support is offered through Discovery—Indigenous Researchers Development. Guidelines are available online.
Requirements: The applicant must have a PhD degree and at least three but no more than eight years of postdoctoral experience at the time of application.
Restrictions: Postgraduate students are not normally eligible to apply for funding: applications seeking support for research training must be submitted by eligible researchers and research teams.
Geographic Focus: All States
Contact: Program Contact; 61-2-6284-6600; fax 61-2-6284-6601; info@arc.gov.au
Internet: http://www.arc.gov.au/info_users/researcher_postgraduate.htm
Sponsor: Australian Research Council
GP.O. Box 2702
Canberra, ACT 2601 Australia

Australian Research Fellowships/Queen Elizabeth II Fellowships **1272**
ARFs and QEIIs provide opportunities for established researchers to undertake research of national and international significance. QEIIs encourage research in Australia by postdoctoral graduates of exceptional promise and proven capacity for original work.
Requirements: Australian Research Fellowships (ARF) and Queen Elizabeth II Fellowships (QEII) are available to researchers with three to eight years of research experience since the award of the PhD or equivalent research doctorate.
Geographic Focus: All States, Australia
Contact: Program Contact, Research Grant and Training Section; 61-2-6284-6600; fax 61-2-6284-6601; info@arc.gov.au
Internet: http://www.arc.gov.au/info_users/researcher_fellowships.htm
Sponsor: Australian Research Council
GP.O. Box 2702
Canberra, ACT 2601 Australia

Aviation Research Grants **1273**
The program objectives are to encourage and support innovative and advanced research in the areas of potential benefit to the long-term technical needs of the National Airspace System; areas related to research on the prevention of catastrophic failures; and areas related to research, development, and implementation of technologies and procedures to counteract terrorist acts against civil aviation. The grants may be used to support research and development projects in the following areas: communications, navigation, and surveillance; capacity and air traffic control technology; aviation weather; airport technology; aircraft safety technology; system security technology; human factors and aviation medicine; environment and energy; and operations research.
Requirements: Colleges, universities, and nonprofit institutions are eligible. Profit-making organizations may be eligible in the area of aviation security.
Restrictions: Discretionary funds are not currently available.
Geographic Focus: All States
Amount of Grant: 25,000-5,000,000 USD
Contact: Program Director; (609) 485-4424; fax (609) 485-6509
Internet: http://www.tc.faa.gov/logistics/grants
Sponsor: U.S. Department of Transportation
800 Independence Avenue SW
Washington, DC 20591

Avon Foundation Breast Care Fund Grants **1274**
The fund awards grants for the development of breast cancer education and outreach activities, particularly breast health education and early detection services for under-served women. AFBCF supports programs that: recruit women for both first time screening and annual screening; develop partnerships between community-based outreach providers and local medical providers; work with health care providers to ensure proper clinical follow-up of abnormal screening results; and educate older women (65+ years old) about Medicare coverage of annual screening mammograms and assist them in obtaining the service from providers who accept Medicare.
Requirements: Private, non-government, nonprofit organizations in the United States and Puerto Rico are eligible. Grants are awarded to community-based programs and/or health care agencies that provide medically under-served women aged 40 and older with direct access to breast cancer education, annual clinical screening services, and prompt follow-up care. All programs must utilize the three-part approach to breast cancer early detection including regular screening mammography, clinical breast examination, and monthly breast self-examination.
Geographic Focus: All States
Date(s) Application is Due: Aug 28
Amount of Grant: 30,000-50,000 USD
Samples: Family Services Center of Coffee County, Enterprise, AL; The Witness Project, University of Arkansas for Medical Sciences, Little Rock, AR; Latino Community Development Agency, Oklahoma City, OK.
Contact: Coordinator; (212) 244-5368; fax (212) 695-3081; admin@avonbreast care.org
Internet: http://www.avonbreastcare.org/fundinginfo.htm
Sponsor: Avon Foundation
505 Eighth Avenue, Suite 1601
New York, NY 10018-6505

Avon Products Foundation Grants **1275**
The foundation's two-fold focus is to support education, community and social services, and arts organizations and programs that provide economic opportunities for women and girls; and to support breast cancer and other women's health organizations and programs. The foundation awards grants in cities and regions with a large concentration of representatives and business operations, with the majority of funds going to U.S.-based institutions. National, international, and New York metropolitan area programs are administered through the foundation's headquarters in New York. For regional funding support, refer to the Avon Foundation website.
Requirements: Applying organizations must be tax-exempt; national and municipal organizations are eligible. Request the guidelines brochure prior to submitting a formal proposal.
Restrictions: Grants do not support individuals; memberships; lobbying organizations; political activities and organizations; religious, veteran, or fraternal organizations; fundraising events; and journal advertisements.
Geographic Focus: All States
Amount of Grant: Up to 23,000,000 USD
Contact: Grants Administrator; (866) 505-2866; info@avonfoundation.org
Internet: http://www.avoncompany.com/women/avonfoundation
Sponsor: Avon Products Foundation
1345 Avenue of the Americas
New York, NY 10105

Award for Encouraging Disadvantaged Students into Careers in the Chemical Sciences **1276**
The purpose of the Award is to recognize individuals who have significantly stimulated or fostered the interest of students, especially minority and/or economically disadvantaged students, in chemistry, thereby promoting their professional development as chemists or chemical engineers, and/or increasing their appreciation of chemistry as the central science. Nominees for the award may come from any professional setting: academia, industry, government, or other independent facility. The award is intended to recognize significant accomplishments in the United States by individuals in stimulating students, especially those currently underrepresented in the profession, to elect careers in the chemical sciences and engineering, and in generating a broader appreciation of chemistry as the central science. The award consists of $5,000 and a certificate. A grant of $10,000

will be made to an academic institution, designated by the recipient, to strengthen its activities in meeting the objectives of the award. Up to $1,500 for travel expenses to the meeting at which the award will be presented will be reimbursed.
Requirements: Any individual, except a member of the award committee, may submit one nomination or seconding letter for the award in any given year. Nominating documents consist of a letter of not more than 1000 words containing an evaluation of the nominee's accomplishments and a specific identification of the work to be recognized, a biographical sketch including date of birth, and a list of publications and patents authored by the nominee.
Restrictions: Self-nominations are not accepted.
Geographic Focus: All States
Date(s) Application is Due: Nov 1
Amount of Grant: 6,500-7,500 USD
Contact: Felicia Dixon, Awards Administrator; (800) 227-5558 or (202) 872-4408; fax (202) 776-8008; f_dixon@acs.org or awards@acs.org
Internet: http://portal.acs.org/portal/acs/corg/content?_nfpb=true&_pageLabel=PP_ARTICLEMAIN&node_id=1319&content_id=CTP_004509&use_sec=true&sec_url_var=region1
Sponsor: American Chemical Society
1155 Sixteenth Street, NW
Washington, DC 20036-4801

Award for Volunteer Service to the American Chemical Society **1277**
The purpose of the Award is to recognize the volunteer efforts of individuals who have served the ACS, contributing significantly to the goals and objectives of the Society. The volunteerism to be recognized may comprise a variety of activities, including, but not limited to, the initiation or sponsorship of a singular endeavor or exemplary performance as a member or chair of a national level ACS committee, or as an elected division or local section officer, or outstanding service in a leadership role in regional meeting(s) or a local section. The award consists of $3,000, a certificate, and an inscription of the recipient's name on a plaque displayed at ACS Headquarters in Washington, DC.
Requirements: A nominee will have been a member of the ACS for at least 15 years, and will have made significant contributions to the ACS.
Restrictions: Past and present members of the ACS Board of Directors and staff are ineligible for the award.
Geographic Focus: All States
Date(s) Application is Due: Nov 1
Amount of Grant: 3,000 USD
Contact: Felicia Dixon, Awards Administrator; (800) 227-5558 or (202) 872-4408; fax (202) 776-8008; f_dixon@acs.org or awards@acs.org
Internet: http://portal.acs.org/portal/acs/corg/content?_nfpb=true&_pageLabel=PP_ARTICLEMAIN&node_id=1319&content_id=CTP_004556&use_sec=true&sec_url_var=region1
Sponsor: American Chemical Society
1155 Sixteenth Street, NW
Washington, DC 20036-4801

AWIS Educational Foundation Predoctoral Awards **1278**
Grants provide funds for women graduate students enrolled in a program in the life, physical, or social sciences or in an engineering program leading to a PhD degree. The award can be used for any aspect of education—tuition, books, housing, research, expenses, equipment, and so forth. U.S. citizens may use the money for study in the United States or abroad; non-U.S. citizens must be enrolled in a U.S. institution of higher education. Each year regular awards and named awards are granted; at times, citations of merit also are awarded.
Requirements: Women students enrolled in any life, physical, or social science or in an engineering program leading to a PhD degree are eligible. Candidates should have passed all qualifying exams and expect to complete the dissertation within two years.
Restrictions: Students pursuing a professional doctorate other than the PhD are not eligible to apply.
Geographic Focus: All States
Date(s) Application is Due: Jan 26
Amount of Grant: 1,000 USD
Contact: Dr. Barbara Filner; (202) 326-8940; fax (202) 326-8960; awisedfd@awis.org
Internet: http://www.awis.org/resource/edfoundation.html
Sponsor: Association for Women in Science Educational Foundation
1200 New York Avenue, Suite 650
Washington, DC 20005

AWM Mentoring Travel Grants for Women **1279**
The objective of the program is to help junior women develop a long-term working and mentoring relationship with a senior mathematician. This relationship should help the junior mathematician to establish her research program and eventually receive tenure. Grants support travel, subsistence, and other required expenses for an untenured woman mathematician to travel to an institute or a department to do research with a specified individual for one month. Any unexpended funds could be used for further travel to work with the same individual during the following year. Applicants may receive up to three grants through their careers, possibly in successive years. Submit five copies of the cover letter; curriculum vita; research proposal of approximately five pages in length that specifies why the proposed travel is particularly beneficial; supporting letter from the proposed mentor (who must promise to be available at the time of the proposed travel and may be either a man or a woman), together with the curriculum vita of the proposed mentor; approximate budget; and information about other sources of funding available.
Requirements: Applicants must be women holding a doctorate or equivalent experience and with a work address in the United States (or home address in case of unemployed). The applicant's research may be in any field that is funded by the Division of Mathematical Sciences of NSF.
Geographic Focus: All States
Date(s) Application is Due: Feb 1
Amount of Grant: Up to 4,000 USD
Contact: Travel Grant Selection Committee; (703) 934-0163; awm@awm-math.org
Internet: http://awm-math.org/travelgrants.html
Sponsor: University of Maryland
4114 Computer and Space Sciences Building
College Park, MD 20742-2461

AWM Travel Grants for Women Researchers **1280**
This program is supported by NSF and the Association for Women in Mathematics. Grants provide full or partial support to women researchers for travel to attend research conferences in their fields of specialization. An applicant should send five copies of a description of her current research and how the proposed travel will benefit her research program; curriculum vita; budget for the proposed travel; and information about all other sources of travel funding. In a cover letter, indicate the conference name, dates, and location (city/state/country). There are three award periods per year.
Requirements: The research conference must be in an area supported by the Division of Mathematical Sciences of NSF. Applicants must be women holding a doctorate (or equivalent experience) and with a work address in the United States.
Restrictions: Individuals who were awarded an AWM-NSF travel grant in the past two years, or who have other sources of external funding, such as any type of NSF grant, are ineligible. Partial support from the applicant's institution or from a nongovernmental agency does not make the applicant ineligible.
Geographic Focus: All States
Date(s) Application is Due: Feb 1; May 1; Oct 1
Amount of Grant: Up to 1,500 USD
Contact: Program Director; (703) 934-0163; awm@awm-math.org
Internet: http://awm-math.org/travelgrants.html
Sponsor: University of Maryland
4114 Computer and Space Sciences Building
College Park, MD 20742-2461

AWRA Richard A. Herbert Memorial Scholarship **1281**
This scholarship is to be used for the enhancement of education in water resources. One scholarship is awarded to a full-time undergraduate student working toward his/her first undergraduate degree and enrolled in a program related to water resources. The second scholarship is awarded to a full-time graduate student enrolled in a program related to water resources. In addition to the scholarships, each winner receives a complimentary membership in AWRA for the year.
Requirements: Full-time graduate and undergraduate students enrolled in a program related to water resources are eligible to apply.
Geographic Focus: All States
Date(s) Application is Due: Apr 29
Amount of Grant: Up to 2,000 USD
Contact: Harriette Bayse, (540) 687-8390; fax (540) 687-8395; harriette@awra.org
Internet: http://www.awra.org/student/herbert.html
Sponsor: American Water Resources Association
4 W Federal Street, P.O. Box 1626
Middleburg, VA 20118

AWWA Abel Wolman Doctoral Fellowships **1282**
The fellowships are designed to encourage promising students to pursue advanced training and research in the field of water supply and treatment. Applicants are evaluated on the basis of the quality of their academic record, the significance of the proposed research to water supply and treatment, and the applicant's potential to perform high-quality research.
Requirements: Applicants must be citizens of Canada, Mexico, or the United States, and anticipate completion of the requirements for the PhD degree within two years of the award.
Geographic Focus: All States
Date(s) Application is Due: Jan 15
Amount of Grant: Up to 20,000 USD
Contact: Linda Moody, (303) 347-6201 or (303) 794-7711; lmoody@awwa.org
Internet: http://www.awwa.org/Membership/Content.cfm?ItemNumber=3501&navItemNumber=13974
Sponsor: American Water Works Association
6666 W Quincy Avenue
Denver, CO 80235

AWWA Academic Achievement Awards **1283**
The awards recognize outstanding academic achievement on a graduate level through the selection of excellent master's theses and doctoral dissertations that relate to the field of potable water. First- and second-place awards are given for master's theses and also for doctoral dissertations. An announcement and form must be obtained from the training programs manager prior to submitting an application.
Requirements: All master's theses and doctoral dissertations that are relevant to the water supply industry are eligible. Applicants must have completed their theses or dissertations within the specified year.
Geographic Focus: All States

Date(s) Application is Due: Oct 1
Amount of Grant: 1,500-3,000 USD
Contact: Linda Moody, (303) 347-6201 or (303) 794-7711; lmoody@awwa.org
Internet: http://www.awwa.org/Membership/Content.cfm?ItemNumber=3501&navItemNumber=13974
Sponsor: American Water Works Association
6666 W Quincy Avenue
Denver, CO 80235

AWWA Holly A. Cornell Scholarship **1284**
The annual scholarship is made to encourage master's-level female and/or minority students to pursue advanced training and careers in the field of water supply and treatment. The recipient is expected to complete all requirements for the MS degree no sooner than December 1 of the year of the award.
Requirements: Female and/or minority (as defined by the Equal Employment Opportunity Commission) U.S. citizens who are currently master's degree students and anticipate completion of the requirements for a master's degree in engineering no sooner than December 1 of the year of the award are eligible to apply.
Geographic Focus: All States
Date(s) Application is Due: Jan 15
Amount of Grant: 5,000 USD
Contact: Linda Moody, (303) 347-6201 or (303) 794-7711; lmoody@awwa.org
Internet: http://www.awwa.org/Membership/Content.cfm?ItemNumber=3501&navItemNumber=13974
Sponsor: American Water Works Association
6666 W Quincy Avenue
Denver, CO 80235

AWWA Larson Aquatic Research Support (LARS) Scholarships **1285**
The Larson Aquatic Research Support scholarship provides support for doctoral and masters students interested in careers in the fields of corrosion control, treatment and distribution of domestic and industrial water supplies, aquatic chemistry, and/or environmental chemistry. Guidelines and application are available online.
Requirements: Applicant must be pursuing a degree (MS or PhD) at an institution of higher learning in the United States, Canada, Puerto Rico, or Mexico. MS candidates must complete the requirements of their degree sometime after August 1 of the year of the award, and PhD candidates must complete degree requirements sometime after December 1 of the year of the award.
Geographic Focus: All States
Date(s) Application is Due: Jan 15
Amount of Grant: 5,000-7,000 USD
Contact: Linda Moody, (303) 347-6201 or (303) 794-7711; lmoody@awwa.org
Internet: http://www.awwa.org/Membership/Content.cfm?ItemNumber=3501&navItemNumber=13974
Sponsor: American Water Works Association
6666 W Quincy Avenue
Denver, CO 80235

AWWA Thomas R. Camp Scholarship **1286**
The annual scholarship is presented to an outstanding graduate student doing applied research in the drinking water field and planning a career in this field. The scholarship is awarded to master's-level students in odd-numbered years and to doctoral-level students in even-numbered years. The recipient should not anticipate completion of his/her degree program before May of the year following the award.
Requirements: Students pursuing a graduate degree at an institution of higher education located in Canada, Guam, Puerto Rico, Mexico, or the United States are eligible.
Geographic Focus: All States
Date(s) Application is Due: Jan 15
Amount of Grant: 5,000 USD
Contact: Linda Moody, (303) 347-6201 or (303) 794-7711; lmoody@awwa.org
Internet: http://www.awwa.org/Membership/Content.cfm?ItemNumber=3501&navItemNumber=13974
Sponsor: American Water Works Association
6666 W Quincy Avenue
Denver, CO 80235

Axe-Houghton Foundation Grants **1287**
The foundation operates in the New York City metropolitan area exclusively for charitable, educational, and scientific purposes to foster and encourage an appreciation of the English language, with major emphasis on the spoken language. Priority is given to projects for the improvement of speech and its uses in the areas of public affairs, education, theater, poetry, debate, and the oral interpretation of literature. A portion of available funds is devoted to speech remediation and to scientific research pertaining to speech. Types of support include program grants and seed money grants.
Requirements: Initial inquiries should be by full proposal and should contain a brief statement of objective, duration of time required to complete this objective, qualifications of personnel responsible for conducting the project, and the general amounts of funds needed. The proposal should also indicate the degree of internal support that would be committed.
Restrictions: Grants are made only to tax-exempt institutions and not to individuals, private foundations or organizations outside of the United States. Grants are not made for scholarships, fellowships, capital, or general support programs.
Geographic Focus: New York
Date(s) Application is Due: Sep 1
Amount of Grant: 1,000-7,500 USD
Contact: Claire Brook, President; (212) 909-8304
Internet: http://www.foundationcenter.org/grantmaker/axehoughton/index.html
Sponsor: Axe-Houghton Foundation
919 3rd Avenue, 2nd Floor
New York, NY 10022

BA Elisabeth Barker Fund **1288**
The fund was established in memory of Elisabeth Barker (1910-1986), diplomatic correspondent and historian of modern Europe. It is intended to support studies in recent European history, particularly the history of central and eastern Europe. Grants may be made to support individual, collective, or institutional projects (including conferences). Awards to individuals may take the form of grants for private study and research in other European countries. British scholars and institutions may also apply for funds to assist scholars from other European countries to visit the United Kingdom, or to extend a stay arranged under other auspices. Application is made by letter, with appropriate supporting documentation, including a curriculum vita, a detailed research proposal describing the project and the scheme of research, a breakdown of the costs, and details of the plans for publication or dissemination of the research. Two referees, from outside the employing institution of the applicant, should be asked to submit references on the applicant and the project direct to the academy. Applicants will be notified of the outcome approximately three months after the closing date.
Requirements: Applicants must be ordinarily resident in the United Kingdom, the Isle of Man, or the Channel Islands. Most awards are made to staff employed in universities and other institutions of higher education, but applicants are not restricted by either academic or employment status.
Restrictions: PhD candidates are not eligible to apply, whether or not the project is related to the topic of their thesis. Awards are not available for the support of courses of study leading to professional qualifications.
Geographic Focus: All States, United Kingdom
Date(s) Application is Due: Jan 15; Apr 15; Oct 15
Amount of Grant: Up to 1,000 GBP
Contact: Grants Department; 020 7969 5217; fax 020 7969 5414; grants@britac.ac.uk
Internet: http://www.britac.ac.uk/funding/guide/intl/ebf.html
Sponsor: British Academy
10 Carlton House Terrace
London, SW1Y 5AH England

Bailey Family Foundation Grants **1289**
The foundation's primary mission is to expand the availability and enhance the quality of postsecondary education. The foundation also conducts research directed toward improving the state of higher education. Financial assistance in the form of scholarships is awarded to students based on their academic record, financial need, and level of community involvement. Scholarship programs include educational scholarships to high school seniors and college students.
Requirements: Applicants must possess a minimum cumulative GPA of 2.5, demonstrate financial need, be a graduating senior from a participating high school/college listed on the application, be U.S. residents, and be pursuing their undergraduate degree.
Geographic Focus: All States
Date(s) Application is Due: Mar 31; Sep 30
Amount of Grant: 5,000 USD
Samples: U of Tampa (FL)—to build an arts complex with classrooms, a gallery, a library, and studio space, **$2 million.**
Contact: Administrator; (703) 971-6203; fax (703) 971-6205; Bailey@Bailey-Family.org
Internet: http://www.bailey-family.org
Sponsor: Bailey Family Foundation
P.O. Box 803
Newington, VA 22122-0803

BA Larger Research Grants **1290**
Grants are available for primary research in the humanities and social sciences for which support is not normally available from other funding agencies. Eligible costs include travel and maintenance, research assistance, workshops, consumables, specialist software, and costs of interpreters in the field. Applications for collaborative or individual research projects are equally welcome. Applications from international groups of scholars are welcome, provided there is a UK-based scholar as lead applicant. The maximum grant period is three years. Guidelines are available online.
Restrictions: The following items are not eligible for support: institutional overheads; computer hardware; books and other permanent resources; the preparation of camera-ready copy; any editorial-related task; publication subventions; costs of publication in electronic media; payment to the principal researcher(s) in lieu of salary, or for personal maintenance at home; replacement teaching costs; travel and maintenance expenses for purposes such as lecture tours, or to write up the results of research; or attendance at conferences or organization of conferences either in the UK or abroad to disseminate the results of research.
Geographic Focus: All States
Date(s) Application is Due: Oct 15
Amount of Grant: 15,000-100,000 GBP
Contact: Grants Department; 020-7969 5217; fax 020-7969 5414; grants@britac.ac.uk
Internet: http://www.britac.ac.uk/funding/guide/lrg.html
Sponsor: British Academy
10 Carlton House Terrace
London, SW1Y 5AH England

Ball Brothers Foundation General Grants 1291

Founded in the name of Edmund B. Ball and his brothers, the Foundation seeks to build and sustain a high quality of life in Indiana by awarding grants to nonprofit organizations in broad subject areas, including elementary, secondary, higher, and adult basic education and literacy skills; cultural activities; community betterment; the environment; and health and human services. Usually, Muncie and Delaware Counties receive a higher priority for funding than requests from across the state. Types of support include general operations, annual campaigns, capital campaigns, building construction/renovation, program development, conferences and seminars, professorships, publication, curriculum development, research, fellowships, matching funds, seed grants, and technical assistance. Preference will be given to catalytic grants that will stimulate others to participate in problem solving or in matching fund programs and to innovative approaches for addressing either traditional or emerging community needs. Applications are reviewed by the board of directors in January, May, and September of each calendar year. Proposals are encouraged to be submitted from February to May. Grant seekers may send a preliminary proposal, complete proposal, or ask for a personal visit to discuss a potential grant request.
Requirements: Indiana 501(c)3 nonprofit institutions and organizations are eligible.
Restrictions: The Foundation will not support: direct assistance to individuals or scholarships; applications coming from outside of Indiana; booster organizations; on-going salary requests of staff personnel to support an organization; services that the community-at-large should normally underwrite (i.e. roads, bus transportation, etc.); capital building projects; research projects (except for philanthropic studies); or unsolicited proposals (all requests must begin with a preliminary proposal).
Geographic Focus: Indiana
Date(s) Application is Due: Apr 1; Sep 1
Amount of Grant: Up to 100,000 USD
Contact: Donna Munchel, Executive Assistant; (765) 741-5500; fax (765) 741-5518; donna.munchel@ballfdn.org or info@ballfdn.org
Internet: http://www.ballfdn.org
Sponsor: Ball Brothers Foundation
222 South Mulberry Street
Muncie, IN 47305

Ball Brothers Foundation Rapid Grants 1292

Ball Brothers Foundation offers funding opportunities for a limited number of Ball Rapid Grants that are designed to provide funding to organizations requiring immediate funding for the following types of needs: to continue or complete a project; to provide professional development; to buy equipment or materials for a project; for travel to meet representatives to advance ideas for a current or future project; to formulate a project idea; to carry out a mandated law or event; or for seed money to begin a new project. Generally requests up to $5,000 are considered. Ball Rapid Grants do not fall within the normal granting period. Requests may be submitted at any time between February 1 and November 30; decisions will be made within four business days and awards sent within 7 to 10 business days.
Requirements: Ball Brothers Foundation is restricted by its charter to grants to nonprofit institutions and organizations within Indiana.
Restrictions: The Foundation will not support: direct assistance to individuals or scholarships; applications coming from outside of Indiana; booster organizations; on-going salary requests of staff personnel to support an organization; services that the community-at-large should normally underwrite (i.e. roads, bus transportation, etc.0; capital building projects; research projects (except for philanthropic studies); or unsolicited proposals (all requests must begin with a preliminary proposal).
Geographic Focus: Indiana
Date(s) Application is Due: Nov 30
Amount of Grant: Up to 5,000 USD
Contact: Donna Munchel, Executive Assistant; (765) 741-5500; fax (765) 741-5518; donna.munchel@ballfdn.org or info@ballfdn.org
Internet: http://www.ballfdn.org/index/staff/generalinfo.asp
Sponsor: Ball Brothers Foundation
222 South Mulberry Street
Muncie, IN 47305

Baltimore Washington Center Adult Psychoanalysis Fellowships 1293

The Adult Psychoanalysis Fellowship Program is a one-year program designed for individuals who have a significant interest in psychoanalysis as a body of knowledge and as a framework with which to understand and carry out therapeutic efforts. Intended for advanced graduate students and residents as well as recent graduates in Psychiatry, Psychology and Social Work, the program at its core consists of monthly seminars presented by psychoanalysts on clinically-related topics. Fellows are also welcome to attend Forums on Psychoanalysis and other conferences offered by the Institute and Society.
Geographic Focus: All States
Date(s) Application is Due: Jul 15
Contact: Elizabeth Manne, Executive Director; (301) 470-3635 or (410) 792-8060; fax (410) 792-4912; admin@bwanalysis.org
Internet: http://www.bwanalysis.org/Fellowship%20Program-Adult.html
Sponsor: Baltimore Washington Center for Psychoanalysis
14900 Sweitzer Lane, Suite 102
Laurel, MD 20707

Baltimore Washington Center Child Psychotherapy Fellowships 1294

The Child Psychotherapy Fellowship program is a one-year program specifically designed for child mental health professionals who are interested in deepening their understanding of working with children from a psychoanalytic or psychodynamic perspective. The program is tuition-free. Detailed discussions of actual clinical work with children and reviews of relevant psychoanalytic literature will be used as a framework to enhance understanding and ability to carry out therapeutic work. Fellowship meetings will be held monthly, September through May, on the second Wednesday of every month, at the homes of Baltimore and Washington faculty members.
Requirements: Professionals, including psychiatrists, psychologists, social workers, and counselors currently working with children in psychotherapy should apply. Recent graduates of training programs should also apply.
Geographic Focus: All States
Date(s) Application is Due: Jul 15
Contact: Elizabeth Manne, Executive Director; (301) 470-3635 or (410) 792-8060; fax (410) 792-4912; admin@bwanalysis.org
Internet: http://www.bwanalysis.org/Fellowship%20Program-Child.html
Sponsor: Baltimore Washington Center for Psychoanalysis
14900 Sweitzer Lane, Suite 102
Laurel, MD 20707

Banfield Charitable Trust Grants 1295

This trust offers grants annually to nonprofit organizations and educational institutions that make life better for Pets and their families. Special considerations will be given to collaborative programs with Pet-related organizations and Banfield team members working together towards a shared goal. At this time, Banfield Charitable Trust is pursuing the following funding priorities: promotion of preventative healthcare for Pets; educating children about veterinary medicine and the Pets they love; programs based on the human-Pet bond and how this relates to longer, healthier lives for Pets and people; and veterinary education programs. The Trust does not generally fund an entire project, but expects to be one of multiple funding sources. Typically, the Trust will not fund more than 50% of the project's entire budget.
Requirements: 501(c)3 organizations and educational institutions are eligible.
Restrictions: Funding will not be considered for the following purposes: Spay and neuter or adoption programs; General operating expenses, deficit reduction, or general administrative overhead expenses; Fundraising campaigns, including special events; Grants to individuals, or to provide support for business enterprise.
Geographic Focus: All States
Date(s) Application is Due: Jun 30; Nov 30
Contact: Darlene Schwartz, (503) 922-5801; info@banfieldcharitabletrust.org
Internet: http://www.banfieldcharitabletrust.org/Grants
Sponsor: Banfield Charitable Trust
8000 North East Tillamook Sreet, P.O. Box 13998
Portland, OR 97213

Banfi Vintners Foundation Grants 1296

The foundation awards general operating grants to nonprofits in its areas of interest, including higher education, civic and public affairs, arts and humanities, wildlife protection, health (hospitals and disease research/prevention), religion, science, social services, and international. Grants are made nationwide, with preference given to requests from Massachusetts and the New York, NY, area.
Geographic Focus: Massachusetts, New York
Amount of Grant: 100-465,000 USD
Samples: Colgate U (Hamilton, NY)—for operating support, $75,000; Friends for Long Island's Heritage (Syosset, NY—for operating support, $25,000; Huntington Hospital Assoc (Huntington, NY)—for operating support, $25,000; Cornell U (Ithaca, NY)—for operating support, $465,000.
Contact: Philip Calderone, Executive Director; (516) 626-9200
Sponsor: Banfi Vintners Foundation
1111 Cedar Swamp Road
Glen Head, NY 11545

Bank of America Charitable Foundation Matching Gifts 1297

The Bank of America Charitable Foundation Matching Gifts program encourages employees to contribute to qualifying charitable organizations. This program supports employee giving by offering a way to double – up to $5,000 (U.S.) per person each calendar year – employees' cash or securities contributions to their favorite charitable organizations and thus improve their communities.
Requirements: Charitable organizations in the United States must be tax-exempt under section 501(c)3 of the Internal Revenue Code and not be classified as a private foundation. Charitable Organizations located in England or Wales must be registered with the Charity Commission. Charitable organizations outside of the United States, England or Wales must be qualified as eligible for donations from CAFAmerica.
Restrictions: The Bank of America Charitable Foundation does not: match charitable gifts to private, family or donor advised funds, or gifts to political or fraternal organizations; or match charitable gifts that benefit students directly or that result in an employee receiving a benefit, including tuition or sponsorships.
Geographic Focus: All States, District of Columbia, Guam, Marshall Islands, Northern Mariana Islands, Puerto Rico, U.S. Virgin Islands, American Samoa, Canada, United Kingdom
Amount of Grant: Up to 5,000 USD
Contact: Anne M. Finucane, Foundation Chairperson/Chief Marketing Officer; (617) 434-9410 or (800) 218-9946; anne.m.finucane@bankofamerica.com
Internet: http://about.bankofamerica.com/en-us/global-impact/matching-gifts-features-and-eligibility.html#fbid=wq9wE7VdEpC
Sponsor: Bank of America Charitable Foundation
100 North Tryon Street
Charlotte, NC 28255

BARD Postdoctoral Fellowships 1298

The United States-Israel Binational Agricultural Research and Development Fund (BARD) was established to promote and support research and development in agriculture for the mutual benefit of both countries. The postdoctoral program was initiated to promote cooperative agricultural research between postdoctoral fellows from one country and senior scientists from the other; to assist young scientists to become professionally established in the scientific community; and to provide BARD with input into new research areas and to enhance scientific competence in these areas. The program emphasizes innovation in agriculture, particularly in the areas of cellular and molecular biology and the use and development of explanatory models in agriculture. Also included in the fellowship is a $5000 allowance for children, and, if necessary, a small allowance not to exceed $3000 to host institution for research-related expenses. The duration of the fellowship is one year. Requests for application kits and information may be obtained from the Maryland address or BARD, P.O. Box 6, Bet Dagan, Israel 50250; phone: 03-968-3230, fax: 03-966-2506, or email: bard@bard-isus.com.

Requirements: Applicants must be U.S. or Israeli citizens. Applicants should have fulfilled the requirements for a PhD degree within the last three years. Candidates must file an application with BARD in English naming only one senior scientist. Application kits may be obtained at either BARD office. Only one application per year for each candidate will be accepted.

Restrictions: Israelis already in the United States and Americans already in Israel (for a period of six months or longer) are not eligible.

Geographic Focus: All States

Date(s) Application is Due: Jan 15

Amount of Grant: 32,000 USD

Contact: BARD, (301) 504-4584; fax (301) 504-4619; lea@bard-isus.com

Internet: http://www.bard-isus.com

Sponsor: United States-Israel Binational Agricultural Research and Development Fund
5601 Sunnyside Avenue
Beltsville, MD 20705

BARD Research Grants 1299

The United States-Israel Binational Agricultural Research and Development Fund (BARD) was established by the governments of the United States and Israel for the purpose of promoting and supporting cooperative research and development projects in agriculture for the mutual benefit of both countries. Proposals may be unidisciplinary or interdisciplinary and may cover any or all phases of research and development. BARD will consider financing research and development projects, exploratory or basic research studies, and small initial feasibility studies and other closely related activities in the areas of soil and water conservation management and utilization; crop production including new technology; crop protection; animal production including aquaculture; veterinary medicine; crop and animal genetic improvement; recycling of wastes; postharvest sciences covering operations from production through processing; agricultural engineering; agricultural economics; and cellular and molecular biology in agriculture. Investigators of either country may request information or mail proposals to Lynn Gipe, USDA-ARS-BARD at the Maryland address, or BARD Executive Director, Volcani Center, P.O. Box 6, Bet Dagan, Israel 50250, 03-9683230. Guidelines and applications are also available on the Web site.

Requirements: Affiliates of public or private nonprofit research institutions that demonstrate the necessary research and development capabilities are eligible to apply for funding. Scientists who wish to apply for grants should submit their proposals through such legally constituted institutions or agencies. The research proposal should be prepared jointly by at least one U.S. and one Israeli investigator and should describe the areas of anticipated cooperative endeavor between them. BARD will assist scientists who are unable to find collaborators; the BARD office in Israel or its liaison office in the United States should be contacted if such assistance is necessary.

Restrictions: BARD will not consider more than one application from the same investigator in a given year or fund the same investigator in more than one concurrent project.

Geographic Focus: All States

Date(s) Application is Due: Sep 1

Amount of Grant: 100,000-350,000 USD

Contact: BARD Program Contact; (301) 504-4522; fax (301) 504-4619; bard@bard-isus.com

Internet: http://www.bard-isus.com

Sponsor: United States-Israel Binational Agricultural Research and Development Fund
5601 Sunnyside Avenue
Beltsville, MD 20705

Barker Award 1300

Open to qualified graduate students registered full-time in the Faculty of Graduate Studies at the University of Calgary.

Requirements: Applicants must be studying in the fields of Business Administration with emphasis on Entrepreneurship, New Venture Development and Marketing. The award is tenable during a student's first year of the MBA program.

Geographic Focus: All States

Date(s) Application is Due: May 30

Amount of Grant: 1,500 CAD

Contact: Connie Baines; (403) 220-5690; cbaines@ucalgary.ca

Internet: https://pr1web.ucalgary.ca/FGS_SAM/public/InternalSearchResults.aspx

Sponsor: University of Calgary
2500 University Drive NW, Earth Sciences Building, Room 720
Calgary, AB T2N 1N4 Canada

Barra Foundation Project Grants 1301

Project grants are one-time grants generally for amounts above $10,000. The Foundation considers grants for innovative projects that aid research in advancing the frontiers of human services, arts and culture, health and education. Grants are not made for ongoing or expanding programs where substantial initial support was previously provided from other sources. Three principal criteria are strictly adhered to in judging the merits of a proposed project. They are: innovation; evaluation; and dissemination. Proposals may be submitted at any time of the year. Initial requests for project grants should be in the form of a preliminary concept paper, generally not to exceed two pages, summarizing: the principal focus and objectives of the proposed project; uniqueness of the concept; the overall methodology; estimated timetable; preliminary budget data; and other sources of support for the project.

Restrictions: The foundation does not provide grants for: ongoing operating budgets; staff salaries; budget deficits; endowments; capital campaigns and projects; international programs; environmental and religious organizations; scholarships and fellowships; or audio/video projects, publications, catalogs, and exhibitions.

Geographic Focus: Pennsylvania

Amount of Grant: 10,000-500,000 USD

Contact: William Harral, III; (215) 233-5115; william.harral@verizon.net

Internet: http://www.barrafoundation.org/grants/index.html

Sponsor: Barra Foundation
8200 Flourtown Avenue, Suite 12
Wyndmoor, PA 19038-7976

Barth Syndrome Foundation Research Grants 1302

Barth syndrome is a serious X-linked recessive condition associated with cardiomyopathy, neutropenia, skeletal muscle weakness, exercise intolerance, growth delay, and diverse biochemical abnormalities (including defects in mitochondrial metabolism and phospholipid biosynthesis). Because many clinical and biochemical abnormalities of Barth syndrome remain poorly understood, the program is seeking proposals for research that will advance knowledge on any aspect of the syndrome. The foundation prefers to award seed grants to experienced investigators for testing of initial hypotheses and collection of preliminary data leading to successful long-term funding by NIH and other major granting institutions. The foundation also encourages investigators new to the field of Barth Syndrome research. Send an electronic version of the full application and all the attachments by the listed application deadline. The Foundation anticipates awarding several one- or two-year grants of up to $40,000 each. Funds will be available as soon as the successful grant applicants have been notified.

Requirements: Principal investigators who are affiliated with nonprofit institutions are eligible.

Geographic Focus: All States

Date(s) Application is Due: Oct 31

Amount of Grant: 10,000-40,000 USD

Contact: Matthew J. Toth; (617) 469-6769; fax (617) 849-5695; mtoth@barthsyndrome.org

Internet: http://www.barthsyndrome.org/english/View.asp?x=1635

Sponsor: Barth Syndrome Foundation
675 VFW Parkway, #372
Chestnut Hill, MA 02467

BA Sir Ernest Cassel Educational Trust Fund Grants 1303

The academy administers funds on behalf of the Sir Ernest Cassel Educational Trust, to promote research in any field falling within the humanities or social sciences. Awards are offered to support travel costs relating to a research project, and are particularly aimed at recent postdoctoral scholars. Awards for travel costs rarely exceed @L1000; however, an application to this fund for travel costs may be combined with an application for a personal research grant for other costs, up to a total of @L7500. Applications should be submitted on the standard research grant application form, available from university research offices, or from the academy. In cases where an applicant seeks both travel and other costs, a single application form should be used. Applicants should not complete two separate forms when submitting a dual application to the Cassel fund for travel and to the research grants fund for other elements of a project. Applicants will be notified of the outcome approximately three months after the closing date.

Requirements: Applicants must be ordinarily resident in the United Kingdom, the Isle of Man, or the Channel Islands. Most awards are made to staff employed in universities and other institutions of higher education, but applicants are not restricted by either academic or employment status. These awards are aimed particularly at recent postdoctoral scholars.

Geographic Focus: All States, United Kingdom

Date(s) Application is Due: Apr 15

Amount of Grant: 1,000-7,500 GBP

Contact: Research Grants Department; 020 7969 5217; fax 020 7969 5414; grants@britac.ac.uk

Internet: http://www.britac.ac.uk/funding/guide/cetf.html

Sponsor: British Academy
10 Carlton House Terrace
London, SW1Y 5AH England

BA Small Research Grants 1304

Small research grants are available to support primary research in the humanities and social sciences. Eligible costs include travel and maintenance, research assistance, workshops, consumables, specialist software, and costs of interpreters in the field. Applications for collaborative or individual projects are equally welcome. Applications from international groups of scholars are welcome, provided there is a UK-based scholar as lead applicant. Grants are tenable for up to 24 months.

Geographic Focus: All States, China

Date(s) Application is Due: Jan 15; Apr 15; Oct 15
Amount of Grant: 500-7,500 GBP
Contact: Grants Department; 020-7969 5217; fax 020-7969 5414; grants@britac.ac.uk
Internet: http://www.britac.ac.uk/funding/guide/srg.html
Sponsor: British Academy
10 Carlton House Terrace
London, SW1Y 5AH England

Batchelor Foundation Grants 1305
The foundation supports food banks and organizations involved with arts and culture, education, the environment, animals and wildlife, health, human services, and economically disadvantaged people. Special emphasis is directed toward programs designed to engage in medical research and provide care for childhood diseases; and promote study, preservation, and public awareness of the natural environment. Funding is available in the form of: capital campaigns; continuing support; endowments; general/ operating support; and program development grants. There are no application forms. Initial approach should be a letter that details the grant proposal.
Requirements: Florida area nonprofits are eligible.
Restrictions: Individuals are not eligible.
Geographic Focus: Florida
Amount of Grant: 1,000-1,000,000 USD
Samples: Community Partnership for Homeless, Miami, FL, $1,056,600—for general operating support and endowment; Fairchild Tropical Botanic Garden, Coral Gables, FL, $540,000— For general operating support; University of Miami, Miami, FL, $525,000— for general operating support.
Contact: Anne Batchelor-Robjohns, Co-C.E.O; (305) 416-9066 or (305) 534-5004; jbatchelor@bellsouth.net
Sponsor: Batchelor Foundation
111 NE 1st Street, Suite 820
Miami, FL 33132

Baxter International Foundation William B. Graham Prize for Health Services Research 1306
Since 1986, The Baxter International Foundation and the Association of University Programs in Health Administration have awarded the William B. Graham Prize for Health Services Research to recognize researchers who have made major contributions to the health of the public through innovative research in health services. The Prize honors the late William B. Graham, longtime chairman and CEO of Baxter International Inc., and is internationally regarded as the premier recognition for individuals conducting health services research. The Prize recognizes individuals who have had a significant impact on the health of the public in one of three primary focus areas: Health Services Management, Health Policy Development and Healthcare Delivery. The Prize includes an award of $25,000 to the individual and $25,000 to a not-for-profit institution that supports the winner's work.
Geographic Focus: All States, All Countries
Amount of Grant: 50,000 USD
Contact: Prize Coordinator; (847) 948-4605; fdninfo@baxter.com
Internet: http://www.baxter.com/about_baxter/sustainability/international_foundation/index.html
Sponsor: Baxter International Foundation
One Baxter Parkway
Deerfield, IL 60015-4633

Bay and Paul Foundations, Inc Grants 1307
General operating and project grants are awarded for support of children's services and precollege educational programs, with an emphasis on technology and enhancement of science, math, and writing curricula; preservation of cultural and natural history collections and collections care training in museums, zoos, libraries, and botanical gardens; advocacy and research programs for preserving biodiversity; and Native American cultural heritage preservation and economic development programs. There are no application forms. Proposals should be directed to the executive director and should include a brief description of the applicant organization and of the program for which funding is requested, including objectives and numbers served; a project budget or financials; other expected sources of support; qualifications of key personnel; and evidence of tax-exempt status. The foundation's directors meet three times a year, usually in January, May, and October, to consider grant proposals. Requests are not accepted via fax. Regular postal service delivery is the preferred method of proposal acceptance.
Requirements: Nonprofits in Connecticut, Massachusetts, Maine, New Hampshire, New Jersey, New York, Rhode Island, and Vermont are eligible.
Restrictions: Grants do not support requests for endowments, building campaigns, building construction or maintenance, sectarian religious programs, books or studies, individual scholarships or fellowships, loans, travel, film, television or video productions, programs consisting primarily of conferences, for annual fund appeals, or to other than publicly recognized charities. First time grants for K-12 arts-in-education programs and K-12 science and math programs are currently geographically restricted to the New York City metropolitan area.
Geographic Focus: Connecticut, Maine, Massachusetts, New Hampshire, New Jersey, New York, Rhode Island, Vermont
Contact: Frederick Bay; (212) 663-1115; info@bayandpaulfoundations.org
Internet: http://www.bayandpaulfoundations.org/areas.html
Sponsor: Bay and Paul Foundations
17 West 94th Street
New York, NY 10025

Bayer Clinical Scholarship Award 1308
This award is intended to facilitate the development of specific clinical expertise in the field of hemophilia for applicants who have completed medical training and have an interest in pursuing a career as a hemophilia treater/researcher. The award will support a mentored physician in training for two years. Clinical duties will encompass diagnosis, evaluation, and the planning of management strategies for patients with hereditary bleeding disorders. In addition to the clinical experience, the applicant may pursue a research project in the field of hemostasis. Clinical scholarships will provide funding for two years. Up to five new awards will be made each year. Guidelines and application are available online.
Requirements: The applicant should have earned his/her medical degree within the previous eight years. The commitment of the mentor to the grantee and the research project (if applicable) is a vital element of the application, as is the quality of the clinical environment at which the applicant will undertake the scholarship.
Geographic Focus: All States
Amount of Grant: 70,000 USD
Contact: Administrator; programadministrator@bayer-hemophilia-awards.com
Internet: http://www.bayer-hemophilia-awards.com/awards.cfm#clinical
Sponsor: Bayer HealthCare Corporation
100 Bayer Road
Pittsburgh, PA 15205-9741

Bayer Foundation Grants 1309
The Bayer Foundation supports programs that enhance the quality of life, provide unique and enriching opportunities that connect diverse groups and ensure preparedness for tomorrow's leaders. The Foundation welcomes proposals from 501(c)3 organizations whose programming matches at least one of the following areas: Civic and Social Service Programs; Education and Workforce Development; Arts and Culture and Health and Human Services.
Requirements: 501(c)3 nonprofit organizations in Bayer operating communities are eligible. Submit proposals to regional offices in California, Connecticut, Georgia, Indiana, Kansas, Massachusetts, New Jersey, New York, North Carolina, Ohio, Pennsylvania, Texas and West Virginia.
Restrictions: The Foundation will not fund: For-profit organizations or those without Internal Revenue Service code 501(c)3 nonprofit, tax-exempt status; Organizations that discriminate on the basis of race, color, creed, gender, sexual orientation or national origin; General operating support for United Way affiliated agencies. Only support for special projects will be considered; Organizations or programs designed to influence legislation or elect candidates to public office; General endowment funds; Deficit reduction or operating reserves; Religious organizations; Charitable dinners, events or sponsorships; Community or event advertising; Individuals; Student trips or exchange programs; Athletic sponsorships or scholarships; Telephone solicitations; Organizations outside of the United States or its territories
Geographic Focus: California, Connecticut, Georgia, Indiana, Kansas, Massachusetts, New Jersey, New York, North Carolina, Ohio, Pennsylvania, Texas, West Virginia
Contact: Bayer Foundation; (800) 422-9374
Internet: http://www.bayerus.com/about/community/i_foundation.html
Sponsor: Bayer Foundation
100 Bayer Road, Building 4
Pittsburgh, PA 15205-9741

Baylor Institute for Faith and Learning Visiting Fellows Program 1310
The Institute for Faith and Learning sponsors a visiting fellows program that brings outstanding scholars to the university for one-year appointments. Fellows pursue research in their academic disciplines informed by Christian intellectual traditions. They serve as participants in the full life of the university, assisting the institute in its aim to encourage academic research that integrates faith and learning, and helping to cultivate a context conducive to the pursuit of such scholarship. Fellows may teach one course per term within a cohosting department, help lead institute-sponsored colloquia, and otherwise further the work of the institute. Applicants should submit a curriculum vita, three recommendation letters, a 750-word description of the project to be undertaken while at the institute, and a published paper.
Requirements: Senior fellows typically hold senior faculty rank at their home institutions, possess records as accomplished, nationally recognized scholars, and pursue research programs of the highest caliber integrative of their academic discipline and the Christian faith. Junior fellows typically hold junior but tenure-track faculty rank at their home institutions, possess records showing clear promise commensurate with their time in rank, and also pursue research programs of the highest caliber integrative of the Christian faith and their area of expertise.
Geographic Focus: All States
Date(s) Application is Due: Nov 1
Amount of Grant: Up to 50,000 USD
Contact: Dr. Douglas Henry, (254) 710-4805; fax (254) 710-4813; IFL@baylor.edu
Internet: http://www.baylor.edu/ifl
Sponsor: Baylor University
1 Bear Place, #97412, Campus Mall
Waco, TX 76798-7412

BBF Fellowships 1311
Barbara Bush Fellows originated at TCALL in 2007, and are funded by The Barbara Bush Texas Fund for Family Literacy. The Barbara Bush Fellows are doctoral students at Texas A&M University-College Station who are engaged in research, writing, and development of resources and presentations in topics relevant to the field of family literacy education. These fellowships consist of a one-year award, which may be used to offset the cost of travel expenses to conferences, dissertation research, and/or tuition, fees, and maintenance while

a student is enrolled at Texas A&M University. During their fellowship year, students will have the opportunity to present their work at national family literacy conferences and publish their research in peer-reviewed journals. Up to four Barbara Bush Fellowships of $25,000 each will be awarded for research on topics relevant to Family Literacy.
Requirements: Applicants for the Barbara Bush Fellowship at TCALL must: be admitted on a full-time basis to Texas A&M University's main College Station campus. You are not required to reside in the Bryan-College Station area, but you must be willing to make yourself available for meetings as needed; be enrolled in a doctoral program. All qualified applicants are encouraged to apply regardless of the discipline or program of study; and submit the complete application packet to TCALL by the deadline.
Geographic Focus: All States
Date(s) Application is Due: Feb 6
Amount of Grant: 25,000 USD
Contact: Ken Appelt; (979) 845-6615; fax (979) 845-0952; kappelt@tamu.edu
Internet: http://www-tcall.tamu.edu/bbfellows/index.html
Sponsor: Barbara Bush Foundation for Family Literacy
1201 15th Street NW, Suite 420
Washington, DC 20005

BCBSM Foundation Excellence in Research Awards for Students **1312**
The annual Excellence in Research Award for Students acknowledges doctoral candidates or medical students enrolled in Michigan universities who have published research papers that represent significant contributions to health policy or clinical care. Entries will be judged based on the subject matter's potential to make a contribution to health policy and medical care as well as the quality and originality of the research. There will be three awards: 1st place-$1,000; 2nd place-$750; 3rd place-$500.
Requirements: Doctoral candidates or medical students enrolled in Michigan universities are eligible. The nomination must be made by a faculty member at the student's university. If the paper has multiple authors, the student must be the first author.
Restrictions: Basic research papers are not eligible.
Geographic Focus: Michigan
Date(s) Application is Due: Jan 1
Amount of Grant: 500-1,000 USD
Contact: Ira Strumwasser; (313) 225-8706; fax (313) 225-7730; foundation@bcbsm.com
Internet: http://www.bcbsm.com/foundation/grant.shtml
Sponsor: Blue Cross Blue Shield of Michigan Foundation
600 East Lafayette Boulevard
Detroit, MI 48226-2998

BCBSM Foundation Investigator Initiated Research Grants **1313**
The program provides grants for applied research that focuses on quality, cost and appropriate access to health care in Michigan. Grants average $75,000 for one year. Multi-year grants or grants in excess of $75,000 are awarded for exemplary projects. Projects that focus on the quality and cost of health care and appropriate access are considered priority and include: Organization and delivery of health care services; Evaluation of new methods or approaches to containing health care costs; Evaluation of new methods or approaches to providing access to high quality health care; Assessment and assurance of quality care; Identification and validation of clinical protocols and evidence-based practice guidelines. Applications are accepted at any time.
Requirements: Applications will be accepted from medical- and doctoral-level researchers based in hospitals, university settings, non-profit health care organizations, health systems, medical or nursing schools, schools of public health, or other relevant academic disciplines such as psychology, sociology and urban studies.
Restrictions: This program does not support basic science or biomedical research, including drug studies or studies using animals, costs of equipment (e.g., personal computers), hardware, software or on-going operating or developmental expenses. The BCBSM Foundation does not provide support to for-profit organizations or individuals associated with organizations not located in Michigan. Blue Cross and Blue Shield of Michigan employees, members of their immediate families, and employees and immediate family members of any Blue Cross and Blue Shield of Michigan affiliate or subsidiary are not eligible to receive BCBSM Foundation grants.
Geographic Focus: Michigan
Amount of Grant: Up to 100,000 USD
Contact: Nora Maloy; (313) 225-8706; fax (313) 225-7730; foundation@bcbsm.com
Internet: http://www.bcbsm.com/foundation/grant.shtml
Sponsor: Blue Cross Blue Shield of Michigan Foundation
600 East Lafayette Boulevard
Detroit, MI 48226-2998

BCBSM Foundation Physician Investigator Research Awards **1314**
The program provides seed money to physicians to explore the merits of a particular research idea for further study. Grants are offered of up to $10,000 for projects that include pilot, feasibility or small research studies in clinical or health services research.
Requirements: Applicants must be a physician interested in research, who is licensed and domiciled in the state of Michigan. Applicants may include physicians working in research environments such as medical schools or university-affiliated hospitals, health care systems, or nonprofit agencies. An application may be submitted at any time.
Geographic Focus: Michigan
Contact: Nora Maloy; (313) 225-8706; fax (313) 225-7730; foundation@bcbsm.com
Internet: http://www.bcbsm.com/foundation/grant.shtml
Sponsor: Blue Cross Blue Shield of Michigan Foundation
600 East Lafayette Boulevard
Detroit, MI 48226-2998

BCBSM Foundation Primary and Clinical Prevention Grants **1315**
The Blue Cross Blue Shield of Michigan Foundation requests letters of interest from Michigan physicians and members of the research community interested in preventing the three leading causes of death in Michigan: heart disease, cancer and stroke. The purpose of this grant initiative is to increase the use of evidence-based prevention efforts to improve the health of Michigan residents. The BCBSM Foundation has allocated $500,000 to fund several projects for up to two years. Funds will be available for salary support, program costs, supplies, office operations and other costs related to the proposed project. Computer equipment expenses, including PC hardware and software, will not be supported. In addition to the overall quality of the project, the feasibility and appropriateness of the budget is a factor in the funding decisions. The required application cover page and the terms and conditions may be downloaded for the BCBSM web site.
Requirements: Applicants must be doctoral level researchers (including physicians) based at Michigan universities, academic medical settings, community hospitals, health systems and community based nonprofit organizations. The principal investigator must have appropriate research and clinical credentials.
Geographic Focus: Michigan
Date(s) Application is Due: Sep 1
Contact: Nora Maloy; (313) 225-8706; fax (313) 225-7730; foundation@bcbsm.com
Internet: http://www.bcbsm.com/foundation/grant.shtml
Sponsor: Blue Cross Blue Shield of Michigan Foundation
600 East Lafayette Boulevard
Detroit, MI 48226-2998

BCBSM Foundation Student Award Program **1316**
The program offers a one year stipend to fund a wide range of health care projects, including applied research, pilot programs, or demonstration and evaluation projects.
Requirements: All doctoral and medical students enrolled in Michigan universities are eligible. For consideration, the proposed project must focus geographically on the state of Michigan and address the BCBSM Foundation's objectives.
Restrictions: Completed or substantially completed dissertations or research projects are not eligible. Students who previously received this award are not eligible. Blue Cross and Blue Shield of Michigan affiliates and subsidiaries are not eligible. Investigation of pharmaceutical efficacy, basic research or research involving non-human subjects is not eligible. Grant monies are not intended to support field placements, practica or internships.
Geographic Focus: Michigan
Date(s) Application is Due: Apr 30
Amount of Grant: 3,000 USD
Contact: Ira Strumwasser; (313) 225-8706; fax (313) 225-7730; foundation@bcbsm.com
Internet: http://www.bcbsm.com/foundation/
Sponsor: Blue Cross Blue Shield of Michigan Foundation
600 East Lafayette Boulevard
Detroit, MI 48226-2998

BCBSM Frank J. McDevitt, DO, Excellence in Research Awards **1317**
The Frank J. McDevitt Excellence in Research Awards for Health Services, Policy and Clinical Care recognize Michigan-based researchers and physicians who have published research in research journals that contributes to improving health and medical care in Michigan. The BCBSM Foundation will award $10,000 for unrestricted research to physician and doctoral level researchers for research on: health policy; health services; clinical care.
Requirements: Applicants must be from Michigan based researchers with terminal research degrees (e.g. PhD, DrPH) and from Michigan based physicians (MD or DO). In the case of multiple authorship, the nominee must be the first author. All authors will be nominated and will constitute one team nomination. The nomination research must have been published, or accepted for publication in a refereed health or medical care journal, within the previous two years. The financial award will be made to the recipient's 501(c)3 non-profit or educational organization.
Restrictions: Only one article per nominee will be accepted for consideration each year.
Geographic Focus: Michigan
Date(s) Application is Due: Jan 1
Contact: Nora Maloy; (313) 225-8706; fax (313) 225-7730; foundation@bcbsm.com
Internet: http://www.bcbsm.com/foundation/grant.shtml
Sponsor: Blue Cross Blue Shield of Michigan Foundation
600 East Lafayette Boulevard
Detroit, MI 48226-2998

BCBSNC Foundation Grants **1318**
The foundation's primary objective is to fund clearly defined, innovative grants that further the foundation mission of improving the health and well-being of North Carolinians. The foundation funds programs typically possessing the following characteristics: programs and/or services designed to produce measurable, long-term impact; programs that are sustainable and designed to be ongoing, rather than one-time or sporadic events; programs and/or services that are replicable. The three primary focus areas include health of vulnerable populations, healthy active communities and community impact through nonprofit excellence.
Requirements: Applicants must meet the following criteria to be eligible for funding: organization is located within North Carolina; organization is a 501(c)3 organization or an educational or governmental entity with tax-exempt status, that is not a private foundation or Type III supporting organization; organization must be able to provide its most recent IRS Form 990. Depending on the type of grant and the size of the organization, an audit may be required as part of the submitted proposal.
Restrictions: The foundation will not provide funding for annual campaigns, political campaigns, religious purposes, individuals, endowments, purchase of advertisements, or for the sole purpose of receiving goods or entitlements from a charitable organization.

Geographic Focus: North Carolina
Contact: Grant Review Committee; (919) 765-7347; foundation@bcbsnc.com
Internet: http://www.bcbsnc.com/foundation/grants.html?previouslyOver=true¤tlyOver=true
Sponsor: Blue Cross Blue Shield of North Carolina Foundation
P.O. Box 2291
Durham, NC 27707-0718

BCBS of Massachusetts Foundation Grants 1319

With a mission to expand access to health care, the foundation awards grants intended to make a significant impact on the health of Massachusetts's low income and uninsured residents. Grant programs include Innovation Fund for the Uninsured; Connecting Consumers with Care; Strengthening the Voice for Access; Pathways to Culturally Competent Health Care; Catalyst Fund; Building Bridges in Children's Mental Health; and Policy Research and Analysis. Grant awards in particular focus areas are made during specific grant cycles. Deadlines and additional information are available online.
Requirements: Massachusetts 501(c)3 nonprofit organizations or government agencies with a mission that includes improving access and removing barriers to healthcare for low-income or uninsured Massachusetts's residents are eligible.
Restrictions: The foundation will not fund individuals; for-profit organizations; capital campaigns, endowments, or building drives; fundraising drives and events; retiring debt or operating deficits; direct political or lobbying activity; projects outside of Massachusetts; or religious organizations for religious purposes (proposals for secular programs of faith-based organizations that meet funding criteria will be considered). Grants are rarely made for curriculum development, conferences, film or video production, or scholarships.
Geographic Focus: Massachusetts
Date(s) Application is Due: Jan 20; Sep 7
Contact: Coordinator; (617) 246-3744; grantinfo@bcbsmafoundation.org
Internet: http://www.bcbsmafoundation.org/foundationroot/en_U.S./grants/focusArea.jsp#aInnovation%20Fund%20for%20the%20Uninsured
Sponsor: Blue Cross Blue Shield of Massachusetts Foundation
401 Park Drive
Boston, MA 02115-3326

BC Center for Retirement Research Dissertation Fellowships 1320

The purpose of the program is to promote the next generation of retirement research scholars and to improve the quality of scholarship in Social Security and income studies. Applicants must demonstrate their dissertation focuses on the Social Security Administration research priorities including Social Security and retirement; macroeconomic analyses of Social Security; wealth and retirement income; program interactions; international research; and demographic research. Doctoral candidates from a wide variety of academic disciplines, including actuarial science, demography, economics, finance, gerontology, political science, public administration, public policy, sociology, social work, and statistics are encouraged to submit a proposal. Awards will be announced in March. Fellowships for successful applicants are renewable for a second-year based on adequate progress. Application and guidelines are available online.
Requirements: Applicants are required to be enrolled in a qualified doctoral program at a U.S. university; have completed all coursework for a PhD by the time funding would start; and have a dissertation advisor and/or committee.
Geographic Focus: All States
Date(s) Application is Due: Jan 31
Contact: Fellowship Administrator; (617) 552-1762; fax (617) 552-0191; crr@bc.edu
Internet: http://www.bc.edu/centers/crr/dissertation.shtml
Sponsor: Boston College
140 Commonwealth Avenue, Fulton Hall 550
Chestnut Hill, MA 02467-3808

BCRF-AACR Grants for Translational Breast Cancer Research 1321

The grants will provide direct support for innovative cancer research projects designed to accelerate the discovery, development, and application of new agents to treat breast cancer and/or for pre-clinical research with direct therapeutic intent. Special emphasis will be placed on research that holds promise for leading to individualized therapeutic options for treatment in the near future.
Requirements: Researchers who are affiliated with any institution involved in cancer research, cancer medicine, or cancer-related biomedical science anywhere in the world may apply. There are no geographic, national, or residency status restrictions. Applicants must have acquired a doctoral degree in a related field. If an applicant has obtained an equivalent degree at a foreign institution, information on the nature of the degree must be provided at the time of application.
Restrictions: Employees or subcontractors of a national government or the for-profit private industry are not eligible to serve as the Principal Investigator for the purpose of these grants. However, collaborations with such individuals are encouraged. Neither the members of the Scientific Review Committee nor the members of their individual laboratories are eligible for these grants.
Geographic Focus: All States
Date(s) Application is Due: May 20; Jul 20
Contact: Hanna Hopfinger, Program Assistant; (267) 646-0665 or (215) 440-9300; fax (215) 440-9372; hanna.hopfinger@aacr.org or awards@aacr.org
Internet: http://www.aacr.org/home/scientists/research-funding—fellowships/bcrf-aacr-grants.aspx
Sponsor: Breast Cancer Research Foundation & the American Assoc for Cancer Research
615 Chestnut Street, 17th Floor
Philadelphia, PA 19106-4404

BCRF Research Grants 1322

The goal of the grant process is a serious peer review of all proposals. Proposals are invited by the Medical Advisory Board, rather than accepted as unsolicited requests. The MAB generally reviews proposals in the late summer. There is no set format, but the BCRF requests a brief narrative (no more than 4-5 pages), lay language summary, annual budget (no more than $225,000 initially, with a maximum of 20% in indirect costs) and budget narrative. Considerable latitude is given to the investigators in terms of the work proposed. Both the Board of Directors and the Medical Advisory Board concur that some of the most important advances in understanding the disease will most likely occur by enabling brilliant minds to pursue some of their most creative theories.
Restrictions: While most of the grants are unrestricted, the Foundation has donors requesting restricted grants for specific purposes. When this occurs, the Medical Advisory Board will meet to determine whether the area of interest merits further exploration. If so, they will identify various research projects meeting the donor's criteria.
Geographic Focus: All States
Amount of Grant: Up to 225,000 USD
Contact: Administrator; (646) 497-2600; fax (646) 497-0890; bcrf@bcrfcure.org
Internet: http://www.bcrfcure.org/action_grantguidelines.html
Sponsor: Breast Cancer Research Foundation
60 East 56th Street, 8th Floor
New York, NY 10022

BC Steven H. Sandell Grants 1323

The program's purpose is to promote research on retirement issues including: Social Security and retirement; macroeconomic analyses of Social Security; wealth and retirement income; program interactions; international research; and demographic research. Within the six broad areas, four topics of particular interest include: retirement planning/risk management decisions of workers and retirees; measures of retirement well-being; trends in labor force participation and their implications for Social Security and for the well-being of the elderly; Social Security reform.
Requirements: The principal investigator must have a PhD in a related discipline or a comparable professional certification; and must be a nontenured junior scholar or senior scholar working in a new area. All applicants must complete the online submission form.
Geographic Focus: All States
Date(s) Application is Due: Mar 31
Amount of Grant: 40,000 USD
Contact: Paige Eppenstein; (617) 552-1092; eppenste@bc.edu
Internet: http://www.bc.edu/centers/crr/sandellguidelines.shtml
Sponsor: Boston College
140 Commonwealth Avenue, Fulton Hall 550
Chestnut Hill, MA 02467-3808

Beatrice Laing Trust Grants 1324

The trust was founded for the relief of poverty and for the advancement of the evangelical faith internationally. Grants are awarded for general charitable purposes at the discretion of the Trustees, with an emphasis on social services and relief, and medical aid and research. Grant recipients typically are charities working in deprived sections of the community in the United Kingdom, to missionary societies and, less frequently, to individuals working in the field of missions in United Kingdom and abroad.
Geographic Focus: All States, Albania, Algeria, Andorra, Angola, Armenia, Austria, Azerbaijan, Belarus, Belgium, Benin, Bosnia & Herzegovina, Botswana, Bulgaria, Burkina Faso, Burundi, Cameroon, Cape Verde, Central African Republic, Chad, Comoros, Congo, Congo, Democratic Republic of, Cote d' Ivoire (Ivory Coast), Croatia, Cyprus, Czech Republic, Denmark, Djibouti, Egypt, Equatorial Guinea, Eritrea, Estonia, Ethiopia, Finland, France, Gabon, Gambia, Georgia, Germany, Ghana, Greece, Guinea, Guinea-Bissau, Hungary, Iceland, Ireland, Italy, Kenya, Kosovo, Latvia, Lesotho, Liberia, Libya, Liechtenstein, Lithuania, Luxembourg, Macedonia, Madagascar, Malawi, Mali, Malta, Mauritania, Mauritius, Moldova, Monaco, Montenegro, Morocco, Mozambique, Namibia, Niger, Nigeria, Norway, Poland, Portugal, Romania, Russia, Rwanda, San Marino, Sao Tome & Principe, Senegal, Serbia, Seychelles, Sierra Leone, Slovakia, Slovenia, Somalia, South Africa, Spain, Sudan, Swaziland, Sweden, Switzerland, The Netherlands, Turkey, Ukraine, United Kingdom, Vatican City
Samples: ActionAid, 10,000 pounds to improve access to education for pastoralist families in Somaliland; OxFam, 10,000 pounds to the You Are Here program; Sense International, 15,000 pounds for work with deaf-blind people in Romania.
Contact: Trust Administrator; 020 8238 8890; fax 020 8238 8897
Sponsor: Beatrice Laing Trust
33 Bunns Lane
London, NW7 2DX England

Beckman Coulter Community Relations Grants 1325

The corporation funds nonprofit organizations, charities and educational institutions that support and promote progress and interest in the areas of science, research-related health care and science education in our plant and customer communities.
Requirements: Organizations seeking funding should submit a letter of request providing the following information: full legal name of the organization; brief description of the organization, including a mission statement and services provided; amount requested; description of how the funds will be used. Additional attachments should include: history of previous support by Beckman Coulter; description of any involvement by Beckman Coulter employees; statement as to why you consider Beckman Coulter an appropriate donor; list of governing board members.
Restrictions: Grants do not support golf tournaments, galas, or religious or political activities.

Geographic Focus: All States
Contact: Cathy Doherty, Community Relations; (714) 961-4478
Internet: http://www.beckman.com/hr/ourcompany/oc_communityRelations.asp#
Sponsor: Beckman Coulter Corporation
Community Relations, 200 S Kraemer Boulevard
Brea, CA 92821

Beckman Scholars Grants 1326

The Program is an invited program for accredited universities and four-year colleges in the United States. It provides scholarships that contribute significantly in advancing the education, research training and personal development of select students in chemistry, biochemistry, and the biological and medical sciences. The sustained, in-depth undergraduate research experiences and comprehensive faculty mentoring are unique in terms of program scope, content and level of scholarship awards. The amount given is $19,300 for two summers and one academic year.
Geographic Focus: All States
Amount of Grant: 19,300 USD
Contact: Program Administrator; (949) 721-2222; fax (949) 721-2225; beckmanscholars@beckman-foundation.com or k.williams@beckman-foundation.com
Internet: http://www.beckman-foundation.com/bsp.html
Sponsor: Arnold and Mabel Beckman Foundation
100 Academy Drive
Irvine, CA 92617

Beckman Young Investigators Grants 1327

The foundation makes grants to nonprofit research institutions to promote research in chemistry and the life sciences, broadly interpreted, and particularly to foster the invention of methods, instruments, and materials that will open up new avenues of research in science. The program is intended to provide research support to the most promising young faculty members in the early stages of academic careers in the chemical and life sciences. The program is open to persons with tenure-track appointments in academic and nonprofit institutions that conduct fundamental research in the chemical and life sciences. The program is intended primarily for U.S. institutions. Only proposals of exceptional merit from foreign institutions will receive consideration. Projects are normally funded for a period of two years. When extraordinary circumstances warrant, support may be provided over a one-year or three-year period. Application materials are available on the Web site.
Requirements: To be eligible, an applicant should not have completed more than three full years in his or her tenure-track or other comparable independent research appointment on the anniversary date of initial appointment in the year in which application is to be made. Individuals may not apply more than three times.
Restrictions: Funding will not be considered for general institutional expenses or general fund-raising campaign expenses such as dinners and mass mailings. The foundation does not provide funds for overhead or indirect costs.
Geographic Focus: All States
Date(s) Application is Due: Sep 28
Amount of Grant: 264,000 USD
Contact: Program Administrator; (949) 721-2222; fax (949) 721-2225; younginvestigators@beckman-foundation.com or k.williams@beckman-foundation.com
Internet: http://www.beckman-foundation.com/byi_guides.html
Sponsor: Arnold and Mabel Beckman Foundation
100 Academy Drive
Irvine, CA 92617

Becton Dickinson and Company Grants 1328

The company supports national and international activities in keeping with its commitment to advance the quality of medical practice and patient care in four areas: the global healthcare fund; the local initiatives fund; the matching gifts program; and the BD product donation program. Call for guidelines.
Geographic Focus: All States
Contact: Community Relations Manager; (201) 847-6800
Internet: http://www.bd.com/responsibility/contributions
Sponsor: Becton Dickinson and Company
1 Becton Drive
Franklin Lakes, NJ 07417

Bedding Plants Foundation Horticulture Scholarships 1329

Scholarships are available to graduate students in the following categories: majors in horticulture or related fields with a specific interest in bedding or flowering potted plants; students interested in international horticulture marketing; work/study abroad programs; students who intern or work for public gardens; and students in horticulture with a business emphasis or business with a horticulture emphasis. Application for all scholarships is available from the BPFI office, or students may apply via Fastweb.com.
Geographic Focus: All States
Date(s) Application is Due: May 1
Amount of Grant: 500-2,000 USD
Contact: William Willbrandt; (517) 333-4617; first@firstinfloriculture.org
Internet: http://www.firstinfloriculture.org/scholarships.htm
Sponsor: Bedding Plants Foundation
P.O. Box 280
Lansing, MI 48826

Bedding Plants Foundation Research Grants 1330

Grants provide funding for scientific research related to the production of bedding and potted plants. Topics include genetic engineering for diseases and insect resistance, management of greenhouse waste, increasing product shelf life and garden life, evaluation of biocontrol methods and techniques, water recirculation and reuse, water quality/alkalinity control, IPM in the greenhouse, insect problems (aphids, whitefly, thrips, fungus gnats), disease problems (botrytis, damping off, root rot, tomato spotted wilt virus, pythium), energy efficient greenhouse designs, and evaluation of alternative media.
Requirements: Work must be carried out by a nonprofit organization with IRS tax-exempt status.
Restrictions: The foundation does not fund overhead or travel expenses.
Geographic Focus: All States
Date(s) Application is Due: Jan 15
Amount of Grant: Up to 100,000 USD
Contact: Executive Director; (517) 333-4617; fax (517) 333-4494; bpfi@aol.com
Sponsor: Bedding Plants Foundation
P.O. Box 280
Lansing, MI 48826

Beim Foundation Grants 1331

The foundation awards grants to eligible nonprofit organizations in its areas of interest, including arts and culture, environmental conservation, education, and social services. Types of support include program grants, general operating grants, capital campaigns, building construction/renovation, land acquisition, equipment acquisition, and seed grants. High priority is given to the following types of projects: capital drives and equipment purchases; innovative start-up programs that require a moderate amount of grant money; intergenerational projects that involve community service; cooperative projects that involve several agencies or volunteers; on-going programs that have proven themselves unique and essential; matching funds drives. Low priority is given to the following types of projects: medical research, debt retirement, national fundraising programs, and requests from public schools and governmental agencies due to the lack of good financial data. Deadlines are February 3 for education, and human services; and July 3 for arts, arts small capital equipment, and environment. Guidelines are available online.
Requirements: 501(c)3 tax-exempt organizations located within Minnesota, as well as the city of Denver, CO; counties of Park and Gallatin in Montana; county of Santa Fe in New Mexico; and county of Cumberland in Maine are eligible.
Restrictions: The foundation does not fund individuals; private foundations; political organizations or campaigns; religious organizations, including schools, except for secular human service activities; memberships, subscriptions, tickets for benefits, conferences, fundraising events, or annual campaigns; organizations that have as a substantial part of their purpose the influencing of legislation; endowment; multi-year commitments; or international efforts.
Geographic Focus: Colorado, Maine, Minnesota, New Mexico
Date(s) Application is Due: Jan 17; Jul 18
Amount of Grant: 2,000-10,000 USD
Contact: Grants Administrator; (612) 605-8192; contact@beimfoundation.org
Internet: http://www.beimfoundation.org/guide.html
Sponsor: Beim Foundation
3109 W 50th Street, Suite 120
Minneapolis, MN 55410-2102

Berea College Appalachian Music Fellowships 1332

The program is designed to support graduate students, faculty, public school teachers, and/or performers for a period of one to six months in order to conduct research in Berea's collection of noncommercial traditional music and to promote the preservation of and access to that music. Proposals should address the applicant's background and interest in Appalachian music, the reason(s) for seeking an opportunity for concentrated study in Berea's traditional music collections, anticipated contribution to the preservation or promotion of these musical resources (e.g., assistance with sound digitization; website development; writing contextual historical summaries, biographies, musical annotations; public performances, publications), and the length of time needed for the project (one month minimum, six months maximum). Contact information for three references should be included with proposals. Guidelines are available online. The fellowships must be taken up between January and August 2006. Fellows are expected to be in residence during the term of the fellowship.
Restrictions: Berea College employees are ineligible.
Geographic Focus: All States
Date(s) Application is Due: Dec 5
Contact: Steve Gowler, (859) 985-3272; steve_gowler@berea.edu
Internet: http://www.berea.edu/hutchinslibrary/specialcollections/MusicFellowship2006.asp
Sponsor: Berea College
Office of Admissions, CP.O. 2220
Berea, KY 40404

Berks County Community Foundation Grants 1333

The foundation supports a broad range of community projects in this Pennsylvania county, including the arts and culture, economic development, education, the environment, and health and human services. Types of support include general operating grants, capital campaigns, demonstration grants, seed grants, and program grants. Although applicants do not have to be located in Berks County, they must provide programs and services within the county.
Requirements: Tax exempt or public benefit organizations, individuals, associations and private or public agencies are eligible to apply. The grant must be used for charitable purposes only. Organizations are eligible to apply to more than one grant program in the same year.

Geographic Focus: Pennsylvania
Contact: Richard Mappin; (610) 685-2223; fax (610) 685-2240; info@bccf.org
Internet: http://www.bccf.org/pages/grants.html
Sponsor: Berks County Community Foundation
501 Washington Street, Suite 801, P.O. Box 212
Reading, PA 19603-0212

Bernard L. Schwartz Fellows Program 1334
The foundation awards a number of fellowships to help individuals establish themselves as credible new voices in the country's public debate. Regular and senior fellowships are available. Fellowships are awarded for one year, subject to renewal. Fellows choosing to work out of the Washington, DC, office will receive fully equipped offices, health benefits, research and editorial assistance, and the opportunity to formulate their ideas in a well-knit community. Applications are available on the Web site.
Requirements: Regular fellows show promise in writing for student, academic, or popular publications, but have yet to establish themselves as leading commentators. Senior fellows are generally established academics, policy analysts, or journalists wishing to make departures in their work.
Geographic Focus: All States
Amount of Grant: 25,000-75,000 USD
Contact: Sherle Schwenninger, Program Director; (202) 986-2700; fax (202) 986-3696; schwenninger@newamerica.net
Internet: http://www.newamerica.net/index.cfm?pg=fellows
Sponsor: New America Foundation
1630 Connecticut Avenue NW, 7th Floor
Washington, DC 20009

Bernard Lowy Fund for the Study of Tropical Botany in Latin America 1335
The Department of Plant Biology of Louisiana State University (LSU) and the family of the late mycologist, Bernard Lowy, established the Bernard Lowy Fund for the Study of Tropical Botany in Latin America. The fund is designed to provide much-needed additional funding for research travel by graduate students and postdoctoral researchers studying diverse aspects of tropical botany in Latin America.
Requirements: Applicant either must be studying or working at a university or research organization in Latin America or be studying, working at, or have received the doctoral degree from LSU at Baton Rouge. Support for applicants from Latin American institutions is for travel and living expenses for research conducted at LSU. These applicants must seek sponsorship from a faculty member in the Plant Biology Department at LSU.
Geographic Focus: All States
Date(s) Application is Due: May 1
Amount of Grant: Up to 1,000 USD
Contact: Dr. Meredith Blackwell, (225) 388-1557; mblackwell@lsu.edu
Sponsor: Bernard Lowy Fund
Department of Plant Biology, Louisiana State University
Baton Rouge, LA 70803-1705

Bernard Van Leer Foundation Grants 1336
The foundation operates in 34 countries in Europe, the Far East, Australia, Central and South America, the United States, and Africa to further the holistic development of young children. Grantmaking focuses on thematic initiatives, which may include partners from countries where the foundation does not otherwise have a grantmaking program. Current themes are: children affected by HIV/AIDS; respect for diversity; and growing up in indigenous societies. The foundation funds projects that promote a holistic approach to early childhood development; the enhancement of parental capacity to support their children's development; a development strategy that is rooted in the local context and is culturally, socially, and economically appropriate; and the building of capacity, local ownership, and working in partnership. Unsolicited proposals are rarely funded. To be considered, submit information about the organization; the project's objectives, strategies, beneficiaries, scope, location, and duration; and the overall budget and the amount required.
Requirements: Grants are made to governmental and nongovernmental organizations located in the list of eligible countries, which include: Brazil, Colombia, the East Caribbean region, Germany, Greece, India, Indonesia, Israel, Kenya, Mexico, Morocco, the Netherlands, Peru, Poland, South Africa, Tanzania, Thailand, Turkey, Uganda, United States of America and Zimbabwe.
Restrictions: The foundation does not fund applications for support to individual children; projects that concentrate solely on one aspect of children's development or learning; projects that specifically focus on children with special needs such as mental and/or physical handicap; proposals for the construction and maintenance of buildings, or the purchase of equipment and materials; isolated requests for scholarships, conferences, media or theatre events; or general support to organisations, or requests to cover recurrent costs or deficits
Geographic Focus: All States
Contact: Peter Laugharn; 31-70-351-20-40; proposal.administration@bvleerf.nl
Internet: http://www.bernardvanleer.org/about/Applications_for_funding
Sponsor: Bernard Van Leer Foundation
Eisenhowerlaan 156, P.O. Box 82334
The Hague, 2508 EH Netherlands

Beryl Henderson Foundation Grants 1337
The Beryl Henderson Foundation and the Australian Federation of University Women will offer a grant of up to $1500 to assist in completion of a research project being undertaken by a female student or researcher in an Australian university. The grant is made in alternate years to honour the memory of Beryl Henderson (1897-1990), a leading women's rights campaigner in Britain and later in Australia. Applications are invited from women who are enrolled either in Honours IV, or programs at an Australian tertiary institution leading to the award of a degree, or embarked on research work with publication in Australia in view. The purpose of the grant is to assist with practical aspects of research, such as intra- or interstate travel to consult library sources, or to consult with research collaborators. All or part of the grant could also be used for obtaining information (e.g. archival material) from overseas which is not available via the internet. It is not envisaged that the grant would be available for equipment such as computer hardware. The award of the grant is not limited to particular disciplines, but research focusing on issues related to the rights and welfare of women and children will be viewed favorably. Selection will be based on merit.
Geographic Focus: All States, Australia
Date(s) Application is Due: Jul 31
Amount of Grant: Up to 1,500 USD
Contact: Gaynor Reeves, Fellowships Convener; (02) 4952 3174; fax (02) 4952 3070; gay.reeves@bigpond.com or gaynor.reeves@gmail.com
Internet: http://www.afuw.org.au/ScholarshipsFS.htm
Sponsor: Beryl Henderson Foundation / Australian Federation of University Women
138 LaPerouse Street
Red Hill, ACT 2603 Australia

Better Way Foundation Grants 1338
Formerly Alpha Omega Foundation, the Better Way Foundation was established in Florida in 1994. Giving is centered geographically in California, Indiana, Minnesota, Washington, and Tanzania. For the most part, the Foundation supports programs designed to provide holistic and cost-effective development opportunities to young children and families. Special emphasis is directed toward programs designed to improve early childhood outcomes. Its primary fields of interest include: Catholic agencies and churches; early childhood education; family services; health care; higher education; human services; and nutrition. Types of funding support includes: capital campaigns; general operating support; program development; research; and scholarship funding. Application forms are not required, and there are no annual deadlines. Funding amounts range up to $200,000.
Requirements: Unsolicited full proposals are not accepted. Organizations interested in presenting an idea for funding must submit a brief letter of inquiry.
Geographic Focus: California, Indiana, Minnesota, Washington, Tanzania
Amount of Grant: Up to 200,000 USD
Contact: Matthew Rauenhorst, (952) 656-4597; info@betterwayfoundation.org
Sponsor: Better Way Foundation
10350 Bren Road West
Minnetonka, MN 55343-9014

Bettina Bahlsen Memorial Graduate Scholarship 1339
The scholarship competition is open to full-time graduate students registered in or eligible to register in the Department of Biological Sciences at the university in the field of cellular, molecular, microbial, or biochemical biology. Selection will be based on academic excellence. Preference will be accorded to a foreign student and, if possible, to a student entering the first year of graduate studies. The scholarship is awarded annually for one-year duration with possibility of renewal in open competition.
Requirements: To be eligible applicants must be graduate students in the department of biological sciences
Geographic Focus: All States
Date(s) Application is Due: Feb 1
Amount of Grant: Up to 17,500 CAD
Contact: Connie Baines; (403) 220-5690; cbaines@ucalgary.ca
Internet: http://www.grad.ucalgary.ca/funding/internal_scholarships/lev_4/bahlsen_bettina.htm
Sponsor: University of Calgary
2500 University Drive NW, Earth Sciences Building, Room 720
Calgary, AB T2N 1N4 Canada

Beverly Willis Architecture Foundation Grants and Fellowships 1340
The foundation offers funding to individuals and institutions to support innovative projects that advance the study and expand the recognition of women in architecture and related professions, and that lead to the dissemination of this knowledge to professional and public audiences alike. Grant funding is divided into the following categories: fellowships for scholarly research, publication, exhibition, or documentary in film or other media; and grants for honoraria to plenary session speakers at professional meetings, conferences, or symposia whose focus matches the mission of BWAF; and travel grants for research trips or professional conferences at which the recipient will be making a presentation related to the purpose of the BWAF. Successful candidates receive their fellowship awards in two equal installments. Applicants are strongly encouraged to send a preliminary inquiry to the director regarding their proposal and its eligibility for funding. Application guidelines are available online. Although not mandatory, applicants are encouraged to send a preliminary inquiry to the Director of BWAF regarding their proposal and its eligibility for funding.
Requirements: Applications should contain, as their cover page, a summary sheet that includes the following: project title; name, address, telephone number, and email address of the applicant; a concise abstract (150 words or less) of the proposed project; a specific amount (in U.S. dollars) sought from the Foundation; a description of the anticipated final product(s) that would result from the proposed project, including plans for distribution of the completed work; names and contact information of the people from who letters of support have been requested. Tax-exempt organizations must include with their grant request a copy of the IRS Determination Letter, which indicates the particular paragraph of the Internal Revenue Code that governs their exempt activities.

Geographic Focus: All States
Date(s) Application is Due: Mar 15
Amount of Grant: Up to 1,500 USD
Contact: Director; (212) 577-1200; director@bwaf.org
Internet: http://www.bwaf.org/grants.html
Sponsor: Beverly Willis Architecture Foundation
2 Columbus Avenue, Suite 3A
New York, NY 10023

BFGoodrich Collegiate Inventors Prize **1341**
The national competition is designed to encourage undergraduate and graduate college students active in science, technology, and creative invention, while stimulating their problem-solving abilities. Prizes are awarded to recognize students and their faculty advisors for research and innovative discoveries that can be patented. Winners are honored alongside inductees of the National Inventors Hall of Fame in September.
Requirements: Students enrolled full-time in any U.S. college or university as of the date of entry submission are eligible to participate.
Geographic Focus: All States
Date(s) Application is Due: Sep 1
Amount of Grant: Up to 30,000 USD
Contact: Ray DePuy, (330) 849-6887; rdepuy@invent.org
Internet: http://www.invent.org/collegiate/overview.html
Sponsor: BFGoodrich Collegiate Inventors Program
221 S Broadway Street
Akron, OH 43308-1505

BIA Native American Graduate Student Fellowships **1342**
Graduate fellowship grants are provided to supplement financial assistance to eligible American Indian/Alaska Native students pursuing a postbaccalaureate degree. OIEP contracts with the American Indian Graduate Center (AIGC). Application requirements and time frames for submitting an application are available with the AIGC by calling (505) 881-4584, writing to 4520 Montgomery Boulevard NE, Suite 1-B, Albuquerque, NM 87109, or visiting their web site at www.aigc.com.
Requirements: Graduate fellowship grants are available to individuals who are pursuing a master's or doctorate degree as a full time student at a U.S. accredited graduate school; able to demonstrate financial need; and an enrolled member of a federally recognized American Indian tribe or Alaskan Native group, or possess one- fourth degree federally recognized Indian blood.
Geographic Focus: All States
Contact: Grants Manager, (202) 208-3478; fax (202) 208-3312
Internet: http://www.aigc.com/fellowship-application/what-is-graduate-education.html
Sponsor: U.S. Department of the Interior
1849 C Street NW, MS-3512 MIB
Washington, DC 20240-0001

Bikes Belong Foundation Research Grants **1343**
The Bikes Belong Foundation, launched in 2006, is a separate, complementary organization to the Bikes Belong Coalition, an organization sponsored by the U.S. bicycle industry. The foundation's focus is on bicycle-safety projects and children's bicycle programs. The foundation offers three distinct grant programs: the Paul David Clark Bicycling Safety Fund; the REI Grant Program; and Bikes Belong Research Grants. The latter focuses on two priority areas of research: projects that examine the economic impact of additional or improved bicycling facilities or bike-related events; and innovative or unique research on the benefits of bicycling (special opportunities). Applicants interested in the special opportunities research area are asked to contact the Research Analyst (see contact information on this page) before submitting an application. Examples of research that has particular interest for Bikes Belong are as follows: economic benefits of bicycling facilities; economic benefits of removing parking and adding bicycle infrastructure; economic benefits of Sunday Streets/Ciclovia events; Quantitative trails to sales studies (bike businesses only); savings for businesses that promote bicycling among employees; academic performance of children who bike to school; and state, regional, or city-level studies of the economic impact of bicycling. Bikes Belong Foundation reviews research applications twice per year. Application guidelines and downloadable applications are available at the website. Organizations should submit their completed applications along with a cover letter and any required supporting materials in one pdf file via email by the deadline date. See application guidelines for more detailed information. Deadlines may vary from year to year. Prospective applicants are encouraged to verify current year deadlines at the grant website.
Requirements: Applicants are limited to U.S. colleges, universities, or other institutions of higher education and non-profit research organizations.
Restrictions: Individuals will not be considered for funding.
Geographic Focus: All States
Date(s) Application is Due: Mar 30; Sep 28
Amount of Grant: 5,000-10,000 USD
Contact: Kate Powlison, Research Analyst and Communications Coordinator; (303) 449-4893 ext. 7; fax (303) 442-2936; kate@bikesbelong.org or grants@bikesbelong.org
Internet: http://www.bikesbelong.org/bikes-belong-foundation/foundation-grants/research-grants
Sponsor: Bikes Belong Foundation
207 Canyon Boulevard. Suite 202
Boulder, CO 80302

Bilateral Scholarships **1344**
Scholarships are offered for three to nine months to foreign students, research workers, and specialists in various fields for studies and research in Finland on a reciprocal basis to nationals of the following countries: Australia, Austria, Belgium, Bulgaria, Canada, China, Cuba, Czech Republic, Denmark, Egypt, France, Germany, Great Britain, Greece, Hungary, Iceland, India, Ireland, Israel, Italy, Japan, Luxembourg, Mexico, Mongolia, the Netherlands, Norway, Poland, Portugal, Republic of Korea, Romania, Slovakia, Spain, Sweden, Switzerland, Turkey, and the United States. Application must be made on a form obtainable from the diplomatic representatives or directly from CIMO.
Requirements: Applicant must have been awarded a degree at a university or institute of higher education in his/her own country; applicant must also have a working knowledge of Finnish, Swedish, English, or German. Selection of scholars is primarily in the hands of the relevant authorities in the candidate's own country who will make the selection and propose the candidates to the Ministry of Education for final approval.
Restrictions: Scholars may not do full-time paid work during the scholarship period.
Geographic Focus: All States
Contact: Ulla-Maija Anttila; 358-9-7747 7680 or 358-1080-6767; cimoinfo@cimo.fi
Internet: http://www.cimo.fi/Resource.phx/cimo/services/scholarships.htx
Sponsor: Center for International Mobility
P.O. Box 343, Hakaniemenkatu 2
Helsinki, 00531 Finland

Bill and Melinda Gates Foundation Agricultural Development Grants **1345**
The Agricultural Development Program supports projects that enable small farmers in developing countries to break the cycle of hunger and poverty, to sell what they grow or raise, increase their incomes, and make their farms more productive and sustainable. Previously funded initiatives include projects that employ a collaborative and comprehensive approach to agricultural development; provide small farmers with the supplies and support they need to succeed; address the needs of women farmers; help small farmers profit from their crops; use science and technology to develop crops that can thrive; gather and analyze data to improve decision-making; encourage greater investment and involvement in agricultural development; and encourage policy and advocacy efforts that accelerate progress against the world's most acute poverty. Additional information on each agricultural development initiative can be found on the website. New proposals are considered, as well as expansion of existing initiatives currently funded by the Foundation.
Requirements: Proposals should aim to help the world's poorest people lift themselves out of hunger and poverty. The Foundation seeks proposals that: are able to produce measurable results; use preventive approaches; promise significant and long-lasting change; leverage support from other sources; and accelerate or are in accordance with work the Foundation already supports.
Restrictions: The majority of funding is made to organizations that are independently identified by Foundation staff. Unsolicited proposals are not accepted. Proposals must be made through 501(c)3 or other tax-exempt organizations. The Foundation is unable to make grants directly to individuals. The Foundation will not fund projects addressing health problems in developed countries; political campaigns and legislative lobbying efforts; building or capital campaigns; or projects that exclusively serve religious purposes.
Geographic Focus: All Countries
Contact: Sam Dryden, Director, Agricultural Development; (206) 709-3400 or (206) 709-3140; info@gatesfoundation.org
Internet: http://www.gatesfoundation.org/agriculturaldevelopment/Pages/default.aspx
Sponsor: Bill and Melinda Gates Foundation
P.O. Box 23350
Seattle, WA 98102

Bill and Melinda Gates Foundation Financial Services for the Poor Grants **1346**
The Gates Foundation seeks to deliver reliable access to a range of safe, affordable financial tools and services to help the world's poorest households build better, healthier lives. The Foundation currently supports distribution channels, saving products, financial systems, and complementary financial systems. To ensure these services benefit the poorest populations will require new models and innovative approaches. New proposals are considered, as well as expansion of existing initiatives currently funded under the Gates Foundation.
Requirements: New technologies and innovative partnerships make it possible to create a "next-generation" banking system. The Foundation seeks to partner with banks, governments, mobile phone companies, retail store chains, and others to make financial services and technology accessible to billions of people throughout the world. The Foundation seeks proposals that are able to produce measurable results; use preventive approaches; promise significant and long-lasting change; leverage support from other sources; and accelerate or are in accordance with work the Foundation already supports.
Restrictions: The majority of funding is made to organizations that are independently identified by Foundation staff. Unsolicited proposals are not accepted. Proposals must be made through 501(c)3 or other tax-exempt organizations. The Foundation is unable to make grants directly to individuals. The Foundation will not fund projects addressing health problems in developed countries; political campaigns and legislative lobbying efforts; building or capital campaigns; or projects that exclusively serve religious purposes.
Geographic Focus: All States, All Countries
Contact: Rodger Vorhies, Director, Financial Services for the Poor; (206) 709-3140; info@gatesfoundation.org
Internet: http://www.gatesfoundation.org/financialservicesforthepoor/Pages/default.aspx
Sponsor: Bill and Melinda Gates Foundation
P.O. Box 23350
Seattle, WA 98102

Bill and Melinda Gates Foundation Water, Sanitation and Hygiene Grants 1347

Poor sanitation causes severe diarrhea, which kills 1.5 million children each year. Smart investments in sanitation can reduce disease, increase family incomes, keep girls in school, help preserve the environment, and enhance human dignity. The Foundation is looking to work with partners in an effort to expand affordable access to sanitation. Detailed information is available at the Foundation website.

Requirements: The Gates Foundation seeks proposals that are able to produce measurable results; use preventive approaches; promise significant and long-lasting change; leverage support from other sources; and accelerate or are in accordance with work the foundation already supports.

Restrictions: The majority of funding is made to organizations that are independently identified by Foundation staff. Unsolicited proposals are not accepted. Proposals must be made through 501(c)3 or other tax-exempt organizations. The Foundation is unable to make grants directly to individuals. The Gates Foundation will not fund projects addressing health problems in developed countries; political campaigns and legislative lobbying efforts; building or capital campaigns; or projects that exclusively serve religious purposes.

Geographic Focus: All States, All Countries

Contact: Kellie Sloan, Interim Director of Water, Sanitation, and Hygiene; (206) 709-3140; info@gatesfoundation.org

Internet: http://www.gatesfoundation.org/watersanitationhygiene/Pages/home.aspx

Sponsor: Bill and Melinda Gates Foundation
P.O. Box 23350
Seattle, WA 98102

Blowitz-Ridgeway Foundation Early Childhood Development Research Award 1348

The foundation supports nonprofit organizations and programs in the areas of medicine, psychology, residential care, and education, and for research in medicine, psychology, social science, and education. Types of support include capital grants, endowments, program development grants, research grants, and scholarships. The Foundation prefers prospective grantees whose programs or services benefit persons who have not yet reached their majority and/or are for the care of individuals or elderly persons who lack sufficient resources to provide for themselves. Although grants may be made to organizations outside the state of Illinois, preference will generally be given to applicants from Illinois. Application information is available on the Web site.

Requirements: Applicants must be classified as 501(c)3 by the IRS.

Restrictions: Grants will not be made for religious or political purposes, nor generally for the production or writing of audio-visual materials.

Geographic Focus: All States

Contact: Laura Romero; (847) 330-1020; fax (847) 330-1028; laura@blowitzridgeway.org

Internet: http://www.blowitzridgeway.org/information/information.html

Sponsor: Blowitz-Ridgeway Foundation
1701 E Woodfield Road, Suite 201
Schaumburg, IL 60173

Blowitz-Ridgeway Foundation Grants 1349

The foundation supports nonprofit agencies that provide medical, psychiatric, and psychological care to economically disadvantaged children and adolescents. Program and capital grants are awarded, primarily in Illinois, in support of medical, psychiatric, psychological, and/or residential care; and research programs in medicine, psychology, social science, and education. The foundation supports operating budgets, and applicants may request commitments that extend beyond one year, but requests for annual funding will not be considered. Applications are accepted throughout the year and are reviewed in the order in which they are received. Guidelines and applications are available online.

Requirements: 501(c)3 nonprofit organizations that offer services to people who lack resources to provide for themselves may apply.

Restrictions: Grants will not be awarded to government agencies or to organizations that subsist mainly on third-party funding and have demonstrated no ability or expended little effort to attract private funding. Grants will not be made for religious or political purposes or for the production or writing of audio-visual materials.

Geographic Focus: Illinois

Amount of Grant: 5,000-30,000 USD

Contact: Serena Moy; (847) 330-1020; fax (847) 330-1028; serena@blowitzridgeway.org

Internet: http://www.blowitzridgeway.org/information/information1.html

Sponsor: Blowitz-Ridgeway Foundation
1701 E Woodfield Road, Suite 201
Schaumburg, IL 60173

Blue Shield of California Grants 1350

Consideration for funding will be given exclusively to organizations that pursue activities directly related to the foundation's program goals, including domestic violence prevention through service provision, education, and outreach; research and education regarding medical best practices and health technologies; and direct or indirect provision of medical insurance or health care to those populations that are uninsured or underinsured, and related policy development.

Requirements: The Foundation funds organizations that are non-profit and tax-exempt under 501(c)3 of the Internal Revenue Service Code (IRC) and defined as a public charity under 509(a)1, 2, or 3 (types I, II, or a functionally integrated type III); accredited schools; units of government/public agencies; tribal governments. The foundation will only support projects that meet the following criteria: the mission of the grantee organization is consistent with the goals and mission of the foundation; the grant is used primarily to serve Californians; the grant seeking organization has a reputation for credibility and integrity; the grant seeking organization is pursuing activities directly related to one of the Foundation's three Program Areas: Health Care and Coverage, Health and Technology and Blue Shield Against Violence.

Restrictions: The foundation does not fund award dinners, athletic events, competitions, special events, or tournaments; conferences or seminars; capital construction; television/film/media production; religious organizations for religious purposes; political causes, candidates, organizations or campaigns; capital projects over $50,000; multi-year projects (generally); grants to individuals (with the exception of the regulated Blue Shield of California Employee Scholarship Program); grants to 509(a) 3, type III supporting organizations that are not functionally integrated.

Geographic Focus: California

Contact: Grants Administrator; (415) 229-5785; fax (415) 229-6268

Internet: http://blueshieldcafoundation.org/grant-center/index.cfm

Sponsor: Blue Shield of California
50 Beale Street
San Francisco, CA 94105-1808

Blum-Kovler Foundation Grants 1351

The Blum-Kovler Foundation was established in 1985 after Everett Kovler retired from his position as President of James Beam Distilling Company. The foundation awards general operating grants to eligible nonprofit organizations in its areas of interest, including social services, Jewish welfare funds, higher education, health services and medical research, and cultural programs. The foundation also supports youth- and child-welfare agencies and public-interest and civic-affairs groups. Grants are awarded primarily in the Chicago metropolitan area and in the Washington, D.C. area. There are no application forms. Applicants should submit a one to two page written proposal with a copy of their IRS determination letter by mid-November to considered for the current year. Typical grant awarded is between $1,000-$5,000.

Requirements: Illinois and District of Columbia nonprofit organizations are eligible.

Geographic Focus: District of Columbia, Illinois

Contact: Hymen Bregar, Secretary; (312) 664-5050
Peter Kovler, Chairperson and Vice President; (312) 664-5050

Sponsor: Blum-Kovler Foundation
875 N Michigan Avenue, Suite 3400
Chicago, IL 60611-1958

BMI Woody Guthrie Fellowships 1352

The foundation offers short-term fellowships to support scholarly use of the Woody Guthrie Archives Research Collection. The program invites applicants who are pursuing research topics or themes related to Woody Guthrie which explore his creative work and contribution to American music and culture. Disciplines may include, but are not limited to: American musicology; historical musicology; ethnomusicology; cultural studies, social sciences, humanities; and American history. A limited number of fellowships will be selected with a value of up to $2,500 per recipient to help defray travel to New York City and residence expenses for the duration of the fellowship. The length of the fellowship will depend on the applicant's research proposal, but is normally limited from one to six months.

Geographic Focus: All States

Date(s) Application is Due: Dec 1

Contact: Administrator; (212) 586-2000; fellowship@woodyguthrie.org

Internet: http://bmi.com/foundation/program/woody_guthrie_fellowship_program/

Sponsor: BMI Foundation
320 W 57th Street
New York, NY 10019

BMW of North America Charitable Contributions 1353

BMW of North America funds charitable programs that benefit society in the areas of education, road-traffic safety, and the environment. The corporation supports education at all levels and specifically focuses on the following: intercultural learning for K-12 students and their teachers; automotive technology, mechanics, and career and repair programs in high schools, technical schools, and community colleges; and research in the areas of safety design, ergonomics, and new materials. In the area of road traffic safety, the corporation supports driver-education programs geared at teenagers and new drivers; basic auto-maintenance programs for women; consumer education on general road-safety issues; and programs to promote the safety of children and young people on the road. In the area of the environment, BMW is committed to sustainable development and focuses grant making on the following: conservation/preservation of natural resources, in particular park lands and waterways; research and promotion of alternative fuels; and environmental education for K-12 students. In general, grants are awarded for specific projects rather than for general operating support, although some operating and capital grants are given consideration. Interested organizations may download application instructions and guidelines at the grant website. Organizations wishing to be considered for a grant must submit an application; telephone solicitations will not considered.

Requirements: 501(c)3 charities or 501(c)9 organizations are eligible to apply.

Restrictions: The corporate giving program does not support non-tax-exempt organizations; individuals; religious organizations for religious purposes; political candidates or lobbying organizations; organizations with a limited constituency, such as fraternal, labor, or veterans groups; travel by groups or individuals; national or local chapters of disease-specific organizations; national conferences, sports events, and other one-time, short-term events; sponsorships or advertising; anti-business groups; team sponsorships or athletic scholarships; or organizations outside the United States or its territories.

Geographic Focus: All States

Contact: Grants Coordinator; (201) 307-4000; fax (201) 307 3607

Internet: http://www.bmwgroupna.com/philanthropy.htm

Sponsor: BMW of North America
300 Chestnut Ridge Road
Woodcliff Lake, NJ 07677-7731

Board of Directors Distinguished Service Award for Senior ACS Administrators 1354
The purpose of the Award is to recognize distinguished service to the Society over a period of years. The award is given at irregular intervals at the discretion of the ACS Board of Directors to recognize outstanding service to the Society by a senior staff member over a period of years. It may be given to the widow, widower, child, or children of the person recognized. The award consists of a cash amount to be determined by the ACS Board of Directors and an appropriate certificate or medallion.
Requirements: Any individual, except a member of the award committee, may submit one nomination or seconding letter for the award in any given year. The nominating documents consist of a letter of not more than 1000 words containing an evaluation of the nominee's accomplishments and a specific identification of the work to be recognized, a biographical sketch including date of birth, and a list of publications and patents authored by the nominee.
Restrictions: Self-nominations are not accepted.
Geographic Focus: All States
Date(s) Application is Due: Nov 1
Contact: Felicia Dixon, Awards Administrator; (800) 227-5558 or (202) 872-4408; fax (202) 776-8008; f_dixon@acs.org or awards@acs.org
Internet: http://portal.acs.org/portal/acs/corg/content?_nfpb=true&_pageLabel=PP_ARTICLEMAIN&node_id=1319&content_id=CTP_004497&use_sec=true&sec_url_var=region1
Sponsor: American Chemical Society
1155 Sixteenth Street, NW
Washington, DC 20036-4801

Bodenwein Public Benevolent Foundation Grants 1355
The Bodenwein Public Benevolent Foundation was established in 1938 under the will of Theodore Bodenwein, owner and publisher of The Day newspaper, to support and promote quality educational, cultural, human-services, and health-care programming for underserved populations. The Foundation specifically serves the people of Greater New London County where The Day has a substantial circulation. The majority of grants from the Bodenwein Public Benevolent Foundation are one year in duration; on occasion, multi-year support is awarded. Applicants must apply online at the grant website. Applicants are strongly encouraged to do the following before applying: review the downloadable state application procedures for additional helpful information, requirements, and restrictions; review the downloadable online-application guidelines at the grant website; review the foundation's funding history (link is available from the grant website); review the online application questions in advance; and review the list of required attachments. These will generally include: a list of board members, financial statements (audited, reviewed, or compiled by independent auditor); an organization summary; a list of other funding sources; an IRS Determination letter; and other required documents. All attachments must be uploaded in the online application as PDF, Word, or Excel files. The annual application deadline is 11:59 p.m. on November 15. Applicants will be notified of grant decisions by letter within three to four months after the proposal deadline.
Requirements: Nonprofit organizations serving Greater New London County (East Lyme, Groton, Ledyard, Lyme, Montville, Mystic, New London, North Stonington, Old Lyme, Salem, Stonington, and Waterford, Connecticut) are eligible to apply. A breakdown of number/percentage of people served by specific towns will be required in the online application.
Restrictions: The foundation does not support requests from individuals, organizations attempting to influence policy through direct lobbying, or any political campaigns.
Geographic Focus: Connecticut
Date(s) Application is Due: Nov 15
Contact: Amy R. Lynch; (860) 657-7015; amy.r.lynch@ustrust.com
Internet: https://www.bankofamerica.com/philanthropic/fn_search.action
Sponsor: Bodenwein Public Benevolent Foundation
200 Glastonbury Boulevard, Suite # 200
Glastonbury, CT 06033-4056

Bodman Foundation Grants 1356
The Bodman Foundation was established by George M. Bodman and his wife Louise Clarke Bodman in 1945. George was born in Toledo, Ohio, in 1882 and died in 1950. Mrs. Bodman was born in Chicago in 1893 and died in 1955. The Bodmans lived for much of their lives in Red Bank, New Jersey, and in New York City, where George Bodman was a senior partner at the investment banking firm of Cyrus J. Lawrence and Sons. The Bodmans were generous supporters of numerous cultural, civic, and service organizations. During World War I Mr. Bodman headed the Intelligence Service of the War Trade Board. During World War II, he served as executive assistant to the Red Cross Commissioner for Great Britain and was regional director in charge of American Red Cross Club operations in England, Scotland, and Ireland. The Bodman Foundation's Certificate of Incorporation states that its funds are to be used for "the aid, support or benefit of religious, educational, charitable, and benevolent objects and purposes for the moral, ethical and physical well-being and progress of mankind." The Bodman Foundation shares trustees, staff, office space, and even a website with the Achelis Foundation which has a similar mission and geographic area of concentration (both foundations give in New York City, while the Bodman Foundation also gives in New Jersey). Funding is concentrated in six program areas: Arts and Culture, Education, Employment, Health, Public Policy, and Youth and Families.
Requirements: 501(c)3 organizations based in New York City that fall within the foundation's areas of interest are welcome to submit an inquiry or proposal letter by regular mail (initial inquiries by email or fax are not accepted, nor are CDs, DVDs, computer discs, or video tapes). An initial inquiry to the foundation should include only the following items: a proposal letter that briefly summarizes the history of the project, need, objectives, time period, key staff, project budget, and evaluation plan; the applicant's latest annual report and complete set of audited financial statements; and the applicant's IRS 501(c)3 tax-exemption letter. Applications may be submitted at any time during the year. Each request is reviewed by staff and will usually receive a written response within thirty days. Those requests deemed consistent with the interests and resources of the foundation will be evaluated further and more information will be requested. Foundation staff may request a site visit, conference call, or meeting. All grants are reviewed and approved by the Trustees at one of their three board meetings in May, September, or December.
Restrictions: The foundation generally does not make grants for the following purposes or program areas: nonprofit organizations outside of New York; annual appeals, dinner functions, and fundraising events; endowments and capital campaigns; loans and deficit financing; direct grants to individuals; individual day-care and after-school programs; housing; organizations or projects based outside the U.S; films or video projects; small art, dance, music, and theater groups; individual K-12 schools (except charter schools); national health and mental health organizations; and government agencies or nonprofit organizations significantly funded or reimbursed by government agencies. Limited resources prevent the foundations from funding the same organization on an ongoing annual basis.
Geographic Focus: New Jersey, New York
Contact: John B. Krieger; (212) 644-0322; main@achelis-bodman-fnds.org
Internet: http://www.achelis-bodman-fnds.org/guidelines.html
Sponsor: Bodman Foundation
767 Third Avenue, 4th Floor
New York, NY 10017-2023

Boeing Company Contributions Grants 1357
The Boeing U.S. contributions program welcomes applications in five focus areas: education; health and human services; arts and culture; civic; and the environment. Primary fields of interest include: arts; elementary and secondary education; the environment; family services, prevention of domestic violence; health care; public affairs; public safety; substance abuse programs; and general human services. The largest single block of charitable contributions goes toward supporting programs and projects related to education. Boeing also looks for innovative initiatives that promote the economic well-being of the community and neighborhood revitalization. Boeing invests in programs that promote participation in arts and cultural activities and experiences, programs that increase public understanding of and engagement in the processes and issues that affect communities and programs that protect and conserve the natural environment. Boeing accepts applications for cash grants, in-kind donations, and services.
Requirements: To apply for support you must be a U.S. based IRS 501(c)3 qualified charitable or educational organization or an accredited K-12 educational institution.
Restrictions: Grants do not support: an individual person or families; adoption services; political candidates or organizations; religious activities, in whole or in part, for the purpose of further religious doctrine; memorials and endowments; travel expenses; nonprofit and school sponsored walk-a-thons, athletic events and athletic group sponsorships other than Special Olympics; door prizes or raffles; U.S. hospitals and medical research; school-affiliated orchestras, bands, choirs, trips, athletic teams, drama groups, yearbooks and class parties; general operating expenses for programs within the United States; organizations that do not follow our application procedures; follow-on applications from past grantees that have not met our reporting requirements or satisfactorily completed the terms of past grants; fundraising events, annual funds, galas and other special-event fundraising activities; advertising, t-shirts, giveaways and promotional items; documentary films, books, etc; debt reduction; dissertations and student research projects; loans, scholarships, fellowships and grants to individuals; for-profit businesses; gifts, honoraria, gratuities; capital improvements to rental properties.
Geographic Focus: Alabama, Arizona, California, Colorado, District of Columbia, Florida, Georgia, Hawaii, Illinois, Kansas, Maryland, Missouri, Nevada, New Mexico, Ohio, Oklahoma, Oregon, Pennsylvania, South Carolina, Texas, Utah, Washington, Australia, Canada
Contact: Antoinette Bailey, (312) 544-2000; fax (312) 544-2082
Internet: http://www.boeing.com/companyoffices/aboutus/community/charitable.htm
Sponsor: Boeing Company Contributions
100 North Riverside
Chicago, IL 60606-1596

Bogliasco Foundation Liguria Study Center Residencies 1358
The fellowships are awarded, without regard to nationality, to qualified persons doing advanced creative work or scholarly research in the following disciplines: archaeology, architecture, classics, dance, film or video, history, landscape architecture, literature, music, philosophy, theater, and visual arts. Applicants are expected to demonstrate significant achievement in their disciplines, commensurate with their age and experience. In addition, they must submit descriptions of the projects that they intend to pursue in Bogliasco. An approved project is presumed to lead to the completion of an artistic, literary, or scholarly work, followed by publication, performance, exhibition, or other public presentation. Approximately 50 fellowships are awarded during two semesters each year—September 8 to December 13 (fall-winter), and February 9 to May 16 (winter-spring). Fellowships usually are for one month (31-32 days) or, in some cases, a half semester (48 days). Fellows may be accompanied by spouses (or spouse-equivalent companions) during their stay at the Liguria Study Center (separate application required).
Geographic Focus: All States
Date(s) Application is Due: Jan 15; Apr 15
Contact: Grants Administrator; info@bfny.org
Internet: http://www.liguriastudycenter.org/english/home.htm
Sponsor: Bogliasco Foundation
885 Second Avenue, Room 3100
New York, NY 10017

Booth-Bricker Fund Grants 1359

The Foundation makes contributions for the purposes of promoting, developing and fostering religious, charitable, scientific, literary and educational programs. Requests are welcomed for capital needs, special projects and other one-time requirements. Applications should be made by letter. There are no forms or deadlines. Requests should include complete information about the applicant organization, including its history, purpose, finances, current operations, governing board and tax status. A detailed explanation of the proposed use of the funds must be provided.
Requirements: Requests are accepted for the funding of projects within the state of Louisiana. Priority is given to the New Orleans area.
Restrictions: The Foundation generally does not provide sustaining (operations and maintenance) funding. No grants are made to individuals.
Geographic Focus: Louisiana
Amount of Grant: 5,000-50,000 USD
Contact: Gray S. Parker, Chairperson; (504) 581-2430
Sponsor: Booth-Bricker Fund
826 Union Street, Suite 300
New Orleans, LA 70112

Bosque Foundation Grants 1360

The foundation gives primarily for higher education, human services, Baptist & Protestant agencies/churches and, medical research in Texas. Types of support include capital campaigns, building/renovation and research grants. There are no application forms or deadlines. Applicants should submit a one-page letter of intent that describes the program and request.
Requirements: Texas nonprofit and for-profit organizations are eligible.
Restrictions: Individuals are not eligible.
Geographic Focus: Texas
Contact: Louis A. Beecherl Jr., Trustee; (214) 956-6732; fax (214) 956-6733
Sponsor: Bosque Foundation
5950 Cedar Springs Boulevard, Suite 210
Dallas, TX 75235-6803

Boston Foundation Grants 1361

The Boston Foundation has a particular concern for low income and disenfranchised communities and residents and supports organizations and programs whose work helps advance the Foundation's high priorities in a variety of subject areas: Arts and Culture; Civic Engagement; Community Safety, Economic Development; Education/Out-of-School Time, Health and Human Services; Housing and Community Economic Development; the Nonprofit Sector, Urban Environment and Workforce Development. The Foundation generally makes the following types of grants: Project or program support for community-based efforts that improve the quality of life in the community, test new models, and promote collaborative and innovative ventures; advocacy and public policy research that is linked to specific action; support for planning to enable organizations and residents to assess community needs, respond to new challenges and opportunities, and provide for the inclusion of new populations; organizational support to develop and build the capacity of nonprofit organizations-support that helps organizations keep pace with the changing requirements and demands of their communities and broader environments; small grants awarded on a rolling basis for one-time organizational development needs through the Vision Fund. In addition, on a very limited basis, the Foundation will consider development grants and strategic alliances.
Requirements: Grants are made only to tax-exempt organizations in Massachusetts.
Restrictions: The committee does not consider more than one proposal from the same organization within a 12-month period. Discretionary grants are generally not made to the following applicants: city or state government agencies or departments; individuals; medical research; endowments; equipment; replacement of lost/expired government funding or gap funding to cover the full cost of providing services; scholarships and fellowships; video and film production; construction and renovation projects and capital campaigns; programs with religious content; travel; summer camps and lobbying. Activities that are generally lower priorities for the Foundation are conferences, lectures, one-time events, programs benefiting only a small number of participants or routine service delivery and/or operating expenses.
Geographic Focus: Massachusetts
Date(s) Application is Due: Jan 5; Jul 1
Contact: Corey Davis; (617) 338-1700; fax (617) 338-1604; info@tbf.org
Internet: http://www.tbf.org
Sponsor: Boston Foundation
75 Arlington Street, 10th Floor
Boston, MA 02116

Boston Psychoanalytic Society and Institute Fellowship in Child Psychoanalytic Psychotherapy 1362

The Child Fellowship Program offers participants an intensive, one year course of study in psychoanalytic approaches to child and adolescent psychotherapy. The curriculum includes both a theoretical course and a clinical course to be taught weekly. The theory courses will introduce the fundamentals of child analytic theories and will review historical as well as modern theoretical approaches. The clinical courses are organized around different developmental stages of childhood. Those who complete the program and wish to continue on in the Child and Adolescent Advanced Training Program will receive credit for their participation and can apply for advanced standing. A printable application form is available for download. Tuition will be $1677 for the academic year.
Requirements: The program is open to mental health clinicians who have completed or are in an advanced phase of their clinical training (residency, graduate internship) and who are interested in further didactic and clinical education in psychoanalytic work with children. Other professionals who work with children and their families such as pediatricians, teachers, and ancillary therapists are also encouraged to apply.
Geographic Focus: All States
Date(s) Application is Due: May 15
Contact: Elizabeth Jordan, Program in Psychoanalytic Studies; (617) 266-0953; fax (617) 266-3466; office@bostonpsychoanalytic.org
Internet: http://www.bostonpsychoanalytic.org/child_fellowship
Sponsor: Boston Psychoanalytic Society and Institute
15 Commonwealth Avenue
Boston, MA 02116

Boston Psychoanalytic Society and Institute Fellowship in Psychoanalytic Psychotherapy 1363

The Fellowship is a one-year program designed to enhance participants' grasp of the principles and practice of psychoanalytic psychotherapy. It is also well suited for clinicians considering more extensive training in psychoanalytic psychotherapy or training in psychoanalysis, but who are either uncertain or not yet ready for a multi-year training commitment. The Fellowship offers two seminars weekly, one theoretical and one clinical, led by outstanding teachers from the Boston Psychoanalytic Society and Institute. The theoretical seminar focuses on Fundamental Concepts of Psychoanalytic Psychotherapy for the first term, on Comparative Theories of Psychotherapeutic Technique for the second term and on Selected Topics in Psychoanalytic Psychotherapy for the third term. The clinical seminar consists of discussion of case material presented by instructors and class members, with senior faculty as discussants. The seminars meet on Thursday evenings for 30 weeks starting in September. Each Fellow is matched with a faculty advisor who is available to discuss individual training goals and options. There is also an opportunity for individual supervision at reduced fees with a wide choice of experienced supervisors from the Boston Psychoanalytic Society and Institute. Tuition will be $1,677 for the academic year. The application fee is $25.
Requirements: The Fellowship is designed especially for mental health clinicians who have completed or are in an advanced phase of their requisite clinical training (psychiatric residency, psychology internship, MSW, masters in psychiatric nursing, etc.) and who desire further didactic and clinical education in psychoanalytic psychotherapy.
Geographic Focus: All States
Date(s) Application is Due: May 15
Amount of Grant: 1,677 USD
Contact: Elizabeth Jordan, Program in Psychoanalytic Studies; (617) 266-0953; fax (617) 266-3466; office@bostonpsychoanalytic.org
Internet: http://www.bostonpsychoanalytic.org/fellowship
Sponsor: Boston Psychoanalytic Society and Institute
15 Commonwealth Avenue
Boston, MA 02116

Bower Award and Prize for Achievement in Science 1364

The Award recognizes outstanding achievement in life, physical, and applied sciences; innovation in the sciences; and training of scientists. The award is given without regard for nationality and includes a gold medal and prize of $250,000. The theme for the 2008 Bower Award for Achievement in Science is Robotics. The Institute seeks to honor an individual who has played a seminal role in either the design and construction of robotic systems or the advancement of enabling technologies as related to robotics such as mechanical structure, sensors, and control algorithms.
Requirements: This is an international competition for individuals whose work has had a significant impact on the field of Robotics. In cases of equal scientific merit, the factor of current economic value of the discovery or application will weigh favorably on behalf of the candidate. Candidates for the Award must be living, and the winner must participate in the April Awards Ceremony in Philadelphia.
Restrictions: This award is for an individual rather than for a group.
Geographic Focus: All States
Date(s) Application is Due: May 31
Amount of Grant: 250,000 USD
Contact: Program Director; (215) 448-1329; fax (215) 448-1364; awards@fi.edu
Internet: http://www.fi.edu/tfi/exhibits/bower/07/bscience_nominate.html
Sponsor: Franklin Institute
222 N 20th Street
Philadelphia, PA 19103-1194

BP Conservation Programme Future Conservationist Awards 1365

The aim of the awards is to develop leadership capacity for biodiversity conservation as a fundamental contribution to sustainable development. The program provides annual grants to passionate people developing innovative projects addressing biodiversity issues of global importance. Projects should address three key areas: development of team capabilities and skills; practical high-priority conservation projects combining research and action; and demonstrate long-term conservation benefits contributing to sustainable development. Application and guidelines are available online.
Requirements: Teams must include only members less than 35 years of age with no more than two years professional conservation and include a minimum of three people.
Restrictions: The program does not fund conference attendance, tuition fees or scholarships, salaries, costly laboratory analyses or gene storage, captive breeding projects, or high school level expeditions. Projects that are specifically for PhD research or master's dissertations will not be supported. Employees from any of the BPCP partner organizations are not eligible to apply.
Geographic Focus: All States

Date(s) Application is Due: Dec 16
Amount of Grant: Up to 12,500 USD
Contact: Marianne Dunn, Program Manager; (44 01223) 277318; fax (44 01223) 277200; bp-conservation-programme@birdlife.org.uk
Internet: http://conservation.bp.com/applications/default.asp
Sponsor: BP Conservation Programme
Wellbrook Court, Griton Road
Cambridge, CB3 ONA England

BPW Career Advancement Scholarships 1366

These scholarships assist mature women who need further education or training to reenter the workforce, enter a new career field, or to improve their chances for advancement. Applicants are strongly encouraged to seek training at the undergraduate or graduate level in computer science, education, science, engineering, paralegal studies, or professional (JD, MD, DDS) degrees. Scholarships are awarded for a one-year period to cover tuition, fees, and school-related expenses such as child care and transportation. Application requests must include a #10 self-addressed double-stamped (first-class) envelope. The listed application deadline date is to request an application.
Requirements: The scholarships are open to women who are U.S. citizens and 25 years of age or older; demonstrate financial need; are officially accepted into a program or course of study in an accredited institution in the United States, Puerto Rico, or the U.S. Virgin Islands; and will graduate within 24 months from date of receipt of funds.
Restrictions: Doctoral studies, study abroad, or correspondence courses are not supported.
Geographic Focus: All States
Date(s) Application is Due: Apr 15
Amount of Grant: 750-1,000 USD
Contact: Tricia Dwyer-Morgan, (202) 777-8932; fax (202) 861-0298; bpwfoundation@act.org or tdwyermorgan@bpwusa.org
Internet: http://www.bpwusa.org/i4a/pages/index.cfm?pageid=4553
Sponsor: Business and Professional Women's Foundation
1900 M Street NW, Suite 310
Washington, DC 20036

Brach Family Foundation Grants 1367

Established in New York in 1991, the Brach Family Foundation offers funding primarily in Brooklyn, New York, to Jewish agencies, temples, and schools. Types of support include research and general operations. Since an application form is not required, applicants should submit the name, address, and phone number of the organization, as well as and argument for financial need. There are no annual deadlines, and grant amounts range from $250 to $10,000.
Geographic Focus: New York
Contact: Zigmond Brach; (718) 236-8000 or (800) 444-5671; fax (718) 236-2400
Sponsor: Brach Family Foundation
1600 63rd Street
Brooklyn, NY 11204-2713

Brainerd Foundation Grants 1368

The foundation is dedicated to protecting the environmental quality of the Pacific Northwest, including Washington, Oregon, Idaho, Montana, Alaska, and British Columbia. Program grants are made in the following areas: endangered ecosystems—conservation biology, conservation assessment, and mining reform and roadless areas; and communications and capacity building—organizational development, and allied voices. Program grants are awarded to cover costs associated with activities such as public education and grassroots outreach, media strategies, litigation, scientific and economic studies, computer networking, and building organizational capacity. Opportunity Fund grants are awarded to organizations for support such as outreach, litigation, applied research, and other unexpected needs. Additional types of support include general operating support, continuing support, equipment acquisition, conferences and seminars, seed money, research, technical assistance, and employee matching gifts. Applications are available online.
Requirements: Nonprofit organizations in the Pacific Northwest are eligible.
Restrictions: The foundation does not favor proposals for school education programs, land acquisition, endowments, capital campaigns, projects sponsored by government agencies, basic research, fellowships, or books or videos that are not part of a broader strategy.
Geographic Focus: Alaska, Idaho, Montana, Oregon, Washington, Canada
Amount of Grant: 250-25,000 USD
Contact: Ann Krumboltz; (206) 448-0676; fax (206) 448-7222; annk@brainerd.org
Internet: http://www.brainerd.org/grants/intro.php
Sponsor: Brainerd Foundation
1601 Second Avenue, Suite 610
Seattle, WA 98101-1541

Breast Cancer Fund Grants 1369

Traditionally, the fund has awarded three types of grants: innovative research grants support projects investigating cutting-edge scientific approaches to the detection, treatment, and prevention of breast cancer; community model grants (non-research grants) to encourage and support the start-up of truly creative, innovative, and replicable U.S. programs that are developing new methods for addressing the support, education, health care access, and advocacy needs of women affected by breast cancer; and discretionary grants that generally fund smaller education, support, and advocacy initiatives. The fund gives preference to projects that address the needs of an ethnically, economically, and geographically diverse representation of women, including persons who are underserved or have low incomes. Initial application for research and community model grants is by letter of intent. If interested, the fund will invite a full proposal. Applicants for discretionary grants may submit an abbreviated application form available from the fund. Contact program staff for project availability.
Requirements: The fund will consider applications from individuals or organizations, whether public or private, for-profit or nonprofit.
Restrictions: No support is available for the direct cost of medical services, and no funds may be used for any lobbying efforts or political expenditures.
Geographic Focus: All States
Contact: Jeanne Rizzo, Executive Director; (415) 346-8223; fax (415) 543-2975; ed@breastcancerfund.org or info@breastcancerfund.org
Internet: http://www.breastcancerfund.org
Sponsor: Breast Cancer Fund
1388 Sutter Street, Suite 400
San Francisco, CA 94109-5400

Brian G. Dyson Foundation Grants 1370

The Foundation, established in Atlanta in 1994 by former Coca-Cola Bottling executive, Brian G. Dyson, offers funding to community foundations, higher education, and federated programs in the Atlanta region and throughout Georgia. Grants typically range up to $20,000, and funding supports general operating costs. There are no specific guidelines, application formats, or deadlines with which to adhere, and initial contact should be made in writing.
Requirements: Applicants must be colleges, public or private schools, or other non-profit organizations located within the state of Georgia.
Geographic Focus: Georgia
Amount of Grant: Up to 20,000 USD
Contact: Brian G. Dyson, Director; (404) 364-2940
Sponsor: Brian G. Dyson Foundation
3060 Peachtree Road NW, Suite 1465
Atlanta, GA 30305-2241

Brico Fund Grants 1371

The mission of the fund is to effect systemic change—to change attitudes, policies and societal patterns. Grants are made to secure full participation in society for women and girls; restore and sustain the earth's natural systems; promote a just and equitable society; and nourish the creative spirit. Types of support include general operating, program, and rarely, capital and endowment grants. Applicants should complete the fund's preliminary application form and a two-page letter of intent, describing the organization's intended project or program.
Requirements: The fund supports organizations with projects and programs within the Greater Milwaukee community. Some funding is done statewide or nationally for programs of broader scope.
Restrictions: Grants do not support conferences and meetings, disease-specific programs, educational institutions, individuals, media projects, medical institutions, religions, or organizations with a focus on animals.
Geographic Focus: Wisconsin
Date(s) Application is Due: Jan 15; Jul 15
Amount of Grant: 1,000-500,000 USD
Contact: Melissa Nimke, Grants Administrator; (414) 272-2747; fax (414) 272-2036; mbn@bricofund.org or bricofund@bricofund.org
Internet: http://www.bricofund.org
Sponsor: Brico Fund
205 E Wisconsin Avenue, Suite 200
Milwaukee, WI 53202

Brinson Foundation Grants 1372

The foundation supports education, public health, and scientific research programs that engage, inform, and inspire committed citizens to confront the challenges that face humanity. Grantmaking priorities are education—awareness and outreach, democracy and citizenship, economically disadvantaged, and libraries and literacy; public health—awareness and outreach, and economically disadvantaged; and scientific research—astrophysics, cosmology, geophysics, medical research (i.e., Alzheimer's disease, cancer, Lou Gehrig's Disease (ALS), and stroke). Types of support include general operating grants and project grants. The foundation does not accept unsolicited grant applications. Grantseekers are asked to review the foundation's mission, vision, beliefs, priorities (accessed from the Who We Are link), and guidelines. If a grantseeker believes the request would match one or more of the foundation's grantmaking priorities, they can make an inquiry by completing the online Grantseeker Information Form. The completed form should be emailed to the office. Further application is by invitation.
Requirements: The foundation will consider inviting grant applications from organizations: whose request matches one or more of the Foundation's grantmaking priorities; located in the United States of America that are exempt from tax under Section 501(c)3 of the Internal Revenue Code and are defined as charitable organizations as described in Section 509(a)1, 2 or 3 or 170(b)1A; located outside the United States of America provided they produce a written legal opinion stating that they are a charitable equivalency to a qualifying U.S. organization and/or a written affidavit containing sufficient information for the Foundation to make a reasonable judgment that the organization is charitable. The Foundation's education and public health grants are generally made to organizations that serve individuals and communities in the greater Chicago area. It also considers leading U.S.-based programs that reach broader populations across the U.S. and internationally or have the potential to have a meaningful impact on best practices at the national or international level. The Foundation's physical science research grants are made to leading organizations across the United States. In this priority area, the location of the program is less critical than the match with the Foundation's grantmaking priorities.

Restrictions: The Foundation will not consider grant inquiries from organizations that: discriminate on the basis of race, gender, religion, ethnicity or sexual orientation. The Foundation will not consider grant inquiries that request funding for: activities that attempt to influence public elections; voter registration; political activity; lobbying efforts; promotion of a specific religious faith; medical research involving human cloning. The Foundation discourages grant inquiries requesting funds for: capital improvements; endowments; fundraising events.
Geographic Focus: All States
Contact: Cheryl A. Heads; (312) 799-4500; fax (312) 799-4310; mail@brinsonfoundation.org
Internet: http://www.brinsonfoundation.org/grant_seekers/grantseekers.shtml
Sponsor: Brinson Foundation
737 N Michigan Avenue, Suite 1850
Chicago, IL 60611

Bristol-Myers Squibb / Meade Johnson Award for Distinguished Achievement in Nutrition Research 1373
The Award for Distinguished Achievement in Nutrition Research was first presented in 1978. The Award includes a $50,000 cash prize and a silver commemorative medallion. The winners are selected by an independent peer-review selection committee whose members are grant administrators of current Bristol-Myers Squibb/Mead Johnson Unrestricted Nutrition Research Grants. The annual application deadline is in late April.
Geographic Focus: All States
Date(s) Application is Due: Apr 24
Amount of Grant: 50,000 USD
Contact: John L. Damonti; (212) 546-4000 or (800) 332-2056; fax (212) 546-9574
Internet: http://www.bms.com/aboutbms/grants/data/nutrit.html
Sponsor: Bristol-Myers Squibb Foundation
345 Park Avenue, Suite 4364
New York, NY 10154

Bristol-Myers Squibb Clinical Outcomes and Research Grants 1374
Bristol-Myers Squibb's mission is to extend and enhance human life. To help achieve that mission, the Company has established programs to support Investigator Sponsored Trials (ISTs). ISTs must be medically appropriate and scientifically valid. While the Company will consider requests for clinical research trials in all clinical and therapeutic areas, it currently gives priority to proposals in the following therapeutic areas: Cardiovascular/Metabolics, Infectious Diseases, Neuroscience, Oncology, Immunology, and Virology. Bristol-Myers Squibb maintains a strict policy of not exercising any influence or control over the design of any investigator initiated clinical research trial supported by BMS.
Requirements: Individuals in the following settings are eligible for support: private practice, hospitals, community health centers, cooperative groups, physician networks, and academic medical centers and universities.
Geographic Focus: All States
Contact: Amit Duggal, (212) 546-4000; fax (212) 546-9574; amit.duggal@bms.com
Internet: http://www.bms.com/responsibility/building_our_communities/Pages/default.aspx
Sponsor: Bristol-Myers Squibb Company
777 Scudders Mill Road
Plainsboro, NJ 08536

Bristol-Myers Squibb Foundation Fellowships 1375
The Program provides support for fellows to gain experience in epidemiological and clinical research as it relates to the care of individuals infected with HIV/AIDS and/or HBV. Goals of the Program include: support of studies that will further strengthen the science and knowledge of HIV/AIDS and HBV; development of a foundation for future prospective and retrospective studies; to provide a forum to share research findings; and support of the development of future clinical researchers. Up to 18 fellows will be selected annually, with grant awards up to $20,000 to support research-related expenses for a one-year research period. Applications must be submitted online.
Requirements: Candidates must: be an active Fellow in good standing in an ACGME-accredited Fellows training program; be a senior Fellow not in the first years of the fellowship; desire to enhance knowledge and skill development in the area of HIV/AIDS, HBV, or Oncology research; and identify a faculty member to serve as the project mentor.
Restrictions: Total grant amount is inclusive of indirect costs and associated IRB fees and is not permitted for use towards travel to conferences or for durable equipment.
Geographic Focus: All States
Date(s) Application is Due: Feb 26
Contact: John L. Damonti; (212) 546-4000 or (800) 332-2056; fax (212) 546-9574
Internet: http://www.bms.com/responsibility/grantsandgiving/Corporate_Giving/Pages/fellowships_scholarships_awards.aspx
Sponsor: Bristol-Myers Squibb Foundation
345 Park Avenue, Suite 4364
New York, NY 10154

Bristol-Myers Squibb Foundation Global HIV/AIDS Initiative Grants 1376
The intent of this initiative is to develop new models in awareness, in medical care, in community development and in prevention and treatment in poor and resource-limited areas of the world, where the need for all such efforts is greatest. Health care infrastructures must be developed and enhanced, stigmatization must be overcome, health care worker capacity must be built and preserved and local people must be empowered to generate and sustain local solutions to this global problem.
Requirements: The Bristol-Myers Squibb Foundation considers requests for support only from tax-exempt organizations that satisfy the requirements of section 501(c)3 of the U.S. Internal Revenue Code.
Restrictions: The Foundation does not award funds to: individuals; political, fraternal, social or veterans' organizations; religious or sectarian organizations unless engaged in a significant project benefiting the entire community; organizations receiving support through United Way or other federated campaigns; endowments; courtesy advertising; or conferences/special events/videos.
Geographic Focus: All States
Contact: John L. Damonti; (212) 546-4000 or (800) 332-2056; fax (212) 546-9574
Internet: http://www.bms.com/responsibility/grantsandgiving/Pages/default.aspx
Sponsor: Bristol-Myers Squibb Foundation
345 Park Avenue, Suite 4364
New York, NY 10154

Bristol-Myers Squibb Foundation Health Disparities Grants 1377
One mission of the Bristol-Myers Squibb Foundation is to reduce health disparities by strengthening community-based health care worker capacity, integrating medical care and community-based supportive services, and mobilizing communities to fight disease. To this end, this Foundation attempts to address health disparities in four strategic disease areas representing major public health burdens and in four highly affected geographies: hepatitis in Asia, HIV/AIDS in Africa, serious mental illness in the U.S., and cancer in Europe. Additional areas of concern include: metabolic diseases, infectious diseases; rheumatoid arthritis; cardiovascular diseases; substance abuse; women's health issues; and overall health care giving.
Requirements: Nonprofit organizations in communities where Bristol-Myers Squibb maintains a facility should submit their requests for company contributions directly to that location. Contact persons are listed at the company website.
Restrictions: The foundation does not support individuals; conferences, special events, or videos; political, fraternal, social, or veterans organizations; religious or sectarian activities, unless they benefit the entire community; organizations funded through federated campaigns; endowments; or courtesy advertising.
Geographic Focus: Connecticut, Indiana, Massachusetts, New Jersey, New York
Contact: John Damonti; (212) 546-4000 or (800) 332-2056; fax (212) 546-9574
Internet: http://www.bms.com/foundation/reducing_health_disparities/Pages/default.aspx
Sponsor: Bristol-Myers Squibb Foundation
345 Park Avenue, Suite 4364
New York, NY 10154

Bristol-Myers Squibb Foundation Health Education Grants 1378
The Program supports the development and replication of novel approaches aimed at individuals in the community to learn more about their health and wellbeing so that they can become more informed decision makers about health care and they can actively participate in disease prevention and management. In many of the grants, the Foundation seeks to be a catalyzing force for changes in government or social policies, supporting programs that can demonstrate the value of new approaches to health care through educational efforts. Nearly all the education programs focus on taking on major diseases, like hepatitis B or cancer and creating model programs for either preventing them or helping treat aspects of them in community settings.
Requirements: The Bristol-Myers Squibb Foundation considers requests for support only from tax-exempt organizations that satisfy the requirements of section 501(c)3 of the U.S. Internal Revenue Code.
Restrictions: The Foundation does not award funds to: individuals; political, fraternal, social or veterans' organizations; religious or sectarian organizations unless engaged in a significant project benefiting the entire community; organizations receiving support through United Way or other federated campaigns; endowments; courtesy advertising; or conferences/special events/videos.
Geographic Focus: All States
Contact: John L. Damonti; (212) 546-4000 or (800) 332-2056; fax (212) 546-9574
Internet: http://www.bms.com/responsibility/grantsandgiving/Pages/default.aspx
Sponsor: Bristol-Myers Squibb Foundation
345 Park Avenue, Suite 4364
New York, NY 10154

Bristol-Myers Squibb Foundation Women's Health Grants 1379
Bristol-Myers Squibb Women's Health grants support projects that enhance women's health with strategies that improve education, prevention, diagnosis, treatment and access to care for women worldwide. Support has been given to projects that test innovative outreach programs, cultivate multi-sectoral partnerships and add new information to the existing body of knowledge to help define and achieve improved health for women around the world. The goal of the program is to generate initiatives that will help enhance women's health through novel interdisciplinary strategies that improve education, prevention, diagnosis, treatment and access to care for women worldwide. Since its inception, significant resources have been invested in programs that educate women about diseases and conditions that particularly threaten them as women.
Requirements: The Bristol-Myers Squibb Foundation considers requests for support only from tax-exempt organizations that satisfy the requirements of section 501(c)3 of the U.S. Internal Revenue Code.
Restrictions: The Foundation does not award funds to: individuals; political, fraternal, social or veterans' organizations; religious or sectarian organizations unless engaged in a significant project benefiting the entire community; organizations receiving support through United Way or other federated campaigns; endowments; courtesy advertising; or conferences/special events/videos.
Geographic Focus: All States
Contact: John L. Damonti; (212) 546-4000 or (800) 332-2056; fax (212) 546-9574
Internet: http://www.bms.com/Documents/foundation/women2002.pdf

Sponsor: Bristol-Myers Squibb Foundation
345 Park Avenue, Suite 4364
New York, NY 10154

Broad Foundation IBD Research Grants **1380**

The program seeks to stimulate innovative research that will lead to both the prevention and successful therapy of inflammatory bowel disease (IBD), including Chrohn's disease and ulcerative colitis. The foundation's goal is to fund basic or clinical research projects that are in the early stages of exploration; propose new directions or ideas; are creative, novel, cutting edge, and imaginative; and are not ready for funding by other more traditional granting agencies. All proposals must be based on sound scientific evidence and careful evaluation of current knowledge in IBD research. Requests must be preceded by a brief (one to three pages)letter of interest, which may be submitted at any time. Funding will be granted for one year; continued funding is based on progress reports and the perceived value of the findings.
Requirements: Nonprofit institutions worldwide, such as universities, hospitals, and research institutes, are eligible.
Restrictions: It is anticipated that the foundation will not fund projects after sufficient progress and maturity have made them fundable by other agencies.
Geographic Focus: All States
Amount of Grant: 100,000 USD
Contact: Heather Kubinec; (310) 954-5091; fax (310) 954-5092; info@broadmedical.org
Internet: http://www.broadmedical.org/funding.htm
Sponsor: Broad Foundation
10900 Wilshire Boulevard, Suite 1200
Los Angeles, CA 90024-6532

Brookdale Foundation Leadership in Aging Fellowships **1381**

The Leadership in Aging Fellowships provide two years of support to junior academics to focus on a project that will help establish them in an area of aging research. Fellowships are open to a broad range of disciplines including, but not limited to, medical, biological and basic sciences, nursing, social sciences, and the arts and humanities. The Fellowship is paid to the candidate's sponsoring institution in support of the candidate's research project. The Fellowship amount of up to $125,000 each year is intended to cover 75% of the fellow's time, base salary and fringe benefits. The award could also be used to include the support of a graduate assistant if necessary as long as the total amount does not exceed $125,000. Additional information is available at the website.
Requirements: Fellowships are open to all professionals in the field of aging. Candidates must meet the following criteria: leadership potential; ongoing commitment to a career in aging; a mentor (or mentors); and a willingness to commit at least 75% of his or her time for career development during each of the two years of the Fellowship.
Geographic Focus: All States
Amount of Grant: Up to 125,000 USD
Contact: Anna Condegni; (212) 308-7355; annaatbrookdale@aol.com
Internet: http://www.brookdalefoundation.org/Leadership/Fellows/fellows.html
Sponsor: Brookdale Foundation
950 Third Avenue, 19th Floor
New York, NY 10022-3668

Brookhaven National Laboratory Sambamurti Memorial Lectureships **1382**

The lectureship is awarded yearly to a young (under 40) high-energy or heavy-ion experimentalist of outstanding achievement. The lecture, which should describe the work for which the lecturer is being honored, is to be delivered to students working at BNL during the summer.
Geographic Focus: All States
Amount of Grant: 500 USD
Contact: Laurence Littenberg; (516) 344-3811; littenbe@bnl.gov
Internet: http://www.phy.bnl.gov/edg/sambamurti.html
Sponsor: Brookhaven National Laboratory
P.O. Box 5000
Upton, NY 11973-5000

Bullard Fellowship in Forest Research **1383**

These fellowships are offered annually to midcareer individuals in biological and physical sciences and social and political studies to promote the advanced study, research, or integration of subjects relating to forested ecosystems. The fellowships are designed to provide an opportunity for candidates to use the facilities at Harvard University and the Harvard Forest to further their scientific and professional growth.
Geographic Focus: All States
Date(s) Application is Due: Feb 1
Amount of Grant: Up to 40,000 USD
Contact: Administrator; (978) 724-3302; fax (978) 724-3595; hfapps@fas.harvard.edu
Internet: http://harvardforest.fas.harvard.edu/education/bullard.html
Sponsor: Harvard University
324 N Main Street, P.O. Box 68
Petersham, MA 01366

Bullitt Foundation Grants **1384**

The foundation functions to protect and restore the environment of the Pacific Northwest, including Washington, Oregon, Idaho, western Montana, coastal rainforests in Alaska, and British Columbia, Canada. Program priorities include aquatic ecosystems; terrestrial ecosystems; conservation and stewardship in agriculture; energy and climate change; growth management and transportation; toxic and radioactive substances; training, organizational development, and unique opportunities (including education and public outreach). Areas of interest include air pollution, climate change, endangered species, energy conservation, environmental education and justice, human health, transportation, and tribal communities. The foundation supports challenge/matching, general operating, project/program, seed money, demonstration, and development grants, as well as requests for conferences/seminars and technical assistance support. Grants will be awarded for one year with possible renewal.
Requirements: Nonprofit organizations in the Pacific Northwest, including Washington, Oregon, Idaho, western Montana, coastal rainforests in Alaska, and British Columbia, Canada are eligible.
Geographic Focus: Alaska, Idaho, Montana, Oregon, Washington, Canada
Date(s) Application is Due: May 1; Nov 1
Contact: Program Officer; (206) 343-0807; fax (206) 343-0822; info@bullitt.org
Internet: http://www.bullitt.org
Sponsor: Bullitt Foundation
1212 Minor Avenue
Seattle, WA 98101-2825

Bunting Fellowship Program **1385**

The Mary Bunting Institute of Radcliffe College is a multidisciplinary research center for women scholars, scientists, artists, and writers and is one of the major centers for advanced study in the United States. One fellowship will be awarded to a professional woman in the field of infant and child development, who is conducting research within the framework of, or contributing to, psychoanalysis. Applications will be judged on the quality and significance of the proposed project, the applicant's record of accomplishment, and the stage in the applicant's career. Office or studio space, auditing privileges, and access to libraries and most other resources of Radcliffe College and Harvard University are provided. Residence in the Boston area and participation in the institute community are required during the one-year fellowship appointment. Fellows are expected to present their work-in-progress at public colloquia, performances, or exhibitions.
Geographic Focus: All States
Date(s) Application is Due: Oct 1
Contact: Coordinator; (617) 496-1324; fax (617) 495-8136; fellowships@radcliffe.edu
Internet: http://www.radcliffe.edu/fellowships/bunting.php
Sponsor: Bunting Institute of Radcliffe College
34 Concord Avenue
Cambridge, MA 02138

Bupa Foundation Medical Research Grants **1386**

The Bupa Foundation is an independent medical research charity that funds medical research to prevent, relieve and cure sickness and ill health. The Foundation aims to produce long-term benefits that have the potential to improve the health of individuals and populations worldwide. The funding will be available for the following areas: achieving sustained behaviour changes in relation to smoking, diet, physical activity and/or alcohol consumption; facilitating wellbeing and preventing mental ill health; improving patient decision-making through, for example, shared decision-making interventions; and, improving the design of community health activities by using new technologies to cost-effectively organise and interpret health outcome data.
Requirements: All projects should develop, apply or test research-based approaches in community settings-with clear intended impacts on public health. The areas for funding as detailed above are not intended to be exclusive of each other. Health professionals working for public or private organizations may apply for research grants for UK-based projects. If you are unsure of your eligibility, go to http://www.bupafoundation.co.uk/About/Grant-Eligibility.htm to see if your project will qualify.
Restrictions: Study for higher/further degrees, medical electives, educational courses, seminars and conferences, although valuable activities, are not eligible for grants. Similarly the Bupa Foundation does not fund the writing of reviews under the research grants scheme.
Geographic Focus: All States, All Countries
Date(s) Application is Due: Nov 30
Contact: Kate Brown, Grants Officer; 020 7656 2591; kate.brown@bupa.com
Internet: http://www.bupafoundation.co.uk/Research-Grants
Sponsor: Bupa Foundation
Bupa House
London, GLONDON WC1A 2BA United Kingdom

Bupa Foundation Multi-Country Grant **1387**

The Bupa Foundation is an independent medical research charity that funds medical research to prevent, relieve and cure sickness and ill health. The Foundation aims to to fund implementation of translational and action orientated research-based solutions, which: drive tangible action to prevent or alleviate chronic disease or the adverse elements of ageing, and to promote wellbeing; lead to the sustained uptake of healthy lifestyles; measurably improve public health to a significant degree. The Foundation also aims to fund research which leads to the development of such solutions. A total grant of up to £600,000 will be awarded to fund one multi-country initiative of one to three years duration. This is defined as a project carried out in at least two of the following countries: United Kingdom, Spain, Australia, New Zealand, the United States or India.
Requirements: Multi-country grant applications are open to clinicians, researchers and health care professionals based in the countries listed above. Applicants are encouraged to include partnerships with universities, healthcare organizations, research institutes, charities, community groups and other appropriate bodies on their application. Projects should contribute to the Foundation's aims and the following objective: Achieving sustained behavior changes in relation to smoking, diet, physical activity and/or alcohol consumption.
Restrictions: Study for higher/further degrees, medical electives, educational courses, seminars, conferences, etc. are not eligible to apply.

Geographic Focus: All States, Australia, India, New Zealand, Spain, United Kingdom
Date(s) Application is Due: Mar 2
Contact: Kate Brown, Grants Officer; 020 7656 2591; kate.brown@bupa.com
Internet: http://www.bupafoundation.co.uk/Multi-country-Grants
Sponsor: Bupa Foundation
Bupa House
London, GLONDON WC1A 2BA United Kingdom

Burden Trust Grants 1388
The trust supports international nonprofit organizations in the fields of medical research and hospitals; schools and training institutions; and care of and homes for the elderly, children, and other individuals in need. Preference will be given to organizations affiliated with the Anglican Church.
Geographic Focus: All States
Date(s) Application is Due: Mar 31
Contact: Patrick O'Conor, 0117 9628611; p.oconor@netgates.co.uk
Sponsor: Burden Trust
51 Downs Park W
Bristol, BS6 7QL England

Burlington Northern Santa Fe Foundation Grants 1389
The foundation is focused on the communities where the company operates and areas where its railways pass. The foundation supports education, including scholarships for Native Americans and scholarships for children of employees in conjunction with the National Merit Scholarships program; the arts, including museums, performing arts, and libraries; and civic and public affairs. Support goes to the Nature Conservancy and for local fire departments and law enforcement. Types of support include general operating support, continuing support, annual campaigns, capital campaigns, and program development. Health and human services funding concentrates on the United Way. Awards are made for a single year and for continuing support. The company also matches employee funds given to public and private colleges and universities, cultural organizations, and hospitals in the United States. Requests for applications should describe the purpose for the grant. Requests are reviewed every six weeks.
Requirements: Any 501(c)3 organization located in Schaumburg, IL, and communities where the corporation operates, including 28 states and two Canadian provinces, are eligible to apply.
Geographic Focus: All States
Contact: Deanna Dugas; (817) 867-6407; fax (817) 352-7924; Deanna.dugas@bnsf.com
Sponsor: Burlington Northern Santa Fe Foundation
2650 Lou Menk Drive, 2nd Floor, P.O. Box 961057
Fort Worth, TX 76131-2830

Burton G. Bettingen Grants 1390
The fields of activity of the corporation are education at all levels, mental health, crime and abuse victims and public protection programs, religion (Christian, Roman Catholic, and Salvation Army), environment, and welfare. Top funding priority is children and youth. The current focus is on child prostitutes, runaways, and abandoned children. Nonprofits servicing the economically disadvantaged also may apply. The corporation provides broad types of support, including operating, capital, research, challenge/matching grants, and endowments. A letter of inquiry stating the applicant's background, goals and objectives, and the specific need for funding is welcome. Unsolicited submissions are considered but receive low priority.
Requirements: IRS 501(c)3 organizations are eligible. Giving primarily, but not limited to, Southern California.
Restrictions: The corporation does not award grants to individuals; for general fund-raising events, dinners, or mass mailings; or to grantmaking organizations.
Geographic Focus: California
Amount of Grant: 5,000-200,000 USD
Contact: Patricia Brown; (323) 938-8478; fax (323) 938-8479; burtonbet@aol.com
Sponsor: Burton G. Bettingen Corporation
134 S Mansfield Avenue
Los Angeles, CA 90036-3019

Bush Foundation Artist Fellowships 1391
The fellowships enable artists to further their work and contribution to their communities. Fellows may decide to take time for solitary work or reflection, engage in collaborative community projects, embark on travel or research, or pursue any other activity that contributes to their lives as artists. Artists may use the fellowship in many ways—to explore new directions, continue work already in progress, or accomplish work not financially feasible otherwise. New to the program, professional development activities have been added including a two-day orientation retreat along with annual workshops combined with artist gatherings. Grant categories rotate on a two-year cycle. Check the website for this year's categories, which could include: Literature (Fiction, Creative Nonfiction, Poetry); Scriptworks (for stage and screen); Film/Video; or, Music Composition. The program also supports visual artists, performance artists, storytellers and traditional folk artists. Fellowships may last from 12 to 24 months.
Requirements: Eligible applicants must be: residents of the funding region: Minnesota, North Dakota or South Dakota, and are U.S. citizens or permanent residents; have lived in the funding region at least 24 months immediately prior to the application deadline; at least 25-years-old on the application deadline.
Restrictions: Artists are not eligible to apply if they are a student enrolled full- or part-time in a degree-granting program after July 1 of the program year; and a director or staff member of the Bush Foundation, or a spouse, parent, child, or grandchild of a director or staff member of the foundation; have already received two BAFs or an Enduring Vision Award.
Geographic Focus: Minnesota, North Dakota, South Dakota
Date(s) Application is Due: Oct 30
Amount of Grant: 48,000 USD
Contact: Julie Gordon Dalgleish, Program Director; (651) 227-0891 or (800) 605-7315; fax (651) 297-6485; jdalgleish@bushfoundation.org
Internet: http://www.bushfellows.org/artist/bush-artist-fellowships
Sponsor: Bush Foundation
332 Minnesota Street, Suite East 900
St. Paul, MN 55101-1315

BWF Ad Hoc Grants 1392
To complement the competitive award programs, the Burroughs Wellcome Fund makes modest grants on an ad hoc basis to nonprofit organizations conducting activities intended to improve the general environment for science. These noncompetitive grants are for activities closely related to the Foundation's focus areas. The Foundation places special priority on working with non-profit organizations, including government agencies, to leverage financial support for targeted areas of research, and on encouraging other foundations to support biomedical research. Proposals should be brief and include a description of the focus of the activity, the expected outcomes, qualifications of the organization or individuals involved; provide certification of the sponsor's Internal Revenue Service tax-exempt status; and give the total budget for the activity, including any financial support obtained or promised. Proposals are given careful preliminary review, and those deemed appropriate are presented for consideration by BWF's board of directors.
Geographic Focus: All States, Canada
Contact: Rolly L. Simpson, Jr., Senior Program Officer; (919) 991-5100; fax (919) 991-5160; info@bwfund.org or rsimpson@bwfund.org
Internet: http://www.bwfund.org/grant-programs/ad-hoc-grants
Sponsor: Burroughs Wellcome Fund
21 TW Alexander Drive, P.O. Box 13901
Research Triangle Park, NC 27709-3901

BWF Career Awards at the Scientific Interface 1393
BWF's Career Awards at the Scientific Interface (CASI) provide $500,000 over five years to bridge advanced postdoctoral training and the first three years of faculty service. These awards are intended to foster the early career development of researchers who have transitioned or are transitioning from undergraduate and/or graduate work in the physical/mathematical/computational sciences or engineering into postdoctoral work in the biological sciences, and who are dedicated to pursuing a career in academic research. Scientific advances such as genomics, quantitative structural biology, imaging techniques, and modeling of complex systems have created opportunities for exciting research careers at the interface between the physical/computational sciences and the biological sciences. Tackling key problems in biology will require scientists trained in areas such as chemistry, physics, applied mathematics, computer science, and engineering. These awards are open to U.S. and Canadian citizens or permanent residents as well as to U.S. temporary residents. Candidates are expected to draw from their training in a scientific field other than biology to propose innovative approaches to answer important questions in the biological sciences. Examples of approaches include, but are not limited to, physical measurement of biological phenomena, computer simulation of complex processes in physiological systems, mathematical modeling of self-organizing behavior, building probabilistic tools for medical diagnosis, developing novel imaging tools or biosensors, developing or applying nanotechnology to manipulate cellular systems, predicting cellular responses to topological clues and mechanical forces, and developing a new conceptual understanding of the complexity of living organisms. Proposals that include experimental validation of theoretical models are particularly encouraged.
Requirements: Citizens and non-citizen permanent residents of the U.S. and Canada are eligible. Candidates must: hold a Ph.D. degree in one of the fields of mathematics, physics, chemistry, computer science, statistics, or engineering (this includes related areas of physical, mathematical, computational, theoretical, and engineering science); demonstrate that their work is truly interdisciplinary; have completed at least 12 months but not more than 48 months of postdoctoral research by the date of the full invited application deadline; be committed to a full-time career in research as an independent investigator at a North American degree-granting institution; have at least one first-author publication; including papers on which first authorship is shared; and be based at a non-profit institution in the U.S. or Canada.
Restrictions: Candidates cannot hold nor have accepted, either in writing or verbally, a faculty appointment as a tenure-track assistant professor at the time of application-both pre-proposal and full application. Candidates must not hold nor have accepted a K99 award from the U.S. National Institutes of Health. Temporary residents of Canada are not eligible.
Geographic Focus: All States, Canada
Date(s) Application is Due: Jan 8
Amount of Grant: Up to 500,000 USD
Contact: Rusty Kelley; (919) 991-5120 or (919) 991-5100; rkelley@bwfund.org
Internet: http://www.bwfund.org/grant-programs/interfaces-science/career-awards-scientific-interface
Sponsor: Burroughs Wellcome Fund
21 TW Alexander Drive, P.O. Box 13901
Research Triangle Park, NC 27709-3901

BWF Career Awards for Medical Scientists (CAMS) 1394
The Career Awards for Medical Scientists (CAMS) is a highly competitive program that provides $700,000 awards over five years for physician-scientists, who are committed to an academic career, to bridge advanced postdoctoral/fellowship training and the early years of faculty service. Awards are made to degree granting institutions in the U.S. or Canada

on behalf of the awardee. The postdoctoral/fellowship portion of the award may last up to two years, and allows the awardee the use of up to $95,000 per year ($65,000 per year salary support). BWF will make up to two additional awards to clinically trained psychiatrists who focus on research at the interface between neuroscience and psychiatry. These proposals must clearly demonstrate evidence of integration of neuroscience and psychiatry in project design.
Requirements: The ideal candidate will be two years away from becoming an independent investigator, have at least two years or more of postdoctoral research experience, and have a significant publication record. Candidates must be nominated by their dean or department chair at the degree-granting institution where they will conduct the postdoctoral/fellowship training under the award. Applications must be approved by an official responsible for sponsored programs (generally from the grants office, office of research, or office of sponsored programs) at the degree-granting institution. Candidates must hold an M.D., D.D.S., or D.V.M. degree. Proposals must be in the area of basic biomedical, disease-oriented, or translational research. Proposals in health services research or involving large-scale clinical trials are ineligible. Candidates must be a fellow, resident, or a postdoctoral researcher and have at least two years of postdoctoral research experience at the time of application.
Restrictions: Proposals in health services research or involving large-scale clinical trials are ineligible.
Geographic Focus: All States, Canada
Date(s) Application is Due: Oct 1
Amount of Grant: 700,000 USD
Contact: Rolly L. Simpson, Jr., Senior Program Officer; (919) 991-5100; fax (919) 991-5160; info@bwfund.org or rsimpson@bwfund.org
Internet: http://www.bwfund.org/grant-programs/biomedical-sciences/career-awards-medical-scientists-2
Sponsor: Burroughs Wellcome Fund
21 TW Alexander Drive, P.O. Box 13901
Research Triangle Park, NC 27709-3901

BWF Collaborative Research Travel Grants 1395

The Collaborative Research Travel Grant (CRTG) program provides up to $15,000 in support for researchers from degree-granting institutions to travel either domestically or internationally to a laboratory to acquire a new research technique or to facilitate a collaboration. To capitalize on what appears to be an opportunity to provide relatively unrestricted travel funds to academic scientists and trainees and to provide a stimulus for those working or contemplating working at the interface of science, this program provides travel grants that can be used both internationally and domestically to acquire new research techniques, and to promote collaborations.
Requirements: Postdoctoral fellows or faculty at degree-granting institutions in the U.S. or Canada are eligible to apply. Applicants must hold a Ph.D. in mathematics, physics, chemistry, computer science, statistics, or engineering at the time of application; and interested in investigating research opportunities in the biological sciences. Biologists holding a doctorate degree at the time of application who are interested in working with physical scientists, mathematicians, engineers, chemists, statisticians, or computer scientists to incorporate their ideas and approaches to answering biological questions are eligible to apply. Applicants must be citizens or permanent residents of the U.S. or Canada at the time of application.
Restrictions: Grants cannot be used for travel to domestic or international conferences, to cover salary support or purchase computers or equipment. Laboratory or lecture courses are not eligible for this grant. No indirect costs or overhead charges can be applied to this grant. This grant is not bi-directional or reciprocal for the collaborator. Public health, health policy, or epidemiology proposals will not be accepted. This grant is not open to researchers at NIH institutions. Previous travel grant recipients are not eligible.
Geographic Focus: All States, Canada
Date(s) Application is Due: Feb 3
Amount of Grant: Up to 15,000 USD
Contact: Rolly L. Simpson, Jr., Senior Program Officer; (919) 991-5110 or (919) 991-5100; fax (919) 991-5160; rsimpson@bwfund.org or info@bwfund.org
Internet: http://www.bwfund.org/grant-programs/biomedical-sciences/collaborative-research-travel-grants
Sponsor: Burroughs Wellcome Fund
21 TW Alexander Drive, P.O. Box 13901
Research Triangle Park, NC 27709-3901

BWF Innovation in Regulatory Science Grants 1396

BWF's Innovation in Regulatory Science Awards provide up to $500,000 over five years to academic investigators developing new methodologies or innovative approaches in regulatory science that will ultimately inform the regulatory decisions the Food and Drug Administration (FDA) and others make. These awards are open to U.S. and Canadian citizens or permanent residents who have a faculty or adjunct faculty appointment at a North American degree-granting institution. Awards are made to degree-granting institutions in the U.S. or Canada on behalf of the awardee. The application process consists of two phases: a preproposal followed by a full proposal invitation. Preproposal applicants selected by the Advisory Committee deemed to meet the goals of this initiative will be invited to submit full proposals.
Requirements: Candidates must: hold an M.D., Ph.D., D.V.M., D.D.S., or M.D.-Ph.D. degree; be based at a non-profit institution [501(c)3 or equivalent] in the U.S. or Canada; hold a faculty position at an accredited, degree-granting institution in the United States or Canada; be an investigator at the adjunct, assistant, associate, or full professor level; be a citizen or permanent resident of the U.S. or Canada at the time of application.
Restrictions: Ineligible applicants include: persons who have applied for permanent resident status but have not received their government documentation by the time of the preproposal application deadline; temporary residents of the U.S. or Canada; postdoctoral students; and candidates who currently hold a BWF award.
Geographic Focus: All States, Canada
Date(s) Application is Due: Apr 1
Amount of Grant: 500,000 USD
Contact: Rusty Kelley, Senior Program Officer; (919) 991-5120 or (919) 991-5100; fax (919) 991-5160; rkelley@bwfund.org or info@bwfund.org
Internet: http://www.bwfund.org/grant-programs/regulatory-science/innovation-regulatory-science
Sponsor: Burroughs Wellcome Fund
21 TW Alexander Drive, P.O. Box 13901
Research Triangle Park, NC 27709-3901

BWF Institutional Unifying Population & Laboratory Based Sciences Grants 1397

The Institutional Program Unifying Population and Laboratory Based Sciences program offers five-year institutional training awards of $500,000 per year to bridge the gap between the population and computational sciences and the laboratory-based biological sciences. The award will support the training of researchers between existing concentrations of research strength in population approaches to human health and in basic biological sciences. The goal is to establish training programs by partnering researchers working in schools of medicine and schools (or academic divisions) of public health. The programs supported by these awards will develop young researchers who will be equally at home with the ideas, approaches, and insights generated at the molecular scale and at the population scale. Proposals that cross institutional boundaries are encouraged.
Requirements: Degree-granting institutions in the U.S. or Canada may submit applications. Proposals must be driven by core components within medical and public health schools, but beyond those required components, departments or centers located within non-medical parts of a university, existing inter-institutional collaboratives, research museums, free-standing research institutes, and other non-profit institutions that provide advanced-level training are all acceptable as potential additional partners. Dental, osteopathic, and veterinary medical schools are appropriate applicants.
Restrictions: For-profit companies may not participate in the application, but could be valuable partners in such training programs. Research groups working at national laboratories and within the federal government are allowable as partners.
Geographic Focus: All States, Canada
Date(s) Application is Due: May 19
Amount of Grant: 500,000 USD
Contact: Victoria McGovern, Ph.D., Senior Program Officer; (919) 991-5112 or (919) 991-5100; fax (919) 991-5160; vmcgovern@bwfund.org or info@bwfund.org
Internet: http://www.bwfund.org/grant-programs/population-sciences/institutional-program-unifying-population-and-laboratory-based
Sponsor: Burroughs Wellcome Fund
21 TW Alexander Drive, P.O. Box 13901
Research Triangle Park, NC 27709-3901

BWF Investigators in the Pathogenesis of Infectious Disease Awards 1398

The Investigators in the Pathogenesis of Infectious Disease Award provides $500,000 over a period of five years to support accomplished investigators at the assistant professor level to study pathogenesis, with a focus on the interplay between human and microbial biology, shedding light on how human and microbial systems are affected by their encounters. The awards are intended to give recipients the freedom and flexibility to pursue new avenues of inquiry and higher-risk research projects that hold potential for significantly advancing the biochemical, pharmacological, immunological, and molecular biological understanding of how microbes and the human body interact. The goal of the program is to provide opportunities for accomplished investigators still early in their careers to study what happens at the points where human and microbial systems connect. The program supports research that sheds light on the fundamentals that affect the outcomes of this encounter: how colonization, infection, commensalism and other relationships play out at levels ranging from molecular interactions to systemic ones.
Requirements: Candidates must: generally have an M.D., D.V.M., or Ph.D. degree; have an established record of independent research and hold a tenure-track position as an assistant professor or equivalent (at the time of application) at a degree-granting institution; and be nominated by accredited, degree-granting institutions in the United States or Canada.
Restrictions: Applications from non-tenure track investigators at tenure-offering, degree-granting institutions will not be accepted.
Geographic Focus: All States, Canada
Date(s) Application is Due: Nov 3
Amount of Grant: 500,000 USD
Contact: Victoria McGovern, Ph.D., Senior Program Officer; (919) 991-5112 or (919) 991-5100; fax (919) 991-5160; vmcgovern@bwfund.org or info@bwfund.org
Internet: http://www.bwfund.org/grant-programs/infectious-diseases-0
Sponsor: Burroughs Wellcome Fund
21 TW Alexander Drive, P.O. Box 13901
Research Triangle Park, NC 27709-3901

BWF Preterm Birth Initiative Grants 1399

The Preterm Birth Initiative was created to increase the understanding of the biological mechanisms underlying parturition and spontaneous preterm birth and will provide up to $600,000 over a four-year period ($150,000 per year). The initiative is designed to stimulate both creative individual scientists and multi-investigator teams to approach the problem of preterm birth using creative basic and translation science methods. Postdoctoral fellows nearing their transition to independent investigator status through senior established investigators are encouraged to apply. Molecular and computational

approaches such genetics/genomics, immunology, microbiology, evolutionary biology, mathematics, engineering, and other basic sciences hold enormous potential for new insights independently or in conjunction with more traditional areas of parturition research such as maternal fetal medicine, obstetrics, and pediatrics. The formation of new connections between reproductive scientists and investigators who are involved in other areas will give preterm birth research a fresh and unique look, and stimulate a new work face to tackle this challenge. Proposals can be submitted by individual investigators or research teams designating a contact principal investigator. Proposals that cross institutional boundaries (partnerships between multiple universities or collaborations within larger universities) are welcomed.
Requirements: Proposals must be submitted from degree-granting institutions in the United States or Canada. The principal investigator must be a postdoctoral fellow in the final 1-2 years of postdoctoral training or hold faculty appointment at a degree-granting institution in the U.S. or Canada. The principal investigator must be a citizen or permanent resident of the U.S. or Canada.
Restrictions: Prior preterm birth full research grant recipients are not eligible to reapply.
Geographic Focus: All States, Canada
Date(s) Application is Due: Dec 1
Amount of Grant: Up to 600,000 USD
Contact: Rolly L. Simpson, Jr., Senior Program Officer; (919) 991-5110 or (919) 991-5100; fax (919) 991-5160; rsimpson@bwfund.org or info@bwfund.org
Internet: http://www.bwfund.org/grant-programs/reproductive-sciences/preterm-birth-initiative
Sponsor: Burroughs Wellcome Fund
21 TW Alexander Drive, P.O. Box 13901
Research Triangle Park, NC 27709-3901

Bydale Foundation Grants **1400**
The foundation emphasizes international understanding, public policy research, environmental quality, cultural programs, the law and civil rights, social services, higher education, and economics. Funding includes support for: conference/seminars, continuing support, general operating support, matching/challenging support, program development, publication, research, and seed money. An application form is not required. Submit an initial approach in the form of a letter or proposal.
Requirements: U.S. 501(c)3 nonprofits are eligible.
Geographic Focus: All States
Date(s) Application is Due: Nov 1
Samples: Clean Water Fund (Washington, DC)—$5000; Greenhouse Crisis Foundation (Washington, DC)—$20,000; Greenpeace U.S.A (Washington, DC)—$2500.
Contact: Milton Solomon, Vice President; (914) 428-3232; fax (914) 428-1660
Sponsor: Bydale Foundation
11 Martine Avenue
White Plains, NY 10606

Byron W. and Alice L. Lockwood Foundation Grants **1401**
The foundation awards grants to Washington nonprofit organizations in its areas of interest, including arts, biomedical research, higher education, hospitals and health care organizations, housing and homelessness, museums, religion and religious welfare programs, and social services. Types of support include capital improvements, continuing support, general operating grants, professorships, projects grants, and research grants. There are no application forms; submit a letter of request.
Requirements: Washington nonprofit organizations are eligible. Grants are awarded primarily in Washington's Seattle and Puget Sound areas.
Geographic Focus: All States
Date(s) Application is Due: Oct 31
Amount of Grant: 900-56,000 USD
Contact: Lee Kraft, Executive Director; (206) 230-8489
Sponsor: Byron W. and Alice L. Lockwood Foundation
P.O. Box 4
Mercer, WA 98040

BYU Religious Studies Center Research Grants **1402**
The mission of the BYU Religious Studies Center is to encourage and sponsor serious, faithful, gospel-related scholarship and the ensuing publication of that scholarship. The mission is accomplished by providing grants to assist worthwhile projects; encouraging faculty members to submit their manuscripts for publication; and publishing and marketing high-quality religious books and periodicals of a scholarly nature.
Geographic Focus: All States
Contact: Richard Neitzel Holzapfel; (801) 422-2332; holzapfel@byu.edu
Internet: http://religion.byu.edu/rsc_about.php
Sponsor: Brigham Young University
Brigham Young University, 167 HGB
Provo, UT 84602

C. Alma Baker Trust Grants **1403**
The trust operates in the fields of agriculture and education and supports students studying for agricultural degrees at Massey University. Types of support include postgraduate scholarships for agricultural and horticultural research, grants for research to further the science of agriculture and horticulture, and travel grants for overseas trips of three to six weeks. Travel grant recipients are expected to contribute to the New Zealand horticultural or agricultural community upon return. Deadlines vary for each program; contact program staff for exact dates.
Geographic Focus: All States, New Zealand
Date(s) Application is Due: Feb 1
Amount of Grant: 10,000-20,000 USD
Contact: Trust Administrator; 01480 411331; fax 01480 459012
Internet: http://www.massey.ac.nz
Sponsor: C. Alma Baker Trust
Private Mail Bag, Massey University
Palmerston North, New Zealand

CAA Professional Development Fellowships for Graduate Students **1404**
CAA's Professional-Development Fellowships support promising artists and art historians who are enrolled in MFA and PhD programs nationwide. Fellows are honored with $5,000 grants to help them with various aspects of their work, whether it be for job-search expenses or purchasing materials for the studio. Fellows also receive a free one-year CAA membership and complimentary registration to the Annual Conference. Honorable mentions, given at the discretion of the jury, also earn a free one-year CAA membership and complimentary conference registration. In Fellowships in the Visual Arts, five student artists in MFA programs will each receive a one-time award of $5,000. In Fellowships in Art History, two doctoral candidates in art history will each receive a one-time award of $5,000.
Requirements: CAA seeks applications from students who are current members; are citizens or permanent residents of the United States; will receive their MFA or PhD degree in the calendar year following the year of application; and have outstanding capabilities and demonstrate distinction in approach, technique, or perspective in their contribution to art history and the visual arts. A jury of artists, curators, and other professionals will review all applications in the fall and announce the recipients in January of the following year. All applications materials must be sent by mail and postmarked by the deadline date.
Geographic Focus: All States
Date(s) Application is Due: Oct 1
Amount of Grant: 5,000 USD
Contact: Michael Fahlund; (212) 392-4401; fax (212) 627-2381; mfahlund@collegeart.org
Internet: http://www.collegeart.org/fellowships/
Sponsor: College Art Association
50 Broadway, 21st Floor
New York, NY 10004

CAA Wyeth Publication Grant **1405**
Since 2005, the Wyeth Foundation for American Art has supported the publication of books on American art through the Wyeth Foundation for American Art Publication Grant, administered by CAA. For this grant program, "American art" is defined as art created in the United States, Canada, and Mexico through 1970. Eligible for the grant are book-length scholarly manuscripts in the history of American art, visual studies, and related subjects that have been accepted by a publisher on their merits but cannot be published in the most desirable form without a subsidy.
Requirements: Application is by the publisher, not the author. Applications are considered only for book-length scholarly manuscripts in the history of American art that are under contract for publication. Awards are open to publishers of all nations; authors need not live or work in the United States. Commercial, university, and museum presses are all eligible. Applicant authors and presses must be institutional CAA members, although on a case-by-case basis CAA may waive that requirement upon request for non-U.S.-based independent presses. Excluded from consideration are excavation or other technical reports, articles, previously published works (including collections of previously published essays), and congress proceedings. Museum exhibition or collection catalogues containing substantial scholarship are eligible. High scholarly and intellectual merit is the sine qua non for an award; however, the jury is also attentive to the following criteria: Topics on art through 1970; Topics with a naturally small market or unusually high expenses; Works by disadvantaged scholars, including those at the earlier stages of a career, or by younger scholars or curators; or issued by smaller museums; or by or about underserved constituencies; Books that break new ground, contribute new scholarship, or publish important primary-source material; and, Beautiful books that increase the audience for American art .
Restrictions: A project that has been rejected for grant may not be resubmitted to the same grant, except in a rare case where substantial revision has been made to the material, and the publisher has so noted in the application. At its discretion, the jury may decline to review the resubmitted application.
Geographic Focus: All States, All Countries
Date(s) Application is Due: Sep 15
Contact: Alex Gershuny; (212) 392-4424; fax (212) 627-2381; agershuny@collegeart.org
Internet: http://www.collegeart.org/wyeth/
Sponsor: College Art Association
50 Broadway, 21st Floor
New York, NY 10004

Cabot Corporation Foundation Grants **1406**
The goal of Cabot Corporation Foundation is to support community outreach objectives, with priority given to science and technology, education, and community and civic improvement efforts in the communities where the company has major facilities or operations. Types of support include capital grants, challenge grants, employee matching gifts, fellowships, general support, professorships, project support, research, scholarships, and seed money. The board meets in January, April, July, and October to consider requests. Applications must be received at least 30 days before a board meeting.
Requirements: The Foundation supports only nonprofit 501(c)3 tax-exempt organizations in areas of company operation. Modest support is available for international organizations that qualify under U.S. tax regulations.

Restrictions: Contributions are not made to individuals; fraternal, political, athletic or veterans organizations; religious institutions; capital and endowment campaigns; sponsorships of local groups/individuals to participate in regional, national, or international competitions, conferences or events; advertising sponsorships; or Tickets or tables at fundraising events.
Geographic Focus: Georgia, Illinois, Louisiana, Massachusetts, New Mexico, Pennsylvania, Texas, West Virginia, Belgium, Canada, China, Switzerland, United Kingdom
Amount of Grant: 2,000-75,000 USD
Contact: Cynthia L. Gullotti, Program Manager; (617) 345-0100; fax (617) 342-6312; Cynthia_Gullotti@cabot-corp.com or cabot.corporation.foundation@cabotcorp.com
Internet: http://www.cabot-corp.com/About-Cabot/Corporate-Giving
Sponsor: Cabot Corporation Foundation
Two Seaport Lane, Suite 1300
Boston, MA 02210-2019

Caesars Foundation Grants **1407**
Founded as Harrah's Foundation in Nevada in 2002, giving is in the area of company operations. The foundation supports programs designed to help older individuals live longer, healthier, and more fulfilling lives; promote a safe and clean environment; and improve the quality of life in communities where Caesars operates. Fields of interest include: aging centers and services; Alzheimer's disease; developmentally disabled services; the environment; food distribution programs; food services; health care; patient services; higher education; hospitals; human services; mental health services; public affairs; public safety; nutrition; and youth services. Types of support being offered include: building and renovation; capital campaigns; continuing support; general operating support; program development; research; scholarship funding; and sponsorships. There are no specific deadlines or application forms. Caesars Foundation Trustees meet on a quarterly basis, typically around the second week of each quarter. Check with your nearest Caesars Entertainment property for deadlines. The foundation generally funds programs and projects of $10,000 or more.
Requirements: Eligible organizations are 501(c)3 nonprofits operating programs in the communities where Caesars employees and their families live and work. Applying organizations must also: demonstrate diversity by providing services and volunteer opportunities to all without regard to race, ethnicity, gender, religion, sexual orientation, identity or disability; illustrate strong leadership that will significantly strengthen communities in which Caesars operates; show sound administrative and financial condition; provide opportunities for Caesars staff involvement as volunteers and/or opportunity to serve on Board of Directors; and, provide branding opportunities and openly support Caesars Foundation in a public forum.
Restrictions: The Foundation is not designed to react to last-minute requests or event sponsorships—plan the timing of your proposal accordingly. Caesars Foundation does not accept requests for in-kind contributions.
Geographic Focus: Arizona, California, Illinois, Indiana, Iowa, Louisiana, Mississippi, Missouri, Nevada, New Jersey, North Carolina, Pennsylvania
Amount of Grant: 10,000 USD
Contact: Gwen Migita, Community Affairs; (702) 880-4728 or (702) 407-6358; fax (702) 407-6520; caesarsfoundation@caesars.com
Internet: http://www.caesarsfoundation.com/
Sponsor: Caesars Foundation
1 Caesars Palace Drive
Las Vegas, NV 89109-8969

California Teachers Association Scholarships **1408**
The CTA Scholarship gives support for study in higher education to active members and/or dependent children of active members and/or dependent children of deceased members. Twenty-five scholarships are awarded annually. Applications are available in October, and may be obtained by contacting the CTA Human Rights Department in Burlingame or any CTA Regional Resource Center office.
Requirements: Applicant must be an active member of California Teachers Association or applicant must be the dependent child of an active, retired-life or deceased member of California Teachers Association depending on which program they are applying for.
Geographic Focus: California
Date(s) Application is Due: Feb 15
Amount of Grant: 2,000 USD
Contact: Manuel Ayan, (650) 552-5204; fax (650) 552-5002; scholarships@cta.org
Internet: http://www.cta.org/InsideCTA/ScholarshipsWorkshops/ScholarshipsWorkshops.htm
Sponsor: California Teachers Association
1705 Murchison Drive, P.O. Box 921
Burlingame, CA 94011-0921

California Wellness Foundation Work and Health Grants **1409**
The foundation focuses its activities on specific priority areas where it has a significant, long-term commitment. Within each priority area, the foundation allocates the majority of its funds toward initiatives. Initiatives are targeted grantmaking programs with distinct objectives and are generally announced through requests for proposals. The foundation awards general grants and project grants. Under general grants, requests for core operating support for organizations that provide direct services to Californians for disease prevention or health promotion are of primary interest. Priority areas under general grants are diversity in the health professions, environmental health, healthy aging, mental health, teenage pregnancy prevention,, violence prevention, women's health, and work and health. Special projects grants are awarded to areas that fall outside the priority areas. Of particular interest are proposals to help California communities respond to cutbacks in federally funded programs. Activities commonly supported under special projects include strengthening traditional safety-net providers, educating consumers about changes in health care systems, advocating for underserved communities in health policy debates, and informing public decision making through policy analysis.
Requirements: Eligible applicants are California 501(c)3 nonprofit organizations, or organizations with a preapproved fiscal sponsor. An organization should first write a succinct letter of interest (one to two pages) that describes the organization, its leadership, the region and population(s) served, and the activities for which funding is needed, including the amount requested.
Geographic Focus: California
Amount of Grant: 5,000-200,000 USD
Contact: Grants Administrator; (818) 702-1900; fax (818) 593-6614
Internet: http://www.tcwf.org/grants_program/index.htm
Sponsor: California Wellness Foundation
6320 Canoga Avenue, Suite 1700
Woodland Hills, CA 91367-7111

Callaway Golf Company Foundation Grants **1410**
The foundation strives to support initiatives in communities where Callaway Golf Company employees live and work. The geographic area of focus is primarily North San Diego County, California. The foundation offers support in the form of: matching funds, special projects, and general operating budgets. Areas of interest include but are not limited to: children and youth; biomedical research, with a special interest in the field of cancer; golf; education; boys & girls clubs of America; drug prevention; American Red Cross; housing; veterans; youth programs; food banks; social services; emergency programs; scholarship program for dependents of Callaway Golf employees; grants for training, competition and equipment needs. The foundation does not require the completion of a formal application document but requests a description of the organization and its history, the project at issue including goals and time lines, the qualifications of the leadership personnel involved in the project, and a detailed project budget. Grants are awarded semiannually. The Callaway Golf Company Foundation does not accept unsolicited requests for grants.
Requirements: IRS 501(c)3 nonprofit organizations in California are eligible.
Restrictions: The foundation will not fund applicants that illegally discriminate on the basis of gender, race, color, religion, national origin, ancestry, age, marital status, medical condition, or physical disability, either in the services they provide or in the hiring of staff; or promote political or particular religious doctrines.
Geographic Focus: California
Amount of Grant: 500-10,000 USD
Contact: Paul Thompson, Executive Director; (760) 930-8686; fax (760) 930-5021; cgcfoundation@callawaygolf.com
Internet: http://www.callawaygolf.com/Global/en-U.S./Corporate/CallawayGolfFoundation.html
Sponsor: Callaway Golf Company Foundation
2180 Rutherford Road
Carlsbad, CA 92008-7328

Calvin Institute of Christian Worship Renewal Grants **1411**
The grants program is designed to foster well-grounded worship renewal in congregations throughout North America. Made possible through support of the Lilly Endowment, Inc., these grants are intended to serve a grass-roots constituency of those concerned for the vitality of the worship life in their local Christian communities in a variety of denominations. Through its grantmaking, the institute intends to stimulate thoughtful and energetic work that will result in worship services that exhibit renewed creativity, theological integrity, and relevance. The institute solicits proposals for grants to support a wide variety of projects. Grants will be awarded both to individuals and to nonprofit organizations, such as churches, colleges, seminaries, or other church-related organizations. Grants to individuals, however, will be awarded through the church or organization with which they are affiliated.
Requirements: Any individual or team of people with leadership role(s) who are directly related to the worship life of a congregation, school, church-related organization, or other ministry is encouraged to apply. This includes pastors, educators, church staff persons, church musicians, directors of worship, liturgists, artists, architects, and scholars.
Geographic Focus: All States
Date(s) Application is Due: Jan 10
Contact: Betty Grit; (616) 526-7168; fax (616) 957-7168; worshipgrants@calvin.edu
Internet: http://webapps.calvin.edu/worship/grants/wrgp.php
Sponsor: Calvin Institute of Christian Worship
3201 Burton SE
Grand Rapids, MI 49546

Camargo Foundation Fellowships **1412**
The program provides, at no cost, furnished apartments and a reference library in the city of Cassis for scholars who wish to pursue projects in the humanities and social sciences related to French and francophone cultures. The foundation also sponsors creative projects by visual artists, photographers, composers, and writers. The term of residence is one semester (early September to mid-December or mid-January to May 31), precise dates being announced each year. An informational brochure and application form are available upon request.
Requirements: Applicants may include members of university and college faculties, including professors emeriti; independent scholars working on specific projects; secondary school teachers; graduate students completing the dissertation required for their degree; and writers, photographers, visual artists, and composers.

Geographic Focus: All States
Date(s) Application is Due: Jan 15
Amount of Grant: 3,500 USD
Contact: Program Contact; (651) 238-8805
Internet: http://www.camargofoundation.org/aboutus.asp
Sponsor: Camargo Foundation
400 Sibley Street, Suite 125
Saint Paul, MN 55101-1928

Cambridge Foundation For Peace Prize 1413
The foundations announces a call for papers that explore the role of diasporas as transnational actors in preventing violence, crisis response, or postconflict peacebuilding. The prize will be awarded for the winning proposal, which will contribute to systematic research on diasporas. In particular, the paper should concentrate on identifying and specifying the mechanisms by which diasporas can be engaged in policy planning and implementation. The regional and/or country focus is open, but preference will be give to Southeastern Europe and/or the Eastern Mediterranean. The CFP will distribute the winning paper to its worldwide audience of policymakers, political leaders, academics, and journalists. The winning paper also will be posted as a working paper on CFP-web.
Requirements: The competition is open to public policy professionals and academics (including graduate students).
Geographic Focus: All States
Date(s) Application is Due: Apr 1
Amount of Grant: 3,000 USD
Contact: Administrator; (617) 254-5437; fax (617) 254-5437; schmitz@cfp-web.org
Internet: http://www.cfp-web.org
Sponsor: Cambridge Foundation for Peace
1770 Massachusetts Avenue, P.O. Box 303
Cambridge, MA 02140-2808

Camille Dreyfus Teacher-Scholar Awards 1414
The Camille Dreyfus Teacher-Scholar Awards Program supports the research and teaching careers of talented young faculty in the chemical sciences. Based on institutional nominations, the program provides discretionary funding to faculty at an early stage in their careers. Criteria for selection include an independent body of scholarship attained within the first five years of their appointment as independent researchers, and a demonstrated commitment to education, signaling the promise of continuing outstanding contributions to both research and teaching. The Camille Dreyfus Teacher-Scholar Awards Program provides an unrestricted research grant of $75,000. Of the total amount, $7,500 is for departmental expenses associated with research and education. Guidelines and required information are available online.
Requirements: The Camille Dreyfus Teacher-Scholar Awards Program is open to academic institutions in the States, Districts, and Territories of the United States of America that grant a bachelor's or higher degree in the chemical sciences, including biochemistry, materials chemistry, and chemical engineering. Nominees must hold a full-time tenure-track academic appointment, and are normally expected to have been appointed no earlier than mid-year 2006. Awardees are from Ph.D. granting departments in which scholarly research is a principal activity. Undergraduate education is an important component of the nominee's activities. Institutions may submit only one Camille Dreyfus nomination annually.
Restrictions: Defrayal of academic-year salary is not permitted.
Geographic Focus: All States
Date(s) Application is Due: Feb 9
Amount of Grant: 75,000 USD
Contact: Mark J. Cordillo, Executive Director; (212) 753-1760; fax (212) 593-2256; mcardillo@dreyfus.org or admin@dreyfus.org
Internet: http://www.dreyfus.org/awards/camille_dreyfus_teacher_award.shtml
Sponsor: Camille and Henry Dreyfus Foundation
555 Madison Avenue, 20th Floor
New York, NY 10022-3301

Campbell Soup Foundation Grants 1415
Since 1953, the Campbell Soup Foundation has provided financial support to local champions that inspire positive change in communities throughout the United States where Campbell Soup Company employees live and work. The Foundation places particular emphasis on Camden, New Jersey, birthplace of Campbell's flagship soup business and world headquarters. The Campbell Soup Foundation focuses its giving on four key areas: hunger relief-supporting food bank organizations in the communities of operation; wellness-addressing the health of consumers in the communities where they live; education-leveraging the Campbell brand portfolio to support educational programs; community revitalization-enhancing the quality of life in the communities that Campbell operates in. The Foundation only considers applications that meet the following criteria: the proposal must fit one of the key focus areas; the organization must display strong and effective leadership; the proposed plan must be clear and compelling, with measurable and sustainable commitments expressed in terms of real results; the proposed activity must be sufficiently visible to leverage additional support from other funding sources. There is no formal deadline. Proposals are accepted and reviewed on a rolling basis.
Requirements: The Foundation limits grants to nonprofit organizations which are tax-exempt under Section 501(c)3 of the Internal Revenue Code. Grants are made to institutions that serve: Camden, New Jersey; Davis, California; Sacramento, California; Stockton, California; Bloomfield, Connecticut; Norwalk, Connecticut; Lakeland, Florida; Downers Grove, Illinois; Marshall, Michigan; Maxton, North Carolina; Camden, New Jersey; South Plainfield, New Jersey; Napoleon, Ohio; Wauseon, Ohio; Willard, Ohio; Denver, Pennsylvania; Downingtown, Pennsylvania; Aiken, South Carolina; Paris, Texas; Richmond, Utah; Everett, Washington; Milwaukee, Wisconsin. Organizations do not need to be located in these communities in order to qualify for funding. However, the programs to be funded must serve these communities. Proposals must be submitted electronically via email to community_relations@campbellsoup.com. Proposals should be prepared in a concise, narrative form, without extensive documentation.
Restrictions: Grants are not made to the following: organizations that are based outside the United States and its territories; individuals; organizations that limit their services to members of one religious group or whose services propagate religious faith or creed; political organizations and those having the primary purpose of influencing legislation of/or promoting a particular ideological point of view; units of government; events and sponsorships; sports related events, activities and sponsorships. Organizations may not submit the same or similar proposals more than once in a Foundation fiscal year (July 1-June 30). Proposals submitted via regular mail will not be reviewed.
Geographic Focus: Connecticut, Florida, Illinois, Michigan, New Jersey, North Carolina, Ohio, Pennsylvania, South Carolina, Texas, Utah, Washington, Wisconsin
Samples: Coriell Institute for Medical Research, $12,500; Scotland County Literacy Council, $15,000; Delaware County Christian School, $500.
Contact: Grant Administrator; (856) 342-6423 or (800) 257-8443; fax (856) 541-8185; community_relations@campbellsoup.com
Internet: http://www.campbellsoupcompany.com/community_center.asp
Sponsor: Campbell Soup Foundation
1 Campbell Place
Camden, NJ 08103-1701

CampusRN/AACN Nursing Scholarship Fund 1416
The scholarship program supports students who are seeking a baccalaureate, master's, or doctoral degree in nursing. Special consideration will be given to students enrolled in a master's or doctoral program with the goal of pursuing a nursing faculty career; completing an RN to baccalaureate program (BSN); or enrolled in an accelerated baccalaureate or master's degree nursing program. Applicants must submit an online application and create a complete profile on CampusRN. All applicants must already be enrolled (not just accepted) at an AACN member institution. Applications are accepted only by email.
Requirements: Nursing students enrolled in a baccalaureate, master's or doctoral degree program in nursing are eligible.
Geographic Focus: All States
Date(s) Application is Due: Jan 1; Mar 1; May 1; Jul 1; Sep 1; Nov 1
Amount of Grant: 2,500 USD
Contact: Administrator; (202) 463-6930; info@campuscareercenter.com
Internet: http://aacn.campusrn.com/scholarships/scholarship_rn.asp
Sponsor: American Association of Colleges of Nursing
1 Dupont Circle NW, Suite 530
Washington, DC 20036

Canada-U.S. Fulbright International Science and Technology Award 1417
Canada-U.S. Fulbright awards offer outstanding students and scholars in Canada and the United States a unique opportunity to participate in one of the world's premier educational exchange programs. The International Fulbright Science and Technology Award Ph.D. offers support for study in science, technology, or engineering at top U.S. institutions. The award carries a value of, on average, U.S.$180,000 over three years.
Requirements: Each applicant is required to submit an online application, along with all supporting documents.
Geographic Focus: All States, Canada
Date(s) Application is Due: May 31
Amount of Grant: 180,000 USD
Contact: Jennifer Regan; (613) 688-5517 or (613) 688-5540; jregan@fulbright.ca
Internet: http://www.fulbright.ca/en/award.asp
Sponsor: Foundation for Educational Exchange between Canada and the USA
350 Albert Street, Suite 2015
Ottawa, ON K1R 1A4 Canada

Canada-U.S. Fulbright Senior Specialists Grants 1418
The Fulbright Senior Specialists Program is designed to provide U.S. and Canadian faculty and professionals with opportunities to collaborate on curriculum and faculty development, institutional planning and a variety of other activities. During the course of their grant, Fulbright Senior Specialists may engage in any of the following activities at their Canadian host institution: conduct needs assessments, surveys, institutional or programmatic research; take part in specialized academic programs and conferences; consult with administrators and instructors of post-secondary institutions on faculty development; present lectures at graduate and undergraduate levels; participate in or lead seminars or workshops at overseas academic institutions; develop and/or assess academic curricula or educational materials; and conduct teacher-training programs at the tertiary level. These are short-term grants of two to six weeks to that are offered on a continuous basis.
Requirements: Eligibility requirements include: Canadian or American citizenship at time of application (permanent resident status or landed immigrant status are not sufficient); a Ph.D. or equivalent professional/terminal degree by December 31 of the application year or equivalent professional experience; and English language proficiency.
Restrictions: Applicants are not eligible for a Canada-U.S. Fulbright Award if he or she: currently holds permanent residency status in the intended host country; is currently residing in or enrolled at a university in the intended host country; has resided abroad for five or more consecutive years in the six year period immediately preceding the date of application; has received a Fulbright Award within the last five years; has had recent substantial experience in the intended host country (recent substantial experience is

defined as study, teaching, research or employment for a period of more than an academic year (nine months) during the past five years); or is a Canadian applicant and you are a dual citizen of Canada and the United States.
Geographic Focus: All States, Canada
Contact: Jennifer Regan; (613) 688-5517 or (613) 688-5540; jregan@fulbright.ca
Internet: http://www.fulbright.ca/en/seniorspecialists.asp
Sponsor: Institute of Int'l Education, Council for International Exchange of Scholars
350 Albert Street, Suite 2015
Ottawa, ON K1R 1A4 Canada

Canada-U.S. Fulbright Traditional Scholar Awards 1419
Grants of four to nine months are available to Canadian and American scholars, post-doctoral researchers and experienced professionals to conduct research and/or teach at a host institution in the United States or Canada, respectively. Canada-U.S. Traditional Fulbright awards offer a unique opportunity to explore important contemporary issues relevant to Canada and the United States and the relationship between the two countries. While the competition is officially field open, applications in the following areas are given preference: comparative public policy; international trade; North American integration; security; communications and culture; ecology and the environment; indigenous issues; law; border issues; public health; Canada-U.S. relations; and Canadian and American studies. The deadline for American scholars is August 1 and the deadline for Canadian scholars is November 16.
Requirements: Eligibility requirements include: Canadian or American citizenship at time of application (permanent resident status or landed immigrant status are not sufficient); a Ph.D. or equivalent professional/terminal degree by December 31 of the application year or equivalent professional experience; and English language proficiency.
Restrictions: Applicants are not eligible for a Canada-U.S. Fulbright Award if he or she: currently holds permanent residency status in the intended host country; is currently residing in or enrolled at a university in the intended host country; has resided abroad for five or more consecutive years in the six year period immediately preceding the date of application; has received a Fulbright Award within the last five years; has had recent substantial experience in the intended host country (recent substantial experience is defined as study, teaching, research or employment for a period of more than an academic year (nine months) during the past five years); is a Canadian applicant and you are a dual citizen of Canada and the United States.
Geographic Focus: All States, Canada
Date(s) Application is Due: Aug 1; Nov 16
Amount of Grant: 25,000 USD
Contact: Jennifer Regan; (613) 688-5517 or (613) 688-5540; jregan@fulbright.ca
Internet: http://www.fulbright.ca/en/seniorscholar.asp
Sponsor: Foundation for Educational Exchange between Canada and the USA
350 Albert Street, Suite 2015
Ottawa, ON K1R 1A4 Canada

Canada Graduate Scholarships and NSERC Postgraduate Scholarships 1420
Canada Graduate Scholarships and NSERC Postgraduate Scholarships provide financial support to high-calibre scholars who are engaged in master's or doctoral programs in the natural sciences or engineering. The Canada Graduate Scholarships will be offered to the top-ranked applicants at each level (master's and doctoral) and the next tier of meritorious applicants will be offered an NSERC Postgraduate Scholarship. This support allows these scholars to fully concentrate on their studies and to seek out the best research mentors in their chosen fields. NSERC encourages interested and qualified Aboriginal students to apply.
Requirements: To be considered eligible for support, applicants must: (1) be a Canadian citizen or a permanent resident of Canada; (2) hold, or expect to hold (at the time you take up the award), a degree in science or engineering from a university whose standing is acceptable to NSERC (if you have a degree in a field other than science or engineering, NSERC may accept your application at its discretion); (3) intend to pursue in the following year full-time graduate studies and research at the master's or doctoral level in an eligible program in one of the areas of the natural sciences and engineering supported by NSERC; and (4) have obtained a first-class average (a grade of 'A-') in each of the last two completed years of study (full-time equivalent). Any exceptions to this requirement must be accompanied by supporting comments that justify the submission.
Restrictions: Scholarships are not available to students entering graduate studies in a qualifying year. Students who, after receiving an NSERC scholarship, change their field of study and research to a field that is not supported by NSERC, cease to be eligible and their awards will be cancelled.
Geographic Focus: All States, Canada
Date(s) Application is Due: Oct 15
Amount of Grant: 17,300-35,000 CAD
Contact: PGS Program Officer; (613) 995-5521; fax (613) 996-2589; schol@nserc.ca
Internet: http://www.nserc-crsng.gc.ca/Students-Etudiants/PG-CS/BellandPostgrad-BelletSuperieures_eng.asp
Sponsor: Natural Sciences and Engineering Research Council of Canada
350 Albert Street, 14th Floor
Ottawa, ON K1A 1H5 Canada

Canadian Patient Safety Institute (CPSI) Research Grants 1421
The Canadian Patient Safety Institute (CPSI) announces its annual Research Competition to continue its strategic mandate of increasing the scope and scale of research and evaluation activities in patient safety. The primary goal for CPSI Research Competition is to develop knowledge about patient safety that can be helpful in a variety of settings and circumstances in organizations across Canada. It is the applicant team's responsibility to identify and secure matching funding from other contributing organization(s). Applicant teams may obtain matching funding from universities, foundations, voluntary health charities, provider associations, provincial government departments, regional health authorities, hospitals, research centers or the private sector.
Requirements: A 1:1 funding ratio is the acceptable minimum required for any project over $20,000.
Geographic Focus: All States, Canada
Date(s) Application is Due: Dec 2
Amount of Grant: 120,000 USD
Contact: Laurel Taylor; (866) 421-6933 or (780) 409-8090; ltaylor@cpsi-icsp.ca
Internet: http://www.patientsafetyinstitute.ca/news/2008researchcompetition.html
Sponsor: Canadian Patient Safety Institute
10235-101 Street
Edmonton, AB T5J 3G1 Canada

Canadian Studies Senior Fellowships 1422
The Senior Fellowship (awarded every other year) provides senior scholars with an opportunity to complete and publish a major study which will significantly benefit the development of Canadian Studies in the United States. A limited number of fellowships are awarded to academics who have a lengthy track record in teaching, researching and publishing on Canada, Canada/United States or Canada/North America; and on the social, cultural, political and economic issues that impact on these relationships. Preference in such instances is to fund a book project after a publisher has indicated an interest.
Requirements: This program is intended for full-time tenured faculty members at accredited four-year U.S. colleges and universities who are fully involved in Canadian Studies. These Canadianists should be in the process of completing research for a book or major monograph. The study must be on a subject of widespread interest to the Canadian Studies community in the United States as well as in Canada. This fellowship is awarded only once to any one recipient and is in recognition of an academic career dedicated to the promotion of Canadian Studies. It is expected that a recipient will be granted a leave of absence or a sabbatical during the award period.
Restrictions: This award is not available two consecutive years to the same recipient. An individual may not receive two individual category Canadian Studies grants during the same grant period.
Geographic Focus: All States
Date(s) Application is Due: Jun 16
Amount of Grant: 3,000-18,000 USD
Contact: Dr. Daniel Abele, Academic Relations Office; (202) 682-7717; fax (202) 682-7791; daniel.abele@dfait-maeci.gc.ca
Internet: http://www.dfait-maeci.gc.ca/can-am/washington/studies/grantguide-en.asp
Sponsor: Canadian Embassy
501 Pennsylvania Avenue NW
Washington, DC 20001

Canon Foundation Grants 1423
The foundation functions to facilitate mutual understanding in all fields of interest and the development of scientific expertise, particularly between Europe and Japan. Types of support include visiting research fellowships and professorships to enable European students to visit Japan and Japanese students to visit Europe. Up to 15 fellowships are awarded each year.
Requirements: Young, highly qualified postgraduate European and Japanese researchers, preferably holding a PhD and not older than 40 years, are eligible.
Geographic Focus: All States, Albania, Andorra, Armenia, Austria, Azerbaijan, Belarus, Belgium, Bosnia & Herzegovina, Bulgaria, Croatia, Cyprus, Czech Republic, Denmark, Estonia, Finland, France, Georgia, Germany, Greece, Hungary, Iceland, Ireland, Italy, Kosovo, Latvia, Liechtenstein, Lithuania, Luxembourg, Macedonia, Malta, Moldova, Monaco, Montenegro, Norway, Poland, Portugal, Romania, Russia, San Marino, Serbia, Slovakia, Slovenia, Spain, Sweden, Switzerland, The Netherlands, Turkey, Ukraine, United Kingdom, Vatican City
Date(s) Application is Due: Sep 15
Amount of Grant: 22,500-27,500 EUR
Contact: Corrie Siahaya-Van Nierop, Director; 31-20-545-8934; fax 31-20-712-8934; foundation@canon-europa.com
Internet: http://www.canonfoundation.org/pages/main.htm
Sponsor: Canon Foundation in Europe
P.O. Box 2262
Amstelveen, 1180 EG Netherlands

CAPAA Mini Grant for Research Related to African Americans in Education 1424
The program awards mini grants to support theoretical and empirical research that examines the current status of, challenges to, and strategies for improving African Americans' access to the University of California system. The College Access Project for African Americans (CAPAA)awards mini grants to support short-term (two years or less) research projects. CAPAA encourages researchers with diverse perspectives to develop ideas and approaches that extend the conventional boundaries of a research question, area, or method. Request for proposals to conduct specific projects are being supported. 1) Survey at the community level and in school districts with students, parents, and teachers designed to elicit perceptions on factors affecting access to and achievement in higher education by African American students. 2) The educational landscape in school districts and institutions of higher education in California, especially the UC and CSU systems, and the extent to which it creates an environment conducive to learning and achievement by African American students. 3) Studies that investigate the role of African American community resources (networks of support, educational enhancement activities/programs, family support) in determining students' choices, references, aspirations, and performance vis-a-vis higher education. Guidelines are available online.

Requirements: Studies must relate to issues of access and equity for African Americans in California's secondary and higher education institutions. Individual as well as collaborative efforts will be considered.
Geographic Focus: All States
Date(s) Application is Due: Nov 12
Amount of Grant: Up to 15,000 USD
Contact: Dr. Ana-Christina Famon, (310) 794-4997; acramon@bunche.ucla.edu
Internet: http://www.bunche.ucla.edu/frames/index.html
Sponsor: UCLA Ralph J. Bunche Center
160 Haines Hall, Box 951545
Los Angeles, CA 90095

Cape Branch Foundation Grants 1425

The foundation awards grants to New Jersey nonprofit organizations in its areas of interest, including education and secondary education, natural resource conservation, and museums. Types of support include general operating support, building construction/renovation, land acquisition, scholarship funds, and research grants. There are no application forms or deadlines. A letter should be submitted outlining purpose and amount of request.
Requirements: New Jersey nonprofit organization are eligible.
Geographic Focus: New Jersey
Amount of Grant: 1,000-329,268 USD
Contact: Dorothy Frank, (609) 987-0300; fax (609) 452-1024
Sponsor: Cape Branch Foundation
P.O. Box 86
Oldwick, NJ 08858

Carl and Eloise Pohlad Family Foundation Grants 1426

The mission of the foundation is to improve the lives of economically disadvantaged children and youth and participate in projects that positively impact the quality of life in the Minneapolis/St.Paul area. The foundation awards grants to Minnesota nonprofits in its areas of interest, including arts and culture, economic development, education, environment, health care, housing, and social services. Types of support include general operating support, continuing support, capital campaigns, building construction/ renovation, endowments, emergency funds, scholarship funds, and research.
Requirements: Minnesota nonprofits are eligible.
Restrictions: Individuals are not eligible. Capital request are considered only for physical plant improvements or significant technology investments. Capital requests for housing construction, endowment, program start-up or expansion or to establish operating reserves are not considered.
Geographic Focus: Minnesota
Amount of Grant: Up to 32,000,000 USD
Contact: Josette Elstad, Grants Manager; (612) 661-3910; fax (612) 661-3715; info@pohladfamilygiving.org
Internet: http://www.pohladfamilyfoundation.org/pff/pff_default.aspx
Sponsor: Carl and Eloise Pohlad Family Foundation
60 South Sixth Street, Suite 3900
Minneapolis, MN 55402

Carl and Lily Pforzheimer Foundation Grants 1427

The foundation awards grants to U.S. nonprofit organizations in its areas of interest, including higher and secondary education, cultural activities and programs, public affairs, and healthcare. Types of support include capital campaigns, fellowships, internship funds, matching grants, professorships, program development, publication, scholarship funds, and seed grants. There are no application forms or deadlines. The board meets in April, June, October, and December to consider requests.
Restrictions: Grants are not awarded to individuals or for building construction/renovation.
Geographic Focus: All States
Amount of Grant: 10,000-150,000 USD
Samples: Bank Street College of Education (New York, NY)—for scholarships and unrestrictd support, $2 million.
Contact: Grants Administrator; (212) 223-6500
Sponsor: Carl and Lily Pforzheimer Foundation
950 Third Avenue, 30th Floor
New York, NY 10022

Carl C. Icahn Foundation Grants 1428

The foundation awards grants to New York and New Jersey nonprofits in the areas of education, arts and culture, health care, child welfare, and Jewish temples and organizations. Types of support include general operating support, annual campaigns, building construction/renovation, and matching funds. There are no application deadlines or forms.
Requirements: New York and New Jersey nonprofits are eligible to apply.
Restrictions: No grants are provided to individuals.
Geographic Focus: New Jersey, New York
Amount of Grant: 500-1,600,000 USD
Samples: Randall's Island Sports Foundation, New York, NY, $1,666,000—to support program/general purposes; Wildlife Conservation Society, Pelham, NY, $10,000—to support program/general purposes; The Ladies' Village Improvement Society, East Hampton, NY, $500—to support program/general purposes.
Contact: Gail Golden-Icahn, Vice Presicent; (212) 702-4300; fax (212) 750-5815
Sponsor: Carl C. Icahn Foundation
767 5th Avenue, 47th Floor
New York, NY 10153-0023

Carl Gellert and Celia Berta Gellert Foundation Grants 1429

The foundation funds religious, charitable, scientific, literary or educational purposes restricted in the nine counties of the greater San Francisco Bay Area (Alameda, Contra Costa, Marin, Napa, San Francisco, San Mateo, Santa Clara, Solano and Sonoma). No grants are made to individuals. Types of support include general operations, annual and capital campaigns, building construction/ renovation, equipment acquisition, debt reduction, program/project development, medical research, publication, and scholarships.
Requirements: California 501(c)3 tax-exempt nonprofit organizations that are not private foundations are eligible.
Restrictions: Grants are not awarded to individuals.
Geographic Focus: California
Date(s) Application is Due: Aug 15
Amount of Grant: 1,000-10,000 USD
Contact: Jack Fitzpatrick, Executive Director; (415) 255-2829
Internet: http://home.earthlink.net/~cgcbg
Sponsor: Carl Gellert and Celia Berta Gellert Foundation
1169 Market Street, Suite 808
San Francisco, CA 94103

Carl M. Freeman Foundation Grants 1430

The Carl M. Freeman Foundation has historically emphasized it's support in the following communities: Montgomery County, Maryland; Sussex County, Delaware; and the Eastern Panhandle of West Virginia . Funding is available for a wide variety of community organizations, having supported everything from arts organizations and hunger centers to educational and health related organizations. To simplify the application process, funding applications are available in five areas of interest: arts/ culture; education/environment; health & human services; housing and; other- this may include anything you feel does not fit in the above categories, for example spaying cats/dogs. Additional guidelines and applications are available at: http://www.freemanfoundation.org/carl/CarlMFreemanFoundation/Grants/GrantGuidelines/ApplyForaGrant/tabid/185/Default.aspx
Requirements: 501(c)3 tax-exempt organizations in Maryland, Delaware and West Virginia are eligible. Nonsectarian religious programs also are eligible.
Restrictions: Grants will not be distributed to: individuals; political associations or candidates; organizations that would disperse the funding to others; organizations that discriminate by race, creed, gender, sexual orientation, age, religion, disability or national origin.
Geographic Focus: Delaware, Maryland, West Virginia
Amount of Grant: 5,000-30,000 USD
Contact: Trish Schechtman; (302) 436-3555; trish@freemanfoundation.org
Internet: http://www.freemanfoundation.org/carl/CarlMFreemanFoundation/Grants/GrantGuidelines/tabid/181/Default.aspx
Sponsor: Carl M. Freeman Foundation
36097 Sand Cove Road
Selbyville, DE 19975

Carlos and Marguerite Mason Trust Grants 1431

The trust awards grants in Georgia with a focus on organ transplants. The trust's interests are patient services—needs relating to the transplantation process for the patient and the immediate family members, as well as donor family members and living donors; donation and transplantation—related education (donor education, professional education, and technical education for transplantation providers; programs to improve consent rates; and programs to promote living organ donations); and research—medical research relating to the area of transplantation. Types of support include capital improvements, start-up grants for new projects, and challenge/matching grants. Grants will be awarded for up to three years. Applicants are encouraged to contact the trustee for an appointment prior to submitting a grant request.
Requirements: Georgia nonprofit organizations are eligible.
Restrictions: Requests for general operating or administrative funds are discouraged. The trust generally does not fund endowment, nor does it fund general goodwill advertising; make grants that would replace existing sources of funding; or pay indirect overhead expenses for projects at colleges, universities, governmental units, or other established organizations.
Geographic Focus: Georgia
Date(s) Application is Due: Jun 1
Amount of Grant: Up to 15,000,000 USD
Contact: Randy Karesh, Grant Manager, Wachovia Bank of Georgia; (404) 332-6677; grantinquiriesga@wachovia.com
Internet: http://www.wachovia.com/corp_inst/charitable_services/0,,4269_3300,00.html
Sponsor: Carlos and Marguerite Mason Trust
3414 Peachtree Road, 5th Floor, MC GA8023
Atlanta, GA 30326

Carnegie Corporation of New York Grants 1432

The foundation provides research, study, and support for projects to improve government at all levels, to increase public understanding of social policy issues, to equalize opportunities for minorities and women, and to increase participation in political and civic life. Also supported are projects that promote electoral reform; education reform from early childhood through higher education; early childhood development; and urban school reform. The foundation will also fund research on the increasing availability and success of after-school and extended service programs for children and teenagers, particularly those in urban areas, that promote high academic achievement. Dissemination of best practices in teacher education will also be emphasized. There is no formal procedure for submitting a proposal. To apply under any of the corporation's grantmaking programs, applicants should submit a full proposal that describes the

project's aims, duration, methods, amount of financial support required, and key personnel. The board meets four times a year, in October, February, April, and June.
Restrictions: Grants are not made for construction or maintenance of facilities or endowments. The Corporation does not generally make grants to individuals except through the Carnegie Scholars Program, that supports the work of select scholars and experts conducting research in the foundation's fields of interest.
Geographic Focus: All States
Amount of Grant: Up to 80,000,000 USD
Contact: Sarina Cipriano, Grants Manager; (212) 371-3200; fax (212) 754-4073
Internet: http://www.carnegie.org/sub/program/grant.html
Sponsor: Carnegie Corporation of New York
437 Madison Avenue
New York, NY 10022

Carnegie Developmental Biology Research Fellowships **1433**
Fellowships are awarded, primarily at the postdoctoral level, for research at the Carnegie Institution Department of Embryology, Baltimore, MD, under the guidance of the department's staff scientists. Topics include the expression of genes through development and growth, chromosome structure and function, and intracellular communication. Facilities include extensive laboratories for biochemistry and molecular biology. Applicants are encouraged to submit materials a year in advance of the proposed starting date. Fellowships are normally awarded for one year with possible renewal for a second year.
Requirements: Applicant must be postdoctoral researcher. Predoctoral candidates must obtain special permission to apply. Application materials should be sent to the director, Carnegie Institution Department of Embryology, 3520 San Martin Drive, Baltimore, MD 21218.
Geographic Focus: All States
Contact: Grants Administrator; (410) 246-3001; fax (410) 243-6311; carnegiecollaborative@ciwemb.edu
Internet: http://www.ciwemb.edu
Sponsor: Carnegie Institution of Washington
1530 P Street NW
Washington, DC 20005

Carnegie Endowment for International Peace Junior Fellowships **1434**
Each year the endowment offers eight to ten one-year fellowships to uniquely qualified graduating college seniors and individuals who have graduated during the past academic year. They are selected from a pool of nominees from close to 400 colleges (a list of participating colleges available online). Fellows work as research assistants to the Endowment's senior associates. Junior Fellows have the opportunity to conduct research for books, co-author journal articles and policy papers, participate in meetings with high-level officials, contribute to congressional testimony and organize briefings attended by scholars, journalists and government officials.
Requirements: The Carnegie Endowment accepts applications only through participating universities via designated nominating officials. Applicants should contact their university for deadlines and a current application.
Restrictions: Students who have started graduate studies are not eligible for nomination. Students need not be U.S. citizens; however, all applicants must be eligible to work in the United States from August 1 through July 31 following graduation. Due to the nominations process and varying on-campus application procedures, the Endowment is not able to send applications directly to students.
Geographic Focus: All States, Canada, Ireland, Spain, United Kingdom
Date(s) Application is Due: Jan 15
Amount of Grant: 35,000 USD
Contact: Fellowship Administrator; (202) 483-7600; fax (202) 483-1840; jrfellowinfo@CarnegieEndowment.org
Internet: http://www.carnegieendowment.org/about/index.cfm?fa=jrFellows
Sponsor: Carnegie Endowment for International Peace
1779 Massachusetts Avenue NW
Washington, DC 20036-2103

Carnegie Plant Biology Fellowships **1435**
Fellowships primarily at the postdoctoral level are available for research at the Carnegie Institution Department of Plant Biology in Stanford, CA, under the guidance of the department's staff scientists. Topics range from biochemical and molecular experiments to broad-ranging ecological research. Facilities include greenhouses, controlled-growth chambers, biochemical and molecular biology laboratories, and varied specialized apparatus. The period of appointment starts July 1. Fellowships are normally awarded for one year and may be renewed for up to two additional years.
Requirements: The successful applicant must have completed, or nearly completed, requirements for the PhD prior to assuming the fellowship. Applications should include educational record, professional positions held, and a list of papers published and submitted. Applicants should write a brief essay describing their interests and what they hope to achieve during the tenure of the fellowship. Applicants should arrange for letters of recommendation from three professional scientists. Materials should be sent to the director, Carnegie Institution Department of Plant Biology, 260 Panama Street, Stanford, CA 94305-4101.
Geographic Focus: All States
Contact: Tina McDowell, Publications Office; (202) 939-1120; fax (202) 387-8092; tmcdowell@pst.ciw.edu
Internet: http://www-ciwdpb.stanford.edu
Sponsor: Carnegie Institution of Washington
1530 P Street NW
Washington, DC 20005

Carnegie Postdoctoral Research Fellowships **1436**
Primarily at the postdoctoral level, fellowships support research at the Geophysical Laboratory and the Department of Terrestrial Magnetism, both in Washington, DC, under the guidance of staff scientists. The Geophysical Laboratory is a private research and educational institution emphasizing high pressure/high temperature research, metamorphic, igneous, and experimental petrology, stable isotope geochemistry, crystallography, mineral physics, theoretical and experimental condensed matter physics, materials research, organic geochemistry, and biogeochemistry. Fellowships are normally awarded for one year and may be renewed for one and occasionally two additional years.
Requirements: The applicant must have completed, or nearly completed, requirements for the PhD prior to assuming the fellowship. Applications should include a curriculum vita, description of thesis research, a list of publications, transcript/grade report, a two- to three-page research proposal for the fellowship period, and three letters of recommendation.
Geographic Focus: All States
Date(s) Application is Due: Dec 31
Contact: Dr. Wesley Huntress Jr.; (202) 478-8900; w.huntress@gl.ciw.edu
Internet: http://www.gl.ciw.edu/employment/postdoc1.php
Sponsor: Carnegie Institution of Washington
5251 Broad Branch Road NW
Washington, DC 20015

Carnegie Postdoctoral Research Fellowships in Astronomy **1437**
Applications are invited for one or more fellowships at the Observatories of the Carnegie Institution for research in theoretical astronomy and/or instrumentation to begin annually in September. Fellows will have access to the observatories' research facilities in Pasadena and on Cerro Las Campanas, Chile. Fellows will be expected to devote a major portion of their time to astronomical research. Fellowships are awarded for one year and may be renewed for up to two additional years. Foreign applicants will be sponsored under the Carnegie Institution's visitor exchange program.
Requirements: The successful applicant must have completed, or nearly completed, requirements for the PhD prior to assuming the fellowship. Persons who have received a PhD degree within the past three years are particularly urged to apply. Applications should include educational record, professional positions held, and a list of papers published and submitted. Applicants also should write a brief essay describing recent astronomical research and indicate what they would hope to accomplish during tenure of the fellowship. Applicants should arrange for separate transmittal of letters of recommendation from three professional scientists. Materials should be sent to The Chair, OCIW Fellowship Committee, The Observatories of the Carnegie Institution, 813 Santa Barbara Street, Pasadena, CA 91101.
Geographic Focus: All States
Date(s) Application is Due: Dec 1
Contact: Dr. John S. Mulchaey, cfellow@ociw.edu
Internet: http://www.ociw.edu/ociw/fellow
Sponsor: Carnegie Institution of Washington
1530 P Street NW
Washington, DC 20005

Carole Fielding Film/Video Production or Research Grants **1438**
Founded in 1947 as the University Film Producers Association, the UFVA has developed into an organization of almost 800 professionals and institutions involved in the production and study of film, video, and other media arts. Few individuals have been as dedicated to UFVA and its members as Carole Fielding. She attended every Association conference for more than twenty-five years, contributed to the Association's activities and made a legion of friends in the organization. Her work to advance the Association and its mission increased further when her husband, Ray, became president of the organization in 1967, and she accompanied him on his trips to Europe when he served as Vice President of the International Congress of Schools of Cinema and Television. Her enthusiastic work for UFVA and affection for its members made her one of the organization's most popular people and she is remembered with fondness and appreciation. The UFVA's annual Carole Fielding Scholarship was created to honor her and has continued to fund outstanding student work for over 20 years.
Restrictions: An applicant must be a student (undergraduate/graduate) at the time the application is made. A faculty member who is an active member of the University Film & Video Association or staff at a UFVA member institution must sponsor the applicant
Geographic Focus: Illinois
Date(s) Application is Due: Dec 15
Contact: Peter J. Bukalski, (866) 647-8382; ufvahome@aol.com
Internet: http://www.ufva.org/content.php?type_id=7&article_id=28
Sponsor: University Film and Video Association
Box 1777
Edwardsville, IL 62026

Carpenter Foundation Grants **1439**
The foundation's primary areas of interest include the arts, education, public interest, and human services. The foundation is deeply concerned with the well-being of children and families and their relationship to their neighborhoods and communities. Also of concern is the health of the web of agencies and organizations which serve them. Grants are awarded for general operating support, program development, capital campaigns, equipment acquisition, scholarship funds, seed money, matching funds, and technical support. Deadlines are generally about six weeks before the quarterly board meetings, held in January, March, June, and September.
Requirements: Tax exempt agencies in the Jackson and Josephine Counties of Oregon may submit proposals.

Restrictions: Grants are not made to individuals. The foundation rarely makes grants for historical applications, hospital construction or equipment, group or individual trips, or activities for religious purposes.
Geographic Focus: Oregon
Amount of Grant: 250-25,000 USD
Contact: Polly Williams; (541) 772-5732; carpfdn@internetcds.com
Internet: http://www.carpenter-foundation.org
Sponsor: Carpenter Foundation
711 E Main Street, Suite 10
Medford, OR 97504

Carrier Corporation Contributions Grants **1440**
Carrier donates approximately $2 million around the world to registered nonprofit organizations. In the United States, Carrier funds only qualified 501(c)3 organizations that meet its eligibility criteria and operate in locations where the company has a significant employee base. Carrier believes in helping people in the communities where they live, work and do business. To better serve those communities and to better align its corporate contributions with mission and values, it focuses giving on the following these areas: environment and sustainability; civic & community; education; arts & culture; health & human services. All U.S. non-profits are required to complete an online grant application. Applications are accepted from March 1 through June 1 of each year, and are reviewed for funding to be paid the following year. Applicants will receive notification in the first quarter of the calendar year in which funding will occur.
Requirements: Carrier funds only qualified 501(c)3 organizations that meet eligibility criteria and operate in locations where it has a significant employee base.
Restrictions: Carrier will not fund: individuals; religious organizations; alumni groups, sororities or fraternities; booster clubs; political groups; any organization determined by Carrier to have a conflict of interest; any organization whose practices are inconsistent with the company's Code of Ethics
Geographic Focus: Alabama, Arizona, Connecticut, Georgia, Illinois, Indiana, Michigan, Nevada, New York, North Carolina, South Carolina, Tennessee, Texas
Date(s) Application is Due: Jun 1
Contact: Rajan Goel, Vice President; (860) 674-3420; fax (860) 622-0488
Internet: http://www.corp.carrier.com/vgn-ext-templating/v/index.jsp?vgnextoid=6afa80757d7e7010VgnVCM100000cb890b80RCRD
Sponsor: Carrier Corporation
One Carrier Place
Farmington, CT 06034-4015

Castle Rock Foundation Grants **1441**
The foundation's grants are awarded to promote a better understanding of the free-enterprise system; preserve principles on which democracy was founded, ensure a limited role for government, and protect individual rights; encourage personal leadership; and uphold traditional American values. Types of support include general operating grants and special project grants.
Requirements: U.S. 501(c)3 organizations are eligible.
Restrictions: Grants requests from human service agencies, museums, organizations primarily supported by tax-derived funding, endowments, scientific or medical research, publications or media projects, churches, debt retirement, special events, or individuals will be denied.
Geographic Focus: All States
Date(s) Application is Due: Mar 15
Amount of Grant: 25,000-75,000 USD
Contact: Sally W. Rippey; (303) 388-1683; generalinfo@castlerockfdn.org
Internet: http://www.castlerockfoundation.org
Sponsor: Castle Rock Foundation
4100 E Mississippi Avenue, Suite 1850
Denver, CO 80246

Catherine Holmes Wilkins Foundation Grants **1442**
The foundation awards grants in Washington's Puget Sound region to nonprofit organizations in its areas of interest. Funding priorities include medical research and education (medical and academic centers conducting research and training in areas such as cancer, heart disease, and mental illness); physically handicapped and mentally ill (community nonprofit agencies providing direct social services to people with physical disabilities or mental illness); and services for the needy (community-based programs providing immediate support to the needy, with particular emphasis on services for abused women and children). Preference will be given to project support, rather than ongoing operating expenses. Proposals are accepted throughout the year. Grants are awarded quarterly.
Requirements: 501(c)3 nonprofits that operate within or significantly affect the residents of the greater Seattle region (Tacoma to Everett, Seattle, and the Eastside) are eligible.
Restrictions: Grants are not made for multi-year projects, debt retirement, operational deficits, or to individuals or for scholarships.
Geographic Focus: Washington
Date(s) Application is Due: Mar 8; Oct 8
Amount of Grant: 3,000-10,000 USD
Contact: Nancy Atkinson, Vice President; (206) 781-3472; OgleFounds@aol.com
Internet: http://fdncenter.org/grantmaker/wilkins
Sponsor: Catherine Holmes Wilkins Foundation
P.O. Box 24565, WA1-501-33-23
Seattle, WA 98124

Catholic University Academic Scholarships, Grants, and Awards **1443**
The university offers a wide range of scholarships that recognize and reward outstanding academic performance in high school, exceptional leadership, and service school, church, and the community. Awards include academic scholarships and grants—Archdiocesan/Trustees Scholarships, University Merit Scholarships, and CUA Incentive Grants; high school nominated scholarships—Award of Excellence, and Leadership/Service Scholarship; alumni grant—CU graduates may recommend new first-time, full-time freshmen for the grant, which is renewable for four years; parish scholarship—parishioners of Catholic churches who are admitted for full-time undergraduate study at CUA; scholarships with special eligibility requirements; and school and departmental scholarships. Deadlines and scholarship amounts vary. Guidelines are available online.
Geographic Focus: All States
Amount of Grant: 1,000-10,000 USD
Contact: Office of University Admissions; (800) 673-2772 or (202) 319-5305; fax (202) 319-6533; cua-admissions@cua.edu
Internet: http://admissions.cua.edu/undergrad/finaid/
Sponsor: Catholic University of America
620 Michigan Avenue NE
Washington, DC 20064

Catholic University Life Cycle Institute Research Fellowship **1444**
The Life Cycle Institute, a multidisciplinary research center for social sciences at the university, offers facilities for visiting research fellows to work for a period of one semester to one year on research projects that develop wider dimensions of topics to which the institute is devoted. The institute is particularly interested in developing global, transnational, and comparative dimensions of current research interests in youth development, gender and education, social movements and civil society, volunteerism and NGO careers including religious vocations, the demographics of professions, and the social frontiers of the information revolution. The institute currently houses specialists in sociology, anthropology, politics, demography, developmental psychology, and religious studies who are on the CUA faculty or doing project-funded research. Fellows will be provided office space, secretarial and other services (including computer, Internet, and library facilities and on-line databases), and administrative support for funded projects within the university's guidelines. Guidelines are available online.
Requirements: Applicants with PhDs in the social sciences and related disciplines, including history and economics, should send a letter of intent outlining a proposed project.
Geographic Focus: All States
Date(s) Application is Due: May 1
Amount of Grant: 6,000 USD
Contact: Dr. John White; (202) 319-5999; fax (202) 319-6267; white@cua.edu
Internet: http://lifecycle.cua.edu
Sponsor: Catholic University of America
620 Michigan Avenue NE
Washington, DC 20064

CBO Economic Policy Fellowships **1445**
The program provides experts in macroeconomics, health economics, financial economics, public finance, and other specialties with a unique opportunity to address complex budgetary and economic policy issues. While in residence, fellows will conduct research based on a submitted proposal, use the agency's data and facilities, and work daily with colleagues at CBO to contribute to analyses and publications. Fellows will also be able to participate in professional development opportunities and will be in an excellent position to draw on the resources of experts at CBO, at other federal agencies, and within the extended community of policy analysts in Washington, DC. Fellowship appointments typically last one year but may vary depending on the project and individual circumstances. Appointment terms are flexible and can be full-time or part-time, and fellows may be employees of CBO or may maintain employment with their permanent employer depending on the availability of such an agreement. Salaries will be commensurate with qualifications and experience. Applicants must submit application materials via email. Guidelines are available online.
Requirements: In addition to expertise in macroeconomics, health economics, financial economics, or public finance, economists or budget policy analysts should have a PhD or equivalent schooling; considerable expertise in their area of proposed research; a commitment to analyze interesting real-world issues in economics and public policy; and a recognized record of research and publications.
Geographic Focus: All States
Date(s) Application is Due: Mar 1
Contact: Nancy Fahey, (202) 226-2628; jobs@cbo.gov
Internet: http://www.cbo.gov/employment/fellowships.shtml
Sponsor: Congressional Budget Office
Second and D Streets SW, Ford House Office Building, 4th Floor
Washington, DC 20515-6925

CCA Alberta Creative Development Initiative: Grants to Individuals **1446**
The Alberta Creative Development Initiative (ACDI) is a partnership between the Alberta Foundation for the Arts and the Canada Council for the Arts, in collaboration with the Calgary Arts Development Authority and the Edmonton Arts Council. The Grant is designed support projects by Alberta professional artists, arts administrators and collectives. The objectives are to: support the development of professional artists and arts organizations in Alberta; foster a dynamic, vibrant arts community in Alberta; and enable the success of Alberta artists and arts organizations locally, provincially, nationally and internationally.

Requirements: Eligible applicants must be: Alberta-based professional artists, arts administrators or collectives working in all artistic disciplines and artistic practices currently supported by the Council or the Alberta Foundation for the Arts. Individuals must be a Canadian citizen or permanent resident, have maintained a regular residence in Alberta in the past year and have been defined as an Alberta resident on their last tax return. Eligible applicants can request funding for projects that will advance their artistic practice and/or professional development. Eligible projects may be research, creation, production/and or dissemination projects or professional development projects including residencies. Aboriginal artists and artists of diverse cultural and regional communities of Canada are eligible.
Restrictions: Full-time undergraduate students at a school, college or university are ineligible. Graduate or post-graduate students are eligible only by meeting the definition of a professional artist or arts administrator, and the proposed project cannot be related to the program of study and cannot be under the direct supervision of a university professor.
Geographic Focus: Canada
Date(s) Application is Due: Nov 1
Amount of Grant: 15,000 CAD
Contact: Marion Vitrac; (800) 263-5588 or (613) 566-4414, ext. 4414;
Internet: http://www.canadacouncil.ca/grants/acdi_idca/tx128351161983158103.htm
Sponsor: Canada Council for the Arts
350 Albert Street, P.O. Box 1047
Ottawa, ON K1P 5V8 Canada

CCA Diefenbaker Award to update 1447
This annual award enables a distinguished German scholar to spend up to 12 months in Canada, which may include brief periods in the United States. While research must be the primary activity during the tenure of the award, the recipient will also be encouraged to participate in the activities of the host institution and to interact more extensively with the research communities in Canada and the United States by visiting other institutions. Candidates may not apply for this award, but instead must be nominated by a host university department or research institute in Canada.
Requirements: Candidates must be German citizens and must have a sound working knowledge of one of Canada's two official languages. German scholars who have demonstrated outstanding ability, especially through a substantial publication record over several years, are eligible to apply.
Geographic Focus: All States, Germany
Date(s) Application is Due: Dec 1
Contact: Nathalie Lauzon, Administrative Assistant; (800) 263-5588 or (613) 566-4414, ext. 4083; fax (613) 566-4430; endowments.prizes@canadacouncil.ca
Internet: http://www.canadacouncil.ca/prizes/john_g_diefenbaker/hn127223038493593750.htm
Sponsor: Canada Council for the Arts
350 Albert Street, P.O. Box 1047
Ottawa, ON K1P 5V8 Canada

CCCF Dora Maclellan Brown Seminary Scholarships 1448
Scholarships are awarded to Christian men and women from the greater Chattanooga area who are theologically and biblically conservative, and are seeking a Masters of Divinity or Masters of Theology degree at a theological Seminary approved by the foundation. The scholarship committee's preference is for candidates for the pulpit ministry, although all applications will be considered. Scholarships are awarded based on merit and need. They are granted usually for three or four years depending on the duration of the degree program. Guidelines and application are available online.
Geographic Focus: Tennessee
Date(s) Application is Due: Feb 15
Contact: Henry Henegar; (423) 266-5257, ext. 101; henryhenegar@cccfdn.org
Internet: http://www.cccfdn.org/page18325.cfm
Sponsor: Chattanooga Christian Community Foundation
736 Market Street, Suite 700
Chattanooga, TN 37402

CCFF Chairman's Distinguished Life Sciences Award 1449
The $10,000 Chairmen's Distinguished Life Sciences Award will be presented to a scientist or researcher who is making or has recently made a significant and positive contribution toward the development of a cutting edge innovation in the field of life sciences. The Chairmen's Award will provide incentive for continuing the research. Individuals will compete and their accomplishments will be judged on how they exemplify excellence in life sciences. Winners must attend the awards ceremony in April at Washington, D.C., in order to receive the award.
Requirements: All Nominees must be living United States citizens. The Nomination must demonstrate a program of ongoing work with a specific outcome suggesting that an important achievement will result. The Nomination's proposed achievement must benefit human health in a unique and effective manner, demonstrating originality from other work in the same or related field(s). The proposed achievement must correlate to a tangible outcome to advance human health.
Restrictions: Nominees must be permitted, by their employer or any other relevant authority, to accept a monetary award bestowed by the Foundation, a Federal government agency. Anyone who cannot attend the award ceremony will automatically be disqualified.
Geographic Focus: All States
Date(s) Application is Due: Feb 26
Contact: Judith M. Shellenberger; (315) 258-0090; judithmscolumbus@cs.com
Internet: http://www.columbusfdn.org/lifesciences/
Sponsor: Christopher Columbus Fellowship Foundation
110 Genesee Street, Suite 390
Auburn, NY 13021

CCFF Distinguished Agriscience Scientist Award 1450
The Christopher Columbus Fellowship Foundation, a Federal government agency, and the American Farm Bureau Federation, have formed a partnership to present four monetary awards and research funding to three groups representing new innovative approaches in the field of agriscience around the country. One $10,000 Distinguished Agriscience Scientist Award will be presented to a scientist who is making or has recently made significant and positive contributions in the field of agriscience.
Requirements: All Candidates must be living United States citizens. The Nomination must demonstrate a program of ongoing work with a specific outcome suggesting that an important achievement will result. The Nomination's proposed achievement must benefit the field of agriscience in a unique and effective manner, demonstrating originality from other work in the same or related field(s). The proposed achievement must correlate to a tangible outcome to advance the field of agriscience.
Geographic Focus: All States
Date(s) Application is Due: Feb 12
Amount of Grant: 10,000 USD
Contact: Judith M. Shellenberger, Executive Director; (315) 258-0090; fax (315) 258-0093; judithmscolumbus@cs.com
Internet: http://www.columbusfdn.org/agriscience/
Sponsor: Christopher Columbus Fellowship Foundation
110 Genesee Street, Suite 390
Auburn, NY 13021

CCH Documentary Project Research and Development Grants 1451
The California Documentary Project (CDP) is a competitive grant program of the California Council for the Humanities (CCH). CDP supports film, radio, and new media projects in which the humanities are used to provide context, depth, and perspective. The humanities spring from a fundamental interest in understanding the values and practices that inform our lives; the need to reflect on the past and the present to make critical choices; and the desire to be moved, delighted, and make sense of the world in which we live. Humanities disciplines include, but are not limited to: history, literature, philosophy, folklore, ethnic studies, religious studies, ethics, jurisprudence, and qualitative approaches in the social sciences. They emphasize analysis, interpretation, and the exchange of ideas. CDP Research and Development grants are designed to strengthen the humanities content and approach of documentary media productions in their earliest stages. Projects must actively involve at least three humanities advisors to help frame and contextualize subject matter throughout the research and development phase. Eligible projects may apply for funding up to $7,000.
Requirements: Eligible applicant organizations/project directors must: have tax-exempt organizational status or a tax-exempt organization as fiscal sponsor; be in good standing with CCH (e.g., without unfulfilled reporting requirements), if a previous grantee; and not have an open grant with CCH, unless a CCH-approved sponsoring organization (see FAQ for more information). Film, radio, and new media projects must meet all of the following criteria: be currently in the research and development stage; document the California experience and explore issues of significance to Californians; approach the subject matter from a humanities perspective—for example, making use of existing or new scholarship and research on the topic, incorporating a variety of perspectives, seeking to foster critical reflection and thoughtful analysis on the part of the audience; actively involve at least three humanities advisors throughout the research and development phase; and be conducted by experienced film, radio, or new media producers and directors. A previously completed work sample must be submitted.
Geographic Focus: California
Date(s) Application is Due: Oct 3
Amount of Grant: Up to 7,000 USD
Contact: Vanessa Whang; (415) 391-1474; fax (415) 391-1312; vwhang@calhum.org
Internet: http://www.calhum.org/guidelines/cdp_2011_RandD_guidelines.htm
Sponsor: California Council for the Humanities
312 Sutter Street, Suite 601
San Francisco, CA 94108-4371

CCIS Visiting Research Fellowships 1452
CCIS will offer a limited number of Visiting Research Fellowships at both the predoctoral and postdoctoral levels for the upcoming academic year. These awards are to support advanced research and writing on any aspect of international migration and refugee flows, in any of the social sciences, history, law, and comparative literature. Visiting Research Fellows are provided with shared office space and a computer, as well as full access to all UCSD academic, institutional, and recreational resources. Stipends are $2,250 per month for predoctoral fellows and $3,000 to $4,000 per month for recent postdoctoral fellows (Ph.D. received within the last 6 years), depending on seniority. Stipends for more senior scholars are negotiable.
Restrictions: Due to funding constraints, CCIS is able to award fellowships only to scholars who have a current or former affiliation with a University of California campus (as a graduate student, faculty member, or researcher).
Geographic Focus: All States
Date(s) Application is Due: Jan 15
Amount of Grant: 27,000-48,000 USD
Contact: John D. Skrentny, Director; (858) 822-4447 or (858) 534-0484; fax (858) 822-4432 or (858) 534-47; jskrentn@ucsd.edu or ccis@ucsd.edu
Internet: http://www.ccis-ucsd.org/Programs/fellowships.htm
Sponsor: Center for Comparative Immigration Studies
University of California-San Diego
La Jolla, CA 92093-0548

CCWH/Berkshire Conference of Women Historians Graduate Student Fellowship 1453

The fellowship is awarded to a woman graduate student completing a dissertation in a history department. History graduate students who apply for the CCWH/Berkshire Award will automatically be considered for the CCWH/Ida B. Wells Award. No additional application is necessary. Request applications from Dr. Ann Le Bar, Department of History, Eastern Washington University, Patterson Hall 200, Cheney, WA 99004; email: alebar@mail.ewu.edu. Information and application form also are available online.
Requirements: The applicant must be a woman graduate student historian in a history department in a U.S. institution; must have passed to ABD status by the time of application; may specialize in any field of history; may hold this award and others simultaneously; and need not attend the award ceremony to receive the award.
Geographic Focus: All States
Date(s) Application is Due: Sep 1
Amount of Grant: 500 USD
Contact: Ann Le Bar; alebar@mail.ewu.edu
Internet: http://theccwh.org/awards.htm
Sponsor: Coordinating Council for Women in History
P.O. Box 9715
Portland, ME 04104-5015

CCWH Catherine Prelinger Award 1454

The award is made to a scholar who has not followed a traditional academic path of uninterrupted and completed secondary, undergraduate, and graduate degrees leading into a tenured faculty position. Although the recipient's degrees do not have to be in history, the recipient's work should clearly be historical in nature. In accordance with the general goals of CCWH, the award is intended to recognize or to enhance the ability of the recipient to contribute significantly to women in history, whether in the profession in the present or in the study of women in the past. Candidates should submit the application package to Carol Gold, Department of History, University of Alaska Fairbanks, P.O. Box 756460, Fairbanks, AK 99775-6460. Telephone: 907-474-6509. Applications may be requested from the same address or on email at ffcg@uaf.edu. Application and guidelines also are available online. Faxed or emailed applications will not be considered.
Requirements: Application is open to scholars with a PhD or ABD degree.
Geographic Focus: All States
Date(s) Application is Due: Mar 13
Amount of Grant: 20,000 USD
Contact: Carol Gold, Chair; (907) 474-6509; ffcg@uaf.edu
Internet: http://theccwh.org/preapp.htm
Sponsor: Coordinating Council for Women in History
P.O. Box 9715
Portland, ME 04104-5015

CCWH Ida B. Wells Graduate Student Fellowship 1455

The fellowship is awarded to an ABD woman graduate student working on a historical dissertation, not necessarily in a history department. History graduate students who apply for the award will automatically be considered for the CCWH/Ida B. Wells award. No additional application is necessary. Request applications from Dr. Ann Le Bar, Department of History, Eastern Washington University, Patterson Hall 200, Cheney, WA 99004; email: alebar@mail.ewu.edu. Application and information also are available online.
Requirements: The applicant must be a woman graduate student in a U.S. institution; must have passed to ABD status by the time of application; may specialize in any field, but must be working on an historical project; may hold this award and others simultaneously; and need not attend the award ceremony to receive the award.
Geographic Focus: All States
Date(s) Application is Due: Sep 1
Amount of Grant: 500 USD
Contact: Ann Le Bar; alebar@mail.ewu.edu
Internet: http://theccwh.org
Sponsor: Coordinating Council for Women in History
P.O. Box 9715
Portland, ME 04104-5015

CDC Cooperative Agreement for Continuing Enhanced National Surveillance for Prion Diseases in the United States 1456

The purpose of the funding is to continue an active surveillance program similar to that conducted by the National Prion Disease Pathology Surveillance Center since 1997 to monitor the occurrence of potentially emerging human TSEs in the United States. This program addresses the Healthy People 2010 focus area(s) of Immunizations and Infectious Diseases. Eligible applicants should have experience in conducting prion disease surveillance and must be capable of fulfilling the ongoing needs for enhancing such surveillance at the national level.
Requirements: Eligible applicants include: public nonprofit organizations; private nonprofit organizations; small, minority, women-owned businesses; universities; colleges; research institutions; hospitals; community-based organizations; faith-based organizations; federally recognized Indian tribal governments; Indian tribes; Indian tribal organizations; state and local governments or their Bona Fide Agents (this includes the District of Columbia, the Commonwealth of Puerto Rico, the Virgin Islands, the Commonwealth of the Northern Marianna Islands, American Samoa, Guam, the Federated States of Micronesia, the Republic of the Marshall Islands, and the Republic of Palau); and political subdivisions of States (in consultation with States).
Restrictions: Recipients may: not use funds for research; not use funds for clinical care; only expend funds for reasonable program purposes, including personnel, travel, supplies, and services, such as contractual. Awardees may not generally use HHS/CDC/ATSDR funding for the purchase of furniture or equipment (any such proposed spending must be identified in the budget). The direct and primary recipient in a cooperative agreement program must perform a substantial role in carrying out project objectives and not merely serve as a conduit for an award to another party or provider who is ineligible.
Geographic Focus: All States
Date(s) Application is Due: Jun 30
Amount of Grant: 100,000,000 USD
Contact: Mattie Jackson, Grants Management; (770) 488-2696; MIJ3@cdc.gov
Internet: http://www.cdc.gov/od/pgo/funding/FOAs.htm
Sponsor: Centers for Disease Control and Prevention
1600 Clifton Road
Atlanta, GA 30333

CDC Cooperative Agreement for the Development, Operation, and Evaluation of an Entertainment Education Program 1457

The purpose of this program is raise awareness and behavioral change concerning public health issues through the use of the television media as means of accessing the television viewing public by providing: accurate public health information and public health issues, as well as accurate depictions of healthy living at all stages of life, to entertainment industry leadership for possible inclusion in television story lines; assistance with public health outreach, including providing additional information concerning public health topics depicted on television story lines, and assistance with public service announcements and informational short videos to be aired in conjunction with drama presentations; and evaluation of public health television story lines and the effect on the viewing public.
Requirements: Eligible applicants include: public nonprofit organizations; private nonprofit organizations; universities; colleges; research institutions; federally recognized Indian tribal governments; Indian tribes; Indian tribal organizations; state and local governments or their Bona Fide Agents (this includes the District of Columbia, the Commonwealth of Puerto Rico, the Virgin Islands, the Commonwealth of the Northern Marianna Islands, American Samoa, Guam, the Federated States of Micronesia, the Republic of the Marshall Islands, and the Republic of Palau); and political subdivisions of States (in consultation with States).
Geographic Focus: All States
Date(s) Application is Due: Jun 7
Contact: Sharon Robertson, Grants Officer; (770) 488-2748; SRobertson1@cdc.gov
Internet: http://www.cdc.gov/od/pgo/funding/FOAs.htm
Sponsor: Centers for Disease Control and Prevention
1600 Clifton Road
Atlanta, GA 30333

CDC Evaluation of the Use of Rapid Testing For Influenza in Outpatient Medical Settings 1458

The purpose of this project is to evaluate how rapid tests for influenza are being implemented and used in clinical practice in outpatient medical settings such as community clinics, solo and group practice physician offices, and hospital emergency rooms across the United States. This evaluation will include a determination of the scope of rapid influenza test use, the types of tests in use and how they are selected, the personnel performing testing, the extent to which good laboratory practices and testing guidelines are being followed, how results are reported and interpreted, how results are used for patient care and antiviral and antibiotic prescribing practices, and the presence of linkages between these outpatient settings and the public health system. Additionally, this project may identify potential opportunities to provide guidance to assist sites in making decisions on the appropriate use of these tests and ways to enhance the connectivity with the public health system. Connectivity with public health is especially important in light of the possibility of an influenza pandemic. This project will identify and evaluate practices used in outpatient settings related to processes such as specimen collection, testing, reporting and referral for influenza.
Requirements: Eligible applicants include: public nonprofit organizations; private nonprofit organizations; small, minority, women-owned businesses; universities; colleges; research institutions; hospitals; community-based organizations; faith-based organizations; federally recognized Indian tribal governments; Indian tribes; Indian tribal organizations; state and local governments or their Bona Fide Agents (this includes the District of Columbia, the Commonwealth of Puerto Rico, the Virgin Islands, the Commonwealth of the Northern Marianna Islands, American Samoa, Guam, the Federated States of Micronesia, the Republic of the Marshall Islands, and the Republic of Palau); and political subdivisions of States (in consultation with States).
Restrictions: Recipients may: not use funds for research; not use funds for clinical care; only expend funds for reasonable program purposes, including personnel, travel, supplies, and services, such as contractual. Awardees may not generally use HHS/CDC/ATSDR funding for the purchase of furniture or equipment (any such proposed spending must be identified in the budget). The direct and primary recipient in a cooperative agreement program must perform a substantial role in carrying out project objectives and not merely serve as a conduit for an award to another party or provider who is ineligible. Reimbursement of pre-award costs is not allowed. Reimbursement of construction costs is not allowed.
Geographic Focus: All States
Date(s) Application is Due: Aug 6
Amount of Grant: 200,000 USD
Contact: Yolanda Sledge, Grants Management; (770) 488-2787; yiso@cdc.gov
Internet: http://www.cdc.gov/od/pgo/funding/FOAs.htm
Sponsor: Centers for Disease Control and Prevention
1600 Clifton Road
Atlanta, GA 30333

CDC Evidence-Based Laboratory Medicine: Quality/Performance Measure Evaluation 1459

The purpose of the program is to evaluate clinical laboratory practice by identifying evidence-based laboratory medicine quality/performance measures associated with the pre- and post-analytic stages of the total testing process, and to identify and address gaps and opportunities for improvement consistent with national health care priorities to improve public health. The primary objectives are to: provide an evidence base which systematically identifies and defines important gaps in laboratory medicine quality related to individuals or populations and/or patient safety health outcomes with demonstrated impacts (i.e., clinical and/or economic); identify, develop, and define laboratory medicine performance/quality measures that can be broadly implemented to evaluate performance associated with patient outcomes; and identify interventions effective in improving performance.
Requirements: Eligible applicants include: public nonprofit organizations; private nonprofit organizations; small, minority, women-owned businesses; universities; colleges; research institutions; hospitals; community-based organizations; faith-based organizations; federally recognized Indian tribal governments; Indian tribes; Indian tribal organizations; state and local governments or their Bona Fide Agents (this includes the District of Columbia, the Commonwealth of Puerto Rico, the Virgin Islands, the Commonwealth of the Northern Marianna Islands, American Samoa, Guam, the Federated States of Micronesia, the Republic of the Marshall Islands, and the Republic of Palau); and political subdivisions of States (in consultation with States).
Restrictions: Recipients may: not use funds for research; not use funds for clinical care; only expend funds for reasonable program purposes, including personnel, travel, supplies, and services, such as contractual. Awardees may not generally use HHS/CDC/ATSDR funding for the purchase of furniture or equipment (any such proposed spending must be identified in the budget). The direct and primary recipient in a cooperative agreement program must perform a substantial role in carrying out project objectives and not merely serve as a conduit for an award to another party or provider who is ineligible. Reimbursement of pre-award costs is not allowed.
Geographic Focus: All States
Date(s) Application is Due: Jul 17
Amount of Grant: 100,000-500,000 USD
Contact: Yolanda Sledge, Grants Specialist; (770) 488-2787; yiso@cdc.gov
Internet: http://www.cdc.gov/od/pgo/funding/FOAs.htm
Sponsor: Centers for Disease Control and Prevention
1600 Clifton Road
Atlanta, GA 30333

CDC Experience Applied Epidemiology Fellowships 1460

The goal of the CDC Experience Applied Epidemiology Fellowship is to increase the number of physicians with a population health perspective by providing medical students with an understanding of applied epidemiology, the role of epidemiology in medicine and health, and the role of physicians in the public health system. The fellowship is held at the Centers of Disease Control and Prevention (CDC) headquarters in Atlanta, Georgia. Fellows start in August and work for ten or twelve months. While at CDC, fellows will: gain training and work experience in applied epidemiology and public health; perform epidemiologic analyses and research; participate in public health field experiences; report findings through written and oral presentations at scientific meetings; write scientific manuscripts for publication; participate in monthly seminars and journal clubs; and attend the annual Epidemic Intelligence Service conference. CDC Program areas available to applicants are chronic diseases, emergency preparedness, environmental health, global health, infectious diseases, injury prevention, and reproductive health. Fellows receive a stipend for living expenses. Applicants must apply online. Supporting documents are required and must be postmarked by the deadline date. Applicants should check the CDC Experience website for further details. Annual deadline dates may vary; applicants are encouraged to check the website or contact program staff to verify current deadline dates. The CDC Experience Applied Epidemiology Fellowship is made possible by a public/private partnership supported by a grant to the CDC Foundation from External Medical Affairs, Pfizer Inc., and is considered a valuable experience to those those seeking career advancement in the areas of clinical medicine, clinical epidemiology, health services research, preventive medicine, and public health.
Requirements: Applicants must: be medical students in their 2nd or 3rd year when they apply; be enrolled in a school accredited by the Liaison Committee on Medical Education or the American Osteopathic Association; be a U.S. citizen or permanent resident; and be willing to relocate to Atlanta, Georgia for the duration of the fellowship. Awardees may start the fellowship while in their 3rd or 4th year.
Geographic Focus: All States
Date(s) Application is Due: Dec 2
Contact: Virginia Watson, MA, Program Contact; (404) 498-6151; VWatson1@cdc.gov or cdcexperience@cdcfoundation.org
Internet: http://www.cdc.gov/CDCExperienceFellowship/index.html
Sponsor: Centers for Disease Control and Prevention
1600 Clifton Road
Atlanta, GA 30333

CDC Foundation Atlanta International Health Fellowships 1461

The Atlanta International Health Fellowships (AIHF) were created in 1984 by Drs. Bob Chen and Katy Irwin to provide financial assistance to international participants in CDC's varied public health training courses. The two young physicians enlisted help from other CDC employees and retirees to raise funds and then formed a partnership with Emory University, which provided free tuition to one individual each year to attend Emory's International Course in Applied Epidemiology. Drs. Chen and Irwin also formed a partnership with Villa International to provide free housing to the AIHF recipient. In 1997, a grant from the Tull Charitable Foundation was added to existing funds, and the Atlanta International Health Fellowship became the first endowed fund at the CDC Foundation. Since the first fellowships were awarded in 1991, 25 individuals from over 22 countries have received stipends to cover costs of travel, tuition and lodging as they enrich their public health expertise at CDC. Interested applicants are encouraged to contact the CDC Foundation or Emory University for information on how to apply for the AIHF fellowship.
Geographic Focus: All States, All Countries
Contact: Verla S. Neslund, J.D., Vice President for Programs; (888) 880-4232 or (404) 653-0790; fax (404) 653-0330; vneslund@cdcfoundation.org
Internet: http://www.cdcfoundation.org/what/program/atlanta-international-health-fellowship-endowment
Sponsor: Centers for Disease Control and Prevention (CDC) Foundation
55 Park Place NE, Suite 400
Atlanta, GA 30303

CDC Foundation Tobacco Network Lab Fellowship 1462

According to a Centers for Disease Control and Prevention (CDC) morbidity and mortality report, tobacco use is the world's leading single preventable cause of death. Worldwide tobacco-related deaths currently exceed 5 million a year and are expected to exceed 8 million a year by 2030 if left unchecked. In the United States, tobacco use is the single leading preventable cause of disease, disability, and death. In 2003, the World Health Organization (WHO) created the Tobacco Laboratory Network (TobLabNet) of government, academic, and independent laboratories worldwide to strengthen national and regional capacity for testing and research of the contents and emissions of tobacco products. The Network's stated goals are: to establish global capacity to test tobacco products for regulatory compliance; to research and develop harmonized standards for contents and emissions testing; to share tobacco research and testing standards and results; to inform risk assessment activities related to the use of tobacco products; and to develop harmonized reporting of results so that data can be transformed into meaningful trend information that can be compared across countries and over time. The Tobacco Network Lab Fellowship is a collaboration between CDC and WHO to place a scientist in CDC's Division of Laboratory Sciences to help TobLabNet address testing and research of tobacco products at the global level in accordance with the tobacco product regulation provisions outlined by the WHO Framework Convention on Tobacco Control. A link to brief information about the fellowship is at the grant website under Chronic Disease & Birth Defects Programs. Interested applicants are encouraged to contact the CDC Foundation or the World Health Organization for more information on how to apply for the fellowship.
Geographic Focus: All States, All Countries
Contact: Verla S. Neslund, J.D., Vice President for Programs; (888) 880-4232 or (404) 653-0790; fax (404) 653-0330; vneslund@cdcfoundation.org
Internet: http://www.cdcfoundation.org/what/programs/list#program-518
Sponsor: Centers for Disease Control and Prevention (CDC) Foundation
55 Park Place NE, Suite 400
Atlanta, GA 30303

CDC Grants for Violence-Related Injury Prevention Research 1463

Funded projects should identify effective ways of preventing violence-related injuries and expand the use of new and current intervention methods for preventing these types of injuries. In the areas of suicide and assaultive behavior, projects may focus on understanding the factors that affect this behavior, such as the nature of suicide among gay and lesbian people as compared to the general population or how unequal access to criminal justice, health care, and education is related to violent behavior. Family and intimate violence prevention projects may examine the intervention strategies that are most effective in preventing injuries. CDC encourages projects to study the needs of mothers and children in families where intimate partner violence takes place and to use population-based research to quantify injury and disability among women as a result of partner violence. Officials are also interested in research that defines the cost of violent injuries and the cost effectiveness of prevention or intervention methods. Contact the program office for deadline dates.
Requirements: Public and private nonprofit and for-profit agencies may apply for grants. This includes state and local agencies, hospitals, and universities.
Geographic Focus: All States
Amount of Grant: Up to 300,000 USD
Contact: Paul Smutz; (770) 488-1508
Internet: http://www.cdc.gov/ncipc/res-opps/VIOLENCE_04045.htm
Sponsor: Centers for Disease Control and Prevention
1600 Clifton Road
Atlanta, GA 30333

CDC National Training Information and Exchange Program 1464

The purpose of this program is to make data and information on trauma care in the United States more accessible to a broad spectrum of individuals and organizations, including trauma care professionals and professional associations, trauma centers and other acute care hospitals, trauma care systems, emergency medical services (EMS) systems, injury researchers and research organizations, public health agencies, health care payers, and the general public.
Requirements: Applications may be submitted by public and private nonprofit organizations and by governments and their agencies.
Geographic Focus: All States
Contact: Cheryl Maddux; (770) 488-2645
Internet: http://www.cdc.gov
Sponsor: Centers for Disease Control and Prevention
1600 Clifton Road
Atlanta, GA 30333

CDC Preventive Medicine Residency and Fellowship 1465

The Preventive Medicine Residency and Fellowship (PMR/F) is an umbrella that includes two preventive medicine programs that provide hands-on experiences in public health agencies at the federal, state, and local levels. The programs consist of a residency (PMR) for physicians only and a fellowship (PMF) for physicians and other health professionals. The PMR is a 24-month nationally accredited graduate medical education program for physicians. It is accredited by the Accreditation Council for Graduate Medical Education (ACGME) and meets the residency requirement of the American Board of Preventive Medicine (ABPM) in the specialty of Public Health and General Preventive Medicine. Select candidates can be considered for one year of residency training. The PMF is a 12-month program open to veterinarians, dentists, registered nurses, physician assistants, and international medical graduates (physicians) who have no ACGME- or American Osteopathic Association (AOA)- accredited postgraduate clinical training. The residency (PMR) and the fellowship (PMF) both provide service, training, and supervised on-the-job learning in public health and preventive medicine practice. PMR/F activities include: designing, implementing, and evaluating public health programs; developing or analyzing health policy; managing public health and preventive medicine projects; bridging medical and public health sectors to improve wellness; and developing and applying leadership skills. The PMR/F program goals are to prepare clinicians for leadership roles in public health and general preventive medicine at the federal, state or local levels. Residents and fellows will receive mentoring by experienced public health and preventive medicine practitioners; hands-on experience in public health and preventive medicine practice; classroom training to enhance professional growth; and M.P.H. coursework for select physicians in the PMR program. Residents and fellows receive a salary and benefits, moving expenses, travel to PMR/F-sponsored training, and tuition support for an M.P.H. degree for residents (as needed). (Qualified applicants interested in the American Board of Preventive Medicine "complementary pathway to board exam certification" must have completed an ACGME accredited residency in another specialty and have an M.P.H. or equivalent degree per ACGME.) Interested applicants may apply online between July 1 and September 15. Additional supporting documents are required. Applications are due by midnight eastern daylight time on the due date. Supporting documents must also be received on the application due date. Applicants are encouraged to mail supporting documents early. All mail delivery to the CDC goes to a central mail facility before being delivered to the PMR/F office. Applicants are encouraged to click on the "Application Information" link at the PMR/F website for additional instructions, guidelines, and information on how to apply and submit the supporting documents.

Requirements: Applicants must: commit to a 1- or 2-year full-time training period (depending on the program) starting in mid-June; be willing to relocate for duration of training; meet the professional and licensing requirements for hiring by the U.S. government according to the U.S. Office of Personnel Management (link provided at the PMR/F website); and have trained in the Epidemic Intelligence Service (EIS) program or have comparable applied epidemiology experience. PMR applicants must be physicians who have completed at least 12 months of ACGME-accredited postgraduate clinical training involving at least 11 months of direct patient care; have a current, full, and unrestricted medical license from a U.S. licensing jurisdiction; and have a Master of Public Health (M.P.H.) or equivalent accredited degree, per ACGME requirements. See the PMR/F website for more information about these requirements. There are some exceptions to the M.P.H. degree requirement. Physicians may be sponsored during the residency to obtain an M.P.H. degree. The completion of necessary courses to earn the degree will be integrated into the training program. M.P.H. sponsored applicants must be Commissioned Corps officers in the U.S. Public Health Service at the time of application with a commitment to an additional 2-year service obligation in the Commissioned Corps following the residency and a willingness to attend the academic institution for the M.P.H. selected by the residency (be willing to relocate). PMF applicants must be veterinarians, dentists, registered nurses, physician assistants, or international medical graduates (physicians who have no ACGME- or American Osteopathic Association- accredited postgraduate clinical training) who have an M.P.H. or equivalent degree from an accredited institution; have a current, full, and unrestricted license from a U.S. licensing jurisdiction in their qualifying clinical specialty; and demonstrate ability in applied epidemiology, public health practice, and written and oral communication for scientific audiences.

Restrictions: Applicants must be U.S. citizens or permanent residents. Physicians who have graduated from a medical school outside the U.S. and do not have a current, full, and unrestricted license from a U.S. licensing jurisdiction may apply.

Geographic Focus: All States

Date(s) Application is Due: Sep 15

Contact: Coordinator; (404) 498-6140; fax (404) 498-6135; PrevMed@cdc.gov

Internet: http://www.cdc.gov/PrevMed/index.html

Sponsor: Centers for Disease Control and Prevention
1600 Clifton Road
Atlanta, GA 30333

CDC Public Health Conference Support Grants 1466

The purpose of the grants is to provide partial support for specific non-federal conferences in the areas of health promotion and disease prevention information/education programs. Funds may be used for direct cost expenditures, such as salaries, speaker fees, rental of necessary equipment, registration fees, and transportation costs (not to exceed economy-class fare) for non-federal individuals. The program continues through March 8, 2009.

Requirements: Public and private organizations, including colleges and universities, and units of state and local government are eligible to apply. Applicants must provide a portion of the conference cost.

Restrictions: Funds may not be used for indirect costs, equipment purchases, honoraria, entertainment, or personal expenses.

Geographic Focus: All States

Contact: Sharon Robertson, Grants Officer; (770) 488-2748; SRobertson1@cdc.gov

Internet: http://www.cdc.gov/od/pgo/funding/FOAs.htm

Sponsor: Centers for Disease Control and Prevention
1600 Clifton Road
Atlanta, GA 30333

CDC Steven M. Teutsch Prevention Effectiveness Fellowships 1467

The Prevention Effectiveness Fellowship (PEF) is a two-year, highly-selective research fellowship for recent doctoral graduates with a background in economics, policy analysis, operations research, decision sciences, and other quantitative areas, and was established to address demand in the field of public health for economics-based inquiry, quantitative policy analysis, and integrative-health-services research. The PEF allows applicants to apply their academic training in quantitative methods to the science of health protection, health promotion, and disease prevention. Prevention effectiveness uses economic evaluation, policy analysis, and decision modeling to systematically assess the impact of public health interventions, policies, programs, and practices on health outcomes. Econometric, decision, simulation, and operations modeling and methods are employed to determine the efficacy, effectiveness, efficiency, and cost of public health services, policies, and interventions in order to facilitate better decision-making about the allocation of resources around programmatic priorities. At CDC, prevention effectiveness methods are used to explain: health-related burden and costs; effectiveness and efficiency of health protection, health promotion, and disease prevention programs; health program prioritization based on optimization modeling; disease transmission and intervention modeling; and health system performance. Fellows have the opportunity to work with leading researchers on issues of critical importance to create the expertise, information, and tools that enable people and communities to protect their health through: health promotion; prevention of disease, injury and disability; and preparedness for new health threats. Fellows are assigned to various CDC Centers, Institute, and Offices (CIOs) where they assess the public health impact of policies, programs, and practices by determining their cost, effectiveness, and overall impact. The PEF provides initial classroom training and hands-on experience to support the development of valuable research skills. After an introduction to basic public health sciences and prevention-effectiveness methods, fellows are assigned to a CDC program area where they take a lead role in prevention-effectiveness studies. Assignment activities include: designing and conducting studies; working closely with national and international experts in public health; presenting and publishing results; and providing technical assistance and service on specific projects or methods. In addition, professional support and career development are provided through didactic training, workshops, research seminars, and peer discussion groups. Fellows are full-time CDC employees during the fellowship and will receive the same benefits (such as medical insurance and vacation) as other full-time CDC employees. During the first year, fellows are paid the equivalent of a GS 12, Step 1 salary. In the second year, they are eligible for salary equivalent to a GS 12, Step 2 with geographic adjustments. Applicants must apply online at the PEF website (click on "Apply Now" on the left hand side of the screen). Applications are due on January 21 before midnight. All required supporting documents must be postmarked by January 19. Deadlines may vary from year to year; applicants should check the PEF website to verify current submission deadlines for proposals and supporting documents. Further essential details on application and supporting document requirements are available from the PEF website.

Requirements: Applicants must hold a doctoral degree in economics or applied economics, decision sciences, health services research or related health sciences, industrial engineering or operations research, public policy or policy analysis, or a related quantitatively-oriented field. A medical degree with additional relevant training is acceptable. Dissertations must be defended by June 17 and degree requirements must be met by July 15 prior to the start date of the fellowship. Written and spoken proficiency in English is essential. Relocation to Atlanta, Georgia is required. As stated in the FAQ at the PEF website, up to $3,000 is provided to cover relocation expenses. Non-U.S. citizens must be legal permanent residents or eligible for J-1 status prior to the program's start date.

Restrictions: Citizens of countries that have been determined by the U.S. Secretary of State to be State Sponsors of Terrorism will not be able to get security clearance to work at CDC and should not apply to the PE Fellowship Program. A list of these may be found at the U.S. Department of State website.

Geographic Focus: All States, All Countries

Date(s) Application is Due: Jan 11

Amount of Grant: 149,198 USD

Contact: Adam G. Skelton, Ph.D., M.P.H., Fellowship Coordinator; (404) 498-6324 or (404) 498-6786; PEF@cdc.gov or adam.skelton@cdc.hhs.gov

Internet: http://www.cdc.gov/PEF/index.html

Sponsor: Centers for Disease Control and Prevention
1600 Clifton Road
Atlanta, GA 30333

CDECD Elizabeth L. Mahaffey Arts Administration Fellowship 1468

The Elizabeth L. Mahaffey Arts Administration Fellowship annually recognizes one exceptionally accomplished Connecticut arts administrator with a $2,500 award to be used for professional development activities. The application is available at the CDECD website.

Requirements: Administrators must be full time state residents employed by Connecticut nonprofit arts organizations, or involved in ongoing arts programs of nonprofit organizations, municipalities, colleges and universities. Applicants must have at least seven years of professional arts administration experience, including three with their current organization.

Restrictions: Administrators employed as consultants, contractors or subcontractors are not eligible to apply. Unpaid and volunteer staff members are also ineligible.

Geographic Focus: Connecticut

Date(s) Application is Due: Feb 1

Amount of Grant: 2,500 USD
Contact: Rhonda Olisky, Special Projects Coordinator; (860) 256-2752; fax (860) 256-2811; rhonda.olisky@ct.gov
Internet: http://www.cultureandtourism.org/cct/cwp/view.asp?a=3933&q=293768
Sponsor: Connecticut Department of Economic and Community Development
One Constitution Plaza, 2nd Floor
Hartford, CT 06103

CDI Postdoctoral Fellowships **1469**

The Children's Discovery Institute (CDI) is a world-class center for pediatric research and innovation created to encourage researchers to ask bold questions and take bold risks to uncover answers. By funding the work of creative scientists and clinicians in collaborative, multi-disciplinary research aimed at some of the most devastating childhood diseases and disorders, CDI will accelerate the realization of better treatments, cures, and preventions. Up to $30,000 per year for up to two years is available for especially promising postdoctoral trainees to complete a research project under the direction of a mentor on an issue relevant to the goals of the CDI.
Requirements: Proposals will be accepted from Washington University faculty and postdoctoral trainees. Preference will be given to faculty members prior to achieving tenure, to teams of investigators from multiple disciplines, and to multi-investigator projects to develop interactive research groups. Preference will also be given to projects with potential for acquisition of new knowledge/translational impact and scientific experience; and to programs synergistic with Washington University and St. Louis Children's Hospital. Letters of intent must be submitted by August 1. For investigators invited to submit a proposal, applications will be due on October 15.
Geographic Focus: Missouri
Date(s) Application is Due: Oct 15
Amount of Grant: Up to 60,000 USD
Contact: Angela Mayer (Corless); (314) 286-2711; cdi@kids.wustl.edu
Internet: http://www.childrensdiscovery.org/Grants/FundingMechanismsDueDates.aspx
Sponsor: Children's Discovery Institute
660 S. Euclid Avenue
St. Louis, MO 63110

Cedar Tree Foundation David H. Smith Conservation Research Fellowship **1470**

The David H. Smith Conservation Research Fellowship program identifies and supports early-career scientists who will shape the field of applied conservation biology. David H. Smith Conservation Research Fellowships are available to post-doctoral researchers (of any nationality) affiliated with a United States institution. In 1998, to help address the need for post-doctoral opportunities for conservation biology graduates, the foundation and founding partner The Nature Conservancy established the David H. Smith Conservation Research Fellowship Program, devoted exclusively to applied conservation research problems. By fostering the development of promising conservation scientists, the Smith Fellowship Program helps encourage this rapidly expanding field of scientific inquiry and link it to the practice of conservation. Each Fellow will receive an annual salary of $50,000 plus benefits, with the post-doctoral position expected to run for two consecutive years. In addition to the stipend, each Fellow receives a travel budget of $8,000 and a research fund of $32,000 over the 2-year fellowship period.
Requirements: To be eligible individuals must have completed their doctorate within the past five years or by the time the award is made.
Geographic Focus: All States
Date(s) Application is Due: Sep 20
Contact: Mike Dombeck; (202) 234-4133, ext. 307; info@cedartreefound.org
Internet: http://www.conbio.org/mini-sites/smith-fellows
Sponsor: Cedar Tree Foundation
100 Franklin Street, Suite 704
Boston, MA 02110-1401

Cen Cal Diving and Aquatic Studies Scholarship **1471**

This scholarship is awarded annually to an outstanding individual who is engaging in studies of underwater habitats. Aquatic-related programs in the disciplines of biology, physical sciences, marine education, maritime archaeology, historical and social aspects of marine resources, or the science of diving are relevant for consideration. Application forms are available on the Web site. Annual deadline dates may vary; contact program staff for exact dates.
Requirements: Applicants must be California residents enrolled in an undergraduate or graduate program at a California college or university, be over 18 years of age, maintain a high academic standing with a minimum GPA of 3.0, and have a national/regional diver certification.
Geographic Focus: California
Date(s) Application is Due: Apr 30
Amount of Grant: 1,000 USD
Contact: Jim Kaller, (415) 362-9134, ext. 12; jameskaller@batnet.com
Internet: http://www.cencal.org/requirements.html
Sponsor: Central California Council of Diving Clubs
P.O. Box 779
Daly City, CA 94017

Center for Advanced Judaic Studies Postdoctoral Fellowships **1472**

The program is designed to engage scholars from within Jewish studies, who specialize in Jewish life under Islamic rule, in fruitful conversation with scholars in Arabic, Syriac, Persian, and Ottoman studies. By bringing together experts from these disparate disciplines who deal with aspects of Jewish, Christian, and Muslim life within pre-modern Islamic polities, the program hopes to encourage a broad view of Islamic societies and to foster new approaches to their religious, ethnic, and linguistic diversity. Guidelines are available online. The 2006-2007 program is Jewish, Christian, and Muslim Life under Caliphs and Sultans. The 2007- 2008 program will be Jewish and Other Imperial Cultures in Late Antiquity: Literary, Social, and Material Histories
Requirements: Any individual or group of scholars in the fields of Jewish studies may submit a proposal for an annual theme.
Geographic Focus: All States
Date(s) Application is Due: Nov 1
Amount of Grant: 33,000 USD
Contact: David B. Ruderman; (215) 238-1290; ruderman@sas.upenn.edu
Internet: http://www.cjs.upenn.edu/program/fellowship.htm
Sponsor: Center for Jewish Studies
420 Walnut Street
Philadelphia, PA 19106

Center for Defense Information Internships **1473**

The center offers an intern program to undergraduate and graduate students and recent graduates who have strong interests in U.S. military issues and related public policy issues. Interns perform a variety of professional support functions but serve mainly as research and outreach assistants. Interns also assist with the production of CDI's weekly television show on military-related current affairs. They work closely with the center staff, usually with considerable responsibility. Intern projects have included such diverse research subjects as U.S. military nuclear wastes, military spending, and U.S. arms transfer policy, as well as media activities. There is no preferred academic background. Fields of study might encompass the social and policy sciences, physical sciences, and preprofessional curricula. Although prior coursework in U.S. military issues and related public policy areas is not required, high academic achievements are important. Experience with video production is a plus. Writing skills are essential. Interns are expected to work on a full-time basis. Internships run from January through May, June through August, and September through December. Application must include personal letter, resume, college transcript, writing samples, and two letters of recommendation.
Requirements: Internships are open to undergraduates, graduates, and graduate students with strong interests in military policy, national security, foreign affairs, and related public policy issues and/or interest in broadcast communications.
Geographic Focus: All States
Date(s) Application is Due: Mar 1; Jul 1; Oct 1
Contact: Coordinator; (202) 332-0600; fax (202) 462-4559; internships@cdi.org
Internet: http://www.cdi.org/about/internships.cfm
Sponsor: Center for Defense Information
1779 Massachusetts Avenue NW
Washington, DC 20036-2109

Center for East Asian Studies Fellowships **1474**

The center was formed to increase interdisciplinary communication among linguists, historians, art historians, political scientists, anthropologists, and others whose research, teaching, or study focuses on China, Korea, or Japan. Some students accepted into the graduate program in East Asian studies at Stanford receive financial assistance through these fellowships. The center also offers a limited number of student awards to Stanford students for summer research and/or study. The most up-to-date information about fellowships is available in the CEAS newsletter which can be accessed on the Web site. Deadlines vary for individual programs; contact program staff for exact dates.
Requirements: Graduate students at Stanford University specializing in East Asia may apply for fellowships at the time of initial application for admission.
Restrictions: Students attending institutions other than Stanford are ineligible.
Geographic Focus: All States
Contact: Connie Chin; (650) 723-3362; fax (650) 725-3350; csquare@stanford.edu
Internet: http://www.stanford.edu/dept/CEAS
Sponsor: Stanford University
Room 51-L, Building 50
Stanford, CA 94305-5013

Center for Education of Women Visiting Scholar Research Grants **1475**

The Center for Education of Women (CEW) invites applications from scholars and practitioners interested in being in residence at the university for a period of one to 12 months to pursue research, writing, and publication interests. Applicants in the following general areas are invited to apply: Women in Higher Education, Women's Work, Women of Color in the Academy, Gender and Education, Career Development, Women in Non-Traditional Fields, Leadership, Gender Equity in Education and Employment, and Gender and Poverty. Other areas of research may also be considered. Scholars will have the opportunity to conduct their research and will be asked to prepare a working paper and/or give a seminar based on this work. CEW will provide office space, access to facilities and programs, modest stipends, and/or research support funds. Applicants should send letter of interest, a one- to two- page outline of proposed project, funding requirements, anticipated dates of stay, and a vita.
Requirements: Applicants should hold an earned PhD, EdD, JD, etc., or have equivalent experience in a relevant field.
Geographic Focus: All States
Date(s) Application is Due: Mar 1; Jun 1
Contact: Beth Sullivan, (734) 998-7225; bsulliva@umich.edu
Internet: http://www.umich.edu/~cew/faculty-staff/vs.htm
Sponsor: University of Michigan
330 E Liberty Street
Ann Arbor, MI 48104-2289

Center for Hellenic Studies Fellowships 1476

Fellowships for research in ancient Greek history, philosophy, literature, language, and religion are offered at the center. Letters of recommendation are due separately at the Director's office by November 1. Application forms are available upon request.
Requirements: Applicants must have PhDs or equivalent, be professionally competent in ancient Greek, and be published writers.
Geographic Focus: All States
Date(s) Application is Due: Oct 15
Amount of Grant: 27,000 USD
Contact: Program Contact; (202) 234-3738; fax (202) 797-3745; chs@fas.harvard.edu
Internet: http://www.chs.harvard.edu/fellowships.sec
Sponsor: Center for Hellenic Studies
3100 Whitehaven Street NW
Washington, DC 20008

Center for Inquiry Science and the Public Visiting Research Fellowships 1477

Each academic year, the center awards visiting research fellowships at the postdoctoral or senior level in the area of science and the public. All relevant disciplines are welcome. Fellowships include an appointment in the appropriate department at State University of New York-Buffalo and a stipend plus accommodations (optional). Candidates should send a research project proposal (1000 to 2000 words), curriculum vita, two references (for recent doctorates), and a writing sample. Guidelines and current research themes are available online.
Geographic Focus: All States
Date(s) Application is Due: Mar 1
Amount of Grant: 15,000 USD
Contact: Austin Dacey; (716) 636-1733; adacey@centerforinquiry.net
Internet: http://www.scienceandthepublic.org/programs.htm
Sponsor: Center for Inquiry
P.O. Box 741
Amherst, NY 14226

Center for Judaic Studies Fellowships 1478

The center invites applications from scholars engaged in all fields of Judaic studies and from scholars in other fields interested in approaching the general topic from a comparative and interdisciplinary perspective. Outstanding graduate students in the final stages of writing their dissertations also may apply. Fellowship recipients also may receive a contribution toward travel expenses. Annual research topics vary; contact the center for current topic. Application materials are available upon written request.
Geographic Focus: All States
Date(s) Application is Due: Nov 15
Amount of Grant: Up to 33,000 USD
Contact: Administrator; (215) 238-1290; fax (215) 238-1540; allenshe@sas.upenn.edu
Internet: http://www.cjs.upenn.edu/program/fellowship.htm
Sponsor: University of Pennsylvania
420 Walnut Street
Philadelphia, PA 19104-6906

Center for National Policy Internships 1479

One intern per semester or summer is employed for a position dealing with the overall general operations of CNP. The program promotes practical opportunities to advance the public policy process, as experts and decisionmakers debate significant national objectives and determine how government can best serve American interests both at home and abroad. The intern will conduct research, write briefings, prepare press materials, participate in fund raising, and perform general office support work. Applications must be received two months prior to the semester.
Requirements: Applicants must be undergraduate students.
Geographic Focus: All States
Date(s) Application is Due: Apr 15; Aug 12
Contact: Alex Sunshine, Intern Coordinator; (202) 682-1800; fax (202) 682-1818; asunshine@cnponline.org or thecenter@cnponline.org
Internet: http://www.cnponline.org/employment.htm
Sponsor: Center for National Policy
1 Massachusetts Avenue NW, Suite 333
Washington, DC 20001

Center for the Study of Books and Media Visiting Research Fellowships 1480

The Center for the Study of Books and Media at the university invites applications for a visiting research fellowship in any field related to the history of books. The fellow is expected to pursue his or her research at Princeton throughout the academic year and is encouraged to participate in the activities of the center. The stipend depends on qualifications.
Requirements: Candidates should have completed a dissertation in book history within the last three years and should propose a project involving research on subjects such as publishing, the book trade, and reading.
Geographic Focus: All States
Date(s) Application is Due: Mar 1
Contact: Thierry Rigogne, trigogne@princeton.edu
Internet: http://web.princeton.edu/sites/english/csbm/postdoc_fellowship.htm
Sponsor: Princeton University
129 Dickinson Hall
Princeton, NJ 08544

Center for the Study of Democratic Politics Fellowships 1481

The Center for the Study of Democratic Politics (CSDP) was created as a research program within the Woodrow Wilson School of Public and International Affairs. The Center supports empirical research on democratic political processes and institutions. The CSDP's research program focuses on the relationship between democratic ideals and democratic practice. Its goal is to encourage rigorous social scientific analysis that informs and is informed by normative theories of democracy. The Center seeks visiting scholars and/or post-doctoral research scholars for research on democratic political processes and institutions. Applications are welcome from political scientists, in addition to related social science disciplines. Each scholar will pursue research and contribute to the intellectual life of the Center, the Department of Politics, and the Woodrow Wilson School of Public and International Affairs. Successful candidates will receive a competitive salary, applicable Princeton University benefits, a taxable moving allowance, and a $5,000 research fund. The application is available at the CSDP website.
Requirements: Center scholars must be in residence at Princeton at least four days each week during the academic terms, and are expected to participate in weekly CSDP seminars in the Department of Politics, as well as occasional lectures, conferences, and other events.
Geographic Focus: All States
Date(s) Application is Due: Nov 30
Amount of Grant: 5,000 USD
Contact: Helene Wood, CSDP Program Assistant; (609) 258-6493 or (609) 258-6980; fax (609) 258-5014; hwood@princeton.edu
Internet: http://www.princeton.edu/csdp/apply/
Sponsor: Princeton University
Center for the Study of Democratic Politics
Princeton, NJ 08544-1013

Center for the Study of Philanthropy International Fellowships 1482

The center provides leadership training through applied research and professional mentorships for young scholar-practitioners in the nonprofit sector. The program offers leadership training to scholar-practitioners outside the U.S. as a means to help build Third-Sector capacity in the fellows' home countries. Specific topical areas are chosen each year. Special attention will also be given to diaspora philanthropy. Fellows participate in a three-month seminar on the U.S. and international voluntary sectors. Fellows are expected to produce a 30-50 page paper on their findings to be presented in the seminar. They will learn about the work of key agencies and meet with foundation and nonprofit representatives. They may also have the opportunity to attend selected conferences. Fellowship includes tuition and monthly stipend, housing, and round-trip air travel to and from the U.S..
Requirements: The program is open to practitioners and researchers under the age of 36 who are citizens of countries other than the U.S.. Applicants must have a college or university degree and speak and write English fluently. Preference will be given to candidates with a strong institutional base and demonstrated research skills.
Geographic Focus: All States
Date(s) Application is Due: Sep 13
Amount of Grant: 1,300-15,600 USD
Contact: Dr. Kathleen McCarthy; (212) 817-2010; info@philanthropy.org
Internet: http://www.philanthropy.org/programs/intnl_fellows_program.html
Sponsor: City University of New York
365 Fifth Avenue
New York, NY 10016-4309

Center for the Study of Philanthropy Senior International Fellowships 1483

The center provides leadership training through applied research and professional mentorships for young scholar-practitioners in the nonprofit sector. The program offers leadership training to scholar-practitioners outside the United States as a means to help build Third-Sector capacity in the fellows' home countries. Specific topical areas are chosen each year. The current year's theme is community foundations. Fellows participate in a three-month seminar on the U.S. and international voluntary sectors. Fellows attend weekly seminars, learn about the work of key agencies and foundations, meet with nonprofit representatives, and study U.S. and international community foundation models. Fellowship includes tuition and monthly stipend, housing, and round-trip air travel to and from the United States.
Requirements: The program is open to senior-level practitioners and researchers under the age of 36 who are citizens of countries other than the United States. Applicants must have a college or university degee and speak and write English fluently. Preference will be given to candidates with a strong institutional base and demonstrated research skills.
Geographic Focus: All States
Date(s) Application is Due: Jun 15; Oct 15
Amount of Grant: 1,300-15,600 USD
Contact: Fellowship Coordinator; (212) 817-1572; csp@gc.cuny.edu
Internet: http://www.philanthropy.org/programs/intnl_fellows_program.html
Sponsor: City University of New York
365 Fifth Avenue
New York, NY 10016-4309

Center for the Study of Professional Military Ethics Resident Fellowship 1484

The goal of this full-time, in-residence program is to help prepare select military officers, career civil servants, academics, and others to teach, apply, and practice ethics in a variety of professions and institutional settings. Each fellow will be required to complete a research project and to participate in a weekly fellows seminar; be encouraged to participate in other activities and programs of the center; and receive office space, computer facilities, library privileges, and a stipend. Applicants should send a letter

describing their interests in ethics in the areas listed; a proposal for a research project to be undertaken during the fellowship; a curriculum vita; and copies of relevant publications. Two letters of reference also should be sent directly to the center. Recipients are announced by the end of April.
Requirements: Ideal candidates from academia will have a background in ethics, international affairs, security studies, and/or public policy, with an interest in pursuing issues of ethics in military affairs, national defense, and international security in their teaching and research.
Geographic Focus: All States
Date(s) Application is Due: Oct 14
Contact: Dr. Albert Pierce, Director, Center for the Study of Professional Military Ethics; (410) 293-6057; acpierce@gwmail.usna.edu
Internet: http://www.usna.edu/Ethics/Programs/Residentfellowship/fellows.htm
Sponsor: U.S. Naval Academy
112 Cooper Road
Annapolis, MD 21402

Center for the Study of Religion Fellowships **1485**
The Center brings a select number of pre-tenure scholars and recent Ph.D. graduates to Princeton University to study religion and religious history. The fellows, who are appointed by the Dean of the Faculty, devote time to servicing the intellectual life of the Center and the University through mentoring graduate and undergraduate students and participating in one of the Center's weekly interdisciplinary seminar. For 2007-2008, fellows will be appointed in two areas: Christian Thought and Practice, special emphasis on the religious life of American Christians, congregations, or clergy; and Public Theology: special emphasis on contemporary issues bridging theology and the social sciences, such as just war theory, pluralism, or social welfare.
Requirements: To apply, submit CV, 3 letters of recommendation, a 5 page proposal, and published or unpublished paper demonstrating scholarly command of topic of one of the three themes. Applications available on Web site.
Restrictions: Princeton University PhDs are not eligible.
Geographic Focus: All States
Date(s) Application is Due: Jan 5
Contact: Anita Klein, Director; (609) 258-5545; fax (609) 258-6940
Internet: http://www.princeton.edu/~csrelig/opportunities/undergrad_ops.html
Sponsor: Princeton University
5 Ivy Lane, 304 Louis Marx Hall
Princeton, NJ 08544

Center of Excellence (COE) in General Aviation Grants **1486**
The FAA has identified a need for a Center of Excellence in general aviation. The center will conduct research, which includes the entire spectrum (i.e., basic research through engineering development, prototyping, and testing) within the scope of general aviation. This scope includes, but is not limited to airport technology; propulsion and structures; aging aircraft; flight safety; fire safety; and training. The FAA intends to provide long-term funding to establish and operate a prestigious partnership with academia, industry, and government. To this end, the FAA encourages offerors to team with organizations that complement their expertise from academia, industry, state/local government, and other governmental agencies. The successful applicant is required to match FAA grant funds with nonfederal funding over the term of the cooperative agreement. Cost sharing (negotiated individually) also is required for any orders placed under the IDIQ contract. The grant minimum per year is $300,000.
Requirements: Colleges and universities are eligible to apply.
Restrictions: Individuals are not eligible for a COE designation and do not qualify for grants under this program.
Geographic Focus: All States
Contact: Pat Watts, Grants Manager; (609) 485-5043; patricia.watts@faa.gov
Internet: http://www.cgar.org/about_program.asp
Sponsor: U.S. Department of Transportation
William J Hughes Technical Center
Atlantic City, NJ 08405

Center on Philanthropy and Civil Society's Emerging Leaders International Fellows Program **1487**
The program provides leadership training through applied research and professional mentorships for young scholar-practitioners in the nonprofit sector. The program is open to scholars and practitioners interested in building Third-Sector capacity in the United States and overseas. This year???s fellows will be selected from abroad and also from communities of color under-represented in the United States grantmaking sector. Fellows are based at the Graduate Center of The City University of New York, where they design and pursue an individualized research project and participate in a seminar with Third-Sector leaders. Specific topical areas are chosen each year.
Requirements: The program is open to practitioners and researchers under the age of 36. Applicants must hold a college or university degree and speak and write English fluently. Preference will be given to candidates with strong ties to a Third-Sector institution and demonstrated research skills.
Geographic Focus: All States
Date(s) Application is Due: Dec 7
Contact: Dr. Kathleen McCarthy; (212) 817-2010; fax (212) 817-1572; csp@gc.cuny.edu
Internet: http://www.philanthropy.org/programs/ifp/application.html
Sponsor: City University of New York
365 Fifth Avenue
New York, NY 10016-4309

Central European University Fellowships **1488**
The university offers full fellowships for master's, doctoral, and doctoral support programs in Budapest, Hungary and Warsaw, Poland. 625 fellowships covering tuition and living expenses are available to individuals from Central and Eastern Europe and the former Soviet Union. A wide variety of other financial aid programs are available to all students on a competitive basis.
Geographic Focus: All States, Austria, Czech Republic, Germany, Hungary, Liechtenstein, Poland, Slovakia, Slovenia, Switzerland
Date(s) Application is Due: Jan 15
Contact: Grants Administrator; (36-1) 327-3000, ext. 2311; finaid@ceu.hu
Internet: http://www.ceu.hu/financial_forms.html
Sponsor: Central European University
Nador u 9
Budapest, 1051 Hungary

Century Foundation Grants **1489**
The foundation sponsors and supervises research on significant economic, social, and political issues. The foundation's research projects usually are designed to produce analytic, book-length manuscripts containing public policy recommendations. They are aimed at an audience that includes the informed public, the press, policymakers, and the academic community. The foundation commissions individuals (project directors) to carry out the projects, and the book-length manuscripts that result are edited and then published by the foundation's press or placed with commercial publishers or university presses. Proposals are welcome for reports and books. Preliminary proposals should summarize in no more than three pages the central argument of the proposed publication, the main themes that would be developed in support of the argument, and examples of evidence that would be explored in making the project director's case. A curriculum vita should be included with the proposal.
Restrictions: The foundation does not award scholarships or support dissertation research, nor does it make grants to individuals or institutions.
Geographic Focus: All States
Contact: Carl Robichaud, Program Officer; (212) 452-7712; fax (212)535-9803; robichaud@tcf.org
Internet: http://www.tcf.org/about.asp
Sponsor: Century Foundation
41 E 70th Street
New York, NY 10021

CFF-NIH Funding Grants **1490**
The Cystic Fibrosis Foundation (CFF) has developed the overall research grant program to complement the awarding mechanism of the NIH. Support from CFF, through various mechanisms, is intended to provide for the development of sufficient preliminary data to make CF-related grant applications highly competitive in the NIH review process. However, as a result of funding constraints on the NIH, coupled with the growing interest in CF research, occasions arise in which highly meritorious grant applications are submitted to the NIH but are not funded. In an effort to assure that all meritorious CF-related research is supported, CFF has developed the CFF/NIH-unfunded Award mechanism to provide funding. The objective of this award is to support excellent CF-related research projects that have been submitted to and approved by the National Institutes of Health (NIH), but cannot be supported by available NIH funds. Applications must fall within the upper 40th percentile. CFF support ranges from $75,000 to $125,000 per year for up to two years.
Restrictions: Indirect costs are not allowed.
Geographic Focus: All States
Amount of Grant: 75,000-125,000 USD
Contact: Grants Management; (800) 344-4823 or (301) 951-4422; grants@cff.org
Internet: http://www.cff.org/research/ForResearchers/FundingOpportunities/Research Grants/#Pilot_and_Feasibility_Awards
Sponsor: Cystic Fibrosis Foundation
6931 Arlington Road, 2nd Floor
Bethesda, MD 20814

CFF Clinical Research Grants **1491**
Cystic Fibrosis Foundation Therapeutics (CFFT), the nonprofit drug discovery and development affiliate of the Cystic Fibrosis Foundation, offers competitive awards to support clinical research projects directly related to CF treatment and care. Projects may address diagnostic or therapeutic methods related to CF or the pathophysiology of cystic fibrosis, and applicants must demonstrate access to a sufficient number of patients from Foundation-accredited care centers and to appropriate controls. Up to $100,000 per year (plus 8% indirect costs) for a maximum of three years may be requested for single-center clinical research grants. For multi-center clinical research, the potential award is up to $225,000 per year (plus 8% indirect costs) for a maximum of three years.
Requirements: Applications will be accepted only after submission and approval of a letter of intent. CFFT strictly adheres to the letter of intent and award application deadlines.
Geographic Focus: All States
Date(s) Application is Due: Mar 6; Sep 11
Amount of Grant: Up to 225,000 USD
Contact: Grants Management; (800) 344-4823 or (301) 951-4422; grants@cff.org
Internet: http://www.cff.org/research/ForResearchers/FundingOpportunities/Clinical ResearchAwards/
Sponsor: Cystic Fibrosis Foundation
6931 Arlington Road, 2nd Floor
Bethesda, MD 20814

CFF First- and Second-Year Clinical Fellowships 1492

The Cystic Fibrosis Foundation offers competitive clinical fellowships for up to five years for physicians interested in cystic fibrosis and other chronic pulmonary and gastrointestinal diseases of children, adolescents, and adults. The intent of this award is to encourage specialized training early in a physician's career and to prepare well-qualified candidates for careers in academic medicine. Training must take place in one of the foundation's accredited centers and must provide thorough grounding in diagnostic and therapeutic procedures, comprehensive care, and cystic fibrosis-related research. All first- and second-year programs must commit a significant portion (at least 30 percent over two years) to research training. Fellows funded by other sources for their first year of training may apply to the foundation for subsequent training support. Third-year fellowships are available for additional basic and/or clinical research training. Research fellowships are also available from the foundation to support physicians beyond the fellowship experience; salary will be commensurate with experience.

Requirements: Applicants must be U.S. citizens or have permanent U.S. resident visas, and must have completed pediatric training and be eligible for board certification in pediatrics by the time the fellowship begins. Candidates with prior training in internal medicine may also apply but must have completed at least two years of an approved adult pulmonary or GI fellowship; be jointly sponsored by the departments of medicine and pediatrics; and be willing to commit at least 75 percent of their time to cystic fibrosis and related problems of young adults.

Geographic Focus: All States

Date(s) Application is Due: Oct 2

Amount of Grant: 47,600-49,250 USD

Contact: Grants Management; (800) 344-4823 or (301) 951-4422; grants@cff.org

Internet: http://www.cff.org/research/ForResearchers/FundingOpportunities/TrainingGrants/

Sponsor: Cystic Fibrosis Foundation
6931 Arlington Road, 2nd Floor
Bethesda, MD 20814

CFF Harry Shwachman Clinical Investigator Award 1493

The three-year CFF Harry Shwachman Clinical Investigator Award provides the opportunity for clinically trained physicians to develop into independent biomedical research investigators who have active involvement in cystic fibrosis-related areas. It is also intended to facilitate the transition from postdoctoral training to a career in academic medicine. Support is available for up to $76,000 per year plus $15,000 for supplies. Support is based on a full-time, twelve-month appointment. Applications are due the second Wednesday in September.

Requirements: U.S. citizenship or permanent resident status is required.

Restrictions: Indirect costs are not allowed.

Geographic Focus: All States

Date(s) Application is Due: Sep 11

Amount of Grant: Up to 91,000 USD

Contact: Grants Management; (800) 344-4823 or (301) 951-4422; grants@cff.org

Internet: http://www.cff.org/research/ForResearchers/FundingOpportunities/ResearchGrants/#Pilot_and_Feasibility_Awards

Sponsor: Cystic Fibrosis Foundation
6931 Arlington Road, 2nd Floor
Bethesda, MD 20814

CFF Leroy Matthews Physician-Scientist Awards 1494

In honor of Dr. LeRoy Matthews' dedication and commitment to cystic fibrosis (CF) research and care, the Cystic Fibrosis Foundation (CFF) announces the Physician Scientist Award named in his memory. Awards will provide up to six years of support for outstanding newly trained pediatricians and internists (MDs and MD/PhDs) to complete sub-specialty training, develop into independent investigators, and initiate research programs. Institutional and individual grants are available. Support ranges from a $48,000 stipend, plus $10,000 for research and development for year one, up to a $76,000 stipend, plus $15,000 for research and development for year six.

Requirements: U.S. citizenship or permanent resident status is required.

Restrictions: Indirect costs are not allowed.

Geographic Focus: All States

Date(s) Application is Due: Sep 11

Amount of Grant: 48,000-86,000 USD

Contact: Grants Management; (800) 344-4823 or (301) 951-4422; grants@cff.org

Internet: http://www.cff.org/research/cystic_fibrosis_foundation_grants/research_grants

Sponsor: Cystic Fibrosis Foundation
6931 Arlington Road, 2nd Floor
Bethesda, MD 20814

CFF Pilot and Feasibility Awards 1495

The Cystic Fibrosis Foundation offers feasibility grants for developing and testing new hypotheses and/or new methods, as well as to support promising new investigators as they establish themselves in research areas relevant to cystic fibrosis. Proposed work must be hypothesis driven and must reflect innovative approaches to critical questions in cystic fibrosis research. Funding priority will be placed on those projects proposing to better understand the mechanisms behind disease pathophysiology and to develop strategies to prevent or treat it. Up to $40,000 per year (plus 8% indirect costs) for two years may be requested.

Requirements: Electronic submission is required.

Restrictions: The award is not meant to support continuation of programs begun under other granting mechanisms.

Geographic Focus: All States

Date(s) Application is Due: Sep 11

Amount of Grant: Up to 40,000 USD

Contact: Grants Management; (800) 344-4823 or (301) 951-4422; grants@cff.org

Internet: http://www.cff.org/research/cystic_fibrosis_foundation_grants/research_grants

Sponsor: Cystic Fibrosis Foundation
6931 Arlington Road, 2nd Floor
Bethesda, MD 20814

CFF Postdoctoral Research Fellowships 1496

Competitive postdoctoral research fellowships in basic or clinical research are made on an annual basis, renewable for a second year. The fellowships are awarded in areas of basic cellular and metabolic research or in other problem areas pertinent to cystic fibrosis and related chronic and recurrent pulmonary and gastrointestinal diseases of childhood. Preference is shown to recent graduates or those just beginning their investigative careers. A third year is offered on a limited basis to highly qualified candidates. Interested individuals are encouraged to contact the foundation for application procedures and/or to discuss the potential relevance of their work to the objectives of the foundation. Stipends are $39,000 (first year), $40,100 (second year), and $42,300 (optional third year). Research expenses of $3,750 per year are available, as well.

Requirements: These awards are offered to MDs, PhDs, and MD/PhDs interested in conducting basic or clinical research related to cystic fibrosis. U.S. citizenship or permanent resident status is required.

Restrictions: Electronic submission only. Indirect costs are not allowed.

Geographic Focus: All States

Date(s) Application is Due: Sep 11

Amount of Grant: 39,000-46,050 USD

Contact: Grants Management; (800) 344-4823 or (301) 951-4422; grants@cff.org

Internet: http://www.cff.org/research/ForResearchers/FundingOpportunities/TrainingGrants/

Sponsor: Cystic Fibrosis Foundation
6931 Arlington Road, 2nd Floor
Bethesda, MD 20814

CFF Research Grants 1497

Grants are offered by the Cystic Fibrosis Foundation in support of high-quality research projects ranging from basic cellular and metabolic mechanisms to therapy of cystic fibrosis and related chronic and recurrent pulmonary and gastrointestinal diseases of childhood. These grants are broadly oriented to the support of projects for developing and initially testing new hypotheses and/or new methods, or those being applied to problems of cystic fibrosis for the first time. The intent of this award is to enable the investigator to collect sufficient data to compete successfully for long-term support from NIH. Interested individuals are encouraged to contact the foundation for guidelines and/or to discuss the potential relevance of their work to the objectives of this program. Support is available for $90,000 per year (plus 8% indirect costs) for a period of two years, at which time a grant may be competitively renewed for an additional year of funding.

Requirements: Investigators who seek support from the Foundation under these funding mechanisms must submit a Letter of Intent (LOI). Applications may be submitted by established investigators or new investigators starting their independent research careers but may not represent support for continuation of a line of research already established by the applicant.

Geographic Focus: All States

Date(s) Application is Due: Sep 11

Amount of Grant: Up to 90,000 USD

Contact: Grants Management; (800) 344-4823 or (301) 951-4422; grants@cff.org

Internet: http://www.cff.org/research/ForResearchers/FundingOpportunities/ResearchGrants/

Sponsor: Cystic Fibrosis Foundation
6931 Arlington Road, 2nd Floor
Bethesda, MD 20814

CFF Student Traineeships 1498

Student Traineeships are offered by the Cystic Fibrosis Foundation to qualified students in order to introduce them to cystic fibrosis research. Each recipient must work with a faculty sponsor on a research project related to cystic fibrosis. Interested individuals are encouraged to contact the foundation for application procedures and/or to discuss the potential relevance of their work to the objectives of the program. Applications are accepted throughout the year. A maximum of $300 of the award may be used for laboratory expenses with the remainder used as a $1,500 stipend for the trainee.

Requirements: Applicants must be students in or about to enter a doctoral program (MD, PhD, or MD/PhD); senior-level undergraduates planning to pursue graduate training may also apply.

Geographic Focus: All States

Contact: Grants Management; (800) 344-4823 or (301) 951-4422; grants@cff.org

Internet: http://www.cff.org/research/ForResearchers/FundingOpportunities/TrainingGrants/

Sponsor: Cystic Fibrosis Foundation
6931 Arlington Road, 2nd Floor
Bethesda, MD 20814

CFF Third-, Fourth-, and Fifth-Year Clinical Fellowships 1499

The Cystic Fibrosis Foundation offers competitive clinical fellowships for up to five years for physicians interested in cystic fibrosis and other chronic pulmonary and gastrointestinal diseases of children, adolescents, and adults to encourage specialized training early in a physician's career and to prepare well-qualified candidates for careers in academic medicine. The third-, fourth-, and fifth-year fellowship award offers support for additional intense basic and/or clinical research training related to cystic fibrosis. Up to $68,250 may be awarded: $58,250 for stipend and $10,000 for research costs.

Requirements: Applicants and sponsors must submit proposals of the research studies to be undertaken and other specialized training that will be offered during this year. Preference will be given to applicants whose training was supported by the foundation.
Restrictions: Recipients who do not enter a career of academic medicine will be subject to payback provisions. Indirect costs are not allowed.
Geographic Focus: All States
Date(s) Application is Due: Sep 11
Contact: Grants Management; (800) 344-4823 or (301) 951-4422; grants@cff.org
Internet: http://www.cff.org/research/ForResearchers/FundingOpportunities/TrainingGrants/
Sponsor: Cystic Fibrosis Foundation
6931 Arlington Road, 2nd Floor
Bethesda, MD 20814

CFR Edward R. Murrow Press Fellowship **1500**
The Council on Foreign Relations (CFR) awards one fellowship annually to a distinguished foreign correspondent or editor for the purpose of promoting the quality of responsible and discerning journalism that distinguished the work of radio and television journalist Edwin R. Murrow. The fellowship was initially launched in 1949 with help from the Carnegie Foundation and was subsequently renamed to honor Mr. Murrow in 1965. The Edward R. Murrow Press Fellow spends nine months full-time-in-residence at CFR's headquarters in New York. The program enables the fellow to engage in sustained analysis and writing, expand his or her intellectual and professional horizons, and extensively participate in CFR's active program of meetings and events. The fellow will be part of the David Rockefeller Studies Program (CFR's think tank) alongside the program's full-time, adjunct, and visiting fellows, whose expertise extends across the broad range of significant foreign-policy issues facing the United States and the international community. Interested candidates who meet the program's eligibility requirements can apply at the CFR website between January 1 and March 1 each year. Official selections and announcement of the award will be made in April. The fellowship should preferably begin in September. The program awards a stipend of $65,000 as well as a modest travel grant. The fellow is considered an independent contractor rather than an employee of CFR, and is not eligible for employment benefits, including health insurance. Founded in 1921, the Council on Foreign Relations is an independent, nonpartisan membership organization, think tank, and publisher dedicated to being a resource for its members, government officials, business executives, journalists, educators and students, civic and religious leaders, and other interested citizens in order to help them better understand the world and the foreign policy choices facing the United States and other countries. The CFR takes no institutional positions on matters of policy.
Requirements: The program is open only to applicants who have distinguished credentials in the field of journalism and who have covered international news as a working journalist for print, broadcast, or online media widely available in the United States. Applicants are limited to those individuals who are authorized to work in the United States and who will continue to be authorized for the duration of the fellowship. Selection as a Murrow Fellow is based on a combination of the following criteria: depth and breadth of professional experience as a foreign correspondent or editor; firm grounding in foreign policy and international relations; and a clear, creative, and original application statement. With respect to the application statement, candidates must submit a detailed description of their research project as well as their expected CFR participation and any planned fellowship travel. The work plan will be judged on its originality, practicality, potential, likelihood of completion during the fellowship period, and the contribution it will make to the applicant's individual career development as well as to CFR.
Restrictions: CFR does not sponsor for visas.
Geographic Focus: All States
Date(s) Application is Due: Mar 1
Contact: Janine Hill, Director, Fellowship Affairs and Studies Strategic Planning; (212) 734-9489 or (212) 434-9740; fax (212) 434-9800; Fellowships@cfr.org
Internet: http://www.cfr.org/thinktank/fellowships/murrow.html
Sponsor: Council on Foreign Relations
58 East 68th Street
New York, NY 10065

CFR International Affairs Fellowships **1501**
The Council on Foreign Relations (CFR) awards ten 12-month International Affairs Fellowships (IAF)s annually. The purpose of this fellowship program, founded in 1967, is to strengthen the mid-career development of outstanding individuals by helping them acquire and apply foreign-policy skills beyond the scope of their professional and scholarly achievements. The program's distinctive character lies in the contrasting professional experiences fellows receive; fellows from academia and the private sector spend their tenures in public service and policy-oriented settings, while government officials spend their tenures in a scholarly atmosphere free from operational pressure. CFR's Fellowship Affairs Office assists all fellows in finding a suitable affiliation for the year. In addition to providing the opportunity to carry out research, the IAF Program integrates all fellows into the intellectual life of CFR. Fellows who are not placed at CFR during their tenure are invited to attend and participate in select CFR meetings and events. Alumni of the program stay connected with CFR and its prestigious network of professionals and leaders, and convene at CFR's annual IAF Conference in New York City each spring. Interested candidates who meet the program's eligibility requirements can apply at the CFR website between June 1 and October 1 each year. Candidates who are selected as IAF finalists will be notified between December and January, with finalist interviews scheduled in Washington, DC, and New York City between January and February. Official selections and announcement of IAF awards will be made between February and March. Fellowships should preferably begin in September. Though deferment is not an option, requests to do so, for up to one year only, will be considered on a case-by-case basis and under special circumstances. The program awards a stipend of $85,000. Fellows are considered independent contractors rather than employees of CFR, and are not eligible for employment benefits, including health insurance. Founded in 1921, the Council on Foreign Relations is an independent, nonpartisan membership organization, think tank, and publisher dedicated to being a resource for its members, government officials, business executives, journalists, educators and students, civic and religious leaders, and other interested citizens in order to help them better understand the world and the foreign policy choices facing the United States and other countries. The CFR takes no institutional positions on matters of policy.
Requirements: The IAF Program is open only to U.S. citizens and permanent residents between the ages of twenty-seven and thirty-five who are eligible to work in the United States. While a Ph.D. is not a requirement, selected fellows generally hold an advanced degree and possess a strong record of work experience as well as a firm grounding in the field of foreign policy. Selection as an IAF is based on a combination of the following criteria: scholarly qualifications, achievements, and promise; depth and breadth of professional experience; firm grounding in foreign policy and international relations; and an application proposal that focuses on solutions to identified problems in U.S. foreign policy. The proposal will be judged on the proposed work's originality, practicality, potential, likelihood of completion during the fellowship period, and the contribution it will make to the applicant's individual career development. Applicants are encouraged to plan a systematic approach to assessing the major substantive and process issues of their planned research.
Restrictions: The program does not fund pre- or postdoctoral research, work toward a degree, or the completion of projects for which substantial progress has been made prior to the fellowship period. CFR does not sponsor for visas.
Geographic Focus: All States
Date(s) Application is Due: Oct 1
Contact: Janine Hill, Director, Fellowship Affairs and Studies Strategic Planning; (212) 434-9489 or (212) 434-9740; fax (212) 434-9800; Fellowships@cfr.org
Internet: http://www.cfr.org/thinktank/fellowships/iaf.html
Sponsor: Council on Foreign Relations
58 East 68th Street
New York, NY 10065

CFR International Affairs Fellowships in Nuclear Security **1502**
The Council on Foreign Relations (CFR) generally awards two International Affairs Fellowships in Nuclear Security (IAF-NS) annually to university-based scholars. The program places academics in U.S. Government positions or with international organizations to work with practitioners. Its goal is to close the gap between research and practice and to enrich the teaching and scholarship of academics while also benefiting policymakers by exposing them to cutting-edge scholarly research. Interested candidates who meet the program's eligibility requirements must submit a cover letter, a CV, and a proposal (maximum 1,000 words in length). Each applicant should arrange to have two letters of recommendation sent that assess the policy relevance of the applicant's proposed project as well as the applicant's qualifications for carrying it out. All application materials must be submitted to the email address given by November 1. The duration of the fellowship is twelve months, preferably beginning in September. Though deferment is not an option, requests to do so, for up to one year only, will be considered on a case-by-case basis and under special circumstances. The program awards a stipend of $125,000. Fellows are considered independent contractors rather than employees of CFR, and are not eligible for employment benefits, including health insurance. The IAF-NS program is supported by the Stanton Foundation. Founded in 1921, CFR is an independent, nonpartisan membership organization, think tank, and publisher dedicated to being a resource for its members, government officials, business executives, journalists, educators and students, civic and religious leaders, and other interested citizens in order to help them better understand the world and the foreign policy choices facing the United States and other countries. The CFR takes no institutional positions on matters of policy.
Requirements: The IAF-NS is open only to faculty members with tenure or on tenure-track lines at accredited universities who propose to conduct policy-relevant research on nuclear security issues. Qualified candidates must be U.S. citizens or permanent residents who are eligible to work in the United States and be between the ages of twenty-nine and forty. Selection as an IAF-NS is based on a combination of the following criteria: scholarly qualifications, achievements and promise; depth and breadth of professional experience; firm grounding in foreign policy and international relations; and an application proposal that focuses on nuclear security issues. The proposal will be judged on the proposed work's originality, practicality, potential, likelihood of completion during the fellowship period, and the contribution it will make to the applicant's individual career development.
Restrictions: CFR does not sponsor for visas.
Geographic Focus: All States
Date(s) Application is Due: Nov 1
Contact: Janine Hill, Director, Fellowship Affairs and Studies Strategic Planning; (212) 434-9489 or (212) 434-9740; fax (212) 434-9800; fellowships@cfr.org
Internet: http://www.cfr.org/thinktank/fellowships/iaf_nuclear.html
Sponsor: Council on Foreign Relations
58 East 68th Street
New York, NY 10065

CFR International Affairs Fellowships in South Korea **1503**
The International Affairs Fellowship in South Korea (IAF-SK), sponsored by the Asan Institute for Policy Studies, was founded in 2010 to strengthen mutual understanding and cooperation between the rising generations of leaders in the United States and South Korea by expanding cultural exchange and enhancing communication between both countries on global issues. The program assists two to three mid-career scholars and professionals from the public and private sectors to advance their analytic capabilities and broaden their foreign-policy experience in South Korea while affiliated with the

Asan Institute. Fellows are drawn from academia, business, government, media, NGOs, and think tanks. Interested candidates who meet the program's eligibility requirements must submit an application form (available at the website as a downloadable PDF), a CV, and a proposal (as specified in the application form). Each applicant should arrange to have two letters of recommendation sent that assess the policy relevance of the applicant's proposed project as well as the applicant's qualifications for carrying it out. All application materials must be submitted to the email address given by October 1. Candidates who are selected as IAF-SK finalists will be notified in November, with finalist interviews scheduled in Washington, DC, and New York City shortly after. Official selections and announcement of IAF-SK awards will be made around December. The duration of the fellowship is between four and twelve months. The program awards a stipend which covers living expenses in Seoul, South Korea plus international travel and necessary research expenses. Fellows are considered independent contractors rather than employees of CFR, and are not eligible for employment benefits, including health insurance. Founded in 1921, the Council on Foreign Relations is an independent, nonpartisan membership organization, think tank, and publisher dedicated to being a resource for its members, government officials, business executives, journalists, educators and students, civic and religious leaders, and other interested citizens in order to help them better understand the world and the foreign policy choices facing the United States and other countries. The CFR takes no institutional positions on matters of policy.
Requirements: The IAF-SK is only open to U.S. citizens. Applicants do not have to be Korea specialists, and knowledge of the Korean language is not a requirement. While a Ph.D. is not a requirement, selected fellows generally hold an advanced degree and possess a strong record of work experience as well as a firm grounding in the field of foreign policy. Applicants should have a capacity for independent work and be eager to undertake serious foreign-policy analysis. Selection as an IAF-SK is based on a combination of the following criteria: scholarly qualifications, achievements and promise; depth and breadth of professional experience; firm grounding in foreign policy and international relations; and an application proposal that focuses on U.S.-South Korea foreign policy issues. The proposed work will be judged on originality, practicality, potential, likelihood of completion during the fellowship period, and the contribution it will make to the applicant's individual career development. With respect to the proposal, applicants are encouraged to plan a systematic approach to assessing the major substantive and process issues of their planned research.
Restrictions: The program does not fund pre- or postdoctoral research, work toward a degree, or the completion of projects for which substantial progress has been made prior to the fellowship period.
Geographic Focus: All States
Date(s) Application is Due: Oct 1
Contact: Janine Hill, Director, Fellowship Affairs and Studies Strategic Planning; (212) 434-9489 or (212) 434-9740; fax (212) 434-9800; fellowships@cfr.org
Internet: http://www.cfr.org/thinktank/fellowships/iaf_southkorea.html
Sponsor: Council on Foreign Relations
58 East 68th Street
New York, NY 10065

CFR International Affairs in Japan Fellowships **1504**
The International Affairs Fellowship in Japan (IAF-J), sponsored by Hitachi, Ltd., was founded in 1997 to strengthen mutual understanding and cooperation between the rising generations of leaders in the United States and Japan. The program provides three to five mid-career U.S. citizens the opportunity to expand their professional horizons by spending a period of research or other professional activity in Japan. Fellows are drawn from academia, business, government, media, NGOs, and think tanks. In cooperation with the Council on Foreign Relations (CFR), Hitachi, Ltd., assists fellows in finding suitable host organizations in Japan. A non-exclusive list of cooperating institutions in Japan can be found on the IAF-J's webpage at the CFR website. Interested candidates who meet the program's eligibility requirements can apply at the CFR website between June 1 and October 1 each year. Candidates who are selected as IAF-J finalists will be notified between December and January, with finalist interviews scheduled in Washington, DC, and New York City between January and February. Official selections and announcement of IAF-J awards will be made between February and March. The duration of the fellowship is between three and twelve months. The program awards a stipend in yen which covers travel and living expenses in Japan. Fellows are considered independent contractors rather than employees of CFR, and are not eligible for employment benefits, including health insurance. Founded in 1921, the Council on Foreign Relations is an independent, nonpartisan membership organization, think tank, and publisher dedicated to being a resource for its members, government officials, business executives, journalists, educators and students, civic and religious leaders, and other interested citizens in order to help them better understand the world and the foreign policy choices facing the United States and other countries. The CFR takes no institutional positions on matters of policy.
Requirements: The IAF-J is open only to U.S. citizens between the ages of twenty-seven and forty-five. The program is intended primarily for those without substantial prior experience in Japan, although the selection committee has made exceptions when it considered that the fellowship would allow an individual to add a significant new dimension to his or her career. Knowledge of the Japanese language is not a requirement. Fellows are selected on the basis of academic and professional accomplishments and promise; the merits and feasibility of their specific research or action proposals; character and personal qualities conducive to promoting cross-cultural communication and cooperation; and the contribution that the proposed research or professional activity will make to the applicant's individual career development.
Geographic Focus: All States
Date(s) Application is Due: Oct 1
Contact: Janine Hill, Director, Fellowship Affairs and Studies Strategic Planning; (212) 434-9489 or (212) 434-9740; fax (212) 434-9800; fellowships@cfr.org
Internet: http://www.cfr.org/thinktank/fellowships/iaf_japan.html
Sponsor: Council on Foreign Relations
58 East 68th Street
New York, NY 10065

CFR Military Fellowships **1505**
Each year, the Chiefs of Staff of the Army and the Air Force, the Chief of Naval Operations, and the Commandants of the Coast Guard and the Marine Corps nominate an outstanding officer from their respective services as a candidate for a Council on Foreign Relations (CFR) Military Fellowship. CFR usually awards four to five such fellowships annually. The program enables the selected officers to broaden their understanding of international relations by spending a year in residence at CFR's headquarters in New York. Fellows expand their knowledge of international relations through a program of individual study, research and reflection, extensive participation in CFR's active program of meetings and events, and interaction with the CFR's diverse membership. They also assist in arranging several military trips for CFR members during the year. Founded in 1921, the Council on Foreign Relations is an independent, nonpartisan membership organization, think tank, and publisher dedicated to being a resource for its members, government officials, business executives, journalists, educators and students, civic and religious leaders, and other interested citizens in order to help them better understand the world and the foreign policy choices facing the United States and other countries. The CFR takes no institutional positions on matters of policy.
Requirements: Interested candidates from the U.S. Army, U.S. Air Force, U.S. Coast Guard, U.S. Marine Corps, and U.S. Navy should contact their human resources officer to learn more about the fellowship application process.
Geographic Focus: All States
Contact: Janine Hill, Director, Fellowship Affairs and Studies Strategic Planning; (212) 434-9489 or (212) 434-9740; fax (212) 434-9800; fellowships@cfr.org
Internet: http://www.cfr.org/thinktank/fellowships/mil.html
Sponsor: Council on Foreign Relations
58 East 68th Street
New York, NY 10065

CFR Stanton Nuclear Security Fellowships **1506**
The Stanton Nuclear Security Fellowship (SNSF) Program, made possible by a generous grant from the Stanton Foundation, offers younger scholars studying nuclear-security issues the opportunity to spend a twelve-month period at the Council on Foreign Relations' (CFR) offices in New York or Washington, DC, conducting policy-relevant research. While in residence full-time at CFR, fellows will be expected to lead a project of their own design, conduct original research, and write at least one policy-relevant document as well as participate fully in CFR's intellectual life. SNSF fellows will be mentored by the fellows of CFR's David Rockefeller Studies Program. Interested candidates must submit an application form (available at the website as a downloadable PDF), a curriculum vitae, and a proposal outlining the work proposed to conduct. Additionally, CFR must receive two letters of recommendation that assess the policy relevance of the applicant's proposed project as well as the applicant's qualifications for carrying it out. All application materials must be submitted to the email address given by the deadline date. The program awards a stipend of $100,000 for junior (non-tenured) faculty; $75,000 for post-doctoral; and $50,000 for pre-doctoral fellows. Payment will be made in twelve equal monthly installments. Fellowships should preferably begin in September. Fellows are considered independent contractors rather than employees of CFR, and are not eligible for employment benefits, including health insurance. Founded in 1921, the Council on Foreign Relations is an independent, nonpartisan membership organization, think tank, and publisher dedicated to being a resource for its members, government officials, business executives, journalists, educators and students, civic and religious leaders, and other interested citizens in order to help them better understand the world and the foreign policy choices facing the United States and other countries. The CFR takes no institutional positions on matters of policy.
Requirements: The SNSF Program seeks to stimulate the development of the next generation of thought leaders in nuclear security. Qualified candidates must be junior (non-tenured) faculty, post-doctoral fellows, or pre-doctoral candidates from any discipline who are working on a nuclear-security related issue. The program is only open to U.S. citizens and permanent residents who are eligible to work in the United States. The fellowships will be awarded on the basis of academic and professional accomplishments and promise, and on the merits of the specific research projects proposed. Fellows may work on a wide range of issues, including nuclear terrorism, nuclear proliferation, nuclear weapons, nuclear force posture, and, as it relates to nuclear security, nuclear energy.
Restrictions: CFR does not sponsor for visas.
Geographic Focus: All States
Date(s) Application is Due: Dec 14
Contact: Janine Hill, Director, Fellowship Affairs and Studies Strategic Planning; (212 434-9489 or (212) 434-9740; fax (212) 434-9800; fellowships@cfr.org
Internet: http://www.cfr.org/thinktank/fellowships/StantonFellowship.html
Sponsor: Council on Foreign Relations
58 East 68th Street
New York, NY 10065

CFUW Beverley Jackson Fellowship **1507**
Founded in 1919, the Canadian Federation of University Women (CFUW) is a voluntary, non-profit, self-funded bilingual organization offering a variety of fellowships to women university graduates. This fellowship is awarded annually to assist a Canadian woman to pursue graduate work at an Ontario university. The applicant must have been accepted into her place of study at time of application. Guidelines and application forms are available from the federation in the spring.

Requirements: Applicant must be a woman over the age of 35 pursuing graduate work in an Ontario university, must hold at least a bachelor's degree or its equivalent from a recognized university, and must be a Canadian citizen or have held landed immigrant status for one year prior to submitting application.
Restrictions: Eligibility for these fellowships and awards is restricted to women.
Geographic Focus: All States, Canada
Date(s) Application is Due: Nov 1
Amount of Grant: 2,000 CAD
Contact: Betty Dunlop, Fellowships Program Manager; (613) 234-8252 or (613) -234-2732; fax (613) 234-8221; cfuwfls@rogers.com
Internet: http://www.cfuw.org/index.php?option=com_content&task=view&id=73&Itemid=159
Sponsor: Canadian Federation of University Women
251 Bank Street, Suite 305
Ottawa, ON K2P 1X3 Canada

CFUW Bourse Georgette LeMoyne Award **1508**
Founded in 1919, the Canadian Federation of University Women (CFUW) is a voluntary, non-profit, self-funded bilingual organization offering a variety of grants and fellowships to women university graduates. Grants awarded by the CFUW in this category are for graduate study at a university where one of the languages of administration and instruction is French. Application forms are available from the federation in the spring.
Requirements: At time of application, applicant must hold a bachelor's degree or its equivalent from a recognized university, must be a Canadian citizen or have held landed immigrant status for one year prior to submitting application, and must have been accepted at her place of study. Finally, the candidate must be studying in French.
Restrictions: Eligibility for these fellowships and awards is restricted to women.
Geographic Focus: All States, Canada
Date(s) Application is Due: Nov 1
Amount of Grant: 7,000 CAD
Contact: Betty Dunlop, Fellowships Program Manager; (613) 234-8252 or (613) -234-2732; fax (613) 234-8221; cfuwfls@rogers.com
Internet: http://www.cfuw.org/index.php?option=com_content&task=category§ionid=7&id=85&Itemid=88&lang=eng
Sponsor: Canadian Federation of University Women
251 Bank Street, Suite 305
Ottawa, ON K2P 1X3 Canada

CFUW Canadian Home Economics Association (CHEA) Fellowship **1509**
Founded in 1919, the Canadian Federation of University Women (CFUW) is a voluntary, non-profit, self-funded bilingual organization offering a variety of fellowships to women university graduates. The candidate for this Fellowship must be studying one or more aspects in the field of home economics, at the masters or doctoral level. Home economics is a field of study that promotes quality home and family life and includes human development, family relations, human sexuality, financial and resource management, consumerism, human ecology, food science, human nutrition, clothing and textiles, housing and shelter and aesthetics. The applicant must also be already accepted into or be enrolled in her postgraduate program in Canada.
Requirements: Candidates must be studying full-time at the master's or doctoral level in Canada or abroad.
Restrictions: Eligibility for these fellowships and awards is restricted to women.
Geographic Focus: All States, Canada
Date(s) Application is Due: Nov 1
Amount of Grant: 6,000 CAD
Contact: Betty Dunlop, Fellowships Program Manager; (613) 234-8252 or (613) -234-2732; fax (613) 234-8221; cfuwfls@rogers.com
Internet: http://www.cfuw.org/index.php?option=com_content&task=view&id=76&Itemid=159
Sponsor: Canadian Federation of University Women
251 Bank Street, Suite 305
Ottawa, ON K2P 1X3 Canada

CFUW Dr. Alice E. Wilson Awards **1510**
These awards are intended to assist women to do refresher work in their chosen field, to do specialized study, or to retrain in new techniques applicable to their field. Special consideration is given to candidates returning to studies after at least three years absence. Work may be undertaken at a Canadian institution or elsewhere. For guidelines and application forms, contact the federation in the spring.
Requirements: Applicant must have a bachelor's degree or equivalent from a recognized university, must be a Canadian citizen or have held landed immigrant status for one year prior to submitting application and documentation, and must have been accepted at her place of study.
Restrictions: Eligibility for these fellowships and awards is restricted to women.
Geographic Focus: All States, Canada
Date(s) Application is Due: Nov 1
Amount of Grant: 6,000 CAD
Contact: Betty Dunlop, Fellowships Program Manager; (613) 234-8252 or (613) -234-2732; fax (613) 234-8221; cfuwfls@rogers.com
Internet: http://www.cfuw.org/index.php?option=com_content&task=view&id=75&Itemid=159
Sponsor: Canadian Federation of University Women
251 Bank Street, Suite 305
Ottawa, ON K2P 1X3 Canada

CFUW Ecole Polytechnique Commemorative Award **1511**
Founded in 1919, the Canadian Federation of University Women (CFUW) is a voluntary, non-profit, self-funded bilingual organization offering a variety of awards and fellowships to women university graduates. This award is made annually for graduate studies in any field, with special consideration given in the study of issues related particularly to women. Application forms are available in August.
Requirements: At the time of application, an applicant must hold at least a bachelor's degree or equivalent from a recognized university, have been accepted into the proposed place of study, and be a Canadian citizen or have held landed immigrant status for at least one year.
Restrictions: Eligibility for this award is restricted to women.
Geographic Focus: All States, Canada
Date(s) Application is Due: Nov 1
Amount of Grant: 7,000 CAD
Contact: Betty Dunlop, Fellowships Program Manager; (613) 234-8252 or (613) -234-2732; fax (613) 234-8221; cfuwfls@rogers.com
Internet: http://www.cfuw.org/index.php?option=com_content&task=view&id=68&Itemid=159
Sponsor: Canadian Federation of University Women
251 Bank Street, Suite 305
Ottawa, ON K2P 1X3 Canada

CFUW Margaret Dale Philp Award **1512**
This award, donated by the Kitchener-Waterloo Club of CFUW, is open to any woman who holds, at the time of application, a bachelor's degree or equivalent from a recognized university and who wishes to enter, or continue, a program leading to an advanced degree in the field of humanities or social sciences. Special consideration will be given to candidates who wish to specialize in Canadian history. Guidelines and application forms are available from the federation in August. These grants are not awarded every academic year. Check with the office or Web site for current offerings.
Requirements: Applicant must: reside in Canada; and must be a Canadian citizen or have held landed immigrant status for one year prior to submitting application and documentation, must reside in Canada, and must have been accepted at her place of study.
Restrictions: Eligibility for these fellowships and awards is restricted to women.
Geographic Focus: All States, Canada
Date(s) Application is Due: Nov 1
Amount of Grant: 3,000 CAD
Contact: Betty Dunlop, Fellowships Program Manager; (613) 234-8252 or (613) -234-2732; fax (613) 234-8221; cfuwfls@rogers.com
Internet: http://www.cfuw.org/index.php?option=com_content&task=view&id=78&Itemid=159
Sponsor: Canadian Federation of University Women
251 Bank Street, Suite 305
Ottawa, ON K2P 1X3 Canada

CFUW Margaret McWilliams Predoctoral Fellowship **1513**
This predoctoral fellowship is open to any woman who is a full-time student and has completed at least one full calendar year in doctoral studies at the time of application. An applicant may be studying abroad when applying, and the fellowship may be used in Canada or elsewhere. Contact the federation for guidelines and application forms in the spring.
Requirements: Applicant must be a Canadian citizen or have held landed immigrant status for one year prior to submitting application and documentation and must be enrolled in a master's degree program.
Geographic Focus: All States, Canada
Date(s) Application is Due: Nov 1
Amount of Grant: 13,000 USD
Contact: Betty Dunlop, Fellowships Program Manager; (613) 234-8252 or (613) -234-2732; fax (613) 234-8221; cfuwfls@rogers.com
Internet: http://www.cfuw.org/index.php?option=com_content&task=view&id=56&Itemid=159
Sponsor: Canadian Federation of University Women
251 Bank Street, Suite 305
Ottawa, ON K2P 1X3 Canada

CFUW Marion Elder Grant Fellowship **1514**
Founded in 1919, the Canadian Federation of University Women (CFUW) is a voluntary, non-profit, self-funded bilingual organization offering a variety of fellowships to women university graduates. At the time of application for this fellowship, the candidate must be enrolled in a full-time course of studies at any level of a second masters or doctoral program. One of her letters of reference must be provided by her graduate supervisor. All else being equal, preference will be given to the holder of an Acadia University degree.
Requirements: Candidates must be studying full-time at the master's or doctoral level in Canada or abroad.
Restrictions: Eligibility for these fellowships and awards is restricted to women.
Geographic Focus: All States, Canada
Date(s) Application is Due: Nov 1
Amount of Grant: 11,000 CAD
Contact: Betty Dunlop, Fellowships Program Manager; (613) 234-8252 or (613) -234-2732; fax (613) 234-8221; cfuwfls@rogers.com
Internet: http://www.cfuw.org/index.php?option=com_content&task=view&id=57&Itemid=159
Sponsor: Canadian Federation of University Women
251 Bank Street, Suite 305
Ottawa, ON K2P 1X3 Canada

CFUW Memorial Fellowship 1515

Founded in 1919, the Canadian Federation of University Women (CFUW) is a voluntary, non-profit, self-funded bilingual organization offering a variety of fellowships to women university graduates. A candidate for the Memorial Fellowship must be enrolled in a masters degree program in science, mathematics, or engineering in the upcoming academic year. She may be studying abroad.
Requirements: Candidates must be studying full-time at the master's or doctoral level in Canada or abroad.
Restrictions: Eligibility for these fellowships and awards is restricted to women.
Geographic Focus: All States, Canada
Date(s) Application is Due: Nov 1
Amount of Grant: 10,000 CAD
Contact: Betty Dunlop, Fellowships Program Manager; (613) 234-8252 or (613) -234-2732; fax (613) 234-8221; cfuwfls@rogers.com
Internet: http://www.cfuw.org/index.php?option=com_content&task=view&id=67&Itemid=159
Sponsor: Canadian Federation of University Women
251 Bank Street, Suite 305
Ottawa, ON K2P 1X3 Canada

CFUW Ruth Binnie Fellowship 1516

Founded in 1919, the Canadian Federation of University Women (CFUW) is a voluntary, non-profit, self-funded bilingual organization offering a variety of fellowships to women university graduates. At the time of application, the candidate must be enrolled in masters or doctoral studies with a focus on one or more aspects of the field of home economics. According to the CFUW, home economics is a field of study that promotes quality home and family life and includes human development, family relations, human sexuality, financial and resource management, consumerism, human ecology, food science, human nutrition, clothing and textiles, housing and shelter and aesthetics. She may be studying abroad.
Requirements: Candidates must be studying full-time at the master's or doctoral level in Canada or abroad.
Restrictions: Eligibility for these fellowships and awards is restricted to women.
Geographic Focus: All States, Canada
Date(s) Application is Due: Nov 1
Amount of Grant: 6,000 CAD
Contact: Betty Dunlop, Fellowships Program Manager; (613) 234-8252 or (613) -234-2732; fax (613) 234-8221; cfuwfls@rogers.com
Internet: http://www.cfuw.org/index.php?option=com_content&task=view&id=77&Itemid=159
Sponsor: Canadian Federation of University Women
251 Bank Street, Suite 305
Ottawa, ON K2P 1X3 Canada

Chapin Hall International Fellowships in Children's Policy Research 1517

The program is designed to increase research and development capacity in the field of child and family policy and to develop leadership for this enterprise. Fellows will work on existing Chapin Hall projects in one of three broadly defined areas of research: developing conceptual bases for and evaluating implementation of community-based supports and services for all children; developing and using data to improve analysis, planning, and management of services for children; and developing and testing new ideas, policies, and programs for children. Some senior fellows may also work on projects of their own design that fit the Chapin Hall research agenda. The two- to 12-month appointments will provide competitive stipends and housing. A form for the fellowship letter of inquiry is available on the Web site.
Requirements: The fellowships are open to individuals from any country with any of a wide range of disciplinary interests and training, including economics, education, history, human development, law, medicine, psychology, public policy, social work, and sociology. All fellows will be expected to conduct work in English.
Geographic Focus: All States
Contact: Christin Glodek, Fellowship Coordinator; (773) 256-5151; fax (773) 256-5351; internationalprogramcoordinator@chapinhall.org
Internet: http://www.about.chapinhall.org/intprograms/intprograms.html
Sponsor: Chapin Hall Center for Children at the University of Chicago
1313 E 60th Street
Chicago, IL 60637

Charles A. King Trust Postdoctoral Research Fellowships 1518

The Charles A. King Trust was established in 1936 to "support and promote the investigation of human disease and the alleviation of human suffering through improved treatment." In keeping with these principles, the King Trust today supports postdoctoral fellows in the basic sciences and clinical/health services research. The program is designed to support postdoctoral scientists in non-profit academic, medical or research institutions in Massachusetts. Two-year grants ranging from $43,500 to $51,000 per year, inclusive of a $2,000 expense allowance, will be awarded. Awards are made to nonprofit academic, medical, or research institutions in the state of Massachusetts on behalf of the Awards Recipient.
Requirements: Applicants must be working in an academic or medical research institution in the state of Massachusetts and have the required minimum/maximum years of experience. Applicants must apply for the fellowships under the guidance of a Mentor who is an established investigator with an active research program. Mentors are expected to be involved in the planning, execution, and supervision of the proposed research. Mentors must confirm that degrees obtained outside the United States are equivalent to the M.D., D.M.D., Ph.D. or other doctoral degree. United States citizenship is not required; visa documentation is not required. Clinical/Health Services Research: By July 1st of each funding cycle, applicants with clinical responsibilities must have completed residency and be enrolled in a postdoctoral fellowship program. Applicants without clinical responsibilities must have completed at least three years of full-time postdoctoral research experience and no more than five years of postdoctoral experience. Basic Science: By July 1st of each funding cycle, applicants holding the Ph.D., M.D., D.M.D., M.D./Ph.D. or equivalent degrees must have completed at least three years of full-time postdoctoral research experience and no more than five years of postdoctoral experience. The online submission deadline is 12:00 Noon, U.S. Eastern time, on December 16, and the mailed submission deadline is 5:00 pm U.S. Eastern time on December 19.
Restrictions: Only one applicant per Mentor may apply per application cycle; there are not institutional limitations on the number of applicants who may submit applications. Funding begins on July 1. A later start date is permitted but the Award must be activated on or before October 1.
Geographic Focus: Massachusetts
Date(s) Application is Due: Dec 16; Dec 19
Amount of Grant: 43,500-51,000 USD
Contact: Erin Johnstone; (617) 279-2240 ext. 710; ejohnstone@hria.org
Internet: http://hria.org/tmfservices/tmfgrants/king.html
Sponsor: Health Resources in Action
95 Berkeley Street
Boston, MA 02116

Charles Abrams Scholarships 1519

This scholarship is awarded annually to a student enrolled in a graduate planning program leading to a master's degree at one of the five schools at which Charles Abrams taught: Division of Urban Planning, Columbia University; Department of City and Regional Planning, Harvard University; Department of Urban Studies and Planning, Massachusetts Institute of Technology; Department for Urban Affairs and Policy Analysis, New School for Social Research; and Department of City and Regional Planning, University of Pennsylvania. The program is designed to aid students who will pursue careers as practicing planners.
Requirements: Applicant must be a U.S. citizen, must be enrolled in a graduate planning program at one of the five eligible schools, must be able to demonstrate a genuine financial need, and must be nominated by his/her school.
Geographic Focus: New York
Date(s) Application is Due: Apr 30
Amount of Grant: 2,000 USD
Contact: Kriss Blank; (312) 786-6722 or (312) 431-9100; kblank@planning.org
Internet: http://www.planning.org/institutions/scholarship.htm
Sponsor: American Planning Association
122 S Michigan Avenue, Suite 1600
Chicago, IL 60603

Charles Edison Fund Grants 1520

The Charles Edison Fund is an endowed philanthropic institution, incorporated in Delaware in 1948, dedicated to the support of worthwhile endeavors generally within the areas of medical research, science education and historic preservation. Funding is available on a national level but, the majority of institutions and, organizations assisted are based principally in the New York-New Jersey Metropolitan area. This concentration of interest facilitates the efforts of the Trustees to evaluate the work of recipient groups, frequently accomplished by personal visitations. The Fund is a seed money organization, primarily starting projects which would not otherwise get off the ground. The Fund meets three times a year, usually in February or March, June and December at which time requests which have been submitted at least three weeks prior to the meeting will be considered.
Requirements: Public and private schools, universities and colleges, and nonprofits are eligible. The Fund recommends that a grant request be submitted on the requesting organization's letterhead and be signed by an official on behalf of the governing board. The Fund does not require or supply application forms. The request should be detailed, complete and include background information about the organization, a full explanation of the project and its costs, and a financial report, current budget and evidence of tax-exempt status of the requesting organization.
Geographic Focus: New Jersey, New York
Contact: John P. Keegan, President; (973) 648-0500; fax (973) 648-0400; info@charlesedisonfund.org
Internet: http://www.charlesedisonfund.org/thefund.html
Sponsor: Charles Edison Fund
One Riverfront Plaza, 4th Floor
Newark, NJ 07102

Charles H. Farnsworth Trust Grants 1521

The Charles H. Farnsworth Trust was established in 1930 to assist older adults to live in dignity and with independence. In describing the purpose of his legacy, Mr. Farnsworth made clear his interest in supporting housing, particularly in developing affordable housing options, and in providing support services to older adults. Program interests include: development of housing, especially housing with support services; services for elderly persons (i.e. health care, homemaker assistance, and nutritional support to enable the elderly to continue living in the community); and research, planning, and communication to better inform individuals, institutions, and the community at large on ways to improve the quality and quantity of housing and support services for seniors. Capital grants related to construction or renovation of housing for older adults is of particular interest. While general operating grants are provided, requests for support for

new or special projects/programs are preferred. Planning grants investigating strategies for the development of supportive housing for the elderly are encouraged. Capital grants generally range between $25,000 and $250,000. General operating grants are usually no more than $10,000. Applicants must apply online at the grant website. Applicants are strongly encouraged to do the following before applying: review the downloadable state application procedures for additional helpful information and clarifications; review the downloadable online-application guidelines at the grant website; review the trust's funding history (link is available from the grant website); review the online application questions in advance; and review the list of required attachments. These will generally include: a list of board members, financial statements (audited, reviewed, or compiled by independent auditor); an organization summary; a list of other funding sources; an IRS Determination letter; and other required documents. All attachments must be uploaded in the online application as PDF, Word, or Excel files. The Farnsworth Trust has biannual deadlines of February 1 and October 15. Grant applicants for the February deadline will be notified of grant decisions by May 31 and applicants for the October deadline will be notified of grant decisions by December 31.
Restrictions: Grant opportunities are restricted to the Greater Boston area.
Geographic Focus: Massachusetts
Date(s) Application is Due: Feb 1; Oct 15
Amount of Grant: 10,000-250,000 USD
Contact: Michealle Larkins; (866) 778-6859; michealle.larkins@baml.com
Internet: https://www.bankofamerica.com/philanthropic/fn_search.action
Sponsor: Charles H. Farnsworth Trust
225 Franklin Street, 4th Floor, MA1-225-04-02
Boston, MA 02110

Charles H. Hood Foundation Child Health Research Awards **1522**
The Charles H. Hood Foundation was incorporated in 1942 to improve the health and quality of life for children through grant support of New England-based pediatric researchers. The intent of the Child Health Research Awards Program is to support newly independent faculty, provide the opportunity to demonstrate creativity, and assist in the transition to other sources of research funding. Two-year grants of $150,000 ($75,000 per year inclusive of 10% indirect costs) are awarded to researchers who are within five years of their first faculty appointment by the funding start date.
Requirements: Applicants must be working in nonprofit academic, medical or research institutions within the six New England states. Grants support hypothesis-driven clinical, basic science, public health, health services research, and epidemiology projects focused on child health. Applicants must hold a doctoral degree with a demonstrated level of independence confirmed by the Department or Division Chair. The Applicant's potential for a lifetime career as an investigator in pediatric research is also critical in the review process. Applicants are required to devote at least 20% effort to the proposed Hood research project. Applicants who have pending R01s or other large applications to the NIH and other agencies are encouraged to submit proposals to the Hood Foundation. Online submissions must be received by April 8, 12:00 Noon, U.S. Eastern time. Printed copies must be received by April 14, 5:00 pm, U.S. Eastern time.
Restrictions: Applicants are ineligible if they have combined federal and non-federal funding totaling $450,000 or more in direct costs at any time during the two years of the Award. This figure refers to external funding only and not an Applicant's start-up package, other intramural support or the Hood Award. Applicants are ineligible if they are currently or have previously been designated as Principal Investigator or Co-P.I. on an R01, P01, Pioneer Award, New Innovator Award or equivalents from federal agencies such as the National Science Foundation (NSF) or Department of Defense (DOD). Applicants who have completed the R00 phase of a K99/R00 are also ineligible for a Hood Award.
Geographic Focus: Connecticut, Maine, Massachusetts, New Hampshire, Rhode Island, Vermont
Date(s) Application is Due: Apr 8; Apr 14
Amount of Grant: 75,000 USD
Contact: Gay Lockwood; (617) 695-9439; fax (617) 423-4619; glockwood@hria.org
Internet: http://hria.org/tmfservices/tmfgrants/hood.html
Sponsor: Health Resources in Action
95 Berkeley Street
Boston, MA 02116

Charles H. Price II Family Foundation Grants **1523**
The Charles H. Price II Family Foundation, established in Kansas City, Missouri, by a prominent American businessman and former Ambassador of the United States, offers support to educational programs and research primarily in Missouri and California (though funding is also occasionally provided outside of these two states). Grants are given for general operating support and scholarships. There are no specific applications or deadlines with which to adhere, and applicants should approach the Foundation in writing via a letter of application.
Geographic Focus: California, Missouri
Amount of Grant: 500-5,000 USD
Contact: Charles H. Price II, Director; (816) 360-6174 or (816) 360-6176
Sponsor: Charles H. Price II Family Foundation
1 W. Armour Boulevard, Suite 300
Kansas City, MO 64111-2004

Charles H. Revson Foundation Grants **1524**
The foundation awards grants nationwide in its areas of interest, including urban affairs and public policy, education and higher education, biomedical research policy, and Jewish education and philanthropy. Types of support include capital campaigns, continuing support, fellowships, internship funds, program development, and research. Preference is given to requests serving New York, NY. There are no application forms or deadlines. The board meets in April, June, October, and December.
Restrictions: Grants do not support local or national health appeals or direct service programs, individuals, building construction/renovation, book projects, charity events, travel expenses, or budgetary support.
Geographic Focus: All States
Contact: Maria Marcantonio, Grants Administrator; (212) 935-3340; fax (212) 688-0633; info@revsonfoundation.org
Internet: http://www.revsonfoundation.org/guidelines.html
Sponsor: Charles H. Revson Foundation
55 E 59th Street, 23rd Floor
New York, NY 10022

Charles Lafitte Foundation Grants **1525**
The foundation is committed to helping groups and individuals foster lasting improvement on the human condition by providing support to education, children's advocacy, medical research, and the arts. Children's advocacy grants support organizations working to improve the quality of life for children, particularly in relation to child abuse, literacy, foster housing, hunger, and after-school programs. Education grants support innovative programs that work to resolve social service issues, address the needs of students with learning disabilities, provide technology and computer-based education, offer leadership skills education, and support at-risk students. Colleges and universities also receive support for research and conferences. The foundation's medical issues and research grants support healthcare studies, with emphasis on cancer research and treatment, children's health, health education, and promoting healthy living and disease prevention. Art grants support emerging artists and educational art programs.
Requirements: 501(c)3 organizations are eligible. The Foundation does not respond to unsolicited submissions, so submit a letter of inquiry prior to preparing an application.
Restrictions: Political organizations or religious-based programs are not supported.
Geographic Focus: All States
Samples: Girl Scouts of the Jersey Shore, $200,000 (2013). Bridge of Books Foundation, $2,500 (2013). Hand in Hand, $10,000 (2013). Jersey Shore University Medical Center, $700,000 (2013).
Contact: Jennifer Vertetis, President; jennifer@charleslafitte.org
Internet: http://charleslafitte.org/grants/overview/
Sponsor: Charles Lafitte Foundation
29520 2nd Ave SW
Federal Way, WA 98023

Charles Lathrop Parsons Award **1526**
This award is given to recognize outstanding public service by a member of the American Chemical Society. The award is normally given every two years; however, the board of directors may at its discretion reduce the interval to one year if, in its judgment, circumstances in a given year warrant such action. Neither the scientific reputation nor the record of scientific achievement of a member affects his or her eligibility for this award, which is not directed toward recognition of scientific accomplishment or stature. The public service to be recognized may be performed either as a part of or completely outside the regular duties and activities of the nominee's employment.
Requirements: A nominee must be a member of ACS and a citizen of the United States, and must have performed outstanding public service. Any individual, except a member of the award committee, may submit one nomination or seconding letter for the award in any given year. Nominating document consists of a letter of not more than 1000 words containing an evaluation of the nominee's accomplishments and a specific identification of the public service to be recognized.
Restrictions: Current members of the ACS board of directors are ineligible.
Geographic Focus: All States
Date(s) Application is Due: Nov 1
Contact: Felicia Dixon, Awards Administrator; (800) 227-5558 or (202) 872-4408; fax (202) 776-8008; f_dixon@acs.org or awards@acs.org
Internet: http://portal.acs.org/portal/acs/corg/content?_nfpb=true&_pageLabel=PP_ARTICLEMAIN&node_id=1319&content_id=CTP_004498&use_sec=true&sec_url_var=region1
Sponsor: American Chemical Society
1155 Sixteenth Street, NW
Washington, DC 20036-4801

Charles M. and Mary D. Grant Foundation Grants **1527**
The Foundation awards grants primarily in the southeast United States in its areas of interest, including community and economic development, health and human services, environment, and education. The Foundation prefers project support, but considers operating support proposals from organizations with budgets of less than $1 million. Grants are made in September. A minimum of three years must elapse between grant awards. Further information is available at the website.
Requirements: Southeast U.S. 501(c)3 tax-exempt organizations are eligible. Specific grant guidelines are available online with the application, which must be filed online.
Restrictions: No grants are made to individuals or for loans. A minimum of three years must elapse between grant awards.
Geographic Focus: Alabama, Florida, Georgia, Kentucky, Mississippi, North Carolina, South Carolina, Tennessee, Virginia, West Virginia
Date(s) Application is Due: Apr 30
Amount of Grant: 20,000-40,000 USD
Samples: Collaborative for the 21st Century Appalachia, Inc., Charleston, West Virginia, first installment of a $60,000 grant to launch the e-Resource and Training

Center, $30,000 (2011); Georgia Legal Services Program, Inc., Atlanta, Georgia, final installment of a $70,000 grant to support affordable housing work in rural Georgia, $30,000 (2011); Louisiana Bucket Brigade, New Orleans, Louisiana, final installment of a $40,000 grant for the Environmental Justice Corps, $20,000 (2011).
Contact: Casey Castaneda, Program Officer; (212) 464-2487; fax (212) 464-2305; casey.b.castaneda@jpmchase.com
Internet: http://fdncenter.org/grantmaker/grant
Sponsor: Charles M. and Mary D. Grant Foundation
J.P. Morgan Private Bank, Philanthropic Services
Dallas, TX 75222-7237

Charles Stewart Mott Foundation Anti-Poverty Program **1528**
This program focuses on improving education, expanding economic opportunity, building organized communities, and special initiatives as pathways out of poverty. The overall goal is to help people vocalize and mobilize around local concerns, grow through participation in educational opportunities, and attain economic self-sufficiency by engaging more fully in the economy. Types of support include challenge/matching grants, conferences and seminars, demonstration grants, general operating grants, program grants, seed money grants, technical assistance, and training grants. The board meets in March, June, September, and December. Organizations should apply at least four months prior to the start date of the project for which they are seeking funds.
Requirements: Nonprofits and K-12 organizations are eligible to apply. A proposal may be submitted by a church-based or similar organization if the project falls clearly within program guidelines and is intended to serve as broad a segment of the population as the program of a comparable nonreligious organization.
Geographic Focus: All States
Amount of Grant: Up to 100,000,000 USD
Contact: Office of Proposal Entry; (810) 238-5651; fax (810) 766-1753; info@mott.org
Internet: http://www.mott.org/programs/poverty.asp
Sponsor: Charles Stewart Mott Foundation
503 S Saginaw Street, Suite 1200
Flint, MI 48502-1851

Charles Stewart Mott Foundation Grants **1529**
The focus of the foundation's grant making is organized in four programs: civil society; environment; Flint, Michigan; and poverty. Flexibility to investigate new opportunities is maintained through an exploratory and special projects program. The civil society program promotes and supports civil society in the United States; Central/Eastern Europe and South Africa. The environment program supports efforts to achieve a healthy global environment capable of sustaining all forms of life. The Flint program seeks to strengthen the capacity of local institutions, including schools and school districts, in the foundation's home community of Flint, Michigan, to respond to economic and social needs. The poverty program addresses issues that contribute to improved life outcomes for children, youth, and families in low-income communities. Programs in low-income communities that connect schools and communities through systemic reform, improved teaching, leadership development, networking, technical assistance, and advocacy are also applicable. In all grant making, particular interest will be given to fresh approaches to solving community problems in the defined program areas; approaches that can generate long-term support from other sources and/or can be replicated in other communities; public policy development and research and development activities to further existing programs as well as to explore new fields of interest; and approaches and activities that lead to systemic change. Although proposals may be submitted at any time, applicants are strongly encouraged to submit during the first quarter of the year for which funding is requested. Grant expenditures are determined by September 1 of each year. The review process takes up to four months from the time the proposal is received. Therefore, proposals should be submitted at least four months prior to the start of the proposed grant period. Funding for unsolicited proposals is limited. It is recommended that letters of inquiry be submitted instead of a full proposal.
Requirements: 501(c)3 organizations are eligible, including schools and school districts.
Restrictions: Grants are not made to/for individuals; religious activities or programs that serve, or appear to serve, specific religious groups or denominations; or local projects outside the Flint area unless the projects are part of a national demonstration or foundation-planned network of grants that have clear and significant implications for replication in other communities.
Geographic Focus: All States
Contact: Mary A. Gailbreath, Director; (810) 238-5651; fax (810) 766-1753; info@mott.org
Internet: http://www.mott.org/about/programs.aspx
Sponsor: Charles Stewart Mott Foundation
503 S Saginaw Street, Suite 1200
Flint, MI 48502-1851

Charles W. Finley Visiting Scholar Education Grant **1530**
The American Academy of Periodontology Foundation established Finley Visiting Scholar Education Grants to honor the memory of Dr. Charles W. Finley and all of the great periodontal leaders of the past, present, and future. Grants are awarded to accredited periodontal programs to offset honoraria, travel, promotion, and seminars by speakers in periodontology. As required, the grants are being matched by funds from each department or institution. Though the speakers will spend the majority of their time with the periodontics department, both students and faculty, the grant requires that the speaker present a formal program to the faculty and students of the institution on a subject of interest to the entire dental institution. Applications are available on the Web site.
Requirements: Accredited periodontal programs are eligible to apply.
Geographic Focus: All States
Date(s) Application is Due: Feb 14
Amount of Grant: 1,500 USD
Contact: Sharon Mellor; (800) 282-4867, ext. 256; fax (312) 573-3272; sharon@perio.org
Internet: http://www.perio.org/foundation/research.html
Sponsor: American Academy of Periodontology Foundation
737 N Michigan Avenue, Suite 800
Chicago, IL 60611

Charlotte W. Newcombe Doctoral Dissertation Fellowships **1531**
This program provides approximately 28 dissertation awards annually to encourage original and significant study of ethical and religious values in all fields. The selection process and all administration for this continuing fellowship program are handled by the Woodrow Wilson National Fellowship Foundation and is supported by the Charlotte W. Newcombe Foundation. The awards are intended to finance the last full year of dissertation writing. Annual deadline dates may vary; contact program staff for exact dates.
Requirements: An applicant must be a candidate for a PhD or ThD degree in a doctoral program at a graduate school in the United States.
Geographic Focus: All States
Date(s) Application is Due: Nov 1
Amount of Grant: 18,000 USD
Contact: Program Director; (609) 452-7007, ext. 131; charlotte@woodrow.org
Internet: http://www.woodrow.org/newcombe/
Sponsor: Charlotte W. Newcombe Fellowships/Woodrow Wilson National Fellowship Foundation
P.O. Box 5281
Princeton, NJ 08543-5281

Chatlos Foundation Grants **1532**
The Chatlos Foundation supports nonprofit organizations in the U.S.A and around the globe. Support is provided to organizations currently exempt by the Internal Revenue Service of the United States. The Foundation's areas of interest are: Bible Colleges/Seminaries, Religious Causes, Medical Concerns, Liberal Arts Colleges and Social Concerns.
Requirements: Applicants must be U.S. tax-exempt, nonprofit organizations that provide services in the following areas: bible colleges, religious causes, medical concerns, liberal arts colleges, and social concerns. Proposals must include cover letter, specific request, tax-exemption letter, and budget. If proposal is to be considered at board level, additional information will be requested.
Restrictions: The foundation will not accept requests from individual church congregations, individuals, organizations in existence for less than two years as indicated by IRS tax-exempt letter of determination, for education below the college level, for medical research projects, or for support of the arts.
Geographic Focus: All States
Contact: C. J. Leff, Administrator; (407) 862-5077; cj@chatlos.org
Internet: http://www.chatlos.org/AppInfo.htm
Sponsor: Chatlos Foundation
P.O. Box 915048
Longwood, FL 32791-5048

Chautauqua Region Community Foundation Grants **1533**
Grants support projects serving communities in Chautauqua County in its areas of interest, including arts and culture, libraries, education, housing and shelters, children and youth, human services and general charitable giving, and government and public administration. Types of support include general operating support, continuing support, building construction/renovation, equipment acquisition, conferences and seminars, publication, seed grants, emergency funds, and undergraduate and graduate scholarships to individuals.
Requirements: Nonprofit organizations may apply for grants in support of projects serving communities in Chautauqua County, excluding the Fredonia/Dunkirk area, which is served by the Northern Chautauqua Community Foundation.
Geographic Focus: New York
Date(s) Application is Due: Jan 31; Mar 1; Nov 1
Contact: June Diethrick, Grants Coordinator; (716) 661-3392; fax (716) 488-0387; jdiethrick@crcfonline.org
Internet: http://www.crcfonline.org/
Sponsor: Chautauqua Region Community Foundation
418 Spring Street
Jamestown, NY 14701

Chest Foundation/LUNGevity Foundation Clinical Research in Lung Cancer Grants **1534**
The goal of this joint clinical research program is, ultimately, to save the lives of those people afflicted with lung cancer by funding innovative research designed to treat and cure lung cancer. A clinical research award of $75,000 will be granted to a single award recipient for clinical research in lung cancer over a period of 2 years. The amount of $37,500 will be granted after July 1. Upon receipt of the Progress Report received before July 15, the remaining $37,500 will be granted.
Requirements: To be considered a candidate for The Chest Foundation/LUNGevity Foundation Clinical Research Award in Lung Cancer, an individual must: be an ACCP member who has completed at least 2 years of pulmonary or oncology fellowship and be within 5 years of fellowship or no more than 5 years in practice; hold a degree of MD, DO, MB,BCh, PharmD, PhD or its equivalent; can be a citizen of the United States, or have a J-1 or H-1 visa, and be working in a certified United States institution or organization, or be a Canadian or international member who is working in a certified institution or organization in his/her country; and be a clinician in lung cancer research.

Geographic Focus: All States
Date(s) Application is Due: Apr 30
Amount of Grant: 75,000 USD
Contact: Sue Ciezadlo; (847) 498-8363; fax (847) 498-5460; sciezadlo@chestnet.org
Internet: http://www.chestfoundation.org/foundation/clinical/LUNGevityAward.php
Sponsor: Chest Foundation
3300 Dundee Road
Northbrook, IL 60062-2348

Chest Foundation Eli Lilly and Company Distinguished Scholar in Critical Care Medicine Award **1535**

The Eli Lilly and Company Distinguished Scholar in Critical Care Medicine will have a 3-year opportunity to examine issues that relate to critically ill patients. The award is intended to permit the investigation of issues that are not easily supported by traditional funding, such as clinical trials or basic science. Rather, the award would support activities such as the development of public policy, patient education models, or economic analysis of treatment or care delivery in this patient group. A stipend of $50,000 annually over the 3-year term of the program will be provided. A fourth year stipend of $10,000 will also be awarded to the individual who will serve in a mentorship role to the fourth Distinguished Scholar position during his/her inaugural year of service.
Requirements: An applicant must be: an FCCP and hold the degree of MD, DO, MBChB, MBBCh, MBBS, DNSc, PharmD, PhD, or EdD; board-certified in critical care medicine; and a recognized clinician and/or scientist with a specialty in sepsis and/or critical care as evidenced by publications, presentations, and/or peers.
Geographic Focus: All States
Date(s) Application is Due: Apr 30
Amount of Grant: Up to 160,000 USD
Contact: Sue Ciezadlo; (847) 498-8363; fax (847) 498-5460; sciezadlo@chestnet.org
Internet: http://www.chestfoundation.org/foundation/clinical/dsCriticalCare.php
Sponsor: Chest Foundation
3300 Dundee Road
Northbrook, IL 60062-2348

Chest Foundation Geriatric Development Research Awards **1536**

The two-year award is intended to provide the impetus required for long-term career development focused on integrating geriatrics into the sub-specialties of internal medicine. The program awards the grant to an academic internist to develop and implement a basic, clinical, or health services research project focused on a geriatric aspect of chest medicine. The funding can support the salary of the award recipient and/or the purchase of supplies, the salaries of technical personnel, and other resources necessary for the completion of the research project. The award also includes a one-time travel grant to attend the meetings of the American Geriatrics Society and the American College of Chest Physicians in the second year of the award.
Requirements: To be eligible for the award, applicants must: have U.S. citizenship or permanent resident status; hold a degree of MD or its equivalent; have completed a sub-specialty internal medicine fellowship leading to certification in his/her sub-specialty by the American Board of Internal Medicine and be within the first three years of his/her faculty appointment; possess a faculty appointment by July 1 (this faculty appointment should be documented in the required letters of recommendation from the applicant's department chair and division director); be a member of the American College of Chest Physicians; commit 75 percent of his/her professional effort to research activities; develop and implement a basic, clinical, or health services research project focused on a geriatric aspect of chest medicine; and generate and implement a career development plan focused on the geriatrics aspects of chest medicine. This plan must include organizing and interacting with a mentorship team made up of a minimum of four members—the applicant's research mentor as the leader of the research team, a sub-specialist in the applicant's field of chest medicine, a geriatrician, and one other member at the applicant's discretion.
Restrictions: The funding cannot be used to acquire administrative or clerical support. Funding is to cover total costs; no indirect cost funds will be provided.
Geographic Focus: All States
Date(s) Application is Due: Apr 30
Amount of Grant: 53,000 USD
Contact: Sue Ciezadlo; (847) 498-8363; fax (847) 498-5460; sciezadlo@chestnet.org
Internet: http://www.chestfoundation.org/foundation/clinical/geriatricAward.php
Sponsor: Chest Foundation
3300 Dundee Road
Northbrook, IL 60062-2348

Chest Foundation Grant in Venous Thromboembolism **1537**

The Chest Foundation is issuing this Request for Proposal (RFP) that will fund two projects with up to a total grant amount of $180,000 each for a 1-year period. The goal of the grant process is to confer awards to projects that include evidence-based replicable programs and tools to address the current gaps in the management and prophylaxis in venous thromboembolism (VTE). The deadline is February 20.
Requirements: To be considered a candidate for The Chest Foundation Grant in Venous Thromboembolism, an individual must: be a current ACCP member and hold the degree of MD, DO, MBChB, MBBCh, MBBS, DNSc, PharmD, PhD, or EdD; and be a recognized scientist focused on an area of VTE or clinical decision making as evidenced by academic productivity and/or publications and presentations.
Geographic Focus: All States
Date(s) Application is Due: Feb 20
Amount of Grant: Up to 180,000 USD
Contact: Sue Ciezadlo; (847) 498-8363; fax (847) 498-5460; sciezadlo@chestnet.org
Internet: http://www.chestfoundation.org/foundation/clinical/VT_Grant.php
Sponsor: Chest Foundation
3300 Dundee Road
Northbrook, IL 60062-2348

Chestnut Hill Charitable Foundation, Inc Grants **1538**

The foundation awards grants to nonprofit organizations in the greater Boston, MA, area in support of medical research, children and youth, and arts and education. Specifically, the foundation's interests include visual arts, museums, humanities, arts and culture, early childhood education and development, education at all levels, hospitals and medical research (cancer, heart and lungs, AIDS, and diabetes), social services, and general charitable giving. Types of support include general operating support, annual campaigns, capital campaigns, building construction/renovations, programs and projects, seed grants, curriculum development, research, and matching funds. There are no application forms; submit a letter of request.
Requirements: Nonprofit organizations in the greater Boston, MA, area may submit letters of request.
Restrictions: Grants are not awarded to support individuals or political or religious organizations.
Geographic Focus: Massachusetts
Amount of Grant: 1,000-15,000 USD
Contact: Kay Kilpatrick, Director Corporate Giving; (617) 630-2415
Sponsor: Chestnut Hill Charitable Foundation
27 Boylston Street
Chestnut Hill, MA 02167-1700

CHF Charles C. Price Fellowship in Polymer History **1539**

Through the Beckman Center, CHF offers long-term fellowships for periods of up to 9 months to support scholars in residence. Generally, research projects must be in an area of the chemical and molecular sciences, technologies, and industries, broadly construed in terms of subject and time period. The Price Fellowship in particular is open to scholars pursuing research in the history of the chemical sciences and technologies. Preference is given to applicants with projects on the history of polymers.
Requirements: To be eligible applicants must either have a Ph.D. (or equivalent) or be a doctoral candidate at the dissertation stage.
Geographic Focus: All States
Date(s) Application is Due: Feb 15
Amount of Grant: Up to 20,000 USD
Contact: Ashley Augustyniak; (215) 873-8269 or (215) 925-2222; fax (215) 925-1954; aaugustyniak@chemheritage.org or fellowships@chemheritage.org
Internet: http://www.chemheritage.org/research/research-nav1-price.html
Sponsor: Chemical Heritage Foundation
315 Chestnut Street
Philadelphia, PA 19106-2702

CHF Glenn E. and Barbara Hodsdon Ullyot Scholarship **1540**

The Ullyot Scholarship sponsors historical research that promotes public understanding of the chemical sciences. Applications are invited from scholars, graduate students, science writers, and journalists. The scholar will spend at least two months in residence at CHF. There is no specific application form to apply for the fellowship. All printed materials and letters of recommendations must be postmarked by the deadline.
Geographic Focus: All States
Date(s) Application is Due: Feb 15
Amount of Grant: Up to 4,500 USD
Contact: Ashley Augustyniak; (215) 873-8269 or (215) 925-2222; fax (215) 925-1954; aaugustyniak@chemheritage.org or fellowships@chemheritage.org
Internet: http://www.chemheritage.org/research/research-nav3-ullyot.html
Sponsor: Chemical Heritage Foundation
315 Chestnut Street
Philadelphia, PA 19106-2702

CHF Long-Term Fellowships **1541**

Through the Beckman Center, CHF offers long-term fellowships for periods of up to 9 months to support scholars in residence. Research projects must be in an area of the chemical and molecular sciences, technologies, and industries, broadly construed in terms of subject and time period. Named fellowships within this group include the following: the Robert W. Allington Fellowship—for general research in the history of chemical and molecular sciences, technologies, and industries; the Gordon Cain Fellowship in Technology, Policy, and Entrepreneurship—for historical research on the development of the chemical industries; the Sydney M. Edelstein Fellowship—for general research in the history of chemical & molecular sciences, technologies, and industries; and the John C. Haas Fellowships—for research that will enhance public understanding of the chemical industries in relation to environmental, societal, health, and safety issues. There is no specific application form to apply.
Requirements: To be eligible applicants must either have a Ph.D. (or equivalent) or be a doctoral candidate at the dissertation stage.
Geographic Focus: All States
Date(s) Application is Due: Feb 15
Contact: Ashley Augustyniak; (215) 873-8269 or (215) 925-2222; fax (215) 925-1954; aaugustyniak@chemheritage.org or fellowships@chemheritage.org
Internet: http://www.chemheritage.org/research/research-nav1.html
Sponsor: Chemical Heritage Foundation
315 Chestnut Street
Philadelphia, PA 19106-2702

CHF Robert W. Gore Fellowship in Materials Innovation 1542

The Fellowship is open to advanced Ph.D. candidates who possess the research expertise, writing skills, and technical background to conduct historical case studies of materials innovation of the past thirty years and to communicate findings to academic, policy, and industrial audiences. The Fellow will undertake one such case study with input from members of the Center for Contemporary History and Policy. The result will be a white paper published by CHF; the fellow will be encouraged (but not required) to publish variants of the study as dissertation chapters or scholarly articles. The fellowship is one component of the larger Gore Innovation Case Studies Program, which will culminate in a conference and an edited volume. In addition to their participation in these and other Case Studies Program activities, the fellow will be given ample opportunity to pursue their own research and academic interests while at CHF. There is no specific application form to apply.
Requirements: To be eligible applicants must either have a Ph.D. (or equivalent) or be a doctoral candidate at the dissertation stage.
Geographic Focus: All States
Date(s) Application is Due: Feb 15
Amount of Grant: Up to 35,000 USD
Contact: Ashley Augustyniak; (215) 873-8269 or (215) 925-2222; fax (215) 925-1954; aaugustyniak@chemheritage.org or fellowships@chemheritage.org
Internet: http://www.chemheritage.org/research/research-nav1-gore.html
Sponsor: Chemical Heritage Foundation
315 Chestnut Street
Philadelphia, PA 19106-2702

CHF Roy G. Neville Short-Term Fellowships 1543

Roy G. Neville Fellowships are open to historians of science, technology, and allied fields, as well as to historians of the book and of print culture, bibliographers, and librarians, who will make use of the Roy G. Neville Historical Chemical Library. The Neville Library contains approximately 6,000 titles dating from the 15th to the 19th centuries and covers all aspects of the history of chemistry and allied fields. Ph.D., Ph.D. candidates, or equivalent preferred but not required.
Geographic Focus: All States
Date(s) Application is Due: Feb 15
Amount of Grant: 3,000-12,000 USD
Contact: Ashley Augustyniak; (215) 873-8269 or (215) 925-2222; fax (215) 925-1954; aaugustyniak@chemheritage.org or fellowships@chemheritage.org
Internet: http://www.chemheritage.org/research/research-nav2.html
Sponsor: Chemical Heritage Foundation
315 Chestnut Street
Philadelphia, PA 19106-2702

CHF Societe de Chimie Industrielle (American Section) Fellowship 1544

The Societe de Chimie Industrielle (American Section) Fellowship is designed to stimulate public understanding of the chemical industries. Applications are encouraged from writers, journalists, educators, and historians of science, technology, or business. Multimedia, popular book projects, and Web-based projects are encouraged. Applicants must specify how the outcomes of their projects will reach a broad audience. The period of fellowship is three months in residence at CHF anytime from September to August.
Geographic Focus: All States
Date(s) Application is Due: Feb 15
Amount of Grant: Up to 15,000 USD
Samples: David Caudill, Professor, Villanova University School of Law, $15,000; Jo Ann Caplin, Temple University, $15,000.
Contact: Ashley Augustyniak; (215) 873-8269 or (215) 925-2222; fax (215) 925-1954; aaugustyniak@chemheritage.org or fellowships@chemheritage.org
Internet: http://www.chemheritage.org/research/research-nav3-chimie.html
Sponsor: Chemical Heritage Foundation
315 Chestnut Street
Philadelphia, PA 19106-2702

CHF Travel Grants 1545

The Beckman Center for the History of Chemistry at CHF offers grants to cover travel and lodging expenses for researchers who wish to use CHF's wealth of resources in the history of the chemical and molecular sciences for short-term research (1 to 4 weeks) at CHF. Travel grant recipients will have access to the collections of the Othmer Library of Chemical History. Researchers are also encouraged to use CHF's oral history collection and its collection of art, artifacts, archives, and images. Philadelphia and the Delaware Valley also boast numerous area resources for scholars. Grants are usually on the order of $750 per week and are intended to help defray the costs of travel and lodging. The application deadline is ongoing.
Requirements: Applicants must reside more than 75 miles from Philadelphia to be eligible. No more than one travel grant per person per fiscal year (1 July to 30 June) will be awarded.
Geographic Focus: All States
Amount of Grant: 750-3,000 USD
Contact: Ashley Augustyniak; (215) 873-8269 or (215) 925-2222; fax (215) 925-1954; aaugustyniak@chemheritage.org or fellowships@chemheritage.org
Internet: http://www.chemheritage.org/research/research-nav5.html
Sponsor: Chemical Heritage Foundation
315 Chestnut Street
Philadelphia, PA 19106-2702

Chicago Community Trust Fellowships 1546

The Chicago Community Trust Fellowship offers professional development opportunities for both emerging and experienced leaders who are open to new learning, have a track record of accomplishments, have potential for significant community impact and demonstrate commitment to building strong nonprofit and public sectors. Fellowship applicants customize their own professional development plans. The Trust offers two fellowship tracks, one for Emerging Leaders and one for Experienced Leaders. Emerging leaders are awarded up to $30,000 and experienced leaders are awarded up to $60,000 to put these plans into practice. The award may fund salary and benefits during a leave of absence from work, costs related to academic courses, coaching or learning from other practitioners.
Requirements: Candidates for the Trust Fellowship must have a demonstrated commitment to, and passion for, their field and must be currently employed full-time in the nonprofit or public sector in the metropolitan Chicago region, which includes Cook, DuPage, Kane, Lake, McHenry and Will counties. For candidates pursuing the Emerging Leaders track, the following are required: program, budget and supervisory responsibility appropriate to the scale of the organization; at least five years of professional working experience and at least two years of management experience; track record of accomplishments which demonstrates leadership ability and potential for professional growth; and, endorsement from his/her organization's leader. For candidates pursuing the Experienced Leaders track, the following are required: membership in an organization's senior leadership team with responsibilities for organizational strategic planning, program/budget and personnel management; at least 15 years of professional working experience and 10 years of management experience; working relationship with the board of directors; and, endorsement from his/her organization's leader or board chair.
Restrictions: Past recipients of Trust Fellowships, professionals from academic institutions and employees or board members of The Chicago Community Trust and its affiliates are not eligible to apply.
Geographic Focus: Illinois
Date(s) Application is Due: Jun 13
Contact: Anne Blanton, (312) 616-8000 or (312) 616-7955; fellowship@cct.org
Internet: http://www.cct.org/about/partnership-initiative/chicago-community-trust-fellowship
Sponsor: Chicago Community Trust
111 East Wacker Drive, Suite 1400
Chicago, IL 60601

Chicago Community Trust Housing Grants: Advancing Affordable Rental Housing 1547

The Trust is committed to support innovation and excellence, and to fund organizations advancing research, policy, advocacy, housing development and renters' rights, making Chicago a national leader in rental solutions for the 21st century. The Trust will directly solicit proposals supporting a coordinated approach to research, policy and advocacy designed to increase resources for rental and supportive housing, remove obstacles to its preservation, rehabilitation and development, and ensure that it meets standards of quality, affordability, location and sustainability across the region; and will directly solicit proposals from groups working to provide information and resources to renters, especially those living in subsidized housing. Finally, the Trust will also issue a request for proposals from nonprofit developers that directly provide quality affordable housing and services to low- and moderate-income families in innovative, cost-effective and impactful ways.
Requirements: The Trust funds nonprofit agencies with evidence of tax-exempt status under Section 501(c)3 of the Internal Revenue Code that are not classified as private foundations. The Trust also accepts applications from agencies that operate under a nonprofit fiscal sponsor. Organizations must be located within and/or primarily serving residents of Cook County, except for regional, statewide or national projects or research that may benefit a substantial portion of Cook County residents. Organizations must also be non-discriminatory in the hiring of staff or in providing services on the basis of race, religion, gender, sexual orientation, age, national origin or disability.
Restrictions: The Trust will not provide grants for: scholarships; individuals; sectarian purposes (programs that promote or require a religious doctrine); support of single-disease oriented research, treatment or care; the sole purpose of writing, publishing, producing or distributing audio, visual or printed material; the sole purpose of conducting conferences, festivals, exhibitions or meetings; or, reducing operating deficits or liquidating existing debt.
Geographic Focus: Illinois
Date(s) Application is Due: Jan 5
Contact: Juanita Irizarry, Program Officer; juanita@cct.org
Internet: http://www.cct.org/apply/funding-priorities/housing#priority2
Sponsor: Chicago Community Trust
111 East Wacker Drive, Suite 1400
Chicago, IL 60601

Chicago Institute for Psychoanalysis Fellowships 1548

This is a program for advanced trainees in psychiatry, social work, and psychology who have significant interest in psychoanalysis as a body of knowledge and as a framework with which to understand and carry out therapeutic work. It is jointly sponsored by the Chicago Psychoanalytic Society and the Institute for Psychoanalysis. A mentor is assigned to each Fellow, who is available throughout the year-long program to discuss cases, papers, research, etc. In addition, Fellows are guests at Society meetings and conferences held locally. Opportunities are provided for the Fellows to discuss presented papers with their mentors. A monthly conference is held to acquaint Fellows with basic psychoanalytic clinical concepts through readings and discussion. Each Fellow receives a copy of the Annual of Psychoanalysis and other publications.
Geographic Focus: All States
Date(s) Application is Due: Aug 15

Contact: Christine Susman, Director; (312) 922-7474, ext. 324; csusman@chicagoanalysis.org or admin@chicagoanalysis.org
Internet: http://www.chicagoanalysis.org/fellow.php
Sponsor: Chicago Institute for Psychoanalysis
122 S Michigan Avenue
Chicago, IL 60603

Children's Advocacy Center Grants **1549**
The program supports the establishment and expansion of children's advocacy centers, with the goal of helping to prevent child abuse. One-year grants are awarded in the following categories: program support, research, member training, prevention, associate member program development, nonmember training, and recognized chapters. Applicants must designate a steering committee that comprises child-welfare workers, law-enforcement officials, medical doctors, and mental-health professionals. Application and guidelines are available online.
Requirements: 501(c)3 tax-exempt organizations are eligible.
Geographic Focus: All States
Date(s) Application is Due: Sep 13
Contact: Grants Administrator; (800) 239-9950, ext. 116
Internet: http://www.nca-online.org
Sponsor: National Children's Alliance
516 C Street NE
Washington, DC 20002

Children's Cardiomyopathy Foundation Research Grants **1550**
The foundation awards funds to support research related to all forms of cardiomyopathy affecting children under the age of 18 years. The goal of CCF's research program is to advance medical knowledge on the disease and develop more accurate diagnostic methods, life-improving therapies and ultimately a cure. The grant program is designed to provide seed funding to investigators for the testing of initial hypotheses and collecting of preliminary data to help secure long-term funding by the National Institutes of Health and other major granting institutions.
Requirements: Proposals will be accepted on an annual basis for innovative basic, clinical or translational research relevant to the cause or treatment of cardiomyopathy in children. Principal investigators must hold a MD, PhD or equivalent degree and reside in the United States or Canada. The investigator must have a faculty appointment at an accredited U.S. or Canadian institution and have the proven ability to pursue independent research as evidenced by original research in peer-reviewed journals. Guidelines and application forms can be downloaded at the website.
Geographic Focus: All States, Canada
Date(s) Application is Due: Oct 5
Contact: Lisa Yue, Executive Director; (866) 808-2873; fax (201) 227-7016; lyue@childrenscardiomyopathy.org or info@childrenscardiomyopathy.org
Internet: http://www.childrenscardiomyopathy.org/site/grants.php
Sponsor: Children's Cardiomyopathy Foundation (CCF)
P.O. Box 547
Tenafly, NJ 07670

Children's Literature Association Research Fellowships **1551**
The association awards a number of fellowships and scholarships annually to support research and original scholarship associated with serious literary criticism of children's literature. In honor of the achievement and dedication of Dr. Margaret P. Esmonde, proposals that deal with critical or original work in the areas of fantasy or science fiction for adolescents or children will be awarded the Margaret P. Esmonde memorial scholarship. It is expected that the research undertaken will lead to publication and make a significant contribution to the field of children's literature. The award may be used for transportation, living expenses, materials, and supplies. Grant recipients should be prepared to submit either a progress report or a summary of the completed project to the Scholarship Committee in February of the year following the award.
Requirements: Recipients must be members of the association. Application consists of a detailed description of the research proposal, a vita that includes a bibliography of major publications and scholarly achievements, and three letters of reference. Applications and supporting materials should be written in or translated into English.
Restrictions: Officers of the association may not apply. Recipients of a scholarship are not eligible to reapply until the third year from the date of the first award. The award may not be used for obtaining advanced degrees, for textbook writing, for pedagogical projects, or for researching or writing a thesis or dissertation.
Geographic Focus: All States
Date(s) Application is Due: Feb 1
Contact: Scholarship Committee; (616) 965-8180; fax (616) 965-3568
Internet: http://ebbs.english.vt.edu/chla/
Sponsor: Children's Literature Association
P.O. Box 138
Battle Creek, MI 49016-0138

Chiles Foundation Grants **1552**
The foundation has a deep concern for and confidence in the future of Oregon and the Pacific Northwest. Although the foundation has made a steady commitment to the improvement of the quality of life for those who live and work in this area, it is not restricted in its grant making to the Pacific Northwest. The foundation has traditionally made grants to certain select institutions of higher education for business schools, scholarships, and athletics; supports basic research in certain select medical institutions; supports religion through divinity schools and religious education; and believes that the arts and cultural activities of a community are important and supports certain select, established institutions. Types of support include building construction/renovation, equipment acquisition, and scholarship funds. Annual deadline dates may vary; contact Foundation by a letter of Inquiry. The Foundation does not accept unsolicited proposals.
Requirements: The preferred initial method of contact is a phone call to the grants administration office to determine whether a prospective proposal is within guidelines; if so, an applicant will be invited to submit a one-page written preliminary proposal. An application form will be sent after approval of the preliminary proposal by the executive committee.
Restrictions: No support for projects involving litigation. Grants are not made to individuals, for deficit financing, mortgage retirement, or projects and conferences already completed.
Geographic Focus: California, Oregon, Germany
Contact: Earle M. Chiles; (503) 222-2143; fax (503) 228-7079; cf@uswest.net
Sponsor: Chiles Foundation
111 SW Fifth Avenue, Suite 4050
Portland, OR 97204-3643

Chiron Foundation Community Grants **1553**
The foundation's focus areas are health and medicine, education, and community. Four imperatives guide health-care giving: accelerating progress toward the prevention and cure or successful management of cancer through research, education, early detection, and public-policy debate; combating infectious disease through prevention-related programs, educational efforts, and therapeutics targeting at-risk populations, with emphasis on the special needs of children and families; ensuring the availability and safety of the blood supply and promoting the highest standards of care for blood donors and recipients; and supporting initiatives in the international medical community to provide vaccines and immunization services to protect at-risk populations, especially children, against the devastation of crippling and lethal diseases. Four imperatives guide education giving: providing training and professional development opportunities to enable classroom teachers to teach science and math more effectively; increasing opportunities for economically disadvantaged students to continue on to postsecondary education; supporting community-based training programs to prepare underrepresented minorities and underprivileged groups for careers in the biosciences; and encouraging and assisting promising scholars to pursue advanced research in the biosciences and medicine. Four imperatives guide community giving: providing early diagnosis, intervention, and support for children with disabilities or other special physical or emotional needs; providing essential social services such as meals and housing to those in greatest need; empowering individuals to achieve self-sufficiency through employment training and job-skill development; and enriching the character and celebrating the distinctives of local communities. Requests are accepted at anytime, and proposals are reviewed quarterly.
Requirements: 501(c)3 tax-exempt organizations in Chiron communities (San Francisco East Bay Area, including Alameda, Contra Costa, and Solano Counties; Seattle; and Philadelphia) are eligible.
Restrictions: In general, Chiron does not support organizations that do not have 501(c)3 tax status; religious, fraternal, service, or veterans' organizations; civic or cultural organizations that do not serve the areas in which Chiron is located; alumni drives and teacher organizations; memorials; municipal and for-profit hospitals; labor unions; city, municipal, or federal government departments; organizations or causes that do not support the company's commitment to non-discrimination and diversity; projects of national scope or from national organizations not related to health care; matching gifts; individuals, including scholarships (other than those awarded as part of the company-sponsored college scholarship program); travel support; fund-raising activities related to individual sponsorship; and fund-raising dinners other than those for health care or medical research organizations aligned with company research and product interests.
Geographic Focus: California, Pennsylvania, Washington
Contact: Community Relations; (510) 601-6952; Chiron_foundation@chiron.com
Internet: http://www.chiron.com/foundation/index.html
Sponsor: Chiron Foundation
4560 Horton Street
Emeryville, CA 94608

Christensen Fund Regional Grants **1554**
The fund (TCF) focuses its grantmaking on maintaining the biological and cultural diversity of the world by focusing on four geographic regions: the greater South West (Southwest United States and Northwest Mexico); Central Asia and Turkey; the African Rift Valley (Ethiopia); and Northern Australia and Melanesia. Grants within these programs are generally directed to organizations based within those regions or, where appropriate, to internationally based organizations working in support of people and institutions on the ground. In general, grants are one year or less; currently grants up to two years are by invitation only.
Requirements: 501(c)3 nonprofit organizations and non-U.S.A institutions with nonprofit or equivalent status in their country of origin are eligible. Partnerships or associations with U.S.A-based nonprofit organizations are preferred.
Restrictions: The fund does not make grants directly to individuals but rather assists individuals through institutions qualified to receive nonprofit support with which such individuals are affiliated.
Geographic Focus: All States
Amount of Grant: Up to 200,000 USD
Contact: Administrator; (415) 644-1600; fax (415) 644-1601; info@christensenfund.org
Internet: http://www.christensenfund.org/index.html
Sponsor: Christensen Fund
260 Townsend Street
San Francisco, CA 94107

Christian Gauss Award 1555

This award is offered for books published in the field of literary scholarship or criticism. An edition of a literary work is not eligible unless it contains an introduction amounting to a substantial critical or historical estimate that might have been published independently. Entries must be submitted by the publisher who will have obtained a copy of the conditions of eligibility prior to submission. By April 10, if the submission is eligible, applicants will receive the shipping list for judges, and should ship one book to each judge by April 30.
Requirements: Entries must be published in the United States during the 12-month period between May 1 (of the past year) and the April 30 (of the current year) deadline date. Entries will ordinarily be the work of a single author. Exceptions may be made for books written by small teams of scholars working in close collaboration. Authors must be U.S. citizens or residents.
Restrictions: Works of fiction and unpublished manuscripts are not eligible.
Geographic Focus: All States
Date(s) Application is Due: Apr 30
Amount of Grant: Up to 2,500 USD
Contact: Sandra Beasley, Awards Coordinator; (202) 265-3808; fax (202) 986-1601; SBeasley@pbk.org or info@pbk.org
Internet: http://staging.pbk.org/AM/Template.cfm?Section=Book_Awards&Template=/CM/HTMLDisplay.cfm&ContentID=2646
Sponsor: Phi Beta Kappa Society
1606 New Hampshire Avenue NW
Washington, DC 20009

Christine Mirzayan Science and Technology Policy Graduate Fellowship 1556

The Christine Mirzayan Science & Technology Policy Graduate Fellowship Program of the National Academies is designed to engage graduate science, engineering, medical, veterinary, business, public policy, and law students in the analytical process that informs the creation of national policy-making with a science/technology element. As a result, students develop basic skills essential to working in the world of science policy.
Requirements: Applications for the fellowships are invited from graduate students through post-doctoral scholars in any physical, biological, or social science field or any field of engineering, medicine/health, or veterinary medicine as well as business, law, education, and other graduate and professional programs. Graduate students and postdoctoral scholars and those who have completed graduate studies or postdoctoral research within the last 5 years are eligible to apply.
Geographic Focus: All States
Date(s) Application is Due: May 1
Amount of Grant: 8,000 USD
Contact: Director; (202) 334-2455; fax (202) 334-1667; policyfellows@nas.edu
Internet: http://www.nationalacademies.org/internship
Sponsor: National Academies
500 5th Street NW, Room 508
Washington, DC 20001

Christine O. Gregoire Youth/Young Adult Award for Outstanding Use of Tobacco Industry Documents 1557

The award recognizes a person 24 years of age or younger who has made a contribution to the health of the public in the recent past through use of tobacco documents. The award also honors innovation in the use and application of tobacco industry documents to improve the public's health and, where applicable, to further the goals of tobacco prevention and control in order to help build a world where young people reject tobacco and anyone can quit. Those nominated should be individuals who have made a notable impact through innovative use of tobacco industry documents as applied to research, policy, or advocacy.
Requirements: At least one of the following two criteria must be met by the nominated individual: Nominees must have made a remarkable research, policy, or advocacy contribution with the use of tobacco industry documents, and/or; Nominees must have employed innovative, creative approaches to the employment of tobacco industry documents that result in an improvement in the health or public awareness of a community or nation. Nominations may be made by individuals such as colleagues, peers, coworkers, instructors/professors, governments, or CBOs that are qualified to make such a nomination because of their familiarity with the nominee's work and contribution.
Restrictions: Nominees must not have any affiliation with the tobacco industry. Members of the awards committee may not nominate potential recipients, although they may provide written support as part of a nomination package. Employees and directors of the American Legacy Foundation and their relatives are not eligible.
Geographic Focus: All States
Date(s) Application is Due: Feb 6
Amount of Grant: 7,500 USD
Contact: Virginia Lockmuller, Grant Coordinator; (202) 454-5555; fax (202) 454-5599; awards@americanlegacy.org
Internet: http://www.legacyforhealth.org/awards/
Sponsor: American Legacy Foundation
1724 Massachusetts Avenue NW
Washington, DC 20036

Christine Wilson Award 1558

The Society for the Anthropology of Food and Nutrition (SAFN) invites students to submit papers in competition for the annual Christine Wilson Awards presented to outstanding undergraduate and graduate student research papers on that examine topics within the perspectives in nutrition, food studies and anthropology. Winners will be recognized and presented with an award at the annual AAA meeting and receive a year's membership in SAFN.
Requirements: Papers may report on research undertaken in whole or in part by the author. Co-authored work is acceptable, provided that submitting student is first author. Papers must have as their primary focus an anthropological approach to the study of food and/or nutrition and must present original, empirical research; literature reviews are not eligible. Papers that propose a new conceptual framework or outline novel research designs or methodological approaches are especially welcome. Eligibility is further restricted to students (undergraduate or graduate) who are currently enrolled or who were enrolled during in the past academic year.
Geographic Focus: All States
Date(s) Application is Due: Oct 5
Contact: Michael R. McDonald, PhD, (239) 590-7212; mmcdonal@fgcu.edu
Internet: http://foodanthro.com/christine-wilson-award/
Sponsor: American Anthropological Association
2200 Wilson Boulevard, Suite 600
Arlington, VA 22201

Chrysalis Scholarships 1559

The association offers scholarships each year to aid women in finishing their theses and completing MS or PhD degree programs in geoscience fields. The support can be used for anything necessary to assist the candidate in completing her thesis, such as typing, drafting expenses, field work, or child care. The applicant should write a letter stating her background, career goals and objectives, involvement in both the geosciences and her community, how she will use the money, and explaining the length and nature of the interruption to her education. The applicant should also submit two letters of reference.
Requirements: Applicants must be women whose educations have been interrupted for at least one year, candidates for advanced degrees in geoscience field, and who need the money to complete their theses during the current academic year.
Geographic Focus: All States
Date(s) Application is Due: Mar 15
Amount of Grant: 2,000 USD
Contact: Program Contact; chrysalis@awg.org
Internet: http://www.awg.org/AWGFoundation/chrysalis.html
Sponsor: Association for Women Geoscientists
P.O. Box 30645
Lincoln, NE 68503-0645

Churchill Fellowships 1560

The fellowships are awarded each year to Australians to undertake overseas study or to conduct an investigative project needing facilities not available in Australia. The value of an applicant's work to the community and the extent to which it will be enhanced by the applicant's overseas study project are considered, as well as the applicant's merit based on past achievements or on demonstrated ability for future achievements in any field. No prescribed qualifications are required. No specific amount of money is attached to a Churchill Fellowship. The average amount awarded is about **$25,000 AUD**.
Requirements: Australian citizens over the age of 18 are eligible. Applicants must be able to complete a minimum of four weeks and a maximum of approximately eight weeks overseas travel to complete their research. Application must be made in the State of residence of the applicant, irrespective of the place of employment. Application can be made in one State or Territory only. Application forms and guidelines can be found at the sponsor's website.
Restrictions: Fellowships will not be awarded to enable the applicant to obtain higher academic or formal qualifications. Proposals for professional or academic investigations will be considered where the project would primarily benefit the Australian community.
Geographic Focus: All States, Australia
Date(s) Application is Due: Feb 28
Amount of Grant: 25,000 USD
Contact: Norman Owens; (02) 9683 9900 or 0418 667 208; norman.owens@abw.org.au
Internet: http://www.churchilltrust.com.au/index.php
Sponsor: Winston Churchill Memorial Trust (Australia)
GP.O. Box 1536
Canberra, ACT 2612 Australia

CIA Undergraduate Co-Op Engineer/Program Management Engineer 1561

The co-op program seeks motivated undergraduate students pursuing degrees in Materials, Chemical, Electrical, Mechanical, Computer Engineering, Telecommunications, or Computer Science to work in the CIA's Directorate of Science and Technology. Students will work in an R&D environment work with experienced Program and Project Managers and will be tasked to help plan and prepare program plans. Program Management Engineers focus on identifying new materials, technologies, and techniques that can be integrated into existing business areas. Students will gain exposure to management of research and development projects in support of materials and technologies, to include: requirements definition, analysis, product design and operational use, fabrication techniques, test procedures, and contract implementation and planning. Candidates must be willing to explore Program Management engineer concepts and procedures. Students selected for this program are often attending academic institutions with established cooperative education programs, though not exclusively. They work on an alternating semester or quarterly basis and are expected to spend a minimum of three semesters or four quarters (this can include a summer tour) on the job prior to graduation. The undergraduate co-op program provides increasingly challenging assignments that are commensurate with the students' academic training and individual ability to assume additional responsibility. To apply for this program, see web page listed and submit resume online.
Requirements: Minimum qualifications: Students should be enrolled in Materials, Chemical, Electrical, Mechanical, Telecommunications, or Computer Engineering

Program, or a Computer Science BS degree program at an accredited college or university with a GPA of 3.0 or better. Applications for a summer co-op will be accepted until November 1st. Applications for winter, spring, and fall co-ops should be sent 6-9 months before the desired start date.
Restrictions: To be considered for Agency employment, applicants must generally not have used illegal drugs within the last 12 months. The issue of illegal drug use prior to 12 months ago is carefully evaluated during the medical and security processing.
Geographic Focus: All States
Date(s) Application is Due: Nov 1
Contact: Office of Public Affairs; (703) 482-0623; fax (703) 482-1739
Internet: https://www.cia.gov/careers/jobs/view-all-jobs/college-students-scientists-engineers-technology/undergraduate-co-op-program-engineer-program-management-engineer.html
Sponsor: Central Intelligence Agency
Office of Public Affairs
Washington, DC 20505

CIA Undergraduate Internship Program-Analysis **1562**
The undergraduate internship program seeks motivated undergraduate students to serve internships in the analytic offices and centers of CIA's Directorate of Intelligence. Undergraduate interns work on teams with our full-time analysts. They research, analyze, write, and brief on international political, military, economic, scientific, technical, and leadership developments. In addition to their analytic responsibilities, undergraduate interns become familiar with the Agency and Intelligence Community by participating in a range of meetings and projects. The program allows participants and the Agency to assess opportunities for a graduate studies internship or permanent employment following the participant's completion of undergraduate school. See website to apply for internship.
Requirements: The CIA is looking for students with a variety of majors, including: international affairs, area studies, economics, geography, physical sciences, or engineering. Students selected for this program must have completed one full year of undergraduate school and be continuing school on a full time basis following this assignment. Interns generally are required to work either a combination of one semester and one summer internship or two 90-day summer internships. A GPA of 3.0 or better is required. All applicants must successfully complete a thorough medical and psychological examination, a polygraph interview, and an extensive background investigation. U.S. citizenship is required. To be considered for Agency employment, applicants must generally not have used illegal drugs within the last 12 months. The issue of illegal drug use prior to 12 months ago is carefully evaluated during the medical and security processing.
Geographic Focus: All States
Contact: Office of Public Affairs; (703) 482-0623; fax (703) 482-1739
Internet: https://www.cia.gov/careers/jobs/view-all-jobs/college-students-analytical-positions/undergraduate-internship-program-analysis.html
Sponsor: Central Intelligence Agency
Office of Public Affairs
Washington, DC 20505

CICF Efroymson Contemporary Arts Fellowships **1563**
The Central Indiana Community Foundation Efroymson Contemporary Arts Fellowships were established to increase public awareness of contemporary art in Indiana, Illinois, Kentucky, and Ohio. The intent of the fellowships is to reward creativity, and to encourage both emerging and established individual artists by supporting their artistic development. Applications are available at the Foundation's website.
Requirements: Applicants must be at least 25 years old and residents of select counties in Indiana, Kentucky, Illinois, or Ohio, with proof that they have lived there for at least one year. They must also make a commitment to reside in their county for the duration of the one year fellowship. Selection is based on: quality and skill of the artist's work; creativity and uniqueness of the artist's work; artist's commitment to developing his or her work; and the impact of the award on the artist's career.
Restrictions: Applicants should refer to the website for a specific list of each state's counties.
Geographic Focus: Illinois, Indiana, Kentucky, Ohio
Date(s) Application is Due: Sep 23
Contact: Beth Barich; (317) 634-2423, ext. 175; fax (317) 684-0943; bethb@cicf.org
Internet: http://www.cicf.org/efroymson-contemporary-arts-fellowships
Sponsor: Central Indiana Community Foundation
615 North Alabama Street, Suite 119
Indianapolis, IN 46204-1498

CIC Smithsonian Institution Fellowships **1564**
The Committee on Institutional Cooperation (CIC) and the Smithsonian Institution (SI) invite fellowship applications for one-year fellowships to support research in residence at Smithsonian Institution facilities. Fellowships carry a stipend of $30,000. All fields of study that are actively pursued by the museums and research organizations of the Smithsonian Institution are eligible. Under this Fellowship, the Smithsonian Institution and the Fellow's university share the cost of support. Students enrolled at CIC member universities apply through their university graduate school, but their application is submitted through the Smithsonian's Online Application System (SOLAA). Each university will nominate one application for the CIC/SI competition. All unfunded applications will be eligible for consideration in the general Smithsonian Institution competition. Fellows are eligible to receive allowances for related expenses and for temporary relocation to the Smithsonian. All applications are submitted through the Smithsonian Online Academic Appointment System, with the link located at the CIC website.
Requirements: In order to qualify, students must have completed all course work for their programs, admitted into doctoral candidacy, and satisfied all requirements except completion of the dissertation. Only students enrolled at CIC institutions are eligible to apply (University of Chicago; University of Illinois; Indiana University; University of Iowa; University of Michigan; Michigan State University; University of Minnesota; University of Nebraska-Lincoln; Northwestern University; Ohio State University; Pennsylvania State University; Purdue University; and University of Wisconsin-Madison).
Geographic Focus: All States
Date(s) Application is Due: Nov 15
Amount of Grant: 30,000 USD
Contact: Amber Marks, Associate Director; (217) 265-8106; aemarks@staff.cic.net
Internet: http://www.cic.net/Home/Students/SmithsonianFellowship.aspx
Sponsor: Committee on Institutional Cooperation
1819 South Neil Street, Suite D
Champaign, IL 61820-7271

CIC Summer Research Opportunities **1565**
The Summer Research Opportunities Program (SROP) is a gateway to graduate education at CIC universities (University of Chicago; University of Illinois at Urbana-Champaign; Indiana University; University of Iowa; University of Michigan; Michigan State University; University of Minnesota; Northwestern University; Ohio State University; Pennsylvania State University; and Purdue University). The goal of the program is to increase the number of underrepresented students who pursue graduate study and research careers. These summer opportunities help prepare undergraduates for graduate study through intensive research experiences with faculty mentors and enrichment activities. The CIC SROP application is accepted from November 1 through February 10. The application is available at the CIC website, in addition to the application guide and videos about the application process, and comments from former participants in the SROP. Applicants must review the application guidelines before applying, where detailed instructions for all required attachments are discussed. Individual campus profiles with program dates, types of applications, stipends, and whether academic credit is available, are posted on the CIC website.
Requirements: Applications must meet all of the following criteria: a cumulative grade point average of 3.0 or higher; citizenship or permanent residency in the U.S; enrolled in a degree-granting program at a college or university in the U.S., Puerto Rico, or other U.S. territory; completed at least two semesters of undergraduate education by the time of the summer experience; at least one semester of undergraduate education remaining after completing the summer research experience; and a strong interest in pursuing a Ph.D. Selected participants are expected to complete 40 hours of research each week. Participants must devote full-time to the program during the eight- to 10-week session. Due to the intense research schedule, SROP participants may not work an additional job during the summer program.
Restrictions: The SROP is not for students pursuing professional degrees such as law, medical, or MBA. Students who have complete an undergraduate degree are not eligible.
Geographic Focus: All States
Contact: Charity Farber, Program Manager; (217) 333-6314; cfarber@staff.cic.net
Internet: http://www.cic.net/Home/Students/SROP/Home.aspx
Sponsor: Committee on Institutional Cooperation
1819 South Neil Street, Suite D
Champaign, IL 61820-7271

CIC Traveling Scholars **1566**
Traveling Scholar allows doctoral students to spend up to a full academic year pursuing specialized courses of study, researching unique library collections, and working in advanced laboratories and facilities at other CIC institutions, with no change in registration procedures from their home university or additional tuition. Doctoral-level students interested in becoming a Scholar must first consult their advisor to determine whether the off-campus opportunity is likely to enhance the student's course of study, and to verify that it is not available on their home campus. The instructor or advisor at the host campus must also approve before a student completes the online application. The application is available at the CIC website with detailed instructions.
Requirements: Students must be enrolled in a doctoral program at their home institution. Advisors at the student's home and host campuses must agree to allow the student to enroll in a course via the Traveling Scholar before the student submits the online application. Scholars remain registered at their home universities, paying the regular fees charged when they are in residence, and must be registered at both institutions. Scholars are not billed by the host institution for registration or fees, but must provide evidence of tuition paid to the home institution. Host universities provide access to libraries, laboratories, recreational facilities, housing, etc., on the same basis as they are made available to resident graduate students. Scholars may be assessed fees for certain services or benefits, and may be required to secure their own health and medical coverage. Credit earned by Scholars is accepted at home universities for the equivalent level credit (undergraduate or graduate) upon receipt of grade reports or transcripts from the host university. Credit toward a degree, however, will be at the discretion of the home university or department. Visits of Scholars may not exceed two semesters or three quarters regardless of the number of courses taken. In most cases, Scholars physically attend courses at another CIC institution. Technology facilitated/distance courses are occasionally permitted, but students must contact the CIC office before submitting a Scholar application.
Restrictions: Candidates must apply at least six weeks prior to the start of the intended course at the host institution.
Geographic Focus: All States
Contact: Charity Farber; (217) 333-6314 or (217) 333-8475; cfarber@staff.cic.net
Internet: http://www.cic.net/Home/Projects/SharedCourses/TScholar/Introduction.aspx
Sponsor: Committee on Institutional Cooperation
1819 South Neil Street, Suite D
Champaign, IL 61820-7271

CICU James C. Ross Research Fellowship 1567

The nine-month research fellowship focuses on higher education and its relationship to state and national priorities. Preference is given to candidates who have a doctorate or have completed all but their dissertation and have a proven track record of communicating research findings to a general audience. The fellow will work in CICU's Albany office, and is expected to produce a final paper by the first Monday in October of next year. Housing is the responsibility of the fellow. To apply, submit a cover letter, 500-1000 word outline of a proposed course of research, resume, recent publication or unpublished manuscript, and two confidential letters of reference. Annual deadlines may vary; contact program staff for exact dates.

Restrictions: Class work is prohibited during the full-time fellowship, though credit arrangements for fellowship participation are permissible.

Geographic Focus: All States

Date(s) Application is Due: Dec 1

Amount of Grant: 20,000 USD

Contact: Sheila Seery, (518) 436-4781; fax (518) 436-0417; sheila@cicu.org

Internet: http://www.cicu.org/CMT/frontpage/04RossApplication.pdf

Sponsor: Commission on Independent Colleges and Universities

17 Elk Street, P.O. Box 7289

Albany, NY 12224

CIGNA Foundation Grants 1568

The Cigna Foundation has identified four areas for grant consideration: health and human services, education, community and civic affairs, and culture and the arts. Health and education are of primary concern and receive priority. Under education, priority is placed on public secondary education, higher education for minorities, and adult basic education/literacy. The foundation also considers requests from U.S. cultural, educational, and public policy organizations that have international components. Requests are accepted and reviewed throughout the year. Consideration will be given to requests for general operating support, program development, annual campaigns, conferences and seminars, fellowships, scholarship funds and employee-related scholarships, and matching gifts and funds.

Requirements: Organizations with 501(c)3 tax-exempt status are eligible.

Restrictions: The foundation will not consider applications for grants to individuals, organizations operating to influence legislation or litigation, political organizations, or religious activities. In general, the foundation will not consider applications from organizations receiving substantial support through the United Way or other CIGNA-supported federated funding agencies; hospitals' capital improvements; or research, prevention, and treatment of specific diseases.

Geographic Focus: All States

Amount of Grant: 5,000-50,000 USD

Contact: Jill Holliday; (860) 226-2094 or (866) 865-5277; jill.holliday@cigna.com

Internet: http://www.cigna.com/aboutus/cigna-foundation

Sponsor: Cigna Foundation

1601 Chestnut Street, TL06B

Philadelphia, PA 19192-1540

CIMO Scholarships for Advanced Studies of the Finnish Language at Finnish Universities 1569

CIMO offers scholarships for advanced studies of the Finnish language at Finnish universities. The scholarships are meant mainly for degree students of Finnish language and literature at universities outside Finland. Preference is given to applicants who are working on their Master's thesis. Scholarships may be applied for one academic semester (3-5 months). The monthly allowance is approximately 725 euros. No additional allowance for housing is paid. The scholarship holder is also responsible for expenses due to international travel to and from Finland.

Requirements: Students apply directly to CIMO for the grant. Application deadlines are in April for the autumn semester and in October for the Spring semester. The applicant must establish contact with the host university department before applying. The application should preferably be made in Finnish. Download the application form from the website.

Geographic Focus: All States

Date(s) Application is Due: Apr 1; Oct 1

Amount of Grant: 2,175-3,625 EUR

Contact: Makela Tarja; +358 (0)207 868 556; makela.tarja@cimo.fi

Internet: http://www.studyinfinland.fi/scholarships/ungergraduate_studies/finnish_studies_and_research.html

Sponsor: Center for International Mobility

P.O. Box 343, Hakaniemenkatu 2

Helsinki, 00531 Finland

Cintas Foundation Fellowships 1570

Cintas Fellowships acknowledge creative accomplishments and encourage excellence in architecture, literature, music composition and the visual arts. Eligibility is limited to artists of Cuban citizenship or direct descent (having a Cuban parent or grandparent). Applications are reviewed by panels of qualified members from each artistic discipline. The Board of the Cintas Foundation, Inc., awards fellowships based on the recommendations of the panels. Cintas Fellows are free to pursue their artistic activities as indicated in their application. The fellowships are not awarded for academic study or research, or to performing artists. Fellowships are not awarded more than twice to the same person. Second awards are granted only to candidates demonstrating outstanding artistic development. Fellowships are awarded annually in the amount of $10,000 each and are paid in quarterly stipends, beginning in September, for twelve consecutive months. The award is currently considered taxable income. The Cintas Foundation Fiscal year starts on September 1st and ends on August 31st of the following year.

Requirements: Creative artists, living outside of Cuba, of Cuban citizenship or direct lineage (having a Cuban parent or grandparent) are eligible.

Restrictions: The fellowships are not awarded for academic study or research, or to performing artists. Fellowships are not awarded more than twice to the same person. Second awards are granted only to candidates demonstrating outstanding artistic development.

Geographic Focus: All States

Date(s) Application is Due: Aug 1

Amount of Grant: 10,000 USD

Contact: Jessica Brodsky, (305) 237-7722; fax (305) 237-7711; jbrodsky@mdc.edu

Internet: http://www.cintasfoundation.org/fellowships.htm

Sponsor: Cintas Foundation

300 Northeast 2nd Avenue

Miami, FL 33132

CIRCLE Civic Education at the High School Level Research Grants 1571

The program seeks research that will help educators and policymakers to improve civic outcomes for U.S. students of high-school age (roughly 14 to 18). Civic outcomes include, but are not limited to knowledge of politics, democracy and civil society; knowledge of social issues; values such as tolerance, trust, patriotism, concern for others' rights and well-being, and efficacy (the belief that one can make a difference); skills and habits of deliberating about public issues and participating in politics and community affairs; volunteering and membership in voluntary and/or nonprofit groups; and intentions to vote or to consider careers in public service (in the government or nonprofit sectors). CIRCLE is interested in research on interventions and reforms that may enhance civic outcomes. These interventions and reforms include, but are not limited to programs of civic education and classes on history, democracy, or law; approaches to the teaching of other disciplines that may have civic benefits; co-curricular activities, including student government and student media; service-learning; games and simulations that involve political or civic issues; student voice or participation in the governance of their schools; the basic structure of high schools (including their size, focus, requirements, climate, admissions criteria, or composition); professional development for teachers, so long as the effects on students can be assessed; and after-school or community-based programs, insofar as these have the potential to reach large numbers of adolescents or to change mainstream education. In most CIRCLE-funded research projects, the outcomes will be civic knowledge, values, skills, or behaviors. However, research that explores whether being civically engaged helps academic outcomes or positive adolescent development also will be considered. The listed application deadline is for letters of inquiry; full proposals are by invitation.

Requirements: CIRCLE welcomes proposals from academics, students (especially PhD candidates at the dissertation stage), independent scholars, practitioners, and research nonprofits and firms. CIRCLE also welcomes proposals from youth of high school age, perhaps working in partnership with adults. Such youth-led research proposals will be evaluated separately and not compared directly to proposals from adults.

Restrictions: CIRCLE funds rigorous research, not advocacy, education, or other forms of practice.

Geographic Focus: All States

Date(s) Application is Due: Dec 15

Amount of Grant: Up to 100,000 USD

Contact: Carrie Donovan, (301) 405-2790; cdonovan@umd.edu

Internet: http://www.civicyouth.org/grants/index.htm

Sponsor: University of Maryland

School of Public Policy

College Park, MD 20742

CIU.S. Neporany Research and Teaching Fellowship 1572

A research and training fellowship in Ukrainian studies will be awarded by the Osyp and Josaphat Neporany Educational Fund. The fellowship will be tenable at any university with research facilities at which the fellow's academic Ukrainian studies specialty may be pursued and the fellow enabled to teach a course related to the specialty. The duration of the fellowship will normally be one term, i.e., half of the academic year. This may be extended where the applicant has been successful in receiving supplemental funding from other sources. Institutions and departments who wish to explore the possibility of hosting a fellow also are encouraged to request information.

Requirements: Applicants must hold doctorates, or have equivalent professional achievement, in Ukrainian studies.

Geographic Focus: All States

Date(s) Application is Due: Mar 1

Amount of Grant: Up to 20,000 CAD

Contact: Coordinator; (780) 492-2973; fax (780) 492-4967; cius@ualberta.ca

Internet: http://www.ualberta.ca/CIU.S./cius-grants.htm

Sponsor: Canadian Institute of Ukrainian Studies

University of Alberta, 450 Athabasca Hall

Edmonton, AB T6G 2E8 Canada

CJ Foundation for SIDS Research Grants 1573

In support of its mission, the CJ Foundation for SIDS (Sudden Infant Death Syndrome) is strongly committed to providing grant support to best-in-class researchers who work toward the ultimate eradication of SIDS and other sleep-related infant and toddler deaths. Applications currently are by invitation only. Preference is given to past grantees and pilot projects that are gathering data for eventual submission to NIH for funding. Applications are reviewed by the Foundation's Medical Advisory Board. Interested applicants should contact the foundation for further information.

Geographic Focus: All States

Contact: Wendy Jacobs, Assistant Executive Director, Programs & Grants; (866) 314-7437 or (551) 996-5111; fax (551) 996-5326; wendy@cjsids.com or info@cjsids.org
Internet: http://www.cjsids.org/grants/grants-overview.html
Sponsor: CJ Foundation for SIDS
30 Prospect Avenue
Hackensack, NJ 07601

Claneil Foundation Grants 1574
The foundation awards grants to nonprofit organizations in its areas of interest, including community development, education, arts and culture, environment, health, women, domestic abuse, and social services. Types of support include capital campaigns, equipment acquisition, continuing support, building construction/renovation, exchange programs, and research. Deadlines listed are for letters of intent. Application forms are required.
Requirements: Nonprofit organizations are eligible. Preference is given to organizations serving Pennsylvania counties including Bucks, Montgomery, Chester, Delaware, and Philadelphia.
Restrictions: Grants are not awarded to governmental organizations, individuals, or religious organizations.
Geographic Focus: Pennsylvania
Date(s) Application is Due: Jun 30; Dec 15
Amount of Grant: 3,000-50,000 USD
Contact: Cathy Weiss, Executive Director; (610) 941-1131; cweiss@claneil.com
Internet: http://www.claneil.org
Sponsor: Claneil Foundation
630 W Germantown Pike, Suite 400
Plymouth Meeting, PA 19462

Clarence E. Heller Charitable Foundation Grants 1575
The charitable foundation supports nonprofit organizations, with priority given to proposals from California, in its areas of interest, including environment and health—to prevent serious risk to human health from toxic substances and other environmental hazards by supporting programs in research, education, and policy development; management of resources—to protect and preserve the earth's limited resources by assisting programs that demonstrate how natural resources can be managed on a sustainable and an ecologically sound basis, and supporting initiatives for sustainable agriculture, and for promoting the long-term viability of communities and regions; music—to encourage the playing, enjoyment, and accessibility of symphonic and chamber music by providing scholarship and program assistance at selected community music organizations and schools, and by helping community-based ensembles of demonstrated quality implement artistic initiatives, diversify and increase audiences, and improve fund-raising capacity; and education—to focus on support for programs that improve the teaching skills of educators and artists in environmental and arts education. Types of support include continuing support, general operating expenses, publications, research, seed money, and special projects. The foundation's board meets typically in March, June, and October. Letters of inquiry are accepted at any time.
Requirements: 501(c)3 tax-exempt organizations are eligible.
Restrictions: Grants are not made to individuals.
Geographic Focus: California
Amount of Grant: 5,000-600,000 USD
Contact: Bruce Hirsch; (415) 989-9839; fax (415) 989-1909; info@cehcf.org
Internet: http://cehcf.org/app_info.html
Sponsor: Clarence E. Heller Charitable Foundation
44 Montgomery Street, Suite 1970
San Francisco, CA 94104

Clark-Huntington Joint Bibliographical Fellowship 1576
Sponsored jointly by the Clark Library and the Huntington Library, this two-month fellowship provides support for bibliographical research in early modern British literature and history as well as other areas where the two libraries have common strengths. The Application form is an online submission, and the applicant should forward a description of research plans, specifying any material required at the Clark and the Huntington Libraries (maximum of 1,000 words, double-spaced).
Requirements: Applicants should hold a PhD degree or have appropriate research experience.
Geographic Focus: All States
Date(s) Application is Due: Feb 1
Contact: Camie Howard-Rock, Fellowship Coordinator; (310) 206-8552; fax (310) 206-8577; camie@humnet.ucla.edu or clarklib@humnet.ucla.edu
Internet: http://www.humnet.ucla.edu/humnet/c1718cs/applic4.htm
Sponsor: UCLA Center for 17th- and 18th-Century Studies
405 Hilgard Avenue, 310 Royce Hall
Los Angeles, CA 90095-1404

Clark Library Short-Term Resident Fellowships 1577
These short-term fellowships facilitate access to the Department of Special Collections for researchers and scholars residing outside of the Los Angeles area. Fellowships are held April through December for periods of one to three months. The department collects primary resources in the humanities and social sciences. Application requirements include a cover letter, a curriculum vita, a brief outline of the research and specific collections to be used (two-page maximum), dates to be spent in residence, and three letters of recommendation. Information on the library's holdings, application, and guidelines are available on the Web site.
Requirements: Scholars holding PhD or equivalent degrees involved in a project suitable to the collection are eligible to apply.
Geographic Focus: All States
Date(s) Application is Due: Feb 1
Amount of Grant: 2,000-6,000 USD
Contact: Coordinator; (323) 731-8529; fax (323) 731-8617; clarklib@humnet.ucla.edu
Internet: http://www.humnet.ucla.edu/humnet/c1718cs/Postd.htm
Sponsor: William Andrews Clark Memorial Library
2520 Cimarron Street
Los Angeles, CA 90018

Claude Pepper Foundation Grants 1578
The foundation makes grants primarily to support the work of the Claude Pepper Center and the Pepper Institute on Aging and Public Policy, both located at Florida State University, Tallahassee, FL. The foundation makes limited grants to other organizations to continue the work and vision of Claude and Mildred Pepper. The foundation also supports a visiting scholars program and an oratory competition for Florida students. Grants are usually made for a period of one year and except in rare instances no grant will be made for longer than a three-year period of time. Guidelines are available online.
Requirements: 501(c)3 tax-exempt organizations are eligible.
Geographic Focus: All States
Contact: John T. Herndon, Executive Director; 850) 644-9309; fax (850) 644-9301; herndon@claudepepperfoundation.org
Internet: http://www.claudepepperfoundation.org/programs_grants.cfm
Sponsor: Claude Pepper Foundation
636 West Call Street
Tallahassee, FL 32306-1122

Clay Foundation Grants 1579
The primary mission of the foundation is to promote and enhance the quality of life of the citizens of West Virginia. Initially, grants will be confined to organizations or projects within the greater Kanawha Valley that significantly affect its residents. The foundation considers a broad range of organizations, though special interest is given to programs in the field of aging; health care, research, and education; vocational education; and services to disadvantaged youth and their families. Applicants are asked to submit a preliminary letter. If foundation priorities and resources permit consideration of the request, a detailed formal proposal may be requested. There are no deadline dates for the submission of preliminary letters. The board meets four times annually, generally in January, April, July, and October, at which times applications are considered and grants made.
Requirements: IRS 501(c)3 tax-exempt organizations in West Virginia are eligible.
Restrictions: Grants are not made for ongoing normal operations, debt retirement or operational deficits, contributions to endowment or scholarship funds, annual appeals or other fund-raising events, national fund-raising campaigns, religious organizations for religious purposes, or grants to conduit organizations.
Geographic Focus: West Virginia
Amount of Grant: 5,000-75,000 USD
Contact: Charles Avampato, President; (304) 344-8656; fax (304) 344-3805
Sponsor: Clay Foundation
1426 Kanawha Boulevard E
Charleston, WV 25301

Cleveland-Cliffs Foundation Grants 1580
Contributions are made to nonprofit organizations to enhance the quality of life of Cleveland-Cliffs Inc employees and in recognition of a corporate responsibility toward educational, health, welfare, civic, and cultural matters within the communities where the company operates. The foundation was formed for the purpose of making contributions to groups organized and operated exclusively for religious, charitable, scientific, literary, or educational purposes and for the prevention of cruelty to children or animals. Types of support include general operating support, annual campaigns, capital campaigns, building/renovation, professorships, scholarship funds, research, and employee matching gifts. The foundation's major emphasis is on supporting education through a matching gift program and direct contributions to educational institutions. Requests for support must be in writing.
Requirements: Nonprofit organizations in the mining communities in which Cleveland-Cliffs Inc operates, including Alabama, Michigan, Minnesota, Ohio and, West Virginia are eligible.
Geographic Focus: Alabama, Michigan, Minnesota, Ohio, West Virginia
Amount of Grant: 250-50,000 USD
Contact: Dana W. Byrne, Vice President; (216) 694-5700; fax (216) 694-4880; publicrelations@cleveland-cliffs.com
Internet: http://www.cliffsnaturalresources.com/Development/CommunityRelations/Pages/Cleveland-CliffsFoundation.aspx
Sponsor: Cleveland-Cliffs Foundation
1100 Superior Avenue, Suite 1500
Cleveland, OH 44114-2589

Clowes ACS/AAST/NIGMS Jointly Sponsored Mentored Clinical Scientist Development Award 1581
The American College of Surgeons and American Association for the Surgery of Trauma announce a program that will provide supplemental salary funding of up to $75,000 per year to an individual who has received a Mentored Clinical Scientist Development Award (K08/K23) from the National Institute for General Medical Science (NIGMS). This award is directed at surgeon-scientists working in the early stages of their research careers. The award supports a three-, four-, or five-year period of supervised research experience that may integrate didactic studies with laboratory or clinical research. The award program offers a means to facilitate the career development of individuals pursuing

careers in trauma surgery research by enhancing salary support over and above that offered by the K08/K23 mechanism.
Requirements: Awardees must be members in good standing of the College and eligible for membership in AAST.
Restrictions: Pre-existing applications and awards are not eligible for consideration.
Geographic Focus: All States
Date(s) Application is Due: Oct 12
Amount of Grant: Up to 75,000 USD
Contact: Kate Early; (312) 202-5281; fax (312) 202-5021; kearly@facs.org
Internet: http://www.facs.org/memberservices/acs-aast-nigms.html
Sponsor: American College of Surgeons
633 N Saint Clair Street
Chicago, IL 60611-3211

CMA Foundation Grants **1582**
The foundation seeks to improve the quality of life for central Ohio residents through the promotion of wellness, prevention of disease, and delivery of services and research. The foundation will review funding requests for matching or challenge grants, multiyear commitments, and capital expenditures on a project-by-project basis. The foundation is the leading local source of funding for health care projects benefiting the local community. Application may be made online.
Requirements: Nonprofit organizations in central Ohio, including Franklin, Delaware, Fairfield, Licking, Madison, Pickaway, and Union Counties, are eligible.
Geographic Focus: Ohio
Contact: Program Contact; (614) 240-7410; yourthoughts@goodhealthcolumbus.org
Internet: http://www.cmaf-ohio.org/cmaf/grants.html
Sponsor: Columbus Medical Association Foundation
431 E Broad Street
Columbus, OH 43215

CMAP Physician Advocacy Fellowship **1583**
This program supports doctors to develop or enhance their advocacy skills by implementing a project in partnership with an advocacy organization. The fellowship seeks to make advocacy a core professional value for physicians by developing a cadre of advocates with expertise in achieving system or policy-level social change at the local, state, and national level. Projects must be focused within the United States and should identify system or policy level changes as the outcomes of the fellowship work. Application guidelines are available online. Deadline listed is for short proposals; full proposals are by invitation.
Requirements: U.S. physicians at all stages of their careers are eligible; however, the most competitive applicants are practicing physicians. Prospective applicants must secure the commitment of an advocacy organization that is prepared to house, mentor, and support them throughout the fellowship period. A list of advocacy organizations that are interested in participating in the fellowship is available online. Applicants also may apply with organizations other than those listed online.
Geographic Focus: All States
Date(s) Application is Due: Oct 7
Contact: Claudia Calhoon; (212) 342-4769; fax cmc100@columbia.edu
Internet: http://www.cmap.columbia.edu/research_fellowship.shtml
Sponsor: Columbia University
630 W 168th Street, P&S Box 11
New York, NY 10032

CMS Hispanic Health Services Research Grants **1584**
The purpose of the grant is to implement Hispanic American health services research activities to meet the needs of diverse CMS beneficiary populations. The program seeks competitive applications for small applied research projects that relate to identifying and evaluating solutions for eliminating health disparities among Hispanic Americans. Funding is available for grants to implement research related to health care delivery and health financing issues affecting Hispanic American communities, including issues of access to health care, utilization of health care services, health outcomes, quality of services, cost of care, health and racial disparities, socio-economic differences, cultural barriers, managed care systems, and active ties related to health screening, prevention, outreach, and education.
Requirements: Investigators should be associated with a university, college, community-based health organization, or a professional association that has a health services research component. Researchers are encouraged to use CMS data as part of their research projects. Researchers are also expected to become involved in the design, implementation, and operation of research projects that address health care issues such as financing, delivery, access, quality, and barriers affecting the Hispanic American community. Applicants must meet one of the following three requirements in order to qualify for funding under this grant program: 1) A health services/disparities researcher at an university or college offering a Ph.D. or Master's Degree Program in one or more of the following disciplines Allied Health, Gerontology, Health Care Administration, Health Education, Health Management, Nursing, Nutrition, Pharmacology, Public Health, Public Policy, Social Work; or 2) a member of a community-based health organization with a Hispanic health services research component; or 3) a member of a professional association focusing on Hispanic health services and health disparities issues. Applications must be submitted electronically (via grants.gov). CMS will accept a hard copy if the applicant is unable to access grants.gov or is having serious problems in sending the application electronically.
Restrictions: Grant funds may not be used for any of the following: to provide direct services to individuals except as explicitly permitted under the grant solicitation; to match any other Federal funds; or, to provide services, equipment, or supports that are already the legal responsibility of another party under Federal law.
Geographic Focus: All States
Date(s) Application is Due: Jul 3
Amount of Grant: 200,000-250,000 USD
Contact: Joi Grymes, (410) 786-7251; Joi.Grymes@cms.hhs.gov
Internet: http://www.cms.hhs.gov/ResearchDemoGrantsOpt/03_Hispanic_Health_Services.asp#TopOfPage
Sponsor: Centers for Medicare and Medicaid Services
7500 Security Boulevard, Room C2-21-15
Baltimore, MD 21244-1850

CMS Historically Black Colleges and Universities (HBCU) Health Services Research Grants **1585**
The purpose of the grant program is to support researchers in implementing health services research activities to meet the needs of diverse CMS beneficiary populations. The goals of the grant program are to: 1) encourage HBCU health services researchers to pursue research issues which impact the Medicare, Medicaid, and SCHIP (State Children's Health Insurance Program) programs, 2) assist CMS in implementing its mission focusing on health care quality and improvement for its beneficiaries, 3) assist HBCU researchers by supporting extramural research in health care capacity development activities for the African American communities, 4) increase the pool of HBCU researchers capable of implementing the research, demonstration, and evaluation activities of CMS, and 5) assist in fostering interuniversity communication and collaboration regarding African American health disparity issues. Funding is available for grants to implement research related to health care delivery and health financing issues affecting African American communities, including issues of access to health care, utilization of health care services, health outcomes, quality of services, cost of care, health and racial disparities, socio-economic differences, cultural barriers, managed care systems, and activities related to health screening, prevention, outreach, and education.
Requirements: To be eligible for grants under this program, an organization must be an HBCU (Historically Black College or University) and meet one of the following three *Requirements:* 1) offer a Ph.D. or Master's Degree Program in one or more of the following disciplines-Allied Health, Gerontology, Health Care Administration, Health Education, Health Management, Nursing, Nutrition, Pharmacology, Public Health, Public Policy, Social Work; or 2) have a School of Medicine; or 3) a member of the National HBCU Network for Health Services and Health Disparities. All proposals should describe research to be conducted with relevance to the CMS Medicare, Medicaid, and SCHIP programs and which area of Healthy People 2010 is served by this project. Applications must be submitted electronically (via grants.gov). CMS will accept a hard copy if the applicant is unable to access grants.gov or is having serious problems in sending the application electronically.
Restrictions: Grant funds may not be used for any of the following: to provide direct services to individuals except as explicitly permitted under the grant solicitation; to match any other Federal funds; to provide services, equipment, or supports that are already the legal responsibility of another party under Federal law.
Geographic Focus: All States
Date(s) Application is Due: Jul 2
Amount of Grant: 200,000-250,000 USD
Contact: Joi Grymes, (410) 786-7251; Joi.Grymes@cms.hhs.gov
Internet: http://www.cms.hhs.gov/ResearchDemoGrantsOpt/02_Historically_Black_Colleges_and_Universities.asp#TopOfPage
Sponsor: Centers for Medicare and Medicaid Services
7500 Security Boulevard, Room C2-21-15
Baltimore, MD 21244-1850

CMS Research and Demonstration Grants **1586**
The general purpose of the Centers for Medicare & Medicaid Services' (CMS) research and demonstration program is to conduct and support projects to develop, test, and implement new health care financing and payment policies and to evaluate the impact of the agency's programs on its beneficiaries, providers, States, and other customers and partners. The scope of the agency's activities embraces all areas of health care: costs, access, quality, service delivery models, and financing and payment approaches.
Requirements: The following themes represent the agency's current priorities in research. Note that all projects must fall into the agency's statutory authorities to operate and improve Medicare, Medicaid, and other CMS programs and activities: (1) Monitoring and Evaluating CMS Programs; (2) Strengthening Medicaid, State Children's Health Insurance Program (SCHIP), and State Programs; (3) Expanding Beneficiaries' Choices and Availability of Managed Care Options; (4) Developing FFS Payment and Service Delivery Systems; (5) Improving Quality of Care and Performance Under CMS Programs; (6) Improving the Health of Our Beneficiary Population; (7) Prescription Drugs; (8) Building Research Capacity. Application packet is available online.
Restrictions: Applicants are expected to contribute towards the project costs. Generally 5 percent of the total project costs is considered acceptable. CMS rarely approves grants or cooperative agreements for research or demonstration projects in which the Federal Government covers 100 percent of the project's costs. The budget may not include costs for construction or remodeling or for project activities that take place before the applicant has received official notification of our approval of the project.
Geographic Focus: All States
Amount of Grant: 25,000-1,000,000 USD
Contact: Grant Officer; (410) 786-5130; Jnorris1@cms.hhs.gov
Internet: http://www.cms.hhs.gov/ResearchDemoGrantsOpt/04_Other_CMS-Grant_Opportunities.asp
Sponsor: Centers for Medicare and Medicaid Services
7500 Security Boulevard, Room C2-21-15
Baltimore, MD 21244-1850

CMU Arts in Society Fellowship 1587

The program awards postdoctoral fellowships to scholars and artists in any field of humanistic inquiry or artistic endeavor. Two fellows will be appointed to the Center for the Arts in Society and will be awarded a stipend plus a small research grant. Fellows will teach one undergraduate course each semester that contributes to the center's arts-histories curriculum. Fellows also participate in the work of the center, which sponsors conferences, lectures, and colloquia on various topics, and give one public lecture. Candidates should submit a statement of current research interests as they relate to the current year's fellowship theme; a one-page proposal for a one-semester undergraduate course related to at least two disciplines—one in the arts and one in the humanities; a full curriculum vita; and at least three letters of recommendation. Applications should be submitted by postal mail.
Requirements: Scholars and artists who have received their terminal degree within the last four years in any field of humanistic inquiry or artistic endeavor are eligible.
Geographic Focus: All States
Date(s) Application is Due: Feb 3
Contact: Judith Schachter, Director; (412) 268-3239; jm1e@andrew.cmu.edu
Internet: http://www.hss.cmu.edu/cas/Content/visiting_fellows.htm
Sponsor: Carnegie Mellon University
Dean's Office, BH 154
Pittsburgh, PA 15213

CNIB Baker Applied Research Fund Grants 1588

The CNIB Baker Applied Research Fund supports research that is focused on the social, educational, cultural and rehabilitative needs of Canadians living with vision loss, and the application of assistive devices to meet these needs. This program is intended both to advance the career development of new investigators, and to encourage senior researchers to develop programs of applied research. Proposals are invited from both. Grants are for a one-year period, with a limit of $35000.
Requirements: Projects must be completed within 12 months. Funds that are unused after 24 months after the start up of the grant must be returned unless a special request for continuance is approved. Applicants must be residents of Canada and research must be conducted primarily in Canada. The applicant must: have an academic appointment at a Canadian university or occupy a supervisory position at a health care facility, or vision rehab facility. All applications will be evaluated with regard to the following criteria: excellence in the health and social sciences as applied to vision care; feasibility of research results translating into improved interventions, programs, and services for Canadians living with vision loss; enhancement of Canada's outstanding research into the needs of Canadians living with vision loss; and creation of training opportunities for students and service providers.
Restrictions: Equipment costs will not generally be covered, but in certain circumstances equipment may be funded on the basis of matching funds from the host facility. Travel and publications costs should not exceed $2,000.
Geographic Focus: Canada
Date(s) Application is Due: Jan 15
Amount of Grant: Up to 35,000 CAD
Contact: Shampa Bose, Executive Assistant and Research Coordinator; (416) 486-2500, ext. 7622 or (800) 563-2642; fax (416) 480-7059; shampa.bose@cnib.ca
Internet: http://www.cnib.ca/en/research/funding/eabaker-applied/
Sponsor: Canadian National Institute for the Blind
1929 Bayview Avenue
Toronto, M4G 3E8 Canada

CNIB Baker New Researcher Fund Grants 1589

The CNIB Baker New Researcher Fund provides one-year grants to encourage new investigations that may lead to the prevention of vision loss. It is intended to benefit new investigators (within 5 years after an academic faculty appointment) by giving them experience and results which can assist them in further grant applications and pilot investigations. A new investigator is someone who is affiliated with a research facility through an academic posting and who has been in the field for less than five years. Grants are for one year, in amounts up to $35,000 (CAD). The application must be submitted online no later than January 15 of each year, with grant decisions made by April 1.
Requirements: Projects should be completed within 12 months. Funds that are unused after 24 months of the start up of the grant must be returned unless a special request for continuance is approved. Applicants must be residents of Canada and research must be conducted primarily in Canada.
Restrictions: Equipment costs will not generally be covered, but in certain circumstances equipment may be funded on the basis of matching funds from the host facility. Travel and publication costs should not exceed $2,000. Applications must be focused and no longer than 10 pages in total.
Geographic Focus: Canada
Date(s) Application is Due: Jan 15
Amount of Grant: Up to 35,000 CAD
Contact: Shampa Bose, Executive Assistant and Research Coordinator; (416) 486-2500, ext. 7622 or (800) 563-2642; fax (416) 480-7059; shampa.bose@cnib.ca
Internet: http://www.cnib.ca/en/research/funding/eabaker-researcher/
Sponsor: Canadian National Institute for the Blind
1929 Bayview Avenue
Toronto, M4G 3E8 Canada

CNIB Barbara Tuck MacPhee Award 1590

The CNIB Barbara Tuck MacPhee Award supports researchers in the field of macular degeneration. The Tuck-MacPhee Award provides a one-year grant of up to $35,000. Projects should be completed within 12 months. The investigator will provide CNIB with a progress report six months following the start up of the grant. A second report including a copy of relevant findings of the research is required at the end of the research and prior to releasing the final quarterly payment. All publications and presentations of this research should acknowledge the support of the CNIB Baker Tuck MacPhee Fund. Applicants must apply online no later than January 15 of each year. The application will be reviewed by a multi-disciplinary review committee, and decisions will be finalized by April 1.
Requirements: Funds that are unused after 24 months must be returned unless a special request for continuance is approved. Equipment costs will not generally be covered, but in certain circumstances equipment may be funded on the basis of matching funds from the host facility. This competition is open to all researchers; however applicants must be residents of Canada and research must be conducted primarily in Canada.
Restrictions: Trainees (residents, fellows, post graduate students) are not eligible to apply.
Geographic Focus: All States, All Countries
Date(s) Application is Due: Jan 15
Contact: Shampa Bose, Executive Assistant and Research Coordinator; (416) 486-2500, ext. 7622 or (800) 563-2642; fax (416) 480-7059; shampa.bose@cnib.ca
Internet: http://www.cnib.ca/en/research/funding/Tuck-MacPhee/
Sponsor: Canadian National Institute for the Blind
1929 Bayview Avenue
Toronto, M4G 3E8 Canada

CNIB Canada Glaucoma Clinical Research Council Grants 1591

The CNIB-CGCRC Award represents a partnership of CNIB and the Canada Glaucoma Clinical Research Council designed to invest in clinical glaucoma research. This partnership contributes about $225,000 over a three year period. Funding of the CNIB-CGCRC award is via an unrestricted grant from Alcon Canada Inc. The application deadline for the CNIB-CGCRC Award is June 30. There are two levels of funding available: $5,000 or less for investigators with training or practice for less than 5 years and $5000 to $25,000 for investigators with training or practice over 5 years.
Geographic Focus: All States
Amount of Grant: Up to 25,000 CAD
Contact: Shampa Bose, Executive Assistant and Research Coordinator; (416) 486-2500, ext. 7622 or (800) 563-2642; fax (416) 480-7059; shampa.bose@cnib.ca
Internet: http://www.cnib.ca/en/research/funding/cnib-cgrc/
Sponsor: Canadian National Institute for the Blind
1929 Bayview Avenue
Toronto, M4G 3E8 Canada

CNIB E. (Ben) & Mary Hochhausen Access Technology Research Grants 1592

The award was established by Ben and Mary Hochhausen to further their long-time volunteer and financial contribution to CNIB. This international award encourages research in the field of access technology for people living with vision loss. Grants up to $10,000 are made each year at the discretion of the E. (Ben) and Mary Hochhausen Fund trustees. Applications are accepted from any country in the world. Awards may be applied to: research projects; study at centres of excellence in technology; fellowships; development of prototypes; or development costs for bringing important new products to market. Initial application should be received by the E. (Ben) and Mary Hochhausen Fund trustees by September 30.
Requirements: Trustees will consider: students enrolled in a recognized College, University or equivalent program; or researchers from industry, academia or applied science.
Geographic Focus: All States, All Countries
Date(s) Application is Due: Sep 30
Amount of Grant: Up to 10,000 CAD
Contact: Shampa Bose, Executive Assistant and Research Coordinator; (416) 486-2500, ext. 7622 or (800) 563-2642; fax (416) 480-7059; shampa.bose@cnib.ca
Internet: http://www.cnib.ca/en/research/funding/hochhausen/
Sponsor: Canadian National Institute for the Blind
1929 Bayview Avenue
Toronto, M4G 3E8 Canada

Coca-Cola Foundation Grants 1593

The foundation has established education as its philanthropic focus and set aside most of its funds to support educational initiatives that address pressing needs. To help prepare youth for life, the foundation gives in three areas: higher education—pipeline programs that connect various levels of education and help students stay in school, scholarships, and minority advancement; classroom teaching and learning—innovative K-12 projects, teacher development, and small projects that deal with classroom activities; and global education—projects that encourage international studies, global understanding, and student exchange. Grants are awarded to both public and private institutions at all levels of education: universities, colleges, and secondary and elementary schools. International educational institutions and health care organizations also receive consideration. Types of support include annual campaigns, donated equipment, employee matching gifts, operating budgets, special projects, capital campaigns, continuing support, fellowships, internships, endowment funds, matching funds, and scholarship funds. Proposals may be submitted at any time.
Requirements: The following are eligible to apply: IRS 501(c)3 non-profits; a foreign organization that has received a ruling from the IRS that it is a section 501(c)3 tax exempt organization; or a foreign organization that is the equivalent of a U.S. charity.
Restrictions: The foundation does not make grants to individuals, religious endeavors, political or fraternal organizations, or organizations without 501(c)3 status.
Geographic Focus: All States
Contact: Helen Smith Price, Executive Director; (404) 676-2568; fax (404) 676-8804
Internet: http://www.thecoca-colacompany.com/citizenship/our_communities.html
Sponsor: Coca-Cola Foundation
P.O. Box 1734
Atlanta, GA 30301

Coeta and Donald Barker Foundation Grants 1594

The foundation awards grants to California and Oregon nonprofit organizations in its areas of interest, including arts, children and youth, community development, disabled, environmental conservation, family services, federated giving, health care and health organizations, heart and circulatory research, higher education, hospitals, mental health, and secondary school education. Types of support include building construction/renovation, equipment acquisition, general operating support, program development, and scholarship funds.

Requirements: California and Oregon nonprofit organizations are eligible.

Restrictions: Grants do not support sectarian religious purposes, federal and tax-dependent organizations, individuals, or endowment funds.

Geographic Focus: California, Oregon

Date(s) Application is Due: Mar 1; Aug 1

Amount of Grant: 100-20,450 USD

Samples: Santa Barbara Zoological Gardens (Santa Barbara, CA), $5000.

Contact: Nancy Harris, Executive Administrator; (760) 324-2656; fax (760) 321-8662

Sponsor: Coeta and Donald Barker Foundation
P.O. Box 936
Rancho Mirage, CA 92270

Coleman Foundation Cancer Care Grants 1595

The Foundation has been an advocate for raising the standards of cancer care in the Midwest region and assuring that direct cancer services are available to cancer patients in the Chicago Metro area. Recently, the Foundation has employed an impact framework which provides a basic outline of the goals for the Foundation's grantmaking. For the Cancer program area, the Foundation has identified strategies that can enable health care providers to gain effective tools and resources in providing supportive oncology (care from diagnosis through end of life). The Cancer Impact Plan (see website) is intended to inform potential grantees as to the particular strategies the Foundation seeks to fund. The Foundation welcomes potential grantees to review the Cancer Impact Plan to assess and determine the best possible strategy to meet the Foundation's intended impact.

Requirements: Grants are made only to 501(c)3 or 509(a)1 nonprofit organizations that are not private foundations. The Foundation's primary geographic focus is the Midwest region, particularly the State of Illinois and the Chicago metropolitan area. Only programs within the United States will be considered, which excludes all international programs. Applicants should submit a letter of inquiry first; LOIs are accepted throughout the calendar year. The Foundation will advise you if a full proposal should be submitted for further review. Proposals are presented by Foundation staff and approved by the Board at quarterly meetings, usually in February, May, August and November.

Restrictions: The program does not fund for-profit businesses, individuals, individual scholarships, advertising books, tickets, equipment purchases (including computer hardware or software), or advertising. General solicitations and annual appeals will not be considered.

Geographic Focus: Illinois, Indiana, Iowa, Michigan, Ohio, Wisconsin

Contact: Rosa Berardi; (312) 902-7120; fax (312) 902-7124; rberardi@colemanfoundation.org

Internet: http://www.colemanfoundation.org/what_we_fund/cancer/

Sponsor: Coleman Foundation
651 West Washington Boulevard, Suite 306
Chicago, IL 60661

College of Human Sciences Scholarships 1596

Human sciences undergraduate and graduate scholarships are available in the College of Human Sciences at Texas Tech University. Programs include human development and family studies, marriage and family therapy, interior design, merchandising, family financial planning, clothing and textiles, restaurant/hotel management, food and nutrition, early childhood, family and consumer sciences education, consumer economics, environmental design, and fashion design. Application deadline is February 1 unless specifically stated otherwise. Check Web site for current details and criteria of available scholarships.

Requirements: Scholarships are awarded only to students who have officially applied to Texas Tech University. Unless the scholarship award is for multiple years, students are required to resubmit scholarship applications during the spring semester to be eligible for renewal of their scholarship.

Geographic Focus: All States

Date(s) Application is Due: Feb 1

Amount of Grant: 200-2,000 USD

Contact: Beverly Pinson, Coordinator; (806) 742-3031; beverly.pinson@ttu.ed

Internet: http://www.hs.ttu.edu/aas/scholarships/default.php

Sponsor: Texas Tech University
Box 41162
Lubbock, TX 79409-1162

Colonel Stanley R. McNeil Foundation Grants 1597

Colonel Stanley R. McNeil and his wife Merna created the Colonel Stanley R. McNeil Foundation in 1993 to support and promote quality educational, human-services, and health-care programming for underserved populations. Special consideration is given to charitable organizations that serve the needs of children. During their lifetimes, the McNeils were actively involved in their local church and community. Colonel McNeil also served on a number of boards, including Lake Bluff Children's Home and Ravenswood Hospital. Although the McNeils had no children, they were strong supporters of children's causes. The foundation is particularly interested in funding programs or organizations that focus on children's causes, start-up initiatives within the human-services or arts and culture arenas, and healthcare. To better support the capacity of nonprofit organizations, multi-year funding requests are considered. Grant requests for naming opportunities that honor the donors are strongly encouraged. Applicants must apply online at the grant website. Applicants are strongly encouraged to do the following before applying: review the downloadable state application procedures for additional helpful information and clarifications; review the downloadable online-application guidelines at the grant website; review the foundation's funding history (link is available from the grant website); review the online application questions in advance; and review the list of required attachments. These will generally include: a list of board members, financial statements (audited, reviewed, or compiled by independent auditor); an organization summary; a list of other funding sources; an IRS Determination letter; and other required documents. All attachments must be uploaded in the online application as PDF, Word, or Excel files. The Colonel Stanley R. McNeil Foundation has biannual deadlines of February 1 and June 1. Applications must be submitted by 11:59 p.m. on the deadline dates. Applicants for the February deadline will be notified of grant decisions by June 30. Applicants for the June deadline will be notified by November 30.

Requirements: Illinois nonprofits serving the Chicago metropolitan area are eligible.

Restrictions: Because requests for support usually exceed available resources, organizations can only apply to either the Lang Burk Fund or the Colonel Stanley McNeil Foundation in the same calendar year. Grant requests to both foundations during the same calendar year will no longer be accepted. The Foundation will consider requests for general operations only if the organization's operating budget is less than $1 million. In general, grant request for individuals, endowment campaigns or capital projects will not be considered.

Geographic Focus: Illinois

Date(s) Application is Due: Feb 1; Jun 1

Samples: Gads Hill Center, Chicago, Illinois, $35,000; Field Museum of Natural History, Chicago, Illinois, $50,000; Organic School Project, Chicago, Illinois, $35,000.

Contact: George Thorn, Vice President, Bank of America; (312) 828-4154; ilgrantmaking@bankofamerica.com

Internet: https://www.bankofamerica.com/philanthropic/fn_search.action

Sponsor: Colonel Stanley R. McNeil Foundation
231 South LaSalle Street, IL1-231-13-32
Chicago, IL 60604

Colorado Bioscience Discovery Evaluation Grants 1598

The program was first created by the Colorado legislature in 2006 with the purpose being to improve and expand the evaluation of new bioscience discoveries at research institutions with the intent of accelerating the development of new products and services-essentially a proof of concept program. 2007 legislation modified the original program by providing additional funding and targeting the new funding to SBIR/STTR recipients and biofuels research. Proof-of-Concept projects can receive a maximum grant of $150,000. SBIR/STTR projects can receive a maximum grant of $100,000.

Requirements: Eligible applicants include any institution located and operating in Colorado that is a public or private, nonprofit institution of higher education, a nonprofit teaching hospital, or a private, nonprofit medical and research center. For-profit entities are not eligible for the program. However, an Office of Technology Transfer must be affiliated with one of these research institutions in order for the Office of Technology Transfer to be eligible to apply under the Program. Application forms and guidelines are available for download at the sponsor's website.

Restrictions: For-profit entities are not eligible for the program.

Geographic Focus: Colorado

Date(s) Application is Due: Jun 30

Contact: Sonya Guram; (303) 892-3840; fax (303) 892-3848; sonya.guram@state.co.us

Internet: http://www.colorado.gov/cs/Satellite?c=Page&childpagename=OEDIT%2FOEDITLayout&cid=1167928017742&p=1167928017742&pagename=OEDITWrapper

Sponsor: Colorado Office of Economic Development and International Trade
1625 Broadway, Suite 2700
Denver, CO 80202

Colorado Clean Energy Fund Solar Innovation Grants (SIG) 1599

The program was established to support innovative programs that can demonstrate a strategy and implementation plan for breaking down financial, educational, political, and technical barriers to greater penetration of solar electric and solar thermal technologies in the residential and commercial sectors. Approximately three hundred and fifty thousand dollars ($350,000) is available during the current competitive cycle. Individual awards will not exceed $50,000.

Requirements: Funding requests will be considered for the following purposes: (a) Education-Applicants will provide educational resources to targeted sectors with the goal of expanding the quantity and quality of solar installations in Colorado. (b) Integrated Design-Applicants will address opportunities to incorporate solar technology into the commercial and production home sectors in both new and existing infrastructure. (c) Utility Programs-Applicants in this category will work to improve the viability of solar technology as a means to offset energy and power demands at the utility level. GEO is soliciting innovative solutions to promoting solar technology at the utility level. (d) Market Analysis & Policy Recommendations-Applicants will identify policy barriers to the wider implementation of solar technology in Colorado and present solutions in the form of a comprehensive policy report. (e) Financing-Applicants will present innovative financing models that will increase the affordability of integrating solar technology in the residential and commercial sectors. Potential recipients include, but are not limited to: Utilities; Home Builders; Installers / System Integrators; Lenders / Financial Institutions; Non-Profits; Trade Associations. Preference will be given to proposals that demonstrate the ability to generate substantial matching funds, however matching funding is not required should the applicant demonstrate sufficient need.

Restrictions: Grant dollars will not be used to buy down the cost of solar systems. Only electronic applications will be accepted.

Geographic Focus: Colorado

Date(s) Application is Due: Apr 20
Amount of Grant: Up to 50,000 USD
Contact: Jeff Lyng, Solar Programs Manager; (303) 866-2264; jeff.lyng@state.co.us
Internet: http://www.colorado.gov/energy/resources/SolarInnovationGrantProgram GovernorsEnergyOfficeColorado.asp
Sponsor: Colorado Governor's Energy Office
1580 Logan Street, Suite 100
Denver, CO 80203

Columbus Foundation Ann Ellis Fund Grants 1600
The Fund is used for eye research, with primary emphasis on sponsoring scientific research on the diagnosis, prevention, and treatment of glaucoma. Once this is accomplished, the fund will assist in the research of other hidden eye problems. For more information, please contact the Community Research and Grant Management Department at the Foundation
Restrictions: Individuals are ineligible. Requests for religious purposes, budget deficits, endowments, conferences, or projects that are normally the responsibility of a public agency are generally not funded.
Geographic Focus: All States
Date(s) Application is Due: Oct 1
Contact: Dottie Henderson, Executive Assistant; (614) 251-4000; fax (614) 251-4009; dhenderson@columbusfoundation.org
Internet: http://www.columbusfoundation.org/gogrants/targeted_needs/specialized_grants.aspx
Sponsor: Columbus Foundation
1234 E Broad Street
Columbus, OH 43205

Commonfund Endowment Management Prize 1601
Commonfund, which manages the world's largest pool of endowment and cash assets for universities, colleges, independent schools, and health care facilities, sponsors this competition for research and scholarship in the advancement of endowment management. The prize is awarded in two installments—upon announcement of the winner and upon acceptance of the final paper. This paper will be published and distributed by Commonfund. The winner is announced in March.
Requirements: Applicants must submit a sample completed paper or working paper written for an academic or other audience, a proposal of 1000 words or less outlining how the applicant would develop a derivative paper written for the endowment management audience, and a curriculum vita.
Geographic Focus: All States
Date(s) Application is Due: Dec 31
Amount of Grant: 50,000 USD
Contact: John Griswold Jr., Senior Vice President; (203) 563-5185; jgriswol@cfund.org
Internet: http://www.commonfund.org/Commonfund/About+Us/pillars_educational_programs_2004.htm
Sponsor: Commonfund
15 Old Danbury Road, P.O. Box 812
Wilton, CT 06897-7717

Commonweal Foundation Community Assistance Grants 1602
The foundation supports educational programs and projects assisting disadvantaged, at-risk youth. The foundation focuses on secondary and, to a lesser extent, elementary education. The foundation also considers grants for educational research and, to a limited extent, health care. Guidelines are available online.
Requirements: 501(c)3 tax-exempt organizations located in the District of Columbia, Maryland, or Northern Virginia are eligible. Applicant organizations should have an annual budget not exceeding $1 million.
Geographic Focus: District of Columbia, Maryland, Virginia
Date(s) Application is Due: Mar 1; Aug 1
Amount of Grant: Up to 25,000 USD
Contact: Gloria Dairsow, (240) 450-0000; fax (240) 450-4115; gdairsow@cweal.org
Internet: http://www.cweal.org/cag.htm
Sponsor: Commonweal Foundation
10770 Columbia Pike, Suite 150
Silver Spring, MD 20901

Commonweal Foundation Grants 1603
The foundation awards grants to eligible nonprofit organizations serving disadvantaged youth in the area comprising the corridor between Baltimore, MD, and Washington, DC. It also offers some assistance to elementary education, educational research, and some health care causes. Programs include Partners in Learning (address low literacy skills of economically disadvantaged children), community assistance grants (i.e., after-school tutoring, parenting classes, and shelters for the homeless or abused women), and learning disabilities support program (special education services to economically disadvantaged children). Contact the office for application procedures.
Requirements: 501(c)3 tax-exempt organizations serving disadvantaged youth in the corridor between Baltimore, MD, and Washington, DC, are eligible.
Geographic Focus: District of Columbia, Maryland, Virginia
Date(s) Application is Due: Feb 1; May 1; Aug 1; Nov 1
Contact: Renee Cottman-Reyes, (240) 450-0000; fax (240) 450-4115; gdairsow@cweal.org
Internet: http://www.commonweal-foundation.org/index.html
Sponsor: Commonweal Foundation
10770 Columbia Pike, Suite 150
Silver Spring, MD 20901

Commonwealth Fund/Harvard University Fellowship in Minority Health Policy 1604
The Commonwealth Fund/Harvard University Fellowship in Minority Health Policy is a one-year, full-time, academic degree-granting program designed to create physician-leaders, particularly minority physician-leaders, who will pursue careers in health policy, public health practice, and academia. It is designed to incorporate the critical skills taught in schools of public health, government, business, and medicine with leadership forums and seminar series conducted by Harvard senior faculty and nationally recognized leaders in minority health and public policy; supervised practicums and shadowing opportunities; and site visits, conferences, and travel. Each fellowship provides: $50,000 stipend; full tuition; health insurance; books; travel; and related program expenses, including financial assistance for a practicum project.
Requirements: Applicants are required to complete applications to both the Commonwealth Fund/Harvard University Fellowship in Minority Health Policy and the Harvard School of Public Health. Applicants must have finished residency and be U.S. citizens.
Geographic Focus: All States
Date(s) Application is Due: Jan 4
Amount of Grant: 50,000 USD
Contact: Joan Y. Reede, Director; (617) 432-2922; mfdp_cfhuf@hms.harvard.edu
Internet: http://www.commonwealthfund.org/Fellowships/Minority-Health-Policy-Fellowship.aspx
Sponsor: Commonwealth Fund
164 Longwood Avenue, 2nd Floor
Boston, MA 02115-5818

Commonwealth Fund Affordable Health Insurance Grants 1605
The Program on Affordable Health Insurance envisions an efficient and equitable health insurance system that makes available to all Americans comprehensive, continuous, and affordable coverage. In support of that vision, the program seeks to: analyze changes in employer-based, private and public insurance coverage for people under age 65, and determine how those changes may affect the affordability and comprehensiveness of coverage, the number of uninsured, the number of under-insured, and churning in and out of coverage; document the consequences of being uninsured, under-insured, and unstably insured, with regard to access to care, health status, personal financial security, and economic productivity; and develop and evaluate federal and state policies to expand and stabilize health insurance, make it more affordable and aligned with incentives to access appropriate and high quality care, and enhance the efficiency with which it is administered.
Requirements: The Commonwealth Fund requests letters of inquiry to initiate the grant application process, and does not wish to review full proposals at this stage. Applicants are encouraged to submit letters of inquiry using our online form.
Restrictions: The Fund makes grants only to tax-exempt organizations and public agencies and does not support: general planning and ongoing activities or existing deficits; endowment or capital costs, including construction, renovation, or equipment; basic biomedical research; conferences, symposia, major media projects, or documentaries, unless they are an outgrowth of one of the Fund's programs; individuals; scholarships; churches or other religious organizations unless the project for which they seek funding is entirely secular in nature; and work for which achievements cannot be measured.
Geographic Focus: All States
Contact: Sara R. Collins, Assistant Vice-President; (212) 606-3838 or (212) 606-3800; fax (212) 606-3500; src@cmwf.org
Internet: http://www.commonwealthfund.org/Content/Program-Areas/Affordable-Health-Insurance.aspx
Sponsor: Commonwealth Fund
1 E 75th Street
New York, NY 10021-2692

Commonwealth Fund Harkness Fellowships in Health Care Policy and Practice 1606
The Harkness Fellowships in Health Care Policy and Practice provide an opportunity for professionals from the Australia, Germany, the Netherlands, New Zealand, Norway, Switzerland, and the United Kingdom to spend four to 12 months in the United States conducting a research study that is relevant to health care policy and practice in both the United States and the fellow's home country, and that is focused on the issues of greatest concern to the fund. Fellowship awards provide up to $107,000, which covers round trip airfare to the United States, a living allowance, funds for project-related travel, research, and conferences, travel to attend the Commonwealth Fund program of fellowship seminars, health insurance, and U.S. and state taxes. A family supplement-including airfare, living allowance, and health insurance-is also provided to Fellows accompanied by a partner and/or children up to age 18.
Requirements: Applicants must be citizens of Australia, Germany, the Netherlands, New Zealand, Norway, Switzerland, and the United Kingdom in their late 20s to early 40s with broad educational backgrounds, not just those in research or academic careers. In order to apply, applicants must be nominated by their institution and submit a formal application, which is available from The Commonwealth Fund in New York City or its representatives in their home country.
Geographic Focus: All States, United Kingdom
Date(s) Application is Due: Sep 15
Amount of Grant: Up to 107,000 USD
Contact: Robin Osborn, Director; (212) 606-3809 or (212) 606-3800; fax (212) 606-3875; ro@cmwf.org
Internet: http://www.commonwealthfund.org/Fellowships/Minority-Health-Policy-Fellowship.aspx
Sponsor: Commonwealth Fund
1 E 75th Street
New York, NY 10021-2692

Commonwealth Fund Health Care Quality Improvement and Efficiency Grants 1607
The goal of the Fund's Program on Health Care Quality Improvement and Efficiency is to improve the quality and efficiency of health care in the United States. To that end, the program supports projects that: promote the development and widespread adoption of health care quality and efficiency measures; assess and enhance the capacity of health care organizations to provide better care more efficiently; and promote the development and adoption of payment and incentive models that encourage providers to improve quality and efficiency.
Requirements: The Commonwealth Fund requests letters of inquiry to initiate the grant application process, and does not wish to review full proposals at this stage. Applicants are encouraged to submit letters of inquiry using our online form.
Restrictions: The Fund makes grants only to tax-exempt organizations and public agencies and does not support: general planning and ongoing activities or existing deficits; endowment or capital costs, including construction, renovation, or equipment; basic biomedical research; conferences, symposia, major media projects, or documentaries, unless they are an outgrowth of one of the Fund's programs; individuals; scholarships; churches or other religious organizations unless the project for which they seek funding is entirely secular in nature; and work for which achievements cannot be measured.
Geographic Focus: All States
Contact: Anne-Marie J. Audet; (212) 606-3800; fax (212) 606-3500; ama@cmwf.org
Internet: http://www.commonwealthfund.org/Content/Program-Areas/Health-Care-Quality-Improvement-and-Efficiency.aspx
Sponsor: Commonwealth Fund
1 E 75th Street
New York, NY 10021-2692

Commonwealth Fund Patient-Centered Coordinated Care Grants 1608
As defined by the Institute of Medicine, patient-centered care is health care that establishes a partnership among practitioners, patients, and their families (when appropriate), to ensure that decisions respect patients' needs and preferences, and that patients have the education and support they need to make decisions and participate in their own care. In primary care, such care is best provided in a medical home-a primary care practice or health center that ensures patients have enhanced access to their clinicians (for example, through the availability of evening or weekend appointments), coordinates care, and engages in continuous quality improvement. The goal of The Commonwealth Fund's Patient-Centered Coordinated Care Program, established in 2005, is to improve the quality of primary care by making it more patient- and family-centered. The initiative supports projects that: promote the collection of information on patient-centered care and the delivery of care to facilitate public reporting and quality improvement; stimulate adoption of effective practices, models, and tools to make primary care practices patient- and family-centered; and improve policy to encourage patient- and family-centered care in medical homes.
Requirements: The Commonwealth Fund requests letters of inquiry to initiate the grant application process, and does not wish to review full proposals at this stage. Applicants are encouraged to submit letters of inquiry using our online form.
Restrictions: The Fund makes grants only to tax-exempt organizations and public agencies and does not support: general planning and ongoing activities or existing deficits; endowment or capital costs, including construction, renovation, or equipment; basic biomedical research; conferences, symposia, major media projects, or documentaries, unless they are an outgrowth of one of the Fund's programs; individuals; scholarships; churches or other religious organizations unless the project for which they seek funding is entirely secular in nature; and work for which achievements cannot be measured.
Geographic Focus: All States
Contact: Melinda Abrams, (212) 606-3800; fax (212) 606-3500; mka@cmwf.org
Internet: http://www.commonwealthfund.org/Content/Program-Areas/Patient-Centered-Coordinated-Care.aspx
Sponsor: Commonwealth Fund
1 E 75th Street
New York, NY 10021-2692

Commonwealth Fund Payment System Reform Grants 1609
To achieve a high performing health system, the U.S. must curb spending growth and improve the way health care is provided. Payment system reform is critical to accomplishing these objectives. As a nation, we need to align incentives so that health care providers are rewarded for high-value care rather than a high volume of services. Rewarding value over volume will also encourage the development of a more integrated health care delivery system. The Commonwealth Fund, through its Program on Payment System Reform, supports analysis and the development of policy options to accomplish these goals. Areas of interest include: reforming the existing payment structure to improve the alignment of incentives and to provide a base for more comprehensive payment reform; modeling and analyzing the potential impact of alternative options for payment reform in Medicare and throughout the health system; using payment reform to encourage the development of new models of health care delivery that provide better and more coordinated care; and using comparative effectiveness research to support better decision-making by providers, payers, and patients.
Requirements: The Commonwealth Fund requests letters of inquiry to initiate the grant application process, and does not wish to review full proposals at this stage. Applicants are encouraged to submit letters of inquiry using our online form.
Restrictions: The Fund makes grants only to tax-exempt organizations and public agencies and does not support: general planning and ongoing activities or existing deficits; endowment or capital costs, including construction, renovation, or equipment; basic biomedical research; conferences, symposia, major media projects, or documentaries, unless they are an outgrowth of one of the Fund's programs; individuals; scholarships; churches or other religious organizations unless the project for which they seek funding is entirely secular in nature; and work for which achievements cannot be measured.
Geographic Focus: All States
Contact: Stuart Guterman, (202) 292-6735 or (212) 606-3800; sxg@cmwf.org
Internet: http://www.commonwealthfund.org/Content/Program-Areas/Payment-System-Reform.aspx
Sponsor: Commonwealth Fund
1 E 75th Street
New York, NY 10021-2692

Commonwealth Fund Quality of Care for Frail Elders Grants 1610
The Commonwealth Fund Program on Quality of Care for Frail Elders aims to transform the nation's nursing homes and other long-term care facilities into resident-centered organizations that are good places to live and work, capable of providing the highest-quality care. The projects it supports aim to: identify, evaluate, and spread models of resident-centered care, or care delivered in accordance with the needs and desires of the people who live in nursing homes; equip nursing home operators to lead transformational change; and promote policy options that support resident-centered care.
Requirements: The Commonwealth Fund requests letters of inquiry to initiate the grant application process, and does not wish to review full proposals at this stage. Applicants are encouraged to submit letters of inquiry using our online form.
Restrictions: The Fund makes grants only to tax-exempt organizations and public agencies and does not support: general planning and ongoing activities or existing deficits; endowment or capital costs, including construction, renovation, or equipment; basic biomedical research; conferences, symposia, major media projects, or documentaries, unless they are an outgrowth of one of the Fund's programs; individuals; scholarships; churches or other religious organizations unless the project for which they seek funding is entirely secular in nature; and work for which achievements cannot be measured.
Geographic Focus: All States
Contact: Mary Jane Koren, (212) 606-3800; fax (212) 606-3500; mjk@cmwf.org
Internet: http://www.commonwealthfund.org/Content/Program-Areas/Quality-of-Care-for-Frail-Elders.aspx
Sponsor: Commonwealth Fund
1 E 75th Street
New York, NY 10021-2692

Commonwealth Fund Small Grants 1611
The fund awards small grants for projects in its major program areas, which include: affordable health insurance; international health care policy and practice; improving the quality of health care services; and improving insurance coverage and access to care. Types of support include employee-matching gifts, program development, program evaluation, and research. Preference is given to projects that seek to solve problems, especially those affecting vulnerable groups; that analyze the effects of policies and trends on well-defined health issues; and that develop and test practical solutions. Prospective grantees should submit a letter of inquiry via regular or electronic mail. Program staff will contact applicants if more detailed information is required. Proposals recommended by fund staff are reviewed and voted upon by the board of directors, which meets in July, April, and November. Letters of inquiry should be brief, no more than two pages.
Requirements: The fund makes grants only to tax-exempt organizations and public agencies.
Restrictions: The fund does not support general planning and ongoing activities or existing deficits; endowment or capital costs, including construction, renovation, or equipment; basic biomedical research; conferences, symposia, major media projects or documentaries, unless they are an outgrowth of one of the fund's programs; individuals; scholarships; churches or other religious organizations, unless the project is entirely secular in nature; or work for which achievements cannot be measured.
Geographic Focus: All States
Contact: Andrea Landes; (212) 606-3800; fax (212) 606-3508; acl@cmwf.org
Internet: http://www.commonwealthfund.org/Grants-And-Programs/Applicant-and-Grantee-Resources/Applying-for-a-Grant.aspx
Sponsor: Commonwealth Fund
1 E 75th Street
New York, NY 10021-2692

Commonwealth Fund State High Performance Health Systems Grants 1612
The Commonwealth Fund's State High Performance Health Systems program is designed to help states develop the infrastructure needed to improve health care quality and outcomes-and to draw out and share lessons of national import from the experience of states that are moving toward comprehensive health care reforms. To enable states to help local health care providers better meet the needs of their patient populations, the program will support approaches that offer providers shared access to the clinical support and practice-management services essential to achieving high performance. Such resources will often be developed though state-initiated, public-private partnerships. To inform and support state health care leaders, additional projects will facilitate information-sharing among states working toward health care reform.
Requirements: The Commonwealth Fund requests letters of inquiry to initiate the grant application process, and does not wish to review full proposals at this stage. Applicants are encouraged to submit letters of inquiry using our online form.
Restrictions: The Fund makes grants only to tax-exempt organizations and public agencies and does not support: general planning and ongoing activities or existing deficits; endowment or capital costs, including construction, renovation, or equipment; basic biomedical research; conferences, symposia, major media projects, or documentaries, unless they are an outgrowth of one of the Fund's programs; individuals; scholarships; churches or other religious organizations unless the project for which they seek funding is entirely secular in nature; and work for which achievements cannot be measured.
Geographic Focus: All States

Contact: Edward L. Schor; (212) 606-3800; fax (212) 606-3500; els@cmwf.org
Internet: http://www.commonwealthfund.org/Content/Program-Areas/State-High-Performance-Health-System.aspx
Sponsor: Commonwealth Fund
1 E 75th Street
New York, NY 10021-2692

Communities Foundation of Texas Grants 1613

The unrestricted funds of the foundation support programs and projects intended to improve the quality of life for the citizens of the Dallas, TX, area. The funds support education, health and hospitals, social services, youth, and cultural programs. The foundation encourages projects developed in consultation with other agencies and planning groups and that promote coordination, cooperation, and sharing among organizations. Types of support include seed grants, emergency funds, building funds, equipment acquisition, matching funds, technical assistance, research, capital campaigns, and operating budgets. Requests for operating funds generally are not granted.
Requirements: 501(c)3 organizations in the Dallas, TX, area may apply.
Restrictions: Grants are not made to or for individuals, endowments, sectarian religious purposes, political or lobbying efforts, deficit financing, media projects, or for operational expenses of established organizations.
Geographic Focus: Texas
Contact: Brent Christopher; (214) 750-4222; fax (214) 750-4210; bchristopher@cftexas.org
Internet: http://www.cftexas.org
Sponsor: Communities Foundation of Texas
5500 Caruth Haven Lane
Dallas, TX 75225-8146

Community Foundation AIDS Endowment Awards 1614

The foundation administers an annual special grants program called the AIDS Endowment Awards. Grants of $1,000 will be awarded to one or two organizations that demonstrate a commitment to working towards the prevention, education and treatment of AIDS so that they may continue their efforts. Services provided could include education programs for the prevention of HIV/AIDS, medical and social services for those living with HIV/AIDS, housing services for those living with HIV/AIDS and medical research for the treatment and prevention of HIV/AIDS.
Requirements: Selected organizations in the metropolitan Richmond area interested in the field of AIDS education, research or community service and care are encouraged to apply. Organizations interested in applying should submit a brief proposal to foundation which should be no more than two single-spaced pages and include: [1] a description of the organization's mission; and [2] a description of the agency's work in the field of AIDS. Include a list of Board of Governors, a copy of the most recent IRS Form 990, an audited financial statement if available, and an IRS Tax Exempt Letter designating your organization a 501(c)3 non-profit.
Geographic Focus: Virginia
Date(s) Application is Due: Feb 15
Amount of Grant: 1,000 USD
Contact: Susan Hallett; (804) 330-7400; fax (804) 330-5992; shallett@tcfrichmond.org
Internet: http://www.tcfrichmond.org/Page2954.cfm
Sponsor: Community Foundation Serving Richmond and Central Virginia
7501 Boulders View Drive, Suite 110
Richmond, VA 23225

Community Foundation for Greater Buffalo Grants 1615

The Community Foundation for Greater Buffalo (CFGB) is a public charity holding more than 800 different charitable funds, large and small, established by individuals, families, nonprofit agencies and businesses to benefit Western New York. Since 1919, the foundation has served the needs of it's community and the wishes of it's donors through personalized service, financial stewardship, local expertise, and community leadership. The Foundation focuse's on four main areas of interest: strengthen the region as a center for arts and culture; natural, historic, and architectural resources; reduce racial and ethnic disparities; increase economic self-sufficiency for low-income individuals and families. Special funding will also be available to support the following: AIDS research and its cure ($5,000); programs that serve the visual, speech or hearing impaired ($5,000). Preference will be given to requests that align with one or more of the four focus areas.
Requirements: Applicants must be 501(c)3 not-for-profit organizations located within the eight counties of Western New York: Allegany, Cattaraugus, Chautauqua, Erie, Genesee, Niagara, Orleans, and Wyoming.
Restrictions: The Foundation will not consider competitive funding for: endowments; religious purposes; schools not registered with the New York State Education Department; attendance at or sponsorship of fundraising events for organizations; annual events or festivals; any partisan political activity. Funds from the foundation cannot be used to support or oppose a candidate for political office. Projects and activities that have occurred. The Foundation will not, except in extraordinary cases, provide payment or reimbursement for expenses incurred prior to the funding decision being communicated to the applicant.
Geographic Focus: New York
Date(s) Application is Due: Mar 1
Amount of Grant: 1,000-25,000 USD
Contact: Jean McKeown, Senior Program Officer; (716) 852-2857, ext. 204; fax (716) 852-2861; jeanm@cfgb.org
Internet: http://www.cfgb.org/page17000.cfm
Sponsor: Community Foundation for Greater Buffalo
712 Main Street
Buffalo, NY 14202

Community Foundation for Greater New Haven Quinnipiac River Fund Grants 1616

The Community Foundation for Greater New Haven (CFFGNH) is a philanthropic institution that was established in 1928. The foundation's mission is to create positive and sustainable change in Connecticut's Greater New Haven region by increasing the amount of and enhancing the impact of community philanthropy. The Quinnipiac River Fund established in 1990 was a result of a settlement between the National Resources Defense Council, Connecticut Fund for the Environment, and the Upjohn Corporation to improve the environmental quality of Quinnipiac River, the New Haven Harbor, and their watersheds, and otherwise benefit the environment of these resources. Consideration given to grant proposals that: research methods of reducing pollution, or otherwise improving the Quinnipiac River's environmental health; address means of reducing both non-point and point sources of pollution to the River; research the permitting process and look at the permits themselves; entail environmental advocacy, except litigation; study the ecology of the Quinnipiac River and the New Haven Harbor; provide public education about the Quinnipiac River and its watershed; purchase land on the Quinnipiac River for conservation purposes, or to reduce pollution and improve public access to the River.
Requirements: IRS 501(c)3 nonprofit organizations are eligible to apply in the greater New Haven area, which includes: Ansonia; Bethany; Branford; Cheshire; Derby; East Haven; Guilford; Hamden; Madison; Milford; New Haven; North Branford; North Haven; Orange; Oxford; Seymour; Shelton; Wallingford; West Haven and; Woodbridge.
Restrictions: Grants are not made to support religious activities, lobbying, or travel.
Geographic Focus: Connecticut
Date(s) Application is Due: Jan 15
Amount of Grant: 5,000-18,000 USD
Contact: Sarah Fabish, Grant/Scholarship Director; (203) 777-2386 or (203) 777-7075; fax (203) 787-6584; sfabish@cfgnh.org
Internet: http://qrwgis.newhaven.edu/QRFprojects.htm
Sponsor: Community Foundation for Greater New Haven
70 Audubon Street
New Haven, CT 06510

Community Foundation for Muskegon County Grants 1617

The community foundation was established to serve the needs of the people of Muskegon County and nearby western Michigan. The foundation's strategic plan for grant making focuses on the prevention of problems rather than the cure; encourages programs that are collaborative, comprehensive, and have the potential to be continuous; encourages leveraging and matching grant opportunities from multiple funders; and supports seed money opportunities for innovative projects. Major grant interest areas include the arts, community development and urban revitalization, education, environment, health/human services, and needs of young children (ages 0-3). Types of support include seed money grants, special projects, matching funds, equipment, scholarship funds, loans, research, publications, conferences and seminars, endowment funds, consulting services, continuing support, emergency funds, internships, professorships, and renovation projects. Although the foundation generally makes one-year grant commitments, it will consider making longer term commitments for new efforts that show strong promise for positive impact. Applications are accepted on specific dates each year; it is recommended that the applicant contact the foundation to discuss their interests and to obtain the next grant application deadline.
Requirements: IRS 501(c)3 organizations and institutions in Muskegon County and nearby western Michigan are eligible.
Restrictions: Support will not be provided for routine operating expenses; capital equipment, computer hardware and software, and motor vehicles; conferences, publications, videos, films, television, or radio programs; endowment campaigns; special fund-raising events; religious programs that serve specific denominations; existing obligations or debts; individual schools or districts; or individuals.
Geographic Focus: Michigan
Amount of Grant: 250-50,000 USD
Contact: Marcy Joy; (231) 722-4538 or (231) 332-4124; mjoy@cffmc.org
Internet: http://www.cffmc.org/grantapply.php
Sponsor: Community Foundation for Muskegon County
425 West Western Avenue, Suite 200
Muskegon, MI 49440

Community Foundation for Southern Arizona Grants 1618

The foundation awards grants in southern Arizona for a broad array of charitable purposes in the areas of arts and humanities, education, the environment, health, social services. Types of support include challenge/matching grants, conferences and seminars, scholarships, equipment acquisition, fellowships, general support, multiyear support, project development, publications, research, seed funding, and technical assistance. Priority is given to proposals that promote collaboration and build on the strengths of individuals and communities. Deadlines are determined on a yearly basis.
Requirements: Nonprofit and grassroots organizations in southern Arizona communities are eligible. Southern Arizona communities include all of Cochise, Santa Cruz, and Pima Counties; and the areas of Yuma, Mariposa, Pinal, Graham, and Greenlee Counties that lie south of the Gila River.
Restrictions: Funds are generally not available for ongoing operating or capital campaigns, debt retirement, endowments, individuals, individual schools, sectarian activities, or underwriting of fund-raising events.
Geographic Focus: Arizona
Contact: Barbara Brown; (520) 770-0800, ext. 107; bbrown@CFSoAZ.org
Internet: http://www.cfsoaz.org/page17480.cfm
Sponsor: Community Foundation for Southern Arizona
2250 E Broadway Boulevard
Tucson, AZ 85719

Community Foundation of East Central Illinois Grants 1619

The community foundation awards grants from its unrestricted funds to Champaign County, IL, nonprofit organizations to use for charitable purposes. Representative categories include arts and humanities, environmental concerns, education, health, human services, research, urban affairs, church, and youth programs. Types of support include continuing support, annual campaigns, building/renovation, equipment, publication, consulting services, and scholarships.

Requirements: Champaign County, IL, nonprofit organizations are eligible.
Geographic Focus: Illinois
Date(s) Application is Due: Aug 31
Contact: Joan Dixon; (217) 359-0125; fax (217) 352-6494; joandixon@cfeci.org
Internet: http://www.cfeci.org/non-profit/grant-application
Sponsor: Community Foundation of East Central Illinois
404 West Church Street
Champaign, IL 61820-3411

Community Foundation of the Fox River Valley 1620

The foundation awards full-time scholarships to graduating high school seniors, undergraduate college students, and graduate school students within its service area, comprising the City of Aurora, Kendall County, and southern Kane County in the State of Illinois. The residency requirement does not apply to graduating high school students whose parent(s) are employed by Nicor Gas; graduating high school students residing within Community High School District 94, West Chicago; and graduating community/junior college students whose parent(s) are employed by a company that maintains membership with the Valley Industrial Association. The majority of scholarships are based on academic ability and financial need. Some awards are renewable. Students who attend the following high schools must obtain applications from their high school guidance office (deadline date for applications determined by the school): Aurora Central Catholic, Aurora Christian, East Aurora, Hinckley-Big Rock, IMSA Kaneland, Marmion, Oswego, Plano, Rosary Waubonsie Valley, West Aurora, and Yorkville High School. All other students may obtain an application by calling the office. The deadline date for receipt of applications is generally mid-February.

Requirements: Eligibility is restricted to students who will be attending school on a full-time basis and whose permanent residence is within the foundation's service area. The foundation's service area includes the City of Aurora, Kendall County, and southern Kane County in the state of Illinois.
Geographic Focus: Illinois
Amount of Grant: 1,000-3,000 USD
Contact: Rhonda Soos, (630) 896-7800; rsoos@communityfoundationFRV.org
Internet: http://www.CommunityFoundationFRV.org
Sponsor: Community Foundation of the Fox River Valley
111 W Downer Place, Suite 312
Aurora, IL 60506-6106

Community Foundation of the Virgin Islands Judith A. Towle Environmental Studies Grants 1621

The Environmental Studies Fund will provide support for internships: sponsored by an institution of higher education, a public sector agency, or an NGO based in the U.S. or British Virgin Islands or in St. Kitts/Nevis; research studies by graduate students: focused on transboundary environmental issues in the U.S. or British Virgin Islands or in St. Kitts/Nevis; workshops, seminars, lectureships: sponsored by an institution of higher education, a public sector agency, or an NGO based in the U.S. or British Virgin Islands or in St. Kitts/Nevis; publications focused on transboundary environmental issues in the U.S. or British Virgin Islands or in St. Kitts/Nevis.

Requirements: Any individual or non-profit organization whose research focuses on the U.S. Virgin Islands, British Virgin Islands, or the Federation of St. Kitts and Nevis may submit a grant application. Preference will be given to graduate students who are residents of the U.S. or British Virgin Islands or of the Federation of St. Kitts/Nevis.
Geographic Focus: U.S. Virgin Islands
Amount of Grant: Up to 3,500 USD
Contact: Dee Baecher-Brown; (340) 774-6031; fax (340) 774-3852; dbrown@cfvi.net
Internet: http://www.cfvi.net/subpages/grants.html
Sponsor: Community Foundation of the Virgin Islands
P.O. Box 11790
St. Thomas, VI 00801-4790

Comprehensive Health Education Foundation Grants 1622

The foundation awards grants to support programs that address health inequities. The initial grantmaking effort will focus on Clark, Pierce, and Spokane Counties in Washington State. One-year grants of up to $20,000 each will be awarded to culturally appropriate, community-led collaborations to test their best idea on how to make it easier for people who suffer from health inequities to move more and eat healthier. Health inequities are defined as differences in the incidence, prevalence, mortality, and burden of diseases that exist for specific populations in the United States. Low-income individuals and people of color within the United States generally have higher rates of poor health and injury than those who are in higher-income groups and are Caucasian.

Requirements: 501(c)3 tax-exempt organizations located in Clark, Pierce, and Spokane Counties in Washington State and units of government that are nondiscriminatory in policy and practice regarding disabilities, age, sex, sexual orientation, race, ethnic origin, or creed are eligible.
Restrictions: Support will not be provided for building or land acquisitions; equipment or furniture purchases; endowment funds; emergency funds; grants to individuals; fellowships/scholarships; research; debt retirement; fundraising activities; general fund drives; indirect overhead; or CHEF programs or products.
Geographic Focus: Washington
Amount of Grant: 500-20,000 USD
Contact: Kari L. Lewis; (800) 323-2433, ext. 1899; KariL@chef.org
Internet: http://www.chef.org/about/grants.php
Sponsor: Comprehensive Health Education Foundation
159 S Jackson Street, Suite 510
Seattle, WA 98104

Compton Foundation Grants 1623

The foundation was founded to address community, national, and international concerns in the fields of peace and world order, population, and the environment. Other concerns of the foundation include equal educational opportunity, community welfare and social justice, and culture and the arts. The foundation makes three kinds of grants. Project grants generally are made to national organizations for projects that fall within the primary areas of peace and world order, population, and the environment. These grants may be for regional (Pacific Coastal states), national, or international activities and are usually for projects of limited duration. Project grants are considered by the board two times a year. Discretionary grants are made at the discretion, and usually at the initiation, of individual board members. Most grants in this area are made for community welfare and social justice, and culture and the arts. Renewal grants provide general support to organizations whose activities have been funded by the foundation for many years and whose work continues to be considered particularly effective by the board. Many of the foundation's grants in the area of equal educational opportunity, and some grants in peace and world order and population, are renewal grants.

Requirements: The foundation makes grants to tax-exempt organizations and institutions.
Restrictions: Grants will not be made to individuals.
Geographic Focus: All States
Date(s) Application is Due: Mar 7; Sep 7
Amount of Grant: 5,000-50,000 USD
Contact: Edith T. Eddy; (650) 508-1181; fax (650) 508-1191; info@comptonfoundation.org
Internet: http://www.comptonfoundation.org/application_procedures.html
Sponsor: Compton Foundation
255 Shoreline Drive, Suite 540
Redwood City, CA 94065

Compton Foundation International Fellowships 1624

The goal of the Compton Foundation's International Fellowship Program is to contribute to the capacity of developing countries, especially in Central America and Sub-Saharan Africa, as well as Mexico, to improve policies and programs relating to Peace and Security, Population and Reproductive Health and Environment and Sustainable Development. The Foundation strives to accomplish this goal by supporting outstanding graduate students who are committed to careers in the program areas of interest to the Foundation within the developing world. Compton Fellows are chosen by selected university-based programs working in partnership with the Foundation.

Geographic Focus: All States
Contact: Edith T. Eddy; (650) 508-1181; fax (650) 508-1191; info@comptonfoundation.org
Internet: http://www.comptonfoundation.org/fellows.html
Sponsor: Compton Foundation
255 Shoreline Drive, Suite 540
Redwood City, CA 94065

Compton Foundation Mentor Fellowships 1625

The fellowship program focuses on graduating college students from the United States. This program is designed to promote creativity and support the commitment of graduating seniors as they move beyond academic preparation to real-world application and contribution. Each year ten fellows are selected from participating universities and awarded a one-year, fellowship. The stipend is to implement a self-directed project, contributing their talents and energy to real-world situations. At the core of the fellowship is the partnership between a fellow and a mentor, who provides guidance, encouragement, and impetus for continued learning and service.

Geographic Focus: All States
Amount of Grant: 36,000 USD
Contact: Edith T. Eddy; (650) 508-1181; fax (650) 508-1191; info@comptonfoundation.org
Internet: http://www.comptonfoundation.org/fellowships.html
Sponsor: Compton Foundation
255 Shoreline Drive, Suite 540
Redwood City, CA 94065

Computing in Chemical Engineering Award 1626

The award, which is sponsored by Mitsubishi Chemical Company and the Dow Chemical Company, recognizes outstanding contributions in the application of computing and systems technology to chemical engineering. All members as well as other interested persons are urged to nominate deserving candidates. All information and supporting documentation are to be included with completed forms. Nominations will remain active for three years. The awardee is invited to deliver an address at a CAST Division luncheon or dinner at the AIChE annual meeting.

Geographic Focus: All States
Date(s) Application is Due: Apr 15
Contact: Director; (212) 591-7107; fax (212) 591-8882; awards@aiche.org
Internet: http://www.aiche.org/awards/awarddtl.asp?AwardID=28
Sponsor: American Institute of Chemical Engineers
3 Park Avenue
New York, NY 10016

Connecticut Community Foundation Grants 1627
The Connecticut Community Foundation serves the people of Greater Waterbury and Northwest Connecticut by supporting public and nonprofit organizations providing programs and services including those for the arts, human services, health care, environment, youth development and education. The Foundation gives priority to efforts that prevent problems, encourage community solutions or improve the organizational capability and financial stability of nonprofit agencies.
Requirements: Nonprofit organizations in Beacon Falls, Bethlehem, Bridgewater, Cheshire, Goshen, Litchfield, Middlebury, Morris, Naugatuck, New Milford, Oxford, Prospect, Roxbury, Southbury, Thomaston, Warren, Watertown, Washington, Wolcott, and Woodbury, CT, may submit applications.
Restrictions: Grants are not awarded for religious purposes, political activities, deficit financing, continuing support, fund-raising events, annual campaigns, newly established arts organizations, commissioning of new works of art, general operating support, or endowments.
Geographic Focus: Connecticut
Date(s) Application is Due: Jan 9; Mar 27; Aug 28
Amount of Grant: 8,000-10,000 USD
Contact: Carol O'Donnell, Associate CEO/Director of Grants; (203) 753-1315; fax (203) 756-3054; info@conncf.org
Internet: http://www.conncf.org/grants/grants.htm
Sponsor: Connecticut Community Foundation
43 Field Street
Waterbury, CT 06702

Connecticut Health Foundation Health Initiative Grants 1628
The foundation awards grants to organizations and institutions that directly respond to its current priority areas and result in improving the health status of Connecticut's underserved and unserved populations. Program priorities include children???s mental health—projects related to children???s mental health, including research, community grants, creating resources for clinical effective practices, and parent advocacy groups; oral health—improving oral health care access, quality, and utilization; and racial and ethnic health disparities—improving the diversity of the health care workforce, and increasing cultural competency in the existing workforce. The foundation awards two major types of grants: strategic and responsive. Application and guidelines are available online.
Requirements: Connecticut state and local units of government, community health centers, health advocacy organizations, community-based organizations, community and cultural groups, schools, and faith-based organizations are eligible. Applicants must have IRS 501(c)3 tax-exempt status or be public entities. Unincorporated organizations may apply through 501(c)3 fiscal agents.
Restrictions: Foundation grants do not support awards to individuals; construction of buildings; capital projects, endowments, or chairs associated with universities, and medical schools; conferences (unless part of a greater project or program); projects that do not benefit Connecticut residents; lobbying or influencing the outcomes of a proposed piece of legislation or election; and indirect cost for discretionary grants.
Geographic Focus: Connecticut
Date(s) Application is Due: Mar 15; Jun 15; Sep 15; Dec 15
Amount of Grant: 50,000-200,000 USD
Contact: Onell Jesus Calderas, Grants Administrator; (860) 224-2200; fax (860) 224-2230; onell@cthealth.org
Internet: http://www.cthealth.org/matriarch
Sponsor: Connecticut Health Foundation
74A Vine Street
New Britain, CT 06052

ConocoPhillips Grants 1629
ConocoPhillips maintains a philanthropic contributions budget for nonprofit, charitable programs closely tied to its corporate goals and focused primarily in locations of strong business interests. Submit an executive summary outlining the purpose of the program or project, how it will be accomplished, expected results, a budget (noting administrative expenses such as: salaries and fees, program expenses and total income), other sources of financial support and a copy of the IRS tax determination letter that confirms 501(c)3 status. The grant must be used in the United States. Requests for grants in non-U.S. locations should be made directly to the ConocoPhillips international office doing business in that part of the world. Education and youth, civic and arts, employee volunteerism, safety and social services, and the environment are focus areas.
Requirements: Applications are accepted from areas where ConocoPhillips has a strong business presence, e.g., Texas and Oklahoma. All contributions are to be used within the United States.
Restrictions: ConocoPhillips does not award funds to individuals, sectarian or religious organization, promotional sponsorship and advertising (marketing related) or an endowment.
Geographic Focus: Alaska, California, Illinois, Louisiana, Montana, New Jersey, Oklahoma, Pennsylvania, Texas
Contact: Program Contact; (281) 293-2685
Internet: http://www.conocophillips.com/about/Contribution+Guidelines/index.htm
Sponsor: ConocoPhillips Corporation
600 N Dairy Ashford, MA 3144
Houston, TX 77079

Conquer Cancer Foundation of ASCO Career Development Award 1630
The Career Development Award (CDA) is a research grant that provides funding to clinical investigators who have received their initial faculty appointment to establish an independent clinical cancer research program. The cancer research must have a patient-oriented focus, including a clinical research study and/or translational research involving human subjects. Proposals with a predominant focus on in vitro or animal studies (even if clinically relevant) are not allowed. CDA is a three-year cancer grant totaling $200,000, paid in three annual increments to the awardee's institution.
Requirements: The applicant must: be a physician (MD, DO, or international equivalent) who is within the first to third year of a full-time, primary faculty appointment in a clinical department at an academic medical institution at the time of grant submission; have a valid, active medical license at the time of application; have completed productive postdoctoral research and demonstrated the ability to undertake independent investigator-initiated clinical research; be an ASCO member (Full Member or International Corresponding) or have submitted a membership application with the grant application; be able to commit more than 50% of full-time effort in research (applies to total research, not just the proposed project) during the award period; have a mentor from the sponsoring institution who must provide a letter of support. If the mentor is not an ASCO Member, a supporting letter from an ASCO Member from the sponsoring institution must be included; be up-to-date and in compliance with all requirements (e.g. progress reports, final reports, budget summaries, IRB approvals, etc.) of any past grants received from Conquer Cancer Foundation. Applicants holding Instructor/Lecturer appointments are eligible to apply; however, the institutional letter of support must include information about the institution's commitment to support the applicant for the duration of the grant period. The full application must be submitted by 11:59PM EDT on the deadline date.
Restrictions: Eligible physicians are allowed to hold only one grant from the Conquer Cancer Foundation at a time. Applicants should not have any current career development awards (i.e., K23, K08, or any other type of career development award) and have not been a Principal Investigator on any large project grants (i.e., R01 or international equivalent, or private foundation grants). Past recipients of training fellowships (i.e., Young Investigator Award or an F32 grant) are eligible. Applicants with institutional KL2/K12 grants are eligible to apply but will be asked to relinquish their institutional grant at the start of the CDA grant period.
Geographic Focus: All States
Date(s) Application is Due: Sep 25
Amount of Grant: 200,000 USD
Contact: Eileen Melnick, Director, Grants and Awards; (571) 483-1700; Grants@conquercancerfoundation.org
Internet: http://www.conquercancerfoundation.org/cancer-professionals/funding-opportunities/career-development-award
Sponsor: Conquer Cancer Foundation
2318 Mill Road, Suite 800
Alexandria, VA 22314

Conquer Cancer Foundation of ASCO Comparative Effectiveness Research Professorship in Breast Cancer 1631
The Translational Research Professorship is designed to provide flexible funding to outstanding translational researchers who have made, and are continuing to make, significant contributions that have changed the direction of cancer research and who provide mentorship to future translational researchers. The award is intended to support qualified individuals who are dedicated to bringing advances in basic sciences into the clinical arena and to mentoring other translational researchers. The award totals $500,000 and is paid in five annual increments of $100,000 made on or about July 1 of each year of the grant term. The award is not designed to cover the total cost of a research project or the investigator's entire compensation, but can be used at the discretion of the awardee to support their overall research activities.
Requirements: Applicants must: have an MD, DO, PhD or equivalent degree; have the rank of full professor (or equivalent); have a full-time faculty appointment at an academic medical center; have made significant contributions that have changed the direction of breast cancer research; be serving as a research mentor to one or more researcher(s) in training, and must be planning to continue to provide leadership in this area throughout the award period; lead a research team in conducting research on comparative effectiveness in breast cancer; be an active member of ASCO or have submitted a membership application with the grant application; commit to spending 75% of time during the award period dedicated to research (applies to total research, not just the proposed project) including leading a team of researchers (in the lab and/or the clinic) and mentoring physician-scientists. The sponsoring institution must be a not-for-profit institution.
Geographic Focus: All States
Date(s) Application is Due: Dec 22
Amount of Grant: 500,000 USD
Contact: Eileen Melnick; (571) 483-1700; Grants@conquercancerfoundation.org
Internet: http://www.conquercancerfoundation.org/cancer-professionals/funding-opportunities/research-professorships
Sponsor: Conquer Cancer Foundation
2318 Mill Road, Suite 800
Alexandria, VA 22314

Conquer Cancer Foundation of ASCO Drug Development Research Professorship 1632
The Drug Development Research Professorship (DDRP) is a five-year award designed to provide flexible cancer research funding to outstanding researchers who have made, and are continuing to make, significant contributions that have changed the direction of cancer research and who provide mentorship to future researchers. This award is intended to support qualified individuals who are dedicated to bringing advances into the clinical arena by exploring new and promising therapeutic compounds that will lead to improved treatments, and to mentoring other researchers in the scientific and regulatory aspects of drug development. The total award amount is $500,000 payable on July 1 in annual increments of $100,000 over five years. The award is not designed to cover the total cost

of a research project or the investigator's entire compensation, but can be used at the discretion of the awardee to support their overall research activities.
Requirements: Applicants must meet the following criteria at the time of grant submission: be a physician (MD, DO, or international equivalent) working in any country with a full-time faculty appointment in a clinical department at an academic medical center; currently hold the rank of full professor (or equivalent) at an academic medical center; made significant contributions to the development of new therapies for cancer; be serving as a research mentor to researcher(s) in training, and must be planning to continue to provide leadership in the area of drug development throughout the award period; lead a research team in the conduct of drug development research that includes clinical trials; be a full member of ASCO in good standing or submit a membership application with grant application; expect to spend 75% of time during the award period dedicated to research and drug development activities, leading a team of researchers (both in the lab and/or in the clinic) and mentoring junior faculty and trainees on research and drug development. Individuals with an endowed professorship from their own institution are eligible to apply for this award. Individuals with administrative responsibilities (such as a Division or Department Chair) must clearly document their time commitment for performing research activities and mentoring activities. The Drug Development Research Professorship is open to international applicants.
Geographic Focus: All States
Date(s) Application is Due: Dec 1
Amount of Grant: 500,000 USD
Contact: Eileen Melnick; (571) 483-1700; Grants@conquercancerfoundation.org
Internet: http://www.conquercancerfoundation.org/cancer-professionals/funding-opportunities/research-professorships
Sponsor: Conquer Cancer Foundation
2318 Mill Road, Suite 800
Alexandria, VA 22314

Conquer Cancer Foundation of ASCO Improving Cancer Care Grants 1633
The Improving Cancer Care Grant (ICCG), funded by Susan G. Komen for the Cure, provides extramural cancer research funding to address important issues regarding access to healthcare, quality of care, and delivery of care, with general applicability to breast cancer. The goal of this cancer grant program is to encourage multi-disciplinary research that will have a major impact on cancer care, with general applicability in the breast cancer arena. Proposals must be focused on developing solutions to current problems, not just describing disparities in care that currently exist. Proposals must be research-focused, with specific aims and hypothesis, not a request for support of a program. ICCG is a three-year cancer research grant totaling $1.35 million, paid in three annual increments of $450,000.
Requirements: Eligible research teams: will focus on implementing and/or evaluating new solutions to existing problems in quality of, access to, and delivery of care, with general applicability to breast cancer; will be led by a single Principal Investigator, who must be an active ASCO member (or have submitted a membership application) with an MD, DO, PhD or equivalent degree; will have a multidisciplinary team of investigators that may include clinicians, nurses, pharmacists, statisticians, epidemiologists, information technologists, and other research experts; will be allowed to obtain expertise not represented in the core team through consultants and/or sub-contracts. All categories of active ASCO members are eligible; will have a multidisciplinary team of investigators that may include clinicians, nurses, pharmacists, statisticians, epidemiologists, information technologists, and other research experts; will be allowed to obtain expertise not represented in the core team through consultants and/or sub-contracts.
Geographic Focus: All States
Date(s) Application is Due: Dec 21
Amount of Grant: 1,350,000 USD
Contact: Eileen Melnick; (571) 483-1700; Grants@conquercancerfoundation.org
Internet: http://www.conquercancerfoundation.org/cancer-professionals/funding-opportunities/improving-cancer-care-grant
Sponsor: Conquer Cancer Foundation
2318 Mill Road, Suite 800
Alexandria, VA 22314

Conquer Cancer Foundation of ASCO International Development and Education Award in Palliative Care 1634
The International Development and Education Award in Palliative Care (IDEA-PC) provides medical education in palliative care, assists with career development, and helps establish strong relationships with leading ASCO members in the field of palliative care who serve as scientific Mentors to each recipient. IDEA-PC Recipients are expected to share the knowledge and training they receive through the program with colleagues in their home countries once they return. The IDEA-PC program enables Recipients to attend sessions at the ASCO Annual Meeting that address symptom management, communication, end-of-life care, and other relevant topics. The program also includes a visit to the Mentor's U.S. or Canadian institution to experience palliative care performed in a multidisciplinary environment, and provides an opportunity to cultivate long-term professional relationships that will help improve palliative care in the Recipient's home countries.
Requirements: Applicants must: have a current passport that does not expire before December; be a current resident of a country classified by the World Bank as Low-Income, Lower-Middle-Income, or Upper-Middle-Income, and have limited resources to attend the ASCO Annual Meeting; be a full Member, Member in Training, or International Corresponding member of ASCO or willing to submit an application for ASCO membership as part of the IDEA application process; be less than ten years past teir oncology program training; be fluent in English (both writing and speaking). Although applicants are not required to submit an abstract for the ASCO Annual Meeting, applicants who do submit an abstract will be given the strongest consideration.
Restrictions: Applicants cannot: have completed more than one academic year of "formal training" in a country classified by the World Bank as High-Income; have previously received the IDEA award or its predecessor, the International Travel Grant (ITG); have previously attended more than one ASCO meeting in the past five (5) years.
Geographic Focus: All States, All Countries
Date(s) Application is Due: Jan 8
Amount of Grant: 1,160 USD
Contact: Eileen Melnick; (571) 483-1700; Grants@conquercancerfoundation.org
Internet: http://www.conquercancerfoundation.org/cancer-professionals/funding-opportunities/international-development-and-education-award
Sponsor: Conquer Cancer Foundation
2318 Mill Road, Suite 800
Alexandria, VA 22314

Conquer Cancer Foundation of ASCO International Development and Education Awards 1635
The International Development and Education Award (IDEA) provides support for early-career oncologists in developing countries and facilitates the sharing of knowledge between these oncologists and ASCO members. The program pairs IDEA Recipients with a leading ASCO member "Mentor" and enables Recipients to attend the ASCO Annual Meeting, participate in a post-meeting visit to their Mentor's institution, and develop long-term relationships to improve cancer care in their countries and inform ASCO programs in developing countries. The award is not intended to completely cover all travel or daily expenses during a Recipient's trip but will cover much of the expense.
Requirements: Applicants must: have a current passport that does not expire before December; be a current resident of a country classified by the World Bank as Low-Income, Lower-Middle-Income, or Upper-Middle-Income, and have limited resources to attend the ASCO Annual Meeting; be a full Member, Member in Training, or International Corresponding member of ASCO or willing to submit an application for ASCO membership as part of the IDEA application process; be less than ten years past teir oncology program training; be fluent in English (both writing and speaking). Although applicants are not required to submit an abstract for the ASCO Annual Meeting, applicants who do submit an abstract will be given the strongest consideration.
Restrictions: Applicants cannot: have completed more than one academic year of "formal training" in a country classified by the World Bank as High-Income; have previously received the IDEA award or its predecessor, the International Travel Grant (ITG); have previously attended more than one ASCO meeting in the past five (5) years.
Geographic Focus: All States, All Countries
Date(s) Application is Due: Jan 8
Amount of Grant: 1,160 USD
Contact: Eileen Melnick; (571) 483-1700; Grants@conquercancerfoundation.org
Internet: http://www.conquercancerfoundation.org/cancer-professionals/funding-opportunities/international-development-and-education-award
Sponsor: Conquer Cancer Foundation
2318 Mill Road, Suite 800
Alexandria, VA 22314

Conquer Cancer Foundation of ASCO International Innovation Grant 1636
The International Innovation Grant provides research funding in support of novel and innovative projects that can have a significant impact on cancer control in low- and middle-income countries. The International Innovation Grant is a one-year research grant of up to $20,000 that is awarded and paid directly to a nonprofit organization or governmental agency in a low-income or middle-income country. The grant may be used by the organization only for the approved, budgeted costs of the research project. Each grant will have a Principal Investigator who is an ASCO member, is affiliated with the Grantee Organization, and is a resident of the low-income or middle-income country. International Innovation Grants are hypothesis-driven research grants that fund a specific research project that may result in the discovery of new knowledge about how to advance cancer control in a low- or middle-income setting. It is anticipated that novel approaches and clinical designs proposed for this grant may differ from what would be considered standard practice within high-income settings. Grantee Organizations and Principal Investigators will be expected to share and disseminate the knowledge gained during their research project.
Requirements: The Applicant Organization must: be an organization with a charitable purpose registered as a not-for-profit with the relevant national authority or must be a government agency; be located in a country categorized by the World Bank as Low-Income, Lower-Middle-Income or Upper-Middle Income; have been operating for at least one full year, have an acceptable management structure and processes in place, and be solvent with or without the support of the International Innovation Grant; should have experience in carrying out activities with tangible outcomes. The Principal Investigator must: be a member of ASCO or have submitted a membership application with the grant application; be a citizen or permanent resident of a country defined by the World Bank as low-income or middle-income, and currently residing in that country; be affiliated with the applicant organization.
Geographic Focus: All States, All Countries
Date(s) Application is Due: Oct 9
Amount of Grant: Up to 20,000 USD
Contact: Eileen Melnick; (571) 483-1700; Grants@conquercancerfoundation.org
Internet: http://www.conquercancerfoundation.org/cancer-professionals/funding-opportunities/international-innovation-grant
Sponsor: Conquer Cancer Foundation
2318 Mill Road, Suite 800
Alexandria, VA 22314

Conquer Cancer Foundation of ASCO Medical Student Rotation Grants 1637

This funding opportunity designed to facilitate the recruitment and retention of individuals from populations underrepresented in medicine to cancer careers and increase access to quality care for underserved communities. The Medical Student Rotation program was initially developed as part of the Conquer Cancer Foundation/ASCO/Susan G. Komen for the Cure Quality of Care Initiative. The program provides 8- to 10-week clinical or clinical research oncology rotations for U.S. medical students from populations underrepresented in medicine who are interested in pursuing oncology as a career. Recipients receive a $5,000 stipend for the rotation plus $1,500 for future travel to the ASCO Annual Meeting, An additional $2,000 will be provided to support the student's mentor.

Requirements: Applicants must be enrolled in an MD or DO, U.S. medical school program and be of an underrepresented population (For the purposes of this program, underrepresented population is defined according to the Association of American Medical Colleges definition, currently specified as those racial and ethnic populations that are underrepresented in the field of medicine relative to their numbers in the general population. Races/ethnicities that have been identified include but are not limited to: American Indian/Alaska Native, Black/African American, Hispanic/Latino, and Native Hawaiian/Pacific Islander). Candidates must be U.S. citizens, U.S. nationals, or permanent residents. Candidates also must demonstrate an interest in pursuing oncology as a career and have a record of good academic standing. Medical students may choose a mentor and rotation setting on their own, or they can request the Conquer Cancer Foundation assistance in pairing them with a mentor. Mentors are responsible for ensuring that the student receives consistent guidance and supervision throughout the rotation, either directly or by another member of the oncology practice or staff able to perform this role. The mentor is responsible for defining the goals and objectives of the rotation for the student, orienting the student to the office or hospital setting, and for providing regular feedback to the student.

Geographic Focus: All States
Date(s) Application is Due: Dec 17
Amount of Grant: Up to 8,500 USD
Contact: Eileen Melnick; (571) 483-1700; Grants@conquercancerfoundation.org
Internet: http://www.conquercancerfoundation.org/cancer-professionals/funding-opportunities/medical-student-rotation-underrepresented-populations
Sponsor: Conquer Cancer Foundation
2318 Mill Road, Suite 800
Alexandria, VA 22314

Conquer Cancer Foundation of ASCO Resident Travel Award for Underrepresented Populations 1638

The Conquer Cancer Foundation of ASCO Resident Travel Award for Underrepresented Populations (RTA) is designed to facilitate the recruitment and retention of individuals from underrepresented populations in medicine to cancer careers and increase access to quality care for underserved communities. RTA provides financial support for residents from underrepresented populations to attend ASCO's Annual Meeting. The intention of this award is to attract residents from underrepresented populations to a possible career in one of the oncology specialties. This travel award will allow the recipients to travel to the ASCO Annual Scientific Meeting, where they will have an opportunity to meet oncologists and to understand the career possibilities in this area. The Resident Travel Award for Underrepresented Populations includes a $1,500 travel advance, complimentary Annual Meeting registration, and access to the Annual Meeting housing block and ASCO's travel agent.

Requirements: Applicants must be enrolled in an ACGME-accredited residency program required for future training in a cancer related subspecialty (i.e. Internal Medicine considering Med/Onc, Surgery considering Surg/Onc) and be of an underrepresented population as defined by the program eligibility criteria. (For the purposes of this program, underrepresented population is defined according to the Association of American Medical Colleges definition, currently specified as those racial and ethnic populations that are underrepresented in the field of medicine relative to their numbers in the general population. Races and ethnicities that have been identified include but are not limited to: American Indiana/Alaska Native, Black/African American, Hispanic/Latino, and Native Hawaiian/Pacific Islander.) Candidates must be U.S. citizens, U.S. nationals or permanent residents. Candidates must have a record of good academic standing.

Geographic Focus: All States
Date(s) Application is Due: Dec 13
Amount of Grant: 1,500 USD
Contact: Eileen Melnick; (571) 483-1700; Grants@conquercancerfoundation.org
Internet: http://www.conquercancerfoundation.org/cancer-professionals/funding-opportunities/resident-travel-award-underrepresented-populations
Sponsor: Conquer Cancer Foundation
2318 Mill Road, Suite 800
Alexandria, VA 22314

Conquer Cancer Foundation of ASCO Translational Research Professorships 1639

The Translational Research Professorship is designed to provide flexible funding to outstanding translational researchers who have made and are continuing to make significant contributions that have changed the direction of cancer research and who provide mentorship to future translational researchers. The award is intended to support qualified individuals who are dedicated to bringing advances in basic sciences into the clinical arena and to mentoring other translational researchers. The award totals $300,000 payable on July 1 in annual increments of $100,000 over three years.

Requirements: The applicant must meet the following criteria at the time of grant award (July 1): be a physician (MD, DO or international equivalent) with a full-time faculty appointment in a clinical department at an academic medical center; currently hold the rank of full professor (or equivalent) at an academic medical center. If applicant does not hold the rank of full professor, but meets all other eligibility criteria, he/she must provide written explanation of why his/her current rank at their institution should be considered equivalent; made significant contributions that have changed the direction of cancer research; be serving as a research mentor to one or more translational researcher(s) in training, and must be planning to continue to provide leadership in this area throughout the award period; lead a research team in the conduct of translational research; be a member of the American Society of Clinical Oncology (ASCO); expect to spend 75% of time during the award period dedicated to translational research activities, leading a team of translational researchers (both in the lab and in the clinic) and mentoring physician-scientists. Individuals with an endowed professorship from their own institution are eligible to apply for this award. Individuals with administrative responsibilities (such as a Division or Department Chair) must clearly document their time commitment for performing research activities and mentoring activities. The sponsoring institution must be a not-for-profit institution.

Geographic Focus: All States
Date(s) Application is Due: Jan 13
Amount of Grant: 300,000 USD
Contact: Eileen Melnick; (571) 483-1700; Grants@conquercancerfoundation.org
Internet: http://www.conquercancerfoundation.org/cancer-professionals/funding-opportunities/research-professorships
Sponsor: Conquer Cancer Foundation
2318 Mill Road, Suite 800
Alexandria, VA 22314

Conseil des arts de Montreal Residency of the Americas 1640

The Residency of the Americas aims to increase the circulation of art and artists across the American continent. It is divided into two parts: a three-month residency for a visual artist and a second, two-month residency for a curator or art critic. Both residencies will enjoy a live-in studio with access to common living areas. They will also have access to high-speed Internet, wood and metal workshops and technical assistance, if needed. The residency includes various activities such as open house events, exhibitions and studio visits by professionals that will enable residents to come in contact with both the public and Montreal's professional arts community. The residencies include: return airfare (max $1,200); a live-in studio for three months; and a stipend of $250 per week. The application and guidelines are available on the Foundry website.

Requirements: Applicants must be artists or curators residing in the United States or Canadian artists living outside of Quebec.

Geographic Focus: All States, Canada
Date(s) Application is Due: Sep 30
Amount of Grant: 1,450 CAD
Contact: Americas Residency Contact; (514) 392-1544; residence@fonderiedarling.org
Internet: http://www.artsmontreal.org/en/programs/residence-des-ameriques
Sponsor: Conseil des arts de Montreal
1210 Sherbrooke Street East
Montreal, QC H2L 1L9 Canada

Conservation, Food, and Health Foundation Grants for Developing Countries 1641

The Conservation, Food, and Health Foundation's geographic focus is the developing world. Through grants to support research and through targeted grants to help solve specific problems, the foundation helps build capacity within developing countries in three areas of interest: conservation, food, and health. The foundation concentrates its grant-making on research, technical assistance, and training projects of benefit to the Third World; favors grants for pilot projects and special programs that have a potential for replication; prefers to support projects that employ and/or train personnel from the developing world; and favors research concerning problems of importance to the developing world. The foundation has two 4-month funding cycles each year. Concept papers received by January 1 will be considered for eligibility for the March 1 full proposal deadline; concept papers received by July 1 will be considered for eligibility for the September 1 full proposal deadline. Full proposals are by invitation. Concept applications must be submitted through the foundation's online application system. Detailed guidelines, instructions, and faqs are provided at the website.

Requirements: The following organizations are eligible to apply: 501(c)3 organizations that are not private foundations under section 509(a) of the United State Internal Revenue Code; "501(c)3 equivalent" foreign or domestic government units; and nongovernmental foreign organizations which can provide secure evidence of their nongovernmental status and charitable purpose.

Restrictions: The foundation does not provide support for buildings, land purchases, or vehicles; quantity purchases of durable medical equipment; endowments or fundraising activities; famine or emergency relief; direct medical care or treatment; feeding or food distribution programs; films, videos, or web-site production; scholarships, fellowships, or travel grants; conferences; re-granting through intermediaries; general operating support; or individuals (however, the foundation may support an individual engaged in research on a problem of significance to the developing world where the research is sponsored by an established, nonprofit organization such as an educational institution and conducted in close partnership with a local nongovernmental organization). The foundation does not consider the states of the former Soviet Union or former Eastern Bloc countries as within its geographic focus.

Geographic Focus: All States, All Countries
Date(s) Application is Due: Jan 1; Jul 1
Amount of Grant: Up to 25,000 USD
Contact: Prentice Zinn; (617) 391-3091; pzinn@gmafoundations.com
Internet: http://cfhfoundation.grantsmanagement08.com/?page_id=5
Sponsor: Conservation, Food, and Health Foundation
77 Summer Street, 8th Floor
Boston, MA 02110-1006

Consortium for Graduate Study in Management Minority Fellowships **1642**
An 11-university consortium, designed to hasten the entry of minorities into managerial positions in business, is offering fellowships to capable young men and women to allow them the opportunity to pursue a high-quality educational program leading to the MBA. Fellowships are made possible through the support and contributions of American business. Summer opportunities are also provided to each student accepted for the program. The 11 participating universities are University of California-Berkley, Indiana University-Bloomington, New York University, University of Michigan-Ann Arbor, University of North Carolina at Chapel Hill, University of Rochester, University of Southern California, University of Texas at Austin, University of Virginia, Washington University-Saint Louis, and University of Wisconsin-Madison. Application forms may be downloaded from the Web site. Applications are accepted between December 1 and January 15.
Requirements: Young African Americans, Puerto Ricans, Cubans, Dominicans, Mexican Americans, or Native Americans who are U.S. citizens or nationals who aspire to managerial careers in business are invited to apply. Applications will be received from college seniors and those holding bachelor's degrees from accredited colleges and universities. Degrees may be in the humanities, life, social and physical sciences, and engineering, as well as business or economics. Those already embarked on other careers but wishing to change careers are also encouraged to apply.
Geographic Focus: All States
Date(s) Application is Due: Jan 15; Dec 1
Amount of Grant: 2,500 USD
Contact: Jackie Olden, (888) 658-6814 or (314) 877-5540; frontdesk@cgsm.org
Internet: http://www.cgsm.org
Sponsor: Consortium for Graduate Study in Management
5585 Pershing, Suite 240
Saint Louis, MO 63112-4621

Cooperative Institute for Research in Environmental Sciences Visiting Fellows **1643**
With support from the Environmental Research Laboratories of the National Oceanic and Atmospheric Administration, the Cooperative Institute for Research in Environmental Sciences (CIRES) at the University of Colorado at Boulder offers up to six one-year visiting fellowships to scientists with research interests in the areas of advanced observing and modeling systems, climate system variability, geodynamics, planetary metabolism, regional processes, and integrating activities. The program provides opportunities for interactions between CIRES scientists and visiting fellows to pursue common research interests. Selections for this program are based in part on the likelihood of interactions between the visiting fellow and the scientists at CIRES and the degree to which both parties will benefit from the exchange of new ideas. To further this goal, priority is given to candidates with research experience at institutions outside the Boulder scientific community. Salary, benefits, and a research budget are provided.
Requirements: Awards may be made to PhD scientists at all levels; faculty planning sabbatical leave and recent PhD recipients are especially encouraged to apply. The program is open to scientists of all countries.
Geographic Focus: All States
Date(s) Application is Due: Dec 31
Contact: Karen Dempsey; (303) 492-1168; fax (303) 492-1149; dempsey@cires.colorado.edu
Internet: http://cires.colorado.edu/visfell
Sponsor: Cooperative Institute for Research in Environmental Sciences
Human Resources Department, 216 UCB
Boulder, CO 80309-0216

Coors Brewing Corporate Contributions Grants **1644**
The company has a firm commitment to giving back to its home-market communities—including Denver, CO; Memphis, TN; and Elkton, VA—and supports grassroots, nonprofit organizations that address community, civic and industry issues. The primary focus is on programs that enhance the quality of life. Vehicles of support include cash grants, Coors products for events, Coors logo items for fund-raisers, volunteer hours by Coors employees or retirees, or used equipment or in-kind services. Preference will be given to groups that focus on issues of national scope. Corporate Contributions usually reviews requests the first Wednesday of every month. A minimum of two months lead time is required before the event or funding need.
Requirements: IRS 501(c)3 nonprofit organizations located in Denver, CO; Memphis, TN; and Elkton, VA; are eligible.
Restrictions: Requests will not be considered for individuals in personal programs; individual scholarships; teams, groups, or races; travel expenses; third-party fund-raisers or sales promotions; political activities; or requests by telephone.
Geographic Focus: Colorado, Tennessee, Virginia
Contact: Buck Boze; (800) 642-6116 or (303) 277-5953; fax (303) 277-6132
Internet: http://www.coors.com
Sponsor: Coors Brewing Company
P.O. Box 4030, Department NH420
Golden, CO 80401

Cord Foundation Grants **1645**
The foundation makes grants to a variety of groups in three broad areas: education, social services, and the arts. Grant recipients have included higher education institutions, research groups, youth organizations, community service agencies, religious organizations, and performing and visual arts centers. Types of support include general operating support, building construction/renovation, equipment acquisition, emergency funds, program development, scholarship funds, research, and matching funds. Grants are awarded nationwide, with preference given to requests from northern Nevada. Application forms are not required.
Requirements: Nonprofits are eligible but giving is primarily in the northern NV area.
Restrictions: The foundation does not support general fund-raising events, memorial campaigns, deficit fundings, conferences, dinners, or mass mailings.
Geographic Focus: Nevada
Amount of Grant: 10,000-250,000 USD
Contact: William Bradley, Trustee; (775) 323-0373
Sponsor: Cord Foundation
418 Flint Street
Reno, NV 89501-2008

Corning Incorporated Foundation Educational Grants **1646**
The grants improve education by supporting selected K-12 school districts, community colleges and four-year institutions of higher learning primarily in communities where the sponsor company has operations. Areas of involvement have included community service programs for students, curriculum enrichment, student scholarships, facility improvement, and instructional technology projects for the classroom.
Requirements: Support goes to institutions that are tax-exempt under Section 501(c)3 of the Internal Revenue Code and which are public charities as defined in Section 509(a) of the Code. All requests to the foundation for support must be made in writing. Grant seekers are advised to submit a two- to three-page letter of inquiry, signed by the senior administrative officer of the organization.
Restrictions: Grants do not support individuals; political parties, campaigns, or causes; labor or veterans' organizations; religious or fraternal groups; volunteer emergency squads; athletic activities; courtesy advertising; or fundraising events.
Geographic Focus: All States
Amount of Grant: 1,000-25,000 USD
Contact: Karen Martin; (607) 974-8722; fax (607) 974-4756; martinkc@corning.com
Internet: http://www.corning.com/inside_corning/foundation.asp
Sponsor: Corning Foundation
1 Riverfront Plaza, MP-BH-07
Corning, NY 14831

Corpus Christi College Research Scholarships **1647**
Successful candidates will pursue, as members of college, a course of study in any field leading to a research-based higher degree in the University of Cambridge. Awards are usually made to those studying for the PhD degree. Subject to satisfactory progress, awards may be extended to a maximum of three years.
Requirements: The awards are open to both men and women who have graduated before October 1 of the year of application. Applicants must first apply for university admission to the Board of Graduate Studies at 4 Mill Lane, Cambridge, CB2 1RZ and must have acceptance as a condition of receiving a scholarship.
Restrictions: Those eligible for United Kingdom State Awards may not apply.
Geographic Focus: All States
Amount of Grant: Up to 500 GBP
Contact: Tutor for Advanced Students; 44 01223-338038; fax 44 01223-338057; graduate-tutor@corpus.cam.ac.uk
Internet: http://www.admin.cam.ac.uk/univ/gsprospectus/colleges/corpuschristi.html
Sponsor: Corpus Christi College
Trumpington Street
Cambridge, CB2 1RH England

Coughlin-Saunders Foundation Grants **1648**
The foundation awards grants to nonprofit organizations in its areas of interest, including arts and arts education, higher education, religion, social services, and youth organizations. Types of support include general operating support, capital campaigns, building construction/renovation, equipment acquisition, emergency funds, program development, professorships, and scholarship funds. Proposals are preferred in January or February. Grants are made primarily for projects which benefit Alexandria, Louisiana, and the surrounding area. New Orleans and the surrounding area will be considered.
Requirements: 501(c)3 tax-exempt organizations in central Louisiana may apply.
Restrictions: Grants are not made to individuals or for fund raisers. Endowments will not be funded.
Geographic Focus: Louisiana
Amount of Grant: 100-40,000 USD
Contact: Ed Crump Jr.; (318) 561-4070; fax (318) 487-7339; csfoundation@kricket.net
Sponsor: Coughlin-Saunders Foundation
2010 Gus Kaplan Drive
Alexandria, LA 71301

Council on Foundations Emerging Philanthropic Leaders Fellowships **1649**
The program is designed to help future foundation leaders become more effective within their communities and the larger field of organized philanthropy. Each year, two two-year fellowships are awarded. Each fellow is matched with a mentor who is a recognized leader in philanthropy. The fellows are asked to identify their professional growth goals and organizational development needs that they would like to address as part of their fellowship. The mentors, drawn from the council's membership, are asked to help develop two-year plans for providing information, advice, and referrals. Applications may be sent by mail, fax, or email.
Requirements: The nominee or applicant must have at least two years experience in the philanthropic area, with at least one year in a leadership position. The nominee or applicant's organization must have a focus on increasing and expanding philanthropic programs within communities that are historically underrepresented in institutional philanthropy. All applicants must submit a simple application letter. The chief executive of the organization also must write a brief letter of support or co-sign the applicant's

submission. Nominees or applicants must be associated with a Council on Foundations member or member-eligible organization.
Geographic Focus: All States
Date(s) Application is Due: Dec 31
Contact: Evelyn Gibson, (703) 879-0691 or (703) 879-0600; Evelyn.Gibson@cof.org
Internet: http://www.cof.org/files/Documents/Diversity/diversitybrochure.pdf
Sponsor: Council on Foundations
2121 Crystal Drive, Suite 700
Arlington, VA 22202

Covidien Advanced MIS Surgical Fellowships 1650

Covidien sponsors fellowships each year at leading institutions that enable individuals to engage in educational and clinical activities that develop their expertise in minimally invasive procedures and techniques. The Foundation for Surgical Fellowship (FSF) is a 501(c)3 organization that handles all aspects of fellowship administration. The Foundation exists to fund high quality fellowships in various areas of surgery, including but not limited to flexible endoscopy, minimal access surgery, and other related disciplines. Through a blinded application and review process, the Foundation identifies and distributes funds to qualified organizations based on criteria of excellence. The Foundation promotes and supports the continued availability of postgraduate surgical training opportunities of the highest quality. Applicants may apply for the Fellowship through the FSF link located at the Covidien website.
Geographic Focus: All States
Date(s) Application is Due: May 18
Contact: Teresa Hacunda; (508) 261-8000; Teresa.Hacunda@covidien.com
Internet: http://www.covidien.com/covidien/pages.aspx?page=Education/Fellowships
Sponsor: Covidien Healthcare Products
15 Hampshire Street
Mansfield, MA 02048

Covidien Surgical Mini-Fellowships 1651

The Covidien Surgical Mini-Fellowship is a three to five day comprehensive educational program meant to more thoroughly train surgeons to perform a specified surgical procedure, and to educate surgeons on procedural specifics and patient management strategies. Fellowships may include detailed review/lecture of surgical techniques, potential intra-operative complications; discussion/lecture about surgeon's role in the care of the patient; discussion/lecture of post-operative patient care via observation of hospital rounds, review of treatment for potential post-operative complications; discussion of reimbursement strategies for laparoscopic procedures; observation of and/or participation in several laparoscopic and/or open cases; observation of operating room procedures with review of patient preparation and optimal patient positioning for the various procedures; and dry skills or animate lab. Applicants may call the contact person for current deadlines and information about the Fellowship.
Requirements: Applicants must have completed at least one master (CME or non CME) and at least one clinical immersion. Additional information about each of these classes is available at the Covidien.
Geographic Focus: All States
Contact: Teresa Hacunda; (508) 261-8000; Teresa.Hacunda@covidien.com
Internet: http://www.covidien.com/covidien/pages.aspx?page=Education/Advanced Training/Mini-Fellowship
Sponsor: Covidien Healthcare Products
15 Hampshire Street
Mansfield, MA 02048

Cowles Charitable Trust Grants 1652

The foundation awards grants, primarily in New York, Florida, and on the East Coast, for the arts and culture, including museums and the performing arts; environment; education, including early childhood education, secondary and higher education, medical school education, adult basic education and literacy, and adult continuing education; hospitals and AIDS programs, including research; social services, including family planning, human services, and federated giving; and community funds, including leadership development, civil rights, and race relations. Types of support include general operating support, capital campaigns, annual campaigns, equipment acquisition, endowment funds, continuing support, seed money, building construction/renovation, matching funds, professorships, and program development. Application forms are required; initial approach should be by letter. The board meets in January, April, July, and October.
Requirements: Nonprofit organizations may apply for grant support. Grants are awarded primarily along the Eastern Seaboard.
Geographic Focus: Florida, New York
Date(s) Application is Due: Mar 1; Jun 1; Sep 1; Dec 1
Amount of Grant: 1,000-40,000 USD
Contact: Mary Croft, Treasurer; (732) 936-9826
Sponsor: Cowles Charitable Trust
P.O. Box 219
Rumson, NJ 07760

CPC Catherine H. Beattie Fellowship 1653

The fellowship was created to promote the conservation of rare and endangered flora in the United States through the programs of the Center for Plant Conservation. The research grant enables a student in either biology or horticulture to conduct field research on rare plants. Preference is given to students whose projects focus on the endangered flora of the Carolinas and the southeastern United States. Additional information and applications may be obtained by contacting the office.
Requirements: Graduate students in biology, horticulture, or a related field are eligible.
Geographic Focus: All States
Date(s) Application is Due: Dec 31
Amount of Grant: 1,000-4,000 USD
Contact: Programs Manager; (314) 577-9450; fax (314) 577-9465; cpc@mobot.org
Internet: http://www.centerforplantconservation.org/beattie.html
Sponsor: Center for Plant Conservation
Missouri Botanical Garden, P.O. Box 299
St. Louis, MO 63166-0299

Credit Union Foundation of British Columbia Grants 1654

The purpose of the foundation is to administer funds for the encouragement, promotion, and advancement of postsecondary education by means of grants-in-aid assistance to British Columbia resident students in need of financial assistance in pursuing vocational, technical, or academic training within British Columbia. Areas of interest include adult and continuing education, education institutions, universities, community colleges, and vocational training. Types of support through the program include scholarships, grants, awards, and bursaries. The majority of student grants are considered in the fall. Contact the office for application forms.
Geographic Focus: All States, Canada
Amount of Grant: 300-2,000 USD
Contact: Program Contact; cufoundation@shaw.ca
Internet: http://www.cufoundation.org
Sponsor: Credit Union Foundation of British Columbia
1441 Creekside Drive
Vancouver, BC V6J 4S7 Canada

CRF Graduate Research Grants 1655

Proposals are welcome from students enrolled in graduate programs in any accredited colleges or universities. Grants will support fully or partially funded projects. Travel expenses may be included if justified by the research. Guidelines are available online.
Geographic Focus: All States
Date(s) Application is Due: Nov 28
Amount of Grant: Up to 15,000 USD
Contact: Fund Coordinator; info@craftcreativitydesign.org
Internet: http://www.craftcreativitydesign.org/research/grants.php
Sponsor: Center for Craft, Creativity, and Design
P.O. Box 1127
Hendersonville, NC 28793

CRF Project Research Grants 1656

Support is provided for research on American studio craft. Research involving international craft must be in relationship to craft in America. The fund seeks proposals that address the goals of the fund through monographs on individual craft artists or themes, historical research, and cross-disciplinary research involving craft. Applicants requesting support for monographs on individual craft artists should refer to www.aaa.si.edu on how the research will compliment or expand the Smithsonian Laitman Documentation Project for Craft and Decorative Arts in America. Guidelines are available online.
Requirements: Proposals are welcome from academic researchers, independent scholars, doctoral students, and museum curators.
Restrictions: General overhead (indirect administrative expenses) is not eligible for university-based projects. No capital equipment purchases are eligible for support.
Geographic Focus: All States
Date(s) Application is Due: Nov 28
Amount of Grant: Up to 15,000 USD
Contact: Fund Coordinator; info@craftcreativitydesign.org
Internet: http://www.craftcreativitydesign.org/research/grants.php
Sponsor: Center for Craft, Creativity, and Design
P.O. Box 1127
Hendersonville, NC 28793

Crohn's Disease and Colitis Research Fellowships 1657

Three-year fellowships are awarded to encourage the development of individuals with research potential to help them prepare for careers of independent research in the areas of Crohn's disease and ulcerative colitis. Yearly renewal is dependent on the receipt of satisfactory progress reports. Application forms must be obtained from the foundation. Letters of intent are due May 1 and November 1. Submission deadlines for full applications are January 14 and July 1 of each year. Annual deadline dates may vary; contact program staff for exact dates.
Requirements: The applicant must hold an MD, PhD, or equivalent degree and be currently employed by an institution engaged in health care and/or health-related research within the United States. Candidates who have obtained MD degrees must have at least two years of postdoctoral experience, one year of which must be documented research experience relevant to inflammatory bowel disease. Candidates holding PhD degrees must have one year of postdoctoral experience and at least one year of documented research experience relevant to IBD.
Geographic Focus: All States
Date(s) Application is Due: Jan 14; May 1; Jul 1; Nov 1
Contact: Carol Cox, Director of Research and Scientific Programs; (212) 685-3440 or (800) 932-2423; fax (212) 779-4098; ccox@ccfa.org
Internet: http://www.ccfa.org/research/?LMI=4
Sponsor: Crohn's and Colitis Foundation of America
386 Park Avenue S, 17th Floor
New York, NY 10016-8804

Crohn's Disease and Ulcerative Colitis Senior Research Grants **1658**

Grants are awarded to individuals for up to two years for research in the basic biomedical and clinical sciences that will increase the understanding of the cause, mechanisms of symptoms, and treatment of inflammatory bowel disease including Crohn's disease and ulcerative colitis. Support for additional one-year periods will be by competitive renewal. Application forms must be secured from the foundation. Submission deadlines for letters of intent are November 1 and May 1. Submission deadlines for full applications are January 14 and July 1 of each year, unless noted otherwise.
Requirements: Applicants must be established researchers in the field of inflammatory bowel disease.
Geographic Focus: All States
Date(s) Application is Due: Jan 14; May 1; Jul 1; Nov 1
Amount of Grant: Up to 115,000 USD
Contact: Carol Cox, Director of Research and Scientific Programs; (212) 685-3440 or (800) 932-2423; fax (212) 779-4098; ccox@ccfa.org
Internet: http://www.ccfa.org/research/?LMI=4
Sponsor: Crohn's and Colitis Foundation of America
386 Park Avenue S, 17th Floor
New York, NY 10016-8804

Croucher Foundation Grants **1659**

The foundation promotes education, learning, and research in the areas of natural science, technology, and medicine. Each year, 20 to 25 scholarships and fellowships, tenable in the United Kingdom, Hong Kong, Canada, Australia, or New Zealand, are awarded.
Requirements: The competition is open to permanent residents of Hong Kong.
Geographic Focus: All States
Contact: Administrator; (852) 2736 6337; fax (852) 2730 0742; cfadmin@croucher.org.hk
Internet: http://www.croucher.org.hk
Sponsor: Croucher Foundation
9 Queen's Road Central, Suite 501 Hong Kong

Cruise Industry Charitable Foundation Grants **1660**

The foundation awards grants to improve the quality of life in U.S. cities and towns where the cruise industry maintains vessel operations, employs a significant number of individuals, and purchases products and services. Areas of interest include civic and community development, educational assistance and training programs, public health programs, and environmental initiatives. There are no application deadlines, see Foundation's website for application guidelines.
Requirements: U.S. 501(c)3 organizations are eligible to apply.
Restrictions: Requests will not be considered from individuals, fraternal organizations, religious organizations, political organizations, or organizations that conduct lobbying activity.
Geographic Focus: All States
Contact: Cynthia Colenda; (703) 522-3160; fax (703) 522-3161; cicf@iccl.org
Internet: http://www.cruisefoundation.org/requirements-and-guidelines.php
Sponsor: Cruise Industry Charitable Foundation
2111 Wilson Boulevard, 8th Floor
Arlington, VA 22201

CSRPC Dissertation Fellowships **1661**

The goal of the fellowship is to enable an outstanding doctoral student interested in the study of race and ethnicity to devote his or her full energies to the completion of the dissertation. The fellowship carries a stipend; a travel and research budget; and will cover advanced residence tuition, fees, and basic university student health insurance, if needed. The successful applicant will be provided with an office and use of a computer at the center. The fellow will be expected to be in residence during the award year, present his or her work at one of the Reproduction of Race and Racial Ideologies Workshop meetings and to actively participate in the workshop and other activities sponsored by the center.
Requirements: University of Chicago doctoral students who have completed all requirements for the PhD but the dissertation, including formal admission to candidacy, and expect to complete all field work by September 26 are eligible to apply. Any dissertation that has as its central focus issues related to race or racialized groups will be considered. Special consideration will be given to projects that attend to the intersection of race or ethnicity with other identities such as gender, class, sexuality, and nationality.
Restrictions: The fellow may not engage in any remunerative activity, including teaching either on or off campus, while holding the award, and will be ineligible for further internal University funding from any source, including teaching appointments, if the degree is not completed within six months of the end of the CSRPC fellowship tenure.
Geographic Focus: All States
Date(s) Application is Due: Mar 4
Amount of Grant: 1,000-18,000 USD
Contact: Center for the Study of Race, Politics, and Culture; (773) 702-8063; fax (773) 834-2200; csrpc@uchicago.edu
Internet: http://csrpc.uchicago.edu/resources_funding.shtml
Sponsor: University of Chicago
5733 S University Avenue
Chicago, IL 60637

CSRPC Postdoctoral Fellowship for Advanced Scholars **1662**

The goal of the fellowship is to support the work of an outstanding advanced scholar whose research focuses on the study of race or ethnicity by allowing the fellow to devote his or her energies to the further development of their research agenda. The fellowship carries a stipend, a travel and research budget, and an allowance for moving expenses. The fellow will be provided with office space, a computer at the center, and full access to University libraries and other facilities. Awardees will be expected to be in full-time residence during the academic year; teach a 10-week course related to race and/or ethnicity (one quarter); give a public lecture; present his or her work at one of the Reproduction of Race and Racial Ideologies Workshop meetings; and actively participate in the workshop and other activities sponsored by the center. Guidelines are available online.
Requirements: Applicants are required to have a PhD; must have been awarded tenure at the time of application; and must be U.S. citizens or permanent residents at the time of application.
Geographic Focus: All States
Date(s) Application is Due: Feb 11
Amount of Grant: 55,000-62,500 USD
Contact: Center for the Study of Race, Politics, and Culture; (773) 702-8063; fax (773) 834-2200; csrpc@uchicago.edu
Internet: http://csrpc.uchicago.edu/resources_funding.shtml
Sponsor: University of Chicago
5733 S University Avenue
Chicago, IL 60637

CSRPC Postdoctoral Fellowship for Junior Scholars **1663**

The goal of the fellowship is to support the work of an outstanding junior scholar whose research focuses on the study of race or ethnicity by allowing the fellow to devote his or her energies to the further development of their research agenda. The fellowship carries a stipend, a travel and research budget, and an allowance for moving expenses. The fellow will be provided with office space, a computer at the center, and full access to university libraries and other facilities. Awardees will be expected to be in full-time residence during the academic year; teach a 10-week undergraduate course related to race and/or ethnicity (one quarter); present his or her work at one of the Reproduction of Race and Racial Ideologies Workshop meetings; and actively participate in the workshop and other activities sponsored by the center. Guidelines are available online.
Requirements: Applicants are required to have a PhD; must be non-tenure track scholars or tenure track faculty who do not expect to receive tenure prior to the start of the fellowship year; and must be U.S. citizens or permanent residents at the time of application.
Geographic Focus: All States
Date(s) Application is Due: Feb 11
Amount of Grant: Up to 2,500 USD
Contact: Center for the Study of Race, Politics, and Culture; (773) 702-8063; fax (773) 834-2200; csrpc@uchicago.edu
Internet: http://csrpc.uchicago.edu/resources_funding.shtml
Sponsor: University of Chicago
5733 S University Avenue
Chicago, IL 60637

CSTE CDC/CSTE Applied Epidemiology Fellowships **1664**

The Council of State and Territorial Epidemiologists (CSTE), in collaboration with the Centers for Disease Control and Prevention (CDC), the Association of Schools of Public Health (ASPH), and the Health Resources and Services Administration (HRSA) has established the Applied Epidemiology Fellowship Program to train recent graduates in the expanding field of applied epidemiology in an effort to meet the nation's ongoing need for applied epidemiology workforce capacity in state and local health departments. In achieving the fellowship's goal which is to provide a high quality training experience and to secure long-term career placement for fellows at the state or local level, the program focuses on balancing three key concepts as identified in the Applied Epidemiology and Training Program (AETP) Development Handbook (McDonnell 2002): providing service to the sponsoring agency; creating and training a core group of public health workers; and strengthening the capacity in applied epidemiology across public health institutions. Participating fellows will receive two years of on-the-job training at a state health agency (starting in June-August of the program year). Fellows are placed under the guidance of highly-trained and experienced primary and secondary mentors and work closely with epidemiologists at the state, local, and federal levels. A primary mentor must have a doctoral-level degree and the secondary mentor is preferred to have doctoral training as well. Mentors must devote four hours per week to spend with the fellow during the first month of the fellowship and two hours per week for the duration. The mentors oversee the training, research and field activities of the fellow, ensure that the fellow is familiar with relevant techniques in a given specialty, help provide day-to-day activities that provide "hands on" epidemiology experience, and encourage the overall professional development of the fellow. Mentors receive an initial orientation conducted by the CSTE National Office prior to the fellow's arrival, and they participate in conference calls at various intervals during the course of the fellowship. Host health agencies are CDC- and CSTE-approved, with a demonstrated capacity to provide an Applied Epidemiology Fellow with technical training, research opportunities, and practical experience in the application of epidemiologic methods. Host sites are selected based on: the career interests of the fellow; the scope, quality, and diversity of the experience offered to the fellow; the experience of supervisors in management; mentors' number of years of experience in epidemiology; academic training of the mentors; the availability of office space; and letters of support for the agency seeking a fellow. Fellows and host agencies are also carefully matched based on the program areas of interest of the fellow and available opportunities of the host agency. Program areas include infectious disease, injury, maternal and child health, environmental heath, chronic disease, substance abuse, and occupational health. Host agencies are expected to integrate fellows into their worksites and to treat them as entry-level permanent employees. If employee programs are offered to regular permanent employees, host agencies are expected to provide comparable programs and financial support for the fellow. The fellowship program will provide M.P.H.-degree fellows with a bimonthly stipend of up to $39,000 annually. Doctoral-level candidates will be paid up to $56,000. Stipends will be set according to location, cost of living, and pay structure at the host agency. The fellowship program will make available up to $3,200 per year to supplement the fellow's individual health insurance coverage. Up to $1,000 will be provided for moving-related expenses. $970 per year is provided each fellow for professional development. Host health agencies and mentors are strongly encouraged to provide financial support and opportunities as well for the fellow to participate in other public health activities that will expand the fellow's scope and depth of epidemiologic knowledge and/or expand his/her job-related capabilities. CDC/CSTE Applied Epidemiology fellows will develop a comprehensive set of core skills through competency-based training. Applicants may view a list of core competencies at the fellowship website by clicking on the competencies tab. Applicants may apply online beginning in November. The link is available from the Application page of the fellowship website (click on "Application" on the left hand side of the screen). Applicants are encouraged to review the fellowship website for complete information on the fellowship and the application process.

Requirements: Applicants must have: an M.P.H., M.S.P.H., M.S., or an equivalent degree or advanced degree in a public-health-related field (ie. Ph.D. in epidemiology, biostatistics, or other pertinent field; an M.D. degree; or a D.V.M. degree); have completed coursework in at least three advanced graduate-level epidemiology courses and one graduate level biostatistics course; and a desire to pursue a long-term career at the state or local level.

Restrictions: United States citizenship

Geographic Focus: All States

Date(s) Application is Due: Feb 1

Amount of Grant: 39,970-56,970 USD

Contact: Ashlyn Beavor, Workforce and Fellowship Administrator; (770) 458-3811; fax (770) 454-8516; abeavor@cste.org

Internet: http://www.cste.org/dnn/ProgramsandActivities/FellowshipProgram/tabid/259/Default.aspx

Sponsor: Council of State and Territorial Epidemiologists
2872 Woodcock Boulevard. Suite 303
Atlanta, GA 30341

CTCRI Idea Grants **1665**

Idea Grants are designed to encourage unique or original research that has the potential to advance knowledge in tobacco control. Grants will allow investigators with innovative ideas and observations to conduct pilot studies, to perform secondary analysis of data sets or to gather new evidence necessary to determine the viability of research directions or hypotheses. Innovative projects are a priority for the Idea Grant program. Preference will be given to proposals that address research priorities identified by the Canadian Tobacco Control Research Summit. Research employing new or unconventional methodologies is encouraged. Proposed methods must be appropriate to the research question(s). Each successful proposal may receive a one-time grant of up to $50,000. Work is expected to be completed in one year's time.

Requirements: Research proposals must meet the following eligibility criteria: The Principal Applicant (PA) is a Canadian citizen or legal resident (Co-applicants may be citizens or residents of other countries); The research proposal demonstrates basic relevance to tobacco abuse / nicotine addiction / tobacco control and addresses research priorities identified by the Canadian Tobacco Control Research Summit; Applicant(s) has disclosed other sources of funding; Commercial interests have been disclosed or applicant(s) has indicated no commercial interests; Applicant(s) has affirmed lack of support from the tobacco industry; Applicant(s) can demonstrate that he/she works in an environment that adequately supports research through ethical review, administration of funds, provision of space and equipment, etc; Research plans include gender analysis, or it has been demonstrated that this is not appropriate; Proposal avoids duplication of previous research, unless it can be demonstrated that replication is of value; Research proposal involves pilot testing of research methods, tools, or hypotheses by a research team in order to strengthen proposals prepared for submission to traditional funding sources; Proposal articulates preliminary plans for development of a full research proposal -OR- Research proposal constitutes a novel idea which falls outside the normal scope of traditional research, either in topic or methodology; Proposal demonstrates that adequate funding is not available from other sources for this purpose.

Restrictions: Members of the staff or Board of Directors of the CTCRI, or a staff member of any CTCRI funding partner organization are not eligible to apply.

Geographic Focus: All States, Canada

Date(s) Application is Due: Apr 1; Oct 1

Amount of Grant: Up to 50,000 USD

Contact: Carol Bishop, Assistant Director; (416) 934-5640; cbishop@cancer.ca

Internet: http://www.ctcri.ca/en/index.php?option=com_content&task=view&id=25&Itemid=44

Sponsor: Canadian Tobacco Control Research Initiative
10 Alcorn Avenue, Suite 200
Toronto, ON M4V 3B1 Canada

CTCRI Knowledge Synthesis Grants **1666**

The goal of the Knowledge Synthesis program is to support interdisciplinary teams of researchers and practitioners / decision-makers to conduct collaborative reviews of evidence for particular tobacco control interventions. The maximum amount awarded for a single grant is $120,000. It is anticipated that at least two applications will be supported in the present competition, contingent upon the availability of funds and the quality of proposals received.

Requirements: Teams are requested to develop proposals to carry out reviews addressing ONE of the following topics: (1) Anti-contraband measures-An issue that undermines the effectiveness of a high tobacco tax strategy is contraband activity that includes the smuggling of tobacco products from lower tax jurisdictions, illicit manufacturing and counterfeiting. Though large price increases may provide incentive for some smokers to quit, it will impel others to seek out lower cost sources. A number of smokers are turning to contraband. (2) The effect of tax and price on prevalence and consumption in subpopulations (defined as tobacco users including 15 to 24 year olds, aboriginal people, persons diagnosed with mental health or substance abuse disorders, and lowest categories of income and formal education). (3) Discount cigarettes (defined as lower-priced brands by small manufacturers and cheaper brands by big tobacco companies)-With the price increase of tobacco products, a number of smokers are turning to price discounted cigarettes. Price discounted cigarettes now comprise more than 40% of market share in cigarette sales. Research proposals must meet the following eligibility criteria: The Principal Applicant (PA) is a Canadian citizen or legal resident (Co-applicants may be citizens or residents of other countries); The research proposal demonstrates basic relevance to tobacco abuse / nicotine addiction / tobacco control and addresses research priorities identified by the Canadian Tobacco Control Research Summit; Applicant(s) has disclosed other sources of funding; Commercial interests have been disclosed or applicant(s) has indicated no commercial interests; Applicant(s) has affirmed lack of support from the tobacco industry; Applicant(s) can demonstrate that he/she works in an environment that adequately supports research through ethical review, administration of funds, provision of space and equipment, etc; Research plans include gender analysis, or it has been demonstrated that this is not appropriate; Proposal avoids duplication of previous research, unless it can be demonstrated that replication is of value; Proposal demonstrates the inclusion of and consultation with end-users (practitioners, policy and decision-makers) throughout the research process; Applicant demonstrates understanding and application of the Better Practices methodology, including definition of the scope of the review.

Restrictions: Members of the staff or Board of Directors of the CTCRI, or a staff member of any CTCRI funding partner organization are not eligible to apply.

Geographic Focus: All States, Canada

Date(s) Application is Due: Apr 1

Amount of Grant: Up to 120,000 USD

Contact: Grants Supervisor; (416) 934-5666; fax (416) 961-4189; info@ctcri.ca

Internet: http://ctcri.ca/en/index.php?option=com_content&task=view&id=212&Itemid=72

Sponsor: Canadian Tobacco Control Research Initiative
10 Alcorn Avenue, Suite 200
Toronto, ON M4V 3B1 Canada

CTCRI Policy Research Grants **1667**

Policy Research grants are intended to stimulate research that will influence, guide or have a direct impact on policy decisions in tobacco control. Suggested research areas and methodologies include: The full policy process from agenda setting and decision making through to development and implementation, enforcement, evaluation and refinement; All areas in which policy might be used to influence tobacco use including fiscal, legislative, regulatory and educational policies; Public and private policy at all levels of government/locale; Preference will be given to projects that address research priorities identified by the Canadian Tobacco Control Research Summit. Research employing new or unconventional methodologies is encouraged, but proposed methods must be appropriate to the research question(s). Each successful proposal may receive a one-time grant of up to $80,000. The project may be completed over the course of two years.

Requirements: Research proposals must meet the following eligibility criteria: The Principal Applicant (PA) is a Canadian citizen or legal resident (Co-applicants may be citizens or residents of other countries); The research proposal demonstrates basic relevance to tobacco abuse / nicotine addiction / tobacco control and addresses research priorities identified by the Canadian Tobacco Control Research Summit; Applicant(s) has disclosed other sources of funding; Commercial interests have been disclosed or applicant(s) has indicated no commercial interests; Applicant(s) has affirmed lack of support from the tobacco industry; Applicant(s) can demonstrate that he/she works in an environment that adequately supports research through ethical review, administration of funds, provision of space and equipment, etc; Research plans include gender analysis, or it has been demonstrated that this is not appropriate; Proposal avoids duplication of previous research, unless it can be demonstrated that replication is of value; Research proposal is relevant to public policy issues surrounding tobacco control in Canada. FastTrack proposals only: Applicants must demonstrate that their proposal is dependent upon policy or legislation which is timely and that the project cannot be delayed until the April 1 or October 1 deadlines.

Restrictions: Members of the staff or Board of Directors of the CTCRI, or a staff member of any CTCRI funding partner organization are not eligible to apply.

Geographic Focus: All States, Canada

Date(s) Application is Due: Apr 1; Oct 1

Contact: Grants Supervisor; (416) 934-5666; fax (416) 961-4189; info@ctcri.ca

Internet: http://ctcri.ca/~ctcri/en/index.php?option=content&task=view&id=206&Itemid=87

Sponsor: Canadian Tobacco Control Research Initiative
10 Alcorn Avenue, Suite 200
Toronto, ON M4V 3B1 Canada

CTCRI Researcher Travel Grants **1668**

Researcher Travel Grants are offered to provide opportunities for graduate and post-doctoral students and individuals affiliated with non-governmental organizations/community groups to attend conferences/meetings related to tobacco abuse and nicotine addiction. Any poster or presentation of research related to tobacco abuse and nicotine addiction is eligible for support. Preference will be given to proposals that address research priorities identified by the Canadian Tobacco Control Research Summit. Research employing new or unconventional methodologies is encouraged, but proposed methods must be appropriate to the research question(s). Grants of up to $3,000 are available to support travel and accommodations. In some cases, the value of the grant may exceed $,3000 (for example, for conferences involving international travel). The number of applications supported annually is contingent upon the availability of funds and the quality of proposals.

Requirements: Research proposals must meet the following eligibility criteria: The Principal Applicant (PA) is a Canadian citizen or legal resident (Co-applicants may be citizens or residents of other countries); The research proposal demonstrates basic relevance to tobacco abuse / nicotine addiction / tobacco control and addresses research priorities identified by the Canadian Tobacco Control Research Summit; Applicant(s) has disclosed other sources of funding; Commercial interests have been disclosed or applicant(s) has indicated no commercial interests; Applicant(s) has affirmed lack of support from the tobacco industry; Applicant(s) can demonstrate that he/she works in an environment that adequately supports research through ethical review, administration of funds, provision of space and equipment, etc; Research plans include gender analysis, or it has been demonstrated that this is not appropriate; Proposal avoids duplication of previous research, unless it can be demonstrated that replication is of value; The Principal Applicant must be conducting graduate or post-doctoral work and must be affiliated with an accredited academic program. OR The Principal Applicant must be affiliated with a non-governmental organization or community group conducting tobacco control research or evaluation; The Principal Applicant's abstract has been accepted for the upcoming conference for which the funding is requested (funding cannot be applied for retroactively); The proposal demonstrates that any current funding will not adequately support the stated objectives of the proposal; The Principal Applicant has not been awarded another CTCRI travel grant to present at a conference or event in the same calendar year as the current application. This does not include attendance at the special tobacco control conferences such as the World Conference on Tobacco or Health (WCOTH) or the Society for Research on Nicotine and Tobacco (SRNT) conference.
Restrictions: Members of the staff or Board of Directors of the CTCRI, or a staff member of any CTCRI funding partner organization are not eligible to apply.
Geographic Focus: All States, Canada
Amount of Grant: Up to 3,000 USD
Contact: Grants Supervisor; (416) 934-5666; fax (416) 961-4189; info@ctcri.ca
Internet: http://ctcri.ca/~ctcri/en/index.php?option=content&task=view&id=208&Itemid=89
Sponsor: Canadian Tobacco Control Research Initiative
10 Alcorn Avenue, Suite 200
Toronto, ON M4V 3B1 Canada

CTCRI Research Planning Grants **1669**

Research Planning Grants are offered for the purpose of bringing together new, multi-sectoral and interdisciplinary research teams to construct research proposals for submission to traditional open funding competitions. The specific objectives of the program are: To facilitate the development of excellent research proposals which are relevant to the priorities of the CTCRI's research agenda; To support the formation of strong, interdisciplinary teams that can compete for research funding through traditional sources. Each successful proposal may receive a one-time grant of up to $15,000. Work is expected to be completed in one year's time. The number of applications supported each year is contingent upon the availability of funds and the quality of proposals.
Requirements: Preference will be given to proposals that address research priorities identified by the Canadian Tobacco Control Research Summit. Suggested methodologies include: Physically gathering members of a team to develop a research plan; Conducting a literature review or other background work to develop a research project; Hiring a research assistant to help write a well grounded proposal (if applicable). Research proposals must meet the following eligibility criteria: The Principal Applicant (PA) is a Canadian citizen or legal resident (Co-applicants may be citizens or residents of other countries); The research proposal demonstrates basic relevance to tobacco abuse / nicotine addiction / tobacco control and addresses research priorities identified by the Canadian Tobacco Control Research Summit; Applicant(s) has disclosed other sources of funding; Commercial interests have been disclosed or applicant(s) has indicated no commercial interests; Applicant(s) has affirmed lack of support from the tobacco industry; Applicant(s) can demonstrate that he/she works in an environment that adequately supports research through ethical review, administration of funds, provision of space and equipment, etc; Research plans include gender analysis, or it has been demonstrated that this is not appropriate; Proposal avoids duplication of previous research, unless it can be demonstrated that replication is of value.
Restrictions: Conducting work to pilot surveys or methods does not qualify for support through this program. Members of the staff or Board of Directors of the CTCRI, or a staff member of any CTCRI funding partner organization are not eligible to apply.
Geographic Focus: All States, Canada
Date(s) Application is Due: Feb 28; May 30; Aug 31; Nov 30
Amount of Grant: Up to 15,000 USD
Contact: Grants Supervisor; (416) 934-5666; fax (416) 961-4189; info@ctcri.ca
Internet: http://ctcri.ca/~ctcri/en/index.php?option=content&task=view&id=209&Itemid=88
Sponsor: Canadian Tobacco Control Research Initiative
10 Alcorn Avenue, Suite 200
Toronto, ON M4V 3B1 Canada

CTCRI Student Research Grants **1670**

The objectives of the program are to support learning opportunities through a one-time grant for students wishing to conduct research projects related to tobacco abuse and nicotine addiction; and to provide opportunities for graduate and post-doctoral students to improve skills in tobacco abuse and nicotine addiction research. The grant program is intended to support a broad range of research topics and areas as long as the proposed research is led by the applicant under professional research supervision. The program is not intended to support the applicant's regular course of study or background work required to prepare a research project, nor is it intended to support the research supervisor's existing project(s). Preference will be given to proposals that address research priorities identified by the Canadian Tobacco Control Research Summit. Each successful proposal may receive a one-time grant of up to $10,000. Work is expected to be completed in one year's time. The number of applications supported each year is contingent upon the availability of funds and the quality of proposals.
Requirements: Graduate and post-doctoral students are eligible for Student Research grant. Research proposals must meet the following eligibility criteria: The Principal Applicant (PA) is a Canadian citizen or legal resident (Co-applicants may be citizens or residents of other countries); The research proposal demonstrates basic relevance to tobacco abuse / nicotine addiction / tobacco control and addresses research priorities identified by the Canadian Tobacco Control Research Summit; Applicant(s) has disclosed other sources of funding; Commercial interests have been disclosed or applicant(s) has indicated no commercial interests; Applicant(s) has affirmed lack of support from the tobacco industry; Applicant(s) can demonstrate that he/she works in an environment that adequately supports research through ethical review, administration of funds, provision of space and equipment, etc; Research plans include gender analysis, or it has been demonstrated that this is not appropriate; Proposal avoids duplication of previous research, unless it can be demonstrated that replication is of value; Proposal demonstrates that any current funding will not adequately support the stated objectives of the proposal; The Research Supervisor has provided a letter of support for the applicant's proposal.
Restrictions: Members of the staff or Board of Directors of the CTCRI, or a staff member of any CTCRI funding partner organization are not eligible to apply.
Geographic Focus: All States, Canada
Date(s) Application is Due: Mar 31; Jun 30; Sep 30; Dec 30
Amount of Grant: Up to 10,000 USD
Contact: Grants Supervisor; (416) 934-5666; fax (416) 961-4189; info@ctcri.ca
Internet: http://ctcri.ca/~ctcri/en/index.php?option=content&task=view&id=211&Itemid=90
Sponsor: Canadian Tobacco Control Research Initiative
10 Alcorn Avenue, Suite 200
Toronto, ON M4V 3B1 Canada

CTCRI Workshop and Learning Opportunities Grants **1671**

Workshop and Learning Opportunity grants are offered to support workshops, meetings and other events to build research capacity and promote collaboration between researchers and end-users. Priority is given to first-time applicants, to regions or populations where development of research is more acutely needed and to events that address research priorities identified by the CTCRI Research Summit: Building capacity for tobacco-related research in the region or population; Providing unique opportunities to bring together individuals from different disciplines and sectors; Supporting the sharing of knowledge, initiation of collaborations or the planning of relevant research activities. Workshops, courses and other educational methods or tools are appropriate for this grant program. Each successful proposal may receive a one-time grant of up to $15,000. Work is expected to be completed in four (4) months time. The number of applications supported annually is contingent upon the availability of funds and the quality of proposals.
Requirements: Proposals to this program must meet the following eligibility criteria: Topic is related to both research and tobacco abuse and/or addiction and intervention; Research could involve identifying research priorities, discussing and/or developing research frameworks and/or disseminating research results; Method involves multiple participants and could be in the form of a workshop or another educational opportunity; Approach is innovative and deals with engaging new topics and methods or involves new populations (e.g., researchers or practitioners new to tobacco control, historically under-represented regions or populations); Proposal demonstrates that any current funding will not adequately support the stated objectives of the proposal. Additionally, the following criteria apply: Research proposals must meet the following eligibility criteria: The Principal Applicant (PA) is a Canadian citizen or legal resident (Co-applicants may be citizens or residents of other countries); The research proposal demonstrates basic relevance to tobacco abuse / nicotine addiction / tobacco control and addresses research priorities identified by the Canadian Tobacco Control Research Summit; Applicant(s) has disclosed other sources of funding; Commercial interests have been disclosed or applicant(s) has indicated no commercial interests; Applicant(s) has affirmed lack of support from the tobacco industry; Applicant(s) can demonstrate that he/she works in an environment that adequately supports research through ethical review, administration of funds, provision of space and equipment, etc; Research plans include gender analysis, or it has been demonstrated that this is not appropriate; Proposal avoids duplication of previous research, unless it can be demonstrated that replication is of value.
Restrictions: Members of the staff or Board of Directors of the CTCRI, or a staff member of any CTCRI funding partner organization are not eligible to apply.
Geographic Focus: All States, Canada
Amount of Grant: Up to 15,000 USD
Contact: Grants Supervisor; (416) 934-5666; fax (416) 961-4189; info@ctcri.ca
Internet: http://ctcri.ca/~ctcri/en/index.php?option=content&task=view&id=210&Itemid=91
Sponsor: Canadian Tobacco Control Research Initiative
10 Alcorn Avenue, Suite 200
Toronto, ON M4V 3B1 Canada

Cudd Foundation Grants **1672**

The foundation awards grants to eligible nonprofit organizations in its areas of interest, including arts, culture, performing arts, children and youth, education, environment, health care, historic preservation, and social services. Types of support include annual campaigns, building construction/renovation, capital campaigns, continuing support, curriculum development, emergency grants, endowments, program development, research, and scholarship funds. The listed application deadline is for letters of intent; full proposals are by invitation.
Requirements: Louisiana and New Mexico nonprofit organizations are eligible.
Geographic Focus: Louisiana, New Mexico

Amount of Grant: 250-222,088 USD
Contact: Amanda Stuermer, (505) 986-8416; fax (505) 986-8427; cuddfdn@aol.com
Sponsor: Cudd Foundation
P.O. Box 2322
Santa Fe, NM 87504

Cushwa Center Research Travel Grants 1673
This program is designed to foster research in the archives and library of the University of Notre Dame. The library collection is particularly rich in the following areas: Catholic newspapers, history of midwestern Catholicism, Catholic literature, and history of Catholicism in the United States. The archives have manuscripts of historical personages, records of 20th-century Catholic organizations, reports of European missionary societies, and other material related to the American Catholic community. Grants to help defray travel and lodging costs are made to scholars of any academic discipline who are engaged in projects which require substantial use of the collections of the library and archives. Submit one copy of the completed application form, current curriculum vita, a 1,000-word description of the project to be undertaken at the center, and a budget estimating travel, lodging, and research expenses. Two letters of recommendation from two people who know your work should be sent directly to the Cushwa Center.
Requirements: The research project must be related to the study of the American Catholicism. Applicants should describe the relevance of the project for American Catholic studies, indicate the specific resources of interest at Notre Dame, and specify any plans for publication of the project.
Geographic Focus: All States
Date(s) Application is Due: Dec 31
Amount of Grant: Up to 2,000 USD
Contact: Director of Cushwa Center for the Study of American Catholicism; (219) 631-5441; fax (219) 631-8471; cushwa.1@nd.edu
Internet: http://www.nd.edu/~cushwa/grants/
Sponsor: University of Notre Dame
1135 Flanner Hall
Notre Dame, IN 46556-5611

Cystic Fibrosis Canada Clinical Fellowships 1674
Cystic Fibrosis Canada (CFC) Clinical Fellowships are intended for those physicians who have already obtained their residency training and who wish to pursue additional clinical training in Cystic Fibrosis (CF) care. The purpose is to train physicians to become CF specialists, so that they can provide ongoing clinical care to individuals with CF in Canada. This training would include developing competence to allow them to participate in clinical trials. This fellowship is not intended to train individuals from other countries who will not stay in Canada. Up to two highly-ranked competitive Clinical Fellowships will be offered by the Foundation each year. The value of a CCFF Clinical Fellowship is dependent upon academic qualifications, and/or research experience. Award levels are reviewed on an annual basis, and will correspond with prevailing Canadian rates.
Requirements: Canadian citizens or permanent residents who have an M.D. degree, have recently completed their clinical training, are exam eligible, and have obtained medical licensure in Canada, are eligible to apply. Clinical Fellowships are tenable in Canada or abroad and it is intended that this experience will directly benefit the Canadian CF community.
Restrictions: Clinical Fellowships will not be awarded to individuals who have not completed residency training.
Geographic Focus: Canada
Date(s) Application is Due: Oct 1
Contact: Administrator; (416) 485-9149; fax (416) 485-0960; info@cysticfibrosis.ca
Internet: http://www.cysticfibrosis.ca/en/treatment/TrainingAwards_ClinicalFellowships.php
Sponsor: Cystic Fibrosis Canada
2221 Yonge Street, Suite 601
Toronto, ON M4S 2B4 Canada

Cystic Fibrosis Canada Clinical Project Grants 1675
Clinical Project grants are intended to bolster the Canadian Cystic Fibrosis Foundation's (CCFF's) commitment to clinical research, and to promote the further development of the Clinical Studies Network. Awarded at the discretion of the leaders of the Medical/Scientific Advisory Committee, in consultation with the Clinical Studies Network, clinical project grants provide a mechanism whereby the most important ideas for clinical studies can be forged into viable protocols, and strategies of urgent clinical relevance can be pursued (i.e. seed money for small, start-up pilot projects). Applications should be made in the form of a letter in which a clear hypothesis is detailed, relevance to CF is demonstrated, and a brief budget is outlined. Applications may be submitted at any time, but the Foundation must be consulted in advance with respect to the availability of funds. The amount of a grant will be determined by the Medical/Scientific Advisory Committee, following a detailed review of the applicant's proposed budget, up to $15,000.
Geographic Focus: Canada
Date(s) Application is Due: Apr 1
Amount of Grant: Up to 15,000 CAD
Contact: Administrator; (416) 485-9149; fax (416) 485-0960; info@cysticfibrosis.ca
Internet: http://www.cysticfibrosis.ca/en/research/clinicalprojectgrants.php
Sponsor: Cystic Fibrosis Canada
2221 Yonge Street, Suite 601
Toronto, ON M4S 2B4 Canada

Cystic Fibrosis Canada Clinic Incentive Grants 1676
Clinic Incentive grants are intended to enhance the standard of clinical care available to Canadians with cystic fibrosis, by providing funds to initiate a comprehensive program for cystic fibrosis patient care, research, and teaching; or to strengthen an existing program. The amount of a grant will be determined by the Medical/Scientific Advisory Committee, following a detailed review of the applicant's proposed budget.
Requirements: Canadian hospitals and/or medical schools are eligible to apply for Clinic Incentive grants.
Restrictions: Basic support for clinical care is the responsibility of provincial governments, and Canadian Cystic Fibrosis Foundation Clinic Incentive grants, which are under the jurisdiction of the cystic fibrosis clinic director, are not intended to supplant public funding.
Geographic Focus: Canada
Date(s) Application is Due: Oct 1
Contact: Administrator; (416) 485-9149; fax (416) 485-0960; info@cysticfibrosis.ca
Internet: http://www.cysticfibrosis.ca/en/treatment/ClinicIncentiveGrants.php
Sponsor: Cystic Fibrosis Canada
2221 Yonge Street, Suite 601
Toronto, ON M4S 2B4 Canada

Cystic Fibrosis Canada Fellowships 1677
A limited number of competitive fellowships are offered by the foundation each year for basic or clinical research training in areas of the biomedical or behavioral sciences pertinent to cystic fibrosis. Fellowships are tenable at approved universities, hospitals, and research institutes in Canada. Initial fellowships are awarded for a two-year period. Fellows may apply for a one-year renewal. No one may receive more than three years of support under a CCFF fellowship. Canadian fellowship applicants of exceptional quality requesting funding to study abroad will be considered. Applicants are expected to demonstrate that comparable training is not available in Canada. Equitable consideration is given to applicants from outside of Canada who intend to return to their own country on completion of the fellowship.
Requirements: Individuals who hold MD or PhD degrees are eligible to apply. Medical graduates should have already completed basic residency training, and must be eligible for Canadian licensure.
Restrictions: Fellowship (Initial) applicants who will have completed four or more years of training as of the October 1st application deadline, following their PhD or MD (post-MD clinical training), are not eligible for Cystic Fibrosis Canada Fellowships. Applicants who have already completed six or more years of postgraduate study or training are not eligible.
Geographic Focus: Canada
Date(s) Application is Due: Oct 1
Contact: Administrator; (416) 485-9149; fax (416) 485-0960; info@cysticfibrosis.ca
Internet: http://www.cysticfibrosis.ca/en/research/TrainingAwards_Fellowships.php
Sponsor: Cystic Fibrosis Canada
2221 Yonge Street, Suite 601
Toronto, ON M4S 2B4 Canada

Cystic Fibrosis Canada Research Grants 1678
Research grants are intended to facilitate the scientific investigation of all aspects of cystic fibrosis. Generally, the Foundation funds research which is carried out in Canada, and the principal investigator must be based at a Canadian institution. While the Foundation will fund a collaboration from outside of Canada, funds must be directed through a Canadian institution and Canadian investigators intending to collaborate with individuals from outside of Canada should contact the Foundation. Applications may be submitted by groups of individuals who plan to collaborate. Grants are awarded for up to three years.
Requirements: A principal investigator should hold a recognized, full-time faculty appointment in a relevant discipline at a Canadian university or hospital. Under exceptional circumstances, and at the discretion of the Research Subcommittee, research grant applications by individuals may be evaluated on a case-by-case basis, with significant emphasis placed on the degree of independence of the applicant, and on the institutional commitments to this individual. Such applications must include a statement from the applicant addressing the issue of salary support during the term of the grant, and the availability of laboratory space; and letters from the Departmental Chair and Dean of Faculty, clarifying the nature and extent of the institutional commitment to the applicant. This same eligibility criterion regarding a faculty appointment applies to any named co-investigator.
Restrictions: Research grants do not provide support for graduate students, postdoctoral fellows, or summer students; construction costs; institutional overheads for laboratory facilities; or purchase of equipment in excess of $10,000.
Geographic Focus: Canada
Date(s) Application is Due: Oct 1
Contact: Administrator; (416) 485-9149; fax (416) 485-0960; info@cysticfibrosis.ca
Internet: http://www.cysticfibrosis.ca/en/index.php
Sponsor: Cystic Fibrosis Canada
2221 Yonge Street, Suite 601
Toronto, ON M4S 2B4 Canada

Cystic Fibrosis Canada Scholarships 1679
Cystic Fibrosis Canada Scholarships provide salary support to a limited number of exceptional investigators, who have made outstanding contributions to, and have demonstrated leadership in, cystic fibrosis research. This award offers new investigators an opportunity to develop outstanding cystic fibrosis research programs, unhampered by heavy teaching or clinical loads. The award will be for a three-year period, renewable for an additional two years on receipt of a satisfactory progress report during the third year

of the award. In no case will an award be for more than five years. The value of a Cystic Fibrosis Canada Scholarship is $60,000 per year.
Requirements: Applications are accepted from candidates who have received their first faculty appointment within the preceding five calendar years.
Geographic Focus: Canada
Date(s) Application is Due: Oct 1
Contact: Administrator; (416) 485-9149; fax (416) 485-0960; info@cysticfibrosis.ca
Internet: http://www.cysticfibrosis.ca/en/research/TrainingAwards_Scholarships.php
Sponsor: Cystic Fibrosis Canada
2221 Yonge Street, Suite 601
Toronto, ON M4S 2B4 Canada

Cystic Fibrosis Canada Senior Scientist Research Training Awards **1680**
The Cathleen Morrison Senior Scientist Research Training Award provides support to a limited number of established senior investigators by offering them an opportunity to obtain additional training that will enhance their capacity to conduct research directly relevant to cystic fibrosis. Cathleen served the CF cause in Canada, and internationally, for thirty years, as Cystic Fibrosis Canada's Executive Director, and then Chief Executive Officer, from 1981 until 2011. Cathleen has been the consummate advocate for CF research – and health research in general. She has recruited, encouraged and guided many researchers from across Canada, throughout their careers. This award can be used for sabbatical support for qualified individuals. Senior Scientist Research Training awards are awarded for a minimum of three months to a maximum of one year. Frequency of application should be reasonable; once every five years. The amount of the award, to a maximum of $15,000, will be determined by the Medical/Scientific Advisory Committee, following a review of the application.
Requirements: Applicants must have held a recognized, full-time faculty appointment in a relevant discipline at a Canadian university or hospital for at least six years. In addition, applicants must hold major research grant funding.
Restrictions: Applicants may hold only one Canadian Cystic Fibrosis Foundation personnel award concurrently.
Geographic Focus: Canada
Date(s) Application is Due: Apr 1; Oct
Amount of Grant: Up to 15,000 CAD
Samples: Dr. Robert E.W. Hancock, The University of British Columbia, $29,116-novel genetic screen employing mouse homozygous knockout ES cells (2011).
Contact: Administrator; (416) 485-9149; fax (416) 485-0960; info@cysticfibrosis.ca
Internet: http://www.cysticfibrosis.ca/en/research/TrainingAwards_SeniorScientistResearch.php
Sponsor: Cystic Fibrosis Canada
2221 Yonge Street, Suite 601
Toronto, ON M4S 2B4 Canada

Cystic Fibrosis Canada Small Conference Grants **1681**
Small conference grants are offered to support medical/scientific conferences which focus on subjects of direct relevance to cystic fibrosis. Such grants will be up to a maximum of $C2500 and are intended to supplement other funding sources. The small conference grants will also support inter-clinic exchanges which are used to facilitate the exchange of special expertise between larger, university-based CF clinics and smaller, more remote clinics. Requests of this type will involve fewer individuals, and will not normally exceed $C1000. Applications may be submitted at any time, but Cystic Fibrosis Canada should be consulted in advance with respect to the availability of funds. Grants are on a first-come, first-served basis.
Requirements: CF clinic directors and CCFF-funded investigators are eligible to apply.
Geographic Focus: Canada
Contact: Administrator; (416) 485-9149; fax (416) 485-0960; info@cysticfibrosis.ca
Internet: http://www.cysticfibrosis.ca/en/research/SmallConferenceGrants.php
Sponsor: Cystic Fibrosis Canada
2221 Yonge Street, Suite 601
Toronto, ON M4S 2B4 Canada

Cystic Fibrosis Canada Special Travel Grants For Fellows and Students **1682**
Cystic Fibrosis Canada (CFC) encourages supported Fellows and Students to attend and participate in scientific meetings related to cystic fibrosis. In addition, allowances may be utilized to acquire new knowledge (i.e. visit another lab). Special Travel allowances may be awarded to CFC-supported Fellows and Students, for each year of their award, upon written application and pending the availability of funds. No CFC-supported Fellow or Student may receive more than one grant per year. Subject to the availability of funds and under certain circumstances, CFC-supported Scholars, physicians who are not CF clinic directors but who specialize in the clinical care of adults with cystic fibrosis at Canadian CF clinics, and recipients of the Skelly and Lackey Summer Studentship awards, may also be eligible. Special Travel allowances are intended to cover travel costs not to exceed the lowest applicable advance booking airfare, and reasonable accommodation and meals, up to a $1,800. Additional guidelines are available at: http://www.cysticfibrosis.ca/en/research/TravelAllowancesStudents.php.
Requirements: Applications should be made in the form of a letter, and must be submitted prior to the proposed travel. The letter should indicate the conference or meeting for which the allowance is requested and must include a proposed travel and accommodation budget, along with the supervisor's signed endorsement. Proof of active participation (i.e. abstract acceptance, etc.) is also required.
Restrictions: An itemized breakdown of expenses incurred and the original receipts must be submitted to the Foundation upon return from the conference or meeting, and any unexpended funds returned to the Foundation.
Geographic Focus: Canada
Amount of Grant: Up to 1,200 CAD
Contact: Grants Administrator; (416) 485-9149; fax (416) 485-0960; ResearchPrograms@cysticfibrosis.ca or info@cysticfibrosis.ca
Internet: http://www.cysticfibrosis.ca/en/research/TravelAllowancesStudents.php
Sponsor: Cystic Fibrosis Canada
2221 Yonge Street, Suite 601
Toronto, ON M4S 2B4 Canada

Cystic Fibrosis Canada Studentships **1683**
A limited number of competitive studentships are offered by the foundation each year to highly qualified graduate students who are registered for a higher degree and who are undertaking full-time research training in areas of the biomedical or behavioral sciences relevant to cystic fibrosis. Students are expected to spend at least 75 percent of their time on the research training described in their application. Equitable consideration will be given to Studentship applicants from outside of Canada, who intend to return to their own country on completion of a Studentship. The value of a Cystic Fibrosis Canada Studentship is reviewed on an annual basis, and will correspond with prevailing Canadian rates. Cystic Fibrosis Canada Studentships are salary awards, and the payment, in addition, of benefits to Cystic Fibrosis Canada students is a matter of host institutional policy.
Requirements: Awards are tenable only at Canadian universities.
Geographic Focus: Canada
Date(s) Application is Due: Oct 1
Contact: Administrator; (416) 485-9149; fax (416) 485-0960; info@cysticfibrosis.ca
Internet: http://www.cysticfibrosis.ca/en/research/TrainingAwards_Studentships.php
Sponsor: Cystic Fibrosis Canada
2221 Yonge Street, Suite 601
Toronto, ON M4S 2B4 Canada

Cystic Fibrosis Canada Summer Studentships **1684**
Cystic Fibrosis Canada Summer studentships are intended to provide support to students engaged in summer research projects in areas of the biomedical or behavioral sciences relevant to cystic fibrosis, under the direction of CF clinic directors or principal investigators. The research project should be attainable within the three-month term of the award. The value of a Cystic Fibrosis Canada Summer Studentship is reviewed on an annual basis, and will correspond with prevailing rates. Most recently, the amount has been established at $5,000 per award.
Requirements: Full-time students pursuing an undergraduate degree in an appropriate discipline are eligible to receive this award. Under certain circumstances, other special academic situations, may be considered for a summer studentship.
Restrictions: Applications must be made by the proposed supervisor on behalf of a student. Applications directly from students cannot be accepted.
Geographic Focus: Canada
Date(s) Application is Due: Feb 1
Amount of Grant: 5,000 CAD
Contact: Administrator; (416) 485-9149; fax (416) 485-0960; info@cysticfibrosis.ca
Internet: http://www.cysticfibrosis.ca/en/research/TrainingAwards_SummerStudentships.php
Sponsor: Cystic Fibrosis Canada
2221 Yonge Street, Suite 601
Toronto, ON M4S 2B4 Canada

Cystic Fibrosis Canada Travel Supplement Grants **1685**
The Cystic Fibrosis Canada has a special travel fund to assist a limited number of individuals who plan to play an active role in European Cystic Fibrosis Society (ECFS) conferences. Up to five travel supplement grants for the upcoming conference will be awarded on a competitive basis, the maximum award will be $2,000 per person. The application deadline each year is February 1 (for Basic Science Conference) and April 1 (for ECFS Conference).
Requirements: The travel supplements will be available to CFC-funded investigators; Cystic Fibrosis Canada clinic directors or clinicians; and Cystic Fibrosis Canada clinic coordinators and allied health professionals (applications from clinicians, coordinators and health professionals must be endorsed by the clinic director); individuals requesting the supplement must show evidence of active participation at the conference, i.e. documentation indicating that you will be presenting a paper or abstract, or participating in a workshop; application is by letter to the Cystic Fibrosis Canada, with a brief budget included; and frequency of application from any particular individual should be reasonable (i.e. once every four years).
Restrictions: An itemized breakdown of expenses incurred and the original receipts must be submitted to the Cystic Fibrosis Canada upon return from the ECFS conference, and any unspent funds returned to the Cystic Fibrosis Canada.
Geographic Focus: Canada
Date(s) Application is Due: Feb 1; Apr 1
Contact: Grants Administrator; (416) 485-9149; fax (416) 485-0960; ResearchPrograms@cysticfibrosis.ca or info@cysticfibrosis.ca
Internet: http://www.cysticfibrosis.ca/en/research/TravelSupplementECFs.php
Sponsor: Cystic Fibrosis Canada
2221 Yonge Street, Suite 601
Toronto, ON M4S 2B4 Canada

Cystic Fibrosis Canada Visiting Allied Health Professional Awards **1686**
A limited number of awards may be made to allied health professionals from abroad who are invited to engage in Cystic Fibrosis clinical observation or activity at a Canadian institution; or Canadian allied health professionals who wish to visit another clinic in Canada or abroad. It is intended that this experience as a Canadian Cystic Fibrosis

Foundation Visiting Allied Health Professional will, in some way, benefit Canadian Cystic Fibrosis clinical care. Applications may be submitted at any time, but the Foundation should be consulted in advance with respect to the availability of funds.
Requirements: To qualify, an allied health professional must be associated with a recognized Cystic Fibrosis clinic, and be an active member in Cystic Fibrosis clinical care. Applications should be made in the form of a letter, with a brief budget included and accompanied by a supporting letter from the head of the appropriate department of the applicant's institution. A supporting letter signed by the department head of the host institution should also be provided.
Geographic Focus: All States, All Countries
Contact: Administrator; (416) 485-9149; fax (416) 485-0960; info@cysticfibrosis.ca
Internet: http://www.cysticfibrosis.ca/en/treatment/TrainingAwards_VisitingAlliedAwards.php
Sponsor: Cystic Fibrosis Canada
2221 Yonge Street, Suite 601
Toronto, ON M4S 2B4 Canada

Cystic Fibrosis Canada Visiting Clinician Awards **1687**
Awards may be made to clinicians from abroad who are invited to engage in Cystic Fibrosis clinical observation or activity at a Canadian institution; or Canadian clinicians who wish to visit another clinic in Canada or abroad. Applications may be submitted at any time, but Cystic Fibrosis Canada should be consulted in advance with respect to the availability of funds.
Requirements: To qualify, a clinician must be associated with a recognized Cystic Fibrosis clinic, and be an active member in Cystic Fibrosis clinical care. It is intended that this experience as a Canadian Cystic Fibrosis Foundation Visiting Clinician will, in some way, benefit Canadian Cystic Fibrosis clinical care. Applications should be made in the form of a letter, with a brief budget included and accompanied by a supporting letter from the head of the appropriate department of the applicant's institution. A supporting letter signed by the department head of the host institution should also be provided.
Geographic Focus: All States, All Countries
Contact: Administrator; (416) 485-9149; fax (416) 485-0960; info@cysticfibrosis.ca
Internet: http://www.cysticfibrosis.ca/en/treatment/TrainingAwards_VisitingClinicianAwards.php
Sponsor: Cystic Fibrosis Canada
2221 Yonge Street, Suite 601
Toronto, ON M4S 2B4 Canada

Cystic Fibrosis Canada Visiting Scientist Awards **1688**
Awards are made to senior investigators from abroad who are invited to engage in CF research at a Canadian institution; or junior and senior Canadian investigators who wish to work in another laboratory in Canada or abroad. It is intended that this experience as a CCFF visiting scientist will, in some way, benefit the Canadian CF research effort. Applications may be submitted at any time, but the foundation should be consulted in advance with respect to the availability of funds.
Requirements: Senior investigators from abroad or junior and senior Canadian investigators who wish to work in another laboratory in Canada or abroad are eligible to apply.
Geographic Focus: Canada
Contact: Administrator; (416) 485-9149; fax (416) 485-0960; info@cysticfibrosis.ca
Internet: http://www.cysticfibrosis.ca/en/research/TrainingAwards_VisitingScientist.php
Sponsor: Cystic Fibrosis Canada
2221 Yonge Street, Suite 601
Toronto, ON M4S 2B4 Canada

Cystic Fibrosis Research Elizabeth Nash Fellowships **1689**
Cystic Fibrosis Research, Inc. (CFRI) is currently accepting applications for postdoctoral research fellowships in basic or clinical research related to cystic fibrosis. Under the Elizabeth Nash Memorial Fellowship Program, $40,000 per year grants will be awarded to a research institution in California to support a principal investigator, who in turn appoints a post-doctoral research fellow. Of particular interests are projects that reflect novel and original research in cystic fibrosis. Fellowships are awarded for one year with a second year's funding contingent on progress reviews. Post-doctoral fellows and Principal Investigators are expected to attend the Spring and Summer research symposia in the San Francisco Bay Area. The annual deadline for application is December 31 for an April 1 start date.
Requirements: Applicants must be California-based research institutions.
Geographic Focus: California
Date(s) Application is Due: Dec 31
Contact: Carroll Jenkins; (855) 237-4669; fax (650) 404-9981; cjenkins@cfri.org
fax (650) 404-9981; dsoohoo@cfri.org
Internet: http://www.cfri.org/ElizabethNashFellowship.shtml
Sponsor: Cystic Fibrosis Research, Inc.
2672 Bayshore Parkway, Suite 520
Mountain View, CA 94043

Cystic Fibrosis Research New Horizons Campaign Grants **1690**
Since 1975, Cystic Fibrosis Rresearch, Inc., has supported cystic fibrosis research at well-established medical facilities through the funding of grants. Its investment in new ideas has enabled researchers to bring new perspectives to the study of this disease. Those involved in varied disciplines of research are encouraged to submit grants to advance the current understanding of cystic fibrosis. The New Horizons Campaign funds two to four research projects at well-established cystic fibrosis laboratories, providing up to $75,000 per year for a maximum of two years for each project. The Research Advisory Committee (RAC) submits vetted proposals to leaders in the field of CF research for their independent review. The top candidates are then selected for funding by CFRI.
Geographic Focus: All States
Amount of Grant: Up to 75,000 USD
Contact: Carroll Jenkins; (855) 237-4669; fax (650) 404-9981; cjenkins@cfri.org
David Soohoo, Director of Programs and Operations; (650) 404-9445 or (855) 237-4669; fax (650) 404-9981; dsoohoo@cfri.org
Internet: http://www.cfri.org/research.shtml
Sponsor: Cystic Fibrosis Research, Inc.
2672 Bayshore Parkway, Suite 520
Mountain View, CA 94043

Cystic Fibrosis Trust Research Grants **1691**
The trust was founded in 1964 to find a complete solution for cystic fibrosis and to improve upon current methods of treatment. Current areas of research include: gene therapy; microbiology; transplantation; preventing lung damage in young children; immunology; pharmacology; and basic science Research grants are awarded in the United Kingdom and internationally in the fields of science and medicine.
Requirements: Only researchers from recognized academic institutions are eligible.
Geographic Focus: All States, All Countries
Date(s) Application is Due: Mar 11
Contact: Grants Officer; 020 8290 8024 or 020 8464 7211; researchgrants@cftrust.org.uk
Internet: http://www.cftrust.org.uk/scope/page/view.go?layout=cftrust&pageid=54
Sponsor: Cystic Fibrosis Trust
11 London Road
Bromley, BR1 1BY United Kingdom

D.F. Halton Foundation Grants **1692**
The foundation awards grants to nonprofit organizations in Charlotte, NC, and San Miguel County, CO, primarily in the areas of youth, education, social services, and the performing arts. Additional areas of interest include historical preservation, education, vocational education, business school education, substance abuse, cancer research, heart/circulatory diseases and research, human and family services, and community development. Grants are awarded for general operating support, annual campaigns, capital campaigns, and scholarship funds. There are no application deadlines or forms.
Requirements: Nonprofit organizations in North Carolina counties, including Mecklenburg, Union, Cleveland, Cabarrus, Stanly, Lincoln, and Gaston, may submit proposals. Nonprofits in San Miguel county, CO. may also submit proposals.
Restrictions: No grants will be awarded to individuals.
Geographic Focus: Colorado, North Carolina
Contact: Dale Halton, President
Sponsor: D.F. Halton Foundation
P.O. Box 834
Ophir, CO 81426

DAAD-AICGS Research Fellowships **1693**
The program is designed to bring scholars and specialists working on Germany, Europe, and/or transatlantic relations to AICGS for research stays of two months each. Project proposals should address a topic closely related to one or more of the Institute's five research and programming areas: Globalization and the German and American Economies; Germany in Europe; Security and Foreign Policy—Changing Agendas and New Challenges; Culture and Politics; or Transnational Issues and German-American Cooperation. Fellowships include a monthly stipend of up to $4,600, depending on the seniority of the applicant, economy class round-trip airfare and transportation to and from Washington (for a maximum of $770), and office space at the Institute.
Requirements: Applicants must be German or American citizens.
Geographic Focus: All States
Date(s) Application is Due: Apr 30; May 31; Sep 30
Amount of Grant: 5,370 USD
Contact: Program Contact; (202) 332-9312; fax (202) 265-9531; kverclas@aicgs.org
Internet: http://www.daad.org/page/98696/
Sponsor: German Academic Exchange Service (DAAD)
871 United Nations Plaza
New York, NY 10017

DAAD-Leo Baeck Institute Fellowships **1694**
Fellowships are awarded for research in New York or Germany on the social, communal, and intellectual history of German-speaking Jewry. Financial assistance is provided to doctoral students for dissertation research and to young academics for the preparation of a scholarly essay or book. The New York fellowship consists of a stipend of $2,000 paid in two installments of $1,000 each. The fellowship for Germany follows the terms of award for the DAAD Research Grant.
Requirements: Applicants must be U.S. citizens and PhD candidates or recent PhDs (degree awarded within the last two years).
Geographic Focus: All States
Date(s) Application is Due: Nov 15
Amount of Grant: 2,000 USD
Contact: Program Contact; (212) 744-6400; fax (212) 988-1305; lbaeck@lbi.cjh.org
Internet: http://www.daad.org/page/48513/
Sponsor: German Academic Exchange Service (DAAD)
871 United Nations Plaza
New York, NY 10017

DAAD Bilateral Programs for Co-Operative Research Grants **1695**

These programs aim to intensify academic collaborative research projects by promoting the project-related exchange of academics. These are bilateral programs agreed between the DAAD in Germany and partner organizations in a number of countries. Support is given to persons participating in specific academic collaborative research projects. Particular importance is attached to the continuing education and training and specialization of young academics working within the framework of research collaboration projects. Financial support is only offered to fund specifically person-related additional costs in as far as these arise through the exchange of participating scientists, scholars, graduates and, for some countries, advanced students. Diploma-level candidates can only be included in the support framework if their final dissertation forms part of the project. The German or respectively the foreign side will pay the travel costs and the stay at the partner institute for the participants of their country on the basis of the generally applicable conditions. Payments by the host institute (for example, for accommodation) will be taken into account. Project-related additional costs (for example, computer time, material purchases, documentation, photocopying, printing) cannot be reimbursed by the DAAD, nor is there any element in the grant to cover bench fees or equipment. To date, programs have been implemented with Argentina, Brazil, Bulgaria, Canada, Chile, China, Croatia, Czech Republic, Finland, France, Greece, Hong Kong, Hungary, India, Italy, Mexico, Norway, Poland, Portugal, Slovakia, Spain, Taiwan, Thailand, the United Kingdom and the United States. Country-specific information, dates and deadlines, and application addresses differ from one country to the next.
Requirements: Applications must be accompanied by a specific academic or scientific research project proposal. Academics from Germany and from the respective partner country must be collaborating in this project. Primary financing for the collaborative research project (personnel and running costs on both sides) must be provided by other sources.
Restrictions: Proposals based purely on study and training are not funded.
Geographic Focus: All States
Contact: Administrator; (212) 758-3223; fax (212) 755-5780; kim@daad.org
Internet: http://www.daad.de/deutschland/foerderung/stipendiendatenbank/00462.en.html?detailid=10&fachrichtung=15&land=44&status=4&seite=1&daad=1
Sponsor: German Academic Exchange Service (DAAD)
871 United Nations Plaza
New York, NY 10017

DAAD Emigre Memorial German Internship Program (EMGIP)-Bundestag **1696**

The Program offers internship opportunities for U.S. and Canadian students in the German parliament, the Bundestag. The internships are two months long in positions matching the student's interest and experience. Interns will be placed with their preferred Fraktion, Ausschuss, with individual members of the Bundestag and their offices. In addition to contributing to the respective offices, interns have the opportunity to study legislative and administrative procedures in the German parliament. The successful applicant will receive compensation of approximately 1,100 Euros per month from the German Bundestag. Subsidized health insurance is available for a monthly fee of about 23 Euros. DAAD can help the interns to obtain housing in Berlin (the average rent for a room is 250 Euros/month) and make contacts with fellow international interns and German students.
Requirements: Applicants for EMGIP Bundestag should possess outstanding academic records and personal integrity as well as some knowledge of the German legislative process. Participants should be advanced undergraduates or graduate students in fields such as political science, international relations, law, history, economics or German. Students must be able to fully communicate in German. U.S. and Canadian citizens and permanent residents are eligible to apply. International students who are enrolled in a full time course of study in the U.S. or Canada may also apply. Applicants must be younger than 32 at the start of the internship. Applicants who are graduating seniors should be prepared to show acceptance to a graduate school or further affiliation with their college to ensure student status in Germany. Travel expenses are the intern's responsibility.
Restrictions: German nationals are not eligible.
Geographic Focus: All States
Date(s) Application is Due: Nov 17
Amount of Grant: 2,200 EUR
Contact: Administrator; (212) 758-3223; fax (212) 755-5780; schenkl@daad.org
Internet: http://www.daad.org/page/53287/
Sponsor: German Academic Exchange Service (DAAD)
871 United Nations Plaza
New York, NY 10017

DAAD Faculty Research Visit Grants **1697**

This program offers grants for one to three months in all academic disciplines to scholars at U.S. and Canadian institutions of higher education to pursue research at universities, libraries, archives, institutes or laboratories in Germany. Grants are awarded for specific research projects. Stipends consist of a monthly maintenance allowance.
Requirements: Candidate must hold a PhD, have been engaged in teaching or research for at least two years after receipt of the doctorate, be a U.S. citizen or permanent resident affiliated with an American institution for at least five years, and have a previous research record in the proposed field. The period for which the grant is awarded must fall entirely within one calendar year.
Restrictions: Grants cannot be used for travel, attendance at conferences or conventions, editorial meetings, lecture tours, or extended guest-professorships. No extra allowance is given for dependents who might accompany the grantee. Three years must elapse before another application by a former study visit grantee will be considered; applicant may not hold a DAAD grant or a grant from another German or American organization consecutively or concurrently.
Geographic Focus: All States
Date(s) Application is Due: Feb 1; Aug 1
Amount of Grant: 1,840-6,720 EUR
Contact: Administrator; (212) 758-3223; fax (212) 755-5780; daadny@daad.org
Internet: http://www.daad.org/page/68573/
Sponsor: German Academic Exchange Service (DAAD)
871 United Nations Plaza
New York, NY 10017

DAAD German Studies Research Grants **1698**

This specialized program offers up to five German Studies Research Grants to highly qualified undergraduate and graduate students who are nominated by their department/program chairs. The grant may be used for short-term research (one to two months) in either North America or Germany. The program is designed to encourage research and promote the study of cultural, political, historical, economic and social aspects of modern and contemporary German affairs from an inter- and multidisciplinary perspective. Research support ranging in value from $1,500 to $2,500 is available to individual scholarship recipients and is intended to offset living and travel costs during the active research phase.
Requirements: Undergraduates with at least junior standing pursuing a German Studies track or minor may be nominated for the grant by their department and/or program chair. Applicants are expected to have completed two years of college German and a minimum of three courses in German Studies (literature, history, politics or other fields) at the time of nomination. Grants are restricted to citizens of the U.S. who are enrolled full time at the university that nominates them. Applicants must be younger than 32 at the start of the grant period.
Restrictions: Support cannot be provided for stays in Germany in the context of study abroad programs.
Geographic Focus: All States
Date(s) Application is Due: May 1; Nov 1
Amount of Grant: 1,500-2,500 USD
Contact: Administrator; (212) 758-3223; fax (212) 755-5780; graaff@daad.org
Internet: http://www.daad.org/page/51532/
Sponsor: German Academic Exchange Service (DAAD)
871 United Nations Plaza
New York, NY 10017

DAAD Intensive Language Courses Grants **1699**

DAAD offers grants to graduate students at North American universities to attend 8-week intensive language courses at leading institutes in Germany. Extensive descriptions of the institutes, their teaching philosophies, course content, as well as course dates are available on the website of each institute. Scholarships are awarded to students currently enrolled full-time in a graduate or PhD program in all fields of study except English, German, or any other modern language or literature. Applicants will be assessed on the basis of their academic record and statements of projected academic and professional future.
Requirements: As a rule, applicants must be citizens of the U.S. or Canada. Foreign nationals may be eligible if they have been full-time graduate students at a U.S. or Canadian University for at least one academic year at the time of application. Applicants must have completed three semesters of college German or have achieved an equivalent level of language proficiency. Applicants must be younger than 32 at the start of the grant period.
Restrictions: Applicants who have received a DAAD summer language grant in the past three years are not eligible to apply.
Geographic Focus: All States
Date(s) Application is Due: Jan 31
Amount of Grant: 2,650 EUR
Contact: Administrator; (212) 758-3223; fax (212) 755-5780; graaff@daad.org
Internet: http://www.daad.org/page/54773/
Sponsor: German Academic Exchange Service (DAAD)
871 United Nations Plaza
New York, NY 10017

DAAD Learn German in Germany Grants **1700**

The Goethe-Institut, through DAAD, offers one to two grants to faculty members in all academic fields except modern languages and literatures to attend intensive language courses at Goethe-Instituts in Germany. Four and eight-week courses are offered year-round. Preference will be given to applicants in the social sciences, the natural sciences, engineering and professional schools, who are in mid-career and are under 46 years of age.
Requirements: Scholars who hold a PhD (or equivalent) and have been working in research or teaching full-time at a United States university or research institution for at least two years after receipt of the doctorate are eligible. Applicants must have a basic knowledge of German and be able to demonstrate a need for acquiring a better proficiency in the German language for their future studies or research.
Restrictions: Applicants must be citizens or permanent residents of the United States. Faculty members who teach in the fields of English, German or any modern languages or literatures are not eligible.
Geographic Focus: All States
Date(s) Application is Due: Jan 31
Amount of Grant: 1,700 EUR
Contact: Administrator; (212) 758-3223; fax (212) 755-5780; graaff@daad.org
Internet: http://www.daad.org/page/47845/
Sponsor: German Academic Exchange Service (DAAD)
871 United Nations Plaza
New York, NY 10017

DAAD Re-invitation Program for Former Scholarship Holders 1701

The Program helps the DAAD maintain contacts with its former one-year scholarship holders and with former scholarship holders who had studied in East Germany (GDR) for at least one year. Former scholarship holders meeting these requirements can apply for re-invitation to Germany to complete a research or work project at a state (public) or state-recognized higher education institution or non-university research institute. Depending on the applicant's work schedule, the research stay can last between one and three months. Depending on the applicant's academic status, the monthly award will amount to 1,840 Euros for assistant lecturers, assistant professors and young lecturers, and 1,990 Euros for professors. In some rare exceptions, 2,240 Euros may be available. In addition to these payments, the DAAD generally will pay an appropriate flat-rate travel allowance, unless these costs are covered by the home country or by another funding source.

Requirements: Applications for the re-invitation program can only be submitted by former one-year scholarship holders who have been back in their home country for at least three years. The most important selection criterion is a convincing and well-planned research or work project to be completed during the stay in Germany.

Geographic Focus: All States

Date(s) Application is Due: Jan 15; Aug 1

Amount of Grant: 1,840-6,720 EUR

Contact: Administrator; (212) 758-3223; fax (212) 755-5780; kim@daad.org

Internet: http://www.daad.de/deutschland/foerderung/stipendiendatenbank/00462.en.html?detailid=38&fachrichtung=15&land=44&status=4&seite=1&daad=1

Sponsor: German Academic Exchange Service (DAAD)
871 United Nations Plaza
New York, NY 10017

DAAD Research Grants 1702

Research grants are awarded to highly qualified undergraduate and graduate students who are nominated by their department/program chairs. The grant may be used for short-term research (one to two months) in either North America or Germany. The program is designed to encourage research and promote the study of cultural, political, historical, economic and social aspects of modern and contemporary German affairs from an inter- and multidisciplinary perspective.

Requirements: Master's level graduate students in the humanities and social services earning a certificate or working on a project in German Studies may be nominated for the grant by their department and/or program chair. Applicants are expected to have completed a minimum of three courses in German Studies (literature, history, politics, or other fields) at the time of nomination. Doctoral degree students in the humanities and social science disciplines in the process of preparing their dissertation proposals on modern German topics may be nominated. Students whose dissertation proposals have already been formally accepted are not eligible for nomination. The intent of the program is to provide an opportunity for short-term exploratory research to determine the viability or to delimit the scope of their proposed dissertations. The program is not intended to supplement or substitute for regular dissertation field work abroad which should lag the short-term research stay by at least one semester. All applicants are expected to have completed two years of college level German language studies. Applicants must be younger than 32 at the start of the grant period.

Restrictions: Grants are restricted to citizens of the U.S. who are enrolled full time at the university that nominates them. Support cannot be provided for stays in Germany in the context of study abroad programs.

Geographic Focus: All States

Date(s) Application is Due: May 1; Nov 1

Contact: Administrator; (212) 758-3223; fax (212) 755-5780; kim@daad.org

Internet: http://www.daad.org/?p=gradresearch

Sponsor: German Academic Exchange Service (DAAD)
871 United Nations Plaza
New York, NY 10017

DAAD Research Grants for Doctoral Candidates, Young Academics & Scientists 1703

The Program provides young foreign academics and scientists with an opportunity to carry out a research project or a course of continuing education and training at a German state (public) or state-recognized higher education institution or non-university research institute. Research grants can be used to carry out; research projects at a German higher education institution for the purpose of gaining a doctorate in the home country; research projects at a German university for the purpose of gaining a doctorate in Germany; or research projects or continuing education and training, but without aiming at a formal degree/qualification. Depending on the project in question and on the applicant's work schedule, grants can be paid generally for between one and ten months, in the case of full doctoral programs in Germany for up to three years, and in exceptions for up to a maximum of four years. Applications for research grants to run for more than six months are decided once a year, and must be submitted by November 15. Applications by musicians, architects and visual artists must be submitted by November 1. Applications for research grants to run for up to six months are decided twice a year, and must be submitted by August 1 or November 15. At the earliest, the grant can begin 6 months after the date of application. Support can only be provided for the completion of a full doctoral program in Germany when special support policy reasons exist. Depending on the award holder's academic level, the program will pay a monthly award of 715 Euros (graduates holding a first degree) or 975 Euros (doctoral candidates). As a rule, the scholarship additionally includes certain payments towards health insurance coverage in Germany. Furthermore, the program generally will pay an appropriate flat-rate travel allowance, unless these costs are covered by the home country or by another funding source.

Requirements: Applications for DAAD research grants are open to excellently-qualified university graduates who hold a Diplom or Master's degree at the time they commence the grant-supported research and, in exceptional cases, graduates holding a Bachelor's degree or already holding a doctorate/PhD (post-docs). It is required that doctoral candidates wishing to take a doctorate/PhD in their home country will already have been admitted to an appropriate course at their home university. Besides previous study achievements, the most important selection criterion is a convincing and well-planned research or continuing education and training project to be completed during the stay in Germany and which has been coordinated and agreed with an academic supervisor at the chosen German host institute. German language skills are generally required, although the required level also depends on the applicant's project and topic, as well as on the available opportunities for learning German in the applicant's home country.

Restrictions: The award of grants is subject to an age limit of 32 years at the time of starting the grant.

Geographic Focus: All States

Date(s) Application is Due: Aug 1; Nov 1; Nov 15

Amount of Grant: 715-9,750 EUR

Contact: Administrator; (212) 758-3223; fax (212) 755-5780; kim@daad.org

Internet: http://www.daad.de/deutschland/foerderung/stipendiendatenbank/00462.en.html?detailid=7&fachrichtung=4&land=44&status=3&seite=1&daad=1

Sponsor: German Academic Exchange Service (DAAD)
871 United Nations Plaza
New York, NY 10017

DAAD Research Internships in Science and Engineering (RISE) 1704

DAAD, in cooperation with science organizations in North America and Germany, offers summer internships in Germany for U.S. and Canadian undergraduate students in the fields of biology, chemistry, physics, earth sciences and engineering. RISE fellows work directly with doctoral students in research groups at top German universities and institutions and can expect to gain serious hands-on research experience. RISE placements provide students the opportunity to live and work in an international context, to gain confidence in their practical and theoretical skills, and to improve their (or begin learning!) German. Last but not least, the research internship should be a source of mutual cultural enrichment for both the interns and their hosts. Every RISE intern receives a pro-rated monthly scholarship of approximately 615 Euros for any period of six weeks to three months between June and August. The program also provides health and accident insurance.

Requirements: To apply for a placement an applicant must: be currently enrolled at an United States or Canadian university/college as a full-time student in the field of Biology, Chemistry, Physics, Earth Sciences or Engineering (or a closely related field); be an undergraduate who will have completed at least 2 years of a degree program by the time of the placement; and prove that he/she will be registered as an undergraduate at a university/college for the academic year.

Geographic Focus: All States

Date(s) Application is Due: Feb 1

Amount of Grant: 615 EUR

Contact: Michaela Gottschling, +49 (0)228-882-567; fax +49 (0)228-882-551; knieps@daad.de or rise@daad.de

Internet: http://www.daad.de/rise/en/index.html

Sponsor: German Academic Exchange Service (DAAD)
Referat 315, Kennedyallee 50
Bonn, 53175 Germany

DAAD Research Stays for University Academics and Scientists 1705

These grants and scholarships aim to provide foreign academics and scientists working in higher education or at research institutes with an opportunity to carry out a research project at a state (public) or state-recognized higher education institution or non-university research institute in Germany. Depending on the applicant's work schedule, the research stay will last between one and three months. Depending on the applicant's academic status, the monthly award will amount to 1,840 Euros for assistant lecturers, assistant professors and young lecturers, and 1,990 Euros for professors. In some rare exceptions, 2,240 Euros may be available. In addition to these payments, the DAAD generally will pay an appropriate flat-rate travel allowance, unless these costs are covered by the home country or by another funding source. Applications for research grants are decided twice a year. Applications must have been submitted by 1 August or 15 January. At the earliest, grants can begin 4 months after the date of application.

Requirements: Applications for DAAD research stays are open to excellently-qualified academics and scientists who should generally hold a doctorate/PhD. All applicants must be working in higher education or at a research institute in their home country. Besides their previous academic achievements (for example, recent publications), the most important selection criterion is a convincing and well-planned research project to be completed during the stay in Germany. The application must provide proof of a workplace being provided at the host institute.

Restrictions: DAAD support for a research stay can only be awarded once in any three-year period. No travel expenses can be paid from the grant.

Geographic Focus: All States

Amount of Grant: 1,840-6,720 EUR

Contact: Scholarship Administrator; (212) 758-3223; fax (212) 755-5780; kim@daad.org or daadny@daad.org

Internet: http://www.daad.de/deutschland/foerderung/stipendiendatenbank/00462.en.html?detailid=39&fachrichtung=15&land=44&status=4&seite=1&daad=1

Sponsor: German Academic Exchange Service (DAAD)
871 United Nations Plaza
New York, NY 10017

DAAD RISE Professional Internships 1706

The Program offers practical, career-building experience in the fields of Biology, Chemistry, Physics, Earth Sciences and Engineering (or closely related subjects). Aimed at Bachelor, Master or PhD students from the U.S. or Canada, the RISE professional internship placements provide students the opportunity to live and work in an international context, to gain confidence in their practical and professional skills, and to improve their German. Further, the internship will be a source of mutual cultural and professional enrichment for both the interns and their host companies. Every intern receives a pro-rated monthly scholarship of approximately 615 Euros for any period of six weeks to three months between June and August. DAAD also provides health and accident insurance.
Requirements: To apply for a placement an applicant must: be currently enrolled at an United States or Canadian university/college as a full-time student in the field of Biology, Chemistry, Physics, Earth Sciences or Engineering (or a closely related field); be an undergraduate who will have completed at least 2 years of a degree program by the time of the placement; and prove that he/she will be registered as an undergraduate at a university/college for the academic year.
Geographic Focus: All States
Date(s) Application is Due: Jan 25
Amount of Grant: 615 EUR
Contact: Martina Ludwig, +49 (0)228) 882-104; rise-pro@daad.de
Internet: http://www.daad.de/rise-pro/en/index.html
Sponsor: German Academic Exchange Service (DAAD)
Referat 315, Kennedyallee 50
Bonn, 53175 Germany

DAAD Scholarships for Artists: Study Visits for Academics 1707

The study visits for academics program aims to enable lecturers from the fields of Fine Art, Design, Film, Music and Architecture as well as Drama, Direction, Dance and Choreography to apply for a supported study visit for the purpose of artistic cooperation with a German host institution. Depending on the applicant's work schedule, the study visit will last between one and three months. Depending on the applicant's academic status, the monthly award will amount to 1,840 euros for assistant lecturers, assistant professors and young lecturers, and 1,990 euros for professors, in exceptions, 2,240 euros. In addition to these payments, the DAAD generally will pay an appropriate flat-rate travel allowance, unless these costs are covered by the home country or by another funding source.
Requirements: To qualify for support under the study visits for artists program, the applicant must be lecturing at a higher education institution in the home country. A well-planned and convincing project is an important selection criterion. The application papers must be accompanied by written confirmation of academic supervision or confirmation of agreement having been reached with a German institution of higher education.
Restrictions: DAAD support for study visits by artists can only be awarded once in any three-year period.
Geographic Focus: All States
Date(s) Application is Due: Jan 15; Aug 1
Amount of Grant: 1,840-6,720 EUR
Contact: Administrator; (212) 758-3223; fax (212) 755-5780; kim@daad.org
Internet: http://www.daad.de/deutschland/foerderung/stipendiendatenbank/00462.en.html?detailid=41&fachrichtung=8&land=44&status=4&seite=1&daad=1
Sponsor: German Academic Exchange Service (DAAD)
871 United Nations Plaza
New York, NY 10017

DAAD Study & Internship Program (SIP) in Germany 1708

A joint program of German Universities of Applied Sciences (UAS7) and DAAD, this program offers students a full academic semester of study abroad at one of the UAS7 universities in Germany followed by a one-semester professional internship experience in a company or research institute in Germany. Scholarships are granted for one semester (6 months). Recipients will be awarded a monthly stipend of about 675 Euro to cover living expenses, including a monthly contribution to cover health and accident liability insurance. In addition, they receive 850 Euro as a contribution to travel expenses. During the internship the company will pay a minimum of 500 Euro/month to help cover the monthly living expenses.
Requirements: Students must: be full-time, in good academic standing, and currently enrolled in an undergraduate degree-granting program at an accredited U.S. or Canadian college or university; be currently enrolled as Sophomores or Juniors in engineering, science, life sciences, business, management, economics, architecture, art, design, journalism, or social work; possess an outstanding academic record and personal integrity, as proven by academic achievement and letters of recommendation; hold U.S. or Canadian citizenship or be a permanent resident thereof (foreign nationals are eligible if they have been full-time students at an accredited U.S. or Canadian university for more than one year at the time of application and intend to return to the U.S. or to Canada after the scholarship period to complete their Bachelor's degree); and meet the age limit of 32 at the start of the grant period.
Geographic Focus: All States
Amount of Grant: 1,525 EUR
Contact: Administrator; (212) 758-3223; fax (212) 755-5780; daadny@daad.org
Internet: http://www.uas7.org/content/programs__projects/study__internship_program/index_en.html
Sponsor: German Academic Exchange Service (DAAD)
871 United Nations Plaza
New York, NY 10017

DAAD Study Visits / Study Seminars and Practicals in Germany by Foreign Students 1709

The program aims to provide students with subject-related knowledge by arranging appropriate visits, tours and information meetings (Study Visits) or by organizing subject-related seminars and practical courses (e.g. specialist courses, block seminars, workshops) at the invitation of a German university. This university is also responsible for organizing practical courses with universities, companies and, possibly, public institutions (Study Seminars and Practicals). The program goals also include: facilitating meetings with German students, academics and researchers to establish and maintain contacts between German and foreign universities; and giving students a greater understanding of and insight (regional and area studies) into economic, political and cultural life in Germany. Cultural events (e.g. concert tours) can be funded when the focus is on meeting students and university teachers from a relevant academic field and this academic relevance is appropriately documented. Study visits, study seminars or practicals should last no less than 7 days. Funding is available for a maximum of 12 days (including travel days), although the visits themselves may last longer. The DAAD takes out health, accident and public/private liability insurance for each funded group.
Restrictions: Funding cannot be provided for annual repeat visits (by applicants, faculties or departments); each applicant, faculty or department can only be considered for a maximum of one application per year. Funding cannot be provided for required/obligatory excursions or for measures that have already been completed. The Program regrets that it is unable to pay any international travel costs.
Geographic Focus: All States
Contact: Katharina Klein, +49 (0)228/882-370; fax +49 (0)228/882-447; k.klein@daad.de
Internet: http://www.daad.de/deutschland/foerderung/stipendiendatenbank/00462.en.html?detailid=15&fachrichtung=15&land=44&status=4&seite=1&daad=1
Sponsor: German Academic Exchange Service (DAAD)
Referat 224, Kennedyallee 50
Bonn, D-53175 Germany

DAAD Undergraduate Scholarships 1710

Highly qualified undergraduate students are invited to apply for scholarships funding study, senior thesis research, and/or internships in Germany. The goal of this program is to support study abroad in Germany and at German universities. Preference will be given to students whose projects or programs are based at and organized by a German university. Scholarships are available either as part of an organized study abroad program or as part of an individual, student-designed study abroad semester or year. The scholarship periods must take place during the German academic year. (short term: October 1 through January 31; long term: October 1 through July 31). Recipients receive a monthly stipend and funds to help defray travel and research expenses, as well as health insurance. Guidelines are available online.
Requirements: Eligible students are current sophomores or juniors of all academic fields; are seeking DAAD support for a four- to ten-month period in Germany during the German academic year; possess outstanding academic records and personal integrity, as evidenced by both their grades and letters of recommendation; are able to receive academic credit at their home institutions or ECTS credits for their activity in Germany; are U.S. or Canadian citizens or permanent residents thereof (Foreign nationals are eligible if they have been full-time students at an accredited U.S. or Canadian university for more than one year at the time of application and will return to the U.S. or Canada after the scholarship period to complete their bachelor's degree.); have well-defined study, research, or internship plans for their stay in Germany; submit the DAAD language evaluation form with their application, although German language competency is not mandatory; demonstrate an interest in contemporary German and European affairs and who explain the significance of their project in Germany to their future studies, research, or professional goals; are enrolled, full-time students in an undergraduate degree-granting program at an accredited North American college or university; and are younger than 32 at the start of the grant period.
Geographic Focus: All States
Date(s) Application is Due: Jan 31
Amount of Grant: 2,460-6,150 EUR
Contact: Administrator; (212) 758-3223; fax (212) 755-5780; daadny@daad.org
Internet: http://www.daad.org/?p=47220
Sponsor: German Academic Exchange Service (DAAD)
871 United Nations Plaza
New York, NY 10017

DAAD Visiting Professorship Grants 1711

This program serves to strengthen the internationalization of the educational experience for scholars, host institutions and students by welcoming educators from abroad to university campuses in Germany for guest teaching assignments. The recent development of international degree programs and traditional curricula looking to infuse an international aspect provide opportunities for professors from other countries to contribute their expertise in particular subjects and teaching methods. Courses need not be taught in German. As the applicant, the German host institution receives the DAAD funding. Visiting Professors are paid by the host institution according to the salary schemes for German university faculty. Funding is available for a period of one semester to two years. Additional allowances can be made for travel, luggage and health insurance.
Requirements: Highly qualified scholars in all academic disciplines, preferably those who hold a doctoral or other terminal degree and have an affiliation with an institution of higher education, are eligible. Candidates must secure an invitation from a German host institution and teach courses integrated into the regular curriculum.
Restrictions: Funds cannot be made available to replace faculty on sabbatical.
Geographic Focus: All States

Date(s) Application is Due: Jan 15; Jul 15
Contact: Grants Administrator; (212) 758-3223; fax (212) 755-5780; daadny@daad.org or gastdozentur@daad.org
Internet: http://www.daad.org/page/50132/
Sponsor: German Academic Exchange Service (DAAD)
871 United Nations Plaza
New York, NY 10017

Dade Community Foundation Grants **1712**
The funding for this program is made available through the Foundation's unrestricted and field of interest funds. This program is designed to honor both the donors interests and address significant community issues such as: education; health; human services; arts and culture; environment; economic development; at-risk youth; abused and neglected children; living with HIV/AIDS; homelessness; social justice; care of animals; heart disease and more.
Requirements: Eligible applicants include nonprofit tax-exempt organizations, as defined by the Internal Revenue Code, which are serving the residents of Miami-Dade County. Preference will be given to organizations based in Miami-Dade County or if located outside the county, are working in partnership with an organization based in Miami-Dade.
Restrictions: The Foundation does not provide grants to individuals, for memberships, fundraising events or memorials. Grants to government agencies are made on a very restricted basis.
Geographic Focus: Florida
Date(s) Application is Due: Nov 15
Amount of Grant: 7,500 USD
Contact: Charisse Grant, Vice President for Programs; (305) 371-2711; fax (305) 371-5342; charisse.grant@dadecommunityfoundation.org
Internet: http://www.dadecommunityfoundation.org/Site/programs/overview.jsp
Sponsor: Dade Community Foundation
200 S Biscayne Boulevard, Suite 505
Miami, FL 33131-5330

Daimler and Benz Foundation Fellowships **1713**
This program is addressed to young postdoctoral scientists in Germany, who are in the early phase of their postgraduate research. The program is, in the sense of the foundation's statutes, open for all scientific disciplines and subject areas. Prerequisites are an own research project and an affiliation with a scientific institution. The amount of the fellowship is 20.000 Euros annually, which is granted for a total of two years. The funds are disbursed by the respective research institution's third-party funds administration, where the fellowship holder is doing research. The funds can be used for equipment, especially for computers, for financing scientific assistants, research travel, participation in conferences, as well as for the organization of own conferences. The fellowship can not, however, be used for living expenses, printing costs, or for expendable items (e.g., chemicals, paper, etc.). The Foundation plans to award approximately ten per year scholarships, the exact number of scholarships may vary depending on the funds available. In accordance with the Foundation's statutes, the program is open to all disciplines and subjects.
Requirements: Postdoctoral scientists and assistant professors in the early phase of their scientific work as well as young researchers in a similar position, such as independent director of junior research groups, are eligible to apply. The applicant must be affiliated with a German university or research institution and carry out their own research project. Applications may be submitted in German and/or English. Each application is reviewed by a committee. The Commission consists of the Board of Management of Daimler and Benz Foundation, the managing director and evaluators. After the deadline initially there will be a pre-selection. The selected applicants will be invited to a seminar in the Daimler and Benz Foundation in which they present their science project in the presence of representatives of the Foundation as well as peer reviewers personally. Participation in this seminar and the presentation of the research project are a prerequisite for the subsequent award of a scholarship. Self-nomination for the Daimler and Benz postdoctoral program is possible and is done without special application forms.
Geographic Focus: Germany
Date(s) Application is Due: Oct 15
Contact: Dr. Jörg Klein; 06 203-10 92-0; info@daimler-benz-stiftung.de
Internet: http://www.daimler-benz-stiftung.de/cms/index.php?page=home_en
Sponsor: Daimler and Benz Foundation
Dr.-Carl-Benz-Platz 2
Ladenburg, 68526 Germany

DaimlerChrysler Corporation Fund Grants **1714**
The fund contributes to organizations grouped under the general categories of education, health and human services, civic and community, religion, and culture and the arts. Within these categories, grants are made available for public welfare or for charitable, scientific, educational, environmental, safety, building, and affirmative action purposes. Higher education grants largely support science and engineering education and business management. A major interest of the corporation is the establishment of national certification standards for elementary and secondary teachers. Another area of concern is the encouragement of early reading skills, and a pilot project has been funded for research in this area. The fund earmarks funds for its future workforce initiatives, which support business and engineering departments, community-based job-skill training, and entry-level work preparation. Types of support include matching gifts, program grants, scholarships, annual campaigns, building construction/renovation, general support, and employee matching gifts. In considering requests, the fund evaluates each applicant organization on its own merits; considered are the programs in which it is engaged, constituencies served, operation procedures, services offered, quality of management, its accountability, finances, and fund-raising practices. Applications are accepted at any time.
Requirements: Eligible for support are nonprofit, tax-exempt educational, health, civic, and cultural organizations primarily in locations where the greatest number of employees of Chrysler and its U.S.-based subsidiaries live and work (Alabama, Delaware, Illinois, Indiana, Michigan, Missouri, New York, Ohio, and Wisconsin). Some support is targeted for national organizations as well.
Restrictions: Grants are not awarded to support endowments, conferences, trips, direct health care delivery, multiyear pledges, capital campaigns, fund-raising activities related to sponsorships, advertising, or debt retirement.
Geographic Focus: All States
Contact: Brian Glowiak; (248) 512-2502; fax (248) 512-2503; mek@dcx.com
Internet: http://www.fund.daimlerchrysler.com
Sponsor: DaimlerChrysler Corporation Fund
1000 Chrysler Drive
Auburn Hills, MI 48326-2766

Damon Runyon Cancer Foundation Fellowships **1715**
The foundation encourages all theoretical and experimental research relevant to the study of cancer and the search for cancer causes, mechanisms, therapies, and prevention. Candidates must apply for the fellowships under the guidance of a sponsor—a senior member of the scientific research community. The sponsor should be actively engaged in the planning, execution, and supervision of the proposed research and should encourage the fellow to report the results of the research in scientific journals and meetings. Awards are made to institutions for the support of the fellow under direct supervision of the sponsor.
Requirements: Applicants must have completed one or more of the following degrees or its equivalent: MD, PhD, DDS, and DVM.
Geographic Focus: All States
Date(s) Application is Due: Mar 15; Aug 15
Amount of Grant: 134,000-174,000 USD
Contact: Fellowship Administrator; (212) 455-0520; awards@drcrf.org
Internet: http://www.drcrf.org/apFellowship.html
Sponsor: Damon Runyon Cancer Research Foundation
675 Third Avenue, 25th Floor
New York, NY 10017

Dana Foundation Science and Health Grants **1716**
The Foundation supports research in neuroscience, immunology, and the effects of arts training on cognition. Areas of focus are: Brain and Immuno-imaging; Human Immunology; Neuroimmunology; Clinical Neuroscience Research; and Arts and Cognition. All other Science and Health Grants are made solely by invitation.
Requirements: The Foundation requires institutions, in many cases, to share the cost of a project or raise matching funds.
Restrictions: The foundation makes no grants directly to individuals; does not support annual operating budgets of organizations, deficit reduction, capital campaigns, or individual sabbaticals; and does not schedule meetings with applicants, other than by specific invitation initiated by the foundation.
Geographic Focus: All States
Contact: Administrator; (212) 223-4040; fax (212) 317-8721; danainfo@dana.org
Internet: http://www.dana.org/grants
Sponsor: Charles A. Dana Foundation
745 Fifth Avenue, Suite 900
New York, NY 10151-0002

Daniel Mendelsohn New Investigator Award **1717**
The Daniel P. Mendelsohn award is intended to encourage new investigators whose work is compatible with this mission. New investigators in the field of research or education whose work relates to health and well being are eligible for the award. Candidates, who may or may not be members of the FRI organization, must have demonstrated scholarly research potential compatible with the mission of FRI. They must show future promise, evidenced by energy and enthusiasm in research that addresses prevention, treatment or education activities. Candidates for the biennial award are reviewed by the Awards Review Committee of FRI that presents their recommendation to the Board of Directors at their October Annual Meeting. Deadline for recommending candidates is May 1 of even numbered years. The Award is a check for $5000 payable to the recipient for her or his personal or professional use. The award will be presented in the fall of the awarding year.
Geographic Focus: All States
Date(s) Application is Due: May 1
Amount of Grant: 5,000 USD
Contact: Janet Klein Brown; (410) 823-5116; fax (410) 823-5131; fri@friendsresearch.org
Internet: http://www.friendsresearch.org/mendelsohn_award.htm
Sponsor: Friends Research Institute
505 Baltimore Avenue, P.O. Box 10676
Baltimore, MD 21285

Danone Institute Belgium Doctoral Scholarship in Nutrition **1718**
The doctoral scholarship, given every four years, is allocated to a candidate who has decided on a career in research on nutrition, and to this end, is enrolled in a doctoral program in a Belgian university in either the French-speaking community or in the Flemish-speaking community. All degrees must be obtained with a minimum grade of distinction.
Requirements: Applicants should be either: holders of a degree of doctor of medicine, pharmacologist, chemical and agricultural engineering, or bioengineering; a graduate in nutrition, physical education and physiotherapy, psychology; or be holders of an equivalent recognized degree obtained in a country of the European Union.
Geographic Focus: All States, Belgium

Amount of Grant: 30,000 EUR
Contact: Fabienne Trinon, + 32 2 770 63 54; fax + 32 2 771 98 97; institute@danone.com
Internet: http://www.danoneinstitute.be/
Sponsor: Danone Institute Belgium
Rue du Duc-100-Hertogstraat
Brussels, 1000 Belgium

Danone Institute Belgium Research in Human Nutrition Grants **1719**
The grants for human nutrition are designed to help promote research in human nutrition, including applied research on an animal model, in Belgium university-level institutions. Candidates should be members of, or an employee with, a research service, research department, research institute or research body within the institution to which he belongs or is affiliated. Three grant awards are given each year.
Requirements: Candidates must be: less than 40 years of age. He/she must also be a member of the permanent academic or scientific staff at a Belgian university-level institution, a permanent researcher at the Fund for Scientific Research, or a research fellow specifically recommended by an university authority.
Geographic Focus: All States, Belgium
Amount of Grant: Up to 12,000 EUR
Contact: Fabienne Trinon, + 32 2 770 63 54; fax + 32 2 771 98 97; institute@danone.com
Internet: http://www.danoneinstitute.be/
Sponsor: Danone Institute Belgium
Rue du Duc-100-Hertogstraat
Brussels, 1000 Belgium

Dart Center Ochberg Fellowship **1720**
The Dart Center provides six or more expense-paid fellowships to midcareer journalists who want to apply knowledge of emotional trauma to improving coverage of violent events. Fellows will attend a two-day seminar on the role emotional trauma plays in coverage of violent events, then will have access to all events and speakers in the annual conference of the International Society for Traumatic Stress Studies (istss.org). Guidelines are available online.
Requirements: Fellowships are open to print and broadcast reporters, photographers, editors, and producers with at least five years of journalism experience.
Geographic Focus: All States
Date(s) Application is Due: Jul 22
Contact: Administrator; (800) 332-0565; fax (206) 543-9285; info@dartcenter.org
Internet: http://www.dartcenter.org/awards_fellowships/index.html
Sponsor: Dart Center for Journalism and Trauma
102 Communications Building, Box 353740
Seattle, WA 98195-3740

Dartmouth College Andrew W. Mellon Postdoctoral Fellowship **1721**
The program offers one two-year postdoctoral teaching/research fellowship in the humanities and related social sciences. Though there are no citizenship requirements, preference will be given to individuals likely to make their careers in the United States. Application information is available online.
Requirements: Applicants should have received the PhD during the past three years.
Geographic Focus: All States
Date(s) Application is Due: Jan 31
Contact: Administrator; (603) 646-0896; fax (603) 646-0998; lhc@dartmouth.edu
Internet: http://www.dartmouth.edu/~lhc/mellon.html
Sponsor: Dartmouth College
6240 Gerry Hall, Room 205
Hanover, NH 03755-3526

Datatel Scholars Foundation Scholarship Program **1722**
The tax-exempt foundation awards undergraduate and graduate scholarships to eligible students to attend a higher learning institution selected from Datatel's more than 600 college, university, and nonprofit client sites. The program reflects the foundation's long-standing commitment to higher education and to give back to the company's client base by focusing Datatel's corporate charitable giving on scholarship. Applicants apply through the institution, which may nominate up to two students. The scholarships are awarded annually on May 1, in conjunction with Datatel's corporate anniversary.
Requirements: To be eligible to apply for a Datatel scholarship, a student must be planning to attend a Datatel client college or university for the year of the award. Students who transfer to another college or university during the award year only maintain their eligibility if the institution to which they transfer is a Datatel client site. Applicants must also be employed at a Datatel non-education client site and attending a college or university of the applicant's choice for the year of the award.
Geographic Focus: All States
Contact: Administrator; (703) 968-9000, ext. 4549; scholars@datatel.com
Internet: http://www.datatel.com/objectViewerBlank.cfm?objectID=EA7DA3E2-F913-11D4-AE900002A5070708
Sponsor: Datatel Scholars Foundation
4375 Fair Lakes Court
Fairfax, VA 22033

Dave Thomas Foundation for Adoption Grants **1723**
The foundation supports advocacy and social service organizations working in the field of adoption. The primary goal of the foundation is to raise public awareness about children in the public welfare system awaiting adoption and to form public and private partnerships to make the adoption process easier and more affordable. Types of support include challenge/matching grants, conferences and seminars, demonstration grants, professorships, program development grants, research grants, seed money grants, and technical assistance. National, regional, or statewide projects will be considered. Priority will be given to projects that request seed money; include matching or other material support from other organizations, government agencies, or funders; coordinate service providers; include a measurement component to evaluate the project's success; and are easily replicable. Proposals are accepted throughout the year. Funding decisions are made four times per year. In order to receive serious consideration, proposals must be received by the submission deadline date for each quarter. Guidelines are available online.
Requirements: Eligible applicants are nonprofit 501(c)3 organizations.
Restrictions: The foundation will not consider funding individual adoption expenses; operating budgets or budget deficits; endowments or capital campaigns; adoption searches or reunions; scholarships; special events; institutions that discriminate on the basis of race, creed, gender, national origin, age, disability, or sexual orientation; or organizations engaged in sectarian religious activities. Unless solicited by the foundation, the following will not be considered: research, educational or promotional videos, publications, television productions, conferences, or public service announcements.
Geographic Focus: All States, Canada
Amount of Grant: 35,000-50,000 USD
Contact: Connie Ackert, Grants Director; (800) 275-3832; fax (614) 766-3871; connie_ackert@davethomasfoundation.org
Internet: http://www.davethomasfoundationforadoption.org
Sponsor: Dave Thomas Foundation for Adoption
525 Metro Place North, Suite 220
Dublin, OH 43017

David Bohnett Foundation Grants **1724**
The foundation's mission is to improve society through social activism. Focus areas include the promotion of the positive portrayal of lesbians and gay men in the media; voter registration activities; animal language research, animal companions, and eliminating rare animal trade; environmental conservation; the reduction and elimination of the manufacture and sale of handguns in the United States; community-based social services that benefit gays and lesbians; and the development of mass transit and non-fossil fuel transportation. Types of support include general operating support, program specific grants, seed money, and multiyear grants. Applicants should email a letter of inquiry to the program officer.
Requirements: Nonprofit organizations are eligible.
Restrictions: Grants do not support individuals, videos or other film productions, or organizations outside the United States.
Geographic Focus: All States
Date(s) Application is Due: Jan 27; Jul 28
Amount of Grant: 5,000-50,000 USD
Contact: Michael Fleming; (310) 277-4611; fax (310) 203-8111; mfpfleming@yahoo.com
Internet: http://www.bohnettfoundation.org/grants/grantapplication.htm
Sponsor: David Bohnett Foundation
2049 Century Park E, Suite 2151
Los Angeles, CA 90067-3123

Daviess County Community Foundation Health Grants **1725**
The Foundation considers proposals for grants on a yearly cycle, which begins each May. At the start of each cycle, a notice is mailed to nonprofit organizations that have applied for grants in the past, have received grants in the past, or have otherwise requested notification of the start of each cycle. Grants in the area of health include activities that: improve and promote health outcomes; provide general and rehabilitative health services; offer mental health services; provide crisis intervention programs; strengthen associations or services associated with specific diseases, disorders, and medical disciplines; and support medical research. Samples of previously funded projects are posted on the website.
Requirements: The Foundation welcomes proposals from nonprofit organizations that are deemed tax-exempt under sections 501(c)3 and 509(a) of the Internal Revenue Code and from governmental agencies serving the County of Daviess, Indiana. Proposals from nonprofit organizations not classified as a 501(c)3 public charity may be considered provided the project is charitable and supports a community need.
Restrictions: Funding is not available for the following: religious organizations for strictly religious purposes; political parties or campaigns; endowment creation or debt reduction; operating costs (not directly related to the proposed project or program); capital campaigns; annual appeals or membership contributions; travel requests for groups or individuals such as bands, sports teams, or classes. Not more than 20% of any grant request may be for personnel costs, office supplies, or other operating costs. Operating costs for any organization must be directly related to the project or program for which funding is being requested.
Geographic Focus: Indiana
Contact: Jeanne Fields; (812) 254-9354; jeanne@daviesscommunityfoundation.org
Internet: http://www.daviesscommunityfoundation.org/program-areas
Sponsor: Daviess County Community Foundation
320 East Main Street, P.O. Box 302
Washington, IN 47501

Davis Family Foundation Grants **1726**
The foundation provides grants primarily to Maine-based educational, medical, and cultural/arts charitable organizations in support of a wide variety of worthwhile projects. Eligible educational organizations include: colleges, universities, and other educational institutions. Eligible medical organizations include: hospitals, clinics and medical research organizations (grant requests will also be considered from other similar health organizations for programs designed to increase the effectiveness or decrease the cost of medical care). Eligible cultural and arts organizations include: those agencies whose

customary and primary activity is to promote music, theater, drama, history, literature, the arts or other similar cultural activities. Further guidelines are available online.
Requirements: Eligible educational organizations include colleges, universities, and other educational institutions (grants are not made to public elementary and secondary schools, nor to schools whose financial support is derived primarily from a church or other religious organization. Trustees will consider grant requests from other educational organizations whose purpose is to promote systemic change in education or to provide innovative programs whose objectives are to improve education). Medical organizations eligible for support include hospitals, clinics, and medical research organizations. Grant requests will also be considered from other similar health organizations for programs designed to increase the effectiveness or decrease the cost of medical care. Eligible cultural/arts organizations include organizations whose customary and primary activity is to promote music, theater, drama, history, literature, the arts, or other similar cultural activities.
Restrictions: The Foundation does not make grants to individuals, religious programs, fellowships, or in the form of loans. The Foundation does not normally provide support for annual giving campaigns or general operating needs. Grants to endowment campaigns have a low priority.
Geographic Focus: Maine
Date(s) Application is Due: Feb 10; May 10; Aug 10; Nov 10
Contact: Anne Vaillancourt; (207) 781-5504; info@davisfoundations.org
Internet: http://www.davisfoundations.org/site/family.asp
Sponsor: Davis Family Foundation
4 Fundy Road
Falmouth, ME 04105

Davis Foundation Postdoctoral Fellowships in Eating Disorders Research **1727**
The Hilda and Preston Davis Foundation's Postdoctoral Fellowship Program is administered and managed by The Medical Foundation. By attracting postdoctoral fellows to the field, dollars allocated to support fellows are leveraged into lifetime contributions to understanding biological causes of eating disorders. The long term goal of the program is to accelerate medical research discoveries that will lead to effective new therapies. Proposals must be focused on relevant aspects of the biological causes of anorexia nervosa and bulimia nervosa as defined by clinical criteria. Up to five three-year fellowships ranging from $43,000-$63,000 per year, inclusive of a $3,000 expense allowance will be awarded to fellows who have completed no more than three years of postdoctoral research training as of the funding start date. The awards will support postdoctoral fellows working in nonprofit academic or research institutions in the United States.
Requirements: Proposals must be focused on relevant aspects of the biological causes of anorexia nervosa and bulimia nervosa as defined by clinical criteria. Examples of funding areas include but are not limited to neural pathways of feeding behavior in animal models; molecular genetic analysis of relevant neural circuit assembly and function; testing of new chemical compounds that might be used in animal models as experimental treatments; and brain imaging technologies that identify neurochemical pathways in patients with these disorders. Applicants must: have a Ph.D. or equivalent awarded from an accredited domestic or foreign institution by July 1 and commit 90% time to research (applicants with clinical responsibilities must commit at least 70% time to research) or, M.D. or equivalent awarded from an accredited domestic or foreign institution by July 1 and commit at least 70% time to research; have completed no more than three years of full-time postdoctoral research experience by the time funding begins on July 1; and, conduct the proposed research project at a hospital, university or other nonprofit research institution where the applicant holds a postdoctoral fellowship appointment. There are no institutional limitations on the number of applicants who may submit applications. However, a mentor may only support one fellow's application. United States citizenship is not required. Online submissions must be received by 12:00 Noon, U.S. Eastern time of January 10. Mailed submissions must be received by 5:00 pm, U.S. Eastern time of January 15.
Restrictions: Clinical psychotherapeutic studies, medication trials, and obesity research are currently outside the scope of this program.
Geographic Focus: All States
Date(s) Application is Due: Jan 10; Jan 15
Amount of Grant: 129,000-189,000 USD
Contact: Jeanne Brown, Program Officer; (617) 279-2240 ext. 709; jbrown@hria.org
Internet: http://hria.org/tmfservices/tmfgrants/davis.html
Sponsor: Health Resources in Action
95 Berkeley Street
Boston, MA 02116

Dean Witter Foundation Grants **1728**
The Foundation supports graduate schools of business and organizations to promote research and higher education in finance. The Foundation makes additional grants, often on a matching basis, to support specific wildlife research and conservation projects in Northern California and seminal opportunities to improve and extend environmental education and to stimulate learning. The Foundation will consider requests for multi-year funding for two or three years into the future to facilitate effective program planning by the institutions supported. Applicants are encouraged to telephone or write the Consultant to determine whether their proposed program falls within the Foundation's areas of interest and grantmaking priorities.
Requirements: The Foundation accepts grant proposals from tax-exempt charitable institutions as defined under Section 501(c)3 of the Internal Revenue Code on a continuing basis. The Foundation does not have a standard application form. Applicants should send one complete proposal to the Consultant with the following elements: cover letter; specific request; personnel information; organizational information; financial information; and addenda.
Restrictions: The Foundation does not: accept funding requests from individuals or award loans, scholarships and grants to specific individuals; assume any obligation to provide continuing support to grantee programs; make grants for annual fundraising events, operating deficits, and capital campaigns; or support sectarian religious activities or sectarian religious facilities.
Geographic Focus: California
Contact: Kenneth Blum; (415) 981-2966; admin@deanwitterfoundation.org
Internet: http://www.deanwitterfoundation.org/FundingGuidelines.html
Sponsor: Dean Witter Foundation
57 Post Street, Suite 510
San Francisco, CA 94104

Deborah Munroe Noonan Memorial Research Grants **1729**
The Deborah Munroe Noonan Memorial Research Fund was established in 1947 by her son, Frank Noonan, to support innovative projects aimed at improving the quality of life for children and adolescents with physical and developmental disabilities. The Noonan Research Fund supports innovative clinical and service system research, demonstration projects, and pilot studies in the Boston area designed to improve the quality of life for children and adolescents with disabilities. The fund provides one-year grants of up to $80,000 (inclusive of 10% indirect costs).
Requirements: Applicants must hold a position within a nonprofit institution or organization within the Fund's geographic area of interest. New investigators are encouraged to submit proposals. Project must address the target age range of birth through 23 years old. Research projects must be conducted within the Fund's geographic area of interest. A complete listing of the geographic eligibility area is contained in the downloadable guidelines available at the grant website. All application information must be completed online, and the proposal uploaded as a PDF by 12:00 Noon U.S. Easter time of the published deadline date.
Restrictions: Proposals for basic science research will not be considered nor will applications for capital costs such as buildings, renovations, or major equipment items. The Noonan Fund does not support direct service, primary prevention projects, primary medical conditions such as obesity or device development. Drug trials are rarely supported by the Noonan Research Fund.
Geographic Focus: Massachusetts
Date(s) Application is Due: Mar 14
Amount of Grant: Up to 80,000 USD
Contact: Jeanne Brown, Program Officer; (617) 279-2240, ext. 709; jbrown@hria.org
Internet: http://hria.org/tmfservices/tmfgrants/noonan.html
Sponsor: Health Resources in Action
95 Berkeley Street
Boston, MA 02116

DEED Grants **1730**
APPA's Demonstration of Energy-Efficient Developments (DEED) program sponsors grants intended for demonstration or early commercialization projects at DEED member utilities that promise either to improve efficiencies or lower costs in the provision of energy services to the consumers of publicly owned electric utilities. DEED grants also are open to all organizations for applied research to early demonstration projects with long-term potential either to improve efficiencies or lower costs in the provision of energy services to the consumers of publicly owned electric utilities; these outside organizations, however, must apply through a DEED member utility. Maximum funding is limited to 25 percent of available DEED funds; amounts greater than $50,000 must be approved by the APPA board after review by the DEED directors. The DEED board of directors meets each year in March and September; therefore, deadlines for receipt of applications are usually in February and August. Call for exact dates and application forms or visit the Web site.
Requirements: Proposals for DEED grants may be submitted only by DEED member utilities that have been DEED members for at least six months prior to grant submission. All proposals shall include a project plan showing tasks, schedule, costs, and decision criteria for termination versus further research.
Geographic Focus: All States
Date(s) Application is Due: Feb 15; Aug 15
Amount of Grant: Up to 75,000 USD
Contact: Michele Ghosh; (202) 467-2960; fax (202) 467-2992; DEED@appanet.org
Internet: http://www.appanet.org/research/index.cfm?ItemNumber=2671&sn.ItemNumber=2415
Sponsor: American Public Power Association
2301 M Street NW
Washington, DC 20037-1484

DEED Student Research Grants and Internships **1731**
DEED (Demonstration of Energy-Efficient Developments) student research grants/internships are intended to promote the involvement of students studying in energy-related disciplines in the public power industry, and to provide host utilities with technical assistance. Successful applicants are expected to conduct research on a project approved by the sponsoring utility and submit a final report on the project, describing activities, cost, bibliography, achievements, problems, results, and recommendations. Additional information is available online.
Requirements: Only graduate or undergraduate students in energy-related disciplines from accredited colleges or universities are eligible. Students must be attending school in a country with at least one DEED member. Applications must be sent from a DEED member utility. Applicants will not be discriminated against on the basis of sex, race, religion, national origin, or citizenship.
Geographic Focus: All States
Date(s) Application is Due: Feb 15; Oct 1
Amount of Grant: 4,000 USD

Contact: DEED Administrator; (202) 467-2960; DEED@APPAnet.org
Internet: http://www.appanet.org/research/index.cfm?ItemNumber=2671&sn.ItemNumber=2415
Sponsor: American Public Power Association
2301 M Street NW
Washington, DC 20037-1484

Delaware Division of the Arts Individual Artist Fellowships **1732**

Individual Artist Fellowships provide funding to Delaware creative artists working in the visual, performing, media, folk, and literary arts. Fellowships are designed to enable recipients to purchase equipment and materials, allocate working time, or fulfill other needs that will allow them to advance in their careers. Emerging Professional Fellowships of $3,000 are available to artists who have not yet established reputations in their fields. Established Professional Fellowships of $6,000 are available to artists who have significant achievements in their fields. A Masters Fellowship in the amount of $10,000 is available on a rotating basis, with category applications changing from year to year. The application and additional guidelines are posted at the website, including guidelines for submitting work samples for each category.
Requirements: Applicants must be at least 18 years of age and residents of Delaware for at least one year at the time of application. Recipients must remain Delaware residents during their fellowship year and not be enrolled in degree or certificate-granting educational programs at the time of application or for the duration of their fellowships. Artists can receive only one fellowship per artistic discipline in each category. Emerging Professional Fellowship recipients may apply in the Established Professional category after a three-year period. Only one fellowship application may be submitted per year. Artists may win only one Masters Fellowship. Visual arts applicants must have a current file in the Delaware Division of the Arts (DDOA) Visual Images Registry.
Geographic Focus: Delaware
Date(s) Application is Due: Aug 1
Amount of Grant: 3,000-10,000 USD
Contact: Paul Weagraff, Director; (302) 577-8278; paul.weagraff.state.de.us
Internet: http://www.artsdel.org/grants/IAFinfo13_DE.pdf
Sponsor: Delaware Division of the Arts
820 North French Street, Carvel State Office Building
Wilmington, DE 19801

Del E. Webb Foundation Grants **1733**

The Foundation applies its resources only for the benefit of the residents of Arizona, California and Nevada for improved and expanded medical services, medical research and education. Within these broad areas of interest, the Foundation draws upon the talent and experience of leaders from many walks of life to select organizations of demonstrated competence which have sound programs that will be able to reach and sustain high levels of performance. In choosing particular projects for support, the Foundation's Board acts on the basis of how the public welfare most effectively may be served. Application information is available online.
Requirements: Grants are confined to the support of nonpartisan, non-profit organizations that are operated in the public interest and have tax-exempt status as charitable organizations granted by the Internal Revenue Service and the states in which the respective organizations are incorporated and/or engage in activities.
Restrictions: Grants are not made to the following: governmental agencies or subdivisions; sectarian or religious organizations whose principle activity is exclusively for the benefit of their own members; organizations soliciting funds in support of projects or programs operated by organizations other than the applicant; expenditures before the recipient incurs and makes such expenditures, and, in the absence of compelling circumstances, the Foundation does not make grants to recipients to liquidate or reduce previously incurred obligations or operating deficits. No grants or loans are awarded or made to individuals for any purpose. Grants are not made for scholarships, student aid or medical assistance. Grants may be made to organizations that provide scholarships, student aid or medical assistance when the organization selects recipients in conformity with accepted standards. The Foundation does not participate in the administration of program(s) that the Foundation funds through the award of the application. The Foundation does not fund or underwrite: galas or gala-like events, testimonial or fund-raising luncheons or dinners, advertising in programs or similar fund-raising activities; organizations that in turn make grants to others; purchase of uniforms, equipment, or trips for school related organizations or amateur sports teams; honoraria for guest speakers or panelists; charities operated by service clubs; or educational seminars.
Geographic Focus: Arizona, California, Nevada
Date(s) Application is Due: Feb 28; May 31; Aug 31; Nov 30
Contact: Program Contact; (928) 684-7223; fax (928) 684-5665
Internet: http://www.dewf.com/DEFAULT.shtml
Sponsor: Del E. Webb Foundation
300 E Willis, Suite C
Prescott, AZ 86301-3110

Dell Foundation Open Grants **1734**

The foundation, the giving arm of Dell Computer, seeks to fund collaborative and innovative solutions to community and children's issues addressing youth (ages newborn to high-school seniors) in Texas, Tennessee, Idaho, and Oregon. Through these efforts, Dell strives to effect change in its local community, while providing lessons and best practices for communities everywhere. Funded areas include arts, education, social services, and health. The foundation supports specific, preventative, and measurable programs with cash, in-kind, and volunteer contributions. Grants are awarded quarterly. An online application form is available on the Web site.
Requirements: Eligibility is open to 501(c)3 nonprofit organizations in Texas—Travis, Williamson County, and McLennan County; Tennessee—Wilson, Davidson County; Idaho—Twin Falls County; and Oregon—Roseburg, Douglas County.
Restrictions: Grants do not support individuals; academic or research projects; civic, religious, or political institutions; school fundraisers; marketing opportunities; or sports events and organizations.
Geographic Focus: Idaho, Oregon, Tennessee, Texas
Date(s) Application is Due: Jan 15; Apr 15; Jul 15; Oct 15
Amount of Grant: Up to 5,000 USD
Contact: Grants Administrator; (512) 338-4400; the_dell_foundation@dell.com
Internet: http://www1.us.dell.com/content/topics/global.aspx/corp/foundation/en/open_grants?c=us&l=en&s=corp
Sponsor: Dell Foundation
1 Dell Way
Round Rock, TX 78682

Deloitte Foundation Grants **1735**

The foundation, funded by Deloitte and Touche U.S.A LLP, supports accounting, business, and related fields of study within the United States. Its national programs benefit university students and faculty and promote excellence in teaching and research, curriculum innovation, and cooperation among practitioners and the academic community. It funds the Deloitte Doctoral Fellowship Program, Trueblood Seminars for Professors, matching gifts program, and many other higher education initiatives. Applications for the doctoral fellowship program must meet the listed deadline; there are no deadlines or application forms for other requests.
Requirements: Doctoral fellowship applicants must be doctoral candidates pursuing their PhD in accounting.
Restrictions: Grants are not awarded to support general purposes, capital campaigns, special programs, publications, or loans.
Geographic Focus: All States
Amount of Grant: 5,000-50,000 USD
Contact: Janet Butchko, Manager; (203) 761-3474
Internet: http://www.deloitte.com/dtt/section_node/0,2332,sid%253D2257,00.html
Sponsor: Deloitte Foundation
10 Westport Road, P.O. Box 820
Wilton, CT 06897-0820

Delta Air Lines Foundation Health and Wellness Grants **1736**

Established in 1968 as Delta's company-managed giving system, the Delta Air Lines Foundation contributes more than $1 million annually in endowed funds to deserving organizations and programs. Delta's Health and Wellness Grants support organizations that dedicate their efforts to the health and well-being of communities with a focus on broad-reaching research for cures and education around diseases that affect all walks of life. Once an application is received, applicants should allow up to three months before review.
Requirements: For proposals which meet the foundation's area of focus, priority will be given to: programs meeting compelling needs in communities where Delta has a presence; proposals that exhibit clear, reasonable goals, and measurable outcomes; distinctive projects where the foundation's involvement will leave a legacy; projects that include collaboration or cooperation with other nonprofit organizations; projects that offer opportunities for Delta employee involvement. The foundation Board of Trustees reviews and approves funding in March, June, September, and November. The deadline for receiving completed proposals is the first day of each of these months.
Restrictions: The foundation will generally not consider: individual applicant's request for support of personal needs; religious activities; political organizations or campaigns; specialized single-issue health organizations; annual or automatic renewal grants; general operating expenses; endowment campaigns; capital campaigns; multi-year commitments; fraternal organizations, professional associations, or membership groups; fundraising events such as benefits; charitable dinners, or sporting events.
Geographic Focus: All States
Date(s) Application is Due: Jun 1; Sep 1; Nov 1
Contact: Administrator; (404) 715-5487 or (404) 715-2554; fax (404) 715-3267
Internet: http://www.delta.com/about_delta/community_involvement/delta_foundation/
Sponsor: Delta Air Lines Foundation
P.O. Box 20706, Department 979
Atlanta, GA 30320-6001

Delta Dental Master's Thesis Award **1737**

The fund invites Master of Science students at the dental schools in Michigan, Ohio, and Indiana to submit proposals for the thesis award program. This program is intended to encourage thesis research that is of direct relevance to the costs or outcomes of dental care. The award provides a stipend to cover costs associated with the conduct of master's thesis research, including materials, supplies, and rental of necessary equipment. A detailed budget with justification for expenditures is required. Partial funding may be approved, dependent on reasonableness of the budget and availability of funds. A copy of the final thesis must be provided to Delta Dental Fund upon its completion. Applications may be submitted at any time. Application and guidelines are available online.
Requirements: Master of Science students at the dental schools in Michigan, Ohio, and Indiana are eligible.
Restrictions: Salaries, wages, indirect costs, and the purchase of equipment will not be covered.
Geographic Focus: Indiana, Michigan, Ohio
Amount of Grant: Up to 3,000 USD
Contact: Program Director; (517) 347-5333; fax (517) 347-5320; ddfund@ddpmi.com
Internet: http://www.ddpmi.com/ddf/masterThesis.htm

Sponsor: Delta Dental Plan of Michigan
P.O. Box 293
Okemos, MI 48805-0293

Delta Gamma Foundation Fellowships for Graduate Study 1738

Fellowships are offered annually for graduate study in any chosen field. Selection is based on scholarship, potential for achievement, and financial need. Annual deadline dates may vary. Contact the committee for application form.
Requirements: Applicants must be Delta Gammas who will have completed their undergraduate study by June 30 of the year in which the fellowship is granted. Application consists of completed application form, official transcripts, two passport-style photos, statement of graduate school acceptance, autobiographical sketch, self-addressed stamped envelope, and sealed letters of recommendation from two persons familiar with applicant's academic work, one alumna of Delta Gamma, and employer.
Geographic Focus: All States
Date(s) Application is Due: Apr 1
Amount of Grant: 2,500 USD
Contact: Jacquelyn Geving Everson, Program Director; (614) 481-8169; fax (614) 481-0133; ScholarFellowandLoans@deltagamma.org
Internet: http://www.deltagamma.org/scholarships_fellowships_and_loans_2.shtml
Sponsor: Delta Gamma Foundation
3250 Riverside Drive, P.O. Box Box 21397
Columbus, OH 43221-0397

Delyte and Dorothy Morris Doctoral Fellowships 1739

Graduate students of the highest caliber are invited to compete for fellowships to support doctoral study in any field at SIU at Carbondale. Fellowships will be awarded for up to three years of full-time study. Application materials and more information are available from the graduate school.
Requirements: Applicant must have an undergraduate grade point average of at least 3.25 overall, or a grade point average for the last two years of bachelor's degree work of 3.5 (A=4.0); if prior graduate study has been undertaken, the applicant must have an overall graduate grade point average of at least 3.7; must have a score in the 75th percentile or higher on a standard test such as the GRE, MAT, GMAT, or ATGSB; and may not already be enrolled in a doctoral program. Preference will be given to citizens or permanent residents of the United States, or immigrants to the United States, and to those who have not previously been enrolled in a graduate program or at SIU.
Geographic Focus: All States
Date(s) Application is Due: Jan 23
Amount of Grant: 18,000 USD
Contact: Fellowship Office; (618) 453-4555; fax (618) 453-4562; gaoffice@siu.edu
Internet: http://www.siu.edu/gradschl/grad_fellowship.htm
Sponsor: Southern Illinois University at Carbondale
Graduate School, Woody Hall B130
Carbondale, IL 62901-4716

DeMatteis Family Foundation Grants 1740

The foundation makes grants in the New York metropolitan area to eligible institutions whose mission involves education, health and human services, medical research, social services, and the arts. Types of support include facilities construction, expansion, renovation; acquisition of capital equipment; scientific/medical research; projects and programs that enable the applicant to expand its mission through new or expanded programs to reach a greater segment of the community served; project-oriented capital campaigns. In general, most grants are made to cover projects that can be accomplished within one year. For construction and longer duration projects, grants may be structured to conform to identified milestones. For major projects, the payment of the grant may be over a number of years.
Requirements: Metropolitan New York 501(c)3 tax-exempt agencies, institutions, and organizations are eligible.
Restrictions: In general, the foundation does not support grants for operating deficits; general operating support; endowments; loans, or financing of any kind; annual appeals, dinner functions, and other special fund raising events; or unrestricted funds.
Geographic Focus: New York
Contact: Grants Administrator; (516) 705-4974
Internet: http://fdncenter.org/grantmaker/dematteis/about.html
Sponsor: DeMatteis Family Foundation
P.O. Box 25
Glen Head, NY 11545

Dennis Leslie Mahony Prize in Legal Theory 1741

The prize will go to the author or authors of an outstanding published work in the field of jurisprudence that best reflects an approach combining legal theory with sociological inquiry, in the tradition of the jurisprudence of the late Professor Julius Stone. A published work need not necessarily be in the form of a traditional book or journal publication. Other types of publication, including reports or papers, are eligible. The recipient of the prize will receive a cash prize, with the offer of an invitation to participate in the activities of the Faculty of Law at the University of Sydney for a period of up to one semester. He or she may also receive an invitation to deliver the prestigious Julius Stone Address in the year following the award of the prize. Entrants are required to submit an application form and five copies of the work, plus five copies of their curriculum vita. Four copies will be returned following judging, and one will be kept in the archive of the Julius Stone Institute. Guidelines are available online.
Requirements: Entries may be directly submitted by the author(s), or on the nomination of a third party.
Geographic Focus: All States
Contact: Helen Irving, 61-2-9351- 0232; fax 61-2-9351-0200; heleni@law.usyd.edu.au
Internet: http://www.law.usyd.edu.au/~jurisprudence
Sponsor: University of Sydney
Julius Stone Institute of Jurisprudence Director, 173-175 Phillip Street, Level 12
Sydney 2000, NSW Australia

DENSO North America Foundation Grants 1742

The Foundation is committed to supporting higher education in engineering and business programs. Priority is given to programs that advance automotive engineering and supply-side business practices. Capital funding is available for equipment, lab development, technological advancements and/or installations, building campaigns and expansion projects. Student projects are supported for university-sanctioned student projects and training competitions. Funding is available for tooling and equipment offering a major, sustaining investment.
Requirements: 501(c)3 nonprofit organizations, educational institutions, and universities located throughout North America are eligible. Proposals should contain: details on how funding will advance student development and training; demonstrate principle(s) of innovation and/or training for efficiency gains in the workplace; contain a clearly articulated desired result.
Restrictions: Grants are not awarded for administrative costs, stipends or trips, conferences and travel expenses.
Geographic Focus: All States
Amount of Grant: 2,000 USD
Contact: Administrator; (248) 372-8233; DENSOfoundation@denso-diam.com
Internet: http://www.densofoundation.org/foundation/foundation.html
Sponsor: DENSO North America Foundation
24777 Denso Drive, MC 4610
Southfield, MI 48086-5047

Denton A. Cooley Foundation Grants 1743

The foundation awards grants in Texas in support of health care, health education, hospitals, and medical research. Types of support include endowments, general operating grants, program grants, and research grants. There are no application forms or deadlines. The board meets quarterly to consider requests.
Restrictions: Grants do not support conferences, loans, individuals, scholarships, fellowships, or publication.
Geographic Focus: Texas
Amount of Grant: 100-250,000 USD
Samples: U of Texas Medical Branch at Galveston (TX)—to establish a professorship in the division of cardiothoracic surgery, $250,000.
Contact: Grants Administrator; (713) 799-2700
Sponsor: Denton A. Cooley Foundation
6624 Fannin, Suite 1640
Houston, TX 77030

Dept of Ed Adult Education—National Leadership Activities Grants 1744

This program supports research, evaluation, information dissemination, and other activities to help states improve adult education and literacy programs. Types of projects include: professional development; national evaluations and surveys; curriculum development; distance learning; data analysis; professional papers; seminars and colloquia.
Requirements: Eligible applicants: Institutions of Higher Education, Local Education Agencies, Nonprofit Organizations, Other Organizations and/or Agencies, State Education Agencies, Postsecondary education institutions, public or private agencies, or consortia of these institutions, agencies, or organizations are eligible for grants, cooperative agreements, and contracts.
Geographic Focus: All States
Contact: Dennis Berry; (202) 245-7814; fax (202) 245-7837; dennis.berry@ed.gov
Internet: http://www.ed.gov/programs/aenla/index.html
Sponsor: U.S. Department of Education
400 Maryland Avenue SW, 11131, PCP
Washington, DC 20202-7240

Dept of Ed International Research and Studies Project Grants 1745

This program supports surveys, studies, and development of instructional materials to improve and strengthen instruction in modern foreign languages, area studies, and other international fields. In addition to surveys and studies, the program provides funds for the development of foreign language materials designed to improve and strengthen foreign language and area and related studies in the U.S. education system.
Requirements: Public and private agencies, organizations, institutions, and individuals may apply.
Restrictions: Funds awarded may not be used for the training of students and teachers.
Geographic Focus: All States
Contact: Ed McDermott, International Education Programs Service; (202) 502-7636; fax (202) 502-7860; ed.mcdermott@ed.gov
Internet: http://www.ed.gov/programs/iegpsirs/index.html
Sponsor: U.S. Department of Education
1990 K Street NW, 6th Floor
Washington, DC 20006-8521

Dept of Ed Jacob K. Javits Gifted and Talented Students Education Program National Research and Development Center 1746

This program conducts research for the purpose of carrying out activities described in Sec. 5464(b) of the statute including research on methods and techniques for identifying and teaching gifted and talented students and for using gifted and talented programs and methods to serve all students. It also conducts program evaluations and surveys. As part of its work, the center collects, analyzes, and develops information about gifted and talented education. Emphasis is given to the identification of and services for students traditionally not included in gifted and talented education, including individuals with limited English proficiency (LEP), individuals with disabilities, and individuals living under economically disadvantaged conditions.
Requirements: Institutions of Higher Education, State Education Agencies and a consortium of IHEs and SEAs may apply.
Geographic Focus: All States
Contact: Anne Sweet; (202) 219-2043; fax (202) 219-2030; anne.sweet@ed.gov
Internet: http://www.ed.gov/programs/nrdcjavits/index.html
Sponsor: U.S. Department of Education
555 New Jersey Avenue NW, Suite 615B
Washington, DC 20208-5573

Dept of Ed Rehabilitation Fellowships 1747

The purpose of this program is to build research capacity by providing support to highly qualified individuals, including those who are individuals with disabilities, to perform research on the rehabilitation of individuals with disabilities. Fellows may conduct original research in any area relating to rehabilitation services and may address any problem faced by persons with disabilities. Two types of fellowships are available. Merit Fellowships are awarded to individuals who have advanced training or experience in independent study. Distinguished Fellowships are available to individuals who have seven or more years of research experience in related fields.
Requirements: Only individuals are eligible to be recipients of fellowships.
Restrictions: Institutions are not eligible to be recipients of fellowships.
Geographic Focus: All States
Date(s) Application is Due: Jun 24
Amount of Grant: 45,000-55,000 USD
Contact: Donna Nangle, (202) 245-7462; donna.nangle@ed.gov
Internet: http://www.ed.gov/legislation/FedRegister/announcements/2005-2/042505c.html
Sponsor: U.S. Department of Education
600 Independence Avenue SW
Washington, DC 20202

Dept of Ed Safe and Drug-Free Schools and Communities State Grants 1748

This program provides support to SEAs for a variety of drug and violence prevention activities focused primarily on school-age youths. Activities may include: developing instructional materials; providing counseling services and professional development programs for school personnel; implementing community service projects and conflict resolution, peer mediation, mentoring and character education programs; establishing safe zones of passage for students to and from school; acquiring and installing metal detectors; and hiring security personnel.
Requirements: State Education Agencies may apply. Local Education Agencies or intermediate education agencies or consortia must apply to the State Education Agency.
Geographic Focus: All States
Contact: Paul Kesner, Office of Safe and Drug-Free Schools; (202) 205-8134; fax (202) 260-7767; paul.kesner@ed.gov
Internet: http://www.ed.gov/programs/dvpformula/index.html
Sponsor: U.S. Department of Education
400 Maryland Avenue SW, Room 3E230, FB-6
Washington, DC 20202-6450

Dept of Ed Star Schools Grants 1749

The purpose of this program is to support distance education projects that: encourage improved instruction in mathematics, science, foreign languages, and other subjects; and serve underserved populations, including disadvantaged, nonreading, and limited English proficient (LEP) populations and individuals with disabilities. Star Schools grants are made to eligible telecommunications partnerships, to enable such partnerships to: develop, construct, acquire, maintain, and operate telecommunications audio and visual facilities and equipment; develop and acquire educational and instructional programming; and obtain technical assistance for the use of such facilities and instructional programming. Annual deadline dates may vary; contact the program office for exact dates.
Requirements: Eligible applicants include either one of the following that is organized on a statewide or multistate basis: a public agency or corporation established for the purpose of developing and operating telecommunications networks to enhance education opportunities provided by education institutions, teacher training centers, and other entities, except that any such agency or corporation shall represent the interests of elementary schools and secondary schools that are eligible to participate in the program under Title I, Part A, of the Elementary and Secondary Education Act of 1965; or a partnership that will provide telecommunications services and that include three or more of the following entities (a-g), at least one of which must be an agency as described in (a) or (b): (a) a local education agency that serves a significant number of elementary and secondary schools that are eligible for assistance under Title I, Part A, of the ESEA, or elementary and secondary schools operated or funded for Indian children by the Department of the Interior; (b) a state education agency; (c) an adult and family education program; (d) an institution of higher education or a state higher education agency; (e) a teacher-training center or academy that provides teacher preservice and in-service training, and receives federal financial assistance or has been approved by a state agency; (f) a public or private entity with experience and expertise in the planning and operation of a telecommunications network, including entities involved in telecommunications through satellite, cable, telephone, or computer; or a public broadcasting entity with such experience; and (g) a public or private elementary or secondary school.
Geographic Focus: All States
Contact: Brian Lekander, Office of Innovation and Improvement; (202) 205-5633; fax (202) 205-5720; brian.lekander@ed.gov
Internet: http://www.ed.gov/programs/starschools/index.html
Sponsor: U.S. Department of Education
400 Maryland Avenue SW, Room 4W226, FB-6
Washington, DC 20202

Dept of Ed Vocational Education National Programs 1750

This program supports research, evaluation, information dissemination, and other activities aimed at improving the quality and effectiveness of vocational and technical education. Projects include: research, development, demonstration, dissemination, identification of best methods, capacity building, technical assistance, evaluation, and assessment activities.
Requirements: Institutions of Higher Education, Nonprofit Organizations, Other Organizations and/or Agencies may apply.
Geographic Focus: All States
Contact: Ricardo Hernandez, U.S. Department of Education, OVAE Supervisor; (202) 245-7818; fax (202) 245-7837; ricardo.hernandez@ed.gov
Internet: http://www.ed.gov/programs/venp/index.html
Sponsor: U.S. Department of Education
400 Maryland Avenue SW, 11137, PCP
Washington, DC 20202-7242

Dept of Ed Vocational Rehabilitation Services Demonstration and Training 1751

This program provides competitive grants to eligible entities to expand and improve the provision of rehabilitation and other services authorized under the Rehabilitation Act of 1973 , as amended. Funding also is provided to further the purposes and policies of the act. More specifically, the program supports activities that increase the provision, extent, availability, scope, and quality of rehabilitation services under the act. Sec. 303 authorizes support of activities serving individuals with disabilities in an array of project types. These diverse projects may include the effective practices that demonstrate methods of service delivery to individuals with disabilities, as well as activities such as technical assistance, systems change, model demonstration, special studies and evaluations, and dissemination and utilization of findings from successful, previously funded projects. The expansion and improvement of rehabilitation and other services will lead to more employment outcomes for individuals with disabilities.
Requirements: State vocational rehabilitation (VR) agencies, community rehabilitation programs, Indian tribes or tribal organizations, or other public or nonprofit agencies or organizations or, as the Rehabilitation Services Administration (RSA) commissioner determines appropriate, for-profit organizations, may apply.
Geographic Focus: All States
Contact: Timothy Muzzio; (202) 245-7458; Timothy.Muzzio@ed.gov
Internet: http://www.ed.gov/programs/demotrain/index.html
Sponsor: U.S. Department of Education
400 Maryland Avenue SW, Room 5052, PCP
Washington, DC 20202

Dept of Ed Women's Educational Equity Grants 1752

This program promotes education equity for women and girls through competitive grants. The program designates most of its funding for local implementation of gender-equity policies and practices. Research, development, and dissemination activities also may be funded. Projects may be funded for up to four years. Examples of allowable activities include: training for teachers and other school personnel to encourage gender equity in the classroom; evaluating exemplary model programs to advance gender equity; school-to-work programs; guidance and counseling activities to increase opportunities for women in technologically demanding workplaces; and, developing strategies to assist LEAs in evaluating, disseminating, and replicating gender-equity programs.
Requirements: Public agencies; private nonprofit agencies; organizations, including community and faith-based organizations; institutions; student groups; community groups; and individuals developing programs that promote gender equity may apply.
Geographic Focus: All States
Contact: Beverly Farrar; (202) 205-3145; fax (202) 205-5630; beverly.a.farrar@ed.gov
Internet: http://www.ed.gov/programs/equity/index.html
Sponsor: U.S. Department of Education
400 Maryland Avenue SW, Room 4W242, FB-6
Washington, DC 20202-5950

Detroit Lions Charities Grants 1753

The organization supports charitable and community causes in Michigan. Funding interests include child abuse and domestic violence prevention, youth recreation, and spinal cord injury research. Grants have been awarded to support the Never, Never Shake a Baby billboard campaign; Pigskin Geography, a learning initiative for youth; Big Brothers/Big Sisters; Grand Rapids Metropolitan YMCA—for swimming lessons for inner-city youth; and Saint Joseph Mercy Hospital—for a domestic violence education program. Requests are accepted between October 1 and December 31.
Geographic Focus: Michigan
Date(s) Application is Due: Dec 31

Samples: Dominion Family Services (MI)— to battle child abuse and domestic violence in Michigan, $210,000.
Contact: Detroit Lions Community Affairs Information Hotline; (313) 216-4056
Internet: http://detroitlions.com/community/index.cfm?cont_id=49715
Sponsor: Detroit Lions Charities
222 Republic Drive
Allen Park, MI 48101

DHHS Comprehensive Community Mental Health Services Grants for Children with Serious Emotional Disturbances 1754

This program provides community-based systems of care for children and adolescents with serious emotional disturbances and their families. The program will ensure that services are provided collaboratively across child-serving systems; that each child or adolescent served through the program receives an individualized service plan developed with the participation of the family; that each individualized plan designates a case manager to assist the child and family; and that funding is provided for mental health services required to meet the needs of youngsters in these systems.
Requirements: States; political subdivisions of a state, such as county or local governments; and federally recognized Native American tribal governments are eligible to apply.
Geographic Focus: All States
Amount of Grant: 200,064-35,000,000 USD
Contact: Gary Blau, Chief, Child Adolescent and Family Branch; (301) 443-1333
Internet: http://mentalhealth.samhsa.gov/publications/allpubs/CA-0013/default.asp
Sponsor: U.S. Department of Health and Human Services
5600 Fishers Lane, Parklawn Building
Rockville, MD 20857

DHHS Emergency Medical Services for Children (EMSC) Program 1755

The purpose of this program is to support demonstration projects for the expansion and improvement of emergency medical services for children who need treatment for trauma or critical care. It is expected that maximum distribution of projects among the states will be made and that priority will be given to projects targeted toward populations with special needs, including Native Americans, minorities, and the disabled. Annual deadline dates may vary; contact program staff for exact dates.
Requirements: States and accredited schools of medicine are eligible to apply.
Geographic Focus: All States
Amount of Grant: 200,000 USD
Contact: Dan Kavanaugh, (301) 443-1321; dkavanaugh@hrsa.gov
Internet: http://mchb.hrsa.gov/programs/emsc
Sponsor: U.S. Department of Health and Human Services
5600 Fishers Lane, Parklawn Building
Rockville, MD 20857

DHHS Emerging Leaders Program Internships 1756

The Emerging Leaders Program (ELP) is a competitive, two-year paid, federal internship with the U.S. Department of Health and Human Services (HHS). The Program provides a unique opportunity to develop enhanced leadership skills in one of the largest federal agencies in the nation. Upon successful completion of the ELP, participants will be eligible for non-competitive conversion to a permanent appointment. The ELP offers participants the following: formal competency-based leadership training and professional development; challenging developmental rotational assignments; mentorship; fast-paced and diverse work environment; and promotion to a permanent career track that targets the following mission-critical occupation specialties: scientific, social science, human resources, administrative, information technology, public health, and law enforcement. To begin the application process, applicants apply to one or more career track(s) that match their education and/or experience specialty. Later applicants may apply to the specific positions under their chosen career track(s). At the date of this writing, the website states that the ELP is not currently accepting applications, but encourages those interested to check back for future updates.
Requirements: Applicants must qualify for a minimum of one career track to move on to the next phase of the selection process.
Geographic Focus: All States
Contact: Grants Coordinator; (877) 696-6775; ELP@hhs.gov
Internet: http://hhsu.learning.hhs.gov/elp/howtoapply.asp
Sponsor: U.S. Department of Health and Human Services
200 Independence Avenue SW
Washington, DC 20201

DHHS Independence Demonstration Program 1757

The objective of the program is to provide for the establishment of demonstration projects designed to determine the social, civic, psychological, and economic effects of providing to individuals and families with limited means an incentive to accumulate assets by saving a portion of their earned income; the extent to which an asset-based policy that promotes saving for postsecondary education, homeownership, and microenterprise development may be used to enable individuals and families with limited means to increase their economic self-sufficiency; and the extent to which an asset-based policy stabilizes and improves families and the community in which the families live. Deadlines are announced in the Federal Register.
Requirements: Nonprofit 501(c)3 tax-exempt organizations and state or local government agencies or tribal governments submitting an application jointly with such a nonprofit organization are eligible to apply.
Geographic Focus: All States
Amount of Grant: 360,000 USD
Contact: James Gatz, (202) 401-4626; AFIProgram@acf.hhs.gov
Internet: http://www.acf.hhs.gov/grants/open/HHS-2004-ACF-OCS-EI-0027.html
Sponsor: U.S. Department of Health and Human Services
370 L'Enfant Promenade SW, Suite 500 W
Washington, DC 20447

DHHS Native American Programs 1758

To provide financial assistance, training and technical assistance, and research, demonstration and evaluation activities to public and private nonprofit organizations including Indian Tribes, urban Indian centers, Alaska Native villages, Native Hawaiian organizations, rural off-reservation groups, and Native American Pacific Island groups for the development and implementation of social and economic development strategies that promote self-sufficiency. These projects are expected to result in improved social and economic conditions of Native Americans within their communities and to increase the effectiveness of Indian Tribes and Native American organizations in meeting their economic and social goals. Additional competitive areas include (1) Environmental Regulatory Enhancement, designed to assist Tribal and Alaska Village governments in developing environmental programs responsive to tribal needs; and (2) Native Languages Preservation and Enhancement, a program to assist Native American tribes and communities in ensuring the survival and continued vitality of their languages.
Requirements: Public and private nonprofit agencies, including but not limited to, governing bodies of Indian tribes on Federal and State reservations, Alaska Native villages and regional corporations established by the Alaska Native Claims Settlement Act, such public and nonprofit private agencies serving Native Hawaiians, Indian and Alaska Native organizations in urban or rural nonreservation areas, and Native American Pacific Islanders (American Samoan Natives, and indigenous peoples of Guam, the Commonwealth of the Northern Mariana and the Republic of Palau) are eligible to apply.
Geographic Focus: All States
Contact: Sheila Cooper, Director of Program Operations; (202) 690-5787
Internet: http://www.acf.hhs.gov/programs/ana/programs/index.html
Sponsor: U.S. Department of Health and Human Services
370 L'Enfant Promenade SW, MS HHH 326-F
Washington, DC 20447

DHHS Oral Health Promotion Research Across the Lifespan 1759

The National Institute of Dental and Craniofacial Research has invited proposals for improving the oral health of people of all ages. The research team must include someone with extensive experience in health promotion, behavioral and/or social science research. The health promotion intervention proposed for funding must be based on a previously conducted assessment of the epidemiology, social, behavioral and/or environmental factors related to the disease or condition under study. Research could focus on maternal and child health, adolescent and young adult health, or health of adults with complex diseases. For example, an applicant might propose to study approaches to involving families, social networks, communities, or neighborhoods in behaviors that promote and improve oral health; improving patient-provider communication related to oral preventive measures; or effective ways to train oral health professional students to communicate with diverse patient populations.
Requirements: Any person with the skills, knowledge, and resources necessary to carry out the proposed research as the project director/principal investigator (PD/PI) is invited to work with his/her organization to develop an application for support. Applications must be submitted electronically through Grants.gov (http://www.grants.gov) using the SF424 research and related forms and the SF424 application guide.
Geographic Focus: All States
Date(s) Application is Due: Mar 5; Jul 5; Nov 5
Contact: Maria Teresa Canto, (301) 594-5497; fax (301) 480-8322; maria.canto@nih.gov
Internet: http://grants.nih.gov/grants/guide/pa-files/PA-07-225.html
Sponsor: National Institute of Dental and Craniofacial Research
45 Center Drive, Natcher Building
Bethesda, MD 20892

DHHS Promoting Safe and Stable Families Grants 1760

The National Center on Child Abuse and Neglect is interested in research on the impact of community-based family support and family preservation programs on child abuse and neglect. Research should focus on expanding the current knowledge base, build on previous research, and provide insights into new approaches to preventing child maltreatment and preserving families through support and preservation services. The center is particularly interested in projects that address specific populations and outcomes. The populations are families who receive family support services but have had no previous contact with child protective services; families who have been referred to child protective services whose cases were unsubstantiated but were found to need services and were referred to family support programs; families who have been in the system whose child abuse or neglect cases were substantiated, who received family preservation or support services, and whose cases are now closed; and families who have open cases whose children have not been removed and who are receiving family preservation services. Outcomes of interest are case finding, which involves families who were not referred to child protective services, and the impact of family support and/or preservation services on prevention, recidivism, and removal of children. Applicants should plan and design the proposed research in collaboration with state and local CPS and Title IV-B agencies as well as community-based entities providing family support services, such as family resource centers. Contact the office for deadline dates.
Requirements: Agencies of state and local governments, public and private nonprofit agencies, and institutions engaged in child and family welfare activities and research are eligible for these grants.

Geographic Focus: All States
Amount of Grant: 194,000-48,000,000 USD
Contact: Joseph Bock, Program Supervisor; (202) 205-8618
Internet: http://www.acf.hhs.gov/programs/cb/programs/fpfs.htm
Sponsor: U.S. Department of Health and Human Services
330 C Street SW, Room 2428
Washington, DC 20447

DHHS Special Programs for the Aging Training, Research, and Discretionary Projects and Grants 1761

Funds are given to provide adequately trained personnel in the field of aging, improve knowledge of the problems and needs of the elderly, and to demonstrate better ways of improving the quality of life for the elderly. Funds may be used to train persons to work in the field of aging, to increase the availability and accessibility of training and education programs in the field of aging, and to conduct activities for the development of knowledge to improve the circumstances of older people. Deadlines are posted in the Federal Register.
Requirements: Grants may be made to any public or nonprofit private agency, organization, or institution.
Restrictions: Grants are not available to individuals.
Geographic Focus: All States
Amount of Grant: 250,000 USD
Contact: Center for Planning and Policy Development; (202) 619-0724
Internet: http://aspe.hhs.gov/SelfGovernance/inventory/Aoa/048.htm
Sponsor: U.S. Department of Health and Human Services
330 Independence Avenue SW
Washington, DC 20201

DHHS Welfare Reform Research, Evaluations, and National Studies Grants 1762

The objectives of the funding are to support research on the benefits, effects, and costs of different welfare reform interventions; and studies such as on the effects of different programs on welfare dependency, illegitimacy, teen pregnancy, employment rates, child well-being, and related areas; and to assist in the development and evaluation of innovative approaches for reducing welfare dependency and increasing the well-being of minority children in welfare families. Grants, cooperative agreements, and contracts are awarded for innovative research, demonstrations, and evaluations that are responsive to the Administration for Children and Families program priorities. Deadlines for grants are announced in the Federal Register.
Requirements: Grants and cooperative agreements may be made to or with governmental entities, colleges, universities, nonprofit, and for-profit organizations (if fee is waived). Contracts may be awarded to nonprofit or for-profit organizations. Grants or cooperative agreements cannot be made directly to individuals.
Geographic Focus: All States
Amount of Grant: 10,000-6,000,000 USD
Contact: Karl Koerper, Office of Planning, Research, and Evaluation, Administration for Children and Families; (202) 401-4535; fax (202) 205-3598; KKoerper@acf.dhhf.gov
Internet: http://www.acf.hhs.gov/programs/opre
Sponsor: U.S. Department of Health and Human Services
370 L'Enfant Promenade SW
Washington, DC 20447

Diaz-Ayala Cuban and Latin American Popular Music Collection Travel Grants 1763

The Diaz-Ayala Cuban and Latin American Popular Music Collection is the most extensive publicly available collection of Cuban music in the United States. The grants will provide graduate students and scholars the opportunity to visit the Diaz-Ayala Music Collection at the FIU Green Library, thereby enhancing its value as a national resource. The program provides a research stipend to offset the costs of a minimum of one-week stay to use the collection. Application and guidelines are available online.
Requirements: Scholars in the humanities and the social sciences whose work will be enhanced by using the resources of the collection are encouraged to apply.
Geographic Focus: All States
Date(s) Application is Due: Feb 15
Amount of Grant: 1,500 USD
Contact: Alma DeRojas, Coordinator; (305) 348-1991; derojasa@fiu.edu
Internet: http://lacc.fiu.edu/centers_institutes/?body=centers_cri_whatsnew&rightbody=centers_cri
Sponsor: Florida International University
11200 SW 8th Street
Miami, FL 33199

Diversifying Higher Education Faculty in Illinois Program (DFI) 1764

The program aims to increase access of underrepresented students to graduate degree programs in Illinois public and private universities and, ultimately, to increase the number of underrepresented faculty and staff in Illinois colleges and universities. Each recipient is awarded an annual stipend and an institutional scholarship that covers tuition and fees. Doctoral recipients may receive awards for up to four years. Those in master's or professional degree programs may receive awards for up to two years.
Requirements: To be considered for an award, an applicant must be an Illinois resident and a United States citizen or permanent resident; a member of an underrepresented group in higher education such as African American, Hispanic, Asian American, or Native American; the recipient of an earned baccalaureate degree; of above-average academic ability as evidenced by admission to a graduate or professional degree program at a participating ICEOP institution; and unable to pursue a graduate or professional degree in the absence of an ICEOP award.
Geographic Focus: Illinois
Date(s) Application is Due: Feb 15
Amount of Grant: 12,500-16,000 USD
Contact: Program Contact; (618) 453-4558; fax (618) 453-5313; fellows@siu.edu
Internet: http://www.dfi.siu.edu
Sponsor: Southern Illinois University at Carbondale
Graduate School, Woody Hall C-224, MC 4723
Carbondale, IL 62901

DOD HBCU/MI Partnership Training Award 1765

This program's goal challenges the scientific community to design innovative research that will foster new directions for, address neglected issues in, and bring new investigators to the field of breast cancer research. The program focuses its funding on innovative projects, particularly those involving multidisciplinary and/or multi-institutional collaborations and alliances that have the potential to make a significant impact on breast cancer. Proposals that address the needs of minority, low-income, rural, and other underrepresented and/or medically underserved populations are strongly encouraged. The listed application deadline is for preproposals. Preproposals must be submitted through the CDMRP eReceipt system. Guidelines are available online.
Requirements: Applicants must be HBCU/MI doctoral-level faculty members. All individuals, regardless of ethnicity, nationality, or citizenship status, may apply as long as they are employed by, or affiliated with, an eligible HBCU/MI institution. Eligible institutions are those approved as HBCU/MIs by the Department of Education.
Geographic Focus: All States
Contact: Commander, U.S. Army Medical Research and Materiel Command; (301) 619-7079; fax (301) 619-7792; cdmrp.pa@det.amedd.army.mil
Internet: http://cdmrp.army.mil/bcrp/default.htm
Sponsor: U.S. Department of Defense
3701 North Fairfax Drive
Arlington, VA 21702-1714

DOE Academies Creating Teacher Scientists (ACTS) Grants 1766

This program is based upon research and best practices in teacher professional development and is aligned with the National Standards for Science Education, the National Teaching Standards, and National Board for Professional Teaching Standards. Instruction is offered to middle and high school teachers in four content areas: biotechnology-genetic engineering; fusion-astrophysics; and energy technologies. Each of these programs teaches knowledge and skills derived from science research and provides a context that when shared with students can help them understand how classroom science is linked to the big science. Teachers will have access to a wealth of mentoring talent that will guide and enrich their understanding of the scientific and technological world. Participants and mentors will create supportive relationships that can follow them into their classrooms. The Teacher Researchers (TR) model provides the teacher with four-weeks of hands-on instruction in a content discipline of their choosing in their initial year followed by two, eight-week summer research internships. Teachers in K-12 classrooms and community college faculty of science, technology, engineering, or mathematics are encouraged to apply. Housing allowances, transportation, and stipends are offered to participants at all the laboratories.
Requirements: All applicants must: be teaching full time in a public or private elementary or secondary school and continue to be full time employees of a school system throughout the duration of the program; have a current teaching assignment with at least 3/4 of classroom contact hours in science, mathematics, and/or technology (applies to secondary school teachers and community college instructors only); be at least 21 years old at the time the appointment starts; be U.S. citizens at the time of application; have current health insurance coverage; and commit to a minimum of 3 years involvement in the program.
Geographic Focus: All States
Date(s) Application is Due: Apr 1
Contact: Todd Clark, (202) 586-7174; fax (202) 586-0019; todd.clark@science.doe.gov
Internet: http://www.scied.science.doe.gov/SciEd/LSTPD/about.htm
Sponsor: U.S. Department of Energy
1000 Independence Avenue SW, Forrestal Building
Washington, DC 20585-1290

DOE Advanced Detector Research Grants 1767

The purpose of this program is to support the development of the new detector technologies needed to perform future high-energy physics experiments. Such experiments will require higher performance detectors to exploit the higher beam energies and intensities of new or upgraded accelerators. Higher performance detectors are also needed to probe for new physical processes in both accelerator- and nonaccelerator-based-experiments. Proposed detector research should be driven by the anticipated needs of experiments to be built within the foreseeable future, as well as upgrades to current experiments. Interesting technologies would include but not be limited to charged particle track detectors, calorimeters, or particle identification detectors that are less sensitive to radiation, have higher resolution, are lower in cost, or can be read out faster than currently available detectors. Applicants are requested to submit a Letter of Intent (LOI) by November 15 which includes the title of the proposal, the name of the principal investigator(s), the requested funding, and a one-page abstract. Failure to submit a letter of intent will not negatively prejudice a responsive formal application submitted in a timely manner. Full applications are due by December 12.
Requirements: Investigators who are currently involved in experimental high-energy physics are eligible to apply and should do so through a U.S. academic institution.
Geographic Focus: All States
Date(s) Application is Due: Nov 15; Dec 12
Amount of Grant: 10,000-600,000 USD

Contact: Dr. Saul Gonzalez, Program Manager; (301) 903-2359; fax (301) 903-2597; Saul.Gonzalez@science.doe.gov
Internet: http://www.science.doe.gov/grants/FAPN07-33.html
Sponsor: U.S. Department of Energy
19901 Germantown Road, MS FE-3
Germantown, MD 20874-1290

DOE Advanced Energy Projects/Laboratory Technology Research Grants 1768
The Advanced Energy Projects program funds research to establish the feasibility of novel, energy-related concepts. These concepts are usually derived from recent advances in basic research, but require additional research to establish their feasibility. A common theme for each concept is the initial linkage of new, or previously neglected, research results to a practical energy payoff. The Laboratory Technology Research program conducts high-risk, energy-related research that advances fundamental science and technology toward innovative applications that could significantly impact the United States' energy economy. Scientists at the Office of Science laboratories enter into cost-shared research partnerships with industry to explore energy applications of research advances in areas of mission relevance to both parties. The partners jointly bring technology research to a point where industry or the department's technology development programs can pursue final development or commercialization. Applications are available on the Web site and may be submitted at any time.
Geographic Focus: All States
Date(s) Application is Due: May 31
Amount of Grant: 10,000-2,000,000 USD
Contact: Martin Rubinstein, Grants and Contracts Division; (301) 903-5212
Internet: http://www.er.doe.gov/production/grants/guide.html
Sponsor: U.S. Department of Energy
19901 Germantown Road, MS FE-3
Germantown, MD 20874-1290

DOE Advanced Nuclear Energy Systems Basic Research Grants 1769
Areas of focus in this program include: understanding of nanoscale interactions under extreme conditions; mastering the behavior of actinides and of fission products; solution behavior under extreme conditions of radiation and temperature; and interfacial behavior under extreme environmental conditions. Research funded under this initiative will pursue breakthroughs in scientific understanding that will advance materials design, will improve characterization of materials and processes, will enhance chemical processes under the extreme conditions present in nuclear energy systems, and will extend interdisciplinary theory-modeling-simulation-experimentation methodology to surmount the existing scientific and technical barriers for nuclear energy systems of the future. Pre-applications are required, and are due no later than November 22. A response to the pre-applications encouraging or discouraging formal applications will be communicated to the applicants by January 4, and formal applications are due March 14.
Requirements: Eligible to apply are institutions of higher education, industry (particularly small and disadvantaged businesses), and nonprofit institutions. Unsolicited research proposals are screened by program officials and, if appropriate, are evaluated by peer review. Successful proposers are awarded one-year or multi-year special research contracts or grants with renewals or incremental funding available, provided performance has been satisfactory. A one- to two-year phase-out period is normally granted to those finishing research projects. Equipment items may be purchased from contract funds with the agreement of DOE.
Geographic Focus: All States
Date(s) Application is Due: Mar 14; Nov 22
Amount of Grant: 10,000-2,000,000 USD
Contact: Tim Fitzsimmons, (301) 903-9830; tim.fitzsimmons@science.doe.gov
Internet: http://www.science.doe.gov/grants/FAPN07-04.html
Sponsor: U.S. Department of Energy
19901 Germantown Road, MS FE-3
Germantown, MD 20874-1290

DOE Albert Einstein Distinguished Educator Fellowship Program 1770
The Albert Einstein Distinguished Educator Fellowship Act was signed into law in November 1994. The law gives the Department of Energy responsibility for administering the program of distinguished educator fellowships for elementary and secondary school mathematics and science teachers. Selected teachers spend up to one year in a Congressional Office or a federal agency. Agencies that have participated include: the Department of Energy (DOE), the National Science Foundation (NSF), the National Aeronautics and Space Administration (NASA), the National Institutes of Health (NIH), the Department of Education (ED), National Institute of Standards and Technology (NIST), the White House Office of Science and Technology Policy (OSTP) and the National Oceanic and Atmospheric Administration (NOAA). The Fellows provide their educational expertise, years of experience and personal insights to these offices. Costs for relocation will be reimbursed based on the distance of current location to Washington DC. During the Fellowship, awardees receive funds for travel to conferences, workshops, and other professional development opportunities.
Requirements: To be eligible, applicants must be U.S. citizens at the time of selection; have a minimum of five years full-time classroom teaching experience; be teaching full-time in a public or private elementary or secondary school; have a current teaching assignment with at least 3/4 of classroom contact hours in science, mathematics, and/or technology (applies to secondary school teachers only); and provide a recommendation from a current school administrator and two additional recommendations.
Geographic Focus: All States
Date(s) Application is Due: Jan 13
Amount of Grant: 5,000-60,000 USD
Contact: Todd Clark, (202) 586-7174; fax (202) 586-0019; todd.clark@science.doe.gov
Internet: http://www.scied.science.doe.gov/scied/Einstein/about.htm
Sponsor: U.S. Department of Energy
1000 Independence Avenue SW, Forrestal Building
Washington, DC 20585-1290

DOE Atomic, Molecular, and Optical Sciences Research Grants 1771
The AMOS Program supports a balanced portfolio of experiment and theory to study the fundamental properties of atoms, ions and small molecules and the interactions between electrons, photons and ions in collisions with atoms, molecules and surfaces. Research is focused on the most complete quantum mechanical description of these properties and interactions and is intended to provide a basic understanding of physical processes. The AMOS program plays an underpinning role in relation to other programs within BES in Chemical and Materials Sciences, in relation to current and future BES facilities in which matter is probed with photons, electrons or heavy ions, and in relation to applied efforts in plasma science. AMOS also contributes at the most fundamental level to the science-based optimization of current energy sources and the development of new ones. Some topics of current interest: studies of the interactions of intense electromagnetic fields, induced by highly charged ions or lasers, with atoms and molecules; coherent control of quantum mechanical processes; development and application of ultrafast x-ray light sources, both laser-based and at BES synchrotron facilities; and theory and experiment on ultracold collisions and quantum condensates.
Requirements: Eligible to apply are institutions of higher education, industry (particularly small and disadvantaged businesses), and nonprofit institutions. Unsolicited research proposals are screened by program officials and, if appropriate, are evaluated by peer review. Successful proposers are awarded one-year or multi-year special research contracts or grants with renewals or incremental funding available, provided performance has been satisfactory. A one- to two-year phase-out period is normally granted to those finishing research projects. Equipment items may be purchased from contract funds with the agreement of DOE.
Geographic Focus: All States
Amount of Grant: 10,000-2,000,000 USD
Contact: Dr. Eric A. Rohlfing, Director; (301) 903-8165; fax (301) 903-4110; Eric.Rohlfing@science.doe.gov
Internet: http://www.science.doe.gov/bes/chm/Programs/programs.html
Sponsor: U.S. Department of Energy
1000 Independence Avenue SW, Forrestal Building
Washington, DC 20585-1290

DOE Basic Energy Sciences (BES) Grants 1772
The mission of the program is to foster and support fundamental research to expand the scientific foundations for new and improved energy technologies and for understanding and mitigating the environmental impacts of energy use. Research is sponsored in a variety of disciplines to broaden the energy supply and technological base knowledge. The disciplines represented are biosciences, chemical sciences, geosciences, engineering, and materials sciences. The program also supports exploratory research on advanced energy projects. Research projects are selected for funding on the basis of scientific merit, the possible relevance to meeting BES long-range research goals, and the contribution toward a balanced, responsive research program. The detailed substance of the program is determined by the selection from among many unsolicited proposals of ideas from the university/scientific community. Applications are accepted at any time.
Requirements: Eligible to apply are institutions of higher education, industry (particularly small and disadvantaged businesses), and nonprofit institutions. Unsolicited research proposals are screened by program officials and, if appropriate, are evaluated by peer review. Successful proposers are awarded one-year or multiyear special research contracts or grants with renewals or incremental funding available, provided performance has been satisfactory. A one- to two-year phase-out period is normally granted to those finishing research projects. Equipment items may be purchased from contract funds with the agreement of DOE.
Geographic Focus: All States
Amount of Grant: 10,000-2,000,000 USD
Contact: Martin R. Rubinstein, Supervisor of Grants and Contracts; (301) 903-4946; fax (301) 903-4194; Martin.Rubinstein@science.doe.gov
Internet: http://www.science.doe.gov/grants/FAPN08-01.html
Sponsor: U.S. Department of Energy
19901 Germantown Road, MS FE-3
Germantown, MD 20874-1290

DOE Catalysis and Chemical Transformations Research Grants 1773
This Program supports basic research to understand the chemical aspects of catalysis, both heterogeneous and homogeneous; the chemistry of fossil resources; and the chemistry the molecules used to create advanced materials. Catalysts are crucial to energy conservation in creating new, less-energy- demanding routes for the production of basic chemical feedstocks and value-added chemicals. Catalysts are also indispensable for processing and manufacturing fuels that are a primary means of energy storage. Results from a fundamental, molecular-level understanding of the syntheses of advanced catalytic materials have the potential of providing new chemicals or materials that can be fabricated with greater energy efficiency or function as energy-saving media themselves. This activity is the Nation's major supporter of catalysis research, and it is the only activity that treats catalysis as a discipline integrating all aspects of homogeneous and heterogeneous catalysis research.
Requirements: Eligible to apply are institutions of higher education, industry (particularly small and disadvantaged businesses), and nonprofit institutions. Unsolicited research proposals are screened by program officials and, if appropriate, are evaluated by peer review. Successful proposers are awarded one-year or multi-year special research contracts

or grants with renewals or incremental funding available, provided performance has been satisfactory. A one- to two-year phase-out period is normally granted to those finishing research projects. Equipment items may be purchased from contract funds with the agreement of DOE.
Geographic Focus: All States
Amount of Grant: 10,000-2,000,000 USD
Contact: Dr. Eric A. Rohlfing, Director; (301) 903-8165; fax (301) 903-4110; Eric.Rohlfing@science.doe.gov
Internet: http://www.science.doe.gov/bes/chm/Programs/programs.html
Sponsor: U.S. Department of Energy
1000 Independence Avenue SW, Forrestal Building
Washington, DC 20585-1290

DOE Chemical Energy and Chemical Engineering Research Grants **1774**
This activity supports research on electrochemistry, thermophysical and thermochemical properties, and physical and chemical rate processes. Emphasis is given to improving and/or developing the scientific base for engineering generalizations and their unifying theories. Also included is fundamental research in areas critical to understanding the underlying limitations in the performance of electrochemical energy storage and conversion systems including anode, cathode, and electrolyte systems and their interactions with emphasis on improvements in performance and lifetime. The program covers a broad spectrum of research including fundamental studies of composite electrode structures; failure and degradation of active electrode materials; thin film electrodes, electrolytes, and interfaces; and experimental and theoretical aspects of phase equilibria, especially of mixtures, including supercritical phenomena.
Requirements: Eligible to apply are institutions of higher education, industry (particularly small and disadvantaged businesses), and nonprofit institutions. Unsolicited research proposals are screened by program officials and, if appropriate, are evaluated by peer review. Successful proposers are awarded one-year or multi-year special research contracts or grants with renewals or incremental funding available, provided performance has been satisfactory. A one- to two-year phase-out period is normally granted to those finishing research projects. Equipment items may be purchased from contract funds with the agreement of DOE.
Geographic Focus: All States
Amount of Grant: 10,000-2,000,000 USD
Contact: Dr. Eric A. Rohlfing, Director; (301) 903-8165; fax (301) 903-4110; Eric.Rohlfing@science.doe.gov
Internet: http://www.science.doe.gov/bes/chm/Programs/programs.html
Sponsor: U.S. Department of Energy
1000 Independence Avenue SW, Forrestal Building
Washington, DC 20585-1290

DOE Chemical Physics Research Grants **1775**
The program supports basic research on fundamental molecular processes related to the mission of the Department in such areas as combustion, catalysis, and environmental restoration. It is the Nation's principal supporter of high temperature chemical kinetics and gas phase chemical physics. Specific areas of research emphasis include, but are not limited to: gas phase chemical reaction theory, computational chemistry, experimental dynamics and spectroscopy, thermodynamics of reaction intermediates, chemical kinetics and reaction mechanisms at high temperatures in the gas phase and at surfaces, combustion diagnostics, and chemical dynamics and kinetics at surfaces and with metal and semiconductor clusters.
Requirements: Eligible to apply are institutions of higher education, industry (particularly small and disadvantaged businesses), and nonprofit institutions. Unsolicited research proposals are screened by program officials and, if appropriate, are evaluated by peer review. Successful proposers are awarded one-year or multi-year special research contracts or grants with renewals or incremental funding available, provided performance has been satisfactory. A one- to two-year phase-out period is normally granted to those finishing research projects. Equipment items may be purchased from contract funds with the agreement of DOE.
Geographic Focus: All States
Amount of Grant: 10,000-2,000,000 USD
Contact: Dr. Eric A. Rohlfing, Director; (301) 903-8165; fax (301) 903-4110; Eric.Rohlfing@science.doe.gov
Internet: http://www.science.doe.gov/bes/chm/Programs/programs.html
Sponsor: U.S. Department of Energy
1000 Independence Avenue SW, Forrestal Building
Washington, DC 20585-1290

DOE Chemical Sciences Research Grants **1776**
The Chemical Sciences program supports a major portion of the nation's fundamental research in the chemical sciences. The research covers those areas of the chemical sciences that impact the Department's energy and environmental missions. Research supported covers atomic, molecular and optical (AMO) sciences; chemical physics; photo- and radiation chemistry; surface chemistry and heterogeneous catalysis; organometallic chemistry and homogeneous catalysis; analytical and separation science; heavy element chemistry; and aspects of chemical engineering sciences. This research provides a foundation for fundamental understanding of the interactions of atoms, molecules, and ions with photons and electrons; the making and breaking of chemical bonds in gas phase, in solutions, at interfaces, and on surfaces; and understanding the energy transfer processes within and between molecules. Applications are accepted at any time.
Requirements: Eligible to apply are institutions of higher education, industry (particularly small and disadvantaged businesses), and nonprofit institutions. Unsolicited research proposals are screened by program officials and, if appropriate, are evaluated by peer review. Successful proposers are awarded one-year or multi-year special research contracts or grants with renewals or incremental funding available, provided performance has been satisfactory. A one- to two-year phase-out period is normally granted to those finishing research projects. Equipment items may be purchased from contract funds with the agreement of DOE.
Geographic Focus: All States
Amount of Grant: 10,000-2,000,000 USD
Contact: Dr. Eric A. Rohlfing, Director; (301) 903-8165; fax (301) 903-4110; Eric.Rohlfing@science.doe.gov
Internet: http://www.science.doe.gov/bes/chm/chmhome.html
Sponsor: U.S. Department of Energy
1000 Independence Avenue SW, Forrestal Building
Washington, DC 20585-1290

DOE Coal Liquefaction Research Grants **1777**
The objectives of the program are to develop advanced technology and an engineering and operations database sufficient for the production of marketable liquid fuels from coal by the year 2015 at a cost competitive with crude oil, between $25 and $30 per barrel. The implementation strategy is based on catalytic staged direct liquefaction, coprocessing of coal/waste and petroleum resid, slurry-phase indirect liquefaction technology, and advanced processes based on exploratory research. Areas covered by the program include direct/indirect coal liquefaction processes, coprocessing of coal-petroleum mixture/coal-waste mixture, refining and chemicals, and supporting research. Proposals are accepted at any time.
Requirements: States, local governments, universities, governmental entities, consortia, nonprofit institutions, commercial corporations, joint federal/industry corporations, U.S. territories, and individuals are eligible to apply.
Geographic Focus: All States
Amount of Grant: 50,000-400,000 USD
Contact: Fred Glaser; (301) 903-2786
Internet: http://www.fe.doe.gov
Sponsor: U.S. Department of Energy
19901 Germantown Road, MS FE-3
Germantown, MD 20874-1290

DOE Combustion Systems Research Grants **1778**
The program emphasizes the development of technologies which would allow more cost-effective and efficient coal use. The objectives of this program include developing the technology necessary for electric power applications with lower cost, increased efficiency, and emission significantly lower than current NSPS. Applications are accepted at any time.
Requirements: States, local governments, universities, governmental entities, consortia, nonprofit institutions, commercial corporations, joint federal/industry corporations, U.S. territories, and individuals are eligible to apply.
Geographic Focus: All States
Amount of Grant: 10,000-25,000,000 USD
Contact: Mary Roland, Department of Energy; (301) 903-3514
Internet: http://www.fe.doe.gov
Sponsor: U.S. Department of Energy
19901 Germantown Road, MS FE-3
Germantown, MD 20874-1290

DOE Competitive Financial Assistance for Renewable Energy and Energy Efficiency Programs **1779**
DOE announces a competitive solicitation for information dissemination, public outreach, training, and related technical analysis and technical assistance activities involving renewable energy and energy efficiency. Areas of interest involving renewable energy include wind, photovoltaic, hydrogen, and bioenergy technologies. Energy efficiency areas of interest include energy efficiency in the transportation, buildings, and industrial sectors.
Requirements: Profit organizations, nonprofit institutions and organizations, state and local governments, universities, individuals, Native American organizations, and Alaskan Native corporations are eligible to apply.
Geographic Focus: All States
Amount of Grant: Up to 6,000,000 USD
Contact: James Damm, Contract Specialist; (303) 275-4788; gostate@nrel.gov
Internet: http://www.tgci.com/fedrgtxt/01-10503.txt
Sponsor: U.S. Department of Energy
1000 Independence Avenue SW, Forrestal Building
Washington, DC 20585-1290

DOE Computational Science Graduate Fellowships **1780**
Sponsored by the DOE Office of Science and Office Defense Programs and administered by the institute, the program supports highly capable students pursuing graduate study at U.S. universities in scientific or engineering disciplines with applications in high-performance computing. Fellows also participate in off-campus research at DOE laboratories. The program offers an annual stipend and payment of tuition and fees for graduate study in scientific and technical disciplines using computational science methods. Appointments are reviewed annually and may be renewed up to a limit of four years.
Requirements: The program is open to U.S. citizens and permanent residents who are in their first or second year of graduate school and working toward a PhD at the time of application. Exceptional senior undergraduates who can meet all of the requirements listed in the application booklet also may apply.
Geographic Focus: All States
Date(s) Application is Due: Jan 11
Amount of Grant: 28,000 USD
Contact: Coodinator; (515) 956-3696; csgf@krellinst.org

Internet: http://www.krellinst.org/csgf
Sponsor: Krell Institute
1609 Golden Aspen Drive, Suite 101
Ames, IA 50010

DOE Control Technology and Coal Preparation Research Grants 1781
The activities of the program emphasize the development of technologies which will allow greater use of coal in an environmentally acceptable manner. These technologies include coal characterization and cleaning, combined flue gas cleanup, processes for cleanup of hot coal combustion and gasification streams, and management of wastes generated in coal conversion and utilization. Applications are accepted at any time.
Requirements: States, local governments, universities, governmental entities, consortia, nonprofit institutions, commercial corporations, joint federal/industry corporations, U.S. territories, and individuals are eligible to apply.
Geographic Focus: All States
Amount of Grant: 10,000-25,000,000 USD
Contact: Mary Roland, Department of Energy; (301) 903-3514
Internet: http://www.fe.doe.gov
Sponsor: U.S. Department of Energy
19901 Germantown Road, MS FE-3
Germantown, MD 20874-1290

DOE Electric Energy Systems Research Grants 1782
The program assists the private sector in the development of advanced technology options for the nation's electric energy networks; methodologies for integrating new technologies into electric utility systems; and analytical procedures to assess and enhance the stability and efficiency of the country's increasingly complex interconnected electric energy system. Supported are innovative research in such areas as high-voltage systems, new materials, and sophisticated control methods. Basic research is supported on increasing the efficiency of electric energy supply and distribution, thus reducing primary energy use per unit of real GNP. Subprograms include reliable electric power delivery, electric field effects, and systems technologies. Proposals are accepted at any time.
Geographic Focus: All States
Amount of Grant: 50,000-500,000 USD
Contact: Office of Utility Technologies; (202) 586-4142
Internet: http://www.eere.energy.gov
Sponsor: U.S. Department of Energy
1000 Independence Avenue SW, Forrestal Building
Washington, DC 20585-1290

DOE Energy-Related Inventions Grants 1783
Grants encourage innovation in developing nonnuclear energy technology by providing assistance in the development of promising energy-related inventions. Assistance provided includes evaluation of energy-related inventions, limited funding assistance where appropriate, advice concerning engineering, marketing, and business planning.
Requirements: There are no restrictions on eligibility. Small businesses, individual inventors, and entrepreneurs are especially invited to participate.
Geographic Focus: All States
Amount of Grant: 50,000-500,000 USD
Contact: Lisa Barnett, Department of Energy; (202) 586-2212
Internet: http://www.eere.energy.gov/inventions
Sponsor: U.S. Department of Energy
1000 Independence Avenue SW, Forrestal Building
Washington, DC 20585-1290

DOE Energy Biosciences Research Grants 1784
To provide financial support for fundamental research in the basic sciences and advanced technology concepts, and assessments in fields related to energy. Financial support, in whole or in part, may be provided for such purposes as the salaries, materials and supplies, equipment, travel, publication costs, and services required for conducting research, related activities, and advanced technology projects or assessments. Restrictions on use of funds depend on grant provisions. Applications may be submitted at any time. Specific grant solicitation notices, issued from time to time, usually contain due dates. Applicants are encouraged to contact the office regarding specific grant due dates and other requirements.
Requirements: Colleges and universities, non-profit organizations, for-profit commercial organizations, State and local governments, and unaffiliated individuals.
Geographic Focus: All States
Amount of Grant: 10,000-2,000,000 USD
Contact: Martin Rubinstein, Grants and Contracts Division; (301) 903-5212
Internet: http://www.cfda.gov/public/viewprog.asp?progid=875
Sponsor: U.S. Department of Energy
19901 Germantown Road, MS FE-3
Germantown, MD 20874-1290

DOE Energy Conservation Research and Development Grants 1785
This program supports scientific and technical research in fields where new knowledge can expand the generic technology base underlying energy conservation. This will allow the private sector to develop ways to make more efficient use of available energy and to facilitate use of nonpetroleum energy sources. Programs include basic and applied research in fields of scientific inquiry that can provide a general technology base that the private sector will be able to use in developing advanced conservation applications and actually getting these technology applications into the marketplace. Applications are accepted at any time.
Requirements: Eligible applicants include profit organizations, private nonprofit institutions/organizations, and state and local governments.
Geographic Focus: All States
Amount of Grant: 50,000-500,000 USD
Contact: Office of Industrial Technologies; (202) 586-0098
Internet: http://www.eere.energy.gov
Sponsor: U.S. Department of Energy
1000 Independence Avenue SW, Forrestal Building
Washington, DC 20585-1290

DOE Energy Conversion Technology Research Grants 1786
Under the Energy Conversion and Utilization Technologies Program, the energy conversion technology subprogram's problem areas that require additional research include thermodynamics and fluid mechanics of closed system energy conversion, combustion processes of internal combustion engines, and physical and chemical aspects of alternative fuels for furnaces and boilers. Potential areas for new research include development of a complete understanding of velocities and turbulence levels at various points during engine operation and how they can be controlled to appropriately tailor combustion processes of advanced engine concepts; complete incorporation of technology base information and development of validated multidimensional large-scale computer models of complete engines for use in solving specific design problems; assemblage of an adequate database of physical and chemical properties for select alternative fuels and experimental evaluation of several combustion parameters of interest to designers of furnaces and boilers; and experimental and analytical investigation of thermodynamic and fluid dynamic aspects of advanced closed-cycle systems and components. Proposals are accepted at any time.
Geographic Focus: All States
Amount of Grant: 50,000-500,000 USD
Contact: Office of Industrial Technologies; (202) 586-0098
Internet: http://www.eere.energy.gov
Sponsor: U.S. Department of Energy
1000 Independence Avenue SW, Forrestal Building
Washington, DC 20585-1290

DOE Energy from Municipal Waste Projects Grants 1787
Grants support research to provide a database on liquids and gas production and to provide a basis for environmental and feedstock preparation. Potential areas for new research initiatives include thermochemical studies to provide a generic technology base for systems producing liquid and gaseous fuels; combustion tasks concerned with kinetics, efficiency, and products formation for corrosion and pollution control; pyrolysis tasks concerned with base data development on effects of various feedstock components and of time, temperature, pressure, and so forth; biochemical studies to provide data on biochemistry and life support systems of anaerobes; water and wastewater studies to provide information to improve energy efficiency in treatment systems; mechanical studies to provide generic data on systems design; and improved densification techniques and binders. Proposals are accepted at any time.
Geographic Focus: All States
Amount of Grant: 50,000-500,000 USD
Contact: Office of Utility Technologies; (202) 586-4142
Internet: http://www.eere.energy.gov
Sponsor: U.S. Department of Energy
1000 Independence Avenue SW, Forrestal Building
Washington, DC 20585-1290

DOE Energy Research Analyses Grants 1788
This program supports energy research analyses of the department's basic and applied research activities. Specific objectives include assessments to identify any duplication or gaps in scientific research activities, and impartial and independent evaluations of scientific and technical research efforts. Applications are accepted at any time and are available on the Web site.
Requirements: Profit organizations, nonprofit institutions, intrastate, interstate, local agencies, and universities are eligible for funding. Preapplication coordination is recommended for unsolicited proposals.
Geographic Focus: All States
Amount of Grant: 10,000-2,000,000 USD
Contact: Martin Rubinstein, Grants and Contracts Division; (301) 903-5212
Internet: http://www.er.doe.gov/production/grants/guide.html
Sponsor: U.S. Department of Energy
1000 Independence Avenue SW, Forrestal Building
Washington, DC 20585-1290

DOE Engineering Research Grants 1789
The objectives of the engineering research program are to extend the body of knowledge underlying current engineering practice in order to open new ways for enhancing energy savings and production, prolonging useful equipment life, and reducing costs while maintaining output and performance quality; and to broaden the technical and conceptual base for solving future engineering problems in the energy technologies. Long-term research topics of current interest include foundations of bioprocessing fuels and energy related wastes; fracture mechanisms; experimental and theoretical studies of multiphase flows; intelligent machines; and diagnostics and control for plasma processing of materials.
Geographic Focus: All States
Date(s) Application is Due: May 31
Amount of Grant: 10,000-2,000,000 USD
Contact: Martin Rubinstein, Grants and Contracts Division; (301) 903-5212

Internet: http://www.er.doe.gov/production/grants/guide.html
Sponsor: U.S. Department of Energy
19901 Germantown Road, MS FE-3
Germantown, MD 20874-1290

DOE Environmental Processes Research Grants 1790

This program addresses global environmental change from increases in atmospheric carbon dioxide and other greenhouse gases. The scope of the global change program encompasses the carbon cycle, climate modeling and diagnostics, atmospheric sciences and meteorology, ecosystem responses, and impacts on resources. The role of clouds and radiation in climate prediction is a particular emphasis. Applications are available on the Web site and are accepted at any time.
Geographic Focus: All States
Amount of Grant: 10,000-2,000,000 USD
Contact: Martin Rubinstein, Grants and Contracts Division; (301) 903-5212
Internet: http://www.er.doe.gov/production/grants/guide.html
Sponsor: U.S. Department of Energy
19901 Germantown Road, MS FE-3
Germantown, MD 20874-1290

DOE Experimental Program to Stimulate Competitive Research (EPSCoR) 1791

The objective of the EPSCoR program is to enhance the capabilities of EPSCoR states to conduct nationally competitive energy-related research and to develop science and engineering manpower to meet current and future needs in energy-related fields. This program addressees research needs across all of the Department of Energy research interests. Research supported by the EPSCoR program is concerned with the same broad research areas addressed by the Office of Science programs that are described above. States eligible for DOE/EPSCoR support include: Alabama, Alaska, Arkansas, Delaware, Hawaii, Idaho, Kansas, Kentucky, Louisiana, Maine, Mississippi, Montana, Nebraska, Nevada, New Mexico, North Dakota, Oklahoma, South Carolina, South Dakota, Tennessee, Vermont, West Virginia, Wyoming, Commonwealth of Puerto Rico, and the Virgin Islands. Applicants are encouraged, but not required, to submit a preliminary proposal to DOE by April.
Geographic Focus: Alabama
Date(s) Application is Due: Sep 21
Amount of Grant: Up to 750,000 USD
Contact: Dr. Matesh Varma, Office of Science; (301) 903-3209; fax (301) 903-9513; matesh.varma@science.doe.gov
Internet: http://www.science.doe.gov/bes/EPSCoR/OVER1.HTM
Sponsor: U.S. Department of Energy
19901 Germantown Road, MS FE-3
Germantown, MD 20874-1290

DOE Faculty and Student Teams (FaST) Grants 1792

The program provides hands-on research opportunities in DOE national laboratories during the summer, and will support a team comprised of one faculty member and 2-3 undergraduate students. The faculty member identifies a mutually beneficial research area amenable to collaboration by the faculty member and the laboratory scientist. Potential areas of collaboration are based upon the Project Descriptions described at the specific DOE Office of Science laboratory. Faculty from colleges and universities with limited research facilities and those institutions serving populations, women, and minorities underrepresented in the fields of science, engineering, and technology are encouraged to apply. The amounts targeted for the student stipends are $4,500 for each student (allocated as ten weekly stipends of $400, and up to $500 for travel), and faculty stipends up to 2/9 academic year salary (up to $12,000) for faculty team members. Both faculty and student team members will receive funding assistance with travel. Student team members will receive funding assistance with housing.
Requirements: Faculty applicants must be United States Citizens or Permanent Resident Aliens. Student team members must: be currently enrolled as an undergraduate student and completed at least one semester of college course work; be 18 years or older at the start of the program; have earned a high school diploma or GED; be United States Citizens or Permanent Resident Aliens; participate in a maximum of two FaST internships; and have coverage under a health insurance plan.
Geographic Focus: All States
Date(s) Application is Due: Feb 1
Contact: Todd Clark, (202) 586-7174; fax (202) 586-0019; todd.clark@science.doe.gov
Internet: http://www.scied.science.doe.gov/scied/PST/about.htm
Sponsor: U.S. Department of Energy
1000 Independence Avenue SW, Forrestal Building
Washington, DC 20585-1290

DOE Fossil Advanced Research and Technology Development Grants 1793

The Advanced Research and Technology Development (AR&TD) Program is directed toward the scientific and technical areas that underlie the development of all fossil energy coal technologies. The AR&TD coal science program conducts fundamental research into coal combustion and conversion and addresses scientific and engineering problems that are barriers to fossil energy technological goals. Areas covered by current projects include materials research; fluids and solids handling; instrumentation and diagnostics; bioprocessing of coal; component development; structure and reactions of coal; multiphase flow and thermodynamics; fundamental problems in combustion of pulverized coal, synfuels, and coal-liquid mixtures; environmental aspects of coal use and conversion; control technology and coal preparation; coal gasification; and coal liquefaction. Applications are accepted at any time.
Requirements: States, local governments, universities, governmental entities, consortia, nonprofit institutions, commercial corporations, joint federal/industry corporations, U.S. territories, and individuals are eligible to apply.
Geographic Focus: All States
Amount of Grant: 10,000-25,000,000 USD
Contact: Mary Roland, Department of Energy; (301) 903-3514
Internet: http://www.fe.doe.gov
Sponsor: U.S. Department of Energy
19901 Germantown Road, MS FE-3
Germantown, MD 20874-1290

DOE Fossil Energy Research and Development Grants 1794

The fossil energy program's role is to support long-term research toward providing an improved capability to convert coal and oil shale to liquid and gaseous fuels; increase domestic production of coal, oil, and gas; ensure that current and new facilities that burn coal can do so in an economically feasible and environmentally acceptable manner; and allow more efficient and more economically attractive utilization of fossil energy resources. Applications are accepted at any time.
Requirements: States, local governments, universities, recognizant industrial governments, U.S. territories, and individuals are eligible to apply.
Geographic Focus: All States
Amount of Grant: 10,000-25,000,000 USD
Contact: Mary Roland, Department of Energy; (301) 903-3514
Internet: http://www.fe.doe.gov
Sponsor: U.S. Department of Energy
19901 Germantown Road, MS FE-3
Germantown, MD 20874-1290

DOE Fusion Energy Science Division Grants 1795

Basic and applied research is carried out in the following areas: (1) basic plasma science research directed at furthering the understanding of fundamental processes in plasmas; (2) improving the theoretical understanding of fusion plasmas necessary for interpreting results from present experiments and the planning and design of future confinement devices; (3) obtaining the critical data on plasma properties, atomic physics, and new diagnostic techniques for support of confinement experiments; (4) supporting exploratory research into concepts that are alternatives to the tokamak; and (5) carrying out research on issues that support the development of inertial fusion energy, for which target development is carried out by the Department of Energy's defense programs. Application forms are available on the Web site. Applications may be submitted at any time.
Requirements: Unsolicited research proposals should be preceded by an informal preproposal of no more than two or three pages that briefly describes the research project and the proposed level of funding.
Geographic Focus: All States
Date(s) Application is Due: May 31
Amount of Grant: 10,000-2,000,000 USD
Contact: Martin Rubinstein, Grants and Contracts Division; (301) 903-5212
Internet: http://www.er.doe.gov/production/grants/guide.html
Sponsor: U.S. Department of Energy
19901 Germantown Road, MS FE-3
Germantown, MD 20874-1290

DOE Fusion Energy Science Technology Division Grants 1796

The technology division's science-oriented goal is to provide the technologies that are required to successfully design, build, and operate near-term experiments aimed at producing, understanding, and optimizing the fusion energy process. The division's energy-oriented goal is to develop the technologies that will be needed in the long term for an economically and environmentally attractive fusion energy source. These goals are pursued through multi-institutional domestic programs and international collaboration partnerships. Applications are accepted at any time and are available on the Web site.
Requirements: Unsolicited research proposals should be preceded by an informal preproposal of no more than two or three pages that briefly describes the research project and the proposed level of funding.
Geographic Focus: All States
Date(s) Application is Due: May 31
Contact: Martin Rubinstein, Grants and Contracts Division; (301) 903-5212
Internet: http://www.science.doe.gov/production/grants/grants.html
Sponsor: U.S. Department of Energy
19901 Germantown Road, MS FE-3
Germantown, MD 20874-1290

DOE Geosciences Research Grants 1797

The Program supports research aimed at developing an understanding of fundamental Earth processes that can be used as a foundation for efficient, effective, and environmentally sound use of energy resources, and provide an improved scientific basis for advanced energy and environmental technologies. The program funds projects that develop a quantitative and predictive understanding of the energy-related aspects of processes within the earth. Emphasis is on the upper levels of the earth's crust and the focus is on geophysics and geochemistry of rock-fluid systems and interactions. Specific topical areas receiving high emphasis include high-resolution geophysical imaging; rock physics; fundamental properties of rocks, minerals, and fluids; scientific drilling; and sedimentary basin systems. The resulting improved understanding and knowledge base are needed to assist efforts in the utilization of the nation's energy resources in an environmentally acceptable fashion. Applications are accepted at any time and are available on the Web site.

Requirements: Eligible to apply are institutions of higher education, industry (particularly small and disadvantaged businesses), and nonprofit institutions. Unsolicited research proposals are screened by program officials and, if appropriate, are evaluated by peer review. Successful proposers are awarded one-year or multi-year special research contracts or grants with renewals or incremental funding available, provided performance has been satisfactory. A one- to two-year phase-out period is normally granted to those finishing research projects. Equipment items may be purchased from contract funds with the agreement of DOE.
Geographic Focus: All States
Amount of Grant: 10,000-2,000,000 USD
Contact: Dr. Eric A. Rohlfing; (301) 903-8165; Eric.Rohlfing@science.doe.gov
Internet: http://www.science.doe.gov/bes/geo/geohome.html
Sponsor: U.S. Department of Energy
19901 Germantown Road
Germantown, MD 20874-1290

DOE Heavy Element Chemistry Research Grants 1798

This Program supports research in actinide and fission product chemistry. Areas of interest include aqueous and non-aqueous coordination chemistry; solution and solid-state speciation and reactivity; measurement of chemical and physical properties; synthesis of actinide-containing materials; chemical properties of the heaviest actinide and transactinide elements; theoretical methods for the prediction of heavy element electronic and molecular structure and reactivity; and the relationship between the actinides, lanthanides, and transition metals. This activity represents the Nation's only funding for basic research in the chemical and physical principles of actinide and fission product materials. The program is primarily based at the national laboratories because of the special licenses and facilities needed to obtain and safely handle radioactive materials. However, research in heavy element chemistry is supported at universities, and collaborations between university and laboratory programs are encouraged. The training of graduate students and postdoctoral research associates is viewed as an important responsibility of this activity.
Requirements: Eligible to apply are institutions of higher education, industry (particularly small and disadvantaged businesses), and nonprofit institutions. Unsolicited research proposals are screened by program officials and, if appropriate, are evaluated by peer review. Successful proposers are awarded one-year or multi-year special research contracts or grants with renewals or incremental funding available, provided performance has been satisfactory. A one- to two-year phase-out period is normally granted to those finishing research projects. Equipment items may be purchased from contract funds with the agreement of DOE.
Geographic Focus: All States
Amount of Grant: 10,000-2,000,000 USD
Contact: Dr. Eric A. Rohlfing, Director; (301) 903-8165; fax (301) 903-4110; Eric.Rohlfing@science.doe.gov
Internet: http://www.science.doe.gov/bes/chm/Programs/programs.html
Sponsor: U.S. Department of Energy
1000 Independence Avenue SW, Forrestal Building
Washington, DC 20585-1290

DOE High-Energy Physics Research Grants 1799

The primary objectives of this program are to understand the nature of and relationships among fundamental forces of nature and to understand the ultimate structure of matter in terms of the properties and interrelations of its basic constituents. The research falls into three broad categories: experimental research, theoretical research, and technology R&D in support of the high-energy physics program. Applications are accepted at any time and are available on the Web site.
Requirements: Unsolicited proposals from universities for research support are analyzed by the program staff and undergo external peer review on an individual basis. Independently, experimental research proposals are submitted to national accelerator laboratories for review. The final decision to make accelerator facilities available for a particular experiment rests with the laboratory management. Successful research proposals usually receive one-year contracts (project grants) with provisions for review and renewal on an annual basis.
Geographic Focus: All States
Date(s) Application is Due: May 31
Contact: Martin Rubinstein, Grants and Contracts Division; (301) 903-5212
Internet: http://www.science.doe.gov/production/grants/grants.html
Sponsor: U.S. Department of Energy
19901 Germantown Road, MS FE-3
Germantown, MD 20874-1290

DOE Industrial Materials for the Future Program 1800

This program is a national effort to research, design, develop, engineer, and test new and improved materials to achieve improvements in energy efficiency, emissions and waste reduction, productivity, product quality, and global competitiveness.
Requirements: Proposals are solicited from universities, and nonprofit research institutes for research and development leading to new materials and processing methods for eventual use in the Industries of the Future. Universities, and nonrofits are required to form partnerships for technology development and to work with industry to ensure that core activities will ultimately lead to successful applications in industry.
Geographic Focus: All States
Date(s) Application is Due: May 31
Contact: Polly Perando, Contract Specialist; polly.perando@ee.doe.gov
Internet: http://www.tgci.com/fedrgtxt/01-11274.txt
Sponsor: U.S. Department of Energy
850 Energy Drive, MS 1221
Idaho Falls, ID 83401-1563

DOE Initial H-Prize Competition for Breakthrough Advances in Materials for Hydrogen Storage 1801

The H-Prize, enacted by Congress, authorized the Secretary of Energy to create a program to competitively award cash prizes that will advance the commercial application of hydrogen energy technologies by dramatizing and incentivizing accelerated research. The H-Prize was originally established by the Energy Independence and Security Act of 2007, in Sec. 654. There are several H-Prize categories, including production, storage, distribution, utilization, and prototypes and transformational technologies. The 2009-11 prize will be awarded in the area of storage materials in mobile systems for light-duty vehicles. The Hydrogen Education Foundation, the charitable, education-focused arm of the National Hydrogen Association, has been chosen to administer the H-Prize. The competition will be a single award for $1 million in the subject area of advanced materials for hydrogen storage—a critical challenge to enable widespread commercialization of hydrogen and fuel cell technologies. (FR Doc. E9-20552)
Requirements: The competition is open to participants, defined as individuals, entities or teams, that are organized or incorporated in the United States, and maintain for the duration of the H-Prize Competition a primary place of business in the United States. Teams may be comprised of any combination of individuals or entities. All individuals (whether participating singly or as part of an entity or team) must be a citizen of, or an alien lawfully admitted for permanent residence into, the United States as of the date of registration in the competition and maintain that status for the duration of the competition. The winner must be present at the award presentation in order to win the H-Prize. If a team or entity is the winner, only the team leader must be present.
Restrictions: The participant, or any member of a participant, shall not be a Federal entity, a Federal employee acting within the scope of his or her employment, or an employee of a National Laboratory acting within the scope of his or her employment. The participant, or any member of a participant, cannot have received Federal funding for research and development of hydrogen storage materials since October 2008. An individual cannot participate on more than one team or compete with multiple entries. Participants cannot be on more than one team or compete with multiple entries.
Geographic Focus: All States
Date(s) Application is Due: Feb 15
Amount of Grant: 1,000,000 USD
Contact: Jeffrey Serfass, Project Director
Internet: http://www.hydrogenprize.org/
Sponsor: U.S. Department of Energy
1000 Independence Avenue SW, Forrestal Building
Washington, DC 20585-1290

DOE Inventions and Innovation Grant Program (I&I) 1802

The goal of the program is to improve energy efficiency through the promotion of innovative ideas and inventions that have a significant potential energy impact and a potential future commercial market. Additionally, DOE provides awardees with nonfinancial support by assisting them with business development and commercialization planning through a network of national and regional resource providers. This assistance is provided at three levels: Up to $50,000 for technologies in early-stage development, up to $250,000 for technologies approaching the point of prototype, and up to $500,000 for technology demonstrations. Cost-share is strongly encouraged to receive a Category 1 or 2 award, and cost-share is required to receive a Category 3 award.
Requirements: U.S. individual inventors, small businesses (profit or nonprofit with less than 500 employees) may apply for category 1 or 2 applications. Universities and nonprofit research institutes may apply for category 1 applications.
Geographic Focus: All States
Date(s) Application is Due: Oct 11
Amount of Grant: 50,000-500,000 USD
Contact: Michael Schledorn, Grant Specialist; (303) 275-4993; goii@go.doe.gov
Internet: http://www.eere.energy.gov/inventions
Sponsor: U.S. Department of Energy
1617 Cole Boulevard
Golden, CO 80401

DOE Materials Sciences and Engineering Research Grants 1803

Research support is directed toward understanding materials properties and phenomena of importance to all energy systems. Emphasis is placed on areas where problems are known to exist or are anticipated. Research is supported in metallurgy, ceramics, solid state physics, materials chemistry, and related disciplines where the emphasis is on the science of materials. Applications are accepted at any time and are available on the Web site.
Requirements: Eligible to apply are institutions of higher education, industry (particularly small and disadvantaged businesses), and nonprofit institutions. Unsolicited research proposals are screened by program officials and, if appropriate, are evaluated by peer review. Successful proposers are awarded one-year or multi-year special research contracts or grants with renewals or incremental funding available, provided performance has been satisfactory. A one- to two-year phase-out period is normally granted to those finishing research projects. Equipment items may be purchased from contract funds with the agreement of DOE.
Geographic Focus: All States
Amount of Grant: 10,000-2,000,000 USD
Contact: Martin R. Rubinstein, Supervisor of Grants and Contracts; (301) 903-4946; fax (301) 903-4194; Martin.Rubinstein@science.doe.gov
Internet: http://www.science.doe.gov/grants/FAPN08-01.html
Sponsor: U.S. Department of Energy
19901 Germantown Road, MS FE-3
Germantown, MD 20874-1290

DOE Mathematical, Information, and Computational Sciences Grants 1804

This program supports a spectrum of fundamental research in applied mathematical sciences, computer science, and networking from basic through prototype development. Results of these efforts are used to form partnerships with users in scientific disciplines to validate the usefulness of the ideas and to develop them into tools. Testbeds on important applications for DOE are supported by this subprogram. Applications are accepted at any time, and application forms can be found on the Web site.

Geographic Focus: All States
Date(s) Application is Due: May 31
Amount of Grant: 10,000-2,000,000 USD
Contact: Martin Rubinstein, Grants and Contracts Division; (301) 903-5212
Internet: http://www.er.doe.gov/production/grants/guide.html
Sponsor: U.S. Department of Energy
19901 Germantown Road, MS FE-3
Germantown, MD 20874-1290

DOE Million Solar Roofs (MSR) Program 1805

The purpose of this initiative is to install solar energy systems on one million U.S. buildings. The department hopes to achieve this goal by establishing state and local MSR partnerships involving states and local governments, businesses, and community-based organizations to promote the use of solar roofs. Matching grants are not required, but will be considered in the application process.

Requirements: Existing and new MSR state and local partnerships are eligible to apply.
Restrictions: Partnerships that received funding under this program or the State Energy Program Special Projects Solicitation in the last fiscal year are ineligible.
Geographic Focus: All States
Amount of Grant: 10,000-50,000 USD
Contact: Sandra Burton, (215) 656-6983; fax (215) 656-6981; sandra.burton@ee.doe.gov
Internet: http://www.millionsolarroofs.com
Sponsor: U.S. Department of Energy
19901 Germantown Road, MS FE-3
Germantown, MD 20874-1290

DOE Nonproliferation and National Security Research 1806

The purpose of the program is to conduct basic and applied research and development on verification technologies needed for effective treaty negotiations and for international agreements on the control of special nuclear materials, nuclear weapons, and weapons of mass destruction. Financial support, in whole or in part, may be provided for salaries, materials, supplies, equipment, travel, publication costs, services required for conducting research, and for developing advanced detection technologies. Contact DOE for deadlines.

Requirements: Eligible to apply are public and private universities and institutions of higher education with postdoctoral programs.
Geographic Focus: All States
Amount of Grant: 100,000-250,000 USD
Contact: Paul Morrison; (202) 586-5751
Internet: http://www.nnsa.doe.gov/na-20
Sponsor: U.S. Department of Energy
19901 Germantown Road, MS FE-3
Germantown, MD 20874-1290

DOE Nuclear Physics Research Grants 1807

The primary objectives of this program are an understanding of the interactions and structures of atomic nuclei and nuclear matter, and an understanding of the fundamental forces of nature as manifested in nuclear matter. Applications are accepted at any time and are available on the Web site.

Requirements: The DOE Nuclear Physics Program supports university-based user groups that plan experiments at the home institution, execute and partially analyze experiments at the national facility, and complete the analyses and publication of results at the home institution. Support of work under Nuclear Theory is almost equally divided between national laboratory-based theorists and university-based theorists. Written proposals for direct funding should be sent to the DOE. Written proposals for access to the accelerator installations should be sent to the director of the facility.
Geographic Focus: All States
Date(s) Application is Due: May 31
Amount of Grant: 10,000-2,000,000 USD
Contact: Martin Rubinstein, Grants and Contracts Division; (301) 903-5212
Internet: http://www.science.doe.gov/production/grants/grants.html
Sponsor: U.S. Department of Energy
19901 Germantown Road, MS FE-3
Germantown, MD 20874-1290

DOE On-Line Temperature Measurement Instrumentation for Gasification Process Control Development Grants 1808

DOE seeks applications for innovative technical approaches to develop an accurate, reliable, robust, and cost-effective real-time temperature-monitoring system capable of measuring temperatures in the high temperature (typically ranging from 2000-2600F) section of pressurized, coal-fired slagging gasifiers. A variety of approaches, including the use of thermocouple or optically based techniques, are acceptable as long as they offer the clear potential to meet the aforementioned objectives. Furthermore, all proposed temperature measurement instrumentation must be suitable for use on large-scale slagging gasification systems and applications must address issues such as laboratory scale-up, potential placement, and method of mounting on actual operating systems. Solicitation DE-PS26-99FT40565.

Geographic Focus: All States
Amount of Grant: 10,000-25,000,000 USD
Contact: John Augustine, Program Director; (412) 386-4524
Internet: http://www.netl.doe.gov
Sponsor: U.S. Department of Energy
19901 Germantown Road, MS FE-3
Germantown, MD 20874-1290

DOE Photochemistry and Radiation Research Grants 1809

The program supports fundamental molecular-level research on interactions of radiation with matter in the condensed phase. The photochemistry research effort emphasizes fundamental processes aimed at the capture and chemical conversion of solar energy. The radiation sciences research effort supports fundamental studies on chemical effects produced by absorption of energy from ionizing radiation.

Requirements: Eligible to apply are institutions of higher education, industry (particularly small and disadvantaged businesses), and nonprofit institutions. Unsolicited research proposals are screened by program officials and, if appropriate, are evaluated by peer review. Successful proposers are awarded one-year or multi-year special research contracts or grants with renewals or incremental funding available, provided performance has been satisfactory. A one- to two-year phase-out period is normally granted to those finishing research projects. Equipment items may be purchased from contract funds with the agreement of DOE.
Geographic Focus: All States
Amount of Grant: 10,000-2,000,000 USD
Contact: Dr. Eric A. Rohlfing, Director; (301) 903-8165; fax (301) 903-4110; Eric.Rohlfing@science.doe.gov
Internet: http://www.science.doe.gov/bes/chm/Programs/programs.html
Sponsor: U.S. Department of Energy
1000 Independence Avenue SW, Forrestal Building
Washington, DC 20585-1290

DOE Program for Ecosystem Research 1810

The mission of this program is to improve the scientific basis for predicting or detecting effects of simultaneous changes in climate and atmospheric composition on terrestrial ecosystems and their component organisms and processes. Ecosystem processes and components of importance to humanity are of special concern. Climatic and atmospheric changes of key interest include (but need not be limited to): warming (and changes in diurnal, seasonal, and interannual temperature cycles), changes in precipitation and evapotranspiration (e.g., intensification of the hydrologic cycle), changes in frequency and/or magnitude of extreme weather events and patterns, and rising atmospheric carbon dioxide and ozone concentrations. Potential applicants are strongly encouraged (but not required) to submit a preapplication for programmatic review. The deadline for preapplications is July 2; deadline for receipt of formal applications is August 13. Annual deadline dates may vary; contact program staff for exact dates.

Geographic Focus: All States
Date(s) Application is Due: Jul 2; Aug 13
Amount of Grant: 10,000-2,000,000 USD
Contact: Martin Rubinstein, Grants and Contracts Division; (301) 903-5212
Internet: http://www.tgci.com/fedrgtxt/01-12539.txt
Sponsor: U.S. Department of Energy
19901 Germantown Road, MS FE-3
Germantown, MD 20874-1290

DOE Programs in Fusion Energy Sciences 1811

DOE announces its interest in receiving grant applications for new research in fusion energy sciences. The specific areas of interest are Magnetic Fusion Concept Exploration Experiments; Inertial Fusion Energy Concept Exploration Research; Inertial Fusion Energy Chamber and Target Research; Magnetic Fusion Liquid Wall Experiments; and Fusion Materials Modeling. Annual deadlines may vary; contact program staff for exact dates. Refer to program notice 00-07.

Requirements: Colleges and universities, nonprofit organizations, for-profit commercial organizations, State and local governments, and unaffiliated individuals.
Geographic Focus: All States
Amount of Grant: 10,000-2,000,000 USD
Contact: Martin Rubinstein, Grants and Contracts Division; (301) 903-5212
Internet: http://www.science.doe.gov/production/grants/grants.html
Sponsor: U.S. Department of Energy
19901 Germantown Road, MS FE-3
Germantown, MD 20874-1290

DOE Reliable Electric Power Delivery Research Grants 1812

Under this program of the Electric Energy Systems Division, recent breakthroughs in new materials and control concepts show the potential for significantly improving our ability to maintain electric power continuity during periods of severe emergencies. New research is needed in such areas as electromagnetic pulse protection, basic mechanisms of insulation failure, emergency detection and control, new systems analysis and equipment research, and system dynamic performance. Proposals are accepted at any time.

Geographic Focus: All States
Amount of Grant: 50,000-500,000 USD
Contact: Office of Utility Technologies; (202) 586-4142
Internet: http://www.eere.energy.gov
Sponsor: U.S. Department of Energy
1000 Independence Avenue SW, Forrestal Building
Washington, DC 20585-1290

DOE Separations and Analysis Research Grants 1813
This Program supports fundamental research covering a broad spectrum of separation concepts, including membrane processes, extraction under both standard and supercritical conditions, adsorption, chromatography, photo-dissociation, and complexation. Also supported is work to improve the sensitivity, reliability, and productivity of analytical determinations and to develop entirely new approaches to analysis. This activity is the Nation's most significant long-term investment in many aspects of separations and analysis, including solvent extraction, ion exchange, and mass spectrometry.
Requirements: Eligible to apply are institutions of higher education, industry (particularly small and disadvantaged businesses), and nonprofit institutions. Unsolicited research proposals are screened by program officials and, if appropriate, are evaluated by peer review. Successful proposers are awarded one-year or multi-year special research contracts or grants with renewals or incremental funding available, provided performance has been satisfactory. A one- to two-year phase-out period is normally granted to those finishing research projects. Equipment items may be purchased from contract funds with the agreement of DOE.
Geographic Focus: All States
Amount of Grant: 10,000-2,000,000 USD
Contact: Dr. Eric A. Rohlfing, Director; (301) 903-8165; fax (301) 903-4110; Eric.Rohlfing@science.doe.gov
Internet: http://www.science.doe.gov/bes/chm/Programs/programs.html
Sponsor: U.S. Department of Energy
1000 Independence Avenue SW, Forrestal Building
Washington, DC 20585-1290

DOE Small Business Innovation Research Commercialization Assistance 1814
The purpose of this program is to accelerate the development, demonstration, and commercialization of products, services, or technology resulting from the research of SBIR Phase II financial assistance award recipients. The major objective of the program is to provide individualized assistance to SBIR Phase II financial assistance recipients that will lead to the successful commercialization of products, services, or technology developed in the SBIR program.
Requirements: Small businesses are eligible to apply.
Geographic Focus: All States
Date(s) Application is Due: Dec 2
Amount of Grant: Up to 750,000 USD
Contact: Grants Administrator; (301) 903-1414; sbir-sttr@science.doe.gov
Internet: http://www.science.doe.gov/sbir
Sponsor: U.S. Department of Energy
9800 S Cass Avenue
Argonne, IL 60439-4899

DOE State Energy Program Special Projects 1815
The goal of this project is to assist states in deploying energy-efficient and renewable energy technologies; to facilitate the acceptance of emerging and underutilized technologies; and to increase the responsiveness of federally funded technology development efforts to private sector energy needs. DOE is providing funding in the areas of Clean Cities/Alternative Fuels, Industrial Technologies, Codes and Standards, Rebuild America, Federal Energy Management Program, Hydrogen Reformer Field Verification, Building America, Wind Energy Case Studies, Biomass Power Projects, and Photovoltaic Projects.
Requirements: All 50 states plus the District of Columbia, the U.S. Virgin Islands, Puerto Rico, Guam, Samoa, and the Commonwealth of the Northern Mariana Islands may apply.
Geographic Focus: All States
Amount of Grant: 20,000-1,000,000 USD
Contact: John Millhone, Program Supervisor; (202) 586-1510
Internet: http://12.46.245.173/pls/portal30/CATALOG.PROGRAM_TEXT_RPT.show
Sponsor: U.S. Department of Energy
1000 Independence Avenue SW, Forrestal Building
Washington, DC 20585-1290

DOE University Coal Research Grants 1816
The grants are awarded to improve scientific and technical understanding of the chemistry and physics involved in the conversion and utilization of coal, furnish technical support for the ongoing and developing coal conversion process, produce clear fuels in an environmentally acceptable manner, and develop new approaches to the design of future coal conversion and utilization technologies. Proposals should be submitted to the National Energy Technology Laboratory, Attn: MS921-107, Department of Energy, P.O. Box 10940, 626 Cochrans Mill Road, Pittsburgh, PA.
Requirements: U.S. institutions of higher education may apply for these grants.
Geographic Focus: All States
Amount of Grant: 50,000-400,000 USD
Contact: Fred Glaser, Office of Advanced Research; (301) 903-2786
Internet: http://www.netl.doe.gov/coal/Advanced%20Research
Sponsor: U.S. Department of Energy
1000 Independence Avenue SW, Forrestal Building
Washington, DC 20585-1290

DOE University Nuclear Science and Reactor Support 1817
The purpose of this program is to provide financial support for research design, analysis, and assessments in science and technology in fields related to nuclear energy. Project grant funds may be used in support of nuclear energy-related research and development financial support, in whole or in part, may be provided for salaries, materials, supplies, equipment, travel, publication costs, supporting costs required for technical activities, market analyses, financing plans, and other activities necessary to achieve the objective.
Requirements: Any individual, partnership, corporation, association, joint venture, institution of higher education, or nonprofit organization may apply.
Geographic Focus: All States
Amount of Grant: 50,000-300,000 USD
Contact: Nancy Hebron-Isreal, Office of Nuclear Energy; (301) 903-1536
Internet: http://nuclear.gov
Sponsor: U.S. Department of Energy
19901 Germantown Road, MS FE-3
Germantown, MD 20874-1290

DOE University Reactor Instrumentation (URI) Program 1818
The purpose of this program is to upgrade and improve the U.S. university nuclear research and training reactors and to contribute to strengthening the academic community's nuclear engineering infrastructure.
Requirements: U.S. colleges and universities having a duly licensed, operating nuclear research or training reactor are eligible to apply.
Geographic Focus: All States
Contact: Suzette Olson, DOE Idaho Operations Office; (208) 526-7385; fax (208) 526-5548; Suzette.Olson@nuclear.energy.gov
Internet: http://www.id.doe.gov/doeid/psd/proc-div.html
Sponsor: U.S. Department of Energy
P.O. Box 1625, MS 3860
Idaho Falls, ID 83415-3860

DOE Vehicle and Engine Research and Development Grants 1819
The grants support research to reduce the transportation sector's vulnerability to petroleum shortages by developing more energy-efficient vehicle propulsion system options and by developing the ability to switch from petroleum to electricity and alternative fuels. Emphasis is placed on advancing heat engine technology through the development of component and materials technologies; supporting long-term, high-risk technology research and development for petroleum savings in the heavy-duty transport sector; advancing electric and hybrid vehicle technology; and developing means of using alternative fuels in vehicles. Identified potential areas for new research initiatives may be requested from one of the program information contacts. Proposals are accepted at any time. For further information on specific programs and prior to submitting formal proposals, contact Advanced Heat Engines, (202) 252-8012; Heavy Duty Transport, (202) 252-8055; Alternative Fuels, (202) 252-8055; and Electric and Hybrid Vehicles, (202) 252-8044.
Geographic Focus: All States
Amount of Grant: 50,000-500,000 USD
Contact: Polly Perando, Program Supervisor; polly.perando@ee.doe.gov
Internet: http://www.eere.energy.gov
Sponsor: U.S. Department of Energy
1000 Independence Avenue SW, Forrestal Building
Washington, DC 20585-1290

DOI African Elephant Conservation Grants 1820
The program objectives are to provide financial assistance to any organization or individual responsible for African elephant conservation and to any organization or individual with experience in African elephant conservation, for approved projects to support research, conservation, management, and protection of African elephants. Funds may be used for approved conservation projects. Funds are matched by nonfederal funds or in-kind support, which must be equal to or exceed the amount of federal funds provided.
Requirements: Individuals and public or private organizations with experience in African elephant conservation may apply. All private sector project proposals must contain evidence of support by governmental entities of countries where the project is to be conducted.
Geographic Focus: All States
Amount of Grant: 50,000 USD
Contact: Program Contact; (703) 358-1754; fax (703) 358-2849
Internet: http://international.fws.gov/grants/grants.html
Sponsor: U.S. Department of the Interior
4401 N Fairfax Drive, Room 730
Arlington, VA 22203

DOI Earthquake Hazards Reduction Research Grants 1821
Applications are being sought for grants for research projects under the National Earthquake Hazards Reduction Program to mitigate earthquake losses that can occur in many parts of the nation by providing earth science data and assessments essential for warning of imminent damaging earthquakes, land-use planning, engineering design, and emergency preparedness decisions. Specific objectives are identified in the annual program announcement.
Requirements: Colleges, universities, for-profit and nonprofit organizations, and state and local governments may submit applications for support by a named principal investigator.
Geographic Focus: All States
Date(s) Application is Due: May 10
Amount of Grant: 5,887-11,000,000 USD
Contact: Michael Blanpied, Program Supervisor; (703) 648-6696; fax (703) 648-6642; gd-erp-coordinator@usgs.gov
Internet: http://aspe.os.dhhs.gov/cfda/p15807.htm
Sponsor: U.S. Department of the Interior
12201 Sunrise Valley Drive
Reston, VA 22092

DOI Geological Research and Data Acquisition Grants 1822

Project grants are awarded to support research in any field of study that helps fulfill the Geological Survey's mission, which is to collect, organize, interpret, and publish information about the nation's energy, minerals, water, and land resources; to determine the geologic structure of the United States; and to develop an understanding of earth processes and hydrologic principles. Applications must be for scientific research projects that are within the survey's area of responsibility. Conferences and symposia will be supported only if it is clear that equivalent results cannot be obtained at regular meetings of professional societies. Cost sharing is encouraged. Applications may be presented at any time.
Requirements: Colleges, universities, profit-making and nonprofit organizations, and state and local governments may make application for support by a named principal investigator.
Restrictions: Office furniture, office equipment, and foreign travel are not normally considered for support.
Geographic Focus: All States
Amount of Grant: 1,000-582,000 USD
Contact: Assistant Director for Research; (703) 648-4460
Internet: http://www.usgs.gov/contracts
Sponsor: U.S. Department of the Interior
12201 Sunrise Valley Drive
Reston, VA 22092

DOI National Spatial Data Infrastructure Cooperative Agreements Grants 1823

The objectives of the program are to help the Federal Geographic Data Committee form partnerships with the nonfederal sector that will further the development and implementation of the National Spatial Data Infrastructure. NSDI means the technology, policies, standards, and human resources necessary to acquire, process, store, distribute, and improve utilization of geospatial data. Partnership activities must be directed towards the creation of the National Geospatial Data Clearinghouse for finding and accessing geospatial data; the development and promulgation of the use of standards in data collection, documentation, transfer, and search and query; the development and creation of a National Digital Geospatial Data Framework; and/or the development and implementation of educational outreach programs to increase awareness and understanding of the major NSDI initiatives.
Requirements: Proposals from federal, state and local government agencies, educational institutions, private firms, private foundations, nonprofit organizations and federally acknowledged or state-recognized Native American tribes or groups are invited. Proposals must demonstrate that more than one organization or agency is involved in the project.
Geographic Focus: All States
Date(s) Application is Due: May 15
Amount of Grant: 20,000-75,000 USD
Contact: David Painter; (703) 648-5513; fax (703) 648-5755; dpainter@fgdc.gov
Internet: http://www.fgdc.gov/funding/cap2002.html
Sponsor: U.S. Department of the Interior
12201 Sunrise Valley Drive, Mail Stop 590
Reston, VA 20192

DOI Rhinoceros and Tiger Conservation Grants 1824

The objectives of the program are to increase conservation of rhinoceros and tigers through strengthening habitat/ecosystem management; surveys and monitoring; conservation education; wildlife inspection, law enforcement, and forensic skills; protected area/reserve management; sustainable development in buffer zones surrounding tiger/rhinoceros habitat; management of human behavior and livestock to decrease conflicts with tigers/rhinoceros; and the use of substitute for tiger/rhinoceros products in oriental medicine. Projects which provide training to strengthen capabilities in these areas will receive priority. Funds may be used for approved conservation projects. Projects proposed should have the support of the local government(s) and have matching funds (cash) or in-kind support (salaries, equipment, etc.) provided by the organization receiving the grant or other partners. Project duration must be one year or less. Proposals may be submitted at any time.
Requirements: U.S. public and private nonprofit institutions and organizations are eligible.
Geographic Focus: All States
Amount of Grant: Up to 50,000 USD
Contact: Program Contact; (703) 358-1754; fax (703) 358-2849
Internet: http://www.fws.gov/international/grants/grants.html
Sponsor: U.S. Department of the Interior
4401 N Fairfax Drive, Room 730
Arlington, VA 22203

DOI Upper Mississippi River System Long-Term Resource Monitoring Grants 1825

The mission of this program is to provide decision makers with information needed to maintain the Upper Mississippi River System (UMRS) as a sustainable large river ecosystem given its multiple-use character. The long-term goals of this program are to: understand the UMRS ecosystem, monitor trends and effects with respect to selected resources, develop resource management alternatives, manage information, and develop useful products. This is a 15-year program. There are no deadlines.
Requirements: Eligible to apply are states, local governments, intrastate/interstate agencies, sponsored organizations, private nonprofit institutions/organizations.
Geographic Focus: Illinois
Contact: Dr. Leslie Holland-Bartels, Center Director; (608) 781-6221; fax (608) 783-6066; Leslie_Holland-Bartels@usga.gov
Internet: http://www.umesc.usgs.gov/ltrmp.html
Sponsor: U.S. Department of the Interior
2630 Fanta Reed Road
La Crosse, WI 54603

DOJ Juvenile Justice and Delinquency Prevention Special Emphasis Grants 1826

Grants support investigative research into and the development and implementation of programs that design, test, and demonstrate effective approaches, techniques, and methods for preventing and controlling juvenile delinquency, such as community-based alternatives to institutional confinement; developing and implementing effective means of diverting juveniles from the traditional juvenile justice and correctional system; developing and supporting programs stressing advocacy activities aimed at improving services to youth impacted by the juvenile justice system; developing model programs to strengthen and maintain the family unit, including self-help programs; developing and implementing special emphasis prevention and treatment programs relating to juveniles who commit serious crimes; developing programs to prevent hate crimes; and developing and implementing further a coordinated, national law-related education program of delinquency prevention. Deadlines are published in program announcements.
Requirements: Public and private nonprofit agencies, organizations, individuals, state and local units of government, and combinations of state and local units are eligible to apply.
Geographic Focus: All States
Contact: Office of Juvenile Justice and Delinquency Prevention; (202) 307-5914; AskDOJ@usdoj.gov
Internet: http://www.usdoj.gov
Sponsor: U.S. Department of Justice
950 Pennsylvania Avenue NW
Washington, DC 20530-0001

DOJ National Institute of Corrections Training and Staff Development Grants 1827

The grants provide assistance to devise and conduct in various geographical locations seminars, workshops, and training programs for law enforcement officers, judges and judicial personnel, probation and parole personnel, correctional personnel, welfare workers, and other personnel, including lay ex-offenders and paraprofessionals, connected with the treatment and rehabilitation of criminal and juvenile offenders. In addition, the grants develop technical training teams to aid in the development of seminars, workshops, and training programs within the several states and with the state and local agencies that work with prisoners, parolees, probationers, and other offenders.
Requirements: States, general units of local government, as well as public and private agencies, educational institutions, organizations, and individuals involved in the development, implementation, or operation of correctional programs and services are eligible to apply.
Geographic Focus: All States
Amount of Grant: 1,500-300,000 USD
Contact: Staff Development Branch, National Institute of Corrections; (202) 307-3106 or (800) 995-6423; fax (202) 307-3361; AskDOJ@usdoj.gov
Internet: http://www.usdoj.gov/10grants/index.html
Sponsor: U.S. Department of Justice
950 Pennsylvania Avenue NW
Washington, DC 20530-0001

DOJ National Institute of Justice Graduate Research Fellowships 1828

The fellowships are intended to improve the quality and quantity of knowledge about crime and the criminal justice system, while helping to increase the number of persons who are qualified to teach in collegiate criminal justice programs; to conduct research related to criminal justice issues; and help such persons perform more effectively within the criminal justice system. A fellowship is funded for up to two years. Applications may be submitted at any time throughout the year. Proposals will be due and subsequently reviewed three times a year, in February, June, and October, with funding decisions made within 60-90 days of the review date.
Requirements: Applicants must be accredited institutions of higher education offering doctoral degree programs. Eligible students are doctoral candidates engaged in dissertation research and writing on a problem relating to law enforcement, crime, or criminal justice. This competitive program provides fellowship stipends, major project costs, and certain university fees.
Geographic Focus: All States
Contact: Sarah Hart, Director; (202) 307-2942; fax (202) 307-6394
Internet: http://www.ojp.gov/nij/funding.htm
Sponsor: U.S. Department of Justice
810 7th Street NW
Washington, DC 20531

DOJ National Institute of Justice Visiting Fellowships 1829

Fellowships provide opportunities for experienced criminal justice practitioners and researchers to pursue projects aimed at improved understanding of crime, delinquency, and criminal justice administration by sponsoring research projects of their own creation and design. Funds may be used to conduct research on crime causation, crime measurements, crime prevention, law enforcement, criminal justice administration, and the effectiveness and efficiency of anti-crime programs. Juvenile delinquency research projects also are eligible for support under this program. Fellows conduct their studies while based at NIJ. Concept papers may be submitted at any time. Applicants should anticipate a decision time frame of six to nine months from concept paper to award. Fellowship guidelines are available at no charge by sending a self-addressed mailing label to Announcement-Visiting Fellowship Program, NCJRS, Box 6000, Rockville, MD 20849-6000, or by phone (800) 851-3420.
Requirements: Fellowships are awarded to individuals or to their parent agencies or organizations. Generally, professionals working in the criminal justice field, including university- or college-based academic researchers and upper-level managers in criminal justice agencies are eligible.

Geographic Focus: All States
Contact: National Institute of Justice Office, Department of Justice; (202) 307-2942
Internet: http://www.usdoj.gov/nij
Sponsor: U.S. Department of Justice
810 7th Street NW
Washington, DC 20531

Donald and Sylvia Robinson Family Foundation Grants 1830
The foundation awards grants to U.S. nonprofit organizations in its areas of interest, including animal and wildlife protection, arms control, arts (general, performing arts, and visual arts), environmental conservation and protection, eye diseases and eye research, family planning and human reproductive health, food distribution, international affairs, Israel, Jewish social services, and social service delivery programs. Types of support include annual campaigns, building construction/renovation, capital campaigns, and general operating support. There are no application deadlines or forms.
Requirements: National nonprofit organizations are eligible to apply.
Restrictions: Individuals are ineligible.
Geographic Focus: Pennsylvania
Amount of Grant: 500-5,000 USD
Contact: Donald Robinson, Treasurer; (412) 661-1200; fax (412) 661-4645
Sponsor: Donald and Sylvia Robinson Family Foundation
6507 Wilkins Avenue
Pittsburgh, PA 15217

Donald Q. Kern Award in Heat Transfer or Energy Conversion 1831
This award is given annually by the Heat Transfer and Energy Conversion Division of AIChE for expertise in a given field of heat transfer or energy conversion. Special emphasis is given to recognition of contributions that have had significant practical applications. A plaque accompanies the award. The recipient will be required to prepare a written review covering an area selected by the recipient.
Requirements: Three to five supporting letters are required that discuss specific contributions by the nominee and the reasons for their value.
Geographic Focus: All States
Date(s) Application is Due: Jun 1
Contact: AIChE Award; (212) 591-7107; fax (212) 591-8882; awards@aiche.org
Internet: http://www.aiche.org/awards/awarddtl.asp?AwardID=40
Sponsor: American Institute of Chemical Engineers
3 Park Avenue
New York, NY 10016

Donald W. Reynolds Foundation Aging and Quality of Life Grants 1832
The Foundation launched its Aging and Quality of Life Program in 1996. Its goal remains improving the quality of life for America's elderly by preparing physicians to provide better care for frail older people. Most physicians today lack adequate training to meet the needs of the frail elderly patient. Such patients typically suffer from interacting physical, social and psychological conditions –both acute and chronic – that limit their independence and threaten their capacity to function in daily life.
Requirements: Proposals are by invitation only. Periodic requests for proposals are announced on the website. Organizations should contact the Foundation to discuss their project before submitting a proposal.
Geographic Focus: All States
Contact: Rani Snyder, Senior Program Officer; (702) 804-6000; fax (702) 804-6099; rani.snyder@dwrf.org or GeneralQuestions@dwrf.org
Internet: http://www.dwreynolds.org/Programs/National/Aging/Aging.htm
Sponsor: Donald W. Reynolds Foundation
1701 Village Center Circle
Las Vegas, NV 89134-6303

Doris Duke Charitable Foundation Clinical Interfaces Award Program 1833
The program seeks to catalyze activity at the interface of clinical and other research disciplines by: supporting the formation of new collaborations and strengthening existing collaborations of outstanding scientists across disciplines; demonstrating successful models for clinical research at the interface of multiple disciplines; and supporting interdisciplinary and inter-institutional endeavors that go beyond the program project mindset. Full grants, awarded over five years, are made to established teams with key investigators from at least three disciplines. Planning grants are awarded to new teams for the development of full proposals over 18 months. The listed application deadline is for pre-proposals; full proposals are by invitation. Though new grants are not being offered continually, potential applicants can sign up to be notified of future competitions.
Requirements: Teams of at least three key investigators whose primary expertise lie in different disciplines are eligible to apply. Key investigators must have advanced degrees (MD, PhD, MD/PhD, or the equivalent), and one of the key investigators must be a clinical researcher. The team leader must work in a U.S. nonprofit institution, such as an academic medical center. The team may include investigators at other institutions in the United States and overseas.
Restrictions: Planning grants will not be awarded in the current competition.
Geographic Focus: All States
Date(s) Application is Due: Nov 2
Contact: Elaine K. Gallin; (212) 974-7104; fax (212) 974-7590; egallin@ddcf.org
Internet: http://www.ddcf.org/page.asp?pageId=299
Sponsor: Doris Duke Charitable Foundation
650 Fifth Avenue, 19th Floor
New York, NY 10019

Doris Duke Charitable Foundation Clinical Research Fellowships for Medical Students 1834
The fellowship program is designed to encourage medical students to pursue careers in clinical research by giving exceptional students the opportunity to take a year to experience clinical research first hand. The CRF program is available at the following medical schools: Columbia University College of Physicians and Surgeons; Harvard Medical School; Mount Sinai School of Medicine; University of California at San Francisco School of Medicine; University of Iowa Roy J. and Lucille A. Carver College of Medicine; University of North Carolina at Chapel Hill School of Medicine; University of Pennsylvania School of Medicine; University of Texas Southwestern Medical Center at Dallas; Washington University Medical School; and Yale University School of Medicine. An additional 11-12 fellowships will be available to medical students interested in conducting clinical research in Africa. Each participating medical school provides medical students with a one-year fellowship experience in clinical research that includes both didactic and research components; matches students to outstanding clinical research mentors; and offers fellowships to at least five students per year.
Requirements: Medical students matriculated at any U.S. medical school who have completed two or more years of medical school prior to the start of the fellowship and who have completed some clinical experience are eligible to apply to any of the participating schools.
Geographic Focus: All States
Date(s) Application is Due: Jan 18
Amount of Grant: 27,000 USD
Contact: Elaine K. Gallin, Program Director; (212) 974-7104; fax (212) 974-7590; egallin@ddcf.org
Internet: http://www.ddcf.org/page.asp?pageId=292
Sponsor: Doris Duke Charitable Foundation
650 Fifth Avenue, 19th Floor
New York, NY 10019

Doris Duke Charitable Foundation Clinical Research Systems Grants 1835
The Program is founded on the premise that clinical research requires not only human and research capital, facilities, and equipment, but it also requires robust support systems, including appropriate regulatory organizations and processes, common language, analytical tools, databases and information systems, as well as mechanisms to distribute findings and interface with the public. The Program is interested in supporting activities that seek to identify ways to strengthen the regulatory processes and systems for clinical research, such as the protection of human subjects. The foundation is willing to consider any exceptional opportunities that may arise.
Geographic Focus: All States
Contact: Elaine Gallin; (212) 974-7104; fax (212) 974-7590; egallin@ddcf.org
Internet: http://www.ddcf.org/page.asp?pageId=302
Sponsor: Doris Duke Charitable Foundation
650 Fifth Avenue, 19th Floor
New York, NY 10019

Doris Duke Charitable Foundation Clinical Scientist Development Award 1836
Grants are awarded to junior physician-scientists to facilitate their transition to independent clinical research careers. The program is designed to help prepare and support new investigators with an MD or MD/PhD as they begin their careers as independent clinical researchers. The program is aimed at conducting clinical research in any disease area. The listed application deadline is for preproposals, which must be submitted electronically.
Requirements: Applicants must: be a physician-scientist conducting clinical research in any disease area; have received an M.D. or a foreign equivalent from an accredited institution; be working in a U.S. degree-granting institution, but do not have to be a U.S. citizen; have a full-time faculty level position not higher than the Assistant Professor level; and have been appointed to their first full-time faculty level position between January 31, 2002 and January 31 of the grant year. (All full-time post-fellowship Instructor level positions will be considered full-time faculty level appointments).
Restrictions: Funds cannot be used on experiments that utilize animals or primary tissues derived from animals. An award will not be made if, prior to the commencement of the award, the applicant becomes the principal investigator on a federal government, peer-reviewed, research or career development award or any non-government award averaging more than $126,000 per year in direct costs and of a duration of three or more years.
Geographic Focus: All States
Date(s) Application is Due: Jan 23
Amount of Grant: 135,000 USD
Contact: Elaine K. Gallin; (212) 974-7104; fax (212) 974-7590; egallin@ddcf.org
Internet: http://www.ddcf.org/page.asp?pageId=291
Sponsor: Doris Duke Charitable Foundation
650 Fifth Avenue, 19th Floor
New York, NY 10019

Doris Duke Charitable Foundation Clinical Scientist Development Award Bridge Grants 1837
The purpose of the bridge grant is to help Doris Duke Clinical Scientist Development Award (CSDA) recipients successfully transition to independent National Institutes of Health (NIH) R01 funding. The Medical Research Program created these grants to bridge the time between when a CSDA investigator learns of a very strong but unfunded score on an NIH application to when s/he is able to successfully obtain an R01 grant. The bridge grants provide protected time and support for CSDA recipients to continue their research projects, collect supportive data, keep patients enrolled if applicable, and strengthen their proposals to resubmit to the NIH. The foundation hopes that the bridge

grants will help keep promising physician-scientists committed to clinical research careers by providing stable support during their critical transition to independent R01 funding. The foundation will award up to four bridge grants of $135,000 each to be spent only on direct costs over 12 to 18 months. Eligible CSDA recipients are invited to submit applications throughout the year up to the deadline of September 1.
Requirements: This competition is open only to former CSDA grantees or current CSDA grantees in the last year of their award (or in a no-cost extension period).
Geographic Focus: All States
Date(s) Application is Due: Sep 1
Amount of Grant: Up to 135,000 USD
Contact: Elaine Gallin; (212) 974-7104; fax (212) 974-7590; egallin@ddcf.org
Internet: http://www.ddcf.org/page.asp?pageId=291
Sponsor: Doris Duke Charitable Foundation
650 Fifth Avenue, 19th Floor
New York, NY 10019

Doris Duke Charitable Foundation Distinguished Clinical Scientist Award 1838
The purpose of the program is to recognize outstanding physician-scientists who are engaged in applying the latest basic science advances to the prevention, diagnosis, treatment, and cure of disease, and to enable the physician-scientist to support and mentor the next generation of physician-scientists conducting translational clinical research. Awards will be granted to physician-scientists conducting translational clinical research in any disease area; this award cycle is not limited to specific disease areas.
Requirements: Grantees must hold an M.D. degree from an accredited institution in the United States (holders of M.D./Ph.D. degrees are also eligible, as are holders of M.D.-equivalent degrees from non-U.S. institutions); hold a full-time university faculty appointment at the level of Associate Professor or above as of the date of nomination; have been appointed to their first full-time, faculty-level position for no more than 15 years; and have an established translational clinical research program.
Restrictions: Experiments that utilize animals or primary tissues derived from animals are not eligible for support through this award program. An award will not be made if, prior to the commencement of the foundation's award, the applicant becomes the principal investigator of a federal government, peer-reviewed, research or career development award or of any nongovernment award averaging $100,000 or more per year and of three years or more duration.
Geographic Focus: All States
Date(s) Application is Due: Feb 14
Amount of Grant: Up to 15,000,000 USD
Contact: Elaine Gallin; (212) 974-7104; fax (212) 974-7590; egallin@ddcf.org
Internet: http://www.ddcf.org/page.asp?pageId=297
Sponsor: Doris Duke Charitable Foundation
650 Fifth Avenue, 19th Floor
New York, NY 10019

Doris Duke Charitable Foundation Operations Research on AIDS Care and Treatment in Africa (ORACTA) Grants 1839
The Program seeks to improve the care and treatment of AIDS patients in resource-limited settings, inform antiretroviral therapy (ART) policy and practice, and improve outcomes of the roll-out and scale-up of ART in Africa. ORACTA grants provide two-year grants of up to $100,000 per year to teams of investigators conducting health operations research on AIDS care and treatment in Africa. Though new grants are offered only periodically, applicants should sign up for the Medical Research Program's mailing list.
Restrictions: Indirect costs are not supported by this award.
Geographic Focus: All States
Amount of Grant: 100,000 USD
Contact: Elaine Gallin; (212) 974-7104; fax (212) 974-7590; egallin@ddcf.org
Internet: http://www.ddcf.org/page.asp?pageId=486
Sponsor: Doris Duke Charitable Foundation
650 Fifth Avenue, 19th Floor
New York, NY 10019

Dorothy Leet Grants 1840
This competition is held in even-numbered years to award grants that assist women graduates from countries with low per capita income, or they may be given to other women graduates who wish to work as experts in these countries or whose research is of value to such countries. Grant is to be used preferably in a country other than that in which the applicant received her education or habitually resides. Grants may be used for obtaining special training essential to research and survey work; training in new techniques in group research and further study; or carrying out independent research or surveys, including completion of projects well advanced at the time of application. Members of the IFUW should apply through their national IFUW unit (in the United States, the American Association of University Women, 1111 16th St NW, Washington, DC 20036; info@aauw.org). Deadline for applications is determined by each affiliate, but normally falls between early September and mid-October of the year preceding the competition. Contact national headquarters for the exact deadline date.
Geographic Focus: All States
Date(s) Application is Due: Oct 1
Contact: Fellowships Officer; 41-22-731-23-80; fax 41-22-738-04-40; info@ifuw.org
Internet: http://www.ifuw.org/fellowships/international.htm
Sponsor: International Federation of University Women
8 rue de l'Ancien Port
Geneva, CH 1201 Switzerland

Dorothy Rider Pool Health Care Grants 1841
The Foundation's intent is to serve as a means to improve the quality of life in the Lehigh Valley community, to build on community strengths and add to its vitality, and to increase the capacity of the community to serve the needs of all its citizens. Within this objective the Foundation's funding program is focused on education, health and welfare, culture and art and community development. Interested applicants should submit a letter of intent of five pages or less.
Requirements: Allentown, PA, nonprofit organizations are eligible.
Restrictions: The Foundation is restricted from providing funds to individuals, legislative or lobbying efforts, political or fraternal organizations or organizations outside the United States and its territories. The Foundation as a policy does not provide operating or capital funds to Sectarian institutions, organizations or programs in which funds will be used primarily for the propagation of religion, hospitals or United Way member agencies. Further, the Foundation does not underwrite charitable or testimonial dinners, fund-raising events or related advertising or the subsidization of books, mailings or articles in professional journals.
Geographic Focus: Pennsylvania
Date(s) Application is Due: Apr 1; Aug 15
Amount of Grant: 100,000 USD
Contact: Ronald Dendas; (610) 770-9346; fax (610) 770-9361; drpool@ptd.net
Internet: http://www.pooltrust.com
Sponsor: Dorothy Rider Pool Health Care Trust
1050 S Cedar Crest Boulevard, Suite 202
Allentown, PA 18103

Dorr Foundation Grants 1842
Dorr Foundation grants are made primarily for programs designed to develop new science curricula from sixth-12th grade. Support is also given to special education projects for youth relating to conservation and the environment if such projects involve the school's curriculum, equipment purchase, program development, emergency funding, and seed money. In addition, some grants are made available to promote research and disseminate information on chemical, metallurgical, and sanitation engineering. Grants are awarded on a national basis, with emphasis in the Northeast states. Types of support include equipment, emergency funds, program development, seed money, curriculum development, scholarship funds, and research. Initial contact should be a phone call. There is no deadline. No response can be expected unless there is interest on the part of the trustees. Applications are accepted at any time.
Requirements: 501(c)3 tax-exempt organizations are eligible.
Restrictions: Grants are not made to individuals or for operation budgets, continuing support, annual campaigns, deficit financing, endowment funds, or conferences and seminars.
Geographic Focus: Connecticut, Maine, Massachusetts, New Hampshire, New York, Vermont
Amount of Grant: 1,000-40,000 USD
Contact: Barbara McMillan, Chairperson; (603) 433-6438
Sponsor: Dorr Foundation
84 Hillside Drive
Portsmouth, NH 03801-5328

Dorr Institute for Arthritis Research and Education 1843
Established in California in 2000, the Dorr Institute for Arthritis Research and Education promotes medical treatment, research and education concerning the causes and treatment of arthritis and joint implant surgery procedures. Its primary fields of interest are arthritis and arthritis research. Recognized as a foundation, it currently funds research for improvement in total hip replacement and total knee replacement. Annually, they support two types of fellows who participate in the research; a research fellow is dedicated to investigations; and a clinical fellow participates in the care of the patients and also participates in research.
Geographic Focus: All States
Contact: Felicia Renty; (310) 695-4400 or (323) 442-5762; felicia@drdorr.com
Jeri Ward, Director; (323) 442-5762 or (213) 977-2280; jeri@drdorr.com
Internet: http://dorrarthritisinstitute.org/research.html
Sponsor: Dorr Institute for Arthritis Research and Education
1520 San Pablo Street, Suite 2000
Los Angeles, CA 90033

Do Something TMI Intern 1844
TMI is the agency of DoSomething.org that works with companies, not-for-profits, and governments to help them in the areas of technology, millennials and impact. The TMI Intern works with the agency team on both the business development front (creating and pitching new ideas for potential clients) and on the client services side (research, writing, etc.) supporting existing projects. Primary responsibilities include: research and development for potential Do Something clients-everything from global education to local teen health; researching companies and brands that are relevant to a teen audience; brainstorming new youth marketing and youth outreach ideas; assisting in the management of our current client and sponsor relations; helping create weekly internal reports with teen insights and helpful data; some administrative work as necessary; helping to manage TMI's social media channels (FB, Twitter, LinkedIn); and helping to develop strategies for client campaigns. Compensation includes: a $1,000 stipend or school credit; tons of fun; a great reference letter; and a chance to work in a super fun and positive work environment. Fall, winter, and summer semester internships are available.
Requirements: General requirements include: ability to travel to the New York City office; commitment to working 12 to 15 hours per week for nine to ten weeks; being a self starter – ability to work well alone and with others; passionate about young people

making an impact and committed to social change; member of DoSomething.org; love of the DoSomething.org culture and can work with a quirky, close-knit team. Specific requirements include: outgoing and enthusiastic personality; confident working independently and taking ownership of full projects; excellent communication skills; solid grammar; ability to convey message clearly; good telephone manner; passion for social change; interest in CSR, marketing, and everything data; knowledge of pop culture and the teen trends of the moment; and the ability to produce client-facing documents and attend meetings with clients.
Geographic Focus: All States
Date(s) Application is Due: May 9; Aug 8; Dec 5
Amount of Grant: 1,000 USD
Contact: Lisa Boyd; (212) 254-2390; lboyd@dosomething.org or intern@tmiagency.org
Internet: https://www.dosomething.org/about/internships#tmi-intern-the-dosomething.org-agency
Sponsor: Do Something
19 West 21st Street, 8th Floor
New York, NY 10010

Do Something User Experience Research Intern **1845**
The Do Something User Experience Research Intern will help get the research needed to make a clean, simple, and intuitive user experience. Responsibilities include: fighting for the user-uncovering what is in the best interest of young people who are looking for ways they can create social change; interviews-recruit, screen, and schedule members and other young people for user interviews; research-field feedback from members and review usability videos, and then consolidate learnings in a way that determines if more research is needed or readiness to implement changes; data digging-be an active member of the quantitative and quantitatively driven data team, helping to lead and report on small and large scale surveys; customer service-use of organizational skills and love of efficiency to come up with better systems that give quick responses to users who need help, or clearly communicate to other team members (tech team, campaigns, social media, etc.); innovation-follow the industry wide conversation on UX innovations and research methodologies. Compensation includes: a $1,000 stipend or school credit; tons of fun; a great reference letter; and a chance to work in a super fun and positive work environment. Fall, winter, and summer semester internships are available.
Requirements: General requirements include: ability to travel to the New York City office; commitment to working 12 to 15 hours per week for nine to ten weeks; being a self starter – ability to work well alone and with others; passionate about young people making an impact and committed to social change; member of DoSomething.org; love of the DoSomething.org culture and can work with a quirky, close-knit team. Specific requirements include: background in a social science (e.g., psychology, sociology, behavioral economics), design (arts), or a related field; curiosity; passion for design or science; and customer service experience.
Geographic Focus: All States
Date(s) Application is Due: May 9; Aug 8; Dec 5
Amount of Grant: 1,000 USD
Contact: Crystal Bell; (212) 254-2390; cbell+intern@dosomething.org
Internet: https://www.dosomething.org/about/internships#user-experience-research-intern
Sponsor: Do Something
19 West 21st Street, 8th Floor
New York, NY 10010

DOS Study of the United States Institutes **1846**
The program seeks grant proposals from U.S. colleges and universities to develop and implement summer institutes. Each institute is a six-week U.S.-based seminar, including up to two weeks of travel, designed for multinational groups of 18 foreign university teaching faculty or 30 secondary school educators. Institute include Study of the United States for Foreign Secondary School Educators; Religious Pluralism in the United States; U.S. Foreign Policy; American Politics and Political Thought; and Contemporary American Literature. The application of the Study of the United States for Foreign Secondary Educators institute is December 13; the January 10 deadline applies to the other institutes. Guidelines are available at the website.
Requirements: U.S. colleges and universities are eligible.
Geographic Focus: All States
Date(s) Application is Due: Jan 10; Dec 13
Contact: Jennifer Phillips, (202) 453-8537; fax (202) 453-8533
Internet: http://exchanges.state.gov/education/rfgps/menu.htm
Sponsor: U.S. Department of State-1
301 4th Street SW, Room 314
Washington, DC 20547

DOT Dwight David Eisenhower Transportation Fellowship Program **1847**
The program's goals are to attract people to the field of transportation, to enhance the careers of transportation professionals by encouraging them to seek advanced degrees, and to retain top talent in the transportation community. The fellowship awards are: the Eisenhower Graduate Fellowships to allow students to pursue master's degrees or doctorates in transportation-related fields; the Eisenhower Historically Black Colleges and Universities Fellowships to provide HBCU students with additional opportunities to enter careers in transportation; the Eisenhower Hispanic Serving Institutions Fellowships to provide HSI students with opportunities to enter transportation careers; the Eisenhower Tribal Colleges Initiatives, which provide fellowships for Native American students and faculty at tribal colleges; and the Eisenhower Faculty Fellowships to provide faculty in transportation fields with opportunities to improve their transportation knowledge by attending conferences, courses, seminars, or workshops.
Geographic Focus: All States
Date(s) Application is Due: Apr 15; Oct 15
Contact: Gwen Sutton, Program Manager; (703) 235-0535; fax (703) 235-0593; transportationedu@fhwa.dot.gov
Internet: http://www.nhi.fhwa.dot.gov/uandg.asp
Sponsor: U.S. Department of Transportation
1200 New Jersey Avenue, SE
Washington, DC 20590

DOT Recreational Boating Safety Grants **1848**
These grants are used to fund projects on various subjects promoting boating safety on the national level. It targets specific boat market segments and recreational boating safety topics.
Requirements: National nongovernmental, nonprofit, public service organizations are eligible to apply.
Geographic Focus: All States
Amount of Grant: Up to 3,000,000 USD
Contact: Vickie Hartberger, (202) 267-0974
Internet: http://www.uscgboating.org/grants/state/rbs.htm
Sponsor: U.S. Department of Transportation
2100 Second Street SW
Washington, DC 20593

DOT Transit Planning and Research Grants **1849**
The objectives of the program are to foster innovation in public transit systems, through local demonstrations of promising, but risky, new technologies and service or operational concepts to provide information that can be used nationally; to address economic and social issues resulting from human impacts on the environment, and develop risk assessment methodologies, integrated assessments, and other analytical tools for effective policy formulation; to develop practical know-how for solving fundamental industry-wide problems, such as how to accommodate the travel needs of persons with disabilities, how to finance transit infrastructure construction and maintenance, and how to meet requirements of the Clean Air Act; and to support development of information and technical assistance to convey results of research, technology development, and innovative demonstrations for adaptation and local implementation.
Requirements: Eligible applicants include public bodies, nonprofit institutions, state and local agencies, universities, and legally constituted public agencies and operators of public transportation services.
Geographic Focus: All States
Contact: Associate Administrator for Research, Demonstration, and Innovation, Federal Transit Administration; (202) 366-4052
Internet: http://www.fta.dot.gov
Sponsor: U.S. Department of Transportation
400 Seventh Street SW
Washington, DC 20590

DOT Transportation Statistics Research Grants **1850**
The purpose of this grant program is to provide financial assistance to eligible organizations to help advance the discipline of transportation statistics. The Bureau of Transportation Statistics (BTS) is an operating administration within the DOT. Its mission is to lead in developing transportation data and information of high quality, and to advance their effective use in public and private transportation decisionmaking. The ultimate goal is to make transportation better—to enhance safety, mobility, economic growth, the human and natural environment, and national security.
Requirements: Public and private nonprofits of higher education are eligible to apply.
Geographic Focus: All States
Amount of Grant: 25,000-200,000 USD
Contact: Mary Hutzler, Acting Director, Bureau of Transportation Statistics; (202) 366-9913; fax (202) 366-3640; answers@bts.gov
Internet: http://www.bts.gov
Sponsor: U.S. Department of Transportation
400 Seventh Street SW
Washington, DC 20590

Douglass Institute Postdoctoral Fellowships **1851**
The Frederick Douglass Institute (FDI) for African and African-American Studies at the University of Rochester brings together scholars working in the humanities and social sciences in the area of Africa and the African Diasporas. Its mission is to further undergraduate and graduate education, faculty research, and intellectual exchange within the University community. The Postdoctoral Fellowship is awarded to scholars who hold a Ph.D. degree in a field that contributes to African and African-American Studies. It carries an annual stipend of $40,000 and a $3,000 stipend for research-related activities. The Fellow will participate in the intellectual life of the Institute, pursue his or her own scholarship, and teach one course in his or her area of specialization. All Fellows receive office space in the Institute, full access to the University's facilities, and opportunities to interact and collaborate with scholars relevant to their projects within the University. Fellows must be in full-time residence during the tenure of their awards, and they are required to be engaged in scholarly activity on a full-time basis. Fellows are expected to be available for consultation with students and colleagues; make at least one formal presentation based upon their research; and contribute generally to the intellectual life of the Institute. Applicants must have their Ph.D. in hand before the fellowship begins. Applications are accepted from September 1 through December 31, with awards announced in early March.

Requirements: Applicants must submit the following: the application form available at the website; a curriculum vitae; a 3-5 page description of the research the applicant plans to carry out during the fellowship year, plus a description of the undergraduate course the applicant would like to teach; a sample of published or unpublished writing on a topic related to the proposal; and three letters of recommendation that comment upon the value and feasibility of the work proposed.
Geographic Focus: All States
Amount of Grant: 43,000 USD
Contact: Ghislaine Radegonde-Eison, Program Manager; (585) 275-9161 or (585) 276-5744; fax (585) 256-2594; fdi@mail.rochester.edu
Internet: http://www.rochester.edu/College/AAS/fellowships/
Sponsor: University of Rochester-Douglass Institute
302 Morey Hall
Rochester, NY 14627-0440

Douglass Institute Predoctoral Fellowships **1852**
The Frederick Douglass Institute for African and African-American Studies at the University of Rochester brings together scholars working in the humanities and social sciences in the area of Africa and the African Diasporas. Its mission is to further undergraduate and graduate education, faculty research, and intellectual exchange within the University community. The Predoctoral Fellowship is awarded to a graduate student of any university whose degree contributes to the scholarship in the field of African and African-American Studies. A principal aim of this Fellowship is to provide a different intellectual environment that will enrich and expedite the completion of the Fellow's dissertation. The Fellow will also gain valuable experience through work with the Institute's director in organizing colloquia, lectures, and other events. There is no teaching obligation. The award carries an annual stipend of $26,000, and the Institute offers research funds to support the Fellow's research. All Fellows receive office space in the Institute, full access to the University's facilities, and opportunities to interact and collaborate with scholars relevant to their projects within the University. Fellows must be in full-time residence during the tenure of their awards and they are required to be engaged in scholarly activity on a full-time basis. Fellows are expected to be available for consultation with students and colleagues; make at least one formal presentation based upon their research; and contribute generally to the intellectual life of the Institute. To qualify for this residential Fellowship, an applicant will need to have completed the following: all required courses; qualifying oral and/or written exams; at least one chapter of the dissertation (this chapter is a required portion of the application package). Application materials and additional information are available at the FDI website. Applications are accepted from September 1 through December 31, with awards announced in early March.
Requirements: Applicants should submit a completed FDI fellowship application form; a curriculum vitae; an official transcript showing completion of all preliminary coursework and qualifying examinations; the dissertation prospectus; completed chapter(s) from the dissertation; and three letters of recommendation, including one from the dissertation supervisor that specifically addresses the candidate's prospects for completing the project within a year.
Geographic Focus: All States
Amount of Grant: 26,000 USD
Contact: Ghislaine Radegonde-Eison, Program Manager; (585) 275-9161 or (585) 276-5744; fax (585) 256-2594; fdi@mail.rochester.edu
Internet: http://www.rochester.edu/College/AAS/fellowships/
Sponsor: University of Rochester-Douglass Institute
302 Morey Hall
Rochester, NY 14627-0440

Dr. Bob and Jean Smith Foundation Grants **1853**
The foundation awards grants to Texas nonprofit organizations in its areas of interest, including arts and culture, higher education, medical education, medical research, and health care. Types of support include annual campaigns, building construction/renovation, capital campaigns, general operating support, matching/challenge grants, and scholarship funds. There are no application deadlines; the board meets quarterly.
Requirements: Texas nonprofit organizations are eligible. Preference is given to Dallas-based organizations.
Restrictions: Individuals are ineligible.
Geographic Focus: Texas
Contact: Sally Smith, Grants Administrator; (214) 521-3461
Sponsor: Dr. Bob and Jean Smith Foundation
3811 Turtle Creek Center, #2150 LB 53
Dallas, TX 75219

Dr. Robert H. Goddard Research Fellowships **1854**
These one-year, nonrenewable fellowships are awarded for full-time graduate study at the institute. In addition to the fellowship stipend, the award includes up to 20 credits of tuition, $1,591/month for 12 months, one year fellowship. Interested persons should contact the graduate admissions office for details on enrolling for graduate study and eligibility for the fellowships. Preference given to doctoral applicants.
Requirements: Applicants must be U.S. citizens with acceptance to a WPI graduate-degree program. GRE or GMAT test scores are required.
Geographic Focus: All States
Date(s) Application is Due: Feb 15
Contact: Graduate Admissions, (508) 831-5301; fax (508) 831-5717; gse@wpi.edu
Internet: http://www.grad.wpi.edu/Financial/fellowships.html
Sponsor: Worcester Polytechnic Institute
100 Institute Road
Worcester, MA 01609-2280

Dreyfus Postdoctoral Program in Environmental Chemistry Grant **1855**
The Camille and Henry Dreyfus Foundation seeks to further the development of scientific leadership in the field of environmental chemistry with a postdoctoral fellowship program. The Postdoctoral Program in Environmental Chemistry provides a principal investigator with an award of $120,000 over two years to appoint a Postdoctoral Fellow in environmental chemistry. Applications most likely to be of interest should describe innovative fundamental research in the chemical sciences or engineering related to the environment. Examples include but are not limited to the chemistry associated with: the climate, the atmosphere, aquatic or marine settings, toxicology, soil or groundwater. Also of interest are chemistry-related energy research (renewable sources, sequestration, etc.), and new or green approaches to chemical synthesis and processing, with a clearly stated relation to the environment.
Requirements: The program is open to all academic and other not-for-profit organizations that have well-established research efforts in environmental science or engineering in the States, Districts, and Territories of the U.S.. These research activities need not be located in traditional departments in the chemical sciences, and collaboration across departments and institutions is encouraged.
Restrictions: No part of the award may be used for institutional administrative purposes.
Geographic Focus: All States
Date(s) Application is Due: Aug 25
Amount of Grant: 120,000 USD
Contact: Mark J. Cordillo, Executive Director; (212) 753-1760; fax (212) 593-2256; mcardillo@dreyfus.org or admin@dreyfus.org
Internet: http://www.dreyfus.org/awards/postdoctoral_program.shtml
Sponsor: Camille and Henry Dreyfus Foundation
555 Madison Avenue, 20th Floor
New York, NY 10022-3301

Dreyfus Prize in the Chemical Sciences **1856**
The Dreyfus Prize in the Chemical Sciences, awarded biennially, consists of a citation, a medal, and a monetary award of $250,000. The prize is awarded to an individual in a selected area of chemistry to recognize exceptional and original research that has advanced the field in a major way. The first Dreyfus Prize was awarded to George Whitesides of Harvard University in the field of materials chemistry, honoring the accomplishments of the Dreyfus brothers, Camille and Henry. All nomination materials and letters of support are due at the Foundation office by the deadline March 1.
Requirements: Any person may nominate a candidate for the Dreyfus Prize in the Chemical Sciences. Self-nominations are not accepted.
Restrictions: The prize is restricted to U.S. residents. Nobel laureates are not eligible. Dreyfus Foundation Advisors and reviewers who serve in the year of the selection are not eligible. There is no restriction on the number of nominees who are at a given institution, nor is institutional approval required.
Geographic Focus: All States
Date(s) Application is Due: Mar 1
Amount of Grant: 250,000 USD
Contact: Mark J. Cordillo, Executive Director; (212) 753-1760; fax (212) 593-2256; mcardillo@dreyfus.org or admin@dreyfus.org
Internet: http://www.dreyfus.org/awards/prize.shtml
Sponsor: Camille and Henry Dreyfus Foundation
555 Madison Avenue, 20th Floor
New York, NY 10022-3301

Dreyfus Senior Scientist Mentor Grants **1857**
The Camille and Henry Dreyfus Foundation supports emeritus faculty who maintain active research programs with undergraduates in the chemical sciences. The Senior Scientist Mentor Program provides an award of $20,000 over two years for undergraduate stipends and modest research support. The program enables undergraduates to conduct research in the chemical sciences in a close working relationship with emeritus faculty members. Only one faculty member may apply per application, although more than one application per department or institution is permitted. A letter of support must be sent directly to the Foundation from a colleague, preferably from outside the institution, who is familiar with the applicant's research and teaching and who can speak to the applicant's experience in mentoring and advising undergraduates. Letters of support cannot be accepted via electronic mail or facsimile.
Requirements: Faculty with emeritus status on or before January of the current year, and who maintain active research programs in the chemical sciences may apply for one of a limited number of awards that will allow undergraduates to do research under their guidance. Successful applicants, who are expected to be closely engaged in a mentoring relationship with the students will receive grants annually for two years for undergraduate stipends and modest research support.
Restrictions: Charges associated with indirect costs or institutional overhead are not allowed. Faculty salary and any expenses associated with graduate students are not permitted.
Geographic Focus: All States
Date(s) Application is Due: Nov 12
Amount of Grant: 20,000 USD
Contact: Mark J. Cordillo, Executive Director; (212) 753-1760; fax (212) 593-2256; mcardillo@dreyfus.org or admin@dreyfus.org
Internet: http://www.dreyfus.org/awards/senior_scientist_mentor.shtml
Sponsor: Camille and Henry Dreyfus Foundation
555 Madison Avenue, 20th Floor
New York, NY 10022-3301

Dreyfus Special Grants Program in the Chemical Sciences 1858
This program offers support to eligible institutions for projects that propose to advance the science of chemistry in innovative ways. Proposals are invited in any area consistent with the Foundation's broad objective of advancing the chemical sciences. For projects that require sustaining support, note that this program is intended to seed the initial phases, with the expectation that grant recipients will find continuing funding from other sources. The Foundation encourages proposals that are judged likely to significantly advance the chemical sciences. Examples of areas of interest include (but are not limited to): the increase in public awareness, understanding, and appreciation of the chemical sciences; innovative approaches to chemistry education at all levels (K-12, undergraduate, and graduate); and efforts to make chemistry careers more attractive. Research proposals are not customarily considered.
Requirements: The Special Grant Program in the Chemical Sciences is open to institutions in the States, Districts, and Territories of the United States of America that have a focus in the chemical sciences. Institutions include schools, colleges and universities, as well as other not-for-profit organizations, such as scientific societies and science museums.
Restrictions: Awards are not made directly to individuals, nor, in general, to private foundations.
Geographic Focus: All States
Contact: Mark J. Cordillo, Executive Director; (212) 753-1760; fax (212) 593-2256; mcardillo@dreyfus.org or admin@dreyfus.org
Internet: http://www.dreyfus.org/awards/special_grant_program_chemical.shtml
Sponsor: Camille and Henry Dreyfus Foundation
555 Madison Avenue, 20th Floor
New York, NY 10022-3301

Drs. Bruce and Lee Foundation Grants 1859
The Foundation's goal is to advance the general welfare and the quality of all life in the Florence, S.C. area by providing economic support to qualified programs and non-profit organizations. The Foundation will support a broad range of charitable purposes including, but not limited to: medical; health; human services; education; arts; religion; civic affairs; and the conservation, preservation and promotion of cultural, historical and environmental resources. There are no application deadlines. Contact the office for application materials and guidelines.
Requirements: Florence, SC, nonprofit organizations are eligible.
Restrictions: The Foundation does not purchase tickets for fundraising events. Grants to individuals will not be considered.
Geographic Focus: South Carolina
Samples: Francis Marion U (Florence, SC)—to construct a facility for the new bachelor's-degree nursing program, $5 million.
Contact: L. Bradley Callicott, Director; (843) 664-2870; blfound@bellsouth.net
Sponsor: Drs. Bruce and Lee Foundation
181 East Evans Street, BTC Box 022
Florence, SC 29506

DSO Cognitive Technology Threat Warning System (CT2WS) Grants 1860
The objective of the CT2WS program is to drive a breakthrough in soldier-portable visual threat warning devices. Recent developments and discoveries in the disparate technology areas of flat-field, wide-angle optics, large pixel-count digital imagers, cognitive visual processing algorithms, neurally-based target detection signatures and ultra-low power analog-digital hybrid signal processing electronics have led DARPA to believe that focused technology development, system design, and system integration efforts may produce revolutionary capabilities for the warfighter. The final objective of the DARPA CT2WS program is the development of prototype soldier-portable digital imaging threat queuing systems capable of effective detection ranges of 1-10 km against dismounts and vehicles while simultaneously surveying a 120-degree or greater field of view (FOV). DARPA requests proposals for the full scope of development (e.g., an end-to-end system designed by a team of multidisciplinary research organizations, plus an integrator for coordination and implementation support). Proposals addressing only individual component-level technologies may be considered non-compliant to this BAA. This BAA affords proposers the choice of submitting proposals for the award of a Grant, Cooperative Agreement, Procurement Contract, Technology Investment Agreement, Other Transaction for Prototype Agreement, or other such appropriate award instrument.
Requirements: The Government anticipates proposals submitted under this BAA will be Unclassified.
Geographic Focus: All States
Date(s) Application is Due: Apr 11
Contact: Amy Kruse, Program Manager; (571) 218-4338; fax (703) 248-1908; amy.kruse@darpa.mil or BAA07-25@darpa.mil
Internet: http://www.darpa.mil/baa/BAA07-25.html
Sponsor: Defense Sciences Office within the Defense Advanced Research Projects Agency
3701 N Fairfax Drive
Arlington, VA 22203-1714

DSO Computer Science Study Group (CSSG) Grants 1861
This Research Announcement (RA) initiates Phase 1 of the 2008 Computer Science Study Group (CSSG). The objective of the CSSG is to rapidly identify ideas in the field of computer science that will provide revolutionary advances, rather than incremental benefit, to the Department of Defense. Phase 1 (12 months) consists of participation on the CS Study Panel. The CS Study Panel anticipates meeting at least four times throughout the 2008 calendar year, totaling approximately twenty days. Two of the meetings will occur during the academic year and will take place in the Washington, D.C. metropolitan area. Two extended meetings, lasting approximately one week each, will take place during the summer, and will involve travel throughout the United States. For successful proposers, host institutions will receive grants or other assistance instruments for up to $100,000. The funding to the host institution will support participant travel expenses for Panel meetings; it will also support participant salary, to include time spent while participating in the CS Study Panel and time to connect the participants' computer science research to critical DoD needs identified during their Panel trips. The funding may also support research assistants' summer salary and costs for administrative support.
Requirements: All responsible sources may submit a proposal that shall be considered by DARPA. Historically Black Colleges and Universities (HBCUs) and Minority Institutions (MIs) are encouraged to submit proposals.
Restrictions: Only U.S. institutions of higher learning are eligible for awards under this RA, and participants must be junior faculty members who are able to receive a U.S. Department of Defense SECRET security clearance.
Geographic Focus: All States
Date(s) Application is Due: Aug 10
Contact: Dr. Benjamin Mann, Program Manager; (571) 218-4246; fax (571) 218-4553; benjamin.mann@darpa.mil or RA07-43@darpa.mil
Internet: http://www.darpa.mil/baa/RA07-43.html
Sponsor: Defense Sciences Office, Defense Advanced Research Projects Agency
3701 N Fairfax Drive
Arlington, VA 22203-1714

DSO Controlling Pathogen Evolution Workshop Grants 1862
The goal of the workshop is to explore: the potential of computational methods to precisely predict pathogen evolution; and vaccination and/or therapeutic strategies to control direction of pathogen evolution. The workshop will investigate tools for predicting protective epitopes, pathogen evolution and protein structure, and then review strategies to force pathogens into an evolutionary trap that prevents escape through mutation. From there, vaccines and therapies can be developed against the particular pathogen strain. Workshop participants are strongly encouraged to prepare posters describing previous or potential research in the area in order to facilitate discussions and/or formation of well-rounded teams.
Requirements: Attendees should have specific expertise in infectious disease, epitope prediction (computational and/or immunomic), vaccinology, mathematical modeling, protein structure, pathogen evolution, drug design, and/or animal modeling.
Geographic Focus: All States
Date(s) Application is Due: Mar 16
Contact: Dr. Michael V. Callahan, Program Manager; (571) 218-4596; fax (571) 218-4553; michael.callahan@darpa.mil
Internet: http://www.darpa.mil/baa/SN07-21.html
Sponsor: Defense Sciences Office, Defense Advanced Research Projects Agency
3701 N Fairfax Drive
Arlington, VA 22203-1714

DSO Defense Sciences Research and Technology Grants 1863
TThe mission of the DSO is to identify and pursue high-risk/high-payoff research initiatives throughout a broad spectrum of the science and engineering disciplines, and to transform these initiatives into important, radically new military capabilities. To carry out this mission, DSO seeks research ideas and areas that might lead to innovations in science and engineering. Therefore, the office is soliciting proposals for advanced research and development in a variety of enabling technical areas. Proposals may be either basic or applied research. However, in all cases, proposers should demonstrate that their proposed effort is aimed at high-risk/high-payoff technologies that have the potential for making, in the 5-10 year time frame, revolutionary rather than incremental improvements to national security, including emerging threats and operational challenges. Multiple awards are anticipated. The amount of resources made available to this BAA will depend on the quality of the proposals received and the availability of funds. While there is no specific requirement for cost and duration of the proposed effort, it is recommended that proposers include a Phase I of 12 to 18 months in length that addresses the most critical issues on the path to success.
Requirements: All responsible sources may submit a proposal that shall be considered by DARPA.
Geographic Focus: All States
Date(s) Application is Due: Feb 14
Contact: Barbara K. McQuiston, Deputy Director; (703) 526-4759; fax (703) 248-1916; barbara.mcquiston@darpa.mil or BAA07-21@darpa.mil
Internet: http://www.darpa.mil/baa/baa07-21mod1.html
Sponsor: Defense Sciences Office, Defense Advanced Research Projects Agency
3701 N Fairfax Drive
Arlington, VA 22203-1714

DSO Radiation Biodosimetry (RaBiD) Grants 1864
The DSO is seeking proposals for new technologies for rapid, high-throughput, portable and low-cost biodosimeters to determine radiation dose to individuals after acute radiation exposure. This technology would provide rapid identification of individuals who have been exposed to high-dose radiation in order to accurately assess radiation exposure levels. The Program is a single phase, 15-month program with a goal of revolutionizing radiation exposure detection. The first 12 months consist of scientific research and development during which the proposer will develop a non- or minimally-invasive radiological biodosimeter and demonstrate accurate radiological detection from biological samples into quartiles of doses for humans with a detection time less than 10 minutes. This will include radiological dose/detection curves, as well as multipoint data regarding decay of biological signal as a function of time after radiation exposure.

A successful proposal will demonstrate a clear path to these deliverables, including a timeline for technology development, experimentation, and delivery of results.
Requirements: All proposers must provide evidence of a competent team capable of developing a new radiation biodosimetry technology meeting all program objectives and milestones. The Program requires that each proposal include a team with demonstrated (or established) capability. At a minimum, proposers are expected to possess expertise or demonstrate collaboration with professionals in the following areas: radiation biologist/radiation oncologist—with expertise in radiation exposure to animal models; and engineering—with developmental expertise needed to translate biological sampling to radiation readout.
Restrictions: Only unclassified proposals will be accepted in response.
Geographic Focus: All States
Date(s) Application is Due: Jul 9
Contact: Dr. Mildred Donlon, Program Manager; (703) 696-2289; fax (703) 741-3896; mildred.donlon@darpa.mil
Internet: http://www.darpa.mil/dso/solicitations/baa07-29.htm
Sponsor: Defense Sciences Office, Defense Advanced Research Projects Agency
3701 N Fairfax Drive
Arlington, VA 22203-1714

DTE Energy Foundation Health and Human Services Grants 1865
The DTE Energy Foundation is at the core of DTE Energy's commitment to the communities and customers it is privileged to serve. The DTE Energy Foundation is dedicated to strengthening the health and human services sector of these communities. Priority will be given to supporting organizations that are in the forefront of addressing the critical, acute human needs brought on by the economic downturn. Grant amounts generally range from $500 to $100,000, and the application process is distinct for each of the following ranges: $500 to $2,000; $2,001 to $10,000; and any amount greater than $10,000. Applications must be submitted electronically by the stated deadlines.
Requirements: Eligible applicants must meet all of the following criteria: be located in or provide services to a community in which DTE Energy does business; and be a nonprofit (i.e. be exempt for federal income tax under section 501(c)3 of the Internal Revenue Code and not a private foundation, as defined in Section 509(a) of the Code).
Restrictions: The Foundation does not provide support to: individuals; political parties, organizations or activities; religious organizations for religious purposes; organizations that are not able to demonstrate commitment to equality and diversity; student group trips; national or international organizations, unless they are providing benefits directly to our service-area residents; projects that may result in undue personal benefit to a member of the DTE Energy Foundation board, or to any DTE Energy employee; conferences unless they are aligned with DTE Energy's business interests; single purpose health organizations; hospitals, for building or equipment needs.
Geographic Focus: Michigan
Date(s) Application is Due: Apr 13; Jul 13; Oct 12; Dec 28
Contact: Karla D. Hall; (313) 235-9271 or (313) 235-9416; foundation@dteenergy.com
Sponsor: DTE Energy Foundation
One Energy Plaza, 1046 WBC
Detroit, MI 48226-1279

DTE Energy Foundation Higher Learning Grants 1866
In the area of Higher Learning grants, the DTE Energy Foundation's focus is on the following priorities: specific academic departments of engineering, science and business that prepare students to enter the professional workforce; programs and institutions that prepare students for technical and skilled trade careers in the energy industry; and programs that increase student retention and success in engineering, the sciences and business with a focus on women and minorities. Grant amounts generally range from $2,000 to $750,000, and the application process is distinct for each of the following ranges: $500 to $2,000; $2,001 to $10,000; and any amount greater than $10,000. Applications must be submitted electronically by the stated deadlines.
Requirements: Eligible applicants must meet all of the following criteria: be located in or provide services to a community in which DTE Energy does business; and be a nonprofit (i.e. be exempt for federal income tax under section 501(c)3 of the Internal Revenue Code and not a private foundation, as defined in Section 509(a) of the Code).
Restrictions: The Foundation does not provide support to: individuals; political parties, organizations or activities; religious organizations for religious purposes; organizations that are not able to demonstrate commitment to equality and diversity; student group trips; national or international organizations, unless they are providing benefits directly to our service-area residents; projects that may result in undue personal benefit to a member of the DTE Energy Foundation board, or to any DTE Energy employee; conferences unless they are aligned with DTE Energy's business interests; single purpose health organizations; hospitals, for building or equipment needs.
Geographic Focus: Michigan
Date(s) Application is Due: Apr 13; Jul 13; Oct 12; Dec 28
Contact: Karla D. Hall; (313) 235-9271 or (313) 235-9416; foundation@dteenergy.com
Internet: http://www.dteenergy.com/dteEnergyCompany/community/foundation/whatWeSupport.html
Sponsor: DTE Energy Foundation
One Energy Plaza, 1046 WBC
Detroit, MI 48226-1279

Duchossois Family Foundation Grants 1867
The foundation focuses its efforts in the Chicago metropolitan area and gives primary consideration to nonprofit organizations that contribute to the community in the area of health. The foundation supports organizations involved with mental health/crisis services, cancer, HIV/AIDS, cancer research, AIDS research and, human services. A one-page summary-request letter should include a description of the organization and its specific needs and purposes, the amount requested, and a list of members of the board of directors and their business/professional affiliations.
Requirements: 501(c)3 tax-exempt public charities serving the Chicago metropolitan area are eligible.
Geographic Focus: Illinois
Contact: Iris Krieg, Executive Director; (312) 641-5765; iriskrieg1@aol.com
Sponsor: Duchossois Family Foundation (Chicago)
203 N Wabash Avenue, Suite 1800
Chicago, IL 60601

Duke Foundation Fellowships 1868
Beginning in the fall semester of 2009, a limited number of graduate fellowships will be available for students in the Ph. D. Program in Public Policy Analysis in the Terry Sanford Institute of Public Policy who intend to focus their dissertation on the theory of foundations, foundation strategy, the measurement of foundation impact, and foundation accountability.
Geographic Focus: All States
Contact: Bruce Kuniholm, Director; (919) 613-7309; bruce.kuniholm@duke.edu
Internet: http://www.pubpol.duke.edu/centers/dfrp/fellowship.php
Sponsor: Duke Foundation, Sanford Institute of Public Policy
Box 90239
Durham, NC 27708-0239

Duke University Adult Cardiothoracic Anesthesia and Critical Care Medicine Fellowships 1869
Nine one year fellowship positions are available each year in the ACGME Accredited Fellowship. Second year research and clinical fellowship opportunities are available; programs are individualized to meet the goals of successful applicants. Generally clinical experience is the focus of the first fellowship year, along with involvement in clinical research. Fellows opting for a two year fellowship continue with one clinical day per week in their second year, in addition to a structured plan involving laboratory and/or clinical research and teaching. Second year clinical specialization in pediatric cardiac anesthesiology or intensive care is also available. The Adult cardiothoracic Anesthesia Fellowship is highly recommended as formal training for specialization in cardiac and/or thoracic anesthesiology. The Fellowship is designed to develop clinical, consultant, and in-depth research expertise. Applications are usually considered starting 18 months prior to the start date
Requirements: Complete application (see Duke University website), in addition to the references required, include a letter from your Chairperson/Program Director and two letters from other faculty members who have worked with you closely.
Geographic Focus: All States
Contact: Mark S. Smith, Director; (919) 681-5046 or (919) 416-3853; fax (919) 681-8993
Internet: http://medschool.duke.edu/modules/som_rt/index.php?id=1
Sponsor: Duke University Medical Center
P.O. Box 3094
Durham, NC 27710

Duke University Ambulatory and Regional Anesthesia Fellowships 1870
The Fellowship in Ambulatory/Regional Anesthesia is designed to provide advanced training in anesthesia in the ambulatory surgery setting. This program will be an extension to the previous training with an emphases on the continued educational component to include research, clinical and teaching responsibilities.
Requirements: The selection process for fellows includes the completion of a Fellowship Application (see website) which is to be substantiated by three letters of recommendation from experts in the field. In addition to the letters of reference the application is to include a current CV and a letter of affirmation stating interest in the program.
Geographic Focus: All States
Date(s) Application is Due: Aug 15
Contact: Stephen M. Klein, Director; (919) 668-2056; klein006@mc.duke.edu
Internet: http://anesthesiology.duke.edu/modules/anes_flwshp/index.php?id=1
Sponsor: Duke University Medical Center
P.O. Box 3094
Durham, NC 27710

Duke University Clinical Cardiac Electrophysiology Fellowships 1871
The Clinical Cardiac Electrophysiology fellowship training program is a comprehensive two-year program including training in catheter ablation of supraventricular tachycardias, ventricular tachycardias, and complex arrhythmias such as atrial fibrillation, ischemic VT, and arrhythmias associated with congenital heart disease. Comprehensive training is also provided in lead extraction and device therapy including permanent pacing, implantable defibrillators, and bi-ventricular devices. Fellows are provided with significant opportunity for research activity through protected block time and are expected to participate in and complete a research project with a faculty mentor. Applicants must have completed at least three years in an ACGME approved Cardiology fellowship program.
Requirements: Applications for training to begin July 1, will be accepted beginning January 1; the application deadline will be September 1. To apply, please include with the completed CCE (EP) Application form (available on the Duke University website), a current curriculum vitae, personal statement (not to exceed 250 words), and the $40 (check or money/order no cash please) processing fee.
Geographic Focus: All States
Date(s) Application is Due: Sep 1
Contact: Dr. Patrick Hranitzky, Program Director; (919) 684-2304; fax (919) 684-4322; Patrick.hranitzky@duke.edu

Internet: http://fellowships.medicine.duke.edu/modules/fellows_cardio/index.php?id=8
Sponsor: Duke University School of Medicine
P.O. Box 3878
Durham, NC 27710

Duke University Hyperbaric Center Fellowships 1872
The clinical Hyperbaric fellow works in the Hyperbaric Center and F.G. Hall Environmental Laboratory, both of which are an integral part of Duke University Medical Center. Fellowships are available for 12 months although they may be extended for additional time. During the period of training the Hyperbaric fellow will be trained in the clinical use of hyperbaric oxygen. This includes the emergency treatment of clostridial myonecrosis, necrotizing fasciitis, arterial gas embolism, decompression sickness, carbon monoxide poisoning and certain other less common conditions such as acute vasculitis with ischemia. The fellow will receive training in the appropriate critical care interventions necessary for the treatment of these patients. This will include hemodynamic monitoring, ventilatory support, fluid resuscitation and the interpretation of arterial blood gases during hyperbaric therapy. The fellow will also receive training in the use of hyperbaric oxygen therapy for non-emergency conditions. The fellow will be responsible for all inpatient and outpatient consults, under the direction of the clinical Hyperbaric faculty. During the training period the fellow will also be required to complete a research project. This will be completed under the guidance of one of the faculty members in the Hyperbaric Center. Part of the didactic program includes regular hyperbaric conferences. The fellow will participate in these and normally be required to present at least one topic during his year of training. The fellow will also have the opportunity to attend the annual two-week course in Hyperbaric and Diving Medicine given by the National Oceanic and Atmospheric Administration in Seattle, Washington. In addition to these activities the applicant may participate in numerous other teaching activities within Duke Medical Center, for example those sponsored by the Departments of Anesthesiology, Medicine, Surgery and Emergency Medicine. Clinical participation under the aegis of those departments can be arranged in some cases, depending upon the training of the individual concerned.
Requirements: Applicants must be eligible for license as a trainee in the state of North Carolina. The Duke Hyperbaric Center Fellowship Application can be downloaded at: http://hyperbaric.mc.duke.edu/education.htm
Geographic Focus: All States
Contact: Tonya Manning, Assistant; (919) 681-1685 or (919) 684-6726
Internet: http://hyperbaric.mc.duke.edu/education.htm
Sponsor: Duke University Medical Center for Hyperbaric Medicine and Environmental Physiology
P.O. Box 3823
Durham, NC 27710

Duke University Interventional Cardiology Fellowships 1873
The Interventional Cardiology Fellowship Training Program at Duke University Medical Center focuses on training leaders in Interventional Cardiology committed to a career in academic cardiology. The length of the fellowship is 12 months. During the fellowship, all fellows will develop both clinical and technical skills in interventional cardiology. All fellows will gain experience in diagnosing, selecting therapies, performing diagnostic and interventional procedures, and judging the effectiveness of treatments for inpatients and outpatients with chronic ischemic heart disease, acute ischemic syndromes, and valvular heart disease. All fellows are expected to participate in research efforts. During the fellowship, each fellow will work with a mentor on a research project which will help foster their long-term goals.
Requirements: Individuals applying for only a 1 year Interventional Cardiology Fellowship position will be required to provide his/her own funding resources for salary support during their training. This funding must be institutional funding (no personal or private funds). A processing fee of $40 (check or money/order made payable to Duke University- no cash please), letter documenting funding support, current curriculum vitae, and personal statement will be required to accompany the completed Interventional Cardiology Application (see Duke University website for application).
Geographic Focus: All States
Date(s) Application is Due: Feb 1
Contact: Sherolyn Patterson, Fellowship Program Coordinator; (919) 684-2304; fax (919) 684-4322; sherolyn.patterson@duke.edu
Internet: http://fellowships.medicine.duke.edu/modules/fellows_cardio/index.php?id=8
Sponsor: Duke University School of Medicine
P.O. Box 3878
Durham, NC 27710

Duke University Libraries Franklin Collection Travel Grants 1874
The Franklin Collection offers travel grants for research involving the use of the collections of the John Hope Franklin Collection or for research with a focus on African and African-American studies. The maximum award per applicant is $1,000. Applicants are strongly encouraged to speak with the public services librarian before applying for the grant.
Requirements: Any faculty member, graduate or undergraduate student, or independent scholar with a research project directly related to the collections held by the Bingham Center or to women's history and culture is eligible to apply.
Restrictions: All applicants must reside beyond a 50 mile radius of Durham, North Carolina.
Geographic Focus: North Carolina
Date(s) Application is Due: Jan 15
Amount of Grant: Up to 1,000 USD
Contact: Karen Jean Hunt, Director; (919) 660-5922; fax (919) 660-5934; franklin-collection@duke.edu
Internet: http://library.duke.edu/specialcollections/franklin/grants.html
Sponsor: Duke University Libraries
Box 90185
Durham, NC 27708-0185

Duke University Libraries Hartman Center Travel Grants 1875
The Grants provide up to $750 to support those individuals whose research focus includes the use of any Hartman Center collections and whose research trips may last any length of time. Grant money may be used for travel to the Rare Book, Manuscript, and Special Collections Library, costs of copying pertinent archival resources, and living expenses while pursuing research here. Travel must originate from at least 60 miles away from Duke University. All applicants must submit a 250-500 word essay summarizing their project and detailing why the Hartman Center collections are essential to their research.
Requirements: Any faculty member, graduate or undergraduate student, or independent scholar with a research project directly related to the collections held by the Bingham Center or to women's history and culture is eligible to apply.
Geographic Focus: North Carolina
Date(s) Application is Due: Jan 15
Amount of Grant: Up to 750 USD
Contact: Jacqueline Reid; (919) 660-5836; fax (919) 660-5934; j.reid@duke.edu
Internet: http://library.duke.edu/specialcollections/hartman/travel-grants/index.html
Sponsor: Duke University Libraries
Box 90185
Durham, NC 27708-0185

Duke University Libraries J. Walter Thompson Research Fellowships 1876
The Fellowships provide $1000 in funds for up to three individuals planning research trips for a minimum of two weeks primarily use the J. Walter Thompson Archives. The JWT archives hold over 5,000 linear feet dating as far back as the 19th century and as recent as the 21st century. The collections document the pioneering role of JWT in the advertising industry and illustrates the changes in the broader industry over the past 100 years through its own growth and diversification. Grant money may be used for travel to the Rare Book, Manuscript, and Special Collections Library, costs of copying pertinent archival resources, and living expenses while pursuing research here. Travel must originate from at least 60 miles away from Duke University. All applicants must submit a 250-500 word essay summarizing their project and detailing why the Hartman Center collections are essential to their research.
Requirements: Any faculty member, graduate or undergraduate student, or independent scholar with a research project directly related to the collections held by the Bingham Center or to women's history and culture is eligible to apply.
Geographic Focus: North Carolina
Date(s) Application is Due: Jan 15
Amount of Grant: Up to 1,000 USD
Contact: Jacqueline Reid; (919) 660-5836; fax (919) 660-5934; j.reid@duke.edu
Internet: http://library.duke.edu/specialcollections/hartman/travel-grants/index.html
Sponsor: Duke University Libraries
Box 90185
Durham, NC 27708-0185

Duke University Libraries Mary Lily Research Grants 1877
The Grants are for research involving the use of their collections or research into other collections with a focus on women's history and culture. Grant money may be used for travel to the Rare Book, Manuscript, and Special Collections Library, costs of copying pertinent archival resources, and living expenses while pursuing research, including accommodations and meals. The maximum award per applicant is $1,000. Applicants are strongly encouraged to speak with the public services librarian before applying for the grant.
Requirements: Any faculty member, graduate or undergraduate student, or independent scholar with a research project directly related to the collections held by the Bingham Center or to women's history and culture is eligible to apply. All applicants must reside beyond a 50 mile radius of Durham, North Carolina.
Geographic Focus: All States
Date(s) Application is Due: Jan 15
Amount of Grant: Up to 1,000 USD
Contact: Kelly Wooten, (919) 660-5967; fax (919) 660-5934; kelly.wooten@duke.edu
Internet: http://library.duke.edu/specialcollections/bingham/grants/index.html
Sponsor: Duke University Libraries
Box 90185
Durham, NC 27708-0185

Duke University Obstetric Anesthesia Fellowships 1878
The Department of Anesthesiology at Duke University Medical Center offers 1-2 positions of a one year Fellowship in Obstetric Anesthesia. Emphasis is on clinical research, advanced clinical training and teaching. Duke University is a tertiary center with approximately 3,000 deliveries per year. The division is currently made up of 8 faculty staff with diverse international pedigree.
Geographic Focus: All States
Contact: Yemi Olufolabi, Fellowship Director; (919) 681-6535 or (919) 668-6266; fax (919) 668-6265; olufo001@mc.duke.edu
Internet: http://anesthesiology.duke.edu/modules/anes_flwshp/index.php?id=1
Sponsor: Duke University Medical Center
P.O. Box 3094
Durham, NC 27710

Duke University Pain Management Fellowships 1879
The Fellowship Program is based in the Duke Health Center at Morreene Road (MRC). Fellows will: cover the Pain Clinic and Durham VAMC, and Acute Pain Service at Duke North; evaluate new patients; perform comprehensive or neurologic studies; formulate diagnosis and plan; become familiar with procedures; learn intervention of the lumbar spine and its application e.g. facet rhizotomy, discography, and intradiscal annuloplasty; implantation therapies; dictation of clinic and procedure notes. Fellows will be asked to participate in research, journal clubs, round table discussions, grand rounds and lectures. There are opportunities for the fellow to become involved in research projects already in progress or to develop an original project. The clinical objectives are to build knowledge, skills and professionalism.
Requirements: Applicants must meet eligibility specified by the ACGME. Acceptance into the program is based on demonstrated academic qualification, interpersonal skills, and evidence of ability to function in a multidisciplinary setting. Fellowship applications may be downloaded from http://anesthesia.duhs.duke.edu/. Submit application, CV and three letters of reference.
Geographic Focus: All States
Date(s) Application is Due: Jun 30
Contact: Dianne L. Scott, Fellowship Director; (919) 684-6736
Internet: http://anesthesiology.duke.edu/modules/anes_flwshp/index.php?id=1
Sponsor: Duke University
932 Morreene Road, Room 232
Durham, NC 27705

Duke University Postdoctoral Research Fellowships in Aging 1880
The goal of the postdoctoral research training program is to produce highly skilled research scientists who have the potential for leadership in gerontological research. In the Duke Research Training Program (RTP), much of the training for each fellow is provided by that person's faculty mentor(s) in a research apprenticeship program. A fellow carries out his/her own research as a junior colleague in the mentor's research program or laboratory. In addition to working in their mentor's programs, all fellows attend a weekly interdisciplinary didactic seminar. The stipend amount is determined by years of relevant postdoctoral experience. The program also provides health insurance for the fellow (using the University's insurance plan), and partial travel support when presenting research at a professional meeting.
Requirements: Applicants must be citizens of the United States, or have been lawfully admitted to the United States for permanent residence. Applicants must have a completed doctoral degree when joining the program. All course work must be completed, and if applying as a Ph.D., final orals must be passed, and the dissertation signed before an applicant can begin the program.
Geographic Focus: All States
Date(s) Application is Due: May 1
Contact: James A. Blumenthal, (919) 660-7517; fax (919) 684-8569; blume003@mc.duke.edu
Internet: http://www.geri.duke.edu/post_doc/
Sponsor: Duke University School of Medicine
P.O. Box 3878
Durham, NC 27710

Durfee Foundation Stanton Fellowship 1881
The Durfee Foundation Stanton Fellowship provides up to six fellows with $75,000 each over a two-year period to think deeply about the intractable problems in their sector, and to tease out solutions that will improve life for the people of Los Angeles. Possible uses for the funds may include: travel to sister organizations in the U.S. or abroad; apprenticing to experts; writing policy papers; enrollment in training programs; or supplementing salaries of co-workers to allow release time for the fellow to concentrate on the project.
Requirements: Individual candidates for this program should: be outstanding leaders who have demonstrated a track record of contribution to the community; work for a nonprofit organization; have worked a minimum of ten years in the nonprofit sector; hold primary or significant responsibility for management of the organization's funds; be willing to participate in a peer network of program alumni; be employed full-time at the applicant organization; and have financial need. The candidate's employing organization should: be located in Los Angeles County; have at least three full-time staff; need financial assistance to underwrite the candidate's leave; and be able to sustain regular operations during the candidate's absence.
Geographic Focus: California
Contact: Executive Director; (310) 899-5120; fax (310) 899-5121; admin@durfee.org
Internet: http://durfee.org/what-we-support/stanton-fellowship/
Sponsor: Durfee Foundation
1453 Third Street, Suite 312
Santa Monica, CA 90401

E.B. Hershberg Award for Important Discoveries in Medicinally Active Substances 1882
This award, supported by Schering-Plough Corporation, is given biennially in odd-numbered years to recognize and encourage outstanding discoveries in the chemistry of medicinally active substances. The discovery for which the award is given should have been made during the past two decades. Nominations are accepted in even-numbered years, and the award consists of $3,000 and a certificate.
Requirements: Any individual, except a member of the award committee, may submit one nomination or seconding letter for the award in any given year. The nominating documents consist of a letter of not more than 1000 words containing an evaluation of the nominee's accomplishments and a specific identification of the work to be recognized, a biographical sketch including date of birth, and a list of publications and patents authored by the nominee.
Restrictions: Self-nominations are not accepted.
Geographic Focus: All States
Date(s) Application is Due: Nov 1
Amount of Grant: 4,000 USD
Contact: Felicia Dixon, Awards Administrator; (800) 227-5558 or (202) 872-4408; fax (202) 776-8008; f_dixon@acs.org or awards@acs.org
Internet: http://portal.acs.org/portal/acs/corg/content?_nfpb=true&_pageLabel=PP_ARTICLEMAIN&node_id=1319&content_id=CTP_004515&use_sec=true&sec_url_var=region1
Sponsor: American Chemical Society
1155 Sixteenth Street, NW
Washington, DC 20036-4801

E. Bright Wilson Award in Spectroscopy 1883
Supported by Rohm and Haas Company, this award recognizes outstanding accomplishments in fundamental or applied spectroscopy in chemistry. The award shall be granted to an individual without regard to age or nationality. The scope of the award is to cover all fields of spectroscopy in chemistry, both fundamental and applied. The recipient will deliver a lecture at the spring awards symposium of the ACS Division of Physical Chemistry.
Requirements: Any individual, except a member of the award committee, may submit one nomination or seconding letter for the award in any given year. Nominating documents consist of a letter of not more than 1000 words containing an evaluation of the nominee's accomplishments and a specific identification of the work to be recognized, a biographical sketch including date of birth, and a list of publications and patents authored by the nominee.
Restrictions: Self-nominations are not accepted.
Geographic Focus: All States
Date(s) Application is Due: Nov 1
Amount of Grant: 6,500 USD
Contact: Felicia Dixon, Awards Administrator; (800) 227-5558 or (202) 872-4408; fax (202) 776-8008; f_dixon@acs.org or awards@acs.org
Internet: http://portal.acs.org/portal/acs/corg/content?_nfpb=true&_pageLabel=PP_ARTICLEMAIN&node_id=1319&content_id=CTP_004511&use_sec=true&sec_url_var=region1
Sponsor: American Chemical Society
1155 Sixteenth Street, NW
Washington, DC 20036-4801

E.J. Sierleja Memorial Fellowship 1884
The fellowship is available annually to a graduate student pursuing advanced studies in the area of transportation, with priority consideration given to, but not limited to, students focusing on rail transportation. The institute supports the advancement of engineering education and research. The program is open to individuals in the United States, Canada, and Mexico.
Requirements: Candidates for these awards must be active institute members, as reflected on the October chapter roster or before. New member applications must be completely processed prior to the end of September in order for students to be eligible. Students must be enrolled full-time in graduate or undergraduate industrial engineering programs for the upcoming year and must have an overall point-hour average of 3.40 on a scale of 4.00; graduate students must have had a 3.40 average as undergraduates to qualify. Students may not apply directly for this scholarship, they must be nominated.
Geographic Focus: All States
Date(s) Application is Due: Nov 15
Contact: Bonnie Cameron; (770) 449-0461, ext. 105; bcameron@iienet.org
Internet: http://www.iienet.org/public/articles/index.cfm?cat=525
Sponsor: Institute of Industrial Engineers
3577 Parkway Lane, Suite 200
Norcross, GA 30092

E.L. Wiegand Foundation Grants 1885
The foundation provides grants to develop and strengthen programs and projects at Arizona, California, the District of Columbia, Idaho, Nevada, New York, Oregon, Utah, and Washington educational institutions in the academic areas of science, business, fine arts, and law; and medicine and health organizations in the areas of heart, eye, and cancer surgery, treatment, and research. The foundation also considers requests for projects that enrich children, communities, public policy, and art. Grants for education, including funds for computers and scientific equipment, are awarded. The board of trustees meets in February, June, and October to choose recipients, but applications may be submitted at any time. Application guidelines are available upon request.
Requirements: Nonprofit organizations in Arizona, California, District of Columbia, Idaho, Nevada, New York, Oregon, Utah, and Washington State are eligible.
Geographic Focus: Arizona, California, District of Columbia, Idaho, Nevada, New York, Oregon, Utah, Washington
Contact: Kristen Avansino, Executive Director; (775) 333-0310; fax (775) 333-0314
Sponsor: E.L. Wiegand Foundation
165 W Liberty Street, Suite 200
Reno, NV 89501

E.V. Murphree Award in Industrial and Engineering Chemistry 1886
This award, sponsored by the ExxonMobile Research and Engineering Company, is given annually to stimulate fundamental research in industrial and engineering chemistry, the development of chemical engineering principles, and their application to industrial processes. A nominee must have accomplished outstanding research of a theoretical or

experimental nature in the fields of industrial chemistry or chemical engineering. The award is granted without regard to age or nationality, and consists of $5,000 and a certificate.
Requirements: Any individual, except a member of the award committee, may submit one nomination or seconding letter for the award in any given year. Nominating documents consist of a letter of not more than 1000 words containing an evaluation of the nominee's accomplishments and a specific identification of the work to be recognized, a biographical sketch including date of birth, and a list of publications and patents authored by the nominee.
Restrictions: Self-nominations are not accepted.
Geographic Focus: All States
Date(s) Application is Due: Nov 1
Amount of Grant: 6,000 USD
Samples: Georges Belfort—award recipient, $5,000; Wolfgang F. Holderich—award recipient, $5,000; Liang-Shih Fan—award recipient, $5,000.
Contact: Felicia Dixon, Awards Administrator; (800) 227-5558 or (202) 872-4408; fax (202) 776-8008; f_dixon@acs.org or awards@acs.org
Internet: http://portal.acs.org/portal/acs/corg/content?_nfpb=true&_pageLabel=PP_ARTICLEMAIN&node_id=1319&content_id=CTP_004520&use_sec=true&sec_url_var=region1
Sponsor: American Chemical Society
1155 Sixteenth Street, NW
Washington, DC 20036-4801

E.W. "Al" Thrasher Awards **1887**
The purpose of the E.W. Al Thrasher Awards is to improve children's health through medical research, with an emphasis on projects that have the potential to translate into clinically meaningful results within a few years. The Fund awards grants three times per year, with no fixed number of awards given in each funding cycle or in each year. There are no deadlines for the awards. Concept papers are accepted on a rolling basis, and are considered by the Fund's Executive Committee approximately once per month. If a full proposal is invited, the applicant will have one year to submit a full proposal. Full proposals recommended by peer reviewers for further consideration are considered by the Fund's Committees in February, June, or October.
Requirements: Principal Investigators must be qualified in terms of education and experience to conduct research. A doctoral-level degree is required. There are no citizenship or residency requirements. The Fund is open to applications from institutions both inside and outside the United States. Ideal applications for the E.W. "Al" Thrasher Award address significant health problems that affect children in large numbers and offer the potential for practical solutions to these problems. Such solutions should be innovative and have the potential for rapid and broad applicability with low financial and/or technical barriers to implementation. Hypothesis-driven research is given priority over exploratory, hypothesis-generating research. The grant amount is based on the actual budgetary needs of the project. The duration of the project can be up to three years. Indirect costs of no more than 7% of direct costs will be paid on E.W. Al Thrasher Awards. Investigators in need of salary support for a specific project may apply for modest support (no more than 20 percent, based on a 40-hour work week). The Fund limits the maximum salary upon which an individual can request salary support. The maximum amount will be based on the NIH guidelines for salary support.
Restrictions: Requests for purchase of major equipment (items costing more than $4,000) are discouraged and rarely funded. The Fund does not award grants for educational programs; general operating expenses; general bridge funding for incomplete projects; construction or renovation of buildings or facilities; loans, student aid, scholarships, tuition; or, support of other funding organizations. Grants generally are not awarded for conferences, workshops, or symposia. Proposals in the areas of research on human fetal tissue or behavioral science research will not be considered. Proposals requesting Thrasher to commit money to a larger pool of funding with no distinct project proposed to Thrasher will not be funded.
Geographic Focus: All States, All Countries
Contact: Megan Duncan; (801) 240-4720; DuncanME@thrasherresearch.org
Internet: http://www.thrasherresearch.org/sites/www_thrasherresearch_org/Default.aspx?page=220
Sponsor: Thrasher Research Fund
68 S. Main Street, Suite 400
Salt Lake City, UT 84101

Earhart Foundation Institutional Grants **1888**
The foundation supports research fellowships for individual projects in economics, history, philosophy, international affairs, and political science. Fellowships are awarded upon direct application to faculty members. Grants for research in these areas also are awarded to educational and research organizations. H.B. Earhart Fellowships are awarded for graduate study (recipients must be nominated). Fields of interest include history and archaeology, philosophy/ethics, economics, political science, and international studies. Types of support include professorships, publication, fellowships, research, grants to individuals, and scholarships to individuals. Application forms are not required.
Restrictions: Grants are not awarded for capital, building, endowment funds, operating budgets, continuing support, annual campaigns, seed money, ermegency funds, deficit financing, matching gifts, or loans.
Geographic Focus: All States
Amount of Grant: 1,000-25,000 USD
Contact: Ingrid Gregg, President; (734) 761-8592
Sponsor: Earhart Foundation
2200 Green Road, Suite H
Ann Arbor, MI 48105

Earle B. Barnes Award for Leadership in Chemical Research Management **1889**
This award, supported by the Dow Chemical Company, is awarded yearly to recognize outstanding achievements in chemical research management. The award is intended to recognize those individuals who have demonstrated outstanding leadership and creativity in promoting the sciences of chemistry and chemical engineering in research management. Nominees should have demonstrated success in research management by exhibiting the proven ability to manage research projects and people. This leadership and creativity must have been shown by a record of successful research projects and by a strong motivation of the researchers on those projects. Recognition of these accomplishments by peers is essential.
Requirements: Any individual, except a member of the award committee, may submit one nomination or seconding letter for the award in any given year. The nominating documents consist of a letter of not more than 1000 words containing an evaluation of the nominee's accomplishments and a specific identification of the work to be recognized, a biographical sketch including date of birth, and a list of publications and patents authored by the nominee. A nominee must be a citizen of the United States.
Restrictions: Self-nominations are not accepted.
Geographic Focus: All States
Date(s) Application is Due: Nov 1
Amount of Grant: 5,000 USD
Contact: Felicia Dixon, Awards Administrator; (800) 227-5558 or (202) 872-4408; fax (202) 776-8008; f_dixon@acs.org or awards@acs.org
Internet: http://portal.acs.org/portal/acs/corg/content?_nfpb=true&_pageLabel=PP_ARTICLEMAIN&node_id=1319&content_id=CTP_004512&use_sec=true&sec_url_var=region1
Sponsor: American Chemical Society
1155 Sixteenth Street, NW
Washington, DC 20036-4801

Earthwatch Institute Student Fellowships **1890**
Earthwatch Student Fellowships offer students with limited exposure to science the opportunity to experience research first hand. Award recipients will: work and learn alongside inspiring professional scientists; discover how science and technology can advance our understanding of the world; interact with talented and committed peers and mentors.
Requirements: High school teachers and counselors are encouraged to nominate one to two students with limited opportunities for enrichment and resources, and no prior experience helping a scientist conduct research. Current juniors are preferred, but sophomores and seniors are welcome to apply. Students with disabilities may be eligible for this program.
Geographic Focus: All States
Contact: Fellowship Program Officer; (800) 776-0188, ext. 118; fax (978) 461-2332; EducationAwards@earthwatch.org
Internet: http://www.earthwatch.org/site/pp.asp?c=dsJSK6PFJnH&b=2221765
Sponsor: Earthwatch Institute
3 Clock Tower Place, Suite 100, Box 75
Maynard, MA 01754-0075

Eastern Europe and the Independent States of the Former Soviet Union Research and Training Grants **1891**
Grants are awarded to organizations to support national programs of advanced research; graduate training; language training; public dissemination of research data, methods, and findings; contact and collaboration among government and private specialists; and/or firsthand experience of the area by U.S. specialists. Organizations that receive the grants act as intermediaries for the federal funds by conducting their own national, open competitions to make awards at the graduate level and above to individual students, scholars, or other institutions. Grant recipients have up to three years to spend their awards. Application deadline will be published in the Federal Register.
Requirements: Applicants must be nonprofit organizations or institutions of higher education with an established track record in conducting research and training programs on the independent states of the former Soviet Union and countries of Eastern Europe. These organizations must run national programs of advanced research; graduate training; language training; public dissemination of research data, methods, and findings; contact and collaboration among Government and private specialists; and/or firsthand experience of the area by U.S. specialists.
Geographic Focus: All States
Amount of Grant: 129,000-12,000,000 USD
Contact: Kenneth Roberts, Executive Director; (202) 736-4572; fax (202) 736-4851
Internet: http://www.state.gov
Sponsor: U.S. Department of State
2201 C Street NW, Room 2251
Washington, DC 20520

Eastman School of Music Graduate Awards Program **1892**
The program offers financial support through the work opportunities that come with the awards. The award includes variable stipends plus approximately one-third to full tuition. The deadline of January 1 is for support of study beginning the following September.
Requirements: Eligibility is restricted to persons who have been accepted for graduate study at the Eastman School of Music of the University of Rochester.
Geographic Focus: All States
Contact: Office of Graduate Studies; (585) 274-1560; cmccamman@esm.rochester.edu
Internet: http://www.esm.rochester.edu/financialaid/grad.php
Sponsor: University of Rochester, Eastman School of Music
26 Gibbs Street
Rochester, NY 14604

Eaton Charitable Fund Grants 1893

The Fund is dedicated to supporting programs that improve the quality of life in communities where the company operates. The Fund gives primary consideration to requests for programs located in an Eaton community, recommended by an Eaton manager and where employees demonstrate leadership involvement. Programs selected for funding will have clearly defined objectives, measurable end results, and provide a positive return on the Funds investment. The Fund's primary interests are in support of community improvement, education, and arts and cultural programs. Program, project and capital grants are awarded. Capital grants are made for special purposes that meet specific community needs within the company's funding focus. On occasion, operating grants are awarded. Proposals should be sent to the manager of the Eaton facility located in an Eaton community.

Requirements: Applicant organizations must be 501(c)3 tax exempt charities and be located in communities where the company has operations.

Restrictions: Eaton does not make contributions to: annual operating budgets of United Way agencies or hospitals; medical research; endowment funds; debt retirement; religious organizations unless they are engaged in a significant program benefiting the entire community; fraternal or labor organizations; individuals or individual endeavors; fund raising benefits, sponsorships or other events.

Geographic Focus: Arkansas, Colorado, Florida, Georgia, Kansas, Kentucky, Louisiana, Maine, Maryland, Michigan, Minnesota, New Jersey, New York, North Carolina, Ohio, Pennsylvania, South Carolina, Texas, Wisconsin

Samples: 81,640 Eaton Multicultural Scholars Program, Cleveland, OH; $15,000 Habitat for Humanity, Sumter, SC; $7,500 United Performing Arts Fund, Milwaukee, WI; and $5,000 YMCA of North Oakland County, Rochester Hills, MI.

Contact: Grants Administrator; (216) 523-4944; fax (216) 479-7013

Internet: http://www.eaton.com/EatonCom/OurCompany/AboutUs/CorporateResponsibility/SocialCommitment/CorporateGiving/index.htm

Sponsor: Eaton Charitable Fund
1111 Superior Avenue
Cleveland, OH 44114-2584

Edmund Niles Huyck Preserve Graduate and Postgraduate Research Grants 1894

Graduate and postgraduate investigators may apply for grants for the support of research utilizing the natural resources of the Biological Station of the Huyck Preserve. Included within the preserve are 2000 acres of natural and reforested woodlands, old fields, Lake Myosotis (100 acres), Lincoln Pond (10 acres), and approximately three miles of permanent and intermittent streams. Housing and laboratory space are provided. Write for instructions on application and additional information.

Requirements: Applicants must be predoctoral and postdoctoral scientists whose research utilizes the natural resources of the Edmund Niles Huyck Preserve.

Geographic Focus: All States

Date(s) Application is Due: Feb 1

Amount of Grant: Up to 2,500 USD

Contact: Richard Wyman; (518) 797-3440; fax (518) 797-3440; rlwyman@capital.net

Internet: http://www.huyckpreserve.org

Sponsor: Edmund Niles Huyck Preserve
P.O. Box 189
Rensselaerville, NY 12147

Educational Communications and Technology Awards 1895

The awards given through the ECT Foundation promote scholarship and leadership in the field of educational communications and technology. Awards are made to defray doctoral dissertation research expenses, for the papers reporting studies addressing a question related to educational technology, and for the papers describing research findings that could be used to improve the process of instructional design, development, and evaluation. Contact the awards program office for official guidelines for presenting a nomination or application.

Geographic Focus: All States

Contact: K.J. Saville; (906) 227-2413; ksaville@nmu.edu or aect@aect.org

Internet: http://www.aect.org/Foundation/Awards/Awards.htm

Sponsor: Association for Educational Communications and Technology
1800 N Stonelake Drive, Suite 2
Bloomington, IN 47408

Education Through Astronomy and Space Science Initiative 1896

This program was developed to establish a comprehensive approach to providing education and public outreach to enhance the publics understanding of space science. The spirit of the program is to provide start-up funding to explore innovative, creative ways to integrate astronomy and space science into U.S. education and public outreach venues through partnerships between the astronomers/space scientists and education professionals.

Requirements: Professionals in astronomy or space science with an affiliation to an institution within the United States are eligible to apply. The proposal must incorporate a partnership with at least one professional educator as coinvestigator.

Geographic Focus: All States

Amount of Grant: Up to 50,000 USD

Contact: Grant Administrator; (410) 338-4798; ideas@stsci.edu

Internet: http://oposite.stsci.edu/pubinfo/edugroup/grants.html#top

Sponsor: Space Telescope Science Institute
3700 San Martin Drive
Baltimore, MD 21218

Educause Jane N. Ryland Fellowships 1897

This fellowship expands opportunities for information technology professionals to attend Educause events, thus helping to build future leaders. Fellowships recognize a combination of past achievement, personal and institutional commitment, potential benefit, and financial need. Awards are made on an annual basis, depending on an applicant's educational goals and the needs of his or her institution.

Requirements: Recipients are selected annually on a competitive basis. Applicants may serve their institutions in any information technology management area: central IT organizations, academic units, or administrative departments.

Geographic Focus: All States

Date(s) Application is Due: Nov 1

Amount of Grant: 500-3,000 USD

Contact: Fellowships Administrator; (303) 449-4430; fax (303) 440-0461; ryland-fellowships@educause.edu

Internet: http://www.educause.edu/awards/fellow

Sponsor: Educause
4772 Walnut Street, Suite 206
Boulder, CO 80301

Edward Bangs Kelley and Elza Kelley Foundation Grants 1898

The Foundation's interest is primarily in Barnstable County, Massachusetts. The Foundation has been the leader in improving the health and welfare of the community. Grants are made to a great variety of health, social and human service agencies, as well as to cultural and environmental organizations. Town libraries, theatre and art groups, and musical organizations have been supported. Grants are sometimes utilized as seed money by young organizations.

Requirements: Grant applicants must be tax-exempt organizations which are not private foundations and must be located in Barnstable County. The proposed project/program must have a direct benefit to the inhabitants of Barnstable County. If the tax-exempt organization is located outside of Barnstable, but the proposed project/program will have a direct and substantial benefit to the inhabitants of Barnstable, the grant application may be eligible for funding.

Restrictions: Applicants must be residents of Barnstable County, Massachusetts, or must demonstrate very significant ties to Barnstable County.

Geographic Focus: Massachusetts

Contact: Henry Murphy Jr.; (508) 775-3117; contact@kelleyfoundation.org

Internet: http://www.kelleyfoundation.org

Sponsor: Edward Bangs Kelley and Elza Kelley Foundation
243 South Street, P.O. Drawer M
Hyannis, MA 02601

Edward W. and Stella C. Van Houten Memorial Fund Grants 1899

Stella C. Van Houten resided in Bergen County, New Jersey. This foundation, providing funding for health and human services, education, education of medical professionals, and the care of children, was established in 1978 in memory of her husband and herself. The Van Houten's had a particular fondness for the Valley Hospital of Ridgewood, New Jersey and for the Rollins College in Florida. The Foundation continues to honor their preferences with grants to these two organizations in addition to grants to other organizations. The Foundation's mission is to: supports agencies, institutions and services in Passaic and Bergen Counties, New Jersey, having to do with the care or cure of sick or disabled persons or for the care of orphaned children or aged persons; educates students in the medical profession; support for educational purposes; support for the care of children. A target of 10% of the grants each year is for medical scholarships.

Requirements: Passaic and Bergen Counties, New Jersey non-profits are eligible to apply. The application form & guidelines are available online at the Wachovia website. The applications must be submitted by January 31 for a March meeting & August 1 for an October meeting.

Geographic Focus: New Jersey

Date(s) Application is Due: Jan 31; Aug 1

Amount of Grant: 6,000-100,000 USD

Contact: Trustee, c/o Wachovia Bank; grantinquiries2@wachovia.com

Internet: https://www.wachovia.com/foundation/v/index.jsp?vgnextoid=00c78689fb0aa110VgnVCM1000004b0d1872RCRD&vgnextfmt=default

Sponsor: Edward W. and Suitella C. Van Houten Memorial Fund
190 River Road, NJ3132
Summit, NJ 07901

Edwin S. Webster Foundation Grants 1900

The policy of the foundation is to support charitable organizations that are well known to the trustees, with emphasis on special projects and capital programs, or operating income for hospitals, medical research, education, youth agencies, cultural activities, and programs addressing the needs of minorities. Types of support include operating budgets, continuing support, annual campaigns, building funds, equipment, land acquisition, endowment funds, matching funds, scholarship funds, professorships, internships, fellowships, special projects, and research.

Requirements: The Foundation confines its grants primarily to the New England area. Grantees must provide evidence of their tax-exempt status. The AGM common proposal format, available on the Internet at http://agmconnect.org is suitable for submission of proposals but not required. There are no set deadlines, but for consideration at the spring meeting, proposals should arrive by May 1 and by November 1 for consideration at the fall meeting. Proposals received after the trustees meet will be held for consideration at the next meeting.

Restrictions: Grants are not made to organizations outside the United States or to individuals.

Geographic Focus: Massachusetts
Date(s) Application is Due: May 1; Nov 1
Amount of Grant: 15,000-50,000 USD
Contact: Michelle Jenney; (617) 391-3087; mjenney@gmafoundations.com
Sponsor: Edwin S. Webster Foundation
GMA Foundations, 77 Summer Street, 8th Floor
Boston, MA 02110-1006

EFWA Accounting Research Grants **1901**
The foundation funds issues papers by faculty, individuals, or business to research technical accounting issues or other issues furthering women in the accounting profession. The foundation also provides research grants for women faculty members to contribute to the field of accounting and assist faculty in their careers.
Geographic Focus: All States
Date(s) Application is Due: May 1
Amount of Grant: 3,000 USD
Contact: Cynthia Hires, (610) 407-9229; fax (610) 644-3713; info@efwa.org
Internet: http://www.efwa.org/research.htm
Sponsor: Educational Foundation for Women in Accounting
P.O. Box 1925
Southeastern, PA 19399-1925

Eisenhower Exchange International Fellowships **1902**
These competitive fellowships are for U.S. citizens to go abroad for a professional exchange program. The program includes travel throughout the selected country/region for research/inquiry purposes. Countries participating have included Argentina, Germany, Hungary, Ireland, Malaysia, Peru, and Taiwan. Fellowships include travel and living expenses and program arrangements/appointment scheduling for a one- to three-month program. Spouse participation is encouraged. Fellowships are awarded in the year preceding departure. Contact the office or visit the Web site for the most current information on participating countries and deadline dates.
Requirements: Midcareer professionals who are U.S. citizens with demonstrated leadership ability and significant contributions to their fields are eligible to apply.
Geographic Focus: All States
Contact: U.S.A Fellowships; (215) 546-1738; fax (215) 546-4567; ike@eisenhowerfellowships.org
Internet: http://eisenhowerfellowships.org
Sponsor: Eisenhower Exchange Fellowships
256 S 16th Street
Philadelphia, PA 19102

Ekhagastiftelsen Foundation Grants **1903**
The foundation operates in Sweden and supports research into agricultural production without the use of poisons and artificial means; and research into the production of natural foods and medical methods based on the capacity of human beings to cure themselves. Types of support include grants to organizations and scholarships and fellowships to individuals.
Geographic Focus: All States, Sweden
Contact: Administrator; 46 70 240 81 81; fax 46-08-21-83-76; info@ekhagastiftelsen.se
Internet: http://www.ekhagastiftelsen.se/eng/stiftelsen.html
Sponsor: Ekhagastiftelsen Foundation
Box 34012
Stockholm, SE-100 26 Sweden

Elias J. Corey Award for Outstanding Original Contribution in Organic Synthesis by a Young Investigator **1904**
The Award, supported by the Pfizer Endowment Fund, is intended to recognize original and insightful work by a young investigator that has had significant impact on the field of synthetic organic chemistry. The award will be granted to an individual under the age of 40 on November 1 of the Award year who has demonstrated outstanding creativity in the field of synthetic organic chemistry. A nominee must have accomplished research that is of exceptional merit and adds value to the field, such as the contribution of a novel and highly useful reagent, reaction, or catalytic process that solves a synthetic problem of considerable importance. Special consideration will be given to the application of such novel chemistry toward the synthesis of natural products and organic molecules of medicinal importance, and to the discovery of novel reactions or processes with broad utility.
Requirements: Any individual, except a member of the award committee, may submit one nomination or seconding letter for the award in any given year. The nominating documents consist of a letter of not more than 1000 words containing an evaluation of the nominee's accomplishments and a specific identification of the work to be recognized, a biographical sketch including date of birth, and a list of publications and patents authored by the nominee.
Restrictions: Self-nominations are not accepted.
Geographic Focus: All States
Date(s) Application is Due: Nov 1
Amount of Grant: 5,000 USD
Contact: Felicia Dixon, Awards Administrator; (800) 227-5558 or (202) 872-4408; fax (202) 776-8008; f_dixon@acs.org or awards@acs.org
Internet: http://portal.acs.org/portal/acs/corg/content?_nfpb=true&_pageLabel=PP_ARTICLEMAIN&node_id=1319&content_id=CTP_004516&use_sec=true&sec_url_var=region1
Sponsor: American Chemical Society
1155 Sixteenth Street, NW
Washington, DC 20036-4801

Elisabeth Severance Prentiss Foundation Grants **1905**
The foundation awards the majority of its support in the field of health and medicine. Scientific and medical research, hospitals, health projects, medical education, and care and support of the elderly are funded. Types of support include building programs, equipment needs, research, operating budgets, continuing support, seed grants, endowment funds, and projects/programs. Decision makers favor specific projects over general operating support. The maximum term of a commitment is five years, with preference for proposals for three years or less. Special consideration will be given to requests from applicants in the Cuyahoga County area. The board meets in June and December.
Requirements: Grants are awarded to promote and improve medical services in the greater Cleveland, OH, area.
Restrictions: Grants are not awarded to individuals for scholarships, fellowships, or grants in aid; or to organizations for fund-raising campaigns, surveys, assessments, studies, or planning activities.
Geographic Focus: Ohio
Date(s) Application is Due: May 15; Nov 15
Amount of Grant: 10,000-100,000 USD
Samples: University Hospitals of Cleveland (OH)—to create an inpatient psychiatric unit for children and adolescents at Rainbow Babies and Children's Hospital, $5 million.
Contact: Michael Galland, National City Bank Contact; (216) 222-2736
Sponsor: Elisabeth Severance Prentiss Foundation
P.O. Box 94651
Cleveland, OH 44104

Elizabeth Glaser International Leadership Awards **1906**
The International Leadership Award (ILA) is a three-year grant focused on international investigations in preventing mother-to-child transmission (PMTCT) of HIV and HIV care and treatment. Recent awards were granted for research into improving the effectiveness of PMTCT services; strengthening maternal and infant diagnosis of HIV; and examining the impact of antiretroviral drugs on children. The Award identifies and supports internationally based physicians or scientists who have the training and potential to develop pediatric HIV prevention and/or treatment programs, but lack the resources to do so. Award recipients are required to mentor a minimum of three additional people who would benefit from the experience and could help them achieve their goals. These are individuals who would be hired and trained to work on the project proposed by the applicant. The Award provides up to $105,000 in total costs per year for three years for a total of $315,000. Funding for the second and third year is dependent on progress in year one, and is not guaranteed. Funds support the salary of the applicant (up to 100% effort), salaries for those hired and mentored as part of the project, and direct costs related to the project itself such as travel, supplies, and equipment. Indirect cost is limited to 5% (included in the total). Salary support per individual investigator may be a maximum of $45,000 per year, and must be commensurate with the level of experience of the investigator, the salary structure in the country and the level of commitment to the project. Awardees must agree to spend at least 50% of their time on the project related to the Award. Final determination as to the amount of the Award is made by the Advisory Board depending on the scope of the project. Interested applicants are asked to submit a Letter of Intent. Only those Letters of the highest caliber will be asked to submit a full grant application. Additional guidelines, instructions, and summaries of previously funded projects are posted on the website.
Requirements: Applicants must be from a developing country, and must be performing work in a developing country in order to apply. Applicants must be living in and performing work in either Africa or India. The Foundation is especially interested in applicants who have specific training or experience with HIV/AIDS. Applicants must have an MD or PhD, and demonstrate commitment to continuing work in a developing country on HIV/AIDS when the award is complete.
Restrictions: U.S. citizens or European citizens living in developing countries are not eligible to apply. Senior Investigators such as full professors are usually not eligible, unless they are changing the direction of their career to include HIV/AIDS research.
Geographic Focus: All Countries
Amount of Grant: 105,000 USD
Contact: Chris Hudnall; (310) 314-3154; fax (310) 314-1469; chris@pedaids.org
Internet: http://www.pedaids.org/What-We-re-Doing/research/grant-programs.aspx
Sponsor: Elizabeth Glaser Pediatric AIDS Foundation
1140 Connecticut Avenue, NW, Suite 200
Washington, DC 20036

Elizabeth Glaser Scientist Award **1907**
The Elizabeth Glaser Scientist Award is the Foundation's highest scientific honor, and is an investment in the most promising HIV/AIDS researchers at a critical stage in their careers. These scientists, who represent the best and brightest investigators from the international medical science community, are selected on the basis of their knowledge, innovation, and dedication. By providing research funding over a five-year period, this award enables recipients to focus their long-term efforts on issues specific to pediatric HIV/AIDS. The Award fosters an unprecedented spirit of collaboration among these scientists. Each year, the Elizabeth Glaser Scientists meet with the Foundation's internationally renowned advisory board to stimulate ideas, report on current programs, and plan future research. Since the program's inception, the Foundation has built an invaluable network of scientists whose work in vaccine development, immune response, breast milk transmission, and other critical areas impacts the entire field of HIV/AIDS research. Interested applicants should contact the Foundation to request a letter of intent cover sheet. Full applications are by invitation.
Requirements: Candidates must have an MD, PhD, DDS, or DVM degree and be at the assistant professor level or above.

Restrictions: The Foundation will not fund large therapeutic trials or the pharmocologic development of drugs.
Geographic Focus: All States, All Countries
Contact: Doug Horner; (202) 296-9165; fax (202) 296-9185; research@pedaids.org
Internet: http://www.pedaids.org/What-We-re-Doing/research/grant-programs
Sponsor: Elizabeth Glaser Pediatric AIDS Foundation
1140 Connecticut Avenue, NW, Suite 200
Washington, DC 20036

Elizabeth McGraw Foundation Grants **1908**
The foundation awards grants nationally, with a focus on the U.S. Northeast, in its areas of interest, including cancer research, elementary and secondary education, art museums, performing arts (opera and theater), and culture. There are no application deadlines or specific guidelines. Send letter containing a detailed description of project and amount of funding requested to: John McGraw, Pres., 157 Eel River Road, Osterville, MA 02655
Geographic Focus: Florida, Massachusetts, New York
Contact: Paul J. Bisset; (212) 454-8607
Sponsor: Elizabeth McGraw Foundation
Deutsche Bank Trust Company of New York, P.O. Box 1297
New York, NY 10008-1297

Elizabeth Nash Foundation Summer Research Awards **1909**
The Elizabeth Nash Foundation is in the sixth year of a partnership with Children's Hospital and Research Center Oakland (CHORI), an internationally renowned biomedical research institute that bridges basic science and clinical research in the prevention and treatment of human disease. The Elizabeth Nash Foundation Summer Research Award is available to high school seniors and undergraduates who have a strong interest in pursuing cystic fibrosis (CF) research. The award is designed to identify and train the next generation of talented CF researchers. Students receive a stipend from the Foundation and CHORI pairs student with a mentor scientist to guide them through a structured CF research project. Previous recipients have co-authored publications in Free Radical Biology and Medicine and Cellular Physiology and Biochemistry involving their summer research. The deadline for applications for summer is 4:00 pm on April 15. The awardee will be provided a one-time stipend of $2,500 for the nine-week period. Announcement of all awards will be made on May 3.
Geographic Focus: All States
Date(s) Application is Due: Apr 15
Amount of Grant: 2,500 USD
Contact: Ann Nash, (408) 399-6919; info@elizabethnashfoundation.org
Internet: http://www.elizabethnashfoundation.org/CHORI.html
Sponsor: Elizabeth Nash Foundation
P.O. Box 1260
Los Gatos, CA 95031-1260

Ellison Medical Foundation/AFAR Julie Martin Mid-Career Award in Aging Research **1910**
The Ellison Medical Foundation and AFAR developed this program for outstanding mid-career scientists who propose novel directions of high importance to biological gerontology. Proposals in areas where NIH awards or other traditional sources are unlikely because the research is high risk, are particularly encouraged if they have the potential for leading to major new advances in our understanding of basic mechanisms of aging. Projects investigating age-related diseases are also supported, but only if approached from the point of view of how basic aging processes may lead to these outcomes. Projects concerning mechanisms underlying common geriatric functional disorders are also encouraged, as long as these include connections to fundamental problems in the biology of aging. Two four-year awards of $500,000 will be given, at the level of $125,000 per year. In addition, up to 10% ($50,000) may be requested for administrative/indirect costs. Recipients of this award are expected to attend the AFAR Grantee Conference. The purpose of the meeting is to promote scientific and personal exchanges among recent AFAR grantees and experts in aging research.
Requirements: The applicant must be an Associate Professor who achieved tenured status after December 1st. Non-tenured Associate Professors at institutions with tenure (even if tenure is only offered at the Full Professor level) are not eligible. Applicants at institutions that do not offer tenure must demonstrate that their appointment is equivalent to that of an Associate Professor who received tenure status after December 1st. The proposed research must be conducted at any type of not-for-profit setting in the United States. The deadline of receipt of applications and all supporting materials is December 16th at 5:00 p.m. EST. Refer to the Julie Martin Mid-Career Award instruction sheet and application at: http://afar.org/grants.html for complete application procedures. Incomplete applications cannot be considered.
Restrictions: Projects that deal strictly with clinical problems such as the diagnosis and treatment of disease, health outcomes, or the social context of aging are not eligible. Applicants who are employees in the NIH Intramural program are not eligible.
Geographic Focus: All States
Date(s) Application is Due: Dec 16
Amount of Grant: 125,000 USD
Contact: Grants Manager; (212) 703-9977 or (888) 582-2327; fax (212) 997-0330; grants@afar.org or info@afar.org
Internet: http://afar.org/Ellison%20Mid-Career.html
Sponsor: American Federation for Aging Research
55 West 39th Street, 16th Floor
New York, NY 10018

Ellison Medical Foundation/AFAR Postdoctoral Fellows in Aging Research **1911**
The Ellison Medical Foundation, in partnership with the American Federation for Aging Research (AFAR), created the Ellison/AFAR Postdoctoral Fellows in Aging Research Program to encourage and further the careers of postdoctoral fellows with outstanding promise in the basic biological and biomedical sciences relevant to understanding aging processes and age related diseases and disabilities. The award is intended to provide significant support to permit these postdoctoral fellows to become established in the field of aging. Projects concerned with understanding the basic mechanisms of aging will be considered. Projects investigating age related diseases are also supported, if approached from the point of view of how basic aging processes may lead to these outcomes. Projects concerning mechanisms underlying common geriatric functional disorders are also considered. It is anticipated that up to 15 one year grants will be awarded, ranging from $44,850 for a first year fellow, up to $58,850 for a fellow with more than 7 years of training. Of the award, up to $7,850 may be requested for expenses such as research supplies, equipment, health insurance and travel to scientific meetings. Refer to the Ellison/AFAR instruction sheet and application on the AFAR website for complete application procedures. Recipients of this award are expected to attend the AFAR Grantee Conference. The purpose of the meeting is to promote scientific and personal exchanges among recent AFAR grantees and experts in aging research.
Requirements: The applicant must be a postdoctoral fellow (MD and/or PhD degree) at the start date of the award (July 1). Any former Ellison/AFAR postdoctoral award recipient may apply for this award. All candidates must submit applications endorsed by their institutions. The proposed research must be conducted at any type of not-for-profit setting in the United States.
Restrictions: Employees in the NIH Intramural program are not eligible. Fellows may not hold any concurrent foundation or not-for-profit funding.
Geographic Focus: All States
Date(s) Application is Due: Dec 16
Amount of Grant: 44,000-58,000 USD
Contact: Grants Manager; (212) 703-9977 or (888) 582-2327; fax (212) 997-0330; grants@afar.org or info@afar.org
Internet: http://www.afar.org/ellisonpostdoc.html
Sponsor: American Federation for Aging Research
55 West 39th Street, 16th Floor
New York, NY 10018

El Pomar Leadership Development and Community Engagement Postgraduate Fellowship **1912**
The program educates young leaders about the role of nonprofit organizations and the foundation within the communities of Colorado. The two-year postgraduate program also focuses on professional development and prepares fellows for positions of leadership in Colorado and the nation. Fellows serve the foundation's operating programs, such as EPYCS, a high school philanthropy initiative. In addition, fellows will be thoroughly educated in the nature of the nonprofit sector and the role of foundations. At a minimum, fellows participate in monthly courses in public speaking, financial and investment management, nonprofit management, fundraising and strategic planning, and an annual Outward Bound course in the central mountain ranges of Colorado. This fellowship has a starting salary of $27,000.
Requirements: Applicant must be a graduate of a four-year university or college, have a Colorado affiliation (e.g., be a state resident, have attended an in-state university or college, or have immediate family who are residents or past residents), and demonstrate a strong leadership capability and potential.
Geographic Focus: Colorado
Amount of Grant: Up to 27,000 USD
Contact: Administrator; (800) 554-7711 or (719) 633-7733; fellowship@elpomar.org
Internet: http://www.elpomar.org/page.asp?pageid=0|6|30&id=0|fellowship
Sponsor: El Pomar Foundation
10 Lake Circle
Colorado Springs, CO 80906

EMBO / EMBL Science Writing Prize **1913**
The Science Writing Prize is given annually for an outstanding piece of science writing designed to communicate a topical issue to a non-scientific audience. There is one winner, and the prize may not be awarded twice to the same person. The theme varies from year to year. The selection jury consists of 15 people, both scientists and non-scientists: the EMBO Science and Society Committee members and EMBO staff. The prize of Euro $1,500 is awarded in June at the EMBO Media Workshop in Heidelberg, where the author is invited to talk about his/her motivation for writing the article and for communicating the topic to the public.
Geographic Focus: All States, Germany
Amount of Grant: 1,500 USD
Contact: Hermann Bujard; 49 0 6221 8891, ext. 101; embo@embo.org
Internet: http://www.embo.org/awards/writing.html
Sponsor: European Molecular Biology Organization
Meyerhofstrasse 1
Heidelberg, D-69117 Germany

EMBO Award for Communication in the Life Sciences **1914**
The Organization launched this award in 2002 in order to give recognition to the huge efforts that some scientists make to communicate their science to the public while remaining fully active in research. The winners of the EMBO Award are nominated for the EU Descartes Prize for science communication.
Geographic Focus: All States, Germany
Amount of Grant: 5,000 USD

Samples: Christian Sardet and Ali Saib, French cell biologists, $5,000 Euro; Armand Marie Leroi, UK scientist, author, and broadcaster, $5,000 Euro.
Contact: Hermann Bujard, Executive Director; 49 0 6221 8891, ext. 101; fax 49 0 6221 8891, ext. 200; embo@embo.org
Internet: http://www.embo.org/awards/communications.html
Sponsor: European Molecular Biology Organization
Meyerhofstrasse 1
Heidelberg, D-69117 Germany

EMBO Installation Grants **1915**
The aim of this scheme is to strengthen science in participating EMBC member states. At present applications are accepted from Croatia, the Czech Republic, Estonia, Hungary, Portugal, Poland and Turkey. Additional member states have expressed an interest to join in the near future. Grants will help scientists to relocate, set up their labs and rapidly establish a reputation in the European scientific community. The program is entirely funded by the participating EMBC member states and successful applicants receive an annual support of Euro $50,000 for three to five years. Each participating country sets an upper limit to the number of grants available per year. Application forms should be completed by the individual scientists and the receiving institute.
Requirements: Eligible scientists should: have an excellent publication record; have spent at least two consecutive years prior to the application deadline, outside the country in which they are applying to establish their lab; and be negotiating a full-time position at an institute/university in participating member states by the date of application. No age limit applies.
Geographic Focus: All States, Albania, Andorra, Armenia, Austria, Azerbaijan, Belarus, Belgium, Bosnia & Herzegovina, Bulgaria, Croatia, Cyprus, Czech Republic, Denmark, Estonia, Finland, France, Georgia, Germany, Greece, Hungary, Iceland, Ireland, Italy, Kosovo, Latvia, Liechtenstein, Lithuania, Luxembourg, Macedonia, Malta, Moldova, Monaco, Montenegro, Norway, Poland, Portugal, Romania, Russia, San Marino, Serbia, Slovakia, Slovenia, Spain, Sweden, Switzerland, The Netherlands, Turkey, Ukraine, United Kingdom, Vatican City
Date(s) Application is Due: Apr 15
Amount of Grant: 150,000-250,000 EUR
Contact: Hermann Bujard, Executive Director; 49 0 6221 8891, ext. 101; fax 49 0 6221 8891, ext. 200; embo@embo.org
Internet: http://www.embo.org/sdig/index.html
Sponsor: European Molecular Biology Organization
Meyerhofstrasse 1
Heidelberg, D-69117 Germany

EMBO Long-Term Fellowships **1916**
Long-Term Fellowships are awarded for prolonged visits (12 to 24 months) and are intended for advanced training through research. As the aim of the EMBO is to promote international research, it follows that mobility is a crucial element in deciding the eligibility of an application. All applications must involve a laboratory of origin or a receiving institute or applicant's nationality from one of the member states. Stipend rates depend on the country being visited, marital status, etc.
Requirements: Applicants must hold a doctorate degree or equivalent before the start of the fellowship but not necessarily when applying. For clarity, it is deemed that this requirement is fulfilled only when the thesis has been stamped as submitted for final examination by the university. As a minimum, applicants must have at least one first author publication in press or published in an international peer reviewed journal at the time of application.
Restrictions: Applications are only considered from candidates with a maximum of 3 years postdoctoral experience.
Geographic Focus: All States, Germany
Date(s) Application is Due: Feb 15; Aug 15
Contact: Liselott Maidment, 49 0 6221 8891, ext. 116 or 49 0 6221 8891 200; fellowships@embo.org
Internet: http://www.embo.org/fellowships/long_term.html
Sponsor: European Molecular Biology Organization
Meyerhofstrasse 1
Heidelberg, D-69117 Germany

EMBO Short-Term Fellowships **1917**
Short-term fellowships are established to advance molecular biology research by helping scientists to visit another laboratory with a view to applying a technique not available in the home laboratory. Such fellowships are not awarded for exchanges between two laboratories within the same country, and are intended for joint research work rather than consultations. The fellowships cover travel plus subsistence of the fellow only and not of any dependents. These fellowships are intended for visits of one week to three months duration (for non-European applicants the short-term fellowships are intended for a fixed duration of 90 days). Stipends vary widely from country to country,
Requirements: Applicants must be either: post-doctoral scientists with less than 10 years of professional experience since finishing their Ph.D. degree (or the equivalent); or pre-doctoral scientists who have not yet completed a Ph.D. degree (or the equivalent).
Restrictions: All applications must involve either a laboratory of origin or a receiving institute from one of the EMBC member states.
Geographic Focus: All States, Germany
Contact: Program Director; 49 0 6221-8891, ext. 115; fax 49 0 6221-8891, ext. 200; fellowships@embo.org
Internet: http://www.embo.org/fellowships/short_term.html
Sponsor: European Molecular Biology Organization
Meyerhofstrasse 1
Heidelberg, D-69117 Germany

Emerson Charitable Trust Grants **1918**
Established in 1944 in Missouri as the Emerson Electric Manufacturing Company Charitable Trust, the Foundation supports: arts and culture—fine arts and cultural institutions to enrich the diversity, creativity, and liveliness of the community; education—programs designed to promote educational systems at all levels; health and human services—programs designed to help individuals and families in times of need, including sickness, old age, family crisis, and natural disasters; civic affairs—programs designed to protect citizenry; further economic health of the community; and build and maintain assets such as parks and zoos; and youth—programs designed to give young people the opportunity to recognize their potential, confidence, and skills to achieve their dreams. The foundation awards college scholarships to children and step-children of employees of Emerson Electric. No specific application form is required, and initial approach should be the complete proposal,
Geographic Focus: All States
Amount of Grant: 5,000-300,000 USD
Contact: Jo Ann Harmon, Senior Vice President; (314) 553-3722; fax (314) 553-1605
Internet: http://www.emerson.com/en-us/about_emerson/company_overview/pages/our_approach_to_corporate_philanthropy.aspx
Sponsor: Emerson Charitable Trust
8000 W Florissant Avenue, P.O. Box 4100
Saint Louis, MO 63136

Emil Buehler Perpetual Trust Grants **1919**
The Trust accepts inquiries from organizations with requests that promote the field of aviation science and technology. Grants are made in support of specific programs or projects. Grants are occasionally made for specific capital purchases if a clear need is demonstrated. Priority is given to cost-effective programs that offer the broadest benefit to the community. The Board of Trustees meets monthly to make grant decisions. Proposals must be received by the 10th day of the month for consideration at that month's meeting.
Requirements: Grants are only made to organizations qualifying for tax exemption under Section 501(c)3 of the Internal Revenue Code.
Restrictions: The Trust does not provide grants for general operating support; endowments; individuals; political activities; loans; or fund-raising events including dinners, benefits and athletic events.
Geographic Focus: All States
Contact: Deborah Janof, Grants Administrator; (201) 967-8040
Sponsor: Emil Buehler Perpetual Trust
305 Route 17 South
Paramus, NJ 07652

Emma B. Howe Memorial Foundation Grants **1920**
The Emma B. Howe Memorial Foundation makes grants through the Minneapolis Foundation. The focus of all grants is to improve: the health and well-being of children, youth and families; opportunities for educational achievement; access to quality affordable housing; and economic vitality throughout the region. Information on the application process is available online.
Requirements: Eligible organizations include 501(c)3 nonprofits, public institutions; and emerging groups organized for nonprofit purposes.
Restrictions: Funds are not available for: individuals; organizations/activities outside of Minnesota; conference registration fees; memberships; direct religious activities; political organizations or candidates; direct fundraising efforts; telephone solicitations; courtesy advertising; financial deficits.
Geographic Focus: Minnesota
Contact: Grants Manager; (612) 672-3836; grants@mplsfoundation.org
Internet: http://www.mplsfoundation.org/partners/emma.htm
Sponsor: Emma B. Howe Memorial Foundation
80 S Eighth Street
Minneapolis, MN 55402

Endangered Language Fund Grants **1921**
The Fund provides grants for language maintenance and linguistic field work. The work most likely to be funded is that which serves both the native community and the field of linguistics. Work which has immediate applicability to one group and more distant application to the other will also be considered. Publishing subventions are a low priority, although they will be considered. Proposals can originate in any country. The language involved must be in danger of disappearing within a generation or two. Endangerment is a continuum, and the location on the continuum is one factor in our funding decisions. Eligible expenses include consultant fees, tapes, films, travel, etc. Grants are normally for a one year period, though extensions may be applied for. The Fund expects grants to be less than $4,000 in size, and to average about $2,000.
Restrictions: Overhead expenses are not allowed.
Geographic Focus: All States
Date(s) Application is Due: Apr 21
Amount of Grant: Up to 4,000 USD
Contact: Grants Coordinator; (203) 865-6163, ext. 265; fax (203) 865-8963; elf@endangeredlanguagefund.org
Internet: http://www.endangeredlanguagefund.org/request.html
Sponsor: Endangered Language Fund
300 George Street, Suite 900
New Haven, CT 06511

Energy Foundation Buildings Grants 1922

The Energy Foundation's mission is to assist the United States and China in the transition to a sustainable energy future by promoting energy efficiency and renewable energy. To this end, grants in the buildings sector hope to: promote the adoption of increasingly stringent building codes, with effective local enforcement; establish stringent appliance standards and complementary consumer labels; create programs to benchmark and rate the energy consumed by buildings and encourage retrofitting the large stock of existing buildings; and develop financial incentives to buy and lease energy-efficient buildings. The Foundation has no fixed format for proposals. Applicants are urged to use the form that best conveys the strengths of their project.
Requirements: The foundation makes grants to nonprofit 501(c)3 charitable organizations. In general, a complete proposal includes the following: the attached application form as a cover sheet; a clear statement of the need(s) or problem(s) to be addressed; strategy; timeline; results you expect from your project; project budget—including a brief explanation of the budget, a list of other sources of actual and potential funding for the project, and a description of plans to secure additional funding; how you will determine whether your project is successful; and a history of organization, including mission and goals.
Restrictions: The Foundation is unable to fund: grants directly to individuals; local projects, unless they have been consciously designed for further replication or have broad regional or national implications; candidates for political office, to influence legislation, or to support sectarian or religious purposes; the research and development of technology (e.g., funds to develop hybrid automobiles or commercialization of an invention); demonstration projects (e.g., model solar homes); community energy projects; endowments or debt reduction; annual fund-raising campaigns or capital construction; the planning, renovation, maintenance, retrofit, or purchase of buildings; the purchase of equipment; or the acquisition of land, even if the intent is to save energy; or unsolicited general operating support.
Geographic Focus: All States
Amount of Grant: 2,000-850,000 USD
Contact: Todd Foland; (415) 561-6700; fax (415) 561-6709; energyfund@ef.org
Internet: http://www.ef.org/programs.cfm?program=buildings
Sponsor: Energy Foundation
301 Battery Street, 5th Floor
San Francisco, 94111

Energy Foundation Climate Grants 1923

The Energy Foundation's mission is to assist the United States and China in the transition to a sustainable energy future by promoting energy efficiency and renewable energy. To this end, grants in the climate sector support work in the following areas: building support for climate protection among diverse allies, such as business associations, veterans' groups, religious organizations, and agricultural interests, among others; policy analysis (such as assessing the effects of climate policy on consumers, business, jobs, and Gross Domestic Product; modeling state-by-state impacts; or evaluating policy design options on issues such as offsets, allowance allocation, or integrating carbon controls with complementary policies for energy efficiency and renewable energy); targeted efforts to educate opinion leaders and stakeholders—nationally or in key states-about climate change and climate policy; supporting implementation of state and regional climate laws and agreements (such as California's AB 32, the Regional Greenhouse Gas Initiative in the Northeast, and the Western Climate Initiative); and improving understanding of the role of the U.S. in an international climate agreement. The Foundation has no fixed deadlines or format for proposals. Applicants are urged to use the form that best conveys the strengths of their project.
Requirements: The foundation makes grants to nonprofit 501(c)3 charitable organizations. In general, a complete proposal includes the following: the attached application form as a cover sheet; a clear statement of the need(s) or problem(s) to be addressed; strategy; timeline; results you expect from your project; project budget—including a brief explanation of the budget, a list of other sources of actual and potential funding for the project, and a description of plans to secure additional funding; how you will determine whether your project is successful; and a history of organization, including mission and goals.
Restrictions: The Foundation is unable to fund: grants directly to individuals; local projects, unless they have been consciously designed for further replication or have broad regional or national implications; candidates for political office, to influence legislation, or to support sectarian or religious purposes; the research and development of technology (e.g., funds to develop hybrid automobiles or commercialization of an invention); demonstration projects (e.g., model solar homes); community energy projects; endowments or debt reduction; annual fund-raising campaigns or capital construction; the planning, renovation, maintenance, retrofit, or purchase of buildings; the purchase of equipment; or the acquisition of land, even if the intent is to save energy; or unsolicited general operating support.
Geographic Focus: All States
Amount of Grant: 2,000-850,000 USD
Contact: Todd Foland; (415) 561-6700; fax (415) 561-6709; energyfund@ef.org
Internet: http://www.ef.org/programs.cfm?program=climate
Sponsor: Energy Foundation
301 Battery Street, 5th Floor
San Francisco, 94111

Energy Foundation Power Grants 1924

The Energy Foundation's mission is to assist the United States and China in the transition to a sustainable energy future by promoting energy efficiency and renewable energy. To this end, grants in the power sector support work in the following areas: policies that yield large-scale purchases of renewable energy; policies that open the electric system to renewable power and associated transmission while supporting resource conservation; policies that yield substantial investments to improve energy efficiency by consumers of electric power and natural gas; policies that remove market and regulatory barriers to renewables, efficiency, and clean distributed generation; resistance to permitting and regulatory approvals for new conventional coal-fired power plants; campaigns to reduce greenhouse gases by directing new utility investments toward clean energy resources; to secure stricter performance standards for emissions from new and existing coal plants, including greenhouse gases; and regulatory controls for carbon capture and sequestration systems. The Foundation has no fixed format for proposals. Applicants are urged to use the form that best conveys the strengths of their project.
Requirements: The foundation makes grants to nonprofit 501(c)3 charitable organizations. In general, a complete proposal includes the following: the attached application form as a cover sheet; a clear statement of the need(s) or problem(s) to be addressed; strategy; timeline; results you expect from your project; project budget—including a brief explanation of the budget, a list of other sources of actual and potential funding for the project, and a description of plans to secure additional funding; how you will determine whether your project is successful; and a history of organization, including mission and goals.
Restrictions: The Foundation is unable to fund: grants directly to individuals; local projects, unless they have been consciously designed for further replication or have broad regional or national implications; candidates for political office, to influence legislation, or to support sectarian or religious purposes; the research and development of technology (e.g., funds to develop hybrid automobiles or commercialization of an invention); demonstration projects (e.g., model solar homes); community energy projects; endowments or debt reduction; annual fund-raising campaigns or capital construction; the planning, renovation, maintenance, retrofit, or purchase of buildings; the purchase of equipment; or the acquisition of land, even if the intent is to save energy; or unsolicited general operating support.
Geographic Focus: All States
Amount of Grant: 2,000-850,000 USD
Contact: Todd Foland; (415) 561-6700; fax (415) 561-6709; energyfund@ef.org
Internet: http://www.ef.org/programs.cfm?program=power
Sponsor: Energy Foundation
301 Battery Street, 5th Floor
San Francisco, 94111

Energy Foundation Transportation Grants 1925

The Energy Foundation's mission is to assist the United States and China in the transition to a sustainable energy future by promoting energy efficiency and renewable energy. To this end, the transportation sector makes grants to promote innovative federal and state policies that reduce global warming pollution from vehicles, encourage low-carbon fuels, and discourage high-carbon fuels. The Foundation focuses on performance-based standards and incentives that guide markets toward cleaner solutions. The Foundation has no fixed deadlines or format for proposals. Applicants are urged to use the form that best conveys the strengths of their project.
Requirements: The foundation makes grants to nonprofit 501(c)3 charitable organizations. In general, a complete proposal includes the following: the attached application form as a cover sheet; a clear statement of the need(s) or problem(s) to be addressed; strategy; timeline; results you expect from your project; project budget-including a brief explanation of the budget, a list of other sources of actual and potential funding for the project, and a description of plans to secure additional funding; how you will determine whether your project is successful; and a history of organization, including mission and goals.
Restrictions: The Foundation is unable to fund: grants directly to individuals; local projects, unless they have been consciously designed for further replication or have broad regional or national implications; candidates for political office, to influence legislation, or to support sectarian or religious purposes; the research and development of technology (e.g., funds to develop hybrid automobiles or commercialization of an invention); demonstration projects (e.g., model solar homes); community energy projects; endowments or debt reduction; annual fund-raising campaigns or capital construction; the planning, renovation, maintenance, retrofit, or purchase of buildings; the purchase of equipment; or the acquisition of land, even if the intent is to save energy; or unsolicited general operating support.
Geographic Focus: All States
Amount of Grant: 2,000-850,000 USD
Contact: Todd Foland; (415) 561-6700; fax (415) 561-6709; energyfund@ef.org
Internet: http://www.ef.org/programs.cfm?program=transportation
Sponsor: Energy Foundation
301 Battery Street, 5th Floor
San Francisco, 94111

Engineering Awards of Excellence 1926

The purpose of this program is to give national recognition to structural engineering excellence and innovation in steel-framed building projects.
Requirements: Projects must be located in the United States, Canada, or Mexico. A significant part of the framing system must be steel wide-flange structural shapes or hollow structural sections. Both new buildings and renovation projects are eligible.
Geographic Focus: All States
Date(s) Application is Due: Mar 1
Contact: Becky LeDonne, Awards Coordinator; (312) 670-5433; fax (312) 670-5403; ledonne@aisc.org
Internet: http://www.aisc.org/Content/NavigationMenu/About_AISC/Competition_and_Awards/Project_Awards/Engineering_Awards_of_Excellence/Engineering_Awards_of_Excellence.htm
Sponsor: American Institute of Suiteel Construction
One E Wacker Drive, Suite 3100
Chicago, IL 60601-2001

Engineering Information Foundation Grants 1927

The Foundation supports developmental projects, instructional projects, and training programs in engineering education and research that fit their fields of interest. These currently include the availability and use of published information, women in engineering, and information access in developing countries. The Foundation has a specific interest in innovative projects with measurable results, projects that promote significant and lasting change, projects that can be successfully replicated elsewhere and methodologies that are specific, well-defined and cost-effective.

Requirements: U.S. 501(c)3 tax-exempt organizations are eligible.

Restrictions: Funds are not available for: operating expenses; requests solely for equipment; general overhead; campaigns; conferences; scholarships, fellowships, assistantships; support for doctoral candidates; loans.

Geographic Focus: All States

Amount of Grant: 5,000-25,000 USD

Contact: Program Contact; (212) 579-7596; fax (212) 579-7517; info@eifgrants.org

Internet: http://www.eifgrants.org

Sponsor: Engineering Information Foundation

180 West 80th Street, Suite 207

New York, NY 10024-6301

Ensign-Bickford Foundation Grants 1928

The corporate foundation gives primarily in areas of company operations, with emphasis on the Simsbury and Avon, CT, areas. Grants are awarded to nonprofit organizations for welfare, education, cultural programs, employee-related scholarships, and community development activities. Types of support include continuing support, annual campaigns, building/renovation, equipment acquisition, land acquisition, programs and projects, conferences and seminars, publication, seed money, internships, scholarships, research, and employee matching gifts. Applications are accepted throughout the year.

Requirements: Nonprofit organizations in company operating areas of Connecticut are eligible to apply.

Restrictions: Grants are not awarded to individuals (except for employee-related scholarships) or for operating budgets, endowment funds, emergency funds, deficit financing, or loans.

Geographic Focus: Connecticut

Amount of Grant: 50-50,000 USD

Contact: Jan Delissio, (860) 843-2388

Internet: http://www.ensign-bickfordind.com/community.html

Sponsor: Ensign-Bickford Foundation

100 Grist Mill Road, P.O. Box 7

Simsbury, CT 06070-0007

Enterprise Community Partners Rose Architectural Fellowships 1929

The Fellowship honors the late Frederick P. Rose, a prominent developer and philanthropist who believed strongly in the value of quality design and the spirit of public service. Under the program's innovative structure, the Enterprise Rose Architectural Fellows work deeply in communities for three years, forging community ties, developing leadership skills, and expanding the capacity of their local host organizations to execute projects with the best possible designs, meaningful community engagement, and the most advanced green building features. Fellows and host organizations promote the principles and best practices of community architecture in a variety of venues, including teaching, conference presentations, and publications. They have initiated the adoption of improved design standards and best practices by their state and local governments and industry organizations. The annual stipend paid to fellows in most communities is $45,000 ($52,000 in high-cost cities, including Los Angeles) an amount that will remain the same for the three years of the fellowship. High cost area may merit a salary adjustment. Adjustments, if any, are described in the application kit for a specific fellowship.

Requirements: Applicants must have a Social Security number or a United States Tax Identification number and must possess a professional degree in architecture from an accredited college or university, or expect to graduate from an accredited school of architecture before the fellowship begins. A professional graduate degree in architecture is preferred. The expected minimum grade point average is "B" or 3.0 on a 4.0 scale for all undergraduate and graduate work. Preference will be given to candidates who have approximately three years professional experience, have completed their Intern Development Program (IDP) requirements, and can work independently within a non-profit environment. However, recent graduates will also be considered, as will candidates who have not yet completed their IDP requirements. Fellows must make a full-time commitment of three years to the fellowship and to the host organization. Applicants with diverse educational, professional, and personal backgrounds, and individuals from disadvantaged communities, as well as minorities are encouraged to apply.

Restrictions: A Fellow may not have any other paid employment or independent consulting work, unless prior written permission is given by the director of the fellowship and an authorized representative of the partner organization. Individuals currently employed by their partner organization are not eligible to apply.

Geographic Focus: All States

Date(s) Application is Due: May 1

Amount of Grant: 45,000-52,000 USD

Contact: Trinity Simons, Program Officer; (781) 235-2006 or (781) 235-4011; tsimons@enterprisecommunity.org or rosefellowship@enterprisecommunity.org

Internet: http://www.enterprisecommunity.com/solutions-and-innovation/design-leadership/rose-architectural-fellowship

Sponsor: Enterprise Community Partners

70 Corporate Center

Columbia, MD 21044

Enterprise Community Partners Terwilliger Fellowship 1930

This fellowship is named for Enterprise Trustee J. Ronald Terwilliger, chairman and CEO of Trammel Crow Residential. This one-year Fellowship plays a leadership role in launching the new Enterprise Innovation Unit, which has primary responsibility for innovation: research, development and planning of potential new Enterprise programs, products and partnerships. The fellow works with the Enterprise executive management team and key partners to develop new solutions to advance Enterprise's mission.

Geographic Focus: All States

Contact: Dave Epley, Program Director; (410) 772-2411; fax (410) 964-1918; depley@enterprisecommunity.org

Internet: http://www.enterprisecommunity.com/about/fellowships-and-awards

Sponsor: Enterprise Community Partners

70 Corporate Center

Columbia, MD 21044

EPA Brownfields Area-Wide Planning Grants 1931

Brownfields Area-Wide Planning is an EPA grant program which provides funding to recipients to conduct research, technical assistance and training that will result in an area-wide plan and implementation strategy for key brownfield sites, which will help inform the assessment, cleanup and reuse of brownfields properties and promote area-wide revitalization. Funding is directed to specific areas, such as a neighborhood, downtown district, local commercial corridor, or city block, affected by a single large or multiple brownfield sites.

Requirements: Eligible governmental entities include a general purpose local unit of government; a land clearance authority or other quasi-governmental entity that operates under the supervision and control of, or as an agent of, a general purpose unit of government; a governmental entity created by a state legislature; a regional council or group of general purpose units of local government; a redevelopment agency that is chartered or otherwise sanctioned by a state; a state; an Indian Tribe (other than in Alaska), or an Alaskan Native Regional Corporation and an Alaska Native Village Corporation as those terms are defined in the Alaska Native Claims Settlement Act (43 U.S.C. 1601 and following); and the Metlakatla Indian Community. Eligible nonprofit organizations include any corporation, trust, association, cooperative, or other organization that is operated mainly for scientific, educational, service, charitable, or similar purpose in the public interest; is not organized primarily for profit; and uses net proceeds to maintain, improve, or expand the operation of the organization.

Geographic Focus: All States

Amount of Grant: Up to 200,000 USD

Contact: Director; (202) 566-2777; fax (202) 566-2757

Internet: http://www.epa.gov/brownfields/areawide_grants.htm

Sponsor: Environmental Protection Agency

1301 Constitution Avenue, NW

Washington, DC 20004

EPA Capacity Building Grants for Compliance Assurance and Enforcement Activities in Indian Country and Other Tribal Areas 1932

The objective of the EPA Capacity Building Grants for Compliance Assurance and Enforcement Activities in Indian Country and Other Tribal Areas is to provide financial resources to build and improve the compliance assurance and enforcement capacity of federally-recognized Indian tribes (tribes), inter-tribal consortia, or tribal organizations. Such capacity building efforts may include economic, social science, statistical research, development, studies, surveys, demonstrations, investigations, public education, training, and fellowships to the extent authorized under the federal environmental laws listed above. This Catalog of Federal Domestic Assistance (CFDA) number covers EPA's Environmental Programs and Management (EPM) resources targeted for compliance assurance and enforcement in Indian country and other tribal areas, including those in Alaska. Project grants can include fellowships, scholarships, research grants, training grants, traineeships, experimental and demonstration grants, evaluation grants, planning grants, technical assistance grants, survey grants, and construction grants. This program has no matching requirements. Awards will range from $3,000 to $40,000 annually.

Requirements: Tribal Organizations, Inter-tribal Consortia, Federal, and State Organizations, Colleges and Universities, and non-for-profit organizations are eligible. Programs can benefit Federally Recognized Indian Tribal Governments; Inter-tribal Consortia and Tribal Organizations; Federal, State, and Multi-jurisdictional State Organizations; colleges and universities; or non-for-profit organizations.

Geographic Focus: All States

Amount of Grant: 3,000-40,000 USD

Contact: Jonathan Binder; (202) 564-2516; fax (202) 564-7083; binder.jonathan@epa.gov

Internet: https://www.cfda.gov/?s=program&mode=form&tab=step1&id=48bd2a8ab477ac80d37743a9bdfeb0f5

Sponsor: Environmental Protection Agency

1200 Pennsylvania Avenue, NW

Washington, DC 20460

EPA Children's Health Protection Grants 1933

The objectives of this program are to catalyze community-based and regional projects and other actions that enhance public outreach and communication; assist families in evaluating risks to children and in making informed consumer choices; build partnerships that increase a community's long-term capacity to advance the protection of children's environmental health and safety; leverage private and public investments to enhance environmental quality by enabling community efforts to continue past EPA's ability to provide assistance to communities; and promote protection of children from environmental threats through lessons learned. There are no deadline dates.

Requirements: Eligible applicants include community groups, public nonprofit institutions/organizations, tribal governments, specialized groups, profit organizations, private nonprofit institutions/organizations, and municipal and local governments. Potential applicants are strongly encouraged to discuss proposed projects with or submit preapplications to program staff prior to the completion of a full proposal.
Geographic Focus: All States
Amount of Grant: 5,000-250,000 USD
Contact: Office of Children's Health Protection; (202) 564-2188; fax (202) 564-2733; fletcher.bettina@epa.gov
Internet: http://yosemite.epa.gov/ochp/ochpweb.nsf/content/grants.htm
Sponsor: Environmental Protection Agency
1200 Pennsylvania Avenue, NW
Washington, DC 20460

EPA Compliance Assistance Support for Services to the Regulated Community and Other Assistance Providers 1934

The EPA has sponsored partnerships with industry, academic institutions, environmental groups, and other agencies to launch sector-specific Compliance Assistance Centers. Each Compliance Assistance Center addresses real world issues in language that is used by the regulated entities. This program's objective is to provide financial assistance to private nonprofit institutions, universities, and public agencies to develop and maintain Compliance Assistance Centers to improve environmental compliance within an identified industrial/ government sector or environmental topic. It is expected that common communications channels used by the industry/government will continue to quickly distribute information to the targeted audience. The sector entry pages will serve as portals to relevant information developed by the Centers and other organizations as appropriate (e.g. EPA). Individual sector entry pages will use navigation structures specific to the sector thereby providing easy access to relevant environmental content. EPA funding must be used only for activities specified in the approved scope of work; discretionary funding will not be provided. EPA generally funds grants and cooperative agreements on a 12-month basis. The grant range is $5,000 to $35,000 per year, with an average of $10,000 annually.
Requirements: Applicants must be nonprofit organizations. Colleges, universities, and community colleges are also eligible to apply.
Restrictions: EPA's Appropriation Act requires that applicants submitting unsolicited research grant proposals share in the cost of conducting research. The amount of the cost share will be based on the mutuality of interest between the Government and the applicant.
Geographic Focus: All States
Amount of Grant: 5,000-35,000 USD
Contact: Tracy Back, National Center for Environmental Research and Quality Assurance; (202) 564-7076; fax (202) 564-0037; back.tracy@epa.gov
Internet: https://www.cfda.gov/?s=program&mode=form&tab=step1&id=556b7138 5864efe15d4605c6ab8f0a72
Sponsor: Environmental Protection Agency
1200 Pennsylvania Avenue, NW
Washington, DC 20460

EPA National Estuary Grants 1935

Project grants and cooperative agreements are awarded to assist qualified organizations and/or persons to develop programs to protect and restore coastal resources in priority estuaries. The program objective is to develop a comprehensive conservation and management plan for each estuary based on established relationships between pollutant loading and impacts on living resources. Actions to be taken under the plan would implement basin-wide reduction of point and non-point sources of pollution adversely impacting the uses of the estuary. Funds are to be used for projects and tasks needed to address priority problems of the estuary as identified within a five-year period and an annual action plan developed through the management conference structure for the priority estuary under study. Each year 12 conferences are formally convened. Funds are available for the priority estuaries at Long Island Sound, Buzzards Bay, Narragansett Bay, New York-New Jersey Harbor, Delaware Bay, Delaware Inland Bays, Sarasota Bay, Galveston Bay, Santa Monica Estuary, Albemarle-Pimlico Sounds, Puget Sound, San Francisco Bay, Casco Bay, Massachusetts Bays, Indian River Lagoon, Tampa Bay, Peconic Bay, San Juan Harbor Bay, Corpus Christi, Tillamook Bay, the Barataria-Terribonne Estuary Complex, New Hampshire Estuaries, Barnegat Bay, Maryland Coastal Bays, Charlotte Harbor, Mobile Bay, Morro Bay, and Columbia Bay. Projects and tasks will include gathering and analyzing existing and historical data, identifying and conducting further research to acquire new and additional data, targeting efforts to increase regional public understanding of the problems and complexities of the estuary and bring public input to the decision-making and management processes, and assessing existing state and local initiatives in water sediment quality and living resource protection programs.
Requirements: State water pollution control agencies, interstate and intrastate agencies, other public or nonprofit private agencies, research institutions, and individuals within the geographic area of a priority estuary are eligible to apply.
Restrictions: Profit-making organizations are not eligible.
Geographic Focus: All States
Date(s) Application is Due: Jun 1
Contact: Darrell Brown, Chief, Coastal Management Branch; (202) 566-1256; fax (202) 566-1366; brown.darrell@epa.gov
Internet: http://www.epa.gov/nep
Sponsor: Environmental Protection Agency
1200 Pennsylvania Avenue, NW
Washington, DC 20460

EPA Pollution Prevention Grants 1936

The EPA Pollution Prevention Grant program applies to the Pollution Prevention (P2) and the Pollution Prevention Information Network (PPIN) grant programs. Both programs were enacted under the Pollution Prevention Act of 1990. The P2 grant program funds grants/cooperative agreements that implement pollution prevention technical assistance services for businesses. The PPIN grant program funds grants/ cooperative agreements through a national network of Pollution Prevention Resource Exchange (P2Rx) Centers. The Centers work to deliver pollution prevention information, resources and tools to state technical assistance providers, businesses and industry sectors. Under both grant programs, EPA is authorized to support projects that utilize pollution prevention techniques to reduce and/or eliminate pollution from air, water and/or land. The objectives include: helping business and industry incorporate P2 techniques into their day-to-day business operations as a means of reducing or eliminating harmful pollutants from entering the air, water or land; to encourage education and training in P2/source reduction techniques and applications; and support of P2 information sharing among State and local technical assistance programs and businesses.
Requirements: Eligible applicants for purposes of funding under this grant program include the 50 states, the District of Columbia, the U.S. Virgin Islands, the Commonwealth of Puerto Rico, and any territory or possession of the United States, any state agency or instrumentality of a state including state colleges and universities, federally-recognized tribes and intertribal consortia. Organizations receiving grant funds through the P2 and PPIN grant programs are required to match federal funds by at least 50 percent.
Restrictions: Local governments, private universities, private nonprofit entities, private businesses, and individuals are not eligible to apply.
Geographic Focus: All States, District of Columbia, Guam, Marshall Islands, Northern Mariana Islands, Puerto Rico, U.S. Virgin Islands, American Samoa
Amount of Grant: 20,500-115,000 USD
Contact: Lee Fiske; (617) 918-1847; fax (617) 918-1810; fiske.lee@epa.gov
Internet: https://www.cfda.gov/?s=program&mode=form&tab=step1&id=53dc0b42 e4cb304e9a54a7ad59905a84
Sponsor: Environmental Protection Agency
1200 Pennsylvania Avenue, NW
Washington, DC 20460

EPA Regional Healthy Indoor Air Projects for Community Outreach and Education 1937

The objective of this program is to support surveys, studies, research, training, outreach, education, investigations or demonstrations performed by organizations that lead to effective outreach strategies to educate key audiences about indoor air pollutants, their associated health risks and encourage effective mitigation and control strategies. Currently, the Program focuses on several critical aspects of indoor air quality that pose significant risks to public health, and in particular, to children and other disproportionately impacted populations. Program focus areas include, but are not limited to reducing the exposure of children and others with asthma to indoor triggers that worsen their health condition, promoting the adoption of operation and maintenance practices in schools, homes and other buildings throughout the nation to reduce the harmful effects of poor indoor air quality on the health of the occupants, promoting voluntary radon testing by homeowners and building operators to identify elevated levels and fix them when they are found, as well as working with homebuilders to incorporate radon resistant and improved indoor air quality construction features into new and remodeled homes, and encouraging adult smokers to protect their children from the adverse health effects of environmental exposure to secondhand smoke by making a conscious decision to keep their homes and cars smokefree. The EPA generally awards grants ranging in value from $5,000 to $750,000 per fiscal year. The average value of each grant is $50,000 per fiscal year.
Requirements: Assistance under this program is generally available to States, local governments, territories, Indian Tribes, and possessions of the U.S., including the District of Columbia, international organizations, public and private universities and colleges, hospitals, laboratories, other public or private nonprofit institutions, which submit applications proposing projects with significant technical merit and relevance to EPA's Office of Air and Radiation's mission.
Geographic Focus: All States, District of Columbia, Guam, Marshall Islands, Northern Mariana Islands, Puerto Rico, U.S. Virgin Islands, American Samoa
Amount of Grant: 5,000-750,000 USD
Contact: Victoria Drew, (202) 343-9049; drew.victoria@epa.gov
Internet: https://www.cfda.gov/?s=program&mode=form&tab=step1&id=d2de8c07 6d885a787769417ed4813ffe
Sponsor: Environmental Protection Agency
Ariel Rios North Building
Washington, DC 20460

EPA Research, Development, Monitoring, Public Education, Training, Demonstrations, and Studies Grants 1938

Grants are awarded to support Research, Development, Monitoring, Public Education, Training, Aid, Demonstrations, and Studies assistance relating to the protection of public health and the environment from pesticides and potential risk from toxic substances. Grants and cooperative agreements are available to support recipients' allowable direct costs incident to approved scopes of work plus allowable indirect costs, in accordance with established EPA policies and regulations. Funding awarded for research does not include research within the purview of EPA's Office of Research and Development. Typically, grants in this category range from $1,000 to $1,500,000, with the average being $500,000.
Requirements: Assistance under this program is generally available to States, U.S. territories or possessions, federally recognized Indian tribal governments and Native American Organizations, public and private universities and colleges, hospitals,

laboratories, other public or private nonprofit institutions, local governments, and individuals and international entities.
Restrictions: For profit organizations are not eligible.
Geographic Focus: All States
Amount of Grant: 1,000-1,500,000 USD
Contact: Sharon Clark, (202) 564- 3810; fax (202) 564-0575; clark.sharon@epa.gov
Internet: https://www.cfda.gov/?s=program&mode=form&tab=step1&id=014d33dd7fea130ebab8d1dca2d45280
Sponsor: Environmental Protection Agency
1200 Pennsylvania Avenue, NW
Washington, DC 20460

EPA Southeastern Multi-Media and Geographic Priority Project Grants 1939

The objective of the Southeastern Multi-Media and Geographic Priority Projects is to fund projects, including surveys, studies, investigations, demonstrations, training, outreach, human health and the environment to: prevent, reduce and eliminate pollution through the promotion of environmental sustainability; build capacity on environmental sustainability and public health issues; support the development and implementation of sustainability strategies and enhance environmental literacy; and promote coordination and collaboration between and among federal, state, and local governmental and non-governmental organizations as well as build environmental education capacity at institutions of higher education. The projects should be centered around providing a forum for academic advisors and sustainability coordinators to collaborate with industry, college and graduate level students to establish and broaden environmental learning opportunities that promote capacity building on environmental sustainability and public health issues, including the creation of hands-on instructional learning of sustainable practices including pollution prevention, water and energy conservation, watershed protection, and environmental stewardship. The project should also serve to foster an environmental literacy plan to ensure that students have an understanding of environmental sustainability and how it interconnects with the green economy. Contact the headquarters or regional office, as appropriate, for application deadlines. The funding range is $15,000 to $100,000, with an average of approximately $90,000 per project.
Requirements: Projects conducted within the geographic boundaries of EPA's Region 4, which comprises the states and tribal lands of Alabama, Georgia, Florida, Kentucky, Mississippi, North Carolina, South Carolina and Tennessee, are eligible for funding when it is available. Interstate, state, tribal, intrastate, and local government agencies, districts, and councils; and public and private universities and colleges, laboratories, and research facilities; and public or private non-governmental, non-profit institutions are eligible to apply, unless restricted by the authorizing statutes. Non-profit organizations must have documentation of non-profit status from the U.S. Internal Revenue Service or their state of incorporation, except that non-profits organizations as defined in Section 501(c)4 of the Internal Revenue Code that engage in lobbying as defined in Section 3 of the Lobbying Disclosure Act 1995 or superseding legislation are ineligible.
Geographic Focus: Alabama, Florida, Georgia, Kentucky, Mississippi, North Carolina, South Carolina, Tennessee
Amount of Grant: 15,000-100,000 USD
Contact: Caroline Robinson, Senior Advisor; (404) 562-8976 or (404) 562-9900; fax (404) 562-8174; robinson.caroline@epa.gov
Internet: https://www.cfda.gov/?s=program&mode=form&tab=step1&id=af6215f79818f81c11793c94fa2fc592
Sponsor: Environmental Protection Agency
Sam Nunn Atlanta Federal Center
Atlanta, GA 30303-8960

EPA Surveys, Studies, Research, Investigations, Demonstrations, and Special Purpose Activities Relating to the Clean Air Act 1940

The primary objective of the program is to support Surveys, Studies, Research, Investigations, Demonstrations and Special Purpose assistance relating to the causes, effects (including health and welfare effects), extent, prevention, and control of air pollution to include such topics as air quality, acid deposition, climate change, global programs, indoor environments, radiation, mobile source technology and community-driven approaches to transportation and emissions reduction. Projects should also focus on addressing environmental justice (EJ) concerns in communities. EJ is the fair treatment and meaningful involvement of all people regardless of race, color, national origin, or income with respect to the development, implementation, and enforcement of environmental laws, regulations, and policies. Grants and cooperative agreements are available to support recipients' allowable direct costs incident to approved Surveys, Studies, Research, Investigations, Demonstrations and Special Purpose plus allowable indirect costs, in accordance with established EPA policies and regulations. Assistance agreement awards under this program may involve or relate to geospatial information. A detailed list of acceptable activities is provided at the grant website. EPA generally award grants ranging in value from $5,000 to $750,000 per fiscal year. The average value of each grant is $150,000 per fiscal year. Contact the headquarters or regional office, as appropriate, for application deadlines.
Requirements: Assistance under this program is generally available to States, local governments, territories, Indian Tribes, and possessions of the U.S., including the District of Columbia, international organizations, public and private universities and colleges, hospitals, laboratories, other public or private nonprofit institutions, which submit applications proposing projects with significant technical merit and relevance to EPA's Office of Air and Radiation's mission.
Geographic Focus: All States
Amount of Grant: 5,000-750,000 USD
Contact: Maureen Hingeley, (202) 564-1306; hingeley.maureen@epa.gov
Internet: https://www.cfda.gov/?s=program&mode=form&tab=step1&id=1d601ed2ad50dabb4dd11eaa76bb7764
Sponsor: Environmental Protection Agency
Ariel Rios North Building
Washington, DC 20460

EPA Targeted Watersheds Grants 1941

Established in 2003, the Targeted Watersheds Grant Program is designed to encourage successful community-based approaches and management techniques to protect and restore the nation's watersheds. The Targeted Watersheds Grant program is a competitive grant program based on the fundamental principles of environmental improvement: collaboration, new technologies, market incentives, and results-oriented strategies. The Targeted Watersheds Grant Program focuses on multi-faceted plans for protecting and restoring water resources that are developed using partnership efforts of diverse stakeholders. There are two distinct funding avenues: Targeted Watersheds Implementation Grants are focused on individual watershed organizations; and Capacity Building Grants are awarded to leading organizations with a national or regional focus that are able to provide training, technical assistance and education to local watershed groups and to teach the critical skills necessary to improve watershed health. The goal of these grants is to better serve the needs of local watershed groups.
Geographic Focus: All States
Contact: Erin Collard; (202) 566-2655; fax (202) 566-1366; initiative.watershed@epa.gov
Internet: http://www.epa.gov/twg/twg_basic.html
Sponsor: Environmental Protection Agency
1200 Pennsylvania Avenue, NW
Washington, DC 20460

EPA Temporally Integrated Monitoring of Ecosystems and Long-Term Monitoring Grants 1942

Clean Air Act section 103(b)(3) authorizes EPA to conduct and promote the coordination and acceleration of research, investigations, experiments, demonstrations, surveys, studies relating to the causes, effects (including health and welfare effects), extent, prevention and control of air pollution. Clean Water Act section 104(b)(3) authorizes EPA to award grants to conduct and promote the coordination and acceleration of research, investigations, experiments, training, demonstrations, surveys and studies relating to the causes, effects, extent, prevention, reduction and elimination of pollution. EPA will award cooperative agreements to study the ecological response to pollution reduction and emissions control programs. The programs will assist in understanding biogeochemical changes in sulfur, nitrogen, mercury, and carbon in streams and lakes in relation to changing pollutant emissions and deposition, as well as other factors, such as potential climate change (e.g., temperature, precipitation, timing of storm events). The TIME/LTM programs will support initiatives that demonstrate the potential for tracking and understanding the ecological response to air pollution reduction programs in acid-sensitive regions of the U.S. Funding will be available for applicants that demonstrate the ability to provide consistent, reliable, quality-assured data from lakes and streams for a variety of surface water chemistry parameters (e.g., pH, sulfate and nitrate concentrations, acid neutralizing capacity, aluminum concentrations, base cations). EPA will implement this grant authority by awarding cooperative agreements to state and local governments, institutions of higher education, and quasi-public and non-profit organizations that demonstrate the capability to perform the sampling, measurement, and parameter characterization of the lakes and streams in this long-term ecological system response program. There is no minimum amount of assistance. EPA anticipates awards ranging in total value of $640,000 to $1,400,000.
Requirements: Assistance under this program is generally available to States, local governments, territories, Indian Tribes, and possessions of the U.S. (including the District of Columbia); public and private universities and colleges; hospitals; laboratories; public or private nonprofit institutions; intertribal consortia; and individuals.
Restrictions: Nonprofit organizations described in Section 501(c)(4) of the Internal Revenue Code that engage in lobbying activities as defined in Section 3 of the Lobbying Disclosure Act of 1995 are not eligible to apply.
Geographic Focus: All States, District of Columbia, Guam, Marshall Islands, Northern Mariana Islands, Puerto Rico, U.S. Virgin Islands, American Samoa
Amount of Grant: 640,000-1,400,000 USD
Contact: Clara Funk, (202) 343-9622; funk.clara@epa.gov
Internet: https://www.cfda.gov/?s=program&mode=form&tab=step1&id=8f43f83e888f5af0a3153751394d2c93
Sponsor: Environmental Protection Agency
Ariel Rios North Building
Washington, DC 20460

Equipment Leasing and Finance Foundation Research Grants 1943

Grants are available to encourage academics in all fields of scholarship to study topics of interest to the equipment leasing and finance industry. Research topics of interest include: transitioning the business model and avoiding the commoditization of a finance product through the bundling of services; best practice case studies; analysis of publicly held lease financing companies—does the current model work?; changes in accounting rules—how does this affect the lessee and lessor in the financial decision making process?; what will finance companies of the future look like?; study of successful industry consolidates; corporate size—when does efficiency of size become inefficient?; limited purpose bank charters—are they the vehicle of the future in leasing?; FASB-and changes to synthetic leases; specialty finance in banking—what is the future?; and maintaining residual values—the challenges and solutions. General topics of research interest are: strategic management/planning, accounting/financial reporting, and finance; tax and legislative/

regulatory; credit/collections; economics; portfolio management; contracts/documentation; technology benchmarking; strategic importance of captives; operations; and best practices. Grants may be used for research studies; booklets, pamphlets and video production; surveys and statistical gathering; and other initiatives that are considered on a case-by-case basis.
Requirements: For-profit and nonprofit organizations and applicants—which may include individuals, universities, foundations, associations, and corporations—are eligible.
Geographic Focus: All States
Date(s) Application is Due: Jan 15; May 15; Sep 1
Amount of Grant: 5,000-50,000 USD
Contact: Lisa Levine, Executive Director; (703) 516-8363; llevine@elamail.com
Internet: http://www.leasefoundation.org/grant
Sponsor: Equipment Leasing and Finance Foundation Grants
4301 N Fairfax Drive, Suite 550
Arlington, VA 22203

Erasmus Institute Fellowships **1944**
The institute will admit six postdoctoral fellows to its center on the campus of the University of Notre Dame. The institute invites applications from academics who by their scholarship and/or faith seek to cultivate Catholic intellectual and cultural traditions. Fellows devote their time in residence to concentrated research and writing on individual projects related to the institute's goals. The fellows will also be expected to teach one course during the academic year, likely in the spring term of next year.
Requirements: Postdoctoral applicants must have received the PhD after May 1, 2001; ABD applicants must fulfill the PhD requirements by June 1 in order to be eligible for the fellowship.
Restrictions: Dissertation and salaried senior and junior faculty fellowships are no longer being offered.
Geographic Focus: All States
Date(s) Application is Due: Jan 27
Amount of Grant: 35,000 USD
Contact: Coordinator; (574) 631-9346; fax (574) 631-3585; erasmus@nd.edu
Internet: http://www.nd.edu/~erasmus
Sponsor: University of Notre Dame
1124 Flanner Hall
Notre Dame, IN 46556-5611

EREF Doctoral Scholarship **1945**
Scholarships are awarded by EREF to recognize excellence in master's, doctoral or post-doctoral waste management research and education. As part of its mission, EREF furthers education in the solid waste management industry. In the past, multiple outstanding candidates have applied for the Fiessinger award in the same year, prompting EREF to periodically utilize foundation funds to award additional scholarships. EREF now regularly awards Ph.D. and postdoctoral scholarships to notable candidates, expanding the scholarship program. At least one EREF Scholar is awarded annually. Award amounts consider the cost of tuition at the recipient's institution and any other funds received. Doctoral scholarships are awarded up to $12,000 per year, paid monthly, and can be extended for up to 3 years from the initial award date.
Requirements: Applications will be considered from full-time PhD students who have a clearly demonstrated interest in environmental research. Applications will be considered from students outside the United States or studying abroad.
Restrictions: Full scholarships may not be awarded to students who will be receiving full-tuition scholarships from other sources (excluding direct university assistance such as tuition remission, or assistantship income) but may still be considered for partial awards.
Geographic Focus: All States, All Countries
Date(s) Application is Due: Aug 1
Amount of Grant: 12,000-36,000 USD
Contact: Bryan F. Staley; (919) 861-6876, ext. 102; fax (919) 861-6878; bstaley@erefdn.org
Internet: http://erefdn.org/index.php/scholarships/
Sponsor: Environmental Research and Education Foundation
3301 Benson Drive, Suite 301
Raleigh, NC 27609

EREF Fiessinger Doctoral Scholarship **1946**
Scholarships are awarded by EREF to recognize excellence in master's, doctoral or post-doctoral waste management research and education. Awards to doctoral students are given in memory of Francois Fiessinger, P.E., Ph.D., a graduate of Rutgers University, who was a founding director of the Environmental Research and Education Foundation. Francois believed in scholarly research and was one of very few people who had a long-term research and development vision for the solid waste industry. In 1997, based on a contribution from Lyonnaise des Eaux, the Francois Fiessinger Memorial Scholarship Fund was established to recognize excellence in Ph.D. or postdoctoral environmental research and education. One Fiessinger Scholar is awarded annually. Award amounts consider the cost of tuition at the recipient's institution and any other funds received. Doctoral scholarships are awarded up to $12,000 per year, paid monthly, and can be extended for up to 3 years from the initial award date.
Requirements: Applications will be considered from full-time PhD students who have a clearly demonstrated interest in environmental research. Applications will be considered from students outside the United States or studying abroad.
Restrictions: Full scholarships may not be awarded to students who will be receiving full-tuition scholarships from other sources (excluding direct university assistance such as tuition remission, or assistantship income) but may still be considered for partial awards.
Geographic Focus: All States, All Countries
Date(s) Application is Due: Aug 1
Amount of Grant: 12,000-36,000 USD
Contact: Bryan F. Staley; (919) 861-6876, ext. 102; fax (919) 861-6878; bstaley@erefdn.org
Internet: http://erefdn.org/index.php/scholarships/
Sponsor: Environmental Research and Education Foundation
3301 Benson Drive, Suite 301
Raleigh, NC 27609

EREF Solid Waste Research Grants **1947**
The Environmental Research and Education Foundation is the only private, grant making institution with a national and international scope whose sole mission is to support solid waste research. Its research grants program is led by a Research Council, a body of volunteers consisting of technical experts in industry, academia and consulting. The work of the Council is guided by a long range strategic plan with the goal to achieve greater sustainability, good environmental stewardship, higher process efficiency and increased knowledge. Council recommended projects are then reviewed by the Board's Projects Committee for a final review and funding allocation. EREF awards several grants each year for research or education in topics pertaining to any aspect of solid waste management. There are two annual deadlines for solicited proposal submissions, including January 5 and July 15. There is no formal application, and submissions should include a description of proposed activity, project timeline, budget, and resumes of principal investigators. Previously awarded grants have ranged from $15,000 to over $500,000 with the average grant being amount being $100,000.
Restrictions: Grants will not be provided for capital campaigns; political contributions; religious causes; operating funds; loans; or support of lobbying activities. The Foundation will not pay overhead in excess of 25%. Proposals submitted in response to this RFP that do not address a specific topic in the research agenda will not be reviewed.
Geographic Focus: All States
Date(s) Application is Due: Jan 5; Jul 15
Amount of Grant: 15,000-500,000 USD
Contact: Bryan F. Staley; (919) 861-6876, ext. 102; fax (919) 861-6878; bstaley@erefdn.org
Internet: http://erefdn.org/index.php/grants/
Sponsor: Environmental Research and Education Foundation
3301 Benson Drive, Suite 301
Raleigh, NC 27609

EREF Sustainability Research Grants **1948**
The Environmental Research and Education Foundation is the only private, grant making institution with a national and international scope whose sole mission is to support solid waste research. Submissions of scientific research proposals related to sustainable solid waste management practices are invited in the following areas: waste minimization; recycling; waste conversion to energy, bio-fuels, chemicals or other useful products, including (but not limited to) the technologies of waste-to-energy, pyrolysis and gasification, oxidation, anaerobic digestion, and composting; and strategies to promote diversion from landfills (e.g. separating organics, market analysis, optimized material management, logistics, etc.). EREF awards several grants each year for research or education in topics pertaining to any aspect of solid waste management. There are two annual deadlines for solicited proposal submissions, including January 5 and July 15. There is no formal application, and submissions should include a description of proposed activity, project timeline, budget, and resumes of principal investigators. Previously awarded grants have ranged from $15,000 to over $500,000 with the average grant being amount being $100,000.
Restrictions: Grants will not be provided for capital campaigns; political contributions; religious causes; operating funds; loans; or support of lobbying activities. The Foundation will not pay overhead in excess of 25%. Proposals submitted in response to this RFP that do not address a specific topic in the research agenda will not be reviewed.
Geographic Focus: All States
Date(s) Application is Due: Jan 5; Jul 15
Contact: Bryan F. Staley; (919) 861-6876, ext. 102; fax (919) 861-6878; bstaley@erefdn.org
Internet: http://erefdn.org/index.php/grants/proposal
Sponsor: Environmental Research and Education Foundation
3301 Benson Drive, Suite 301
Raleigh, NC 27609

Erikson Institute Child Development Doctoral Fellowships **1949**
The PhD program, offered jointly with the Department of Psychology at Loyola University Chicago, prepares students for careers in college teaching and research, administration, program design and evaluation, and child advocacy. Named fellowships include the Fellowship in Applied Research in Child Development and the Irving B. Harris Leadership Fellowship in Child Development. The fellowships include tuition plus a research/teaching assistantship or living stipend and are awarded on the basis of academic promise, experience, and research interest.
Geographic Focus: All States
Date(s) Application is Due: Apr 1
Contact: Dr. Robert Halpern, (312) 893-7142; rhalpern@erikson.edu
Internet: http://www.erikson.edu/students.asp?file=applyfinaid&highlight=fellowships
Sponsor: Erikson Institute
420 N Wabash Avenue
Chicago, IL 60611

Ernest Guenther Award in the Chemistry of Natural Products **1950**
This award, supported by Givaudan, is given annually to recognize and encourage outstanding achievements in analysis, structure elucidation, and chemical synthesis of natural products. Special consideration will be given to the independence of thought and originality shown. The award consists of $6,000, a medallion, and a certificate.

Requirements: Any individual, except a member of the award committee, may submit one nomination or seconding letter for each award in any given year. Nominating documents consist of a letter of not more than 1000 words containing an evaluation of the nominee's accomplishments and a specific identification of the work to be recognized, a biographical sketch including date of birth, and a list of publications and patents authored by the nominee.
Restrictions: Self-nominations are not accepted.
Geographic Focus: All States
Date(s) Application is Due: Nov 1
Amount of Grant: 8,500 USD
Contact: Felicia Dixon, Awards Administrator; (800) 227-5558 or (202) 872-4408; fax (202) 776-8008; f_dixon@acs.org or awards@acs.org
Internet: http://chemistry.org/portal/Chemistry?PID=acsdisplay.html&DOC=awardsguenther.html
Sponsor: American Chemical Society
1155 Sixteenth Street, NW
Washington, DC 20036-4801

ESF Latsis Prize **1951**
The foundation invites nominations for the European Latsis Prize. Presented each year at the ESF's Annual Assembly, the European Latsis Prize is awarded to an individual scholar or research group in recognition of outstanding and innovative contributions in a selected field of European research. Nomination information and nomination forms are available on the Web site.
Requirements: Nominations may be received from institutions and individuals for an individual scholar, or from institutions generally for research groups.
Restrictions: Self-nominations are not accepted.
Geographic Focus: All States
Amount of Grant: 100,000 CHF
Contact: Secretary General; 33-0-3-88-76-71-00; fax 33-0-3-88-37-05-32; prize@esf.org
Internet: http://www.esf.org/prize
Sponsor: European Science Foundation
1 quai Lezay Marnesia, BP 90015
Strasbourg Cedex, F-67080 France

ESRI Conservation Grants **1952**
The conservation grants program provides donations and discounts of GIS software, data, books, and training. There are no application deadlines. The program does not grant hardware or cash, but partners with many other groups who do (list available online). General grants and basic grants are available. Grants are awarded to organizations that can demonstrate a strong commitment to conservation; compelling reasons to use GIS in obtaining their objectives; and the organizational capacity to use GIS effectively over a long period of time. All applicants must present a credible case for how GIS and the support from ESRI will advance their cause of conservation and environmental protection, broadly defined. Successful applicants will effectively make their case about how they propose to integrate and analyze disparate spatial data to accomplish their goals. To request application materials, send a blank email message (no subject, no content) to ecpgrant@esri.com, or visit the Web site.
Requirements: Eligible applicants are U.S.-based 501c(3) nonprofit organizations that actively engage the public in resource conservation and environmental protection. This includes, but is not limited to, grassroots conservation and environmental organizations, community action groups, economic development organizations, sustainable development groups, community-based conservation groups, growth management organizations, environmental justice groups, and tribal conservation organizations. International groups must have a U.S.-based tax-exempt organization as their sponsor.
Restrictions: The following types of organizations are not eligible to apply: colleges and universities; public schools or school districts; local, state, and federal government agencies or natural resource conservation districts; association and coalitions organized to fund-raise for or to benefit local, state, and federal government agencies; natural resource conservation districts; and CTSP's nonprofit sponsors and their associated members.
Geographic Focus: All States
Contact: Charles Convis; (909) 793-2853, ext. 2488; fax (909) 307-3025; ecp@esri.com
Internet: http://www.conservationgis.org/aaesrigrants.html
Sponsor: Environmental Systems Research Institute
380 New York Street
Redlands, CA 92373-8100

Ethics and Excellence in Journalism Foundation Grants **1953**
This Foundation was founded for charitable, scientific and educational purposes, including the improvement of the quality of the practice of journalism among various media. Support includes funding for creative projects and research that promotes excellence in journalism and instills and encourages high ethical standards in journalism. Funding is available to organizations via an application and review process.
Requirements: U.S. nonprofits, including universities and colleges, are eligible.
Restrictions: Individuals are not eligible.
Geographic Focus: All States
Date(s) Application is Due: Apr 15; Oct 15
Contact: Nancy Hodgkinson, Senior Program Officer; (405) 604-5388; fax (405) 604-0297; nancy.hodgkinson@journalismfoundation.org
Internet: http://www.journalismfoundation.org
Sponsor: Ethics and Excellence in Journalism Foundation
210 Park Avenue, Suite 3150
Oklahoma City, OK 73102

ETS Predoctoral Summer Program **1954**
Students chosen for the eight-week program will participate in ongoing independent research with access to a mentor in one of the following areas: education; educational technology; learning; linguistics; literacy; minority issues; new constructs; policy research; psychology; psychometrics; statistics; teaching; and testing issues (including alternate forms of assessment for special populations and new forms of assessment). An explicit goal of the program is to increase the number of women and minority professionals in educational measurement and related fields. Applicants must submit their application materials electronically. Guidelines are available online.
Requirements: Graduate students who are currently enrolled in a doctoral program and have completed a minimum of two years of full-time graduate study in a program emphasizing one of the areas specified are eligible. A letter of recommendation from the major advisor or committee chair and transcripts, undergraduate and graduate, must be included with completed application form.
Geographic Focus: All States
Date(s) Application is Due: Feb 1
Amount of Grant: 5,000 USD
Contact: Grants Administrator; (609) 734-5543; internfellowships@ets.org
Internet: http://www.ets.org/research/fellowships.html
Sponsor: Educational Testing Service
Rosedale Road, MS 07R
Princeton, NJ 08541-0001

ETS Sylvia Taylor Johnson Minority Fellowship in Educational Measurement **1955**
The two-year fellowship, renewable after the first year by mutual agreement, is designed to provide talented minority scholars an opportunity to carry out independent research under the mentorship of ETS senior researchers. The stipend will be set in relation to the successful applicant's compensation at the home institution. Applicants may obtain the center's brochure before they apply.
Requirements: Applicants must have received their doctoral degree within the past ten years and must be U.S. citizens or permanent residents.
Geographic Focus: All States
Date(s) Application is Due: Feb 1
Contact: Linda DeLauro, (609) 734-1806; fax (609) 734-5410; internfellowships@ets.org
Internet: http://www.ets.org/portal/site/ets/menuitem.c988ba0e5dd572bada20bc47c3921509/?vgnextoid=1dadaf5e44df4010VgnVCM10000022f95190RCRD&vgnextchannel=49f5be3a864f4010VgnVCM10000022f95190RCRD
Sponsor: Educational Testing Service
Rosedale Road, MS 07R
Princeton, NJ 08541-0001

EU Center for Humanistic Inquiry Junior and Postdoctoral Fellowships **1956**
The purpose of the program is to stimulate and support humanistic research by providing scholars in early stages of their careers with the necessary time, space, and other resources. Research projects must be humanistic, but fellows may hold the PhD in any discipline. Fellows are expected to offer an upper-level undergraduate course on a subject of their choice during the spring of their fellowship year. The program provides stipends, research budgets, office space, and access to the university libraries. Application forms and further information are available from the office and online.
Requirements: Junior fellows are scholars who, at the beginning of the fellowship year, will be at least three years beyond receipt of the PhD, and normally no more than 10. Postdoctoral fellows, who must have the PhD before the submission of their applications, are awarded to those who have held the PhD for no more than three years before receiving the fellowship.
Geographic Focus: All States
Date(s) Application is Due: Feb 23
Amount of Grant: 38,500 USD
Contact: Humanistic Inquiry; (404) 727-6424; fax (404) 727-1669; chi@emory.edu
Internet: http://www.chi.emory.edu/Fellowships/index.html
Sponsor: Emory University
1715 N Decatur Road
Atlanta, GA 30307

Eugene M. Lang Foundation Grants **1957**
The foundation awards grants in New York and Pennsylvania in its areas of interest, including education (early childhood education and higher education), medical and health programs, arts, health organizations, medical research, minorities, and performing arts. Types of support include annual campaigns, conferences and seminars, continuing support, fellowships, general operating support, internship funds, professorships, program development, scholarship funds, and seed money. The foundation favors social services such as those helping homeless or single mothers. Locally based groups wanting support must involve a Lang family member. There are no application deadlines; initial approach should be by letter.
Requirements: Organizations in New York and Pennsylvania are eligible to apply.
Restrictions: Grants are not made to individuals, or for building funds, equipment and materials, capital or endowment funds, deficit financing, publications, or matching gifts.
Geographic Focus: New York, Pennsylvania
Amount of Grant: 500-50,000 USD
Contact: Program Contact; (212) 949-4100
Sponsor: Eugene M. Lang Foundation
535 5th Avenue, Suite 906
New York, NY 10017

Eugene McDermott Foundation Grants 1958
The foundation awards grants to Texas nonprofit organizations in its areas of interest, including children and youth, community development, education (early childhood through higher education), health care/organizations, international human rights, medical research, minorities, and social service delivery programs. Types of support include annual campaigns, building construction/renovation, capital campaigns, equipment/land acquisition, general operating grants, matching/challenge grants, professorships, programs/project support, research grants, scholarship funds, and seed grants. There are no application deadlines or forms. The board meets quarterly to consider requests.
Requirements: Texas nonprofit organizations are eligible.
Geographic Focus: Texas
Amount of Grant: 1,000-25,000 USD
Contact: Grants Administrator; (214) 521-2924
Sponsor: Eugene McDermott Foundation
3808 Euclid Avenue
Dallas, TX 75205

EU Robert W. Woodruff Library Research Fellowships 1959
Short-term fellowships support scholarly use of the library's research collections in the areas of modern literature and African American studies. The Special Collections Department has extensive holdings related to the Irish literary renaissance and the finest collection outside of Ireland for the study of contemporary Irish poetry. The library also holds the literary archive of the late poet laureate of England, Ted Hughes, and related British literary collections. The department also houses extensive collections focusing on black print culture, the civil rights and post-civil rights movements, communism and the Left, and African American religion and culture. Fellowship support helps defray expenses in traveling to and residing in Atlanta during the fellowship tenure, usually one month.
Geographic Focus: All States
Date(s) Application is Due: May 31
Amount of Grant: Up to 2,000 USD
Contact: Fellowship Coordinator; (404) 727-0360; speccollref@emory.edu
Internet: http://web.library.emory.edu/services/hr/fellows.html
Sponsor: Emory University
Robert W Woodruff Library, 540 Asbury Circle, Emory University
Atlanta, GA 30322-2870

Evan and Susan Bayh Foundation Grants 1960
Established in 2002 by Senator Evan Bayh and his wife Susan, the foundation gives to various education and human services programs mostly in Indiana.
Requirements: There are no deadlines and no specific forms to use. It is suggested that you contact the foundation prior to submitting an application.
Geographic Focus: Indiana
Amount of Grant: 2,500-5,000 USD
Contact: G. Frederick Glass, (317) 237-0300; fred.glass@bakerd.com
Sponsor: Evan and Susan Bayh Foundation
300 N Meridian Street, Suite 2700
Indianapolis, IN 46204-1750

Evan Frankel Foundation Grants 1961
The foundation awards grants in its areas of interest, including higher education in the humanities and the environment. Giving is primarily in Manhattan and Suffolk County, NY and Los Angeles, CA. Submit a letter to request guidelines and deadline dates.
Restrictions: Individuals are not eligible.
Geographic Focus: All States
Amount of Grant: 500-100,000 USD
Contact: Nancy Wendell, (631) 329-2833; fax (631) 329-7102; frankelfound@hamptons.com
Sponsor: Evan Frankel Foundation
P.O. Box 5072
East Hampton, NY 11937

Evanston Community Foundation Grants 1962
Grants awarded by the Foundation: encourage and support new initiatives and innovative approaches to addressing community needs; build the capacity of local nonprofit organizations to fulfill their missions more effectively; encourage collaborative ventures that will strengthen the community; provide initial support of projects that will have impact beyond the scope and timeline of the proposed project; build community partnerships and resources; and strengthen the area's nonprofit community. Grants typically provide initial seed money to launch new projects, capstone dollars for a larger project that will grow over a longer period, and support a one-time activity or initial phase of a new program. The Foundation's area of interests include: arts and culture; basic human needs; community development; education; environment; health; women and girls and youth and families. The Foundation's RFP and application instructions are available online.
Requirements: IRS 501(c)3 organizations serving the Evanston community are eligible.
Geographic Focus: Illinois
Date(s) Application is Due: Feb 26
Samples: Evanston Youth Bank (Evanston, IL)—for Northwestern University Undergraduate Leadership Program, $1000.
Contact: Sara Schastok, Executive Director; (847) 492-0990; fax (847) 492-0904; schastok@evcommfdn.org or info@evcommfdn.org
Internet: http://www.evcommfdn.org/grant_making.htm
Sponsor: Evanston Community Foundation
1007 Church Street, Suite 108
Evanston, IL 60201

Evjue Foundation Grants 1963
The Foundation contributes each year to worthy educational, cultural and charitable organizations that contribute to the quality of life in Madison and Dane County. The Foundation's interests include: theater, hunger, troubled youth, and universities. The Foundation looks favorably upon projects responding to overall community needs and priorities that do not duplicate existing services. Application and guidelines are available online.
Requirements: Grants are only made to nonprofit organizations.
Restrictions: Grants are not made to individuals. All grants for scholarships are given to educational institutions, which have full responsibility for selecting the individual recipients. Grants are not usually made to establish or add to endowment funds. Grants are not ordinarily made to fund specific medical or scientific research inquiries, nor to support operating expenses and general administrative expenses of organizations.
Geographic Focus: Wisconsin
Date(s) Application is Due: Mar 5
Contact: Arlene Hornung; (608) 252-6401; ahornung@madison.com
Internet: http://host.madison.com/ct/about/evjue/article_36a7a66e-a46a-11de-ac74-001cc4c002e0.html
Sponsor: Evjue Foundation
1901 Fish Hatchery Road
Madison, WI 53713

Ewing Halsell Foundation Grants 1964
The foundation awards grants to eligible Texas nonprofit organizations in its areas of interest, including education, environment, health care and health organizations, medical research, social services, and youth services. Types of support include annual campaigns, building construction/renovation, equipment acquisition, land acquisition, publication, research, seed grants, and technical assistance. There are no application deadlines or forms.
Requirements: Texas nonprofit organizations are eligible. Preference is given to requests from southwestern Texas, particularly San Antonio.
Restrictions: The foundation's grants do not support individuals or requests for deficit financing, emergency funds, general endowments, matching gifts, scholarships, fellowships, demonstration projects, general purposes, conferences, or loans.
Geographic Focus: Texas
Contact: Grants Administrator; (210) 223-2640
Sponsor: Ewing Halsell Foundation
711 Navarro Street, Suite 537
San Antonio, TX 78205

Expect Miracles Foundation Grants 1965
The MMLC Funding Award is given on an annual basis on behalf of the Miracle Maker Leadership Council to cancer fighting causes and patient care programs where such support is not always readily available. The proposed program should have the ability to materially impact an organization or a specific program of a qualified organization.
Requirements: Funding Award recipients must have 501(c)3 status as defined by the IRS. Applicants must provide service in at least one of the following areas: cancer patient care support programs and initiatives; cancer awareness and educational programs; or cancer prevention/health projects. Priority will be given to initiatives demonstrating sustainable benefits to the communities where the Foundation events are held and where Foundation supporters and the Miracle Maker Leadership Council (MMLC) members live.
Restrictions: Although there are no specific guidelines on the size, annual budget, or annual fundraising of a recipient organization, the program is not designed to assist in funding large institutional/national/society entities. The goal of the MMLC is to select programs and causes where funding is not always readily available. MMLC guidelines prohibit funding for the following: private pursuits; political parties, associations and representatives of advocacy groups; organizations that discriminate by race, creed, gender, sexual orientation, age, religion, or national origin; religious causes; advertising, promotion, or sponsorship; and donation to an individual or team fundraising initiative or program.
Geographic Focus: California, Connecticut, District of Columbia, Illinois, Maine, Maryland, Massachusetts, New Hampshire, New Jersey, New York, Pennsylvania, Rhode Island, Vermont
Date(s) Application is Due: Mar 5
Amount of Grant: Up to 27,000 USD
Contact: Alana Chin, (617) 391-9235 or (617) 827-7463; achin@expectmiraclesfoundation.org
Internet: http://www.expectmiraclesfoundation.org/
Sponsor: Expect Miracles Foundation
6 Quail Run
Hingham, MA 02043

Explorers Club Exploration Fund Grants 1966
The Explorers Club is an international multidisciplinary professional society dedicated to the advancement of field research and the ideal that it is vital to preserve the instinct to explore. Since its inception in 1904, the Club has served as a meeting point and unifying force for explorers and scientists worldwide. The Exploration Fund of the Explorers Club provides grants in support of exploration and field research. Grants are made primarily to graduate students. Awards grants range from $500 to $1,500.
Requirements: Individuals at the advanced graduate and professional levels are eligible to apply for support. Applicants need not be members of the Explorers Club.
Geographic Focus: All States
Date(s) Application is Due: Mar 31
Amount of Grant: 500-1,500 USD
Contact: Matthew Williams, Administrative Assistant; (212) 628-8383; fax (212) 288-4449; mwilliams@explorers.org

Internet: http://explorers.org/resources/funding/funding.php
Sponsor: Explorers Club
46 E 70th Street
New York, NY 10021

Explorers Club Scott Pearlman Field Awards 1967
The Scott Pearlman Field Awards provide grants for professional artists, writers, photographers, filmmakers, and journalists in support of scientific expeditions sponsored by The Explorers Club. Grants in amounts up to $1,500 are made to professionals who are recommended by the leader of an expedition. Applications will be judged solely on the merit of the proposal, the competence and ability of the candidate, and the appropriateness of the budget.
Requirements: Samples of the candidates work must be submitted with this application, along with two references from individuals familiar with the candidate's skills, personality, and ability to provide reproduction-quality documentation of the specified expedition.
Geographic Focus: All States
Amount of Grant: Up to 1,500 USD
Contact: Matthew Williams, Administrative Assistant; (212) 628-8383; fax (212) 288-4449; mwilliams@explorers.org
Internet: http://www.explorers.org/resources/funding/pearlmanawards/pearlmanawards.php
Sponsor: Explorers Club
46 E 70th Street
New York, NY 10021

ExxonMobil Foundation Malaria Initiative Grants 1968
As a major employer and investor in Africa, ExxonMobil is committed to working with partners — local institutions, international NGOs and governments — to stop the spread of malaria on the continent. Since 2000, the corporation's funding has reached nearly 83 million people through support exceeding $110 million. Funding supports partners who help improve the delivery and use of prevention tools such as bed nets; train healthcare workers to ensure malaria interventions are reaching those who need them most; provide technical assistance to help countries increase their capacity to control malaria; and facilitate the monitoring and promotion of progress through integrated communications programs.
Requirements: Grants are made to tax-exempt organizations within principal company-operating areas.
Restrictions: Grants are not made to individuals for scholarships, fellowships, research, or travel. Organizations that are primarily religious in nature are ineligible.
Geographic Focus: All States, United Kingdom
Contact: Suzanne McCarron, President; (972) 444-1100 or (972) 444-1007; fax (972) 444-1405; contributions@exxonmobil.com
Internet: http://www.exxonmobil.com/Corporate/community_malaria_initiative.aspx
Sponsor: ExxonMobil Foundation
5959 Las Colinas Boulevard
Irving, TX 75039-2298

Eye-Bank for Sight Restoration and Fight for Sight Summer Student Research Fellowship 1969
Beginning in 2011, The Eye-Bank for Sight Restoration and Fight for Sight co-sponsored a Summer Student Research Fellowship. Annual grants of $2,100 will support full-time research conducted from June to August. The goal of this award is to advance the skills needed to initiate and carry out work in a scientific environment.
Requirements: Applicants must be currently enrolled undergraduates, medical students or graduate students studying in the United States who wish to explore ophthalmology or eye research as a career. Students are expected to complete a short, independent project during the summer months under the guidance of a senior scientist or clinician. Candidates are invited to apply by visiting www.fightforsight.org for an application.
Restrictions: Lab employees are not eligible for this fellowship.
Geographic Focus: All States
Date(s) Application is Due: Feb 1
Amount of Grant: 2,100-2,100 USD
Contact: Janice Benson; (212) 679-6060; fax (212) 679-4466; janice@fightforsight.org
Internet: http://www.eyedonation.org/scholarship_program.html#fight_sight
Sponsor: Eye-Bank for Sight Restoration
120 Wall Street
New York, NY 10005-3902

Ezra Jack Keats/Kerlan Collection Memorial Fellowship 1970
This fellowship provides a talented writer and/or illustrator of children's books with funds from the Children's Literature Research Collections at the University of Minnesota-Twin Cities Campus to use the Kerlan Collection for artistic development. Special consideration will be given to someone who would find it difficult to finance the visit to the Kerlan Collection. The fellow will receive transportation and a per diem allotment.
Requirements: Grants are awarded to individuals based on need. Candidates may be of any nationality.
Geographic Focus: All States
Date(s) Application is Due: Dec 31
Amount of Grant: 1,500 USD
Contact: Fellowship Committee; (612) 624-4576; fax (612) 625-5525; clrc@tc.umn.edu
Internet: http://special.lib.umn.edu/clrc/kerlan/awards.php
Sponsor: University of Minnesota
222 21st Avenue South, 113 Andersen Library
Minneapolis, MN 55455

F. Albert Cotton Award in Synthetic Inorganic Chemistry 1971
The award recognizes outstanding synthetic accomplishment in the field of inorganic chemistry and will be granted without regard to age or nationality. Creativity and imagination will be especially valued. The award consists of $5,000 and a certificate. Reasonable travel expenses to the meeting at which the award will be presented will be reimbursed.
Requirements: Any individual, except a member of the award committee, may submit one nomination or seconding letter for the award in any given year. The nominating documents consist of a letter of not more than 1000 words containing an evaluation of the nominee's accomplishments and a specific identification of the work to be recognized, a biographical sketch including date of birth, and a list of publications and patents authored by the nominee.
Restrictions: Self-nominations are not accepted.
Geographic Focus: All States
Date(s) Application is Due: Nov 1
Amount of Grant: 5,000 USD
Contact: Felicia Dixon, Awards Administrator; (800) 227-5558 or (202) 872-4408; fax (202) 776-8008; f_dixon@acs.org or awards@acs.org
Internet: http://portal.acs.org/portal/acs/corg/content?_nfpb=true&_pageLabel=PP_ARTICLEMAIN&node_id=1319&content_id=CTP_004489&use_sec=true&sec_url_var=region1
Sponsor: American Chemical Society
1155 Sixteenth Street, NW
Washington, DC 20036-4801

F.W. Olin Foundation Construction Grant 1972
Each year, the foundation accepts applications for a grant to pay the entire cost of constructing and equipping a new academic building or library on a U.S. independent college or university campus with a full-time undergraduate enrollment of 500 or more; the cost will not be shared with any other donor. Submit the original application to the New York office and one copy to the Minneapolis office c/o William Horn, 1500 Forshay Tower, Minneapolis, MN 55402; (612) 341-2581; fax: (612) 341-3801.
Geographic Focus: All States
Contact: Dr. Richard Miller, President; (212) 832-0508; richard.miller@olin.edu
Internet: http://www.olin.edu/about_olin/olin_foundation.asp
Sponsor: F.W. Olin Foundation
780 Third Avenue, Suite 3403
New York, NY 10017-7090

Facebook Fellowships 1973
Every day Facebook confronts the most complex technical problems and believes that close relationships with the academy will enable them to address many of these problems at a fundamental level and solve them. As part of its ongoing commitment to academic relations, Facebook has created the Facebook Fellowship program to support graduate students. Interest is in a wide range of academic topics, including the following topical areas: Internet Economics—auction theory and algorithmic game theory relevant to online advertising auctions; Cloud Computing—storage, databases, and optimization for computing in a massively distributed environment; Social Computing—models, algorithms and systems around social networks, social media, social search and collaborative environments; Data Mining and Machine Learning—learning algorithms, feature generation, and evaluation methods to produce effective online and offline models of behavioral signals; Systems—hardware, operating system, runtime, and language support for fast, scalable, efficient data centers; Information Retrieval—search algorithms, information extraction, question answering, cross-lingual retrieval and multimedia retrieval. Tuition and fees will be paid for the academic year: $30K stipend (paid over 9 months of the academic year); $5K per year toward conference attendance and travel; $5K toward a personal computer. Fellows will have an opportunity to apply for a paid summer internship. Fellowships will begin in August or September, depending on the start of the award recipient's academic year.
Requirements: Full-time Ph.D. students in topical areas represented by these fellowships who are currently involved in on-going research are eligible to apply. Non-U.S. citizens are also eligible but must be enrolled at a U.S. university and meet the eligibility requirements and have a valid Visa. Students must be studying Computer Science, Computer Engineering, Electrical Engineering, System Architecture, or a related area. Students must be enrolled during the academic year that the Fellowship is awarded and must be nominated by a faculty member.
Restrictions: Recipients are responsible for the taxation on any funds.
Geographic Focus: All States
Date(s) Application is Due: Feb 15
Contact: Greg Badros; (650) 543-4811; fellowships@facebook.com
Internet: http://www.facebook.com/careers/fellowship.php
Sponsor: Facebook
1601 South California Avenue
Palo Alto, CA 94304-1111

Fairlawn Foundation Grants 1974
The Foundation is dedicated to supporting excellence and innovation in health care delivery, education and research in the Worcester area. The Foundation was established to: provide financial support to organizations, both public and private, that have specific, well-defined proposals to improve and/or expand the practice and delivery of medical and allied health care services in the Worcester area; and to support the education of Worcester area residents in health care related areas.
Requirements: Funding preference is given to institutions involved in higher education, medical or scientific research, hospitals and health care providers, and other organizations capable of

making a long-term impact on health care through innovation or research. Applicants must be tax-exempt, nonprofit organizations as recognized by IRS code 501(c)3.
Restrictions: Except in unusual circumstances, the Foundation generally will not approve grants: to any program or individual in excess of $120,000; to programs that require more than five years to complete; for projects supported by other funds in which the Fairlawn Foundation contribution would comprise less than half of total funding; or for pre-existing projects as a minor participant.
Geographic Focus: Massachusetts
Date(s) Application is Due: Mar 15; Sep 15
Amount of Grant: Up to 120,000 USD
Contact: Lois Smith; (508) 755-0980, ext. 107; lsmith@greaterworcester.org
Internet: http://www.greaterworcester.org/grants/Fairlawn.htm
Sponsor: Greater Worcester Community Foundation
370 Main Street, Suite 650
Worcester, MA 01608-1738

FAMRI Clinical Innovator Awards 1975
The purpose of the Clinical Innovator Awards (CIA) from the Flight Attendant Medical Research Institute (FAMRI) is to stimulate novel medical and clinical scientific research studies on the topic of second hand tobacco smoke. Of high priority to FAMRI are new technologies to detect second hand tobacco-caused illnesses and novel therapies to treat these diseases. FAMRI is interested in clinical and translational research, from basic molecular mechanisms of disease to occupational health. These areas include: studies on the etiology and pathogenesis of tobacco-caused respiratory diseases in humans; therapeutic interventions; clinical trials; and research to establish a protocol for inquiry regarding second hand tobacco smoke exposure by all health care professionals. FAMRI plans to provide awards of up to $100,000 per year for direct costs plus no more than 8.5% for indirect costs, subject to the availability of funds (FAMRI cannot guarantee annual and total levels of support). Awards usually will be made for three years with a possibility of a no-cost extension for one year upon request. Applicants must first submit an online letter of intent by the due date. Upon invitation, they may then submit online proposals. Instructions and links for submitting the letter of intent and proposal are available under "Current and Upcoming Awards" at the FAMRI website. Deadlines may vary from year to year. To be kept up to date of all current and future funding opportunities, interested applicants are encouraged to sign up for the FAMRI ListServ (click on "listserv signup" under the researchers tab at the FAMRI website).
Requirements: Applicants must be certain that their institution can agree to the terms of the FAMRI grant agreement. A copy of the agreement is available for review under "Current and Upcoming Awards" at the FAMRI website. Grantees must devote a minimum of 50% of professional time to research during the period of this award; at least 15% of the grantee's time (or a total of 15% of the combined times of the co-grantees) should be devoted to the research project described in the application. Researchers with an advanced degree (MD, PhD, or MD/PhD, or the equivalent) are eligible to apply. Individuals who apply to FAMRI for funding need not be U.S. citizens. They normally apply through and with an institution or organization that is qualified under United States tax laws to receive grants from FAMRI, a U.S.-based private foundation. The applying institution receives and administers FAMRI's funds and supervises the work of the Principal Investigator. If the applying institution is organized under the laws of a country other than the United States, it must provide FAMRI any and all documentation that FAMRI may require, in form and content acceptable to FAMRI, to ensure that any grant from FAMRI shall be a "qualified distribution" under Internal Revenue Service laws and regulations, including but not limited to affiliation with a qualified U.S.-based institution, a current affidavit from the applicant or an opinion of qualified legal counsel stating the facts and circumstances of the applicant's organization and activities and certifying that any grant from FAMRI shall be a "qualified distribution." Research projects that involve human subjects must follow Public Health Service guidelines (or its equivalent in non-U.S. institutions) and must have the approval of the appropriate Institutional Review Boards (IRB) or foreign equivalent prior to the start of the research. The IRB (or foreign equivalent) approval must be maintained and be in effect throughout the research project. All research must be performed in accordance with all relevant institutional and federal policies and guidelines (or foreign equivalent) relating to human subject protection, radiation and environmental health and safety. All FAMRI grantees are encouraged to collaborate with fellow grantees and may be required to participate periodically in FAMRI's collaborative efforts.
Restrictions: The Clinical Innovator Award must not exceed $100,000 per year plus 8.5 % indirect costs for a total of $108,500 per year. FAMRI will only support animal research that utilizes mice or rats. Research involving other animals will not be considered unless the animals will not be sacrificed as a result of the research. Pending Association for Assessment of Laboratory Animal Care (AALAC) approval must be submitted prior to funding of a grant involving animal subjects. To avoid all conflicts, any principal investigator or other participating applicant who is currently accepting tobacco industry funding, has recently applied for tobacco funding, or has any other association or consulting position with a tobacco company or affiliate of the tobacco industry is not eligible for FAMRI funding. All prior funding and/or association with tobacco companies must be disclosed and will be considered on a case-by-case basis. An award will not be made or will be withdrawn if the applicant becomes the Principal Investigator on a peer-reviewed grant with the same general aims as those proposed in the application to FAMRI. FAMRI reserves the right to announce the award publicly in a manner and at a time of its choosing.
Geographic Focus: All States, All Countries
Date(s) Application is Due: Sep 26
Contact: Elizabeth Kress; (305) 379-7007; fax (305) 577-0005; ekress@famri.org
Internet: http://www.famri.org/current_awards/index.php
Sponsor: Flight Attendant Medical Research Institute
201 S Biscayne Boulevard, Suite 1310
Miami, FL 33131

FAMRI Young Clinical Scientist Awards 1976
The Young Clinical Scientist Award (YSCA) is intended to prepare and support new clinical investigators with an MD or PhD as they begin their careers as independent researchers in the field of smoking-related disorders. Awards are offered to two groups of scientists: those individuals who are currently faculty members, or who expect to be before the start date of the award (YCSA faculty), and those individuals who do not expect to become faculty until after the start date of the award (YCSA fellow). Applicants must initially submit an online letter of intent (LOI) by 5 p.m. on the LOI due date, and upon subsequent invitation submit a full proposal by 5 p.m. on the proposal due date. Instructions and links for submitting the letter of intent and proposal are available under "Current and Upcoming Awards" at the FAMRI website. YCSA awards can be extended to a total of no more than five years. Renewal forms will be sent to current grantees in early February of the appropriate year and will be due late March. Grantees who are fellows should notify both the Flight Attendant Medical Research Institute (FAMRI) and the American Institute of Biological Sciences (AIBS) of any change in status. Grantees are encouraged to call or email the FAMRI Administrative Assistant to obtain current AIBS contact information. Deadlines may vary from year to year. To be kept up to date of all current and future funding opportunities, interested applicants are encouraged to sign up for the FAMRI ListServ (click on "listserv signup" under the researchers tab at the FAMRI website).
Requirements: Applicants must be certain that their institution can agree to the terms of the FAMRI grant agreement. A copy of the agreement is available for review under "Current and Upcoming Awards" at the FAMRI website. Each individual other than the institutional official must submit a tobacco disclosure form. A tobacco disclosure form is also required from the applicant's institution. Tobacco disclosure forms may be downloaded from the FAMRI website and after they are signed, they may be scanned and uploaded. Researchers with an advanced degree (M.D., Ph.D., or M.D./Ph.D., or the equivalent) are eligible to apply. A YCSA grantee must devote a minimum of 75% of professional time to research during the period of this award and at least 50% of the grantee's time should be devoted to the research project described in the application. Individuals who apply to FAMRI for funding need not be U.S. citizens. They normally apply through and with an institution or organization that is qualified under United States tax laws to receive grants from FAMRI, a U.S.-based private foundation. The applying institution receives and administers FAMRI's funds and supervises the work of the Principal Investigator. If the applying institution is organized under the laws of a country other than the United States, it must provide FAMRI any and all documentation that FAMRI may require, in form and content acceptable to FAMRI, to ensure that any grant from FAMRI shall be a "qualified distribution" under Internal Revenue Service laws and regulations, including but not limited to affiliation with a qualified U.S.-based institution, a current affidavit from the applicant or an opinion of qualified legal counsel stating the facts and circumstances of the applicant's organization and activities and certifying that any grant from FAMRI shall be a "qualified distribution." Research projects that involve human subjects must follow Public Health Service guidelines (or its equivalent in non-U.S. institutions) and must have the approval of the appropriate Institutional Review Boards (IRB) or foreign equivalent prior to the start of the research. The IRB (or foreign equivalent) approval must be maintained and be in effect throughout the research project. All research must be performed in accordance with all relevant institutional and federal policies and guidelines (or foreign equivalent) relating to human subject protection, radiation and environmental health and safety. All FAMRI grantees are encouraged to collaborate with fellow grantees and may be required to participate periodically in FAMRI's collaborative efforts.
Restrictions: The program is limited to the development of young researchers in smoking-related disorders. The YCSA/fellow award must not exceed $75,000 per year plus 8.5 % indirect costs for a total of $81,375 per year. The YCSA/faculty award must not exceed $100,000 per year plus 8.5 % indirect costs for a total of $108,500 per year. FAMRI will only support animal research that utilizes mice or rats. Research involving other animals will not be considered unless the animals will not be sacrificed as a result of the research. Pending Association for Assessment of Laboratory Animal Care (AALAC) approval must be submitted prior to funding of a grant involving animal subjects. To avoid all conflicts, any principal investigator or other participating applicant who is currently accepting tobacco industry funding, has recently applied for tobacco funding, or has any other association or consulting position with a tobacco company or affiliate of the tobacco industry is not eligible for FAMRI funding. All prior funding and/or association with tobacco companies must be disclosed and will be considered on a case-by-case basis. An award will not be made or will be withdrawn if the applicant becomes the Principal Investigator on a peer-reviewed grant with the same general aims as those proposed in the application to FAMRI. FAMRI reserves the right to announce the award publicly in a manner and at a time of its choosing.
Geographic Focus: All States, All Countries
Date(s) Application is Due: Sep 26
Amount of Grant: 162,750-542,500 USD
Contact: Elizabeth Kress; (305) 379-7007; fax (305) 577-0005; ekress@famri.org
Internet: http://www.famri.org/current_awards/index.php
Sponsor: Flight Attendant Medical Research Institute
201 S Biscayne Boulevard, Suite 1310
Miami, FL 33131

Fannie and John Hertz Foundation Graduate Fellowships 1977
The foundation supports graduate students working towards the PhD degree in the applied physical sciences. Fellows must have proposed fields of graduate study in engineering, applied physics, computer science, math, chemistry, or other area in the applied physical sciences; have excellent previous scholastic performance, typically including at least an A- grade average during the last two years of undergraduate work; be a U.S. citizen or a permanent resident of the United States; and be ready to make themselves, their skills, and abilities available for the common defense in the event of national emergency.

Requirements: Applicant must be a U.S. citizen and have a bachelor's degree and propose to complete a program of graduate study leading to an advanced degree, but not in a professional degree program or a joint PhD/professional degree program.
Restrictions: Joint-professional degree programs are not funded.
Geographic Focus: All States
Date(s) Application is Due: Oct 28
Amount of Grant: 31,000-36,000 USD
Contact: Jay Davis, PhD; (510) 373-1642; fax (510) 373-6329; askhertz@aol.com
Internet: http://www.hertzfoundation.org/dx/fellowships/fellowshipaward.aspx
Sponsor: Fannie and John Hertz Foundation
2300 First Street, Suite 250
Livermore, CA 94550

Fannie E. Rippel Foundation Grants **1978**
The foundation aids, assists, funds, equips, and provides maintenance for corporations, institutions, associations, organizations, or societies maintained for the relief and care of aged women; provides funds for the building, equipping, and maintenance of hospitals; and provides funds for corporations, institutions, and other organizations existing for treatment of and/or research on heart disease or cancer. The foundation gives emphasis to the equipment and programmatic needs of major teaching medical centers and local rural hospitals, particularly where opportunities exist for leveraging the expertise or capabilities of the medical centers/rural hospitals. Programs should reach underserved rural and urban groups, advocate preventive care, present strategies to change behaviors of the people served, and promote humanistic medicine and mind-body-spirit connections in the healing process. Preference also is given to proposed projects where the benefits can be leveraged through challenge grants.
Requirements: Organizations, associations, institutions, and hospitals in the Northeast are eligible.
Restrictions: Grants are not awarded to individuals.
Geographic Focus: Connecticut, Maine, Massachusetts, New Hampshire, Rhode Island, Vermont
Amount of Grant: 50,000-300,000 USD
Contact: Leigh Scherrer, Foundation Associate; (973) 540-0101, ext. 305; fax (973) 540-0404; lscherrer@rippelfoundation.org
Internet: http://rippelfoundation.org/
Sponsor: Fannie E. Rippel Foundation
14 Maple Avenue, Suite 200
Morristown, NJ 07960

Fannie Mae Foundation Kennedy School of Government Fellowships **1979**
The foundation offers up to 35 fellowships annually as part of its partnership with the John F. Kennedy School of Government at Harvard University. The program enhances the management and decision-making skills of senior public and nonprofit officials committed to improving affordable housing opportunities in the United States. The program engages accomplished leaders experienced in managing housing and related community development programs. The curriculum focuses on organizational strategy, political management, policy development, management control and operations, and management of human resources. Fellowship funds cover the cost of the admission deposit, program tuition, and room and board for the session. Applicants must be able to attend the entire three-week session and are responsible for their own transportation and other incidental costs.
Requirements: Admissions decisions are made solely by the Kennedy School of Government. Applicants must satisfy the requirements for admission to be accepted by the Kennedy School to attend one of the state and local program sessions.
Geographic Focus: All States
Date(s) Application is Due: Apr 1
Contact: Fellowship Administrator; (202) 274-8000; fax (202) 274-8100
Internet: http://www.fanniemaefoundation.org/grants/kennedy_school.shtml
Sponsor: Fannie Mae Foundation
4000 Wisconsin Avenue NW, N Tower, Suite 1
Washington, DC 20016-2804

FAR Fund Grants **1980**
The FAR Fund Project is a New Orleans-based program exploring Hurricane Katrina's effects on New Orleans therapists and therapeutic practice. It was designed by and for clinicians. The project's mission is two fold: to offer support and concrete help to local clinicians, and to develop a model for better understanding how shared trauma affects therapists and therapy. Through this project, the FAR Fund wishes to unite and revitalize clinician communities following large-scale disasters wherever they occur, starting in New Orleans.
Requirements: The FAR Fund Project is open to New Orleans psychotherapists of all disciplines and theoretical orientations. Their mission is two-fold: to offer support and concrete help to local mental health clinicians, and to develop a psychodynamic model to better understand how shared trauma affects therapists and therapy.
Geographic Focus: Louisiana
Contact: Shirlee Taylor; (212) 982-8400; fax (212) 982-8477; FARFund@mac.com
Sponsor: Far Fund
928 Broadway, Suite 902
New York, NY 10010

Farmers Insurance Corporate Giving Grants **1981**
The corporate community relations program awards grants in the areas of education, public safety, arts and culture, civic improvement, and health and human services. Education giving focuses on literacy programs, mentoring programs, adopt-a-school programs, employee matching grants, and aid-to-education undergraduate scholarships. Public safety awards support tougher laws against drunk driving, drug- and alcohol-free graduation night parties, neighborhood crime prevention, highway safety, and earthquake relief. Arts and culture funding supports children's programs and public television. Civic improvement focuses on recognizing exemplary youth, voter registration drives, adopt-a-highway programs, and community paint-a-thons. Health and human services giving supports March of Dimes, United Way, aid for families with cancer, aid to migrant farmworkers, and feeding the hungry. There are no application deadlines. Requests for contributions should be in the form of a letter outlining the purpose of the organization or program. The letter also should include the amount requested, its intended use, and a description of how Farmers' support will be recognized. Additional information should include a budget, annual report, proof of tax-exempt status, and a roster of the board of directors.
Requirements: 501(c)3 tax-exempt organizations are eligible.
Restrictions: Farmers does not make charitable contributions to individuals, political candidates, or religious groups or for sports events, advertising or raffle tickets, construction projects, or international programs.
Geographic Focus: All States
Contact: Doris Dunn, Director of Community Relations; (888) 327-6335
Internet: http://www.farmers.com/corporate_giving.html
Sponsor: Farmers Insurance Group of Companies
4680 Wilshire Boulevard
Los Angeles, CA 90010

FASSE Demonstration Project Grant **1982**
This research award is to engage and support educators in strengthening and advocating social studies, where social studies is defined as the integrated study of the social sciences and humanities to promote civic competence. This annual award includes a commemorative gift, annual conference session for research presentation and publicity.
Requirements: Nomination *Requirements:* no self nominations; cover page; letter of nomination supporting, rationalizing the nomination; vitae or professional resume prepared by nominee; nominated study.
Geographic Focus: All States
Date(s) Application is Due: May 1
Contact: Prema Parmar, Programs Assistant; (301) 588-1800, ext. 106; fax (301) 588-2049; pparmar@ncss.org or excellence@ncss.orgexcellence@ncss.orgexcellence@n
Internet: http://www.socialstudies.org/awards/research/exemplary/
Sponsor: Exemplary Research in Social Studies Award
8555 Sixteenth Street, Suite 500
Silver Spring, MD 20910

Father James B. Macelwane Annual Awards **1983**
To stimulate interest in meteorology among college students through the encouragement of original student papers concerned with some phase of the atmospheric sciences, these awards, supported by Weather Corporation of America and administered by the American Meteorological Society, are given for three papers submitted that are judged to be the best in the previous 12-month period; also considered may be papers from the previous 12-month period if they did not receive any award during the previous competition. To be considered, the paper must be accompanied by a letter of application from the author and a letter from the department head or other faculty member of the major department confirming that the author was an undergraduate student at the time the paper was written and indicating the elements of the paper that represent original contributions by the student.
Requirements: All registered undergraduates in a college or university in the Americas are eligible to participate.
Restrictions: Participating students must be enrolled as undergraduates at the time the paper is written, and no more than two students from any one institution may enter papers in any one contest.
Geographic Focus: All States
Amount of Grant: 100-300 USD
Contact: Stephanie Armstrong; (617) 227-2426, ext. 235; dfernand@ametsoc.org
Internet: http://www.ametsoc.org/amsstudentinfo/scholfeldocs
Sponsor: American Meteorological Society
45 Beacon Street
Boston, MA 0002108-3693

FCD Child Development Grants **1984**
The foundation awards grants to support programs for children, particularly the disadvantaged, and promote their well-being through basic and policy-relevant research about the factors that promote optimal development of children and adolescents; policy analysis, advocacy, services, and public education to enhance the discussion and adoption of social policies that support families in their important child-raising responsibilities; and leadership development activities linked to the programmatic focus of the foundation. Grants focus on the integration of research, policy, and advocacy in two areas: the availability of and access to early childhood education programs and health care for children. Most grants support research, but a small number of direct service grants are made for New York City-based projects that advance the foundation's research and policy analysis efforts. There are no application deadlines; submit a brief letter of inquiry. Full proposals are by invitation, following a strict preproposal process.
Requirements: Nonprofit organizations are eligible.
Restrictions: The foundation does not consider requests for scholarships or support for individuals, capital campaigns, building purchase or renovation, or equipment purchase. The foundation does not make grants outside the United States.
Geographic Focus: All States
Amount of Grant: Up to 200,000 USD

Contact: Mark Bogosian, Communications and Grants Officer; (212) 867-5777; fax (212) 867-5844; mark@fcd-us.org or info@fcd-us.org
Internet: http://fcd-us.org/grants
Sponsor: Foundation for Child Development
295 Madison Avenue, 40th Floor
New York, NY 10017

FCD Young Scholars Fellowships 1985

The Foundation for Child Development has endured for over 100 years as a result of the constancy of its commitment to promoting the well-being of children and the flexibility with which it evolves with the times, embracing new strategies as children's needs, our society, and research knowledge have changed. The Foundation's Young Scholars Program (YSP) is no exception to this process of renewal. We remain steadfastly committed to this flagship program aimed at developing the next generation of developmental scientists who are equipped to advance policy and practice on behalf of the nation's children. But it is also an opportune moment to refresh how we address this goal in light of decision-makers' knowledge needs and scientific advancements. FCD is committed to the following issues in developing the next generation of the YSP: empirical and professional mentoring; recruiting scholars from underrepresented groups, notably scholars of color; creating a community of interdisciplinary scholars and supporting collaboration among them; ensuring that pressing policy and practice challenges inform the work of the scholars; fostering the scholars' capacity to translate their work for policy and practice audiences; and expanding opportunities for the young scholars to engage with other FCD grantees by, in part, creating greater thematic coherence across all of FCD's strategic initiatives.
Requirements: Fellowships will be available in the Spring of 2014 to scholars who have earned their PhD within the last 15 years. Applicants must hold a PhD or its equivalent in one of the behavior or social sciences or in an allied professional field (i.e., public policy, public health, education, social work, nursing). Applicants must hold a position as a full-time, tenure-track faculty member of a U.S. college or university.
Restrictions: The foundation does not consider requests for: capital campaigns, the purchase, construction or renovation of buildings, grants for projects outside the United States, the direct provision of preschool education or child care, or health care; or under the foundation's health focus, research, policy, or direct-service projects concerned with specific illnesses.
Geographic Focus: All States
Contact: Mark Bogosian, Communications and Grants Officer; (212) 213-8337, ext. 203; fax (212) 213-5897; inforequest@ffcd.org
Internet: http://fcd-us.org/our-work/new-american-children/young-scholars-program-ysp
Sponsor: Foundation for Child Development
295 Madison Avenue, 40th Floor
New York, NY 10017

FDHN Bridging Grants 1986

Awards of $50,000 each are available to assist investigators were not awarded their first R01 or the first renewal of their R01 to continue their gastroenterology related research. Two awards will be made following each of the three annual NIH review cycles and availability of Summary Statements. The primary objective of the award is to provide interim support to AGA Members who have submitted grants to NIH that were approved on the basis of scientific merit, but received priority scores out of the funding range.
Requirements: Applicants must: be AGA members who are applying for their first R01 or their second R01 either as a competitive renewal or a new R01 (this includes investigators going from a K award to their first R01); hold an MD, PhD, or equivalent degree (e.g., MB, ChB, MBBS, DO); hold full-time faculty positions at North American universities or professional institutes at the time of application; and have submitted a proposal to NIH that has undergone the peer review process and was either approved and not funded or received high commendation but not funded.
Restrictions: Applicants cannot receive more than $75,000 of any additional extramural research support. Up to 25% of total amount the award may be used for salary support of the Principal Investigator. Indirect costs, including travel are not allowed.
Geographic Focus: All States
Date(s) Application is Due: Feb 14; Jun 14; Oct 14
Contact: Awards Manager; (301) 222-4012; fax (301) 652-3890; awards@fdhn.org
Internet: http://www.fdhn.org/wmspage.cfm?parm1=119
Sponsor: Foundation for Digestive Health and Nutrition
4930 Del Ray Avenue
Bethesda, MD 20814

FDHN Centocor International Research Fellowship in Gastrointestinal Inflammation & Immunology 1987

This award provides $50,000 to enable promising young investigators from outside the United States to spend a year at a U.S. institution engaged full-time in research related to the fundamental processes in gastrointestinal inflammation and immunology. The primary objectives of the award are: to provide the opportunity for young investigators from outside the U.S. to participate in basic research on inflammatory disease processes at prominent institutions in the U.S. (the award provides salary support for fulltime research); to initiate future international collaborative research efforts in inflammatory digestive diseases; and to address an important question in such a way as to provide a meaningful answer.
Requirements: Candidates must hold an MD or equivalent degree (eg, MB. ChB, MBBS). Applicants should currently be on the faculty of an academic institution outside the U.S. Applicants must be AGA members or be eligible for and submit an application for membership.
Restrictions: At the time of application, candidates are not allowed to hold any other similar research grant. These awards may not be renewed. No more than one application will be accepted from any institution in a given year.
Geographic Focus: All States
Date(s) Application is Due: Jan 14
Amount of Grant: 50,000 USD
Contact: Awards Manager; (301) 222-4012; fax (301) 652-3890; awards@fdhn.org
Internet: http://www.fdhn.org/wmspage.cfm?parm1=105
Sponsor: Foundation for Digestive Health and Nutrition
4930 Del Ray Avenue
Bethesda, MD 20814

FDHN Designated Outcomes Award in Geriatric Gastroenterology 1988

Two awards in the amount of $35,000 for one year are available to support investigator-initiated outcomes research in geriatric gastroenterology. In general, outcomes studies examine clinical outcomes, patient satisfaction, quality of life, economic evaluation, quality of care, functional status, appropriateness of care, conformance of recommended/desirable standards of performance, or change in practice patterns. The objective of this award is to promote research by young investigators in the area of outcomes, broadly defined above, as it relates to geriatric gastroenterology. Please review the AGA Future Trends Committee Report: Effects of Aging of the Population on Gastroenterology Practice, Education and Research to learn more about research topics encouraged for study. Funds may be used for salary support of personnel and technicians only, supplies and/or equipment/services. Women and minorities are strongly encouraged to apply.
Requirements: Investigators must possess an MD, PhD or equivalent and must hold faculty positions at accredited North American academic institutions by the time of the start date of the award. MD applicants: No more than five years should elapse following the completion of your clinical training (GI fellowship or equivalent) and the start date of this award. Applicants must be AGA Members.
Restrictions: The award is intended for junior faculty; therefore, established investigators are not eligible. Candidates may not hold awards on a similar topic from other agencies. Indirect costs are not allowed.
Geographic Focus: All States
Date(s) Application is Due: Sep 5
Amount of Grant: 35,000 USD
Contact: Awards Manager; (301) 222-4012; fax (301) 652-3890; awards@fdhn.org
Internet: http://www.fdhn.org/wmspage.cfm?parm1=234
Sponsor: Foundation for Digestive Health and Nutrition
4930 Del Ray Avenue
Bethesda, MD 20814

FDHN Designated Research Award in Geriatric Gastroenterology 1989

This award provides $75,000 per year for three years (total $225,000) for young investigators working toward independent careers in academic research related to geriatric gastroenterology. The overall objective is to enable young investigators to develop independent and productive research careers, with a focus on research related to geriatric gastroenterology, by ensuring that a major proportion of their time is protected for research. Non-recipient applicants for this award will be considered for the Research Scholar Awards. Applicants should review the AGA Future Trends Committee Report: Effects of Aging of the Population on Gastroenterology Practice, Education and Research to learn more about research topics encouraged for study. The award is not intended for fellows, but for young faculty who have demonstrated unusual promise and have some record of accomplishment in research.
Requirements: Candidates must hold an MD, PhD, or equivalent degree (e.g., MB, ChB, MBBS, DO). Applicants must hold full-time faculty positions at North American universities or professional institutes at the time award begins. Applicants must be Members of the AGA. # MD applicants: no more than five years shall have elapsed following the completion of your clinical training (GI fellowship or its equivalent) and the start date of this award. PhD applicants: no more than five years shall have elapsed from the completion of your postdoctoral training and the start date of this award. Candidates must devote at least 70 percent of their efforts to research related to geriatric gastroenterology.
Restrictions: Candidates should be in the beginning years of their careers; therefore, established investigators are not appropriate candidates.
Geographic Focus: All States
Date(s) Application is Due: Sep 15
Amount of Grant: 75,000 USD
Contact: Awards Manager; (301) 222-4012; fax (301) 652-3890; awards@fdhn.org
Internet: http://www.fdhn.org/wmspage.cfm?parm1=224
Sponsor: Foundation for Digestive Health and Nutrition
4930 Del Ray Avenue
Bethesda, MD 20814

FDHN Designated Research Award in Research Related to Pancreatitis 1990

The overall objective of this Award is to enable young investigators to develop independent and productive research careers, with a focus on pancreatic disease, by ensuring that a major proportion of their time is protected for research. The Award provides $75,000 per year for three years (total $225,000) for young investigators working toward independent careers in academic research related to understanding, improving treatments or curing Pancreatitis.
Requirements: Candidates must hold an MD, PhD, or equivalent degree (e.g., MB, ChB, MBBS, DO). Applicants must hold full-time faculty positions at North American universities or professional institutes at the time award begins. Applicants must also be members of the AGA. The award is not intended for fellows, but for young faculty who have demonstrated unusual promise and have some record of accomplishment in research. Candidates should be in the beginning years of their careers; therefore, established investigators are not appropriate candidates. MD applicants: No more than five years shall have elapsed following the completion of your clinical training (GI fellowship or its equivalent) and the start date of

this award. PhD applicants: No more than five years shall have elapsed from the completion of your postdoctoral training and the start date of this award.
Geographic Focus: All States
Date(s) Application is Due: Apr 17
Amount of Grant: 75,000 USD
Contact: Awards Manager; (301) 222-4012; fax (301) 652-3890; awards@fdhn.org
Internet: http://www.fdhn.org/wmspage.cfm?parm1=100
Sponsor: Foundation for Digestive Health and Nutrition
4930 Del Ray Avenue
Bethesda, MD 20814

FDHN Fellow Abstract Prizes **1991**
Two awards of $1000 each will be given to fellows who have submitted abstracts chosen to be presented during Digestive Disease Week. Awards will be presented at a ceremony during DDW. The primary objective of this award is to stimulate interest in GI research careers through competition and recognition
Requirements: Qualified candidates are MD or PhD postdoctoral fellows who are trainee members the AGA. Women and minority investigators are strongly encouraged to apply. Applicants must: be sponsored by an AGA member; and be the first author of an abstract accepted for presentation at DDW and provide evidence of abstract acceptance. Individuals with faculty appointments are not eligible. Applicants may only submit one abstract for consideration.
Restrictions: Fellows who have been awarded must present the abstract. No substitute presenters are allowed. Abstracts must have been selected for presentation at DDW. A letter of recommendation from the sponsor is required.
Geographic Focus: All States
Date(s) Application is Due: Mar 21
Amount of Grant: 1,000 USD
Contact: Awards Manager; (301) 222-4012; fax (301) 652-3890; awards@fdhn.org
Internet: http://www.fdhn.org/wmspage.cfm?parm1=124
Sponsor: Foundation for Digestive Health and Nutrition
4930 Del Ray Avenue
Bethesda, MD 20814

FDHN Fellowship to Faculty Transition Awards **1992**
This award provides $40,000 per year for two years for current trainees in gastroenterology related fields so they may gain additional research training in gastrointestinal, liver function or related diseases. The objective is to prepare physicians for independent research careers in digestive diseases. The award provides salary support for additional full-time research training in basic science to acquire modern laboratory skills. The additional two years of research training provided by this award should broaden and expand the scope of investigative tools available to the recipient, generally in basic disciplines such as cell or molecular biology or immunology.
Requirements: Applicants must be MDs or MD/PhDs currently in a gastroenterology-related fellowship at an accredited North American institution, committed to academic careers. They will have completed two years of research training at the start of this award. Applicants must be AGA Trainee Members and be sponsored by an AGA Member.
Restrictions: Individuals who hold a PhD are ineligible. Although the institution may supplement the award, the applicant may not concurrently hold a similar training award or grant from another organization, such as the NIH, ALF, CCFA, or the Glaxo Institute of Digestive Health.
Geographic Focus: All States
Date(s) Application is Due: Sep 5
Amount of Grant: 40,000 USD
Contact: Awards Manager; (301) 222-4012; fax (301) 652-3890; awards@fdhn.org
Internet: http://www.fdhn.org/wmspage.cfm?parm1=102
Sponsor: Foundation for Digestive Health and Nutrition
4930 Del Ray Avenue
Bethesda, MD 20814

FDHN Funderburg Research Scholar Award in Gastric Biology Related to Cancer **1993**
This grant of $25,000 per year for two years (total $50,000) is awarded to an established investigator working on novel approaches in gastric cancer, including the fields of gastric mucosal regeneration and regulation of cell growth (not as they relate to peptic ulcer disease inflammation (including Helicobacter pylori) as precancerous lesions; genetics of gastric oncogenes in gastric epithelial malignancies; epidemiology of gastric cancer; etiology of malignancies; or clinical research in the diagnosis or treatment of gastric carcinoma. The primary objective of the award is to support an active, established investigator in the field of gastric biology who enhances the fundamental understanding of gastric cancer pathobiology in order to ultimately develop a cure for the disease.
Requirements: Applicants must hold faculty positions at accredited North American institutions and must have established themselves as independent investigators in the field of gastric biology. Women and minority investigators are strongly encouraged to apply. Applicants must be Members of the AGA.
Geographic Focus: All States
Date(s) Application is Due: Sep 5
Amount of Grant: 25,000 USD
Contact: Awards Manager; (301) 222-4012; fax (301) 652-3890; awards@fdhn.org
Internet: http://www.fdhn.org/wmspage.cfm?parm1=90
Sponsor: Foundation for Digestive Health and Nutrition
4930 Del Ray Avenue
Bethesda, MD 20814

FDHN Graduate Student Awards **1994**
Two awards of $20,000 a year for two years are offered to fund graduate students undertaking research in the biology and epidemiology of diseases of the gastrointestinal tract, liver or pancreas. The award includes $18,000 for stipend and $2,000 to be used towards fringe benefits such as medical insurance and travel to a national meeting. The primary objective of the award is to provide a salary stipend for graduate students performing doctoral research related to the gastrointestinal tract, liver or pancreas.
Requirements: Applicants should have completed at least one year and no more than three years of training towards the doctoral degree and have selected and confirmed the laboratory or department in which they will conduct doctoral research. Research is to be conducted at an accredited academic institution within North America and the research advisor must be a Member of the AGA. Applicants should be U.S. citizens, permanent residents or overseas students who have a current visa to pursue education within North America.
Restrictions: Indirect costs, excluding travel, are not allowed. Tuition costs are expected to be covered by the institution or host department and laboratory.
Geographic Focus: All States
Date(s) Application is Due: Mar 14
Amount of Grant: 20,000 USD
Contact: Awards Manager; (301) 222-4012; fax (301) 652-3890; awards@fdhn.org
Internet: http://www.fdhn.org/wmspage.cfm?parm1=154
Sponsor: Foundation for Digestive Health and Nutrition
4930 Del Ray Avenue
Bethesda, MD 20814

FDHN Jon I. Isenberg International Research Scholar Award **1995**
This award provides a total of $50,000 for one year of study, $25,000 from the AGA and $25,000 in matching funds from the applicant's national GI society. The award provides for non-U.S. citizen young investigators to spend one year performing GI-related research at an American institution under the tutelage of an AGA member. Four awards will be given annually. The primary objective of this award is to promote international scholarship, increase AGA's involvement within the international GI community and foster international collaboration in training GI investigators.
Requirements: An eligible candidate must be nominated by his or her national GI Society. The national GI society must match the $25,000 contribution from the AGA. Before submitting an application, a candidate must identify a sponsoring American institution and a research preceptor who agrees in writing to supervise his or her training and research. Qualified candidates must possess a doctoral degree, either an MD, or equivalent degree, and/or a PhD degree. In keeping with the intent to support the career development of young investigators, the candidate must be within five (5) years of completing GI training or, if a PhD, within five (5) years of the receipt of the degree, at the time of initiation of the award. A documented parental leave of absence will not be counted towards the five (5) years of eligibility.
Geographic Focus: All States
Date(s) Application is Due: Mar 1
Amount of Grant: 50,000 USD
Contact: Awards Manager; (301) 222-4012; fax (301) 652-3890; awards@fdhn.org
Internet: http://www.fdhn.org/wmspage.cfm?parm1=101
Sponsor: Foundation for Digestive Health and Nutrition
4930 Del Ray Avenue
Bethesda, MD 20814

FDHN June & Donald O. Castell MD, Esophageal Clinical Research Award **1996**
One award of $35,000 is made annually to provide research and/or salary support for junior faculty involved in clinical research in esophageal diseases. This award is funded by the June and Donald O. Castell, MD, Gastroenterology Research and Education Trust. The primary objective of the award is to support investigators who have demonstrated high potential to develop independent, productive research careers.
Requirements: Candidates must hold a MD or PhD or equivalent. Applicants must hold a full-time faculty position at a North American universities or professional institute. Applicants must be members of the AGA. The recipient must be at or below the level of assistant professor, and his/her initial appointment to the faculty position must have been within seven (7) years of the time of application. This award is not intended for fellows, but for junior faculty who have demonstrated unusual promise; have some record of accomplishment in research; and have established independent research programs at the time of the award. Candidates must devote at least 50 percent of their efforts to research related to esophageal function or diseases.
Restrictions: If an award recipient receives notification of another award with overlapping scientific objectives, prior to the start date of an AGA or FDHN award, the applicant must choose between the two awards.
Geographic Focus: All States
Date(s) Application is Due: Jan 14
Contact: Awards Manager; (301) 222-4012; fax (301) 652-3890; awards@fdhn.org
Internet: http://www.fdhn.org/wmspage.cfm?parm1=104
Sponsor: Foundation for Digestive Health and Nutrition
4930 Del Ray Avenue
Bethesda, MD 20814

FDHN Moti L. & Kamla Rustgi International Travel Awards **1997**
This program awards grants to young basic, translational and clinical investigators to support their travel and related expenses to attend Digestive Disease Week. Two awards of $500 each will be given to selected individuals residing outside North America. The primary objective of the award is to enable young investigators outside of North American (U.S. or

Canada) institutions to attend Digestive Disease Week and encourage international trainees to become more involved in digestive tract, pancreatic and liver disease research.
Requirements: Candidates must: be MD or PhD or MD PhD postdoctoral fellows who are international trainee members of the AGA; be sponsored by an International member of AGA; be 35 years of age or younger at the time of the meeting; be fluent in English; and have a sufficient number of scientific papers (impact factor) published and/or poster presentations.
Geographic Focus: All States
Date(s) Application is Due: Mar 21
Amount of Grant: 500 USD
Contact: Awards Manager; (301) 222-4012; fax (301) 652-3890; awards@fdhn.org
Internet: http://www.fdhn.org/wmspage.cfm?parm1=126
Sponsor: Foundation for Digestive Health and Nutrition
4930 Del Ray Avenue
Bethesda, MD 20814

FDHN Non-Career Research Awards **1998**
This award provides approximately $10,000 (per symposia) for travel support for young investigators and selected established investigators to participate in symposia on gastrointestinalrelated topics. The primary objective of the award is to foster interactions and enhance the exchange of information between clinical and basic science investigators, and established and junior investigators working in gastrointestinal research. Eligibility women and minority organizers are strongly encouraged to apply. Travel support may be provided for: junior investigators who are within 5 years of completing their clinical training in GI or from completion of their PhD doctoral thesis and are at or below the rank of assistant professor; and up to two established investigators who are invited speakers. Biosketches of these individuals and a letter addressing their contribution to the scientific content and atmosphere of the meeting should be submitted with the application.
Requirements: Within 60 days of the conclusion of the meeting, a one-page summary of the highlights of the meeting and an accounting of the use of funds must be submitted. The names of participants and speakers in the symposium, addresses, dates of birth and academic ranks of the attending scientists and their itemized expenses, must be sent to the Foundation. Any unexpended funds must be returned.
Restrictions: Indirect costs are not allowed.
Geographic Focus: All States
Date(s) Application is Due: Feb 1; May 1; Oct 1
Amount of Grant: 10,000 USD
Contact: Awards Manager; (301) 222-4012; fax (301) 652-3890; awards@fdhn.org
Internet: http://www.fdhn.org/wmspage.cfm?parm1=123
Sponsor: Foundation for Digestive Health and Nutrition
4930 Del Ray Avenue
Bethesda, MD 20814

FDHN Non-Career Research Grants **1999**
A research initiative grant of $25,000 for one year is offered to investigators to support pilot research projects in gastroenterology- or hepatology-related areas. The primary objective of the award is to provide non-salary funds for new investigators to help them establish their research careers or to support pilot projects that represent new research directions for established investigators. The intent is to stimulate research in gastroenterology- or hepatology-related areas by permitting investigators to obtain new data that can ultimately provide the basis for subsequent grant applications of more substantial funding and duration. Women and minorities are strongly encouraged to apply.
Requirements: Investigators must possess an MD or PhD degree or equivalent and must hold faculty positions at accredited North American institutions. Candidates may not hold awards on a similar topic from other agencies. Applicants must be AGA Members. Applicants for this award may not simultaneously apply for the AGA/Miles and Shirley Fiterman Foundation Basic Research Award or the AGA June and Donald O. Castell, MD, Esophageal Clinical Research Award.
Restrictions: If an award recipient receives notification of another award with overlapping scientfic objectives, prior to the start date of an AGA or FDHN award, the applicant must choose between the two awards.
Geographic Focus: All States
Date(s) Application is Due: Jan 14
Contact: Awards Manager; (301) 222-4012; fax (301) 652-3890; awards@fdhn.org
Internet: http://www.fdhn.org/wmspage.cfm?parm1=121
Sponsor: Foundation for Digestive Health and Nutrition
4930 Del Ray Avenue
Bethesda, MD 20814

FDHN Research Scholar Awards **2000**
These awards provide salary support for young investigators working in any area of gastrointestinal, liver function, or related diseases. The primary intent of the program is to support physician-investigators who have a high potential to develop independent, productive research careers in gastroenterology and hepatology. Candidates must devote at least 70 percent of their effort to research related to the gastrointestinal tract or liver. There must be a strong commitment from the candidate's division and department to support the candidate by protecting time for research and providing adequate laboratory space and facilities.
Requirements: Applicants must hold full-time faculty positions at North American universities or professional institutes. Nonphysician candidates with a PhD will also be considered. Candidates should be early in their research careers and commonly will have recently completed their fellowship training.
Restrictions: Indirect costs are not allowed. Candidates who have been at the assistant professor level or equivalent for more than five years are not eligible. Nor can applicants hold, or have held, an RO1, R29, K11, K08, or VA research award or any award with similar objectives from nonfederal sources (such as ALF, CCFA, or Glaxo Institute of Digestive Health). However, awards or grants obtained after receipt of this award need not be surrendered.
Geographic Focus: All States
Date(s) Application is Due: Sep 5
Amount of Grant: 75,000 USD
Contact: Awards Manager; (301) 222-4012; fax (301) 652-3890; awards@fdhn.org
Internet: http://www.fdhn.org/wmspage.cfm?parm1=103
Sponsor: Foundation for Digestive Health and Nutrition
4930 Del Ray Avenue
Bethesda, MD 20814

FDHN Student Research Fellowships **2001**
This program offers financial support for students to spend a minimum of 10 weeks performing research in digestive diseases or nutrition and is intended to stimulate interest in research careers in these areas. The work may take place at any time during the year. Up to 20 fellowships will be available for full-time research with a preceptor, who must be a faculty member who directs a research project in a gastroenterology-related area at an accredited North American institution. A complete financial statement and scientific progress report are required upon completion of the program. The AGA has recognized the need to attract and encourage minority individuals to enter and pursue gastroenterology research careers. In response to this concern, seven of the student research fellowship awards will be reserved for underrepresented minority students. For the purpose of this award, minorities have been defined as African American, Hispanic, Native American/Alaskan Native, and Pacific Islander.
Requirements: Candidates may be high school, undergraduate, medical, or graduate students (not yet engaged in thesis research) in accredited North American institutions. Women and minority students are strongly encouraged to apply. The preceptor must be a faculty member who directs a research project in a gastroenterology-related area at an accredited North American institution.
Restrictions: Candidates may not hold similar salary support from other agencies: e.g., American Liver Foundation, Crohn's and Colitis Foundation.
Geographic Focus: All States
Date(s) Application is Due: Mar 5
Amount of Grant: 2,000-3,000 USD
Contact: Awards Manager; (301) 222-4012; fax (301) 652-3890; awards@fdhn.org
Internet: http://www.fdhn.org/wmspage.cfm?parm1=115
Sponsor: Foundation for Digestive Health and Nutrition
4930 Del Ray Avenue
Bethesda, MD 20814

FDHN TAP Endowed Designated Research Award in Acid-Related Diseases **2002**
This award provides $75,000 per year for three years (total $225,000) for young investigators working toward independent careers in acid-related diseases. The overall objective is to enable young investigators to develop independent and productive careers in acid-related research by ensuring that a major proportion of their time is protected for research.
Requirements: Candidates must hold an MD, PhD, or equivalent degree (e.g., MB, ChB, MBBS, DO). Applicants must hold full-time faculty positions at North American universities or professional institutes at the time award begins. Applicants must be Members of the AGA. The award is not intended for fellows, but for young faculty who have demonstrated unusual promise and have some record of accomplishment in research. Candidates should be in the beginning years of their careers; therefore, established investigators are not appropriate candidates. Candidates must devote at least 70 percent of their efforts to research related to geriatric gastroenterolgy.
Restrictions: MD applicants: no more than five years shall have elapsed following the completion of your clinical training (GI fellowship or its equivalent) and the start date of this award. PhD applicants: no more than five years shall have elapsed from the completion of your postdoctoral training and the start date of this award.
Geographic Focus: All States
Date(s) Application is Due: Sep 5
Contact: Awards Manager; (301) 222-4012; fax (301) 652-3890; awards@fdhn.org
Internet: http://www.fdhn.org/wmspage.cfm?parm1=132
Sponsor: Foundation for Digestive Health and Nutrition
4930 Del Ray Avenue
Bethesda, MD 20814

FDHN Translational Research Awards **2003**
One award of $100,000 per year for two years will be made annually to support translational research in gastroenterology and/or hepatology. Translational research will be defined as the process of applying ideas, insights and discovery generated through basic science research to the diagnosis, treatment or prevention of human disease. The primary objective of the award is to enhance interaction between researchers with basic science and clinical backgrounds with the goal of accelerating the pace of discovery that is directly applicable to patient care. The creation of teams including both a Ph.D. and an M.D. researcher is particularly encouraged.
Requirements: This award must be applied for jointly by a team of researchers (typically consisting of two members), including at least one researcher with significant training and experience in a basic science discipline (including areas outside the traditional biomedical sciences, such as physical sciences and engineering) and at least one researcher who is qualified to provide direct clinical care to patients. Junior investigators in either or both categories are particularly encouraged to apply. Applicants must be members of the AGA. Research must be conducted at an accredited North American institution.
Restrictions: Candidates may not hold support for the same project from another agency. If a proposed award recipient receives notification of another award with overlapping

scientific objectives prior to the start date of an AGA or FDHN award, the applicant must choose only one of the awards to accept.
Geographic Focus: All States
Date(s) Application is Due: Mar 14
Amount of Grant: 100,000 USD
Contact: Awards Manager; (301) 222-4012; fax (301) 652-3890; awards@fdhn.org
Internet: http://www.fdhn.org/wmspage.cfm?parm1=148
Sponsor: Foundation for Digestive Health and Nutrition
4930 Del Ray Avenue
Bethesda, MD 20814

Fermilab Postdoctoral Fellowship **2004**
Fermilab, a Department of Energy laboratory, offers postdoctoral fellowships for work in experimental physics. The objective of the programs is to broaden or intensify the research experience of recent PhDs through involvement in experimental or theoretical research. Recent PhDs receive research associate appointments generally for two or more years.
Geographic Focus: All States
Contact: Dr. Michael Albrow, Head-Experimental Physics Projects Department, Particle Physics Division; (630) 840-3922; albrow@fnal.gov
Internet: http://www.fnal.gov
Sponsor: Fermi National Accelerator Laboratory
P.O. Box 500, MS 116
Batavia, IL 60510-0500

Fetzer Institute Research Grants **2005**
The Foundation uses the bulk of its income to actively run its own programs or services. The Institute's mission is to foster awareness of the power of love and forgiveness through research, education, and service programs. Current focus includes scientific research on altruistic love, compassionate love, and forgiveness. Individuals and organizations invited by the Institute to participate in the creation and implementation of its programs can receive funding. Requests are posted on the Foundation's website. The Foundation does not accept unsolicited proposals.
Geographic Focus: All States
Contact: Heidi Ihrig, (269) 375-2000; educationrfp@fetzer.org
Internet: http://www.fetzer.org/Programs.aspx?PageID=Programs&NavID=1
Sponsor: Fetzer Institute
9292 W KL Avenue
Kalamazoo, MI 49009-9398

FIC Research Collaboration Award **2006**
Project grants facilitate collaborative research efforts between U.S. and foreign scientists that will expand and enhance the NIH-supported research program of the U.S. principal investigator, while at the same time benefiting the scientific interests of the collaborating foreign scientist. These small grants will provide funds for up to three years to purchase supplies, materials, and small equipment items necessary to conduct the collaborative research in the foreign scientist's laboratory. Travel support for both the U.S. principal investigator and the foreign collaborator also is provided. Applications will be reviewed on the basis of scientific and technical merit of the research proposal. Collaborative potential, significance of the proposed research, and the competence of the foreign collaborator are also prime factors. All biomedical and behavioral research topics supported by the NIH are eligible for inclusion under this program. Contact the office for deadline dates. PAR: PA-02-057
Requirements: The grant applicant is the U.S. institution where the U.S. principal investigator is employed. The foreign collaborator must hold a position at a public or nonprofit private institution.
Geographic Focus: All States
Date(s) Application is Due: Jan 21; May 21; Sep 21
Contact: Dr. Kathleen Michels; (301) 496-1653; fax (301) 402-0779; FIRCA@nih.gov
Internet: http://www.fic.nih.gov/programs/firca.html
Sponsor: Fogarty International Center
31 Center Drive, Building 31
Bethesda, MD 20892-2220

Fidelity Foundation Grants **2007**
Fidelity Investments Chairman Edward C. Johnson 3d and his father, the founder of the company, established the Fidelity Foundation in 1965 with several operating principles in mind. These principles, still current today, guide the Foundation's decisions and grantmaking. The Foundation grant program was designed to strengthen the long-term effectiveness of nonprofit institutions. The types of projects it funds, and the way in which it funds them, are specifically intended to help nonprofits build the organizational capabilities they need to better fulfill their missions and serve their constituencies. The Fidelity Foundation considers Letters of Inquiry from organizations with current IRS 501(c)3 public charity status only. Grants are made to fund only significant, transformative projects usually budgeted at $50,000 or more. The Foundation's primary philanthropic investments are allocated to the following sectors: arts and culture; community development and social services; health; and education.
Requirements: Grants are generally made only to organizations with operating budgets of $500,000 or more. The Fidelity Foundation considers projects from organizations of regional or national importance throughout the United States.
Restrictions: Grants are not awarded to support sectarian organizations, disease-specific associations, or public school systems. Support does not go to individuals or for scholarships, civic or start-up organizations, corporate memberships, operating support, or participation in benefit events, film, or video projects.
Geographic Focus: All States
Amount of Grant: 50,000-1,000,000 USD
Contact: Kathleen Ward; (617) 563-6806; info@FidelityFoundation.org
Internet: http://www.fidelityfoundation.org/index.html
Sponsor: Fidelity Foundation
82 Devonshire Street, S2
Boston, MA 02109

Field Museum Graduate Student Fellowships **2008**
A limited number of Field Museum fellowships are available for graduate students engaged in dissertation research associated with the museum. These fellowships provide stipend support.
Requirements: Candidates will be expected to have formal involvement with the museum; a curator should serve on the student's academic committee, and the research should rely heavily upon the collections and facilities of the museum. Students must be in residence in the Chicago area and are expected to spend a significant portion of their research time at the museum. The period of appointment is one year and starts on September 1st. Applicants must contact the sponsoring curator well in advance of the application deadline.
Restrictions: Only electronic applications using the museum's web site will be accepted.
Geographic Focus: Illinois
Date(s) Application is Due: Feb 1
Contact: Scholarship Coordinator; (312) 922-9410
Internet: http://www.fieldmuseum.org/research_collections/scholarships/default.htm#grad_fellow
Sponsor: Field Museum
1400 S Lake Shore Drive
Chicago, IL 60605-2496

Field Museum High School Internships **2009**
A few salaried internships per year are available for high school students or recent graduates to work directly with collections and research staff at the Museum. Internships are 10 weeks long and are usually taken in the summer. Other internships, including those in administrative areas, are also available.
Requirements: Applicants for these internships must contact the museum curator who seems most appropriate, since the internship project and work schedule will be arranged with her/him (see website for list of curators and contact information). The museum curators travel frequently and should be contacted well in advance of the application deadline.
Geographic Focus: All States
Date(s) Application is Due: Feb 1
Contact: Sam Burns, Human Resources; (312) 665-7279; sburns@fieldmuseum.org
Internet: http://www.fieldmuseum.org/research_collections/scholarships/default.htm#ungrad_intern
Sponsor: Field Museum
1400 S Lake Shore Drive
Chicago, IL 60605-2496

Field Museum Undergraduate Internships **2010**
A limited number of internships per year are available for undergraduate students or recent graduates to work directly with collections and research staff at the Museum. Internships are 10 weeks long and are usually taken in the summer. Other internships, including those in administrative areas, are also available.
Requirements: Applicants for these internships must contact the museum curator who seems most appropriate, since the internship project and work schedule will be arranged with her/him (see website for list of curators and contact information). The museum curators travel frequently and should be contacted well in advance of the application deadline.
Geographic Focus: All States
Date(s) Application is Due: Feb 1
Contact: Sam Burns, Human Resources; (312) 665-7279; sburns@fieldmuseum.org
Internet: http://www.fieldmuseum.org/research_collections/scholarships/default.htm#ungrad_intern
Sponsor: Field Museum
1400 S Lake Shore Drive
Chicago, IL 60605-2496

Field Museum Visiting Scholarships **2011**
Providing opportunities for scientists who wish to use the Field Museum's collections, funds are earmarked for travel and for subsistence while visitors are conducting their research. Young professionals and graduate students can be funded for periods of a few days to several weeks. Owing to the limited availability of these funds, awards are typically less than $1,500 per scholar. Short or extended-term visits by distinguished national and international scientists can be funded for periods of several weeks up to one year. These stipends are negotiable.
Requirements: Applications must be endorsed by a Field Museum curator (see website for list of curators responsible for the collection you wish to study). Contact the appropriate curator beforehand to communicate your background, goals and objectives. Applications require statements regarding: (a) purpose and significance of the proposed research, (b) relevance of the Field Museum's collections to the project, (c) collaboration(s) with Field Museum curators, if any, (d) procedures and methods used in the project.
Restrictions: Only electronic applications using the museum's web site will be accepted.
Geographic Focus: All States
Date(s) Application is Due: May 1; Nov 1
Contact: Scholarship Coordinator; (312) 922-9410

Internet: http://www.fieldmuseum.org/research_collections/scholarships/default.htm#visiting_scholar
Sponsor: Field Museum
1400 S Lake Shore Drive
Chicago, IL 60605-2496

Fight for Sight-Streilein Foundation for Ocular Immunology Research Award 2012
Beginning in 2011, Fight for Sight and the Streilein Foundation for Ocular Immunology co-sponsored a new annual Summer Student Research Fellowship, Fight for Sight-Streilein Foundation for Ocular Immunology Research Award, for the study of ocular immunology. Candidates for the award, an annual $2,100 grant for fulltime research conducted from June to August, are invited to apply by visiting www.fightforsight.org for an application.
Requirements: The award is open to undergraduates, graduate and medical students studying in the United States. Applications will be reviewed by the Fight for Sight Scientific Review Committee and Streilein Foundation for Ocular Immunology and a grantee will be selected based upon the committee's recommendations to the Boards of both organizations.
Geographic Focus: All States
Date(s) Application is Due: Feb 1
Contact: Janice Benson; (212) 679-6060; fax (212) 679-4466; janice@fightforsight.org
Internet: http://www.streilein-foundation.org/fellowships.html
Sponsor: Streilein Foundation for Ocular Immunology
P.O. Box 6104
Boston, MA 02114

Fight for Sight Grants-in-Aid 2013
Fight for Sight was founded in 1946 by Mildred Weisenfeld, a young woman with retinitis pigmentosa, to encourage and fund research in ophthalmology, vision and related sciences. The goal of Fight for Sight is to encourage and facilitate research in detection, understanding, prevention, treatment and cures of visual disorders especially those diseases leading to impaired sight or blindness. To that end, FFS funds Grants-in-Aid to academic researchers within the first three years of their faculty appointments. FFS primarily supports new investigators, promoting the development of scientific skills and enabling the development of preliminary scientific findings and pilot studies necessary to successfully apply for more substantial federal and private funding such as that provided by the National Eye Institute and other divisions of the NIH. Support may be used to defray costs of personnel (but not the applicant), equipment and consumable supplies needed for the specific research project. One year awards of $20,000 are provided and may start between July 1 and September 1.
Requirements: Applications will only be considered from researchers who have received their first faculty or research appointment in eye/vision within the previous three years. Fringe benefits are not included and institutional overhead charges are not covered. Travel costs are generally not supported.
Geographic Focus: All States
Date(s) Application is Due: Jan 1
Contact: Janice Benson; (212) 679-6060; fax (212) 679-4466; janice@fightforsight.org
Internet: https://www.fightforsight.org/Grants/Research-Award-Types/
Sponsor: Fight for Sight (U.S.A)
381 Park Avenue S, Suite 809
New York, NY 10016

Fight for Sight Post-Doctoral Awards 2014
Post-Doctoral Awards support individuals with a doctorate (Ph.D., M.D., O.D., Dr.PH, or D.V.M.) who are interested in academic careers in basic or clinical research in ophthalmology, vision or related sciences. This funding is intended to offer those interested in an academic career the opportunity to spend a year engaged in vision and eye research under the supervision of a senior scientist/clinician mentor. Clinical post-doctoral researchers are required to spend sufficient time on the funded research project to carry out the proposed objectives while basic researchers are expected to work full-time. One year grants of $20,000 are awarded for start dates between July 1 and September 1.
Requirements: Applications are considered from individuals who are within three years of their doctoral degrees or clinical residency training and have not received a previous Fight for Sight fellowship award. If at the time of application a doctorate has not yet been obtained, a cover letter must be submitted by the conferring institution advising when such degree is expected to be awarded. Recipients may supplement their awards with institutional or other funds however any anticipated supplemental support must be disclosed at the time of application. Total combined salary support must not exceed the annual stipend level set by the NIH for National Research Service Award recipients.
Restrictions: Fringe benefits are not provided by Fight for Sight.
Geographic Focus: All States
Date(s) Application is Due: Feb 1
Contact: Janice Benson; (212) 679-6060; fax (212) 679-4466; janice@fightforsight.org
Internet: https://www.fightforsight.org/Grants/Research-Award-Types/
Sponsor: Fight for Sight (U.S.A)
381 Park Avenue S, Suite 809
New York, NY 10016

Fight for Sight Summer Student Fellowships 2015
Fight for Sight was founded in 1946 by Mildred Weisenfeld, a young woman with retinitis pigmentosa, to encourage and fund research in ophthalmology, vision and related sciences. The goal of Fight for Sight is to encourage and facilitate research in detection, understanding, prevention, treatment and cures of visual disorders especially those diseases leading to impaired sight or blindness. Unrestricted awards of $2,100 are given for two to three months of full-time research, usually during June-August.
Requirements: Fight for Sight's primary mission is to support and encourage promising scientists early in their careers. FFS primarily supports new investigators, promoting the development of scientific skills and enabling the development of preliminary scientific findings and pilot studies necessary to successfully apply for more substantial federal and private funding such as that provided by the National Eye Institute and other divisions of the NIH. Summer Student Fellowships are available to undergraduates, graduate and medical students who are interested in pursuing eye-related clinical or basic research. For most students, this is their first exposure to eye or vision research and the experience has resulted in many students choosing academic ophthalmology or eye research as a full-time career.
Restrictions: Students receiving stipends from other sources are generally not eligible.
Geographic Focus: All States
Date(s) Application is Due: Feb 1
Contact: Janice Benson; (212) 679-6060; fax (212) 679-4466; janice@fightforsight.org
Internet: https://www.fightforsight.org/Grants/Research-Award-Types/
Sponsor: Fight for Sight (U.S.A)
381 Park Avenue S, Suite 809
New York, NY 10016

FINRA Investor Education Foundation General Grants 2016
The Foundation funds research and educational projects that support its mission of providing underserved Americans with the knowledge, skills and tools necessary for financial success throughout life. The Foundation seeks proposals that advance its mission through high-impact investor education or investor protection projects that are efficiently delivered and achieve significant and tangible benefits for the target audience; research that expands the body of knowledge and offers solutions in the fields of investor education and investor protection; and initiatives that lead with a research element and follow with a high-impact investor education or investor protection project based on the results of the research. Of particular interest are projects that better prepare older Americans for handling their finances during retirement; encourage Americans to implement steps to achieve basic retirement security and avoid investing pitfalls; encourage women to take control of their financial future through investor education; advance practice, policy, and thought in the fields of investor education and investor protection; and encourage investors to check the background of financial professionals prior to doing business with them. There is no set minimum or maximum grant amount. Projects should be completed within 24 months of grant award. Those exceeding 24 months will be considered if justification for the additional time is included in the proposal.
Requirements: 501(c)3 tax-exempt organizations are eligible.
Restrictions: The foundation will not award grants to individuals; organizations affiliated with a current member of the foundation board of directors or FINRA board of governors; securities firms regulated by FINRA; organizations affiliated with a securities firm or individual regulated by FINRA, such as a foundation established by a securities firm; securities regulators, self-regulatory organizations, or securities industry trade associations; organizations that are termed disqualified persons pursuant to Article III, Section 3(d) of the FINRA by-laws; foreign organizations; or entities that discriminate on the basis of age, color, disability, marital status, national origin, race, religion, sex, sexual orientation, or veteran status. The Foundation will generally not consider proposals to fund: international programs or projects; expenses that are not directly related to the project for which funding is sought; salaries of permanent staff (for example, prorated salaries of administrative and executive personnel, or oversight and coordination activities of a project principal); capital costs, such as building and construction or equipment such as computer hardware and office furniture; pass-through funding (for example, if the 501(c)3 organization plans to turn over the funding to a proprietary organization or consultant); projects with a potential conflict of interest (for example where funded technical support or expertise might be provided by a board member of the 501(c)3 organization); conferences and similar activities that fail to provide a long-term solution or sufficiently broad outreach; distribution methodologies that require ongoing maintenance when the ability to perform upkeep without continued funding is questionable (for example, materials with a short "shelf life" that would require ongoing funding for frequent updating); projects with proprietary elements, such as for-profit activities, use or purchase of copyrighted or trademarked materials, and proprietary research; lobbying, political contributions, fund-raising events, or other similar activities designed to influence legislation or intervene in political campaigns; donations, endowments, challenge grants, matching funds, and other similar programs; or direct or matching payments to members of the public, such as scholarships, assistance with personal and family financial difficulties, registration fees for conferences and training, or similar activities.
Geographic Focus: All States
Amount of Grant: Up to 750,000 USD
Contact: Susan M. Sarver, (202) 728-6948; susan.sarver@finra.com
Internet: http://www.finrafoundation.org/grants/general/
Sponsor: FINRA Investor Education Foundation
1735 K Street, NW
Washington, DC 20006-1506

Fischelis Grants for Research in the History of American Pharmacy 2017
The institute supports a wide range of scholarly activities as long as they promise to contribute significantly to historical understanding and are clearly related to the modern practice of pharmacy in the United States. Projects must be connected directly to American pharmacy practice. Preference will be given to 20th-century topics, although studies of earlier periods will be considered. Historical discussions of current practice issues are also eligible. An application received after the deadline date will be evaluated individually if any funds remain unawarded; otherwise it will be considered in the next year's program.
Requirements: Each application is referred for evaluation to a committee of the institute. Principal criteria used in evaluating an application are a candidate's qualifications and

record relevant to the intended purpose, relevance to the history of modern American pharmacy, and the adequacy of resources to fulfill the stated purpose.
Geographic Focus: All States
Date(s) Application is Due: Mar 1
Amount of Grant: Up to 5,000 USD
Contact: Gregory Higby, Director; (608) 262-5378; grants@aihp.org
Internet: http://www.pharmacy.wisc.edu/aihp
Sponsor: American Institute of the History of Pharmacy
777 Highland Avenue
Madison, WI 53705-2222

FishAmerica Foundation Marine and Anadromous Fish Habitat Restoration Grants **2018**
The FishAmerica Foundation Marine and Anadromous Fish Habitat Restoration program requests proposals for local efforts to accomplish meaningful on-the-ground restoration of marine, estuarine and riparian habitats, including salt marshes, seagrass beds, mangrove forests, and freshwater habitats important to anadromous fish species (fish like salmon and striped bass that spawn in freshwater and migrate to the sea). Emphasis is on using a hands-on, grassroots approach to restore fisheries habitat across coastal America and the Great Lakes.
Requirements: Projects must result in the implementation of locally-driven habitat restoration projects that emphasize stewardship and yield ecological and socioeconomic benefits. These projects must clearly demonstrate significant benefits to marine, estuarine or anadromous fisheries resources, particularly sportfish, and should involve community participation through an educational or volunteer component tied to the restoration activities. Nonprofit organizations such as sporting clubs, civic organizations, conservation groups, and to a lesser extent state agencies, in the United States and Canada are eligible. Projects must result in on-the-ground habitat restoration, clearly demonstrate significant benefits to marine, estuarine or anadromous fisheries resources, particularly sportfish, and must involve community participation through an educational or volunteer component tied to the restoration activities.
Geographic Focus: All States
Date(s) Application is Due: Apr 30
Amount of Grant: 10,000-75,000 USD
Contact: Patrick Egan, Grants Manager; (703) 519-9691, ext. 247; fax (703) 519-1872; pegan@asafishing.org or fishamerica@asafishing.org
Internet: http://www.fishamerica.org/grants.html
Sponsor: FishAmerica Foundation
1001 North Fairfax Street, Suite 501
Alexandria, VA 22314

FishAmerica Foundation Research Grants **2019**
Overall, the FishAmerica Foundation will fund research projects that further the National Fish Habitat Plan. Specifically, the program funds research projects that have regional or national implication, not local. Grants support research in the following areas: fisheries management, water quality, habitat studies, stock enhancement, economic impact studies related to sport fishing, and tagging.
Requirements: Nonprofit organizations such as sporting clubs, civic organizations, conservation groups, and to a lesser extent state agencies, in the United States and Canada are eligible.
Restrictions: Grants do not support the following projects or purposes: donations of fishing tackle; salaries, administration, overhead, or travel for conservation projects; individuals; local stream monitoring programs; political activities that attempt to influence political campaigns or legislation; access projects such as road construction, boat ramps, or fishing piers; loans, endowments, trips, tours, tickets, advertising, or publications; or permitting and related costs.
Geographic Focus: All States
Date(s) Application is Due: Jul 31
Amount of Grant: Up to 25,000 USD
Contact: Patrick Egan, Grants Manager; (703) 519-9691, ext. 247; fax (703) 519-1872; pegan@asafishing.org or fishamerica@asafishing.org
Internet: http://www.fishamerica.org/grants.html
Sponsor: FishAmerica Foundation
1001 North Fairfax Street, Suite 501
Alexandria, VA 22314

Fishman Family Foundation Grants **2020**
The Foundation considers grants for: research, education, and cultural development of and for the community; scholarships related to Jewish services, education, social, and community activities; medical and scientific research; providing resources to meet critical needs in Israel; and educational grants and scholarships. Proposals are reviewed in April and October. Application information is available online.
Requirements: 501(c)3 nonprofits are eligible.
Geographic Focus: All States
Date(s) Application is Due: Mar 31; Sep 30
Contact: Betty Fishman, President; info@fishman.org
Internet: http://www.fishman.org/apply.html
Sponsor: Fishman Family Foundation
730 E Cypress Avenue
Monrovia, CA 91016

Fitzpatrick and Francis Family Business Continuity Foundation Grants **2021**
The Family Business Continuity Foundation is dedicated to providing financial support for the training and education of family business advisors either through grants directly to individuals or to institutions that provide education and training to individuals. The Foundation wants to ensure that family businesses receive the best possible advice as they create Family Councils, select a competent board of directors, and plan to successfully transfer ownership and leadership of their enterprise from one generation to the next. Primary fields of interest include: the arts, community and economic development, and education. A formal application is required, though there are no specified annual deadlines. Funding is primarily offered in Kansas, Missouri, and Vancouver, British Columbia, though the Foundation does provide grants outside of its geographic designated areas. Grants range from $500 to $10,000.
Geographic Focus: All States, Canada
Amount of Grant: 500-10,000 USD
Contact: John J. Fitzpatrick; (785) 273-8596 or (785) 273-8500; fax (785) 273-8768
Internet: http://familybizfoundation.org/
Sponsor: Fitzpatrick and Francis Family Business Continuity Foundation
P.O. Box 4815
Topeka, KS 66604-0815

Five College Women's Research Center Ford Associateships **2022**
The center invites applications from international applicants for two one-semester Ford Associateships (fall and spring), which offer a stipend plus a housing/travel allowance in return for teaching (in English) one undergraduate course in the women's studies program at the University of Massachusetts or the Women and Gender Studies Department at Amherst College. UMass and Amherst seek a total of two experienced researchers and teachers to strengthen their undergraduate curriculum and, in the case of the university, to complement their graduate program. For the fall, the University of Massachusetts seeks a researcher with expertise in the Middle East or Latin America with an emphasis on globalization, transnational studies, or postcolonial studies. Research should focus on gender, race, ethnicity, class, and sexuality within the context of globalization. For fall or spring 2006, Amherst College seeks a researcher who is also an experienced teacher and who works on the Middle East, Latin America, Asia, or Africa with expertise in the field of gender in the media especially in the context of war and civil unrest. Ford Associates need not be studying their own region of origin. Ford Associates are expected to be in residence in the Five College area, working full-time at the center on the project outlined in their application, during the period of appointment. Each Ford Associate is expected to give a public presentation on her work, lead a faculty seminar, and participate actively in the daily life of the center. Application and guidelines are available online.
Requirements: Applications are welcomed worldwide from writers, visual artists, film and video makers, as well as academic scholars. Ford applicants may work on any and all regions of the world and need not be studying their own region of origin.
Geographic Focus: All States
Date(s) Application is Due: Feb 13
Amount of Grant: 15,000 USD
Contact: Ford Associateships Administrator; (413) 538-2275; fax (413) 538-3121; fcwsrc@wscenter.hampshire.edu
Internet: http://www.fivecolleges.edu/sites/fcwsrc/about
Sponsor: Five Colleges
50 College Street
South Hadley, MA 01075-6406

Fleishhacker Foundation Eureka Fellowships **2023**
The Foundation offers an unrestricted fellowship program for visual artists. Artists must be nominated by a local nonprofit arts organization to qualify for the Fellowship. Awards of $25,000 per artist are determined by a review panel of non-Bay Area arts experts. There are no particular deadlines with which to adhere, and nominating individuals or organizations should begin by contacting the Foundation.
Requirements: Nominees must reside in the San Francisco Bay area.
Geographic Focus: California
Contact: Christine Elbel; (415) 561-5350; info@fleishhackerfoundation.org
Internet: http://www.fleishhackerfoundation.org/eureka.html
Sponsor: Fleishhacker Foundation
P.O. Box 29918
San Francisco, CA 94129-0918

Flinn Foundation Grants **2024**
The foundation's grantmaking programs, limited to Arizona, include enhancing community-based solutions to local health care needs, especially those for children and youth; strengthening medical education and biomedical research programs in Arizona; strengthening Arizona's universities through an undergraduate scholarship program for outstanding Arizona high school students; and enhancing the visibility and long-term artistic mission of Arizona's principal visual and performing arts organizations. Types of support include program development, seed money, scholarship funds, and research grants.
Requirements: Arizona-based institutions or organizations whose programs are operated for the benefit of Arizona institutions and individuals are eligible to apply. Applications are accepted at any time. There is no application form, but a preliminary letter of inquiry or phone call is requested to determine the appropriateness of a full submission.
Restrictions: The foundation rarely provides grants to individuals, building projects (capital campaigns), purchase of equipment, endowment projects, annual fund-raising campaigns, ongoing operating expenses, or deficit needs. Requests to support conferences and workshops, publications, or the production of films and video are considered only when these activities are an integral component of a larger foundation initiative.

Geographic Focus: Arizona
Amount of Grant: 2,500-150,000 USD
Contact: JoAnn Fazio; (602) 744-6800; fax (602) 744-6815; info@flinn.org
Internet: http://www.flinn.org/about/grants.cms
Sponsor: Flinn Foundation
1802 N Central Avenue
Phoenix, AZ 85004-1506

Florence Tan Moeson Fellowships 2025

This fellowship is made possible by the generous donation by Mrs. Florence Tan Moeson, a former cataloger in the Chinese Team of the Regional and Cooperative Cataloging Division for 45 years. The purpose of the fellowship is to provide individuals with the opportunity to pursue research on East, Southeast, and/or South Asia (including the overseas Asian communities), using the unparalleled collections of the Library of Congress in Washington, DC. The grants are for a minimum of five business days of research and are to be used to cover travel to and from Washington, overnight accommodations, and photocopying; or, to cover expenses incurred while traveling to the Library of Congress to engage in scholastic research in the area of Asian studies. Up to 15 awards are available from $300 to $2,500. Awards are announced in December.
Requirements: Graduate students, independent scholars, community college teachers, researchers without regular teaching appointments, and librarians with a need for fellowship support are especially encouraged to apply. Applications are accepted online only.
Restrictions: Non-U.S. citizens and permanent residents who do not have a U.S. Social Security number must submit a Form W-7 to the Internal Revenue Service in order to obtain an individual taxpayer identification number (ITIN) prior to commencing their fellowship at the Library, regardless of the taxability of their income under this program or exemption under a treaty with the U.S.
Geographic Focus: All States
Date(s) Application is Due: Sep 30
Amount of Grant: 300-2,500 USD
Contact: Anchi Hoh; (202) 707-5673; fax (202) 707-1724; adia@loc.gov
Internet: http://www.lcasianfriends.org/ftm-fellowship
Sponsor: Library of Congress
101 Independence Avenue SE
Washington, DC 20540-4810

Florida Division of Cultural Affairs Individual Artist Fellowships 2026

The Individual Artist Fellowship program is designed to recognize practicing, professional, and creative artists residing in Florida through monetary fellowship awards. The program provides support for artists of exceptional talent and demonstrated ability to improve their artistic skills and enhance their careers. For purposes of eligibility to apply to the program, a professional creative artist is defined as a person who creates, on an ongoing basis, original works of art within an artistic discipline, and is pursuing it as a means of livelihood or for the highest level of professional recognition. Applicants are eligible to submit only one application and only one artist may apply to the Fellowship program even if there is collaboration between one or more artists.
Requirements: To be eligible for a fellowship the applicant must: submit a complete application and sample(s) of work on or before the postmark deadline; be a legal resident of Florida, as defined by section 196.015, Florida Statutes or section 22.17, Florida Statutes and agree to maintain Florida residency for the duration of the fellowship period; be a professional creative artist at least 18 years of age; not be enrolled in any undergraduate or graduate degree-seeking program during the fellowship period; not have received a monetary fellowship award from the Florida Department of State, Division of Cultural Affairs during or after fiscal year 2002; not be an individual serving as a fellowship review panelist within the same discipline; not be identified by name on sample(s) of work or Form 3 of the application, if the application is reviewed under a blind review process; submit sample(s) of work completed within the last three years prior to the application deadline.
Restrictions: Funds must be spent on activities directly related to the advancement of the fellowship recipient's artistic career.
Geographic Focus: Florida
Date(s) Application is Due: Jul 1
Amount of Grant: 5,000 USD
Contact: Morgan Barr Lewis, Arts Administrator; (850) 245-6356; fax (850) 245-6497; morgan.lewis@dos.myflorida.com
Internet: http://www.florida-arts.org/programs/fellowship/
Sponsor: Florida Division of Cultural Affairs
500 South Bronough Street, 3rd Floor
Tallahassee, FL 32399-0250

Florida Sea Turtle Grants 2027

Funded by a portion of revenues from Florida's Sea Turtle Specialty License Plate, the grants program distributes funds each year to support sea turtle research, conservation, and education programs throughout Florida. Types of grants include competitive grants, funding for ongoing activities, and emergency grants.
Requirements: Florida coastal local governments, Florida-based nonprofit organizations, and education and research institutions that actively participate in marine turtle research, conservation, and educational activities within the state of Florida are eligible to apply.
Geographic Focus: Florida
Contact: Administrator; (352) 373-6441; fax (352) 375-2449; stgp@helpingseaturtles.org
Internet: http://www.helpingseaturtles.org/stgp.htm
Sponsor: Caribbean Conservation Corporation
4424 NW 13th Street, Suite B11
Gainesville, FL 32609

Fluor Foundation Grants 2028

The Fluor Corporation achieves its contribution objectives through the Fluor Foundation and corporate giving. The Foundation's areas of interest are: education; human services; cultural outreach; and public/civic affairs. The Foundation considers requests for operating, program, capital or endowment support. Priority is given to funding organizations with employee volunteer participation. Application information is available online.
Requirements: Fluor's giving programs focus on community organizations in those locations around the world where the company has a presence. See website for local contact information: http://www.fluor.com/sustainability/community/fluor_giving/Pages/applying_for_fluor_foundation_grants.aspx
Restrictions: Funding is not available for: film production/publishing activities; individuals; sports organizations/programs; veterans, fraternal, labor or religious organizations; or lobbying/political organizations or campaigns.
Geographic Focus: Alaska, California, Louisiana, New Jersey, New York, North Carolina, Pennsylvania, South Carolina, Tennessee, Texas, Virginia, Washington, Albania, Andorra, Armenia, Australia, Austria, Azerbaijan, Belarus, Belgium, Bosnia & Herzegovina, Bulgaria, Canada, Caribbean, China, Croatia, Cyprus, Czech Republic, Denmark, Estonia, Finland, France, Georgia, Germany, Greece, Hungary, Iceland, Ireland, Italy, Japan, Kosovo, Latvia, Liechtenstein, Lithuania, Luxembourg, Macedonia, Malta, Mexico, Moldova, Monaco, Montenegro, New Zealand, Norway, Peru, Philippines, Poland, Poland, Portugal, Romania, Russia, Russia, San Marino, Serbia, Slovakia, Slovenia, Spain, Sweden, Switzerland, The Netherlands, Turkey, Ukraine, United Kingdom, Vatican City, Venezuela
Contact: Suzanne Esber, Executive Director of Community Relations; (949) 349-7847; fax (949) 349-7694; suzanne.esber@fluor.com
Internet: http://www.fluor.com/sustainability/community/fluor_giving/Pages/applying_for_fluor_foundation_grants.aspx
Sponsor: Fluor Foundation
3 Polaris Way
Aliso Viejo, CA 92698

FMC Foundation Grants 2029

The foundation supports education, community improvement, urban affairs, health and human services, and public issues/economic education. Higher education is supported through scholarships and employee matching gifts. Education support tends to be in business, engineering, chemistry, and some minority education programs. Eligible applicants need to contact the FMC in their geograhic area. Each individual FMC location determines how their contributions will be given.
Requirements: Grants are awarded to 501(c)3 organizations in FMC-plant communities. U.S.-based organizations with an international focus are also eligible.
Restrictions: Grants are not made to individuals.
Geographic Focus: Arizona, California, Delaware, Florida, Illinois, Louisiana, Maine, Maryland, Missouri, New Jersey, New York, North Carolina, Pennsylvania, Tennessee, Texas, West Virginia, Wyoming, Canada
Contact: Program Contact; (215) 299-6000; fax (215) 299-6140
Internet: http://www.fmc.com
Sponsor: FMC Foundation
1735 Market Street
Philadelphia, PA 19103

Ford Foundation International Fellowships 2030

The program supports advanced study by exceptional individuals who will use this education to become leaders in their respective fields, furthering development in their own countries and greater economic and social justice worldwide. Eligible fields of study are any academic discipline or field of study related to the foundation's three grant-making areas: asset building and community development—children, youth and families, sexuality and reproductive health, work-force development, development finance and economic security, environment and development, community development; education, media, and arts and culture—education reform; higher education and scholarship; religion, society and culture; media; and arts and culture; peace and social justice—human rights; international cooperation; governance; and civil society. The international fellowships provide support for up to three years of formal graduate-level study leading to a masters or doctoral degree. Fellows are selected from countries where the foundation maintains active overseas programs. Once selected, fellows may enroll in an appropriate university program anywhere in the world, including their country of residence. The program also enables fellows to undertake short-term language study and training in research and computer skills prior to graduate school enrollment.
Requirements: Applicants must be resident nationals or residents of Asia, Africa, the Middle East, Latin America, and Russia. Successful candidates will demonstrate superior achievement in their undergraduate studies and hold a baccalaureate degree or its equivalent; have substantial experience in community service or development-related activities; possess leadership potential evidenced by their employment and academic experience; propose to pursue a post-baccalaureate degree that will directly enhance their leadership capacity in a practical, policy, academic, or artistic discipline or field corresponding to one or more of the foundation's areas of endeavor; present a plan specifying how they will apply their studies to social problems or issues in their own countries; and commit themselves to working on these issues following the fellowship period.
Restrictions: U.S. nationals are not eligible, although fellows may study in the United States.
Geographic Focus: All States
Contact: Fellowship Administrator; (212) 984-5558 or (212) 573-5000; fax (212) 351-3677; question@fordifp.org
Internet: http://www.fordifp.net/

Sponsor: Ford Foundation
320 E 43rd Street
New York, NY 10017

Ford Foundation Peace and Social Justice Grants 2031
This program area supports two categories: human rights and international cooperation, and governance and civil society. Grants under human rights and international cooperation support projects involving women's rights and civil rights in the United States, international human rights, international law, multilateral conflict prevention, and U.S. foreign policy. Governance and civil society grants support projects concerning civic participation, electoral reform, and innovations in state and local government. Most grants are given to organizations. The foundation also makes grants to individuals, though they are few in number relative to demand and are limited to research, training, and other activities related to the foundation's program interests. Support for graduate fellowships is generally provided through grants to universities and other organizations, which select recipients. Applications are considered throughout the year; a letter of inquiry is the first step of the application process.
Requirements: The letter of inquiry should include the purpose of the project, problems and issues addressed, information about the applicant organization, estimated overall budget, time period for which funds are requested, and qualifications of those engaged in the project.
Restrictions: Support is not normally given for routine operating costs of institutions or for religious activities. Except in rare cases, funding is not available for the construction or maintenance of buildings. The foundation does not award undergraduate scholarships or make grants for purely personal or local needs.
Geographic Focus: All States
Amount of Grant: 75,000-250,000 USD
Contact: Luis Antonio Ubinas, President; (212) 573-5000; fax (212) 599-4584; office-secretary@fordfound.org
Internet: http://www.fordfound.org/grants
Sponsor: Ford Foundation
320 E 43rd Street
New York, NY 10017

Fordham Prizes 2032
The foundation awards two prizes annually. The Thomas B. Fordham Prize for Distinguished Scholarship is awarded to a scholar who has made major contributions to education reform via research, analysis, and successful engagement in the war of ideas. The Thomas B. Fordham Prize for Valor is awarded to a leader who has made major contributions to education reform via noteworthy accomplishments at the national, state, local, and/or school levels. Nominations from individuals or organizations familiar with the work of outstanding education reformers and scholars are welcome. Self-nominations will not be considered. Nomination guidelines are available online.
Requirements: Anyone can be nominated whose work has had a profound impact on education in the United States. Candidates may be nominated either for cumulative lifetime achievement or for extraordinary one-time accomplishments.
Restrictions: Employees and trustees of the foundation are not eligible, nor are members of the prize committee.
Geographic Focus: Ohio
Contact: Jennifer Leischer; (202) 223-5452; fax (202) 223-9226; jleischer@edexcellence.net
Internet: http://www.edexcellence.net/foundation/global/page.cfm?id=305
Sponsor: Thomas B. Fordham Foundation
1701 K Street NW, Suite 1000
Washington, DC 20006

Ford Motor Company Fund Grants 2033
The fund supports not-for-profit organizations in three major areas: innovation and education; community development and American legacy, and auto-related safety education. The fund seeks to build partnerships with organizations that have a well-defined sense of purpose, a demonstrated commitment to maximizing available resources, and a reputation for meeting objectives and delivering quality programs and services. Priority is placed on the support and development of organizations that promote diversity and inclusion. Requests for support are accepted and reviewed throughout the year. The fund now implements an online application system. Details are available online.
Restrictions: Ford does not fund: advocacy-directed programs; animal-rights organizations; beauty or talent contests; day-to-day business operations; debt reduction; donation of vehicles; efforts to influence legislation, or the outcome of any elections or any specific election of candidates to public office or to carry on any voter registration drive; endowments; fraternal organizations; general operating support to hospitals and health care institutions; individual sponsorship related to fundraising activities; individuals; labor groups; loans for small businesses; loans to program-related investments; organizations that do not have 501(c)3 status; organizations that unlawfully discriminate in their provision of goods and services based on race, color, religion, gender, gender identity or expression, ethnicity, sexual orientation, national origin, physical challenge, age, or status as a protected veteran; political contributions; private K-12 schools; profit-making enterprises; religious programs or sectarian programs for religious purposes; species-specific organizations; sports teams.
Geographic Focus: All States
Contact: Ford Fund Coordinator; (888) 313-0102; Fordfund@ford.com
Internet: http://www.ford.com/en/goodWorks/fundingAndGrants/fordMotorCompanyFund/default.htm
Sponsor: Ford Motor Company Fund
P.O. Box 1899
Dearborn, MI 48121-1899

Forest Products Society Wood Award 2034
The Forest Products Society (FPS), Borden Inc, and Dynea U.S.A Inc, sponsor the annual wood awards. Two awards are made for the best 2000- to 4000-word paper on the forestry industry. Topics may cover any of a wide range of problems from harvesting trees, to development and manufacture of products, to distribution and marketing of such products. Judges will rate the essays on the basis of subject, skill of treatment, conclusions, applicability, and conformance to competition rules. Winners will make a poster presentation at the annual meeting of the FPS. Tentative titles are due by the listed application deadline.
Requirements: The competition is open to all graduate students who are registered for a full quarter or full semester at any university or college the year previous to application.
Geographic Focus: All States
Date(s) Application is Due: Feb 6
Amount of Grant: 500-1,000 USD
Contact: Vickie Bruce, (608) 231-1361, ext. 210; fax (608) 231-2152; vbruce@forestprod.org or info@forestprod.org
Internet: http://www.forestprod.org/awa-wood.html
Sponsor: Forest Products Society
2801 Marshall Court
Madison, WI 53705-2295

Forrest C. Lattner Foundation Grants 2035
The Foundation's primary objectives are in six areas of interest: arts and humanities; education; environment; health and social services; historic preservation; and medical research. The Foundation also wishes to encourage the development of innovative model programs. Application guidelines are available online.
Restrictions: The foundation is unable to fulfill grant requests to individuals, or to support programs that are the primary responsibility of the public sector.
Geographic Focus: Florida, Georgia, Kansas, Rhode Island, Texas
Date(s) Application is Due: Mar 1; Sep 1
Contact: Susan Lattner Lloyd; (561) 278-3781; fax (561) 278-3167; lattner@bellsouth.net
Internet: http://www.lattnerfoundation.org
Sponsor: Forrest C. Lattner Foundation
777 E Atlantic Avenue, Suite 317
Delray Beach, FL 33483

Foundation Fighting Blindness Marjorie Carr Adams Women's Career Development Awards 2036
The Program's primary purpose is to support female clinical research scientists of superior dedication and talent to pursue vigorous research programs to find the cures for retinal degenerative diseases research (RDD) i.e. inherited orphan retinal degenerative diseases and dry age-related macular degeneration. The goal of this program is to facilitate advances in the laboratory and clinical research, to elucidate the mechanisms for the etiology and pathogenesis of RDDs and to develop innovative strategies to prevent, treat and cure these diseases. This award provides support for mentored study and research for clinically trained biomedical research professionals who have the potential to assume leadership roles as clinician-scientists in RDD research.
Requirements: Female clinical scientists possessing an, M.D., D.O., or recognized equivalent foreign degrees and who are in their first, second or third year of a junior faculty appointment are eligible to apply for the Adams WCDA. Applicants do not have to be U.S. citizens.
Geographic Focus: All States
Date(s) Application is Due: Nov 14
Amount of Grant: 65,000-325,000 USD
Contact: Cindy Settar, (410) 568-0150 or (800) 683-5555; csettar@Fightblindness.org or grants@Fightblindness.org
Internet: http://www.blindness.org/index.php?option=com_content&view=section&id=9&Itemid=107#career-development
Sponsor: Foundation Fighting Blindness
7168 Columbia Gateway Drive, Suite 100
Columbia, MD 21046

Foundation for Endangered Languages Grants 2037
The Foundation for Endangered Languages exists to support, enable and assist the documentation, protection and promotion of endangered languages. In order to do this, it aims to: raise awareness of endangered languages, both inside and outside the communities where they are spoken, through all channels and media; support the use of endangered languages in all contexts: at home, in education, in the media, and in social, cultural and economic life; monitor linguistic policies and practices, and to seek to influence the appropriate authorities where necessary; support the documentation of endangered languages, by offering financial assistance, training, or facilities for the publication of results; collect together and make available information of use in the preservation of endangered languages; and disseminate information on all of the above activities as widely as possible.
Geographic Focus: All States
Amount of Grant: 1,000-2,000 USD
Samples: John Hobson, University of Sydney—Banjalang: Northern Rivers district, NSW, Australia, $1450; Chun Huang, University of Florida—Siraya (Taiwan), $1000; Molly Babel, Mono Lake Northern Paiute: Bridgeport, California, U.S.A, $550.
Contact: Nicholas Ostler, nostler@chibcha.demon.co.uk
Internet: http://www.ogmios.org/grant.htm
Sponsor: Foundation for Endangered Languages
Batheaston Villa, 172 Bailbrook Lane
Bath, BA1 7AA England

Foundation for Enhancing Communities Grants 2038
The Foundation awards grants to Pennsylvania nonprofit organizations in the greater Harrisburg area. Areas of interest include art/cultural programs, education, health care and health organizations, human services, and community development. Types of support include program development, publication, seed money, and scholarships. Deadlines vary according to funding availability.
Requirements: Tax exempt 501(c)3 organizations in Pennsylvania's Dauphin, Cumberland, Franklin, and Perry Counties are eligible.
Restrictions: Grants do not support individuals, operating budgets, or capital expenses.
Geographic Focus: Pennsylvania
Contact: Jennifer Kuntch; (717) 236-5040; fax (717) 231-4463; jkuntch@tfec.org
Internet: http://www.ghf.org
Sponsor: Foundation for Enhancing Communities
200 North 3rd Street, P.O. Box 678
Harrisburg, PA 17108-0678

Foundation for Evangelism Doctoral Scholarships 2039
The Foundation for Evangelism sponsors scholarships for those pursuing doctoral studies in evangelism. Currently these programs are led by professors supported by The Foundation for Evangelism and administered through a ThD program at Duke Divinity School and a PhD program at Asbury Theological Seminary. The overall goal in establishing these scholarships is to equip scholars in the Wesleyan tradition who express an interest in being considered for seminary teaching positions in evangelism. These students will help fulfill the Foundation's vision to raise up generations of leaders who have a passion for evangelism by not only being a leader but also training others to lead.
Geographic Focus: All States
Contact: Dr. Stephen Gunter, Professorship Director; (828) 454-6800 or (800) 737-8333; fax (828) 454-6810; sgunter@foundationforevangelism.org
Internet: http://foundationforevangelism.org/grants/doctoral-education/
Sponsor: Foundation for Evangelism
125 North Lakeshore Drive, P.O. Box 985
Lake Junaluska, NC 28745

Foundation for Evangelism Harry Denman Fellowships 2040
The Foundation for Evangelism currently funds the Harry Denman Fellowships which are given to scholars in the United Methodist tradition who are pursuing doctoral studies in evangelism or related disciplines. The Fellowships are to encourage and equip people who might one day teach evangelism in United Methodist schools of theology or institutions of higher education. Candidates are chosen for their commitment to Christ and the church as well as for their academic excellence. A Foundation for Theological Education (AFTE) administers these fellowships through their John Wesley Fellowship Program.
Geographic Focus: All States
Contact: Dr. Stephen Gunter, Professorship Director; (828) 454-6800 or (800) 737-8333; fax (828) 454-6810; sgunter@foundationforevangelism.org
Internet: http://foundationforevangelism.org/grants/doctoral-education/
Sponsor: Foundation for Evangelism
125 North Lakeshore Drive, P.O. Box 985
Lake Junaluska, NC 28745

Foundation for Seacoast Health Grants 2041
The mission of the foundation is to invest its resources to improve the health and well being of Seacoast residents. The foundation considers a very limited number of new grant initiatives which address one or more of the following prioritized health needs; access to affordable mental health services; access to preventative and restorative dental services; access to affordable child care and after school care; access to affordable primary medical care; and coordination and dissemination of health information related to identified priority needs. The deadline for the submission of the letter of intent is March 1 for Infants, Children and Adolescence and June 1 for Promoting Health and Preventing Disease. Annual deadline dates may vary; contact program staff for exact dates.
Requirements: Nonprofit organizations in cities and towns in the New Hampshire/Maine Seacoast area, including Greenland, New Castle, Newington, North Hampton, Portsmouth, and Rye, NH; and Eliot, Kittery, and York, ME, may submit applications.
Restrictions: Grants do not support ongoing general operating expenses, deficit elimination, political activities, travel, conferences, or lodging.
Geographic Focus: Maine, New Hampshire
Date(s) Application is Due: Mar 1; Jun 1
Contact: Susan Bunting, President; (603) 422-8200; ffsh@communitycampus.org
Internet: http://www.ffsh.org/grants.cfm
Sponsor: Foundation for Seacoast Health
100 Campus Drive, Suite 1
Portsmouth, NH 03801

Foundation for the Advancement of Mesoamerican Studies Grants 2042
The foundation awards grants to support scholarly contributions to the understanding of ancient Mesoamerican cultures (Mexico, Belize, Guatemala, Honduras, and El Salvador). The foundation supports projects in the disciplines of archaeology, art history, epigraphy, linguistics, ethnohistory, ethnography, and sociology. Interdisciplinary projects are encouraged, especially those that combine disciplines in novel and potentially productive ways. The program provides research grants and contingency grants. Contingency grants provide emergency funds for unforeseen situations encountered during initial research, secondary analysis, and/or the final dissemination of findings.
Restrictions: FAMSI does not provide funds for equipment purchases, salaries, stipends, or honoraria.
Geographic Focus: All States
Amount of Grant: 500-10,000 USD
Contact: Sandra Noble; (352) 795-5990; fax (352) 795-1970; sandra@famsi.org
Internet: http://research.famsi.org/grants/
Sponsor: Foundation for the Advancement of Mesoamerican Studies
268 South Suncoast Boulevard
Crystal River, FL 34429-5498

Foundation for the Future Grants 2043
The program awards grants to support scholars undertaking research that is directly related to a better understanding of the factors affecting the quality of life of the long-term future of humanity. Preliminary Grant Applications must be submitted between January 1 and April 30 of the year in which grant funding is sought. An application form is available on the Web site.
Requirements: Individuals and organizations worldwide may apply.
Geographic Focus: All States
Date(s) Application is Due: Apr 30
Amount of Grant: 5,000-25,000 USD
Contact: Sesh Velamoor; (425) 451-1333; seshvelamoor@futurefoundation.org
Internet: http://www.futurefoundation.org/grants/index.html
Sponsor: Foundation for the Future
123-105th Avenue SE
Bellevue, WA 98004-6265

Foundation of the American Thoracic Society Research Grants 2044
The American Thoracic Society established the ATS Foundation Research program to foster the development and training of future leaders in science and biomedical research. Through collaborations with non-profit and corporate partners the research program is able to provide research and career development grants and mentored fellowships to encourage young investigators to devote their talents and time to generating novel scientific ideas leading to new life-saving therapies for patients and their families. Since its founding in 2002, the Foundation of the ATS has awarded $10 million in grants to 105 young researchers investigating a wide spectrum of lung diseases, ranging from asthma and COPD to pulmonary fibrosis and alpha-1 antitrypsin. The program has also funded the career development of 24 pulmonary and critical care fellows. The grants amounts vary and range from $5,000 a year for 1 year to $50,000 per year for two years.
Restrictions: Grants are not awarded to individuals.
Geographic Focus: All States
Amount of Grant: 5,000-50,000 USD
Contact: Alyssa Chase; (212) 315-8600; fax (212) 315-6498; achase@thoracic.org
Internet: http://foundation.thoracic.org/programs/research-program.php
Sponsor: Foundation of the American Thoracic Society
25 Broadway, 18th Floor
New York, NY 10004

Frances C. and William P. Smallwood Foundation Grants 2045
The foundation awards grants in support of education, higher education, and social services. Applications will be received from the following organizations; educational institutions, both public and private, primarily at the college or university level, as well as other non-profit organizations involved in American education and is confined to the area within a 100 mile radius of the Dallas/Fort Worth Metroplex, the area within a 100 mile radius of the city of Gainesville, FL, and the area within a 100 mile radius of Minden, Nevada. Potential applicants should request guidelines prior to applying.
Requirements: Contributions are limited to non-profit, tax-exempt organizations which have obtained IRS status under Section 501(c)3 of the IRS code, and, where appropriate, under Section 170(c).
Restrictions: Funding will not be considered for: general endowment funds of an organization; fund-raising events, i.e. tickets, dinner, telethons; corporate memberships or contributions to Chambers of Commerce, taxpayer associations, and other similar bodies; and contributions to political organizations, campaigns and candidates.
Geographic Focus: Florida, Nevada, Texas
Date(s) Application is Due: Feb 28
Contact: Suzanne Stockdale, (775) 782-3678
Sponsor: Frances C. and William P. Smallwood Foundation
P.O. Box 2050
Fort Worth, TX 76113

Frances L. and Edwin L. Cummings Memorial Fund Grants 2046
The foundation supports nonprofits in New York City and northern New Jersey in the areas of education, especially programs that serve public school children from disadvantaged backgrounds; social welfare concerns; and campaigns to build endowments through establishment of challenge grants. Other areas of interest include elementary education, secondary education, vocational education, adult basic education and literacy, and higher education; hospitals, medical care, AIDS, and cancer; children, youth, and human services; and community development. Support is provided for endowment funds, seed money, consulting services, matching funds, technical assistance, professorships, and program development. The board meets in June and December. Application forms are not required.
Requirements: Grants are awarded to nonprofit organizations in the metropolitan New York, NY, area, with emphasis on New York City, southern Westchester County, and northern New Jersey.
Restrictions: Grants are not awarded to individuals or for capital building campaigns, general operating support, moving expenses, conferences, surveys, annual fund-raising campaigns, or research conducted by individuals or private institutions.

Geographic Focus: New Jersey, New York
Amount of Grant: 10,000 USD
Contact: Elizabeth Costas; (212) 286-1778; fax (212) 682-9458
Sponsor: Frances L. and Edwin L. Cummings Memorial Fund
501 Fifth Avenue, Suite 708
New York, NY 10017-6103

Francis C. Wood Scholar in Residence Program **2047**
Short-term grants will be awarded to scholars engaged in projects requiring personal use of the historical collections of the library and/or Mutter Museum. Letters of application outlining the proposed project (proposal should not exceed five pages), necessary length of residence, historical materials to be used and a budget with specific information on travel, lodging, and research expenses should be submitted, along with a curriculum vitae and two letters of recommendation by January 31.
Requirements: The program is restricted to individuals pursuing advanced research. Candidates must have a PhD or an equivalent record of professional experience and scholarly publication.
Restrictions: The award will not be made to degree candidates or to persons seeking support for work leading to a degree.
Geographic Focus: All States
Amount of Grant: 30,000 USD
Contact: Ed Morman; (215) 563-3737, ext. 265; emorman@collphyphil.org
Internet: http://www.collphyphil.org/woodpg1.shtml
Sponsor: College of Physicians of Philadelphia
19 S 22nd Street
Philadelphia, PA 19103

Francis P. Garvan-John M. Olin Medal **2048**
This award, sponsored by the the Francis P. Garvan-John M. Olin Medal Endowment and administered by the American Chemical Society, is given annually to recognize distinguished service to chemistry by women chemists who are U.S. citizens. The award consists of $5,000, a medallion with a presentation box, and a certificate. Up to $2,500 for travel expenses to the meeting at which the award will be presented will be reimbursed.
Requirements: A nominee must be a citizen of the United States and have performed distinguished service to chemistry. Any individual, except a member of the awards committee, may submit one nomination or seconding letter for the award in any given year. The nominating documents consist of a letter of not more than 1000 words containing an evaluation of the nominee's accomplishments and a specific identification of the work to be recognized, a biographical sketch including date of birth, and a list of publications and patents authored by the nominee.
Restrictions: Self-nominations are not accepted.
Geographic Focus: All States
Date(s) Application is Due: Nov 1
Contact: Awards Administrator; (202) 872-4575; fax (202) 776-8008; awards@acs.org
Internet: http://portal.acs.org/portal/acs/corg/content?_nfpb=true&_pageLabel=PP_ARTICLEMAIN&node_id=1319&content_id=CTP_004521&use_sec=true&sec_url_var=region1
Sponsor: American Chemical Society
1155 Sixteenth Street, NW
Washington, DC 20036-4801

Frank H. Field and Joe L. Franklin Award for Outstanding Achievement in Mass Spectrometry **2049**
This award, sponsored by Bruker Daltonics, is given annually to recognize outstanding achievement in the development or application of mass spectrometry. The award is granted without regard to age, nationality, or the date of the achievement recognized by the award. In odd-numbered years the award will be presented for advances in techniques or fundamental processes in mass spectrometry. Recognition will be given in even-numbered years to development of the applications of mass spectrometry. The award consists of $5,000 and a certificate.
Requirements: Any individual, except a member of the award committee, may submit one nomination or seconding letter for each award in any given year. The nominating documents consist of a letter of not more than 1000 words containing an evaluation of the nominee's accomplishments and a specific identification of the work to be recognized, a biographical sketch including date of birth, and a list of publications and patents authored by the nominee.
Restrictions: Self-nominations are not accepted.
Geographic Focus: All States
Date(s) Application is Due: Nov 1
Contact: Felicia Dixon; (800) 227-5558 or (202) 872-4408; fax (202) 776-8008; f_dixon@acs.org or awards@acs.org
Internet: http://portal.acs.org/portal/acs/corg/content?_nfpb=true&_pageLabel=PP_ARTICLEMAIN&node_id=1319&content_id=CTP_004522&use_sec=true&sec_url_var=region1
Sponsor: American Chemical Society
1155 Sixteenth Street, NW
Washington, DC 20036-4801

Franklin Mosher Baldwin Fellowships **2050**
The Baldwin fellowship is intended for scholars with citizenship in an African country who seek to obtain an advanced degree or specialized training in an area of study related to human origins research. This award is for a program of approved special training and/or advanced training towards an MA, PhD, or equivalent and is limited to two years of support.
Requirements: Scholars with citizenship in an African country are eligible to apply.
Geographic Focus: All States
Date(s) Application is Due: Feb 15
Amount of Grant: Up to 12,000 USD
Contact: Grants Officer; (415) 561-4646; fax (415) 561-4647; grants@leakeyfoundation.org
Internet: http://www.leakeyfoundation.org/grants/g3.jsp
Sponsor: L.S.B. Leakey Foundation
1002A O'Reilly Avenue, P.O. Box 29346
San Francisco, CA 94129-0346

Frank Stanley Beveridge Foundation Grants **2051**
The foundation welcomes proposals in the areas of: animal care; arts, culture and humanities; civil rights, social action, advocacy; education; employment/jobs; environmental quality, protection and beautification; food, nutrition and agriculture; health; housing; human services; medical research; mental health; philanthropy; safety; recreation; religion; science; social services; and youth development. The board meets in October and April to consider requests. Multiyear grants are rare. Contact the foundation via the Web site only. No phone or written inquiries will be accepted.
Requirements: Applicants must be 501(c)3 nonprofit organizations or foundations in Massachusetts's Hampden and Hampshire Counties.
Restrictions: The Foundation prefers not to support: awards or prizes; commissioning of new artistic work; conferences/seminars; curriculum development; debt reduction; employee matching gifts; employee-related scholarships; endowment funds; exhibitions; faculty/staff development; fellowship funds; fellowships to individuals; film/video/radio production; foundation administered programs; general operating support; grants to individuals; income development; internship funds; management development; performance/production costs; professorships; program-related investment/loans; publications; scholarships to individuals; student aid; technical assistance.
Geographic Focus: Massachusetts
Date(s) Application is Due: Feb 1; Aug 1
Amount of Grant: 50,000 USD
Contact: Philip Caswell, President; (800) 229-9667; fax (561) 748-0644; administrator@beveridge.org or caswell@beveridge.org
Internet: http://www.beveridge.org/
Sponsor: Frank Stanley Beveridge Foundation
1340 U.S. Highway 1, Suite 102
Jupiter, FL 33469

Frederick Douglass Summer Fellowships **2052**
Clarion University has established the Frederick Douglass Summer Fellowships to offer opportunities to three graduate students who are pursuing a career in college teaching and are in their final year of a terminal degree program. This program is part of a Pennsylvania State System of Higher Education Collaborative that is intended to commemorate Douglass' life-long contributions to equality and public service. Fellowships are designed to attract advanced graduate students primarily from HBCU's and HACU's. Selected scholars will team and/or co-teach one course during a five-week Summer Session II, which typically runs from the first week of June through the first week of July. The stipend for classroom instruction, community presentations, and other duties as assigned (for the full duration of Summer Session II) is $3,000. The University will also provide housing assistance, academic support including office space, computer and library privileges, and an allowance of $1,000 for research-related expenses.
Requirements: Applicants must be U.S. citizens, and must have completed all doctoral work, except the dissertation, by the end of the next academic year. In addition to the application, candidates must submit two full sets of the following materials in hard copy: full curriculum vitae and letter of interest; graduate school transcript (a copy of the original transcript is acceptable); letter from the candidate's major or dissertation adviser regarding their qualifications for this program; one page summary of the dissertation topic; and at least two reference letters that address the candidate's qualifications.
Restrictions: Electronic applications are not accepted.
Geographic Focus: All States
Date(s) Application is Due: Apr 30
Amount of Grant: 4,000 USD
Contact: Stacy Boyer-Davis; (814) 393-2109; fax (814) 393-1618; sboyerdavis@clarion.edu
Internet: http://www.clarion.edu/48794/
Sponsor: Clarion University-Office of Social Equity
207 Carrier Building
Clarion, PA 16214

Frederick Douglass Summer Scholars Fellowship **2053**
In keeping with Frederick Douglass life of public service and commitment to equity and access, the Frederick Douglass Institute for Intercultural Research offers teaching opportunities to graduate students pursuing careers as university faculty members and entering their final year of terminal degree and/or doctoral programs. As a Frederick Douglass Fellow, scholars will teach and/or co-teach one course during the summer sessions, continue their research, and contribute to the life of the university. Dates of employment typically run from May through August. Douglass Scholar applicants may also be considered for full- or partial-year appointments at the discretion of individual departments. The Douglass Fellowships provide professional experiences, mentoring, and potential employment opportunities within universities that are strongly committed to cultural diversity. Applicants from diverse cultural backgrounds, especially those from historically underrepresented groups, are encouraged to apply. Scholars will be compensated as adjunct faculty, according to the university's collective bargaining agreement. Additionally, Fellows may be provided housing and travel. Faculty benefits

(office space, computer and network access, and library privileges) are also provided. The application is available at the Douglass Institute website.
Requirements: Applicant must be U.S. citizens or permanent residents; M.A. degree completed and enrollment in a terminal degree program (Ph.D., M.F.A., Ed.D., etc.); possess academic background in one of the disciplines taught at the university. Teaching experience is preferred, with A.B.D. or doctorate within the last two years. Priority consideration will be given to completed applications postmarked by December 16.
Restrictions: The appointment does not cover families, and scholars are responsible for their own transportation.
Geographic Focus: All States
Date(s) Application is Due: Dec 16
Contact: Veronica Watson; (724) 357-3299 or (724) 357-4940; Veronica.Watson@iup.edu
Internet: http://www.iup.edu/page.aspx?id=39215
Sponsor: Indiana University of Pennsylvania-Douglass Institute
Stright Hall, Room 108
Indiana, PA 15705

Frederick Gardner Cottrell Foundation Grants **2054**
Giving primarily in Woods Hole, Massachusetts, Hunt Valley, Maryland, Brunswick, Maine, and Houston, Texas. The foundation awards grants to educational and scientific organizations to advance research and chair teaching positions. The foundation's areas of interest include eye research, higher education, marine science, and science. There are no application deadline. Submit a written request for application.
Geographic Focus: Maine, Maryland, Massachusetts, Texas
Amount of Grant: 100,000-300,000 USD
Contact: Gary Munsinger; (520) 748-4400; fax (520) 748-0025; munsinger@rctech.com
Sponsor: Frederick Gardner Cottrell Foundation
5210 East Williams Circle, Suite 240
Tucson, AZ 85711-4410

Frederic Stanley Kipping Award in Silicon Chemistry **2055**
This award, supported by Dow Corning Corporation, is given biennially in even-numbered years to recognize distinguished achievement in research in silicon chemistry and to stimulate the creativity of others toward further advancement of this field of chemistry. A nominee must have accomplished distinguished achievement in research in silicon chemistry during the preceding 10 years. The measure of this achievement should focus primarily on the nominee's significant publications in the field of silicon chemistry but may include consideration of contributions to the related field of organometallic chemistry. There are no limits on age or nationality. Applications are accepted in odd-numbered years. The award consists of $5,000 and a certificate.
Requirements: Any individual, except a member of the award committee, may submit one nomination or seconding letter for the award in any given year. Nominating documents consist of a letter of not more than 1000 words containing an evaluation of the nominee's accomplishments and a specific identification of the work to be recognized, a biographical sketch including date of birth, and a list of publications and patents authored by the nominee.
Restrictions: Self-nominations are not accepted.
Geographic Focus: All States
Date(s) Application is Due: Nov 1
Amount of Grant: 5,000 USD
Contact: Felicia Dixon, Awards Administrator; (800) 227-5558 or (202) 872-4408; fax (202) 776-8008; f_dixon@acs.org or awards@acs.org
Internet: http://portal.acs.org/portal/acs/corg/content?_nfpb=true&_pageLabel=PP_ARTICLEMAIN&node_id=1319&content_id=CTP_004523&use_sec=true&sec_url_var=region1
Sponsor: American Chemical Society
1155 Sixteenth Street, NW
Washington, DC 20036-4801

Freshwater Future Project Grants **2056**
The Project Grants program provides grant awards ranging from $500 to $5,000 (USD) to be used for specific project expenses. Projects to protect aquatic habitats in the following strategic program areas will be considered: projects to protect and restore wetlands that will result in ongoing or positive permanent change in the habitat; on-the-ground restoration activities that include an action component to create permanent change (a small amount of funding for this area); advocating for low-impact solutions to community storm water issues; participation in land use planning and zoning; participation in watershed planning; watch-dogging and participating in the development, implementation, and enforcement of local, state, provincial, and federal aquatic habitat protection regulations; non-partisan voter education, voter registration, and candidate forums; advocating for naturally functioning river and coastal ecosystems; reducing polluted runoff; launching special initiatives or creating unique opportunities to strengthen citizen involvement in aquatic habitat protection and restoration; and communicating how local wetlands, lakes and rivers are interconnected in the Great Lakes Basin. The spring cycle deadline for project grants is March 31, while the fall cycle deadline is September 30.
Geographic Focus: All States
Date(s) Application is Due: Mar 31; Sep 30
Amount of Grant: 500-5,000 USD
Contact: Jill Ryan; (231) 348-8200; jill@freshwaterfuture.org or info@freshwaterfuture.org
Internet: http://www.freshwaterfuture.org/grant-programs/project-grant-program.html
Sponsor: Freshwater Future
P.O. Box 2479
Petoskey, MI 49770

Fritz B. Burns Foundation Grants **2057**
The foundation supports nonprofit organizations primarily in southern California by awarding grants for education, with an emphasis on buildings, equipment, endowments (except for ordinary operating expenses), student scholarship and loan funds, and faculty fellowships; to/for hospitals, hospital equipment, and medical research; and religious organizations (Christian, Jewish, Latter-day Saints, nondenominational, Presbyterian, Protestant, Roman Catholic, and Salvation Army). Proposals should be concise, containing a brief description of what is planned, with a clear statement of the objective sought; IRS letter certifying tax exemption; financial statements; and a list of officers and directors. No formalized application or proposal format is required. Proposals are considered on a quarterly basis.
Requirements: Nonprofits in southern California may apply.
Restrictions: Grant requests are not considered from individuals nor from tax-supported entities.
Geographic Focus: California
Date(s) Application is Due: Sep 30
Amount of Grant: 10,000-250,000 USD
Contact: Joseph Rawlinson, President; (818) 840-8802
Sponsor: Fritz B. Burns Foundation
4001 West Alameda Avenue, Suite 203
Burbank, CA 91505-4338

FTE Dissertation Fellowships **2058**
The fellowships support African American PhD or ThD students in their final year of dissertation work in religious or theological studies. The objectives of the program are to increase completion rates of African American students in the field of theological studies; collaborate with graduate programs in identifying and supporting students at this juncture of their studies; and offer doctoral students special opportunities for scholarly exchange and peer contacts, as well as assistance in the enhancement of writing and research skills. The fellowship includes a stipend for living expenses and other educational costs; full tuition grants; and attendance at a summer conference of workshops to assist in the development of research, writing, and teaching skills. Applicants must be able to write full time during the fellowship year. Application materials are available on the Web site.
Requirements: U.S. African American PhD or ThD students at the final writing stage of their graduate work in religious or theological studies are eligible. The applicant's dissertation research proposal and writing plan must have been approved by the dissertation committee.
Geographic Focus: All States
Date(s) Application is Due: Feb 1
Amount of Grant: 15,000 USD
Contact: Sharon Watson Fluker, Director; (404) 727-1450; sfluker@thefund.org
Internet: http://www.thefund.org/programs/fellowships/dissertation/index.html
Sponsor: Fund for Theological Education
825 Houston Mill Road, Suite 250
Atlanta, GA 30329-4211

FTE Doctoral Fellowships **2059**
The fellowships support African American students entering a PhD or ThD program in religious or theological studies. The objectives of the program are to increase the number of qualified African American students in the field of theological studies; establish partnerships with graduate programs and theological schools to strengthen their efforts toward the recruitment and support of these students; provide opportunities for intellectual stimulation and encouragement through conferences, mentoring, and peer relationships; and undertake assessment of how racial/ethnic minority doctoral candidates are best supported in their work and the factors that contribute to successful completion of their programs. The fellowship includes a stipend for living expenses and other educational costs; full tuition grants; and attendance at a summer conference of workshops to assist in the development of research, writing, and teaching skills. Application materials are available on the Web site.
Requirements: African American students entering their first year of an accredited graduate program leading to a PhD or ThD degree in religious or theological studies are eligible. Applicants must be committed to providing leadership within theological education and strongly considering a career in seminary teaching and research.
Geographic Focus: All States
Date(s) Application is Due: Mar 1
Amount of Grant: Up to 15,000 USD
Contact: Ellen Echols Purdum; (404) 727-1450; fax (404) 727-1490; epurdum@thefund.org
Internet: http://www.thefund.org/programs/fellowships/doctoral/index.html
Sponsor: Fund for Theological Education
825 Houston Mill Road, Suite 250
Atlanta, GA 30329-4211

FTE Expanding Horizons Partnership Grants **2060**
The program supports African American students in PhD and ThD programs in religious and theological studies. It links educators, academic institutions, and those pursuing doctoral work for the purpose of promoting quality in the preparation for and practice of scholarly research and teaching in theological seminaries. The program supports doctoral fellows entering their first year of a PhD or ThD program in religious or theological studies; and dissertation support to fellows at the final stage of their doctoral work in religious or theological studies. Dissertation fellowship applications are due February 1; doctoral fellowship applications are due March 1.
Requirements: African American students in PhD and ThD programs in religious and theological studies are eligible.
Geographic Focus: All States

Date(s) Application is Due: Feb 1; Mar 1
Amount of Grant: 15,000 USD
Contact: The Rev. Ellen Echols Purdum, Director; (404) 727-1450; fax (404) 727-1490; sfluker@thefund.org or fte@thefund.org
Internet: http://www.thefund.org/programs/horizons/index.html
Sponsor: Fund for Theological Education
825 Houston Mill Road, Suite 250
Atlanta, GA 30329-4211

FTE North American Doctoral Fellowships **2061**
The fellowships support talented racial or ethnic minority students currently enrolled in a PhD or ThD program in religious or theological studies. Candidates must pursue full-time study with high academic performance; be committed to teaching and scholarship; and demonstrate capacity for leadership in theological education. Preference is given to students nearing the end of their studies. Application materials are available online.
Requirements: U.S. or Canadian PhD or ThD students who are members of racial or ethnic groups traditionally underrepresented in graduate education are eligible.
Restrictions: DMin students are ineligible.
Geographic Focus: All States
Date(s) Application is Due: Mar 1
Amount of Grant: Up to 5,000 USD
Contact: Sharon Watson Fluker, Director; (404) 727-1450; sfluker@thefund.org
Internet: http://www.thefund.org/programs/fellowships/doctoral/northamerican.html
Sponsor: Fund for Theological Education
825 Houston Mill Road, Suite 250
Atlanta, GA 30329-4211

Fulbright-Hays Seminars Abroad Bilateral Projects Grants **2062**
The program provides short-term study and travel seminars abroad for U.S. educators in the social sciences and humanities for the purpose of improving their understanding and knowledge of the peoples and cultures of other countries. Support is generally made available through inter-agency agreements. There are approximately seven to ten seminars with fourteen to sixteen participants in each seminar annually. Seminars are four to six weeks in duration. All seminars are held in countries outside of Western Europe. Terms of the award include a round-trip economy airfare, room and board, tuition and fees, and program-related travel within the host country.
Requirements: Applicants must: be a citizen of the U.S. or a permanent resident; hold at least a bachelor's degree from an accredited college or university; have at least three years of full-time experience by the time of departure for the seminar; be currently employed full-time in a U.S. school system, institution of higher education, local education agency (LEA), state education agency (SEA), library, or museum; meet the general language requirement; and be physically and psychologically able to participate in all phases of the seminar.
Geographic Focus: All States
Contact: Gail Holdren, International Education Graduate Programs contact; (202) 502-7691; fax (202) 502-7859; gale.holdren@ed.gov
Internet: http://www.ed.gov/programs/iegpssap/index.html
Sponsor: Institute of International Education
1400 K Street NW, 7th Floor
Washington, DC 20005-2403

Fulbright/Garcia Robles Grants **2063**
The program is especially interested in attracting students pursuing fieldwork or internships in public administration or public policy. Other fields of interest are projects on Mexican culture, society and politics to deepen the understanding of Mexico in the United States. Areas of collaboration go far beyond border issues, which are also of relevance, and can include projects on migration, ecological issues, public health, education, public policy, environmental protection, human rights, and trade policy. The program will also support projects in the arts that further dialogue and contact between U.S. and Mexican arts communities.
Requirements: Applicants must: be U.S. citizens at the time of application (permanent residents are not eligible); be in good health (grantees will be required to submit a satisfactory Medical Certificate of Health from a physician); and hold a B.A. degree or the equivalent before the start of the grant (applicants who have not earned a B.A. degree or the equivalent, but who have extensive professional study and/or experience in business, may be considered). Excellent spoken and written Spanish is required at the time of application.
Restrictions: Fields of study not recommended include medicine and dentistry.
Geographic Focus: All States
Contact: Jody Dudderar; (212) 984-5565; fax (212) 984-5325; jdudderar@iie.org
Internet: http://us.fulbrightonline.org/program_country.html?id=70#full
Sponsor: Institute of International Education
1400 K Street, NW, 7th Floor
Washington, DC 20005-2403

Fulbright Alumni Initiatives Awards Grants **2064**
The objective of the AIA program is to help translate the individual Fulbright experience into long-term institutional impact. To this end, the program provides small institutional grants to Fulbright alumni to develop innovative projects that will foster institutionally supported linkages and sustainable, mutually beneficial relationships between the Fulbright scholar's home and host institutions. Just about any activity resulting in the creation or fostering of a sustainable institutional relationship that will have an impact on both the Fulbright alum's home institution and the Fulbright host institution abroad and which both institutions are prepared to support in both the long and short terms will be considered. The program has been temporarily suspended, and applicants should check back for updates.
Requirements: Applicants must be U.S. citizens. The program is open to eligible alumni whose grants occurred at any time from the 1998/1999 academic year through the present academic year.
Restrictions: This program is limited to application by a Fulbright alum (whether U.S. or Visiting Scholar) in partnership with his/her host institution colleague. However, other faculty members from both applicants' institutions may be part of the proposal and participate in the project.
Geographic Focus: All States
Contact: Stacey Bustillos, Program Officer; (202) 686-6252; sbustillos@cies.iie.org
Internet: http://www.cies.org/aia/
Sponsor: Institute of Int'l Education, Council for International Exchange of Scholars
3007 Tilden Street NW, Suite 5L
Washington, DC 20008-3009

Fulbright Beginning Professional Journalism Awards **2065**
The program provides five Professional Journalism Awards for beginning professional journalists or recent graduates in journalism or related fields to conduct independent research and serve as interns or guest journalists at newspapers, publishers or television and radio stations in Germany. The grantee will spend the first few months studying and conducting independent research, followed by one or more internships or guest journalist positions at German newspapers, publishers or television and radio stations. All Journalism areas including photo, print, and broadcast journalism as well as electronic media will be considered. Research and internship affiliation will be arranged by the Fulbright Commission in Berlin; grantee preferences will be accommodated as far as possible. Research may be conducted at a German university or other relevant institution. University matriculation in Germany (actual coursework is optional) will be required.
Requirements: Solid proficiency in German commensurate with the project is required.
Geographic Focus: All States
Contact: Jon Adler, Program Manager; (212) 984-5326; jadler@iie.org
Internet: http://us.fulbrightonline.org/program_country.html?id=41#journ
Sponsor: Institute of International Education
1400 K Street, NW, 7th Floor
Washington, DC 20005-2403

Fulbright Binational Business Grants in Mexico **2066**
This program is designed to enhance the knowledge, expertise, and understanding of business in Mexico for U.S. students in business, law, or engineering. It combines graduate courses (a minimum of 3 and a maximum of 6) in the pertinent academic area (e.g., business, finance, international trade, or comparative law among others) with an internship with a Mexico-based company, firm, or NGO dedicated to international business. Candidates with a BA degree and recent M.B.A., J.D. or master's degree in business administration, finance, economics, international relations, engineering, or accounting, and some work experience are preferred. A Master's degree is recommended but not required. Significant work experience in a business environment is strongly recommended for these candidates and for candidates who do not have a business-related degree.
Requirements: Applicants must: be U.S. citizens at the time of application (permanent residents are not eligible); be in good health (grantees will be required to submit a satisfactory Medical Certificate of Health from a physician); and hold a B.A. degree or the equivalent before the start of the grant (applicants who have not earned a B.A. degree or the equivalent, but who have extensive professional study and/or experience in business, may be considered). Excellent spoken and written Spanish is required at the time of application.
Geographic Focus: All States, Mexico
Contact: Jody Dudderar; (212) 984-5565; fax (212) 984-5325; jdudderar@iie.org
Internet: http://us.fulbrightonline.org/program_country.html?id=70#binational
Sponsor: Institute of International Education
1400 K Street, NW, 7th Floor
Washington, DC 20005-2403

Fulbright Business Grants in Spain **2067**
These awards are co-sponsored by the Instituto de Empresa Foundation in Madrid. Instituto de Empresa is one of Europe's leading business schools and has hosted nearly 28,000 students, representing 71 different nationalities. Awards are for full-time study in the bilingual English/Spanish International MBA program which offers the basic MBA core, two components of elective courses, and a Global Affairs Seminar in which students analyze business problems focusing on five different geographic areas.
Requirements: Applicants must: be U.S. citizens at the time of application (permanent residents are not eligible); be in good health (grantees will be required to submit a satisfactory Medical Certificate of Health from a physician); and hold a B.A. degree or the equivalent before the start of the grant (applicants who have not earned a B.A. degree or the equivalent, but who have extensive professional study and/or experience in business, may be considered). Excellent spoken and written Spanish is required at the time of application.
Geographic Focus: All States, Albania, Andorra, Armenia, Austria, Azerbaijan, Belarus, Belgium, Bosnia & Herzegovina, Bulgaria, Croatia, Cyprus, Czech Republic, Denmark, Estonia, Finland, France, Georgia, Germany, Greece, Hungary, Iceland, Ireland, Italy, Kosovo, Latvia, Liechtenstein, Lithuania, Luxembourg, Macedonia, Malta, Moldova, Monaco, Montenegro, Norway, Poland, Portugal, Romania, Russia, San Marino, Serbia, Slovakia, Slovenia, Spain, Sweden, Switzerland, The Netherlands, Turkey, Ukraine, United Kingdom, Vatican City
Contact: Jon Adler, Program Manager; (212) 984-5326; jadler@iie.org
Internet: http://us.fulbrightonline.org/program_country.html?id=98#MBA
Sponsor: Institute of International Education
1400 K Street, NW, 7th Floor
Washington, DC 20005-2403

Fulbright Critical Language Enhancement Awards 2068

The Award, a component of the National Security Language Initiative, is available to students who have been awarded a Fulbright U.S. Student full grant and who intend to use the eligible languages in their Fulbright project. Students must also demonstrate the equivalent of an academic year of study in that language by the grant start date. Application for an Award is made in conjunction with the Fulbright Program application. Awards will be granted based on recommendations made by the selection committees, and final selections by Fulbright Commissions or U.S. Embassies in the corresponding countries. The purpose of the Award is to cultivate higher levels of language proficiency prior to and during the Fulbright grant period and beyond. Ultimately, awardees will achieve a high level of proficiency in a targeted language and will go on to careers or further study which will incorporate the use of this and/or related languages. The languages available for the Award are: all dialects of Arabic, Azeri, Bengali, Chinese (Mandarin only), Farsi, Gujarati, Hindi, Korean, Marathi, Pashto, Punjabi, Russian, Tajik, Turkish, Urdu, and Uzbek. It is preferred that language training take place in the same country as that of the award.

Requirements: The grant includes pre- and post-testing in the relevant language, a minimum of 20 hours of formal study per week in-country, and a commitment to study beyond the Critical Language Enhancement Award training period. Applicants must: be U.S. citizens at the time of application (permanent residents are not eligible); be in good health (grantees will be required to submit a satisfactory Medical Certificate of Health from a physician); and hold a B.A. degree or the equivalent before the start of the grant (applicants who have not earned a B.A. degree or the equivalent, but who have extensive professional study and/or experience in fields in which they wish to pursue a project, may be considered). In the creative and performing arts area, four years of professional study and/or experience meets the basic eligibility requirement.

Restrictions: Due to scheduling constraints, Fulbright English Teaching Assistants (ETAs) are ineligible for this award. The following persons are ineligible: anyone who has already held a U.S. Department of State-funded Fulbright student grant of any type; anyone who has previously received a Department of Education-funded Doctoral Dissertation Research Abroad (Fulbright Hays grant); employees of the U.S. Department of State, and their immediate families, for a period ending one year following termination of such employment; employees of private and public agencies (excluding educational institutions) under contract to the U.S. Department of State to perform administrative or screening services on behalf of the U.S. Department of State's exchange program, for a period ending one year following the termination of their services for the U.S. Department of State provided such employees have been directly engaged in performing services related to the exchange programs; applicants holding a doctoral degree at the time of application; applicants seeking enrollment in a medical degree program abroad; or applicants currently residing in the countries of Australia, Belgium/Luxembourg, Canada, Chile, Finland, Hungary, Mexico, Netherlands, New Zealand, Sweden, or Switzerland.

Geographic Focus: All States, Algeria, Angola, Benin, Botswana, Burkina Faso, Burundi, Cameroon, Cape Verde, Central African Republic, Chad, Comoros, Congo, Congo, Democratic Republic of, Cote d' Ivoire (Ivory Coast), Djibouti, Egypt, Equatorial Guinea, Eritrea, Ethiopia, Gabon, Gambia, Ghana, Guinea, Guinea-Bissau, Kenya, Lesotho, Liberia, Libya, Madagascar, Malawi, Mali, Mauritania, Mauritius, Morocco, Mozambique, Namibia, Niger, Nigeria, Rwanda, Sao Tome & Principe, Senegal, Seychelles, Sierra Leone, Somalia, South Africa, Sudan, Swaziland

Contact: Jermaine Jones, Program Manager; (212) 984-5341; jjones@iie.org

Internet: https://us.fulbrightonline.org/thinking_clea.html

Sponsor: Institute of International Education
1400 K Street, NW, 7th Floor
Washington, DC 20005-2403

Fulbright Distinguished Chairs Grants 2069

The Fulbright Distinguished Chairs Program comprises approximately forty distinguished lecturing, distinguished research and distinguished lecturing/research awards ranging from three to 12 months. Awards in the Fulbright Distinguished Chairs Program are viewed as among the most prestigious appointments in the Fulbright Scholar Program. Candidates should be senior scholars and have a significant publication and teaching record. Applicants should submit hard copies of the Distinguished Chairs Application Form (one page), a letter of interest (about three pages), a curriculum vitae (maximum eight pages) and, if required, a sample syllabus (maximum four pages). Chairs are available in Australia, Austria, Brazil, Canada, Denmark, Finland, France, Germany, Hungary, Ireland, Israel, Italy, Netherlands, Poland, Portugal, Russia, and Sweden. Because an objective of the Fulbright Program is to provide an educational exchange experience for those not previously afforded such an opportunity, preference will usually be given to candidates who have not had substantial recent experience in the country to which they are applying.

Requirements: Applicants must meet all of the following eligibility *Requirements:* U.S. citizenship at the time of application (permanent resident status is not sufficient); possess a Ph.D. or equivalent professional terminal degree at the time of application (for professionals and artists outside academe, recognized professional standing and substantial professional accomplishments); have college or university teaching experience at the level and in the field of the proposed lecturing activity as specified in the award description (for Distinguished Chairs awards, candidates should be senior scholars with a significant publication and teaching record); have foreign language proficiency only if specified in the award description or required for the completion of the proposed project; be of sound physical and mental health; and disclose prior conviction or current indictment for commission of a felony.

Restrictions: Previous Fulbright scholar grantees are eligible to apply only if five years will have elapsed between the ending date of one scholar award and the beginning date of the new scholar award. This rule does not apply if the previous grant was for less than two months. Employees, spouses or dependent children of the United States Department of State or public and private organizations under contract to the United States Department of State are ineligible to apply for a Fulbright grant until one year after the employee's termination.

Geographic Focus: All States

Date(s) Application is Due: Aug 3

Contact: Maria Bettua, Program Officer; (202) 686-6245; mbettua@cies.iie.org

Internet: http://www.cies.org/Chairs/

Sponsor: Institute of Int'l Education, Council for International Exchange of Scholars
3007 Tilden Street NW, Suite 5L
Washington, DC 20008-3009

Fulbright English Teaching Assistantships (ETAs) 2070

The Fulbright program has offered opportunities for U.S. students to serve as English teachers and teaching assistants at schools, colleges and universities abroad for many years. Since an applicant may only apply to one country and one program, it is vital to select carefully, based on educational and career goals, academic background and preparation, language proficiency, and geographic interests. ETAs in Asia are placed primarily in elementary and secondary schools and knowledge of the host country language at the time of application is not required; however, ETAs in South America will be working with university and adult students and must have proficiency in the host country language. In most cases, ETAs: are placed in schools or universities outside of capital cities; are assigned various activities designed to improve their students' language abilities and knowledge of the United States; are fully integrated into the host community, increasing their own language skills and knowledge of the host country; may pursue individual study/research plans in addition to ETA responsibilities; and have some background or training in education and/or teaching English. Check the guidelines on line for the ever-changing list of host countries.

Requirements: Applicants must: be U.S. citizens at the time of application (permanent residents are not eligible); be in good health (grantees will be required to submit a satisfactory Medical Certificate of Health from a physician); and have sufficient proficiency in the written and spoken language of the host country (where required) to communicate with the people and to carry out the proposed ETA.

Restrictions: The following persons are ineligible: anyone who has already held a U.S. Department of State-funded Fulbright student grant of any type; anyone who has previously received a Department of Education-funded Doctoral Dissertation Research Abroad (Fulbright Hays grant); employees of the U.S. Department of State, and their immediate families, for a period ending one year following termination of such employment; employees of private and public agencies (excluding educational institutions) under contract to the U.S. Department of State to perform administrative or screening services on behalf of the U.S. Department of State's exchange program, for a period ending one year following the termination of their services for the U.S. Department of State provided such employees have been directly engaged in performing services related to the exchange programs; or applicants holding a doctoral degree at the time of application.

Geographic Focus: All States

Date(s) Application is Due: Oct 20

Contact: Walter Jackson, Program Manager; (212) 984-5327; wjackson@iie.org

Internet: http://us.fulbrightonline.org/thinking_teaching.html

Sponsor: Institute of International Education
1400 K Street, NW, 7th Floor
Washington, DC 20005-2403

Fulbright European Union Scholar-in-Residence (EU.S.IR) Grants 2071

The primary objective of the European Union Scholar-in-Residence (EU.S.IR) program is to strengthen expertise in European Union affairs. Through an arrangement with the EU, grants will be available to bring European scholars specializing in EU affairs to U.S. campuses as resident scholars for one term of the academic year. The scholars will give guest lectures and conduct seminars as appropriate, consult with faculty and students on research, engage in collaborative study and provide outreach to neighboring institutions and the local community. The resident scholars are not expected to teach regular course offerings. Detailed program guidelines, proposal forms, and further information are available upon request; materials are available in May prior to the deadline date.

Geographic Focus: All States

Date(s) Application is Due: Oct 15

Contact: Alma Ford; (202) 686-6252; fax (202) 362-3442; aford@cies.iie.org

Internet: http://www.cies.org/sir/eusir/

Sponsor: Institute of Int'l Education, Council for International Exchange of Scholars
3007 Tilden Street NW, Suite 5L
Washington, DC 20008-3009

Fulbright German Studies Seminar Grants 2072

The program allows participation in a group seminar on current German society and culture. The program will begin in Berlin and include visits to other cities in Germany. The focus will be on the formation of policies in current issues at the core of modern society such as climate change, food technology, gene technology, stem cell research and the broad scope of education. The seminar will explore how Germany and its European neighbors view the role of science in their societies, examining the factors that result in differing approaches to these issues and the challenges of harmonizing different standards under a uniform EU-policy umbrella. With an eye to the global marketplace and worldwide competition in science, the seminar will further examine the multiple interests which structure relations between national governments, economic corporations, political and supranational bodies as well as research and development institutions. Scholars from U.S. universities, colleges, and community colleges who hold full-time teaching appointments and meet other academic requirements (Ph.D., Ph.D. candidacy or other equivalent degree or qualifications) are eligible. The seminar lasts two weeks, typically during the month of June.

Requirements: Applicants must be U.S. citizens with permanent residence in the United States at the time of application.

Geographic Focus: All States, Germany
Date(s) Application is Due: Nov 1
Contact: Maria Bettua, Assistant Director; (202) 686-6245; mbettua@cies.iie.org
Internet: http://www.iie.org/cies/us_scholars/us_awards/GSS/index.html
Sponsor: Institute of Int'l Education, Council for International Exchange of Scholars
3007 Tilden Street NW, Suite 5L
Washington, DC 20008-3009

Fulbright International Education Administrators (IEA) Seminar Grants **2073**
The IEA seminars are designed to introduce participants to the society, culture and higher education systems of these countries through campus visits, meetings with foreign colleagues and government officials, attendance at cultural events and briefings on education. Participants in the International Education Administrators Program gain a firsthand look into the host country's academic infrastructure and culture. They gain new perspective on the need to internationalize U.S. campuses and insight into how it can be done. The deadline for the German program is February 1, while the deadline for both Japan and Korea is November 1.
Requirements: To be eligible, applicants must: be U.S. citizens; be international education professionals and senior university administrators (e.g., deans, provosts, vice presidents) with significant responsibility for international programs and activities; have an affiliation with an accredited college or university or nonprofit international exchange organization administering postsecondary student or faculty exchange; and have a minimum of three years of work experience in international education. Applicants for the Japan program must be affiliated with a four-year college or university, while Germany and Korea will consider applicants from both two- and four-year institutions.
Restrictions: Employees, spouses or dependent children of the United States Department of State or public and private organizations under contract to the United States Department of State are ineligible to apply for a Fulbright grant until one year after the employee's termination. TEFL administrators are ineligible for these seminar programs, but they are encouraged to view a listing of other Fulbright opportunities.
Geographic Focus: All States
Date(s) Application is Due: Feb 1; Nov 1
Contact: Maria Bettua; (202) 686-6245; mbettua@cies.iie.org
Internet: http://www.cies.org/IEA/
Sponsor: Institute of Int'l Education, Council for International Exchange of Scholars
3007 Tilden Street NW, Suite 5L
Washington, DC 20008-3009

Fulbright mtvU Awards **2074**
Up to four awards will be available to pursue projects around an aspect of international contemporary or popular music as a cultural force for expression. Preference will be given to creative projects that are conveyed in a dynamic fashion and are accompanied by a feasible plan. In addition to presenting unique projects on music as a global force for mutual understanding, applicants must submit a documentation and outreach plan describing how they intend to share their activities with their peers during their Fulbright year abroad through mtvU print, broadcast and/or online mediums.
Requirements: Applicants must: be U.S. citizens at the time of application (permanent residents are not eligible); be in good health (grantees will be required to submit a satisfactory Medical Certificate of Health from a physician); have sufficient proficiency in the written and spoken language of the host country (where required) to communicate with the people; and hold a B.A. degree or the equivalent before the start of the grant (applicants who have not earned a B.A. degree or the equivalent, but who have extensive professional study and/or experience in fields in which they wish to pursue a project, may be considered). In the creative and performing arts area, four years of professional study and/or experience meets the basic eligibility requirement. All candidates for the Europe and Eurasia region are required to obtain their own affiliations, generally established with an educational and/or research institution in the host country or countries. Candidates are responsible for securing their own research clearance as required by the host country.
Restrictions: The following persons are ineligible: anyone who has already held a U.S. Department of State-funded Fulbright student grant of any type; anyone who has previously received a Department of Education-funded Doctoral Dissertation Research Abroad (Fulbright Hays grant); employees of the U.S. Department of State, and their immediate families, for a period ending one year following termination of such employment; employees of private and public agencies (excluding educational institutions) under contract to the U.S. Department of State to perform administrative or screening services on behalf of the U.S. Department of State's exchange program, for a period ending one year following the termination of their services for the U.S. Department of State provided such employees have been directly engaged in performing services related to the exchange programs; or applicants currently residing in the countries of Australia, Belgium/Luxembourg, Canada, Chile, Finland, Hungary, Mexico, Netherlands, New Zealand, Sweden, or Switzerland.
Geographic Focus: All States
Date(s) Application is Due: Mar 1
Contact: Walter Jackson, Program Manager; (212) 984-5327; wjackson@iie.org
Internet: https://us.fulbrightonline.org/overview_typesgrants.html
Sponsor: Institute of International Education
1400 K Street, NW, 7th Floor
Washington, DC 20005-2403

Fulbright New Century Scholars (NCS) Grants **2075**
The NCS Program brings 30 top academics and professionals from around the world together each year to collaborate on an issue of global importance. Of the thirty, approximately one-third will be U.S. citizens while the remaining two thirds will be visiting scholars from countries with an operational Fulbright Scholar Program. NCS will provide a platform for scholars from the U.S. and around the world to engage in debate and dialogue based on multidisciplinary research and to develop new global models for understanding the social context within which nations and communities shape their responses to the many challenges of the 21st century. This particular aspect of the New Century Scholars program is a unique feature that distinguishes it from the core Fulbright Scholar Program.
Requirements: Applicants must be conducting current research relevant to the program's theme and objectives, be open to exploring and incorporating comparative, interdisciplinary approaches in their investigations, and interested in developing collaborative activities with other NCS Scholars. U.S. applicants must have U.S. citizenship and be residing permanently in the United States. For academic applicants, a Ph.D. or equivalent terminal degree in a relevant field is required. For applicants in the professional fields, the appropriate terminal degree in a relevant field is required. Non-U.S. applicants must be citizens or permanent residents of and residing in the country from which they are applying at the time of application. All applicants must have fluency in English.
Restrictions: Non-U.S. applicants holding permanent residency green cards, whether or not they reside in the U.S., are not eligible.
Geographic Focus: All States
Contact: Jonathan Looper; (202) 686-6235; fax (202) 362-3442; jlooper@cies.iie.org
Internet: http://www.cies.org/NCS/
Sponsor: Institute of Int'l Education, Council for International Exchange of Scholars
3007 Tilden Street NW, Suite 5L
Washington, DC 20008-3009

Fulbright Occasional Lecturer Program (OLP) Grants **2076**
The Council for International Exchange of Scholars provides travel awards through the Occasional Lecturer Program (OLP), that enable Fulbright Visiting Scholars to accept guest lecturing invitations at colleges and universities throughout the United States. OLP Travel Awards are granted at three different levels ranging from $250 to $750, which CIES determines based on distances involved.
Geographic Focus: All States
Amount of Grant: 250-750 USD
Contact: Jonathan Looper; (202) 686-6235; fax (202) 362-3442; jlooper@cies.iie.org
Internet: http://www.cies.org/sir/olp/
Sponsor: Institute of Int'l Education, Council for International Exchange of Scholars
3007 Tilden Street NW, Suite 5L
Washington, DC 20008-3009

Fulbright Public Policy Initiative Program Grants in Mexico **2077**
This program is designed to enhance the knowledge, expertise, and understanding of public policy in Mexico for U.S. students in business, law, or engineering. Pursuant to this interest in public policy, the Fulbright Commission will fund three grants specifically for students conducting fieldwork or research in the areas of public administration or public policy.
Requirements: Applicants must: be U.S. citizens at the time of application (permanent residents are not eligible); be in good health (grantees will be required to submit a satisfactory Medical Certificate of Health from a physician); and hold a B.A. degree or the equivalent before the start of the grant (applicants who have not earned a B.A. degree or the equivalent, but who have extensive professional study and/or experience in business, may be considered). Excellent spoken and written Spanish is required at the time of application.
Geographic Focus: All States
Contact: Jody Dudderar; (212) 984-5565; fax (212) 984-5325; jdudderar@iie.org
Internet: http://us.fulbrightonline.org/program_country.html?id=70#public
Sponsor: Institute of International Education
1400 K Street, NW, 7th Floor
Washington, DC 20005-2403

Fulbright Scholar-in-Residence (SIR) Grants **2078**
Grants strengthen the institutional commitment of U.S. higher education to international education and scholarship by enabling colleges and universities to invite visiting scholars from abroad in the humanities and social sciences or in professional specializations with strong international focus. A Fulbright Scholar in Residence may teach regular courses from a foreign area perspective, serve as a resource person in interdisciplinary courses, assist in developing new courses, or participate in special seminars. Host institutions are expected to share the scholar's expertise with a wide range of departments and with neighboring institutions, involve him/her in community activities and professional organizations, and provide opportunities for the visitor to pursue personal research interests. Detailed program guidelines, proposal forms, and further information are available upon request; materials are available in May prior to deadline date.
Geographic Focus: All States
Date(s) Application is Due: Oct 15
Contact: Alma Ford; (202) 686-6252; fax (202) 362-3442; aford@cies.iie.org
Internet: http://www.cies.org/sir/
Sponsor: Institute of Int'l Education, Council for International Exchange of Scholars
3007 Tilden Street NW, Suite 5L
Washington, DC 20008-3009

Fulbright Specialists Grants **2079**
The Program is designed to provide short-term academic opportunities (two to six weeks) for U.S. faculty and professionals. Shorter grant lengths give specialists greater flexibility to pursue a grant that works best with their current academic or professional commitments. Applications for the Program are accepted on a rolling basis, and peer review of applications is conducted eight times per year. Program goals include: to increase the participation of leading U.S. scholars and professionals in Fulbright academic exchanges; to encourage new activities that go beyond the traditional Fulbright

activities of lecturing and research; and to promote increased connections between U.S. and non-U.S. post-secondary academic institutions.
Requirements: Applicants must: be U.S. citizen at the time of application (permanent resident status is not sufficient; if a naturalized citizen, applicant must provide actual date of naturalization); possess a Ph.D. or equivalent professional/terminal degree at the time of application; have a minimum of five years of post-doctoral teaching or professional experience in the field in which you are applying (for professionals and artists outside academe, recognized professional standing and substantial professional accomplishments plus a minimum of five years of professional experience in the field in which you are applying); disclose prior conviction or current indictment for commission of a felony; and be residing in the United States at the time they are approved for a grant and intend to return to their U.S. institution after the grant's completion.
Restrictions: Employees, spouses or dependent children of the United States Department of State or public and private organizations under contract to the United States Department of State exchange programs are ineligible to apply for a Fulbright grant until one year after the employee's termination.
Geographic Focus: All States
Date(s) Application is Due: Jan 28; Feb 24; Mar 25; Apr 21; May 20; Jul 15; Sep 9; Nov 4; Dec 30
Contact: Ryan Hathaway, Coordinator; (202) 686-4026; rhathaway@cies.iie.org
Internet: http://www.cies.org/specialists/#top
Sponsor: Institute of Int'l Education, Council for International Exchange of Scholars
3007 Tilden Street NW, Suite 5L
Washington, DC 20008-3009

Fulbright Traditional Scholar Program in Europe and Eurasia **2080**
The traditional Fulbright Scholar Program sends 800 U.S. faculty and professionals abroad each year. Grantees lecture and conduct research in a wide variety of academic and professional fields. Distribution of awards to countries in the region will vary annually according to the caliber of the applicants. Grants are available to: Albania, Andorra, Armenia, Austria, Azerbaijan, Belarus, Belgium, Bosnia and Herzegovina, Bulgaria, Croatia, Cyprus, Czech Republic, Denmark, Estonia, European Union, Finland, France, Georgia, Germany, Greece, Hungary, Iceland, Ireland, Italy, Latvia, Lithuania, Luxembourg, Macedonia, Moldova, Netherlands, Norway, Poland, Portugal, Romania, Russia, Slovakia (Slovak Republic), Slovenia, Spain, Sweden, Switzerland, Turkey, Ukraine, and United Kingdom. Only countries Considered to be in the Eastern Europe or Eurasia regions may be part of a multi-country application (2 to 3 countries). Multi-country proposals are not permitted for Western Europe. Language requirements vary by country, and prior knowledge of the local language may not be required, particularly where language is not commonly taught in the U.S.
Requirements: Applicants must: be U.S. citizens at the time of application (permanent residents are not eligible); be in good health (grantees will be required to submit a satisfactory Medical Certificate of Health from a physician); have sufficient proficiency in the written and spoken language of the host country (where required) to communicate with the people and to carry out the proposed study; and hold a B.A. degree or the equivalent before the start of the grant (applicants who have not earned a B.A. degree or the equivalent, but who have extensive professional study and/or experience in fields in which they wish to pursue a project, may be considered). In the creative and performing arts area, four years of professional study and/or experience meets the basic eligibility requirement. All candidates for the Europe and Eurasia region are required to obtain their own affiliations, generally established with an educational and/or research institution in the host country or countries. Candidates are responsible for securing their own research clearance as required by the host country.
Restrictions: The following persons are ineligible: anyone who has already held a U.S. Department of State-funded Fulbright student grant of any type; anyone who has previously received a Department of Education-funded Doctoral Dissertation Research Abroad (Fulbright Hays grant); employees of the U.S. Department of State, and their immediate families, for a period ending one year following termination of such employment; employees of private and public agencies (excluding educational institutions) under contract to the U.S. Department of State to perform administrative or screening services on behalf of the U.S. Department of State's exchange program, for a period ending one year following the termination of their services for the U.S. Department of State provided such employees have been directly engaged in performing services related to the exchange programs; applicants holding a doctoral degree at the time of application; applicants seeking enrollment in a medical degree program abroad; or applicants currently residing in the countries of Australia, Belgium/Luxembourg, Canada, Chile, Finland, Hungary, Mexico, Netherlands, New Zealand, Sweden, or Switzerland.
Geographic Focus: All States, Albania, Andorra, Armenia, Austria, Azerbaijan, Belarus, Belgium, Bosnia & Herzegovina, Bulgaria, Croatia, Cyprus, Czech Republic, Denmark, Estonia, Finland, France, Georgia, Germany, Greece, Hungary, Iceland, Ireland, Italy, Kosovo, Latvia, Liechtenstein, Lithuania, Luxembourg, Macedonia, Malta, Moldova, Monaco, Montenegro, Norway, Poland, Portugal, Romania, Russia, San Marino, Serbia, Slovakia, Slovenia, Spain, Sweden, Switzerland, The Netherlands, Turkey, Ukraine, United Kingdom, Vatican City
Date(s) Application is Due: Aug 1
Contact: Rachel Holskin, IIE Program Manager; (212) 984-5326; rholskin@iie.org
Internet: http://us.fulbrightonline.org/program_regions_countries.php?id=3
Sponsor: Institute of International Education
1400 K Street, NW, 7th Floor
Washington, DC 20005-2403

Fulbright Traditional Scholar Program in Sub-Saharan Africa **2081**
The traditional Fulbright Scholar Program sends 800 U.S. faculty and professionals abroad each year. Grantees lecture and conduct research in a wide variety of academic and professional fields. Selection for countries in sub-Saharan Africa, with the exception of South Africa, will be made based on the quality of the applications, rather than per-country quotas. Distribution of awards to countries in the region will vary annually according to the caliber of the applicants. Grants are available to: Benin, Botswana, Burkina Faso, Cameroon, Chad, Eritrea, Ethiopia, Ghana, Guinea, Ivory Coast, Kenya, Madagascar, Malawi, Mali, Mauritius, Mozambique, Namibia, Niger, Nigeria, Senegal, South Africa, Swaziland, Tanzania, Togo, Uganda, Zambia, and Zimbabwe. All countries within the sub-Saharan Africa Region, with the exception of South Africa, may be part of a multi-country application (2 to 3 countries). For unlisted countries, applications may be considered on a case-by-case basis, but for dissertation research only.
Requirements: Applicants must: be U.S. citizens at the time of application (permanent residents are not eligible); be in good health (grantees will be required to submit a satisfactory Medical Certificate of Health from a physician); have sufficient proficiency in the written and spoken language of the host country to communicate with the people and to carry out the proposed study; and hold a B.A. degree or the equivalent before the start of the grant (applicants who have not earned a B.A. degree or the equivalent, but who have extensive professional study and/or experience in fields in which they wish to pursue a project, may be considered). In the creative and performing arts area, four years of professional study and/or experience meets the basic eligibility requirement. All candidates for Africa are required to obtain their own affiliations, generally established with an educational and/or research institution in the host country or countries.
Restrictions: Grants are not available to: Burundi, Central African Republic, Guinea-Bissau, Liberia, Somalia, or Sudan. The following persons are ineligible: anyone who has already held a U.S. Department of State-funded Fulbright student grant of any type; anyone who has previously received a Department of Education-funded Doctoral Dissertation Research Abroad (Fulbright Hays grant); employees of the U.S. Department of State, and their immediate families, for a period ending one year following termination of such employment; employees of private and public agencies (excluding educational institutions) under contract to the U.S. Department of State to perform administrative or screening services on behalf of the U.S. Department of State's exchange program, for a period ending one year following the termination of their services for the U.S. Department of State provided such employees have been directly engaged in performing services related to the exchange programs; applicants holding a doctoral degree at the time of application; applicants seeking enrollment in a medical degree program abroad; or applicants currently residing in the countries of Australia, Belgium/Luxembourg, Canada, Chile, Finland, Hungary, Mexico, Netherlands, New Zealand, Sweden, or Switzerland.
Geographic Focus: All States, Algeria, Angola, Benin, Botswana, Burkina Faso, Burundi, Cameroon, Cape Verde, Central African Republic, Chad, Comoros, Congo, Congo, Democratic Republic of, Cote d' Ivoire (Ivory Coast), Djibouti, Egypt, Equatorial Guinea, Eritrea, Ethiopia, Gabon, Gambia, Ghana, Guinea, Guinea-Bissau, Kenya, Lesotho, Liberia, Libya, Madagascar, Malawi, Mali, Mauritania, Mauritius, Morocco, Mozambique, Namibia, Niger, Nigeria, Rwanda, Sao Tome & Principe, Senegal, Seychelles, Sierra Leone, Somalia, South Africa, Sudan, Swaziland
Date(s) Application is Due: Aug 1
Contact: Jermaine Jones, Program Manager; (212) 984-5341; jjones@iie.org
Internet: http://us.fulbrightonline.org/program_regions_countries.php?id=1
Sponsor: Institute of International Education
1400 K Street, NW, 7th Floor
Washington, DC 20005-2403

Fulbright Traditional Scholar Program in the East Asia/Pacific Region **2082**
The traditional Fulbright Scholar Program sends 800 U.S. faculty and professionals abroad each year. Grantees lecture and conduct research in a wide variety of academic and professional fields. Distribution of awards to countries in the region will vary annually according to the caliber of the applicants. Grants are available to: Australia, Cambodia, China, Hong Kong, Indonesia, Japan, Korea, Laos, Macau, Malaysia, Mongolia, New Zealand, Philippines, Singapore, Taiwan, Thailand, and Vietnam. All countries within the East Asia/Pacific region, with the exception of China, may be part of a multi-country application (2 to 3 countries). For unlisted countries, applications may be considered on a case-by-case basis, but for dissertation research only. Language requirements vary by country, and prior knowledge of the local language may not be required, particularly where language is not commonly taught in the U.S.
Requirements: Applicants must: be U.S. citizens at the time of application (permanent residents are not eligible); be in good health (grantees will be required to submit a satisfactory Medical Certificate of Health from a physician); have sufficient proficiency in the written and spoken language of the host country to communicate with the people and to carry out the proposed study; and hold a B.A. degree or the equivalent before the start of the grant (applicants who have not earned a B.A. degree or the equivalent, but who have extensive professional study and/or experience in fields in which they wish to pursue a project, may be considered). In the creative and performing arts area, four years of professional study and/or experience meets the basic eligibility requirement. All candidates for the East Asia/Pacific region are required to obtain their own affiliations, generally established with an educational and/or research institution in the host country or countries. Candidates are responsible for securing their own research clearance as required by the host country.
Restrictions: Grants are not available to: Brunei, the Cook Islands, East Timor, Fiji, Myanmar, the Pacific Island Nations, Papua New Guinea, or Western Samoa. The following persons are ineligible: anyone who has already held a U.S. Department of State-funded Fulbright student grant of any type; anyone who has previously received a Department of Education-funded Doctoral Dissertation Research Abroad (Fulbright Hays grant); employees of the U.S. Department of State, and their immediate families,

for a period ending one year following termination of such employment; employees of private and public agencies (excluding educational institutions) under contract to the U.S. Department of State to perform administrative or screening services on behalf of the U.S. Department of State's exchange program, for a period ending one year following the termination of their services for the U.S. Department of State provided such employees have been directly engaged in performing services related to the exchange programs; applicants holding a doctoral degree at the time of application; applicants seeking enrollment in a medical degree program abroad; or applicants currently residing in the countries of Australia, Belgium/Luxembourg, Canada, Chile, Finland, Hungary, Mexico, Netherlands, New Zealand, Sweden, or Switzerland.
Geographic Focus: All States, Abkhazia, Afghanistan, Armenia, Australia, Azerbaijan, Bahrain, Bangladesh, Bhutan, British Indian Ocean Territory, Brunei, Burma (Myanmar), Cambodia, China, Christmas Island, Cocos, Cyprus, Georgia, Hong Kong, India, Indonesia, Iran, Iraq, Israel, Japan, Jordan, Kazakhstan, Kuwait, Kyrgyzstan, Laos, Lebanon, Macau, Malaysia, Maldives, Mongolia, Nagorno-Karabakh, Nepal, North Korea, Northern Cyprus, Oman, Pakistan, Palestinian Authority, Philippines, Qatar, Russia, Saudi Arabia, Singapore, South Korea, South Ossetia, Sri Lanka, Syria, Taiwan, Tajikistan, Thailand, Timor-Lester, Turkey, Turkmenistan, United Arab Emirates, Uzbekistan, Vietnam, Yemen
Date(s) Application is Due: Aug 1
Contact: Jonathan Akeley, Program Manager; (212) 984-5487; jakeley@iie.org
Internet: http://us.fulbrightonline.org/program_regions_countries.php?id=2
Sponsor: Institute of International Education
1400 K Street, NW, 7th Floor
Washington, DC 20005-2403

Fulbright Traditional Scholar in the Near East and North Africa Region 2083
The traditional Fulbright Scholar Program sends 800 U.S. faculty and professionals abroad each year. Grantees lecture and conduct research in a wide variety of academic and professional fields. Selection for countries in the Near East and North Africa region will be made based on the quality of the applications, rather than per-country quotas. Distribution of awards to countries in the region will vary annually according to the caliber of the applicants. Grants are available to: Bahrain, Egypt, India, Jordan, Kuwait, Morocco, Oman, Qatar, Syria, Tunisia, United Arab Emirates, and Yemen. Multi-country applications (2 to 3 countries) are available, except in Egypt, India, Jordan, and Morocco. For unlisted countries, applications may be considered on a case-by-case basis.
Requirements: Applicants must: be U.S. citizens at the time of application (permanent residents are not eligible); be in good health (grantees will be required to submit a satisfactory Medical Certificate of Health from a physician); have sufficient proficiency in the written and spoken language of the host country to communicate with the people and to carry out the proposed study; and hold a B.A. degree or the equivalent before the start of the grant (applicants who have not earned a B.A. degree or the equivalent, but who have extensive professional study and/or experience in fields in which they wish to pursue a project, may be considered). In the creative and performing arts area, four years of professional study and/or experience meets the basic eligibility requirement. Candidates are responsible for securing research clearance, as required. In countries with Fulbright Commissions, assistance may be provided.
Restrictions: Grants are not available to: Algeria, Iran, Iraq, Lebanon, Saudi Arabia, the West Bank, or Gaza. The following persons are ineligible: anyone who has already held a U.S. Department of State-funded Fulbright student grant of any type; anyone who has previously received a Department of Education-funded Doctoral Dissertation Research Abroad (Fulbright Hays grant); employees of the U.S. Department of State, and their immediate families, for a period ending one year following termination of such employment; employees of private and public agencies (excluding educational institutions) under contract to the U.S. Department of State to perform administrative or screening services on behalf of the U.S. Department of State's exchange program, for a period ending one year following the termination of their services for the U.S. Department of State provided such employees have been directly engaged in performing services related to the exchange programs; applicants holding a doctoral degree at the time of application; applicants seeking enrollment in a medical degree program abroad; or applicants currently residing in the countries of Australia, Belgium/Luxembourg, Canada, Chile, Finland, Hungary, Mexico, Netherlands, New Zealand, Sweden, or Switzerland.
Geographic Focus: All States, Algeria, Angola, Benin, Botswana, Burkina Faso, Burundi, Cameroon, Cape Verde, Central African Republic, Chad, Comoros, Congo, Congo, Democratic Republic of, Cote d' Ivoire (Ivory Coast), Djibouti, Egypt, Equatorial Guinea, Eritrea, Ethiopia, Gabon, Gambia, Ghana, Guinea, Guinea-Bissau, Israel, Kenya, Lesotho, Liberia, Libya, Madagascar, Malawi, Mali, Mauritania, Mauritius, Morocco, Mozambique, Namibia, Niger, Nigeria, Rwanda, Sao Tome & Principe, Senegal, Seychelles, Sierra Leone, Somalia, South Africa, Sudan, Swaziland, Uzbekistan
Date(s) Application is Due: Aug 1
Contact: Jermaine Jones, Program Manager; (212) 984-5341; jjones@iie.org
Internet: http://us.fulbrightonline.org/program_regions_countries.php?id=4
Sponsor: Institute of International Education
1400 K Street, NW, 7th Floor
Washington, DC 20005-2403

Fulbright Traditional Scholar Program in the South and Central Asia Region 2084
The traditional Fulbright Scholar Program sends 800 U.S. faculty and professionals abroad each year. Grantees lecture and conduct research in a wide variety of academic and professional fields. Selection for countries in the South and Central Asia region will be made based on the quality of the applications, rather than per-country quotas. Distribution of awards to countries in the region will vary annually according to the caliber of the applicants. Grants are available to: Bangladesh, India, Kazakhstan, Kyrgyz Republic, Nepal, Sri Lanka, Tajikistan, and Uzbekistan. Multi-country applications (2 to 3 countries) are not available. Language requirements vary by country. Applicants to India who are recommended for final review will need to submit research visa applications in January or February.
Requirements: Applicants must: be U.S. citizens at the time of application (permanent residents are not eligible); be in good health (grantees will be required to submit a satisfactory Medical Certificate of Health from a physician); have sufficient proficiency in the written and spoken language of the host country to communicate with the people and to carry out the proposed study; and hold a B.A. degree or the equivalent before the start of the grant (applicants who have not earned a B.A. degree or the equivalent, but who have extensive professional study and/or experience in fields in which they wish to pursue a project, may be considered). In the creative and performing arts area, four years of professional study and/or experience meets the basic eligibility requirement. Candidates are responsible for securing research clearance, as required. In countries with Fulbright Commissions, assistance may be provided.
Restrictions: Grants are not available to: Afghanistan, Bhutan, Pakistan, Republic of Maldives, or Turkmenistan. The following persons are ineligible: anyone who has already held a U.S. Department of State-funded Fulbright student grant of any type; anyone who has previously received a Department of Education-funded Doctoral Dissertation Research Abroad (Fulbright Hays grant); employees of the U.S. Department of State, and their immediate families, for a period ending one year following termination of such employment; employees of private and public agencies (excluding educational institutions) under contract to the U.S. Department of State to perform administrative or screening services on behalf of the U.S. Department of State's exchange program, for a period ending one year following the termination of their services for the U.S. Department of State provided such employees have been directly engaged in performing services related to the exchange programs; applicants holding a doctoral degree at the time of application; applicants seeking enrollment in a medical degree program abroad; or applicants currently residing in the countries of Australia, Belgium/Luxembourg, Canada, Chile, Finland, Hungary, Mexico, Netherlands, New Zealand, Sweden, or Switzerland.
Geographic Focus: All States, Kazakhstan, Uzbekistan
Date(s) Application is Due: Aug 1
Contact: Jonathan Akeley, Program Manager; (212) 984-5487; jakeley@iie.org
Internet: http://us.fulbrightonline.org/program_regions_countries.html?id=6
Sponsor: Institute of International Education
1400 K Street, NW, 7th Floor
Washington, DC 20005-2403

Fulbright Traditional Scholar Program in the Western Hemisphere 2085
The traditional Fulbright Scholar Program sends 800 U.S. faculty and professionals abroad each year. Grantees lecture and conduct research in a wide variety of academic and professional fields. Selection for countries in Central America and the Caribbean will be made based on the quality of the applications, rather than per-country quotas. Distribution of awards to countries in the region will vary annually according to the caliber of the applicants. Grants are available to: Argentina, Barbados and the Eastern Caribbean, Bolivia, Brazil, Canada, Chile, Colombia, Costa Rica, Dominican Republic, Ecuador, El Salvador, Guatemala, Honduras, Jamaica, Mexico, Nicaragua, Panama, Paraguay, Peru, Trinidad and Tobago, Uruguay, and Venezuela. English Teaching Assistantships are available in Argentina, Brazil, Chile, Uruguay and Venezuela. Language proficiency may be preferred or required. Applicants for English Teaching Assistantships can apply to only one country. Multi-country applications (2 to 3 countries) are available in all other disciplines throughout the Western Hemisphere. For unlisted countries, applications may be considered on a case-by-case basis, but for dissertation research only.
Requirements: Applicants must: be U.S. citizens at the time of application (permanent residents are not eligible); be in good health (grantees will be required to submit a satisfactory Medical Certificate of Health from a physician); have sufficient proficiency in the written and spoken language of the host country to communicate with the people and to carry out the proposed study; and hold a B.A. degree or the equivalent before the start of the grant (applicants who have not earned a B.A. degree or the equivalent, but who have extensive professional study and/or experience in fields in which they wish to pursue a project, may be considered). In the creative and performing arts area, four years of professional study and/or experience meets the basic eligibility requirement. Candidates are responsible for securing research clearance, as required. In countries with Fulbright Commissions, assistance may be provided.
Restrictions: Grants are not available to: Bahamas, Belize, Cuba, French Guiana, Guyana, Haiti, Martinique, or Suriname. The following persons are ineligible: anyone who has already held a U.S. Department of State-funded Fulbright student grant of any type; anyone who has previously received a Department of Education-funded Doctoral Dissertation Research Abroad (Fulbright Hays grant); employees of the U.S. Department of State, and their immediate families, for a period ending one year following termination of such employment; employees of private and public agencies (excluding educational institutions) under contract to the U.S. Department of State to perform administrative or screening services on behalf of the U.S. Department of State's exchange program, for a period ending one year following the termination of their services for the U.S. Department of State provided such employees have been directly engaged in performing services related to the exchange programs; applicants holding a doctoral degree at the time of application; applicants seeking enrollment in a medical degree program abroad; or applicants currently residing in the countries of Australia, Belgium/Luxembourg, Canada, Chile, Finland, Hungary, Mexico, Netherlands, New Zealand, Sweden, or Switzerland.
Geographic Focus: All States, Antigua & Barbuda, Argentina, Bahamas, Barbados, Belize, Bolivia, Brazil, Canada, Chile, Colombia, Costa Rica, Cuba, Dominica, Dominican Republic, Ecuador, El Salvador, Grenada, Guatemala, Guyana, Haiti, Honduras, Jamaica, Mexico, Nicaragua, Paraguay, Peru
Date(s) Application is Due: Aug 1

Contact: Jody Dudderar, Program Manager in South America, Mexico and Canada; (212) 984-5565; fax (212) 984-5325; jdudderar@iie.org
Internet: http://us.fulbrightonline.org/program_regions_countries.php?id=5
Sponsor: Institute of International Education
1400 K Street, NW, 7th Floor
Washington, DC 20005-2403

Fulbright Travel Grants 2086

Travel grants are available only to Germany, Hungary, or Italy. They are available to supplement an award from a non-IIE source that does not provide funds for travel or to supplement a student's own funds for study. Travel grants provide round-trip transportation to the country where the student will pursue studies for an academic year, supplemental health and accident insurance, and the cost of all orientation courses abroad, if applicable.
Requirements: Applicants must be U.S. citizens with at least the bachelor's degree and capable of meeting the eligibility criteria for Fulbright grants.
Geographic Focus: All States, Germany, Hungary
Contact: Walter Jackson, Program Manager; (212) 984-5327; wjackson@iie.org
Internet: https://us.fulbrightonline.org/overview_typesgrants.html
Sponsor: Institute of International Education
1400 K Street, NW, 7th Floor
Washington, DC 20005-2403

Fulbright Visiting Specialists Program: Direct Access to the Muslim World 2087

The program aims to promote American understanding of Islamic civilization and the history, politics, and culture of today's Muslim world. The program further aims to strengthen resources for enriching understanding of relevant global issues. The program offers a unique opportunity for U.S. colleges and universities to host specialists from countries with significant Muslim populations for short-term programs of intensive lecturing, public outreach, and consultation. The Fulbright Visiting Specialists will lecture on and off campus in a variety of humanities and social science disciplines, including the arts, with a focus on Islamic society or civilization. The Program encourages inter-religious dialog through on-campus and off-campus activities. The Visiting Specialists will also work together with the U.S. host institutions on projects with lasting benefits such as curriculum consultation, faculty development, and institutional linkage-building.
Requirements: Institutions with few or no programs/courses related to Islam or the Muslim world are encouraged to apply. Institutions that have well-established Islam-related or area studies programs should clearly indicate that their proposed program for the Fulbright Visiting Specialists Program is in a field/subject that is not currently supported at the institution. Community colleges and minority-serving institutions, such as Historically Black Colleges and Universities, Hispanic Serving Institutions, and Tribal Colleges and Universities, are encouraged to apply and will receive preference for an award. At the time of program participation, the Fulbright Visiting Specialist should be a resident of his/her home country.
Restrictions: The program does not support the teaching of languages or research projects. Institutions that have already hosted a Fulbright Visiting Specialist are no longer eligible. While more than one application may be submitted by the same institution, only one award to the same institution will be offered. Persons holding U.S. citizenship or permanent residency in the U.S. are not eligible.
Geographic Focus: All States
Date(s) Application is Due: Apr 1
Contact: Alma Ford; (202) 686-6252; fax (202) 362-3442; aford@cies.iie.org
Internet: http://www.cies.org/Visiting_Specialists/
Sponsor: Institute of Int'l Education, Council for International Exchange of Scholars
3007 Tilden Street NW, Suite 5L
Washington, DC 20008-3009

Fund for Folk Culture Artists Support Grants 2088

The Artists Support Program, a program of the Fund for Folk Culture (FFC) underwritten through generous support from the Ford Foundation, The William and Flora Hewlett Foundation and The San Francisco Foundation, offers individual folk and traditional artists in the United States the opportunity to pursue activities that will help them to grow artistically and professionally, to better connect with their cultural communities or to develop new audiences for their work. The program offers grants of up to $5,000 for specific projects or for general expenses necessary to an artist's work.
Requirements: All folk and traditional artists who are at least 18 years of age, and either citizens or permanent residents of the United States. Examples of eligible proposals include (but are not limited to) purchase of materials, supplies or equipment; studio, rehearsal or performance space rental; development of portfolio or marketing materials; CD production; new work development; collaborations with other artists or cultural organizations; international and out-of-state travel for participation in artist gatherings and retreats or conferences; work on collaborative projects; research related to artistic growth; and other creative or professional development activities. Individual folk and traditional artists will be invited to apply through a nomination process. Do not submit a proposal unless you have been invited to do so. Questions about the nomination process or the Artists Support Program can be directed to Betsy Peterson.
Restrictions: In-state apprenticeship activity (frequently supported through state arts agencies or other state folk arts organizations) is not eligible for support.
Geographic Focus: All States
Contact: Elizabeth E. Peterson; (512) 472-8757, ext. 2; bpeterson@folkculture.org
Internet: http://www.folkculture.org/about/asp.html
Sponsor: Fund for Folk Culture
P.O. Box 303366
Austin, TX 78703-0057

Fund for Scientific Research—Flanders Grants 2089

The foundation awards pre- and postdoctoral fellowships and grants for research in Flemish universities and research institutions in all fields of science, including medicine, technology, environmental studies, social science, law, and the humanities. National and international research is supported.
Geographic Focus: All States
Contact: Secretary General; 32 (2) 512-91-10; fax 32 (2) 512-58-90; post@fwo.be
Internet: http://sun.fwo.be/page1en.php?lang=en
Sponsor: Fund for Scientific Research-Flanders
Egmontstraat 5
Brussels, 1000 Belgium

Fund for the Advancement of the Discipline 2090

The American Sociological Association invites submissions for the Fund for the Advancement of the Discipline (FAD) awards. Supported by the American Sociological Association through a matching grant from the National Science Foundation, the goal of this award is to nurture the development of scientific knowledge by funding small, ground breaking research initiatives and other important scientific research activities. FAD awards provide scholars with venture capital for innovative research that has the potential for challenging the discipline, stimulating new lines of research, and creating new networks of scientific collaboration. The award is intended to provide opportunities for substantive and methodological breakthroughs, broaden the dissemination of scientific knowledge, and provide leverage for acquisition of additional research funds.
Requirements: Proposals are reviewed for scientific merit and the importance of the proposed research project. Within this context, specific evaluation criteria include the following elements: innovativeness and promise of the research idea; originality and significance of research goals; the potential of the study as a building block in the development of future research; appropriateness and significance of the research hypothesis; feasibility and adequacy of project design; plans for analysis and evaluation of data; plans for dissemination of results; appropriateness of requested budget. The awards are limited to individuals with PhD degrees or the equivalent. Preference is given to applicants who have not previously received a FAD award.
Geographic Focus: All States
Date(s) Application is Due: Jun 15; Dec 15
Amount of Grant: Up to 7,000 USD
Contact: Roberta Spalter-Roth; (202) 383-9005, ext. 317; spalter-roth@asanet.org
Internet: http://asanet.org/page.ww?section=Funding&name=Fund+for+the+Advancement+of+the+Discipline
Sponsor: American Sociological Association
1430 K Street NW, Suite 600
Washington, DC 20005

Furthermore Grants in Publishing 2091

The program is concerned with nonfiction book publishing about the city; natural and historic resources; art, architecture, and design; cultural history; and civil liberties and other public issues of the day. Grants apply to writing, research, editing, design, indexing, photography, illustration, and printing and binding. Work is sought that appeals to an informed general audience; gives evidence of high standards in editing, design, and production; promises a reasonable shelf life; might not otherwise achieve top quality or even come into being; and represents a contribution without which we would be the poorer. Book proposals to which a university press or trade publisher is already committed and for which there is a feasible distribution plan are usually preferred. In geographical reach, the program is drawn, but in no way limited, to New York City and to New York State and its Hudson Valley.
Requirements: 501(c)3 tax-exempt organizations are eligible to apply. Trade publishers and public agencies may apply for Furthermore grants in partnership with an eligible nonprofit project sponsor. Grants from $500 to roughly $15,000 are awarded in spring and fall with March 1 and September 1 deadlines.
Restrictions: Applications from individuals cannot be accepted
Geographic Focus: All States
Date(s) Application is Due: Mar 1; Sep 1
Amount of Grant: 500-15,000 USD
Contact: Joan K. Davidson; (518) 828-8900; fax (518) 828-8901; info@furthermore.org
Internet: http://www.furthermore.org
Sponsor: J.M. Kaplan Fund
261 Madison Avenue, 19th Floor
New York, NY 10016

G. Harold and Leila Y. Mathers Charitable Foundation Grants 2092

The foundation awards grants to support research in the basic life sciences. Requests for general operating support also will be considered. There are no application forms or deadlines. The board meets two or three times each year.
Requirements: U.S. research organizations are eligible.
Geographic Focus: All States
Amount of Grant: 10,000-200,000 USD
Contact: James Handelman; (914) 242-0465; bcheikin@mathersfoundation.org
Internet: http://www.mathersfoundation.org
Sponsor: G. Harold and Leila Y. Mathers Charitable Foundation
118 N Bedford Road, Suite 203
Mount Kisco, NY 10549-2555

G. Unger Vetlesen Foundation Grants 2093

The foundation's mission is to aid and contribute to religious, charitable, scientific, literary, and educational purposes throughout the world. Grants have been concentrated in the fields of oceanography, climate studies, earth sciences, cultural programs emphasizing Norwegian-American relations, environmental programs, libraries, international affairs, and public policy. A letter of inquiry, 3 pages or less, should be submitted at any time during the year. Proposals are to be submitted only upon invitation. Unsolicited proposals will not be reviewed.
Requirements: 501(c)3 nonprofit organizations are eligible.
Restrictions: The Foundation does not make grants to individuals.
Geographic Focus: All States
Date(s) Application is Due: Apr 30; Oct 31
Contact: George Rowe Jr; (212) 586-0700; fax (212) 245-1863; info@monellvetlesen.org
Internet: http://www.monellvetlesen.org/vetlesen/appguide.htm
Sponsor: G. Unger Vetlesen Foundation
Fulton, Rowe & Hart, One Rockefeller Plaza, Suite 301
New York, NY 10020-2002

Gabor A. Somorjai Award for Creative Research in Catalysis 2094

The Award, supported by the Gabor A. Somorjai Endowment Fund, is intended to recognize outstanding theoretical, experimental, or developmental research resulting in the advancement of understanding or application of catalysis. U.S. residents will be given preference. Outstanding contributions to catalysis in any discipline will be considered. The award consists of $5,000 and a certificate.
Requirements: Any individual, except a member of the award committee, may submit one nomination or seconding letter for the award in any given year. Nominating documents consist of a letter of not more than 1000 words containing an evaluation of the nominee's accomplishments and a specific identification of the work to be recognized, a biographical sketch including date of birth, and a list of publications and patents authored by the nominee.
Restrictions: Self-nominations are not accepted.
Geographic Focus: All States
Date(s) Application is Due: Nov 1
Contact: Felicia Dixon, Awards Administrator; (800) 227-5558 or (202) 872-4408; fax (202) 776-8008; f_dixon@acs.org or awards@acs.org
Internet: http://portal.acs.org/portal/acs/corg/content?_nfpb=true&_pageLabel=PP_ARTICLEMAIN&node_id=1319&content_id=CTP_004525&use_sec=true&sec_url_var=region1
Sponsor: American Chemical Society
1155 Sixteenth Street, NW
Washington, DC 20036-4801

Gaius Charles Bolin Fellowships for Minority Graduate Students 2095

Gaius Charles Bolin was the first black graduate of Williams. Fellowships were founded in 1985 on the centennial of his admission to the College. The Gaius Charles Bolin Fellowships are designed to promote diversity on college faculties by encouraging students from underrepresented groups to complete a terminal graduate degree and to pursue careers in college teaching. Bolin Fellowships are two-year residencies, and up to three scholars or artists are appointed each year. Fellows devote the bulk of the first year to the completion of dissertation work, or in the case of Master of Fine Arts (MFA) applicants, building their professional portfolios, while also teaching one course as a faculty member in one of the College's academic departments or programs. The second year of residency (ideally with degree in hand) is spent on academic career development while again teaching just one course. The annual stipend is $35,000 with a $4,000 allowance for research-related expenses.
Requirements: Applicants must be United States citizens or permanent residents who intend to pursue a professorial career in the United States. Ph.D. candidates must have completed all doctoral work except the dissertation by the end of the current academic year. MFA candidates must be recent recipients of the degree. Fellowships are awarded to applicants from underrepresented groups, including ethnic minorities, those who are first-generation college graduates, women in predominantly male fields, or disabled scholars.
Geographic Focus: All States
Date(s) Application is Due: Nov 15
Amount of Grant: 35,000-35,000 USD
Contact: Peter T. Murphy, Dean of Faculty; (413) 597-3131; gburda@williams.edu
Internet: http://dean-faculty.williams.edu/graduate-fellowships/gaius-charles-bolin-dissertation-and-post-mfa-fellowships/
Sponsor: Williams College
880 Main Street
Williamstown, MA 01267

Gas Technology Institute Basic Research Program 2096

The institute supports a broad program of basic research relevant to the U.S. natural gas industry. Research topics include solid oxide fuel cell stacks and new drilling concepts. Contracts may run as long as three years. Research topics may change in content or timing during the year; changes will be posted on the Internet. Only written, fax, and Internet requests for individual proposal requests (RFPs) will be accepted.
Requirements: Potential applicants must contact the office for specific individual requests for proposals.
Geographic Focus: All States
Contact: GTI Managers; (847) 768-0500; fax (847) 768-0501; vp_rd@gastechnology.org
Internet: http://www.gastechnology.org/webroot/app/xn/xd.aspx?it=enweb&xd=gtihome.xml
Sponsor: Gas Technology Institute
1700 S Mount Prospect Road
Des Plaines, IL 60018-1804

Gates Family Foundation Fellowships 2097

The John F. Kennedy School of Government at Harvard University and the Gates Family Foundation have partnered since 1980, each year sending up to five individuals to participate in the Senior Executives in State and Local Government Program. The three-week program operates as an interactive classroom, where case studies of real issues are analyzed to find solutions to complex problems, and faculty and participants work together and learn from each other. Gates Fellowships will be awarded to residents of Colorado by an admissions committee at the Kennedy School based upon the recommendations of a nominating panel of distinguished Colorado residents. The program fee is $10,900 which includes tuition, housing, curricular materials, and most meals. The fellowship is $10,000 and covers most program costs. Participants or their employers must fund the balance.
Geographic Focus: Colorado
Date(s) Application is Due: Jan 15; Apr 1; Jul 15; Oct 1
Contact: Karen White Mather; (303) 722-1881; info@gatesfamilyfoundation.org
Internet: http://www.gatesfamilyfoundation.org/www/gates.php?section=gates_fellowship&p=
Sponsor: Gates Family Foundation
3575 Cherry Creek North Drive, Suite 100
Denver, CO 80209-3600

GCA Interchange Fellowship 2098

The program provides a reciprocal exchange of British and American students in horticulture, landscape architecture, and related fields. The intent of the program is to foster cultural understanding, as well as to promote horticultural studies and the exchange of information in this field. GCA offers an academic year of graduate work for a British student in America. The Martin McLaren Scholarship provides a noncredit work/study program for an American recipient at universities such as Reading, Wye College of University of London, Gloucester College, and the Royal Botanic Gardens, Kew; Royal Horticultural Society Gardens, Wisley; and Royal Botanic Garden, Edinburgh. The program is jointly sponsored with the Institute of Horticulture in the British Isles.
Requirements: Recent college graduates who have earned the bachelor's degree in related fields are eligible. Because of a 26-year-old age limit on student travel vouchers in Great Britain and Europe, it is strongly advised that the applicant be 26 or younger.
Geographic Focus: All States
Date(s) Application is Due: Nov 15
Contact: Connie Sutton; (212) 753-8287; fax (212) 753-0134; csutton@gcamerica.org
Internet: http://www.gcamerica.org/scholarship/gcainterchange.html
Sponsor: Garden Club of America
14 E 60th Street
New York, NY 10022-1002

GCA Katharine M. Grosscup Scholarships 2099

Scholarships provide assistance to college juniors, seniors, and graduate students pursuing the study of horticulture, agriculture, and related subjects. Preference is given to students from Ohio, Pennsylvania, West Virginia, Michigan, Kentucky, and Indiana.
Geographic Focus: All States
Date(s) Application is Due: Feb 1
Contact: Nancy Stevenson, Grosscup Scholarship Committee; (216) 721-2056
Internet: http://www.gcamerica.org/scholarship/grosscup.html
Sponsor: Garden Club of America
11030 East Boulevard
Cleveland, OH 44106

GCA Summer Environmental Studies Scholarships 2100

Each year the GCA Awards for Summer Environmental Studies provide financial aid toward summer studies doing field work or research in the environmental field. The awards offer students who have demonstrated a keen interest in the betterment of the environment an opportunity for further study in the field of ecology. With these scholarships, young men and women can pursue summer programs beyond the regular course of study to gain additional knowledge and experience. Work may award academic credit but should be in addition to required courses. College students may apply for these awards to pursue study following their freshman, sophomore, or junior year. Guidelines and application are available online.
Geographic Focus: All States
Date(s) Application is Due: Feb 10
Contact: Connie Sutton; (212) 753-8287; fax (212) 753-0134; csutton@gcamerica.org
Internet: http://www.gcamerica.org/scholarship/summeraward_org.html
Sponsor: Garden Club of America
14 E 60th Street
New York, NY 10022-1002

GCA Tropical Botany Research Awards 2101

The program was established to promote the preservation of tropical forests by enlarging the body of botanists with field experience. The awards enable recipients to pursue independent field research in the tropics. Two awards will be offered annually.
Requirements: PhD candidates are eligible to apply.
Geographic Focus: All States
Date(s) Application is Due: Dec 31
Contact: Judith Mashinya; (202) 861-8316; judith.mashinya@wwfus.org
Internet: http://www.gcamerica.org/scholarship/tropical.html
Sponsor: Garden Club of America
1250 24th Street NW
Washington, DC 20037

GCSAA Essay Contest 2102

The contest is open to undergraduate and graduate students pursuing degrees in turfgrass science, agronomy, or any field related to golf course management. Essays should be seven to 12 pages and should focus on the relationship between golf courses and the environment.
Requirements: Undergraduate and graduate students pursuing degrees in turfgrass science, agronomy, or any field related to golf course management are eligible to apply. Applicants must be a member of GCSAA.
Geographic Focus: All States
Date(s) Application is Due: Mar 31
Amount of Grant: 1,000-2,000 USD
Contact: Scholarship Coordinator; (800) 472-7878, ext. 4424; ahoward@gcsaa.org
Internet: http://www.gcsaa.org/Students/scholarships/essay/default.asp
Sponsor: Golf Course Superintendents Association of America Foundation
1421 Research Park Drive
Lawrence, KS 66049

GCSAA Watson Fellowships 2103

These awards are available to candidates for master's and doctoral degrees in fields related to golf course management. The goal of this program is to identify tomorrow's leading teachers and researchers.
Requirements: To be eligible applicants must be in the second year of a recognized graduate program in turfgrass science or a related field. Typically, this includes candidates for masters degrees or doctorates in appropriate fields of study. Applicants must also be planning to pursue a career in research, instruction, or extension in a university setting.
Geographic Focus: All States
Date(s) Application is Due: Oct 1
Amount of Grant: 5,000 USD
Contact: Amanda Howard, Scholarship Coordinator; (800) 472-7878, ext. 4424 or (785) 832-3678; ahoward@gcsaa.org
Internet: http://www.gcsaa.org/students/scholarships/default.asp
Sponsor: Golf Course Superintendents Association of America Foundation
1421 Research Park Drive
Lawrence, KS 66049

GEM MS Engineering Fellowships 2104

The objective of the master's program is to increase the pool of minority MS engineering graduates. Each participant is provided with a summer internship experience for a GEM corporate member and a portable academic fellowship of tuition, fees, and stipend that must be used at a GEM university member. Fellows are supported for three semesters or four quarters.
Requirements: Eligibility includes U.S. citizenship; ethnic identification of Native American, African American, Mexican American, or Puerto Rican; academic classification of junior, senior, or baccalaureate degree recipient in an accredited engineering discipline; and an undergraduate record that indicates the ability to successfully pursue graduate studies in engineering.
Geographic Focus: All States
Date(s) Application is Due: Nov 1
Amount of Grant: 10,000 USD
Contact: Fellowship Office; (219) 631-7771; fax (219) 287-1486; gem@nd.edu
Internet: http://was.nd.edu/gem/gemwebapp/public/gem_01_100.htm
Sponsor: National GEM Center
P.O. Box 537
Notre Dame, IN 46556

GEM PhD Engineering Fellowships 2105

The objective of this program is to offer doctoral fellowships to underrepresented minority students. Fellowships may be used at any participating GEM member university. Tuition, fees, and a yearly stipend are provided. Fellows may be required to accept a research or teaching assistantship. A paid summer internship may be required.
Requirements: Eligibility includes U.S. citizenship and ethnic identification of Native American, African American, Mexican American, or Puerto Rican. Applicants must have or be in the process of attaining a master's degree in engineering and have an academic record that indicates the ability to successfully pursue doctoral studies in engineering.
Geographic Focus: All States
Date(s) Application is Due: Nov 1
Amount of Grant: 19,900 USD
Contact: Fellowship Office; (219) 631-7771; fax (219) 287-1486; gem@nd.edu
Internet: https://ws4.nd.edu/gem/gemwebapp/public/gem_01_100.htm
Sponsor: National GEM Center
P.O. Box 537
Notre Dame, IN 46556

GEM PhD Science Fellowships 2106

The goal of this program is to increase the number of minority students pursuing doctoral degrees in the natural, physical, and life sciences (chemistry, physics, mathematics, and computer science, as well as the earth, biological, and pharmaceutical sciences). Fellowships are portable to any participating GEM member university, and include a stipend, tuition, and fees. The award is renewable through the university for up to four additional years. Fellows may be required to accept a research or teaching assistantship. A paid summer internship is required prior to entering a doctoral program.
Requirements: Eligibility includes U.S. citizenship; ethnic identification of Native American, African American, Hispanic, or Puerto Rican; academic classification of junior, senior, or baccalaureate degree recipient in an accredited science discipline; and an undergraduate record that indicates the ability to successfully pursue doctoral studies in the natural sciences.
Geographic Focus: All States
Date(s) Application is Due: Nov 1
Amount of Grant: 19,900 USD
Contact: Fellowship Office; (219) 631-7771; fax (219) 287-1486; gem@nd.edu
Internet: http://was.nd.edu/gem/gemwebapp/public/gem_01_100.htm#03
Sponsor: National GEM Center
P.O. Box 537
Notre Dame, IN 46556

General Motors Foundation Grants Support Program 2107

With a strong commitment to diversity in all areas, the targeted areas of focus for the Foundation are: education; health and human services; civic and community; public policy; arts and culture; and environment and energy. Primary consideration is given to requests that meet the following criteria: exhibit a clear purpose and defined need in one of the foundation's areas of focus; recognize innovative approaches in addressing the defined need; demonstrate an efficient organization and detail the organization's ability to follow through on the proposal; and, explain clearly the benefits to the foundation and the plant city communities. Paper applications are no longer accepted. Completion of an online eligibility quiz is the first step in the application process.
Requirements: Nonprofit, tax-exempt organizations and institutions are eligible to apply. Applications must be made online.
Restrictions: The Foundation not not support organizations that discriminate on the basis of race, religion, creed, gender, age, veteran status, physical challenge or national origin. Contributions are generally not provided for: individuals; religious organizations; political parties or candidates; U.S. hospitals and health care institutions (general operating support); capital campaigns; endowment funds; conferences, workshops or seminars not directly related to GM's business interests.
Geographic Focus: All States
Contact: Grant Coordinator; (313) 556-5000
Internet: http://www.gm.com/company/gmability/community/guidelines/index.html
Sponsor: General Motors Foundation
300 Renaissance Center, P.O. Box 300
Detroit, MI 48265-3000

General Service Foundation Human Rights and Economic Justice Grants 2108

The goal of the General Service Foundation's Human Rights and Economic Justice program is to support efforts that protect, promote and create good jobs with living wages for workers, including low-wage workers in the United States and Mexico. The Foundation seeks to: strengthen worker voices; promote public policies that protect labor rights; and democratize corporate power and promote corporate accountability. In assessing potential grantees, the Foundation will apply the following criteria: supporting organizations that work to address needs identified by the underrepresented and low-income communities that area directly impacted, connecting at the local and national level; and concentrating on organizations that take risks and try new ideas to enhance their skills, and reflect the diversity of their constituency. Samples of previously funded projects are posted on the Foundation website.
Requirements: All applicants, regardless of their prior grant history with the Foundation, start their application process by submitting a letter of inquiry in the spring and fall. Organizations should determine if their project is an appropriate fit for the Foundation, then send a letter of inquiry via the Foundation's online application. Letters of inquiry should include: a political and situational analysis, statement of the issues to be addressed under the proposed project, the history and goals of the organization, and an explanation of why the organization or coalition is the entity that is most likely to achieve success; a brief summary of the project, short and long term goals and anticipated outcomes; the approximate starting date and duration of the proposed activities; the total amount of funding needed, the amount requested from the Foundation, a budget, and information about other sources of support; and a copy of IRS tax exempt 501(c)3 letter. If the project is judged appropriate, the organization will be offered an application to submit to the Foundation.
Restrictions: The Foundation does not fund: organizations based outside the United States or Mexico; projects without significant promise of impact beyond a local or state level; research and publications not directly linked to policy outcomes; direct service delivery; and development or relief projects.
Geographic Focus: All States, Mexico
Date(s) Application is Due: Feb 1; Sep 1
Amount of Grant: 5,000-35,000 USD
Contact: Holly Bartling, Program Contact-Mexico grants; (202) 232-1005 or (970) 920-6834, ext 4; fax (970) 920-4578; holly@generalservice.org
Internet: http://www.generalservice.org/International%20Peace.htm
Sponsor: General Service Foundation
557 North Mill Street, Suite 201
Aspen, CO 81611

Genuardi Family Foundation Grants 2109

The Foundation supports direct providers of services in the areas of education, health, human services, and culture. The Foundation gives preference to projects which receive broad-based community support and provides for reasonable costs associated with conducting the proposed project or program. Grant proposals may include the allocation of a reasonable percentage of grant monies towards general support of the recipient organization. On a limited basis, the Foundation will also consider grants in support of capital campaigns for facilities or equipment, and organizational capacity building.
Requirements: The Foundation only considers grant requests from organizations designated as tax exempt under Section 501(c)3 of the Internal Revenue Code. Preference will be given to non-profit organizations based in the Greater Philadelphia Area.

Restrictions: The Foundation's funding will not process grant requests from: individuals; fraternal and/or civic organizations; political candidates or to influence legislation; other foundations; general fund raising and endowment drives; debt reduction; environmental issues or initiatives; annual appeals or letters of solicitation; or public, private or parochial schools that serve the general public.
Geographic Focus: Pennsylvania
Date(s) Application is Due: Aug 1
Amount of Grant: 5,000-50,000 USD
Contact: Meredith A. Huffman, Executive Director; (610) 834-2030; info@genuardifamilyfoundation.org
Internet: http://www.genuardifamilyfoundation.org/genuardi_priorities.html
Sponsor: Genuardi Family Foundation
Blue Bell Executive Campus, 470 Norristown Road, Suite 300
Auburn, NY 13021

Geographic Education Dissertation Award **2110**
Awards are made for outstanding doctoral research to encourage research in geography teaching and learning. Applications for the award must be made through a 15-page paper based on the applicant's doctoral dissertation.
Requirements: Scholars eligible for the award must have received the doctoral degree within the previous two years. All applicants are expected to present their research at the annual meeting.
Geographic Focus: All States
Date(s) Application is Due: Mar 15
Amount of Grant: 100-300 USD
Contact: Executive Director; (256) 782-5293; fax (256) 782-5336; ncge@ncge.org
Internet: http://www.ncge.org/resources/awards/dissertation.cfm
Sponsor: National Council for Geographic Education
Jacksonville State University, 206A Martin Hall
Jacksonville, AL 36265-1602

George A. Olah Award in Hydrocarbon or Petroleum Chemistry **2111**
The award is conferred to recognize, encourage, and stimulate outstanding research achievements in hydrocarbon or petroleum chemistry. Special consideration will be given to independence of thought and originality. The award consists of $5,000 and a certificate. Reasonable travel expenses to the meeting at which the award will be presented will be reimbursed.
Requirements: A nominee must be a resident of the United States or Canada and have accomplished outstanding research in the chemistry of hydrocarbons or petroleum and its products. Any individual, except a member of the award committee, may submit one nomination or seconding letter for the award in any given year. The nominating documents consist of a letter of not more than 1000 words, containing an evaluation of the nominee's accomplishments and a specific identification of the work to be recognized, a biographical sketch including date of birth, and a list of publications and patents authored by the nominee.
Geographic Focus: All States
Date(s) Application is Due: Nov 1
Amount of Grant: 6,500 USD
Contact: Felicia Dixon, Awards Administrator; (800) 227-5558 or (202) 872-4408; fax (202) 776-8008; f_dixon@acs.org or awards@acs.org
Internet: http://portal.acs.org/portal/acs/corg/content?_nfpb=true&_pageLabel=PP_ARTICLEMAIN&node_id=1319&content_id=CTP_004526&use_sec=true&sec_url_var=region1
Sponsor: American Chemical Society
1155 Sixteenth Street, NW
Washington, DC 20036-4801

George A Ohl Jr. Foundation Grants **2112**
The purpose of the George A Ohl Jr. Foundation is to improve the well-being of the citizens of the State of New Jersey through science, health, recreation, education and increased good citizenship. Grants are made to organizations engaged in such work whether through research, publications, health, school or college activities. The Foundation's mission is the relief of the poor; the improvement of living conditions; the care of the sick, the young, the aged, the homeless, the incompetent and the helpless. The foundation will target: 35% of its grants to Community Redevelopment; 35% to Health and Human Services organizations; 15% for Arts and culture; 15% for Educational requests. Application deadlines are: January 22 for a March meeting and, June 15 for an August meeting. Application forms are available online. Applicants will receive notice acknowledging receipt of the grant request, and subsequently be notified of the grant declination or approval.
Requirements: New Jersey 501(c)3 nonprofit organizations are eligible to apply. Proposals should be submitted in the following format: completed Common Grant Application Form; an original Proposal Statement; an audited financial report and a current year operating budget; a copy of your official IRS Letter with your tax determination; a listing of your Board of Directors. Proposal Statements (second item in the above Format) should answer these questions: what are the objectives and expected outcomes of this program/project/request; what strategies will be used to accomplish your objective; what is the timeline for completion; if this is part of an on-going program, how long has it been in operation; what criteria will you use to measure success; if the request is not fully funded, what other sources can you engage; an Itemized budget should be included; please describe any collaborative ventures. Prior to the distribution of funds, all approved grantees must sign and return a Grant Agreement Form, stating that the funds will be used for the purpose intended. Progress reports and Completion reports must also be filed as required for your specific grant. All current grantees must be in good standing with required documentation prior to submitting new proposals to any foundation.
Restrictions: Grants are not made for political purposes, nor to organizations which discriminate on the basis of race, ethnic origin, sexual or religious preference, age or gender
Geographic Focus: New Jersey
Date(s) Application is Due: Jan 22; Jun 15
Amount of Grant: 3,000-35,000 USD
Contact: Wachovia Bank, N.A., Trustee; grantinquiries2@wachovia.com
Internet: https://www.wachovia.com/foundation/v/index.jsp?vgnextoid=e0f78689fb0aa110VgnVCM1000004b0d1872RCRD&vgnextfmt=default
Sponsor: George A Ohl Jr. Foundation
190 River Road, NJ3132
Summit, NJ 07901

George B. Storer Foundation Grants **2113**
The foundation awards grants, primarily in Florida, in its areas of interest, including higher education, social services—particularly for the blind, youth organizations, conservation, hospitals, and cultural programs. Types of support include building construction/renovation, capital campaigns, general operating support, matching/challenge grants, and research grants. There are no application forms. Applications should be submitted between October 15 and the listed deadline.
Geographic Focus: All States
Date(s) Application is Due: Nov 15
Contact: Grants Administrator, c/o Thomas McDonald
Sponsor: George B. Storer Foundation
P.O. Box 1040
Tavernier, FL 33070

George C. Pimentel Award in Chemical Education **2114**
This award, supported by the Dow Chemical Company, is given annually to recognize outstanding contributions to chemical education. A nominee must have made outstanding contributions to chemical education considered in its broadest meaning, including the training of professional chemists; the dissemination of reliable information about chemistry to prospective chemists, to members of the profession, to students in other fields, and to the general public; and the integration of chemistry into our educational system. The activities recognized by the award may lie in the fields of teaching (at any level), organization and administration, influential writing, educational research, the methodology of instruction, establishment of standards of instruction, and public enlightenment. Preference shall be given to U.S. citizens. The Award includes $5,000, a certificate and travel expenses to award meeting.
Requirements: Any individual, except a member of the award committee, may submit one nomination or seconding letter for the award in any given year. The nominating documents consist of a letter of not more than 1000 words containing an evaluation of the nominee's accomplishments and a specific identification of the work to be recognized, a biographical sketch including date of birth, and a list of publications and patents authored by the nominee.
Restrictions: Self-nominations are not accepted.
Geographic Focus: All States
Date(s) Application is Due: Nov 1
Amount of Grant: 5,000 USD
Contact: Felicia Dixon, Awards Administrator; (800) 227-5558 or (202) 872-4408; fax (202) 776-8008; f_dixon@acs.org or awards@acs.org
Internet: http://portal.acs.org/portal/acs/corg/content?_nfpb=true&_pageLabel=PP_ARTICLEMAIN&node_id=1319&content_id=CTP_004528&use_sec=true&sec_url_var=region1
Sponsor: American Chemical Society
1155 Sixteenth Street, NW
Washington, DC 20036-4801

George Foundation Grants **2115**
The George Foundation strive to support organizations and programs that assist in developing strong, stable families across Fort Bend County region of Texas. The Foundation's areas of interest include: family stability; scholarships; foundation initiated programs, current programs include, Youth in Philanthropy (YIP), Leadership Excellence for Non-Profits, Integrated Mental Health Care, and Transportation. The Foundation prefers to fund the following types of grants to support Fort Bend organizations in their delivery of services to the community (listed in order of priority): program/project support; foundation initiated; general operating; capital. Proposal deadlines for making grant applications to the Foundation are January 15, April 15, July 15 and October 15 of each year. All proposals for capital support will be grouped together for review. The deadline for capital proposals will be October 15 of each year.
Requirements: Non-profits in Fort Bend County, Texas may submit grant proposals.
Restrictions: The Foundation does not fund: grants to organizations that do not have a current 501(c)3 determination letter; churches or other organized religious bodies; grants to another organization that distributes money to recipients of its own selection, i.e., a regranting organization; regional, national or international programs; grants for research or studies; grants for travel, conferences, conventions, group meetings, or seminars; the purchase of event tickets, tables, ads or sponsorships; support to fairs and festivals; religious or private schools; request for funds to develop films, videos, books or other media projects; direct mail campaigns; loans of any kind; grants to individuals; grants to fraternal organizations; political interests of any kind; and institutions that discriminate on the basis of race, creed, gender, national origin, age, disability or sexual orientation in policy or in practice.
Geographic Focus: Texas
Date(s) Application is Due: Jan 15; Apr 15; Jul 15; Oct 15

Amount of Grant: 1,000-4,000,000 USD
Contact: Dee Koch; (281) 342-6109; fax (281) 341-7635; dkoch@thegeorgefoundation.org
Internet: http://www.thegeorgefoundation.org
Sponsor: George Foundation
310 Morton Street, PMB Suite C
Richmond, TX 77469

George Frederick Jewett Foundation Grants **2116**

The foundation is concerned primarily with people and values. The grants program focuses on the future and on stimulating and supporting activities and projects of established, voluntary, nonprofit organizations that are of importance to human welfare. Grants are made in the fields of arts and humanities, conservation and preservation, education, health care and medical services, population, religion, and social welfare. The foundation may support research on and studies of important problems of public concern solely for the purpose of aiding in the gathering and presenting of facts that may assist the public to better understand such problems and to arrive at realistic and effective solutions to them. From time to time, support may be given to the scholarship, fellowship, and research programs of established institutions when sufficient evidence is available to establish clearly that the applicant organization is awarding such grants in accordance with the regulations established by the IRS. Grants are awarded to support activities in progress, research into potential projects, building and equipment, general operations, program development, seed funding, research, technical assistance, and matching funds. Inquiries for clarification of the foundation's policy and program emphasis are encouraged.
Requirements: Preference is given to public charities or nonprivate foundations. The foundation confines its grants largely to requests from eastern Washington and the San Francisco Bay area.
Restrictions: Grants do not support advertising; advocacy, athletic, international, religious, political, or veterans organizations; or individuals.
Geographic Focus: California, Washington
Amount of Grant: 5,000-50,000 USD
Contact: Ann Gralnek, (415) 421-1351; fax (415) 421-0721; ADGjewettf@aol.com or tfbjewettf@aol.com
Sponsor: George Frederick Jewett Foundation
235 Montgomery Street, Suite 612
San Francisco, CA 94104

George Gund Foundation Grants **2117**

The Foundation's long-standing interests include: arts; economic development; community revitalization; education; environment; and human services. The Foundation considers global climate change an urgent issue that cuts across all of the Foundation's programs. The Foundation takes seriously it's own responsibility and wants to hear from grant applicants what they are doing or considering to reduce or to eliminate their organizational impact on climate change. The Foundation also supports special projects grants, which currently include: Retinal Degenerative Diseases research grants, making an annual commitment for research on the causes, nature and prevention of inherited retinal degenerative diseases and; philanthropic services grants, offering support to organizations that strengthen the infrastructure of the nonprofit and philanthropic communities. The George Gund Foundation also supports capital requests but only for projects that are clearly aligned with their program priorities and that meet Green Building Council LEED (Leadership in Energy and Environmental Design) certification. The Foundation's green building policy covers both planning and construction grants. In addition, the Foundation supports opportunities that cross program boundaries and that integrate elements of the Foundation's interests. Although the Foundation's focus is centered in the Greater Cleveland, Ohio region, a portion of their grantmaking will continue to support state and national policy making that bolsters their work. Proposals should be mailed directly to the George Gund Foundation. All proposals are screened and evaluated by the staff before presentation at Trustee Meetings. Receipt of proposals will be acknowledged by mail.
Requirements: 501(c)3 nonprofit organizations are eligible to apply for funding, with a special interest in Greater Cleveland, Ohio region. Proposals are accepted three times: March 15, July 15 and November 15. Proposals are due the next business day if a deadline falls on a weekend. The grant application form is available on the George Gund Foundation website. All proposals must include a climate change statement, the Foundation's website includes resources to assist grantees with this task. Applicants also must include a completed Grant Application Cover Sheet, which is signed by the organization's board chair and executive director. Proposals should also include: organizational background; history; mission; types of programs offered; constituencies served; project description; justification of need; specific goals and objectives; activities planned to meet goals and objectives; project time line; qualifications of key personnel; methods of evaluation; project budget; anticipated expenses, including details about how Foundation funds would be used; anticipated income, including information about other sources approached for funding; organizational budget; previous and current year budget and proposed budget for project year(s) showing both income and expenses; the organization's most recent audited financial statement, do not include IRS 990 forms; supporting documents; list of current trustees; letters of support; readily available printed material about organization such as annual reports and brochures; IRS letter confirming Internal Revenue Code 501(c)3 status and classification as a public charity or information confirming status as a government unit or agency. The Foundation also will accept the Ohio Common Grant Form, available at www.ohiograntmakers.org, if organizations are using it to apply to multiple funders. Faxed or electronic proposals are not accepted.
Restrictions: Do not submit proposals in notebooks, binders or plastic folders and print proposals on both sides of each sheet of paper. The Foundation normally does not consider grants for endowments. Capital requests must meet the Foundation's program goals and also adhere to green building standards of environmental sustainability. Details on these requirements are available from the Foundation. Grants are not made for debt reduction or to fund benefit events. The Foundation does not make grants to individuals, nor does it administer programs it supports. Grants are limited to organizations located in the United States.
Geographic Focus: All States
Date(s) Application is Due: Mar 15; Jul 15; Nov 15
Contact: David Abbott; (216) 241-3114; fax (216) 241-6560; info@gundfdn.org
Internet: http://www.gundfdn.org/what.asp
Sponsor: George Gund Foundation
1845 Guildhall Building, 45 Prospect Avenue, West
Cleveland, OH 44115

George H. Hitchings New Investigator Award in Health Research **2118**

Provides awards to provide flexible support for young researchers in their second year of graduate school. Students planning to follow a career path in academic teaching and research that is closely tied to the health of North Carolinians and the strength of North Carolina Science are particularly encouraged to apply.
Requirements: Applications/nominations must be submitted by the institution on behalf of the student(s). Nomination applications are accepted from Duke, NCCU, NCSU, and UNC-Chapel Hill.
Restrictions: Only one nomination/application per graduate program/department within the university will be accepted.
Geographic Focus: North Carolina
Date(s) Application is Due: Mar 5
Amount of Grant: 10,000 USD
Contact: Libby Long; (919) 474-8370, ext. 134; fax (919) 941-9208; libby@trialglecf.org
Internet: http://www.trianglecf.org/DatasetRecord.cfm?recordID=10001417&returnURL=%2Fpage10000237%2Ecfm&returntoname=View%20all%20Scholarships&sidepageid=10000237&thetitle=George%20H.%20Hitchings%20New%20Investigator%20Award%20in%20Health%20Research&Ds_PagepropId=1
Sponsor: Triangle Community Foundation
324 Blackwell Street, Suite 1220
Durham, NC 27701

George Kress Foundation Grants **2119**

The foundation awards grants to eligible Wisconsin nonprofit organizations in its areas of interest, including arts and culture; boys and girls clubs; children and youth services; Christian agencies and churches; community/economic development; education; family services; health organizations; higher education; historic preservation/historical societies; hospitals; human services; libraries; recreation; United Ways and Federated Giving Programs; and YM/YWCAs and YM/YWHAs. Types of support include annual campaigns; building/renovation; capital campaigns; continuing support; professorships; program development; and research. Preference is given to nonprofit organizations that benefit Green Bay and Madison.
Requirements: There are no application forms or deadlines. Interested applicants should submit a letter of inquiry describing their proposed project.
Geographic Focus: Wisconsin
Contact: John Kress, President; (920) 433-3109
Sponsor: George Kress Foundation
c/o Green Bay Packaging Company
Green Bay, WI 54301

George S. and Dolores Dore Eccles Foundation Grants **2120**

The foundation awards grants to eligible Utah organizations in its areas of interest, including arts, children and youth, economics, higher education, hospitals, medical research, performing arts, visual arts, and social services. Types of support include building construction/renovation, capital campaigns, equipment acquisition, general operating grants, matching/challenge grants, professorships, program development, research grants, and scholarship funds. A request for application is available online.
Requirements: Giving primarily in Utah.
Restrictions: Funding requests will not be considered from the following types of organizations: those that have not received a tax exemption letter establishing 501(c)3 status from the Internal Revenue Service, unless they are a unit of government, in which case such a letter is not required; other private foundations; those of a political nature that attempt to influence legislation and/or candidacy of persons for elected public office; conduit organizations, unified funds, or those that use funds to make grants to support other organizations; those that do not have fiscal responsibility for the proposed project. Funds will also not be considered for: contingencies, deficits, or debt reduction; general endowment funds; direct aid to individuals; conferences, seminars, or medical research; requests which do not fall within the Foundation's specified areas of interest.
Geographic Focus: Utah
Contact: Director; (801) 246-5340; fax (801) 350-3510; gseg@gseccles.org
Internet: http://www.gsecclesfoundation.org
Sponsor: George S. and Dolores Dore Eccles Foundation
79 South Main Street, 12th Floor
Salt Lake City, UT 84111

Georgetown University Chemistry Department Graduate Research Fellowships **2121**

Fellowships enable qualified students to pursue graduate work leading to PhD degrees while gaining valuable experience assisting in undergraduate courses during at least one of the years of graduate study. The graduate program provides a firm foundation in all areas of chemistry as well as opportunities for research under 17 faculty members on a considerable variety of topics.
Requirements: Applicant must have a BS degree in chemistry or its equivalent.

Geographic Focus: All States
Contact: Chairperson; (202) 687-6073; fax (202) 687-6209; chemad@georgetown.edu
Internet: http://www.georgetown.edu/departments/chemistry/graduate.htm
Sponsor: Georgetown University
Department of Chemistry, Box 571227
Washington, DC 20057-2222

George W. Codrington Charitable Foundation Grants **2122**
The George W. Codrington Charitable Foundation gives primarily to nonprofit organizations in Ohio, but may consider other areas. The Foundation funds higher education, hospitals, museums, arts groups and performing arts, and youth programs. Types of support include annual and capital campaigns, continuing support, equipment, general/operating support, program development, and research.
Requirements: Application forms are not required. Applicants should submit three copies of the following: their IRS determination letter; a brief history of the organization and description of its mission; the geographic area to be served; a list of the board of directors, trustees, officers, and other key individuals with their affiliations; and a detailed description of the project and amount of funding requested. Proposals should be submitted one month before the board meets in April, June, September, and December. Organizations are notified of funding promptly after the board meeting.
Restrictions: Funding is not available for individuals, endowment funds, or loans.
Geographic Focus: Ohio
Contact: Craig Martahus; (216) 566-8674; tommie.robertston@thomasonhine.com
Sponsor: George W. Codrington Charitable Foundation
127 Public Square, 39th Floor
Cleveland, OH 44114-1216

Georgia-Pacific Foundation Environment Grants **2123**
Since a healthy environment is crucial to any town, city or neighborhood, nowhere is Georgia-Pacific's commitment to building strong communities better exemplified than in its stewardship of natural resources and the environment. The Foundation believes that adhering to sound science is the best way to keep and protect our natural resources. Consequently, it works with scientific institutions to develop the solutions, expertise and scientific data necessary to sustain our natural resources and help resolve environmental concerns. To support these efforts, the Georgia-Pacific Foundation funds environmental programs, such as the Keystone Institute Science School, Keep America Beautiful, and The Nature Conservancy.
Requirements: Nonprofit organizations in Georgia-Pacific communities are eligible to apply for funding. To find out whether a program qualifies for consideration for a Georgia-Pacific Foundation grant, applicants should complete the Eligibility Survey available online. The Georgia-Pacific Foundation will accept proposals for grants and in-kind donations from January 1 through October 31. It is recommended that applicants submit proposals early in the grant cycle.
Restrictions: No support is given for discriminatory organizations, political candidates, churches or religious denominations, religious or theological schools, social, labor, veterans', alumni, or fraternal organizations not of direct benefit to the entire community, athletic associations, national organizations with local chapters already receiving support, medical or nursing schools, or pass-through organizations. No grants to individuals (except for scholarships), or for emergency needs for general operating support, political causes, legislative lobbying, or advocacy efforts, goodwill advertising, sporting events, general operating support for United Way member agencies, tickets or tables for testimonials or similar benefit events, named academic chairs, social sciences or health science programs, fundraising events, or trips or tours.
Geographic Focus: Alabama, Arkansas, California, Florida, Georgia, Illinois, Indiana, Iowa, Kentucky, Louisiana, Massachusetts, Michigan, Mississippi, New Jersey, New York, North Carolina, Ohio, Oklahoma, Oregon, Pennsylvania, South Carolina, Tennessee, Texas, Virginia, Washington, West Virginia, Wisconsin
Date(s) Application is Due: Oct 31
Contact: Curley M. Dossman, Jr.; (404) 652-4182; cmdossma@gapac.com
Internet: http://gp.com/gpfoundation/environment.html
Sponsor: Georgia-Pacific Foundation
133 Peachtree Street, NE, 39th Floor
Atlanta, GA 30303

Georgia Institute of Technology Postdoctoral Fellowship **2124**
The 12-month postdoctoral fellowship starts in mid-August. Duties include individual/group counseling, outreach, crisis intervention, consultation, and testing/evaluation. The fellowship provides a stipend plus benefits. Applicants should send resumes to the office.
Requirements: Individuals with the PhD in psychology and experience in counseling/clinical psychology are eligible.
Geographic Focus: All States
Date(s) Application is Due: Mar 9
Amount of Grant: 22,000 USD
Contact: Dr. Irene Dalton, (404) 894-2575; fax (404) 894-1804
Internet: http://www.ohr.gatech.edu
Sponsor: Georgia Institute of Technology
500 Tech Parkway
Atlanta, GA 30332-0286

Georgia Power Foundation Grants **2125**
Giving is focused on issues that directly affect customers, employees, business and shareholders. These include: improving the quality of education by partnering with organizations to assist students with personal development, mentoring and career exploration; protecting the environment by promoting programs to improve air and water quality, preserve natural resources and protect endangered species; preventing cancer; and promoting diversity. The Foundation gives strong preference to Georgia-based organizations and programs that seek to improve the quality of life for the state's residents. Applicants may apply online or with a written proposal.
Requirements: The foundation makes grants to tax-exempt organizations that seek to improve the quality of life for Georgia's residents.
Restrictions: The Foundation does not provide grants to individuals, private elementary or secondary schools, and religious organizations, nor political campaigns or causes. The Foundation does not provide multi-year funding commitments.
Geographic Focus: Georgia
Date(s) Application is Due: Feb 15; May 15; Aug 15; Nov 15
Amount of Grant: 10,000 USD
Contact: Grants Administrator; (404) 506-6784; gpfoundation@southernco.com
Internet: http://www.georgiapower.com/community/apply.asp
Sponsor: Georgia Power Foundation
241 Ralph McGill Boulevard NE, Bin 10131
Atlanta, GA 30308-3374

Gerber Foundation Grants **2126**
The foundation awards grants to support national programs that have a significant impact on issues facing infants and young children. Areas of interest include pediatric health—promoting health and preventing disease, including projects geared toward research or interventions that will reduce the incidence of serious chronic illnesses (e.g., diabetes, heart disease, obesity, or cancer) or improve cognitive, social, and emotional aspects of development; pediatric nutrition—assuring adequate nutrition for infants and young children through projects of research or interventions; and the effects of environmental hazards. The listed deadlines are for letters of inquiry. Guidelines and application are available online.
Requirements: 501(c)3 nonprofit organizations are eligible. Priority is given to projects that improve infant and young children nutrition, care, and development from the first year before birth to three years of age.
Restrictions: The foundation does not make grants or loans to individuals. Outside the West Michigan area, the foundation does not support capital campaigns, operating support, national child welfare programs, international based programs, or food/baby products giveaway programs.
Geographic Focus: All States
Date(s) Application is Due: Jun 1; Dec 1
Amount of Grant: Up to 1,000,000 USD
Contact: Program Contact; (231) 924-3175; fax (231) 924-7906; cobits@ncresa.org
Internet: http://www.gerberfoundation.org
Sponsor: Gerber Foundation
4747 W 48th Street, Suite 153
Fremont, MI 49412-8119

Gerda Henkel Foundation Grants **2127**
The Foundation funds activities concentrating on German and foreign academia in the fields of historical humanities. In particular, research projects in the following fields are supported: history, prehistory and early history; archaeology; art history; history of Islam; and legal history. Funds are provided for: specific temporary research projects through grants for personnel, travel and material expenses; research and doctoral scholarships for German and foreign scholars; innovative academic conferences; publications of particularly successful projects supported by the Foundation. Application information is available online.
Restrictions: Support is not provided for: permanent positions; academic institutions through endowments; festschrifts or dictionaries; annual meetings and academic anniversaries; publications which are not related to projects already being supported by the Foundation; excursions; exhibition projects which are not related to projects already being supported by the Foundation.
Geographic Focus: All States, Germany
Date(s) Application is Due: Jun 29
Contact: Grants Administrator; 49-(0)211-359853; fax 49-(0)211-357137; info@gerda-henkel-stiftung.de
Internet: http://www.gerda-henkel-stiftung.de/02_foerderung/en_foerderung.htm
Sponsor: Gerda Henkel Foundation
Malkastenstrasse 15
Dusseldorf, 40211 Germany

Germanistic Society of America Fellowships **2128**
Usually up to six fellowships are awarded annually to enable prospective specialists to study for an academic year in Germany, primarily in art history; economics and banking; German language and literature; history; international law; philosophy; political science; and public affairs. Candidates selected for these awards, administered in the United States by the IIE, will be considered for Fulbright travel grants.
Requirements: A master's degree is desirable, but candidates must have a bachelor's degree and be U.S. citizens by November 1 of the year preceding the award.
Geographic Focus: All States
Amount of Grant: 12,000 USD
Contact: Programs Division; (212) 984-5330; fax (212) 984-5325; info@iie.org
Internet: http://www.iie.org/Content/NavigationMenu/Fulbright_Demo_Site/U_S__Student_Program/Fulbright_Grant_Opportunities/Germany.htm
Sponsor: Institute of International Education
1400 K Street, NW, 7th Floor
Washington, DC 20005-2403

Gerrish/FTE Technology Education Graduate Fellowship 2129
The fellowship is available to technology teachers at the K-12 grade level. Candidates must be accepted into a graduate degree program in technology education and beginning or continuing full-time graduate work. All applicants must be members of the International Technology Education Association. Additional information is available upon request.
Geographic Focus: All States
Date(s) Application is Due: Dec 1
Amount of Grant: 5,000 USD
Contact: Kendall Starkweather; (703) 860-2100; fax (703) 860-0353; iteaordr@iris.org
Internet: http://www.doe.mass.edu/tgfa/tech.html#2
Sponsor: Foundation for Technology Education
1914 Association Drive, Suite 201
Reston, VA 20191-1539

Gertrude B. Elion Mentored Medical Student Research Awards 2130
An annual award granted to support women medical students interested in pursuing health-related research projects.
Requirements: Students must have the support of a faculty mentor and must conduct their research at one of the four medical schools in North Carolina. Applications are accepted from Duke University Medical Center, East Carolina University School of Medicine, UNC-Chapel Hill School of Medicine or Wake Forest University Baptist Medical Center only.
Restrictions: Candidates must be women, enrolled as full-time students, and will have completed at least one year of medical school prior to the start of the award. They must conduct their research at the applying institution. Candidates must be citizens or permanent residents of the United States or Canada at the time of application. Documentation of permanent residency status must be provided with the application. Persons who have applied for permanent residency but have not received their government documentation by the time of application are not eligible.
Geographic Focus: North Carolina
Date(s) Application is Due: Mar 5
Amount of Grant: 10,000 USD
Contact: Libby Long; (919) 474-8370, ext. 134; fax (919) 941-9208; libby@trianglecf.org
Internet: http://www.trianglecf.org/page10000237.cfm
Sponsor: Triangle Community Foundation
324 Blackwell Street, Suite 1220
Durham, NC 27701

Getty Research Grants for Institutions-Archival Grants 2131
These grants support the basic arrangement and description of important archival collections of art-historical material in order to make them more accessible to scholars. Most projects result in the creation of an electronic finding aid. These grants do not support digitization of already catalogued archival materials, nor do they support archival research or transcription projects. Applicants should submit a preliminary letter of inquiry.
Requirements: Nonprofit institutions that own significant archival collections (including both documentary and photographic archives) relevant to the study of art history are eligible to apply for support. The collections must be open for use by interested scholars.
Restrictions: Requests for support to preserve archival collections are not eligible; however, certain preservation costs may be considered as a part of a larger arrangement and description project.
Geographic Focus: All States
Contact: Int'l Research Grants Supervisor; (310) 440-7320; fax (310) 440-7703
Internet: http://www.getty.edu/grants/research/institutions/archival.html
Sponsor: J. Paul Getty Trust
1200 Getty Center Drive, Suite 800
Los Angeles, CA 90049-1685

Getty Research Grants for Institutions-Art History Research Grants 2132
Grants in this category support projects that advance the scholarly mission of Art History Research Centers. Projects that enhance scholarly access to library and archival resources are particularly encouraged. Applicants should submit a preliminary letter of inquiry to determine eligibility.
Requirements: Grants are reserved for independent advanced research centers for the history of art that have significant library holdings and residential fellowship programs, and are open to scholars worldwide.
Geographic Focus: All States
Contact: Art History Research Center; (310) 440-7320; fax (310) 440-7703
Internet: http://www.getty.edu/grants/research/institutions/art_historical.html
Sponsor: J. Paul Getty Trust
1200 Getty Center Drive, Suite 800
Los Angeles, CA 90049-1685

Getty Research Grants for Institutions-Cataloguing of Museum Collections Grants 2133
These grants support full scholarly research and documentation of one or more distinct groups of objects in a museum's permanent collection by experienced scholars. Support is also available for the publication costs of catalogues. Applicants should submit a preliminary letter of inquiry to determine eligibility.
Requirements: Museums and other nonprofit organizations that own and house art collections and are open to the public on a regular basis are eligible to apply.
Restrictions: These grants do not support electronic collection catalogues, special exhibition catalogues, registrars' inventories, checklists, or summary catalogues.
Geographic Focus: All States
Contact: Int'l Research Grants Supervisor; (310) 440-7320; fax (310) 440-7703
Internet: http://www.getty.edu/grants/research/institutions/cataloguing.html
Sponsor: J. Paul Getty Trust
1200 Getty Center Drive, Suite 800
Los Angeles, CA 90049-1685

Getty Research Grants for Institutions-Critical Reference Resources Grants 2134
These grants support the preparation and publication of scholarly databases, reference works, innovative electronic resources, and other research tools that provide critical resource materials for art-historical scholarship. Priority is given to projects of international importance for which resources are otherwise limited. Applicants should submit a preliminary letter of inquiry to determine eligibility.
Requirements: Nonprofit institutions are eligible to apply. Grants are intended to support a crucial phase or portion of a larger project, not to provide long-term, ongoing support.
Geographic Focus: All States
Contact: Int'l Research Grants Supervisor; (310) 440-7320; fax (310) 440-7703
Internet: http://www.getty.edu/grants/research/institutions/critical_reference.html
Sponsor: J. Paul Getty Trust
1200 Getty Center Drive, Suite 800
Los Angeles, CA 90049-1685

Getty Research Institute Nonresidential Grants-Collaborative Research Grant 2135
These grants provide opportunities for teams of scholars to pursue interpretive research projects that offer new explanations of art and its history. Collaborations that foster a cross-fertilization of ideas and methodologies are particularly encouraged. Grants also fund the research in preparation for scholarly exhibitions. Grants provide support for projects during a determined time frame. Grant periods vary according to the needs of the individual projects, but are generally available for research periods of one to two years. Although team members may alternate their period of leave to work on the project, the proposed plan for the project's completion must include a portion of time dedicated to joint study; such periods of joint study may include travel. Applications are welcome from scholars of all nationalities. Potential applicants are strongly encouraged to send a brief letter of inquiry. Grant amounts may vary. Application information is available online.
Requirements: These grants are intended to support established scholars who have attained distinction in their fields. Teams may consist of two or more art historians, or of one or more art historians together with one or more scholars from other disciplines. Teams for exhibition projects should include scholars from both museums and universities.
Restrictions: Individual scholars may not apply as a member of more than one team. Applications for the development of basic research tools, such as computer databases or art-historical reference works, are not eligible in this grant category.
Geographic Focus: All States
Date(s) Application is Due: Nov 1
Contact: Manager; (310) 440-7374; fax (310) 440-7703; researchgrants@getty.edu
Internet: http://www.getty.edu/grants/research/scholars/research_collaborative.html
Sponsor: J. Paul Getty Trust
1200 Getty Center Drive, Suite 800
Los Angeles, CA 90049-1685

Getty Research Institute Nonresidential Grants-Curatorial Research Fellowships 2136
This fellowship supports the professional scholarly development of curators by providing them with time off from regular museum duties to undertake short-term research or study projects that advance the understanding of art and its history. Applicants may apply for a research period of one to three consecutive months. Fellowships provide stipends toward salary replacement and travel expenses. Eligible projects may include research for an exhibition or publication, travel to visit sites or collections, revision of a dissertation for publication, an intensive period of reading or writing, or other projects that support professional scholarly development. Application information is available online.
Requirements: Fellowships are reserved for full-time curators of all nationalities who have a minimum of three years' professional experience and are employed at museums with art collections. Applicants must obtain authorization from their museums for the proposed fellowship term.
Restrictions: Funding is not available to cover institutional overhead costs or for carrying out the planning requirements of an exhibition apart from research. Fellowships are not intended to fund research for the completion of an academic degree.
Geographic Focus: All States
Date(s) Application is Due: Nov 1
Amount of Grant: 3,500-5,000 USD
Contact: Coordinator; (310) 440-7374; fax (310) 440-7703; researchgrants@getty.edu
Internet: http://www.getty.edu/grants/research/scholars/research_curatorial.html
Sponsor: J. Paul Getty Trust
1200 Getty Center Drive, Suite 800
Los Angeles, CA 90049-1685

Getty Research Institute Nonresidential Grants-Postdoctoral Fellowships 2137
These fellowships provide support for outstanding scholars in the early stages of their careers, allowing them the flexibility to travel and study wherever necessary to undertake their work. Grants support interpretive research projects that make a substantial and original contribution to the understanding of art and its history. Fellowships provide a stipend for a twelve-month period. Application information is available online.
Requirements: Scholars of all nationalities whose doctoral degrees in art history (or the equivalent in countries outside the United States) have been or will be officially conferred

between 1/1/01 and 1/1/07 are eligible to apply. Candidates who hold doctoral degrees in fields outside art history are eligible to apply if they can demonstrate that their work promises to make a substantial and original contribution to the understanding of art and its history.
Restrictions: During the tenure of the fellowship, fellows may not accept other awards or grants, nor may they undertake any form of employment. These awards may not be deferred, and they are not renewable.
Geographic Focus: All States
Date(s) Application is Due: Nov 1
Amount of Grant: 40,000 USD
Contact: Fellowship Coordinator; (310) 440-7374; researchgrants@getty.edu
Internet: http://www.getty.edu/grants/research/scholars/research_postdoc.html
Sponsor: J. Paul Getty Trust
1200 Getty Center Drive, Suite 800
Los Angeles, CA 90049-1685

Getty Scholars and Visiting Scholars Grants **2138**
These grants provide a unique research experience. Recipients are in residence at the Getty Research Institute or the Getty Villa in Malibu where they are able to pursue their own projects free from academic obligations, make use of Getty collections, and participate in the intellectual life of the Getty. Scholars are in residence for the entire academic year. A salary replacement stipend is awarded. The grant also includes an office, research assistance, airfare to and from Los Angeles, an apartment and health benefits. Application information is available online.
Requirements: These grants are for established scholars, artists, or writers who have attained distinction in their fields. Applications are welcome from researchers of all nationalities who are working in the arts, humanities, or social sciences.
Geographic Focus: All States
Date(s) Application is Due: Nov 1
Contact: Grants Manager; (310) 440-7374; fax (310) 440-7703; researchgrants@getty.edu
Internet: http://www.getty.edu/grants/research/scholars/research_grischolars.html
Sponsor: J. Paul Getty Trust
1200 Getty Center Drive, Suite 800
Los Angeles, CA 90049-1685

Gil and Dody Weaver Foundation Grants **2139**
Established in Texas in 1980, the Gil and Dody Weaver Foundation offers support throughout the States of Texas, Oklahoma, and Louisiana, with some emphasis on the Dallas-Fort worth area. The Foundation's primary fields of interest include: cancer; children and youth services; education; health organizations; human services; social services; and recreational camps. Major types of support come in the form of: annual campaigns; general operating/continual support; and scholarship funds. Although no formal application is required, the Foundation does provide specific application guidelines. These guidelines require a history of the organization, detailed information about the proposed project, and budgetary needs. The annual deadline is May 31, with final notifications by September 30. Recent grants have ranged from $1,000 to $20,000, with occasional higher amounts for special circumstances.
Requirements: 501(c)3 organization serving the residents of Texas, Oklahoma, and Louisiana, are welcome to apply.
Restrictions: No grants are given to individuals. No applications are accepted from organizations located in states other than Texas, Oklahoma, or Louisiana.
Geographic Focus: California, Louisiana, Mississippi, New Mexico, Oklahoma, Texas
Date(s) Application is Due: May 31
Amount of Grant: 1,000-20,000 USD
Contact: William R. Weaver, (214) 999-9497; fax (214) 999-9496
Sponsor: Gil and Dody Weaver Foundation
1845 Woodall Rodgers Freeway, Suite 1275
Dallas, TX 75201-2299

Gilbert F. White Postdoctoral Fellowships **2140**
The fellowships support professionals who wish to devote a year to scholarly work on social science or policy problems in the areas of natural resources, energy, or the environment. Two fellowships per year are awarded for a minimum of nine and a maximum of 12 months. The amount of the stipend is commensurate with experience. Fellows will reside in one of three RFF units: the energy and natural resources division, the quality of the environment division, or the center for risk management. Applications via fax are not accepted.
Requirements: The award is open to individuals in any discipline who will have completed their doctoral requirements by the beginning of the academic year for which the fellowship is requested. Faculty members who will be on sabbatical leave during the fellowship are also encouraged to apply.
Restrictions: Individuals holding positions in government as well as at academic institutions are eligible.
Geographic Focus: All States
Date(s) Application is Due: Feb 28
Amount of Grant: 1,000 USD
Contact: Coordinator; (202) 328-5155; fax (202) 939-3460; white-award@rff.org
Internet: http://www.rff.org/rff/About/Fellowships_and_Internships/White/Gilbert-F-White-Postdoctoral-Fellowship-Program.cfm
Sponsor: Resources for the Future
1616 P Street NW
Washington, DC 20036-1400

Gilbreth Memorial Fellowships **2141**
These fellowships are awarded to recognize graduate students pursuing advanced degrees in industrial engineering in accredited industrial engineering programs at colleges and universities in the United States, Canada, or Mexico. Consideration is given to proven scholastic ability, character, leadership qualities, potential service to industrial engineering, and the need for financial assistance. Nominations are solicited from department heads at the end of August for candidates for the following academic year. Nomination packets are then sent to the nominees.
Requirements: Institute members who are enrolled full time in graduate industrial engineering programs and have had a 3.4 to 4.0 GPA as undergraduates are eligible. Nominations are accepted only from department heads of accredited industrial engineering programs.
Geographic Focus: All States
Date(s) Application is Due: Nov 15
Amount of Grant: Up to 4,000 USD
Contact: Coordinator; (800) 494-0460 or (770) 449-0460; info@iienet.org
Internet: http://www.iienet.org/public/articles/index.cfm?cat=525
Sponsor: Institute of Industrial Engineers
3577 Parkway Lane, Suite 200
Norcross, GA 30092

Gilroy and Lillian P. Roberts Charitable Foundation Grants **2142**
The foundation awards grants to eligible nonprofit organizations in its areas of interest, including arts and fine arts, health care and health organizations, higher education, Jewish temples and organizations, and social services. Types of support include annual campaigns, capital campaigns, continuing support, fellowships, general operating grants, professorships, program development, and scholarships. Most grants are awarded in Montgomery and Delaware Counties, PA. There are no application deadlines or forms.
Restrictions: Individuals are not eligible.
Geographic Focus: Pennsylvania
Amount of Grant: Up to 408,750 USD
Contact: Stanley Merves, Treasurer; (610) 668-1998
Sponsor: Gilroy and Lillian P. Roberts Charitable Foundation
101 W Elm Street, Suite 500
Conshohocken, PA 19428

Girl's Best Friend Foundation Grants **2143**
The foundation supports and promotes programs by and for girls and young women in Illinois (ages eight through 21). The foundation is dedicated to effecting change at the grassroots level by funding community-based organizations statewide. Types of grants made include general operating support, project specific support, planning or start-up support, technical assistance, and collaborative action research (research that lays the groundwork for policy changes conducted by a partnership between two nonprofits or a nonprofit contracting with a researcher or academic institution). When multiyear funding is requested, the foundation will consider making two-year grants if the requesting organization has received at least two previous grants from GBF, is seeking two years of funding for the same purpose, and has clearly demonstrated the need for a two-year grant. Applications are due the first Monday in August.
Requirements: Letters of intent will be accepted from 501(c)3 nonprofits for projects that serve or have a direct impact on girls living in the Chicago metropolitan area including Cook, DuPage, Kane, Lake, McHenry, and Will counties.
Restrictions: Organizations with budgets that exceed $650,000 may apply only for technical assistance and/or collaborative action research grants. The foundation generally does not fund individuals, capital campaigns, debt reduction, scholarships, or government or religious organizations.
Geographic Focus: Illinois
Amount of Grant: 5,000-20,000 USD
Contact: Robin Dixon, Senior Program Officer; (312) 266-2842; fax (312) 266-2972; robin@girlsbestfriend.org or contact@girlsbestfriend.org
Internet: http://www.girlsbestfriend.org/apply/index.html
Sponsor: Girl's Best Friend Foundation
900 N Franklin, Suite 210
Chicago, IL 60610

Gladys Krieble Delmas Foundation Grants for Venetian Research **2144**
Predoctoral and postdoctoral grants will be awarded for travel to and residence in Venice and Veneto for research on Venice and the former Venetian empire and for study of contemporary Venetian society and culture. Disciplines of the humanities and social sciences are eligible areas of study, including, but not limited to, archaeology, architecture, art, bibliography, economics, history, history of science, law, literature, music, political science, religion, and theater. Prospective applicants should write to the foundation for application forms.
Requirements: Applicants must be U.S. citizens or permanent residents, have some experience in advanced research, and, if graduate students, have completed all doctoral requirements except for the dissertation at the time of application.
Geographic Focus: All States
Date(s) Application is Due: Dec 15
Amount of Grant: Up to 19,900 USD
Contact: Program Contact; (212) 687-0011; fax (212) 687-8877; info@delmas.org
Internet: http://www.delmas.org/guidelines/v_ir_a.html
Sponsor: Gladys Krieble Delmas Foundation
521 5th Avenue, Suite 1612
New York, NY 10175-1699

Glaucoma Foundation Grants **2145**

The foundation funds research initiatives to determine the causes of glaucoma, to improve methods of treatment, and to develop cures for the various kinds of glaucoma. Two areas of particular focus are: optic nerve rescue and restoration—research into new approaches designed to protect the optic nerve against glaucomatous damage, to restore vision lost to glaucoma and eventually reverse blindness by restoring or regenerating the function of the optic nerve cells, and to explore the feasibility of achieving transplantation of optic nerve cells; and molecular genetics—research into the genetic causes of the various forms of glaucoma, particularly the identification of the responsible genes, with the long-term goal of finding ways to reverse these genetic defects. Grants are awarded for a one-year period and are renewable. Guidelines and application are available online.
Requirements: Applicants must have a full time faculty position or the equivalent.
Restrictions: The foundation does not provide funds for investigator salaries, travel, overhead, or other indirect costs.
Geographic Focus: All States
Date(s) Application is Due: Mar 1; Sep 1
Amount of Grant: 40,000-50,000 USD
Contact: Kira A. Zmuda, Director; (212) 651-2509; kzmuda@glaucomafoundation.org
Internet: http://www.glaucomafoundation.org/Grant_Application.htm
Sponsor: Glaucoma Foundation Grants
80 Maiden Lane, Suite 700
New York, NY 10038

Glaucoma Research Pilot Project Grants **2146**

These grants provide funds to encourage innovative and pilot research, as well as to aid ongoing studies that seek to protect and restore the optic nerve, accurately monitor glaucoma's progression, find the genes responsible for the glaucoma, understand the intraocular pressure system and develop better treatments, and determine the risk factors for glaucoma damage.
Requirements: Applicants must have a graduate degree.
Restrictions: Funding is not granted for equipment purchases, overhead commercial applications, or indirect costs.
Geographic Focus: All States
Date(s) Application is Due: Aug 1
Amount of Grant: 40,000 USD
Contact: Jennifer Rulon, (800) 826-6693 or (415) 986-3162; research@glaucoma.org
Internet: http://www.glaucoma.org/research/index.php
Sponsor: Glaucoma Research Foundation
251 Post Street, Suite 600
San Francisco, CA 94108

Glenn/AFAR Breakthroughs in Gerontology Awards **2147**

The goal of the program is to provide timely support to a small number of pilot research programs that may be of high risk but offer significant promise of yielding transforming discoveries in the fundamental biology of aging. The hope is that one or more of the funded research projects will lead to major new insights into the molecular factors that coordinate aging in multiple cells and tissues, and the ways in which the aging process is differentially timed in long-lived species. Projects that focus on genetic controls of aging and longevity, on delay of aging by pharmacological agents or dietary means, or that elucidate the mechanisms by which alterations in hormones, anti-oxidant defenses, or repair processes promote longevity are all well within the intended scope of this competition. Projects that focus instead on specific diseases or on assessment of health care strategies will receive much lower priority, unless the research plan makes clear and direct connections to fundamental issues in the biology of aging. Studies of invertebrates, mice, human clinical materials, or cell lines are all potentially eligible for funding. Although preliminary data are always helpful for evaluating the feasibility of the experiments proposed, the emphasis in review will be on creativity and the likelihood that the findings will open new vistas and approaches to aging research that might merit intensive follow-up studies. Applications from individuals not previously engaged in aging research are particularly encouraged, as long as the research proposals show high promise for leading to important new discoveries in biological gerontology. The proposed research must be conducted at any type of nonprofit setting in the United States. Guidelines are available online.
Requirements: Applicants must at the time they submit their proposal be full-time faculty members at the rank of assistant professor or higher. A strong record of independent publication beyond the postdoctoral level is a requirement.
Restrictions: Applicants who are employees in the NIH Intramural program are not eligible.
Geographic Focus: All States
Date(s) Application is Due: Dec 16
Amount of Grant: 200,000 USD
Contact: Grants Manager; (212) 703-9977 or (888) 582-2327; fax (212) 997-0330; grants@afar.org or info@afar.org
Internet: http://afar.org/grants.html
Sponsor: American Federation for Aging Research
55 West 39th Street, 16th Floor
New York, NY 10018

Glenn/AFAR Research Grants for Postdoctoral Fellows **2148**

AFAR funds research projects concerned with understanding the basic mechanisms of aging. Projects investigating age-related diseases are also supported, especially if approached from the point of view of how basic aging processes may lead to these outcomes. Projects concerning mechanisms underlying common geriatric functional disorders are also encouraged. Examples of promising areas of research include aging and immune function; genetic control of longevity; neurobiology and neuropathology of aging; invertebrate or vertebrate animal models; cardiovascular aging; aging and cellular stress resistance; metabolic and endocrine changes; age-related changes in cell proliferation; caloric restriction and aging; DNA repair and control of gene expression; biology of the menopause; aging and apoptosis; and biodemographic analysis of aging. Funding begins July 1. Guidelines and application are available online.
Requirements: Postdoctoral fellows who will have had at least two and not more than five years of prior postdoctoral training at the time of the award are eligible.
Restrictions: Funds may not be requested for overhead or indirect costs. Fellows may not hold any concurrent foundation or not-for-profit funding.
Geographic Focus: All States
Date(s) Application is Due: Dec 15
Amount of Grant: 60,000 USD
Contact: Grants Manager; (212) 703-9977; fax (212) 997-0330; amfedaging@aol.com
Internet: http://www.afar.org/grants.html
Sponsor: American Federation for Aging Research
55 West 39th Street, 16th Floor
New York, NY 10018

Glenn T. Seaborg Award for Nuclear Chemistry **2149**

The purpose of the Award is to recognize and encourage research in nuclear and radiochemistry or their applications. A nominee must have made outstanding contributions to nuclear or radiochemistry or to their applications. The award will be granted without regard to age or nationality. The award consists of $3,000 and a certificate. Reasonable travel expenses to the meeting at which the award will be presented will be reimbursed.
Requirements: Any individual, except a member of the award committee, may submit one nomination or seconding letter for the award in any given year. Nominating documents consist of a letter of not more than 1000 words containing an evaluation of the nominee's accomplishments and a specific identification of the work to be recognized, a biographical sketch including date of birth, and a list of publications and patents authored by the nominee.
Restrictions: Self-nominations are not accepted.
Geographic Focus: All States
Date(s) Application is Due: Nov 1
Amount of Grant: 3,000 USD
Contact: Felicia Dixon, Awards Administrator; (800) 227-5558 or (202) 872-4408; fax (202) 776-8008; f_dixon@acs.org or awards@acs.org
Internet: http://portal.acs.org/portal/acs/corg/content?_nfpb=true&_pageLabel=PP_ARTICLEMAIN&node_id=1319&content_id=CTP_004529&use_sec=true&sec_url_var=region1
Sponsor: American Chemical Society
1155 Sixteenth Street, NW
Washington, DC 20036-4801

Global Center for Dispute Resolution Research Grants **2150**

The center is dedicated to research that provides insight, information, and guidance on the growth and practice of alternative dispute resolution (ADR) methods throughout the world, particularly in the area of cross-border commerce. The grants program funds research papers. Areas of interest include pattern and trend analysis of the collected data, such as global trends in ADR by industry, by country, by process type, and by forces in the marketplace that lead to caseload fluctuations in different sectors; and designed experiment research that will examine specific aspects of dispute resolution processes, such as process effectiveness, cost savings, time savings, quality of result, party satisfaction, and optimization of results. Research findings and their application will inform policy development in the field and increase process knowledge of practitioners.
Geographic Focus: All States
Contact: Grants Administrator; (212) 716-3900; fax (212) 716-5913; woodw@adr.org
Internet: http://www.globalcenteradr.org/english/research.jsp
Sponsor: Global Center for Dispute Resolution Research
335 Madison Avenue, 10th Floor
New York, NY 10017

Global Change Education Program **2151**

This program provides research opportunities in areas related to global change such as atmospheric sciences, ecology, global carbon cycles, climatology, and terrestrial processes. Program funding includes: $475 weekly, plus travel for undergraduates; $1500 monthly stipend, plus tuition and fees for graduates.Types of support include graduate research environmental fellowships (GREF) and summer undergraduate research experiences (SURE).
Requirements: College juniors, seniors, and graduate students are eligible to apply.
Geographic Focus: All States
Contact: Dr. Milton Constantin; (865) 576-7009; constanm@orau.gov
Internet: http://see.orau.org/ProgramDescription.aspx?Program=10125
Sponsor: Oak Ridge Institute for Science and Education
P.O. Box 117
Oak Ridge, TN 37831-0117

GMFU.S. Black Sea Trust for Regional Corporation Grants **2152**

The Black Sea Trust (BST) for Regional Cooperation invites proposals from organizations and institutions in the countries of the Wider Black Sea Region on the Trust's three programs: civic participation, cross-border initiatives, and Eastern Links. The Trust promotes regional cooperation and good governance in the Wider Black Sea region; accountable, transparent, and open governments; strong, effective civic sectors; and independent and professional media. To respond to the rapid shifts in the region, BST staff regularly consult with regional experts and aim to sharpen the program's grantmaking

strategy in order to more effectively achieve the Trust's goals. With the complexity and diversity of the region, BST priorities are revised regularly to respond to the region's changing needs. Adjustments are made in consultation with the BST Advisory Board, the German Marshall Fund's network of offices and internal expertise, and in coordination with other donors active in the region. The maximum duration of projects is 12 months. As an exception, outstanding projects of a longer duration may be considered. Grants generally range from $5,000 and $75,000, with most grants falling between $15,000 and $25,000. BST can support exceptional multi-year projects, renewable on an annual basis contingent upon satisfactory interim reports and performance. There are no application deadlines, and grant decisions are made monthly.
Requirements: Non-governmental organizations, governmental entities, community groups, policy institutes, other associations legally registered in Armenia, Azerbaijan, Bulgaria, Georgia, Moldova, Romania, Turkey, Ukraine and Russia. The projects have to be implemented in one or more of the countries mentioned. An online application and budget form are required, and must be completed in English.
Restrictions: Individuals and political parties may not apply. Non-indigenous organizations may not apply. Cooperative projects between indigenous and non-indigenous organizations are considered under the Eastern Links component, yet non-indigenous administrative costs will not receive BST support. The BST does not support scholarly research, academic fellowships and scholarships, one-off events, humanitarian aid, refugee or IDP return, religious activities, or the arts or sciences. BST only supports travel and website creation/maintenance as components of larger activities.
Geographic Focus: Armenia, Azerbaijan, Bulgaria, Moldova, Romania, Russia, Turkey, Ukraine
Amount of Grant: 15,000-25,000 USD
Contact: Program Contact; (202) 683-2650; fax (202) 265-1662; info@gmfus.org
Internet: http://www.gmfus.org/grants-fellowships/grantmaking-programs/black-sea-trust
Sponsor: German Marshall Fund of the United States
1744 R Street NW
Washington, DC 20009

GNOF Gulf Coast Oil Spill Grants **2153**
Shortly after the tragic explosion of BP's Deepwater Horizon drilling rig in 2010, the Greater New Orleans Foundation (GNOF) responded to the crisis by establishing the Gulf Coast Oil Spill Fund to serve as a conduit between the donor community and GNOF's nonprofit partners. The Gulf Coast Oil Spill Fund (GCOS) supports work that helps fisherman and their families in the parishes most affected by the oil spill: Plaquemines, St. Bernard, lower Jefferson, Terrebonne, and Lafourche parishes. The long-term goal of GCOS is to help strengthen coastal communities against future environmental catastrophes such as the Deepwater spill. Specifically GCOS embraces the following objectives: the creation of a more resilient commercial fishing industry that is better able to respond to future crises (whether natural or man-made) through the introduction of new technologies and practices; the development and implementation of job training opportunities for fishers and charter-boat workers temporarily or permanently displaced from their chosen vocation because of the spill; the development of innovative marketing and distribution plans for local seafood; the development of a more comprehensive and accessible system for monitoring both water quality and seafood quality; a mechanism for more effectively organizing the region's commercial fishers which focuses on transcending geographic, racial, ethnic, and linguistic barriers that have traditionally divided commercial fishing populations; the development of laws and/or policies that will be of direct benefit to commercial fishers and others who derive the majority of their income from water-related activities; the provision of free or reduced-cost legal advice and mental-health services to commercial fishers and individuals whose income has been curtailed or interrupted by the oil disaster; and a mechanism for better coordinating the provision of emergency assistance to aquatic and other wildlife affected by "mini-landfalls" or other contact with lingering oil from the Deep Water Horizon spill. GNOF accepts GCOS applications through its RFP process. Interested organizations are encouraged to subscribe to GNOF's email newsletter so they can be notified of future RFPs. The subscription link is available from GNOF's Apply-for-a-Grant web page which contains a comprehensive listing of GNOF's current and past funding opportunities. Interested organizations are also encouraged to contact the GNOF Director for Regional Initiatives with any questions.
Requirements: Applicant organizations must meet all of the following criteria: be classified by the Internal Revenue Sercie as 501(c)3 nonprofit entities or make use of a fiscal agent that has a 501(c)3 designation; be based within GNOF's thirteen-parish service area (Assumption, Jefferson, Lafourche, Orleans, Plaquemines, St. Bernard, St. Charles, St. James, St. John the Baptist, St. Tammany, Tangipahoa, Terrebonne, and Washington); and demonstrate close collaboration with commercial fishing organizations or populations in affected communities in the GNOF service area.
Restrictions: The GCOS fund will not support film or documentary projects, art projects, or academic research (except in the case of developing a more comprehensive and accessible system for monitoring both water quality and seafood quality).
Geographic Focus: Louisiana
Contact: Dr. Marco Cocito-Monoc; (504) 598-4663; marco@gnof.org
Internet: http://www.gnof.org/gnof-oil-spill-fund/
Sponsor: Greater New Orleans Foundation
1055 St. Charles Avenue, Suite 100
New Orleans, LA 70130

GNOF IMPACT Grants for Education **2154**
Through the IMPACT Program, the Greater New Orleans Foundation (GNOF) makes grants to organizations serving the Greater New Orleans region. The ultimate goal of the IMPACT program is to create a resilient, sustainable, vibrant, and equitable region in which individuals and families flourish and in which the special character of the New Orleans region and its people is preserved, celebrated, and given the means to develop. Specifically GNOF hopes to accomplish the following objectives through its IMPACT grants: provide a much needed source of financial and other support to nonprofit organizations that are struggling in the current financial environment and that are important to the health and vibrancy of the region; develop a better sense of the nonprofit organizations serving the region so GNOF can more effectively match donor desires with effective charitable work; identify and nurture promising new leaders and initiatives, especially in those communities that are in greatest need; and gain knowledge that will help nonprofit leaders and GNOF staff develop better long-term strategies for addressing regional needs and taking best advantage of important opportunities. IMPACT grants are awarded in four categories: Arts and Culture, Youth Development, Education, and Health and Human Services. In the category of Education support is available to organizations that seek to ensure that all young people in public K-12 schools in GNOF's 13-parish service area attend high-performing schools. Priority will be given to work that has the following goals: to advance public education reforms in Orleans parish and other under served areas in GNOF's service area; to hold public agencies accountable for the success of our public schools; to improve student achievement by training and engaging key stakeholders in education (parents, school leaders, etc.) in the skills of organizing, policy research and analysis, policy advocacy, and the use of data in making policy recommendations; and conduct research and/or work with school leaders and other financial decision makers to address the financial sustainability issues within the charter school system. Interested organizations must submit a letter of intent along with all attachments via one email by 5 p.m. on July 30. GNOF program staff will review all letters of intent and will contact those organizations that are invited to submit a full application for funding. Awards are announced in November. Deadlines may vary from year to year. Interested organizations should verify the current deadline at the GNOF website where they can also obtain complete guidelines and requirements as well as a downloadable application form and cover sheet. Prospective applicants can also subscribe to GNOF's email newsletter for announcements of future funding opportunities. The subscription link is available from GNOF's Apply-for-a-Grant web page which contains a comprehensive listing of GNOF's current and past funding opportunities.
Requirements: Nonprofit, tax-exempt organizations that serve the thirteen parishes of Greater New Orleans are eligible to apply for funding. Organizations that are not tax-exempt but have a fiscal agent relationship with a 501(c)3 organization are also eligible.
Restrictions: Through its IMPACT program, the Greater New Orleans Foundation is unable to fund the following types of requests: requests for individual support, either through scholarships or other forms of financial assistance; special events or conferences; programs that promote religious doctrine; endowments; and scientific or medical research.
Geographic Focus: All States
Date(s) Application is Due: Jul 30
Amount of Grant: Up to 20,000 USD
Contact: Roy Williams; (504) 598-4663; grants@gnof.org or roy@gnof.org
Internet: http://www.gnof.org/programs/impact/
Sponsor: Greater New Orleans Foundation
1055 St. Charles Avenue, Suite 100
New Orleans, LA 70130

Goddess Scholars Grants **2155**
The purpose of these grants is to develop young researchers at the entry faculty level (Instructor or Assistant Professor) who will focus a research career on studying the unique aspects of stroke in women. Such research may be basic or clinical in orientation, including areas such as mechanisms of ischemic brain injury, neuroprotective treatments, primary or secondary stroke prevention, epidemiological studies, educational programs for patients or health care providers, rehabilitation, and outcome research. The grant provides funding for two years, with potential non-competitive renewal for a third year should satisfactory progress be demonstrated. These grants are intended to provide bridging funds for individuals who are finishing or have completed residency or postdoctoral training and require additional support before their subsequent establishment as independent investigators.
Requirements: Goddess Scholar grants are targeted at individuals who: hold an M.D., Ph.D., or equivalent degree; are at the fellow, instructor, or assistant professor rank; and are within 5 years of completion of residency or post-doctoral fellowship at the time of grant activation. Individuals who are currently enrolled in a fellowship program may apply if there is assurance of a faculty position at the time of the grant's activation.
Geographic Focus: All States
Amount of Grant: 65,000 USD
Contact: Erin Tower, Program/Operations Assistant; (212) 713-6789; fax (212) 288-2160; erin@thegoddessfund.org
Internet: http://www.thegoddessfund.org/grants.html
Sponsor: Hazel K. Goddess Fund for Stroke Research in Women
785 Park Avenue
New York, NY 10021-3552

Goldhirsh Foundation Brain Tumor Research Grants **2156**
The Goldhirsh Foundation is interested in providing strategic investment in both pediatric and adult brain tumor research to accelerate progress toward more effective treatment for malignant diffuse glioma tumors. Applications are encouraged from investigators working in the continuum between basic research and clinical application, integrating and translating knowledge in various disciplines into meaningful progress for patients. The Initial Proposal Application must be submitted by Thursday, January 7, 12:00 Noon, U.S. and Canada Eastern Time (GMT-5), to the online application system. Applicants who are invited to submit Full Proposals will be notified by email in early

March. Up to three investigators will receive grants of $400,000 USD (inclusive of 10% indirect costs) over two years at $200,000 USD per year. The Foundation also makes awards of $100,000 USD (inclusive of 10% indirect costs) for one-year pilot studies. Please note that The Goldhirsh Foundation is not offering the three-year $600,000 USD award for the 2010 Grant Cycle.
Requirements: The Goldhirsh Foundation is accepting Initial Proposal Applications from applicants who meet the following eligibility *Requirements:* have an M.D. and/or Ph.D. degree(s) or equivalent degree; hold a faculty appointment at a nonprofit academic, medical or research institution in the United States, Canada or Israel. The applicant may collaborate with investigators from other institutions and these institutions may include for-profit companies; preference will be given to originality of ideas, regardless of applicant seniority; eligibility is not limited to those investigators currently working in brain tumor research. Investigators from other fields are encouraged to apply with proposals directly relevant to malignant diffuse glioma tumors; projects must be relevant to tumors within the category of malignant diffuse gliomas, i.e., glioblastomas, diffuse astrocytomas (WHO grade III), oligoastrocytomas (WHO grade III), and oligodendrogliomas (WHO grade III).
Restrictions: Projects focusing on WHO grade I-II gliomas, non-glial neuroepithelial tumors, meningeal tumors, nerve sheath tumors, CNS lymphomas, germ-cell tumors and tumors of the sella region are beyond the scope of the grant program. Projects that characterize genes and gene products of normal cellular development are not eligible, nor are epidemiological studies.
Geographic Focus: All States
Date(s) Application is Due: Jan 7
Amount of Grant: Up to 600,000 USD
Contact: Sally E. McNagny; (617) 279-2254; smcnagny@goldhirshfoundation.org
Internet: http://www.goldhirshfoundation.org/application_information.htm
Sponsor: Goldhirsh Foundation
95 Berkeley Street, Suite 201
Boston, MA 02116

Goldman Philanthropic Partnerships Program **2157**

Since its founding, the Program has solicited, validated and helped donors co-fund innovative research with the potential to quickly find cures for life-altering diseases, and supported these donors using for-profit business tools to insure that the research and philanthropic goals are met through proper project and resource management. The Program provides funding research for cures for a range of catastrophic diseases by exploring groundbreaking research in all disciplines of medicine, including those areas of research outside of conventional medicine such as alternative medicine. The Program usually co-funds a venture with at least one organizational, institutional, or corporate partner. The Partnerships maintains a strong interest in diseases of children and young adults, although it will fund research into any patient population where it believes there is a high likelihood of return on disease prevention, treatment or cure.
Geographic Focus: All States
Contact: Dr. Bruce E. Bloom, President; (312) 601-8856 or (312) 780-3440; fax (312) 780-3459; bruce@goldmanpartnerships.org
Internet: http://www.goldmanpartnerships.org/aboutus.html
Sponsor: Goldman Philanthropic Partnerships
70 West Madison Street, Suite 1500
Chicago, IL 60602

Goldsmith Awards Program **2158**

The program, funded by the Goldsmith-Greenfield Foundation, is sponsored by the Joan Shorenstein Center on the Press, Politics, and Public Policy at Harvard University's John F. Kennedy School of Government. The program annually awards the Goldsmith Prize for Investigative Reporting and two Goldsmith Book Prizes. The Goldsmith Prize is given annually to the journalist(s) whose investigative reporting in a story or series of related stories best promotes more effective and ethical conduct of government, the making of public policy, or the practice of politics. The Goldsmith Book Prizes are given annually to the author(s) of the best book that aims at improving the quality of government or politics through an examination of the press and government or the intersection of press and politics in the formation of public policy. Both prizes have a submission deadline of December 31, or a pstmark deadline of January 2. The program also grants Goldsmith Research Awards on an ongoing basis to scholars, graduate students, and journalists in the field of press/politics. Applications are available on the Web site.
Requirements: Publication must have occurred within 12 months (January 1 to December 31) preceding the submission deadline, and all submissions must be in English.
Restrictions: Edited volumes will not be accepted for the book prize. Books will not be accepted as submissions for the investigative reporting prize.
Geographic Focus: All States
Date(s) Application is Due: Jan 2
Amount of Grant: 2,000-25,000 USD
Contact: Alison Kommer, (617) 495-1329; fax (617) 495-8696; alison_kommer@harvard.edu
Internet: http://www.hks.harvard.edu/presspol/prizes_lectures/goldsmith_awards/
Sponsor: Harvard University
79 John F Kennedy Street
Cambridge, MA 02138

GPP Charles E. Culpeper Foundation Biomedical Pilot Projects Initiative Grants **2159**

The initiative is designed to encourage the investigation of novel ideas in cures for disease, particularly in the areas of molecular genetics, bio-engineering, and molecular pharmacology. Research into complimentary and alternative medicine will also be considered. Grants will be made on a one-time basis with the possibility for renewal for a second year upon reapplication. No more than 8 percent of the grant may be allocated to cover indirect costs. The purpose of these grants is to explore new and even untested hypotheses, thus substantial preliminary information is not required. Proposals are accepted throughout the year and each successful
Requirements: Nonprofit health care organizations, accredited medical schools, and universities in the United States are eligible.
Geographic Focus: All States
Date(s) Application is Due: Nov 16
Amount of Grant: Up to 25,000 USD
Contact: Grants Administrator; (847) 948-5512; fax (847) 948-5516
Internet: http://www.goldmanpartnerships.org/Culpeper/CulpeperBiomedicalPilotInitiativegrants.htm
Sponsor: Goldman Philanthropic Partnerships
155 N Pfingsten Road, Suite 109
Deerfield, IL 60015

Graduate Institute Hodson Trust Teacher Fellowships **2160**

The Hodson Trust Teacher Fellowship Program is funded through the generosity of the Hodson Trust, and offers a grant of 70% of cost during the summer for eligible full-time primary or secondary school teachers. This Fellowship is open to students from the states of Maryland, Virginia, Washington D.C., Delaware, Pennsylvania, New Jersey, Connecticut, and New York. Applicants must complete the online form and attach a letter from the principal of their school verifying their full time employment as a teacher in the academic year prior to enrollment to the institute.
Requirements: Full-time public or private primary and secondary school teachers with a minimum of one year of teaching experience are eligible for this fellowship.
Geographic Focus: Connecticut, Delaware, District of Columbia, Maryland, New Jersey, New York, Pennsylvania, Virginia
Contact: Janet Huang, (410) 626-2502; fax (410) 626-2885; janet.huang@sjca.edu
Internet: http://www.stjohnscollege.edu/GI/finaid.shtml
Sponsor: Saint John's College
P.O. Box 2800
Annapolis, MD 21404-2800

Grammy Foundation Grants **2161**

The foundation awards grants in two program areas. Research Projects grants are made to organizations and individuals to support efforts that advance the research and/or broad- reaching implementations of original research projects related to the impact of music study on early childhood development, the human development, and the medical and occupational well-being of music professionals. Archiving and Preservation Projects grants are made to organizations and individuals to support efforts that advance the archiving and preservation of the music and recorded sound heritage of the Americas. The foundation funds preservation of original, preexisting media and source material; preservation projects that follow the recommended methodology; projects of historical, artistic, cultural, and or/national significance; and archiving projects including the rescue, organization of, and access to preexisting media and materials. Application and guidelines are available online.
Restrictions: The foundation does not fund Recording Academy chapters, trustees, governors, officers, or staff; organizations that discriminate on the basis of race, sex, religion, national origin, disability, or age; projects promoting advocacy issues; a single organization or individual for more than three consecutive years; organizations or individuals not based in the Americas; purchase of collections; recording projects, demo tapes, or performance events; proposals for commercial purposes (i.e., CD reissue or textbook/ A/V package); purchase of repairs of equipment; purchase or repairs of musical instruments; maintenance or upgrading of computer systems; competitions or any expense associated with competitions; work toward academic degrees; music education or in-residence programs; documentaries; endowments and fundraising; buildings and facilities; marketing, publicity, design costs; or projects where copyright status is unknown.
Geographic Focus: All States
Date(s) Application is Due: Oct 1
Contact: Kristin Murphy, (310) 392-3777, ext. 8662; grants@grammy.com
Internet: http://www.grammy.com/GRAMMY_Foundation/Grants/
Sponsor: Grammy Foundation
3030 Olympic Boulevard
Santa Monica, CA 90405

Grass Foundation Marine Biological Laboratory Advanced Imaging Fellowships **2162**

The Grass Foundation Advanced Imaging Fellowships at Marine Biological Laboratory will support individuals working in any area of neuroscience at any stage of their career, ranging from graduate student to professor. This opportunity is intended to allow individuals to bring their preparations to the MBL for up to four weeks, learning and utilizing the available microscopy resources, gathering data and interacting with other imaging specialists. Grass Fellows in Advanced Imaging are integrated into the traditional Grass Fellows program during the summer, providing a rich environment and access to many activities at the MBL. As such, Imaging Fellows will receive bench space in the Grass Laboratory for sample preparation, and are expected to attend Grass Laboratory lab meetings. In addition, Imaging Fellows are encouraged to audit the many classes available including the neurobiology course, which offers lectures on the theory and practice of advanced microscopy. Support will include research supplies, travel, housing and meals at the Marine Biological Laboratory for the individual Fellow. Applicant instructions with detailed information of all attachments needed, such as proposal and vitae, are available at the Grass Foundation website.

Requirements: Applicants at any stage in their academic career are eligible. Priority is given to applicants with an experimental need to use high-end state of the art imaging systems. Prior research experience at the MBL, including previous Grass Fellowships, is neither required nor disqualifying. IRS regulations require that international Fellows (i.e., not U.S. citizens or resident aliens) hold a J-1 visa for the duration of the Fellowship. The duration of the program is two to four weeks during July. Fellows are expected to be in residence at the MBL the entire time. Up to three Fellows will be accepted.
Restrictions: Holders of H1-B visas are not eligible to apply. Funding is available for the individual Fellow only.
Geographic Focus: All States, All Countries
Date(s) Application is Due: Feb 20
Contact: Ann Woolford; (508) 289-7521; fax (508) 289-7931; awoolford@mbl.edu
Internet: http://www.grassfoundation.org/grass-fellowships-in-advanced-imaging
Sponsor: Grass Foundation
P.O. Box 241458
Los Angeles, CA 90024

Grass Foundation Marine Biological Laboratory Fellowships 2163
Grass Fellowships at the Marine Biological Laboratory (MBL) in Woods Hole, Massachusetts, support investigator-designed, independent research projects by scientists early in their career. Supported approaches include neurophysiology, biophysics, integrative neurobiology, neuroethology, neuroanatomy, neuropharmacology, systems neuroscience, cellular and developmental neurobiology, and computational approaches to neural systems. Grass Fellowships provide research support including laboratory space, animals, equipment, and supplies for one summer at the MBL. Additionally, the investigator, his/her spouse or legal domestic partner, and dependent children are provided housing, a daily meal allowance and round-trip travel to the MBL. Fellows function as an intellectual and social group within the MBL scientific community while sharing space in the Grass Laboratory. In a weekly private seminar series, eminent investigators at the MBL discuss their work with the Fellows. In addition, a yearly Forbes Lecturer will spend a portion of the summer in the Grass Lab interacting with Fellows. From 8 to 10 Fellowships are awarded annually. The Fellowship last from late May to early August. The application and supplemental material are available at the Foundation website.
Requirements: Early investigators (late stage predoctoral trainees and beyond) are eligible to apply. This includes applicants with prior experience at MBL or with the Grass Foundation. MBL course alumni are encouraged to apply. Priority is given to applicants with a demonstrated commitment to pursuing a research career. International Fellows must hold either a J-1 or H-1B visa for the entire duration of the Fellowship.
Geographic Focus: All States, All Countries
Date(s) Application is Due: Dec 5
Contact: Ann Woolford; (508) 289-7521; fax (508) 289-7931; awoolford@mbl.edu
Internet: http://www.grassfoundation.org/grass-fellowship-program
Sponsor: Grass Foundation
P.O. Box 241458
Los Angeles, CA 90024

Greater Saint Louis Community Foundation Grants 2164
The Greater Saint Louis Community Foundation (GSLCF) was founded in 1915, one year after the first community foundation was established in Cleveland, Ohio. Currently GSLCF administers over 400 individual charitable funds that total $170 million in assets. These funds annually make over $17 million in grants that shape the greater Saint Louis region, touch communities across the nation, and reach across the globe. Historically, the mission of GSLCF has been two-fold: to serve donors and ensure that their dollars work in line with the goals that are important to them; and to promote charitable giving through community investment in nonprofit organizations capable of addressing community issues in measurable ways. To this end GSLCF maintains an online database YOURGivingLink to help donors find deserving nonprofit organizations that match their giving interests. Nonprofits are encouraged to register their organizations with the database; the link is available at the GSLCF website.
Requirements: While the Foundation predominately supports St. Louis area nonprofits, grants are also made to national and international charities.
Geographic Focus: Illinois, Missouri
Contact: Amy Basore Murphy, Director of Scholarships and Donor Services; (314) 588-8200, ext. 139 or (314) 880-4965; fax (314) 588-8088; amurphy@stlouisgives.org
Internet: http://www.stlouisgives.org/charities/
Sponsor: Greater Saint Louis Community Foundation
319 North Fourth Street, Suite 300
Saint Louis, MO 63102-1906

Greater Worcester Community Foundation Jeppson Memorial Fund for Brookfield Grants 2165
The Fund provides money to civic and community projects that help improve the lives of residents and enrich the cultural environment. The Fund provides support for: cultural or artistic performances; public seminars; festivals or exhibitions; services that help frail or vulnerable citizens, or that contribute to public health and safety; opportunities for educational enrichment; youth involvement in recreation, sports and the arts; and projects that foster community awareness and connections among different groups. Application information is available online.
Requirements: Any nonprofit or civic organization that serves the residents of Brookfield may apply.
Restrictions: Grant funds may not be used for expenses already incurred by the applicant. Fund awards are not intended to replace municipal funds.
Geographic Focus: Massachusetts
Date(s) Application is Due: Jul 15
Contact: Pamela B. Kane, (508) 755-0980; pkane@greaterworcester.org
Internet: http://www.greaterworcester.org/grants/Jeppson.htm
Sponsor: Greater Worcester Community Foundation
370 Main Street, Suite 650
Worcester, MA 01608-1738

Great Lakes Fishery Commission Pilot Research Project Grants 2166
Great Lakes Fishery Commission Pilot Research Project grants are designed to explore new hypotheses and concepts and to acquire preliminary data to guide the develpment of full research proposals. Pilot project proposals may be submitted at any time and must be less than $10,000. Funding decisions will be made after the March and September SLRB meetings. Deadlines for submission of pilot project proposals are January 15th for consideration in March or July 15th for consideration in September.
Geographic Focus: All States
Date(s) Application is Due: Jan 15; Jul 15
Amount of Grant: Up to 10,000 USD
Contact: Mike Siefkes, Sea Lamprey Program Specialist; (734) 662-3209, ext. 22; fax (734) 741-2010; msiefkes@glfc.org or info@glfc.org
Internet: http://www.glfc.org/research/SRpp.php
Sponsor: Great Lakes Fishery Commission
2100 Commonwealth Boulevard, Suite 100
Ann Arbor, MI 48105

Great Lakes Fishery Commission Sea Lamprey Research Grants 2167
The purpose of the Sea Lamprey Research program is to conduct research that will provide the information needed to further the integrated management of sea lampreys, including the development of novel, effective strategies for their control. Biological, ecological, and management-related research on sea lampreys and their effects on Great Lakes fish communities and fisheries is supported. Theme areas include: assessment; barriers and trapping; lampricides; pheromones, and sterile males. Research projects that do not fit into an existing theme area are supported under the category Non-Theme Research. Highly meritorious concepts, not connected to existing theme areas may be funded. Projects range from $10,000 to $100,000 per year. The average award ranges from $35,000 to $50,000 per year. Most projects are two to three years in duration. Annual funding available for Sea Lamprey Research program projects is typically $500,000.
Geographic Focus: All States
Date(s) Application is Due: Jun 1
Amount of Grant: 10,000-100,000 USD
Contact: Mike Siefkes, Sea Lamprey Program Specialist; (734) 662-3209, ext. 22; fax (734) 741-2010; msiefkes@glfc.org or info@glfc.org
Internet: http://www.glfc.org/research/sldesc.php
Sponsor: Great Lakes Fishery Commission
2100 Commonwealth Boulevard, Suite 100
Ann Arbor, MI 48105

Great Lakes Fishery Trust Access Grants 2168
The Great Lakes Fishery Trust's (GLFT) Access to the Great Lakes Fishery grant category focuses on significantly increasing access to the Great Lakes fishery for shore-based angling and tribal fishing. The goal of this funding category is to improve or create opportunities for shore-based access to fishing for Great Lakes species for use by tribal and/or recreational users. The GLFT supports its access goals through a combination of competitive and funder-directed grantmaking. Supported activities may include: new construction of access sites; upgrades and renovations to existing access sites; engineering and feasibility studies for proposed access sites; land acquisition to support subsequent access site development; and communication and outreach efforts regarding existing shore-based angling opportunities. Fishery Access proposals are due on August 23 each year.
Requirements: Organizations that are eligible to apply for GLFT grants include nonprofit organizations with a 501(c)3 designation from the IRS (or nongovernmental organizations that hold charitable status in their country), as well as educational and governmental (including tribal) organizations.
Geographic Focus: Illinois, Indiana, Iowa, Michigan, Ohio, Wisconsin
Date(s) Application is Due: Jan 25
Contact: Jonathon Beard; (517) 371-7468; fax (517) 484-6549; jbeard@glft.org
Internet: http://www.glft.org/site/fishing_access_grant.shtml
Sponsor: Great Lakes Fishery Trust
600 W. Saint Joseph, Suite 10
Lansing, MI 48933-2265

Great Lakes Fishery Trust Ecological and Biological Fisheries Research to Inform Management Grants 2169
The Great Lakes Fishery Trust pursues its fisheries research efforts through competitive grantmaking within established research themes. Fundable activities include hypothesis-driven research as well as tools, resources, and capacity-building efforts that support the development or use and application of research. The GLFT will support research that: builds research capacity and management expertise needed to understand and manage the Great Lakes ecosystem for sustainable production of valuable species; enhances the ability of managers and agencies to respond to changes in the fishery and ecosystem; and/or reduces the cost of fishery management in relation to benefits provided. he main theme areas for the GLFT's gran tmaking in Ecological and Biological Fisheries Research are: fish health; recruitment; lake sturgeon rehabilitation; ecosystem modeling; and emerging issues. Pre-proposals for Ecological and Biological Fisheries research projects are due by January 22 of each year. The online application will be available in November.

Requirements: Organizations that are eligible to apply for GLFT grants include nonprofit organizations with a 501(c)3 designation from the IRS (or nongovernmental organizations that hold charitable status in their country), as well as educational and governmental (including tribal) organizations.
Geographic Focus: Illinois, Indiana, Iowa, Michigan, Ohio, Wisconsin
Contact: Jonathon Beard; (517) 371-7468; fax (517) 484-6549; jbeard@glft.org
Internet: http://www.glft.org/site/dam_management.shtml
Sponsor: Great Lakes Fishery Trust
600 W. Saint Joseph, Suite 10
Lansing, MI 48933-2265

Great Lakes Protection Fund Grants 2170

The Great Lakes Protection Fund welcomes preproposals for projects that enhance the health of the Great Lakes ecosystem. Applicants should propose projects that will return the greatest ecosystem benefits. Current funding interests include preventing biological pollution, restoring natural flow regimes; and using market mechanisms for environmental improvement. Additional projects are sought to add to, and expand fund-supported work in these areas. The fund also welcomes projects that are designed to identify and explore other master variables that if acted upon, will result in tangible improvements to the health of the Great Lakes ecosystem. Proposals may be submitted at any time. The board makes grant decisions at its March, June, September, and December meetings.
Requirements: Nonprofit organizations (including environmental organizations, trade associations, and universities), for-profit businesses, government agencies, and individuals are eligible for fund support.
Restrictions: The fund does not give to general operating funds, environmental education, and groups from Indiana.
Geographic Focus: All States
Amount of Grant: 25,000-420,000 USD
Contact: J. David Rankin, Program Director; (847) 425-8150; fax (847) 424-9832; drankin@glpf.org or preproposals@glpf.org
Internet: http://www.glpf.org/about-the-fund
Sponsor: Great Lakes Protection Fund
1560 Sherman Avenue, Suite 880
Evanston, IL 60201

Green Diamond Charitable Contributions 2171

Green Diamond Resource Company is a fifth-generation, family-owned forest-products company that has operations in the states of California and Washington. The mission of Green Diamond's contributions program is to improve the quality of life in communities where the company has a significant number of employees living and working; and to serve as a catalyst for employees to become involved and to provide leadership in their communities. The company has supported a broad range of organizations in its operating communities that address key community needs such as education, health services, economic development, the arts, and more. The company also provides support to environmental and conservation projects, such as environmental education, habitat restoration, and research relevant to the timber industry, as well as other projects that impact the timber business. Contributions are generally made in locations where the company has operations. To the extent possible, contributions will support organizations of interest to, or recommended by, Green Diamond employees. Green Diamond prefers to make capital contributions that will benefit the operating communities for the long term as opposed to contributing operating funds. Generally, support is committed for one year at a time and in amounts less than $5,000. The company supports the United Way in all of its operating communities. The company accepts applications twice a year, generally in May and in late summer/early fall. Interested applicants should call or mail the contact person given for their state to request application materials. More detailed information about application deadlines will be communicated at that time.
Requirements: Organizations that serve Pierce, Mason, Thurston, and Grays Harbor counties in Washington or Del Norte and Humboldt counties in California are eligible to apply. Criteria taken into account in determining the amount of any contributions are as follows: degree of support from company employees; relative size and importance of company operations in the community (balance among Green Diamond communities); needs of organization or program for which funding is requested; amount of previous company contributions to the organization; amount committed by other companies, foundations, and/or governments (projects should demonstrate broad-based community support); and proximity of the requesting organization to Green Diamond operations or administrative offices.
Geographic Focus: California, Washington
Amount of Grant: Up to 5,000 USD
Contact: Jacki Deuschle, Public Affairs Manager; (707) 668-4488; fax (707) 668-4402; jdeuschle@greendiamond.com
Internet: http://www.greendiamond.com/charitable-giving/
Sponsor: Green Diamond Resource Company
1301 Fifth Avenue, Suite 2700
Seattle, WA 98101-2613

Green Foundation Human Services Grants 2172

The Foundation's mission is to uncover new opportunities, encourage growth and ultimately effect positive change within those institutions that best reflect the core focus areas and the communities it serves. Within the human services area, the Foundation focuses on institutions that provide hope and support to those least able to help themselves as well as the general community; including children, adolescents, the elderly, the homeless, and families who struggle with domestic abuse.
Requirements: 501(c)3 nonprofits (as per the IRS Service Code of 1986) are eligible. Most grant making is limited to institutions that serve the Los Angeles community; however the Foundation will consider requests beyond this geographic boundary for those institutions with the potential to impact communities statewide or nationally.
Restrictions: The Foundation does not provide funds for: those with net assets or fund balances of less than $100,000; multi-year commitments; annual meetings, conferences, and/or seminars; religious programs; capital campaigns; direct mail campaigns; conduit institutions, unified funds, fiscal agents, or institutions using grant funds from donors to support other institutions or individuals; private foundations; or individuals.
Geographic Focus: All States
Amount of Grant: 1,000-150,000 USD
Contact: Kylie Wright; (626) 793-6200, ext. 1; fax (626) 793-6201; kylies@ligf.org
Internet: http://www.ligf.org/humanservices.php
Sponsor: Green Foundation
225 South Lake Avenue, Suite 1410
Pasadena, CA 91101

Greenspun Family Foundation Grants 2173

The Greenspun Family Foundation supports many causes with an emphasis on education, health, children, Jewish issues, and the greater Las Vegas community. Preference is given to requests from the Las Vegas area. There are no application forms or deadlines. Applicants should submit a letter of inquiry. Information about current programs supported is available at the Foundation website.
Geographic Focus: Nevada
Amount of Grant: 50,000-2,000,000 USD
Contact: Dr. Brian Cram, Director; (702) 259-2323 or (702) 259-4023; fax (702) 259-4019; brian.cram@lasvegassun.com
Internet: http://www.thegreenspuncorp.com/philanthropy.php
Sponsor: Greenspun Family Foundation
901 North Green Valley Parkway, Suite 210
Henderson, NV 89074

Greenwall Foundation Bioethics Grants 2174

The Foundation provides funding for physicians, lawyers, philosophers, theologians and other professionals to address micro and macro issues in bioethics, providing guidance for those engaged in decision-making at the bedside as well as those responsible for shaping institutional and public policy. The Foundation is especially interested in the work of junior investigators and pilot projects that may lead to NIH support, and it is prepared to address issues regarded by some as sensitive or potentially controversial.
Restrictions: The Foundation is not normally interested in proposals to support equipment purchase, facility construction or renovation, or general operating expenses, and will not normally consider grants to private foundations, endowment funds, or individual applicants.
Geographic Focus: All States
Date(s) Application is Due: Feb 1; Aug 1
Amount of Grant: 5,000-50,000 USD
Contact: Sam Teigen; (212) 679-7266; fax (212) 679-7269; steigen@greenwall.org
Internet: http://www.greenwall.org/guidebio.htm
Sponsor: Greenwall Foundation
420 Lexington Avenue, Suite 2500
New York, NY 10170

Greenwall Foundation Faculty Scholars Program in Bioethics 2175

The Program is a career development program that enables junior faculty members to carry out original research on policy and moral dilemmas at the intersection of ethics and the life sciences. To maximize scholarly development, three years of support are provided, requiring a 50% time commitment (carefully monitored) in each of the three years. Each year the Scholars participate in a learner-centered educational program that includes required attendance at two workshops (spring and fall). Continued involvement in this activity, after each Scholar's support period is over, provides Scholar-alumni/ae with ongoing professional development and engages them in the training of successor cohorts. Preliminary applications are due on December 3. Approximately 15 to 20 applicants will be invited to submit full proposals due in February.
Requirements: Applicants must be junior faculty members holding at least a 60% appointment at a university or non-profit research institute in the U.S. Priority will be given to applicants who are below the rank of Associate Professor, who have not received a comparable career development award, and whose work will have an impact on public policy or clinical practice. Within this group, priority will be given to applicants whose research addresses innovative and emerging topics.
Geographic Focus: All States
Date(s) Application is Due: Dec 3
Contact: Sam Teigen, Grants Administrator; (415) 476-6241 or (212) 679-7266; fax (212) 679-7269; steigen@greenwall.org or greenwall@medicine.ucsf.edu
Internet: http://www.greenwallfsp.org/
Sponsor: Greenwall Foundation
420 Lexington Avenue, Suite 2500
New York, NY 10170

Greenwall Foundation Fellowship Program in Bioethics and Health Policy 2176

The Fellowship program and its faculty cover much of the waterfront of issues and methodologies in bioethics, from ethics and advances in biomedical science, to ethics and public health and health policy, to ethics and clinical care. The defining focus of the Program, however, is bioethics and public policy. The Program is intended for people who are early in their careers. Greenwall applicants are subdivided by discipline (medicine, philosophy and law) and reviewed by a selection committee composed of program faculty with relevant disciplinary backgrounds. Typically, only one Fellow per year is accepted from each discipline.

Geographic Focus: All States
Date(s) Application is Due: Dec 1
Amount of Grant: 111,650 USD
Contact: Maria Merritt, (410) 614-6335; mmerritt@jhsph.edu
Internet: http://www.greenwall.org/guideaffil.htm
Sponsor: Greenwall Foundation
420 Lexington Avenue, Suite 2500
New York, NY 10170

Gruber Foundation Cosmology Prize **2177**

The Gruber Foundation honors and encourages educational excellence, social justice and scientific achievements that better the human condition. Foundation is a private, United States-based philanthropic organization established in 1993 under the 501(c)3 section of U.S. Corporate Law. It is funded entirely by Peter and Patricia Gruber, who serve as its Chairman and President, respectively. The Foundation is headquartered at Yale University. A major focus of the Foundation's philanthropy is its International Prize Program, created to recognize excellence in science and humanities by highlighting five fields that create a better world: Cosmology, Genetics, Neuroscience, Justice, and Women's Rights. The Cosmology Prize, cosponsored by the International Astronomical Union since 2001, honors a leading cosmologist, astronomer, astrophysicist or scientific philosopher for theoretical, analytical, conceptual or observational discoveries leading to fundamental advances in our understanding of the universe. The Prize acknowledges and encourages further exploration in a field that shapes the way we perceive and comprehend our universe. In doing so, The Gruber Foundation seeks to extend the pioneering legacy of, among others, Plato and Aristotle; Ptolemy and Copernicus; Brahe, Kepler, and Galileo; Newton and Halley; Einstein and Hubble. Individuals whose achievements have produced fundamental, advances in our understanding of the structure and evolution of the universe may be nominated. A gold medal and unrestricted $500,000 cash prize are presented annually.
Requirements: Nominations for the Cosmology Prize are invited annually from the fields of astronomy, cosmology, mathematics, and the philosophy of science. Individuals, organizations, and institutions that are active in or have an appreciation for contemporary cosmological research and study may submit nominations.
Geographic Focus: All States, All Countries
Date(s) Application is Due: Dec 15
Amount of Grant: 500,000 USD
Contact: Bernetia Akin, Foundation Contact; (203) 432-6231 or (340) 775-4430; nominations@gruberprizes.org or info@gruberprizes.org
Internet: http://www.gruberprizes.org/GruberPrizes/Cosmology.php
Sponsor: Gruber Foundation
157 Church Street, 9th Floor
New Haven, CT 06510

Gruber Foundation Gentics Prize **2178**

The Gruber Foundation honors and encourages educational excellence, social justice and scientific achievements that better the human condition. Foundation is a private, United States-based philanthropic organization established in 1993 under the 501(c)3 section of U.S. Corporate Law. It is funded entirely by Peter and Patricia Gruber, who serve as its Chairman and President, respectively. The Foundation is headquartered at Yale University. A major focus of the Foundation's philanthropy is its International Prize Program, created to recognize excellence in science and humanities by highlighting five fields that create a better world: Cosmology, Genetics, Neuroscience, Justice, and Women's Rights. The Genetics Prize, established in 2001, is presented to a leading scientist, or up to three, in recognition of groundbreaking contributions to any realm of genetics research. A gold medal and unrestricted $500,000 cash award is awarded for fundamental insights in the field of genetics. These may include original discoveries in genetic function, regulation, transmission, and variation, as well as in genomic organization. Nominations for the Prize close on December 15.
Requirements: Nominations may be submitted by individuals, organizations, and institutions that are active in or have an appreciation for contemporary genetic research or problems. Individuals who have made original discoveries in the fields of genetic function, regulation, transmission, or variation or in genomic organization are eligible.
Geographic Focus: All States
Date(s) Application is Due: Dec 15
Contact: Bernetia Akin, Foundation Contact; (203) 432-6231 or (340) 775-4430; nominations@gruberprizes.org or info@gruberprizes.org
Internet: http://www.gruberprizes.org/GruberPrizes/Genetics.php
Sponsor: Gruber Foundation
157 Church Street, 9th Floor
New Haven, CT 06510

Gruber Foundation International Astronomical Union Fellowships **2179**

With the aim to promote the science of cosmology and other branches of astronomy, the Gruber Foundation has created the TGF Fellowship Program. One Fellowship will be awarded every year in the amount of $50,000. The award can cover travel, subsistence, and research expenses during a postdoctoral appointment for a period which is typically of one year duration starting no later than 1st October of the award year, but may be extended to cover up to two years. The Fellowship may be held in conjunction with any other fellowship or award. The Fellowship will be awarded to an extremely promising, young astrophysicist, working in any field of astrophysics, either theoretical, observational or experimental. There are no limitations on nationality, but preference will be given to applicants from countries in difficult economic conditions.
Geographic Focus: All States, All Countries
Amount of Grant: 50,000 USD
Contact: Bernetia Akin, Foundation Contact; (203) 432-6231 or (340) 775-4430; nominations@gruberprizes.org or info@gruberprizes.org
Internet: http://www.gruberprizes.org/GruberPrizes/YoungScientists_Astronomical.php
Sponsor: Gruber Foundation
157 Church Street, 9th Floor
New Haven, CT 06510

Gruber Foundation International Research Award **2180**

The Gruber Foundation International Research Award in Neuroscience recognizes two young neuroscientists for outstanding research and educational pursuit in an international setting. If the nominee is a U.S. citizen or permanent resident, the nominee must be studying or working at an institution located outside the U.S; if a non-U.S. citizen or permanent resident, the nominee must be studying or working at an institution that is located in the U.S. Nominees must be active in research at the time the award is given. Successful candidates are graduate students, postdoctoral fellows, or research associates. Each recipient receives a $25,000 prize, in addition to air or ground transportation to the SfN annual meeting with two nights hotel accommodations, and complimentary registration. All nominating materials must be submitted using the SfN's award nomination site.
Requirements: Although nominees do not need to be SfN members, they must be nominated or endorsed by an SfN member.
Restrictions: Self-nomination is not permitted. No person may nominate more than one candidate. Current SfN officers and councilors are ineligible for nomination.
Geographic Focus: All States, Puerto Rico, Argentina, Australia, Brazil, Canada, China, Denmark, India, Israel, Malaysia, Mexico, New Zealand, Nigeria, Norway, Singapore, South Korea, Spain, Switzerland, Turkey, Ukraine, United Kingdom
Date(s) Application is Due: Jun 1
Amount of Grant: 25,000 USD
Contact: Pat R. Levitt, PhD, Chairperson; (323) 442-1509; fax (323) 442-2145; awards@sfn.org or plevitt@med.usc.edu
Internet: http://www.sfn.org/index.aspx?pagename=FellowshipAndAwards_gruberinternational
Sponsor: Gruber Foundation
157 Church Street, 9th Floor
New Haven, CT 06510

Gruber Foundation Neuroscience Prize **2181**

The Gruber Foundation honors and encourages educational excellence, social justice and scientific achievements that better the human condition. Foundation is a private, United States-based philanthropic organization established in 1993 under the 501(c)3 section of U.S. Corporate Law. It is funded entirely by Peter and Patricia Gruber, who serve as its Chairman and President, respectively. The Foundation is headquartered at Yale University. A major focus of the Foundation's philanthropy is its International Prize Program, created to recognize excellence in science and humanities by highlighting five fields that create a better world: Cosmology, Genetics, Neuroscience, Justice, and Women's Rights. The Neuroscience Prize honors scientists for major discoveries that have advanced the understanding of the nervous system. The Prize, established in 2004, is an unrestricted cash award of $500,000, a gold medal inscribed with the recipient's name, and a citation describing the achievement for which the recipient is being honored. It is awarded each year to a person or persons chosen by a distinguished advisory board of neuroscience experts from nominations that are received from around the world. Nominations for the Prize close on December 15.
Requirements: Nominations for the Neuroscience Prize are invited annually and may be submitted by individuals, organizations, and institutions that are active in or have an appreciation for contemporary neuroscience research and study. Individuals from anywhere in the world who have conducted highly distinguished research in the field of the brain, spinal cord, or peripheral nervous system are eligible.
Geographic Focus: All States, All Countries
Date(s) Application is Due: Dec 15
Amount of Grant: 500,000 USD
Contact: Bernetia Akin, Foundation Contact; (203) 432-6231 or (340) 775-4430; nominations@gruberprizes.org or info@gruberprizes.org
Internet: http://www.gruberprizes.org/GruberPrizes/Neuroscience.php
Sponsor: Gruber Foundation
157 Church Street, 9th Floor
New Haven, CT 06510

Gruber Foundation Prize for Global Justice and Women's Rights **2182**

The Gruber Foundation honors and encourages educational excellence, social justice and scientific achievements that better the human condition. Foundation is a private, United States-based philanthropic organization established in 1993 under the 501(c)3 section of U.S. Corporate Law. It is funded entirely by Peter and Patricia Gruber, who serve as its Chairman and President, respectively. The Foundation is headquartered at Yale University. A major focus of the Foundation's philanthropy is its International Prize Program, created to recognize excellence in science and humanities by highlighting five fields that create a better world: Cosmology, Genetics, Neuroscience, Justice, and Women's Rights. Since 2001, the Justice Prize has honored individuals who have advanced the cause of justice as delivered through the legal system. And, since 2003, the Women's Rights Prize has honored individuals who have made significant contributions to human rights that advance the rights of women and girls around the world. In 2012, the two prizes have been combined into the Prize for Global Justice and Women's Rights. It includes a gold medal and unrestricted $500,000 cash award, and is an international award presented annually. Recipients are selected by a distinguished panel of international legal experts from nominations that are received from around the world. The annual deadline is December 15.

Geographic Focus: All States, All Countries
Date(s) Application is Due: Dec 15
Amount of Grant: 500,000 USD
Contact: Bernetia Akin, Foundation Contact; (203) 432-6231 or (340) 775-4430; nominations@gruberprizes.org or info@gruberprizes.org
Internet: http://www.gruberprizes.org/GruberPrizes/Justice.php
Sponsor: Gruber Foundation
157 Church Street, 9th Floor
New Haven, CT 06510

Gruber Foundation Rosalind Franklin Young Investigator Award **2183**
The Gruber Foundation funds the Rosalind Franklin Young Investigator Award – a career development research award of $75,000 over three years-to 2 young woman geneticists from anywhere in the world. The award, which honors the groundbreaking contributions of Dr. Rosalind Franklin, is intended to inspire and support new generations of women in the field of genetics. One award will fund genetics research in human and non-human mammals, and the other will fund genetics research in model organisms. This unrestricted cash award is presented every three years.
Requirements: The successful candidate must be in her first one to three years of an independent faculty-level position in any area of genetics.
Geographic Focus: All States, All Countries
Date(s) Application is Due: Dec 15
Amount of Grant: 75,000 USD
Contact: Bernetia Akin, Foundation Contact; (203) 432-6231 or (340) 775-4430; nominations@gruberprizes.org or info@gruberprizes.org
Internet: http://www.gruberprizes.org/GruberPrizes/YoungScientists_Rosalind Franklin.php
Sponsor: Gruber Foundation
157 Church Street, 9th Floor
New Haven, CT 06510

Gruber Foundation Weizmann Institute Awards **2184**
The Weizmann Institute is a center of basic multidisciplinary scientific research and graduate study, addressing crucial problems in technology, medicine and health, energy, agriculture, and the environment. Supported by personal funds donated by Peter and Patricia Gruber, the Gruber Awards are used to fund the research of new scientists at the Weizmann Institute and the scientific travel of selected postdoctoral fellows and doctoral students through an award of $30,000 per year over a three-year period. A faculty committee advises the President of the Institute as to a worthy Gruber Award recipient. The inaugural award was presented to Elad Schneidman,a senior scientist in the Institute's Department of Neurobiology, in May 2007. One new scientist will receive the award annually. The Award provides vital funding for gifted new scientists at the beginning of their research activities, and helps enrich their work.
Geographic Focus: All States, All Countries
Date(s) Application is Due: Dec 15
Contact: Bernetia Akin, Foundation Contact; (203) 432-6231 or (340) 775-4430; nominations@gruberprizes.org or info@gruberprizes.org
Internet: http://www.gruberprizes.org/GruberPrizes/YoungScientists_Weizmann.php
Sponsor: Gruber Foundation
157 Church Street, 9th Floor
New Haven, CT 06510

GSA Research Grants **2185**
The society offers research grants to provide partial support of master's and doctoral thesis research in earth science for graduate students at universities in the United States, Canada, Mexico, and Central America. Application forms are available on the Web site. Faxed or email applications will not be accepted.
Requirements: Eligibility is restricted to GSA members. Applicant must be enrolled in a graduate program in the United States, Canada, Mexico, or Central America and must outline the proposed research. Application must be on current GSA forms, which are available from the geology departments of most colleges and universities in the United States or upon request from the Research Grants Administrator of the society. Completed forms must be accompanied by two faculty evaluations of the applicant. Applicants must use forms for the current grant year.
Restrictions: The program does not offer scholarships to be used for tuition, housing, etc., nor are students in seeking acceptance into graduate school given assistance.
Geographic Focus: All States
Date(s) Application is Due: Feb 1
Contact: Program Officer; (303) 357-1028; awards@geosociety.org
Internet: http://www.geosociety.org/profdev/grants/index.htm
Sponsor: Geological Society of America
P.O. Box 9140
Boulder, CO 80301-9140

GUITS Grants for the Publication of Scholarly Books and Journals **2186**
Grants for the Publication of Scholarly Books and Journals are intended to cover part of the publication costs of scholarly books and journals in the field of Turkish Studies to be published in the U.S. Most grants are disbursed as reimbursements and require the presentation of documentation before payment is made. Grant announcements are sent out in mid-May.
Requirements: To be eligible for a grant, applicants should be: faculty in any field of the social sciences and/or humanities; and U.S. citizens or permanent residents at the time of the application.
Geographic Focus: All States
Date(s) Application is Due: Mar 7
Contact: David C. Cuthell; (202) 687-0295; fax (202) 687-3780; dcc@turkishstudies.org
Internet: http://www.turkishstudies.org/grantinfo.html
Sponsor: Georgetown University Institute of Turkish Studies
Intercultural Center, Box 571033, Room 305R
Washington, DC 20057-1033

GUITS Grants for Undergraduate Study **2187**
The Institute of Turkish Studies is a non-profit, private educational institution devoted solely to the advancement of training, research, and teaching in the field of Ottoman and modern Turkish Studies. Grants for Undergraduate Study are for students who wish to participate in study abroad programs in Turkey. These awards will be made to institutions, which will be responsible for their distribution to qualified applicants. ITS welcomes applications from institutions seeking to either expand or develop overseas programs in Turkey. The maximum award is $10,000.
Requirements: To be eligible for a grant, applicants should be: faculty in any field of the social sciences and/or humanities; and U.S. citizens or permanent residents at the time of the application.
Geographic Focus: All States
Date(s) Application is Due: Mar 7
Amount of Grant: 1,000-3,000 USD
Contact: David C. Cuthell; (202) 687-0295; fax (202) 687-3780; dcc@turkishstudies.org
Internet: http://www.turkishstudies.org/grantinfo.html
Sponsor: Georgetown University Institute of Turkish Studies
Intercultural Center, Box 571033, Room 305R
Washington, DC 20057-1033

GUITS Matching Seed-Money Grants for New Positions in Turkish Studies **2188**
Matching Seed-Money Grants for New Positions in Turkish Studies are intended for institutions. ITS offers matching seed money grants to universities for the establishment of these new positions in any discipline in the field of Turkish Studies. Universities are expected to make commitment to establish a tenure-track position for this position. The maximum award is .50 FTE of a junior faculty appointment.
Requirements: To be eligible for a grant, applicants should be: faculty in any field of the social sciences and/or humanities; and U.S. citizens or permanent residents at the time of the application.
Geographic Focus: All States
Date(s) Application is Due: Mar 7
Contact: David C. Cuthell; (202) 687-0295; fax (202) 687-3780; dcc@turkishstudies.org
Internet: http://www.turkishstudies.org/grantinfo.html
Sponsor: Georgetown University Institute of Turkish Studies
Intercultural Center, Box 571033, Room 305R
Washington, DC 20057-1033

GUITS Post-Doctoral Summer Travel-Research Grants **2189**
The Institute of Turkish Studies is a non-profit, private educational institution devoted solely to the advancement of training, research, and teaching in the field of Ottoman and modern Turkish Studies. The grant is intended to provide partial support for travel and research to Turkey for those who hold a Ph.D. in social sciences or humanities discipline. The time period for the grant varies with each individual application. Normally, the recipients of the grants are expected to spend a minimum of four weeks in Turkey. Maximum award is round-trip airfare to Turkey. Application information is available online.
Requirements: Eligible recipients must be U.S. citizens or U.S. permanent residents and currently live/work in the U.S.
Geographic Focus: All States
Date(s) Application is Due: Mar 7
Contact: David C. Cuthell; (202) 687-0295; fax (202) 687-3780; dcc@turkishstudies.org
Internet: http://www.turkishstudies.org/grantinfo.html
Sponsor: Georgetown University Institute of Turkish Studies
Intercultural Center, Box 571033, Room 305R
Washington, DC 20057-1033

GUITS Research Grants in Comparative Studies of Modern Turkey **2190**
The Institute of Turkish Studies is a non-profit, private educational institution devoted solely to the advancement of training, research, and teaching in the field of Ottoman and modern Turkish Studies. A significant portion of the project should be devoted comparatively to one or more states or political entities in Europe, Latin America, the Middle East, and Asia in addition to the Turkish Republic. The grants are primarily, but not exclusively, in the field of Political Science. The grant is for the academic year, and permission to postpone the tenure of the grant beyond this period will normally not be given. Application information is available online.
Requirements: Grants are awarded to graduate students at the dissertation research stage and for post-doctoral scholars in the U.S. who study aspects of the Republic of Turkey (post 1922) in a comparative context. Applicants must be U.S. citizens or permanent residents in the U.S. and affiliated with a university at the U.S. at the time of the application.
Geographic Focus: All States
Date(s) Application is Due: Mar 7
Contact: David C. Cuthell; (202) 687-0295; fax (202) 687-3780; dcc@turkishstudies.org
Internet: http://www.turkishstudies.org/grantinfo.html
Sponsor: Georgetown University Institute of Turkish Studies
Intercultural Center, Box 571033, Room 305R
Washington, DC 20057-1033

GUITS Sabbatical Research Grants 2191

The Institute of Turkish Studies is a non-profit, private educational institution devoted solely to the advancement of training, research, and teaching in the field of Ottoman and modern Turkish Studies. These grants are for the academic year and permission to postpone the tenure of the grant beyond this period will normally not be given. Application information is available online.

Requirements: To be eligible for a grant, applicants should be: faculty in any field of the social sciences and/or humanities; and U.S. citizens or permanent residents at the time of the application. The grants will be awarded only to applicants who certify that they are taking sabbatical leave and will be conducting research in the field of Turkish Studies.

Geographic Focus: All States

Date(s) Application is Due: Mar 7

Amount of Grant: 25,000 USD

Contact: David C. Cuthell; (202) 687-0295; fax (202) 687-3780; dcc@turkishstudies.org

Internet: http://www.turkishstudies.org/grantinfo.html

Sponsor: Georgetown University Institute of Turkish Studies
Intercultural Center, Box 571033, Room 305R
Washington, DC 20057-1033

GUITS Summer Research Grants in Turkey for Graduate Students 2192

The Institute of Turkish Studies is a non-profit, private educational institution devoted solely to the advancement of training, research, and teaching in the field of Ottoman and modern Turkish Studies. These grants are awarded for summer travel to Turkey to carry out research projects. The time period for grants varies with each individual application. Normally, the recipients of the Summer Research Grants are expected to spend a minimum of two months in Turkey. Application information is available online.

Requirements: Eligible recipients must be: graduate students in any field of the social sciences and/or humanities in the United States; U.S. citizens or permanent residents at the time of the application; and currently not engaged in dissertation writing.

Geographic Focus: All States

Date(s) Application is Due: Mar 7

Amount of Grant: 1,000-2,000 USD

Contact: David C. Cuthell; (202) 687-0295; fax (202) 687-3780; dcc@turkishstudies.org

Internet: http://www.turkishstudies.org/grantinfo.html

Sponsor: Georgetown University Institute of Turkish Studies
Intercultural Center, Box 571033, Room 305R
Washington, DC 20057-1033

H.A. and Mary K. Chapman Charitable Trust Grants 2193

H. Allen Chapman was born in Colorado in 1919. In 1976, he established the H.A. and Mary K. Chapman Charitable Trust, a perpetual charitable private foundation that maintains endowments to fund charitable grants to public charities. The trustees and staff that administer the foundation also provide public stewardship through service to charitable organizations and causes. A major charitable focus of H. A. Chapman during his life, and the lives of his philanthropic parents, James A. and Leta Chapman, was education and medical research. Though not limited geographically, most grants and public service are within Oklahoma. Grants to human services and civic and community programs and projects are primarily focused in the area of Tulsa. There are two steps in the process of applying for a grant. The first is a Letter of Inquiry from the applicant. This letter is used to determine if the applicant will be invited to take the second step of submitting a formal Grant Proposal.

Requirements: IRS 501(c)3 non-profits are eligible to apply.

Restrictions: Grant requests for the following purposes are not favored: endowments, except as a limited part of a capital project reserved for maintenance of the facility being constructed; deficit financing and debt retirement; projects or programs for which the Chapman Trusts would be the sole source of financial support; travel, conferences, conventions, group meetings, or seminars; camp programs and other seasonal activities; religious programs of religious organizations; project or program planning; start-up ventures are not excluded, but organizations with a proven strategy and results are preferred; purposes normally funded by taxation or governmental agencies; requests made less than nine months from the declination of a previous request by an applicant, or within nine months of the last payment made on a grant made to an applicant; requests for more than one project.

Geographic Focus: Oklahoma

Amount of Grant: Up to 300,000 USD

Contact: Andrea Doyle; (918) 496-7882; fax (918) 496-7887; andie@chapmantrusts.com

Internet: http://www.chapmantrusts.org/grants_programs.html

Sponsor: H.A. and Mary K. Chapman Charitable Trust
6100 South Yale, Suite 1816
Tulsa, OK 74136

H.B. Earhart Fellowships 2194

These fellowships are awarded to move talented individuals through graduate study in the fields of international studies, economics, political science, and history in optimum time to embark upon careers in college or university teaching or in research. Awards are made to graduate students nominated by faculty sponsors whose participation is invited annually. Sponsors also monitor performance. Direct applications from candidates or from noninvited sponsors are not accepted.

Geographic Focus: All States

Amount of Grant: 1,000-25,000 USD

Contact: Ingrid Ann Gregg, President; (734) 761-8592

Sponsor: Earhart Foundation
2200 Green Road, Suite H
Ann Arbor, MI 48105

H.J. Heinz Company Foundation Grants 2195

The Foundation is committed to promoting the health and nutritional needs of children and families. Priority is given to programs in communities where Heinz operates with a special focus given to southwestern Pennsylvania. The Foundation proactively donates funds to develop and strengthen organizations that are dedicated to nutrition and nutritional education, youth services and education, diversity, healthy children and families, and quality of life. Application information is available online.

Requirements: Only organizations that have 501(c)3 tax status under the U.S. Internal Revenue Code are eligible for support domestically. International organizations are encouraged to submit a letter of inquiry prior to preparing a full proposal to determine eligibility. International organizations will need to provide a 501(c)3 determination letter from the United States Internal Revenue Service or sufficient documentation to demonstrate that the non-U.S. grantee is the equivalent of a U.S. public charity. Documentation should be provided in English. The Foundation will also accept proposals in the Common Grant Application Format. All proposals must be submitted in writing. Electronic submissions are not accepted at this time.

Restrictions: The Foundation will not provide grants to individuals nor make multi-year pledges except for major capital or grant campaigns. Generally, the Foundation does not make loans and does not provide grants for individuals, equipment, conferences, travel, general scholarships, religious programs, political campaigns, and unsolicited research projects.

Geographic Focus: All States

Contact: Tammy B. Aupperle, Program Director; (412) 456-5773; fax (412) 442-3227; heinz.foundation@hjheinz.com

Internet: http://www.heinz.com/sustainability.aspx/social/heinz-foundation.aspx

Sponsor: H.J. Heinz Company Foundation
P.O. Box 57
Pittsburgh, PA 15230-0057

H.W. Wilson Foundation Grants 2196

The foundation awards grants nationwide in its areas of interest, including libraries, library schools, and library science; higher education; and cultural outreach. Types of support include program grants, scholarships, and research grants. There are no application forms or deadlines. The board meets in January, March, May, August, and October.

Restrictions: Grants do not support building construction/renovation, endowment funds, or operating expenses.

Geographic Focus: All States

Contact: William Stanton, President; (718) 588-8400

Sponsor: H.W. Wilson Foundation
950 University Avenue
Bronx, NY 10452

HAF Riley Frazel Memorial Scholarship 2197

The Riley Frazel Memorial Scholarship is offered for graduating high school senior from Arcata or McKinleyville high schools. Recipients may be planning to attend any form of higher education (college, university, vocational or trade school), full or part time, and are not limited on where they may enroll. Any major is eligible. Financial need is considered, but grade point average is not. Community service and extra curricular activities are taken into consideration. The scholarship is worth a one time award of $1,000. The application is available at the website.

Requirements: Applicants must be graduating seniors of either Arcata or McKinleyvile high schools. They must also have attended Fieldbrook Elementary School at any time from kindergarten through eighth grade. Candidates will submit the application, a personal statement, their official transcript, and two letter of recommendation.

Geographic Focus: California

Date(s) Application is Due: Mar 15

Amount of Grant: 1,000 USD

Contact: Cassandra Wagner; (707) 442-2993, ext. 323; cassandraw@hafoundation.org

Internet: http://www.hafoundation.org/index.php?option=content&task=view&id=61

Sponsor: Humboldt Area Foundation
373 Indianola Road
Bayside, CA 95524

Hague Academy of International Law Summer Scholarships 2198

The Hague Academy of International Law awards scholarships for one of two summer sessions of study in public and private international law taught by selected authorities. Sessions are for three weeks of intensive study in French or in English; diplomas are available if arrangements for taking the tests are made prior to acceptance of the application deadline of March 1 of each year. Forms may be obtained by writing directly to the academy or to the heads of diplomatic missions and other authorities and scientific institutions connected with international law.

Requirements: Application must be submitted by the applicant and must be accompanied by a recommendation from a professor of international law, a curriculum vita, and a statement of evidence that the applicant considers to be of value in support of his/her candidacy. Applicants must be under 40 years of age.

Restrictions: No more than two scholarships will be awarded to applicants from any one country during the same year; an applicant may obtain only one scholarship.

Geographic Focus: All States

Date(s) Application is Due: Mar 1

Contact: Secretariat; 070-302-42-42

Internet: http://www.hagueacademy.nl/eng-home.html

Sponsor: Hague Academy of International Law
Peace Palace, Carnegieplein 2
The Hague, 2517 KJ Netherlands

Hampton Roads Community Foundation Mental Health Research Grants **2199**
The Foundation's mission is to inspire philanthropy and transform the quality of life in southeastern Virginia. Funding for Mental Health Research Grants is provided by the Benjamin R. Brown and Charles G. Brown Funds. Applications are available on the website. Applicants from the Eastern Shore of Virginia should speak with a Program Officer before submitting a proposal.
Requirements: Nonprofit organizations serving residents of south Hampton Roads (Chesapeake, Franklin, Norfolk, Portsmouth, Suffolk, Virginia Beach and Isle of Wight County) and the Eastern Shore of Virginia are eligible.
Geographic Focus: Virginia
Date(s) Application is Due: Oct 15
Contact: Program Officer; (757) 622-7951; grants@hamptonroadscf.org
Internet: http://www.hamptonroadscf.org/nonprofits/specialInterestGrants.html
Sponsor: Hampton Roads Community Foundation
One Commercial Plaza, Suite 1410
Norfolk, VA 23510

Harold Alfond Foundation Grants **2200**
The foundation awards grants to eligible organizations in its areas of interest, including secondary education, higher education, medical research, health care, and general charities. Grants support: education—public and private colleges and universities, as well as private secondary schools, to fund athletically oriented capital projects and scholarship endowments; medical research—individual research projects are considered when sponsored by a recognized medical research center; health—community support for capital campaigns and endowment funds; and general charities—community organizations that support youth, the arts, persons with disabilities, underprivileged, substance abuse rehabilitation, and annual fund drives of national organizations focusing on the above areas. There are no application deadlines or forms.
Restrictions: Grants are not made to individuals.
Geographic Focus: Florida, Maine
Amount of Grant: Up to 2,000,000 USD
Contact: Gregory Powell, (207) 828-7999
Sponsor: Harold Alfond Foundation
Two Monument Square
Portland, ME 04101-4093

Harold R. Bechtel Charitable Remainder Uni-Trust Grants **2201**
Established in Davenport, Iowa, in 1987, the Harold R. Bechtel Charitable Remainder Uni-Trust supports programs in Scott County, Iowa, with giving primarily for health associations, education, and for children, youth, and social services. The Foundation's primary interest areas include: children and youth services; community and economic development; education; family services; health organizations and associations; heart and circulatory research; higher education; human services; museums, music; and performing arts. Types of support building construction and renovation, as well as program-related funding. There are no specific deadlines, and applicants should contact the office in writing or via telephone to request application forms.
Requirements: 501(c)3 organizations serving the residents of Rock County, Iowa, are eligible to apply.
Restrictions: No grants are given to individuals, or for endowment funds, debt retirement, past operating deficit, general or continuing support, or for scholarly research in an established discipline.
Geographic Focus: Iowa
Amount of Grant: Up to 150,000 USD
Contact: R. Richard Bittner; (563) 328-3333; fax (563) 328-3352; loseband@blwlaw.com
Sponsor: Harold R. Bechtel Charitable Remainder Uni-Trust
201 West 2nd Street, Suite 1000
Davenport, IA 52801-1817

Harold R. Bechtel Testamentary Charitable Trust Grants **2202**
Established in Davenport, Iowa, in 1989, the Harold R. Bechtel Testamentary Charitable Trust Grants gives primarily to higher education and youth services. Major fields of interest include: children and youth services; community development; economic development; education; human services; and medical research. Applicants should contact the office in writing or via telephone to request application forms.
Requirements: 501(c)3 organizations serving the residents of Davenport, Iowa, Rock Island, Illinois, and Moline, Illinois, are eligible to apply.
Restrictions: No grants are given to individuals, or for endowment funds, debt retirement, past operating deficit, general or continuing support, or for scholarly research in an established discipline.
Geographic Focus: Iowa
Contact: R. Richard Bittner; (563) 328-3333; fax (563) 328-3352; loseband@blwlaw.com
Sponsor: Harold R. Bechtel Testamentary Charitable Trust
201 West 2nd Street, Suite 1000
Davenport, IA 52801-1817

Harold Simmons Foundation Grants **2203**
The foundation awards grants to Texas nonprofit organizations, with emphasis on social services, religion, health, the arts, and youth. Grants also support community programs and projects, child development, and adult basic education/literacy programs. The foundation also supports international development and relief efforts in Third World countries. Grants are awarded for general operating support, annual campaigns, capital campaigns, building construction/renovation, continuing support, seed money, and program development. Application forms are not required, and there are no deadline dates.
Requirements: Dallas, TX, nonprofits are eligible.
Restrictions: Grants are not awarded to support individuals or for endowment funds.
Geographic Focus: Texas
Contact: Lisa Simmons Epstein, President; (972) 233-2134
Sponsor: Harold Simmons Foundation
5430 LBJ Freeway, Suite 1700
Dallas, TX 75240-2697

Harris Graduate School of Public Policy Studies Research Development Grants **2204**
The Harris School supports this program for social science scholars interested in food assistance research. Awards will be made to scholars who propose research including, but not limited to interactions between food assistance programs and other welfare programs with respect to participation, administration, budget exposure, and the role of food assistance as a personal and fiscal stabilizer; the effects of the macroeconomic environment on the need for food assistance, level of participation, and food assistance program costs; and the well-being of current and former food assistance recipients. Other topics related to welfare reform and macroeconomic interactions with food assistance will be considered. This program is designed to encourage experienced researchers in other areas to start projects in the area of food assistance; research on food assistance using innovative approaches and research methods; smaller, start-up projects with the potential to make a significant contribution to food assistance research; and younger and junior scholars to develop research agendas in the area of food assistance. Funding may include compensation for the principal investigator's time; research assistance; travel; and purchase of data, computers, or other research related items. Application guidelines are available online.
Requirements: Applicants must hold a PhD degree.
Geographic Focus: All States
Date(s) Application is Due: May 1
Amount of Grant: Up to 40,000 USD
Contact: Development Grants Program; (773) 702-2028; spopa@uchicago.edu
Internet: http://harrisschool.uchicago.edu/Research/funding.asp
Sponsor: University of Chicago
1155 E 60th Street
Chicago, IL 60637

Harry A. and Margaret D. Towsley Foundation Grants **2205**
In 1959, Margaret Towsley created the Harry A. and Margaret D. Towsley Foundation with an initial gift of $4 Million in Dow Chemical Company stock. While the Foundation's initial goals were typical of general family foundations, its mission later became focused on programs promoting education, health care, shelter, and nutrition for children. As its assets grew, its areas of concentration expanded into college and university education, medical education, planned parenthood, and interdisciplinary programs with the schools of law and social work. These areas reflected Dr. and Mrs. Towsley's common interest in teaching. The foundation currently awards grants to Michigan organizations in its areas of interest, including environment, medical and preschool education, social services, continuing education, and research in the health sciences. Types of support include annual campaigns, building construction and renovation, capital campaigns, continuing support, employee matching gifts, endowments, general operating support, matching/challenge support, professorships, program development, research, and seed grants. There are no application forms; submit a letter of inquiry between January and the listed application deadline.
Restrictions: Grants are not awarded to individuals or for travel, scholarships, fellowships, conferences, books, publications, films, tapes, audio-visual or communication media, or loans.
Geographic Focus: Michigan
Date(s) Application is Due: Mar 31
Contact: Lynn Towsley White, President; (989) 837-1100; fax (989) 837-3240
Sponsor: Harry A. and Margaret D. Towsley Foundation
140 Ashman Street, P.O. Box 349
Midland, MI 48640

Harry B. and Jane H. Brock Foundation Grants **2206**
The Harry B. and Jane H. Brock Foundation honors a former Fort Payne, Alabama, native who changed the structure of the banking industry in Alabama. Today, the Foundation awards grants to eligible Alabama organizations to support community development and higher education. Primary areas of interest include community service, community funds, education, the environment, higher education, volunteerism, women's services, social services, and cancer treatment and research. Types of support include: annual campaigns, capital campaigns, endowments, general operating support, program development, and research. Written proposals should describe the project and organization and include a copy of the IRS tax-determination letter. The annual deadline is November 1.
Requirements: 501(c)3 organizations serving the residents of Alabama are eligible.
Restrictions: No grants to individuals are awarded. Giving primarily in Birmingham and Huntsville, Alabama.
Geographic Focus: Alabama
Date(s) Application is Due: Nov 1
Amount of Grant: 1,000-25,000 USD
Contact: Harry B. Brock, Jr.; (205) 939-0236 or (205) 918-0833; fax (205) 939-0806
Sponsor: Harry B. and Jane H. Brock Foundation
2101 Highland Avenue, Suite 250, P.O. Box 11643
Birmingham, AL 35202-1643

Harry Frank Guggenheim Foundation Dissertation Fellowships 2207

The Foundation awards ten or more dissertation fellowships each year to individuals who will complete their dissertations within the award year. These fellowships are designed to help doctoral candidates finish writing their dissertations rather than to support dissertation research. Questions that interest the Foundation concern violence and aggression in relation to social change, intergroup conflict, war, terrorism, crime, and family relationships, among other subjects. Priority will also be given to areas and methodologies not receiving adequate attention and support from other funding sources. Applicants may be citizens of any country and studying at colleges or universities in any country. The application deadline is February 1, with final decisions made by the Board of Directors at its June meeting for funding that fall.

Requirements: Applicants must submit the following to apply: title page; abstract page and survey; advisor's letter and abbreviated CV; applicant's CV and graduate school transcript; project description; protection of subjects; and their other support for funding. Applicants should refer to the guidelines for additional information and specific instructions for their abstract page and survey (based on their individual discipline), and contact the Foundation if they have questions.

Restrictions: Dissertations with no relevance to understanding human violence and aggression will not be supported. Applicants should only apply if they are entering the dissertation stage of graduate school (fieldwork or other research is complete and writing has begun). Recipients of the dissertation fellowship must submit a copy of the dissertation, approved and accepted by their institution, within six months after the end of the award year. Any papers, books, articles, or other publications based on the research should also be sent to the foundation.

Geographic Focus: All States, All Countries
Date(s) Application is Due: Feb 1
Amount of Grant: 20,000 USD
Contact: Program Officer; (646) 428-0971; fax (646) 428-0981; info@hfg.org
Internet: http://www.hfg.org/df/guidelines.htm
Sponsor: Harry Frank Guggenheim Foundation
25 West 53rd Street
New York, NY 10019-5401

Harry Frank Guggenheim Foundation Research Grants 2208

The Foundation invites proposals in any of the natural sciences, social sciences or humanities that promise to increase understanding of the causes, manifestations, and control of human violence and aggression. Highest priority is given to research that can increase understanding and improve urgent problems of violence and aggression in the modern world. Questions that interest the foundation concern violence and aggression in relation to social change, intergroup conflict, war, terrorism, crime, and family relationships, among other subjects. Priority will also be given to areas and methodologies not receiving adequate attention and support from other funding sources. Applicants may be citizens of any country. Most awards fall within the range of $15,000 to $40,000 per year for periods of one or two years. Applications for larger amounts and longer durations must be very strongly justified. Applications must be received by August 1, with a decision given in December.

Requirements: Applicants must mail two copies of the typed application in English to the Foundation. Applications may not be faxed or emailed. Along with two copies of the application, applicants must include the following: title page; abstract and survey; budget and its justification; personnel; research plan; other support; protection of subjects; and referee comments. Applicants should refer to the Foundation website for specific instructions on how to submit each of the attachments, and contact the Foundation if they have any questions.

Restrictions: Research with no relevance to understanding human problems will not be supported, nor will proposals to investigate urgent social problems where the Foundation cannot be assured that useful, sound research can be done.While almost all recipients of the Foundation grants possess a Ph.D., M.D., or equivalent degree, there are no formal degree requirements for the grant. The grant, however, may not be used to support research undertaken as part of the requirements for a graduate degree. Applicants need not be affiliated with an institution of higher learning, although most are college or university professors. The foundation awards research grants to individuals (or a few principal investigators at most) for individual projects, but does not award grants to institutions for institutional programs. Individuals who receive research grants may be subject to taxation.

Geographic Focus: All States, All Countries
Date(s) Application is Due: Aug 1
Contact: Program Officer; (646) 428-0971; fax (646) 428-0981; info@hfg.org
Internet: http://www.hfg.org/rg/guidelines.htm
Sponsor: Harry Frank Guggenheim Foundation
25 West 53rd Street
New York, NY 10019-5401

Harry Ransom Humanities Research Center Research Fellowships 2209

The center offers a number of fellowships on an international basis to scholars for the purpose of conducting research based on the center's collections. The center is the principal rare book and manuscript library of the University of Texas at Austin and noted for its collections of British, American, and French literary materials, with major strengths in photography, music, film, and theater arts. Most fellowships are for one month, but a limited number of fellowships may be awarded for periods of residency up to four months to scholars with longer-term projects. Recipients are expected to be in continuous residence for the duration of their awards. The program also awards domestic and foreign travel stipends to scholars with research projects that require less than one month's research at the center (stipends may not be combined with other Ransom Center fellowships); and dissertation fellowships, which are designated for graduate students who are working on doctoral dissertations. Guidelines are available online.

Requirements: Applicants must have received the PhD and be engaged in postdoctoral or equivalent research based on the center's collections. A two-page proposal, one-page curriculum vita, and two letters of recommendation are required for application.

Restrictions: Individuals who have received the HRHRC Research Fellowship are ineligible to reapply for one year. Fellowships do not support work on the dissertation.

Geographic Focus: All States
Date(s) Application is Due: Feb 1
Amount of Grant: 4,000-14,700 USD
Contact: Administrator; (512) 471-8944; fax (512) 471-9646; fellows@hrc.utexas.edu
Internet: http://www.hrc.utexas.edu/about/fellowships/application
Sponsor: University of Texas at Austin
P.O. Box 7219
Austin, TX 78713-7219

Harry W. Bass, Jr. Foundation Grants 2210

The Dallas-based Harry W. Bass, Jr. Foundation seeks to enrich the lives of the citizens of Texas by providing support to qualified organizations in the areas of education, health, human services, civic & community, science, research, arts and culture. Grant applications for specific programs or projects, capital projects or, less often, general operations are considered. Endowment gifts are rare. The Harry W. Bass, Jr. Foundation also considers program-related investments as part of its grant-making activities. The foundation strives to be responsive to the needs of all eligible organizations and considers requests of any amount. There is no formal application form. Grant requests are accepted at any time throughout the year. Each organization is limited to one application within a twelve-month period. Requests are usually processed within three to four months. Electronic grant applications should be submitted to dcalhoun@hbrf.org (no file attachments, please). For more information, contact the Harry W. Bass, Jr. Foundation.

Requirements: Texas 501(c)3 tax-exempt organizations are eligible.

Restrictions: In general, grants are not made for purposes of: church or seminary construction; annual fundraising events or general sustentation drives; professional conferences and symposia; out-of-state performances or competition expenses; to other private foundations. Unsolicited grant requests are restricted to organizations based in the Greater Dallas area. In the event a grant is approved, the recipient organization will be unable to submit another grant request for two years due to the foundation policy of not providing subsequent year grants.

Geographic Focus: Texas
Contact: F. David Calhoun; (214) 599-0300; fax (214) 599-0405; dcalhoun@hbrf.org
Internet: http://www.harrybassfoundation.org/about.asp
Sponsor: Harry Bass Foundation
4809 Cole Avenue, Suite 252
Dallas, TX 75205

Harvard Postdoctoral Fellowships on Education 2211

The graduate school invites applications from scholars for a two-year residential postdoctoral fellowship. Applications are encouraged from a broad range of scholarly traditions and backgrounds whose research links theory, practice, and policy in education. Special consideration will be given to applicants committed to careers in university research and teaching related to the improvement of education, particularly K-12 education. Fellows will teach one course during each of the two academic years and will give an annual lecture to the graduate school community about their research. Fellows receive an annual stipend plus a small housing supplement, are eligible for a travel supplement for participation in professional conferences, and receive additional support for course development or research. Applicants should send a letter of application, including a proposal for a two-year plan of research and teaching that identifies a member of the HGSE faculty who might serve as a mentor and provides a rationale for the selection of the mentor. Contact the office for additional information.

Requirements: Applicants must be within two years of receiving the PhD or EdD. Their degree must be completed by July and cannot have been awarded prior to May 2003.

Geographic Focus: All States
Date(s) Application is Due: Feb 11
Amount of Grant: 80,000-84,000 USD
Contact: Kenje Ogata, (617) 496-2805; kenje_ogata@gse.harvard.edu
Internet: http://www.gse.harvard.edu/~finaid/fellowship.html
Sponsor: Harvard University
122 Longfellow Hall
Cambridge, MA 02138

Harvard University Administrative Fellowships 2212

The program seeks to attract candidates, especially those from underrepresented ethnic minority groups and those committed to addressing the underrepresentation of ethnic minorities in university administration, to administrative careers in higher education. The fellowship offers a 12-month management experience complemented by a professional development program. Fellows from business, government, education, and the professions will be selected. Each participant receives a salary from Harvard University and is placed in a full-time work assignment in Harvard schools or central administration departments for the duration of the program. All costs for the required educational seminar will be covered by Harvard University. Participants must secure and cover the cost of their housing. Guidelines and application are available online.

Requirements: Applicants must have a bachelors degree and at least five years of work experience. The program seeks applicants with degree and backgrounds in the fields of athletics/sports management, information technology, alumni affairs and development, finance, human resources, and library science and management.

Geographic Focus: All States

Date(s) Application is Due: Feb 10
Contact: Teresa Malonzo; (617) 495-8919; teresa_malonzo@harvard.edu
Internet: http://www.oap.harvard.edu/afp
Sponsor: Harvard University
1350 Massachusetts Avenue
Cambridge, MA 02138

Harvard University Graduate School of Education Fellowships and Awards 2213
Harvard's graduate school of education (HGSE) offers several merit-based fellowships and awards. Presidential Fellowships, awarded to entering doctoral students, provide full tuition and fees for four years of study and an annual stipend for the first three years of study. Recipients must be full-time and are determined by the admissions committee based on the strength of the individual's application for admission. Entering Awards are awarded to first-year doctoral students who are either highly promising future professors or school leaders. These awards are for one year only and provide full tuition and fees plus a stipend. Roy E. Larsen Awards are awarded to first-year doctoral students who exhibit superior achievement and outstanding research potential. These awards are for one year only and provide full tuition and fees plus a stipend. Conant Fellowships support the professional growth of outstanding teachers and administrators in the Boston and Cambridge public schools. To be eligible, applicants must be teachers or administrators with professional status, under contract with either the Boston or the Cambridge public schools, and admitted to a degree program at HGSE. These fellowships cover the cost of full tuition are pro-rated for part-time study. Awards for doctoral Conant Fellowships carry the possibility of a one-year renewal. Guidelines for these programs are available online.
Geographic Focus: All States
Date(s) Application is Due: Feb 4
Amount of Grant: 7,000 USD
Contact: Office; (617) 496-2805; fax (617) 496-0840; kenje_ogata@harvard.edu
Internet: http://www.gse.harvard.edu/~finaid/fellowship.html#merit
Sponsor: Harvard University
108 Longfellow Hall, 13 Appian Way
Cambridge, MA 02138

Harvard University John M. Olin Prizes 2214
The Harvard University John M. Olin Prizes, supported by a grant from the John M. Olin Foundation are awarded annually for the best papers written in the area of law and economics. The amount of a prize is $1,000. A professor must submit the paper on behalf of the student with a statement describing why the professor believes the paper is meritorious. Submission should be made before the end of classes in the Spring semester by email to the Program Administrator.
Geographic Focus: All States
Date(s) Application is Due: Nov 30
Contact: Irina Goldina; (617) 496-1670; fax (617) 384-5653; igoldina@law.harvard.edu
Internet: http://www.law.harvard.edu/programs/olin_center/
Sponsor: Harvard University
1563 Massachusetts Avenue
Cambridge, MA 02138

Harvard University Weatherhead Center for International Affairs Fellowships 2215
The Harvard University Weatherhead Center for International Affairs Fellows program is the principal forum for the interaction of practitioners with the faculty, students, and visiting scholars of the Weatherhead Center. The program was inaugurated in 1958, simultaneously with the establishment of the Center itself, in the belief that the Center's research activities would benefit from the contributions of non-academic professionals with broad practical experience in international affairs, and that the Center could broaden the practitioners' horizons and deepen their understanding of the contemporary issues in which they are professionally involved. While on campus, Fellows are exposed to social science theories and research intended to place their experiences in an academic perspective. Their primary activity is to work on a research project of their choosing. They may also organize and lead study groups for Harvard undergraduates, participate in seminars at the Center and elsewhere, provide advice to undergraduate and graduate students, and otherwise interact with the faculty, visiting scholars, and other members of the University's community. For all successful applicants, the Center hopes to establish an agreement with a Harvard faculty member associated with the Weatherhead Center that results in the formation of an active scholarly relationship. Past Fellows have been senior diplomats, military officers, politicians, journalists, international civil servants, officials from nongovernmental organizations, and business leaders from around the world. The single characteristic they share is leadership in international affairs, as demonstrated by both a record of past achievement and the promise of future accomplishments.
Geographic Focus: All States
Contact: Kathleen Molony, Director, Fellows Program; (617) 495-2644 or (617) 495-4420; fax (617) 496-5642; kmolony@wcfia.harvard.edu
Internet: http://wcfia.harvard.edu/programs/fellows
Sponsor: Harvard University
61 Kirkland Street, Room 110
Cambridge, MA 02138

Harvard University Women's Studies in Religion Research Associate and Visiting Faculty 2216
The Harvard University Women's Studies in Religion Research Associate and Visiting Faculty program is designed to explore women's distinctive religious experience and perspectives in their cultural and racial diversity, and to examine male-centeredness and sexism in religious traditions of the world. The program encourages and guides the development of women's studies in religion to help transform the study of religion so it more adequately reflects the study of human experience. Each year, the school appoints as visiting lecturers and research associates in the program five scholars whose work provides leadership in women's studies in the various fields of religion. These scholars conduct individual research projects on topics related to the history and function of gender in the symbolization of religious traditions; the institutionalization of roles in religious communities; and the dynamics of the interaction between religious systems of belief and the personal, social, and cultural conditions of women. Each is also a member of one of the school's departments and offers in one of the three curricular areas a semester course related to this research project. Full-time residence at Harvard Divinity School is required while carrying out proposed research projects. In the spring semester each associate gives a public lecture on her/his research. Application guidelines and deadlines are available from the Harvard Divinity School. Salary for each position will be $55,000. The appointment is full-time, lasting ten months, and includes health benefits and reimbursement of some expenses.
Requirements: Candidates with doctorates in the fields of religion and those with primary competence in other humanities, social science, and public policy fields who have serious investment in religion are eligible. Applicants must have received their PhD by October 1 of the application year.
Geographic Focus: All States
Date(s) Application is Due: Oct 15
Amount of Grant: 55,000 USD
Contact: Divinity School Office; (617) 495-5705 or (617) 495-5761; fax (617) 496-8564; admin@wsrp.harvard.edu
Internet: http://www.hds.harvard.edu/wsrp/apply
Sponsor: Harvard University
45 Francis Avenue, Andover Hall
Cambridge, MA 02138

Harvard Women and Public Policy Postdoctoral Fellowship 2217
The program will award a postdoctoral residential fellowship to an outstanding scholar in security affairs from the United States and elsewhere working to promote basic research in the broad area of international security with a particular focus on gender, prevention, peace building, and sustainable security. The fellow will hold his/her fellowship jointly with the International Security Program of the Belfer Center for Science and International Affairs. The fellow will be expected to participate actively in the weekly ISP seminar and in the intellectual life of the Women and Public Policy Program; take an active organizational role in the monthly Boston Consortium meetings on Gender Security and Human Rights and devote some portion of his/her time to collaborative endeavors; and complete a book, monograph, scholarly article, or other significant publication during the period of residence. Residential fellowships also provide for individual health insurance, institutional support, and some research-related travel stipends.
Requirements: The fellow must hold a PhD by September.
Geographic Focus: All States
Date(s) Application is Due: Jan 31
Amount of Grant: 30,000 USD
Contact: Theresa Lund, John F. Kennedy School of Government Office; (617) 496-6609; fax (617) 496-6154; theresa_lund@harvard.edu
Internet: http://www.ksg.harvard.edu/wappp/research/fellowship.html
Sponsor: Harvard University
79 John F Kennedy Street
Cambridge, MA 02138

Harvey Randall Wickes Foundation Grants 2218
The Foundation awards grants to eligible Michigan nonprofit organizations in its areas of interest, including the arts; children/youth services; education; hospitals; human services; libraries; and recreation. Types of support include annual campaigns, building/renovation, equipment, and seed money. There are no application deadline dates. Applications must be received two weeks prior to board meetings in March, June, September, and December. Award notification is given within two weeks of the board meeting.
Requirements: Michigan 501(c)3 organizations in Saginaw County are eligible.
Restrictions: Funding is not available for government where support is forthcoming from tax dollars, individuals, endowments, travel, conferences, loans, or film or video projects.
Geographic Focus: Michigan
Contact: Hugo Braun, Jr.; (989) 799-1850; fax (989) 799-3327; hrwickes@att.net
Sponsor: Harvey Randall Wickes Foundation
4800 Fashion Square Boulevard, Plaza N., Suite 472
Saginaw, MI 48604-2677

Hatton W. Sumners Foundation for the Study & Teaching of Self Government Grants 2219
The Foundation's purpose is to encourage the study, teaching and research in the science and art of self-government so that citizens understand the fundamental principles of democracy in shaping governmental policies. Working through qualified, tax-exempt organizations, the Foundation seeks to reach, educate, and motivate the general public and the current and future leaders of American society. The Foundation gives to youth organizations and higher education. Types of support include: conferences and seminars; continuing support; curriculum development; endowments; fellowships; general operating support; internship funds; matching and challenge support; and research. Application information is available at the Foundation website. Grant applications will be accepted from January 1st through August 1st of each year. Final decisions on grant proposals are made by the Trustees in October of each year.
Requirements: Grants are made only to 501(c)3 tax-exempt organizations in Texas, New Mexico, Oklahoma, Louisiana, Arkansas, Kansas, Nebraska and Missouri.

Restrictions: The Foundation does not fund religious organizations or individual grants.
Geographic Focus: Arkansas, Kansas, Louisiana, Missouri, Nebraska, New Mexico, Oklahoma, Texas
Date(s) Application is Due: Aug 1
Contact: Hugh Akin, Executive Director; (214) 220-2128; fax (214) 953-0737; hugh@hattonsumners.org or info@hattonsumners.org
Internet: http://www.hattonsumners.org
Sponsor: Hatton W. Sumners Foundation for Study and Teaching of Self Government
325 North St. Paul Street, Suite 3920
Dallas, TX 75201

Haverford College Mellon Postdoctoral Fellowships 2220

The John B. Hurford '60 Center for the Arts and Humanities of Haverford College invites applications for a two-year Mellon Postdoctoral Fellowship in the Humanities to begin in the Fall Semester. The program seeks scholars interested in the anarchist tradition and its historical and contemporary influence on art, politics, and culture. Women and members of underrepresented minority groups are strongly encouraged to apply. Area of specialization is open, but might include history, history of art, music, political science, political economy, anthropology, literature, sociology, philosophy, and cultural studies, among other possibilities. Scholars with broad historical and interdisciplinary interests are encouraged to apply.
Requirements: Candidates who received the PhD within the last five years as of the listed application deadline date are eligible.
Geographic Focus: All States
Date(s) Application is Due: Jan 31
Contact: Emily Carey Cronin; (610) 896-1336 or (610) 896-1000; ecronin@haverford.edu
Internet: http://www.haverford.edu/HCAH/mellon_post-doctoral_program/index.php
Sponsor: Haverford College
370 Lancaster Avenue
Haverford, PA 19041

Hawaii Community Foundation Health Education and Research Grants 2221

The program provides support from three distinct funds: Tobacco Prevention & Control Trust Fund; the Leahi Fund Research Fund; and the Medical Research Program. Funding will be provided for medical education and research in the fields of: cancer; heart disease; lung disease and research. Priority is given to projects that: demonstrate a foreseeable benefit to the people of Hawaii; support new investigators in Hawaii; and support collaborative efforts. Proposal submission information is available online.
Requirements: To be eligible for consideration an organization must be a 501(c)3 organization or a unit of government. The Principal Investigator must be based in Hawaii and conducting the research in Hawaii.
Geographic Focus: Hawaii
Date(s) Application is Due: Feb 27; Jul 17; Aug 14; Sep 10
Amount of Grant: 25,000-50,000 USD
Contact: Christel Wuerfel, Philanthropic Services Assistant; (808) 537-6333 or (888) 731-3863; cwuerfel@hcf-hawaii.org
Internet: http://www.hawaiicommunityfoundation.org/index.php?id=71&categoryID=23
Sponsor: Hawai'i Community Foundation
1164 Bishop Street, Suite 800
Honolulu, HI 96813

Healthcare Foundation of New Jersey Grants 2222

The Foundation's funding priorities are the vulnerable populations of the greater Newark New Jersey community; the emergent health needs of serving at-risk individuals and families in the MetroWest Jewish community; and clinical research/medical education initiatives that significantly and directly impact these populations. The Foundation seeks grant proposals that promise innovation and change or a significant enhancement of services. Proposals are accepted on a rolling basis throughout the calendar year. Proposal submission instructions are available online.
Requirements: Nonprofits in the Greater Newark area are eligible to apply. The Foundation strongly suggests that application documents be submitted electronically.
Restrictions: Grants are made only to private nonprofit organizations that have tax-exempt status under Section 501(c)3 of the Internal Revenue Code and that are not private foundations. The Foundation does not make grants to individuals or government agencies. The Foundation does not typically fund the following: organizations outside of Essex, Morris or Union County, New Jersey; programs not related to health care; direct support of an individual's healthcare needs; fundraising events or endowment campaigns; advertising campaigns; lobbying; or scholarships.
Geographic Focus: New Jersey
Contact: Program Contact; (973) 921-1210; fax (973) 921-1274; info@hfnj.org
Internet: http://www.hfnj.org
Sponsor: Healthcare Foundation of New Jersey
60 East Willow Street, 2nd Floor
Millburn, NJ 07041

Health Foundation of Southern Florida Responsive Grants 2223

The foundation awards Responsive Grants through two grant cycles per year. With exceptions, the foundation focuses on providing one to three year grants that do not exceed $300,000 annually. The majority of grants are funded in the $50,000 to $150,000 range over one or two years. Funding is provided in four categories: Project Planning; Health Services; Organizational Capacity Building and Health System/Health Policy Development.
Requirements: Though the foundation welcomes proposal applications anytime, the applications are reviewed on a semi-annual basis. Applicant organizations must be tax-exempt nonprofit under section 501(c)3 of the Internal Revenue Code or a local or state governmental agency. The project must serve exclusively the residents of Broward, Miami-Dade and/or Monroe counties. Initially, a preliminary proposal is required (see the sponsor's website for specific details). If approved, the sponsor will then invite a full proposal. Download the sponsor's grant guide from the website.
Restrictions: The foundation does not fund: Biomedical research or other research that will not impact local residents within the immediate future (1-3 years) or that does not have a direct application to implementing a community-driven health intervention; Capital campaigns of over $1 million (versus grants toward specific health-related equipment or the 'build out' of a specific health-focused space); Secondary and tertiary services (versus preventive and primary medical, oral and behavioral health care services); Health promotion and/or health care with a high per capita cost (this figure will vary depending upon the type of intervention, but over $1,000 per person/year cost may be a rule of thumb); Service expansion or new projects without viable sustainability (unable to reach sustainability without the foundation's resources within a four-year period).
Geographic Focus: Florida
Date(s) Application is Due: Mar 13; Apr 24
Amount of Grant: Up to 300,000 USD
Contact: Eliane Morales, (305) 374-7200; fax (305) 374-7003; emorales@hfsf.org
Internet: http://www.hfsf.org/ORIGHTML/responsive.html
Sponsor: Health Foundation of South Florida
2 South Biscayne Boulevard
Miami, FL 33131

Health Management Scholarships and Grants for Minorities 2224

This financial assistance program provides scholarships and grants to ethnic minority students for graduate study in health care administration/management. Annual tuition scholarships and grants are made to outstanding students in health services management and related programs. Grants are provided to cover the costs of books, tutorial assistance, and other nontuition educational aids. Students are eligible to receive a combination of awards during the same time period.
Requirements: To be eligible, applicants must be members of an ethnic minority group; must be U.S. citizens or have permanent resident status (birth certificate, green card, or current passport); they must be college juniors, seniors or graduate students; they must be health care administration/management majors; they must have a strong academic background (2.5 GPA on a 4.0 scale); and they must have strong extracurricular and community service activities.
Geographic Focus: All States
Amount of Grant: 500-2,500 USD
Contact: Program Contact; (800) 233-0996 or (312) 422-2680; fax (312) 422-4566
Internet: http://www.diversityconnection.org/diversityconnection_app/career-center/internships/Standard-Internship-page.jsp?fl=S4
Sponsor: Institute for Diversity in Health Management
1 N Franklin, 30th Floor
Chicago, IL 60606

Hearst Foundations Health Grants 2225

The Hearst Foundations fund non-profit organizations working in the fields of culture, education, health, and social services. The Foundations have two offices, one in New York which manages funding for organizations headquartered east of the Mississippi River and one in San Francisco which manages funding for organizations to the west. About 80% of the Foundations' total funding goes to prior grantees; the Foundations receive approximately 1,200 grant requests annually. The Foundations' health funding comprises 30% of their total giving; 80% of the Foundations' health funding goes to organizations having budgets over ten-million dollars. In the area of health, the Foundations assist leading regional hospitals, medical centers, and specialized medical institutions providing access to high-quality healthcare for low-income populations. In response to the shortage of healthcare professionals necessary to meet the country's evolving needs, the Foundations also fund programs designed to enhance skills and increase the number of practitioners and educators across roles in healthcare. Because the Foundations seek to use their funds to create a broad and enduring impact on the nation's health, support for medical research and the development of young investigators is also considered. Preference is given to the following types of programs; professional development programs; programs improving access to high-quality healthcare for low-income populations; programs developing and providing specialized care for the complex needs of elderly populations; programs scaling innovative healthcare delivery systems to provide efficient, coordinated care; and research, particularly related to finding new cures and treatments for prevalent diseases, such as cancer. The Foundations provide program, capital, and, on a limited basis, endowment support. Requests are accepted year round. These must be submitted via the Foundations' online application portal. Each request goes through an evaluation process that generally spans four to six weeks. The Foundations conduct a site visit of semi-finalists and may also consult with experts in a given field. Applicants will receive an email confirmation of receipt of submission and can follow the status of their request through the online system. Instructions for using the system, guidelines (in the form of an FAQ), and the link to the Foundations' online portal are at the Foundations' website.
Requirements: Grants are made only to 501(c)3 organizations.
Restrictions: Organizations must wait one year from the date of their notice of decline before the Foundations will consider another request. Grantees must wait a minimum of three years from their grant award date before the Foundations will consider another request. The Foundations do not fund individuals or the following types of requests: those from organizations operating outside the United States; those from organizations with operating budgets under one million dollars; those from organizations involved in publishing, radio, film, or television; those from local chapters of organizations; those from organizations lacking demonstrable long-term impact on populations served; requests to fund tours, conferences, workshops, or seminars; requests to fund advocacy

or public-policy research; requests to fund special events, tickets, tables, or advertising for fundraising events; requests for seed money or to fund start-up projects; and request to fund program-related investments.
Geographic Focus: All States
Amount of Grant: Up to 600,000 USD
Contact: Mason Granger; (212) 649-3750; fax (212) 586-1917; hearst.ny@hearstfdn.org
Internet: http://www.hearstfdn.org/funding-priorities/
Sponsor: Hearst Foundations
300 West 57th Street, 26th Floor
New York, NY 10019-3741

Hebrew University Lady Davis Trust Fellowships 2226
The trust offers fellowships to visiting professors, postdoctoral researchers, and doctoral students at the Hebrew University of Jerusalem and at the Technion, Israel Institute of Technology in Haifa. Distinguished scholars spend from three months to one year at one of these institutions developing research networks, teaching and enjoying the world-class academic environment.
Requirements: Scholars of any age, in any field, and from any region may apply.
Geographic Focus: All States
Contact: Program Coordinator; (800) 404-8622 or (212) 472-2288; fax (212) 517-4548; admission@roth.mscc.huji.ac.il
Internet: http://www.huji.ac.il/unew/main.html
Sponsor: Hebrew University of Jerusalem
Givat Ram, Jerusalem 91904
Israel

Heineman Foundation for Research, Education, Charitable and Scientific Purposes 2227
The purpose of the Heineman Foundation is to provide seed money to start-up projects and new projects within existing organizations for a maximum of three to five years. Preference will be given to organizations that we have not previously funded. The average range of our donations is $20,000.00 to $50,000.00, per annum. An organization must have 501(c)3 status and upload copies of corresponding IRS documents to the online application form in order for the application to be considered. The Foundation's general areas of interest are the following (in no particular order): programs that enable economically challenged women to enter and remain in the workplace; on site day care centers for women in the workplace; job training programs for women; language and leadership skills for women; environmental research that will help prevent, reduce and/or eliminate water degradation; music as education and preserver of culture; research into prevention of and treatment for childhood illnesses; programs that enable youth to think, create and communicate effectively; and programs that support and promote high achievement in music, science, and literature. Applications/proposals must be submitted online no later than September 1st.
Geographic Focus: All States
Date(s) Application is Due: Sep 1
Amount of Grant: 20,000-50,000 USD
Contact: Simon Rose, President; (212) 493-8000; info@heinemanfoundation.org
Internet: http://www.heinemanfoundation.org/guidelines
Sponsor: Heineman Foundation
140 Broadway
New York, NY 10005-1108

Helen Bader Foundation Grants 2228
Throughout her life, Helen Bader sought to help others. She played many roles-student, mother, businesswoman, and social worker-believing that everyone should have the opportunity to reach their fullest potential. Growing up in the railroad town of Aberdeen, South Dakota, Helen learned the value of hard work and self-reliance. The Great Depression and the sacrifices of World War II also taught her the importance of reaching out to those in need. Helen attended Downer College in Milwaukee, earning a degree in botany. She married Alfred Bader, a chemist from Austria, and together they started a family and created a business, the Aldrich Chemical Company. From the 1950s to the 1970s, their hard work helped build one of Wisconsin's most successful start-up enterprises of the era. The Baders' eventual divorce led Helen to again become self-reliant. She subsequently finished her Master of Social Work at the University of Wisconsin-Milwaukee. While doing her field work with the Legal Aid Society of Milwaukee, Helen met and helped many people in need, including single mothers and adults with mental illness. In the process, she gained a deeper appreciation for their everyday struggles. After graduation, she worked at the Milwaukee Jewish Home, where working with older adults brought home the many issues of aging. At a time when Alzheimer's disease was almost a complete mystery, she helped open the resident' minds and hearts through dance and music. Helen felt that the residents' quality of life depended upon the small details, so she was happy to run errands or escort them to the symphony. She found herself touched by the arts and studied the violin and guitar at the Wisconsin Conservatory of Music. Helen eventually faced cancer. As the illness began to sap her physical strength, she shared a wish with her family: to continue to aid those in need. She died in 1989. After her death, patterns of Helen's quiet style of philanthropy became more apparent. When she had come across an organization that impressed her, she would just pull out her checkbook without a lot of fanfare. In her name, the Helen Bader Foundation (HBF) supports worthy organizations working in key areas affecting the quality of life in Milwaukee, the state of Wisconsin, and Israel. The foundation also seeks to inspire the generosity in others, as every individual can make a difference through gifts of time, talent, and resources. The foundation will consider multiple-year requests with 24 or 36 month terms. Multi-year grants are subject to annual review before funds for subsequent years are released. The application deadline for the online preliminary proposal is January 5. The application deadline for organizations invited to complete a full proposal is February 2. The link to the online application system is available at the grant website. Application deadlines may vary from year to year. Prospective applicants are encouraged to visit the grant website to verify current deadline dates.
Requirements: Grants are awarded for projects consistent with one or more of the Helen Bader Foundation's program areas: Alzheimer's and aging (national in scope, with priority given to Wisconsin); economic development (restricted to the city of Milwaukee); community partnerships for youth (restricted to the city of Milwaukee); community initiatives (restricted to greater Milwaukee); arts (restricted to the city of Milwaukee); and directed grants and initiatives such as aid and support to Israel (for which proposals must be staff-solicited). Grants are given only to U.S. organizations which are tax exempt under Section 501(c)3 of the Internal Revenue Code or to government entities; grants will only be approved for foreign entities which meet specific charitable status requirements.
Restrictions: The Foundation does not provide direct support for individuals, such as individual scholarships.
Geographic Focus: All States, All Countries
Date(s) Application is Due: Jan 5; Feb 25
Amount of Grant: 10,000-100,000 USD
Contact: Tamara Hogans; (414) 224-6464; tammy@hbf.org or info@hbf.org
Internet: http://www.hbf.org/apply.htm
Sponsor: Helen Bader Foundation
233 North Water Street, Fourth Floor
Milwaukee, WI 53202

Helen Carr Minority Fellowships 2229
These fellowships provide support to African American engineering students or faculty at Hampton University, Morgan State University, Howard University, North Carolina A&T State University, Prairie View A&M University, Tennessee State University, Tuskegee University, and Southern University who have not yet completed their doctorates and wish to do so. The fellowships are awarded for one year and may be renewed. The amount awarded is dependent on the educational requirements and financial needs of the recipients.
Requirements: Eligible to apply are African American students or engineering faculty who are interested in earning doctoral degrees and returning to one of the historically black engineering colleges to teach.
Geographic Focus: All States
Date(s) Application is Due: May 1
Contact: Supervisor; (202) 331-3525 or (202) 331-3500; projects@asee.org
Internet: http://www.asee.org/fellowships/hbecc.cfm
Sponsor: American Society for Engineering Education
1818 N Street NW, Suite 600
Washington, DC 20036

Helen Kellogg Institute for International Studies Residential Fellowships 2230
The Kellogg Institute awards Faculty Residential Fellowships to enable Kellogg faculty fellows to undertake significant projects consistent with the Institute's research priorities. The Faculty Residential Fellowship, which normally covers the recipient's salary for one semester, allows a fellow the opportunity to apply for an academic leave from his or her department or college. While not a given, leave is usually granted. Consequently, during the fellowship period the fellow is relieved of teaching and/or administrative responsibilities.
Requirements: Only members of the regular teaching and research faculty who are fellows of the Kellogg Institute are eligible for this program. Assistant professors may apply for faculty residential fellowships, but awards are contingent on their having passed third-year review before beginning the fellowship.
Restrictions: Applicants may not have had a paid leave in the previous two years, or a Kellogg Faculty Residential Fellowship in the preceding six years. Projects funded by this program must be conducted in residence at Notre Dame. Proposals for field research outside the University and/or for projects that would entail other than brief absences of the faculty member during the fellowship period are not eligible.
Geographic Focus: All States
Date(s) Application is Due: Nov 2
Contact: Sharon Schierling; (574) 631-8524 or (574) 631-6580; sschierl@nd.edu
Internet: http://kellogg.nd.edu/faculty/grants/resfellowship.shtml
Sponsor: University of Notre Dame
130 Hesburgh Center
Notre Dame, IN 46556-5677

Helen V. Brach Foundation Grants 2231
Established in 1974 in Illinois by the wife of Frank Brach, principal owner of the E.J. Brach and Sons Candy Company of Chicago, the foundation operates to prevent cruelty to animals or to children; for religious, charitable, scientific, literary, and education purposes; and for public safety testing through support of Midwest 501(c)3 tax-exempt organizations carrying out programs and activities in these areas. Brach provides grants nationally and has wide-ranging interests. For example, it supports homeless and women's emergency shelters, teen pregnancy prevention programs, parenting education, summer school for disadvantaged children, job training for welfare mothers, orphanages, and scholarships for economically disadvantaged students. Types of support include annual campaigns, building construction and renovation, equipment, general operating support, publications, research, and special projects. The foundation ordinarily does not make multi-year grants or commitments. Applicants are required to complete in full a brief application form, which may be obtained from the office. The board of directors gives final consideration to all applications received in a given year at the board's meeting, which is usually held in March.
Requirements: Although 501(c)3 nonprofits from across the nation are eligible, giving is primarily made in the Midwest, as well as California, Massachusetts, Ohio, Pennsylvania, and South Carolina.

Restrictions: Grants are not made to individuals or to organizations outside the United States. Typically grants are not made in excess of 10 percent of a group's operating budget, which automatically excludes start-up grants.
Geographic Focus: All States
Date(s) Application is Due: Dec 31
Amount of Grant: 5,000-25,000 USD
Contact: John P. Hagnell, Associate Director; (312) 372-4417; fax (312) 372-0290
Sponsor: Helen V. Brach Foundation
55 W Wacker Drive, Suite 701
Chicago, IL 60601-1609

Hendrick Foundation for Children Grants **2232**
The foundation supports programs designed to provide health, medical, social welfare, and educational services to benefit children with illness, disease, injury, pain, disability, incapacity, or other disadvantages; and improve quality of life for children with life-threatening or chronic injuries, illness, and disabilities. It is also contemplated that the Foundation will honor grants to other charitable and educational organizations sponsoring programs which are community oriented and focus on improving the quality of life of children with life-threatening or chronic injuries, illnesses and disabilities. All requests shall be given careful and thorough consideration by the Hendrick Foundation for Children Grant Committee. Grant awards will be determined on the merits of the organization, its objectives and mission. All grant requests should be submitted in writing on the organization's letterhead
Geographic Focus: All States
Amount of Grant: 3,500-250,000 USD
Contact: Charles V. Ricks, President; (704) 568-5550
Internet: http://www.thehendrickfoundation.org/grants.html
Sponsor: Hendrick Foundation for Children
P.O. Box 240070
Charlotte, NC 28224-0070

Henry Dreyfus Teacher-Scholar Awards **2233**
The Henry Dreyfus Teacher-Scholar Awards Program supports the research and teaching careers of talented young faculty in the chemical sciences. Based on institutional nominations, the program provides discretionary funding to faculty at an early stage in their careers. The Henry Dreyfus Teacher-Scholar Awards Program is based on accomplishment in scholarly research with undergraduates, as well as a compelling commitment to teaching. The Henry Dreyfus Teacher-Scholar Awards Program provides an unrestricted research grant of $60,000.
Requirements: The Henry Dreyfus Teacher-Scholar Awards Program is open to academic institutions in the States, Districts, and Territories of the United States of America that grant a bachelor's or master's degree in the chemical sciences, including biochemistry, materials chemistry, and chemical engineering. If the department has a modest Ph.D. program, the nominee's research accomplishments must be almost exclusively with undergraduates. Nominees must hold a full-time tenure-track academic appointment, be between the fourth and twelfth years of their independent academic careers, and be engaged in research and teaching primarily with undergraduates. Awardees are typically in departments that do not grant a doctoral degree. Institutions may submit only one Henry Dreyfus nomination annually and renominations are accepted.
Restrictions: Awards are not made directly to individuals.
Geographic Focus: All States
Date(s) Application is Due: Aug 6
Amount of Grant: 60,000 USD
Contact: Mark J. Cordillo, Executive Director; (212) 753-1760; fax (212) 593-2256; mcardillo@dreyfus.org or admin@dreyfus.org
Internet: http://www.dreyfus.org/awards/henry_dryfus_teacher_award.shtml
Sponsor: Camille and Henry Dreyfus Foundation
555 Madison Avenue, 20th Floor
New York, NY 10022-3301

Henry J. Kaiser Family Foundation Media Fellowships **2234**
The purpose of the fellowship is to help journalists and commentators do the best possible job of keeping the public informed about health issues at this critical time in the evolution of the health care system. The goal of the fellowship is to provide a highly flexible range of opportunities tailored to each fellow's interests; help journalists improve the quality of the work they do; provide time to research specific topics; and deepen participants' commitment to becoming specialists in health reporting. Fellows may want to participate in the fellowship program on a part-time basis and continue writing or reporting during their fellowship. Fellows are encouraged to do reporting based on their fellowship research and to experiment with different media or forms of writing. Most fellows work from home, or base themselves at a local academic or research institution. There is no obligation to pursue academic coursework, although that is an option open to all fellows. There is no application form. Guidelines are available online.
Requirements: Fellowships will be awarded to U.S. journalists interested in U.S. health policy issues.
Geographic Focus: All States
Date(s) Application is Due: Mar 16
Amount of Grant: Up to 50,000 USD
Contact: Penny Duckham; (650) 234-9220; fax (650) 854-4800; pduckham@kff.org
Internet: http://www.kff.org/about/mediafellowships.cfm
Sponsor: Henry J. Kaiser Family Foundation
2400 Sand Hill Road
Menlo Park, CA 94025

Henry Luce Foundation/ACLS Dissertation Fellowships in American Art **2235**
The ACLS Dissertation Fellowships in American Art support graduate students in any stage of Ph.D. dissertation research or writing. The Fellowships are for a one-year non-renewable term beginning in the summer.
Requirements: Applicants must have completed all requirements for the Ph.D. except the dissertation before beginning the one-year tenure. A student must be a Ph.D. candidate in a department of art history, and the candidate's dissertation must be focused on a topic in the history of the visual arts of the United States. Students whose degrees will be granted by another department may be eligible if their principal dissertation adviser is in a department of art history and their dissertation topic is object-oriented. Specific guidelines and the online Fellowship application are available at the website.
Restrictions: Fellowship funds may not be used to defray tuition costs. Students who received grants of $5,000 or more from the Henry Luce Program previously administered through university art departments are not eligible, nor are students preparing theses for the Master of Fine Arts degree.
Geographic Focus: All States
Date(s) Application is Due: Nov 9
Amount of Grant: 27,000 USD
Contact: Nicole Stahlmann, Director of Fellowship Programs; (212) 697-1505, ext. 134; fax (212) 949-8058; nstahlmann@acls.org
Internet: http://www.acls.org/grants/Default.aspx?id=516
Sponsor: American Council of Learned Societies
633 Third Avenue
New York, NY 10017-6795

Henry Luce Foundation Special Projects in Higher Education Grants **2236**
This award is a responsive grant made to colleges and universities for special projects which fall outside the context of the Foundation's funding categories. Higher education is an underlying theme for most of the Foundation's programs, through support for research, training, and the creation of new resources. The current primary focus of the program has reflected the Foundation's commitment to strengthen America's capacity for international understanding. No special forms are required for the application process. An initial letter of inquiry should be addressed to the appropriate program contact. Requests are accepted at any time. Awards are determined by the board of directors, which meets three times a year.
Restrictions: The Foundation does not support health care or medical projects and does not fund development assistance work overseas. It does not normally assist journalism, media and film projects or the performing arts.
Geographic Focus: All States
Contact: Supervisor; (212) 489-7700; fax (212) 581-9541; hlf1@hluce.org
Internet: http://www.hluce.org/4heddefm.html
Sponsor: Henry Luce Foundation
111 W 50th Street, Room 4601
New York, NY 10020

Henry W. Bull Foundation Grants **2237**
Established in 1960 in California, the foundation was established by Maud Bull in memory of her husband, Henry W. Bull. The foundation grants awards on a nationwide basis, giving primarily in the areas of: arts/performing arts; education; health and; human services. Types of support include annual campaigns, building construction/renovation, capital campaigns; program development, equipment acquisition, matching/challenge grants; general operating support, continuing support, and research. There is no formal grant application form. Submit a simple, concise statement of needs and objectives with pertinent supportive data, c/o Janice Gibbons, Santa Barbara Bank & Trust, P.O. Box 2340, Santa Barbara, CA, 93120-2340.
Requirements: 501(c)()3 nonprofits are eligible.
Restrictions: No grants to individuals or private foundations.
Geographic Focus: All States
Date(s) Application is Due: Apr 1; Sep 1
Amount of Grant: 500-20,000 USD
Contact: Janice Gibson; (805) 899-8405; fax (805) 884-1404
Sponsor: Henry W. Bull Foundation
P.O. Box 2340
Santa Barbara, CA 93120-2340

HERA Foundation Ovarian Cancer OSB1 Seed Grant **2238**
The Foundation is a non-profit organization whose mission is to prevent ovarian cancer and improve prognosis through innovative scientific research and education. The HERA Foundation will fund two basic, translational and/or clinical research projects: one grant of $25,000 is for projects that include, but are not limited to, those that study high grade ovarian carcinoma etiology and pathogenesis, and those that explore novel approaches to early ovarian cancer detection and targeted therapy; a second grant of $20,000 is earmarked for low grade/micropapillary serous carcinoma/low malignant potential ovarian tumor research for research leading to a better understanding and/or more effective treatment. The goal is to encourage outside the box thinking and solutions and provide seed money for such projects. Chemotherapy has been in use for many years and to date we have not improved overall survival by 1%. New approaches to understanding and treating this disease are needed and are what is of interest to the Foundation.
Requirements: Applicants must hold either a Ph.D. and/or an M.D. degree, and be either a post-doctoral research fellow or clinical fellow. A previous research background in ovarian cancer is not essential, as the HERA Foundation seeks to encourage new young investigators to enter the field. However, candidates should demonstrate the relevance of

their previous research background to the proposed work, and indicate a strong long-term commitment to research in the field of ovarian cancer.
Geographic Focus: All States
Date(s) Application is Due: Jan 15
Amount of Grant: 20,000-25,000 USD
Samples: Dr. Brant Wang, Johns Hopkins University; Dr. Hiroyuki Yoshida, Anderson Cancer Center; Dr. Martina Bazzaro, Johns Hopkins University.
Contact: Sheridia Daniels, (970) 948-7360; sdaniel6@jhmi.edu
Internet: http://www.herafoundation.org/research.cfm
Sponsor: HERA Foundation
P.O. Box 664
Carbondale, CO 81623

Herbert A. and Adrian W. Woods Foundation Grants 2239
The Herbert A. and Adrian W. Woods Foundation was established on June 9, 1999 upon the death of Mrs. Adrian W. Woods. Mrs. Woods had a long history of charitable giving in the St. Louis community and wanted to establish this foundation to continue that legacy of giving. The Foundation supports charitable organizations primarily in the greater St. Louis, Missouri area. The Trustees will consider the donor's past giving history and any special needs of any of the charities she has given to in the past. The Trustees will also consider requests from charitable organizations that fall into one or more of the following categories: abused, neglected, or troubled children; the poor; the Episcopal Church and affiliates, including outreach programs; arts and culture in the Metropolitan St. Louis area; animal welfare (in Missouri); and victims of illness or disability, including research in this area. The following types of requests may be submitted: special projects; capital campaign requests (capital grants are awarded only as a source of support among a broad community of funders); challenge or matching grants; and general operation funding. Applicants must apply online at the grant website. Applicants are strongly encouraged to do the following before applying: review the downloadable state application procedures for additional helpful information and clarifications; review the downloadable online-application guidelines at the grant website; review the foundation's funding history (link is available from the grant website); review the online application questions in advance; and review the list of required attachments. These will generally include: a list of board members, financial statements (audited, reviewed, or compiled by independent auditor); an organization summary; a list of other funding sources; an IRS Determination letter; and other required documents. All attachments must be uploaded in the online application as PDF, Word, or Excel files. The application deadline for this foundation is 11:59 p.m. on September 1. Final decisions about grant rewards will be made by November 30.
Requirements: Applicants must have 501(c)3 tax-exempt status.
Restrictions: The fund will not consider requests for multi-year grants (pledges) or endowment creation and funding. The fund does not support requests from individuals, organizations attempting to influence policy through direct lobbying, or any political campaigns.
Geographic Focus: Missouri
Date(s) Application is Due: Sep 1
Contact: Shanise Evans, Vice President; (314) 466-8027; shanise.evans@baml.com
Internet: https://www.bankofamerica.com/philanthropic/fn_search.action
Sponsor: Herbert A. and Adrian W. Woods Foundation
100 North Broadway MO2-100-07-15
Saint Louis, MO 63102-2728

Herbert C. Brown Award for Creative Research in Synthetic Methods 2240
This award, sponsored by Purdue Borane Research Fund and the Herbert C. Brown Award Endowment, is to recognize and encourage outstanding and creative contributions to research in synthetic methods. A nominee must have accomplished outstanding and creative research that involved the discovery and development of novel and useful methods for chemical synthesis. The synthetic method that is the basis for the award may be related to any area of chemistry. The award consists of $5,000, a medallion with a presentation box, and a certificate.
Requirements: Any individual, except a member of the award committee, may submit one nomination or seconding letter for the award in any given year. The nominating documents consist of a letter of not more than 1000 words containing an evaluation of the nominee's accomplishments and a specific identification of the work to be recognized, a biographical sketch including date of birth, and a list of publications and patents authored by the nominee.
Restrictions: Self-nominations are not accepted.
Geographic Focus: All States
Date(s) Application is Due: Nov 1
Amount of Grant: 5,000 USD
Contact: Felicia Dixon, Awards Administrator; (800) 227-5558 or (202) 872-4408; fax (202) 776-8008; f_dixon@acs.org or awards@acs.org
Internet: http://portal.acs.org/portal/acs/corg/content?_nfpb=true&_pageLabel=PP_ARTICLEMAIN&node_id=1319&content_id=CTP_004530&use_sec=true&sec_url_var=region1
Sponsor: American Chemical Society
1155 Sixteenth Street, NW
Washington, DC 20036-4801

Herbert H. and Grace A. Dow Foundation Grants 2241
The Foundation has charter goals to improve the educational, religious, economic and cultural lives of Michigan's people. Priority is given to organizations that: have clearly stated objectives, strong and purposeful management and are publicly accountable; have needs which are in areas not normally funded by governmental or public financing; are not hesitant to explore, initiate, volunteer, or execute original ideas or concepts; are willing to collaborate with other persons or organizations to give synergy to a common objective or goal; have purposes which tend to advance private enterprise and the preservation of a free, open and self-resourceful society. Application information is available online.
Requirements: Only organizations in Michigan are eligible to apply.
Restrictions: The Foundation does not make grants directly to individuals. It cannot legally support: organizations to which contributions are not tax deductible, according to Internal Revenue Service regulations; organizations that practice discrimination by race, sex, creed, age or national origin; political organizations or organizations whose purposes are to influence legislation.
Geographic Focus: Michigan
Contact: Margaret Ann Riecker; (989) 631-3699; info@hhdowfoundation.org
Internet: http://www.hhdowfdn.org
Sponsor: Herbert H. and Grace A. Dow Foundation
1018 West Main Street
Midland, MI 48640-4292

Herbert Hoover Presidential Library Association Travel Grants 2242
Travel grants are awarded to encourage scholarly use of the Herbert Hoover Presidential Library in West Branch, Iowa. The program is intended to promote the study of subjects of interest and concern to Herbert Hoover, Lou Henry Hoover, their associates, and other public figures as reflected in the library's collections. It is strongly suggested that the applicant contact the archival staff to determine whether library holdings are pertinent to the applicant's research. In recent years, awards have ranged from $500 to $1,500 per applicant. The Association will consider requests for extended research at the library. An independent committee of distinguished scholars from Iowa colleges and universities evaluates the research proposals.
Requirements: Current graduate students, post-doctoral scholars, and independent researchers are eligible to apply. Three separately mailed references must be submitted.
Geographic Focus: All States
Date(s) Application is Due: Mar 1
Amount of Grant: 500-1,500 USD
Contact: Delene McConnaha; (800) 828-0475 or (319) 643-5327; fax (319) 643-2391; delene.mcconnaha@hooverassociation.org or info@hooverassoc.org
Internet: http://www.hooverassociation.org/grantsawards/travel_grant.php
Sponsor: Herbert Hoover Presidential Library Association
302 Parkside Drive
West Branch, IA 52358

Herb Society of America Research and Education Grants 2243
Grants are meant to further the knowledge and use of herbs and contribute the results of the study and research to the records of horticulture, science, literature, history, art, or economics. The research could involve such varied topics as history, medicinal research and usage from a historical perspective, fragrance, flavor, horticulture, landscaping, economic importance, chemurgy, taxonomy, and political and social influence. Grants are made for periods up to one year.
Requirements: Applicants should be graduate students with proposed programs of scientific or academic investigation of herbal plants, or persons capable of demonstrating in-depth knowledge of herbs.
Restrictions: The grant is not given for financial aid to individuals, rather for specific research on herbal projects. The grant does not provide for indirect costs or personal travel for the benefit of any individual.
Geographic Focus: All States
Date(s) Application is Due: Jan 31
Amount of Grant: Up to 5,000 USD
Contact: Administrator; (440) 256-0514; fax (440) 256-0541; herbs@herbsociety.org
Internet: http://www.herbsociety.org/research.php
Sponsor: Herb Society of America
9019 Kirtland Chardon Road
Kirtland, OH 44094

Hereditary Disease Foundation John J. Wasmuth Postdoctoral Fellowships 2244
Support is offered for research projects that contribute to identifying and understanding the basic defect of Huntington's disease. Areas of interest include trinucleotide expansions, animal models, gene therapy, neurobiology and development of the basal ganglia, cell survival and death, and intercellular signaling in striatal neurons. Awards may be renewed for a third year. To receive an application, submit a letter of intent. Letter of intent forms may be submitted electronically via the Web site.
Geographic Focus: All States
Date(s) Application is Due: Feb 15; Jun 15; Oct 15
Amount of Grant: 45,000 USD
Contact: Dr. Carl D. Johnson, (212) 928-2121; carljohnson@hdfoundation.org
Internet: http://www.hdfoundation.org/funding/postdoct.php
Sponsor: Hereditary Disease Foundation
3960 Broadway, 6th Floor
New York, NY 10032

Hereditary Disease Foundation Lieberman Award 2245
The Hereditary Disease Foundation announces a special Lieberman Award, to catalyze innovative proposals leading to the treatment and cure of Huntington's disease. A Lieberman Award can be funded for two years for up to $75,000 per year. Areas of interest include trinucleotide expansions, animal models, gene therapy, neurobiology and development of the basal ganglia, cell survival and death, and intercellular signaling in striatal neurons.

Geographic Focus: All States
Amount of Grant: 75,000-150,000 USD
Contact: Dr. Carl D. Johnson, (212) 928-2121; carljohnson@hdfoundation.org
Internet: http://www.hdfoundation.org/funding/lieberman.php
Sponsor: Hereditary Disease Foundation
3960 Broadway, 6th Floor
New York, NY 10032

Hereditary Disease Foundation Research Grants 2246

The focus of the Hereditary Disease Foundation is on Huntington's disease. Support will be for research projects that contribute to identifying and understanding the basic defect in Huntington's disease. Grants are usually for one year, with the possibility of renewal for up to three years. Areas of interest include trinucleotide expansions, animal models, gene therapy, neurobiology and development of the basal ganglia, cell survival and death, and intercellular signaling in striatal neurons. In addition to its regular grants, the foundation also offers the Lieberman Award for innovative proposals accelerating the discovery of a treatment and cure of Huntington's disease. Grants that receive funding are considered seed money. If the project shows promise, it is hoped that other institutions will fund it thereafter. To obtain an application, submit a one-page letter of intent. Letter of intent forms may be submitted electronically via the Web site.
Geographic Focus: All States
Date(s) Application is Due: Feb 15; Jun 15; Oct 15
Amount of Grant: Up to 50,000 USD
Contact: Dr. Carl D. Johnson, (212) 928-2121; carljohnson@hdfoundation.org
Internet: http://www.hdfoundation.org/funding/grants.php
Sponsor: Hereditary Disease Foundation
3960 Broadway, 6th Floor
New York, NY 10032

Herman Kahn Fellowship 2247

One fellowship is given annually in the area of national security studies, domestic or international economics, education policy, or the study of political institutions to support PhD candidates who have completed their coursework and have only their dissertations remaining. Studies will be carried out in Indianapolis, IN, for one academic year. The fellow's time will be equally divided between dissertation work and policy research assigned by the institute. Application requirements are available upon request. Selected candidates will be interviewed in Indianapolis.
Requirements: Eligible to apply are graduate students who have completed all PhD coursework within the last five years and who can provide evidence that they have been formally admitted to candidacy for the PhD. Consideration will be given only to applicants with outstanding academic records who are strongly recommended by faculty members and have demonstrated interest in and capability for policy-oriented research.
Geographic Focus: All States
Date(s) Application is Due: Apr 1
Amount of Grant: 18,000 USD
Contact: Director of Research; (202) 974-2400; fax (202) 974-2410; rsvp@hudson.org
Internet: http://www.hudson.org/learn/index.cfm?fuseaction=mission_statement
Sponsor: Hudson Institute
1015 15th Street NW, 6th Floor
Washington, DC 20005

HHMI-NIBIB Interfaces Initiative Grants 2248

HHMI and the National Institute of Biomedical Imaging and Bioengineering have formed a partnership to support biomedical research institutions in developing graduate-level research training programs in emerging interdisciplinary fields. The primary goal of this initiative is to train a cadre of PhD scientists who possess the knowledge and skills to conduct interdisciplinary research at the interface between the biomedical sciences and the physical science, computational, engineering, or mathematical disciplines. These fields may include, but are not limited to, chemistry, imaging, science, materials science, nanotechnology, and physics. Another goal is to reduce barriers to interdisciplinary graduate science education. Proposals should reflect the unique educational and scientific capabilities and strengths of the applicant institution(s) as well as address the specific goals of the initiative. The initiative consists of two phases: Phase I supports the establishment of new interdisciplinary training programs; Phase II will sustain the training programs through their critical years. Registration letters of intent to apply are due by January 20, with the proposal submission deadline June 15.
Requirements: All U.S. institutions that grant PhD degrees in appropriate science or engineering disciplines are eligible to apply. Collaborative programs between two or more institutions are acceptable.
Geographic Focus: All States
Date(s) Application is Due: Jan 20; Jun 15
Amount of Grant: Up to 1,000,000 USD
Contact: Maryrose Franko; (800) 448-4882, ext. 8880; interdisc@hhmi.org
Internet: http://www.hhmi.org/grants/institutions/nibib.html
Sponsor: Howard Hughes Medical Institute
4000 Jones Bridge Road
Chevy Chase, MD 20815-6789

HHMI-NIH Cloister Research Scholars Program 2249

The goal of this program, a joint venture of HHMI and NIH, is to expand the pool of medically trained researchers by encouraging medical students to pursue research careers. The program provides support for an intensive nine months to one year of full-time research, with a possible one-year extension, at NIH in Bethesda, MD. Most students enter after their second year of medical school. The Cloister, a residential facility for scholars, is available on the NIH campus. The research project is selected upon arrival at NIH, after a round of laboratory visits. Application forms may be downloaded from the Web site or requested from the program officer.
Requirements: Applicant must be in good standing at a medical or dental school in the U.S. or Puerto Rico and must receive permission from the school to participate in the program. The school does not have to be a major academic medical center or research-oriented school.
Restrictions: Enrollees in an M.D./Ph.D. or D.D.S./Ph.D. program are not eligible.
Geographic Focus: All States
Date(s) Application is Due: Jan 10
Amount of Grant: 17,800 USD
Contact: Office of Grants and Special Programs; (800) 424-9924 or (301) 215-8873; fax (301) 215-8888; research_scholars@hhmi.org
Internet: http://www.hhmi.org/research/cloister/program.html
Sponsor: Howard Hughes Medical Institute
4000 Jones Bridge Road
Chevy Chase, MD 20815-6789

HHMI/EMBO Start-up Grants for Central Europe 2250

The program encourages young scientists to obtain independent faculty positions in select countries that have ongoing HHMI and EMBO science programs. Applications are encouraged from scientists to set up their first independent laboratories in a Central European EMBC member state (Croatia, Czech Republic, Estonia, Hungary, Poland, and Slovenia). Application forms should be completed by the individual scientist and the receiving institute. It is expected that the institute make the applicant an offer that goes beyond that of a position and laboratory space. HHMI will contribute $50,000, while the participating member state must provide the remaining $25,000. Application and guidelines are available online.
Requirements: Eligible scientists should have an excellent publication record; have a maximum of 10 years postdoctoral experience; be negotiating a position at an institute/university in Central Europe by the date of the application; and be located outside Central Europe at the time of application.
Geographic Focus: All States, Austria, Czech Republic, Germany, Hungary, Liechtenstein, Poland, Slovakia, Slovenia, Switzerland
Date(s) Application is Due: Aug 1
Amount of Grant: 75,000 USD
Contact: Grants Administrator; (301) 215-8500
Internet: http://www.embo.org/projects/yip/embo_hhmi_startup_grants.html
Sponsor: Howard Hughes Medical Institute
4000 Jones Bridge Road
Chevy Chase, MD 20815-6789

HHMI Biomedical Research Grants for International Scientists: Infectious Diseases and Parasitology 2251

The Program provides five-year grants to support promising basic biomedical research scientists working in the fields of infectious diseases and parasitology. The Institute seeks to build the strongest foundation of scientific knowledge and capability by creating a global network of scientists who make important contributions outside the U.S. research community. It support these leaders in their home countries to facilitate the application and dissemination of their knowledge and to strengthen research environments and educational opportunities.
Requirements: Applicants must hold a full-time appointment or have a pending appointment at a nonprofit scientific research organization in any country other than the United States and the United Kingdom. Researchers may not be citizens or permanent residents of the United States.
Restrictions: Only scientists working in the research areas specified in the program announcement are eligible to apply for grant support. The Program does not consider unsolicited grant proposals or award grants for clinical trials and research on health education, health-care delivery, or health services.
Geographic Focus: All States
Date(s) Application is Due: Sep 15
Amount of Grant: 50,000-100,000 USD
Contact: Jill G. Conley; (301) 215-8873; fax (301) 215-8888; parasite@hhmi.org
Internet: http://www.hhmi.org/grants/individuals/idap.html
Sponsor: Howard Hughes Medical Institute
4000 Jones Bridge Road
Chevy Chase, MD 20815-6789

HHMI Biomedical Research Grants for International Scientists in Canada and Latin America 2252

HHMI's International Research Scholars Program provides five-year grants to support promising basic biomedical research scientists working in Canada and Latin America. Eligible epidemiology research includes that directed toward an understanding of disease distribution in populations or of associations that may suggest causality or preventive strategies.
Requirements: The supported research must be conducted in one of the following eight eligible countries: Argentina, Brazil, Canada, Chile, Mexico, Peru, Uruguay, or Venezuela. Applicants must hold a full-time appointment or have a pending full-time appointment at a nonprofit scientific research organization in any one of the eight eligible countries.
Restrictions: Clinical trials and research on health education, health care delivery, or health services are not eligible fields.
Geographic Focus: All States

Date(s) Application is Due: Sep 14
Amount of Grant: 50,000-100,000 USD
Contact: Jill G. Conley; (301) 215-8873; fax (301) 215-8888; canlatam@hhmi.org
Internet: http://www.hhmi.org/grants/individuals/canlatam.html
Sponsor: Howard Hughes Medical Institute
4000 Jones Bridge Road
Chevy Chase, MD 20815-6789

HHMI Biomedical Research Grants for International Scientists in the Baltics, Central and Eastern Europe, Russia, and Ukrain 2253

The Program will award five-year grants to support promising basic biomedical research scientists working in the Baltics, Central and Eastern Europe, Russia, and Ukraine. Eligible epidemiology research includes that directed toward an understanding of disease distribution in populations or of associations that may suggest causality or preventive strategies. Allowable expense categories include salaries and stipends for award recipients, graduate students, postdoctoral fellows, and technicians, equipment and supplies, travel, and publication costs. Up to 10 percent of the grant may be used to pay for indirect costs (including administrative costs) of the scientist's institution.
Requirements: The supported research must be conducted in one of the following eligible countries: Bulgaria, Croatia, Czech Republic, Estonia, Hungary, Latvia, Lithuania, Poland, Romania, Russia, Slovak Republic, Slovenia, or Ukraine.
Restrictions: Researchers may not be citizens or permanent residents of the United States. Grant funds may not be used for laboratory renovations.
Geographic Focus: All States, Albania, Andorra, Armenia, Austria, Azerbaijan, Belarus, Belgium, Bosnia & Herzegovina, Bulgaria, Croatia, Cyprus, Czech Republic, Denmark, Estonia, Finland, France, Georgia, Germany, Greece, Hungary, Hungary, Iceland, Ireland, Italy, Kosovo, Latvia, Liechtenstein, Lithuania, Luxembourg, Macedonia, Malta, Moldova, Monaco, Montenegro, Norway, Poland, Portugal, Romania, Russia, Russia, San Marino, Serbia, Slovakia, Slovenia, Spain, Sweden, Switzerland, The Netherlands, Turkey, Ukraine, Ukraine, United Kingdom, Vatican City
Date(s) Application is Due: Nov 17
Amount of Grant: 50,000-100,000 USD
Contact: Jill G. Conley; (301) 215-8873; fax (301) 215-8888; bceeru@hhmi.org
Internet: http://www.hhmi.org/grants/individuals/bceeru.html
Sponsor: Howard Hughes Medical Institute
4000 Jones Bridge Road
Chevy Chase, MD 20815-6789

HHMI Gilliam Fellowships for Advanced Study 2254

The Fellowships provide full support for up to five years of study toward a Ph.D. for outstanding students who are from groups underrepresented in the sciences or from disadvantaged backgrounds. Gilliam fellows attend the university of their choice and work alongside distinguished scientists. Chosen for their academic excellence and scientific potential, they will become the leaders of a new generation of biomedical researchers. The Fellowship is available up to 5 years at $44,000 per year.
Requirements: Must be: past participants in the HHMI Exceptional Research Opportunities Program (EXROP); graduating seniors or recent baccalaureate recipients (within two years) who have completed no more than one year of Ph.D. training. Former EXROP students pursuing an M.D./Ph.D. degree may apply, but the fellowship supports only the Ph.D. portion of the studies. There is no citizenship requirement.
Geographic Focus: All States
Date(s) Application is Due: Oct 6
Amount of Grant: 44,000 USD
Contact: Maryrose Franko, Senior Program Office; (800) 424-9924 or (301) 215-8880; fax (301) 215-8888; gilliam@hhmi.org
Internet: http://www.hhmi.org/grants/individuals/gilliam.html
Sponsor: Howard Hughes Medical Institute
4000 Jones Bridge Road
Chevy Chase, MD 20815-6789

HHMI Grants and Fellowships Programs 2255

The institute administers a grants program that focuses on improving science education from preschool through postdoctoral training, enhancing the science literacy of the general public, and supporting the research of biomedical scientists in selected foreign countries. Currently, grants are administered through five programs: an international program that supports biomedical scientists outside the United States and provides funding for selected courses and workshops; a precollege science-education program that offers grants for educational activities to science museums, botanical and zoological gardens, etc; a biomedical-research program that supports medical schools and research organizations; a graduate science-education program that provides fellowships for graduate students, medical students, and physicians, and supports special courses; and a biological-sciences education program that awards grants to selected undergraduate institutions.
Geographic Focus: All States
Samples: United Way of New York City (NY)—for the September 11th Fund, established to support emergency-assistance to organizations and other nonprofit health and human-service groups providing disaster-relief services to victims of the terrorist attacks in New York, Pennsylvania, and Washington, $50,000.
Contact: Jill G. Conley; (301) 215-8873; fax (301) 215-8888; grantswww@hhmi.org
Internet: http://www.hhmi.org/grants
Sponsor: Howard Hughes Medical Institute
4000 Jones Bridge Road
Chevy Chase, MD 20815-6789

HHMI Initiative for Colleges Grants 2256

HHMI awards four-year grants to colleges (baccalaureate and master's institutions) to support science education at the undergraduate, K-12, and community college levels. Grants support a range of activities that engage students in research and broaden their access to science; create courses that convey the excitement of contemporary science; develop new, current, and future faculty members; and establish or expand outreach programs that extend to elementary, high school, and community college faculty and students. The registration deadline is May 16, with final proposals due by October 17.
Requirements: Only invited institutions are eligible to apply. Invitations are based on an institution's classification by the Carnegie Foundation for the Advancement of Teaching as a baccalaureate and master's institution and on the institution's record of postbaccalaureate student achievement in the sciences.
Geographic Focus: All States
Date(s) Application is Due: May 16; Oct 17
Amount of Grant: 800,000-16,000,000 USD
Contact: Grants Officer; (301) 215-8870 or (301) 215-8500; grantugr@hhmi.org
Internet: http://www.hhmi.org/grants/institutions/colleges.html
Sponsor: Howard Hughes Medical Institute
4000 Jones Bridge Road
Chevy Chase, MD 20815-6789

HHMI Initiative for Universities Grants 2257

HHMI awards four-year grants to research and doctoral universities to support science education at the undergraduate and K-12 levels. The program supports initiatives that strengthen education in the biological sciences and other scientific disciplines as they relate to biology. Program objectives include: to strengthen research and teaching in undergraduate education and to support student involvement in faculty research; to prepare undergraduates for graduate studies and for careers in biomedical research, medicine, and science education; to bring fresh perspectives to established scientific disciplines, and develop novel programs and courses in emerging areas; to foster linkages between universities and elementary and secondary schools; and to develop approaches that encourage teamwork among faculty, postdoctoral fellows, graduate students, and undergraduates. The registration deadline is May 12, with final proposals due by October 18.
Requirements: Only invited institutions are eligible to apply. Invitations are based on an institution's classification by the Carnegie Foundation for the Advancement of Teaching as a baccalaureate and master's institution and on the institution's record of postbaccalaureate student achievement in the sciences.
Geographic Focus: All States
Date(s) Application is Due: May 14; Oct 1
Amount of Grant: 12-22,000,000 USD
Contact: Grants Officer; (301) 215-8870 or (301) 215-8500; grantugr@hhmi.org
Internet: http://www.hhmi.org/grants/institutions/universities.html
Sponsor: Howard Hughes Medical Institute
4000 Jones Bridge Road
Chevy Chase, MD 20815-6789

HHMI International Research Scholars Program 2258

This five year grant program supports non-U.S. biomedical scientists, scientific meetings for grant-supported scientists, and other international educational activities. Grants are awarded to promising biomedical scientists who have made significant contributions to fundamental research. Application information is available online.
Requirements: To be considered, individuals must hold a full-time academic or research appointment at a university, medical school, or other nonprofit scientific institution. They are not permitted to have major administrative responsibilities.
Geographic Focus: All States
Contact: Jill G. Conley, Program Director; (301) 215-8873; fax (301) 215-8888
Internet: http://www.hhmi.org/grants/for_grantees/
Sponsor: Howard Hughes Medical Institute
4000 Jones Bridge Road
Chevy Chase, MD 20815-6789

HHMI Med into Grad Initiative Grants 2259

HHMI will award grants to institutions to improve the understanding of medicine and pathobiology by scientists conducting biomedical research. Grants will be used to modify existing graduate training or initiate new programs to develop a cadre of Ph.D. researchers who understand pathophysiology and are committed to working at the interface of the basic sciences and clinical medicine. Applications will be accepted from any university in the United States that offers Ph.D.-level training in an appropriate science or engineering discipline. It is expected that as much as $10 million will be awarded in this competition. Awards will range from $400,000 to $1 million, and the grant funds will be allocated over four consecutive years.
Requirements: U.S. institutions that grant PhD degrees in the appropriate science and engineering disciplines are eligible to apply.
Geographic Focus: All States
Date(s) Application is Due: Apr 27
Amount of Grant: 400,000-1,000,000 USD
Contact: Maryrose Franko, Senior Program Officer; (800) 448-4882, ext. 8880; fax (301) 215-8888; medintograd@hhmi.org
Internet: http://www.hhmi.org/grants/institutions/medintograd.html
Sponsor: Howard Hughes Medical Institute
4000 Jones Bridge Road
Chevy Chase, MD 20815-6789

HHMI Physician-Scientist Early Career Award **2260**
Through this competitive grant initiative, HHMI awards five-year grants to selected alumni of the HHMI-NIH Research Scholars Program and the HHMI Research Training Fellowships for Medical Students Program to support these individuals as they begin careers as independent physician-scientists. The award provides $375,000 over a five-year period for direct research costs.
Requirements: Only alumni of the HHMI-NIH Research Scholars Program and the HHMI Research Training Fellowship for Medical Students Program who have received an M.D., M.D./Ph.D., D.D.S, or equivalent degree are eligible to apply.
Restrictions: The funds may not be used for the salary of the awardee or institutional indirect costs.
Geographic Focus: All States
Amount of Grant: 375,000 USD
Contact: Office of Grants and Special Programs; (800) 448-4882, ext. 8889 or (800) 424-9924; fax (301) 215-8888; earlycareer@hhmi.org
Internet: http://www.hhmi.org/grants/individuals/earlycareer.html
Sponsor: Howard Hughes Medical Institute
4000 Jones Bridge Road
Chevy Chase, MD 20815-6789

HHMI Precollege Outreach Initiative for Biomedical Research Institutions Grants **2261**
HHMI will award five-year grants to support pre-K to 12th-grade science education. Funding will be provided to medical schools, dental schools, veterinary schools, public health schools, hospitals, academic health centers, and independent research institutions to enable these science-rich organizations to work with school systems, museums, and other partners skilled in delivering science content to students. Letters of Intent to submit a proposal are due by June 13, and proposal submissions are due by September 12.
Requirements: HHMI does not make grants directly to individuals, school systems, or to pre-K to 12th-grade schools under this initiative.
Geographic Focus: All States
Date(s) Application is Due: Jun 13; Sep 12
Amount of Grant: 200,000-500,000 USD
Contact: Grants Officer; (301) 215-8870 or (301) 215-8500; grantugr@hhmi.org
Internet: http://www.hhmi.org/grants/institutions/biomedical.html
Sponsor: Howard Hughes Medical Institute
4000 Jones Bridge Road
Chevy Chase, MD 20815-6789

HHMI Research Training Fellowships **2262**
Medical fellowships are intended to strengthen and expand the nation's pool of medically trained researchers. These fellowships provide funds to help meet fellows' research-related expenses and education costs during the research training period. Fellowships are awarded annually to provide support for one year of full-time training in fundamental biomedical, clinical, translational, or applied research. The fellowship includes a stipend, a research allowance to meet research-related expenses, and a fellow's allowance to be used on behalf of the fellow for health care, tuition and fees, and other research-related expenses. Application information is available online.
Requirements: Fellowships are awarded to students enrolled in medical and dental programs in the U.S.
Geographic Focus: All States
Date(s) Application is Due: Jan 11
Contact: Program Officer; (301) 215-8500 or (800) 424-9924; fax (301) 215-8888; medfellows@hhmi.org
Internet: http://www.hhmi.org/grants/individuals/medfellows.html
Sponsor: Howard Hughes Medical Institute
4000 Jones Bridge Road
Chevy Chase, MD 20815-6789

HHMI Undergraduate Science Education Grants for Professors **2263**
The Program awards four-year grants to research-active undergraduate faculty. The grants are intended to empower leading scientists at doctoral and research universities to work more closely with undergraduates at their home institutions and provide other institutions with innovative models for transmitting the excitement and values of scientific research to undergraduate education.
Requirements: Institutions must be invited by HHMI to nominate faculty members to compete for the HHMI professors awards. Invitations are based on the institutions' classifications by the Carnegie Foundation for the Advancement of Teaching as doctoral and research universities and the institutions' records of postbaccalaureate student achievement in the sciences. Invitees are listed in the program announcement.
Geographic Focus: All States
Contact: Office of Grants; (800) 424-9924 or (301) 215-8873; grantugr@hhmi.org
Internet: http://www.hhmi.org/grants/individuals/professors.html
Sponsor: Howard Hughes Medical Institute
4000 Jones Bridge Road
Chevy Chase, MD 20815-6789

HIAS Scholarship Awards **2264**
The annual competition offers scholarships to deserving students in the United States and Israel. The awarding of scholarships is based on academic excellence, financial need, and involvement within the Jewish community. Guidelines and application are available online.
Geographic Focus: All States
Date(s) Application is Due: Mar 15; Jun 30
Amount of Grant: 1,500 USD
Contact: Scholarship Committee; (212) 613-1358; scholarship@hias.org
Internet: http://www.hias.org/Scholarships/apply.html
Sponsor: HIAS Scholarship Program
333 Seventh Avenue, 16th Floor
New York, NY 10001-5004

Higher Education Graduate Scholarships, Fellowships, and Grants **2265**
The program in higher education at New York University offers graduate assistantships and fellowships to graduate students. Degree programs in higher education include the MA, EdD (community college leadership focus), and PhD. Information and applications are available on the Web site.
Requirements: These opportunities are open specifically to students enrolled in the master's and doctoral degree programs in higher education at New York University.
Geographic Focus: All States
Amount of Grant: 1,000-15,000 USD
Contact: Office of Graduate Admissions; (212) 998-5030; ed.gradadmissions@nyu.edu
Internet: http://www.nyu.edu/financial.aid
Sponsor: New York University
82 Washington Square E, 2nd Floor
New York, NY 10003-6680

Hilda and Preston Davis Foundation Grants **2266**
The Foundation provides funds to charitable organizations whose programs advance the development of all areas of the lives of children and young adults. The Foundation places special emphasis on, and channels most of its financial resources toward, those organizations whose attention is concentrated on eating disorders and education for the underprivileged. Grants may range from $10,000 up to $100,000. In unique circumstances, the Foundation does consider a more significant grant for a program having a major impact in one or more of its areas of interest. The Foundation encourages pilot initiatives that test new program models. Of particular interest to the Foundation are organizations that promote partnerships and collaborative efforts among multiple groups and organizations. The Foundation almost always limits grant durations to three years or less. Applications are accepted throughout the year.
Requirements: ll applicants must be an approved tax-exempt non-profit organization as defined by the Internal Revenue Service. Areas of greatest importance to the Foundation are funded on a national basis. However, a good deal of the Foundation's grant making in other program areas is focused in areas of geographic importance to the donors–the northeastern United States (southern New England & the middle Atlantic states), as well as California. Priority will be given to requests that show specific plans for funding beyond the present.
Restrictions: The Foundation will not allow any funds to be earmarked for indirect costs or institutional overhead in cases where the grant relationship was developed independent of that institution's direct involvement. The Foundation generally will not provide grants to the following: organizations not determined to be tax-exempt under section 501(c)3 of the Internal Revenue Code; individuals; general fundraising drives; endowments; government agencies; or organizations that subsist mainly on third party funding and have demonstrated no ability or expended little effort to attract private funding.
Geographic Focus: California, Connecticut, Delaware, District of Columbia, Maryland, Massachusetts, New Hampshire, New Jersey, New York, Pennsylvania, Rhode Island, Vermont, Virginia, West Virginia
Amount of Grant: 10,000-100,000 USD
Contact: Grants Administrator; (203) 629-8552; fax (203) 547-6112; davis@fsllc.net
Internet: http://www.hpdavis.org/application.htm
Sponsor: Hilda and Preston Davis Foundation
640 West Putnam Avenue, 3rd Floor
Greenwich, CT 06830

Hill Crest Foundation Grants **2267**
The Hill Crest Foundation awards grants to Alabama nonprofits, primarily health associations and human services and education, with some funding for the arts. Types of support include: building and renovation; capital campaigns; endowments; equipment; matching and challenge support; professorships; program development; publication; research; scholarship funds; seed money; and technical assistance. There are no deadlines or applications. The Board of Directors meets quarterly.
Requirements: Application forms are not required. Before submitting a proposal, organizations should submit a letter, detailing the project and the amount requested, along with descriptive literature about their organization.
Restrictions: Grantmaking is limited to Alabama. The Foundation does not fund individuals.
Geographic Focus: Alabama
Contact: Charles Terry, Chairperson; (205) 425-5800
Sponsor: Hill Crest Foundation
P.O. Box 530507
Mountain Brook, AL 35253-0507

Hilton Hotels Corporate Giving Grants **2268**
Hilton makes charitable contributions to nonprofit organizations involved with K-12 education, youth development, public policy, homelessness, and civic affairs. Support is given primarily in areas of company operations, with emphasis on California, including Los Angeles and San Francisco, and Tennessee, including Memphis; giving also to national organizations. Contact Ellen Gonda for additional information at: corporate_communications@hilton.com.

Requirements: 501(c)3 tax-exempt organizations are eligible.
Restrictions: No support for sport teams, religious organizations not of direct benefit to the entire community, government-supported organizations (over 20 percent of budget), hospitals, private schools, pre-schools, or day care facilities, film and video production, or promotional materials. No grants to individuals (except for employee-related scholarships), or for fellowships, sports activities, debt reduction, capital campaigns or endowments, film or video projects, or promotional merchandise.
Geographic Focus: California, Tennessee
Contact: Ellen Gonda, Senior Vice President; (310) 278-4321; fax (310) 205-7678; corporate_communications@hilton.com
Internet: http://www.hiltonworldwide.com
Sponsor: Hilton Hotels Corporation
9336 Civic Center Drive
Beverly Hills, CA 90210-3604

Hirtzel Memorial Foundation Grants 2269
The foundation awards grants to eligible organizations in New York and Pennsylvania in its areas of interest, including medical research and health care; community and neighborhood improvement; higher education; and social and human services. Types of support include building construction and renovation, capital campaigns, scholarships, equipment acquisition, general operating support, research, and scholarships. There are no application deadlines; contact the office for appropriate forms.
Requirements: New York nonprofit organizations in Ripley, Chautauqua County, and Pennsylvania nonprofit organizations in North East, Erie County, are eligible.
Geographic Focus: New York, Pennsylvania
Amount of Grant: 750-250,000 USD
Contact: Laurie Moritz, Grants Administrator; (412) 234-0023
Sponsor: Orris C. Hirtzel and Beatrice Dewey Hirtzel Memorial Foundation
P.O. Box 185
Pittsburgh, PA 15230

Hispanic Scholarships 2270
Scholarships are intended for undergraduate and graduate students of Hispanic American background. Successful candidates are chosen on the basis of academic achievement, personal strengths, leadership, and financial need. For more information and application requirements, write to the selection committee at the Hispanic Scholarship Fund. Applications can be downloaded from Web site.
Requirements: Applicants must be U.S. citizens or legal permanent residents who come from Mexican American, Puerto Rican, Cuban, Caribbean, Central American, or South American parentage and who attend college in one of the 50 states or Puerto Rico on a full-time basis, must have completed a minimum of 12 units of college work prior to submission of an application, and must have a GPA of 2.7 on a 4.0 scale. Community college applicants must be enrolled in majors transferable to a four-year institution offering a baccalaureate degree.
Restrictions: HSF is unable to make awards to fully employed persons or practicing professionals.
Geographic Focus: All States
Amount of Grant: 2,000-10,000 USD
Contact: Sara Martinez Tucker, President; (877) 473-4636 or (415) 445-9930; fax (415) 808-2302; info@hsf.net
Internet: http://www.hsf.net/scholarship/Special.htm#nhfa
Sponsor: Hispanic Scholarship Fund
55 Second Street, Suite 1500
San Francisco, CA 94105

Hoglund Foundation Grants 2271
The foundation supports nonprofits working in the areas of education, health science and services, social services, and children's health and development. Grants are made primarily in Dallas, and Houston, Texas, with a limited number of grants awarded outside of this area. Types of support include project support, capital support, and general operating support. Application guidelines are available on the foundation's Web site.
Requirements: To be eligible for consideration, an organization must provide proof that they have received a determination letter from the Internal Revenue Service indicating that it is a tax exempt organization as described in Section 501(c)3 of the Internal Revenue Code of 1986 and is treated as other than a private foundation as within the meaning of Section 509(a) of the Code. An organization may also qualify under Section 170(c)(1) if the grant requested is to be used exclusively for public purposes as described in the Code.
Restrictions: Grants are not made to individuals.
Geographic Focus: Texas
Date(s) Application is Due: Mar 15; Jul 15; Nov 15
Amount of Grant: 1,000-200,000 USD
Contact: Kelly Compton; (214) 987-3605; fax (214) 363-6507; info@hoglundfoundation.org
Internet: http://www.hoglundfoundation.org/grants.html
Sponsor: Hoglund Foundation
5910 North Central Expressway, Suite 255
Dallas, TX 75206

Holberg Memorial Fund International Prize 2272
The Holberg International Memorial Prize is named after Ludvig Holberg, who was born in Bergen in 1684, and held the Chairs of Metaphysics and Logic, Latin Rhetoric and History at the University of Copenhagen. Holberg played an important part in bringing the Enlightenment to the Nordic countries. Through his interdisciplinary and internationally-oriented efforts, Holberg sought to modernize academic subjects and teaching methods. The main purpose of the Holberg Fund is to increase the profile of the arts and humanities, the social sciences, law and theology. The Prize is awarded to scholars who have made outstanding contributions to research in the arts and humanities, social science, law, or theology, either within these fields or through interdisciplinary work. The prize winner must have had a decisive influence on international research. The Prize is worth 4.5 million NOK (approximately EUR 570,000 or USD 800,000 USD). Specific instructions and a current nomination form are available at the website.
Requirements: Scholars holding positions at universities and other research institutions, including academies, are entitled to nominate candidates for the Prize. The letter of nomination should be written in English and state the reasons for the nomination in two to three pages. The nomination should also include the candidate's CV and suggest referees who know the scholar's work. The function of the nomination is to make the Holberg Prize Academic Committee aware of the candidate's work.
Restrictions: Nominations are strictly confidential. They should not be disclosed to the nominee or to others at any time.
Geographic Focus: All States, All Countries
Date(s) Application is Due: Sep 15
Amount of Grant: 775,000 USD
Contact: Trine Kleven; 47 55 58 69 92; Trine.kleven@holbergprisen.no
Internet: http://www.holbergprisen.no/en/holberg-international-memorial-prize.html
Sponsor: University of Bergen
Thormohlensgate 51, 2nd Floor, P.O. Box 7800
Bergen, N-5020 Norway

Holberg Memorial Fund Nils Klim Prize 2273
The Nils Klim Prize is awarded to a young Nordic researcher in the academic fields of the arts and humanities, social sciences, law, and theology. The Prize is awarded to researchers who have made an outstanding contribution, either within one of the relevant fields alone or of an interdisciplinary nature. Particular importance is attached to the researcher's independence and innovative ability. Consideration is also given to whether the research contribution in question can be deemed particularly outstanding or promising in light of the researcher's age, the nature of the research project, and the research traditions in the academic field in question. The Board of the Holberg Memorial Fund makes the award on the basis of the recommendation of the Nils Klim Academic Committee, which consists of outstanding Nordic scholars in the relevant academic fields. The prize is worth NOK 250,000 (approximately $32,000 EUR or $42,000 USD). Specific guidelines and a current nomination form are available at the Prize website.
Requirements: Candidates must be under the age of 35 on the closing date of the nomination. Scholars holding a senior position at universities and other research institutions within the above mentioned academic fields are entitled to nominate candidates. The letter of nomination must be written in English, no more than five pages, and state the reasons for the nomination. The letter should also contain a CV for the candidate including list of publications and copies of the candidate's doctoral thesis and any another significant publications. The letter should also suggest referees who know the researcher and his/her work.
Geographic Focus: Denmark, Finland, Iceland, Norway, Sweden
Date(s) Application is Due: Sep 15
Amount of Grant: 250,000 NOK
Contact: Trine Kleven; 47 55 58 69 92; Trine.kleven@holbergprisen.no
Internet: http://www.holbergprisen.no/en/nils-klim-prize.html
Sponsor: University of Bergen
Thormohlensgate 51, 2nd Floor, P.O. Box 7800
Bergen, N-5020 Norway

Holberg Memorial Fund School Project 2274
Every year students at upper secondary schools in Norway are invited to carry out a research project in cooperation with established researchers in an effort to stimulate young people to take an interest in the arts, social sciences, law, and theology. These projects span a wide range of topics such as different views on alternative medicine or clothes as an expression. Three projects are awarded prizes of NOK 15,000, 10,000, and 5,000 respectively. The prize winning students are also invited to attend the award ceremony. The Project also awards a prize of NOK 15,000 to the teachers who present the best educational program. Specific information is available at the website.
Geographic Focus: Norway
Amount of Grant: 5,000-15,000 NOK
Contact: Barbara Wendelbo; +47 55 58 90 45; barbara.wendelbo@holbergprisen.no
Internet: http://www.holbergprisen.no/en/holberg-prize-school-project/info.html
Sponsor: University of Bergen
Thormohlensgate 51, 2nd Floor, P.O. Box 7800
Bergen, N-5020 Norway

HomeBanc Foundation Grants 2275
Giving primarily in Florida and Georgia, the HomeBanc Foundation also, gives to national organizations. The Foundation has three main goals: provide support to cancer related causes through funding education, advocacy, and research; support the American dream of homeownership for those who might not reach it on their own; and provide college funding to students who strive to achieve their goals through education beyond high school. Scholarship forms are available online. Send a Letter of Inquiry to the Foundation, unsolicited applications are not excepted. The Foundation will contact you if your proposal fits their criteria.
Requirements: Nonprofits providing services in Georgia and Florida are eligible.
Geographic Focus: Florida, Georgia
Contact: Amanda Albertelli; (404) 497-1000; aalbertelli@homebanc.com

Sponsor: HomeBanc Foundation
2002 Summit Boulevard, Suite 100
Atlanta, GA 30319-1497

Honeybee Health Improvement Project Grants 2276
The mysterious disappearance of bees, called Colony Collapse Disorder (CCD), is a growing threat to Honey Bees, the mainstay of pollination services in agriculture. The North American Pollinator Protection Campaign (NAPPC), a tri-national coalition dedicated to promoting the health of all pollinators partners with different organizations to perform research for improving the health of honey bees and reversing the threats they face. In partnership with Burt's Bees, the Honeybee Health Improvement Project was created. It awards annual research grants to scientists seeking solutions to improve honeybee health—including genetic stock improvements, commercial beekeeping practices, and forage opportunities on public and private land. NAPPC will fund multiple projects up to $8,000 each.
Requirements: Review and selection of proposals will be conducted by members of the Honey Bee Health Improvement Task Force (HBH). Funds must be used within a one-year period. Focused, targeted projects with a high likelihood of providing tangible results that can be applied to improving honey bee health are preferred. Proposals providing valuable extensions of previously funded projects will be considered. The HBH Task Force has identified six priority areas for funding, though other areas will be considered as well: (1) Effects of nutrition on honey bee behavior, physiology and/or colony health; (2) Effects of pesticides on honey bee behavior, physiology and/or colony health; (3) Effects of pathogens and pests on honey bee behavior, physiology and/or colony health; including the development of novel methods to mitigate these effects, such as RNAi technology; (4) Development of methods to improve genetic stocks of managed honey bee populations; (5) Effects of climate or environmental variables on: a) plants, especially nectar and pollen quantity and quality; and/or b) honey bee physiology and/or colony health; (6) The development of diagnostics or indicators for the presence of stressors that effect honey bee health, particularly those that can be used by beekeepers. Proposals that address multiple priority areas or have implications for the health of other managed or native bee species are encouraged.
Restrictions: As a nonprofit organization, the Pollinator Partnership/NAPPC does not pay overhead on funded research grants.
Geographic Focus: All States
Date(s) Application is Due: Dec 19
Amount of Grant: Up to 8,000 USD
Contact: Jennifer Tsang, (510) 643-7430; jt@pollinator.org
Internet: http://www.pollinator.org/honeybee_health.htm
Sponsor: North American Pollinator Protection Campaign
423 Washington Street, 5th Floor
San Francisco, CA 94111

Honeywell Corporation Habitat and Conservation Grants 2277
In partnership with environmental organizations, Honeywell provides students with unique learning opportunities and educators with new teaching tools to promote green science in the classroom. The Corporation believes that energy, water, waste management and air quality are among the most prominent issues around the world today. Operated in partnership with the New Jersey Audubon Society, Montezuma Audubon Center, Onondaga Audubon Society, and Maryland Science Center, the Honeywell Institute for Ecosystems Education takes science and social studies teachers through an explorative tour of their local ecosystems. Educators learn new teaching techniques and discover creative methods to integrate environmental science into their existing curriculum. Applications are required, and potential applicants should begin by contacting the office by email. There are no specific deadlines.
Geographic Focus: All States
Contact: Michael Holland; (973) 455-2728; michael.holland@honeywell.com
Internet: http://honeywell.com/Citizenship/Pages/habitat-conservation.aspx
Sponsor: Honeywell Corporation
101 Columbia Road
Morristown, NJ 07962

Honeywell Corporation Sustainable Opportunities Grants 2278
The Honeywell Corporation is dedicated to protecting the environment with a comprehensive and contemporary commitment to address some of the world's toughest challenges. Giving in this area is on a national basis within areas of company operations, with emphasis on New Jersey and New York. Types of support include: education, program development, sponsorships, and environmental protection programs. Applications are required, and potential applicants should begin by contacting the office by email. There are no specific deadlines.
Geographic Focus: All States
Contact: Michael Holland, Hometown Solutions Media Contact; (973) 455-2728 or (973) 727-6891; michael.holland@honeywell.com
Internet: http://honeywell.com/Citizenship/Pages/our-commitments.aspx
Sponsor: Honeywell Corporation
101 Columbia Road
Morristown, NJ 07962

Honor the Earth Grants 2279
As a unique national Native initiative, Honor the Earth works to raise public awareness, and raise and direct funds to grassroots Native environmental groups. Honor the Earth's Board encourages proposals on Native environmental justice, sustainable development, and cultural preservation, with a grant limit of $5,000. The organization's focus will remain on sustainable Indigenous communities support. Their work also focuses on opposition to fossil fuels extraction and destructive mining practices. Additional information, guidelines, and information on previously funded projects is available at the website. Applicants are encouraged to check back periodically for grant funding through the Building Resilience program. Proposals may be submitted in hard copy or by email. Faxed proposals cannot be accepted.
Requirements: Grants are awarded solely to organizations that are led and managed by Native peoples. Priority is given to grassroots, community-based organizations and groups with a lack of access to federal and/or tribal funding resources.
Restrictions: Funding is not available for individuals.
Geographic Focus: All States, Canada
Date(s) Application is Due: Nov 9
Amount of Grant: 1,000-5,000 USD
Contact: Winona LaDuke, Executive Director; (218) 375-3200; HonorGrants@honorearth.org or info@honortheearth.org
Internet: http://www.honorearth.org/grantmaking
Sponsor: Honor the Earth
607 Main Avenue
Callaway, MN 56521

Horowitz Foundation for Social Policy Grants 2280
The Horowitz Foundation approves approximately fifteen grants each year, in the amount of $7,500 per grant-$5,000 initially and an additional $2,500 upon receipt of a final report or a copy of the product of the research. The foundation supports and advances research and understanding in the major fields of the social sciences, specifically in the fields of psychology, anthropology, sociology, economics, urban affairs, area studies, political science, and other disciplines. Grants provide direct assistance to individual scholars worldwide who require small grants to further their research with emphasis on policy-oriented studies. Preference will be given to projects that deal with contemporary issues in the social sciences or issues of policy relevance, as well as to scholars in the initial stages of work. The annual deadline is January 31.
Geographic Focus: All States
Date(s) Application is Due: Jan 31
Amount of Grant: 3,000-7,500 USD
Contact: Mary E. Curtis, Chairman; (732) 445-2280; fax (732) 445-3138; applications@horowitz-foundation.org or info@horowitz-foundation.org
Internet: http://www.horowitz-foundation.org
Sponsor: Horowitz Foundation for Social Policy
P.O. Box 7
Rocky Hill, NJ 08553-0007

Horowitz Foundation for Social Policy Special Awards 2281
The Horowitz Foundation provides seven special awards that may be granted for certain projects. These often carry an additional modest stipend beyond the customary amounts of a grant. The eight areas are as follows: the Donald R. Cressey Award for work done at the empirical level that has direct implication for changes in criminal justice and penology practices; the Joshua Feigenbaum Award for empirical research on policy aspects of the arts and popular culture, with special reference to mass communication; the Eli Ginzberg Award for a project involving solutions to major health and welfare problems in urban settings; the Irving Louis Horowitz Higher Education Award for research that supports the development of organizational leadership in higher education (specific to Rutgers University doctoral students); the Harold D. Lasswell Award for policy related projects in international relations and foreign affairs; the Robert K. Merton Award for studies in the relation between social theory and public policy; the Martinus Nijhoff Award for policy implications of scientific, technological and medical research; and the John L. Stanley Award for a work that seeks to expand our understanding of the political and ethical foundations of policy research.
Geographic Focus: All States
Date(s) Application is Due: Jan 31
Contact: Mary E. Curtis, Chairman; (732) 445-2280; fax (732) 445-3138; applications@horowitz-foundation.org or info@horowitz-foundation.org
Internet: http://www.horowitz-foundation.org/special-awards.html
Sponsor: Horowitz Foundation for Social Policy
P.O. Box 7
Rocky Hill, NJ 08553-0007

Horton Research Grants 2282
One or more grants are awarded annually in support of a research project in hydrology and water resources, with the objective of fostering graduate student research leading to the completion of doctoral dissertations. Appropriate topics may be in hydrology including its physical, chemical, or biological aspects, or in water resources policy sciences including economic systems analysis, sociology, and law. A proposal must be signed by the student and by the faculty supervisor; a formal application form must accompany the proposal.
Requirements: Applicant must be a PhD candidate.
Geographic Focus: All States
Date(s) Application is Due: Apr 15
Amount of Grant: 10,000 USD
Contact: Grant Coordinator; (800) 966-2481 or (202) 462-6900; service@agu.org
Internet: http://www.agu.org/inside/honors.html#SH
Sponsor: American Geophysical Union
2000 Florida Avenue NW
Washington, DC 20009

Hospital Libraries Section / MLA Professional Development Grants 2283

The purpose of this Medical Libraries Association (MLA) award which is sponsored by the association's Hospital Libraries Section (HLS) is to provide librarians working in hospital and similar clinical settings with the support needed for educational or research activities. These include developing and acquiring the knowledge and skills delineated in "Competencies for Lifelong Learning and Professional Success: The Educational Policy Statement of the Medical Library Association" and "The Research Imperative: The Research Policy Statement of the Medical Library Association." The grant may also be used to support reimbursement for expenses incurred in conducting scientific research, such as professional assistance in survey-research design, statistical analyses, etc. The application form and guidelines are available for download at the website. Applications and supporting documents must be received by MLA via email, fax, or mail by December 1. email applications will be accepted as PDF or MS Word files only. Founded in 1898, the Medical Library Association (MLA) is a nonprofit, educational organization of more than 1,100 institutions and 3,600 individual members in the health sciences information field committed to educating health-information professionals, supporting health-information research, promoting access to the world's health-sciences information, and working to ensure that the best health information is available to all.
Requirements: Applicants must have been employed as a health-sciences librarian within the last year in either a hospital or other clinical-care institution. It is preferred that applicants be a member of the Hospital Libraries Section of the MLA.
Restrictions: Previous recipients of the HLS/MLA Professional Development Grant are not eligible. Applicants must not have received an MLA grant, scholarship, or other award within the last year. An applicant can only receive one award per year. An award will be made to no more than one employee per institution per year. Awards will not be given to support work toward a degree or certificate program.
Geographic Focus: All States, Canada
Date(s) Application is Due: Dec 1
Amount of Grant: Up to 800 USD
Contact: Carla Funk; (312) 419-9094, ext. 14; fax (312) 419-8950; mlapd2@mlahq.org
Internet: http://www.mlanet.org/awards/grants/index.html
Sponsor: Medical Library Association
65 East Wacker Place, Suite 1900
Chicago, IL 60601-7246

Houghton Library Visiting Fellowships 2284

Short-term fellowships are available to scholars pursuing research in literature, history, philosophy, religion, history of science, theater, dance, and printing and graphic arts. Fellows also have access to collections in Widener as well as throughout the world's largest university library.
Geographic Focus: All States
Date(s) Application is Due: Jan 13
Amount of Grant: 2,750 USD
Contact: Dennis Marnon; (617) 495-2441; dmarnon@fas.harvard.edu
Internet: http://hcl.harvard.edu/libraries/houghton/public_programs.html#fellowships
Sponsor: Harvard University
Houghton Library
Cambridge, MA 02138

HRF Graduate Fellowships 2285

The Hudson River Foundation (HRF) will award up to six full-time research Fellowships to advanced graduate students conducting research on the Hudson River system. Fellowships awarded to doctoral students will include a stipend of up to $15,000 for one year, with a research budget of up to $1,000. Fellowships awarded to master's level students will include a stipend of up to $11,000 for one year, and research budget of up to $1,000. In special cases, applicants can apply for a 15-month Fellowship in order to extend the proposed project through an additional summer. Application documents are available at the grant website. Applications must be emailed in a single PDF document by 5:00 p.m. on the deadline date. One original copy, postmarked by the deadline date, must be mailed to the address given. Applications are generally due in March. Decisions will be made in June.
Requirements: Applicants must be enrolled in an accredited doctoral or master's program, have a thesis advisor and advisory committee (if appropriate to the institution), and have a thesis research plan approved by the student's institution or department. The student's home university will be expected to cover any indirect costs associated with the project, and to be the primary source of support for materials and expenses required to do the research.
Restrictions: The Foundation will not accept applications for extensions of existing fellowships. The award is conditional upon a full tuition waiver or reimbursement by the University.
Geographic Focus: All States
Date(s) Application is Due: Mar 21
Amount of Grant: 12,000-16,000 USD
Contact: Dr. Dennis Suszkowski, Science Director; (212) 483-7667; fax (212) 924-8325; dennis@hudsonriver.org or info@hudsonriver.org
Internet: http://www.hudsonriver.org/graduate_fellow.htm
Sponsor: Hudson River Foundation
17 Battery Place, Suite 915
New York, NY 10004

HRF Hudson River Improvement Grants 2286

The HRF Improvement Grants were created to support projects that promote the enhancement of public use and enjoyment of the natural, scenic and cultural resources of the Hudson River and its shores, with an emphasis on physical projects that require capital construction, development or improvement. Examples of projects that may be considered for funding include: development or improvement of facilities that increase public physical or visual access to the Hudson River, including but not limited to docks, boats, piers and shore front access points; repair, restoration or creation of habitat; development or improvement and equipping of facilities suitable for Hudson River education programs, such as interpretive centers, marsh boardwalks or waterfront classrooms. The Improvement Fund encourages proposals that focus on environmentally beneficial shoreline enhancement and education projects. Samples of previously funded projects are available at the website. Application guidelines and a the application form are available at the website. Proposals should be assembled electronically and emailed. Additionally, applicants must mail six complete hard copies (one signed) to the contact person.
Requirements: Applicants must include a copy of the organization's tax exemption certificate under Section 501(c)3 of the Internal Revenue Code or an indication that the applicant is a subdivision of a government unit.
Geographic Focus: New York
Date(s) Application is Due: Nov 19
Amount of Grant: Up to 10,000 USD
Contact: Dr. Dennis Suszkowski, Science Director; (212) 483-7667; fax (212) 924-8325; dennis@hudsonriver.org or info@hudsonriver.org
Internet: http://www.hudsonriver.org/hrif/
Sponsor: Hudson River Foundation
17 Battery Place, Suite 915
New York, NY 10004

HRF Mark B. Bain Graduate Fellowships 2287

The Foundation awards up to six full-time research Fellowships to advanced graduate students conducting research on the Hudson River system. Doctoral student Fellowships include a stipend of up to $15,000 for one year, and research budget of up to $1,000. Master's student Fellowships include a stipend of up to $11,000 for one year, and a research budget of up to $1,000. Additional information about the Fellowship and application process can be found at the Call for Proposals attachment posted on the HRF call for proposals site.
Requirements: The award is conditional upon a full tuition waiver or reimbursement by the University. Applicants must be enrolled in an accredited doctoral or master's program, must have a thesis advisor and advisory committee (if appropriate to the institution), and must have a thesis research plan approved by the student's institution or department. The student's home university will be expected to be the primary source of support for materials and expenses required to do the thesis research and will also be expected to cover any indirect costs associated with the project. In special cases, applicants can apply for a 15-month Fellowship in order to extend the proposed project through an additional summer.
Geographic Focus: All States
Date(s) Application is Due: Mar 18
Amount of Grant: Up to 28,000 USD
Contact: Dennis J. Suszkowski, Science Director; (212) 483-7667; fax (212) 924-8325; dennis@hudsonriver.org or info@hudsonriver.org
Internet: http://www.hudsonriver.org/hrfund/
Sponsor: Hudson River Foundation
17 Battery Place, Suite 915
New York, NY 10004

HRF New York City Environmental Grants for Newton Creek 2288

The New York City Environmental Fund (NYCEF) was formed to encourage active community stewardship of the waterways, shorelines, parklands and open spaces of the City of New York. NYCEF was created to support projects that (a) protect, preserve, improve or restore the environment; (b) facilitate the public's enjoyment of natural resources; or (c) enhance the public's awareness, knowledge, and understanding of ecological principles, natural resources and other environmental issues. The NYCEF Newtown Creek Fund requests proposals for environmental education, public access and stewardship projects that will benefit communities along Newtown Creek, one of New York City's six designated Maritime and Industrial Zones. In accordance with the settlement agreement creating the Fund, priority will be given to projects in and around Greenpoint, Brooklyn, and along Newtown Creek and the East River (both Brooklyn and Queens). NYCEF will provide financial support for qualifying projects that promote the environmental health and revitalization of natural resources and that increase public understanding of, access to and enjoyment of the environment. The NYCEF Newtown Creek Environmental Benefits Program seeks proposals for projects that protect and improve the environmental health of natural areas and the environment; educate people about the value of natural resources, biodiversity, conservation, and environmental protection; enhance access to natural areas and encourage stewardship activities to protect them; promote community involvement and encourage stewardship (personal and community responsibility and guardianship) for natural areas within the Newton Creek target area; and provide a framework for community stewardship and programming related to broader capital projects and improvements. The typical grant size will be between $5,000 and $25,000.
Requirements: Any nonprofit group or organization that qualifies as tax-exempt under Section 501(c)3 of the Internal Revenue Code is eligible to apply for support from the NYCEF Newtown Creek Environmental Benefits Program. Additionally, schools and government agencies with qualifying projects may apply. Groups without 501(c)3 certification may, for the purposes of administration of a grant, associate with an organization that does qualify.
Geographic Focus: New York
Amount of Grant: 5,000-25,000 USD
Contact: Dr. Dennis Suszkowski, Science Director; (212) 483-7667; fax (212) 924-8325; dennis@hudsonriver.org or info@hudsonriver.org
Internet: http://www.hudsonriver.org/nycef/ncf.html
Sponsor: Hudson River Foundation
17 Battery Place, Suite 915
New York, NY 10004

HRF Tibor T. Polgar Fellowships 2289

The Polgar Fellowships are conducted jointly by the Hudson River Foundation and the New York State Department of Environmental Conservation. The Fellowship provides summer funding, and funds for up to eight undergraduate and graduate students to conduct research on the Hudson River. Researchers gather important information on all aspects of the river, and train students in conducting estuarine studies and public policy research. Polgar Fellowships may be awarded for studies anywhere within the tidal Hudson estuary from New York Harbor to the Federal Dam at Troy, New York, including the four marshes of the National Estuarine Research Reserve (Stockport Flats, the Tivoli Bays, Iona Island Marsh, and Piermont Marsh).

Requirements: Graduate and undergraduate students are eligible. Because of the training and educational aspects of this program, each potential Fellow must be sponsored by a primary advisor. The advisor must be willing to commit sufficient time for supervision of the research, and to attend two meetings (orientation and final reports) with their students. Advisors will receive a $500 stipend.

Geographic Focus: All States

Date(s) Application is Due: Feb 20

Amount of Grant: 3,800-4,800 USD

Contact: Dr. Dennis Suszkowski, Science Director; (212) 483-7667; fax (212) 924-8325; dennis@hudsonriver.org or info@hudsonriver.org

Internet: http://www.hudsonriver.org/polgar.htm

Sponsor: Hudson River Foundation
17 Battery Place, Suite 915
New York, NY 10004

HRSA Nurse Education, Practice, Quality and Retention (NEPQR) Grants 2290

Through its Nurse Education, Practice, Quality and Retention grants program, the Division of Nursing will solicit three-year cooperative agreements that propose to develop and implement innovative career ladder programs that will increase the enrollment, progression, and graduation of Veterans in Bachelor of Science in Nursing programs. This VBSN program supports HRSA's strategic plan to improve access to quality health care and services; strengthen, the nation's healthcare workforce; build healthy communities; and improve health equity. Awarded VBSN applications will complement the collaborative efforts of the Health Resources and Services Administration (HRSA), the Department of Defense (DoD), and the Department of Veteran¿s Affairs (VA) that seek to: reduce barriers that prevent veterans from transitioning into nursing careers; develop BSN career ladder programs targeted to the unique needs of veterans; explore innovative educational models to award academic credit for prior health career experience/training or other relevant military training; improve employment opportunities for veterans through high demand careers training, and address the growing national demand for BSN prepared Registered Nurses. The intermediate program goals are to facilitate the transition of veterans into the field of professional nursing, while building upon skills, knowledge, and training acquired during their military service in order to increase employment opportunities. Sub-goals include increasing and diversifying the health workforce, and ensuring that healthcare providers are trained to provide high quality care that is culturally and linguistically aligned with the communities they will serve. The VBSN project will: provide program participants with the opportunity to receive academic credit for prior military medical training and experience; provide participants with knowledge, skills, and support(s) needed to successfully matriculate through innovative BSN career ladder training programs; and prepare program participants for the National Council Licensing Examination for Registered Nurses (NCLEX- RN).

Requirements: Eligible applicants include accredited schools of nursing as defined in section 801(2), of the Public Health Service Act, a health care facility as defined in section 801(11) of the Public Health Service Act, or partnership of such a school and facility.

Geographic Focus: All States

Date(s) Application is Due: Feb 18

Contact: Rebecca Spitzgo; (301) 443-5794 or (888) 275-4772; fax (301) 443-8586

Internet: http://www.grants.gov/view-opportunity.html?oppId=249594

Sponsor: Health Resources and Services Administration
5600 Fishers Lane
Rockville, MD 20857

HSBC Contributions Community Grants 2291

HSBC believes that banks play a positive role in the economy and society by providing individuals and businesses with the financial services they need to meet their ambitions. Focusing on education and the environment, HSBC builds strong, enduring relationships with a wide range of organizations, providing resources and tools to sustain long-term results. HSBC supports the communities in which it operates both through the involvement of its employees and monetary contributions. On a limited basis, HSBC provides support for initiatives that fall outside of its focus areas of education and the environment. This support is primarily driven by its Community Reinvestment Act activities and employee volunteer efforts.

Requirements: HSBC initiates the majority of its grant relationships. Eligible non-profits (those with 501(c)3 tax-deductible status located in areas in which HSBC serves) will be invited to submit their proposal for funding to HSBC via the online HSBC Contribution Application. Proposals must be received by November 1 of each year; all organizations will receive a response in writing. If approved, the correspondence will outline the grant terms along with any requirements for evaluation and reporting. For Education proposals, the goal is to support initiatives that help give young people hope, a positive attitude, opportunities for success and a better future. Providing access to education, particularly for children from disadvantaged backgrounds or for schools in economically deprived areas, is the central aim of HSBC's focus. HSBC Contributions support primary and secondary education programs for children ages 3-25, including environmental education. Additionally, preference will be given to efforts aimed at raising levels of financial and business literacy in the community, especially among young people. Note that general operating grants are not provided to elementary and secondary schools. For Environment proposals, HSBC supports organizations/programs that focus on environmental policy development, public awareness-building, education of teachers or other adults (i.e. over 25), scientific research and projects to improve the sustainability or resilience of eco-systems. An important component in any contribution is the opportunity for HSBC employees to be engaged and have opportunities to volunteer in relevant environmental projects, most importantly where HSBC has a significant local presence.

Restrictions: Support is prohibited to oOrganizations outside of the United States and/or organizations that do not hold 501(c)3 tax deductible status under the Internal Revenue Code; fraternal, veteran, labor, or athletic organizations; for-profit student aid or scholarship programs, aside from those already established by the company; political, lobbying or voter registration programs, or those supporting the candidacy of a particular individual; funds to support travel-group or individual; organizations that might in any way pose a conflict with HSBC's corporate values, products, customers or employees; religious organizations that serve a limited constituency; advertising; or, individuals.

Geographic Focus: All States

Date(s) Application is Due: Nov 1

Contact: Jan Garner; (224) 880-7128; janice.e.garner@us.hsbc.com

Internet: http://www.us.hsbc.com/1/2/home/about/corporate-sustainability/community-partners

Sponsor: HSBC Bank U.S.A
P.O. Box 2013
Buffalo, NY 14240

HUD Community Outreach Partnership Centers Grants 2292

The program provides funds to Institutes of Higher Learning (IHLs) to establish and operate Community Outreach Partnership Centers. These centers carry out research and administer outreach programs in the surrounding communities that focus on addressing the problems of urban areas. Centers should conduct not only outreach but technical assistance, applied research, and empowerment activities for the local neighborhood and community-based organizations. Centers will be required to address at least three urban problems in their outreach programs; these can include issues relating to housing, economic development, neighborhood revitalization, infrastructure, health care, job training, education, crime prevention, planning, and community organization. Outreach activities may include technical assistance, counseling, workshops, and assistance to community and neighborhood organizations.

Requirements: Institutions of higher education with demonstrated ability to carry out eligible activities are eligible. A 50 percent match of the total cost of research activities is required, as well as a 25 percent match of the total cost of establishing and operating outreach activities.

Geographic Focus: All States

Amount of Grant: 200,000-400,000 USD

Contact: Kinnard Wright, Office of University Partnerships, Policy Development and Research; (202) 708-3061, ext. 7495; Kinnard_D._Wright@hud.gov

Internet: http://www.hud.gov/grants

Sponsor: U.S. Department of Housing and Urban Development
451 Seventh Street SW
Washington, DC 20410

HUD Urban Scholars Postdoctoral Fellowships 2293

The program is intended to expand urban research and ensure a continued pipeline of new researchers who will be vitally important to both research and public policy. The fellowships enable recent doctoral recipients to make a positive impact on the scholarship relating to housing and urban development. Applicants must have an academic appointment that will extend beyond the 15-month duration of the fellowship. The institution of higher education where the fellow is employed must designate a faculty adviser with relevant technical and subject expertise to monitor progress on the research project; provide office space and computer access; and waive indirect costs above the 8 percent allowed by the fellowship. Applicants must identify a mentor at their institution or elsewhere who is a respected scholar in the area of the research topic. (The mentor and faculty adviser can be, but do not need to be, the same person.) Fellows will also have a mentor at HUD who conducts research on a related topic and who will keep the fellow advised of policy developments.

Requirements: Applicants must have earned a PhD degree no earlier than January 1, 1997, in a field that relates to the research topic eligible for support.

Restrictions: A fellow may not hold another comparable major fellowship concurrently.

Geographic Focus: All States

Amount of Grant: Up to 55,000 USD

Contact: Fellowship Program; (202) 334-2872; fax (202) 334-3419; infofell@nas.edu

Internet: http://www7.nationalacademies.org/fellowships

Sponsor: U.S. Department of Housing and Urban Development
2101 Constitution Avenue
Washington, DC 20418

Huebner Foundation Fellowships 2294

The purpose of the fellowships is to increase the number of qualified professors specializing in insurance and risk management by providing financial assistance to graduate students aspiring to academic careers at the college or university level. Postdoctoral fellowships also are available. A successful candidate for a Foundation grant must certify that it is his or her intention to (1) follow a career as a full-time college or university faculty member with a teaching or research specialization in insurance economics and risk management; (2) specialize

in insurance economics and risk management, and in a related discipline such as finance, for a graduate degree; and (3) not engage in any outside work for pay or profit without the consent of the Executive Director of the Foundation during the period of the fellowship.
Requirements: Applicant must be a citizen of the United States or Canada and have a bachelor degree from an accredited college or university. Candidates for Foundation grants must apply on a special form available from the Foundation. Candidates for Foundation grants must apply separately and directly for admission to the Wharton Doctoral Program. They must take the Graduate Management Examination Test (GMAT) or the Graduate Record Examination (GRE) administered by the Educational Testing Service in Princeton, New Jersey. A copy of the GMAT or GRE results must be sent by ETS to the Wharton Doctoral Program office, located at 1150 Steinberg Hall-Dietrich Hall at the University of Pennsylvania.
Geographic Focus: All States
Amount of Grant: 15,000 USD
Contact: Executive Director; (215) 898-9631; fax (215) 573-2218
Internet: http://www.huebnergeneva.org
Sponsor: S.S. Huebner Foundation
3733 Spruce Street, Wharton School, 430 Vance Hall
Philadelphia, PA 19104-6301

Huie-Dellmon Trust Grants **2295**
The trust awards grants to Louisiana nonprofits in its areas of interest, including hospitals, higher and secondary education, libraries, and Protestant churches and organizations. Types of support include general operating support, capital campaigns, building construction/renovation, equipment acquisition, program development, scholarship funds, research, and matching funds. There are no application forms or deadlines.
Requirements: Central Louisiana nonprofit organizations are eligible.
Geographic Focus: Louisiana
Amount of Grant: 1,000-100,000 USD
Samples: Christus Cabrini Foundation, Alexandria, LA, $100,000; Calvary Baptist Church, Alexandria, LA, $12,500; Rapides Parish Library, Alexandria, LA, $30,000.
Contact: Richard Crowell Jr., Trustee; (318) 748-8141
Sponsor: Huie-Dellmon Trust
P.O. Box 330
Alexandria, LA 71309-0330

Humanities Institute Visiting Scholars Fellowships **2296**
The institute offers residential fellowships each academic year to one or two visiting scholars to pursue advanced work in the humanities. Projects may contribute to scholarly knowledge or to the general public's understanding of the humanities. Recipients might eventually produce scholarly articles, a monograph on a specialized subject, a book on a broad topic, an archaeological site report, a translation, an edition, or other scholarly tools. Fellowships support projects that can be completed during the tenure of an award or those that are part of a long-term endeavor. Applicants may be faculty or staff members of colleges or universities, or scholars and independent writers. Guidelines are available online.
Requirements: Applicants should have held the PhD for five years or more or possess a record of professional accomplishment.
Restrictions: Fellowships do not support projects to study teaching methods or theories. Neither do they support surveys of courses and programs or the preparation of institutional curricula.
Geographic Focus: All States
Date(s) Application is Due: Jan 15
Amount of Grant: 40,000 USD
Contact: UCHI; (860) 486-9057 or (860) 486-9058; fax (860) 486-9136; UCHI@uconn.edu
Internet: http://www.humanities.uconn.edu
Sponsor: University of Connecticut
215 Glenrook Road, U-4234
Storrs, CT 06269-4234

Human Source Foundation Grants **2297**
The Human Source Foundation is organized to support charitable and educational initiatives that improve the human condition. The Foundation provides assistance and financial support for programs involved with education, human services, and youth development. Giving is restricted to the state of Texas, primarily in Denton and Fort Worth counties. There are no specific deadlines or application forms, and applicants should begin by forwarding a letter of request.
Requirements: Tax-exempt 501(c)3 organizations serving Texas, primarily Denton and Fort Worth counties, are eligible to apply.
Geographic Focus: Texas
Contact: Mary G. Palko; (817) 926-2799; fax (817) 926-5202; mary@ftw.com
Sponsor: Human Source Foundation
2409 Winton Terrace West, P.O. Box 100423
Fort Worth, TX 76185-0423

Hunter's Hope Foundation Grants **2298**
Hunter's Hope funds projects directly related to Krabbe disease or other leukodystrophies. Eligible studies may investigate new treatment approaches, or study basic mechanisms related to these diseases. Areas of relevant research include, among others, gene therapy, myelin development, neurodegeneration and transplantation, developmental neurobiology, and molecular biology. Applications are accepted in three project categories: postdoctoral fellowships provide stipend support for individuals who are within five years of receiving their terminal degree (MD, DVM, PhD, or equivalent degree). Fellows will be working with senior investigators within their sponsoring institution. Fellowships are awarded for two years. Pilot studies are one-year projects designed to test new concepts or ideas and can serve as a source of preliminary data for future studies. Major research grants provide support to senior investigators conducting major research studies. Projects may be funded for one to three years. Applications are reviewed twice annually.
Geographic Focus: All States
Date(s) Application is Due: Apr 15; Sep 15
Amount of Grant: 30,000-100,000 USD
Contact: Executive Director; (877) 984-4673 or (716) 667-1200; fax (716) 667-1212; info@huntershope.org
Internet: http://www.huntershope.org/research/default.asp
Sponsor: Hunter's Hope Foundation
P.O. Box 643
Orchard Park, NY 14127

Huntington's Disease Society of America Research Fellowships **2299**
Fellowships are designed to assist promising young postdoctoral investigators in the early stages of their careers who are engaged in research at the basic and clinical levels relating to the cause and treatment of Huntington's disease. The amount of the award will depend upon the policy of the sponsoring accredited medical school or university and the training and experience of the applicant. Fellowships are awarded for one year with the possibility of renewal. Application forms and requested information on curriculum vita, publications, research proposal, sponsorship, and other sources of funding are provided by the society.
Requirements: All applicants must have an MD or PhD degree or the equivalent and work must in some way be related to Huntington's disease. There is no citizenship requirement.
Geographic Focus: All States
Amount of Grant: Up to 80,000 USD
Contact: Robert Graze; (212) 242-1968 or (800) 345-4372, ext. 227; rgraze@hdsa.org
Internet: http://www.hdsa.org/research/grant-applications.html
Sponsor: Huntington's Disease Society of America
505 8th Avenue, Suite 902
New York, NY 10018

Huntington's Disease Society of America Research Grants **2300**
The Society awards research grants for one year for support of basic or clinical research related to Huntington's disease. Grant awards are provided as seed monies for new or innovative research projects in the hope that they will develop sufficiently to attract funding from other sources. Application forms are provided by the society and generally follow the format of NIH grant applications.
Requirements: Applicants may be doctorate or graduate students or equivalent professionals. There is no citizenship requirement.
Geographic Focus: All States
Amount of Grant: Up to 100,000 USD
Contact: Robert Graze; (212) 242-1968 or (800) 345-4372, ext. 227; rgraze@hdsa.org
Internet: http://www.hdsa.org/research/grant-applications.html
Sponsor: Huntington's Disease Society of America
505 8th Avenue, Suite 902
New York, NY 10018

Huntington Library Babara Thom Postdoctoral Fellowships **2301**
The Huntington is an independent research center with holdings in British and American history, literature, art history, and the history of science and medicine. Babara Thom Postdoctoral Fellowships are designed to support non-tenured faculty members who are revising a manuscript for publication. The amount of each award is $50,000 for a nine- to twelve-month tenure. Applications should be postmarked by December 15.
Requirements: Applicants must: be pursuing scholarship in a field appropriate to the Huntington's collections; and be non-tenured faculty. Recipients of all fellowships are expected to be in continuous residence at the Huntington and to participate in and make a contribution to its intellectual life.
Geographic Focus: All States
Date(s) Application is Due: Dec 15
Amount of Grant: 450,000 USD
Contact: Dr. Steve Hindle; (626) 405-2194; fax (626) 449-5703; cpowell@huntington.org
Internet: http://www.huntington.org/huntingtonlibrary.aspx?id=566
Sponsor: Huntington Library and Art Gallery
1151 Oxford Road
San Marino, CA 91108

Huntington Library Clark-Huntington Joint Bibliographical Fellowship **2302**
The Huntington is an independent research center with holdings in British and American history, literature, art history, and the history of science and medicine. The Clark-Huntington Joint Bibliographical Fellowship has a two-month duration, with one month at the William Andrews Clark Library; one month at the Huntington. The amount of the award is $5,000 Applications should be postmarked by February 1.
Requirements: Applicant scholars must have the PhD or equivalent degree.
Geographic Focus: All States
Date(s) Application is Due: Feb 1
Amount of Grant: 5,000 USD
Contact: Dr. Steve Hindle; (626) 405-2194; fax (626) 449-5703; cpowell@huntington.org
Internet: http://www.huntington.org/huntingtonlibrary.aspx?id=566
Sponsor: Huntington Library and Art Gallery
1151 Oxford Road
San Marino, CA 91108

Huntington Library Dana and David Dornsife Fellowships 2303

The Huntington Library and Art Gallery is an independent research center with holdings in British and American history, literature, art history, the history of science, and photography. The Dana and David Dornsife fellowships support researchers pursuing scholarship in a field appropriate to the Huntington's collections. Preference will be given to scholars who have not held major awards in the three years preceding the year of the award. Recipients are expected to be in continuous residence at the Huntington during the nine- to 12-month fellowship tenure and to participate in the intellectual life of the center. There is no application form. Application consists of cover sheet, description of the project, curriculum vita, and three letters of recommendation that must be received by the application deadline.
Requirements: Application is open to individuals with the PhD or equivalent degree who have received their degree by June of the previous year.
Geographic Focus: All States
Date(s) Application is Due: Dec 15
Amount of Grant: 50,000 USD
Contact: Dr. Steve Hindle; (626) 405-2194; fax (626) 449-5703; cpowell@hungtington.org
Internet: http://huntington.org/huntingtonlibrary.aspx?id=566
Sponsor: Huntington Library and Art Gallery
1151 Oxford Road
San Marino, CA 91108

Huntington Library Dibner History of Science Long-Term Fellowships 2304

The Huntington is an independent research center with holdings in British and American history, literature, art history, and the history of science and medicine. The Dibner History of Science Program offers historians of science and technology the opportunity to study in the Burndy Library and the other history of science and technology resources at the Huntington. Long-term fellowship awards are of a nine- to twelve month duration, and are in the amount of $50,000. Applicants can be conducting research or already be at the writing stage and need reference materials only. The annual deadline is a postmark of December 15.
Requirements: Eligible candidates must have a PhD or equivalent as of June of the previous year.
Geographic Focus: All States
Date(s) Application is Due: Dec 15
Amount of Grant: 50,000 USD
Contact: Dr. Steve Hindle; (626) 405-2194; fax (626) 449-5703; cpowell@hungtington.org
Internet: http://huntington.org/huntingtonlibrary.aspx?id=566
Sponsor: Huntington Library and Art Gallery
1151 Oxford Road
San Marino, CA 91108

Huntington Library Dibner History of Science Short-Term Fellowships 2305

The Huntington is an independent research center with holdings in British and American history, literature, art history, and the history of science and medicine. The Dibner program offers long-term fellowships which are designed to further study in the Burndy Library and the other history of science and technology resources at the Huntington. Short-term awards are $2,500 per month for a one- to five-month period. Applications must be postmarked by December 15.
Requirements: Applicants for short-term awards must: have a PhD or equivalent; or be a doctoral candidate at the dissertation stage.
Geographic Focus: All States
Date(s) Application is Due: Dec 15
Amount of Grant: 2,500-12,500 USD
Contact: Dr. Steve Hindle; (626) 405-2194; fax (626) 449-5703; cpowell@hungtington.org
Internet: http://www.huntington.org/huntingtonlibrary.aspx?id=566
Sponsor: Huntington Library and Art Gallery
1151 Oxford Road
San Marino, CA 91108

Huntington Library Mellon Fellowships 2306

The Huntington is an independent research center with holdings in British and American history, literature, art history, and the history of science and medicine. This Fellowship is designed to support applicants pursuing scholarship in a field appropriate to the Huntington's collections. The amount of each award is $50,000 for a nine- to twelve-month tenure. Applications must be postmarked by December 15.
Requirements: Applicants must: be pursuing scholarship in a field appropriate to the Huntington's collections; and have received the PhD in June of the application year. Recipients of all fellowships are expected to be in continuous residence at the Huntington and to participate in and make a contribution to its intellectual life.
Geographic Focus: All States
Date(s) Application is Due: Dec 15
Amount of Grant: 50,000 USD
Contact: Dr. Steve Hindle; (626) 405-2194; fax (626) 449-5703; cpowell@hungtington.org
Internet: http://www.huntington.org/huntingtonlibrary.aspx?id=566
Sponsor: Huntington Library and Art Gallery
1151 Oxford Road
San Marino, CA 91108

Huntington Library National Endowment for the Humanities Fellowships 2307

The Huntington Library and Art Gallery, an independent research center with holdings in British and American history, literature, art history, and the history of science, will award a minimum of three fellowships funded by National Endowment for the Humanities. These fellowships are designed to permit scholars to spend from four to twelve months at the Huntington. Fellowships may be combined with sabbaticals or other stipendiary support. Preference will be given to scholars who have not held major fellowships during the three years preceding the year for which the center is awarding the fellowships. Fellows will be able to pursue their own research and writing while participating in the intellectual life of the Huntington. The amount of each fellowship is $50,400.
Requirements: Applicants must be established scholars at the postdoctoral level or its equivalent, must be U.S. citizens or foreign nationals who have resided in the United States for at least three years, and must be pursuing scholarship in a field appropriate to the Huntington's collections.
Geographic Focus: All States
Date(s) Application is Due: Dec 15
Amount of Grant: Up to 50,400 USD
Contact: Dr. Steve Hindle; (626) 405-2194; fax (626) 449-5703; cpowell@hungtington.org
Internet: http://www.huntington.org/huntingtonlibrary.aspx?id=566
Sponsor: Huntington Library and Art Gallery
1151 Oxford Road
San Marino, CA 91108

Huntington Library Short-Term Fellowships 2308

The Huntington is an independent research center with holdings in British and American history, literature, art history, and the history of science and medicine. Within this category, The Huntington awards a number of specialized fellowships, including: Francis Bacon Foundation Fellowships in Renaissance England; Haynes Foundation Fellowships in Los Angeles History; Reese Fellowship in American Bibliography and the History of the Book in the Americas; Trent R. Dames Civil Engineering History Fellowship; Francis J. Weber Research Fellowship in Roman Catholic History; and Christopher Isherwood Foundation Fellowships. The amount of each award is $2,500 per month for a one- to five-month tenure. Applications must be postmarked by December 15.
Requirements: Applicant scholars must have the PhD or equivalent degree or be doctoral candidates at the dissertation stage. Recipients of all fellowships are expected to be in continuous residence at the Huntington and to participate in and make a contribution to its intellectual life.
Geographic Focus: All States
Date(s) Application is Due: Dec 15
Amount of Grant: 2,500-12,500 USD
Contact: Dr. Steve Hindle; (626) 405-2194; fax (626) 449-5703; cpowell@hungtington.org
Internet: http://www.huntington.org/huntingtonlibrary.aspx?id=566
Sponsor: Huntington Library and Art Gallery
1151 Oxford Road
San Marino, CA 91108

Huntington Library Travel Grants and Exchange Fellowships for Study in Great Britain 2309

The Huntington Library and Art Gallery is an independent research center with holdings in British and American history, literature, art history, the history of science, and photography. The Huntington offers several travel grants in any of the fields in which the Huntington collections are strong and where the research will be carried out in Great Britain. The program also offers exchange fellowships with Linacre College and Lincoln College, Oxford. Duration is for one month, and applications should be postmarked by December 15.
Requirements: Eligibility is restricted to PhD or equivalent candidates or doctoral candidate at the dissertation stage.
Geographic Focus: All States
Date(s) Application is Due: Dec 15
Contact: Dr. Steve Hindle; (626) 405-2194; fax (626) 449-5703; cpowell@hungtington.org
Internet: http://huntington.org/huntingtonlibrary.aspx?id=566
Sponsor: Huntington Library and Art Gallery
1151 Oxford Road
San Marino, CA 91108

Hutchinson Community Foundation Grants 2310

The Hutchinson Community Foundation provides funding in the counties of Hutchinson and, Reno, Kansas, inluding the immediate area surrounding area. Areas of interest include, but are not, limited to: community needs in areas such as early childhood, youth, education, arts and culutre, and neighborhood development.
Requirements: Nonprofits in Kansas are eligible to apply. Special purpose units of government can apply for support of innovative projects located in Reno County.
Restrictions: Grant proposals from individuals or non-qualifying organizations will not be considered.
Geographic Focus: Kansas
Contact: Audrey Abbott Patterson; (620) 663-5293; aubrey@hutchcf.org
Internet: http://www.hutchcf.org/default.asp?sm=2&si=1
Sponsor: Hutchinson Community Foundation
One North Main, Suite 501, P.O. Box 298
Hutchinson, KS 67504-0298

Hydrogen Student Design Contest 2311

The annual Hydrogen Student Design Contest challenges university students to design hydrogen energy applications for real-world use. Established in 2004 by the Hydrogen Education Foundation, the Contest showcases the talents of students in many disciplines, including engineering, architecture, marketing, and entrepreneurship. Although the formal rules and guidelines are still under development, this Contest plans to address three key areas: 1) Hydrogen Production—Students will develop a strategy for supplying hydrogen for the community. At least 33% of the community's hydrogen must be produced from renewable sources. In their plans, students must include at least one hydrogen

energy station that co-produces power, heat, and hydrogen (also known as a CHHP system). 2) Early Markets—Students will conduct a market analysis to determine near-term customers in the community for their hydrogen supply. Early markets for hydrogen include back-up power, portable power, and materials handling vehicles. Students will identify at least five businesses or other organizations in the community that could be customers for hydrogen in these early markets. 3) Fueling Stations—Students will plan and design a cluster of hydrogen fueling stations for vehicles in the community. Teams will be required to create a plan for the cluster that identifies the number of fueling stations needed to meet passenger vehicle demand; the approximate locations for each station; the station type (retail, private fleet, mixed use, etc.); and total cost estimates for deploying the plan. Students will be required to design the main components of three fueling stations. The Grand Prize Winner will receive the opportunity to present their design in a Keynote Presentation slot in the NHA Annual Hydrogen Conference and Expo, a stipend of up to $5,000 to cover airfare, meals, hotel accommodations, NHA conference registrations, and incidental trip expenses for up to eight team members and their faculty representative, and extensive media recognition. Honorable Mentions will receive the opportunity to give poster presentations of their design at the NHA Annual Hydrogen Conference and Expo, complimentary hotel rooms and registrations to the NHA Conference for up to four team members and their faculty representative.
Requirements: The contest is open to current college and university undergraduate and graduate students in the U.S. or abroad. Team members must be enrolled in a college or university at the time of the Contest but do not have to be enrolled full-time. Given the multi-disciplinary nature of the competition, teams are encouraged to include members from any field of study relevant to the team's design. Common disciplines include: engineering, architecture/planning, industrial design, economics, business, environmental science, policy, chemistry, marketing, and education. Registration must be completed by January 22; Entries are due March 24.
Geographic Focus: All States
Date(s) Application is Due: Jan 22; Mar 24
Amount of Grant: 5,000 USD
Contact: Kyle Gibeault; (202) 223-5547; gibeaultk@hydrogenassociation.org
Internet: http://www.hydrogencontest.org/contest.asp
Sponsor: Hydrogen Education Foundation
1211 Connecticut Avenue NW, Suite 600
Washington, DC 20036-2701

IAFF Burn Foundation Research Grants **2312**

The IAFF Burn Foundation Research Grant Program continues to lead to new knowledge and innovative approaches to the prevention and treatment of the physical and psychological problems that impair the quality of life for a burn patient. Research grant proposals are accepted annually and awards are distributed during the American Burn Association Annual Meeting. Topics of interest include by are not limited to pain management, physical and psychological rehabilitation, and wound healing and scarring, especially in children. Clinical projects are recommended. The foundation's Board of Medical Advisors oversees the selection process, reviewing each of the submissions. While all of candidates are deserving, only seven to ten grants are chosen and submitted to the IAFF Burn Foundation for consideration. Typically, five to seven are funded.
Requirements: The foundation welcomes research grant applications from members of the American Burn Association. The areas of funded research fall into three categories: (1) Quality of Life-This area of research is the most common, focusing on issues such as pain management, nutrition, and physical and psychological rehabilitation. The major focus in this area is the care of burned children. (2) Burn Prevention-Research in this area is especially important in shaping prevention programs. (3) Basic Science-Research grants funded in this category surround nutrition, wound healing and scar formation. No animal studies are funded. The required application and detailed guidelines can be found at the sponsor's website.
Restrictions: No animal studies are funded. The grant does not include indirect costs.
Geographic Focus: All States, Canada
Amount of Grant: 25,000 USD
Contact: Patrick Morrison; (202) 824-8620; fax (202) 637-0839; burnfoundation@iaff.org
Internet: http://burn.iaff.org/grants.shtml
Sponsor: International Association of Fire Fighters
1750 New York Avenue NW
Washington, DC 20006

Ian Axford (New Zealand) Fellowships in Public Policy **2313**

The Ian Axford (New Zealand) Fellowships in Public Policy are for outstanding mid-career professionals to research, travel and gain practical experience in public policy in New Zealand for seven months. The program has three goals: to reinforce New Zealand-United States links by enabling Americans of high intellectual ability and leadership potential to come to New Zealand to gain experience and build contacts in the field of public policy development; to help improve the practice of public policy in New Zealand and the United States by the cross-fertilization of ideas and experience in the two countries; and to build a network of public policy experts on both sides of the Pacific, and encourage ongoing policy exchange between New Zealand and the United States. Fellows will be based at a relevant government agency in Wellington and carefully paired with a leading policy expert who will serve as their mentor. In addition, they will be expected to spend a substantial part of their Fellowship in contact with other relevant organizations to gain practical experience in their field in New Zealand. These contacts may include an extended or part-time placement.
Requirements: Applicants must be: a U.S. citizen with at least five years of experience in your profession; a mid-career professional active in any part of the public, business or non-profit sector; a potential leader and opinion-former in your chosen field; and have an interest in learning from your experience in New Zealand and capable of putting to effective use in the U.S., any policy lessons learned.
Restrictions: Fellowships are not awarded to support basic research or to enable study for an academic degree.
Geographic Focus: All States
Date(s) Application is Due: Mar 1
Contact: Martin Boswell, +64 4 494 1507 or +64 4 472 2065; fax +64 4 499 5364; martin@fulbright.org.nz
Internet: http://www.fulbright.org.nz/awards/am-ian-axford.html
Sponsor: Fullbright New Zealand
120 Featherston Street, P.O. Box 3465, Level 8
Wellington, 6140 New Zealand

Ian N. McKinnon Memorial Fellowship **2314**

This four-month fellowship is open to qualified graduates of any recognized university who are registered in or admissible to a full-time program leading to a master's or doctoral degree at the University of Calgary. Fields of study supported are all areas relevant to the effective development and utilization of energy resources with special emphasis on economics, engineering, and geology.
Requirements: Candidates must be Canadian citizens or landed immigrants at the time of taking up the award.
Geographic Focus: All States, Canada
Date(s) Application is Due: Apr 15
Amount of Grant: 3,500 CAD
Contact: Faculty of Graduate Studies; (403) 220-5417; graduate@ucalgary.ca
Internet: http://www.grad.ucalgary.ca/funding/internal_scholarships/lev_4/mckinnon.htm
Sponsor: University of Calgary
2500 University Drive NW, Earth Sciences Building, Room 720
Calgary, AB T2N 1N4 Canada

IARC Expertise Transfer Fellowship **2315**

The Fellowship is intended to enable an established investigator to spend normally from six to twelve months in an appropriate host institute in a low- to medium-resource country in order to transfer knowledge and expertise in a research area relevant for the host country and related to epidemiology, biostatistics, environmental chemical carcinogenesis, cancer etiology and prevention, infection and cancer, molecular cell biology, molecular genetics, molecular pathology and mechanisms of carcinogenesis. Applications should include a proposed collaborative research project, specifying the link to IARC's on-going activities and a letter of support from the host lab giving details of feasibility and anticipated benefit to the receiving institute. Priority will be given to projects directly linked to IARC's on-going research program, involving at least one contact at IARC. There will be an annual remuneration of up to $70,000, which will take into account the on-going salary of the Fellow. This amount may include limited support for the project. The cost of travel will also be met.
Requirements: Applicants should be established cancer researchers actively engaged in the field with appropriate scientific or medical qualifications and an excellent publications' record. They must also belong to the staff of a university or a research institution.
Geographic Focus: All States
Date(s) Application is Due: Nov 30
Amount of Grant: Up to 70,000 USD
Contact: Dr. Paolo Boffetta, +33-(0)472-73-84-48; fax +33 (0)472-73-80-80; fel@iarc.fr or vsa@iarc.fr
Internet: http://www.iarc.fr/en/education-training/expertisetransfer.php
Sponsor: International Agency for Research on Cancer
150 cours Albert-Thomas
Lyon Cedex 08, 69372 France

IARC Postdoctoral Fellowships for Training in Cancer Research **2316**

Applications for training fellowships are invited from junior scientists from low- or medium-resource countries wishing to complete their training in those aspects of cancer research related to coordinate and conduct both epidemiological and laboratory research into the causes of cancer. Disciplines covered include: epidemiology, biostatistics, environmental chemical carcinogenesis, cancer etiology and prevention, infection and cancer, molecular cell biology, molecular genetics, molecular pathology and mechanisms of carcinogenesis, with emphasis on interdisciplinary projects. The fellowship is for a period of one year, with the possibility of an extension for a second year subject to satisfactory appraisal.
Requirements: Candidates should have spent less than five years abroad (including doctoral studies) and have finished their doctoral degree within five years of the closing date for application or be in the final phase of completing their doctoral degree (M.D. or Ph.D.). They must provide evidence of their ability to return to their home country and keep working in cancer research. The working languages at IARC are English and French. Candidates must be proficient in English at a level sufficient for scientific communication.
Restrictions: Candidates already working as a postdoctoral fellow at the Agency at the time of application or who have had any contractual relationship with IARC during the 6 months preceding the application deadline or who have already spent more than one year at IARC cannot be considered.
Geographic Focus: All States
Date(s) Application is Due: Nov 30
Contact: Dr. Paolo Boffetta, +33-(0)472-73-84-48; fel@iarc.fr or vsa@iarc.fr
Internet: http://www.iarc.fr/en/education-training/postdoc.php
Sponsor: International Agency for Research on Cancer
150 cours Albert-Thomas
Lyon Cedex 08, 69372 France

IARC Visiting Scientist Award for Senior Scientists 2317

The IARC is offering this Award for a qualified and experienced investigator with recent publications in international peer-reviewed scientific journals who wishes to spend from six to twelve months at the IARC working on a collaborative project in a research area related to the Agency's programs: epidemiology, biostatistics, environmental chemical carcinogenesis, cancer etiology and prevention, infection and cancer, molecular cell biology, molecular genetics, molecular pathology and mechanisms of carcinogenesis. There will be an annual remuneration of up to $80,000, which will take into account the on-going salary of the visiting scientist plus the cost of travel.
Requirements: Applicants must belong to the staff of a university or a research institution and should provide written assurance of a post to return to at the end of the period of award.
Geographic Focus: All States
Date(s) Application is Due: Nov 30
Amount of Grant: Up to 80,000 USD
Contact: Dr. Paolo Boffetta, +33-(0)472-73-84-48; fel@iarc.fr or vsa@iarc.fr
Internet: http://www.iarc.fr/en/education-training/vsa.php
Sponsor: International Agency for Research on Cancer
150 cours Albert-Thomas
Lyon Cedex 08, 69372 France

IAS Mellon Fellowships for Assistant Professors 2318

The School of Historical Studies at the institute, with the support of the Andrew Mellon Foundation, has established a program of one-year memberships for assistant professors at universities and colleges in the United States and Canada to support promising young scholars who have embarked on professional careers. While at the institute they will be expected to engage exclusively in scholarly research and writing. Appointments will be for one full year (July 1 through June 30 with the option of staying through the second summer until August 15) and will carry all the privileges of membership at the Institute for Advanced Study. The stipend will match the combined salary and benefits at the member's home institution at the time of application. Application and instructions are available online.
Requirements: Assistant professors in areas represented in the School of Historical Studies (Greek and Roman civilization, the history of Europe, the Islamic world, East Asian studies, the history of art, and modern international relations) may apply, provided at the time of their arrival they will have served at least two and not more than four years as assistant professors in institutions of higher learning in the United States or Canada and provided they can return to their institution.
Geographic Focus: All States
Date(s) Application is Due: Nov 15
Contact: Administrative Manager; mzelazny@ias.edu
Internet: http://www.hs.ias.edu/mellon.htm
Sponsor: Institute for Advanced Study
Einstein Drive
Princeton, NJ 08540

IATA Research Grants 2319

In order to enhance the National Athletic Trainer's Association's (NATA) Research and Education Foundation, the Illinois Athletic Trainers' Association has provided limited funding for athletic training related research in the state of Illinois. Grant monies from the IATA are awarded for projects concerning research in the domains relevant to athletic training within the state of Illinois. Grant applications will be reviewed and monies will be awarded during the Spring and Fall IATA BOD meetings.
Requirements: Applicants must be Certified/Licensed, members in good standing not only with the NATA but also the IATA. Applications must be submitted 3 months prior to the beginning of the research study, and the research must be relevant to the profession of athletic training within the state of Illinois. All applications will be analyzed bi-annually and as stated above must be submitted prior to the beginning of the project. Please note that there is limited funding for these awards, and you may be turned down due to lack of funding.
Restrictions: There will be no awards granted for recovery of cost involved with research that has already been initiated.
Geographic Focus: Illinois
Contact: Mike Sullivan; (630) 853-0820; webmaster@illinoisathletictrainers.org
Internet: http://www.illinoisathletictrainers.org/Researchgrts.htm
Sponsor: Illinois Athletic Trainers' Association
701 Thomas Road
Wheaton, IL 60187

IBM Center for the Business of Government Grants 2320

Through grants for research and forums, the organization stimulates research and facilitates discussion on new approaches to improving the effectiveness of government at the federal, state, local, and international levels. Individuals may apply for grants to produce a 30-40 page research report in one of the areas presented in the Research Announcement. The manuscript must be submitted no later than six months after the start of the project. Interested individuals should submit a three-page description of the proposed research and a resume, including a list of publications. Application and guidelines are available online.
Requirements: Individuals working in universities, nonprofit organizations, and journalism are eligible.
Geographic Focus: All States
Date(s) Application is Due: Mar 1; Oct 1
Amount of Grant: 20,000 USD
Contact: Jonathan D. Breul; (202) 515-4382; jonathan.d.breul@us.ibm.com
Internet: http://www.businessofgovernment.org/main/apply/index.asp
Sponsor: IBM Center for the Business of Government
1301 K Street NW, 4th Floor, West Tower
Washington, DC 20005

IBRO/SfN International Travel Grants 2321

The Society for Neuroscience offers travel fellowships to their annual meeting for neuroscientists under the age of 35 who are currently working and living in a resource-restricted country – those defined by the World Bank as low, lower-middle, or upper-middle income. Registration fees are waived for successful candidates attending the meeting. Awardees will be notified prior to the SfN abstract deadline of May 15 of the current year as to whether they have been selected for the SfN travel grant. However, this award is contingent upon the awardee's abstract being accepted by SfN and proof of attendance at the meeting (in addition to the requirements). Candidates must apply through the IBRO online application system.
Requirements: Candidates must be a citizen of and reside in a country that the World Bank has classified low, lower-middle, or upper-middle income. The applicant should be the first author of an abstract to be presented at the SfN annual meeting. A copy of the abstract submitted to SfN for a poster or a platform presentation is required.
Restrictions: Candidates from the U.S. and Canada are not eligible to apply.
Geographic Focus: All Countries
Date(s) Application is Due: Feb 1
Amount of Grant: Up to 2,000 USD
Contact: Robynn Rockstad-Rex, Director of Programmes and Communications; 33 (0) 1 46 47 92 92; fax 33 (0) 1 46 47 42 50; robynn@ibro.info
Internet: http://funding.ibro.info/Pub/Pub_Main_Display.asp?LC_Docs_ID=3169
Sponsor: International Brain Research Organization
255 rue Saint-Honore
Paris, 75001 France, Metropolitan

IBRO Albert Aguayo Fellowship 2322

The Aguayo Fellowship enables students from developing countries to spend a three-month stay in a Quebec laboratory. One Fellowship is awarded each year, funded jointly and selected by the Institute of Neurosciences, Mental Health and Addition (INMHA) and Le Fonds de la recherche en sante du Quebec (FRSQ). Fellowships must begin in Quebec, Canada within six to twelve months of being awarded. A list of prospective host laboratories and research centers is provided at the IBRO website. Candidates may also identify their own host and contact them directly. Applications must be submitted through the IBRO online application process.
Requirements: Candidates should be citizens of and reside in one of the following IBRO regions at the time of their application: Africa; Asia-Pacific (excluding Australia and New Zealand); Central/Eastern Europe; and Latin America. Candidates must have a pre-doctoral or doctoral degree, PhD, MD, DrPH, MBBS or equivalent qualification. Candidates must provide evidence that they will return to a suitable position in their home country at the completion of the Fellowship.
Restrictions: The Fellowship is not renewable. emailed or hard copy applications are not accepted.
Geographic Focus: All Countries
Date(s) Application is Due: Dec 15
Amount of Grant: 8,000 CAD
Contact: Robynn Rockstad-Rex, Director of Programmes and Communications; 33 (0) 1 46 47 92 92; fax 33 (0) 1 46 47 42 50; robynn@ibro.info
Internet: http://funding.ibro.info/Pub/Pub_Main_Display.asp?LC_Docs_ID=5443
Sponsor: International Brain Research Organization
255 rue Saint-Honore
Paris, 75001 France, Metropolitan

IBRO Asia Regional APRC Exchange Fellowships 2323

The IBRO-APRC Exchange Fellowship aims to allow high-quality junior scientists at the level of postdoctoral fellow or junior faculty (under the age of 45) from diverse geographic and scientific areas to broaden the scope of their training in neuroscience by working four to six months abroad in established laboratories. Senior Ph.D. students with good publication records and strong justification for the exchange program will also be considered. Both the applicant and the host laboratory must be within the Asia Pacific Region. Up to $8,500 per Fellowship is available, although funding may vary depending on the country in which the candidate studies. The Fellowship is used along with matching funds from the host laboratory to defray the cost of travel and subsidize living costs. Applications must be submit through the IBRO online application system. emailed or hard copy applications are not acceptable.
Requirements: Selection criteria include academic credentials of the applicant and the host, as well as the quality of the research proposal with a structured research plan. Priority is given to applicants from less developed and less well-funded countries. Strong justification must be made by the applicant that he/she will return home after the exchange, bringing with them new knowledge and skills to advance neuroscience in their home countries. A firm commitment from his/her parent institution to this effect will be viewed favorably by the selection committee.
Geographic Focus: Afghanistan, Bahrain, Bangladesh, Bhutan, Brunei, Cambodia, China, East Timor, India, Indonesia, Iran, Iraq, Israel, Japan, Jordan, Kazakhstan, Kuwait, Kyrgyzstan, Laos, Lebanon, Malaysia, Maldives, Mongolia, Myanmar (Burma), Nepal, North Korea, Oman, Pakistan, Philippines, Qatar, Russia, Saudi Arabia, Singapore, South Korea, Sri Lanka, Syria, Taiwan, Tajikistan, Thailand, Turkey, Turkmenistan, United Arab Emirates, Uzbekistan, Vietnam, Yemen
Date(s) Application is Due: Oct 15
Contact: Robynn Rockstad-Rex, Director of Programmes and Communications; 33 (0) 1 46 47 92 92; fax 33 (0) 1 46 47 42 50; robynn@ibro.info

Internet: http://funding.ibro.info/Pub/Pub_Main_Display.asp?LC_Docs_ID=4272
Sponsor: International Brain Research Organization
255 rue Saint-Honore
Paris, 75001 France, Metropolitan

IBRO Europe Regional Grants for Western European Workshops & Symposia 2324
With the aim to foster cooperation among regional young neuroscientists, IBRO's Western Europe Regional Committee (WERC) offers partial funding of courses, workshops and symposia on important topics in neuroscience organized in West European countries. Preference will be given to activities that include younger scientists and offer training for scientists from countries with limited resources for research or teaching. Applications must be submitted through the IBRO online application system. emailed or hard copy applications are not acceptable.
Geographic Focus: Austria, Belgium, Denmark, Finland, France, Germany, Greece, Iceland, Ireland, Italy, Luxembourg, Netherlands, Norway, Portugal, Spain, Sweden, Switzerland, United Kingdom
Contact: Juan Lerma; 965 91 92 39; jlerma@umh.es
Internet: http://funding.ibro.info/Pub/Pub_Main_Display.asp?LC_Docs_ID=5373
Sponsor: International Brain Research Organization
255 rue Saint-Honore
Paris, 75001 France, Metropolitan

IBRO InEurope Short Stay Grants 2325
As a result of the successful launch of the FENS/IBRO Programme of European Neuroscience Schools, IBRO's Central and Eastern Europe and Western Europe regional committees decided that the collaboration would be further strengthened by the activation of interregional networking and mobility of young scientists. InEurope Short Stay Grants fund visits to European institutions by researchers living and working within Europe, so that they can acquire new methods or specific techniques that are necessary for their work. The aim of the programme is to increase intra-European mobility of young researchers by providing grants for short, goal-directed exchanges within European laboratories. Exchanges are limited to no more than four weeks. Funding can be used to cover travel and local expenses. Applications are accepted at any time, but must be submitted through the IBRO online application process. Selected applicants will be notified at the end of February, May, August and November. Applicants who are not selected in one round of evaluations will remain in competition for the next round.
Requirements: Applicants must be Ph.D. students or postdoc Fellows, 35 years of age or less.
Restrictions: emailed or hard copy applications are not acceptable.
Geographic Focus: Albania, Andorra, Armenia, Austria, Azerbaijan, Belarus, Belgium, Bosnia & Herzegovina, Bulgaria, Croatia, Cyprus, Czech Republic, Denmark, Estonia, Finland, France, Germany, Greece, Hungary, Iceland, Ireland, Italy, Latvia, Liechtenstein, Lithuania, Luxembourg, Macedonia, Malta, Moldova, Monaco, Montenegro, Netherlands, Norway, Poland, Portugal, Romania, Russia, San Marino, Serbia, Slovakia, Slovenia, Spain, Sweden, Switzerland, Turkey, Ukraine, United Kingdom, Vatican City
Amount of Grant: Up to 3,000 EUR
Contact: Robynn Rockstad-Rex, Director of Programmes and Communications; 33 (0) 1 46 47 92 92; fax 33 (0) 1 46 47 42 50; robynn@ibro.info
Internet: http://funding.ibro.info/Pub/Pub_Main_Display.asp?LC_Docs_ID=4112
Sponsor: International Brain Research Organization
255 rue Saint-Honore
Paris, 75001 France, Metropolitan

IBRO International Travel Grants 2326
The IBRO International Travel Grants aim to foster neuroscience research especially in less well-funded countries by providing support to high quality neuroscientists from diverse geographic and scientific areas (U.S./Canada Region excluded) who wish to participate at international neuroscience meetings. Priority is given to those who have not obtained funding from this programme within the past three years. Applications should be submitted by February 1 for travel to conferences occurring from July to December of that year. Applications should be submitted by September 1 for travel occurring from January to June of the following year. Candidates must apply through the IBRO online application system.
Restrictions: Applicants cannot apply for more than one category of funding within travel grants.
Geographic Focus: All Countries
Date(s) Application is Due: Feb 1; Sep 1
Contact: Robynn Rockstad-Rex, Director of Programmes and Communications; 33 (0) 1 46 47 92 92; fax 33 (0) 1 46 47 42 50; robynn@ibro.info
Internet: http://funding.ibro.info/Pub/Pub_Main_Display.asp?LC_Docs_ID=3169
Sponsor: International Brain Research Organization
255 rue Saint-Honore
Paris, 75001 France, Metropolitan

IBRO Latin America Regional Funding for Neuroscience Schools 2327
The Latin America Regional Funding for Neuroscience Schools provides financial support for the training activities of advanced students and young researchers in the field of neuroscience in order to increase the quality of neuroscience education in the region; increase the links between young neuroscientists; create a network of former awardees and senior researchers; and establish a mentorship programme of continuous support of alumni. Candidates must submit applications and all supporting materials through the IBRO online application process.
Requirements: The Latin America Regional Committee (LARC) will review applications and grant financial support for Schools of Neuroscience with the following characteristics based on the IBRO School Guidelines (posted on the website): uniform high-quality education with an active involvement of students and teachers during the entire event; three levels of schools (basic, advanced/specialized, and inter-regional); cover broad and topical issues, but also include subjects of general interest, such as guidance on writing papers or grant proposals, and ethical issues; and a well balanced series of lectures and laboratory-type practical activities. Neuroscientists from any country in the region are eligible as prospective school directors. Each school should have no less than 15 and not more than 100 students and young scientists. The level of the school should match the level of selected students. The application should contain the scientific program of the school, a list of local and invited faculty, and the budget. Additional information is available at the IBRO website.
Restrictions: Hard copy or emailed applications are not accepted.
Geographic Focus: Puerto Rico, Antigua & Barbuda, Argentina, Bahamas, Barbados, Belize, Bolivia, Brazil, Chile, Colombia, Costa Rica, Cuba, Dominica, Dominican Republic, Ecuador, El Salvador, Grenada, Guatemala, Guyana, Haiti, Honduras, Jamaica, Mexico, Nicaragua, Panama, Paraguay, Peru, Saint Kitts And Nevis, St. Lucia, St. Vincent and the Grenadines, Suriname, Trinidad and Tobago, Uruguay, Venezuela
Contact: Robynn Rockstad-Rex, Director of Programmes and Communications; 33 (0) 1 46 47 92 92; fax 33 (0) 1 46 47 42 50; robynn@ibro.info
Internet: http://funding.ibro.info/Pub/Pub_Main_Display.asp?LC_Docs_ID=4661
Sponsor: International Brain Research Organization
255 rue Saint-Honore
Paris, 75001 France, Metropolitan

IBRO Latin America Regional Funding for PROLAB Collaborations 2328
The PROLAB Program seeks to foster scientific collaboration and development of human resources among Latin American and Caribbean neuroscience research groups. PROLAB uses the neuroscience scientific network of the region, and provides economic support to joint research projects. Its main goal is to diminish the asymmetries in the development of the neurosciences in the continent, thus allowing established research groups to set up collaboration with emerging groups and open a space for horizontal collaboration among established groups within the region. The scientific collaborations will be supported through grants for research missions (exchange trips by doctoral and post doctoral students from the participating institutions) planned to work on scientific questions specifically proposed in the submitted project. Financial support from PROLAB/IBRO/LARC is exclusively aimed at supporting research missions (three to six months) by doctoral and post-doctoral student members of the research groups for activities within the approved project. Applications must be submitted through the IBRO online application process. There are no hard copy application forms.
Requirements: Projects should aim at scientific investigations and training of human resources in research. They should be structured around a scientific research project in neuroscience focused on relevant scientific topics in the forefront of this area of human knowledge. Applications should have a joint coordination by the leaders of the research groups involved. The participant groups will choose one of the research leaders to be the responsible liaison with LARC. The projects should be planned for a two year period, with possibility of extension of financial support for another two years. The application should describe how the collaboration will be organized, and the chronogram of activities for the first two years of collaboration. At the end of each year, the project's liaison to LARC must present a joint report of the collaborative activities that took place among the research groups during that period and an update of the chronogram for the next year. Approval by LARC IBRO is necessary for continuation of the support.
Geographic Focus: Puerto Rico, Antigua & Barbuda, Argentina, Bahamas, Barbados, Belize, Bolivia, Brazil, Chile, Colombia, Costa Rica, Cuba, Dominica, Dominican Republic, Ecuador, El Salvador, Grenada, Guatemala, Guyana, Haiti, Honduras, Jamaica, Mexico, Nicaragua, Panama, Paraguay, Peru, Saint Kitts And Nevis, St. Lucia, St. Vincent and the Grenadines, Suriname, Trinidad and Tobago, Uruguay, Venezuela
Contact: Robynn Rockstad-Rex, Director of Programmes and Communications; 33 (0) 1 46 47 92 92; fax 33 (0) 1 46 47 42 50; robynn@ibro.info
Internet: http://funding.ibro.info/Pub/Pub_Main_Display.asp?LC_Docs_ID=5357
Sponsor: International Brain Research Organization
255 rue Saint-Honore
Paris, 75001 France, Metropolitan

IBRO Latin America Regional Travel Grants 2329
The International Brain Research Organization's Latin America Regional Committee (LARC) evaluates funding for attending courses/workshops or for presenting results at meetings within the Latin American region. Candidates should not be older than 40 years of age. Priority will be given to doctoral students and postdocs who are also planning a short stay in a lab. When the travel grant is linked to a short stay, an application for LARC Short Research Stay Grants must also be submitted. Applications must be submitted through IBRO's online application process.
Restrictions: emailed applications are not accepted. Applications for activities outside the region or within the applicant's own country will be considered only if adequately justified.
Geographic Focus: Puerto Rico, Antigua & Barbuda, Argentina, Bahamas, Barbados, Belize, Bolivia, Chile, Colombia, Costa Rica, Cuba, Dominica, Dominican Republic, Ecuador, El Salvador, Grenada, Guatemala, Guyana, Haiti, Honduras, Jamaica, Mexico, Nicaragua, Panama, Paraguay, Peru, Saint Kitts And Nevis, St. Lucia, St. Vincent and the Grenadines, Suriname, Trinidad and Tobago, Uruguay, Venezuela
Contact: Robynn Rockstad-Rex, Director of Programmes and Communications; 33 (0) 1 46 47 92 92; fax 33 (0) 1 46 47 42 50; robynn@ibro.info
Internet: http://funding.ibro.info/Pub/Pub_Main_Display.asp?LC_Docs_ID=4618
Sponsor: International Brain Research Organization
255 rue Saint-Honore
Paris, 75001 France, Metropolitan

IBRO Regional Grants for Int's Fellowships to U.S. Laboratory Summer Neuroscience Courses 2330

The U.S.-Canada Regional Committee for IBRO (U.S.CRC) provides fellowship support for students from developing nations who are accepted into the eligible high level neuroscience-related summer courses at the Marine Biological Laboratory in Woods Hole, Massachusetts and at the Cold Spring Harbor Laboratory in Cold Spring Harbor, New York. Substantial financial support is available to enable the participation of students from resource-restricted countries in these programs. The Fellowships are an effort to allow the most worthy alumni of IBRO Neuroscience Schools and other outstanding young scientists to gain outstanding experiences in leading training programs. These experiences will benefit scientists who will continue to study and work in less financially advantaged regions such as Latin America, Africa, Central and Eastern Europe, and Asia/Pacific. Candidates should apply directly to MBL or CSHL through their online application process. Each lab will inform candidates if they have been selected. Information about courses offered at each lab is available at the IBRO website.

Restrictions: Applicants should contact each lab directly about whether their chosen courses are eligible for the Fellowship.

Geographic Focus: All Countries

Date(s) Application is Due: Feb 1

Contact: Robynn Rockstad-Rex, Director of Programmes and Communications; 33 (0) 1 46 47 92 92; fax 33 (0) 1 46 47 42 50; robynn@ibro.info

Internet: http://funding.ibro.info/Pub/Pub_Main_Display.asp?LC_Docs_ID=4572

Sponsor: International Brain Research Organization
255 rue Saint-Honore
Paris, 75001 France, Metropolitan

IBRO Research Fellowships 2331

The IBRO Funding Programme aims to foster neuroscience, especially in less well-funded countries, by providing support to high quality neuroscientists from diverse geographic and scientific areas (U.S./Canada region excluded) who wish to broaden the scope of their training in neuroscience by working abroad. IBRO Research Fellowships, which include the John G. Nicholls and Rita Levi-Montalcini Fellowships, support post-doctoral laboratory training to applicants under the age of 45 for up to one year overseas. The funding for a 12-month Fellowship is 35,000 euros. Priority is given to those who have not obtained funding from this programme within the past three years and who, after completion of the Fellowship, are willing to return to their home countries, bringing with them new knowledge and skills to advance neuroscience in their regions.

Requirements: Applicants must meet the following *Requirements:* submit a letter of acceptance from the proposed supervisor (host scientist) indicating that he/she will accept the candidate and agrees on the proposed project; a letter of reference from the present supervisor; and, if available, evidence of an offer that the applicant will have a position to return to his/her home country after the Fellowship. Only applications completed through IBRO's online system will be considered.

Geographic Focus: All Countries

Date(s) Application is Due: Feb 1

Amount of Grant: 35,000 EUR

Contact: Robynn Rockstad-Rex , Director of Programmes and Communications; 33 (0) 1 46 47 92 92; fax 33 (0) 1 46 47 42 50; robynn@ibro.info

Internet: http://funding.ibro.info/Pub/Pub_Main_Display.asp?LC_Docs_ID=4254

Sponsor: International Brain Research Organization
255 rue Saint-Honore
Paris, 75001 France, Metropolitan

IBRO Return Home Fellowships 2332

The IBRO Fellowships Programme aims to foster neuroscience research, especially in less well-funded countries, by providing support to high quality neuroscientists from diverse geographic and scientific areas (U.S./Canada Region excluded) who wish to broaden the scope of their laboratory training in neuroscience by working abroad. Priority is given to those who have not obtained funding from this programme within the past three years and who, after completion of the fellowship, are willing to return to their home countries, bringing with them new knowledge and skills to advance neuroscience in their regions. The objective of the Return Home Programme of IBRO is to coordinate efforts with other organizations to both improve the opportunities for productive neuroscience research within the less advantaged regions of the world and to provide more aid to those researchers trained overseas who wish return to their home countries. Three profiles of scientists will be the main target of this Fellowship: postdoctoral Fellows who have finished research training in neurosciences (including clinical research) in a center of excellence of a developed country; research students who have been trained in a Centre of Excellence in Brain Research (CEBR); and scientists who are developing a successful basic/clinical research career in a developed country and wish to return to their country of origin or to a less developed country for personal or cultural reasons. Only applications completed through the IBRO application system will be considered.

Geographic Focus: All Countries

Date(s) Application is Due: Sep 1

Amount of Grant: Up to 20,000 EUR

Contact: Robynn Rockstad-Rex, Director of Programmes and Communications; 33 (0) 1 46 47 92 92; fax 33 (0) 1 46 47 42 50; robynn@ibro.info

Internet: http://funding.ibro.info/Pub/Pub_Main_Display.asp?LC_Docs_ID=5169

Sponsor: International Brain Research Organization
255 rue Saint-Honore
Paris, 75001 France, Metropolitan

ICC Faculty Fellowships 2333

Based on Ernest Boyer's concept that community engagement provides an excellent means by which the three components of faculty professional development of teaching, research, and professional service can support and strengthen each other, Indiana Campus Compact awards Faculty Fellowships as a model for faculty professional development through the scholarship of engagement. Up to 7 fellowships are competitively awarded in each academic year, in addition to one Senior Fellow. The fellowship amount is $3,750. Institutions must match at least $1,250. The fellowship fiscal period is twelve months, from June 1 to May 31.

Requirements: Faculty from all disciplines and professional schools at Indiana Campus Compact (ICC) member campuses are eligible to apply. Former Fellows are eligible to apply; however, preference is given to new applicants. Fellows are expected to model the integration of community engagement into all three aspects of their professional development in the following ways: (1) Teaching-Each Fellow will teach a service-learning course within the program period; (2) Research-Each Fellow will design a research, scholarly, or artistic project relating to the service performed with a community organization or to the practice of service-learning, to be initiated during the program period; (3) Professional Service-Each Fellow will provide direct service to a nonprofit organization during the program year to assist in addressing a particular community issue that is related to his/her academic discipline or expertise as an educator. Note: ALL proposals must be properly routed through the fiscal routing system at an applicant's institution prior to electronic submission to Indiana Campus Compact. However, if the institution is an Indiana University campus, routing through Research & Sponsored Programs or Sponsored Research Services is not required.

Geographic Focus: Indiana

Contact: J.R. Jamison; (317) 274-6500; fax (317) 274-6510; jrjamiso@iupui.edu

Internet: http://www.indianacampuscompact.org/Default.aspx?tabid=86

Sponsor: Indiana Campus Compact
620 Union Drive, Suite 203
Indianapolis, IN 46202

ICC Scholarship of Engagement Faculty Grants 2334

Eighteen faculty per year will be selected to choose one of three options related to the Scholarship of Engagement: develop or revise a service-learning course to be taught by the recipient, provide scholarly research on service-engagement, or provide a professional service to a local community agency. The selected faculty members will also encourage the development of service-engagement infrastructure on their respective campuses through trainings and workshops of other faculty. Award amounts are $2,250. Institutions are required to cash match at least $750.

Requirements: Adjunct or full-time faculty from all disciplines and professional schools at Indiana Campus Compact (ICC) member campuses are eligible to apply. An adjunct or full-time faculty member applicant must propose a project that focuses on one of three areas: (1) Teaching-Design a new course or revise an existing course to include a service-learning component; (2) Research-Develop and initiate a research project related to his/her discipline that addresses a community issue and advances the field of engagement through service-learning. Preference will be given to projects that are participatory action research; (3) Professional Service-Provide direct service to a nonprofit organization to assist in addressing a particular community issue. This professional service must be related to the faculty member's academic discipline or his/her expertise as an educator. Note: ALL proposals must be properly routed through the fiscal routing system at an applicant's institution prior to electronic submission to Indiana Campus Compact. However, if the institution is an Indiana University campus, routing through Research & Sponsored Programs or Sponsored Research Services is not required.

Restrictions: Grants are awarded on a reimbursement basis. Grant funds cannot be used for indirect costs, items of durable equipment costing more than $100 (unless approved in writing by ICC), payment for participants to do service, fundraising activities, entertainment costs, tickets to recreational events, ball games, zoos, etc., clothing such as T-shirts or hats, or cash incentives.

Geographic Focus: Indiana

Date(s) Application is Due: Nov 6

Contact: Lindsay Doucette, Program Director; (317) 274-6500; lgdoucet@iupui.edu

Internet: http://www.indianacampuscompact.org/Default.aspx?tabid=86

Sponsor: Indiana Campus Compact
620 Union Drive, Suite 203
Indianapolis, IN 46202

ICCS Graduate Students Scholarships 2335

The scholarships enable successful candidates to spend four to six weeks at a Canadian university or research site other than their own doing research related to their thesis or dissertation in the field of Canadian studies. Scholarships are not intended to initiate a thesis or dissertation, but rather to provide access to crucial scholarly information and resources in Canada in support of a thesis/dissertation that is close to or at the point of being written.

Requirements: Each applicant must be nominated by his/her national Canadian studies association The student must obtain in writing the support of a faculty member at a Canadian university who has agreed to act as the student's academic sponsor during the tenure of his/her award.

Geographic Focus: All States, Canada

Date(s) Application is Due: Dec 31

Amount of Grant: Up to 5,000 CAD

Contact: Administrator; (613) 789-7834, ext. 242; csppec@iccs-ciec.ca

Internet: http://www.iccs-ciec.ca/pages/4_ICCSprogs/a_gradstu.html

Sponsor: International Council for Canadian Studies
75 Albert Street, Suite 908
Ottawa, ON K1P 5E7 Canada

ICCS Publishing Fund 2336

The fund assists with the publication and distribution in Canada of scholarly monographs on Canada written by foreign Canadianists who are members of a Canadian Studies Association or associate member belonging to the International Council for Canadian Studies. This fund assists foreign Canadianists by granting financial aid to a recognized scholarly press once the work is published. The fund may also grant financial assistance for the translation from English or French into a third language and from a third language into French or English. The funds allocated will apply to the printing or translating costs of a manuscript/book and not to the costs of research, typing, reproduction or any other stage in the manuscript preparation process. Applications must be presented to the ICCS in either of Canada's official languages. Guidelines are available online.
Requirements: Only scholarly manuscripts on Canada written by foreign Canadianists are eligible. Manuscripts must be previously unpublished. Manuscripts consisting of previously published texts, such as excerpts of books or articles, are eligible if a third of the texts have not previously been published. The author must be a member in good standing of an ICCS member association or associate member. A letter of support from a Canadian Studies Association or associate member belonging to the ICCS must accompany the application.
Restrictions: Collections of poetry, novels, plays, magazine articles or conference proceedings are not eligible.
Geographic Focus: All States
Date(s) Application is Due: Oct 31
Contact: Guy LeClair, (613) 789-7834, ext. 228; fax (613) 789-7830; gleclair@cs-ciec.ca
Internet: http://www.iccs-ciec.ca/pages/4_ICCSprogs/f_publishfund.html
Sponsor: International Council for Canadian Studies
75 Albert Street, Suite 908
Ottawa, ON K1P 5E7 Canada

Ida Smedley MacLean International Fellowship 2337

The fellowship shall be awarded to encourage advanced scholarship by enabling university women to undertake original research in some country other than that in which the holder has received her education or habitually resides. Fellowships are intended to cover at least eight months of work and should be taken up within a reasonable period from the date of the award. Members of the IFUW should apply through their national affiliates (in the United States, the American Association of University Women, 1111 16th St NW, Washington, DC 20036). The competition is held in even-numbered years. The deadline for receipt of applications is determined by each affiliate but normally falls between early September and mid-October of the year preceding the competition. Contact the national headquarters for the exact deadline date.
Requirements: An applicant must be a woman graduate who is a member of IFUW and must be well started on the research program to which the application relates.
Restrictions: A fellowship will not be awarded for the first year of a PhD program.
Geographic Focus: All States
Date(s) Application is Due: Oct 1
Amount of Grant: 8,000 CHF
Contact: Fellowships Officer; 41-22-731-23-80; fax 41-22-738-04-40; info@ifuw.org
Internet: http://www.ifuw.org/fellowships/international.htm
Sponsor: International Federation of University Women
8 rue de l'Ancien Port
Geneva, CH 1201 Switzerland

IDEM Section 319(h) Nonpoint Source Grants 2338

This program funds projects that will work on a watershed level to reduce nonpoint source pollution in Indiana's lakes, rivers, and streams. Nonpoint source pollution does not come from a pipe. It results when water (rain or snowmelt) moves across land, such as city streets, agricultural fields and residential backyards, and picks up dirt, fertilizers, pesticides, animal wastes, road salt, motor oil and other pollutants. Nonpoint source pollution is also caused by wind, which like rain, can pick up soil particles and deposit them in lakes and streams. These pollutants have harmful effects on drinking water supplies, recreation, fisheries and wildlife. Nonpoint source pollution is the leading cause of water quality problems in Indiana and is responsible for many of the impairments identified on the 303(d) List of Impaired Waterbodies. Funding for selected projects will be provided by Clean Water Act Section 319 grant funds and match is provided by grant recipients and partners.
Requirements: To be considered for funding, the project sponsor (the entity responsible for the project and its overall success), must be one of the following: 1) Municipality; 2) County Government; 3) State Government; 4) Federal Government; 5) College/University; 6) Nonprofit 501(c)3. The program provides funding and technical assistance to groups that work on the watershed level with citizens to develop locally-based solutions to nonpoint source pollution. Specific ways to address nonpoint source water pollution include education/outreach on watershed management, information gathering activities such as conducting watershed inventories and water quality assessments for the purpose of developing comprehensive watershed management plans and implementing those plans, including implementation of best management practices that directly reduce sources of nonpoint source pollution. IDEM provides sixty percent (60%) of an approved project's total cost with Section 319 funds. A grant recipient must provide the remaining forty percent (40%) of the total project cost as match. Match may be in-kind services or cash. Match cannot come from any federal funding sources. Guidelines and application forms are available for download at the website. Do not use forms from previous years-they will not be accepted by IDEM.
Restrictions: The following is a list of activities that cannot be funded with Section 319 funds and cannot be counted as matching funds for a Section 319 grant: Permit fees; Food for meetings or other events; Purchase of agricultural equipment, or other large pieces of equipment (equipment modifications and leasing are allowable); Purchase of land or land easements (these activities can be counted as matching funds in some cases); Any project which is directed at water quantity rather than water quality, such as dredging, drainage, or flood control; Any practices, equipment, or supplies used to fulfill the requirements of any federal permit (NPDES permit, Section 401 Water Quality Certification, permits from the U.S. Army Corps of Engineers, as examples) or to comply with IDEM's Confined Feeding Operation rule or permit requirements, or to meet enforcement requirements; Wetland mitigation sites; Incentive payments or yield losses; Nonpoint source best management practices not sanctioned by IDEM or not sanctioned by a partner agency of IDEM; Practices not installed in accordance with standards and specifications developed by NRCS, IDNR or other recognized standards; Office furniture; Sales tax. Additionally, the following will not be funded by this grant program: Septic system pump outs, repairs, rehabilitations, or demonstrations of alternative septic systems; Projects whose sole purpose is data collection, research, demonstration of best management practices, or education/outreach. However, these activities may be incorporated as elements into a proposal that meets one of the three priorities.
Geographic Focus: Indiana
Date(s) Application is Due: Sep 1
Contact: Laura Bieberich, Team Leader; (317) 233-1863
Internet: http://www.in.gov/idem/resources/grants_loans/319h/index.html
Sponsor: Indiana Department of Environmental Management
Indiana Government Center North, 100 N Senate Avenue, Mail Code 50-01
Indianapolis, IN 46204-2251

IEEE Electrical History Fellowship 2339

This fellowship supports one year of full-time graduate study in the history of electrical engineering and technology at an engineering school of recognized standing in the United States or Canada.
Requirements: Individuals doing graduate or postgraduate work in the history of electrical technology are eligible.
Geographic Focus: All States
Date(s) Application is Due: Feb 1
Amount of Grant: 15,000 USD
Contact: Director; (732) 932-1066; fax (732) 932-1193; history@ieee.org
Internet: http://www.ieee.org/portal/index.jsp?pageID=corp_level1&path=membership/students/awards&file=sc_ehistory.xml&xsl=generic.xsl
Sponsor: Institute of Electrical and Electronics Engineers
39 Union Street
New Brunswick, NJ 08901-8538

IERF Sepmeyer Research Grants 2340

The Foundation's mission is to conduct research and disseminate information on world educational systems and to facilitate the integration of individuals educated outside the United States into the U.S. educational environment and work force. The Foundation conducts and supports comprehensive, quality research on world education systems, shares research findings with the international community, and provides research-based credentials, evaluations, and related services. The purpose of the research grant is to aid the Foundation in carrying out its mission. There is no application form. Applicants should submit a research project summary. Application information is available online.
Requirements: Research grants are available to all persons, irrespective of citizenship and country of residence.
Geographic Focus: All States
Date(s) Application is Due: Feb 1
Amount of Grant: Up to 5,000 USD
Contact: Susan Bedil; (310) 258-9451; fax (310) 342-7086; grants@ierf.org
Internet: http://www.ierf.org/grants.asp
Sponsor: International Education Research Foundation
P.O. Box 3665
Culver City, CA 90231-3665

IES Faculty Fellowships 2341

Founded in 1950, the Institute for the International Education of Students (IES Abroad) is a Chicago-based, not-for-profit academic consortium that offers financial aid to students studying for a semester or full year at one of its centers abroad. IES Abroad operates study-abroad programs on the continents of Europe, Africa, Asia, South America, Australia, and New Zealand. In cooperation with the Newberry Library, the institute is very pleased to offer a one-month research fellowship for an IES Abroad faculty member at the Newberry Library in Chicago, Illinois (U.S.A) during the summer. Recipients will have full access to the Newberry Library's extensive collections in the areas of European discovery, exploration, and settlement of the Americas; American literature and history; the Renaissance; the French Revolution; Portuguese and Brazilian history; American Indian history and literature; the history of cartography; the history and theory of music; the history of printing; and early philology and linguistics. IES Abroad will cover all international transportation costs, provide a monthly stipend of $1600, and pay housing costs for the duration of the fellowship. The recipient is responsible for any additional expenses. Prospective applicants should contact their respective IES Abroad Center Directors for application instructions and further information. A contact list has been provided at the IES website (click on the "Contact Us" and then on "Advisors and Faculty" links).
Requirements: All eligible IES Abroad faculty whose research would benefit from access to the Newberry Library's collections are invited to apply for this fellowship. At the end of the one-month research period, recipients will be asked to present their findings and to report on their experience to the IES Abroad Chicago staff.
Geographic Focus: All States, All Countries

Amount of Grant: 1,600 USD
Contact: Michael Steinberg; (800) 995-2300 or (312) 944-1750; fax (312) 944-1448; msteinberg@IESabroad.org or info@IESabroad.org
Internet: https://www.iesabroad.org/IES/Advisors_and_Faculty/Faculty_Development/fellowships.html
Sponsor: Institute for the International Education of Students
33 North LaSalle Street, 15th Floor
Chicago, IL 60602

IFP Minnesota McKnight Filmmaking Fellowships 2342
IFP Minnesota in partnership with the McKnight Foundation awards $25,000 fellowships to two Minnesota filmmakers annually. These fellowships recognize Minnesota artists for talent in working with film/video as demonstrated by two examples of completed, original works in any of the genres and formats of narrative, documentary, experimental, or animation in feature or short lengths. The program is intended to support mid-career artists by providing financial assistance, professional encouragement, and industry recognition. Work examples and application material are judged by a panel of industry professionals from outside the state of Minnesota. Judges look for consistent artistic excellence and merit, clarity and uniqueness of vision, professional quality in the technical aspects of production, and demonstrable, sustained growth in the artist's career. Additionally, judges assess the ability of the artist to present their application in an articulate and professional manner.
Geographic Focus: Minnesota
Date(s) Application is Due: Mar 3
Amount of Grant: 25,000 USD
Contact: Lu Lippold, (651) 644-1912, ext.106; fax (651) 644-5708; llippold@ifpmn.org
Internet: http://www.ifpmn.org/mcknight.html
Sponsor: Independent Feature Project Minnesota
2446 University Avenue West, Suite 100
St. Paul, MN 55114

IFP Minnesota McKnight Screenwriters Fellowships 2343
IFP Minnesota in partnership with the McKnight Foundation awards fellowships to two Minnesota screenwriters annually. These fellowships recognize Minnesota artists for talent in writing for the screen as demonstrated by one completed feature-length screenplay. The program is intended to support mid-career writers by providing financial assistance, professional encouragement, and industry recognition. Screenplays and application material are judged by two panels of industry professionals from outside the state of Minnesota. In addition to the monetary award, the winners receive entrance into the IFP Market in New York, as well as a live reading on stage at the Jungle Theater in Minneapolis with professional actors. Screenplays and applicants are judged on originality of ideas and uniqueness of writer's voice, skillful story structure, talent with creating dialog, and talent with character development.
Restrictions: The following are ineligible: organizations and companies; full-time and part-time students in any film-, video- or screenwriting-related degree program; screenplays associated with a degree-granting program; and employees of IFP/MSP or the McKnight Foundation. Fellows cannot attend school full time during the fellowship year. The screenplay may not be an adaptation of any previously written performed or filmed work by another artist; or have been submitted to the fellowship more than three times previously, under any title.
Geographic Focus: Minnesota
Date(s) Application is Due: Feb 4
Amount of Grant: 25,000 USD
Contact: Lu Lippold, (651) 644-1912, ext.106; fax (651) 644-5708; llippold@ifpmn.org
Internet: http://www.ifpmn.org/mcknight.html
Sponsor: Independent Feature Project Minnesota
2446 University Avenue West, Suite 100
St. Paul, MN 55114

IFP New York Anthony Radziwill Documentary Fund Grants 2344
The fund provides grants to emerging and established documentary filmmakers in the form of development funds (seed money) for specific new projects. Administered by IFP New York, the fund seeks to provide funding for independent nonfiction filmmakers at the earliest stage of new work, traditionally a difficult point at which to secure funding. Grants will be awarded to documentary projects that are seeking funds needed for research, treatment and script development, initial interviews and shooting, the production of trailers/clips for further funding needs, etc. Typically a grant will be seed money, though in some cases it could be a supplement to another initial small grant.
Requirements: Grants are given to individuals with creative and financial control on the project. The grantee must be a U.S. legal resident who is 18 years of age or older. Grants are for nonfiction projects intended to be: feature length (over 50 minutes); for general audiences; and primarily for wide distribution via theatrical, television, and festival markets.
Restrictions: The following are not eligible: international projects for which neither the producer nor director is a legal U.S. resident; projects by students in undergraduate or graduate degree programs, or for fulfillment of any course requirement by nonmatriculated students; projects intended primarily for the educational or industrial markets or for promotional purposes; projects for which film or video is not the primary artistic component (work designed to accompany a larger theater piece or multimedia installation); individuals and immediate family members who are employees of any IFP chapter, or currently serving as evaluators, panelists, or on the fund advisory committee; and grantees who have received funding from the fund for any grant cycle within the previous two-year period.
Geographic Focus: All States
Date(s) Application is Due: Mar 1; Sep 1
Contact: Administrator; (212) 465-8200, ext. 830; fax (212) 465-8525; info@ifp.org
Internet: http://www.ifp.org/common/page.php?ref=grants
Sponsor: Independent Feature Project New York
104 W 29th Street, 12th Floor
New York, NY 10001

IFT Foundation Grants 2345
The foundation awards funding in science communications and career guidance. Science communications delivers the scientific perspective on food issues to media, policy makers, allied societies, and members. Priority topics include food safety, biotechnology, diet and health, food labeling, food and the environment, and support for research in food science and technology. Career guidance includes educational videos on food science and food science careers. Funds available for distribution by the foundation vary yearly. It is recommended that applicants contact the foundation prior to preparing a proposal. The intent of the foundation is to serve as a catalyst to establish programs rather than a permanent funding source for programs.
Geographic Focus: All States
Contact: Tekla Syers, Director of Development; (312) 782-8424; tasyers@ift.org
Internet: http://www.ift.org/cms/?pid=1000260
Sponsor: Institute of Food Technologists Foundation
525 W Van Buren Street, Suite 1000
Chicago, IL 60601

IFUW Dr. A. Vibert Douglas International Fellowship 2346
This is a triennial fellowship funded by Canadian Federation of University Women (CFUW) and administered by International Federation of University Women (IFUW). This competition is open to women graduates for research for a period of eight months abroad. Members of the IFUW should apply through their national affiliate (in the United States, the American Association of University Women, 1111 16th St NW, Washington, DC 20036). The deadline for receipt of applications is determined by each affiliate but normally falls between early September and mid-October of the year preceding the competition; contact national headquarters for exact date.
Requirements: An applicant must be a woman graduate and must be well started on the research program to which the application relates.
Restrictions: A fellowship will not be awarded for the first year of a PhD program.
Geographic Focus: All States, Canada
Amount of Grant: 12,000 CAD
Contact: Betty Dunlop, Fellowships Program Manager; (613) 234-8252 or (613) -234-2732; fax (613) 234-8221; cfuwfls@rogers.com
Internet: http://www.cfuw.org/index.php?option=com_content&task=view&id=499&Itemid=159
Sponsor: International Federation of University Women
8 rue de l'Ancien Port
Geneva, CH 1201 Switzerland

IFUW Winifred Cullis Grants 2347
This competition is held in even-numbered years and is open to women graduates who are members of the International Federation of University Women (IFUW). Grants are awarded for research and/or training in a country other than the one in which the candidate has received her education or habitually resides. The grants are to assist women graduates to obtain specialized training essential to their research; to train in new techniques in the humanities, social sciences, and natural sciences; or to carry out independent research including completion of a piece of well-advanced research. Members of the IFUW should apply through the IFUW affiliate in their country (in the United States, the American Association of University Women, IFUW Liaison Executive Office, 1111 16th St NW, Washington, DC 20036). Application deadlines for members applying through on of IFUWs 71 national affiliates vary by country, but normally fall between early September and early October of the year preceding the competition; contact national headquarters for exact dates. Application deadline for independent members and international individual members is November 1.
Geographic Focus: All States
Date(s) Application is Due: Oct 1; Nov 1
Amount of Grant: 5,000 CHF
Contact: Anamaria Vere; 41-22-731-23-80; fax 41-22-738-04-40; info@ifuw.org
Internet: http://www.ifuw.org/i_fell.htm
Sponsor: International Federation of University Women
10 rue du Lac
Geneva, CH 1207 Switzerland

IHI Quality Improvement Fellowships 2348
The Institute for Healthcare Improvement (IHI), an independent not-for-profit organization based in Cambridge, Massachusetts, is an innovator in health and health care improvement worldwide. The IHI Fellowship Program is a year-long immersion program for mid-career health professionals. The program has two overarching goals, to: (1) develop health care leaders with the drive, skills, and experience to spread improvement in the United States and globally; and (2) build capability within health care organizations to reach dramatically higher levels of performance. Fellows spend one year at IHI's office in Cambridge, Massachusetts, and return to their home organization to lead transformative change. Fellowships are one year in duration, July through June.
Requirements: The IHI Quality Improvement Fellowship supports fellows who live and work in the United States. The fellowship is designed for mid-careers professionals who will return to senior leadership positions in their home organizations. IHI will, however, consider a very strong application from a candidate earlier in his/her career. All successful

candidates must have full-time employment with an organization who will sponsor them for the fellowship (salary and benefits for the year are paid by the home organization) and who shows commitment to leveraging the fellow's learning appropriately post-fellowship. Applicants are not required to have a M.D; IHI encourages applications from nurse leaders, quality improvement professionals, and allied health professionals.
Geographic Focus: All States
Contact: Joelle Baehrend, (617) 301-4816; fax (617) 301-4848; jbaehrend@ihi.org
Internet: http://www.ihi.org/offerings/Training/Fellowships/Pages/default.aspx
Sponsor: Institute for Healthcare Improvement
20 University Road, 7th Floor
Cambridge, MA 02138

IIE Central Europe Summer Research Institute Summer Research Fellowship 2349
The Central Europe Summer Research Institute (CESRI) is a fellowship opportunity for U.S. graduate students in science and engineering. With support from the National Science Foundation (NSF) and the German Academic Exchange Service (DAAD), CESRI will provide eight U.S. graduate students with a high-quality international research experience in Austria, the Czech Republic, Germany, Hungary, Poland or Slovakia. Two CESRI grants are specifically for CESRI Fellowships to Germany. The remaining six CESRI grants are funded by NSF for the entire Central Europe region, with preference for grants to countries other than Germany. The program is intended for scientific research projects. The CESRI program will be eight weeks in length, with the first four days spent as a group in Budapest, Hungary in a specially designed cultural and academic orientation to the region. Participants will spend the remaining 7.5 weeks working in individually-arranged placements in university labs or other appropriate sites where they can participate in creative research activities under the supervision of European mentors. In addition to the scholarship, recipients receive round trip airfare to Central Europe, room and board for 8 weeks, health and accident insurance, and a four day academic and cultural orientation in Budapest.
Requirements: All applicants must be U.S. citizens or permanent residents in the U.S., and be current Master's or Ph.D. students at a U.S. university in one of the following fields: Biology, Chemistry, Computer Science, Engineering, Environmental Science, or Mathematics. Candidates should submit the online application, along with the online self-evaluation form and two completed professional reference forms.
Restrictions: Projects involving field work, policy, ethics, or clinical research on human objects are not eligible for consideration.
Geographic Focus: All States
Date(s) Application is Due: Feb 1
Amount of Grant: 2,000 USD
Contact: Fellowship Contact; (212) 883-8200; fax (212) 984-5452; cesri@iie.org
Internet: http://www.iie.org/Programs/CESRI
Sponsor: Institute of International Education
1400 K Street, NW, 7th Floor
Washington, DC 20005-2403

IIE David L. Boren Fellowships 2350
Boren Fellowships provide up to $30,000 funding to U.S. graduate students to add an international and language component to their graduate education through specialization in area study, language study, or increased language proficiency. Boren Fellowships support study and research in areas of the world that are critical to U.S. interests, including Africa, Asia, Central & Eastern Europe, Eurasia, Latin America, and the Middle East. Boren Fellowships are funded by the National Security Education Program (NSEP), which focuses on geographic areas, languages, and fields of study that are critical to U.S. national security. Applicants should identify how their projects, as well as their future academic and career goals, will contribute to U.S. national security.
Requirements: Boren Fellowships promote long term linguistic and cultural immersion. Therefore, applicant preference will be given to those proposing overseas programs of 6 months or longer. However, applicants proposing overseas programs of 3-6 months, especially those in the STEM (science, technology, engineering, and mathematics) fields are encouraged to apply. Boren Fellowships are awarded with preference for countries, languages, and fields of study critical to U.S. national security. Preference is also given to students who will study abroad for longer periods of time, and who are highly motivated by the opportunity to work in the federal government. Applicants should refer to the website for current deadlines and detailed criteria for application.
Restrictions: Applicants must commit to a length of study of at last 12 weeks.
Geographic Focus: All States
Date(s) Application is Due: Jan 31
Amount of Grant: Up to 30,000 USD
Contact: Michael Saffle; (800) 618-6737; fax (202) 326-7672; boren@iie.org
Internet: http://www.borenawards.org/boren_fellowship
Sponsor: Institute of International Education
1400 K Street, NW, 7th Floor
Washington, DC 20005-2403

IIE David L. Boren Scholarships 2351
Boren Scholarships provide undergraduates with up to $20,000 funding to acquire skills and experiences in areas of the world critical to U.S. national security, in exchange for their commitment to seek work in the federal government. Boren Scholarships are funded by the National Security Education Program (NSEP), which focuses on geographic areas, languages, and fields of study critical to U.S. national security. NSEP draws on national security issues such as sustainable development, environmental degradation, global disease and hunger, population growth and migration, and economic competitiveness. Boren Scholarships allow students to study in areas of the world that are underrepresented in study abroad programs, including Africa, Asia, Central and Eastern Europe, Eurasia, Latin America, and the Middle East. Scholars represent a variety of academic backgrounds, but are interested in studying less commonly taught languages, including but not limited to Arabic, Chinese, Korean, Portuguese, Russian, and Swahili. Individual on-campus deadlines will be earlier than February 13.
Requirements: Students are eligible to apply if they are U.S. citizens enrolled as undergraduates in a degree-seeking program at a college or university for the duration of the scholarship (they may not graduate until the scholarship is complete); and applying to a study abroad program that meets their home institution standards in a country outside of Western Europe, Canada, Australia, or New Zealand. Boren Scholarships are awarded with preference for countries, languages, and fields of study critical to U.S. national security. Preference is also given to students who will study abroad for longer periods of time, and who are highly motivated by the opportunity to work in the federal government. In exchange for scholarship funding, all Boren scholars must agree to the NSEP Service Requirement posted on the website. Applicants should refer to the website for current information and their individual campus representative for their on-campus deadline.
Restrictions: Boren Scholarships are not for study in the U.S.
Geographic Focus: All States
Date(s) Application is Due: Feb 13
Amount of Grant: 8,000-20,000 USD
Contact: Lisa Mayorga; (800) 618-6737; fax (202) 326-7672; boren@iie.org
Internet: http://www.borenawards.org/boren_scholarship
Sponsor: Institute of International Education
1400 K Street NW, 7th Floor
Washington, DC 20005-2403

IIE General Electric Foundation TECH Awards 2352
The GE Foundation TECH Award is designed to stimulate innovation in new technology, design, research methods and applications among university students in China. The Award encourages market-oriented technology innovations in the following areas: energy, water, electronics and electrical engineering, machinery and mechanical processing, chemistry and material science. The Award recognizes innovation and excellence in top Chinese university students in the fields of engineering and the sciences. It also encourages recipients to continue in these fields in China after graduation, and helps to foster the development of local talent and human resources. Recipients also receive a trip to the GE China Technology Center in Shanghai to participate in a recognition ceremony and workshops, as well as the opportunity to intern at GE China.
Requirements: Applicants must attend one of the following universities: Shanghai Jiao Tong University; Zhejiang University; China University of Technology; Tsinghua University; Xi'an Jiaotong University; Xi'an University of Electronic Science and Technology; University of Electronic Science and Technology; Harbin Institute of Technology; or Huazhong University of Science and Technology. Along with the online application, students must submit a detailed Chinese plan, its summary, two letters of recommendation, college transcripts, and a copy of the applicant's Master's and/or Ph.D. degree.
Geographic Focus: China
Date(s) Application is Due: Oct 17
Amount of Grant: 800-2,000 USD
Contact: Award Contact; (8) 52 2103-1505; fax 852 2603-5765; china_awards@iiehongkong.org
Internet: http://www.iie.org/Programs/GE-Foundation-TECH-Award
Sponsor: Institute of International Education
1400 K Street, NW, 7th Floor
Washington, DC 20005-2403

IIE KAU.S.T Graduate Fellowships 2353
The King Abdullah University of Science and Technology (KAU.S.T) engages students, faculty, and researchers in advancing science and technology through collaborative inquiry focused on issues of regional and global significance. Their studies focus on the continued advancement of the global quality of life in all issues associated with energy, water, and food. They also explore environment and the Red Sea, and computational science and engineering (high-performance computing, computing for a small planet) as an enabling technology for their research activities. A KAU.S.T Fellowship includes full tuition, a monthly stipend, housing and a travel benefit.
Requirements: Eligibly fields of study include: applied mathematics and computational science; bioscience; chemical and biological engineering; chemical science; computer science; earth science and engineering; electrical engineering; environmental science and engineering; marine science and engineering; materials science and engineering; and mechanical engineering.To qualify for the KAU.S.T fellowship, applicants must meet the following requirements before completing the online application: a bachelor's degree in a KAU.S.T-relevant field of study, pursuing a graduate degree in a similar field; minimum GPA of 3.5; a TOEFL score of 79 or 6.0 on the IELTS (International English Language Testing System; see website for specific requirements where TOEFL is not required); must submit GRE scores; a statement of purpose; three letters of recommendation; and an official university transcript.
Geographic Focus: All States, All Countries
Date(s) Application is Due: Jan 15
Contact: KAU.S.T Fellowship Contact; +966 (0) 2 803 3428
Internet: http://www.kaust.edu.sa/admissions/tokaust/fellowship.html
Sponsor: King Abdullah University of Science and Technology
Graduate Affairs, Building 9, Suite 4328
Thuwal, Jeddah, 23955-6900 Saudi Arabia

IIE Leonora Lindsley Memorial Fellowships 2354

The Lindsley Scholarship was established for Leonora Lindsley, an American member of an international all-female combat unit on the Western Front during World War II, who was killed the day before the war officially ended. The scholarship was created in her honor by French citizens who are descendants from Resistance fighters to fund those pursuing educational studies in the United States. An award is granted to one or two students for graduate study for a twelve month period, and is renewable for a second year. The winner(s) are chosen every other year. The amount of the award varies from year to year, but ranges from $10,000 to $12,000.

Requirements: The Fulbright Commission in Paris gathers applications and sends them to IIE where they are reviewed for completeness and eligibility. Once this is done, a selection panel is convened and finalists chosen. Any major is eligible.

Restrictions: Only descendants of French Resistance fighters are eligible. Grantees must be at an accredited U.S. university.

Geographic Focus: All States, France

Amount of Grant: Up to 12,000 USD

Contact: Leonara Lindsley Fellowship Contact; (212) 984-5552

Internet: http://www.iie.org/Programs/Leonora-Lindsley-Memorial-Fellowship

Sponsor: Institute of International Education

1400 K Street, NW, 7th Floor

Washington, DC 20005-2403

IIE Rockefeller Foundation Bellagio Center Residencies 2355

The Bellagio Center Residency program offers scholars, artists, thought leaders, policymakers and practitioners a serene setting conducive to focused, goal-oriented work, and the opportunity to establish new connections with fellow residents across an array of disciplines and geographies. The Center sponsors three kinds of residencies—for scholars, creative artists and practitioners. The Center offers one-month residency for each category and is able to accommodate a group working on a particular project. The Center is particularly interested in applicants whose work connects in some way to the Foundation's issues areas (basic survival safeguards, global health, climate and environment, urbanization, and social and economic security).

Requirements: Applicants must submit an online application, a detailed project proposal of no more than 1,250 words, an abbreviated curriculum vitae, three reference letters, and work samples.

Geographic Focus: All States

Date(s) Application is Due: Dec 1

Contact: Bellagio Center Committee

Internet: http://www.rockefellerfoundation.org/bellagio-center

Sponsor: Institute of International Education

1400 K Street, NW, 7th Floor

Washington, DC 20005-2403

IIE Whitaker International Fellowships and Scholarships 2356

The goal of the Whitaker funding is to assist in the development of outstanding professional engineers and scientists who will lead and serve the profession with an international outlook. The term of the Fellowship awards will be one academic year (as defined by the academic calendar of the host country). The term of the Scholarship awards can be for as little as one academic semester or as long as a full academic year. Benefits include maintenance allowance, airfare, accident and sickness insurance, partial tuition reimbursement (Fellows only), and access to grantee events, as well as membership in an elite alumni resource network. Applicants should consult the website for specific information about host institutions in Australia, Brazil, France, Germany, Ireland, Italy, The Netherlands, South Africa, Spain, Switzerland, and the United Kingdom.

Requirements: U.S. citizens and non-citizen permanent residents may apply. Fellows must have a BS or MS degree in biomedical engineering or bioengineering, and they will not have a doctorate at the time they start the grant. The Fellows applicant's most recent degree may not have been obtained more than two years prior to the start of the grant. Scholars must have either a degree in biomedical engineering and a doctorate which may not have been obtained more than two years prior to the start of the grant, or must be advanced Ph.D. candidates who have been admitted to candidacy and who will receive the Ph.D. prior to departure.

Geographic Focus: All States

Date(s) Application is Due: Oct 8

Amount of Grant: Up to 35,000 USD

Contact: Sabeen Altaf, Program Manager; (212) 984-5442; whitaker@iie.org

Internet: http://www.iie.org/Programs/Whitaker-International-Fellows-and-Scholars-Program

Sponsor: Institute of International Education

1400 K Street, NW, 7th Floor

Washington, DC 20005-2403

IIPP Graduate Fellowship 2357

The program supports graduate education toward a masters degree in international affairs or a related field. Fellows apply to and enroll in the nation's leading graduate programs, most members of the Association of Professional Schools of International Affairs (APSIA). Participating graduate schools provide monetary support over the course of a two-year program. IIPP, in turn, provides each of its fellows with a matching grant to be applied in accordance with an agreement between IIPP and the institution the fellow attends.

Requirements: To date, IIPP fellows have enrolled at the following institutions: American University, Columbia University, Georgetown University, George Washington University, Harvard University, Johns Hopkins University, Princeton University, Syracuse University, Tufts University, University of Denver, University of Maryland, University of Michigan, University of Washington, and Yale University. Non-APSIA schools include Cornell University, Pepperdine University, New York University, and the Bush School at Texas A&M.

Geographic Focus: All States

Date(s) Application is Due: Mar 15

Amount of Grant: 30,000 USD

Contact: Administrator; (800) 530-6232 or (703) 205-7623; fax (703) 205-7645

Internet: http://161.58.87.106/content/program.cfm#grad

Sponsor: Institute for International Public Policy

2750 Prosperity Avenue, Suite 600

Fairfax, VA 22031

Illinois Clean Energy Community Foundation Renewable Energy Grants 2358

The Illinois Clean Energy Community Foundation helps non-profit groups and local government organizations to purchase and install renewable energy technologies, including photovoltaic panels, solar thermal systems, wind turbines and biomass systems. Renewable energy provides heating, cooling, hot water, steam or electricity, in an environmentally sound manner while also lowering or stabilizing utility costs. Before considering renewable energy, make the building as energy efficient as possible so that the installed renewable energy system can be smaller, less costly and able to offset a larger percentage of the energy consumed. The Foundation's goal, through its support of renewable energy technologies, is to improve the environment, help grow the renewable energy sector in Illinois, create jobs and lower energy costs. We also work with colleges and universities, as well as K-12 schools and teachers around the state to help educate and train Illinois' next generation about the benefits and potential of renewable energy.

Requirements: 501(c)3 tax-exempt organizations, educational institutions, and state and local government agencies serving Illinois residents are eligible.

Restrictions: The foundation will not provide funding for remediation of environmentally impaired properties; technology research; promotion of proprietary products; reoccurring operating costs; political campaigns or lobbying; capital campaigns or support for an organization's endowment; or projects undertaken by individuals.

Geographic Focus: Illinois

Date(s) Application is Due: Jan 15; Jul 15

Contact: Bob Romo; (312) 372-5191; fax (312) 372-5190; info@illinoiscleanenergy.org

Internet: http://www.illinoiscleanenergy.org/renewable-energy/

Sponsor: Illinois Clean Energy Community Foundation

2 N LaSalle Street, Suite 1140

Chicago, IL 60602

Illinois DCEO Coal Research Grants 2359

The Illinois Clean Coal Institute (ICCI) Coal Research and Development (R&D) Program is the technical component of the Illinois Department of Commerce and Economic Opportunity (DCEO) Office of Coal Development (OCD). The DCEO / OCD administers the Illinois Coal Research Program and is dedicated to the development and use of Illinois' extensive coal resources as a fuel source for the 21st century. The main funding cycle of the ICCI is through the annual Request for Proposals (RFP) solicitation. The RFP solicits proposals for research projects up to the commercialization stage. The program provides partial funding to research clean coal technology, coal chemistry, mining productivity and coal combustion byproduct utilization. Subgrants are typically between $60,000 and $250,000 and are issued by Southern Illinois University-Carbondale. Grant periods are limited to 24 months. The annual program cycle begins in April of each year with a Request for Proposals issued by the ICCI. Proposal review and project selection is typically completed by the end of August. Unsolicited proposals may be submitted at any time. All projects under this solicitation are capped at a maximum of $200,000. Project funds requested near the upper range should involve pilot-scale work or other commercial demonstrations, with extensive cost sharing by industrial entities and research partners. ICCI funds may be requested as cost sharing for larger-scale demonstration projects with an Illinois coal emphasis. Allocation of ICCI funds to such projects is of special interest.

Requirements: Any entity may apply. Proposals undergo a rigorous peer review. Preference is given to Illinois applicants. Researchers are encouraged to meet with ICCI staff at any time to discuss concepts that may have the potential to further the Illinois coal industry. If the concept appears sound and has the potential to significantly benefit the Illinois coal industry, the ICCI staff will assist the group in the formulation and submission of a Statement of Interest (SOI) document rather than submitting a formal proposal. This document will be sent to researchers after initial discussions with the ICCI. The SOI does not contain all the detail required in a formal proposal, and the scope of the SOI can be as narrow or broad as desired. Another difference between the SOI and a formal proposal is that the SOI allows communications between the ICCI and the prospective proposer prior to preparation of a formal proposal. Usually, the SOI will be between two and four pages. After review of the SOI, the applicant will be notified in writing as to whether or not their project was selected for further consideration. If reviews of the SOI indicate that a formal proposal should be submitted, the proposer will be guided by the ICCI staff to submit an unsolicited proposal. Unsolicited proposals use a similar format as the RFP. Although this program has a coal research emphasis, there are numerous opportunities that do not require experience in hands-on coal research, such as carbon management, materials, catalysis, fundamental and applied chemistry, business practices, mining, and engineering studies.

Restrictions: Proposals dealing with the following types of R&D are not of interest: any projects involving the use, classification, and upgrading of coal combustion residues (CCR's). This includes the recovery of special materials or elements from the ash; any projects involving plasma gasification, either as a stand-alone technology or as a supplemental technology; laboratory-scale and small pilot-scale research on CO2 capture and/or separation; or, any projects focusing on fuel cell research. Proposals dealing with these subjects will be deemed nonresponsive and will not be reviewed or considered.

Geographic Focus: Illinois
Amount of Grant: Up to 250,000 USD
Contact: Dr. Francois Botha, (618) 985-3500
Internet: http://www.icci.org/funding.php
Sponsor: Illinois Department of Commerce and Economic Opportunity
500 East Monroe Street
Springfield, IL 62701-1643

Illinois DCEO Coal Revival Grants **2360**
The Illinois Coal Revival Program provides financial assistance in the form of grants to assist with the development of new, coal-fired electric generation capacity or coal gasification facilities in Illinois. Financial assistance through the program will be provided in the form of a grant based on State Retail Occupation Taxes that will be paid on Illinois coal purchases for new facilities. Qualifying facilities may be eligible for grants roughly equal to the present value of future sales taxes paid on Illinois-mined coal over a 25-year period, up to a maximum amount of $100 million. Unsolicited proposals may be submitted at any time during the year. Funding of selected proposals may be substantially delayed due to the need for legislative action to establish a funding appropriation.
Requirements: Businesses must propose to construct a new electric generating or expand at an existing electric generating facility, including transmission lines and associated equipment to provide baseload electric power or to construct a coal gasification facility. The proposed new facility or facility expansion must: 1) have an aggregate nameplate generating capacity of 400 megawatts or more for all units at one site, will use coal or gases derived from coal as its primary fuel source at the proposed facility, and will support the creation of at least 150 new Illinois coal-mining jobs, or 2) use coal gasification or IGCC to generate chemical feedstocks, transportation fuels or electricity.
Geographic Focus: Illinois
Amount of Grant: Up to 100,000,000 USD
Contact: Dan Wheeler, (217) 785-1671; fax (217) 558-2647; Dan.Wheeler@illinois.gov
Internet: http://www.illinoisbiz.biz/dceo/Bureaus/Coal/Grants/
Sponsor: Illinois Department of Commerce and Economic Opportunity
500 East Monroe Street
Springfield, IL 62701-1643

Illinois DNR Youth Recreation Corps Grants **2361**
The Illinois Youth Recreation Corps was established for making grants to local sponsors to provide wages to youth operating and instructing in recreational and conservation programs for the benefit of other youth. Such programs shall provide recreational opportunities for children of all age levels and shall include, but not be limited to, the coordination and teaching of physical activities, arts and handicrafts, and learning activities.
Requirements: Potential applicants include local sponsors who can provide necessary facilities, materials and management for summer recreational and conservation activities for youth within the community and who desire a grant for hiring eligible youth as supervisors, instructors, instructional aides or maintenance personnel. Local sponsors must be units of local government or not?for?profit entities. Enrollment (hiring) is limited to citizens of the State of Illinois who, at the time of hiring, are 16, 17, 18 or 19 years of age, and who have skills that can be used in the summer recreation or conservation program.
Restrictions: Youth currently employed in any manner by the local sponsor are not eligible for inclusion in the program.
Geographic Focus: Illinois
Date(s) Application is Due: Jun 30
Contact: Valerie Keener, Administrator; (217) 524-4126 or (217) 782-7481; fax (217) 782-9599; valerie.keener@illinois.gov or dnr.grants@illinois.gov
Internet: http://www.dnr.illinois.gov/grants/Pages/default.aspx
Sponsor: Illinois Department of Natural Resources
One Natural Resources Way
Springfield, IL 62702-1271

IMLS 21st Century Museum Professionals Grants **2362**
The 21st Century Museum Professionals program supports a range of activities, including professional training in all areas of museum operations and leadership development. This program provides the museum community with support for a variety of training and personnel development activities for museum staff members across all types of museums, as well as the collection and dissemination of information to museum professionals and the public. Project design could include direct dissemination of information through workshops, seminars, and courses, or indirect communication through publications and Web sites. Projects should benefit multiple institutions or diverse constituencies.
Requirements: Museums that fulfill the Eligibility Criteria for Museums may apply. Private not-for-profit museum services organizations or associations that engage in activities designed to advance the well-being of museums and the museum profession also may apply. In addition, institutions of higher education, including public and not-for-profit universities, are eligible. Please see Program Guidelines for specific eligibility criteria.
Geographic Focus: All States
Date(s) Application is Due: Mar 15
Amount of Grant: 15,000-500,000 USD
Contact: Twinet G. Kimbrough; (202) 653-4703; tkimbrough@imls.gov
Internet: http://www.imls.gov/applicants/grants/21centuryMuseums.shtm
Sponsor: Institute of Museum and Library Services
1800 M Street NW, 9th Floor
Washington, DC 20036-5802

IMLS Conservation Project Support Grants **2363**
The grants provide matching funds to help museums identify conservation needs and priorities and perform activities to ensure the safekeeping of their collections. Projects may assess four types of collections: nonliving, systematics/natural history, living plants, and living animals. Grants are available for different types of conservation projects, such as environmental improvements, research, surveys and long-range plans, training, and treatment. Projects may not exceed two years. IMLS will pay up to one-half the cost of the project. A museum's matching funds must come from cash contributions, earned income, in-kind contributions, or non-federal funds. This program also funds Exceptional Projects. These are projects with broad-reaching effects that benefit multiple institutions. The deadline apples to programs for museums in counties of Alabama, Florida, Louisiana, and Mississippi that have been declared disaster areas by the federal government.
Requirements: Museums located in one of the 50 states, the Commonwealth of Puerto Rico, American Samoa, the Virgin Islands, the Northern Mariana Islands, the Trust Territory of the Pacific Islands, Guam, or the District of Columbia may apply. Applicant organizations must have provided museum services, including exhibiting objects to the general public, on a regular basis for at least two years prior to application. A public or private nonprofit organization such as a municipality, college, or university that is responsible for the operation of a museum may, if necessary, apply on behalf of the museum.
Restrictions: A museum operated by a department or agency of the federal government is not eligible to apply. Funds may not be used for the acquisition of objects, contributions to endowment funds, costs of social activities, major renovations, new construction, or preaward costs.
Geographic Focus: All States
Date(s) Application is Due: Oct 1
Amount of Grant: 1-150,000 USD
Contact: Mark Feitl, Program Specialist; (202) 653-4635; mfeitl@imls.gov
Internet: http://www.imls.gov/applicants/grants/conservProject.shtm
Sponsor: Institute of Museum and Library Services
1800 M Street NW, 9th Floor
Washington, DC 20036-5802

IMLS Grants to State Library Administrative Agencies **2364**
Through the program, the Institute of Museum and Library Services provides funds to State Library Administrative Agencies (SLAAs) using a population-based formula. State libraries may use the appropriation to support statewide initiatives and services. They also may distribute the funds through subgrant competitions or cooperative agreements to public, academic, research, school, and special libraries in their state.
Requirements: State library administrative agencies located in one of the 50 states of the United States, the District of Columbia, the Commonwealth of Puerto Rico, Guam, American Samoa, the U.S. Virgin Islands, the Commonwealth of the Northern Mariana Islands, the Republic of the Marshall Islands, the Federated States of Micronesia, and the Republic of Palau are eligible to submit five-year plans.
Geographic Focus: All States
Date(s) Application is Due: Apr 1
Contact: Laurie C. Brooks; (202) 653-4650; stateprograms@imls.gov
Internet: http://www.imls.gov/programs/programs.shtm
Sponsor: Institute of Museum and Library Services
1800 M Street NW, 9th Floor
Washington, DC 20036-5802

IMLS Laura Bush 21st Century Librarian Grants **2365**
The program supports efforts to recruit and educate the next generation of librarians and the faculty who will prepare them for careers in library science. It also supports grants for research related to library education and library staffing needs, curriculum development, and continuing education and training. Program priorities are masters-level programs, doctoral programs, pre-professional programs, research, programs to build institutional capacity, and continuing education and training. The grant period is up to three years, except doctoral programs, which may be up to four years. Applications, guidelines, and examples of successful proposals are available online.
Requirements: All applicants for the Laura Bush 21st Century Librarian program are required to file their applications, including all attachments, online through Grants.gov. All types of libraries, except federal and for-profit libraries, may apply. Eligible libraries include public, school, academic, special, private (not-for-profit), archives, library agencies, library consortia, and library associations. In addition, research libraries that give the public access to services and materials suitable for scholarly research not otherwise available to the public and that are not part of a university or college are eligible. Institutions of higher education, including public and not-for-profit universities and colleges, also are eligible. Graduate schools of library and information science may apply as part of an institution of higher education.
Geographic Focus: All States
Date(s) Application is Due: Dec 15
Contact: Kevin Cherry, Senior Program Officer; (202) 653-4662; kcherry@imls.gov
Internet: http://www.imls.gov/applicants/grants/21centuryLibrarian.shtm
Sponsor: Institute of Museum and Library Services
1800 M Street NW, 9th Floor
Washington, DC 20036-5802

IMLS National Leadership Grants (NLG) **2366**
National Leadership Grants enable libraries and museums to help people gain the knowledge, skills, attitudes, behaviors, and resources that enhance their engagement in community, work, family, and society. Projects should enable libraries and museums to address current problems in creative ways, develop and test innovative solutions, and expand the boundaries within which cultural heritage institutions operate. The results

of these projects will help equip tomorrow's libraries and museums to better meet the needs of a Nation of Learners. Successful proposals will show evidence that they will have national impact and generate results-new tools, research, models, services, practices, or alliances-that can be widely adapted or replicated to extend the benefit of federal support. Proposals will reflect an understanding of current issues and needs, showing the potential for far-reaching impact throughout the museum and/or library community. Projects will provide creative solutions to issues of national importance and provide leadership for other organizations.
Requirements: All types of libraries, except federal and for-profit libraries, may apply. Eligible libraries include public, school, academic, special, private (not-for-profit), archives, library agencies, library consortia, and library associations. Research libraries and archives that give the public access to services and materials suitable for scholarly research not otherwise available and that are not part of a university or college also are eligible. Digital libraries that make library materials publicly available and provide services including selection, organization, description, reference, and preservation under the supervision of at least one permanent professional staff librarian are eligible to apply. Institutions of higher education, including public and not-for-profit universities and colleges, also are eligible. An academic unit, such as a graduate school of library and information science, may apply as part of an institution of higher education. Library applicants may apply individually or as partners. Additionally, all types of museums, large and small, are eligible for funding. Eligible museums include aquariums, arboreta and botanical gardens, art museums, youth museums, general museums, historic houses and sites, history museums, nature centers, natural history and anthropology museums, planetariums, science and technology centers, specialized museums, and zoological parks. Private nonprofit museum services organizations or associations that engage in activities designed to advance the well-being of museums and the museum profession also may apply. In addition, institutions of higher education, including public and nonprofit universities, are eligible.
Geographic Focus: All States
Date(s) Application is Due: Feb 1
Amount of Grant: 50,000-1,000,000 USD
Contact: Rachel Frick, Program Officer, Libraries; (202) 653-4667; rfrick@imls.gov
Internet: http://www.imls.gov/applicants/grants/nationalLeadership.shtm
Sponsor: Institute of Museum and Library Services
1800 M Street NW, 9th Floor
Washington, DC 20036-5802

Inamori Foundation Kyoto Prize **2367**
The Kyoto Prize is awarded annually to persons who have made significant contributions in the three categories of Advanced Technology, Basic Sciences, and Arts and Philosophy. Through this Prize, the and groups worldwide Foundation seeks not only to recognize outstanding achievements but also to promote academic and cultural development and to contribute to mutual international understanding.
Geographic Focus: All States, Japan
Samples: U.S. immunologist and geneticist Dr. Leonard A. Herzenberg, of Stanford, California; Japanese statistical mathematician Dr. Hirotugu Akaike, of Tokyo; and Japanese designer Issey Miyake, an artist whose innovative creations transcend time, culture and social status, Tokyo.
Contact: Program Director; (075) 353-7272; fax (075) 353-7270
Internet: http://www.inamori-f.or.jp/e_fd_out_out.html
Sponsor: Inamori Foundation
620 Suiginya-cho
Shimogyo-ku, Koyoto 600-8411 Japan

Inamori Foundation Research Grants **2368**
Through the Inamori Grants and other research funding, the Foundation supports a wide variety of activities by young researchers in Japan in the fields of natural sciences, humanities, and social sciences. It aims to cultivate and nurture human resourses who will contribute to the future of our global society, to promote academic and cultural development, and to contribute to international understanding. In the founder's own words, it is my sincere hope that the activities of the Foundation contribute to the progress and development of humankind and thereby allow me to repay, in some way, my indebtedness to all those in my local community, in my country, and throughout the world who have helped make me and my company what they are today.
Requirements: Must be a resident researcher living in Japan.
Geographic Focus: All States, Japan
Contact: Research Program Director; (075) 353-7272; fax (075) 353-7270
Internet: http://www.inamori-f.or.jp/e_fd_out_out.html
Sponsor: Inamori Foundation
620 Suiginya-cho
Shimogyo-ku, Koyoto 600-8411 Japan

Independent Institute John M. Templeton Fellowships Essay Contest **2369**
The Templeton Fellowships Essay Contest gives monetary awards to junior faculty and students who submit an essay based on a particular topic. Essays are submitted from all over the world. First price junior faculty are awarded $10,000, with students awarded first ($2,500), second ($1,500), and third ($1,000) prizes. Applicants should consult the website for current information, as fellowships are not available every year.
Requirements: College or university students who are part-time, full-time, undergraduate, graduate, and from all nations and academic disciplines are eligible. (Student must be enrolled on date of essay submission.) Students must be no older than 35 years. Junior faculty members must hold a position of assistant professor or higher, not yet tenured, and may not be older than 35 years.
Restrictions: Previous fellowship recipients, judges, and current or past institute employees are ineligible.
Geographic Focus: All States, All Countries
Amount of Grant: 1,000-10,000 USD
Contact: Carl Close, Academic Affairs Director; (510) 632-1366; fax (510) 568-6040
Internet: http://www.independent.org/students/essay/winners2010.asp
Sponsor: Independent Institute
100 Swan Way
Oakland, CA 94621-1428

Indiana 21st Century Research and Technology Fund Awards **2370**
The Indiana 21st Century Research and Technology Fund of the Indiana Economic Development Corporation (IEDC) is open to proposals from all public and private entities for technology-based commercialization activities encompassing science/technology creation, innovation, and transfer intended to have commercial impacts. The fund intends to increase the numbers, and rates of development, of new and expanding technology-based companies by funding promising opportunities that, in some cases, the financial markets might find too risky. The Fund makes awards in two broad categories: Science and Technology Commercialization and Centers of Excellence. In addition, the Fund provides cost-share on behalf of Federal proposals submitted by Indiana-based entities. Generally awards are made in multiples of $50,000 up to $2,000,000. Support for awards in excess of $2,000,000 will be rare.
Requirements: The IEDC defines a technology-based company as one that is involved in transferring advanced technology into products, developing technologies with the near-term intention of creating products, or using new or advanced technologies in its design, development, and/or manufacturing of products. The Fund emphasizes the creation of academic-sector-commercial-sector partnerships. In making awards, the Fund expects significant leverage from the partners involved in the projects. Important: before applying, contact Fund staff (email preferred) to discuss your interest in submitting a proposal and to discuss your technology and commercialization goals. While not a review criterion, the fund encourages the inclusion of interns from any academic institution, or participating commercial sector partner, in order to increase project-related involvement of students at all levels.
Restrictions: Only direct costs will be supported. Institutions will not be provided indirect (overhead) cost support. Entities with previous Fund awards that are not current with regard to financial or technical reporting requirements will be disqualified from making new submissions to the Fund. Resubmissions of previously declined proposals will be considered only if substantive changes have been made to the proposal. Fund staff will determine whether to review resubmissions.
Geographic Focus: Indiana
Amount of Grant: 500,000-2,000,000 USD
Contact: Carla Phelps, Financial Manager; (317) 233- 4336; cphelps@21fund.org
Internet: http://www.21fund.org/
Sponsor: Indiana Economic Development Corporation
1 North Capitol Avenue, Suite 900
Indianapolis, IN 46204

Indiana SBIR/STTR Commercialization Enhancement Program (ISCEP) **2371**
ISCEP provides funds to enhance commercialization activities of Indiana-based SBIR/STTR (Small Business Innovation Research/Small Business Technology Transfer) awardees. Specific ISCEP goals include: Support of thoughtfully structured commercialization plans of SBIR/STTR Phase II awardees; Accelerate and enhance commercial impacts of SBIR/STTR technologies, and; Establish and enhance successful technology-based businesses in Indiana. Proposals will be accepted from a small business that has received a federal Phase II SBIR or STTR and at least 50% of the federal program dollars must be expended prior to submission of a proposal in response to the RFP. Review considers both the technology development stage and the related business plan. The final stage of review involves a presentation to the IEDC. Between two and four awards of up to $350,000 may be made per funding cycle. The awards will be made with performance periods of up to 2 years and are contingent on the availability of funds.
Requirements: Eligible applicants must have a principal place of business in Indiana and the benefits from commercialization must accrue to an Indiana small business. Applicants must be a small business that has received a federal Phase II SBIR or STTR and at least 50% of the federal program dollars must be expended prior to submission of a proposal in response to the RFP, because an essential metric used in judging suitability for an award will involve assessing current progress toward achievement of Phase II objectives. The technology being commercialized must be directly related to the technology funded under the SBIR or STTR Phase II award. There is no limit as to the number of proposals that a single company can submit. However, only one award per company will be made per RFP cycle. The RFP, proposal tips and application are available at the website.
Restrictions: Awardees must maintain a principal place of business in Indiana for a term of 10 years beginning on the Effective Date of the agreement. Otherwise forfeiture and repayment the award plus interest will be required. Successful applicants must adhere to specific reporting requirements, including: quarterly progress reports, annual reports, annual site visits and a final report.
Geographic Focus: Indiana
Date(s) Application is Due: Jul 20
Amount of Grant: Up to 350,000 USD
Contact: Brooke Pyne, Program Director; (812) 384-3078 or (812) 381-0350; fax (812) 384-3487; bpyne@iedc.in.gov
Internet: http://www.in.gov/iedc/sbir/index.html
Sponsor: Indiana Economic Development Corporation
SBIR/STTR Program Office, 32 E Main Street
Bloomfield, IN 47424

Indiana Space Grant Consortium Discovery Program Crosscutting Research Team Grants 2372

The mission of the Indiana Space Grant Consortium is to promote science, technology, engineering and mathematics (STEM) workforce development, formal and informal education, and research by the dissemination of NASA related activities, content, and opportunities to the residents of the State of Indiana. The Vision for Space Exploration lays out a plan of action to support this mission. As NASA moves forward to develop the science and technology to fully implement the Vision, the Discovery Program for Research offers a means for INSGC members to address that broad mission. These funding opportunities seek to expand the realm of knowledge through discovery of new ideas and new techniques. These specific awards are intended to provide funds for developing new synergies for research collaboration, or a significant inter-institutional/ inter-organizational funding initiative meeting INSGC and NASA priorities. Strong student participation in STEM workforce development type activities is encouraged. Amount of up to $25,000 are available for multiple institutions, and up to $15,000 for multiple departments / units at a single institution. The recommended match is 1.5 to 1 for all-academic partnerships, and 2 to 1 if at least one partner is from industry.
Requirements: Faculty and students receiving NASA/INSGC funds for direct support (salary or travel) must be U.S. citizens. At least two participating organizations must be named with cost share amounts coming from each of the participants.
Geographic Focus: Indiana
Date(s) Application is Due: Feb 13
Amount of Grant: Up to 25,000 USD
Contact: Dr. Barrett Caldwell, Director; (765) 494-5873; fax (765) 496-3449; bscaldwell@purdue.edu or director@insgc.org
Internet: http://www.insgc.org/?q=node/108
Sponsor: Indiana Space Grant Consortium
203 S Martin Jischke Drive, Gerald D. and Edna E Mann Hall, Room 160
West Lafayette, IN 47907-1971

Indiana Space Grant Consortium Discovery Program Research Grants 2373

The mission of the Indiana Space Grant Consortium is to promote science, technology, engineering and mathematics (STEM) workforce development, formal and informal education, and research by the dissemination of NASA related activities, content, and opportunities to the residents of the State of Indiana. The Vision for Space Exploration lays out a plan of action to support this mission. As NASA moves forward to develop the science and technology to fully implement the Vision, the Discovery Program for Research offers a means for INSGC members to address that broad mission. These funding opportunities seek to expand the realm of knowledge through discovery of new ideas and new techniques. These awards are intended to support early career faculty (less than 6 years experience) members as they establish their research agenda on the national level, or for faculty with established careers attempting to develop new lines for research prior to requesting external funding. For these research grants, the amount available for early career faculty and research initiation is up to $10,000; faculty salary is limited to $3,000, with a match of 1:1 required.
Requirements: Faculty and students receiving NASA/INSGC funds for direct support (salary or travel) must be U.S. Citizens. Significant opportunities for student research experiences are required, and must be described prominently in the proposal.
Geographic Focus: Indiana
Date(s) Application is Due: Feb 13
Amount of Grant: Up to 10,000 USD
Contact: Dr. Barrett Caldwell, Director; (765) 494-5873; fax (765) 496-3449; bscaldwell@purdue.edu or director@insgc.org
Internet: http://www.insgc.org/?q=node/108
Sponsor: Indiana Space Grant Consortium
203 S Martin Jischke Drive, Gerald D. and Edna E Mann Hall, Room 160
West Lafayette, IN 47907-1971

Indiana Space Grant Consortium Discovery Program Student Hands-On Research Grants 2374

The mission of the Indiana Space Grant Consortium is to promote science, technology, engineering and mathematics (STEM) workforce development, formal and informal education, and research by the dissemination of NASA related activities, content, and opportunities to the residents of the State of Indiana. The Vision for Space Exploration lays out a plan of action to support this mission. As NASA moves forward to develop the science and technology to fully implement the Vision, the Discovery Program for Research offers a means for INSGC members to address that broad mission. These funding opportunities seek to expand the realm of knowledge through discovery of new ideas and new techniques. These types of project support are geared toward workforce development, with opportunities supporting faculty and student research by providing assistance with funded activities. Up to $3,500 per semester is available, including summer semester.
Requirements: A minimum 1 to 1 match is required. A higher match or leverage is recommended, particularly if there is support from ongoing sponsored research or contract projects. All students receiving NASA/INSGC funds for direct support (salary or travel) must be U.S. Citizens. Student activities must be research driven, and must focus on actual research rather than clerical or janitorial support. A student must be the project planner and active researcher.
Geographic Focus: Indiana
Date(s) Application is Due: Feb 13
Amount of Grant: Up to 3,500 USD
Contact: Dr. Barrett Caldwell, Director; (765) 494-5873; fax (765) 496-3449; bscaldwell@purdue.edu or director@insgc.org
Internet: http://www.insgc.org/?q=node/108
Sponsor: Indiana Space Grant Consortium
203 S Martin Jischke Drive, Gerald D. and Edna E Mann Hall, Room 160
West Lafayette, IN 47907-1971

Indiana Space Grant Consortium Fellowships 2375

The mission of the Indiana Space Grant Consortium is to promote science, technology, engineering and mathematics (STEM) workforce development, formal and informal education, and research by the dissemination of NASA related activities, content, and opportunities to the residents of the State of Indiana. Available fellowships include: doctoral fellowships in Science Education of $15,000; doctoral fellowships in science and engineering disciplines of $15,000; and MA/MS fellowships of $7,500. Funds for fellowships may be used to offset costs for project expenses or student stipends. It is the applicant's responsibility to work with the institution to make the necessary arrangement for the type of funds being sought. INSGC Fellowships do not include tuition or fee remission. If the student receives a fee remission or any other form or support from their institution as part of their fellowship award, that is considered matching funds. Doctoral and masters level students may receive an award for up to two consecutive years, but must reapply separately for each year of fellowship support.
Requirements: To be eligible for a fellowship, an applicant must be: a U.S. citizen; enrolled in good standing as an active full-time undergraduate student at an INSGC affiliate institution throughout the scholarship period; and major in a science, technology, engineering, or mathematics (STEM) discipline or STEM education (or a first-year program leading into a STEM discipline).
Restrictions: Indirect costs may not be charged to fellowships or scholarships. Administrative costs associated with fellowship and scholarship programs may be provided as matching funds by the institution.
Geographic Focus: Indiana
Date(s) Application is Due: Feb 13
Amount of Grant: 7,500-15,000 USD
Contact: Dr. Barrett Caldwell, Director; (765) 494-5873; fax (765) 496-3449; bscaldwell@purdue.edu or director@insgc.org
Internet: http://www.insgc.org/?q=node/109
Sponsor: Indiana Space Grant Consortium
203 S Martin Jischke Drive, Gerald D. and Edna E Mann Hall, Room 160
West Lafayette, IN 47907-1971

Indiana Waste Tire Fund (WTF) Grants 2376

The Indiana Waste Tire Fund (WTF) Program was created to help Indiana businesses undertake research, development, and/or commercial manufacturing projects that develop markets for use of scrap tires. Funds are made available through grants of up to $100,000 with a 50% match requirement are available for innovative research and development projects. Key elements include project partnerships and technology transfer. Provisions will be made for confidentiality of proprietary information.
Requirements: Eligible applicants are limited to Indiana businesses and not-for-profits. Interested applicants must contact OPPTA (Office of Pollution Prevention and Technical Assistance) to receive a grant application. Activities that may be funded through the program include the following: Civil engineering applications using shredded tires; Asphalt-rubber applications using ground rubber; Development of advanced technology and processes for production of products from scrap tires and; Tire-derived-fuel (TDF).
Restrictions: Projects must take place within the state of Indiana.
Geographic Focus: Indiana
Amount of Grant: Up to 100,000 USD
Contact: Regional Grant Representative; (800) 988-7901; recycling@idem.in.gov
Internet: http://www.in.gov/recycle/funding/wtf.html
Sponsor: Indiana Department of Environmental Management
100 North Senate Avenue, MC 64-01
Indianapolis, IN 46204-2251

Indian Council of Social Science Research Grants 2377

The council funds research projects of scientists in the social sciences and sponsors projects on its own initiatives (e.g., research on the northeastern region of India, women's studies, and entrepreneurship). Types of support include research fellowships, contingency grants, grants for study and publications, and conferences and seminars. The council collaborates in research and exchange programs with a number of countries and evaluates research proposals submitted by foreign nationals intending to undertake research in India.
Geographic Focus: All States
Contact: Professor Andre Beteille, Chair; (0) 26179679; chairman@icssr.org
Internet: http://www.icssr.org/rp_main.htm
Sponsor: Indian Council of Social Science Research
35 Ferozeshah Road
New Delhi, 110 001 India

Infinity Foundation Grants 2378

The foundation awards grants in a broad array of areas including charitable, scientific, religious, educational, and holistic healing activities by organizations and public agencies that work to better the lives of people. Grants also support individuals' research and development of educational materials to improve the authenticity of the portrayal of Indic traditions in the educational system. Proposed projects could result in one or more of the following: books, curriculum development, articles, conferences, CD-Roms, digital

slide shows, Internet presentations, and audio/video materials. Topics covered may include philosophy, history, religion, science, art, and sociology, as they pertain to the educational curricula on Indic traditions. Proposals should be submitted by email.
Requirements: Grantee may be a scholar, teacher, visionary, or spiritual leader whose work in the designated topics would be enhanced by a foundation grant.
Geographic Focus: All States
Contact: Rajiv Malhotra; (609) 683-0548; fax (609) 683-0478; rm.infinity@gmail.com
Internet: http://www.infinityfoundation.com/callforgrantproposals.htm
Sponsor: Infinity Foundation
66 Witherspoon Street, Suite 400
Princeton, NJ 08542

Institute Award for Excellence in Industrial Gases Technology 2379
Sponsored by Praxair Inc and administered by AIChE, the award is given to recognize sustained excellence in the contribution to the advancement of technology for the production, distribution, and application of industrial gases. The recipient of the award shall have a record of sustained contribution recognized for its impact on advancing the frontier of industrial gases technology. The contribution can be characterized by a sustained record of important fundamental research, innovation, technological development, or the novel application of technology, either fostering or leading to important commercial results. Recipients need not be members of the institute.
Requirements: All members of AIChE as well as other interested persons are urged to nominate deserving candidates. All information and supporting documents are to be included with the completed form; nomination forms and instructions may be obtained from the institute.
Geographic Focus: All States
Date(s) Application is Due: Feb 15
Amount of Grant: 3,500 USD
Contact: Programs Director; (212) 591-7107; fax (212) 591-8890; awards@aiche.org
Internet: http://www.aiche.org/awards/awarddtl.asp?AwardID=59
Sponsor: American Institute of Chemical Engineers
3 Park Avenue
New York, NY 10016

Institute for Advanced Studies in Culture Postdoctoral Fellowships 2380
The Institute supports research addressing the complex and dynamic relationship between religion and democratic ideals, institutions, and practices. Residential and nonresidential fellowships are available. Application materials and information are available online.
Requirements: Applicants may be from any discipline, but must be working on a project concerned with religion and public life, and must have defended their dissertation by May 1.
Geographic Focus: All States
Date(s) Application is Due: Jan 15
Amount of Grant: 35,000 USD
Contact: James Davison Hunter; (434) 924-7705; iasc@virginia.edu
Internet: http://www.virginia.edu/iasc/programs.html
Sponsor: University of Virginia
P.O. Box 400816
Charlottesville, VA 22904-4816

Institute for Agriculture and Trade Policy Food and Society Fellowships 2381
The IATP Food and Society Fellows Program provides fellowships to professionals in food and agriculture from across North America, enabling them to use mass media channels to inform and shape the public agenda. The goal of the program is to create sustainable food systems that promote good health, vibrant communities, environmental stewardship, worker justice and accessibility for all. Fellows come from many disciplines: chefs, farmers, nutritionists, activists, public health professionals, fishers, policy experts and academics. Together they form an interdisciplinary team that works to: use communication to influence the issues that reach the public agenda, thereby creating policy changes at the personal, organizational and public policy levels that advance sustainable food and farming systems; increase the mass media communications on issues around sustainable food and farming systems that produce healthy, green, fair and affordable foods; raise the profile of the fellows as food system experts among media and policymakers; and build capacity, leadership, and cohesiveness in a group of experts who collaborate and communicate using mass media channels to bring sustainable food system issues to the public agenda.
Requirements: Applicants should be U.S. residents actively involved in a professional career that involves two or more of the following: agriculture, health promotion, youth development, food production, and/or policy analysis.
Geographic Focus: All States
Contact: Mark Muller; (612) 870-3420; fax (612) 870-4846; mmuller@iatp.org
Internet: http://www.foodandsocietyfellows.org/about/page/about-us
Sponsor: Institute for Agriculture and Trade Policy
2105 First Avenue South
Minneapolis, MN 55404

Institute for Palestine Studies, Publishing, and Research Organization Grants 2382
The institute operates in the United States, Lebanon, France, and Jerusalem and awards research grants on all aspects of the Palestine problem, including the history and development of Palestine, the Palestinian problem and the Arab-Israeli conflict, and possible ways of arriving at a peaceful resolution.
Geographic Focus: All States
Contact: Linda Butler; (202) 342-3990; fax (202) 342-3927; ipsdc@palestine-studies.org
Internet: http://www.palestine-studies.org/final/en/intro/research.php
Sponsor: Institute for Palestine Studies, Publishing, and Research Organization
3501 M Street NW
Washington, DC 20007

Institute of American Cultures Graduate and Predoctoral Fellowships 2383
The UCLA Center for Asian American Studies, American Indian Studies Center, Center for African American Studies, and the Chicano Studies Research Center, in conjunction with the Institute of American Cultures, offer a limited number of predoctoral fellowships to support scholarly work in relevant ethnic studies. The awards are for one year. Reapplication for a second year of support is permitted. The deadline for African American and Chicana/o studies is January 13 and the deadline for American Indian and Asian American studies is March 30. On occasion, centers have chosen to divide the graduate fellowship between two or among three students. Application and guidelines are available online.
Requirements: The fellow must be a full-time UCLA graduate student and make a contribution to the ongoing activities of the center. Application for the fellowship in African American Studies is open only to doctoral students who will have advanced to candidacy by the beginning of the fellowship year.
Geographic Focus: All States
Date(s) Application is Due: Jan 13; Mar 30
Contact: Program Contact; (310) 825-1233; IACcoordinator@gdnet.ucla.edu
Internet: http://www.gdnet.ucla.edu/iacweb/preweber.htm
Sponsor: Institute of American Cultures
1237 Murphy Hall, Box 951419
Los Angeles, CA 90095-1419

Institute of American Cultures Postdoctoral and Visiting Scholar Fellowships 2384
The UCLA Center for Asian American Studies, American Indian Studies Center, Center for African American Studies, and the Chicano Studies Research Center, in conjunction with the Institute of American Cultures, offers a limited number of postdoctoral fellowships. Currently, the IAC, in cooperation with UCLA's four Ethnic Studies Research Centers and the UCLA Library's Center for Oral History Research offers four postdoctoral fellowships (one per Center) that focus on conducting and theorizing oral history research in African American, American Indian, Asian American, and Chicano communities. Awards will depend on experience for those who have recently received a PhD. Senior scholars will receive a supplement to their sabbatical salaries to support work in ethnic studies. Application and guidelines are available online.
Requirements: Applicants must be citizens or permanent residents of the United States and hold a PhD from an accredited college/university (or, in the case of the arts, a terminal degree) in the appropriate field at the time of appointment. Fellows must teach or do research in the programs of the center.
Restrictions: UCLA faculty, staff, and currently enrolled students are not eligible to apply.
Geographic Focus: All States
Date(s) Application is Due: Jan 13
Amount of Grant: 36,000 USD
Contact: IAC Coordinator; (310) 825-1233; iaccoordinator@gdnet.ucla.edu
Internet: http://www.gdnet.ucla.edu/iacweb/pstweber.htm
Sponsor: UCLA Institute of American Cultures
1237 Murphy Hall, Box 951419
Los Angeles, CA 90095-1419

Institute of Global Conflict and Cooperation Research, Conference, and Teaching Grants 2385
Institute of Global Conflict and Cooperation (IGCC) research and teaching grants are awarded to University of California (UC) faculty to stimulate independent research and education projects on international issues of contemporary importance. Grants support research and teaching on the nine UC campuses on the causes of international conflict and opportunities to promote international cooperation. IGCC research, research conference, and teaching grants are designed to stimulate independent research and education projects on international issues of contemporary importance. Topics of interest are posted on the Web site, as are application procedures.
Requirements: Applications are accepted from UC faculty in all disciplines. Collaborative work is encouraged; however, the principal investigator must have an integral role in the project. Collaborative projects between UC faculty and the scientific and the technical staff of the national laboratories are eligible. Research projects may be either theoretical or policy analytical.
Geographic Focus: All States
Date(s) Application is Due: Mar 7
Amount of Grant: Up to 35,000 USD
Contact: Kim Newin; (858) 534-8602; ktnewin@ucsd.edu
Internet: http://www-igcc.ucsd.edu/cprograms/fac_grant_menu.php
Sponsor: University of California, San Diego
9500 Gilman Drive, Department 0518
La Jolla, CA 92093-0518

Institute of Noetic Sciences Research Internships 2386
As part of its objective to advance the science of consciousness, IONS is dedicated to training emerging noetic scientists. IONS hosts a limited number of research, education, and communications interns each year. The Institute's interns have been high school and college students wishing to gain research and nonprofit experience, graduate and post-graduate students who engage in more formal training experiences, and those returning to the workforce or making a career change. Internships are unpaid, and interns cannot live on campus.
Geographic Focus: All States
Contact: Cassandra Vieten, Director of Research; (707) 779-8211 or (707) 779-8212; cvieten@noetic.org or research@noetic.org
Internet: http://www.noetic.org/research/participate/internships/

Sponsor: Institute of Noetic Sciences
625 2nd Street, Suite 200
Petaluma, CA 94952-5120

Institute of Paper Science and Technology Fellowships 2387

Fellowships are awarded in the interdisciplinary program in chemistry, chemical engineering, mechanical engineering, physics, biology, and mathematics leading to the MS and PhD degrees. Alumni have obtained responsible positions in research, development, production, and management in the pulp and paper industry and in the chemical and related industries, with a small but increasing number in advanced education.
Requirements: Applicants generally have BA or BS degrees in chemistry, paper science, chemical engineering, or mechanical engineering, but well-qualified individuals with degrees in other physical sciences and engineering areas will be considered. Citizens or permanent legal residents of the United States, Canada, or Mexico enrolled at the institute are eligible.
Geographic Focus: All States
Amount of Grant: 16,000-18,000 USD
Contact: Project Manager; (800) 558-6611; energy.challenge@ipst.gatech.edu
Internet: http://www.ipst.gatech.edu/energy_challenge/website/challenger/ipst.html
Sponsor: Institute of Paper Science and Technology
500 10th Street NW
Atlanta, GA 30318

Institute of Textile Technology Graduate Fellowships 2388

Up to 10 students will each be awarded a tuition scholarship and a nonservice fellowship. Each ITT fellowship provides the student with a stipend for the nine months of the academic year. The Academic Committee will decide which students receive these financial awards. Students also may apply for Federal aid in the form of student loans. Fellowships require no service, such as assistance with laboratory work or teaching, and are continued for the second year, provided the student's performance during the first year is satisfactory. All students are expected to maintain a minimum average of B in the course work.
Requirements: Fellowships are intended for U.S. citizens holding bachelor's degrees and interested in careers in the textile industry.
Restrictions: If at the end of the first year the required average is not attained, the fellowship cannot be renewed for the second year.
Geographic Focus: All States
Date(s) Application is Due: Mar 15
Amount of Grant: 12,000 USD
Contact: Program Officer; (919) 513-7583; fax (919) 882-9410
Internet: http://www.itt.edu/Education/FinancialAid.asp
Sponsor: Institute of Textile Technology
2401 Research Drive, Box 8301
Raleigh, NC 27695-8301

Institute of Turkish Studies Dissertation Writing Grants 2389

The institute is devoted solely to the advancement of training, research, and teaching in the field of Ottoman and modern Turkish studies. These grants are intended for advanced students who have finished the research stage of their dissertation, and they may not be used for dissertation research. The dissertation writing grants will be awarded only to applicants who certify that they will not be involved in teaching beyond the half-time level. Application and guidelines are available online.
Requirements: Graduate students in any field of the social sciences and/or humanities, who are U.S. citizens or permanent residents at the time of application; currently enrolled in a PhD degree program in the United States; and expecting to complete all PhD requirements except their dissertation by March are eligible.
Geographic Focus: All States
Contact: David Cuthell, (202) 687-0295; fax (202) 687-3780; dcc@turkishstudies.org
Internet: http://www.turkishstudies.org/grand_app.html
Sponsor: Institute of Turkish Studies
Intercultural Center, Box 571033-Room 305R
Washington, DC 20057-1033

Institute of Turkish Studies Research Grants in Comparative Studies of Modern Turkey 2390

The program supports and encourages the development of research, scholarship, and learning in the field of Turkish Studies in the United States. This grant is for graduate students at the dissertation research stage and for post-doctoral scholars in the U.S. who study aspects of the Republic of Turkey (post-1922) in a comparative context. A significant portion of the project should be devoted comparatively to one or more states or political entities in Europe, Latin America, the Middle East, and Asia in addition to the Turkish Republic. These grants are primarily, but not exclusively, in the field of Political Science. Applications and further information are available online.
Requirements: Applicants must be U.S. citizens or permanent residents in the U.S. and affiliated with a university in the U.S at the time of the application.
Geographic Focus: All States
Amount of Grant: 10,000 USD
Contact: David Cuthell, (202) 687-0295; fax (202) 687-3780; dcc@turkishstudies.org
Internet: http://www.turkishstudies.org/grand_app.html
Sponsor: Institute of Turkish Studies
Intercultural Center, Box 571033-Room 305R
Washington, DC 20057-1033

Institute of Turkish Studies Summer Language Study Grants in Turkey 2391

These grants are for summer travel to Turkey for language study in preparation for graduate research. The time period for these grants varies with each individual application. Normally, recipients are expected to spend a minimum of two months in Turkey at an established Ottoman or Turkish language training facility. Application guidelines are available online.
Requirements: Applicants must be graduate students in any field of the social sciences and/or humanities; U.S. citizens or permanent residents at the time of the application; and currently enrolled in a university in the United States.
Geographic Focus: All States
Date(s) Application is Due: Mar 11
Amount of Grant: 1,000-2,000 USD
Contact: David Cuthell, (202) 687-0295; fax (202) 687-3780; dcc@turkishstudies.org
Internet: http://www.turkishstudies.org/grantsprogram.html
Sponsor: Institute of Turkish Studies
Intercultural Center, Box 571033-Room 305R
Washington, DC 20057-1033

Institute of Turkish Studies Summer Research Grants in Turkey 2392

These grants are for summer travel to Turkey to carry out projects. The time period for these grants varies with each individual application. Normally, grant recipients are expected to spend a minimum of two months in Turkey. Application and guidelines are available online.
Requirements: Applicants must be graduate students in any field of the social sciences and/or humanities in the United States; U.S. citizens or permanent residents at the time of the application; and currently not engaged in dissertation writing.
Geographic Focus: All States
Amount of Grant: 1,000-2,000 USD
Contact: David Cuthell, (202) 687-0295; fax (202) 687-3780; dcc@turkishstudies.org
Internet: http://www.turkishstudies.org/grand_app.html
Sponsor: Institute of Turkish Studies
Intercultural Center, Box 571033-Room 305R
Washington, DC 20057-1033

Intel Bachelor's and Master's Degree Internships 2393

Several Intel manufacturing facilities, in cooperation with participating universities, offer paid internships for students enrolled full-time in specific undergraduate and graduate-level electronics and technology programs. These internships offer hands-on technical education and experience and valuable networking opportunities. Undergraduate and graduate level internships are available at all our major U.S. sites. One of the great advantages of internships is the flexible scheduling, designed to work around the typical school schedule. Interns receive: competitive salaries; vacation and holiday time; access to Intel University classes for professional and personal development; networking with Intel managers and executives; relocation assistance; and earned credit toward sabbaticals. Application details are available through the career centers of participating colleges.
Requirements: In order to be eligible applicants must: be enrolled in bachelor's, master's, or doctoral programs in engineering, science, and business-related fields with a 3.0 or higher GPA on a 4.0 scale.
Geographic Focus: All States
Contact: Internship Coordinator; (408) 765-8080; fax (408) 765-9904
Internet: http://www.intel.com/jobs/usa/students/internships/degree.htm
Sponsor: Intel Corporation
2200 Mission College Boulevard
Santa Clara, CA 95054-1549

Intel Foundation PhD Fellowships 2394

The Intel Foundation PhD Fellowship Program awards fellowships to PhD candidates doing work in fields related to Intel's business and research interests. These fellowships, only available at select U.S. universities, include a cash award (tuition/fees/stipend), an Intel PC, introduction to an Intel Mentor, and the opportunity to conduct research or an internship at Intel. Approximately 35 fellowships are awarded annually. In addition, Intel Foundation collaborates with The National Consortium for Graduate Degrees for Minorities in Engineering and Science (GEM) to help support graduate education for underrepresented minorities at the master's and PhD levels. Contact your school's engineering or computer science department to receive additional information.
Geographic Focus: All States
Date(s) Application is Due: Nov 15
Contact: Fellowship Coordinator; (408) 765-8080; fax (408) 765-9904
Internet: http://www.intel.com/jobs/usa/students/internships/fellowship.htm
Sponsor: Intel Corporation Foundation
2200 Mission College Boulevard
Santa Clara, CA 95054-1549

Intel Technician Internships 2395

Several Intel manufacturing facilities, in cooperation with participating colleges, offer paid internships for students enrolled full-time in specific two-year electronics and technology programs. These internships offer hands-on technical education and experience and valuable networking opportunities. There are a limited number of internships available, so early application is encouraged. One of the great advantages of technician internships is the flexible scheduling, designed to work around your school schedule. The Compressed Work Week offers a three or four day week that gives an Intern time for all other interests and activities. Interns receive: competitive salaries; vacation and holiday time; access to Intel University classes for professional and personal development; networking with Intel managers and executives; relocation assistance; and earned credit toward sabbaticals. Application details are available through the career centers of participating colleges.

Requirements: In order to be eligible for Intel's technical college internship, applicants must: have declared in a specific full-time degree program such as semiconductor manufacturing, electronics technology or automation, etc. at a participating college; maintain a 3.0 or higher GPA on a 4.0 scale; be able to work a minimum number of hours as set by their Intel manager; and be enrolled in classes and making significant progress toward a degree.
Geographic Focus: All States
Contact: Internship Coordinator; (408) 765-8080; fax (408) 765-9904
Internet: http://www.intel.com/jobs/usa/students/internships/technician.htm
Sponsor: Intel Corporation
2200 Mission College Boulevard
Santa Clara, CA 95054 1549

International Crane Foundation Internships 2396
This is a formal stipend program with internships being offered in the fields of aviculture and restoration. In the aviculture program, interns will receive intensive hands-on training in the care and management of endangered cranes, including husbandry, handling techniques, behavior, stimulating reproduction, incubation, chick rearing, artificial insemination, health care, and genetic management. The associate also manages the volunteer chick parent program. Each intern is responsible for developing and completing a research project with assistance from the staff. Internships usually run from March through May, June through August, September through November, and December through February. Application deadlines are for spring, summer, fall, and winter internships. Under the restoration program, interns will be involved in the ecosystems restoration program, which directs the restoration of prairie, savanna, and wetland communities on the ICF property. Interns will develop important ecological, hydrological, botanical, and horticultural skills. Depending on the season, interns may be involved in prescribed burns, field studies of wild sandhill care populations; seed collecting, planting, and site preparation for community restorations; and interaction with foreign scientists. The associate also will manage the seed collection program. Internships usually run for six months from March through August and from June through November. Application deadlines are January 1 and April 1 for the spring-summer and summer-fall internships, respectively. The program runs for nine months from March through November.
Requirements: Applications may be made by recent college sophomores through college graduates in biology, zoology, or botany who have a willingness to work at manual tasks and can work in a self-directed manner. Individuals seeking graduate training are especially encouraged.
Geographic Focus: All States
Date(s) Application is Due: Jan 1; Apr 1; Jul 1; Oct 1
Amount of Grant: 350 USD
Contact: Korie Klink, (608) 356-9462, ext. 127; fax (608) 356-9465; korie@savingcranes.org
Internet: http://www.savingcranes.org/about/get_involved/intern.cfm
Sponsor: International Crane Foundation
P.O. Box 447
Baraboo, WI 53913-0447

International Federation of University Women Research Fellowships 2397
This competition is open to women graduates for research for a period of at least eight months, preferably in a country other than the one in which the candidate has received her education or habitually resides. In addition, the Study and Action Program Fellowship is awarded for research related to the IFUW Study and Action Program—Woman's Role in Changing Society. Members of the IFUW should apply through their national affiliates (in the United States, the American Association of University Women, 1111 16th St NW, Washington, DC 20036). Competitions are held in even-numbered years. The deadline for receipt of applications is determined by each affiliate but normally falls between early September and November of the year preceding the competition.
Requirements: An applicant must be a woman graduate who is an IFUW member and must be started on the research program to which the application relates.
Restrictions: A fellowship will not be awarded for the first year of a PhD program.
Geographic Focus: All States
Date(s) Application is Due: Nov 1
Amount of Grant: 3,000-6,000 CHF
Contact: Fellowships Officer; (022) 731 23 80; fax (022) 738 04 40; info@ifuw.org
Internet: http://www.ifuw.org/fellowships/index.htm
Sponsor: International Federation of University Women
8 rue de l'Ancien Port
Geneva, CH 1201 Switzerland

International Fellowships in Medical Education 2398
The program provides opportunities for faculty from schools of medicine outside the United States to study aspects of medical education in the United States that have the potential to improve medical education in their home country institutions. Mentoring will be provided by preceptors in U.S. basic and clinical science departments, medical education departments, and health system institutions. Eligible study areas include educational methodology, curriculum design, evaluation systems, medical school governance, development of basic and clinical science departments, and the design and operation of health care and public health system programs linked to medical education. Fellows may concurrently pursue collaborative research interests with U.S. basic science faculty or observe some clinical activity with clinical faculty. Fellowship allowance generally includes a monthly stipend, round-trip economy-class air fare for the fellow, and travel to one scientific meeting in the United States. Programs may range from six months to one year. The interactive, web-based application is now available online; paper applications are no longer accepted.
Requirements: Candidates must reside and work in their home countries at the time of application; have a graduate or professional degree in medicine or in a basic medical science; have not less than three years of work experience in their chosen field in the home country; hold an academic appointment as a faculty member in a school of medicine; have the ability to communicate effectively in English; have the endorsement of a home country medical school or organization; and have a position to return to in the home country medical school or organization.
Restrictions: Fellowships are not provided for programs in basic or clinical research; degree-granting educational programs that require acceptance to an institution and tuition payments; tuition grants for short-term courses; specialty training in residency programs, or solely for training in clinical procedures.
Geographic Focus: All States
Amount of Grant: 14,400-28,800 USD
Contact: Grants Administrator; (215) 386-5900; fax (215) 386-9196; info@ecfmg.org
Internet: http://www.ecfmg.org/about.html
Sponsor: Educational Commission for Foreign Medical Graduates
3624 Market Street
Philadelphia, PA 19104-2685

International Human Rights Funders Grants 2399
Human rights grantmaking supports a wide range of efforts to ensure that all people have the opportunity to enjoy a genuinely human existence. As affirmed in the Universal Declaration of Human Rights (UDHR) and the international and domestic laws to which it gave rise, the people of the world are endowed with inalienable civil, cultural, economic, political and social rights. What gives human rights work such power is that all governments, including the United States, have a recognized obligation to respect, protect and fulfill these rights, both domestically and internationally. Increasingly, non-state actors corporations and individuals are also being held accountable. Funders contribute to the advancement of human rights around the world through their grantmaking. Grants varying in type and amount support a broad range of approaches, including: public education to inform people about their human rights and how to exercise them; documenting, reporting and fact-finding to expose human rights violations; litigation to uphold human rights and hold abusers accountable; policy advocacy to ensure that states and non-state actors conform to human rights standards; research and scholarship to define the content of rights; networking and coalition building to further the effectiveness of a global human rights movement; and capacity building for organizations engaged in the above work, locally and internationally.
Requirements: The applicant must be an organization that works to defend or promote human rights; and is based in India, Bangladesh, Pakistan, and be working, in whole or in part, on human rights issues in the country in which the organization is based, or is based in Algeria, Tunisia, or Morocco, and be working, in whole or in part, on human rights issues in the country in which the organization is based.
Restrictions: The fund does not support stand-alone conferences, individuals, businesses, scholarships, fundraising events, university-based research, government agencies, or activities directly or indirectly intended to support candidates for political office.
Geographic Focus: All States
Contact: Michael Hirschhorn, Director; (202) 609-2631; fax (202) 609-2633; MichaelH@ihrfg.org or info@IHRFG.org
Internet: http://www.hrfunders.org/hrfunding/index.html
Sponsor: International Human Rights Funders Group
1410 Broadway, 23rd Floor
New York, NY 10018

International League of Antiquarian Booksellers Bibliographical Prize 2400
Every four years (next in 2006) the ILAB awards a prize to the author of the best work, published or unpublished, of learned bibliography or of research into the history of books or of typography, and books of general interest on the subject. Two copies of each work to be considered must be deposited at the office of the secretary at the very latest 16 months before date of award. The award period covers the four years prior to the year of deadline date.
Requirements: Entries must be submitted in a language that is universally used. A work already published is eligible only if publication occurred within the four years immediately preceding the closing date for submission or if it has an imprint bearing a date within those four years.
Restrictions: Entries in the form of a specialized catalog of one or more books destined for sale are not eligible, nor are periodicals or public library catalogs.
Geographic Focus: All States
Date(s) Application is Due: Dec 31
Amount of Grant: 10,000 USD
Contact: Raymond Kilgarriff; +44 (0)1424 426146; rmkilgarriff@btinternet.com
Internet: http://www.ilab-lila.com/english/prize.htm
Sponsor: International League of Antiquarian Booksellers
Hauptstrasse 19 A
Bad Honnef, D-53604 Germany

International Max Planck Research School Mathematics in the Sciences Fellowships 2401
The research school offers PhD fellowships for qualified students with an excellent background in mathematics or related fields. It will lead interested graduate students towards research problems in the physical and life sciences. This involves a broad range of mathematical fields, including geometry, partial differential equations and functional analysis, stochastics, and discrete mathematics. Specific subject areas are: a) Partial

Differential Equations and Material Sciences; b) Numerical Analysis and Scientific Computing; c) Riemannian and Symplectic Geometry and Hamiltonian Systems; d) Quantum Field Theory, Particle Physics; e) Algebraic Geometry, String Theory; f) Geometric and Functional Analytic Methods in Mathematical Physics; g) Stochastic Processes, Many Particle Systems; h) Complex Systems in Evolutionary Processes and Neurobiology. The fellowships will include a generous stipend, and there will be no tuition fees. Teaching will be in English. In addition, German language courses will be provided free of charge.
Requirements: There is no application form to fill out. However, interested applications should submit the following in paper form: CV; transcript of examination results; a copy of the master thesis (if applicable); and two letters of recommendation by academic teachers, to reach the school in sealed envelopes or to be sent directly.
Restrictions: For students who did not write a master thesis there are qualifying grants available as an entrance level to the research school.
Geographic Focus: All States
Date(s) Application is Due: Mar 25; Jun 25; Sep 25; Dec 25
Contact: Hayk Mikayelyan; +49 341 9959 50; fax +49 341 9959 658; imprsmis.mpg.de
Internet: http://www.imprs-mis.mpg.de/positions.html
Sponsor: Max Planck Institute for Social Anthropology
Max-Planck-Institut fur Mathematik in den Naturwissenschaften; IMPRS Mathematics in the Sciences; InselstraBe 22
Leipzig, 04103 Germany

International Peace Research Institute, Oslo Grants **2402**
PRIO offers scholarships for graduate students, mainly at Norwegian universities. In addition to a student stipend, the scholarship includes supervision by PRIO's researchers and office space at the institute's building in Oslo. Residence in Oslo during the scholarship period is essential. Guidelines are available online. The scholarship lasts for an agreed period of time, up to one year. If the thesis is finalized within the agreed period, PRIO may offer an extension of the stipend period to secure academic publishing of the results, e.g. by writing an article for an international journal.
Geographic Focus: All States
Date(s) Application is Due: Oct 1
Amount of Grant: Up to 42,000 NOK
Contact: Stein Tonnesson; 47-22-54-77-31; stein@prio.no or info@prio.no
Internet: http://www.prio.no/page/preview/preview/9346/21744.html
Sponsor: International Peace Research Institute, Oslo
Fuglehauggta 11
Oslo, 0260 Norway

International Studies Graduate Fellowships **2403**
The Graduate Programs in International Studies department at the university offers MA and PhD fellowships to train graduate students and promote scholarship on transnational issues. The program focuses on traditional problems of foreign and security policy, international business and economics, as well as post-Cold-War priorities such as gender, human rights, and migration.
Geographic Focus: All States
Contact: Program Coordinator; (757) 683-5700; fax (757) 683-5701; isgpd@odu.edu
Internet: http://al.odu.edu/gpis
Sponsor: Old Dominion University
Batten Arts and Letters Building, Room 620
Norfolk, VA 23529-0086

Ipatieff Prize **2404**
This award is given every three years to recognize outstanding chemical experimental work in the field of catalysis or high pressure, carried out by individuals of any nationality and not over 40 years of age (by date of presentation). If experimental investigations in these fields shall have been abandoned to such a degree that no outstanding results have been achieved, then the award may be given for highly meritorious work in a closely allied field of chemistry. Special weight shall be given to the independence of thought and the originality shown. The award may be made for investigations carried out in any country and without consideration of the nationality of the recipient; however, preference will be given to American chemists.
Requirements: Any individual, except a member of the award committee, may submit one nomination or seconding letter for the award in the award year. Nominating documents consist of a letter of not more than 1000 words containing an evaluation of the nominee's accomplishments and a specific identification of the work to be recognized, a biographical sketch including date of birth, and a list of publications and patents authored by the nominee.
Restrictions: Self-nominations are not accepted.
Geographic Focus: All States
Date(s) Application is Due: Nov 1
Amount of Grant: 5,000 USD
Contact: Felicia Dixon, Awards Administrator; (800) 227-5558 or (202) 872-4408; fax (202) 776-8008; f_dixon@acs.org or awards@acs.org
Internet: http://portal.acs.org/portal/acs/corg/content?_nfpb=true&_pageLabel=PP_ARTICLEMAIN&node_id=1319&content_id=CTP_004533&use_sec=true&sec_url_var=region1
Sponsor: American Chemical Society
1155 Sixteenth Street, NW
Washington, DC 20036-4801

IRA Albert J. Harris Awards **2405**
The Albert J. Harris Award is given for a recently published journal article or monograph that makes an outstanding contribution to the understanding of prevention or assessment of reading or learning disabilities. Publications may be submitted by the author or anyone else. Copies may be duplicated from the actual publication. Reprints are also acceptable. Nomination for the Albert J. Harris Award is open to all literacy professionals. Specific guidelines and criteria for submission are at the IRA website.
Requirements: Publications that have appeared in refereed professional journals the previous submission year are eligible for consideration. Eligible works include reports of empirical studies that involve the collection of original data and/or articles that develop theory or synthesize knowledge. Submissions must include the cover sheet located with the guidelines at the website.
Geographic Focus: All States
Date(s) Application is Due: Sep 1
Amount of Grant: 800 USD
Contact: Marcella Moore, Program Supervisor; (302) 731-1600 or (800) 336-7323; fax (302) 731-1057; research@reading.org
Internet: http://www.reading.org/Resources/AwardsandGrants/albert_j_harris_award.aspx
Sponsor: International Reading Association
800 Barksdale Road, P.O. Box 8139
Newark, DE 19714-8139

IRA Elva Knight Research Grant **2406**
The Elva Knight Research Grant provides up to $8,000 for research in reading and literacy. Projects should be completed within two years and may be carried out using any research method or approach as long as the focus of the project is on research in reading or literacy. A sample of an accepted proposal is available at the IRA website.
Requirements: To be eligible, all applicants must be members of the International Reading Association. Specific guidelines and application forms are available at the IRA website. The applicants can complete and submit their applications online using a web-based grant management system. Candidates should not include their name, institutional affiliation, or any other personal information that will identify them other than the Profile page, as the applications are conducted as a blind review.
Restrictions: Activities such as developing new programs or instructional materials are not eligible for funding, unless these activities are necessary procedures for the conduct of the research.
Geographic Focus: All States, All Countries
Date(s) Application is Due: Nov 1
Amount of Grant: Up to 8,000 USD
Contact: Marcella Moore, Program Contact, Research Division; (302) 731-1600; fax (302) 731-1057; research@reading.org
Internet: http://www.reading.org/Resources/AwardsandGrants/research_knight.aspx
Sponsor: International Reading Association
800 Barksdale Road, P.O. Box 8139
Newark, DE 19714-8139

IRA Esther Zolt Academic Research Award **2407**
The Esther Zolt Academic Research Award is given annually for a completed research study on the effective classroom implementation of In2Books that inspires others to improve teaching and learning in innovative ways. Specific guidelines and a current application are available at the website.
Requirements: Applicants must be IRA members, and submit the application electronically. In preparing the proposal, the following guidelines must be met: do not include the applicant's name, institutional affiliation, or any other personal information that will identify the applicant on any part of the application other than the Cover Sheet (provided in the guidelines). The completed Cover Sheet must be sent as a separate document to ensure a blind review of all proposals.
Geographic Focus: All States, All Countries
Date(s) Application is Due: Nov 15
Amount of Grant: 2,500 USD
Contact: Marcella Moore, Award Contact; (302) 731-1600 or (800) 336-7323; fax (302) 731-1057; research@reading.org
Internet: http://www.reading.org/Resources/AwardsandGrants/award_erwin_zolt.aspx
Sponsor: International Reading Association
800 Barksdale Road, P.O. Box 8139
Newark, DE 19714-8139

IRA Esther Zolt Teacher Research Award **2408**
The International Reading Association (IRA) Esther Zolt Teacher Research Award is given annually for a completed action research study. The study should be conducted by a teacher who has implemented In2Books with exceptional innovation and rigor, leading to the development of curious and independent learners. Specific guidelines and a current application are available at the website.
Requirements: Applicants must be members of the IRA. In preparing the proposal, these guidelines must be followed: do not include the applicant's name, institutional affiliation, or any other personal information that will identify the applicant on any part of the application other than the Cover Sheet (provided in the guidelines). The completed Cover Sheet should be emailed as a separate document from the proposal to endure a blind review of all proposals.
Geographic Focus: All States, All Countries
Date(s) Application is Due: Nov 15
Amount of Grant: 2,500 USD
Contact: Marcella Moore, Program Contact; (302) 731-1600 or (800) 336-7323; fax (302) 731-1057; research@reading.org

Internet: http://www.reading.org/Resources/AwardsandGrants/award_erwin_zolt.aspx
Sponsor: International Reading Association
800 Barksdale Road, P.O. Box 8139
Newark, DE 19714-8139

IRA Helen M. Robinson Dissertation Research Award **2409**
The Helen R. Robinson Dissertation Research Award supports doctoral students at the early stages of their dissertation research in the area of reading and literacy. Applicants must be members of the International Reading Association (IRA).
Requirements: Applicants must adhere to the following guidelines: complete and submit a proposal online using a web-based grant management system (Nimble Net) from the IRA website; do not include the applicant's name, institutional affiliation, or other personal information that identifies the applicant on any part of the proposal other than the Profile page (grant competitions will be conducted as a blind review); do not exceed the maximum character requirement (spaces and punctuation are included in the total count; and complete all editing before submitting the proposal, as editing cannot be done after submission. Specified instructions, criteria for proposal review, and the current application is available at the website.
Restrictions: Applicants may apply for more than one research grant, but are limited to one research award per year.
Geographic Focus: All States, All Countries
Date(s) Application is Due: Nov 1
Amount of Grant: 1,200 USD
Contact: Program Contact; (302) 731-1600,; fax (302) 731-1057; research@reading.org
Internet: http://www.reading.org/Resources/AwardsandGrants/research_robinson.aspx
Sponsor: International Reading Association
800 Barksdale Road, P.O. Box 8139
Newark, DE 19714-8139

IRA Jeanne S. Chall Research Fellowship **2410**
The Chall Fellowship was established to encourage and support reading research. The Fellowship supports research efforts in the following areas: beginning reading (theory, research, and practice that improves the effectiveness of learning to read); readability (methods of predicting the difficulty of texts); reading difficulty (diagnosis, treatment, and prevention); stages of reading development; the relation of vocabulary to reading; and diagnosing and teaching adults with limited reading ability. Additional guidelines, including instructions for the web-based application system and review criteria, are available at the website.
Requirements: All applicants must be members of the International Reading Association. Open to doctoral students who are planning or beginning their dissertations and to university based graduate students who are already pursuing reading research.
Geographic Focus: All States, All Countries
Date(s) Application is Due: Nov 1
Amount of Grant: 6,000 USD
Contact: Marcella Moore, Program Supervisor; (302) 731-1600 or (800) 336-7323; fax (302) 731-1057; research@reading.org
Internet: http://www.reading.org/Resources/AwardsandGrants/research_chall.aspx
Sponsor: International Reading Association
800 Barksdale Road, P.O. Box 8139
Newark, DE 19714-8139

IRA Outstanding Dissertation of the Year Award **2411**
The Outstanding Dissertation of the Year Award is intended for doctoral students who have focused their research in the reading field or who have conducted related research having implications for reading. Studies using any research approach (ethnographic, experimental, historical, survey, etc.) are encouraged. Each study will be assessed by the scholarly qualification of its report and its significant contributions to knowledge within the reading field.
Requirements: Dissertations must have been approved by May of the previous application year. Applicants must be current members of the International Reading Association. A 40-page monograph based on the dissertation should be emailed to research@reading.org or mailed to the IRA's address by October 1. The application package should include: title page, abstract, table of contents, dissertation monograph, and references. Additional guidelines and a current application are available at the IRA website.
Geographic Focus: All States, All Countries
Date(s) Application is Due: Oct 1
Amount of Grant: 1,000 USD
Contact: Marcella Moore, Program Supervisor; (302) 731-1600, ext. 423; fax (302) 731-1057; research@reading.org
Internet: http://www.reading.org/Resources/AwardsandGrants/research_outstanding.aspx
Sponsor: International Reading Association
800 Barksdale Road, P.O. Box 8139
Newark, DE 19714-8139

IRA Pearson Foundation-IRA-Rotary Literacy Awards **2412**
The Pearson Foundation-IRA-Rotary Literacy Awards will fund literacy projects with IRA and Rotarian members working together to improve the literacy needs of the local community or anywhere in the world. In addition to the cash award, each honoree will identify a literacy organization to receive up to 1,000 books through the digital literacy initiative, We Give Books, created by the Pearson Foundation and Penguin. For each book read at wegivebooks.org, a book will be donated to the predetermined literacy organization. Two awards of $2,500 each are given. Specific guidelines and a current application are available at the website.
Requirements: Applicants must be IRA members or Rotary members in good standing.
Geographic Focus: All States, All Countries
Date(s) Application is Due: Jun 15
Amount of Grant: 2,500 USD
Contact: Awards Contact; (302) 731-1600 or (800) 336-7323; fax (302) 731-1057; irawash@reading.org
Internet: http://www.reading.org/Resources/AwardsandGrants/award_pearsonfoundation.aspx
Sponsor: International Reading Association
800 Barksdale Road, P.O. Box 8139
Newark, DE 19714-8139

IRA Steven A. Stahl Research Grants **2413**
The Steven A. Stahl Research Grant encourages and supports promising graduate students conducting classroom research in reading. Applicants must have at least three year of pre-K-12 teaching experience. Specific guidelines and instructions for using the web-based application process are available at the website.
Restrictions: The research must be conducted in the classroom(s), focused on improving reading instruction and children's reading achievement, empirically rigorous, and completed within two years.
Geographic Focus: All States, All Countries
Date(s) Application is Due: Nov 1
Amount of Grant: 1,000 USD
Contact: Marcella Moore, Program Supervisor; (302) 731-1600, ext. 423 or (800) 336-7323; fax (302) 731-1057; research@reading.org
Internet: http://www.reading.org/Resources/AwardsandGrants/research_stahl.aspx
Sponsor: International Reading Association
800 Barksdale Road, P.O. Box 8139
Newark, DE 19714-8139

IRA Teacher as Researcher Grants **2414**
The Teacher as Researcher Grants support classroom teachers who undertake action research inquiries about literacy and instruction. Grants can be awarded up to $4,000, although priority will be given to smaller grants in order to support as many teacher researchers as possible. Studies may be carried out using any research method or approach so long as the focus is on reading/writing or literacy. Additional guidelines, samples of winning proposals, and the web-based application is available at the website.
Requirements: All applicants must be members of the International Reading Association and practicing pre-K-12 teachers with full time or permanent half time teaching responsibilities (includes librarians, Title I teachers, classroom teachers, and resource teachers). Classroom teachers will be given preference. Applicants may apply as a collaborative group or individually.
Restrictions: There must be a span of three years before past grant recipients can apply for another grant. Activities such as developing new programs or instructional materials are not eligible for funding unless these activities are necessary for conducting the research.
Geographic Focus: All Countries
Date(s) Application is Due: Nov 1
Amount of Grant: Up to 4,000 USD
Contact: Marcella Moore, Program Supervisor; (302) 731-1600 or (800) 336-7323; fax (302) 731-1057; research@reading.org
Internet: http://www.reading.org/Resources/AwardsandGrants/research_teacher_as_researcher.aspx
Sponsor: International Reading Association
800 Barksdale Road, P.O. Box 8139
Newark, DE 19714-8139

Ireland Family Foundation Grants **2415**
Established in 2000, the Ireland Family Foundation givings primarily in North Carolina with some funding available in California. The Foundations supports non-profit organizations involved with autism research, human services and education.
Requirements: The initial approach should be in the form of a letter, followed up with a proposal containing : a copy of IRS Determination Letter; a detailed description of project and amount of funding requested.
Restrictions: No grants to individuals.
Geographic Focus: California, North Carolina
Amount of Grant: 10,000-60,000 USD
Samples: ARC of Orange County, Chapel Hill, NC., $35,000;
Contact: Lori Ireland, President; (919) 932-3556
Sponsor: Ireland Family Foundation
1434 Arboretum Drive
Chapel Hill, NC 27517-9161

IREX ECA Alumni Small Grants **2416**
The Program is a professional and community development program open to alumni of the Eurasian Undergraduate Program (UGRAD) formerly known as FSAU and Edmund S. Muskie Graduate Fellowship Program. Alumni must complete their small grant activities within Eurasia. Currently, Eurasia is defined as Armenia, Azerbaijan, Belarus, Georgia, Kazakhstan, Kyrgyz Republic, Moldova, Russia, Tajikistan, Turkmenistan, Ukraine and Uzbekistan. Possible project ideas include: initiating a public or community service program; launching a pilot program at an NGO or academic institution; organizing training programs or conferences for professional colleagues and/or other alumni; funding travel of U.S. colleagues to eligible Eurasian countries for conferences, training programs and professional collaboration; development and publication of curricula, textbooks, or related reference or educational materials; publication of public information

pamphlets or brochures; and conducting any other projects judged by the selection committee and/or ECA to be in the interest of program goals.
Geographic Focus: All States
Date(s) Application is Due: Jun 1
Amount of Grant: Up to 3,000 USD
Contact: Program Contact; (202) 628-8188; fax (202) 628-8189; asgp@irex.org
Internet: http://www.irex.org/programs/asgp/index.asp
Sponsor: International Research and Exchanges Board
2121 K Street NW, Suite 700
Washington, DC 20037

IREX Edmund S. Muskie Graduate Fellowship Program (Muskie) 2417
The program provides opportunities for graduate students and professionals from Armenia, Azerbaijan, Belarus, Georgia, Kazakhstan, Kyrgyzstan, Moldova, Russia, Tajikistan, Turkmenistan, Ukraine and Uzbekistan for one-year non-degree, one-year degree or two-year degree study in the United States. Eligible fields of study are: business administration, economics, education, environmental management, international affairs, journalism and mass communication, law, library and information science, public administration, public health, and public policy.
Geographic Focus: All States
Date(s) Application is Due: Jan 31
Contact: Grants Administrator; (202) 628-8188; fax (202) 628-8189; muskie@irex.org
Internet: http://www.irex.org/programs/muskie/index.asp
Sponsor: International Research and Exchanges Board
2121 K Street NW, Suite 700
Washington, DC 20037

IREX Eurasian Undergraduate Exchange Program (UGRAD) Grants 2418
Established by the U.S. Congress in 1992 to foster democratization and economic development in Eurasia, the program was originally called the FREEDOM Support Act Undergraduate Program (FSAU). Key objectives are to: foster democratization and economic development; and promote cultural understanding. Project activities include: encouraging youth leadership; promoting community service in the United States and Eurasia; and supporting practical experience for professional development. IREX opened its first field office in Moscow, Russia and since has expanded its field presence to Armenia, Azerbaijan, Belarus, Georgia, Kazakhstan, Kyrgyzstan, Moldova, Turkmenistan, Tajikistan, Turkey, Ukraine, and Uzbekistan. Programs in Eurasia span academic exchanges, educational advising, alumni programming, independent media assistance and development, Internet training and access, professional training, NGO development, and partnership building.
Requirements: Mid-level university faculty, researchers, scholars, and undergraduate students from the United States are eligible.
Geographic Focus: All States, New Independent States
Date(s) Application is Due: Jan 31; Nov 6
Contact: Program Officers; (202) 628-8188; fax (202) 628-8189; ugrad@irex.org
Internet: http://www.irex.org/programs/ugrad/index.asp
Sponsor: International Research and Exchanges Board
2121 K Street NW, Suite 700
Washington, DC 20037

IREX Individual Advanced Research Opportunities (IARO) Grants 2419
The IARO program seeks to attract, select, and support in-depth field research by U.S. students, scholars and experts in policy-relevant subject areas related to Southeast Europe and Eurasia, as well as to disseminate knowledge about these regions to a wide network of constituents in the United States and abroad. The IARO Program provides fellows with the means and support necessary to conduct incountry research on contemporary political, economic, historical, or cultural developments relevant to U.S. foreign policy*. The IARO Program plays a vital role in supporting the emergence of a dedicated and knowledgeable cadre of U.S. scholars and experts who can enrich the U.S. understanding of developments in Southeast Europe and Eurasia. In addition to conducting their research, IARO fellows are asked to make themselves available as temporary consultants or experts to schools, local NGOs or the U.S. Embassy in their host country. Fellows with grants of 4 months or less are asked to perform 10 hours of service during their grant period, and fellows with grants of 5 or more months are asked perform 20 hours. Grants will be awarded in EACH of the four categories: Master's Student—must be enrolled in a Master's program during the grant period; Predoctoral Student—must be enrolled in a PhD program during the grant period; Professional—must have one of the following degrees (MA, MS, MFA, MBA, MPA, MLIS, MPH, JD, MD) and must not currently be enrolled as a student; Postdoctoral Scholar—must hold a PhD by the application deadline.
Requirements: Applicants must be U.S. citizens who have a command of the host country language sufficient for the completion of their research. Normally, applicants are required to have a full-time affiliation with a college or university and to be faculty members or doctoral candidates who will have completed all requirements for the PhD except the dissertation by the time of participation.
Geographic Focus: All States
Date(s) Application is Due: Nov 15
Contact: Program Officers; (202) 628-8188; fax (202) 628-8189; iaro@irex.org
Internet: http://www.irex.org/programs/iaro
Sponsor: International Research and Exchanges Board
2121 K Street NW, Suite 700
Washington, DC 20037

IREX International Leadership in Education Program (ILEP) Grants 2420
The Program brings outstanding secondary teachers from the Near East, South Asia and Southeast Asia to the United States to further develop expertise in their subject areas, enhance their teaching skills, and increase their knowledge about the United States. ILEP is a program of the Bureau of Educational and Cultural Affairs of the U.S. Department of State. Key issues and objectives are to: contribute to improving the quality of secondary education in participating countries; strengthen the ability of women and under-served populations to play a part in national development; and develop professional and personal relationships between American and international teachers. Project activities include: teacher training; providing leadership programs for women and under-served populations; and forming international partnerships for professional development and education,
Requirements: Applicant must be secondary teachers from the Near East, South Asia and Southeast Asia
Geographic Focus: All States, New Independent States
Contact: Program Officer; (202) 628-8188; fax (202) 628-8189; iep@irex.org
Internet: http://www.irex.org/programs/iep/index.asp
Sponsor: International Research and Exchanges Board
2121 K Street NW, Suite 700
Washington, DC 20037

IREX MENA Media Emerging Leaders Fellowships 2421
The Middle East and North Africa (MENA) program is open to media managers and supervisors based in the Middle East and North Africa with the desire to enhance their media management skills and increase their exposure to alternate media business practices. Fellows will attend sessions on leadership skills and media management trends at Northwestern University's Media Management Center followed by a fellowship placement with a U.S. media outlet. The key objective is to support independent media in the Middle East and North Africa.
Requirements: Applicants must have at least ten years experience in either broadcast or print outlets in the MENA region and be proficient in English.
Geographic Focus: All States
Contact: Program Officer; (202) 628-8188; fax (202) 628-8189; mena@irex.org
Internet: http://www.irex.org/programs/MENAmedia/index.asp
Sponsor: International Research and Exchanges Board
2121 K Street NW, Suite 700
Washington, DC 20037

IREX Policy-Connect Collaborative Research Grants 2422
The program provides fellowships to U.S. scholars and professionals for overseas research on contemporary political, economic, historical, or cultural developments relevant to U.S. foreign policy. Fellowships support collaborative teams of two or three U.S. scholars and professionals for up to 12 months. The program seeks to attract, select, and support advanced research by U.S. experts in policy-relevant subject areas related to Southeast Europe and Eurasia, facilitate collaboration among and between U.S. and international scholars, and disseminate knowledge about Europe and Eurasia to a wide network of constituents in the United States and abroad. Upon completion of the project, scholars will be requested to present their research findings at a Policy Forum at the U.S. Department of State and to write a short policy paper.
Requirements: The principal investigator must be a U.S. citizen or permanent resident. A PhD or equivalent terminal degree is required. Collaborative research programs involving international colleagues are strongly encouraged.
Geographic Focus: All States
Date(s) Application is Due: Apr 1
Amount of Grant: Up to 30,000 USD
Contact: Program Officers; (202) 628-8188; fax (202) 628-8189; iaro@irex.org
Internet: http://www.irex.org/programs/policy-connect/index.asp
Sponsor: International Research and Exchanges Board
2121 K Street NW, Suite 700
Washington, DC 20037

IREX Regional Policy Symposium Grants 2423
The Program provides U.S. students, scholars, and professionals with a forum to examine and discuss current policy research on the countries of Eurasia and Central and East Europe from multi-disciplinary and multi-regional approaches. The research ultimately results in the development and dissemination of policy recommendations to academic and policy communities. The program has three primary goals: to enable U.S. junior and senior scholars to work together in analyzing complex issues affecting the countries of Eurasia and Central and East Europe from multi-disciplinary and multi-regional approaches; to encourage the cross-fertilization of ideas and networking opportunities among scholars with similar regional interests; and to provide policymaking communities with knowledge of current research on evolving regions and valuable policy conclusions drawn from intensive interaction among scholars.
Geographic Focus: All States
Date(s) Application is Due: Dec 15
Contact: Program Officers; (202) 628-8188; fax (202) 628-8189; iaro@irex.org
Internet: http://www.irex.org/programs/symp/index.asp
Sponsor: International Research and Exchanges Board
2121 K Street NW, Suite 700
Washington, DC 20037

IREX Short-Term Travel Grants 2424
Grants provide support for scholarly projects focusing on Central and Eastern Europe, Eurasia, and Mongolia in the humanities and social sciences disciplines. Projects should demonstrate academic merit and relevance for the American academic community studying

these regions, as well as a positive impact on public, cultural, and historical knowledge of these regions through dissemination of research results. Funding is available for individual scholarly research visits to archives, libraries, museums, etc., or for conducting interviews; presentations at scholarly conferences focused on and located in Central and Eastern Europe or Eurasia; and collaborative projects such as joint publications or comparative surveys. The award provides; travel from the United States to the host country; a stipend to cover in-country costs for meals, lodging, and local transportation; and miscellaneous research expenses directly related to the project incidental expenses, such as conference registration fees, visa fees, and research expenses.
Requirements: Applicants must be U.S. citizens or permanent residents (green card holder) of the U.S. for three consecutive years prior to application; submit a research proposal on a topic in one of the academic disciplines listed; and hold a PhD or other terminal degree.
Restrictions: Applicants cannot be a current IREX employee or consultant or their immediate family members (spouses, parents, children, and siblings). Projects must be in the humanities and social science disciplines only. Travel may not exceed a total of 60 days. Individuals may apply for only one trip and one project per application deadline. IREX-funded project activity must be completed within one year of the application deadline.
Geographic Focus: All States
Date(s) Application is Due: Feb 1
Amount of Grant: Up to 5,000 USD
Contact: Program Officer; (202) 628-8188; fax (202) 628-8189; stg@irex.org
Internet: http://www.irex.org/programs/stg/index.asp
Sponsor: International Research and Exchanges Board
2121 K Street NW, Suite 700
Washington, DC 20037

IREX University Administration Support Program (UASP) 2425
The Program benefits university administrators at select state universities in Russia by providing access to pertinent university management skills and models of administration. The Program is further designed to provide training to university administrators in the region while providing more concentrated funding to those university administrators who demonstrate they have the initiative, skills, and institutional support necessary to turn their institutions into models of administrative excellence. To this end, IREX implements a two-tiered program. First, fellowships in University Administration are offered that provide funding for senior level administrators to spend approximately 10 weeks at host universities in the United States. Second, after completion of their fellowships in the United States, the university administrators apply for small grants to jump-start reform projects at their home universities.
Geographic Focus: All States, Russia
Contact: Program Officer; (202) 628-8188; fax (202) 628-8189; uasp@irex.org
Internet: http://www.irex.org/programs/uasp/index.asp
Sponsor: International Research and Exchanges Board
2121 K Street NW, Suite 700
Washington, DC 20037

IREX U.S. Embassy Policy Specialist (EPS) Grants 2426
The Program (EPS) was established to support U.S. embassies and consulates overseas by providing policy specialists-in-residence. While serving at the embassy or consulate the specialists also conduct their own research. EPS fellows have served missions in: Baku, Azerbaijan; Bishkek, Kyrgyzstan; Dushanbe, Tajikistan; Ekaterinburg, Russia; and Vladivostok, Russia. EPS fellowships are one to two months. Grant length and dates are determined in consultation with the specific embassy or consulate. Examples of embassy service can include conducting field research, writing policy papers, consulting embassy staff, and assisting with grant panels.
Geographic Focus: All States
Contact: Program Officer; (202) 628-8188; fax (202) 628-8189; eps@irex.org
Internet: http://www.irex.org/programs/eps/index.asp
Sponsor: International Research and Exchanges Board
2121 K Street NW, Suite 700
Washington, DC 20037

Irving and Rose Crown Scholarships and Fellowships in American History 2427
These fellowships are awarded for a program of graduate study and research leading to the PhD degree in American History at Brandeis University. The primary goal of the program is to train historical scholars, teachers, and researchers. Full tuition is provided. Fellowships are normally renewable for four years; additional research grants are available for fellows.
Requirements: Applicants must have a BA, MA, or professional training, preferably with a strong background in history, American studies, or related fields. Applicants must be accepted into the PhD program in American History at Brandeis University.
Geographic Focus: All States
Amount of Grant: 43,345 USD
Contact: Program Contact; (781) 736-2270; fax (781) 736-2273; historia@brandeis.edu
Internet: http://www.brandeis.edu/departments/history/grad/amst-crown.html
Sponsor: Brandeis University-History Department
Department of History, MS 036
Waltham, MA 02454-9110

ISA Hyland R. Johns Grants 2428
This program provides funds to researchers for projects of interest and benefit to the arboricultural industry. Research must focus on the biology, management, and care of trees, and their relation to environmental, social, and economic benefits. The society encourages research partnerships to enhance industry involvement, increase interdisciplinary interaction, and provide broader funding potential. Write or fax the society for a list of priority areas. Phone calls will not be accepted.
Restrictions: Grant funds cannot be used to pay overhead expenses.
Geographic Focus: All States
Date(s) Application is Due: May 1
Amount of Grant: 7,500-25,000 USD
Contact: ISA Research Trust; (888) 472-8733 or (217) 355-9411; fax (217) 355-9516; isa@isa-arbor.com
Internet: http://www.treefund.org/GrantSumm.asp#Johns
Sponsor: International Society of Arboriculture
1400 W Anthony Drive, P.O. Box 3129
Champaign, IL 61826

ISA John Z. Duling Grants 2429
The goal of the program is to provide seed money or partial support for research and technology transfer projects addressing topics that have the potential of benefiting the everyday work of arborists. Proposals in the following priority areas are more likely to be funded: root and soil management; planting and establishment; plant health care; and risk assessment and worker safety. Application information is available online.
Requirements: Proposals must be submitted on the application form or an exact duplicate. No faxed or reduced copies of the original will be accepted. Applications sent electronically will not be accepted.
Restrictions: Funds cannot be used to pay for overhead expenses or student tuition and fees.
Geographic Focus: All States
Amount of Grant: Up to 7,500 USD
Contact: Director; (630) 221-8127; fax (630) 690-0702; treefund@treefund.org
Internet: http://www.treefund.org/grants/Grants.aspx
Sponsor: International Society of Arboriculture
711 East Roosevelt Road
Wheaton, IL 60187

ISGC Graduate Fellowships 2430
The ISGC Graduate Fellowships support outstanding graduate students pursuing aerospace, space science, earth system science and other interdisciplinary space-related science, engineering or mathematics fields. Up to ten fellowships of $10,000 each are awarded every academic year. Award winners will be chosen based on student qualifications, scientific and educational relevance of the proposed activity, faculty member's expressed commitment to the project, and the application. Women, under-represented minorities, and persons with disabilities are strongly encouraged to apply. Candidates may apply at the website's online application system. Applications are due by April 12, with award notifications in early May.
Requirements: Applicants must meet the following *Requirements:* U.S. citizenship; full-time current enrollment at one of the near twenty ISGC institutions posted on the website; a minimum 3.0 to 4.0 GPA; a good academic record; and research or design experience in a field related to astronomy, aerospace, cosmology, astrophysics, or earth science.
Geographic Focus: Illinois
Date(s) Application is Due: Apr 12
Amount of Grant: 10,000 USD
Contact: Diane Jeffers; (217) 244-8048; fax (217) 244-0720; dejeffer@illnois.edu
Internet: http://www.ae.illinois.edu/isgc/funding/grad-fellowship.html
Sponsor: Illinois Space Grant Consortium
306 Talbot Laboratory, 104 South Wright Street
Urbana, IL 61801-2935

ISGC Informal Education Grants 2431
The Informal Education Grants are designed to increase learning, educate students, educators, and the general public on specific science, technology, engineering, or math (STEM) content areas. Innovative activities associated with aerospace, space sciences, and earth system sciences are preferred. The grants are available to faculty members or educators at all ISGC institutions posted at the website. Funds may be used for salary, supplies, and minimal travel. A total of $8,000 has been allocated for one-year programs beginning in May. Programs targeted to women, underrepresented minorities, and persons with disabilities are strongly encouraged. The application is available at the website, and is due February 15.
Requirements: Grants must support programs in the STEM content areas. Activities associated with aerospace, physics, astronomy, cosmology, or earth system sciences are preferred. Projects must also include program staff and supplemental materials/handouts. Grants must be matched (at least one-to-one) with funds from non-federal sources. The matching fund's source must be identified and confirmed by a letter from the appropriate authority. Waived indirect costs and faculty effort can qualify as matching funds. Any beneficiary of direct support (salary, wages, travel) from this grant must be a U.S. citizen.
Restrictions: Funds cannot be used for the purchase of equipment. Proposal submission for this program is limited to faculty/professional staff (research and professorial) at ISGC affiliate and lead institutions.
Geographic Focus: Illinois
Date(s) Application is Due: Feb 15
Amount of Grant: 8,000 USD
Contact: Diane Jeffers; (217) 244-8048; fax (217) 244-0720; dejeffer@illinois.edu
Internet: http://www.ae.illinois.edu/ISGC/funding/informal-ed.html
Sponsor: Illinois Space Grant Consortium
306 Talbot Laboratory, 104 South Wright Street
Urbana, IL 61801-2935

ISGC Multiple Undergraduate Research Grants 2432

The Multiple Undergraduate Research Grants are designed to fund programs of undergraduate research in aerospace, physics, astronomy, cosmology, earth system science and other interdisciplinary space-related science or engineering fields at ISGC institutions. The program should involve multiple students working with multiple research advisors. The grants are available to faculty members or research scientists at all ISGC institutions posted on the website. Emphasis will be placed on hands-on activities and projects in areas of interest to NASA. One-year grant periods begin in May. Women, under-represented minorities and persons with disabilities are encouraged to apply. The application is available at the website.

Requirements: The grant must support research in an area of interest to NASA, such as aerospace engineering, physics, astronomy, cosmology or earth system science. Grants must be matched (at least one-to-one) with funds from non-federal sources. Prior to the execution of a sub-award agreement, the source of matching funds must be identified and confirmed by a letter from the appropriate authority. Waived indirect costs and faculty effort can qualify as matching funds. Proposals can include requests for salary, supplies, and travel support. Any beneficiary of direct support (salary, wages, travel) from this grant must be U.S. citizen.

Restrictions: Funds cannot be used for the purchase of equipment. Participation in this program is limited to faculty (research and professorial) at ISGC affiliate and lead institutions

Geographic Focus: Illinois

Date(s) Application is Due: Feb 15

Amount of Grant: Up to 124,000 USD

Contact: Diane Jeffers; (217) 244-8048; fax (217) 244-0720; dejeffer@illinois.edu

Internet: http://www.ae.illinois.edu/ISGC/funding/multi-undergrad.html

Sponsor: Illinois Space Grant Consortium
306 Talbot Laboratory, 104 South Wright Street
Urbana, IL 61801-2935

ISGC Research Seed Grants 2433

The Research Seed Grants are designed to facilitate the development of research expertise in aerospace engineering/space science at ISGC institutions, especially with smaller research institutions and/or assistant professors. Grants will allow the award recipients to obtain preliminary results needed to support larger proposals to other federal or non-federal funding agencies. Emphasis is placed on research in an area of interest to NASA and new research ideas. A total of $40,000 is allocated for a one-year period beginning in May. The funding level of each seed grant is $10,000. Proposals will be reviewed by ISGC Directors, with selection based on adherence to scientific merit and equitable distribution of resources across the Consortium. Women, under-represented minorities and persons with disabilities are encouraged to apply. The application is available through the online processing system.

Requirements: The seed grant must support research in an area of interest to NASA, such as aerospace engineering, physics, astronomy, cosmology or earth system science. Grants must be matched (at least one-to-one) with funds from non-federal sources. Prior to the execution of a sub-award agreement, the source of matching funds must be identified and confirmed by a letter from the appropriate authority. Waived indirect costs and faculty effort can qualify as matching funds. Any beneficiary of direct support (salary, wages, travel) from the grant must be a U.S. citizen. Proposals can include requests for salary, supplies, and travel support.

Restrictions: Participation in this program is limited to faculty (research and professorial) at ISGC affiliate and lead institutions that are not participating in an ISGC-funded multiple undergraduate research program. Funds cannot be used for the purchase of equipment.

Geographic Focus: Illinois

Date(s) Application is Due: Feb 15

Amount of Grant: 10,000 USD

Contact: Diane Jeffers; (217) 244-8048; fax (217) 244-0720; dejeffer@illinois.edu

Internet: http://www.ae.illinois.edu/ISGC/funding/seedgrant.html

Sponsor: Illinois Space Grant Consortium
306 Talbot Laboratory, 104 South Wright Street
Urbana, IL 61801-2935

ISUD Jacobsen Prizes 2434

ISUD is a group of scholars and philosophers from many parts of the world who have joined together to discuss and debate leading questions concerning the theory and practice of global community. The Jacobsen Prizes recognize continuation of Jens A.B. Jacobsen's quest to look beyond the individual ego and seek universal meaning that can be shared by all. Submitted papers are eligible for the awards, which will be made during the Congress. All papers presented at the Congress wil be published in the ISUD bi-annual proceedings. Papers must be submitted by the listed deadline. Detailed information is contained on the Web site.

Requirements: Participation in the Congresses, which take place every two years (odd-numbered years), is open to anyone with a philosophical bent and with the love of wisdom in his or her heart.

Geographic Focus: All States

Date(s) Application is Due: Nov 15

Amount of Grant: 1,500 USD

Contact: Dr. Daniel Shannon, Department of Philosophy; deshan@depauw.edu

Internet: http://www.isud.org/papers/jacobsen.shtml

Sponsor: International Society for Universal Dialogue
DePauw University, 212 Asbury Hall
Greencastle, IN 46135,

ITEA Maley/FTE Scholarship 2435

The Foundation for Technology Education established the Maley/FTE Technology Teacher Scholarship in honor of Dr. Donald Maley. Its purpose is to support teachers in their preparation to increase the positive outcomes of technology education. Criteria include: (1) evidence of teaching success, (2) plans for action research, (3) recommendations, (4) plans for professional development, and (5) the applicant's need. The scholarship is provided directly to the teacher and will be presented during the ITEA Annual Conference.

Requirements: Applicant must be a member of the International Technology Education Association. (Membership may be enclosed with scholarship application). Applicant must be a technology teacher at any grade level who is beginning or continuing graduate study. See the sponsor's website for the list of required items for the application package.

Geographic Focus: All States

Date(s) Application is Due: Dec 1

Contact: Kendall N. Starkweather; (703) 860-2100; itea@iteaconnect.org

Internet: http://www.iteaconnect.org/Awards/scholarshipmaley.htm

Sponsor: International Technology Education Association
1914 Association Drive, Suite 201
Reston, VA 20191-1539

Ittleson Foundation AIDS Grants 2436

In regards to AIDS, the foundation is particularly interested in new model, pilot, and demonstration efforts: addressing the needs of underserved at-risk populations and especially those programs recognizing the overlap between such programs; responding to the challenges facing community-based AIDS service organizations and those organizations addressing systemic change; providing meaningful school-based sex education; making treatment information accessible, available and easily understandable to those in need of it; or, addressing the psycho-social needs of those infected and affected by AIDS, especially adolescents.

Requirements: Tax-exempt organizations may apply.

Restrictions: The foundation generally does not provide funds for capital building projects, endowments, grants to individuals, scholarships or internships (except as part of a program), direct service programs (especially outside New York City), projects that are local in focus and unlikely to be replicated, continuing or general support, projects and organizations that are international in scope or purpose, or biomedical research.

Geographic Focus: All States

Date(s) Application is Due: Sep 1

Contact: Anthony C. Wood, Executive Director; (212) 794-2008; fax (212) 794-0351

Internet: http://www.ittlesonfoundation.org/aids.html

Sponsor: Ittleson Foundation
15 E 67th Street, 5th Floor
New York, NY 10021

Ittleson Foundation Environment Grants 2437

The foundation supports innovative pilot, model and demonstration projects that will help move individuals, communities, and organizations from environmental awareness to environmental activism by changing attitudes and behaviors. This program seeks to encourage and nurture environmental action through: supporting the present generation of environmental activists, whether professionals or volunteers through education, training and other activities; educating and engaging the next generation of environmentalists with a special interest in supporting the training of those who are teaching that generation; strengthening the infrastructure of the environmental movement with a particular focus on efforts at the grassroots and statewide levels; activating new constituencies, particularly those focused on environmental equity issues.

Requirements: Tax-exempt organizations may apply.

Restrictions: The foundation generally does not provide funds for capital building projects, endowments, grants to individuals, scholarships or internships (except as part of a program), direct service programs (especially outside New York City), projects that are local in focus and unlikely to be replicated, continuing or general support, projects and organizations that are international in scope or purpose, or biomedical research.

Geographic Focus: All States

Date(s) Application is Due: Sep 1

Contact: Anthony C. Wood, Executive Director; (212) 794-2008; fax (212) 794-0351

Internet: http://www.ittlesonfoundation.org/enviro.html

Sponsor: Ittleson Foundation
15 E 67th Street, 5th Floor
New York, NY 10021

Ittleson Foundation Mental Health Grants 2438

For this program, the foundation is interested in innovative, pilot, model and demonstration projects that are: fighting the stigma associated with mental illness and working to change the public's negative perception of people who have mental illness; utilizing new knowledge and current technological advances to improve programs and services for people who have mental illness; bringing the full benefits of this new knowledge and technology to those who presently do not have access to them; or, advancing preventative mental health efforts, especially those targeted to youth and adolescents, with a special focus on strategies that involve parents, teachers, and others in close contact with these populations.

Requirements: Tax-exempt organizations may apply.

Restrictions: The foundation generally does not provide funds for capital building projects, endowments, grants to individuals, scholarships or internships (except as part of a program), direct service programs (especially outside New York City), projects that are local in focus and unlikely to be replicated, continuing or general support, projects and organizations that are international in scope or purpose, or biomedical research.

Geographic Focus: All States
Date(s) Application is Due: Sep 1
Contact: Anthony C. Wood, Executive Director; (212) 794-2008; fax (212) 794-0351
Internet: http://www.ittlesonfoundation.org/mental.html
Sponsor: Ittleson Foundation
15 E 67th Street, 5th Floor
New York, NY 10021

ITVS Commissioned Funding 2439

ITVS accepts proposals on an ongoing basis for documentary projects that fall outside of the standing initiatives (DDF, LINCS, Open Call). For development funding, activities may include travel, research, script development, preliminary production for fundraising/ work-in-progress reels, or other early-phase activities. For production funding, all production and post-production activities are eligible. Producers should note that ITVS's capacity to fund projects outside of its standing initiatives is extremely limited and highly competitive, but will consider programs for special support and funding on a case-by-case basis. For development funding, ITVS accepts: proposals for documentaries that have not yet begun production; and programs that can become eligible for ITVS production funding initiatives as single programs of standard broadcast length (half-hour or one-hour). For production funding, ITVS accepts proposals for limited series (two to four episodes). There is no deadline for applications. Interested producers may submit their materials at any time. The entire proposal review process takes six to eight weeks.
Requirements: Independent producers who are at least 18 years old and are citizens or legal residents of the United States or its external territories are eligible. Applicants must have some previous film or television production experience in a principal role (producer, co-producer, director or co-director) as demonstrated by submitting a sample tape of previously completed work.
Restrictions: As with other ITVS initiatives, applicants may only have one project under consideration at any one time. Applicants not eligible to apply include: individuals who do not meet ITVS' definition of an independent producer; students; current signatories of ITVS production contracts who have not completed delivery; producers representing foreign-based production entities; producers regularly employed by a public or commercial broadcast entity, including public access stations; or organizations such as universities, foundations and non-profit media organizations.
Geographic Focus: All States
Contact: Jonathan Archer, Programming Coordinator; (415) 356-8383, ext. 284; fax (415) 356-8391; jonathan_archer@itvs.org
Internet: http://www.itvs.org/funding/edf
Sponsor: Independent Television Service
651 Brannan Street, Suite 410
San Francisco, CA 94107

IUCP Archives Program Fellowship 2440

Grants support research in any discipline at Indiana University's Ruth Lilly Philanthropy Archives. To apply, submit a resume or vita and cover letter including the following information: name, contact information (including telephone and email address); title and brief description of the project; description of how the Philanthropy Archival collection will be utilized in the research project; dates and duration of the project; and amount requested and budget summary. Allowable expenses include round-trip travel, temporary lodging, food, photocopying, and other research expenses. Applications are accepted on an ongoing basis. Contact staff prior to proposal submission to ensure that the required research documents are available in the library.
Requirements: Individual scholars and researchers are eligible.
Geographic Focus: All States
Amount of Grant: Up to 4,000 USD
Contact: Kathy Steinberg, (317) 684-8957; fax (317) 684-8900; ksteinbe@iupui.edu
Internet: http://www.philanthropy.iupui.edu/research.html
Sponsor: Indiana University Center on Philanthropy
550 W North Street, Suite 301
Indianapolis, IN 46202-3162

IUCP Dissertation Fellowship 2441

One-year grants are awarded for doctoral research relating to issues facing philanthropy and nonprofit organizations. Participants are encouraged to obtain matching funds from their home institutions.
Requirements: Applicants from Indiana institutions must be conducting dissertation research relevant to the center's priority research areas. Applicants from universities outside of Indiana may be asked to contact the Aspen Institute's Nonprofit Research Fund.
Geographic Focus: All States
Contact: Kathy Steinberg, (317) 684-8957; fax (317) 684-8900; ksteinbe@iupui.edu
Internet: http://www.philanthropy.iupui.edu
Sponsor: Indiana University Center on Philanthropy
550 W North Street, Suite 301
Indianapolis, IN 46202-3162

IUCP Indiana Research Fund 2442

Grants are awareded for research relating to issues facing philanthropy and nonprofit organizations. Priority will be given to well-designed projects consistent with the center's strategic emphasis on research and practice that may be translated into potential solutions for problems facing philanthropy, fund raising, and the nonprofit sector. Proposals involving collaboration among nonprofit practitioners, academic researchers, independent scholars, and policy analysts are welcome. Annual deadlines dates may vary; contact program staff for exact dates.
Requirements: Grants will be awarded to faculty members affiliated with any of the branch campuses of Indiana University.
Geographic Focus: All States
Contact: Maggie Bowden, (317) 278-8984; mtbowden@iupui.edu
Internet: http://www.philanthropy.iupui.edu/giving_opportunities.html
Sponsor: Indiana University Center on Philanthropy
550 W North Street, Suite 301
Indianapolis, IN 46202-3162

IUVSTA Prize for Science 2443

The Prize is given for experimental and/or theoretical research in vacuum science, technique or its applications. The purpose is to recognize and encourage outstanding internationally-acclaimed research in the fields of interest to the IUVSTA. The award will normally be given to an individual; however, in exceptional cases involving team research, multiple awards may be given. The Prize consists of a cash award, a struck medal and a certificate setting forth the reasons for the award. The Prize will be conferred at intervals of not less than three years. Reasonable travel expenses of the recipient to the meeting at which the Prize is presented shall be agreed upon and reimbursed.
Requirements: The nominee must have accomplished outstanding experimental and/or theoretical research in vacuum science, technique or its applications within the ten years preceding the year in which the award is made. Special consideration will be given to nominees currently engaged in an active career of research. Neither nominees nor nominators are required to be members of IUVSTA member societies.
Restrictions: Current officers and members of the Executive Council and Standing Committees, as well as Division Officers of IUVSTA, are not eligible.
Geographic Focus: All States
Date(s) Application is Due: Apr 2
Contact: Dr. Masatoshi Ono, +81 44 366 6331; ono.m@funai-atri.co.jp
Internet: http://www.iuvsta.org/prizeinfo.html
Sponsor: International Union for Vacuum Science, Technique, and Applications
1-12 Minami-Wataridi-Cho
Kawasaki-shi, Kanagawa 210-0855 Japan

IUVSTA Prize for Technology 2444

The purpose of the Award is to recognize and encourage outstanding internationally-acclaimed achievements in technology and instrumentation in the fields of interest to the IUVSTA. The award will normally be given to an individual; however, in exceptional cases involving team research, multiple awards may be given. The Prize consists of a cash award, a struck medal and a certificate setting forth the reasons for the award. The Prize will be conferred at intervals of not less than three years. Reasonable travel expenses of the recipient to the meeting at which the Prize is presented shall be agreed upon and reimbursed.
Requirements: The nominee must have accomplished outstanding results in the area of technology and instrumentation within the ten years preceding the year in which the award is made. Special consideration will be given to nominees currently actively involved in this work. Neither nominees nor nominators are required to be members of IUVSTA member societies.
Restrictions: Current officers and members of the Executive Council and Standing Committees, as well as Division Officers of IUVSTA, are not eligible.
Geographic Focus: All States
Date(s) Application is Due: Apr 2
Contact: Dr. Masatoshi Ono, +81 44 366 6331; ono.m@funai-atri.co.jp
Internet: http://www.iuvsta.org/prizeinfo.html
Sponsor: International Union for Vacuum Science, Technique, and Applications
1-12 Minami-Wataridi-Cho
Kawasaki-shi, Kanagawa 210-0855 Japan

IUVSTA Welch Foundation Scholarship 2445

The scholarship is offered for a one-year period, starting in September, to a promising scholar who wishes to contribute to the study of vacuum science techniques or their application in any field. Because of the international nature of the scholarship, strong preference is given to applicants who propose study in a foreign lab in which they have not yet studied. Payment of the scholarship is in three installments totalling $15,000; one of $7500 at the beginning, another of $7000 after six months, and a third of $500 upon delivery of a final report after completion of work. The scholarship holder is encouraged to seek funds in addition to the scholarship, but should obtain authorization of the chairperson of the Welch Committee of the IUVSTA before accepting any additional funds.
Requirements: Candidates should have at least a bachelor's degree; a doctoral degree is preferred. Application should include a curriculum vita; a photocopy of, or attestation of, all diplomas; the name and address of laboratory chosen and a letter indicating that this laboratory's facilities will be available to the applicant; a 200-word abstract describing the proposed research; a declaration that the candidate will not violate any laws of his/her own country during the tenure of the scholarship, as well as a declaration that the candidate will not violate any laws or engage in any political activity in the country where he/she intends working; and two recommendations from present or past professors or research directors. Additionally, candidates must produce satisfactory evidence of reasonable fluency either in the language of the country where he/she will work during the tenure of the scholarship or in English.
Restrictions: If the candidate cannot begin work as scheduled, he or she may begin within three months after September 1. In the case of a delay of more than three months, another candidate may be chosen. The laboratory where the candidate wishes to work must approve any delay in the commencement of work.
Geographic Focus: All States
Date(s) Application is Due: Apr 15

Amount of Grant: 15,000 USD
Contact: Dr. Frank R. Shepherd, (613) 763-3285; frank_shepherd@avs.org
Internet: http://www.iuvsta.org/welchann.html
Sponsor: International Union for Vacuum Science, Technique, and Applications
30 Avenue de la Renaissance
Brussels, B-1000 Belgium

IWMF Public Health Fellowship **2446**

The goals of the fellowship are to offer women newspaper editors and radio producers the opportunity to enhance their ability to cover public health issues, build their journalism skills, and learn first-hand about standards and practices in the U.S. media. The six-month fellowship includes four months in the United States and two months in the fellows' home countries, where the fellows will implement a public health reporting project. Each applicant must clearly identify and describe a public health reporting project that she will focus on when she returns to her home country for that segment of the fellowship. Those demonstrating original approaches to covering HIV/AIDS will be given top consideration. The fellowship includes application costs for a U.S. visa; round-trip economy airfare from the fellows' home countries to Washington, DC; round-trip economy airfare from Washington, DC, to the fellowship site and health insurance; and a fixed stipend to cover lodging, meals, and ground transportation during the U.S. segment of the fellowship. For the home country segment, a fixed stipend will be provided to support the fellow and cover all costs associated with implementing her project. Application and guidelines are available online.
Requirements: The fellowship is open to women journalists who are working in one of the following countries: Asia (China and Indonesia), West Africa (Benin, Burkina Faso, Cameroon, Cape Verde, Chad, Gambia, Ghana, Guinea, Guinea Bissau, Ivory Coast, Liberia, Mali, Mauritania, Niger, Nigeria, Senegal, Sierra Leone, and Togo), and Southern Africa (Angola, Botswana, Lesotho, Malawi, Mozambique, Namibia, Swaziland, Zambia, and Zimbabwe). A successful applicant will be dedicated to a career in journalism and have five or more years of full-time experience in the profession, with at least three of those years having been focused on public health journalism. She must be currently working as a newspaper editor or radio producer and show a strong commitment to sharing knowledge and skills with colleagues upon returning home. The fellow must also have excellent written and spoken English skills.
Geographic Focus: All States
Date(s) Application is Due: Apr 15
Contact: Administrator; (202) 496-1992; fax (202) 496-1977; fellowship@iwmf.org
Internet: http://iwmf.org/programs/8702
Sponsor: International Women's Media Foundation
1625 K Street NW, Suite 1275
Washington, DC 20006

Izaak Walton Killam Memorial Postdoctoral Fellowships at the University of British Columbia **2447**

These awards are offered to candidates who have shown superior ability in research and who possess doctorates or who are deemed to have similar qualifications. The candidate may pursue any field of research other than the arts. The basis of the award will be special distinction of intellect, with due regard for sound character and personal qualities. Subject to review at the end of the first year, Killam Memorial Postdoctoral Fellowships are awarded for two years. Fellowships are tenable at the University of British Columbia.
Requirements: Applicants must have obtained the PhD no more than two academic years prior to the anticipated commencement date of the fellowship. The doctorate must be from a university other than The University of British Columbia. Application is open to citizens of any country.
Geographic Focus: All States
Date(s) Application is Due: Nov 15
Amount of Grant: 44,000 CAD
Contact: Killam Secretary; (604) 822-2848; graduate.awards@ubc.ca
Internet: http://www.grad.ubc.ca/awards/index.asp?menu=015,000,000,000
Sponsor: University of British Columbia
180-6371 Crescent Road
Vancouver, BC V6T 1Z2 Canada

Izaak Walton Killam Memorial Predoctoral Fellowships at the University of British Columbia **2448**

These fellowships are awarded to highest ranked doctoral candidates in the university's graduate fellowships competitions. Approximately 25 awards are made each year. Applications are available on the Web site.
Requirements: Applicants must be graduate students with a first-class standing in the last two years of full-time study. Students from any country and from all fields are eligible.
Geographic Focus: All States
Amount of Grant: 26,500 CAD
Contact: Killam Secretary; (604) 822-2848; graduate.awards@ubc.ca
Internet: http://www.grad.ubc.ca/awards/index.asp?menu=003,000,000,000
Sponsor: University of British Columbia
180-6371 Crescent Road
Vancouver, BC V6T 1Z2 Canada

J.B. Reynolds Foundation Grants **2449**

The foundation makes grants in the local Kansas City, MO, area, up to a 150-mile radius of the city. Grants are awarded for building and equipment, community development, medical research, social welfare, and the arts and humanities. Some support is given to colleges and universities. Additional types of support include general operating support, continuing support, annual campaigns, endowment funds, publications, and research. The board meets in April and December of each year to consider letters of requests, and only invited proposals are reviewed. All grants are awarded in December.
Requirements: 501(c)3 organizations in the Kansas City, MO, area may submit applications.
Geographic Focus: Missouri
Amount of Grant: 5,000-50,000 USD
Contact: Richard L. Finn, Secretary-Treasurer; (816) 753-7000; fax (816) 753-1354
Sponsor: J.B. Reynolds Foundation
P.O. Box 219139
Kansas City, MO 64141-6139

J.H. Stewart Reid Memorial Fellowship for Doctoral Studies **2450**

This fellowship is awarded annually for doctoral study in a Canadian university in any field of study. Fellows must reapply for continuation.
Requirements: Applicants must be Canadian citizens or have held permanent resident (landed immigrant) status for one year prior to application; must hold registration in a doctoral program at a Canadian university; must have completed comprehensive exams and have doctoral thesis proposal accepted by the listed deadline date of the year applying; must have a first-class academic record; and must not hold scholarships that exceed in total $25,000 inclusive of the J. H. Stewart Reid Memorial Fellowship Trust.
Geographic Focus: All States, Canada
Date(s) Application is Due: Apr 30
Amount of Grant: 5,000 CAD
Contact: Johanne Smith; (613) 820-2270; fax (613) 820-7244; stewartreid@caut.ca
Internet: http://stewartreid.caut.ca/English/policy.htm
Sponsor: Canadian Association of University Teachers
2675 Queensview Drive
Ottawa, ON K2B 8K2 Canada

J.S. Mill Fellowships in Economics for Exceptional Students **2451**

The fellowship assists students who are enrolled in the graduate program in economics at Washington University. Application materials may be obtained from the department.
Requirements: Applicants must be admitted to the graduate program in the department of economics at Washington University.
Geographic Focus: All States
Date(s) Application is Due: Jan 15
Amount of Grant: 16,000 USD
Contact: Karen Rensing, Secretary, Department of Economics; (314) 935-5670; fax (314) 935-4156; karenr@wuecona.wustl.edu or gradsec@wueconc.wustl.edu
Internet: http://economics.wustl.edu/graduate/brochure/brochure.pdf
Sponsor: Washington University
Campus Box 1208, 205 Eliot Hall
Saint Louis, MO 63130-4899

J.W. Kieckhefer Foundation Grants **2452**

The Kieckhefer Foundation awards grants to nation-wide 501(c)3 organizations in support of medical research, hospices, and health agencies; family planning services; social services; higher education; youth and child welfare agencies; ecology and conservation; community funds; and cultural programs. Types of support include the following: annual campaigns; building renovation; conferences and seminars; continuing support; emergency funds; endowments; equipment; general operating support; land acquisition; matching and challenge support; program development; publication; and research.
Requirements: Applications are not accepted. Organizations should submit a letter of inquiry to the Foundation, with a description of their project and amount requested.
Restrictions: Grants are not awarded to individuals.
Geographic Focus: All States
Contact: John I. Kieckhefer, Trustee; (928) 445-4010
Sponsor: J.W. Kieckhefer Foundation
116 East Gurley Street
Prescott, AZ 86301-3821

Jackson Foundation Grants **2453**

The Jackson Foundation was established October 1960 pursuant to the last will of Maria C. Jackson. Its purpose is to respond to the requests deemed appropriate to promote the welfare of the public of the City of Portland or the State of Oregon, or both. The Foundation considers projects located outside the Portland metropolitan area only if the project is of statewide appeal, rather than of local concern.
Requirements: Grants are awarded to nonprofit 501(c)3 tax-exempt agencies located within the state of Oregon.
Restrictions: Funding is not available for individuals or private businesses. Grants are generally not made to a K-12 school. Do not contact the Foundation by phone to check the status of an application.
Geographic Focus: Oregon
Date(s) Application is Due: Mar 31; Jun 30; Sep 30; Dec 31
Contact: Robert H. Depew, c/o U.S. Bank, (503) 275-4414
Internet: http://www.thejacksonfoundation.com
Sponsor: Jackson Foundation
P.O. Box 3168
Portland, OR 97208

Jacob and Hilda Blaustein Foundation Grants **2454**

The Foundation promotes social justice and human rights through its five program areas: Jewish life; strengthening Israeli democracy; health and mental health; educational opportunity; and human rights. Support is provided to organizations in the United States and abroad, with a preference to the Baltimore, Maryland area. The Foundation supports organizations that promote systemic change; involve constituents in planning and decision-making; encourage volunteer and professional development; and engage in ongoing program evaluation. Additional guidelines and application information is available online.
Requirements: Nonprofit organizations are eligible to apply.
Restrictions: Support is unavailable for the following: individuals; scholarships to individuals; unsolicited proposals for academic, scientific, or medical research; direct mail; annual giving; membership campaigns; fundraising; commemorative events.
Geographic Focus: Maryland
Contact: Betsy Ringel; (410) 347-7103; fax (410) 347-7210; info@blaufund.org
Internet: http://www.blaufund.org/foundations/jacobandhilda_f.html
Sponsor: Jacob and Hilda Blaustein Foundation
10 E Baltimore Street, Suite 1111
Baltimore, MD 21202

Jacob and Valeria Langeloth Foundation Grants **2455**
The Foundation's grantmaking program is centered on the concepts of health and well-being. The Foundation's purpose is to promote and support effective and creative programs, practices and policies related to healing from illness, accident, physical, social or emotional trauma and to extend the availability of programs that promote healing to underserved populations. The Foundation is particularly interested in funding programs that address the health of individuals who, because of barriers to difficulty accessing care. Preference is given in the areas of: caregivers and; correctional health care.
Requirements: 501(c)3 organizations that promote physical and emotional healing, especially to underserved populations, such as: community-based organizations, health care providers and research institutions in the state of New York are eligible to apply.
Restrictions: The Foundation will not consider proposals for annual or capital campaigns, for building or renovation projects, for budgetary relief, or preventive medicine. The Foundation does not support projects that focus on children or end-of-life issues. Neither will the Foundation make grants to individuals or for sectarian or religious purposes, or for political activities or lobbying. The Foundation will not fund any organization that discriminates on the basis of age, gender, national origin, race, or sexual preference.
Geographic Focus: New York
Date(s) Application is Due: Feb 1; Aug 1
Contact: Scott Moyer; (212) 687-1133; fax (212) 681-2628; smoyer@langeloth.org
Internet: http://www.langeloth.org/apply.php
Sponsor: Jacob and Valeria Langeloth Foundation
521 Fifth Avenue, Suite 1612
New York, NY 10175-1699

Jacobs Research Funds Small Grants **2456**
Grants are given for anthropological research on living Native Americans on cultural, social, psychological, aesthetic, and linguistic problems. While the fund has a Pacific Northwest focus, research projects within Canada, Mexico, the rest of the continental United States, and Alaska are eligible. The fund emphasizes field research rather than the analysis of previously collected materials. Grant funds may cover fees for consultants, supplies (notebooks, audio- and videotapes, etc.), transportation to and from the field, and lodging in the field. Grants are made for a maximum of one year and are renewed only on the basis of a new application. Application forms are available upon request.
Requirements: Grants are made to individuals. Applicants need not have completed degrees. Formal institutional affiliation is not required; however, in the absence of formal research training or institutional affiliation, cooperation with or supervision by an experienced research scholar is strongly recommended. Applications from Native Americans are encouraged.
Restrictions: Projects in archaeology, physical anthropology, applied anthropology, and applied linguistics are not included in the program. Research on archive materials or museum collections is not ordinarily considered for funding. Funds are not provided for the salary of the researcher, major equipment, or living expenses.
Geographic Focus: All States
Date(s) Application is Due: Feb 15
Amount of Grant: 1,200 USD
Contact: Administrator; (360) 676-6981; fax (360) 738-7409; museuminfo@cob.org
Internet: http://www.whatcommuseum.org/pages/info/info.htm
Sponsor: Whatcom Museum of History and Art
121 Prospect Street
Bellingham, WA 98225

James and Sylvia Thayer Short -Term Research Fellowships **2457**
The James and Sylvia Thayer Short-Term Research Fellowships support the use of special collections materials by visiting scholars and UCLA graduate students. Collections that are administered by the newly integrated UCLA Library Special Collections and available for Thayer Fellowship-supported research include materials in the humanities and social sciences; medicine; life and physical sciences; visual and performing arts; and UCLA history. Research residencies may last up to three months between March 1 and December 17. Recipients receive stipends ranging from $500 to $2500. (Awards vary yearly). Those receiving fellowships are expected to provide a report on the results of their research that can be mounted on the UCLA Library Web site. Applications are due December 11 and should include: cover letter; curriculum vitae; outline of research and special collections to be used (two pages maximum); brief budget for travel, living, and research expenses; dates to be spent in residence; two letters of recommendation from faculty or other scholars familiar with the research project.
Requirements: United States citizens and permanent residents with the legal right to work in the U.S. who are engaged in graduate-level, post-doctoral, or independent research are invited to apply.
Geographic Focus: All States
Date(s) Application is Due: Dec 11
Amount of Grant: 500-2,500 USD
Contact: Eunice MacGill, Administrative Assistant; (310) 794-4408; emacgill@library.ucla.edu or lib_thayer@library.ucla.edu
Internet: http://www.library.ucla.edu/special/thayer.cfm
Sponsor: UCLA Library
A1713 Charles E Young Research Library, P.O. Box 951575
Los Angeles, CA 90095-1575

James Flack Norris Award in Physical Organic Chemistry **2458**
This award, sponsored by The Northeastern Section of ACS, is given annually to encourage and reward outstanding contributions to physical organic chemistry. The award consists of $5,000 and a certificate. Reasonable travel expenses to the meeting at which the award will be presented will be reimbursed.
Requirements: Any individual, except a member of the award committee, may submit one nomination or seconding letter for the award in any given year. Nominating documents consist of a letter of not more than 1000 words containing an evaluation of the nominee's accomplishments and a specific identification of the work to be recognized, a biographical sketch including date of birth, and a list of publications and patents authored by the nominee.
Restrictions: Self-nominations are not accepted.
Geographic Focus: All States
Date(s) Application is Due: Nov 1
Contact: Felicia Dixon; (800) 227-5558 or (202) 872-4408; f_dixon@acs.org
Internet: http://portal.acs.org/portal/acs/corg/content?_nfpb=true&_pageLabel=PP_ARTICLEMAIN&node_id=1319&content_id=CTP_004536&use_sec=true&sec_url_var=region1
Sponsor: American Chemical Society
1155 Sixteenth Street, NW
Washington, DC 20036-4801

James Graham Brown Foundation Brown Fellowships **2459**
The Brown Fellows Program is unique in that it is a single initiative that spans two campuses. In partnership with Centre College and the University of Louisville, Brown Fellows receive full tuition, room and board and will participate in an extraordinary summer program featuring international travel as a group before freshman year. Fellows also receive an enhanced academic experience consisting of individual study, field-based experiential learning and community service. Each student will have an on campus mentor, engage in enrichment opportunities and manage a customized Leadership Project. In successive years the program will focus on the individual student to provide an extraordinary summer enrichment experience focused in thematic areas of service, international research and study, and leadership.
Requirements: Eligibility and selection criteria include scholastic attainment, intellectual leadership, visionary thinking, and exemplary community involvement. Additional consideration will be given for a demonstrated ability to overcome adversity, as well as a willingness and ability to serve as an ambassador for Kentucky.
Geographic Focus: Kentucky
Contact: Dodie L. McKenzie; (502) 896-2440 or (866) 896-5423; grants@jgbf.org
Internet: http://www.brownfellows.org/
Sponsor: James Graham Brown Foundation
4350 Brownsboro Road, Suite 200
Louisville, KY 40207

James H. Cummings Foundation Grants **2460**
The Foundation, gives exclusively for charitable purposes in advancing medical science, research, and education in selected cities in the U.S. and Canada, and for charitable work among underprivileged boys and girls, and aged and infirm persons in designated areas. Priority is given to medical proposals. The funding is limited to: Toronto, Ontario; Canada; Hendersonville, North Carolina and; Buffalo, New York. Grants are available, in the following types of support: building/renovation; capital campaigns; equipment; land acquisition; matching/challenge support; research and seed money.
Requirements: The giving program is limited to the vicinity of the cities of Buffalo, NY; Hendersonville, NC; and Toronto, ON, Canada.
Restrictions: No support for national health organizations. Grants are not awarded to individuals, nor for: loans, annual campaigns, program support, endowment funds, operating budgets, emergency funds, deficit financing, scholarships, fellowships, publications, conferences, or continuing support.
Geographic Focus: New York, North Carolina, Canada
Contact: William McFarland; (716) 874-0040; cummings.foundation@verizon.net
Sponsor: James H. Cummings Foundation
1807 Elmwood Avenue, Room 112
Buffalo, NY 14207

James M. Collins Foundation Grants **2461**
The foundation awards grants to Texas nonprofit organizations in its areas of interest, including: arts, economic development, health organizations, association, higher education, human services, museums, Salvation Army, secondary school/education.

Types of support include research grants, program support, and social services. There are no application deadlines or forms.
Requirements: Texas nonprofit organizations are eligible.
Restrictions: Individuals are not eligible.
Geographic Focus: Texas
Contact: Dorothy Dann Collins Torbert, President; (214) 691-2032
Sponsor: James M. Collins Foundation
8115 Preston Road, Suite 680
Dallas, TX 75225

James McKeen Cattell Fund Fellowships **2462**

The nonrenewable fellowships advance the science of psychology and its useful applications by enabling faculty members to take a full year sabbatical leave for independent study and research. Annual deadline dates may vary; contact program staff for exact dates.
Requirements: Eligibility is limited to residents of North America. The fellowships are intended for tenured faculty of a psychology department who are eligible for sabbatical leave. Matching institutions' support required.
Geographic Focus: All States
Date(s) Application is Due: Dec 1
Amount of Grant: Up to 32,000 USD
Contact: Dr. Christina Williams, Program Contact; williams@psych.duke.edu
Internet: http://www.cattell.duke.edu
Sponsor: James McKeen Cattell Fund
9 Flowers Drive, Box 90086, Duke University, Department of Psychological and Brain Sciences
Durham, NC 27708-0086

James R. Dougherty Jr. Foundation Grants **2463**

The James R. Dougherty, Jr. Foundation, established in 1950, gives to organizations primarily in Texas. Areas of interest include: family services, domestic violence, human services and women. Support is offered in the form of: annual campaigns, building/renovation, capital campaigns, continuing support, curriculum development, endowments, equipment, general/operating support, income development, management development/capacity building, matching/challenge support, program development, program evaluation, research, scholarship funds, seed money and, technical assistance grants. The board meets twice a year, in the Spring and Fall. No application form is required, contact the Foundation, before submitting a proposal.
Requirements: Nonprofit organizations in Texas are eligible to apply.
Geographic Focus: Texas
Date(s) Application is Due: Mar 1; Sep 1
Contact: Daren Wilder, Grants Administrator; (512) 358-3560
Sponsor: James R. Dougherty Jr. Foundation
P.O. Box 640
Beeville, TX 78104-0640

James S. McDonnell Foundation Brain Cancer Research Collaborative Activity Awards **2464**

The Foundation offers Collaborative Activity Awards to initiate interdisciplinary discussions on problems or issues, to help launch interdisciplinary research networks, or to fund communities of researchers/practitioners dedicated to developing new methods, tools, and applications of basic research to applied problems. All proposals submitted to the foundation must clearly link the experimental models and questions to human disease. Proposals primarily intending to characterize basic mechanisms of growth and development that may plausibly but are not yet known to be contributory to human brain cancer are not encouraged. Proposals testing molecules as possible treatment interventions should consider including tests designed to uncover unintended biological effects of such molecules that would disqualify future clinical usefulness. The Foundation is particularly interested in supporting novel research that will generate new knowledge leading to increased rates of survival and improve functional recovery for individuals with brain cancer.
Restrictions: The Foundation does not fund: undergraduate tuition, stipends, scholarships, fellowships, research or travel expenses, or other educational expenses; graduate or postdoctoral stipends, scholarships, fellowships, research or travel expenses, or other educational expenses (exceptions include requests that qualify as allowable budget items as part of specific Research Award applications); expenses tied to projects whose explicit goal is the publication of a book or other bound volume (although publication of a book or special issue of a journal may be one goal or outcome of work funded by the Foundation); expenses tied to the establishment or day-to-day running of a journal or small press; scientific meetings or workshops other than those JSMF puts together or which are affiliated with Collaborative Activity Awards; ongoing operational support for university-based centers, programs, or institutes; professional society meetings or specific sessions at such meetings; charitable functions, museum exhibitions, or similar causes or events; or charitable donations to individuals and organizations.
Geographic Focus: All States
Contact: Cheryl Washington; (314) 721-1532; washington@jmsf.org or info@jmsf.org
Internet: http://www.jsmf.org/programs/bc/index.htm
Sponsor: James S. McDonnell Foundation
1034 S Brentwood Boulevard, Suite 1850
Saint Louis, MO 63117

James S. McDonnell Foundation Complex Systems Collaborative Activity Awards **2465**

The Foundation offers Collaborative Activity Awards to initiate interdisciplinary discussions on problems or issues, to help launch interdisciplinary research networks, or to fund communities of researchers/practitioners dedicated to developing new methods, tools, and applications of basic research to applied problems. The Program supports scholarship and research directed toward the development of theoretical and mathematical tools that can be applied to the study of complex, nonlinear systems. It is anticipated that research funded in this program will address issues in fields such as biology, biodiversity, climate, demography, epidemiology, technological change, economic Complex Systems development, governance, or computation. While the program's emphasis is on the development and application of theoretical models used in these research fields and not on particular fields per se, JSMF is particularly interested in projects attempting to apply complex systems approaches to real world problems. Proposals attempting to apply tools and models to problems where such approaches are not yet considered usual or mainstream (for example, differentiating normal physiology from disease) are encouraged.
Restrictions: The Foundation does not fund: undergraduate tuition, stipends, scholarships, fellowships, research or travel expenses, or other educational expenses; graduate or postdoctoral stipends, scholarships, fellowships, research or travel expenses, or other educational expenses (exceptions include requests that qualify as allowable budget items as part of specific Research Award applications); expenses tied to projects whose explicit goal is the publication of a book or other bound volume (although publication of a book or special issue of a journal may be one goal or outcome of work funded by the Foundation); expenses tied to the establishment or day-to-day running of a journal or small press; scientific meetings or workshops other than those JSMF puts together or which are affiliated with Collaborative Activity Awards; ongoing operational support for university-based centers, programs, or institutes; professional society meetings or specific sessions at such meetings; charitable functions, museum exhibitions, or similar causes or events; or charitable donations to individuals and organizations.
Geographic Focus: All States
Contact: Cheryl Washington, Grants Manager; (314) 721-1532; fax (314) 721-7421; washington@jmsf.org or info@jmsf.org
Internet: http://www.jsmf.org/programs/cs/index.htm
Sponsor: James S. McDonnell Foundation
1034 S Brentwood Boulevard, Suite 1850
Saint Louis, MO 63117

James S. McDonnell Foundation Research Grants **2466**

The Awards are designed to support research projects with a high probability of generating new knowledge and insights. Projects submitted for funding consideration should be at an early, even preliminary stage of development that intend to break new ground or to challenge commonly-held assumptions. Projects submitted should be sufficiently novel, cross-disciplinary, or heterodox so that they have a strong likelihood of influencing the development of new ways of thinking about important problems. Awards provide adequate, flexible funding over a sufficient time period to allow investigators to pursue and develop innovative directions to their research programs. Funds can be expended over a minimum of 3 years or a maximum of 6 years. Smaller amounts of money expended over shorter amounts of time may be requested to help investigators pursue pilot projects or test the feasibility of an experimental approach. The applicant can apply the grant funds towards any research-based expense, including travel, equipment, and supplies. Funds can be used to support collaborative projects. A percentage of the funds can also be used to support small workshops organized by the applicant where the goal of the workshop is to gather expertise in support of the research objective.
Requirements: Applications for grants are considered only from 501(c)3 organizations which are not private foundations. Applications from foreign organizations will be considered only if they provide written legal opinion that they qualify as tax-exempt under these sections of the U.S. IRS code or if a qualifying American organization is authorized to receive funds for them under this code. Individuals also are eligible.
Geographic Focus: All States
Amount of Grant: Up to 450,000 USD
Contact: Cheryl Washington, Grants Manager; (314) 721-1532; fax (314) 721-7421; washington@jmsf.org or info@jmsf.org
Internet: http://www.jsmf.org/apply/research/index.htm
Sponsor: James S. McDonnell Foundation
1034 S Brentwood Boulevard, Suite 1850
Saint Louis, MO 63117

James S. McDonnell Foundation Scholar Awards **2467**

The Awards program derives from and is consistent with the foundation's commitment to supporting high quality research and scholarship leading to the generation of new knowledge and its responsible application. Currently, the Awards are only available in the Brain, Mind & Behavior program area, and provide largely unrestricted funding over a sufficient time period to allow investigators to pursue and develop new directions to their research programs.
Requirements: Applications must be sponsored by a nonprofit institution as defined by Section 501(c)3 of the United States Internal Revenue Tax Code. Eligible Scholar Award nominees must have completed all doctoral, postdoctoral, or fellowship training and hold an independent research position. Nominees in the earlier career stages are encouraged.
Restrictions: Researchers with current grant support from JSMF are not eligible to apply for a Scholar Award until all prior JSMF grant funds have been fully expended and a final grant report has been received and approved by the foundation office.
Geographic Focus: All States
Amount of Grant: Up to 600,000 USD
Contact: Cheryl Washington, Grants Manager; (314) 721-1532; fax (314) 721-7421; washington@jmsf.org or info@jmsf.org
Internet: http://www.jsmf.org/apply/scholar/index.htm
Sponsor: James S. McDonnell Foundation
1034 S Brentwood Boulevard, Suite 1850

Saint Louis, MO 63117

James S. McDonnell Foundation Understanding Human Cognition Awards 2468
The long-standing interest of the James S. McDonnell Foundation on human mind/brain is reflected in changes made to the 2008 RFA, including the explicit emphasis on understanding human cognition. The program is intended to help investigators pursue experiments designed to answer well-articulated questions. JSMF Scholar Awards support research studying how neural systems are linked to and support cognitive functions and how cognitive systems are related to an organism's (preferably human) observable behavior. Studies with model organisms should justify why such models were selected and how data obtained from models advances our understanding of human cognition. Proposals proposing to use functional imaging to identify the neural correlates of cognitive or behavioral tasks (for example, mapping the parts of the brain that light up when different groups of subjects play chess, solve physics problems, or choose apples over oranges) are not funded through this program. In general, JSMF and its expert advisors have taken an unfavorable view of projects attempting too wide a leap in a single bound. Functional imaging studies using poorly characterized tasks as proxies for complex behavioral issues involving empathy, moral judgments, or social decision-making are generally not appropriate responses to this call for proposals. In past competitions, proposals structured along such lines were eliminated from funding consideration early in the review process.
Requirements: Aspects of proposals appropriate to the JSMF UHC program would include, but are not limited to: characterizing the cognitive operations involved in performing a task; studying how the brain identifies, extracts and uses relevant information from complicated environments; examining how manipulations and/or perturbations at one spatial or temporal scale are meaningful at finer or coarser levels of organization (e.g. does a synaptic change account for a change in network function and vice versa); re-examining common wisdom assumptions (such as the existence of critical periods in human learning); evaluating the usefulness of methodologies or improving the usefulness of methodologies commonly used in mind/brain research; applying approaches and knowledge from cognitive psychology or cognitive science to important problems in education, training, or rehabilitation; taking a comparative, evolutionary approach to characterizing the uniqueness of the human brain and of human cognition.
Restrictions: The Foundation does not fund: undergraduate tuition, stipends, scholarships, fellowships, research or travel expenses, or other educational expenses; graduate or postdoctoral stipends, scholarships, fellowships, research or travel expenses, or other educational expenses (exceptions include requests that qualify as allowable budget items as part of specific Research Award applications); expenses tied to projects whose explicit goal is the publication of a book or other bound volume (although publication of a book or special issue of a journal may be one goal or outcome of work funded by the Foundation); expenses tied to the establishment or day-to-day running of a journal or small press; scientific meetings or workshops other than those JSMF puts together or which are affiliated with Collaborative Activity Awards; ongoing operational support for university-based centers, programs, or institutes; professional society meetings or specific sessions at such meetings; charitable functions, museum exhibitions, or similar causes or events; or charitable donations to individuals and organizations.
Geographic Focus: All States
Contact: Cheryl Washington, Grants Manager; (314) 721-1532; fax (314) 721-7421; washington@jmsf.org or info@jmsf.org
Internet: http://www.jsmf.org/programs/uhc/index.htm
Sponsor: James S. McDonnell Foundation
1034 S Brentwood Boulevard, Suite 1850
Saint Louis, MO 63117

James T. Grady-James H. Stack Award for Interpreting Chemistry for the Public 2469
This Award is given annually to recognize, encourage, and stimulate outstanding reporting directly to the public, which materially increases the knowledge and understanding of chemistry, chemical engineering, and related fields. This information may have been disseminated through the press, radio, television, films, the lecture platform, or books or pamphlets for the lay public. The award consists of $3,000, a medallion with a presentation box, and a certificate.
Requirements: Any individual, except a member of the award committee, may submit one nomination or seconding letter for the award in any given year. The nominating documents consist of a letter of not more than 1000 words containing an evaluation of the nominee's accomplishments and a specific identification of the work to be recognized, a biographical sketch including date of birth, and a list of publications authored by the nominee.
Restrictions: Application or self-nomination is not acceptable.
Geographic Focus: All States
Date(s) Application is Due: Nov 1
Amount of Grant: 3,000 USD
Contact: Felicia Dixon, Awards Administrator; (800) 227-5558 or (202) 872-4408; fax (202) 776-8008; f_dixon@acs.org or awards@acs.org
Internet: http://portal.acs.org/portal/acs/corg/content?_nfpb=true&_pageLabel=PP_ARTICLEMAIN&node_id=1319&content_id=CTP_004537&use_sec=true&sec_url_var=region1
Sponsor: American Chemical Society
1155 Sixteenth Street, NW
Washington, DC 20036-4801

Jane Beattie Memorial Scholarship 2470
The purpose of the fund is to provide scholarships to subsidize travel to the U.S. for purposes of scholarly activity by a foreign scholar in the area of judgment and decision research, broadly defined. Attendance at the annual SJDM meetings is one example of an activity that would be appropriate for support, but by no means the only one. In most years, the Fund awards one or two scholarships in amounts of $400-$700 each.
Requirements: Applicants should be scholars living and working in a country other than the U.S. who will use the award to help pay for travel to the U.S. for scholarly activities associated with research in judgment and decision making. It is anticipated that most awards will be granted to faculty or graduate students at colleges and universities, especially recent and soon-to-be Ph.D.'s, but others will also be considered. The required application form can be downloaded from: http://gsbwww.uchicago.edu/fac/joshua.klayman/more/BeattieForm.rtf
Restrictions: U.S. Scholars are not eligible to apply.
Geographic Focus: All States
Date(s) Application is Due: Jul 26
Amount of Grant: 400-700 USD
Contact: Joshua Klayman, (773) 834-9134; joshk@uchicago.edu
Internet: http://www.sjdm.org/content/beattie-award
Sponsor: Society for Judgment and Decision Making
College of Business, Florida State University, P.O. Box 3061110
Tallahassee, FL 32306-1110

Jane Bradley Pettit Foundation Health Grants 2471
The Jane Bradley Pettit Foundation awards grants to initiate and sustain projects in the Greater Milwaukee community, with a focus on programs and projects that serve low-income and disadvantaged individuals, women, children, and the elderly. In the area of Health, the Foundation gives priority to community-based health care and prevention programs which address the physical and mental health needs of families, children, persons at-risk and the elderly. Support and advocacy for victims of abuse and neglect are also a priority. An initial application should be in the form of a letter of intent.
Requirements: Milwaukee-area charitable organizations are eligible.
Geographic Focus: Wisconsin
Date(s) Application is Due: Jan 15; May 15; Sep 15
Amount of Grant: 5,000-100,000 USD
Contact: Heidi Jones, Director of Administration; (414) 982-2880 or (414) 982-2874; fax (414) 982-2889; hjones@staffordlaw.com
Internet: http://www.jbpf.org/guidelines/index.html
Sponsor: Jane Bradley Pettit Foundation
1200 N. Mayfair Road, Suite 430
Wauwatosa, WI 53226-3282

Japan Foundation Center for Global Partnership (CGP) Grants 2472
Program supports Japan-U.S. collaborative policy-oriented projects in the target areas listed below. In addition, CGP recognizes the value of including a multinational dimension within projects, and therefore gives priority to those that incorporate Asia into its agenda. Prospective grantees must submit letters of inquiry for CGP consideration, 2-3 pages minimum, no later than March 1. Completed applications invited by CGP must be received by May 1.
Requirements: Proposals from U.S. organizations with 501(c)3 status only will be accepted by CGP New York. Appropriate Japan-based institutional collaboration must be demonstrated and secured throughout the course of the project.
Geographic Focus: All States
Date(s) Application is Due: Mar 1; May 1
Contact: Tomoki Akazawa, Program Director; (212) 489-1255; fax (212) 489-1344; tomoki_akazawa@cgp.org
Internet: http://www.cgp.org/index.php?option=section&id=3
Sponsor: Japan Foundation Center for Global Partnership
152 West 57th Street, 17th Floor
New York, NY 10019-3101

Japan Foundation Center for Global Partnership (CGP) Institutional Grants 2473
With emphasis on academic/intellectual excellence and national and international public consequence, CGP's institutional grant program supports policy-oriented projects conducted by an array of institutions from the United States and Japan, including non-profit organizations, universities, and think tanks as well as policymakers, scholars, and educators, with a view to creating new networks and providing opportunities for advancing research and candid discussion of these issues. Within the mandate of enhancing Japan-U.S. collaboration, CGP recognizes the value of including a multinational dimension within projects, and gives priority to those that incorporate Asia into its agenda. Applicants must submit letters of inquiry for consideration, 2-3 pages minimum, no later than March 3. Full proposals must be received by May 1.
Requirements: Applicants can include: non-profit organizations; universities; think tanks; policymakers; scholars; and educators.
Geographic Focus: All States
Date(s) Application is Due: Mar 3; May 1
Contact: Yukiko Ono; (212) 489-1255; fax (212) 489-1344; yukiko_ono@cgp.org
Internet: http://www.jpf.go.jp/cgp/e/grant/index.html
Sponsor: Japan Foundation Center for Global Partnership
152 West 57th Street, 17th Floor
New York, NY 10019-3101

Japan Foundation Center for Global Partnership (CGP) NP.O. Fellowships 2474
The Fellowship supports practical training in the United States for professionals committed to working on social and global issues through the activities of a Japanese nonprofit organization. The program seeks to build long-lasting relationships between fellows and their counterparts in the United States, fostering information exchange between the

nonprofit sectors in both countries and global perspectives on the role of civil society in addressing pressing social issues. It is particularly focused on nurturing new nonprofit leaders who have the ability to initiate meaningful change within Japan and lead future collaborative efforts between the United States and Japan. Fellows pursue self-directed training agendas from the home base of a U.S. nonprofit organization (host organization) that is well-aligned with the interests and objectives of the fellowship. Training agendas enable the fellow to develop management skills for nonprofit organizations and knowledge around a selected theme, such as program development and implementation, development (fund raising), staff/volunteer/board management, and communications. The fellow's specialty field provides a context for the development of skills and knowledge. In the past, fellows have come from fields such as volunteer promotion, at-risk youth, domestic violence, social entrepreneurship, and capacity-building for nonprofits.

Requirements: All fellows are required to participate in orientations in Japan. Fellows have a minimum of three years of experience at Japanese nonprofits. Ability to communicate in English and capacity to pro-actively carry out fellowship activities in a foreign culture are also considered.

Geographic Focus: All States

Contact: Yukiko Ono; (212) 489-1255; fax (212) 489-1344; yukiko_ono@cgp.org

Internet: http://www.cgp.org/index.php?option=article&task=default&articleid=156&id=14

Sponsor: Japan Foundation Center for Global Partnership

152 West 57th Street, 17th Floor

New York, NY 10019-3101

Japan Foundation Center for Global Partnership Security Studies Fellowships 2475

The Fellowship program, administrated by the Research Institute for Peace and Security (RIPS) and operated in collaboration with RIPS and CGP, aims to cultivate excellent young researchers with a global viewpoint in the field of security in Japan. The program endeavors to rear experts who will contribute to security studies, domestically and internationally, by providing the young researchers interested in this area with a research scholarship and by holding research seminars to provide support for specialized research. At the same time, the program aims to heighten academic interest relating to international security, arms control, strategic studies, Japanese security issues and other such fields which were late to develop in Japan's spheres of research and education.

Geographic Focus: All States

Contact: Yoshinobu Yamamoto; +81-(0)3-5562-3542; fax +81-(0)3-5562-3504

Internet: http://www.jpf.go.jp/cgp/e/fellow/ssfp/index.html

Sponsor: Japan Foundation Center for Global Partnership

Ark Mori Building, 20 Floor, 1-12-32, Akasaka

Minato-ku, Tokyo, 107-6021 Japan

Japan Foundation Japanese Studies Fellowships 2476

In order to promote Japanese Studies overseas, the Japan Foundation's Japanese Studies Fellowships offer preeminent foreign scholars in Japanese studies the opportunity to conduct research in Japan. The foundation offers three types of fellowships: Scholars and Researchers-Long Term (2-12 months); Scholars and Researchers-Short Term (21-59 days); and Doctoral Candidates (4-12 months). Long-term and Doctoral fellows receive the following benefits: a monthly stipend of either 370,000 or 430,000 yen for long-term fellows (amount is determined according to fellow's professional status) or 310,000 yen for Doctoral fellows; a settling-in allowance of 250,000 yen (for fellows not already in Japan); a departure allowance of 50,000 yen; a dependent allowance of 50,000 yen per month (for dependents who reside in Japan continuously for more than six months); coverage of enrollment fees (up to 50,000 yen per month); and a research allowance for Long-term fellows of up to 40,000 yen per month. Short-term fellows receive a per-diem allowance of 17,000 yen per day. All fellows who are not already in Japan receive economy-class airfare to and from Japan and traveler's insurance. U.S. citizens (except for those who are residing in other countries as permanent residents) and permanent U.S. residents should submit their applications online no later than November 1 at the Japan Foundation New York website-and should also follow the application guidelines at that site. All other applicants have a deadline of December 1 and may download their application guidelines and forms from the Japan Foundation website. Applicants who are citizens or permanent residents in countries other than the U.S. or Taiwan should submit their application materials to the nearest Foundation Office or Japanese diplomatic mission. Taiwanese applicants, including permanent residents in Taiwan, should submit their application materials to the Taipei office of the Interchange Association, Japan. Applicants residing in countries in which they hold neither nationality nor permanent residency (including residents in Japan) should submit their applications to the Foundation Headquarters in Tokyo. A list of contact information is included in the downloadable program guidelines (WORD) at the foundation website. Please note that applications received after the deadline will not be considered and that multiple applications from the same applicant will make all applications invalid. The Japan Foundation was established in 1972 by the Japanese legislature and became one of Japan's "Independent Administrative Institutions" in October 2003. The foundation maintains its headquarters in Tokyo and operates through a network of 21 overseas offices in 20 countries worldwide. The foundation's mission is to promote international cultural exchange and mutual understanding between Japan and other countries. To do so it offers three major categories of programming: Arts & Culture; Japanese Studies; and Japanese-Language Education.

Requirements: Applicants for long- and short-term fellowships must hold a Ph.D. or equivalent professional experience in research, teaching, and writing in the fields of the humanities or social sciences. Doctoral Candidates must have achieved All-But-Dissertation (ABD) status by the time the fellowship begins. Applicants must hold nationality or permanent residency in countries that have diplomatic relations with Japan (Taiwanese applicants, including permanent residents of Taiwan, are eligible to apply for the long-term and doctoral-candidate fellowships). Applicants must also satisfy the following criteria: be in good health; be proficient in either Japanese or English; secure all affiliation arrangements by the application deadline (excluding the short-term fellowship); and stay continuously in Japan for the term of fellowship.

Restrictions: Previous recipients of Japan Foundation long-term or doctoral fellowships are eligible to reapply only if a full three years have elapsed since the termination of their previous fellowship. Previous short-term fellows may reapply after one full year from the termination of their most recent fellowship. Applicants who have received Japan Foundation fellowships in the past on two or more occasions will be given lower priority. Individuals meeting any of the following conditions are not eligible to apply: scholars or researchers in the natural sciences, medicine, or engineering; undergraduates; master's degree candidates; those intending to enroll in undergraduate or master's courses at universities in Japan; graduates intending to pursue Ph.D. courses in Japan; those who intend to use the fellowship for political or religious purposes; those who primarily intend to enhance their Japanese-language ability or technical skills; those who intend to develop and/or produce Japanese-language teaching materials; those who intend to acquire artistic training in the area of Japanese traditional culture (tea ceremony, flower arrangement, etc.); those who plan to receive other grants concurrently with the fellowship; or those who have been in Japan continuously for more than a year at the time of application. Japanese citizens must hold permanent residency in another country to be eligible to apply. Applicants holding nationality of countries not maintaining diplomatic relations with Japan are not eligible to apply, even if they have permanent residency in a country with diplomatic relations with Japan.

Geographic Focus: All Countries

Date(s) Application is Due: Nov 1; Dec 1

Amount of Grant: 357,000-4,880,000 JPY

Contact: Program Director; 03-5369-6051 or 03-5369-6069; fax 03-5369-6041

Internet: http://www.jpf.go.jp/e/program/intel.html

Sponsor: Japan Foundation

4-4-1 Yotsuya, Shinjuku-ku

Tokyo, 160-0004 Japan

Japan Foundation New York Institutional Support for Japanese Studies Grants 2477

To encourage the sustained and innovative growth of Japanese Studies in national as well as transnational contexts, the Japan Foundation is expanding its Institutional Support Program (ISP) in the United States. Individual grants will be made annually to colleges, universities (or consortia of colleges and universities), and academic organizations that play a central role in Japanese studies in the U.S. and that demonstrate backing for those proposals through matching funds. Proposals should address the needs, opportunities, and visions of particular institutions. Applicants are encouraged to think practically and inventively about their discrete projects. Projects may engage specific or comprehensive goals and may entail smaller or larger grants. Priority will be given to projects that contribute significantly to infrastructure building for the field, provide opportunities for students and faculty to travel to Japan, and, most desirably, enhance the quality of teaching and research in Japanese Studies as well as the quality of equipment and materials (including those in libraries). Links to application guidelines and forms are available at the website. Application packets must be mailed to the Japanese Foundation New York and must be received by the due date. As application deadlines may vary from year, interested applicants are advised to verify current deadline dates. The Japan Foundation was established in 1972 by the Japanese legislature and became one of Japan's "Independent Administrative Institutions" in October 2003. The foundation maintains its headquarters in Tokyo and operates through a network of 21 overseas offices in 20 countries worldwide including offices in New York and Los Angeles, and a Center for Global Partnership, also in New York. The foundation's mission is to promote international cultural exchange and mutual understanding between Japan and other countries. To do so, it offers three major categories of programming: Arts & Culture; Japanese Studies; and Japanese-Language Education.

Requirements: United States colleges, universities (or consortia of colleges and universities), and academic organizations playing a central role in Japanese Studies are eligible to apply. Small institutions, newly-established institutions, and institutions without an established program of Japanese studies at the graduate level are encouraged to apply and to submit proposals that emphasize lasting impact and sustainability (for example, creation of new faculty positions in Japanese studies, etc.). The foundation requires submission of multi-year proposals (up to three years in length). Applicants may request a maximum of $150,000 per year. Applicants must be prepared to match the grant equally, and optimally, at higher levels. Matching funds can be both cash and in-kind contributions.

Restrictions: Previous ISP grantees may not apply through a project whose grant duration overlaps with their current grant project. The ten universities that received sizable Japanese-government donations in 1973 are eligible to apply, but will be given lower priority except in cases in which their proposals are involved in outreach, consortium building, or partnering with institutions and organizations in Japan. Proposed projects must not be used for religious, political, or commercial purposes, or to support specific doctrines or claims. The foundation will not consider the following types of requests: projects in the field of medicine, manual training, or natural sciences; organization of or participation in sporting events and training; projects in the field of art; goodwill exchanges; social-welfare or international-development projects; tourist-oriented programs; and language-education programs (however, professional language training for Japanese-Studies scholars is eligible). The applicant may not request funding for the following items: indirect costs or costs that are not directly needed to carry out the applied projects; overhead costs of the administrative body of the institution; salaries of administrative staff (funds may be requested for assistants directly related to the project, up to a limit of thirty days); equipment such as computers, photocopiers, fax machines, electric appliances, office furniture, or computer supplies; real estate; construction fees; rental fees for facilities that would be used long-term; rental fees for facilities within the

applying institution; international airfare that is higher than the estimated economy-class fare for the shortest route from the city of boarding to the destination; or IATA Special Excursion Fares (PEX discount fare).
Geographic Focus: All States
Date(s) Application is Due: Dec 9
Contact: Noriko Yamamoto; (212) 489-0299; fax (212) 489-0409; info@jfny.org
Internet: http://www.jfny.org/japanese_studies/institutionalsupport.html
Sponsor: Japan Foundation in New York
152 W 57th Street, 17th Floor
New York, NY 10019

Japan Foundation New York Japanese Studies Fellowships **2478**

The Japan Foundation New York's Japanese Studies Fellowships give outstanding scholars in the humanities and social sciences in the U.S. the opportunity to conduct research in Japan. The foundation offers three types of fellowships: Scholars and Researchers-Long Term (2-12 months); Scholars and Researchers-Short Term (21-59 days); and Doctoral Candidates (4-12 months). Long-term fellows and Doctoral fellows receive the following benefits: a monthly stipend of either 370,000 or 430,000 yen for long-term fellows (amount is determined according to fellow's professional status) or 310,000 yen for graduate fellows; a settling-in allowance of 250,000 yen (for fellows not already in Japan); a departure allowance of 50,000 yen; a dependent allowance of 50,000 yen per month (for dependents who reside in Japan continuously for more than six months); coverage of enrollment fees (up to 50,000 yen per month); and a research allowance for long-term fellows of up to 40,000 yen per month. Short-term fellows receive a per-diem allowance of 17,000 yen per day. All fellows who are not already in Japan receive economy-class airfare to and from Japan and traveler's insurance. Links to the application guidelines (program instructions) and an online submission system are available at the website. American citizens residing in the U.S. or permanent U.S. residents must submit their applications via this on-line system by the November 1 deadline date. (American citizens with permanent residency abroad have different application requirements and should visit the website and/or contact the foundation for information on how to apply.) All candidates are encouraged to carefully peruse the website and/or email the foundation for more detailed requirements and guidelines. The Japan Foundation was established in 1972 by the Japanese legislature and became one of Japan's "Independent Administrative Institutions" in October 2003. The foundation maintains its headquarters in Tokyo and operates through a network of 21 overseas offices in 20 countries worldwide including offices in New York and Los Angeles, and a Center for Global Partnership, also in New York. The foundation's mission is to promote international cultural exchange and mutual understanding between Japan and other countries. To do so it offers three major categories of programming: Arts & Culture; Japanese Studies; and Japanese-Language Education.
Requirements: Applicants must be American citizens, American citizens temporarily residing abroad, or permanent residents of the United States. Non-U.S. citizens are encouraged to visit the Japan Foundation Headquarters website (www.jpf.go.jp/e/) to apply for similar fellowship opportunities. Applicants for long- and short-term fellowships must hold a Ph.D. or equivalent professional experience in research, teaching, and writing in the fields of the humanities or social sciences. Doctoral Candidates must have achieved All-But-Dissertation (ABD) status by the time the fellowship begins. Applicants must also satisfy the following criteria: be in good health; be proficient in either Japanese or English; secure all affiliation arrangements by the application deadline (excluding the short-term fellowship); and stay continuously in Japan for the term of fellowship.
Restrictions: The duration of the fellowship may be adjusted by the Foundation. Previous recipients of Japan Foundation fellowships are eligible to reapply only if a full three years have elapsed since the termination of their previous fellowship. Previous Short-Term Research Fellows may reapply after one full year from the termination of their most recent fellowship. Individuals meeting any of the following conditions are not eligible to apply: scholars or researchers in the natural sciences, medicine, or engineering; undergraduates; master's degree candidates; those intending to enroll in undergraduate or master's courses at universities in Japan; graduates intending to pursue Ph.D. courses in Japan; those who intend to use the fellowship for political or religious purposes; those who primarily intend to enhance their Japanese-language ability or technical skills; those who intend to develop and/or produce Japanese-language teaching materials; those who intend to acquire artistic training in the area of Japanese traditional culture (tea ceremony, flower arrangement, etc.); those who plan to receive other grants concurrently with the fellowship; or those who have been in Japan continuously for more than a year at the time of application.
Geographic Focus: All States
Date(s) Application is Due: Nov 1
Amount of Grant: 357,000-4,880,000 JPY
Contact: Noriko Yamamoto, Program Director, Japanese Studies; (212) 489-0299; fax (212) 489-0409; info@jfny.org
Internet: http://www.jfny.org/japanese_studies/fellowship.html
Sponsor: Japan Foundation in New York
152 W 57th Street, 17th Floor
New York, NY 10019

Japan Foundation Six-Month Grants for Specialists in Cultural and Academic Fields **2479**

The Japan Foundation provides non-native specialists (researchers, postgraduate students, librarians, museum curators, etc.) with an opportunity to take a six-month Japanese-language course specifically designed for their research needs. The program focuses on Japanese-language proficiency necessary for specialized work or research activities and assists participants to acquire language skills necessary to collect and present information in their major fields as well as network with people in their fields. The program also coaches participants as they carry out their research and work activities (visits to other institutions, interviewing professionals, etc.). Training takes place from September through March at the foundation's Japanese-Language Institute (JFJLI, Kansai) in Japan. Exact dates may vary from year to year and are posted at the foundation website. The following expenses will be covered by the training grant: round-trip airfare (discount economy class), airport tax, and foreign-travel tax; accommodation at JFJLI, Kansai; overseas-travel insurance; meals during the training program; and other expenditures necessary for participation in the program. Nationals from the following regions may also receive a set allowance to cover necessary expenses such as public-transportation fees, communication fees, etc.: Asia and Oceania; Central and South America; Eastern Europe; Middle East and North Africa; and Africa. (Applicants should check the guidelines for specifics.) Detailed program information and application guidelines as well as official application forms are downloadable from the foundation website. Applications should be submitted to the applicant's nearest Japan Foundation overseas office or Japanese diplomatic mission no later than December 1. In Taiwan, applications should be submitted to the Interchange Association, Japan, Taipei office. Applicants must notify their respective overseas office, mission, or interchange of their intention to apply prior to submitting an application. Contact information for both the foundation's Japan and overseas offices is available under the "JF Worldwide" tab at the foundation website. Deadlines may vary from year to year. Applicants are encouraged to verify current deadline dates. The Japan Foundation was established in 1972 by the Japanese legislature and became one of Japan's "Independent Administrative Institutions" in October 2003. The foundation maintains its headquarters in Tokyo and operates through a network of 21 overseas offices in 20 countries worldwide. The foundation's mission is to promote international cultural exchange and mutual understanding between Japan and other countries. To do so it offers three major categories of programming: Arts & Culture; Japanese Studies; and Japanese-Language Education.
Requirements: This grant is an intensive residential-training course for specialists (e.g. scholars, researchers, full-time postgraduate students, and full-time librarians and museum curators) outside Japan who need a good command of the Japanese language in order to conduct field-related work and research. Applicants must be postgraduate students in institutions outside Japan or other specialists who are employed by (and upon their return continue to work for) applicable organizations outside Japan (e.g. institutions of higher education, research centers, public-sector organizations, research or public libraries, cultural-exchange organizations, or museums which are currently working for international-exchange programs that concern Japan). Permission to participate from the applicant's affiliated institution is an eligibility requirement. Additionally applicants must satisfy the following criteria: be in the fields of Japan-related social sciences or humanities (excluding Japanese-language study and Japanese-language education); have a need for Japanese-language proficiency to successfully carry out their work; have a specific and concrete theme related to Japan in their study or work and a plan to make a presentation on the achievements; have a minimum proficiency score of Level 4 on the Japanese-Language Proficiency Test (JLPT); be under the age of 40 at time of application; and be in good physical and mental health. All else being equal, priority will be given to the following applicants: specialists who are currently engaged in their profession; postgraduate students who plan to take a job related to Japanese studies; students officially registered in a Ph.D.-level program at a graduate school; applicants who have affiliations with major institutions; and those whose applications are related to major cultural-exchange events announced by the Ministry of Foreign Affairs of Japan.
Restrictions: Participants are not permitted to bring family members. The following individuals are not eligible to apply: those who have participated in the training programs for teachers of Japanese language at JFJLI, Urawa in the last five years; those who have participated in the Japanese Language Programs at JFJLI, Kansai for librarians or for specialists in cultural and academic fields; those who have participated in the Japanese Language programs at JFJLI, Kansai for researchers and postgraduate students for more than four-months in the past five years; those who completed their compulsory education in Japan; those who are scheduled for study or training under other programs in Japan; and those who hold nationality in a country that does not have diplomatic relations with Japan (Taiwanese are eligible to apply for this program).
Geographic Focus: All States, All Countries
Date(s) Application is Due: Dec 1
Contact: Program Director, Japanese-Language Teacher and Institutional Support Dept; 048-834-1181-2; fax 048-834-1170
Internet: http://www.jpf.go.jp/e/program/japanese.html#1
Sponsor: Japan Foundation
4-4-1 Yotsuya, Shinjuku-ku
Tokyo, 160-0004 Japan

Jefferson Science Fellows Program **2480**

This program establishes a new model for engaging the American academic science, technology, and engineering communities in the formulation and implementation of U.S. foreign policy. Each fellow will spend one year at the U.S. Department of State for an on-site assignment in Washington, DC, that may also involve extended stays at U.S. foreign embassies and/or missions. All JSF assignments will be designed in consultation with regional and/or functional bureaus within the U.S. Department of State. Following the fellowship year, the fellow will return to his/her academic career, but will remain available to the U.S. government as an experienced consultant for short-term projects over the following five years. Application and guidelines are available online.
Requirements: Nominees and applicants must be U.S. citizens and must hold a tenured faculty position at a U.S. degree granting academic institution of higher learning. Nominations will be accepted from U.S. academic institutions that have signed a JSF/MOU with the U.S. Department of State.
Geographic Focus: All States
Date(s) Application is Due: Dec 1
Contact: Fellowships Office; (202) 334-2872; fax (202) 334-3419; jsf@nas.edu
Internet: http://www7.nationalacademies.org/fellowships/Jefferson_Science_Fellows.html

Sponsor: National Academies
500 5th Street NW, Room 508
Washington, DC 20001

Jeffress Trust Awards in Interdisciplinary Research **2481**
The Thomas F. and Kate Miller Jeffress Memorial Trust was established in 1981 to support basic research in chemical, medical, and other scientific research through grants to educational and research organizations in the Commonwealth of Virginia. The trust was established under the will of Mr. Robert M. Jeffress, a business executive and philanthropist of Richmond, Virginia. Grants are made to assist scientists in educational and research institutions to conduct investigations in the natural sciences, generally considered to include chemistry, biology (with the exception of field studies, classification, or other largely observational studies), and the basic medical sciences, such as biochemistry, microbiology and others. The principal evaluation criteria will be the scientific significance of the proposed work and the competence of the investigator. The trust is particularly interested in supporting fundamental research by scientists early in their careers and new areas of research or more speculative projects by established investigators. In general, grants will cover direct-expense items essential to the successful completion of the proposed project. Grants from the Jeffress Memorial Trust are one year in duration. Funding may be renewed at the discretion of the allocations committee. Application materials are available for download at the grant website. Applicants are strongly encouraged to review the application guidelines for additional helpful information and clarifications before applying.
Requirements: Eligible research areas for Jeffress support include astronomy, biosciences, chemistry, computer sciences, engineering, environmental sciences, material science, mathematics and physics. Full-time faculty at institutions in Virginia that are within seven years of their first faculty appointment are eligible to apply as Principal Investigators. Student participation is a requirement of the proposed research plan. Online submissions must be received by January 14, 12:00 Noon, U.S. Eastern time. Mailed hard copies must be received by The Medical Foundation division by January 16, 5:00 pm, U.S. Eastern time.
Restrictions: Research in social, economic and behavioral sciences, including psychology, are currently outside the scope of Jeffress funding. Faculty from medical schools and schools of Osteopathic Medicine may not apply as Principal Investigators; however, they are encouraged to collaborate as Co-Investigators.
Geographic Focus: Virginia
Date(s) Application is Due: Jan 14; Jan 16
Amount of Grant: 100,000 USD
Contact: Jeanne Brown, Program Officer; (617) 279-2240 ext. 709; jbrown@hria.org
Internet: http://www.wm.edu/grants/OPPS/jeffress.htm
Sponsor: Thomas F. and Kate Miller Jeffress Memorial Trust
1111 E. Main Street, VA2-300-12-92
Richmond, VA 23219

Jeffrey Campbell Graduate Fellow in Anthropology **2482**
The university's Department of Anthropology invites applications for a fellowship in biological (physical) anthropology or paleoanthropology for next year's academic year. The successful candidate will be a specialist and qualified to teach in the area of human origins. The Campbell Fellow will receive a stipend per academic year with the possibility of additional funds for travel to conferences and professional meetings. The fellow will also be afforded office space and a personal computer. It is expected that fellows will reside at the university for one academic year and to teach one course per semester in an area of the candidate's research interest (in consultation with the department). Fellows will also present a research-based paper in the fellows' lecture series each year. Send a detailed letter of application, curriculum vita, three letters of reference, and a short (one to two pages) statement of pedagogy to the office. Guidelines are available online.
Requirements: Applicants must have completed coursework and preliminary examinations for the PhD in anthropology. Candidates should be members of racial or ethnic groups historically underrepresented in higher education in the United States.
Restrictions: Geographic area is open, except for South Asia.
Geographic Focus: New York
Amount of Grant: 28,500 USD
Contact: Dr. Alice Pomponio; (315) 229-5797; apomponio@stlawu.edu
Internet: http://web.stlawu.edu/resources/positions/faculty/fy%202005-2006/anthrojcg f0805.htm
Sponsor: Saint Lawrence University
Department of Anthropology
Canton, NY 13617

Jeffrey Thomas Stroke Shield Foundation Research Grants **2483**
The Dr. Jeffrey Thomas Stroke Shield Foundation (JTSSF) welcomes submissions of grant requests for funding covering all aspects of stroke prevention, treatment and awareness. JTSSF is actively seeking to provide seed money grants for projects that are important, achievable, and innovative that will ultimately lead to stroke prevention and a better treatment and quality of life for stroke sufferers. From the Foundation's perspective within clinical stroke treatment and research, it understands the importance of directing targeted research funds to the scientists and investigators who can move the best ideas in stroke research prevention and treatment forward. Letters of Intent (LOIs) are required and must be postmarked by February 28. All LOI submissions will undergo a review and selection process by the JTSSF Scientific Advisory Board. Only those projects deemed to be of suitable scientific merit will be invited to submit a full application. If invited to submit an application, applicants will receive additional instructions and information necessary for submission. Final grant applications are due by May 16.
Requirements: Principal investigators who hold appointments/positions in accredited hospitals and clinics, healthcare institutions, academic medical centers, professional associations, member societies and other not?for?profit entities in the U.S. are welcome to apply.
Geographic Focus: All States
Date(s) Application is Due: May 16
Contact: Nancy Weber, Executive Administrator; (415) 830-6031; info@strokeshieldfoundation.org
Internet: http://strokeshieldfoundation.org/researchers/grant-program/
Sponsor: Jeffrey Thomas Stroke Shield Foundation
3053 Fillmore Street, #268
San Francisco, CA 94123

Jenifer Altman Foundation Grants **2484**
The foundation is a small foundation working for a socially just and ecologically sustainable future, innovative research and demonstration projects in mind-body health, and improved child care prospects for at-risk children. Small grants are made in two areas: initiatives concerning the impact of endocrine disrupting chemicals on human health and the preservation of the environment and initiatives supporting the global citizens' movement for a just and sustainable future in relation to international governmental decision-making forums. The foundation awards a few grants for innovative research and demonstration programs in mind-body health, especially those focused on cancer. Local initiatives that contribute to the quality of life in Bolinas, CA, are also of interest. Applicants should send a concept letter to determine if their projects align with the foundations areas of interest.
Requirements: Creative nonprofits are eligible to apply for one-time grants.
Restrictions: Grants are not made to individuals. The foundation rarely funds programs for the conservation of specific species or habitats and rarely funds research at universities on environmental issues. Grants are not made for civic programs, projects in other countries, or projects that are focused entirely on one city or region in the United States.
Geographic Focus: All States
Amount of Grant: 500-50,000 USD
Contact: Marni Rosen, Executive Director; (415) 561-2182 or (415) 561-2187; fax (415) 561-6480; info@jaf.org or mrosen@jaf.org
Internet: http://www.jaf.org/apply/index.html
Sponsor: Jenifer Altman Foundation
P.O. Box 29209
San Francisco, CA 94129

Jessie Smith Noyes Foundation Grants **2485**
The foundation is committed to preventing damage to the natural systems upon which all life depends and to strengthening individuals and institutions committed to protecting natural systems and ensuring a sustainable society. The foundation makes grants primarily in the interrelated areas of environment and reproductive rights. The program components include toxics, sustainable agriculture (both with emphasis on southern and Rocky Mountain states), and U.S. reproductive rights. In addition, a few grants are made in four areas of special concern: sustainable communities, U.S. environmental justice, strengthening the U.S. nonprofit sector, and environmental issues in the metropolitan New York region. Letters of inquiry are received at any time; proposals will be requested from the foundation after review.
Requirements: 501(c)3 tax-exempt organizations are eligible.
Restrictions: Normally, the foundation will not consider requests for direct service, endowments, loans or scholarships to individuals, capital construction funds, conferences, media events, production of media and TV programming, or general fund-raising drives. General research projects are not funded per se.
Geographic Focus: All States
Contact: Millie Buchanan; (212) 684-6577; fax (212) 689-6549; noyes@noyes.org
Internet: http://www.noyes.org
Sponsor: Jessie Smith Noyes Foundation
6 E 39th Street, 12th Floor
New York, NY 10016-0112

Jewish Federation of Metropolitan Chicago Scholarships **2486**
Scholarship funds are available for Jewish college students. The scholarships support full-time students, predominantly those legally domiciled in the metropolitan Chicago area, who are identified as having promise for significant contributions in their chosen careers. Assistance is available for those in the helping professions, i.e., medicine, education, Jewish communal service, the rabbinate, social service, law, and communications at the University of Illinois (Champaign-Urbana).
Requirements: Applicants must be Jewish; meet one or both of the following criteria—born and raised in either Cook County, Chicago metropolitan area or Northwest Indiana, or prior to starting professional education, worked full-time for at least 12 months in Cook County or Chicago metropolitan area; intend to remain in the Chicago metropolitan area after completing school; and be entering a professional postgraduate education program on a full-time basis (school acceptance is not required at time of scholarship application). Undergraduate applicants must be entering the junior or senior year of a professional education program that will allow them to pursue employment in their chosen helping professions at a bachelor's degree level.
Geographic Focus: Illinois
Date(s) Application is Due: Feb 15
Amount of Grant: Up to 500,000 USD
Contact: Program Administrator; (312) 673-3457; jvsscholarship@jvschicago.org
Internet: http://www.jvschicago.org/scholarship
Sponsor: Jewish Vocational Service
216 W Jackson Boulevard, Suite 700
Chicago, IL 60606

Jewish Social Entrepreneurs Fellowships 2487
The fellowship supports women and men who are developing projects that address current social issues in the Jewish community and the United States. Projects can be an idea in an advanced stage of planning, a young independent nonprofit organization, or an original transformative program within an existing organization. Selected programs can focus on a wide variety of social arenas, including education, social justice, the arts, technology, the environment, and spirituality. The fellowship provides seed capital, entrepreneurial training, mentorships, technical assistance, and Jewish learning. Each fellow is encouraged and supported in raising additional funds for his or her project. Guidelines and application forms are available online.
Requirements: Jewish innovators between the ages of 21 and 35 are eligible.
Geographic Focus: All States
Amount of Grant: 5,000-30,000 USD
Contact: Administrator; (415) 777-4500; fellowship@joshuaventure.org
Internet: http://www.joshuaventure.org
Sponsor: Joshua Venture
28 Second Street, Suite 500
San Francisco, CA 94105

JHMI Patrick C. Walsh Prostate Cancer Research Fund 2488
Funding is available to support multidisciplinary research in prostate cancer through the Patrick C. Walsh Prostate Cancer Research Fund. Awards of $50,000 to $100,000 for up to 2 years are available to fund career development and developmental research programs (pilot projects). This fund will attract and support the best and brightest scientists throughout Johns Hopkins to join in an effort to defeat the number one cancer in men. 1) Awards will be for pilot projects, focused on prostate cancer, that develop a new research direction, explore an innovative idea, test an unconventional, but potentially important, new hypothesis, or ascertain the feasibility of a new research approach. These grants will be reviewed by a scientific advisory board which will award funding to projects that hold the most promise.
Requirements: All faculty of the Johns Hopkins University may apply, whether or not they are members of the James Buchanan Brady Urological Institute or of the Sidney Kimmel Comprehensive Cancer Center. However, an investigator may hold only one pilot project award at a time.
Geographic Focus: All States
Date(s) Application is Due: Jan 8
Amount of Grant: 50,000-100,000 USD
Contact: Angela Sciuto; (410) 614-0257; asciuto2@jhmi.edu
Internet: http://prostatecancerprogram.onc.jhmi.edu/
Sponsor: Johns Hopkins Medical Institute
733 N Broadway, Suite 117
Baltimore, MD 21205

JILA Visiting Fellowships and Postdoctoral Research Associateships 2489
One-year postdoctoral research associateships are intended to provide advanced research training in the years immediately after the PhD degree. These research associateships are generally renewable for a second year. Visiting fellowships are awarded for independent research and study, usually to those with extensive research experience. For those in doubt as to which type of appointment is more suitable, the institute advises application for a research associateship. If all materials for the completed application are received by November 1, the applicant may request to be considered for both the visiting fellowship program and the postdoctoral research associateship program. Transportation to Boulder may be paid for the appointee and his/her spouse and minor children, and an allowance for shipment of personal effects and an allowance for appropriate professional travel within the United States during the period of the appointment are usually made available.
Geographic Focus: All States
Date(s) Application is Due: Nov 1
Contact: Secretary, Visiting Scientists Program; (303) 492-7796; fax (303) 492-5235; jilavf@jila.colorado.edu
Internet: http://jilawww.colorado.edu/www/programs/pd.html
Sponsor: Joint Institute for Laboratory Astrophysics
University of Colorado, 440 UCB
Boulder, CO 80309-0440

JM Foundation Grants 2490
The foundation awards grants to eligible nonprofit organizations in its areas of interest, including education and research that fosters market-based policy solutions; developing state and national organizations that promote free enterprise, entrepreneurship, and private initiative; and identifying and educating young leaders. Types of support include internships, matching/challenge grants, program grants, publication, research grants, seed grants, and technical assistance. The foundation's board of directors meets bi-annually, usually in May and October. There are no formal proposal deadlines. Inquiries and proposals are processed on an ongoing basis.
Requirements: Public charities, including 501(c)3, 509(a)1, and 170(b)1(a)(vi) nonprofit organizations, that shares the foundation's priority interests is invited to submit a proposal.
Restrictions: Grants do not support individuals, the arts, government agencies, public schools, and international agencies; or requests for operating expenses, annual fundraising campaigns, capital campaigns, equipment, endowment funds, and loans.
Geographic Focus: All States
Amount of Grant: 5,000-100,000 USD
Contact: Carl Helstrom, Executive Director; (212) 687-7735; fax (212) 697-5495
Internet: http://foundationcenter.org/grantmaker/jm/guide_jm.html
Sponsor: JM Foundation
654 Madison Avenue, Suite 1605
New York, NY 10065

JMO Law and Legal System Grants 2491
Grants are given to deepen the understanding of the American judicial system and to preserve the rule of law as the bedrock of American constitutional government. The foundation supports public interest law and studies related to the judicial system, jurisprudence, and the relationship between law and economics. Financial support is awarded for research, institutional support, fellowships, professorships, lectures, conferences/seminars, books, scholarly journals and journals of opinion, and sometimes for television and radio programs.
Requirements: A 501(c)3 nonprofit may apply by submitting a concise written proposal describing the project and grant request and offering background on the organization and its current sources of funding.
Geographic Focus: All States
Contact: Caroline Hemphill; (212) 661-2670; fax (212) 661-5917; chemphill@jmof.org
Internet: http://www.jmof.org/grant_programs.html
Sponsor: John M. Olin Foundation
330 Madison Avenue, 22nd Floor
New York, NY 10017

JMO Public Policy Research Grants 2492
Grants support research on the formulation, implementation, and evaluation of public policy in the social and economic fields. Grants are made in such areas as regulatory policy, tax policy, fiscal policy, monetary policy, and welfare policy. The larger grants are made to graduate schools at universities in these areas. Financial support is awarded for research, institutional support, fellowships, professorships, lectures, conferences/seminars, books, scholarly journals and journals of opinion, and sometimes for radio and television programs.
Requirements: A 501(c)3 nonprofit may apply by submitting a concise written proposal describing the project and grant request and offering background on the organization and its current sources of funding.
Restrictions: Grants do not support endowments, building campaigns, annual giving programs, or individuals.
Geographic Focus: All States
Amount of Grant: 20,000-300,000 USD
Contact: Caroline Hemphill; (212) 661-2670; fax (212) 661-5917; chemphill@jmof.org
Internet: http://www.jmof.org/grant_programs.html
Sponsor: John M. Olin Foundation
330 Madison Avenue, 22nd Floor
New York, NY 10017

JMO Strategic and International Studies Grants 2493
Grants support projects that address the relationship between American institutions and the international context in which they operate. Such projects include studies of national security affairs, strategic issues, American foreign policy, and the international economy. Financial support is awarded for research, institutional support, fellowships, professorships, lectures, conferences/seminars, books, scholarly journals and journals of opinion, and sometimes for television and radio programs.
Requirements: A 501(c)3 nonprofit may apply by submitting a concise written proposal describing the project and grant request and offering background on the organization and its current sources of funding.
Geographic Focus: All States
Amount of Grant: 10,000-300,000 USD
Contact: Caroline Hemphill, Program Officer; (212) 661-2670; fax (212) 661-5917; chemphill@jmof.org or inquiry@jmof.org
Internet: http://www.jmof.org/grant_programs.html
Sponsor: John M. Olin Foundation
330 Madison Avenue, 22nd Floor
New York, NY 10017

Joel Henry Hildebrand Award in the Theoretical and Experimental Chemistry of Liquids 2494
This award, supported by the ExxonMobil Research and Engineering Company, is given annually to recognize distinguished contributions to the understanding of the chemistry and physics of liquids. The award shall be granted without regard to age or nationality. The award consists of $5,000 and a certificate.
Requirements: Any individual, except a member of the award committee, may submit one nomination or seconding letter for the award in any given year. Nominating documents consist of a letter of not more than 1000 words containing an evaluation of the nominee's accomplishments and a specific identification of the work to be recognized, a biographical sketch including date of birth, and a list of publications and patents authored by the nominee.
Restrictions: Self-nominations are not accepted.
Geographic Focus: All States
Date(s) Application is Due: Nov 1
Contact: Felicia Dixon, Awards Administrator; (800) 227-5558 or (202) 872-4408; fax (202) 776-8008; f_dixon@acs.org or awards@acs.org
Internet: http://portal.acs.org/portal/acs/corg/content?_nfpb=true&_pageLabel=PP_ARTICLEMAIN&node_id=1319&content_id=CTP_004539&use_sec=true&sec_url_var=region1
Sponsor: American Chemical Society
1155 Sixteenth Street, NW
Washington, DC 20036-4801

Joel L. Fleishman Civil Society Fellowships **2495**

The Fellowship provides a select group of leaders from domestic non-profit organizations, international non-governmental organizations, foundations, government, socially responsible businesses, and other civil society groups in the United States and internationally with the opportunity to come in residence at the Sanford Institute for a four-week mini-sabbatical. While at Duke, fellows will perform research and work with institute faculty and other Duke affiliates on issues relating to the development of civil society. Applicants will be selected based upon their proposed research project and how they intend to utilize Duke's resources to benefit their professional work in civil society. The fellowship provides housing and program expenses; a stipend; and access to the university's library, research centers, and recreational facilities. Download the application online.

Requirements: Employees of nonprofit and nongovernmental groups worldwide are eligible.

Restrictions: Individuals employed at government agencies and academic institutions are not eligible. Full-time academics are not eligible to apply.

Geographic Focus: All States

Date(s) Application is Due: May 1

Amount of Grant: 6,000 USD

Samples: Frehiwot Alebachew, Saves Lives Ethiopia (SaLE), Ethiopia; Anthony Ehrenreich, Congress of South African Trade Unions (COSATU), South Africa; Liza Lim, Institute of Social Order, Philippines.

Contact: Melynn Glusman; (919) 613-7432; melynn.glusman@duke.ed

Internet: http://www.pubpol.duke.edu/centers/civil/

Sponsor: Duke University

P.O. Box 90239

Durham, NC 27708

John B. Bene Fellowships in Social Forestry **2496**

This fellowship covers graduate studies that focus on the relationship of forest resources to the social, economic, and environmental welfare of people, especially those in developing countries. The award must be used, at least in part, to help fund field research in a developing country. The fellowships are awarded for one year and are renewable for up to an additional two years. Applications can be downloaded from Web site.

Requirements: Applicants must be Canadian citizens or permanent residents of Canada enrolled in a Canadian university at the master's or doctoral level and have an academic background that combines forestry or agroforestry with social sciences. Research proposal must be approved by the research/thesis committee. Applicants must provide evidence of affiliation with an institution or organization in the region in which the research will take place, and have completed course work and passed comprehensive examinations by the time of award tenure.

Geographic Focus: All States, Canada

Date(s) Application is Due: Mar 1

Amount of Grant: 15,000 CAD

Contact: International Development Research Center; (613) 236-6163, ext. 2098; fax (613) 563-0815; cta@idrc.ca

Internet: http://web.idrc.ca/en/ev-23378-201-1-DO_TOPIC.html

Sponsor: International Development Research Center

P.O. Box 8500

Ottawa, ON K1G 3H9 Canada

John Ben Snow Memorial Trust Grants **2497**

The mission of the Foundation is to make grants within specific focus areas to enhance the quality of life in Central and Northern New York State. Historically, the Foundation has made grants in the following program areas: arts and culture, community development, education, environment, historic preservation, and journalism. The Foundation responds to the ever-changing needs of various segments of the population, especially to the needs of young people and people who are disadvantaged either physically or economically. It is the Foundation's general policy to give preference to proposals seeking funds for new or enhanced programs, one-time, short-term grants to sustain a program until funding is stabilized, matching grants used to encourage the participation of other donors, and last dollars towards a capital campaign. There are no minimums or maximum grant amounts; however, most grants range from $5,000 to $15,000.

Requirements: Giving is primarily focused in Maryland, Nevada, and central New York.

Restrictions: The Foundation will not accept proposals from individuals or for-profit organizations. Additionally, the Foundation does not encourage proposals from religious organizations or proposals for endowments, contingency funding, or debt reduction.

Geographic Focus: All States

Date(s) Application is Due: Apr 1

Amount of Grant: 5,000-25,000 USD

Contact: Jonathan L. Snow, (315) 471-5256; fax (315) 471-5256

Internet: http://www.johnbensnow.com/jbsmt

Sponsor: John Ben Snow Memorial Trust

50 Presidential Plaza, Suite 106

Syracuse, NY 13202

John D. and Catherine T. MacArthur Foundation Global Challenges Grants **2498**

The global challenges area focuses on emerging opportunities and threats in a period of rapid globalization—the development of complex political, social, and economic interconnections that result from the increased capacity for people, goods, capital, and information to move freely across national borders. Grants are made in the areas of human rights and economic governance. Grants are made throughout the year. Guidelines are available on the Web site.

Requirements: Grants are provided to nonprofit organizations, public agencies, universities, research institutes, and networks of such institutions. Only rarely are grants awarded for individual research projects, preference being given to multifaceted programs.

Restrictions: The foundation will not provide funding for publications and conferences, political activities, attempts to influence legislation, foundation development campaigns, or for programs that are the responsibility of governments.

Geographic Focus: All States

Amount of Grant: 50,000-2,000,000 USD

Contact: Office of Grants; (312) 726-8000; fax (312) 920-6258; 4answers@macfound.org

Internet: http://www.macfound.org/programs/gss/GC.htm

Sponsor: John D. and Catherine T. MacArthur Foundation

140 S Dearborn Street, Suite 1100

Chicago, IL 60603-5285

John Dana Archbold Fellowships **2499**

The fellowship program supports educational exchange between the United States and Norway. It is administered by the Nansen Fund in cooperation with the Norway-American Association in Oslo and the University of Oslo. Each fellowship provides for a year of graduate or postdoctoral study and research in Norway. The University of Oslo International Summer School offers Norwegian language courses and orientation for six weeks before the regular academic year begins. University of Oslo waives tuition costs; attendance is required. The maintenance stipend is sufficient to meet ordinary expenses of a single person. Travel allowance is provided to cover overseas airfare to and from Oslo. Application must be complete, including references and transcripts. Successful candidates will be notified before March 15. Fellowships are for one year starting in even-numbered years only.

Requirements: Eligibility is limited to individuals who are citizens and residents of the United States, 20 to 35 years old, in good health, and of good character. Applicants are judged competitively, with emphasis on proven ability in their field, seriousness of purpose, and a record of social adaptability. A high degree of proficiency in the Norwegian language is needed only in such fields as Norwegian literature and law; otherwise it is not necessary particularly for persons doing independent research. Undergraduate applicants must have a BA or BS degree before departure for Norway.

Geographic Focus: All States

Date(s) Application is Due: Jan 15

Amount of Grant: 5,000 USD

Contact: Anne Brith Berge, (713) 526-6222; fax (713) 521-9473; nacc@net1.net

Sponsor: Nansen Fund

77 Saddlebrook Lane

Houston, TX 77024

John Deere Foundation Grants **2500**

The foundation invests in programs in education, health/human services, community improvement, arts and culture. Types of support include annual campaigns, building construction/renovation, continuing support, fellowships, general operating support, scholarship funds, and seed money grants. Foundation interest also includes support for Third World development through U.S.-based nonprofits with international building funds, research grants, general operating purposes, and continuing support. There are no application deadlines.

Requirements: Nonprofit organizations in communities with major John Deere operating units, and employee presence are eligible. Eligible U.S. locations are: Augusta, GA; Quad City Region, IL; Des Moines, IA, Dubuque, IA, Iowa Quad Cities, IA, Ottumwa, IA, Waterloo, IA, Coffeyville, KS, Lenexa, KS, Thibodaux, LA, Springfield, MO, Cary, NC, Fuquay-Varina, NC, Fargo, ND, Greeneville, TN, Madison, WI, Horicon, WI. Exceptions include: accredited colleges and universities; organizations focused on international development initiatives related to John Deere Solutions for World Hunger initiative. Because John Deere dealerships are owned and operated independently, their communities are not included in this geographic scope. Also eligible to apply are organizations and institutions of national or international scope that reflect the foundation's concerns.

Restrictions: Funds are not available for the following organizations or purposes: individual initiatives, including scholarships; sports teams, racing teams, athletic endeavors or scholarships designated for athletes; faith-based organizations for sectarian purposes; political candidates, campaigns or organizations; private clubs, fraternities or sororities; other foundations for purposes of building endowment; tax-supported entities.

Geographic Focus: Georgia, Illinois, Iowa, Kansas, Louisiana, Missouri, North Carolina, Tennessee, Wisconsin, Belarus, Brazil, Canada, Estonia, Latvia, Lithuania, Moldova, Ukraine

Contact: Amy Nimmer, Director; (309) 765-8000

Internet: http://www.deere.com/en_U.S./globalcitizenship/socialinvestment/index.html

Sponsor: John Deere Foundation

1 John Deere Place

Moline, IL 61265-8098

John G. Duncan Trust Grants **2501**

The trust awards grants in the following areas of interest: arts; education; health care; human services; religion; and support: building/renovation; capital campaigns; emergency funds; equipment; program development; seed money; and research. Contact the foundation, for the application guidelines and application form.

Requirements: IRS 501(c)3 organizations in Colorado are eligible.

Restrictions: No support for other grantmaking organizations, individuals, endowments, or for general operating expenses.

Geographic Focus: Colorado

Date(s) Application is Due: Jan 31; Apr 30; Jul 31; Oct 31

Amount of Grant: 1,000-10,000 USD

Contact: Jason Craig, c/o Wells Fargo Bank, N.A., (720) 947-6820

Sponsor: John G. Duncan Trust
1740 Broadway, MAC C7300-483
Denver, CO 80274-0001

John Merck Scholars Awards 2502

Scholars are chosen from the ranks of the most promising assistant professors currently working, or planning to work, in neurobiological and cognitive sciences relating to the biology of mental disability and developmental disabilities, including developmental studies of cognition, perception, language, reading, learning and motor performance. The foundation will accept one application for a neurobiologist and one for a cognitive scientist from major universities and other research centers. Scholars will receive $75,000 per year for a four-year period, subject to an annual review of research progress. The sponsoring institution is responsible for fiscal management.
Requirements: The foundation will fund the most promising young researchers whose work illuminates neurodevelopmental disorders from the perspectives of (i) synapse formation and synaptic plasticity; (ii) learning and memory, and synaptic plasticity; (iii) perception, cognition and behavior; (iv) neurogenesis and pattern formation; and (v) genetics and early development. Also encouraged are proposals that (i) investigate the possible role of environmental chemicals in the origins of developmental disabilities, or that (ii) aim to distinguish subgroups within accepted diagnostic categories through the use of sophisticated behavioral and neuroimaging tests of perception, cognition, and emotions based on concepts from modern cognitive neuroscience. In all cases, they seek proposals from young scientists conducting research that is of the highest quality and that has the greatest chance of increasing understanding of neurodevelopmental disorders. Applicants must have the following: Academic rank in a university or medical school, or equivalent standing in a research institute or medical center; A record of research in areas relating to the foundation's interest in the underlying causes of developmental disabilities; Not more than four years of experience in an independent faculty position; Evidence of a commitment to a career in neuroscience or cognitive science. Request application forms by mail, email or phone.
Restrictions: Applicants may not: hold tenured positions or their equivalents; apply in more than two rounds of competition; apply for continued postdoctoral support. Holding other fellowships concurrently with the John Merck Scholars Award is discouraged. Prior approval by the foundation for an overlapping fellowship is required and will be given only in unusual circumstances.
Geographic Focus: All States
Date(s) Application is Due: Sep 15
Amount of Grant: 75,000 USD
Contact: Jason Bentsman; (617) 556-4120; fax (617) 556-4130; j bentsman@jmfund.org
Internet: http://www.jmfund.org/jm_scholars_program.html
Sponsor: John Merck Fund
2 Oliver Street, 8th Floor
Boston, MA 02108

John O. Galloway Memorial Scholarship in Geology and Geophysics 2503

One scholarship is awarded annually for one year of graduate study in the Department of Geology and Geophysics at the University of Calgary. Students interested in the award should apply first to the department.
Requirements: Candidates must be graduate students enrolled in the master's program in the Department of Geology and Geophysics who have completed a minimum of two academic terms of full-time graduate work prior to the commencement of the award.
Geographic Focus: All States
Date(s) Application is Due: Feb 1
Amount of Grant: 5,000 CAD
Contact: Connie Baines; (403) 220-5690; cbaines@ucalgary.ca
Internet: http://www.grad.ucalgary.ca/funding/internal_scholarships/lev_4/galloway.htm
Sponsor: University of Calgary
2500 University Drive NW, Earth Sciences Building, Room 720
Calgary, AB T2N 1N4 Canada

John P. McGovern Foundation Grants 2504

The John P. McGovern Foundation, established in 1961, supports the charitable interests of the donor to support the activities of established nonprofit organizations, which are of importance to human welfare with special focus on children and family health education and promotion, treatment and disease prevention. The Foundation gives primarily in Texas with an emphasis on the Houston area, but also provides funding in the Southwest region as well. The types of support offered are: building/renovation; conferences/seminars; continuing support; curriculum development; emergency funds; endowments; general/operating support; matching/challenge support; professorships; publication; research; and scholarship funds.
Requirements: Non-profits in Texas are eligible to apply.
Geographic Focus: Texas
Date(s) Application is Due: Aug 31
Amount of Grant: 10,000-100,000 USD
Contact: Kathrine G. McGovern, President; (713) 661-4808; fax (713) 661-3031
Sponsor: John P. McGovern Foundation
2211 Norfolk Street, Suite 900
Houston, TX 77098-4044

John R. Oishei Foundation Grants 2505

The foundation's primary mission is to support medical research and care and education, as well as cultural and social needs existing in the Buffalo Niagara region. The foundation favors creative programs that attempt to advance from the status quo, are strategically sound, and are strongly focused on excellence. Programs that provide opportunities for foundation support to be leveraged into greater support from other sources will be especially favored. The foundation generally does not fund operating expenses, though occasional exceptions may be made for organizations that address basic human needs. Requests for capital funds will not be considered unless they are an integral part of an otherwise eligible proposal. Contact the foundation to discuss eligibility before submitting an application. Grant applications are accepted throughout the year.
Requirements: It is the general policy of the foundation to confine its support to activities located in the Buffalo, NY, metropolitan region.
Restrictions: Grants do not support endowments; capital requests (buildings or equipment); deficit funding or loans; individual scholarships or fellowships (except within specific foundation programs); travel, conferences, seminars, or workshops; fundraising events; or 509(a) private foundations.
Geographic Focus: New York
Amount of Grant: 100,000-200,000 USD
Contact: Blythe T. Merrill, Senior Program Officer; (716) 856-9490, ext. 3; fax (716) 856-9493; btmerrill@oisheifdt.org
Internet: http://www.oisheifdt.org/Home/Fund/WhatWeFundOverview
Sponsor: John R. Oishei Foundation
1 HSBC Center, Suite 3650
Buffalo, NY 14203-2805

John S. Dunn Research Foundation Grants and Chairs 2506

Established in 1985, the foundation funding area is limited to the Texas, providing support to health and medical-related organizations, especially hospitals; support also for cancer, other medical research, and freestanding clinics.
Requirements: Grants are made to nonprofit organizations within Texas having 501(c)3 tax-exempt status.
Restrictions: No grants to individuals, or for multi-year or seed money grants.
Geographic Focus: Texas
Amount of Grant: 10,000-100,000 USD
Contact: Dr. Lloyd Gregory; (713) 626-0368; fax (713) 626-3866
Sponsor: John S. Dunn Research Foundation
3355 West Alabama street, Suite 720
Houston, TX 77098-1718

John Simon Guggenheim Memorial Foundation Fellowships 2507

Often characterized as mid-career awards, Guggenheim Fellowships are intended for men and women who have already demonstrated exceptional capacity for productive scholarship or exceptional creative ability in the arts. Fellowships are awarded through two annual competitions: one open to citizens and permanent residents of the United States and Canada, and the other open to citizens and permanent residents of Latin America and the Caribbean. Candidates must apply to the Guggenheim Foundation in order to be considered in either of these competitions. The Foundation receives between 3,500 and 4,000 applications each year. Although no one who applies is guaranteed success in the competition, there is no prescreening: all applications are reviewed. Approximately 220 Fellowships are awarded each year.
Requirements: Fellowships are awarded to individuals who have already demonstrated exceptional capacity for productive scholarship or exceptional creative ability in the arts.
Restrictions: Guggenheim Fellowships are not scholarships, and they are not available to those who seek to complete their training either as undergraduate, graduate, or part-time students. They are also not intended for immediate postgraduate work.
Geographic Focus: All States, Puerto Rico, Anguilla, Antigua & Barbuda, Argentina, Aruba, Bahamas, Barbados, Bolivia, Brazil, British Virgin Islands, Canada, Cayman Islands, Chile, Colombia, Costa Rica, Cuba, Dominica, Dominican Republic, Ecuador, El Salvador, French Guiana, Grenada, Guadeloupe, Guatemala, Haiti, Honduras, Jamaica, Martinique, Mexico, Montserrat, Nicaragua, Panama, Paraguay, Peru, Saint Kitts And Nevis, St. Lucia, St. Vincent and the Grenadines, Trinidad and Tobago, Turks and Caicos Islands, Uruguay, Venezuela, Virgin Islands (U.S.)
Date(s) Application is Due: Oct 1; Dec 1
Contact: Associate Secretary; (212) 687-4470; fax (212) 697-3248; fellowships@gf.org
Internet: http://www.gf.org/about-the-foundation/the-fellowship/
Sponsor: John Simon Guggenheim Memorial Foundation
90 Park Avenue
New York, NY 10016

Johnson & Johnson Corporate Contributions Grants 2508

Johnson & Johnson and its many operating companies support community-based programs that improve health and well-being. The Company works with community-based partners that have the greatest insight into the needs of local populations and the strategies that stand the greatest chances of success. Giving focuses on: saving and improving the lives of women and children; building on the skills of people who serve community health needs, primarily through education; and preventing diseases and reducing stigma and disability in underserved communities where we have a high potential for impact.
Requirements: Grants are awarded to nonprofit and tax-exempt local, national, and international organizations and institutions.
Restrictions: Grants are not awarded to individuals, for deficit funding, capital or endowment funds, demonstration projects, or publications.
Geographic Focus: All States
Contact: Shaun Mickus, (732) 524-2086; smickus@corus.jnj.com
Internet: http://www.jnj.com/connect/caring/corporate-giving/
Sponsor: Johnson and Johnson
1 Johnson & Johnson Plaza
New Brunswick, NJ 08933-0001

John Stauffer Charitable Trust Grants 2509

The trust awards grants to California higher education institutions and hospitals. Types of support include building construction/renovation, challenge/matching funds, endowments, equipment acquisition, fellowships, professorships, and scholarship funds. There are no application forms or deadlines.
Requirements: California nonprofits are eligible.
Geographic Focus: California
Amount of Grant: 100,000-500,000 USD
Contact: H. Jess Senecal; (626) 793-9400; fax (626) 793-5900; jess@lagerlof.com
Sponsor: John Stauffer Charitable Trust
301 N Lake Avenue, 10th Floor
Pasadena, CA 91101-4108

John W. Alden Trust Grants 2510

The trust awards grants to Massachusetts nonprofit organizations in its areas of interest, including education and therapy for children who are blind, disabled, retarded, or mentally or physically ill. Grants will normally be made for specific projects rather than general operating purposes. There are no application deadlines. Grant applications should be submitted online at www.cybergrants.com/alden. Detailed instructions on the web page will guide you through the process. All questions should be directed to the grants coordinator.
Requirements: Eastern Massachusetts 501(c)3 tax-exempt organizations are eligible.
Restrictions: Grants are not made to individuals.
Geographic Focus: Massachusetts
Date(s) Application is Due: Jan 15; Apr 15; Jul 15; Oct 15
Amount of Grant: Up to 15,000 USD
Contact: Susan Monahan; (617) 951-1108; fax (617) 542-7437; smonahan@rackemann.com
Internet: http://www.cybergrants.com/alden
Sponsor: John W. Alden Trust
160 Federal Street, 13th Floor
Boston, MA 02110-1700

Joint Center for Poverty Research Grants 2511

The center seeks research proposals on the causes and consequences of poverty. Research priorities include the changing labor market, substance abuse and teen pregnancy, concentrated urban poverty, and the effects of recent policy changes on immigrants. Grants also focus on the disabled, the well-being of children, and urban populations.
Requirements: Preference is given to researchers without tenures but who have full-time academic employment.
Restrictions: Applicants are not required to have a PhD.
Geographic Focus: All States
Date(s) Application is Due: May 1
Amount of Grant: Up to 30,000 USD
Samples: Julia Henly, University of Chicago, School of Social Service Administration (Chicago, IL)—for the Study on Work-Child Care Fit.
Contact: Poverty Research; (847) 491-3395; fax (847) 491-9916; ipr@northwestern.edu
Internet: http://www.jcpr.org/researchfunding.html
Sponsor: University of Chicago
2040 Sheridan Road
Evanston, IL 60208

Joint Center for Poverty Research Visiting Scholars Grants 2512

The Northwestern University/University of Chicago Joint Center for Poverty Research sponsors this program of visiting scholars in residence. Each of the two campuses will host one visiting scholar. Applicants must be actively involved in research directly relevant to poverty in the United States. Preference will be given to scholars who can spend an entire academic year in residence and who can particularly benefit from interaction with the affiliated faculty in one or both of the two universities involved. Candidates are invited to express a preference for one of the two universities. The award includes a salary for the academic year, plus benefits and some research support. There are no teaching requirements. Applicants with partial funding from other sources are welcome.
Requirements: The program is intended for untenured or junior researchers who have completed their PhDs.
Restrictions: PhD graduates of Northwestern University or the University of Chicago are not eligible.
Geographic Focus: All States
Amount of Grant: Up to 50,000 USD
Contact: Poverty Research; (847) 491-3395; fax (847) 491-9916; ipr@northwestern.edu
Internet: http://www.jcpr.org/researchfunding.html
Sponsor: Northwestern University
2046 Sheridan Road
Evantson, IL 60208-4105

Joseph Alexander Foundation Grants 2513

Established in 1960 in New York, the Foundation supports primarily in the New York region. Fields of interest include the following: higher education; health organizations; medical research, particularly optic nerve research; social services; and Jewish organizations. Types of support also include the following: annual campaigns; building/renovation; capital campaigns; conferences/seminars; curriculum development; endowments; equipment; exchange programs; general/operating support; program development; research; and scholarship funds. Applicants should submit a letter requesting application guidelines before submitting a proposal. There is no deadline, but the board meets in January, April, July, and October.
Requirements: Nonprofit organizations are eligible to apply.
Restrictions: Grants are not made to individuals.
Geographic Focus: New York
Amount of Grant: 5,000-50,000 USD
Contact: Robert Weintraub, President; (212) 355-3688
Sponsor: Joseph Alexander Foundation
110 East 59th Street
New York, NY 10022-1304

Joseph C. Wey Fellowship in Medieval Studies 2514

The Basilian Fathers of Toronto fund the fellowship for study at the Pontifical Institute of Mediaeval Studies during the forthcoming academic year. As the Basilians have for many years been engaged in teaching and pastoral activity in Mexico and Columbia, a special fellowship named for the former Superior General, Joseph C. Wey, CSB is intended to extend their mission to graduate study in some aspect of the Middle Ages. Medievalists in Latin America are therefore encouraged to apply for this new fellowship, as are scholars from outside of North America whose intentions are to return to their original country in pursuance of their teaching careers. The fellow will be given an office and full access to the faculty and the library of the institute, and will also have access to the libraries and faculty of the University of Toronto. The holder must make a firm commitment to return to his or her country of origin to pursue teaching there.
Requirements: Any medievalist, including those presently holding academic positions, may apply. All candidates must have at least one postsecondary qualification and show a strong commitment to teaching. A doctorate is highly desirable, as is entry into the institute's programme for the Licence. Candidates will be expected to come prepared with the languages needed for their research. English, in which all candidates should be proficient, will be the language of instruction and discussion.
Geographic Focus: All States
Date(s) Application is Due: May 1
Amount of Grant: 35,000 USD
Contact: Jeanette Jardine; (416) 926-7142; fax (416) 926-7292; jeanettejardine@utoronto.ca
Internet: http://www.pims.ca
Sponsor: Pontifical Institute of Mediaeval Studies
59 Queen's Park Crescent E
Toronto, ON M5S 2C4 Canada

Joseph L. Fisher Dissertation Awards 2515

Resources for the Future will award several fellowships for the coming year for support of doctoral dissertation research in economics on issues related to the environment, natural resources, or energy. It is expected that fellowship recipients will engage in no employment during the period of fellowship tenure and that RFF will be notified immediately of financial assistance made available from any other source for support of doctoral work. Application materials and additional information may be obtained from the office. Applications sent by fax or email will not be accepted.
Requirements: This fellowship is intended to support graduate students in the final year of their dissertation research. Candidates must have completed the preliminary examinations for the doctorate not later than February 1. Awards will be announced in May. Women and members of minority groups are especially encouraged to apply.
Geographic Focus: All States
Date(s) Application is Due: Feb 28
Amount of Grant: 12,000 USD
Contact: Coordinator for Academic Programs; (202) 328-5155; fisher-award@rff.org
Internet: http://www.rff.org/rff/About/Fellowships_and_Internships/Fisher/Joseph-L-Fisher-Doctoral-Dissertation-Fellowships.cfm
Sponsor: Resources for the Future
1616 P Street NW
Washington, DC 20036-1400

Josiah Macy Jr. Foundation Grants 2516

The foundation awards grants to improve the education of physicians and other health care professionals. Grant making focuses on improving medical and health professionals' education in the context of the changing health care system, increasing the number of minority physicians, boosting teamwork between health care professionals, and using educational strategies to increase care for under-served populations. Current initiatives cover the fields of pediatric residency, inter-professional curriculum, improved clinical training, increasing diversity among health care professions, Macy Scholars program, women's health in medical curricula, women in medicine, and funding for a New York post-baccalaureate program. Applications may be submitted at any time. A preliminary letter of inquiry is useful in assisting the foundation staff in determining whether submission of a full proposal is appropriate.
Requirements: Grants are made to colleges, universities, and other 501(c)3 nonprofit organizations for activities within established program areas. Grant proposals, addressed to the president, should include the name of the sponsoring agency or institution; description of the project; names and qualifications of the person who will be responsible for the project; expected cost and duration of the project including an itemized budget; documents substantiating the tax-exempt status of the sponsoring institution; and a letter of endorsement from the sponsoring institution.
Restrictions: Grants are not awarded to individuals or for general undesignated support or for construction/renovation projects.
Geographic Focus: All States
Amount of Grant: Up to 5,000,000 USD
Contact: George E. Thibault; (212) 486-2424; jmacyinfo@josiahmacy
Internet: http://www.josiahmacyfoundation.org/index.php?section=grant_guidelines
Sponsor: Josiah Macy Jr. Foundation

44 East 64th Street
New York, NY 10021

Josiah W. and Bessie H. Kline Foundation Grants **2517**
The foundation was established for charitable, scientific, literary, and educational purposes. The objectives for its grantmaking activities are to aid blind or incapacitated persons or crippled children. Grants are made to Pennsylvania colleges and universities and to hospitals, to institutions for crippled children, or to any other benevolent or charitable institution. Generally, grants are made to such institutions that are located in south-central Pennsylvania. Grants also are made for scientific and medical research. Types of support include continuing support, annual campaigns, capital campaigns, equipment acquisition, building/renovation, curriculum development, scholarship funds, matching funds, research, and emergency funds. The foundation does not have an application form. Instead, a letter of request must be submitted with the following information: a description of the need and purpose of the applicant organization, a project budget, amount requested from the foundation and the dates of the need, and a copy of the 501(c)3 tax-exemption letter. Requests may be submitted at any time. The board generally meets in March and November to consider requests.
Requirements: 501(c)3 organizations serving south-central Pennsylvania communities are eligible.
Restrictions: The foundation does not make loans and does not make grants to individuals or to normal operational phases of established programs, endowments, campaigns or national organizations, or religious programs.
Geographic Focus: Pennsylvania
Amount of Grant: 500-150,000 USD
Contact: John Obrock; (717) 561-0820 or (717) 561-4373; fax (717) 561-0826
Sponsor: Josiah W. and Bessie H. Kline Foundation
515 S 29th Street
Harrisburg, PA 17104

Jovid Foundation Grants **2518**
The foundation partners with nonprofits in Washington, DC, to help low-income residents of the district learn how to become self-sufficient. Particularly interest is given to funding neighborhood-based efforts that provide programs and services to DC adults. The foundation has a modest budget for the support of the arts. Mini grants are awarded to help increase capacity for small groups through board training, staff development, membership dues, conferences, or Web site development. Potential applicants should initiate contact with the foundation by writing a one- to two-page letter describing the proposed project. Submit a two-page letter of inquiry; full proposals are by request.
Requirements: Washington, DC, nonprofits are eligible.
Geographic Focus: District of Columbia
Date(s) Application is Due: Jan 10; Apr 4; Jul 11; Oct 3
Contact: Bob Wittig; (202) 686-2616; fax (202) 686-2621; jovidfoundation@yahoo.com
Internet: http://fdncenter.org/grantmaker/jovid
Sponsor: Jovid Foundation
5335 Wisconsin Avenue NW, Suite 440
Washington, DC 20015-2003

Joyce Foundation Education Grants **2519**
The Joyce Foundation is committed to improving education in Midwest cities, especially by eliminating the barriers that prevent low-income students and children of color from reaching their full educational potential. The Foundation is a long-time supporter of Chicago school reform, and has also worked in Milwaukee, Detroit, Cleveland, and currently in Indianapolis and Minneapolis. Over the years, Joyce has explored a variety of strategies, including school finance, parent organizing, educational technology, charter schools, and early childhood education; it has also consistently supported research identifying and evaluating promising new strategies. Because research shows that having a first-rate teacher has an enormous impact on student achievement, the Foundation now concentrates the bulk of its grant making on improving teacher quality. It also continues to support other strategies for addressing the achievement gap, including making sure children learn to read by third grade, and supporting high quality charter schools and other educational innovations.
Requirements: Education grant making in K-12 focuses on Chicago, Indianapolis, and Minneapolis. The Education Program works to close the achievement gaps that separate low-income students and communities of color from their peers by improving the quality of teachers they encounter in school, enhancing early reading policies, and exploring such innovations as charter schools. Program priorities are Teacher Quality, Early Reading and Innovation. A small portion of program funds is reserved for other outstanding opportunities, Innovation Grants, to close the achievement gap, especially policy-oriented efforts to expand the supply of high-quality charter schools in Chicago, Indianapolis, and Minneapolis. A letter of interest (LOI) should be submitted first. If acceptable, applicants will be requested to submit a full proposal. Applicants are strongly encouraged to plan their application and proposal submission process for the April or July meetings, since most grant funds will be distributed at those times.
Restrictions: The Foundation does not generally support capital proposals, endowment campaigns, religious activities, commercial ventures, direct-service programs, or scholarships.
Geographic Focus: Illinois, Indiana, Minnesota
Date(s) Application is Due: Apr 8; Aug 14; Dec 3
Contact: Veronica Salter, Grants Manager; (312) 782-2464; fax (312) 595-1350; info@joycefdn.org or vsalter@joycefdn.org
Internet: http://www.joycefdn.org/programs/education/
Sponsor: Joyce Foundation
321 North Clark Street, Suite 1500
Chicago, IL 60654

Joyce Foundation Employment Grants **2520**
The Joyce Foundation believes that equipping workers to succeed in good-paying jobs benefits individuals, their families and communities, and the Midwest economy as a whole. In the last two decades, the Foundation has tested strategies to boost workers' skills and help women on welfare, low-skilled workers, ex-offenders, and others who face significant barriers in joining the workforce. Through it all, Joyce has consistently funded research to evaluate promising strategies and learn from the results; Joyce-supported research on welfare-to-work and transitional jobs were major contributions to the field. Current evaluation research focuses on adult education programs that prepare adults without high school diplomas or the equivalent with skills needed for careers.
Requirements: The Employment Program primarily focuses on federal and state policy grants, but will make some grants to support targeted metro-level progress in Chicago, Indianapolis, and Minneapolis/St. Paul. The current areas of interest are: Basic Foundational Skills-preparing adults with the skills required for jobs in today's market. In particular, the Foundation supports the evaluation and expansion of promising adult education programs that prepare people for post-secondary education and careers; Industry Training Partnerships-The Foundation supports expanding partnerships between industry associations or groups of businesses and educational organizations to create and promote career advancement opportunities. Grants encourage development and promotion of industry-recognized credentials that help low-skilled workers advance in the labor market. The Foundation also supports research into the return on investment relating to employer-sponsored skill development programs; and, Innovation. A letter of interest (LOI) should be submitted first. If acceptable, applicants will be requested to submit a full proposal. Applicants are strongly encouraged to plan their application and proposal submission process for the April or July meetings, since most grant funds will be distributed at those times.
Restrictions: The Foundation does not generally support capital proposals, endowment campaigns, religious activities, commercial ventures, direct-service programs, or scholarships.
Geographic Focus: Illinois, Indiana, Minnesota
Date(s) Application is Due: Apr 8; Aug 14; Dec 3
Contact: Veronica Salter; (312) 782-2464; fax (312) 595-1350; info@joycefdn.org
Internet: http://www.joycefdn.org/programs/employment/
Sponsor: Joyce Foundation
321 North Clark Street, Suite 1500
Chicago, IL 60654

JSA JLab Graduate Fellowships **2521**
The JLab Graduate Fellowships fund doctoral students at SURA member universities (listed at the website) conducting research related to the theoretical and experimental programs at Jefferson Lab, including the following: nuclear and related particle physics; accelerator physics; associated science and engineering; and free electron laser program. Each Fellowship awards $10,000 toward the normal academic year research assistant stipend, plus a $2,000 supplement. Conference travel support is available with each Fellowship. Selection is based on the following criteria: merit and quality of proposed research; utility and relevance of proposed research to Jefferson Lab and proposed utilization of Lab resources; academic qualifications; letters of recommendation; and the likelihood of accomplishment of proposed research objectives by graduation.
Requirements: Eligible students must be enrolled full-time in a relevant doctoral program at a SURA member university. Students may apply at any stage of their graduate careers or while undergraduate seniors. Awards are limited to two years and students seeking second-year support must resubmit a complete application.
Restrictions: Awardees may not concurrently hold another major full-time fellowship or internship.
Geographic Focus: All States
Amount of Grant: 12,000 USD
Contact: Hugh Montgomery; (757) 269-7100; fax (757) 269-7398; mont@jlab.org
Internet: http://www.jsallc.org/IF/IndexGF.html
Sponsor: Jefferson Science Associates
c/o Southeastern Universities Research Association, Inc.
Washington, DC 20005

JSA JLab Sabbatical Research Leave Support Grants **2522**
The JLab Sabbatical Research Leave Support Grants fund faculty from universities conducting research at the Jefferson Lab during approved sabbatical leave from their home institutions. The Program provides $1,500 per month living expense support for up to twelve months for researchers to relocate to the Lab during their sabbatical. Additional information is available at the website.
Geographic Focus: All States
Amount of Grant: 18,000 USD
Contact: Hugh Montgomery; (757) 269-7100; fax (757) 269-7398; mont@jlab.org
Internet: http://www.jsallc.org/IF/IndexSS.html
Sponsor: Jefferson Science Associates
c/o Southeastern Universities Research Association, Inc.
Washington, DC 20005

JSA Junior Scientist Travel Support **2523**
Jefferson Science Associates awards travel support to graduate students and post docs working at the Jefferson Lab for attendance and/or presentations at national and international conferences and workshops. This support enables junior researchers to receive recognition for their work, and helps participants to improve communication skills as they further their careers in the field. Travel funds may cover up to 50% of the travel cost. The traveler is expected to find grants or other support for the remaining cost of travel. Generally, up

to $400-$500 may be awarded for local/national meetings and up to $800-$1,000 for international meetings. Advisors/supervisors should complete the request form at the website for matching travel funds. Requests for only attendance to meetings (no presentation or talk expected) are also considered. Requests may be submitted at any time.
Requirements: Eligible participants are undergraduate, graduate, or Ph.D. students and post docs who are not employees of the Jefferson Lab and are not receiving other JSA Initiatives Fund support.
Geographic Focus: All States, All Countries
Amount of Grant: 400-1,000 USD
Contact: Marco Battaglieri; (757) 269-7100; fax (757) 269-7363; battagli@jlab.org
Internet: http://www.jsallc.org/IF/IndexJSTF.html
Sponsor: Jefferson Science Associates
c/o Southeastern Universities Research Association, Inc.
Washington, DC 20005

JSPS Bilateral Exchange Programs with Foreign Academic Institutions 2524
JSPS administers programs for scientific cooperation and exchange through memoranda of understanding with 75 foreign academic institutions in 42 countries. Cooperation primarily involves exchange visits by senior and junior scientists. The purpose of the program is to develop scientific understanding between Japan and the cooperating countries. Types of support include research, fellowships, and international exchange programs. Awards cover travel expenses, maintenance, and other necessary expenses. Contact the program office for more information regarding the JSPS Bilateral Exchange Programs with Foreign Academic Institutions and the JSPS Cooperation Programs with Southeast Asian Countries.
Requirements: Eligibility depends on the memoranda of understanding between the cooperating countries and the JSPS.
Geographic Focus: All States
Contact: Program Administrator; 81-03- 3263-2365; fax 81-03-3263-1673
Internet: http://www.jsps.go.jp/english/e-bilat/main.html
Sponsor: Japan Society for the Promotion of Science
Nihon Gakujutsu Shinko-kai, 5-3-1, Kojimachi, Chiyoda-ku
Tokyo, 102-8471 Japan

JSPS Fellowships for Foreign Scientists 2525
JSPS offers these fellowships to facilitate cooperative research between visiting foreign scientists and hosting Japanese scientists. There are two types of fellowships: short-term, granted to eminent scientists, such as Nobel laureates, and other outstanding senior scientists to visit Japan; and long-term, granted to foreign researchers who wish to conduct research in cooperation with, or under the supervision of, scientists at Japanese universities or research institutions. Awards for short-term fellowships include round-trip airfare, a living allowance of @Y18,000 per day, and an additional allowance for travel within Japan. The long-term fellowship includes round-trip airfare, a monthly stipend of @ Y240,000-@Y300,000, and additional allowances for housing and travel within Japan.
Requirements: Applications must be submitted by the inviting scientist through the head of the institution concerned. Candidates eligible for short-term fellowships include eminent or senior scientists. Candidates for long-term fellowships are those who want to conduct joint research with Japanese scientists.
Geographic Focus: All States
Contact: Fellowship Office; (81-03) 263-1721; fax (81-03) 222-1986
Internet: http://www.jsps.go.jp/english/e-fellow/fellow.html
Sponsor: Japan Society for the Promotion of Science
Sumitomo-Ichibancho Building, 6 Ichibancho, Chiyoda-ku
Tokyo, 102-8471 Japan

JSPS International Joint Research Projects and Scientific Seminars 2526
JSPS provides support for Japanese scientists to undertake cooperative research projects with foreign scientists. This program funds joint international ventures in scientific research with Japanese scientists in a wide variety of disciplines.
Requirements: Japanese scientists engaged in joint research projects and seminars with foreign scientists are eligible to apply.
Geographic Focus: All States, Japan
Contact: Program Administrator; 81-03-3263-1724; fax 81-03-3263-1673
Internet: http://www.jsps.go.jp/english/site_map.html
Sponsor: Japan Society for the Promotion of Science
Sumitomo-Ichibancho Building, 6 Ichibancho, Chiyoda-ku
Tokyo, 102-8471 Japan

JSPS Postdoctoral Fellowship for Foreign Researchers 2527
To facilitate cooperative research between young foreign researchers and hosting Japanese researchers, fellowships are awarded to enable postdoctoral researchers to undertake research at Japanese universities and institutions. The program also supports short-term fellowships of 15 days to 11 months. The fellowship includes round-trip airfare, a monthly stipend of @Y270,000, and an additional settling-in allowance. In addition, JSPS may provide a special research grant up to @Y1.5 million, depending on need of applicants.
Requirements: Fellowships are available to citizens of countries that have diplomatic relations with Japan. Candidates must hold the doctorate degree when the fellowship goes into effect; the degree must have been received within six years prior to April of the fiscal year in which the fellowship commences. The candidate must have arranged in advance a research plan with his/her Japanese host researcher, who must be employed at a Japanese university/research institution.
Geographic Focus: All States
Amount of Grant: 5,000-13,000 USD
Contact: Program Administrator; 81-3-3263-1721; fax 81-3-3263-1854
Internet: http://www.jsps.go.jp/english/e-fellow/postdoctoral.html#short
Sponsor: Japan Society for the Promotion of Science
Nihon Gakujutsu Shinko-kai, 5-3-1, Kojimachi, Chiyoda-ku
Tokyo, 102-8471 Japan

Judith and Jean Pape Adams Charitable Foundation ALS Grants 2528
Jean Pape Adams was interested in supporting charitable organizations actively engaged in discovering the causes of Amyotrophic Lateral Sclerosis (ALS) and in finding a cure. The Foundation encourages research that is collaborative, innovative and aggressive. These initiatives may fill gaps not currently being investigated and may involve novel approaches. The research must be of high scientific merit in finding a cure for ALS. Proposals may be longer than two pages. A lay summary is also required. Applications are reviewed by an Advisory Committee comprised of scientific researchers, neurologists and others who have had close relationships with friends and family members afflicted with ALS. The Committee makes funding recommendations to the Trustees.
Requirements: Public agencies and organizations that are classified as a charitable organization described in Section 501(c)3 of the Internal Revenue Code and as a public charity under Section 509(a) of the Internal Revenue Code may apply.
Restrictions: Support groups, education and medical equipment are not supported.
Geographic Focus: All States
Date(s) Application is Due: Sep 13
Amount of Grant: 1,000-250,000 USD
Contact: Shelley L. Carter; (830) 997-7347; fax (918) 591-2433; scarter@jjpafoundation.com
Internet: http://www.jjpafoundation.com/guidelines.html
Sponsor: Judith and Jean Pape Adams Charitable Foundation
7030 South Yale Avenue, Suite 600
Tulsa, OK 74136

Judith Graham Pool Postdoctoral Research Fellowship in Hemophilia 2529
A limited number of fellowships are awarded through professional and graduate schools or research institutions. The Judith Graham Pool (JGP) program is designed to support research studies of high scientific merit and relevance to hemophilia. Permissible research topics include clinical or basic research on the biochemical, genetic, hematologic, orthopedic, psychiatric, or dental aspects of the hemophilias or von Willebrand disease. Other acceptable topics include rehabilitation, therapeutic modalities, psychosocial issues, or AIDS/HIV as they pertain to hemophilias or von Willebrand disease. Awards are made for a one- or two-year period, with the second-year renewal contingent upon demonstrated progress in the fellow's research project. The fellowships will be awarded for the period between July 1 of one year and June 30 of the next. Applications are available on the Web site.
Requirements: Applicants must be citizens or permanent residents of the United States who have completed doctoral training and must enter the JGP fellowship program from a doctoral, postdoctoral, internship, or residency training program.
Restrictions: Established investigators or faculty members are not eligible.
Geographic Focus: All States
Date(s) Application is Due: Dec 1
Amount of Grant: 42,000 USD
Contact: Dr. Rita Barsky, Assistant Director of Research; (800) 424-2634, ext. 3730 or (212) 328-3730; fax (212) 328-3788; rbarsky@hemophilia.org
Internet: http://www.hemophilia.org/research/jgp.htm
Sponsor: National Hemophilia Foundation
116 W 32nd Street, 11th Floor
New York, NY 10001

Juvenile Justice Clearing House Grants 2530
The Juvenile Justice Clearing House, of the Department of Justice, awards grants under its Juvenile Justice and Delinquency Prevention Program. Areas of interest include delinquency prevention research, juvenile gangs, serious violent juvenile offenders, and juvenile justice statistics. Types of support include conferences and seminars, development grants, fellowships, internships, project grants, residencies, visiting scholars grants, and research grants. There are no matching requirements. Awardees generally receive from one to three years of support.
Requirements: Public and private organizations, including institutions of higher education, and individuals may submit applications for grant support.
Geographic Focus: All States
Amount of Grant: Up to 100,000,000 USD
Contact: Director; (800) 851-3420; fax (301) 519-5600; askjj@ncjrs.org
Internet: http://ojjdp.ncjrs.org/funding/FundingList.asp
Sponsor: Juvenile Justice Clearing House
P.O. Box 6000
Rockville, MD 20849-6000

Karst Research Fellowships and Grants 2531
The foundation annually awards graduate fellowships and grants for graduate research in any karst-related discipline anywhere in the world. Proposals involving surface or subsurface research are acceptable. The stipend is administered directly between the awardee and the foundation; the study itself is to be supervised at the graduate school of the awardee's choice by the thesis advisor. Application information is available upon request.
Requirements: Applicants must be graduate students at the master's or doctoral level and be in good standing with the graduate school.
Geographic Focus: All States
Date(s) Application is Due: Apr 15
Amount of Grant: Up to 10,000 USD
Contact: Dr. Tom Poulson, Grants & Fellowships Chairman; (561) 630-3643; pnkambesis@juno.com

Internet: http://www.cave-research.org/grants/grants.html
Sponsor: Cave Research Foundation
345 Middlefield Road, MS 975
Menlo Park, CA 94025

Kate Neal Kinley Memorial Fellowship 2532
The fellowships support students of fine and applied arts who have received degrees in art, architecture, or music. Three major fellowships will be awarded, two each in art, architecture, or music and an additional one in music. The fellowships are to be used toward defraying the expenses of advanced study of the fine arts in America or abroad. Two or three additional fellowships of lesser amounts may also be granted. Contact the committee chair for an application form and instructions.
Requirements: The fellowship is open to graduates of the College of Fine and Applied Arts at the university and to graduates of similar national and international institutions of equal educational standing whose principal or major studies have been in the fields of architecture, art, or music. Preference is given to applicants under 25 years of age.
Geographic Focus: All States
Date(s) Application is Due: Dec 3
Contact: Dr. Kathleen Conlin; (217) 333-1661; fax (217) 244-8381; nacox.3@uiuc.edu
Internet: http://www.faa.uiuc.edu/about_faa/funding_opportunities.html
Sponsor: University of Illinois
100 Architecture Building, 608 East Lorado Taft Drive
Champaign, IL 61820

Kavli Foundation Research Grants 2533
The Kavli Foundation supports research in the areas of astrophysics, nanoscience, neuroscience, and theoretical physics three scientific frontiers that promise great discoveries in the 21st century and beyond. This mission is implemented through an international program of research institutes, professorships, symposia and workshops, as well as prizes in the fields of astrophysics, nanoscience, and neuroscience. Dedicated to advancing science for the benefit of humanity and promoting increased public understanding and support for scientists and their work, the Foundation also makes available materials on these sciences at no cost to educational institutions.
Geographic Focus: All States
Contact: Grants Administrator; (805) 983-6000; fax (805) 988-4800
Internet: http://www.kavlifoundation.org/areasofscience/
Sponsor: Kavli Foundation
1801 Solar Drive, Suite 250
Oxnard, CA 93031

Kavli Foundation Research Prizes 2534
The Kavli Prize includes three separate prizes, which are awarded for outstanding scientific research in the fields of astrophysics, neuroscience and nanoscience, and each consists of $1,000,000 in cash, a medal and a diploma. The Kavli Prize will be awarded in Oslo every two years. The Prize is intended not only as an accolade for the individual scientists concerned, but also as a recognition of research efforts that are of the utmost importance for the future of mankind. It aims to promote international cooperation between researchers and to increase awareness of the importance of research among the general public. The call for nominations will be open September 1-December 15.
Geographic Focus: All States
Amount of Grant: 1,000,000 USD
Contact: Prize Administrator; (805) 983-6000; fax (805) 988-4800
Internet: http://www.kavlifoundation.org/kavli-prize
Sponsor: Kavli Foundation
1801 Solar Drive, Suite 250
Oxnard, CA 93031

Kellogg Institute for International Studies Andrónico Luksic Grants 2535
The objective of the Kellogg Institute for International Studies Andrónico Luksic Grants program is to encourage the Notre Dame community to find new ways to engage with colleagues from PUC (on research, exchanges, conferences, etc.) and in particular to increase the number of Notre Dame faculty and staff going to Santiago. Notre Dame faculty and staff may apply for grants of up to $10,000. Applicants are encouraged to demonstrate other sources of funding or cost sharing which will complement the Luksic Collaboration fund. Funding is available for a variety of collaborative ventures, including but not limited to the following: initiating collaborative research projects (travel, research costs, etc.); sponsoring or attending research conferences and workshops hosted at PUC; sponsoring educational ventures (faculty and staff exchanges, workshops, etc.); Notre Dame Graduate research assistant and postdoctoral fellow stipends and benefits.
Restrictions: Awards will not support clerical or secretarial help. Awards may support expenses incurred only at Notre Dame or by Notre Dame faculty and staff (e.g. hiring of research assistants employed by and working at PUC will generally not be funded through the Notre Dame Luksic grant.
Geographic Focus: All States
Date(s) Application is Due: Apr 15; Jul 15; Oct 14
Amount of Grant: Up to 10,000 USD
Contact: Anne Hayes, Associate Director; (574) 631-5527 or (574) 631-6580; fax (574) 631-6717; ahayes2@nd.edu
Internet: http://international.nd.edu/faculty-resources/#Funding
Sponsor: Kellogg Institute for International Studies
130 Hesburgh Center
Notre Dame, IN 46556-5677

Kellogg Institute for International Studies Collaborative Faculty Grants 2536
Collaborative Faculty Grants promote innovative, collaborative, and interdisciplinary research on issues closely aligned with the Kellogg Institute core research themes, democracy and human development. Collaborative projects might include the expansion of faculty members' current projects or the instigation of new collaborative ventures. The Institute seeks to support scholarly projects that have significant interdisciplinary and collaborative components and an ambitious mix of programmatic elements. Funded projects will transcend those eligible for Kellogg funding in the categories of individual faculty research, working group, or conference support. By providing a strong financial incentive for fellows to undertake major research projects, especially collaborative and longer-term ones, the Institute hopes to foster more, and more significant, innovative research that will make major contributions to scholarship on its core themes. Among other things, Kellogg especially seeks to support research that: crosses disciplinary boundaries and leads to crossfertilization across regions and themes; links important normative and theoretical work with high-level empirical research; unites its core themes of democracy and human development; uses a multidisciplinary approach to address complex problems of development; integrates research on religion and democracy/human development as a transversal theme; and/or contributes to relevant contemporary debates on public policy. Programmatic elements may include but are not limited to: working groups; workshops, roundtables, short-term visitors, and conferences; course buy-outs; team-taught seminars or courses; and graduate student assistance. Project funding may typically run between $10,000 and $70,000 for one or two years.
Requirements: Project leaders must be Kellogg faculty fellows, although all participants in the collaborative project need not be. Applicants are strongly encouraged to seek and priority is given to projects with financial support from other sources, both internal and external to Notre Dame.
Geographic Focus: All States
Date(s) Application is Due: Feb 11; Apr 1; Sep 17; Nov 19
Amount of Grant: 10,000-70,000 USD
Contact: Sharon Schierling, Associate Director; (574) 631-8524 or (574) 631-6580; fax (574) 631-6717; sschierl@nd.edu
Internet: http://kellogg.nd.edu/faculty/grants/collaborative.shtml
Sponsor: Kellogg Institute for International Studies
130 Hesburgh Center
Notre Dame, IN 46556-5677

Kellogg Institute for International Studies Course Development Grants 2537
New courses that address the questions and concerns of the Kellogg Institute research agenda can serve to enrich and help internationalize the undergraduate curriculum at the University, as well as to foster the intellectual engagement of undergraduates with Kellogg research themes. Funding is available to develop courses during the summer before offering them during the subsequent calendar year. Recipients are expected to offer the new course at least twice within five years following the grant. A maximum of $6,000 is available per grant, including a summer stipend of $2,500 for an individual faculty member, or $1,500 each for two faculty members developing a team-taught course. Up to $3,500 may be budgeted for supplementary funding, which may include but is not limited to teaching materials, books, research assistance, faculty or visitor travel expenses, honoraria, and payment for services.
Requirements: All full-time members of the regular faculty are eligible.
Restrictions: Course proposals with an in-country experiential learning component will be considered but Kellogg course development grants may not be used to fund student travel.
Geographic Focus: All States
Date(s) Application is Due: Feb 11; Apr 1; Sep 17; Nov 19
Amount of Grant: Up to 6,000 USD
Contact: Sharon Schierling, Associate Director; (574) 631-8524 or (574) 631-6580; fax (574) 631-6717; sschierl@nd.edu
Internet: http://kellogg.nd.edu/faculty/grants/cdev.shtml
Sponsor: Kellogg Institute for International Studies
130 Hesburgh Center
Notre Dame, IN 46556-5677

Kellogg Institute for International Studies Dissertation Year Fellowships 2538
The Kellogg Institute offers Dissertation Year Fellowships in International Studies to advanced Notre Dame graduate students in the social sciences, history, philosophy, theology, and international human rights law. The fellowships are intended for students to do field research or to complete their dissertations. The dissertation topic must pertain to international studies, addressing one or more of the thematic priorities of the Kellogg Institute: democratization and the quality of democracy; growth and development; religion and society; social movements and organized civil society; public policies for social justice. The awards will be equal to the respective department's academic year stipend for graduate student assistants.
Requirements: Applicants must be enrolled and in good standing in the Departments of Economics, Political Science, History, Philosophy, Sociology, or Theology, or in the J.S.D. program in International Human Rights.
Restrictions: Students who have other dissertation support equal to a full departmental graduate student stipend from Notre Dame sources for the time period being considered are not eligible. Students applying for this fellowship are not eligible for the Institute's Graduate Research Grant competition.
Geographic Focus: All States
Date(s) Application is Due: Mar 10
Contact: Denise Wright; (574) 631-8523 or (574) 631-6580; dwright1@nd.edu
Internet: http://kellogg.nd.edu/students/grad/grants/dissert.shtml
Sponsor: Kellogg Institute for International Studies

130 Hesburgh Center
Notre Dame, IN 46556-5677

Kellogg Institute for International Studies Experiencing The World Fellowships 2539
The Kellogg Institute offers the opportunity for freshmen and sophomores to engage in initial exploratory projects in Africa, Asia or Latin America. The award funds exceptionally qualified and committed undergraduates who seek to undertake innovative projects in Africa, Asia or Latin America. These may include research, non-profit work, study, or other activities that will increase their commitment to and knowledge of one of the regions. Award recipients may receive up to $5000 to pursue field projects for up to three months in the chosen region. Students interested in this award must justify their project as something that will substantially deepen and enrich their undergraduate experience and must show that this project is something that is otherwise not possible within existing Notre Dame programs. Student projects might include exploratory research for senior theses, area studies essays, or international scholars research; collaborative research with a professor; internship work for a non-profit organization or NGO; enrollment in an academic or language program that would not be possible through existing Notre Dame arrangements; or other creative forms of experiential education.
Requirements: Students interested in this award must meet the following criteria: have a GPA of 3.0 or higher; have taken at least one course at Notre Dame related to the proposed region of the project and/or demonstrate prior interest in the region; demonstrate adequate language skills to carry out the proposed project; and have not previously received an ETW Fellowship.
Geographic Focus: All States
Date(s) Application is Due: Mar 5
Amount of Grant: Up to 5,000 USD
Contact: Holly Rivers; (574) 631-6023 or (574) 631-6580; hrivers@nd.edu
Internet: http://kellogg.nd.edu/students/etw/index.shtml
Sponsor: Kellogg Institute for International Studies
130 Hesburgh Center
Notre Dame, IN 46556-5677

Kellogg Institute for International Studies Faculty Fellows' Research Grants 2540
The Kellogg Institute for International Studies offers individual and collaborative research grants to support summer and academic-year research by Kellogg faculty fellows in all disciplines. Funding is available for a variety of project costs, including but not limited to the following: travel, research assistance, or acquisition of research materials, including electronic databases; travel to research meetings and workshops in cases in which other Notre Dame funding does not exist or is not sufficient; communication with colleagues and institutions in other countries for research or other professional purposes; computer software and databases (but not hardware); and collaboration with coauthors and, in collective research projects, with other individuals involved in the project. Proposals will be accepted for most research-related expenses; proposals should justify the relevance and importance of proposed expenses. The committee will consider requests that involve multiple dimensions (e.g., two field trips, one field trip plus research assistance, etc.). Individual grants will not normally exceed $10,000 and proposals will be considered by the Grants Committee for any amount up to this figure.
Requirements: Applicants must notify the Kellogg Institute of any funding received from another source; the Institute may adjust the award accordingly. No duplicate funding (that is, funding for the same purpose) is allowed.
Geographic Focus: All States
Date(s) Application is Due: Feb 11; Apr 1; Sep 17; Nov 19
Amount of Grant: Up to 10,000 USD
Contact: Sharon Schierling; (574) 631-8524 or (574) 631-6580; sschierl@nd.edu
Internet: http://kellogg.nd.edu/faculty/grants/research.shtml
Sponsor: Kellogg Institute for International Studies
130 Hesburgh Center
Notre Dame, IN 46556-5677

Kellogg Institute for Int'l Studies Global Collaboration Initiative Grants 2541
The Office of the Vice President and Associate Provost for Internationalization, Notre Dame International (NDI), announces the Global Collaboration Initiative (GCI). Under the auspices of the GCI, several grants between $10,000 and $20,000 (with the possibility of renewal for up to three years) will be awarded to fund initiatives in international research, scholarship and collaboration with major foreign universities and research organizations. The GCI is intended to enable through international cooperation the establishment or continuation of programs of research, scholarship and sustained collaboration that will make major contributions in any field of study. These initiatives should contribute to increasing the internationalization of Notre Dame by elevating the university's visibility in the international academic and research community. Successful proposals will demonstrate the potential to create meaningful and sustainable exchange and/or research collaborations with colleagues and students at major foreign research universities and other research institutions, and to elevate the University's visibility in the international academic and research community.
Requirements: Eligible applicants include full-time regular members of the Notre Dame faculty, including tenured and tenure-track faculty, research faculty, library faculty and special professional faculty, are eligible to apply. Centers (e.g. NDnano, cSEND) are also eligible to apply provided that at least one faculty member is affiliated with the project.
Restrictions: Faculty receiving funding from the GCI Grant Program during the previous fiscal year may not apply for a new grant but may apply for a continuation of the current grant for up to three years. Ineligible project requests include: programs that do not involve collaboration with faculty/researchers outside the U.S; faculty summer salary; faculty academic year course reduction; projects in the final stages of completion; and ongoing research projects funded by another sponsor.
Geographic Focus: All States
Date(s) Application is Due: Oct 1
Contact: Geraldine Meehan, (574) 631-7742 or (574) 631-6580; fax (574) 631-6717; Geraldine.Meehan.19@nd.edu
Internet: http://international.nd.edu/global-collaboration-initiative-gci/
Sponsor: Kellogg Institute for International Studies
130 Hesburgh Center
Notre Dame, IN 46556-5677

Kellogg Institute for International Studies Graduate Research Grants 2542
The purpose of the Kellogg Institute for International Studies Graduate Research Grants is to support Notre Dame graduate students involved in overseas (or, if demonstrably appropriate, in the United States) research in international studies, addressing the thematic priorities of the Kellogg Institute. The grant is not intended to be a substitute for seeking external support for dissertation research but instead to fill gaps in other funding or to enable exploratory trips that will strengthen students' ability to prepare externally competitive dissertation proposals. Preference is given to students in a doctoral program and with the equivalent of two years of full-time graduate study prior to the beginning date of their proposed research. Funding is available for a variety of project costs, including research materials and assistance as well as travel. Individual grants will not normally exceed $7,000, and proposals will be considered for any amount below this figure.
Requirements: Student applicants must be enrolled in graduate school at the University of Notre Dame and be in good standing.
Restrictions: Students applying for this award are not eligible to apply for the Institute's Dissertation Year Fellowship in the same academic year.
Geographic Focus: All States
Date(s) Application is Due: Feb 26; Oct 16
Amount of Grant: Up to 7,000 USD
Contact: Denise Wright, Program Coordinator; (574) 631-8523 or (574) 631-6580; fax (574) 631-6717; dwright1@nd.edu
Internet: http://kellogg.nd.edu/students/grad/grants/gradresearch.shtml
Sponsor: Kellogg Institute for International Studies
130 Hesburgh Center
Notre Dame, IN 46556-5677

Kellogg Institute for Int'l Studies Grants for Conferences and Workshops 2543
The Kellogg Institute for International Studies makes grants of up to $35,000 for research conferences organized by its faculty fellows, giving preference to conferences that match major Institute priorities. On top of this amount, the Institute may match up to $15,000 from non-Institute sources for conference funding. Grants up to $10,000 will go through the Grants Committee. The Faculty Committee can supplement those awards with additional amounts for major research conferences addressing Kellogg priorities. Workshops, on the other hand, generally are small and informal, often addressing research issues at an exploratory stage of a project. Workshop grants may also be used to invite scholars and experts to critique a faculty member's manuscript. Grants of up to $5,000 are available for such workshops.
Requirements: Faculty members are strongly encouraged to seek other funding sources to supplement and leverage Kellogg support. Kellogg-supported conferences ordinarily are cosponsored by other Notre Dame units or external entities, and need not be held at the University. Conferences often result in published volumes. Applicants' proposals and budgets must include information about other sources of funding and sponsorship.
Geographic Focus: All States
Date(s) Application is Due: Feb 11; Apr 1; Sep 17; Nov 19
Amount of Grant: Up to 35,000 USD
Contact: Sharon Schierling; (574) 631-8524 or (574) 631-6580; sschierl@nd.edu
Internet: http://kellogg.nd.edu/faculty/grants/events.shtml
Sponsor: Kellogg Institute for International Studies
130 Hesburgh Center
Notre Dame, IN 46556-5677

Kellogg Institute for Int'l Studies Grants for Distinguished Lectureships 2544
The Kellogg Institute for International Studies will consider proposals from faculty fellows to fund especially distinguished lecturers or performers for periods of up to two weeks. Distinguished lecturers are expected to give a major public lecture or a series of presentations. The maximum award is $10,000. Preference is given to proposals for candidates whose visit would have maximum impact at Notre Dame through such activities as participating in faculty seminars or courses; workshops or other research collaborations; events at residence halls; or extracurricular activities sponsored by student groups.
Geographic Focus: All States
Date(s) Application is Due: Feb 11; Apr 1; Sep 17; Nov 19
Amount of Grant: Up to 10,000 USD
Contact: Sharon Schierling; (574) 631-8524 or (574) 631-6580; sschierl@nd.edu
Internet: http://kellogg.nd.edu/faculty/grants/events.shtml
Sponsor: Kellogg Institute for International Studies
130 Hesburgh Center
Notre Dame, IN 46556-5677

Kellogg Institute for International Studies Grants for Events 2545
The Kellogg Institute for International Studies makes available to faculty fellows small grants to support group events such as lectures, seminars, and films that advance understanding of international studies and of Kellogg's major themes in particular. The best proposals will bring together different elements of the Kellogg and/or Notre Dame communities in innovative ways. Proposals may be submitted at any time, but well in

advance of the proposed activity. Proposals will be evaluated by the Grants Committee on a rolling basis. The Grants Committee usually meets only during the academic year, so applicants should plan accordingly. The maximum award is $5,000.
Geographic Focus: All States
Date(s) Application is Due: Feb 11; Apr 1; Sep 17; Nov 19
Amount of Grant: Up to 5,000 USD
Contact: Sharon Schierling; (574) 631-8524 or (574) 631-6580; sschierl@nd.edu
Internet: http://kellogg.nd.edu/faculty/grants/events.shtml
Sponsor: Kellogg Institute for International Studies
130 Hesburgh Center
Notre Dame, IN 46556-5677

Kellogg Institute for International Studies Intramural Grants 2546
Kellogg Intramural Grants primarily fund research and educational activities. Requests to support student activities must make explicit the connection between the proposed activity and academic work or educational goals. Proposals by student groups must have a faculty or staff sponsor. The grants are intended to support projects of an institutional or group nature rather than individuals' research. The Institute will provide a maximum of $5,000 to support research conferences organized by Notre Dame faculty members who are not Institute fellows.
Requirements: Intramural grants are open to any faculty or staff member, student group, or unit within the ND community. Projects must be within the field of international studies or have a clear international dimension, and preferably relate to one of the Institute's five research themes.
Restrictions: Grants are restricted to the University of Notre Dame academic community and are intended primarily (though not exclusively) for on-campus activities and events.
Geographic Focus: All States
Date(s) Application is Due: Feb 11; Apr 1; Sep 17; Nov 19
Amount of Grant: Up to 5,000 USD
Contact: Sharon Schierling; (574) 631-8524 or (574) 631-6580; sschierl@nd.edu
Internet: http://kellogg.nd.edu/faculty/grants/intramural.shtml
Sponsor: Kellogg Institute for International Studies
130 Hesburgh Center
Notre Dame, IN 46556-5677

Kellogg Institute for Int'l Studies Kroc Undergraduate Research Grants 2547
The Kellogg and Kroc Institutes announce a Research Grants Program for undergraduate students at Notre Dame. The grants are intended for juniors whose interests include a clear international dimension preferably related to the Kellogg Institute's themes. Several grants of up to $5,000 each will be awarded for research abroad or, if demonstrably appropriate, in the United States. Applicants are required to notify the Kellogg Institute of any funding received from another source. If duplicate funding is received, the Institute may adjust the award.
Requirements: Students must be in their junior year of studies at Notre Dame and be in good academic standing. They must have the endorsement and pledge of supervision from a faculty member.
Geographic Focus: All States
Date(s) Application is Due: Mar 1
Amount of Grant: Up to 5,000 USD
Contact: Holly Rivers; (574) 631-6023 or (574) 631-6580; hrivers@nd.edu
Internet: http://kellogg.nd.edu/students/grants/index.shtml
Sponsor: Kellogg Institute for International Studies
130 Hesburgh Center
Notre Dame, IN 46556-5677

Kellogg Institute for Int'l Studies Quechua Summer Language Fellowships 2548
The Latin American Studies Program through the Kellogg Institute for International Studies will offer several summer fellowships for the summer to Notre Dame undergraduate and graduate students who would like to pursue the study of Quechua, the language of the Incas and of millions of indigenous peoples living in Bolivia, Colombia, Ecuador, and Peru. Students will attend classes at the Centro Tinku in Cusco, Peru. Classes are available at the beginning, intermediate and advanced levels. Students will participate in a seven week study of the language and culture. Recipients of the award will receive a small stipend as well as funding to cover airfare, program fees, room and board, and health insurance.
Requirements: Applicants must meet the following criteria: be able to attend the seven week program; demonstrate interest in Latin American Studies through coursework, work, or living experience, internships, etc; and be proficient in Spanish.
Geographic Focus: All States
Date(s) Application is Due: Mar 15
Contact: Holly Rivers; (574) 631-6023 or (574) 631-6580; hrivers@nd.edu
Internet: http://kellogg.nd.edu/students/lasp/fellowships/quechua.shtml
Sponsor: Kellogg Institute for International Studies
130 Hesburgh Center
Notre Dame, IN 46556-5677

Kellogg Institute for International Studies Summer Internships 2549
Through the Kellogg Institute for International Studies Summer Internship program, Notre Dame undergraduates are offered a variety of opportunities to gain hands-on experience with policy institutes, non-governmental organizations and various businesses that will better equip them to function in an increasingly globalized world. Students may work with organizations within Africa, Asia, Latin America and the United States. Participation in the program allows students to improve their language skills, learn about major policies and issues facing these regions and gain real world experience in their chosen field. Students spend a minimum of eight weeks with their host institution. Dates of internships vary. Some programs have set requirements while dates for other organizations are arranged between the supervisor and the intern.
Requirements: The program seeks students who have the academic interests, skills and initiative that best match the needs of the host organization, ensuring a mutually beneficial relationship for both the students and the organization. Applicants must be able to clearly articulate the relationship between their academic and career goals and the work needed at the host organization. Preference is given to students completing their junior year but sophomores and freshmen may apply. Kellogg internships are available to Notre Dame undergraduate students.
Restrictions: Students are not eligible for more than one award.
Geographic Focus: All States
Date(s) Application is Due: Feb 25
Contact: Sharon Schierling; (574) 631-8524 or (574) 631-6580; sschierl@nd.edu
Internet: http://kellogg.nd.edu/students/internship/index.shtml
Sponsor: Kellogg Institute for International Studies
130 Hesburgh Center
Notre Dame, IN 46556-5677

Kellogg Institute for Int'l Studies Supplemental Awards for PhD Students 2550
The Kellogg Institute for International Studies Supplemental Awards for PhD Students aims to attract outstanding graduate students who have already been offered departmental fellowships in the Political Science, Sociology, Economics, or History departments at Notre Dame. The Kellogg Institute will offer the supplemental fellowships for up to five years to qualified students. Recipients will receive $5,000 per year to supplement their departmental stipends. These funds will be disbursed in the fall and spring semesters. Fellowships are renewable annually for four additional academic years, contingent upon satisfactory progress toward the PhD. The annual deadline in in early March.
Requirements: To be eligible, a student must intend to devote their major doctoral field to the study of Latin America, Africa, and/or Asia.
Geographic Focus: All States
Date(s) Application is Due: Mar 3
Amount of Grant: Up to 5,000 USD
Contact: Denise Wright, Program Coordinator; (574) 631-8523 or (574) 631-6580; fax (574) 631-6717; dwright1@nd.edu
Internet: http://kellogg.nd.edu/students/grad/grants/ccaa.shtml
Sponsor: Kellogg Institute for International Studies
130 Hesburgh Center
Notre Dame, IN 46556-5677

Kellogg Institute for International Studies Visiting Fellowships 2551
The Kellogg Institute seeks both accomplished and promising scholars whose work and presence will promote interdisciplinary international research on the Institute's core themes. These visiting fellowships afford the time to move a research project forward in an intellectually stimulating environment, share findings, and develop linkages with renowned U.S. and international scholars. As a visiting fellow, you pursue research related to Kellogg Institute themes of democracy and human development, share your research with the Notre Dame scholarly community, and have the opportunity to publish in Kellogg's peer-reviewed Working Paper Series. Visiting Fellowships include: a stipend; travel expenses; medical insurance benefits; and a partial housing subsidy for the Hesburgh Center Residences or other on-campus housing.
Requirements: Fellowships are open to candidates from the U.S. and abroad who hold a PhD or equivalent degree in a social science discipline or history (exceptions considered), distinguished senior candidates who do not hold a PhD, and advanced graduate students who will complete their PhD before beginning the fellowship.
Geographic Focus: All States
Date(s) Application is Due: Nov 1
Contact: Denise Wright, Program Coordinator; (574) 631-8523 or (574) 631-6580; fax (574) 631-6717; dwright1@nd.edu
Internet: http://kellogg.nd.edu/vfellowships/index.shtml
Sponsor: Kellogg Institute for International Studies
130 Hesburgh Center
Notre Dame, IN 46556-5677

Kennedy Center Fellowship at the DeVos Institute of Arts Management 2552
The Fellowship at the DeVos Institute of Art Management is an immersive program in arts management strategy designed to prepare mid-career arts managers for executive positions in today's complex arts environment. Hosted at the nation's cultural center and one of the largest and most dynamic performing arts institutions, the Fellowship offers a structured blend of academic training, hands-on work experience, independent and collective learning opportunities, and personalized mentoring. Fellows attend weekly classes throughout the Fellowship taught by highly experienced performing arts professionals that include case study, independent research, and collaborative group work. During their course of study, Fellows gain hands-on experience in three different Kennedy Center departments working closely with senior Kennedy Center staff. These rotations extend and complement Fellows' experience in their courses and project work. The Seminar Series provides participants with contemporary knowledge essential to leaders of performing arts organizations. Seminar topics include, but are not limited to: accessibility; arts education; human resource management; law and the arts; media relations; programming; technology and arts management; and union negotiations. Participants receive a $20,000 living allowance, program materials, and reimbursement

for health insurance coverage. They also receive tickets to performance at the Kennedy Center and relocation assistance. The application is available at the website.
Requirements: Applicants must have a minimum of five years of professional arts management experience, a bachelor's degree in the arts or related discipline, and demonstrated knowledge of issues in arts management.
Geographic Focus: All States
Date(s) Application is Due: Apr 15
Amount of Grant: 20,000 USD
Contact: Syrah Gunning; (202) 416-8897; artsmanager@kennedy-center.org
Internet: http://www.kennedy-center.org/education/artsmanagement/fellowships.cfm
Sponsor: John F. Kennedy Center for the Performing Arts
2700 F Street NW
Washington, DC 20566

Kennedy Center Summer Fellowships for International Arts Managers **2553**
The DeVos Institute's Summer Fellowships for International Arts Managers offer international arts managers an immersive program of study in arts management strategy. For four weeks each summer, over three consecutive years, international Fellows participate in seminars, group projects, individual mentorships, and practical assignments supervised by the Institute's faculty and Center senior staff. The Fellowship includes the following: seminars and coursework in strategic planning; board development; fundraising; marketing; artistic planning; individual advising from Institute faculty and Center senior staff; air transportation to and from the U.S; lodging during the Fellowship; per diem to cover living expenses; visa sponsorship; and program materials. The application, a list of frequently asked questions, and informational brochure with endorsement comments from previous participants are available at the website.
Requirements: Applicants must meet the following criteria: a minimum of five years working experience in an administrative capacity at an arts organization outside North America; currently working as an executive or department head with decision-making authority; ability to commit to the full term of the Fellowship (four weeks in late June/July for three consecutive summers); and excellence verbal and writing skills in business English.
Geographic Focus: All Countries
Contact: Syrah Gunning, Summer Fellowship Coordinator; (202) 416-8897; segunning@kennedy-center.org or artsmanager@kennedy-center.org
Internet: http://www.kennedy-center.org/education/artsmanagement/international.cfm
Sponsor: John F. Kennedy Center for the Performing Arts
2700 F Street NW
Washington, DC 20566

Kenneth T. and Eileen L. Norris Foundation Grants **2554**
The foundation supports Los Angeles County nonprofits in the areas of: medicine—to improve access to health care, increase knowledge through research, and provide facilities for those activities to take place; youth—to provide constructive activities, positive role models, and opportunities for disadvantaged, disabled, and misguided children; community—to support law enforcement agencies, good citizenship, and environmental conservation; culture—to support museums, symphony orchestras, and dance and theater companies; and education and science—to focus on private education, especially secondary and college levels. Types of support include general operating support, continuing support, building construction and/or renovation, equipment acquisition, endowment funds, program development, professorships, scholarship funds, research, and matching funds. Education/science and medicine projects are accepted between May 1 and June 30; youth requests are accepted between February 15 and March 31; cultural (the arts) and community requests are accepted between December 1 and January 31; and medicine proposals are due between May 1 and June 30.
Requirements: Grants are awarded to organizations in southern California.
Geographic Focus: California
Date(s) Application is Due: Jan 31; Mar 31; Jun 30
Amount of Grant: 5,000-25,000 USD
Contact: Lisa D. Hansen; (562) 435-8444; fax (562) 436-0584; grants@ktn.org
Internet: http://www.norrisfoundation.org/grant.html
Sponsor: Kenneth T. and Eileen L. Norris Foundation
11 Golden Shore, Suite 450
Long Beach, CA 90802

Kettering Fund Grants **2555**
The mission of the Kettering Fund is to support scientific, medical, social and educational studies and research conducted by nonprofit, charitable organizations that are located and/or provide service in Ohio. The Distribution Committee meets on a semi-annual basis to make funding decisions, generally in May and November. The committee will consider support for the following purposes: capital needs (such as the purchase of land, new construction, renovations and equipment); seed grants for new organizations or programs; project support; endowments; research; and scholarship funds.
Requirements: 501(c)3 nonprofit organizations that provide services in or are located in Ohio may apply.
Restrictions: The fund does not support: religious organizations for religious purposes; activities of specific individuals; public elementary or secondary schools; loans; travel; deficit reduction; benefit events; or efforts to carry on propaganda or otherwise attempt to influence legislation.
Geographic Focus: Ohio
Amount of Grant: 5,000-200,000 USD
Contact: Judith M. Thompson; (973) 228-1021; fax (973) 228-2399; info@ketteringfund.org
Internet: http://www.ketteringfund.org/main.html
Sponsor: Kettering Fund
1480 Kettering Tower
Dayton, OH 45423

KFF Barbara Jordan Health Policy Scholars Program **2556**
The program brings talented African American, Latino, American Indian/Alaska Native, and Asian/Pacific Islander college seniors and recent graduates to Washington, DC, where they are placed in Congressional offices and learn about health policy. Through the nine-week program (May 23 through July 28), scholars gain knowledge about federal legislative procedure and health policy issues, while further developing their critical thinking and leadership skills. Scholars also participate in seminars and site visits to augment their knowledge of health care issues, and write and present a health policy research memo. Scholars receive a stipend upon completion of the program; a daily expense allowance for meals and local transportation; transportation/airfare to and from Washington, DC; and lodging at Howard University in Washington, DC. Application and guidelines are available online.
Requirements: Eligible candidates must be U.S. citizens who will be seniors or recent graduates of an accredited U.S. college or university in the fall. Successful applicants will show evidence of academic achievement and will demonstrate the following: be a member of a population that is adversely affected by racial or ethnic disparities in health; and have experience working in or with programs that address health disparities (e.g., local health departments, community health centers, inner city/rural hospitals, or organizations that conduct research on or develop policy related to health or health care disparities) or have done previous academic work (e.g., taken courses, written papers, or participate in research projects) related to health disparities.
Restrictions: Currently enrolled law, medical, and graduate students are not eligible.
Geographic Focus: All States
Date(s) Application is Due: Jan 9
Amount of Grant: 7,000 USD
Contact: Grants Administrator; (650) 854-9400; fax (650) 854-4800
Internet: http://www.kff.org/about/jordanscholars.cfm
Sponsor: Henry J. Kaiser Family Foundation
2400 Sand Hill Road
Menlo Park, CA 94025

Kiplinger Program in Public Affairs Journalism Fellowships **2557**
The program in public affairs journalism is dedicated to giving journalists time to produce meaningful work and deepen their knowledge of public affairs issues so they can better inform citizens and contribute to public life. Fellows will have six months to work on an independent public affairs print project of their choice. Fellows who successfully complete the program will receive a Certificate in Public Affairs Journalism from the John Glenn Institute, as well as a transcript reflecting credits for the courses they have taken. Fellows must either be U.S. citizens with five years of relevant professional experience in the United States or abroad, or foreign nationals with five years of professional experience within the United States or working abroad for U.S. organizations. Application and guidelines are available online.
Requirements: Applicants must have at least five years of professional experience in journalism. Those eligible include reporters, editors, and employees from newspapers, wire services, magazines, online publications, and freelancers. Television and radio reporters will also be considered if they seek to produce a print project.
Geographic Focus: All States
Date(s) Application is Due: Jul 20
Amount of Grant: 20,000 USD
Contact: Betsy Hubbard; (614) 247-8845; fax (614) 292-4868; hubbard.160@osu.edu
Internet: http://www.kiplingerprogram.org/kiplinger
Sponsor: Ohio State University
350 Page Hall, 1810 College Road
Columbus, OH 43210

Klarman Family Foundation Grants in Eating Disorders Research **2558**
The Klarman Family Foundation is interested in providing strategic investment in translational research that will accelerate progress in developing effective treatments for anorexia nervosa, bulimia nervosa and binge eating disorder. The Program's short-term goal is to support the most outstanding science and expand the pool of scientists whose research explores the basic biology of feeding, anorexia nervosa, bulimia nervosa, and/or binge eating disorder. The long-term goal is to improve the lives of patients suffering from these conditions.
Requirements: At the time of application, investigators must hold a faculty appointment at a nonprofit academic, medical or research institution in the United States, Canada or Israel. Investigators who previously submitted applications or have concluded their Klarman Family Foundation Award remain eligible to apply. Applicants do not need to be nominated by their institutions. Investigators working outside the field of eating disorders research are encouraged to apply. Examples of funding areas include but are not limited to molecular genetic analysis of relevant neural circuit assembly and function; genetic and epigenetic research; animal models created by genetically altering neural circuits; testing of new chemical entities that might be used in animal models as exploratory treatments; and clinical trials involving medications or other non-psychotherapeutic interventions. Applications must be received by The Medical Foundation by February 11, 1:00 pm U.S. Eastern time.
Restrictions: Only one application may be submitted per applicant. Clinical psychotherapeutic studies, imaging studies involving humans, and research in the medical complications of these disorders are outside the scope of this program.
Geographic Focus: All States, Canada, Israel
Date(s) Application is Due: Feb 11
Amount of Grant: 400,000 USD

Contact: Gay Lockwood, Senior Program Officer; (617) 279-2240 ext. 702; fax (617) 423-4619; glockwood@hria.org
Internet: http://hria.org/tmfservices/tmfgrants/klarman.html
Sponsor: Klarman Family Foundation
P.O. Box 171627
Boston, MA 02117

Kluge Center David B. Larson Fellowship in Health and Spirituality 2559
The Library of Congress invites qualified scholars to apply for a postdoctoral fellowship in the field of health and spirituality. It seeks to encourage the pursuit of scholarly excellence in the scientific study of the relation of religiousness and spirituality to physical, mental, and social health. The fellowship provides an opportunity for a period of six to 12 months of concentrated use of the collections of the Library of Congress, through full-time residency in the Library's John W. Kluge Center. If necessary, special arrangements may be made with the National Library of Medicine for access to its materials as well. Applicants must submit a formal application packet, including an application form, a two-page curriculum vita that should indicate major prior scholarship, a one-paragraph project summary, a bibliography of basic sources, a research proposal of no more than 1500 words, and three letters of reference (in English) from people who have read the research proposal.
Requirements: Applicants must by U.S. citizens or permanent residents and must possess a doctoral degree (PhD, MD, ScD, DrPH, DSW, PPsy, DST, ThD, and JD) awarded by the deadline date.
Geographic Focus: All States
Date(s) Application is Due: Apr 17
Amount of Grant: 25,200-50,400 USD
Contact: Larson Fellowship; (202) 707-3302; fax (202) 707-3595; scholarly@loc.gov
Internet: http://www.loc.gov/loc/kluge/fellowships/larson.html
Sponsor: Library of Congress
101 Independence Avenue SE
Washington, DC 20540-4860

Kluge Center Distinguished Chairs 2560
The Kluge Scholars-occupants of the Kluge Chairs-are free to pursue their own research, with access to all the multi-lingual, multi-medial, multi-disciplinary resources of the Library. Appointed for a limited period of time, the only obligations during their residency will be to help craft and participate in some meetings or conversations open to Members of Congress and congressional staff, and to offer at least one public presentation for the broader public policy community in Washington. There are five broadly defined Kluge Chairs: American Law and Governance; Countries and Cultures of the North; Countries and Cultures of the South; Technology and Society; and Modern Culture. Taken together, these broadly defined Chairs bring a critical mass of the world's leading thinkers to Washington from all over America and the world. Their presence provides an opportunity for a new type of dialog with political leaders, in an authentically scholarly atmosphere.
Geographic Focus: All States
Contact: Program Director; (202) 707-3302; fax (202) 707-3595; scholarly@loc.gov
Internet: http://www.loc.gov/loc/kluge/fellowships/chairs.html
Sponsor: Library of Congress
101 Independence Avenue SE
Washington, DC 20540-4860

Kluge Center Fellowships 2561
The fellowships are awarded to qualified scholars worldwide, and especially encourages humanistic and social science research that makes use of the Library's large and varied collections. Fellowships are funded up to 12 months at a rate of $4,200 per month. Interdisciplinary, cross-cultural, or multi-lingual research is particularly welcome. Among the collections available to researchers are the world's largest law library and outstanding multi-lingual collections of books and periodicals. Deep special collections of manuscripts, maps, music, films, recorded sound, prints and photographs are also available.
Requirements: Scholars worldwide who have received a terminal advanced degree within the past seven years in the humanities, social sciences, or in a professional field such as architecture or law are eligible. Exceptions may be made for individuals without continuous academic careers. Applicants may be U.S. citizens or foreign nationals. Upon selection, and in accordance with relevant visa regulations, foreign nationals will be assisted in obtaining the appropriate visa. To meet the minimum eligibility requirements, the degree must be formally awarded by the deadline date.
Geographic Focus: All States
Date(s) Application is Due: Jul 15
Amount of Grant: 4,200-50,400 USD
Contact: Coordinator; (202) 707-3302; fax (202) 707-3595; scholarly@loc.gov
Internet: http://www.loc.gov/loc/kluge/fellowships/kluge.html
Sponsor: Library of Congress
101 Independence Avenue SE
Washington, DC 20540-4860

Kluge Center Kislak Fellowships in American Studies 2562
The Program supports scholarly research that contributes significantly to a greater understanding of the cultures and history of the Americas. It provides an opportunity for a period of up to 8 months of concentrated use of materials from the Kislak Collection and other collections of the Library of Congress, through full-time residency at the Library. The program supports research projects in the disciplines of archaeology, history, cartography, epigraphy, linguistics, ethno-history, ethnography, bibliography and sociology, with particular emphasis on Florida, the circum-Caribbean region and Mesoamerica. We encourage interdisciplinary projects that combine disciplines in novel and productive ways.
Requirements: Applicants may be of any nationality and must possess a PhD degree, or equivalent terminal degree, awarded by the application deadline date.
Geographic Focus: All States
Date(s) Application is Due: Feb 28
Contact: Program Director; (202) 707-3302; fax (202) 707-3595; scholarly@loc.gov
Internet: http://www.loc.gov/loc/kluge/fellowships/kislak.html
Sponsor: Library of Congress
101 Independence Avenue SE
Washington, DC 20540-4860

Kluge Center Kislak Short-Term Fellowships in American Studies 2563
The Program offers short-term fellowships for independent scholars, undergraduate and graduate students, and college and university faculty to conduct research based on items from the Kislak Collection. The Program supports scholarly research that contributes significantly to a greater understanding of the cultures and history of the Americas. It provides an opportunity for a period of up to four months of concentrated use of materials from the Kislak Collection and other collections of the Library of Congress, through full-time residency at the Library. The program supports research projects in the disciplines of archaeology, history, cartography, epigraphy, linguistics, ethno-history, ethnography, bibliography and sociology, with particular emphasis on Florida, the circum-Caribbean region and Mesoamerica. We encourage interdisciplinary projects that combine disciplines in novel and productive ways.
Requirements: Applicants may be of any nationality and must possess a PhD degree, or equivalent terminal degree, awarded by the application deadline date.
Geographic Focus: All States
Date(s) Application is Due: Jan 31
Amount of Grant: 4,200-16,800 USD
Contact: Program Director; (202) 707-3302; fax (202) 707-3595; scholarly@loc.gov
Internet: http://www.loc.gov/loc/kluge/fellowships/kislakshort.html
Sponsor: Library of Congress
101 Independence Avenue SE
Washington, DC 20540-4860

Kluge Center Kissinger Chair in Foreign Policy and International Relations 2564
The Kissinger Scholar is a distinguished senior research position in residence at the Library for a period of ten months. Using research facilities and services at the Library of Congress, the Scholar is expected to engage in research on foreign policy and international affairs that will lead to publication. The Scholar may be of any nationality. A stipend of $135,000 during the term of appointment supports the Scholar.
Geographic Focus: All States
Date(s) Application is Due: Nov 1
Amount of Grant: 135,000 USD
Contact: Program Director; (202) 707-3302; fax (202) 707-3595; scholarly@loc.gov
Internet: http://www.loc.gov/loc/kluge/fellowships/kissinger.html
Sponsor: Library of Congress
101 Independence Avenue SE
Washington, DC 20540-4860

Kluge Center Kissinger Lecturer Award 2565
Another part of the Kissinger Program is the Kissinger Lecturer, who is appointed annually to deliver the Kissinger Lecture. Like the Kissinger Scholar, the Lecturer may be of any nationality. The Lecturer will have achieved distinction in the field of foreign affairs. Each lecture will be published and, every five years, the lectures will be aggregated into an edited volume that may contain additional material. The Lecturers, who will receive a $25,000 honorarium and round- trip transportation, will be appointed upon recommendation of the Steering Committee. There will be no direct applications, but nominations are welcome and may be submitted at any time. Selection will be based upon an achieved level of distinction.
Geographic Focus: All States
Contact: Program Director; (202) 707-3302; fax (202) 707-3595; scholarly@loc.gov
Internet: http://www.loc.gov/loc/kluge/fellowships/kissinger.html
Sponsor: Library of Congress
101 Independence Avenue SE
Washington, DC 20540-4860

Knoll-MacDonald Foundation Grants 2566
Established in 1999, the Knoll-MacDonald Foundation provides grant funding to institutions of higher education in Virginia, Maryland, New York, Rhode Island, and Michigan. The Foundation's primary field of interest is engineering education and biomedical research. Grant amounts are relatively small, most recently ranging from $350 to $2,000. Though there are no annual deadlines, applications should submit a letter of application, which includes: contact information, a detailed description of the project, and amount requested.
Geographic Focus: Maryland, Michigan, New York, Rhode Island, Virginia
Contact: Geraldine MacDonald, President; (703) 435-2122
Sponsor: Knoll-MacDonald Foundation
12172 Tryton Way
Reston, VA 20190-3281

Kobe College Corporation Graduate Fellowships 2567
The institute sponsors graduate fellowships for one year of study or research in Japan. The program is open to U.S. citizens and supports teaching or research at the master's or doctoral level. Graduate students who have a record of teaching effectively about Japan, or who show

promise to do so in the future, are favorable candidates. Fellowship recipients must plan to teach in the U.S. after completing their degree, either at the secondary or higher education levels, and should have good public speaking skills. One fellowship will be awarded to a female graduate student. Currently, the Graduate Fellowship is only be offered in alternate years.
Requirements: Applicants must be U.S. citizens, enrolled, in good standing, in a graduate program at an accredited higher education institution in the U.S..
Restrictions: Students enrolled in professional graduate degree programs are not eligible (i.e., M.B.A., J.D., M.D., etc.).
Geographic Focus: All States
Date(s) Application is Due: Jan 10
Contact: Administrator; (312) 364-9336; kobecollegecorp@prodigy.net
Internet: http://www.kccjee.org/pages/program_pages/gradfellowshippage.htm
Sponsor: Kobe College Corporation
188 W Randolph, Suite 1809
Chicago, IL 60601

Komen Greater NYC Clinical Research Enrollment Grants **2568**
Women, especially racial and ethnic minorities, are not adequately represented in cancer research efforts. Multiple barriers — cultural, linguistic, financial, and systematic — hinder their participation in clinical research, and particularly in clinical trials. Through Clinical Research Enrollment Grants, Komen Greater NYC seeks to fund projects that employ effective strategies to overcome these barriers to enrollment and retention in breast cancer clinical research. The funding period is two years, based on the Komen Greater NYC fiscal year — April 1st through March 31st.. Programs applying for the CRE grant may request up to $200,000 over this two-year period. Organizations that have previously received Komen Greater NYC funding are welcome to apply for additional funding.
Requirements: To apply, the organization must be: a not-for-profit organization (a charitable or educational tax-exempt organization), a government agency, or an Indian tribe; conducting National Cancer Institute (NCI) or Department of Defense (DOD) approved breast cancer clinical research (this includes studies through the Clinical Trials Cooperative Group Program including but not limited to ACOSOG, ECOG, SWOG, NSABP, ACRIN, and RTOG; and located in the Komen Greater NYC service area, which includes the five boroughs of New York City, Long Island, Westchester County, and Rockland County.
Geographic Focus: New York
Date(s) Application is Due: Oct 9
Amount of Grant: Up to 200,000 USD
Contact: Michelle Marquez, Director of Development; (212) 461-6186 or (212) 560-9590; fax (212) 560-9598; mmarquez@komennyc.org
Internet: http://www.komennyc.org/site/PageServer?pagename=grants_clinicaltrials
Sponsor: Susan G. Komen for the Cure Foundation
470 Seventh Avenue, 7th Floor
New York, NY 10018

Komen Greater NYC Community Breast Health Grants **2569**
The mission of Komen Greater NYC is to eradicate breast cancer as a life-threatening disease by advancing research, education, screening, and treatment. The Greater New York City Affiliate is, therefore, soliciting applications from non-profit organizations that have current or planned breast health programs within the five boroughs of New York City, Long Island, Westchester County, and Rockland County. The purpose of the Community Breast Health Grants Program (formerly known as Screening, Treatment, and Education Program- STEP) is to support community-based programs that provide the medically under-served with access to breast health education, screening coordination, diagnostic, treatment, and support services. The funding period is twelve months, based on the Komen Greater NYC fiscal year — April 1st through March 31st. Programs applying for the CBH grant may request up to $75,000 over this one year period. Organizations that have previously received Komen Greater NYC funding are welcome to apply for additional funding. However, all CBH grant awards are for one year only.
Requirements: Applicant organizations must be: a not-for-profit organization (a charitable or educational tax-exempt organization), a government agency, or an Indian tribe; located in the Komen Greater NYC service area, which includes the five boroughs of New York City, Long Island, Westchester County, and Rockland County; and a current or planned provider of breast health services.
Geographic Focus: New York
Date(s) Application is Due: Oct 9
Amount of Grant: Up to 75,000 USD
Contact: Michelle Marquez, Director of Development; (212) 461-6186 or (212) 560-9590; fax (212) 560-9598; mmarquez@komennyc.org
Internet: http://www.komennyc.org/site/PageServer?pagename=grants_breasthealthgrants
Sponsor: Susan G. Komen for the Cure Foundation
470 Seventh Avenue, 7th Floor
New York, NY 10018

Kosciuszko Foundation Fellowships for Polish Citizens **2570**
The Kosciuszko Foundation annually awards a number of Non-Teaching Fellowships to Poles for advanced study and research at universities and other institutions of higher learning in the United States. The Foundation provides a cost-of-living stipend which includes: transatlantic travel, housing allowance, health and accident insurance coverage, and, when warranted, domestic travel. Non-Teaching Fellowships are awarded to postgraduate scholars, professionals or artists with doctoral degrees. In determining awards, in addition to overall excellence, consideration is given to number of other factors such as scholarly affiliation or geographic location in order to achieve fair and reasonable diversity among awardees. There are no restrictions as to fields of study or research.
Requirements: Only Polish citizens residing permanently in Poland are eligible to apply. This program is not for Poles residing outside of Poland, nor those residing temporarily outside of Poland, whether or not engaged in research or study. Dual citizens (United States and Poland) as well as individuals applying or holding United States permanent residency status (green card) are not eligible. Applicants must have strong English language proficiency to communicate and to carry out proposed projects. The level of English proficiency will be determined during the personal interviews in Warsaw.
Restrictions: There is no allowance for dependents (spouse or child) support. Awardees are solely responsible for all expenses of accompanying dependents.
Geographic Focus: Poland
Date(s) Application is Due: Oct 15
Amount of Grant: 7,650-25,500 USD
Contact: Addy Tymczyszyn; (212) 734-2130, ext. 210; addy@thekf.org
Internet: http://www.thekf.org/scholarships/exchange-us/
Sponsor: Kosciuszko Foundation
15 East 65th Street
New York, NY 10065

Kosciuszko Foundation Graduate Study and Research in Poland Scholarships **2571**
This scholarship supports graduate level research at universities in Poland by American graduate students and university faculty members with funding from the Polish Ministry of Education and the Kosciuszko Foundation. Research projects may be conducted from October through June of the upcoming academic year at Polish universities and institutions which fall under the jurisdiction of the Polish Ministry of Education. The scholarship provides a Ministry stipend in the amount of 1,350 zloty per month for housing and living expenses. Additional funding of $250 per month of approved study/research is awarded by the Kosciuszko Foundation.
Requirements: Applicants must have working knowledge of the Polish language and are reviewed based on academic background, motivation for pursuing graduate studies and/or research in Poland, and proposals for studies and research in Poland. Applicants must be U.S. citizens or permanent residents.
Restrictions: Transportation to and from Poland is not included. This scholarship does not include tuition to attend classes at Polish universities.
Geographic Focus: All States
Date(s) Application is Due: Jan 5
Amount of Grant: 400-4,000 USD
Contact: Addy Tymczyszyn; (212) 734-2130, ext. 210; addy@thekf.org
Internet: http://www.thekf.org/scholarships/exchange-poland/research/
Sponsor: Kosciuszko Foundation
15 East 65th Street
New York, NY 10065

Kosciuszko Foundation Grants for Polish Citizens **2572**
The Kosciuszko Foundation annually awards a number of grants to Poles for study and research at universities and other institutions of higher learning in the United States. The Foundation provides a cost-of-living stipend which includes: transatlantic travel, housing allowance, health and accident insurance coverage, and, when warranted, domestic travel. Grants are awarded to those without doctoral degrees. In determining awards, in addition to overall excellence, consideration is given to number of other factors such as scholarly affiliation or geographic location in order to achieve fair and reasonable diversity among awardees. There are no restrictions as to fields of study or research supported, and each proposal is evaluated individually.
Requirements: Only Polish citizens residing permanently in Poland are eligible to apply. This program is not for Poles residing outside of Poland, nor those residing temporarily outside of Poland, whether or not engaged in research or study. Dual citizens (United States and Poland) as well as individuals applying or holding United States permanent residency status (green card) are not eligible. Applicants must have strong English language proficiency to communicate and to carry out proposed projects. The level of English proficiency will be determined during the personal interviews in Warsaw.
Restrictions: There is no allowance for dependents (spouse or child) support. Awardees are solely responsible for all expenses of accompanying dependents.
Geographic Focus: Poland
Date(s) Application is Due: Oct 15
Amount of Grant: 7,650-25,500 USD
Contact: Addy Tymczyszyn; (212) 734-2130, ext. 210; addy@thekf.org
Internet: http://www.thekf.org/scholarships/exchange-us/
Sponsor: Kosciuszko Foundation
15 East 65th Street
New York, NY 10065

Kosciuszko Foundation Teaching Fellowships for Polish Citizens **2573**
The Kosciuszko Foundation annually awards a number of Teaching Fellowships to Poles for teaching assignments at universities and other institutions of higher learning in the United States. The Foundation provides a cost-of-living stipend which includes: transatlantic travel, housing allowance, health and accident insurance coverage, and, when warranted, domestic travel. Teaching Fellowships are for university faculty members with teaching assignments at the United States universities in addition to individual research project. In determining awards, in addition to overall excellence, consideration is given to number of other factors such as scholarly affiliation or geographic location in order to achieve fair and reasonable diversity among awardees. There are no restrictions as to fields of teaching.
Requirements: Only Polish citizens residing permanently in Poland are eligible to apply. This program is not for Poles residing outside of Poland, nor those residing temporarily outside of Poland, whether or not engaged in research or study. Dual citizens (United States and Poland) as well as individuals applying or holding United States permanent residency status (green card) are not eligible. Applicants must have strong English

language proficiency to communicate and to carry out proposed projects. The level of English proficiency will be determined during the personal interviews in Warsaw.
Restrictions: There is no allowance for dependents (spouse or child) support. Awardees are solely responsible for all expenses of accompanying dependents.
Geographic Focus: Poland
Date(s) Application is Due: Oct 15
Amount of Grant: 7,650-25,500 USD
Contact: Addy Tymczyszyn; (212) 734-2130, ext. 210; addy@thekf.org
Internet: http://www.thekf.org/scholarships/exchange-us/
Sponsor: Kosciuszko Foundation
15 East 65th Street
New York, NY 10065

Kosciuszko Foundation Tuition Scholarships **2574**
Domestic scholarships foster knowledge and appreciation of Polish culture, history, and traditions and serve to strengthen understanding and friendship between citizens of the United States and Poland through educational and cultural ties. Tuition scholarships support full-time study in law, medicine, education, Slavic studies, literature, political science, chemistry, marketing, international relations, journalism, dentistry, nursing, and many other fields. Certain scholarships are available to assist eligible students of law and engineering at DePaul University, Chicago; Bayonne, New Jersey, residents studying nursing, teaching or business; and to eligible students who are residents of Amsterdam, New York, Chicopee, Massachusetts, and the state of New Hampshire. The Tuition Scholarship program also supports certain programs including the one-year Masters program at the Center for European Studies at the Jagiellonian University in Cracow and English Schools of Medicine in Poland. The period of study is one academic year with possible renewal for one additional year. Scholarships range from $1,000 to $7,000.
Requirements: Eligible students must be United States citizens of Polish descent, Poles who are permanent residents of the United States, and Americans of non-Polish descent who are pursuing courses of study or research concerning Polish subjects. United States citizens of non-Polish descent are considered when their studies are primarily focused on Polish studies. Only one member per immediate family may receive a Tuition Scholarship during any given academic year.
Geographic Focus: All States
Date(s) Application is Due: Jan 5
Amount of Grant: 1,000-7,000 USD
Contact: Addy Tymczyszyn; (212) 734-2130, ext. 210; addy@thekf.org
Internet: http://www.thekf.org/scholarships/tuition/ts/
Sponsor: Kosciuszko Foundation
15 East 65th Street
New York, NY 10065

Kovler Family Foundation Grants **2575**
The Kovler Family Foundation awards grants in the areas of the arts, children/youth services, medical research (particularly diabetes), education, human services, higher education, human services, and Jewish federated giving programs. General operating or research grants are awarded primarily in the Chicago metropolitan area. There are no application forms. Applicants should submit a one to two page written proposal letter with a copy of their IRS determination letter by mid-November. Typical grant awarded is between $1,000-$5,000.
Requirements: Illinois nonprofit organizations are eligible to apply.
Restrictions: The Foundation does not award grants to individuals.
Geographic Focus: Illinois
Contact: Jonathan Kovler, President and Treasurer; (312) 664-5050
Sponsor: Kovler Family Foundation
875 North Michigan Avenue
Chicago, IL 60611-1958

Kress Foundation Art History Pre-Doctoral Fellowships **2576**
The predoctoral fellowship in the history of European art supports a two-year research appointment in association with one of the following institutes: Florence, Kunsthistorishches Institut; Jerusalem, Nelson Glueck School of Biblical Archaeology, and Hebrew Union College; Leiden, Prentenkabinet/Kunsthistorisch Instituut der Rijksuniversiteit; London, Courtauld Institute of Art, University of London Warburg Institute, and University of London; Munich, Zentralinstitut Kunstgeschichte; Nicosia, Cyprus American Archaeological Research Institute; Paris, American University in Paris; Rome, Bibliotheca Hertziana; and Zurich, Swiss Institute for Art Research.
Requirements: Nominees must be U.S. citizens or individuals matriculated at an American university, and must be nominated by their art history departments.
Geographic Focus: All States
Date(s) Application is Due: Nov 30
Amount of Grant: 22,500 USD
Contact: Lisa Ackerman; (212) 861-4993; fax (212) 628-3146
Internet: http://www.kressfoundation.org/prog.html
Sponsor: Samuel H. Kress Foundation
174 E 80th Street
New York, NY 10021

L'Oreal U.S.A Women in Science Fellowship **2577**
The program supports the aspirational goals of young women interested in pursuing studies and careers in the natural sciences (biological and physical), engineering, computer science, and mathematics. Applicants must be planning to pursue their field of study or advanced research at a U.S. nonprofit institution. Five one-year grants will be offered. The fellowships may not be deferred or renewed. Candidates may obtain matching grants or supplemental fellowship grants. Guidelines and application are available online.
Requirements: The fellowship is available to postdoctoral associates only. U.S. citizens and individuals with permanent resident status are eligible.
Geographic Focus: All States
Date(s) Application is Due: Oct 15
Amount of Grant: 20,000 USD
Contact: Fellowship Administrator; InfoOffice@us.loreal.com
Internet: http://www.lorealusa.com/fwis/2006_fellowship.aspx
Sponsor: L'Oreal U.S.A
575 Fifth Avenue
New York, NY 10017

Lady Davis Fellowships **2578**
Fellowships of three to nine months make the cultural heritage of ancient and modern Israel widely known to people from both technologically advanced and evolving societies. The program enables scholars of various disciplines and nationalities to study, perform research, or teach at the Hebrew University of Jerusalem or at the Technion-Israel Institute of Technology, Haifa. Types of support include postdoctoral fellowships visiting professorships, and graduate studies abroad. Prospective candidates for Hebrew University should contact the university to ensure that there is a faculty member willing to sponsor the application. Fellows will be selected on the basis of demonstrated excellence in their studies, promise of distinction in their chosen fields of specialization, as well as on qualities of mind, intellect, and character. The graduate fellowship includes tuition, travel, and a stipend to meet reasonable living expenses; the postdoctoral fellowship includes travel and a larger stipend to meet reasonable living expenses; and the visiting professorship includes travel and stipend.
Requirements: Citizenship is unrestricted. Only students who are enrolled in a PhD program overseas are eligible applicants for the graduate fellowship at the Hebrew University. Applicants for the Technion must have completed their studies with excellent marks.
Geographic Focus: All States
Contact: M. Mark Sopher, 972-2-658-4723; fax 972-2-566-3848; LDFT@vms.huji.ac.il
Internet: http://ldft.huji.ac.il
Sponsor: Lady Davis Fellowship Trust
Hebrew University, Givat Ram
Jerusalem, 91904 Israel

Lady Gregory Postgraduate Fellowship **2579**
The Faculty of Arts at NUI Galway awards three postgraduate fellowships to support PhD research within the Arts Faculty. Fellowships are awarded on a competitive basis, one in each of the general academic categories of humanities, languages and literatures, and social and behavioral sciences. Areas of study include: archeology, classics, education, English, French, Gaeilge, geography, German, history, Italian, philosophy, political science and sociology, psychology, Spanish and the Centers for Human Settlement and Historical Change, Irish studies, and the Houston School of Film and Digital Media. The fellowship, which begins in the fall, is tenable for a maximum of three years. Guidelines and application are available online.
Requirements: To be considered for a fellowship, the applicant must be the holder of, or expect to attain before July, a first-class honors degree (or equivalent) from a recognized institution in a relevant discipline. Fellowships will normally be awarded to individuals embarking on the first year of their PhD program. The applicant must also have an established proficiency in English or Irish languages. For applicants whose first language is not English or Irish, a language proficiency assessment may be applied.
Geographic Focus: All States
Date(s) Application is Due: Mar 3
Amount of Grant: 20,000 EUR
Contact: Mairead Ni Dhomhnaill, Office of Faculty of Arts; 353 91 493622; fax 353 91 495508; mairead.nidhomhnaill@nuigalway.ie
Internet: http://www.nuigalway.ie/arts/ladygregory.html
Sponsor: National University of Ireland Galway
University Road
Galway, Ireland

Lalor Foundation Postdoctoral Fellowships **2580**
The Lalor Foundation postdoctoral fellowship program supports promising new researchers in establishing scientific and teaching careers. The mission of the program is to support these researchers early in their work so that they can become independently funded in the field of mammalian reproductive biology as related to the regulation of fertility. The individual nominated by the applicant institution for the postdoctoral fellowship for conduct of the work may be a citizen of any country. Fellowships will be $42,000 per year for coverage of fellowship stipend, fringes and institutional overhead.
Requirements: U.S. institutions must be exempt from federal income taxes under Section 501(c)3 of the U.S. Internal Revenue Code and must submit a determination letter from the Internal Revenue Service stating that it is not a private foundation. The individual nominated should have training and experience at least equal to the Ph.D. or M.D. level and should not have a faculty appointment (i.e., instructor, lecturer or higher). Potential fellows should not have held the doctoral degree more than two years from receipt of the degree.
Restrictions: Institutional overhead may not exceed 10 percent of the total fellowship award.
Geographic Focus: All States
Date(s) Application is Due: Jan 15
Amount of Grant: 42,000 USD

Contact: Susan Haff, (617) 426-7080, ext. 323; fax (617) 426-5441; shaff@gmafoundationst.com or fellowshipmanager@gmafoundations.com
Internet: http://www.lalorfound.org/?page_id=13
Sponsor: Lalor Foundation
77 Summer Street, 8th Floor
Boston, MA 02110-1006

LAM Foundation Established Investigator Awards **2581**
The Established Investigator Awards support research directly related to lymphangioleiomyomatosis (LAM) and is targeted at established investigators who focus on the abnormal proliferation of smooth muscle that occurs in the disease. Awards provide a maximum of $50,000 per year, renewable for up to two additional years. The structure and terms of this award are identical to the LAM Fellowship Award except that with the LAM Established Award, faculty level investigators are eligible to receive funding for technician support and supplies.
Requirements: U.S. citizens and foreign nationals with appropriate immigrant visa status are eligible. At the time of application, an applicant must hold a doctoral degree and faculty appointment at the level of assistant or associate professor and be undertaking a project related to LAM.
Geographic Focus: All States
Date(s) Application is Due: Sep 30
Contact: Francis X. McCormack; (513) 558-4831; frank.mccormack@uc.edu
Internet: http://www.thelamfoundation.org/research/apply-for-lam-funding
Sponsor: LAM Foundation
231 Albert Sabin Way, MSB Room 6053
Cincinnati, OH 45267-0564

LAM Foundation Pilot Project Grants **2582**
Pilot project awards are granted for one year to enable investigators to gather sufficient preliminary data for more substantial LAM funding. Funding is intended to support faculty-level investigators, or postdoctoral-level investigators who will be performing LAM research in the laboratory of an established scientist who is an expert in areas that are directly pertinent to LAM. Letters of intent (LOI) are required prior to submission of grant proposals to the LAM Foundation. Application materials and details are available online.
Requirements: Applications are accepted from investigators of all nationalities who have MDs or PhDs, but candidates must possess visas that allow for completion of the proposed project in the original laboratory.
Geographic Focus: All States
Date(s) Application is Due: Sep 30
Amount of Grant: Up to 25,000 USD
Contact: Francis X. McCormack; (513) 558-4831; frank.mccormack@uc.edu
Internet: http://www.thelamfoundation.org/research/apply-for-lam-funding
Sponsor: LAM Foundation
231 Albert Sabin Way, MSB Room 6053
Cincinnati, OH 45267-0564

LAM Foundation Postoctoral Fellowships **2583**
The foundation offers postdoctoral fellowships for the study of the cellular and molecular basis of the abnormal smooth muscle proliferation that occurs in the disease, Lymphangioleiomyomatosis (LAM). Fellowships are renewable for up to two additional years. The fellowship is intended to support postdoctoral level investigators who will be performing LAM research in the laboratory of an established scientist who is an expert in areas that are directly pertinent to LAM. Letters of intent (LOI) are required prior to submission of grant proposals to the LAM Foundation. Application materials and details are available online.
Requirements: Applications are accepted from investigators of all nationalities who have MDs or PhDs, but candidates must possess visas that allow for completion of the proposed project in the original laboratory.
Restrictions: The award is generally not available to individuals who have attained faculty rank, or who are seeking transition funds for the establishment of a laboratory in a new setting.
Geographic Focus: All States
Date(s) Application is Due: Sep 30
Amount of Grant: Up to 50,000 USD
Contact: Francis X. McCormack; (513) 558-4831; frank.mccormack@uc.edu
Internet: http://www.thelamfoundation.org/research/apply-for-lam-funding
Sponsor: LAM Foundation
231 Albert Sabin Way, MSB Room 6053
Cincinnati, OH 45267-0564

LAM Foundation Research Grants **2584**
The program provides seed money to investigators seeking to understand the abnormal smooth muscle proliferation that occurs in lymphangioleiomyomatosis (LAM). The grant supports clinical, laboratory, epidemiological, or any other kind of research that directly relates to LAM. Grants are subject to annual review and may be granted for up to two years. Scientists who which to apply for LAM Foundation funding must submit a Letter of Intent (LOI). Letters of Intent must reach The LAM Foundation by July 30th.
Requirements: U.S. citizens and foreign nationals with appropriate immigrant visa status are eligible. At the time of application, an applicant must hold a doctoral degree and faculty appointment with an academic institution and have completed two years of research training.
Restrictions: Residents, interns, postdoctoral fellows, students enrolled in degree-granting programs, and established investigators are ineligible.
Geographic Focus: All States
Date(s) Application is Due: Sep 30
Amount of Grant: Up to 40,000 USD
Contact: Francis X. McCormack; (513) 558-4831; frank.mccormack@uc.edu
Internet: http://www.thelamfoundation.org/research/apply-for-lam-funding
Sponsor: LAM Foundation
231 Albert Sabin Way, MSB Room 6053
Cincinnati, OH 45267-0564

Lance Armstrong Foundation Research Grants **2585**
The foundation funds and disseminates research that builds on the body of knowledge and services focused on improving the quality of life for cancer survivors across physical, emotional, and practical challenges. The research program also supports research focused on testicular cancer. The foundation funds research that is not readily fundable from traditional sources and that encourages and supports the efforts of both young and established investigators. Young investigators are defined as within eight years of completing a terminal degree or within five years of initiating independent research within a mentored laboratory, while general research awards support new research projects initiated by established investigators. Proposals are accepted from all countries. The listed application deadline is for letters of intent; full RFPs are available online.
Geographic Focus: All States
Date(s) Application is Due: May 13
Amount of Grant: Up to 75,000 USD
Contact: Amy Nunn; (512) 236-8820, ext. 181; amy.nunn@laf.org
Internet: http://www.livestrong.org/site/c.jvKZLbMRIsG/b.739079/k.9EDA/Research_Grants.htm
Sponsor: Lance Armstrong Foundation
P.O. Box 161150
Austin, TX 78716-1150

Landon Foundation-AACR Innovator Award for Cancer Prevention Research **2586**
This Award was established and first given in 2008 to recognize the outstanding achievement of an early career Assistant Professor, and to provide support for cancer prevention research of significant scientific merit in any discipline across the continuum of research. The goals of the program are to: encourage younger investigators to pursue cancer prevention research of significant scientific merit; provide the support necessary to sustain and enhance highly meritorious cancer prevention research; foster interactions between and among cancer scientists and disseminate the scientific knowledge about cancer prevention research; and contribute to a global impact against cancer. A two-year grant of $100,000 ($50,000 per year) will be provided to the recipient's institution to support direct research expenses; salary and benefits; and attendance at the AACR Annual Meeting, Cancer Prevention Conference, or other AACR meeting relevant to cancer prevention for the purpose of participating in scholarly exchange.
Requirements: Candidates must be scientists at the level of Assistant Professor who completed postdoctoral studies or clinical fellowships on or after July 1. Candidates cannot be tenured or under consideration for tenured academic positions at the time of application. A candidate whose title is not Assistant Professor but who believes that he or she meets the eligibility requirements for this Award must attach a letter from the institution's dean certifying that the position held is equivalent to Assistant Professor. Both AACR members and nonmembers are eligible to apply. However, nonmembers must submit a satisfactory application for AACR Active membership on or before October 31. Applications must be submitted by 12:00 noon (United States Eastern Daylight Time) on November 20, using the proposalCENTRAL website at http://proposalcentral.altum.com. One paper copy with original signatures and all supporting documents must be mailed to the AACR office, and be postmarked no later then November 25.
Restrictions: Employees or subcontractors of a national government or private industry are not eligible for this grant.
Geographic Focus: All States
Date(s) Application is Due: Nov 20; Nov 25
Amount of Grant: 100,000 USD
Contact: Julia Laurence, Program Assistant; (215) 440-9300, ext. 102 or (267) 646-0655; fax (215) 440-9372; julia.laurence@aacr.org or awards@aacr.org
Internet: http://www.aacr.org/home/scientists/research-funding—fellowships/landon-aacr-innovator-award-for-cancer-prevention-research.aspx
Sponsor: American Association for Cancer Research
615 Chestnut Street, 17th Floor
Philadelphia, PA 19106-4404

Landon Foundation-AACR Innovator Award for International Collaboration **2587**
This Award will support highly meritorious research that is being conducted collaboratively by investigators in different countries around the world. The goals of the program are to: promote international cancer research collaboration as an effective means to accelerate progress against cancer; provide the support necessary to sustain and enhance highly meritorious international cancer research collaborations; foster interactions between and among cancer scientists and disseminate the scientific knowledge gained from international collaboration; and contribute to a global impact against cancer. The Award of $100,000 will be provided to the recipients' institutions in the form of a two-year grant to support the collaborative research project, and may be used for such purposes including, but not limited to: direct research expenses attributable to the proposed research, which may include the salary of research assistants or technicians; expenses for travel to the institutions of the collaborators relevant to the research project; expenses for training in new techniques that will benefit the research project, such as training received through a scholar exchange/visiting professor program; and expenses related to

the presentation of research data at scientific meetings or through other means that will contribute to the dissemination of the scientific knowledge gained.
Requirements: Two or more independent researchers who working within an established international cancer research collaboration involving institutes in multiple countries may apply. Proposals to initiate a collaborative project will not be accepted. The co-investigators may be affiliated with any institution involved in cancer research, cancer medicine, or cancer-related biomedical science anywhere in the world. There is no limit to the number of co-investigators within each of the collaborating institutes. There are no national or residency status restrictions. Applicants need not be members of the AACR. Applications must be completed online using the proposalCENTRAL website, with one paper copy submitted to the AACR office. Online applications must be completed by 12:00 noon (United States Eastern Daylight Time) on November 20. The paper copy with original signatures and all supporting documents must be postmarked no later than November 25.
Restrictions: Scientific investigators or health professionals who are funded by the tobacco industry for any project, or whose named mentors in the case of mentored grants are funded by the tobacco industry for any project, may not apply and will not be eligible for AACR grants.
Geographic Focus: All States
Date(s) Application is Due: Nov 20; Nov 25
Amount of Grant: 100,000 USD
Contact: Hanna Hopfinger, Grants Program Assistant; (267) 646-0665; fax (215) 440-9372; hanna.hopfinger@aacr.org
Internet: http://www.aacr.org/home/scientists/research-funding—fellowships/landon-aacr-innovator-award-for-international-collaboration.aspx
Sponsor: American Association for Cancer Research
615 Chestnut Street, 17th Floor
Philadelphia, PA 19106-4404

Law and Public Affairs (LAPA) Fellowships at Princeton University 2588
The program invites outstanding faculty, independent scholars, lawyers, and judges to apply for appointments as fellows for the academic year. LAPA Fellows devote the major portion of their time to their own research and writing on law-related subjects of empirical, interpretive and/or normative significance. In addition, LAPA Fellows are expected to participate in LAPA programs, including a biweekly faculty-graduate seminar and a weekly discussion group, as well as some public events and conferences. Salaries vary according to individual circumstances but will not exceed a maximum that is set each fall. One of the general Fellows who will teach an undergraduate course through LAPA will be named the Martin and Kathleen Crane Fellow in Law and Public Affairs.
Requirements: The Fellows program is open to all regardless of citizenship, but it does not support work toward the completion of a degree. All applicants should have received a doctorate, juris doctor, or an equivalent professional degree by the beginning of the fellowship. In evaluating an application, the selection committee looks particularly closely at the proposal outlining work the applicant proposes to do while in residence at Princeton. Successful LAPA applicants should demonstrate substantial expertise in law-related matters, but in explaining research projects, applicants would be well advised to write for an audience of academic generalists. LAPA is particularly interested in potential Fellows with interests in policy or in normative or interpretive inquiry. The selection committee, made up of Princeton faculty associated with the program, will evaluate applicants on the basis of (1) the quality of their achievements in their field of specialization and their ability to benefit from the activities of the program; (2) the quality and significance of their proposed projects; (3) the contributions they are likely to make in the future to legal scholarship and practice; and (4) their ability to contribute to intellectual life in legal studies at Princeton. In any given year, the program tries to get a mix of senior and junior scholars, domestic and international scholars, and those based in law schools or in the practice of law on the one hand and those whose homes are in other disciplines on the other. LAPA does not announce annual themes for its Fellows so applications on all law-related subjects are welcome for the general Fellows program. Applications should be submitted using the online application.
Restrictions: The program does not normally support, as a primary activity, off-site fieldwork, development of course materials, work in legal practice, direct advocacy of causes or residence elsewhere.
Geographic Focus: All States
Date(s) Application is Due: Nov 6
Contact: Paul Frymer; (609) 258-5626; fax (609) 258-0922; pfrymer@princeton.edu
Internet: http://lapa.princeton.edu/fellowships.php
Sponsor: Princeton University
Wallace Hall
Princeton, NJ 08544

Law and Social Science Doctoral Dissertation Fellowships 2589
The purpose of this fellowship program is to encourage original and significant research on law and legal institutions. Applications are invited from outstanding students who are candidates for PhD degrees in the social sciences. Proposed dissertation research must be in the general area of socio-legal studies or in social scientific approaches to law, the legal profession, or legal institutions. The dissertation must address critical issues in the field and show promise of a major contribution to social scientific understanding of law and legal processes.
Requirements: Candidates must have completed all doctoral requirements except the dissertation by September 1 of the application year.
Geographic Focus: All States
Amount of Grant: 30,000-34,500 USD
Contact: Grants Administrator; (312) 988-6513; fellowships@abfn.org
Internet: http://www.abf-sociolegal.org/abffellow.html
Sponsor: American Bar Foundation
750 N Lake Shore Drive
Chicago, IL 60611-4403

Lawrence C. Gallen Teaching Fellowship 2590
Gallen Teaching fellows will teach three courses a semester of the Augustine and Culture Villanova Seminar in the College of Arts and Sciences Core Humanities Program. The Core Humanities Program is a writing-intensive interdisciplinary seminar program required of all first-year students. Besides teaching responsibilities, Gallen fellows are expected to share in the life of the department and participate in the department's many activities, including curriculum and faculty development, as well as other committee and/or administrative responsibilities. In addition, all Gallen fellows are expected to contribute to Villanova's residential learning communities. Candidates must submit unofficial undergraduate and graduate school transcripts, three letters of recommendation, a writing sample, and a curriculum vita. Each fellowship is a two-year renewable fellowship. Three fellowships are available. Guidelines are available online.
Requirements: Candidates for these positions must have a PhD in a Humanities discipline, a strong commitment to undergraduate teaching in the humanities, and an interest in teaching the thought of Augustine and its influence. Candidates with an interest in Hebrew Scriptures and the New Testament are also encouraged to apply.
Geographic Focus: All States
Contact: John A. Doody; (610) 519-4691 or (610) 519-7325; john.doody@villanova.edu
Internet: http://www.villanova.edu/artsci/vcle/fellowships/gallen.htm
Sponsor: Villanova University
St. Augustine Center Liberal Arts Room 104, 800 Lancaster Avenue
Villanova, PA 19085

Lawrence K. Cecil Award in Environmental Chemical Engineering 2591
This award is given to recognize outstanding chemical engineering contributions and achievements toward the preservation or improvement of the natural environment. There is no age limit. The award may be based on new discoveries, research, or the development of new processes or equipment relating to chemical engineering applications in the protection of the environment; outstanding contribution of a chemical engineering nature in the design, construction, operation, or management of environmental protection facilities or enterprises; or distinguished service of a chemical engineering nature in the protection of the environment as a professional engineer, industrial leader, or educator.
Requirements: All members of AIChE as well as other interested persons are urged to nominate deserving candidates. All information and supporting documents are to be included with the completed form; nominating forms are available from the institute.
Geographic Focus: All States
Date(s) Application is Due: Feb 15
Amount of Grant: 2,500 USD
Contact: Program Director; (212) 591-7107; fax (212) 591-8890; awards@aiche.org
Internet: http://www.aiche.org/awards/awarddtl.asp?AwardID=32
Sponsor: American Institute of Chemical Engineers
3 Park Avenue
New York, NY 10016

LBI-L Leo Baeck/Studienstiftung Fellowships 2592
The Leo Baeck Fellowship is an international Fellowship for doctoral candidates pursuing research into the history and culture of German-speaking Jewry. Fellowships fund doctoral students who wish to carry out research in more than one country. Every year, funding is offered for up to 12 junior researchers to spend a year working on research at the location of their choice. In addition to financial support for one year in the form of a monthly stipend and a travel allowance, Fellowships offer regular workshops and a common intranet to encourage Fellows to present their research and discuss it with eminent academics. Students enrolled at a university in Germany may apply for an extension after the first year. The online application form with guidelines is available at the grant website.
Requirements: The program is open to students of any nationality at any institution of higher education. Applicants must have one or more university degree(s) with formal qualification for doctoral studies, and a doctoral research project focusing on the history and culture of German-speaking Jewry.
Restrictions: Financial support for tuition costs is not included.
Geographic Focus: All States, All Countries
Date(s) Application is Due: Feb 1
Amount of Grant: 13,800 EUR
Contact: Prof. Raphael Gross, Director; +44(0) 20 7882 5690 or +44(0) 20 7882 5689; fax +44(0) 20 7882 6901; info@leobaeck.co.uk
Internet: http://www.studienstiftung.de/en/leo-baeck.html
Sponsor: Leo Baeck Institute London
2nd Floor, Arts Two Building
London, E14NS Great Britain

LBI-NY/DAAD Fellowships in Germany 2593
The German Academic Exchange Service (DAAD) and the Leo Baeck Institute (LBI) award one six-month, or two three-month graduate fellowship(s) annually for the purpose of providing financial assistance to doctoral students doing dissertation research or to academics preparing a scholarly essay or book that requires a period of research in libraries, archives or research institutions in the Federal Republic of Germany. The research must be in the field of study served by the LBI (social, communal and intellectual history of German-speaking Jewry). A link to application guidelines is available at the grant website.
Requirements: Applicants must be U.S. citizens and PhD candidates or recent PhDs (degree awarded within the last two years).
Geographic Focus: All States

Date(s) Application is Due: Nov 1
Amount of Grant: Up to 2,000 EUR
Contact: Carol Kahn Strauss, Executive Director; (212) 744-6400 or (212) 294-8340; fax (212) 988-1305; carol@lbi.cjh.org
Internet: http://www.lbi.org/about/fellowships/lbidaad-fellowships/
Sponsor: Leo Baeck Institute New York
15 West 16th Street
New York, NY 10011

LBI-NY/DAAD Fellowships in New York **2594**
The German Academic Exchange Service (DAAD) and the Leo Baeck Institute (LBI) award two fellowships annually for research at the LBI in New York on the social, communal, and intellectual history of German-speaking Jewry. The LBI is an international research, study, and lecture center whose library and archives in New York offer comprehensive documentation for the study of the subject area. The DAAD is a German national agency that provides programs and funding for students, faculty, researchers and others in higher education; represents the German higher education system as an academic and research destination; and helps build ties between institutions around the world. The fellowships provide financial assistance to doctoral students for dissertation research and to young academics for the preparation of a scholarly essay or book. Each fellowship consists of a stipend of $2,000, paid in two installments of $1,000, and is normally used within one calendar year. A link to application guidelines and to the application form is provided at the grant website.
Requirements: Applicants must be U.S. citizens and Ph.D. candidates or recent Ph.D.s (degree awarded within two years preceding application for the fellowship).
Restrictions: Support for travel or family members is not available.
Geographic Focus: All States
Date(s) Application is Due: Nov 1
Amount of Grant: 2,000 USD
Contact: Carol Kahn Strauss, Executive Director; (212) 744-6400 or (212) 294-8340; fax (212) 988-1305; carol@lbi.cjh.org or lbaeck@lbi.cjh.org
Internet: http://www.lbi.org/about/fellowships/lbidaad-fellowships/
Sponsor: Leo Baeck Institute New York
15 West 16th Street
New York, NY 10011

LBI-NY David Baumgardt Memorial Fellowships **2595**
The David Baumgardt fellowship is intended to assist scholars whose research projects are connected with the writings of Professor David Baumgardt or his scholarly interests in ethics, Wissenschaft des Judentums, and the modern intellectual history of German-speaking Jewry. The fellowship, in principal, requires extensive use of the facilities of the Leo Baeck Institute (LBI) in New York and particularly of the David Baumgardt Collection. The fellowship pays a stipend based on the requirements of the particular project being conducted. Support for travel or family members is not available. The stipend will be paid in monthly installments. Application guidelines and a link to a downloadable application form are available at the grant website. Applications must be submitted in writing no later than November 1.
Requirements: Doctoral candidates and academics affiliated with an accredited institution of higher education are eligible to apply. The fellowship should be used by the end of the year in which it was granted.
Geographic Focus: All Countries
Date(s) Application is Due: Nov 1
Amount of Grant: Up to 3,000 USD
Contact: Dr. Frank Mecklenburg, Director of Research, Chief Archivist; (212) 744-6400 or (212) 294-8340; fax (212) 988-1305; fmecklenburg@lbi.cjh.org
Internet: http://www.lbi.org/about/fellowships/david-baumgardt-memorial-fellowship/
Sponsor: Leo Baeck Institute New York
15 West 16th Street
New York, NY 10011

LBI-NY Fred Grubel Fellowship **2596**
The Fred Grubel Fellowship is a paid summer internship program for a graduate student who will participate in work on a specific research topic (jointly determined by the candidate and the Leo Baeck Institute) related to Leo Baeck Institute (LBI) collections, which can include archives, and library, photo, and art collections. The research project should pertain to the lives of refugees of the 1930s and 1940s in New York. The fellow will be supervised by the Director of Research and will work on a day-to-day basis with archives and library staff. The deadline for applications is the fall of the year prior to the summer internship. Decisions will be made in early spring. The compensation is $1,500 per month for three summer months. The application form is available online.
Requirements: Ph.D. candidates from history, sociology, literature, or Jewish studies programs are eligible.
Geographic Focus: All Countries
Date(s) Application is Due: Nov 1
Contact: Dr. Frank Mecklenburg, Director of Research, Chief Archivist; (212) 744-6400 or (212) 294-8340; fax (212) 988-1305; fmecklenburg@lbi.cjh.org
Internet: http://www.lbi.org/about/fellowships/fred-grubel-fellowship/
Sponsor: Leo Baeck Institute New York
15 West 16th Street
New York, NY 10011

LBI-NY Fritz Halbers Fellowships **2597**
The Leo Baeck Institute awards the Fritz Halbers Fellowship to assist scholars whose projects are connected with the culture and history of German-speaking Jewry. The amount of the award will be determined according to the requirements of the project and the availability of funds. The fellowship consists of an award, not exceeding $3,000, to be determined according to the requirements of the project. Application guidelines and an application form are available at the grant website.
Requirements: Applicants must be students enrolled in a Ph.D. program at an accredited institution of higher education. The fellowship should be used by the end of the year in which it is granted; support for travel or family members is not available.
Geographic Focus: All Countries
Date(s) Application is Due: Nov 1
Contact: Dr. Frank Mecklenburg, Director of Research, Chief Archivist; (212) 744-6400 or (212) 294-8340; fax (212) 988-1305; fmecklenburg@lbi.cjh.org
Internet: http://www.lbi.org/about/fellowships/fritz-halbers-fellowship/
Sponsor: Leo Baeck Institute New York
15 West 16th Street
New York, NY 10011

LBI-NY Gerald Westheimer Career Development Fellowship **2598**
The Leo Baeck Institute offers a Career Development Fellowship to a scholar or professional in an early career stage (e.g. before gaining tenure in an academic institution or its equivalent). Proposed work should deal with topics within the Leo Baeck Institute's mission, namely historical or cultural issues of the Jewish experience in German-speaking lands. Funding of up to $20,000 will cover one year and, at the discretion of the reviewing board, may be renewed for a second year. Funding is intended to provide for the following activities and expenses: obtaining scholarly material (e.g. publications); temporary help in research and production needs; membership in scholarly organizations; travel; computer, copying, and communication charges; and a summer stipend for non-tenured academics. Applications outlining the nature and scope of the proposed project including a budget should be submitted, in no more than two pages, by the specified deadline date to the Director of Research. A curriculum vitae, names of three references, and supporting material (e.g. outline of proposed work, draft of chapters, previous publications) should be appended. Submission by email is encouraged.
Restrictions: Funding is limited to applicants in early in their career.
Geographic Focus: All States, All Countries
Date(s) Application is Due: Mar 1
Amount of Grant: Up to 20,000 USD
Contact: Dr. Frank Mecklenburg, Director of Research, Chief Archivist; (212) 744-6400 or (212) 294-8340; fmecklenburg@lbi.cjh.org
Internet: http://www.lbi.org/about/fellowships/career-development-fellowship/
Sponsor: Leo Baeck Institute New York
15 West 16th Street
New York, NY 10011

Leahi Fund **2599**
The purpose of the Fund is to support programs of research and education in, and the prevention of, pulmonary disease. Highest priority will be given to: research and education with foreseeable clinical application to pulmonary problems that exist in Hawai'i; scientific research in the prevention or treatment of pulmonary diseases and infirmities. Preference will be given to supporting new investigators and to new areas of investigation for established researchers. Application information is available online.
Requirements: Research grants can only be approved for individuals affiliated with tax-exempt organizations classified by the Internal Revenue Services as 501(c)3 type, or with units of government.
Restrictions: Funding is not available for: salary for the principal investigator; patient care; travel; major equipment purchases; printing or publication costs.
Geographic Focus: Hawaii
Date(s) Application is Due: Jun 21
Amount of Grant: 30,000 USD
Contact: Christel Wuerfel, Philanthropic Services Assistant; (808) 537-6333 or (888) 731-3863; cwuerfel@hcf-hawaii.org
Internet: http://www.hawaiicommunityfoundation.org/index.php?id=71&categoryID=23
Sponsor: Hawai'i Community Foundation
1164 Bishop Street, Suite 800
Honolulu, HI 96813

Leakey Foundation Paleoanthropology Award **2600**
The grant is intended for potentially long-term, multidisciplinary research programs that seek to recover the physical and/or cultural remains of early humans and their hominid ancestors. All applicants must make an explicit argument for how their research is related to understanding human evolution. Usually awarded to senior scientists, these grants are potentially renewable for up to three years.
Requirements: Usually awarded to established scientists.
Geographic Focus: All States
Date(s) Application is Due: Jul 15; Dec 15
Amount of Grant: Up to 22,000 USD
Contact: Grants Officer; (415) 561-4646; fax (415) 561-4647; grants@leakeyfoundation.org
Internet: http://www.leakeyfoundation.org/grants.htm
Sponsor: L.S.B. Leakey Foundation
1002A O'Reilly Avenue, P.O. Box 29346
San Francisco, CA 94129-0346

Leo Baeck Essay Prize **2601**

The Leo Baeck Essay Prize has been established to stimulate new research on the history and culture of German-speaking Jewry and to promote young researchers in the field. The essay can be on any topic on the history and culture of German-speaking, Central-European Jewry from early modern times through the post-war period and up to the present. Along with a cash prize, the winning essayist will receive a free year's print and online subscription to the Leo Baeck Institute Year Book, with the winning essay published in the Year Book's upcoming volume. While there is no runner-up prize, the editors may also publish any entries that are highly commended by the judges. The online application is located at the Baeck website, along with entry guidelines.
Requirements: The Leo Baeck Essay Prize is open to researchers who have completed their postgraduate studies no more than five years earlier than the current deadline date. All entrants must grant an exclusive license to the Year Book.
Restrictions: Essays submitted for the Leo Baeck Essay Prize must not be under consideration for publication elsewhere.
Geographic Focus: All Countries
Date(s) Application is Due: Oct 5
Amount of Grant: 500 GBP
Contact: Professor Raphael Gross, Editor; +44(0) 20 7882 5690 or +44(0) 20 7882 5689; fax +44(0) 20 7882 6901; info@leobaeck.co.uk
Internet: http://www.oxfordjournals.org/our_journals/lbaeck/essay_prize.html
Sponsor: Leo Baeck Institute London
2nd Floor, Arts Two Building
London, E14NS Great Britain

Leonard and Helen R. Stulman Charitable Foundation Grants **2602**
The Leonard and Helen R. Stulman Charitable Foundation supports work in four areas of interest. Ninety five percent (95%) of funding from The Foundation is reserved for programs in greater Baltimore and the State of Maryland. The foundation will make a limited number of new grants each year to projects that address one or more of the following: research as to causes and treatment of mental illness and services for people with mental illness and their families; initiatives for an older adult and aging population and programs for geriatric medicine; programs engaged in health care treatment and prevention as well as those involved in medical research; and higher educational institutions. To apply to the Foundation for funding, applicants should submit a two-page letter of inquiry, with basic background on your organization, the identified needs your project proposes to address, an overview of the proposed project, and the amount you intend to request.
Requirements: Organizations (or their fiscal agents) that qualify as public charities under section 501(c)3 of the Internal Revenue Code and do not discriminate on the basis of race, creed, national origin, color, physical handicap, gender or sexual orientation are eligible to apply.
Geographic Focus: All States
Amount of Grant: 50,000 USD
Contact: Cathy Brill; (410) 332-4171; fax (410) 837-4701; cathybrill@comcast.net
Internet: http://www.bcf.org/ourgrants/ourgrantsdetail.aspx?grid=4
Sponsor: Leonard and Helen R. Stulman Charitable Foundation
2 East Read Street, 9th Floor
Baltimore, MD 21202

Leon Lowenstein Foundation Grants **2603**
The foundation awards grants to eligible organizations in its areas of interest, including education, medical research, and disaster relief. Types of support include research grants, general operating support, program development, and seed funding. There are no application deadlines or forms. The board meets twice each year to consider requests.
Requirements: Nonprofit organizations are eligible. Preference will be given to requests from the New York City metropolitan area.
Geographic Focus: All States
Contact: Executive Director; (212) 319-0670; fax (212) 688-0134
Sponsor: Leon Lowenstein Foundation
126 E 56th Street
New York, NY 10022

LI American Religion Dissertation Fellowship **2604**
The program is designed to support the final year of writing on promising PhD and ThD dissertation projects dealing with aspects of American religious life that are related to the concerns of the Louisville Institute. Preference will be given to proposals that attempt to describe more fully how the Christian faith is actually lived by contemporary persons and to bring the resources of the Christian faith into closer relation to their daily lives; and to help us understand more adequately the institutional reconfiguration of American religion. Proposals on certain other issues of importance to the churches are also welcome. Proposed projects may employ a variety of methodological perspectives, including, but not limited to, history; ethics; the social sciences; and historical, systematic, and practical theology. They may also be interdisciplinary in nature. Guidelines and application forms are available online.
Requirements: Applicants must be candidates for the PhD or ThD degree who have fulfilled all predissertation requirements, including approval of the dissertation proposal, by January 15 of the award year and expect to complete the dissertation by the end of the following academic year.
Restrictions: The institute is interested in funding projects that can contribute in significant ways to the understanding of contemporary religious communities. Consequently, very few dissertation projects funded in recent years have dealt with American religion prior to the mid-nineteenth century.
Geographic Focus: All States
Date(s) Application is Due: Jan 15
Amount of Grant: Up to 18,000 USD
Contact: Fellowships Director; (502) 895-3411, ext. 251; fax (502) 894-2286; info@louisville-institute.org
Internet: http://www.louisville-institute.org/secondary/DFdetail.asp
Sponsor: Louisville Institute
1044 Alta Vista Road
Louisville, KY 40205-1798

Library Company of Philadelphia and The Historical Society of Pennsylvania Visiting Research Fellowships **2605**
The Library Company of Philadelphia and The Historical Society of Pennsylvania will jointly award approximately thirty one-month fellowships for research in residence in either or both collections during the academic year. These two independent research libraries, adjacent to each other in Center City Philadelphia, have complementary collections capable of supporting research in a variety of fields and disciplines relating to the history of America and the Atlantic world from the 17th through the 19th centuries, as well as Mid-Atlantic regional history to the present. The Stipend is $2,000. Fellowships are tenable for any one-month period between June and May. They support advanced, post doctoral, and dissertation research. Candidates are strongly encouraged to inquire about the appropriateness of the proposed topic. Deadline for receipt of applications is March 1 with a decision to be made by April 15.
Geographic Focus: All States
Date(s) Application is Due: Mar 1
Amount of Grant: 2,000 USD
Contact: James Green, (215) 546-3181; fax (215) 546-5167; jgreen@librarycompany.org
Internet: http://www.librarycompany.org/fellowships/american.htm
Sponsor: Library Company of Philadelphia
1314 Locust Street
Philadelphia, PA 19107-5698

Library Company of Philadelphia Barra Foundation International Research Fellowships **2606**
The Library Company of Philadelphia and the Historical Society of Pennsylvania each year offer two one-month fellowships to support research in residence in their collections by foreign national scholars of early American history and culture living outside the United States. The fellowships support both post-doctoral and dissertation research. The project proposal should demonstrate that the Library Company and the Historical Society have primary sources central to the research topic. Candidates are encouraged to inquire about the appropriateness of a proposed topic before applying. The fellowships are tenable for one month at any time from June to May. The stipend is $2,500, plus an allowance for travel expenses. Deadline for receipt of application is March 1, and appointments will be made by April 15. There are no application forms. To apply send a curriculum vitae, a two to four page description of the project, and one letter of reference.
Geographic Focus: All States
Date(s) Application is Due: Mar 1
Amount of Grant: 2,500 USD
Contact: James Green, (215) 546-3181; fax (215) 546-5167; jgreen@librarycompany.org
Internet: http://www.librarycompany.org/fellowships/barra.htm
Sponsor: Library Company of Philadelphia
1314 Locust Street
Philadelphia, PA 19107-5698

Library Company of Philadelphia NEH Long-term Post-Doctoral Fellowships **2607**
The fellowship supports research in residence at the Library Company on any subject relevant to its collections, which are capable of supporting research in a variety of fields and disciplines relating to the history of America and the Atlantic world from the 17th through the 19th centuries. The fellowships are tenable from September through May, but the awards may both be divided between two applicants, each of whom would spend a semester in residence. Candidates are strongly encouraged to inquire about the appropriateness of the proposed topic before applying.
Requirements: Applicants must hold a PhD. Midcareer and senior scholars are particularly encouraged to apply.
Geographic Focus: All States
Date(s) Application is Due: Nov 1
Amount of Grant: 40,000-20,000 USD
Contact: James Green, (215) 546-3181; fax (215) 546-5167; jgreen@librarycompany.org
Internet: http://www.librarycompany.org/fellowships/postdoc.htm
Sponsor: Library Company of Philadelphia
1314 Locust Street
Philadelphia, PA 19107-5698

Library Company of Philadelphia Prizes **2608**
The Program in Early American Economy and Society makes two awards annually to recognize the best book and journal article published in the previous year relating to an aspect of early American economic history, broadly defined, up to 1850. Submissions are welcomed in such fields as the history of finance, business, commerce, agriculture, manufacturing, technology, and economic policymaking. The author of the winning book receives a cash award and presents his or her work during an awards ceremony. The author of the winning article receives a cash award. Complete entries consist of three copies of the work and a nominating letter explaining the work's importance to the field of early American economic history. Forward nominations and entries to the Program in Early American Economy and Society.
Requirements: Nominations should be made by someone other than the author.

Geographic Focus: All States
Date(s) Application is Due: Mar 15
Amount of Grant: 1,000-2,500 USD
Contact: Cathy Matson; (215) 546-3181; fax (215) 546-5167; cmatson@udel.edu
Internet: http://www.librarycompany.org/economics/earticleprize.htm
Sponsor: Library Company of Philadelphia
1314 Locust Street
Philadelphia, PA 19107-5698

Library Company of Philadelphia Program in Early American Economy and Society (PEAES) Dissertation-Level Fellowships **2609**

The fellowships are designed to promote scholarship in early American economy and society, broadly defined, from its colonial beginnings to roughly the 1850s. Some of the possible topics of research include the history of commerce, finance, technology, manufacturing, agriculture, internal improvements, and political economy. One dissertation-level fellowship, carrying a stipend of $20,000, is tenable for nine consecutive months of residency from September 1 to May 31. It may also be divided between two scholars, who would each receive $10,000 for the periods between September 1 to December 15, or December 15 to May 31. Deadline for receipt of the applications is March 1, with reply date being March 31.
Geographic Focus: All States
Date(s) Application is Due: Mar 1
Amount of Grant: 20,000 USD
Contact: Cathy Matson; (215) 546-3181; fax (215) 546-5167; cmatson@udel.edu
Internet: http://www.librarycompany.org/Economics/EFellowships.htm
Sponsor: Library Company of Philadelphia
1314 Locust Street
Philadelphia, PA 19107-5698

Library Company of Philadelphia Program in Early American Economy and Society (PEAES) One-Month Fellowships **2610**

These fellowships are designed to promote scholarship in early American economy and society, broadly defined, from its colonial beginnings to roughly the 1850s. Some of the possible topics of research include the history of commerce, finance, technology, manufacturing, agriculture, internal improvements, and political economy. Available to scholars at all levels, four one-month fellowships, carrying stipends of $2,000 each, are tenable for a month of research at the Library Company between June 1 and May 31. Deadline for receipt of one-month and dissertation fellowship applications is March 1, and reply date is March 31.
Geographic Focus: All States
Date(s) Application is Due: Mar 1
Amount of Grant: 20,000 USD
Contact: Cathy Matson; (215) 546-3181; fax (215) 546-5167; cmatson@udel.edu
Internet: http://www.librarycompany.org/Economics/EFellowships.htm
Sponsor: Library Company of Philadelphia
1314 Locust Street
Philadelphia, PA 19107-5698

Library Company of Philadelphia Program in Early American Economy and Society (PEAES) Post-Doctoral Research Fellowships **2611**

This fellowship is designed to promote scholarship in early American economy and society, broadly defined, from its colonial beginnings to roughly the 1850s. Some of the possible topics of research include the history of commerce, finance, technology, manufacturing, agriculture, internal improvements, and political economy. One post-doctoral research fellowship, carrying a stipend of $40,000, is tenable for nine consecutive months of residency from September 1 to May 31. It may also be divided between two scholars, who would each receive $20,000 for the periods between September 1 to December 15 or December 15 to May 31. Deadline for receipt of applications is November 1, and reply date is December 15.
Geographic Focus: All States
Date(s) Application is Due: Nov 1
Amount of Grant: 40,000 USD
Contact: Cathy Matson; (215) 546-3181; fax (215) 546-5167; cmatson@udel.edu
Internet: http://www.librarycompany.org/Economics/EFellowships.htm
Sponsor: Library Company of Philadelphia
1314 Locust Street
Philadelphia, PA 19107-5698

Library of Congress Fellowships in International Studies **2612**

The program supports postdoctoral research in all disciplines of the humanities and social sciences using the foreign language collections of the Library of Congress. Proposals in multidisciplinary and cross-disciplinary studies are also welcome, as are proposals focused on single or multiple geographical areas. Up to 10 fellowships will be available for four to nine months each. During the fellowship period, scholars will be expected to be engaged in full-time research at the Library. The fellowships may be initiated at any time between July 1 and February 1. The fellowships may be combined with sabbatical and other fellowship funds to extend the research period up to a total of 12 months. Completed applications must be submitted through the ACLS online fellowship application system.
Requirements: Applicants must hold a PhD conferred prior to the listed application deadline, and strong preference will be given to scholars at an early stage of their career—those within seven years of the degree. Applicants must be U.S. citizens or permanent residents as of the application deadline date, and may be affiliated with any academic institution. Independent scholars are also welcome to apply.
Geographic Focus: All States
Date(s) Application is Due: Nov 3
Amount of Grant: 16,000-36,000 USD
Contact: Mary Lou Reker, Programs Manager; (202) 707-5027
Internet: http://www.loc.gov/kluge
Sponsor: Library of Congress
101 Independence Avenue SE, LJ-120
Washington, DC 20540-4860

LI General Grants **2613**

The institute offers a modest general grant program that supports a limited number of individual and collaborative research and leadership education projects related to the priorities of the Louisville Institute. Such grants, for example, might cover an individual scholar's research expenses or the costs of convening a research team or group of religious leaders. Applicants who believe their request fits within institute guidelines should submit a preliminary letter of no more than two pages, in which they tell about their project and the amount of support they are requesting. The board meets three times per year (April, August, and December) to evaluate proposals.
Requirements: Academics, pastors, and other religious leaders are encouraged to apply.
Geographic Focus: All States
Date(s) Application is Due: Mar 1; Jul 1; Nov 1
Amount of Grant: 25,000-75,000 USD
Contact: Administrator; (502) 992-5432; fax (502) 894-2286; info@louisville-institute.org
Internet: http://www.louisville-institute.org/secondary/generalgrant.asp
Sponsor: Louisville Institute
1044 Alta Vista Road
Louisville, KY 40205-1798

Lillian S. Wells Foundation Grants **2614**

The Lillian S. Wells Foundation established in 1976, primarily supports the Fort Lauderdale, Florida, and Chicago, Illinois regions with funding interests that include: medical research, with emphasis on brain cancer research, women's health, substance abuse, and at-risk youth. Additional funding for education and the arts. Contact the Foundation for an application form, the Board meets quarterly in January, April, July, and October.
Requirements: Nonprofit organizations in Chicago, Illinois and Fort Lauderdale, Florida area are eligible to apply.
Geographic Focus: Florida, Illinois
Date(s) Application is Due: Mar 15; Jun 15; Sep 15; Dec 15
Amount of Grant: 1,000-1,900,000 USD
Contact: Patricia F. Mulvaney; patricia.mulvaney@thewellsfamily.net
Sponsor: Lillian S. Wells Foundation
600 Sagamore Road
Fort Lauderdale, FL 33301-2215

Lillian Sholtis Brunner Summer Fellowships for Historical Research in Nursing **2615**

These summer fellowships for historical research in nursing are offered by the Center for the Study of the History of Nursing in the School of Nursing at the University of Pennsylvania. Selection of the scholars will be based on evidence of preparation and/or productivity in historical research related to nursing. It is expected that the research and new materials produced by Brunner scholars will help ensure the growth of scholarly work focused on the history of nursing. The fellowship will support up to eight weeks of residential study and use of the center's collections. The application should be sent via email to either Patricia Antonio, PhD, RN, FAAN (dantonio@nursing.upenn.edu) or Barbra Mann Wall, PhD, RN (wallbm@nursing.upenn.edu).
Geographic Focus: All States
Date(s) Application is Due: Dec 31
Amount of Grant: 2,500 USD
Contact: Dr. Karen Buhler-Wilkerson, Director; (215) 898-4725; fax (215) 573-2168; karenwil@nursing.upenn.edu
Internet: http://www.nursing.upenn.edu/history/Pages/Brunner_Fellowship.aspx
Sponsor: University of Pennsylvania
School of Nursing, 418 Curie Boulevard
Philadelphia, PA 19104-6020

Lilly Endowment Teacher Creativity Fellowships **2616**

One of the longest-standing programs of Lilly Endowment is the Teacher Creativity Fellowship Program. It has evolved to expand eligibility to teachers in private and parochial schools and to principals and assistant principals. The commitment behind this competitive program is to help Indiana become an even more rewarding arena for teachers and other education professionals. The Fellowships have supported creative projects that are personally renewing and intellectually revitalizing. While personal renewal is the main goal of this program, applicants are encouraged to demonstrate how their proposed projects will engage their students in the learning process. The Endowment encourages teams of eligible individuals to submit collaborative proposals. If a collaborative project is funded, each participant may receive an $8,000 fellowship. Proposals will be judged on the substance, clarity, originality and feasibility of the project. Preference will be given to proposals demonstrating that substantial thought and creativity has been given to this renewal opportunity. Proposal criteria and requirements are available at the website, along with tips on writing a proposal. Applicants should also consult the website for detailed information regarding time commitments and reporting for the project.
Requirements: Licensed teachers in Indiana public, private and parochial schools (including guidance counselors and library/media specialists), principals and assistant principals in such schools are eligible to apply. They must be currently employed full-time and must be Indiana-certified in grades K-12.

Restrictions: Proposals to develop new course outlines or lesson units typically do not receive funding.
Geographic Focus: Indiana
Date(s) Application is Due: Nov 4
Amount of Grant: 8,000 USD
Contact: Barbara S. DeHart, Director; (317) 916-7345; teachercreativity@lei.org
Internet: http://www.lillyendowment.org/ed_tc.html
Sponsor: Lilly Endowment
2801 N Meridian Street, P.O. Box 88068
Indianapolis, IN 46208-0068

Link Foundation Energy Doctoral Fellowships 2617
The objective of the program is to foster energy research, to enhance both the theoretical and practical knowledge and application of energy research, and to disseminate the results of that research through lectures, seminars, and publications. On the basis of applications to the foundation in the form of research proposals, awards will be made to doctoral students in academic institutions. Preference will be shown to proposals dealing directly with energy and exploring ideas not yet fully tested, rather than to developed programs already in progress. Ideas that can be implemented in the relatively near term will be given first priority. Application guidelines are available online.
Requirements: Preference will be given to candidates who have a well-defined thesis direction but who still have enough time remaining working toward their PhD that receipt of a Link Foundation fellowship could make a difference.
Restrictions: The foundation does not support student tuition, institutional overhead, or pooled costs.
Geographic Focus: All States
Date(s) Application is Due: Dec 1
Amount of Grant: 21,500-24,000 USD
Contact: Dr. Lee Lynd, Thayer School of Engineering Manager; (603) 646-2674
Internet: http://www.linkenergy.org
Sponsor: Link Foundation
Dartmouth College
Hanover, NH 03755

LITA/Gaylord Award for Achievement in Library and Information Technology 2618
The purpose of the award is to recognize distinguished leadership in technology, notable developments of applications of technology, superior accomplishment in research or education or original contribution to literature in the field.
Requirements: Nominations for the award may be made by any member of the American Library Association. Nominations are invited and judged by the LITA/Gaylord Awards Committee.
Geographic Focus: All States
Date(s) Application is Due: Mar 1
Amount of Grant: 1,000 USD
Contact: Megan Griffin, (800) 545-2433, ext. 4270; fax (312) 280-3257; mgriffin@ala.org
Internet: http://www.ala.org/ala/mgrps/divs/lita/litaresources/litascholarships/litascholarships.cfm#oclchttp://www.ala.org/ala/lita/litaresources/litascholarships/litascholarships.htm#oclc
Sponsor: American Library Association
50 East Huron Street
Chicago, IL 60611-2795

Littlefield-AACR Grants in Metastatic Colon Cancer Research 2619
The grants will provide support for innovative cancer research projects designed to accelerate the discovery and development of new agents to treat metastatic colon cancer and/or for pre-clinical research with direct therapeutic intent. Special emphasis will be placed on research that holds promise for leading to individualized therapeutic options for treatment in the near future or for developing promising new cancer therapeutics for metastatic colon cancer, which will translate into clinical applications within a one- to two-year period.
Requirements: The prizes are open to all cancer researchers who are affiliated with any institution involved in cancer research, cancer medicine, or cancer-related biomedical science anywhere in the world. Such institutions include those in academia, industry, or government. Candidates must be active researchers and have a record of recent publications.
Geographic Focus: All States
Date(s) Application is Due: Feb 1
Amount of Grant: 500,000-1,000,000 USD
Contact: Julia Laurence, Program Assistant; (215) 440-9300, ext. 102 or (267) 646-0655; fax (215) 440-9372; julia.laurence@aacr.org or laurence@aacr.org
Internet: http://www.aacr.org/home/scientists/research-funding—training-grants/research-funding/littlefield-aacr-grants-for-metastatic-colon-cancer-research.aspx
Sponsor: Littlefield 2000 Trust and the American Association for Cancer Research
615 Chestnut Street, 17th Floor
Philadelphia, PA 19106-4404

Lloyd A. Fry Foundation Health Grants 2620
The Lloyd A. Fry Foundation supports organizations with the strength and commitment to address persistent problems of urban Chicago resulting from poverty, violence, ignorance, and despair. The Foundation's Health Program is committed to increasing access to high quality primary care and reducing health care disparities for Chicago's low-income residents. To accomplish these goals, the Foundation is interested in supporting: efforts to implement medical-home models of care which provide comprehensive integrated primary care services across multi-disciplinary team members in single or multiple settings; high quality primary care services that are not widely available to low-income populations; community outreach to connect hard-to-reach individuals with high-quality primary care; and policy advocacy focused on improving the quality of health care and increasing access to health care for low-income populations in Chicago. Typically, grants in this area range from $18,000 to $100,000.
Requirements: Grants are made only to tax-exempt organizations and are rarely made to organizations outside the Chicago metropolitan area.
Restrictions: The foundation does not fund government entities, political or religious activities, fundraising events, or medical research.
Geographic Focus: Illinois
Date(s) Application is Due: Mar 1; Sep 1; Dec 1
Amount of Grant: 18,000-100,000 USD
Contact: Unmi Song; (312) 580-0310; fax (312) 580-0980; usong@fryfoundation.org
Internet: http://www.fryfoundation.org/grants/health/
Sponsor: Lloyd A. Fry Foundation
120 South LaSalle Street, Suite 1950
Chicago, IL 60603-3419

Loeb Fellowships in Advanced Environmental Design Studies 2621
Loeb Fellowships provide part-time and full-time residencies in environmental studies in Harvard's Graduate School of Design. Part-time fellowships are limited to persons living within a two-hour airline distance of Cambridge, MA; it is expected that they will spend at least two days per week at Harvard during an academic semester or the academic year. Full-time fellows are expected to obtain a leave of absence for the duration of the appointment. Experienced professionals and academicians in the field are requested to nominate those they consider eligible for the fellowships.
Requirements: Midcareer professionals in design, planning, and related fields, normally with at least five years of professional experience, may apply. While residency/citizenship are not required, fellowships are limited to individuals whose work directly impacts the built environment of the United States.
Geographic Focus: All States
Date(s) Application is Due: Jan 4
Contact: Administrator; (617) 495-9345; loeb_fellowship@gsd.harvard.edu
Internet: http://www.gsd.harvard.edu/loebfell
Sponsor: Harvard University
48 Quincy Street, Room 420
Cambridge, MA 02138

Lowe Foundation Grants 2622
The foundation awards grants to organizations in Texas, primarily for programs that support the critical needs of women and children. The arts and higher education also receive support. Grants are awarded to support capital campaigns, building construction/renovation, and general operations. The board meets in April of each year. The December 1 deadline listed is for pre-proposals, with full proposals by request due on December 31.
Requirements: 501(c)3 organizations in Texas may apply for grant support.
Geographic Focus: Texas
Date(s) Application is Due: Dec 1; Dec 31
Contact: Clayton Maebius; (512) 322-0041; info@thelowefoundation.org
Internet: http://www.thelowefoundation.org/guidelines.htm
Sponsor: Lowe Foundation
1005 Congress Avenue, Suite 895
Austin, TX 78701

LSAC Research Grants 2623
Law School Admission Council (LSAC) Research Grants fund research on a wide variety of topics related to the mission of LSAC. Specifically included are projects investigating precursors to legal training, selection into law schools, legal education, and the legal profession. LSAC welcomes proposals for research proceeding from any of a variety of methodologies, a potentially broad range of topics, and varying time frames. Proposals will be judged on the importance of the questions addressed, their relevance to the mission of LSAC, the quality of the research designs, and the capacity of the researchers to carry out the project. LSAC's membership includes law schools in the United States, Canada, and Australia. Comparative proposals about topics outside the United States, Canada, and Australia are welcome, but they must include some explicit connection to legal education or the legal profession within those countries. A list of possible topics and topics of particular interest can be found on the website.
Requirements: Eligible research projects must inform either the process of selecting law students or legal education itself in a demonstrable way. The program welcomes proposals for research proceeding from any of a variety of methodologies, a potentially broad range of topics, and varying time frames. Proposals will be judged on the importance of the questions addressed, their relevance to the mission of LSAC, the quality of the research designs, and the capacity of the researchers to carry out the project. Eligible investigators need not be members of law school faculties. Proposals from interdisciplinary teams of law faculty and researchers from outside law schools are strongly encouraged. Applicants from all countries are eligible. Principal investigators need not be based in law schools; proposals are welcome from social, behavioral, and educational researchers of all kinds. Collaborations are encouraged between those who know legal education most intimately and those who know most about how to design and conduct empirical research.
Restrictions: Grants must be made to an institution or organization, not to individuals. Not eligible to receive grants are members of the LSAC board, members of the LSAC test development and research committee or its grants subcommittee, and persons who had been members of one or more of those bodies within one year prior to application. Grants do not fund legal education tuition.
Geographic Focus: All Countries

Date(s) Application is Due: Feb 1; Sep 1
Contact: Ann Gallagher; (215) 968-1258; fax (215) 504-1408; AGallagher@LSAC.org
Internet: http://www.lsac.org/LSACResources/Grants/lsac-legal-education-grant-program.asp
Sponsor: Law School Admission Council
662 Penn Street, P.O. Box 40
Newtown, PA 18940-0040

Lucile Horton Howe and Mitchell B. Howe Foundation Grants **2624**
The foundation supports youth organizations, social services, medical research, hospitals, religion, child welfare, drug abuse, education, and family services. Types of support include continuing support, general operating funds, and research. Requests from qualified organizations must be received prior to the second Tuesday of March for the grant year. There is only one consideration meeting a year.
Requirements: Only nonprofit organizations in the Prox-Pasadena area and the San Gabriel Valley of California are eligible. A brief letter and a copy of the organizations 501(c)3 form are the requirements for application.
Restrictions: No restricted grants will be funded by the foundation.
Geographic Focus: California
Date(s) Application is Due: Feb 28
Contact: Mitchell B. Howe, President; (626) 792-2771; lhmbhowefoun@earthlink.net
Sponsor: Lucile Horton Howe and Mitchell B. Howe Foundation
180 South Lake Avenue
Pasadena, CA 91101-4932

Ludwick Family Foundation Grants **2625**
The foundation is a California nonprofit public benefit corporation established exclusively for charitable, scientific, literary, and educational purposes. Grants are awarded to United States or U.S. based international organizations in adherence to the mission of Ludwick Family Foundation and its founding documents. The foundation tends to fund tangible types of items that will remain with and can be used repeatedly by the organization.
Requirements: Eligible applicants include U.S. or U.S.-based international organizations, 501(c)3 nonprofit public charities, and government agencies (any level). All grants are to be used exclusively for charitable, public benefit purposes.
Restrictions: The foundation does not grant requests for salaries, general operating expenses, scholarships, endowment funds, fundraising events or capital campaigns, feasibility studies, consulting fees, or advertising. The foundation will no longer accept any unsolicited requests for research or from public/private schools (K-12), universities/colleges, child daycare/development centers, hospitals, or libraries. Grant consideration will be given only to those organizations that have been invited to submit formal proposals and that have completed the application process.
Geographic Focus: All States
Amount of Grant: 5,000-50,000 USD
Contact: Trista Campbell; (626) 852-0092; fax (626) 852-0776; ludwickfndn@ludwick.org
Internet: http://www.ludwick.org
Sponsor: Ludwick Family Foundation
203 South Glendora Avenue, Suite B, P.O. Box 1796
Glendora, CA 91740

Ludwig Vogelstein Foundation Grants **2626**
The foundation provides one-time grants to individuals working in the fine arts (excluding photography, film, and video) and in the literary arts, including playwriting. It also supports certain unaffiliated scholarly and/or research projects. Grants ranging from $1,000 to $3,000 are awarded on the criteria of merit and need; the foundation is especially concerned with those who have no other source of funding.
Requirements: Send email or SASE to request current guidelines (email is preferable and quick). The foundation prefers not to receive phone calls. Currently, the sponsor does not have a website.
Geographic Focus: All States
Amount of Grant: 1,000-3,000 USD
Contact: Diana Braunschweig, Executive Director; lvf@earthlink.net
Sponsor: Ludwig Vogelstein Foundation
4001 Inglewood Avenue, Suite 101-309
Redondo Beach, CA 90278

Lumina Foundation for Education Grants **2627**
The Lumina Foundation for Education believes that education provides the basis for individual opportunity, economic vitality and social stability. With its partners, Lumina strives to meet workforce demands and close gaps in attainment for groups not historically well-served by higher education. Lumina's overarching goal is to increase the higher education attainment rate of the United States to 60 percent by 2025. This will represent an increase of 23 million graduates above current levels of production. The Foundation's proactive portfolio focuses on: increasing awareness of the benefits of higher education; improving student access to and preparedness for college; improving student success in college; and productivity across the higher education system. Note that Foundation grants vary in size by their scope but that grants resulting from unsolicited inquiries typically range from $50,000-$250,000. The usual duration for a grant is one to three years.
Requirements: IRS 501(c)3 or 509(a)(1, 2, or 3) tax-exempt organizations and public school organizations are eligible.
Restrictions: The foundation will not award grants that support corporate sponsorships and fundraising events; fund partisan political or lobbying efforts; or provide direct support for individuals, including scholarships or institutional scholarship funds.
Geographic Focus: All States
Amount of Grant: 50,000-250,000 USD
Contact: Candace Brandt; (800) 834-5756 or (317) 951-5300; fax (317) 951-5063
Internet: http://www.luminafoundation.org/grants.html
Sponsor: Lumina Foundation for Education
30 South Meridian Street, Suite 700, P.O. Box 1806
Indianapolis, IN 46204

Lumosity Human Cognition Grant **2628**
Lumosity invites researchers to submit proposals for studies that use functional neuroimaging techniques to investigate mechanisms underlying cognitive processes implicated in Lumosity's games and assessments. These studies would ideally involve both neuroimaging and behavioral methods with healthy adult populations. Examples of types of projects that would be prioritized include, but are not limited to demonstrations of: task-related neural activity within and across brain regions; changes in neural activity that accompany training-related changes in cognitive performance; and neural specificity of cognitive training effects. All applications are encouraged to focus explicitly on the use of neuroimaging as a tool for studying Lumosity's cognitive training platform. A single proposal will be awarded in an amount not to exceed $150,000 for a maximum period of 12 months beginning early next year. The deadline for submission is September 27.
Requirements: Individuals from academic, medical, or research institutions based in the United States are eligible to apply. Institutions may be public or private, and must not be commercial.
Restrictions: Funds are to be used to cover research expenses only and cannot be used for equipment/supplies, travel, conference or publication costs.
Geographic Focus: All States
Date(s) Application is Due: Sep 27
Amount of Grant: Up to 150,000 USD
Contact: Faraz Farzin, Research Scientist; (570) 498-9018; faraz@lumoslabs.com
Internet: http://hcp.lumosity.com/research/from_the_lab/lumosity-announces-human-cognition-grant-for-fall-2013
Sponsor: Lumosity
153 Kearny Street
San Francisco, CA 94108

Lumpkin Family Foundation Healthy People Grants **2629**
Historically, the Lumpkin Family Foundation has supported entities that provide services that help keep people in its geographic area healthy. Requests will be evaluated on how well they accomplish one or more of the following: support the creativity of nonprofit organizations by seeding new projects and encouraging experimentation and innovation; support organizations demonstrating outstanding leadership in their field or community; promote the effectiveness of organizations and the nonprofit sector by supporting planning, learning and the professional development of staff and board leaders; facilitate collaboration across traditional organization or sector boundaries for community benefit; and develop public understanding of issues and promote philanthropic support necessary to address issues of community importance.
Requirements: Grants are awarded to nonprofit organizations that serve the community without discrimination on the basis of race, sex, or religion. Special consideration will be given to organizations and programs in East Central Illinois.
Restrictions: Proposals will not be considered from organizations that are not 501(c)3 tax-exempt; organizations whose primary purpose is to influence legislation; political causes, candidates, organizations or campaigns; individuals; or religious organizations, unless the particular program will benefit a large portion of the community and does not duplicate the work of other agencies in the community.
Geographic Focus: All States
Contact: Bruce Karmazin, (217) 234-5915 or (217) 235-3361; fax (217) 258-8444; Bruce@lumpkinfoundation.org
Internet: http://www.lumpkinfoundation.org/bWHATbwefund/HealthyPeople.aspx
Sponsor: Lumpkin Family Foundation
121 South 17th Street
Mattoon, IL 61938

Lustgarten Foundation for Pancreatic Cancer Research Grants **2630**
The foundation aims to make rapid advances in the battle against pancreatic cancer through a combination of cutting-edge technology and novel ideas. Grants can be submitted in one of three areas: novel technologies for pancreatic cancer genetics; screening for the early detection of pancreatic cancer; and novel therapies in pancreatic cancer. Proposals that exhibit leveraging of institutional resources and the attainment of major new programs are encouraged. This initiative encourages, but does not require, the development of integrative and collaborative teams of investigators from within a single institution or among several institutions. Letters of intent must be received on or before 5:00 p.m. Eastern Time on July 31, with full applications due no later than 5:00 p.m. Eastern Time on August 18. Grants will be awarded for a one-year period for a maximum amount of $100,000, of which no more than 10% can be used for indirect costs.
Requirements: Applications will be accepted from individual investigators as well as collaborating investigators. Applicants must have an MD or a PhD degree to become principal investigators. Eligible organizations include nonprofit organizations; public or private institutions, such as universities, colleges, hospitals, and laboratories; and domestic or foreign institutions/organizations.
Geographic Focus: All States
Amount of Grant: Up to 100,000 USD
Contact: Kerri Kaplan, Executive Director; (516) 803-2304 or (866) 789-1000; fax (516) 803-2303; kkaplan@cablevision.com or lsasso@cablevision.com
Internet: http://www.lustgarten.org/Page.aspx?pid=666

Sponsor: Lustgarten Foundation for Pancreatic Cancer Research
1111 Stewart Avenue
Bethpage, NY 11714

Luther I. Replogle Foundation Grants 2631
The foundation focuses its giving on the following areas: programs addressing the needs of youth and children living in, or at risk of, long-term poverty (especially children of inner-city residents and migrant workers); programs to improve educational opportunities for inner- city children, including enrichment programs in the arts and sciences, alternative schools, after-school tutoring and mentoring, and scholarship programs; programs for affordable and supportive housing that reach groups of people frequently left out of traditional shelter programs, including single mothers and families with children, the elderly, ex-offenders, and youth; projects, lectures, and fellowships in classical archaeology; projects and institutions working for the conservation of maps and globes, and dissemination and education in the area of geography; and a modest budget for support of the arts. Types of support include direct service efforts; advocacy at the local, state, and/or federal levels; general operating support, new projects, and capital campaigns (limited). The board meets twice a year to make funding decisions, once in the autumn and once in the spring. Guidelines and application are available online.
Requirements: 501(c)3 nonprofits in Chicago, IL; Minneapolis, MN; Palm Beach County, FL; and Washington, DC, are eligible. Preference is given to organizations with small or modest operating budgets
Geographic Focus: District of Columbia, Florida, Illinois, Minnesota
Date(s) Application is Due: Mar 15; Sep 15
Amount of Grant: 500-45,000 USD
Contact: Gwenn Gebhard; (202) 296-3686; fax (202) 296-3948; info@lirf.org
Internet: http://www.lirf.org
Sponsor: Luther I. Replogle Foundation
1111 19th Street NW, Suite 900
Washington, DC 20036

Lymphatic Education & Research Network Additional Support Grants for NIH funded F32 Postdoctoral Fellows 2632
The Lymphatic Education & Research Network (LE&RN) is a 501(c)3 not?for profit organization whose mission is to fight lymphatic disease and lymphedema through education, research and advocacy. LE&RN has retained The Medical Foundation, a division of Health Resources in Action (The Medical Foundation division), to manage the administrative aspects of the LE&RN Awards Program. LE&RN seeks to accelerate the prevention, treatment and cure of the disease while bringing patients and medical professionals together to address the unmet needs surrounding lymphatic disorders. The goal of the Additional Support Awards Program is to help foster career interest in the field of lymphatic research by offering additional funds for the F32 postdoctoral research projects. The additional support may be used for research supplies, equipment, health insurance and travel to scientific meetings.
Requirements: Postdoctoral fellows who are supported by the NIHF32 Postdoctoral Fellowship Program, have at least one year remaining of F32 support as of July 1, and conduct research that clearly advances the mission of LE&RN are eligible to apply. Award recipients with an active F32 Fellowship will receive $10,000 annually for up to two years. The additional support award may be used for salary support of personnel who are assisting in the research project, as well as research supplies, equipment, health insurance and travel to scientific meetings. A budget proposal is not required for the application. Online submissions must be received by The Medical Foundation by January 10, 1:00 pm, U.S. Eastern time.
Geographic Focus: All States
Date(s) Application is Due: Jan 10
Amount of Grant: 20,000 USD
Contact: Erin Johnstone; (617) 279-2240 ext. 710; ejohnstone@hria.org
Internet: http://www.hria.org/tmfservices/tmfgrants/lern32.html
Sponsor: Lymphatic Education & Research Network
261 Madison Avenue
New York, NY 10016

Lymphatic Education & Research Network Postdoctoral Fellowships 2633
The Lymphatic Education & Research Network (LE&RN) is a 501(c)3 not-for profit organization whose mission is to fight lymphatic disease and lymphedema through education, research and advocacy. LE&RN seeks to accelerate the prevention, treatment and cure of the disease while bringing patients and medical professionals together to address the unmet needs surrounding lymphatic disorders. The LE&RN Postdoctoral Fellowship Program is designed to support postdoctoral scientists in academic, medical or research institutions throughout the world. The awards will support investigators who have recently received their doctorates, a critical point in career development when young scientists choose their lifelong research focus. The goal of the Fellowship Program is to expand and strengthen the pool of outstanding junior investigators in the field of lymphatic research. Two-year fellowships ranging from $82,000 to $95,000 will be awarded to fellows who have completed no more than three years of postdoctoral training by July 1st of the LE&RN funding cycle.
Requirements: Each applicant must be working under the supervision of an established investigator who is the designated Mentor. Examples of funding areas include but are not limited to basic, translational or clinical investigations into normal and aberrant lymphatic biology; mechanisms of normal and pathological lymphangiogenesis; imaging and understanding the mechanisms that regulate lymphatic structure and function; animal models of lymphatic system disorders; and the cross-talk between the lymphatic and the immune systems. Applications will be accepted worldwide from postdoctoral fellows who will meet the following eligibility requirements by July 1: received a Ph.D. or equivalent and/or an M.D. or equivalent from an accredited institution; commit at least 90% time for research or, at least 70% time for research if the applicant also has clinical responsibilities; have completed no more than three years of full-time postdoctoral research experience; be a postdoctoral fellow under the supervision of a faculty member; and, conduct the proposed research project at a nonprofit hospital, university or other research institution (there are not institutional limitations on the number of applicants who may submit). Eligibility is not limited to those investigators with past experience in the field. Completed applications must be received by The Medical Foundation by January 10, 1:00 pm, U.S. Eastern time.
Restrictions: Projects in lymphocyte biology, leukemia, lymphoma and conditions secondary to lymphedema such as cutaneous infections are outside the scope of this program.
Geographic Focus: All States, All Countries
Date(s) Application is Due: Jan 10
Contact: Erin Johnstone; (617) 279-2240 ext. 710; ejohnstone@hria.org
Internet: http://hria.org/tmfservices/tmfgrants/lern.html
Sponsor: Lymphatic Education & Research Network
261 Madison Avenue
New York, NY 10016

Lynde and Harry Bradley Foundation Fellowships 2634
The Bradley Fellowship program is a major part of the Foundation's program interest in strengthening America's intellectual infrastructure at the higher-education level. Begun in 1986, the program is a model for influencing the intellectual framework of national life. It helps fulfill the two basic prerequisites for liberal education — good teaching and good curriculum — by enabling colleges and universities to offer students in the advanced stages of graduate work the opportunity to pursue serious studies with excellent teachers. For the program, the Foundation chooses institutions of higher education each year to support graduate and post-graduate students, who are designated as Bradley Fellows. The Fellows are chosen by each institution upon the recommendation of one or more distinguished professors in the humanities, social sciences, and the law.
Geographic Focus: All States
Contact: Daniel P. Schmidt, Vice President; (414) 291-9915; fax (414) 291-9991
Internet: http://www.bradleyfdn.org/bradley_fellows.asp
Sponsor: Lynde and Harry Bradley Foundation
1241 North Franklin Place
Milwaukee, WI 53202-2901

Lynde and Harry Bradley Foundation Grants 2635
The foundation aims to encourage projects that focus on cultivating a renewed, healthier, and more vigorous sense of citizenship among the American people and among peoples of other nations. Projects likely to be supported will generally treat free people as self-governing, personally responsible citizens, not as victims or clients; aim to restore the intellectual and cultural legitimacy of common sense, the wisdom of experience, everyday morality, and personal character; seek to reinvigorate and re-empower the traditional, local institutions—families, schools, churches, and neighborhoods—that provide training in and room for the exercise of genuine citizenship, that pass on everyday morality to the next generation, and that cultivate personal character; and encourage decentralization of power and accountability away from centralized, bureaucratic, national institutions back to states, localities, and revitalized mediating structures. Eligible projects may address any arena of public life—economics, politics, culture, or civil society; the problem of citizenship at home or abroad; Milwaukee and Wisconsin community and state projects that aim to improve the life of the community through increasing cultural and educational opportunities, grassroots economic development, and social and health services; the resuscitation of citizenship in the economic, political, cultural, or social realms; policy research and writing about approaches encouraging that resuscitation; academic research and writing that explore the intellectual roots of citizenship; and popular writing and media projects that illustrate for a broader public audience the themes of citizenship. The foundation supports programs that research the needs of gifted children and techniques of providing education for students with superior skills and/or intelligence; research programs investigating how learning occurs in gifted children; and demonstration programs of instruction.
Requirements: As an initial step, tax-exempt and nonprofit organizations should prepare a brief letter of inquiry presenting a concise description of their project, its objectives and significance, and the qualifications of the organizations and individuals involved. If the project appears to fall within the foundation's mandate, the applicant will be invited to submit a formal proposal. If invited to submit a formal proposal, the applicant should submit another letter. It should include a more-thorough, yet still concise description of the project, its objectives and significance, and the qualifications of the groups and individuals involved in it. It should also include a project budget, the specific amount being sought from Bradley, and a list of its other sources of support, philanthropic or otherwise.
Restrictions: The foundation favors projects that are normally not financed by public funds and will consider requests from religious organizations that are not denominational in character. Grants without significant importance to the foundation's areas of interest will only under special conditions be considered for endowment or deficit financing proposals. Grants will not be made to individuals, for overhead costs, or for fund-raising counsel.
Geographic Focus: All States
Date(s) Application is Due: Feb 1; May 1; Aug 1; Nov 1
Contact: Daniel P. Schmidt; (414) 291-9915; fax (414) 291-9991
Internet: http://www.bradleyfdn.org/program_interests.asp
Sponsor: Lynde and Harry Bradley Foundation
1241 North Franklin Place
Milwaukee, WI 53202-2901

M. Bastian Family Foundation Grants 2636
The foundation supports nonprofit organizations in its areas of interest, including music, the arts, higher education, health care and health organizations, religion (Christian and

Latter-day Saints), social services, and wildlife conservation. Types of support include general operating support and scholarship funds. Contact the office for application forms. There are no application deadlines.
Requirements: Funding focus is primarily in Utah.
Restrictions: Grants are not made to individuals.
Geographic Focus: All States
Contact: McKay Matthews, Program Contact; (801) 225-2455
Sponsor: M. Bastian Family Foundation
51 W Center Street, Suite 305
Orem, UT 84057

M.D. Anderson Foundation Grants **2637**
The Anderson Foundation funds projects for the improvement of working class conditions among workers, and for the establishment, support and maintenance of hospitals, homes and institutions for the care of the young, sick, the aged, and the helpless. The Foundation also gives for the improvement of general living conditions and for the promotion of health, science, education, and the advancement of knowledge. Funding is given for aging center and services; education; employment; government/public administration; health care; human services; medical specialties public policy and research; and youth services. Types of support include building and/or renovation, equipment, matching/challenge support, research, and seed money.
Requirements: Organizations should submit a letter of inquiry with the following information: a copy of their IRS determination letter; a detailed description of their project and amount of funding requested; a copy of the current year's organizational budget and/or project budget; and a listing of additional sources and amount of support. Applicants should submit five copies of the proposal. There are no deadlines. The Board meets once a month, with organizations contacted within four weeks.
Restrictions: Funding is given primarily in Texas, with emphasis on the Houston area. Grants are not available to individuals, operating funds, or endowments.
Geographic Focus: Texas
Amount of Grant: 25,000-200,000 USD
Contact: Karen Jenkins, Grant Contact; (713) 216-1095
Sponsor: M.D. Anderson Foundation
P.O. Box 2558
Houston, TX 77252-8037

M.J. Murdock Charitable Trust College Research Program for Life Sciences **2638**
The Murdock Trust's College Research Program for Life Science is designed to support research at private, predominantly undergraduate colleges and universities in the Trust's funding region. The primary criterion used in evaluating proposals will be the perceived scientific quality of the research as measured by its likelihood to contribute significantly to fundamental knowledge in the field. A clear commitment of the college or university to support the research will also strengthen a proposal, as will substantial participation of undergraduate students as research associates, rather than as technical assistants. Eligible costs include: minor equipment and supplies needed (with institutional cost sharing); student stipends of up to $3,500 for ten weeks of full time participation; faculty stipends of up to $7,500 for up to ten weeks of full time participation; travel costs; and special/unusual expenses that are essential in the performance of the research. Additional detailed information is available at the Trust website.
Requirements: To be eligible, the institution must offer the baccalaureate degree, but no doctoral degree in the sciences. The principal investigator must either hold a faculty appointment in a department of biology, be a member of a department of environmental science doing research on biological systems or organisms, or be a member of a psychology department or its equivalent, doing laboratory neuroscience research. Application forms may be downloaded, but cannot be submitted electronically. The application and all supporting materials must be mailed to the Trust.
Restrictions: Faculty members in departments of biochemistry are not eligible. The following are also ineligible for funding: faculty and student academic year stipends, postdoctoral stipends, and graduate student stipends; indirect costs and fringe benefits, except as noted above; secretarial assistance, telephone, routine office supplies; publication costs, books, and journals; service or use charges for equipment housed at the home institution; and travel expenses to scientific meetings.
Geographic Focus: Alaska, Idaho, Montana, Oregon, Washington
Date(s) Application is Due: Oct 1
Contact: Dana Miller, Senior Program Director; (360) 694-8415; fax (360) 694-1819
Internet: http://www.murdock-trust.org/grants/college-research-life-sciences.php
Sponsor: M.J. Murdock Charitable Trust
703 Broadway, Suite 710
Vancouver, WA 98660

M.J. Murdock Charitable Trust College Science Research Program **2639**
The Murdock College Science Research Program is designed exclusively for private four-year liberal arts colleges in the Trust's grantmaking region. The goal of the program is to strengthen faculty and undergraduate student research efforts in the natural sciences at these colleges. Grants are given to institutions with the capacity for developing division-wide, faculty-student research programs that can be sustained with institutional resources and external funding. To date, the following colleges have been admitted into the program: Gonzaga University, Lewis & Clark College, University of Puget Sound, Whitman College, Linfield College, Northwest Nazarene University, Pacific Lutheran University, Pacific University, Seattle University, Willamette University, and the University of Portland.
Requirements: Colleges prepared to enter into and benefit from this program are invited to contact Murdock Trust.
Restrictions: Funding is exclusively for private liberal arts colleges and universities in Alaska, Idaho, Montana, Oregon, and Washington State.
Geographic Focus: Alaska, Idaho, Montana, Oregon, Washington
Contact: Dana Miller, Senior Program Director; (360) 694-8415; fax (360) 694-1819
Internet: http://www.murdock-trust.org/grants/college-science-research.php
Sponsor: M.J. Murdock Charitable Trust
703 Broadway, Suite 710
Vancouver, WA 98660

M.J. Murdock Charitable Trust Exceptional Opportunity Grants **2640**
The Murdock Charitable Trust Exceptional Opportunity Grants meets the special program needs of regional graduate research universities. It is appropriate for projects that have special merit, but require special funding due to time constraints, no immediate source of funding, or some other unusual circumstance. The Vice President for Research, or the equivalent, at each of the universities with which arrangements have been made, may authorize an investigator or team to submit a proposal for a maximum amount of $50,000.
Restrictions: This grant is open to applicants only upon invitation or special arrangements with the Trust. Any potential applicant or organization interested in this programs should confer with the Trust before serious consideration is given to making application. Proposals may not be submitted to the Trust except at the invitation of the Vice President for Research at the principal investigator's institution. Applicants should contact the Trust for additional information.
Geographic Focus: Alaska, Idaho, Montana, Oregon, Washington
Amount of Grant: Up to 50,000 USD
Contact: Dana Miller, Senior Program Director; (360) 694-8415; fax (360) 694-1819
Internet: http://www.murdock-trust.org/grants/exceptional-opportunity-grants.php
Sponsor: M.J. Murdock Charitable Trust
703 Broadway, Suite 710
Vancouver, WA 98660

M.J. Murdock Charitable Trust General Grants **2641**
The M.J. Murdock Charitable Trust General Grants provides support for education, arts and culture, and health and human services. The Trust considers educational projects in formal and informal settings, emphasizing enhancement or expansion, as well as new educational approaches. Of special interest to the Trust are the following: performance and visual arts projects which enrich the cultural environment of the region and educational outreach efforts; and programs that emphasize preventative efforts which address physical, spiritual, social, and psychological needs, with a focus on youth. Grants are awarded for capital projects, program initiation, expansion, or for increased organizational capacity. Organizations are encouraged to view the grants awarded link for examples of previously funded grants.
Requirements: Before proceeding, interested parties should review the General Grant Application Guidelines to see if their organization is eligible or their project is appropriate for application to the Trust. After determining eligibility and appropriateness, organizations must submit a letter of inquiry before proceeding with the application. Upon approval, the organization will use the Trust's General Grant Application form and procedures.
Restrictions: The Trust only funds programs in the Pacific Northwest region: Alaska, Idaho, Montana, Oregon, and Washington. The following requests for funding are not considered: specific individuals and/or their personal benefit; individuals unauthorized to act on behalf of a qualified tax-exempt organization; funds that will ultimately be passed through to other organizations; propagandizing or for influencing legislation and elections; institutions that in policy or practice unfairly discriminate against race, ethnic, origin, sex, creed, or religion; sectarian or religious organizations whose principal activity is for the primary benefit of their own members; or for long-term loans, debt retirement, or operational deficits. The following funding requests are rarely considered: normal ongoing operations or the continuation of existing projects; endowments or revolving funds that act as such; continuation of programs previously financed from other external sources; urgent needs, emergency, or gap funding; organizations organized or operating outside any state or territory of the United States.
Geographic Focus: Alaska, Idaho, Montana, Oregon, Washington
Contact: Marybeth Stewart Goon; (360) 694-8415; fax (360) 694-1819
Internet: http://www.murdock-trust.org/grants/general-grants.php
Sponsor: M.J. Murdock Charitable Trust
703 Broadway, Suite 710
Vancouver, WA 98660

M.J. Murdock Charitable Trust Partners in Science Grants **2642**
The Murdock Partners in Science Grants are designed to help high school science teachers work with a mentor doing cutting-edge research over the course of two summers. The purpose of this grant is to bring the knowledge from the research lab back into the high school science classroom, promoting hands-on science education. Partners in Science grants provide research opportunities for teachers to work with investigators in academic labs or in labs associated with other non-profit institutions. Teachers are expected to arrange their own partnerships, although the Trust may provide some guidance. Approximately 25 grants are awarded each year to teacher-mentor partnerships in the Pacific Northwest. After completion of the Partners in Science Grants, recipients may then apply for the Partners in Science Supplemental Award of up to $7,000. These awards are made directly to the teacher's high school, and are for the purpose of implementing ideas from the research partnership in their classrooms. The application deadline for these awards is November 1. More information about Supplemental Awards and application forms is available from the Trust.
Requirements: Applications are accepted from researchers in the natural sciences at institutions located in areas where the Partners in Science program is available.

Applications must be submitted through a college, university or other qualified research institution which accepts responsibility for the administration of the award. To apply, an investigator must hold a Ph.D. and must have an active research program as evidenced by grants and/or publications and hold an appointment in a natural science department at a college, university, or other qualified institution in the Pacific Northwest (Alaska, Idaho, Montana, Oregon, and Washington). High school teachers should possess academic qualifications in appropriate disciplines, in most cases a four year degree or the equivalent in chemistry, physics, biology, or earth science. Applicants must be active in classroom teaching and have a high school appointment. The teacher and high school would normally be within convenient commuting distance of the research host institution so that a research partnership may continue beyond the 8-10 week summer research periods. Qualified faculty members/research scientists (mentors) should request application packets from the Trust (as previously noted awards are made through, and must be endorsed by, a qualified institution of higher education or research institution), or download it from the Trust's website. Applications are in two parts: a short three-page application form to be completed by the mentor; and a two-page teacher candidate statement to be completed by the teacher. Applications must include both parts, for the contributions of each partner will be considered in the review process. There are no restrictions with regard to age, rank, or length of service of either teachers or mentors, but preference may be given to teachers who have not had a previous opportunity to do research, those who have had three or more years of teaching experience, and those who plan to continue teaching for several years, so the benefits of the program can be transferred back to the classroom. Applicants should refer to the website for more information about how mentor/teacher partnerships can be formed.
Geographic Focus: Alaska, Idaho, Montana, Oregon, Washington
Date(s) Application is Due: Dec 1
Amount of Grant: Up to 15,000 USD
Contact: Dana Miller, Senior Program Director; (360) 694-8415; fax (360) 694-1819
Internet: http://www.murdock-trust.org/grants/partners-science.php
Sponsor: M.J. Murdock Charitable Trust
703 Broadway, Suite 710
Vancouver, WA 98660

M.J. Murdock Charitable Trust Research Start-up Grants for New Science Faculty **2643**

The Murdock Start-up Grants offer support to supplement start-up research packages offered to new faculty in the natural sciences departments at private educational institutions. Grants are primarily targeted at undergraduate colleges and universities in the Trust's five-state region (Alaska, Idaho, Montana, Oregon, and Washington).
Requirements: Applicants should prepare a two page cover letter that addresses the following items: what is the vision for the department, extending at least over the next five years; how are teaching loads counted in the department, and what is considered a normal load (address at least classroom and also laboratory duties); how many majors have graduated in the department in each of the last five years; what are institutional and departmental expectations for the new hire regarding peer reviewed publications and external funding; how will the new hire fit into the future of the department. The application form is available online and should include the following information: institutional information; departmental information; consultant information; budget summary; department signatures; and a start-up package budget to be offered.
Geographic Focus: Alaska, Idaho, Montana, Oregon, Washington
Date(s) Application is Due: Aug 1
Amount of Grant: Up to 30,000 USD
Contact: Dana Miller, Senior Program Director; (360) 694-8415; fax (360) 694-1819
Internet: http://www.murdock-trust.org/grants/start-up-new-science-faculty.php
Sponsor: M.J. Murdock Charitable Trust
703 Broadway, Suite 710
Vancouver, WA 98660

M.J. Murdock Charitable Trust Scientific Research Grants **2644**

The M.J. Murdock Charitable Trust Scientific Research Grants fund select public research universities and medical institutes located within the Pacific Northwest region. The Trust prefers requests for projects in the natural sciences where the main objective is the acquisition of new knowledge. However, requests for research in engineering and medicine are also eligible. Training students in conducting research is an important consideration. Additional information and the application are located at the Trust website.
Requirements: A letter of inquiry is required unless the organization has received a Trust grant for scientific research within the last five years.
Restrictions: Only applications from scientific research institutions or major universities will be considered. Applications from individuals will not be considered.
Geographic Focus: Alaska, Idaho, Montana, Oregon, Washington
Contact: Dana Miller, Senior Program Director; (360) 694-8415; fax (360) 694-1819
Internet: http://www.murdock-trust.org/grants/general-scientific-research-grants.php
Sponsor: M.J. Murdock Charitable Trust
703 Broadway, Suite 710
Vancouver, WA 98660

MAA-Tensor Foundation Women and Mathematics Grants **2645**

MAA and the Tensor Foundation award grants to advance the study of math among women and girls, especially those in middle school, high school, and postsecondary grades. Grants support project directors and encourage mathematics faculty to develop projects that increase the participation of women in mathematics. Grants are made to the institution of the project director. An institution is expected to supply matching funds or in-kind support as an indication of commitment to the project. Some grants may be renewed for a second year. Possible projects are to create a network of women professional mentors who will direct mathematics projects for girls, organize a club for women interested in math, bring high school women onto a college campus for a Math Day, and form partnerships with industry to acquaint women students with real-life applications of mathematics. Proposals can be up to three single-spaced pages.
Requirements: College, university, and high school mathematics faculty (in conjunction with college or university faculty) and their departments and institutions may submit proposals.
Geographic Focus: All States
Date(s) Application is Due: Feb 25
Amount of Grant: Up to 5,000 USD
Contact: Dr. Florence Fasanelli, (202) 966-5591; fax (202) 966-5591; ffasanelli@juno.com
Internet: http://www.maa.org/projects/tensor_solic.html
Sponsor: Mathematical Association of America
1529 18th Street NW
Washington, DC 20036

Mabel H. Flory Charitable Trust Grants **2646**

The Mabel H. Flory Charitable Trust was established in 1979 and funds organizations primarily in Washington, DC., and Baltimore, MD. Support is available for research designed to develop new cures and methods of treatment: for children with hearing and sight problems; arthritis in both children and adults; and aphasia related problems.
Requirements: Your initial approach should be in the form of a letter, prior to the deadline date. Mail letter to the following address: c/o BB&T Co. of Virginia, Attn.: Trustee, 6400 Arlington Boulevard, Falls Church, VA, 22042.
Restrictions: No grants to individuals.
Geographic Focus: District of Columbia, Maryland
Date(s) Application is Due: Aug 31
Contact: BB&T Co. Trustee, (703) 531-2053
Sponsor: Mabel H. Flory Charitable Trust c/o BBandT Co. of Virginia
P.O. Box 2907
Wilson, NC 27894-2907

Mabel Y. Hughes Charitable Trust Grants **2647**

The Trust awards grants to Denver-area nonprofit organizations in its areas of interest, including children and youth services; education; family services; health care; higher education; human services; children's and art museums; performing arts centers; performing arts, opera; and reproductive health and family planning. Types of support include: annual campaigns; continuing support; emergency funds; endowments; equipment; general/operating support; program development; research; and seed money.
Requirements: Organizations should submit a letter of inquiry to the Trust, and if their project is appropriate for funding, they will be asked to submit a proposal.
Restrictions: The Trust does not support funding for individuals, deficit financing, scholarships, fellowships, or loans.
Geographic Focus: Colorado
Amount of Grant: 10,000-30,000 USD
Contact: Peggy Toal, c/o Wells Fargo Bank, Private Client Services; (720) 947-6725
Sponsor: Mabel Y. Hughes Charitable Trust
1740 Broadway
Denver, CO 80274

MacArthur Foundation Conservation and Sustainable Development Grants **2648**

The MacArthur Foundation is concerned about the deteriorating condition of the biosphere that supports life on earth. For the decade 2000-2010, the Foundation pursued a strategy that focused on eight ecological hotspots (places of high biodiversity under threat) in Madagascar, Melanesia, the Andes, Insular Caribbean, Albertine Rift, Eastern Himalayas, and Lower Mekong. In 2011, the Foundation launched a new strategy that is focused on one of the most compelling environmental challenges of the 21st century: the conservation of ecosystems. Ecosystems and their biodiversity underpin human well-being. They provide food and water; regulate floods, drought, and disease; and support soil formation and pollination. They also have intangible value as places of spiritual significance. All these are vital services, essential to economies and nation states. Hence, with the current Conservation and Sustainable Development program, the objective is to slow this degradation of ecosystems and, eventually, to reverse it. The geographical focus will be on three regions: the Great Lakes of East Central Africa, the Greater Mekong and its headwaters, and the watersheds of the Andes. Each is a place of high biodiversity, important freshwater service, and carbon-storage value. MacArthur's conservation and sustainable development program makes grants on a three-year recurring cycle, with one portfolio of grants prepared annually within each region and for the coastal marine initiative.
Geographic Focus: All States
Date(s) Application is Due: Feb 3; Apr 20; Jun 15
Amount of Grant: 50,000-2,000,000 USD
Contact: Douglas G. Siegel, Grants Administrator; (312) 726-8000; fax (312) 920-6258; 4answers@macfound.org
Internet: http://www.macfound.org/programs/conservation/
Sponsor: John D. and Catherine T. MacArthur Foundation
140 S Dearborn Street, Suite 1100
Chicago, IL 60603-5285

MacDonald-Peterson Foundation Grants **2649**

Established in 1995 in Texas, the MacDonald-Peterson Foundation offers funding for health and human services programs in Texas. The Foundation's primary fields of interest

include: children's services; Christian agencies and churches; community and economic development; food banks; health organizations; higher education; hospitals; human services; medical research; mental health counseling and support groups; performing arts (orchestras); and youth development and community service clubs. There is no specified application format or annual deadlines, so applicants should begin the process by contacting the Foundation directly. Awards generally range from $5,000 to $20,000, and are given for general operations.
Requirements: 501(c)3 organization in, or serving the residents of, Texas are eligible.
Geographic Focus: Texas
Samples: 100 Club of Houston, Houston, Texas, $10,000-general operations (2012); Art and Environmental Architecture, Houston, Texas, $20,000-general operations (2012); Asian Relief, Riverdale, Maryland, $10,000-general operations (2012).
Contact: Guy Tabor, (713)579-2355
Sponsor: MacDonald-Peterson Foundation
1 Riverway, Suite 1000
Houston, TX 77056-1944

Macquarie Bank Foundation Grants **2650**
The Foundation is one of Australia's oldest and largest corporate foundations contributing more than $150 million to more than 1500 community organizations world-wide since 1985. The Foundation focuses its resources in five core areas-the arts, education, environment, health, and welfare. The Foundation is also committed to projects specifically aimed at supporting indigenous communities. The Foundation's funding criteria is flexible and open. It welcomes applications from a diverse range of community organizations that are working in innovative ways to provide long-term benefits. Funding levels are flexible and are dictated by the needs of the organization and funding availability. Each application is assessed on its individual merit, with priority given to programs which support a broad section of the community at a regional, state or national level; have the involvement or potential for involvement of Macquarie Bank staff through volunteering, fundraising, pro bono work and board and/or management committee involvement; are located in cities/countries where Macquarie Bank staff are located; and deliver long-term benefits and build community sustainability. Prospective applicants are encouraged to check the Foundation website or contact Foundation Staff for more information on how to apply.
Geographic Focus: All States, All Countries
Amount of Grant: 100-500,000 USD
Contact: Heather Matwejev, Macquarie Bank Foundation, Asia; +61 2 8232 6951; fax +61 2 8232 0019; heather.matwejev@macquarie.com or foundation@macquarie.com
Internet: http://www.macquarie.com/mgl/com/foundation/about/application-guidelines
Sponsor: Macquarie Bank Foundation
G.P.O. Box 4294
Sydney, NSW 1164 Australia

Maine Community Foundation Keepers Preservation Education Grants **2651**
The Keepers Preservation Education Fund was established to support aspiring or established historic preservation professionals who wish to increase their professional knowledge or career potential. Awards may be used for tuition, professional meeting attendance, special book or other types of media purchases, domestic and foreign study travel, and other purposes as deemed appropriate. Applications are reviewed on a rolling basis.
Requirements: Eligible individuals must be: students enrolled full or part-time in institutions of higher learning with majors in historic preservation or allied fields including American studies, anthropology, archeology, architecture, landscape architecture, art history, architectural history, history, planning, law, economics, or building conservation; volunteer or paid staff of organizations or entities involved in historic preservation; and practicing professionals in historic preservation-related fields. Applicants may be from anywhere in the United States.
Restrictions: Individuals must apply through a sponsoring institution or organization, which reviews and forwards the application to the Foundation for funding review. Applicants may not be members of the awards committee or the immediate family of a committee member.
Geographic Focus: All States
Contact: Cherie Galyean; (877) 700-6800; fax (207) 667-0447; cgalyean@mainecf.org)
Internet: http://www.mainecf.org/keepersfund.aspx
Sponsor: Maine Community Foundation
245 East Main Street
Ellsworth, ME 04605

Manitoba Arts Council Aboriginal Arts Creative Development Pilot Grants **2652**
The Aboriginal Arts Creative Development Pilot Projects initiative is designed to enable Aboriginal artists in all arts disciplines, including dance/music, drumming, singing, carving, painting, storytelling, new media, writing, theatre or crafts, to create or produce new works over a three month period. Only one Creative Development grant may be received in any given fiscal year (April 1 to March 31). The maximum award is $7,500.
Requirements: Professional Aboriginal artists are eligible who: are Canadian citizens or permanent residents of Canada; have lived in Manitoba for at least one year immediately prior to submitting an application; have completed basic training (not necessarily in a formal school, college or university); view the practice of their art as a career and are recognized as professionals in their art form by their peers; have completed at least three full years of professional activity; are able to devote a sufficient time over a six-month period to complete the project for which they are requesting assistance.
Geographic Focus: Canada
Date(s) Application is Due: Sep 25
Amount of Grant: Up to 7,500 CAD
Contact: Tracey Longbottom; (204) 945-3180; tlongbottom@artscouncil.mb.ca
Internet: http://artscouncil.mb.ca/2010/02/aboriginal-arts-creative-development/
Sponsor: Manitoba Arts Council
525-93 Lombard Avenue
Winnipeg, MB R3B 3B1 Canada

Manitoba Arts Council Craft Grants **2653**
The Craft Grant is designed to provide assistance to professional Manitoba craft artists in the production of new creative work. The Grant supports contemporary or traditional craft projects based on the artist's original idea. Artists may apply for new work or for work that is part of a series. Craft artists contemplating a major project involving concentrated work over several years may define their proposed project as one phase of this work. Artists may apply for living and project expenses to the following maximums: established artists can receive a maximum of $10,000; and emerging artists can receive a maximum of $5,000. Eligible costs include living costs and direct project costs for new work: research; travel; supplies and materials; the purchase or rental of relevant equipment; documentation; technical assistance; temporary rental of studio space or facilities; and apprenticeship or craft development workshops. Artists are eligible for a maximum of two Manitoba Arts Council Craft Grants every forty-eight months. Artists are encouraged to discuss their eligibility with the program consultant prior to applying.
Requirements: Established artists must have a significant record of professional craft activity and have established a national or international profile. Emerging artists must have a modest record of professional craft activity following basic training, including the exhibition or public presentation of their work.
Restrictions: The program does not provide for: exhibition expenses which are the responsibility of the exhibition space, such as opening costs, invitations, mailing, printing of catalogs or posters; and the costs of exhibiting existing work.
Geographic Focus: Canada
Date(s) Application is Due: Sep 15
Amount of Grant: Up to 10,000 CAD
Contact: Rekha Ramachandran, Program Consultant; (204) 945-2237; fax (204) 945-5925; rramachandran@artscouncil.mb.ca or info@artscouncil.mb.ca
Internet: http://artscouncil.mb.ca/2010/02/craft-grant/
Sponsor: Manitoba Arts Council
525-93 Lombard Avenue
Winnipeg, MB R3B 3B1 Canada

Manitoba Arts Council Dance Creation/Production Grants **2654**
The Dance Creation/Production Grant is available to support the creation of original works by Manitoba choreographers, as well as for dance production projects. Individuals, professional dance artist collectives and professional dance organizations may be eligible. The maximum grant available through this program is $15,000. Assistance may be requested for: research and creation projects (creating new work or developing an existing work); a production project or a series of productions (the public presentation of a choreographic work involving a majority of professional Manitoban artists); a combined creation and production project; and creation of choreography by independent Manitoban choreographers for organizations receiving annual funding by the Manitoba Arts Council (wherein the choreographic fee is the only eligible expense). Eligible costs include: fees to artists (performer, choreographer, mentor, designer, composers, etc.); a monthly living allowance of $2,000 for the primary artist/choreographer (this period of time may include research, creation and/or production); research material; studio expenses; audio-visual expenses; production expenses; and marketing and communication expenses.
Requirements: Eligible applicants include professional dance artists, professional dance artist collectives, and professional dance organizations who: do not receive annual/operating support from the Manitoba Arts Council; employ a majority of artists who are Canadian citizens, permanent residents or landed immigrants; employ a majority of artists who have lived in Manitoba for at least one year immediately prior to making an application (exceptional circumstances require a written explanation); and employ a majority of artists who view the discipline as a vocation and are recognized as professionals in their artistic discipline by their peers.
Restrictions: This program does not fund: rehearsal costs, dancers' fees or production costs for the creation of choreography by independent Manitoban choreographers for organizations receiving annual funding by the Manitoba Arts Council; projects for Artistic Directors and Choreographers in-residence of organizations receiving annual funding by the Manitoba Arts Council; or amateur or school productions.
Geographic Focus: Canada
Date(s) Application is Due: Sep 15
Amount of Grant: Up to 15,000 CAD
Contact: Cathleen Enns, Dance and Music; (204) 945-8631 or (204) 945-2237; fax (204) 945-5925; cenns@artscouncil.mb.ca or info@artscouncil.mb.ca
Internet: http://artscouncil.mb.ca/2010/02/dance-creationproduction-grant/
Sponsor: Manitoba Arts Council
525-93 Lombard Avenue
Winnipeg, MB R3B 3B1 Canada

MAP International Medical Fellowships **2655**
The MAP International Medical Fellowship encourages lifelong involvement in global health issues by providing selected medical students firsthand exposure in a Christian context to the health, social and cultural characteristics of a developing world community. MAP International, through a grant from the founders of Reader's Digest, DeWitt and Lila Wallace, sponsors fourth year medical students by funding travel arrangements for medical students to their chosen destination. The field experience must be designed to provide the student exposure to community health in a Christian context. Students select a mission agency or hospital that has an outreach among the poor in a rural or urban setting.

Students must spend at least eight weeks in the field (six weeks for residents and interns). The fellowship provides 100 percent of the approved round trip airfare to one destination. In most instances, students pay room and board as well as any in-country travel expenses. To obtain an application, check with the dean of your school or the Student Affairs Office. Applications are sent to all medical schools annually. You may also download the application in PDF or Microsoft Word format on the Map International website.
Requirements: Fourth-year medical students, residents and interns are eligible. Applications should be submitted during the academic year prior to travel.
Geographic Focus: All States
Date(s) Application is Due: Mar 1
Contact: Margaret Stevenson, Medical Fellowship Coordinator; (404) 880-0540; fax (912) 265-6170; mstevenson@map.org
Internet: http://www.map.org/site/PageServer?pagename=what_Medical_Fellowship
Sponsor: Map International
4700 Glynco Parkway
Brunswick, GA 31525-6800

Marcia and Otto Koehler Foundation Grants **2656**
Grants are made to support arts and culture, medical, education, and social organizations in San Antonio, Texas. Proposals must demonstrate leadership in effecting positive change; encourage collaborative effort; serve large and diverse sectors of the population; and demonstrate vision, effectiveness, and good fiscal management. Types of support include general operating support, building construction/renovation, and research. Grants are made for one time only. Applicants must send a letter of request for application by March 1 to receive an application.
Requirements: Grants are awarded only to organizations in Bexar County, Texas.
Restrictions: Grants will not be made to individuals or to support other foundations or endowments; salaries; operating deficits; political organizations; or churches, synagogues, or parishes.
Geographic Focus: Texas
Date(s) Application is Due: Jun 1
Amount of Grant: 2,500-50,000 USD
Contact: Thomas K. Killion, Senior Vice President, Bank of America; (210) 270-5422; fax (210) 270-5520; thomas.k.killion@ustrust.com
Sponsor: Marcia and Otto Koehler Foundation
P.O. Box 121
San Antonio, TX 78291-0121

Marcus and Theresa Levie Educational Fund Scholarships **2657**
Graduate scholarships in the helping professions are awarded to individual students for the academic year; eligible applicants may receive application forms from the Jewish Vocational Service. Personal interviews are required.
Requirements: Applicants must be Jewish, legally domiciled in Cook County, IL, demonstrate financial need, and be enrolled or entering an accredited college or university in a full-time vocational or professional training program, preferably graduate level or above, in the helping professions; at least a C-plus average is required. The scholarship supervisor may be reached between December 1 and February 15 at the fund.
Geographic Focus: Illinois
Date(s) Application is Due: Mar 1
Amount of Grant: 1,000-5,000 USD
Contact: Marla Levine; (312) 346-6700; fax (312) 855-3282; jvschicago@jon.cjfny.org
Internet: http://www.free-4u.com/levie_educational_fund_scholarship1.htm
Sponsor: Marcus and Theresa Levie Educational Fund
1 S Franklin Street
Chicago, IL 60606

Margaret and James A. Elkins Jr. Foundation Grants **2658**
The Elkins Foundation Grants support nonprofit programs and organizations in the metropolitan Houston, Texas area. Grants are given primarily for charitable, religious, scientific, or educational and literacy programs, including public safety testing, and the prevention of cruelty to children and animals. Fields of interest include: biology/life sciences; child development, education; child development, services; children and youth services; Christian agencies and churches; elementary and secondary education; engineering and technology; health organizations, association; higher education; hospitals (general); medical research, institute; medical education; religion; science; and safety/disasters. The Foundation also supports: building renovation; capital campaigns; emergency funds; endowments; equipment; program development; and research.
Requirements: Applications are not required. Applicants should submit a letter of inquiry, along with a copy of their most recent annual report, audited financial statement, or 990 form. Board meetings vary, and there is no deadline.
Restrictions: Grants are not made to individuals, or to support annual fundraising campaigns or operating deficits.
Geographic Focus: Texas
Contact: Larry Medford, Secretary-Treasurer; (713) 652-2052
Sponsor: Margaret and James A. Elkins Jr. Foundation
1001 Fannin Street, 1166 First City Tower
Houston, TX 77002

Margaret T. Morris Foundation Grants **2659**
The Margaret T. Morris Foundation awards grants, primarily in Arizona, in its areas of interest, including: animal welfare; arts; children and youth, services; education; environment; higher education; homeless service; human services; marine science; medical research and education; mental health and crisis services; museums; performing arts; reproductive health and family planning; and hospices. The Foundation's types of support include: building renovation; capital campaigns; debt reduction; endowments; general operating support; land acquisition; matching and challenge support; and program development. The Board of Directors meets in August, December, and as needed.
Requirements: Applications are not accepted. Applicants should submit a letter of inquiry with their request for funding and a description of the project.
Geographic Focus: Arizona
Contact: Thomas Polk, Trustee; (928) 445-4010
Sponsor: Margaret T. Morris Foundation
P.O. Box 592
Prescott, AZ 86302-0592

Margaret Yardley Fellowship **2660**
This fellowship is awarded toward payment of tuition only for a woman from New Jersey who is a college graduate and plans to continue her education. She must be enrolled as a full-time student in graduate work or a doctoral program for an uninterrupted year of study in an American college of her choice. Selection is based on scholastic record, financial need, ability, character, and potential service to the chosen field.
Requirements: Request an application form by the listed deadline date; a self-addressed, stamped envelope must be enclosed.
Geographic Focus: New Jersey
Date(s) Application is Due: Mar 1
Amount of Grant: 1,000 USD
Contact: Deen Meloro; (732) 249-5485 or (732) 249-5474; dmeloro@njsfwc.org
Internet: http://www.njsfwc.org/MYFellowship.html
Sponsor: New Jersey State Federation of Women's Clubs
55 Labor Center Way
New Brunswick, NJ 08901

Marijuana Policy Project Grants **2661**
The program supports efforts that promote the legalization and regulation of marijuana. Projects may encompass any aspect of marijuana-law reform, such as: ballot initiatives, grassroots organizing, litigating for the right to use marijuana for religious purposes, lobbying the American Medical Association, marijuana research, and revising the definition of marijuana abuse and dependence in the American Psychiatric Association's Diagnostic and Statistical Manual. Submit letters of inquiry; full proposals are by invitation.
Requirements: Preference is given to 501(c)3 tax-exempt organizations, but such status is not an eligibility requirement.
Restrictions: The grants program does not fund political parties or candidates for office, state ballot initiatives, or hemp-related projects.
Geographic Focus: All States
Date(s) Application is Due: Jan 2; May 1; Sep 1
Contact: Stephanie Vogel, (202) 462-5747, ext.104; stephanie@mpp.org
Internet: http://www.mpp.org/grants/index.html
Sponsor: Marijuana Policy Project
P.O. Box 77492
Washington, DC 20013

Marine Corps Historical Center Beeler-Raider Fellowship **2662**
The annual fellowship encourages graduate-level and advanced study of the combat contributions of enlisted marines. Applicants may either suggest a study topic or ask the staff at the center for guidance in selecting an appropriate topic. Preference is given to projects covering the pre-1991 period, for which records are made available to scholars. Fellowship recipients are encouraged to do part of their research in the main federal archival centers in Washington, DC. Preliminary application involves submitting a letter that outlines the applicant's qualifications and the topic (or requesting a topic); applicants will be invited to submit formal applications.
Requirements: The program concentrates on graduate students but considers other qualified persons. Applicants must be U.S. citizens or nationals.
Geographic Focus: All States
Date(s) Application is Due: May 1
Contact: Grants Coordinator; (703) 432-4874
Internet: http://hqinet001.hqmc.usmc.mil/HD/General/Grants_Internships_Grants.htm#Grant3
Sponsor: United States Marine Corps
Marine Corps University, 3079 Moreell Avenue
Quantico, VA 22134

Marine Corps Historical Center Dissertation Fellowship **2663**
One dissertation fellowship per academic year is awarded to a qualified graduate student working on a doctoral dissertation pertinent to Marine Corps history. Topics in U.S. military and naval history, as well as history and history-based studies in the social and behavioral sciences, with a direct relationship to the history of the U.S. Marine Corps will be considered. Topics may encompass wars, institutions, organization and administration, policy, biography, civil affairs and civic action, civil-military relations, weaponry and technology, manpower, training and education, strategy, tactics, and logistics; as well as diplomatic, political, economic, social, and intellectual trends affecting the Marine Corps during peace and war. The topic must have the approval of the applicant's dissertation director and contribute to the study and understanding of Marine Corps history. Fellows are expected to do a portion of their research in Washington, DC.
Requirements: The applicant must be a U.S. citizen who is enrolled in a recognized graduate school; have completed by September all requirements for the doctoral degrees except the dissertation; and have an approved, pertinent dissertation topic.

Restrictions: A fellowship will not be awarded to an applicant who has held or accepted an equivalent fellowship from any other Department of Defense agency; however, recipients of the Marine Corps's master's thesis fellowships may apply.
Geographic Focus: All States
Date(s) Application is Due: May 1
Contact: Grants Coordinator; (703) 432-4874
Internet: http://hqinet001.hqmc.usmc.mil/HD/General/Grants_Internships_Grants.htm
Sponsor: United States Marine Corps
Marine Corps University, 3079 Moreell Avenue
Quantico, VA 22134

Marion Gardner Jackson Charitable Trust Grants 2664
The Marion Gardner Jackson Charitable Trust funds the arts and humanities, education, health, and human service organizations in the Quincy, Illinois area and surrounding communities in Adams County. The trust supports capital, program, and operating grants. The application is available at the trust website through Bank of America.
Restrictions: For program support grants, the yearly request may not be more than 50% of the program's budget. Organizations can submit one application per year and will not receive more than one award from the trust in any given year. Organizations receiving a multi-year award from the trust that continues into the next grant year are not eligible to apply for an additional grant until the end of the grant cycle. The trust does not support requests from individuals, organizations attempting to influence policy through direct lobbying, or any political campaigns.
Geographic Focus: Illinois
Date(s) Application is Due: Jul 31
Contact: Debra Grand; (312) 828-4154; ilgrantmaking@bankofamerica.com
Internet: https://www.bankofamerica.com/philanthropic/fn_search.action
Sponsor: Marion Gardner Jackson Charitable Trust
231 South LaSalle Street, IL1-231-13-32
Chicago, IL 60604

Marjorie C. Adams Charitable Trust Grants 2665
The Marjorie C. Adams Charitable Trust, established in New York in 1987, is administered by the JP Morgan Chase Bank of Newark and supports the care and treatment of deaf and blind people, research into the causes and the treatment of deafness and blindness, and K-12 education. Types of support for these programs include: building and renovation projects, capital campaigns, and endowments. There are no specific application forms required or deadlines with which to adhere, and applicants should begin by contacted the Foundation office directly.
Requirements: 501(c)3 organizations, primarily serving residents of New York and Maryland, are eligible to apply.
Geographic Focus: Maryland, New York
Amount of Grant: Up to 150,000 USD
Contact: Frank Lemma, Vice President; (212) 464-2439 or (212) 648-1477
Sponsor: Marjorie C. Adams Charitable Trust
270 Park Avenue, 16th Floor
New York, NY 10017

Marshall Scholarships 2666
The scholarships enable U.S. citizens who are graduates of U.S. colleges and universities to study for degrees in any subject at a university in the United Kingdom for a period of at least two academic years, extendible for one more year. Up to 40 new awards are offered each year. Application guidelines may be obtained after June 1 each year from British Information Services in New York City; the British Consulate-General in Boston, Chicago, Atlanta, and San Francisco; or from the British Embassy (Cultural Department). The deadline for application falls in mid-October of the year preceding the scholarship tenure. The award's amount includes University fees, cost of living expenses, annual book grant, thesis grant, research and daily travel grants, fares to and from the United States and, where applicable, a contribution towards the support of a dependent spouse.
Requirements: Marshall scholarships are open only to citizens of the United States 25 years of age or under who must, by the time of taking up residence in a British university, have obtained a first degree requiring at least three years of study from an accredited degree-granting university or college in the United States. A minimum grade point average of 3.7 for academic courses after the freshman year will normally be required. Married persons are eligible for the scholarships, but among candidates of equal merit, preference will be given to those who intend to remain unmarried until at least the end of their first-year tenure.
Geographic Focus: All States
Date(s) Application is Due: Oct 1
Contact: Scholarship Supervisor; (202) 588-7800; fax (202) 588-7918; enquiries@britishcouncil-usa.org
Internet: http://www.acu.ac.uk/marshall/index.html
Sponsor: British Council
3100 Massachusetts Avenue NW
Washington, DC 20008

Martin E.P. Seligman Award for Outstanding Research in Positive Psychology 2667
The competition awards a cash grant to a scholar who has completed a PhD dissertation in any area of research related to positive psychology in the last three years and who plans to continue research in this area. Interested candidates should submit six copies of their curriculum vita and a summary of their dissertation research (not to exceed eight single-spaced pages, including tables and figures, but excluding references). The summary should contain a description of the problem investigated; an explanation of the methods used to investigate the problem; a summary of the findings; a discussion of the significance of the findings in relation to positive psychology; and an outline of how the work might be continued in the future. The award is presented at the APA convention. Travel to the convention and one day's lodging and expenses are also included with the award.
Requirements: The dissertation must have been completed after March 30, 2002.
Geographic Focus: All States
Date(s) Application is Due: Apr 1
Amount of Grant: 1,000 USD
Contact: Administrator; (610) 941-2828; fax (610) 825-1730; info@templeton.org
Internet: http://www.templeton.org/SeligmanAward/index.html
Sponsor: John Templeton Foundation
300 Conshohocken State Road, Suite 500
West Conshohocken, PA 19428

Martin Luther King Jr. Graduate Memorial Scholarships 2668
The scholarships provide graduate study aid to racial/ethnic minority group CTA members and/or their dependent children, and student CTA members. This scholarship is supported by voluntary contributions from the membership and has the goal of promoting professional growth for minority teachers. Scholarships vary each year depending on the amount of contributions and on the financial need of individual applicants. Applications are available in January each year and may be obtained by contacting CTA regional resource center offices or the CTA Human Rights Department in Burlingame.
Geographic Focus: All States
Date(s) Application is Due: Mar 15
Contact: Human Rights; (650) 697-1400; fax (650) 552-5001; scholarships@cta.org
Internet: http://www.cta.org/InsideCTA/TrainingHR/Scholarships/ScholarMLK.htm
Sponsor: California Teachers Association
1705 Murchison Drive, P.O. Box 921
Burlingame, CA 94011-0921

Mary Baker Eddy Library Summer Research Fellowships 2669
The library awards fellowships to scholars and other researchers for research in its collections in the fields of women's history, spirituality and health, religious studies, 19th-century history, and journalism—themes for which the library's collections hold special relevance and value. Consideration will be given to applicants with outstanding writing skills, evidence of having previous publications, and a definite plan for disseminating the research through publication, lecturing, etc. Fellowship recipients are expected to be in regular and continuous use of the collections at the library in Boston during the period of the fellowship (between June 13 and September 2). Fellows are responsible for making their own arrangements for housing in the Boston area. Application and guidelines are available online.
Requirements: Fellowships are open to doctoral and postdoctoral scholars, faculty members, and independent researchers.
Geographic Focus: All States
Date(s) Application is Due: Feb 13
Amount of Grant: Up to 8,400 USD
Contact: Administrator; (617) 450-7206; fellowships@marybakereddylibrary.org
Internet: http://www.marybakereddylibrary.org/programs/fellowshipapplications.jhtml
Sponsor: Mary Baker Eddy Library
200 Massachusetts Avenue
Boston, MA 02115

Mary Gates Research Scholarships 2670
Mary Gates Research Scholarships are competitive scholarships intended to enhance the educational experiences of undergraduate students at the University of Washington while they are engaged in research guided by faculty. Students may pursue research in a variety of disciplines. The application and additional information about the selection criteria are available at the Mary Gates website.
Requirements: Applicants must be full time students enrolled at the University of Washington in a four year program. Candidates need not be U.S. citizens. Candidates must already be engaged in research with faculty at the time of application, and continue with their research while full time students.
Restrictions: Applicants are not eligible for more than one Mary Gates Scholarship at the same time. Candidates should consult the Gates Endowment office to discuss if the Research Scholarship may interfere with other financial aid they receive.
Geographic Focus: All States, All Countries
Amount of Grant: 4,000 USD
Contact: Jodene Davis; (206) 616-3925; fax (206) 616-4389; mgates@u.washington.edu
Internet: http://www.washington.edu/uaa/mge/apply/research/index.htm
Sponsor: Mary Gates Endowment for Students
University of Washington
Seattle, WA 98195-2803

Mary Ingraham Bunting Institute Peace Fellowship 2671
One fellowship will be awarded to support a woman committed to domestic or international policy work, or actively engaged in finding peaceful and just solutions to conflict, either real or potential, among groups or nations. Involvement with peace issues may be of an activist or scholarly nature.
Requirements: Women who have demonstrated effectiveness in senior-level work related directly to peace and justice are eligible.
Geographic Focus: All States
Date(s) Application is Due: Oct 3
Amount of Grant: 32,000 USD
Contact: Coordinator; (617) 495-8212; fax (617) 495-8136; fellowships@radcliffe.edu

Internet: http://www.radcliffe.edu/fellowships/bunting.php
Sponsor: Bunting Institute of Radcliffe College
34 Concord Avenue
Cambridge, MA 02138

Mary Isabel Sibley Fellowship **2672**

The fellowship is awarded annually for advanced research in Greek language, literature, history, or archaeology (odd-numbered years); or French language or literature (even-numbered years). Applicants must have shown ability to carry on original research and must be planning to devote full-time work to research during the fellowship year. Application forms may be obtained from the society.
Requirements: Applicants must be unmarried women between 25 and 35 years of age who hold doctorates or have completed all requirements except the dissertation.
Restrictions: Eligibility is not restricted to Phi Beta Kappa members or to U.S. citizens.
Geographic Focus: All States
Date(s) Application is Due: Jan 15
Amount of Grant: 20,000 USD
Contact: Sandra Beasley; (202) 265-3808; fax (202) 986-1601; SBeasley@pbk.org
Internet: http://staging.pbk.org/AM/Template.cfm?Section=Sibley_Fellowship&Template=/CM/HTMLDisplay.cfm&ContentID=2647
Sponsor: Phi Beta Kappa Society
1606 New Hampshire Avenue NW
Washington, DC 20009

Mary K. Chapman Foundation Grants **2673**

Mary K. Chapman was born in Oklahoma in 1920. She graduated from the University of Tulsa and worked as a nurse before her marriage to Allen Chapman in 1960. After the death of her husband in 1979, Mary Chapman maintained her own personal charitable giving program. Before her death in 2002, she established The Mary K. Chapman Foundation, a charitable trust founded to perpetuate her own charitable giving program. This foundation was fully funded with a bequest from her estate in 2005. Mary K. Chapman was very interested in supporting education, but as a former nurse and a very compassionate person, much of her charity was directed to health, medical research, and educating and caring for the less fortunate and disadvantaged. There are two steps in the process of applying for a grant. The first is a Letter of Inquiry from the applicant. This letter is used to determine if the applicant will be invited to take the second step of submitting a formal Grant Proposal.
Requirements: IRS 501(c)3 non-profits are eligible to apply.
Restrictions: Grant requests for the following purposes are not favored: endowments, except as a limited part of a capital project reserved for maintenance of the facility being constructed; deficit financing and debt retirement; projects or programs for which the Chapman Trusts would be the sole source of financial support; travel, conferences, conventions, group meetings, or seminars; camp programs and other seasonal activities; religious programs of religious organizations; project or program planning; start-up ventures are not excluded, but organizations with a proven strategy and results are preferred; purposes normally funded by taxation or governmental agencies; requests made less than nine months from the declination of a previous request by an applicant, or within nine months of the last payment made on a grant made to an applicant; requests for more than one project.
Geographic Focus: Oklahoma
Amount of Grant: Up to 300,000 USD
Contact: Andie Doyle; (918) 496-7882; fax (918) 496-7887; andie@chapmantrusts.com
Internet: http://www.chapmantrusts.org/grants_programs.php
Sponsor: Mary K. Chapman Foundation
6100 South Yale, Suite 1816
Tulsa, OK 74136

Mary Kay Foundation Cancer Research Grants **2674**

In May 2010, the Foundation awarded $1.3 million in grants to select doctors and medical scientists focusing on curing cancers that affect women. These 13 recipients from across the United States received a $100,000 grant to conduct cutting-edge research. Since 1996, the Foundation has given more than $14 million to support this effort. Grants are awarded each year to researchers at medical schools recommended by the The Mary Kay Foundation Research Review Committee, which is composed of prominent doctors who volunteer their time to help the Foundation select the best recipients across the United States. After reviewing these recommendations, the Board of Directors at the Foundation selects the grant recipients. The Mary Kay Foundation accepts grant applications from November to mid-February each year.
Geographic Focus: All States
Amount of Grant: 100,000 USD
Contact: Lana Rowe, (972) 687-4822 or (877) 652-2737; Lana.Rowe@mkcorp.com or MKCares@marykayfoundation.org
Internet: http://www.marykayfoundation.org/Pages/CancerGrantProgram.aspx
Sponsor: Mary Kay Foundation
P.O. Box 799044
Dallas, TX 75379-9044

Mary S. and David C. Corbin Foundation Grants **2675**

The Corbin Foundation gives primary consideration to charitable organizations and/or local chapters of national charities located in Akron and Summit County, Ohio, although extremely worthy causes outside of this preferred area may be considered. Areas of interest include arts and culture; civic and community; education; health care; housing; social services; medical research; and youth. The Foundation meets in May and November to consider requests. Grant requests must be received no later than March 1 for consideration in May and September 1 for consideration in November. The application and guidelines are available at the Foundation website.
Requirements: The Foundation has a general application cover sheet which applicants must complete. Organizations should also send a brief letter on their letterhead, submitting one original and one copy of all application materials.
Restrictions: Only written applications will be considered. Telephone and personal interviews are discouraged unless requested by the Foundation. The Foundation does not fund individuals; annual fundraising campaigns; ongoing requests for general operating support (although some repeat grants are made); operating deficits; or organizations which in turn make grants to others.
Geographic Focus: Ohio
Date(s) Application is Due: Mar 1; Sep 1
Contact: Erika J. May; (330) 762-6427; fax (330) 762-6428; corbin@nls.net
Internet: http://foundationcenter.org/grantmaker/corbin/guide.html
Sponsor: Mary S. and David C. Corbin Foundation
Akron Central Plaza
Akron, OH 44308-1830

Massachusetts Bar Foundation Legal Intern Fellowships **2676**

The program was established to encourage careers in the law that further the goals of social justice, while contributing valuable legal support to organizations serving the underrepresented in Massachusetts. To assist law students in gaining practical experience in the public sector, the foundation awards four stipends to law students who intern during the summer months at nonprofit organizations providing civil legal services to low-income clients in Massachusetts. Guidelines are available online.
Requirements: Applications are welcome from law students who have secured a volunteer internship with a qualified nonprofit organization in Massachusetts; are currently enrolled in U.S. law schools, with preference given to second- and third-year students; demonstrate a commitment to public interest law, including experience working with low-income clients; and commit to work no less than 10 continuous, full-time weeks between June 1 and October 1.
Geographic Focus: Massachusetts
Date(s) Application is Due: Mar 17
Amount of Grant: 6,000 USD
Contact: Administrator; (617) 338-0534; fax (617) 338-0550; foundation@massbar.org
Internet: http://www.massbarfoundation.org/grant_programs/legal_intern
Sponsor: Massachusetts Bar Foundation
20 West Street
Boston, MA 02111-1204

Massachusetts Cultural Council Artist Fellowships **2677**

MCC's Artist Fellowships recognize exceptional work by Massachusetts artists across a range of disciplines. These highly competitive awards provide artists crucial validation among their peers and the public. They catalyze artistic advancement and pave the way for creative innovation of enduring cultural value. The Fellowships provide direct, unrestricted support to Massachusetts artists in recognition of exceptional original work, to foster the creation of new art in the Commonwealth. Fellowships in a range of disciplines are awarded to primary creators of original works of art. The Artist Fellowships provide competitive grants of $7,500 and finalist awards of $500. The number of awards per discipline category varies according to the number of applications reviewed, the recommendations of each panel, and program funds available. Drawing, Painting, and Traditional Arts applications are accepted August 15 through October 7. Choreography, Fiction/Creative Nonfiction, and Poetry applications are accepted December 15 through January 30.
Requirements: Any artist 18 years or older who has been a legal resident of Massachusetts for the last two years at the time of application is eligible. All Artist Fellows/Finalists must be legal residents of Massachusetts when grants are awarded.
Geographic Focus: Massachusetts
Date(s) Application is Due: Jan 30; Oct 7
Contact: Kelly Bennett, Program Coordinator; (617) 727-3668, ext. 349 or (800) 232-0960; fax (617) 727-0044; kelly.bennett@art.state.ma.us
Internet: http://www.massculturalcouncil.org/programs/artistfellows.asp
Sponsor: Massachusetts Cultural Council
10 St. James Avenue, 3rd Floor
Boston, MA 02116-3803

Massachusetts Foundation for the Humanities Project Grants **2678**

The Foundation makes project grants to support lecture series, professional development for teachers, public forums, post or pre-performance discussions, film screenings with discussions, reading and discussion programs, oral history projects, and other public humanities activities in Massachusetts. In most cases, the maximum award for a grant for a public program is $5,000, and most proposals the Foundation receives fit under the general category of 'public program'. The current thematic initiative-Liberty and justice for all-gives priority to public humanities programs that explore two fundamental principles in American political life, and their interplay, past and present. Projects that address this theme are eligible for a maximum of $10,000.
Requirements: Proposals are due on or before the first business day of February, May, August, and November. Any non-profit organization, whether or not it has 501(c)3 status, is eligible for MFH funding.
Geographic Focus: Massachusetts
Contact: Hayley Wood; (413) 584-8440; fax (413) 584-8454; hwood@mfh.org
Internet: http://www.mfh.org/grants/index.htm
Sponsor: Massachusetts Foundation for the Humanities
66 Bridge Street

Northampton, MA 01060

Massage Therapy Foundation Practitioner Case Report Contest 2679

The Foundation has chosen to encourage the writing of case reports to provide an opportunity for massage therapists and bodyworkers to develop research skills and enhance their ability to provide knowledge-based massage to the public. This contest is intended to enhance professional development skills of the practitioner: writing case reports help develop communication skills, critical thinking skills, and could contribute to future research and clinical practice. Cash and publication recognition will be awarded to practitioners submitting the top reports.

Requirements: The Foundation requires that HIPAA guidelines for confidentiality are followed.
Geographic Focus: All States
Date(s) Application is Due: Oct 8
Contact: Alison Pittas, Program Manager of Research and Grants; (847) 869-5019, ext. 167 or (847) 905-1667; fax (847) 864-1178; apittas@massagetherapyfoundation.org
Internet: http://www.massagetherapyfoundation.org/practitionercontest.html
Sponsor: Massage Therapy Foundation
500 Davis Street, Suite 900
Evanston, IL 60201

Massage Therapy Foundation Research Grants 2680

One of the primary purposes of the Foundation is to fund solid research studies investigating the many beneficial applications of massage therapy. Foundation research grants are awarded to individuals or teams conducting studies that promise to advance our understanding of specific therapeutic applications of massage, public perceptions of and attitudes toward massage therapy, and the role of massage therapy in health care delivery. The Research Grant deadline is normally March 1 annually (or the following Monday if this falls on a weekend). The normal award for 12 months is $1,000 to $30,000, and must be used in the specific time period for which it has been awarded. This grant supports high quality, independent research that contributes to the basic science of massage therapy application, including applied research investigating massage therapy as a health/mental health treatment and/or prevention modality.

Requirements: Research Grants are available to investigators who: have experience in the relevant field of research; are presently associated with, or have secured the cooperation of a university, independent research organization, health center, or other institution qualified and willing to function as a Sponsoring Organization for the purpose of this project; or new investigators without prior research experience must document support from an experience investigator willing to act as a collaborator.

Restrictions: Research Grants will not be awarded to spouses, domestic partners, children, descendant, spouses of descendants or any other individual related to any officers or trustees of the Massage Therapy Foundation, or to members of the Research Proposal Review Committee.
Geographic Focus: All States
Date(s) Application is Due: Mar 1
Amount of Grant: Up to 30,000 USD
Contact: Alison Pittas, Program Manager of Research and Grants; (847) 869-5019, ext. 167 or (847) 905-1667; fax (847) 864-1178; apittas@massagetherapyfoundation.org
Internet: http://www.massagetherapyfoundation.org/grants_research.html
Sponsor: Massage Therapy Foundation
500 Davis Street, Suite 900
Evanston, IL 60201

Massage Therapy Foundation Student Case Report Contest 2681

The program is a creative way for the foundation and massage therapy schools to work together to advance the development of research literacy and evidence-based practice habits in massage practitioners. The contest encourages students at massage-therapy schools to conduct research and produce case reports on their findings. Guidelines are available online.
Geographic Focus: All States
Date(s) Application is Due: Mar 1
Amount of Grant: 1,500-2,500 USD
Contact: Alison Pittas, Program Manager of Research and Grants; (847) 869-5019, ext. 167 or (847) 905-1667; fax (847) 864-1178; apittas@massagetherapyfoundation.org
Internet: http://www.massagetherapyfoundation.org/contest.html
Sponsor: Massage Therapy Foundation
500 Davis Street, Suite 900
Evanston, IL 60201

Massage Therapy Research Fund (MTRF) Grants 2682

The purpose of the MTRF is to fund high quality research that investigates the efficacy, and clinical effectiveness of massage therapy and that contributes to our understanding of how massage therapy achieves its effects. A secondary goal is to increase research knowledge and capacity among massage therapists through their engagement in research practices at all levels. The MTRF invites applications in areas indicated above. For both competition cycles, a total funding amount of $50,000 is offered. Grant applications up to $15,000 will be considered.

Requirements: All properly submitted applications will be considered. Studies may be qualitative, quantitative, or mixed method. Initially, funding preference will be given to smaller, pilot, or creative study designs. In this way, the MTRF aims to assist the scientific and massage communities in building a foundation for the eventual construction of more complicated research projects. The MTRF encourages applications that also seek funding from other sources. The principal investigator must be affiliated with a Canadian charitable institution (university or hospital) that is willing to administer the finances of the grant. Funds are awarded to the charitable institution, not to an individual. The research team may include members affiliated with a non-Canadian institution. One member of the research team should be a massage therapist. Students at the masters level and above are encouraged to apply.
Geographic Focus: All States, Canada
Date(s) Application is Due: Mar 1; Sep 30
Amount of Grant: 15,000 USD
Contact: Keren Brown, Executive Director; (416) 778-4443 or (866) 778-4443; fax (416) 778-5438; info@holistichealthresearch.ca
Internet: http://www.holistichealthresearch.ca/massagefund.gk
Sponsor: Holistic Health Research Foundation of Canada
80 Carlton Street
Toronto, ON M5B 1L6 Canada

Materials Research Society Medalist Awards 2683

This program constitutes public and professional recognition of outstanding recent achievements in materials research. A medal and cash prize will be awarded annually for a specific discovery or advancement that is expected to have a major impact on the progress of any materials-related field. A medal will normally be awarded to one individual for a cited achievement; however, it may be awarded to the collaboration of two or three individuals. In that case, each recipient will receive a personal medal trophy, registration, and travel, and the cash award will be divided equally. Recipients are selected from nominations prepared by peers of the recipients, not the recipients themselves. A nominee need not be a member of the society and may be of any nationality or citizenship. A nomination for the medal remains under consideration for three years, and may be updated by the nominator during that time. The award includes an engraved medal, registration fee for MRS meeting, and reasonable travel expenses reimbursement.

Requirements: Nominations are open to scientists and engineers who have in recent years been responsible for major advances in any materials-related field of research. Nomination package consists of statement by nominator supporting the candidate's suitability with particular reference to specific research achievement, impact of the achievement within the relevant materials field and its effect on other sciences, and innovative and perceptive quality of the candidate's work; supporting information and documents such as curriculum vita including relevant publications and copies of important papers relevant to the research achievements; letters of support from two established scientists familiar with the nominee's qualifications and area of research; and a list of supporting documents submitted. Package should not exceed 10 pages (excluding reprints).

Restrictions: Current members of the MRS awards committee and MRS officers are not eligible.
Geographic Focus: All States
Date(s) Application is Due: Jun 1
Amount of Grant: 5,000 USD
Contact: John Ballance; (724) 779-3003; fax (724) 779-8313; ballance@mrs.org
Internet: http://www.mrs.org/awards/index.html
Sponsor: Materials Research Society
506 Keystone Drive
Warrendale, PA 15086-7573

Matilda R. Wilson Fund Grants 2684

Matilda Rausch Dodge Wilson died on September 19, 1967, leaving most of her wealth to the Matilda R. Wilson Fund, a charitable trust she had established in Detroit in 1944. Today, the Matilda R. Wilson Fund awards grants, primarily in southeast Michigan, in support of the arts, youth agencies, higher education, hospitals, and social services. Types of support include building construction/renovation, endowments, equipment acquisition, general operating support, matching/challenge grants, program development, research, and scholarship funds. There are no application forms; initial approach should be a letter of request. The board considers requests at board meetings in January, April, and September.

Requirements: Michigan tax-exempt organizations are eligible.
Restrictions: Grants or loans are not made to individuals.
Geographic Focus: Michigan
Amount of Grant: 10,000-100,000 USD
Contact: David P. Larsen; (313) 259-7777; fax (313) 393-7579; roosterveen@bodmanllp.com
Sponsor: Matilda R. Wilson Fund
1901 Saint Antoine Street, 6th Floor
Detroit, MI 48226-2310

Mattel Children's Foundation Grants 2685

Charitable organizations that demonstrate that they directly serve children in need may be eligible for grants of $5,000 up to $20,000. Such organizations must have a mission that focuses on the direct service of children ages zero to 12 years. Applicants or programs must show creative and/or innovative methods to address locally defined need directly impacting children, and must be aligned with Mattel's philanthropic priorities: Learnin—increasing access to education for underserved children and in particular, innovative strategies to promote and address literacy; health—supporting the physical health and well-being of children, with particular emphasis on promoting healthy, active lifestyles; and Girl Empowerment—promoting self-esteem in young girls, up to age 12. Organizations may submit under a fiscal agent as long as the program has been in existence for at least two years.

Requirements: Applicants must be a 501(c)3 Public Charity organizations. If an organization is a project under a fiscal sponsor, that sponsor must have a valid tax exemption status. Organizations must serve children in communities within the U.S. Organizations must not discriminate against a person or a group on the basis of age, political affiliation, race, national origin, ethnicity, gender, disability, sexual orientation or religious beliefs. Organizations must have an annual operating budget of less than $1,000,000.

Restrictions: Organizations may not be affiliated in any way with national organizations, regardless of whether funds are received from the national entity. Examples of organizations

that are ineligible to apply include Boys & Girls Clubs, United Ways, YMCA's, and disease-affiliated organizations such as the American Heart Association.
Geographic Focus: All States
Amount of Grant: 5,000-20,000 USD
Contact: Administrator; (310) 252-2000; fax (310) 252-2180; foundation@mattel.com
Internet: http://corporate.mattel.com/about-us/philanthropy/programs.aspx
Sponsor: Mattel Children's Foundation
333 Continental Boulevard, M1-1418
El Segundo, CA 90245-5012

Mattel International Grants **2686**
The corporation and its foundation award grants to charitable organizations that directly serve children in need. Funding priorities include international organizations or programs that creatively address a defined need directly impacting children in need, and international organizations or programs that align with Mattel's philanthropic priorities, including health—supporting the physical and mental health and well-being of children, and increasing access to health care services for children in need; education—increasing access to education, promoting literacy to children in need, and resources that promote after-school educational achievement; and girls empowerment—promoting self-esteem of girls and increasing access to education, health, and community resources for girls. Types of support include program-specific grants—funding for new programs or expansion of existing programs; and core operating support—support of organizations to sustain their programs. Applications must be submitted online. Applications will not be accepted by fax or mail. Submit questions via email.
Requirements: Organizations must serve children in non-U.S. communities; and must not discriminate against a person or a group on the basis of age, political affiliation, race, national origin, ethnicity, gender, disability, sexual orientation, or religious belief. Pilot projects and new organizations may be considered as long as all eligibility criteria are met; however, preference will be given to organizations that have at least two years of experience. Preference is given to organizations that have an annual operating budget of less than $1 million and are not affiliated with a national organization.
Restrictions: The program does not not fund capital funding for physical property purchase, renovation, or developments; individuals; political parties, candidates, or partisan political organizations; labor organizations, fraternal organizations, athletic clubs, or social clubs; sectarian or denominational religious organizations, except for programs that are available to anyone, broadly promoted, and free from religious orientation; schools and school districts; fundraising events (e.g., dinners, tournaments); advertising or marketing sponsorships; or research overhead and/or indirect costs (including fiscal sponsor fees) that exceed 15 percent of the direct project costs.
Geographic Focus: All States
Amount of Grant: 5,000-25,000 USD
Contact: Administrator; (310) 252-2000; fax (310) 252-2180; foundation@mattel.com
Internet: http://corporate.mattel.com/about-us/philanthropy/internationalgrants.aspx
Sponsor: Mattel Children's Foundation
333 Continental Boulevard, M1-1418
El Segundo, CA 90245-5012

Max and Victoria Dreyfus Foundation Grants **2687**
The Foundation awards grants for: museums; cultural, performing, and visual arts programs, schools; hospitals, educational and skills training projects and programs for youth, seniors, and the disabled. Areas of interest include arts, education, health care, hospitals, social services, and medical research. Submit an inquiry letter (three-page maximum) that describes the nonprofit organization, its annual operating budget, the project to be funded, and the project budget; include proof of tax-exempt status.
Requirements: U.S. nonprofit organizations are eligible.
Restrictions: Foreign charitable organizations and individuals are ineligible.
Geographic Focus: All States
Date(s) Application is Due: Mar 10; Jul 10; Nov 10
Amount of Grant: 5,000-20,000 USD
Contact: John W. Hager, (202) 337-3300; fax (202) 337-4925
Sponsor: Max and Victoria Dreyfus Foundation
2233 Wisconsin Avenue NW, Suite 414
Washington, DC 20007-4122

Maximilian E. and Marion O. Hoffman Foundation **2688**
The foundation has distributed grants in the areas of medicine, education, the arts, and to various charitable organizations in Connecticut and throughout the country. The principal focus of its philanthropy has been worthy activities in the State of Connecticut.
Geographic Focus: Connecticut
Amount of Grant: 1,000-300,000 USD
Contact: Marion Barrak, President; (860) 521-2949
Sponsor: Hoffman Foundation
970 Farmington Avenue, Suite 203
West Hartford, CT 06107-2134

Mayo Clinic Administrative Fellowship **2689**
The Mayo Clinic Administrative Fellowship Program is a postgraduate fellowship that fosters the development of outstanding individuals committed to careers in health care administration. Since 1983, the Administrative Fellowship Program has provided participants with practical, skill-building experience through a broad variety of administrative rotations. The Administrative Fellowship Program prepares fellows for health care administration leadership opportunities within the Mayo Clinic system through direct participation in activities designed to build and strengthen essential administrative skills. Mayo Clinic Administrative Fellows will build the following skills: acquire and refine managerial competencies, including analytical, persuasive and human relations skills; develop an awareness and understanding of Mayo Clinic organizational structure and the interrelation between departments and personnel; receive individualized mentoring from a wide array of administrators and physicians; and establish a performance record through project work and departmental staff responsibilities. Those interested in the fellowship should review the website for available positions. Deadlines vary, but a general timeline for the fellowship recruitment and selection process is as follows: early August-job posted; mid-September-application deadline; October-phone interviews; early November-candidate on-site interviews; mid-November fellow offers extended; December-acceptance letters mailed; and July-fellowship begins. The application and additional information, including a detailed description of a variety of rotations, informational webinars, and a list of frequently asked questions, are available at the website.
Requirements: Diverse candidates from strong master's-level accredited programs are sought for the Administrative Fellowship Program. Qualified candidates are master's-prepared individuals who have completed the degree requirements or are degree candidates in health care administration, business administration, public health, health services administration or related fields of study. Applicants should have experience within the health care industry, such as an internship, externship, fellowship, or applicable employment. Applicants should also possess broad knowledge of current and historical perspectives of the health care industry that can include business management and administration, clinical practice management, clinical research and education, hospital administration and management, finance, human resources and organizational development, information systems, and managed care. Ideal candidates should possess the following traits and behaviors: the need to be challenged at a high level; energetic, hard-working, ambitious and a self-starter; flexibility and willingness to assume a variety of role assignments and deal effectively with ambiguity; objective and insightful; service, team and learning orientation; and understanding and commitment to cultural competency and diversity in the workplace and in the practice of health care.
Restrictions: Mayo Clinic is not able to provide visa sponsorship.
Geographic Focus: All States
Date(s) Application is Due: Sep 15
Contact: Stacy M. Fuhrman; (507) 266-6599; fuhrman.stacy@mayo.edu
Internet: http://www.mayoclinic.org/afp
Sponsor: Mayo Clinic
200 First Street SW
Rochester, MN 55905

Mayo Clinic Administrative Fellowship-Eau Claire **2690**
The Administrative Fellowship Program is a postgraduate fellowship that fosters the development of outstanding individuals committed to careers in health care administration. The Mayo Clinic Health System in Eau Claire has provided participants with practical experience through a broad variety of administrative interactions, committee involvement, and projects leadership opportunities. The Fellowship Program in Healthcare Administration at Mayo Clinic Health System in Eau Claire is designed to foster the development of master's-prepared individuals committed to a career in healthcare administration. Under the guidance of the chief administrative officer, this one-year management training experience is tailored to meet the needs and interests of the individual candidate. This program allows the fellow to experience the major aspects of managing an integrated healthcare organization. The Fellowship Program offers exposure to all levels of management, allowing fellows to gain an applied understanding of the operations of an integrated healthcare system. Fellows are provided with opportunities to attend various executive-level meetings, and encouraged to participate in committees and other team projects. Examples of previous fellowship experiences include a physician need analysis, a community benefit assessment and a supply chain management project. The Fellow will receive a competitive salary and benefits. The application and additional information, including a detailed list of the objectives and expectations of the Fellowship, are posted at the website.
Requirements: Diverse candidates from strong master's-level accredited programs are sought for the Administrative Fellowship Program. Qualified candidates are master's-prepared individuals who have completed the degree requirements or are degree candidates in health care administration, business administration, public health, health services administration or related fields of study. Applicants should have experience within the health care industry, such as an internship, externship, fellowship, or applicable employment. Applicants should also possess broad knowledge of current and historical perspectives of the health care industry that can include business management and administration, clinical practice management, clinical research and education, hospital administration and management, finance, human resources and organizational development, information systems, and managed care. Ideal candidates should possess the following traits and behaviors: the need to be challenged at a high level; energetic, hard-working, ambitious and a self-starter; flexibility and willingness to assume a variety of role assignments and deal effectively with ambiguity; objective and insightful; service, team and learning orientation; and understanding and commitment to cultural competency and diversity in the workplace and in the practice of health care.
Restrictions: Lodging is the responsibility of the individual. There is no commitment of employment on behalf of the individual or organization beyond the one-year program.
Geographic Focus: All States
Date(s) Application is Due: Oct 1
Contact: Nicole Egan; (715) 838-5081; egan.nicole@mayo.edu
Internet: http://mayoclinichealthsystem.org/locations/eau-claire/careers/for-students/administrative-fellowship
Sponsor: Mayo Clinic
200 First Street SW
Rochester, MN 55905

Mayo Clinic Business Consulting Fellowship 2691
The Business Consulting Fellowship Program at Mayo Clinic is a one-year learning experience designed to develop outstanding individuals committed to a career in health care management. Through this program, recipients will gain practical experience and actively participate in projects within Mayo Clinic. Fellows will assist experienced analysts on large and small projects designed to improve the efficiency, quality, and effectiveness of Mayo Clinic. Sample project topics include the following: business planning; quality improvement; information technology systems implementation; forecasting and access; simulation; patient flow; facilities design; electronic systems usability; and staffing analysis. The Clinic offers competitive compensation and comprehensive benefits for Fellow participants. The program typically begins in June and December. Candidates may apply at the website.
Requirements: Applicants must have recently completed a master's degree and have at least one year of relevant internship or work experience, or have a bachelor's degree and at least four years of work or internship experience. Previous health care experience is preferred. Degrees should be in business or health care administration, industrial engineering, operations research, or a closely related field. Successful candidates will have skills and experience in several of these areas: consulting; workflow and operations analysis; staffing analysis; process design, improvement or re-engineering; simulation; electronic system implementation; and project management. Strong candidates should also have the following skills: objectivity; analytical, innovative and results oriented; comfortable with ambiguity; curious and willing to ask questions; highly skilled communicators; and able to work both independently and as a part of team.
Geographic Focus: All States
Contact: Chad Musolf, Human Resources; (507) 284-4851; musolf.chad@mayo.edu
Internet: http://www.mayoclinic.org/train-sp-rst/
Sponsor: Mayo Clinic
200 First Street SW
Rochester, MN 55905

MBL Albert and Ellen Grass Summer Fellowships 2692
Marine Biological Laboratory (MBL) summer researchers come from all over the world. Visiting researchers find an infrastructure and an informal, interactive scientific community that allows them to launch into research as soon as they arrive. Research awards are available to investigators wishing to do visiting research at the MBL. The Albert and Ellen Grass Fellowship promotes innovative research collaborations, ideally between investigators from different institutions. Preference will be given to junior investigators, who are a required member of any team seeking support from this Fellowship. The MBL Research Awards provide costs for research and housing, and also enable awardees to benefit from the intellectual and interactive environment of the scientific community at the MBL. Funding requires a minimum stay of six weeks at the MBL. Awards can be used to support independent research, collaborative research, or in affiliation with the discovery courses. Proposals for research award support will be considered in, but are not limited to, the following fields of investigation: cell biology; development biology; ecology; evolution; microbiology; neurobiology; physiology; and tissue engineering. Candidates may apply at the MBL website.
Requirements: Research proposed in the resident program or courses requires pre-approval of the host lab and program and the course directors, in addition to the CASO office.
Geographic Focus: All States, All Countries
Date(s) Application is Due: Dec 15
Contact: Julie Early; (508) 289-7173; researchawards@mbl.edu or jearly@mbl.edu
Internet: http://hermes.mbl.edu/research/summer/awards_general.html
Sponsor: Marine Biological Laboratory
7 MBL Street
Woods Hole, MA 02543

MBL Ann E. Kammer Memorial Summer Fellowship 2693
Marine Biological Laboratory (MBL) summer researchers come from all over the world. Visiting researchers find an infrastructure and an informal, interactive scientific community that allows them to launch into research as soon as they arrive. Research awards are available to these visiting researchers. In memory of renowned zoologist, Ann E. Kammer, this Fellowship is available to women investigators working in the neurosciences. The MBL Research Awards provide costs for research and housing, and also enable awardees to benefit from the intellectual and interactive environment of the scientific community at the MBL. Funding requires a minimum stay of six weeks at the MBL. Awards can be used to support independent research, collaborative research, or in affiliation with the discovery courses. Proposals for Research award support will be considered in, but are not limited to, the following fields of investigation: cell biology; development biology; ecology; evolution; microbiology; neurobiology; and physiology. Candidates may fill out applications at the MBL online process.
Requirements: Research proposed in the resident program or courses requires pre-approval of the host lab and program and the course directors, in addition to the CASO office.
Geographic Focus: All States, All Countries
Date(s) Application is Due: Dec 15
Contact: Julie Early; (508) 289-7173; researchawards@mbl.edu or jearly@mbl.edu
Internet: http://hermes.mbl.edu/research/summer/awards_general.html
Sponsor: Marine Biological Laboratory
7 MBL Street
Woods Hole, MA 02543

MBL Associates Summer Fellowships 2694
Marine Biological Laboratory (MBL) summer researchers come from all over the world. Visiting researchers find an infrastructure and an informal, interactive scientific community that allows them to launch into research as soon as they arrive. Research awards are available to these visiting researchers. The MBL Associates were formed to provide a connection to the Laboratory for friends, both scientists and non-scientists who have an interest in the research, educational and cultural activities of the MBL. The Associates provide fellowship funds to help support the costs of summer research at the MBL. The MBL Research Awards provide costs for research and housing, and also enable awardees to benefit from the intellectual and interactive environment of the scientific community at the MBL. Funding requires a minimum stay of six weeks at the MBL. Awards can be used to support independent research, collaborative research, or in affiliation with the discovery courses. Proposals for Research award support will be considered in, but are not limited to, the following fields of investigation: cell biology; development biology; ecology; evolution; microbiology; neurobiology; physiology; and tissue engineering. Candidates may fill out applications at the MBL online process.
Requirements: Research proposed in the resident program or courses requires pre-approval of the host lab and program and the course directors, in addition to the CASO office.
Geographic Focus: All States, All Countries
Date(s) Application is Due: Dec 15
Contact: Julie Early; (508) 289-7173; researchawards@mbl.edu or jearly@mbl.edu
Internet: http://hermes.mbl.edu/research/summer/awards_general.html
Sponsor: Marine Biological Laboratory
7 MBL Street
Woods Hole, MA 02543

MBL Baxter Postdoctoral Summer Fellowship 2695
Marine Biological Laboratory (MBL) summer researchers come from all over the world. Visiting researchers find an infrastructure and an informal, interactive scientific community that allows them to launch into research as soon as they arrive. Research awards are available to these visiting researchers. Baxter International, Inc. has established a Fellowship for postdoctoral investigators in the biological and biomedical sciences. This Fellowship supports a scientist undertaking independent research at the MBL. The MBL Research Awards provide costs for research and housing, and also enable awardees to benefit from the intellectual and interactive environment of the scientific community at the MBL. Funding requires a minimum stay of six weeks at the MBL. Awards can be used to support independent research, collaborative research, or in affiliation with the discovery courses. Candidates may fill out applications at the MBL online process.
Requirements: Research proposed in the resident program or courses requires pre-approval of the host lab and program and the course directors, in addition to the CASO office.
Geographic Focus: All States, All Countries
Date(s) Application is Due: Dec 15
Contact: Julie Early; (508) 289-7173; researchawards@mbl.edu or jearly@mbl.edu
Internet: http://hermes.mbl.edu/research/summer/awards_general.html
Sponsor: Marine Biological Laboratory
7 MBL Street
Woods Hole, MA 02543

MBL Burr and Susie Steinbach Summer Fellowship 2696
Marine Biological Laboratory (MBL) summer researchers come from all over the world. Visiting researchers find an infrastructure and an informal, interactive scientific community that allows them to launch into research as soon as they arrive. Research awards are available to these visiting researchers. The Steinbach Fellowship honors the memory of H. "Burr" and Eleanor "Susie" Steinbach. Dr. Steinbach was a scientist and director of the MBL who made important contributions toward the understanding of bioelectric generators, ionic equilibria across cell membranes, and the enzymology of cell organelles. This Fellowship helps cover laboratory fees, housing expenses, and other appropriate personal expenses. The MBL Research Awards provide costs for research and housing, and also enable awardees to benefit from the intellectual and interactive environment of the scientific community at the MBL. Funding requires a minimum stay of six weeks at the MBL. Awards can be used to support independent research, collaborative research, or in affiliation with the discovery courses. Proposals for Research award support will be considered in, but are not limited to, the following fields of investigation: cell biology; development biology; ecology; evolution; microbiology; neurobiology; physiology; and tissue engineering. Candidates may fill out applications at the MBL online process.
Requirements: Research proposed in the resident program or courses requires pre-approval of the host lab and program and the course directors, in addition to the CASO office.
Geographic Focus: All States, All Countries
Date(s) Application is Due: Dec 15
Contact: Julie Early; (508) 289-7173; researchawards@mbl.edu or jearly@mbl.edu
Internet: http://hermes.mbl.edu/research/summer/awards_general.html
Sponsor: Marine Biological Laboratory
7 MBL Street
Woods Hole, MA 02543

MBL E.E. Just Summer Fellowship for Minority Scientists 2697
Marine Biological Laboratory (MBL) summer researchers come from all over the world. Visiting researchers find an infrastructure and an informal, interactive scientific community that allows them to launch into research as soon as they arrive. Research awards are available to these visiting researchers. The Just Fellowship honors Dr. Just's contributions to developmental biology and experimental embryology. Dr. Just was the first African American to be professionally recognized for his scientific work, much of it accomplished during the early 1900s. This Fellowship will underwrite an outstanding minority scientist's participation in research at the MBL. The MBL Research Awards provide costs for research and housing, and also enable awardees to benefit from the intellectual and interactive environment of the scientific community at the MBL. Funding requires a minimum stay of

six weeks at the MBL. Awards can be used to support independent research, collaborative research, or in affiliation with the discovery courses. Proposals for Research award support will be considered in, but are not limited to, the following fields of investigation: cell biology; development biology; ecology; evolution; microbiology; neurobiology; physiology; and tissue engineering. Candidates may fill out applications at the MBL online process.
Requirements: Research proposed in the resident program or courses requires pre-approval of the host lab and program and the course directors, in addition to the CASO office.
Geographic Focus: All States, All Countries
Date(s) Application is Due: Dec 15
Contact: Julie Early; (508) 289-7173; researchawards@mbl.edu or jearly@mbl.edu
Internet: http://hermes.mbl.edu/research/summer/awards_general.html
Sponsor: Marine Biological Laboratory
7 MBL Street
Woods Hole, MA 02543

MBL Erik B. Fries Summer Fellowships **2698**
Marine Biological Laboratory (MBL) summer researchers come from all over the world. Visiting researchers find an infrastructure and an informal, interactive scientific community that allows them to launch into research as soon as they arrive. Research awards are available to these visiting researchers. The Erik B. Fries Fellowships defray laboratory or library expenses for independent investigators at the MBL. Fries, a long time Woods Hole resident, was a biology professor at City College of New York, and MBL's botany course in the early 1900s. Preference is given to investigator members from Fries' former department at City College. The MBL Research Awards provide costs for research and housing, and also enable awardees to benefit from the intellectual and interactive environment of the scientific community at the MBL. Funding requires a minimum stay of six weeks at the MBL. Awards can be used to support independent research, collaborative research, or in affiliation with the discovery courses. Proposals for Research award support will be considered in, but are not limited to, the following fields of investigation: cell biology; development biology; ecology; evolution; microbiology; neurobiology; physiology; and tissue engineering. Candidates may fill out applications at the MBL online process.
Requirements: Research proposed in the resident program or courses requires pre-approval of the host lab and program and the course directors, in addition to the CASO office.
Geographic Focus: All States, All Countries
Date(s) Application is Due: Dec 15
Contact: Julie Early; (508) 289-7173; researchawards@mbl.edu or jearly@mbl.edu
Internet: http://hermes.mbl.edu/research/summer/awards_general.html
Sponsor: Marine Biological Laboratory
7 MBL Street
Woods Hole, MA 02543

MBL Eugene and Millicent Bell Summer Fellowships **2699**
Marine Biological Laboratory (MBL) summer researchers come from all over the world. Visiting researchers find an infrastructure and an informal, interactive scientific community that allows them to launch into research as soon as they arrive. Research awards are available to these visiting researchers. The Eugene and Millicent Bell Fellowships in Tissue Engineering provide funding to visiting graduate students, postdoctoral fellows, or faculty pursuing research in a variety of biomedical questions that advance knowledge of biological composition and function at levels ranging from molecules to cells, to tissue, to organs and whole organisms. This work is grounded in a solid base of chemistry, physics, math and biology and aims to expand the possibilities for tissue and organ reconstitution and replacement. The MBL Research Awards provide costs for research and housing, and also enable awardees to benefit from the intellectual and interactive environment of the scientific community at the MBL. Funding requires a minimum stay of six weeks at the MBL. Awards can be used to support independent research, collaborative research, or in affiliation with the discovery courses. Candidates may fill out applications at the MBL online process.
Requirements: Research proposed in the resident program or courses requires pre-approval of the host lab and program and the course directors, in addition to the CASO office.
Geographic Focus: All States, All Countries
Date(s) Application is Due: Dec 15
Contact: Julie Early; (508) 289-7173; researchawards@mbl.edu or jearly@mbl.edu
Internet: http://hermes.mbl.edu/research/summer/awards_general.html
Sponsor: Marine Biological Laboratory
7 MBL Street
Woods Hole, MA 02543

MBL Evelyn and Melvin Spiegal Summer Fellowship **2700**
Marine Biological Laboratory (MBL) summer researchers come from all over the world. Visiting researchers find an infrastructure and an informal, interactive scientific community that allows them to launch into research as soon as they arrive. Research awards are available to these visiting researchers. The Spiegal Fellowship support summer research at the MBL in any relevant biomedical discipline. The MBL Research Awards provide costs for research and housing, and also enable awardees to benefit from the intellectual and interactive environment of the scientific community at the MBL. Funding requires a minimum stay of six weeks at the MBL. Awards can be used to support independent research, collaborative research, or in affiliation with the discovery courses. Proposals for Research award support will be considered in, but are not limited to, the following fields of investigation: cell biology; development biology; ecology; evolution; microbiology; neurobiology; physiology; and tissue engineering. Candidates may fill out applications at the MBL online process.
Requirements: Research proposed in the resident program or courses requires pre-approval of the host lab and program and the course directors, in addition to the CASO office.
Geographic Focus: All States, All Countries
Date(s) Application is Due: Dec 15
Contact: Julie Early; (508) 289-7173; researchawards@mbl.edu or jearly@mbl.edu
Internet: http://hermes.mbl.edu/research/summer/awards_general.html
Sponsor: Marine Biological Laboratory
7 MBL Street
Woods Hole, MA 02543

MBL Frank R. Lillie Summer Fellowship **2701**
Marine Biological Laboratory (MBL) summer researchers come from all over the world. Visiting researchers find an infrastructure and an informal, interactive scientific community that allows them to launch into research as soon as they arrive. Research awards are available to these visiting researchers. The Lillie Fellowship was established in memory of Frank R. Lillie, a distinguished developmental biologist and the laboratory's director for many years. The Fellowship provides a research laboratory and funds for supplies. The MBL Research Awards provide costs for research and housing, and also enable awardees to benefit from the intellectual and interactive environment of the scientific community at the MBL. Funding requires a minimum stay of six weeks at the MBL. Awards can be used to support independent research, collaborative research, or in affiliation with the discovery courses. Proposals for Research award support will be considered in, but are not limited to, the following fields of investigation: cell biology; development biology; ecology; evolution; microbiology; neurobiology; physiology; and tissue engineering. Candidates may fill out applications at the MBL online process.
Requirements: Research proposed in the resident program or courses requires pre-approval of the host lab and program and the course directors, in addition to the CASO office.
Geographic Focus: All States, All Countries
Date(s) Application is Due: Dec 15
Contact: Julie Early; (508) 289-7173; researchawards@mbl.edu or jearly@mbl.edu
Internet: http://hermes.mbl.edu/research/summer/awards_general.html
Sponsor: Marine Biological Laboratory
7 MBL Street
Woods Hole, MA 02543

MBL Frederik B. and Betsy G. Bang Summer Fellowships **2702**
Marine Biological Laboratory (MBL) summer researchers come from all over the world. Visiting researchers find an infrastructure and an informal, interactive scientific community that allows them to launch into research as soon as they arrive. Research awards are available to these visiting researchers. The Bang Fellowships are offered for the study of the immune capability of marine animals and for the use of marine models for research in molecular biology or biomedicine. Funds primarily support laboratory space and research expenses. Funding requires a minimum stay of six weeks at the MBL. Candidates may fill out applications at the MBL online process.
Requirements: Research proposed in the resident program or courses requires pre-approval of the host lab and program and the course directors, in addition to the CASO office.
Geographic Focus: All States, All Countries
Date(s) Application is Due: Dec 15
Contact: Julie Early; (508) 289-7173; researchawards@mbl.edu or jearly@mbl.edu
Internet: http://hermes.mbl.edu/research/summer/awards_general.html
Sponsor: Marine Biological Laboratory
7 MBL Street
Woods Hole, MA 02543

MBL Fred Karush Library Readership **2703**
Marine Biological Laboratory (MBL) summer researchers come from all over the world. Visiting researchers find an infrastructure and an informal, interactive scientific community that allows them to launch into research as soon as they arrive. A variety of research awards are available to these visiting researchers. The Karush Library Readership provides the fees for desk rental, and full privileges in the MBL/WHOI Library on a monthly basis during the summer. The MBL Research Awards provide costs for research and housing, and also enable awardees to benefit from the intellectual and interactive environment of the scientific community at the MBL. Funding requires a minimum stay of six weeks at the MBL. Awards can be used to support independent research, collaborative research, or in affiliation with the discovery courses. Proposals for Research award support will be considered in, but are not limited to, the following fields of investigation: cell biology; development biology; ecology; evolution; microbiology; neurobiology; physiology; and tissue engineering. Candidates may fill out applications at the MBL online process.
Requirements: Research proposed in the resident program or courses requires pre-approval of the host lab and program and the course directors, in addition to the CASO office.
Geographic Focus: All States, All Countries
Date(s) Application is Due: Feb 15
Contact: Julie Early; (508) 289-7173; researchawards@mbl.edu or jearly@mbl.edu
Internet: http://hermes.mbl.edu/research/summer/awards_general.html
Sponsor: Marine Biological Laboratory
7 MBL Street
Woods Hole, MA 02543

MBL Gruss Lipper Family Foundation Summer Fellowship **2704**
Marine Biological Laboratory (MBL) summer researchers come from all over the world. Visiting researchers find an infrastructure and an informal, interactive scientific community that allows them to launch into research as soon as they arrive. Research awards are available to these visiting researchers. The Gruss Lipper Fellowship

underwrites expenses and travel to bring Israeli investigators and advanced students to the MBL to work independently, collaborate with MBL scientists, or to participate in the MBL summer research and short courses. The MBL Research Awards provide costs for research and housing, and also enable awardees to benefit from the intellectual and interactive environment of the scientific community at the MBL. Funding requires a minimum stay of six weeks at the MBL. Awards can be used to support independent research, collaborative research, or in affiliation with the discovery courses. Proposals for Research award support will be considered in, but are not limited to, the following fields of investigation: cell biology; development biology; ecology; evolution; microbiology; neurobiology; physiology; and tissue engineering. Candidates may fill out applications at the MBL online process.
Requirements: Research proposed in the resident program or courses requires pre-approval of the host lab and program and the course directors, in addition to the CASO office.
Geographic Focus: Israel
Date(s) Application is Due: Dec 15
Contact: Julie Early; (508) 289-7173; researchawards@mbl.edu or jearly@mbl.edu
Internet: http://hermes.mbl.edu/research/summer/awards_general.html
Sponsor: Marine Biological Laboratory
7 MBL Street
Woods Hole, MA 02543

MBL H. Keffer Hartline and Edward F. MacNichol, Jr. Fellowships **2705**
Marine Biological Laboratory (MBL) summer researchers come from all over the world. Visiting researchers find an infrastructure and an informal, interactive scientific community that allows them to launch into research as soon as they arrive. Research awards are available to these visiting researchers. The Hartline and MacNichol Fellowships were established in memory of Dr. Hartline and his impact on sensory science. Dr. Hartline came to the MBL as a young scientist, initiated his pioneering studies of the eye, and continued this research at the MBL for many years. These Fellowships support scientists carrying out research in sensory neurobiology. The MBL Research Awards provide costs for research and housing, and also enable awardees to benefit from the intellectual and interactive environment of the scientific community at the MBL. Funding requires a minimum stay of six weeks at the MBL. Awards can be used to support independent research, collaborative research, or in affiliation with the discovery courses. Candidates may fill out applications at the MBL online process.
Requirements: Research proposed in the resident program or courses requires pre-approval of the host lab and program and the course directors, in addition to the CASO office.
Geographic Focus: All States, All Countries
Date(s) Application is Due: Dec 15
Contact: Julie Early; (508) 289-7173; researchawards@mbl.edu or jearly@mbl.edu
Internet: http://hermes.mbl.edu/research/summer/awards_general.html
Sponsor: Marine Biological Laboratory
7 MBL Street
Woods Hole, MA 02543

MBL Herbert W. Rand Summer Fellowship **2706**
Marine Biological Laboratory (MBL) summer researchers come from all over the world. Visiting researchers find an infrastructure and an informal, interactive scientific community that allows them to launch into research as soon as they arrive. A variety of research awards are available to these visiting researchers. The Rand Fellowship includes a summer stipend, and funds or assists with such expenses as travel, supplies, and laboratory fees. The MBL Research Awards provide costs for research and housing, and also enable awardees to benefit from the intellectual and interactive environment of the scientific community at the MBL. Funding requires a minimum stay of six weeks at the MBL. Awards can be used to support independent research, collaborative research, or in affiliation with the discovery courses. Proposals for Research award support will be considered in, but are not limited to, the following fields of investigation: cell biology; development biology; ecology; evolution; microbiology; neurobiology; physiology; and tissue engineering. Candidates may fill out applications at the MBL online process.
Requirements: Research proposed in the resident program or courses requires pre-approval of the host lab and program and the course directors, in addition to the CASO office.
Geographic Focus: All States, All Countries
Date(s) Application is Due: Dec 15
Contact: Julie Early; (508) 289-7173; researchawards@mbl.edu or jearly@mbl.edu
Internet: http://hermes.mbl.edu/research/summer/awards_general.html
Sponsor: Marine Biological Laboratory
7 MBL Street
Woods Hole, MA 02543

MBL James E. and Faith Miller Memorial Summer Fellowship **2707**
Marine Biological Laboratory (MBL) summer researchers come from all over the world. Visiting researchers find an infrastructure and an informal, interactive scientific community that allows them to launch into research as soon as they arrive. Research awards are available to these visiting researchers. The Miller Fellowship provides funds for summer research at the MBL in any relevant field of medical research. The MBL Research Awards provide costs for research and housing, and also enable awardees to benefit from the intellectual and interactive environment of the scientific community at the MBL. Funding requires a minimum stay of six weeks at the MBL. Awards can be used to support independent research, collaborative research, or in affiliation with the discovery courses. Proposals for Research award support will be considered in, but are not limited to, the following fields of investigation: cell biology; development biology; ecology; evolution; microbiology; neurobiology; physiology; and tissue engineering. Candidates may fill out applications at the MBL online process.
Requirements: Research proposed in the resident program or courses requires pre-approval of the host lab and program and the course directors, in addition to the CASO office.
Geographic Focus: All States, All Countries
Date(s) Application is Due: Dec 15
Contact: Julie Early; (508) 289-7173; researchawards@mbl.edu or jearly@mbl.edu
Internet: http://hermes.mbl.edu/research/summer/awards_general.html
Sponsor: Marine Biological Laboratory
7 MBL Street
Woods Hole, MA 02543

MBL John M. Arnold Award **2708**
Marine Biological Laboratory (MBL) summer researchers come from all over the world. Visiting researchers find an infrastructure and an informal, interactive scientific community that allows them to launch into research as soon as they arrive. A variety of research awards are available to these visiting researchers. The John M. Arnold Awards funds a retired or senior researcher in any field of study compatible with the mission of the MBL. The MBL Research Awards provide costs for research and housing, and also enable awardees to benefit from the intellectual and interactive environment of the scientific community at the MBL. Funding requires a minimum stay of six weeks at the MBL. Awards can be used to support independent research, collaborative research, or in affiliation with the discovery courses. Proposals for Research award support will be considered in, but are not limited to, the following fields of investigation: cell biology; development biology; ecology; evolution; microbiology; neurobiology; physiology; and tissue engineering. Candidates may fill out applications at the MBL online process.
Requirements: Research proposed in the resident program or courses requires pre-approval of the host lab and program and the course directors, in addition to the CASO office.
Geographic Focus: All States, All Countries
Date(s) Application is Due: Dec 15
Contact: Julie Early; (508) 289-7173; researchawards@mbl.edu or jearly@mbl.edu
Internet: http://hermes.mbl.edu/research/summer/awards_general.html
Sponsor: Marine Biological Laboratory
7 MBL Street
Woods Hole, MA 02543

MBL Laura and Arthur Colwin Summer Fellowships **2709**
Marine Biological Laboratory (MBL) summer researchers come from all over the world. Visiting researchers find an infrastructure and an informal, interactive scientific community that allows them to launch into research as soon as they arrive. Research awards are available to these visiting researchers. The Colwin Summer Fellowships provide support for independent investigators conducting research in the fields of cell and developmental biology. The MBL Research Awards provide costs for research and housing, and also enable awardees to benefit from the intellectual and interactive environment of the scientific community at the MBL. Funding requires a minimum stay of six weeks at the MBL. Candidates may fill out applications at the MBL online process.
Requirements: Research proposed in the resident program or courses requires pre-approval of the host lab and program and the course directors, in addition to the CASO office.
Geographic Focus: All States, All Countries
Date(s) Application is Due: Dec 15
Contact: Julie Early; (508) 289-7173; researchawards@mbl.edu or jearly@mbl.edu
Internet: http://hermes.mbl.edu/research/summer/awards_general.html
Sponsor: Marine Biological Laboratory
7 MBL Street
Woods Hole, MA 02543

MBL Lucy B. Lemann Memorial Summer Fellowship **2710**
Marine Biological Laboratory (MBL) summer researchers come from all over the world. Visiting researchers find an infrastructure and an informal, interactive scientific community that allows them to launch into research as soon as they arrive. Research awards are available to these visiting researchers. Established in memory of Lucy B. Lemann, a Woods Hole resident, this Fellowship helps support the costs of summer research at the MBL. The MBL Research Awards provide costs for research and housing, and also enable awardees to benefit from the intellectual and interactive environment of the scientific community at the MBL. Funding requires a minimum stay of six weeks at the MBL. Awards can be used to support independent research, collaborative research, or in affiliation with the discovery courses. Proposals for Research award support will be considered in, but are not limited to, the following fields of investigation: cell biology; development biology; ecology; evolution; microbiology; neurobiology; physiology; and tissue engineering. Candidates may fill out applications at the MBL online process.
Requirements: Research proposed in the resident program or courses requires pre-approval of the host lab and program and the course directors, in addition to the CASO office.
Geographic Focus: All States, All Countries
Date(s) Application is Due: Dec 15
Contact: Julie Early; (508) 289-7173; researchawards@mbl.edu or jearly@mbl.edu
Internet: http://hermes.mbl.edu/research/summer/awards_general.html
Sponsor: Marine Biological Laboratory
7 MBL Street
Woods Hole, MA 02543

MBL M.G.F. Fuortes Summer Fellowships **2711**
Marine Biological Laboratory (MBL) summer researchers come from all over the world. Visiting researchers find an infrastructure and an informal, interactive scientific community that allows them to launch into research as soon as they arrive. Research awards are available to these visiting researchers. Established in memory of M. G. F. Fuortes, a renowned

neurophysiologist, the Fuortes Summer Fellowships are available to young investigators working in the neurosciences. The MBL Research Awards provide costs for research and housing, and also enable awardees to benefit from the intellectual and interactive environment of the scientific community at the MBL. Funding requires a minimum stay of six weeks at the MBL. Awards can be used to support independent research, collaborative research, or in affiliation with the discovery courses. Proposals for Research award support will be considered in, but are not limited to, the following fields of investigation: cell biology; development biology; ecology; evolution; microbiology; neurobiology; physiology; and tissue engineering. Candidates may fill out applications at the MBL online process.
Requirements: Research proposed in the resident program or courses requires pre-approval of the host lab and program and the course directors, in addition to the CASO office.
Geographic Focus: All States, All Countries
Date(s) Application is Due: Dec 15
Contact: Julie Early; (508) 289-7173; researchawards@mbl.edu or jearly@mbl.edu
Internet: http://hermes.mbl.edu/research/summer/awards_general.html
Sponsor: Marine Biological Laboratory
7 MBL Street
Woods Hole, MA 02543

MBL Nikon Summer Fellowship **2712**
Marine Biological Laboratory (MBL) summer researchers come from all over the world. Visiting researchers find an infrastructure and an informal, interactive scientific community that allows them to launch into research as soon as they arrive. Research awards are available to these visiting researchers. A summer Fellowship at the MBL is available from Nikon, Inc. to a young investigator doing research in an area of biology in which she or he can make extensive use of advanced microscopy or micro-manipulation systems. This Fellowship includes a summer laboratory, housing, and a budget for incidental expenses, equipment rental, and supplies. Requests will also be considered for collaborative research projects at the MBL during the off-season period. The MBL Research Awards provide costs for research and housing, and also enable awardees to benefit from the intellectual and interactive environment of the scientific community at the MBL. Funding requires a minimum stay of six weeks at the MBL. Awards can be used to support independent research, collaborative research, or in affiliation with the discovery courses. Proposals for Research award support will be considered in, but are not limited to, the following fields of investigation: cell biology; development biology; ecology; evolution; microbiology; neurobiology; physiology; and tissue engineering. Candidates may fill out applications at the MBL online process.
Requirements: Research proposed in the resident program or courses requires pre-approval of the host lab and program and the course directors, in addition to the CASO office.
Geographic Focus: All States, All Countries
Date(s) Application is Due: Dec 15
Contact: Julie Early; (508) 289-7173; researchawards@mbl.edu or jearly@mbl.edu
Internet: http://hermes.mbl.edu/research/summer/awards_general.html
Sponsor: Marine Biological Laboratory
7 MBL Street
Woods Hole, MA 02543

MBL Plum Foundation John E. Dowling Fellowships **2713**
Marine Biological Laboratory (MBL) summer researchers come from all over the world. Visiting researchers find an infrastructure and an informal, interactive scientific community that allows them to launch into research as soon as they arrive. Research awards are available to these visiting researchers. The Dowling Fellowships help underwrite the costs of room and board, supplies, and laboratory space rental for outstanding applicants pursuing research on the neurobiology of vision and retinal research. The MBL Research Awards provide costs for research and housing, and also enable awardees to benefit from the intellectual and interactive environment of the scientific community at the MBL. Funding requires a minimum stay of six weeks at the MBL. Awards can be used to support independent research, collaborative research, or in affiliation with the discovery courses. Proposals for Research award support will be considered in, but are not limited to, the following fields of investigation: cell biology; development biology; ecology; evolution; microbiology; neurobiology; physiology; and tissue engineering. Candidates may fill out applications at the MBL online process.
Requirements: Research proposed in the resident program or courses requires pre-approval of the host lab and program and the course directors, in addition to the CASO office.
Geographic Focus: All States, All Countries
Date(s) Application is Due: Dec 15
Contact: Julie Early; (508) 289-7173; researchawards@mbl.edu or jearly@mbl.edu
Internet: http://hermes.mbl.edu/research/summer/awards_general.html
Sponsor: Marine Biological Laboratory
7 MBL Street
Woods Hole, MA 02543

MBL Robert Day Allen Summer Fellowship **2714**
Marine Biological Laboratory (MBL) summer researchers come from all over the world. Visiting researchers find an infrastructure and an informal, interactive scientific community that allows them to launch into research as soon as they arrive. Research awards are available to these visiting researchers. The Robert Day Allen Fellowship honors the memory of Professor Allen, a distinguished investigator of cell motility, and one of the developers of video-enhanced optical microscopy with differential interference contrast. The Fellowship funds a young investigator committed to research on cell motility and cytoarchitecture. The MBL Research Awards provide costs for research and housing, and also enable awardees to benefit from the intellectual and interactive environment of the scientific community at the MBL. Funding requires a minimum stay of six weeks at the MBL. Candidates may fill out applications at the MBL online process.
Requirements: Research proposed in the resident program or courses requires pre-approval of the host lab and program and the course directors, in addition to the CASO office.
Geographic Focus: All States, All Countries
Date(s) Application is Due: Dec 15
Contact: Julie Early; (508) 289-7173; researchawards@mbl.edu or jearly@mbl.edu
Internet: http://hermes.mbl.edu/research/summer/awards_general.html
Sponsor: Marine Biological Laboratory
7 MBL Street
Woods Hole, MA 02543

MBL Stephen W. Kuffler Summer Fellowships **2715**
Marine Biological Laboratory (MBL) summer researchers come from all over the world. Visiting researchers find an infrastructure and an informal, interactive scientific community that allows them to launch into research as soon as they arrive. Research awards are available to investigators wishing to do visiting research at the MBL. The Stephen W. Kuffler Fellowships are intended to encourage the career development of promising young investigators by helping to support them in the intense intellectual atmosphere of the MBL for the summer. It covers part of the costs of laboratory rental, housing, and other personal expenses. The MBL Research Awards provide costs for research and housing, and also enable awardees to benefit from the intellectual and interactive environment of the scientific community at the MBL. Funding requires a minimum stay of six weeks at the MBL. Awards can be used to support independent research, collaborative research, or in affiliation with the discovery courses. Proposals for Research award support will be considered in, but are not limited to, the following fields of investigation: cell biology; development biology; ecology; evolution; microbiology; neurobiology; physiology; and tissue engineering. The application is available at the MBL website.
Requirements: Research proposed in the resident program or courses requires pre-approval of the host lab and program and the course directors, in addition to the CASO office.
Geographic Focus: All States, All Countries
Date(s) Application is Due: Dec 15
Contact: Julie Early; (508) 289-7173; researchawards@mbl.edu or jearly@mbl.edu
Internet: http://hermes.mbl.edu/research/summer/awards_general.html
Sponsor: Marine Biological Laboratory
7 MBL Street
Woods Hole, MA 02543

MBL William Townsend Porter Summer Fellowships for Minority Investigators **2716**
Marine Biological Laboratory (MBL) summer researchers come from all over the world. Visiting researchers find an infrastructure and an informal, interactive scientific community that allows them to launch into research as soon as they arrive. Research awards are available to these visiting researchers. The Porter Foundation offers fellowship support for young scientists (undergraduates, senior graduate students, and postdoctoral trainees) who are from an under-represented minority group, and are U.S. citizens or permanent residents to do summer research with senior investigators at the MBL. The Fellowship contributes toward the expenses at the MBL during the summer. The MBL Research Awards provide costs for research and housing, and also enable awardees to benefit from the intellectual and interactive environment of the scientific community at the MBL. Funding requires a minimum stay of six weeks at the MBL. Awards can be used to support independent research, collaborative research, or in affiliation with the discovery courses. Proposals for Research award support will be considered in, but are not limited to, the following fields of investigation: cell biology; development biology; ecology; evolution; microbiology; neurobiology; physiology; and tissue engineering. Candidates may fill out applications at the MBL online process.
Requirements: Research proposed in the resident program or courses requires pre-approval of the host lab and program and the course directors, in addition to the CASO office.
Geographic Focus: All States, All Countries
Date(s) Application is Due: Dec 15
Contact: Julie Early; (508) 289-7173; researchawards@mbl.edu or jearly@mbl.edu
Internet: http://hermes.mbl.edu/research/summer/awards_general.html
Sponsor: Marine Biological Laboratory
7 MBL Street
Woods Hole, MA 02543

McCarthy Family Foundation Grants **2717**
The McCarthy Family Foundation is a small private foundation established in 1988. Organized as a California public benefit nonprofit corporation, it operates from San Diego, California. The foundation's current program interest areas have been funded throughout its almost two decades history. These funding areas are: K-12 science education; HIV/AIDS research, education and direct services; assistance to homeless persons; and child abuse prevention and services for victims and families. The board will occasionally fund a special project outside these categorical areas and/or for regional, national or international charitable purposes. A small portion of the grantmaking budget is provided for director matching grants to encourage and amplify personal philanthropy by the foundation's board of directors. Proposals will be considered in the above general areas or for special project areas established by the Board. Proposals will only be accepted for programs within San Diego County. Multiple year proposals may be considered but it is not the foundation's intention to provide annual support. It is not normally the foundation's desire to be the sole source of funds for a project. Most grants are expected to be small, $5,000-$15,000, reflecting the foundation's limited budget although there are no fixed minimum or maximum amounts. Grant proposals

received by March 15 will be considered for decision/funding in June. Grant proposals received by September 15 will be considered for decision/funding in December.
Requirements: Proposals will only be accepted for programs within San Diego County, California. All applicants should first submit a letter of inquiry, which can be submitted directly from the Foundation's web site. The letter (1-2 pages) must contain a brief statement describing the applicant and the need for funds with enough information for the foundation to determine whether or not the application falls within its program areas. Proposals submitted without an initial letter of inquiry will not be reviewed by the foundation. Applicants should briefly and clearly provide a statement of their needs and the specific request to the foundation, taking into account other possible sources of funding. Letters of inquiry will be acknowledged upon their receipt, but because the foundation operates without a professional staff, a more detailed response may be delayed. Applicants who receive a favorable response to their initial inquiry will be invited to submit a grant application. The foundation accepts the Common Grant Application of San Diego Grantmakers, which can be downloaded from the Foundation's web site. Five copies of the application (proposal) should be submitted. Only one copy of required attachments need be submitted. The application should be signed by the organization's board chair or the executive director (or equivalent individuals).
Restrictions: The foundation does not consider grants for individuals, scholarship funds, or sectarian religious activities. Normally the foundation does not consider requests for general fundraising drives. It does not make grants intended directly or indirectly to support political candidates or to influence legislation.
Geographic Focus: California
Date(s) Application is Due: Mar 15; Sep 15
Amount of Grant: 500-250,000 USD
Contact: Rachel McCarthy Bender; (858) 485-0129; mail@mccarthyfamilyfdn.org
Internet: http://www.mccarthyfamilyfdn.org/guide.html
Sponsor: McCarthy Family Foundation
P.O. Box 27389
San Diego, CA 92198-1389

McCombs Foundation Grants **2718**
The McCombs Foundation, established in 1981, makes grants to Texas nonprofit charitable, philanthropic, educational, and benevolent organizations. Areas of interest include arts, athletics, education and higher education, historic preservation, medical research, philanthropy, recreation and sports, youth services, and voluntarism. There are no application deadlines or forms. Applicants should submit a letter of application stating a brief history of the organization, any available printed support materials, and a budget detail. Final notification takes approximately two weeks.
Requirements: Texas nonprofits are eligible. Application form not required.
Restrictions: No grants are given to individuals.
Geographic Focus: Texas
Amount of Grant: 1,000-6,000,000 USD
Samples: University of Texas, Houston, Texas, $5,000,000-Anderson Cancer Center; Saint Mary's Hall, San Antonio, Texas, $300,000-general operations.
Contact: Gary Woods, Treasurer; (210) 821-6523
Sponsor: McCombs Foundation
755 East Mulberry, Suite 600
San Antonio, TX 78212

McCune Charitable Foundation Grants **2719**
The Marshall L. and Perrine D. McCune Charitable Foundation is dedicated to enriching the health, education, environment, and cultural and spiritual life of New Mexicans. The Foundation engages in proactive grantmaking that seeks to foster positive social change. Specifically, the Foundation funds projects that benefit New Mexico in the areas of arts, economic development, education, environment, health, and social services. It is working to stimulate economic diversity, nurture sustainability, and bridge the economic gaps that exist in our communities with the aim of creating wealth for all New Mexicans.
Requirements: Grants can be awarded to qualified 501(c)3 nonprofit organizations, federally recognized Indian tribes, public schools, and governmental agencies. Applications are available online via the Foundation's website.
Restrictions: Grants are not awarded to individuals or to support endowments.
Geographic Focus: New Mexico
Date(s) Application is Due: Sep 30
Amount of Grant: Up to 150,000 USD
Contact: Norty Kalishman; (505) 983-8300; fax (505) 983-7887; mccune@nmmccune.org
Internet: http://www.nmmccune.org/
Sponsor: McCune Charitable Foundation
345 East Alameda Street
Santa Fe, NM 87501

McCune Foundation Civic Grants **2720**
The McCune Foundation's grants are assigned to one of four program areas, including: education, human services, humanities, and civic. In the area of civic affairs, the Foundation focuses new job growth and generating community wealth, support of core community assets, and quality of life. Community development, public amenities, conservation, and overall regional grants are made in this program area. The civic agenda supports capital projects, research and development, organizational capacity building, and programming with the following strategic priorities: spur investment in Pittsburgh's urban core and metropolitan area; leverage public and private dollars for broad-based support of community assets; revitalize Pittsburgh's neighborhoods by changing blighted market conditions; support broad based community assets improving the livability of the region; and develop a New Economy in Pittsburgh that will return greater prosperity to the region.
Requirements: The foundation supports 501(c)3 organizations in southwestern Pennsylvania and throughout the country, with emphasis on the Pittsburgh area. This area includes the following counties: Allegheny, Beaver, Butler, Armstrong, Westmoreland, and Washington. To apply, an organization should send a brief (2 to 3 page) initial inquiry, preferably using the Foundation's website. The letter should contain: project overview-describe what the proposed efforts are intended to achieve for the region as well as for the organization; what activities/actions are planned to meet the stated goals; project timeline; resources required-total cost of the project; anticipated income, including private and public funders; amount of funding requested; IRS 501(c)3 determination letter-attach a copy (either scanned via email or hard copy via regular mail); and a copy of the organization's latest audit (either scanned via email or hard copy via regular mail).
Restrictions: Grants are not awarded to individuals or for general operating purposes or loans. Unsolicited proposals from outside the funding area are not accepted.
Geographic Focus: All States
Amount of Grant: 1,000-1,000,000 USD
Contact: Henry S. Beukema; (412) 644-8779; fax (412) 644-8059; info@mccune.org
Internet: http://www.mccune.org/foundation:Website,mccune,grants
Sponsor: McCune Foundation
750 Sixth PPG Place
Pittsburgh, PA 15222

McCune Foundation Education Grants **2721**
The McCune Foundation's grants are assigned to one of four program areas, including: education, human services, humanities, and civic. In the area of education, the Foundation promotes excellence in the institutions of higher education, independent elementary and secondary schools, and ancillary education programs (general education). The education program supports capital projects, research and development, financial aid, programming, and endowment to accomplish the following strategic priorities: leverage university research capacity to enhance the region's competitive advantage in economic development; use universities as anchors for community revitalization strategies; increase access to independent schools and higher education; recruit and retain exceptional faculty; and develop and deliver quality enrichment programs for school aged children, particularly with a focus on math, science and reading.
Requirements: The foundation supports 501(c)3 organizations in southwestern Pennsylvania and throughout the country, with emphasis on the Pittsburgh area. This area includes the following counties: Allegheny, Beaver, Butler, Armstrong, Westmoreland, and Washington. To apply, an organization should send a brief (2 to 3 page) initial inquiry, preferably using the Foundation's website. The letter should contain: project overview-describe what the proposed efforts are intended to achieve for the region as well as for the organization; what activities/actions are planned to meet the stated goals; project timeline; resources required-total cost of the project; anticipated income, including private and public funders; amount of funding requested; IRS 501(c)3 determination letter-attach a copy (either scanned via email or hard copy via regular mail); and a copy of the organization's latest audit (either scanned via email or hard copy via regular mail).
Restrictions: Grants are not awarded to individuals or for general operating purposes or loans. Unsolicited proposals from outside the funding area are not accepted.
Geographic Focus: All States
Amount of Grant: 1,000-1,000,000 USD
Contact: Henry S. Beukema; (412) 644-8779; fax (412) 644-8059; info@mccune.org
Internet: http://www.mccune.org/foundation:Website,mccune,grants
Sponsor: McCune Foundation
750 Sixth PPG Place
Pittsburgh, PA 15222

McCune Foundation Human Services Grants **2722**
The McCune Foundation's grants are assigned to one of four program areas, including: education, human services, humanities, and civic. In the area of human services, the Foundation focuses on health, social services, and community improvement. The interdisciplinary approach to grantmaking in this program area works with community-based and regional institutions to address pressing community needs by supporting the critical work of existing programs, as well as new initiatives and programs that seek to find and ameliorate the root causes of community distress and disinvestment. The human services area supports capital projects, research and development, organizational capacity building, and programming with the following strategic priorities: transfer education and research assets into economic opportunities for the region; leverage public and private dollars for broad-based support of community assets; promote self-sufficiency of residents; increase social and economic stability of communities and the region; and test and support effective prevention programs.
Requirements: The foundation supports 501(c)3 organizations in southwestern Pennsylvania and throughout the country, with emphasis on the Pittsburgh area. This area includes the following counties: Allegheny, Beaver, Butler, Armstrong, Westmoreland, and Washington. To apply, an organization should send a brief (2 to 3 page) initial inquiry, preferably using the Foundation's website. The letter should contain: project overview-describe what the proposed efforts are intended to achieve for the region as well as for the organization; what activities/actions are planned to meet the stated goals; project timeline; resources required-total cost of the project; anticipated income, including private and public funders; amount of funding requested; IRS 501(c)3 determination letter-attach a copy (either scanned via email or hard copy via regular mail); and a copy of the organization's latest audit (either scanned via email or hard copy via regular mail).
Restrictions: Grants are not awarded to individuals or for general operating purposes or loans. Unsolicited proposals from outside the funding area are not accepted.
Geographic Focus: All States
Amount of Grant: 1,000-1,000,000 USD

Contact: Henry S. Beukema; (412) 644-8779; fax (412) 644-8059; info@mccune.org
Internet: http://www.mccune.org/foundation:Website,mccune,grants
Sponsor: McCune Foundation
750 Sixth PPG Place
Pittsburgh, PA 15222

MCEAS Dissertation Fellowships **2723**
The center was established to marshal the vast scholarly and archival resources of 10 area institutions in an interdisciplinary research program focused on the history and culture of the United States to 1850, to increase scholarly and public awareness of the heritage of the Delaware Valley. At least six new fellows will be appointed to nine-month terms beginning September 1st, with a stipend of at least $18,000 payable in monthly installments. While a limited number of shorter-term awards may be made, all candidates should apply for a nine-month appointment. Fellows receive office space in the Center's magnificent new building on the University of Pennsylvania's historic campus and library, computer, and other privileges at the University. Limited travel funds for research are also available. While no teaching is required for most fellowships, all McNeil Center fellows are expected to be in residence in Philadelphia during the academic year and to participate regularly in the Center's program of seminars and other activities.
Requirements: Fellowships are available to advanced graduate students from any PhD-granting institution who are in the dissertation research or writing stage.
Geographic Focus: All States
Date(s) Application is Due: Mar 1
Amount of Grant: 18,000 USD
Contact: Daniel K. Richter, (215) 898-9251; drichter@history.upenn.edu
Internet: http://www.mceas.org/dissertationfellowships.htm
Sponsor: McNeil Center for Early American Studies
3355 Woodland Walk
Philadelphia, PA 19104-4531

McGraw-Hill Companies Community Grants **2724**
McGraw-Hill focuses its charitable efforts on its goal of financial capability for all. The Corporation partners with nonprofit organizations to help individuals gain the necessary knowledge to make smart savings, credit and spending decisions. McGraw-Hill also extends its support to organizations that share the Corporation's dedication to the arts and culture, education and health and human services. The company gives priority consideration to organizations and projects that: promote and support excellence in education and learning, with a primary emphasis on financial literacy; further financial literacy in the communities and markets where the many diverse businesses of The McGraw-Hill Companies operate; utilize unique applications of innovative and developing technologies; extend their reach globally; can be evaluated and can serve as models elsewhere; are staffed and administered by people with demonstrated competence and experience in their fields; have been determined tax-exempt 501(c)3 organizations and qualify as public charities under IRS rules, or are the local country-specific equivalent of a charitable nonprofit organization.
Requirements: Proposals must include the following: a program narrative (no more than three pages); a detailed organization overview; a concise description of the grant's purpose; and all required attachments. If the request is considered eligible, a meeting may be arranged with the Corporate Responsibility and Sustainability staff. On-site visits may also be made. Proposals are accepted at any time, with funding decisions made quarterly.
Restrictions: McGraw-Hill does not support courtesy advertising, pledges for a walk-a-thon, or similar activities. Funding is also not available for the following: institutions and agencies clearly outside the company's primary geographic concerns and interests; libraries and schools of higher education and K-12; political activities or organizations established to influence legislation; sectarian or religious organizations; member-based organizations, i.e., fraternities, labor, veterans, athletic, social clubs; individuals; endowment funds; or loans. Grants are not renewed automatically. Requests for renewed support must be submitted each year.
Geographic Focus: All States
Contact: Susan Wallman, Manager, Corporate Contributions; (212) 512-6480; fax (212) 512-3611; susan_wallman@mcgraw-hill.com
Internet: http://www.mcgraw-hill.com/site/cr/community/giving#section_2
Sponsor: McGraw-Hill Companies
1221 Avenue of the Americas, 20th Floor
New York, NY 10020-1095

McGregor Fund Grants **2725**
The fund makes grants to organizations in the following areas: human services—emphasis on activities in southeastern Michigan addressing emergency needs in housing, food, clothing, and other direct aid; education—early child education and development, colleges and universities in the Detroit metropolitan area, liberal arts colleges and universities in Michigan and Ohio; arts and culture in southeastern Michigan; health care—improving access to primary care for underserved populations in Detroit; and public benefit—civic and community organizations that improve the quality of life in southeastern Michigan. Types of support include project support, operating support, special projects, and capital support. Grant decisions are made by the board of trustees four times annually, in February, May, September, and November. Grant requests may be submitted at any time, but requests may take up to three months for staff review. Application guidelines are available online.
Requirements: Organizations in the metropolitan Detroit, MI area, are eligible. Requests will be considered from organizations located elsewhere for programs or projects that significantly benefit the metropolitan Detroit area (city of Detroit and Wayne, Oakland, and Macomb Counties).
Restrictions: The fund discourages proposals for student scholarships, travel, seminars, conferences, workshops, film or video projects, as well as disease-specific organizations and their local affiliates.
Geographic Focus: Michigan
Amount of Grant: 25,000-50,000 USD
Contact: C. David Campbell; (313) 963-3495; fax (313) 963-3512; info@mcgregfund.org
Internet: http://www.mcgregorfund.org/guideline.html
Sponsor: McGregor Fund
333 W Fort Street, Suite 2090
Detroit, MI 48226

McKnight Doctoral Fellowships **2726**
The fellowship is awarded only to African Americans who intend to seek the PhD degree in any field in the arts and sciences, mathematics, business, or engineering. The fellowship award includes tuition and a stipend for up to five years of support. Additional information and applications are available on the Web site.
Requirements: Applicants must be African American, U.S. citizens, and hold a minimum of a bachelor's degree from a regionally accredited college or university.
Geographic Focus: All States
Date(s) Application is Due: Jan 15
Amount of Grant: 17,000 USD
Contact: Program Officer; (813) 272-2772; fax (813) 272-2784; mdf@fl-educ-fd.org
Internet: http://www.fefonline.org/mdf.html
Sponsor: Florida Education Fund
201 E Kennedy Boulevard, Suite 1525
Tampa, FL 33602

McLean Contributionship Grants **2727**
Originally established in 1951 as The Bulletin Contributionship for charitable, educational and scientific purposes, the Contributionship became The McLean Contributionship on May 1, 1980 when the association of the McLean family with the Bulletin ended. Independent Publications, also owned by the McLean family, continues as the main financial supporter of the Contributionship. The Contributionship favors projects that stimulate a better understanding of the natural environment and encourage the preservation of its important features; encourage more compassionate and cost-effective care for the ill and aging in an atmosphere of dignity and self-respect; or promote education, medical, scientific, or on occasion, cultural developments that enhance quality of life. In addition, the Trustees from time to time support projects which motivate promising young people to assess and develop their talents despite social and economic obstacles or encourage those in newspaper and related fields to become more effective and responsible in helping people better understand how events in their communities and around the world affect them. The Trustees meet several times a year. The Contribution accepts and processes applications for grants throughout the year. Applications must be received at least six weeks before a meeting date. Interested organizations may view the annual-meeting schedule on the Application Procedure page at the grant website. The Contributionship accepts the common-grant-application form of Delaware Valley Grantmakers Association (link is provided on the Application Procedure web page). Application may also be made by letter (guidelines are included on the Application Procedure web page).
Requirements: Applicants must show evidence of tax-exempt status.
Restrictions: Geographic area is limited to the following locations: Greater Philadelphia area; Nashua, New Hampshire; Dubois Pennsylvania; and Central Florida. The Contributionship does not fund the costs or expenses of existing staff allocated to a project it is asked to support.
Geographic Focus: Florida, New Hampshire, Pennsylvania
Amount of Grant: 2,000-100,000 USD
Contact: Sandra McLean, Executive Director; (610) 527-6330; fax (610) 527-9733
Internet: http://fdncenter.org/grantmaker/mclean
Sponsor: McLean Contributionship
945 Haverford Road, Suite A
Bryn Mawr, PA 19010-3814

MDA Development Grant **2728**
The association will consider an application for a research grant from a candidate who may be a member of a research team in the laboratory of a senior investigator under whose guidance the applicant will be given flexibility to conduct a neuromuscular disease research project. Awards are for one to three years. Applications must be requested by the listed application deadlines. Guidelines are available online. Development grants are a maximum of $60,000 per year.
Requirements: To be eligible for a Development Grant, an applicant must: hold a Doctor of Medicine (M.D.), Doctor of Philosophy (Ph.D.), Doctor of Science (D.Sc.) or equivalent degree (i.e. D.O.); be a member of a research team at an appropriate institution; be qualified to conduct a program of original research under the supervision of a Principal Investigator; have an acceptable research plan for a specific disease in MDA's program; have access to institutional resources necessary to conduct the proposed research project; and have eighteen (18) months of post-doctoral research laboratory training at the time of application, but no more than 5 years (60 months). A pre-proposal form must be submitted through proposalCENTRAL to formally request an application for an MDA research grant.
Geographic Focus: All States
Date(s) Application is Due: Jan 15; Jul 15
Amount of Grant: Up to 60,000 USD
Contact: Grants Manager; (520) 529-2000; fax (520) 529-5454; grants@mdausa.org
Internet: http://www.mdausa.org/research/guidelines.html#developmentGrant
Sponsor: Muscular Dystrophy Association

3300 E. Sunrise Drive
Tucson, AZ 85718

MDA Neuromuscular Disease Research Grants **2729**

MDA supports research aimed at developing treatments for the muscular dystrophies and related diseases of the neuromuscular system. These are the muscular dystrophies (among which are Duchenne and Becker); motor neuron diseases (including ALS and SMA); the peripheral nerve disorders (CMT and Friedreich's ataxia); inflammatory myopathies; disorders of the neuromuscular junction; metabolic diseases of muscle as well as other myopathies. Proposals from applicants outside the United States will be considered for projects of highest priority to MDA and when, in addition to the applicant's having met the requirements noted above the applicant's country of residence may not have adequate sources of financial support for biomedical research. Funding levels for primary Research Grants are unlimited. Awards are for either one, two or three years for all grant types.
Requirements: To be eligible to apply for an MDA research grant, an applicant must: hold a Doctor of Medicine (M.D.), Doctor of Philosophy (Ph.D.), Doctor of Science (D.Sc.) or equivalent degree (i.e. D.O.); be a professional or faculty member (Professor, Associate Professor or Assistant Professor) at an appropriate educational, medical or research institution; be qualified to conduct and mentor a program of original research within their own laboratory; assume both administrative and financial responsibility for the grant; and have access to institutional resources necessary to conduct the proposed research project. A pre-proposal form must be submitted through proposalCENTRAL to formally request an application for an MDA research grant.
Geographic Focus: All States
Date(s) Application is Due: Jan 15; Jul 15
Contact: Grants Manager; (520) 529-2000; fax (520) 529-5454; grants@mdausa.org
Internet: http://static.mda.org/research/guidelines.html
Sponsor: Muscular Dystrophy Association
3300 E. Sunrise Drive
Tucson, AZ 85718

MDARD Specialty Crop Block Grant Program-Farm Bill **2730**

The Specialty Crop Block Grant Program-Farm Bill (SCBGP-FB) provides grants annually to assist in enhancing the competitiveness of specialty crops (fruits, vegetables, tree nuts, and nursery crops). Specialty crops are defined as fruits, vegetables, tree nuts, dried fruits, horticulture, and nursery crops (including floriculture). Examples of project areas that would qualify for funds include, but are not limited to: food safety; food security; nutrition; trade enhancement; education; research; promotion; marketing; plant health programs; "Buy Local" programs; increased consumption; enhanced innovation; improved efficiency of distribution system; environmental concerns and conservation; product development; and cooperative development. This is a cost reimbursement program.
Requirements: Eligible applicants must be non-profit organizations, local, state, and federal government entities, for-profit organizations, and universities. Organizations must be legal entities recognized by the IRS and must reside, and/or conduct their business or organization in Michigan and must be in good standing. Universities submitting applications for funding (research) must have the application submitted by the commodity group for which the research is being performed. Applications for grant funds should demonstrate how the project potentially impacts and produces measurable outcomes for the specialty crop industry rather than a single organization, institution, or individual. Multi-state organizations proposals are permitted.
Restrictions: Grant funds will not be awarded for projects that directly benefit a particular commercial product or provide a profit to a single organization, institution, or individual, as these projects do not enhance specialty crop industry competitiveness.
Geographic Focus: Michigan
Date(s) Application is Due: Apr 12
Contact: Nancy Nyquist; (517) 241-4381; fax (517) 335-0628; nyquistn@michigan.gov
Internet: http://www.michigan.gov/mdard/0,1607,7-125-1568_51684—-,00.html
Sponsor: Michigan Department of Agriculture and Rural Development
P.O. Box 30017
Lansing, MI 48909

MDEQ Beach Monitoring Grants-Inland Lakes **2731**

The Beach Monitoring Grants Program provides funding for more effective monitoring of public bathing beaches. With this funding the goals of the program are to: determine levels of bacteria in select bathing beach waters of the state; notify the public of the results; evaluate the effectiveness of state programs in preventing bacterial contamination of surface waters; and develop methods that can differentiate sources of bacterial contamination. Deadlines may vary from year to year based on funding priorities. Applicants will usually have 60 days to prepare and submit a grant application. Grant awards will be announced following department review. The department reserves 90 days for the review process. In the past, a total of $200,000 was available for grants of up to $25,000.
Requirements: Local county health departments, universities, and non-profit organizations are eligible to apply. The grants require a minimum of 25 percent local match (can include in-kind services). Grant applications are selected for funding by the department based on program priorities and the following criteria: location and frequency of beach use; history of beach monitoring; history of bacterial contamination; access to lab facilities to analyze samples promptly; ability to communicate results to the public in an efficient manner; ability to respond and take appropriate action in an event of beach contamination; proximity of the beach to known contamination source; availability of matching funds; innovativeness and feasibility of the proposed project; and, ability to reduce time delay between sampling and results.
Geographic Focus: Michigan
Amount of Grant: Up to 25,000 USD
Contact: Shannon Briggs, (517) 335-1214; briggss4@michigan.gov
Internet: http://www.michigan.gov/documents/deq/deq-essd-grantsloans-catalog_210643_7.pdf
Sponsor: Michigan Department of Environmental Quality
525 West Allegan Street
Lansing, MI 48909

MDEQ Local Water Quality Monitoring Grants **2732**

The water quality monitoring grants are meant to fund water quality monitoring activities to address local issues of concern and to identify new chemicals/issues that may be impacting the quality of Michigan's surface waters. Monitoring activities are limited to surface waters (i.e., rivers, streams, public access lakes, wetlands, the Great Lakes, etc.) and may include ambient chemical, biological, or physical monitoring activities, as well as the development of tools to help with the assessment of such data. Approximately $250,000 is available for water quality monitoring projects (non-beach) under the Clean Michigan Initiative-Clean Water Fund (CMI-CWF). Inland beach grants are available biennially. Disbursement of grant funds is done on a cost-reimbursement basis.
Requirements: Local units of government and nonprofit entities are eligible to receive grant funding. Nonprofit entities are those exempt from taxation under Section 501(c)3 of the Internal Revenue Code. Eligible entities generally include county, city, township, and village agencies, watershed and environmental action councils, universities, regional planning agencies, and other incorporated not-for-profit organizations. Applications must be received no later than 5:00 p.m. of the published deadline date. Eligible groups must have undergone a successful financial audit within the 24-month period immediately preceding the application; and, must not have had a grant revoked or terminated, or demonstrated an inability to manage a grant or meet the obligations in a project contract with the Department of Environmental Quality (DEQ), or its predecessor agencies, within the 24-month period immediately preceding the application. There is no maximum dollar amount for these grants. The dollar amount requested should be based upon what is needed to carry out the tasks identified in the project proposal. This grant requires a minimum 25 percent (%) match of public or private funds. Match may not include other CMI funds or Clean Water Act funds awarded as grants by the state. Project contracts cannot exceed two years.
Restrictions: Individuals, private lake associations (that are not tax exempt under Section 501(c)3 of the Internal Revenue Code), for-profit organizations, and federal and state agencies are not eligible to apply. Additionally, restoration efforts/education only projects and volunteer monitoring activities are ineligible for funding.
Geographic Focus: Michigan
Date(s) Application is Due: Mar 22
Contact: Dawn Roush, (517) 335-3307; roushd@michigan.gov
Internet: http://www.michigan.gov/deq/0,1607,7-135-3313_3686_3728—-,00.html
Sponsor: Michigan Department of Environmental Quality
525 West Allegan Street
Lansing, MI 48909

MDEQ Scrap Tire Market Development Grants **2733**

The primary purpose of the MDEQ Scrap Tire Market Development Grants program is to issue grants for projects that will result in the development of increased markets for scrap tires. Overall, the goal is to use available funding to promote increased markets for scrap tire material in order to ensure appropriate markets for this material exist in an effort to reduce the public health and environmental concerns associated with scrap tires. The applicant should submit a Letter of Interest (LOI) for approval prior to submitting a full application. LOIs are due by February 15, with full grants due by April 15.
Requirements: The applicant's project must demonstrate new or increased uses of scrap tires in manufactured products, such as modified asphalt, molded or extruded rubber products, or aggregate replacement materials; or the project must be for the research and development of methods to increase the use of scrap tires. The project must be located in Michigan, and the applicant must be registered to do business in Michigan. Scrap tires must be purchased from a Michigan processor or other Michigan generator of scrap tires. The applicant must submit documentation of the property owner's permission for the applicant to do business at that location. The applicant must not be in litigation with the state or any other public entity concerning compliance with the Part 169, Scrap Tires, or Part 115, Solid Waste Management, of the Natural Resources and Environmental Protection Act, 1994, PA 451, as amended. Typical projects must be completed within 18 months of the start date.
Restrictions: Grants are available for up to 50% of the cost of purchasing equipment, or for research and development costs. The applicant must commit to at least 50% of the total project amount, in cash. In-kind services are not allowed as part of the match, but will be considered during the application evaluation process.
Geographic Focus: Michigan
Date(s) Application is Due: Apr 15
Contact: Rhonda Oyer, (517) 373-4750; oyerr@michigan.gov
Internet: http://www.michigan.gov/documents/deq/deq-essd-grantsloans-catalog_210643_7.pdf
Sponsor: Michigan Department of Environmental Quality
525 West Allegan Street
Lansing, MI 48909

Mead Johnson Nutritionals Medical Education Grants **2734**

Mead Johnson Nutritionals has established a Grant program to support medical education and philanthropic initiatives. The program will consider supporting Independent Medical Education (IME) programs and activities designed to enhance the professional skills and knowledge of health care professionals involved in pediatrics, when the program is independently developed and conducted by a qualified medical education provider. Supported programs must be objective, balanced, and scientifically rigorous. Health care provider

educational programming can be given for: medical education programs; scientific conferences; creation of health care publications; and other educational and scientific activities. In addition to the U.S. Medical Education Grants, Mead Johnson also funds International Medical Education Grants. Organizations outside the U.S. that wish to request an Independent Medical Education (IME) grant should refer to Mead Johnson contact information within their country or contact Mead Johnson at meadjohnson.grants1@bms.com.
Requirements: Eligible recipients for medical education grants include: academic medical centers; accredited medical education companies; community health centers; hospitals; managed care organizations; medical/professional societies; medical universities; patient advocacy groups; pharmacies; and similar entities.
Restrictions: Grants may not be tied, in any way, to past, present, or future prescribing, purchasing, or recommending (including formulary recommendations) of any product.
Geographic Focus: All States
Contact: Administrator; (812) 429-7831; meadjohnson.grants1@bms.com
Internet: http://www.mjn.com/app/iwp/MJN/Content2.do?dm=mj&id=/MJN_Home2/mjnBtnSocialResponsibility2/mjnBtnMedicalEducationApplications&iwpst=B2C&ls=0&csred=1&r=3435836001
Sponsor: Mead Johnson Nutritionals
2400 West Lloyd Expressway
Evansville, IN 47721-0001

Mead Johnson Nutritionals Scholarships **2735**
Mead Johnson Nutritionals provides scholarship grants to medical students, residents, fellows, and other healthcare professionals in training to attend classes and major educational or scientific meetings of a national medical or professional association. The selection of the individuals to receive such support and the grant application must be made by the institution at which they are being trained or the organization hosting the event. Allow at least three months for the processing of your funding application. Independent Medical Education (IME) grant applications submitted less than three months prior to an event may be rejected, due to insufficient time to process the application.
Requirements: Eligible recipients for such grants: academic medical centers; hospitals; medical/professional societies; and medical universities.
Geographic Focus: All States
Contact: Administrator; (812) 429-7831; meadjohnson.grants1@bms.com
Internet: http://www.mjn.com/app/iwp/MJN/Content2.do?dm=mj&id=/MJN_Home2/mjnBtnSocialResponsibility2/mjnBtnMedicalEducationApplications/MedScholarships&iwpst=B2C&ls=0&csred=1&r=3435836794
Sponsor: Mead Johnson Nutritionals
2400 West Lloyd Expressway
Evansville, IN 47721-0001

Meadows Foundation Grants **2736**
The Meadows Foundation seeks to assist people and institutions of Texas improve the quality and circumstances of life for themselves and future generations. Grants are made in five areas of interest: arts and culture, human services, health, education, and civic and public affairs. Within those three areas, the Foundation seeks to fund programs that: improve educational outcomes of Texas students; preserve and enhance the natural environment; and address the detection and treatment of mental illness.Projects should show the ability to continue and expand after the grant period.
Requirements: Grants are made to qualified organizations in Texas or programs benefiting Texas residents. Applicants must prove one or more of the following conditions in their proposal: Foundation support would be vital or catalytic to a proposed project's success; the project is well planned and the agency has the capacity to execute the plan; and financial support from other sources exists to ensure that the project will be implemented and continue after the grant period. Applications can be found at the website and should be submitted online.
Restrictions: Grants are not made to: individuals; for church or seminary construction; annual fund-raising events or drives; biomedical research; out-of-state performances or competition expenses; single artistic events or performances; or professional conferences and symposia.
Geographic Focus: Texas
Contact: Cynthia Cass; (214) 826-9431 or (800) 826-9431; grants@mfi.org
Bruce H. Esterline, Vice President for Grants; (214) 826-9431
Internet: http://www.mfi.org
Sponsor: Meadows Foundation of Texas
3003 Swiss Avenue
Dallas, TX 75204-6090

MedImmune Charitable Grants **2737**
The program supports educational and charitable activities that ultimately improve patient healthcare and further understanding of disease states and treatment options within the community. While the sponsor will consider initiatives in many areas, priority is given to those in the following areas of interest: (a) health and science education for young people; (b) influenza; (c) immunology; (d) pediatric infectious disease; (e) pediatric respiratory disease.
Requirements: Eligible programs include: (a) Community outreach designed to increase public awareness (e.g, fundraising walk or run, nonprofit community hospital or hospital foundation, etc.); (b) Fellowships; (c) Independent educational/scientific programs, accredited and non-accredited; (d) Medical or Scientific education for Healthcare Professionals; (e) Scholarships; or, (f) Academic Chair Endowments. Applications must be completed online at https://web.medimmune.com/corporategrants/, and must be submitted at least 6 weeks prior to the program date or event.
Restrictions: Applications will not be considered from: (1) individuals, physicians, or physician practices; (2) retroactive support; (3) exhibit or display booths at conventions/conferences; (4) textbooks; (5) operating expenses; (6) fraternal, social, leisure, labor or political organizations; (7) corporate foundations; (8) public or private schools or scholarship funds, except where a grant is specific to a science education initiative.
Geographic Focus: All States
Contact: Grants Office; (866) 396-6235; corporatefunding@medimmune.com
Internet: http://www.medimmune.com/culture_grants.aspx
Sponsor: MedImmune
One MedImmune Way
Gaithersburg, MD 20878

Medtronic Foundation Fellowships **2738**
The U.S. based Medtronic Fellows Program works to prepare the next generation of scientific innovators. Through the Foundation's partnerships with preselected universities such as Johns Hopkins, Duke University, Stanford University, and Massachusetts Institute of Technology, it supports college and university scholars and fellows in science, engineering and health-related fields. Applicants should review the website for specific university participants and contact the individual universities for more information.
Geographic Focus: All States, United Kingdom
Contact: Kris Fortman; (763) 514-4000 or (800) 633-8766; kris.fortman@medtronic.com
Internet: http://www.medtronic.com/foundation/programs_msp.html
Sponsor: Medtronic Foundation
710 Medtronic Parkway
Minneapolis, MN 55432-5604

Mellon Post-Doctoral Fellowship in Philosophy **2739**
The Philosophy Department invites applications for a two-year post-doctoral fellowship. Applications from candidates who will have their Ph.D. by the start date or who have completed their Ph.D. within the last two years may apply. The areas of specialty and competence are open. The successful candidate will be appointed for a term of two years and will carry a teaching load of one course each semester. Mellon Fellows are included in Department activities as regular junior faculty members. They work closely with a faculty mentor; they participate in Williams' Project in Effective Teaching; and they receive feedback on pedagogical skills and teaching effectiveness through standard evaluation procedures. The baseline stipend is $42,000 per year plus an allowance of $2,250 for research and professional travel expenses. Full benefits including health and dental insurance are provided.
Requirements: Applicants must be citizens or permanent residents of the United States, or expect to pursue a teaching career in the United States.
Geographic Focus: All States
Contact: Steven Gerrard; (413) 497-4351; communication@williams.edu
Internet: http://dean-faculty.williams.edu/graduate-fellowships/mellon-post-doctoral-fellowship-in-philosophy/
Sponsor: Williams College
880 Main Street
Williamstown, MA 01267

Mellon Post-Doctoral Fellowship in Sustainable Agriculture **2740**
Williams College Program in Environmental Studies invites applications for a two-year post-doctoral fellowship in Sustainable Agriculture. Candidates who will have their Ph.D. by the start date or who have completed their Ph.D. within the last three years may apply. Candidates with training in the natural or social sciences who can contribute to the curriculum and vitality of an undergraduate program are sought. The position requires vision and initiative and a desire to join Williams' vibrant multidisciplinary community of environmental scholars, practitioners and activists. Relevant fields include the traditional disciplines of agriculture, biology, geography, anthropology, sociology, and economics. Applicants with interdisciplinary training in agrobiodiversity, soil science, agroecology, or food studies are encouraged. An international focus is highly desirable. Williams places a high value on undergraduate teaching; consequently, strong preference is given to applicants who can demonstrate a capacity for excellent teaching. The successful candidate will be appointed for a term of two years and will carry a teaching load of one course each semester. The baseline stipend is $42,000 per year plus an allowance of $2,250 for research and professional travel expenses. Full benefits including health and dental insurance are provided.
Requirements: Applicants may send a cover letter, curriculum vitae, a two-page statement of research project and teaching interests, and three letters of reference.
Geographic Focus: All States
Date(s) Application is Due: Dec 1
Contact: Jennifer L. French; (413) 497-4351; fax (413) 597-3553; jfrench@williams.edu
Internet: http://dean-faculty.williams.edu/graduate-fellowships/mellon-post-doctoral-fellowship-in-sustainable-agriculture/
Sponsor: Williams College
880 Main Street
Williamstown, MA 01267

Melville Charitable Trust Grants **2741**
The Melville Trust was organized in 1990 with the desire to understand the growing problem of homelessness and help eradicate it. The foundation decided to focus its study on Connecticut and to create programs that would alleviate and reduce the impact of homelessness on children, adults and communities. The trust also funds educational, research and advocacy initiatives in the state and on the national level.
Requirements: The Trust does not use a formal application form. Instead, applicants are asked to submit a concept paper of not more than three pages describing the project. If there is interest in pursuing the request after the concept paper has been reviewed, the applicant organization will be contacted to either submit a more formal proposal and/or provide more detailed information to the Board. Applications must be from an

organization that has been determined to be tax-exempt under section 501(c)3 of the Internal Revenue Code and not a private foundation under section 509(a) of the Code.
Restrictions: Grants will not be awarded for: direct support to individuals or scholarships; budget deficits; religious organizations for religious purposes; or general fund raising drives or events.
Geographic Focus: All States
Contact: Aimee Hendrigan; (617) 236-2244; ahendrigan@melvilletrust.org
Internet: http://www.melvilletrust.org/funding_guidelines.aspx
Sponsor: Melville Charitable Trust
240 Newbury Street, 2nd Floor
Boston, MA 02116-2580

Mericos Foundation Grants **2742**
The Mericos Foundation primarily awards grants to organizations based in California with emphasis on Santa Barbara. Fields of interest include aging; animals/wildlife; environment; arts and arts education; child development; children and youth; elementary and secondary education; higher education; hospitals; medical care; medical research; libraries; and museums (art and natural history). Types of support include building/renovation, equipment, fellowships, general/operating support, matching/challenge support, and program development. Grants are usually initiated by the foundation. Interested organizations should contact the Vice-President to discuss their project.
Requirements: Nonprofit organizations in California are eligible to apply.
Restrictions: Grants are not made to individuals.
Geographic Focus: California
Amount of Grant: 15,000-200,000 USD
Contact: Linda Blinkenberg, Vice President; (626) 441-5188; fax (626) 441-3672
Sponsor: Mericos Foundation
625 South Fair Oaks Avenue, Suite 360
South Pasadena, CA 91030-2630

Mesothelioma Applied Research Foundation Grants **2743**
The program seeks to stimulate translational research for the treatment of pleural or peritoneal mesothelioma. Grants support projects that apply cutting-edge general cancer research concepts specifically to mesothelioma, and translate directly to improved mesothelioma treatment. Eligible projects may relate to either benchwork research or clinical research, must not be presently funded or pending review, and may be conducted through any not-for-profit academic, medical, or research institution, in the United States or abroad. Grants are awarded for 12 months, renewable for 12 more months.
Requirements: The investigator must be affiliated with a nonprofit academic, medical, or research institution.
Geographic Focus: All States
Date(s) Application is Due: Aug 15
Amount of Grant: 50,000 USD
Contact: Danele Mangone; (805) 563-8400, ext. 7273; dmangone@curemeso.org
Internet: http://www.curemeso.org/site/c.kkLUJ7MPKtH/b.4097447/k.E6A3/Mesothelioma_Research_Grants__Call_for_Applications.htm
Sponsor: Mesothelioma Applied Research Foundation
3944 State Street, Suite 340, P.O. Box 91840
Santa Barbara, CA 93105

Metanexus Institute Templeton Research Lectures **2744**
The goal of the program is to promote the constructive engagement and original research between the physical, biological, and human sciences and those modes of inquiry and understanding generally found within the domains of theology, religious studies, and philosophy. This is a three-year program, with the possibility of a fourth-year extension, that supports scholars who want to develop on-going, interdisciplinary programs within their universities. The program seeks to create vibrant and long-term networks of scholars engaged in vigorous dialog, careful research, and significant publication. The grant offers an honorarium for an individual scholar selected each year of the program to give the annual Templeton Research Lectures at the host institution. This is an opportunity to bring a distinguished scholar from another institution for sustained interaction or possibly in specialized situations the elevation of an internal faculty to give the annual lectures. These lectures must result in a book- length publication whose target audience is either the academic community or the educated layperson. Guidelines and application procedures are available online.
Requirements: Applications for the grant should be submitted by an interdisciplinary committee of scholars employed by their host university and forming the core of a governing committee to oversee the program. One person should be identified as both the chair of the group and the contact person for the grant applicant.
Geographic Focus: All States
Date(s) Application is Due: Jan 1
Amount of Grant: 90,000-130,000 USD
Contact: Dr. William Grassie, (215) 789-2200; fax (215) 789-2222; lectures@metanexus.net
Internet: http://www.metanexus.net/lectures/request/overview.html
Sponsor: Metanexus Institute on Religion and Science
3624 Market Street, Suite 301
Philadelphia, PA 19104

MetLife/TCG Aha! Think it Do it Grants **2745**
The program supports creative thinking and action in two ways: Think It grants (up to $25,000) give theater professionals the time and space for research and development, Do It grants (up to $50,000) support the implementation and testing of new ideas. The intent of the program is to enable theaters to dare to try new approaches to problem-solving artistic, managerial, production and/or technological challenges to try things the organization doesn't and couldn't normally do. Proposals from organizations seeking to test an idea that is new to them, but not to the field will be accepted. Emphasis is on identifying strategies that will be major catalysts for change to the organization, rather than strategies which will bring minimal change.
Requirements: To be eligible, the applying theater must be a member theater of TCG in good standing and may apply in only one category. The Project Leader must be a key artistic, administrative or production staff member who holds a salaried position at the theater. Theaters receiving THINK IT grants are eligible to apply for DO IT grants in a future round, provided that their original grant was completed. Preference will be given to projects that break from the organization's previous practice and have the best chance at long-term, field-wide impact over a longer span of time.
Restrictions: Strategies and projects that have already been tested or implemented at the applicant organization are not eligible.
Geographic Focus: All States
Date(s) Application is Due: May 15
Amount of Grant: 25,000-50,000 USD
Contact: Sarah McLellan, Artistic Programs Associate
Internet: http://www.tcg.org/grants/aha/aha_guidelines.cfm
Sponsor: Theatre Communications Group
520 8th Avenue, 24th Floor
New York, NY 10018-4156

Meyer and Pepa Gold Family Foundation Grants **2746**
Established in 1986 and formerly known as the Gold Family Foundation, the revamped Meyer and Pepa Gold Family Foundation offers grant support in New Jersey and New York. Its primary fields of interest include the arts, health organizations, and medical research agencies. The are no specified application forms required or annual deadlines. Interested parties should begin by contacting the Foundation directly, offering an overview of their program and budgetary needs. Recent grants, which predominately support general operations, range up to $35,000.
Requirements: Any 501(c)3 located in, or serving the residents of, New Jersey and/or New York are eligible to apply.
Restrictions: No grants are offered to individuals.
Geographic Focus: New Jersey, New York
Amount of Grant: Up to 35,000 USD
Contact: Pepa Gold, Trustee; (732) 229-0569 or (732) 660-1770
Sponsor: Meyer and Pepa Gold Family Foundation
P.O. Box 777
Oakhurst, NJ 07755

Meyer and Stephanie Eglin Foundation Grants **2747**
The Eglin Foundation supports programs in the Philadelphia area. Fields of interest include: cancer research; higher education; hospitals; Jewish agencies, synagogues, and federated giving programs; museums; music and the performing arts; and United Way programs. Applications are not required, and there are no deadlines. Organizations should submit a letter of intent.
Requirements: Along with the letter of intent, organizations should submit an annual report/audited financial statement/990; a 501(c)3 tax-exempt letter; descriptive literature about their organization; and a detailed description of the project, with amount requested.
Geographic Focus: Pennsylvania
Contact: Stephanie Eglin, President; (215) 496-9381
Sponsor: Meyer and Stephanie Eglin Foundation
Eglin Square Garage
Philadelphia, PA 19102

Meyer Memorial Trust Responsive Grants **2748**
Responsive Grants are awarded in the areas of human services; health; affordable housing; community development; conservation and environment; public affairs; arts and culture; and education. Funding ranges from $40,000 to $300,000, with grants periods from one to three years in length. Responsive grants help support many kinds of projects, including core operating support, building and renovating facilities, and strengthening organizations. There are two stages of consideration before Responsive Grants are awarded. Initial Inquiries are accepted at any time through MMT's online grants application. Applicants that pass initial approval are invited to submit full proposals. The full two-step proposal investigation usually takes five to seven months. Final decisions on Responsive Grants are made by trustees monthly, except in January, April and August. Additional information about the application process, along with the online application, is available at the website.
Requirements: Support is for 501(c)3 nonprofits in Oregon and Clark County, Washington.
Restrictions: Funding is not available for sectarian or religious organizations for religious purposes, or for animal welfare organizations, or projects that primarily benefit students of a single K-12 school (unless the school is an independent alternative school primarily serving low-income and/or special needs populations). Funding is also not available for individuals or for endowment funds, annual campaigns, general fund drives, special events, sponsorships, direct replacement funding for activities previously supported by federal, state, or local public sources, deficit financing, acquisition of land for conservation purposes (except through Program Related Investments), or hospital capital construction projects (except through Program Related Investments).
Geographic Focus: Oregon, Washington
Amount of Grant: 40,000-300,000 USD
Contact: Maddelyn High, Grants Administrator; (503) 228-5512; maddelyn@mmt.org
Internet: http://www.mmt.org/program/responsive-grants

Sponsor: Meyer Memorial Trust
425 NW 10th Avenue, Suite 400
Portland, OR 97209

MfA Newton Fellowships **2749**
The fellowships are available to mathematically sophisticated individuals who are interested in using their talents to make a difference in the lives of young people. The program seeks college graduates and mid-career professionals who can demonstrate a strong interest in teaching, have a bachelor's degree with substantial coursework in mathematics, and are willing to commit to a five-year program that includes one year of full-time graduate study and four years of teaching mathematics in New York City high schools. Fellows receive a stipend, full tuition to master's in education program, mentoring and coaching, and professional development. Applicants are required to take the ETS Praxis II Math subject tests. Application and guidelines are available online.
Requirements: U.S. citizens and permanent residents are eligible.
Restrictions: Individuals who are currently teaching, are certified to teach, or have completed an education degree program are not eligible.
Geographic Focus: New York
Date(s) Application is Due: Feb 10
Amount of Grant: 90,000 USD
Contact: Fellowship Administrator; (212) 206-0053; fax (212) 206-1622; info@mathforamerica.org
Internet: http://www.mathforamerica.org
Sponsor: Math for America
101 Fifth Avenue, 5th Floor
New York, NY 10003

MfA Newton Master Teacher Fellowships **2750**
The fellowship is a four-year award available to outstanding teachers of mathematics in the New York City public high schools. The program's purpose is to recognize mathematics teachers who demonstrate solid mathematical knowledge and outstanding performance records. The program provides opportunities for professional development and collaboration with other teachers. Guidelines are available online.
Requirements: Candidates must be current New York City public high school math teachers, hold permanent certification in mathematics from New York State, and possess a license to teach in New York City.
Geographic Focus: New York
Date(s) Application is Due: Mar 24
Amount of Grant: 12,500-50,000 USD
Contact: Fellowship Administrator; (212) 206-0053; fax (212) 206-1622; info@mathforamerica.org
Internet: http://www.mathforamerica.org
Sponsor: Math for America
101 Fifth Avenue, 5th Floor
New York, NY 10003

MGFA Post-Doctoral Research Fellowships **2751**
The Myasthenia Gravis Foundation of America, Inc. announces a competitive twelve-month post-doctoral research fellowship for clinical or basic research pertinent to myasthenia gravis or related neuromuscular disorders. Research may be concerned with neuromuscular transmission, immunology, molecular or cell biology of the neuromuscular synapse, or the etiology/pathology or diagnosis of myasthenia gravis. See the Foundations website for application and, additional guidelines.
Requirements: The applicant must be: a permanent resident of the United States or Canada who has been accepted to work in the laboratory of an established investigator at an institution in the United States, Canada or abroad deemed appropriate by the Medical/Scientific Advisory Board of the Myasthenia Gravis Foundation of America, Inc; or a foreign national who has been accepted to work in the laboratory of an established investigator at an institution in the United States or Canada deemed appropriate by the Medical/Scientific Advisory Board of the Myasthenia Gravis Foundation of America, Inc.
Geographic Focus: All States
Date(s) Application is Due: Oct 1
Amount of Grant: Up to 50,000 USD
Contact: Tor Holtan, Chief Executive; (800) 541-5454 or (212) 297-2156; fax (212) 370-9047; Tor.Holtan@myasthenia.org or mgfa@myasthenia.org
Internet: http://www.myasthenia.org/hp_fellowships.cfm#nursing
Sponsor: Myasthenia Gravis Foundation of America
355 Lexington Avenue, 15th Floor
New York, NY 10017

MGFA Student Fellowships **2752**
These fellowships are awarded annually to current medical students or graduate students interested in the scientific basis of myasthenia gravis or related neuromuscular conditions, serving both to further scientific inquiries into the nature of these disorders and to encourage more research. The stipend is up to $5,000. See the Foundation's website for application form and guidelines.
Requirements: The fellowships are awarded to current medical students or graduate students. Briefly describe, in abstract form, the question that you propose to study, its association to myasthenia gravis or related neuromuscular conditions, and how you will approach the project. Applicants and sponsoring institutions must comply with policies governing the protection of human subjects, the humane care of laboratory animals and the inclusion of minorities in study populations.
Geographic Focus: All States
Date(s) Application is Due: Mar 31
Contact: Tor Holtan, Chief Executive; (800) 541-5454 or (212) 297-2156; fax (212) 370-9047; Tor.Holtan@myasthenia.org or mgfa@myasthenia.org
Internet: http://www.myasthenia.org/hp_fellowships.cfm#nursing
Sponsor: Myasthenia Gravis Foundation of America
355 Lexington Avenue, 15th Floor
New York, NY 10017

MGM Resorts Foundation Community Grants **2753**
The MGM Resorts Foundation invites proposals from nonprofit agencies providing direct services to people living in its communities. All of the funds allocated through the Foundation come from employee contributions and their desire to make a difference in the communities where they live and work. Foundation grant allocations are 100% employee-driven. The Foundation empowers MGM Resorts employees to choose to make direct contributions to the agency of their choice, or to contribute to the Community Grant Funds, which provides grants to nonprofits through an annual Request for Proposal (RFP) process. MGM Resorts Foundation grants are for a one-year period and do not automatically renew. Continued or expanded projects and programs (your organization is currently providing these services) can request a per year maximum grant of $65,000 in Southern Nevada; $10,000 in Northern Nevada, Mississippi and the Detroit, Michigan area. New projects and programs (your organization is not currently providing these services) can request a per year maximum grant of $35,000 in Southern Nevada; $5,000 in Northern Nevada, Mississippi and the Detroit, Michigan area.
Requirements: To receive a grant from the Foundation, your agency must meet the following
Requirements: Operate as an IRS 501(c)3 organization and have been doing so for a minimum of 36 months; provide service within the regions MGM Resorts employees live, work and care for their families (Nevada, Mississippi, and the greater Detroit, Michigan area); your organization's administrative costs must be 25% or under; provide a human service; and, meet the MGM Resorts diversity policy: open to all people, without regard to race, color, creed, sex, sexual orientation, religion, disability, or national origin. Agencies must request funding for projects/programs that provide services in the following focus areas: Strengthening Neighborhoods (self-sufficiency, revitalization of communities); Strengthening Children (early childhood development, success in school, prevention / intervention); and, Strength in Difficult Times (recovery and counseling services). Proposals must be received by the Foundation by 5:00 pm of the deadline date.
Restrictions: The program does not support the following types of organizations or activities: projects/programs that are exclusively for medical research; public schools or privately funded / tuition-based schools; governmental entities; religious organizations that do not have 501(c)3 status; pass-through agencies (organizations whose staff does not provide direct client services but who allocate funding to subsequent organizations to provide projects/programs and services); sponsorship of special events and/or fundraising activities; capital campaigns or endowment funds; political issues, such as, election campaigns, issue endorsements, bill drafts or legislation reform; organizations that require clients to embrace specific beliefs or traditions; projects/programs that are exclusively recreational or athletic sponsorships; membership-based organizations without a sliding fee scale and scholarship system already in place.
Geographic Focus: Michigan, Mississippi, Nevada
Date(s) Application is Due: May 2
Amount of Grant: 5,000-65,000 USD
Contact: Shelley Gitomer, Vice President of Philanthropy & Community Engagement; (702) 692-9643; foundation@mgmresorts.com
Internet: http://www.mgmmirage.com/csr/community/foundation.aspx
Sponsor: MGM Resorts Foundation
3260 Industrial Road
Las Vegas, NV 89109

MGN Family Foundation Grants **2754**
The MGN Family Foundation makes grants to qualified 501(c)3 organizations specializing in the following areas: education; health care and medical research; children in need; armed service personnel. Areas of particular interest include: colleges, universities and private schools, examples would be: to fund a chair to provide lecturers in literature, philosophy or the arts, and provide scholarships; hospitals and clinics that specialize in excellent patient care and continuing medical research such as Memorial Sloan-Kettering Cancer Center and the Mayo and Cleveland Clinics are other examples, also under consideration would be hospice organizations that provide palliative care to the dying; organizations that support children in need whether due to emotional, physical abuse, neglect or disadvantaged circumstances. Support can be for basic necessities such as food, shelter, education, medical as well as, spiritual and emotional counseling; in light of recent events, this Foundation would like to offer assistance to our servicemen/ servicewomen and their families through those charities which give support to their unique needs. The examples provided above reflect the true mission and goals of the MGN Foundation. There should be no exclusions due to race or creed, provided all applicants have a strong moral base and core values in the areas of education, health care, welfare of children and service personnel and their families. The Foundation Board meets semi-annually in May and November. Applications are due by April 1st and October 1st. Application form is available online.
Requirements: Qualified 501(c)3 organizations specializing in the following areas: education; health care and medical research; children in need; armed service personnel. To apply, submit seven (7) sets of the following items: Grant application form completed, dated, and signed by the Chief Executive Officer or Chairman of the Board of the organization; list of Board of Directors; Financial Statement (audited if available), for the most recent complete fiscal year; copy of IRS 501(c)3 Determination Letter. If you wish, you may submit your proposal in a narrative format of not more than two pages in addition

to the completed application form. Optional materials may be submitted but are not required (such as brochures discussing or depicting the activities of the organization).
Restrictions: Do not staple materials or place them in a bound notebook.
Geographic Focus: All States
Date(s) Application is Due: Apr 1; Oct 1
Amount of Grant: 1,000-10,000 USD
Contact: Pamela Nothstein; (843) 937-4614; grantinquiries7@wachovia.com
Internet: https://www.wachovia.com/foundation/v/index.jsp?vgnextoid=61078689fb0a a110VgnVCM1000004b0d1872RCRD&vgnextfmt=default
Sponsor: MGN Family Foundation
16 Broad Street (SC1000)
Charleston, SC 29401

MHS Long Term Research Fellowships 2755

The Society will award two long-term grants of six to twelve months, although either fellowship may be divided into two awards of a maximum of five months. The stipend, governed by an NEH formula, will be no more than $40,000 for a term of six to twelve months and smaller amounts for shorter terms. Tenure must be continuous. Within the constraints of the NEH's guidelines, the Society will also supplement each stipend with a housing allowance of up to $500 per month.
Requirements: MHS-NEH fellowships are open to U.S. citizens and to foreign nationals who have lived in the United States for at least the three years immediately preceding the application deadline. Applicants must have completed their professional training.
Restrictions: NEH-sponsored fellowships are not available to graduate students.
Geographic Focus: All States
Date(s) Application is Due: Jan 15
Amount of Grant: 43,000-46,000 USD
Contact: Jane Becker, (617) 646-0518 or (617) 536-1608; fellowships@masshist.org
Internet: http://www.masshist.org/fellowships/long_term.cfm
Sponsor: Massachusetts Historical Society
1154 Boylston Street
Boston, MA 02215

MHS New England Regional Fellowships 2756

The New England Regional Fellowship Consortium, a collaboration of 18 major cultural agencies, provides a stipend for eight weeks of research at participating institutions. Participants include Baker Library, Harvard Business School; Baker Library; Boston Athenaeum; Bostonian Society; Colonial Society of Massachusetts; Connecticut Historical Society; Francis A. Countway Library of Medicine; Harvard Law School, Special Collections; Historic Deerfield; Houghton Library, Harvard University; Maine Historical Society; Massachusetts Historical Society; Mystic Seaport; New England Historic Genealogical Society; New Hampshire Historical Society; Rhode Island Historical Society; Schlesinger Library; and Vermont Historical Society. The grants are designed to encourage projects that draw on the resources of several agencies. Each award will be for research at a minimum of three different institutions. Fellows must work at each of these organizations for at least two weeks. Grants are for June of one year through May 31 of the next. Application and guidelines are available online.
Requirements: Applications are welcome from anyone with a serious need to use the collections and facilities of the organizations.
Geographic Focus: All States
Date(s) Application is Due: Feb 1
Amount of Grant: 5,000 USD
Contact: Jane Becker, (617) 646-0518 or (617) 536-1608; fellowships@masshist.org
Internet: http://www.nerfc.org/
Sponsor: Massachusetts Historical Society
1154 Boylston Street
Boston, MA 02215

MHS Short Term Research Fellowships 2757

The society offers approximately 20 fellowships each year for researchers who need to use the collections to complete a major project. Fellowships for the current cycle include: African American Studies Fellowship; Boston Marine Society Fellowship; W.B.H. Dowse Fellowships; Marc Friedlaender Fellowship; Massachusetts Society of the Cincinnati Fellowship; Andrew W. Mellon Fellowships; Ruth R. Miller Fellowship; Andrew Oliver Research Fellowship; Paul Revere Memorial Association Fellowship; and Benjamin F. Stevens Fellowship. Each fellowship includes a stipend for four weeks of research, which must be conducted in the society's archives sometime between July 1 of one year June 30 of the next. Candidates who live 50 or more miles from Boston receive preference. Guidelines and application are available online.
Requirements: Recipients must be U.S. citizens or foreign nationals holding the appropriate U.S. government documents. Typical applicants include but are not limited to independent scholars, advanced graduate students, and college and university faculty in all disciplines.
Geographic Focus: All States
Date(s) Application is Due: Mar 1
Amount of Grant: 1,500 USD
Contact: Jane Becker, (617) 646-0518 or (617) 536-1608; fellowships@masshist.org
Internet: http://www.masshist.org/fellowships/short_term.cfm
Sponsor: Massachusetts Historical Society
1154 Boylston Street
Boston, MA 02215

MHS Suzanne and Caleb Loring Research Fellowship 2758

The Massachusetts Historical Society and the Boston Athenaeum will award one Suzanne and Caleb Loring Fellowship on the Civil War, Its Origins, and Consequences this year. The recipient will conduct research for at least four weeks at each institution. The Athenaeum's Civil War collections are anchored by its holdings of Confederate states imprints, the largest in the nation, consisting of books, maps, broadsides, sheet music, government documental publications, and other materials organized according to the Parrish & Willingham bibliography. The Society's manuscript holdings on the Civil War are particularly strong. They include, for instance, diaries, photographs, correspondence from the battlefield and the home front, papers of political leaders, materials on black regiments raised in Massachusetts, and extensive holdings on the U.S. Sanitary Commission. The Athenaeum and the Society are especially interested in projects for which both repositories' resources are vital. The fellowship carries a stipend of $4,000. Each institution will automatically refer unsuccessful proposals to its short-term fellowship competition.
Geographic Focus: All States
Date(s) Application is Due: Feb 15
Amount of Grant: 4,000 USD
Contact: Jane Becker, (617) 646-0518 or (617) 536-1608; fellowships@masshist.org
Internet: http://www.masshist.org/fellowships/loring.cfm
Sponsor: Massachusetts Historical Society
1154 Boylston Street
Boston, MA 02215

MHS Teacher Fellowships 2759

The Massachusetts Historical Society (MHS) is offering fellowships to public and/or parochial/independent schoolteachers and library media specialists during the summer months. The fellowships carry a stipend of $4,000 for four weeks of on-site research at the MHS. Two distinct fellowships are available: the Adams Teacher Fellowship—projects that make use of the Society's Adams Family Papers; and the Swensrud Teacher Fellowship—any project for which the Society's collections offer sufficient primary sources.
Requirements: Applications are welcome from any K-12 teacher who has a serious interest in using the collections at the MHS to prepare primary-source-based curricula, supported by documents and visual aids, in the fields of American history, world history, or English/language arts.
Geographic Focus: All States
Date(s) Application is Due: Mar 5
Amount of Grant: 4,000 USD
Contact: Jane Becker, (617) 646-0518 or (617) 536-1608; fellowships@masshist.org
Internet: http://www.masshist.org/fellowships/teacher_fellowships.cfm
Sponsor: Massachusetts Historical Society
1154 Boylston Street
Boston, MA 02215

Michael J. Fox Foundation Clinical Discovery Grants 2760

The goal of this program is to encourage translational research by stimulating well-designed clinical research projects focused on potentially high-impact approaches to the field of Parkinson's disease. Through the program, researchers will be invited to submit grant applications to conduct new or novel clinical research in humans relevant to the cure, cause, prevention, or improved treatment of Parkinson's disease and its complications. The program will support a variety of innovative research studies that are directly related to patients or patient populations affected by the disease. Appliciants may request a project period of up to three years. The foundation employs an expedited, abbreviated granting mechanism to accelerate discoveries and hopefully translate findings to Parkinson's disease patients more quickly. A mandatory letter of intent is due on the listed application deadline; full applications are by invitation.
Requirements: Both U.S. and non-U.S. organizations are eligible. Applications may be submitted by public and private nonprofit entities, units of state and local governments, eligible agencies of the federal government, and for- profit entities. Collaborative efforts are welcome.
Restrictions: Postdoctoral students or fellows are not eligible to apply as principal investigators.
Geographic Focus: All States
Date(s) Application is Due: Nov 21
Contact: Grants Administrator; research@michaeljfox.org
Internet: http://www.michaeljfox.org/research/index.php
Sponsor: Michael J. Fox Foundation for Parkinson's Research
P.O. Box 4777, Grand Central Station
New York, NY 10163

Michael J. Fox Foundation for Parkinson's Research LEAPS Program 2761

Investigators are invited to apply for funding through a novel mechanism called LEAPS (Linked Efforts to Accelerate Parkinson's Solutions). LEAPS are multi-year, potentially multi-million dollar grant awards for collaborative, milestone-driven efforts to address questions that will have significant practical impact on the understanding and treatment of Parkinson's disease. LEAPS grants are goal-oriented and focused on answering a broad research question, one whose answer would represent a major, practical step forward in Parkinson's disease—a solution to a Parkinson's problem. Examples of such questions could be categorized under broad aims of: repairing the PD brain; slowing or halting the progression of PD; preventing or controlling levodopa-induced dyskinesias; and identifying the cause of PD. LEAPS are not intended to be conventional research grants. LEAPS investigators will be asked to interact regularly with foundation staff and advisors to discuss elements of the project, research directions, progress against stated milestones,

and possible ways (other than through additional financial resources) the foundation may offer support or assistance. The RFP and preproposal cover sheet are available online.
Requirements: Applications may be submitted by U.S. and non-U.S. entities; public and private nonprofit entities, such as universities, colleges, hospitals, laboratories, units of state and local governments, and eligible agencies of the federal government; and for-profit entities.
Restrictions: Postdoctoral students or fellows are not eligible to apply as coordinating or principal investigators.
Geographic Focus: All States
Date(s) Application is Due: Apr 29
Contact: Grants Administrator; (800) 708-7644
Internet: http://www.michaeljfox.org
Sponsor: Michael J. Fox Foundation for Parkinson's Research
P.O. Box 4777, Grand Central Station
New York, NY 10163

Michael J. Fox Foundation Parkinson's Cell Line Research Grants **2762**
Researchers are invited to submit investigator-initiated proposals using embryonic stem cells targeted at achieving optimized transplantation materials and protocols promoting graft survival, stability of dopaminergic phenotype, and long-term functional recovery in animal models of Parkinson's disease. The intent is to use the acquired knowledge to develop cell replacement therapy for PD patients. Work in animal or human cell lines is acceptable; however, a primary focus on human cell lines is required, and research on human embryonic stem cells is considered appropriate. Mandatory letters of intent are due August 9. Guidelines are available online.
Requirements: Investigators with scientific interests and demonstrated success in the development, culture, and manipulation of individual cell lines are invited to apply.
Geographic Focus: All States
Date(s) Application is Due: Aug 9; Sep 20
Samples: Ernest Arenas of the Karolinska Institute (Stockholm Sweden), Jun Takahashi of the Kyoto U Graduate School of Medicine (Japan), Xuejun Wen of Clemson U (SC), and Su-Chung Zhang of the U of Wisconsin (Madison, WI)—for stem-cell research related to finding a cure for Parkinson's disease, $1 million divided.
Contact: Grants Administrator; (800) 708-7644; research@michaeljfox.org
Internet: http://www.michaeljfox.org/research/fundingopps.php?id=9
Sponsor: Michael J. Fox Foundation for Parkinson's Research
P.O. Box 4777, Grand Central Station
New York, NY 10163

Microscopy Society of America Student Scholarships **2763**
The scholarship program is designed to further the interest in microscopy and facilitate information exchange for students working in the field of microscopy. Recipient must attend the annual meeting and present a research paper at the regular scientific session.
Requirements: Recipient must be a full-time student at a recognized college or university at the time of the annual meeting of MSA, must submit an abstract of research or project, and must be sponsored with a signed statement by a member of MSA, preferably a research advisor.
Geographic Focus: All States
Date(s) Application is Due: Dec 15
Amount of Grant: Up to 3,000 USD
Contact: Business Office; (800) 538-3672 or (312) 644-1527; fax (312) 644-8557
Internet: http://www.microscopy.org
Sponsor: Microscopy Society of America
230 E Ohio Street, Suite 400
Chicago, IL 60611

Microsoft Research A. Richard Newton Breakthrough Research Awards **2764**
Microsoft invites proposals for breakthrough research projects. Breakthrough, in this sense, refers to projects which clearly leverage innovative computational techniques and advantages at the heart of the research, and which may integrate approaches from multiple disciplines. Additionally, projects considered breakthrough demonstrate potentially high impact by solving problems of great importance or by inventing promising new approaches for enabling their solution. The total amount available under this request for proposals (RFP) is U.S. $1,000,000. Microsoft Research anticipates making approximately ten to twelve awards, with a maximum of $100,000 for any single award. Awards are made for the purpose of seed-funding larger initiatives, proofs of concept, or demonstrations of feasibility. It is important to understand that funding will continue after the first year only in exceptional circumstances, and that the principal investigators should therefore make every effort to leverage Microsoft Research funds as one component of a diverse funding base in a larger or longer-running project.
Requirements: All qualifying institutions are eligible without regard for geographic location. The proposing institution must be either: an accredited degree-granting college or university (or international equivalent) with non-profit status and awarding degrees at the baccalaureate level or above; or a research institution with non-profit status.
Restrictions: For all awards, payment of indirect costs (overhead) is not permitted.
Geographic Focus: All States
Date(s) Application is Due: Dec 3
Amount of Grant: Up to 100,000 USD
Contact: Rick Rashid; (800) 642-7676; fax (425) 936-7329; erpinq@microsoft.com
Internet: http://research.microsoft.com/ur/us/fundingopps/rfps/ARichardNewtonAward.aspx
Sponsor: Microsoft Research
One Microsoft Way
Redmond, WA 98052-6399

Microsoft Research Beyond Search Semantic Computing and Internet Economics Grants **2765**
Microsoft would like to enable cross-disciplinary research in the area of semantic computing, Internet economics and online advertising, and is looking for innovative proposals which explore the technical, research, societal, and commercial issues around the topics of semantic computing and Internet economics. The total amount available under this request for proposals (RFP) is $1,000,000. Microsoft Research anticipates making approximately 20 awards averaging $50,000, with a maximum of $100,000 for any single award. Awards are made for the purpose of seed-funding larger initiatives, proofs of concept, or demonstrations of feasibility.
Requirements: All qualifying institutions are eligible without regard for geographic location. The proposing institution must be either: an accredited degree-granting college or university (or international equivalent) with non-profit status and awarding degrees at the baccalaureate level or above; or a research institution with non-profit status.
Restrictions: For all awards, payment of indirect costs (overhead) is not permitted.
Geographic Focus: All States
Date(s) Application is Due: Nov 1
Amount of Grant: Up to 100,000 USD
Contact: Rick Rashid; (800) 642-7676; fax (425) 936-7329; erpinq@microsoft.com
Internet: http://research.microsoft.com/ur/us/fundingopps/rfps/BeyondSearch_RFP.aspx
Sponsor: Microsoft Research
One Microsoft Way
Redmond, WA 98052-6399

Microsoft Research Biomedical Computing for Genome Wide Association Studies Grants **2766**
Microsoft Research encourages researchers to develop tools that can facilitate better data usage and analysis for genomewide association studies to provide a stronger framework for enabling personalized treatment methods. It anticipates making approximately 8-10 awards averaging $70,000, with a maximum of $100,000 for any single award. Awards are made for the purpose of seed-funding larger initiatives, proofs of concept, or demonstrations of feasibility. It is important to understand that funding will continue after the first year only in exceptional circumstances, and that the principal investigators should therefore make every effort to leverage Microsoft Research funds as one component of a diverse funding base in a larger or longer-running project.
Requirements: All qualifying institutions are eligible without regard for geographic location. The proposing institution must be either: an accredited degree-granting college or university (or international equivalent) with non-profit status and awarding degrees at the baccalaureate level or above; or a research institution with non-profit status.
Restrictions: For all awards, payment of indirect costs (overhead) is not permitted.
Geographic Focus: All States
Amount of Grant: Up to 100,000 USD
Contact: Rick Rashid; (800) 642-7676; fax (425) 936-7329; erpinq@microsoft.com
Internet: http://www.microsoft.com/presspass/press/2007/jul07/07-16FacultySummitPR.mspx
Sponsor: Microsoft Research
One Microsoft Way
Redmond, WA 98052-6399

Microsoft Research Cell Phone as a Platform for Healthcare Grants **2767**
Microsoft Research will support selected academic research with the goal of advancing the state-of-the-art in smart cell phones for health care applications. Successful projects must seek to increase the capacity of cell phone-based health care solutions in under-served rural and urban communities. They must also take into consideration the social context of application deployment. Topics to consider include: medical applications that are relevant, worldwide, for smart mobile phones (application and Web-enabled) in rural, and urban, communities; and appropriate services and infrastructures needed to provide affordable and accessible health care services. The total amount available under this request for proposals (RFP) is $1,000,000. Microsoft Research anticipates making approximately 10-12 awards averaging $80,000, with a maximum of $100,000 for any single award. Awards are made for the purpose of seed-funding larger initiatives, proofs of concept, or demonstrations of feasibility. It is important to understand that funding will continue after the first year only in exceptional circumstances, and that the principal investigators should therefore make every effort to leverage Microsoft Research funds as one component of a diverse funding base in a larger or longer-running project.
Requirements: All qualifying institutions are eligible without regard for geographic location. The proposing institution must be either: an accredited degree-granting college or university (or international equivalent) with non-profit status and awarding degrees at the baccalaureate level or above; or a research institution with non-profit status.
Restrictions: For all awards, payment of indirect costs (overhead) is not permitted.
Geographic Focus: All States
Date(s) Application is Due: Oct 29
Amount of Grant: Up to 100,000 USD
Contact: Rick Rashid; (800) 642-7676; fax (425) 936-7329; erpinq@microsoft.com
Internet: http://research.microsoft.com/en-us/um/redmond/about/collaboration/awards/cellphone-healthcare_awards.aspx
Sponsor: Microsoft Research
One Microsoft Way
Redmond, WA 98052-6399

Microsoft Research Human-Robot Interaction (HRI) Grants **2768**
Microsoft Research encourages research to take human-robot interaction to the next level through development of tools and methods that lead to practical applications

with realistic commercial potential within five to 10 years. It anticipates making approximately 8-10 awards averaging $50,000, with a maximum of $100,000 for any single award. Awards are made for the purpose of seed-funding larger initiatives, proofs of concept, or demonstrations of feasibility. It is important to understand that funding will continue after the first year only in exceptional circumstances, and that the principal investigators should therefore make every effort to leverage Microsoft Research funds as one component of a diverse funding base in a larger or longer-running project.
Requirements: All qualifying institutions are eligible without regard for geographic location. The proposing institution must be either: an accredited degree-granting college or university (or international equivalent) with non-profit status and awarding degrees at the baccalaureate level or above; or a research institution with non-profit status.
Restrictions: For all awards, payment of indirect costs (overhead) is not permitted.
Geographic Focus: All States
Amount of Grant: Up to 100,000 USD
Contact: Rick Rashid; (800) 642-7676; fax (425) 936-7329; erpinq@microsoft.com
Internet: http://www.microsoft.com/presspass/press/2007/jul07/07-16FacultySummitPR.mspx
Sponsor: Microsoft Research
One Microsoft Way
Redmond, WA 98052-6399

Microsoft Research Intelligent Web 3.0 Grants **2769**

This Program encourages research to help find, discover, extract, publish, and share information, at a desk or on the go, safely, making the Web meaningful (from string manipulation to meaning computation) and enabling a human-centric, context-aware model of information access. It anticipates making approximately 8-10 awards averaging $50,000, with a maximum of $100,000 for any single award. Awards are made for the purpose of seed-funding larger initiatives, proofs of concept, or demonstrations of feasibility. It is important to understand that funding will continue after the first year only in exceptional circumstances, and that the principal investigators should therefore make every effort to leverage Microsoft Research funds as one component of a diverse funding base in a larger or longer-running project.
Requirements: All qualifying institutions are eligible without regard for geographic location. The proposing institution must be either: an accredited degree-granting college or university (or international equivalent) with non-profit status and awarding degrees at the baccalaureate level or above; or a research institution with non-profit status.
Restrictions: For all awards, payment of indirect costs (overhead) is not permitted.
Geographic Focus: All States
Amount of Grant: Up to 100,000 USD
Contact: Rick Rashid; (800) 642-7676; fax (425) 936-7329; erpinq@microsoft.com
Internet: http://www.microsoft.com/presspass/press/2007/jul07/07-16FacultySummitPR.mspx
Sponsor: Microsoft Research
One Microsoft Way
Redmond, WA 98052-6399

Microsoft Research Mechanisms for Safe and Scalable Multicore Computing Grants **2770**

Microsoft Research will encourage research in how operating systems and runtimes can evolve to enable safe and scalable concurrent programs. It anticipates making approximately 8-10 awards averaging $50,000, with a maximum of $100,000 for any single award. Awards are made for the purpose of seed-funding larger initiatives, proofs of concept, or demonstrations of feasibility. It is important to understand that funding will continue after the first year only in exceptional circumstances, and that the principal investigators should therefore make every effort to leverage Microsoft Research funds as one component of a diverse funding base in a larger or longer-running project.
Requirements: All qualifying institutions are eligible without regard for geographic location. The proposing institution must be either: an accredited degree-granting college or university (or international equivalent) with non-profit status and awarding degrees at the baccalaureate level or above; or a research institution with non-profit status.
Restrictions: For all awards, payment of indirect costs (overhead) is not permitted.
Geographic Focus: All States
Amount of Grant: Up to 100,000 USD
Contact: Rick Rashid; (800) 642-7676; fax (425) 936-7329; erpinq@microsoft.com
Internet: http://www.microsoft.com/presspass/press/2007/jul07/07-16FacultySummitPR.mspx
Sponsor: Microsoft Research
One Microsoft Way
Redmond, WA 98052-6399

Microsoft Research New Faculty Fellowship Program **2771**

The program seeks nominees who are advancing computing research in novel directions with the potential for high impact, and who demonstrate the likelihood of becoming thought leaders of the field. The program identifies, recognizes, and supports exceptional new faculty members engaged in innovative computing research. The objective is to stimulate and support creative research by promising researchers who have the potential of making a profound impact to the state of the art in their research discipline. Examples of research areas may include, but are not limited to, security, scientific computing, interdisciplinary research, bioinformatics, computational biology, and software engineering. Each fellowship award includes a cash gift as well as Microsoft resources. The awardees will have the option of exploring collaborations with MSR researchers. Guidelines are available online.
Requirements: The potential fellow must be nominated by the university provost. No more than one nomination from the university will be accepted. The new faculty member must hold a tenure-track faculty position in a U.S. research university. The applicant must be in the first, second, or third year as a new faculty member in his/her first faculty appointment. Five letters of recommendation are required from established researchers familiar with the applicant's research; no more than two letters of recommendation can come from within the nominating institution. Applications may only be submitted through the online application system.
Restrictions: Direct applications are not accepted from a new faculty candidate. The applicant must be no more than six years from the completion of his/her most recent PhD.
Geographic Focus: All States
Date(s) Application is Due: Nov 19
Amount of Grant: 200,000 USD
Samples: Magdalena Balazinska, University of Washington, $200,000; Josh Bongard, University of Vermont, $200,000; Adam Siepel, Cornell University, $200,000.
Contact: Fellowship Program Director; (800) 642-7676
Internet: http://research.microsoft.com/ur/us/nff/default.aspx
Sponsor: Microsoft Research
One Microsoft Way
Redmond, WA 98052-6399

Microsoft Research Sustainable Computing Grants **2772**

Microsoft Research encourages research in innovative approaches toward power-optimized system architectures, and adaptive power management solutions for maximizing the energy efficiency of computing infrastructure. It anticipates making approximately 8-10 awards averaging $50,000, with a maximum of $100,000 for any single award. Awards are made for the purpose of seed-funding larger initiatives, proofs of concept, or demonstrations of feasibility. It is important to understand that funding will continue after the first year only in exceptional circumstances, and that the principal investigators should therefore make every effort to leverage Microsoft Research funds as one component of a diverse funding base in a larger or longer-running project.
Requirements: All qualifying institutions are eligible without regard for geographic location. The proposing institution must be either: an accredited degree-granting college or university (or international equivalent) with non-profit status and awarding degrees at the baccalaureate level or above; or a research institution with non-profit status.
Restrictions: For all awards, payment of indirect costs (overhead) is not permitted.
Geographic Focus: All States
Amount of Grant: Up to 100,000 USD
Contact: Rick Rashid; (800) 642-7676; fax (425) 936-7329; erpinq@microsoft.com
Internet: http://www.microsoft.com/presspass/press/2007/jul07/07-16FacultySummitPR.mspx
Sponsor: Microsoft Research
One Microsoft Way
Redmond, WA 98052-6399

Miles of Hope Breast Cancer Foundation Grants **2773**

The Fund provides programs and support services for women and families in the Hudson Valley, New York, affected by breast cancer. All funds raised for the Foundation are used in the Hudson Valley. The Miles of Hope Breast Cancer Foundation is currently offering grants for projects in the areas of breast health and breast cancer education, outreach, screening, treatment and support projects.
Requirements: Services must be provided in the eight counties served by the Miles of Hope Breast Cancer Foundation, which include: Columbia, Dutchess, Greene, Orange, Putnam, Rockland, Westchester and Ulster.
Restrictions: Project must be specific to breast health and/or breast cancer. Applicants must be a U.S. nonprofit (federally tax-exempt) organization. Nonprofit organizations, known educational institutions (i.e. college or university), government agencies, and Indian tribes are eligible. Equipment costs, if applicable, may not exceed 30% of direct costs and should be used exclusively on this project. Salaries, if requested, are for personnel related to this project only and not the general work of the employee. Funds will not be awarded to capital campaigns
Geographic Focus: New York
Contact: Administrator; (845) 527-6884 or (845) 264-2005; info@milesofhopebcf.org
Internet: http://www.milesofhopebcf.org/funds.html#grants
Sponsor: Miles of Hope Breast Cancer Foundation
P.O. Box 405
La Grangeville, NY 12540

Miller Brewing Corporate Contributions Grants **2774**

The corporate contributions program funds programs of nonprofit organizations in corporate operating communities. Applicants should call the local Miller community affairs manager to discuss an idea before submitting a proposal. The corporation also supports the Thurgood Marshall Scholarship Fund, which provides four-year merit scholarships to students attending the 44 historically black public colleges and universities, and two-year scholarships to law schools at Texas Southern, North Carolina Central, Southern, and Howard Universities; and ?Adelante! U.S. Education Leadership Fund, which awards scholarships to Hispanic-American students.
Requirements: Nonprofits must be near Miller operating locations, in Irwindale, CA; Albany, GA; Trenton, OH; Eden, NC; Fort Worth, TX; and Milwaukee, WI.
Geographic Focus: California
Amount of Grant: 1,000-15,000 USD
Contact: Grants Administrator; (414) 931-3110; fax (414) 931-6352
Internet: http://www.millerbrewing.com/inthecommunity/default.asp
Sponsor: Miller Brewing Corporation
3939 W Highland Boulevard
Milwaukee, WI 53201-0482

Millipore Foundation Grants **2775**

The foundation supports nonprofit organizations in company operating communities in five main areas: education and research, social services, culture, health care, and public policy. Types of support include general operating support, employee matching gifts, and employee-related scholarships. Applications are accepted throughout the year.
Requirements: 501(c)3 tax-exempt organizations are eligible.
Geographic Focus: All States
Amount of Grant: 2,500-5,000 USD
Contact: Charleen Johnson; (978) 715-1268; Charleen_Johnson@millipore.com
Internet: http://www.millipore.com/corporate/milliporefoundation.nsf/foundationhome
Sponsor: Millipore Foundation
290 Concord Road
Billerica, MA 01821-7037

Millstone Foundation Grants 2776
The foundation awards grants to eligible Missouri nonprofit organizations in its areas of interest, including education and higher education, Jewish services, and Israel. Types of support include annual campaings, continuing support, deficit financing, emergency funds, general operating support, and research grants. There are no application forms or deadlines.
Requirements: Missouri tax-exempt organizations are eligible. Preference is given to requests from Saint Louis.
Geographic Focus: Missouri
Contact: Colleen Millstone, Director; (314) 961-8500
Sponsor: Millstone Foundation
7701 Forsyth Boulevard, Suite 925
Saint Louis, MO 63105-1842

Mineralogical Society of America Research Grants 2777
This is a grant for research in crystallography. The award selection will be based on the qualifications of the applicant, the quality, innovativeness, and scientific significance of the proposed research, and the likelihood of success of the project. There are no restrictions on how the grant funds may be spent, as long as they are used in support of research.
Requirements: The applicant must have reached his or her 25th birthday but not yet have reached his or her 36th birthday on the date the grant is awarded, and that the person is not a MSA Counsellor.
Geographic Focus: All States
Date(s) Application is Due: Jun 1
Contact: J. Alex Speer; (703) 652-9950; fax (703) 652-9951; j_a_speer@minsocam.org
Internet: http://www.minsocam.org/MSA/Awards/Crystallography_Award.html
Sponsor: Mineralogical Society of America
1015 18th Street NW, Suite 601
Washington, DC 20036-5212

Minerva Fellowships 2778
The program enables German and Israeli scientists and researchers to spend longer-term research residencies at institutions in the host country from six to thirty-six months (for graduates) or from six to twenty-four months (for post docs). Some fifty scientists receive fellowships each year. Young scientists and researchers (post docs) are given priority in the selection process. In addition to the monthly stipend, dependency, children and travel allowance will be given along with payment in kind for books, travel to symposia, etc.
Requirements: For the graduate fellowship, eligible candidates must have completed a university degree ('Diplom', MA, M.Sc., or equivalent qualification). For the post-doctoral fellowship applications will be accepted from candidates who provide proof that they have submitted their doctoral thesis prior to the start of the fellowship year, or who have completed their doctorate no more than seven years prior to the beginning of the fellowship. Maternal leave will be taken into account. Applicants who are presently residing in Israel/Germany but are not nationals of either of the two countries must show a proven record of integration into the Israeli or German scientific community. An affiliation to a research institution in Israel or Germany is required together with a residency of 5 years or longer in either of the two countries. Conversely, if applicants are citizens of one country but have been living in the other for an extended period of time (more than 4 years) a scholarship in the second country cannot be granted since it would not truly support German-Israeli exchange. A questionnaire and list of required application documents are available at the website.
Restrictions: Scientists with tenure are not eligible to apply. BA students are not eligible to apply.
Geographic Focus: All States, Germany, Israel
Date(s) Application is Due: Jun 15
Contact: Frau Sieglinde; +49-(0)89-2108-1242; reichardt@gv.mpg.de
Internet: http://www.minerva.mpg.de/fellowships/fellowships.html
Sponsor: Minerva Foundation
HofgartenstraBe 8
Munich, D-80539 Germany

Minerva Short-Term Research Grants 2779
The primary goal of grants is to give young German and Israeli scientists the opportunity to visit research partners or to establish contacts with scientists in the respective guest country. The grants also allow the scientists to participate in seminars and workshops, particularly as guest speakers. The short-term grants are for stays of a minimum of 1 week and up to 8 weeks. Grants are available for all subject areas and are open to applicants from all research facilities (in Israel-public or governmental research institutes) and universities in Germany and Israel. Graduate students will receive EUR 300/week, while Postdocs will receive EUR 425/week. Flights are covered by a lump sum of up to EUR 700.
Requirements: Applicants should not be older than 38 years. (Maternal leave will be taken into account with regard to the age limit). Application forms may be downloaded from http://www.minerva.mpg.de. Both German and Israeli scientists must submit their applications to the Minerva Office in Munich. In addition, Israeli applicants are requested to send copies of their applications by electronic mail to the Israeli Ministry of Science (nurit@most.gov.il). Applications must be written in English and sent by electronic mail. Selection criteria are: scientific excellence; scientifically promising and innovative research topics; relevance of the research topic to German-Israeli cooperation.
Restrictions: Scientists presently holding a Minerva Fellowship are not eligible to apply. In addition, principal investigators in projects within the German Federal Ministry for Education and Research and the Israeli Ministry of Science and Technology programs are not eligible to apply. Only complete applications and applications sent by electronic mail will be processed.
Geographic Focus: All States, Germany, Israel
Date(s) Application is Due: May 1
Contact: Michael Nagel, Head Office Contact; 49 89 2108 1258; nagel@gv.mpg.de
Internet: http://www.minerva.mpg.de/fellowships/research.html
Sponsor: Minerva Foundation
HofgartenstraBe 8
Munich, D-80539 Germany

Minnesota War Orphans Educational Assistance Grants 2780
This program serves children of Minnesota veterans who died while on active duty military service or as a result of injuries or disease incurred while on active duty military service. Funding may be used for vocational/occupational, community college, undergraduate, graduate, professional, or postdoctoral education. Funds cover tuition, fees, room and board, or books and supplies. Awards are renewable. An application form must be obtained from the VA, which includes necessary instructions for completing the form and supporting documents.
Requirements: Applicant must be resident of Minnesota for at least two years prior to application; the deceased parent-veteran must have been a Minnesota resident at time of entry to active duty service.
Restrictions: Scholarships may not be used at the University of Minnesota.
Geographic Focus: Minnesota
Amount of Grant: 750 USD
Contact: Service Officer; (612) 296-2562; fax (612) 296-3954; lee.peterson@state.mn.us
Internet: http://www.mdva.state.mn.us/education.htm
Sponsor: Minnesota State Department of Veterans Affairs
20 W 12th Street, Veterans Service Building
Saint Paul, MN 55155-2079

Minority Faculty Fellowship 2781
Indiana University (IU) annually seeks minorities for participation in this fellowship program. The purpose of the program is to introduce to the Bloomington campus outstanding African American, Hispanic American, and Native American scholars. Each year summer fellows are invited to teach one class during the first (six-week) or second (eight-week) summer session, and academic-year fellows teach one or more classes during the fall and spring semesters. This program provides fellows the opportunity to interact with the Bloomington faculty and to conduct research in the IU libraries and academic facilities. In the past, some IU departments and schools have offered fellows permanent tenure-track positions.
Requirements: U.S. citizens or permanent residents within one year of completion of the doctorate and those who have completed the PhD within the last four years are eligible to apply.
Geographic Focus: All States
Date(s) Application is Due: Nov 30
Amount of Grant: 4,000 USD
Contact: Rebecca Jones; (812) 855-0542; fax (812) 856-5477; mffp@indiana.edu
Internet: http://www.iub.edu/~shs/ffp_overview.shtml
Sponsor: Indiana University
1021 E Third Street, Memorial Hall W 108
Bloomington, IN 47405-7005

MIT Knight Science Journalism Fellowships 2782
The fellowships provide nine-month academic residency at the Massachusetts Institute of Technology, auditing courses, visiting laboratories, and attending twice-weekly seminars with leading scientists. The goal of the program is to increase exposure of the fellows to front-rank researchers and to deepen the fellows' knowledge of current issues in science and technology. The annual online application deadline is March 1. Fellows are selected by a panel of judges-usually five or six judges-after they send a fully completed application.
Requirements: Knight Science Journalism Fellowships are for mid-career journalists who demonstrate a high level of journalistic excellence and accomplishment, as well as a long-term commitment to their craft. Journalists from all countries compete on an equal basis.
Restrictions: Part-time writers or producers are not eligible.
Geographic Focus: All States, All Countries
Date(s) Application is Due: Mar 1
Amount of Grant: 55,000 USD
Contact: Bianca Sinausky; (617) 253-3442; fax (617) 258-8100; singleta@mit.edu
Internet: http://ksj.mit.edu/fellowships
Sponsor: Massachusetts Institute of Technology
77 Massachusetts Avenue, Room E19-623
Cambridge, MA 02139-4307

MLA Cunningham Memorial International Fellowship 2783
The Medical Library Association (MLA) Cunningham Memorial International Fellowship was established in 1967 with a bequest from the estate of Eileen R. Cunningham. The fellowship supports the education and training of a health-science

librarian from a country outside the United States or Canada and includes a stipend plus approved travel for four months in either country. The fellowship also pays for the cost of attending MLA's Annual Meeting. Guidelines and application materials are available at the website. All materials and letters of support must be received by MLA by December 1. Founded in 1898, the Medical Library Association (MLA) is a nonprofit, educational organization of more than 1,100 institutions and 3,600 individual members in the health sciences information field, committed to educating health-information professionals, supporting health-information research, promoting access to the world's health-sciences information, and working to ensure that the best health information is available to all.
Requirements: Candidates must have both an undergraduate degree and a master's-level library degree (latter requirement may be waived) and must either be working in or preparing to work in a medical library in their country. The fellowship recipient must agree to return to his or her country and work in a medical library for a period of two years. In addition, a satisfactory score must be achieved on the TOEFL English competency examination.
Restrictions: Fellowship is not available to U.S. or Canadian citizens. Travel arrangements to and from the United States or Canada must be paid by the fellow. Previous recipients of the Cunningham fellowship are ineligible.
Geographic Focus: All Countries
Date(s) Application is Due: Dec 1
Contact: Carla Funk, Program Contact; (312) 419-9094, ext. 14; fax (312) 419-8950; mlapd2@mlahq.org or grants@mlahq.org
Internet: http://www.mlanet.org/awards/grants/cunningham.html
Sponsor: Medical Library Association
65 East Wacker Place, Suite 1900
Chicago, IL 60601-7246

MLA David A. Kronick Traveling Fellowship **2784**

Established in 2002, the David A. Kronick Traveling Fellowship grants one $2,000 award each year to cover the expenses involved in traveling to three or more medical libraries in the United States or Canada, for the purpose of studying a specific aspect of health information management. Activities that are operational in nature or have only local usefulness will be considered. Guidelines and application materials are available at the Medical Library Association (MLA) website. Applications and supporting documents must be received by MLA via email, fax, or mail by December 1. email applications and documents will be accepted as PDF or MS Word files only. Founded in 1898, the Medical Library Association (MLA) is a nonprofit, educational organization of more than 1,100 institutions and 3,600 individual members in the health sciences information field committed to educating health-information professionals, supporting health-information research, promoting access to the world's health-sciences information, and working to ensure that the best health information is available to all.
Requirements: Eligible candidates should satisfy the following *Requirements:* hold a graduate degree in library science; currently be a practicing health-sciences librarian with at least five years of professional experience; be an individual member of MLA; and be a citizen of or permanent resident in either the United States or Canada. Consideration will be given in exceptional cases to an outstanding candidate not meeting the criteria above. The award winner is required to submit a report to MLA on the results of the project, that may be published in appropriate health information journals.
Geographic Focus: All States
Date(s) Application is Due: Dec 1
Amount of Grant: 2,000 USD
Contact: Carla Funk, Executive Director; (312) 419-9094, ext. 14; fax (312) 419-8950; mlapd2@mlahq.org or grants@mlahq.org
Internet: http://www.mlanet.org/awards/grants/index.html
Sponsor: Medical Library Association
65 East Wacker Place, Suite 1900
Chicago, IL 60601-7246

MLA Donald A.B. Lindberg Research Fellowship **2785**

The Medical Library Association (MLA) believes that access to high-quality information improves decision-making by health professionals, scientists, and consumers and is a major determinant in improved health and quality of care both nationally and internationally. MLA awards a $10,000 fellowship annually to support an individual whose research will extend the underlying knowledge-base of health-sciences information management or enhance the practice of the information professions, particularly health-sciences librarianship. The MLA's areas of interest include the organization, delivery, and use of information and knowledge and their impact on health-care access and delivery, public-health services, consumers' use of health information, biomedical research, and education for the health professions. The Lindberg Fellowship is an unrestricted grant that is awarded to the applicant, not to the sponsoring organization, and may be used for salaries, supplies, equipment, travel, fees, insurance, salaries for research assistants, and other research-related costs as specified in the grant application. The award may be used to supplement or extend other awards, including other private or government-supported fellowships, but is not contingent on receiving other awards. The award is not restricted to disbursement in a single year; funding may be disbursed over a period of up to five years depending on the needs of the research fellow. The application form and guidelines are available at the website. The application and supporting documents must be received by MLA via email, fax, or mail by November 15. Applications and documents sent via email will be accepted as PDF or MS Word files only. The Lindberg Research Fellowship is named in honor of Donald A.B. Lindberg, M.D., Director of the National Library of Medicine (NLM), in recognition of his significant national and international achievements at the NLM and its National Center for Biotechnology Information. The Lindberg Research Fellow program of MLA is administered through the Lindberg Research Fellow Jury.
Requirements: Health-sciences librarians, health professionals, researchers, educators, and administrators are eligible to apply. Applicants must be sponsored by an institution or organization; one or a combination of the following institutions or organizations may sponsor applicants: MLA's board, committees, sections, and chapters; graduate schools of library and information science; university deans or department chairs in health-professions schools or health-care organizations; library organizations; or scientific academies and societies. Applicants must have a bachelor's, master's, or doctor's degree or be enrolled in a program leading to such a degree and demonstrate a commitment to the health sciences; additionally applicants must be a citizen or permanent resident of either the United States or Canada, or have been lawfully admitted for permanent residence at the time of appointment.
Restrictions: The grant may not be used for institutional overhead, other indirect costs, income tax payments, or tuition. Acceptance of the grant may be subject to institutional rules and regulations and to all applicable tax laws. The fellowship is not designed to support research for a doctoral dissertation or master's thesis. An applicant may resubmit an application to reapply for the Lindberg Fellowship for up to a two-year period.
Geographic Focus: All States, Canada
Date(s) Application is Due: Nov 15
Amount of Grant: 10,000 USD
Contact: Carla Funk, Executive Director; (312) 419-9094, ext. 14; fax (312) 419-8950; mlapd2@mlahq.org or grants@mlahq.org
Internet: http://www.mlanet.org/awards/grants/index.html
Sponsor: Medical Library Association
65 East Wacker Place, Suite 1900
Chicago, IL 60601-7246

MLA Graduate Scholarship **2786**

The Medical Library Association (MLA) annually awards a $5,000 scholarship to a student who shows excellence in scholarship and potential for accomplishment in health-sciences librarianship. The scholarship is announced at the annual meeting of the association, where the recipient will also receive a one-year student membership in the Medical Library Association and free inclusive student registration at the association's annual meeting. The recipient will assume the other costs of attending the meeting and the ceremony at which the presentation is made. Guidelines and application materials are available at the website. All materials and letters of support must be received by MLA via email, fax, or mail by December 1. email applications and documents will be accepted as PDF or MS Word files only. The recipient will be notified in March before the annual meeting. Founded in 1898, the Medical Library Association is a nonprofit, educational organization of more than 1,100 institutions and 3,600 individual members in the health sciences information field, committed to educating health-information professionals, supporting health-information research, promoting access to the world's health-sciences information, and working to ensure that the best health information is available to all.
Requirements: To be eligible applicants must be either citizens or permanent residents of the United States or Canada. Additionally, applicants must either be entering or have completed no more than one-half of the academic requirements of a Masters program at an American Library Association (ALA)-accredited graduate school of library science at the time of the granting.
Restrictions: Past recipients of the MLA Scholarship or the MLA Scholarship for Minority Students are not eligible to apply.
Geographic Focus: All States, Canada
Date(s) Application is Due: Dec 1
Amount of Grant: Up to 5,000 USD
Contact: Carla Funk, Executive Director; (312) 419-9094, ext. 14; fax (312) 419-8950; mlapd2@mlahq.org or grants@mlahq.org
Internet: http://www.mlanet.org/awards/grants/scholar.html
Sponsor: Medical Library Association
65 East Wacker Place, Suite 1900
Chicago, IL 60601-7246

MLA Graduate Scholarship for Minority Students **2787**

The Medical Library Association (MLA) annually awards a $5,000 scholarship to a student who shows excellence in scholarship and potential for accomplishment in health-sciences librarianship. The scholarship is announced at the annual meeting of the association, where the recipient will also receive a one-year student membership in the Medical Library Association and free inclusive student registration at the association's annual meeting. The recipient will assume the other costs of attending the meeting and the ceremony at which the presentation is made. Guidelines and application materials are available at the website. All materials and letters of support must be received by MLA via email, fax, or mail (in order of preference) by December 1. email applications and documents will be accepted as PDF or MS Word files only. The recipient will be notified in March before the annual meeting. Founded in 1898, the Medical Library Association (MLA) is a nonprofit, educational organization of more than 1,100 institutions and 3,600 individual members in the health sciences information field, committed to educating health-information professionals, supporting health-information research, promoting access to the world's health-sciences information, and working to ensure that the best health information is available to all.
Requirements: To be eligible the applicant must be a member of a minority group (Black or African-American, Hispanic or Latino, Asian, Aboriginal, North-American Indian or Alaskan Native, or Native Hawaiian or other Pacific Islander) and be a citizen or permanent resident of the United States or Canada. Additionally, the applicant must either be entering or have completed no more than one-half of the academic requirements of a Masters program at an American Library Association (ALA)-accredited graduate school of library science at the time of the granting.

Restrictions: Past recipients of the MLA Scholarship or the MLA Scholarship for Minority Students are not eligible to apply.
Geographic Focus: All States, Canada
Date(s) Application is Due: Dec 1
Amount of Grant: Up to 5,000 USD
Contact: Carla Funk; (312) 419-9094, ext. 14; fax (312) 419-8950; mlapd2@mlahq.org
Internet: http://www.mlanet.org/awards/grants/minstud.html
Sponsor: Medical Library Association
65 East Wacker Place, Suite 1900
Chicago, IL 60601-7246

MLA Murray Gottlieb Prize Essay Award 2788
The Murray Gottlieb Prize was established in 1956 by Ralph and Jo Grimes of the Old Hickory Bookshop (Brinklow, Maryland) in memory of Murray Gottlieb, a New York antiquarian-book dealer. The purpose of the award is to stimulate the health-sciences librarian's interest in the history of medicine. Currently the award is sponsored by the Medical Library Association (MLA) History of the Health Sciences Section and is awarded annually for the best unpublished essay on the history of medicine and allied sciences written by a health-sciences librarian. The author of the winning essay receives complimentary registration to the annual meeting, a certificate at the association's annual meeting, and a cash award of $100 afterwards. The recipient assumes all costs of attending the meeting and the ceremony at which the presentation is made. Guidelines and a submission form are available at the website. All materials must be received by MLA via email, fax, or mail (in order of preference) by November 1. The recipient will be notified in March before the annual meeting. Founded in 1898, the Medical Library Association is a nonprofit, educational organization of more than 1,100 institutions and 3,600 individual members in the health sciences information field committed to educating health-information professionals, supporting health-information research, promoting access to the world's health-sciences information, and working to ensure that the best health information is available to all.
Requirements: Authors of papers submitted must be health-sciences librarians. The submitted paper must treat some aspect of the history of medicine or allied sciences. Entries will be judged according to quality of bibliographic research, quality of the experimental design or arguments developed to support a particular hypothesis, contribution to the study of the history of the health sciences, and originality and style (clarity, appearance, conciseness, and structure).
Restrictions: Papers may be under consideration for publication at the time of submission but cannot have been published.
Geographic Focus: All States, All Countries
Date(s) Application is Due: Nov 1
Amount of Grant: 100 USD
Contact: Carla Funk; (312) 419-9094, ext. 14; fax (312) 419-8950; mlapd2@mlahq.org
Internet: http://www.mlanet.org/awards/honors/gottlieb.html
Sponsor: Medical Library Association
65 East Wacker Place, Suite 1900
Chicago, IL 60601-7246

MLA Research, Development, and Demonstration Project Grant 2789
The purpose of the Medical Library Association (MLA)'s Research, Development, and Demonstration Project Grant is to provide support for research, development, or demonstration projects that will help to promote excellence in the field of health-sciences librarianship and information sciences. Grants range from $100 to $1000. The application form and guidelines are available for download at the website. Applications and supporting documents must be received by MLA via email, fax, or mail by December 1. emailed applications and documents will be accepted as PDF or MS Word files only. MLA may offer Research, Development, and Demonstration Project Grants more than once a year. Applicants are encouraged to visit the website or contact the association for information about additional deadlines. Founded in 1898, the Medical Library Association (MLA) is a nonprofit, educational organization of more than 1,100 institutions and 3,600 individual members in the health sciences information field committed to educating health-information professionals, supporting health-information research, promoting access to the world's health-sciences information, and working to ensure that the best health information is available to all.
Requirements: Eligible applicants must satisfy the following criteria: hold a graduate degree in library science; be a practicing health-sciences librarian with at least two years of professional experience; be a citizen of or have permanent residence status in either the United States or Canada; and preferably be an individual member of the MLA. Consideration will be given in exceptional cases to an outstanding candidate(s) not meeting the criteria above.
Restrictions: Grants will not be given to support an activity that is operational in nature or that has only local usefulness.
Geographic Focus: All States, Canada
Date(s) Application is Due: Dec 1
Amount of Grant: 100-1,000 USD
Contact: Carla Funk; (312) 419-9094, ext. 14; fax (312) 419-8950; mlapd2@mlahq.org
Internet: http://www.mlanet.org/awards/grants/index.html
Sponsor: Medical Library Association
65 East Wacker Place, Suite 1900
Chicago, IL 60601-7246

MLA Rittenhouse Award 2790
The Rittenhouse Award, initiated in 1967 and sponsored by Rittenhouse Book Distributors, Inc., is presented annually by the Medical Library Association (MLA) for the best unpublished paper or web-based project on health-sciences librarianship or medical informatics submitted by a student in an ALA-accredited program of library and information studies or by an internship trainee in medical informatics or health-sciences librarianship. The author of the winning essay or project receives complimentary student registration to MLA's annual meeting, a certificate at the meeting, and a cash award of $500 after the meeting. Winning entries will be submitted to the Journal of the Medical Library Association for consideration. The recipient assumes all costs of attending the meeting and the ceremony at which the presentation is made. Guidelines and a submission form are available at the website. All materials must be received by MLA via email, fax, or mail (in order of preference) by November 1. The recipient will be notified in March before the annual meeting. Founded in 1898, the Medical Library Association is a nonprofit, educational organization of more than 1,100 institutions and 3,600 individual members in the health sciences information field committed to educating health-information professionals, supporting health-information research, promoting access to the world's health-sciences information, and working to ensure that the best health information is available to all.
Requirements: Authors of papers or web-based projects submitted for the Rittenhouse Award must be enrolled in a course for credit in an ALA-accredited graduate program of library and information studies or have recently graduated from such a program and be currently enrolled as an internship trainee in medical informatics or health-sciences librarianship. Submissions must treat some aspect of health-sciences librarianship or medical informatics and may be bibliographical, address a health-sciences issue or topic, or report the results of research. Authors of web-based projects should submit a needs assessment, information on development of the website, evaluation of the website, and a summary outlining the project plan and evaluation results. Entries will be judged according to the following standards: thoroughness and relevance of bibliographic research or needs assessment; importance of topic that would warrant a bibliographic treatment or web-based project; quality of experimental design or arguments developed to support a particular hypothesis or question; originality of argument and positions taken; methodological rigor; quality of website solution; impact on the field of health-sciences librarianship or medical informatics; and style (clarity, appearance, conciseness, organization, and structure).
Restrictions: Papers must have been written or web-based project completed during the 18 months preceding the submission deadline and during the student's course work toward a graduate degree in library and information studies or medical informatics. Papers that have been published or that are under consideration for publication are not eligible.
Geographic Focus: All States
Date(s) Application is Due: Nov 1
Amount of Grant: 500 USD
Contact: Carla Funk; (312) 419-9094, ext. 14; fax (312) 419-8950; mlapd2@mlahq.org
Internet: http://www.mlanet.org/awards/honors/ritten.html
Sponsor: Medical Library Association
65 East Wacker Place, Suite 1900
Chicago, IL 60601-7246

MLA Thomas Reuters / Frank Bradway Rogers Information Advancement Award 2791
The Thomson Reuters/Frank Bradway Rogers Information Advancement Award is presented annually by the Medical Library Association in recognition of outstanding contributions in the following areas: the use of technology to deliver health sciences information; the science of information; and facilitating the delivery of health-sciences information. Originally sponsored by the Institute for Scientific Information (ISI), the award was presented for the first time in 1983 to Frank Bradway Rogers, M.D., for whom it was subsequently renamed. The recipient receives a certificate at the association's annual meeting and a cash award of $500 after the annual meeting. Guidelines and a nomination form are available at the website. In order for the nomination to be considered, all materials and letters of support must be received by MLA via email, fax, or mail (in order of preference) by November 1. The recipient will be notified in March before the annual meeting. Founded in 1898, the Medical Library Association (MLA) is a nonprofit, educational organization of more than 1,100 institutions and 3,600 individual members in the health sciences information field, committed to educating health-information professionals, supporting health-information research, promoting access to the world's health-sciences information, and working to ensure that the best health information is available to all.
Requirements: It is preferred that an individual nominee be a member of MLA, or if a small group is nominated for a combined effort, that at least one member of the group be a member of MLA. Nominations should be for innovative contributions to improved information handling and use. Preference will be given to pioneering efforts (i.e., new theories or applications).
Geographic Focus: All States, All Countries
Date(s) Application is Due: Nov 1
Amount of Grant: 500 USD
Contact: Carla Funk; (312) 419-9094, ext. 14; fax (312) 419-8950; mlapd2@mlahq.org
Internet: http://www.mlanet.org/awards/honors/rogers.html
Sponsor: Medical Library Association
65 East Wacker Place, Suite 1900
Chicago, IL 60601-7246

MoEF Dr. Salim Ali National Wildlife Fellowship 2792
In order to commemorate the memory of the great wildlife conservationist of the country, Dr. Salim Ali to inspire and promote, particularly the younger generation of wildlife managers and scientists for taking up research and experimental projects aimed at conservation of the rich wildlife heritage of the Government of India. The Ministry of Environment and Forests has decided to award the fellowship alternatively each year. The fellowship will be awarded normally for a period of two years at the rate of Rs. 4000 per month. In addition, an amount of Rs. 18,000 per annum will be paid for meeting

the contingent expenditure. Preference will be given to the officials of the State and/or Forest and Wildlife Department.
Requirements: Only Indian citizens engaged in wildlife management or research are eligible.
Geographic Focus: India
Amount of Grant: 132,000 INR
Contact: Munjurul Hannan Khan, Deputy Secretary; +880 2 71 67 472; fax +880 1 71 344 31 79; dsenv1@moef.gov.bd or munjurulkhan@gmail.com
Internet: http://moef.nic.in/modules/fellowships-and-awards/awards/index.php?f=1
Sponsor: Ministry of Environment and Forests
Paryavaran Bhawan, CGO Complex, Lodhi Road
New Delhi, 110 003 India

MOEF Indira Gandhi Paryavaran Puraskar Award 2793

The MOEF Indira Gandhi Paryavaran Puraskar Award was instituted in 1987 by the Ministry of Environment and Forests (MoEF), Government of India. This national prize on environment is named after the former Prime Minister of India late Shrimati Indira Gandhi and shall be known as Indira Gandhi Paryavaran Puraskar. The prize will be given each year on any environment related occasion for significant contributions in the field of environment, which are recognized as having measurable impact on the protection and overall improvement of the environment. From the year 2009 onwards, two prizes of Rs. 5.00 lakh each along with silver lotus trophy under the organization category, and three prizes of Rs. 5.00 lakh, Rs. 3.00 lakh and Rs. 2.00 lakh respectively to individuals in individual category along with silver lotus trophy shall be given annually. Following is an indicative but not exhaustive list of work areas concerned: prevention of pollution; conservation of natural resources; rational use of depletable resources; environmental planning and management; environmental impact assessment; outstanding field work; environmental education; and creating awareness of environmental issues.
Requirements: Any citizen of India or Indian organization engaged in the Scientific Work or Field Work for the cause of protection of environment is eligible. There is no age limit for individuals.
Geographic Focus: India
Amount of Grant: 200,000-500,000 INR
Contact: S M Munjurul Hannan Khan, Deputy Secretary; +880 2 71 67 472; fax +880 1 71 344 31 79; dsenv1@moef.gov.bd or munjurulkhan@gmail.com
Internet: http://moef.nic.in/citizen/award/igpp.html
Sponsor: Ministry of Environment and Forests
Paryavaran Bhawan, CGO Complex, Lodhi Road
New Delhi, 110 003 India

Monfort Family Foundation Grants 2794

The foundation awards general operating grants to Colorado nonprofit organizations in its areas of interest, including education, health and medical research, arts and culture, agriculture, and general charitable giving. Application forms must be obtained.
Requirements: Colorado nonprofits, primarily in Weld County, are eligible.
Restrictions: Grants are not made to individuals.
Geographic Focus: Colorado
Date(s) Application is Due: May 1; Oct 1
Amount of Grant: 5,000-100,000 USD
Contact: Dave Evans, Program Contact; (970) 454-1357
Sponsor: Monfort Family Foundation
Box 337300
Greeley, CO 80633

Morgan Adams Foundation Grants 2795

The Morgan Adams Foundation supports laboratory and clinical research in the area of pediatric cancer, with an emphasis on cancers of the brain and spine. The Foundation's particular interest is to provide seed and bridge grants for viable investigations not yet ripe enough for total funding by larger organizations. Additionally, the Foundation is able to partially fund translational studies, experimental therapeutics studies and/or Phase I and II clinical trials on a case by case basis. The Foundation is seeking out projects that are highly collaborative and/or multi-institutional in nature, hoping to facilitate through the strategic placement of funding, more rapid trials development in order to get the best possible treatment options into circulation with the greatest expediency possible. Through careful consideration and placement of its funding, the Foundation encourages and supports research intended to improve treatment effectiveness, improve treatment outcomes and improve the quality of life for children battling cancer.
Geographic Focus: All States
Contact: Joan Slaughter, Executive Director; (303) 758-2130; fax (303) 758-2134; joan@morganadamsfoundation.org or info@morganadamsfoundation.org
Internet: http://morganadamsfoundation.hotpressplatform.com/funding-research/how-funds-become-hope
Sponsor: Morgan Adams Foundation
5303 E. Evans Avenue, Suite 202
Denver, CO 80222

Motorola Foundation Grants 2796

Motorola concentrates its giving in the following areas: engineering, technical, science or mathematics programs in K-12 schools or in universities and colleges; programs reaching traditionally under-represented groups in the areas of math, science, engineering and business; programs providing technical assistance, research and statistical information on the state of science and engineering education, or; progams that support the protection and preservation of the environment.
Requirements: All grants are made to 501(c)3 tax-exempt organizations.
Restrictions: Grants may not be made to individuals (scholarships, travel, business loans), religious groups, fraternity or sorority programs, political campaigns, private foundations, benefit events or ads, single disease causes, national health organizations or their local chapters, trade schools, product donations, media projects, capital fund drives, sports sponsorships, or endowment funds.
Geographic Focus: Arizona, California, Florida, Georgia, Illinois, Massachusetts, New Jersey, Pennsylvania, Texas
Amount of Grant: 10,000-100,000 USD
Contact: Matthew Blakely, (847) 576-7895; fax (847) 576-9440; giving@motorola.com
Internet: http://www.motorola.com/content.jsp?globalObjectId=8152
Sponsor: Motorola Foundation
1303 E Algonquin Road
Schaumburg, IL 60196

Motorola Foundation Innovation Generation Collaborative Grants 2797

The Innovation Generation Collaborative Grants Support medium to large-scale STEM education collaborations between two or more non-profit organizations, schools and/or school districts, with joint requests ranging from $150,000 to $500,000. Programming must have nationwide and/or online reach or take place near communities where Motorola has an employee presence: Tempe, Arizona; San Diego, California; Miami, Florida; Chicago, Illinois; Lowell, Massachusetts; New York City and Long Island, New York; Philadelphia, Pennsylvania; or Fort Worth, Texas. Program must demonstrate measurable outcomes & impact on at least 250 primary participants. Grants may be for up to two years of project work starting after June of this year. Grant applications are accepted at the Motorola Foundation website, with a deadline date of April 1st.
Requirements: Any U.S. non-profit organization, school or school district may apply. Programming must have nationwide and/or online reach or take place near communities where Motorola has an employee presence: Tempe, Arizona; San Diego, California; Miami, Florida; Chicago, Illinois; Lowell, Massachusetts; New York City and Long Island, New York; Philadelphia, Pennsylvania; or Fort Worth, Texas. All requests must be partnerships between two or more non-profit organizations or schools. Only prior Motorola Foundation Innovation Generation grant recipients may apply. Organizations will be expected to interact and share program attributes with other grantees through conferences and online forums. Applications must be accompanied by letters of support from each member of the partnership. Funding priority will be placed on partnerships that: focus on a key STEM interest area such as teacher training, girls programming, out-of-school programming or online programs; engage students and teachers in innovative, hands-on activities; teach not only STEM, but also innovative thinking and creative problem-solving skills; focus on girls and underrepresented minorities; engage Motorola employees as volunteers.
Restrictions: Only prior Motorola Foundation Innovation Generation grant recipients may apply.
Geographic Focus: All States
Amount of Grant: 150,000-500,000 USD
Contact: Matthew Blakely, (847) 576-7895; fax (847) 576-9440; giving@motorola.com
Internet: http://www.motorola.com/staticfiles/Business/Corporate/U.S.-EN/corporate-responsibility/society/community-investment-education-more-about-innovation-generation-grants.html?globalObjectId=8153
Sponsor: Motorola Foundation
1303 E Algonquin Road
Schaumburg, IL 60196

MPI for Social Anthropology Department II Postdoctoral Grants 2798

The postdoctoral grants are awarded with a specific theme in mind (see the Society's website for the chosen theme). All studies will gather standardized data, and individuals will be expected to expand their research based on specific topics related to the theme. The postdoctoral grants are awarded initially for two years, with the possibility of a one-year extension. Individuals with disabilities will be given priority, assuming equal qualifications.
Requirements: Participants must have a PhD and will undertake comparative fieldwork in light of the project's theme. PhD must be in anthropology or a related discipline to be submitted before commencement. There is no application to fill out; individuals should send the following documentation either by mail or email: a cover letter; a CV including a list of publications; a project proposal (one page outline); photocopies of university degrees; names of 2-3 referees.
Restrictions: Postdoctoral grants can only be awarded to scientists who have received their doctorate within the last ten years.
Geographic Focus: All States
Date(s) Application is Due: Jun 30
Contact: Ms. Janka Diallo, 49 345 29 27 404; fax 49 345 29 27 402; jdiallo@eth.mpg.de
Internet: http://www.mpg.de/english/careerOpportunities/jobOffers/extern/ethnologische_forschung/stellenangebot29/
Sponsor: Max Planck Institute for Social Anthropology
P.O. Box 11 03 51
Halle/Saale, 06017 Germany

MPI for Social Anthropology Max Planck Fellow Research Group Law, Organizations, Science, and Technology (LOST) 2799

This fellowship is open to scientists who have completed their doctorate within the last ten years. The primary regional focus is on Africa. Special attention is given to medical practice and argumentation in juridical contexts such as in the control of epidemics, the legitimization and legalization of diagnostic and healing practices, intellectual property rights, medical evidence in various forms of courts of justice etc. (for more details see http://www.eth.mpg.de/). The PhD grants are generally awarded for 2 years, with the

possibility of two six-month extensions. The postdoctoral grants are for two years (a one-year extension is possible).
Requirements: Applicants should have an anthropological background with at least some experience in Science & Technology Studies, or an S&TS background with a least some anthropological experience. Good field research skills, the ability to work in a team, communication skills with good written and spoken English are required. PhD students are expected to complete a year-long period of fieldwork as part of their studies. No application form is required, however applicants should mail or email the following: a cover letter; a CV including a list of publications; a project resume (five pages only); photocopies of university degrees; names of 2-3 referees, who can be contacted.
Restrictions: Postdoctoral grants can only be awarded to scientists who have received their doctorate within the last ten years.
Geographic Focus: All States
Date(s) Application is Due: Apr 30
Contact: Ms. Wagenbrett, Personnel Administration; +49 345/ 29 27 404; fax +49 345/ 29 27 402; wagenbrett@eth.mpg.de
Internet: http://www.mpg.de/english/careerOpportunities/jobOffers/extern/ethnologische_forschung/stellenangebot23/
Sponsor: Max Planck Institute for Social Anthropology
P.O. Box 11 03 51
Halle/Saale, 06017 Germany

MSI Alden G. Clayton Dissertation Proposal Competition **2800**
This annual competition will grant up to five awards of $5,000 each for dissertation proposals. Some preference will be given to proposals that address the current MSI Research Priority Topics. (See the website for current research priorities.) Each winner will receive an unrestricted cash prize of $5,000 designed to aid in the completion of the dissertation. Winners are asked to submit the results to MSI's working paper series.
Requirements: Entrants must be candidates for the doctorate in marketing or a related field (e.g., organizational behavior, consumer psychology, economics, strategy, management, finance etc.) in a fully accredited university. The dissertation must be no more than fifty percent completed at the time of submission, and the primary thesis advisor must approve submission of the proposal for the competition. Submissions must be accompanied by a letter from the primary thesis advisor certifying that the dissertation is no more than fifty percent completed at the time of submission.
Restrictions: For blind reviewing purposes, please omit any identifying information (name, school, etc.) from the proposal itself. Please include with your submission a separate file/hard copy sheet containing your name, the title of the dissertation, school, and detailed contact information for both the entrant and his/her thesis advisor(s).
Geographic Focus: All States
Date(s) Application is Due: Jul 31
Amount of Grant: 5,000 USD
Contact: Ross Rizley, Research Director; (617) 491-2060; rizley@msi.org
Internet: http://www.msi.org/awards/index.cfm?id=60
Sponsor: Marketing Science Institute
1000 Massachusetts Avenue
Cambridge, MA 02138-5396

MSI Marketing Strategy Meets Wall Street Research Competition **2801**
The Marketing Science Institute (MSI) and the Emory Brand Institute (EmoryBI) have allocated resources to fund high-quality empirical research on several critical questions in this area. These research topics were prioritized in a multi-disciplinary discussion session of leading accounting, finance, and marketing faculty, as well as representatives from corporations and investment banks. Priority Research Topics: How informed are stock prices relative to important customer, brand, and marketing developments?; Understanding brand valuation; Marketing spending and firm value; The investor community as a customer.
Requirements: The co-sponsors invite proposals for original, scientifically rigorous research projects on these priority topics. Proposals should follow the MSI guidelines (available at the website).
Restrictions: The following expense categories are acceptable for funding: research assistance, data acquisition and analysis, travel among co-authors. Awarded funds may be used only for out-of-pocket expenses of the research (i.e., they may not be used for university overhead, researcher's salary or course relief, tuition remission, etc.).
Geographic Focus: All States
Date(s) Application is Due: Sep 1
Amount of Grant: 5,000-15,000 USD
Contact: Ross Rizley, Research Director; (617) 491-2060; rizley@msi.org
Internet: http://www.msi.org/research/index.cfm?id=146
Sponsor: Marketing Science Institute
1000 Massachusetts Avenue
Cambridge, MA 02138-5396

MSI Research Support Standard Grants **2802**
MSI funds high-quality research that deals with topics of importance to member companies. The organization supports research with the potential for application by managers as well as more basic or exploratory work. No one approach or methodology is favored over another as long as the form is appropriate to the objectives of the research. Studies may be conceptual or empirical and may involve literature reviews, comparative studies, field or laboratory experiments, model building, or theory development. Cross-disciplinary work building on theories, research results, and methods from disciplines of relevance to marketing are encouraged. MSI and its member companies strongly endorse using actual consumers, customers, and executives rather than student subjects in research projects.
Requirements: Academic researchers (faculty members, or doctoral students working collaboratively with faculty advisors) can apply for financial and/or non-financial support for research projects. Most MSI grants are made to cover researchers' out-of-pocket costs for data collection, respondent fees, research assistants, and similar expenses. Both proposals and pre-proposals are accepted, though there is no required standard format. MSI operates on a monthly review cycle, and proposals may be submitted at any time. See website for detailed guidelines.
Restrictions: Grants do not provide salary replacement for the principal researcher(s), funds for the purchase of equipment or software, university overhead, tuition, or funds for travel to non-MSI conferences.
Geographic Focus: All States
Amount of Grant: 5,000-20,000 USD
Contact: Research Director; (617) 491-2060; fax (617) 491-2065; msi@msi.org
Internet: http://www.msi.org/research/index.cfm?id=48
Sponsor: Marketing Science Institute
1000 Massachusetts Avenue
Cambridge, MA 02138-5396

Mt. Sinai Health Care Foundation Academic Medicine and Bioscience Grants **2803**
The Foundation seeks to strengthen Cleveland's prominence as a national leader in medical education and biomedical science, perpetuating Mt. Sinai's leadership role in these areas. Emerging collaborations among Cleveland's academic medical centers present opportunities to build organizational capacity and advance progress. The Board of Directors of the Foundation meets on a quarterly basis to review proposals.
Requirements: 501(c)3 nonprofits serving greater Cleveland, Ohio may submit proposals for grant support. The foundation welcomes and encourages an informal conversation with program staff prior to the submission of a grant request.
Restrictions: In general, the foundation does not support general operating expenses, direct provision of health services, building or equipment expenses, fund-raising events, projects outside of greater Cleveland, endowment funds, lobbying, program advertising, grants for individuals, or scholarships.
Geographic Focus: Ohio
Date(s) Application is Due: Jan 1; Apr 1; Jul 1; Oct 1
Contact: Shelly Galvin; (216) 421-5500; fax (216) 421-5633; sgg2@case.edu
Internet: http://www.mtsinaifoundation.org/whatwefund_academic.html
Sponsor: Mount Sinai Health Care Foundation
11000 Euclid Avenue
Cleveland, OH 44106-1714

Munby Fellowship in Bibliography **2804**
One fellowship is offered annually to cover bibliographical research, normally at the postdoctoral level, based on the collections of the Cambridge University libraries. There is no restriction on the choice of topic within the fields of historical bibliography, the history of the book trade, and book collecting. Research may be concerned with printed or manuscript material, Western or Oriental, in any language. Application consists of a curriculum vita, a statement outlining the research proposed, a list of publications, and names and addresses of two references. There are no interviews. Election will take place in autumn for the fellowship beginning the following October and running until September 30 of the next year.
Requirements: Eligible to apply are graduates of any university, who may be of any nationality. Preference will be given to promising younger scholars already established in their careers.
Geographic Focus: All States
Date(s) Application is Due: Sep 9
Amount of Grant: 20,540 GBP
Contact: Deputy Librarian; 01223-333047; fax 01223-333160; library@lib.cam.ac.uk
Internet: http://www.lib.cam.ac.uk/munby.html
Sponsor: Cambridge University Library
West Road
Cambridge, CB3 9DR England

Murphy Institute Center for Ethics and Public Affairs Faculty Fellowships **2805**
The Center for Ethics and Public Affairs supports residential faculty fellowships to support outstanding faculty whose teaching and research focus on questions of ethics and moral choice in such areas as architecture, business, government, law, medicine, urban design and planning, and engineering. Fellows will participate in conferences and seminars and devote the majority of time to conducting their own research. All Fellows will receive a stipend, an office, library privileges, a research allowance and a monthly housing supplement. The fellowships are open to all, regardless of citizenship. See website for application form.
Requirements: Applicants should hold a doctorate in philosophy, political theory, or related disciplines; or a professional degree in architecture, business, engineering, law, medicine, urban design and planning, or public policy. Preference will be given to applicants at an early stage of their careers, normally no more than 10 years from the terminal degree in their field.
Geographic Focus: Louisiana
Date(s) Application is Due: Nov 17
Amount of Grant: Up to 35,000 USD
Contact: Meg Keenan; (504) 862-3236; fax (504) 862-8360; mkeenan@tulane.edu
Internet: http://www.tulane.edu/~murphy/Faculty%20F.htm
Sponsor: Murphy Institute, Tulane University
108 Tilton Hall
New Orleans, LA 70118

Murphy Institute Center for Ethics and Public Affairs Graduate Fellowships **2806**
The Center for Ethics and Public Affairs encourages teaching and research on ethics, in moral and political philosophy, and in public life and the professions. The Center's Graduate Fellowships support outstanding Tulane graduate students in these fields. Graduate Fellows receive a stipend for one academic year, which varies according to individual circumstance.
Requirements: Student applicants should be enrolled in a doctoral program at Tulane University or in one of Tulane's professional schools; and should have completed all course requirements and general examinations by September prior to the fellowship year. The Center seeks applicants who have excelled in their fields of specialization, have demonstrated an interest in ethical questions that cut across disciplinary boundaries, and who are likely to make significant contributions to teaching and scholarship in the fields of ethics, moral and political philosophy, political theory, or professional ethics.
Geographic Focus: Louisiana
Date(s) Application is Due: Mar 3
Contact: Meg Keenan; (504) 862-3236; fax (504) 862-8360; mkeenan@tulane.edu
Internet: http://murphy.tulane.edu/center/graduate-fellowships.php
Sponsor: Murphy Institute, Tulane University
108 Tilton Hall
New Orleans, LA 70118

Musgrave Foundation Grants **2807**
The foundation awards grants to eligible Missouri tax-exempt organizations in its areas of interest, including civic and public affairs, community development, elementary education, secondary education, higher education, and social services. Types of support include annual campaigns, building construction/renovation, capital campaigns, continuing support, equipment acquisition, general operating support, matching grants, and scholarship funds.
Requirements: 501(c)3 tax-exempt organizations and colleges, universities, and schools in Missouri may apply.
Restrictions: Grants are not made to individuals.
Geographic Focus: Missouri
Date(s) Application is Due: May 31
Amount of Grant: 500-61,000 USD
Contact: Jerry Redfern; (417) 841-4698 or (417) 882-9090; jredfern@musgravefoundation.org
Internet: http://www.musgravefoundation.org
Sponsor: Musgrave Foundation
P.O. Box 10327
Springfield, MO 65804

NAAB Coordinated Research Grants **2808**
This program is designed to sustain growth and improve efficiency of the artificial insemination industry by fostering research and development in animal breeding and reproductive physiology with special emphasis on artificial insemination. Research priorities include disease transmittability, improvement of health testing procedures as they relate to artificial insemination, male and female fertility, semen quality, indicators of genetic merit, genetic factors affecting semen production, genetic evaluation, and marketing methods and strategies. The program supports research primarily related to dairy and beef cattle. However, research on other species may be considered if applicable to the bovine species. Grants are awarded for one year, July 1 to June 30. Preproposals are due on the listed application deadline date.
Requirements: Applicants should hold the doctorate or be of equivalent professional level in applicable discipline fields.
Geographic Focus: All States
Date(s) Application is Due: Oct 1
Amount of Grant: 1,000-15,000 USD
Contact: Jere Mitchell, Technical Director; (573) 445-4406; fax (573) 446-2279; jmitchel@naab-css.org or naab-css@naab-css.org
Internet: http://www.naab-css.org/membership/research_grant.html
Sponsor: National Association of Animal Breeders
P.O. Box 1033
Columbia, MO 65205

NAACP Agnes Jones Jackson Graduate Scholarships **2809**
This scholarship is awarded annually to NAACP members who are or plan to be full-time graduate students at accredited colleges or universities and who can demonstrate financial need. The scholarship is awarded for one year and a recipient may apply for renewal at deadline date. Application forms and required documents must be postmarked on or before the last Friday in the month of March.
Requirements: Current regular members of the NAACP who are U.S. citizens under the age of 25 with a 2.5 (undergraduate) or 3.0 (graduate) GPA are eligible to apply.
Geographic Focus: All States
Date(s) Application is Due: Mar 31
Amount of Grant: 1,500-2,500 USD
Contact: Program Supervisor; (410) 358-8900; fax (410) 764-7357
Internet: http://www.naacp.org/departments/education/scholarship_jones.html
Sponsor: National Association for the Advancement of Colored People
8260 Willow Oaks Corporate Drive
Fairfax, VA 22031

NAACP Willems Scholarships **2810**
The scholarship is awarded to U.S. citizens who is male and majoring in engineering, chemistry, physics, or computer and mathematical sciences. Applicant must possess a cumulative grade point average of at least 3.0. Both undergraduate and graduate students may apply. Membership and participation in NAACP is highly desired. To request a scholarship packet, your request must be in writing. Application forms and required documents must be postmarked on or before the last Friday in the month of March.
Geographic Focus: All States
Date(s) Application is Due: Mar 31
Amount of Grant: 2,000-3,000 USD
Contact: National Education Coordinator; (410) 580-5760
Internet: http://www.naacp.org/inc/docs/education/education_scholarship_instruction_app.pdf
Sponsor: National Association for the Advancement of Colored People
4805 Mount Hope Drive
Baltimore, MD 21215

NAA Norwegian Marshall Fund Grants **2811**
The program awards up to 15 grants each year to support Americans invited to conduct postgraduate study or research in Norway on areas of mutual importance to Norway and the United States. The goal of the program is to increase knowledge and understanding and strengthen the ties of friendship between the two countries.
Requirements: Citizens of the United States, who have arranged with a Norwegian sponsor or research institution to pursue a research project or program in Norway are encouraged to apply for an award. Under special circumstances, the awards can be extended to Norwegians for study or research in the United States.
Geographic Focus: All States
Contact: Gunnar Gran; 47 22 447716; fax 47 22 447831; namerika@online.no
Internet: http://www.noram.no/index_am.html
Sponsor: Norway-America Association
Drammensveien 20c
Oslo, N-0255 Norway

NACC David Stevenson Fellowships **2812**
The goals of the David Stevenson fellowship are to acknowledge David Stevenson's many contributions to the development of data and information about the nonprofit sector and the important role he played in teaching and working with undergraduate students; and to encourage junior faculty members of color to provide leadership for the field and to assist them in gaining tenure in their respective academic institutions. Two fellowships are awarded annually. Fellowships may be used to support research, travel to present research at conferences, and professional education seminars related to research and teaching. The fellowship must be completed between August 1 and July 31. The award will be made payable to the award recipient's affiliate institution. Guidelines are available online. The Nonprofit Academic Centers Council (NACC) is a membership association comprised of academic centers or programs at accredited colleges and universities that focus on the study of nonprofit organizations, voluntarism and/or philanthropy. Established in 1991, NACC is the first group entirely dedicated to the promotion and networking of centers that provide research and education in philanthropy and the nonprofit sector.
Requirements: Applicants must meet the following eligibility criteria: be junior faculty in a tenure track position; be affiliated with an accredited institution of higher education in the United States or Canada (the institution need not be a member of NACC); be members of the African-American, Latino, Asian/Pacific Islander, or American-Indian/Alaskan-Native communities; be committed to excellence in research and teaching in the field of philanthropic or nonprofit-sector studies or nonprofit management from any one of a variety of academic disciplines; and be willing and able to participate in the annual ARNOVA conference (travel expenses will be paid by NACC). Applications will be evaluated by a selection committee of distinguished scholars in the field and colleagues of David Stevenson using the following criteria: significance and quality of proposed research to the field of philanthropic and nonprofit-sector studies; relationship of the topic to the education of undergraduate students or to the development of relevant data and information systems; merit of the proposed project and use of fellowship funds; ability of the applicant to carry out the proposed activity; and the applicant's ability to present a well-written, thoughtfully-prepared application. Priority consideration will be made for nontenured junior faculty. An emphasis on quality and the potential for a candidate to provide leadership in the field will also be a considering factor.
Restrictions: Travel expenses to attend the ARNOVA conference are not transferable as additional award money. Travel to attend the ARNOVA conference is limited to the year of the award and up to two years after the award to present fellowship-related research results.
Geographic Focus: All States
Date(s) Application is Due: Mar 31
Amount of Grant: 15,000 USD
Contact: Sean W. Shacklett; (216) 368-0969; sshacklett-nacc@case.edu
Internet: http://www.naccouncil.org/stevenson.asp
Sponsor: Case Western Reserve University
10900 Euclid Avenue
Cleveland, OH 44106-7167

NACC William Diaz Fellowships **2813**
The goals of the William Diaz Fellowship are to acknowledge William Diaz's contributions to the development of minority leadership for nonprofit organizations, foundations, and the field of philanthropic and nonprofit management studies and to encourage a better understanding of philanthropy from diverse communities. Two fellowships are awarded annually. Fellowships may be used to support research, travel to present research at conferences, professional education seminars related to research and teaching, and course development. All funds must be expended and accounted for within

one year after the end of the fellowship. The award will be made payable to the award recipient's affiliate institution. Guidelines are available online. The Nonprofit Academic Centers Council (NACC) is a membership association comprised of academic centers or programs at accredited colleges and universities that focus on the study of nonprofit organizations, voluntarism and/or philanthropy. Established in 1991, NACC is the first group entirely dedicated to the promotion and networking of centers that provide research and education in philanthropy and the nonprofit sector.
Requirements: Applicants must meet the following eligibility criteria: be faculty of any rank (including senior faculty new to philanthropic and nonprofit management studies) whose work focuses on philanthropy, nonprofit organizations, and diversity; be affiliated with an accredited institution of higher education in the United States or Canada (the institution need not be a member of NACC); be members of the African-American, Latino, Asian/Pacific Islander, or American-Indian/Alaskan-Native communities; be committed to excellence in research and teaching in the field of philanthropic or nonprofit-sector studies or nonprofit management from any one of a variety of academic disciplines; be willing and able to participate in the annual ARNOVA Conference (travel expenses will be paid by NACC). Applications will be evaluated by a selection committee of distinguished scholars in the field and colleagues of William Diaz using the following criteria: significance and quality of proposed research; relationship of the proposed research, if any, to the development of minority leadership for nonprofit organizations, foundations, and the field of philanthropic and nonprofit-sector studies; merit of the proposed project and use of fellowship funds; ability of the applicant to carry out the proposed activity; and applicant's ability to present a well-written, thoughtfully-prepared application.
Restrictions: Travel expenses to attend the ARNOVA conference are not transferable as additional award money. Travel to attend the ARNOVA conference is limited to the year of the award and up to two years after the award to present fellowship-related research results.
Geographic Focus: All States
Date(s) Application is Due: Mar 30
Amount of Grant: 15,000 USD
Contact: Sean W. Shacklett, M.P.A., Executive Director; (216) 368-0969; fax (216) 368-0969; sshacklett-nacc@case.edu
Internet: http://www.naccouncil.org/diaz.asp
Sponsor: Case Western Reserve University
10900 Euclid Avenue
Cleveland, OH 44106-7167

NAEA Research Grants **2814**
The National Art Education Foundation invites proposals to support research in art education that advances knowledge in the field of art education and that promulgates the goals outlined in 'Creating a Visual Arts Research Agenda Toward the 21st Century.' Funds are awarded to selected art educators whose proposals specifically focus on issues relating to one of the recommendations identified in this document. NAEF invites proposals to support research in art education that advances knowledge in the field of art education. Grants are awarded to art educators to pursue a broad range of research topics that are aligned with the NAEA Strategic Goals: advocacy, learning, research and knowledge, and organizational vibrancy. NAEF encourages the submissions of proposals that conduct research that supports the impact and importance of arts education in student learning and provides hard data to support the findings of the research. The project year runs from July 1 through June 30.
Requirements: NAEF Grants are open to active NAEA members who have been members for at least one year prior to the date of application; to state/province associations of NAEA; and to recognized special issues groups of NAEA. All eligible applicants are encouraged to apply. NAEF encourages submissions of the following proposals: proposals that support the creation of communities of learners, including both researchers and practitioners, working together to explore a research question and/or project; and, proposals that support the identification of best practice and research that leads to further understanding of the impact and importance of arts education to student learning in and through the visual arts in a variety of settings, with an interest in research that provides quantitative data to support its findings. These priorities do not preclude other submissions meeting the eligibility of the current Research Guidelines.
Restrictions: The following applicants will not be considered: non-members; nonprofit organizations or institutions; members applying on behalf of the nonprofit organization or institution where they work; relatives of current members of the NAEF Board of Trustees. Applications from individuals who are members of only the state/province association are not considered NAEA members. Individuals must have an NAEA ID# in order to apply.
Geographic Focus: All States
Date(s) Application is Due: Oct 1
Amount of Grant: Up to 10,000 USD
Contact: Kathi R. Levin, (703) 860-8000; naef@arteducators.org
Internet: http://www.arteducators.org/grants/national-art-education-foundation
Sponsor: National Art Education Foundation
1806 Robert Fulton Drive, Suite 300
Reston, VA 20191

NAEF Teacher Incentive Grants **2815**
The National Art Education Foundation (NAEF) invites proposals for scholarship support of projects that promote the improvement of the teaching of art. Funds are awarded to classroom-based action research projects that facilitate development of teachers' capacities to provide visual art instruction that reflects the vision promulgated by the National Art Education Association in areas such as curriculum; student learning; student assessment; classroom behavior, management, or discipline; or other practices relating to instructional interaction and the achievement of student learning. The project year runs from July 1 through June 30. Grants are limited to a maximum of $2,500 per award.
Requirements: NAEF Grants are open to active NAEA members who have been members for at least one year prior to the date of application; to state/province associations of NAEA; and to recognized special issues groups of NAEA. All eligible applicants are encouraged to apply.
Restrictions: The following applicants will not be considered: non-members; nonprofit organizations or institutions; members applying on behalf of the nonprofit organization or institution where they work; relatives of current members of the NAEF Board of Trustees. Applications from individuals who are members of only the state/province association are not considered NAEA members. Individuals must have an NAEA ID# in order to apply.
Geographic Focus: All States
Date(s) Application is Due: Oct 1
Amount of Grant: Up to 2,500 USD
Contact: Kathi R. Levin; (703) 860-8000; naef@arteducators.org
Internet: http://www.arteducators.org/grants/national-art-education-foundation
Sponsor: National Art Education Foundation
1806 Robert Fulton Drive, Suite 300
Reston, VA 20191

NAF Fellowships **2816**
The foundation supports new and innovative studies that are relevant to the cause, pathogenesis, or treatment of the hereditary or sporadic ataxias. Postdoctoral fellowship awards serve as a bridge from postdoctoral positions to junior faculty positions. The award will permit individuals to spend an additional third year in a postdoctoral position and increase chances to establish an independent ataxia research program. Letters of Intent should arrive by August 15, with full proposals due by September 15. Application and guidelines are available online.
Requirements: Applicants should have completed at least one year of postdoctoral training, but not more than two at the time of application, and should have shown a commitment to research in the field of ataxia.
Restrictions: Indirect costs are not allowed.
Geographic Focus: All States
Date(s) Application is Due: Aug 15; Sep 15
Amount of Grant: 35,000 USD
Contact: Grants Administrator; (763) 553-0020; fax (763) 553-0167; naf@ataxia.org
Internet: http://www.ataxia.org/research/ataxia-research-grants.aspx
Sponsor: National Ataxia Foundation
2600 Fernbrook Lane, Suite 119
Minneapolis, MN 55447-4752

NAF Kyle Bryant Translational Research Award **2817**
The Friedreich's Ataxia Research Alliance (FARA) and the National Ataxia Foundation (NAF) invite proposals, under a competitive Request for Applications (RFA) process, to award a grant focusing on pre-clinical investigations that will facilitate clinical trials for Friedreich's Ataxia. Funding will be for one year. The total award is limited to $120,000 (direct costs only). The Foundation is requesting that applicants submit Letters of Intent by July 15, and full applications by July 31.
Requirements: Each application must include a project objectives and lay summary that will be used as FARA and NAF wishes and should not contain confidential information. The other parts of the grant application are considered confidential and will only be released to review committee, FARA and NAF staff and Board of Directors. Applications will be accepted from for-profit organizations, non-profit organizations, public or private institutions, and foreign institutions.
Geographic Focus: All States
Date(s) Application is Due: Jul 15; Jul 31
Amount of Grant: Up to 120,000 USD
Contact: Susan Hagen, (763) 553-0020; susan@ataxia.org
Internet: http://www.ataxia.org/research/ataxia-research-grants.aspx
Sponsor: National Ataxia Foundation
2600 Fernbrook Lane, Suite 119
Minneapolis, MN 55447-4752

NAF Research Grants **2818**
NAF research grants are for new and innovative studies that are relevant to the cause, pathogenesis or treatment of the hereditary or sporadic ataxias. Due to the larger availability of funding for Ataxia-Telangiectasia (A-T), those research proposals will receive a lower priority. However, a higher ranking will be given to those Ataxia-Telangiectasia research studies which lend themselves to an overall better understanding of the ataxia process. Research grants are offered primarily as seed monies to assist investigators in the early or pilot phase of their studies and as additional support for ongoing investigations on demonstration of need. It is hoped that these studies will be further developed to attract future funding from other sources. Grants are awarded for one year only. If funding for a second year is desired then another grant application should be submitted for that funding period. The average funds granted will be up to $15,000, but funding may be considered for up to $30,000 for projects deserving special consideration.
Restrictions: Indirect costs will not be funded.
Geographic Focus: All States
Date(s) Application is Due: Aug 15
Amount of Grant: 5,000-15,000 USD
Contact: Grants Administrator; (763) 553-0020; fax (763) 553-0167; naf@ataxia.org
Internet: http://www.ataxia.org/research/ataxia-research-grants.aspx
Sponsor: National Ataxia Foundation

2600 Fernbrook Lane, Suite 119
Minneapolis, MN 55447-4752

NAF Young Investigator Awards **2819**
The purpose of this award is to encourage young clinical and scientific investigators to pursue a career in the field of ataxia research. It is the foundation's hope that ataxia research will be invigorated by the work of young, talented individuals supported by stable multi-year funding. This award will be made to the sponsoring institution and will be used as salary support or for direct research expenses. The sponsoring institution must agree in writing to the following provision: a clinical investigator (MD) must be free to allocate a minimum of half time for a project; a non-clinician investigator (PhD) must allocate a minimum of three-fourths time for the project. Periodic progress reports will be required. Applicants are asked to submit a brief email with the intentions to apply for NAF funding, including a full title of the research proposal, on or before August 1. Deadline for full application is September 1.
Requirements: Candidates must have attained the MD or PhD degree, and have an appointment as a junior faculty member. Clinicians must have finished their residency no more than five years prior to applying, and investigators must have completed their postdoctoral training no more than five years prior to applying.
Restrictions: Individuals at the associate professor level are not eligible. Awards are for direct costs only, not for indirect costs or institutional overhead.
Geographic Focus: All States
Date(s) Application is Due: Aug 1; Sep 1
Amount of Grant: 50,000 USD
Contact: Grants Administrator; (763) 553-0020; fax (763) 553-0167; naf@ataxia.org
Internet: http://www.ataxia.org/research/ataxia-research-grants.aspx
Sponsor: National Ataxia Foundation
2600 Fernbrook Lane, Suite 119
Minneapolis, MN 55447-4752

NAGC Hollingworth Award **2820**
The annual Hollingworth Award competition is sponsored by the National Association for Gifted Children (NAGC) to encourage educational and psychological research studies of potential benefit to gifted and talented children. The award is named for Leta Stetter Hollingworth (1886-1939), a pioneer in the field of gifted education. Long associated with Teachers College of Columbia University.
Requirements: This international competition is open to individuals, educational institutions, and other organizations that present proposals for publishable research projects concerning gifted and talented youth. You may be a professor, educational administrator, psychologist, teacher, graduate student, or other professional individual. The research projects may be sponsored by universities, schools systems, individual schools, public agencies, or private nonprofit organizations.
Geographic Focus: All States
Date(s) Application is Due: Apr 20
Amount of Grant: 2,500 USD
Contact: Jeff Danielian; (202) 785-4268; fax (202) 785-4248; danielian@nagc.org
Internet: http://www.nagc.org/index.aspx?id=1120
Sponsor: National Association for Gifted Children
1707 L Street NW, Suite 550
Washington, DC 20036

NAGC Masters and Specialists Award **2821**
The National Association for Gifted Children (NAGC) presents a maximum of three awards to those who have demonstrated at least four of the following: teaching; writing; advocacy; curriculum development; community service in gifted education; juried presentations (local, state, or national); facilitating the social and emotional development of gifted and talented students; mentoring/tutoring of children and adolescents; publication of papers in juried and/or non-juried publications; participation in graduate student research colloquia; teacher in-service/parent workshop.
Requirements: Viable candidates will have a proven record of at least four of the following: teaching; writing; advocacy; curriculum development; community service in gifted education; juried presentations (local, state, or national); facilitating the social and emotional development of gifted and talented students; mentoring/tutoring of children and adolescents; publication of papers in juried and/or non-juried publications; participation in graduate student research colloquia; teacher in-service/parent workshop
Restrictions: An individual may receive this award once. The award must be given within a year of the candidate's graduation. A maximum of one candidate and award can be given to one institution.
Geographic Focus: All States
Date(s) Application is Due: Apr 20
Amount of Grant: 200 USD
Contact: Jeff Danielian; (202) 785-4268; fax (202) 785-4248; danielian@nagc.org
Internet: http://www.nagc.org/index.aspx?id=696
Sponsor: National Association for Gifted Children
1707 L Street NW, Suite 550
Washington, DC 20036

NAGPRA Museum Consultation/Documentation Grants **2822**
The Native American Graves Protection and Repatriation Act (NAGPRA) addresses the rights of lineal descendants, Indian tribes, Alaska Native villages and corporations, and Native Hawaiian organizations to certain Native American human remains and cultural items with which they are affiliated. Section 10 of the statute authorizes the Secretary of the Interior to make grants to museums and to Indian tribes, Alaska Native villages and corporations, and Native Hawaiian organizations for the purposes of assisting in the inventory, documentation, and repatriation of Native American human remains and cultural items.
Requirements: NAGPRA grants are available to any institution or state or local government agency (including any institution of higher learning) that has possession of, or control over, Native American human remains or cultural items.
Restrictions: Museums that have previously received a NAGPRA grant, but whose grant expired without successfully completing major elements of the proposed work, or without meeting the conditions of the grant award will not be considered for funding.
Geographic Focus: All States
Date(s) Application is Due: Mar 6
Amount of Grant: 5,000-90,000 USD
Contact: Sangita Chari; (202) 354-2203; fax (202) 371-5197; Nagpra_Grants@nps.gov
Internet: http://www.cr.nps.gov/nagpra/GRANTS/INDEX.HTM
Sponsor: National Park Service
1849 C Street NW, MS 2253
Washington, DC 20240

NAGPRA Museum Repatriation Grants **2823**
The Native American Graves Protection and Repatriation Act (NAGPRA) addresses the rights of lineal descendants, Indian tribes, Alaska Native villages and corporations, and Native Hawaiian organizations to certain Native American human remains and cultural items with which they are affiliated. Section 10 of the statute authorizes the Secretary of the Interior to make grants to museums and to Indian tribes, Alaska Native villages and corporations, and Native Hawaiian organizations for the purposes of assisting in the inventory, documentation, and repatriation of Native American human remains and cultural items. Repatriation awards will be granted to museums to repatriate culturally affiliated Native American human remains and cultural items.
Requirements: NAGPRA grants are available to any institution or state or local government agency (including any institution of higher learning) that has possession of, or control over, Native American human remains or cultural items.
Restrictions: Applicants for repatriation awards must collaborate with the tribes(s), Alaska Native village(s) and corporation(s), or Native Hawaiian organization(s) to whom the human remains or cultural items are being repatriated.
Geographic Focus: All States
Date(s) Application is Due: Jun 30
Amount of Grant: Up to 15,000 USD
Contact: Sangita Chari; (202) 354-2203; fax (202) 371-5197; Nagpra_Grants@nps.gov
Internet: http://www.cr.nps.gov/nagpra/GRANTS/INDEX.HTM
Sponsor: National Park Service
1849 C Street NW, MS 2253
Washington, DC 20240

Nakanishi Prize **2824**
To Prize is intended to recognize and stimulate significant work that extends chemical and spectroscopic methods to the study of important biological phenomena, and was established by the students and colleagues of Koji Nakanishi in the United States and Japan and by the members of the scientific community of both countries. Consequently, the prize is given in the United States and Japan in alternate years through the designated societies and separate endowments administered accordingly. Presentations in odd-numbered years are administered by ACS and in even-numbered years by the Kanagawa Academy of Science and Technology. The prize shall be awarded to an individual without regard to age or nationality for outstanding scientific achievement. Special consideration will be given for work that has contributed broadly on an international scope. In addition to the prize, a bronze medal and certificate are awarded.
Restrictions: Self-nominations are not accepted.
Geographic Focus: All States
Date(s) Application is Due: Nov 1
Amount of Grant: 3,000 USD
Contact: Felicia Dixon, Awards Administrator; (800) 227-5558 or (202) 872-4408; fax (202) 776-8008; f_dixon@acs.org or awards@acs.org
Internet: http://portal.acs.org/portal/acs/corg/content?_nfpb=true&_pageLabel=PP_ARTICLEMAIN&node_id=1319&content_id=CTP_004540&use_sec=true&sec_url_var=region1
Sponsor: American Chemical Society
1155 Sixteenth Street, NW
Washington, DC 20036-4801

NAMM Foundation Music Research Grants **2825**
The Foundation supports scientific research exploring the relationship between music and physical and emotional wellness, with particular attention to the elderly population, the impact of music making on at-risk youth, and music education and the effects of music and music making. The research division also convenes scientists, educators and others around critical issues in music research and disseminates research through various on-line research referral services and archives. Interested applicants should submit a pre-proposal by email to info@nammfoundation.org. Pre-proposal information is available online.
Geographic Focus: All States
Contact: Administrator; (760) 438-8001; fax (760) 438-7327; info@nammfoundation.org
Internet: http://www.music-research.org/Grants/guidelines.html
Sponsor: NAMM Foundation
5790 Armada Drive
Carlsbad, CA 92008

Nanyang Technological University Lee Kuan Yew Postdoctoral Fellowship **2826**
The three-year fellowship will be held concurrently with the candidate's appointment as a research staff in the school working on any of the following identified research areas: College of Engineering, School of Biological Sciences, School of Civil and Environmental Engineering, School of Computer Engineering, School of Electrical and Electronic Engineering, School of Mechanical and Aerospace and Engineering, School of Materials Science and Engineering, and School of Natural Sciences. The program is open to nationals of all countries. Applications must be made on the NTU application form, available at http://www.ntu.edu.sg/personnel/adminres/StaffRec/Researchform.doc. Application and guidelines are available online.
Requirements: Applications are invited from young candidates (below 35 years of age) who have obtained their PhD degrees in the last few years and with excellent records and research potential.
Geographic Focus: All States
Date(s) Application is Due: Mar 31
Contact: Vice President, Human Resources; (65) 67919340; lklim@ntu.edu.sg
Internet: http://www.ntu.edu.sg/personnel/adminres/StaffRec/AdLKY2005.html
Sponsor: Nanyang Technological University
50 Nanyang Avenue
Singapore, 639798
Singapore

NAPM Doctoral Dissertation Grant **2827**
Four yearly fellowships are available to produce useful research that can be applied to the purchasing and materials management functions, and to help develop more high-potential academicians who will teach and conduct research in the purchasing and materials management fields. Grants are to be paid directly to the doctoral candidate in three equal installments: upon receipt of the final research proposal approved by the candidate's dissertation committee; upon receipt of a written statement from the dissertation committee verifying that the research is approximately 50 percent completed; and upon receipt of completed dissertation.
Requirements: Doctoral candidates who are pursuing PhDs or DBAs in purchasing, business, management, logistics, economics, industrial engineering, or related fields in accredited U.S. universities are eligible for the grant.
Geographic Focus: All States
Date(s) Application is Due: Jan 31
Amount of Grant: Up to 12,000 USD
Contact: Dr. Joseph Cavinato, Senior Vice President; (480) 752-6276, ext. 3029 or (800) 888-6276; fax (480) 752-7890; jcavinato@ism.ws
Internet: http://www.napm.org/OnlineGuides/doctoralgrant.cfm
Sponsor: National Association of Purchasing Management
2055 E Centennial Circle, P.O. Box 22160
Tempe, AZ 85285-2160

NAPNAP Foundation Graduate Student Research Grant **2828**
The program supports research by graduate students currently enrolled in a pediatric nurse practitioner program. The purposes of the award are to: encourage graduate students to participate in the research process; and stimulate interest in research regarding children and their families. The recipient of the award will be invited to participate at the research presentation session at the annual conference in the spring following the award. Application and guidelines are available online.
Requirements: The graduate student must be a current member of NAPNAP.
Geographic Focus: All States
Date(s) Application is Due: Apr 1
Amount of Grant: 1,000 USD
Contact: Dolores Jones; (856) 857-9700; fax (856) 857-1600; djones@napnap.org
Internet: http://www.napnap.org/PNPResources/research/AvailGrantsScholarships.aspx
Sponsor: National Association of Pediatric Nurse Associates and Practitioners
20 Brace Road, Suite 200
Cherry Hill, NJ 08034-2634

NAPNAP Foundation Nursing Research Grants **2829**
The foundation supports nursing research efforts that contribute to the improvement of the quality of life for children and their families. The foundation is primarily interested in the following areas of research: impact of nurse practitioner role in healthcare delivery; development/implementation of innovative methods for delivery of health care services in the areas of school health, community health, behavior and development, adolescent health, and chronic disease management; health care problems of children and their families; advanced nursing practice problems and issues; exploratory/evaluative research of PNP education and credentialing; and educational programs and materials for healthcare providers and/or children and families. Up to two grants will be awarded annually.
Geographic Focus: All States
Date(s) Application is Due: Apr 1
Amount of Grant: 2,500 USD
Contact: Dolores Jones; (856) 857-9700; fax (856) 857-1600; djones@napnap.org
Internet: http://www.napnap.org/index.cfm?page=13&sec=102
Sponsor: National Association of Pediatric Nurse Associates and Practitioners
20 Brace Road, Suite 200
Cherry Hill, NJ 08034-2634

NARSAD Colvin Prize for Outstanding Achievement in Mood Disorders Research **2830**
The Colvin Prize for Outstanding Achievement in Mood Disorders Research, is an award of $50,000 given to an outstanding scientist carrying out work on the causes, pathophysiology, treatment, or prevention of affective disorders. The scientist to be recognized is one who gives particular promise for advancing the understanding of affective illness or its basic brain mechanisms that will lead to new treatment approaches. The Prize is given at the annual New York City Awards Dinner held each October. Nominations are due by June 1.
Geographic Focus: All States
Date(s) Application is Due: Jun 1
Amount of Grant: 50,000 USD
Contact: Sho Tin Chen; (516) 829-0091; fax (516) 487-6930; schen@bbrfoundation.org
Internet: http://bbrfoundation.org/outstanding-achievement-prizes#Lieber
Sponsor: Brain and Behavior Research Foundation
60 Cutter Mill Road, Suite 404
Great Neck, NY 11021

NARSAD Distinguished Investigator Grants **2831**
The Brain and Behavior Research Foundation (formerly National Alliance for Research on Schizophrenia and Depression, or NARSAD) distributes funds for psychiatric brain and behavior disorder research. NARSAD Distinguished Investigator Grants provide support for experienced investigators (full professor or equivalent) conducting neurobiological and behavioral research. Areas of particular interest include patient populations with unique or unusual characteristics and central nervous system developments. A one-year grant of up to $100,000 is provided for established scientists pursuing innovative projects in diverse areas of neurobiological research.
Requirements: Applicants must be a full professor (or equivalent), and maintain peer reviewed competitively funded scientific programs. Guidelines and application deadlines are at the Foundation's website.
Restrictions: Previous grantees may not receive the grant for a second time until five years have elapsed since the beginning date of the prior award. Grant requirements from all previous awards must be met. Only one applicant may apply per application. Applicants may submit only one application per cycle.
Geographic Focus: All States
Amount of Grant: Up to 100,000 USD
Contact: Sho Tin Chen, Associate Manager, Research Grants; (800) 829-8289 or (516) 829-5576; fax (516) 487-6930; grants@narsad.org
Internet: http://bbrfoundation.org/narsad-distinguished-investigator-grant
Sponsor: Brain and Behavior Research Foundation
60 Cutter Mill Road, Suite 404
Great Neck, NY 11021

NARSAD Goldman-Rakic Prize for Cognitive Neuroscience **2832**
The Goldman-Rakic Prize for Cognitive Neuroscience Research is given in recognition of a research scientist who has made distinguished contributions to the understanding of cognitive neuroscience. The Prize includes $40,000 and an honorary lecture at Yale University to honor the memory and accomplishments of Patricia Goldman-Rakic. The Prize is specifically for excellence in neurobiological research at the cellular, physiological, or behavioral levels that may lead to a greater understanding of underlying psychiatric or neurological disease. The Prize is given at the annual New York City Awards Dinner held each October. Nominations for the Award are due by June 1.
Geographic Focus: All States
Date(s) Application is Due: Jun 1
Amount of Grant: 40,000 USD
Contact: Sho Tin Chen, Associate Manager, Research Grants; (516) 829-0091; fax (516) 487-6930; schen@bbrfoundation.org
Internet: http://bbrfoundation.org/outstanding-achievement-prizes#Lieber
Sponsor: Brain and Behavior Research Foundation
60 Cutter Mill Road, Suite 404
Great Neck, NY 11021

NARSAD Independent Investigator Grants **2833**
The Brain and Behavior Research Foundation (formerly National Alliance for Research on Schizophrenia and Depression, or NARSAD) distributes funds for psychiatric brain and behavior disorder research. The NARSAD Independent Investigator Grants provide support for investigators during the critical period between the initiation of research and the receipt of sustained funding. Basic and/or clinical investigators are supported, but research must be relevant to schizophrenia, major affective disorders, or other serious mental illnesses. A two-year grant up to $100,000, or $50,000 per year, is provided to scientists at the associate professor level or equivalent, who are clearly independent and have won national competitive support as a principal investigator.
Requirements: Applicants must have a doctoral-level degree, be an associate professor (or equivalent), and have received, as Principal Investigator, competitive research support at a national level. Independent Investigators are eligible for a maximum of two grants. Previous awardees at the associate professor level are encouraged to apply for a second grant. Applications must be submitted through the online application system. Additional guidelines can be found at the Foundation's website.
Restrictions: The total budget request may not exceed $100,000. Awards cannot overlap in time, all prior grant requirements (including NARSAD Young Investigator Grants) must be met, and an applicant cannot submit more than one application per cycle. Investigators that have been promoted to associate professor, but have currently active NARSAD Young Investigator Grants, are not eligible.
Geographic Focus: All States
Date(s) Application is Due: Nov 15

Amount of Grant: Up to 50,000 USD
Contact: Sho Tin Chen, Associate Manager, Research Grants; (800) 829-8289 or (516) 829-5576; fax (516) 487-6930; grants@bbrfoundation.org
Internet: http://bbrfoundation.org/narsad-independent-investigator-grants
Sponsor: Brain and Behavior Research Foundation
60 Cutter Mill Road, Suite 404
Great Neck, NY 11021

NARSAD Lieber Prize for Schizophrenic Research 2834

The Lieber Prize for Schizophrenia Research is given in recognition of a research scientist who has made distinguished contributions to the understanding of schizophrenia. The $50,000 cash rewards past achievement and provides further incentive for an outstanding working scientist to continue to do exceptional research into the causes, prevention, and treatment of schizophrenia. The prize is bestowed at the annual New York City Awards Dinner held each October. The presentation of the prize helps build public understanding of the importance of schizophrenia research, and acts as a goal towards which researchers might strive. Nominations are due by June 1. The Lieber Prize winner will select the winner of the the Sidney R. Baer, Jr. Prize for Schizophrenia Research, worth $40,000. The winner is typically a NARSAD Young Investigator at the Lieber Prize winner's institution, conducting work in schizophrenia research.
Geographic Focus: All States
Date(s) Application is Due: Jun 1
Amount of Grant: 50,000 USD
Contact: Sho Tin Chen, Associate Manager, Research Grants; (516) 829-0091 or (800) 829-8289; fax (516) 487-6930; schen@bbrfoundation.org or info@bbrfoundation.org
Internet: http://bbrfoundation.org/outstanding-achievement-prizes#Lieber
Sponsor: Brain and Behavior Research Foundation
60 Cutter Mill Road, Suite 404
Great Neck, NY 11021

NARSAD Ruane Prize for Child and Adolescent Psychiatric Research 2835

The Ruane Prize is an award of $50,000 given to an outstanding scientist carrying out research on the causes, pathophysiology, treatment, or prevention of severe mental illness in children. The scientist to be recognized is one who gives particular promise for advancing the understanding of psychotic, affective or other severe brain and behavior disorders having their onset in childhood or adolescence. Contributions may be for clinical research or relevant basic science. The Prize is given at the annual New York City Awards Dinner held each October. Nominations are due by June 1.
Geographic Focus: All States
Date(s) Application is Due: Jun 1
Amount of Grant: 50,000 USD
Contact: Sho Tin Chen, Associate Manager, Research Grants; (516) 829-0091; fax (516) 487-6930; schen@bbrfoundation.org
Internet: http://bbrfoundation.org/outstanding-achievement-prizes#Lieber
Sponsor: Brain and Behavior Research Foundation
60 Cutter Mill Road, Suite 404
Great Neck, NY 11021

NARSAD Young Investigator Grants 2836

The Brain and Behavior Research Foundation (formerly National Alliance for Research on Schizophrenia and Depression, or NARSAD) distributes funds for psychiatric brain and behavior disorder research. The Foundation's NARSAD Young Investigator Grants offer up to $30,000 a year for up to two years to enable promising investigators to either extend their research fellowship training or to begin careers as independent research faculty. Grants are intended to facilitate innovative research opportunities and support basic, as well as translational and/or clinical investigators. Research must be relevant to the understanding, treatment and prevention of serious psychiatric disorders such as schizophrenia, bipolar, mood and anxiety disorders, or child and adolescent psychiatric disorders. Awardees become eligible for the Freedman Prize for Outstanding Research, with a cash prize of $1,000. They are also eligible for the Klerman Prize for Outstanding Clinical Research, also with a cash prize of $1,000. Young Investigator winners also become eligible for the Baer Prize for Schizophrenia Research, worth $40,000.
Requirements: Applicants must have a doctoral level degree and already be employed in research training, or be in a faculty or independent research position. Applicants must have an on-site mentor or senior collaborator who is an established investigator in areas relevant to psychiatric disorders. Applications must be submitted online from the Foundation's website.
Restrictions: Pre-doctoral students or investigators at the rank of associate professor or equivalent are not eligible. Institutional overhead is not allowable.
Geographic Focus: All States
Date(s) Application is Due: Sep 15
Amount of Grant: Up to 60,000 USD
Contact: Sho Tin Chen, Associate Manager, Research Grants; (800) 829-8289 or (516) 829-5576; fax (516) 487-6930; grants@bbrfoundation.org
Internet: http://bbrfoundation.org/narsad-young-investigator-grant
Sponsor: Brain and Behavior Research Foundation
60 Cutter Mill Road, Suite 404
Great Neck, NY 11021

NASA Cooperative Agreement Notice for Technology Development Grants 2837

The objective of this opportunity is to establish partnerships with the goal of developing a technology to meet a specific NASA need. This goal will be accomplished by selecting partners who will cooperatively share in the development cost of the technology that meets the specified NASA need. The partnership selection is a two step process with an initial Notice of Intent followed with the request for the Project Proposal on selected technology concepts. Specific areas of interest include: Rocket Propulsion Test Technologies and Integrated System Health Management. NASA plans to fund projects at approximately $20,000 to $500,000 per project with a period of performance of approximately 12 months.
Requirements: The partner is expected to contribute at least 50 percent of the total resources required to accomplish the objectives of the project. However, special situations may arise in which NASA will allow the partner contribution to be less than 50 percent. Partner contributions may be cash, non-cash or both.
Geographic Focus: All States
Date(s) Application is Due: Sep 30
Amount of Grant: 20,000-500,000 USD
Contact: James D. Huk; (228) 688-1045; fax (228) 688-1141; james.huk@ssc.nasa.gov
Internet: http://nspires.nasaprs.com/external/solicitations/summary.do?method=init&solId={993EDEC9-8D62-6A05-EEB7-DF2EDF97F902}&path=open
Sponsor: National Aeronautics and Space Administration
NASA Business Management Division, Acquisition Management Office, Code BA30
Stennis Space Center, MS 39529-6000

NASA Goddard Visiting Fellowships 2838

The program seeks highly qualified PhD scientists for collaboration in scientific research in one of the following research theme areas of the Division: aerosols; carbon cycle; clouds and precipitation; polar climate change; solar impacts on climate; terrestrial water; or weather and short-term climate forecasting. Selected candidates will be offered appointments as Goddard Visiting Fellows to pursue independent research in collaboration with scientists in the Laboratories within the Earth Sciences Division, and will have full access to NASA computing facilities and other resources at either the Goddard Space Flight Center in Greenbelt, Maryland, or at the Goddard Institute for Space Studies in New York, New York. Candidates will be selected, by competition, amongst the above theme areas of research but awards may not be made in all theme areas in any one competition year.
Requirements: This solicitation is aimed at exceptional post-doctoral candidates and university faculty interested in sabbatical arrangements.
Geographic Focus: All States
Date(s) Application is Due: Mar 1
Contact: Dr. Raymond M. Hoff; (410) 455-1610; fax (410) 455-1291; hoff@umbc.edu
Internet: http://gest.umbc.edu/employment_opp/goddard_visiting_fellows.html
Sponsor: Goddard Earth Sciences and Technology Center / NASA Goddard Earth Sciences Division
5523 Research Park Drive, Suite 320
Baltimore, MD 21228

NASA Graduate Student Researchers Program (GSRP) Fellowships 2839

The Program awards fellowships for graduate study leading to masters or doctoral degrees in the fields of science, mathematics, and engineering related to NASA research and development. This twelve month award includes a required internship at the NASA Center affiliated with the NASA sponsored research. The goal of the Program is to cultivate research ties to the academic community, to help to meet the continuing needs of the Nation's aeronautics and space requirements by increasing the number of highly trained scientists and engineers in aeronautics and space-related disciplines, and to broaden the base of students pursuing advanced degrees in science, mathematics, and engineering. Research opportunities are assessed and updated annually to complement the mission requirements of NASA. Research areas are in disciplines that lead to aeronautics and space careers. Training Grants are awarded for one in the amount of $30,000. This amount includes a $21,000 student stipend, a student travel allowance of $4,000, up to $1,000 for health insurance, and a $4,000 university allowance, which typically goes to the Research Adviser, who becomes the Principal Investigator for the Training Grant. The student stipend of $21,000 may cover tuition, room and board, books, software, meal plans, school and laboratory supplies, and other related expenses. The $4,000 student allowance may be used for national and international conferences and data collection. Awards are renewable up to three years based on satisfactory academic advancement, research progress, and available funding. The program supports approximately 300 graduate students annually.
Requirements: Applicants must be currently enrolled or accepted as a full-time graduate student in an accredited U.S. college or university. Applicants must be U.S. citizens. Students may apply at any time during their graduate program, or prior to receiving their baccalaureate degree, provided they have been accepted to an accredited graduate program at a U.S. college or university. All applicants must have a Faculty Adviser (who will be the Principal Investigator of the GSRP Training Grant) from the institution where they plan to receive their graduate degree.
Geographic Focus: All States
Date(s) Application is Due: Feb 1
Amount of Grant: 30,000 USD
Contact: Dr. Katie Blanding, (202) 358-0402; katie.blanding@nasa.gov
Internet: http://fellowships.hq.nasa.gov/gsrp/program/index.cfm
Sponsor: National Aeronautics and Space Administration
Office of Education, NASA Headquarters
Washington, DC 20546-0001

NASA Lewis Educational Research and Collaborative Internships 2840

This is an educational program that provides internships for students in science, engineering, professional administrative, and technical areas. Internships are available with various starting dates during the summer at the NASA Glenn Research Center in Cleveland, Ohio. The internships provide students with introductory professional experiences to complement their academic programs. Interns receive assignments in research and development as well as technical and administrative projects under the personal guidance of NASA professional staff members. Each assignment is commensurate with the academic level and field of study of the student. Interns are integrated into the day-to-day activities of the Center to the greatest extent possible. Time is devoted to a program of scheduled educational, professional, and social events. Internships are available for college students, secondary teachers, and high school students.
Requirements: Teachers must be U.S. citizens employed full-time as secondary school teachers. College students must be U.S. citizens and full-time students pursuing a baccalaureate or higher degree in science, engineering, technology, or professional administration at an accredited college or university in the United States. College-bound high school graduates or two-year college students are also eligible. Students must have cumulative GPA of 3.0 on a 4.0 scale.
Geographic Focus: All States
Date(s) Application is Due: Jan 15; Jan 31
Contact: Mary Roberts; (440) 962-3170, ext. 5006; internships@oai.org
Internet: http://www.oai.org/education/student/lercip.html
Sponsor: Ohio Aerospace Institute
22800 Cedar Point Road
Cleveland, OH 44142

NASA Marslink Initiative **2841**
This program is an effort to bring one of the most current NASA missions into the classrooms of various U.S. schools across the country. This initiative is designed to impact one school per congressional district. Schools will be chosen on a first-come, first-served basis. The interactive program allows students to simulate a mission to Mars in three phases. The project is funded by $75,000 in matching grant money from NASA. Schools must be able to provide $225 toward the cost of the $500 curriculum.
Requirements: K-12 public schools with sufficient technological capability and ability to meet the required match are eligible to apply.
Geographic Focus: All States
Amount of Grant: 500 USD
Contact: Tim Lewaren, (800) 965-3763, ext. 262; tim@space-explorers.com
Internet: http://www.state.nj.us/cgi-bin/education/grants/grants_active2.pl?string=recnum=00050&maxhits=500&fed=1
Sponsor: National Aeronautics and Space Administration
1825 Nimitz Drive
De Pere, WI 54115

NASA New Aeronautics Competition for University Students **2842**
University students are invited to design the next generation, 21st Century, DC-3 type aircraft. Some of the design considerations include reduced environmental impact, reduced noise, daily operations on short runways, cost analysis for production and operation, passenger and cargo limits, structure and materials, engines. Students should also briefly describe three valid scenarios for potential use of this vehicle in the year 2058. Entries will be limited to a 25-page research paper, excluding ancillary material, and adhere to a standard format. Teams or individuals may enter. Winning entries may be invited to a student forum sponsored by ARMD and/or industry, may receive offers of student internships, and may receive other prizes, including cash, depending on available funds. Cash prize amounts will be on the order of $5000 for first place, $3500 for second place, $2000 for third place. In the case of a tie, prizes will be determined based on funds available.
Requirements: Any U.S. citizen enrolled in an accredited college or university in the U.S. or its territories is eligible to enter the competition. International students, either in the U.S. or elsewhere, may enter but will not be eligible for certain prizes. Trophies and certificates will be awarded to each winner, regardless of citizenship. Students who are receiving NASA funds for research projects or who are working on a NASA funded project for a faculty member should disclose this information in their entry packet.
Restrictions: Only U.S. citizens are eligible for prizes or NASA funded internships.
Geographic Focus: All States
Date(s) Application is Due: Apr 30
Amount of Grant: 2,000-5,000 USD
Contact: Dr. Elizabeth Ward; fax (757) 864-7638; elizabeth.b.ward@nasa.gov
Internet: http://aero.larc.nasa.gov/competitions_univ.htm
Sponsor: National Aeronautics and Space Administration
NASA Langley Research Center, Mail Stop 218, Building 1212, Room 128
Hampton, VA 23681

NASA Undergraduate Student Researchers Project (U.S.RP) Grants **2843**
The Program offers undergraduate students across the United States internship experiences at NASA Centers. Students may apply for a 10-week summer session or a 15-week spring or fall session. The purpose of the Program is three-fold: to attract undergraduate students from the widest array of backgrounds, who are fully representative of America's racial, ethnic, and cultural diversity and to provide them with hands-on, challenging research experiences that stimulate continued student interest in the fields/disciplines aligned with NASA's research and development mission; to build a national program bridge-from existing NASA K-12 Education Program activities to NASA Higher Education Program options-that encourages and facilitates student interest in future professional opportunities with NASA and its partner organizations; and to extend and strengthen NASA's commitment to educational excellence and university research, highlighting the critical need to increase the Nation's undergraduate and graduate science, engineering, mathematics, and technology skill base. Students will receive a $6,000 (10-week summer session) or $9,000 (15-week spring or fall session) stipend for the research experience plus one round-trip airfare or ground transportation allowance to and from the NASA Host Center.
Requirements: Applicants must meet the following eligibility requirements to participate: U.S. citizenship; classified as a Junior or Senior after Spring semester/quarter of the previous year; enrolled full-time in an accredited U.S. college or university (community college students may apply, however placement is contingent on matriculation into a four-year institution as a rising Junior); academic major or demonstrated coursework concentration in engineering, mathematics, computer science, or physical/life sciences; demonstrated strong interest/commitment to one of the above career fields or disciplines (e.g., science/math/engineering fairs, clubs, or awards; tutoring/mentoring; internships or other related experience); and minimum cumulative grade point average of 3.0 on a scale of 4.0.
Geographic Focus: All States
Date(s) Application is Due: Jan 31; Feb 28
Amount of Grant: 6,000 USD
Contact: Nick Koltun; (757) 766-5210; fax (757) 766-5205; nkoltun@odu.edu
Internet: http://www.vsgc.odu.edu/Menu3_1_3.htm
Sponsor: National Aeronautics and Space Administration
600 Butler Farm Road, Suite 200
Hampton, VA 23666

National Academies Postdoctoral and Senior Research Awards **2844**
The program administers postdoctoral and senior research awards through its associateship programs, part of the Policy and Global Affairs Division. The Research Associateship Programs (RAP) are sponsored by federal laboratories and NASA research centers at over 100 locations in the United States and overseas. Awards are given for the purpose of conducting research at a specific laboratory chosen by the applicant; and are made to doctoral level scientists and engineers who can apply their special knowledge and research talents to research areas that are of interest to them and to the host laboratories and centers. Guidelines and application are available online.
Requirements: Postdoctoral research associateships are awarded to persons who have held the doctorate less than five years at the time of application and are made initially for one year. Senior research associateships are awarded to applicants who have held the doctorate five years or more at the time of application. Senior research associateship applicants should have research experience that has resulted in significant contributions and recognition as established investigators in their specialized fields.
Geographic Focus: All States
Date(s) Application is Due: Feb 1; May 1; Aug 1; Nov 1
Amount of Grant: 45,000 USD
Contact: Awards Coordinator; (202) 334-2760; fax (202) 334-2759; rap@nas.edu
Internet: http://www4.nationalacademies.org/pga/rap.nsf/WebDocuments/Home+Page
Sponsor: National Academy of Sciences
500 Fifth Street NW, GR 322A
Washington, DC 20001

National Academy of Education Spencer Postdoctoral Fellowships **2845**
Fellowships are designed to promote scholarly inquiry and discussion concerning the ends and means of education, in all of its forms, in the United States and abroad. Each year the National Academy of Education solicits applicants in education, the humanities, and the social and the behavioral sciences; their interests must describe research whose relevance to education is apparent. Fellows are expected to devote one academic year to full-time research or two contiguous academic years to half-time research. Applications may be made directly by the individual. Up to 30 fellowships will be awarded each year. This fellowship is open to citizens of all countries; the grant is expected to be used for salary replacement or to pay related research costs. Guidelines and application packet are available online.
Requirements: Applicants must have received the PhD, EdD, or equivalent degrees between January 1, 2000, and December 31 of the current year.
Restrictions: Concurrent funding for the proposed project is not permitted.
Geographic Focus: All States
Date(s) Application is Due: Nov 10
Amount of Grant: 55,000-110,000 USD
Contact: Spencer Fellowship Program Director; (202) 334-2341; nae.info@nyu.edu
Internet: http://www.nae.nyu.edu
Sponsor: National Academy of Education
500 Fifth Street NW, #1049
Washington, DC 20001

National Assessment of Educational Progress Visiting Scholar **2846**
ETS sponsors the visiting scholar program to enable scholars to conduct their own independent research using the NAEP database with access to senior NAEP staff and ETS research staff and have an opportunity to contribute to the effectiveness of the program, especially as it pertains to African American, Hispanic, and other minority students by identifying productive areas of investigation for future NAEP activities relevant to the education of those minority students. The goals of this program are to provide research opportunities for recent awardees of the doctorate and to increase the number of women and minority professionals working in the areas specified. The term of visit is normally from September 1 through June 30. Stipend is set in relation to compensation at the guest scholar's home institution. In addition, limited relocation expenses consistent with ETS guidelines will be reimbursed upon presentation of receipts.
Requirements: The program is open to any individual who holds a doctorate in a relevant discipline and provides evidence of prior research relevant to the education of minorities.
Geographic Focus: All States

Contact: Linda DeLauro; (609) 734-1806; fax (609) 683-2800; internfellowships@ets.org
Internet: http://www.ets.org/portal/site/ets/menuitem.22f30af61d34e9c39a77b13bc3921509/?vgnextoid=49f5be3a864f4010VgnVCM10000022f95190RCRD
Sponsor: Educational Testing Service
Rosedale Road, MS 07R
Princeton, NJ 08541-0001

National Black MBA Association Annual Scholarships **2847**
NBMBAA is a nonprofit organization of minorities that hold or are currently pursuing graduate business degrees. Students applying for the national scholarships (27 awarded each year) must complete the official NBMBAA scholarship application, provide verification of grade point average and enrollment, complete a financial need assessment provided, and write a two-page essay that responds to a topic specified by the association. There are also 30 state chapters, many of which offer local scholarships in addition to supporting the national scholarships. Scholarships are offered at the undergraduate, master's, and doctoral levels. Students wishing to pursue chapter scholarships must contact the chapter president in their area for details; a list of chapters, presidents, and their mailing addresses is available upon request.
Requirements: Minority students actively pursuing graduate degrees in business are eligible to apply. They must be enrolled full time in accredited graduate business programs in either the fall or winter term of the year in which they apply.
Geographic Focus: All States
Date(s) Application is Due: Apr 30
Amount of Grant: 7,000-12,500 USD
Contact: Lori Johnson, NBMBAA National Office; (312) 236-2622, ext. 8086; fax (312) 236-4131; scholarship@nbmbaa.org
Internet: http://nbmbaa.org/scholarship.cfm
Sponsor: National Black MBA Association
180 N Michigan Avenue
Chicago, IL 60601

National Blood Foundation Research Grants **2848**
The National Blood Foundation (NBF), established in 1983, has a history of supporting research and education that advances transfusion medicine and cellular therapies by funding scientific research that benefits patients and donors. Funds are raised annually from corporations, blood centers, foundations and individuals by the NBF for the National Blood Foundation Research and Education Trust Fund (NBFRET) and the NBF. The NBF and NBFRET, which are 501(c)3 organizations, provide grants for scientific research in the field of transfusion medicine and cellular therapies and support educational initiatives that benefit this community. The National Blood Foundation annually awards grants for one or two-year research projects, with a maximum award per grant of $75,000.
Requirements: NBF has also funded research and studies on critical issues that benefit patients and donors in the fields of transfusion medicine and cellular therapies, and are in addition to scientific research. Preference is given to new investigators in the field. There is a $150 application fee.
Geographic Focus: All States
Date(s) Application is Due: Dec 13
Amount of Grant: Up to 65,000 USD
Contact: Amy Quiggins, Manager; (301) 215-6552 or (301) 215-6551; fax (301) 215-5751; aquiggins@aabb.org or nbf@aabb.org
Internet: http://www.aabb.org/programs/nbf/Pages/default.aspx
Sponsor: National Blood Foundation
8101 Glenbrook Road
Bethesda, MD 20814-2749

National Book Scholarship Fund **2849**
The purpose of the National Book Scholarship Fund (NBSF) is to distribute direct in-kind assistance to local literacy providers throughout the United States. NBSF grants are intended to provide the New Readers Press books and other materials essential to begin a new literacy outreach program or to significantly expand an existing effort. First priority will be given to programs that focus on women and intergenerational or family literacy concerns. Other organizations working in economically disadvantaged urban and rural areas, among homeless populations, in refugee resettlement efforts, in workplace literacy efforts, and English-as-a-second-language initiatives will also receive consideration for support. Types of support include program development grants, research grants, scholarships, technical assistance, and training grants. Applications are accepted from September until December. applicants are encouraged to apply early because NBSF applications are reviewed and awards are made on a first-come, first-served basis.
Requirements: Programs applying for a grant must agree to provide, from local sources, a cash payment equal to 20% of the total amount being requested in the grant application to defray the costs of NBSF administration.
Geographic Focus: All States
Date(s) Application is Due: Apr 14
Amount of Grant: 500-8,000 USD
Contact: Michele Diecuch, Program Administrator; (888) 528-2224 or (315) 422-9121, ext. 321; fax (315) 422-6369; mdiecuch@proliteracy.org
Internet: http://www.servicelearning.org/funding-sources#category310
Sponsor: ProLiteracy Worldwide
1320 Jamesville Avenue
Syracuse, NY 13210

National Center for Responsible Gaming Centers of Excellence in Gambling Research Grants **2850**
The purpose of the Centers of Excellence in Gambling Research Grants program is to: advance the understanding of gambling disorders through innovative, multi-disciplinary research investigations; and enhance the dissemination of research findings to the public. The center grants are intended to provide funding for a stable, long-term institutional focus on a complex set of gambling-related problems requiring an innovative, multi-disciplinary approach. Recipients are expected to provide leadership in the field by: conducting cutting-edge investigations of gambling-related disorders; translating research findings for non-academic audiences; and cultivating the next generation of gambling researchers by mentoring young investigators. Investigative efforts should be broadly based, encompassing a variety of areas, including biological, biomedical, social, behavioral, and/or clinical sciences. The proposed research investigation may focus on a broad range of research that develops and tests psychosocial or pharmacological approaches for prevention, intervention, treatment or relapse prevention of gambling disorders. The NCRG is especially interested in brief interventions targeted at underrepresented populations, such as minorities, young adults and persons with sub-clinical gambling disorders. Other priorities include the following topics: impact of Indian gaming, gambling and minorities, secondary data analysis,and technology and gambling. Applicants may request up to a total of $350,000 in direct costs for a period not to exceed 36 months.
Requirements: Both public and private nonprofit organizations are eligible to apply. Profit-making organizations should contact Christine Reilly (creilly@ncrg.org). Foreign institutions are required to collaborate with a U.S. institution. Applications involving a non-U.S. institution must have a principal investigator and fiscal agent based at a U.S. institution.
Geographic Focus: All States, All Countries
Date(s) Application is Due: Jul 2
Amount of Grant: Up to 350,000 USD
Contact: Christine Reilly; (978) 338-6610; fax (978) 522-8452; creilly@ncrg.org
Internet: http://www.ncrg.org/research-center/apply-ncrg-funding/ncrg-centers-excellence-gambling-research
Sponsor: National Center for Responsible Gaming
900 Cummings Center, Suite 418-U
Beverly, MA 01915

National Center for Responsible Gaming Conference Scholarships **2851**
The National Center for Responsible Gaming (NCRG) is offering the need-based scholarship program for the NCRG Conference on Gambling and Addiction. Insofar it is able, it is the NCRG's wish to provide fair access to the NCRG Conference to all interested stakeholders, especially those in academia and the clinical and public health sectors. Last year, the NCRG supported almost thirty NCRG Conference attendees. This year, the NCRG will award scholarships to more than fifteen individuals, and a portion of these scholarships may also include travel assistance. Additional scholarship funds may be available at a later time. To apply for a scholarship, interested applicants must submit their application and registration form by June 15.
Geographic Focus: All States
Date(s) Application is Due: Jun 15
Contact: Christine Reilly; (978) 338-6610; fax (978) 522-8452; creilly@ncrg.org
Internet: http://www.ncrg.org/public-education-and-outreach/conference/conference-scholarships
Sponsor: National Center for Responsible Gaming
900 Cummings Center, Suite 418-U
Beverly, MA 01915

National Center for Responsible Gaming Early Stage Investigator Grants **2852**
The NCRG Early Stage Investigator Grants for Gambling Researchers is a career development award that will provide scientists with the mentoring and research experiences necessary for research independence. Applications for this award must propose a research plan that has: intrinsic research importance in the area of prevention and treatment of gambling disorders; and will serve as a suitable vehicle for learning the methodology, theories and concepts needed for a well-trained, independent researcher in the area of gambling research. Applicants may request up to $65,000 in direct costs for a period not to exceed 24 months.
Requirements: The Early Stage Investigator Grant is intended for individuals who are within 10 years of completing their terminal research degree or within 10 years of completing medical residency (or the equivalent). The candidate for this award may not have been a recipient of an NIH Career Development Award or an NCRG grant. Candidates who are not U.S. citizens but have an appointment at a U.S. institution are eligible to apply. Candidates based at foreign institutions are not eligible to apply.
Geographic Focus: All States
Date(s) Application is Due: May 1
Amount of Grant: Up to 65,000 USD
Contact: Christine Reilly; (978) 338-6610; fax (978) 522-8452; creilly@ncrg.org
Internet: http://www.ncrg.org/research-center/apply-ncrg-funding/early-stage-investigator-grants
Sponsor: National Center for Responsible Gaming
900 Cummings Center, Suite 418-U
Beverly, MA 01915

National Center for Responsible Gaming Exploration Grants **2853**
Exploration Grants provide quick access to funding for researchers focused on gambling disorders. Funds can be used to support a broad range of activities including: pilot data to complete an initial but unfunded project, pilot data needed for preliminary results,new direction on a current project not otherwise supported, and funded projects that did not build costs into budgets for services previously free or subsidized which now carry a cost. The proposed research investigation may focus on a broad range of research that develops and tests psychosocial or pharmacological approaches for prevention, intervention, treatment or relapse prevention of gambling disorders. The NCRG is

especially interested in brief interventions targeted at underrepresented populations, such as minorities, young adults and persons with sub-clinical gambling disorders. Other priorities include the following topics: impact of Indian gaming; gambling and minorities; secondary data analysis; and technology and gambling.
Requirements: Both public and private nonprofit organizations are eligible to receive grants from the NCRG. Profit-making organizations should contact Christine Reilly (creilly@ncrg.org). Foreign institutions are required to collaborate with a U.S. institution. Applications involving a non-U.S. institution must have a principal investigator and fiscal agent based at a U.S. institution.
Geographic Focus: All States, All Countries
Amount of Grant: Up to 10,000 USD
Contact: Christine Reilly; (978) 338-6610; fax (978) 522-8452; creilly@ncrg.org
Internet: http://www.ncrg.org/research-center/apply-ncrg-funding/exploration-grants
Sponsor: National Center for Responsible Gaming
900 Cummings Center, Suite 418-U
Beverly, MA 01915

National Center for Responsible Gaming Independent Scientist Grants 2854
The Independent Scientist Award provides support for newly independent scientists who can demonstrate the need for a period of intensive research focus as a means of enhancing their research careers. The grant is intended to foster the development of outstanding scientists and to enable them to expand their potential to make significant contributions to the field of gambling disorders. Eligible applicants have independent peer-reviewed research support. Applicants may request up to $40,000 in direct costs per year for a period not to exceed 24 months.
Requirements: Both public and private nonprofit organizations are eligible to apply. Profit-making organizations should contact Christine Reilly (creilly@ncrg.org). Applications involving a non-U.S. institution must have a principal investigator and fiscal agent based at a U.S. institution. Candidates must have a doctoral degree and independent, peer reviewed research support at the time the award is made and must be willing to commit a minimum of one-third of full-time professional effort conducting research and relevant career development activities during the period of the award.
Geographic Focus: All States, All Countries
Date(s) Application is Due: May 1
Amount of Grant: Up to 40,000 USD
Contact: Christine Reilly; (978) 338-6610; fax (978) 522-8452; creilly@ncrg.org
Internet: http://www.ncrg.org/research-center/apply-ncrg-funding/independent-scientist-grant
Sponsor: National Center for Responsible Gaming
900 Cummings Center, Suite 418-U
Beverly, MA 01915

National Center for Responsible Gaming Large Grants 2855
Large Grants provide up to $75,000 for up to two years for discrete, specified, circumscribed research projects related to gambling disorders. The proposed research investigation may focus on a broad range of research that develops and tests psychosocial or pharmacological approaches for prevention, intervention, treatment or relapse prevention of gambling disorders. The NCRG is especially interested in brief interventions targeted at underrepresented populations, such as minorities, young adults and persons with sub-clinical gambling disorders. Other priorities include the following topics: impact of Indian gaming; gambling and minorities; secondary data analysis; and technology and gambling. Applicants may request up to $75,000 in direct costs per year for a period not to exceed 24 months.
Requirements: Both public and private nonprofit organizations are eligible to apply. Profit-making organizations should contact Christine Reilly, senior research director (creilly@ncrg.org). Foreign institutions are required to collaborate with a U.S. institution. Applications involving a non-U.S. institution must have a principal investigator and fiscal agent based at a U.S. institution.
Geographic Focus: All States
Date(s) Application is Due: Jun 3
Amount of Grant: Up to 75,000 USD
Contact: Christine Reilly; (978) 338-6610; fax (978) 522-8452; creilly@ncrg.org
Internet: http://www.ncrg.org/research-center/apply-ncrg-funding/large-grants
Sponsor: National Center for Responsible Gaming
900 Cummings Center, Suite 418-U
Beverly, MA 01915

National Center for Responsible Gaming Postdoctoral Fellowships 2856
The NCRG Postdoctoral Fellowship Grant is intended to help ensure that a pool of highly trained scientists is available to address the research needs of the field of gambling disorders. Eligible applicants are within five years of completing their terminal research degree or within five years of completing medical residency. The proposed research investigation may focus on a broad range of research that develops and tests psychosocial or pharmacological approaches for prevention, intervention, treatment or relapse prevention of gambling disorders. The NCRG is especially interested in brief interventions targeted at underrepresented populations, such as minorities, young adults and persons with sub-clinical gambling disorders. Other priorities include the impact of Indian gaming; gambling and minorities; secondary data analysis; and technology and gambling. Applicants may request up to $85,000 in direct costs for a period not to exceed 12 months.
Requirements: The Postdoctoral Fellowship is intended for individuals who are within five years of completing their terminal research degree or within five years of completing medical residency (or the equivalent). The candidate for this award may not have been a recipient of an NIH Career Development Award or an NCRG grant. Candidates who are not U.S. citizens but have an appointment at a U.S. institution are eligible to apply. Candidates based at foreign institutions are not eligible to apply.
Geographic Focus: All States
Date(s) Application is Due: Jun 3
Amount of Grant: Up to 85,000 USD
Contact: Christine Reilly; (978) 338-6610; fax (978) 522-8452; creilly@ncrg.org
Internet: http://www.ncrg.org/research-center/apply-ncrg-funding/addiction-fellowship
Sponsor: National Center for Responsible Gaming
900 Cummings Center, Suite 418-U
Beverly, MA 01915

National Center for Responsible Gaming Seed Grants 2857
NCRG Seed Grants support a variety of research activities, including: pilot and feasibility studies; secondary analysis of existing data; small, self-contained research projects; development of research methodology; development of new research technology. The proposed research investigation may focus on a broad range of research that develops and tests psychosocial or pharmacological approaches for prevention, intervention, treatment or relapse prevention of gambling disorders. The NCRG is especially interested in brief interventions targeted at underrepresented populations, such as minorities, young adults and persons with sub-clinical gambling disorders. Other priorities include the following topics: impact of Indian gaming, gambling and minorities, secondary data analysis and technology and gambling.
Requirements: Both public and private nonprofit organizations are eligible to receive grants from the NCRG. Applicants from profit-making companies should contact Christine Reilly, senior research director (creilly@ncrg.org). Foreign institutions are required to collaborate with a U.S. institution. Applications involving a non-U.S. institution must have a principal investigator and fiscal agent based at a U.S. institution.
Geographic Focus: All States
Contact: Christine Reilly; (978) 338-6610; fax (978) 522-8452; creilly@ncrg.org
Internet: http://www.ncrg.org/research-center/apply-ncrg-funding/seed-grants
Sponsor: National Center for Responsible Gaming
900 Cummings Center, Suite 418-U
Beverly, MA 01915

National Center for Responsible Gaming Travel Grants 2858
Travel Grants are intended to support the attendance of postdoctoral investigators at scientific meetings at which they will present a paper or a poster on gambling research. Applicants may request up to $1,500 in support of conference registration fees, airfare, ground transportation, meals and lodging. Eligible meetings include academic conferences that employ rigorous peer review in the selection of presentations.
Requirements: Travel Grants are available to scientists who are not more than ten years beyond their doctorate (Ph.D.) or medical residence. The applicant must be the first author of the paper or poster to be presented. Eligible candidates must be based at a U.S. institution.
Restrictions: The Principal Investigator may apply for only one Travel Grant per cycle.
Geographic Focus: All States
Amount of Grant: Up to 1,500 USD
Contact: Christine Reilly; (978) 338-6610; fax (978) 522-8452; creilly@ncrg.org
Internet: http://www.ncrg.org/research-center/apply-ncrg-funding/travel-grants
Sponsor: National Center for Responsible Gaming
900 Cummings Center, Suite 418-U
Beverly, MA 01915

National Center on Philanthropy and the Law Fellowship in Nonprofit Law 2859
The NCPL Fellowship permits one Fellow each year to spend one year in residence at the New York University Office of Legal Counsel, working with the General Counsel of NYU and the staff of eleven attorneys on legal issues facing the University. The Fellow will delve into an unusually wide variety of legal and organizational issues and will participate in the legal work conducted by the Office of Legal Counsel, including conducting factual and legal research, drafting litigation papers, transactional and policy documents, and correspondence, and participating in the lawyering enterprises of the University.
Requirements: Fellows are selected for their scholarship, leadership, and commitment to practicing in the field of nonprofit law.
Geographic Focus: All States
Date(s) Application is Due: Dec 5
Amount of Grant: 47,000 USD
Contact: Erin Bainbridge Ortiz; (212) 998-6168; fax (212) 995-3149; ncpl.info@nyu.edu
Internet: http://www.law.nyu.edu/ncpl/carrframe.html
Sponsor: National Center on Philanthropy and the Law
110 W Third Street, Room 205
New York, NY 10012

National Council of Farmer Cooperatives Graduate Awards 2860
The awards support one doctoral dissertation and two master's theses with themes dealing with some aspect of economics, finance, operation, or structure of American agricultural cooperatives. The Edwin G. Nourse award is for the best doctoral dissertation. The Kenneth D. Naden award and E.A. Stokdyk award are for the two best master's theses. Application forms are required.
Requirements: Any graduate student in economics, business, communications, sociology, or other relevant field who has an agricultural vocational objective is eligible. All entries must be completed between April 1 of the previous year and March 31 of the year of the award and must be accompanied by a registration form available from NCFC.
Geographic Focus: All States
Date(s) Application is Due: Apr 15
Amount of Grant: 600-1,500 USD
Contact: Catherine Wilson, Executive Assistant; (202) 626-8700; fax (202) 626-8722
Internet: http://www.ncfc.org

Sponsor: National Council of Farmer Cooperatives Education Foundation
50 F Street NW, Suite 900
Washington, DC 20001

National Education Reporting Fellowships 2861
The association offers a study/travel fellowship for a two-month period for reporters to pursue stories of their own design. Fellows will receive half-salary for the two-month study period; travel expenses; access to expert sources; editing assistance; the opportunity to share work with other reporters at EWA regional meetings and the national seminar; and consultation with EWA staff about progress and problems during the project. A distinguished panel of journalists will select the fellows on the basis of the quality of their proposals. Annual deadlines fall in May; contact the office for specific dates.
Requirements: The fellowships are open to full-time print or broadcast journalists who have been covering education for at least two years, have the endorsement of their employers, and can show that they have likely outlets for the products of their studies. Freelance writers may apply if they write about education for a substantial portion of their time and if they have agreements with media organizations to publish or air stories resulting from the fellowships.
Geographic Focus: All States
Contact: Lisa Walker; (202) 452-9830; fax (202) 637-9707; lwalker@ewa.org
Internet: http://www.ewa.org/desktopdefault.aspx?page_id=12
Sponsor: Education Writers Association
2122 P Street NW, Suite 201
Washington, DC 20037

National Endowment for Democracy Discretionary Grants 2862
The objective of the worldwide grant program is to strengthen democratic institutions throughout the world through private, nongovernmental efforts. The endowment is currently funding programs abroad in the areas of pluralism—development of strong, independent, private sector organizations, especially trade unions and business associations; democratic governance and political processes—promotion of strong, stable political parties committed to the democratic process; education, culture, and communications—programs that nourish a strong democratic civic culture including support for publications and other communications media and training programs; the production and dissemination of books and other materials to strengthen popular understanding and intellectual advocacy of democracy; and programs of democratic education. The Endowment is especially interested in proposals that originate with indigenous democratic groups. It is also interested in nonpartisan programs seeking to strengthen democratic values among all sectors of the democratic political spectrum. To a much lesser extent, the endowment also funds programs in international cooperation and research. The endowment staff is willing to discuss proposal ideas and welcomes preliminary letters of inquiry prior to submission of formal proposals. Applications are accepted any time, but are due six weeks prior to the board meeting. The board meets quarterly to approve grants; call to confirm deadlines.
Restrictions: The endowment does not make grants to individuals or for projects within the domestic United States.
Geographic Focus: All States
Date(s) Application is Due: Jan 18; Mar 28; Jun 13; Sep 12
Contact: Public Info Officer; (202) 378-9700; fax (202) 378-9407; info@ned.org
Internet: http://www.ned.org/grants/grants.html
Sponsor: National Endowment for Democracy
1025 F Street NW, Suite 800
Washington, DC 20004

National Endowment for Democracy Reagan-Fascell Democracy Fellowships 2863
The program enables democracy activists, practitioners, scholars, and journalists from around the world to deepen their understanding of democracy and enhance their ability to promote democratic change. Fellows maintain full-time residence at the International Forum for Democratic Studies, the research arm of the Endowment. The Forum hosts 12 to 15 Reagan-Fascell fellows per year for periods ranging from three to ten months. The program offers two tracks, a practitioner track and a scholarly track. Fellowships on the practitioner track tend to be short-term (three to five months), typically culminating in a strategy memorandum, short article, or op-ed and a presentation of the fellow's analysis and ideas. The scholarly track is principally for professors and researchers from emerging and aspiring democracies. Accomplished scholars from established democracies are also eligible to apply. During their three- to 10-month stay at the Forum, scholars make at least one presentation and complete a substantial piece of writing (a monograph or book) for publication. Each fellow receives a monthly stipend for living expenses, plus health insurance and reimbursement for travel to and from Washington, DC. Fellows also receive a fully equipped office and support services. Application and guidelines are available online.
Requirements: There are no specific degree requirements for applicants selecting the practitioner track; candidates are expected to have substantial practical experience working on behalf of some aspect of democracy building—such as human rights advocacy, political party activism, voter education, election monitoring, journalism and press freedom, ethnic pluralism, civic education, etc. Individuals applying in the scholarly track are expected to possess a PhD or academic equivalent at the time of application, to have published in their field of expertise, and to have developed a detailed research outline for their fellowship project. Examples of eligible candidates for the scholarly track include college and university professors, scholars, analysts at research centers, and independent writers.
Geographic Focus: All States
Date(s) Application is Due: Nov 10
Contact: Coordinator; (202) 378-9700; fax (202) 378-9407; fellowships@ned.org
Internet: http://www.ned.org/forum/reagan-fascell.html
Sponsor: National Endowment for Democracy
1025 F Street NW, Suite 800
Washington, DC 20004

National Endowment for the Arts Jazz Masters Fellowships 2864
The National Endowment for the Arts Jazz Masters Fellowship is the highest honor bestowed upon a jazz musician. Each year, the program elevates to its ranks a select number of living legends who have made exceptional contributions to the advancement of jazz. For the NEA Jazz Masters, the selection criteria are artistic excellence and significance of the nominees' contributions to the development and performance of jazz. The Arts Endowment will honor musicians who represent a range of styles and instruments. In addition, one fellowship (A.B. Spellman NEA Jazz Masters Fellowship for Jazz Advocacy) will be given to an individual who has made major contributions to the appreciation, knowledge, and advancement of the American jazz art form. Fellowships are awarded to living individuals on the basis of nominations from the public, including the jazz community. The nomination form is available at the NEA website.
Requirements: Nominees must have nationally recognized and publicly accessible bodies of work in the field of jazz. The recipients must be citizens or permanent residents of the United States.
Restrictions: Posthumous nominations are not eligible for nomination.
Geographic Focus: All States
Date(s) Application is Due: Oct 1
Amount of Grant: 25,000 USD
Contact: Katja von Schuttenbach; (202) 682-5711; vonschuttenbach@arts.gov
Internet: http://www.nea.gov/national/jazz/Award.html
Sponsor: National Endowment for the Arts
1100 Pennsylvania Avenue NW
Washington, DC 20506-0001

National Endowment for the Arts Literature Fellowships: Creative Writing 2865
The Literature Fellowships for Creative Writing offer $25,000 grants in prose (fiction and creative nonfiction) and poetry to published creative writers that enable the recipients to set aside time for writing, research, travel, and general career advancement. The Fellowships operate on a two-year cycle with prose and poetry available in alternating years. Applications are reviewed through an anonymous process in which the only criteria for review are artistic excellence and artistic merit. To review the applications, the Endowment assembles a different advisory panel every year, each diverse with regard to geography, race and ethnicity, and artistic points of view. Deadlines vary each year according to the writing category. Applications must be submitted through Grants.gov.
Requirements: Applicants must be citizens or permanent residents of the U.S. Poetry applicants may apply if they have published within the past ten years. They may also apply if they have published a volume of 48 or more pages; or twenty or more different poems or pages of poetry in five or more literary journals, anthologies, or publications which regularly include poetry as a portion of their format. Up to 16 poems may be in a single volume of poetry of fewer than 48 pages. This volume, however, may count as only one of the required five places of publication. Applicants may use digital, audio, or online publications to establish eligibility, provided that such publications have competitive selection processes and stated editorial policies. If the online publication or website no longer exists, applicants must provide, upon request, sufficient evidence that the work once appeared online. The following may not be used to establish eligibility: pre-publication material, such as galleys, proofs, and advance reader's copies; self-publication including work that has appeared in a publication where the applicant is the editor, publisher, or staff; collaborative work; scholarly writing; instructional writing; journalism; book reviews; editorials/letters to the editor; interviews; student publications and publications which primarily print work by persons who are affiliated with a particular academic institution. In addition, eligibility may not be established by publication in presses that require individual writers to pay for part or all of the publications costs; asks writers to buy or sell copies of the publication; publish the work of anyone who subscribes to the publication or joins the organization through membership fees; publish the work of anyone who buys an advertisement in the publication; publish work without competitive selection; or publish work without professional editing.
Restrictions: Individuals may apply only once each year. Applicants must be published authors. The Fellowship is also not available for the following: individuals who previously have received two or more Literature Fellowships (in fiction, creative nonfiction, or poetry) or Translation Fellowships from the NEA; individuals who have received any Literature Fellowship (in fiction, creative nonfiction, or poetry) or Translation Fellowship from the NEA within the past ten years; news reporting; scholarly writing; or work toward academic degrees.
Geographic Focus: All States
Amount of Grant: 25,000 USD
Contact: Amy Stolls, Literature Program Officer; (202) 682-5771 or (202) 682-5034; stollsa@arts.gov or LitFellowships@arts.gov
Internet: http://www.nea.gov/grants/apply/Lit/index.html
Sponsor: National Endowment for the Arts
1100 Pennsylvania Avenue NW
Washington, DC 20506-0001

National Endowment for the Arts Literature Fellowships: Translation Projects 2866
Through Fellowships to exceptionally talented, published translators, the National Endowment for the Arts supports projects for the specific translation of prose, poetry, or drama from other languages into English. The work to be translated should be of interest for its literary excellence and value. Priority will be given to projects that involve

work that has not yet been translated into English. Applications must be submitted electronically through Grants.gov.
Requirements: Applicants must be U.S. citizens or permanent residents. Individuals are eligible to apply if, alone or in collaboration, they have published a translation into English of a novel or a volume of 48 or more pages of fiction or poetry, or a book-length translation of a work of literary merit; or published a total of 48 pages of translations of creative literature into English in literary magazines, anthologies, or books; or translated into English at least one full-length play that has been produced by a professional theater company.
Restrictions: The following is not available for funding: individuals who previously have received three or more Literature Fellowships (in fiction, creative nonfiction, or poetry) or Translation Fellowships; individuals who have received any Literature Fellowship (in fiction, creative nonfiction, or poetry) or Translation Fellowship from the National Endowment for the Arts within the past five years; scholarly writing; or work toward an academic degree.
Geographic Focus: All States
Date(s) Application is Due: Jan 3
Amount of Grant: 12,500-25,000 USD
Contact: Amy Stolls, Literature Program Officer; (202) 682-5771; stollsa@arts.gov
Internet: http://arts.endow.gov/grants/apply/LitTranslation/index.html
Sponsor: National Endowment for the Arts
1100 Pennsylvania Avenue NW
Washington, DC 20506-0001

National Endowment for the Humanities Fellowships at Winterthur **2867**
Founded almost 60 years ago by collector and horticulturist Henry Francis du Pont, Winterthur consists of a prestigious museum of American decorative arts, a research library that serves scholars around the world, and a 1,000 acre preserve located on the grounds of du Pont's childhood home. Winterthur has a National Endowment for the Humanities (NEH)-funded program that offers four- to twelve-month fellowships for scholars wishing to pursue advanced research in the following areas: American studies; American cultural, social, and intellectual history; art and design history; material culture; decorative arts; consumer culture; garden and landscape studies; Shaker studies, travel and tourism; the Atlantic World; and objects in literature. The program encourages the exchange of ideas among research fellows, the Winterthur staff, and graduate students through colloquia, conferences, discussions, and informal gatherings. Housing is available on the Winterthur grounds and is within easy walking distance of the library and museum where fellows enjoy many staff privileges, including 24-hour access to the library. Fellowships stipends range up to $40,000. Detailed guidelines and a downloadable application form are available at Winterthur's website. The preferred method of submission is email; however, applicants may also mail the completed application and required supporting documentation (essay, bibliography, vita, etc.) to the address given. All applicants are strongly encouraged to search Wintercat (the library's online catalog), visit Winterthur, and contact staff members to discuss potential research projects. The suitability of a project to Winterthur's collections is the primary consideration of the fellowship award committee.
Requirements: Academic, museum, and independent scholars are invited to apply.
Restrictions: This fellowship may not be used in the pursuit of a degree, although fellows with doctorate in hand may revise a dissertation for publication. Applicants must be U.S. citizens or residents for three years prior to application.
Geographic Focus: All States
Date(s) Application is Due: Jan 15
Amount of Grant: Up to 40,000 USD
Contact: Rosemary Krill; (302) 888-4637; fax (302) 888-4870; rkrill@winterthur.org
Internet: http://www.winterthur.org/?p=535
Sponsor: Winterthur Museum, Garden, and Library
5105 Kennett Pike
Wilmington, DE 19735

National Foundation for Catholic Youth Grants **2868**
In an effort to provide support for the field of Catholic youth ministry, the foundation awards financial grants to parish, diocesan, and national initiatives. Grant projects are activities, events, strategies, or campaigns designed to train youth and coordinators of youth ministry, to develop new approaches to comprehensive youth ministry, and to respond to the needs of today's young church. Guidelines are available online.
Requirements: The foundation awards grants to parish, diocesan, and national initiatives.
Restrictions: The foundation does not provide scholarships for undergraduate or postsecondary education.
Geographic Focus: All States
Date(s) Application is Due: Dec 9
Amount of Grant: 500-4,000 USD
Contact: Grants Administrator; (202) 636-3825; fax (202) 526-7544; info@cyfusa.org
Internet: http://www.cyfusa.org/grants.html
Sponsor: National Foundation for Catholic Youth
415 Michigan Avenue NE, Suite 40
Washington, DC 20017-4502

National Gallery of Art Kress Interpretive Fellowship **2869**
The Kress Interpretive Fellowship aims to cultivate students and young professionals interested in museum careers, focusing on curatorial and educational collaboration. These two-year fellowships, renewable for a third year, will provide curatorial training and support scholarly research related to the gallery's collections of European and American art. Fellows will be fully integrated into a specific curatorial department with duties, privileges, and status equivalent to an assistant curator; divide their time between specific research projects and more general curatorial work within the department, including research on the collection and new acquisitions, work on the presentation of the collection, participation in aspects of special exhibition projects, and opportunities to give public lectures and gallery talks; and, in consultation with the supervising curator, develop a concrete project intended to complement their own research interests. Fellowships include stipend, annual travel allowance, and eligibility for medical and term life insurance through the federal government. The stipend for the 12-month fellowship period will be $25,000 and is subject to all applicable taxes. The Fellow will work full time and may begin the fellowship term at a mutually agreed upon date (as early as September 17 and not later than October 1).
Requirements: Graduate art history students who are currently enrolled or have recently graduated are encouraged to apply. The ideal candidate will have a broad knowledge of European art history, with excellent communications skills, both oral and written, and a demonstrated interest in teaching and interpretation of premodern European art.
Geographic Focus: All States
Date(s) Application is Due: Aug 1
Amount of Grant: 25,000 USD
Contact: Grants Administrator; (202) 842-6257; intern@nga.gov
Internet: http://www.nga.gov/education/fellowed.htm
Sponsor: National Gallery of Art
2000B South Club Drive
Landover, MD 20785

National Geographic Society/Waitt Grants **2870**
The National Geographic Society/Waitt Grants Program funds projects that require venture capital, supporting exceptional projects while foregoing a time-consuming peer-review process. NGS/Waitt grants are able to fund proof of concept research for applicants at an earlier stage in their careers than other NGS grant programs. Special emphasis is placed on expedited grant processing and turnaround. The selection committee endeavors to have funding decisions made within eight weeks of application submission. Applications are accepted on a rolling basis.
Requirements: Funding is not restricted to United States citizens. Applicants planning work in foreign countries should include at least one local collaborator as part of their research teams. Applicants are not required to have advanced degrees to be eligible for funding, though they will be required to show a commensurate level of expertise and experience. Applicants must have an affiliation with an educational organization or other institution.
Restrictions: This grant program does not pay educational tuition, nor does it offer scholarships or fellowships of any kind. Society grants may not be used for indirect costs, overhead, and other expenses not directly related to the project. Fringe benefits are also excluded, as are salaries. Funds may not be used for travel to scientific/professional meetings or conferences, legal actions, land acquisition, endowments, construction of permanent field stations, or to publish research results.
Geographic Focus: All States
Amount of Grant: 5,000-15,000 USD
Contact: Grants Administrator; (800) 647-5463; waitt@ngs.org
Internet: http://www.nationalgeographic.com/field/grants-programs/waitt-grants.html
Sponsor: National Geographic Society
1145 Seventeenth Street NW
Washington, DC 20036-4688

National Geographic Society Conservation Trust Grants **2871**
The objective of the trust is to support conservation activities around the world. Grants support projects that contribute significantly to the preservation and sustainable use of the Earth's biological, cultural, and historical resources. Researchers planning work in foreign countries should include at least one local collaborator as part of their research teams. The strength of the program is to identify cutting edge programs that might be overlooked due to inherent risks associated with new investigators and fields. Funds are intended to function as complementary support; the trust strongly encourages applicants to seek additional, concurrent funding from other funding agencies. Before receiving an application form, each principal investigator must submit a pre-application form online. If the pre-application is approved, the principal investigator will be sent an electronic application and instructions.
Requirements: Applicants are not expected to have advanced degrees (PhD or equivalent). However, applicants must provide a record of prior research or conservation action as it pertains to the proposed project.
Restrictions: Society grants may not be used for indirect costs, overhead, and other expenses not directly related to the project. Fringe benefits are also excluded, as are salaries. Funds may not be used for travel to scientific/professional meetings or conferences, legal actions, land acquisition, endowments, construction of permanent field stations, or to publish research results.
Geographic Focus: All States
Amount of Grant: 15,000-20,000 USD
Contact: Grants Administrator; (800) 647-5463; conservationtrust@ngs.org
Internet: http://www.nationalgeographic.com/conservation
Sponsor: National Geographic Society
1145 Seventeenth Street NW
Washington, DC 20036-4688

National Geographic Society Expeditions Council (EC) Grants **2872**
The Expeditions Council consists of representatives from National Geographic editorial divisions (magazines, television, books, and so on) who review and vote on grant applications and an advisory board of external consultants. The Expeditions Council is editorially driven; projects must have the potential to yield compelling stories and images. Applications are also judged on the qualifications of applicants and their teams and on the merit and uniqueness of the project. In addition to financial support, the Expeditions Council offers its grantees the opportunity to work effectively with the National Geographic's many

divisions. Grantees are therefore able to share the results of their expeditions with the National Geographic global audience. Grants generally range from $5,000 to $35,000 and are to be used for direct field expenses: transportation, supplies, subsistence, and permit costs as well as other related fees (e.g., interpreters, guides, and porters).
Requirements: International applicants are encouraged. However, submissions must be made in English if they are to receive timely consideration by the Expeditions Council. Applicants are expected to have qualifications and experience pertinent to the expedition or project they propose, and advanced academic degrees are not required. Those planning work in countries other than their own should consider including at least one local collaborator as part of their expedition team.
Restrictions: The Expeditions Council does not provide fees for photography, videography, and writing. Society grants may not be used for indirect costs, overhead, and other expenses not directly related to the project. Fringe benefits are also excluded, as are salaries. Funds may not be used for travel to scientific/professional meetings or conferences, legal actions, land acquisition, endowments, construction of permanent field stations, or to publish research results.
Geographic Focus: All States
Amount of Grant: 5,000-35,000 USD
Contact: Grants Administrator; (800) 647-5463; ecouncil@ngs.org
Internet: http://www.nationalgeographic.com/field/grants-programs/expeditions-council.html
Sponsor: National Geographic Society
1145 Seventeenth Street NW
Washington, DC 20036-4688

National Geographic Society Field Research and Exploration Grants **2873**
The National Geographic Society awards grants for scientific field research and exploration through its Committee for Research and Exploration. The program provides grants throughout the world for basic research in the sciences pertinent to geography including, but not limited to, anthropology, archaeology, astronomy, biology, botany, geography, geology, oceanography, paleontology, and zoology. Particular emphasis is placed on multidisciplinary projects of an environmental nature. Priorities favor research that addresses pressing environmental concerns. All projects must have geographic relevance or significance. Types of support include grants in aid, research grants, and seed money grants. Prior to applying, a potential investigator must submit a pre-proposal (500 words maximum length) outlining the significance of the research and describing who will conduct the research and where it will occur.
Requirements: Applicants are expected to have advanced degrees (Ph.D. or equivalent) and be associated with an educational organization or institution. Independent researchers or those pursuing a Ph.D.-level degree may apply, but awards to non-Ph.D. applicants are rare. As a general rule, all applicants are expected to have published a minimum of three articles in peer-reviewed scientific journals.
Restrictions: Society grants may not be used for expenses that are not directly related to the project. Fringe benefits for any individual or institution are excluded, as are salaries for those with faculty status, or any other paid professional affiliation. Funds may not be used for travel to scientific/professional meetings or conferences, nor to publish research results.
Geographic Focus: All States
Amount of Grant: 15,000-20,000 USD
Contact: Grants Administrator; (800) 647-5463; cre@ngs.org
Internet: http://www.nationalgeographic.com/research/grant/rg1.html
Sponsor: National Geographic Society
1145 Seventeenth Street NW
Washington, DC 20036-4688

National Geographic Society Young Explorers Grants **2874**
Young Explorers Grants (YEG) offer opportunities to individuals between the ages of 18 and 25 to pursue research, conservation, and exploration-related projects consistent with National Geographic's existing grant programs, including: the Committee for Research and Exploration (CRE), the Expeditions Council (EC), and the Conservation Trust (CT). Grants vary in amount depending on significance of the project, though most range between $2,000 and $5,000.
Requirements: Applicants are not expected to have advanced degrees (PhD or equivalent). However, applicants must provide a record of prior research or conservation action as it pertains to the proposed project.
Restrictions: Society grants may not be used for indirect costs, overhead, and other expenses not directly related to the project. Fringe benefits are also excluded, as are salaries. Funds may not be used for travel to scientific/professional meetings or conferences, legal actions, land acquisition, endowments, construction of permanent field stations, or to publish research results.
Geographic Focus: All States
Amount of Grant: 2,000-5,000 USD
Contact: Administrator; (800) 647-5463; cre@ngs.org or conservationtrust@ngs.org
Internet: http://www.nationalgeographic.com/field/grants-programs/yeg-application.html
Sponsor: National Geographic Society
1145 Seventeenth Street NW
Washington, DC 20036-4688

National Headache Foundation Research Grants **2875**
The National Headache Foundation is interested in research protocols that are objectively sound and whose results can, when published in the medical literature, contribute to the better understanding and treatment of headache and pain. Grants may be used for the purchasing of supplies needed for the study (chemical or pharmacological reagents, laboratory animals, tissue culture materials. forms, etc.) and for costs related to the recruiting of patients, data analysis, interpreting or reading results, etc. Applications are available online, and are due by December 1.
Requirements: Funds for a research project approved by the National Headache Foundation are awarded to the individual who is the principal investigator of the study. The funds are sent to the institution that employs the PI but designated for the PI and the approved study.
Restrictions: Grants from the NHF will not cover costs including overhead, salaries (payment for technicians, others collecting data, administrative services, etc.), purchasing of equipment, rent or other indirect expenses.
Geographic Focus: All States
Date(s) Application is Due: Dec 1
Contact: Carolyn Smith, Executive Assistant; (312) 274-2652 or (312) 274-2650; fax (312) 640-9049; info@headaches.org
Internet: http://www.headaches.org/For_Professionals/Research_Grants
Sponsor: National Headache Foundation
820 N Orleans, Suite 217
Chicago, IL 60610-3132

National Headache Foundation Seymour Diamond Clinical Fellowship in Headache Education **2876**
The one-year fellowship is intended to encourage, provide for, and engage physicians in the clinical management of headache patients. The program is an individualized hands-on approach at a headache clinic mutually selected by the committee and the applicant. The objectives are to develop competence in the management of common and complex headache problems, to publish a research paper keyed to the treatment of headache, and to prepare the applicant for a career in treating headache patients. The participating institution will be a specialized facility dedicated primarily to headache management, with facilities for inpatient and outpatient treatment. Up to 80 percent of the fellow's time will be clinical, 20 percent study/research.
Requirements: Applicant will be a graduate of an approved U.S. medical school and be four years or more postgraduate.
Geographic Focus: All States
Date(s) Application is Due: Feb 1
Amount of Grant: 40,000 USD
Contact: Fellowship Selection Committee; (312) 274-2650; fax (312) 640-9049; info@headaches.org
Internet: http://www.headaches.org/For_Professionals/Fellowship_Grants
Sponsor: National Headache Foundation
820 N Orleans, Suite 217
Chicago, IL 60610-3132

National Housing Endowment Challenge/Build/Grow Grant **2877**
The National Housing Endowment is the philanthropic arm of the National Association of Home Builders (NAHB). Working with organizations within the NAHB federation, the National Housing Endowment is offering eligible applicants up to $5,000 in matching challenge grant funds (dollar for dollar) for programs and projects addressing NAHB's highest priority areas of Education, Land Use and Labor Shortage/Worker Training.
Requirements: The program/project must address the National Housing Endowment mission of furthering the education, training, and research goals of the home building industry. The program/project can initially be at the local level, but should have the potential to be used as a national model to assist in addressing the education, land use and workforce training/recruitment issue(s). In addition, the program/project should adhere to the IRS regulation definition of ????serving the public interest???? and not exist to serve the sole interests of the home building industry.
Restrictions: Programs and projects are not eligible for support over a multi-year period. HBA projects that have already received funding in 2007 may not apply. Preference will be given to newly created programs/projects especially those that create partnerships in their communities.
Geographic Focus: All States
Date(s) Application is Due: Apr 16
Amount of Grant: Up to 3,000 USD
Contact: Tracy Hensler, (800) 368-5242, ext. 8483; fax (202) 266-8177; nhe@nahb.com
Internet: http://www.nationalhousingendowment.com/CBG.htm
Sponsor: National Housing Endowment
1201 15th Street NW
Washington, DC 20005

National Leadership Grant Competition **2878**
The Sons of Italy awards merit-based grants for Italian American students enrolled full time in undergraduate or graduate programs at accredited colleges or universities. Criteria for the award include academic excellence and leadership potential.
Requirements: Applicants must be U.S. citizens who are full-time undergraduate or graduate students of Italian heritage.
Geographic Focus: All States
Date(s) Application is Due: Feb 28
Amount of Grant: 4,000-25,000 USD
Contact: Administrator; (202) 547-5106; fax (202) 547-8168; scholarships@osia.org
Internet: http://www.osia.org/public/scholarships/grants.asp
Sponsor: Sons of Italy Foundation
219 E Street NE
Washington, DC 20002

National LGBT Scholarship Fund **2879**

The mission of the foundation is to provide financial support, mentoring, and hope to meritorious gay, lesbian, bisexual, and transgender students, who are marginalized because of their sexual orientation or gender identity. Submit applications online by the listed application deadline.
Requirements: The application process is open to all LGBT students nationwide regardless of level of education.
Geographic Focus: All States
Date(s) Application is Due: Mar 1
Samples: To be distributed among 10 individuals—for scholarships for gay, lesbian, bisexual, and transgender college students, $132,000.
Contact: Scholarship Administrator; (866) 337-6468; fax (866) 397-6468; info@thepointfoundation.org
Internet: http://www.thepointfoundation.org/scholarships.html
Sponsor: Point Foundation
P.O. Box 11210
Chicago, IL 60611

National MPS Society Grants **2880**
The society awards research grants and fellowships to qualified medical researchers to promote medical research in the fields of mucopolysaccharidosis (MPS) and mucolipidosis (ML). The program supports MPS (I, II, III IV, V, and VI) research, ML II/III research, and general MPS research. In February of each year the Society announces the grants offered, and funding begins July 1. Occasionally a second grant offering is made later in the year. All applications are reviewed and ranked by members of the National MPS Society Scientific Advisory Board. The final decision of award recipients is determined by the Board of Directors of the National MPS Society. Funding for the grants comes from the annual National MPS Society 5k walk/runs in addition to donations and fundraisers. Grants and fellowships are typically awarded for a two-year period.
Requirements: Qualified medical researchers are eligible.
Geographic Focus: All States
Date(s) Application is Due: May 1
Amount of Grant: Up to 50,000 USD
Contact: Barbara Wedehase, Executive Director; (877) 677-1001 or (919) 806-0101; barbara@mpssociety.org or info@mpssociety.org
Internet: http://www.mpssociety.org/research/
Sponsor: National MPS Society
P.O. Box 14686
Durham, NC 27709-4686

National Parkinson Foundation Clinical Research Fund Grants **2881**
The National Parkinson Foundation (NPF)'s Clinical Research Fund (CRF) is designed to support cutting-edge clinical studies of new therapeutic options for Parkinson's patients as well as research comparing the effectiveness of existing treatment options. Conducted by or in collaboration with NPF's Centers of Excellence, CRF-funded research aims to discover the best treatment and care options for people living with Parkinson's disease. Broad clinical-research areas funded by the CRF program are as follows: care quality; imaging; cognition; physical therapy; care models; exercise; and biomarkers. CRF Requests for Application are issued as funding becomes available, but no less than once per year. Applicants may request two-year funding (or less) for their projects. An example of the CRF application process is as follows: an RFP is released in August; NPF schedules pre-submission conference calls for September and October; applicants submit their proposals in December; peer-review is conducted January through March; awards are announced in April. Specific deadline dates will vary from year to year and will be detailed in the RFP. More detailed information is generally posted in a listserv to which all applicants are invited to subscribe (address given in the RFP). Applications must be submitted electronically through NPF's online grant submission system. Registration is required.
Requirements: Principal investigators must be clinical investigators affiliated with a currently-certified NPF Center of Excellence. Management of the grant budget and reporting must be done at the Center of Excellence. Further, each application must be cosigned by the institutional officer responsible for grants and contracts. Principal investigators must meet their institution's criteria for eligibility to receive grants and oversee expenditures. Individual investigators at non-certified institutions may apply as collaborators. Applicants must reasonably expect to have Institutional Review Board approval shortly after the award date.
Restrictions: There are no limits on the number of grants an institution may request but each grant is limited to no more than $250,000 over a two-year period. Postdoctoral fellows, fellows, and students cannot serve as a principal investigator. Grants cannot be used to purchase equipment or to support conference fees or travel (except for travel to a Center of Excellence collaborating on the research project). Indirect costs will not be funded. NPF no longer funds basic laboratory research.
Geographic Focus: All States, All Countries
Amount of Grant: Up to 250,000 USD
Contact: Jorge Zamudio, (305) 243-4830; fax (305) 243-8144; jzamudio@parkinson.org or iiresearchgrants@parkinson.org
Internet: http://www.parkinson.org/Professionals/Grants—-Research.aspx
Sponsor: National Parkinson's Foundation
1501 N.W. Ninth Avenue / Bob Hope Road
Miami, FL 33136-1494

National Psoriasis Foundation Research Grants **2882**
The foundation will award up to five $50,000 one year pilot project grants to support advancement of both basic and clinical research into the cause and cure of psoriasis and psoriatic arthritis. The purpose of pilot project grants is to provide support for innovative early stage ideas. These grants are designed to allow the investigator to determine whether their idea is worth pursuing and then to secure longer-term funding.
Requirements: Applicants should use their NIH biographical sketches. Download the required application form from the website.
Geographic Focus: All States
Date(s) Application is Due: Oct 20
Amount of Grant: Up to 50,000 USD
Contact: Bruce F. Bebo, Jr., Director of Research and Medical Programs; (503) 244-7404 or (800) 723-9166; research@psoriasis.org
Internet: http://www.psoriasis.org/research/grants/index.php
Sponsor: National Psoriasis Foundation
6600 SW 92nd Avenue, Suite 300
Portland, OR 97223-7195

National Rising Scholars Award to Advance Research on Higher Education for the Public Good **2883**
The program makes awards for pretenured faculty, young practitioners, and advanced graduate students in any discipline in U.S. higher education institutions who engage in research that explores higher education's role in serving the public good. The award will be made to individuals in order to support outstanding research and encourage publication. Recipients receive waiver for conference registration fees to attend one of the national conferences of a collaborating association (Association for Institutional Research, American Association of Higher Education, American Education Research Association, Association for the Study of Higher Education, and American College Personnel Association); participation in the Kellogg Forum Higher Educational Intergenerational Scholars Symposium; and inclusion of the manuscript from their project in an annual Kellogg Forum edited monograph and consideration for publication in the peer review journals of the collaborating associations. Applicants may apply online.
Requirements: All recipients should hold a doctorate (no more than five years post degree) or be at the dissertation stage of their doctorate studies; they must also be affiliated with a higher educational institution.
Geographic Focus: All States
Date(s) Application is Due: Apr 1
Contact: Kellogg Forum; (734) 615-8882; fax (734) 615-9777; Rising.Scholars@umich.edu
Internet: http://www.thenationalforum.org/projects_risingscholars.shtml
Sponsor: National Forum on Higher Education for the Public Good
610 E University Avenue, 2339 SEB
Ann Arbor, MI 48109-1259

National Security Education Program David L. Boren Graduate Fellowships **2884**
Each year, NSEP publishes a list of geographic areas, languages, and fields of study identified as critical to U.S. national security. In addition, it emphasizes a diverse list of fields of study, including business, economics, history, international affairs, law, applied sciences and engineering, health and biomedical sciences, political science, and other social sciences. Applications will be considered for study in other countries and world regions where increased understanding and appreciation represents an important contribution to U.S. national security. Fellowship recipients are required to seek employment with an agency or office of the federal government involved in national security affairs. Awards are made for a minimum of one and a maximum of six academic semesters (24 months). Applicants design their own programs and may combine domestic language and cultural study with overseas study. All fellowships must include study of a modern language other than English and the study of an area and culture.
Requirements: NSEP welcomes applications from U.S. citizens enrolled in or applying to a graduate degree program in an accredited U.S. college or university located within the United States.
Restrictions: Support may not be used for study of Western Europe, Canada, Australia, or New Zealand.
Geographic Focus: All States
Date(s) Application is Due: Jan 30
Contact: Contact; (800) 498-9360 or (202) 884-8285; nsep@aed.org
Internet: http://www.aed.org/nsep/application.html
Sponsor: Academy for Educational Development
1825 Connecticut Avenue NW, Suite 744
Washington, DC 20009-5721

National Security Education Program Grants to Institutions of Higher Education **2885**
The National Security Education Program (NSEP) encourages proposals that address programs in specific foreign languages, countries, and areas generally underrepresented in U.S. programs; and/or address professional, disciplinary, and/or interdisciplinary opportunities to incorporate international education into the curriculum. Proposals are sought that improve language acquisition and cultural knowledge; improve study and work opportunities abroad, stimulate faculty interest and involvement in international aspects of curriculum, improve information dissemination and linkages across institutions, and/or broaden the base of interdisciplinary and institutional relationships. Evaluation of proposals will be based on significance of the problem, responsiveness to identified needs, cost effectiveness, prospects for wider impact, applicant's capacity, and evaluation plans. Prospective applicants should obtain the complete program announcement, which contains final deadline dates and proposal requirements. The grant solicitation may be downloaded from the listed NSEP Web site.
Requirements: Accredited two-year and four-year U.S. institutions of higher education.
Restrictions: Federal government schools are ineligible.
Geographic Focus: All States

Contact: Grants Manager; (202) 326-7697; fax (202) 326-7698; nsep@iie.org
Internet: http://www.ndu.edu/nsep
Sponsor: National Security Education Program
1400 K Street NW, Suite 650
Washington, DC 20005-2403

National Sleep Foundation Pickwick Postdoctoral Fellowship **2886**
The fellowships enable young researchers to devote full-time professional effort to research related to the study of sleep or sleep disorders. Fellows receive funding for one or two years. The second year of support is dependent upon satisfactory progress in the first year. Applicants must conduct research in recognized U.S. or Canadian programs of study or laboratories with strong mentorship in the appropriate area. Fellowships are available for basic, applied, or clinical research. Minorities are encouraged to apply.
Requirements: Candidates must hold either an MD, DVM, PhD, or DO degree in which the degree or subsequent training has been completed within the past five years.
Restrictions: Faculty positionS or who have received an NIH grant are ineligible.
Geographic Focus: All States
Date(s) Application is Due: Dec 1
Contact: Jessica Steinitz, (202) 347-3471, ext. 209; jsteinitz@sleepfoundation.org
Internet: http://www.sleepfoundation.org/nsfalert/index.php?id=31&secid=159#item159
Sponsor: National Sleep Foundation
1522 K Street NW, Suite 500
Washington, DC 20005

National Turkey Federation Research Award **2887**
This award is given each even-numbered year for the outstanding record of turkey research in the United States during the past five years. The work must be original and deal specifically with turkeys. The recipient will be announced at the annual meeting of the Poultry Science Association. The award includes a cash amount and a plaque.
Requirements: The work, specifically with turkeys, must be original in one of a wide range of fields such as, but not limited to, economics, food technology, genetics, merchandising, pathology, physiology, and nutrition. Any administrator or researcher may nominate a qualified candidate. Nominations can be made by any individual knowledgeable about the nominee's accomplishments. The nomination should include the following items: nomination letter, biography, pertinent bibliography, reprints of appropriate and supporting data. Material must be submitted electronically via the PSA awards website no later than March 1 prior to the annual meeting at which the award will be given.
Geographic Focus: All States
Date(s) Application is Due: Mar 1
Amount of Grant: 1,000 USD
Contact: Mary Swenson; (217) 356-5285; fax (217) 398-4119; marys@assochq.org
Internet: http://www.poultryscience.org/awards.asp
Sponsor: Poultry Science Association
1111 North Dunlap Avenue
Savoy, IL 61874

National Urban Fellows **2888**
The National Urban Fellows leadership development program is a rigorous, 14-month, full-time graduate degree program comprising two semesters of academic course work and a nine-month mentorship assignment. Fellows receive a $25,000 stipend, health insurance, a book allowance, relocation and travel reimbursement, and full payment of tuition, in addition to their ongoing personal and professional development. The program culminates in a Master of Public Administration (MPA) degree from the City of New York's Bernard M. Baruch College, School of Public Affairs. During the mentorship, Fellows complete course work via distance learning. An MPA degree is awarded to Fellows upon completion of all academic and program requirements. Application materials are available at the website.
Requirements: United States citizens who have bachelor's degrees; have three to five years' full-time work experience in an administrative or management capacity; have demonstrated exceptional ability, maturity, leadership potential, and commitment to the solutions of urban and rural problems; and are willing to relocate for the duration of the fellowship year are eligible. An official transcript of all undergraduate and graduate education will be required.
Geographic Focus: All States
Amount of Grant: 26,000 USD
Contact: Miguel A. Garcia, Jr.; (212) 730-1700; fax (212) 730-1823
Internet: http://www.nuf.org/fellows-overview
Sponsor: National Urban Fellows
989 Avenue of the Americas, Suite 400
New York, NY 10018

National Wetlands Awards **2889**
Cosponsored by the Environmental Law Institute, U.S. Environmental Protection Agency, U.S. Fish and Wildlife Service, USDA Forest Service, Natural Resources Conservation Service, and National Marine Fisheries Service, the awards recognize individuals who have demonstrated excellence in wetland conservation. Awards will be made in the the following categories: education and outreach; science research; conservation and restoration; landowner stewardship; state, tribal, and local program development; and wetland community leader. Awardees are chosen by a committee of wetland experts representing the conservation and business communities, as well as federal, state, and local governments. Guidelines and nomination form are available online.
Requirements: Nominations must be for individuals—not organizations or groups of individuals—whose programs or projects operate at the regional, state, or local level.
Restrictions: Federal employees are not eligible. Projects not currently in compliance with the Clean Water Act or similar state or local statutes, as well as projects compelled by enforcement actions under these statutes, are not eligible.
Geographic Focus: All States
Date(s) Application is Due: Dec 15
Contact: Roxanne Thomas, Wetlands Program Director; (202) 939-3827; fax (202) 939-3868; thomas@eli.org or wetlandsawards@eli.org
Internet: http://www2.eli.org/nwa/nwaprogram.htm
Sponsor: Environmental Law Institute
2000 L Street NW, Suite 620
Washington, DC 20036

National Wildlife Federation Campus Ecology Fellowships **2890**
National Wildlife Federation's Campus Ecology Fellows confront global warming on their campuses and help to educate and engage the campus community on global warming impacts and solutions. Federation Fellowships allow students to pursue their vision of an ecologically sustainable future through tangible projects to confront global warming on campus and in the community. Fellows gain practical experience in the conservation field and first-hand knowledge of the challenges and opportunities inherent in successful conservation efforts. Fellowship guidelines and the application are available at the Federation website.
Requirements: Candidates may apply directly through the Federation's job database.
Geographic Focus: All States
Contact: Fellowship Coordinator; (800) 822-9919
Internet: http://www.nwf.org/Global-Warming/Campus-Solutions/Get-Involved/Apply-for-a-Fellowship.aspx
Sponsor: National Wildlife Federation
11100 Wildlife Center Drive
Reston, VA 20190

National Wildlife Federation Internships **2891**
Through the internship program, the National Wildlife Federation (NWF) brings students and recent graduates to its offices nationwide, where they can experience many aspects of NWF's conservation mission, including work on endangered species, land stewardship, and protecting our nation's water. Departments benefiting from the skills interns provide include conservation, education, fundraising, communications, legal and publications. Working side by side with knowledgeable professionals, interns receive support from the internship coordinator, while developing a variety of professional and personal life skills. Internship projects support commonsense conservation in communities throughout the nation. Interns receive medical, dental, and holiday benefits. Interns must work a minimum of 20 hours per week and are expected to complete a 24-week internship to be eligible for these benefits. Some departments substitute a paid salary with college credit. Students apply for the internship through the Federation's job page.
Requirements: Graduates and graduate students from all fields of study may apply. To ensure a rigorous and meaningful professional, educational, and personal experience, interns must: document and evaluate their experience for the internship coordinator at scheduled intervals throughout the internship. They must also complete the work objectives agreed to at the beginning of the internship. Applicants are selected based on a variety of measures, including professional goals, course work, job experience, resume and cover letter, and grade point average, along with professional/academic references. Candidates should demonstrate commitment to and experience working with multi-cultural communities, environmental justice and diversity. Due to the high volume of applications, only candidates selected for an interview will be contacted.
Restrictions: For-credit interns are not eligible for benefits.
Geographic Focus: All States
Amount of Grant: 500 USD
Contact: Internship Coordinator; (800) 822-9919; internopp@nwf.org
Internet: http://www.nwf.org/About/Jobs-at-NWF.aspx
Sponsor: National Wildlife Federation
11100 Wildlife Center Drive
Reston, VA 20190

National Wildlife Refuge System Centennial Scholarship Program **2892**
The scholarship supports students whose research will contribute toward improved management and understanding of the diverse fish, wildlife, and plant resources found on U.S. National Wildlife Refuges. Under this competitive program, scholarships will be awarded to support students' research expenses, tuition, fees, books, and room and board. The one-time, one-year awards are based on merit and will take into consideration the value of the research toward improving management and understanding of fish, wildlife, and plant resources on or related to National Wildlife Refuges as well as the student's academic achievements.
Requirements: To be eligible for consideration, a student must be a U.S. citizen or legal resident enrolled in an accredited institution of higher education in the United States pursuing a graduate or undergraduate degree (sophomores and juniors in the current academic year only) in environmental science, natural resource management, biology, public policy, education, geography, political science, or related disciplines. Only those students working on a National Wildlife Refuge or whose work is directly applicable to resource management on a National Wildlife Refuge(s) are eligible.
Geographic Focus: All States
Date(s) Application is Due: Apr 15
Amount of Grant: 5,000-15,000 USD
Contact: Kevin Kilcullen, (703) 358-2029; Kevin_Kilcullen@fws.gov

Internet: http://www.nfwf.org/programs/centennialscholarship/index.cfm
Sponsor: National Fish and Wildlife Foundation #6
1133 15th Street, NW, Suite 1100
Washington, DC 97205

National Wild Turkey Federation Research Grants **2893**
The federation awards grants to promote research that will maximize wild turkey populations in all suitable habitats and assure the optimum sustained consumptive and nonconsumptive recreational use. The research program is also intended to promote public awareness and support for the conservation and wise management of the American wild turkey. State Super Fund grants deal with priority projects as determined by the NWTF state chapters and their respective wildlife agencies. These projects may include habitat management (planting wildlife openings), education (creating private landowner awareness of turkey management concepts), hunter safety, and restoration. National projects, which are broader in scope than state projects, may take the form of research grants to support university research or as a publication that can be used by government and corporate biologists as a management reference tool.
Geographic Focus: All States
Contact: Director of Research and Management; (800) 843-6983 or (803) 637-3106; fax (803) 637-0034; kmorris@nwtf.net
Internet: http://www.nwtf.com/conservation
Sponsor: National Wild Turkey Federation
770 Augusta Road, P.O. Box 530
Edgefield, SC 29824

NATO Advanced Research Workshops (ARW) Program **2894**
The aims of an ARW are to contribute to the critical assessment of existing knowledge on new important topics, to identify directions for future research, and to promote close working relations between scientists from different countries and with different professional experience. Workshops host from 20 to 50 experts and last from two to five days. Funding is provided to ARW directors to cover organizational expenses and the travel and living expenses of key speakers, and to contribute to the travel and living expenses of other participants of whom 40-50 percent are from partner countries. ARWs are codirected by a scientist from a partner country and a scientist from a NATO country; ARWs may be held in either a NATO or a partner country. Areas of support and deadlines are as follows: physical and engineering science, March 1, August 1, November 1; life sciences and environmental and earth sciences, April 1, September 1, December 1; and security-related civil sciences, February 1, May 1, October 1. Applications are available on the Web site.
Requirements: Scientists and engineers working permanently in NATO member countries and cooperating Euro-Atlantic Partnership Council (EAPC) countries are eligible to apply.
Geographic Focus: All States
Date(s) Application is Due: Feb 1; Mar 1; Apr 1; May 1; Aug 1; Sep 1; Oct 1; Nov 1; Dec 1
Contact: Scientific Affairs; 011 322 707 41 11; fax 011 322 707 42 32; natodoc@hq.nato.int
Internet: http://www.nato.int/science-old/e/arw.htm
Sponsor: North Atlantic Treaty Organization
Boulevard Leopold III
Brussels, B-1110 Belgium

NATO Advanced Study Institutes (ASI) Program **2895**
Grants are offered to organize Advanced Study Institutes, which are high-level tutorial activities where a carefully defined subject is treated in depth by lecturers of international standing, and which include new advances in the subject not yet in university courses. Guidelines are available online.
Requirements: Support is available for ASIs only in priority research topics in the areas of defense against terrorism or countering other threats to security and/or in a partner-country priority.
Geographic Focus: All States
Date(s) Application is Due: Mar 1; Sep 1; Dec 1
Contact: Scientific Affairs; 011 322 707 41 11; fax 011 322 707 42 32; natodoc@hq.nato.int
Internet: http://www.nato.int/science/support/asi/asi-nfa.htm
Sponsor: North Atlantic Treaty Organization
Boulevard Leopold III
Brussels, B-1110 Belgium

NATO Collaborative Linkage Grants (CLG) **2896**
The aim of the program is to facilitate collaboration between research scientists or research teams in order to stimulate scientific research through the pooling of capabilities and resources. CLGs are available for collaboration between researchers in partner and NATO countries. Funding is available for reciprocal visits abroad of up to five members of each research team collaborating on a joint research project. Awards are for one or two years. Computer networking supplements are available to upgrade the computer networking capabilities of the partner-country laboratories, upon additional application once a CLG has been awarded. Grants are given for projects that rely for basic costs on national funding but where the costs for the international collaboration cannot be met from other sources. The grants support travel and living expenses of investigators for short visits to partner institutions abroad. Areas of support and deadlines are as follows: physical and engineering science, March 1, August 1, November 1; life sciences and environmental and earth sciences, April 1, September 1, December 1; and security-related civil sciences, February 1, May 1, October 1. Applications are available on the Web site. Grants will be in Euro. Payment of the grant, however, may be in any currency, as specified by the grantholder at the time of payment.
Requirements: To be eligible for support a project must be specific and the collaboration should be between scientists in NATO countries and NATO's Partner countries of the Euro-Atlantic Partnership Council. Scientists of the Mediterranean Dialogue countries may also apply for support of a CLG, and should draw up applications in cooperation with scientists from NATO countries. Applicants should follow the guidelines given for NATO Partner countries.
Restrictions: No salaries or stipends for any of the collaborators, students, technical assistants, etc., will be offered under these grants. No institutional overhead expenses may be charged to the grant. No support can be supplied for scientists while on sabbatical or other extended leave abroad, or to allow attendance at conferences, symposia, workshops, etc. Scientists may not be involved in more than one collaborative linkage grant at the same time.
Geographic Focus: All States
Date(s) Application is Due: Feb 1; Mar 1; Apr 1; May 1; Aug 1; Sep 1; Oct 1; Nov 1; Dec 1
Amount of Grant: 5,000 USD
Contact: Scientific Affairs; 011 322 707 41 11; fax 011 322 707 42 32; natodoc@hq.nato.int
Internet: http://www.nato.int/science-old/e/clg.htm
Sponsor: North Atlantic Treaty Organization
Boulevard Leopold III
Brussels, B-1110 Belgium

NATO Expert Visits (EV) Grants **2897**
The objective of expert visits program is to allow high-level specialists from NATO and partner countries to benefit from one another's expertise. Grants support the travel and living expenses of experts from NATO countries visiting research laboratories in partner countries, and vice versa, for consultation on current research projects. The duration of visits may range from a few days to a maximum of one month. Applications are available on the Web site.
Requirements: To be eligible for support the visits must enable senior scientists to provide expertise on a specific project, and must be for visits either of a partner-country expert to institutions in a NATO country or a NATO-country expert to institutions in a Euro-Atlantic Partnership Council (EAPC) country.
Geographic Focus: All States
Date(s) Application is Due: Mar 1; Sep 1; Dec 1
Contact: Scientific Affairs; 011 322 707 41 11; fax 011 322 707 42 32; natodoc@hq.nato.int
Internet: http://www.nato.int/science/support/ev/ev-info.htm
Sponsor: North Atlantic Treaty Organization
Boulevard Leopold III
Brussels, B-1110 Belgium

NATO Science for Peace Grants **2898**
The objective of the Science for Peace subprogram is to offer support to Euro-Atlantic Partnership Council (EAPC) partner countries in their transition towards a market-oriented, environmentally sound economy. This subprogram supports applied science and technology projects in partner countries that relate to industrial or environmental problems, when such problems involve collaboration between science and industry or between science and other end-users. Science for Peace helps partner scientists to increase contacts in the NATO science community, while building a stronger science infrastructure in their home countries. Support is for three- to five-year funding of applied R&D projects having well-defined objectives, well-defined work plans, and well-defined budgets. Projects are of a cooperative nature, jointly carried out among NATO and partner countries; they take place in partner countries. Projects involving more than one partner country are encouraged. Funds are provided for partner-country participants (and Greece and Turkey) to cover mainly scientific equipment, computers, software, travel, training of project personnel and project-specific consumables. Funds are provided for NATO-country participants mainly to cover travel and project-specific consumables. Calls for proposals, application forms, and guidelines are posted on the Web site when available, generally in March.
Geographic Focus: All States
Contact: Scientific Affairs; 011 322 707 41 11; fax 011 322 707 42 32; natodoc@hq.nato.int
Internet: http://www.nato.int/science/sfp/index.htm
Sponsor: North Atlantic Treaty Organization
Boulevard Leopold III
Brussels, B-1110 Belgium

NBME Stemmler Medical Education Research Fund Grants **2899**
The purpose of the fund is to provide support for research or development of innovative evaluation methodology or techniques with the potential to advance assessment in medical education or practice, at eligible medical schools. Collaborative investigations within or among institutions are eligible, particularly as they strengthen the likelihood of the project's contribution and success. Proposed projects may request up to two years of support. Applicants for both new proposals and competing renewals may request up to $150,000 of NBME funding support for a project period of up to two years.
Requirements: Medical schools accredited by the Liaison Committee on Medical Education or the American Osteopathic Association are eligible.
Geographic Focus: All States
Date(s) Application is Due: Oct 3
Amount of Grant: Up to 150,000 USD
Contact: Officer of Research Support; (215) 590-9657; stemmlerfund@nbme.org
Internet: http://www.nbme.org/research/stemmler/index.html
Sponsor: National Board of Medical Examiners
3750 Market Street
Philadelphia, PA 19104

NCAA Postgraduate Scholarships **2900**
The NCAA awards up to 174 postgraduate scholarships annually, 87 for men and 87 for women. The scholarships are awarded to student-athletes who excel academically

and athletically and who are in their last year of intercollegiate athletics competition. The one-time grants are awarded for fall sports, winter sports, and spring sports. Each sports season (fall, winter and spring), there are 29 scholarships available for men and 29 scholarships available for women. The scholarships are one-time, nonrenewable grants.
Requirements: Eligibility is restricted to student-athletes at NCAA member institutions who, in their final year of eligibility, have performed with distinction as members of varsity teams in the sport for which they are nominated. Nominees must have a minimum grade-point average of 3.0 (based on a 4.0 scale), be nominated by the faculty athletics representative or athletics director, and intend to continue academic work beyond the baccalaureate degree as a full-time graduate student.
Geographic Focus: All States
Date(s) Application is Due: Feb 24; May 5; Dec 9
Amount of Grant: 7,500 USD
Contact: Committee Coordinator; (317) 917-6222; fax (317) 917-6888; info@ncaa.org
Internet: http://www1.ncaa.org/membership/ed_outreach/prof_development/postgrad_scholarships.html
Sponsor: National Collegiate Athletic Association
P.O. Box 6222
Indianapolis, IN 46206-6222

NCAR Graduate Research Fellowships 2901

The NCAR Advanced Study Program offers a small number of graduate fellowships that support university PhD thesis research conducted in collaboration with NCAR scientists. Usually, these fellowships are restricted to research projects that require facilities or expertise available at NCAR but not at the collaborating university. Fellowships may also be supported when the NCAR scientist contributes scientific or technical expertise in an area outside the research interests of the faculty at the collaborating university. Research topics at NCAR include atmospheric dynamics, climate, mesoscale weather, atmospheric chemistry, studies of the upper atmosphere, solar physics, turbulence, cloud physics, oceanography, the planetary boundary layer, societal impacts of atmospheric processes, mathematical and numerical methods applied to modeling of the atmosphere or sun, development of instrumentation for field and laboratory investigations, and many other areas of atmospheric research. The fellowship may support benefits and some travel costs. Fellowships are not available every year. Contact the organization or check the Web site for more information.
Requirements: The student must intend to work on a PhD thesis in cooperation with an NCAR program and must be in residence at NCAR for the thesis research.
Geographic Focus: All States
Amount of Grant: 18,070 USD
Contact: Barbara Hansford; (303) 497-1601; fax (303) 497-1646; barbm@ucar.edu
Internet: http://www.asp.ucar.edu/asp/arnaud.html
Sponsor: National Center for Atmospheric Research
P.O. Box 3000
Boulder, CO 80307-3000

NCAR Postdoctoral Fellowships in the Atmospheric Sciences 2902

The program provides an opportunity for talented atmospheric scientists to continue to pursue their research interests and to offer an opportunity to scientists from other disciplines to apply their training to research in the atmospheric sciences. Each fellow is encouraged to pursue his/her own research objectives while taking maximum advantage of the opportunities available at NCAR, including collaboration with staff and use of research facilities. NCAR concentrates most of its research efforts on coordinated projects in atmospheric dynamics (on all scales) and models, climate science, cloud physics, atmospheric chemistry and radiation, turbulence, upper-atmosphere physics, solar and solar-terrestrial physics (including ionospheric studies and aeronomy), oceanography, and atmospheric technology. The NCAR program also includes studies of the interaction of the atmosphere with the oceans, the cryosphere, the earth's surface, and human society. Scientists interested in applying such skills as biology, ecology, geology, economics, and political science to atmospheric issues are encouraged to apply.
Requirements: An applicant must have a PhD in an area related to NCAR science during the first four years preceding the application date or must be able to complete such a PhD before starting the NCAR appointment.
Restrictions: Scientists with more experience should apply to the director of the division in which they would like to work for positions as visiting scientists.
Geographic Focus: All States
Date(s) Application is Due: Jan 5
Amount of Grant: 47,000-49,000 USD
Contact: Paula Fisher, Administrator; (303) 497-1646; asp-apply@ucar.edu
Internet: http://www.asp.ucar.edu/asp/pdann.html
Sponsor: National Center for Atmospheric Research
P.O. Box 3000
Boulder, CO 80307-3000

NCCAM Exploratory Developmental Grants for Complementary and Alternative Medicine (CAM) Studies of Humans 2903

The National Center for Complementary and Alternative Medicine (NCCAM), the National Cancer Institute (NCI), the National Institute on Aging (NIA), and the Office of Dietary Supplements (ODS) invites high quality exploratory/developmental research grant applications of humans in all domains of complementary and alternative medicine (CAM), in order to obtain preliminary data that can be used as a foundation for a larger clinical study. NCCAM groups CAM practices into four domains: biologically-based practices; energy medicine; manipulative and body-based therapies, and mind-body medicine. In addition, NCCAM funds studies of whole medical systems that employ practices drawn from the four domains. The total project period for an application submitted in response to this funding opportunity may not exceed three years. Direct costs are limited to $400,000 over a three-year period, with no more than $250,000 in direct costs allowed in any single year.
Requirements: The following organizations and institutions may apply: public/state controlled institutions of higher education; private institutions of higher education; Hispanic-serving institutions; historically Black colleges and universities (HBCUs); tribally controlled colleges and universities (TCCUs); Alaska native and native Hawaiian serving institutions; nonprofits with 501(c)3 IRS status; nonprofits without 501(c)3 IRS status; small businesses; for-profit organizations; State governments; regional organizations; U.S. territories or possessions; Indian/Native American tribal governments; Indian/Native American tribally designated organizations; county governments; city or township governments; special district governments; independent school districts; public housing authorities/Indian housing authorities; eligible agencies of the Federal government; and faith-based or community based organizations.
Geographic Focus: All States
Date(s) Application is Due: Feb 16; Jun 16; Oct 16
Amount of Grant: Up to 400,000 USD
Contact: April Bower, (301) 451-3560; fax (301) 480-3621; bowera@mail.nih.gov
Internet: http://grants.nih.gov/grants/guide/pa-files/PAR-08-135.html#SectionVII
Sponsor: National Center for Complementary and Alternative Medicine
9000 Rockville Pike
Bethesda, MD 20892-5475

NCCAM Ruth L. Kirschstein National Research Service Awards for Postdoctoral Training in Complementary and Alternative Medicine 2904

The National Center for Complementary and Alternative Medicine (NCCAM) awards individual postdoctoral research training fellowships to promising applicants with the desire and potential to become productive, independent investigators in the field of complementary and alternative medicine (CAM) research. The primary objective of this funding opportunity is to help ensure that diverse pools of highly trained scientists will be available in adequate numbers and in appropriate research areas to carry out the Nation's biomedical, behavioral, and clinical CAM research agendas. Applicants may come from a range of career backgrounds, including biomedical research, conventional allopathic medicine, or CAM healing practices. The number of awards and the total amount of funding that NCCAM expects to award through this announcement will depend on the scientific merit of applications received, relevance to NCCAM's program priorities, and the availability of funds.
Requirements: Eligible organizations include for-profit and non-profit organizations, public or private institutions such as universities, colleges, hospitals and laboratories, units of State and local governments, eligible agencies and labs of the Federal government including NIH intramural labs, and domestic or foreign institutions/organizations. Applicant fellows must be citizens or non-citizen nationals of the United States, or have been lawfully admitted to the United States for permanent residence and have in their possession an Alien Registration Receipt Card.
Restrictions: Individuals on temporary or student visas are not eligible.
Geographic Focus: All States
Date(s) Application is Due: Apr 8; Aug 8; Dec 8
Contact: Nancy J. Pearson, (301) 594-0519; fax (301) 480-3621; pearsonn@mail.nih.gov
Internet: http://grants.nih.gov/grants/guide/pa-files/PAR-07-319.html
Sponsor: National Center for Complementary and Alternative Medicine
9000 Rockville Pike
Bethesda, MD 20892-5475

NCCAM Translational Tools for Clinical Studies of CAM Interventions Grants 2905

This Funding Opportunity Announcement (FOA) issued by the National Center for Complementary and Alternative Medicine (NCCAM) of the National Institutes of Health (NIH) encourages investigator(s)-initiated applications that propose to develop, enhance, and validate translational tools to facilitate rigorous study of complementary and alternative medicine (CAM) approaches that are in wide use by the public. Multidisciplinary studies and collaboration among investigators with expertise in appropriate disciplines are encouraged. NCCAM intends to award up to $3 million (total costs) per year, to support approximately 4 to 6 new grants. A maximum of $300,000 (direct costs) per year may be requested and the total project period for an application submitted in response to this FOA will be five years.
Requirements: The following organizations and institutions are eligible to apply: individuals; public/state controlled institutions of higher education; private institutions of higher education; Hispanic-serving institutions; historically Black colleges and universities (HBCUs); tribally controlled colleges and universities (TCCUs); Alaska Native and Native Hawaiian serving institutions; nonprofits with 501(c)3 IRS status; nonprofits without 501(c)3 IRS status; for-profit organizations; state governments; Indian/Native American tribal governments (federally recognized); Indian/Native American tribally designated organizations; county governments; city or township governments; special district governments; independent school districts; public housing authorities/Indian housing authorities; U.S. territory or possession; Indian/Native American tribal governments (other than federally recognized); regional organizations; non-domestic (non-U.S.) entities (foreign organizations); other eligible agencies of the federal government; and faith-based or community-based organizations.
Geographic Focus: All States
Date(s) Application is Due: Jul 17
Amount of Grant: 300,000 USD
Contact: Dr. Partap S. Khalsa, Division of Extramural Research Contact; (301) 594-3462; fax (301) 480-1587; khalsap@mail.nih.gov

Internet: http://grants.nih.gov/grants/guide/rfa-files/RFA-AT-09-002.html
Sponsor: National Center for Complementary and Alternative Medicine
9000 Rockville Pike
Bethesda, MD 20892-5475

NCFR Affiliate Councils Outstanding Student Research Papers 2906
The National Council on Family Relations, founded in 1938, is the oldest multidisciplinary, nonpartisan professional organization focused solely on family research, practice and education. There are two awards each year, one for a graduate student paper and one for an undergraduate student paper. The graduate student paper winner will receive $250 and a plaque. The undergraduate student paper winner will receive $100 and a certificate.
Requirements: While students may submit their papers to NCFR on their own, if multiple submissions are received from within one affiliate, the submissions may be sent back to the local affiliate's board/officers to choose the strongest entry/ies (regional affiliates will be permitted one paper per state in the region). For the Graduate Student Outstanding Paper, authors must be a graduate student, a member of NCFR, and a member of a student, state, or regional affiliate council. For the Undergraduate Student Outstanding Paper, authors must be an undergraduate student and a member of a student, state, or regional affiliate council.
Restrictions: Papers already published or presented at a national conference are not eligible.
Geographic Focus: All States
Date(s) Application is Due: Jun 1
Amount of Grant: 100-250 USD
Contact: Allison Wickler, (763) 231-2881; allisonwickler@ncfr.org
Internet: http://www.ncfr.org/about/awards/affiliate-councils-awards/affiliate-councils-outstanding-student-research-papers-gradua
Sponsor: National Council on Family Relations
1201 West River Parkway, Suite 200
Minneapolis, MN 55454-1115

NCFR Family Therapy Section Best Research Paper Awards 2907
The National Council on Family Relations, founded in 1938, is the oldest multidisciplinary, nonpartisan professional organization focused solely on family research, practice and education. Family Therapy Section paper awards are intended to foster scholarly activity and assist in identifying future leaders. Winners receive a $250 cash award and a certificate. Two awards will be presented-one to a student and one to a new professional. The awards are presented annually at the Family Therapy Section Annual Meeting at the NCFR Annual Conference. The deadline is concurrent with the NCFR conference submission deadline.
Requirements: Upon acceptance of paper for presentation, students are contacted by a section representative and invited to apply and submit their papers for award consideration. The winner must be: a student or new professional within three years of graduation; a current member of NCFR and the FT Section when proposal is submitted; the first author on paper submitted to section; and, ready to present paper at the NCFR Annual Conference.
Geographic Focus: All States
Date(s) Application is Due: Mar 3
Amount of Grant: 250 USD
Contact: Jeanne Strand; (763) 781-9331; jeannestrand@ncfr.org
Internet: http://www.ncfr.org/awards/section-awards/family-therapy/best-research-paper-awards-one-student-award-one-new-profession
Sponsor: National Council on Family Relations
1201 West River Parkway, Suite 200
Minneapolis, MN 55454-1115

NCFR Innovation Grants 2908
The National Council on Family Relations, founded in 1938, is the oldest multidisciplinary, nonpartisan professional organization focused solely on family research, practice and education. NCFR will support up to two $10,000 innovation grants over the next year. The grants program is designed to enhance interdisciplinary collaborations around issues of importance to family well-being. A range of activities could be supported by these funds such as: a) research pilot projects, b) theory development and enhancement, c) policy initiatives, and d) outreach, engagement, and training opportunities. NCFR encourages innovative and groundbreaking proposals that will further the goals of NCFR and, in so doing, strengthen and support families.
Requirements: Proposals must involve significant interdisciplinary collaboration and demonstrate how the proposed activities and contributions will be strengthened by the interdisciplinary approach. The proposal must demonstrate how it meets the NCFR Global Ends. The primary applicant (or applicants, in the case of co-PIs) must be an NCFR member who is an academic scholar or other family professional. Postdoctoral fellows may apply as PIs. Students may be involved in the proposed project as collaborators but not as PIs. Applicants may be included on more than one proposal but should be listed as the PI on only on proposal. Priority will be given to proposals with NCFR members from two or more different disciplines. Projects that aim to bridge the areas of research and practice and that build on a stong theoretical framework or advance theory are encouraged. The call for proposals is intentionally broad to encourage a variety of innovative proposals that meet these objectives. Proposals must be received by NCFR by 11:59PM CDT of the deadline date.
Restrictions: Funding will not be granted for supporting conferences or workshops.
Geographic Focus: All States
Date(s) Application is Due: Sep 1
Amount of Grant: Up to 10,000 USD
Contact: Jeanne Strand; (763) 781-9331; jeannestrand@ncfr.org
Internet: http://www.ncfr.org/awards/ncfr-innovation-grants
Sponsor: National Council on Family Relations
1201 West River Parkway, Suite 200
Minneapolis, MN 55454-1115

NCFR Ruth Hathaway Jewson Award 2909
The National Council on Family Relations, the NCFR Sections, CFLE Advisory Board, and the Affiliate Councils have established awards to recognize those individuals and groups who further the interests of the organization and/or exemplify the standards to which members aspire. Award winners are distinguished colleagues who have dedicated themselves to working in the family field, and have given outstanding service. Members nominate other members for the awards, send supporting evidence, and list other people who can verify the nominee's qualifications. A selection committee evaluates the applicants and selects the winner(s). This award is given in honor of Ruth Hathaway Jewson, NCFR's second executive director. It funds the the best dissertation proposal submitted by a doctoral candidate in the field of family studies. Dr. Jewson was responsible for expanding NCFR's membership from a small organization to over 4,000 members when she retired. The award winner receives a $500 cash award and a plaque, which are presented at the beginning of a plenary session at the NCFR Annual Conference. This award is presented biannually (even years).
Requirements: Awardees must be nominated by NCFR members. The nominee must be a doctoral student and have been a member of NCFR for at least six months. All materials must be received by NCFR by May 1. The winner will be notified by July 31.
Geographic Focus: All States
Date(s) Application is Due: May 1
Contact: Jeanne Strand; (763) 781-9331; jeannestrand@ncfr.org
Internet: http://www.ncfr.org/awards/ncfr-sponsored-awards/ruth-hathaway-jewson-award-snp
Sponsor: National Council on Family Relations
1201 West River Parkway, Suite 200
Minneapolis, MN 55454-1115

NCHS National Center for Health Statistics Postdoctoral Research Appointments 2910
The National Center for Health Statistics (NCHS) is the Nation's principal health statistics agency and provides the information needed to develop the programs and policies that will improve the health of the American people. The objective of the NCHS Postdoctoral Research Program is to provide opportunities for postdoctoral candidates of unusual promise and ability to conduct research in problems of their choosing that are compatible with the interests of NCHS. A postdoctoral researcher is a temporary salaried employee of NCHS analogous to fellows or similar temporary researchers at the postdoctoral level in universities and other organizations. Appointments are two years in length, but extensions are possible with special consideration. Benefits include sick leave, annual leave, thrift savings, and health and life insurance. Flex time and alternate work schedules are available if desired, contingent on the supervisor's approval. Cost of relocation will be determined on an individual basis. An evaluation is conducted after one year to ensure that the postdoctoral researcher is making suitable progress. General areas of interest for research at NCHS include statistical theory, survey methodology, statistical computing, economics, demography, and social and behavioral science. Specific areas of interest are: confidentiality protection and related topics; data linkage; quality control, quality improvement, and its measurement; adjustment for nonresponse in sample surveys; telephone survey methods; data collection in a changing health care system; small-area estimation; cognitive processes and question response; data editing and imputation; statistical computing; sampling and estimation; questionnaire design and measurement research; human factors and usability research; and data mining. To apply, applicants must submit a statement of research interest, a curriculum vitae, official transcripts, and three reference letters. The statement of research interest should describe the purpose and context of the research project, the research approach and methods to be used, key activities to be undertaken during the fellowship period, and the expected contributions of the proposed research. For fullest consideration, applicants should submit the application by the deadline date to the address given. However, applications will be accepted throughout the year until positions are filled. Evaluation of proposals will be conducted annually in March or April by a special panel convened for this purpose. Notification of awards will be made beginning in mid-April. The date on which an appointment would begin is negotiable on an individual basis but is expected to occur between June 1 and December 31. Applicants can find complete information on the postdoctoral research program and on application submission requirements at the program website. Applicants are encouraged to contact NCHS to obtain specific information on current research and available technical facilities and to speak to NCHS researchers working in the specific area of interest.
Requirements: Awardees must hold a Ph.D. or other earned research degree recognized in the United States as equivalent to the Ph.D. or must present acceptable evidence of having completed all the formal academic requirements for the degree before appointment. Applicants must have demonstrated ability for creative research. An applicant's training and research experience may be in any appropriate discipline or combination of disciplines required for the proposed research. Postdoctoral researchers must devote their full-time effort to the research program and must be in residence at NCHS during the program. No period of tenure may be spent in residence at another agency or institution.
Restrictions: Applicants must be U.S. citizens or legal permanent residents with a work authorization. Postdoctoral research positions are awarded only to persons who have held doctorates for less than 3 years or are in the process of receiving doctorate degrees at the time of application. No additional monetary aid or other remuneration may be accepted from another appointment, fellowship, or similar grant during the period of the program.
Geographic Focus: All States
Date(s) Application is Due: Feb 28

Amount of Grant: 149,744-231,484 USD
Contact: Cynthia Link , (800) 232-4636 or (301) 458-4344; cynthia.link@cdc.hhs.gov
Internet: http://www.cdc.gov/nchs/about/postdoc.htm
Sponsor: National Center for Health Statistics
3311 Toledo Road
Hyattsville, MD 20782

NCI/NCCAM Quick-Trials for Novel Cancer Therapies 2911

This PA sponsored by NCI and NCCAM expands the initiative to all cancer sites and provides investigators with rapid access to support for pilot, phase I, and phase II cancer clinical trials testing new agents and patient monitoring and laboratory studies to ensure the timely development of new therapeutic approaches. QUICK-TRIAL will provide a new approach designed to simplify the grant application process and provide a rapid turnaround from application to funding. Features include a modular grant application and award process, inclusion of the clinical protocol within the grant application, and accelerated peer review with the goal of issuing new awards within five months of application receipt. Inclusion of the complete clinical protocol within the PHS 398 grant application is intended to simplify the application process by eliminating the need to duplicate protocol details in the research plan section. Investigators may apply for a maximum of two years of funding support using the exploratory/developmental (R21) grant mechanism for up to $250,000 direct costs per year. Therapeutic trials in human subjects employing new agents and therapeutic approaches, including CAM treatments, whether used as a single agent/modality or in combination, are appropriate. Preference will be given to clinical trials that include laboratory studies to validate mechanistic hypotheses or clinical correlates that can meaningfully guide further clinical development. A letter of intent is due one month prior to application receipt date. PA: PAR-04-155
Requirements: Applications may be submitted by foreign and domestic, for-profit and nonprofit organizations, both public and private; units of state and local governments; and eligible agencies of the federal government.
Geographic Focus: All States
Date(s) Application is Due: Apr 9; Aug 9; Dec 9
Amount of Grant: Up to 250,000 USD
Contact: Amy Connolly; (301) 496-8786; fax (301) 496-8601; amyconnolly@nih.gov
Internet: http://grants.nih.gov/grants/guide/pa-files/par-04-155.html
Sponsor: National Cancer Institute
6130 Executive Boulevard, Executive Plaza N, Suite 3157, MSC 7328
Bethesda, MD 20892-7328

NCI/NIDA Transdisciplinary Tobacco Use Research Centers Grants 2912

Transdisciplinary tobacco use research centers (TTURCs) are expected to conduct research that will lead to major scientific advances in knowledge of tobacco use, nicotine addiction, prevention and treatment interventions, and the translation of this knowledge into practice. There are several overarching goals: increase the number of investigators from relevant disciplines who focus on the study of tobacco use as part of transdisciplinary teams; generate basic research evidence to improve understanding of the etiology and natural history of tobacco use; produce evidence-based tobacco use interventions that can translate to the community and specific understudied or underserved populations; increase the number of evidence-based interventions that are novel, including the development, testing and dissemination of innovative behavioral treatments and prevention strategies based upon findings from basic research; train transdisciplinary investigators capable of conducting cutting-edge tobacco use research; and to increase the number of peer-reviewed publications in the area of tobacco use, nicotine addiction, and treatment. RFA-CA-04-012
Requirements: Applications may be submitted by public and private, domestic, for-profit and nonprofit organizations, units of state and local governments, and eligible agencies of the federal government. Domestic applications with foreign components are permitted. To be considered, applicant organizations must have: (1) an established investigator as a strong principal investigator who can oversee and conduct planning activities, provide direction to the TTURC, ensure a transdisciplinary research emphasis, and build a career development program; full institutional commitment from the applicant institution; clearly delineated and formally confirmed, subcontracted, collaborative scientific arrangements with scientists from other institutions, including foreign institutions, may be included.
Geographic Focus: All States
Contact: Crystal Wolfrey, Grants Administration Branch, NCI; (301) 496-8634; fax (301) 496-8601; crystal.wolfrey@nih.gov or wolfreyc@mail.nih.gov
Internet: http://grants1.nih.gov/grants/guide/rfa-files/RFA-CA-04-012.html
Sponsor: National Cancer Institute
6130 Executive Boulevard, Executive Plaza N
Bethesda, MD 20892

NCI Academic-Industrial Partnerships for Development and Validation of In Vivo Imaging Systems and Methods for Cancer Investigations 2913

The overall goal of this Funding Opportunity Announcement (FOA) is to accelerate the transition of in vivo spectroscopic and imaging systems and methods into cancer research, clinical trials, and/or clinical practice. A related goal is to advance new technologies and methods for animal imaging. To promote these goals, the Cancer Imaging Program (CIP) of the National Cancer Institute (NCI) solicits applications from multi-institutional research partnerships formed by academic, industrial, and other investigators. The partners will establish an inter-disciplinary, multi-institutional research team for the purpose of developing imaging technology that is applicable to cancer research involving human subjects and/or animal models. The academic-industrial partnerships must be dedicated to the complete research and development cycle, culminating in the product prototype stage. The partnership model is expected to more readily overcome various barriers to the development of new imaging devices and technologies faced by either academia or industry working alone.
Requirements: Eligible institutions include: public and state controlled institutions of higher education; private institutions of higher education; nonprofits with 501(c)3 IRS status (other than institutions of higher education); nonprofits without 501(c)3 IRS status (other than institutions of higher education); small businesses; for-profit organizations (other than small businesses); state governments; U.S. territories or possessions; non-domestic (non-U.S.) entities (foreign organizations); and eligible agencies of the federal government.
Geographic Focus: All States
Date(s) Application is Due: Feb 5; Jun 5; Oct 5
Contact: Guoying Liu, (301) 496-9531; fax (301) 480-3507; guoyingl@mail.nih.gov
Internet: http://grants.nih.gov/grants/guide/pa-files/PAR-07-214.html
Sponsor: National Cancer Institute
6130 Executive Plaza, EPN Suite 6000, MSC 7412
Bethesda, MD 20892-7412

NCI Application and Use of Transformative Emerging Technologies in Cancer Research 2914

This Funding Opportunity Announcement (FOA) solicits grant applications proposing exploratory research projects on evaluating the performance of emerging molecular and cellular analysis technologies utilizing an appropriate cancer-relevant biological system. Technologies proposed for evaluation may have broad applicability but must ultimately be suitable for cancer-relevant analyses such as the detection (at various molecular and cellular levels) of cancer-related characteristics and/or structural and/or functional alterations. Letters of intent are requested by January 23, April 27, and August 30; full proposal deadlines are February 23, May 27, and September 30. Annual deadline dates may vary; contact program staff for exact dates.
Requirements: Eligible institutions include: public and state controlled institutions of higher education; private institutions of higher education; nonprofits with 501(c)3 IRS status (other than institutions of higher education); nonprofits without 501(c)3 IRS status (other than institutions of higher education); small businesses; for-profit organizations (other than small businesses); state governments; U.S. territories or possessions; non-domestic (non-U.S.) entities (foreign organizations); and eligible agencies of the federal government.
Geographic Focus: All States
Date(s) Application is Due: Jan 23; Feb 23; Apr 27; May 27; Aug 30; Sep 30
Amount of Grant: Up to 275,000 USD
Contact: Richard Aragon, (301) 496-1550; fax (301) 496-7807; raragon@mail.nih.gov
Internet: http://grants.nih.gov/grants/guide/rfa-files/RFA-CA-09-006.html
Sponsor: National Cancer Institute
31 Center Drive, Building 31, Room 10A52, MSC 2580
Bethesda, MD 20892-2580

NCI Application of Metabolomics for Translational and Biological Research Grants 2915

The purpose of this Funding Opportunity Announcement (FOA), sponsored by multiple NIH Institutes/Centers (ICs), is to promote the use of metabolomic technologies in translational research in human health and disease for the purposes of enabling and improving disease detection, diagnosis, risk assessment, prognosis, and prediction of therapeutic responses. In addition to this overall goal, specific scopes and objectives of the participating ICs are presented below. Using the NIH Research Project Grant (R01) funding mechanism, this FOA focuses on well developed projects supported by preliminary data.
Requirements: Application can be made by the following institutions and organizations: public and state controlled institutions of higher education; private institutions of higher education; small businesses; for-profit organizations (other than small businesses); and non-domestic (non-U.S.) entities (foreign organizations).
Geographic Focus: All States
Date(s) Application is Due: Feb 5; Feb 16; Jun 5; Jun 16; Oct 5; Oct 16
Contact: Padma Maruvada, (301) 496-3893; fax (301) 402-8990; maruvadp@mail.nih.gov
Internet: http://grants.nih.gov/grants/guide/pa-files/PA-07-301.html
Sponsor: National Cancer Institute
6130 Executive Boulevard, Executive Plaza N, Suite 3144, MSC 7362
Bethesda, MD 20892-7362

NCI Basic Cancer Research in Cancer Health Disparities Grants 2916

Through this Funding Opportunity Announcement (FOA), the Center to Reduce Cancer Health Disparities (CRCHD) and the Division of Cancer Biology (DCB), at the National Cancer Institute (NCI), invite cooperative agreement research (U01) grant applications from investigators interested in conducting basic research studies into the causes and mechanisms of cancer health disparities. These awards will support pilot and feasibility studies, development and testing of new methodologies, secondary data analyses, and innovative mechanistic studies that investigate biological/genetic bases of cancer health disparities. This FOA is also designed to aid and facilitate the growth of a nationwide cohort of scientists with a high level of basic research expertise in cancer health disparities research who can develop resources and tools, such as biospecimens, cell lines and methods that are necessary to conduct basic research in cancer health disparities. The total project period for an application submitted in response to this funding opportunity may not exceed five years. Direct costs are limited to $250,000/year and may not exceed 5 years.
Requirements: The following organizations and institutions are eligible to apply: public/state controlled institutions of higher education; private institutions of higher education; Hispanic-serving institutions; historically Black colleges and universities (HBCUs); tribally controlled colleges and universities (TCCUs); Alaska Native and Native Hawaiian serving institutions; nonprofits with 501(c)3 IRS status; nonprofits without 501(c)3 IRS status; small businesses; for-profit organizations; state governments; Indian/Native

American tribal governments (federally recognized); Indian/Native American tribally designated organizations; U.S. territories or possessions; Indian/Native American tribal governments (other than federally recognized); regional organizations; eligible agencies of the federal government; and units of local governments.
Geographic Focus: All States
Date(s) Application is Due: Jun 23; Nov 23
Amount of Grant: Up to 250,000 USD
Contact: Phillip J. Daschner, (301) 496-1951; fax (301) 496-2025; PD93U@nih.gov
Internet: http://grants.nih.gov/grants/guide/pa-files/PAR-09-161.html
Sponsor: National Cancer Institute
6130 Executive Boulevard, Room 5014, MSC 7344
Bethesda, MD 20892-7344

NCI Cancer Education and Career Development Program 2917
This Funding Opportunity Announcement (FOA) represents the continuation of the Cancer Education and Career Development Program (CECDP) established by the National Cancer Institute (NCI). The purpose of the CECDP is to support the development and implementation of institutional curriculum-dependent pre-doctoral and/or postdoctoral programs with specific core didactic and research requirements to train a cadre of scientists in interdisciplinary and collaborative cancer research settings. Examples of interdisciplinary/collaborative areas of cancer research that are particularly applicable to intent of the CECDP are cancer prevention and control, behavioral and population sciences, nutrition, imaging, and molecular diagnosis. This FOA makes an NCI-specific use of the NIH Education R25 grant mechanism.
Requirements: Applications may be submitted on behalf of the principal investigator by domestic, nonfederal, public or private organizations. Applications may include more than one institution to create a program through consortium agreements. Predoctoral and postdoctoral candidates must be citizens or noncitizen nationals of the U.S., or must have been lawfully admitted to the U.S. for permanent residence.
Restrictions: No application may request more than $500,000 in direct costs per annum without written approval from NCI.
Geographic Focus: All States
Amount of Grant: Up to 500,000 USD
Contact: Catherine Blount; (301) 496-3179; fax (301) 496-8601; blountc@mail.nih.gov
Internet: http://grants.nih.gov/grants/guide/pa-files/PAR-06-511.html
Sponsor: National Cancer Institute
6130 Executive Boulevard, Executive Plaza N, Suite 3144, MSC 7362
Bethesda, MD 20892-7362

NCIC Canadian Breast Cancer Research Initiative Grants 2918
The initiative is a broad alliance of partners, comprising the Canadian Cancer Society, Health Canada through the National Health Research and Development Program, the Medical Research Council, the National Cancer Institute of Canada, the Breast Cancer International Centre, the Canadian Breast Cancer Foundation, and the Canadian Breast Cancer Network. The grants provide research support for up to three years to individuals formally affiliated with Canadian universities or other recognized Canadian institutions. Proposals must clearly state their relevance to the problem of breast cancer. Grantees are required to submit annual progress reports. Obtain application form 801 from the institute.
Restrictions: Applications may not be submitted in duplicate to sponsoring agencies. Applications for support of infrastructure will not be accepted.
Geographic Focus: All States, Canada
Contact: Dr. Pascale Macgregor, Research Program Director; (416) 596-6598, ext. 360; fax (416) 596-1714; pfmacgregor@cbcf.org
Internet: http://www.breast.cancer.ca/cbcra_research_program
Sponsor: National Cancer Institute of Canada
10 Alcorn Avenue, Suite 200
Toronto, ON M4V 3B1 Canada

NCIC Canadian Cancer Society Research Grants 2919
The grants are intended to stimulate Canadian investigators in the design and execution of clinical and psychosocial/behavioral research in cancer and in the exploration of all the areas of the basic sciences having relevance to cancer. Grants for research will be awarded to projects deemed worthy of support, provided that the basic equipment and research facilities are available in the institution concerned and that it will provide the necessary administrative services. Such awards will be granted for the purchase and maintenance of animals, for expendable supplies, minor items of equipment, payment of graduate students and other assistants, and for research travel. Researchers focusing on breast cancer are strongly encouraged to submit applications to the Canadian Breast Cancer Research Initiative.
Requirements: A researcher who is designated as the principal investigator must be based in, or formally affiliated with (but not necessarily receive salary support from), a Canadian host institution such as a university, research institute, or healthcare agency.
Restrictions: Funds may not be used for personal salary support of the grantee.
Geographic Focus: All States, Canada
Date(s) Application is Due: Oct 17
Amount of Grant: Up to 300,000 CAD
Contact: Director of Research; (416) 961-7223; fax (416) 961-4189; research@cancer.ca
Internet: http://www.ncic.cancer.ca/ncic/internet/standard/0,3621,84658243_85805597__langId-en,00.html
Sponsor: National Cancer Institute of Canada
10 Alcorn Avenue, Suite 200
Toronto, ON M4V 3B1 Canada

NCIC Canadian Cancer Society Travel Award for Senior Level PhD Students 2920
The purpose of this program is to provide financial assistance to senior-level PhD or MD/PhD students by helping to defray the travel costs associated with making a scientific presentation at a conference, symposium, or other appropriate professional gathering. The travel award will be made to defray actual costs of travel, registration, accommodation, etc. Three competitions are held annually. Conference dates for which funding is being requested must occur within six months of the competition deadline. Individuals may receive this award only once. A maximum of one award per supervisor will be given in the current calendar year. Applications must be completed using form 405/96. Letters announcing the results of eligible applications will be issued approximately one month following the deadline date.
Requirements: Applicants must be students enrolled in a PhD or MD/PhD program at a Canadian institution and be in their final phase of study. Candidates must be attending a conference for the purpose of presenting data (paper or poster presentation) from a cancer-related project on a first-author basis. It is not a requirement that the student and/or the supervisor be supported by an NCIC grant.
Geographic Focus: All States, Canada
Date(s) Application is Due: Apr 1; Sep 1; Dec 1
Contact: Program Contact; (416) 961-7223; fax (416) 961-4189; research@cancer.ca
Internet: http://www.ncic.cancer.ca/ncic/internet/standard/0,3621,84658243_85813951_91005239_langId-en,00.html
Sponsor: National Cancer Institute of Canada
10 Alcorn Avenue, Suite 200
Toronto, ON M4V 3B1 Canada

NCI Centers of Excellence in Cancer Communications Research 2921
NCI invites applications for Centers of Excellence in Cancer Communications Research (CECCRs). The purpose is to identify areas of discovery that build upon important recent developments in knowledge and technology and that hold promise for making significant progress against all cancers. The centers must include three or more individual hypothesis-driven research projects, pilot or developmental research projects, shared resources, and a plan for career development. To be effective, the centers' research should integrate cancer communications appropriately into one or more contexts of the cancer continuum—from prevention through treatment to survivorship and end-of-life research. Communications research also is needed about challenging topics such as cancer information seeking, decision making under uncertainty, and genetic testing. Centers' research also should provide insight into mechanisms underlying how people process information. Annual deadline dates may vary; contact program staff for exact dates. RFA: CA-03-007
Requirements: Applications may be submitted by domestic for-profit and nonprofit organizations, both public and private; units of state and local governments; and eligible agencies of the federal government.
Geographic Focus: All States
Contact: Crystal Wolfrey, Grants Administration Branch; (301) 496-8634; fax (301) 496-8601; crystal.wolfrey@nih.gov or wolfreyc@mail.nih.gov
Internet: http://grants.nih.gov/grants/guide/rfa-files/RFA-CA-08-004.html
Sponsor: National Cancer Institute
6130 Executive Boulevard, Executive Plaza N, Suite 3144, MSC 7362
Bethesda, MD 20892-7362

NCIC Equipment Grants 2922
The grants provide for the purchase of items of permanent equipment necessary for research projects supported by the institute, if such equipment is not otherwise available. Application for minor items of apparatus and equipment with a total cost for each item of less than $C10,000 should be included as part of the application for the research grant. Application for funds to purchase each item with a total cost of less than $C100,000 but more than $C5000 may be made either as part of an application for a grant for research or as a separate request if the applicant is the recipient of a continuing award. For items of specialized major equipment having a value of $C100,000 or more, the institute may have to contemplate alternative approaches to funding; therefore, a prior consultation with the institute is required.
Geographic Focus: All States
Date(s) Application is Due: Oct 17
Contact: Administrator; (416) 961-7223; fax (416) 961-4189; research@cancer.ca
Internet: http://www.ncic.cancer.ca/ncic/internet/standard/0,3621,84658243_85808828__langId-en,00.html
Sponsor: National Cancer Institute of Canada
10 Alcorn Avenue, Suite 200
Toronto, ON M4V 3B1 Canada

NCIC Terry Fox Postdoctoral Research Fellowships 2923
Fellowships are designed to provide training in cancer research for outstanding candidates who plan careers in Canada in this field of investigation. Post-MD fellowships are made to provide support for specialized training in the research aspects of the broad field of oncology (including epidemiology, medical statistics, clinical trials, and other forms of biomedical research). Initial tenure of a fellowship will normally be two years. One or more years of additional support may be available to individuals who demonstrate a serious commitment to cancer research, e.g., by registration in a degree program. Post-MD fellows may wish to maintain their clinical skills by means of service to the host institution during the period of the award. Such activities are acceptable provided the service is directly related to the cancer-oriented plan of the fellow and neither exceeds 10 working hours per week nor adversely affects the quantity or quality of the research training.
Requirements: Candidates must be graduates of universities recognized by NCIC and must be accepted for postdoctoral training in the field of cancer research in an academic environment also recognized by NCIC.

Restrictions: In general, applicants will not be eligible for funding beyond five years of postdoctoral training.
Geographic Focus: All States, Canada
Date(s) Application is Due: Feb 1
Contact: Director of Research; (416) 961-7223; fax (416) 961-4189; research@cancer.ca
Internet: http://www.ncic.cancer.ca/ncic/internet/standard/0,3621,84658243_85811512_90904221_langId-en,00.html
Sponsor: National Cancer Institute of Canada
10 Alcorn Avenue, Suite 200
Toronto, ON M4V 3B1 Canada

NCIC Terry Fox Research Studentships 2924

Research studentships provide training to PhD students in cancer research and are offered to outstanding individuals who plan a career in cancer research in Canada. Awards are made for up to a maximum of four years and are not renewable. Research studentship award recipients must devote full time to the research and training program during the tenure of the studentship. If the supervisor considers a certain amount of teaching to be part of the training, such teaching and duties must be limited to not more than 10 hours per week. At the completion of each award year, a letter summarizing the progress to date, signed by the research student and the research supervisor, must be submitted to the institute.
Requirements: Applicants must be residents of Canada and be enrolled in a PhD program at a Canadian institution at the time of application. Candidates must have completed at least two years of research training at the graduate level or have equivalent research experience.
Geographic Focus: All States, Canada
Contact: Michael Wosnick; (416) 934-5672; fax (416) 961-4189; mwosnick@cancer.ca
Internet: http://www.ncic.cancer.ca/ncic/internet/standard/0,3621,84658243_85813949_90911828_langId-en,00.html
Sponsor: National Cancer Institute of Canada
10 Alcorn Avenue, Suite 200
Toronto, ON M4V 3B1 Canada

NCI Diet, Epigenetic Events, and Cancer Prevention Grants 2925

The purpose of this Funding Opportunity Announcement (FOA) is to promote innovative preclinical and clinical research to determine how diet and dietary factors, including dietary supplements, impact epigenetic processes involved in cancer prevention. Although much evidence exists that dietary components are linked to cancer prevention, the specific nutrients and sites of action remain elusive. Diet, in fact, has been implicated in many of the pathways of cancer, including apoptosis, cell cycle control, differentiation, inflammation, angiogenesis, DNA repair, and carcinogen metabolism. These are also processes that are likely regulated by epigenetic events to impact gene function and chromatin stability. Thus, research supported by this initiative could address one of the following issues: How bioactive food components regulate epigenetic events for cancer prevention, how bioactive food components might alter aberrant epigenetic patterns or events and restore gene function, and how these components might circumvent or compensate for genes and pathways that are altered by epigenetic events. Another important aim of this FOA is to encourage collaborations between nutrition scientists and experts in epigenetics to study bioactive food components with cancer-preventative properties and to examine key epigenetic events in cancer processes (e.g., carcinogen metabolism, cell division, differentiation, and apoptosis) in order to begin to establish linkages between epigenetic events and patterns, and tumor incidences/behaviors. NIDDK is particularly interested in receiving applications that address normal and neoplastic conditions focusing on the impact of epigenetics on cell division, differentiation, and/or apoptosis in the gastrointestinal (GI) tract and related organs. It is also expected that each application demonstrate experience in nutrition and cancer prevention as well as in epigenetic research. Because the nature and scope of the proposed research will vary from application to application, it is anticipated that the size and duration of each award will also vary.
Requirements: Application can be made by the following institutions and organizations: public/state controlled institutions of higher education; private institutions of higher education; nonprofits with 501(c)3 IRS status (other than institutions of higher education); nonprofits without 501(c)3 IRS status (other than institutions of higher education); small businesses; for-profit organizations (other than small businesses); state governments; U.S. territories or possessions; Indian/Native American tribal governments (federally recognized); Indian/Native American tribal governments (other than Federally recognized); Indian/Native American tribally designated organizations; non-domestic (non-U.S.) entities (foreign organizations); Hispanic-serving institutions; historically Black colleges and universities (HBCUs); tribally controlled colleges and universities (TCCUs); Alaska Native and Native Hawaiian serving institutions; regional organization; eligible agencies of the Federal government; and faith-based or community based organizations.
Geographic Focus: All States
Date(s) Application is Due: Feb 5; Jun 5; Oct 5
Contact: Vishnudutt Purohit, (301) 443-2689; fax (301) 594-0673; vpurohit@mail.nih.gov
Internet: http://grants.nih.gov/grants/guide/pa-files/PA-07-175.html
Sponsor: National Cancer Institute
6130 Executive Boulevard, Executive Plaza N, Suite 3157, MSC 7328
Bethesda, MD 20892-7328

NCI Engineered Isogenic Cell Lines with Relevant Cancer Targets Research Grants 2926

This PA encourages the small business community to develop isogenic cell lines in which functions of a molecular activity or pathway found in human tumors can be analyzed in pair-wise fashion. Such cell lines could be used for drug screening with the intent of identification of compounds that influence the action of the identified target. The end result of the PA-supported projects would be a series of genetically engineered screening models that the small businesses could market directly or utilize in a license agreement or other type of collaboration. PAR-98-068 may be obtained through the NIH grant line (301-402-2221), the NIH Web site, mail, and email.
Geographic Focus: All States
Contact: Toby Friedberg, Extramural Activities Division Conact; (301) 496-3428; fax (301) 402-0275; friedbet&7dea.nci.nih.gov
Internet: http://www.grants.nih.gov/grants/guide/pa-files/PAR-98-068.html
Sponsor: National Cancer Institute
6130 Executive Boulevard, Executive Plaza N
Bethesda, MD 20892

NCI Exploratory/Developmental Grants for Diagnostic Cancer Imaging 2927

The Cancer Imaging Program (CIP) of the Division of Cancer Treatment and Diagnosis solicits exploratory/developmental (R21) grants that articulate highly innovative research concepts in diagnostic cancer imaging. Within each area of importance in imaging, there exists a need for innovative and creative approaches leading to new avenues of research. One way to encourage research into high-risk/high-impact areas is to provide investigators with the initial resources required to accomplish feasibility and pilot testing of innovative ideas. Contact program staff for deadline dates. PA-04-045
Requirements: Applications may be submitted by foreign and domestic, for-profit and nonprofit organizations, public and private.
Geographic Focus: All States
Amount of Grant: Up to 275,000 USD
Contact: Kathryn Dunn, Grants Management Specialist, National Cancer Institute; (301) 846-6829; fax (301) 846-5720; dunnkath@mail.nih.gov
Internet: http://grants.nih.gov/grants/guide/pa-files/PA-04-045.html
Sponsor: National Cancer Institute
6120 Executive Boulevard, Executive Plaza S
Bethesda, MD 20892

NCI Exploratory Grants for Behavioral Research in Cancer Control 2928

This FOA will support innovative pilot projects or feasibility studies, which will facilitate the growth of research science in the cancer control continuum from a behavioral perspective. This FOA includes and incorporates the research interests of the Behavioral Research program, the Office of Cancer Survivorship, and the Community Oncology and Prevention Trials Research Group. This FOA is appropriate for testing timely interventions in pilot studies for feasibility or using rigorous qualitative research methods to assess the potential efficacy of an intervention. It is also appropriate for the psychometric evaluation of new measures or culturally appropriate ones to be adapted for use in populations where measures have not yet been developed or validated. This FOA encourages applications that include small cross-disciplinary teams of investigators who bring perspectives from the behavioral and social sciences, as well as other fields of public health.
Requirements: Eligible institutions and agencies include: public/state controlled institutions of higher education; private institutions of higher education; nonprofits with 501(c)3 IRS status (other than institutions of higher education); nonprofits without 501(c)3 IRS status (other than institutions of higher education); small businesses; for-profit organizations (other than small businesses); state governments; U.S. territories or possessions; regional organizations; non-domestic (non-U.S.) entities (foreign organizations); eligible agencies of the federal government; and units of local governments.
Geographic Focus: All States
Date(s) Application is Due: Feb 16; Jun 16; Oct 16
Amount of Grant: Up to 275,000 USD
Contact: Sabra F. Woolley, (301) 435-4589; fax (301) 594-0673; woolleys@mail.nih.gov
Internet: http://grants.nih.gov/grants/guide/pa-files/PA-09-130.html
Sponsor: National Cancer Institute
6130 Executive Boulevard, Room 4084, MSC 7365
Bethesda, MD 20850-7365

NCI Improving Diet and Physical Activity Assessment Grants 2929

This Funding Opportunity Announcement (FOA) is aimed at advancing the quality of measurements of dietary intake and physical activity pertinent to cancer and/or other pathologies through support of research on improved instruments, technologies, and/or statistical/analytical techniques. Studies proposed in the grant applications should be aimed at optimizing the combined use of objective and self-report measures of physical activity and/or dietary intake for testing in both general and diverse populations. Applicants responding to this FOA, which uses the R01 grant mechanism, should have already tested and validated new approaches and have obtained feasibility data in the context of cancer or other relevant diseases. Because the nature and scope of the proposed research will vary from application to application, it is anticipated that the size and duration of each award will also vary.
Requirements: Application can be made by the following institutions and organizations: public/state controlled institutions of higher education; private institutions of higher education; nonprofits with 501(c)3 IRS status (other than institutions of higher education); nonprofits without 501(c)3 IRS status (other than institutions of higher education); small businesses; for-profit organizations (other than small businesses); state governments; U.S. territories or possessions; Indian/Native American tribal governments (federally recognized); Indian/Native American tribal governments (other than Federally recognized); Indian/Native American tribally designated organizations; non-domestic (non-U.S.) entities (foreign organizations); Hispanic-serving institutions; historically Black colleges and universities (HBCUs); tribally controlled colleges and universities (TCCUs); Alaska Native and Native Hawaiian serving institutions; regional organization; eligible agencies of the Federal government; and faith-based or community based organizations.

Restrictions: Applicants without sufficiently extensive preliminary data or those who wish to explore the utility of new untested dietary or physical activity assessment methods in pilot studies are advised to consider submitting applications under the parallel FOA, PAR-06-103, which uses the Exploratory/Developmental Grant (R21) award mechanism.
Geographic Focus: All States
Date(s) Application is Due: Feb 5; Jun 5; Oct 5
Contact: Richard Troiano, Ph.D; (301) 435-6822; fax (301) 435-3710; troianor@mail.nih.gov
Internet: http://grants.nih.gov/grants/guide/pa-files/PAR-09-225.html
Sponsor: National Cancer Institute
6130 Executive Boulevard, Executive Plaza N, Room 4005, MSC 7344
Bethesda, MD 20892-7344

NCI Mentored Career Development Award to Promote Diversity **2930**
This mechanism establishes a pathway of recruiting, training, and retaining underrepresented investigators in research fields that address problems pertinent to the biology, etiology, pathogenesis, prevention, diagnosis, control, and treatment of human cancer and who can conduct independent competitive cancer research programs. The Mentored Career Development Award to Promote Diversity (K01) provides support for a sustained period of protected time for intensive research career development under the guidance of an experienced mentor, or sponsor, in the biomedical, behavioral or clinical sciences leading to research independence. The expectation is that through this sustained period of research career development and training, awardees will launch independent research careers and become competitive for new research project grant funding.
Requirements: The following organizations and institutions may apply: for-profit and non-profit organizations; public or private institutions, such as universities, colleges, hospitals, and laboratories; and domestic institutions.
Restrictions: Foreign institutions and organizations are not eligible to apply.
Geographic Focus: All States
Date(s) Application is Due: Feb 12; Mar 12; Jun 12; Jul 12; Oct 12; Nov 12
Amount of Grant: 75,000 USD
Contact: Belinda Locke, (301) 496-7344; fax (301) 402-4551; Lockeb@mail.nih.gov
Internet: http://grants.nih.gov/grants/guide/pa-files/PAR-06-220.html
Sponsor: National Cancer Institute
6116 Executive Boulevard, Suite 7031, MSC 8350
Bethesda, MD 20892-8350

NCI Ruth L. Kirschstein National Research Service Award Institutional Research Training Grants **2931**
The National Institutes of Health (NIH) will award Ruth L. Kirschstein National Research Service Award (NRSA) Institutional Research Training Grants (T32) to eligible institutions as the primary means of supporting predoctoral and postdoctoral research training to help ensure that a diverse and highly trained workforce is available to assume leadership roles related to the Nation's biomedical, behavioral and clinical research agenda. The primary objective of the T32 program is to prepare qualified individuals for careers that have a significant impact on the health-related research needs of the Nation. This program supports predoctoral, postdoctoral and short term research training programs at domestic institutions of higher education with the T32 funding mechanism. Stipends are provided as a subsistence allowance for trainees to help defray living expenses during the research training experience and are based on a 12-month appointment period.
Requirements: The following organizations are eligible to apply: public/state controlled institutions of higher education; private institutions of higher education; nonprofits with 501(c)3 IRS status (other than institutions of higher education); nonprofits without 501(c)3 IRS status (other than institutions of higher education); for-profit organizations (other than small businesses); state governments; U.S. territories or possessions; Indian/Native American tribal governments (Federally recognized); Hispanic-serving institutions; historically Black colleges and universities (HBCUs); tribally controlled colleges and universities (TCCUs); Alaska Natives and Native Hawaiians serving institutions; regional organizations; eligible agencies of the federal government; faith-based or community based organizations.
Geographic Focus: All States
Date(s) Application is Due: Apr 8; Aug 8; Dec 8
Contact: Crystal Wolfrey, Grants Administration Branch, NCI; (301) 496-8634; fax (301) 496-8601; crystal.wolfrey@nih.gov or wolfreyc@mail.nih.gov
Internet: http://grants.nih.gov/grants/guide/pa-files/PA-08-226.html
Sponsor: National Cancer Institute
6120 Executive Boulevard, Room 243, MSC 7150
Bethesda, MD 20892-7150

NCI Stages of Breast Development: Normal to Metastatic Disease Grants **2932**
The National Cancer Institute (NCI), National Institute of Child Health and Human Development (NICHD), National Institute of Diabetes and Digestive and Kidney Diseases (NIDDK), National Institute on Aging (NIA), and National Institute of Environmental Health Sciences (NIEHS) invite investigator-initiated research grant applications to study the molecular, cellular, endocrine, and other physiological influences on the development and maturation of the normal mammary gland and alterations involved in early malignant and metastatic breast cancer. Multi-disciplinary collaborations, for example between cell biologists, molecular endocrinologists, bioengineers, geneticists, and mammary pathologists, are encouraged. Appropriate studies include, but are not limited to, documenting the role of dynamic hormonal influences and determining the role of cell growth, apoptosis, and differentiation in mammary gland maturation; integrating knowledge of cell signaling in breast tissue with whole organ biology; developing models of breast differentiation; and studies focusing on the characteristics of breast tumor physiology particularly in relation to metastasis. It is expected that a complete understanding of mammary gland development will form critical underpinnings for continued advances in detecting, preventing and treating breast cancer.
Requirements: Applications may be submitted by domestic and foreign, for-profit and non-profit organizations, public and private, such as universities, colleges, hospitals, laboratories, units of State or local governments, and eligible agencies of the Federal government.
Geographic Focus: All States
Date(s) Application is Due: Feb 5; Jun 5; Oct 5
Amount of Grant: Up to 250,000 USD
Contact: Padma Maruvada, (301) 496-3893; fax (301) 402-8990; maruvadp@mail.nih.gov
Internet: http://grants.nih.gov/grants/guide/pa-files/PA-99-162.html
Sponsor: National Cancer Institute
6130 Executive Boulevard, Executive Plaza N, Suite 3144, MSC 7362
Bethesda, MD 20892-7362

NCI Technologies and Software to Support Integrative Cancer Biology Research (SBIR) Grants **2933**
This Funding Opportunity Announcement (FOA), issued by the National Cancer Institute (NCI), invites Small Business Innovation Research (SBIR) grant applications from small business concerns (SBCs) that propose the development of commercial software tools, computational/mathematical methods, and technologies that will enable integrative cancer biology research. Integrative cancer biology focuses on understanding cancer as a complex biological system by utilizing both computational and experimental biology to integrate heterogeneous data sources and ultimately to generate predictive computational models of cancer processes. Phase I applications may request up to 2 years of support with total costs of up to $300,000 for the duration of the award (not to exceed $200,000 total costs in any one year). Budgetary requests of Phase II applications will conform to standard SBIR guidelines (up to $750,000 total costs for up to two years).
Requirements: Only United States small business concerns (SBCs) are eligible to submit SBIR applications.
Geographic Focus: All States
Date(s) Application is Due: Apr 5; Aug 5; Dec 5
Amount of Grant: 200,000-750,000 USD
Contact: Greg Evans, (301) 594-8807; fax (301) 480-4082; evansgl@mail.nih.gov
Internet: http://grants.nih.gov/grants/guide/pa-files/PA-09-188.html
Sponsor: National Cancer Institute
31 Center Drive, Building 31, Room 10A19, MSC 2580
Bethesda, MD 20892-2580

NCRR Networks and Pathways Collaborative Research Projects Grants **2934**
This announcement solicits applications for research project grants that will leverage and complement the ongoing technology development being pursued in the National Technology Centers for Networks and Pathways (TCNPs), a program of the NIH Roadmap for Medical Research. These collaborative projects should focus either on addressing a challenging biological problem using the technology developed in one or more of the TCNPs, or on the development of technology that will complement that which is being developed in the centers. Applicants may request support for their own work as well as supplemental support for components pursued in the participating TCNP. While the nature and scope of these applications will vary, it is anticipated that the annual direct costs for these projects will typically range from $150,000 to $250,000.
Requirements: Application can be made by the following institutions and organizations: public/state controlled institutions of higher education; private institutions of higher education; nonprofits with 501(c)3 IRS status (other than institutions of higher education); nonprofits without 501(c)3 IRS status (other than institutions of higher education); small businesses; for-profit organizations (other than small businesses); state governments; U.S. territories or possessions; non-domestic (non-U.S.) entities (foreign organizations); historically Black colleges and universities (HBCUs); Tribally controlled colleges and universities (TCCUs); and other eligible agencies of the Federal government.
Geographic Focus: All States
Date(s) Application is Due: Feb 5; Jun 5; Oct 5
Contact: Douglas M. Sheeley, (301) 594-9762; sheeleyd@mail.nih.gov
Internet: http://grants.nih.gov/grants/guide/pa-files/PA-07-266.html
Sponsor: National Center for Research Resources
6701 Democracy Boulevard, MSC 4874
Bethesda, MD 20892-4874

NCRR Novel Approaches to Enhance Animal Stem Cell Research **2935**
The purpose of this funding opportunity announcement (FOA) is to encourage the submission of applications for research to enhance animal stem cells as model biological systems. Innovative approaches to isolate, characterize and identify toti-potent and multi-potent stem cells from nonhuman biomedical research animal models, as well as to generate reagents and techniques to characterize and separate those stem cells from other cell types is encouraged.
Requirements: The following organizations and institutions may apply: public/state controlled institutions of higher education; private institutions of higher education; Hispanic-serving institutions; historically Black colleges and universities (HBCUs); tribally controlled colleges and universities (TCCUs); Alaska Native and Native Hawaiian serving institutions; nonprofits with 501(c)3 IRS status (other than institution of higher education); nonprofits without 501(c)3 IRS status (other than institution of higher education); non-domestic (non-U.S.) entities (foreign organizations); small businesses; for-profit organizations (other than small businesses).
Restrictions: Studies involving human subjects are not allowed.
Geographic Focus: All States

Date(s) Application is Due: Feb 5; Jun 5; Oct 5
Contact: John D. Harding, (301) 435-0744; fax (301) 480-3819; hardingj@mail.nih.gov
Internet: http://grants.nih.gov/grants/guide/pa-files/PA-07-303.html
Sponsor: National Center for Research Resources
6701 Democracy Boulevard, MSC 4874
Bethesda, MD 20892-4874

NCSS/FASSE/CUFA Inquiry Grant **2936**
The Fund for the Advancement of Social Studies Education (FASSE) and the College and University Faculty Assembly (CUFA) of the National Council for the Social Studies (NCSS) have established a grant to support inquiry in citizenship education. Grant proposals should affirm social, cultural, and racial diversity and address issues of equality, equity, and social justice. Proposals that address aims for citizen action are preferred. Proposals should be relevant to school, university or community-based educational settings. FASSE and CUFA will award a $10,000 grant to the successful applicant who presents a proposal for a research project that demonstrates potential to inform the educational field about justice oriented, citizenship education. The original and four copies of the proposal must be submitted no later than June 30.
Requirements: Assistant, Associate or a full professor, or a graduate student with the demonstrated support of a university mentor/advisor are eligible to apply. Applicant(s) must be an NCSS member. If a graduate student, then the mentor must also be an NCSS member.
Restrictions: No fax or email submissions will be accepted.
Geographic Focus: All States
Date(s) Application is Due: Jun 30
Contact: Prema Cordeiro, Program Manager; (301) 588-1800, ext. 106; fax (301) 588-2049; pcordeiro@ncss.org or excellence@ncss.orgexcellence@ncss.org
Internet: http://www.socialstudies.org/fasse
Sponsor: National Council for the Social Studies
8555 Sixteenth Street, Suite 500
Silver Spring, MD 20910

NCSS Exemplary Research in Social Studies Award **2937**
It is the mission of National Council for the Social Studies to engage and support educators in strengthening and advocating social studies, where social studies is defined as the integrated study of the social sciences and humanities to promote civic competence. In support of this mission, NCSS and the Research Community sponsor this annual research award designed to recognize an exemplary single-study research conception, model, design, procedure and analysis in social studies and social education. For the purpose of this award, social education includes work in the individual social sciences; women's studies and gender studies; multicultural, citizenship, global, economic, and geographic education; decision making; reflective inquiry; moral education; critical theory; teacher education; teacher thinking; and teaching-learning phenomena. Research is broadly constructed to mean any and all forms of recognized, formal research methodologies, paradigms, and perspectives. Detailed instructions about the nomination process are available at the website.
Restrictions: No self-nominations are accepted.
Geographic Focus: All States
Date(s) Application is Due: May 1
Contact: Prema Cordeiro, Program Manager; (301) 588-1800, ext. 106; fax (301) 588-2049; pcordeira@ncss.org or excellence@ncss.org
Internet: http://www.socialstudies.org/awards/research/exemplary/
Sponsor: National Council for the Social Studies
8555 Sixteenth Street, Suite 500
Silver Spring, MD 20910

NCSS Jean Dresden Grambs Distinguished Career Research in Social Studies Award **2938**
The Jean Dresden Grambs Distinguished Career Research Award recognizes professionals who have made extensive contributions to knowledge concerning significant areas of social studies education through meritorious research. The use of the term research in reference to this award shall be broadly construed to mean any and all forms of recognized formal research-methodologies, paradigms. and perspectives. Eligible research is that which has: added significantly to the understanding of the status, impact, characteristics, and activities of social studies education or social education, including several separate and distinct contributions related to curriculum, instruction, evaluation, and the learning and teaching of and in the social studies; stimulated a sense of responsibility among and fostered other professionals to engage in additional research efforts; added significantly to the information bases to those working in the social studies/social education field as evidenced by citations of their work within the social studies literature; and been used by professionals as the basis for significant decisions, rationales, and practices related to one or more areas of social studies curriculum instruction, evaluation, and teaching or learning.
Requirements: Applicants are eligible for consideration and receipt of this award when they have met the following conditions: published documentation of research endeavors and findings covering ten or more years; published documentation that research reveals a focus or direction in a significant segment of the research effort, without intending that the individual has spent the entire time on a single or a limited range of research topics; documentation exists that cites in positive ways the person's research contributions; and letters of support on behalf of the researcher are received by not less than five recognized researchers in the field of social studies/social science education. The person does not need to be currently involved in active research and publications in the social studies/social education areas or fields or a current member of NCSS.
Restrictions: No self-nominations are accepted.
Geographic Focus: All States
Date(s) Application is Due: Jul 18
Contact: Prema Cordeiro, Program Manager; (301) 588-1800, ext. 106; fax (301) 588-2049; pcordeiro@ncss.org or excellence@ncss.orgexcellence@ncss.orgexcellence@n
Internet: http://www.socialstudies.org/awards/research/grambs/
Sponsor: National Council for the Social Studies
8555 Sixteenth Street, Suite 500
Silver Spring, MD 20910

NCSS Larry Metcalf Exemplary Dissertation Award **2939**
The annual Metcalf Dissertation Award is designed to recognize substantive scholarly inquiry in social education completed in pursuit of the doctoral degree. Social education includes work in the individual social sciences; women studies and gender studies; multicultural, citizenship, global, economic, and geographic education; decision making; reflective theory; moral education; critical theory; teacher education; teacher thinking; and teaching-learning phenomena. Research is broadly defined to include experimental, qualitative, historical, and philosophical work. The award recipient must be willing to present the research at a one-hour session during the annual conference.
Requirements: Research in pursuit of the doctoral degree must be completed between March of the year preceding the award and February of the year in which the award is presented.
Restrictions: Self nominations are not accepted.
Geographic Focus: All States
Date(s) Application is Due: Apr 1
Amount of Grant: 250 USD
Contact: Prema Cordeiro, Program Manager; (301) 588-1800, ext. 106; fax (301) 588-2049; pcordeiro@ncss.org or excellence@ncss.org
Internet: http://www.socialstudies.org/awards/research/metcalf/
Sponsor: National Council for the Social Studies
8555 Sixteenth Street, Suite 500
Silver Spring, MD 20910

NCTE Teacher Researcher Award **2940**
The council regards as appropriate research any organized inquiry, theoretical or applied, into the subject of English, including the language arts or related fields, that has significance for the teaching or learning of English. Proposals are ecouraged that focus on underrepresented populations, curriculum changes and the effect the changes have on students, school policies, home-school literacy reationships, after-school programs, student literacy practices, etc. Applications are available online.
Requirements: An applicant must be a member of NCTE. Applicant is urged to define the purpose clearly and limit the project so that it may be completed in a reasonable time, include a valid method of analyzing data, demonstrate that both the question and related research have been explored, and select persons as references who can speak specifically of applicant's ability to complete the project.
Restrictions: Award may not be used for travel costs or subsidies for loss of salary, nor for indirect or overhead costs.
Geographic Focus: All States
Date(s) Application is Due: Aug 15
Amount of Grant: 500-12,500 USD
Contact: Dale Allender, Program Officer; (510) 642-5791; fax (510) 642-4799
Internet: http://www.ncte.org/groups/research/awards/120596.htm
Sponsor: National Council of Teachers of English
1111 W Kenyon Road
Urbana, IL 61801

NCTM E. Glenadine Gibb 7-12 Classroom Research Grants **2941**
The purpose of this grant is support and encourage classroom-based research in pre-college mathematics education in collaboration with college or university mathematics educators. For the upcoming year, grants with a maximum of $6,000 each will be awarded to classroom teachers currently teaching mathematics at the grades 7–12 level. The research must be a significant collaborative effort involving a college or university mathematics educator (a mathematics education researcher or a teacher of mathematics learning, teaching, or curriculum) and one or more grades 7–12 classroom teachers. The proposal may include, but is not restricted to, research on the following topics: curriculum development and implementation; involvement of at-risk or minority students; students' thinking about a particular mathematics concept or set of concepts; connection of mathematics to other disciplines; focused learning and teaching of mathematics with embedded use of technology (any acquisition of equipment must support the proposed plan but not be the primary focus of the grant); and innovative assessment or evaluation strategies. Involvement of preservice teachers is encouraged but not required. It is hoped that this research will lead to an article suitable for publication in the "Journal for Research in Mathematics Education" or in one of the NCTM school journals.
Requirements: Proposals must address the following: research design, the plan for collecting and analyzing data, and the anticipated impact on students' learning. The applicant must be a current Full Individual or E-Member of NCTM or teach at a school having a current NCTM PreK-8 school membership. The college or university mathematics educator must be a member of the NCTM. All grant activities must be completed between June 1 and May 31.
Restrictions: Past grantees are not eligible to apply. No person(s) may receive more than one award administered by the Mathematics Education Trust in the same academic year. Past recipients of this grant are not eligible to reapply.
Geographic Focus: All States
Date(s) Application is Due: Nov 11
Amount of Grant: Up to 6,000 USD

Contact: Nihad Ziad; (703) 620-9840, ext. 2112; infocentral@nctm.org
Internet: http://www.nctm.org/resources/content.aspx?id=22418
Sponsor: National Council of Teachers of Mathematics
1906 Association Drive
Reston, VA 20191-1502

NCTM Edward G. Begle PreK-6 Classroom Research Grants 2942

The purpose of this grant is to support and encourage classroom-based research in precollege mathematics education in collaboration with college or university mathematics educators. Grants with a maximum of $6,000 each will be awarded to classroom teachers currently teaching mathematics at the grades PreK–6 level. The research must be a significant collaborative effort involving a college or university mathematics educator (a mathematics education researcher or a teacher of mathematics learning, teaching, or curriculum) and one or more grades PreK–6 classroom teachers. The proposal may include, but is not restricted to, research on the following topics: curriculum development and implementation; involvement of at-risk or minority students; students' thinking about a particular mathematics concept or set of concepts; connection of mathematics to other disciplines; focused learning and teaching of mathematics with embedded use of technology (any acquisition of equipment must support the proposed plan but not be the primary focus of the grant); or innovative assessment or evaluation strategies. Involvement of preservice teachers is encouraged but not required. It is hoped that this research will lead to an article suitable for publication in the Journal for Research in Mathematics Education or in one of the NCTM school journals.
Requirements: Proposals must address the following: research design, the plan for collecting and analyzing data, and the anticipated impact on students' learning. The applicant must be a current Full Individual or E-Member of NCTM or teach at a school having a current NCTM PreK-8 school membership. The college or university mathematics educator must be a member of the NCTM.
Restrictions: No person(s) may receive more than one award administered by the Mathematics Education Trust in the same academic year. Past recipients of this grant are not eligible to reapply.
Geographic Focus: All States
Date(s) Application is Due: Nov 11
Amount of Grant: Up to 6,000 USD
Contact: Nihad Ziad, Grants Administrator; (703) 620-9840, ext. 2112; fax (703) 476-2970; infocentral@nctm.org or affiliates@nctm.org
Internet: http://www.nctm.org/resources/content.aspx?id=1330
Sponsor: National Council of Teachers of Mathematics
1906 Association Drive
Reston, VA 20191-1502

NEA/TCG Career Development Grants for Theatre Directors 2943

The program offers six exceptionally talented, early-career theatre directors the opportunity to spend six months over a two-year period expanding their artistic boundaries and increasing their knowledge of the field. Each experience is hand-tailored, matching recipients' goals to appropriate and challenging assignments. Under the program, recipient directors will receive $25,000, which represents six stipend payments of $3,900 and a final payment of $1,600. these stipends cannot be used to underwrite fees for directing jobs, unless those assignments are related to the goals the recipient articulates in their application and interview. In addition, each recipient will receive up to $2,000 to support ongoing life needs such as health insurance, child care or elder care and up to $1,500 in supplementary travel.
Requirements: Applicants of diverse backgrounds representing the full spectrum of theatrical activity, including those who have worked in non-traditional venues, are encouraged to apply to this program. All applicants: must have submitted online registration and mailed completed application packet by the postmark deadline; must be a citizen or permanent resident of the United States at the time of the application postmark deadline; must have directed three fully staged professional productions. Productions are considered professional if all artists are compensated at least a minimal amount for their work. Productions at educational institutions or community theaters do not meet this requirement. It is not necessary for the production to have operated on a union contract.
Restrictions: Applicants must not be enrolled in, or on leave from, a university or conservatory training program at the commencement of the program period; and, must not work exclusively in other disciplines (e.g. opera, dance). Previous recipients of this program, the National Endowment for the Arts Director Fellowship or the NEA/TCG Director Fellowship are not eligible to apply.
Geographic Focus: All States
Date(s) Application is Due: Jan 30
Amount of Grant: 25,000 USD
Contact: Casey Baltes; (212) 609-5900, ext.281; cbaltes@tcg.org
Internet: http://www.tcg.org/grants/cdpdirect/cdp_dir_index.cfm
Sponsor: Theatre Communications Group
520 8th Avenue, 24th Floor
New York, NY 10018-4156

NEA National Heritage Fellowships in the Folk and Traditional Arts 2944

To honor and preserve our nation's diverse cultural heritage, the National Endowment for the Arts (NEA) annually awards eight National Heritage Fellowships to master folk and traditional artists. These fellowships recognize lifetime achievement, artistic excellence, and contributions to our nation's traditional arts heritage. Nominees must be worthy of national recognition and have a record of continuing artistic accomplishment. They must be actively participating in their art form, either as practitioners or as teachers. Fellows are selected according to criteria of artistic excellence, significance within the particular artistic tradition, and contributions to living cultural heritage. In addition, one NEA National Heritage Fellowship will be given to an individual who has made major contributions to the excellence, vitality, and public appreciation of the folk and traditional arts. Nominations are accepted by using the online nomination form located at the NEA website. A list of previous fellowship recipients is also located at the website.
Requirements: Recipients are selected on the basis of nominations from the public. Nominations may be individuals or a group of individuals. Nominees must be citizens or permanent residents of the United States. Additional detailed instructions about the nomination process are available at the website.
Restrictions: Posthumous or self-nominations are not eligible. Individuals who have previously received a NEA lifetime honor award (National Heritage Fellowship, Jazz Masters Fellowship, or an Opera Honor) are not eligible.
Geographic Focus: All States
Date(s) Application is Due: Oct 1
Amount of Grant: Up to 25,000 USD
Contact: Cheryl Schiele; (202) 682-5587; schielec@arts.gov
Internet: http://www.nea.gov/honors/heritage/index.html
Sponsor: National Endowment for the Arts
1100 Pennsylvania Avenue NW
Washington, DC 20506-0001

NEDA/AED Charron Family Research Grant 2945

Committed to promoting study into the causes, cures and prevention of eating disorders, NEDA first established the Research Grants Program in 2002 in partnership with the Academy for Eating Disorders (AED). The aim of these grants is to expand eating disorders research while drawing promising new scientists into the field. In 2004 Shelley Charron-Bahniuk died from complications of her eating disorder while she was in recovery. The Charron Family Research Grant was created as a direct result of this illness that impacted both her immediate and extended family.
Requirements: Individuals working in the field of eating disorders who have completed their training, and have the doctorate degree or beyond are eligible.
Geographic Focus: All States
Amount of Grant: 10,000 USD
Contact: Dee Christoff; (206) 382-3587, ext. 15; info@nationaleatingdisorders.org
Internet: http://www.nationaleatingdisorders.org/p.asp?WebPage_ID=712
Sponsor: National Eating Disorders Association / Academy for Eating Disorders
603 Stewart Street, Suite 803
Seattle, WA 98101

NEDA/AED Joan Wismer Research Grant 2946

Committed to promoting study into the causes, cures and prevention of eating disorders, NEDA first established the Research Grants Program in 2002 in partnership with the Academy for Eating Disorders (AED). The aim of these grants is to expand eating disorders research while drawing promising new scientists into the field.
Requirements: Individuals working in the field of eating disorders who have completed their training, and have the doctorate degree or beyond are eligible.
Geographic Focus: All States
Amount of Grant: 10,000 USD
Contact: Dee Christoff, Development Director; (206) 382-3587, ext. 15; fax (206) 829-8501; info@nationaleatingdisorders.org
Internet: http://www.nationaleatingdisorders.org/p.asp?WebPage_ID=818
Sponsor: National Eating Disorders Association / Academy for Eating Disorders
603 Stewart Street, Suite 803
Seattle, WA 98101

NEDA/AED Tampa Bay Eating Disorders Task Force Award 2947

The National Eating Disorders Association sponsors a small grants program to support research in the areas of eating disorders etiology, prevention, and treatment. Through the grants, the organization seeks to expand innovative eating disorders research while supporting investigators in the early stages of their careers in the eating disorders field. Support can be included for: salary for technicians, research assistants and other ancillary personnel; laboratory, radiological and pharmaceutical costs in clinical research projects; rating instruments and questionnaires; equipment and supplies; acquisition and maintenance of laboratory animals; subject payment; and advertising.
Requirements: Any junior investigator worldwide who works in the field of eating disorders and has completed their terminal graduate level degree and training is eligible.
Restrictions: Trainees (e.g. students, grad students, residents) are not eligible and investigators must be within seven years following the end of their training. Individuals who previously served or currently serve as a principal investigator on a substantial externally funded grant (e.g. RO1, RO3, or K award in the U.S.) are ineligible. Funds cannot be used for the following purposes: salary support for residents, fellows or other trainees; faculty or investigator salary; secretarial salary and benefits; tuition for residents, graduate students or post-doctoral fellows; publication costs or purchase of reprints or books; room and board for clinical subjects; studies that are also sponsored by pharmaceutical firms; computers; or travel.
Geographic Focus: All States
Amount of Grant: 10,000 USD
Contact: Dee Christoff; (206) 382-3587, ext. 15; info@nationaleatingdisorders.org
Internet: http://www.nationaleatingdisorders.org/p.asp?WebPage_ID=816
Sponsor: National Eating Disorders Association / Academy for Eating Disorders
603 Stewart Street, Suite 803
Seattle, WA 98101

NEFE Grants 2948

NEFE seeks projects whose outcomes can improve the public's ability to achieve personal and household financial well-being. Of particular interest are pro-active research projects initiated from one of a broad spectrum of scholarly disciplines whose findings may cultivate critical thinking in the financial literacy community. Also of interest are development projects that put research recommendations into action. Project outcomes must be capable of achieving traction and measurable impact with audiences such as financial education intermediaries, researchers, practitioners, decision makers, and others who can achieve effective outreach to a target population with an unmet financial literacy need or to the general public. Funding requests are assessed within the parameters of the following three key grant themes: understanding financial behavior; advance innovative behavior; and assure significance to society. Recent grant awards for research projects have ranged from $100,000 to $150,000. Grant durations generally range from 12 to 36 months.
Requirements: 501(c)3 tax-exempt organizations are eligible.
Restrictions: NEFE does not fund: individuals; foreign organizations; international programs or projects; an organization's operating and administrative expenses; salaries of permanent staff; capital costs; pass-through funding; budget deficits; projects with a potential conflict of interest; conferences, workshops, seminars and similar activities that fail to provide a long-term solution; distribution methodologies that require ongoing maintenance; projects with commercial or proprietary elements; lobbying, political contributions, fund-raising events, or other similar activities designed to influence legislation; donations, endowments, challenge grants, matching funds, sponsorships, and other similar programs; or direct or matching payments to members of the public.
Geographic Focus: All States
Amount of Grant: 100,000-150,000 USD
Contact: Greta Zwickey; (303) 741-6333; fax (303) 220-0838; gnz@nefe.org
Internet: http://www.nefe.org/pages/innovative.html
Sponsor: National Endowment for Financial Education
1331 17th Street, Suite 1200
Denver, CO 80202

NEH Awards for Faculty at Historically Black Colleges and Universities 2949

Grants awarded under this program support individual faculty members at Historically Black Colleges and Universities (HBCUs) pursuing research of value to humanities scholars, students, or general audiences. Awards are designed to be flexible, allowing applicants to define the audience, type of research, award periods, and administrative arrangements that best fit their projects. Awards can be used for a wide range of projects that are based on humanities research. Eligible projects include pursuing research in primary and secondary materials; producing articles, monographs, books, digital materials, archaeological site reports, translations, editions, or other scholarly resources; and conducting basic research leading to the improvement of an existing undergraduate course or the achievement of institutional or community research goals.
Requirements: This program accepts applications from faculty members and retired faculty members affiliated with Historically Black Colleges or Universities. All U.S. citizens who teach at or have retired from Historically Black Colleges and Universities are eligible to apply. Foreign nationals who have been living in the United States or its jurisdictions for at least the three years prior to the application deadline are also eligible, provided that they teach at or have retired from any of these institutions.
Restrictions: NEH Awards for Faculty may not be used for: curricular or pedagogical tools, methods, theories, or surveys; preparation or revision of textbooks; projects that seek to promote a particular political, religious, or ideological point of view; projects that advocate a particular program of social action; works in the creative and performing arts, i.e., painting, writing fiction or poetry, dance performance, etc; or research leading to the improvement of graduate courses.
Geographic Focus: All States
Date(s) Application is Due: Apr 14
Amount of Grant: Up to 50,400 USD
Contact: Jane Aikin, Director; (202) 606-8212 or (202) 606-8200; jaikin@neh.gov
Internet: http://www.neh.gov/grants/guidelines/AF_HBCU.html
Sponsor: National Endowment for the Humanities
400 7th Street SW
Washington, DC 20506

NEH Awards for Faculty at Institutions with High Hispanic Enrollment 2950

Grants awarded under this program support individual faculty members at Institutions with High Hispanic Enrollment pursuing research of value to humanities scholars, students, or general audiences. Awards are designed to be flexible, allowing applicants to define the audience, type of research, award periods, and administrative arrangements that best fit their projects. Awards can be used for a wide range of projects that are based on humanities research. Eligible projects include pursuing research in primary and secondary materials; producing articles, monographs, books, digital materials, archaeological site reports, translations, editions, or other scholarly resources; and conducting basic research leading to the improvement of an existing undergraduate course or the achievement of institutional or community research goals.
Requirements: This program accepts applications from faculty members and retired faculty members affiliated with Institutions with High Hispanic Enrollment. All U.S. citizens who teach at or have retired from Institutions with High Hispanic Enrollment are eligible to apply. Foreign nationals who have been living in the United States or its jurisdictions for at least the three years prior to the application deadline are also eligible, provided that they teach at or have retired from any of these institutions.
Restrictions: NEH Awards for Faculty may not be used for: curricular or pedagogical tools, methods, theories, or surveys; preparation or revision of textbooks; projects that seek to promote a particular political, religious, or ideological point of view; projects that advocate a particular program of social action; works in the creative and performing arts, i.e., painting, writing fiction or poetry, dance performance, etc; or research leading to the improvement of graduate courses.
Geographic Focus: All States
Date(s) Application is Due: Apr 14
Amount of Grant: Up to 50,400 USD
Contact: Jane Aikin, Director; (202) 606-8212 or (202) 606-8200; jaikin@neh.gov
Internet: http://www.neh.gov/grants/guidelines/AF_IHHE.html
Sponsor: National Endowment for the Humanities
400 7th Street SW
Washington, DC 20506

NEH Awards for Faculty at Tribal Colleges and Universities 2951

Grants awarded under this program support individual faculty members at Tribal Colleges and Universities pursuing research of value to humanities scholars, students, or general audiences. Awards are designed to be flexible, allowing applicants to define the audience, type of research, award periods, and administrative arrangements that best fit their projects. Awards can be used for a wide range of projects that are based on humanities research. Eligible projects include pursuing research in primary and secondary materials; producing articles, monographs, books, digital materials, archaeological site reports, translations, editions, or other scholarly resources; and conducting basic research leading to the improvement of an existing undergraduate course or the achievement of institutional or community research goals.
Requirements: This program accepts applications from faculty members and retired faculty members affiliated with Tribal Colleges or Universities. All U.S. citizens who teach at or have retired from Tribal Colleges and Universities are eligible to apply. Foreign nationals who have been living in the United States or its jurisdictions for at least the three years prior to the application deadline are also eligible, provided that they teach at or have retired from any of these institutions.
Restrictions: NEH Awards for Faculty may not be used for: curricular or pedagogical tools, methods, theories, or surveys; preparation or revision of textbooks; projects that seek to promote a particular political, religious, or ideological point of view; projects that advocate a particular program of social action; works in the creative and performing arts, i.e., painting, writing fiction or poetry, dance performance, etc; or research leading to the improvement of graduate courses.
Geographic Focus: All States
Date(s) Application is Due: Apr 14
Amount of Grant: Up to 50,400 USD
Contact: Jane Aikin, Director; (202) 606-8212 or (202) 606-8200; jaikin@neh.gov
Internet: http://www.neh.gov/grants/guidelines/AF_TCU.html
Sponsor: National Endowment for the Humanities
400 7th Street SW
Washington, DC 20506

NEH Collaborative Research Grants 2952

Collaborative Research Grants support interpretive research undertaken by a team of two or more scholars, for full-time or part-time activities for periods of a minimum of one year up to a maximum of three years. Support is available for various combinations of scholars, consultants, and research assistants; project-related travel; field work; applications of information technology; and technical support and services. All grantees are expected to communicate the results of their work to the appropriate scholarly and public audiences. Eligible projects include: research that significantly adds to knowledge and understanding in the humanities; conferences on topics of major importance in the humanities that will benefit scholarly research; archaeological projects that include the interpretation and communication of results (projects may encompass excavation, materials analysis, laboratory work, field reports, and preparation of interpretive monographs); and research that uses the knowledge and perspectives of the humanities and historical or philosophical methods to enhance understanding of science, technology, medicine, and the social sciences. These grants support full-time or part-time activities for periods of one to three years.
Requirements: Eligibility is limited to: institutional applicants, i.e., U.S. nonprofit organizations with IRS 501(c)3 tax-exempt status, and state and local governments and federally recognized Indian tribal governments; and project directors who are not affiliated with an eligible institution, i.e., U.S. citizens or foreign nationals who have been living in the United States or its jurisdictions for at least the three years immediately preceding the application deadline.
Restrictions: Collaborative Research Grants may not be used for: projects undertaken by individuals (without collaboration); specific policy studies or educational or technical impact assessments; projects that seek to promote a particular political, religious, or ideological point of view; projects that advocate a particular program of social action; editions or translations (see the note below); support for the preparation or publication of textbooks intended primarily for classroom use; development of tools, including digital tools, or any materials analysis whose primary goal is refinement of a method, unless the development contributes to the larger interpretive goals of the project; inventories of collections; empirical social science research; projects in cognitive psychology; creation and enhancement of databases unless part of an interpretive humanities project; collection of oral history interviews, unless part of interpretive humanities research; research in educational methods, tests, or measurements, or any projects primarily intended for students in formal learning environments; recurrent meetings of professional organizations, societies, and scholarly organizations; planning or putting on exhibitions; works in the creative and performing arts (e.g., painting, writing fiction or poetry, dance performance); or support for publication costs or publication subvention.
Geographic Focus: All States
Date(s) Application is Due: Dec 8

Contact: Jane Aikin, Director; (202) 606-8212 or (202) 606-8200; jaikin@neh.gov
Internet: http://www.neh.gov/grants/guidelines/collaborative.html
Sponsor: National Endowment for the Humanities
400 7th Street SW
Washington, DC 20506

NEH Digital Humanities Implementation Grants **2953**
The NEH Digital Humanities Implementation Grants program is designed to fund the implementation of innovative digital-humanities projects that have successfully completed a start-up phase and demonstrated their value to the field. Such projects might enhance our understanding of central problems in the humanities, raise new questions in the humanities, or develop new digital applications and approaches for use in the humanities. The program can support innovative digital-humanities projects that address multiple audiences, including scholars, teachers, librarians, and the public. Unlike NEH's start-up grant program, which emphasizes basic research, prototyping, experimentation, and potential impact, the Digital Humanities Implementation Grants program seeks to identify projects that have successfully completed their start-up phase and are well positioned to have a major impact. Proposals are welcome for digital initiatives in any area of the humanities. Digital Humanities Implementation Grants may involve: implementation of computationally-based methods or techniques for humanities research; implementation of new digital tools for use in humanities research, public programming, or educational settings; efforts to ensure the completion and long-term sustainability of existing digital resources (typically in conjunction with a library or archive); studies that examine the philosophical or practical implications of the use of emerging technologies in specific fields or disciplines of the humanities, or in interdisciplinary collaborations involving several fields or disciplines; or implementation of new digital modes of scholarly communication that facilitate peer review, collaboration, or the dissemination of humanities scholarship for various audiences. Awards range from $100,000 to $325,000.
Requirements: Eligibility is limited to: U.S. nonprofit organizations with IRS 501(c)3 tax-exempt status; and state and local governmental agencies and federally recognized Indian tribal governments.
Restrictions: Individuals are not eligible to apply. Digital Humanities Implementation Grants may not be used for: projects that mainly involve digitization; support for recurring publications (such as the creation of a journal); conferences, professional meetings, or workshops; acquisition of computer equipment or software in excess of 20 percent of the grant total; creative or performing arts; work undertaken in the pursuit of an academic degree; projects that seek to promote a particular political, religious, or ideological point of view; or projects that advocate a particular program of social action.
Geographic Focus: All States
Date(s) Application is Due: Feb 18
Amount of Grant: 100,000-325,000 USD
Contact: Brett Bobley; (202) 606-8401 or (202) 606-8364; bbobley@neh.gov o
Internet: http://www.neh.gov/grants/guidelines/digitalhumanitiesimplementation.html
Sponsor: National Endowment for the Humanities
400 7th Street SW
Washington, DC 20506

NEH Digital Humanities Start-Up Grants **2954**
The National Endowment for the Humanities (NEH) invites applications to the Digital Humanities Start-Up Grants program. This program is designed to encourage innovations in the digital humanities. By awarding relatively small grants to support the planning stages, NEH aims to encourage the development of innovative projects that promise to benefit the humanities. Proposals should be for the planning or initial stages of digital initiatives in any area of the humanities. Digital Humanities Start-Up Grants may involve: research that brings new approaches or documents best practices in the study of the digital humanities; planning and developing prototypes of new digital tools for preserving, analyzing, and making accessible digital resources, including libraries' and museums' digital assets; scholarship or studies that examine the philosophical or practical implications and impact of the use of emerging technologies in specific fields or disciplines of the humanities, or in interdisciplinary collaborations involving several fields or disciplines; innovative uses of technology for public programming and education utilizing both traditional and new media; and new digital modes of publication that facilitate the dissemination of humanities scholarship in advanced academic as well as informal or formal educational settings at all academic levels. Two levels of awards will be made in this program: Level I awards are small grants designed to fund brainstorming sessions, workshops, early alpha-level prototypes, and initial planning; and Level II awards are larger grants that can be used for more fully-formed projects that are ready to begin implementation or demonstrate proofs of concept.
Requirements: All applicants must propose an innovative approach, method, tool, or idea that has not been used before in the humanities. Eligibility is limited to: U.S. nonprofit organizations or institutions with IRS 501(c)3 tax-exempt status; and state and local governmental agencies and federally recognized Indian tribal governments.
Restrictions: Individuals are not eligible to apply. Digital Humanities Start-Up Grants cannot be used for: projects that mainly involve digitization, unless the applicant is proposing an innovative method for digitization; the creation or conversion of a scholarly journal (however, the exploration of or planning for new models of scholarly publication is allowed); the implementation or assessment of existing digital applications in the humanities (however, exploration of or planning for a new direction or tool for an established project is allowed); recurring or established conferences or professional meetings; acquisition of computer equipment or software in excess of 20 percent of the grant total; creative or performing arts; empirical social scientific research; work undertaken in the pursuit of an academic degree; the preparation or publication of textbooks; projects that seek to promote a particular political, religious, or ideological point of view; or projects that advocate a particular program of social action.
Geographic Focus: All States
Date(s) Application is Due: Apr 2; Sep 11
Amount of Grant: 5,000-60,000 USD
Contact: Brett Bobley; (202) 606-8401 or (202) 606-8364; bbobley@neh.gov o
Internet: http://www.neh.gov/grants/guidelines/digitalhumanitiesstartup.html
Sponsor: National Endowment for the Humanities
400 7th Street SW
Washington, DC 20506

NEH Documenting Endangered Languages Grants **2955**
Documenting Endangered Languages is a joint, multi-year funding program of the National Science Foundation and the National Endowment for the Humanities to develop and advance scientific and scholarly knowledge concerning endangered human languages. Made urgent by the imminent death of approximately half of the 6000-7000 currently used human languages, DEL seeks not only to document these endangered languages but to integrate, systematize, and make knowledge concerning them widely available by exploiting advances in information technology. Approximately 6-10 Standard or Continuing Grants ranging from $12,000 to $150,000 per year for one to three years are available. Approximately 12 Fellowships of $4,200 per month for awards lasting from six to twelve months are available; the maximum stipend is $50,400 for a twelve-month tenure period.
Requirements: Academic institutions and non-profit, non-academic organizations located in the United States are eligible. U.S. citizens are eligible to apply for fellowships. Foreign nationals who have been living in the United States or its jurisdictions for at least the three years prior to the proposal deadline are also eligible to apply for fellowships.
Restrictions: For-profit organizations are not eligible to apply.
Geographic Focus: All States
Date(s) Application is Due: Sep 20
Amount of Grant: 62,400-212,400 USD
Contact: Barbara Bays, Senior Program Officer; (202) 606-8290 or (202) 606-8269; bbays@neh.gov or publicpgms@neh.gov
Internet: http://www.neh.gov/grants/guidelines/del.html
Sponsor: National Endowment for the Humanities
1100 Pennsylvania Avenue NW, Room 426
Washington, DC 20506

NEH Fellowships **2956**
Fellowships support individuals pursuing advanced research that is of value to humanities scholars, general audiences, or both. Recipients usually produce articles, monographs, books, digital materials, archaeological site reports, translations, editions, or other scholarly resources in the humanities. Projects may be at any stage of development. Fellowships support continuous full-time work for a period of six to twelve months. NEH encourages submission of Fellowships applications from faculty at Historically Black Colleges and Universities, Institutions with High Hispanic Enrollment, and Tribal Colleges and Universities. Fellowships cover periods lasting from six to twelve months at a stipend of $4,200 per month in outright funds. The maximum stipend is $50,400 for a twelve-month period.
Requirements: The Fellowships program accepts applications from researchers, teachers, and writers, whether they have an institutional affiliation or not. All U.S. citizens, whether they reside inside or outside the United States, are eligible to apply. Foreign nationals who have been living in the United States or its jurisdictions for at least the three years prior to the application deadline are also eligible.
Restrictions: Fellowships may not be used for: projects that seek to promote a particular political, religious, or ideological point of view; projects that advocate a particular program of social action; specific policy studies; research for doctoral dissertations or theses by students enrolled in a degree program; the preparation or revision of textbooks; curriculum development; the development of pedagogical tools (including teaching methods or theories); educational or technical impact assessments; the creation or enhancement of databases, unless part of a larger interpretive project; empirical social science research, unless part of a larger humanities project; inventories of collections; or works in the creative and performing arts, i.e., painting, writing fiction or poetry, dance performance, etc.
Geographic Focus: All States
Date(s) Application is Due: Apr 30
Amount of Grant: 4,200-50,400 USD
Contact: Jane Aikin, Director; (202) 606-8212 or (202) 606-8200; jaikin@neh.gov
Internet: http://www.neh.gov/grants/guidelines/fellowships.html
Sponsor: National Endowment for the Humanities
400 7th Street SW
Washington, DC 20506

NEH Fellowships at Independent Research Institutions **2957**
Grants for Fellowship Programs at Independent Research Institutions (FPIRI) support fellowships at institutions devoted to advanced study and research in the humanities. NEH fellowships provide scholars with research time and access to resources that might not be available at their home institutions. Fellowship programs may be administered by independent centers for advanced study, libraries, and museums in the United States; American overseas research centers; and American organizations that have expertise in promoting research on foreign cultures. Individual scholars must apply directly to the institutions themselves. FPIRI grants provide funding to programs for humanities fellowships of four to twelve months. Awards are generally made to support fellowships for three years. Recent grants have ranged from $78,600 to $480,600.

Requirements: Any U.S. nonprofit institution with IRS 501(c)3 tax-exempt status is eligible, as are state and local governmental agencies and federally recognized Indian tribal governments. In addition, to apply for a FPIRI grant, institutions must: be financed, governed, and administered independently of institutions of higher education, and have established and maintained fellowship programs with their own or other funding.
Restrictions: Individuals are not eligible to apply.
Geographic Focus: All States
Date(s) Application is Due: Aug 17
Amount of Grant: 78,600-480,600 USD
Contact: Jane Aikin, Director; (202) 606-8212 or (202) 606-8200; jaikin@neh.gov
Internet: http://www.neh.gov/grants/guidelines/fpiri.html
Sponsor: National Endowment for the Humanities
400 7th Street SW
Washington, DC 20506

NEH Fellowships for Advanced Social Science Research on Japan **2958**
The Fellowship Program for Advanced Social Science Research on Japan is a joint activity of the Japan-U.S. Friendship Commission (JU.S.FC) and the National Endowment for the Humanities. Awards support research on modern Japanese society and political economy, Japan's international relations, and U.S.-Japan relations. The program encourages innovative research that puts these subjects in wider regional and global contexts and is comparative and contemporary in nature. Research should contribute to scholarly knowledge or to the general public's understanding of issues of concern to Japan and the United States. Appropriate disciplines for the research include anthropology, economics, geography, history, international relations, linguistics, political science, psychology, public administration, and sociology. The fellowships are designed for researchers with advanced language skills whose research will require use of data, sources, and documents in their original languages or whose research requires interviews onsite in direct one-on-one contact. Fellows may undertake their projects in Japan, the United States, or both, and may include work in other countries for comparative purposes. Projects may be at any state of development. Awards usually result in articles, monographs, books, digital materials, archaeological site reports, translations, editions, or other scholarly resources. As tax-supported federal agencies, JU.S.FC and NEH endeavor to make grant products available to the broadest possible audience. Our goal is for scholars, educators, students, and the American public to have ready and easy access to the wide range of grant products. For projects that lead to the development of websites, all other considerations being equal, preference is given to those that provide free access to the public. Fellowships cover uninterrupted periods lasting from six to twelve months at a stipend of $4,200 per month. The maximum stipend is $50,400 for a twelve-month award period.
Requirements: The Fellowships for Advanced Social Science Research on Japan program accepts applications from researchers, teachers, and writers, whether they have an institutional affiliation or not. All U.S. citizens, whether they reside inside or outside the United States, are eligible to apply. Foreign nationals who have been living in the United States or its jurisdictions for at least the three years prior to the application deadline are also eligible.
Restrictions: Advanced Social Science Research on Japan Fellowships may not be used for: projects that seek to promote a particular political, religious, or ideological point of view; projects that advocate a particular program of social action; specific policy studies; research for doctoral dissertations or theses by students enrolled in a degree program; the preparation or revision of textbooks; curriculum development; the development of pedagogical tools (including teaching methods or theories); educational or technical impact assessments; the creation or enhancement of databases, unless part of a larger interpretive project; inventories of collections; or works in the creative and performing arts, i.e., painting, writing fiction or poetry, dance performance, etc.
Geographic Focus: All States
Date(s) Application is Due: May 3
Amount of Grant: 25,200-50,400 USD
Contact: Jane Aikin, Director; (202) 606-8212 or (202) 606-8200; jaikin@neh.gov
Internet: http://www.neh.gov/grants/guidelines/fellowships-japan.html
Sponsor: National Endowment for the Humanities
400 7th Street SW
Washington, DC 20506

NEI Clinical Study Planning Grant **2959**
The National Eye Institute (NEI) supports large-scale clinical vision research projects, including randomized clinical trials and epidemiologic studies. At the time of submission, applications requesting support for these activities are expected to provide detailed information regarding the study's rationale, design, analytic techniques, protocols and procedures, facilities and environment, organizational structure, and collaborative arrangements. This information is best conveyed in a well-documented Manual of Procedures (MOP), the development of which represents a costly and time-consuming activity. The Clinical Study Planning Grant is designed to facilitate activities central to the refinement of a study's protocol and procedures and the development of a detailed MOP.
Requirements: Applications may be submitted by domestic for-profit and nonprofit organizations, both public and private; units of state and local governments; and eligible agencies of the federal government; domestic and foreign institutions.
Restrictions: Foreign institutions are not eligible.
Geographic Focus: All States
Amount of Grant: Up to 150,000 USD
Contact: Donald F. Everett., (301) 451-2020; fax (301) 402-0528; dfe@nei.nih.gov
Internet: http://grants.nih.gov/grants/guide/pa-files/PAR-05-115.html
Sponsor: National Eye Institute
5635 Fishers Lane, Suite 1300, MSC 9300
Bethesda, MD 20892-9300

NEI Clinical Vision Research Development Award **2960**
This award is intended to strengthen interactions among clinicians, biostatisticians, epidemiologists, statistical geneticists, and other clinical trial specialists to facilitate the design and conduct of clinical research projects, such as the development of coordinating center capabilities. The intent of this projects is to advance the understanding, prevention, or clinical management of visual system disorders. A number of experienced investigators and research groups are needed to provide leadership in the design and implementation of such clinical vision research projects. Therefore, NEI wishes to help institutions with NEI-funded investigators augment staff expertise and acquire the resources necessary to enhance their clinical vision research programs. PA: PAR-00-050
Requirements: Applications may be submitted by domestic for-profit and non-profit organizations, public and private, such as universities, colleges, hospitals, laboratories, units of State and local governments, and eligible agencies of the Federal government.
Restrictions: Foreign institutions and institutions having current or past NEI Core Grant for Vision Research (P30) support for a biostatistics module are not eligible to apply.
Geographic Focus: All States
Amount of Grant: Up to 100,000 USD
Contact: Mary Frances Cotch, (301) 496-5983; fax (301) 402-0528; mfcotch@nei.nih.gov or deverett@nei.nih.gov
Internet: http://grants.nih.gov/grants/guide/pa-files/PAR-00-050.html
Sponsor: National Eye Institute
5635 Fishers Lane, Suite 1300, MSC 9300
Bethesda, MD 20892-9300

NEI Mentored Clinical Scientist Development Program Award (K12) **2961**
The NEI Mentored Clinician Scientist Development Award (K12) is an award to an educational institution or professional organization to facilitate and support career development experiences which lead clinician scientists to research independence. It is expected that scholars appointed to the program will subsequently apply for their own individual research support. The program should be designed to accommodate candidates with varying levels of research experience who require an individualized career development program ranging from two to five years. Applications for this award should propose a structured, phased development plan overseen by a Program Director and a Program Advisory Committee. A designated period of didactic training should be followed by a period of supervised research experience. Candidates should be exposed to both research that is typically undertaken with a few investigators and to research approaches that require a multidisciplinary team to address complex research questions as appropriate. It is expected that at the end of the career development period, appointees will transition successfully into positions as independent investigators. This award provides five years of potentially renewable support. The subsequent continuation will depend in part on the progress made by the Scholars in achieving independent research support. Renewal applications must contain a progress report documenting the accomplishments and current career activities of former Scholars. PA# PAR-09-083
Requirements: The following organizations and institutions are eligible to apply: public/state controlled institutions of higher education; private institutions of higher education; hispanic-serving Institutions; historically black colleges and universities; tribally controlled colleges and universities; Alaska Native and Native Hawaiian serving institutions; nonprofits with 501(c)3 IRS Status (other than institutions of higher education); nonprofits without 501(c)3 IRS Status (other than institutions of higher education). Any individual with the skills, knowledge, and resources necessary to carry out the proposed research is invited to work with his/her institution to develop an application for support. Individuals from underrepresented racial and ethnic groups as well as individuals with disabilities are always encouraged to apply for NIH support.
Restrictions: Foreign institutions are not eligible to apply.
Geographic Focus: All States
Date(s) Application is Due: May 13
Amount of Grant: 1,125,000 USD
Contact: Neeraj Agarwal, (301) 451-2020; fax (301) 402-0528; agarwalnee@nei.nih.gov
Internet: http://grants.nih.gov/grants/guide/pa-files/PAR-09-083.html
Sponsor: National Eye Institute
5635 Fishers Lane, Suite 1300, MSC 9300
Bethesda, MD 20892-9300

NEI Research Grant For Secondary Data Analysis Grants **2962**
The NEI Research Grant for Secondary Data Analysis (R21) program is designed to provide investigators with the support necessary to conduct secondary data analyses utilizing existing database resources. Applications may be related to, but must be distinct from, the specific aims of the original data collection. The total project period for an application submitted in response to this funding opportunity may not exceed two years. Direct costs are limited to $275,000 over an R21 two-year period, with no more than $200,000 in direct costs allowed in any single year. PA # PAR-06-326
Requirements: Eligible organizations: for-profit and non-profit organizations; public and private institutions (such as universities, colleges, hospitals, and laboratories); units of state and local governments; eligible agencies of the federal government; units of state; local tribal governments; domestic and foreign institutions. Eligible Project Directors/Principal Investigators (PD/PIs): any individual with the skills, knowledge, and resources necessary to carry out the proposed research. Individuals from underrepresented racial and ethnic groups as well as individuals with disabilities are always encouraged to apply for NIH support.
Geographic Focus: All States
Date(s) Application is Due: May 8
Amount of Grant: 200,000-275,000 USD
Contact: Natalie Kurinij, (301) 451-2020; fax (301) 402-0528; kurinijn@mail.nih.gov

Internet: http://grants.nih.gov/grants/guide/pa-files/PAR-06-326.html
Sponsor: National Eye Institute
5635 Fishers Lane, Suite 1300, MSC 9300
Bethesda, MD 20892-9300

NEI Ruth L. Kirschstein National Research Service Award Short-Term Institutional Research Training Grants 2963

The National Institutes of Health (NIH) will award Ruth L. Kirschstein National Research Service Award (NRSA) Short-Term Institutional Research Training Grants (T35) to eligible institutions to develop or enhance research training opportunities for individuals interested in careers in biomedical, behavioral and clinical research. Many of the NIH Institutes and Centers (ICs) use this grant mechanism exclusively to support intensive, short-term research training experiences for students in health professional schools during the summer. In addition, the Short-Term Institutional Research Training Grant may be used to support other types of predoctoral and postdoctoral training in focused, often emerging scientific areas relevant to the mission of the funding IC. The proposed training must be in either basic, behavioral or clinical research aspects of the health-related sciences. This program is intended to encourage graduate and/or health professional students to pursue research careers by exposure to and short-term involvement in the health- related sciences. The training should be of sufficient depth to enable the trainees, upon completion of the program, to have a thorough exposure to the principles underlying the conduct of research. PA # PA-08-227
Requirements: The following organizations/institutions are eligible to apply: public/state controlled institutions of higher education; private institutions of higher education; hispanic-serving institutions; historically black colleges and universities (HBCUs); tribally controlled colleges and universities (TCCUs); Alaska Native and Native Hawaiian serving institutions; nonprofits with 501(c)3 IRS Status (Other than Institutions of Higher Education); nonprofits without 501(c)3 IRS Status (Other than institutions of higher education); indian/native american tribal governments (federally recognized); indian/native american tribally designated organizations; U.S. Territory or possession; indian/native american tribal governments (other than federally recognized); faith-based or community-based organizations. Only domestic, non-profit, private or public institutions may apply for grants to support National Research Service Award (NRSA) short-term research training programs. Foreign institutions are not eligible to apply. The applicant institution must have a strong and high quality research program in the area(s) proposed for the research training and must have the requisite staff and facilities on site to conduct the proposed research training program. Any individual with the skills, knowledge, and resources necessary to organize and implement a high-quality research training program is invited to work with their institution to develop an application for support. Individuals from underrepresented racial and ethnic groups, individuals with disabilities, and individuals from disadvantaged backgrounds are always encouraged to apply for NIH support.
Restrictions: Foreign institutions are not eligible to apply.
Geographic Focus: All States
Date(s) Application is Due: Sep 8
Contact: Neeraj Agarwal, (301) 451-2020; fax (301) 402-0528; agarwalnee@mail.nih.gov
Internet: http://grants.nih.gov/grants/guide/pa-files/PA-08-227.html
Sponsor: National Eye Institute
5635 Fishers Lane, Suite 1300, MSC 9300
Bethesda, MD 20892-9300

NEI Scholars Program 2964

This program provides an opportunity for outstanding individuals to obtain laboratory or clinical research training within NEI intramural environment and to facilitate the successful transition to continue their research career at an extramural institution as independent vision researchers. This is accomplished by first providing individuals with the necessary resources to receive high quality research training for three to four years at NEI. This is followed by providing extramural funding to support the research program for two years at the extramural institution to which NEI scholar is recruited. It is anticipated that NEI scholars will subsequently compete for independent funding to continue their research. During the intramural phase, the scholar is expected to spend full time on research. During the extramural phase, the scholar must spend a minimum of 75 percent of a full-time professional effort conducting research and engaging in research career development activities for the two years of the award. Applicants are encouraged to contact NEI extramural staff regarding their eligibility for this award and NEI intramural laboratory scientist regarding sponsorship information, prior to submitting an application. PA: PAR-98-107
Requirements: Individuals must have a research or health professional doctoral level degree or its equivalent and must have demonstrated the potential for a highly productive research career during predoctoral training and the immediate postdoctoral period, if applicable. Individuals at NEI or other NIH intramural laboratories who meet the eligibility requirements are eligible to apply. Before submitting an application for the program, the candidate must identify an individual in the NEI Division of Intramural Research who will serve as the sponsor and will be committed to supervise the training and research project. The sponsor must be an active investigator in the area of the proposed research and have research training experience and resources needed to support the NEI scholar.
Restrictions: Individuals who have had more than five years of postdoctoral research training at the time of application are not eligible to apply; however, clinical training does not count against the five years. Former principal investigators on NIH research project grants R01; SBIR/STTR awards R43, R44/R41, R42; subprojects of program project grants P01; center grants P30, P50; K08, K23, K24, or the equivalent are not eligible.
Geographic Focus: All States
Date(s) Application is Due: Feb 1; Jun 1; Oct 1
Amount of Grant: Up to 175,000 USD
Contact: Dr. Maria Giovanni, Division of Extramural Research; (301) 496-0484; fax (301) 402-0528; myg@nei.nih.gov
Internet: http://www.grants.nih.gov/grants/guide/pa-files/PAR-98-107.html
Sponsor: National Eye Institute
5635 Fishers Lane, Suite 1300, MSC 9300
Bethesda, MD 20892-9300

NEI Translational Research Program On Therapy For Visual Disorders 2965

This program continues support of collaborative, multidisciplinary research programs focused on new therapeutic approaches to restore or prevent the loss of function due to visual system diseases and disorders. The rapid and efficient translation of laboratory research findings into clinical development requires a comprehensive and highly integrated approach involving collaborative teams of scientists and clinicians with expertise in multiple disciplines. Such a collaborative approach would be particularly appropriate for research focused on pathways that will likely be targeted by biological intervention, such as gene therapy, cell-based therapy, pharmacological approaches, or development of appropriate delivery systems. The intention of this program is to make resources available to scientists from several disciplines to form research teams to address scientific and technical questions that would be beyond the capabilities of any one research group. It is anticipated that applications funded under this program will lead to therapies that can be tested in a clinical trial.
Requirements: Eligible Institutions, you may submit an application (s) if your organization has any of the following characteristics: for-profit organizations; non-profit organizations; public or private institutions, such as universities, colleges, hospitals, and laboratories; units of State government; units of local government; eligible agencies of the Federal government; domestic Institutions Foreign components to domestic applications will be allowed if the foreign component provides significant scientific input that is critical to the program and for which there is no similar resource in the United States. Eligible Individuals: any individual with the skills, knowledge, and resources necessary to carry out the proposed research is invited to work with their institution to develop an application for support. Individuals from underrepresented racial and ethnic groups as well as individuals with disabilities are always encouraged to apply for NIH programs.
Geographic Focus: All States
Amount of Grant: Up to 2,000,000 USD
Contact: Dr. Peter Dudley, (301) 451-2020; fax (301) 402-0528; pad@nei.nih.gov
Internet: http://grants.nih.gov/grants/guide/pa-files/PAR-05-110.html
Sponsor: National Eye Institute
5635 Fishers Lane, Suite 1300, MSC 9300
Bethesda, MD 20892-9300

NEI Vision Research Core Grants 2966

The primary objective of Core Grants for Vision Research is to provide groups of investigators who have achieved independent National Eye Institute (NEI) funding with additional, shared support to enhance their own and their institution's capability for conducting vision research. Secondary objectives of this program include facilitating collaborative studies and attracting other scientists to research on the visual system.
Requirements: Institutions applying for a Core Grant must hold, on the receipt date, a minimum of eight NEI research project grants (R01), Bioengineering Research Partnership Grants (R24), Mentored Clinical Scientist Development Awards (K08), Mentored Patient-Oriented Research Career Development Award (K23), or Midcareer Investigator Award in Patient-Oriented Research (K24), including any noncompeting extensions of these awards made with or without additional funds. No other mechanisms or source of research support will be considered in determining eligibility. Only one Core Grant will be made to any single applicant organization. For multicampus institutions, no more than one Core Grant will be made to each campus. Joint applications may be submitted by investigators at neighboring, independent institutions.
Geographic Focus: All States
Date(s) Application is Due: Sep 25
Amount of Grant: 2,000,000-2,500,000 USD
Contact: Hemin R. Chin, (301) 451-2020; fax (301) 402-0528; hemin@nei.nih.gov
Internet: http://www.nei.nih.gov/funding/p30.asp
Sponsor: National Eye Institute
5635 Fishers Lane, Suite 1300, MSC 9300
Bethesda, MD 20892-9300

Nelda C. and H.J. Lutcher Stark Foundation Grants 2967

The foundation awards grants in its areas of interest, including education, health and social services, medical and dental, community enrichment, arts and culture. Giving is primarily in Texas, with preference to programs that directly impact Southeast Texas and limited giving in Southwest Louisiana. There are no application forms. Submit a letter of proposal containing the following information: description of the project signed by the president or CEO; brief history, description, programs, and mission of the organization; project time line; specific amount requested; list of other funding sources and amounts; list of board of directors and occupations; form 990 for most recent year-end; most recent audited financial statements or organization prepared statements; and current copy of IRS determination letter certifying tax-exempt status. Guidelines are available online. All grants from the third quarter of 2008 and for 2009 have been suspended due to recovery efforts in which the Foundation must engage for its own operations as a result of damages from Hurricane Ike, together with the negative impact of the downturn in financial markets on the Foundation's corpus, the Stark Foundation will continue the prior suspension of the grant-making aspect of its operations through the entirety of Year 2009. Therefore, the Stark Foundation will

not accept any grant applications for Year 2009, nor will be able to consider any grant applications that may be submitted notwithstanding this suspension.
Requirements: To be eligible for a grant, an organization must: be exempt from taxation under Section 501(c)3 of the Internal Revenue Code; not be a private foundation within the meaning of Section 509(a) of the Code. No grants are made for endowment purposes, to individuals, or to supporting organizations as described in Section 509(a)(3) of the Internal Revenue Code.
Geographic Focus: All States, Canada
Date(s) Application is Due: Mar 1; Jun 1; Oct 1
Contact: Grant Department
Internet: http://www.starkfoundation.org/Grants/Grant-Guidelines.aspx
Sponsor: Nelda C. and H.J. Lutcher Stark Foundation
P.O. Box 909
Orange, TX 77631-0909

Nellie Mae Education Foundation District-Level Change Grants **2968**
The Nellie Mae Education Foundation District-Level Systems Change Grants support the shift in high school-level education in New England to prepare students to thrive in a global, complex, fast changing society. Grant programs promote student centered models of schooling that offer a path away from a one size fits all approach to a more customized approach to maximize learning for all students. The initiative focuses on three priority areas: developing school and district designs and practices that enable all learners to achieve high standards; creating sustainable policy change to support these new approaches; and generating public will and increasing demand for changes in practice. Examples of allowable activities for funding include: engaging central office, school administrators, and teachers in identifying current initiatives and how they compete with or support a student centered learning frame; surveying student needs and challenges and what motivates them to achieve academically and personally; creating greater building level autonomy and funding streams based on student needs rather than programs; researching feasibility, designing and planning for the appropriate blended learning model including visits to blended learning schools; using technology as an essential tool in learning environments and professional development; or administrator visits from student centered schools. The Foundation will support self-assessment activities, efforts to build, forge or nurture a common vision and mission for the work, and integration of this work with other school/district initiatives. Funds can support purchasing online and hardcopy literacy materials to support complex Common Core State Standards (CCSS) text demands; engaging a team of district stakeholders to conduct an analysis of culture of the system's various dimensions; or administrative visits from student centered schools. Proposal information, including the request for proposal, a list of supplemental materials and instructions, questions to address in the proposal, a list of frequently asked questions and previously funded projects, and an instructional webinar, is located at the Foundation website.
Requirements: The Foundation's activities include both making grants to the public charities it supports and providing services to those organizations. The Foundation operates exclusively to promote the charitable and educational purposes of nonprofit educational organizations, including universities, colleges, secondary schools, elementary schools and other educational organizations that are described in the IRS Code Section 501(c)3.
Restrictions: Funding for this project is concentrated on urban school districts and learning centers in Connecticut, Massachusetts, and Rhode Island. The Foundation does not fund capital campaigns, endowments, scholarships or fellowships, debt reduction or cash reserves, building construction or renovation, and certain indirect costs.
Geographic Focus: Connecticut, Massachusetts, Rhode Island
Date(s) Application is Due: Oct 18
Amount of Grant: 25,000-1,000,000 USD
Contact: Stephanie Cheney; (781) 348-4240; fax (781) 348-4299; scheney@nmefdn.org
Internet: http://www.nmefoundation.org/grants/district-level-systems-change
Sponsor: Nellie Mae Education Foundation
1250 Hancock Street, Suite 205N
Quincy, MA 02169

Nellie Mae Education Foundation Public Understanding Grants **2969**
The Nellie Mae Education Foundation Public Understanding Grants work to promote awareness of the student-centered approaches that lead to deeper learning for all students, and create the public will to demand greater adoption of these approaches. Grants focus on three areas: educating communities on student-centered approaches to learning, what they are, and why they are critical; transforming public perception of how students learn and why it must change; and building public demand for student-centered learning approaches in communities, districts and states. Funding for projects is posted on the Foundation website when it becomes available. A list of frequently asked questions and previously funded projects are located on the website.
Requirements: The Foundation's activities include both making grants to the public charities it supports and providing services to those organizations. The Foundation operates exclusively to promote the charitable and educational purposes of nonprofit educational organizations, including universities, colleges, secondary schools, elementary schools and other educational organizations as described in the IRS Code Section 501(c)3. Organizations located in the New England area are eligible to apply (Maine, New Hampshire, Vermont, Massachusetts, Rhode Island, and Connecticut). Additionally, the Foundation supports programs, organizations, research and conferences in other states in order to influence policy, and advance knowledge of public opinion toward education.
Restrictions: The Foundation does not fund capital campaigns, endowments, scholarships or fellowships, debt reduction or cash reserves, building construction or renovation, and certain indirect costs. The Foundation also does not support Section 509(a) organizations or individuals.
Geographic Focus: All States
Amount of Grant: 7,500-250,000 USD
Contact: Stephanie Cheney; (781) 348-4240; fax (781) 348-4299; scheney@nmefdn.org
Internet: http://www.nmefoundation.org/grants/public-understanding
Sponsor: Nellie Mae Education Foundation
1250 Hancock Street, Suite 205N
Quincy, MA 02169

Nellie Mae Education Foundation Research and Development Grants **2970**
The Nellie Mae Education Foundation Research and Development Grants work to build and deepen understanding of student-centered learning approaches in order to foster more widespread adoption and implementation. Grants focus on three areas: gathering and promoting existing knowledge of student-centered approaches to learning; creating new research to promote student-centered approaches in order to support overall system change, policy change and community engagement; and developing new tools that support student-centered teaching, learning and assessment. Funding opportunities for this grant are posted on the Foundation website when funding is available. A list of frequently asked questions and additional list of previously funded projects are listed on the website.
Requirements: The Foundation's activities include making grants to the public charities it supports and providing services to those organizations. The Foundation operates exclusively to promote the charitable and educational purposes of nonprofit educational organizations, including universities, colleges, secondary schools, elementary schools and other educational organizations that are described in IRS Code Section 501(c)3. Funding is available primarily for the New England area, but the Foundation occasionally supports programs, organizations, research, and conferences outside of the New England area in order to influence policy, and advance knowledge of public opinion toward education.
Restrictions: The Foundation does not fund capital campaigns, endowments, scholarships or fellowships, debt reduction or cash reserves, building construction or renovation, and certain indirect costs.
Geographic Focus: All States
Contact: Stephanie Cheney; (781) 348-4240; fax (781) 348-4299; scheney@nmefdn.org
Internet: http://www.nmefoundation.org/grants/research-development
Sponsor: Nellie Mae Education Foundation
1250 Hancock Street, Suite 205N
Quincy, MA 02169

Nellie Mae Education Foundation State Level Systems Change Grants **2971**
The Nellie Mae Education Foundation State-Level Systems Change (SLSC) initiative works to remove barriers and shape state and federal policies that promote deeper learning through student-centered approaches. The SLSC focuses on four areas of priority: state accountability systems that include evaluation of new basic skills such as creative thinking and problem solving; high school graduation requirements based more on students demonstrating competency in target areas than on time spent in the classroom; teacher certification requirements that support teachers and help students develop skills and knowledge through relevant, interest-driven experiences; and state funding formulas that enable and support student-centered learning environments. Request for proposals, with supplementary material and instructions, should be posted on the website if funding is currently available and proposals may be submitted. Previously funded projects and a list of frequently asked questions are located at the Foundation website.
Requirements: The Foundation's activities include both making grants to the public charities it supports and providing services to those organizations. The Foundation operates exclusively to promote the charitable and educational purposes of nonprofit educational organizations, including universities, colleges, secondary schools, elementary schools and other educational organizations that are described in IRS Code Section 501(c)3. Funding is concentrated in Connecticut, Maine, New Hampshire, Rhode Island, Vermont, and Massachusetts, but is also available outside the area to support programs, organizations, research, and conference in order to influence policy, and advance knowledge of public opinion toward education.
Restrictions: The Foundation does not fund capital campaigns, endowments, scholarships or fellowships, debt reduction or cash reserves, building construction or renovation, and certain indirect costs.
Geographic Focus: All States
Amount of Grant: 50,000-700,000 USD
Contact: Stephanie Cheney; (781) 348-4240; fax (781) 348-4299; scheney@nmefdn.org
Internet: http://www.nmefoundation.org/grants/state-level-systems-change
Sponsor: Nellie Mae Education Foundation
1250 Hancock Street, Suite 205N
Quincy, MA 02169

Nell Warren Elkin and William Simpson Elkin Foundation **2972**
The foundation awards grants to charitable organizations in its areas of interest, including universities, cancer clinics, specific disease associations, religious groups, and Bible studies. Application form and guidelines are available online.
Requirements: Nonprofit, charitable organizations are eligible.
Geographic Focus: All States
Date(s) Application is Due: Feb 28; Jul 31
Amount of Grant: 2,000-10,000 USD
Contact: Dale Welch; (404) 724-3773; rs.foundationinfo@suntrust.com
Internet: http://www.elkinfoundation.org
Sponsor: Nell Warren Elkin and William Simpson Elkin Foundation
P.O. Box 4655, MC 221
Atlanta, GA 30302

Neotropical Grassland Conservancy Derald G. Langham Memorial Research Grant **2973**

Memorial Research Grants directly promote conservation research. Grants are reviewed by outside scientists and the NGC board. Grants are one-year, unrestricted funds that can be used for direct or indirect costs, including salary. The Derald G. Langham Memorial Research Grant provides a $2,000 grant to scientists in the American tropics featuring grassland plants. The grant is awarded in memory of Derald G. Langham. Derald received the first Ph.D. from Cornell University in plant genetics in 1937. He and his family spent the next 30 years in Venezuela as Derald established plant nurseries and headed the national plant genetics program. His passion for plants, charisma, and support of new ways of thinking of the world are memorialized in this award.
Geographic Focus: All States
Date(s) Application is Due: Mar 1
Contact: Valentina Ferretti; (916) 967-3223; info@conservegrassland.org
Internet: http://www.conservegrassland.org/memorial_research_grants.htm
Sponsor: Neotropical Grassland Conservancy
6274 Heathcliff Drive
Carmichael, CA 95608-0909

Neotropical Grassland Conservancy Memorial Grants **2974**
Memorial Research Grants directly promote conservation research. Grants are reviewed by outside scientists and the NGC board. Grants are one-year, unrestricted funds that can be used for direct or indirect costs, including salary. Specific grants may be restricted to certain taxa or new research initiatives. See individual grants for details. Unless specifically stated, these grants are not restricted to conservation-oriented projects, but you will need to address the issue of how this grant may impact conservation efforts in your area under the section conservation implications of your project. Applications are accepted until the 1st of March each year and are reviewed by outside reviewers and members of the board.
Geographic Focus: All States
Date(s) Application is Due: Mar 1
Contact: Valentina Ferretti; (916) 967-3223; info@conservegrassland.org
Internet: http://www.conservegrassland.org/memorial_research_grants.htm
Sponsor: Neotropical Grassland Conservancy
6274 Heathcliff Drive
Carmichael, CA 95608-0909

Neotropical Grassland Conservancy Remembrance Fund Grants **2975**
The Remembrance Fund honors special people and special events. Many people can contribute to the fund in any amount. Remembrance Fund donations are acknowledged to the person you designate, and are listed on the NGC Web site and in its E-newsletter. A donation on behalf of a person or an occasion is an open expression of thoughtfulness and generosity. Remembrance fund donations have no minimum and can be set-up as one-time or annual contributions. Examples of remembrance events, to which many people can contribute, are graduations, anniversaries, birthdays, memorials, etc. Annual awards are $500 or more, and are named in honor of the person or event (such as the Christopher Miller Award) or can be added to the Remembrance Fund. Grants from the Remembrance Fund are awarded every year and are listed on the web site with a short description.
Geographic Focus: All States
Amount of Grant: 500 USD
Contact: Gary M. Langham; (916) 967-3223; info@conservegrassland.org
Internet: http://www.conservegrassland.org/remembrance_fund.htm
Sponsor: Neotropical Grassland Conservancy
6274 Heathcliff Drive
Carmichael, CA 95608-0909

Neotropical Grassland Conservancy Student Grants **2976**
The Student Grant Program (SGP) was designed in response to feedback from Venezuelan professors. Graduate student research projects are often hindered by lack of funds for basic needs such as transportation, lodging, meals and consumable field supplies. As little as $500 can launch a lifelong career in conservation. Student researchers in the Neotropical Grassland Conservancy's growing network outline their project and describe its impact on grassland conservation. Applications are accepted on the 1st of April, August, and December and are reviewed by members of the Student Grant Program committee and two outside reviewers. Award decisions are based on the student's demonstrated commitment to conservation, feasibility and merit of the proposed project, and how closely the project meets the mission of the Neotropical Grassland Conservancy.
Requirements: Latin American students enrolled in a M.S. or Ph.D. program in a U.S. or South American university are eligible to apply.
Geographic Focus: All States
Date(s) Application is Due: Apr 1; Aug 1; Dec 1
Amount of Grant: 500-2,500 USD
Contact: Peter B. McIntyre; (916) 967-3223; info@conservegrassland.org
Internet: http://www.conservegrassland.org/student_grant.htm
Sponsor: Neotropical Grassland Conservancy
6274 Heathcliff Drive
Carmichael, CA 95608-0909

Neotropical Grassland Conservancy W. J. and Virginia W. Moorhouse Research Grant **2977**
Memorial Research Grants directly promote conservation research. Grants are reviewed by outside scientists and the NGC board. Grants are one-year, unrestricted funds that can be used for direct or indirect costs, including salary. The W. J. and Virginia W. Moorhouse Research Grant provides a $2,000 grant to scientists doing ornithological research in the grasslands of the American tropics. This grant is awarded in memory of Bill and Jinny Moorhouse.
Geographic Focus: All States
Date(s) Application is Due: Mar 1
Amount of Grant: 2,000 USD
Contact: Valentina Ferretti; (916) 967-3223; info@conservegrassland.org
Internet: http://www.conservegrassland.org/memorial_research_grants.htm
Sponsor: Neotropical Grassland Conservancy
6274 Heathcliff Drive
Carmichael, CA 95608-0909

Neotropical Grassland Conservancy Wijdenes Memorial Research Grant **2978**
Memorial Research Grants directly promote conservation research. Grants are reviewed by outside scientists and the NGC board. Grants are one-year, unrestricted funds that can be used for direct or indirect costs, including salary. The Wijdenes Memorial Research Grant provides a $5,000 grant to scientists in the American tropics initiating new grassland conservation research. The grant is awarded in memory of Sue Wijdenes. Sue was an avid birder, traveler, and advocate of conservation.
Geographic Focus: All States
Date(s) Application is Due: Mar 1
Amount of Grant: 5,000 USD
Contact: Valentina Ferretti; (916) 967-3223; info@conservegrassland.org
Internet: http://www.conservegrassland.org/memorial_research_grants.htm
Sponsor: Neotropical Grassland Conservancy
6274 Heathcliff Drive
Carmichael, CA 95608-0909

Nestle Foundation Large Research Grants **2979**
The Nestlé Foundation supports research in human nutrition with public health relevance in low-income and lower middle-income countries. The results of the research projects should ideally provide a basis for implementation and action which will lead to sustainable effects in the studied populations as generally applicable to the population at large. They should also enable institutional strengthening and capacity building in a sustainable manner in the host country and further cooperation and collaboration between institutions in developed and developing countries. Categories include: training grants, pilot grants, research grants (small and large), and re-entry grants. Large research grants include a full grant application of a complete research proposal. Applications are accepted all year and evaluated twice a year.
Requirements: The Foundation's work is primarily concerned with human nutrition research issues dealing with: maternal and child nutrition, including breastfeeding and complementary feeding; macro- and micronutrient deficiencies and imbalances; interactions between infection and nutrition; and nutrition education and health promotion. Interested scientists should first submit a letter of intent briefly describing the project and the estimated budget. Instructions are available on the Foundation's website. If approved applicants will receive an invitation to submit a full grant proposal. Guidelines and forms are available at the website. The letter of intent and the grant application should include detailed, evidence-based information about the public health relevance of the project as well as its immediate impact and sustainability. Funding is primarily based on the scientific quality, public health relevance in the short and long term, sustainability, capacity-building component, and budget considerations.
Restrictions: The Foundation expects research proposals to be primarily the initiative of local researchers from developing countries but considers applications jointly made by scientists from developed countries, provided it is clear that the initiative will result in capacity building and human resource development, and the bulk of the budget is spent in the developing country.
Geographic Focus: All Countries
Date(s) Application is Due: Jan 10; May 10
Amount of Grant: Up to 100,000 USD
Contact: Catherine Loeb; +41 21 320 33 51; fax +41 21 320 33 92; nf@nestlefoundation.org
Internet: http://www.nestlefoundation.org/e/research.html
Sponsor: Nestle Foundation
4 Place de la Gare
Lausanne, CH-1001 Switzerland

Nestle Foundation Pilot Grants **2980**
The Nestlé Foundation supports research in human nutrition with public health relevance in low-income and lower middle-income countries. The results of the research projects should ideally provide a basis for implementation and action which will lead to sustainable effects in the studied populations as generally applicable to the population at large. They should also enable institutional strengthening and capacity building in a sustainable manner in the host country and further cooperation and collaboration between institutions in developed and developing countries. Categories include: training grants, pilot grants, research grants (small and large), and re-entry grants. Pilot grants provide support for research with a high potential to lead to a subsequent full research grant. Additional guidelines are available at the Foundation website. Applications are accepted all year and evaluated twice a year.
Requirements: The Foundation's work is primarily concerned with human nutrition research issues dealing with: maternal and child nutrition, including breastfeeding and complementary feeding; macro- and micronutrient deficiencies and imbalances; interactions between infection and nutrition; and nutrition education and health promotion. Interested scientists should first submit a letter of intent briefly describing the project and the estimated budget. Instructions are available on the Foundation's website. If approved, applicants will receive an invitation to submit a full grant proposal. Guidelines and forms are available at the website. The letter of intent and the grant application should include detailed, evidence-based information about the public health relevance of the project as well as its immediate impact and sustainability. Funding is

primarily based on the scientific quality, public health relevance in the short and long term, sustainability, capacity-building component, and budget considerations.
Restrictions: The Foundation expects research proposals to be primarily the initiative of local researchers from developing countries but considers applications jointly made by scientists from developed countries with those from developing countries provided it is clear that the initiative will result in capacity building and human resource development in the latter and the bulk of the budget is spent in the developing country.
Geographic Focus: All Countries
Date(s) Application is Due: Jan 10; May 10
Amount of Grant: Up to 20,000 USD
Contact: Catherine Loeb; +41 21 320 33 51; fax +41 21 320 33 92; nf@nestlefoundation.org
Internet: http://www.nestlefoundation.org/e/research.html
Sponsor: Nestle Foundation
4 Place de la Gare
Lausanne, CH-1001 Switzerland

Nestle Foundation Re-entry Grants **2981**

The Nestlé Foundation supports research in human nutrition with public health relevance in low-income and lower middle-income countries. The results of the research projects should ideally provide a basis for implementation and action which will lead to sustainable effects in the studied populations as generally applicable to the population at large. They should also enable institutional strengthening and capacity building in a sustainable manner in the host country and further cooperation and collaboration between institutions in developed and developing countries. Categories include: training grants, pilot grants, research grants (small and large), and re-entry grants. Re-entry grants encourage the return and re-establishment of post-graduate students into careers in their own countries. Applications are accepted all year and evaluated twice a year.
Requirements: The Foundation's work is primarily concerned with human nutrition research issues dealing with: maternal and child nutrition, including breastfeeding and complementary feeding; macro- and micronutrient deficiencies and imbalances; interactions between infection and nutrition; and nutrition education and health promotion. Interested scientists should first submit a letter of intent briefly describing the project and the estimated budget. Instructions are available on the Foundation's website. If approved applicants will receive an invitation to submit a full grant proposal. Guidelines and forms are available at the website. The letter of intent and the grant application should include detailed, evidence-based information about the public health relevance of the project as well as its immediate impact and sustainability. Funding is primarily based on the scientific quality, public health relevance in the short and long term, sustainability, capacity-building component, and budget considerations.
Restrictions: The host institution will need to guarantee a post for the returnee and ensure career development within the host institution. Contribution of support to the eligible candidate from the host institution is essential, while support and collaboration from the overseas institution where the candidate trained is helpful.
Geographic Focus: All Countries
Date(s) Application is Due: Jan 10; May 10
Amount of Grant: Up to 50,000 USD
Contact: Catherine Loeb; +41 21 320 33 51; fax +41 21 320 33 92; nf@nestlefoundation.org
Internet: http://www.nestlefoundation.org/e/research.html
Sponsor: Nestle Foundation
4 Place de la Gare
Lausanne, CH-1001 Switzerland

Nestle Foundation Small Research Grants **2982**

The Nestlé Foundation supports research in human nutrition with public health relevance in low-income and lower middle-income countries. The results of the research projects should ideally provide a basis for implementation and action which will lead to sustainable effects in the studied populations as generally applicable to the population at large. They should also enable institutional strengthening and capacity building in a sustainable manner in the host country and further cooperation and collaboration between institutions in developed and developing countries. Categories include: training grants, pilot grants, research grants (small and large), and re-entry grants. Small research grants provide support of a small research study that represents a continuation of a training grant or a pilot grant. Applications are accepted all year and evaluated twice a year.
Requirements: The Foundation's work is primarily concerned with human nutrition research issues dealing with: maternal and child nutrition, including breastfeeding and complementary feeding; macro- and micronutrient deficiencies and imbalances; interactions between infection and nutrition; and nutrition education and health promotion. Interested scientists should first submit a letter of intent briefly describing the project and the estimated budget. Instructions are available on the Foundation's website. If approved applicants will receive an invitation to submit a full grant proposal. Guidelines and forms are available at the website. The letter of intent and the grant application should include detailed, evidence-based information about the public health relevance of the project as well as its immediate impact and sustainability. Funding is primarily based on the scientific quality, public health relevance in the short and long term, sustainability, capacity-building component, and budget considerations.
Restrictions: The Foundation expects research proposals to be primarily the initiative of local researchers from developing countries but considers applications jointly made by scientists from developed countries, provided it is clear that the initiative will result in capacity building and human resource development, and the bulk of the budget is spent in the developing country.
Geographic Focus: All Countries
Date(s) Application is Due: Jan 10; May 10
Amount of Grant: Up to 50,000 USD
Contact: Catherine Loeb; +41 21 320 33 51; fax +41 21 320 33 92; nf@nestlefoundation.org
Internet: http://www.nestlefoundation.org/e/research.html
Sponsor: Nestle Foundation
4 Place de la Gare
Lausanne, CH-1001 Switzerland

Nestle Foundation Training Grant **2983**

The Nestlé Foundation supports research in human nutrition with public health relevance in low-income and lower middle-income countries. The results of the research projects should ideally provide a basis for implementation and action which will lead to sustainable effects in the studied populations as generally applicable to the population at large. They should also enable institutional strengthening and capacity building in a sustainable manner in the host country and further cooperation and collaboration between institutions in developed and developing countries. Categories include: training grants, pilot grants, research grants (small and large), and re-entry grants. Training grants support a small research project such as a master's degree in science, Ph.D. thesis project, or a training endeavor. Applications are accepted all year and evaluated twice a year.
Requirements: The Foundation's work is primarily concerned with human nutrition research issues dealing with: maternal and child nutrition, including breastfeeding and complementary feeding; macro- and micronutrient deficiencies and imbalances; interactions between infection and nutrition; and nutrition education and health promotion. Interested scientists should first submit a letter of intent briefly describing the project and the estimated budget. Instructions are available on the Foundation's website. If approved applicants will receive an invitation to submit a full grant proposal. Guidelines and forms are available on our website. The letter of intent and the grant application should include detailed, evidence-based information about the public health relevance of the project as well as its immediate impact and sustainability. Funding is primarily based on the scientific quality, public health relevance in the short and long term, sustainability, capacity-building component, and budget considerations.
Restrictions: The Foundation expects research proposals to be primarily the initiative of local researchers from developing countries but considers applications jointly made by scientists from developed countries with those from developing countries provided it is clear that the initiative will result in capacity building and human resource development in the latter and the bulk of the budget is spent in the developing country.
Geographic Focus: All Countries
Date(s) Application is Due: Jan 10; May 10
Amount of Grant: Up to 20,000 USD
Contact: Catherine Loeb; +41 21 320 33 51; fax +41 21 320 33 92; nf@nestlefoundation.org
Internet: http://www.nestlefoundation.org/e/research.html
Sponsor: Nestle Foundation
4 Place de la Gare
Lausanne, CH-1001 Switzerland

Nevada Arts Council Artists Fellowships **2984**

Artists Fellowships of $5,000 and Honorable Mention grants of $500 are awarded awarded in each of three categories – literary, performing and visual arts. They are given to professional individual artists with outstanding talent and demonstrated ability and commitment. The funds may be used for support for the creation of new works, works in progress, and/or professional development. Fellowship support for the categories alternate every other year between creative writing/visual arts and performing arts. Interested artists should first obtain application forms and guidelines from the council.
Requirements: Applicants must have been Nevada residents for a full year prior to the application deadline.
Restrictions: Individuals who are degree-candidate students are not eligible. Artists who received the fellowships in the past two years are also ineligible.
Geographic Focus: Nevada
Date(s) Application is Due: May 25
Amount of Grant: 5,000 USD
Contact: Fran Morrow; (775) 687-7106; fax (702) 486-3887; fkmorrow@clan.lib.nv.us
Internet: http://nac.nevadaculture.org/index.php?option=com_content&view=article&id=1043&Itemid=412
Sponsor: Nevada Arts Council
716 N Carson Street, Suite A
Carson City, NV 89701

New Hampshire Charitable Foundation Grants **2985**

The foundation seeks to improve the quality of life for New Hampshire citizens by distributing grants to qualifying organizations and scholarship assistance to the state's residents. Areas of support include arts, humanities, environment and conservation, health, and social and community services. Types of support include seed grants, scholarships and scholarship funds, program development, technical assistance, demonstration grants, development grants, endowments, and training grants. Four regional divisions of the foundation exist to benefit particular areas of the state: Greater Piscataqua Community Foundation, Lakes Region Charitable Foundation, Northern New Hampshire Foundation, and Upper Valley Community Foundation. Although a limited number of multi-year grants are made, awards are not usually repeated or renewed. Grantmaking meetings are held three times each year by the foundation.
Requirements: New Hampshire 501(c)3 tax-exempt organizations are eligible.
Restrictions: Grants are not awarded for capital campaigns, the purchase of major equipment, as operating support for ongoing or annual programs, to replace public funding, for sectarian or religious purposes, for deficit reduction, to support academic research, or for travel out of state by school or youth groups or by individuals for scholarly or professional purposes.

Geographic Focus: New Hampshire
Amount of Grant: 5,000-15,000 USD
Contact: Jennifer P. Hopkins; (800) 464-6641, ext. 1239; jph@nhcf.org
Internet: http://www.nhcf.org/page16898.cfm
Sponsor: New Hampshire Charitable Foundation
37 Pleasant Street
Concord, NH 03301-4005

NFID Advanced Vaccinology Course Travel Grant **2986**

These grants are intended to assist recent postdoctoral graduates who have a demonstrated interest in a career in vaccinology. This program provides participants with a comprehensive overview of vaccinology, from immunology to the social, economic, political, and ethical issues of vaccination. Practical approaches to the implementation of effective vaccination programs suitable for particular population groups will be discussed. To be eligible for this award the applicant must have made separate application to the course. Further information about the course can be obtained from: Betty Dodet, Fondation Merieux, 17, rue Bourgelat, BP 2021, Lyon Cedex 02, France. Annual deadline dates vary; contact program staff for exact dates. Note: Please do not contact NFID for information about the course. Contact the listed contact person.
Requirements: Individuals who are U.S. citizens who reside in the United States at the time of the course and who are within three years of obtaining a doctoral degree, or in the case of physicians, within two years of completing specialty training or a subspecialty fellowship, are eligible to apply. Candidates must have a demonstrated commitment to vaccinology as a career, and must be conducting research or working in a recognized and accredited U.S. institution of higher learning or in a governmental agency.
Geographic Focus: All States
Date(s) Application is Due: Nov 14
Amount of Grant: 4,000 USD
Contact: Betty Dodet; (33-4) 72 40 79 50; betty.dodet@fond-merieux.org
Internet: http://www.nfid.org/fellow/grantsfellow.pdf
Sponsor: National Foundation for Infectious Diseases
4733 Bethesda Avenue, Suite 750
Bethesda, MD 20814-5278

NFID Colin L. Powell Minority Postdoctoral Fellowship in Tropical Disease Research **2987**

The program, sponsored by NFID and GlaxoSmithKline, invites applications from qualified minority researchers to become specialists and investigators in the field of tropical diseases. The one-year fellowship includes a stipend, $3000 of which may be used for travel and supplies. Fellowships commence on or about July 1, with notification on or about April 15. Annual deadline dates may vary; contact program staff for exact dates.
Requirements: The applicant must hold a doctorate from a recognized U.S. university and be a U.S. citizen or permanent resident or spouse of a U.S. citizen or permanent resident.
Geographic Focus: All States
Date(s) Application is Due: Jan 6
Amount of Grant: 30,000 USD
Contact: Executive Director; (301) 656-0003; fax (301) 907-0878; info@nfid.org
Internet: http://www.nfid.org/fellow/grantsfellow.pdf
Sponsor: National Foundation for Infectious Diseases
4733 Bethesda Avenue, Suite 750
Bethesda, MD 20814-5278

NFID John P. Utz Medical Mycology Postdoctoral Fellowship **2988**

The fellowship is intended to encourage and assist a qualified physician to become a specialist and investigator in medical mycology. Selection criteria for the one-year fellowship will include, but not be limited to, the applicant's record of scholarship performance and professional qualifications; scientific merit of the clinical research project proposed; and validity of research rationale and adequacy of facilities available to the applicant, host department, and host institution. The recipient will be required to submit two single-spaced, one-page progress reports during his/her year of clinical research. One report will be due after six months and the second after 12 months or upon completion of the fellowship year. Notification of award will be made on or about April 15.
Requirements: The applicant must be a physician who is a U.S. citizen. He/she must demonstrate aptitude and accomplishment in research and must confirm arrangements for conduct of the proposed research in a recognized host laboratory. A letter from the chairperson of the department of infectious diseases expressing a willingness to assume responsibility for training the applicant must accompany the application.
Restrictions: The fellowship may not be awarded if the applicant has received or will receive a major fellowship, research grant, or traineeship from the federal government or another foundation in excess of the amount of this award.
Geographic Focus: All States
Date(s) Application is Due: Jan 6
Amount of Grant: 41,000 USD
Contact: Executive Director; (301) 656-0003; fax (301) 907-0878; info@nfid.org
Internet: http://nfid.org/fellow/grantsfellow.pdf
Sponsor: National Foundation for Infectious Diseases
4733 Bethesda Avenue, Suite 750
Bethesda, MD 20814-5278

NFID New Investigator Matching Grants **2989**

Matching grants are intended to assist new investigators who are beginning their research. The proposal must be based on research relevant to infectious diseases, microbiology, clinical medicine, epidemiology, nursing, etc. Priority will be given to applications proposing research on HIV and AIDS; tuberculosis, influenza, and other respiratory diseases; hospital-acquired and opportunistic infections; viral hepatitis; gastrointestinal, diarrheal, and food-borne diseases; sexually transmitted diseases; antibiotic resistance; tropical, travelers', and parasitic diseases; diseases transferred from animals to humans; and new and reemerging infectious diseases. Each application must be accompanied by an agreement from the applicant's sponsoring institution to match the award with equal funds.
Requirements: A new investigator is defined as an individual with full-time junior faculty (instructor or assistant professor) status at a recognized and accredited institution of higher learning. Priority will be given to those who do not have research grants and whose studies represent pilot work with intent to further develop their research.
Geographic Focus: All States
Date(s) Application is Due: Feb 14
Amount of Grant: 2,000 USD
Contact: Executive Director; (301) 656-0003; fax (301) 907-0878; info@nfid.org
Internet: http://www.nfid.org/fellow/grantsfellow.pdf
Sponsor: National Foundation for Infectious Diseases
4733 Bethesda Avenue, Suite 750
Bethesda, MD 20814-5278

NFID Nosocomial Infection Research and Training Postdoctoral Fellowship **2990**

The fellowship is intended to encourage and assist a qualified physician researcher to become a specialist and investigator in the field of nosocomial infections. Priority will be given to the study of gram positive nosocomial infections. Each application must be accompanied by a letter from the host laboratory and signed by the laboratory director or appropriate department chairperson indicating a willingness to accept the applicant and to provide guidance, research space, and necessary research equipment. Overhead to the sponsoring institution will not be paid. One thousand dollars of the stipend may be used for travel and supplies.
Requirements: The applicant must be a physician and a citizen of the United States. He/she must demonstrate aptitude and accomplishment in research and must confirm arrangements for conduct of the proposed research in a recognized host laboratory. An applicant with no grant support will receive priority consideration.
Restrictions: Applicants receiving or to be awarded grants in the same academic year in excess of the amount of this award are ineligible.
Geographic Focus: All States
Date(s) Application is Due: Jan 6
Amount of Grant: Up to 40,000 USD
Contact: Executive Director; (301) 656-0003; fax (301) 907-0878; info@nfid.org
Internet: http://nfid.org/fellow/grantsfellow.pdf
Sponsor: National Foundation for Infectious Diseases
4733 Bethesda Avenue, Suite 750
Bethesda, MD 20814-5278

NFID Postdoctoral Fellowships in Emerging Infectious Diseases **2991**

This one-year fellowship is intended to encourage and assist a qualified physician to become a recognized authority on emerging infectious diseases and epidemiology. The fellow will be assigned to the National Center for Infectious Diseases, CDC, in Atlanta, GA, and will participate in analysis and reporting of the national burden of infectious diseases and potential threats of emerging infectious diseases.
Requirements: The applicant must be a physician who is a U.S. citizen who preferably will have completed an infectious diseases fellowship.
Restrictions: The fellowship may not be awarded if the applicant has received or will receive a major fellowship, research grant, or traineeship from the federal government or another foundation in excess of $20,000.
Geographic Focus: All States
Date(s) Application is Due: Jan 6
Amount of Grant: 50,000 USD
Contact: Executive Director; (301) 656-0003; fax (301) 907-0878; info@nfid.org
Internet: http://nfid.org/fellow/grantsfellow.pdf
Sponsor: National Foundation for Infectious Diseases
4733 Bethesda Avenue, Suite 750
Bethesda, MD 20814-5278

NFIE Foundation Learning and Leadership Grants **2992**

Grants will fund recipients to participate in a high-quality professional development experience or organize a collegial study group that leads to improvements in practice, curriculum, and student achievement in the school or university. One-shot professional growth experiences such as attending a national conference or engaging a professional speaker are discouraged. Decisions regarding the content of the professional growth activities must be based on an assessment of student work undertaken with colleagues and must be embedded in the institutional planning process. Grant funds may be used for fees, travel expenses, books, or other materials that enable applicants to learn subject matter, instructional approaches, and skills. Recipients are expected to exercise professional leadership by sharing their new learning with their colleagues. Applications may be submitted at any time; notification will be made within seven months of receipt of the application.
Requirements: Practicing public school teachers, public school education support personnel, and public higher education faculty and staff are eligible to apply. Preference will be given to members of the National Education Association.
Restrictions: Grant funds may not be used to pursue degrees, to pay indirect costs or grant administration fees, or to pay salaries. With the exception of study groups, funds may not be used by applicants to pay themselves stipends. Grant funds may not be used to serve K-12 students in after-school or weekend settings. Funds may not be used for lobbying or religious purposes. Identical applications will not be considered.
Geographic Focus: All States

Date(s) Application is Due: Feb 1; Jun 1; Sep 15
Amount of Grant: 2,000-5,000 USD
Contact: Grants Coordinator; (202) 822-7840; fax (202) 822-7779
Internet: http://www.nfie.org/grants.htm
Sponsor: NEA Foundation for the Improvement of Education
1201 16th Street NW, Suite 416
Washington, DC 20036

NFJC Gantz-Zahler Grant **2993**
The grant is awarded every other year to support the publication of a nonfiction work reflecting on Jewish life, history, tradition, thought, or culture. Proposals must include a contract or letter from an established publisher indicating intent to publish the completed work. Grants can cover any publication cost, including research, editorial, printing, marketing, or distribution. Guidelines and application are available online.
Requirements: The author must be a U.S. citizen or permanent resident of the United States. The work must be nonfiction. The work must have a commitment to be published from a publisher.
Geographic Focus: All States
Date(s) Application is Due: Apr 21
Amount of Grant: 2,500 USD
Contact: Kristen Runk, (212) 629-0500, ext. 215; fax (212) 629-0508; krunk@jewishculture.org or nfjc@jewishculture.org
Internet: http://www.jewishculture.org/grants/gantz_zahler_grant.html
Sponsor: National Foundation for Jewish Culture
330 Seventh Avenue, 21st Floor, P.O. Box 489
New York, NY 10001

NFL Charities Medical Grants **2994**
NFL Charities supports medical research and enhancing scientific knowledge in order to benefit all who are actively involved in competitive sports and recreational athletic activities. These grants support research that addresses some of the risk factors that exist for all athletes and citizens with active lifestyles. The NFL Charities Board considers Medical Grant proposals that place emphasis on scientific merit, clinical relevance and significance to the NFL, especially projects concerning concussion and traumatic brain injury, cardiovascular research and MRSA. Grants will be available in the end of June or early July and will be open for approximately 8 weeks.
Requirements: Funding is available for non-profit educational and research institutions within the United States. Applications must be submitted online via the NFL Charities Grant Application Management System (G.A.M.S.) website. Every field in the application must be populated with requested information. Indirect costs should not be included in the project's total budget, and only direct research costs are funded by the grant. Salaries for Principal Investigators (PI's) and Co-Principal Investigators (Co-PI's) (Ph.D's and M.D.'s) must follow NIH guidelines. PI's and Co-PI's must disclose all previous and/or current grant support and any potential overlap must be submitted. Postdoctoral fellows serving as PI or Co-PI must include assurances that they will remain at the same institution for the duration of the grant period; however, PI's and co-PI's may work at different institutions, but the grant must be submitted as associated with the institution of the primary PI.
Restrictions: For-profit groups are not eligible for funding. Only submission per PI or Co-PI will be funded, and only one submission per institution can be considered for NFL Charities Medical Grant funding. Prospective applicants first should be vetted through their institutional administrative offices in order to ensure that an institution only endorses and submits its top two proposals. Administrative contacts at the associated institution must sign off on the Compliance Certificate that accompanies each application(s) from a PI's institution, thereby ensuring that such vetting has occurred.
Geographic Focus: All States
Date(s) Application is Due: Jun 1
Amount of Grant: Up to 100,000 USD
Contact: Clare Graff, (212) 450-2435 or (917) 816-2885; clare.graff@nfl.com
Internet: http://www.nflcharities.org/grants/medical
Sponsor: NFL Charities
280 Park Avenue, 17th Floor
New York, NY 10017

NFWF/Exxon Save the Tiger Fund Grants **2995**
The fund supports a variety of efforts aimed at the conservation of tigers in the wild, including on-the-ground tiger conservation and habitat/prey protection projects in tiger-range countries, as well as education activities both in tiger-range countries and internationally. Applications should demonstrate: anticipated benefits to wild tiger populations; improved management/protection of occupied and/or suitable tiger habitats; enhanced prey protection measures for in situ efforts; cooperation and coordination with other tiger conservation organizations; integration of tiger-range governments and local communities into project; lack of duplication and conflict with other ongoing conservation efforts; and methods for resolving conflicts between tiger conservation and the economic health and day-to-day lives of surrounding communities.
Requirements: U.S. or international nonprofit organizations, academic institutions, and government agencies, excluding U.S. federal agencies, are eligible.
Restrictions: The program will not fund political advocacy or litigation activities; captive tiger projects that are not presently part of approved Species Survival Plan; and general administrative overhead of sponsoring agency.
Geographic Focus: All States
Date(s) Application is Due: Jan 21; Nov 5
Contact: Mahendra Shrestha; (202) 857-0166; Mahendra.Shrestha@nfwf.org
Internet: http://www.savethetigerfund.org/Content/NavigationMenu2/Research/Grantees/Applyforagrant/default.htm
Sponsor: National Fish and Wildlife Foundation #6
1133 15th Street, NW, Suite 1100
Washington, DC 97205

NFWF Acres for America Grants **2996**
Acres for America is a partnership between Walmart Stores, Inc. and the National Fish and Wildlife Foundation. The Acres for America program was established to provide funding for projects that conserve important habitat for fish, wildlife, and plants through acquisition of interest in real property. The goal of the Acres for America program is to offset the footprint of Walmart's domestic facilities on at least an acre by acre basis through these acquisitions. Applicants are strongly urged to contact the National Fish and Wildlife Foundation regional director in their area to discuss project ideas prior to submitting pre-proposals. All grant awards require a minimum 1:1 match of cash or contributed goods and services. Federal funds may be considered as match. Higher ratios of matching funds will at times aid in making applications more competitive. All applications to the Acres for America Program will be made through a two-stage process, which includes a pre- and then a full-proposal. Only electronic pre-proposals submitted through the on-line pre-proposal form will be considered. Following a review period of about four weeks, appropriate pre-proposals will be selected and applicants will be invited to submit full proposals.
Requirements: Nonprofit conservation organizations, counties and cities, state government agencies, tribes, federal agencies, and schools are eligible.
Restrictions: The Foundation does not fund political advocacy or litigation of any kind, shortfalls in government agency budgets, general administrative overhead, or indirect costs.
Geographic Focus: All States
Date(s) Application is Due: Aug 1
Amount of Grant: 200,000 USD
Contact: Thomas E. Kelsch; (503) 417-8700, ext. 6008; Tom.Kelsch@nfwf.org
Internet: http://www.nfwf.org/AM/Template.cfm?Section=Charter_Programs_List&Template=/TaggedPage/TaggedPageDisplay.cfm&TPLID=60&ContentID=17798
Sponsor: National Fish and Wildlife Foundation #6
1133 15th Street, NW, Suite 1100
Washington, DC 97205

NFWF Bird Conservation Initiative Grants **2997**
The National Fish and Wildlife Foundation and its partners have keystone initiatives which are the core portfolio of multi-year initiatives through which they seek to achieve measurable outcomes. The Bird Conservation Initiative focuses on migratory and resident species and habitats that occur in the United States or its territories and that have been identified as high priorities for the nation. The goal is to find the best conservation investments, fund the best solutions, and deliver measurable results for bird conservation. Under this mission statement the Foundation has selected initiatives that focus on making a measurable impact on the status of specific species and their habitats. Individual initiatives are described on the website. Application is made online from the website.
Requirements: Applicants should review the business plans for this initiative and prepare application materials accordingly. Prospective applicants should contact the Director to discuss project ideas prior to submitting proposals to help ensure the relevance of funding requests. The Foundation will consider multi-year funding for important projects, subject to annual performance review of the grantee. A minimum 1:1 match is required for most Foundation grants, and higher ratios of match to Foundation funds are desired and improve the competitiveness of applications.
Geographic Focus: All States
Date(s) Application is Due: Aug 1; Dec 1
Amount of Grant: 50,000-375,000 USD
Contact: Daniel Petit; (202) 857-0166; fax (202) 857-0162; daniel.petit@nfwf.org
Internet: http://www.nfwf.org/am/template.cfm?section=Bird_
Sponsor: National Fish and Wildlife Foundation #6
1133 15th Street, NW, Suite 1100
Washington, DC 97205

NFWF Budweiser Renewable Energy and Wildlife Conservation Research Prize **2998**
This research prize will be awarded under a competitive process and will recognize and support a single innovative project that advances new technologies or practices that contribute to making renewable energy a practical energy alternative. The $100,000 total prize will be awarded in consideration of a project's ability to develop or evaluate new cost-effective renewable energy technologies for industrial application while demonstrating the measurable benefits such practices can have in advancing wildlife conservation. Strong preference will be given to projects that demonstrate additional financial commitment in the form of cash or in-kind matching funds at a ratio equal to or exceeding 1:1. Funds may be used for items such as: salaries and benefits, student stipends, travel, publication expenses, and supplies. Applications will be accepted through August 15.
Requirements: Eligibility for the research prize is limited to accredited post-secondary academic institutions with preference given to those working collaboratively among different departments within an institution(s), and with public agencies and nonprofit organizations.
Restrictions: Funds may not be used for: political advocacy or litigation of any kind, shortfalls in government agency budgets, or general administrative overhead or indirect costs.
Geographic Focus: All States
Date(s) Application is Due: Aug 15
Amount of Grant: 100,000 USD
Contact: Ellen Gabel, Program Director; (202) 857-0166; fax (202) 857-0162; ellen.gabel@nfwf.org or info@nfwf.org

Internet: http://www.nfwf.org/AM/Template.cfm?Section=Grants&Template=/CM/ContentDisplay.cfm&ContentID=3634
Sponsor: National Fish and Wildlife Foundation #6
1133 15th Street, NW, Suite 1100
Washington, DC 97205

NFWF Freshwater Fish Conservation Initiative Grants **2999**
The National Fish and Wildlife Foundation and its partners have keystone initiatives which are the core portfolio of multi-year initiatives through which they seek to achieve measurable outcomes. The Freshwater Fish Conservation Initiative focuses on species and habitats species that occur in the United States or its territories and that have been identified as high priorities for the nation. Freshwater fish (anadromous and catadromous) and associated aquatic life such as mussels, crayfish, and other invertebrates are included. The goal is to find the best conservation investments, fund the best solutions, and deliver measurable results for fish conservation. Under this mission statement there are selected initiatives that focus on making a measurable impact on the status of specific species and their habitat. Individual initiatives are listed on the website and are focused on conserving freshwater fish, aquatic organisms, and their habitats. Application is made online from the website.
Requirements: Applicants should review the business plans for this initiative and prepare application materials accordingly. Prospective applicants should contact the Director to discuss project ideas prior to submitting proposals to help ensure the relevance of funding requests. The Foundation will consider multi-year funding for important projects, subject to annual performance review of the grantee. A minimum 1:1 match is required for most Foundation grants, and higher ratios of match to Foundation funds are desired and improve the competitiveness of applications.
Geographic Focus: All States
Date(s) Application is Due: Aug 1; Dec 1
Amount of Grant: 50,000-375,000 USD
Contact: Jim Sedell, Director, Fish Conservation; (503) 417-8700; fax (503) 417-8787
Internet: http://www.nfwf.org/AM/Template.cfm?Section=Fish_Conservation2
Sponsor: National Fish and Wildlife Foundation #6
1133 15th Street, NW, Suite 1100
Washington, DC 97205

NFWF International Sea Turtle Conservation Fund Grants **3000**
The National Fish and Wildlife Foundation announces the availability of matching grant funding for international sea turtle conservation projects in the Western Hemisphere. Priority will be given to projects along the Atlantic and Pacific Coasts of Central and South America and in the wider Caribbean (including the U.S. Virgin Islands and Puerto Rico) and Canada. Priority consideration will be given to projects demonstrating clear and immediate benefit to the conservation of hawksbill (Eretmochelys imbricata) and leatherback (Dermochelys coriacea) turtles in the wider Caribbean and East Pacific, and east Pacific green turtles (Chelonia mydas). A limited amount of funds are available to support single or multi-year projects focused on addressing the incidental capture of sea turtles in artisanal and commercial gillnet fisheries. Proposals must be submitted in English to the International Sea Turtle Conservation Fund via the Foundation's online system by 11:59PM EST on February 4.
Geographic Focus: All States
Date(s) Application is Due: Feb 4
Amount of Grant: Up to 25,000 USD
Contact: Tom Barry; (202) 857-0166; fax (202) 857-0162; Tom.Barry@nfwf.org
Internet: http://www.nfwf.org/AM/Template.cfm?Section=Browse_All_Programs&TEMPLATE=/CM/ContentDisplay.cfm&CONTENTID=7639
Sponsor: National Fish and Wildlife Foundation #6
1133 15th Street, NW, Suite 1100
Washington, DC 97205

NFWF Marine & Coastal Conservation Initiative Grants **3001**
The National Fish and Wildlife Foundation and its partners have keystone initiatives which are the core portfolio of multi-year initiatives through which they seek to achieve measurable outcomes. The Marine and Coastal Conservation Initiative focuses on building conservation partnerships to overcome the most important challenges to the health of the marine and coastal environment. The goal is to find the best conservation investments, fund the best solutions, and deliver measurable results for a broad cross-section of fish and wildlife representing marine and coastal ecosystem health. Under this mission statement, the Foundation has selected initiatives that are focused on making a measureable impact on specific species and their habitat. Individual initiatives are listed on the website but are focused under three themes that are critical for marine and coastal wildlife in America. The themes are promoting sustainable fishers, protecting marine natural heritage, and climate change and adaptation. Application is made online from the website.
Requirements: Applicants should review the business plans for this initiative and prepare application materials accordingly. Prospective applicants should contact the Director to discuss project ideas prior to submitting proposals to help ensure the relevance of funding requests. The Foundation will consider multi-year funding for important projects, subject to annual performance review of the grantee. A minimum 1:1 match is required for most Foundation grants, and higher ratios of match to Foundation funds are desired and improve the competitiveness of applications.
Geographic Focus: All States
Date(s) Application is Due: Aug 1; Dec 1
Contact: Anthony C. Chatwin; (202) 857-0166; anthony.chatwin@nfwf.org
Internet: http://www.nfwf.org/am/template.cfm?section=Marine_and_Coastal
Sponsor: National Fish and Wildlife Foundation #6
1133 15th Street, NW, Suite 1100
Washington, DC 97205

NFWF Marine Debris Research and Technology Grants **3002**
The Marine Debris Research and Technology Grants Program (Program) is a partnership between the NOAA Marine Debris Program and National Fish and Wildlife Foundation. The Program seeks to support projects that have a strong likelihood of reducing marine debris caused by derelict fishing gear (both commercial and recreational), to increase our understanding of the impacts, location and best methods for removal, and projects that engage marinas, ports and the fishing industry. The Program provides competitive grants to finance innovative proposals that seek to work with marinas, ports, and the fishing industry to significantly reduce the occurrence of debris in these areas. Technology proposals to test new innovations in gear development to prevent loss or make the gear stop fishing or entangling if lost are also of high priority and strongly encouraged. In addition, the Program is accepting research proposals that address the biological, social, or economic impact of marine debris on species, habitat, and coastal businesses. Approximately $800,000 is available for calendar year. The majority of grants are anticipated to fall between $20,000 and $200,000. Larger grants will be reserved for those projects that are structured to definitively answers key research questions or implement projects over a broad scale. Pre-proposals must be received by May 5, and final proposals are due June 27.
Requirements: Eligible applicants are institutions of higher education, other non-profits, commercial organizations, and state (including Commissions), local, and tribal governments whose projects have the potential to benefit NOAA trust resources through marine debris research and prevention projects.
Restrictions: Applications from federal agencies or employees of federal agencies will not be considered. Funds cannot be used for political advocacy, boycotts, litigation expenses, terrorist activities or activities conducted in violation of the Foreign Corrupt Practices Act.
Geographic Focus: All States
Date(s) Application is Due: May 5; Jun 27
Contact: Tom Barry; (202) 857-0166; fax (202) 857-0162; Tom.Barry@nfwf.org
Internet: http://www.nfwf.org/AM/Template.cfm?Section=Home&Template=/TaggedPage/TaggedPageDisplay.cfm&TPLID=32&ContentID=9272
Sponsor: National Fish and Wildlife Foundation #6
1133 15th Street, NW, Suite 1100
Washington, DC 97205

NFWF More Fish Partnership Fund Grants **3003**
The goal of the Fund is to improve awareness about aquatic resources by supporting both grassroots capacity building with the ability to measure that performance and successfully plan and implement habitat conservation projects. Under this program, grants of up to $50,000 will be awarded competitively. Proposed activities eligible for funding will include: aquatic habitat restoration; enhancement and protection that protect and improve water quantity/quality, aquatic vegetation; landowner/water user conservation, outreach and stewardship training; and conservation watershed planning and management. The program's primary focus is smaller, community-based projects for up to $50,000, but larger requests for funds will be considered on a competitive basis. The following costs are eligible: protection, enhancement, and restoration of fish habitat within and along rivers and streams, lakes, reservoirs, ponds, coastal areas, etc. and the upland areas affecting them; volunteer training workshops, stakeholder roundtables, and community outreach; and project design and development that is anticipated to lead to an on-the-ground project within 18 months. Matching funds will be required on a 1:1 basis or higher.
Requirements: Applicants representing B.A.S.S. Clubs, local non-profits, educational institutions, angling clubs, tribes, conservation districts, or local governments are eligible. Community groups without non-profit status are encouraged to seek an eligible sponsor/fiscal agent.
Restrictions: Full fee acquisition, less-than-fee acquisition, free-standing educational projects, mitigation or corrective actions, general planning and studies, and legal or lobbying efforts are ineligible.
Geographic Focus: All States
Date(s) Application is Due: Nov 1
Amount of Grant: Up to 50,000 USD
Contact: Barrett Bohnengel; (503) 417-8700; Barrett.Bohnengel@nfwf.org
Internet: http://www.nfwf.org/AM/Template.cfm?Section=Home&Template=/CM/ContentDisplay.cfm&ContentID=3937
Sponsor: National Fish and Wildlife Foundation
806 SW Broadway, Suite 750
Portland, OR 97205

NFWF National Whale Conservation Fund Grants **3004**
The fund supports research, management, conservation, and education/outreach activities related to the conservation and recovery of whales, particularly projects relating to great whales that are the most endangered (e.g., North Atlantic and North Pacific right whales). Priority will be given to projects that address anthropogenic threats to endangered whales of U.S. waters (i.e., fishing gear entanglement, ship/whale collisions, noise, habitat degradation, contaminants, etc.). Proposals that address tasks or needs identified in endangered whale recovery plans or conservation plans adopted by the National Marine Fisheries Service are encouraged. Applicants are encouraged to show leverage of a one-to-one or greater match to the proposal request. NWCF actively seeks to form new partnerships with corporations, organizations, and individuals to leverage NWCF???s resources sufficiently to meet its aggressive conservation goals. Applications should clearly describe anticipated benefits to whale conservation; project goals, objectives, and methodology, including provisions for monitoring and evaluation; qualifications of key project personnel; detailed budget

illustrating use of NWCF grant; cooperation and coordination with other appropriate federal, state, and private organizations; and ability to leverage NWCF grant award.
Restrictions: NWCF will not fund political advocacy, boycotts, litigation, or indirect or unallocated expenses.
Geographic Focus: All States
Contact: Stephanie Pendergrass, Assistant Director; (202) 857-0166; fax (202) 857-0162; Stephanie.Pendergrass@nfwf.org
Internet: http://www.nfwf.org/AM/Template.cfm?Section=Who_We_Are&Template=/TaggedPage/TaggedPageDisplay.cfm&TPLID=30&ContentID=9725
Sponsor: National Fish and Wildlife Foundation #6
1133 15th Street, NW, Suite 1100
Washington, DC 97205

NFWF Nature Restoration Trust Grants **3005**

The Nature Restoration Trust seeks to fund projects that benefit fish, wildlife, and plants, as well as provide educational opportunities and encourage community involvement in conservation efforts. The Nature Restoration Trust will fund conservation projects within the Pacific Gas and Electric Company's service area in northern and central California through the Five Star Restoration Program. Note that the Nature Restoration Trust program especially encourages schools and school-based projects to apply.
Requirements: Partners, funders, and/or participants necessary to complete the project include schools or youth organizations; local or tribal governments; local businesses or corporations; conservation organizations or local citizens groups; state and federal resource management agencies; and foundations or other funders. Conservation projects within the Pacific Gas and Electric Company's service area in northern and central California are eligible.
Restrictions: Projects involving only research, monitoring, or planning; and projects that are part of a mitigation requirement are not eligible for funding.
Geographic Focus: California
Date(s) Application is Due: Feb 15
Amount of Grant: 24,000 USD
Contact: Claire Thorp; (415)243-3104 or (415) 778-0999; Claire.Thorp@nfwf.org
Internet: http://www.nfwf.org/AM/Template.cfm?Section=Home&Template=/TaggedPage/TaggedPageDisplay.cfm&TPLID=30&ContentID=8301
Sponsor: National Fish and Wildlife Foundation
90 New Montgomery Street, Suite 700
San Francisco, CA 94105

NFWF Puget Sound Marine Conservation Fund Grants **3006**

Together with the U.S. Fish and Wildlife Service, the Foundation will use funds to establish the Puget Sound Marine Conservation Fund to receive monies and begin a grant-making program to finance projects in the State of Washington within a geographic region that includes Puget Sound, and the Straits of Juan de Fuca, and waters subject to tidal influence within that geographic area for any of the following purposes: habitat conservation, protection, restoration and management projects to benefit the fish and wildlife resources and the habitats upon which they depend; landowner outreach and incentive programs for restoration and management of natural resources; public use and natural/cultural projects that benefit the National Wildlife Refuge System and other federal and state refuges, reserves and protected natural areas; and collaborative projects from local coastal communities seeking environmentally and economically sustainable solutions to natural resource problems. Pre-proposals can address a multitude of program purposes (listed at the web site). The people living in or near such habitats are intended to be the beneficiaries of the Fund, on behalf of such species and habitats.
Requirements: The following organizations are eligible to apply: 501(c)3 registered non-profit conservation organizations; counties; local, state or federal government agencies; tribes; schools; soil and water conservation districts; and other special districts.
Restrictions: Funds cannot be used for: political advocacy, litigation expenses, mitigation or terrorist activities; equipment purchases not primarily related to the direct completion of the project; videos; improvement of natural resources for commercial or private profit required mitigation; required Habitat Conservation Plan activities; or straight percentage of administrative overhead/indirect costs (a portion of the grant may be used for direct project-related administrative expenses such as salaries, travel, supplies, telephone, but these must be identified specifically in the budget as line items).
Geographic Focus: All States
Contact: Cara Rose; (503) 417-8700, ext. 21; fax (503) 417-8787; cara.rose@nfwf.org
Internet: http://www.nfwf.org/AM/Template.cfm?Section=Who_We_Are&Template=/CM/ContentDisplay.cfm&ContentID=3998
Sponsor: National Fish and Wildlife Foundation
806 SW Broadway, Suite 750
Portland, OR 97205

NFWF Pulling Together Initiative Grants **3007**

The program aims to create partnerships between local agencies, private landowners, and other interested parties in developing long-term weed management projects as part of an integrated pest management strategy. The goals of PTI are: to prevent, manage, or eradicate invasive and noxious plants through a coordinated program of public/private partnerships; and to increase public awareness of the adverse impacts of invasive and noxious plants. Preproposals should be submitted inline by the October 26 due date, with full proposals due in January.
Requirements: Eligible applicants are partnerships of federal, state, and/or local government agencies, nonprofit organizations, private businesses, and landowners. The foundation makes challenge grants based on a two-to-one nonfederal funding match; awardees are required to supply remaining funds.
Geographic Focus: All States
Date(s) Application is Due: Jan 11; Oct 26
Contact: Ellen Gabel; (202) 857-0166; fax (202) 857-0162; ellen.gabel@nfwf.org
Internet: http://www.nfwf.org/AM/Template.cfm?Section=Who_We_Are&Template=/CM/ContentDisplay.cfm&ContentID=4790
Sponsor: National Fish and Wildlife Foundation #6
1133 15th Street, NW, Suite 1100
Washington, DC 97205

NFWF Shell Marine Habitat Grants **3008**

NFWF and Shell Oil Company Foundation fund natural resource conservation projects in the Gulf of Mexico. Projects will be funded that protect, conserve, restore, or enhance the Gulf coasts of Florida, from Tampa north; Mobile Bay, AL; coastal Louisiana, with an emphasis on the La Branche wetlands; and Port Arthur, Galveston Bay, and the Deer Park Ship Channel in Texas. Projects of interest include those that protect or restore estuarine habitat in the above geographic areas. Projects that include opportunities for hands-on participation by volunteers are of special interest. Focal species include brown pelicans, with an emphasis on habitat restoration and conservation of breeding sites. In addition, education programs intended to minimize harm from monofilament entanglement and other human impacts are of particular interest. Priority will be given to projects that build new or enhance existing partnerships and that address conservation needs of species characteristic of southeastern U.S. marine ecosystems. Preproposals must be submitted online by the listed application deadline. Application and guidelines are available online.
Requirements: Nonprofit IRS 501(c)3 tax-exempt organizations are eligible. Entities that are subdivisions of government (federal, state, county/parish, or city) are also eligible. Applicants must be based in the United States.
Restrictions: Individuals are not eligible for grants under this program. The program will not fund international projects, with very limited exceptions. The foundation will not fund political advocacy, lobbying, or litigation of any kind; shortfalls in government agency budgets; general administrative overhead or indirect costs of any kind; multiyear grants (applicant may reapply); or basic research.
Geographic Focus: All States
Date(s) Application is Due: Apr 1; Jun 1
Contact: Suzanne Sessine; (202) 857-0166; fax (202) 857-0162; Suzanne.Sessine@nfwf.org
Internet: http://www.nfwf.org/AM/Template.cfm?Section=Home&Template=/TaggedPage/TaggedPageDisplay.cfm&TPLID=32&ContentID=9173
Sponsor: National Fish and Wildlife Foundation #6
1133 15th Street, NW, Suite 1100
Washington, DC 97205

NFWF Wildlife and Habitat Conservation Initiative Grants **3009**

The National Fish and Wildlife Foundation and its partners have keystone initiatives which are the core portfolio of multi-year initiatives through which they seek to achieve measurable outcomes. The Wildlife and Habitat Conservation Initiative addresses conservation needs for mammals, reptiles, amphibians, invertebrates, and plants, as well as landscape-level and issue-based conservation. The goal is to find the best conservation investments, fund the best solutions, and deliver measurable results for wildlife. Under this mission statement, there are selected initiatives focused on making a measureable impact on specific species and their habitats. Individual initiatives are listed on the website and are focused under three themes which are imperiled wildlife species, wildlife movement and migration, and climate change and adaptation. Application is made online from the website.
Requirements: Applicants should review the business plans for this initiative and prepare application materials accordingly. Prospective applicants should contact the Director to discuss project ideas prior to submitting proposals to help ensure the relevance of funding requests. The Foundation will consider multi-year funding for important projects, subject to annual performance review of the grantee. A minimum 1:1 match is required for most Foundation grants, and higher ratios of match to Foundation funds are desired and improve the competitiveness of applications.
Geographic Focus: All States
Date(s) Application is Due: Aug 1; Dec 1
Amount of Grant: 50,000-375,000 USD
Contact: Carly Vynne; (202) 857-0166; fax (202) 857-0162
Internet: http://www.nfwf.org/am/template.cfm?section=Wildlife_and_Habitat12
Sponsor: National Fish and Wildlife Foundation #6
1133 15th Street, NW, Suite 1100
Washington, DC 97205

NFX Patient-Oriented Research Grants **3010**

The foundation supports a broad range of research endeavors that will lead to better recognition, treatment, and an eventual cure of children and adults with fragile X syndrome. The following forms of investigation are potentially appropriate for the current round of funding: educational or therapy interventions, medical treatment interventions, studies of family support and counseling, and diagnostic and reproductive studies. Basic research and patient-oriented research grants are awarded on a six-month alternating cycle. The January deadline is for basic science and clinical research applications. Applications should be submitted both as email attachments and as hard-copy documents.
Requirements: The program is open to all investigators with demonstrated ability to carry out research on fragile X syndrome.
Geographic Focus: All States
Date(s) Application is Due: Jan 31
Amount of Grant: Up to 35,000 USD
Contact: Administrator; (800) 688-8765 or (925) 938-9300; NATLFX@FragileX.org
Internet: http://www.fragilex.org/html/request_for_application.htm
Sponsor: National Fragile X Foundation
P.O. Box 190488
San Francisco, CA 94119

NFX Summer Research Fellowships 3011

The foundation awards summer research fellowships in the area of fragile X syndrome. Proposals should be no longer than four pages, and should describe the research project in depth and the specific roles and activities of the summer fellow and the mentor. The curriculum vitas of the mentor and the fellow should be included as an addendum. Guidelines are available online.

Requirements: The fellow can be either an undergraduate or graduate student, and the mentor should be experienced in research concerning fragile X syndrome.

Geographic Focus: All States

Date(s) Application is Due: Apr 17

Amount of Grant: 2,500 USD

Contact: Administrator; (800) 688-8765 or (925) 938-9300; NATLFX@FragileX.org

Internet: http://www.fragilex.org/html/request_for_application.htm

Sponsor: National Fragile X Foundation
P.O. Box 190488
San Francisco, CA 94119

NHC Summer Institutes in Literary Studies Grants 3012

The center offers three summer institutes designed to provide junior faculty members in departments of literature and other disciplines the opportunity for intensive critical study of a single text or small group of texts. The institutes will seek to engage literary works deeply through close reading. Each institute will include only 12 participants and will be led by a distinguished scholar. The center will supply all texts and cover travel, meal, and lodging costs. Participants also receive a stipend. Information and application are available online.

Requirements: The institutes are open to scholars who have received a PhD within the last 10 years and who teach in departments of literature or other relevant disciplines at U.S. colleges or universities.

Geographic Focus: All States

Date(s) Application is Due: Feb 25

Amount of Grant: 1,500 USD

Contact: Richard Schramm, (877) 271-7444; rschramm@ga.unc.edu

Internet: http://www.nhc.rtp.nc.us/siliterarystudies/index.htm

Sponsor: National Humanities Center
7 Alexander Drive, P.O. Box 12256
Research Triangle Park, NC 27709-2256

NHGRI Mentored Research Scientist Development Award 3013

These awards are intended to foster the career development of individuals with expertise in scientific disciplines that would further technological developments critical to the success of the Human Genome Project and the understanding of the genetic basis of diseases. Up to three scientists may be selected and appointed to this program by the grantee institutions. Up to the Federal salary cap, plus fringe benefits, is available. Up to $20,000 of research support per year is also available.

Requirements: Eligibility is limited to individuals with degrees in computer sciences, mathematics, chemistry, engineering, physics, and closely related scientific disciplines, such as bioinformatics, computational biology, statistics, biomathematics, bioengineering. Applications will be accepted from domestic, nonfederal organizations of higher education that have strong, well-established research and training programs. The following organizations and institutions may apply: public/state controlled institutions of higher education; private institutions of higher education; Hispanic-serving institutions; historically Black colleges and universities (HBCUs); tribally controlled colleges and universities (TCCUs); Alaska native and native Hawaiian serving institutions; nonprofits with 501(c)3 IRS status; nonprofits without 501(c)3 IRS status; small businesses; for-profit organizations; State governments; regional organizations; U.S. territories or possessions; Indian/Native American tribal governments; Indian/Native American tribally designated organizations; eligible agencies of the Federal government; and faith-based or community based organizations.

Geographic Focus: All States

Date(s) Application is Due: Feb 12; Mar 12; Jun 12; Jul 12; Oct 12; Nov 12

Contact: Bettie J. Graham, (301) 496-7531; bettie_graham@nih.gov

Internet: http://grants.nih.gov/grants/guide/pa-files/PAR-98-062.html

Sponsor: National Human Genome Research Institute
5635 Fishers Lane, Suite 4076
Rockville, MD 20852

NHLBI Airway Smooth Muscle Function and Targeted Therapeutics in Human Asthma Grants 3014

The purpose of this funding opportunity announcement, issued by the NHLBI, NIH, is to solicit Research Project Grant applications from institutions and organizations that propose to investigate the complex role of airway smooth muscle (ASM) functions in the development of human asthma and identify novel therapeutic targets. Budgets for direct costs of up to $350,000 per year (exclusive of indirect costs associated with consortia) and a project duration of up to four years may be requested for a maximum of $1,400,000 direct costs over a four-year project period. Letters of Intent should be received by December 8, and full proposals are due by January 6.

Requirements: The following organizations and institutions are eligible to apply: public/state controlled institutions of higher education; private institutions of higher education; Hispanic-serving institutions; historically Black colleges and universities (HBCUs); tribally controlled colleges and universities (TCCUs); Alaska Native and Native Hawaiian serving institutions; nonprofits with 501(c)3 IRS status; nonprofits without 501(c)3 IRS status; for-profit organizations; state governments; Indian/Native American tribal governments (federally recognized); Indian/Native American tribally designated organizations; county governments; city or township governments; special district governments; independent school districts; public housing authorities/Indian housing authorities; U.S. territory or possession; Indian/Native American tribal governments (other than federally recognized); regional organizations; non-domestic (non-U.S.) entities (foreign organizations); other eligible agencies of the federal government; and faith-based or community-based organizations.

Geographic Focus: All States

Date(s) Application is Due: Jan 6; Dec 8

Amount of Grant: Up to 350,000 USD

Contact: Susan Banks-Schlegel, (301) 435-0202; Schleges@nhlbi.nih.gov

Internet: http://grants1.nih.gov/grants/guide/rfa-files/RFA-HL-09-007.html

Sponsor: National Institutes of Health
Building 31, Room 5A48
Bethesda, MD 20892

NHLBI Ancillary Studies in Clinical Trials 3015

This funding opportunity invites research grant applications to conduct time-sensitive ancillary studies related to heart, lung, and blood diseases and sleep disorders in conjunction with ongoing clinical trials and other large clinical studies supported by NIH or non-NIH entities. The program establishes an accelerated review/award process to support the crucial time frame in which these ancillary studies must be performed. Time-sensitive ancillary studies include those that require active longitudinal data collection and thus need to begin recruiting subjects as close as possible to the start of the parent study. The ancillary study can address any research questions related to the mission of NHLBI for which the parent study can provide participants, infrastructure, and data. The parent studies most often will be a clinical trial, but also can be an observational study or registry that can provide a sufficient cohort of well-characterized patients. Each ancillary study application must demonstrate the time-sensitive nature of the application and must explicitly address why an expedited review is essential to its feasibility. Applications are due May 24, September 24, and January 24 by 5:00 pm local time of applicant.

Requirements: The following may apply: public/state controlled and private institutions of higher education; nonprofits with or without 501(c)3 IRS status (other than institutions of higher education); small businesses; for-profit organizations (other than small businesses); state, county, city or township governments; special district governments; Indian/Native American Tribal governments; eligible agencies of the Federal government; U.S. territory or possession; independent school districts; public housing authorities/Indian housing authorities; faith-based or community-based organizations; regional organizations; non-domestic (non-U.S.) entities (Foreign Institutions). The following types of Higher Education Institutions are always encouraged to apply for NIH support as public or private institutions of higher education: Hispanic-serving institutions; historically Black colleges and universities; Tribally controlled colleges and universities; Alaska Native and Native Hawaiian serving institutions. Non-domestic (non-U.S.) components of U.S. Organizations are eligible to apply. Any individual(s) with the skills, knowledge, and resources necessary to carry out the proposed research as the Program Director/Principal Investigator (PD/PI) is invited to work with his/her organization to develop an application for support. Individuals from underrepresented racial and ethnic groups as well as individuals with disabilities are always encouraged to apply for NIH support. An application may request a budget for direct costs up to $250,000 (10 modules) per year, excluding subcontractor or consortium facilities and administrative (F&A) costs. The scope of the proposed project should determine the project period. The maximum project period is four years.

Geographic Focus: All States, Guam, Marshall Islands, Northern Mariana Islands, Puerto Rico, U.S. Virgin Islands, All Countries, American Samoa

Date(s) Application is Due: Jan 24; May 24; Sep 24

Contact: Suzanne Goldberg, R.N., M.S.N., (301) 435-0532; fax (301) 480-3667; goldbergsh@mail.nih.gov

Internet: http://grants.nih.gov/grants/guide/rfa-files/RFA-HL-14-004.html

Sponsor: National Institutes of Health
Building 31, Room 5A48
Bethesda, MD 20892

NHLBI Bioengineering and Obesity Grants 3016

This funding opportunity solicits Research Project Grant (R01) applications from institutions/organizations that propose to solicit applications to develop and validate new and innovative engineering approaches to address clinical problems related to energy balance, intake, and expenditure. Novel sensors, devices, imaging, and other technologies, including technologies to detect biochemical markers of energy balance are expected to be developed and evaluated by collaborating engineers, physical scientists, mathematicians, and scientists from other relevant disciplines with expertise in obesity and nutrition. Because the nature and scope of the proposed research will vary from application to application, it is anticipated that the size and duration of each award will also vary. The total amount awarded and the number of awards will depend upon the mechanism numbers, quality, duration, and costs of the applications received.

Requirements: The following are eligible to apply: public/state controlled and private institutions of higher education; nonprofit with or without 501(c)3 IRS status (other than institution of higher education); small business; for-profit organization (other than small business); state government; U.S. Territory or Possession; Indian/Native American tribal government (federally recognized); Indian/Native American tribal government (other than federally recognized); Indian/Native American tribally designated organization; Non-domestic (non-U.S.) entity (foreign organization); Hispanic-serving institution; historically Black colleges and universities (HBCUs); tribally controlled colleges and universities (TCCUs); Alaska Native and Native Hawaiian serving institutions; regional organizations. Individuals with the skills, knowledge, and resources necessary to carry out the proposed research are invited to work with their institution/organization to develop an

application for support. Individuals from underrepresented racial and ethnic groups as well as individuals with disabilities are always encouraged to apply for NIH support. More than one PD/PI, or multiple PDs/PIs, may be designated on the application. Applicants may submit more than one application, provided each application is scientifically distinct.
Geographic Focus: All States, Guam, Marshall Islands, Northern Mariana Islands, Puerto Rico, U.S. Virgin Islands, American Samoa
Date(s) Application is Due: Jan 25
Contact: Abby G. Ershow, (301) 435-0550; fax (301) 480-2858; ErshowA@mail.nih.gov
Internet: http://grants.nih.gov/grants/guide/pa-files/PA-07-354.html
Sponsor: National Institutes of Health
Building 31, Room 5A48
Bethesda, MD 20892

NHLBI Bioengineering Approaches to Energy Balance and Obesity Grants for SBIR 3017

This funding opportunity will develop and validate new and innovative bioengineering technology to address clinical problems related to energy balance, intake, and expenditure. Novel sensors, devices, imaging, and other approaches are expected to be developed and evaluated by collaborating engineers, physical scientists, and scientists from other relevant disciplines with expertise in obesity and nutrition. The goal is to increase the number of useful technologies and tools available to scientists to facilitate their research in energy balance and health. Eventually these research tools should facilitate therapeutic advances and behavioral changes to address such problems as weight control and obesity. The total amount awarded and the number of awards will depend upon the mechanism numbers, quality, duration, and costs of the applications received. This FOA will utilize the SBIR (R43/R44) grant mechanisms for Phase I, Phase II, and Fast-Track applications and runs in parallel with a FOA of identical scientific scope, PA-07-436, that solicits applications under the Small Business Technology Transfer (STTR) (R41/R42) grant mechanisms. The total amount awarded and the number of awards will depend upon the mechanism numbers, quality, duration, and costs of the applications received. Program Announcement (PA) Number: PA-07-435.
Requirements: Only United States small business concerns (SBCs) are eligible to submit SBIR applications. A small business concern is one that, at the time of award for both Phase I and Phase II SBIR awards, meets all of the following criteria: (1) is independently owned and operated, is not dominant in the field of operation in which it is proposing, has a place of business in the United States and operates primarily within the United States or makes a significant contribution to the U.S. economy, and is organized for profit; (2) is (a) at least 51% owned and controlled by one or more individuals who are citizens of, or permanent resident aliens in, the United States, or (b) for SBIR only, it must be a for-profit business concern that is at least 51% owned and controlled by another for-profit business concern that is at least 51% owned and controlled by one or more individuals who are citizens of, or permanent resident aliens in, the United States; (3) has, including its affiliates, an average number of employees for the preceding 12 months not exceeding. Any individual with the skills, knowledge, and resources necessary to carry out the proposed research is invited to work with his/her organization to develop an application for support. Individuals from underrepresented racial and ethnic groups as well as individuals with disabilities are always encouraged to apply for NIH support.
Geographic Focus: All States
Date(s) Application is Due: Apr 5; Aug 5; Dec 5
Contact: J. Timothy Baldwin; (301) 435-0513; BaldwinT@mail.nih.gov
Abby G. Ershow, Sc.D., (301) 435-0550; fax (301) 480-1338; ErshowA@mail.nih.gov
Internet: http://grants.nih.gov/grants/guide/pa-files/PA-07-435.html
Sponsor: National Institutes of Health
Building 31, Room 5A48
Bethesda, MD 20892

NHLBI Bioengineering Approaches to Energy Balance and Obesity Grants for STTR 3018

This funding opportunity will develop and validate new and innovative bioengineering technology to address clinical problems related to energy balance, intake, and expenditure. Novel sensors, devices, imaging, and other approaches are expected to be developed and evaluated by collaborating engineers, physical scientists, and scientists from other relevant disciplines with expertise in obesity and nutrition. The goal is to increase the number of useful technologies and tools available to scientists to facilitate their research in energy balance and health. Eventually these research tools should facilitate therapeutic advances and behavioral changes to address such problems as weight control and obesity. The total amount awarded and the number of awards will depend upon the mechanism numbers, quality, duration, and costs of the applications received. Program Announcement (PA) Number: PA-07-436.
Requirements: Only United States small business concerns (SBCs) are eligible to submit STTR applications. Individuals with the skills, knowledge, and resources necessary to carry out the proposed research are invited to work with their organization to develop an application for support. Individuals from underrepresented racial and ethnic groups as well as individuals with disabilities are always encouraged to apply for NIH support. On an STTR application, the PD/PI may be employed with the SBC or the participating non-profit research institution as long as he/she has a formal appointment with or commitment to the applicant SBC, which is characterized by an official relationship between the small business concern and that individual. Applicant SBCs may submit more than one application, provided each application is scientifically distinct. More than one PD/PI, or multiple PDs/PIs, may be designated on the application.
Geographic Focus: All States
Date(s) Application is Due: Apr 5; Aug 5; Dec 5
Contact: Abby G. Ershow, (301) 435-0550; fax (301) 480-1338; ErshowA@mail.nih.gov
Internet: http://grants.nih.gov/grants/guide/pa-files/PA-07-436.html
Sponsor: National Institutes of Health
Building 31, Room 5A48
Bethesda, MD 20892

NHLBI Biomedical Research Training for Individuals from Underrepresented Groups 3019

The NHLBI established the Biomedical Research Training Program for Individuals from Underrepresented Groups (BRTPUG) to offer opportunities for underrepresented post-baccalaureate individuals to receive training in basic, translational, and clinical research. The purpose of the program is to enhance career opportunities in biomedical sciences, including clinical and laboratory medicine, epidemiology, and biostatistics as applied to the etiology and treatment of heart, blood vessel, lung, and blood diseases. BRTPUG offers each participant the opportunity to work closely with leading research scientists in the Division of Intramural Research and extramural scientists in the Division of Cardiovascular Sciences -Prevention and Population Sciences Program. The program provides participants with hands-on training in a research environment, which will prepare them to continue their studies and advance their careers in clinical and basic research. The trainee's research internship is a one-time appointment of 1-2 year's beginning the summer of selection. Each trainee is assigned to a mentor who is responsible for designing a carefully planned training program.
Requirements: The program supports recently completed post-baccalaureates, who have completed academic training in course work relevant to biomedical, behavioral, or statistical research. Applicants must have a cumulative grade point average (GPA) or science course GPA of 3.3 or better on a 4.0 scale, or 4.3 or better on a 5.0 scale, and be a U.S. citizen or permanent resident. Underrepresented post-baccalaureate individuals planning to pursue advanced degrees in the biomedical sciences, who are planning to apply to graduate or professional (medical/dental/veterinary, pharmacy) school are eligible to apply.
Geographic Focus: All States
Date(s) Application is Due: Jan 31
Contact: Dr. Helena O. Mishoe, (301) 451-5081; mishoeh@nhlbi.nih.gov
Internet: http://www.nhlbi.nih.gov/funding/training/redbook/pbbrtu.htm
Sponsor: National Institutes of Health
Building 31, Room 5A48
Bethesda, MD 20892

NHLBI Cardiac Development Consortium Grants 3020

The National Heart, Lung, and Blood Institute (NHLBI) and the Canadian Institutes of Health Research (CIHR) invite applications to participate in a new NHLBI Cardiac Development Consortium, a cooperative investigative group that will drive an integrated approach to the investigation of cardiovascular development. The purpose of this initiative is to support basic collaborative research leading to a comprehensive understanding of the regulatory networks controlling cardiovascular development. The consortium will consist of up to four Research Centers, scientific Cores as necessary, and a Steering Committee. Research Center applicants will propose projects that will select key regulatory pathways, identify the components of the pathways and targets, and rapidly disseminate data to the scientific community. The Research Centers will collaborate to maximize interoperability of the data. This Consortium is part of a new NHLBI translational program in pediatric cardiovascular disease, which includes the companion NHLBI Pediatric Cardiac Genomics Consortium and Administrative Coordinating Center. These consortia will interact with each other and the NHLBI Pediatric Heart Network to encourage translation of results from basic science to clinical research, and to provide clinical input on pressing needs for basic research. Maximum allowable direct costs for the research centers are $160,000 in the Year 1 planning phase, and up to $1,070,000 per year in years two through six. Letters of Intent should be received by January 6, and full proposals are due by February 6.
Requirements: The following organizations and institutions are eligible to apply: public/state controlled institutions of higher education; private institutions of higher education; Hispanic-serving institutions; historically Black colleges and universities (HBCUs); tribally controlled colleges and universities (TCCUs); Alaska Native and Native Hawaiian serving institutions; nonprofits with 501(c)3 IRS status; nonprofits without 501(c)3 IRS status; small businesses; for-profit organizations; state governments; and non-domestic (non-U.S.) entities in North America.
Geographic Focus: All States
Date(s) Application is Due: Jan 6; Feb 6
Amount of Grant: Up to 5,510,000 USD
Contact: Charlene Schramm, (301) 435-0510; SchrammC@nhlbi.nih.gov
Internet: http://grants.nih.gov/grants/guide/rfa-files/RFA-HL-09-002.html
Sponsor: National Institutes of Health
Building 31, Room 5A48
Bethesda, MD 20892

NHLBI Career Transition Awards 3021

The purpose of the NHLBI Career Transition Award (K22) program is to provide highly qualified postdoctoral fellows with an opportunity to receive mentored research experience in the NHLBI Division of Intramural Research and then to provide them with funding to facilitate the transition of their research programs as new investigators to extramural institutions. To achieve these objectives, the NHLBI Career Transition Award will support two phases of research: a mentored intramural phase (two years) and an extramural phase (three years), for a total of five years of combined support. Transition from the intramural phase of support to the extramural phase is not automatic. Approval of the transition will be based on the success of the awardees research program as determined

by an NHLBI progress review, which will include an evaluation of a research plan to be carried out at the extramural institution. The total period of support is five years (two years intramural, plus three years extramural). Awards are not renewable. Applications are due by 5:00 pm local time of the applicant organization (see dates below).
Requirements: The following organizations are eligible to apply: public/state controlled and private institutions of higher education; Hispanic-serving institutions; historically black colleges and universities; tribally controlled colleges and universities; Alaska Native and Native Hawaiian serving institutions; nonprofits with or without 501(c)3 IRS status (other than institutions of higher education); small businesses; for-profit organizations (other than small businesses); state, county, and city/township governments; special district governments; Indian/Native American tribal governments (federally recognized and other than federally recognized); eligible agencies of the Federal government; U.S. territory or possession; independent school districts; public or Indiana housing authorities; Native American tribal organizations (other than federally recognized tribal governments); faith-based or community-based organizations; and regional organizations. Applicant organizations may submit more than one application, provided that each application is scientifically distinct. By the time of award, the individual must be a citizen or a non-citizen national of the United States or have been lawfully admitted for permanent residence. Candidates for this award must have earned a terminal clinical or research doctorate (including PhD, MD, DO, DC, ND, DDS, DVM, ScD, DNS, PharmD., or equivalent doctoral degree), or a combined clinical and research doctoral degree. The candidate must have postdoctoral research experience, during which the potential for highly productive basic or clinical research was demonstrated. Individuals from underrepresented racial and ethnic groups as well as individuals with disabilities are always encouraged to apply for NIH support. During the intramural phase of the award, the candidate will spend full time on research. The required research experience must be completed in an intramural NIH laboratory. To obtain support for the extramural phase, at the time of the award candidates must have a full time formal tenure-track (or equivalent) appointment offer at the academic institution that is the applicant institution. Candidates who have VA appointments may not consider part of the VA effort toward satisfying the full time requirement at the applicant institution. Candidates with VA appointments should contact the NHLBI staff prior to preparing an application to discuss their eligibility. Support for the extramural phase will be provided to the extramural institution. Total direct costs for the extramural phase cannot exceed $249,000, including fringe benefits, per year. The total costs cannot exceed $747,000 for the three-year period.
Restrictions: NIH will not accept any application that is essentially the same as one already reviewed. An individual may not have two or more competing NIH career development applications pending review concurrently. Non-domestic (non-U.S.) entities (foreign institutions) and non-domestic (non-U.S.) components of U.S. organizations are not eligible to apply. Multiple Principal Investigators are not allowed.
Geographic Focus: All States, Guam, Marshall Islands, Northern Mariana Islands, Puerto Rico, U.S. Virgin Islands, American Samoa
Date(s) Application is Due: Feb 12; Jun 11; Oct 12
Amount of Grant: Up to 747,000 USD
Contact: Herbert M. Geller; (301) 451-9440; direducation@nhlbi.nih.gov
Internet: http://grants.nih.gov/grants/guide/pa-files/PAR-12-137.html
Sponsor: National Institutes of Health
Building 31, Room 5A48
Bethesda, MD 20892

NHLBI Characterizing the Blood Stem Cell Niche Grants 3022

This FOA issued by the National Heart, Lung, and Blood Institute, National Institutes of Health, solicits Research Project Grant applications in a specific area of stem cell research, the blood stem cell niche, an area critical to advancing stem cell biology and its applications to cellular therapeutics including hematopoietic stem cell transplantation. This FOA is being initiated to foster collaborative research projects and insightful approaches to dissect the cellular components and factors involved in the hematopoietic stem cell niche. Development of conditional genetic knock-out models to test the role of factors in specific cell lineages and imaging technology to facilitate following stem cell engraftment in the niche in vivo are integral to this initiative. An individual research grant application may not request more than $250,000 direct cost a year unless collaborative in vivo cell imaging studies are included, in which case up to $350,000 direct cost may be requested in a year. An applicant may request a project period of up to 4 years. Letters of Intent should be received by December 8, and full proposals are due by January 6.
Requirements: The following organizations and institutions are eligible to apply: public/state controlled institutions of higher education; private institutions of higher education; Hispanic-serving institutions; historically Black colleges and universities (HBCUs); tribally controlled colleges and universities (TCCUs); Alaska Native and Native Hawaiian serving institutions; nonprofits with 501(c)3 IRS status; nonprofits without 501(c)3 IRS status; for-profit organizations; state governments; Indian/Native American tribal governments (federally recognized); Indian/Native American tribally designated organizations; county governments; city or township governments; special district governments; independent school districts; public housing authorities/Indian housing authorities; U.S. territory or possession; Indian/Native American tribal governments (other than federally recognized); regional organizations; non-domestic (non-U.S.) entities (foreign organizations); other eligible agencies of the federal government; and faith-based or community-based organizations.
Geographic Focus: All States
Date(s) Application is Due: Jan 6; Dec 8
Amount of Grant: 250,000-350,000 USD
Contact: John Thomas, (301) 435-0065; fax (301) 451-5453; ThomasJ@NHLBI.NIH.Gov
Internet: http://grants.nih.gov/grants/guide/rfa-files/RFA-HL-09-010.html
Sponsor: National Institutes of Health
Building 31, Room 5A48
Bethesda, MD 20892

NHLBI Circadian-Coupled Cellular Function in Heart, Lung, and Blood Tissue Grants 3023

The purpose of this FOA is to stimulate Phase I translation (T1) of fundamental genome-based circadian biology to improve our understanding of heart, lung, and blood disease pathogenesis in non-neural tissues. For the purpose of this initiative, T1 research starts after gene discovery and has as its goal the development of molecular models with application to an improved understanding of human health and disease pathogenesis, predictive testing for disease susceptibility, prognostic testing, or the selection of therapeutic strategies. Budget for direct costs up to $325,000 per year for project duration of up to four years may be requested. Letters of Intent should be received by December 8, and full proposals are due by January 6.
Requirements: The following organizations and institutions are eligible to apply: public/state controlled institutions of higher education; private institutions of higher education; Hispanic-serving institutions; historically Black colleges and universities (HBCUs); tribally controlled colleges and universities (TCCUs); Alaska Native and Native Hawaiian serving institutions; nonprofits with 501(c)3 IRS status; nonprofits without 501(c)3 IRS status; for-profit organizations; state governments; Indian/Native American tribal governments (federally recognized); Indian/Native American tribally designated organizations; county governments; city or township governments; special district governments; independent school districts; public housing authorities/Indian housing authorities; U.S. territory or possession; Indian/Native American tribal governments (other than federally recognized); regional organizations; non-domestic (non-U.S.) entities (foreign organizations); other eligible agencies of the federal government; and faith-based or community-based organizations.
Geographic Focus: All States
Date(s) Application is Due: Jan 6; Dec 8
Amount of Grant: Up to 325,000 USD
Contact: Aaron Laposky; (301) 435-0199; fax (301) 480-3451; laposkya@nhlbi.nih.gov
Internet: http://grants.nih.gov/grants/guide/rfa-files/RFA-HL-09-012.html
Sponsor: National Institutes of Health
Building 31, Room 5A48
Bethesda, MD 20892

NHLBI Clinical Centers for the NHLBI Asthma Network (AsthmaNet) Grants 3024

The purpose of this funding opportunity, issued by NHLBI, NIH, is to invite applications to participate in the NHLBI Asthma Network (AsthmaNet), a clinical research network that will develop and conduct multiple clinical trials to address the most important asthma management questions and new treatment approaches in pediatric and adult populations. AsthmaNet is designed to promote cooperation and coordination, facilitate scientific exchange, provides training opportunities, and leverage resources. AsthmaNet will include multiple Clinical Centers and one Data Coordinating Center. The protocols will include clinical trials to evaluate and/or compare existing or new therapeutic approaches to asthma management as well as a limited number of proof-of-concept studies to advance the development of novel therapies and studies that investigate the mechanistic bases for interventions examined in AsthmaNet. Letters of Intent should be received by December 30, and full proposals are due by January 30.
Requirements: The following organizations and institutions are eligible to apply: public/state controlled institutions of higher education; private institutions of higher education; Hispanic-serving institutions; historically Black colleges and universities (HBCUs); tribally controlled colleges and universities (TCCUs); Alaska Native and Native Hawaiian serving institutions; nonprofits with 501(c)3 IRS status; nonprofits without 501(c)3 IRS status; for-profit organizations; state governments; Indian/Native American tribal governments (federally recognized); Indian/Native American tribally designated organizations; county governments; city or township governments; special district governments; independent school districts; public housing authorities/Indian housing authorities; U.S. territory or possession; Indian/Native American tribal governments (other than federally recognized); regional organizations; non-domestic (non-U.S.) entities (foreign organizations); other eligible agencies of the federal government; and faith-based or community-based organizations.
Geographic Focus: All States
Date(s) Application is Due: Jan 30; Dec 30
Contact: Virginia S. Taggart, (301) 435-0202; taggartv@nhlbi.nih.gov
Internet: http://grants.nih.gov/grants/guide/rfa-files/RFA-HL-08-010.html
Sponsor: National Institutes of Health
Building 31, Room 5A48
Bethesda, MD 20892

NHLBI Developmental Origins of Altered Lung Physiology and Immune Function Grants 3025

This FOA issued by the NHLBI, National Institutes of Health, solicits Research Project Grant applications from institutions and organizations that propose to perform research that will enhance the understanding of how the pre- and postnatal environments affect the interplay of the lung and immune system during development resulting in sustained changes in lung physiology and immune function that compromise respiratory health and outcomes. Budgets for direct costs of up to $350,000 per year (exclusive of indirect costs associated with consortia) and a project duration of up to five years may be requested for a maximum of $1,750,000 direct costs over a five-year project period. Letters of Intent should be received by September 19, and full proposals are due by October 21.
Requirements: The following organizations and institutions are eligible to apply: public/state controlled institutions of higher education; private institutions of higher education;

Hispanic-serving institutions; historically Black colleges and universities (HBCUs); tribally controlled colleges and universities (TCCUs); Alaska Native and Native Hawaiian serving institutions; nonprofits with 501(c)3 IRS status; nonprofits without 501(c)3 IRS status; for-profit organizations; state governments; Indian/Native American tribal governments (federally recognized); Indian/Native American tribally designated organizations; county governments; city or township governments; special district governments; independent school districts; public housing authorities/Indian housing authorities; U.S. territory or possession; Indian/Native American tribal governments (other than federally recognized); regional organizations; non-domestic (non-U.S.) entities (foreign organizations); other eligible agencies of the federal government; and faith-based or community-based organizations.
Geographic Focus: All States
Date(s) Application is Due: Sep 19; Oct 21
Amount of Grant: Up to 350,000 USD
Contact: Patricia Noel, (301) 435-0202; noelp@nhlbi.nih.gov
Internet: http://grants.nih.gov/grants/guide/rfa-files/RFA-HL-08-009.html
Sponsor: National Institutes of Health
Building 31, Room 5A48
Bethesda, MD 20892

NHLBI Exploratory Studies in the Neurobiology of Pain in Sickle Cell Disease 3026

The goal of this initiative is to foster novel basic and translational research of the neurobiology of pain in sickle cell disease. Priority will be given to the application of investigational techniques that have been utilized in other pain syndromes in both human and nonhuman studies. A multidisciplinary collaboration, between neurobiologists, clinical specialists in pain, and hematologists will be necessary to maximize the research effort.
Requirements: The following organizations and institutions are eligible to apply: public/state controlled institutions of higher education; private institutions of higher education; Hispanic-serving institutions; historically Black colleges and universities (HBCUs); tribally controlled colleges and universities (TCCUs); Alaska Native and Native Hawaiian serving institutions; nonprofits with 501(c)3 IRS status; nonprofits without 501(c)3 IRS status; for-profit organizations; state governments; Indian/Native American tribal governments (federally recognized); Indian/Native American tribally designated organizations; county governments; city or township governments; special district governments; independent school districts; public housing authorities/Indian housing authorities; U.S. territory or possession; Indian/Native American tribal governments (other than federally recognized); regional organizations; non-domestic (non-U.S.) entities (foreign organizations); other eligible agencies of the federal government; and faith-based or community-based organizations.
Geographic Focus: All States
Date(s) Application is Due: Jan 5; Feb 3
Amount of Grant: Up to 300,000 USD
Contact: Harvey Luksenburg, (301) 435-0050; luksenburgh@mail.nih.gov
Internet: http://grants.nih.gov/grants/guide/rfa-files/RFA-HL-09-008.html
Sponsor: National Institutes of Health
Building 31, Room 5A48
Bethesda, MD 20892

NHLBI Immunomodulatory, Inflammatory, and Vasoregulatory Properties of Transfused Red Blood Cell Units as a Function of Preparation and Storage 3027

The goal of this Funding Opportunity Announcement (FOA) in blood banking and transfusion medicine is to solicit novel basic and translational research projects which will identify the molecular and cellular changes that occur during red blood cell unit preparation and storage; and evaluate the immunomodulatory, inflammatory, and vasoactive effects of storage lesion elements from red blood cell units on the blood vessel wall, host cells, and tissue oxygenation. Only research applications involving human red blood cell units (and their components) suitable for transfusion in the U.S. and that are collected and stored according to methods approved by the U.S. FDA will be considered responsive to this FOA. Applications are encouraged that propose collaborations between investigators from different disciplines such as hematology, blood banking, transfusion medicine, immunology, cell biology, rheology, nanotechnologies, molecular biology, and other relevant disciplines. This program will promote interactions and resource sharing among the award recipients to maximize their research efforts. Letters of Intent should be received by April 21 (1st cycle) or December 19 (2nd cycle), and full proposals are due by either May 21 (1st cycle) or January 21 (2nd cycle).
Requirements: The following organizations and institutions are eligible to apply: public/state controlled institutions of higher education; private institutions of higher education; Hispanic-serving institutions; historically Black colleges and universities (HBCUs); tribally controlled colleges and universities (TCCUs); Alaska Native and Native Hawaiian serving institutions; nonprofits with 501(c)3 IRS status; nonprofits without 501(c)3 IRS status; small businesses; for-profit organizations; state governments; Indian/Native American tribal governments; Indian/Native American tribally designated organizations; and U.S. territories or possessions.
Geographic Focus: All States
Date(s) Application is Due: Jan 21; Apr 21; May 21; Dec 19
Contact: Simone Glynn, (301) 435-0065; glynnsa@nhlbi.nih.gov
Internet: http://grants.nih.gov/grants/guide/rfa-files/RFA-HL-08-005.html
Sponsor: National Institutes of Health
Building 31, Room 5A48
Bethesda, MD 20892

NHLBI Independent Scientist Award 3028

The award is a special salary-only grant designed to provide protected time for newly independent scientists who currently have non-research obligations such as heavy teaching loads, clinical work, committee assignments, service, and administrative duties that prevent them from having a period of intensive research focus. The award is targeted to persons with doctoral degrees who have completed their research training, have independent peer-reviewed research support, and who need a period of protected research time in order to foster their research career development. New applications are due February 12, June 12, and October 12, while renewal applications are due March 12, July 12, and November 12. Up to $75,000 per year plus fringe benefits is available. No other research development support funds are provided.
Requirements: The following organizations and institutions are eligible to apply: public/state controlled institutions of higher education; private institutions of higher education; nonprofits with 501(c)3 IRS status; nonprofits without 501(c)3 IRS status; small businesses; for-profit organizations; State governments; U.S. territories or possessions; Indian/Native American tribal governments (Federally recognized and other than Federally recognized); Indian/Native American tribally designated organizations; Hispanic-serving institutions; historically Black colleges and universities (HBCUs); tribally controlled colleges and universities (TCCUs); Alaska Native and Native Hawaiian serving institutions; regional organizations; and faith-based or community based organizations.
Restrictions: The award is not intended for investigators who already have full time to perform research, or have substantial publication records or considerable research support indicating that they are well established in their fields. Foreign institutions are not eligible to apply.
Geographic Focus: All States
Date(s) Application is Due: Feb 12; Mar 12; Jun 12; Jul 12; Oct 12; Nov 12
Amount of Grant: 75,000 USD
Contact: Traci Heath Mondoro, (301) 435-0052; mondorot@nhlbi.nih.gov
Internet: http://grants.nih.gov/grants/guide/pa-files/PA-09-038.html
Sponsor: National Institutes of Health
Building 31, Room 5A48
Bethesda, MD 20892

NHLBI Intramural Research Training Awards 3029

The Office of Education of the Division of Intramural Research coordinates and assists the training program and provides administrative support and resources for all research training activities within the Division. Programs are available within the Laboratories and Branches within the Division to provide specialized research training for high school students, college students, graduate and medical students, postdoctoral fellows and medical residents and fellows. The Office of Education provides assistance to applicants at each of these levels. NHLBI is committed to improving the representation of underrepresented groups, including minorities, women, and scientists with disabilities in the mainstream of basic and clinical research.
Geographic Focus: All States
Contact: Herbert M. Geller, (301) 451-9440; direducation@nhlbi.nih.gov
Internet: http://www.training.nih.gov/student/pre-irta/previewpostbac.asp
Sponsor: National Heart, Lung, and Blood Institute
10 Center Drive, Building 10, Room 2N242, MSC 1754
Bethesda, MD 20892-1754

NHLBI Investigator Initiated Multi-Site Clinical Trials 3030

The support of multi-site clinical trials is one strategy NHLBI uses to improve the understanding of the clinical mechanisms of disease and to improve prevention, diagnosis, and treatment. The purpose of this opportunity is to provide a vehicle for submitting grant applications for investigator-initiated multi-site randomized controlled clinical trials. The trials may address any research question related to the mission and goals of the NHLBI and may test clinical or behavioral interventions. The funding opportunity is appropriate for applications to conduct phase II and phase III randomized clinical trials where participants are recruited from multiple sites. Large-scale pragmatic trials (such as comparative effectiveness trials) as well as trials designed to test efficacy of an intervention are appropriate. The trials may randomize at the individual (patient) level or at a group level (e.g., randomization of clinics, schools, worksites, etc.). Clinical trials involving NHLBI mission-related rare diseases that require coordination across multiple clinical sites are also suitable for submission to this funding opportunity. In any case, the trial should propose the most efficient study design to complete the specific aims. The maximum project period is 5 years.
Requirements: The following organizations are eligible to apply: public/state controlled and private institutions of higher education; Hispanic-serving institutions; historically black colleges and universities; tribally controlled colleges and universities; Alaska Native and Native Hawaiian serving institutions; Asian American Native American Pacific Islander serving institutions; nonprofits with or without 501(c)3 IRS status (other than institutions of higher education); small businesses; for-profit organizations (other than small businesses); state, county, and city/township governments; special district governments; Indian/Native American tribal governments (federally recognized and other than federally recognized); eligible agencies of the Federal government; U.S. territory or possession; independent school districts; public or Indiana housing authorities; Native American tribal organizations (other than federally recognized tribal governments); faith-based or community-based organizations; regional organizations; and non-domestic (non-U.S.) entities (foreign institutions) and non-domestic (non-U.S.) components of U.S. organizations. Clinical Coordination Center (CCC) and Data Coordination Center (DCC) applications are required to be submitted together when the proposed costs of the clinical trial exceed $500,000 (minus F&A for subcontracts) in any given year. Any individual(s) with the skills, knowledge, and resources necessary to carry out the proposed research as the Program Director(s)/Principal Investigator(s) (PD(s)/

PI(s)) is invited to work with his/her organization to develop an application for support. Individuals from underrepresented racial and ethnic groups as well as individuals with disabilities are always encouraged to apply for NIH support.
Geographic Focus: All States, All Countries
Date(s) Application is Due: Feb 12; Jun 12; Oct 12
Amount of Grant: Up to 500,000 USD
Contact: David J. Gordon, (301) 435-0564; fax (301) 480-7971; gordond@nhlbi.nih.gov
Internet: http://grants.nih.gov/grants/guide/pa-files/PAR-13-128.html
Sponsor: National Institutes of Health
Building 31, Room 5A48
Bethesda, MD 20892

NHLBI Lymphatics in Health and Disease in the Digestive, Cardiovascular and Pulmonary Systems 3031

This program encourages Exploratory/Developmental Grant (R21) applications for research into aspects of lymphatic vessel physiology and pathophysiology related to health and disease of digestive, cardiovascular and pulmonary system organs and resolution of thromboembolic events, and inflammation and immune responses as they relate to these diseases. However, studies with the major focus on immune mechanisms will not be considered responsive. Studies to understand the factors that control local lymphatic vessel functional anatomy and physiology during health or disease in these organs and systems, and the mechanisms by which alterations of lymphatic vessel function affect organ function, are of interest. Specific areas of interests for the National Heart, Lung, and Blood Institute (NHLBI) include approaches that will identify the genetic, molecular, and cellular defects that contribute to congenital malformation of the lymphatic system; congenital-lymphatic-malformation-induced pulmonary dysfunction; whether and how lymphangiogenesis affects cardiovascular and pulmonary diseases; whether and how lymphatic drainage affects transplant of heart or lungs; whether and how lymphatic vessel hyperplasia in the dermal interstitium is involved in salt-sensitive hypertension; and whether and how the manipulation of platelet-lymphatic endothelial cell interaction may provide clinical benefit during thromboembolic events.
Requirements: The following organizations and institutions are eligible to apply: public/state controlled and private institutions of higher education; Hispanic-serving institutions; historically Black colleges and universities (HBCUs); Tribally controlled colleges and universities (TCCUs); Alaska Native and Native Hawaiian serving institutions; Nonprofits with or without 501(c)3 IRS status; small businesses; for-profit organizations; state, county, or city/township governments; special district governments; Indian/Native American tribal governments; Indian/Native American tribally designated organizations; independent school districts; public housing authorities/Indian housing authorities; U.S. territory or possession; Indian/Native American tribal governments (other than federally recognized); regional organizations; non-domestic (non-U.S.) entities (foreign organizations); other eligible agencies of the federal government; and faith-based or community-based organizations. Non-domestic (non-U.S.) components of U.S. Organizations are also eligible to apply. Applicant organizations may submit more than one application, provided that each application is scientifically distinct. Any individual(s) with the skills, knowledge, and resources necessary to carry out the proposed research as the Program Director(s)/Principal Investigator(s) (PD(s)/PI(s)) is invited to work with his/her organization to develop an application for support. Individuals from underrepresented racial and ethnic groups as well as individuals with disabilities are always encouraged to apply for NIH support.
Restrictions: NIH will not accept any application that is essentially the same as one already reviewed.
Geographic Focus: All States, Guam, Marshall Islands, Northern Mariana Islands, Puerto Rico, U.S. Virgin Islands, All Countries
Date(s) Application is Due: Feb 16; Jun 16; Oct 16
Internet: http://grants.nih.gov/grants/guide/pa-files/PAR-12-260.html
Sponsor: National Institutes of Health
Building 31, Room 5A48
Bethesda, MD 20892

NHLBI Lymphatics in Health and Disease in the Digestive, Urinary, Cardiovascular and Pulmonary Systems 3032

This program is to encourage Research Project Grant (R01) applications for research into aspects of lymphatic vessel physiology and pathophysiology related to health and disease of digestive system and urinary tract organs, and cardiovascular and pulmonary systems; in resolution of thromboembolic events; and inflammation and immune responses as they relate to these diseases. However, studies with the major focus on immune mechanisms will not be considered responsive. Studies to understand the factors that control local lymphatic vessel functional anatomy and physiology during health or disease in these organs/systems, and the mechanisms by which alterations of lymphatic vessel function affect organ function, are of interest. Applications must be received by 5:00 PM local time of applicant organization.
Requirements: The following organizations and institutions are eligible to apply: public/state controlled and private institutions of higher education; Hispanic-serving institutions; historically Black colleges and universities (HBCUs); Tribally controlled colleges and universities (TCCUs); Alaska Native and Native Hawaiian serving institutions; Nonprofits with or without 501(c)3 IRS status; small businesses; for-profit organizations; state, county, or city/township governments; special district governments; Indian/Native American tribal governments; Indian/Native American tribally designated organizations; independent school districts; public housing authorities/Indian housing authorities; U.S. territory or possession; Indian/Native American tribal governments (other than federally recognized); regional organizations; non-domestic (non-U.S.) entities (foreign organizations); other eligible agencies of the federal government; and faith-based or community-based organizations. Non-domestic (non-U.S.) components of U.S. Organizations are also eligible to apply. Applicant organizations may submit more than one application, provided that each application is scientifically distinct. Any individual(s) with the skills, knowledge, and resources necessary to carry out the proposed research as the Program Director(s)/Principal Investigator(s) (PD(s)/PI(s)) is invited to work with his/her organization to develop an application for support. Individuals from underrepresented racial and ethnic groups as well as individuals with disabilities are always encouraged to apply for NIH support.
Restrictions: NIH will not accept any application that is essentially the same as one already reviewed.
Geographic Focus: All States, Guam, Marshall Islands, Northern Mariana Islands, Puerto Rico, U.S. Virgin Islands, All Countries
Date(s) Application is Due: Feb 5; Jun 5; Oct 5
Contact: H. Eser Tolunay; (301) 435-0560; Eser.Tolunay@nih.gov
Internet: http://grants.nih.gov/grants/guide/pa-files/PAR-07-420.html
Sponsor: National Institutes of Health
Building 31, Room 5A48
Bethesda, MD 20892

NHLBI Mentored Career Award for Faculty at Institutions that Promote Diversity (K01) 3033

The National Institutes of Health (NIH) recognizes a unique and compelling need to promote diversity in the biomedical, behavioral, clinical and social sciences research workforce. The NIH expects efforts to diversify the workforce to lead to the recruitment of the most talented researchers from all groups, to improve the quality of the educational and training environment, to balance and broaden the perspective in setting research priorities, to improve the ability to recruit subjects from minority and other health disparity populations into clinical research protocols, and to improve the Nation's capacity to address and eliminate health disparities. The objectives of this program are to: (1) Advance the awardee's career development trajectory by strengthening research capacity, publishing and other scholarly activities; (2) Improve success and retention in a research career; (3) Promote scientific collaborations that lead to acquisition of new skills or research in other fields of scholarly interest; and (4) Increase the number of highly trained investigators at institutions that promote diversity whose basic and clinical research interests are grounded in the advanced methods and experimental approaches needed to solve problems related to cardiovascular, pulmonary, and hematologic diseases in general and in populations that suffer disproportionately from these conditions. An award made to a successful applicant to this FOA will be a three-to-five-year non-renewable career development award to support faculty at a non-research intensive institution with an institutional mission focused on serving diverse communities that are not well represented in NIH-funded research, or institutions that have been identified by federal legislation as having specially focused institutional missions. The candidates must have research experience and be committed to developing into independent biomedical investigators in research areas relevant to the mission of the NHLBI (i.e., cardiovascular, pulmonary, hematologic, or sleep disorders research).
Requirements: Higher Education Institutions (Public/State Controlled Institutions of Higher Education and Private Institutions of Higher Education) are eligible to apply. The following types of Higher Education Institutions are always encouraged to apply for NIH support as Public or Private Institutions of Higher Education: Hispanic-serving Institutions;Historically Black Colleges and Universities (HBCUs); Tribally Controlled Colleges and Universities (TCCUs); Alaska Native and Native Hawaiian Serving Institutions. The Institution must be a domestic college or university with an institutional mission focused on serving minority and other health disparity populations underrepresented in scientific research, or institutions that have been identified by federal legislation as having specifically focused institutional missions. To demonstrate need for research capacity development in heart, lung, and blood diseases and sleep disorders, applicant institutions should not have received over $1,515,000 in NIH funding (direct costs) from the NHLBI per year in any of the previous five years for research training and education grants in the above NHLBI mission areas. Applicant organizations may submit more than one application, provided that each application is scientifically distinct. Any candidate with the skills, knowledge, and resources necessary to carry out the proposed research as the Program Director/Principal Investigator (PD/PI) is invited to work with his/her mentor and organization to develop an application for support. Individuals from underrepresented racial and ethnic groups as well as individuals with disabilities are always encouraged to apply for NIH support. Multiple Principal Investigators are not allowed. By the time of award, the individual must be a citizen or a non-citizen national of the United States or have been lawfully admitted for permanent residence. Individuals on temporary or student visas are not eligible.
Restrictions: Non-domestic (non-U.S.) Entities (Foreign Institutions) and non-domestic (non-U.S.) components of U.S. Organizations are not eligible to apply. Foreign components are not allowed.
Geographic Focus: All States
Date(s) Application is Due: Feb 7
Amount of Grant: Up to 100,000 USD
Contact: Henry Chang, M.D., (301) 435-0067; changh@nhlbi.nih.gov
Internet: http://www.nhlbi.nih.gov/funding/training/redbook/newrek01.htm
Sponsor: National Institutes of Health
Building 31, Room 5A48
Bethesda, MD 20892

NHLBI Mentored Career Award For Faculty At Minority Institutions 3034

The NHLBI Mentored Career Award for Faculty at Minority Institutions is a three-to-five-year non-renewable career development award made to faculty at a minority institution. The candidates are faculty members at a minority institution, must have research experience (length of time may vary), and be committed to developing into

independent biomedical investigators in research areas relevant to the mission of the NHLBI (i.e, cardiovascular, pulmonary, hematologic, or sleep disorders research). The award will enable suitable faculty members holding doctoral degrees, such as the Ph.D., M.D., D.O., D.V.M., or an equivalent degree, to undertake special study and supervised research under a mentor who is an accomplished investigator in the research area proposed and has experience in developing independent investigators. Letters of Intent are due on June 21, while full proposals are due by July 19.
Requirements: Minority scientists and physicians with limited research experience needing guided coursework and supervised laboratory experiences, as well as minority faculty needing only an intensive research experience under the guidance of an established scientist, are eligible to apply. At the time of award, it is required that at least two years have elapsed since the receipt of the doctoral degree and that the candidate have at least one year of prior documented research experience.
Restrictions: Current or past principal investigators of an NIH grant or its equivalent, including the Clinical Investigator Award, Physician Scientist Award, Clinical Investigator Development Award, or Mentored Clinical Scientist Development Award are not eligible for the Mentored Research Scientist Development Award for Minority Faculty. Similarly, individuals serving as responsible investigators or project leaders on large grants, such as a Program Project Grant, are not eligible for the Mentored Research Scientist Development Award for Minority Faculty.
Geographic Focus: All States
Date(s) Application is Due: Jul 19
Amount of Grant: Up to 200,000 USD
Contact: Traci Heath Mondoro, (301) 435-0052; mondorot@nhlbi.nih.gov
Internet: http://grants.nih.gov/grants/guide/rfa-files/RFA-HL-05-016.html#SectionVII
Sponsor: National Institutes of Health
Building 31, Room 5A48
Bethesda, MD 20892

NHLBI Mentored Career Development Award to Promote Faculty Diversity/ Re-Entry in Biomedical Research 3035
This program invites applications to increase the number of highly trained investigators, from diverse backgrounds underrepresented in research areas of interest to the NHLBI or from those individuals who wish to re-enter their research careers (e.g., after a hiatus due to family circumstances). Minorities are not the only group underrepresented in the scientific community. Individuals with disabilities are also underrepresented. It is geared toward individuals whose basic and clinical research interests are grounded in the advanced methods and experimental approaches needed to solve problems related to cardiovascular, pulmonary, and hematologic diseases in the general and health disparities populations. NHLBI encourages research training and career development crossing disciplinary boundaries (e.g., biophysics, biostatistics, bioinformatics, bioengineering) to develop a new interdisciplinary work force. Award budgets are composed of salary and other program-related expenses. The total project period may not exceed 5 years. Applications are due by the posted deadline date by 5:00 PM local time of applicant organization.
Requirements: The following organizations are eligible to apply: Public/state controlled and private institutions of higher education; nonprofits with and without 501(c)3 IRS Status (other than institutions of higher education); small businesses; for-profit organizations (other than small businesses); state governments; county governments; city or township governments; special district governments; Indian/Native American Tribal governments (federally recognized); Indian/Native American Tribal governments (other than federally recognized); U.S. territory or possession. The following types of higher education institutions are always encouraged to apply for NIH support as public or private institutions of higher education: Hispanic-serving institutions; historically Black colleges and universities; Tribally controlled colleges and universities; Alaska Native and Native Hawaiian serving institutions. Any candidate with the skills, knowledge, and resources necessary to carry out the proposed research as the Program Director/Principal Investigator (PD/PI) is invited to work with his/her mentor and organization to develop an application for support. Individuals from underrepresented racial and ethnic groups as well as individuals with disabilities are always encouraged to apply for NIH support. Multiple Principal Investigators are not allowed. By the time of award, the individual must be a citizen or a non-citizen national of the United States or have been lawfully admitted for permanent residence.
Restrictions: Non-domestic (non-U.S.) Entities (Foreign Institutions) are not eligible to apply. Non-domestic (non-U.S.) components of U.S. Organizations are not eligible to apply. Foreign components, as defined in the NIH Grants Policy Statement, are not allowed. Individuals on temporary or student visas are not eligible.
Geographic Focus: All States, Guam, Marshall Islands, Northern Mariana Islands, Puerto Rico, U.S. Virgin Islands, American Samoa
Date(s) Application is Due: Feb 7
Amount of Grant: Up to 625,000 USD
Contact: Lorraine M. Silsbee, M.H.S., (301) 435-0709; silsbeeL@nhlbi.nih.gov
Internet: http://www.nhlbi.nih.gov/funding/training/redbook/newsck01.htm
Sponsor: National Institutes of Health
Building 31, Room 5A48
Bethesda, MD 20892

NHLBI Mentored Clinical Scientist Research Career Development Awards 3036
This award enables candidates holding professional degrees (e.g., M.D., D.O., D.V.M., or equivalent degrees) to undertake 3 to 5 years of special study and supervised research with the goal of becoming independent investigators. The award also allows awardees to pursue a research career development program suited to their experience and capabilities under a mentor who is competent to provide guidance in the chosen research area. Institutions may submit applications on behalf of candidates who hold professional degrees. At least 2 years must have elapsed since the health professional degree was granted. Candidates can have varying levels of clinical training and research experience. New applications are due February 12, June 12, and October 12, while resubmitted applications are due March 12, July 12, and November 12. Up to $75,000 per year plus fringe benefits is available, and up to $25,000 per year for research support for new awards.
Requirements: All candidates must be U.S. citizens, non-citizen nationals, or legal permanent residents of the U.S. Persons with temporary or student visas are not eligible. The grantee institution must have a strong, well-established research and research training program in the chosen area, accomplished faculty in the basic and clinical sciences, and a commitment to the candidate's research development. The proposed program should include an appropriate mentor.
Geographic Focus: All States
Date(s) Application is Due: Feb 12; Mar 12; Jun 12; Jul 12; Oct 12; Nov 12
Amount of Grant: Up to 100,000 USD
Contact: Lorraine M. Silsbee, (301) 435-0709; Lorraine_Silsbee@nih.gov
Internet: http://www.nhlbi.nih.gov/funding/training/redbook/newrek08.htm
Sponsor: National Institutes of Health
Building 31, Room 5A48
Bethesda, MD 20892

NHLBI Mentored Patient-Oriented Research Career Development Awards 3037
This award supports the career development of investigators who are committed to patient-oriented research. It provides support for supervised study and research for clinically trained professionals who have the potential to develop into productive, clinical investigators focusing on patient-oriented research. This program provides research development opportunities for clinicians with varying levels of research experience. Support is provided for a minimum of 3 years and a maximum of 5 years. New applications are due February 12, June 12, and October 12. Resubmissions are due March 12, July 12, and November 12.
Requirements: The following organizations are eligible to apply: Public/state controlled and private institutions of higher education; nonprofits with and without 501(c)3 IRS Status (other than institutions of higher education); small businesses; for-profit organizations (other than small businesses); state governments; county governments; city or township governments; special district governments; Indian/Native American Tribal governments (federally recognized); Indian/Native American Tribal governments (other than federally recognized); U.S. territory or possession. The following types of higher education institutions are always encouraged to apply for NIH support as public or private institutions of higher education: Hispanic-serving institutions; historically Black colleges and universities; Tribally controlled colleges and universities; Alaska Native and Native Hawaiian serving institutions. Applicant organizations may submit more than one application, provided that each application is scientifically distinct. Any candidate with the skills, knowledge, and resources necessary to carry out the proposed research as the Program Director/Principal Investigator (PD/PI) is invited to work with his/her mentor and organization to develop an application for support. Individuals from underrepresented racial and ethnic groups as well as individuals with disabilities are always encouraged to apply for NIH support. Multiple Principal Investigators are not allowed. By the time of award, the individual must be a citizen or a non-citizen national of the United States or have been lawfully admitted for permanent residence. Candidates for this award must have a health-professional doctoral degree. Such degrees include but are not limited to the M.D., D.O., D.D.S., D.M.D., O.D., D.C., Pharm. D., N.D. (Doctor of Naturopathy), as well as a doctoral degree in nursing research or practice. Candidates with Ph.D. degrees are eligible for this award if the degree is in a clinical field and they usually perform clinical duties. Individuals with the Ph.D. or other doctoral degree in clinical disciplines such as clinical psychology, clinical genetics, speech-language pathology, audiology or rehabilitation are also eligible. Individuals holding the Ph.D. in a non-clinical discipline but who are certified to perform clinical duties should contact the appropriate Institute concerning their eligibility for a K23 award. Candidates also must have completed their clinical training, including specialty and, if applicable, subspecialty training prior to receiving an award. However, candidates may submit an application prior to the completion of clinical training.
Restrictions: Non-domestic (non-U.S.) Entities (Foreign Organizations) are not eligible to apply. Foreign (non-U.S.) components of U.S. Organizations are not allowed. Individuals on temporary or student visas are not eligible.
Geographic Focus: All States, Guam, Marshall Islands, Northern Mariana Islands, Puerto Rico, U.S. Virgin Islands, American Samoa
Date(s) Application is Due: Feb 12; Mar 12; Jun 12; Jul 12; Oct 12; Nov 12
Amount of Grant: Up to 620,000 USD
Contact: Sandra Colombini Hatch; (301) 435-0222; hatchs@nhlbi.nih.gov
Internet: http://www.nhlbi.nih.gov/funding/training/redbook/newk23.htm
Sponsor: National Institutes of Health
Building 31, Room 5A48
Bethesda, MD 20892

NHLBI Mentored Quantitative Research Career Development Awards 3038
The goal of this program is to foster interdisciplinary collaboration in biomedical and behavioral research by supporting supervised research experiences for scientists with quantitative and engineering backgrounds. This award provides research and career development opportunities for scientists and engineers with little or no biomedical or behavioral research experience who are committed to establishing themselves in careers as independent biomedical or behavioral investigators. Examples of quantitative scientific and technical backgrounds outside of biology and medicine considered appropriate for this award include, but are not limited to: mathematics, statistics, computer science, informatics, physics, chemistry, and engineering. This mechanism is intended for research-oriented investigators from the postdoctoral level to the level of senior faculty. New applications are due February 12, June 12, October 12. Resubmissions are due

March 12, July 12, and November 12. Award budgets are composed of salary and other program-related expenses. The total project period may not exceed 5 years.
Requirements: Institutions may submit applications on behalf of candidates who hold an advanced degree in a quantitative area of science or engineering. The following organizations are eligible to apply: Public/state controlled and private institutions of higher education; nonprofits with and without 501(c)3 IRS Status (other than institutions of higher education); small businesses; for-profit organizations (other than small businesses); state governments; county governments; city or township governments; special district governments; Indian/Native American Tribal governments (federally recognized); Indian/Native American Tribal governments (other than federally recognized); U.S. territory or possession. The following types of higher education institutions are always encouraged to apply for NIH support as public or private institutions of higher education: Hispanic-serving institutions; historically Black colleges and universities; Tribally controlled colleges and universities; Alaska Native and Native Hawaiian serving institutions. Individuals with an advanced degree in a quantitative area of science or engineering are eligible candidates. Multiple Principal Investigators are not allowed. All candidates must be U.S. citizens, non-citizen nationals, or legal permanent residents of the U.S. The grantee institution must have a strong, well-established research and research training program in the chosen area, accomplished faculty in the basic and clinical sciences, and a commitment to the candidate's research development. The proposed program should include an appropriate mentor.
Restrictions: NHLBI has a 6 year limit of cumulative support on institutional and mentored Ks (e.g., K12 or KL2 plus the K25). A candidate for the K25 may not concurrently apply for or have an award pending for any other NIH career development award. Non-domestic (non-U.S.) Entities (Foreign Organizations) are not eligible to apply. Foreign (non-U.S.) components of U.S. Organizations are not allowed.
Geographic Focus: All States, Guam, Marshall Islands, Northern Mariana Islands, Puerto Rico, U.S. Virgin Islands, American Samoa
Date(s) Application is Due: Feb 12; Mar 12; Jun 12; Jul 12; Oct 12; Nov 12
Contact: Drew E. Carlson, (301) 435-0535; fax (301) 480-7404; carlsonde@nhlbi.nih.gov
Internet: http://www.nhlbi.nih.gov/funding/training/redbook/newk25.htm
Sponsor: National Institutes of Health
Building 31, Room 5A48
Bethesda, MD 20892

NHLBI Microbiome of the Lung and Respiratory Tract in HIV-Infected Individuals and HIV-Uninfected Controls **3039**

The National Heart, Lung, and Blood Institute (NHLBI) solicits grant applications under this Funding Opportunity Announcement (FOA) to characterize the microbiome of the lung (the airways and airspaces below the glottis and the lung parenchyma) alone or in combination with the nasal and/or oropharyngeal cavities in HIV-infected individuals and matched HIV-uninfected controls, including normal healthy controls, using molecular techniques to identify bacteria and if possible other organisms, e.g., viruses, cell-wall deficient organisms, protozoa, and fungi. Investigators should use high-throughput technology platforms to create a data set of sufficient quality and depth to allow analysis of how changes of microbiota relate to HIV lung disease progression/complications. These data will be used to examine the impact of changes in the respiratory microbiome on the pathogenesis and progression of HIV disease, on HIV-related respiratory complications, and the effects of anti-HIV therapies. The NHLBI also invites applications for four Clinical/Sequencing research sites and one supporting Data Coordinating Center (DCC). Each application for a clinical/sequencing research site should propose approaches to characterize the lung microbiome and applicants are encouraged to include at least one hypothesis-driven mechanistic aim. Direct costs up to $525,000 per year for project duration of up to five years may be requested.
Requirements: The following organizations/institutions are eligible to apply: Public/State Controlled Institutions of Higher Education; Private Institutions of Higher Education; Hispanic-serving Institutions; Historically Black Colleges and Universities (HBCUs); Tribally Controlled Colleges and Universities (TCCUs); Alaska Native and Native Hawaiian Serving Institutions; Nonprofits with 501(c)3 IRS Status (Other than Institutions of Higher Education); Nonprofits without 501(c)3 IRS Status (Other than Institutions of Higher Education); Small Businesses; For-Profit Organizations (Other than Small Businesses); State Governments; Indian/Native American Tribal Governments (Federally Recognized); Indian/Native American Tribally Designated Organizations; County Governments; City or Township Governments; Special District Governments; Independent School Districts; Public Housing Authorities/Indian Housing Authorities; U.S. Territory or Possession; Indian/Native American Tribal Governments (Other than Federally Recognized); Regional Organizations; Other Eligible Agencies of the Federal Government; and, Faith-based or Community-based Organizations. Any individual with the skills, knowledge, and resources necessary to carry out the proposed research as the PD/PI is invited to work with his/her institution to develop an application for support. Individuals from underrepresented racial and ethnic groups as well as individuals with disabilities are always encouraged to apply for NIH support. More than one PD/PI, or multiple PDs/PIs, may be designated on the application for projects that require a "team science" approach and therefore clearly do not fit the single-PD/PI model. All PDs/PIs must be registered in the NIH eRA Commons prior to the submission of the application.
Restrictions: Foreign organizations are not eligible, however, domestic institutions may form consortia and enter into subcontracts with Non-domestic (non-U.S.) Entities (Foreign Organizations) if need can be demonstrated – foreign subcontract use the same criteria as are used for justifying foreign grants.
Geographic Focus: All States
Date(s) Application is Due: Feb 5; Jun 5; Oct 5
Amount of Grant: Up to 525,000 USD
Contact: Sandra Colombini Hatch, (301) 435-0222; Hatchs@nhlbi.nih.gov
Internet: http://grants.nih.gov/grants/guide/rfa-files/RFA-HL-09-006.html
Sponsor: National Institutes of Health
Building 31, Room 5A48
Bethesda, MD 20892

NHLBI Midcareer Investigator Award in Patient-Oriented Research **3040**

This award provides protected time to mid-career clinical investigators who are typically at the Associate Professor level or the equivalent and who have their own independent peer-reviewed research support to provide mentoring to junior clinical investigators, particularly K23 grantees, in patient-oriented research and to stabilize the careers of these investigators so that they could continue to conduct patient-oriented research and be available as mentors in patient-oriented research. It is expected that the K24 recipients will obtain new or additional independent peer-reviewed funding for patient-oriented research as a Program Director/Principal Investigator and establish and assume leadership roles in collaborative patient-oriented research programs. In addition, it is expected that there will be an increased effort and commitment to act as a mentor to beginning clinician investigators in patient-oriented research to enhance the research productivity of both the K24 investigator and increase the pool of well-trained clinical researchers of the future. New applications are due on February 12, June 12 and October 12. Renewal applications are due March 12, July 12, and November 12.
Requirements: Candidates must be U.S. citizens, non citizen nationals, or legal permanent residents of the U.S. Candidates for this award must have a health-professional doctoral degree or its equivalent. Such degrees include but are not limited to the M.D., D.O., D.D.S., D.M.D., O.D., D.C., Pharm.D., N.D. (Doctor of Naturopathy), as well as a doctoral degree in nursing. Candidates with Ph.D. degrees are eligible for this award if the degree is in a clinical field or they perform patient-oriented research (P.O.R). This may include clinical psychologists, clinical geneticists, speech and language pathologists. Candidates should typically be in the midcareer stage at the Associate Professor level or functioning at that rank in an academic setting or equivalent non-academic setting and must have an established record of independent, peer-reviewed patient-oriented research grant funding including at the time of application and record of publications. This award is intended for individuals who have a record of supervising and mentoring patient-oriented researchers.
Geographic Focus: All States
Date(s) Application is Due: Feb 12; Mar 12; Jun 12; Jul 12; Oct 12; Nov 12
Contact: Sandra Colombini Hatch, M.D., Medical Officer; (301) 435-0222; fax (301) 480-3557; hatchs@nhlbi.nih.gov
Internet: http://www.nhlbi.nih.gov/funding/training/redbook/estk24.htm
Sponsor: National Institutes of Health
Building 31, Room 5A48
Bethesda, MD 20892

NHLBI National Research Service Award Programs in Cardiovascular Epidemiology and Biostatistics **3041**

The National Research Service Award (NRSA) is one of many training programs sponsored by the National Heart, Lung, and Blood Institute (NHLBI). This program enables research institutions to support pre- and post-doctoral training in areas of particular interest to the Institute and to prepare individuals for careers in biomedical and behavioral research. The cardiovascular epidemiology and biostatistics research training programs offer opportunities in the fields of epidemiology of cardiovascular diseases and statistical analysis. These awards do not support study leading to the M.D., D.D.S. or other similar professional degrees, nor do they support residency training.
Requirements: Trainees are selected through review procedures established by the program director at the grantee institution. Students are required to pursue their research training full-time and must be U.S. citizens, noncitizen nationals or legal permanent residents of the United States. At the time of appointment, trainees must have received a baccalaureate degree and must be training at the post-baccalauareate level in a program leading to a Ph.D., Sc.D. or equivalent degree. Students who wish to interrupt their medical, dental, or other professional school studies to engage in full-time research training before completing their professional degrees are also eligible. Attention is given to recruiting individuals from minority groups that are under-represented nationally in the biomedical sciences (African Americans, Hispanics, Native Americans, Alaskan Natives, and Pacific Islanders).
Restrictions: Persons on temporary or student visas are not eligible.
Geographic Focus: All States
Date(s) Application is Due: Jan 25
Contact: Lorraine M. Silsbee, (301) 435-0709; fax (301) 480-1667; silsbeeL@nih.gov
Internet: http://www.nhlbi.nih.gov/funding/training/epi-bio/nrsa.htm
Sponsor: National Institutes of Health
Building 31, Room 5A48
Bethesda, MD 20892

NHLBI New Approaches to Arrhythmia Detection and Treatment Grants for SBIR **3042**

The purpose of this initiative is to improve our ability to detect, treat and prevent cardiac arrhythmias using new or improved methods, tools, and technologies. Applications are invited for the development and significant improvements of innovative monitoring tools, diagnostic methods, and therapeutic approaches for arrhythmia detection, treatment and prevention.
Requirements: Only United States small business concerns (SBCs) are eligible to submit SBIR applications.
Geographic Focus: All States
Date(s) Application is Due: Apr 5; Aug 5; Dec 5

Contact: David A. Lathrop, PhD; (301) 435-0507; LathropD@NHLBI.NIH.GOV
Internet: http://grants.nih.gov/grants/guide/pa-files/PA-07-032.html
Sponsor: National Institutes of Health
Building 31, Room 5A48
Bethesda, MD 20892

NHLBI New Approaches to Arrhythmia Detection and Treatment Grants for STTR 3043

The purpose of this initiative is to improve our ability to detect, treat and prevent cardiac arrhythmias using new or improved methods, tools, and technologies. Applications are invited for the development and significant improvements of innovative monitoring tools, diagnostic methods, and therapeutic approaches for arrhythmia detection, treatment and prevention.
Requirements: Only United States small business concerns (SBCs) are eligible to submit STTR applications.
Geographic Focus: All States
Date(s) Application is Due: Apr 5; Aug 5; Dec 5
Contact: David A. Lathrop, PhD; (301) 435-0507; LathropD@NHLBI.NIH.GOV
Internet: http://grants.nih.gov/grants/guide/pa-files/PA-07-031.html
Sponsor: National Institutes of Health
Building 31, Room 5A48
Bethesda, MD 20892

NHLBI Nutrition and Diet in the Causation, Prevention, and Management of Heart Failure Research Grants 3044

The purpose of this Funding Opportunity Announcement (FOA) is to encourage submission of investigator-initiated research applications on the role of nutrition and diet in the causation, prevention, and treatment of cardiomyopathies and heart failure. Basic, translational, and applied interdisciplinary research applications with rigorous hypothesis-testing designs for projects in animals or humans are of interest. The overall goal is to develop a satisfactory science base for preventive approaches in high-risk individuals and for rational nutritional management of patients in various stages of heart failure.
Requirements: The following groups are eligible to apply: for-profit organizations on-profit organizations; public or private institutions, such as universities, colleges, hospitals, and laboratories; units of State government; units of local government; eligible agencies of the Federal government; foreign institutions; domestic institutions; faith-based or community-based organizations; units of State tribal government; and units of local Tribal government.
Geographic Focus: All States
Date(s) Application is Due: Feb 16; Jun 16; Oct 16
Amount of Grant: Up to 200,000 USD
Contact: Abby G. Ershow, (301) 435-0550; fax (301) 480-1338; ErshowA@mail.nih.gov
Internet: http://grants.nih.gov/grants/guide/pa-files/PA-06-136.html
Sponsor: National Institutes of Health
Building 31, Room 5A48
Bethesda, MD 20892

NHLBI Pathway to Independence Awards 3045

The Pathway to Independence (PI) Award is designed to facilitate a timely transition from a mentored postdoctoral research position to a stable independent research position with independent NIH or other independent research support at an earlier stage than is currently the norm. The PI award will provide up to 5 years of support consisting of two phases. The initial phase will provide 1-2 years of mentored support for postdoctoral research scientists. This phase will be followed by up to 3 years of independent support contingent on securing an independent tenure-track or equivalent research position. New applications are due February 12, June 12, and October 12. Resubmitted applications are due March 12, July 12, and November 12.
Requirements: The following organizations are eligible to apply: Public/state controlled and private institutions of higher education; nonprofits with and without 501(c)3 IRS Status (other than institutions of higher education); small businesses; for-profit organizations (other than small businesses); state governments; county governments; city or township governments; special district governments; Indian/Native American Tribal governments (federally recognized); Indian/Native American Tribal governments (other than federally recognized); U.S. territory or possession. The following types of higher education institutions are always encouraged to apply for NIH support as public or private institutions of higher education: Hispanic-serving institutions; historically Black colleges and universities; Tribally controlled colleges and universities; Alaska Native and Native Hawaiian serving institutions. The applicant institution will be the mentored phase K99 institution. All institution/organization types listed above are eligible for both the mentored and independent phase, with one exception: eligible agencies of the Federal government, such as the NIH intramural program, are eligible only for the mentored phase. Any candidate with the skills, knowledge, and resources necessary to carry out the proposed research as the Program Director/Principal Investigator (PD/PI) is invited to work with his/her mentor and organization to develop an application for support. Individuals from underrepresented racial and ethnic groups as well as individuals with disabilities are always encouraged to apply for NIH support. Multiple Principal Investigators are not allowed. At the time of application submission (or resubmission) candidates for this award must have earned a terminal clinical or research doctorate (including Ph.D., M.D., D. O., D.C., N.D., D.D.S., D.V.M., Sc.D., D.N.S., Pharm. D., or equivalent doctoral degree, or a combined clinical and research doctoral degree); and have no more than 5 years (60 months) of research experience since completing the requirements of the doctoral degree (resubmissions must also comply with this requirement). Individuals whose terminal or research doctorate was awarded more than 60 months before submission of the K99/R00 application are strongly urged to contact one of the NHLBI Program staff designated below before submission to determine whether they are eligible to apply. This confirmation of eligibility should be documented in a cover letter to accompany the application. Applications from ineligible individuals will be returned without review. U.S. citizens and non-U.S. citizens are eligible.
Restrictions: Non-domestic (non-U.S.) Entities (Foreign Organizations) are not eligible to apply. Foreign (non-U.S.) components of U.S. Organizations are not allowed.
Geographic Focus: All States, Guam, Marshall Islands, Northern Mariana Islands, Puerto Rico, U.S. Virgin Islands, American Samoa
Date(s) Application is Due: Feb 12; Mar 12; Jun 12; Jul 12; Oct 12; Nov 12
Contact: Herbert Geller, Ph.D., Director, Office of Education; (301) 451-9440; fax (301) 594-8133; direducation@nhlbi.nih.gov
Internet: http://grants.nih.gov/grants/guide/pa-files/PA-11-197.html
Sponsor: National Institutes of Health
Building 31, Room 5A48
Bethesda, MD 20892

NHLBI Pediatric Cardiac Genomics Consortium Grants 3046

The purpose of this initiative is to support collaborative genetic and genomic research leading to a comprehensive understanding of the causes and outcomes of congenital heart disease in the human population. The Consortium will consist of up to 6 Research Centers, Cores, and a Steering Committee. Research Center applicants will propose detailed investigation of the genetics and genomics of one or more human cardiac malformations, with the goal of identifying genetic influences on outcome, as well as causative genes. Individual Research Centers will collaborate in recruitment of individuals with congenital heart disease and their families for all Research Center projects. Maximum allowable direct costs for the Research Centers per year are $160,000 in Year 1, and $535,000 per year in years 2-6. Letters of Intent should be received by January 6, and full proposals are due by February 6.
Requirements: The following organizations and institutions are eligible to apply: public/state controlled institutions of higher education; private institutions of higher education; Hispanic-serving institutions; historically Black colleges and universities (HBCUs); tribally controlled colleges and universities (TCCUs); Alaska Native and Native Hawaiian serving institutions; nonprofits with 501(c)3 IRS status; nonprofits without 501(c)3 IRS status; for-profit organizations; small businesses; state governments; non-domestic (non-U.S.) entities in North America; and other eligible agencies of the Federal Government.
Geographic Focus: All States
Date(s) Application is Due: Jan 6; Feb 6
Amount of Grant: Up to 2,835,000 USD
Contact: Gail D. Pearson, (301) 435-0510; pearsong@mail.nih.gov
Internet: http://grants.nih.gov/grants/guide/rfa-files/RFA-HL-09-003.html
Sponsor: National Institutes of Health
Building 31, Room 5A48
Bethesda, MD 20892

NHLBI Phase II Clinical Trials of Novel Therapies for Lung Diseases 3047

The purpose of this FOA is to solicit research grant applications to conduct Phase II clinical therapeutic trials that have the potential to advance development of novel therapies for a lung disease or a cardiopulmonary disorder of sleep. Each application will propose one Phase II interventional trial that will most likely use physiological or biochemical rather than clinical endpoints along with at least one smaller basic ancillary research study that is tightly related to the clinical question. Although definitive Phase III trials will not be supported, the proposed studies must provide proof of concept for a novel intervention that has high potential for modifying current treatments and could be disease modifying.
Requirements: The following organizations and institutions are eligible to apply: public/state controlled institutions of higher education; private institutions of higher education; Hispanic-serving institutions; historically Black colleges and universities (HBCUs); tribally controlled colleges and universities (TCCUs); Alaska Native and Native Hawaiian serving institutions; nonprofits with 501(c)3 IRS status; nonprofits without 501(c)3 IRS status; for-profit organizations; state governments; Indian/Native American tribal governments (federally recognized); Indian/Native American tribally designated organizations; county governments; city or township governments; special district governments; independent school districts; public housing authorities/Indian housing authorities; U.S. territories or possessions; Indian/Native American tribal governments (other than federally recognized); regional organizations; non-domestic (non-U.S.) entities (foreign organizations); other eligible agencies of the federal government; and faith-based or community-based organizations.
Geographic Focus: All States
Date(s) Application is Due: Jan 29; May 18; Jun 18; Aug 15; Sep 15; Dec 30
Amount of Grant: Up to 1,515,000 USD
Contact: Andrea L. Harabin, Ph.D; (301) 435-0222; harabina@nhlbi.nih.gov
Internet: http://grants.nih.gov/grants/guide/rfa-files/RFA-HL-10-003.html
Sponsor: National Institutes of Health
Building 31, Room 5A48
Bethesda, MD 20892

NHLBI Prematurity and Respiratory Outcomes Program Grants 3048

This RFA intends to fill the need for understanding the underlying pathophysiology by supporting the investigation of the molecular mechanisms that contribute to acute and chronic respiratory morbidity of the premature infant. The well phenotyped populations developed in this program will significantly increase the chance of studying groups with common pathophysiology and lay the groundwork for novel biomarker discovery. The program will use the cooperative agreement mechanism and require multidisciplinary expertise including neonatologists/pulmonologists, and physiological, molecular, biological, or bioinformatics scientists. Participating sites will plan and coordinate

their own research and will develop and implement shared protocols for respiratory phenotyping and respiratory outcomes of the premature infant NICU graduate. Letters of Intent should be received by May 4, and full proposals are due by June 2 each year.
Requirements: The following organizations and institutions are eligible to apply: public/state controlled institutions of higher education; private institutions of higher education; Hispanic-serving institutions; historically Black colleges and universities (HBCUs); tribally controlled colleges and universities (TCCUs); Alaska Native and Native Hawaiian serving institutions; nonprofits with 501(c)3 IRS status; nonprofits without 501(c)3 IRS status; for-profit organizations; state governments; Indian/Native American tribal governments (federally recognized); Indian/Native American tribally designated organizations; county governments; city or township governments; special district governments; independent school districts; public housing authorities/Indian housing authorities; U.S. territories or possessions; Indian/Native American tribal governments (other than federally recognized); regional organizations; non-domestic (non-U.S.) entities (foreign organizations); other eligible agencies of the federal government; and faith-based or community-based organizations.
Geographic Focus: All States
Date(s) Application is Due: May 4; Jun 2
Amount of Grant: 150,000-350,000 USD
Contact: Carol J. Blaisdell, (301) 435-0222; blaisdellcj@nhlbi.nih.gov
Internet: http://grants.nih.gov/grants/guide/rfa-files/RFA-HL-10-007.html
Sponsor: National Institutes of Health
Building 31, Room 5A48
Bethesda, MD 20892

NHLBI Progenitor Cell Biology Consortium Administrative Coordinating Center Grants 3049

The goal of the Consortium is to identify and characterize progenitor cell lineages, to direct the differentiation of stem and progenitor cells to desired cell fates, and to develop new strategies to address the unique challenges presented by the transplantation of these cells. The Consortium will assemble multiple independent research projects, each with a multi-disciplinary team of Principal Investigators, core research support facilities, and skills development components to establish synergistic virtual hubs focused on progenitor cell biology. The total project period for an application submitted in response to this FOA is 7 years. Maximum allowable direct costs are $600,000 per year for ACC operations. Letters of Intent are dur by April 1, while full applications should arrive by May 1.
Requirements: The following organizations and institutions are eligible to apply: public/state controlled institutions of higher education; private institutions of higher education; Hispanic-serving institutions; historically Black colleges and universities (HBCUs); tribally controlled colleges and universities (TCCUs); Alaska Native and Native Hawaiian serving institutions; nonprofits with 501(c)3 IRS status; nonprofits without 501(c)3 IRS status; small businesses; for-profit organizations; and state governments.
Geographic Focus: All States
Date(s) Application is Due: Apr 1; May 1
Amount of Grant: Up to 4,200,000 USD
Contact: Denis Buxton, (301) 435-0513; Buxtond@nhlbi.nih.gov
Internet: http://grants.nih.gov/grants/guide/rfa-files/RFA-HL-09-005.html
Sponsor: National Institutes of Health
Building 31, Room 5A48
Bethesda, MD 20892

NHLBI Protein Interactions Governing Membrane Transport in Pulmonary Health and Disease Grants 3050

This Funding Opportunity Announcement (FOA) issued by the National Heart, Lung, and Blood Institute, National Institutes of Health (NIH), solicits research grant applications to delineate the protein interactions and pathways governing membrane trafficking pathways operative in pulmonary health and disease and develop novel therapeutic interventions. Because the nature and scope of the proposed research will vary from application to application, it is anticipated that the size and duration of each award will also vary. The total amount awarded and the number of awards will depend upon the mechanism numbers, quality, duration, and costs of the applications received.
Requirements: The following organizations and institutions are eligible to apply: public/state controlled institutions of higher education; private institutions of higher education; Hispanic-serving institutions; historically Black colleges and universities (HBCUs); tribally controlled colleges and universities (TCCUs); Alaska Native and Native Hawaiian serving institutions; nonprofits with 501(c)3 IRS status; nonprofits without 501(c)3 IRS status; for-profit organizations; state governments; Indian/Native American tribal governments (federally recognized); Indian/Native American tribally designated organizations; county governments; city or township governments; special district governments; independent school districts; public housing authorities/Indian housing authorities; U.S. territory or possession; Indian/Native American tribal governments (other than federally recognized); regional organizations; non-domestic (non-U.S.) entities (foreign organizations); other eligible agencies of the federal government; and faith-based or community-based organizations.
Geographic Focus: All States
Date(s) Application is Due: Feb 5; Jun 5; Oct 5
Contact: Susan Banks-Schlegel, (301) 435-0202; Schleges@nhlbi.nih.gov
Internet: http://grants.nih.gov/grants/guide/pa-files/PA-07-137.html
Sponsor: National Institutes of Health
Building 31, Room 5A48
Bethesda, MD 20892

NHLBI Research Demonstration and Dissemination Grants 3051

The purpose of this Funding Opportunity Announcement (FOA) is to encourage the scientific community to conduct Demonstration and Dissemination (D&D) studies to test the effectiveness of interventions in children, adolescents, and/or adults to: promote healthful behaviors; reduce risk factors for heart, lung, and blood diseases, and sleep disorders; improve the prevention or management of heart, lung, and blood diseases, and sleep disorders, including the delivery of health care services; and enhance understanding of the processes of intervention implementation and diffusion, or sustainability in a defined population, or defined clinical or community setting. The total project period for an application submitted in response to this funding opportunity may not exceed 5 years. Direct costs are limited to less than $500,000 in any single year, unless pre-approval is obtained from the Institute.
Requirements: The following organizations and institutions are eligible to apply: public/state controlled institutions of higher education; private institutions of higher education; Hispanic-serving institutions; historically Black colleges and universities (HBCUs); tribally controlled colleges and universities (TCCUs); Alaska Native and Native Hawaiian serving institutions; nonprofits with 501(c)3 IRS status; nonprofits without 501(c)3 IRS status; for-profit organizations; state governments; Indian/Native American tribal governments (federally recognized); Indian/Native American tribally designated organizations; county governments; city or township governments; special district governments; independent school districts; public housing authorities/Indian housing authorities; U.S. territory or possession; Indian/Native American tribal governments (other than federally recognized); regional organizations; non-domestic (non-U.S.) entities (foreign organizations); other eligible agencies of the federal government; and faith-based or community-based organizations.
Geographic Focus: All States
Date(s) Application is Due: Jan 25; May 25; Sep 25
Amount of Grant: Up to 500,000 USD
Contact: Charlotte Pratt, (301) 435-0382; fax (301) 480-5158; PrattC@nhlbi.nih.gov
Internet: http://grants.nih.gov/grants/guide/pa-files/PA-07-017.html
Sponsor: National Institutes of Health
Building 31, Room 5A48
Bethesda, MD 20892

NHLBI Research Grants on the Relationship Between Hypertension and Inflammation 3052

This Funding Opportunity Announcement (FOA) solicits grant applications that propose the study of the sequence of events in which the vascular inflammatory state contributes to the development and maintenance of hypertension. Ample evidence suggests that inflammation may play a role in the pathogenesis of hypertension or that it may characterize a functional state of the vessel wall as a consequence of high blood pressure. Angiotensin II (Ang II), a widely recognized vasoconstrictor and anti-natriuretic involved in blood pressure regulation, also acts as a pro-inflammatory factor in the cardiovascular system. Ang II stimulates the expression of several inflammatory cytokines, which in turn affect blood pressure. A potential linkage between Ang II and immuno-cytokines is their shared ability to induce the level of reactive oxygen species (ROS), which serve as second messengers for many intracellular signaling pathways. The production of ROS not only decreases bioavailability of nitric oxide (NO), a vasodilator, but also initiates the functional and morphological alterations, such as remodeling, in the vascular wall that accompany the hypertensive state over time. This FOA would provide an opportunity to bring focus on the potential causal relationship between hypertension and inflammation in a cohesive, integrated manner. A new understanding of hypertension and inflammation would provide novel opportunities to prevent and treat the disease.
Requirements: The following organizations and institutions are eligible to apply: public/state controlled institutions of higher education; private institutions of higher education; Hispanic-serving institutions; historically Black colleges and universities (HBCUs); tribally controlled colleges and universities (TCCUs); Alaska Native and Native Hawaiian serving institutions; nonprofits with 501(c)3 IRS status; nonprofits without 501(c)3 IRS status; for-profit organizations; state governments; Indian/Native American tribal governments (federally recognized); Indian/Native American tribally designated organizations; county governments; city or township governments; special district governments; independent school districts; public housing authorities/Indian housing authorities; U.S. territory or possession; Indian/Native American tribal governments (other than federally recognized); regional organizations; non-domestic (non-U.S.) entities (foreign organizations); other eligible agencies of the federal government; and faith-based or community-based organizations.
Geographic Focus: All States
Date(s) Application is Due: Feb 5; Jun 5; Oct 5
Contact: Eser Tolunay, (301) 435-0560; tolunaye@nhlbi.nih.gov
Internet: http://grants.nih.gov/grants/guide/pa-files/PA-07-138.html
Sponsor: National Institutes of Health
Building 31, Room 5A48
Bethesda, MD 20892

NHLBI Research on the Role of Cardiomyocyte Mitochondria in Heart Disease: An Integrated Approach 3053

The National Heart, Lung, and Blood Institute (NHLBI) invites applications for collaborative research projects to develop an integrated understanding of cardiomyocyte mitochondria and its contributions to myocardial adaptations and heart disease progression by combining functional data with information derived from powerful new technologies. Budgets for direct costs of up to $500,000 per year and a project duration of up to four years may be requested for a maximum of $2,000,000 direct costs over a four-year project period. Because the nature and scope of the proposed research will vary from application to application, it is anticipated that the size and duration of each award will also vary.

Requirements: The following organizations/institutions are eligible to apply: Public/State Controlled Institutions of Higher Education; Private Institutions of Higher Education; Hispanic-serving Institutions; Historically Black Colleges and Universities (HBCUs); Tribally Controlled Colleges and Universities (TCCUs); Alaska Native and Native Hawaiian Serving Institutions; Nonprofits with 501(c)3 IRS Status (Other than Institutions of Higher Education); Nonprofits without 501(c)3 IRS Status (Other than Institutions of Higher Education); Small Businesses; For-Profit Organizations (Other than Small Businesses); State Governments; Indian/Native American Tribal Governments (Federally Recognized); Indian/Native American Tribally Designated Organizations; County Governments; City or Township Governments; Special District Governments; Independent School Districts; Public Housing Authorities/Indian Housing Authorities; U.S. Territory or Possession; Indian/Native American Tribal Governments (Other than Federally Recognized); Regional Organizations; Non-domestic (non-U.S.) Entities (Foreign Organizations); Other Eligible Agencies of the Federal Government; and, Faith-based or Community-based Organizations. Any individual(s) with the skills, knowledge, and resources necessary to carry out the proposed research as the PD/PI is invited to work with his/her organization to develop an application for support. Individuals from underrepresented racial and ethnic groups as well as individuals with disabilities are always encouraged to apply for NIH support. More than one PD/PI (i.e., multiple PDs/PIs) may be designated on the application for projects that require a "team science" approach and therefore clearly do not fit the single-PD/PI model. All PDs/PIs must be registered in the NIH electronic Research Administration (eRA) Commons prior to the submission of the application. Applicants may submit more than one application, provided each application is scientifically distinct. Request for Applications (RFA) Number: RFA-HL-10-002
Restrictions: Applicants are not permitted to submit a resubmission application. Renewals will not be allowed.
Geographic Focus: All States, Guam, Marshall Islands, Northern Mariana Islands, Puerto Rico, U.S. Virgin Islands, All Countries
Date(s) Application is Due: Feb 5; Jun 5; Oct 5
Amount of Grant: Up to 2,000,000 USD
Contact: Isabella Liang, (301) 435-0504; fax (301) 451-5458; Liangi@mail.nih.gov
Internet: http://grants.nih.gov/grants/guide/rfa-files/RFA-HL-10-002.html
Sponsor: National Institutes of Health
Building 31, Room 5A48
Bethesda, MD 20892

NHLBI Research Supplements to Promote Diversity in Health-Related Research **3054**

This supplement enables principal investigators with eligible NHLBI research grants/contracts to include high school students in their projects. Priority is given to applications requesting support for individuals from underrepresented racial and ethnic groups, individuals with disabilities, and individuals from disadvantaged backgrounds. Nationally, underrepresented groups in biomedical research careers include but are not limited to African Americans, Hispanic Americans, American Indians/Alaska Natives, Native Hawaiians and Pacific Islanders. The research activity proposed for the student must be part of the approved research for the parent grant or contract. The student should be encouraged to participate in ongoing team discussions of research findings and directions. The purpose of this program is to provide high school students with an opportunity to obtain a meaningful experience in various aspects of health-related research to stimulate their interest in careers in biomedical, behavioral, biometric, clinical, nursing and social sciences. Students are expected to devote sufficient effort to the research project and related activities during the period of support to gain insight into the process of scientific discovery. Support for at least three months is encouraged during any one year. Principal Investigators are encouraged to seek high school students who will devote at least two years to this program (i.e., equivalent to two three-month, full-time periods). More than one high school research supplements can be awarded per research grant, subproject of a program project grant (P01), or contract. Applications will be accepted at any time, however, applications should arrive at least three months before the requested start date, to allow time for review.
Requirements: All principal investigators at U.S. institutions with eligible NHLBI research grants or contracts may apply. Grants/contracts with adequate time for a summer research experience or one year remaining at the time of award are eligible to apply for high school students. This announcement is for supplements to existing projects. To be eligible, the parent award must be active and the research proposed in the supplement must be accomplished within the competitive segment. All additional costs must be within the scope of the peer-reviewed and approved project. Before submitting an application for a research supplement, applicants are strongly encouraged to contact their program official to discuss the program. Supplemental awards under this program are limited to citizens or non-citizen nationals of the United States or to individuals who have been lawfully admitted for permanent residence in the United States (i.e., in possession of an Alien Registration Receipt Card or some other legal evidence of admission for permanent residence at the time of application). This program may not be used to provide technical support to NIH-supported investigators. Application budgets are limited to no more than the amount of the current parent award, and must reflect actual needs of the proposed project. Direct costs for individual administrative supplements vary from less than $5,000 to more than $100,000 depending on the career level of the candidate. Administrative supplements end with the competitive cycle of the parent grant.
Restrictions: Non-domestic (non-U.S.) Entities (Foreign Institutions) and non-domestic (non-U.S.) components of U.S. Organizations are not eligible to apply. Foreign components, as defined in the NIH Grants Policy Statement, are not allowed.
Geographic Focus: All States
Contact: Nara Gavini, (301) 451-5081; fax (301) 480-0862; gavininn@nhlbi.nih.gov
Internet: http://grants.nih.gov/grants/guide/pa-files/PA-12-149.html
Sponsor: National Institutes of Health
Building 31, Room 5A48
Bethesda, MD 20892

NHLBI Right Heart Function in Health and Chronic Lung Diseases Grants **3055**

This program will promote research on cellular, molecular and physiological determinants of right ventricular function in health, and function/ dysfunction in association with chronic lung diseases. The goal is to encourage research which concentrates on building a foundation of basic knowledge of right ventricular physiology and the basic molecular and biomechanical factors contributing to right ventricular dysfunction associated with pulmonary diseases. This knowledge base will be used in determining the best methods of preventing, diagnosing and treating right heart failure. This FOA seeks to fill the current gaps that exist in our understanding of the right heart and how right heart function affects lung function in both health and disease. The emphasis on interaction of the heart and the lung during chronic lung disease in this program is intended to promote collaboration between the heart and lung research communities.
Requirements: The following organizations and institutions are eligible to apply: public/state controlled institutions of higher education; private institutions of higher education; Hispanic-serving institutions; historically Black colleges and universities (HBCUs); tribally controlled colleges and universities (TCCUs); Alaska Native and Native Hawaiian serving institutions; nonprofits with 501(c)3 IRS status; nonprofits without 501(c)3 IRS status; for-profit organizations; state governments; Indian/Native American tribal governments (federally recognized); Indian/Native American tribally designated organizations; county governments; city or township governments; special district governments; independent school districts; public housing authorities/Indian housing authorities; U.S. territory or possession; Indian/Native American tribal governments (other than federally recognized); regional organizations; non-domestic (non-U.S.) entities (foreign organizations); other eligible agencies of the federal government; and faith-based or community-based organizations.
Geographic Focus: All States
Date(s) Application is Due: Feb 5; Jun 5; Oct 5
Contact: Elizabeth M. Denholm, (301) 435-0222; denholme@mail.nih.gov
Internet: http://grants.nih.gov/grants/guide/pa-files/PA-07-043.html
Sponsor: National Institutes of Health
Building 31, Room 5A48
Bethesda, MD 20892

NHLBI Ruth L. Kirschstein National Research Service Award Institutional Research Training Grants **3056**

This award enables research institutions to support predoctoral research training in specific areas and fields of shortage. Selection of institutions is by national competition; trainees are selected through local review procedures established by the program director at the grantee institution. Trainees are required to pursue their research training full-time, and trainees in clinical areas are expected to confine their clinical duties to those that are part of their research training. Institutional grants are renewable, through competition, for up to 5 years. The NHLBI will accept all types of competing T32 applications (new, renewal/competing continuation, resubmission/amended) until January 25. Only resubmission/amended T32 applications will be accepted until September 25.
Requirements: Trainees must be U.S. citizens, noncitizen nationals, or legal permanent residents of the U.S. Persons on temporary or student visas are not eligible. At the time of appointment, trainees must have received a baccalaureate degree and must be training at the graduate level in a program leading to the Ph.D., Sc.D., or other equivalent degree. Trainees who wish to interrupt their medical, dental, veterinary, or other professional school studies to engage in full-time research training before completing their professional degrees are also eligible. Applications requesting $500,000 or more in direct costs in any given year must receive prior approval from the NHLBI before submitting an application.
Restrictions: These awards do not support study leading to the M.D., D.O., D.D.S., or other similar professional degrees, nor do they support residency training.
Geographic Focus: All States
Date(s) Application is Due: Jan 25; Sep 25
Contact: Jane Scott, ScD, MSN, Director; (301) 435-0535; scottj2@nhlbi.nih.gov
Internet: http://www.nhlbi.nih.gov/funding/training/redbook/gradt32.htm
Sponsor: National Institutes of Health
Building 31, Room 5A48
Bethesda, MD 20892

NHLBI Ruth L. Kirschstein National Research Service Awards for Individual Postdoctoral Fellows **3057**

The National Heart, Lung, and Blood Institute is one of many Institutes in the National Institutes of Health that participate in the Ruth L. Kirschstein National Research Service Awards (NRSA) for Individual Postdoctoral Fellows. This program offers health scientists the opportunity to receive full-time research training for up to 3 years in areas that reflect the national need for biomedical, clinical and behavioral research in cardiovascular, pulmonary, hematologic, and sleep disorders. These grants are not intended for study leading to the M.D., D.O., D.D.S., or equivalent professional degrees, nor do they support residency training. The proposed postdoctoral training must offer an opportunity to enhance the applicant's understanding of the health-related sciences, and must be within the broad scope of biomedical, behavioral, or clinical research or other specific disciplines relevant to the research mission of the participating NIH Institutes and Centers. Applicants with a health professional doctoral degree may use the proposed

postdoctoral training to satisfy a portion of the degree requirements for a master's degree, a research doctoral degree or any other advanced research degree program.
Requirements: Any applicant fellow with the skills, knowledge, and resources necessary to carry out the proposed research as the Project Director/Principal Investigator (PD/PI) is invited to work with his/her sponsor and organization to develop an application for support. Individuals from underrepresented racial and ethnic groups as well as individuals with disabilities are always encouraged to apply for NIH support. Applicants must be U.S. citizens, noncitizen nationals, or legal permanent residents of the U.S. Training can be conducted abroad if the site provides opportunities that are not available in this country. Applicants are cautioned that not all NIH Institutes and Centers (ICs) participate in this program, and that consultation with relevant IC staff prior to submission of an application is strongly encouraged. The participating ICs have different emphases and program requirements for this program. Before submitting a fellowship application, the applicant fellow must identify a sponsoring institution. The sponsoring institution must have staff and facilities available on site to provide a suitable environment for performing high-quality research. Eligible sponsoring institutions include: Public/State Controlled and Private Institutions of Higher Education; Nonprofits with or without 501(c)3 IRS Status (Other than Institutions of Higher Education); Small Businesses; For-Profit Organizations (Other than Small Businesses); Eligible Agencies of the Federal Government; and, Non-domestic (non-U.S.) Entities (Foreign Organizations). The following types of Higher Education Institutions are always encouraged to apply for NIH support as Public or Private Institutions of Higher Education: Hispanic-serving Institutions; Historically Black Colleges and Universities (HBCUs); Tribally Controlled Colleges and Universities (TCCUs); and, Alaska Native and Native Hawaiian Serving Institutions. When the fellowship begins, the applicant must have received a doctoral degree and have arranged to work with a sponsor affiliated with an institution that has the staff and facilities needed for the proposed training.
Restrictions: A Kirschstein-NRSA fellowship may not be used to support the clinical years of residency training. However, these awards are appropriate for the research fellowship years of a residency program. Research clinicians must devote full-time to their proposed research training and confine clinical duties to those activities that are part of the research training program.
Geographic Focus: All States, Guam, Marshall Islands, Northern Mariana Islands, Puerto Rico, U.S. Virgin Islands, All Countries
Date(s) Application is Due: Apr 8; Aug 8; Dec 8
Contact: Sandra Colombini Hatch, (301) 435-0222; hatchs@nhlbi.nih.gov
Internet: http://www.nhlbi.nih.gov/funding/training/redbook/phdf32.htm
Sponsor: National Institutes of Health
Building 31, Room 5A48
Bethesda, MD 20892

NHLBI Ruth L. Kirschstein National Research Service Awards for Individual Predoctoral Fellowships to Promote Diversity in Health-Related Research 3058

NHLBI is one of many Institutes in the NIH that participate in the Ruth L. Kirschstein National Research Service Awards for Individual Predoctoral Fellowships to Promote Diversity in Health-Related Research. This program encourages students from underrepresented racial and ethnic groups, individuals with disabilities, and individuals from disadvantaged backgrounds to seek research doctoral degrees in the biomedical and behavioral sciences to help increase the number of well-trained scientists from underrepresented groups. The fellowship provides up to 5 years of support for research training leading to the Ph.D. or equivalent research degree, the combined M.D./Ph.D. degree, or other combined degrees in the biomedical or behavioral sciences.
Requirements: Individuals with disabilities, or from underrepresented racial and ethnic groups, or individuals from disadvantaged backgrounds pursuing advanced degrees in the biomedical and behavioral sciences are eligible for fellowship awards. The Fellowship applicant must have a baccalaureate degree and be currently enrolled in an eligible doctoral program. At the time of appointment, students must be U.S. citizens, noncitizen nationals, or lawfully admitted to the U.S. for permanent residence. Individuals on temporary or student visas are not eligible. Before submitting a fellowship application, the applicant fellow must identify a sponsoring institution. The sponsoring institution must have staff and facilities available on site to provide a suitable environment for performing high-quality research. Eligible sponsoring institutions include Public/State Controlled and Private Institutions of Higher Education; Nonprofits with or without 501(c)3 IRS Status (Other than Institutions of Higher Education); Small Businesses; For-Profit Organizations (Other than Small Businesses); Eligible Agencies of the Federal Government; Non-domestic (non-U.S.) Entities (Foreign Organizations). The following types of Higher Education Institutions are always encouraged to apply for NIH support as Public or Private Institutions of Higher Education: Hispanic-serving Institutions; Historically Black Colleges and Universities (HBCUs); Tribally Controlled Colleges and Universities (TCCUs); Alaska Native and Native Hawaiian Serving Institutions.
Restrictions: NIH will not accept any application that is essentially the same as one already reviewed. An individual may not have two or more competing NIH fellowship applications pending review concurrently.
Geographic Focus: All States
Date(s) Application is Due: Apr 8; Aug 8; Dec 8
Contact: Xenia J. Tigno; (301) 435-0202; tignoxt@mail.nih.gov
Internet: http://www.nhlbi.nih.gov/funding/training/redbook/gradf31.htm
Sponsor: National Institutes of Health
Building 31, Room 5A48
Bethesda, MD 20892

NHLBI Ruth L. Kirschstein National Research Service Awards for Individual Predoctoral MD/PhD Fellows and Other Dual Degree Fellows 3059

The NHLBI is interested in supporting individual predoctoral fellowships for combined MD/PhD training in research areas relevant to the NHLBI mission. The purpose of the Ruth L. Kirschstein National Research Service Awards (Kirschstein-NRSA) is to provide support to individuals for combined MD/PhD and other dual doctoral degree training (e.g. DO/PhD, DDS/PhD, AuD/PhD). The participating Institutes award this Kirschstein-NRSA individual fellowship (F30) to qualified applicants with the potential to become productive, independent, highly trained physician-scientists and other clinician-scientists, including patient-oriented researchers in their scientific mission areas. This funding opportunity supports individual predoctoral F30 fellowships with the expectation that these training opportunities will increase the number of future investigators with both clinical knowledge and skills in basic, translational or clinical research. Award budgets are composed of stipends, tuition and fees, and institutional allowance.
Requirements: Domestic for-profit or non-profit institutions/organizations, or public or private institutions, such as universities, colleges, hospitals and laboratories, eligible agencies of the federal government, and NIH intramural laboratories are eligible to apply. The sponsoring institution must have staff and facilities available on site to provide a suitable environment for performing high-quality research training. The PhD phase of the program may be conducted outside the sponsoring institution, e.g. a Federal laboratory including the NIH intramural laboratories. This training, however, must be part of a combined MD/PhD program. Applications to the program will be accepted from students currently enrolled in a formally combined MD/PhD program. By the time of award, all candidates for the Kirschstein-NRSA F30 award must be citizens or non-citizen nationals of the United States, or must have been lawfully admitted to the United States for Permanent Residence.
Restrictions: Foreign institutions are not eligible to apply.
Geographic Focus: All States
Date(s) Application is Due: Apr 8; Aug 8; Dec 8
Contact: Drew E. Carlson, Ph.D., Program Director, Office of Research Training and Career Development; (301) 435-0535; fax (301) 480-7404; carlsonde@nhlbi.nih.gov
Internet: http://www.nhlbi.nih.gov/funding/training/redbook/gradmdphdf30.htm
Sponsor: National Institutes of Health
Building 31, Room 5A48
Bethesda, MD 20892

NHLBI Ruth L. Kirschstein National Research Service Awards for Individual Senior Fellows 3060

The National Heart, Lung, and Blood Institute is one of many Institutes in the National Institutes of Health that participate in the Ruth L. Kirschstein National Research Service Awards (NRSA) for individual Senior Fellows (Parent F33). This program enables experienced scientists to change the direction of their research careers, broaden their scientific background, acquire new research capabilities, or enlarge their command of an allied research field. The purpose of the senior fellowship (F33) award is to provide senior fellowship support to experienced scientists who wish to make major changes in the direction of their research careers or who wish to broaden their scientific background by acquiring new research capabilities as independent research investigators in scientific health-related fields relevant to the missions of the participating NIH Institutes and Centers. These awards will enable individuals with at least seven years of research experience beyond the doctorate (individuals with PhD, MD, OD, associate professors, full professors, etc.), and who have progressed to the stage of independent investigator, to take time from regular professional responsibilities for the purpose of receiving training to increase their scientific capabilities. In most cases, this award is used to support sabbatical experiences for established independent scientists seeking support for retraining or additional career development. This program is not designed for postdoctoral level investigators seeking to enhance their research experience prior to independence. Award budgets are composed of stipends, tuition and fees, and institutional allowance.
Requirements: Applicants must be U.S. citizens, non-citizen nationals, or permanent residents of the U.S. When the fellowship begins, an applicant must have received the Ph.D., M.D., D.O., D.D.S., D.V.M., O.D., D.P.M., Sc.D., D.Eng., DNSc., or equivalent degree. The applicant must arrange to work full-time with a particular sponsor affiliated with an institution that has the staff and facilities needed for the proposed training. Training can be conducted abroad if the site provides opportunities not available in the U.S. The following are eligible sponsoring organizations: public/state controlled and private institutions of higher education; nonprofits with or without 501(c)3 IRS status (other than institutions of higher education); small businesses; for-profit organizations (other than small businesses); eligible agencies of the Federal government; non-domestic (non-U.S.) entities (foreign organizations). The following types of Higher Education Institutions are always encouraged to apply for NIH support as public or private institutions of higher education: Hispanic-serving institutions; historically Black colleges and universities; Tribally controlled colleges and universities; Alaska Native and Native Hawaiian serving institutions.
Restrictions: Multiple Principal Investigators are not allowed.
Geographic Focus: All States, Guam, Marshall Islands, Northern Mariana Islands, Puerto Rico, U.S. Virgin Islands, American Samoa
Date(s) Application is Due: Apr 8; Aug 8; Dec 8
Contact: Charlotte A. Pratt; (301) 435-0382; fax (301) 480-1864; prattc@nhlbi.nih.gov
Internet: http://www.nhlbi.nih.gov/funding/training/redbook/estf33.htm
Sponsor: National Institutes of Health
Building 31, Room 5A48
Bethesda, MD 20892

NHLBI Ruth L. Kirschstein National Research Service Award Short-Term Institutional Research Training Grants 3061

This program provides funds to research institutions to make awards to individuals in health professional schools for research opportunities that would not be available through their regular course of study. Awards are made to training institutions by national competition. Many of the NIH Institutes and Centers (ICs) use this grant mechanism exclusively to support intensive, short-term research training experiences for

students in health professional schools during the summer. In addition, the Short-Term Institutional Research Training Grant may be used to support other types of predoctoral and postdoctoral training in focused, often emerging scientific areas relevant to the mission of the funding IC. Grants may be for project periods up to five years in duration and are renewable. Trainees selected for short-term training are required to pursue research training for 2-3 months on a full-time basis.
Requirements: Public and private institutions of higher education; nonprofits with or without 501(c)3 IRS status (other than Institutions of Higher Education); Indian/Native American Tribal Governments; U.S. Territories or Possessions; Native American tribal organizations; and, Faith-based or Community-based Organizations are eligible to apply. The following types of Higher Education Institutions are always encouraged to apply for NIH support as Public or Private Institutions of Higher Education: Hispanic-serving Institutions; Historically Black Colleges and Universities (HBCUs); Tribally Controlled Colleges and Universities (TCCUs); and, Alaska Native and Native Hawaiian Serving Institutions. The proposed training must be in basic, behavioral or clinical research aspects of the health-related sciences. This program is intended to encourage graduate and/or health professional students to pursue research careers by exposure to and short-term involvement in the health-related sciences. The training should be of sufficient depth to enable the trainees, upon completion of the program, to have a thorough exposure to the principles underlying the conduct of research. Trainees are selected by the grantee institution and must be U.S. citizens, noncitizen nationals, or legal permanent residents of the U.S. Trainees should have successfully completed at least one semester at an accredited school of medicine, optometry, osteopathy, dentistry, veterinary medicine, pharmacy, or public health before entering the program. The award can be used to support individuals who already have an M.S. or Ph.D. and have been accepted to health professional schools.
Restrictions: The award cannot be used to support courses that are required for the M.D., D.O., D.D.S., D.V.M., or similar professional degrees. Non-domestic (non-U.S.) Entities (Foreign Organizations) are not eligible to apply. Foreign (non-U.S.) components of U.S. Organizations are not allowed.
Geographic Focus: All States, Guam, Marshall Islands, Northern Mariana Islands, Puerto Rico, U.S. Virgin Islands
Date(s) Application is Due: Jan 25
Contact: Charlotte Pratt; (301) 435-0382; fax (301) 480-1864; PrattC@nhlbi.nih.gov
Internet: http://www.nhlbi.nih.gov/funding/training/redbook/gradt35.htm
Sponsor: National Institutes of Health
Building 31, Room 5A48
Bethesda, MD 20892

NHLBI Short-Term Research Education Program to Increase Diversity in Health-Related Research 3062

The National Heart, Lung, and Blood Institute, National Institutes of Health invites Research Education (R25) applications to promote diversity in undergraduate and health professional participant populations by providing short-term research education support to stimulate career development in cardiovascular, pulmonary, hematologic, and sleep disorders research. The overall goal of the program is to provide research opportunities for individuals from backgrounds underrepresented in biomedical science, including individuals from disadvantaged backgrounds, individuals from underrepresented racial and ethnic groups, and individuals with disabilities that will significantly contribute to a diverse research workforce in the future. The research opportunities should be of sufficient depth to enable the participants, upon completion of the program, to have a thorough exposure to the principles underlying the conduct of research, and help prepare participants interested in research to pursue competitive fellowships, or other research training or career development awards. NHLBI intends to fund three to five new awards corresponding to a total of $900,000 for new grants per fiscal year.
Requirements: The following types of organizations are eligible to apply: Higher Education Institutions (Public/State Controlled Institutions of Higher Education, Private Institutions of Higher Education); Nonprofits other than Institutions of Higher Education, with or without 501(c)3 IRS Status; For-Profit Organizations; Small Businesses; State Governments; County Governments; City or Township Governments; Special District Governments; Indian/Native American Tribal Governments (Federally Recognized); Indian/Native American Tribal Governments (Other than Federally Recognized); Eligible Agencies of the Federal Government; U.S. Territory or Possession; Independent School Districts; Public Housing Authorities/Indian Housing Authorities; Native American Tribal Organizations (other than Federally recognized tribal governments); Faith-based or Community-based Organizations; Regional Organizations. The following types of Higher Education Institutions are always encouraged to apply for NIH support as Public or Private Institutions of Higher Education: Hispanic-serving Institutions; Historically Black Colleges and Universities (HBCUs); Tribally Controlled Colleges and Universities (TCCUs); Alaska Native and Native Hawaiian Serving Institutions. Some institutions provide unique opportunities for access to participants and faculty from diverse backgrounds. These institutions can assist NIH in its efforts to recruit the most talented researchers from all groups, to improve the ability to recruit subjects from diverse backgrounds into clinical research protocols, and improve the Nation's capacity to address and eliminate health disparities. These institutions are therefore encouraged to apply for NIH support. The applicant institution must have a strong research program in the area(s) proposed for research training and must have the requisite staff and facilities to carry out the proposed program. The applicant institution must have the available research facilities, personnel, and support for the program in the areas of cardiovascular, pulmonary, or hematologic diseases or sleep disorders. Institutions with adequate staff and resources in these areas are encouraged to apply. Although the size of the award may vary within the scope of the research education program proposed, it is expected that applications will stay within the following budgetary guidelines: the total institutional annual direct cost should not exceed $319,000. The total project period for an application submitted in response to this funding opportunity may not exceed 5 years.
Restrictions: Non-domestic (non-U.S.) Entities (Foreign Institutions) and non-domestic (non-U.S.) components of U.S. Organizations are not eligible to apply. Foreign components, as defined in the NIH Grants Policy Statement, are not allowed.
Geographic Focus: All States, Guam, Marshall Islands, Northern Mariana Islands, Puerto Rico, U.S. Virgin Islands
Date(s) Application is Due: Feb 7
Amount of Grant: Up to 2,700,000 USD
Contact: Drew Carlson, Ph.D., (301) 435-0535; carlsonde@nhlbi.nih.gov
Internet: http://grants.nih.gov/grants/guide/rfa-files/RFA-HL-13-020.html
Sponsor: National Institutes of Health
Building 31, Room 5A48
Bethesda, MD 20892

NHLBI Summer Institute for Training in Biostatistics II 3063

The National Heart, Lung, and Blood Institute (NHLBI) and the National Center for Research Resources (NCRR) invite applications for grants to develop, conduct, and evaluate summer courses in the basic principles and methods of biostatistics as employed in biomedical research. The courses would introduce participants to the field of biostatistics for the purpose of attracting new students into the field. The course will attract students from the entire U.S.A and will cover the fundamental concepts of probability, statistical reasoning and inferential methods motivated, in part, by examples that include data collected in studies of heart, lung, blood, and sleep disorders. The course will be taught during the summer, with appropriate modifications or refinements following each of the first two summers' sessions. Programs may choose an area of emphasis, such as clinical trials, statistical genetics, bioinformatics or epidemiology or take a general approach. Recent college graduates as well as advanced undergraduates should be eligible to participate provided they have not participated in SIBS previously. Direct costs are limited to $248,000 per year for a three-year period. Letters of Intent should be received by December 5, and full proposals are due by January 6.
Requirements: The following organizations and institutions are eligible to apply: public/state controlled institutions of higher education; private institutions of higher education; Hispanic-serving institutions; historically Black colleges and universities (HBCUs); tribally controlled colleges and universities (TCCUs); Alaska Native and Native Hawaiian serving institutions; nonprofits with 501(c)3 IRS status; nonprofits without 501(c)3 IRS status; for-profit organizations; state governments; Indian/Native American tribal governments (federally recognized); Indian/Native American tribally designated organizations; county governments; city or township governments; special district governments; independent school districts; public housing authorities/Indian housing authorities; U.S. territory or possession; Indian/Native American tribal governments (other than federally recognized); regional organizations; non-domestic (non-U.S.) entities (foreign organizations); other eligible agencies of the federal government; and faith-based or community-based organizations.
Geographic Focus: All States
Date(s) Application is Due: Jan 6; Dec 5
Amount of Grant: Up to 248,000 USD
Contact: Song Yang, (301) 435-0431; fax (301) 480-1862; yangso@nhlbi.nih.gov
Internet: http://grants.nih.gov/grants/guide/rfa-files/RFA-HL-09-009.html
Sponsor: National Institutes of Health
Building 31, Room 5A48
Bethesda, MD 20892

NHLBI Summer Internship Program in Biomedical Research 3064

The goal of the Summer Internship Program in Biomedical Research at NHLBI is to expose students to research investigation in a highly enriched environment that is devoted exclusively to biomedical research and training. The Program is open to high school and college, graduate or Medical and Dental Students. Participants join a research laboratory for a minimum of ten weeks between June and August and conduct research in selected areas of investigation under the guidance of an NHLBI intramural research scientist. To augment the training experience, students participate in a Seminar Series at the NHLBI, as well as the National Institutes of Health Summer Seminar Series and Summer Research Program Poster Day. While applications may be filed between November 15 and March 1, early applications are encouraged.
Requirements: Applicants must be either U.S. citizens or permanent residents and be currently enrolled full-time in high school, college, professional or graduate school with a minimum GPA of 3.0 on a 4.0 scale. College graduates are eligible for the summer between graduation and beginning either graduate or professional school. High school students should be at least 16 years of age prior to June 1. Applications must be submitted online using the link for provided for Summer Internship Applicants.
Geographic Focus: All States
Date(s) Application is Due: Mar 1
Contact: Aurora Taylor, Summer Program Coordinator; (301) 451-9440; fax (301) 594-8133; taylora2@nhlbi.nih.gov or direducation@nhlbi.nih.gov
Internet: http://www.nhlbi.nih.gov/research/intramural/education/summer.htm
Sponsor: National Institutes of Health
Building 31, Room 5A48
Bethesda, MD 20892

NHLBI Targeted Approaches to Weight Control for Young Adults Grants 3065

The National Heart, Lung, and Blood Institute (NHLBI) of the National Institutes of Health (NIH) invites applications for cooperative agreements (U01) to conduct two-phase clinical research studies to develop, refine, and test innovative behavioral

and/or environmental approaches for weight control in young adults at high risk for weight gain. Weight control interventions can address weight loss, prevention of weight gain, or prevention of excessive weight gain during pregnancy. The first phase will consist of formative research to refine the proposed intervention, recruitment, retention, and adherence strategies targeted to young adults. The second phase will consist of a randomized controlled trial to test the efficacy of the intervention. For the purpose of this FOA, young adults are defined as 18-35 years of age. The maximum award for the research component will be approximately $3,000,000 in direct costs, including direct costs of up to $500,000 for the first year, $650,000 for the second year, $650,000 for the third year, $650,000 for the fourth year, and $550,000 for the fifth year. Letters of Intent should be received by September 10, and full proposals are due by October 10.
Requirements: The following organizations and institutions are eligible to apply: public/state controlled institutions of higher education; private institutions of higher education; Hispanic-serving institutions; historically Black colleges and universities (HBCUs); tribally controlled colleges and universities (TCCUs); Alaska Native and Native Hawaiian serving institutions; nonprofits with 501(c)3 IRS status; nonprofits without 501(c)3 IRS status; for-profit organizations; Indian/Native American tribal governments; Indian/Native American Tribally Designated Organizations; eligible agencies of the federal government; faith-based or community-based organizations; and regional organizations.
Geographic Focus: All States
Date(s) Application is Due: Sep 10; Oct 10
Contact: Catherine (Cay) Loria, (301) 435-0702; loriac@nhlbi.nih.gov
Internet: http://grants.nih.gov/grants/guide/rfa-files/RFA-HL-08-007.html
Sponsor: National Institutes of Health
Building 31, Room 5A48
Bethesda, MD 20892

NHLBI Targeting Calcium Regulatory Molecules for Arrhythmia Prevention Grants 3066

The purpose of this Funding Opportunity Announcement (FOA) is to stimulate the development of therapeutic strategies for cardiac arrhythmias that target Ca2+ handling and Ca2+ signaling molecules. The overall goal is to exploit present knowledge of Ca2+ regulatory molecules and the role they play in arrhythmogenesis to develop novel therapeutic targets. Critical components of successful responses to this program announcement are: preclinical or early clinical testing of promising new therapeutic strategies to terminate and prevent cardiac arrhythmias in established models of human arrhythmogenic disease; and the feasibility to move positive findings to Phase I or II clinical trials. Grantees must budget for and attend two Project Director/Principal Investigator (PD/PI) meetings in Bethesda, MD, usually scheduled during the second and fourth year of the award.
Requirements: The following organizations and institutions are eligible to apply: public/state controlled institutions of higher education; private institutions of higher education; Hispanic-serving institutions; historically Black colleges and universities (HBCUs); tribally controlled colleges and universities (TCCUs); Alaska Native and Native Hawaiian serving institutions; nonprofits with 501(c)3 IRS status; nonprofits without 501(c)3 IRS status; for-profit organizations; state governments; Indian/Native American tribal governments (federally recognized); Indian/Native American tribally designated organizations; county governments; city or township governments; special district governments; independent school districts; public housing authorities/Indian housing authorities; U.S. territory or possession; Indian/Native American tribal governments (other than federally recognized); regional organizations; non-domestic (non-U.S.) entities (foreign organizations); other eligible agencies of the federal government; and faith-based or community-based organizations.
Geographic Focus: All States
Date(s) Application is Due: Feb 5; Jun 5; Oct 5
Contact: Dennis A. Przywara; (301) 435-0506; przywarad@nhlbi.nih.gov
Internet: http://grants.nih.gov/grants/guide/pa-files/PA-07-101.html
Sponsor: National Institutes of Health
Building 31, Room 5A48
Bethesda, MD 20892

NHLBI Training Program Grants for Institutions that Promote Diversity 3067

The purpose of this program is to increase the participation of individuals from diverse backgrounds in cardiovascular, pulmonary, hematologic and sleep disorders research across the career development continuum. The NHLBI's T32 Training Program for Institutions That Promote Diversity is a Ruth L. Kirschstein National Research Service Award Program intended to support training of predoctoral and health professional students and individuals in postdoctoral training at non-research intensive institutions with an institutional mission focused on serving diverse communities that are not well represented in NIH-funded research, or identified federal legislation of same, with the potential to develop meritorious training programs in cardiovascular, pulmonary, hematologic, and sleep disorders. The Training Program is designed to expand the capability for biomedical research by providing grant support to institutions that have developed successful programs that promote diversity and that offer doctoral degrees in the health professions or in health-related sciences. These institutions are uniquely positioned to engage health disparities populations in research and translation of research advances that impact health outcomes, as well as provide health care for these populations. The primary goals of the Training Program are to: (1) contribute to the expansion of the future pool of individuals from diverse backgrounds underrepresented in the biomedicine science enterprise, (2) enable trainees to increase their competitiveness for peer-review research funding, (3) strengthen publication record of trainees, and (4) foster institutional environments conducive to professional development in the biomedical science. The application is due by 5:00 pm local time of the applicant organization.
Requirements: The following organizations/institutions are eligible to apply: Public/State Controlled Institutions of Higher Education and Private Institutions of Higher Education. The following types of Higher Education Institutions are always encouraged to apply for NIH support as Public or Private Institutions of Higher Education: Hispanic-serving Institutions; Historically Black Colleges and Universities (HBCUs); Tribally Controlled Colleges and Universities (TCCUs); Alaska Native and Native Hawaiian Serving Institutions. The Institution must be a domestic college or university with an institutional mission focused on serving diverse communities or with federal recognition of the same. To demonstrate need for research capacity development in heart, lung, and blood diseases and sleep disorders, applicant institutions should not have received over $1,515,000 in funding (direct costs) from the NHLBI per year in each of the previous five years for research training and education grants in the above NHLBI mission areas. Applicant organizations may submit more than one application, provided that each application is programmatically distinct. Any individual(s) with the skills, knowledge, and resources necessary to carry out the proposed research as the Program Director/Principal Investigator (PD/PI) is invited to work with his/her organization to develop an application for support. Individuals from underrepresented racial and ethnic groups as well as individuals with disabilities are always encouraged to apply for NIH support. The PD/PI should be an established investigator in the scientific area in which the application is targeted and capable of providing both administrative and scientific leadership to the development and implementation of the proposed program. The PD/PI will be expected to monitor and assess the program and submit all documents and reports as required. Although a letter of intent is not required, is not binding, and does not enter into the review of a subsequent application, the information that it contains allows IC staff to estimate the potential review workload and plan the review.
Restrictions: Non-domestic (non-U.S.) Entities (Foreign Organizations) are not eligible to apply. Foreign (non-U.S.) components of U.S. Organizations are not allowed.
Geographic Focus: All States, Guam, Marshall Islands, Northern Mariana Islands, Puerto Rico, U.S. Virgin Islands
Date(s) Application is Due: Feb 7
Amount of Grant: Up to 250,000 USD
Contact: Sandra Colombini Hatch, M.D., (301) 435-0222; hatchs@nhlbi.nih.gov
Internet: http://grants.nih.gov/grants/guide/rfa-files/RFA-HL-12-032.html
Sponsor: National Institutes of Health
Building 31, Room 5A48
Bethesda, MD 20892

NHLBI Translating Basic Behavioral and Social Science Discoveries into Interventions to Reduce Obesity: Centers for Behavioral Intervention Development Grants 3068

This FOA solicits cooperative agreement (U01) applications from institutions/organizations that propose to translate findings from basic research on human behavior into more effective clinical, community, and population interventions to reduce obesity and improve obesity-related behaviors. This FOA will support Centers for Behavioral Intervention Development (CBIDs) in which interdisciplinary teams of basic and applied behavioral and social science researchers develop and refine novel interventions to reduce obesity and alter obesity-related health behaviors (e.g., diet, physical activity). The interventions to be developed include any of a wide range of strategies aimed at promoting weight loss and/or preventing weight gain. Interventions can be targeted to the individual, family, social network, community, environmental, clinical or population level, or combinations of these, and to any age group. Investigators at each CBID will conduct several types of studies during the funding period, including laboratory experimental studies, formative research, early phase trials and pilot/feasibility studies, in order to create promising new avenues for reducing obesity and improving obesity-related behaviors. The maximum award for the research component will be approximately $3,750,000 in direct costs across five years, with a limit of $750,000 in direct cost for any single year. Letters of Intent should be received by December 16, and full proposals are due by January 13.
Requirements: The following organizations and institutions are eligible to apply: public/state controlled institutions of higher education; private institutions of higher education; Hispanic-serving institutions; historically Black colleges and universities (HBCUs); tribally controlled colleges and universities (TCCUs); Alaska Native and Native Hawaiian serving institutions; nonprofits with 501(c)3 IRS status; nonprofits without 501(c)3 IRS status; for-profit organizations; state governments; Indian/Native American tribal governments (federally recognized); Indian/Native American tribally designated organizations; county governments; city or township governments; special district governments; independent school districts; public housing authorities/Indian housing authorities; U.S. territory or possession; Indian/Native American tribal governments (other than federally recognized); regional organizations; non-domestic (non-U.S.) entities (foreign organizations); other eligible agencies of the federal government; and faith-based or community-based organizations.
Geographic Focus: All States
Date(s) Application is Due: Jan 13; Dec 16
Amount of Grant: Up to 750,000 USD
Contact: Susan M. Czajkowski, (301) 435-0406; czajkows@mail.nih.gov
Internet: http://grants.nih.gov/grants/guide/rfa-files/RFA-HL-08-013.html
Sponsor: National Institutes of Health
Building 31, Room 5A48
Bethesda, MD 20892

NHLBI Translational Programs in Lung Diseases 3069

The purpose of this program is to encourage interdisciplinary approaches to lung translational research, allowing for potentially high-impact research, which will translate basic observations to applied clinical research. An application to NHLBI requires a minimum of three interrelated research projects that focus a number of scientific disciplines on investigations of a complex biomedical theme or research question. Applicants will be required to propose research that will become progressively more clinical and translational during the life of the program project, and they are expected to assemble interactive, multidisciplinary teams that have the combined expertise to formulate a plan for successful translation. Research teams are not required to have prior collaborative experience but must be able to demonstrate an integrated, practical approach that will result in the effective progression of mechanistic/basic concepts toward application in the clinic. This program allows for a maximum of two 5-year awards (a new application and one competing renewal). There are two application cycles. Letters of Intent should be received by April 25 and August 25 (respectively), and full proposals are due by May 25 and September 25 (respectively).
Requirements: The following organizations and institutions are eligible to apply: public/state controlled institutions of higher education; private institutions of higher education; Hispanic-serving institutions; historically Black colleges and universities (HBCUs); tribally controlled colleges and universities (TCCUs); Alaska Native and Native Hawaiian serving institutions; nonprofits with 501(c)3 IRS status; nonprofits without 501(c)3 IRS status; for-profit organizations; state governments; Indian/Native American tribal governments (federally recognized); Indian/Native American tribally designated organizations; county governments; city or township governments; special district governments; independent school districts; public housing authorities/Indian housing authorities; U.S. territories or possessions; Indian/Native American tribal governments (other than federally recognized); regional organizations; non-domestic (non-U.S.) entities (foreign organizations); other eligible agencies of the federal government; and faith-based or community-based organizations.
Geographic Focus: All States
Date(s) Application is Due: Apr 25; May 25; Aug 25; Sep 25
Contact: Patricia Noel, (301) 435-0202; noelp@nhlbi.nih.gov
Internet: http://grants.nih.gov/grants/guide/pa-files/PAR-09-185.html
Sponsor: National Institutes of Health
Building 31, Room 5A48
Bethesda, MD 20892

NHPRC Archives-Basic Projects Grants **3070**
The National Historical Publications and Records Commission seeks proposals for archives and records projects to carry out fundamental archival activities that promote the preservation and use of America's documentary heritage essential to understanding our democracy, history, and culture. The Commission is particularly concerned that some historical collections in archives and repositories are difficult for most researchers to find because they are not processed or represented in national catalogs. It hopes to encourage repositories to reveal these hidden collections by concentrating on materials in their backlogs. In addition, the NHPRC wishes to continue support for planning for archives, collection development, and appropriate preservation. As a result, the Commission seeks proposals for projects that support backlog processing and collections cataloging, phased preservation, and/or collections development. A grant normally is for one or two years and for up to $150,000. The Commission expects to make up to 20 grants in this category for a total of up to $1,000,000.
Requirements: Although projects focused exclusively on professional development are not eligible in this grant category, the Commission strongly encourages each application to include professional development components necessary for the success of the project.
Geographic Focus: All States
Date(s) Application is Due: Oct 6
Amount of Grant: Up to 150,000 USD
Contact: Daniel A. Stokes; (202) 357-5487 or (202) 357-5010; daniel.stokes@nara.gov
Internet: http://www.archives.gov/nhprc/announcement/basic.html
Sponsor: National Historical Publications and Records Commission
700 Pennsylvania Avenue NW, Room 106
Washington, DC 20408-0001

NHPRC Archives-Detailed Processing Projects Grants **3071**
The National Historical Publications and Records Commission seeks proposals for detailed processing and preservation of collections of national significance. The collections should also have high research demand or substantial preservation challenges. Detailed Processing Projects may include reformatting records onto microfilm or other media in order to ensure their permanence in the face of high demand. Applicants may process and create detailed descriptions at the series or file level. Such descriptions will improve user access to historical records and help preserve collections. If parts of collections deserve processing to the item level, applicants must provide specific justifications for this detailed degree of work and provide estimates of the percentage of collections to be processed to the item level. A grant normally is for one to three years and ranges between $40,000-$200,000. The Commission expects to make up to 5 grants in this category, for a total of up to $500,000.
Requirements: Eligible projects include archives and other repositories of historical documents that are part of: nonprofit organizations or institutions; colleges and universities and other academic institutions; state or local government agencies; and federally-acknowledged or state-recognized Native American tribes or groups.
Geographic Focus: All States
Date(s) Application is Due: Oct 6
Amount of Grant: 40,000-200,000 USD
Contact: Daniel A. Stokes; (202) 357-5487 or (202) 357-5010; daniel.stokes@nara.gov
Internet: http://www.archives.gov/nhprc/announcement/detailed.html
Sponsor: National Historical Publications and Records Commission
700 Pennsylvania Avenue NW, Room 106
Washington, DC 20408-0001

NHPRC Digitizing Historical Records Grants **3072**
The Commission seeks proposals that use cost-effective methods to digitize nationally significant historical record collections and make the digital versions freely available online. Projects must make use of existing holdings of historical repositories and consist of entire collections or series. The materials should already be available to the public at the archives and described so that projects can re-use existing information to serve as metadata for the digitized collection. A grant normally is for 1 to 3 years and up to $150,000. The Commission expects to make up to 5 grants in this category, for a total of up to $400,000.
Requirements: Eligible applicants include: nonprofit organizations or institutions with IRS 501(c)3 tax exempt status; colleges, universities, and other academic institutions; state or local government agencies; and federally-acknowledged or state-recognized Native American tribes or groups. Cost sharing is required. It is the financial contribution the applicant pledges to the cost of a project. Cost sharing can include both direct and indirect expenses, in-kind contributions, non-Federal third-party contributions, and any income earned directly by the project. The NHPRC will provide up to 50 percent of the total project costs.
Restrictions: Projects may not use grant funds to create descriptive metadata or edited transcriptions of the digitized materials.
Geographic Focus: All States
Date(s) Application is Due: Jun 5
Amount of Grant: Up to 150,000 USD
Contact: Daniel A. Stokes; (202) 357-5487 or (202) 357-5010; daniel.stokes@nara.gov
Internet: http://www.archives.gov/nhprc/announcement/digitizing.html
Sponsor: National Historical Publications and Records Commission
700 Pennsylvania Avenue NW, Room 106
Washington, DC 20408-0001

NHPRC Electronic Records Projects Grants **3073**
The National Historical Publications and Records Commission seeks proposals that will increase the capacity of archival repositories to create electronic records archives that preserve records of enduring historical value. The NHPRC supports efforts by archivists and records managers to meet the challenges of electronic records. Projects must involve institutions that have already established archives and records management programs. A grant normally is for 1 to 3 years and up to $300,000. The Commission expects to make up to 6 grants in this category, for a total of up to $600,000. The NHPRC will provide up to 50 percent of the total project costs for Electronic Records Archives projects.
Requirements: Eligible applicants include: nonprofit organizations or institutions with IRS 501(c)3 tax exempt status; colleges, universities, and other academic institutions; state or local government agencies; and federally-acknowledged or state-recognized Native American tribes or groups. Cost sharing is required. It is the financial contribution the applicant pledges to the cost of a project. Cost sharing can include both direct and indirect expenses, in-kind contributions, non-Federal third-party contributions, and any income earned directly by the project.
Restrictions: Projects in this category cannot digitize historical records.
Geographic Focus: All States
Date(s) Application is Due: Jun 5
Amount of Grant: Up to 300,000 USD
Contact: Daniel A. Stokes; (202) 357-5487 or (202) 357-5010; daniel.stokes@nara.gov
Internet: http://www.archives.gov/nhprc/announcement/electronic.html
Sponsor: National Historical Publications and Records Commission
700 Pennsylvania Avenue NW, Room 106
Washington, DC 20408-0001

NHPRC Historical Editing Fellowships **3074**
The National Historical Publications and Records Commission seeks proposals from active NHPRC-supported publications projects to serve as a host institution for a Historical Documentary Editing Fellowship. Applicants should demonstrate the capability to provide strong post-graduate training in documentary editing, including document collection, accessioning, and control; selection; transcription; annotation; proofreading; indexing; and project management. Awards are for one year grants of $55,000 each. The basic stipend may be up to $42,000. A 25 percent allowance for fringe benefits supplements the stipend. In addition, up to $2,500 may be used by the institution to cover the costs of interviewing individual applicants and/or sending fellows to professional conferences or training institutes. Depending on the quality of proposals and the availability of funding, the Commission expects to fund one to two fellowship projects, each with a single fellow.
Requirements: Proposals may be submitted only by active NHPRC-supported historical documentary editing projects that are affiliated with: nonprofit organizations and institutions; colleges, universities, and other academic institutions; state or local government agencies; federally-acknowledged or state-recognized Native American tribes or groups.
Geographic Focus: All States
Date(s) Application is Due: Oct 6
Amount of Grant: Up to 55,000 USD
Contact: Daniel A. Stokes; (202) 357-5487 or (202) 357-5010; daniel.stokes@nara.gov
Internet: http://www.archives.gov/nhprc/announcement/fellowship.html
Sponsor: National Historical Publications and Records Commission
700 Pennsylvania Avenue NW, Room 106
Washington, DC 20408-0001

NHPRC Publishing Historical Records Grants **3075**
The National Historical Publications and Records Commission seeks proposals to publish historical records of national significance. Projects may focus on the papers of major

figures from American life or cover broad historical movements in politics, military, business, reform movements, the arts, and other aspects of the national experience. The historical value of the records and their expected usefulness to broad audiences must justify the costs of the project. Grants are awarded for collecting, describing, preserving, compiling, editing, and publishing documentary source materials. Eligible Activities Include: scholarly documentary editions in printed and bound volumes; scholarly documentary editions in online, compact disc, and other formats; image editions in online, compact disc, microfilm, and other formats; conversion of existing print and microfilm editions to electronic publications; and combinations of the above. Award amounts ordinarily range from $20,000 to $250,000 annually. The Commission expects to make as many as 35-40 grants in this category. Cost sharing can include both direct and indirect expenses, in-kind contributions, non-Federal third-party contributions, and any income earned directly by the project.
Requirements: Eligibility includes: individuals; nonprofit organizations or institutions with IRS 501(c)3 tax exempt status; colleges, universities, and other academic institutions; state or local government agencies; and federally-acknowledged or state-recognized Native American tribes or groups. Cost sharing is required.
Restrictions: The NHPRC does not fund proposals to purchase historical records; it also does not fund proposals to publish the papers of anyone who has been deceased for fewer than ten years.
Geographic Focus: All States
Date(s) Application is Due: Jun 5
Amount of Grant: 20,000-250,000 USD
Contact: Daniel A. Stokes; (202) 357-5487 or (202) 357-5010; daniel.stokes@nara.gov
Internet: http://www.archives.gov/nhprc/announcement/publishing.html
Sponsor: National Historical Publications and Records Commission
700 Pennsylvania Avenue NW, Room 106
Washington, DC 20408-0001

NHPRC State and National Archival Partnership Grants 3076
The National Historical Publications and Records Commission seeks proposals to strengthen archives and historical records programs in each of the states and build a national archival network. The Commission awards grants to State Historical Records Advisory Boards to: provide statewide archival services, including professional education, public information about records and archival programs, and other activities to implement state plans; operate grant programs for eligible archives, manuscript repositories, and other organizations within a state; collaborate on projects with other organizations to address common problems or shared opportunities within a state or among a consortium of state archives; assess the health of archival and records programs, the conditions of records, and the challenges and opportunities facing historical records keepers and users; and to implement strategies and programs to address pressing archival issues; hold meetings and public forums on statewide or national archival issues; and participate in national archival organizations. The NHPRC expects to award up to 56 grants in this program. Applicants may apply for outright basic grants of between $5,000 and $20,000 for 18 month projects. For State and National Archival Partnership Projects, the Commission will supply up to 50 percent of the total project costs.
Requirements: These grants are awarded only to State Historical Records Advisory Boards (SHRABs) in each state, or to the state agency responsible for the SHRAB, ordinarily the state archives. Another state agency, or a non-profit organization, such as a foundation or university, acting on behalf of the designated state agency may apply.
Geographic Focus: All States
Date(s) Application is Due: Mar 6
Amount of Grant: 5,000-20,000 USD
Contact: Daniel A. Stokes; (202) 357-5487 or (202) 357-5010; daniel.stokes@nara.gov
Internet: http://www.archives.gov/nhprc/announcement/partnership.html
Sponsor: National Historical Publications and Records Commission
700 Pennsylvania Avenue NW, Room 106
Washington, DC 20408-0001

NHPRC Strategies and Tools for Archives and Historical Publishing Grants 3077
The National Historical Publications and Records Commission seeks proposals to develop new strategies and tools that can improve the preservation, public discovery, or use of historical records. Projects may also focus on techniques and tools that will improve the professional performance and effectiveness of those who work with such records, such as archivists, documentary editors, and records managers. Projects concerning records may focus on methods of working with any format including born-digital records. Publishing of historical records must focus on methods of presenting archival records as primary sources. The Commission does not fund projects focused on artifacts or books. A grant normally is for one to three years. The Commission expects to make one to three grants of between $50,000 and $150,000. The total amount allocated to this category is up to $350,000 during the fiscal year.
Requirements: Eligibility includes: nonprofit organizations or institutions with IRS 501(c)3 tax exempt status; colleges, universities, and other academic institutions; state or local government agencies; and federally-acknowledged or state-recognized Native American tribes or groups. Cost sharing is required.
Geographic Focus: All States
Date(s) Application is Due: Jun 5
Amount of Grant: 50,000-150,000 USD
Contact: Daniel A. Stokes; (202) 357-5487 or (202) 357-5010; daniel.stokes@nara.gov
Internet: http://www.archives.gov/nhprc/announcement/strategies.html
Sponsor: National Historical Publications and Records Commission
700 Pennsylvania Avenue NW, Room 106
Washington, DC 20408-0001

NIAAA Independent Scientist Award 3078
The award is a special salary-only grant designed to provide protected time for newly independent scientists who currently have non-research obligations such as heavy teaching loads, clinical work, committee assignments, service, and administrative duties that prevent them from having a period of intensive research focus. The award is targeted to persons with doctoral degrees who have completed their research training, have independent peer-reviewed research support, and who need a period of protected research time in order to foster their research career development. New applications are due February 12, June 12, and October 12, while renewal applications are due March 12, July 12, and November 12. Up to 75% per year of base salary plus fringe benefits is available. Research support is not provided in most cases. Individuals engaged in predominantly theoretical work, such as modeling or computer simulation, may request up to $25,000 per year for research expenses.
Requirements: The following organizations and institutions are eligible to apply: public/state controlled institutions of higher education; private institutions of higher education; nonprofits with 501(c)3 IRS status; nonprofits without 501(c)3 IRS status; small businesses; for-profit organizations; State governments; U.S. territories or possessions; Indian/Native American tribal governments (Federally recognized and other than Federally recognized); Indian/Native American tribally designated organizations; Hispanic-serving institutions; historically Black colleges and universities (HBCUs); tribally controlled colleges and universities (TCCUs); Alaska Native and Native Hawaiian serving institutions; regional organizations; and faith-based or community based organizations. The candidate must have a doctoral degree and peer-reviewed, independent research support at the time the award is made. The candidate must spend a minimum of 75 percent effort conducting research during the period of the award.
Restrictions: The award is not intended for investigators who already have full time to perform research, or have substantial publication records or considerable research support indicating that they are well established in their fields. Foreign institutions are not eligible to apply.
Geographic Focus: All States
Date(s) Application is Due: Feb 12; Mar 12; Jun 12; Jul 12; Oct 12; Nov 12
Amount of Grant: Up to 100,000 USD
Contact: Michael Hilton, (301) 402-9402; mhilton@mail.nih.gov
Internet: http://grants.nih.gov/grants/guide/pa-files/PA-09-038.html
Sponsor: National Institute on Alcohol Abuse and Alcoholism
5635 Fishers Lane, MSC 9304
Bethesda, MD 20892-9304

NIAAA Mechanisms of Alcohol and Drug-Induced Pancreatitis Grants 3079
This Funding Opportunity Announcement (FOA) issued by the National Institute on Alcohol Abuse and Alcoholism (NIAAA), the National Institute of Diabetes and Digestive and Kidney Diseases (NIDDK), and the National Institute on Drug Abuse (NIDA), National Institutes of Health (NIH), solicits grant applications from institutions/organizations that propose to investigate the underlying molecular, biochemical, and cellular mechanisms by which long-term alcohol ingestion and drugs lead to the development of pancreatitis. Research is also encouraged to understand the role of various predisposing factors, including substance abuse, that make the pancreas susceptible to alcoholic injury. Understanding the mechanisms as well as the role of predisposing factors may help in developing strategies for the prevention or treatment of the disease. Because the nature and scope of the proposed research will vary from application to application, it is anticipated that the size and duration of each award will also vary.
Requirements: Application can be made by the following institutions and organizations: public/state controlled institutions of higher education; private institutions of higher education; nonprofits with 501(c)3 IRS status (other than institutions of higher education); nonprofits without 501(c)3 IRS status (other than institutions of higher education); small businesses; for-profit organizations (other than small businesses); state governments; U.S. territories or possessions; Indian/Native American tribal governments (federally recognized); Indian/Native American tribal governments (other than Federally recognized); Indian/Native American tribally designated organizations; non-domestic (non-U.S.) entities (foreign organizations); Hispanic-serving institutions; historically Black colleges and universities (HBCUs); tribally controlled colleges and universities (TCCUs); Alaska Native and Native Hawaiian serving institutions; regional organization; eligible agencies of the Federal government; and faith-based or community based organizations.
Geographic Focus: All States
Date(s) Application is Due: Feb 5; Jun 5; Oct 5
Contact: Vishnudutt Purohit, (301) 443-2689; fax (301) 443-0673; vpurohit@mail.nih.gov
Internet: http://grants.nih.gov/grants/guide/pa-files/PA-07-067.html
Sponsor: National Institute on Alcohol Abuse and Alcoholism
5635 Fishers Lane, MSC 9304
Bethesda, MD 20892-9304

NIAAA Mentored Clinical Scientist Research Career Development Awards 3080
This award enables candidates holding professional degrees (e.g., M.D., D.O., D.V.M., or equivalent degrees) to undertake 3 to 5 years of special study and supervised research with the goal of becoming independent investigators. The award also allows awardees to pursue a research career development program suited to their experience and capabilities under a mentor who is competent to provide guidance in the chosen research area. Institutions may submit applications on behalf of candidates who hold professional degrees. At least 2 years must have elapsed since the health professional degree was granted. Candidates can have varying levels of clinical training and research experience. New applications are due February 12, June 12, and October 12, while resubmitted applications are due March

12, July 12, and November 12. Up to Up to 75 percent of base salary plus fringe benefits is available, and up to $50,000 per year for research support for new awards.
Requirements: All candidates must be U.S. citizens, non-citizen nationals, or legal permanent residents of the U.S. Persons with temporary or student visas are not eligible. The grantee institution must have a strong, well-established research and research training program in the chosen area, accomplished faculty in the basic and clinical sciences, and a commitment to the candidate's research development. The proposed program should include an appropriate mentor.
Geographic Focus: All States
Date(s) Application is Due: Feb 12; Mar 12; Jun 12; Jul 12; Oct 12; Nov 12
Amount of Grant: 100,000 USD
Contact: Judith A. Arroyo, (301) 402-0717; fax (301) 443-6077; jarroyo@mail.nih.gov
Internet: http://grants.nih.gov/grants/guide/pa-files/pa-06-512.html
Sponsor: National Institute on Alcohol Abuse and Alcoholism
5635 Fishers Lane, MSC 9304
Bethesda, MD 20892-9304

NIA Aging, Oxidative Stress and Cell Death Grants 3081
The purpose of this Program Announcement is to encourage the submission of applications to support research on the relationship between oxidative stress and apoptosis, and how these biological processes are involved in aging and/or change with age. Although the National Institute of General Medical Sciences is not participating in this program announcement, that Institute continues to be interested in receiving applications on basic mechanisms of programmed cell death and oxidative stress when these applications do not focus on aging mechanisms. Modular Grant applications will request direct costs in $25,000 modules, up to a total direct cost request of $250,000 per year.
Requirements: The following organizations and institutions may apply: public/state controlled institutions of higher education; private institutions of higher education; Hispanic-serving institutions; historically Black colleges and universities (HBCUs); tribally controlled colleges and universities (TCCUs); Alaska native and native Hawaiian serving institutions; nonprofits with 501(c)3 IRS status; nonprofits without 501(c)3 IRS status; small businesses; for-profit organizations; State governments; regional organizations; U.S. territories or possessions; Indian/Native American tribal governments; Indian/Native American tribally designated organizations; county governments; city or township governments; special district governments; independent school districts; public housing authorities/Indian housing authorities; eligible agencies of the Federal government; and faith-based or community based organizations.
Geographic Focus: All States
Date(s) Application is Due: Feb 5; Jun 5; Oct 5
Amount of Grant: Up to 250,000 USD
Contact: Chyren Hunter, (301) 496-9322; fax (301) 402-0528; Hunterc@nia.nih.gov
Internet: http://grants.nih.gov/grants/guide/pa-files/PA-00-081.html
Sponsor: National Institute on Aging
Gateway Building, Room 2C218, 7201 Wisconsin Avenue, MSC 9205
Bethesda, MD 20892-9205

NIA Aging Research Dissertation Awards to Increase Diversity 3082
This Funding Opportunity Announcement (FOA) issued by the National Institute on Aging (NIA) of the National Institutes of Health (NIH) publicizes the re-issuance of an FOA that publicizes the availability of dissertation awards in aging research to increase the diversity of the research-on-aging workforce. These awards are available to all qualified Predoctoral students in accredited research doctoral programs in the United States (including Puerto Rico and other U.S. territories or possessions).
Requirements: Public or private institutions, such as universities, colleges, hospitals, and laboratories, that offer accredited research doctoral programs may submit an application.
Restrictions: Domestic institutions only are eligible.
Geographic Focus: All States
Date(s) Application is Due: Feb 16; Jun 16; Oct 16
Amount of Grant: 50,000 USD
Contact: Michael-David, (301) 402-7713; fax (301) 402-2945; michael-david.kerns@nih.hhs.gov
Internet: http://grants.nih.gov/grants/guide/pa-files/PAR-08-250.html
Sponsor: National Institute on Aging
Gateway Building, Room 2C218, 7201 Wisconsin Avenue, MSC 9205
Bethesda, MD 20892-9205

NIA AIDS and Aging: Behavioral Sciences Prevention Research Grants 3083
NIA, NIMH, and NINR invite qualified researchers to submit applications to investigate prevention issues relevant to AIDS in middle-aged and older populations. This announcement solicits AIDS prevention research proposals to study primary prevention of disease transmission as well as secondary and tertiary prevention of negative behavioral and social consequences of HIV/AIDS for persons with AIDS, their families, and communities. Thus, the primary goals are to identify social and behavioral factors associated with HIV transmission and disease progression in later life; examine behavioral and social consequences of HIV infection/AIDS across the life course; develop and evaluate age-appropriate behavioral and social interventions for preventing AIDS in middle-aged and older adults and/or ameliorating problems associated with older adult caregiver responsibilities and burdens; explore health care issues surrounding AIDS care; and strengthen existing research and evaluation methods.
Requirements: Applications may be submitted by domestic and foreign for-profit and non-profit organizations, public and private, such as universities, colleges, hospitals, laboratories, units of State and local governments, and eligible agencies of the Federal government. Applications may be submitted by single institutions or by a consortia of institutions.
Geographic Focus: All States
Date(s) Application is Due: Feb 5; Feb 16; Jun 5; Jun 16; Oct 5; Oct 16
Contact: Ying Tian, (301) 496-6761; fax (301) 402-1784; tiany@nia.nih.gov
Internet: http://www.grants.nih.gov/grants/guide/pa-files/PA-97-069.html
Sponsor: National Institute on Aging
Gateway Building, Suite 3C307, 7201 Wisconsin Avenue, MSC 9205
Bethesda, MD 20892-9205

NIA Alzheimer's Disease Core Centers Grants 3084
The National Institute on Aging (NIA) invites applications from qualified institutions for support of Alzheimer's Disease Core Centers (ADCCs). These centers are designed to support and conduct research on Alzheimer's disease (AD), to serve as shared research resources that will facilitate research in AD and related disorders, distinguish them from the processes of normal brain aging and mild cognitive impairment (MCI), provide a platform for training, develop novel techniques and methodologies, and translate these research findings into better diagnostic, prevention and treatment strategies. New applications should request a project period of five years and a budget for direct costs of up to $750,000 per year. Competing renewal applications may request direct costs for all cores (both required and optional), and the other listed functions, i.e., satellites and pilot grants at a level not exceeding the combined direct costs of all funded activities awarded during the final year of the present funding period plus a 3% increase, or $750,000 per year, whichever is larger. Letters of Intent should arrive by the April 6 deadline, with full proposals due by May 5
Requirements: The following organizations and institutions are eligible to apply: public/state controlled institutions of higher education; private institutions of higher education; Hispanic-serving institutions; historically Black colleges and universities (HBCUs); tribally controlled colleges and universities (TCCUs); Alaska Native and Native Hawaiian serving institutions; nonprofits with 501(c)3 IRS status (other than institutions of higher education); and nonprofits without 501(c)3 IRS status (other than institutions of higher education).
Restrictions: At the time of award, the applicant institution cannot have another Alzheimer's Disease Center funded by the NIA.
Geographic Focus: All States
Date(s) Application is Due: Apr 6; May 5
Amount of Grant: 750,000 USD
Contact: Creighton H. Phelps, (301) 496-9350; fax (301) 496-1494; phelpsc@nia.nih.gov
Internet: http://grants.nih.gov/grants/guide/rfa-files/RFA-AG-10-001.html
Sponsor: National Institute on Aging
Gateway Building, Suite 350, 7201 Wisconsin Avenue, MSC 9205
Bethesda, MD 20892-9205

NIA Alzheimer's Disease Drug Development Grants 3085
The NIA and other NIH Institutes currently support research projects for the study of the epidemiology, etiology, diagnosis, and treatment of Alzheimer's disease (AD). Investigators in all relevant scientific disciplines and organizations are invited to develop and pre-clinically test compounds which can slow the disease progression, reverse the disease process, or delay the onset of or prevent AD, mild cognitive impairment (MCI), or age-related cognitive decline.
Requirements: The following organizations and institutions may apply: public/state controlled institutions of higher education; private institutions of higher education; nonprofits with 501(c)3 IRS status; nonprofits without 501(c)3 IRS status; small businesses; for-profit organizations; State governments; and non-domestic (non-U.S.) entities (foreign organizations).
Geographic Focus: All States
Date(s) Application is Due: Jan 2; Feb 2; Sep 1; Oct 0
Amount of Grant: 300,000-800,000 USD
Contact: Suzana Petanceska, (301) 496-9350; petanceskas@nia.nih.gov
Internet: http://grants.nih.gov/grants/guide/pa-files/PAR-08-266.html
Sponsor: National Institute on Aging
Gateway Building, Suite 350, 7201 Wisconsin Avenue, MSC 9205
Bethesda, MD 20892-9205

NIA Alzheimer's Disease Pilot Clinical Trials 3086
The objective of the Alzheimer's Disease Pilot Clinical Trials initiative is to improve the quality of clinical research designed to evaluate interventions for the prevention and treatment of Alzheimer's disease (AD), mild cognitive impairment (MCI), and age-associated cognitive decline by stimulating applications for pilot clinical trials to test interventions aimed at delaying the onset of or preventing AD, MCI, and age-associated cognitive decline; slowing, halting, or, if possible, reversing the progressive decline in cognitive function; and modifying the cognitive and behavioral symptoms in AD and MCI. The goal is not to duplicate or compete with pharmaceutical companies but to encourage, complement, and accelerate the process of testing new, innovative, and effective treatments. The pilot research project application should directly address how it will inform and advance the design of a subsequent full-scale clinical trial. The application should also address the intrinsic scientific merit of the pilot trial itself. The application may include the following: studies to refine the intervention strategy (e.g. drug dosage, duration, delivery system; behavioral intensity and duration); studies to define and refine the target population and ensure adequate enrollment, protocol adherence and subject retention; collection of preliminary data for establishing measures of clinical efficacy including clinical/neuropsychological/behavioral measures, neuroimaging measures, and other biological measures in blood and cerebrospinal fluid; assessment of efficacy (including primary and secondary outcome and assessment measures for both patients and caregivers); and studies of safety.
Requirements: The following organizations and institutions may apply: public/state controlled institutions of higher education; private institutions of higher education;

Hispanic-serving institutions; historically Black colleges and universities (HBCUs); tribally controlled colleges and universities (TCCUs); Alaska native and native Hawaiian serving institutions; nonprofits with 501(c)3 IRS status; nonprofits without 501(c)3 IRS status; small businesses; for-profit organizations; State governments; regional organizations; U.S. territories or possessions; Indian/Native American tribal governments; Indian/Native American tribally designated organizations; county governments; city or township governments; special district governments; independent school districts; public housing authorities/Indian housing authorities; eligible agencies of the Federal government; and faith-based or community based organizations.
Geographic Focus: All States
Date(s) Application is Due: Feb 5; Jun 5; Oct 5
Amount of Grant: 450,000-1,350,000 USD
Contact: Laurie M. Ryan, (301) 496-9350; fax (301) 496-1494; ryanl@nia.nih.gov
Internet: http://grants.nih.gov/grants/guide/pa-files/PAR-08-062.html
Sponsor: National Institute on Aging
Gateway Building, Suite 350, 7201 Wisconsin Avenue, MSC 9205
Bethesda, MD 20892-9205

NIA Alzheimer's Disease Research Centers Grants 3087
The National Institute on Aging (NIA) invites applications from qualified institutions for support of Alzheimer's Disease Research Centers (ADRCs). These centers are designed to support and conduct research on Alzheimer's disease (AD), to serve as shared research resources that will facilitate research in AD and related disorders, distinguish them from the processes of normal brain aging and mild cognitive impairment (MCI), provide a platform for training, develop novel techniques and methodologies, and translate these research findings into better diagnostic, prevention and treatment strategies. New applications should request a project period of five years and a budget for direct costs of up to $1 million per year. Letters of Intent are due by April 6 annually, with full applications received by May 5.
Requirements: The following organizations and institutions are eligible to apply: public/state controlled institutions of higher education; private institutions of higher education; Hispanic-serving institutions; historically Black colleges and universities (HBCUs); tribally controlled colleges and universities (TCCUs); Alaska native and native Hawaiian serving institutions; nonprofits with 501(c)3 IRS status (other than institutions of higher education); and nonprofits without 501(c)3 IRS status (other than institutions of higher education).
Geographic Focus: All States
Date(s) Application is Due: Apr 5; May 6
Amount of Grant: Up to 1,000,000 USD
Contact: Creighton H. Phelps, (301) 496-9350; fax (301) 496-1494; phelpsc@nia.nih.gov
Internet: http://grants.nih.gov/grants/guide/rfa-files/RFA-AG-10-002.html
Sponsor: National Institute on Aging
Gateway Building, Suite 350, 7201 Wisconsin Avenue, MSC 9205
Bethesda, MD 20892-9205

NIA Archiving and Development of Socialbehavioral Datasets in Aging Related Studies Grants 3088
This Funding Opportunity Announcement (FOA), issued by the National Institute on Aging (NIA) is seeking small grant applications to stimulate and facilitate data archiving and development related to cognitive psychology, behavioral interventions in the context of randomized controlled trials (RCTs), demography, economics, epidemiology, behavioral genetics and other behavioral research on aging for secondary analysis.
Requirements: The following organizations and institutions may apply: public/state controlled institutions of higher education; private institutions of higher education; Hispanic-serving institutions; historically Black colleges and universities (HBCUs); tribally controlled colleges and universities (TCCUs); Alaska native and native Hawaiian serving institutions; nonprofits with 501(c)3 IRS status; nonprofits without 501(c)3 IRS status; small businesses; for-profit organizations; State governments; regional organizations; U.S. territories or possessions; Indian/Native American tribal governments; Indian/Native American tribally designated organizations; county governments; city or township governments; special district governments; independent school districts; public housing authorities/Indian housing authorities; eligible agencies of the Federal government; and faith-based or community based organizations.
Geographic Focus: All States
Date(s) Application is Due: Feb 16; Jun 16; Oct 16
Amount of Grant: Up to 100,000 USD
Contact: Partha Bhattacharyya, (301) 496-3136; bhattacharyyap@mail.nih.gov
Internet: http://grants.nih.gov/grants/guide/pa-files/PA-08-252.html
Sponsor: National Institutes of Health
Building 31, Room 5A48
Bethesda, MD 20892

NIA Awards to Support Research on the Biology of Aging in Invertebrates Grants 3089
This Funding Opportunity Announcement (FOA) issued by the National Institute on Aging (NIA), National Institutes of Health, invites applications that propose the identification and development of new invertebrate models for use in biomedical research that is highly relevant to the biology of aging. Applications should focus on the identification, development and characterization of new invertebrate models that have short lifespans. Invertebrate models with tractable genetics and genome sequence available or in progress, negligible or induced senescence, and tissue regeneration in adults are of particular interest to NIA. Examples of invertebrate models of interest include, but are not limited to planaria, hydra, rotifers, and tunicates. These invertebrate models are currently not well developed for aging research. Within the scope of this FOA, invertebrate models other than those mentioned above currently used by the biomedical research community, but not yet fully developed or realized for aging research, will be considered responsive to this FOA. In particular, applications that focus on the characterization of these novel models for aging research utilizing molecular, genetic, cellular or physiological tools and approaches to identify genes involved in longevity, senescence and cellular pathways in tissue homeostasis during aging are encouraged. Letters of Intent are due by September 30, with final applications due by October 30.
Requirements: The following organizations and institutions are eligible to apply for support: public/state controlled institutions of higher education; private institutions of higher education; Hispanic-serving institutions; historically Black colleges and universities (HBCUs); tribally controlled colleges and universities (TCCUs); Alaska Native and Native Hawaiian serving institutions; nonprofits with 501(c)3 IRS status (other than institutions of higher education); nonprofits without 501(c)3 IRS status (other than institutions of higher education); Indian/Native American tribal governments (federally recognized); Indian/Native American tribally designated organizations; U.S. territories or possessions; and eligible agencies of the federal government.
Geographic Focus: All States
Date(s) Application is Due: Sep 30; Oct 30
Amount of Grant: Up to 250,000 USD
Contact: Anna M. McCormick, Ph.D; (301) 496-6402; mccormia@nia.nih.gov
Internet: http://grants.nih.gov/grants/guide/rfa-files/RFA-AG-10-004.html
Sponsor: National Institute on Aging
Gateway Building, Suite 2C231, 7201 Wisconsin Avenue, MSC 9205
Bethesda, MD 20892

NIA Behavioral and Social Research Grants on Disasters and Health 3090
The purpose of this FOA is to stimulate research in the behavioral and social sciences on the consequences of natural and man-made disasters for the health of children, the elderly, and vulnerable groups, with an ultimate goal of preventing and mitigating harmful consequences and health disparities. Disasters include severe weather-related events, earthquakes, large-scale attacks on civilian populations, technological catastrophes or perceived catastrophes, and influenza pandemics. For the elderly and for children and youth, the health outcomes of greatest interest include mortality, disability and resilience, severe distress and clinically significant morbidity (as opposed to mild or transient symptoms and dysphoria), and economic hardship sufficient to harm health. For children and youth, long-term effects on development are also of interest.
Requirements: The following organizations and institutions may apply: for-profit organizations; non-profit organizations; public or private institutions, such as universities, colleges, hospitals, and laboratories; units of State government; units of Local government; units of State Tribal government; units of Local Tribal government; eligible agencies of the Federal government; foreign institutions; domestic institutions; faith-based or community-based organizations; Indian/Native American tribal governments (federally recognized); Indian/Native American tribal governments (other than federally recognized); and Indian/Native American tribally designated organizations.
Geographic Focus: All States
Date(s) Application is Due: Feb 16; Jun 16; Oct 16
Contact: Rebecca L. Clark, (301) 496-1175; rclark@mail.nih.gov
Internet: http://grants.nih.gov/grants/guide/pa-files/PA-06-452.html
Sponsor: National Institute on Aging
Gateway Building, Room 533, 7201 Wisconsin Avenue, MSC 9205
Bethesda, MD 20892

NIA Bioenergetics, Fatigability, and Activity Limitations in Aging 3091
This FOA encourages Research Project Grant applications proposing to study bioenergetic factors underlying increased fatigability and activity limitations in aging. Increased fatigability is a significant cause of restricted physical and cognitive activity in older adults. Alterations in bioenergetics-the production and utilization of energy, and the regulation of these processes-may contribute significantly to increased fatigability. This FOA encourages applications that propose to: elucidate specific alterations in bioenergetics related to increased fatigability and activity limitations; develop and evaluate improved measures of fatigability related to bioenergetics; and evaluate interventions for increased fatigability and activity limitations that target alterations in bioenergetics and lead to improved quality of life.
Requirements: The following organizations and institutions may apply: public/state controlled institutions of higher education; private institutions of higher education; Hispanic-serving institutions; historically Black colleges and universities (HBCUs); tribally controlled colleges and universities (TCCUs); Alaska native and native Hawaiian serving institutions; nonprofits with 501(c)3 IRS status; nonprofits without 501(c)3 IRS status; small businesses; for-profit organizations; State governments; regional organizations; U.S. territories or possessions; Indian/Native American tribal governments; Indian/Native American tribally designated organizations; county governments; city or township governments; special district governments; independent school districts; public housing authorities/Indian housing authorities; eligible agencies of the Federal government; and faith-based or community based organizations.
Geographic Focus: All States
Date(s) Application is Due: Feb 5; Feb 16; Jun 5; Jun 16; Oct 5; Oct 16
Contact: Cheryl L. McDonald, (301) 435-0560; mcdonalc@nhlbi.nih.gov
Internet: http://grants.nih.gov/grants/guide/pa-files/PA-09-190.html
Sponsor: National Institute on Aging
Building 31, Room 5C27, 31 Center Drive, MSC 2292
Bethesda, MD 20892-7710

NIA Diversity in Medication Use and Outcomes in Aging Populations Grants 3092

The National Institute on Aging (NIA) in collaboration with the National Center for Complementary and Alternative Medicine, the Office of Research on Minority Health and the Office of Research on Women's Health invites qualified researchers to submit applications to investigate issues relevant to medication use and outcomes among older people. This program announcement solicits both basic and applied research proposals to investigate the multiple factors that influence medication use by the elderly. Goals of this announcement are to: identify the epidemiological, psychosocial, health care and clinical factors associated with medication use by older people, including the use of unconventional medical products; assess social, behavioral, psychological and cognitive factors that play a role in older people's understanding of and adherence to medication regimens; determine the role of medical and pharmaceutical professionals in facilitating or hindering proper use of prescribed and over-the-counter medications and unconventional medications including herbs and dietary supplements; refine methodologies to link these factors to medical or psychosocial outcomes; investigate interventions to improve medication adherence; and increase our understanding of biological factors which contribute to therapeutic outcomes in the use of medications by the elderly.
Requirements: Applications may be submitted by foreign and domestic for-profit and non-profit organizations, public and private, such as universities, colleges, hospitals, laboratories, units of state and local governments, and eligible agencies of the Federal government.
Geographic Focus: All States
Contact: Laurie M. Ryan, (301) 496-9350; fax (301) 496-1494; ryanl@nia.nih.gov
Internet: http://grants.nih.gov/grants/guide/pa-files/PA-99-097.html
Sponsor: National Institute on Aging
Gateway Building, Suite 350, 7201 Wisconsin Avenue, MSC 9205
Bethesda, MD 20892-9205

NIA Effects of Gene-Social Environment Interplay on Health and Behavior in Later Life Grants 3093

The National Institute on Aging (NIA) invites applications for the development of multidisciplinary collaborations among existing longitudinal twin and family studies, with a focus on social and behavioral factors associated with aging outcomes. This FOA is intended to lay the foundation for future studies of the role of gene-environment interplay in accounting for links between social experiences and physical health, functionality, and psychological well-being in midlife and older age. Of particular interest are applications that can embed this foundation within a lifespan perspective. Applicants are encouraged to use this FOA as a starting point for: developing collaborations among existing data sets; harmonizing social phenotypes across a number of twin and family studies; conducting analyses aimed at further refining social phenotypes for genetic analysis; using animal models of gene-social environment interplay to inform work using twin studies; and/or establishing preliminary data on the feasibility of genetic/genomic approaches that could be applied to questions of how social experiences are biologically embedded and influence middle and later life outcomes. Use of existing data is strongly encouraged; funds from this award are not intended for new data collection, but a small percentage of funds may be used for adding modules to ongoing data collection efforts to augment phenotypic or genetic data. NIA expects to make 1-2 awards totaling $750,000 total costs in the first year, and $3,750,000 in total costs over the 5 year project period. Direct costs are expected to range from $200,000 -$500,000 in the first year. Letters of Intent are due by October 9, with final applications due by November 9.
Requirements: The following organizations and institutions may apply: public/state controlled institutions of higher education; private institutions of higher education; Hispanic-serving institutions; historically Black colleges and universities (HBCUs); tribally controlled colleges and universities (TCCUs); Alaska native and native Hawaiian serving institutions; nonprofits with 501(c)3 IRS status; nonprofits without 501(c)3 IRS status; small businesses; for-profit organizations; State governments; regional organizations; U.S. territories or possessions; Indian/Native American tribal governments; Indian/Native American tribally designated organizations; county governments; city or township governments; special district governments; independent school districts; public housing authorities/Indian housing authorities; eligible agencies of the Federal government; and faith-based or community based organizations.
Geographic Focus: All States
Date(s) Application is Due: Oct 9; Nov 9
Contact: Erica L. Spotts, (301) 451-4503; fax (301) 402-0051; spottse@mail.nih.gov
Internet: http://grants.nih.gov/grants/guide/rfa-files/RFA-AG-10-006.html
Sponsor: National Institute on Aging
Gateway Building, Room 533, 7201 Wisconsin Avenue, MSC 9205
Bethesda, MD 20892

NIA Factors Affecting Cognitive Function in Adults with Down Syndrome Grants 3094

The purpose of this solicitation is to encourage investigator initiated research to focus on various factors that maximize and maintain cognitive function in adults (>21 years of age) with Down syndrome. This solicitation encourages applications from investigators from very diverse backgrounds and interdisciplinary and transdisciplinary proposals, as well as those that involve systematic longitudinal study rather than study of samples of convenience. These studies could serve to form an evidence base that will inform not only health care management and treatment, but also issues such as quality of life, access to services, education, and ultimately affect public policy. This solicitation encourages applications that would inform several areas of inquiry: the natural history of medical and mental health issues in adults with Down syndrome, their incidence and onset; health disparities that occur, and their prevention or effective treatment; the development of better diagnostic tools to assess functional and cognitive impairment in adults with Down syndrome; improved quality of life measures; community-based interventions that improve individual educational, occupational, and social outcomes; the impact of early entry to services, type of intervention, and/or family history on quality of life for adults with Down Syndrome; and the impact of family structure and ancestry on critical life transitions, co-occurring conditions, and developmental changes of adults with Down syndrome through longitudinal studies rather than studies of samples of convenience. The requested budget is limited to $500,000 total costs per year for a maximum of five years. Letters of Intent are due by May 29, and full applications should arrive by June 30.
Requirements: The following organizations and institutions are eligible to apply: public/state controlled institutions of higher education; private institutions of higher education; nonprofits with 501(c)3 IRS status (other than institutions of higher education); nonprofits without 501(c)3 IRS status (other than institutions of higher education); for-profit organizations (other than small businesses); State governments; county governments; eligible agencies of the Federal government; and faith-based or community-based organizations.
Restrictions: Foreign institutions may not submit applications, but collaborators at foreign institutions are allowed.
Geographic Focus: All States
Date(s) Application is Due: May 29; Jun 30
Contact: Mary Lou Oster-Granite, (301) 435-6866; fax (301) 496-3791; mo96o@nih.gov
Internet: http://grants.nih.gov/grants/guide/rfa-files/RFA-HD-09-028.html
Sponsor: Eunice Kennedy Shriver National Institute of Child Health and Human Development
6100 Executive Boulevard, Room 4B05L, MSC 7510
Bethesda, MD 20892-7510

NIAF Fellowship Grants 3095

The National Italian American Foundation's mission is to promote, preserve and protect Italian American heritage and culture. The NIAF supports grants to organizations, individuals, schools, and communities that provide opportunities to encourage collaboration among students and academics in the United States and Italy and improve the quality of Italian language teaching in the United States. This includes NIAF's existing Teacher Fellowship Program, the Fulbright Fellowship at Georgetown University, the Mastrobuono Academic Fellowship and any other academic fellowship program either in the U.S. or Italy.
Geographic Focus: All States
Date(s) Application is Due: Jun 30
Contact: Alexandra Dall, Scholarships and Grants Manager; (202) 939-3118 or (202) 387-0600; fax (202) 387-0800; adall@niaf.org or grants@niaf.org
Internet: http://www.niaf.org/grants/About.asp
Sponsor: National Italian American Foundation
1860 19th Street NW
Washington, DC 20009-5501

NIA Grants for Alzheimer's Disease Drug Discovery 3096

The objective of this solicitation is to stimulate preclinical research in the discovery, design, development and testing of novel compounds aimed at slowing, halting, or, if possible, reversing the progressive decline in cognitive function and modifying the behavioral symptoms in Alzheimer's disease as well as delaying the onset of or preventing AD. This initiative is intended to stimulate basic research and development efforts. The goal is not to duplicate or compete with pharmaceutical companies but to encourage, complement, and accelerate the process of discovering new, innovative, and effective compounds for the prevention and treatment of the cognitive impairment and behavioral symptoms associated with Alzheimer's disease. The development of compounds for ameliorating, modifying, or improving potential aberrations in neuronal cellular communication mechanisms is encouraged. These compounds should be designed to affect fundamental processes such as the neuronal dysfunction, death, and loss of connectivity associated with the disease by targeting molecules and mechanisms such as neurotransmitters, neuromodulators, and neurotrophins; receptors and ion channels; second and third messenger systems; all facets of relevant amyloid precursor protein, amyloid beta protein, and tau neurobiology; protein synthesis, aggregation, and degradation; energy utilization; and oxidative, immunological, and inflammatory mechanisms.
Requirements: The following organizations are eligible to apply: for-profit organizations; non-profit organizations; public or private institutions, such as universities, colleges, hospitals, and laboratories; units of State government; units of local government; eligible agencies of the Federal government; foreign institutions; and domestic institutions.
Geographic Focus: All States
Date(s) Application is Due: Feb 16; Jun 16; Oct 16
Amount of Grant: Up to 275,000 USD
Contact: Neil S. Buckholtz, (301) 496-9350; fax (301) 496-1494; Buckholn@nia.nih.gov
Internet: http://grants.nih.gov/grants/guide/pa-files/PAS-06-261.html
Sponsor: National Institute on Aging
Gateway Building, Suite 350, 7201 Wisconsin Avenue, MSC 9205
Bethesda, MD 20892-9205

NIA Harmonization of Longitudinal Cross-National Surveys of Aging Grants 3097

The National Institute on Aging (NIA) will provide support through the R21 mechanism to facilitate the ex-ante and ex-post harmonization of international longitudinal surveys of aging populations to the Health and Retirement Study to maximize cross-national comparability of data. Facilitation will include, but is not limited to, meetings, pilots, and the development of methods to improve cross-national comparability of international aging surveys to the Health and Retirement Study and existing international data files that are harmonized with the HRS: the English Longitudinal Study on Ageing (ELSA); and the Survey of Health, Ageing, and Retirement in Europe (SHARE).
Requirements: The following organizations and institutions may apply: public/state controlled institutions of higher education; private institutions of higher education; nonprofits with 501(c)3 IRS status; nonprofits without 501(c)3 IRS status; for-profit

organizations; State governments; and non-domestic (non-U.S.) entities (foreign organizations).
Geographic Focus: All States
Date(s) Application is Due: Feb 16; Jun 16; Oct 16
Amount of Grant: Up to 275,000 USD
Contact: John W. R. Phillips, (301) 496-3138; fax (301) 402-0051; phillipj@mail.nih.gov
Internet: http://grants.nih.gov/grants/guide/pa-files/PAS-07-387.html
Sponsor: National Institute on Aging
Gateway Building, Room 533, 7201 Wisconsin Avenue, MSC 9205
Bethesda, MD 20892

NIA Health Behaviors and Aging Grants **3098**
The National Institute on Aging (NIA) invites qualified researchers to submit applications for research and research training on those health-related behaviors and attitudes of older adults, their families, and significant others, that can affect health and functioning as people grow older. Studies are sought that extend scientific understanding of how older adults health behaviors and attitudes develop under varying social conditions; how they relate to health promotion and disease prevention, care and treatment of disease, rehabilitation or death; and how they can be modified as relevant new scientific knowledge is developed.
Requirements: Applications may be submitted by foreign and domestic, for-profit and non-profit organizations, public and private, such as universities, colleges, hospitals, laboratories, units of State and local governments, and eligible agencies of the Federal government. Applications from minority individuals and women are encouraged.
Geographic Focus: All States
Date(s) Application is Due: Feb 5; Jun 5; Oct 5
Amount of Grant: Up to 150,000 USD
Contact: Chyren Hunter, (301) 496-9322; fax (301) 402-0528; Hunterc@nia.nih.gov
Internet: http://grants.nih.gov/grants/guide/pa-files/PA-93-064.html
Sponsor: National Institute on Aging
Gateway Building, Room 2C218, 7201 Wisconsin Avenue, MSC 9205
Bethesda, MD 20892-9205

NIA Healthy Aging through Behavioral Economic Analyses of Situations Grants **3099**
The National Institute on Aging (NIA), National Institutes of Health, solicits Research Project Grant applications that propose to translate basic findings from Behavioral Economics into behavior change interventions targeting health behaviors associated with chronic health conditions of mid-life and older age. Applications should propose small pilot clinical trials or demonstration projects, ideally based on collaborations between individuals with expertise in behavioral economics and psychologists, psychiatrists, clinicians, or others with expertise in aging or implementing behavioral interventions. Depending on the availability of funding, the NIA expects to make 4 or 5 awards totaling $1,250,000 total costs in the first year, and $3,750,000 over the 3 year project period. The direct cost amount for each individual award will be no greater than $525,000 in total for three years, and no more than $175,000 in any year. Letters of Intent are due by October 2, with final applications due by November 2.
Requirements: The following organizations and institutions may apply: public/state controlled institutions of higher education; private institutions of higher education; Hispanic-serving institutions; historically Black colleges and universities (HBCUs); tribally controlled colleges and universities (TCCUs); Alaska native and native Hawaiian serving institutions; nonprofits with 501(c)3 IRS status; nonprofits without 501(c)3 IRS status; small businesses; for-profit organizations; State governments; regional organizations; U.S. territories or possessions; Indian/Native American tribal governments; Indian/Native American tribally designated organizations; county governments; city or township governments; special district governments; independent school districts; public housing authorities/Indian housing authorities; eligible agencies of the Federal government; and faith-based or community based organizations.
Geographic Focus: All States
Date(s) Application is Due: Oct 2; Nov 2
Amount of Grant: Up to 175,000 USD
Contact: Jonathan W. King, (301) 402-4156; fax (301) 402-0051; kingjo@nia.nih.gov
Internet: http://grants.nih.gov/grants/guide/rfa-files/RFA-AG-10-008.html
Sponsor: National Institute on Aging
Gateway Building, Room 533, 7201 Wisconsin Avenue, MSC 9205
Bethesda, MD 20892

NIA Higher-Order Cognitive Functioning and Aging Grants **3100**
The National Institute on Aging (NIA) invites qualified researchers to submit new applications for research projects that focus on adulthood and aging related changes in the higher-order processes and strategies required for judgment, decision-making, reasoning, problem-solving, and processing complex information. As in earlier periods of life older adults continue to make decisions related to everyday life, but with advanced age, new, and sometimes even more complex, decision-making is required of them. Recent research indicates that age-related limitations in cognitive processing resources (e.g., speed and working memory) may impact decision-making. Research also indicates that some older adults experience growth in specific areas of cognitive functioning (e.g., expertise, semantic knowledge, emotional regulation) and continue to use adaptive intelligence, demonstrating multi-directionality in adult cognitive change. It is generally recognized that research on higher-order processing is underdeveloped in the field of aging. Research proposals are needed that examine the actual processes that are engaged when older adults make important decisions, how these processes change with age and context, and what environmental supports, interventions, and training may be necessary for optimal functioning. Research may investigate either individual or collaborative, or social processes. Modular Grant applications will request direct costs in $25,000 modules, up to a total direct cost request of $250,000 per year.
Requirements: Applications may be submitted by foreign and domestic, for-profit and non-profit, public and private organizations such as universities, colleges, hospitals, laboratories, units of State and local governments, and eligible agencies of the Federal government. Racial/ethnic minority individuals, women, and persons with disabilities are encouraged to apply as principal investigators.
Geographic Focus: All States
Date(s) Application is Due: Feb 16; Jun 16; Oct 16
Contact: Chyren Hunter, (301) 496-9322; fax (301) 402-0528; Hunterc@nia.nih.gov
Internet: http://grants.nih.gov/grants/guide/pa-files/PA-00-052.html
Sponsor: National Institute on Aging
Gateway Building, Room 2C218, 7201 Wisconsin Avenue, MSC 9205
Bethesda, MD 20892-9205

NIA Human Biospecimen Resources for Aging Research Grants **3101**
The purpose of this announcement is to disseminate information about available human biospecimen resources for age-related studies, and promote independent use of existing human biospecimen resources through collaborative and other arrangements with the studies that collect biospecimens. Websites for the participating studies are accessed via the Virtual Repository. NIA is interested in aging-related research that integrates biomarker analyses with the study of normal aging, age-associated diseases, and behavioral and psychosocial data. Budgets for direct costs of up to $50,000 per year and a project duration of up to two years may be requested for a maximum of $100,000 direct costs over a two-year project period.
Requirements: The following organizations and institutions may apply: for-profit organizations; non-profit organizations; public or private institutions, such as universities, colleges, hospitals, and laboratories; units of State government; units of Local government; units of State Tribal government; units of Local Tribal government; eligible agencies of the Federal government; foreign institutions; domestic institutions; faith-based or community-based organizations; Indian/Native American tribal governments (federally recognized); Indian/Native American tribal governments (other than federally recognized); and Indian/Native American tribally designated organizations.
Geographic Focus: All States
Date(s) Application is Due: Feb 16; Jun 16; Oct 16
Contact: Nancy L. Nadon, (301) 402-7744; fax (301) 402-0010; nadonn@nia.nih.gov
Internet: http://grants.nih.gov/grants/guide/pa-files/PA-06-443.html
Sponsor: National Institute on Aging
Gateway Building, Suite 2C231, 7201 Wisconsin Avenue, MSC 9205
Bethesda, MD 20892

NIAID Independent Scientist Award **3102**
The award is a special salary-only grant designed to provide protected time for newly independent scientists who currently have non-research obligations such as heavy teaching loads, clinical work, committee assignments, service, and administrative duties that prevent them from having a period of intensive research focus. The award is targeted to persons with doctoral degrees who have completed their research training, have independent peer-reviewed research support, and who need a period of protected research time in order to foster their research career development. New applications are due February 12, June 12, and October 12, while renewal applications are due March 12, July 12, and November 12. Up to $75,000 plus fringe benefits per year is available. Additional research is not generally provided.
Requirements: The following organizations and institutions are eligible to apply: public/state controlled institutions of higher education; private institutions of higher education; nonprofits with 501(c)3 IRS status; nonprofits without 501(c)3 IRS status; small businesses; for-profit organizations; State governments; U.S. territories or possessions; Indian/Native American tribal governments (Federally recognized and other than Federally recognized); Indian/Native American tribally designated organizations; Hispanic-serving institutions; historically Black colleges and universities (HBCUs); tribally controlled colleges and universities (TCCUs); Alaska Native and Native Hawaiian serving institutions; regional organizations; and faith-based or community based organizations. The candidate must have a doctoral degree and peer-reviewed, independent research support at the time the award is made. The candidate must spend a minimum of 75 percent effort conducting research during the period of the award.
Restrictions: The award is not intended for investigators who already have full time to perform research, or have substantial publication records or considerable research support indicating that they are well established in their fields. Foreign institutions are not eligible to apply.
Geographic Focus: All States
Date(s) Application is Due: Feb 12; Mar 12; Jun 12; Jul 12; Oct 12; Nov 12
Contact: Milton J. Hernandez, (301) 496-3775; mh35c@nih.gov
Internet: http://grants.nih.gov/grants/guide/pa-files/PA-09-038.html
Sponsor: National Institute of Allergy and Infectious Diseases
6610 Rockledge Drive
Bethesda, MD 20892

NIA Independent Scientist Award **3103**
The award is a special salary-only grant designed to provide protected time for newly independent scientists who currently have non-research obligations such as heavy teaching loads, clinical work, committee assignments, service, and administrative duties that prevent them from having a period of intensive research focus. The award is targeted to persons with doctoral degrees who have completed their research training, have independent peer-reviewed research support, and who need a period of protected research time in order to foster their research career development. New applications are

due February 12, June 12, and October 12, while renewal applications are due March 12, July 12, and November 12. Up to 75% per year of base salary plus fringe benefits is available. No other research development support funds are provided.
Requirements: The following organizations and institutions are eligible to apply: public/state controlled institutions of higher education; private institutions of higher education; nonprofits with 501(c)3 IRS status; nonprofits without 501(c)3 IRS status; small businesses; for-profit organizations; State governments; U.S. territories or possessions; Indian/Native American tribal governments (Federally recognized and other than Federally recognized); Indian/Native American tribally designated organizations; Hispanic-serving institutions; historically Black colleges and universities (HBCUs); tribally controlled colleges and universities (TCCUs); Alaska Native and Native Hawaiian serving institutions; regional organizations; and faith-based or community based organizations. The candidate must have a doctoral degree and peer-reviewed, independent research support at the time the award is made. The candidate must spend a minimum of 75 percent effort conducting research during the period of the award.
Restrictions: The award is not intended for investigators who already have full time to perform research, or have substantial publication records or considerable research support indicating that they are well established in their fields. Foreign institutions are not eligible to apply.
Geographic Focus: All States
Date(s) Application is Due: Feb 12; Mar 12; Jun 12; Jul 12; Oct 12; Nov 12
Contact: Chyren Hunter, (301) 496-9322; fax (301) 402-0528; Hunterc@nia.nih.gov
Internet: http://grants.nih.gov/grants/guide/pa-files/PA-09-038.html
Sponsor: National Institute on Aging
Gateway Building, Room 2C218, 7201 Wisconsin Avenue, MSC 9205
Bethesda, MD 20892-9205

NIA Malnutrition in Older Persons Research Grants 3104

NIA continues to encourage research on the full spectrum of issues related to the causes, prevention, and treatment as well as sociobehavioral and economic aspects related to malnutrition in older persons. Support for this program will be through the research project grant and mentored research scientist development award. Applications will compete for available funds on the basis of scientific merit with all other applications assigned to the institute.
Requirements: Applications may be submitted by foreign and domestic, for-profit and non-profit organizations, public and private, such as universities, colleges, hospitals, laboratories, unit of State and local governments, and eligible agencies of the Federal government. Applications from minority individuals and women are encouraged.
Geographic Focus: All States
Date(s) Application is Due: Feb 12; Mar 12; Jun 12; Jul 12; Oct 12; Nov 12
Contact: Chyren Hunter, (301) 496-9322; fax (301) 402-0528; Hunterc@nia.nih.gov
Internet: http://grants.nih.gov/grants/guide/pa-files/PA-94-088.html
Sponsor: National Institute on Aging
Building 31, Room 5C27, 31 Center Drive, MSC 2292
Bethesda, MD 20892-7710

NIA Mechanisms, Measurement, and Management of Pain in Aging: from Molecular to Clinical 3105

This FOA encourages Research Project Grant applications from institutions and organizations that propose to: study biological, neurobiological, psychosocial, and clinical mechanisms and processes by which aging and/or age-related diseases affect the experience of pain; examine biological, neurobiological, psychosocial, and clinical factors that impact pain experience and prevalence in older people; evaluate existing pain assessment and/or management approaches in older adults; or develop new assessment methods and/or management strategies for pain with particular attention to the needs of older adults. Studies involving animal models or human subjects are appropriate under this program announcement.
Requirements: The following organizations and institutions may apply: public/state controlled institutions of higher education; private institutions of higher education; Hispanic-serving institutions; historically Black colleges and universities (HBCUs); tribally controlled colleges and universities (TCCUs); Alaska native and native Hawaiian serving institutions; nonprofits with 501(c)3 IRS status; nonprofits without 501(c)3 IRS status; small businesses; for-profit organizations; State governments; regional organizations; U.S. territories or possessions; Indian/Native American tribal governments; Indian/Native American tribally designated organizations; county governments; city or township governments; special district governments; independent school districts; public housing authorities/Indian housing authorities; eligible agencies of the Federal government; and faith-based or community based organizations.
Geographic Focus: All States
Date(s) Application is Due: Feb 5; Feb 16; Jun 5; Jun 16; Oct 5; Oct 16
Contact: Wen G. Chen, (301) 496-9350; fax (301) 496-1494; chenw@nia.nih.gov
Internet: http://grants.nih.gov/grants/guide/pa-files/PA-09-193.html
Sponsor: National Institute on Aging
Gateway Building, Suite 350, 7201 Wisconsin Avenue, MSC 9205
Bethesda, MD 20892-9205

NIA Mechanisms Underlying the Links between Psychosocial Stress, Aging, the Brain and the Body Grants 3106

This FOA encourages multidisciplinary and interdisciplinary research to elucidate the mechanistic links between psychosocial stress and health in aging, as well as how the aging process and age-related diseases affect the responses to psychosocial stressors. Generally, research should be focused on: aging and how neural mechanisms respond to psychosocial stress and affect other body systems; characterizing the behavioral, psychological and social mechanisms and pathways involved in transducing psychosocial stressors into health outcomes; how stressors modulate physiological process underlying life-span, immune mechanisms, and metabolism; and how psychosocial stress contributes to the development or progression of geriatric syndromes, chronic medical conditions, and disabilities in later life. Research is strongly encouraged that aims to identify appropriate targets for intervention, at any level of analysis, from societal to molecular. Research spanning multiple levels of analysis is particularly encouraged.
Requirements: The following organizations and institutions are eligible to apply: public/state controlled institutions of higher education; private institutions of higher education; Hispanic-serving institutions; historically Black colleges and universities (HBCUs); tribally controlled colleges and universities (TCCUs); Alaska native and native Hawaiian serving institutions; nonprofits with 501(c)3 IRS status (other than institutions of higher education); nonprofits without 501(c)3 IRS status (other than institutions of higher education); small businesses; for-profit organizations (other than small businesses); state governments; Indian/Native American tribal governments (federally recognized); Indian/Native American tribally designated organizations; county governments; city or township governments; special district governments; independent school districts; public housing authorities/Indian housing authorities; U.S. territories or possessions; Indian/native American tribal governments (other than federally recognized); regional organizations; and non-domestic (non-U.S.) entities (foreign organizations).
Restrictions: Research focused on oxidative stress or on environmental or physical stressors of a non-psychosocial nature is not appropriate for this FOA.
Geographic Focus: All States
Date(s) Application is Due: Mar 5; Jul 5; Nov 5
Contact: Suzana S. Petanceska, (301) 496-9350; petanceskas@mail.nih.gov
Internet: http://grants.nih.gov/grants/guide/pa-files/PA-09-216.html
Sponsor: National Institute on Aging
Gateway Building, Suite 350, 7201 Wisconsin Avenue, MSC 9205
Bethesda, MD 20892-9205

NIA Medical Management of Older Patients with HIV/AIDS Grants 3107

The National Institute on Aging (NIA), the National Institute of Allergy and Infectious Diseases (NIAID), the Center for Mental Health Research on AIDS of the National Institute of Mental Health (NIMH), and the National Institute of Nursing Research (NINR), invite Research Project Grant applications addressing clinical and translational health issues in the diagnosis and/or management of HIV infection and its consequences in older persons. The goal of this funding opportunity is to improve medical outcomes, functional status and quality of life in older patients with HIV/AIDS through improved understanding of interactions among aging processes, HIV viral infection, treatment effects and toxicities, and multiple morbidities commonly occurring in older persons.
Requirements: The following organizations and institutions may apply: public/state controlled institutions of higher education; private institutions of higher education; Hispanic-serving institutions; historically Black colleges and universities (HBCUs); tribally controlled colleges and universities (TCCUs); Alaska native and native Hawaiian serving institutions; nonprofits with 501(c)3 IRS status; nonprofits without 501(c)3 IRS status; small businesses; for-profit organizations; State governments; regional organizations; U.S. territories or possessions; Indian/Native American tribal governments; Indian/Native American tribally designated organizations; county governments; city or township governments; special district governments; independent school districts; public housing authorities/Indian housing authorities; eligible agencies of the Federal government; and faith-based or community based organizations.
Geographic Focus: All States
Date(s) Application is Due: Jan 7; May 7; Sep 7
Contact: Susan G. Nayfield, (301) 496-6761; fax (301) 402-1784; nayfiels@nia.nih.gov
Internet: http://grants.nih.gov/grants/guide/pa-files/PA-09-017.html
Sponsor: National Institute on Aging
Gateway Building, Suite 3C307, 7201 Wisconsin Avenue, MSC 9205
Bethesda, MD 20892-9205

NIA Mentored Clinical Scientist Research Career Development Awards 3108

The overall objective of the NIH Research Career Development Award program is to prepare qualified individuals for careers that have a significant impact on the health-related research needs of the Nation. The objective of the NIH Mentored Clinical Scientist Research Career Development Award (K08) program is to support didactic study and mentored research for individuals with clinical doctoral degrees (e.g., M.D., D.D.S., D.M.D., D.O., D.C., O.D., N.D., D.V.M., Pharm.D., or Ph.D. in clinical disciplines). The purpose of this Funding Opportunity Announcement (FOA) is to continue the long-standing NIH support of the K08 program. This award provides support and protected time for an intensive, mentored research career development experience in biomedical or behavioral research, including translational research. For the purpose of this award, translational research is defined as application of basic research discoveries toward the diagnosis, management, and prevention of human disease.
Requirements: The following organizations and institutions may apply: public/state controlled institutions of higher education; private institutions of higher education; Hispanic-serving institutions; historically Black colleges and universities (HBCUs); tribally controlled colleges and universities (TCCUs); Alaska native and native Hawaiian serving institutions; nonprofits with 501(c)3 IRS status; nonprofits without 501(c)3 IRS status; small businesses; for-profit organizations; State governments; regional organizations; U.S. territories or possessions; Indian/Native American tribal governments; Indian/Native American tribally designated organizations; county governments; city or township governments; special district governments; independent school districts; public housing authorities/Indian housing authorities; eligible agencies of the Federal government; and faith-based or community based organizations.
Restrictions: Individuals on temporary or student visas and organizations described in section 501(c)4 of the Internal Revenue Code that engage in lobbying are not eligible.

Geographic Focus: All States
Date(s) Application is Due: Feb 12; Mar 12; Jun 12; Jul 12; Oct 12; Nov 12
Contact: Chyren Hunter, (301) 496-9322; fax (301) 402-0528; Hunterc@nia.nih.gov
Internet: http://grants.nih.gov/grants/guide/pa-files/PA-09-042.html
Sponsor: National Institute on Aging
Gateway Building, Room 2C218, 7201 Wisconsin Avenue, MSC 9205
Bethesda, MD 20892-9205

NIA Mentored Research Scientist Development Award 3109
The award is for research scientists who have established careers in biomedical, behavioral, or social research and wish to change career direction toward aging research; for more junior researchers with training in aging research who need additional periods of mentored research experience prior to becoming fully independent; or for researchers with training and experience in some aspect of aging research who wish to gain complementary training to expand their research interests in aging. The primary purpose of the program is to provide support and protected time (three, four, or five years) for an intensive, supervised career development experience in the biomedical, behavioral, or clinical sciences leading to research independence.
Requirements: The candidate must have a research or health-professional doctorate or its equivalent and must have demonstrated the capacity or potential for highly productive independent research in the period after the doctorate. The candidate must identify a mentor with extensive research experience and must spend a minimum of 75 percent of full-time professional effort conducting research for the period of the award. The following organizations and institutions may apply: public/state controlled institutions of higher education; private institutions of higher education; Hispanic-serving institutions; historically Black colleges and universities (HBCUs); tribally controlled colleges and universities (TCCUs); Alaska native and native Hawaiian serving institutions; nonprofits with 501(c)3 IRS status; nonprofits without 501(c)3 IRS status; small businesses; for-profit organizations; State governments; regional organizations; U.S. territories or possessions; Indian/Native American tribal governments; Indian/Native American tribally designated organizations; eligible agencies of the Federal government; and faith-based or community based organizations.
Geographic Focus: All States
Date(s) Application is Due: Feb 12; Mar 12; Jun 12; Jul 12; Oct 12; Nov 12
Contact: Chyren Hunter, (301) 496-9322; fax (301) 402-0528; Hunterc@nia.nih.gov
Internet: http://grants.nih.gov/grants/guide/pa-files/PAR-09-060.html
Sponsor: National Institute on Aging
Gateway Building, Room 2C218, 7201 Wisconsin Avenue, MSC 9205
Bethesda, MD 20892-9205

NIA Minority Dissertation Research Grants in Aging 3110
Small grants to support doctoral dissertation research will be available for underrepresented minority doctoral candidates. Grant support is designed to aid the research of such investigators and to encourage underrepresented minority individuals from a variety of academic disciplines and programs to conduct research related to aging. It is anticipated that a total of five to six new applications will be funded in each year. The maximum budget request should be limited to $25,000 in direct costs for the initial budget period. No more than $30,000 in total direct costs will be provided across the two year period.
Requirements: This research initiative is to provide underrepresented minority students assistance to complete their dissertation research on an aging-related topic and thereby increase their representation in aging research.
Geographic Focus: All States
Date(s) Application is Due: Mar 15; Nov 15
Amount of Grant: Up to 30,000 USD
Contact: Laurie M. Ryan, (301) 496-9350; fax (301) 496-1494; ryanl@nia.nih.gov
Internet: http://grants.nih.gov/grants/guide/pa-files/PAR-98-110.html
Sponsor: National Institute on Aging
Gateway Building, Suite 350, 7201 Wisconsin Avenue, MSC 9205
Bethesda, MD 20892-9205

NIAMS Pilot and Feasibility Clinical Research Grants in Arthritis and Musculoskeletal and Skin Diseases 3111
The goal of this initiative is to encourage novel clinical research aimed at improving the understanding of the causes, treatment or prevention of arthritis and musculoskeletal and skin diseases. This Pilot and Feasibility Clinical Research Grants Program is designed to allow initiation of exploratory, short-term clinical studies, intended to facilitate the development of new ideas which may be investigated without stringent requirements for preliminary data. Research aims appropriate for an investigator-initiated R01 grant, such as understanding underlying biological mechanisms of disease, would not be expected in an application under this FOA. Proposed studies should focus on research questions that are likely to gather critical preliminary data in support of a future, planned clinical trial or to benefit clinical research and trials more broadly. Overall goals of these studies are intended to generate data critical to future clinical research or trials such as by demonstrating the feasibility of a recruitment target or approach, obtaining data in support of potential inclusion and exclusion criteria, or to gather preliminary evidence of efficacy, tolerability and/or toxicity of an available drug, biologic or device in a new population relevant to the NIAMS mission. Examples include testing of: new prevention strategies; a new application of existing pharmacologic or non-pharmacologic interventions or their unique combination; a new cost effective or adaptive trial design in a specific population or across populations; and studies intended to expand the science of clinical trials overall. Results generated from this program may lead to critical input into a new clinical trial; trials that inform decision-making in urgent or high priority clinical practice needs for arthritis and musculoskeletal and skin diseases; or preliminary data for trials in understudied and rare diseases. A high priority is the use of such studies to help stimulate the translation of promising research developments from the laboratory into clinical practice. The total project period for an application submitted in response to this funding opportunity may not exceed 2 years. Although the size of award may vary with the scope of research proposed, it is expected that applications will stay within the budgetary guidelines for an exploratory/developmental project; direct costs are limited to $275,000 over an R21 two-year period, with no more than $200,000 in direct costs allowed in any single year.
Requirements: The following organizations/institutions are eligible to apply: public/state controlled institutions of higher education; private institutions of higher education; Hispanic-serving institutions; historically Black colleges and universities; tribally controlled colleges and universities; Alaska native and native Hawaiian serving institutions; nonprofits with 501(c)3 IRS status (other than institutions of higher education); nonprofits without 501(c)3 IRS status (other than institutions of higher education); small businesses; for-profit organizations (other than small businesses); state governments; Indian/Native American tribal governments (federally recognized); Indian/Native American tribally designated organizations; county governments; city or township governments; special district governments; independent school districts; public housing authorities/Indian housing authorities; U.S. territory or possession; Indian/Native American tribal governments (other than federally recognized); regional organizations; non-domestic (non-U.S.) entities (foreign organizations); eligible agencies of the federal government; and faith-based or community-based organizations.
Geographic Focus: All States
Date(s) Application is Due: Mar 1; Jul 1; Nov 1
Amount of Grant: Up to 275,000 USD
Contact: James Witter, (301) 594-1963; fax (301) 480-4543; witterj@mail.nih.gov
Internet: http://grants.nih.gov/grants/guide/pa-files/PAR-10-282.html#SectionVII
Sponsor: National Institute of Arthritis and Musculoskeletal and Skin Diseases
1 AMS Circle
Bethesda, MD 20892-3675

NIA Multicenter Study on Exceptional Survival in Families: The Long Life Family Study (LLFS) Grants 3112
This limited competition FOA is to continue the Long Life Family Study (LLFS). The LLFS was designed to determine the degree and patterns of familial transmission and aggregation of exceptional longevity and healthy survival to advanced age as characterized by a variety of phenotypic measures, with a further goal of also understanding potential genetic factors that contribute to exceptional survival. Five awards, to four study centers and one data management and coordinating center, were made in response to RFA-AG-03-004.
Geographic Focus: All States
Date(s) Application is Due: Aug 30; Sep 30
Contact: Winifred K. Rossi, (301) 496-3836; fax (301) 402-1784; winnie_rossi@nih.gov
Internet: http://grants.nih.gov/grants/guide/rfa-files/RFA-AG-10-005.html
Sponsor: National Institute on Aging
Gateway Building, Suite 3C307, 7201 Wisconsin Avenue, MSC 9205
Bethesda, MD 20892-9205

NIA Nathan Shock Centers of Excellence in Basic Biology of Aging Grants 3113
The NIA invites applications for support of centers, known as Nathan Shock Centers of Excellence in Basic Biology of Aging. These center grants will provide funding for research and training activities related to basic biology of aging. They are intended for institutions with a substantial investment in and commitment to aging research, but they are not intended to directly support clinical research or clinical trials. The goal of this program is to enhance the ability of institutions with well-developed research programs in basic research on aging to utilize state-of-the-art research resources to provide the strongest environment for the conduct of research on aging. Thus, this RFA is intended to enhance the quality of research in the basic biology of aging, facilitate the planning and coordination of aging research activities, provide support and a suitable environment for investigators new to aging research to acquire research skills and experience at institutions that have demonstrated commitment to, and expertise in, basic biology of aging research, and to develop potential regional and/or national resource centers. Annual deadline dates may vary.
Requirements: Applications may be submitted by domestic nonprofit organizations and institutions, state and local governments and their agencies, and authorized federal institutions. The applicant institution must currently support a minimum of 15 peer-reviewed, externally funded research projects on aging to be eligible. In the case of currently funded program projects (P01s), or similar multiproject research grants, each research project will be deemed to be a separate project. (Core components do not qualify.)
Geographic Focus: All States
Date(s) Application is Due: Apr 16; May 20
Contact: Dr. Huber Warner, (301) 496-4996; fax (301) 402-0010; warnerh@nia.nih.gov
Internet: http://grants.nih.gov/grants/guide/rfa-files/RFA-AG-04-010.html
Sponsor: National Institute on Aging
Gateway Building, Suite 2C231, 7201 Wisconsin Avenue, MSC 9205
Bethesda, MD 20892

NIA Network Infrastructure Support for Emerging Behavioral and Social Research Areas in Aging 3114
The purpose of this FOA is to provide infrastructure support in specific emerging interdisciplinary areas of behavioral and social research in aging using the NIH Resource-Related Research Project mechanism. The infrastructure support will facilitate research networks through meetings, conferences, small scale pilots, training, and dissemination to encourage growth and development in specified emerging areas and resources. Budgets for direct costs of up to $150,000 per year and project duration of 3-5 years may be requested with a maximum of $750,000 direct costs over a five-year project period.

Letters of Intent are due by September 30, December 28, and April 26, respectively. Final applications due by October 30, January 25, and May 25, respectively..
Requirements: The following organizations and institutions may apply: public/state controlled institutions of higher education; private institutions of higher education; Hispanic-serving institutions; historically Black colleges and universities (HBCUs); tribally controlled colleges and universities (TCCUs); Alaska native and native Hawaiian serving institutions; nonprofits with 501(c)3 IRS status; nonprofits without 501(c)3 IRS status; small businesses; for-profit organizations; State governments; regional organizations; U.S. territories or possessions; Indian/Native American tribal governments; Indian/Native American tribally designated organizations; county governments; city or township governments; special district governments; independent school districts; public housing authorities/Indian housing authorities; eligible agencies of the Federal government; and faith-based or community based organizations.
Geographic Focus: All States
Date(s) Application is Due: Jan 25; Apr 26; May 25; Sep 30; Oct 30; Dec 28
Amount of Grant: 150,000-750,000 USD
Contact: Lis Nielsen, (301) 402-4156; fax (301) 402-0051; nielsenli@nia.nih.gov
Internet: http://grants.nih.gov/grants/guide/pa-files/PAR-09-233.html
Sponsor: National Institute on Aging
Building 31, Room 5C27, 31 Center Drive, MSC 2292
Bethesda, MD 20892-7710

NIA Pilot Research Grants **3115**
NIA is seeking small grant applications in specific areas to stimulate and facilitate the entry of promising new investigators into aging research or encourage established investigators to enter new targeted, high-priority areas in this research field. This small grant program provides support for pilot research that is likely to lead to subsequent individual research project grants and/or a significant advancement of aging research. Annual deadline dates may vary; contact program staff for exact dates.
Requirements: Applications may be submitted by domestic for-profit and nonprofit organizations, both public or private; units of state and local governments; and eligible agencies of the federal government. The program supports new and established investigators. For a new investigator to be eligible, the individual should be in the first five years of his or her independent research career. If the applicant is in the final stages of training it is permissible to apply for an R03, but no award will be made to individuals who are still in training or fellowship status at the time of award. For an established investigator to be eligible, the individual must propose research that is unrelated to a currently funded research project in which the investigator participates.
Restrictions: Foreign organizations and institutions are not eligible.
Geographic Focus: All States
Amount of Grant: Up to 50,000 USD
Contact: Dr. David Finkelstein, Biology of Aging Program Contact; (301) 496-6402; fax (301) 402-0010; BAPquery@extramur.nia.nih.gov
Internet: http://grants.nih.gov/grants/guide/pa-files/PA-99-049.html
Sponsor: National Institute on Aging
Gateway Building, Suite 2C231, 7201 Wisconsin Avenue, MSC 9205
Bethesda, MD 20892

NIA Postdoctoral Training in Research on Aging in Canada **3116**
The objective of the National Research Service Award (NRSA) F32 program is to provide support to promising postdoctoral applicants who have the potential to become productive and successful independent research investigators in scientific health-related fields relevant to the missions of the participating NIH Institutes and Centers (ICs). A primary purpose of this initiative is to offer training in research on aging that exposes developing American postdoctoral investigators to diverse combinations of expertise and resources singularly available in Canada through the Canadian Institutes of Health-Institute on Aging (CIHR-IA), combinations that also take into account the unique cultural and ethnic composition of Canada and its aging population.
Requirements: Applicants must choose Canadian institutions and mentors selected by CIHR-IA. These institutions and mentors can be located through the CIHR Funding Database. Any individual with the skills, knowledge, and resources necessary to carry out the proposed research training is invited to work with his/her sponsor and institution to develop an application for support. Individuals from underrepresented racial and ethnic groups as well as individuals with disabilities are always encouraged to apply for NIH support.
Restrictions: Candidates for the postdoctoral fellowship award must be citizens or non-citizen nationals of the United States, or must have been lawfully admitted to the United States for Permanent Residence.
Geographic Focus: All States
Date(s) Application is Due: Apr 8; Aug 8; Dec 8
Contact: Michael-David A.R.R. Kerns, (301) 402-7713; fax (301) 402-2945; michael-david.kerns@nih.hhs.gov
Internet: http://grants.nih.gov/grants/guide/pa-files/PA-06-469.html
Sponsor: National Institute on Aging
Gateway Building, Room 2C218, 7201 Wisconsin Avenue, MSC 9205
Bethesda, MD 20892-9205

NIA Promoting Careers in Aging and Health Disparities Research Grants **3117**
The Mentored Research Scientist Development Award provides support for a sustained period of protected time for intensive research career development under the guidance of an experienced mentor, or sponsor, in the biomedical, behavioral or clinical sciences leading to research independence. The expectation is that through this sustained period of research career development and training, awardees will launch independent research careers and become competitive for new research project grant funding.
Requirements: The following organizations and institutions may apply: public/state controlled institutions of higher education; private institutions of higher education; Hispanic-serving institutions; historically Black colleges and universities (HBCUs); tribally controlled colleges and universities (TCCUs); Alaska native and native Hawaiian serving institutions; nonprofits with 501(c)3 IRS status; nonprofits without 501(c)3 IRS status; small businesses; for-profit organizations; State governments; regional organizations; U.S. territories or possessions; Indian/Native American tribal governments; Indian/Native American tribally designated organizations; county governments; city or township governments; special district governments; independent school districts; public housing authorities/Indian housing authorities; eligible agencies of the Federal government; and faith-based or community based organizations.
Geographic Focus: All States
Date(s) Application is Due: Feb 12; Jun 12; Oct 12
Amount of Grant: 450,000-750,000 USD
Contact: Dr. J. Taylor Harden, (301) 496-0765; Taylor_Harden@nih.gov
Internet: http://grants.nih.gov/grants/guide/pa-files/PAR-09-136.html
Sponsor: National Institute on Aging
Building 31, Room 5C27, 31 Center Drive, MSC 2292
Bethesda, MD 20892-7710

NIA Renal Function and Chronic Kidney Disease in Aging Grants **3118**
This Funding Opportunity Announcement (FOA) issued by the National Institute on Aging (NIA) and the National Institute of Diabetes and Digestive and Kidney Diseases (NIDDK), National Institutes of Health, invites applications that propose basic, clinical, and translational research on chronic kidney disease (CKD) and its consequences in aging and in older persons. Applications should focus on: biology and pathophysiology of CKD in animal models; etiology and pathophysiology of CKD in the elderly; epidemiology and risk factors for the development of CKD with advancing age; and/or diagnosis, medical management and clinical outcomes of CKD in this population. Research supported by this initiative should enhance knowledge of CKD and its consequences in the elderly and provide evidence-based guidance in the diagnosis, prevention, and treatment of CKD in older persons.
Requirements: The following organizations and institutions may apply: public/state controlled institutions of higher education; private institutions of higher education; Hispanic-serving institutions; historically Black colleges and universities (HBCUs); tribally controlled colleges and universities (TCCUs); Alaska native and native Hawaiian serving institutions; nonprofits with 501(c)3 IRS status; nonprofits without 501(c)3 IRS status; small businesses; for-profit organizations; State governments; regional organizations; U.S. territories or possessions; Indian/Native American tribal governments; Indian/Native American tribally designated organizations; county governments; city or township governments; special district governments; independent school districts; public housing authorities/Indian housing authorities; eligible agencies of the Federal government; and faith-based or community based organizations.
Geographic Focus: All States
Date(s) Application is Due: Feb 5; Feb 16; Jun 5; Jun 16; Oct 5; Oct 16
Contact: Susan G. Nayfield, (301) 496-6761; fax (301) 402-1784; nayfiels@nia.nih.gov
Internet: http://grants.nih.gov/grants/guide/pa-files/PA-09-165.html
Sponsor: National Institute on Aging
Gateway Building, Suite 3C307, 7201 Wisconsin Avenue, MSC 9205
Bethesda, MD 20892-9205

NIA Role of Apolipoprotein E, Lipoprotein Receptors and CNS Lipid Homeostasis in Brain Aging and Alzheimer's Disease **3119**
This FOA encourages multidisciplinary and interdisciplinary research to elucidate how Apolipoprotein E, lipoprotein receptors and CNS lipid homeostasis influence brain aging and the transition to neurodegeneration in Alzheimer's disease (AD). The ultimate goal is to gain an in depth understanding of the mechanisms by which the Apolipoprotein E e4 allele confers increased AD risk for the purpose of advancing the overall search for efficacious AD treatments and Apolipoprotein E e4-directed therapeutics in particular. To this end we encourage research spanning multiple levels of analysis in multiple species (from mice to man). Studies aimed at identifying new therapeutic targets for the treatment of AD and other age-related neurodegenerative conditions associated with lipid neurobiology are strongly encouraged. Also of great interest are projects that aim to use lipidomics, various types of imaging and other cutting edge technologies to identify and develop early biomarkers of neurodegeneration associated lipid dyshomeostasis. The estimated amount of funds available to support projects under this FOA is $2,500,000.
Requirements: The following organizations and institutions are eligible to apply: public/state controlled institutions of higher education; private institutions of higher education; Hispanic-serving institutions; historically Black colleges and universities (HBCUs); tribally controlled colleges and universities (TCCUs); Alaska native and native Hawaiian serving institutions; nonprofits with 501(c)3 IRS status (other than institutions of higher education); nonprofits without 501(c)3 IRS status (other than institutions of higher education); small businesses; for-profit organizations (other than small businesses); state governments; Indian/Native American tribal governments (federally recognized); Indian/Native American tribally designated organizations; county governments; city or township governments; special district governments; independent school districts; public housing authorities/Indian housing authorities; U.S. territories or possessions; Indian/native American tribal governments (other than federally recognized); regional organizations; and non-domestic (non-U.S.) entities (foreign organizations).
Geographic Focus: All States
Date(s) Application is Due: Mar 5; Jul 5; Nov 5
Contact: Suzana S. Petanceska, (301) 496-9350; petanceskas@mail.nih.gov
Internet: http://grants.nih.gov/grants/guide/pa-files/PA-09-217.html

Sponsor: National Institute on Aging
Gateway Building, Suite 350, 7201 Wisconsin Avenue, MSC 9205
Bethesda, MD 20892-9205

NIA Role of Nuclear Receptors in Tissue and Organismal Aging Grants 3120
This Funding Opportunity Announcement (FOA) solicits grant applications proposing to conduct research into underlying biologic mechanisms involving nuclear receptors, their co-regulators and intracellular signaling systems in the process of aging and the connections of the aging process with pathophysiology in middle- and old-age. The focus of the proposed research must be on processes of aging and/or age-related changes, i.e., research proposed in applications responding to this FOA must have clear relevance to aging. Potential topic areas of interest to the NIA include, but are not limited to, the following: age-changes in levels, sensitivity, and gene expression activation profiles of NRs and co-regulators, and their effects on processes of aging (e.g., age-changes in tissue function, inflammation, glucose and lipid metabolism, adipogenesis, muscle, and brain function) and connections with disease processes for diseases common in older individuals; modification of lifespan in model organisms (e.g., the NR homologue DAF-12 in the insulin-IGF-like pathway in the nematode, and mammalian homologues) and the health-protecting, life-extending effects of caloric restriction or other life-span altering genetic mutations or transgenic models in mammalian organisms; and age-changes in steroid metabolism or in tissue sensitivity and/or response to exogenous sex steroid (or tissue-selective sex steroid receptor modulators) therapy in middle- and old-age.
Requirements: The following organizations and institutions may apply: public/state controlled institutions of higher education; private institutions of higher education; Hispanic-serving institutions; historically Black colleges and universities (HBCUs); tribally controlled colleges and universities (TCCUs); Alaska native and native Hawaiian serving institutions; nonprofits with 501(c)3 IRS status; nonprofits without 501(c)3 IRS status; small businesses; for-profit organizations; State governments; regional organizations; U.S. territories or possessions; Indian/Native American tribal governments; Indian/Native American tribally designated organizations; eligible agencies of the Federal government; and faith-based or community based organizations.
Geographic Focus: All States
Date(s) Application is Due: Feb 5; Jun 5; Oct 5
Contact: Ronald Margolis, (301) 594-8819; fax (301) 435-6047; rm76f@nih.gov
Internet: http://grants.nih.gov/grants/guide/pa-files/PAS-07-267.html
Sponsor: National Institute on Aging
Gateway Building, Suite 2C231, 7201 Wisconsin Avenue, MSC 9205
Bethesda, MD 20892

NIA Support of Scientific Meetings as Cooperative Agreements 3121
The purpose of the program announcement is to inform the scientific community that the National Institute on Aging (NIA) will now support scientific meetings as cooperative agreements in addition to the current practice of supporting them through the traditional grant mechanism. This program provides guidelines for when it is appropriate to request support of a meeting as a cooperative agreement and explains procedures for preparing and submitting such applications. The Principal Investigator will have the primary authority and responsibility to define objectives and approaches; plan, publicize, and conduct the scientific meeting; and publish the results of the meeting.
Requirements: Applications may be submitted by U.S. institutions, including scientific or professional societies eligible to receive grants from Public Health Service (PHS) agencies. In the case of an international conference, the U.S. representative organization of an established international scientific or professional society is the eligible applicant.
Restrictions: Foreign institutions are not eligible.
Geographic Focus: All States
Contact: Chyren Hunter, (301) 496-9322; fax (301) 402-0528; Hunterc@nia.nih.gov
Internet: http://grants.nih.gov/grants/guide/pa-files/PA-00-128.html
Sponsor: National Institute on Aging
Gateway Building, Suite 2C231, 7201 Wisconsin Avenue, MSC 9205
Bethesda, MD 20892

NIA Thyroid in Aging Grants 3122
The purpose of this funding opportunity announcement (FOA) is to encourage submission of investigator-initiated research applications on the thyroid in aging. This FOA is intended to promote basic, translational, and clinical studies leading to increased understanding of the physiology of the aging thyroid and improved diagnosis and management of thyroid disease in the elderly.
Requirements: The following organizations and institutions may apply: public/state controlled institutions of higher education; private institutions of higher education; Hispanic-serving institutions; nonprofits with 501(c)3 IRS status; nonprofits without 501(c)3 IRS status; small businesses; for-profit organizations; State governments; regional organizations; U.S. territories or possessions; Indian/Native American tribal governments; Indian/Native American tribally designated organizations; non-domestic (non-U.S.) entities (foreign organizations); Hispanic-serving institutions; historically Black colleges and universities (HBCUs); tribally controlled colleges and universities (TCCUs); Alaska native and native Hawaiian serving institutions; and regional organizations.
Geographic Focus: All States
Date(s) Application is Due: Feb 5; Feb 16; Jun 5; Jun 16; Oct 5; Oct 16
Contact: Nancy J. Emenaker; (301) 496-0116; emenaken@mail.nih.gov
Internet: http://grants.nih.gov/grants/guide/pa-files/PA-08-037.html
Sponsor: National Institute on Aging
Gateway Building, Suite 3C307, 7201 Wisconsin Avenue, MSC 9205
Bethesda, MD 20892-9205

NIA Transdisciplinary Research on Fatigue and Fatigability in Aging Grants 3123
The purpose of this Funding Opportunity Announcement (FOA) is to encourage submission of research applications on fatigue and fatigability in aging. This FOA is intended to promote research studies employing transdisciplinary approaches that could lead to increased understanding of mechanisms contributing to, assessment of, or potential interventions for, increased fatigue or fatigability in older persons. Both animal models and humans are appropriate for study under this FOA.
Requirements: The following organizations and institutions may apply: public/state controlled institutions of higher education; private institutions of higher education; Hispanic-serving institutions; historically Black colleges and universities (HBCUs); tribally controlled colleges and universities (TCCUs); Alaska native and native Hawaiian serving institutions; nonprofits with 501(c)3 IRS status; nonprofits without 501(c)3 IRS status; small businesses; for-profit organizations; State governments; regional organizations; U.S. territories or possessions; Indian/Native American tribal governments; Indian/Native American tribally designated organizations; county governments; city or township governments; special district governments; independent school districts; public housing authorities/Indian housing authorities; eligible agencies of the Federal government; and faith-based or community based organizations.
Geographic Focus: All States
Date(s) Application is Due: Feb 5; Feb 16; Jun 5; Jun 16; Oct 5; Oct 16
Contact: Kathleen Jett, PhD; (301) 594-2154; fax (301) 480-8260; jettk@mail.nih.gov
Internet: http://grants.nih.gov/grants/guide/pa-files/PA-08-161.html
Sponsor: National Institute on Aging
Gateway Building, Suite 3C307, 7201 Wisconsin Avenue, MSC 9205
Bethesda, MD 20892-9205

NIA Translational Research at the Aging/Cancer Interface (TRACI) Grants 3124
The purpose of this Funding Opportunity Announcement (FOA) is to enhance translational research in the overlapping areas of human aging and cancer by: integrating knowledge of basic processes in cancer biology and aging into clinical care of older patients with cancer (bench to bedside); and exploring clinical observations from the patient care setting at more basic and molecular levels (bedside to bench). Ultimately, information from the research supported by this initiative should lead to further improvements in prevention, diagnosis and disease management, improving the health and well-being of elderly patients at risk for, or diagnosed with, cancer and decreasing the functional impairment and morbidity associated with cancer in this population.
Requirements: The following organizations and institutions may apply: public/state controlled institutions of higher education; private institutions of higher education; Hispanic-serving institutions; historically Black colleges and universities (HBCUs); tribally controlled colleges and universities (TCCUs); Alaska native and native Hawaiian serving institutions; nonprofits with 501(c)3 IRS status; nonprofits without 501(c)3 IRS status; small businesses; for-profit organizations; State governments; regional organizations; U.S. territories or possessions; Indian/Native American tribal governments; Indian/Native American tribally designated organizations; county governments; city or township governments; special district governments; independent school districts; public housing authorities/Indian housing authorities; eligible agencies of the Federal government; and faith-based or community based organizations.
Geographic Focus: All States
Date(s) Application is Due: Feb 5; Feb 16; Jun 5; Jun 16; Oct 5; Oct 16
Contact: Susan G. Nayfield, (301) 496-6761; fax (301) 402-1784; nayfiels@nia.nih.gov
Internet: http://grants.nih.gov/grants/guide/pa-files/PA-08-230.html
Sponsor: National Institute on Aging
Gateway Building, Suite 3C307, 7201 Wisconsin Avenue, MSC 9205
Bethesda, MD 20892-9205

NIA Vulnerable Dendrites and Synapses in Aging and Alzheimer's Disease Grants 3125
The purpose of this initiative is to solicit research proposals that will investigate the factors regulating synaptic plasticity and dysfunction with a particular emphasis on the age-dependent changes in the functions of dendrites, spines and synapses of key cell types in regions of brain that are selectively vulnerable in certain neurodegenerative disorders such as AD, or with aging using, for example, in vitro and in vivo models.
Requirements: The following organizations and institutions may apply: public/state controlled institutions of higher education; private institutions of higher education; Hispanic-serving institutions; historically Black colleges and universities (HBCUs); tribally controlled colleges and universities (TCCUs); Alaska native and native Hawaiian serving institutions; nonprofits with 501(c)3 IRS status; nonprofits without 501(c)3 IRS status; small businesses; for-profit organizations; State governments; regional organizations; U.S. territories or possessions; Indian/Native American tribal governments; Indian/Native American tribally designated organizations; county governments; city or township governments; special district governments; independent school districts; public housing authorities/Indian housing authorities; eligible agencies of the Federal government; and faith-based or community based organizations.
Geographic Focus: All States
Date(s) Application is Due: Feb 5; Jun 5; Oct 5
Contact: D. Stephen Snyder, (301) 496-9350; fax (301) 496-1494; ss82f@nih.gov
Internet: http://grants.nih.gov/grants/guide/pa-files/PA-09-061.html
Sponsor: National Institute on Aging
Gateway Building, Suite 350, 7201 Wisconsin Avenue, MSC 9205
Bethesda, MD 20892-9205

NICHD Innovative Therapies and Clinical Studies for Screenable Disorders 3126

Grants

This Funding Opportunity Announcement (FOA) issued by the National Institute of Child Health and Human Development (NICHD), the National Institute on Deafness and Other Communication Disorders (NIDCD), and the National Institute of Diabetes and Digestive and Kidney Diseases (NIDDK), National Institutes of Health (NIH), solicit applications for research relevant to the basic understanding and development of therapeutic interventions for currently screened conditions and high priority genetic conditions for which screening could be possible in the near future. Because the nature and scope of the proposed research will vary from application to application, it is anticipated that the size and duration of each award will also vary. The total amount awarded and the number of awards will depend upon the numbers, quality, duration, and costs of the applications received.
Requirements: Application can be made by the following institutions and organizations: public/state controlled institutions of higher education; private institutions of higher education; nonprofits with 501(c)3 IRS status (other than institutions of higher education); nonprofits without 501(c)3 IRS status (other than institutions of higher education); small businesses; for-profit organizations (other than small businesses); state governments; U.S. territories or possessions; Indian/Native American tribal governments (federally recognized); Indian/Native American tribal governments (other than Federally recognized); Indian/Native American tribally designated organizations; non-domestic (non-U.S.) entities (foreign organizations); Hispanic-serving institutions; historically Black colleges and universities (HBCUs); tribally controlled colleges and universities (TCCUs); Alaska Native and Native Hawaiian serving institutions; regional organization; eligible agencies of the Federal government; and faith-based or community based organizations.
Geographic Focus: All States
Date(s) Application is Due: Feb 5; Mar 5; Jun 5; Jul 5; Oct 5; Nov 5
Contact: Gilian Engelson, (301) 451-2137; fax (301) 451-5512; engelsong@mail.nih.gov
Internet: http://grants.nih.gov/grants/guide/pa-files/PAR-07-184.html
Sponsor: National Institute of Child Health and Human Development
31 Center Drive, Building 31 Room 2A32, MSC 2425
Bethesda, MD 20892-2425

NICHD Ruth L. Kirschstein National Research Service Award Institutional Predoctoral Training in Systems Biology & Birth Defects 3127

The objective of this NRSA T32 program is to provide research training to predoctoral students interested in establishing research careers that use systems biology approaches to study developmental biology and the formation of structural birth defects. Stipends are provided as a subsistence allowance to help defray trainees' living expenses during the research training experience. Funds are provided to offset the cost of tuition and fees. The NIH provides 60% of the combined requested costs, up to $16,000 annually per Ph.D. student, and up to $21,000 annually for students in formal programs leading to combined professional and Ph.D. degrees. Letters of Intent are due by April 25, with full applications due May 25.
Requirements: The following organizations are eligible to apply: public/state controlled institutions of higher education; private institutions of higher education; nonprofits with 501(c)3 IRS status (other than institutions of higher education); nonprofits without 501(c)3 IRS status (other than institutions of higher education); for-profit organizations (other than small businesses); state governments; U.S. territories or possessions; Indian/Native American tribal governments (Federally recognized); Hispanic-serving institutions; historically Black colleges and universities (HBCUs); tribally controlled colleges and universities (TCCUs); Alaska Natives and Native Hawaiians serving institutions; regional organizations; eligible agencies of the federal government; faith-based or community based organizations.
Geographic Focus: All States
Date(s) Application is Due: Apr 25; May 25
Contact: James Coulombe, (301) 496-5541; fax (301) 480-0303; coulombej@mail.nih.gov
Internet: http://grants.nih.gov/grants/guide/pa-files/PAR-08-054.html
Sponsor: National Institute of Child Health and Human Development
31 Center Drive, Building 31 Room 2A32, MSC 2425
Bethesda, MD 20892-2425

Nick Traina Foundation Grants 3128

The Foundation supports organizations involved in the diagnosis, research, treatment, and/or family support of manic-depression, and other forms of mental illness, suicide prevention, child abuse and children in jeopardy, and provides assistance to struggling musicians in the areas of mental illness. The Foundation may give special consideration to proposals that address manic-depression in children and young adults. There are no deadlines and there is no formal application form. The Board meets four times a year to review proposals.
Requirements: Applicants must be 501(c)3 organizations. Proposals should be kept to three pages in addition to attachments and should include: the organization's purpose, including a description of its history and mission and the population served; the reason for the grant request, including the amount requested and how the funds will be used; whether the request is for general organizational support or for a specific project (including the duration of a specific project); a copy of the IRS determination letter; and a copy of the most recent annual report and the last two years' financial statements.
Restrictions: Organizations may apply for one grant each calendar year. Requests from individuals are ineligible. Grant funding focuses on the San Francisco Bay area.
Geographic Focus: California
Contact: Danielle Steel, President; (415) 771-4224; info@nicktrainafoundation.org
Internet: http://nicktrainafoundation.com/grants.htm
Sponsor: Nick Traina Foundation
P.O. Box 470427
San Francisco, CA 94147-0427

NIDA Mentored Clinical Scientist Research Career Development Awards 3129

The overall objective of the NIH Research Career Development Award program is to prepare qualified individuals for careers that have a significant impact on the health-related research needs of the Nation. The objective of the NIH Mentored Clinical Scientist Research Career Development Award (K08) program is to support didactic study and mentored research for individuals with clinical doctoral degrees (e.g., M.D., D.D.S., D.M.D., D.O., D.C., O.D., N.D., D.V.M., Pharm.D., or Ph.D. in clinical disciplines). The purpose of this Funding Opportunity Announcement (FOA) is to continue the long-standing NIH support of the K08 program. This award provides support and protected time for an intensive, mentored research career development experience in biomedical or behavioral research, including translational research. For the purpose of this award, translational research is defined as application of basic research discoveries toward the diagnosis, management, and prevention of human disease.
Requirements: The following organizations and institutions may apply: public/state controlled institutions of higher education; private institutions of higher education; Hispanic-serving institutions; historically Black colleges and universities (HBCUs); tribally controlled colleges and universities (TCCUs); Alaska native and native Hawaiian serving institutions; nonprofits with 501(c)3 IRS status; nonprofits without 501(c)3 IRS status; small businesses; for-profit organizations; State governments; regional organizations; U.S. territories or possessions; Indian/Native American tribal governments; Indian/Native American tribally designated organizations; county governments; city or township governments; special district governments; independent school districts; public housing authorities/Indian housing authorities; eligible agencies of the Federal government; and faith-based or community based organizations.
Geographic Focus: All States
Date(s) Application is Due: Feb 12; Mar 12; Jun 12; Jul 12; Oct 12; Nov 12
Amount of Grant: Up to 140,000 USD
Contact: Mimi M. Ghim, (301) 443-6071; ghimm@mail.nih.gov
Internet: http://grants.nih.gov/grants/guide/pa-files/PA-09-042.html#SectionIII
Sponsor: National Institute on Drug Abuse
6001 Executive Boulevard, Room 5213, MSC 9561
Bethesda, MD 20892-9561

NIDA Pilot and Feasibility Studies in Preparation for Drug Abuse Prevention Trials 3130

This FOA for R34 applications seeks to support: pilot and/or feasibility testing of new, revised, or adapted preventive intervention approaches targeting the initiation of drug use, the progression to abuse or dependence, and the acquisition or transmission of HIV infection among diverse populations and settings; and pre-trial feasibility testing for prevention services and systems research.
Requirements: The following organizations and institutions may apply: public/state controlled institutions of higher education; private institutions of higher education; Hispanic-serving institutions; historically Black colleges and universities (HBCUs); tribally controlled colleges and universities (TCCUs); Alaska native and native Hawaiian serving institutions; nonprofits with 501(c)3 IRS status; nonprofits without 501(c)3 IRS status; small businesses; for-profit organizations; State governments; regional organizations; U.S. territories or possessions; Indian/Native American tribal governments; Indian/Native American tribally designated organizations; county governments; city or township governments; special district governments; independent school districts; public housing authorities/Indian housing authorities; eligible agencies of the Federal government; and faith-based or community based organizations.
Restrictions: The NIDA R34 mechanism does not support the development of intervention protocols, manuals, or the standardization of protocols.
Geographic Focus: All States
Date(s) Application is Due: Feb 16; Jun 16; Oct 16
Amount of Grant: 45,000-225,000 USD
Contact: Aleta Meyer, (301) 402-1725; fax (301) 443-2636; meyera2@nida.nih.gov
Internet: http://grants.nih.gov/grants/guide/pa-files/PA-09-146.html
Sponsor: National Institute on Drug Abuse
6001 Executive Boulevard, Room 5213, MSC 9561
Bethesda, MD 20892-9561

NIDCD Mentored Clinical Scientist Research Career Development Award 3131

This award enables candidates holding professional degrees (e.g., M.D., D.O., D.V.M., or equivalent degrees) to undertake 3 to 5 years of special study and supervised research with the goal of becoming independent investigators. The award also allows awardees to pursue a research career development program suited to their experience and capabilities under a mentor who is competent to provide guidance in the chosen research area. Institutions may submit applications on behalf of candidates who hold professional degrees. At least 2 years must have elapsed since the health professional degree was granted. Candidates can have varying levels of clinical training and research experience. New applications are due February 12, June 12, and October 12, while resubmitted applications are due March 12, July 12, and November 12. Up to $105,000 per year plus fringe benefits is available, and up to $85,000 per year for research support for new awards.
Requirements: A candidate must have a clinical degree or its equivalent, must have initiated postgraduate clinical training, must identify a mentor with extensive research experience, and must be willing to spend a minimum of 75 percent of full-time professional effort conducting research. Applications may be submitted on behalf of candidates by domestic, nonfederal organizations, public or private, such as medical, dental, or nursing schools or other institutions of higher education. Candidates must be U.S. citizens, noncitizen nationals, or permanent residents.
Geographic Focus: All States
Date(s) Application is Due: Feb 12; Mar 12; Jun 12; Jul 12; Oct 12; Nov 12
Amount of Grant: Up to 185,000 USD

Contact: Daniel Sklare, (301) 496-1804; fax (301) 402-6251; sklared@nidcd.nih.gov
Internet: http://grants.nih.gov/grants/guide/pa-files/pa-06-512.html
Sponsor: National Institute on Deafness and Other Communication Disorders
6120 Executive Boulevard, Executive Plaza S
Bethesda, MD 20892

NIDCD Mentored Research Scientist Development Award 3132

The NIDCD utilizes the K01 Program for the retooling and transition of junior- and midcareer-level scientists and clinicians (generally below the rank of full professor, or the equivalent in nonacademic settings) into two research domains: translational research, i.e., the application of basic research discoveries toward advancing the diagnosis, management and prevention of deafness and other communication disorders; and clinical research. Junior-level clinically-trained individuals should apply for the Mentored Clinical Scientist Development Award or the Mentored Patient-Oriented Research Career Development Award, as appropriate.
Requirements: The candidate must have a research or health-professional doctorate or its equivalent, and must have demonstrated the capacity or potential for highly productive independent research in the period after the doctorate. The candidate must identify a mentor with extensive research experience, and must be willing to spend a minimum of 75 percent of full-time professional effort conducting research and research career development activities for the period of the award. Applications may be submitted on behalf of candidates by domestic, public or private, nonfederal organizations. Candidates must be U.S. citizens or noncitizen nationals.
Geographic Focus: All States
Date(s) Application is Due: Feb 12; Mar 12; Jun 12; Jul 12; Oct 12; Nov 12
Amount of Grant: Up to 105,000 USD
Contact: Daniel A. Sklare, (301) 496-1804; sklared@nidcd.nih.gov
Internet: http://www.grants.nih.gov/grants/guide/pa-files/PAR-99-027.html
Sponsor: National Institute on Deafness and Other Communication Disorders
6120 Executive Boulevard, Executive Plaza S
Bethesda, MD 20892

NIDDK Advances in Polycystic Kidney Disease Grants 3133

The National Institute of Diabetes and Digestive and Kidney Diseases (NIDDK) through its Division of Kidney, Urologic and Hematologic Diseases (DKUHD) invites experienced and new investigators to submit research grant applications to pursue basic and applied investigations in order to better understand the etiology and pathogenesis of Polycystic Kidney Disease (PKD), in both its autosomal dominant and autosomal recessive forms. Such applications may examine the genetic determinants, and cellular and molecular mechanisms, which disrupt normal kidney function; mechanisms of cyst formation and growth; development of experimental model systems; development of markers of disease progression; and the identification of innovative therapeutic interventions and gene targeted strategies to prevent progressive renal insufficiency due to this disorder. The intent of this funding opportunity is to intensify investigator-initiated research, to attract new investigators to the field, and to increase interdisciplinary research. The ultimate aim is to facilitate PKD-related research studies, which will provide the basis for new therapeutic approaches.
Requirements: The following organizations may apply: for-profit organizations; non-profit organizations; public or private institutions, such as universities, colleges, hospitals, and laboratories; units of State government; units of local government; eligible agencies of the Federal government; foreign institutions; domestic institutions; and faith-based or community-based organizations.
Geographic Focus: All States
Date(s) Application is Due: Feb 5; Jun 5; Oct 5
Contact: Catherine M. Meyers, (301) 594-7717; fax (301) 480-3510; cm420i@nih.gov
Internet: http://grants.nih.gov/grants/guide/pa-files/PA-06-158.html
Sponsor: National Institute of Diabetes and Digestive and Kidney Diseases
6707 Democracy Boulevard
Bethesda, MD 20892-5460

NIDDK Adverse Metabolic Side Effects of Second Generation Psychotropic Medications Leading to Obesity and Increased Diabetes Risk Grants 3134

This Funding Opportunity invites investigator-initiated research grant applications for studies examining the adverse metabolic effects (i.e., obesity and diabetes) of psychotropic medications in animal models and across the human lifespan (including pediatric, adult and geriatric populations). Applications responsive to this FOA should focus on: increasing the understanding of the nature, rates, and pathophysiology of adverse metabolic effects of psychotropic medications; elucidating biomedical and psychosocial risk factors for the development of metabolic adverse effects of psychiatric therapeutics; and develop interventions to prevent and/or mitigate metabolic adverse effects across the lifespan.
Requirements: The following organizations and institutions may apply: public/state controlled institutions of higher education; private institutions of higher education; Hispanic-serving institutions; historically Black colleges and universities (HBCUs); tribally controlled colleges and universities (TCCUs); Alaska native and native Hawaiian serving institutions; nonprofits with 501(c)3 IRS status; nonprofits without 501(c)3 IRS status; small businesses; for-profit organizations; State governments; regional organizations; U.S. territories or possessions; Indian/Native American tribal governments; Indian/Native American tribally designated organizations; eligible agencies of the Federal government; and faith-based or community based organizations.
Geographic Focus: All States
Date(s) Application is Due: Feb 22; Jun 22; Oct 22
Amount of Grant: Up to 500,000 USD
Contact: Christine Hunter, (301) 594-4728; hunterchristine@niddk.nih.gov
Internet: http://grants.nih.gov/grants/guide/pa-files/PAR-08-160.html
Sponsor: National Institute of Diabetes and Digestive and Kidney Diseases
6707 Democracy Boulevard
Bethesda, MD 20892-5460

NIDDK Ancillary Studies of Kidney Disease Accessing Information from Clinical Trials, Epidemiological Studies, and Databases Grants 3135

The National Institute of Diabetes and Digestive and Kidney Diseases (NIDDK) invites investigator-initiated research project applications for ancillary studies to ongoing or completed clinical trials and epidemiological studies of kidney disease as well as clinical trials and epidemiological studies for other diseases or populations that lend themselves to the study of kidney disease. These studies may range from new analyses of existing datasets of completed studies to additional collection of data and biological specimens in ongoing investigations. The goal of these studies should be to extend our understanding of the risk factors for developing kidney disease and their associated co-morbid illnesses such as malnutrition and cardiovascular disease, factors associated with rapid decline in kidney function among persons with chronic kidney disease, and the impact of these diseases on quality of life and mental and physical functioning. Studies of biomarkers for patients with acute renal failure and other kidney diseases, for example, are also appropriate topics for further investigation. Studies ancillary to both government and non-government supported clinical trials and epidemiological studies are encouraged. Analysis of large public access databases and other databases is also encouraged.
Requirements: The following organizations and institutions may apply: public/state controlled institutions of higher education; private institutions of higher education; Hispanic-serving institutions; historically Black colleges and universities (HBCUs); tribally controlled colleges and universities (TCCUs); Alaska native and native Hawaiian serving institutions; nonprofits with 501(c)3 IRS status; nonprofits without 501(c)3 IRS status; small businesses; for-profit organizations; State governments; regional organizations; U.S. territories or possessions; Indian/Native American tribal governments; Indian/Native American tribally designated organizations; eligible agencies of the Federal government; and faith-based or community based organizations.
Geographic Focus: All States
Date(s) Application is Due: Feb 5; Jun 5; Oct 5
Contact: John W. Kusek, (301) 594-7735; fax (301) 480-3510; jk61x@nih.gov
Internet: http://grants.nih.gov/grants/guide/pa-files/PA-07-050.html
Sponsor: National Institute of Diabetes and Digestive and Kidney Diseases
6707 Democracy Boulevard
Bethesda, MD 20892-5460

NIDDK Ancillary Studies to Major Ongoing NIDDK and NHLBI Clinical Research Studies 3136

This Funding Opportunity Announcement (FOA) solicits grant applications from qualified investigators to conduct ancillary studies of selected ongoing major clinical research studies, including clinical trials, epidemiological studies and disease databases, supported by the National Institute of Diabetes and Digestive and Kidney Diseases and the National Heart, Lung, and Blood Institute, National Institutes of Health (NIH). The objective is to introduce NIDDK-oriented ancillary studies into either NIDDK or NHLBI-funded parent studies, or NHLBI-oriented ancillary studies into NIDDK-funded parent studies.
Requirements: The following organizations and institutions may apply: public/state controlled institutions of higher education; private institutions of higher education; Hispanic-serving institutions; historically Black colleges and universities (HBCUs); tribally controlled colleges and universities (TCCUs); Alaska native and native Hawaiian serving institutions; nonprofits with 501(c)3 IRS status; nonprofits without 501(c)3 IRS status; small businesses; for-profit organizations; State governments; regional organizations; U.S. territories or possessions; Indian/Native American tribal governments; Indian/Native American tribally designated organizations; eligible agencies of the Federal government; and faith-based or community based organizations.
Geographic Focus: All States
Date(s) Application is Due: Feb 5; Jun 5; Oct 5
Contact: John W. Kusek, (301) 594-7735; fax (301) 480-3510; jk61x@nih.gov
Internet: http://grants.nih.gov/grants/guide/pa-files/PAR-07-024.html
Sponsor: National Institute of Diabetes and Digestive and Kidney Diseases
6707 Democracy Boulevard
Bethesda, MD 20892-5460

NIDDK Basic and Clinical Studies of Congenital Urinary Tract Obstruction Grants 3137

The purpose of this funding opportunity is to address the numerous scientific and clinical uncertainties related to the development, treatment and prognosis of congenital obstructive uropathy, by encouraging and facilitating research in diverse areas. These areas include: the development of objective prognostic markers; the genetic determinants of this congenital disorder; the development of reliable animal models of the disorder; and, evaluation of the long-term effectiveness of various treatment strategies. Because the nature and scope of the proposed research will vary from application to application, it is anticipated that the size and duration of each award will also vary. The total amount awarded and the number of awards will depend upon the mechanism numbers, quality, duration, and costs of the applications received.
Requirements: The following organizations and institutions may apply: public/state controlled institutions of higher education; private institutions of higher education; Hispanic-serving institutions; historically Black colleges and universities (HBCUs); tribally controlled colleges and universities (TCCUs); Alaska native and native Hawaiian serving institutions; nonprofits with 501(c)3 IRS status; nonprofits without 501(c)3 IRS status; small businesses; for-profit organizations; State governments;

regional organizations; U.S. territories or possessions; Indian/Native American tribal governments; Indian/Native American tribally designated organizations; eligible agencies of the Federal government; and faith-based or community based organizations.
Geographic Focus: All States
Date(s) Application is Due: Feb 5; Jun 5; Oct 5
Contact: Marva Moxey-Mims, (301) 594-7717; fax (301) 480-3510; mm726k@nih.gov
Internet: http://grants.nih.gov/grants/guide/pa-files/PA-07-059.html
Sponsor: National Institute of Diabetes and Digestive and Kidney Diseases
6707 Democracy Boulevard
Bethesda, MD 20892-5460

NIDDK Basic Research Grants in the Bladder and Lower Urinary Tract 3138
This funding opportunity encourages basic cellular, molecular, developmental and genetic research relevant to the bladder and lower urinary tract. Basic research studies that address age and gender differences in bladder and lower urinary tract function are also encouraged. New and established investigators from related fields of study are encouraged to apply their expertise to these problem areas. Investigators with diverse basic science and clinical backgrounds are encouraged to develop collaborative research relationships. Discoveries resulting from these studies may serve as the basis for the development of new agents, techniques, and strategies for detecting, preventing, and treating diseases of the bladder and lower urinary tract, as well as dealing with bladder complications resulting from other diseases, such as diabetes.
Requirements: Application can be made by the following institutions and organizations: for-profit organizations; non-profit organizations; public or private institutions (such as universities, colleges, hospitals, and laboratories); units of State government; units of local government; eligible agencies of the Federal government; foreign institutions; and domestic institutions.
Geographic Focus: All States
Date(s) Application is Due: Feb 5; Jun 5; Oct 5
Contact: Chris Mullins, (301) 594-7717; fax (301) 480-3510; cm419z@nih.gov
Internet: http://grants.nih.gov/grants/guide/pa-files/PA-06-254.html
Sponsor: National Institute of Diabetes and Digestive and Kidney Diseases
6707 Democracy Boulevard
Bethesda, MD 20892-5460

NIDDK Calcium Oxalate Stone Diseases Grants 3139
The purpose of this Funding Opportunity Announcement (FOA) is to increase investigator interest in research into the genetics and heritability of oxalate regulation and the oxalate stone diseases, and to develop new treatments for the disorder. This initiative solicits research as well as pilot and feasibility studies that utilize new and innovative approaches to study the diagnosis, treatment and prevention of these disorders. Because the nature and scope of the proposed research will vary from application to application, it is anticipated that the size and duration of each award will also vary. The total amount awarded and the number of awards will depend upon the mechanism numbers, quality, duration, and costs of the applications received.
Requirements: The following organizations and institutions may apply: public/state controlled institutions of higher education; private institutions of higher education; Hispanic-serving institutions; historically Black colleges and universities (HBCUs); tribally controlled colleges and universities (TCCUs); Alaska native and native Hawaiian serving institutions; nonprofits with 501(c)3 IRS status; nonprofits without 501(c)3 IRS status; small businesses; for-profit organizations; State governments; regional organizations; U.S. territories or possessions; Indian/Native American tribal governments; Indian/Native American tribally designated organizations; eligible agencies of the Federal government; and faith-based or community based organizations.
Geographic Focus: All States
Date(s) Application is Due: Feb 5; Jun 5; Oct 5
Contact: Rebekah S. Rasooly, (301) 594-6007; fax (301) 480-3510; rr185i@nih.gov
Internet: http://grants.nih.gov/grants/guide/pa-files/PA-07-051.html
Sponsor: National Institute of Diabetes and Digestive and Kidney Diseases
6707 Democracy Boulevard
Bethesda, MD 20892-5460

NIDDK Cell-Specific Delineation of Prostate and Genitourinary Development 3140
This RFA is intended to encourage the development of new research tools and methods to study the development and biology of the prostate and genitourinary tract. Strategies to be supported include systematic assessment of gene expression in specific cell types in the developing prostate and genitourinary tract; methods to tag individual cell types for purification, analysis, and characterization; tools to manipulate gene expression in vivo in individual cell types; and development of clinically relevant cell lines. The major goal of this initiative is to ensure that research tools are available so that a broad range of innovative methods can be applied to studying prostate development. Annual deadline dates may vary; contact program staff for exact dates.
Requirements: Application can be made by the following institutions and organizations: public/state controlled institutions of higher education; private institutions of higher education; nonprofits with 501(c)3 IRS status (other than institutions of higher education); nonprofits without 501(c)3 IRS status (other than institutions of higher education); small businesses; for-profit organizations (other than small businesses); state governments; U.S. territories or possessions; Indian/Native American tribal governments (federally recognized); Indian/Native American tribal governments (other than Federally recognized); Indian/Native American tribally designated organizations; non-domestic (non-U.S.) entities (foreign organizations); Hispanic-serving institutions; historically Black colleges and universities (HBCUs); tribally controlled colleges and universities (TCCUs); Alaska Native and Native Hawaiian serving institutions; regional organization; eligible agencies of the Federal government; and faith-based or community based organizations.
Geographic Focus: All States
Date(s) Application is Due: Feb 5; Mar 5; Jun 5; Jul 5; Oct 5; Nov 5
Amount of Grant: Up to 350,000 USD
Contact: Dr. Robert Star, (301) 594-7715; StarR@extra.niddk.nih.gov
Internet: http://grants.nih.gov/grants/guide/rfa-files/RFA-DK-00-015.html
Sponsor: National Institute of Diabetes and Digestive and Kidney Diseases
31 Center Drive, Room 9A-35, MSC 2560
Bethesda, MD 20892-2560

NIDDK Co-Activators and Co-Repressors in Gene Expression Grants 3141
This initiative stems from an NIDDK workshop in molecular endocrinology and is designed to stimulate research that addresses the fundamental underlying mechanisms by which nuclear accessory proteins mediate signaling through hormone receptors at the level of the regulation of gene expression. Staff contact prior to submission is highly recommended.
Requirements: Applications may be submitted by domestic and foreign, for-profit and nonprofit organizations, both public and private; units of state and local governments; and eligible agencies of the federal government.
Geographic Focus: All States
Date(s) Application is Due: Feb 16; Jun 16; Oct 16
Amount of Grant: 100,000-250,000 USD
Contact: Dr. Ronald Margolis, (301) 594-8819; fax (301) 435-6047; rm76f@nih.gov
Internet: http://grants.nih.gov/grants/guide/pa-files/PA-99-111.html
Sponsor: National Institutes of Health
6707 Democracy Boulevard, Building 2DEM, Room 693
Bethesda, MD 20892

NIDDK Collaborative Interdisciplinary Team Science in Diabetes, Endocrinology and Metabolic Diseases Grants 3142
The purpose of the Collaborative Interdisciplinary Team Science Program described in this announcement is to provide support to enable strong investigative teams to do inter- and/or trans-disciplinary research on a complex problem in biomedical science relevant to Diabetes, Endocrinology and Metabolic Diseases. It is anticipated that 1-3 projects will be funded per year, but because the nature and scope of the proposed research will vary from application to application, it is anticipated that the size and duration of each award will also vary. The total amount awarded and the number of awards will depend upon the quality, duration, and costs of the applications received, as well as on the availability of funds.
Requirements: The following organizations and institutions are eligible to apply: public/state controlled institutions of higher education; private institutions of higher education; Hispanic-serving institutions; historically Black colleges and universities (HBCUs); tribally controlled colleges and universities (TCCUs); Alaska Native and Native Hawaiian serving institutions; and non-domestic (non-U.S.) entities (foreign organizations).
Geographic Focus: All States
Date(s) Application is Due: Feb 24; Mar 24
Contact: Karen Salomon, (301) 594-7733; fax (301) 435-6047; ks495c@nih.gov
Internet: http://grants.nih.gov/grants/guide/pa-files/PAR-08-182.html
Sponsor: National Institute of Diabetes and Digestive and Kidney Diseases
6707 Democracy Boulevard
Bethesda, MD 20892-5460

NIDDK Developmental Biology and Regeneration of the Liver Grants 3143
The purpose of this FOA is to invite qualified scientific investigators to submit applications on liver development and regeneration to fully define the molecular and cellular mechanisms underlying these processes in health and disease and to apply these findings to developing improved therapies for liver disease. Because the nature and scope of the proposed research will vary from application to application, it is anticipated that the size and duration of each award will also vary. The total amount awarded and the number of awards will depend upon the mechanism numbers, quality, duration, and costs of the applications received.
Requirements: The following organizations and institutions may apply: public/state controlled institutions of higher education; private institutions of higher education; Hispanic-serving institutions; historically Black colleges and universities (HBCUs); tribally controlled colleges and universities (TCCUs); Alaska native and native Hawaiian serving institutions; nonprofits with 501(c)3 IRS status; nonprofits without 501(c)3 IRS status; small businesses; for-profit organizations; State governments; regional organizations; U.S. territories or possessions; Indian/Native American tribal governments; Indian/Native American tribally designated organizations; eligible agencies of the Federal government; and faith-based or community based organizations.
Geographic Focus: All States
Date(s) Application is Due: Feb 5; Jun 5; Oct 5
Contact: Samir Zakhari, (301) 443-0799; fax (301) 594-0673; szakhari@mail.nih.gov
Internet: http://grants.nih.gov/grants/guide/pa-files/PA-07-026.html
Sponsor: National Institute of Diabetes and Digestive and Kidney Diseases
6707 Democracy Boulevard
Bethesda, MD 20892-5460

NIDDK Development of Assays for High-Throughput Drug Screening Grants 3144
The overall goal of the Molecular Libraries Initiative, part of the NIH Roadmap for Medical Research, is to offer public sector researchers opportunities to screen small molecules with the high-throughput chemical screening (HTS) methods commonly used by the private sector to develop therapeutic agents. The purpose of this FOA is to support the development of innovative assays that may ultimately be adapted for

automated screening. The assay should aim to identify new tools for basic research or promising new avenues for therapeutics development, especially in areas related to the missions of the participating institutes. Awards issued under this FOA are contingent upon the availability of funds and the submission of a sufficient number of meritorious applications. The total amount awarded and the number of awards will depend upon the quality and costs of the applications received.
Requirements: The following organizations and institutions may apply: public/state controlled institutions of higher education; private institutions of higher education; Hispanic-serving institutions; historically Black colleges and universities (HBCUs); tribally controlled colleges and universities (TCCUs); Alaska native and native Hawaiian serving institutions; nonprofits with 501(c)3 IRS status; nonprofits without 501(c)3 IRS status; small businesses; for-profit organizations; State governments; regional organizations; U.S. territories or possessions; Indian/Native American tribal governments; Indian/Native American tribally designated organizations; eligible agencies of the Federal government; and faith-based or community based organizations.
Geographic Focus: All States
Date(s) Application is Due: Feb 5; Jun 5; Oct 5
Contact: Laura K. Moen, (301) 594-4748; fax (301) 480-3510; lm232f@nih.gov
Internet: http://grants.nih.gov/grants/guide/pa-files/PA-07-320.html
Sponsor: National Institute of Diabetes and Digestive and Kidney Diseases
6707 Democracy Boulevard
Bethesda, MD 20892-5460

NIDDK Development of Disease Biomarkers Grants **3145**
The goal of this Funding Opportunity Announcement (FOA) is to validate biomarkers for well-defined human diseases of liver, kidney, urological tract, digestive and hematologic systems, and endocrine and metabolic disorders, diabetes and its complications, and obesity, for which there are no or very few biomarkers, or for which standard biomarkers are currently prohibitively invasive or expensive. New biomarkers will stimulate bench to bedside translation by providing measures of the biological effects of potential new treatments. The ideal biomarker can be measured in a minimally invasive way, can be measured repeatedly over time, identifies early stages of disease, is indicative of disease prognosis, and correlates well with progression and response to therapy. Especially of interest would be studies designed to test the validity of candidate biomarkers or new technologies to monitor candidate biomarkers in patient tissue samples or small groups of well-characterized patients. For biomarkers already validated in human subjects, the larger effort might be to establish a reliable assay prior to pursuing larger patient studies. Priority will be given to those projects with high promise for improving clinical care in the relatively near future. Studies of potential biomarkers conducted in animal models of disease, and extended prospective clinical trials for marker validation against hard clinical endpoints, are outside the scope of this initiative. Because the nature and scope of the proposed research will vary from application to application, it is anticipated that the size and duration of each award will also vary.
Requirements: Application can be made by the following institutions and organizations: public/state controlled institutions of higher education; private institutions of higher education; nonprofits with 501(c)3 IRS status (other than institutions of higher education); nonprofits without 501(c)3 IRS status (other than institutions of higher education); small businesses; for-profit organizations (other than small businesses); state governments; U.S. territories or possessions; Indian/Native American tribal governments (federally recognized); Indian/Native American tribal governments (other than Federally recognized); Indian/Native American tribally designated organizations; non-domestic (non-U.S.) entities (foreign organizations); Hispanic-serving institutions; historically Black colleges and universities (HBCUs); tribally controlled colleges and universities (TCCUs); Alaska Native and Native Hawaiian serving institutions; regional organization; eligible agencies of the Federal government; and faith-based or community based organizations.
Geographic Focus: All States
Date(s) Application is Due: Feb 5; Jun 5; Oct 5
Contact: Maren R. Laughlin, (301) 594-8802; fax (301) 480-0475; ml33q@nih.gov
Internet: http://grants.nih.gov/grants/guide/pa-files/PA-07-052.html
Sponsor: National Institute of Diabetes and Digestive and Kidney Diseases
6707 Democracy Boulevard
Bethesda, MD 20892-5460

NIDDK Diabetes, Endocrinology, and Metabolic Diseases NRSAs—Individual **3146**
Individual National Research Service Awards (NRSAs) are given only at the postdoctoral level for research training in diabetes, endocrine, or metabolic diseases. To receive one, an individual must submit an application that describes a specific research project, which is guided and sponsored by a specific preceptor at a particular institution. The award may be for up to 36 months of training and consists of a stipend and an allowance to help meet expenses incurred during the course of fellowship.
Requirements: All awardees must be citizens or have been admitted to the United States for permanent residence. Individuals and public and private institutions, nonprofit and for-profit, that propose to establish, expand, and improve research activities in health sciences and related fields may apply for research grants. Individuals must be nominated and sponsored by public or nonprofit private institutions having staff and facilities appropriate to the proposed research training programs. SBIR grants can be awarded only to domestic small businesses. STTR grants can be awarded only to domestic small business concerns that partner with research institutions in cooperative research and development.
Geographic Focus: All States
Date(s) Application is Due: Apr 5; Aug 5; Dec 5
Contact: Dr. Judith Fradkin, Director; (301) 496-7348
Internet: http://grants.nih.gov/grants/guide/pa-files/PA-07-107.html
Sponsor: National Institute of Diabetes and Digestive and Kidney Diseases
45 Center Drive, Natcher Building
Bethesda, MD 20892-2560

NIDDK Diabetes Research Centers **3147**
The mission of the Diabetes Centers is to serve as a key component of the NIDDK-supported research effort to develop new therapies and improve the health of Americans with, or at risk for, diabetes and related endocrine and metabolic disorders. The Centers promote new discoveries and enhance scientific progress through support of cutting-edge basic and clinical research related to the etiology and complications of diabetes, with the goal of rapidly translating research findings into novel strategies for the prevention, treatment and cure of diabetes and related conditions. This Funding Opportunity Announcement (FOA) solicits new and competing continuation applications for Diabetes Endocrinology Research Centers (DERCs) and Diabetes Research and Training Centers (DRTCs). Both types of Centers are designed to support and enhance the national research effort in diabetes and related endocrine and metabolic diseases. DERCs support three primary research-related activities: biomedical research cores, a Pilot and Feasibility (P&F) program, and an Enrichment program. DRTCs possess all elements of a DERC, with additional dedicated core services and P&F awards to support research in diabetes prevention and control. All activities pursued by Diabetes Centers are designed to enhance the efficiency, productivity, effectiveness and multidisciplinary nature of research in Diabetes Center topic areas.
Requirements: The following organizations and institutions are eligible to apply: public/state controlled institutions of higher education; private institutions of higher education; Hispanic-serving institutions; historically Black colleges and universities (HBCUs); tribally controlled colleges and universities (TCCUs); Alaska Native and Native Hawaiian serving institutions; nonprofits with 501(c)3 IRS status (other than institutions of higher education); nonprofits without 501(c)3 IRS status (other than institutions of higher education); for-profit organizations (other than small businesses); and state governments.
Geographic Focus: All States
Date(s) Application is Due: Jun 22; Jul 15
Contact: James F. Hyde, (301) 594-7692; fax (301) 480-3503; jh486z@nih.gov
Internet: http://grants.nih.gov/grants/guide/rfa-files/RFA-DK-08-008.html
Sponsor: National Institute of Diabetes and Digestive and Kidney Diseases
6707 Democracy Boulevard
Bethesda, MD 20892-5460

NIDDK Diet Composition and Energy Balance Grants **3148**
The goal of this funding opportunity announcement (FOA) is to invite Research Project Grant applications investigating the role of diet composition in energy balance, including studies in both animals and humans. Both short and longer-term studies are encouraged, ranging from basic studies investigating the impact of micro-or macronutrient composition on appetite, metabolism, and energy expenditure through clinical studies evaluating the efficacy of diets differing in micro- or macronutrient composition, absorption, dietary variety, or energy density for weight loss or weight maintenance. Because the nature and scope of the proposed research will vary from application to application, it is anticipated that the size and duration of each award will also vary. The total amount awarded and the number of awards will depend upon the mechanism numbers, quality, duration, and costs of the applications received.
Requirements: The following organizations and institutions may apply: public/state controlled institutions of higher education; private institutions of higher education; Hispanic-serving institutions; historically Black colleges and universities (HBCUs); tribally controlled colleges and universities (TCCUs); Alaska native and native Hawaiian serving institutions; nonprofits with 501(c)3 IRS status; nonprofits without 501(c)3 IRS status; small businesses; for-profit organizations; State governments; regional organizations; U.S. territories or possessions; Indian/Native American tribal governments; Indian/Native American tribally designated organizations; eligible agencies of the Federal government; and faith-based or community based organizations.
Geographic Focus: All States
Date(s) Application is Due: Feb 5; Jun 5; Oct 5
Contact: Susan Z. Yanovski, (301) 594-8882; fax (301) 480-8300; sy29f@nih.gov
Internet: http://grants.nih.gov/grants/guide/pa-files/PA-07-218.html
Sponsor: National Institute of Diabetes and Digestive and Kidney Diseases
6707 Democracy Boulevard
Bethesda, MD 20892-5460

NIDDK Education Grants **3149**
This funding opportunity announcement (FOA) solicits Research Education grant applications from applicant organizations that propose to create educational opportunities to attract undergraduate students, graduate students, and postdoctoral fellows to careers in areas of biomedical or behavioral research of particular interest to the NIDDK while fostering the career development of these students and fellows. The NIDDK is especially interested in attracting students and postdoctoral fellows from scientific disciplines underrepresented in disease-oriented biomedical research such as engineering, informatics, computer science, and computational sciences, to encourage them to apply their expertise to research relevant to diabetes and other endocrine and metabolic diseases, digestive and liver diseases, nutrition, obesity research and prevention, and kidney, urologic and hematologic diseases. The total project period for an application submitted in response to this funding opportunity may not exceed five years. Direct costs are limited to $100,000 per year.
Requirements: The following organizations and institutions may apply: public/state controlled institutions of higher education; private institutions of higher education; Hispanic-serving institutions; historically Black colleges and universities (HBCUs); tribally controlled colleges and universities (TCCUs); Alaska native and native Hawaiian

serving institutions; nonprofits with 501(c)3 IRS status; nonprofits without 501(c)3 IRS status; and Indian/Native American tribally designated organizations.
Geographic Focus: All States
Date(s) Application is Due: Jan 25; May 25; Sep 25
Amount of Grant: Up to 100,000 USD
Contact: James F. Hyde, (301) 594-7692; fax (301) 480-3503; jh486z@nih.gov
Internet: http://grants.nih.gov/grants/guide/pa-files/PAR-06-554.html
Sponsor: National Institute of Diabetes and Digestive and Kidney Diseases
6707 Democracy Boulevard
Bethesda, MD 20892-5460

NIDDK Endoscopic Clinical Research Grants In Pancreatic And Biliary Diseases 3150

These small grants (R03) may be used as planning grants for full-scale multi-center clinical trials or for pilot studies that could lead to full-scale multi-center clinical trials designed to provide evidence for or against changes in the current standard of care. Pilot epidemiological studies are encouraged that could lead to more extended research that would provide evidence for or against changes in health policy, especially as related to disease and cancer prevention. It is expected that these R03 grants will serve as a basis for planning future multi-center research project grant applications (R01) or cooperative agreement (U01) awards. New and experienced investigators in relevant fields and disciplines may apply for these small grants. Investigators are encouraged to take advantage of recent endoscopic and laboratory developments. In addition, the small grant is a good mechanism for new and experienced investigators to become better equipped to perform clinical and epidemiological research. A project period of up to two years and a budget for direct costs of up to two $25,000 modules or $50,000 per year may be requested.
Requirements: The following organizations and institutions may apply: for-profit organizations; non-profit organizations; public or private institutions, such as universities, colleges, hospitals, and laboratories; units of State government; units of local government; eligible agencies of the Federal government; domestic institutions; and foreign institutions.
Geographic Focus: All States
Date(s) Application is Due: Feb 16; Jun 16; Oct 16
Contact: Jose Serrano, (301) 594-8871; fax (301) 480-8300; js362q@nih.gov
Internet: http://grants.nih.gov/grants/guide/pa-files/PAR-06-171.html
Sponsor: National Institute of Diabetes and Digestive and Kidney Diseases
6707 Democracy Boulevard
Bethesda, MD 20892-5460

NIDDK Enhancing Zebrafish Research with Research Tools and Techniques 3151

This FOA encourages investigator-initiated applications designed to exploit the power of the zebrafish as a vertebrate model for biomedical and behavioral research. Applications proposing to develop new research tools or techniques that are of high priority to the zebrafish community and that will advance the detection and characterization of genes, pathways, and phenotypes of interest in development and aging, organ formation, neural processes, behavior, sensory processing, physiological processes, and disease processes are welcome. Budgets for direct costs under $500,000 (direct costs) per year and a project duration of up to five years may be requested. Letters of Intent should arrive by August 17, with full proposals due by September 17.
Requirements: The following organizations and institutions may apply: public/state controlled institutions of higher education; private institutions of higher education; Hispanic-serving institutions; historically Black colleges and universities (HBCUs); tribally controlled colleges and universities (TCCUs); Alaska native and native Hawaiian serving institutions; nonprofits with 501(c)3 IRS status; nonprofits without 501(c)3 IRS status; small businesses; for-profit organizations; State governments; regional organizations; U.S. territories or possessions; Indian/Native American tribal governments; Indian/Native American tribally designated organizations; eligible agencies of the Federal government; and faith-based or community based organizations.
Geographic Focus: All States
Date(s) Application is Due: Aug 17; Sep 17
Contact: Dr. Rebekah S. Rasooly, (301) 594-6007; fax (301) 480-3510; rr185i@nih.gov
Internet: http://grants.nih.gov/grants/guide/pa-files/PAR-08-139.html
Sponsor: National Institute of Diabetes and Digestive and Kidney Diseases
6707 Democracy Boulevard
Bethesda, MD 20892-5460

NIDDK Erythroid Lineage Molecular Toolbox Grants 3152

The ultimate goals of this Funding Opportunity Announcement (FOA) will be to use erythroid cells at different stages of development and differentiation to assemble the complete collection of genes expressed, describe how they are expressed and elucidate physical parameters, including subcellular localization of the proteins translated during erythropoiesis. These reagents may then be used to discover the structure-function relationships that exist in erythroid cells with possible application to other cell types. Because the nature and scope of the proposed research will vary from application to application, it is anticipated that the size and duration of each award will also vary.
Requirements: Application can be made by the following institutions and organizations: public/state controlled institutions of higher education; private institutions of higher education; nonprofits with 501(c)3 IRS status (other than institutions of higher education); nonprofits without 501(c)3 IRS status (other than institutions of higher education); small businesses; for-profit organizations (other than small businesses); state governments; U.S. territories or possessions; Indian/Native American tribal governments (federally recognized); Indian/Native American tribal governments (other than Federally recognized); Indian/Native American tribally designated organizations; non-domestic (non-U.S.) entities (foreign organizations); Hispanic-serving institutions; historically Black colleges and universities (HBCUs); tribally controlled colleges and universities (TCCUs); Alaska Native and Native Hawaiian serving institutions; regional organization; eligible agencies of the Federal government; and faith-based or community based organizations.
Geographic Focus: All States
Date(s) Application is Due: Feb 5; Jun 5; Oct 5
Contact: Terry Rogers Bishop, (301) 594-7726; tb232j@nih.gov
Internet: http://grants.nih.gov/grants/guide/pa-files/PA-07-011.html
Sponsor: National Institute of Diabetes and Digestive and Kidney Diseases
6707 Democracy Boulevard
Bethesda, MD 20892-5460

NIDDK Erythropoiesis: Components and Mechanisms Grants 3153

This FOA issued by the National Institute of Diabetes and Digestive and Kidney Diseases (NIDDK) the National Institute of Aging (NIA), and the National Heart, Lung, and Blood Institute (NHLBI), National Institutes of Health, encourages investigator-initiated R01 applications that propose hypothesis-driven research using erythroid cells. The aim of this program is to support research efforts towards a complete description of the molecular and cellular components of erythropoiesis and how these components contribute to erythropoiesis. Components include genes that are expressed (transcriptome) in erythroid cells, either during development or during differentiation, and the proteins (proteome) that are translated in erythroid cells, especially with post-translational modifications or subcellular localizations that are unique to erythroid cells. A long range goal of this program is to generate a concise description of erythropoiesis that unifies genetics, molecular processes and cytokine determinants in the erythroid lineages so that new therapeutics may be developed to measure and combat anemia. Budgets for direct costs of up to $500,000 per year and a project duration of up to five years may be requested.
Requirements: The following organizations and institutions may apply: public/state controlled institutions of higher education; private institutions of higher education; Hispanic-serving institutions; historically Black colleges and universities (HBCUs); tribally controlled colleges and universities (TCCUs); Alaska native and native Hawaiian serving institutions; nonprofits with 501(c)3 IRS status; nonprofits without 501(c)3 IRS status; small businesses; for-profit organizations; State governments; regional organizations; U.S. territories or possessions; Indian/Native American tribal governments; Indian/Native American tribally designated organizations; county governments; city or township governments; special district governments; independent school districts; public housing authorities/Indian housing authorities; eligible agencies of the Federal government; and faith-based or community based organizations.
Geographic Focus: All States
Date(s) Application is Due: Feb 5; Jun 5; Oct 5
Amount of Grant: Up to 500,000 USD
Contact: Terry Rogers Bishop, (301) 594-7726; tb232j@nih.gov
Internet: http://grants.nih.gov/grants/guide/pa-files/PA-09-255.html
Sponsor: National Institute of Diabetes and Digestive and Kidney Diseases
6707 Democracy Boulevard
Bethesda, MD 20892-5460

NIDDK Exploratory/Developmental Clinical Research Grants in Obesity 3154

The goal of this initiative is to encourage exploratory/developmental clinical research that will accelerate the development of effective interventions for prevention or treatment of overweight or obesity in adults and/or children. The goal of this mechanism is to provide flexibility for initiating exploratory, short-term studies, thus allowing new ideas to be investigated in a more expeditious manner without stringent requirements for preliminary data. Such support is needed to encourage investigators to pursue new approaches, underdeveloped topics, or more creative avenues for research including new partnerships. Epidemiological research with a goal of informing translational/clinical research on prevention or treatment of obesity or overweight in adults and/or children is encouraged. The emphasis is thus on the development of exploratory clinical studies, pilot and feasibility studies, or small randomized clinical trials that will provide preliminary data for intervention and epidemiological studies that will inform translational/clinical research.
Requirements: The following organizations and institutions may apply: public/state controlled institutions of higher education; private institutions of higher education; Hispanic-serving institutions; historically Black colleges and universities (HBCUs); tribally controlled colleges and universities (TCCUs); Alaska native and native Hawaiian serving institutions; nonprofits with 501(c)3 IRS status; nonprofits without 501(c)3 IRS status; small businesses; for-profit organizations; State governments; regional organizations; U.S. territories or possessions; Indian/Native American tribal governments; Indian/Native American tribally designated organizations; eligible agencies of the Federal government; and faith-based or community based organizations.
Geographic Focus: All States
Date(s) Application is Due: Feb 16; Jun 16; Oct 16
Amount of Grant: 200,000-275,000 USD
Contact: Carolyn W. Miles, (301)451-3759; fax (301) 480-8300; cm294e@nih.gov
Internet: http://grants.nih.gov/grants/guide/pa-files/PA-09-124.html
Sponsor: National Institute of Diabetes and Digestive and Kidney Diseases
6707 Democracy Boulevard
Bethesda, MD 20892-5460

NIDDK Grants for Basic Research in Glomerular Diseases 3155

This FOA invites applications from investigators with diverse scientific interests to apply their expertise in basic research to enhance the understanding of the pathogenesis

of the various primary or secondary forms of glomerular disease. Recent observations regarding intrinsic glomerular cell biology, particularly in the podocyte, have provided exciting new insights into potential pathogenic mechanisms of human glomerular disease. Although both immune and nonimmune mechanisms of glomerular injury have been studied previously, experimental models of disease and recent techniques that provide tools for molecular and proteomic profiling show great promise for identifying glomerular disease biomarkers. Despite these recent advances, additional basic studies are needed to facilitate a better understanding of glomerular disease pathogenesis and ultimately impact on the treatment of human glomerulopathies. It is anticipated that applications submitted in response to this FOA could address a number of different aspects concerning the pathogenesis, natural history, therapy pre-emption or prevention of the various morphologic forms of experimental glomerular disease.
Requirements: The following organizations and institutions may apply: public/state controlled institutions of higher education; private institutions of higher education; Hispanic-serving institutions; historically Black colleges and universities (HBCUs); tribally controlled colleges and universities (TCCUs); Alaska native and native Hawaiian serving institutions; nonprofits with 501(c)3 IRS status; nonprofits without 501(c)3 IRS status; small businesses; for-profit organizations; State governments; regional organizations; U.S. territories or possessions; Indian/Native American tribal governments; Indian/Native American tribally designated organizations; eligible agencies of the Federal government; and faith-based or community based organizations.
Geographic Focus: All States
Date(s) Application is Due: Feb 5; Jun 5; Oct 5
Contact: Marva Moxey-Mims, (301) 594-7717; fax (301) 480-3510; mm726k@nih.gov
Internet: http://grants.nih.gov/grants/guide/pa-files/PA-07-367.html
Sponsor: National Institute of Diabetes and Digestive and Kidney Diseases
6707 Democracy Boulevard
Bethesda, MD 20892-5460

NIDDK Health Disparities in NIDDK Diseases **3156**
It is recognized that there are many diseases and disorders that disproportionately affect the health of racial and ethnic minority populations in the United States. It is evident that African-Americans, Hispanic Americans, American Indians, Alaska Natives, some Asian-Americans, and Native Hawaiians and other Pacific Islanders experience much higher risks and poorer health status than the general population. Several of the diseases that disproportionately afflict minorities are high priority research areas for NIDDK, including diabetes, obesity, nutrition-related disorders, hepatitis C, gallbladder disease, H. Pylori infection, sickle cell disease, kidney diseases, and metabolic, gastrointestinal, hepatic, and renal complications from infection with HIV. NIDDK encourages efforts health disparity research to reduce the human and economic costs resulting from inequities in health care and health outcomes. Because the nature and scope of the proposed research will vary from application to application, it is anticipated that the size and duration of each award will also vary. The total amount awarded and the number of awards will depend upon the mechanism numbers, quality, duration, and costs of the applications received.
Requirements: The following organizations and institutions may apply: public/state controlled institutions of higher education; private institutions of higher education; Hispanic-serving institutions; historically Black colleges and universities (HBCUs); tribally controlled colleges and universities (TCCUs); Alaska native and native Hawaiian serving institutions; nonprofits with 501(c)3 IRS status; nonprofits without 501(c)3 IRS status; small businesses; for-profit organizations; State governments; regional organizations; U.S. territories or possessions; Indian/Native American tribal governments; Indian/Native American tribally designated organizations; eligible agencies of the Federal government; and faith-based or community based organizations.
Geographic Focus: All States
Date(s) Application is Due: Feb 5; Jun 5; Oct 5
Contact: Myrlene Staten, (301) 402-7886; fax (301) 480-3503; ms808k@nih.gov
Internet: http://grants.nih.gov/grants/guide/pa-files/PA-07-027.html
Sponsor: National Institute of Diabetes and Digestive and Kidney Diseases
6707 Democracy Boulevard
Bethesda, MD 20892-5460

NIDDK Identifying and Reducing Diabetes and Obesity Related Health Disparities within Healthcare Systems **3157**
This FOA requests applications designed to identify or address factors or barriers that result in disparate outcomes within a healthcare system. All applications should measure the impact of identified factors or interventions on health outcomes. Research is sought that examines at least one of the following factors in a healthcare system and/or the interaction between these factors; healthcare professionals, healthcare organizations, and the patients and communities they serve.
Requirements: The following organizations and institutions may apply: public/state controlled institutions of higher education; private institutions of higher education; Hispanic-serving institutions; historically Black colleges and universities (HBCUs); tribally controlled colleges and universities (TCCUs); Alaska native and native Hawaiian serving institutions; nonprofits with 501(c)3 IRS status; nonprofits without 501(c)3 IRS status; small businesses; for-profit organizations; State governments; regional organizations; U.S. territories or possessions; Indian/Native American tribal governments; Indian/Native American tribally designated organizations; eligible agencies of the Federal government; and faith-based or community based organizations.
Geographic Focus: All States
Date(s) Application is Due: Feb 5; Jun 5; Oct 5
Contact: Christine Hunter, (301) 594-4728; fax (301) 480-3503; ch514c@.nih.gov
Internet: http://grants.nih.gov/grants/guide/pa-files/PA-07-388.html
Sponsor: National Institute of Diabetes and Digestive and Kidney Diseases
6707 Democracy Boulevard
Bethesda, MD 20892-5460

NIDDK Insulin Signaling and Receptor Cross Talk Grants **3158**
This Funding Opportunity Announcement (FOA) solicits investigator-initiated research projects that will define the mechanistic basis of crosstalk between the insulin receptor signaling pathway and other signaling pathways, with the goal of defining how insulin action is influenced by the complex microenvironment found in insulin responsive tissues.
Requirements: The following organizations and institutions may apply: public/state controlled institutions of higher education; private institutions of higher education; Hispanic-serving institutions; historically Black colleges and universities (HBCUs); tribally controlled colleges and universities (TCCUs); Alaska native and native Hawaiian serving institutions; nonprofits with 501(c)3 IRS status; nonprofits without 501(c)3 IRS status; small businesses; for-profit organizations; State governments; regional organizations; U.S. territories or possessions; Indian/Native American tribal governments; Indian/Native American tribally designated organizations; eligible agencies of the Federal government; and faith-based or community based organizations.
Geographic Focus: All States
Date(s) Application is Due: Feb 5; Jun 5; Oct 5
Contact: Kristin Abraham, (301) 451-8048; fax (301) 480-3503; ka136s@nih.gov
Internet: http://grants.nih.gov/grants/guide/pa-files/PA-07-058.html
Sponsor: National Institute of Diabetes and Digestive and Kidney Diseases
6707 Democracy Boulevard
Bethesda, MD 20892-5460

NIDDK Intestinal Failure, Short Gut Syndrome and Small Bowel Transplantation Grants **3159**
The National Institute of Diabetes and Digestive and Kidney Diseases (NIDDK) seeks grant applications to study the pathogenesis, natural history, treatment and complications of intestinal failure and its therapies, including parenteral nutrition and small bowel transplantation. The overall objective of this funding opportunity is to encourage basic and clinical research into intestinal failure, short gut syndrome and intestinal transplantation. Because the nature and scope of the proposed research will vary from application to application, it is anticipated that the size and duration of each award will also vary.
Requirements: Application can be made by the following institutions and organizations: public/state controlled institutions of higher education; private institutions of higher education; nonprofits with 501(c)3 IRS status (other than institutions of higher education); nonprofits without 501(c)3 IRS status (other than institutions of higher education); small businesses; for-profit organizations (other than small businesses); state governments; U.S. territories or possessions; Indian/Native American tribal governments (federally recognized); Indian/Native American tribal governments (other than Federally recognized); Indian/Native American tribally designated organizations; non-domestic (non-U.S.) entities (foreign organizations); Hispanic-serving institutions; historically Black colleges and universities (HBCUs); tribally controlled colleges and universities (TCCUs); Alaska Native and Native Hawaiian serving institutions; regional organization; eligible agencies of the Federal government; and faith-based or community based organizations.
Geographic Focus: All States
Date(s) Application is Due: Feb 5; Jun 5; Oct 5
Contact: Dr. Michael K. May, (301) 594-8884; fax (301) 480-8300; mm102i@nih.gov
Internet: http://grants.nih.gov/grants/guide/pa-files/PA-07-010.html
Sponsor: National Institute of Diabetes and Digestive and Kidney Diseases
6707 Democracy Boulevard
Bethesda, MD 20892-5460

NIDDK Intestinal Stem Cell Consortium Grants **3160**
This FOA invites new applications to participate in the Intestinal Stem Cell Consortium (ISCC). The goals of this initiative are: to establish a network of individual research projects and a Coordinating Center focused on research on stem cells of the small intestine; to isolate, culture, characterize, functionally validate, and compare stem cell populations from the small intestine epithelium in vivo and in vitro; to share data, biomaterials, models, reagents, resources and methods between projects in the ISCC; and make these publicly available through a web site to be developed by the Coordinating Center. The objective of support for this consortium is to accelerate research on stem cells of the intestinal epithelium in order to facilitate understanding of the biology and function of the intestine and aid development of therapies for intestinal diseases and conditions where damage and replacement of this epithelium are important components. Direct costs are limited to $250,000 per year for a maximum five-year period for research projects and $350,000 per year for a maximum five-year period for the Coordinating Center.
Requirements: Application can be made by the following institutions and organizations: public/state controlled institutions of higher education; private institutions of higher education; nonprofits with 501(c)3 IRS status (other than institutions of higher education); nonprofits without 501(c)3 IRS status (other than institutions of higher education); small businesses; for-profit organizations (other than small businesses); state governments; U.S. territories or possessions; Indian/Native American tribal governments (federally recognized); Indian/Native American tribal governments (other than Federally recognized); Indian/Native American tribally designated organizations; non-domestic (non-U.S.) entities (foreign organizations); Hispanic-serving institutions; historically Black colleges and universities (HBCUs); tribally controlled colleges and universities (TCCUs); Alaska Native and Native Hawaiian serving institutions; regional organization; eligible agencies of the Federal government; and faith-based or community based organizations.
Geographic Focus: All States
Date(s) Application is Due: Mar 18
Amount of Grant: Up to 250,000 USD

Contact: Ronald Margolis, (301) 594-8819; fax (301) 435-6047; rm76f@nih.gov
Internet: http://grants.nih.gov/grants/guide/rfa-files/RFA-DK-08-010.html
Sponsor: National Institute of Diabetes and Digestive and Kidney Diseases
6707 Democracy Boulevard
Bethesda, MD 20892-5460

NIDDK Mentored Research Scientist Development Award **3161**
The overall objective of the NIH Research Career Development Award program is to prepare qualified individuals for careers that have a significant impact on the health-related research needs of the Nation. The objective of the NIDDK Mentored Research Scientist Development Award (K01) is to provide support for a sustained period of protected time for intensive research career development under the guidance of an experienced mentor, or sponsor, in the biomedical, behavioral or clinical sciences leading to research independence, thereby ensuring a future cadre of well-trained Ph.D. scientists working in research areas supported by NIDDK. The expectation is that through this sustained period of research career development and training, awardees will launch independent research careers and become competitive for new research project grant (R01) funding. By providing support for the critical transition period between postdoctoral training and independent R01 funding for non-clinical investigators, the NIDDK hopes to foster the careers of these investigators who are vital for the future excellence of the NIDDK research endeavor. Applicants must justify the need for a period of mentored research experience and provide a convincing case that the proposed period of support will substantially enhance their careers as independent investigators.
Requirements: The following organizations and institutions may apply: public/state controlled institutions of higher education; private institutions of higher education; Hispanic-serving institutions; historically Black colleges and universities (HBCUs); tribally controlled colleges and universities (TCCUs); Alaska native and native Hawaiian serving institutions; nonprofits with 501(c)3 IRS status; nonprofits without 501(c)3 IRS status; small businesses; for-profit organizations; State governments; regional organizations; U.S. territories or possessions; Indian/Native American tribal governments; Indian/Native American tribally designated organizations; eligible agencies of the Federal government; and faith-based or community based organizations.
Geographic Focus: All States
Date(s) Application is Due: Mar 12; Jul 12; Nov 12
Contact: James F. Hyde, (301) 594-7692; fax (301) 480-3503; jh486z@nih.gov
Internet: http://grants.nih.gov/grants/guide/pa-files/PAR-09-060.html
Sponsor: National Institute of Diabetes and Digestive and Kidney Diseases
6707 Democracy Boulevard
Bethesda, MD 20892-5460

NIDDK Multi-Center Clinical Study Cooperative Agreements **3162**
This PAR provides for grant applications for investigator-initiated, multi-center clinical studies. NIDDK supports investigator-initiated, multi-center clinical studies through a two-part process that includes an implementation planning (U34) grant. Beginning with the receipt date, NIDDK will accept, peer review, and consider for funding applications for investigator-initiated, multi-center clinical studies only from NIDDK Multi-Center Clinical Study Implementation Planning (U34) grant awardees. Because the nature and scope of the proposed research will vary from application to application, it is anticipated that the size and duration of each award will also vary. The total amount awarded and the number of awards will depend upon the mechanism numbers, quality, duration, and costs of the applications received.
Requirements: The following organizations and institutions may apply: public/state controlled institutions of higher education; private institutions of higher education; Hispanic-serving institutions; historically Black colleges and universities (HBCUs); tribally controlled colleges and universities (TCCUs); Alaska native and native Hawaiian serving institutions; nonprofits with 501(c)3 IRS status; nonprofits without 501(c)3 IRS status; small businesses; for-profit organizations; State governments; regional organizations; U.S. territories or possessions; Indian/Native American tribal governments; Indian/Native American tribally designated organizations; eligible agencies of the Federal government; and faith-based or community based organizations.
Geographic Focus: All States
Date(s) Application is Due: Feb 5; Jun 5; Oct 5
Contact: Catherine Meyers, (301) 451-4901; fax (301) 480-3510; meyersc@mail.nih.gov
Internet: http://grants.nih.gov/grants/guide/pa-files/PAR-08-058.html
Sponsor: National Institute of Diabetes and Digestive and Kidney Diseases
6707 Democracy Boulevard
Bethesda, MD 20892-5460

NIDDK Multi-Center Clinical Study Implementation Planning Grants **3163**
NIDDK supports investigator-initiated, multi-center clinical studies through a two-part process that includes an implementation planning grant. The planning grant is designed to: permit early peer review of the rationale for the proposed clinical study; permit assessment of the design/protocol of the proposed study; provide support for the development of a complete study protocol and associated documents including a manual of operations; and support the development of other essential elements required for the conduct of a clinical study. Completion of the required products of a U34 grant is a prerequisite for submission of a multi-center clinical study cooperative agreement application, which will support the actual conduct of the study. Because the nature and scope of the proposed research will vary from application to application, it is anticipated that the size and duration of each award will also vary. The total project period for an application submitted in response to this funding opportunity may not exceed two years. Applicants may request up to $250,000 in direct costs per year.
Requirements: The following organizations and institutions may apply: public/state controlled institutions of higher education; private institutions of higher education; Hispanic-serving institutions; historically Black colleges and universities (HBCUs); tribally controlled colleges and universities (TCCUs); Alaska native and native Hawaiian serving institutions; nonprofits with 501(c)3 IRS status; nonprofits without 501(c)3 IRS status; small businesses; for-profit organizations; State governments; regional organizations; U.S. territories or possessions; Indian/Native American tribal governments; Indian/Native American tribally designated organizations; eligible agencies of the Federal government; and faith-based or community based organizations.
Geographic Focus: All States
Date(s) Application is Due: Mar 18; Jun 23
Contact: Catherine Meyers, (301) 451-4901; fax (301) 480-3510; meyersc@mail.nih.gov
Internet: http://grants.nih.gov/grants/guide/pa-files/PAR-08-057.html
Sponsor: National Institute of Diabetes and Digestive and Kidney Diseases
6707 Democracy Boulevard
Bethesda, MD 20892-5460

NIDDK Neurobiology of Diabetic Complications Grants **3164**
NIDDK invites investigator-initiated research grant applications to study the mechanisms by which diabetes results in painful and disabling neuropathies and other neurological complications and to apply this information to the development of interventions to prevent, limit, or reverse these complications. The intent of this RFA is to attract basic neuroscientists to the study of diabetic neuropathy and neurobiology relevant to diabetes, and enhance interdisciplinary approaches to research in this area. Annual deadline dates may vary; contact the appropriate NIH institute for exact dates. Letters of intent are due by March 24, with full applications due by April 25 each year.
Requirements: Applications may be submitted by domestic and foreign, for-profit and nonprofit organizations, both public and private; units of state and local governments; and eligible agencies of the federal government.
Geographic Focus: All States
Date(s) Application is Due: Mar 24; Apr 25
Amount of Grant: Up to 3,000,000 USD
Contact: Dr. Paul Nichols, (301) 496-9964; fax (301) 402-2060; pn13w@nih.gov
Internet: http://grants.nih.gov/grants/guide/rfa-files/RFA-NS-00-002.html
Sponsor: National Institute of Diabetes and Digestive and Kidney Diseases
45 Center Drive, MSC 6600
Bethesda, MD 20892-6600

NIDDK Neuroimaging in Obesity Research Grants **3165**
This FOA issued by the National Institute of Diabetes and Digestive and Kidney Diseases, the National Institute on Drug Abuse, and the National Institute of Biomedical Imaging and Bioengineering, of the National Institutes of Health, solicits Research Project Grant (R01) applications from institutions/ organizations that propose to use neuroimaging approaches in obesity research in human subjects and animal models. Many areas of the brain interact or communicate with other organs to control eating behavior, physical activity and energy metabolism, and functional neuroimaging holds enormous promise for expanding our understanding of how food intake and energy expenditure are mismatched in a setting of abundantly available nutrients, leading to excessive fat storage. The total amount of funding that the ICs expect to award through this announcement is up to $4.5 Million, to fund approximately 9-15 awards. The duration and requested direct costs will vary depending on the scope of the proposed research, but cannot exceed 5 years and $500,000 per year. Letters of intent are due by February 18, with full applications due by 5:00 p.m. on March 18.
Requirements: Application can be made by the following institutions and organizations: public/state controlled institutions of higher education; private institutions of higher education; nonprofits with 501(c)3 IRS status (other than institutions of higher education); nonprofits without 501(c)3 IRS status (other than institutions of higher education); small businesses; for-profit organizations (other than small businesses); state governments; U.S. territories or possessions; Indian/Native American tribal governments (federally recognized); Indian/Native American tribal governments (other than Federally recognized); Indian/Native American tribally designated organizations; non-domestic (non-U.S.) entities (foreign organizations); Hispanic-serving institutions; historically Black colleges and universities (HBCUs); tribally controlled colleges and universities (TCCUs); Alaska Native and Native Hawaiian serving institutions; regional organization; eligible agencies of the Federal government; and faith-based or community based organizations.
Geographic Focus: All States
Date(s) Application is Due: Feb 18; Mar 18
Amount of Grant: Up to 500,000 USD
Contact: Maren R. Laughlin, (301) 594-8802; fax (301) 480-0475; ml33q@nih.gov
Internet: http://grants.nih.gov/grants/guide/rfa-files/RFA-DK-08-009.html
Sponsor: National Institute of Diabetes and Digestive and Kidney Diseases
6707 Democracy Boulevard
Bethesda, MD 20892-5460

NIDDK New Technologies for Liver Disease SBIR **3166**
The purpose of this Funding Opportunity Announcement (FOA) is to solicit Small Business Innovation Research (SBIR) grant applications from small business concerns (SBCs) that propose to develop resources, research tools, instrumentations, biomarkers, devices, drugs or new and innovative approaches to diagnosis, monitoring, management, treatment and prevention of liver diseases. Areas of interest include development of reliable and practical means of diagnosis of liver diseases; biomarkers for disease activity and stage; noninvasive tests for inflammation, fibrosis and fat in the liver; and drugs, complementary and alternative modalities, biologics or molecular reagents for the therapy or prevention of liver diseases. The

goal of this announcement is to enlist members of the small business research community in advancing means of diagnosis, treatment and prevention of liver disease and facilitate the goals outlined in the trans-NIH Action Plan for Liver Disease Research.
Requirements: Only United States small business concerns (SBCs) are eligible to submit SBIR applications.
Geographic Focus: All States
Date(s) Application is Due: Apr 5; Aug 5; Dec 5
Contact: Christine L. Densmore, (301) 402-8714; fax (301) 480-8300; cd121z@nih.gov
Internet: http://grants.nih.gov/grants/guide/pa-files/PA-09-095.html
Sponsor: National Institute of Diabetes and Digestive and Kidney Diseases
6707 Democracy Boulevard
Bethesda, MD 20892-5460

NIDDK New Technologies for Liver Disease STTR 3167
The objective of this FOA is to encourage and enable scientists at small businesses to develop and evaluate new technologies, drugs, devices, and approaches to diagnosis, management, and prevention of liver disease. Development of new technologies as well as application of existing technologies may be proposed. Studies may include use of animal models or human participants or both. If appropriate, plans for manufacturing and clinical evaluation of developed technologies, drugs, devices and innovative approaches should be included in the application. However, clinical trials beyond Phase I studies will not be considered appropriate to this announcement.
Requirements: Only United States small business concerns (SBCs) are eligible to submit STTR applications.
Geographic Focus: All States
Date(s) Application is Due: Apr 5; Aug 5; Dec 5
Contact: Christine L. Densmore, (301) 402-8714; fax (301) 480-8300; cd121z@nih.gov
Internet: http://grants.nih.gov/grants/guide/pa-files/PA-09-094.html
Sponsor: National Institute of Diabetes and Digestive and Kidney Diseases
6707 Democracy Boulevard
Bethesda, MD 20892-5460

NIDDK Non-Invasive Methods for Diagnosis and Progression of Diabetes, Kidney, Urological, Hematological and Digestive Diseases 3168
This funding opportunity is a call for the application of imaging and other non- or minimally-invasive technologies to detect, characterize, diagnose, identify persons with predisposition to, or monitor treatment of diseases of interest to the National Institute of Diabetes and Digestive and Kidney Diseases. Also needed are new, robust surrogate markers for clinical trial endpoints, and new ways to characterize normal and pathological tissues in vivo. Diseases of interest include type 1 and 2 diabetes, obesity, and kidney, liver, urologic, hematologic, digestive, endocrine and metabolic diseases and their complications. Applicable techniques include molecular imaging and functional imaging approaches, imaging methods with high spatial, chemical or time resolution, and new spectroscopic or sensor array technologies for monitoring metabolic or physiological events.
Requirements: The following organizations and institutions may apply: for-profit organizations; non-profit organizations; public or private institutions, such as universities, colleges, hospitals, and laboratories; units of State government; units of local government; eligible agencies of the Federal government; foreign institutions; domestic institutions; and faith-based or community-based organizations.
Geographic Focus: All States
Date(s) Application is Due: Feb 5; Jun 5; Oct 5
Contact: Maren R. Laughlin, (301) 594-8802; fax (301) 480-0475; ml33q@nih.gov
Internet: http://grants.nih.gov/grants/guide/pa-files/PA-06-143.html
Sponsor: National Institute of Diabetes and Digestive and Kidney Diseases
6707 Democracy Boulevard
Bethesda, MD 20892-5460

NIDDK Pancreatic Development and Regeneration Grants: Toward Cellular Therapies for Diabetes 3169
This Funding Opportunity Announcement (FOA) is intended to promote basic research in pancreatic development, stem cell biology and regeneration that will provide the foundation for the rationale design of future cell replacement therapies for diabetes. Appropriate topics for investigation include, but are not limited to: prospective isolation, purification and characterization of pancreatic stem/progenitor cells; identification of the signals, signaling pathway components and transcriptional factors that could direct the formation of pancreatic endoderm and progenitors, islet cell differentiation, morphogenesis, growth, function and survival in vitro; use of existing NIH-approved human ES cell lines or mouse embryonic cells, pancreatic stem cells, non-pancreatic adult stem cells such as those from liver or gut, human amniotic cells or bone marrow-derived cells to efficiently generate endoderm, pancreatic progenitor cells, or differentiated beta cells in culture; development of methods and reagents to assess quantitatively, the survival, migration, fate and function of transplanted animal or human pancreatic progenitors and their progeny in the living animal; determination of the factors important for beta cell regeneration and designing strategies to foster regeneration; development of new animal models that can used to identify exogenous or endogenous factors that might promote beta cell regeneration or neogenesis; characterization the lineage of cells that may contribute or participate in beta cell regeneration by genetic marking; and development of assays to screen chemical libraries for modifiers of beta cell regeneration, differentiation, and or expansion.
Requirements: The following organizations and institutions may apply: public/state controlled institutions of higher education; private institutions of higher education; Hispanic-serving institutions; historically Black colleges and universities (HBCUs); tribally controlled colleges and universities (TCCUs); Alaska native and native Hawaiian serving institutions; nonprofits with 501(c)3 IRS status; nonprofits without 501(c)3 IRS status; small businesses; for-profit organizations; State governments; regional organizations; U.S. territories or possessions; Indian/Native American tribal governments; Indian/Native American tribally designated organizations; eligible agencies of the Federal government; and faith-based or community based organizations.
Geographic Focus: All States
Date(s) Application is Due: Feb 5; Jun 5; Oct 5
Contact: Sheryl M. Sato, (301) 594-8811; fax (301) 480-0475; ss68z@nih.gov
Internet: http://grants.nih.gov/grants/guide/pa-files/PA-07-056.html
Sponsor: National Institute of Diabetes and Digestive and Kidney Diseases
6707 Democracy Boulevard
Bethesda, MD 20892-5460

NIDDK Pilot and Feasibility Clinical Research Grants in Diabetes, Endocrine and Metabolic Diseases 3170
The goal of this initiative is to encourage exploratory/developmental clinical research related to the prevention or treatment of diabetes, obesity and endocrine and genetic metabolic diseases. The Pilot and Feasibility Clinical Research Grants Program is designed to allow initiation of exploratory, short-term clinical studies, so that new ideas may be investigated without stringent requirements for preliminary data. Such support can be used by experienced investigators, as well as new investigators, to pursue new approaches and investigate underdeveloped research areas. The short-term studies should focus on research questions that are likely to have high clinical impact. They can include testing a new prevention strategy, a new intervention, or unique combinations of therapies. A high priority is the use of such studies to help stimulate the translation of promising research developments from the laboratory into clinical practice in diabetes, endocrine diseases and genetic metabolic diseases, including cystic fibrosis.
Requirements: The following organizations and institutions may apply: public/state controlled institutions of higher education; private institutions of higher education; Hispanic-serving institutions; historically Black colleges and universities (HBCUs); tribally controlled colleges and universities (TCCUs); Alaska native and native Hawaiian serving institutions; nonprofits with 501(c)3 IRS status; nonprofits without 501(c)3 IRS status; small businesses; for-profit organizations; State governments; regional organizations; U.S. territories or possessions; Indian/Native American tribal governments; Indian/Native American tribally designated organizations; eligible agencies of the Federal government; and faith-based or community based organizations.
Geographic Focus: All States
Date(s) Application is Due: Feb 16; Jun 16; Oct 16
Contact: Karen Salomon, (301) 594-7733; fax (301) 435-6047; ks495c@nih.gov
Internet: http://grants.nih.gov/grants/guide/pa-files/PA-09-133.html
Sponsor: National Institute of Diabetes and Digestive and Kidney Diseases
6707 Democracy Boulevard
Bethesda, MD 20892-5460

NIDDK Pilot and Feasibility Clinical Research Grants in Kidney or Urologic Diseases 3171
This funding opportunity announcement (FOA) is a re-issuance of PAR-06-113 and specifically encourages the submission of applications for small scale or pilot and feasibility clinical and translational research studies, including epidemiological studies or clinical trials related to kidney or urologic disease research that address important clinical and translational questions and are potentially of high impact. It is anticipated that some applications for pilot and feasibility studies may lead to full-scale clinical studies including evaluation of diagnostic strategies, epidemiologic studies, or trials in the diagnosis, prevention, or treatment of kidney or urologic diseases. These grants may be used to plan, evaluate the feasibility, or implement clinical trials that assess pharmacological, dietary, surgical, or behavioral interventions for the prevention or treatment of kidney or urologic disease. Small scale or pilot epidemiological studies are also encouraged.
Requirements: The following organizations and institutions may apply: public/state controlled institutions of higher education; private institutions of higher education; Hispanic-serving institutions; historically Black colleges and universities (HBCUs); tribally controlled colleges and universities (TCCUs); Alaska native and native Hawaiian serving institutions; nonprofits with 501(c)3 IRS status; nonprofits without 501(c)3 IRS status; small businesses; for-profit organizations; State governments; regional organizations; U.S. territories or possessions; Indian/Native American tribal governments; Indian/Native American tribally designated organizations; eligible agencies of the Federal government; and faith-based or community based organizations.
Geographic Focus: All States
Date(s) Application is Due: Feb 16; Jun 16; Oct 16
Amount of Grant: 200,000-275,000 USD
Contact: Catherine M. Meyers, (301) 594-7717; fax (301) 480-3510; cm420i@nih.gov
Internet: http://grants.nih.gov/grants/guide/pa-files/PAR-09-077.html
Sponsor: National Institute of Diabetes and Digestive and Kidney Diseases
6707 Democracy Boulevard
Bethesda, MD 20892-5460

NIDDK Pilot and Feasibility Clinical Research Studies in Digestive Diseases and Nutrition 3172
The goal of this pilot and feasibility mechanism is to provide flexibility for initiating preliminary, short-term clinical studies, thus allowing new ideas to be investigated in a more expeditious manner without stringent requirements for preliminary data. Such support is needed to encourage experienced investigators as well as new investigators to pursue new approaches, underdeveloped topics, or more creative avenues of research. If successful, these awards should lead to significant scientific advances in digestive disease research and should facilitate translation into clinical practice to improve patient outcomes.

Requirements: The following organizations and institutions are eligible to apply: public/state controlled institutions of higher education; private institutions of higher education; Hispanic-serving institutions; historically Black colleges and universities (HBCUs); tribally controlled colleges and universities (TCCUs); Alaska Native and Native Hawaiian serving institutions; nonprofits with 501(c)3 IRS status (other than institutions of higher education); nonprofits without 501(c)3 IRS status (other than institutions of higher education); small businesses; for-profit organizations (other than small businesses); state governments; Indian/Native American tribal governments (federally recognized); Indian/Native American tribally designated organizations; county governments; city or township governments; special district governments; independent school districts; public housing authorities; Indian housing authorities; U.S. territory or possession; Indian/Native American tribal governments (other than federally recognized); regional organizations; non-domestic (non-U.S.) entities (foreign organizations); other eligible agencies of the federal government; and faith-based or community-based organizations.
Geographic Focus: All States
Date(s) Application is Due: Feb 16; Jun 16; Oct 16
Contact: Mary Evans, (301) 594-4578; fax (301) 480-8300; me189d@nih.gov
Internet: http://grants.nih.gov/grants/guide/pa-files/PA-09-151.html
Sponsor: National Institute of Diabetes and Digestive and Kidney Diseases
6707 Democracy Boulevard
Bethesda, MD 20892-5460

NIDDK Planning Grants For Translational Research For The Prevention and Control Of Diabetes And Obesity 3173

The National Institute of Diabetes and Digestive and Kidney Diseases (NIDDK) seeks to develop cost effective and sustainable interventions, that can be adopted in real world settings, for the prevention and control of diabetes and obesity. Research should be based on interventions already proven efficacious in clinical trials to prevent and reverse obesity and type 2 diabetes, to improve care of type 1 and type 2 diabetes and to prevent or delay its complications. The interventions proposed under this solicitation should have the potential to be widely disseminated to clinical practice, individuals and communities at risk. Relevant topics include: methods to improve health care delivery to patients with or at risk of diabetes; to enhance diabetes self management; methods to develop strategies to promote healthy lifestyles to reduce the risk of diabetes; and methods to promote lifestyle change that will prevent or reverse overweight and obesity, and 5) new cost effective ways to identify people with pre-diabetes and undiagnosed diabetes. This FOA will use the National Institutes of Health (NIH) clinical trial planning grant (R34) award mechanism, which will provide up to $150,000 in direct costs per year.
Requirements: Applications may be submitted by: for-profit organizations; non-profit organizations; public or private institutions, such as universities, colleges, hospitals, and laboratories; units of State government; units of local government; eligible agencies of the Federal government; domestic institutions; faith-based or community-based organizations; Indian/Native American tribal governments; and Indian/Native American tribally designated organizations.
Geographic Focus: All States
Date(s) Application is Due: Feb 16; Jun 16; Oct 16
Amount of Grant: Up to 150,000 USD
Contact: Sanford Garfield, (301) 594-8803; fax (301) 480-3503; sg50o@nih.gov
Internet: http://grants.nih.gov/grants/guide/pa-files/PAR-06-358.html
Sponsor: National Institute of Diabetes and Digestive and Kidney Diseases
6707 Democracy Boulevard
Bethesda, MD 20892-5460

NIDDK Proteomics: Diabetes, Obesity, And Endocrine, Digestive, Kidney, Urologic, And Hematologic Diseases 3174

This Funding Opportunity Announcement (FOA) issued by the National Institute of Diabetes and Digestive and Kidney Diseases (NIDDK), National Institutes of Health (NIH), solicits projects that advance research to identify and quantitate protein expression patterns, post-translational modification of proteins, and protein-protein interactions on cells, tissues, organ systems relevant to diabetes, obesity, endocrine and metabolic diseases, nutritional function and diseases of the alimentary tract, exocrine pancreas, liver, kidney, bladder and prostate and normal biological processes related to the function of these systems. The development and improvement of innovative proteomic technologies is also encouraged through their application to relevant biological questions related to the pathophysiology of endocrine glands, gastrointestinal tract, liver and kidney, bladder and prostate.
Requirements: The following organizations and institutions may apply: public/state controlled institutions of higher education; private institutions of higher education; Hispanic-serving institutions; historically Black colleges and universities (HBCUs); tribally controlled colleges and universities (TCCUs); Alaska native and native Hawaiian serving institutions; nonprofits with 501(c)3 IRS status; nonprofits without 501(c)3 IRS status; small businesses; for-profit organizations; State governments; regional organizations; U.S. territories or possessions; Indian/Native American tribal governments; Indian/Native American tribally designated organizations; eligible agencies of the Federal government; and faith-based or community based organizations.
Geographic Focus: All States
Date(s) Application is Due: Feb 5; Jun 5; Oct 5
Contact: Salvatore Sechi, (301) 594-8814; fax (301) 480-2688; ss24q@nih.gov
Internet: http://grants.nih.gov/grants/guide/pa-files/PA-07-016.html
Sponsor: National Institute of Diabetes and Digestive and Kidney Diseases
6707 Democracy Boulevard
Bethesda, MD 20892-5460

NIDDK Research Grants for Studies of Hepatitis C in the Setting of Renal Disease 3175

It is the intent of this Funding Opportunity Announcement (FOA) to invite applications from investigators with diverse scientific interests, who wish to apply their expertise into basic and applied research to enhance the understanding of the pathogenesis, natural history, therapy,and prevention of hepatitis C in the setting of renal disease or renal transplantation. Because the nature and scope of the proposed research will vary from application to application, it is anticipated that the size and duration of each award will also vary. The total amount awarded and the number of awards will depend upon the mechanism numbers, quality, duration, and costs of the applications received.
Requirements: The following organizations and institutions may apply: public/state controlled institutions of higher education; private institutions of higher education; Hispanic-serving institutions; historically Black colleges and universities (HBCUs); tribally controlled colleges and universities (TCCUs); Alaska native and native Hawaiian serving institutions; nonprofits with 501(c)3 IRS status; nonprofits without 501(c)3 IRS status; small businesses; for-profit organizations; State governments; regional organizations; U.S. territories or possessions; Indian/Native American tribal governments; Indian/Native American tribally designated organizations; eligible agencies of the Federal government; and faith-based or community based organizations.
Geographic Focus: All States
Date(s) Application is Due: Feb 5; Jun 5; Oct 5
Contact: Catherine M. Meyers, (301) 594-7717; fax (301) 480-3510; cm420i@nih.gov
Internet: http://grants.nih.gov/grants/guide/pa-files/PA-07-015.html
Sponsor: National Institute of Diabetes and Digestive and Kidney Diseases
6707 Democracy Boulevard
Bethesda, MD 20892-5460

NIDDK Research Grants on Improving Health Care for Obese Patients 3176

This funding opportunity announcement (FOA) solicits Research Project Grant applications from institutions and organizations that propose to conduct research to determine the barriers to optimal health care for obese patients, and to test innovations or modifications in care delivery to improve health outcomes for obese patients independent of weight loss. Awards issued under this FOA are contingent upon the availability of funds and the submission of a sufficient number of meritorious applications. Applications can be renewed by competing for additional project periods.
Requirements: The following organizations and institutions may apply: public/state controlled institutions of higher education; private institutions of higher education; Hispanic-serving institutions; historically Black colleges and universities (HBCUs); tribally controlled colleges and universities (TCCUs); Alaska native and native Hawaiian serving institutions; nonprofits with 501(c)3 IRS status; nonprofits without 501(c)3 IRS status; small businesses; for-profit organizations; State governments; regional organizations; U.S. territories or possessions; Indian/Native American tribal governments; Indian/Native American tribally designated organizations; eligible agencies of the Federal government; and faith-based or community based organizations.
Geographic Focus: All States
Date(s) Application is Due: Feb 5; Jun 5; Oct 5
Contact: Susan Z. Yanovski, (301) 594-8882; fax (301) 480-8300; sy29f@nih.gov
Internet: http://grants.nih.gov/grants/guide/pa-files/PA-07-013.html
Sponsor: National Institute of Diabetes and Digestive and Kidney Diseases
6707 Democracy Boulevard
Bethesda, MD 20892-5460

NIDDK Role of Gastrointestinal Surgical Procedures in Amelioration of Obesity-Related Insulin Resistance & Diabetes Independent of Weight Loss 3177

This Funding Opportunity Announcement (FOA) issued by the National Institute of Diabetes and Digestive and Kidney Diseases, National Institutes of Health encourages Research Project Grant (R01) applications from institutions and organizations that propose to explain the underlying mechanism(s) by which various gastrointestinal surgical procedures ameliorate obesity-related insulin resistance and diabetes independent of the resultant weight loss. Awards issued under this FOA are contingent upon the availability of funds and the submission of a sufficient number of meritorious applications.
Requirements: The following organizations and institutions may apply: public/state controlled institutions of higher education; private institutions of higher education; Hispanic-serving institutions; historically Black colleges and universities (HBCUs); tribally controlled colleges and universities (TCCUs); Alaska native and native Hawaiian serving institutions; nonprofits with 501(c)3 IRS status; nonprofits without 501(c)3 IRS status; small businesses; for-profit organizations; State governments; regional organizations; U.S. territories or possessions; Indian/Native American tribal governments; Indian/Native American tribally designated organizations; eligible agencies of the Federal government; and faith-based or community based organizations.
Geographic Focus: All States
Date(s) Application is Due: Feb 5; Jun 5; Oct 5
Contact: Carolyn W. Miles, (301) 451-3759; fax (301) 480-8300; cm294e@nih.gov
Internet: http://grants.nih.gov/grants/guide/pa-files/PA-08-014.html
Sponsor: National Institute of Diabetes and Digestive and Kidney Diseases
6707 Democracy Boulevard
Bethesda, MD 20892-5460

NIDDK Secondary Analyses in Obesity, Diabetes and Digestive and Kidney Diseases Grants 3178

The specific objectives of this announcement on Secondary Analyses In Obesity, Diabetes, Digestive and Kidney Diseases are to support the following: secondary analyses of data related to the epidemiology of disease areas of NIDDK; important and/or innovative hypotheses explored through analysis of existing data sets; secondary analyses designed

to inform and support subsequent applications for individual research awards; rapid analyses of new databases and experimental modules to inform the design and content of future studies; the archiving of datasets to be made publicly available for research purposes related to disease areas of NIDDK, including both epidemiological studies and multi-center clinical trials. Research that employs analytic techniques that demonstrate or promote methodological advances in patient oriented and epidemiologic research is also of interest. International comparative analyses are encouraged. Applications that are innovative and high risk with the likelihood for high impact are especially encouraged.
Requirements: The following organizations and institutions may apply: public/state controlled institutions of higher education; private institutions of higher education; Hispanic-serving institutions; historically Black colleges and universities (HBCUs); tribally controlled colleges and universities (TCCUs); Alaska native and native Hawaiian serving institutions; nonprofits with 501(c)3 IRS status; nonprofits without 501(c)3 IRS status; small businesses; for-profit organizations; State governments; regional organizations; U.S. territories or possessions; Indian/Native American tribal governments; Indian/Native American tribally designated organizations; eligible agencies of the Federal government; and faith-based or community based organizations.
Geographic Focus: All States
Date(s) Application is Due: Feb 16; Jun 16; Oct 16
Amount of Grant: 200,000-275,000 USD
Contact: James E. Everhart, (301) 594-8878; fax (301) 480-8300; je17g@nih.gov
Internet: http://grants.nih.gov/grants/guide/pa-files/PA-09-131.html
Sponsor: National Institute of Diabetes and Digestive and Kidney Diseases
6707 Democracy Boulevard
Bethesda, MD 20892-5460

NIDDK Seeding Collaborative Interdisciplinary Team Science in Diabetes, Endocrinology and Metabolic Diseases Grants 3179
The purpose of the Seeding Collaborative Interdisciplinary Team Science Program described in this announcement is to provide initial support to enable strong new investigative teams to form, and to foster preliminary research activities. It is anticipated that research teams receiving seeding support under this program will be well positioned to compete for funds through the parent NIDDK Collaborative Interdisciplinary Research Program in Diabetes, Endocrinology and Metabolic Diseases solicitation. It is anticipated that 5-10 projects will be funded, but because the nature and scope of the proposed research will vary from application to application, it is anticipated that the size and duration of each award will also vary. Budgets for direct costs of up to $300,000 per year with a project duration of up to one year may be requested.
Requirements: The following organizations and institutions are eligible to apply: public/state controlled institutions of higher education; private institutions of higher education; Hispanic-serving institutions; historically Black colleges and universities (HBCUs); tribally controlled colleges and universities (TCCUs); Alaska Native and Native Hawaiian serving institutions; and non-domestic (non-U.S.) entities (foreign organizations).
Geographic Focus: All States
Date(s) Application is Due: Jan 25; May 25; Sep 25
Contact: Karen Salomon, (301) 594-7733; fax (301) 435-6047; ks495c@nih.gov
Internet: http://grants.nih.gov/grants/guide/pa-files/PAR-08-181.html
Sponsor: National Institute of Diabetes and Digestive and Kidney Diseases
6707 Democracy Boulevard
Bethesda, MD 20892-5460

NIDDK Silvio O. Conte Digestive Diseases Research Core Centers Grants 3180
The objective of the Silvio O. Conte Digestive Diseases Research Core Centers is to bring together, on a cooperative basis, basic science and clinical investigators to enhance the effectiveness of their research related to digestive and/or liver diseases and their complications. A core center must be an identifiable unit within a single university medical center or a consortium of cooperating institutions, including an affiliated university. An existing program of excellence in biomedical research in the area of digestive and/or liver diseases is a prerequisite for applying. This research must be in the form of NIH research projects, program projects, or other peer-reviewed research that is already funded at the time of submission of a center grant application. Close cooperation, communication, and collaboration among all involved personnel of various professional disciplines are the ultimate objectives.
Requirements: The following organizations and institutions are eligible to apply: public/state controlled institutions of higher education; private institutions of higher education; nonprofits with 501(c)3 IRS status (other than institutions of higher education); nonprofits without 501(c)3 IRS status (other than institutions of higher education); and for-profit organizations (other than small businesses).
Geographic Focus: All States
Date(s) Application is Due: Oct 12; Nov 9
Contact: Judith Podskalny, (301) 594-8876; fax (301) 480-8300; jp53s@nih.gov
Internet: http://grants.nih.gov/grants/guide/rfa-files/RFA-DK-09-005.html
Sponsor: National Institute of Diabetes and Digestive and Kidney Diseases
6707 Democracy Boulevard
Bethesda, MD 20892-5460

NIDDK Small Business Innovation Research to Develop New Therapeutics & Monitoring Technologies for Type 1 Diabetes Towards an Artificial Pancreas 3181
A promising therapeutic option for the treatment of diabetes is a system (termed an artificial pancreas or closed-loop) that can mimic normal pancreatic beta cell function thereby restoring normal metabolic homeostasis without causing hypoglycemia. The design of any system capable of achieving this goal is complex and raises scientific, clinical and regulatory challenges. This FOA solicits Small Business Innovation Research (SBIR) grant applications from small business concerns (SBCs) for funding to perform research leading to the development of innovative technologies that may advance progress toward integrated long term glucose regulated insulin delivery system (artificial pancreas). Letters of Intent are due by March 16, with full proposals to be received by April 15.
Requirements: Only United States small business concerns (SBCs) are eligible to submit SBIR applications.
Geographic Focus: All States
Date(s) Application is Due: Mar 16; Apr 15
Contact: Dr. Guillermo Arreaza, (301) 594-4724; fax (301) 480-3503; ga96b@nih.gov
Internet: http://grants.nih.gov/grants/guide/rfa-files/RFA-DK-09-001.html
Sponsor: National Institute of Diabetes and Digestive and Kidney Diseases
6707 Democracy Boulevard
Bethesda, MD 20892-5460

NIDDK Small Grant Program for K08/K23 Recipients 3182
The NIDDK invites recipients of its K08/K23 awards to apply for grant support during the final two years of their awards through the NIDDK Small Grant Program. This additional support will enable K awardees to either expand their current research objectives or to branch out to a closely related pilot study, thus demonstrating their growth as investigators and their independence. This should facilitate the transition to fully independent investigator status. The added grant support in the latter years of an award is expected to have the following benefits: increased fiscal independence for the K08/K23 award recipient as a precursor to complete independence; an opportunity for the recipient to generate additional publications and data to form the basis for an R01 application; an opportunity for a Scientific Review Group (SRG) to evaluate accomplishments made during the first two years of the K08/K23 award; an opportunity for the applicant to provide more detailed research plans for the last two years of the K award; and an opportunity for the applicant to demonstrate additional success in the peer review process during the course of their K08/K23 award. A project period of up to two years and a budget for direct costs of up to two $25,000 modules or $50,000 per year may be requested.
Requirements: This award is available only to those currently holding the K08 award from NIDDK. Details of the K08 award are given in NIH Program Announcement PA-00-003. Applications may be submitted on behalf of candidates by domestic non-federal nonprofit organizations, public or private.
Geographic Focus: All States
Date(s) Application is Due: Feb 16; Jun 16; Oct 16
Contact: Dr. Judith Podskalny, (301) 594-8876; fax (301) 480-8300; jp53s@nih.gov
Internet: http://grants.nih.gov/grants/guide/pa-files/PAR-06-172.html
Sponsor: National Institute of Diabetes and Digestive and Kidney Diseases
6707 Democracy Boulevard
Bethesda, MD 20892-5460

NIDDK Small Grants for Underrepresented Minority Scientists in Diabetes and Digestive and Kidney Diseases 3183
The goal of this program is to enable the applicant to accept a tenure-earning position, gain additional research experience and obtain preliminary data on which to base a subsequent research grant application in an area of diabetes, endocrinology, metabolism, digestive diseases, and nutrition, kidney, urology, or hematology. Annual deadline dates may vary; contact program staff for exact dates.
Requirements: Awards will be limited to minority citizens or non-citizen nationals of the United States or to individuals who have been lawfully admitted for permanent residence at the time of application. Applicants must also have a doctoral degree.
Restrictions: Applicants may not hold, nor apply concurrently for any other PHS research project grant at the time of this application.
Geographic Focus: All States
Date(s) Application is Due: Mar 22
Amount of Grant: Up to 250,000 USD
Contact: Dr. Charles H. Rodgers, (301) 594-7717; rodgersc@extra.niddk.nih.gov
Internet: http://grants.nih.gov/grants/guide/rfa-files/RFA-DK-00-007.html
Sponsor: National Institute of Diabetes and Digestive and Kidney Diseases
45 Center Drive, MSC 6600
Bethesda, MD 20892-6600

NIDDK Traditional Conference Grants 3184
Conference Grants are awarded to institutions and organizations (not individuals) to provide partial support for international or domestic meetings, conferences, and workshops to coordinate, exchange, and disseminate research information. NIDDK is especially interested in conference grants used to cover travel expenses for young and minority investigators, although meeting publications, salaries, consultant services, equipment rental, and supplies may be requested. Indirect costs normally are not allowed.
Requirements: Applicants must submit a letter of intent to apply at least 8 weeks before the application receipt date. This letter should include the names of the organizing committee, a preliminary agenda, a list of accepted or invited speakers, a proposed budget and other planned sources of funding, and a description of how the conference will address each of the above criteria.
Geographic Focus: All States
Contact: Dr. M. Ken May; (301) 594-8884; fax (301) 480-8300; maym@mail.nih.gov
Internet: http://www2.niddk.nih.gov/Funding/Grants/ApplicantGuidelines/ConferenceSupport.htm
Sponsor: National Institute of Diabetes and Digestive and Kidney Diseases
6707 Democracy Boulevard
Bethesda, MD 20892-5460

NIDDK Training in Clinical Investigation in Kidney and Urology Grants 3185

The Division of Kidney, Urology, and Hematology (DKUH) has long been interested in stimulating more training in clinical research methods in programs related to nephrology and urology. At the same time, there is a dearth of clinicians that have advanced training in clinical trial methodology. The purpose of this program announcement is to encourage applications for new training programs and to encourage program directors of existing Institutional Training Grants supported by DKUH to collaborate with schools of public health and departments of epidemiology, statistics, and biostatistics to offer didactic and practical training in clinical research methodologies. The goal of this program announcement is to increase the number of well-trained investigators who are capable of conceiving, designing, conducting, analyzing, and presenting the results of clinical research in areas pertaining to nephrology or urology.
Requirements: For new training programs, only domestic, nonprofit, private or public institutions may apply. The applicant institution must have the staff and facilities required for the proposed program. The research training program director at the institution will be responsible for the selection and appointment of trainees to receive National Research Service Award (NRSA) support and for the overall direction of the program.
Restrictions: Positions on NRSA institutional grants may not be used for study leading to the MD, DDS or other clinical health professional degrees except when those studies are a part of a formal combined research degree program such as the MD/PhD. Similarly, trainees may not accept NRSA support for studies which are a part of residency training leading to a medical specialty or subspecialty except when the residency program credits a period of full-time, postdoctoral research training toward board certification and the trainee intends to pursue a research career.
Geographic Focus: All States
Date(s) Application is Due: Mar 10
Contact: Dr. Charles Rodgers, (301) 594-7717; rodgersc@extra.niddk.nih.gov
Internet: http://grants.nih.gov/grants/guide/pa-files/PAR-99-012.html
Sponsor: National Institute of Diabetes and Digestive and Kidney Diseases
45 Center Drive, MSC 6600
Bethesda, MD 20892-6600

NIDDK Translational Research for the Prevention and Control of Diabetes and Obesity **3186**
The National Institute of Diabetes and Digestive and Kidney Diseases (NIDDK), the National Institute of Nursing Research and the Office of Behavior and Social Sciences Research seek to develop cost effective and sustainable interventions that can be adopted in real world settings, for the prevention and control of diabetes and obesity. Research should be based on interventions already proven efficacious in clinical trials to prevent and reverse obesity and type 2 diabetes, to improve care of type 1 and type 2 diabetes and to prevent or delay its complications. The interventions proposed under this solicitation should have the potential to be widely disseminated to clinical practice, individuals and communities at risk. Relevant topics include: methods to improve health care delivery to patients with or at risk of diabetes; strategies to enhance diabetes self management; methods to develop strategies to promote healthy lifestyles to reduce the risk of diabetes; methods to promote lifestyle change that will prevent or reverse overweight and obesity; and new cost effective ways to identify people with pre-diabetes and undiagnosed diabetes.
Requirements: The following organizations and institutions may apply: public/state controlled institutions of higher education; private institutions of higher education; Hispanic-serving institutions; historically Black colleges and universities (HBCUs); tribally controlled colleges and universities (TCCUs); Alaska native and native Hawaiian serving institutions; nonprofits with 501(c)3 IRS status; nonprofits without 501(c)3 IRS status; small businesses; for-profit organizations; State governments; regional organizations; U.S. territories or possessions; Indian/Native American tribal governments; Indian/Native American tribally designated organizations; eligible agencies of the Federal government; and faith-based or community based organizations.
Geographic Focus: All States
Date(s) Application is Due: Jan 25; May 25; Sep 25
Contact: Sanford Garfield, (301) 594-8803; fax (301) 480-3503; sg50o@nih.gov
Internet: http://grants.nih.gov/grants/guide/pa-files/PAR-06-532.html
Sponsor: National Institute of Diabetes and Digestive and Kidney Diseases
6707 Democracy Boulevard
Bethesda, MD 20892-5460

NIDDK Transmission of Human Immunodeficiency Virus (HIV) In Semen **3187**
The National Institute of Diabetes and Digestive and Kidney Diseases (NIDDK) through its Division of Kidney, Urologic and Hematologic Diseases (DKUHD) and the National Institute of Child Health and Human Development, Center for Population Research invite investigators to submit research grant applications that will increase the basic and clinical knowledge of the biology of HIV in semen. Direct exposure to semen of HIV seropositive men is a major route for transmission of HIV type I. This Funding Opportunity Announcement (FOA) focuses on studies that will elucidate the factors that determine HIV shedding in the male genital tract. This includes studies that elucidate the infectivity of HIV in semen fractions, the effect of antiretroviral therapy on HIV infectivity in semen fractions, the relationship between the immunobiology of the male genital tract and HIV replication and infectivity, and factors such as genital tract inflammation which influence HIV transmission through semen. Of particular interest are studies of the anatomic sites of virus infected cells and virus production in the male genital tract. The intent of this Funding opportunity is to encourage investigator-initiated research, to attract new investigators to the field, and to increase interdisciplinary research to enhance the scope and effectiveness of research in this area. Because the nature and scope of the proposed research will vary from application to application, it is anticipated that the size and duration of each award will also vary. The total amount awarded and the number of awards will depend upon the mechanism numbers, quality, duration, and costs of the applications received.
Requirements: The following organizations and institutions are eligible to apply: public/state controlled institutions of higher education; private institutions of higher education; Hispanic-serving institutions; historically Black colleges and universities (HBCUs); tribally controlled colleges and universities (TCCUs); Alaska Native and Native Hawaiian serving institutions; nonprofits with 501(c)3 IRS status (other than institutions of higher education); nonprofits without 501(c)3 IRS status (other than institutions of higher education); small businesses; for-profit organizations (other than small businesses); state governments; Indian/Native American tribal governments (federally recognized); Indian/Native American tribally designated organizations; county governments; city or township governments; special district governments; independent school districts; public housing authorities; Indian housing authorities; U.S. territory or possession; Indian/Native American tribal governments (other than federally recognized); regional organizations; non-domestic (non-U.S.) entities (foreign organizations); other eligible agencies of the federal government; and faith-based or community-based organizations.
Geographic Focus: All States
Date(s) Application is Due: Feb 5; Jun 5; Oct 5
Contact: Leroy M. Nyberg, Jr., (301) 594-7717; ln10f@nih.gov
Internet: http://grants.nih.gov/grants/guide/pa-files/PA-07-044.html
Sponsor: National Institute of Diabetes and Digestive and Kidney Diseases
6707 Democracy Boulevard
Bethesda, MD 20892-5460

NIDDK Type 2 Diabetes in the Pediatric Population Research Grants **3188**
NIDDK and NICHD invite investigator-initiated research grant applications to study the epidemiology, natural history, pathophysiology, prevention, and treatment of type 2 diabetes in children in the United States. Type 2 diabetes has traditionally been viewed as a disease of adults. More recently, however, it has become apparent that an increasing number of cases of type 2 diabetes are being reported in children. New hypothesis-driven studies are needed to describe the epidemiology, refine the diagnosis, define the metabolic abnormalities, and formulate treatment options for type 2 diabetes in children. Annual deadline dates may vary; contact program staff for exact dates.
Requirements: Applications may be submitted by domestic for-profit and nonprofit organizations, both public and private; units of state and local governments; and eligible agencies of the federal government.
Geographic Focus: All States
Contact: Dr. Gilman Grave, (301) 496-5593; fax (301) 480-9791
Internet: http://www2.niddk.nih.gov/Research/ScientificAreas/Diabetes/Type2Diabetes/DM2K.htm
Sponsor: National Institute of Diabetes and Digestive and Kidney Diseases
6707 Democracy Boulevard
Bethesda, MD 20892-5460

NIDRR Field-Initiated Projects **3189**
The goals of this project include: developing methods and technologies that maximize the integration of individuals with disabilities into society; employment and family environments; and improving the effectiveness of services authorized under the Rehabilitation Act. Applicants who receive awards under this program may carry out either research or development activities, and applications for each type of project will be reviewed separately. The applicant is responsible for indicating which type of project is being proposed. Projects may last up to three years. Annual deadline dates may vary; contact program staff for exact dates.
Requirements: Proposals will be accepted from any public or private agency or organization.
Geographic Focus: All States
Contact: Donna Nangle, (202) 245-7462; fax (202) 245-7323; Donna.Nangle@ed.gov
Internet: http://www.ed.gov/fund/grant/apply/nidrr/index.html
Sponsor: National Institute on Disability and Rehabilitation Research
550 12th Street SW, 6th Floor, Switzer Building
Washington, DC 20202-6510

NIEHS Hazardous Materials Worker Health and Safety Training Grants **3190**
This FOA issued by the National Institute of Environmental Health Sciences (NIEHS) invites applications for cooperative agreements to support the development of model programs for the training and education of workers engaged in activities related to hazardous materials and waste generation, removal, containment, transportation and emergency response. The major objective of this solicitation is to prevent work-related harm by assisting in the training of workers in how best to protect themselves and their communities from exposure to hazardous materials encountered during hazardous waste operations, hazardous materials transportation, environmental restoration of contaminated facilities or chemical emergency response. A variety of sites, such as those involved with chemical waste clean up and remedial action and transportation-related chemical emergency response may pose severe health and safety concerns to workers and the surrounding communities. These sites contain a multiplicity of hazardous substances, sometimes unknown substances, and often the site is uncontrolled. A major goal of the Worker Education and Training Program (WETP) is to provide assistance to organizations in developing their institutional competency to provide appropriate model training and education programs. The total project period for an application submitted in response to this funding opportunity must be 5 years. A new applicant may request a budget for direct costs of up to $700,000 for the first year. Letters of Intent (LOIs) are due by October 23, with full proposals due by November 23.
Requirements: The following organizations and institutions are eligible to apply: public/state controlled institutions of higher education; private institutions of higher education; Hispanic-serving institutions; historically Black colleges and universities (HBCUs); tribally controlled colleges and universities (TCCUs); Alaska Native and

Native Hawaiian serving institutions; and nonprofits with 501(c)3 IRS status (other than institutions of higher education).
Geographic Focus: All States
Date(s) Application is Due: Oct 23; Nov 23
Amount of Grant: Up to 700,000 USD
Contact: Joseph Hughes; (919) 541-0217; fax (919) 541-0462; hughes3@niehs.nih.gov
Internet: http://grants.nih.gov/grants/guide/rfa-files/RFA-ES-09-004.html
Sponsor: National Institute of Environmental Health Sciences
Keystone Building, Room 3039, P.O. Box 12233, MD K3-14
Research Triangle Park, NC 27709

NIEHS Hazmat Training at Doe Nuclear Weapons Complex Grants 3191

The National Institute of Environmental Health Sciences (NIEHS) invites applications for cooperative agreements to support the development of model programs for the training and education of workers engaged in activities related to hazardous materials and waste generation, removal, containment, transportation and emergency response within the Department of Energy (DOE) Nuclear Weapons Complex. The major objective of this solicitation is to prevent work related harm by assisting in the training and education of workers in the DOE nuclear weapons complex. Safety and health training will transmit skills and knowledge to workers in how best to protect themselves and their communities from exposure to hazardous materials encountered during hazardous waste operations, facility decommissioning and decontamination, hazardous materials transportation, environmental restoration of contaminated facilities or chemical emergency response. The total project period for an application submitted in response to this funding opportunity must be 5 years. A new applicant may request a budget for direct costs of up to $700,000 for the first year. Letters of Intent (LOIs) are due by October 23, with full proposals due by November 23.
Requirements: The following organizations and institutions are eligible to apply: public/state controlled institutions of higher education; private institutions of higher education; Hispanic-serving institutions; historically Black colleges and universities (HBCUs); tribally controlled colleges and universities (TCCUs); Alaska Native and Native Hawaiian serving institutions; and nonprofits with 501(c)3 IRS status (other than institutions of higher education).
Geographic Focus: All States
Date(s) Application is Due: Oct 23; Nov 23
Amount of Grant: Up to 700,000 USD
Contact: Joseph Hughes; (919) 541-0217; fax (919) 541-0462; hughes3@niehs.nih.gov
Internet: http://grants.nih.gov/grants/guide/rfa-files/RFA-ES-09-003.html
Sponsor: National Institute of Environmental Health Sciences
Keystone Building, Room 3039, P.O. Box 12233, MD K3-14
Research Triangle Park, NC 27709

NIEHS Independent Scientist Award 3192

The award is a special salary-only grant designed to provide protected time for newly independent scientists who currently have non-research obligations such as heavy teaching loads, clinical work, committee assignments, service, and administrative duties that prevent them from having a period of intensive research focus. The award is targeted to persons with doctoral degrees who have completed their research training, have independent peer-reviewed research support, and who need a period of protected research time in order to foster their research career development. Strong preference is given to candidates with active research project grant support from the NIEHS. New applications are due February 12, June 12, and October 12, while renewal applications are due March 12, July 12, and November 12. Up to $75,000 per year of base salary plus fringe benefits is available. Research support of up to $25,000 is provided in some cases. Individuals engaged in predominantly theoretical work, such as modeling or computer simulation, may request up to $25,000 per year for research expenses.
Requirements: The following organizations and institutions are eligible to apply: public/state controlled institutions of higher education; private institutions of higher education; nonprofits with 501(c)3 IRS status; nonprofits without 501(c)3 IRS status; small businesses; for-profit organizations; State governments; U.S. territories or possessions; Indian/Native American tribal governments (Federally recognized and other than Federally recognized); Indian/Native American tribally designated organizations; Hispanic-serving institutions; historically Black colleges and universities (HBCUs); tribally controlled colleges and universities (TCCUs); Alaska Native and Native Hawaiian serving institutions; regional organizations; and faith-based or community based organizations. The candidate must have a doctoral degree and peer-reviewed, independent research support at the time the award is made. The candidate must spend a minimum of 75 percent effort conducting research during the period of the award.
Restrictions: The award is not intended for investigators who already have full time to perform research, or have substantial publication records or considerable research support indicating that they are well established in their fields. Foreign institutions are not eligible to apply.
Geographic Focus: All States
Contact: Carol Shreffler, (919) 541-1445; fax (919) 541-5064; shreffl1@niehs.nih.gov
Internet: http://grants.nih.gov/grants/guide/pa-files/PA-09-038.html
Sponsor: National Institute of Environmental Health Sciences
Keystone Building, Room 3039, P.O. Box 12233, MD K3-14
Research Triangle Park, NC 27709

NIEHS Mentored Clinical Scientist Research Career Development Awards 3193

This award enables candidates holding professional degrees (e.g., M.D., D.O., D.V.M., or equivalent degrees) to undertake 3 to 5 years of special study and supervised research with the goal of becoming independent investigators. The award also allows awardees to pursue a research career development program suited to their experience and capabilities under a mentor who is competent to provide guidance in the chosen research area. Institutions may submit applications on behalf of candidates who hold professional degrees. At least 2 years must have elapsed since the health professional degree was granted. Candidates can have varying levels of clinical training and research experience. New applications are due February 12, June 12, and October 12, while resubmitted applications are due March 12, July 12, and November 12. Up to $75,000 per year plus fringe benefits is available, and up to $25,000 per year for research support for new awards.
Requirements: All candidates must be U.S. citizens, non-citizen nationals, or legal permanent residents of the U.S. Persons with temporary or student visas are not eligible. The grantee institution must have a strong, well-established research and research training program in the chosen area, accomplished faculty in the basic and clinical sciences, and a commitment to the candidate's research development. The proposed program should include an appropriate mentor.
Geographic Focus: All States
Date(s) Application is Due: Feb 12; Mar 12; Jun 12; Jul 12; Oct 12; Nov 12
Amount of Grant: Up to 100,000 USD
Contact: Carol Shreffler, (919) 541-1445; fax (919) 541-5064; shreffl1@niehs.nih.gov
Internet: http://grants.nih.gov/grants/guide/contacts/pa-06-512_contacts.htm
Sponsor: National Institute of Environmental Health Sciences
Keystone Building, Room 3039, P.O. Box 12233, MD K3-14
Research Triangle Park, NC 27709

NIEHS Mentored Research Scientist Development Award 3194

The MRSDA is for a research scientist who needs an additional period of sponsored research experience as a way to gain expertise in a research area new to the candidate or in an area that would demonstrably enhance the candidate's scientific career. Research projects proposed in response to this Announcement will be expected to have a defined impact on the environmental health sciences and be directly responsive to the mission of the NIEHS, which is distinguished from that of other Institutes by its focus on research programs seeking to link the effects of environmental exposures to the cause, mechanisms, moderation, or prevention of a human disease or disorder or relevant pathophysiologic process. Applications must include as part of the justification the specific human disease, dysfunction, pathophysiologic condition, or relevant human biologic process altered, and the specific environmentally relevant toxicant. Examples of environmentally relevant toxicants include industrial chemicals or manufacturing byproducts, metals, pesticides, herbicides, air pollutants and other inhaled toxicants, particulates or fibers, fungal, bacterial, or biologically derived toxins. Agents considered non-responsive to this announcement include, but are not limited to: alcohol, chemotherapeutic agents, radiation which is not the result of an ambient environmental exposure, drugs of abuse, and infectious or parasitic agents. Ecologic, biomonitoring, biotransformation or biodegradation studies are also not responsive. Applicants must demonstrate a significant career shift or career enhancement activity which will achieve research independence upon completion of the program. Preference will be given to candidates whose mentor has NEIHS research grant support.
Requirements: The candidate must have a research or health-professional doctorate or its equivalent and must have demonstrated the capacity or potential for highly productive independent research in the period after the doctorate. The candidate must identify a mentor with extensive research experience and must spend a minimum of 75 percent of full-time professional effort conducting research for the period of the award. Applications may be submitted on behalf of candidates by domestic, nonfederal organizations, public or private, such as medical, dental, or nursing schools or other institutions of higher education. Candidates must be U.S. citizens, noncitizen nationals, or permanent residents.
Geographic Focus: All States
Date(s) Application is Due: Feb 12; Mar 12; Jun 12; Jul 12; Oct 12; Nov 12
Amount of Grant: 75,000 USD
Contact: Carol Shreffler, (919) 541-1445; fax (919) 541-5064; shreffl@niehs.nih.gov
Internet: http://grants.nih.gov/grants/guide/pa-files/PA-09-040.html
Sponsor: National Institute of Environmental Health Sciences
Keystone Building, Room 3039, P.O. Box 12233, MD K3-14
Research Triangle Park, NC 27709

NIEHS Small Business Innovation Research (SBIR) Grants 3195

The focus of these grants is on understanding how chemical and physical agents cause pathological changes in molecules, cells, tissues, and organs and become manifested as respiratory disease; neurological, behavioral, and developmental abnormalities; cancer; and other disorders. By understanding the relationship between environmental exposures and the subsequent development of disease or biological injury, human health may be protected. SBIR program objectives are to stimulate technological innovation, use small businesses to meet federal research and development needs, increase private sector commercialization of innovations derived from federal research and development, and foster and encourage participation by minority and disadvantaged persons in technological innovation. Phase I awards, generally for six months, establish the technical merit and feasibility of proposed research efforts that may lead to commercial products or processes. Phase II awards are for the continuation of research initiated in Phase I.
Requirements: SBIR grants can be awarded only to domestic small businesses (entities that are independently owned and operated for profit, are not dominant in the field in which research is proposed, and have no more than 500 employees). Primary employment (more than one-half) of the principal investigator must be with the small business at the time of award and during the conducting of the proposed research. In both Phase I and II, the research must be performed in the United States.
Geographic Focus: All States
Date(s) Application is Due: Apr 15; Aug 15; Dec 15

Amount of Grant: 100,000-750,000 USD
Contact: Jo Anne Goodnight, SBIR/STTR; (301) 435-2688; fax (301) 480-0146; jg128w@nih.gov or sbir@od.nih.gov
Internet: http://grants.nih.gov/grants/funding/sbir.htm#sbir
Sponsor: National Institute of Environmental Health Sciences
Keystone Building, Room 3039, P.O. Box 12233, MD K3-14
Research Triangle Park, NC 27709

NIGMS/NIDCR Evolution of Infectious Diseases Research Grants **3196**

The goal of this joint NIGMS/NIDCR request for applications is to encourage development of a predictive science of infectious diseases by applying the perspectives, theories, and methods of population and evolutionary biology to important issues of disease emergence, prevention, and treatment. To achieve this goal, this RFA seeks collaborations between two major groups of scientists: those with expertise in population and evolutionary biology and molecular phylogenetics, including mathematical modeling and complexity theory; and infectious disease experts such as clinicians, epidemiologists, immunologists, microbiologists, veterinarians, or plant pathologists. PA: PA-99-147
Requirements: Applications may be submitted by domestic and foreign, for-profit and nonprofit organizations, units of state and local governments, and eligible agencies of the federal government.
Restrictions: Foreign institutions are not eligible for program project grants.
Geographic Focus: All States
Amount of Grant: Up to 250,000 USD
Contact: Dr. Irene Anne Eckstrand, Division of Genetics and Developmental Biology; (301) 594-0943; fax (301) 480-2228; Irene_Eckstrand@nih.gov
Internet: http://grants2.nih.gov/grants/guide/pa-files/PA-99-147.html
Sponsor: National Institute of General Medical Sciences
45 Center Drive, Natcher Building
Bethesda, MD 20892

NIGMS Mentored Clinical Scientist Research Career Development Awards **3197**

This award enables candidates holding professional degrees (e.g., M.D., D.O., D.V.M., or equivalent degrees) to undertake 3 to 5 years of special study and supervised research with the goal of becoming independent investigators. The award also allows awardees to pursue a research career development program suited to their experience and capabilities under a mentor who is competent to provide guidance in the chosen research area. Institutions may submit applications on behalf of candidates who hold professional degrees. At least 2 years must have elapsed since the health professional degree was granted. Candidates can have varying levels of clinical training and research experience. New applications are due February 12, June 12, and October 12, while resubmitted applications are due March 12, July 12, and November 12. Up to $75,000 per year plus fringe benefits is available, and up to $20,000 per year for research support for new awards.
Requirements: The candidate must have an MD degree or its equivalent, must have completed postgraduate clinical training and have secured a faculty appointment in an appropriate research-intensive environment, must identify a mentor with extensive research experience, and must be willing to spend a minimum of 75 percent of full-time professional effort conducting research and research career development. Applications may be submitted on behalf of candidates by domestic, nonfederal organizations, public or private, such as medical schools. Minorities and women are encouraged to apply. Candidates must be U.S. citizens or noncitizen nationals, or must have been lawfully admitted for permanent residence and possess Alien Registration Receipt Card (I-551) or some other verification of legal admission as permanent residents.
Restrictions: Former principal investigators on NIH research project (R01), sub-projects of program project (P01), or center grants (P50), or the equivalent, are not eligible. A candidate for the award may not concurrently apply for any other PHS award that duplicates the provisions of this award nor have another application pending award. K08 recipients are encouraged to apply for independent research grant support during the period of this award. The K08 recipient would be allowed to maintain the award if other PHS support is procured, as long as the new support does not interfere with the ability to meet the MCSDA requirements.
Geographic Focus: All States
Date(s) Application is Due: Feb 12; Mar 12; Jun 12; Jul 12; Oct 12; Nov 12
Amount of Grant: Up to 95,000 USD
Contact: Dr. Alison Cole; (301) 594-1826; fax (301) 480-2802; colea@nigms.nih.gov
Internet: http://www.grants.nih.gov/grants/guide/pa-files/PAR-98-084.html
Sponsor: National Institute of General Medical Sciences
45 Center Drive, Natcher Building, Room 2AS-49, MSC 6200
Bethesda, MD 20892-6200

NIGMS Pharmacogenomics Research Network Grants **3198**

The purpose of this funding opportunity announcement is to allow new and renewal applications for research groups to compete to become part of the Pharmacogenomics Research Network (PGRN), and to create network resources that serve the entire PGRN. The PGRN is a consortium of interdisciplinary groups with a core mission to conduct research into understanding the genetic basis of variable drug responses, both therapeutic and adverse. The scientific objectives of the research network include definition of the pressing problems for safe and effective use of medications, establishment of truly cross-disciplinary research groups, application of the most appropriate genetic and genomic methods to important sample collections, discovery and replication studies that identify variation in critical genes and biological pathways, elucidation of the functional consequences of genetic variation for the drugs being studied, and understanding the clinical implications of genotyping to predict drug responses. The expected direct costs amount for each individual research group award is $1 to $1.5 million per year, and for each add-on network resource is $0.1 to $0.5 million per year. Letters of Intent are due by May 2, with full proposals received by June 2.
Requirements: The following organizations and institutions are eligible to apply: public/state controlled institutions of higher education; private institutions of higher education; Hispanic-serving institutions; historically Black colleges and universities (HBCUs); nonprofits with 501(c)3 IRS status (other than institutions of higher education); nonprofits without 501(c)3 IRS status (other than institutions of higher education); small businesses; and for-profit organizations (other than small businesses).
Geographic Focus: All States
Date(s) Application is Due: May 2; Jun 2
Contact: Rochelle M. Long, (301) 594-3827; rochelle.long@nih.gov
Internet: http://grants.nih.gov/grants/guide/rfa-files/RFA-GM-10-001.html
Sponsor: National Institute of General Medical Sciences
45 Center Drive, Building 45
Bethesda, MD 20892

NIGMS Predoctoral Training in Bioinformatics and Computation Biology Grants **3199**

The purpose of this program is to develop a mechanism to train a cadre of scientists whose primary professional identification and disciplinary affiliation is bioinformatics or computational biology. For the purpose of this announcement, bioinformatics and computational biology are defined broadly to include the use of theory, computer implementation, and application to the full spectrum of basic research in the biomedical sciences. The terms thus include analysis of molecular sequence and structure, molecular function, cellular function, physiology, genomics, and genetics, as well as computational modeling of complex phenomena such as neural circuits and equilibrium phenomena, population biology, theoretical and mathematical biology, and the analysis of complex systems. The goal of this program is to train PhD students in the background theory, computational implementation, and biological application of information sciences (including computer science, statistics, mathematics and others), and to use this training to study problems relevant to biomedical research. PA: PAR-99-146
Requirements: Applications may be submitted by domestic public and private institutions with established programs leading to a PhD degree.
Geographic Focus: All States
Amount of Grant: Up to 16,488 USD
Contact: Dr. Marion Zatz; (301) 594-0943; fax (301) 480-2228
Internet: http://grants.nih.gov/grants/guide/pa-files/PAR-99-146.html
Sponsor: National Institute of General Medical Sciences
45 Center Drive, Natcher Building
Bethesda, MD 20892

NIGMS Ruth L. Kirschstein National Research Service Awards for Individual Predoctoral Fellows in PharmD/PhD Grants **3200**

NIGMS is committed to increasing the number of pharmacy students who wish to pursue doctoral research in the pharmaceutical sciences. It recognizes that current career paths for Doctor of Pharmacy graduates have reduced the number of individuals who pursue academic research training. NIGMS's intent in issuing this Funding Opportunity Announcement (FOA) is to provide individual research training fellowships to promising predoctoral applicants who are enrolled in PharmD/PhD programs in Schools of Pharmacy who desire to become productive and successful independent investigators whose research meets the mission of NIGMS. This fellowship program will provide support for the dissertation stage of predoctoral students enrolled in joint PharmD/PhD programs. It is expected that the length of time for an award will be 3 years or less.
Requirements: The following organizations and institutions are eligible to apply: for-profit organizations; non-profit organizations; domestic institutions; public or private institutions, such as universities, colleges, and hospitals; units of State and local governments; eligible agencies of the Federal government.
Restrictions: Individuals on temporary or student visas are not eligible.
Geographic Focus: All States
Date(s) Application is Due: Apr 8; Aug 8; Dec 8
Contact: Richard Okita, (301) 594-3827; fax (301) 480-2802; okitar@nigms.nih.gov
Internet: http://grants.nih.gov/grants/guide/pa-files/PA-09-029.html
Sponsor: National Institute of General Medical Sciences
45 Center Drive, Natcher Building, Room 2AS-49, MSC 6200
Bethesda, MD 20892-6200

NIGMS Short Courses on Mathematical, Statistical, and Computational Tools for Studying Biological Systems **3201**

The goal of this FOA is to prepare researchers at all professional levels for multidisciplinary research on complex biological systems and phenotypes by improving integration of mathematical, statistical, and computational approaches into biological and/or behavioral research. Progress in contemporary biological and behavioral disciplines depends heavily on investigators who are skilled in the use of mathematics, computation, and statistics (collectively referred to as quantitative methods). This FOA issued by the National Institute of General Medical Sciences, the National Center for Complementary and Alternative Medicine, the National Institute on Aging and the National Institute of Mental Health encourages applications for Research Education Grants (R25) from institutions and organizations to conduct workshops and short courses to improve integration of mathematical, statistical, and computational approaches into biological and/or behavioral research. Support will be limited to activities that reach a wide audience of researchers.
Requirements: Application can be made by the following institutions and organizations: public/state controlled institutions of higher education; private institutions of higher

education; nonprofits with 501(c)3 IRS status (other than institutions of higher education); nonprofits without 501(c)3 IRS status (other than institutions of higher education); small businesses; for-profit organizations (other than small businesses); state governments; U.S. territories or possessions; Indian/Native American tribal governments (federally recognized); Indian/Native American tribal governments (other than Federally recognized); Indian/Native American tribally designated organizations; non-domestic (non-U.S.) entities (foreign organizations); Hispanic-serving institutions; historically Black colleges and universities (HBCUs); tribally controlled colleges and universities (TCCUs); Alaska Native and Native Hawaiian serving institutions; regional organization; eligible agencies of the Federal government; and faith-based or community based organizations.
Restrictions: The program announcement is not intended for university course or curriculum development.
Geographic Focus: All States
Date(s) Application is Due: Jan 25; May 25; Sep 25
Amount of Grant: Up to 200,000 USD
Contact: Matthew E. Portnoy, (301) 594-0943; mportnoy@nigms.nih.gov
Internet: http://grants.nih.gov/grants/guide/pa-files/PA-09-002.html
Sponsor: National Institute of General Medical Sciences
45 Center Drive, Building 45, 2AS.25P
Bethesda, MD 20892-6200

NIGMS Support of Competitive Research (SCORE) Pilot Project Awards 3202
The objective of the SCORE program is to foster the development of faculty at minority serving institutions (MSIs) in order to increase their research competitiveness and promote their transition to non-SCORE external sources of funding. This objective is expected to translate into an increase in the number of individuals from groups underrepresented in biomedical and behavioral research professionally engaged in these areas of research, and an enhancement of an institution's research base. SCORE SC grants are offered to eligible MSIs. For the purposes of this program, eligible MSIs are those with more than 50% student enrollment of individuals from groups underrepresented in biomedical and behavioral research. The SC2 awards may not exceed $300,000 (direct costs) for the entire length of the award which may be one to three years maximum.
Requirements: Eligible applicants include: public or private post secondary educational institution, such as a university or college awarding associate, undergraduate or graduate degrees located in the United States of America or its territories including Puerto Rico, Guam and Virgin Islands with more than 50 percent student enrollment from groups underrepresented in the biomedical and behavioral sciences (such as African American, Hispanic American, Native American, Alaska Natives or natives of the U.S. Pacific Islands); an Indian tribe that has a recognized governing body and that performs substantial governmental functions, or an Alaska Regional Corporation (ARC), as defined in the Alaska Native Claims Settlement Act (43 U.S.C. 1601 et. Seq.); or an institution that has a significantly high student enrollment from groups underrepresented in the biomedical or behavioral sciences but less than 50% (if the Secretary of the Department of Health and Human Services, through the MBRS Chief, determines that the institution has demonstrated special commitment to the retention and graduation of students from groups underrepresented in the biomedical and behavioral sciences and to the hiring and retention of science faculty from underrepresented groups).
Restrictions: Foreign institutions are not eligible to apply.
Geographic Focus: All States
Date(s) Application is Due: Jan 25; May 25; Sep 25
Amount of Grant: 50,000-100,000 USD
Contact: Hinda Zlotnik, (301) 594-3900; fax (301) 480-2753; zlotnikh@nigms.nih.gov
Internet: http://grants.nih.gov/grants/guide/pa-files/PAR-08-027.html
Sponsor: National Institute of General Medical Sciences
45 Center Drive, Room 2AS.37, MSC 6200
Bethesda, MD 20892-6200

NIGMS Support of Competitive Research Research Advancement Awards 3203
The objective of the SCORE program is to foster the development of faculty at minority serving institutions (MSIs) in order to increase their research competitiveness and promote their transition to non-SCORE external sources of funding. This objective is expected to translate into an increase in the number of individuals from groups underrepresented in biomedical and behavioral research professionally engaged in these areas of research, and an enhancement of an institution's research base. SCORE SC grants are offered to eligible MSIs. For the purposes of this program, eligible MSIs are those with more than 50% student enrollment of individuals from groups underrepresented in biomedical and behavioral research. The anticipated amount of individual SC1 awards is expected to range between $125,000 to $250,000 maximum direct costs per year.
Requirements: Eligible applicants include: public or private post secondary educational institution, such as a university or college awarding associate, undergraduate or graduate degrees located in the United States of America or its territories including Puerto Rico, Guam and Virgin Islands with more than 50 percent student enrollment from groups underrepresented in the biomedical and behavioral sciences (such as African American, Hispanic American, Native American, Alaska Natives or natives of the U.S. Pacific Islands); an Indian tribe that has a recognized governing body and that performs substantial governmental functions, or an Alaska Regional Corporation (ARC), as defined in the Alaska Native Claims Settlement Act (43 U.S.C. 1601 et. Seq.); or an institution that has a significantly high student enrollment from groups underrepresented in the biomedical or behavioral sciences but less than 50% (if the Secretary of the Department of Health and Human Services, through the MBRS Chief, determines that the institution has demonstrated special commitment to the retention and graduation of students from groups underrepresented in the biomedical and behavioral sciences and to the hiring and retention of science faculty from underrepresented groups).
Restrictions: Foreign institutions are not eligible to apply.
Geographic Focus: All States
Date(s) Application is Due: Jan 25; May 25; Sep 25
Amount of Grant: 125,000-250,000 USD
Contact: Hinda Zlotnik, (301) 594-3900; fax (301) 480-2753; zlotnikh@nigms.nih.gov
Internet: http://grants1.nih.gov/grants/guide/pa-files/PAR-08-026.html
Sponsor: National Institute of General Medical Sciences
45 Center Drive, Room 2AS.37, MSC 6200
Bethesda, MD 20892-6200

NIH Academic Research Enhancement Awards (AREA) 3204
NIH makes a special effort to stimulate research in educational institutions that provide baccalaureate training for a significant number of the nation's research scientists but that historically have not been major recipients of NIH support. The AREA funds are intended to support new research projects or expand ongoing research activities proposed by faculty members of these institutions in areas related to the health sciences. Funding decisions will be based on the proposed research project's scientific merit and relevance to NIH programs and the institution's contribution to the undergraduate preparation of doctoral-level health professionals. Annual deadline dates vary; contact program staff for exact dates. PA: PA-99-062
Requirements: All health professional schools/colleges and other academic components of domestic institutions offering baccalaureate or advanced degrees in the sciences related to health are eligible to apply.
Restrictions: Applicant principal investigators must not have active research grant support from either NIH or ADAMHA at the applicant institution at the time of award of an AREA grant; may not submit a regular NIH or ADAMHA research grant application for essentially the same project as a pending AREA application; are expected to conduct the majority of their research at their own institutions, although limited access to special facilities or equipment other institutions is permitted; and may not be awarded more than one AREA grant at a time nor be awarded a second AREA grant to continue research initiated under the first AREA grant.
Geographic Focus: All States
Date(s) Application is Due: May 7; Jun 25; Sep 7
Amount of Grant: Up to 150,000 USD
Contact: Denise Russo; (301) 451-7972; fax (301) 480-0146; drusso1@mail.nih.gov
Internet: http://grants.nih.gov/grants/guide/pa-files/PA-06-042.html
Sponsor: National Institutes of Health
6705 Rockledge Drive, Room 3524
Bethesda, MD 20892

NIH Biobehavioral Research for Effective Sleep 3205
The goal of this program announcement is to stimulate clinical and applied research on behavioral, psychosocial, and physiological consequences of acute and chronic partial sleep deprivation in either chronically ill or healthy individuals and to develop environmental, clinical management, and other interventions with the potential to reduce sleep disturbances and significantly improve the health of large numbers of people. Although sleep disorders are a cause of sleep loss in affected individuals, the questions to be addressed under this solicitation should focus on causes and consequences of sleep deprivation, apart from any sleep pathology. Inquiries may be directed to program staff at the appropriate NIH institute.
Requirements: Applications may be submitted by domestic and foreign, for-profit and non-profit organizations, public and private, such as universities, colleges, hospitals, laboratories, units of State and local governments, and eligible agencies of the Federal government. Racial/ethnic minority individuals, women, and persons with disabilities are encouraged to apply as Principal Investigators.
Geographic Focus: All States
Date(s) Application is Due: Feb 16; Jun 16; Oct 16
Amount of Grant: Up to 250,000 USD
Contact: Susan Grove; (301) 435-0714; sg16d@nih.gov or grantsinfo@od.nih.gov
Internet: http://grants.nih.gov/grants/guide/pa-files/PA-00-046.html
Sponsor: National Institutes of Health
6701 Rockledge Drive
Bethesda, MD 20892

NIH Biology, Development, and Progression of Malignant Prostate Disease Research Grants 3206
Investigator-initiated research grants will be made to examine a range of fundamental biological issues considered critical for progress in defeating prostate cancer. The purpose of funding is to encourage new projects focusing on the biology that underlies the development and progression of malignant prostatic disease. Multidisciplinary collaborations between basic and clinical scientists in projects ranging from exploratory basic biology studies to disease progression in appropriate biological systems or models are encouraged.
Requirements: Applications may be submitted by foreign or domestic for-profit and nonprofit organizations, both public and private; units of state and local government; and eligible agencies of the federal government.
Geographic Focus: All States
Date(s) Application is Due: Feb 5; Feb 16; Jun 5; Jun 16; Oct 5; Oct 16
Amount of Grant: Up to 250,000 USD
Contact: Frank Bellino, (301) 496-6402; fax (301) 402-0010; fb12a@nih.gov
Internet: http://grants.nih.gov/grants/guide/pa-files/PA-99-081.html
Sponsor: National Institutes of Health

Gateway Building, 7201 Wisconsin Avenue, Suite 2C231
Bethesda, MD 20892

NIH Biomedical Research on the International Space Station 3207

The National Institutes of Health (NIH) and the National Aeronautics and Space Administration (NASA) are cooperating to facilitate biomedical research in space for better understanding of human physiology and human health on Earth. Applications to this FOA should propose innovative biomedical research on the molecular or cellular level that is directly relevant to the NIH mission and can be carried out on the ISS. Awards made through this FOA will initially support milestone-driven, ground based preparatory studies (UH2 ground feasibility phase), with possible rapid transition to the second, ISS-based research phase (UH3 ISS experimental phase). The ground feasibility phase (UH2) will allow investigators to focus on ground-based preparatory work to meet scientific milestones and technical requirements leading to the ISS experimental phase (UH3). The UH3 phase will include preparing the experiments for launch, conducting them on the ISS, and the subsequent data analyses on Earth. UH3s will be awarded after administrative review of the eligible UH2s that have met the scientific milestones and feasibility requirements necessary to conduct research on the ISS. Applicants should contact individuals at the specific departments listed on the web site. Letters of Intent are due by August 31, and full applications should arrive by September 30.

Requirements: The UH2/UH3 application must be submitted as a single application, and applicants should note specific instructions for each phase in this FOA. The following organizations and institutions may apply: public/state controlled institutions of higher education; private institutions of higher education; Hispanic-serving institutions; historically Black colleges and universities (HBCUs); tribally controlled colleges and universities (TCCUs); Alaska native and native Hawaiian serving institutions; nonprofits with 501(c)3 IRS status; nonprofits without 501(c)3 IRS status; small businesses; for-profit organizations; State governments; regional organizations; U.S. territories or possessions; Indian/Native American tribal governments; Indian/Native American tribally designated organizations; county governments; city or township governments; special district governments; independent school districts; public housing authorities/Indian housing authorities; eligible agencies of the Federal government; and faith-based or community based organizations.

Geographic Focus: All States
Date(s) Application is Due: Aug 31; Sep 30
Amount of Grant: 300,000 USD
Contact: Susan Grove; (301) 435-0714; sg16d@nih.gov or grantsinfo@od.nih.gov
Internet: http://grants.nih.gov/grants/guide/pa-files/PAR-09-120.html
Sponsor: National Institutes of Health
6701 Rockledge Drive
Bethesda, MD 20892

NIH Bone and the Hematopoietic and Immune Systems Research Grants 3208

The participating institutes seek investigator-initiated projects that have the potential to illuminate functional interactions between bone and the hematopoietic and immune systems. Recent observations underscore the linkage between endochondral bone formation and the establishment of hematopoietic marrow, and suggest that interactions between bone, marrow, and the immune system persist in the mature skeleton. Marrow stromal cells include the precursors of the osteochondrogenic lineage, exert important influences on osteoclastogenesis and lymphopoiesis, and mediate the effects of some systemic factors on bone turnover. Recent evidence indicates that hematopoietic cells can influence the differentiation of osteogenic cells, and suggests that mature lymphocytes can influence osteoclastic and osteoblastic functions. In order to explore the mechanisms that underlie these interactions, the participating institutes support outstanding projects that have the potential to either clarify the importance of specific cell types and effector molecules or identify previously unrecognized cellular and molecular agents that influence bone physiology. Collaborations among bone biologists, hematologists, and immunologists, and between basic scientists and clinical investigators, are particularly encouraged. Inquiries may be directed to the staff at the appropriate NIH institute: Dr. William Sharrock, Musculoskeletal Diseases Branch, NIAMS, 45 Center Drive, Bethesda, MD 20892; Dr. Frank Bellino, (301) 496-6402, Biology of Aging Program, NIA, Gateway Building, Bethesda, MD 20892; Dr. Kenneth Gruber, (301) 584-4836, Chronic and Disabling Diseases Branch, NIDCR, 45 Center Drive, Bethesda, MD 20892; and Dr. Charles Peterson, (301) 435-0050, Division of Blood Diseases and Resources, NHLBI, 6701 Rockledge Drive, Bethesda, MD 20892. PA: PA-99-085

Requirements: Applications may be submitted by domestic and foreign, for-profit and nonprofit, public and private organizations, units of state and local governments, and eligible agencies of the federal government.

Restrictions: Foreign institutions are not eligible for program project grant support.
Geographic Focus: All States
Amount of Grant: Up to 250,000 USD
Contact: Frank Bellino, (301) 496-6402; fax (301) 402-0010; fb12a@nih.gov
Internet: http://grants.nih.gov/grants/guide/pa-files/PA-99-085.html
Sponsor: National Institutes of Health
Gateway Building, 7201 Wisconsin Avenue, Suite 2C231
Bethesda, MD 20892

NIH Building Interdisciplinary Research Careers in Women's Health Grants 3209

The NIH Office of Research on Women's Health (ORWH) and its cosponsors invite institutional career development award applications for Building Interdisciplinary Research Careers in Women's Health (BIRCWH) Career Development Programs, hereafter termed Programs. Programs will support mentored research career development of junior faculty members, known as BIRCWH Scholars, who have recently completed clinical training or postdoctoral fellowships, and who will be engaged in interdisciplinary basic, translational, behavioral, clinical, and/or health services research relevant to women's health or sex/gender factors. Direct costs are limited to $500,000 total costs for a five-year period. Letters of Intent are due by September 23, with final applications due by October 22.

Requirements: The following organizations and institutions are eligible to apply for support: public/state controlled institutions of higher education; private institutions of higher education; Hispanic-serving institutions; historically Black colleges and universities (HBCUs); tribally controlled colleges and universities (TCCUs); Alaska Native and Native Hawaiian serving institutions; nonprofits with 501(c)3 IRS status (other than institutions of higher education); nonprofits without 501(c)3 IRS status (other than institutions of higher education); Indian/Native American tribal governments (federally recognized); Indian/Native American tribally designated organizations; U.S. territories or possessions; and eligible agencies of the federal government.

Geographic Focus: All States
Date(s) Application is Due: Sep 23; Oct 22
Contact: Joan Davis Nagel, (301) 496-9186; fax (301) 402-1798; joandav@mail.nih.gov
Internet: http://grants.nih.gov/grants/guide/rfa-files/RFA-OD-09-006.html
Sponsor: National Institutes of Health
6707 Democracy Boulevard, Suite 400
Bethesda, MD 20892

NIHCM Foundation Health Care Research Awards 3210

To encourage outstanding work from researchers furthering innovation in health care policy and management, the National Institute for Health Care Management (NIHCM) Research and Educational Foundation recognizes excellence in original and creative research with a $10,000 award. Selection criteria for the award includes the originality and creativity of the research. Impact that the research may have in advancing the public policy debate, improving health care delivery, management and the relevance to one or more of the subject areas. The award will be presented to the winner at a dinner in Washington, DC in May.

Requirements: Articles must have been published in a peer-reviewed journal or similar quality publication during calendar year 2009. Applications are encouraged from authors in diverse fields, including economics, health policy, political science, public health, etc. and from individuals in non-academic settings, such as research firms and policy organizations. All research articles must address some aspect of health care financing, delivery, organization or the implementation of health care policy.

Geographic Focus: All States
Date(s) Application is Due: Feb 26
Contact: Nancy Chockley; (202) 296-4426; fax (202) 296-4319; nihcm@nihem.org
Internet: http://nihcm.org/awards/research
Sponsor: National Institute for Health Care Management Foundation
1225 19th Street NW, Suite 710
Washington, DC 20036

NIHCM Foundation Health Care Television and Radio Journalism Awards 3211

The Third Annual NIHCM Foundation Health Care Television and Radio Journalism Award recognizes excellence in television and radio reporting on health care issues and policy. Selection criteria will include, new insights generated by the reporting, impact on health care policy and quality of writing and production. A $10,000 prize will be presented to the winner at a dinner in Washington, DC in May.

Requirements: Entries must originally have aired during calendar year 2009. Entries will be accepted from staff employees of national or local news organizations, nationally-distributed cable programs, or from freelance journalists who have produced reports disseminated by same. Online stories will be accepted if produced for and disseminated originally on the web in affiliation with a television or radio news organization. Material repurposed from other media will not be accepted.

Geographic Focus: All States
Date(s) Application is Due: Feb 26
Amount of Grant: 10,000 USD
Contact: Nancy Chockley; (202) 296-4426; fax (202) 296-4319; nihcm@nihem.org
Internet: http://nihcm.org/awards/televison_and_radio_journalism
Sponsor: National Institute for Health Care Management Foundation
1225 19th Street NW, Suite 710
Washington, DC 20036

NIH Dietary Supplement Research Centers: Botanicals Grants 3212

ODS, NCCAM and NCI invite new and renewal applications to support research centers which will: promote collaborative integrated interdisciplinary study of botanicals, particularly those found as ingredients in dietary supplements; and conduct research of high potential for being translated into practical benefits for human health. This initiative is intended to advance the spectrum of botanical research activities, ranging from plant identification and characterization to early phase clinical studies. Preclinical research that will inform future clinical studies is encouraged as the primary research focus of this program. ODS has a particular interest in botanicals as part of health care for conditions relating to health maintenance or primary prevention. NCCAM is interested in the study of botanicals broadly used by the American public. The NCI Division of Cancer Prevention is interested in supporting research focused on mechanisms by which botanically derived bioactive food components might influence cancer risk and tumor behavior. An applicant may request a project period of up to five years; an annual budget in direct costs up to $1.0 million.

Requirements: The following organizations and institutions are eligible to apply: individuals; public/state controlled institutions of higher education; private institutions of higher education; Hispanic-serving institutions; historically Black colleges and universities (HBCUs); tribally controlled colleges and universities (TCCUs); Alaska Native and Native Hawaiian serving institutions; nonprofits with 501(c)3 IRS status; nonprofits without 501(c)3 IRS status; for-profit organizations; state governments;

Indian/Native American tribal governments (federally recognized); Indian/Native American tribally designated organizations; county governments; city or township governments; special district governments; independent school districts; public housing authorities/Indian housing authorities; U.S. territory or possession; Indian/Native American tribal governments (other than federally recognized); regional organizations; non-domestic (non-U.S.) entities (foreign organizations); other eligible agencies of the federal government; and faith-based or community-based organizations.
Geographic Focus: All States
Date(s) Application is Due: Dec 3
Contact: Christine A. Swanson; (301) 435-2920; Swansonc@od.nih.gov
Internet: http://grants.nih.gov/grants/guide/pa-files/PAR-09-091.html
Sponsor: National Institutes of Health
6100 Executive Boulevard, Room 3B01, MSC 7517
Bethesda, MD 20892

NIH Earth-Based Research Relevant to the Space Environment 3213
The purpose of this Program Announcement (PA) is to stimulate ground-based research on basic, applied, and clinical biomedical and behavioral problems that are relevant to human space flight or that could use the space environment as a laboratory. Although none of the research supported under this initiative would be conducted in space, it is anticipated that it would form a basis for future competitively reviewed studies which could be conducted on the International Space Station, or other space flight opportunities, by skilled on-board specialists. Modular Grant applications will request direct costs in $25,000 modules, up to a total direct cost request of $250,000 per year.
Requirements: The following organizations and institutions may apply: public/state controlled institutions of higher education; private institutions of higher education; Hispanic-serving institutions; historically Black colleges and universities (HBCUs); tribally controlled colleges and universities (TCCUs); Alaska native and native Hawaiian serving institutions; nonprofits with 501(c)3 IRS status; nonprofits without 501(c)3 IRS status; small businesses; for-profit organizations; State governments; regional organizations; U.S. territories or possessions; Indian/Native American tribal governments; Indian/Native American tribally designated organizations; county governments; city or township governments; special district governments; independent school districts; public housing authorities/Indian housing authorities; eligible agencies of the Federal government; and faith-based or community based organizations.
Geographic Focus: All States
Date(s) Application is Due: Feb 5; Jun 5; Oct 5
Contact: Susan Grove; (301) 435-0714; sg16d@nih.gov or grantsinfo@od.nih.gov
Internet: http://grants.nih.gov/grants/guide/pa-files/PA-00-088.html
Sponsor: National Institutes of Health
6701 Rockledge Drive
Bethesda, MD 20892

NIH Enhancing Adherence to Diabetes Self-Management Behaviors 3214
This PA solicits applications for investigator-initiated research related to sociocultural, environmental, and behavioral mechanisms and biological/technological factors that contribute to successful and ongoing self-management in diabetes. Self-management is defined as client strategies and behaviors that contribute to blood glucose normalization, improved health, and prevention or reduction of complications. It is broader than adherence to specific regimen components and incorporates deliberate problem solving and decision-making processes. Applications are encouraged for both type 1 and type 2 diabetes; representative and minority populations; and all age groups. Inquiries may be directed to program staff at the appropriate NIH institute. Modular Grant applications will request direct costs in $25,000 modules, up to a total direct cost request of $250,000 per year.
Requirements: Applications may be submitted by domestic and foreign, for-profit and non-profit organizations, public and private, such as universities, colleges, hospitals, laboratories, units of State and local governments, and eligible agencies of the Federal government. Racial/ethnic minority individuals, women, and persons with disabilities are encouraged to apply as principal investigators.
Geographic Focus: All States
Date(s) Application is Due: Feb 16; Jun 16; Oct 16
Amount of Grant: Up to 250,000 USD
Contact: Susan Grove; (301) 435-0714; sg16d@nih.gov or grantsinfo@od.nih.gov
Internet: http://grants.nih.gov/grants/guide/pa-files/PA-00-049.html
Sponsor: National Institutes of Health
6701 Rockledge Drive
Bethesda, MD 20892

NIH Exceptional, Unconventional Research Enabling Knowledge Acceleration Grants 3215
The purpose of the EUREKA (Exceptional Unconventional Research Enabling Knowledge Acceleration) initiative is to foster exceptionally innovative research that, if successful, will have an unusually high impact on the areas of science that are germane to the mission of one or more of the participating NIH Institutes. Before submitting an application, it is extremely important to verify that the proposed research is of interest to at least one of the NIH Institutes that is participating in the EUREKA FOA, since applications that are not germane to the mission of one or more of the participating Institutes will be withdrawn without review.
Requirements: The following organizations and institutions are eligible to apply: public/state controlled institutions of higher education; private institutions of higher education; Hispanic-serving institutions; historically Black colleges and universities (HBCUs); tribally controlled colleges and universities (TCCUs); Alaska Native and Native Hawaiian serving institutions; nonprofits with 501(c)3 IRS status (other than institutions of higher education); nonprofits without 501(c)3 IRS status (other than institutions of higher education); small businesses; for-profit organizations (other than small businesses); non-domestic (non-U.S.) entities (foreign organizations).
Restrictions: EUREKA is for new projects, not for continuation of existing projects. Nor is EUREKA for support of pilot projects, i.e., projects of limited scope that are designed primarily to generate data that will enable the PI to seek other funding opportunities.
Geographic Focus: All States
Date(s) Application is Due: Nov 24
Amount of Grant: Up to 800,000 USD
Contact: Laurie Tompkins, (301) 594-0943; tompkinL@nigms.nih.gov
Internet: http://grants.nih.gov/grants/guide/rfa-files/RFA-GM-10-009.html
Sponsor: National Institutes of Health
45 Center Drive, MSC 6200
Bethesda, MD 20892-6200

NIH Exploratory/Developmental Research Grant 3216
The Exploratory/Developmental Grant (R21) mechanism is intended to encourage exploratory and developmental research projects by providing support for the early and conceptual stages of these projects. These studies may involve considerable risk but may lead to a breakthrough in a particular area, or to the development of novel techniques, agents, methodologies, models, or applications that could have a major impact on a field of biomedical, behavioral, or clinical research.
Requirements: Eligible organizations: for profit organizations; non-profit organizations; public or private institutions, such as universities, colleges, hospitals and laboratories; units of state government; units of local government; eligible institutions of the federal government; domestic institutions; foreign institutions; faith-based or community-based organizations; units of state tribal government; and units of local tribal government. Eligible Project Directors/Principal Investigators: any individual with the skills, knowledge and resources necessary to carry out the proposed research, is invited to work with his/her institution to develop an application for support; individuals from underrepresented racial and ethnic groups; individuals with disabilities
Geographic Focus: All States
Date(s) Application is Due: Feb 16; Jun 16; Oct 16
Amount of Grant: Up to 275,000 USD
Contact: Ellen S. Liberman, (301) 451-2020; esl@nei.nih.gov
Internet: http://grants.nih.gov/grants/guide/pa-files/PA-06-181.html
Sponsor: National Eye Institute
5635 Fishers Lane, Suite 1300, MSC 9300
Bethesda, MD 20892-9300

NIH Exploratory Innovations in Biomedical Computational Science and Technology (BISTI) Grants 3217
The NIH is interested in promoting research and developments in computational science and technology that will support rapid progress in areas of scientific opportunity in biomedical research. As defined here, biomedical computing or biomedical information science and technology includes database design, graphical interfaces, querying approaches, data retrieval, data visualization and manipulation, data integration through the development of integrated analytical tools, and tools for electronic collaboration, as well as computational and mathematical research including the development of structural, functional, integrative, and analytical models and simulations. This particular FOA is intended to support exploratory biomedical informatics and computational biology research-applications should be innovative, with high risk/high impact in new areas that are lacking preliminary data or development. Because the nature and scope of the proposed research will vary from application to application, it is anticipated that the size and duration of each award will also vary.
Requirements: Application can be made by the following institutions and organizations: for-profit organizations; non-profit organizations; public or private institutions (such as universities, colleges, hospitals, and laboratories); units of state government; units of local government; eligible agencies of the Federal government; foreign institutions; domestic institutions; faith-based or community-based organizations; Indian/Native American tribal governments (federally recognized); Indian/Native American tribal governments (other than federally recognized); and Indian/Native American tribally designated organizations.
Geographic Focus: All States
Date(s) Application is Due: Feb 16; Jun 16; Oct 16
Contact: Peter Lyster, (301) 451-6446; lysterp@mail.nih.gov
Internet: http://grants.nih.gov/grants/guide/pa-files/PAR-06-411.html
Sponsor: National Institutes of Health
45 Center Drive, Room 2AS.55k, MSC 6200
Bethesda, MD 20892-6200

NIH Fine Mapping and Function of Genes for Type 1 Diabetes Grants 3218
The purpose of this initiative is to bring together investigators with experience in genetics, immunology, and biochemistry to perform fine mapping of the loci discovered and study the function of the genes being identified and to identify the mechanisms by which newly discovered genes predispose to T1D. This would lead to elucidation of the mechanisms whereby changes in the function or regulation of these genes are likely to provide crucial new insights into disease pathogenesis. The discovery of the genes would be relevant to developing a predictive strategy for individuals who may develop diabetes. In the future, genetic testing may lead to personalized treatment regimens by identifying the most appropriate class of drugs for particular patients. Total funding available for this five year program is $20 million to fund 4-10 applications, contingent

upon the availability of funds and the submission of a sufficient number of meritorious applications. The maximum direct costs are $5 million that can be spent on a project period of up to 5 years. Letters of Intent are due by February 27, with final applications due by March 30. The office strongly urges communication from applicants via email.
Requirements: The following organizations and institutions are eligible to apply: public/state controlled institutions of higher education; private institutions of higher education; Hispanic-serving institutions; historically Black colleges and universities (HBCUs); tribally controlled colleges and universities (TCCUs); Alaska Native and Native Hawaiian serving institutions; nonprofits with 501(c)3 IRS status (other than institutions of higher education); nonprofits without 501(c)3 IRS status (other than institutions of higher education); small businesses; for-profit organizations (other than small businesses); non-domestic (non-U.S.) entities (foreign organizations); and other eligible agencies of the federal government.
Geographic Focus: All States
Date(s) Application is Due: Feb 27; Mar 30
Amount of Grant: 1,500,000-5,000,000 USD
Contact: Beena Akolkar, PhD; (301) 594-8812; ba92i@nih.gov
Internet: http://grants.nih.gov/grants/guide/rfa-files/rfa-dk-08-006.html
Sponsor: National Institutes of Health
6707 Democracy Boulevard, Room 6105
Bethesda, MD 20892-5460

NIH Health Promotion Among Racial and Ethnic Minority Males 3219

This Funding Opportunity Announcement solicits Exploratory/Developmental grant applications from applicants that propose to stimulate and expand research in the health of minority men. Specifically, this initiative is intended to: enhance our understanding of the numerous factors (e.g., sociodemographic, community, societal, personal) influencing the health promoting behaviors of racial and ethnic minority males and their sub-populations across the life cycle; and solicit applications focusing on the development and testing of culturally and linguistically appropriate health-promoting interventions designed to reduce health disparities among racially and ethnically diverse males and their sub-populations age 21 and older. The total project period for an application submitted in response to this funding opportunity may not exceed two years. Direct costs are limited to $275,000 over an R21 two-year period, with no more than $200,000 in direct costs allowed in any single year.
Requirements: Application can be made by the following institutions and organizations: public/state controlled institutions of higher education; private institutions of higher education; nonprofits with 501(c)3 IRS status (other than institutions of higher education); nonprofits without 501(c)3 IRS status (other than institutions of higher education); small businesses; for-profit organizations (other than small businesses); state governments; U.S. territories or possessions; Indian/Native American tribal governments (federally recognized); Indian/Native American tribal governments (other than Federally recognized); Indian/Native American tribally designated organizations; non-domestic (non-U.S.) entities (foreign organizations); Hispanic-serving institutions; historically Black colleges and universities (HBCUs); tribally controlled colleges and universities (TCCUs); Alaska Native and Native Hawaiian serving institutions; regional organization; eligible agencies of the Federal government; and faith-based or community based organizations.
Geographic Focus: All States
Date(s) Application is Due: Feb 16; Jun 16; Oct 16
Amount of Grant: 200,000-275,000 USD
Contact: Dr. Charlotte Pratt; (301) 435-0382; fax (301) 480-5158; prattc@nhlbi.nih.gov
Internet: http://grants.nih.gov/grants/guide/pa-files/PA-07-421.html
Sponsor: National Institutes of Health
6701 Rockledge Drive, Suite 10118, MSC 7936
Bethesda, MD 20892-7936

NIH Human Connectome Project Grants 3220

This Funding Opportunity Announcement is issued as an initiative of the NIH Blueprint for Neuroscience Research. The Neuroscience Blueprint is a collaborative framework through which 16 NIH Institutes, Centers and Offices jointly support neuroscience-related research, with the aim of accelerating discoveries and reducing the burden of nervous system disorders. The overall purpose of this five year Human Connectome Project (HCP) is to develop and share knowledge about the structural and functional connectivity of the human brain. This purpose will be pursued through the following specific efforts: existing, but cutting-edge, non-invasive imaging technologies will be optimized and combined to acquire structural and functional in vivo data about axonal projections and neural connections from brains of hundreds of healthy adults. Demographic data and data regarding sensory, motor, cognitive, emotional, and social function will also be collected for each subject, as will DNA samples and blood (to establish cell lines). Models to better understand and use these data will be developed. Connectivity patterns will be linked to existing architectonic data. Data and models will be made available to the research community immediately via a user-friendly system to include tools to query, organize, visualize and analyze data. Outreach activities will be conducted to engage and educate the research community about the imaging tools, data, models, and informatics tools. After five years, these specific efforts are expected to deliver: a set of integrated, non-invasive imaging tools to obtain connectivity data from humans in vivo; a high quality and well characterized, quantitative set of human connectivity data linked to behavioral and genetic data as well as to general, existing architectonic data, and associated models, from up to hundreds of healthy adult female and male subjects; and rapid, user-friendly dissemination of connectivity data, models, and tools to the research community via outreach activities and an informatics platform. Letters of Intent are due by October 24, with final applications due by November 24.
Requirements: The following organizations and institutions are eligible to apply for support: public/state controlled institutions of higher education; private institutions of higher education; Hispanic-serving institutions; historically Black colleges and universities (HBCUs); tribally controlled colleges and universities (TCCUs); Alaska Native and Native Hawaiian serving institutions; nonprofits with 501(c)3 IRS status (other than institutions of higher education); nonprofits without 501(c)3 IRS status (other than institutions of higher education); Indian/Native American tribal governments (federally recognized); Indian/Native American tribally designated organizations; U.S. territories or possessions; and eligible agencies of the federal government.
Geographic Focus: All States
Date(s) Application is Due: Oct 24; Nov 24
Amount of Grant: Up to 6,000,000 USD
Contact: Michael F. Huerta, (301) 443-1815; fax (301) 443-1731; mhuertl@mail.nih.gov
Internet: http://grants.nih.gov/grants/guide/rfa-files/RFA-MH-10-020.html
Sponsor: National Institutes of Health
6001 Executive Boulevard, Room 7202, MSC 9645
Bethesda, MD 20892-9645

NIH Independent Scientist Award 3221

The award is a special salary-only grant designed to provide protected time for newly independent scientists who currently have non-research obligations such as heavy teaching loads, clinical work, committee assignments, service, and administrative duties that prevent them from having a period of intensive research focus. The award is targeted to persons with doctoral degrees who have completed their research training, have independent peer-reviewed research support, and who need a period of protected research time in order to foster their research career development. New applications are due February 12, June 12, and October 12, while renewal applications are due March 12, July 12, and November 12.
Requirements: The following organizations and institutions are eligible to apply: public/state controlled institutions of higher education; private institutions of higher education; nonprofits with 501(c)3 IRS status; nonprofits without 501(c)3 IRS status; small businesses; for-profit organizations; State governments; U.S. territories or possessions; Indian/Native American tribal governments (Federally recognized and other than Federally recognized); Indian/Native American tribally designated organizations; Hispanic-serving institutions; historically Black colleges and universities (HBCUs); tribally controlled colleges and universities (TCCUs); Alaska Native and Native Hawaiian serving institutions; regional organizations; and faith-based or community based organizations. The candidate must have a doctoral degree and peer-reviewed, independent research support at the time the award is made. The candidate must spend a minimum of 75 percent effort conducting research during the period of the award.
Restrictions: The award is not intended for investigators who already have full time to perform research, or have substantial publication records or considerable research support indicating that they are well established in their fields. Foreign institutions are not eligible to apply.
Geographic Focus: All States
Date(s) Application is Due: Feb 12; Mar 12; Jun 12; Jul 12; Oct 12; Nov 12
Amount of Grant: 75,000-100,000 USD
Contact: Susan Grove; (301) 435-0714; sg16d@nih.gov or grantsinfo@od.nih.gov
Internet: http://grants.nih.gov/grants/guide/contacts/PA-09-038_contacts.html
Sponsor: National Institutes of Health
6701 Rockledge Drive
Bethesda, MD 20892

NIH Innovations in Biomedical Computational Science and Technology Grants 3222

The NIH is interested in promoting research and developments in computational science and technology that will support rapid progress in areas of scientific opportunity in biomedical research. As defined here, biomedical computing or biomedical information science and technology includes database design, graphical interfaces, querying approaches, data retrieval, data visualization and manipulation, data integration through the development of integrated analytical tools, and tools for electronic collaboration, as well as computational and mathematical research including the development of structural, functional, integrative, and analytical models and simulations. Because the nature and scope of the proposed research will vary from application to application, it is anticipated that the size and duration of each award will also vary.
Requirements: Application can be made by the following institutions and organizations: for-profit organizations; non-profit organizations; public or private institutions (such as universities, colleges, hospitals, and laboratories); units of state government; units of local government; eligible agencies of the Federal government; foreign institutions; domestic institutions; faith-based or community-based organizations; Indian/Native American tribal governments (federally recognized); Indian/Native American tribal governments (other than federally recognized); and Indian/Native American tribally designated organizations.
Geographic Focus: All States
Date(s) Application is Due: Feb 5; Jun 5; Oct 5
Amount of Grant: Up to 500,000 USD
Contact: Peter Lyster, (301) 451-6446; lysterp@mail.nih.gov
Internet: http://grants.nih.gov/grants/guide/pa-files/PAR-09-218.html
Sponsor: National Institutes of Health
45 Center Drive, Room 2AS.55k, MSC 6200
Bethesda, MD 20892-6200

NIH Innovations in Biomedical Computational Science and Technology Initiative Grants for SBIR 3223

This funding opportunity announcement (FOA) solicits Small Business Innovation Research (SBIR) grant applications from small business concerns (SBCs) that propose innovative research in biomedical computational science and technology to promote the progress of biomedical research. There exists an expanding need to speed the progress of biomedical research through the power of computing to manage and analyze data and to model biological processes. The NIH is interested in promoting research and developments in biomedical computational science and technology that will support rapid progress in areas of scientific opportunity in biomedical research. As defined here biomedical computing or biomedical information science and technology includes database design, graphical interfaces, querying approaches, data retrieval, data visualization and manipulation, data integration through the development of integrated analytical tools, and tools for electronic collaboration, as well as computational research including the development of structural, functional, integrative, and analytical models and simulations.
Requirements: Only United States small business concerns (SBCs) are eligible to submit SBIR applications.
Geographic Focus: All States
Date(s) Application is Due: Apr 5; Aug 5; Dec 5
Amount of Grant: 150,000-750,000 USD
Contact: Peter Lyster, (301) 451-6446; lysterp@mail.nih.gov
Internet: http://grants.nih.gov/grants/guide/pa-files/PAR-07-160.html
Sponsor: National Institutes of Health
45 Center Drive, Room 2AS.55k, MSC 6200
Bethesda, MD 20892-6200

NIH Innovations in Biomedical Computational Science and Technology Initiative Grants for STTR 3224
This funding opportunity announcement (FOA) solicits Small Business Technology Transfer (STTR) grant applications from small business concerns (SBCs) that propose innovative research in biomedical computational science and technology to promote the progress of biomedical research. There exists an expanding need to speed the progress of biomedical research through the power of computing to manage and analyze data and to model biological processes. The NIH is interested in promoting research and developments in biomedical computational science and technology that will support rapid progress in areas of scientific opportunity in biomedical research. As defined here biomedical computing or biomedical information science and technology includes database design, graphical interfaces, querying approaches, data retrieval, data visualization and manipulation, data integration through the development of integrated analytical tools, and tools for electronic collaboration, as well as computational research including the development of structural, functional, integrative, and analytical models and simulations.
Requirements: Only United States small business concerns (SBCs) are eligible to submit STTR applications.
Geographic Focus: All States
Date(s) Application is Due: Apr 5; Aug 5; Dec 5
Contact: Peter Lyster, (301) 451-6446; lysterp@mail.nih.gov
Internet: http://grants.nih.gov/grants/guide/pa-files/PAR-07-161.html
Sponsor: National Institutes of Health
45 Center Drive, Room 2AS.55k, MSC 6200
Bethesda, MD 20892-6200

NIH Mentored Clinical Scientist Development Award 3225
This award enables candidates holding professional degrees (e.g., M.D., D.O., D.V.M., or equivalent degrees) to undertake 3 to 5 years of special study and supervised research with the goal of becoming independent investigators. The award also allows awardees to pursue a research career development program suited to their experience and capabilities under a mentor who is competent to provide guidance in the chosen research area. Institutions may submit applications on behalf of candidates who hold professional degrees. At least 2 years must have elapsed since the health professional degree was granted. Candidates can have varying levels of clinical training and research experience. New applications are due February 12, June 12, and October 12, while resubmitted applications are due March 12, July 12, and November 12.
Requirements: All candidates must be U.S. citizens, non-citizen nationals, or legal permanent residents of the U.S. Persons with temporary or student visas are not eligible. The grantee institution must have a strong, well-established research and research training program in the chosen area, accomplished faculty in the basic and clinical sciences, and a commitment to the candidate's research development. The proposed program should include an appropriate mentor.
Geographic Focus: All States
Date(s) Application is Due: Feb 12; Mar 12; Jun 12; Jul 12; Oct 12; Nov 12
Contact: Susan Grove; (301) 435-0714; sg16d@nih.gov or grantsinfo@od.nih.gov
Internet: http://grants.nih.gov/grants/guide/pa-files/pa-06-512.html
Sponsor: National Institutes of Health
6701 Rockledge Drive
Bethesda, MD 20892

NIH Mentored Patient-Oriented Research Career Development Award 3226
The purpose of this program is to support the career development of investigators who have made a commitment to focus their research endeavors on patient-oriented research. This mechanism provides support for three to five years of supervised study and research for clinically trained professionals who have the potential to develop into productive, clinical investigators focusing on patient-oriented research. Inquiries may be directed to program staff at the appropriate NIH institute. See the following web page for specific contact information: http://grants.nih.gov/grants/guide/contacts/PA-09-043_contacts.html
Requirements: Applicants must have a clinical doctoral degree or its equivalent and must have completed their clinical training, specialty and, if applicable, sub-specialty training prior to receiving an award. However, candidates may submit an application prior to the completion of clinical training. Candidates must identify a mentor with extensive research experience, and must be willing to spend a minimum of 75 percent of full-time professional effort conducting research career development and clinical research.
Restrictions: Current and former principal investigators on NIH research project (R01), FIRST Awards (R29), comparable career development awards (K01, K07, or K08), sub-projects of program project (P01) or center grants (P50), and the equivalent are not eligible for funding under this grant.
Geographic Focus: All States
Contact: Susan Grove; (301) 435-0714; sg16d@nih.gov or grantsinfo@od.nih.gov
Internet: http://grants.nih.gov/grants/guide/pa-files/PA-09-043.html#SectionVII
Sponsor: National Institutes of Health
6701 Rockledge Drive
Bethesda, MD 20892

NIH Mentored Research Scientist Development Awards 3227
The overall objective of the NIH Research Career Development Award program is to prepare qualified individuals for careers that have a significant impact on the health-related research needs of the Nation. The objective of the NIH Mentored Research Scientist Development Award is to provide support for a sustained period of protected time for intensive research career development under the guidance of an experienced mentor, or sponsor, in the biomedical, behavioral or clinical sciences leading to research independence. The expectation is that through this sustained period of research career development and training, awardees will launch independent research careers and become competitive for new research project grant funding. Inquiries should be directed to program staff at the appropriate NIH institute. See the website for specific contact information.
Requirements: Applications may be submitted, on behalf of candidates, by domestic, non-federal organizations, public or private, such as medical, dental, or nursing schools or other institutions of higher education. Candidates must be U.S. citizens or non-citizen nationals, or must have been lawfully admitted for permanent residence by the time of award. Candidates must have a research or a health-professional doctorate or its equivalent; have demonstrated the capacity or potential for highly productive independent research in the period after the doctorate; must identify a mentor with extensive research experience; and must be willing to spend a minimum of 75 percent of full-time professional effort conducting research and research career development during the entire award period.
Restrictions: Individuals on temporary or student visas are not eligible for this award.
Geographic Focus: All States
Date(s) Application is Due: Feb 12; Mar 12; Jun 12; Jul 12; Oct 12; Nov 12
Contact: Susan Grove; (301) 435-0714; sg16d@nih.gov or grantsinfo@od.nih.gov
Internet: http://grants.nih.gov/grants/guide/pa-files/PA-09-040.html
Sponsor: National Institutes of Health
6701 Rockledge Drive
Bethesda, MD 20892

NIH Nonhuman Primate Immune Tolerance Cooperative Study Group Grants 3228
NIAID and NIDDK invite applications from single institutions or consortia of institutions to participate in a multi-center, cooperative research program to evaluate existing and new tolerance induction treatment regimens and to elucidate the underlying mechanisms of the induction, maintenance and/or loss of tolerance in nonhuman primate models of kidney and islet transplantation. The goals of this research program are to evaluate further the safety, toxicity, and efficacy of existing tolerance induction regimens; define the underlying mechanisms of action of the therapeutic approaches under investigation; and develop and validate immune and/or surrogate markers of the induction, maintenance, and loss of tolerance, graft function, graft acceptance, and graft survival. The deadline for Letters of Intent is August 22, with full applications due September 22.
Requirements: Applications may be submitted by domestic for-profit and nonprofit organizations and public and private institutions.
Restrictions: Foreign institutions are not eligible to apply. This RFA will not support: studies in animal models other than non-human primates; human studies or trials; xenotransplantation; preliminary development of a non-human primate transplantation model; studies to improve the isolation, preservation, or supply of organs, tissues or cells; hematopoietic stem cell transplantation, unless in the context of kidney or islet transplantation (i.e. bone marrow chimerism experiments); heart or lung transplantation models; or studies of non-immunological side effects of an immuno-modulatory agent.
Geographic Focus: All States
Date(s) Application is Due: Aug 22; Sep 22
Amount of Grant: Up to 67,000,000 USD
Contact: Kristy Kraemer; (301) 496-5598; fax (301) 480-0693; kkraemer@niaid.nih.gov
Internet: http://grants.nih.gov/grants/guide/rfa-files/RFA-AI-06-018.html
Sponsor: National Institutes of Health
6610 Rockledge Drive, Room 3043, MSC-6601
Bethesda, MD 20892

NIH Receptors and Signaling in Bone in Health and Disease 3229
The objective of this initiative is to elicit grant submissions that focus on systemic hormones, local growth factors, and bone-active cytokines, their receptors and mechanisms of signaling in bone. While the primary focus is on basic research, the long-term emphasis is on identifying mechanisms or processes related to hormone action with potential applicability as targets for therapeutic agents that may have efficacy in the treatment of diseases that adversely affect bone, such as osteoporosis and primary hyperparathyroidism. Modular Grant applications will request direct costs in $25,000 modules, up to a total direct cost request of $250,000 per year.

Requirements: Applications may be submitted by domestic and foreign for-profit and nonprofit organizations, public and private, such as universities, colleges, hospitals, laboratories, units of State and local governments, and eligible agencies of the Federal Government. Racial/ethnic minority individuals, women, and persons with disabilities are encouraged to apply as principal investigators.
Geographic Focus: All States
Date(s) Application is Due: Feb 5; Jun 5; Oct 5
Amount of Grant: Up to 250,000 USD
Contact: Susan Grove; (301) 435-0714; sg16d@nih.gov or grantsinfo@od.nih.gov
Internet: http://grants.nih.gov/grants/guide/pa-files/PA-00-017.html
Sponsor: National Institutes of Health
6701 Rockledge Drive
Bethesda, MD 20892

NIH Recovery Act Limited Competition: Academic Research Enhancement Awards 3230

The purpose of the Academic Research Enhancement Award (AREA) program is to stimulate research in educational institutions that provide baccalaureate or advanced degrees for a significant number of the Nation's research scientists, but that have not been major recipients of NIH support. These AREA grants create opportunities for scientists and institutions otherwise unlikely to participate extensively in NIH programs, to contribute to the Nation's biomedical and behavioral research effort. AREA grants are intended to support small-scale health-related research projects proposed by faculty members of eligible, domestic institutions. The requested budget is limited to $300,000 total costs for the entire three-year project (maximum time period). Applicants should contact individuals at the specific departments listed on the web site. No Letters of Intent are necessary, though full applications should arrive by September 24.
Requirements: The following institutions and organizations may apply: public/state controlled institutions of higher education; private institutions of higher education; Hispanic-serving institutions; historically Black colleges and universities (HBCUs); tribally controlled colleges and universities (TCCUs); Alaska Native and Native Hawaiian serving institutions; public or private institutions and components of institutions defined below as health professional schools or colleges; and other academic components.
Restrictions: Foreign organizations/institutions are not permitted as the applicant organization.
Geographic Focus: All States
Date(s) Application is Due: Sep 24
Amount of Grant: Up to 300,000 USD
Contact: Susan Grove; (301) 435-0714; sg16d@nih.gov or grantsinfo@od.nih.gov
Internet: http://grants.nih.gov/grants/guide/rfa-files/RFA-OD-09-007.html
Sponsor: National Institutes of Health
6701 Rockledge Drive
Bethesda, MD 20892

NIH Recovery Act Limited Competition: Biomedical Research, Development, and Growth to Spur the Acceleration of New Technologies Pilot Grants 3231

This Funding Opportunity Announcement (FOA), supported by funds provided to the NIH under the American Recovery & Reinvestment Act of 2009 (Recovery Act or ARRA), Public Law 111-5, solicits grant applications for a new initiative called Biomedical Research, Development, and Growth to Spur the Acceleration of New Technologies (BRDG-SPAN) Pilot Program. The purpose of this pilot program is to address the funding gap between promising research and development (R&D) and transitioning to the market — often called the Valley of Death — by contributing to the critical funding needed by applicants to pursue the next appropriate milestone(s) toward ultimate commercialization; i.e., to carry out later stage research activities necessary to that end. This program aims to accelerate the transition of research innovations and technologies toward the development of products or services that will improve human health, help advance the mission of NIH and its Institutes and Centers (ICs), and create significant value and economic stimulus. This program also aims to foster partnerships among a variety of research and development (R&D) collaborators working toward these aims. The requested budget is limited to $1 million total costs per year for a maximum of three years. Applicants should contact individuals at the specific departments listed on the web site. Letters of Intent are due by August 3, and full applications should arrive by September 1.
Requirements: The applicant United States institution and organization must be located in the 50 states, territories and possessions of the U.S., Commonwealth of Puerto Rico, Trust Territory of the Pacific Islands, or District of Columbia.
Restrictions: Foreign organizations/institutions are not permitted as the applicant organization.
Geographic Focus: All States
Date(s) Application is Due: Aug 3; Sep 1
Amount of Grant: Up to 1,000,000 USD
Contact: Susan Grove; (301) 435-0714; sg16d@nih.gov or grantsinfo@od.nih.gov
Internet: http://grants.nih.gov/grants/guide/rfa-files/RFA-OD-09-008.html
Sponsor: National Institutes of Health
6701 Rockledge Drive
Bethesda, MD 20892

NIH Recovery Act Limited Competition: Building Sustainable Community Linked Infrastructure to Enable Health Science Research 3232

Applications are invited from domestic (United States) institutions/organizations proposing to develop or expand needed infrastructures that will fundamentally transform collaboration and communication between academic health centers and local communities. Such collaborative infrastructures are essential to advance the health science research enterprise while ensuring that important research findings are effectively disseminated and implemented to improve public health and advance health care delivery. The primary intent is to fund projects to develop infrastructure for productive and sustainable academic-community research partnerships that can be leveraged in the future for efficiently conducting research that includes and is relevant to affected communities, and through which research findings can be disseminated in a manner that maximizes impact on public health. It is expected that each grantee will be able to leverage this Recovery Act funding into future research grants using the infrastructure partnership between itself and the participating community. Letters of Intent are due by November 12, with final applications due by December 11.
Requirements: The following organizations and institutions may apply: public/state controlled institutions of higher education; private institutions of higher education; Hispanic-serving institutions; historically Black colleges and universities (HBCUs); tribally controlled colleges and universities (TCCUs); Alaska native and native Hawaiian serving institutions; nonprofits with 501(c)3 IRS status; nonprofits without 501(c)3 IRS status; small businesses; for-profit organizations; State governments; regional organizations; U.S. territories or possessions; Indian/Native American tribal governments; Indian/Native American tribally designated organizations; county governments; city or township governments; special district governments; independent school districts; public housing authorities/Indian housing authorities; eligible agencies of the Federal government; and faith-based or community based organizations.
Restrictions: This RFA is not intended to fund research or evaluation projects, clinical trials, or public health campaigns. Foreign organizations/institutions are not permitted as the applicant organization.
Geographic Focus: All States
Date(s) Application is Due: Nov 12; Dec 11
Contact: Dr. Suzanne Heurtin-Roberts, (301) 402-2277; sheurtin@mail.nih.gov
Internet: http://grants.nih.gov/grants/guide/rfa-files/RFA-OD-09-010.html
Sponsor: National Institutes of Health
1 Center Drive, Room B3-11
Bethesda, MD 20892

NIH Recovery Act Limited Competition: Small Business Catalyst Awards for Accelerating Innovative Research Grants 3233

This NIH Funding Opportunity Announcement, supported by funds provided to the NIH under the American Recovery and Reinvestment Act of 2009 (Recovery Act or ARRA), Public Law 111-5, invites grant applications from small business concerns that propose to accelerate innovation through high risk, high reward research and development (R&D) that has commercial potential and is relevant to the mission of the NIH. The Small Business Catalyst Award is further expected to support entrepreneurs of exceptional creativity, drawn from scientific and technological environments beyond NIH, who propose pioneering and possibly transformative approaches to addressing major biomedical or behavioral challenges with the potential for downstream commercial development. The Small Business Catalyst Award for Accelerating Innovative Research funding opportunity seeks to encourage fresh research perspectives and approaches to serve the mission of NIH. In particular, applications from small business concerns without a history of NIH Small Business Innovation Research (SBIR) or Small Business Technology Transfer (STTR) support may receive funding priority. Solicited are applications for support for projects that have the potential to generate high impact results (e.g., products, processes or services) and/or innovative research applications, research tools, techniques, devices, inventions, or methodologies. The outcomes of the research supported should have potential to lead to products that will improve public health and create significant value and economic stimulus. This FOA solicits early-stage ideas that promise to lead to major leaps forward in capabilities important to serving the mission of NIH rather than incremental improvements of existing technologies. In accord with the funding priority of this initiative to attract applicants without a history of SBIR/STTR support from NIH, the focus of the projects solicited by this FOA is on early stage technology development. Budget requests are limited to $200,000 total costs for a maximum project period of one year. Applicants should contact individuals at the specific departments listed on the web site. Letters of Intent are due by August 3, and full applications should arrive by September 1.
Requirements: Only United States small business concerns (SBCs) are eligible to submit SBIR applications. A small business concern is one that, at the time of award of SBIR Phase I and Phase II, meets all of the following criteria: is organized for profit, with a place of business located in the United States, which operates primarily within the United States or which makes a significant contribution to the United States economy through payment of taxes or use of American products, materials or labor; is in the legal form of an individual proprietorship, partnership, limited liability company, corporation, joint venture, association, trust or cooperative, except that where the form is a joint venture, there can be no more than 49 percent participation by foreign business entities in the joint venture; is at least 51 percent owned and controlled by one or more individuals who are citizens of, or permanent resident aliens in, the United States, or it must be a for-profit business concern that is at least 51% owned and controlled by another for-profit business concern that is at least 51% owned and controlled by one or more individuals who are citizens of, or permanent resident aliens in, the United States, except in the case of a joint venture, where each entity to the venture must be 51 percent owned and controlled by one or more individuals who are citizens of, or permanent resident aliens in, the United States; and has, including its affiliates, not more than 500 employees.
Geographic Focus: All States
Date(s) Application is Due: Aug 3; Sep 1
Amount of Grant: 200,000 USD
Contact: Susan Grove; (301) 435-0714; sg16d@nih.gov or grantsinfo@od.nih.gov
Internet: http://grants.nih.gov/grants/guide/rfa-files/RFA-OD-09-009.html

Sponsor: National Institutes of Health
6701 Rockledge Drive
Bethesda, MD 20892

NIH Recovery Act Limited Competition: Supporting New Faculty Recruitment to Enhance Research Resources through Biomedical Research Core Centers Grants 3234

This initiative, which is supported by funds provided to the NIH under the Recovery Act, is designed to provide the necessary resources for U.S. academic institutions/organizations to enhance their biomedical research efforts through the development of Biomedical Research Core Centers. For this announcement, Biomedical Core Centers are defined as a community of multidisciplinary researchers focusing on areas of biomedical research relevant to NIH, such as centers, departments, programs, and/or trans-departmental collaborations or consortia. These Core Centers are designed to provide scientific and programmatic support for promising research faculty and their areas of research. The NIH invites applications that include plans to recruit and hire investigators to conduct biomedical research in all scientific disciplines, including the field of bioethics. Applicants should contact individuals at the specific departments listed on the web site. Letters of Intent are due by April 29, and full applications should arrive by May 29.
Requirements: The following institutions and organizations may apply: public/state controlled institutions of higher education; private institutions of higher education; Hispanic-serving institutions; historically Black colleges and universities (HBCUs); tribally controlled colleges and universities (TCCUs); Alaska Native and Native Hawaiian serving institutions; nonprofits with 501(c)3 IRS status (other than institutions of higher education); nonprofits without 501(c)3 IRS status (other than institutions of higher education); and Indian/Native American tribally designated organizations .
Restrictions: Foreign organizations/institutions are not permitted as the applicant organization.
Geographic Focus: All States
Date(s) Application is Due: Apr 29; May 29
Contact: Susan Grove; (301) 435-0714; sg16d@nih.gov or grantsinfo@od.nih.gov
Internet: http://grants.nih.gov/grants/guide/rfa-files/RFA-OD-09-005.html
Sponsor: National Institutes of Health
6701 Rockledge Drive
Bethesda, MD 20892

NIH Research On Ethical Issues In Human Subjects Research 3235

The purpose of this funding opportunity announcement is to solicit research addressing the ethical challenges of human subjects research in order to optimize the protection of human subjects and enhance the ethical conduct of human subjects research. The research design for studies on ethical issues in human subjects research should be appropriate to the nature of the project(s) proposed and the disciplines involved. Given the conceptual and methodological complexity of many of these research questions, interdisciplinary and collaborative projects are encouraged, particularly those involving clinical researchers, ethicists, and behavioral/social scientists.
Requirements: The following organizations and institutions may apply: public/state controlled institutions of higher education; private institutions of higher education; nonprofits with 501(c)3 IRS status (other than institution of higher education); nonprofits without 501(c)3 IRS status (other than institution of higher education); small businesses; for-profit organizations (other than small business); State governments; U.S. territories or possessions; Indian/Native American tribal governments (Federally recognized); Indian/Native American tribal governments (other than Federally recognized); Indian/Native American tribally designated organizations; non-domestic (non-U.S.) entities (foreign organization); eligible agencies of the Federal government; and faith-based or community based organizations.
Geographic Focus: All States
Date(s) Application is Due: Feb 5; Jun 5; Oct 5
Contact: Kim Witherspoon, (301) 496-8866; withersk@ctep.nci.nih.gov
Internet: http://grants.nih.gov/grants/guide/pa-files/PA-07-277.html
Sponsor: National Institutes of Health
6130 Executive Boulevard, Executive Plaza N, Room 7009, MSC 7432
Bethesda, MD 20892-7432

NIH Research on Sleep and Sleep Disorders 3236

This Funding Opportunity Announcement (FOA) solicits grant applications proposing research to advance biomedical knowledge related to sleep or sleep disorders, improve understanding of the neurobiology or functions of sleep over the life-span, enhance timely diagnosis and effective treatment for individuals affected by sleep-related disorders, or implement and evaluate innovative community-based public health education and intervention programs. Because the nature and scope of the proposed research will vary from application to application, it is anticipated that the size and duration of each award will also vary. The total amount awarded and the number of awards will depend upon the mechanism numbers, quality, duration, and costs of the applications received.
Requirements: The following organizations and institutions may apply: public/state controlled institutions of higher education; private institutions of higher education; nonprofits with 501(c)3 IRS status; nonprofits without 501(c)3 IRS status; small businesses; State governments; non-domestic (non-U.S.) entities (foreign organizations); eligible agencies of the Federal government; and faith-based or community based organizations.
Geographic Focus: All States
Date(s) Application is Due: Feb 5; Jun 5; Oct 5
Contact: Susan Grove, (301) 435-0714; sg16d@nih.gov or grantsinfo@od.nih.gov
Internet: http://grants.nih.gov/grants/guide/pa-files/pa-07-140.html
Sponsor: National Institutes of Health
6701 Rockledge Drive
Bethesda, MD 20892

NIH Research Project Grants 3237

The Research Project Grant (R01) is an award to support a discrete, specified, circumscribed project to be performed by named Project Directors/Principal Investigators (PDs/PIs) in areas representing the investigators' specific interests and competencies, based on the mission of the NIH. The R01 is the original and historically the oldest grant mechanism used by the NIH to support health-related research and development. The NIH awards R01 grants to institutions/organizations of all types. This mechanism allows the PDs/PIs to define the scientific focus or objective of the research based on particular areas of interest and competence. Although the PDs/PIs write the grant application and are responsible for conducting and supervising the research, the actual applicant is the research institution/organization. Research grant applications are assigned to an NIH IC based on receipt and referral guidelines and many applications are assigned to multiple ICs with related research interests. Each IC maintains a Web site with funding opportunities and areas of interest. Contact with an IC representative may help focus the research plan based on an understanding of the mission of the IC. For specific information about the mission of each NIH IC, see http://www.nih.gov/icd, which provides a brief summary of the research interests in each IC and access to individual IC home pages.
Requirements: The following organizations and institutions may apply: public/state controlled institutions of higher education; private institutions of higher education; Hispanic-serving institutions; historically Black colleges and universities (HBCUs); tribally controlled colleges and universities (TCCUs); Alaska native and native Hawaiian serving institutions; nonprofits with 501(c)3 IRS status; nonprofits without 501(c)3 IRS status; small businesses; for-profit organizations; State governments; regional organizations; U.S. territories or possessions; Indian/Native American tribal governments; Indian/Native American tribally designated organizations; eligible agencies of the Federal government; and faith-based or community based organizations.
Geographic Focus: All States
Date(s) Application is Due: Feb 5; Jun 5; Oct 5
Contact: Grants Coordinator; (301) 435-0714; grantsinfo@od.nih.gov
Internet: http://grants.nih.gov/grants/guide/pa-files/PA-07-070.html
Sponsor: National Institutes of Health
6100 Executive Boulevard
Bethesda, MD 20892

NIH Ruth L. Kirschstein National Research Service Awards for Individual Senior Fellows 3238

The objective of the National Research Service Award (NRSA) F33 program is to provide senior fellowship support to experienced scientists who wish to make major changes in the direction of their research careers or who wish to broaden their scientific background by acquiring new research capabilities as independent research investigators in scientific health-related fields relevant to the missions of the participating NIH Institutes and Centers (ICs). Senior fellowship applicants for the F33 award must include a research training proposal that offers an opportunity for individuals to broaden their scientific background or to extend their potential for research in health-related areas as independent researchers. These awards will enable individuals with at least seven years of research experience beyond the doctorate, and who have progressed to the stage of independent investigator, to take time from regular professional responsibilities for the purpose of receiving training to increase their scientific capabilities. In most cases, this award is used to support sabbatical experiences for established independent scientists seeking support for retraining or additional career development.
Requirements: The following organizations are eligible to apply: for-profit organizations; non-profit organizations; public or private institutions, such as universities, colleges, hospitals, and laboratories; eligible agencies and labs of the Federal government, including NIH intramural labs; domestic institutions; and foreign institutions.
Restrictions: This program is not designed for postdoctoral level investigators seeking to enhance their research experience prior to independence.
Geographic Focus: All States
Date(s) Application is Due: Apr 8; Aug 8; Dec 8
Contact: Nancy C. Lohrey, (301) 496-8580; fax (301) 402-4472; lohreyn@mail.nih.gov
Internet: http://grants.nih.gov/grants/guide/pa-files/PA-07-172.html
Sponsor: National Institutes of Health
6116 Executive Boulevard, Suite 7019, MSC 8346
Bethesda, MD 20892-8346

NIH Ruth L. Kirschstein National Research Service Awards for Individual Postdoctoral Fellows 3239

The purpose of the National Research Service Award Act (NRSA) is to help ensure that diverse pools of highly trained scientists will be available in adequate numbers and in appropriate research areas to carry out the Nation's biomedical, behavioral and clinical research agendas. The National Institutes of Health (NIH) awards individual postdoctoral fellowships (F32) to promising applicants who have the potential to become productive, independent investigators in fields related to the mission of participating NIH Institutes and Centers. The proposed postdoctoral training must be within the broad scope of biomedical, behavioral, or clinical research or other specific disciplines relevant to the research mission of the participating NIH Institutes and Centers. The proposed training must offer an opportunity to enhance the fellow's understanding of the health-related sciences and extend his/her potential for a productive research career.
Requirements: Applicant organizations must have any of the following characteristics: for-profit organizations; non-profit organizations; public or private institutions, such as universities, colleges, hospitals, and laboratories; units of State and local governments; eligible agencies and labs of the Federal government including NIH intramural labs; domestic institutions; and foreign institutions.

Geographic Focus: All States
Date(s) Application is Due: Apr 8; Aug 8; Dec 8
Contact: Dr. Walter Schaffer, (301) 435-2687; fax (301) 480-0146; ws11q@nih.gov
Internet: http://grants.nih.gov/grants/guide/pa-files/PA-06-373.html
Sponsor: National Institutes of Health
45 Center Drive, Natcher Building
Bethesda, MD 20892

NIH Ruth L. Kirschstein National Research Service Awards for Individual Predoctoral Fellows 3240

The objective of this funding opportunity announcement is to help ensure that highly trained scientists will be available in adequate numbers and in appropriate research areas to carry out the Nation's biomedical, behavioral, and clinical research agenda. The participating Institutes of the National Institutes of Health (NIH) provide individual predoctoral research training fellowship awards to promising doctoral candidates who have the potential to become productive, independent investigators in research fields relevant to the missions of these participating NIH Institutes and Centers. Each participating NIH Institute has a unique scientific purview and different program goals and initiatives that evolve over time. Prior to preparing an application, it is critical that all applicants consult the appropriate Institute website for details of research areas supported by that Institute. Applicants should also contact the appropriate Institute scientific/research contact to obtain current information about specific program priorities and policies.
Requirements: The following organizations are eligible to apply: for-profit organizations; non-profit organizations; public or private institutions, such as universities, colleges, hospitals, and laboratories; eligible agencies and labs of the Federal government, including NIH intramural labs; domestic institutions; and foreign institutions.
Geographic Focus: All States
Date(s) Application is Due: Apr 8; Aug 8; Dec 8
Contact: Michael-David Kerns, (301) 496-9322; kernsmd@mail.nih.gov
Internet: http://grants.nih.gov/grants/guide/pa-files/PA-07-002.html
Sponsor: National Institutes of Health
6120 Executive Boulevard
Bethesda, MD 20892-7180

NIH Ruth L. Kirschstein National Research Service Awards for Individual Predoctoral Fellowships to Promote Diversity in Health-Related Research 3241

The purpose of the F31 predoctoral fellowship to promote diversity in health-related research is to provide up to five years of support for research training leading to the PhD or equivalent research degree, the combined MD/PhD degree; or another formally combined professional degree and research doctoral degree in biomedical, behavioral, health services, or clinical sciences. These fellowships will enhance the diversity of the biomedical, behavioral, health services, and clinical research labor force in the United States by providing opportunities for academic institutions to identify and recruit students from diverse population groups to seek graduate degrees in health-related research and apply for this fellowship. The goal of this program is to increase the number of scientists from diverse population groups who are prepared to pursue careers in biomedical, behavioral, social, clinical, or health services research.
Requirements: The following organizations and institutions are eligible to apply: public/state controlled institutions of higher education; private institutions of higher education; Hispanic-serving institutions; historically Black colleges and universities (HBCUs); tribally controlled colleges and universities (TCCUs); Alaska Native and Native Hawaiian serving institutions; nonprofits with 501(c)3 IRS status (other than institutions of higher education); nonprofits without 501(c)3 IRS status (other than institutions of higher education); for-profit organizations (other than small businesses); non-domestic (non-U.S.) entities (foreign organizations); and other eligible agencies of the Federal Government.
Geographic Focus: All States
Date(s) Application is Due: Apr 8; Aug 8; Dec 8
Contact: Susan Grove, (301) 435-0714; sg16d@nih.gov or grantsinfo@od.nih.gov
Internet: http://grants.nih.gov/grants/guide/pa-files/PA-09-209.html
Sponsor: National Institutes of Health
6701 Rockledge Drive
Bethesda, MD 20892

NIH Ruth L. Kirschstein National Research Service Awards for Individual Predoctoral MD/PhD and Other Dual Doctoral Degree Fellows 3242

The purpose of the Ruth L. Kirschstein National Research Service Awards (Kirschstein-NRSA) is to provide support to individuals for combined MD/PhD and other dual doctoral degree training (e.g. DO/PhD, DDS/PhD, AuD/PhD). The participating Institutes award this Kirschstein-NRSA individual fellowship to qualified applicants with the potential to become productive, independent, highly trained physician-scientists and other clinician-scientists, including patient-oriented researchers in their scientific mission areas. This funding opportunity supports individual predoctoral F30 fellowships with the expectation that these training opportunities will increase the number of future investigators with both clinical knowledge and skills in basic, translational or clinical research.
Requirements: The following organizations and institutions are eligible to apply: public/state controlled institutions of higher education; private institutions of higher education; Hispanic-serving institutions; historically Black colleges and universities (HBCUs); tribally controlled colleges and universities (TCCUs); Alaska Native and Native Hawaiian serving institutions; nonprofits with 501(c)3 IRS status (other than institutions of higher education); nonprofits without 501(c)3 IRS status (other than institutions of higher education); for-profit organizations (other than small businesses); and other eligible agencies of the Federal Government.
Restrictions: Non-domestic (non-U.S.) entities (foreign organizations) are not eligible to apply.
Geographic Focus: All States
Date(s) Application is Due: Apr 8; Aug 8; Dec 8
Contact: Nancy C. Lohrey, (301) 496-8580; fax (301) 402-4472; lohreyn@mail.nih.gov
Internet: http://grants.nih.gov/grants/guide/pa-files/PA-09-207.html
Sponsor: National Institutes of Health
6116 Executive Boulevard, Suite 7019, MSC 8346
Bethesda, MD 20892-8346

NIH Sarcoidosis: Research into the Cause of Multi-Organ Disease and Clinical Strategies for Therapy Grants 3243

The purpose of this Funding Opportunity Announcement (FOA) is to stimulate research on the etiology and management of sarcoidosis, an immune-mediated granulomatous inflammatory disorder. Studies supported by this FOA would include investigations to find the etiology(ies) and related host factors that might enhance susceptibility to sarcoidosis, especially development of symptomatic multi-organ disease that has a propensity to involve critical organs (such as lungs, heart, eyes, central/peripheral nervous system, liver, kidneys and other abdominal viscera) that create serious illness. Study of the still problematic management and the biological, behavioral, or psychosocial burden on individuals, families, and community is encouraged. Moreover, there is interest for approaches to risk reduction, psychological coping, and management of complications and side effects of treatment.
Requirements: The following organizations and institutions may apply: public/state controlled institutions of higher education; private institutions of higher education; Hispanic-serving institutions; historically Black colleges and universities (HBCUs); tribally controlled colleges and universities (TCCUs); Alaska native and native Hawaiian serving institutions; nonprofits with 501(c)3 IRS status; nonprofits without 501(c)3 IRS status; small businesses; for-profit organizations; State governments; regional organizations; U.S. territories or possessions; Indian/Native American tribal governments; Indian/Native American tribally designated organizations; eligible agencies of the Federal government; and faith-based or community based organizations.
Geographic Focus: All States
Date(s) Application is Due: Feb 5; Jun 5; Oct 5
Contact: Herbert Y. Reynolds, (301) 435-0222; Reynoldh@nhlbi.nih.gov
Internet: http://grants.nih.gov/grants/guide/pa-files/PA-07-136.html
Sponsor: National Institutes of Health
6701 Rockledge Drive, Room 10018, MSC 7936
Bethesda, MD 20892-7952

NIH School-based Interventions to Prevent Obesity 3244

This Funding Opportunity Announcement (FOA) encourages the formation of partnerships between academic institutions and school systems in order to develop and implement controlled, school-based intervention strategies designed to reduce the prevalence of obesity in childhood. This FOA also encourages evaluative comparisons of different intervention strategies, as well as the use of methods to detect synergistic interactions between different types of interventions. Because the nature and scope of the proposed research will vary from application to application, it is anticipated that the size and duration of each award will also vary. The total amount awarded and the number of awards will depend upon the mechanism numbers, quality, duration, and costs of the applications received.
Requirements: The following organizations and institutions may apply: public/state controlled institutions of higher education; private institutions of higher education; Hispanic-serving institutions; historically Black colleges and universities (HBCUs); tribally controlled colleges and universities (TCCUs); Alaska native and native Hawaiian serving institutions; nonprofits with 501(c)3 IRS status; nonprofits without 501(c)3 IRS status; small businesses; for-profit organizations; State governments; regional organizations; U.S. territories or possessions; Indian/Native American tribal governments; Indian/Native American tribally designated organizations; eligible agencies of the Federal government; and faith-based or community based organizations.
Geographic Focus: All States
Date(s) Application is Due: Feb 5; Jun 5; Oct 5
Contact: Gilman Grave, (301) 496-5593; fax (301) 480-9791; graveg@mail.nih.gov
Internet: http://grants.nih.gov/grants/guide/pa-files/PA-07-180.html
Sponsor: National Institutes of Health
6100 Executive Boulevard, 4B-11, MSC 7510
Bethesda, MD 20892-7510

NIH Self-Management Strategies Across Chronic Diseases Grants 3245

The purpose of this Program Announcement (PA) is to solicit applications to expand research on established self-management interventions to multiple chronic diseases across the life-course. Interventions aimed at chronic disease self-management are numerous and many are well described in the literature. They are often presented as specific to a particular chronic disease. This PA encourages applicants to investigate the applicability of effective self-management interventions to a broader spectrum of chronic diseases. Chronic disease, for this announcement, is defined as illnesses that are prolonged, are rarely cured completely, and require self-management behaviors by affected individuals and/or their caretakers. Modular Grant applications will request direct costs in $25,000 modules, up to a total direct cost request of $250,000 per year.
Requirements: The following organizations and institutions may apply: public/state controlled institutions of higher education; private institutions of higher education; Hispanic-serving institutions; historically Black colleges and universities (HBCUs); tribally controlled colleges and universities (TCCUs); Alaska native and native Hawaiian serving institutions; nonprofits with 501(c)3 IRS status; nonprofits without 501(c)3 IRS status; small businesses; for-profit organizations; State governments; regional organizations; U.S. territories or possessions; Indian/Native American tribal

governments; Indian/Native American tribally designated organizations; county governments; city or township governments; special district governments; independent school districts; public housing authorities/Indian housing authorities; eligible agencies of the Federal government; and faith-based or community based organizations.
Geographic Focus: All States
Date(s) Application is Due: Feb 5; Jun 5; Oct 5
Amount of Grant: Up to 250,000 USD
Contact: Susan Grove, (301) 435-0714; sg16d@nih.gov or grantsinfo@od.nih.gov
Internet: http://grants.nih.gov/grants/guide/pa-files/PA-00-109.html
Sponsor: National Institutes of Health
6701 Rockledge Drive
Bethesda, MD 20892

NIH Solicitation of Assays for High Throughput Screening in the Molecular Libraries Probe Production Centers Network (MLPCN) 3246

he NIH Molecular Libraries Roadmap Initiative wishes to encourage HTS assay applications from investigators who have the interest and capability to work with the Molecular Libraries Probe Production Centers Network (MLPCN) for chemical probe development. This Funding Opportunity Announcement (FOA) promotes discovery and development of new chemical probes as research tools for use by scientists in both the public and private sectors to advance the understanding of biological functions and disease mechanisms. This initiative is one of the integrated components of the NIH Molecular Libraries Roadmap initiative that offers biomedical researchers access to large-scale automated high throughput screening (HTS) centers in the MLPCN, diverse compound libraries in the Small Molecule Repository (MLSMR) and information on biological activities of small molecules in the PubChem BioAssay public database. Letters of Intent (LOIs) are due on December 4, April 4, and August 3 respectively. Full application must be received by January 4, May 4, and September 3 respectively.
Requirements: The following organizations and institutions are eligible to apply: public/state controlled institutions of higher education; private institutions of higher education; Hispanic-serving institutions; historically Black colleges and universities (HBCUs); tribally controlled colleges and universities (TCCUs); Alaska native and native Hawaiian serving institutions; nonprofits with 501(c)3 IRS status (other than institutions of higher education); nonprofits without 501(c)3 IRS status (other than institutions of higher education); small businesses; for-profit organizations (other than small businesses); state governments; Indian/Native American tribal governments (federally recognized); Indian/Native American tribally designated organizations; county governments; city or township governments; special district governments; independent school districts; public housing authorities/Indian housing authorities; U.S. territories or possessions; Indian/native American tribal governments (other than federally recognized); regional organizations; non-domestic (non-U.S.) entities (foreign organizations); eligible agencies of the federal government; and faith-based or community-based organizations.
Geographic Focus: All States
Date(s) Application is Due: Jan 4; May 4; Sep 3
Amount of Grant: Up to 50,000 USD
Contact: Yong Yao, (301) 443-6102; yyao@mail.nih.gov
Internet: http://grants.nih.gov/grants/guide/pa-files/PAR-09-129.html#SectionIII
Sponsor: National Institutes of Health
6001 Executive Boulevard, Room 7175, MSC 9641
Bethesda, MD 20892-9641

NIH Structural Biology of Membrane Proteins Grants 3247

This PA solicits applications to develop research and methods to enhance the rate of membrane protein structure determination and to determine specific membrane protein structures. Innovative methods for expression, oligomerization, solubilization, stabilization, purification, characterization, crystallization, isotopic labeling, and structure determination of unique and biologically significant membrane proteins by x-ray diffraction, nuclear magnetic resonance (NMR), electron microscopic, mass spectrometry, and other biophysical techniques are encouraged. Projects that will lead in the near term to determining the structures of biologically important membrane proteins are also encouraged.
Requirements: Applicant organizations must be: for-profit organizations; non-profit organizations; public or private institutions, such as universities, colleges, hospitals, and laboratories; units of State government; units of local government; eligible agencies of the Federal government; foreign institutions; domestic institutions; faith-based agencies; or community-based organizations.
Geographic Focus: All States
Date(s) Application is Due: Feb 5; Jun 5; Oct 5
Contact: Peter C. Preusch, (301) 594-3827; preuschp@nigms.nih.gov
Internet: http://grants.nih.gov/grants/guide/pa-files/PA-06-119.html
Sponsor: National Institutes of Health
45 Center Drive, Building 45, 2AS.55E, MSC 6200
Bethesda, MD 20892

NIH Summer Internship Program in Biomedical Research 3248

Summer programs at the National Institutes of Health (NIH) provide an opportunity to spend a summer working side-by-side with some of the leading scientists in the world, in an environment devoted exclusively to biomedical research. The NIH consists of the 240-bed Mark O. Hatfield Clinical Research Center and more than 1,200 laboratories located on the main campus in Bethesda, Maryland, as well as in Baltimore and Frederick, Maryland; Research Triangle Park, North Carolina; Phoenix, Arizona; Hamilton, Momtana; and Detroit, Michigan. Awards cover a minimum of eight weeks, with students generally arriving at the NIH in May or June. The stipends for trainees are adjusted yearly, with supplements for prior experience. Prospective candidates must apply online. The application is available from mid-November to March 1.
Requirements: The Summer Internship Program is for students who will be sixteen years of age or older at the time they begin the program and who are currently enrolled at least half-time in high school or an accredited U.S. college or university. Students who have been accepted into a college or university program may also apply.
Restrictions: To be eligible, candidates must be U.S. citizens or permanent residents.
Geographic Focus: All States
Date(s) Application is Due: Mar 1
Contact: Arnetta Courtney, (301) 827-3724; arnetta.courtney@fda.hhs.gov
Internet: http://www.training.nih.gov/student/sip/info.asp
Sponsor: National Institutes of Health
31 Center Drive, Building 31, Room 9A06, MSC 2560
Bethesda, MD 20892-2560

NIMH AIDS and Aging: Behavioral Sciences Prevention Research Grants 3249

NIA, NIMH, and NINR invite qualified researchers to submit applications to investigate prevention issues relevant to AIDS in middle-aged and older populations. This announcement solicits AIDS prevention research proposals to study primary prevention of disease transmission as well as secondary and tertiary prevention of negative behavioral and social consequences of HIV/AIDS for persons with AIDS, their families, and communities. Thus, the primary goals are to identify social and behavioral factors associated with HIV transmission and disease progression in later life; examine behavioral and social consequences of HIV infection/AIDS across the life course; develop and evaluate age-appropriate behavioral and social interventions for preventing AIDS in middle-aged and older adults and/or ameliorating problems associated with older adult caregiver responsibilities and burdens; explore health care issues surrounding AIDS care; and strengthen existing research and evaluation methods.
Requirements: Applications may be submitted by domestic and foreign for-profit and non-profit organizations, public and private, such as universities, colleges, hospitals, laboratories, units of State and local governments, and eligible agencies of the Federal government. Applications may be submitted by single institutions or by a consortia of institutions.
Geographic Focus: All States
Date(s) Application is Due: Feb 5; Feb 16; Jun 5; Jun 16; Oct 5; Oct 16
Contact: Dr. Willo Pequegnat, Division of Mental Disorders, Behavioral Research and AIDS; (301) 443-6100; fax (301) 443-9719; wpequegn@nih.gov
Internet: http://www.grants.nih.gov/grants/guide/pa-files/PA-97-069.html
Sponsor: National Institute of Mental Health
5600 Fishers Lane, Parklawn Building
Rockville, MD 20857

NIMH Basic and Translational Research in Emotion Grants 3250

This Funding Opportunity Announcement encourages Research Project Grant applications to expand basic and translational research on the processes and mechanisms involved in the experience, expression, and regulation of emotion. The study of emotion encompasses a wide range of physiological, psychological, social, cognitive, and developmental phenomena. Emotional processes can be studied in human or animal subjects. Important objects of study include, but are not limited to: central and peripheral nervous system activity in the origins, expression, regulation and modulation of emotion; the contribution of emotional and motivational systems to cognitive faculties such as perception, attention, learning, and memory; investigations of overt behaviors, interpersonal relationships, communication and decision making; and the environmental circumstances and experiences that evoke and modulate different emotions.
Requirements: The following organizations and institutions may apply: public/state controlled institutions of higher education; private institutions of higher education; Hispanic-serving institutions; historically Black colleges and universities (HBCUs); tribally controlled colleges and universities (TCCUs); Alaska native and native Hawaiian serving institutions; nonprofits with 501(c)3 IRS status; nonprofits without 501(c)3 IRS status; small businesses; for-profit organizations; State governments; regional organizations; U.S. territories or possessions; Indian/Native American tribal governments; Indian/Native American tribally designated organizations; county governments; city or township governments; special district governments; independent school districts; public housing authorities/Indian housing authorities; eligible agencies of the Federal government; and faith-based or community based organizations.
Geographic Focus: All States
Date(s) Application is Due: Feb 5; Jun 5; Oct 5
Contact: Wendy Nelson, (301) 435 4590; fax (301) 435 7547; nelsonw@mail.nih.gov
Internet: http://grants.nih.gov/grants/guide/pa-files/PA-09-137.html
Sponsor: National Institute of Mental Health
6001 Executive Boulevard, Room 7179, MSC 9637
Bethesda, MD 20892-9637

NIMH Curriculum Development Award in Neuroinformatics Research and Analysis 3251

The purpose of this program is to encourage and support applications from individuals, with the requisite scientific expertise and leadership, for the development of courses and curricula designed to train interdisciplinary neuroinformatics scientists at U.S. educational institutions. The field of neuroinformatics combines neuroscience research with informatics research developed from the computer sciences, mathematics, physics, engineering or closely related sciences. It is anticipated that these courses or curricula would be useful to students and scientists who wish: to develop new conceptual approaches to basic and/or clinical neuroscientific research and analysis; or to acquire, store, retrieve, organize, manage, analyze, visualize, manipulate, integrate, synthesize, disseminate, and share data about

the brain and behavior. Development of courses at the graduate and undergraduate level is encouraged. As part of this program, awardees will be expected to develop and implement the courses or curricula in their institution. It is expected that such courses and curricula will be models that could be transferable to other institutions in whole or in part.
Requirements: The principal investigator must be engaged in neuroscience research or research in computer science, mathematics, physics, engineering, or a related informatics field. Collaborator(s) who will contribute to the interdisciplinary nature of the courses or curricula must be identified. The principal investigator must be willing to spend at least 20 percent of full-time professional effort on course(s) and curricula development during the period of the award. The principal investigator must also identify appropriate researcher(s) who will agree to collaborate on the development of course(s) and curricula.
Geographic Focus: All States
Date(s) Application is Due: Feb 16; Jun 16; Oct 16
Contact: Stephen H. Koslow, (301) 443-1815; fax (301) 443-1867; koz@helix.nih.gov
Internet: http://grants.nih.gov/grants/guide/pa-files/PAR-99-135.html
Sponsor: National Institute of Mental Health
6001 Executive Boulevard, Room 6167, MSC 9613
Bethesda, MD 20892

NIMH Early Identification and Treatment of Mental Disorders in Children and Adolescents Grants 3252

This program supports research on early identification and treatment of mental disorders in children and adolescents. In particular, this announcement intends to encourage research on disorders such as schizophrenia, schizoaffective disorder, bipolar disorder, major depression, obsessive-compulsive disorder, and anorexia nervosa, alone or comorbid with other common mental or substance abuse disorders. PA: PA-00-094
Requirements: Applications may be submitted by domestic for-profit and nonprofit organizations, both public and private; units of state or local governments; and eligible agencies of the federal government.
Geographic Focus: All States
Amount of Grant: 125,000-250,000 USD
Contact: Shelli Avenevoli, (301) 443-5944; avenevos@mail.nih.gov
Internet: http://grants.nih.gov/grants/guide/notice-files/NOT-MH-08-006.html
Sponsor: National Institute of Mental Health
6001 Executive Boulevard
Bethesda, MD 20892

NIMH Jointly Sponsored Ruth L. Kirschstein National Research Service Award Institutional Predoctoral Training Program in the Neurosciences 3253

The Jointly Sponsored NIH Predoctoral Training Program in the Neurosciences supports broad and fundamental, early-stage graduate research training in the neurosciences via institutional NRSA research training grants (T32) at domestic institutions of higher education. Trainees are supported during years 1 and 2 of their graduate training when they are typically not committed to a dissertation laboratory. The primary objective is to prepare qualified individuals for careers in neuroscience that have a significant impact on the health-related research needs of the Nation. Because the nature and scope of the proposed research training will vary from application to application, it is anticipated that the size and duration of each award will also vary. The total amount awarded and the number of awards will depend upon the number, quality, duration, and costs of the applications received.
Requirements: The following organizations are eligible to apply: non-profit organizations; public or private institutions, such as universities, colleges, hospitals, and laboratories; and domestic institutions.
Geographic Focus: All States
Date(s) Application is Due: Apr 25; May 25
Contact: Lindsey Grandison, (301) 443-0606; fax (301) 443-1650; lgrandis@mail.nih.gov
Internet: http://grants.nih.gov/grants/guide/pa-files/PAR-08-101.html
Sponsor: National Institute of Mental Health
5635 Fishers Lane, MSC 9304
Bethesda, MD 20892-9304

NIMH Short Courses in Neuroinformatics 3254

The purpose of this program is to encourage and support short-term research for the development of short courses, seminars, and workshops on interdisciplinary neuroinformatics education . This short-term training will be provided to scientists seeking to combine knowledge about the various subdisciplines of neuroscience and behavioral science research with expertise in informatics research. It is anticipated that these short courses will allow the participants to acquire new conceptual approaches to basic neuroscience research and analyses; and learn to develop unique strategies for acquiring, storing, retrieving, organizing, managing, analyzing, visualizing, manipulating, integrating, synthesizing, disseminating, and sharing data about the brain and behavior.
Requirements: Any nonprofit or for-profit organization engaged in health-related education or research and located in the United States, its possessions, or territories is eligible to apply.
Geographic Focus: All States
Date(s) Application is Due: Feb 16; Jun 16; Oct 16
Amount of Grant: Up to 150,000 USD
Contact: Stephen H. Koslow, (301) 443-1815; fax (301) 443-1867; koz@helix.nih.gov
Internet: http://grants.nih.gov/grants/guide/pa-files/PAR-99-137.html
Sponsor: National Institute of Mental Health
6001 Executive Boulevard, Room 6167, MSC 9613
Bethesda, MD 20892

NINDS Optimization of Small Molecule Probes for the Nervous System Grants 3255

This Funding Opportunity Announcement (FOA) encourages Small Business Innovation Research (SBIR) grant applications from small business concerns (SBCs) that propose to develop new small molecule probes for investigating biological function in the nervous system via the application of advanced medicinal chemistry and the biological testing of compounds. Eligible SBCs will have identified probe candidates via screening of small molecule collections, using in vitro assays of biological activity developed to interrogate these collections, and be able to show that the structural features of these small molecules are related to their biological activity. Applications should nominate small molecule probe candidates from distinct structural series for the further, iterative design and testing of analogues in structure-activity relationship studies, using in vitro assays of biological function adapted to the medium throughput screening requirements of this work. These studies should have the goal of developing a small molecule probe possessing the attributes (eg: affinity, selectivity, activity) required for its use in future pharmacological studies proposed by the SBC. Applicants are strongly encouraged to utilize publicly available cheminformatic capabilities for the acquisition of compounds, and semi-custom synthesis of analogues.
Requirements: Only United States small business concerns (SBCs) are eligible to submit SBIR applications.
Geographic Focus: All States
Date(s) Application is Due: Apr 5; Aug 5; Dec 5
Amount of Grant: Up to 150,000 USD
Contact: Mark Scheideler, (301) 496-1779; fax (301) 402-1501; scheidelerm@ninds.nih.gov
Internet: http://grants.nih.gov/grants/guide/pa-files/PAR-09-260.html
Sponsor: National Institute of Neurological Disorders and Stroke
6001 Executive Boulevard, NSC 2107
Bethesda, MD 20892-9527

NINDS Optimization of Small Molecule Probes for the Nervous System Grants 3256

This Funding Opportunity Announcement (FOA) encourages Small Business Technology Transfer (STTR) grant applications from small business concerns (SBCs) that propose to develop new small molecule probes for investigating biological function in the nervous system via the application of advanced medicinal chemistry and the biological testing of compounds. Eligible SBCs will have identified probe candidates via screening of small molecule collections, using in vitro assays of biological activity developed to interrogate these collections, and be able to show that the structural features of these small molecules are related to their biological activity. Applications should nominate small molecule probe candidates from distinct structural series for the further, iterative design and testing of analogues in structure-activity relationship studies, using in vitro assays of biological function adapted to the medium throughput screening requirements of this work. These studies should have the goal of developing a small molecule probe possessing the attributes (eg: affinity, selectivity, activity) required for its use in future pharmacological studies proposed by the SBC. Applicants are strongly encouraged to utilize publicly available cheminformatic capabilities for the acquisition of compounds, and semi-custom synthesis of analogues, which is required of these studies.
Requirements: Only United States small business concerns (SBCs) are eligible to submit STTR applications.
Geographic Focus: All States
Date(s) Application is Due: Apr 5; Aug 5; Dec 5
Amount of Grant: Up to 150,000 USD
Contact: Mark Scheideler, (301) 496-1779; fax (301) 402-1501; scheidelerm@ninds.nih.gov
Internet: http://grants.nih.gov/grants/guide/pa-files/PAR-09-260.html
Sponsor: National Institute of Neurological Disorders and Stroke
6001 Executive Boulevard, NSC 2107
Bethesda, MD 20892-9527

NINDS Optimization of Small Molecule Probes for the Nervous System Grants 3257

This FOA issued by participating institutes of the National Institutes of Health, encourages research grant applications from institutions/ organizations that propose to develop new small molecule probes for investigating biological function in the nervous system via the application of advanced medicinal chemistry and the biological testing of compounds. Eligible investigators will have identified probe candidates via screening of small molecule collections, using in vitro assays of biological activity developed to interrogate these collections, and be able to show that the structural features of these small molecules are related to their biological activity. Proposals should nominate small molecule probe candidates from distinct structural series for the further, iterative design and testing of analogues in structure-activity relationship studies, using in vitro assays of biological function adapted to the medium throughput screening requirements of this work. These studies should have the goal of developing a small molecule probe possessing the attributes (eg: affinity, selectivity, activity) required for its use in future pharmacological studies proposed by the investigator. Applicants are strongly encouraged to utilize publicly available cheminformatic capabilities for the acquisition of compounds, and semi-custom synthesis of analogues.
Requirements: The following organizations and institutions are eligible to apply: public/state controlled institutions of higher education; private institutions of higher education; Hispanic-serving institutions; historically Black colleges and universities (HBCUs); tribally controlled colleges and universities (TCCUs); Alaska native and native Hawaiian serving institutions; nonprofits with 501(c)3 IRS status (other than institutions of higher education); nonprofits without 501(c)3 IRS status (other than institutions of higher education); small businesses; for-profit organizations (other than small businesses); state governments; Indian/Native American tribal governments (federally recognized); Indian/Native American tribally designated organizations; county governments; city or township governments; special district governments; independent school districts; public housing authorities/Indian housing authorities; U.S. territories or possessions; Indian/native American tribal governments (other than federally recognized); regional

organizations; non-domestic (non-U.S.) entities (foreign organizations); eligible agencies of the federal government; and faith-based or community-based organizations.
Geographic Focus: All States
Date(s) Application is Due: Feb 16; Jun 16; Oct 16
Amount of Grant: Up to 150,000 USD
Contact: Mark Scheideler, (301) 496-1779; fax (301) 402-1501; scheidelerm@ninds.nih.gov
Internet: http://grants.nih.gov/grants/guide/pa-files/PAR-09-251.html
Sponsor: National Institute of Neurological Disorders and Stroke
6001 Executive Boulevard, NSC 2107
Bethesda, MD 20892-9527

NINDS Support of Scientific Meetings as Cooperative Agreements 3258
The purpose of the program announcement is to inform the scientific community that the National Institute of Neurological Disorders and Stroke (NINDS) will now support scientific meetings as cooperative agreements in addition to the current practice of supporting them through the traditional grant mechanism. This program provides guidelines for when it is appropriate to request support of a meeting as a cooperative agreement and explains procedures for preparing and submitting such applications. The Principal Investigator will have the primary authority and responsibility to define objectives and approaches; plan, publicize, and conduct the scientific meeting; and publish the results of the meeting.
Requirements: Applications may be submitted by U.S. institutions, including scientific or professional societies eligible to receive grants from Public Health Service (PHS) agencies. In the case of an international conference, the U.S. representative organization of an established international scientific or professional society is the eligible applicant.
Restrictions: Foreign institutions are not eligible.
Geographic Focus: All States
Contact: Danilo A. Tagle, (301) 496-5745; fax (301) 402-2060; tagled@ninds.nih.gov
Internet: http://grants.nih.gov/grants/guide/notice-files/NOT-NS-02-003.html
Sponsor: National Institute of Neurological Disorders and Stroke
6001 Executive Boulevard, Neuroscience Center, Room 2114
Bethesda, MD 20892-9525

NINR Acute and Chronic Care During Mechanical Ventilation Research Grants 3259
NINR seeks research applications for investigator-initiated research to improve the clinical management and care delivery of patients who require mechanical ventilation. For the purpose of this program announcement, the time course of mechanical ventilation is defined as beginning with endotracheal or tracheal tube intubation, maintenance of full and/or partial ventilatory support, discontinuation of mechanical ventilation via weaning techniques, and concluding with care of patients who require chronic mechanical ventilation. Two different, but related, types of basic science and/or clinical research on acute and/or chronic mechanical ventilation are encouraged: studies examining the effects of physiological, environmental, and/or psychological approaches to facilitate the care of patients during any phase of mechanical ventilation, and studies directed toward examining the physical and psychological ramifications of technology dependency on patients and their families/significant others. Priority research areas include optimum ways to deliver patient care during mechanical ventilatory support; ethical issues related to initiation, maintenance, and withdrawal of life support technology; emotional support and communication during mechanical ventilation; methods of optimally managing the artificial airway; impact of care on family dynamics; and decision making regarding withdrawal of ventilatory support. Research is also needed to describe and intervene with care-giving resources, family functioning, and quality of life for patients on mechanical ventilation particularly chronically ill children and adults being maintained on home ventilators; optimum methods and techniques to deliver nursing care while critically ill patients receive mechanical ventilation; and determine optimal measures for promoting recovery as patients move through levels of care in acute, intermediate, and long-term care. This PA covers a wide range of topics relevant to clinical management and clinical problem solving issues for patients who require mechanical ventilation.
Requirements: Applications may be submitted by public and private, domestic and foreign, for-profit and nonprofit organizations; units of state and local governments; and eligible agencies of the federal government.
Restrictions: The total project period for an application may not exceed five years.
Geographic Focus: All States
Date(s) Application is Due: Feb 5; Feb 16; Jun 5; Jun 16; Oct 5; Oct 16
Contact: Josephine Boyington, (301) 594-2542; boyingtonje@mail.nih.gov
Internet: http://grants.nih.gov/grants/guide/pa-files/PA-99-003.html
Sponsor: National Institute of Nursing Research
31 Center Drive, Room 5B10
Bethesda, MD 20892-2178

NINR AIDS and Aging: Behavioral Sciences Prevention Research Grants 3260
NIA, NIMH, and NINR invite qualified researchers to submit applications to investigate prevention issues relevant to AIDS in middle-aged and older populations. This announcement solicits AIDS prevention research proposals to study primary prevention of disease transmission as well as secondary and tertiary prevention of negative behavioral and social consequences of HIV/AIDS for persons with AIDS, their families, and communities. Thus, the primary goals are to identify social and behavioral factors associated with HIV transmission and disease progression in later life; examine behavioral and social consequences of HIV infection/AIDS across the life course; develop and evaluate age-appropriate behavioral and social interventions for preventing AIDS in middle-aged and older adults and/or ameliorating problems associated with older adult caregiver responsibilities and burdens; explore health care issues surrounding AIDS care; and strengthen existing research and evaluation methods.
Requirements: Applications may be submitted by domestic and foreign for-profit and non-profit organizations, public and private, such as universities, colleges, hospitals, laboratories, units of State and local governments, and eligible agencies of the Federal government. Applications may be submitted by single institutions or by a consortia of institutions.
Geographic Focus: All States
Date(s) Application is Due: Feb 5; Feb 16; Jun 5; Jun 16; Oct 5; Oct 16
Contact: Josephine Boyington, (301) 594-2542; boyingtonje@mail.nih.gov
Internet: http://www.grants.nih.gov/grants/guide/pa-files/PA-97-069.html
Sponsor: National Institute of Nursing Research
31 Center Drive, Room 5B10
Bethesda, MD 20892-2178

NINR Chronic Illness Self-Management in Children and Adolescents Grants 3261
The purpose of this Funding Opportunity Announcement (FOA) issued by the NINR, NICHD, NIDDK, NCI, and the ODS is to solicit research to improve self-management and quality of life in children and adolescents with chronic illnesses. Biobehavioral studies of children within the context of family and family-community dynamics are encouraged. Children with a chronic disease and their families have a long-term responsibility for self-management. The child with the illness will have a life-long responsibility to maintain and promote health and prevent complications. Research related to sociocultural, environmental, and behavioral mechanisms as well as biological/technological factors that contribute to successful and ongoing self-management of particular chronic illnesses in children is encouraged. Because the nature and scope of the proposed research will vary from application to application, it is anticipated that the size and duration of each award will also vary.
Requirements: Application can be made by the following institutions and organizations: public/state controlled institutions of higher education; private institutions of higher education; nonprofits with 501(c)3 IRS status (other than institutions of higher education); nonprofits without 501(c)3 IRS status (other than institutions of higher education); small businesses; for-profit organizations (other than small businesses); state governments; U.S. territories or possessions; Indian/Native American tribal governments (federally recognized); Indian/Native American tribal governments (other than Federally recognized); Indian/Native American tribally designated organizations; non-domestic (non-U.S.) entities (foreign organizations); Hispanic-serving institutions; historically Black colleges and universities (HBCUs); tribally controlled colleges and universities (TCCUs); Alaska Native and Native Hawaiian serving institutions; regional organization; eligible agencies of the Federal government; and faith-based or community based organizations.
Restrictions: This FOA is restricted to studies of chronic illnesses in children and adolescents ages 8 to 21 grouped by developmental stages according to the discretion of the investigator. Studies of chronic mental illness or serious cognitive disability are beyond the scope of this FOA.
Geographic Focus: All States
Date(s) Application is Due: Feb 5; Jun 5; Oct 5
Contact: Dr. Linda S. Weglicki, (301) 594-6908; weglickils@mail.nih.gov
Internet: http://grants.nih.gov/grants/guide/pa-files/PA-07-097.html
Sponsor: National Institute of Nursing Research
31 Center Drive, Room 5B10
Bethesda, MD 20892-2178

NINR Clinical Interventions for Managing the Symptoms of Stroke Research Grants 3262
NINR, NINDS, NIA, NIDCD, and NICHD seek research applications that address effective nonpharmacological approaches to managing the initial events and subsequent symptoms of a stroke. This research would focus on reducing the impairments, preventing secondary complications, and improving the functional independence and the quality of life for the individual following a stroke. Research areas that need to be explored include, but are not limited to: techniques to improve assessment during the initial stroke event and at each stage of recovery to guide nonpharmacologic treatment decisions and monitor patient progress; interventions to manage common symptoms, such as dysphagia, aphasia, bowel and bladder problems, sensorimotor deficits, depression, cognitive and perceptual deficits as well as difficulties in nutrition, hydration, skin integrity, and sleep; strategies to determine optimal timing, intensity, and duration of nonpharmacologic treatments; methods to prevent secondary complications, improve function, and address the sequelae of common comorbid conditions; strategies to promote self-care in both activities of daily living and instrumental activities of daily living and to enhance or maintain health promotion activities; measures to enhance family caregiving during all stages of recovery; methods to improve care during transition from one setting to another and care in specialized nurse managed units; and studies on the extent to which age and pre-existing conditions, such as specific diseases, sensory, motor, and cognitive declines, influence treatment outcome and management after stroke.
Requirements: Applications may be submitted by domestic and foreign, for-profit and nonprofit organizations, public and private.
Geographic Focus: All States
Amount of Grant: Up to 250,000 USD
Contact: Karin F. Helmers, (301) 594-2177; fax (301) 480-8260; karin_helmers@nih.gov
Internet: http://www.grants.nih.gov/grants/guide/pa-files/PA-99-088.html
Sponsor: National Institute of Nursing Research
31 Center Drive, Room 5B10
Bethesda, MD 20892-2178

NINR Diabetes Self-Management in Minority Populations Grants 3263
This Program Announcement (PA) solicits applications for investigator-initiated research related to sociocultural, environmental, and behavioral mechanisms and biological/

technological factors that contribute to successful and ongoing self-management of diabetes in minority populations. Applications that expand accepted intervention strategies in majority populations to minority populations are encouraged. Testing new interventions designed to promote self-management in minority diabetes populations will also be responsive to the PA. Self-management is defined as client strategies and behaviors that contribute to blood glucose normalization, improved health, and prevention or reduction of complications. The concept is broader than adherence to specific regimen components and incorporates deliberate problem solving and decision making processes. Applications are encouraged for both type 1 and type 2 diabetes and all age groups. Modular Grant applications will request direct costs in $25,000 modules, up to a total direct cost request of $250,000 per year.
Requirements: Applications may be submitted by domestic and foreign, for-profit and non-profit organizations, public and private, such as universities, colleges, hospitals, laboratories, units of State and local governments, and eligible agencies of the Federal government. Racial/ethnic minority individuals, women, and persons with disabilities are encouraged to apply as principal investigators.
Geographic Focus: All States
Date(s) Application is Due: Feb 5; Jun 5; Oct 5
Amount of Grant: Up to 250,000 USD
Contact: Dr. Karen Huss, (301) 594-5970; fax (301) 451-5649; husk@mail.nih.gov
Internet: http://grants.nih.gov/grants/guide/pa-files/PA-00-113.html
Sponsor: National Institute of Nursing Research
31 Center Drive, Room 5B10
Bethesda, MD 20892-2178

NINR Mechanisms, Models, Measurement, & Management in Pain Research Grants 3264
The purpose of this Funding Opportunity Announcement (FOA), Mechanisms, Models, Measurement, and Management in Pain Research, is to inform the scientific community of the pain research interests of the various Institutes and Centers (ICs) at the National Institutes of Health (NIH) and to stimulate and foster a wide range of basic, clinical, and translational studies on pain as they relate to the missions of these ICs. New advances are needed in every area of pain research, from the micro perspective of molecular sciences to the macro perspective of behavioral and social sciences. Although great strides have been made in some areas, such as the identification of neural pathways of pain, the experience of pain and the challenge of treatment have remained uniquely individual and unsolved. Furthermore, our understanding of how and why individuals transition to a chronic pain state after an acute insult is limited. Research to address these issues conducted by interdisciplinary and multidisciplinary research teams is strongly encouraged, as is research from underrepresented, minority, disabled, or female investigators. Direct costs are limited to $275,000 over a two-year period, with no more than $200,000 in direct costs allowed in any single year.
Requirements: The following organizations and institutions are eligible to apply: public/state controlled institutions of higher education; private institutions of higher education; Hispanic-serving institutions; historically Black colleges and universities (HBCUs); tribally controlled colleges and universities (TCCUs); Alaska native and native Hawaiian serving institutions; nonprofits with 501(c)3 IRS status (other than institutions of higher education); nonprofits without 501(c)3 IRS status (other than institutions of higher education); small businesses; for-profit organizations (other than small businesses); state governments; Indian/Native American tribal governments (federally recognized); Indian/Native American tribally designated organizations; county governments; city or township governments; special district governments; independent school districts; public housing authorities/Indian housing authorities; U.S. territories or possessions; Indian/native American tribal governments (other than federally recognized); regional organizations; non-domestic (non-U.S.) entities (foreign organizations); eligible agencies of the federal government; and faith-based or community-based organizations.
Geographic Focus: All States
Date(s) Application is Due: Feb 16; Jun 16; Oct 16
Amount of Grant: Up to 275,000 USD
Contact: Dr. Susan Marden, (301) 496-9623; fax (301) 480-8260; mardens@mail.nih.gov
Internet: http://grants.nih.gov/grants/guide/pa-files/PA-10-007.html
Sponsor: National Institute of Nursing Research
31 Center Drive, Room 5B10
Bethesda, MD 20892-2178

NINR Mentored Research Scientist Development Award 3265
The objective of the NIH Mentored Research Scientist Development Award is to provide support for a sustained period of protected time for intensive research career development under the guidance of an experienced mentor, or sponsor, in the biomedical, behavioral or clinical sciences leading to research independence. The expectation is that through this sustained period of research career development and training, awardees will launch independent research careers and become competitive for new research project grant funding. NINR encourages mentored research scientist development in areas of symptom management, pulmonary, critical care, trauma, reproductive health, end-of-life and palliative care. NINR limits the length of the K01 award to a 3-year period and generally considers career development applications only from doctorally prepared applicants who will advance the science of nursing.
Requirements: The candidate must have a research or health-professional doctorate or its equivalent and must have demonstrated the capacity or potential for highly productive independent research in the period after the doctorate. The candidate must identify a mentor with extensive research experience and must spend a minimum of 75 percent of full-time professional effort conducting research for the period of the award. Applications may be submitted on behalf of candidates by domestic, nonfederal organizations, public or private, such as medical, dental, or nursing schools or other institutions of higher education. Candidates must be U.S. citizens, noncitizen nationals, or permanent residents.
Geographic Focus: All States
Date(s) Application is Due: Feb 12; Mar 12; Jun 12; Jul 12; Oct 12; Nov 12
Amount of Grant: 50,000 USD
Contact: David Banks, (301) 496-9558; fax (301) 480-8260; banksd@mail.nih.gov
Internet: http://grants.nih.gov/grants/guide/pa-files/PA-09-040.html
Sponsor: National Institute of Nursing Research
31 Center Drive, Room 5B10
Bethesda, MD 20892-2178

NINR Quality of Life for Individuals at the End of Life Grants 3266
The National Institute of Nursing Research (NINR) and 6 other ICs seek research grant applications that will generate scientific knowledge to improve the quality of life for individuals who are facing end-of-life issues and for their families. Research applications may include basic, clinical or care delivery studies focused on management of physical and psychological symptoms, patient-provider and patient-family communication, ethics and clinical decision-making, caregiver support, or the context of care delivery for those facing life-limiting illnesses. In a broad sense the purpose of this program announcement is to enhance the quality of life remaining for individuals who are nearing the end of their lives. Modular Grant applications will request direct costs in $25,000 modules, up to a total direct cost request of $250,000 per year.
Requirements: Applications may be submitted by domestic and foreign, for-profit and non-profit organizations, public and private, such as universities, colleges, hospitals, laboratories, units of State and local governments, and eligible agencies of the Federal government. Racial/ethnic minority individuals, women, and persons with disabilities are encouraged to apply as principal investigators.
Geographic Focus: All States
Date(s) Application is Due: Feb 5; Jun 5; Oct 5
Amount of Grant: Up to 250,000 USD
Contact: Nancy C. Lohrey, (301) 496-8580; fax (301) 402-4472; lohreyn@mail.nih.gov
Internet: http://grants.nih.gov/grants/guide/pa-files/PA-00-127.html
Sponsor: National Institute of Nursing Research
31 Center Drive, Room 5B10
Bethesda, MD 20892-2178

NINR Ruth L. Kirschstein National Research Service Award (NRSA) for Individual Predoctoral Fellows in Nursing Research 3267
This grant program will provide predoctoral training support for doctoral students. The applicant must propose a research training program and dissertation research that is consistent with the scientific mission of the NINR. Research topics and skills that will serve as a foundation for an ongoing program of research are of particular interest. The research training experience must enhance the applicant's conceptualization of research problems and research skills, under the guidance and supervision of a committed mentor who is an active and established investigator in the area of the applicant's proposed research. The research training program should be carried out in a research environment that includes appropriate human and technical resources and is demonstrably committed to the research training of the applicant in the program he/she proposes in the application. Up to five years of aggregate NRSA support may be provided. Fellowship awardees are required to pursue their research training on a full-time basis, devoting at least 40 hours per week to the training program.
Requirements: The following organizations are eligible to apply: for-profit organizations; non-profit organizations; public or private institutions, such as universities, colleges, hospitals, and laboratories; and domestic institutions.
Geographic Focus: All States
Date(s) Application is Due: Apr 8; Aug 8; Dec 8
Contact: David Banks, (301) 496-9558; fax (301) 480-8260; banksd@mail.nih.gov
Internet: http://grants.nih.gov/grants/guide/pa-files/PAR-09-227.html
Sponsor: National Institute of Nursing Research
31 Center Drive, Room 5B10
Bethesda, MD 20892-2178

NIST Fire Research Grants 3268
This program supports research into the dynamics of fire, causes of fire, and a better understanding of fire prevention and safety measures. It aims to advance fire science and engineering practices by developing a better understanding of dynamic fire phenomena. This includes performing research into the heat and mass transfer processes of fires, in order to improve predictions of growth, suppression, and emissions from fires. This program also seeks to develop a more definitive understanding of the behavior, prevention, and control of large fires.
Requirements: The Program is open to institutions of higher education; hospitals; non-profit organizations; commercial organizations; state, local, and Indian tribal governments; foreign governments; organizations under the jurisdiction of foreign governments; and international organizations. Joint programs with participation by more than one eligible entity are possible.
Geographic Focus: All States
Amount of Grant: 10,000-100,000 USD
Contact: Wanda J. Duffin-Ricks; (301) 975-6863; wanda.duffin-ricks@nist.gov
Internet: http://www.nist.gov/public_affairs/grants.htm
Sponsor: National Institute of Standards and Technology
100 Bureau Drive, Stop 8660
Gaithersburg, MD 20899-8660

NIST Hollings Manufacturing Extension Partnerships 3269
The purpose of the program is to establish, maintain, and support manufacturing extension centers and services to accelerate the usage of appropriate manufacturing

technology by smaller U.S.-based manufacturing firms; and partner with the states in developing such technical assistance programs and services for their manufacturing base. Funds shall be used for the creation and support of regional manufacturing extension services, or used by the states to plan for and pilot test statewide extension services. Funds may also be used to plan for and pilot test services within a multi-state region which has sufficient regional linkages to justify such services. Extension service providers shall be affiliated with a U.S.-based nonprofit institution or organization or group thereof. Funds may be used for the purpose of demonstrations, technology deployment, active transfer and dissemination of research findings and extension service expertise to a wide range of companies and enterprises, especially small manufacturers with fewer than 500 employees. Extension service planning and pilot testing agreements require a state to provide cost share by an amount at least equal to the amount of federal assistance requested. Deadline dates are published in the Federal Register.
Requirements: Eligible to apply for extension services are U.S.-based nonprofit institutions or organizations or groups thereof. Applicants must contribute 50 percent or more of the proposed service's capital, annual operating and maintenance costs. At least 50 percent of the applicant's share must be as cash or full-time personnel.
Geographic Focus: All States
Amount of Grant: 200,000-65,000,000 USD
Contact: Roger D. Kilmer; (301) 975-4676; fax (301) 963-6556; roger.kilmer@nist.gov
Internet: http://www.nist.gov/public_affairs/guide/mep.htm
Sponsor: National Institute of Standards and Technology
100 Bureau Drive, Stop 4800
Gaithersburg, MD 20899-4800

NIST Materials Science and Engineering Grants 3270
The Program supports work in polymers, ceramics, metallurgy, and neutron scattering and spectroscopy research at academic, industrial, and other non-federal institutions. Projects are usually funded for one year and are subject to availability of funds.
Requirements: Universities, colleges, professional institutes and associations, nonprofit organizations, and state and local governments are eligible for these grants.
Geographic Focus: All States
Contact: Tanya Burke, (301) 975-4711; tanya.burke@nist.gov
Internet: http://www.nist.gov/public_affairs/grants.htm
Sponsor: National Institute of Standards and Technology
100 Bureau Drive, Room A120, Stop 3580
Gaithersburg, MD 20899-3580

NIST National Standard Reference Data System Research Grants 3271
These grants support research for data evaluation that will aid the scientist or engineer by providing reliable evaluated information on the chemical and physical properties of substances. Evaluations are carried out through a network of data centers and projects. Data compilations published and databases established by the program are for sale without restriction. Contact the program for deadline information. Project support is provided for one to two years.
Requirements: Research grants, contracts, and cooperative agreements for data evaluation are made to technical groups in universities, industrial laboratories, and other research establishments.
Geographic Focus: All States
Contact: Office of Standard Reference Data; (301) 975-2208 or (301) 975-2200; fax (301) 926-0416
Internet: http://www.nist.gov/srd/
Sponsor: National Institute of Standards and Technology
100 Bureau Drive, Stop 2300
Gaithersburg, MD 20899-2300

NIST Precision Measurement Grants 3272
The Program supports researchers in U.S. colleges and universities for experimental and theoretical studies of fundamental physical phenomena. It is seeking proposals for significant, primarily experimental, research in the field of fundamental measurement or the determination of fundamental constants. The purpose of the program is to encourage basic, measurement-related research and to foster contacts between NIST scientists and those faculty members of U.S. academic institutions who are actively engaged in such work. Applicants should propose multi-year projects for up to three years at no more than $50,000 per year. Applicants should begin by submitting an abbreviated proposal, which must be received by the end of business on February 2. Finalists will be selected by March 23, and will be required to submit full proposals to NIST by close of business on May 4.
Requirements: Universities and colleges in the United States are eligible to apply.
Geographic Focus: All States
Date(s) Application is Due: Feb 2; May 4
Amount of Grant: Up to 50,000 USD
Contact: Dr. Peter J. Mohr, fax (301) 975-3217; mohr@nist.gov
Internet: http://physics.nist.gov/ResOpp/grants/grants.html
Sponsor: National Institute of Standards and Technology
100 Bureau Drive, Stop 8420
Gaithersburg, MD 20899-8420

NIST Small Business Innovation Research (SBIR) Grants 3273
Ten federal agencies set aside a portion of their extramural research and development budget each year to fund research proposals from small science and technology-based firms. This Program supports creative advanced research in important scientific and engineering areas and is designed to encourage the conversion of government-funded R&D into technological innovation and commercial application. The Program goals are: to increase private sector commercialization of innovations derived from federal R&D; to use small business to meet federal research and development (R&D) needs; to stimulate small business innovation in technology; and to foster and encourage participation by minority and disadvantaged persons in technological innovation. The program consists of three phases: In Phase 1, a small business can receive up to $75,000 to establish the scientific or technical merit or feasibility of ideas that appear to have commercial potential. The successful completion of Phase 1 enables the awardee to compete for a Phase 2I contract for up to $300,000 to continue research in the same technical area. Phase 1I awardees have 6 months to complete their work; Phase 2 awardees have up to 2 years. Phase 2 is the principal R&D effort. Under Phase 3, no SBIR funds are available to pursue commercial applications of the technology.
Requirements: Each organization submitting a proposal for both Phase 1 and Phase 2 must qualify as a small business concern (Section 2.10) for research or R&D purposes (Section 2.7) at the time of award. In addition, the primary employment of the principal investigator must be with the small business at the time of the award and during the conduct of the proposed research. More than one-half of the principal investigator's time must be spent with the small business for the period covered by the award. Primary employment with a small business precludes full-time employment with another organization.
Geographic Focus: All States
Amount of Grant: 50,000-200,000 USD
Contact: Program Director; (301) 975-3085; fax (301) 975-3482; sbir@nist.gov
Internet: http://patapsco.nist.gov/ts_sbir/
Sponsor: National Institute of Standards and Technology
100 Bureau Drive, Stop 2200
Gaithersburg, MD 20899-2200

NKF Clinical Scientist Awards 3274
The award supports investigators who have demonstrated outstanding clinical or basic research potential to promote their continued success as independent investigators. Scientists are expected to devote a minimum of 75 percent of their time to research. This undertaking includes the establishment of a new Study Section which will review all proposals involving clinical research for all funding mechanisms, based on the following definition from the Nathan Report, which describes three categories of clinical research: Patient-Oriented Research, defined as research conducted with human subjects (or on material of human origin such as tissues, specimens and cognitive phenomena) for which an investigator (or colleague) directly interacts with human subjects in either an outpatient or inpatient setting; Epidemiologic and Biobehavioral Studies; and Outcomes Research and Health Services Research. Applications must be submitted electronically no later than November 30.
Requirements: Applicants must be citizens or permanent residents of the United States at the time of application. Applicants must hold an M.D. or equivalent domestic or foreign degree and must be full-time staff members of a department within an institution in the United States at the time of the award. Applicants will generally have at least five, but not more than 10 years of postdoctoral research experience in nephrology by the beginning of the award date.
Geographic Focus: All States
Date(s) Application is Due: Nov 30
Amount of Grant: 50,000 USD
Contact: Emily B. Newell, Scientific Communications Director; (212) 889-2210, ext. 288 or (800) 622-9010; fax (212) 689-9261; emilyn@kidney.org or research@kidney.org
Internet: http://www.kidney.org/professionals/research/index.cfm
Sponsor: National Kidney Foundation
30 East 33rd Street
New York, NY 10016

NKF Franklin McDonald/Fresenius Medical Care Clinical Research Grants 3275
The National Kidney Foundation is soliciting applications for the annual Franklin McDonald/Fresenius Medical Care Clinical Research Grant. Applicants should be able to spend a minimum of 50% of their time conducting research and 25% of effort in the direct care of maintenance dialysis patients and/or in teaching medical students, residents or nephrology fellows the pathophysiology and treatment of CKD, ESRD, or maintenance dialysis. Proposals are due on December 1.
Requirements: Applicants must hold a full-time faculty appointment at a university or an equivalent position as a scientist on the staff of a research organization.
Geographic Focus: All States
Date(s) Application is Due: Dec 1
Contact: Emily B. Newell, Scientific Communications Director; (212) 889-2210, ext. 288 or (800) 622-9010; fax (212) 689-9261; emilyn@kidney.org or research@kidney.org
Internet: http://www.kidney.org/professionals/research/index.cfm
Sponsor: National Kidney Foundation
30 East 33rd Street
New York, NY 10016

NKF Professional Councils Research Grants 3276
The purpose of the CNNT Research Grants Program is to further knowledge of nursing and technician issues in the management of kidney failure. The grants may cover basic or applied research on nursing and technical issues in the area of kidney failure; early intervention and treatment of chronic kidney disease; development and evaluation of education programs to enhance patient/family understanding of kidney failure treatment; demonstration projects related to kidney failure and rehabilitation. Letter of intent due on October 15, and full proposals must be received by December 1.
Requirements: Applicants must hold a regular membership in CNNT, a minimum of two years work experience as a nephrology nurse or technician, residence in the U.S. or its territories, written approval from the department head or facility director at the

institution/facility where the research is to be conducted, and have prior research training or evidence of support from an individual with research experience.
Restrictions: Project support and stipend not to exceed 20% of requested amount; consultant fees not to exceed 15% of total requested funds.
Geographic Focus: All States
Date(s) Application is Due: Oct 15; Dec 1
Amount of Grant: Up to 20,000 USD
Contact: Beverly Sneed, Grant Contact; (214) 514-3846 or (800) 622-9010; fax (212) 689-9261; bevyann1@sbcglobal.net or research@kidney.org
Internet: http://www.kidney.org/professionals/research/profcouncil.cfm
Sponsor: National Kidney Foundation
30 East 33rd Street
New York, NY 10016

NKF Research Fellowships **3277**
The purpose of National Kidney Foundation (NKF) Research Fellowships is: to foster training of young and new investigators with the potential of making contributions to the understanding and cure of kidney diseases; to foster, through training fellows, kidney-related research that is of high merit; and to encourage high quality applicants who want to make a career change into academic Nephrology. The fellowships are open to qualified investigators of any nationality who are interested in a career in kidney research and who will not have had more than four and one-half years of research training prior to the commencement of the award. The one-year awards may be renewed.
Requirements: Applicants must have completed no more than 4.5 years of post-doctoral research training (after receipt of M.D., Ph.D, D.O., or equivalent degree) at time of the activation of the award. For those applying for a second year (competitive renewal), applicants must have completed no more than 5.5 years of post-doctoral training.
Restrictions: Anyone who has ever held a faculty appointment at the level of Assistant Professor is ineligible.
Geographic Focus: All States
Date(s) Application is Due: Nov 30
Contact: Emily B. Newell, Scientific Communications Director; (212) 889-2210, ext. 288 or (800) 622-9010; fax (212) 689-9261; emilyn@kidney.org or research@kidney.org
Internet: http://www.kidney.org/professionals/research/awards.cfm
Sponsor: National Kidney Foundation
30 East 33rd Street
New York, NY 10016

NKF Young Investigator Grants **3278**
The purpose of this program is to support research in the fields of nephrology, urology, and related disciplines by junior faculty now starting their investigative work, who are often hampered in their progress by difficulty in obtaining suitable financial support for their research during the formative periods of their careers. The applicant's research and career goals must be directed to the study of normal or abnormal kidney function or of diseases of the kidney or urinary tract.
Requirements: Applicants must have completed research fellowship training in nephrology, urology or closely related field prior to the start of the grant award, intend to pursue directly related research, and hold a full-time appointment to a faculty position at a university or an equivalent position as a scientist on the staff of a research oriented institution for no more than four years at the start of the award. Recipients typically hold an appointment as Assistant Professor.
Geographic Focus: All States
Date(s) Application is Due: Nov 30
Amount of Grant: Up to 50,000 USD
Contact: Emily B. Newell, Scientific Communications Director; (212) 889-2210, ext. 288 or (800) 622-9010; fax (212) 689-9261; emilyn@kidney.org or research@kidney.org
Internet: http://www.kidney.org/professionals/research/awards.cfm
Sponsor: National Kidney Foundation
30 East 33rd Street
New York, NY 10016

NL American Society for Eighteenth-Century Studies Fellowships **3279**
The Newberry, open to the public without charge, is an independent research library dedicated to the advancement and dissemination of knowledge, especially in the humanities. The library acquires and preserves a broad array of special-collections research materials (spanning six centuries of European and American civilizations) and promotes and provides for these collections' effective use, fostering research, teaching, publication, life-long learning, and civic engagement. Fellowships at the Newberry provide support for researchers wishing to use the Newberry's collections and are either short-term (one to two months) or long-term (four to twelve months). Short-term fellowships are intended to geographically facilitate access for distant scholars to materials essential to a project. Long-term fellowships are intended to support significant works of scholarship that draw on the strengths of the Newberry's collections. The Newberry American Society for Eighteenth-Century Studies (ASECS) Fellowships are short-term fellowships offered to scholars who wish to pursue studies from the period of 1660 to 1815. These fellowships provide a stipend of $2,000 per month. At the time of this writing, the current application submission date for most Newberry short-term fellowships is February 6 and awardees are usually notified near the end of April; however deadlines may vary from year to year. Applications must be submitted as email attachments by 11:59 p.m. on the deadline date. Applicants can find guidelines and forms at the grant website. Applicants are also encouraged to verify current deadline dates.
Requirements: Newberry ASECS Fellowship recipients must be members of the ASECS at the time of the award. Most short-term fellowships are for either Ph.D. candidates or postdoctoral scholars. All fellowships at the Newberry must be conducted in-residence. The Newberry collections must be necessary to the short-term fellowship project since the purpose of these fellowships is to enable access to particular materials. Applications from former recipients of any fellowship at the Newberry are welcome, but preference is given to applicants who have not held a major fellowship within the past three years. In general, all fellowships at the Newberry must be completed by the end of the academic year for which they are awarded.
Restrictions: Most short-term fellowships are restricted to individuals who live outside the Chicago area who need to examine specific items in the Newberry's collections. Applicants may apply for more than one fellowship at the Newberry by designating those for which they wish to be considered on the cover sheet (available at the grant website). Applicants may also apply for both short- and long-term fellowships at the Newberry; however individuals may accept only one fellowship at the Newberry per year. Only scholars who have previously received a short-term fellowship at the Newberry may apply for a long-term fellowship for the same project. No other project extensions are allowable.
Geographic Focus: All States, All Countries
Date(s) Application is Due: Feb 6
Amount of Grant: 2,000-6,000 USD
Contact: Daniel Greene, Vice President for Research and Academic Programs; (312) 255-3666 or (312) 255-3535; research@newberry.org or greened@newberry.org
Internet: http://www.newberry.org/short-term-fellowships
Sponsor: Newberry Library
60 West Walton Street
Chicago, IL 60610-3380

NL Arthur and Lila Weinberg Fellowships for Independent Scholars and Researchers **3280**
The Newberry, open to the public without charge, is an independent research library dedicated to the advancement and dissemination of knowledge, especially in the humanities. The library acquires and preserves a broad array of special-collections research materials (spanning six centuries of European and American civilizations) and promotes and provides for these collections' effective use, fostering research, teaching, publication, life-long learning, and civic engagement. Fellowships at the Newberry provide support for researchers wishing to use the Newberry's collections and are either short-term (one to two months) or long-term (four to twelve months). Most short-term fellowships are intended to geographically facilitate access for distant scholars to materials essential to a project; however residents in the Chicago area are eligible to apply for Weinberg Fellowships. Long-term fellowships are intended to support significant works of scholarship that draw on the strengths of the Newberry's collections. The Arthur and Lila Weinberg Fellowship is a short-term fellowship awarded to independent scholars or researchers who work outside of traditional academic settings (e.g. scholars, journalists, writers, filmmakers, visual and performing artists, and other humanists) who wish to use the Newberry's collections in the creation of a publication or comparable product. Preference is given to scholars working on historical issues related to social justice or reform. The Weinberg fellowship is intended to support one month-in-residence at the Newberry. Fellows will receive a total stipend of $2,000. At the time of this writing, the current application submission date for most Newberry short-term fellowships is February 6 and awardees are usually notified near the end of April; however deadlines may vary from year to year. Applications must be submitted as email attachments by 11:59 p.m. on the deadline date. Applicants can find guidelines and forms at the grant website. Applicants are also encouraged to verify current deadline dates.
Requirements: Both local and non-local residents of the Chicago area are eligible to apply for this fellowship. The Newberry collections must be necessary to the short-term fellowship project since the purpose of these fellowships is to enable access to particular materials. Applications from former recipients of any fellowship at the Newberry are welcome, but preference is given to applicants who have not held a major fellowship within the past three years. In general, all fellowships must be completed by the end of the academic year for which they are awarded.
Restrictions: This fellowship is intended to support one month-in-residence at the Newberry; however, local residents who are awarded a Weinberg Fellowship may choose to distribute their time over a period of 9-12 months on a flexible basis to be determined in conversation with Newberry staff.
Geographic Focus: All States, All Countries
Date(s) Application is Due: Feb 6
Amount of Grant: 2,000 USD
Contact: Daniel Greene, Vice President for Research and Academic Programs; (312) 255-3666 or (312) 255-3535; research@newberry.org or greened@newberry.org
Internet: http://www.newberry.org/short-term-fellowships
Sponsor: Newberry Library
60 West Walton Street
Chicago, IL 60610-3380

NL Associated Colleges of the Midwest/Great Lakes College Association Faculty Fellowships **3281**
The Newberry, open to the public without charge, is an independent research library dedicated to the advancement and dissemination of knowledge, especially in the humanities. The library acquires and preserves a broad array of special-collections research materials (spanning six centuries of European and American civilizations) and promotes and provides for these collections' effective use, fostering research, teaching, publication, life-long learning, and civic engagement. Fellowships at the Newberry provide support for researchers wishing to use the Newberry's collections and are either short-term (one to two months) or long-term (four to twelve months). The Associated Colleges of the Midwest (ACM)/Great Lakes College Association (GLCA) Faculty Fellowships support ACM- and GLCA-college faculty as they teach a small group of

select undergraduate students in an advanced research seminar while doing their own research. These fellowships are considered long-term fellowships. Fellows can come from any of the ACM or GLCA colleges and from any discipline. At the time of this writing, the current application-submission date for most Newberry long-term fellowships is December 12 and awardees are usually notified in mid- to late-March; however deadlines may vary from year to year. Potential applicants should call or email the ACM contact given for information on how to apply.
Restrictions: Applicants may apply for more than one fellowship at the Newberry by designating those for which they wish to be considered on the cover sheet (available at the grant website). Applicants may also apply for both short- and long-term fellowships at the Newberry; however individuals may accept only one fellowship at the Newberry per year. Only scholars who have previously received a short-term fellowship at the Newberry may apply for a long-term fellowship for the same project. No other project extensions are allowable.
Geographic Focus: Colorado, Illinois, Indiana, Iowa, Michigan, Minnesota, Ohio, Pennsylvania, Wisconsin
Amount of Grant: Up to 40,000 USD
Contact: John Ottenhoff, Vice President; (312) 263-5000; jottenhoff@acm.edu
Internet: http://www.newberry.org/long-term-fellowships
Sponsor: Newberry Library
60 West Walton Street
Chicago, IL 60610-3380

NL Audrey Lumsden-Kouvel Fellowships 3282
The Newberry, open to the public without charge, is an independent research library dedicated to the advancement and dissemination of knowledge, especially in the humanities. The library acquires and preserves a broad array of special-collections research materials (spanning six centuries of European and American civilizations) and promotes and provides for these collections' effective use, fostering research, teaching, publication, life-long learning, and civic engagement. Fellowships at the Newberry provide support for researchers wishing to use the Newberry's collections and are either short-term (one to two months) or long-term (four to twelve months). Short-term fellowships are intended to geographically facilitate access for distant scholars to materials essential to a project. Long-term fellowships are intended to support significant works of scholarship that draw on the strengths of the Newberry's collections. Long-term fellowships support individual research and promote serious intellectual exchange through active participation in the Newberry's scholarly activities, including a biweekly fellows' seminar. Applicants may combine long-term fellowship awards with sabbatical or other stipendiary support. The Audrey Lumsden-Kouvel Fellowship is a long-term fellowship for post-doctoral scholars who wish to use the Newberry's extensive holdings in late medieval and early modern history and literature. Preference will be given to projects focusing on Romance cultures, including work that draws on sources from the colonial Americas. Topics in Portuguese, Spanish, and Latin American studies are especially welcome, as are translation projects. The fellowship carries a monthly stipend of $4,200. At the time of this writing, the current application-submission date for most Newberry long-term fellowships is December 12 and awardees are usually notified in mid- to late-March; however deadlines may vary from year to year. Applications must be submitted as email attachments by 11:59 p.m. on the deadline date. Applicants can find guidelines and forms at the grant website. Applicants are also encouraged to verify current deadline dates.
Requirements: All long-term fellowships are restricted to postdoctoral scholars. Recent Ph.D. recipients are encouraged to apply. Applicants must request at least four and no more than six months of support. All fellowships at the Newberry must be in-residence. Applications from former recipients of any fellowship at the Newberry are welcome, but preference is given to applicants who have not held a major fellowship within the past three years. In general, all fellowships must be completed by the end of the academic year for which they are awarded.
Restrictions: Applicants may apply for more than one fellowship at the Newberry by designating those for which they wish to be considered on the cover sheet (available at the grant website). Applicants may also apply for both short- and long-term fellowships at the Newberry; however individuals may accept only one fellowship at the Newberry per year. Only scholars who have previously received a short-term fellowship at the Newberry may apply for a long-term fellowship for the same project. No other project extensions are allowable.
Geographic Focus: All States, All Countries
Amount of Grant: 16,800-25,200 USD
Contact: Daniel Greene, Vice President for Research and Academic Programs; (312) 255-3666 or (312) 255-3535; research@newberry.org or greened@newberry.org
Internet: http://www.newberry.org/long-term-fellowships
Sponsor: Newberry Library
60 West Walton Street
Chicago, IL 60610-3380

NL British Academy Fellowship for Study in the United Kingdom 3283
The Newberry, open to the public without charge, is an independent research library dedicated to the advancement and dissemination of knowledge, especially in the humanities. The library acquires and preserves a broad array of special-collections research materials (spanning six centuries of European and American civilizations) and promotes and provides for these collections' effective use, fostering research, teaching, publication, life-long learning, and civic engagement. In cooperation with the British Academy, the Newberry Library offers an exchange fellowship for three months of study in the United Kingdom in any field in humanities in which the Newberry's collections are strong. Deadlines may vary from year to year. For more information on the program, current deadline dates, and application guidelines, those interested in applying should call or email the Newberry Library using the contact information given on this page.
Requirements: Applicants must be established scholars at the postdoctoral level or its equivalent who are on the staffs of universities, museums, or libraries. Preference is given to readers and staff of the Newberry and to scholars who have previously used the Newberry.
Geographic Focus: All States
Date(s) Application is Due: Dec 6
Amount of Grant: 1,350-16,200 GBP
Contact: Daniel Greene, Vice President for Research and Academic Programs; (312) 255-3666 or (312) 255-3535; research@newberry.org or greened@newberry.org
Internet: http://www.newberry.org/
Sponsor: Newberry Library
60 West Walton Street
Chicago, IL 60610-3380

NL Frances C. Allen Fellowships 3284
The Newberry, open to the public without charge, is an independent research library dedicated to the advancement and dissemination of knowledge, especially in the humanities. The library acquires and preserves a broad array of special-collections research materials (spanning six centuries of European and American civilizations) and promotes and provides for these collections' effective use, fostering research, teaching, publication, life-long learning, and civic engagement. Fellowships at the Newberry provide support for researchers wishing to use the Newberry's collections and are either short-term (one to two months) or long-term (four to twelve months). Short-term fellowships are intended to geographically facilitate access for distant scholars to materials essential to a project. Long-term fellowships are intended to support significant works of scholarship that draw on the strengths of the Newberry's collections. First awarded in 1983, the Frances C. Allen Fellowships are considered to be short-term fellowships, however their tenure may extend from one month to a full year. Their particular goal is to encourage Native-American women in their studies of any field related to the Newberry's collections. The fellowships provide a stipend of $2,000 per month; however supplemental funding may be available on a case by case basis. At the time of this writing, the current application submission date for most Newberry short-term fellowships is February 6 and awardees are usually notified near the end of April; however deadlines may vary from year to year. Applications must be submitted as email attachments by 11:59 p.m. on the deadline date. Applicants can find guidelines and forms at the grant website. Applicants are also encouraged to verify current deadline dates.
Requirements: These fellowships are available to women of Native-American heritage who are pursuing an academic program at any stage beyond the undergraduate degree. Allen fellows are expected to spend a significant part of their tenure in residence at the Newberry's D'Arcy McNickle Center for American Indian History. Applications from former recipients of any fellowship at the Newberry are welcome, but preference is given to applicants who have not held a major fellowship within the past three years. In general, all fellowships must be completed by the end of the academic year for which they are awarded.
Restrictions: Most short-term fellowships are restricted to individuals who live outside the Chicago area who need to examine specific items in the Newberry's collections. Applicants may apply for more than one fellowship at the Newberry by designating those for which they wish to be considered on the cover sheet (available at the grant website). Applicants may also apply for both short- and long-term fellowships at the Newberry; however individuals may accept only one fellowship at the Newberry per year. Only scholars who have previously received a short-term fellowship at the Newberry may apply for a long-term fellowship for the same project. No other project extensions are allowable.
Geographic Focus: All States, All Countries
Date(s) Application is Due: Feb 6
Amount of Grant: 2,000-24,000 USD
Contact: Daniel Greene, Vice President for Research and Academic Programs; (312) 255-3666 or (312) 255-3535; research@newberry.org or greened@newberry.org
Internet: http://www.newberry.org/short-term-fellowships
Sponsor: Newberry Library
60 West Walton Street
Chicago, IL 60610-3380

NL Graduate Scholar-in-Residence Opportunities 3285
The Newberry, open to the public without charge, is an independent research library dedicated to the advancement and dissemination of knowledge, especially in the humanities. The library acquires and preserves a broad array of special-collections research materials (spanning six centuries of European and American civilizations) and promotes and provides for these collections' effective use, fostering research, teaching, publication, life-long learning, and civic engagement. The Graduate Scholar-in-Residence program at the Newberry encourages Ph.D. candidates in the humanities to conduct research in the library's collections and to join its community of scholars. Selected scholars are provided with reserve space, access to the Newberry during extended hours, and opportunities to present their work-in-progress to the Newberry's scholarly community. Applications to become a Graduate Scholar-in-Residence are accepted each year in May. Candidates must submit via email a project description along with a filled-in application form (available for download from the website) by 11:59 p.m. on the deadline date. Applicants will be notified of their acceptance in June and will begin their residence in September so they may join the Newberry's new long-term fellows in fall-orientation activities. More detailed guidelines on how to apply are given at the grant website.
Requirements: Graduate students with advanced to Ph.D. candidacy are invited to apply for this status, with preference to those whose dissertation projects are well underway. Candidates' need for and intensive use of the Newberry's collections will be a crucial factor in the library's consideration of applications. Graduate scholars are expected to be in residence at least ten hours per week and should be wiling to provide a small amount of service to the Newberry.
Geographic Focus: Illinois

Date(s) Application is Due: May 1
Contact: Daniel Greene, Vice President for Research and Academic Programs; (312) 255-3666 or (312) 255-3535; research@newberry.org or greened@newberry.org
Internet: http://www.newberry.org/graduate-scholar-residence-program
Sponsor: Newberry Library
60 West Walton Street
Chicago, IL 60610-3380

NL Herzog August Bibliothek Wolfenbuttel Fellowship **3286**

The Newberry, open to the public without charge, is an independent research library dedicated to the advancement and dissemination of knowledge, especially in the humanities. The library acquires and preserves a broad array of special-collections research materials (spanning six centuries of European and American civilizations) and promotes and provides for these collections' effective use, fostering research, teaching, publication, life-long learning, and civic engagement. Fellowships at the Newberry provide support for researchers wishing to use the Newberry's collections and are either short-term (one to two months) or long-term (four to twelve months). Short-term fellowships are intended to geographically facilitate access for distant scholars to materials essential to a project. Long-term fellowships are intended to support significant works of scholarship that draw on the strengths of the Newberry's collections. Applicants for either long- or short-term fellowships at the Newberry may also apply for the Herzog August Bibliothek Wolfenbuttel fellowship which provides an additional two months at the Herzog August library in Wolfenbuttel, Germany. The Herzog award will pay 1,050 EUR per month plus up to 600 EUR for travel expenses. At the time of this writing, the current application submission date for most Newberry short-term fellowships is February 6, and for long-term fellowships, December 12. Short-term awards are usually announced near the end of April; long-term awards, mid- to late-March; however deadlines may vary from year to year. Applications must be submitted as email attachments by 11:59 p.m. on the deadline dates. Applicants can find guidelines and forms at the grant website. Applicants are also encouraged to verify current deadline dates.
Requirements: The applicant's proposed project must link the collections of both libraries (the Newberry and the Herzog August); the applicant should plan to hold both fellowships sequentially to ensure continuity of research. Further requirements will depend on the particular Newberry short- or long-term fellowship to which the applicant plans to jointly apply.
Restrictions: Restrictions will depend on the particular Newberry short- or long-term fellowship to which the applicant plans to jointly apply.
Geographic Focus: All States, All Countries
Date(s) Application is Due: Feb 6; Dec 12
Amount of Grant: 2,100-2,700 EUR
Contact: Daniel Greene, Vice President for Research and Academic Programs; (312) 255-3666 or (312) 255-3535; research@newberry.org or greened@newberry.org
Internet: http://www.newberry.org/short-term-fellowships
Sponsor: Newberry Library
60 West Walton Street
Chicago, IL 60610-3380

NL Lawrence Lipking Fellowship at the Newberry **3287**

The Newberry, open to the public without charge, is an independent research library dedicated to the advancement and dissemination of knowledge, especially in the humanities. The library acquires and preserves a broad array of special-collections research materials (spanning six centuries of European and American civilizations) and promotes and provides for these collections' effective use, fostering research, teaching, publication, life-long learning, and civic engagement. Fellowships at the Newberry provide support for researchers wishing to use the Newberry's collections and are either short-term or long-term. Short-term fellowships are intended to geographically facilitate access for distant scholars to materials essential to a project. Long-term fellowships are intended to support significant works of scholarship that draw on the strengths of the Newberry's collections. The annual Lipking-Newberry Fellowship enables a qualified Northwestern graduate student in English to spend a quarter conducting research at the Newberry. A stipend equivalent to Northwestern's University Fellowship stipend ($5,130 in 2009-10) and tuition will be provided. Candidates for the fellowship should supply a short-term cover sheet, a curriculum vitae, one letter of recommendation, and a research plan of no more than 500 words detailing intended use of Newberry resources. The cover sheet and further application guidelines are available at the grant website. Applications must be submitted as email attachments by 11:59 p.m. on the deadline date to the Newberry (and copied to Northwestern). The email addresses to use for both organizations are specified at the grant website. At the time of this writing, the current application submission date for most Newberry short-term fellowships is February 6 and awardees are usually notified near the end of April; however, deadlines may vary from year to year. Candidates are encouraged to verify current deadline dates.
Requirements: Applicants must be qualified Northwestern graduate students in the English department. Full-time residence at the Newberry is required.
Restrictions: In the unlikely event that no English graduate student should qualify, the fellowship will be opened to other Northwestern departments. Applicants may apply for more than one fellowship at the Newberry by designating those for which they wish to be considered on the cover sheet (available at the grant website). Applicants may also apply for both short- and long-term fellowships at the Newberry; however individuals may accept only one fellowship at the Newberry per year. Only scholars who have previously received a short-term fellowship at the Newberry may apply for a long-term fellowship for the same project. No other project extensions are allowable.
Geographic Focus: Illinois
Date(s) Application is Due: Feb 6
Contact: Daniel Greene; (312) 255-3666 or (312) 255-3535; research@newberry.org
Internet: http://www.newberry.org/short-term-fellowships
Sponsor: Newberry Library
60 West Walton Street
Chicago, IL 60610-3380

NL Lester J. Cappon Fellowship in Documentary Editing **3288**

The Newberry, open to the public without charge, is an independent research library dedicated to the advancement and dissemination of knowledge, especially in the humanities. The library acquires and preserves a broad array of special-collections research materials (spanning six centuries of European and American civilizations) and promotes and provides for these collections' effective use, fostering research, teaching, publication, life-long learning, and civic engagement. Fellowships at the Newberry provide support for researchers wishing to use the Newberry's collections and are either short-term (generally one to two months) or long-term (four to twelve months). Short-term fellowships are intended to geographically facilitate access for distant scholars to materials essential to a project. Long-term fellowships are intended to support significant works of scholarship that draw on the strengths of the Newberry's collections. The Lester J. Cappon Fellowship in Documentary Editing is considered a short-term fellowship to support scholarly editing projects based on Newberry materials and carried out in residence. Cappon fellowships provide a stipend of $2,000 per month; applicants may request up to three months of support for this fellowship. At the time of this writing, the current application submission date for most Newberry short-term fellowships is February 6 and awardees are usually notified near the end of April; however deadlines may vary from year to year. Applications must be submitted as email attachments by 11:59 p.m. on the deadline date. Applicants can find guidelines and forms at the grant website. Applicants are also encouraged to verify current deadline dates.
Requirements: The Cappon Fellowship is restricted to postdoctoral scholars. While most short-term fellowships at the Newberry are restricted to scholars who live outside the Chicago area, applicants for the Cappon fellowship may be local. All fellowships at the Newberry must be conducted in-residence. Applications from former recipients of any fellowship at the Newberry are welcome, but preference is given to applicants who have not held a major fellowship within the past three years. In general, all fellowships at the Newberry must be completed by the end of the academic year for which they are awarded.
Restrictions: Applicants may apply for more than one fellowship at the Newberry by designating those for which they wish to be considered on the cover sheet (available at the grant website). Applicants may also apply for both short- and long-term fellowships at the Newberry; however individuals may accept only one fellowship at the Newberry per year. Only scholars who have previously received a short-term fellowship at the Newberry may apply for a long-term fellowship for the same project. No other project extensions are allowable.
Geographic Focus: All States, All Countries
Date(s) Application is Due: Feb 6
Contact: Daniel Greene; (312) 255-3666 or (312) 255-3535; research@newberry.org
Internet: http://www.newberry.org/short-term-fellowships
Sponsor: Newberry Library
60 West Walton Street
Chicago, IL 60610-3380

NL Lloyd Lewis Fellowships in American History **3289**

The Newberry, open to the public without charge, is an independent research library dedicated to the advancement and dissemination of knowledge, especially in the humanities. The library acquires and preserves a broad array of special-collections research materials (spanning six centuries of European and American civilizations) and promotes and provides for these collections' effective use, fostering research, teaching, publication, life-long learning, and civic engagement. Fellowships at the Newberry provide support for researchers wishing to use the Newberry's collections and are either short-term (one to two months) or long-term (four to twelve months). Short-term fellowships are intended to geographically facilitate access for distant scholars to materials essential to a project. Long-term fellowships are intended to support significant works of scholarship that draw on the strengths of the Newberry's collections. Long-term fellowships support individual research and promote serious intellectual exchange through active participation in the Newberry's scholarly activities, including a biweekly fellows' seminar. Applicants may combine long-term fellowship awards with sabbatical or other stipendiary support. The Lloyd Lewis Fellowships in American History are long-term fellowships for post-doctoral scholars pursuing projects in any area of American history appropriate to the Newberry's collections. The fellowship carries a monthly stipend of $4,200. At the time of this writing, the current application-submission date for most Newberry long-term fellowships is December 12 and awardees are usually notified in mid- to late-March; however deadlines may vary from year to year. Applications must be submitted as email attachments by 11:59 p.m. on the deadline date. Applicants can find guidelines and forms at the grant website. Applicants are also encouraged to verify current deadline dates.
Requirements: Applicants must have their Ph.D. or its equivalent and must request at least four but no more than 12 months of support. All fellowships at the Newberry must be in-residence. Applications from former recipients of any fellowship at the Newberry are welcome, but preference is given to applicants who have not held a major fellowship within the past three years. In general, all fellowships must be completed by the end of the academic year for which they are awarded.
Restrictions: Applicants may apply for more than one fellowship at the Newberry by designating those for which they wish to be considered on the cover sheet (available at the grant website). Applicants may also apply for both short- and long-term fellowships at the Newberry; however individuals may accept only one fellowship at the Newberry per year. Only scholars who have previously received a short-term fellowship at the Newberry may apply for a long-term fellowship for the same project. No other project extensions are allowable.

Geographic Focus: All States, All Countries
Date(s) Application is Due: Dec 12
Amount of Grant: 16,800-50,400 USD
Contact: Daniel Greene, Vice President for Research and Academic Programs; (312) 255-3666 or (312) 255-3535; research@newberry.org or greened@newberry.org
Internet: http://www.newberry.org/long-term-fellowships
Sponsor: Newberry Library
60 West Walton Street
Chicago, IL 60610-3380

NLM Academic Research Enhancement Awards (AREA) 3290

The purpose of the Academic Research Enhancement Award (AREA) program is to stimulate research in educational institutions that provide baccalaureate or advanced degrees for a significant number of the Nation's research scientists, but that have not been major recipients of support. NLM's participation is confined to applications that address research in biomedical informatics and bioinformatics. Only electronic applications are accepted. Deadlines for new applications are February 25, June 25 and October 25 each year.
Geographic Focus: All States
Date(s) Application is Due: Feb 25; Jun 25; Oct 25
Contact: Dr. Hua-Chuan Sim, Clinical and Public Health Informatics; (301) 594-5983; fax (301) 402-1384; simh@mail.nih.gov
Internet: http://www.nlm.nih.gov/ep/area.html
Sponsor: National Library of Medicine
6705 Rockledge Drive, Suite 301
Bethesda, MD 20892

NLM Advancing Novel Science in Women's Health Research Exploratory Developmental Research Grants 3291

The purpose of this funding opportunity announcement (FOA), issued by the Office of Research on Women's Health (ORWH), is to promote innovative, interdisciplinary research that will advance new concepts in women's health research and the study of sex/gender differences. Recent research reports have established the importance of studying issues specific to women, including the scientific and clinical importance of analyzing data separately for females and males. ORWH is particularly interested in encouraging extramural investigators to undertake new interdisciplinary research to advance studies on how sex and gender factors affect women's health; however, applications in all areas of women's health and/or sex/gender research are invited. The deadline for new applications is October 16, and the deadline for revised applications is November 16.
Geographic Focus: All States
Date(s) Application is Due: Oct 16; Nov 16
Contact: Dr. Hua-Chuan Sim, Clinical and Public Health Informatics; (301) 594-5983; fax (301) 402-1384; simh@mail.nih.gov
Internet: http://www.nlm.nih.gov/ep/answhr.html
Sponsor: National Library of Medicine
6705 Rockledge Drive, Suite 301
Bethesda, MD 20892

NLM Advancing Novel Science in Women's Health Research Small Project Grants 3292

The purpose of this funding opportunity announcement (FOA), issued by the Office of Research on Women's Health (ORWH), is to promote innovative, interdisciplinary research that will advance new concepts in women's health research and the study of sex/gender differences. Recent research reports have established the importance of studying issues specific to women, including the scientific and clinical importance of analyzing data separately for females and males. ORWH is particularly interested in encouraging extramural investigators to undertake new interdisciplinary research to advance studies on how sex and gender factors affect women's health; however, applications in all areas of women's health and/or sex/gender research are invited. The deadline for new applications is October 16, and the deadline for revised applications is November 16.
Geographic Focus: All States
Date(s) Application is Due: Oct 16; Nov 16
Contact: Dr. Hua-Chuan Sim, Clinical and Public Health Informatics; (301) 594-5983; fax (301) 402-1384; simh@mail.nih.gov
Internet: http://www.nlm.nih.gov/ep/answhr.html
Sponsor: National Library of Medicine
6705 Rockledge Drive, Suite 301
Bethesda, MD 20892

NLM Behavioral and Social Science Research on Understanding and Reducing Health Disparities Research Grants 3293

The goal of this funding opportunity announcement (FOA) is to encourage behavioral and social science research on the causes and solutions to health and disability disparities in the U.S. population. Health disparities between, on the one hand, racial/ethnic populations, lower socioeconomic classes, and rural residents and, on the other hand, the overall U.S. population are major public health concerns.
Geographic Focus: All States
Date(s) Application is Due: Sep 18
Contact: Dr. Hua-Chuan Sim, Clinical and Public Health Informatics; (301) 594-5983; fax (301) 402-1384; simh@mail.nih.gov
Internet: http://www.nlm.nih.gov/ep/healthdisp.html
Sponsor: National Library of Medicine
6705 Rockledge Drive, Suite 301
Bethesda, MD 20892

NL Mellon Postdoctoral Research Fellowships 3294

The Newberry, open to the public without charge, is an independent research library dedicated to the advancement and dissemination of knowledge, especially in the humanities. The library acquires and preserves a broad array of special-collections research materials (spanning six centuries of European and American civilizations) and promotes and provides for these collections' effective use, fostering research, teaching, publication, life-long learning, and civic engagement. Fellowships at the Newberry provide support for researchers wishing to use the Newberry's collections and are either short-term (one to two months) or long-term (four to twelve months). Short-term fellowships are intended to geographically facilitate access for distant scholars to materials essential to a project. Long-term fellowships are intended to support significant works of scholarship that draw on the strengths of the Newberry's collections. Long-term fellowships support individual research and promote serious intellectual exchange through active participation in the Newberry's scholarly activities, including a biweekly fellows' seminar. Applicants may combine long-term fellowship awards with sabbatical or other stipendiary support. The Mellon Postdoctoral Research Fellowships are long-term fellowships that support postdoctoral scholars in any field relevant to the Newberry's collections for six to ten months of residential research and writing. The fellowship carries a monthly stipend of $4,200. Submission deadlines may vary from year to year. For more information on the program, current deadline dates, and application guidelines, those interested in applying should call or email the Newberry Library using the contact information given on this page.
Requirements: Postdoctoral scholars at any stage in their careers are eligible. Applications from former recipients of any fellowship at the Newberry are welcome, but preference is given to applicants who have not held a major fellowship within the past three years. In general, all fellowships must be completed by the end of the academic year for which they are awarded.
Geographic Focus: All States, All Countries
Date(s) Application is Due: Dec 12
Amount of Grant: 25,200-42,000 USD
Contact: Daniel Greene, Vice President for Research and Academic Programs; (312) 255-3666 or (312) 255-3535; research@newberry.org or greened@newberry.org
Internet: http://www.newberry.org/
Sponsor: Newberry Library
60 West Walton Street
Chicago, IL 60610-3380

NLM Exceptional, Unconventional Research Grants Enabling Knowledge Acceleration 3295

This RFA is for exceptionally innovative research on novel hypotheses or difficult problems, solutions to which would have an extremely high impact on biomedical or biobehavioral research that is germane to the mission of the National Library of Medicine. The RFA supports new projects, not continuation of projects that have already been initiated. For the EUREKA program, NLM seeks innovative applications in the following two discovery science areas: integrated discovery mining for biology and medicine; and integrated hypothesis testing for biology and medicine. Only electronic applications are accepted. The deadline for Letter of Intent is September 29, and the deadline for new applications is October 28.
Restrictions: This grant does not support pilot projects, i.e., projects of limited scope that are designed primarily to generate data that will enable the PI to seek other funding opportunities.
Geographic Focus: All States
Date(s) Application is Due: Sep 29; Oct 28
Contact: Dr. Valerie Florance, (301) 594-5983; florancev@mail.nih.gov
Internet: http://www.nlm.nih.gov/ep/eureka.html
Sponsor: National Library of Medicine
6705 Rockledge Drive, Suite 301
Bethesda, MD 20892

NLM Exploratory/Developmental Grants 3296

The purpose of this program is to support for early stage conceptual work and feasibility tests in biomedical informatics and bioinformatics. These studies may involve considerable risk but may lead to a breakthrough in a particular area, or to the development of novel techniques or approaches that could have major impact on a field of biomedical, behavioral, or clinical research. Preliminary data are not required, but applicants are urged to provide milestones by which progress on the project can be measured. Only electronic applications are accepted. Deadlines for new applications are February 16, June 16 and October 16 each year. Deadlines for revised applications are March 16, July 16 and November 16 each year.
Geographic Focus: All States
Date(s) Application is Due: Feb 16; Mar 16; Jun 16; Jul 16; Oct 16; Nov 16
Contact: Dr. Jane Ye, Bioinformatics and Computational Biology; (301) 594-5983; fax (301) 402-1384; yej@mail.nih.gov
Internet: http://www.nlm.nih.gov/ep/GrantExDev.html
Sponsor: National Library of Medicine
6705 Rockledge Drive, Suite 301
Bethesda, MD 20892

NLM Express Research Grants in Biomedical Informatics and Bioinformatics 3297

The purpose of the program is to provide support for rigorous scientific research in biomedical informatics and bioinformatics. NLM Express Research Grant applications are limited to a 15-page research plan and to a modular budget. The scope of interests is outlined in the funding opportunity announcement. Investigators whose needs are not met by this announcement can also apply through the NIH Parent R01 (PA 07-070), with a cover letter requesting assignment to NLM. Those who do so should use PAR 08-

080 as a source of guidance about NLM's research grant priorities and scope of interest. Deadlines for new applications are February 5, June 5 and October 5 each year. Deadlines for revised applications are March 5, July 5 and November 5 each year.
Geographic Focus: All States
Date(s) Application is Due: Feb 5; Mar 5; Jun 5; Jul 5; Oct 5; Nov 5
Contact: Dr. Jane Ye; (301) 594-5983; fax (301) 402-1384; yej@mail.nih.gov
Internet: http://www.nlm.nih.gov/ep/GrantResearch.html
Sponsor: National Library of Medicine
6705 Rockledge Drive, Suite 301
Bethesda, MD 20892

NLM Grants for Scholarly Works in Biomedicine and Health 3298
Small grants are awarded for up to three years to provide short-term assistance for the preparation of book-length manuscripts and, in some cases, the publication of important scientific information needed by U.S. health professionals. Grants are awarded for major critical reviews and analyses of current developments in important areas of the health sciences, historical studies, works about health sciences informatics, librarianship, and certain kinds of secondary information and literature tools in the health sciences. Publication may be in formats other than print-on-paper (e.g., electronic, film, etc.), and may involve new and innovative ways of organizing and presenting information. Only electronic applications are accepted.
Requirements: Grants may be made to public or private, nonprofit institutions on behalf of a principal investigator; or to unaffiliated individuals to support salaries, consultant fees, equipment and supplies, travel, and other justified costs.
Restrictions: The program does not support textbooks, the production of curriculum materials, the initial reporting of original biomedical scientific research findings, proceedings of annual meetings, projects highly likely to have substantial commercial sales (best-seller type publications), or projects of local interest only. The program does not provide support for the free distribution of publications resulting from grants.
Geographic Focus: All States
Date(s) Application is Due: Feb 2
Amount of Grant: 50,000-150,000 USD
Contact: Dr. Hua-Chuan Sim, Clinical and Public Health Informatics; (301) 594-5983; fax (301) 402-1384; simh@mail.nih.gov
Internet: http://www.nlm.nih.gov/ep/GrantPubs.html
Sponsor: National Library of Medicine
6705 Rockledge Drive, Suite 301
Bethesda, MD 20892

NL Midwest Modern language Association Fellowship 3299
The Newberry, open to the public without charge, is an independent research library dedicated to the advancement and dissemination of knowledge, especially in the humanities. The library acquires and preserves a broad array of special-collections research materials (spanning six centuries of European and American civilizations) and promotes and provides for these collections' effective use, fostering research, teaching, publication, life-long learning, and civic engagement. Fellowships at the Newberry provide support for researchers wishing to use the Newberry's collections and are either short-term (one to two months) or long-term (four to twelve months). Short-term fellowships are intended to geographically facilitate access for distant scholars to materials essential to a project. Long-term fellowships are intended to support significant works of scholarship that draw on the strengths of the Newberry's collections. A short-term fellowship is available for Ph.D. candidates or post-doctoral scholars who are members of the Midwest Modern Language Association (MMLA) for up to one month's work-in-residence at the Newberry. The fellowship provides a stipend of $2,000 per month. At the time of this writing, the current application submission date for most Newberry short-term fellowships is February 6 and awardees are usually notified near the end of April; however deadlines may vary from year to year. Applications must be submitted as email attachments by 11:59 p.m. on the deadline date. Applicants can find guidelines and forms at the grant website. Applicants are also encouraged to verify current deadline dates.
Requirements: MMLA membership must be current at the time of application and through the period of the fellowship. Most short-term fellowships at the Newberry are for Ph.D. candidates and postdoctoral scholars. All fellowships at the Newberry must be conducted in-residence. The Newberry collections must be necessary to the short-term fellowship project since the purpose of these fellowships is to enable access to particular materials. Applications from former recipients of any fellowship at the Newberry are welcome, but preference is given to applicants who have not held a major fellowship within the past three years. In general, all fellowships at the Newberry must be completed by the end of the academic year for which they are awarded.
Restrictions: Most short-term fellowships are restricted to individuals who live outside the Chicago area who need to examine specific items in the Newberry's collections. Applicants may apply for more than one fellowship at the Newberry by designating those for which they wish to be considered on the cover sheet (available at the grant website). Applicants may also apply for both short- and long-term fellowships at the Newberry; however individuals may accept only one fellowship at the Newberry per year. Only scholars who have previously received a short-term fellowship at the Newberry may apply for a long-term fellowship for the same project. No other project extensions are allowable.
Geographic Focus: All States
Date(s) Application is Due: Feb 6
Amount of Grant: Up to 2,000 USD
Contact: Daniel Greene, Vice President for Research and Academic Programs; (312) 255-3666 or (312) 255-3535; research@newberry.org or greened@newberry.org
Internet: http://www.newberry.org/short-term-fellowships
Sponsor: Newberry Library
60 West Walton Street
Chicago, IL 60610-3380

NLM Internet Connection for Medical Institutions Grants 3300
NLM encourages the development of a communications infrastructure to promote the rapid interchange of medical information nationally and throughout the world. Internet access provides health professionals engaged in education, research, clinical care, and administration with a means of accessing remote databases and libraries, transferring files and images, and interacting with colleagues throughout the world. To accelerate the pace with which health-related institutions become part of the electronic information web, NLM offers grants to support Internet connections. Funds may be used for the purchase and installation of a gateway system and associated connection hardware. Funds may also be used to defray the cost of installation and leasing of communication circuits to connect the network. Annual deadline dates may vary; contact the program office for exact dates. RFA: LM-00-001
Requirements: Domestic public and private nonprofit institutions working in health sciences administration, education research, and/or clinical care services can apply for these grants. Groups that are health-related institutions are also encouraged to apply.
Geographic Focus: All States
Amount of Grant: 30,000-50,000 USD
Contact: Dr. Jane Ye, Bioinformatics and Computational Biology; (301) 594-5983; fax (301) 402-1384; yej@mail.nih.gov
Internet: http://grants.nih.gov/grants/guide/rfa-files/RFA-LM-00-001.html
Sponsor: National Library of Medicine
8600 Rockville Pike, Building 38, Room B1E08
Bethesda, MD 20894

NLM Limited Competition for Continuation of Biomedical Informatics Bioinformatics Resource Grants 3301
This limited competition funding opportunity announcement (FOA) solicits competitive renewal applications from investigators currently supported under PAR-04-142, NLM Resource Grant in Biomedical Informatics/Bioinformatics (P41) or its predecessors. NLM's Biomedical Informatics/Bioinformatics Resource Grant program provides support for five (5) resource centers that support activities, databases or software tools that advance research or practice in the biomedical sciences, clinical medicine or health services research. Direct costs for the first year may not exceed 5% above the current award of the active year of NLM funding at the time of application, or a maximum of $500,000, whichever is smaller. Applicants may include cost-of-living increases of up to 3% per year for direct costs in future years. The deadline for Letters of Intent is July 28, while the deadline for Renewal Applications is August 27.
Geographic Focus: All States
Date(s) Application is Due: Jul 28; Aug 27
Contact: Dr. Valerie Florance, (301) 594-5983; florancev@mail.nih.gov
Internet: http://www.nlm.nih.gov/ep/grantp41.html
Sponsor: National Library of Medicine
6705 Rockledge Drive, Suite 301
Bethesda, MD 20892

NLM Manufacturing Processes of Medical, Dental, & Biological Technologies (SBIR) Grants 3302
NLM is encouraging eligible United States small business concerns to submit SBIR and STTR Phase I, Phase II, and Fast-Track grant applications whose biomedical research is related to advanced processing, manufacturing processes, equipment and systems, and manufacturing workforce skills and protection. Only electronic applications are accepted.
Geographic Focus: All States
Date(s) Application is Due: Apr 5; Aug 5; Dec 5
Contact: Dr. Jane Ye, Bioinformatics and Computational Biology; (301) 594-5983; fax (301) 402-1384; yej@mail.nih.gov
Internet: http://www.nlm.nih.gov/ep/grantsbmanu.html
Sponsor: National Library of Medicine
6705 Rockledge Drive, Suite 301
Bethesda, MD 20892

NLM New Technologies for Liver Disease Grants 3303
The goal of this announcement is to enlist members of the small business research community in advancing means of diagnosis, treatment and prevention of liver disease and facilitate the goals outlined in the trans-NIH Action Plan for Liver Disease Research. Annual deadline dates are April 5, August 5 and December 5.
Geographic Focus: All States
Date(s) Application is Due: Apr 5; Aug 5; Dec 5
Contact: Dr. Jane Ye, Bioinformatics and Computational Biology; (301) 594-5983; fax (301) 402-1384; yej@mail.nih.gov
Internet: http://www.nlm.nih.gov/ep/grantsbliver.html
Sponsor: National Library of Medicine
6705 Rockledge Drive, Suite 301
Bethesda, MD 20892

NL Monticello College Foundation Fellowship for Women 3304
The Newberry, open to the public without charge, is an independent research library dedicated to the advancement and dissemination of knowledge, especially in the humanities. The library acquires and preserves a broad array of special-collections research materials (spanning six centuries of European and American civilizations) and promotes and provides for these collections' effective use, fostering research, teaching, publication, life-long learning, and civic engagement. Fellowships at the Newberry provide support

for researchers wishing to use the Newberry's collections and are either short-term (one to two months) or long-term (four to twelve months). Short-term fellowships are intended to geographically facilitate access for distant scholars to materials essential to a project. Long-term fellowships are intended to support significant works of scholarship that draw on the strengths of the Newberry's collections. Long-term fellowships support individual research and promote serious intellectual exchange through active participation in the Newberry's scholarly activities, including a biweekly fellows' seminar. Applicants may combine long-term fellowship awards with sabbatical or other stipendiary support. The Monticello College Foundation Fellowship for Women is considered a long-term fellowship and is designed primarily for women at early stages in their professional careers whose work gives clear promise of scholarly productivity and whose careers would be significantly enhanced by research, writing, and participation in the intellectual life of the Newberry. The fellowship carries a monthly stipend of $4,200. At the time of this writing, the current application-submission date for most Newberry long-term fellowships is December 12 and awardees are usually notified in mid- to late-March; however deadlines may vary from year to year. Applications must be submitted as email attachments by 11:59 p.m. on the deadline date. Applicants can find guidelines and forms at the grant website. Applicants are also encouraged to verify current deadline dates.
Requirements: Applicants' topics should be related to the Newberry's collections; preference will be given to proposals particularly concerned with the study of women. Applicants must have their Ph.D. at the time of application and must request at least four but no more than six months of support. All fellowships at the Newberry must be in-residence. Applications from former recipients of any fellowship at the Newberry are welcome, but preference is given to applicants who have not held a major fellowship within the past three years. In general, all fellowships must be completed by the end of the academic year for which they are awarded.
Restrictions: Applicants may apply for more than one fellowship at the Newberry by designating those for which they wish to be considered on the cover sheet (available at the grant website). Applicants may also apply for both short- and long-term fellowships at the Newberry; however individuals may accept only one fellowship at the Newberry per year. Only scholars who have previously received a short-term fellowship at the Newberry may apply for a long-term fellowship for the same project. No other project extensions are allowable.
Geographic Focus: All States
Date(s) Application is Due: Dec 12
Contact: Daniel Greene, Vice President for Research and Academic Programs; (312) 255-3666 or (312) 255-3535; research@newberry.org or greened@newberry.org
Internet: http://www.newberry.org/long-term-fellowships
Sponsor: Newberry Library
60 West Walton Street
Chicago, IL 60610-3380

NLM Predictive Multiscale Models of the Physiome in Health and Disease Grants **3305**
The goal of this funding opportunity announcement (FOA) is to move the field of biomedical computational modeling forward through the development of more realistic and predictive models of health and disease. NIH recognizes the need for sophisticated, predictive, computational models of development and disease that encompass multiple biological scales. These models may be designed to uncover biological mechanisms or to make predictions about clinical outcome and may draw on a variety of data sources including relevant clinical data. Ultimately the models and the information derived from their use will enable researchers and clinicians to better understand, prevent, diagnose and treat the diseases or aberrations in normal development. Specifically this FOA solicits the development of predictive multiscale models of health and disease states that must include higher scales of the physiome. The specific objectives are to develop multiscale models that are physiologically mechanistic and biomedically relevant, to bring together modeling and biomedical expertise to collaborate on building models, to validate and test models with standard datasets, and to develop models that can be explicitly shared with other modelers.
Geographic Focus: All States
Date(s) Application is Due: Jan 14; May 14; Sep 15
Contact: Dr. Jane Ye, Bioinformatics and Computational Biology; (301) 594-5983; fax (301) 402-1384; yej@mail.nih.gov
Internet: http://www.nlm.nih.gov/ep/predictive.html
Sponsor: National Library of Medicine
6705 Rockledge Drive, Suite 301
Bethesda, MD 20892

NLM Research Project Grants **3306**
The primary purpose of this program is to provide support for rigorous scientific research in biomedical informatics and bioinformatics. Medical informatics is an interdisciplinary field that combines health sciences specialty areas with the information and computer sciences, computational linguistics, and decision theory, among others. NLM's program emphasizes knowledge issues without restriction as to specific categorical diseases. Proposed research must be clearly health relevant. Areas of interest include knowledge representation, more effective human-machine interfaces, and more efficient information storage and retrieval. The cognitive aspects of information-related behaviors are of particular interest. This area includes problems of medical decision analysis, medical decision support systems, and retrieval from image databases. Deadlines for new applications are February 5, June 5 and October 5 each year. Deadlines for revised applications are March 5, July 5 and November 5 each year.
Requirements: Institutions or organizations with research capabilities in the health information fields or in medical informatics are eligible to apply.
Geographic Focus: All States
Date(s) Application is Due: Feb 5; Mar 5; Jun 5; Jul 5; Oct 5; Nov 5
Amount of Grant: 25,000-2,000,000 USD
Contact: Dr. Jane Ye; (301) 594-5983; fax (301) 402-1384; yej@mail.nih.gov
Internet: http://www.nlm.nih.gov/ep/GrantResearchParent.html
Sponsor: National Library of Medicine
6705 Rockledge Drive, Suite 301
Bethesda, MD 20892

NLM Research Supplements to Promote Reentry in Health-Related Research **3307**
Principal Investigators holding NIH research grants of the specified types (see announcement) can apply for administrative supplements to support individuals with high potential to reenter an active research career after taking time off to care for children or attend to other family responsibilities. It is anticipated that at the completion of the supplement, the reentry scientist will be in a position to apply for a career development award, a research award, or some other form of independent research support. The grantee institution must submit the application for supplemental funds directly to the awarding component that supports the parent grant. The application must not be submitted to the NIH Center for Scientific Review.
Geographic Focus: All States
Contact: Dr. Hua-Chuan Sim, Clinical and Public Health Informatics; (301) 594-5983; fax (301) 402-1384; simh@mail.nih.gov
Internet: http://www.nlm.nih.gov/ep/grantsupreentry.html
Sponsor: National Library of Medicine
6705 Rockledge Drive, Suite 301
Bethesda, MD 20892

NLM Small Business Innovation Research (SBIR) Grants **3308**
The NLM supports research and development projects in biomedical informatics and bioinformatics. NLM defines biomedical informatics as the intersection of basic informational and computing sciences with a wide range of application domains in biomedicine and public health. Bioinformatics is the intersection of basic informational and computing sciences with the biological sciences. Innovative technologies and methodologies fuel progress in biomedical and behavioral research and represent an increasingly important area of the economy. The SBIR and STTR programs provide support for research or development of new or improved technologies and methodologies that have the potential to succeed as commercial products or services. Only electronic applications are accepted.
Geographic Focus: All States
Date(s) Application is Due: Apr 5; Aug 5; Dec 5
Contact: Dr. Jane Ye, Bioinformatics and Computational Biology; (301) 594-5983; fax (301) 402-1384; yej@mail.nih.gov
Internet: http://www.nlm.nih.gov/ep/grantsbir.html
Sponsor: National Library of Medicine
6705 Rockledge Drive, Suite 301
Bethesda, MD 20892

NLM Technological Innovations for Interdisciplinary Research Incorporating the Behavioral and Social Sciences Grants **3309**
The intent of the program is to solicit Small Business Innovation Research (SBIR) and Small Business Technology Transfer (STTR) grant applications from small business concerns (SBCs) for development of new, innovative technologies for research integrating human social and/or behavioral science with other disciplines. These announcements cover Phase I, Phase II, and Fast-Track applications. Only electronic applications are accepted.
Geographic Focus: All States
Date(s) Application is Due: Jan 15; May 1; Sep 17
Contact: Dr. Jane Ye, Bioinformatics and Computational Biology; (301) 594-5983; fax (301) 402-1384; yej@mail.nih.gov
Internet: http://www.nlm.nih.gov/ep/grantsbsoc.html
Sponsor: National Library of Medicine
6705 Rockledge Drive, Suite 301
Bethesda, MD 20892

NL National Endowment for the Humanities Fellowships **3310**
The Newberry, open to the public without charge, is an independent research library dedicated to the advancement and dissemination of knowledge, especially in the humanities. The library acquires and preserves a broad array of special-collections research materials (spanning six centuries of European and American civilizations) and promotes and provides for these collections' effective use, fostering research, teaching, publication, life-long learning, and civic engagement. Fellowships at the Newberry provide support for researchers wishing to use the Newberry's collections and are either short-term (one to two months) or long-term (four to twelve months). Short-term fellowships are intended to geographically facilitate access for distant scholars to materials essential to a project. Long-term fellowships are intended to support significant works of scholarship that draw on the strengths of the Newberry's collections. Long-term fellowships support individual research and promote serious intellectual exchange through active participation in the Newberry's scholarly activities, including a biweekly fellows' seminar. Applicants may combine long-term fellowship awards with sabbatical or other stipendiary support. The National Endowment for the Humanities Fellowships are long-term fellowships to support post-doctoral scholars in any field appropriate to the Newberry's collections. The fellowships carries a monthly stipend of $4,200. At the time of this writing, the current application-submission date for most Newberry long-term fellowships is December 12 and awardees are usually notified in mid- to late-March; however deadlines may vary

from year to year. Applications must be submitted as email attachments by 11:59 p.m. on the deadline date. Applicants can find guidelines and forms at the grant website. Applicants are also encouraged to verify current deadline dates.
Requirements: All long-term fellowships are restricted to postdoctoral scholars. Applicants must be United States citizens or foreign nationals with three years' residence. Applicants must request at least four and no more than twelve months of support. All fellowships at the Newberry must be in-residence. Applications from former recipients of any fellowship at the Newberry are welcome, but preference is given to applicants who have not held a major fellowship within the past three years. In general, all fellowships must be completed by the end of the academic year for which they are awarded.
Restrictions: Applicants may apply for more than one fellowship at the Newberry by designating those for which they wish to be considered on the cover sheet (available at the grant website). Applicants may also apply for both short- and long-term fellowships at the Newberry; however individuals may accept only one fellowship at the Newberry per year. Only scholars who have previously received a short-term fellowship at the Newberry may apply for a long-term fellowship for the same project. No other project extensions are allowable.
Geographic Focus: All States
Date(s) Application is Due: Dec 12
Amount of Grant: 16,800-50,400 USD
Contact: Daniel Greene, Vice President for Research and Academic Programs; (312) 255-3666 or (312) 255-3535; research@newberry.org or greened@newberry.org
Internet: http://www.newberry.org/long-term-fellowships
Sponsor: Newberry Library
60 West Walton Street
Chicago, IL 60610-3380

NL Newberry Consortium on American Indian Studies Faculty Fellowships 3311
The Newberry, open to the public without charge, is an independent research library dedicated to the advancement and dissemination of knowledge, especially in the humanities. The library acquires and preserves a broad array of special-collections research materials (spanning six centuries of European and American civilizations) and promotes and provides for these collections' effective use, fostering research, teaching, publication, life-long learning, and civic engagement. Fellowships at the Newberry provide support for researchers wishing to use the Newberry's collections and are either short-term (one to two months) or long-term (four to twelve months). Short-term fellowships are intended to geographically facilitate access for distant scholars to materials essential to a project. Long-term fellowships are intended to support significant works of scholarship that draw on the strengths of the Newberry's collections. Long-term fellowships support individual research and promote serious intellectual exchange through active participation in the Newberry's scholarly activities, including a biweekly fellows' seminar. Applicants may combine long-term fellowship awards with sabbatical or other stipendiary support. The Newberry Consortium in American Indian Studies (NCAIS) offers long-term fellowships for faculty at participating institutions. The fellowships provide support for one semester in residence at the Newberry and carry a monthly stipend of $4,200. Fellows will work on a project in American Indian Studies requiring research in the Newberry's collections and will receive a research carrel and be accorded the same privileges as other Newberry long-term fellows. At the time of this writing, the current application submission date for most Newberry short-term fellowships is December 12 and awardees are usually notified in mid- to late-March; however deadlines may vary from year to year. Applications must be submitted as email attachments by 11:59 p.m. on the deadline date. Applicants can find guidelines and forms at the grant website. Applicants are also encouraged to verify current deadline dates.
Requirements: Fellowship recipients must be faculty at institutions participating in NCAIS. At this writing those institutions include Cornell University, Harvard University, Michigan State University, Princeton University, The University of British Columbia, The University of Illinois at Urbana-Champaign, University of Manitoba, University of Minnesota, University of Montana, University of New Mexico, University of North Carolina at Chapel Hill, University of Winnipeg, University of Wisconsin Milwaukee, University of Wyoming, Vanderbilt University, and Yale University. Preference is given to scholars at an early career stage. Faculty Fellows will present research, participate in both the McNickle Center Seminar in American Indian and Indigenous Studies and the Newberry Library Fellows' Seminar, and be available to NCAIS Graduate Student Fellows for consultation. Applicants must request at least four and no more than six months of support.
Restrictions: Applicants may apply for more than one fellowship at the Newberry by designating those for which they wish to be considered on the cover sheet (available at the grant website). Applicants may also apply for both short- and long-term fellowships at the Newberry; however individuals may accept only one fellowship at the Newberry per year. Only scholars who have previously received a short-term fellowship at the Newberry may apply for a long-term fellowship for the same project. No other project extensions are allowable.
Geographic Focus: All States
Date(s) Application is Due: Dec 12
Amount of Grant: 16,800-25,200 USD
Contact: Daniel Greene, Vice President for Research and Academic Programs; (312) 255-3666 or (312) 255-3535; research@newberry.org or greened@newberry.org
Internet: http://www.newberry.org/long-term-fellowships
Sponsor: Newberry Library
60 West Walton Street
Chicago, IL 60610-3380

NL Newberry Consortium on American Indian Studies Graduate Student Fellowships 3312
The Newberry, open to the public without charge, is an independent research library dedicated to the advancement and dissemination of knowledge, especially in the humanities. The library acquires and preserves a broad array of special-collections research materials (spanning six centuries of European and American civilizations) and promotes and provides for these collections' effective use, fostering research, teaching, publication, life-long learning, and civic engagement. The Newberry Consortium on American Indian Studies (NCAIS) offers short-term graduate-student fellowships for doctoral candidates at participating institutions. The fellowships offer support for one to two months of dissertation research and carry stipends of $2,500 per month. NCAIS graduate-student fellows will be offered individual research space at the Newberry and accorded the same privileges as other Newberry short-term fellows. These NCAIS fellowships may also be used to fund research in other libraries, archives, or in the field. At the time of this writing, the current application submission date for most Newberry short-term fellowships is February 6 and awardees are usually notified near the end of April; however deadlines may vary from year to year. Applications must be submitted as email attachments by 11:59 p.m. on the deadline date. Applicants can find guidelines and forms at the grant website. Applicants are also encouraged to verify current deadline dates.
Requirements: Fellowship recipients must be doctoral candidates at institutions participating in NCAIS. At this writing those institutions include Cornell University, Harvard University, Michigan State University, Princeton University, The University of British Columbia, The University of Illinois at Urbana-Champaign, University of Manitoba, University of Minnesota, University of Montana, University of New Mexico, University of North Carolina at Chapel Hill, University of Winnipeg, University of Wisconsin Milwaukee, University of Wyoming, Vanderbilt University, and Yale University. Fellowship recipients are expected to present their research at the NCAIS's annual graduate student conference or at a Newberry-sponsored seminar in American Indian and indigenous studies.
Restrictions: Applicants may apply for more than one fellowship at the Newberry by designating those for which they wish to be considered on the cover sheet (available at the grant website). Applicants may also apply for both short- and long-term fellowships at the Newberry; however individuals may accept only one fellowship at the Newberry per year. Only scholars who have previously received a short-term fellowship at the Newberry may apply for a long-term fellowship for the same project. No other project extensions are allowable.
Geographic Focus: All States, Canada
Date(s) Application is Due: Feb 6
Amount of Grant: 2,500-5,000 USD
Contact: Daniel Greene, Vice president of Research and Education; (312) 255-3666 or (312) 255-3535; research@newberry.org or greened@newberry.org
Internet: http://www.newberry.org/short-term-fellowships
Sponsor: Newberry Library
60 West Walton Street
Chicago, IL 60610-3380

NL Newberry Library/Ecole Nationale des Chartes Exchange Fellowship 3313
The Newberry, open to the public without charge, is an independent research library dedicated to the advancement and dissemination of knowledge, especially in the humanities. The library acquires and preserves a broad array of special-collections research materials (spanning six centuries of European and American civilizations) and promotes and provides for these collections' effective use, fostering research, teaching, publication, life-long learning, and civic engagement. Fellowships at the Newberry provide support for researchers wishing to use the Newberry's collections and are either short-term or long-term. Short-term fellowships are intended to geographically facilitate access for distant scholars to materials essential to a project. Long-term fellowships are intended to support significant works of scholarship that draw on the strengths of the Newberry's collections. Long-term fellowships support individual research and promote serious intellectual exchange through active participation in the Newberry's scholarly activities, including a biweekly fellows' seminar. Applicants may combine long-term fellowship awards with sabbatical or other stipendiary support. The Newberry Library/École Nationale des Chartes Exchange Fellowship (considered a long-term fellowship) has existed since 1979 and provides three months of auditor tuition and a monthly stipend of 2,000 Euros for an American or Canadian Ph.D. student to study at the École Nationale des Chartes in Paris, preferably during the fall semester. The École des Chartes is the oldest institution in Europe specializing in the archival sciences, including paleography, codicology, bibliography, diplomatics, textual editing, and the history of the book. Applications must be submitted as email attachments by 11:59 p.m. on the deadline date. At the time of this writing, the current application-submission date for most Newberry long-term fellowships is December 12 and awardees are usually notified in mid- to late-March; however deadlines may vary from year to year. Applicants are encouraged to verify current deadline dates at the grant website. Application forms and more specific guidelines are available at the grant website.
Requirements: Applications are especially encouraged from students in the early stages of preparing their dissertations. Applicants must provide specific reference to their need for the training available at the École des Chartes and how it will contribute to their dissertation or future projects. Preference is given to students attending member institutions of the Newberry's Center for Renaissance Studies Consortium which are as follows: DePaul University; Eastern Illinois University; Georgia Institute of Technology; Illinois State University; Indiana University; Kent State University; Loyola University; Marquette University; McGill University; Miami University; Michigan State University; Northern Illinois University; Northwestern University, Ohio University; Oklahoma State University; Purdue University; Queen Mary, University of London; Saint Louis University; Southern Illinois University; Tulane University; University of Aberdeen; University of Akron; University of Arizona; University of Chicago; University of Illinois at Chicago; University of Illinois at Urbana-Champaign; University of Iowa; University of Kansas Lawrence; University of Louisville; University of Michigan; University of Minnesota; University of Missouri; University of Nebraska—Lincoln; University of Nevada; University of North Texas; University of Notre Dame; University of Oklahoma;

University of Pittsburgh; University of Tennessee—Knoxville; University of Texas; University of Warwick; University of Waterloo; University of Western Ontario; University of Wisconsin—Madison; University of Wisconsin—Milwaukee; Vanderbilt University; Washington University; Wayne State University.
Geographic Focus: All States, All Countries
Date(s) Application is Due: Dec 12
Amount of Grant: 6,000 USD
Contact: Daniel Greene, Vice President for Research and Academic Programs; (312) 255-3666 or (312) 255-3535; research@newberry.org or greened@newberry.org
Internet: http://www.newberry.org/long-term-fellowships
Sponsor: Newberry Library
60 West Walton Street
Chicago, IL 60610-3380

NL Northeast Modern Language Assocation Fellowship **3314**
The Newberry, open to the public without charge, is an independent research library dedicated to the advancement and dissemination of knowledge, especially in the humanities. The library acquires and preserves a broad array of special-collections research materials (spanning six centuries of European and American civilizations) and promotes and provides for these collections' effective use, fostering research, teaching, publication, life-long learning, and civic engagement. Fellowships at the Newberry provide support for researchers wishing to use the Newberry's collections and are either short-term (one to two months) or long-term (four to twelve months). Short-term fellowships are intended to geographically facilitate access for distant scholars to materials essential to a project. Long-term fellowships are intended to support significant works of scholarship that draw on the strengths of the Newberry's collections. A short-term fellowship is available for Ph.D. candidates or post-doctoral scholars who are members of the Northeast Modern Language Association (NeMLA) for up to one month's work-in-residence at the Newberry. The fellowship provides a stipend of $2,000 per month. Preference will be given to projects focusing on materials written in French, German, Italian, or Spanish. At the time of this writing, the current application submission date for most Newberry short-term fellowships is February 6 and awardees are usually notified near the end of April; however deadlines may vary from year to year. Applications must be submitted as email attachments by 11:59 p.m. on the deadline date. Applicants can find guidelines and forms at the grant website. Applicants are also encouraged to verify current deadline dates.
Requirements: Applicants must be members of the NeMLA at the time of the award. Most short-term fellowships at the Newberry are for Ph.D. candidates and postdoctoral scholars. All fellowships at the Newberry must be conducted in-residence. The Newberry collections must be necessary to the short-term fellowship project since the purpose of these fellowships is to enable access to particular materials. Applications from former recipients of any fellowship at the Newberry are welcome, but preference is given to applicants who have not held a major fellowship within the past three years. In general, all fellowships at the Newberry must be completed by the end of the academic year for which they are awarded.
Restrictions: Most short-term fellowships are restricted to individuals who live outside the Chicago area who need to examine specific items in the Newberry's collections. Preference is given to residents of the Northeastern United States or Eastern Canada; Applicants may apply for more than one fellowship at the Newberry by designating those for which they wish to be considered on the cover sheet (available at the grant website). Applicants may also apply for both short- and long-term fellowships at the Newberry; however individuals may accept only one fellowship at the Newberry per year. Only scholars who have previously received a short-term fellowship at the Newberry may apply for a long-term fellowship for the same project. No other project extensions are allowable.
Geographic Focus: All States, Canada
Date(s) Application is Due: Feb 6
Amount of Grant: Up to 2,000 USD
Contact: Daniel Greene, Vice President for Research and Academic Programs; (312) 255-3666; research@newberry.org or greened@newberry.org
Internet: http://www.newberry.org/short-term-fellowships
Sponsor: Newberry Library
60 West Walton Street
Chicago, IL 60610-3380

NL Scholar-in-Residence Opportunity **3315**
The Newberry, open to the public without charge, is an independent research library dedicated to the advancement and dissemination of knowledge, especially in the humanities. The library acquires and preserves a broad array of special-collections research materials (spanning six centuries of European and American civilizations) and promotes and provides for these collections' effective use, fostering research, teaching, publication, life-long learning, and civic engagement. The Scholar-in-Residence program at the Newberry encourages postdoctoral scholars in the humanities to conduct research in the library's collections and to join its community of scholars. Selected scholars are provided with reserve space, access to the Newberry during extended hours, and opportunities to present their work-in-progress to the Newberry's scholarly community. Applications to become a Scholar-in-Residence are accepted each year in May. Candidates must submit via email a project description along with a filled-in application form (available for download from the website) by 11:59 p.m. on the deadline date. Applicants will be notified of their acceptance in June and will begin their residence in September and will join the Newberry's new long-term fellows in fall-orientation activities. Candidates can find more detailed guidelines and information at the grant website.
Requirements: Applicants must be postdoctoral scholars in the humanities.
Geographic Focus: Illinois
Date(s) Application is Due: May 1
Contact: Daniel Greene, Executive Vice President for Research and Academic Programs; (312) 255-3666 or (312) 255-3535; research@newberry.org or greened@newberry.org
Internet: http://www.newberry.org/scholar-residence-program
Sponsor: Newberry Library
60 West Walton Street
Chicago, IL 60610-3380

NL Short-Term Fellowships in the History of Cartography **3316**
The Newberry, open to the public without charge, is an independent research library dedicated to the advancement and dissemination of knowledge, especially in the humanities. The library acquires and preserves a broad array of special-collections research materials (spanning six centuries of European and American civilizations) and promotes and provides for these collections' effective use, fostering research, teaching, publication, life-long learning, and civic engagement. Fellowships at the Newberry provide support for researchers wishing to use the Newberry's collections and are either short-term (one to two months) or long-term (four to twelve months). Short-term fellowships are intended to geographically facilitate access for distant scholars to materials essential to a project. Long-term fellowships are intended to support significant works of scholarship that draw on the strengths of the Newberry's collections. The Short-Term Fellowship in the History of Cartography supports work-in-residence at the Newberry. Projects must be related to the history of cartography and focus on cartographic materials in the Newberry's collection. The fellowship provides a stipend of $2,000 per month. At the time of this writing, the current application submission date for most Newberry short-term fellowships is February 6 and awardees are usually notified near the end of April; however deadlines may vary from year to year. Applications must be submitted as email attachments by 11:59 p.m. on the deadline date. Applicants can find guidelines and forms at the grant website. Applicants are also encouraged to verify current deadline dates.
Requirements: Most short-term fellowships are for either Ph.D. candidates or postdoctoral scholars. All fellowships at the Newberry must be conducted in-residence. The Newberry collections must be necessary to the short-term fellowship project since the purpose of these fellowships is to enable access to particular materials. Applications from former recipients of any fellowship at the Newberry are welcome, but preference is given to applicants who have not held a major fellowship within the past three years. In general, all fellowships at the Newberry must be completed by the end of the academic year for which they are awarded.
Restrictions: Most short-term fellowships are restricted to individuals who live outside the Chicago area who need to examine specific items in the Newberry's collections. Applicants may apply for more than one fellowship at the Newberry by designating those for which they wish to be considered on the cover sheet (available at the grant website). Applicants may also apply for both short- and long-term fellowships at the Newberry; however individuals may accept only one fellowship at the Newberry per year. Only scholars who have previously received a short-term fellowship at the Newberry may apply for a long-term fellowship for the same project. No other project extensions are allowable.
Geographic Focus: All States, All Countries
Date(s) Application is Due: Feb 6
Amount of Grant: 2,000-4,000 USD
Contact: Daniel Greene, Vice President for Research and Academic Programs; (312) 255-3666 or (312) 255-3535; research@newberry.org or greened@newberry.org
Internet: http://www.newberry.org/short-term-fellowships
Sponsor: Newberry Library
60 West Walton Street
Chicago, IL 60610-3380

NL Short-Term Resident Fellowships for Individual Research **3317**
The Newberry, open to the public without charge, is an independent research library dedicated to the advancement and dissemination of knowledge, especially in the humanities. The library acquires and preserves a broad array of special-collections research materials (spanning six centuries of European and American civilizations) and promotes and provides for these collections' effective use, fostering research, teaching, publication, life-long learning, and civic engagement. Fellowships at the Newberry provide support for researchers wishing to use the Newberry's collections and are either short-term (one to two months) or long-term (four to twelve months). Short-term fellowships are intended to geographically facilitate access for distant scholars to materials essential to a project. Long-term fellowships are intended to support significant works of scholarship that draw on the strengths of the Newberry's collections. Newberry's Short-Term Resident Fellowships provide access to the Newberry's collections for Ph.D. candidates or post-doctoral scholars who live and work outside the Chicago area. Normally, these fellowships are awarded to individual scholars. However, the Newberry will award a small number of fellowships for collaborative projects. The fellowships provide a stipend of $2,000 per month (for each scholar). At the time of this writing, the current application submission date for most Newberry short-term fellowships is February 6 and awardees are usually notified near the end of April; however deadlines may vary from year to year. Applications must be submitted as email attachments by 11:59 p.m. on the deadline date. Applicants can find guidelines and forms at the grant website. Applicants are also encouraged to verify current deadline dates.
Requirements: Ph.D. candidates and post-doctoral scholars who live and work outside the Chicago area are eligible to apply. Teams of two or three scholars who plan to collaborate on a single, substantive project are also eligible to apply. All fellowships at the Newberry must be conducted in-residence. The Newberry collections must be necessary to the short-term fellowship project since the purpose of these fellowships is to enable access to particular materials. Applications from former recipients of any fellowship at the Newberry are welcome, but preference is given to applicants who have not held a major fellowship within the past three years. In general, all fellowships at the Newberry must be completed by the end of the academic year for which they are awarded.

Restrictions: Most short-term fellowships are restricted to individuals who live outside the Chicago area who need to examine specific items in the Newberry's collections. Applicants may apply for more than one fellowship at the Newberry by designating those for which they wish to be considered on the cover sheet (available at the grant website). Applicants may also apply for both short- and long-term fellowships at the Newberry; however individuals may accept only one fellowship at the Newberry per year. Only scholars who have previously received a short-term fellowship at the Newberry may apply for a long-term fellowship for the same project. No other project extensions are allowable.
Geographic Focus: All States, All Countries
Date(s) Application is Due: Feb 6
Amount of Grant: 2,000-6,000 USD
Contact: Daniel Greene, Vice President for Research and Academic Programs; (312) 255-3666 or (312) 255-3535; research@newberry.org or greened@newberry.org
Internet: http://www.newberry.org/short-term-fellowships
Sponsor: Newberry Library
60 West Walton Street
Chicago, IL 60610-3380

NL South Central Modern Languages Association Fellowship **3318**

The Newberry, open to the public without charge, is an independent research library dedicated to the advancement and dissemination of knowledge, especially in the humanities. The library acquires and preserves a broad array of special-collections research materials (spanning six centuries of European and American civilizations) and promotes and provides for these collections' effective use, fostering research, teaching, publication, life-long learning, and civic engagement. Fellowships at the Newberry provide support for researchers wishing to use the Newberry's collections and are either short-term (one to two months) or long-term (four to twelve months). Short-term fellowships are intended to geographically facilitate access for distant scholars to materials essential to a project. Long-term fellowships are intended to support significant works of scholarship that draw on the strengths of the Newberry's collections. A one-month short-term fellowship is available for postdoctoral scholars who are members of the South Central Modern Languages Association (SCMLA) for work-in-residence at the Newberry. The fellowship provides a stipend of $2,000 per month. At the time of this writing, the current application submission date for most Newberry short-term fellowships is February 6 and awardees are usually notified near the end of April; however deadlines may vary from year to year. Applications must be submitted as email attachments by 11:59 p.m. on the deadline date. Applicants can find guidelines and forms at the grant website. Applicants are also encouraged to verify current deadline dates.
Requirements: Applicants must be postdoctoral scholars and at the time of the award, be members of SCMLA. All fellowships at the Newberry must be conducted in-residence. The Newberry collections must be necessary to the short-term fellowship project since the purpose of these fellowships is to enable access to particular materials. Applications from former recipients of any fellowship at the Newberry are welcome, but preference is given to applicants who have not held a major fellowship within the past three years. In general, all fellowships at the Newberry must be completed by the end of the academic year for which they are awarded.
Restrictions: Most short-term fellowships are restricted to individuals who live outside the Chicago area who need to examine specific items in the Newberry's collections. Applicants may apply for more than one fellowship at the Newberry by designating those for which they wish to be considered on the cover sheet (available at the grant website). Applicants may also apply for both short- and long-term fellowships at the Newberry; however individuals may accept only one fellowship at the Newberry per year. Only scholars who have previously received a short-term fellowship at the Newberry may apply for a long-term fellowship for the same project. No other project extensions are allowable.
Geographic Focus: All States
Date(s) Application is Due: Feb 6
Amount of Grant: 2,000 USD
Contact: Daniel Greene, Vice President for Research and Academic Programs; (312) 255-3666 or (312) 255-3535; research@newberry.org or greened@newberry.org
Internet: http://www.newberry.org/
Sponsor: Newberry Library
60 West Walton Street
Chicago, IL 60610-3380

NL Susan Kelly Power and Helen Hornbeck Tanner Fellowship **3319**

The Newberry, open to the public without charge, is an independent research library dedicated to the advancement and dissemination of knowledge, especially in the humanities. The library acquires and preserves a broad array of special-collections research materials (spanning six centuries of European and American civilizations) and promotes and provides for these collections' effective use, fostering research, teaching, publication, life-long learning, and civic engagement. Fellowships at the Newberry provide support for researchers wishing to use the Newberry's collections and are either short-term (one to two months) or long-term (four to twelve months). Short-term fellowships are intended to geographically facilitate access for distant scholars to materials essential to a project. Long-term fellowships are intended to support significant works of scholarship that draw on the strengths of the Newberry's collections. The Susan Kelly Power and Helen Hornbeck Tanner Fellowship is a short-term fellowship for Ph.D. candidates and post-doctoral scholars of American Indian heritage. The fellowship supports up to two months of residential research in any field in the humanities using the collections of the Newberry at a stipend of $2,000 per month. At the time of this writing, the current application submission date for most Newberry short-term fellowships is February 6 and awardees are usually notified near the end of April; however deadlines may vary from year to year. Applications must be submitted as email attachments by 11:59 p.m. on the deadline date. Applicants can find guidelines and forms at the grant website. Applicants are also encouraged to verify current deadline dates.
Requirements: Most short-term fellowships at the Newberry are for Ph.D. candidates and postdoctoral scholars. All fellowships at the Newberry must be conducted in-residence. The Newberry collections must be necessary to the short-term fellowship project since the purpose of these fellowships is to enable access to particular materials. Applications from former recipients of any fellowship at the Newberry are welcome, but preference is given to applicants who have not held a major fellowship within the past three years. In general, all fellowships at the Newberry must be completed by the end of the academic year for which they are awarded.
Restrictions: While most short-term fellowships at the Newberry are restricted to individuals who live outside the Chicago area and who need to examine specific items in the Newberry's collections, applicants for the Susan Kelly Power and Helen Hornbeck Tanner Fellowship may be local. Applicants may apply for more than one fellowship at the Newberry by designating those for which they wish to be considered on the cover sheet (available at the grant website). Applicants may also apply for both short- and long-term fellowships at the Newberry; however individuals may accept only one fellowship at the Newberry per year. Only scholars who have previously received a short-term fellowship at the Newberry may apply for a long-term fellowship for the same project. No other project extensions are allowable.
Geographic Focus: All States
Date(s) Application is Due: Jan 16
Amount of Grant: Up to 4,000 USD
Contact: Daniel Greene, Vice President for Research and Academic Programs; (312) 255-3666 or (312) 255-3535; research@newberry.org or greened@newberry.org
Internet: http://www.newberry.org/short-term-fellowships
Sponsor: Newberry Library
60 West Walton Street
Chicago, IL 60610-3380

NL Weiss/Brown Publication Subvention Award **3320**

The Newberry, open to the public without charge, is an independent research library dedicated to the advancement and dissemination of knowledge, especially in the humanities. The library acquires and preserves a broad array of special-collections research materials (spanning six centuries of European and American civilizations) and promotes and provides for these collections' effective use, fostering research, teaching, publication, life-long learning, and civic engagement. With support from the Roger W. Weiss and Howard Mayer Brown Fund, the Newberry will award up to $15,000 to subsidize the publication of a scholarly book, monograph, or edition on European civilization before 1700 in one of the following fields: music, theater, French or Italian literature, or cultural studies. This award aspires to enable the publication of works of the highest quality either by publishing them in a particularly appropriate way (with special typography, plates, or appendices that would otherwise be prohibitively expensive, for example) or by significantly reducing the cover price, thereby allowing the publication to reach a wider audience. Once these criteria are met, preference will be given to publications that are unique, that are unusual in concept or execution, that bring into print previously unpublished source material, that represent a departure from the normal habits of a given publishing house or entity, or that promise to reach the broadest possible audience for the type of book envisioned. A link to application guidelines is available at the grant website. Applications are due via email by the deadline date. Awardees will be notified six to eight weeks after the deadline. Deadlines may vary from year to year; applicants are encouraged to verify current deadline dates.
Requirements: Applications for the Weiss/Brown Award will require the following documents: the author's curriculum vitae; a copy of the publication contract or other evidence of commitment by the publisher to publish the book; copies of the reader's reports on the manuscript, or equivalent outside evaluations; and a production budget and schedule. The award will be paid to the publisher; as a condition of accepting funds, publishers must include an acknowledgement of the award in the publication and provide the Newberry Library with a gratis copy of the book.
Restrictions: Faxed applications will not be accepted.
Geographic Focus: All States, All Countries
Date(s) Application is Due: Feb 6
Amount of Grant: Up to 15,000 USD
Contact: Daniel Greene, Vice President for Research and Academic Programs; (312) 255-3666 or (312) 255-3535; research@newberry.org or greened@newberry.org
Internet: http://www.newberry.org/short-term-fellowships
Sponsor: Newberry Library
60 West Walton Street
Chicago, IL 60610-3380

NMF California Community Service Scholarships **3321**

The program offers scholarships to third- and fourth-year students attending medical schools in California. Scholarships are awarded based on demonstrated commitment to practice in California, interest in community-based primary care, satisfactory academic performance, leadership, and financial need. Students are encouraged to apply for the scholarship program during their second or third year of medical school. The goal of the program is to expose students who are interested in receiving first-hand exposure to community-based primary care for the equivalent of six weeks; or conducting research for the equivalent of at least eight to 12 weeks in areas of critical need (e.g., HIV/AIDS care and research, hypertension, tuberculosis, cardiovascular disease, diabetes, asthma, substance abuse, women's health research, etc.). It also fosters their commitment to practicing in California. The scholarships will be awarded on the basis of personal commitment to practice in California, demonstrated interest in community-based primary care or research, satisfactory academic performance, and leadership. Priority will be given to NMF scholars.
Requirements: African-Americans, mainland Puerto Ricans, Mexican-Americans, Native Hawaiians, Alaska Natives, and American Indians who are U.S. citizens attending

California medical schools. Eligibility is limited to rising third-year or fourth-year students. The required application is available at the sponsor's website.
Restrictions: Only students attending California medical schools are ebigible for the program.
Geographic Focus: California
Amount of Grant: 15,000 USD
Contact: Mary Wade; (415) 397-2526; marywade@nmfonline.org
Internet: http://www.nmfonline.org/Programs/Scholarships/CaliforniaCSSP/Overview.htm
Sponsor: National Medical Fellowships
5 Hanover Square
New York, NY 10004

NMF General Electric Medical Scholars Fellowships **3322**
The sponsor has established a partnership with General Electric (GE) to provide need-based scholarship support to historically underrepresented minority medical students, and to allow them to engage in clinical work in underserved communities. The program also offers eligible students an opportunity to complete a two-month elective in Ghana, West Africa focusing on areas of critical need such as HIV/AIDS, diabetes, tuberculosis, heart disease, cancer and infant mortality. Five fellowships are awarded annually. Students will be matched with a mentor for the period of the program. Travel, hosing and a stipend will be provided for students.
Requirements: Students must be nominated by their medical school deans. Eligible candidates must be underrepresented minority students enrolled in their fourth year in an accredited U.S. medical school. Candidates must demonstrate outstanding academic achievement, leadership, and potential for community-based work or initiation of innovative projects in school or community. Additionally, candidates must also demonstrate potential for distinguished contributions to medicine. Complete application requirements can be found at the sponsor's website.
Geographic Focus: All States
Contact: Dr. Stephen Keith, President; (240) 243-8000; fax (240) 465-0450
Internet: http://www.nmfonline.org/index.html
Sponsor: National Medical Fellowships
5 Hanover Square
New York, NY 10004

NMF National Medical Association Awards for Medical Journalism **3323**
In general, these awards recognize academic achievement, leadership, and potential for distinguished contributions to medicine. Specifically, these awards recognize demonstrated skills in journalism. Deadlines vary-contact the sponsor for dates.
Requirements: Eligibility is limited to African-American medical students. Each medical school may submit up to two nominations. Candidates must be in the third or fourth year of medical school and must have published articles and photographs in, or been writers, editors, photographers on the staffs of: medical school newspapers; medical students' journals; recognized professional journals; or, other accredited scientific journal. Students who have written, produced or directed health-related films, commercials or videos are also encouraged to submit their applications and samples of their work. Applicants must also submit samples of their journalistic work (articles, photographs, audio tapes, video tapes, etc.) for review by the selection committee. Entries must have been published or produced within the last three years. The application form and more detailed guidelines can be found at the sponsor's website.
Geographic Focus: All States
Amount of Grant: 2,500 USD
Contact: Programs; (212) 483-8880; fax (212) 239-9718; nmf2@nmfonline.org
Internet: http://www.nmfonline.org/index.html
Sponsor: National Medical Fellowships
5 Hanover Square
New York, NY 10004

NMF National Medical Association Emerging Scholar Awards **3324**
The National Medical Association and NMF sponsor need-based award programs that recognize and reward African American medical students for special achievements, academic excellence, leadership, and potential for outstanding contributions to medicine. The NMA Emerging Scholar Awards are presented to current first, second, and third-year medical students.
Requirements: Eligibility for these awards is limited to African-American medical students. Each school may submit two nominations. Applicants must demonstrate and document financial need. Deadlines vary-contact the sponsor for specific dates.
Geographic Focus: All States
Amount of Grant: 2,250 USD
Contact: Programs; (212) 483-8880; fax (212) 483-8897; nmf1@nmfonline.org
Internet: http://www.nmfonline.org/index.html
Sponsor: National Medical Fellowships
5 Hanover Square
New York, NY 10004

NMF National Medical Association Patti LaBelle Award **3325**
This award is one of the highest honors presented to students by the National Medical Association. It recognizes academic achievement, leadership and potential for distinguished contributions to medicine. This award is presented to current first-, second-, and third-year medical students with demonstrated financial need.
Requirements: Eligibility is limited to African-American medical students. Application materials must include a letter of recommendation that fully discusses the candidate's academic performance, extracurricular activities and other qualifications; a complete scholarship application form; an essay of at least 500 words discussing the applicant's motivation for a career in medicine and plans over the next ten years; and, an official academic transcript.
Geographic Focus: All States
Amount of Grant: 5,000 USD
Contact: Programs; (212) 483-8880; fax (212) 483-8897; nmf1@nmfonline.org
Internet: http://www.nmfonline.org/index.html
Sponsor: National Medical Fellowships
5 Hanover Square
New York, NY 10004

NMF Need-Based Scholarships **3326**
The program provides financial assistance to United States citizens from groups currently underrepresetned in the medical profession; specifically, African-Americans, Mexican-Americans, Native Americans, Alaska Natives, Native Hawaiians, and mainland Puerto Ricans who permanently reside within the 50 U.S. states. These awards are offered to first- and second-year medical students. Scholarships are granted on the basis of financial need as determined by the student's total resources (including parental and spousal support), cost of education, and receipt of other scholarships and grants. Awards have ranged from $500 to $10,000.
Requirements: Applicants must be U.S. citizens by the application deadline and must have been accepted to AAMC or AOA-accredited U.S. medical schools for study leading to M.D. or D.O. degrees. The application form and FAQs can be found at the sponsor's website. Applicants must demonstrate and document financial need by submitted complete copies of their parents', spouse's and their own most recent 1040, 1040A or 1040EZ tax forms. All nontaxable income (e.g. AFDC, ADC, Social Security benefits, etc.) must also be documented by the appropriate agency. Applicants with neither taxable nor nontaxable income must provide verification of means of support. All parental financial data must be submitted in order to evaluate dependent status.
Geographic Focus: All States
Date(s) Application is Due: Aug 31
Amount of Grant: 500-10,000 USD
Contact: Programs Department; (212) 483-8880, ext. 304; fax (212) 483-8897; NMF1@NMFonline.org or info@NMFonline.org
Internet: http://www.nmfonline.org/programs.html
Sponsor: National Medical Fellowships
5 Hanover Square
New York, NY 10004

NMF Ralph W. Ellison Prize **3327**
The prize is presented to a graduating underrepresented medical student who has demonstrated outstanding academic achievement, leadership, and potential to make significant contributions to medicine. One prize is presented each year (nonrenewable) and includes a certificate of merit and a stipend. Annual deadline date may vary; contact program staff for exact dates.
Requirements: African-Americans, mainland Puerto Ricans, Mexican-Americans, Native Hawaiians, Alaska Natives, and American Indians who are U.S. citizens attending M.D. degree-granting institutions accredited by the Liaison Committee on Medical Education of the Association of American Medical Colleges, or in D.O. degree-granting colleges of osteopathic medicine accredited by the Bureau of Professional Education of the American Osteopathic Association are eligible.
Geographic Focus: All States
Amount of Grant: 500 USD
Contact: Programs; (212) 483-8880; fax (212) 483-8897; nmf1@nmfonline.org
Internet: http://www.nmfonline.org/index.html
Sponsor: National Medical Fellowships
5 Hanover Square
New York, NY 10004

NMSF Ernest F. Hollings Ocean Awareness Trust Fund Grants **3328**
The Foundation builds partnerships to promote conservation and develop an ocean literate public who is able to make informed and responsible decisions regarding the ocean and its resources. This fund was established in 2005 as a legacy to the longstanding commitment of former U.S. Senator Ernest F. Hollings to promoting public involvement and understanding of the benefits of the oceans to current and future generations. The Ocean Awareness Trust Fund seeks to continue this important work by ensuring that all Americans understand and value these precious resources, and by promoting opportunities for the people of all ages to become actively interested and involved in ocean stewardship. Special consideration will be given to projects that address marine debris, human interaction with marine mammals, runoff and pollution or public health issues such as red tides or marine pharmaceuticals.
Requirements: Eligible applicants are formal or informal education institutions, nonprofits, commercial organizations, and state, local and Indian tribal governments.
Geographic Focus: All States
Date(s) Application is Due: Aug 17
Amount of Grant: 10,000-25,000 USD
Contact: Allison Alexander, Grants Manager; (301) 608-3040; fax (301) 608-3044; Allison@nmsfocean.org or grants@NMSFocean.org
Internet: http://nmsfocean.org/grants.asp
Sponsor: National Marine Sanctuary Foundation
8601 Georgia Avenue, Suite 501
Silver Spring, MD 20910

NOAA Abrupt Climate Change Studies (ARCHES) Grants **3329**

The office invites proposals that build on the ARCHES program model-a multidisciplinary, single-institution effort that advances NOAA's abrupt climate change research interests (including paleo-drought) along three timescales: the era of anthropogenic influence, which includes the modern period of instrumental data as well as the future; the last 2000 years; and the Holocene. It is expected that proposals will describe an integrated research program that incorporates a variety of paleoproxy and modern datasets, has strong modeling, diagnostics and synthesis components and also incorporates a research thrust to study the past and future impacts of abrupt climate change. Due to the level of funding available, new observations are not encouraged, unless they are highly-leveraged and greatly enhance the overall effort. It is anticipated that the successful proposal will be from a single institution that can demonstrate a strong, collocated multi-disciplinary research program. Consideration will be given to an institution wishing to link its effort with one of the several NOAA Cooperative Institutes as well as the NOAA Regional Integrated Science and Assessments program. While it is in the best interest of the applicants and their institutions to submit a Letter of Intent (LOI), it is not a requirement; applicants who do not submit an LOI are allowed to submit a full proposal. For those that do submit an LOI, full proposals will be encouraged only for LOIs deemed relevant.
Requirements: Eligible applicants are institutions of higher education, other nonprofits, commercial organizations, international organizations, and state, local and Indian tribal governments.
Restrictions: Federal agencies or institutions are not eligible to receive Federal assistance under this notice.
Geographic Focus: All States
Date(s) Application is Due: Sep 24
Amount of Grant: 2,000,000 USD
Contact: James Todd; (301) 734-1258; fax (301) 713-0518; james.todd@noaa.gov
Internet: http://www.climate.noaa.gov/index.jsp?pg=/opportunities/opp_index.jsp&opp=2008/acchange_info.jsp
Sponsor: National Oceanic and Atmospheric Administration
1315 East-West Highway, 12th Floor
Silver Spring, MD 20910

NOAA Arctic Research Grants 3330

The goal of the Arctic Research Program is to provide climate-relevant observations and analysis of the broader Arctic region, with an emphasis on the Pacific sector of the Arctic, to support improved climate projections and assessment of impacts in the Arctic and regions influenced by the Arctic. The Program is soliciting proposals in the following area: participation in the Russian-American Long-term Census of the Arctic (RU.S.ALCA) program in FY2008 and beyond. All proposals should display proposed start dates of April 1, 2008, and extend to March 31, 2013 for a total of 5 years.
Requirements: Eligible applicants are institutions of higher education, other nonprofits, commercial organizations, international organizations, and state, local and Indian tribal governments. Proposals will be accepted only from investigators that responded to the previous call for proposals for RU.S.ALCA issued by CIFAR and who were selected for future funding, or from the existing RU.S.ALCA logistics providers.
Restrictions: Federal agencies or institutions are not eligible to receive Federal assistance under this notice.
Geographic Focus: All States
Date(s) Application is Due: Sep 24
Contact: Kathy Crane, (301) 734-1212; fax (301) 713-0517; Kathy.Crane@noaa.gov
Internet: http://www.climate.noaa.gov/index.jsp?pg=/opportunities/opp_index.jsp&opp=2008/arp_info.jsp
Sponsor: National Oceanic and Atmospheric Administration
1315 East-West Highway, 12th Floor
Silver Spring, MD 20910

NOAA Assessing Atlantic Meridional Overturning Circulation Grants 3331

The AMOC near-term priority is designed to take advantage of rapidly advancing observing and assimilation capabilities as well as leverage substantial international investment. The program will include activities in the following areas: fundamental research (e.g. description of the AMOC, its variability, critical processes); observing and monitoring the AMOC (e.g. design and development, stewardship of data); ocean analyses (e.g. improving ocean models and historical ocean analyses); development of nowcasting capabilities and experimental products; and assessment of potential impacts of rapid AMOC changes. Letters of Intent (LOI) will be due November 2, with responses from program management by November 7. Full proposals will be due December 7, with funding to begin August 1.
Requirements: Eligible applicants are institutions of higher education, other nonprofits, commercial organizations, international organizations, and state, local and Indian tribal governments.
Restrictions: Federal agencies or institutions are not eligible to receive Federal assistance under this notice.
Geographic Focus: All States
Date(s) Application is Due: Nov 2; Dec 7
Amount of Grant: 50,000-200,000 USD
Contact: James Todd; (301) 734-1258; fax (301) 713-0518; james.todd@noaa.gov
Internet: http://www.climate.noaa.gov/index.jsp?pg=/opportunities/opp_index.jsp&opp=2008/amoc_info.jsp
Sponsor: National Oceanic and Atmospheric Administration
1315 East-West Highway, 12th Floor
Silver Spring, MD 20910

NOAA Atlantic Salmon Conservation Grants 3332

NOAA Atlantic Salmon Conservation Grants provide funding to catalyze the implementation of locally driven projects that address priority needs for Atlantic salmon recovery and restoration. Through these grants, NOAA provides funding and technical assistance to support restoration of habitat connectivity and function for the benefit of Atlantic salmon within their current and historical range in New England. Funded projects have strong on-the-ground habitat restoration or protection components that lead to long-term ecological habitat improvements for Atlantic salmon, and also provide educational and social benefits related to Atlantic salmon conservation for citizens and their communities. Proposals selected for funding through this solicitation will be implemented through a cooperative agreement. Funding of up to $2,500,000 is expected to be available for Atlantic Salmon Conservation Grants. Typical awards will range from $50,000 to $250,000. Although a select few may fall outside of this typical award range, project proposals requesting less than $30,000 or more than $500,000 will not be accepted.
Requirements: Eligible applicants are institutions of higher education, other non-profits, industry and commercial (for profit) organizations, organizations under the jurisdiction of foreign governments, international organizations, and state, local and Indian tribal governments whose projects have the potential to benefit NOAA trust resources.
Restrictions: Applications from federal agencies or employees of federal agencies will not be considered.
Geographic Focus: All States
Date(s) Application is Due: Nov 14
Amount of Grant: 30,000-500,000 USD
Contact: Craig Woolcott, (301) 713-0174; fax (301) 713-0184; Craig.Woolcott@noaa.gov
Internet: http://www.grants.gov/search/search.do;jsessionid=LP0hqWfp11PVPKSr9yfL1LLhNQTn2gsNBmYzZtzGKTQhJRdRBYxW!-2005900814?oppId=42303&flag2006=false&mode=VIEW
Sponsor: U.S. Department of Commerce
1401 Constitution Avenue, NW, Room 5128
Washington, DC 20230

NOAA Atmospheric Composition and Climate Program (ACC) Grants 3333

The program pursues two overall research objectives: to improve the predictive understanding of the radiative forcing of the climate system by aerosols and by chemically active greenhouse gases, and to better characterize the recovery of the stratospheric ozone layer, including its role in climate change. The integrated research activities that address these objectives involve instrument 'developmentegional' to global observations, laboratory studies, and theoretical modeling by NOAA and extramural partners. While it is in the best interest of the applicants and their institutions to submit a Letter of Intent (LOI), it is not a requirement; applicants who do not submit an LOI are allowed to submit a full proposal. For those that do submit an LOI, full proposals will be encouraged only for LOIs deemed relevant.
Requirements: Eligible applicants are institutions of higher education, other nonprofits, commercial organizations, international organizations, and state, local and Indian tribal governments.
Restrictions: Federal agencies or institutions are not eligible to receive Federal assistance under this notice.
Geographic Focus: All States
Date(s) Application is Due: Sep 24
Amount of Grant: 50,000-200,000 USD
Contact: Kea Duckenfield; (301) 734-1216; kea.duckenfield@noaa.gov
Internet: http://www.climate.noaa.gov/index.jsp?pg=/opportunities/opp_index.jsp&opp=2008/acc_info.jsp
Sponsor: National Oceanic and Atmospheric Administration
1315 East-West Highway, 12th Floor, Room 12707
Silver Spring, MD 20910

NOAA Ballast Water Management Demonstration Program (RDTE Facility) Grants 3334

The Program supports projects to develop, test, and demonstrate ballast water technologies and practices that reduce the threat of introduction of aquatic invasive species to U.S. waters through the discharge of ballast water. The ultimate mission of the program is the implementation and use of effective treatments. The Program's priority is to support establishment of RDTE facilities that demonstrate: a clear statement of the purpose and goals of the facility; appropriate physical and organizational infrastructure to accomplish their purpose; access to appropriate expertise in multiple disciplines; ability to provide fair and unbiased decisions and advice concerning what technologies or ballast water management strategies will be addressed and what activities will be undertaken; active support of state regulatory and maritime private sector interests; adequate quality control and accountability procedures; and the financial ability to maintain the facility even in the event that NOAA funding is less than expected or unavailable after the grant period.
Requirements: Only those applicants who submit Letters of Intent by the preliminary proposal deadline and who submit LOIs that meet the other requirements are eligible to submit full proposals. Eligible applicants are individuals, institutions of higher education, nonprofit organizations, for-profit organizations, Federal, State, local and Indian tribal governments, foreign governments, organizations under the jurisdiction of foreign governments, and international organizations.
Restrictions: Applications from non-Federal and eligible Federal applicants (including NOAA and MARAD employees) will be evaluated in the same selection process.
Geographic Focus: All States
Date(s) Application is Due: Apr 24
Amount of Grant: 500,000 USD
Contact: Dorn Carlson, Grants Coordinator; (301) 734-1080; ballast.water@noaa.gov

Internet: http://www07.grants.gov/search/search.do;jsessionid=HWbpMDzmZ104MfvCGSbHhsGGPnb8dGTRG2hM0vfdTvpFRTZdkPyK!-796138783?oppId=16384&flag2006=true&mode=VIEW
Sponsor: National Oceanic and Atmospheric Administration
1315 East-West Highway, 12th Floor
Silver Spring, MD 20910

NOAA Ballast Water Management Demonstration Program (Technologies and Practices) Grants 3335

The Ballast Water Management Demonstration Program supports projects to develop, test, and demonstrate ballast water technologies and practices that reduce the threat of introduction of aquatic invasive species to U.S. waters through the discharge of ballast water. The ultimate mission of the program is the implementation and use of effective treatments. This grant program supports projects to develop, test, and demonstrate ballast water technologies and practices in order to reduce the threat of introduction of aquatic invasive species to U.S. waters through the discharge of ballast water. This FFO is for projects dealing with technologies and practices necessary for successful ballast water management. For example, this could include technologies or practices that treat ballast water, or that make it possible to test, evaluate, regulate or use ballast water treatment technologies or practices. Letters of Intent must be received by the National Sea Grant Office by 4:00 PM EST February 21. Full proposals must be received by 4:00 PM EDT April 3. Proposals selected for funding from non-Federal applicants will be funded through project grants or cooperative agreements.
Requirements: Only those applicants who submit Letters of Intent by the preliminary proposal deadline and who submit LOIs that meet the other requirements are eligible to submit full proposals. Eligible applicants are individuals, institutions of higher education, nonprofit organizations, for-profit organizations, Federal, State, local and Indian tribal governments, foreign governments, organizations under the jurisdiction of foreign governments, and international organizations.
Restrictions: Applications from non-Federal and eligible Federal applicants (including NOAA and MARAD employees) will be evaluated in the same selection process.
Geographic Focus: All States
Date(s) Application is Due: Apr 3
Amount of Grant: 200,000 USD
Contact: Dorn Carlson, Grants Coordinator; (301) 734-1080; ballast.water@noaa.gov
Internet: http://www07.grants.gov/search/search.do;jsessionid=HWbpMDzmZ104MfvCGSbHhsGGPnb8dGTRG2hM0vfdTvpFRTZdkPyK!-796138783?oppId=16385&flag2006=true&mode=VIEW
Sponsor: National Oceanic and Atmospheric Administration
1315 East-West Highway, 12th Floor
Silver Spring, MD 20910

NOAA Ballast Water Technology Demonstration Program (Research, Development, Testing and Evaluation Facility) Grants 3336

The Program supports projects to develop, test, and demonstrate technologies that treat ships' ballast water in order to reduce the threat of introduction of aquatic invasive species to U.S. waters through the discharge of ballast water. The purpose of the Program is to foster the development of technologies that: are effective at removing, inactivating, or preventing the transfer of aquatic organisms in the ballast tanks of ships, including in the ballast water or associated sediments; are practicable from the standpoint of ship operations, safety, and environmental protection; have the ability to meet all applicable regulatory requirements; and have the potential to be a commercially viable product. The Program has funded, and continues to fund, grants for one- and two-year ballast water treatment technology demonstration projects. Preliminary proposals are due on October 27th, with final proposals due on February 10th.
Requirements: Eligible applicants include: individuals; institutions of higher education; nonprofit organizations; for-profit organizations; Federal, State, local and Indian tribal governments; foreign governments; organizations under the jurisdiction of foreign governments; and international organizations. Applications from non-Federal and eligible Federal applicants (including NOAA employees) will be evaluated in the same selection process.
Geographic Focus: All States
Date(s) Application is Due: Feb 10; Oct 27
Amount of Grant: 950,000 USD
Contact: Dorn Carlson, Grants Coordinator; (301) 734-1080; ballast.water@noaa.gov
Internet: http://www07.grants.gov/search/search.do;jsessionid=HXXNv3Mt6rGXD7y5CX8dvQqGkVgGF8H6Hw5BGwTw1BFBfBQs3fdn!1282164016?oppId=1562&flag2006=true&mode=VIEW
Sponsor: National Oceanic and Atmospheric Administration
1315 East-West Highway, 12th Floor
Silver Spring, MD 20910

NOAA Ballast Water Technology Demonstration Program (Treatment Technology Demonstration Project) Grants 3337

The Program supports projects to develop, test, and demonstrate technologies that treat ships' ballast water in order to reduce the threat of introduction of aquatic invasive species to U.S. waters through the discharge of ballast water. This Program is only for the technology demonstration grants competition described. There is a separate FFO for the competition to develop and operate a regional ballast water technology research, development, testing and evaluation facility. Depending on annual appropriations, NOAA and the U.S. Fish and Wildlife Service expect to make available up to about $2 million, and the U.S. Maritime Administration (MARAD) expects to make available several vessels for use as test platforms, to support ballast water treatment technology demonstration projects. The maximum amount of award will vary with the scale of the proposed project. Depending on the funding available and the number and quality of proposals received, approximately 10 grants with a median value of about $200,000 are anticipated to be awarded. Preliminary proposals must be received by 4 p.m. EDT on August 26, with full proposals due by 4 p.m. EST on December 2. Only those who submit preliminary proposals meeting the preliminary proposal deadline and other requirements are eligible to submit full proposals.
Requirements: Eligible applicants include: individuals; institutions ofhigher education; nonprofit organizations; for-profit organizations; Federal, State, local and Indian tribal governments; foreign governments; organizations under the jurisdiction of foreign governments; and international organizations. Applications from non-Federal and eligible Federal applicants (including NOAA employees) will be evaluated in the same selection process.
Geographic Focus: All States
Date(s) Application is Due: Aug 26; Dec 2
Contact: Dorn Carlson, Grants Coordinator; (301) 734-1080; ballast.water@noaa.gov
Internet: http://www07.grants.gov/search/search.do;jsessionid=HXXNv3Mt6rGXD7y5CX8dvQqGkVgGF8H6Hw5BGwTw1BFBfBQs3fdn!1282164016?oppId=1414&flag2006=true&mode=VIEW
Sponsor: National Oceanic and Atmospheric Administration
1315 East-West Highway, 12th Floor
Silver Spring, MD 20910

NOAA California Bay Watershed Education and Training (B-WET) Grants 3338

The California B-WET grant program is a competitively based program that supports existing environmental education programs, fosters the growth of new programs, and encourages the development of partnerships among environmental education programs throughout the San Francisco Bay, Monterey Bay, and Santa Barbara Channel watersheds. Projects support organizations that provide students meaningful watershed educational experiences and teachers professional development opportunities in the area of environmental education related watersheds.
Requirements: Eligible applicants are K-through-12 public and independent schools and school systems, institutions of higher education, nonprofit organizations, state or local government agencies, and Indian tribal governments. No cost sharing is required under this program; however, the National Marine Sanctuary Program strongly encourages applicants applying for either area of interest to share as much of the costs of the award as possible.
Geographic Focus: All States
Date(s) Application is Due: Oct 6
Amount of Grant: 50,000-60,000 USD
Contact: Seaberry Nachbar, (831) 647-4201; seaberry.nachbar@noaa.gov
Internet: http://www.grants.gov/search/search.do;jsessionid=LT5QC6LgrvSTzGVTkBfmSkJyDJJs2yy1yRRxhSQppp55y4bm5Hbl!-1724161176?oppId=42316&flag2006=false&mode=VIEW
Sponsor: National Oceanic and Atmospheric Administration
299 Foam Street
Monterey, CA 93940

NOAA Chesapeake Bay Office B-WET (Bay Watershed Education and Training) Grants 3339

The grant is a competitively based program that supports existing environmental education programs, fosters the growth of new programs, and encourages the development of partnerships among environmental education programs throughout the entire Chesapeake Bay watershed. Funded projects assist in meeting the Stewardship and Community Engagement goals of the Chesapeake 2000 Agreement. Specifically, projects support organizations that provide meaningful watershed educational experiences for students or related professional development for teachers. NCBO is encouraging applications that include innovative technologies in the delivery of these experiences.
Requirements: Eligible applicants are K-12 public and independent schools and school systems, institutions of higher education, community-based and nonprofit organizations, state or local government agencies, interstate agencies, and Indian tribal governments in the Chesapeake Bay watershed. The Department is strongly committed to broadening the participation of historically black colleges and universities, Hispanic serving institutions, tribal colleges and universities, and institutions that work in undeserved areas.
Restrictions: Federal agencies or institutions are not eligible to receive federal assistance under this funding program.
Geographic Focus: Delaware, District of Columbia, Maryland, New York, Pennsylvania, Virginia, West Virginia
Date(s) Application is Due: Oct 19
Contact: Shannon Sprague; (410) 267-5664 or (410) 267-5660; Shannon.Sprague@noaa.gov
Internet: http://noaa.chesapeakebay.net/educationgrants.aspx
Sponsor: National Oceanic and Atmospheric Administration
410 Severn Avenue
Annapolis, MD 21403

NOAA Chesapeake Bay Office Cooperative Research Grants 3340

The Program is intended to formalize and expand collaborative research among the Chesapeake Bay's commercial fishing industry, marine scientists, and fishery management communities. The goal of this initiative is to enhance the data upon which fishery management decisions are made as well as to facilitate communication and collaboration among commercial and collaboration among commercial fishermen, scientists, and fishery managers. Through this initiative, the NCBO will develop a collaborative and cooperative program to set research priorities that meet management and fishing industry needs. Priority areas are modified slightly each year but may include stock assessment research and management, and multi-species research, management and fisheries ecosystem plan implementation, including ecosystem understanding, retrospective analyses, and ecosystem modeling. $100,000 to $150,000 in funding may be available for awards; there are no restrictions on minimum or maximum funding requests. Award periods may be up

to a maximum of three years. Letters of Intent (LOI) are requested within 30 days and full proposals 30-45 days after LOI. Applicants must submit a full proposal by the deadline, regardless of whether they have heard back from NOAA following the LOI submission. The deadline for proposal submission is usually within the March/April time frame.
Requirements: Eligible applicants are institutions of higher education, other nonprofits, commercial organizations, foreign governments, organizations under the jurisdiction of foreign governments, international organizations, state, local and Indian tribal governments.
Restrictions: Federal agencies or institutions are not eligible to receive federal assistance under this funding program.
Geographic Focus: Delaware, District of Columbia, Maryland, New York, Pennsylvania, Virginia, West Virginia
Contact: Shannon Sprague; (410) 267-5664 or (410) 267-5660; Shannon.Sprague@noaa.gov
Internet: http://noaa.chesapeakebay.net/CooperativeResearch.aspx
Sponsor: National Oceanic and Atmospheric Administration
410 Severn Avenue
Annapolis, MD 21403

NOAA Chesapeake Bay Office Fisheries Science Grants 3341
The Fisheries component of the Program is competitively based and supports research, monitoring, modeling and management addressing various aspects of Chesapeake Bay fisheries. Funded projects foster knowledge and understanding of the Chesapeake Bay ecosystem by providing biological information and life history characteristics for many individual Chesapeake Bay fisheries stocks, and by broadening the multispecies knowledge base for development of Fisheries Ecosystem Planning. Priority areas are modified slightly each year but may include stock assessment research and management, and multi-species research, management and fisheries ecosystem plan implementation, including ecosystem understanding, retrospective analyses, and ecosystem modeling. Letters of Intent (LOI) are requested within 30 days and full proposals 30-45 days after LOI. Applicants must submit a full proposal by the deadline, regardless of whether they have heard back from NOAA following the LOI submission. The deadline for proposal submission is usually within the March/April time frame.
Requirements: Eligible applicants are institutions of higher education, other nonprofits, commercial organizations, foreign governments, organizations under the jurisdiction of foreign governments, international organizations, state, local and Indian tribal governments.
Restrictions: Federal agencies or institutions are not eligible to receive federal assistance under this funding program.
Geographic Focus: Delaware, District of Columbia, Maryland, New York, Pennsylvania, Virginia, West Virginia
Contact: Shannon Sprague; (410) 267-5664 or (410) 267-5660; Shannon.Sprague@noaa.gov
Internet: http://noaa.chesapeakebay.net/fisheriesgrants.aspx
Sponsor: National Oceanic and Atmospheric Administration
410 Severn Avenue
Annapolis, MD 21403

NOAA Chesapeake Bay Office Non-Native Oyster Research Grants 3342
The NCBO seeks proposals for projects that will provide scientific information to support a programmatic Environmental Impact Statement (EIS) on the proposed introduction of a non-native oyster species. The EIS is being prepared by federal and state agencies in response to a proposal by the states of Maryland and Virginia to introduce the non-native oyster, Crassostrea ariakensis, to the Chesapeake Bay and other tidal waters of Maryland and Virginia. It is expected that approximately $1.8M will be available through this research program. Funds will be used to support multi-year projects currently underway, as well as new projects that can be completed within a 12-18 month project period to fill remaining information gaps. The NCBO requests proposals for projects that will provide data and information in the following three areas: biological research and analysis; economic research and analysis; and evaluation of EIS alternatives. Proposals should exhibit familiarity with related work that is completed or ongoing. Proposals may be submitted for up to 3 years.
Requirements: Eligible applicants are institutions of higher education, other nonprofits, commercial organizations, foreign governments, organizations under the jurisdiction of foreign governments, international organizations, state, local and Indian tribal governments. No cost sharing is required under this program, however, the NCBO strongly encourages applicants to share as much of the project costs as possible. Funds from other Federal awards may not be considered matching funds.
Restrictions: Federal agencies or institutions are not eligible to receive Federal assistance under this notice.
Geographic Focus: Maryland, Virginia
Date(s) Application is Due: Mar 15
Contact: Jamie King; (410) 267-5655; fax (410) 267-5666; Jamie.King@noaa.gov
Internet: http://noaa.chesapeakebay.net/nonnativeoystergrants.aspx
Sponsor: National Oceanic and Atmospheric Administration
410 Severn Avenue
Annapolis, MD 21403

NOAA Chesapeake Bay Office Submerged Aquatic Vegetation (SAV) Culture and Restoration Grants 3343
The Submerged Aquatic Vegetation (SAV) component of the Chesapeake Bay Integrated Science Program is competitively-based and designed to enhance and increase the important fisheries habitat in the Chesapeake Bay and its tidal tributaries. This program focuses on areas including large-scale SAV planting, site assessments for future SAV planting, applied research on SAV seed ecology, enhanced capacity for SAV propagation, and integrated restoration (e.g., with oysters). Availability of funding for this program is included in the December NOAA Omnibus notice published in the Federal Register. Pre-proposals are typically due in January or February.
Requirements: Eligible applicants are institutions of higher education, other nonprofits, commercial organizations, foreign governments, organizations under the jurisdiction of foreign governments, international organizations, state, local and Indian tribal governments.
Restrictions: Federal agencies or institutions are not eligible to receive federal assistance under this funding program.
Geographic Focus: Delaware, District of Columbia, Maryland, New York, Pennsylvania, Virginia, West Virginia
Date(s) Application is Due: Mar 26
Contact: Shannon Sprague; (410) 267-5664 or (410) 267-5660; Shannon.Sprague@noaa.gov
Internet: http://noaa.chesapeakebay.net/aquaticvegetationgrants.aspx
Sponsor: National Oceanic and Atmospheric Administration
410 Severn Avenue
Annapolis, MD 21403

NOAA Climate Change Data and Detection (CCDD) Grants 3344
The goal of the program is to provide data and information management support to assure the availability of critical data sets for a variety of international programs and assessments. The data and resulting products extend the existing long-term climate record and, also, serve as essential input to predictive models. For the upcoming fiscal year, the CCDD program is soliciting proposals for research in the following areas: development of climate reference data sets; climate change detection and attribution studies, in joint sponsorship with the Department of Energy's Climate Change Prediction Program; and paleoclimate. While it is in the best interest of the applicants and their institutions to submit a Letter of Intent (LOI) by the July 23 due date, it is not a requirement; applicants who do not submit an LOI are allowed to submit a full proposal. For those that do submit an LOI, full proposals will be encouraged only for LOIs deemed relevant.
Requirements: Eligible applicants are institutions of higher education, other nonprofits, commercial organizations, international organizations, and state, local and Indian tribal governments.
Restrictions: Federal agencies or institutions are not eligible to receive Federal assistance under this notice.
Geographic Focus: All States
Date(s) Application is Due: Sep 24
Amount of Grant: 50,000-200,000 USD
Contact: Chris Miller, (301) 734-1241; fax (301) 713-0518; christopher.d.miller@noaa.gov
Internet: http://www.climate.noaa.gov/cpo_pa/ccdd/
Sponsor: National Oceanic and Atmospheric Administration
1315 East-West Highway, 12th Floor
Silver Spring, MD 20910

NOAA Climate Dynamics and Experimental Prediction (CDEP) Grants 3345
NOAA is soliciting proposals to initiate Climate Test Bed (CTB) projects under the Climate Dynamics and Experimental Prediction (CDEP) program. Due to limited available funding, the priority areas are focused on enhancing multi-model ensemble based forecasts, anchored by the NCEP Climate Forecast System, and on improving monthly-to-seasonal climate forecast products and applications, especially for drought. Priority areas include: climate forecast system improvements; generation and evaluation of multi-model ensemble forecasts; improving drought prediction; and enhancing operational drought forecast products and applications. Prospective investigator is strongly encouraged to submit a Letter of Intent (LOI) by the published deadline of July 23.
Requirements: Eligible applicants are institutions of higher education, other nonprofits, commercial organizations, international organizations, and state, local and Indian tribal governments.
Restrictions: Federal agencies or institutions are not eligible to receive Federal assistance under this notice.
Geographic Focus: All States
Date(s) Application is Due: Sep 24
Amount of Grant: 50,000-200,000 USD
Contact: Kenneth Mooney, (301) 734-1242; Kenneth.Mooney@noaa.gov
Internet: http://www.climate.noaa.gov/index.jsp?pg=/opportunities/opp_index.jsp&opp=2008/cdep_info.jsp
Sponsor: National Oceanic and Atmospheric Administration
1315 East-West Highway, 12th Floor
Silver Spring, MD 20910

NOAA Climate Prediction Program for the Americas (CPPA) Grants 3346
The Program is a competitive research program with a goal to improve operational intraseasonal to interannual climate and hydrologic predictions for the Americas with quantified uncertainties sufficient for making informed decisions. To achieve its goal, CPPA has the following major objectives: to improve the predictive understanding and model simulation of ocean, atmosphere and land-surface processes; to quantify the sources and limits of predictability of climate variations on intra-seasonal to interannual time scale; to advance NOAA's operational climate forecasts, monitoring; and analysis systems and to develop climate-based hydrologic forecasting capabilities for decision support and water resource applications. Currently, CPPA is soliciting proposals under the following research areas: VOCALS (VAMOS Ocean-Cloud-Atmosphere-Land-Studies) Field Experiment; Hydrologic applications of NCEP multi-model seasonal hindcasts; and predictability and mechanisms of drought. While it is in the best interest of the applicants and their institutions to submit a Letter of Intent (LOI) by the July 23 due date, it is not a requirement; applicants who do not submit an LOI are allowed to submit a full proposal. For those that do submit an LOI, full proposals will be encouraged only for LOIs deemed relevant.
Requirements: Eligible applicants are institutions of higher education, other nonprofits, commercial organizations, international organizations, and state, local and Indian tribal governments.

Restrictions: Federal agencies or institutions are not eligible to receive Federal assistance under this notice.
Geographic Focus: All States
Date(s) Application is Due: Sep 24
Amount of Grant: 50,000-200,000 USD
Contact: Jin Huang, (301) 734-1226; fax (301) 713-0517; Jin.Huang@noaa.gov
Internet: http://www.climate.noaa.gov/cpo_pa/cppa/
Sponsor: National Oceanic and Atmospheric Administration
1315 East-West Highway, 12th Floor
Silver Spring, MD 20910

NOAA Climate Variability and Human Health Program (CVHH) Grants 3347
The Program is designed to take a problem-oriented approach to the use of climate and environmental information for social benefit. The goal of the program is to provide scientifically sound, socially relevant and technically usable climate and environmental information for the public health community. The CVHH program is comprised of four main components: research; application; capacity building and training; and community building. In collaboration with other agencies, academia, international institutions, and regional partners, the CVHH program supports training and other activities that build capacity to use climate information for public health purposes. The program also focuses on building a robust climate and health community, helping to bring together various communities and disciplines to foster an iterative definition of the problem and collective development of approaches and solutions.
Requirements: All institutions eligible to receive support from NOAA, EPA, NASA, and NSF are able to apply. Universities, nonprofit organizations, for profit organizations, state and local governments, and Indian tribes are included among entities eligible for funding under this announcement.
Restrictions: Funding for foreign institutions is not available. Federal agencies or institutions are not eligible to receive Federal assistance under this notice.
Geographic Focus: All States
Date(s) Application is Due: Feb 27; Apr 23
Amount of Grant: 50,000-250,000 USD
Contact: Diane Brown; (301) 734-1206; fax (301) 713-0518; diane.brown@noaa.gov
Internet: http://www.ogp.noaa.gov/mpe/csi/cvhh/index.htm
Sponsor: National Oceanic and Atmospheric Administration
1315 East-West Highway, 12th Floor
Silver Spring, MD 20910

NOAA Climate Variability and Predictability (CVP) Grants 3348
The program seeks to understand, through modeling and diagnostic studies, large-scale patterns of climate variability and their teleconnections on seasonal to decadal time scales. The program aims to address climate variability and predictability over the United States, recognizing that this may require knowledge of climate variability in other regions of the globe. Goals include: identifying and understanding the major patterns of climate variability on seasonal, decadal and longer time scales and evaluating their predictability; expanding capacity in short term (seasonal to interannual) climate predictability and searching for ways to predict decadal variability; better documenting the record of rapid climate changes and the mechanisms for these events, and evaluating the potential for abrupt climate changes in the future; evaluating and enhancing the models used to project climate change due to human activity, including anthropogenically induced changes in atmospheric composition; and detecting and describing any global climate changes that may occur. Currently, CVP is soliciting proposals that specifically address strategies for correcting tropical biases in coupled climate models to advance the understanding and prediction of climate. While it is in the best interest of the applicants and their institutions to submit a Letter of Intent (LOI) by the July 23 due date, it is not a requirement; applicants who do not submit an LOI are allowed to submit a full proposal. For those that do submit an LOI, full proposals will be encouraged only for LOIs deemed relevant.
Requirements: Eligible applicants are institutions of higher education, other nonprofits, commercial organizations, international organizations, and state, local and Indian tribal governments.
Restrictions: Federal agencies or institutions are not eligible to receive Federal assistance under this notice.
Geographic Focus: All States
Date(s) Application is Due: Sep 24
Amount of Grant: 50,000-200,000 USD
Contact: James Todd; (301) 734-1258; fax (301) 713-0518; james.todd@noaa.gov
Internet: http://www.climate.noaa.gov/index.jsp?pg=./opportunities/opp_index.jsp&opp=info_cvp.jsp
Sponsor: National Oceanic and Atmospheric Administration
1315 East-West Highway, 12th Floor
Silver Spring, MD 20910

NOAA Coastal and Estuarine Land Conservation Grants 3349
The purpose of this document is to advise eligible coastal states and territories (requirements described below) that OCRM is soliciting coastal and estuarine land conservation project proposals for competitive funding under the CELCP. This announcement solicits proposals for land acquisition projects (fee simple interest or conservation easements) that can be completed within 18 months from the start date of the award and that have the purpose of protecting important coastal and estuarine areas. NOAA may extend project grants up to an additional 18 months (for a total project period of 3 years maximum) if circumstances warrant.
Requirements: Only states and territories with Coastal Zone Management Programs or National Estuarine Research Reserves approved under the CZMA that have submitted a draft CELCP plan to NOAA on or before May 1, 2008, are eligible to participate in this funding competition. The eligible states and territories are: Alabama, Alaska, American Samoa, California, Connecticut, Delaware, Florida, Georgia, Guam, Hawaii, Indiana, Louisiana, Maine, Maryland, Massachusetts, Michigan, Mississippi, New Hampshire, New Jersey, New York, North Carolina, Northern Mariana Islands, Ohio, Oregon, Pennsylvania, Puerto Rico, Rhode Island, South Carolina, Texas, Virginia, Washington, and Wisconsin.
Geographic Focus: All States
Date(s) Application is Due: Jul 23
Amount of Grant: Up to 3,000,000 USD
Contact: Elisabeth Morgan, (301) 713-3155, ext. 166; Elisabeth.Morgan@noaa.gov
Internet: http://www07.grants.gov/search/search.do;jsessionid=LQCZGLyvwpgGPznL6jnLRyjK1vpGyTRsfyCJKLZtfJnqhLHphMnJ!280595828?oppId=41800&flag2006=false&mode=VIEW
Sponsor: U.S. Department of Commerce
1401 Constitution Avenue, NW, Room 5128
Washington, DC 20230

NOAA Coastal Ecosystem Research Project 3350
This program solicits 1-year and 2-year proposals for modeling, monitoring, and retrospective studies of coastal ecosystem research in the Northern Gulf of Mexico. Modeling studies are needed to provide a framework for studies in the northern Gulf of Mexico. Modeling activities will be used to guide further program development and identify important processes for the extensive fieldwork anticipated to follow this preliminary phase. Modeling studies may include: models that simulate impacts of varying nutrient flux on productivity and trophic response in the northern Gulf of Mexico ecosystem, including impacts to and responses of commercially and recreationally important fisheries; NPZF models; physical-biological coupled models of processes in the Gulf ecosystem influenced by the Mississippi River discharge, including transport and population dynamics of key zooplankton and fishery populations; models of oceanographic and climate influences on nutrients and their impact on Gulf productivity; models of biogeochemical cycling of nutrients within the Gulf and its relationship to the dynamics of organic carbon flux in the Gulf; and models of water column stability and hypoxic zone dynamics. This announcement requests full proposals only.
Geographic Focus: All States
Date(s) Application is Due: Mar 14
Amount of Grant: 25,000-1,000,000 USD
Contact: Rob Magnien; (301) 713-3338 or (301) 713-3020; rob.magnien@noaa.gov
Internet: http://frwebgate2.access.gpo.gov/cgi-bin/getdoc.cgi?dbname=2001_register&docid=fr17ja01-56
Sponsor: National Oceanic and Atmospheric Administration
1305 East-West Highway, N/SCI2 SSMC4, Room 8243
Silver Spring, MD 20910

NOAA Coastal Hypoxia Research Program (CHRP) Grants 3351
The purpose of this document is to advise the public that NOAA is soliciting proposals for projects of 2 to 5 years in duration that advance understanding, predicting, and managing the causes and ecological and economic impacts of hypoxia in representative coastal ecosystems. Proposals for funding under this announcement will be accepted for hypoxia research in U.S. Great Lakes, estuaries, and coastal ocean regions, exclusive of the dead zone in the Gulf of Mexico over the Louisiana continental shelf. Funding is contingent upon the availability of annual Federal appropriations.
Requirements: Institutions of higher education, not-for- profit institutions, State, commercial organizations, local and Indian tribal governments and Federal agencies are eligible to apply.
Geographic Focus: All States
Date(s) Application is Due: Sep 11
Contact: Rob Magnien; (301) 713-3338 or (301) 713-3020; rob.magnien@noaa.gov
Internet: http://www.cop.noaa.gov/opportunities/grants/fundingarchive/fy2007/CHRP_FFO_0606.html
Sponsor: National Oceanic and Atmospheric Administration
1305 East-West Highway, N/SCI2 SSMC4, Room 8243
Silver Spring, MD 20910

NOAA Coastal Management Fellowships 3352
The Coastal Management Fellowship was established in 1996 to provide on-the-job education and training opportunities in coastal resource management and policy for postgraduate students and to provide project assistance to state coastal zone management programs. The program matches postgraduate students with state coastal zone programs to work on projects proposed by the state and selected by the fellowship sponsor, the National Oceanic and Atmospheric Administration (NOAA) Coastal Services Center. This two-year opportunity offers a competitive salary, medical benefits, and travel and relocation expense reimbursement. Proposals due from the states on October 17, and applications are due from candidates on January 26.
Geographic Focus: All States
Date(s) Application is Due: Jan 26
Contact: Margaret VanderWilt, Fellowship Coordinator; (843) 740-1273; fax (843) 740-1224; margaret.vanderwilt@noaa.gov or csc.fellowships@noaa.gov
Internet: http://www.csc.noaa.gov/cms/fellows.html
Sponsor: National Oceanic and Atmospheric Administration
2234 South Hobson Avenue
Charleston, SC 29405

NOAA Coastal Ocean Grants 3353

This program is designed to provide predictive capability for managing coastal ecosystems. Grants may be used to fund research and inter-agency initiatives under specific COP programs related to NOAA's mission in coastal ecosystem oceanography, cumulative coastal impacts, and coastal forecasting/natural hazards. The program supports research on critical issues which exist in the nation's estuaries, coastal waters, and the Great Lakes; and translates its findings into accessible information for coastal managers, planners, lawmakers, and the public. Grant periods may last from one to three years with funding approved annually. Deadlines are announced each year.

Requirements: Institutions of higher education, not-for- profit institutions, State, commercial organizations, local and Indian tribal governments and Federal agencies are eligible to apply.

Geographic Focus: All States

Amount of Grant: 25,000-1,000,000 USD

Contact: Rob Magnien, Director; (301) 713-3338 or (301) 713-3020; fax (301) 713-4044; rob.magnien@noaa.gov

Internet: http://12.46.245.173/cfda/CFDA_Static/Programs/11_478.html

Sponsor: National Oceanic and Atmospheric Administration

1305 East-West Highway, N/SCI2 SSMC4, Room 8243

Silver Spring, MD 20910

NOAA Collaborative Science, Technology, and Applied Research Grants 3354

Through the Collaborative Science, Technology, and Applied Research (CSTAR) program, the NWS provides funding to the university community for highly collaborative applied research activities. The CSTAR program creates a cost-effective transition from basic and applied research to operations and services. This service-science linkage provides a foundation for the ongoing infusion of science and technology into NWS operations. The long-term objective of the CSTAR program is to improve the overall forecast and warning capabilities of the operational hydrometeorological community by addressing the following national science priorities: quantitative precipitation estimation (QPE) and forecasting (QPF), including precipitation type and probabilistic QPF (PQPF); flash flood and probabilistic river prediction; prediction of seasonal-to-interannual and decadal climate variability, and the impacts of these variabilities on extreme weather events; prediction of tropical cyclones near landfall, including track, intensity, and associated precipitation, and hazardous weather; prediction of marine conditions, including fog, winds, coastal ocean, and open ocean waves; the effect of topography and other surface forcing on local weather regimes; locally hazardous weather, especially severe convection, winter weather, and phenomena that affect aviation; and conditions conducive for the rapid development of wildfires and the dispersion of smoke and other air-quality hazards. Individual NWS regions and centers have a subset of these science priorities due to differences in factors such as topography, weather regimes, and mission.

Requirements: All accredited U.S. colleges and universities, including federally funded educational institutions, such as the Naval Postgraduate School, are eligible.

Geographic Focus: All States

Amount of Grant: 100,000-125,000 USD

Contact: Sam Contorno, Federal Program Officer; (301) 713-3557, ext. 150; fax (301) 713-1253; samuel.contorno@noaa.gov

Internet: http://www.epa.gov/EPA-IMPACT/2003/January/Day-07/i224.htm

Sponsor: National Oceanic and Atmospheric Administration

1315 East-West Highway, Room 15330

Silver Spring, MD 20910

NOAA Communicating Hurricane Information Grants 3355

This solicitation aims to advance basic research in the social, behavioral and economic sciences related to the communication of hurricane outlooks, forecasts, watches, and warnings both to decision makers (i.e., emergency managers, elected officials) and to the general public. Although all proposals addressing this topic will be considered, investigators are urged to consider specific communication concerns: how messages are received and understood; and what the linkages are between communications and behavior. The role of the physical scientist or engineer on each project is to provide expertise about the state of hurricane forecast technologies and what is likely to be available in the near future. This interaction between the social scientists and the physical scientist or engineer should continue throughout the duration of the project. The Program anticipates making approximately a total of 6-12 awards for research projects. These awards will be for 1 to 3 years.

Requirements: Eligible research proposals must have at least one investigator from the social, behavioral, or economic sciences, and at least one from the physical sciences or engineering.

Geographic Focus: All States

Date(s) Application is Due: Jun 3

Amount of Grant: Up to 400,000 USD

Contact: Robert E. O'Connor, Directorate for Social, Behavioral and Economic Sciences; (703) 292-7263; fax (703) 292-9068; roconnor@nsf.gov

Internet: http://www07.grants.gov/search/synopsis.do;jsessionid=HYTSYzGB611pwvpt7Dx3cqnZjD2lJsFhsGLMd6v2XS8TJcVsWY6j!1282164016

Sponsor: National Oceanic and Atmospheric Administration

4201 Wilson Boulevard

Arlington, VA 22230

NOAA Cooperative Institute: Eastern U.S. Continental Shelf Frontier Exploration, Research, and Technology Development Grants 3356

The purpose of this announcement is to invite the submission of proposals to establish a CI for the eastern U.S. Continental Shelf frontier exploration, research, and technology development, and to provide details on the application, review, and selection process. This CI will give NOAA the benefit of working with complementary capabilities at one or more research institutions that contribute to NOAA's ocean exploration, research, and technology priorities on the eastern U.S. Continental Shelf. CI Concept/Program Background A CI is a NOAA-supported, non-Federal organization that has established an outstanding research program in one or more areas that are relevant to the NOAA mission to understand and predict changes in the Earth's environment and conserve and manage coastal and marine resources to meet our Nation's economic, social, and environmental needs.

Requirements: Eligibility is limited to non-Federal public and private non-profit universities, colleges and research institutions that offer accredited graduate level degree-granting programs in NOAA- related sciences.

Geographic Focus: All States

Date(s) Application is Due: Oct 6

Amount of Grant: 150,000-25,000,000 USD

Contact: Dr. John Cortinas, (301) 734-1090; John.Cortinas@noaa.gov

Internet: http://www.grants.gov/search/search.do;jsessionid=LP0hqWfp11PVPKSr9yfL1LLhNQTn2gsNBmYzZtzGKTQhJRdRBYxW!-2005900814?oppId=42235&flag2006=false&mode=VIEW

Sponsor: National Oceanic and Atmospheric Administration

1315 East-West Highway, Room 11326

Silver Spring, MD 20910

NOAA Cooperative Institute for Applied Meteorological Studies Grants 3357

CIAMS is affiliated with the Fort Worth based headquarters office of the Southern Region of the National Weather Service. The Institute employs research scientists and graduate students in a broad program of applied research and service to Texas and surrounding states, in agricultural meteorology, marine meteorology and air-sea interactions over the Gulf of Mexico, lightning and severe weather, and Doppler radar studies from the installations in the Texas A&M Department of Atmospheric Sciences and at the Houston Weather Forecast Office. The NWS is responsible for the overall identification of tasks needed to be addressed and funds (through cooperative agreements) for faculty and student work towards those goals. Texas A&M University also provides some funding, as well as computer and observational facilities, to help support faculty and students' work on projects. Awards are for a 12-month period. In some instances, award periods may extend beyond 12 months. There are no deadline dates.

Requirements: Eligible applicants are Sole Source programs with Texas A&M University and public- and state-supported institutions.

Geographic Focus: All States

Contact: Sam Contorno, Federal Program Officer; (301) 713-3557, ext. 150; fax (301) 713-1253; samuel.contorno@noaa.gov

Internet: http://atmo.tamu.edu/ciams/

Sponsor: National Oceanic and Atmospheric Administration

1315 East-West Highway, Room 15330

Silver Spring, MD 20910

NOAA Cooperative Institute for Tropical Meteorology Research Grants 3358

The NOAA Cooperative Institute for Tropical Meteorology (CITM) is affiliated with the Department of Meteorology at Florida State University. The Program offers cooperative research opportunities for graduate level students. This cooperative research has an emphasis on hurricane studies, flash floods, tropical convection, and local weather. Student participation is a large component of the program. Florida State University also provides some funding, as well as computer and observational facilities, to help support faculty and students' work on projects. Awards are for a 12-month period. In some instances, award periods may extend beyond 12 months. There are no deadline dates.

Requirements: Eligible applicants are Sole Source programs with Florida State University and public- and state-supported institutions.

Geographic Focus: All States

Amount of Grant: 100,000-125,000 USD

Contact: Sam Contorno; (301) 713-3557, ext. 150; samuel.contorno@noaa.gov

Internet: http://www.met.fsu.edu/index.pl/acprograms/gradprogram

Sponsor: National Oceanic and Atmospheric Administration

1315 East-West Highway, Room 15330

Silver Spring, MD 20910

NOAA Cooperative Research Grants 3359

The CRP program provides financial assistance for projects that seek to increase and improve the working relationship between researchers from the NMFS, state fishery agencies, universities, and fishermen in the Gulf of Mexico and off the South Atlantic states of North Carolina, South Carolina, Georgia, and Florida involving the U.S. fishing industry (recreational and commercial). The program is a means of involving commercial and recreational fishermen in the collection of fundamental fisheries information. Collection efforts support the development and evaluation of management and regulatory options.

Requirements: Eligible applicants may be institutions of higher education, nonprofits, commercial organizations, individuals, and state, local, and Indian tribal governments.

Restrictions: Federal agencies or institutions are not eligible. Foreign governments, organizations under the jurisdiction of foreign governments, and international organizations are also excluded for purposes of this solicitation.

Geographic Focus: All States

Date(s) Application is Due: Sep 9

Amount of Grant: 25,000-400,000 USD

Contact: Robert Sadler; (727) 824-5324; Robert.Sadler@noaa.gov

Internet: http://www07.grants.gov/search/search.do;jsessionid=LQCZGLyvwpgGPznL6jnLRyjK1vpGyTRsfyCJKLZtfJnqhLHphMnJ!280595828?oppId=42299&flag2006=false&mode=VIEW

Sponsor: National Oceanic and Atmospheric Administration

263 13th Avenue South
Saint Petersburg, FL 33702

NOAA Cooperative Unit for Fisheries Education and Research Internships 3360

The Cooperative Unit for Fisheries Education and Research (CUFER) was established in 1992 within the Cooperative Institute for Marine and Atmospheric Studies (CIMAS) of the Rosenstiel School of Marine and Atmospheric Sciences (RSMAS) of the University of Miami (UM) in response to a need for the development of methods for improved quantitative assessment of fish populations and as a source of advice for resource sustainability. These activities are envisioned to complement NOAA Fisheries mission of resource stewardship. CUFER is designed to attract bright and skilled people to a program that encourages students and scientists with quantitative interests to apply themselves to collaborative research projects on quantitative fisheries science and marine resource modeling. Check the web site for available internships.
Geographic Focus: All States
Contact: David Die; (305) 421-4607; fax (305) 361-4457; ddie@rsmas.miami.edu
Internet: http://cufer.rsmas.miami.edu/wordpress/?page_id=77
Sponsor: National Oceanic and Atmospheric Administration
4600 Rickenbacker Causeway
Miami, FL 33149

NOAA Coral Reef Ecosystem Studies (CRES) Grants 3361

The purpose of this program is to advise the public that NOAA/NOS/NCCOS/CSCOR is soliciting proposals for projects of 3-5 years in duration for the Coral Reef Ecosystem Studies Program (CRES) and up to 3 years in duration for the Deep Coral Reef Ecosystem Studies Program (Deep-CRES). The development of predictive models (e.g., bio-physical models to investigate larval transport of reef organisms and their recruitment to reef systems in the context of variable oceanographic conditions; water quality models to investigate the relationship between watershed-based pollutant inputs and effects on reef ecosystems; economic models to investigate the relationship between coral reef health and local economies) as a component of these studies is required. Scientific information, syntheses, models, and ecosystem forecasting capabilities from this multi-disciplinary, long-term effort should be clearly usable by resource managers to make more informed decisions on managing U.S. coral reef ecosystems. Funding is contingent upon the availability of Fiscal Year Federal appropriations.
Requirements: Institutions of higher education, not-for- profit institutions, State, commercial organizations, local and Indian tribal governments and Federal agencies are eligible to apply.
Geographic Focus: All States
Date(s) Application is Due: Nov 13
Contact: Rob Magnien; (301) 713-3338 or (301) 713-3020; rob.magnien@noaa.gov
Internet: http://www.cop.noaa.gov/opportunities/grants/fundingarchive/fy2007/CRES_FFO_0606.html
Sponsor: National Oceanic and Atmospheric Administration
1305 East-West Highway, N/SCI2 SSMC4, Room 8243
Silver Spring, MD 20910

NOAA CRCP Coral Reef Ecosystem Monitoring Grants 3362

The Grant Program, as authorized under the Coral Reef Conservation Act of 2000, provides matching grants to Governor-appointed point of contact agencies for the jurisdictions of Puerto Rico, the U.S. Virgin Islands, Florida, Hawaii, American Samoa, Guam, the Commonwealth of the Northern Mariana Islands, the Republic of Palau, the Federated States of Micronesia (including Chuuk, Yap, Kosrae, and Pohnpei), and the Republic of the Marshall Islands to support State and territory coral reef monitoring activities. The current closing date for applications is February 22, with pre-applications due on November 6. The expected number of awards is 15, with total funding equaling $1,100,000.
Requirements: The island jurisdictions of Puerto Rico, the U.S. Virgin Islands, Florida, Hawaii, American Samoa, Guam, the Commonwealth of the Northern Mariana Islands, the Republic of Palau, the Federated States of Micronesia (including Chuuk, Yap, Kosrae, and Pohnpei), and the Republic of the Marshall Islands are eligible to apply.
Geographic Focus: All States
Date(s) Application is Due: Feb 22; Nov 6
Amount of Grant: 73,333 USD
Contact: Steve Drescher, Policy Advisor; (301) 713-0942, ext.122; fax (301) 713-0947; Steve.J.Drescher@noaa.gov
Internet: http://www07.grants.gov/search/search.do;jsessionid=HWbpMDzmZ104MfvCGSbHhsGGPnb8dGTRG2hM0vfdTvpFRTZdkPyK!-796138783?oppId=14663&flag2006=true&mode=VIEW
Sponsor: National Oceanic and Atmospheric Administration
1315 East-West Highway, 12th Floor
Silver Spring, MD 20910

NOAA CSCOR Regional Ecosystem Prediction Program: Cumulative Impacts of Stressors at the Land-Water Interface in the Mid-Atlantic 3363

Proposals are solicited for projects of up to 5 years in duration for a regional-scale ecosystem research study in the Mid-Atlantic investigating the cumulative impacts of multiple stressors at the land-water interface of estuaries and bays on recreationally, economically or ecologically important living resource populations and communities. This research is needed to support management efforts to protect and restore submerged and wetland plant communities at the land-water interface of Mid-Atlantic estuaries and bays and species that depend upon these critical habitats including finfish, shellfish, amphibians, reptiles and birds. Generally, costs that are allowable include salaries, equipment, supplies, and training, as long as these are necessary and reasonable.
Requirements: Eligible applicants are: institutions of higher education; other non-profits; states; local governments; commercial organizations; and Federal agencies that possess the statutory authority to receive financial assistance.
Restrictions: The Program will not fund any Federal Full Time Employee (FTE) salaries.
Geographic Focus: All States
Date(s) Application is Due: Oct 15
Amount of Grant: 500,000-1,000,000 USD
Contact: Elizabeth Turner; (603) 862-4680; elizabeth.turner@noaa.gov
Internet: http://www07.grants.gov/search/search.do;jsessionid=HXXNv3Mt6rGXD7y5CX8dvQqGkVgGF8H6Hw5BGwTw1BFBfBQs3fdn!1282164016?oppId=14667&flag2006=true&mode=VIEW
Sponsor: National Oceanic and Atmospheric Administration
1315 East-West Highway, 12th Floor
Silver Spring, MD 20910

NOAA CSCOR Regional Ecosystem Prediction Program: Ecosystem Goal Setting in Coastal Waters and Reefs of South Florida 3364

Two- to three-year proposals are solicited to develop, undertake and conclude a consensus-building process that results in scientifically-based quantifiable goals for aquatic resources and habitats of the Florida Bay and Keys that are of concern to or fall within the legal mandates of management agencies at the federal, state and local level. Funding instruments available are project grants and cooperative agreements. Generally, costs that are allowable include salaries, equipment, supplies, and training, as long as these are necessary and reasonable.
Requirements: Eligible applicants are: institutions of higher education; other non-profits; states; local governments; commercial organizations; and Federal agencies that possess the statutory authority to receive financial assistance.
Restrictions: The Program will not fund any Federal Full Time Employee (FTE) salaries.
Geographic Focus: All States
Date(s) Application is Due: Oct 29
Amount of Grant: Up to 500,000 USD
Contact: Larry Pugh; (301) 713-3338, ext. 160; larry.pugh@noaa.gov
Internet: http://www07.grants.gov/search/search.do;jsessionid=HXXNv3Mt6rGXD7y5CX8dvQqGkVgGF8H6Hw5BGwTw1BFBfBQs3fdn!1282164016?oppId=14667&flag2006=true&mode=VIEW
Sponsor: National Oceanic and Atmospheric Administration
1315 East-West Highway, 12th Floor
Silver Spring, MD 20910

NOAA CSCOR Regional Ecosystem Prediction Program: Invasive Species in the Great Lakes—A Regional Scale Approach 3365

Within an ecosystem framework impacted by invasive species, proposals are solicited for projects up to 5 years in duration for a regional-scale ecosystem research study to develop environmental forecasting tools for assessing and predicting changes in water quality, habitat, living resources and consequences to the food web of the Great Lakes. Proposals must address this issue in a regional context (all or most of the Great Lakes) and be of obvious utility to managing natural resources in the face of current invasive populations. Proposals must include justification for the questions addressed, in terms of environmental management importance, linkage to invasive species and need for a regional ecosystem construct. Proposals must also provide detailed descriptions of the improved quantitative understanding to be achieved and predictive capability to be developed, identify management issue(s) that are being addressed and specify how results could change current management practices. Funding instruments available are project grants and cooperative agreements. Generally, costs that are allowable include salaries, equipment, supplies, and training, as long as these are necessary and reasonable.
Requirements: Eligible applicants are: institutions of higher education; other non-profits; states; local governments; commercial organizations; and Federal agencies that possess the statutory authority to receive financial assistance.
Restrictions: The Program will not fund any Federal Full Time Employee (FTE) salaries.
Geographic Focus: All States
Date(s) Application is Due: Oct 1
Amount of Grant: 500,000-1,000,000 USD
Contact: Felix Martinez; (301) 713-3338, ext. 153; felix.martinez@noaa.gov
Internet: http://www07.grants.gov/search/search.do;jsessionid=HXXNv3Mt6rGXD7y5CX8dvQqGkVgGF8H6Hw5BGwTw1BFBfBQs3fdn!1282164016?oppId=14667&flag2006=true&mode=VIEW
Sponsor: National Oceanic and Atmospheric Administration
1315 East-West Highway, 12th Floor
Silver Spring, MD 20910

NOAA Cumulative Impacts of Multiple Stressors (MultiStress) Grants 3366

The purpose of this program is to advise the public that NOAA/NOS/NCCOS/CSCOR is soliciting proposals for projects of up to 5 years in duration to investigate the impacts of multiple stressors in coastal ocean ecosystems, including estuaries and the Great Lakes. These projects should be interdisciplinary, multiple investigator, and well integrated studies designed to develop capabilities for understanding, predicting, and managing the effects of multiple stressors (both anthropogenic and natural) in coastal ecosystems. Funding is contingent upon the availability of annual Federal appropriations.
Requirements: Institutions of higher education, not-for- profit institutions, State, commercial organizations, local and Indian tribal governments and Federal agencies are eligible to apply.
Geographic Focus: All States
Date(s) Application is Due: Oct 23

Contact: Rob Magnien; (301) 713-3338 or (301) 713-3020; rob.magnien@noaa.gov
Internet: http://www.cop.noaa.gov/opportunities/grants/fundingarchive/fy2007/Multistress_FFO_0606.html
Sponsor: National Oceanic and Atmospheric Administration
1305 East-West Highway, N/SCI2 SSMC4, Room 8243
Silver Spring, MD 20910

NOAA Dr. Nancy Foster Scholarships **3367**
The Scholarship Program provides support for outstanding scholarship and encourages independent graduate-level research in oceanography, marine biology, or maritime archaeology, particularly by women and members of minority groups. Dr. Nancy Foster Scholarships carry a 12-month stipend for each student of $20,000 and an annual cost-of-education allowance of up to $12,000 and up to $20,000 support for a four to six week research collaboration at a NOAA facility. Masters students may be supported for up to two years, and doctoral students for up to four years. About four scholarships are awarded each year. The annual stipend is paid directly to the scholar.
Requirements: Those eligible to apply are United States citizens currently pursuing or intending to pursue a masters or doctoral level degree in oceanography, maritime archaeology (including the curation, preservation, and display of maritime artifacts) or marine biology, (including all science, engineering, and resource management of ocean and coastal areas). Studies must be conducted on a full-time basis.
Geographic Focus: All States
Date(s) Application is Due: Feb 8
Amount of Grant: Up to 52,000 USD
Contact: Pritti Brahma, (301) 713-9437, ext. 118; priti.brahma@noaa.gov
Internet: http://fosterscholars.noaa.gov/aboutscholarship.html
Sponsor: National Oceanic and Atmospheric Administration
1315 East-West Highway, Room 10725
Silver Spring, MD 20910

NOAA Ecological Forecasting (ECOFORE) Grants **3368**
The NOAA/NOS/NCCOS/CSCOR is soliciting proposals for projects of 2 to 5 years in duration for the development of new ecological forecasting capabilities and the transition of existing ecological forecasts to operational status. These ecological forecasts are to support critical management decisions for the management of the Nation's Great Lake , estuarine, coastal and ocean ecosystems. The EcoFore program seeks to develop capabilities to provide coastal managers with state-of-the-art science in a form that they can utilize to make proactive management decisions in an ecosystem context. Proposals should provide detailed descriptions of the predictive/forecasting capability to be developed, management issue(s) to be addressed, the target audience or users of developed forecasts, and how forecasts will be provided to users. Explicit identification of the end user group(s) is required.
Requirements: Institutions of higher education, not-for- profit institutions, State, commercial organizations, local and Indian tribal governments and Federal agencies are eligible to apply.
Geographic Focus: All States
Date(s) Application is Due: Nov 18
Contact: Elizabeth Turner, (603) 862-4680 or (301) 713-3020; fax (301) 713-4044; Elizabeth.Turner@noaa.gov
Internet: http://www.cop.noaa.gov/opportunities/grants/fundingarchive/fy2006.html
Sponsor: National Oceanic and Atmospheric Administration
1305 East-West Highway, N/SCI2 SSMC4, Room 8243
Silver Spring, MD 20910

NOAA Ecology and Oceanography of Harmful Algal Blooms Grants **3369**
The Program is seeking applications proposing targeted research projects of up to 3 years duration and multi-disciplinary regional studies for 3 to 5 years duration for the Ecology and Oceanography of Harmful Algal Blooms (ECOHAB) program. This program provides support for research on algal species whose populations may cause or result in deleterious effects on ecosystems, coastal economies, and human health. Studies of the causes of such blooms, their detection, effects, mitigation, and control in U.S. coastal waters (including estuaries and Great Lakes) are solicited.
Requirements: Institutions of higher education, not-for- profit institutions, State, commercial organizations, local and Indian tribal governments and Federal agencies are eligible to apply.
Geographic Focus: All States
Date(s) Application is Due: Oct 4
Contact: Rob Magnien; (301) 713-3338 or (301) 713-3020; rob.magnien@noaa.gov
Internet: http://www.cop.noaa.gov/opportunities/grants/fundingarchive/fy2008/Ecohab_08.html
Sponsor: National Oceanic and Atmospheric Administration
1305 East-West Highway, N/SCI2 SSMC4, Room 8243
Silver Spring, MD 20910

NOAA Ernest F. Hollings Scholarships **3370**
The purposes of the Scholarship are to: increase undergraduate training in oceanic and atmospheric science, research, technology, and education and foster multidisciplinary training opportunities; increase public understanding and support for stewardship of the ocean and atmosphere and improve environmental literacy; recruit and prepare students for public service careers with the National Oceanic and Atmospheric Administration and other natural resource and science agencies at the federal, state and local levels of government; and recruit and prepare students for careers as teachers and educators in oceanic and atmospheric science and to improve scientific and environmental education in the United States. The award provides financial assistance (up to a maximum of $8,000) for two academic years of full-time study and a 10-week, full-time internship position ($650/week) during the summer at a NOAA facility.
Requirements: Applicants must be U.S. citizens; have full-time junior academic status in an accredited college or university (fall semester) within the United States or U.S. territories; have a cumulative GPA of 3.0 (based on a 4.0 scale) in all completed undergraduate courses and major field of study; be majoring in a discipline area related to oceanic and atmospheric science, research, technology, and education; and be supportive of the purposes of NOAA???s programs and mission.
Geographic Focus: All States
Date(s) Application is Due: Feb 28
Amount of Grant: Up to 22,500 USD
Contact: Chantell Haskins, (301) 713-9437, ext. 125; chantell.haskins@noaa.gov
Internet: http://www.oesd.noaa.gov/Hollings_info.html
Sponsor: Oak Ridge Institute for Science and Education
1315 East-West Highway
Silver Spring, MD 20910

NOAA Faculty and Student Intern Research Program (FSIRP) **3371**
The primary focus of the program is to provide work experiences and research opportunities to faculty, as well as graduate and undergraduate students who are pursuing degrees in the NOAA-related sciences. Appointments are also available for high school students planning to enroll in and attend an institution of higher education upon graduation. Disciplines include: business, computer sciences, engineering, life sciences, physical sciences. Appointments are made available on a year-round basis. Summer internships are generally for ten weeks. A weekly stipend of $440 is provided for undergraduate students, $550 for graduate students, and $1,200 for faculty. Limited travel reimbursement (round-trip transportation expenses between facility and home or campus)and housing allowance are also available.
Geographic Focus: All States
Contact: Alfred A. Corea, (301) 713-0500, ext. 114; Alfred.A.Corea@noaa.gov
Internet: http://see.orau.org/ProgramDescription.aspx?Program=10049
Sponsor: U.S. Department of Commerce
1401 Constitution Avenue, NW, Room 5128
Washington, DC 20230

NOAA Financial Assistance for National Centers for Central Coastal Ocean Science **3372**
Research grants are awarded to determine the long-term consequences of human activities that adversely affect the coastal and marine environments; to assess the consequences of these activities in terms of ecological, economic, and social impacts; and to define and evaluate management alternatives that minimize adverse consequences of human use of the coastal and marine environments and resources. There are no specific deadlines for proposal submission. Proposals will be reviewed several times a year. Support for projects may be provided for periods up to five years.
Requirements: Universities, colleges, technical schools, institutes, laboratories, state and local government agencies, profit or nonprofit entities or individuals are eligible to apply.
Geographic Focus: All States
Contact: National Ocean Service, Office of Ocean Resources Conservation and Assessment; (301) 713-3102
Internet: http://www.nccos.noaa.gov
Sponsor: National Oceanic and Atmospheric Administration
1305 East-West Highway, N/SCI2 SSMC4, Room 8243
Silver Spring, MD 20910

NOAA General Coral Reef Conservation Grants **3373**
The NOAA Coral Reef Conservation Program/ General Coral Reef Conservation Grants Program (GCRCGP) provides funding to institutions of higher education, non-profit organizations, commercial organizations, Freely Associated State government agencies, and local and Indian tribal governments to support coral reef conservation projects in the United States and the Freely Associated States in the Pacific, as authorized under the Coral Reef Conservation Act of 2000. Projects funded through the GCRCGP support on-the ground efforts that: help preserve, sustain and restore the condition of coral reef ecosystems; promote the wise management and sustainable use of coral reef resources; increase public knowledge and awareness of coral reef ecosystems and issues regarding their conservation; and develop sound scientific information on the condition of coral reef ecosystems and the threats to such ecosystems. Projects should complement and fill gaps in state, territorial and commonwealth coral reef programs, emphasize community- based conservation, or address local action strategy priorities.
Requirements: Institutions of higher education, non-profit organizations, commercial organizations, local and Indian tribal governments and Freely Associated State Government Agencies can apply for funding under the GCRCGP.
Restrictions: U.S. federal, state, territory, and commonwealth governments and Regional Fishery Management Councils are not eligible for this program.
Geographic Focus: All States
Date(s) Application is Due: Nov 3
Amount of Grant: 15,000-50,000 USD
Contact: Liz Fairey, (301) 713-3459; Liz.fairey@noaa.gov
Internet: http://www07.grants.gov/search/search.do;jsessionid=LQCZGLyvwpgGPznL6jnLRyjK1vpGyTRsfyCJKLZtfJnqhLHphMnJ!280595828?oppId=42306&flag2006=false&mode=VIEW
Sponsor: U.S. Department of Commerce
1401 Constitution Avenue, NW, Room 5128
Washington, DC 20230

NOAA Global Carbon Cycle (GCC) Grants 3374

The goal of the program is to improve the ability to predict the fate of anthropogenic carbon dioxide and future atmospheric carbon dioxide concentrations using a combination of atmospheric and oceanic global observations, process-oriented field studies and modeling. GCC research is integrated across four over-riding themes: global location, magnitude and dynamics of carbon sources and sinks; northern hemisphere carbon sinks; causes of variability in sources and sinks; and future atmospheric carbon dioxide concentrations.
Requirements: Eligible applicants are institutions of higher education, other nonprofits, commercial organizations, international organizations, and state, local and Indian tribal governments.
Restrictions: Federal agencies or institutions are not eligible to receive Federal assistance under this notice.
Geographic Focus: All States
Amount of Grant: 50,000-200,000 USD
Contact: Kathy Tedesco, (301) 427-2382; fax (301) 713-0518; kathy.tedesco@noaa.gov
Internet: http://www.climate.noaa.gov/index.jsp?pg=./cpo_pa/cpo_pa_index.jsp&pa=gcc&sub=1
Sponsor: National Oceanic and Atmospheric Administration
1315 East-West Highway, 12th Floor
Silver Spring, MD 20910

NOAA Graduate Sciences Program (GSP) Grants 3375

The Graduate Sciences Program (GSP) is aimed primarily at increasing opportunities for students in NOAA-related fields to pursue research and educational training in atmospheric, environmental, remote sensing and oceanic sciences at minority serving institutions (MSI) when possible. The GSP offers at least two years of NOAA-related research and training opportunities. The program's primary objective is to provide college graduates with degrees in math, science, economics, law, social science, and engineering entry-level employment and hands-on research and work experience at NOAA. The program is designed to improve NOAA's outreach and recruitment efforts of under-represented individuals in the scientific mission-related occupational fields, i.e., fishery biology, environmental law, meteorology, cartography, oceanography, hydrology, computer science, environmental economics, and engineering. The award is equal to 640 hours paid employment, full tuition, housing allowance, meal allowance, and travel.
Requirements: Qualified candidates must have and maintain a minimum GPA of 3.0 or higher every semester/quarter.
Geographic Focus: All States
Date(s) Application is Due: Jan 31
Contact: Pritti Brahma, (301) 713-9437, ext. 118; priti.brahma@noaa.gov
Internet: http://epp.noaa.gov/grad_science/welcome.html
Sponsor: National Oceanic and Atmospheric Administration
1315 East-West Highway, Room 10725
Silver Spring, MD 20910

NOAA Great Lakes Habitat Restoration Partnership Grants 3376

The NOAA Great Lakes Habitat Restoration Program invites applications requesting funding to establish one or more regional habitat restoration partnership(s) for 1 to 3 years. The partnership(s) is expected to catalyze the implementation of habitat restoration projects that will benefit coastal resources through improved Great Lakes habitat quality. The centerpiece of the program will be one or more restoration projects in an Area of Concern that: are based on strong science and data availability; are ecosystem focused; and, involve significant problems and lake-wide improvements. Project areas should include locations where: maximum use can be made of on-going restoration efforts and partnerships; availability of matching funds are met; the problem is significant to the Great Lakes region, NOAA's mission and established priorities; and there is a scientific merit in restoration. NOAA envisions working jointly on such a partnership(s) through its Great Lakes Habitat Restoration Program (GLHRP) to fund and administer projects that support community-identified priorities such as: restoring and enhancing critical, nearshore areas, tributaries and connecting channels; remediating basin-wide sources of stress; protecting healthy functioning areas; and monitoring ecosystem health.
Requirements: Eligible applicants are institutions of higher education, hospitals, other non-profits, commercial (for-profit) organizations, organizations under the jurisdiction of foreign governments, international organizations, state, local and Indian tribal governments whose applications propose to benefit Great Lakes coastal and open-lake habitats.
Restrictions: Applications from federal agencies or employees of federal agencies will not be considered.
Geographic Focus: All States
Date(s) Application is Due: Aug 31
Amount of Grant: 250,000 USD
Contact: Jenni Wallace, (301) 713-0174, ext. 191; GLHRP.GLERL@NOAA.gov
Internet: http://www07.grants.gov/search/search.do;jsessionid=LQCZGLyvwpgGPznL6jnLRyjK1vpGyTRsfyCJKLZtfJnqhLHphMnJ!280595828?oppId=14637&flag2006=true&mode=VIEW
Sponsor: U.S. Department of Commerce
1401 Constitution Avenue, NW, Room 5128
Washington, DC 20230

NOAA International Coral Reef Conservation Project Grants 3377

The NOAA Coral Reef Conservation Grant Program, as authorized under the Coral Reef Conservation Act of 2000, provides matching grants of financial assistance for international coral reef conservation projects. The Program solicits proposals under four funding categories: Promote Watershed Management in the Wider Caribbean, Brazil, and Bermuda; Support Planning for Effective Marine Protected Area Management; Encourage the Development of National Networks of Marine Protected Areas in the Wider Caribbean, Bermuda, Brazil, Southeast Asia, and the South Pacific; and Promote Regional Socio-Economic Training and Monitoring in Coral Reef Management in the Wider Caribbean, Brazil, Bermuda, the Western Indian Ocean, the Red Sea, the South Pacific, South Asia, and Southeast Asia. Each funding category has specific applicant and project eligibility criteria. For specific country eligibility per category, refer to individual category descriptions online. Pre-applications are due on November 3, with full applications received by February 13.
Requirements: Eligible applicants include institutions of higher education, U.S. and international non-profit organizations, and commercial organizations. The proposed work must be conducted at a non-U.S. site.
Restrictions: U.S. federal agencies and individuals are not eligible.
Geographic Focus: All States
Date(s) Application is Due: Feb 13; Nov 3
Amount of Grant: 15,000-80,000 USD
Contact: Scot Frew, (301) 713-3078, ext. 220; Scot.Frew@noaa.gov
Internet: http://www07.grants.gov/search/search.do;jsessionid=LQCZGLyvwpgGPznL6jnLRyjK1vpGyTRsfyCJKLZtfJnqhLHphMnJ!280595828?oppId=42317&flag2006=false&mode=VIEW
Sponsor: U.S. Department of Commerce
1401 Constitution Avenue, NW, Room 5128
Washington, DC 20230

NOAA Joint Hurricane Testbed Grants 3378

The Office of Oceanic and Atmospheric Research (OAR), National Oceanic and Atmospheric Administration (NOAA), is soliciting LOIs (Letters of Intent) under the United States Weather Research Program (U.S.WRP), as administrated by the U.S.WRP Joint Hurricane Testbed (JHT). This notice: provides guidelines for the submission of full proposals; describes opportunities and application procedures for the transfer of relevant research and technology advances into tropical cyclone analysis and forecast operations; and calls for researchers to submit proposals to test and evaluate, and modify if necessary, in a quasi-operational environment, their own scientific and technological research applications. Projects satisfying metrics for success and operational constraints may be selected for operational implementation by the operational center(s) after the completion of the JHT-funded work. The period of the award is from one to two years. LOIs must be submitted by July 31, and full proposals are due by October 30.
Requirements: Eligible applicants are institutions of higher education; other nonprofits; commercial organizations; foreign governments; organizations under the jurisdiction of foreign governments; international organizations; state, local and Indian tribal governments; and Federal agencies. Applications from non-Federal and Federal agencies will be competing against each other.
Geographic Focus: All States
Date(s) Application is Due: Jul 31; Oct 30
Amount of Grant: 50,000-200,000 USD
Contact: Dorothy Fryar, Program Contact; (301) 734-1179; Dorothy.Fryar@noaa.gov
Internet: http://www07.grants.gov/search/search.do;jsessionid=LQCZGLyvwpgGPznL6jnLRyjK1vpGyTRsfyCJKLZtfJnqhLHphMnJ!280595828?oppId=42328&flag2006=false&mode=VIEW
Sponsor: National Oceanic and Atmospheric Administration
1315 East-West Highway, Room 11209
Silver Spring, MD 20910

NOAA Joint Research Institutes Grants 3379

The objective of this grants program is to promote collaboration between university scientists and NOAA scientists so that academic and government researchers can work together to learn about and apply their knowledge to environmental processes. The NOAA Joint Research Institutes are centers of excellence in research relevant to understanding the Earth's oceans, atmosphere, Great Lakes, Arctic regions, and regions similar to the U.S. intermountain west, as well as the solar-terrestrial environment. Cutting-edge research is being conducted under a broad umbrella of environmental themes, such as climate variability, air quality, sea-ice dynamics, coastal ecosystems, fisheries oceanography, hydrology, and weather research and forecasting. Contact the office for further details.
Requirements: Major state universities located near the NOAA environmental research laboratories are eligible to apply.
Geographic Focus: All States
Amount of Grant: 3,000-150,000 USD
Contact: Dr. John Cortinas Jr., Joint Institute Program Manager; (301) 713-2465, ext. 206; fax (301) 713-0158; john.cortinas@noaa.gov
Internet: http://www.oar.noaa.gov/organization/backgrounders/joint.html
Sponsor: National Oceanic and Atmospheric Administration
1315 East-West Highway, Room 11554
Silver Spring, MD 20910

NOAA Living Marine Resources Cooperative Science Center Internships/ Fellowships 3380

The Mission of the Center is to conduct research congruent with the interests of NOAA Fisheries and to prepare students for careers in research, management, and public policy that support the sustainable harvest and conservation of our nation's living marine resources. Research activities of the Center fall into four thematic areas and is a collaborative effort among researchers from the partner institutions and NOAA Fisheries staff. The thematic areas of research are: Fisheries Socioeconomics, Quantitative Fisheries, Essential Fish Habitat, and Aquaculture. The Internship/Fellowship program

is a NOAA partnership with the University of Maryland Eastern Shore, Delaware State University, Savannah State University, Hampton University, the UMBI Center of Marine Biotechnology, and the University of Miami.
Geographic Focus: All States
Contact: Dionne Hoskins; (312) 352-8925; dionne.hoskins@noaa.gov
Internet: http://www.ssufisheries.com/
Sponsor: National Oceanic and Atmospheric Administration
P.O. Box 20228
Savannah, GA 31404

NOAA Living Marine Resources Cooperative Science Center Mini Grants 3381
The program has been established by the Living Marine Resources Cooperative Science Center (LMRCSC) to award and distribute funds up to $1,000.00 for research projects and/or travel directly related to the goals, objectives, and Mission of the LMRCSC. Requests for fee payment are accepted up to $500. Suggested activities include travel to NOAA laboratories or partner institutions, student centered projects, resource enhancement that encourages and sustains research and collaboration. Travel or projects with clear NOAA partnerships will be given preferential consideration. Applications are received and awarded on a continuing basis as funding allows.
Requirements: Eligible research and travel applicants are SSU faculty or fulltime staff, and students enrolled at Savannah State University. Fee applicants must be graduate students matriculating at SSU. All students must have and maintain a 3.0 gpa.
Geographic Focus: Georgia
Amount of Grant: Up to 1,000 USD
Contact: Dionne Hoskins; (312) 352-8925; dionne.hoskins@noaa.gov
Internet: http://www.ssufisheries.com/
Sponsor: National Oceanic and Atmospheric Administration
P.O. Box 20228
Savannah, GA 31404

NOAA Miami Lab Summer Internships 3382
The Miami Lab Summer Internship Program provides educational opportunities for both high school and undergraduate level students with research positions offering on-the-job training experience. Students receive direct supervision from faculty members at the Rosenstiel School of Marine and Atmospheric Science (RSMAS) and from staff scientists at the Atlantic Oceanographic Marine Laboratory (AOML) and the Southeast Fisheries Center (SEFC). This program provides students an opportunity to work in a scientific environment and to appraise career opportunities in oceanographic science. The annual deadline is the end of May, and grant awards vary depending upon the experience of the individual student. Interested applicants should contact the office directly.
Geographic Focus: All States
Date(s) Application is Due: May 31
Contact: Essie C. Duffie, (305) 361-4237; essie.c.duffie@noaa.gov
Sponsor: U.S. Department of Commerce
1401 Constitution Avenue, NW, Room 5128
Washington, DC 20230

NOAA Mid-Atlantic Bight (MAB) National Undersea Research Center Grants 3383
The MAB Center supports undersea research to improve knowledge of processes governing change and stability in ecosystems of coastal and oceanic waters and seabed south of Montauk Point, New York and off New York, New Jersey, Pennsylvania, Delaware, Maryland and Virginia. The Center provides undersea technology with an emphasis on the REMU.S. Autonomous Undersea Vehicle (AUV) and the LEO-15 Observatory and associated sensors in order to address research questions at a range of scales, and is now requesting research proposals related to these priorities. Qualified proposals are eligible to receive support in the way AUV and LEO-15 Node access, including operators and divers, data download and processing, and AUV shipment and operator travel costs. Modest funds will be available for development project supplies. Proposals should provide a summary of the proposed research demonstration or development, describe research goals and facilities/ equipment requirements, outline time or logistic constraints, specify area of operations including depths, and estimate the level of support required. Demonstration proposals have a limit of $15,000 per project, to include platform costs, travel, transportation, and salaries for NURP technical support (such as REMU.S. operators). Proposals received after May 21 will be reviewed on a rotational basis dependent on funds.
Requirements: Demonstration or development proposals are required and must be submitted by email by May 21.
Restrictions: Proposals for support of science beyond technical demonstrations or developments are discouraged during this funding cycle.
Geographic Focus: All States
Date(s) Application is Due: May 21
Amount of Grant: 15,000 USD
Contact: Michael P. De Luca, Director; (732) 932-6555, ext. 512; fax (732) 932 8578; deluca@marine.rutgers.edu
Internet: http://marine.rutgers.edu/nurp/proposals.html
Sponsor: U.S. Department of Commerce
1401 Constitution Avenue, NW, Room 5128
Washington, DC 20230

NOAA Mid-Atlantic Research Set-Aside Grants 3384
The RSA program provides a mechanism to fund research and compensate vessel owners through the sale of fish harvested under the research quota. Vessels participating in an approved research project to harvest and land fish in excess of any imposed trip limit or during fishery closures. Landings from such trips are sold to generate funds that help defray the costs associated with the approved research projects. No Federal funds are provided for research under this notification. NMFS is soliciting proposals for research activities concerning the summer flounder, scup, black sea bass, Loligo squid, Illex squid, Atlantic mackerel, butterfish, Atlantic bluefish, and tilefish fisheries. The closing date for applications is March 24.
Requirements: Eligible applicants include institutions of higher education, hospitals, other nonprofits, commercial organizations, individuals, and state, local, and Native American tribal governments.
Restrictions: Federal agencies and institutions are not eligible to receive Federal assistance under this notice. Additionally, employees of any Federal agency or Regional Fishery Management Council are ineligible to submit an application under this program.
Geographic Focus: All States
Date(s) Application is Due: Mar 24
Contact: Cheryl A. Corbett, Cooperative Programs Specialist; (508) 495-2070; fax (508) 495-2004; cheryl.corbett@noaa.gov
Internet: http://www07.grants.gov/search/search.do;jsessionid=HWbpMDzmZ104MfvCGSbHhsGGPnb8dGTRG2hM0vfdTvpFRTZdkPyK!-796138783?oppId=16824&flag2006=true&mode=VIEW
Sponsor: National Oceanic and Atmospheric Administration
166 Water Street
Woods Hole, MA 02543

NOAA Monitoring and Event Response for Harmful Algal Blooms Grants 3385
The purpose of this document is to advise the public that NCCOS/CSCOR is soliciting proposals for two types of research projects: MERHAB-targeted and MERHAB-regional. MERHAB-targeted proposals will incorporate tools, approaches and technologies from HAB research programs into existing harmful algal bloom (HAB) monitoring programs. MERHAB regional proposals will create partnerships to enhance existing and initiate new HAB monitoring capabilities that will provide managers with timely information needed to mitigate HAB impacts on coastal communities.
Requirements: Institutions of higher education, not-for- profit institutions, State, commercial organizations, local and Indian tribal governments and Federal agencies are eligible to apply.
Geographic Focus: All States
Date(s) Application is Due: Oct 2
Contact: Rob Magnien; (301) 713-3338 or (301) 713-3020; rob.magnien@noaa.gov
Internet: http://www.cop.noaa.gov/opportunities/grants/fundingarchive/fy2007/MERHAB_FFO_0606.html
Sponsor: National Oceanic and Atmospheric Administration
1305 East-West Highway, N/SCI2 SSMC4, Room 8243
Silver Spring, MD 20910

NOAA Monkfish Research Set-Aside Grants 3386
The intent of this program is for fishing vessels to utilize these research DAS to conduct monkfish related research, rather than their allocated monkfish DAS, thereby eliminating any cost to the vessel associated with using a monkfish DAS. Landings from such research trips may be sold to generate funds to help defray research costs. No Federal funds are provided for research under this notification. Rather, projects funded under the Program would be provided with additional opportunity to harvest monkfish, and the catch sold to generate income to offset research costs. Projects funded under an RSA DAS award must enhance the knowledge of the monkfish fishery resource or contribute to the body of information on which management decisions are made. The Councils and NMFS will give priority to funding research proposals in the following general areas: research concerning monkfish bycatch and discards; research to minimize bycatch and interactions with sea turtles and other protected species; research to minimize impacts of monkfish fishery on essential fish habitat (EFH), or other sensitive habitats; research to establish an exempted monkfish trawl fishery in the Northern Fishery Management Area; research on the biology or population structure and dynamics of monkfish; tagging studies; mesh and gear selectivity studies, including studies on gear efficiency; and cooperative stock assessment surveys. Note that the research subject areas listed above are not listed in order of priority.
Requirements: Eligible applicants are institutions of higher education, hospitals, other nonprofits, commercial organizations, individuals, state, local and Indian tribal governments.
Restrictions: Federal agencies or institutions are not eligible to receive Federal assistance under this notice. Also, a person is not eligible to submit an application under this program if he/she is an employee of any Federal agency. Fishery Management Council members who are not Federal employees may submit an application.
Geographic Focus: All States
Date(s) Application is Due: Aug 31
Contact: Steve Drescher; (301) 713-0942, ext. 122; Steve.J.Drescher@noaa.gov
Internet: http://www07.grants.gov/search/search.do;jsessionid=HV8GpydVhXQylKWtj8sRt2JWS1dB9H223h3VT8tly7LpfYB52vmB!2066172073?oppId=14650&flag2006=true&mode=VIEW
Sponsor: National Oceanic and Atmospheric Administration
1315 East-West Highway, 12th Floor
Silver Spring, MD 20910

NOAA National Climatic Data Center (NCDC) Internships 3387
Students from across the Nation are invited to participate in the NCDC Internship Program. Topics for investigation will cover a variety of scientific areas. Previous student research has ranged from hurricanes to configuration management to precipitation trends. Any climate variable holds possibilities for an opportunity of study. Each student is expected to prepare a presentation. The presentation can be completed using a team approach. Several students are permitted to work together if the research topics are

related. All students hired will be employees of The National Oceanic and Atmospheric Administration or the designated program. Atmospheric Administration or the designated program. Salary is commensurate with experience.
Geographic Focus: All States
Date(s) Application is Due: Feb 28
Contact: Carmella Davis Watkins, (828) 271-4105; carmella.davis.watkins@noaa.gov
Internet: http://www.ncdc.noaa.gov/oa/about/about.html
Sponsor: National Oceanic and Atmospheric Administration
151 Patton Avenue
Asheville, NC 28801

NOAA National Coastal and Estuarine Research and Technology Grants 3388
The National Oceanic and Atmospheric Administration (NOAA) seeks to establish a national estuarine research and technology program which operates in partnership with the National Estuarine Research Reserve System (NERRS). Funds will be used to conduct collaborative research and transform the best available science into practical innovative tools that coastal managers can use to detect, prevent, and reverse the impacts of coastal pollution and habitat degradation. Additionally, the program will provide coastal and estuarine managers a better understanding of what tools are available, how well they work, and how best to apply them to detect, prevent, and reverse the impacts of coastal pollution and habitat degradation.
Requirements: Eligible applicants are non-Federal institutions of higher education, other non-profits, commercial organizations, and state and local governments that possess the statutory authority to receive financial assistance.
Geographic Focus: All States
Date(s) Application is Due: Sep 9
Amount of Grant: Up to 5,232,000 USD
Contact: Erica Seiden; (301) 563-1172; Erica.Seiden@noaa.gov
Internet: http://www07.grants.gov/search/search.do;jsessionid=LQCZGLyvwpgGPznL6jnLRyjK1vpGyTRsfyCJKLZtfJnqhLHphMnJ!280595828?oppId=42319&flag2006=false&mode=VIEW
Sponsor: U.S. Department of Commerce
1401 Constitution Avenue, NW, Room 5128
Washington, DC 20230

NOAA National Estuarine Research Reserve Land Acquisition and Construction Grants 3389
he National Estuarine Research Reserve System consists of estuarine areas of the United States and its territories which are designated and managed for research and educational purposes. Each reserve within the system is chosen to represent different bio-geographic regions and to include a variety of ecosystem types in accordance with the classification scheme of the national program. Through the funding of designated reserve agencies and universities to undertake land acquisition and construction projects that support the NERRS purpose, NOAA will strengthen protection of key land and water areas; enhance long-term protection of the area for research and education; and provide for facility and exhibit construction. It is anticipated that the awards will run for up to two years. In the past, funding for land acquisition/construction awards has ranged in amount from approximately $50,000 to $3 million.
Requirements: Eligible applicants are NERR lead state agencies or universities in coastal states.
Geographic Focus: All States
Date(s) Application is Due: Nov 30
Amount of Grant: 50,000-3,000,000 USD
Contact: Nina Garfield, (301) 563-1171, ext. 171; Nina.Garfield@noaa.gov
Internet: http://www07.grants.gov/search/search.do;jsessionid=LQCZGLyvwpgGPznL6jnLRyjK1vpGyTRsfyCJKLZtfJnqhLHphMnJ!280595828?oppId=42321&flag2006=false&mode=VIEW
Sponsor: U.S. Department of Commerce
1401 Constitution Avenue, NW, Room 5128
Washington, DC 20230

NOAA National Estuarine Research Reserve System Graduate Research Fellowships 3390
The National Estuarine Research Reserve System's Graduate Research Fellowship provides master's degree students and Ph.D. candidates with an opportunity to conduct research of local and national significance that focuses on enhancing coastal zone management. Fellows conduct their research within a National Estuarine Research Reserve and gain hands-on experience by participating in the host reserve's research and monitoring programs. Graduate Research Fellowship projects are based on the reserves' local needs, the reserve system's national priorities and the students' interest. Fellows receive a stipend of $20,000 per year, and Fellowships may be funded for up to three years.
Requirements: Students admitted to or enrolled in a full-time master's or doctoral program at U.S. accredited universities are eligible to apply.
Restrictions: Electronically transmitted letters of support are not acceptable.
Geographic Focus: All States
Date(s) Application is Due: Nov 1
Contact: Alison Krepp, (301) 713-3155, ext. 105; Alison.Krepp@noaa.gov
Internet: http://nerrs.noaa.gov/Fellowship/welcome.html
Sponsor: National Oceanic and Atmospheric Administration
1305 East-West Highway, N/SCI2 SSMC4, Room 8243
Silver Spring, MD 20910

NOAA National Height Modernization Grants 3391
The purpose of this notice is to solicit proposals for cooperative agreements and/or grants between NOAA and partnering entities in the United States, implementing NOAA's National Height Modernization Program (NHMP) Plan. Proposals submitted in response to this announcement should contribute to the beneficial public outcomes associated with the five priority issues in this plan which consists of enhancing the vertical component of the National Spatial Reference System (NSRS); enabling users to access the vertical component of the NSRS; outreach and education regarding geospatial issues and activities as they relate to NHMP; capacity building and technology transfer as they relate to NHMP; coordination, cooperation, and collaboration with other entities to accomplish common goals as they relate to NHMP. This competition is focused on the geography of the United States and its territories in response to NOAA's NHMP Regional Implementation Plan and subsequent congressional appropriations. Pre-proposals are required and due by August 1. Full proposals are due by September 2.
Requirements: Eligible funding applicants are institutions of higher education, state, local and Indian tribal governments.
Geographic Focus: All States
Date(s) Application is Due: Aug 1; Sep 2
Contact: Gilbert Mitchell, (301) 713-3228, ext. 114; Glbert.Mitchell@NOAA.gov
Internet: http://www.grants.gov/search/search.do;jsessionid=LP0hqWfp11PVPKSr9yfL1LLhNQTn2gsNBmYzZtzGKTQhJRdRBYxW!-2005900814?oppId=42318&flag2006=false&mode=VIEW
Sponsor: National Oceanic and Atmospheric Administration
1305 East-West Highway, N/SCI2 SSMC4, Room 8243
Silver Spring, MD 20910

NOAA National Sea Grant Gulf of Mexico Oyster Industry Grants 3392
The goal of this program is to encourage multi-disciplinary research and extension projects that contribute directly to the efficiency and profitability of oyster-related businesses and to the safety of oyster products. This competition generally runs every two years. Annual deadlines vary. Applicants should visit the web site for further details.
Requirements: Institutions of higher education, not-for- profit institutions, State, commercial organizations, local and Indian tribal governments and Federal agencies are eligible to apply.
Geographic Focus: All States
Contact: Megan Agy; (301) 734-1066 or (301) 734-1077; Megan.Agy@noaa.gov
Internet: http://www.seagrant.noaa.gov/funding/rfp.html#Gulf%20Oyster
Sponsor: National Oceanic and Atmospheric Administration
1305 East-West Highway, N/SCI2 SSMC4, Room 8243
Silver Spring, MD 20910

NOAA National Sea Grant Industry Fellowships 3393
The Sea Grant Industry Fellowship provides support for graduate students who are pursuing research and development projects in topics of interest to a particular industry/company. In a true partnership, the student, the faculty advisor, the Sea Grant college or institute, and the industry representative work together on a project from beginning to end.
Requirements: The Fellowship is available to U.S. citizens who are graduate students enrolled in PhD degree programs in academic institutions in the United States and its territories.
Geographic Focus: All States
Date(s) Application is Due: Jan 20
Contact: Miguel Lugo; (301) 734-1077, ext. 1075; miguel.lugo@noaa.gov
Internet: http://www.seagrant.noaa.gov/funding/fellowships.html#industry
Sponsor: National Oceanic and Atmospheric Administration
1305 East-West Highway, N/SCI2 SSMC4, Room 8243
Silver Spring, MD 20910

NOAA National Sea Grant John A. Knauss Marine Policy Fellowships 3394
The program is designed to provide a unique educational experience to students who have an interest in marine/ocean/Great Lakes resources and in the national policy decisions affecting those resources. The program matches highly qualified graduate students with hosts in the legislative branch, the executive branch, or appropriate associations/institutions located in the Washington, DC, area for a one-year paid fellowship.
Requirements: A student who, at the time of application, is in a graduate or professional degree program in a marine-related field at an accredited institution of higher education in the United States may apply. All applicants must submit an application to one of the Sea Grant College Programs in their state. If there is no program in the applicant's state, he or she should apply through the closest state Sea Grant College Program. A list of the programs may be obtained from the Web site http://www.mdsg.umd.edu/NSGO/SGDirectors.html.
Geographic Focus: All States
Date(s) Application is Due: Feb 20
Amount of Grant: 41,500 USD
Contact: Miguel Lugo, Program Manager; (301) 734-1077, ext. 1075; fax (301) 713-0799; miguel.lugo@noaa.gov
Internet: http://www.seagrant.noaa.gov/funding/rfp.html#knauss
Sponsor: National Oceanic and Atmospheric Administration
1315 East-West Highway, Room 8243
Silver Spring, MD 20910

NOAA National Sea Grant Oyster Disease Research Grants 3395
This extensive program of ongoing research coupled with outreach and management efforts aims to better serve the restoration of healthy populations of oysters in the nation's coastal waters through research related to oyster disease. This competition generally runs every two years. Annual deadlines vary. Applicants should visit the web site for further details.

Requirements: Institutions of higher education, not-for- profit institutions, State, commercial organizations, local and Indian tribal governments and Federal agencies are eligible to apply.
Geographic Focus: All States
Contact: Megan Agy; (301) 734-1066 or (301) 734-1077; Megan.Agy@noaa.gov
Internet: http://www.seagrant.noaa.gov/funding/rfp.html#oyster
Sponsor: National Oceanic and Atmospheric Administration
1305 East-West Highway, N/SCI2 SSMC4, Room 8243
Silver Spring, MD 20910

NOAA Northern Gulf of Mexico Ecosystems and Hypoxia Assessment Grants 3396
NGOMEX has two distinct components. The Modeling the Causes of Hypoxia component takes a regional ecosystem prediction approach to advance model development assessing the association between the northern Gulf hypoxic zone and causative factors. The proposed research for this competition should be 3-5 years in duration. The Modeling the Impacts of Hypoxia component takes an ecosystem stressors approach to advance understanding of hypoxia on ecologically and commercially important living resource populations of the northern Gulf of Mexico ecosystem. These projects should be 3 to 5 years of duration. Funding is contingent upon the availability of annual Federal appropriations. It is anticipated that final recommendations for funding under this announcement will be made by February and that projects funded under this announcement will have a July 1 through August 1 start date.
Geographic Focus: All States
Date(s) Application is Due: Oct 20
Contact: Rob Magnien; (301) 713-3338 or (301) 713-3020; rob.magnien@noaa.gov
Internet: http://www.cop.noaa.gov/opportunities/grants/fundingarchive/fy2009.html
Sponsor: National Oceanic and Atmospheric Administration
1315 East-West Highway, Room 8243
Silver Spring, MD 20910

NOAA Ocean Exploration Omnibus: Marine Archaeology Grants 3397
Proposals for exploration and discovery of significant maritime heritage sites are the priority. Submerged, previously subaerial, landscapes, shipwrecks, and other maritime cultural sites are typical focus subjects of the program. Proposals will emphasize the early phases of field archaeology: searching, locating, evaluating or inventorying sites. All applicants must convincingly describe: the archaeological significance of their site(s), and their importance to human history; how their research fits within the realm of exploration; and why their methodologies are innovative and make the most economical use of current marine technology. Pre-proposals are required and due by August 29. Full proposals are due by October 21.
Requirements: Eligible applicants are: institutions of higher education; other nonprofits; commercial organizations; foreign governments; organizations under the jurisdiction of foreign governments; international organizations; state, local and Indian tribal governments; and Federal agencies.
Restrictions: The Announcement does not invite proposals to support later phases of archaeological research, such as intensive site excavations, and major conservation projects.
Geographic Focus: All States
Date(s) Application is Due: Aug 29; Oct 21
Amount of Grant: Up to 400,000 USD
Contact: Nicolas Alvarado; (301) 734-1015; OAR.OE.FAQ@noaa.gov
Internet: http://www.grants.gov/search/search.do;jsessionid=LP0hqWfp11PVPKSr9yfL1LLhNQTn2gsNBmYzZtzGKTQhJRdRBYxW!-2005900814?oppId=42332&flag2006=false&mode=VIEW
Sponsor: National Oceanic and Atmospheric Administration
1315 East-West Highway, SSMC 3, Room 10244
Silver Spring, MD 20910

NOAA Ocean Exploration Omnibus: Ocean Exploration Grants 3398
OE is seeking pre-proposals and full proposals to support its mission, consistent with NOAA's Strategic Plan, to search, investigate, and document poorly-known and unknown areas of the ocean and Great Lakes through interdisciplinary exploration, and to advance and disseminate knowledge of the ocean environment and its physical, chemical, and biological resources. Competitive OE proposals will be bold, innovative and interdisciplinary in their approach to Ocean Exploration. NOAA OE anticipates a total of approximately $1,400,000 including costs for ship and submersible assets will be available through this announcement. Pre-proposals are required and due by August 29. Full proposals are due by October 21.
Requirements: Eligible applicants are: institutions of higher education; other nonprofits; commercial organizations; foreign governments; organizations under the jurisdiction of foreign governments; international organizations; state, local and Indian tribal governments; and Federal agencies.
Restrictions: Only exploratory proposals will be funded, any other kind of project will not be reviewed.
Geographic Focus: All States
Date(s) Application is Due: Aug 29; Oct 21
Amount of Grant: 350,000 USD
Contact: Nicolas Alvarado; (301) 734-1015; OAR.OE.FAQ@noaa.gov
Internet: http://www.grants.gov/search/search.do;jsessionid=LP0hqWfp11PVPKSr9yfL1LLhNQTn2gsNBmYzZtzGKTQhJRdRBYxW!-2005900814?oppId=42333&flag2006=false&mode=VIEW
Sponsor: National Oceanic and Atmospheric Administration
1315 East-West Highway, SSMC 3, Room 10244
Silver Spring, MD 20910

NOAA Pacific Islands Assistantships 3399
The NOAA Pacific Services Center (PSC) and the NOAA Coastal Services Center have supported the Pacific Islands Assistantship (PIA) since 2001. The PIA program matches highly qualified recipients of professional degrees with hosts from the four Pacific Island coastal resource management organizations of American Samoa, the Commonwealth of the Northern Mariana Islands, Guam, and Hawaii. New assistants are placed every two years for two-year terms. The focus of each assistant depends on the needs of the programs and partners.
Geographic Focus: All States
Date(s) Application is Due: Jan 26
Contact: NOAA Pacific Services; (808) 532-3202; fax (808) 532-3224; psc@noaa.gov
Internet: http://www.csc.noaa.gov/cms/fellows/pacific-island.html
Sponsor: National Oceanic and Atmospheric Administration
737 Bishop Street, Suite 2250
Honolulu, HI 96813

NOAA Population Dynamics Recruiting Grants 3400
The Population Dynamics Recruiting Program (PDRP) was developed to increase the number of highly qualified applicants for population dynamics positions with the Southeast Fisheries Science Center. The PDRP provides unique opportunities for top undergraduate students in the southeastern U.S. to learn about the field of marine resources population dynamics first-hand, all expenses paid, from some of the top experts in the field. Students interested must enter the PDRP through a one week Undergraduate Workshop, held in early January, each year, in the Florida Keys . The highly competitive application process seeks students with a strong quantitative background and a basic understanding of ecology. The second phase of the PDRP is the Summer Program where 4-8 of the top, PDRP Workshop participants spend part of their summers learning about current, important marine resource issues. The top students participating in the summer program are offered M.S. Graduate Assistantships with the NMFS RTR Unit at Virginia Tech.
Geographic Focus: All States
Date(s) Application is Due: Oct 15
Contact: Dr. Jim Berkson, (540) 231-5910 or (540) 552-2510; Jim.Berkson@noaa.gov
Internet: http://www.nmfs.vt.edu/sumprograms.htm
Sponsor: National Oceanic and Atmospheric Administration
100 Cheatham Hall
Blacksburg, VA 24061

NOAA Regional Ecosystem Prediction Program on Coral Reef Ecosystem Studies (CRES) Grants 3401
The goal of this funding opportunity is to utilize existing scientific tools and approaches (e.g., biophysical models; coupled watershed and hydrodynamic models) within a social, cultural, and economic framework to develop and implement effective coastal ecosystem management practices in the Pacific Islands. Proposals should be regional in scale, interdisciplinary, comprehensive, integrated, and include multiple investigators to develop capabilities for innovative forecasts and predictions for improved management and control capabilities. Projects under this program will be 3 to 5 years in duration. Funding is contingent upon the availability of annual Federal appropriations. It is anticipated that final recommendations for funding under this announcement will be made by February and that projects funded under this announcement will have a July 1 through August 1 start date.
Geographic Focus: All States
Date(s) Application is Due: Oct 9
Contact: Rob Magnien; (301) 713-3338 or (301) 713-3020; rob.magnien@noaa.gov
Internet: http://www.cop.noaa.gov/opportunities/grants/fundingarchive/fy2009.html
Sponsor: National Oceanic and Atmospheric Administration
1305 East-West Highway, N/SCI2 SSMC4, Room 8243
Silver Spring, MD 20910

NOAA Regional Integrated Sciences and Assessments (RISA) Grants 3402
The program supports integrated, place-based research across a range of social, natural, and physical science disciplines to expand decision-makers' options in the face of climate change and variability at the regional level. RISA teams are comprised of researchers from the physical, natural, engineering and social sciences who work together and partner with stakeholders in a region to determine how climate impacts key resources and how climate information could aid in decision making and planning for those stakeholders. This effort often includes analyses of adaptation options in the face of a varying and changing climate. Currently, the RISA program is soliciting proposals to launch one new RISA effort in a region of the U.S. not currently covered by the program. The region needs to be one where drought is a serious concern to decision makers and public policy officials. Proposed projects should build on progress already achieved by integrated climate-society research and assessments projects in the region in terms of working with stakeholders and advancing regional climate impacts science. NOAA plans to have funding on the order of up to $500,000 to $600,000 per year available for this new RISA. While it is in the best interest of the applicants and their institutions to submit a Letter of Intent (LOI) by the July 23 due date, it is not a requirement; applicants who do not submit an LOI are allowed to submit a full proposal. For those that do submit an LOI, full proposals will be encouraged only for LOIs deemed relevant.
Requirements: Eligible applicants are institutions of higher education, other nonprofits, commercial organizations, international organizations, and state, local and Indian tribal governments.
Restrictions: Federal agencies or institutions are not eligible to receive Federal assistance under this notice.
Geographic Focus: All States
Date(s) Application is Due: Sep 24
Amount of Grant: 500,000-600,000 USD
Contact: Hannah Campbell, (301) 734-1208; hannah.campbell@noaa.gov

Internet: http://www.climate.noaa.gov/index.jsp?pg=/opportunities/opp_index.jsp&opp=2008/risa_info.jsp
Sponsor: National Oceanic and Atmospheric Administration
1315 East-West Highway, 12th Floor
Silver Spring, MD 20910

NOAA Scientific Data Stewardship (SDS) Grants **3403**
The Program seeks to support the routine, operational production of Climate Data Records (CDRs) for data from the atmosphere, oceans, and land surface. This production requires collaboration between experts in the climate community and experts in data management, and must be informed by user feedback on the accessibility and usability of the produced CDRs. Applicants to this program must demonstrate that proposed CDRs have reached a critical level of maturity in science and data preservation and are of demonstrable societal benefit to be considered for funding under this Program.
Requirements: Eligible applicants are institutions of higher education, other nonprofits, commercial organizations, international organizations, and state, local and Indian tribal governments.
Restrictions: Federal agencies or institutions are not eligible to receive Federal assistance under this notice.
Geographic Focus: All States
Date(s) Application is Due: Sep 24
Amount of Grant: 50,000-200,000 USD
Contact: Diane Brown; (301) 734-1206; fax (301) 713-0518; diane.brown@noaa.gov
Internet: http://www.climate.noaa.gov/index.jsp?pg=/opportunities/opp_index.jsp&opp=2008/sds_info.jsp
Sponsor: National Oceanic and Atmospheric Administration
1315 East-West Highway, 12th Floor
Silver Spring, MD 20910

NOAA Sea Scallop Research Set-Aside Grants **3404**
NMFS requests research proposals for fishing to utilize portions of the total allowable catch (TAC) and Days-at-Sea (DAS) allowance in the Atlantic sea scallop fishery that have been set aside by the New England Fishery Management Council (Council) for sea scallop research endeavors under a research set-aside (RSA) program. The program provides a mechanism to fund research and compensate vessel owners through the sale of fish harvested under the research quota. Vessels participating in an approved research project may be authorized by the Administrator, Northeast Region, NMFS (Regional Administrator), to harvest and land species in excess of any imposed trip limit or to harvest and land during fishery closures. Landings from such trips would be sold to generate funds that would help defray the costs associated with research projects. The DOC/NOAA encourages applications from members of the fishing community and applications that involve fishing community cooperation and participation.
Requirements: Eligible applicants are institutions of higher education, other nonprofits, commercial organizations, international organizations, and state, local and Indian tribal governments.
Restrictions: Federal agencies or institutions are not eligible to receive Federal assistance under this notice. Also, a person is not eligible to submit an application under this program if he/she is an employee of any Federal agency. Fishery Management Council members who are not Federal employees may submit an application.
Geographic Focus: All States
Date(s) Application is Due: Nov 18
Contact: Steve Drescher; (301) 713-0942, ext.122; Steve.J.Drescher@noaa.gov
Internet: http://www07.grants.gov/search/search.do;jsessionid=HV8GpydVhXQylKWtj8sRt2JWS1dB9H223h3VT8tly7LpfYB52vmB!2066172073?oppId=16903&flag2006=true&mode=VIEW
Sponsor: National Oceanic and Atmospheric Administration
1315 East-West Highway, 12th Floor
Silver Spring, MD 20910

NOAA Sector Applications Research Program (SARP) Grants **3405**
The Program supports the overarching goals of the NOAA Climate Program by developing the knowledge base, decision support tools, capacities and partnerships in sectors affected by climate in a substantial and increasingly visible way. SARP is designed to catalyze and support interdisciplinary research, innovative outreach and education activities that enhance the capacity of key socio-economic sectors to respond to and plan for climate variability and change through the use of climate information and related decision support resources. The program is designed to systematically build an interdisciplinary and expressly applicable knowledge base and mechanism for the creation, dissemination and exchange of climate-related research findings and decision support resources critical for understanding and addressing resource management challenges in vital social and economic sectors (e.g., coastal resources, water, agriculture, health, etc.). While it is in the best interest of the applicants and their institutions to submit a Letter of Intent (LOI) by the July 23rd due date, it is not a requirement; applicants who do not submit an LOI are allowed to submit a full proposal. For those that do submit an LOI, full proposals will be encouraged only for LOIs deemed relevant.
Requirements: Eligible applicants are institutions of higher education, other nonprofits, commercial organizations, international organizations, and state, local and Indian tribal governments.
Restrictions: Federal agencies or institutions are not eligible to receive Federal assistance under this notice.
Geographic Focus: All States
Date(s) Application is Due: Sep 24
Amount of Grant: 50,000-200,000 USD
Contact: Nancy Beller-Simms, (301) 734-1205; Nancy.Beller-Simms@noaa.gov
Internet: http://www.climate.noaa.gov/index.jsp?pg=/opportunities/opp_index.jsp&opp=2008/sarp_info.jsp
Sponsor: National Oceanic and Atmospheric Administration
1315 East-West Highway, 12th Floor
Silver Spring, MD 20910

NOAA South Florida Grants **3406**
NCCOS/CSCOR is soliciting 1-year and 2-year proposals to support coastal ecosystem studies in South Florida including Florida Bay , Florida Keys, the Florida Keys National Marine Sanctuary (FKNMS), and adjacent coastal waters. It will provide support for the NOAA South Florida Program (SFP) and the FKNMS. The overall goal of this announcement is to fund high priority research and long term observational data collection needed to model and predict the impacts of Everglades restoration on the South Florida coastal ecosystem and to fulfill NOAA commitments to the South Florida Ecosystem Restoration effort and the Comprehensive Everglades Restoration Plan (CERP).
Requirements: Institutions of higher education, not-for- profit institutions, State, commercial organizations, local and Indian tribal governments and Federal agencies are eligible to apply.
Geographic Focus: All States
Date(s) Application is Due: Sep 29
Contact: Rob Magnien, Director; (301) 713-3338 or (301) 713-3020; fax (301) 713-4044; rob.magnien@noaa.gov
Internet: http://www.cop.noaa.gov/opportunities/grants/fundingarchive/fy2006/sfl06.html
Sponsor: National Oceanic and Atmospheric Administration
1305 East-West Highway, N/SCI2 SSMC4, Room 8243
Silver Spring, MD 20910

NOAA Spherical Display Systems for Earth System Science-Installations and Content Development Grants **3407**
NOAA seeks partners to further promote the inclusion of Earth system science concepts in informal science education venues through the development of exhibits involving spherical display systems. There are two priorities for this competition: priority 1 installations of spherical display systems into public exhibits and, priority 2 development and evaluation of Earth system science modules. A single institution may apply to both priorities, but with separate applications, one to each priority. Applications will be reviewed independently. Funds awarded cannot necessarily pay for all the costs that the recipient might incur in the course of carrying out the project. NOAA anticipates the availability of approximately $4,000,000 of Federal financial assistance
Requirements: Eligible applicants are institutions of higher education, other nonprofits, and state, local and Indian tribal governments in the United States.
Restrictions: Applications to priority 1 requesting Federal support from NOAA of more than $100,000 total will not be considered for funding through this announcement. Applications for priority 2 requesting Federal support from NOAA of more than $300,000 total will not be considered for funding. For profit organizations, foreign institutions, foreign organizations and foreign government agencies are not eligible to apply.
Geographic Focus: All States
Date(s) Application is Due: Oct 30
Amount of Grant: 100,000-300,000 USD
Contact: Carrie McDougall, (202) 482-0875; carrie.mcdougall@noaa.gov
Internet: http://www.oesd.noaa.gov/Spherical_Display_awards.html
Sponsor: National Oceanic and Atmospheric Administration
1401 Constitution Avenue NW, Room 6863
Washington, DC 20230

NOAA Transition of Research Applications to Climate Services Grants **3408**
The TRACS Program transitions experimentally mature climate information tools, methods, and processes, including computer related applications (e.g. web interfaces, visualization tools), from research mode into settings where they may be applied in an operational and sustained manner. TRACS primary goal is to generate sustained delivery of useful climate information products and services to local, regional, national, and international decision and policy makers. TRACS seeks not only to support implementation of these transitions, but also to learn from partners how to better accomplish technology transition processes for public goods applications and improved risk management. The title TRACS, besides evoking a focus of being on-track, should also call attention to three key elements and their interplay. These activities include, transitions (i.e. a focus on partnerships where technology hand-offs occur), research applications (i.e. experimentally developed and tested, end-user-friendly information to support decision making), and climate services (i.e. the routine and timely delivery of that information, including via partnerships).
Requirements: Eligible applicants are institutions of higher education, other nonprofits, commercial organizations, international organizations, and state, local and Indian tribal governments.
Restrictions: Federal agencies or institutions are not eligible to receive Federal assistance under this notice.
Geographic Focus: All States
Contact: Josh Foster; (301) 734-1218; fax (301) 713-0518; josh.foster@noaa.gov
Internet: http://www.climate.noaa.gov/cpo_pa/nctp/
Sponsor: National Oceanic and Atmospheric Administration
1315 East-West Highway, 12th Floor
Silver Spring, MD 20910

NOAA Undersea Research Center Coral Reef Conservation Program Coral Research Grants **3409**
The NOAA Undersea Research Center at the University of North Carolina Wilmington seeks proposals for undersea research projects to be conducted off the southeast U.S. from North Carolina to Texas. This funding opportunity is for one to two year targeted research projects that address one or more of the following threats facing U.S. coral reefs in Florida (includes the Florida Keys, Southeast Florida, and the Eastern Gulf

of Mexico) and the Flower Garden Banks: fishing, pollution, coastal uses (including invasive species), climate change, and extreme events. Proposals should be geared towards management-oriented research to: develop a better understanding of the role of the threats listed above in coral reef ecosystem decline; predict the impacts from and responses of reef communities to anthropogenic and natural stressors; and design, test, and evaluate the effectiveness of specific reef management measures, including marine protected areas and habitat restoration techniques. Pre-proposals are required and must be submitted by email by January 14 to: pottst@uncw.edu.
Requirements: Proposals submitted under the CRCP funding opportunity require 100% non-federal match. Non-federal matching funds may be comprised of a variety of public and private sources and may include in-kind contributions and other non-cash support.
Geographic Focus: All States
Date(s) Application is Due: Jan 14
Amount of Grant: 75,000 USD
Contact: Thomas Potts; (910) 962-2442; fax (910) 962-2444; pottst@uncw.edu
Internet: http://www.uncw.edu/nurc/funding/opportunities.htm
Sponsor: U.S. Department of Commerce
1401 Constitution Avenue, NW, Room 5128
Washington, DC 20230

NOAA Undersea Research Center Ocean Acidification Research at Aquarius Reef Base Grants 3410
The NOAA Undersea Research Center at the University of North Carolina Wilmington seeks proposals for undersea research projects to be conducted off the southeast U.S. from North Carolina to Texas. Private foundation funds in the amount of $75,000 for direct science support are available for a research mission focused on the impacts of ocean acidification on coral reefs at Aquarius Reef Base. Proposals may be written for one year (opportunity 1) or two years (opportunity 2) of direct science support of $75,000 per year and should contain an in situ field component. Undersea system and support vessel time is provided by the Center at no cost to the principal investigator. Pre-proposals are required and must be submitted by email by January 14 to: pottst@uncw.edu.
Requirements: Proposals submitted under the CRCP funding opportunity require 100% non-federal match. Non-federal matching funds may be comprised of a variety of public and private sources and may include in-kind contributions and other non-cash support.
Geographic Focus: All States
Date(s) Application is Due: Jan 14
Amount of Grant: 75,000 USD
Contact: Karen Kohanowich, Acting Director; (301) 734-1006 or (301) 713-1000; fax (301) 713-1967; karen.kohanowich@noaa.gov
Internet: http://www.uncw.edu/nurc/funding/opportunities.htm
Sponsor: U.S. Department of Commerce
1401 Constitution Avenue, NW, Room 5128
Washington, DC 20230

NOAA Undersea Research Grants 3411
NOAA's Undersea Research Program (NURP) is a unique comprehensive research program that provides scientists with the tools and expertise to conduct cutting-edge underwater research using advanced underwater technologies (e.g., human occupied submersibles, remotely operated vehicles, an undersea laboratory, mixed gas and technical diving). Research priorities are identified on an annual basis by NURP Headquarters in its Science Guidance. The research priorities are based upon the NOAA Strategic Plan and research needs as identified by NOAA resource managers. Highest priority is given to proposals for studies within the U.S. Exclusive Economic Zone and waters under U.S. jurisdiction, including the Great Lakes, territorial seas, and adjacent waters of the United States; U.S. Territories; and the Freely-Associated States. Funding opportunities will be announced through the individual NURP Center or NIU.S.T.
Requirements: Eligible applicants are U.S. Institutions of higher education, not-for-profit institutions, and federal, state, and local governments.
Restrictions: Federal agencies may not charge salary or overhead.
Geographic Focus: All States
Contact: Karen Kohanowich, Acting Director; (301) 734-1006 or (301) 713-1000; fax (301) 713-1967; karen.kohanowich@noaa.gov
Internet: http://www.nurp.noaa.gov/Funding.htm
Sponsor: U.S. Department of Commerce
1401 Constitution Avenue, NW, Room 5128
Washington, DC 20230

NOAA Undersea Research Program Project Grants 3412
NOAA's Undersea Research Program (NURP) is a unique comprehensive research program that provides scientists with the tools and expertise to conduct cutting-edge underwater research using advanced underwater technologies (e.g., human occupied submersibles, remotely operated vehicles, an undersea laboratory, mixed gas and technical diving). Some funds (less than $10,000 per project) are available for developmental projects or small exploratory research grants. These funds are reserved for late-breaking, high priority issues and may be awarded at the discretion of the Center Director. Funding opportunities will be announced through the individual NURP Center or NIU.S.T.
Requirements: Eligible applicants are U.S. Institutions of higher education, not-for-profit institutions, and federal, state, and local governments.
Restrictions: Federal agencies may not charge salary or overhead.
Geographic Focus: All States
Amount of Grant: Up to 10,000 USD
Contact: Karen Kohanowich, Acting Director; (301) 734-1006 or (301) 713-1000; fax (301) 713-1967; karen.kohanowich@noaa.gov
Internet: http://www.nurp.noaa.gov/Funding.htm
Sponsor: U.S. Department of Commerce
1401 Constitution Avenue, NW, Room 5128
Washington, DC 20230

Nobel Laureate Signature Award for Graduate Education in Chemistry 3413
This award, sponsored by Mallinckrodt Baker Inc. and administered by the American Chemical Society, is awarded annually to recognize an outstanding graduate student and his or her preceptor(s) in the field of chemistry as broadly defined. The graduate student nominee must have completed a PhD dissertation in chemistry within the 12-month period before the deadline for receipt of nominations. The award will recognize only work done while the nominee was a graduate student; it is granted without regard to age or nationality. The award includes a stipend, plaque, and travel expenses for the student and preceptor(s).
Requirements: A nominating document shall consist of a letter of nomination, a brief biographical sketch of the graduate student nominee and the preceptor(s), a synopsis of the nominee's PhD dissertation not longer than 10 typewritten pages (double-spaced), and one or two letters from experts in the field of the dissertation independently appraising its significance. Nominees may subsequently be asked to submit five copies of the complete dissertation. All documents must be in English. Six copies of all items to be included in the nomination must be submitted.
Restrictions: Self-nominations are not accepted.
Geographic Focus: All States
Date(s) Application is Due: Feb 1
Amount of Grant: 3,000-6,000 USD
Contact: Awards Administrator; (202) 872-4408; fax (202) 872-6317; awards@acs.org
Internet: http://portal.acs.org/portal/acs/corg/content?_nfpb=true&_pageLabel=PP_ARTICLEMAIN&node_id=1319&content_id=CTP_004541&use_sec=true&sec_url_var=region1
Sponsor: American Chemical Society
1155 Sixteenth Street, NW
Washington, DC 20036-4801

NOCSAE Research Grants 3414
Due to the diversity and complexity of potential research proposals, NOCSAE has instituted a two-phase application procedure. Those interested in seeking funding are required to first submit a Preliminary Grant Application. These brief, one page proposals are reviewed by the NOCSAE Board of Directors. The Board votes to invite Full Grant Applications from these preliminary proposals after consideration of the current goals and the available funds. The invited full proposals are assigned and reviewed by an external scientific study section. Applications should describe projects based upon basic and/or applied research bearing a rational relationship toward increasing our understanding of sports injury mechanisms and injury prevention through the use of protective sports equipment. Priority will be given to projects that demonstrate evidence of recurring injury where the injury is either catastrophic, serious, or costly.
Requirements: Applications should describe projects based upon basic and/or applied research bearing a rational relationship towards increasing our understanding of sports injury mechanisms and injury prevention through the use of protective sports equipment. Projects related specifically to equipment design are not appropriate. Projects analyzing athletic performance are not appropriate. Any individual and institution residing in North America is eligible. All applications regardless of origin will be evaluated by the same strict scientific guidelines.
Restrictions: Projects related specifically to equipment design and development are not appropriate and will not be accepted. Projects analyzing athletic performance are also not appropriate.
Geographic Focus: All States
Date(s) Application is Due: Oct 1
Amount of Grant: 50,000 USD
Contact: Mike Oliver; (913) 888-1340; fax (913) 498-8817; mike.oliver@nocsae.org
Internet: http://www.nocsae.org/research/funding.html
Sponsor: National Operating Committee on Standards for Athletic Equiipment
11020 King Street, Suite 215
Overland, KS 66210

Norcliffe Foundation Grants 3415
The Norcliffe Foundation is a private nonprofit family foundation established in 1952 by Paul Pigott for the purpose of improving the quality of life for all people in the Puget Sound region. The foundation provides grants in the areas of health, education, social services, civic improvement, religion, culture and the arts, the environment, historic preservation and youth programs. Foundation funding types include capital campaigns for building and equipment, certain operating budgets, endowments, challenge/matching grants, land acquisition, new projects, start-up funds, renovation, and research. Requests for publication, videos/films, and website production may be considered as may scholarships, fellowships / chairs, conferences / seminars, social enterprise development, and technical assistance. Multi-year and renewable funding may be considered. Applications are accepted year-round. One copy of a letter proposal and/or common grant application form should be directed to the President in care of the Foundation Manager at the contact information given. An initial phone call is optional. Guidelines, instructions, and a copy of the common grant proposal form are provided at the website. Funding decisions and notification generally occur three to six months after receipt of request.
Requirements: 501(c)3 organizations in the Puget Sound region in and around Seattle, Washington are eligible to apply.
Restrictions: The Norcliffe Foundation does not provide the following types of assistance: deficit financing, emergency funds, grants to individuals, matches to employee-giving,

Program-Related Investments / Loans (PRIs), in-kind services, volunteer / loaned executive, and internships. Applicants may only submit one request per year from date of funding or denial. Applicants who have previously received grants of $50,000 or more from the Foundation must wait two years from final payment before submitting a new application.
Geographic Focus: Washington
Amount of Grant: 1,000-25,000 USD
Contact: Arline Hefferline, Foundation Manager; (206) 682-4820; fax (206) 682-4821; arline@thenorcliffefoundation.com
Internet: http://www.thenorcliffefoundation.com/
Sponsor: Norcliffe Foundation
999 Third Avenue, Suite 1006
Seattle, WA 98104-4001

Nordson Corporation Foundation Grants 3416

The corporate foundation awards grants to nonprofit organizations in geographic areas where Nordson facilities and employees are located: Cuyahoga and Lorain counties in Ohio; the Greater Atlanta, Georgia area; California, Rhode Island and Southeastern Massachusetts. The foundation provides a source of stable funding for community programs and projects in the areas of education, human welfare, civic, and arts and culture. Educational support is generally limited to improving elementary and secondary schools and certain programs for public and private higher education. Human welfare grants focus on children and youth. The foundation considers other funding areas, including urban affairs, volunteerism, public policy, health and health organizations, and literacy. Grants are awarded for general operating support, continuing support, annual campaigns, capital campaigns, building construction/renovation, equipment acquisition, emergency funds, seed money, scholarship funds, employee matching gifts, and technical assistance. The board meets four times each year. Deadlines vary by state; contact program staff for appropriate deadlines. Application forms are available on the website and may only be submitted online.
Requirements: Private, nonprofit organizations in California, Georgia, Ohio, Rhode Island, and southeastern Massachusetts may apply.
Restrictions: Funding is not provided for organizations whose services are not provided within the foundation's geographic areas of interest; direct grants or scholarships to individuals; organizations not eligible for tax-deductible support; organizations not exempt under Section 501(c)3 of the Internal Revenue Code; political causes, candidates, organizations or campaigns; organizations that discriminate on the basis of race, sex, or religion; or special occasion, goodwill advertising, i.e., journals or dinner programs.
Geographic Focus: California, Georgia, Massachusetts, Ohio, Rhode Island
Date(s) Application is Due: Feb 15; May 15; Aug 15; Nov 15
Amount of Grant: Up to 15,000,000 USD
Contact: Cecilia H. Render; (440) 892-1580, ext. 5172; crender@nordson.com
Internet: http://www.nordson.com/Corporate/Community/Foundation
Sponsor: Nordson Corporation Foundation
28601 Clemens Road
Westlake, OH 44145-1119

North Carolina Arts Council Artist Fellowships 3417

The program operates on a two-year cycle. Composers of jazz, chamber, electronic, experimental, and symphonic music; songwriters; playwrights and screenwriters; spoken-word artists; and writers of fiction, poetry, literary nonfiction, and literary translation are eligible to apply during even years. Choreographers, film/video artists, craft artists, and visual artists are eligible to apply in odd years.
Requirements: Applicants must: be at least 18 years old; be a U.S. citizen or holder of permanent resident alien status; have been a year-round resident of North Carolina for at least one year immediately prior to the application deadline; remain a North Carolina resident during the grant period and be physically present in the state for the majority of the grant period; not be enrolled at the time of application and will not be enrolled during the grant period in an academic or degree-granting program.
Geographic Focus: North Carolina
Date(s) Application is Due: Nov 3
Amount of Grant: 10,000 USD
Contact: Wayne Martin; (919) 807-6506; wayne.martin@ncmail.net
Internet: http://www.ncarts.org/grants_category.cfm?ID=35
Sponsor: North Carolina Arts Council
Department of Cultural Resources, 109 East Jones Street, MSC #4632
Raleigh, NC 27601-2807

North Carolina Arts Council Arts Folklife Internship 3418

This three-month internship provides intensive, directed training with the Arts Council's Folklife staff. The intern receives a $4,000 stipend for the three-month period. The internship takes place for twelve weeks between April and August. Applicants must have completed at least one year of graduate study in folklife or related fields (ethnomusicology, anthropology, history, etc.) or have experience with traditional arts and culture. The next application deadline is March 2.
Requirements: Applicants must have completed at least one year of graduate study in folklife or related fields (ethnomusicology, anthropology, history, etc.) or have otherwise gained significant experience with traditional arts and culture. Applicants must be North Carolina residents.
Geographic Focus: North Carolina
Date(s) Application is Due: Mar 2
Amount of Grant: 4,000 USD
Contact: Wayne Martin; (919) 807-6506; wayne.martin@ncmail.net
Internet: http://www.ncarts.org/freeform_scrn_template.cfm?ffscrn_id=119
Sponsor: North Carolina Arts Council
Department of Cultural Resources, 109 East Jones Street, MSC #4632
Raleigh, NC 27601-2807

North Carolina Arts Council Folklife Public Grants 3419

Public programs support: exhibitions, performances and workshops, publications, and radio and television programs that promote public knowledge and appreciation of the state's traditional arts; or salary assistance to organizations that create a staff position for a professional folklorist. Grant funds also support expenses directly related to the project including production costs, stipends, honoraria, marketing, travel, supplies, telephone, and postage.
Requirements: North Carolina nonprofit organizations are eligible.
Geographic Focus: North Carolina
Date(s) Application is Due: Mar 2
Amount of Grant: Up to 10,000 USD
Contact: Sally Peterson, Folklife Specialist; (919) 807-6507; sally.peterson@ncdcr.gov
Internet: http://www.ncarts.org/grants_programs.cfm?ID=6
Sponsor: North Carolina Arts Council
Department of Cultural Resources, 109 East Jones Street, MSC #4632
Raleigh, NC 27601-2807

North Carolina Arts Council Residency Center Grants 3420

The N.C. Arts Council sponsors one- and two-month residencies for artists in a variety of disciplines at residency centers in California and Vermont. The purpose of the N.C. Arts Council's residency program is to provide both emerging and established North Carolina artists with time for their work. Two artists will be selected for residencies at Headlands Center for the Arts and for residencies at Vermont Studio Center as part of this program. The deadline for Headlands is June 6, and the deadline for Vermont is June 16.
Requirements: North Carolina poets, fiction writers, playwrights, storytellers, and visual artists may apply.
Geographic Focus: North Carolina
Date(s) Application is Due: Jun 6; Jun 16
Amount of Grant: 1,000-7,500 USD
Contact: Jeff Pettus, Visual Arts Director; (919) 807-6513; jeff.pettus@ncmail.net
Internet: http://www.ncarts.org/grants_category.cfm?ID=36&CFID=4457402&CFTOKEN=25653586&jsessionid=2830289859123l777988333
Sponsor: North Carolina Arts Council
Department of Cultural Resources, 109 East Jones Street, MSC #4632
Raleigh, NC 27601-2807

North Carolina Biotechnology Center Collaborative Funding Grants 3421

The grant financially supports a postdoctoral fellow or technician in a university research laboratory who will perform a project of commercial interest under the guidance of a principal investigator. The Kenan Institute for Engineering, Technology and Science and the Biotechnology Center jointly sponsor this grant. The Grant aims to: encourage collaboration between academia and industry; help to move technology from the lab to commercialization; and leverage Biotechnology Center money with the Kenan Institute.
Requirements: Proposed programs and activities must clearly relate to biotechnology research, development, commercialization, education, or training. Priority will be giving to proposals requesting support for innovative programs or activities in biotechnology. Only North Carolina academic institutions, non profit research organizations, or businesses may apply.
Restrictions: Awards are not made to individuals.
Geographic Focus: North Carolina
Date(s) Application is Due: Feb 25; Aug 6
Amount of Grant: Up to 50,000 USD
Contact: Ginny DeLuca, (919) 549-8842; virginia_deluca@ncbiotech.org
Internet: http://www.ncbiotech.org/services_and_programs/grants_and_loans/collaborative_funding/index.html
Sponsor: North Carolina Biotechnology Center
15 T.W. Alexander Drive, P.O. Box 13547
Research Triangle Park, NC 27709-3547

North Carolina Biotechnology Center Institutional Development Grants 3422

The purpose of the Institutional Development Grant program is to provide research equipment or core facilities that serve multiple investigators. Research Intensive universities require at least six investigators; Non-research intensive universities require at least three. If the proposal has any of the following features, it will be more competitive: shows that the request is occurring within the context of an institutional initiative; involves new ideas that may lead to biotechnology breakthroughs; involves collaboration between academic and industrial scientists; and strengthens regional and statewide biotechnology capabilities.
Requirements: Awards are made to any North Carolina academic or nonprofit institution.
Geographic Focus: North Carolina
Date(s) Application is Due: Sep 10
Amount of Grant: Up to 250,000 USD
Contact: Ginny DeLuca, (919) 549-8842; virginia_deluca@ncbiotech.org
Internet: http://www.ncbiotech.org/services_and_programs/grants_and_loans/institutional_development/index.html
Sponsor: North Carolina Biotechnology Center
15 T.W. Alexander Drive, P.O. Box 13547
Research Triangle Park, NC 27709-3547

North Carolina Biotechnology Center Oliver Smithies Faculty Recruitment Grant 3423

The goal of this program is to aid recruitment of outstanding research faculty into tenure-track positions at the senior level to non-Research Extensive colleges and universities. It is expected that the recruit will significantly impact an academic institution's overall biotechnology research initiative. These awards are intended to increase an institution's competitive advantage in negotiations with the recruit.
Requirements: Proposed programs and activities must clearly relate to biotechnology research, development, commercialization, education, or training. Priority will be giving to proposals requesting support for innovative programs or activities in biotechnology. Only North Carolina academic institutions, non profit research organizations, or businesses may apply.
Restrictions: Awards are not made to individuals. Duke, North Carolina State University and University of North Carolina at Chapel Hill are ineligible for this program.
Geographic Focus: North Carolina
Amount of Grant: 250,000 USD
Contact: Ginny DeLuca, (919) 549-8842; virginia_deluca@ncbiotech.org
Internet: http://www.ncbiotech.org/services_and_programs/grants_and_loans/faculty_recruitment/index.html
Sponsor: North Carolina Biotechnology Center
15 T.W. Alexander Drive, P.O. Box 13547
Research Triangle Park, NC 27709-3547

North Carolina Biotechnology Center Research Grants 3424

Biotechnology Research Grants support novel research projects that have the potential to strengthen biotechnology research and development at academic and non-profit research institutions in the early stages of building their research capacity. Awards are intended to support investigators at these institutions allowing for the generation of preliminary data required to attract funding from other sources, with the ultimate goal of technology commercialization. The preproposal application deadline is October 1. For applicants invited, the deadline for full proposals is February 4.
Requirements: Proposed programs and activities must clearly relate to biotechnology research, development, commercialization, education, or training. Priority will be giving to proposals requesting support for innovative programs or activities in biotechnology. Only North Carolina academic institutions, non profit research organizations, or businesses may apply.
Restrictions: Awards are not made to individuals. Duke, North Carolina State University and University of North Carolina at Chapel Hill are ineligible for this program.
Geographic Focus: North Carolina
Date(s) Application is Due: Feb 4; Oct 1
Amount of Grant: Up to 75,000 USD
Contact: Ginny DeLuca, (919) 549-8842; virginia_deluca@ncbiotech.org
Internet: http://www.ncbiotech.org/services_and_programs/grants_and_loans/biotech_research.html
Sponsor: North Carolina Biotechnology Center
15 T.W. Alexander Drive, P.O. Box 13547
Research Triangle Park, NC 27709-3547

North Carolina Biotechnology Centers of Innovation Grants 3425

The Centers of Innovation Program is bringing together North Carolina's best scientific and technical minds in the life sciences. This program is designed to focus the state's efforts in biotechnology research, development and commercialization in targeted industrial sectors important to economic development and job creation. The North Carolina Biotechnology Center (NCBC) will work with university researchers, technology transfer offices, industrial partners, non-profit stake holders as well as regional and state-wide community leaders to establish nine Centers of Innovation (COI). Initial Centers of Innovation will complement efforts already under way in the state to align academic and industrial resources. The Phase I award is a 12-month $100,000 planning grant that will allow the development of a business plan and organization plan. The Phase II award is funding for the implementation of the business plan. Only invited applicants may apply. Potential applicants will need to demonstrate that they have established a cohesive academic-industry consortium in order to receive an invitation to submit an application.
Geographic Focus: North Carolina
Amount of Grant: Up to 2,400,000 USD
Contact: Mary Beth Thomas, (919) 541-9366; marybeth_thomas@ncbiotech.org
Internet: http://www.ncbiotech.org/services_and_programs/COI/index.html
Sponsor: North Carolina Biotechnology Center
15 T.W. Alexander Drive, P.O. Box 13547
Research Triangle Park, NC 27709-3547

North Dakota Council on the Arts Individual Artist Fellowships 3426

The Individual Fellowship Program was established to recognize practicing artists residing in North Dakota with a monetary fellowship award. This program is designed to support professional artists with outstanding talent and ability to improve their artistic skills and enhance their career opportunities. Two fellowships of $2,500 are annually awarded. An applicant will be considered professional if they are pursuing their original artwork on an ongoing basis within an artistic discipline, and if they are pursuing it as a means of livelihood, either on a partial or total basis. Works in progress are acceptable with an explanation of the work's goal in the Artist's Statement.
Requirements: To be eligible for a fellowship the applicant must: be at least 18 years of age and a resident of the state of North Dakota as of January of the grant year and agree to maintain that residency during the fellowship period; submit a completed application and samples of work on or before the deadline; not be enrolled as an undergraduate or graduate degree-seeking student during the fellowship period; submit samples of work completed within the last three years prior to the application deadline.
Restrictions: Collaborative works will not be accepted. Artists may submit collaborative work as evidence of artistic skill only if the applicant's role in the creation of the work is clearly defined.
Geographic Focus: North Dakota
Date(s) Application is Due: Feb 15
Amount of Grant: 2,500 USD
Contact: Rebecca Engelman; (701) 328-7593; fax (701) 328-7595; rengelman@nd.gov
Internet: http://www.nd.gov/arts/grants/IAFguidelines.html
Sponsor: North Dakota Council on the Arts
1600 E Century Avenue, Suite 6
Bismarck, ND 58503-0649

North Dakota Indian Scholarships 3427

The intent of the program is to assist Native American students in obtaining a basic college education. To that end, priority in making awards will be given to undergraduate full-time students. In order to provide a reward as well as an incentive to students with high academic achievements, those applicants with a cumulative grade point average of 3.50 or above, shall be given priority in funding.
Requirements: Applicants must be an enrolled member of a federally recognized Indian tribe and a resident of North Dakota. Residency for each student shall be determined by the institution that student is attending. All applicants must have been accepted for admission at an institution of higher learning or state vocational education program within North Dakota. Scholarship recipients must be enrolled full-time and may not have a grade point average below 2.00. Students participating in internships, student teaching, teaching assistance, or cooperative education programs may be eligible for a scholarship award only if participation in that program is required for the degree and only if tuition must be paid for the credits earned.
Geographic Focus: North Dakota
Date(s) Application is Due: Jul 15
Amount of Grant: 500-2,000 USD
Contact: Rhonda Schauer, Coordinator of Multicultural Education; (701) 224-2497; fax (701) 328-2961; rhonda.schauer@ndus.nodak.edu
Internet: http://www.ndus.nodak.edu/students/financial-aid/details.asp?id=111
Sponsor: North Dakota University System
10th Floor, State Capitol 600 East Boulevard Avenue, Department 215
Bismarck, ND 58505-0230

Northrop Grumman Corporation Grants 3428

Northrop Grumman is committed to supporting communities throughout the U.S., especially those where its employees live and work. The Contributions Program seeks to address critical issues and needs by providing financial assistance to accredited schools and 501(c)3 nonprofit organizations. The majority of these contributions address education, services for veterans and the military, health and human services, and the environment.
Requirements: The following information must be included in the grant request: Tax Exempt Number; non-profit contact name, title, address, phone number, fax number; brief history of the organization, mission statement, goals and objectives; type and scope of services offered, and the geographical area served; specific details as to how requested funds would be used; project budget and amount requested; demographic impact; current operating budget including latest financial statement; list of other corporate funders; list of directors and/or officers, and their affiliations; Northrop Grumman employee sponsor, if applicable (name and phone number); and contact information.
Restrictions: Grants are not awarded to: religious organizations; political groups; fraternal organizations; individuals; athletic groups or activities, including charity-benefit sporting events; charter schools, unless they have open enrollment and hold the same standards as public schools; bands or choirs; capital campaigns; organizations providing services primarily to animals; communities outside of the United States; or organizations whose programs discriminate based on race, color, age, sex, religion, national origin, sexual orientation, disability, veteran status or any other characteristic protected by law.
Geographic Focus: All States
Date(s) Application is Due: Apr 30; Sep 30
Contact: Cheryl Horn, cheryl.horn@ngc.com
Internet: http://www.northropgrumman.com/corporate-responsibility/corporate-citizenship/contribution-guidelines.html
Sponsor: Northrop Grumman Corporation
1840 Century Park East
Los Angeles, CA 90067

Northrop Grumman Foundation Grants 3429

Northrop Grumman's priority is to provide opportunities related to science, technology, engineering and mathematics (STEM) for students and teachers. The foundation supports diverse and sustainable NATIONAL-LEVEL programs that enhance the education experience for students and provide teachers with the training and tools they need to be successful in the classroom.
Requirements: Grant requests must be submitted online via the Foundation's website. Grant requests must include: tax exempt number; non-profit or institution contact name, title, address, phone number, fax number; brief history of the organization, mission statement, goals and objectives; type and scope of services offered, and the geographical area serviced; specific details as to how requested funds would be used; project budget and amount requested; demographic impacted; current operating budget including latest financial statement; list of other corporate

funders; list of directors and/or officers, and their affiliations; Northrop Grumman employee sponsor, if applicable (name and phone number); and contact information.
Restrictions: The Foundation will not consider the following requests: individuals; fundraising events such as raffles, walk-a-thons, banquets or dinners; campus student organizations, fraternities, sororities, and honor societies; religious schools or colleges whose primary focus is to promote religious beliefs; athletic teams, support organizations; advertising or underwriting expenses; capital campaigns; endowments; tuition; choirs, bands, or drill teams.
Geographic Focus: All States
Amount of Grant: 500-200,000 USD
Contact: Carleen Beste, (888) 478-5478; carleen.beste@ngc.com
Internet: http://www.northropgrumman.com/corporate-responsibility/corporate-citizenship/foundation-grant-guidelines.html
Sponsor: Northrop Grumman Foundation
1840 Century Park E, MS 131/CC
Los Angeles, CA 90067

Northwestern Mutual Foundation Grants 3430
The people of Northwestern Mutual have a long history of community involvement. To carry out its corporate contributions program, Northwestern Mutual created a private foundation, The Northwestern Mutual Foundation. The Northwestern Mutual Foundation focuses its funding in Milwaukee, where the company is headquartered. Therefore, applying for a grant from the Northwestern Mutual Foundation is limited to Milwaukee-area nonprofit organizations with programs centered on Education, Health & Human Services or Arts & Culture.
Requirements: Northwestern Mutual Foundation does not accept unsolicited grant requests. Groups new to the Foundation are asked to submit a written letter of intent, describing the organization, funding request, and provide information about the program for which you are seeking support. The Foundations team will review your letter of intent and, if they are interested in learning more about your project, you will receive an application form. Groups that have an active relationship with the Foundation and currently receive funding are allowed to apply annually for support. Once an organization submits an application, the review process takes about 90 days. The Foundation review's Milwaukee-area grants on a monthly basis.
Restrictions: Milwaukee-area nonprofits are eligible to apply. The Foundation defines the Milwaukee-area as Milwaukee County, with limited programmatic grants in Waukesha, Racine and Ozaukee Counties.
Geographic Focus: Wisconsin
Amount of Grant: 10,000-800,000 USD
Contact: Edward J. Zore , President; (414) 271-1444
Internet: http://www.nmfn.com/tn/aboutus—fd_intro
Sponsor: Northwestern Mutual Foundation
720 East Wisconsin Avenue
Milwaukee, WI 53202-4797

Northwest Fund for the Environment Grants 3431
The Northwest Fund for the Environment is designated to be spent promoting change in the uses of natural resources to increase their protection and preservation in the state of Washington. Emphasis is placed on the protection of wild fish, native wildlife, natural forests, wetlands and shorelines, and the preservation of pure and free-flowing waters. Special program areas of interest include aquatic ecosystem protection of freshwater and saltwater; growth management to promote smart growth land use policies and to prevent damage to environmentally sensitive areas; and community response in support of grassroots work that feeds into the fund's goals or supports emergent issues and opportunities not covered by previous funding. The Foundation requires grantees to propose a scope of work lasting up to 16 months, complete the entire 16 month grant period, and submit a final report prior to applying for a new standard grant. Standard grants range from $3,000 to $20,000. The Community Response Fund Grants range from $500 to $3,000. Application deadlines vary according to particular funding requested. Samples of previous funding are available at the website.
Requirements: Applicants must follow the online application process, which includes calling the Foundation to discuss whether it may be eligible for funding. Unsolicited applications or letters of inquiry will not be accepted. Only nonprofit organizations classified as 501(c)3 public charities serving Washington state are eligible. First time applicants may only apply for project support.
Restrictions: The following are not eligible for funding: individuals or for profit organizations; endowments or capital campaigns; debt reduction or loans to organizations; capital projects, such as building acquisition or construction, large equipment purchases or vehicle acquisition; land acquisition; structural restoration, such as tree planting or culvert removal; purely educational activities (except when an integral part but only one component of your application); organization of, scholarships for, or support of public conferences, as distinct from invited meetings or convening of groups engaged in related efforts; youth group support; production of publications and videos for the general public; web page development; museum displays; art projects, including performances; visual art projects and arts education; partisan political activities; academic scholarships or fellowships; academic research; university overhead costs; or field research (except when an integral part but only one component of your application).
Geographic Focus: Washington
Date(s) Application is Due: Feb 1
Contact: Pam Fujita-Yuhas; (206) 386-7220; fax (206) 386-7223; staff@nwfund.org
Internet: http://www.nwfund.org
Sponsor: Northwest Fund for the Environment
1904 Third Avenue, Suite 615
Seattle, WA 98101

Northwood University/Alden B. Dow Creativity Center Summer Residency 3432
The program provides a 10-week residency at the university from mid-June through mid-August to allow each successful applicant to pursue innovative ideas having the potential for impact in his or her field. Project ideas must be specific with accomplishable goals for the 10-week period. Selection is based on newness and uniqueness of idea. Fellows are expected to be capable of independent research on their proposed projects. From education and literature to health care and chemistry, physics, engineering, and architecture and the arts, no creative area will be overlooked when a project idea may result in a significant advancement. Foreign citizens are also invited, provided they can communicate in written and spoken English. Project budgets are approved on an individual basis; however, ideas involving heavy developmental costs are discouraged. The grant includes travel within the United States (foreign applicants must provide transportation to the United States), room and board, and a stipend for personal expenses and/or project materials. Application fee is $10; forms and instructional brochures are available upon request.
Geographic Focus: All States
Date(s) Application is Due: Dec 31
Amount of Grant: 750 USD
Contact: Dr. Grover Proctor, Executive Director; (989) 837-4478; fax (989) 837-4468; creativity@northwood.edu
Internet: http://www.northwood.edu/abd
Sponsor: Northwood University
4000 Whiting Drive
Midland, MI 48640-2398

Norwegian Academy of Science and Letters Abel Prize 3433
The Abel Prize is an international prize for outstanding scientific work in the field of mathematics, including mathematical aspects of information technology, mathematical physics, probability theory, numerical analysis and computational science, statistics and applications of mathematics in other sciences. The prize is meant to recognize contributions of extraordinary depth and influence to pure and applied mathematics. Such work may have resolved fundamental problems, created powerful new techniques, introduced unifying principles or opened up major new fields of research. The prize amount is 6 million NOK (about 750,000 Euro or 1,000,000USD) and was awarded for the first time in 2003.
Requirements: Candidates may send the committee a nomination letter. The nomination should be accompanied by a description of the work and impact of the nominee/nominees, together with names of distinguished specialists in the field of the nominee/nominees who can be contacted for an independent opinion. Further information and a current online submission form is available online.
Geographic Focus: All States, All Countries
Date(s) Application is Due: Sep 15
Amount of Grant: 6,000,000 NOK
Contact: Oyvind Anderson; + 47 22 12 10 90; fax + 47 22 12 10 99; dnva@online.no
Internet: http://www.abelprisen.no/en/abelprisen/nomination.html
Sponsor: Norwegian Academy of Science and Letters
Drammensveien 78
Oslo, N-0271 Norway

Notsew Orm Sands Foundation Grants 3434
The Notsew Orm Sands Foundation was established in 1995 via an initial donation from Charles Burnett, III, a British-born race car driver who set the steam-powered land speed record in 2009. Giving is on a national and international basis, with some emphasis on Houston, Texas, and the United Kingdom. Primary fields of interest are: higher education, hospitals, human services, medical research, and Protestant agencies and churches. There are no specific deadlines, and applicants should submit a detailed description of project and amount of funding requested, along with a copy of current year's organizational budget and/or project budget.
Restrictions: No grants are given to individuals.
Geographic Focus: All States
Amount of Grant: 250-75,000 USD
Samples: Baylor College of Medicine, Houston, Texas, $75,000; University of Saint Thomas, Houston, Texas, $61,000; Yale University, New Haven, Connecticut, $55,636; Centurion Ministries, Princeton, New Jersey, $20,000.
Contact: Charles Burnett, III, President and Secretary; (281) 497-0744
Sponsor: Notsew Orm Sands Foundation
2470 S. Dairy Ashford Street, Suite 802
Houston, TX 77077-5716

NPS American Battlefield Protection Grants 3435
The American Battlefield Protection Program (ABPP) promotes the preservation of significant historic battlefields associated with wars on American soil. The goals of the program are: to protect battlefields and sites associated with armed conflicts that influenced the course of our history; to encourage and assist all Americans in planning for the preservation, management, and interpretation of these sites; and to raise awareness of the importance of preserving battlefields and related sites for future generations. The ABPP focuses primarily on land use, cultural resource and site management planning, and public education.
Requirements: Federal, tribal, state, and local governments as well as private and nonprofit organizations are eligible to apply.
Geographic Focus: All States
Date(s) Application is Due: Jan 21
Amount of Grant: 1,000-115,000 USD
Contact: Kristen Stevens, ABPP Grants Manager; (202) 354-2037; fax (202) 371-1616; kristen_stevens@nps.gov

Internet: http://www.cr.nps.gov/hps/abpp/
Sponsor: National Park Service
1849 C Street NW, MS 2253
Washington, DC 20240

NPSC Graduate Fellowships in the Physical Sciences 3436

The consortium awards graduate fellowships in the physical sciences (astronomy, chemistry, computer science, geology, materials science, mathematical sciences, physics, and their subdisciplines, and related engineering fields—chemical, computer, electrical, environmental, and mechanical) at any participating NPSC member university or college (list available on request). This is a full-time study, six-year, renewable, doctoral track fellowship program. The fellowship covers tuition and fees and provides a stipend. NPSC welcomes applications from all qualified students with continued emphasis on recruiting underrepresented minorities: African American; Hispanic; Native American Indian, Eskimo, Aleut; Pacific Islander (Polynesian); and/or women, the underrepresented groups in the physical sciences and related engineering fields. Application and guidelines are available online.
Requirements: Applicant must be: a U.S. citizen and have the ability to pursue graduate work at an NPSC member institution; in the senior year with at least a 3.0/4.0 grade-point average, or in the first year of a graduate program pursuing a PhD, or in a terminal master's program, or returning from the workforce with no more than a master's degree.
Restrictions: Persons who already possess a doctoral degree in any field are ineligible.
Geographic Focus: All States
Date(s) Application is Due: Nov 5
Amount of Grant: 16,000 USD
Contact: Executive Director; (800) 854-6772 or (213) 821-2409; npschq@npsc.org
Internet: http://www.npsc.org/students/info.html
Sponsor: National Physical Science Consortium
3716 S Hope, Suite 348
Los Angeles, CA 90007

NRA Foundation Grants 3437

Grant requests must conform to, and foster the purposes set forth in The NRA Foundation's Articles of Incorporation. These purposes are as follows: to promote, advance and encourage firearms and hunting safety; to educate individuals, including the youth of the United States, with respect to firearms and firearms history and hunting safety and marksmanship, as well as with respect to other subjects that are of importance to the well-being of the general public; to conduct research in furtherance of improved firearms safety and marksmanship facilities and techniques; to support activities of the National Rifle Association of America, but only to the extent that such activities are in furtherance of charitable, educational or scientific purposes within the meaning of section 501(c)3 of the Internal Revenue Code; to engage in any other activity that is incidental to, connected with, or in advancement of the foregoing purposes and that is within the scope of allowable purposes under 26 U.S.C. 501(c)3.
Requirements: Any organization, association or other entity, whether formally incorporated or not, that has, as a minimum, a unique federal employer identification number (EIN) issued by the Internal Revenue Service. Eligible organizations may apply for an NRA Foundation grant for a qualifying project. Additionally, allowable projects must qualify under IRS 501(c)3 regulations in one of the following categories: charitable; scientific; testing for public safety; literary; educational; fostering national/international amateur sports competition (cannot include the provision of athletic facilities or equipment.)
Restrictions: The following organizations are not eligible for grants: political candidates or organizations; labor organizations; state fund committees; friends of NRA committees; private business/private enterprise; other organizations or groups that have not been assigned a federal employer identification number by the Internal Revenue Service. The following activities are not eligible for funding: projects which confer private benefit upon the applying organization/group; deficit financing; projects for commercial ventures; projects which require membership in the NRA or in the applying organization/group; applications from organizations or groups that have not submitted a final report for a previously awarded grant. In addition, the following limitations/restrictions apply to grants which are otherwise eligible for funding: no funding will be awarded to an applicant for payment of administrative fees, office overhead, or other similar charges; the Foundation does not approve multi-year funding of projects; requests must be submitted for consideration each year. Funding cannot be given for competitions requiring NRA or other club or association membership. In addition, although grants may be sought for the purpose of fostering national or international amateur sports competition, grant awards cannot be made for the purpose of providing facilities or equipment to be used in such competitions.
Geographic Focus: All States
Amount of Grant: 5,000-9,000,000 USD
Contact: Wayne Sheets; (800) 423-6894; fax (703) 267-3985; nraf@nrahq.org
Internet: http://www.friendsofnra.org/National.aspx?cid=9
Sponsor: National Rifle association Foundation
11250 Waples Mill Road
Fairfax, VA 22030

NSERC Brockhouse Canada Prize for Interdisciplinary Research in Science and Engineering Grant 3438

The Brockhouse Canada Prize for Interdisciplinary Research in Science and Engineering recognizes outstanding Canadian teams of researchers from different disciplines who have combined their expertise to produce achievements of outstanding international significance in the natural sciences and engineering. The prize, accompanied by a team research grant of $250,000, reflects NSERC's commitment to supporting Canadian research through strategic investments in people, discovery and innovation. The grant supports the direct costs of university-based research and/or the enhancement of research facilities. The grant may be distributed in one lump sum or up to five instalments, depending on the needs of the recipients.
Requirements: Research teams nominated for the Brockhouse Canada Prize must have at least two members who are independent researchers, one of whom must hold an NSERC grant. The team can be part of an international effort, but the majority of the nominated team members must be employed at a Canadian university or public or private organization. NSERC recognizes that teams may change between the time of the specific research achievements and the time of nomination. Nominations will be accepted when changes have occurred but only as long as the core of the team remains intact. Contributions must be primarily in the natural sciences and engineering, and of an interdisciplinary and collaborative nature.
Restrictions: Any Canadian citizen may nominate a research team for the Brockhouse Canada Prize. Self nominations will not be accepted. Current NSERC Council members cannot be nominated. Since the prize is for interdisciplinary research, nominators should consult NSERC's Guidelines for the Preparation and Review of Applications in Interdisciplinary Research. Nominators should also consult NSERC's Guidelines for the Preparation and Review of Applications in Engineering and the Applied Sciences. See website for deadline & hyper links to above mentioned guidelines.
Geographic Focus: All States, Canada
Date(s) Application is Due: Jun 1
Amount of Grant: 250,000 USD
Contact: Administrator; (613) 995-5829; brockhouse@nserc-crsng.gc.ca
Internet: http://www.nserc-crsng.gc.ca/Prizes-Prix/Brockhouse-Brockhouse/Index-Index_eng.asp
Sponsor: Natural Sciences and Engineering Research Council of Canada
350 Albert Street
Ottawa, ON K1A 1H5 Canada

NSERC E.W.R. Steacie Memorial Fellowships 3439

Every year, NSERC awards up to six Steacie Fellowships that are held for a two-year period. Successful Fellows are relieved of teaching and administrative duties, so that they can devote all their time and energy to research. The Fellowships are held at a Canadian university or affiliated research institution.
Requirements: Candidates should have obtained their doctorate within the last 12 years and hold a grant from NSERC.
Restrictions: Candidates may not apply on their own behalf. They must be nominated by senior members of the Canadian science and engineering community, and the nominations must be endorsed by the executive head of the candidate's university.
Geographic Focus: All States, Canada
Date(s) Application is Due: Jun 30
Amount of Grant: 90,000 USD
Contact: Administrator; (613) 996-7042; inquiries.steacie@nserc-crsng.gc.ca
Sponsor: Natural Sciences and Engineering Research Council of Canada
350 Albert Street
Ottawa, ON K1A 1H5 Canada

NSERC Gerhard Herzberg Canada Gold Medal for Science and Engineering Grants 3440

The NSERC Herzberg Medal is awarded annually to an individual who has demonstrated sustained excellence and influence in research, for a body of work conducted in Canada that has substantially advanced the natural sciences or engineering fields. The award is meant to celebrate Canada's most outstanding scientists and engineers and to raise public awareness about the major contributions that Canada's top researchers make to international science and technology, and to bettering people's lives. In addition to the medal, the winner is guaranteed $1 million to use for personal university-based research or to direct in some related way such as the establishment of research scholarships, fellowships or chairs in his or her name at Canadian universities. The monetary award will be distributed over a five-year period.
Requirements: To be eligible, a candidate must be a scientist or engineer from a Canadian university, government lab or private firm. Candidates may be nominated by any individual or group; however, the nominators must be Canadian citizens or permanent residents of Canada. Posthumous or self nominations will not be accepted. Current NSERC Council members are not eligible for nomination.
Restrictions: An individual may win the NSERC Herzberg Medal only once.
Geographic Focus: All States, Canada
Date(s) Application is Due: May 1
Amount of Grant: 1,000,000 USD
Samples: John C. Polanyi, University of Toronto; Chemistry
Contact: Administrator; (613) 996-7042; inquiries.steacie@nserc-crsng.gc.ca
Internet: http://www.nserc-crsng.gc.ca/Prizes-Prix/Herzberg-Herzberg/About-Apropos_eng.asp
Sponsor: Natural Sciences and Engineering Research Council of Canada
350 Albert Street
Ottawa, ON K1A 1H5 Canada

NSERC Howard Alper Postdoctoral Prize 3441

The NSERC Howard Alper Postdoctoral Prize is awarded to the most outstanding candidate in NSERC's postdoctoral fellowship competition. Valued at $20,000, the prize recognizes academic excellence, existing and potential research contributions, interpersonal and communication skills, and leadership abilities. It was established by Howard Alper, recipient of the 2000 Gerhard Herzberg Canada Gold Medal for Science and Engineering.
Requirements: Award candidates must be in the NSERC's postdoctoral fellowship competition.
Geographic Focus: All States, Canada
Amount of Grant: 20,000 USD

Contact: Grants Administrator; (613) 947-3789; scholfin@nserc-crsng.gc.ca
Internet: http://www.nserc-crsng.gc.ca/Prizes-Prix/Alper-Alper/Index-Index_eng.asp
Sponsor: Natural Sciences and Engineering Research Council of Canada
350 Albert Street
Ottawa, ON K1A 1H5 Canada

NSERC Industrial R&D Fellowships 3442
The IRDF Program provides financial contributions that support the most promising recent doctoral graduates to engage in research and development in the private sector and with not-for-profit and non-governmental organizations. Its objective is to: encourage excellent recent Ph.D. graduates in science and engineering to gain experience and seek careers in the Canadian private, not-for-profit and non-governmental sectors; promote awareness in the Canadian private, not-for-profit and non-governmental sectors of the capabilities of Canadian universities and university research; facilitate the transfer of expertise and technology; and provide an opportunity for Ph.D. holders seeking university careers to gain experience in the private, not-for-profit and non-governmental sectors; research and development. The value is $30,000 per year for two years plus a minimum contribution of $10,000 per year from the host organization. See website for further detail.
Requirements: Industrial R&D Fellowships are intended primarily for recent doctoral graduates seeking employment in the Canadian private, not-for-profit and non-governmental sectors for the first time. To be nominated for a fellowship, you must: be a Canadian citizen or permanent resident of Canada at the time of your nomination;hold a doctoral degree in a discipline of science or engineering that NSERC supports or expect to have completed all the requirements for such a degree, including your thesis defense, by the proposed date of appointment. In addition, candidates must have completed their doctoral degree within the last five years. Awards to candidates who completed their doctoral degree more than five years but no more than seven years prior to the nomination will be made only in exceptional circumstances and will have to be justified by the nominating organization. If you have withdrawn from the workforce and active research for at least one year after you received your doctorate for maternity leave or to raise a child, NSERC will extend this period to nine years.
Geographic Focus: All States, Canada
Amount of Grant: 30,000 USD
Contact: Director of Programs; (613) 947-0428; schol@nserc-crsng.gc.ca
Internet: http://www.nserc-crsng.gc.ca/Students-Etudiants/PD-NP/Industrial-Industrielle_eng.asp
Sponsor: Natural Sciences and Engineering Research Council of Canada
350 Albert Street
Ottawa, ON K1A 1H5 Canada

NSERC Industrial Research Chairs 3443
Industrial Research Chairs (IRC) are intended to assist universities in building on existing strengths to achieve the critical mass required for a major research endeavour in science and engineering of interest to industry and/or assist in the development of research efforts in fields that have not yet been developed in Canadian universities but for which there is an important industrial need. Description NSERC offers three types of Industrial Research Chairs: Senior Industrial Research Chairs (for distinguished senior researchers), Associate Industrial Research Chairs (for promising junior faculty researchers), and Executive Industrial Research Chairs (for outstanding R&D professionals). Emphasis is on the development and transfer of new knowledge from the university to industry, and on the training of highly qualified personnel in fields of importance to industry. A reasonable level of industry participation in terms of a monetary, equipment, or personnel commitment is a prerequisite for consideration of NSERC funding. University authorities should contact NSERC concerning application procedures. Applications are accepted at any time.
Requirements: Candidates must be researchers of international stature who are at or approaching the most productive period of their careers, who have the necessary qualifications to hold the rank of professor or associate professor and are recognized experts in their fields. The applicant must hold a senior administrative position in a Canadian university, such as department head, dean, vice president or president, with direct or ultimate responsibility for the proposed chair.
Restrictions: The applicant cannot be the candidate for the chair.
Geographic Focus: All States, Canada
Contact: Secretary; (613) 943-3087; fax (613) 943-3282; information@chairs.gc.ca
Internet: http://www.nserc-crsng.gc.ca/Professors-Professeurs/CFS-PCP/IRC-PCI_eng.asp
Sponsor: Natural Sciences and Engineering Research Council of Canada
350 Albert Street, 16th Floor, P.O. Box 1610
Ottawa, ON K1A 1H5 Canada

NSERC Industrial Undergraduate Student Research Awards 3444
These awards are meant to stimulate your interest in research in the natural sciences and engineering. They are also meant to encourage you to undertake graduate studies and to pursue a research career in these fields. If you would like to gain research work experience in a non-university setting that complements your studies, these awards can provide you with financial support through your host organization.
Requirements: To be eligible to apply for an award, you must: be a Canadian citizen or permanent resident of Canada; be registered at the time you apply in a bachelor's degree program at an eligible university; and have obtained, over the previous years of study, a cumulative average of at least second class (a grade of B or B-, if applicable) as defined by your university.
Geographic Focus: All States, Canada
Amount of Grant: 4,500 USD
Contact: Undergraduate Cordinator; (613) 947-0428; schol@nserc-crsng.gc.ca
Internet: http://www.nserc-crsng.gc.ca/Students-Etudiants/UG-PC/U.S.RAI-BRPCI_eng.asp
Sponsor: Natural Sciences and Engineering Research Council of Canada
350 Albert Street
Ottawa, ON K1A 1H5 Canada

NSERC Innovation Challenge Awards 3445
The Innovation Challenge Awards honor graduate students in the natural sciences or engineering who have demonstrated an entrepreneurial spirit by identifying ways in which their research thesis results can be developed into products and processes to benefit Canadians. Since 2004, these awards have recognized and rewarded Canada's brightest minds. The awards are aimed at fostering, in master's or PhD students who are in their final year or have recently graduated, an appreciation of real-world applications of their research. The students review their thesis work and submit an essay identifying potential products or services that could result from applying their research. Since universities play a crucial role in the innovation cycle in Canada, university Industrial Liaison Offices (ILOs), or their equivalent, are charged with selecting and submitting their institution's nominations. There are three awards: one $10,000 grand prize and two $5,000 runner-up prizes.
Requirements: Eligible nominees must have conducted research within the natural sciences or engineering at the masters or PhD level at an eligible Canadian university & be Canadian citizens or permanent residents in Canada at the time of nomination.
Geographic Focus: All States, Canada
Date(s) Application is Due: May 15
Amount of Grant: 5,000-10,000 USD
Contact: Supervisor; (613) 947-9452; fax (613) 992-5337; innovation@nserc-crsng.gc.ca
Internet: http://www.nserc-crsng.gc.ca/Prizes-Prix/Ica-Pdi/About-Apropos_eng.asp
Sponsor: Natural Sciences and Engineering Research Council of Canada
350 Albert Street
Ottawa, ON K1A 1H5 Canada

NSERC John C. Polanyi Awards 3446
In tribute to the excellence in research that Dr. John C. Polanyi continues to exemplify, NSERC offers a prestigious research prize in his name. Although Dr. Polanyi's most significant research accomplishments have been in chemistry, the NSERC John C. Polanyi Award recognizes a recent outstanding advance in any field of the natural sciences or engineering.
Requirements: To be considered for the NSERC John C. Polanyi Award, candidates must: be nominated by two active researchers from the Canadian science and engineering research community; currently hold an NSERC grant. If a team is being nominated, one team member must hold an NSERC grant and have received NSERC support for the research that led to the advance for which the individual or team is being nominated.
Geographic Focus: All States, Canada
Date(s) Application is Due: Mar 2
Amount of Grant: 250,000 USD
Contact: Coordinator; (613) 995-5829; fax (613) 992-5337; polanyi@nserc-crsng.gc.ca
Internet: http://www.nserc-crsng.gc.ca/Prizes-Prix/Polanyi-Polanyi/Index-Index_eng.asp
Sponsor: Natural Sciences and Engineering Research Council of Canada
350 Albert Street
Ottawa, ON K1A 1H5 Canada

NSERC Northern Research Internships Program 3447
Northern Research Internships Program supports senior undergraduate students, graduate students and postdoctoral fellows pursuing studies in the Canadian North1 by facilitating extended stays and subsidizing the costs associated with activities in the North. Up to $10,000 for a 16-week internship, including a logistics allowance of up to $4,000.
Requirements: See website for extensive list of requirements.
Geographic Focus: All States, Canada
Amount of Grant: 10,000 USD
Contact: Scholarships and Fellowships Division; (613) 947-0428; fax (613) 996-2589; schol@nserc-crsng.gc.ca
Internet: http://www.nserc-crsng.gc.ca/Students-Etudiants/NorthernIntern-NordiqueStage_eng.asp
Sponsor: Natural Sciences and Engineering Research Council of Canada
350 Albert Street
Ottawa, ON K1A 1H5 Canada

NSERC Postdoctoral Fellowships Program 3448
The Postdoctoral Fellowships (PDF) Program provides support to a core of the most promising researchers at a pivotal time in their careers. The fellowships are also intended to secure a supply of highly qualified Canadians with leading-edge scientific and research skills for Canadian industry, government, and universities. The value of his fellowship is $40,000 per year for two years.
Requirements: You must be a Canadian citizen or a permanent resident of Canada and hold or expect to hold a doctorate in one of the fields of research that NSERC supports. In addition: You must expect to complete all your doctoral requirements by January 1 of the year following the award announcement. You must have completed all requirements of your first doctoral degree no more than two years before the deadline date of the year in which you apply. If you have acquired at least six months of full-time relevant work experience in industry or government (experience in a university or its affiliated hospitals, research centers, and other laboratories will not be considered) after you received your doctorate, NSERC will extend this eligibility period to three years. If you have withdrawn from the workforce and active research for maternity leave or to raise

a child for at least one year after you received your doctorate, NSERC will extend this eligibility period to six years. NSERC uses the date of degree completion to determine applicant eligibility. This is the date on which all requirements of your degree have been met, including successful defense and submission of the corrected copy of your thesis. You can hold only one NSERC Postdoctoral Fellowship. If NSERC does not award you a fellowship after your first application, you may apply again, provided you meet the eligibility criteria. You can apply only twice for a Postdoctoral Fellowship. You may submit only one application per year to either NSERC or CIHR or SSHRC.
Geographic Focus: All States, Canada
Date(s) Application is Due: Oct 15
Amount of Grant: 40,000 USD
Contact: Director of Programs; (613) 947-0428; schol@nserc-crsng.gc.ca
Internet: http://www.nserc-crsng.gc.ca/Students-Etudiants/PD-NP/PDF-BP_eng.asp
Sponsor: Natural Sciences and Engineering Research Council of Canada
350 Albert Street
Ottawa, ON K1A 1H5 Canada

NSERC Research Tools and Instruments (RTI) Grants **3449**
Research Tools and Instruments (RTI) Grants foster and enhance the discovery, innovation and training capability of university researchers in the natural sciences and engineering by supporting the purchase of research equipment and installations. RTI Grants are one-year awards that assist in buying or developing research equipment that costs more than $7,000. In exceptional cases, multiple-year requests will be accepted. The onus is on the applicant to clearly explain the rationale. Items costing less than $7,000 each can be purchased with Discovery Grant funds or be included as a complement to a main piece of equipment being requested.RTI grants are divided into three categories according to the total net cost of the equipment, i.e., excluding taxes, shipping, and handling. These categories are: RTI Category 1: $7,001 to $150,000; RTI Category 2: $150,001 to $325,000; and RTI Category 3: more than $325,000.
Requirements: To apply, submit an Application for a Grant (Form 101) and a Personal Data Form (Form 100) for the applicant and each co-applicant. Only applications from applicants and co-applicants who currently hold or are applying for an NSERC research grant (e.g., Discovery, Strategic, Research Partnership, Canada Research Chairs) at the time of application will be accepted.
Geographic Focus: All States, Canada
Date(s) Application is Due: Oct 1; Oct 25
Amount of Grant: 7,000 USD
Contact: Grants Administrator; (613) 995-5829; fax (613) 992-5337; resgrant@nserc.ca
Internet: http://www.nserc-crsng.gc.ca/Professors-Professeurs/RTII-OIRI/RTI-OIR_eng.asp
Sponsor: Natural Sciences and Engineering Research Council of Canada
350 Albert Street
Ottawa, ON K1A 1H5 Canada

NSERC Synergy Awards for Innovation **3450**
The annual Synergy Awards for Innovation recognize examples of university-industry collaboration that stand as a model of effective partnership. By working together, award-winning companies and universities have proven that effective partnerships are the foundation of achievement. Their success has enriched the academic and research programs within Canadian universities while providing tangible benefits to Canadians. Synergy Awards are offered in four categories: small and Medium-Sized Companies a partnership with a single company of up to 500 employees; large Companies a partnership with a single company of more than 500 employees; two or More Companies a partnership with two or more companies of any size; Leo Derikx Award an established innovative model of long-standing university-industry partnership in pre-competitive R&D that has improved the general well-being of an industry. Up to four awards may be presented annually, one in each category.
Requirements: The nominated partnership must be between a Canadian university professor or group of professors and a Canadian-based company or companies. Nominations must be submitted by a senior official from one of the participating organizations. Nominations will be accepted only from the President, CEO or Vice-President of Research (or equivalent) of the industrial partner, or from the President, Rector, Vice-President of Research or Vice-Rector of Research (or equivalent) from the university. Letters of support from the other partner organization(s) must be from a senior executive in the company or executive head of the university, and must be included with the nomination. Refer to Nomination Checklist to ensure that your nomination package is complete. The checklist will be found on the NSERC website.
Geographic Focus: All States, Canada
Date(s) Application is Due: May 1
Amount of Grant: 200,000 USD
Contact: Administrator; (613) 947-9452; synergy@nserc-crsng.gc.ca
Internet: http://www.nserc-crsng.gc.ca/Prizes-Prix/Synergy-Synergie/Index-Index_eng.asp
Sponsor: Natural Sciences and Engineering Research Council of Canada
350 Albert Street
Ottawa, ON K1A 1H5 Canada

NSERC Undergraduate Student Research Awards **3451**
Undergraduate Student Research Awards (U.S.RA) are meant to stimulate your interest in research in the natural sciences and engineering. They are also meant to encourage you to undertake graduate studies and pursue a research career in these fields. If you would like to gain research work experience that complements your studies in an academic setting, these awards can provide you with financial support through your host university.
Requirements: To be eligible to apply for an award, you must: be a Canadian citizen or permanent resident of Canada; be registered at the time you apply in a bachelor's degree program at an eligible university and have obtained, over the previous years of study, a cumulative average of at least second class (a grade of B or B-, if applicable) as defined by your university.
Geographic Focus: All States, Canada
Amount of Grant: 4,500 USD
Contact: Undergraduate Cordinator; (613) 947-0428; schol@nserc-crsng.gc.ca
Internet: http://www.nserc-crsng.gc.ca/Students-Etudiants/UG-PC/U.S.RA-BRPC_eng.asp
Sponsor: Natural Sciences and Engineering Research Council of Canada
350 Albert Street
Ottawa, ON K1A 1H5 Canada

NSF-CONACyT Collaborative Research Opportunities **3452**
The purpose of the initiative is to advance scientific and engineering knowledge through joint research efforts of investigators from the United States of America and Mexico having complementary talents and interests, and capitalizing on the international character of modern scientific research and its support by computer network infrastructures. Projects supported in this collaborative initiative must include researchers and educators in institutions from both countries in areas relating to either CISE or ENG. Projects may range in scope from supplements to an existing award through joint efforts to a new joint-effort proposal up to $100,000 with a duration from one to three years; larger grants may be awarded under special circumstances if appropriate. Refer to brochure NSF 96-145.
Requirements: Scientists and engineers from universities and other nonprofit research institutions in the United States of America and Mexico conducting research in areas relating to either CISE or ENG may submit proposals for support under this initiative. All research proposals submitted under this initiative must be bilateral, collaborative, and related to a corresponding proposal from the other country.Proposals must be received by NSF no later than the first Wednesday in February.
Restrictions: Research proposals submitted to NSF under this collaborative initiative should avoid requests for equipment. Nonetheless, under special circumstances, the latter may be requested with appropriate justification. CONACyT has its own guidelines with respect to equipment.
Geographic Focus: All States, Mexico
Amount of Grant: Up to 100,000 USD
Contact: Jeannette M. Wing, Assistant Director; (703) 292-8900; jwing@nsf.gov
Internet: http://www.nsf.gov/pubs/1998/nsf96145/nsf96145.htm
Sponsor: National Science Foundation
4201 Wilson Boulevard
Arlington, VA 22230

NSF-NIST Interaction in Chemistry, Materials Research, Molecular Biosciences, Bioengineering, and Chemical Engineering **3453**
This program is intended to facilitate interactions between faculty and students supported by the National Science Foundation (NSF) and scientists at the National Institute of Standards and Technology's (NIST) Chemical Science and Technology Laboratory (CSTL) and Materials Science and Engineering Laboratory (MSEL), including the NIST Center for Neutron Research (NCNR). Chemistry, materials research, molecular biology, bioengineering, and chemical engineering are centralized at NIST in these laboratories. Support may be requested for supplements to existing NSF awards to provide the opportunity for faculty and students to participate in research at NIST facilities. Support may be requested as supplements to existing NSF awards for travel expenses and per diem associated with work at these NIST laboratories for faculty, students and other personnel associated with the NSF/NIST activity. It is estimated that 10-20 supplements will be made annually, up to $20,000 each.
Requirements: Support may be requested only by organizations that have currently active awards in any of the participating NSF Divisions, viz. the Division of Bioengineering and Environmental Systems (BES), the Division of Chemical and Transport Systems (CTS), the Division of Chemistry (CHE), the Division of Materials Research (DMR), and the Division of Molecular and Cellular Biosciences (MCB). Only faculty who are principal investigators on current NSF awards from BES, CHE, CTS, DMR, or MCB are eligible to apply for supplements. Standard NSF Grant Proposal Guidelines apply (see http://www.nsf.gov/publications/pub_summ.jsp?ods_key=gpg).
Restrictions: Indirect costs (F&A) are not allowed.
Geographic Focus: All States
Amount of Grant: Up to 20,000 USD
Contact: Dr. William F. Koch; (301) 975-8301; william.koch@nist.gov
Internet: http://www.nsf.gov/publications/pub_summ.jsp?ods_key=nsf03568
Sponsor: National Science Foundation
4201 Wilson Boulevard
Arlington, VA 22230

NSF/DOE Partnership in Basic Plasma Science and Engineering Grants **3454**
The goal of this three year (FY09-FY11) program initiative is to enhance plasma research and education in this broad, multidisciplinary field by coordinating efforts and combining resources of the two agencies. The initiative will address fundamental issues in plasma science and engineering that can have impact in other areas or disciplines in which improved basic understanding of the plasma state is needed.
Requirements: Proposals may be submitted by individual investigators or small groups from universities and nonprofit organizations.
Restrictions: An applicant may submit only one proposal as principal or co-principal investigator.
Geographic Focus: All States
Date(s) Application is Due: Oct 22
Amount of Grant: 25,000-200,000 USD

Contact: Richard L. Berger, (703) 292-2183; fax (703) 292-9078; rlberger@nsf.gov
Internet: http://www.nsf.gov/funding/pgm_summ.jsp?pims_id=5602
Sponsor: National Science Foundation
4201 Wilson Boulevard
Arlington, VA 22230

NSF/NEH Documenting Endangered Languages Project Grants & Fellowships: data, infrastructure and computational methods 3455

This funding partnership between the National Science Foundation (NSF) and the National Endowment for the Humanities (NEH) supports projects to develop and advance knowledge concerning endangered human languages. Made urgent by the imminent death of roughly half of the approximately 7000 currently used languages, this effort aims to exploit advances in information technology to build computational infrastructure for endangered language research. The program supports projects that contribute to data management and archiving, and to the development of the next generation of researchers. Funding can support fieldwork and other activities relevant to the digital recording, documenting, and archiving of endangered languages, including the preparation of lexicons, grammars, text samples, and databases. Funding will be available in the form of one- to three-year project grants as well as fellowships for up to twelve months and doctoral dissertation research improvement grants for up to 24 months. Approximately 18 to 22 awards will be chosen including up to 12 Fellowships.
Requirements: Proposals may only be submitted by: Universities and Colleges [Universities and two- and four-year colleges (including community colleges) accredited in, and having a campus located in the U.S., acting on behalf of their faculty members. Such organizations also are referred to as academic institutions.]; Non-profit, non-academic organizations (Independent museums, observatories, research labs, professional societies and similar organizations in the U.S. associated with educational or research activities); Unaffiliated Individuals (Scientists, engineers or educators in the U.S. who are U.S. citizens); Tribal organizations and other American Indian, Alaska Native, and Hawaiian serving organizations.
Geographic Focus: All States
Date(s) Application is Due: Sep 15
Contact: Shobhana Chelliah, Program Director; (703) 292-4381; schellia@nsf.gov
Internet: http://www.nsf.gov/funding/pgm_summ.jsp?pims_id=12816
Sponsor: National Science Foundation
4201 Wilson Boulevard
Arlington, VA 22230

NSF Academic Research Infrastructure Program: Recovery and Reinvestment 3456

As a result of the American Recovery and Reinvestment Act of 2009, NSF will invest $200 million in the Nation's research facilities and research training infrastructure. This investment will advance the science and engineering research enterprise at many institutions. The purpose of this program is to enhance the Nation's existing research facilities where sponsored and/or unsponsored research activities and research training take place to enable next-generation research infrastructure that integrates shared resources across user communities. Consistent with NSF's mission to strengthen the U.S. science and engineering enterprise, the ARI-R^2 program will: update existing research facilities at institutions of higher education (including graduate and undergraduate institutions, among which are included community colleges) and other non-profit research organizations (e.g., independent research museums, independent research laboratories, and research consortia) in order to support research that can address the challenges of the 21st century; enable academic departments, disciplinary and cross-disciplinary units, or multi-organization consortia to renovate research facilities through the addition or augmentation of cyberinfrastructure, other than general-purpose computing systems or data storage systems, to create environments that enhance research and integrate research with education; improve access to and increase use of next-generation research facilities for researchers, educators and students; and assist research organizations, including those that have historically received limited Federal research and development funds, to improve their science and engineering research environments.
Requirements: Proposals may only be submitted by the following: universities and two- and four-year colleges (including community colleges) accredited in, and having a campus located in, the United States, its territories and possessions (distinct academic campuses (that award their own degrees, have independent administrative structures, admissions policies, alumni associations, etc.) within multi-campus systems qualify as separate institutions); and independent non-profit research museums, independent non-profit research institutions and laboratories, and non-profit research consortia located in the United States, its territories and possessions (these organizations must have an independent administrative and financial structure. Non-profit research consortia must also have 501(c)3 tax status. Any organization can collaborate in a consortium; however, only eligible organizations can receive NSF funds.
Restrictions: Organizations that may not receive funds include foreign organizations (e.g., foreign academic institutions, museums and laboratories); research arms of for-profit firms; and Federally Funded Research and Development Centers (FFRDCs). Construction, engineering, architecture and other facility service divisions of for-profit firms that will carry out the work of renovation and repair may receive NSF funds as sub-awards. The American Recovery and Reinvestment Act of 2009 states that funds may not be used by any State or local government, or any private entity, for any casino or other gambling establishment, aquarium, zoo, golf course, or swimming pool.
Geographic Focus: All States
Date(s) Application is Due: Aug 24
Amount of Grant: 250,000-10,000,000 USD
Contact: Stephen Meacham, Senior Staff Associate; (703) 292-8040; fax (703) 292-9040; smeacham@nsf.gov
Internet: http://www.nsf.gov/funding/pgm_summ.jsp?pims_id=503380
Sponsor: National Science Foundation
4201 Wilson Boulevard
Arlington, VA 22230

NSF Accelerating Discovery In Science & Engineering Through PETASCALE Simulations and Analysis (PetaApps) 3457

The primary purpose of this program is to support projects that will enable researchers to capitalize on emerging petascale computing architectures and thus catalyze progress in science and engineering beyond the current state-of-the-art. The emphasis is on implementation and exploitation of forefront techniques. Proposals from or including junior researchers are encouraged as one of the goals of this solicitation is to build a community capable of using petascale computing. Proposers must be prepared to demonstrate that they have a science or engineering research challenge requiring petascale computing and that they have a path to exploiting such computation capabilities effectively.
Requirements: Proposals are encouraged in the following areas, although work in other areas will also be considered: enhancing algorithmic scalability using techniques that better exploit multi-threaded, highly parallel, hierarchical architectures; improving and creating data sampling, analysis, and clustering algorithms for massive data sets, including scalable interactive visualization tools; developing innovative modeling, simulation or optimization algorithms suitable for petascale systems; developing platform-independent software to solve forefront scientific problems on petascale systems; optimizing software for specific petascale hardware or predicted 'best guess' extrapolations to future hardware; exploring innovative computational techniques that were not previously considered viable because of limited hardware capability; conducting performance analysis and profiling of software that is heavily used but may never have been analyzed for scalability, bottle-necks, and optimization; changing functioning problem solvers by substituting algorithmic implementations known from CS research to scale more effectively to computing systems of very large scale; adapting a research code that currently uses MPI on supercomputers, to one or more of Co-Array Fortran, UPC or Titanium, in a form that will be useful for specific petascale problems. Estimated program budget, number of awards and average award size/duration are subject to the availability of funds. Proposal budgets should not exceed $2M in size and the requested duration should not exceed five years.
Restrictions: An individual may be the Primary Investigator in no more than one proposal that responds to this solicitation.
Geographic Focus: All States
Date(s) Application is Due: Oct 30
Amount of Grant: 900,000-2,000,000 USD
Contact: Hans Kaper, (703) 292-4859; hkaper@nsf.gov
Internet: http://nsf.gov/funding/pgm_summ.jsp?pims_id=501015&org=NSF&sel_org=NSF&from=fund
Sponsor: National Science Foundation
4201 Wilson Boulevard
Arlington, VA 22230

NSF Accelerating Innovation Research 3458

To accelerate the process of innovation, NSF is undertaking two related, new activities. The first will encourage the translation of the numerous, technologically-promising, fundamental discoveries made by NSF researchers, while drawing upon and building the entrepreneurial spirit of the researchers and students. The second activity will foster connections between an existing NSF innovation research alliance (including consortia such as Engineering Research Centers (ERC), Industry University Cooperative Research Centers (I/UCRC), Partnerships for Innovation (PFI), Science and Technology Centers (STC), Nanoscale Science and Engineering Centers (NSEC), Materials Research Science and Engineering Centers (MRSEC) grantees) and other institutions, whose complementary focus will spur the development of discoveries into innovative technologies through collaboration. Both of these activities are designed to strengthen the U.S. innovation ecosystem.
Requirements: For the Technology Translation Competition, the Principal Investigator or a co-PI must be a current or prior NSF awardee and a faculty member at a U.S. college or university at the time of award in the current competition. Subject technology must be derived from a discovery research project already conducted or initiated by the National Science Foundation. For the Research Alliance Competition, one of the partners must be an NSF funded innovation research alliance (including Centers for Analysis and Synthesis, Centers for Chemical Innovation, Engineering Research Centers, Industry/University Collaborative Research Centers, Materials Research Science and Engineering Centers, Nanoscale Science and Engineering Centers, Science and Technology Centers, and Science of Learning Centers). The research alliance must be active at the time of award in the current competition. The other partner(s) may be another research entity (either NSF-funded, other government agency funded, or privately funded), a small business consortium, or a local or regional innovation entity.
Geographic Focus: All States
Date(s) Application is Due: Feb 1
Amount of Grant: 350,000-1,000,000 USD
Contact: Rathindra DasGupta; (703) 292-8353; fax (703) 292-9057; rdasgupt@nsf.gov
Internet: http://www.nsf.gov/funding/pgm_summ.jsp?pims_id=503553
Sponsor: National Science Foundation
4201 Wilson Boulevard
Arlington, VA 22230

NSF Active Nanostructures and Nanosystems Grants 3459

The National Science Foundation announces a program on collaborative research and education in the area of active nanostructures, nanosystems, and on the long-term societal change associated with these innovations. Active nanostructures change or evolve

their structure, property, or function during their operation. The goal of this program is to support fundamental research and catalyze synergistic science and engineering research and education in several emerging areas of nanoscale science and engineering, including: active nanostructures; nanosystems with improved functionality and new architectures; hierarchical nanomanufacturing; and long-term societal and educational implications of scientific and technological advances on the nanoscale. This solicitation provides support for Nanoscale Interdisciplinary Research Teams (NIRT) and Nanoscale Exploratory Research (NER). Other research and education projects in nanoscale science and engineering will continue to be supported in the relevant programs and divisions. The anticipated funding amount is $29,000,000 total, of which $24,500,000 is for approximately 20 NIRT awards and $4,500,000 is for approximately 40 NER awards. Full proposals should be submitted via FastLane or Grants.gov.
Requirements: Proposals may be submitted by a single organization or a group of organizations consisting of a lead organization in partnership with one or more partner organizations. Principal investigators are encouraged to form synergistic collaborations among researchers.
Restrictions: Only U.S. academic institutions with significant research and degree-granting education programs in disciplines normally supported by NSF are eligible.
Geographic Focus: All States
Date(s) Application is Due: Nov 15
Amount of Grant: Up to 29,000,000 USD
Contact: Matthew Realff; (703) 292-7081; fax (703) 292-9056; mrealff@nsf.gov
Internet: http://www.nsf.gov/funding/pgm_summ.jsp?pims_id=13633
Sponsor: National Science Foundation
4201 Wilson Boulevard
Arlington, VA 22230

NSF Activities in Science, Engineering, and Mathematics for Persons With Disabilities 3460

The program funds projects that enhance the academic and career development of persons with disabilities in fields related to science, math, technology, and engineering (SMET). Demonstration and intervention projects may last from one to eight years, and funding for this category is to help grantees design and carry out innovative projects that reduce barriers to science, mathematics, engineering, and technology education for persons with disabilities. Information dissemination projects support innovative activities designed to disseminate and promote use of program results, to promote replication of successful strategies used to increase the recruitment, education, and retention in SMET education and career development of persons with disabilities, and to provide the Nation's SMET teaching force information on strategies that permit full and successful participation of students with disabilities in SMET educational environments. Capacity building within community colleges projects encourage proposals designed to establish and test models by which students with disabilities are recruited and trained in SMET disciplines in community college environments. All applicants must submit a preliminary proposal and are encouraged to discuss their proposals with NSF staff before doing so. Full proposals should be submitted via the FastLane system at NSF or Grants.gov. Refer to NSF brochure NSF 00-69.
Requirements: Organizations eligible to submit proposals under this program include universities and colleges; nonprofit, nonacademic institutions, including museums, professional associations, and community-based organizations; state and local governments, including school districts; and for-profit research institutions and businesses.
Geographic Focus: All States
Date(s) Application is Due: May 15; Aug 15
Amount of Grant: Up to 45,000,000 USD
Contact: Lawrence Scadden; (703) 306-1635, ext. 6865; lscadden@nsf.gov
Internet: http://www.nsf.gov/pubs/2000/nsf0069/nsf0069.htm
Sponsor: National Science Foundation
4201 Wilson Boulevard
Arlington, VA 22230

NSF ADVANCE: Increasing the Participation and Advancement of Women in Academic Science and Engineering Careers Grants 3461

The goal of the ADVANCE program is to develop systemic approaches to increase the representation and advancement of women in academic science, technology, engineering and mathematics (STEM) careers, thereby contributing to the development of a more diverse science and engineering workforce. Creative strategies to realize this goal are sought from women and men. Members of underrepresented minority groups and individuals with disabilities are especially encouraged to apply. Proposals that address the participation and advancement of women with disabilities and women from underrepresented minority groups are particularly encouraged. Creative strategies to realize this goal are sought from women and men. Members of underrepresented minority groups and individuals with disabilities are especially encouraged to apply. Proposals that address the participation and advancement of women with disabilities and women from underrepresented minority groups are particularly encouraged. Letters of Intent are due by January 20, with full proposals due by February 24.
Requirements: Proposals from primarily undergraduate institutions, teaching intensive colleges, community colleges, minority-serving institutions (e.g. Tribal Colleges and Universities, Historically Black Colleges and Universities, Hispanic-Serving Institutions), women's colleges, and institutions primarily serving persons with disabilities are encouraged. Institutional Transformation proposals may be submitted by academic institutions of higher learning in the U.S., its territories or possessions, or the Commonwealth of Puerto Rico, that award degrees in a field supported by NSF. Partnerships involving industry, government, professional societies and other not-for-profit organizations are encouraged but not required; however, in the case of partnerships, the lead partner organization must be an academic institution of higher learning and must accept overall management responsibility for the activity. Government organizations (other than academic institutions of higher learning) are not eligible to apply, but may participate in partnerships when the lead partner organization is an academic institution of higher learning with overall management responsibility for the activity. Partner organizations and institutions must be based in the U.S., its territories or possessions, or the Commonwealth of Puerto Rico. Partnerships for Adaptation, Implementation and Dissemination proposals may be submitted by: academic institutions of higher learning that award degrees in a field supported by NSF; professional societies; or other not-for-profit organizations. Submitting institutions and organizations must be based in the U.S., its territories or possessions, or the Commonwealth of Puerto Rico. Government organizations (other than academic institutions of higher learning) are not eligible to apply, but may participate in partnerships when the lead partner organization is an academic institution of higher learning with overall management responsibility for the activity. Partner institutions and organizations must be based in the U.S., its territories or possessions, or the Commonwealth of Puerto Rico.
Restrictions: Organizations that have received NSF ADVANCE Institutional Transformation awards are not eligible to apply for another institutional transformation award or for an IT-Start award. Organizations may submit only one Institutional Transformation proposal or one IT-Start proposal. There is no limitation on the number of PAID proposals.
Geographic Focus: All States
Date(s) Application is Due: Jan 20; Feb 24
Amount of Grant: 500,000 USD
Contact: Jessie DeAro; (703) 292-5350; fax (703) 292-9018; jdearo@nsf.gov
Internet: http://www.nsf.gov/funding/pgm_summ.jsp?pims_id=5383
Sponsor: National Science Foundation
4201 Wilson Boulevard
Arlington, VA 22230

NSF Advanced Learning Technologies (ALT) Grants 3462

Through this program, the NSF supports research that: enables radical improvements in learning through innovative computer and information technologies; and advances research in computer science, information technology, learning, and cognitive science through the unique challenges posed by learning environments and learning technology platforms. Integrative research approaches that build across disciplines and establish tight linkages among theory, experiment, and design are strongly encouraged. Technology goals may include systems for tutoring or assessment, modeling and sensing of cognitive or emotional states, context awareness, natural language interfaces, collaboration, knowledge management, and non-traditional goals that redefine the roles of technology in learning. Educational foci for ALT projects must include an area of science, technology, engineering, or mathematics (STEM), or general cross-cutting skills directly relevant to STEM. The anticipated funding amount is $2,800,000 total, to be divided into approximately 6 to 10 awards annually. Full proposals should be submitted via FastLane or Grants.gov.
Requirements: In response to this solicitation, an investigator may participate as PI or Co-PI in no more than one proposal.
Geographic Focus: All States
Date(s) Application is Due: Apr 25
Amount of Grant: 280,000-470,000 USD
Contact: John C. Cherniavsky; (703) 292-5136; fax (703) 292-9046; jchernia@nsf.gov
Internet: http://www.nsf.gov/funding/pgm_summ.jsp?pims_id=12834
Sponsor: National Science Foundation
4201 Wilson Boulevard
Arlington, VA 22230

NSF Advanced Technologies and Instrumentation (ATI) Grants 3463

The Advanced Technologies and Instrumentation (ATI) program provides grants to support the development and construction of state-of-the-art detectors and instruments for the visible, infrared, and radio regions of the spectrum, including interferometric imaging instrumentation and adaptive optics, and the application of new hardware and software technology and innovative techniques in astronomical research. Instrumentation projects should be driven by scientific considerations and clearly articulated scientific goals. Proposals to the ATI program should include a brief task implementation plan with milestones, schedules, and costs. Full proposals should be submitted via the FastLane system or Grants.gov.
Geographic Focus: All States
Date(s) Application is Due: Nov 1
Contact: Jeffrey R. Pier; (703) 292-2977; fax (703) 292-9034; jpier@nsf.gov
Internet: http://www.nsf.gov/funding/pgm_summ.jsp?pims_id=5660
Sponsor: National Science Foundation
4201 Wilson Boulevard
Arlington, VA 22230

NSF Advances in Biological Informatics Grants 3464

The Advances in Biological Informatics (ABI) program seeks to encourage new approaches to the analysis and dissemination of biological knowledge for the benefit of both the scientific community and the broader public. The ABI program is especially interested in the development of informatics tools and resources that have the potential to advance- or transform- research in biology supported by the Directorate for Biological Sciences at the National Science Foundation. The ABI program accepts three major types of proposals: Innovation awards that seek to pioneer new approaches to the application of informatics to biological problems, Development awards that seek to provide robust cyberinfrastructure that will enable transformative biological research, and Sustaining awards that seek to support ongoing operations and maintenance of existing cyberinfrastructure that is critical for continued advancement of priority biological research. The annual deadline the second Tuesday in August.
Geographic Focus: All States

Contact: Julie A. Dickerson; (703) 292-8167; fax (703) 292-9063; jdickers@nsf.gov
Internet: http://nsf.gov/funding/pgm_summ.jsp?pims_id=5444
Sponsor: National Science Foundation
4201 Wilson Boulevard
Arlington, VA 22230

NSF Advancing Theory in Biology (ATB) Grants 3465
The Biological Sciences Directorate invites submission of proposals that advance our conceptual and theoretical understanding of living systems. The Advancing Theory in Biology (ATB) solicitation supports the development of new theoretical approaches that will improve our understanding of fundamental biological principles that integrate phenomena across levels of biological organization. The anticipated funding amount is $5,000,000 total, of which 10 to 15 grants will be awarded. Full proposals should be submitted via the FastLane system at NSF or Grants.gov.
Geographic Focus: All States
Date(s) Application is Due: Feb 17
Amount of Grant: 330,000-500,000 USD
Contact: Kamal Shukla; (703) 292-7131; fax (703) 292-9061; kshukla@nsf.gov
Internet: http://www.nsf.gov/funding/pgm_summ.jsp?pims_id=501066
Sponsor: National Science Foundation
4201 Wilson Boulevard
Arlington, VA 22230

NSF Aeronomy Grants 3466
The Aeronomy program supports research on upper and middle atmosphere phenomena of ionization, recombination, chemical reaction, photo emission, and transport; the transport of energy, and momentum. This program also supports research into mass in the mesosphere-thermosphere-ionosphere system, including the processes involved and the coupling of this global system to the stratosphere below and magnetosphere above and the plasma physics of phenomena manifested in the coupled ionosphere-magnetosphere system, including the effects of high-power radio wave modification. Full proposals should be submitted via the FastLane system at NSF or Grants.gov.
Geographic Focus: All States
Contact: Cassandra Fesen; (703) 292-8529; fax (703) 292-9022; cfesen@nsf.gov
Internet: http://www.nsf.gov/funding/pgm_summ.jsp?pims_id=11686
Sponsor: National Science Foundation
4201 Wilson Boulevard
Arlington, VA 22230

NSF Alan T. Waterman Award 3467
The annual award recognizes an outstanding young researcher in any field of science or engineering supported by the National Science Foundation. In addition to a medal, the awardee receives a grant of $1,000,000 over a five year period for scientific research or advanced study in the mathematical, physical, biological, engineering, social, or other sciences at the institution of the recipient's choice.
Requirements: Institutions may nominate an unlimited number of individuals. Nomination packages consist of a nomination and four letters of reference submitted via FastLane. Candidates must be U.S. citizens, normally 35 years of age or younger, or not more than seven years beyond receipt of the PhD degree by December 31 of the year nominated. Candidates should have demonstrated exceptional individual achievements in scientific or engineering research of sufficient quality to place them at the forefront of their peers. Criteria include originality, innovation, and significant impact on the field.
Geographic Focus: All States
Date(s) Application is Due: Dec 31
Amount of Grant: 1,000,000 USD
Contact: Mayra N. Montrose; (703) 292-8040; fax (703) 292-9040; mmontros@nsf
Internet: http://www.nsf.gov/od/waterman/waterman.jsp
Sponsor: National Science Foundation
4201 Wilson Boulevard
Arlington, VA 22230

NSF Algebra, Number Theory and Combinatorics (ANTC) Grants 3468
The Program supports research in algebra, including algebraic structures, general algebra, and linear algebra; number theory, including algebraic, analytic number theory, arithmetic geometry, quadratic forms, and automorphic forms; combinatorics, including graph theory; and algebraic geometry. Full proposals should be submitted via the FastLane system at NSF or Grants.gov.
Geographic Focus: All States
Date(s) Application is Due: Oct 6
Contact: Tie Luo; (703) 292-8448; fax (703) 292-9032; tluo@nsf.gov
Internet: http://www.nsf.gov/funding/pgm_summ.jsp?pims_id=5431&org=DMS&from=home
Sponsor: National Science Foundation
4201 Wilson Boulevard
Arlington, VA 22230

NSF Analysis Grants 3469
The Program supports basic research in that area of mathematics whose roots can be traced to the calculus of Newton and Leibniz. Given its centuries-old ties to physics, analysis has influenced developments from Newton's mechanics to quantum mechanics and from Fourier's study of heat conduction to Maxwell's equations of electromagnetism to Witten's theory of supersymmetry. More generally, research supported by analysis provides the theoretical underpinning for the majority of applications of the mathematical sciences to other scientific disciplines. Current areas of significant activity include: nonlinear partial differential equations; dynamical systems and ergodic theory; real, complex and harmonic analysis; operator theory and algebras of operators on Hilbert space; mathematical physics; and representation theory of Lie groups/algebras. Emerging areas include random matrix theory and its ties to classical analysis, number theory, quantum mechanics, and coding theory, and development of noncommutative geometry with its applications to modeling physical phenomena. It should be stressed, however, that the underlying role of the Analysis Program is to provide support for research in mathematics at the most fundamental level. Although this is often done with the expectation that the research will generate a payoff in applications at some point down the road, the principal mission of the Program is to tend and replenish an important reservoir of mathematical knowledge, maintaining it as a dependable resource to be drawn upon by engineers, life and physical scientists, and other mathematical scientists, as need arises. Full proposals should be submitted via the FastLane system at NSF or Grants.gov.
Geographic Focus: All States
Date(s) Application is Due: Oct 6
Contact: Joe W. Jenkins; (703) 292-4870; fax (703) 292-9032; jjenkins@nsf.gov
Internet: http://www.nsf.gov/funding/pgm_summ.jsp?pims_id=5434&org=DMS&from=home
Sponsor: National Science Foundation
4201 Wilson Boulevard
Arlington, VA 22230

NSF Analytical & Surface Chemistry (ASC) Grants 3470
Supports fundamental chemical research directed toward the characterization and analysis of all forms of matter. Studies of elemental and molecular composition and of the microstructure of both bulk and surface domains are included. The program supports projects that develop the fundamentals of measurement science, new sensors and new instruments, and innovative approaches to data processing and interpretation. Investigations designed to probe the chemical structure and reactivity of the interface between different forms of matter also are supported. The program is linked to several other chemistry research programs within NSF, including Solid State Chemistry (Materials Research Division, MPS Directorate); Biochemistry and Biophysics (Molecular and Cellular Biosciences Division, BIO Directorate); and Chemical Reaction Processes and Interfacial, Transport, and Separation Processes (Chemical and Transport Systems Division, ENG Directorate). Full proposals should be submitted via the FastLane system at NSF or Grants.gov.
Geographic Focus: All States
Date(s) Application is Due: Jul 31; Nov 30
Contact: Zeev Rosenzweig; (703) 292-7719; fax (703) 292-9037; zrosenzw@nsf.gov
Internet: http://www.nsf.gov/funding/pgm_summ.jsp?pims_id=5677
Sponsor: National Science Foundation
4201 Wilson Boulevard
Arlington, VA 22230

NSF Animal Developmental Mechanisms Grants 3471
Animal Developmental Systems programmatic area supports research that seeks to understand the processes that result in the complex phenotype of animals. Because different organisms may be more amenable to certain approaches than others, analyses of development in a wide range of different species are encouraged. The Developmental Systems Cluster is also particularly interested in understanding how emergent properties result in the development of complex phenotypes and lead to the evolution of developmental mechanisms. Proposals may be submitted at any time; however, receipt by target dates will ensure review at the regular panel meeting. Full proposals should be submitted via the FastLane system at NSF or Grants.gov.
Requirements: The most frequent recipients of support are academic institutions and nonprofit research groups. In special circumstances, grants also are awarded to other types of institutions and to individuals.
Geographic Focus: All States
Date(s) Application is Due: Jan 12; Jul 12
Contact: James Deshler, Program Director; (703) 292-8417; jdeshler@nsf.gov
Internet: http://www.nsf.gov/funding/pgm_summ.jsp?pims_id=501087
Sponsor: National Science Foundation
4201 Wilson Boulevard
Arlington, VA 22230

NSF Antarctic Aeronomy and Astrophysics Grants 3472
The Program supports studies of three major domains: the stratosphere and the mesosphere; the thermosphere, the ionosphere, and the magnetosphere; and astronomy and astrophysical studies of the Universe, including solar astronomy and cosmic ray physics. Major goals are to sponsor research that requires or would benefit from the unique conditions of the Antarctic, to contribute to the understanding of the role of the Antarctic in global environmental change, to participate in interdisciplinary studies of the middle and upper atmosphere, and to improve understanding of the coupling of the Earth's polar atmosphere with the magnetosphere and of the ways in which both are affected by solar activity. A total of $20,000,000 is available, plus up to $35,000,000 in outyear increments for continuing awards. Deadlines are: May 2 for research proposals on the Oden during November through December; and June 6 for all proposals other than work proposed on Oden for November through December. Full proposals should be submitted via the FastLane system at NSF or Grants.gov.
Geographic Focus: All States
Amount of Grant: 400,000 USD
Contact: Vladimir Papitashvili, Program Director; (703) 292-7425; vpapita@nsf.gov
Internet: http://www.nsf.gov/funding/pgm_summ.jsp?pims_id=13420
Sponsor: National Science Foundation
4201 Wilson Boulevard

Arlington, VA 22230

NSF Antarctic Artists and Writers Grants 3473

The purpose of the Program is to enable serious writings and works of art that exemplify the Antarctic heritage of humankind. In particular, the program seeks to increase public understanding of the Antarctic region, including the continent and the surrounding oceans, as well as the associated research and education endeavors. The Program provides opportunities for professional artists and writers to travel to Antarctica — at research stations, field camps, and aboard ships — to make the observations necessary to complete their proposed projects. While the majority of award recipients are established artists and writers, the program also seeks to support early career artists and writers in an effort to broaden participation. Full proposals should be submitted via the FastLane system at NSF or Grants.gov.

Geographic Focus: All States
Date(s) Application is Due: Jun 8
Contact: Winifred M. Reuning, (703) 292-8033; fax (703) 292-9079; wreuning@nsf.gov
Internet: http://www.nsf.gov/funding/pgm_summ.jsp?pims_id=12783
Sponsor: National Science Foundation
4201 Wilson Boulevard
Arlington, VA 22230

NSF Antarctic Earth Sciences Grants 3474

The Program supports research in both terrestrial and marine settings to interpret Antarctica's rich history and to understand the processes that shape it today. Current projects address such diverse topics as: understanding the ice sheets using sediment records to reconstruct their history and determine the geologic controls of their formation and stability; deciphering paleoenvironmental and paleobiological records to understand global climate, ocean circulation, and the evolution of life; exploring Antarctica's tectonic evolution, from its central role in Gondwana's breakup to the present-day deformation driving volcanism, rifting, and orogenesis; and investigating unique processes, such as the formation of subglacial lakes or the aeolian and permafrost sculpting of the Dry Valleys. All of these problems require a better understanding of Antarctica's geology, including the structure and composition of its crust and underlying mantle, and its connection to global earth science questions. The program therefore supports and encourages field, laboratory, and theoretical work in geology, geophysics, and any other area of earth sciences, including the intersections between these fields and biology, glaciology, and oceanography. The program also strongly encourages investigators new to the program, international collaboration, and research-related education and outreach. Work on previously collected samples and data is also encouraged. A total of $20,000,000 is available, plus up to $35,000,000 in outyear increments for continuing awards. Deadlines are: May 2 for research proposals on the Oden during November through December; and June 6 for all proposals other than work proposed on Oden for November through December. Full proposals should be submitted via the FastLane system at NSF or Grants.gov.

Geographic Focus: All States
Date(s) Application is Due: May 2; Jun 6
Amount of Grant: 400,000 USD
Contact: Thomas P. Wagner; (703) 292-4746; fax (703) 292-9079; twagner@nsf.gov
Internet: http://www.nsf.gov/funding/pgm_summ.jsp?pims_id=8173
Sponsor: National Science Foundation
4201 Wilson Boulevard
Arlington, VA 22230

NSF Antarctic Glaciology Grants 3475

The Program is concerned with the study of the history and dynamics of all naturally occurring forms of snow and glacial ice, including floating ice shelves, glaciers, and continental and marine ice sheets. Program emphases include paleoenvironments from ice cores, ice dynamics, numerical modeling, glacial geology, and remote sensing of ice sheets. Some specific objectives are: correlation of climatic fluctuations evident in antarctic ice cores with data from arctic and lower-latitude ice cores, and integration of the ice record with the terrestrial and marine record; documentation of the geographic extent of climatic events noted in paleoclimatic records; extension of the ice core time series to provide information on astronomical forcing of climate; establishment of more precise dating methodologies for deep ice cores; determination of the Cenozoic history of antarctic ice sheets and their interaction with global climate and uplift of the Transantarctic Mountains; response of the antarctic ice sheets to the Pliocene warming; investigation of the physics of fast glacier flow with emphasis on processes at glacier beds; investigation of ice-shelf stability; and identification and quantification of the feedback between ice dynamics and climate change. A total of $20,000,000 is available, plus up to $35,000,000 in outyear increments for continuing awards. Deadlines are: May 2 for research proposals on the Oden during November through December; and June 6 for all proposals other than work proposed on Oden for November through December. Full proposals should be submitted via the FastLane system at NSF or Grants.gov.

Geographic Focus: All States
Date(s) Application is Due: May 2; Jun 6
Amount of Grant: 400,000 USD
Contact: Julie M. Palais; (703) 292-8033; fax (703) 292-9079; jpalais@nsf.gov
Internet: http://www.nsf.gov/funding/pgm_summ.jsp?pims_id=12798
Sponsor: National Science Foundation
4201 Wilson Boulevard
Arlington, VA 22230

NSF Antarctic Integrated System Science (AISS) Grants 3476

In general terms, the AISS program administers projects that transcend disciplinary boundaries, are highly integrated and address questions broader in scope than those typically supported by the disciplinary programs. AISS projects must have compelling intellectual merit, broad impact and expand the frontiers of our knowledge. Projects must not be so broad in scope that tractable research strategies are not practical. It is recognized that integrated system proposals can be challenging to review. At this juncture in accordance with existing NSF guidelines, proposers may choose to submit single collaborative proposals or multiple related proposals that share some common text. Proposals will be reviewed by both ad-hoc mail reviews and a combination of panelists from the disciplinary panels as appropriate. Those considering submission to AISS are encouraged to contact the program director in advance. A total of $20,000,000 is available, plus up to $35,000,000 in outyear increments for continuing awards. Deadlines are: May 2 for research proposals on the Oden during November through December; and June 6 for all proposals other than work proposed on Oden for November through December. Full proposals should be submitted via the FastLane system at NSF or Grants.gov.

Restrictions: AISS does not fund programs that recast disciplinary questions into a form requiring minimal expertise from other disciplines when progress is possible within a discipline.
Geographic Focus: All States
Date(s) Application is Due: May 2; Jun 6
Amount of Grant: 400,000 USD
Contact: Kelly Falkner; (703) 292-8033; fax (703) 292-9079; kfalkner@nsf.gov
Internet: http://www.nsf.gov/pubs/2008/nsf08535/nsf08535.htm
Sponsor: National Science Foundation
4201 Wilson Boulevard
Arlington, VA 22230

NSF Antarctic Ocean and Atmospheric Sciences Grants 3477

The Program is intended to foster advances in understanding of the oceanic and tropospheric environments at high southern latitudes and their links to each other and global scale processes. Innovative approaches involving surface and/or satellite-based field observations and/or modeling studies are particularly encouraged. Major program elements include but are not limited to: physical oceanographic studies that probe the dynamics and kinematics of the polar oceans, such as the effects of interface driving forces like wind, solar radiation, tides, and heat exchange, water mass production and modification processes, ocean dynamics at the pack ice edge, and the effect of polynyas on ventilation and relationships to the distribution of marine biota; chemical oceanographic studies that investigate the chemical composition of and reactions in sea water, track fluxes of material within ocean basins and at their boundaries, and apply chemical tracers to study time and space scales of physical and biogeochemical oceanic processes; sea ice studies that encompass material characteristics at the individual crystal level to large-scale patterns of freezing, deformation, and melting and the relationships of sea ice dynamics with the ocean and atmosphere over a variety of temporal and spatial scales; and meteorological studies that investigate atmospheric circulation systems and dynamics entailing aspects such as local and/or global exchange of heat, momentum and trace constituents and atmospheric chemistry. A total of $20,000,000 is available, plus up to $35,000,000 in outyear increments for continuing awards. Deadlines are: May 2 for research proposals on the Oden during November through December; and June 6 for all proposals other than work proposed on Oden for November through December. Full proposals should be submitted via the FastLane system at NSF or Grants.gov.

Geographic Focus: All States
Date(s) Application is Due: May 2; Jun 6
Amount of Grant: 400,000 USD
Contact: Kelly Falkner; (703) 292-8033; fax (703) 292-9079; kfalkner@nsf.gov
Internet: http://www.nsf.gov/funding/pgm_summ.jsp?pims_id=13422
Sponsor: National Science Foundation
4201 Wilson Boulevard
Arlington, VA 22230

NSF Antarctic Organisms and Ecosystems Grants 3478

The goal of the Program is to improve understanding of antarctic organisms and their interactions within the biosphere and geosphere. The program supports projects directed at all levels of biological organization from molecular, cellular, and organismal, to communities and ecosystems over regional and global scales. Investigators are encouraged to develop and apply theory and innovative technologies to understand how organisms adapt to and live in high-latitude environments and how populations and ecosystems may respond to global change. Particular emphases include: marine ecosystems; terrestrial and freshwater ecosystems; population dynamics physiological ecology, and adaptation; and genomics. A total of $20,000,000 is available, plus up to $35,000,000 in outyear increments for continuing awards. Deadlines are: May 2 for research proposals on the Oden during November through December; and June 6 for all proposals other than work proposed on Oden for November through December. Full proposals should be submitted via the FastLane system at NSF or Grants.gov.

Geographic Focus: All States
Date(s) Application is Due: May 2; Jun 6
Amount of Grant: 400,000 USD
Contact: Roberta L. Marinelli; (703) 292-7448; fax (703) 292-9079; rmarinel@nsf.gov
Internet: http://www.nsf.gov/funding/pgm_summ.jsp?pims_id=13421
Sponsor: National Science Foundation
4201 Wilson Boulevard
Arlington, VA 22230

NSF Antarctic Research Grants 3479

Scientific research and operational support of that research are the principal activities supported by the United States Government in Antarctica. The goals are to expand

fundamental knowledge of the region, to foster research on global and regional problems of current scientific importance, and to use Antarctica as a platform from which to support research. For projects involving fieldwork, the U.S. Antarctic Program supports only research that can be done exclusively in Antarctica or that is best done from Antarctica. The program also supports antarctic-related analytical research performed at home organizations. A total of $20,000,000 is available, plus up to $35,000,000 in outyear increments for continuing awards. Deadlines are: May 2 for research proposals on the Oden during November through December; and June 6 for all proposals other than work proposed on Oden for November through December. Full proposals should be submitted via the FastLane system at NSF or Grants.gov.
Geographic Focus: All States
Date(s) Application is Due: May 2; Jun 6
Amount of Grant: 400,000 USD
Contact: Kelly Falkner; (703) 292-8033; fax (703) 292-9079; kfalkner@nsf.gov
Internet: http://www.nsf.gov/funding/pgm_summ.jsp?pims_id=5519
Sponsor: National Science Foundation
4201 Wilson Boulevard
Arlington, VA 22230

NSF Antarctic Special Opportunity for Science aboard the Swedish Ice Breaker Oden Grants 3480
The 5-year U.S.-Swedish cooperative agreement for use of the icebreaker Oden provides the opportunity for research proposals to support participation on cruises aboard the Class 1A icebreaker Oden in the Southern Ocean. The Oden will lead the annual break-in of the McMurdo Ship Channel in the Ross Sea in the January-to-February time period. Prior to that time, and in conjunction with the transit of the Oden from South America to the Ross Sea, there will be approximately 20 dedicated science days co-sponsored by Sweden and the United States. The expeditions provide the possibility to collect data in sectors of Antarctic marine regions that are rarely visited and data-sparse.
Requirements: Proposals that include well-developed collaborations among U.S. and Swedish scientists are particularly encouraged. Scientists from Sweden may apply for support from the Swedish Research Council (http://www.vr.se/). The Swedish Research Council is offering a similar schedule for cruise participation, with review and decisions compatible with the U.S. timetable outlined above. Scientists from third countries can be part of U.S. or Swedish Oden-programs.
Geographic Focus: All States
Date(s) Application is Due: May 2
Contact: Roberta L. Marinelli; (703) 292-7448; fax (703) 292-9079; rmarinel@nsf.gov
Internet: http://www.nsf.gov/pubs/2008/nsf08535/nsf08535.htm
Sponsor: National Science Foundation
4201 Wilson Boulevard
Arlington, VA 22230

NSF Applied Mathematics Grants 3481
These grants support research motivated by or having an effect on problems arising in science and engineering. Mathematical merit and novelty, as well as breadth and quality of impact on applications, are important factors. Proposals to develop critical mathematical techniques from individual investigators as well as interdisciplinary teams are encouraged. The proposal window is November 1 through November 15 each year. Full proposals should be submitted via the FastLane system at NSF or Grants.gov. PD 07-1266.
Geographic Focus: All States
Date(s) Application is Due: Nov 15
Contact: Michael Steuerwalt; (703) 292-4860; fax (703) 292-9032; msteuerw@nsf.gov
Internet: http://www.nsf.gov/funding/pgm_summ.jsp?pims_id=5664&org=DMS&from=home
Sponsor: National Science Foundation
4201 Wilson Boulevard
Arlington, VA 22230

NSF Archaeology and Archeaometry Research Grants 3482
The program provides support for anthropologically relevant archaeological research at both a senior and doctoral dissertation level. It also funds anthropologically significant archaeometric research. High risk exploratory research proposals are accepted for consideration and a description of these competitions is provided in the Archaeology Program Overview. Deadline dates for senior research grants are July 1 and December 1; other archaeometry grants are due October 31. Doctoral dissertation research and high risk research in anthropology proposals are accepted at any time. Full proposals should be submitted via the FastLane system at NSF or Grants.gov. Refer to brochure PD 98-1391 for program announcement.
Geographic Focus: All States
Date(s) Application is Due: Jul 1; Oct 31; Dec 1
Contact: Dr. John E. Yellen; (703) 292-8759; fax (703) 292-9068; jyellen@nsf.gov
Internet: http://www.nsf.gov/funding/pgm_summ.jsp?pims_id=11690
Sponsor: National Science Foundation
4201 Wilson Boulevard
Arlington, VA 22230

NSF Arctic Natural Science Grants in Atmospheric Sciences 3483
Research in arctic atmospheric sciences focuses on arctic climate and meteorology, including atmosphere-sea and atmosphere-ice interactions. Research on past climates and atmospheric gases preserved in snow and ice is encouraged. The program also supports research on stratospheric and tropospheric processes. For upper atmospheric and space physics, proposals often are considered jointly with the Division of Atmospheric Sciences in the Geosciences Directorate and OPP's Antarctic Aeronomy and Astrophysics Program when conjugate studies are proposed. Full proposals should be submitted via the FastLane system at NSF or Grants.gov.
Requirements: Only U.S. organizations are eligible to submit proposals under this solicitation. There is no limit on PI eligibility, and there is no limit on the number of proposals that may be submitted.
Restrictions: Proposals that treat generic processes that could be adequately studied outside the Arctic are more appropriate to other programs within the Foundation.
Geographic Focus: All States
Date(s) Application is Due: Nov 10
Amount of Grant: 400,000 USD
Contact: Jane Dionne; (703) 292-7427; fax (703) 292-9082; jdionne@nsf.gov
Internet: http://www.nsf.gov/funding/pgm_summ.jsp?pims_id=13424
Sponsor: National Science Foundation
4201 Wilson Boulevard
Arlington, VA 22230

NSF Arctic Natural Science Grants in Biological Sciences 3484
The Program supports projects that emphasize understanding of the adaptation of organisms to the arctic environment. Biological studies in the Arctic include research in freshwater, marine, and terrestrial biology, and ecology. Full proposals should be submitted via the FastLane system at NSF or Grants.gov.
Requirements: Only U.S. organizations are eligible to submit proposals under this solicitation. There is no limit on PI eligibility, and there is no limit on the number of proposals that may be submitted.
Restrictions: Proposals that treat generic processes that could be adequately studied outside the Arctic are more appropriate to other programs within the Foundation.
Geographic Focus: All States
Date(s) Application is Due: Nov 10
Amount of Grant: 400,000 USD
Contact: Jane Dionne; (703) 292-7427; fax (703) 292-9082; jdionne@nsf.gov
Internet: http://www.nsf.gov/funding/pgm_summ.jsp?pims_id=13424
Sponsor: National Science Foundation
4201 Wilson Boulevard
Arlington, VA 22230

NSF Arctic Natural Science Grants in Earth Sciences 3485
This Program supports a range of terrestrial and marine geology and geophysics project, but of greatest interest are projects that will bring a better understanding of the arctic geological processes, which will improve our ability to interpret the geologic record of environmental change in the polar regions, particularly in the Late Cretaceous and Cenozoic with special emphasis on the Quaternary. Also a priority is understanding the processes responsible for the evolution of permafrost and its consequent effects and a better understanding and reconstruction of the plate tectonic history of the Arctic Ocean. Full proposals should be submitted via the FastLane system at NSF or Grants.gov. NSF 06-603.
Requirements: Only U.S. organizations are eligible to submit proposals under this solicitation. There is no limit on PI eligibility, and there is no limit on the number of proposals that may be submitted.
Restrictions: Proposals that treat generic processes that could be adequately studied outside the Arctic are more appropriate to other programs within the Foundation.
Geographic Focus: All States
Date(s) Application is Due: Nov 10
Amount of Grant: 400,000 USD
Contact: Jane Dionne; (703) 292-7427; fax (703) 292-9082; jdionne@nsf.gov
Internet: http://www.nsf.gov/funding/pgm_summ.jsp?pims_id=13424
Sponsor: National Science Foundation
4201 Wilson Boulevard
Arlington, VA 22230

NSF Arctic Natural Science Grants in Glaciology 3486
Glaciology projects generally focus on the history and dynamics of all naturally-occurring forms of snow and ice, including seasonal snow, glaciers, and the Greenland ice sheet. Modeling of mass balance, glacial geology, and remote sensing studies of ice sheets are also supported. Full proposals should be submitted via the FastLane system at NSF or Grants.gov. NSF 06-603
Requirements: Only U.S. organizations are eligible to submit proposals under this solicitation. There is no limit on PI eligibility, and there is no limit on the number of proposals that may be submitted.
Restrictions: Graduate students are not encouraged to initiate proposals but are encouraged to participate as research assistants to principal investigators. Proposals that treat generic processes that could be adequately studied outside the Arctic are more appropriate to other programs within the Foundation.
Geographic Focus: All States
Date(s) Application is Due: Nov 10
Amount of Grant: 400,000 USD
Contact: Jane Dionne; (703) 292-7427; fax (703) 292-9082; jdionne@nsf.gov
Internet: http://www.nsf.gov/funding/pgm_summ.jsp?pims_id=13424
Sponsor: National Science Foundation
4201 Wilson Boulevard
Arlington, VA 22230

NSF Arctic Natural Science Grants in Ocean Sciences 3487
The Program supports ocean science projects that advance knowledge of the structure and processes of the Arctic Ocean and adjacent seas and their interactions with their

boundaries, including the arctic sea-ice cover. Areas of special interest are: low-temperature life processes; the formation, movement, and mixing of arctic water masses; the dynamics of sea ice; and the role of the Arctic Ocean and adjacent seas in global climate. Proposals concerned with the interdependencies of chemical and physical processes and marine organisms and productivity are encouraged. Full proposals should be submitted via the FastLane system at NSF or Grants.gov. Brochure NSF 06-603
Requirements: Only U.S. organizations are eligible to submit proposals under this solicitation. There is no limit on PI eligibility, and there is no limit on the number of proposals that may be submitted.
Restrictions: Graduate students are not encouraged to initiate proposals but are encouraged to participate as research assistants to principal investigators. Proposals that treat generic processes that could be adequately studied outside the Arctic are more appropriate to other programs within the Foundation.
Geographic Focus: All States
Date(s) Application is Due: Nov 10
Contact: Jane Dionne; (703) 292-7427; fax (703) 292-9082; jdionne@nsf.gov
Internet: http://www.nsf.gov/funding/pgm_summ.jsp?pims_id=13424
Sponsor: National Science Foundation
4201 Wilson Boulevard
Arlington, VA 22230

NSF Arctic Natural Sciences Grants 3488

The Program provides core support for disciplinary and interdisciplinary research on arctic processes and coordinates its support of arctic research with the Directorates for Geosciences, Mathematical and Physical Sciences, Social, Behavioral, and Economic Sciences, and Biological Sciences. Upper atmospheric and space physics proposals often are considered jointly with the Division of Atmospheric Sciences in the Geosciences Directorate and with OPP's Antarctic Aeronomy and Astrophysics Program when conjugate studies are proposed. Areas of special interest include marine and terrestrial ecosystems, arctic atmospheric and oceanic dynamics and climatology, arctic geological and glaciological processes, and their connectivity to lower latitudes. The Program particularly encourages proposals that treat uniquely arctic processes and that provide hypothesis testing required to produce the understanding needed to develop predictive tools based on first principles. Proposals to perform monitoring per se are discouraged. ANS supports projects that emphasize understanding the adaptation of organisms to the unique aspects of the arctic environment. Terrestrial and marine geology and geophysics projects of greatest interest are those that will improve our ability to interpret the geologic record of environmental change in the Arctic, particularly during the Quaternary. Understanding the processes responsible for the evolution of permafrost and consequences of changing permafrost remains a priority, as well. Projects that focus on the history and dynamics of all naturally occurring forms of arctic snow and ice, including seasonal snow, glaciers, and the Greenland ice sheet, are encouraged. The Program supports ocean science projects that advance knowledge of the processes of the Arctic Ocean and adjacent seas and their interactions with their boundaries. The development of sensors necessary to observe these processes is also supported by ANS. Full proposals should be submitted via the FastLane system at NSF or Grants.gov.
Requirements: Only U.S. organizations are eligible to submit proposals under this solicitation. There is no limit on PI eligibility, and there is no limit on the number of proposals that may be submitted.
Restrictions: Proposals that treat generic processes that could be adequately studied outside the Arctic are more appropriate to other programs within the Foundation.
Geographic Focus: All States
Date(s) Application is Due: Nov 10
Contact: William J. Wiseman; (703) 292-4750; fax (703) 292-9082; wwiseman@nsf.gov
Internet: http://www.nsf.gov/funding/pgm_summ.jsp?pims_id=13424
Sponsor: National Science Foundation
4201 Wilson Boulevard
Arlington, VA 22230

NSF Arctic Research and Education Grants 3489

This program supports activities that bridge research and education in concert with funded research grants and agreements through supplement requests or as separate proposal requests to support new ventures. Proposals to this program may include formal or informal education or outreach for students K-12 and higher or to the broader public. Most commonly awards are made as supplements to research grants or as small grants. The program seeks to collaborate with other directorates at NSF to promote the integration of research and education. Arctic research spans the major fields of science, technology, engineering and mathematics and is often multi- or interdisciplinary. Research in the Arctic has clear applications for education and outreach at many levels. The region itself is an interesting hook for teaching about life, physical, and social sciences, and concepts such as ocean and atmosphere circulation, climate, earth system science, animal migrations, and life in extreme environments. Projects bridging research and education that have received support from this program include: teacher enhancement experiences through field research and mentoring by researchers; curriculum development infusing inquiry-based research into K-12 science; travel for arctic experts (e.g., researcher, Alaska Natives, educators) to share their knowledge through seminars and lectures to K-12 and university classrooms, community groups, and other audiences; opportunities for minorities and underrepresented groups to participate in science; projects that involve people of the arctic in science and education, and journalist/science writer immersion in arctic science. Full proposals should be submitted via the FastLane system at NSF or Grants.gov.
Geographic Focus: All States
Date(s) Application is Due: Nov 10
Amount of Grant: 400,000 USD
Contact: Renee D. Crain, Research Support and Logistics Manager; (703) 292-4482; fax (703) 292-9082; rcrain@nsf.gov
Internet: http://www.nsf.gov/funding/pgm_summ.jsp?pims_id=13448
Sponsor: National Science Foundation
4201 Wilson Boulevard
Arlington, VA 22230

NSF Arctic Research Opportunities 3490

The goal of the NSF Arctic Sciences Section is to gain a better understanding of the Earth's physical, biological, geological, chemical, social and cultural processes, and the interactions of ocean, land, atmosphere, biological, and human systems in the Arctic. The Arctic Sciences Section and other NSF programs support projects that contribute to the development of the next generation of researchers and scientific literacy for all ages through education, outreach and broadening participation in science, technology, engineering and mathematics. Program representatives from the Office of Polar Programs (OPP) and other NSF programs that support arctic research coordinate across NSF, including joint review and funding of arctic proposals and mutual support of special projects with high logistical costs. Pending availability of funds, $16,000,000 may be available for proposals to this program. This does not include logistics support that may be provided through the Arctic Research Support and Logistics program. NSF estimates 40 awards per year as standard or continuing grants, or cooperative agreements. The number of awards and average award size and duration are subject to the availability of funds. NSF 06-603
Requirements: Only U.S. organizations are eligible to submit proposals to this program. There is no limit on PI eligibility, and there is no limit on the number of proposals that may be submitted. Emphasis areas and special research opportunities include: (1) Arctic Natural Sciences Program; (2) Arctic Social Sciences Program; (3) Arctic System Science Program; (4) Arctic Research Support and Logistics Program; (5) Arctic Research and Education Program; (6) International Polar Year (IPY). Detailed descriptions for each category can be found in the program solicitation.
Geographic Focus: All States
Date(s) Application is Due: Nov 10
Contact: Renee D. Crain; (703) 292-4482; fax (703) 292-9082; rcrain@nsf.gov
Internet: http://nsf.gov/funding/pgm_summ.jsp?pims_id=5521
Sponsor: National Science Foundation
4201 Wilson Boulevard
Arlington, VA 22230

NSF Arctic Research Support and Logistics Grants 3491

The Program supports the field component of research projects funded through science programs in the Arctic Sciences Section of the Office of Polar Programs and through other programs at National Science Foundation (NSF). The program accepts proposals that: support long-term observations of the Arctic; support the acquisition of data sets useful to a broad segment of the arctic research community; will lead to Cooperative Agreements to operate multi-use arctic research facilities; or provide services that broadly support the arctic research community, such as facilitating communication, developing research ideas in an arctic-wide community setting, and cooperating with arctic communities. The types of proposals considered by RSL are: long-term observations; data acquisitions or collection; facility operations; and field work. Full proposals should be submitted via the FastLane system at NSF or Grants.gov. Detailed information pertaining to proposal submission is included in the Arctic Research Opportunities solicitation (NSF 05-514).
Requirements: Proposals may only be submitted by U.S. organizations. Academic institutions and academically related nonprofit research organizations may submit proposals on behalf of any U.S. scientist; nonprofit, nonacademic research institutions; private profit organizations; and state and local governments are also eligible for support. Under certain circumstances unaffiliated scientists may be eligible. All proposals involving fieldwork must include sufficient information to assure safety and feasibility. Proposals must identify all proposed field personnel and describe their qualifications for the work they are to do.
Geographic Focus: All States
Date(s) Application is Due: Nov 10
Amount of Grant: 400,000 USD
Contact: Renee D. Crain; (703) 292-4482; fax (703) 292-9082; rcrain@nsf.gov
Internet: http://www.nsf.gov/funding/pgm_summ.jsp?pims_id=13437
Sponsor: National Science Foundation
4201 Wilson Boulevard
Arlington, VA 22230

NSF Arctic Social Sciences Grants 3492

The Arctic Social Sciences Program (ASSP) encompasses all social sciences. These include, but are not limited to anthropology, archaeology, economics, geography, linguistics, political science, psychology, science and technology studies, sociology, traditional knowledge and related subjects. Although unsolicited proposals in any of the social sciences mentioned above are welcome, areas of particular interest include culture and environment, resources and economic change, development of social and political institutions, ethnic (cultural) and regional identities, and knowledge systems. Full proposals should be submitted via the FastLane system at NSF or Grants.gov.
Requirements: Proposals may only be submitted by U.S. organizations. Academic institutions and academically related nonprofit research organizations may submit proposals on behalf of any U.S. scientist; nonprofit, nonacademic research institutions; private profit organizations; and state and local governments are also eligible for support. Under certain circumstances unaffiliated scientists may be eligible. All proposals involving fieldwork must include sufficient information to assure safety and feasibility. Proposals must identify all proposed field personnel and describe their qualifications for the work they are to do.
Geographic Focus: All States

Date(s) Application is Due: Nov 10
Amount of Grant: 400,000 USD
Contact: Anna M. Kerttula de Echave, Program Director; (703) 292-7432; fax (703) 292-9082; akerttul@nsf.gov
Internet: http://www.nsf.gov/funding/pgm_summ.jsp?pims_id=13425
Sponsor: National Science Foundation
4201 Wilson Boulevard
Arlington, VA 22230

NSF Arctic System Science (ARCSS) Grants 3493
The program goals are to understand the physical, chemical, biological, and sociocultural processes of the arctic system that interact with the total earth system and thus contribute to or are influenced by global change. ARCSS has four linked and ongoing components: ocean-ice-atmosphere interactions; land-atmosphere-ice interactions; human dimensions of the arctic system; and paleoenvironmental arctic sciences. Specific program opportunities are announced annually for single and multidisciplinary investigations.
Requirements: Any U.S. scientist interested in doing research may initiate a proposal, which is submitted formally by his or her institution. Nonprofit, nonacademic research institutions; private profit-making organizations; and local and state governments also are eligible for support.
Restrictions: Graduate students are not encouraged to initiate proposals but are encouraged to participate as research assistants to principal investigators.
Geographic Focus: All States
Date(s) Application is Due: Nov 10
Amount of Grant: 400,000 USD
Contact: Neil R. Swanberg; (703) 292-8029; fax (703) 292-9082; nswanber@nsf.gov
Internet: http://www.nsf.gov/funding/pgm_summ.jsp?pims_id=13426
Sponsor: National Science Foundation
4201 Wilson Boulevard
Arlington, VA 22230

NSF Astronomy and Astrophysics Postdoctoral Fellowships 3494
NSF Astronomy and Astrophysics Postdoctoral Fellowships provide an opportunity for highly qualified young investigators within 3 years of obtaining their PhD to carry out an integrated program of independent research and education. Fellows may engage in research of observational, instrumental, or theoretical nature, in combination with a coherent educational plan for the duration of the fellowship. The program supports researchers for a period of up to 3 years with fellowships that can be taken to the institution or national facility of their choice. The program is intended to recognize young investigators of significant potential, and provide them with experience in research and education that will establish them in positions of distinction and leadership in the community.
Geographic Focus: All States
Date(s) Application is Due: Oct 9
Contact: Edward A. Ajhar; (703) 292-5039; fax (703) 292-9034; eajhar@nsf.gov
Internet: http://nsf.gov/funding/pgm_summ.jsp?pims_id=5291
Sponsor: National Science Foundation
4201 Wilson Boulevard
Arlington, VA 22230

NSF Astronomy and Astrophysics Research Grants 3495
The Astronomy and Astrophysics Research Grants program provides individual investigator and collaborative research grants for observational, theoretical, laboratory and archival data studies in all areas of astronomy and astrophysics, including but not limited to the following areas of study: Planetary Astronomy-studies of the detailed structure and composition of the surfaces, interiors and atmospheres of the planets and satellites in the Solar System, the nature of small bodies (asteroids and comets), the inter-planetary medium, and the origin and development of the Solar System; Stellar Astronomy and Astrophysics-studies of the structure and activity of the Sun and other stars, the physical properties and composition of all types of single and multiple stars, compact objects and their interactions, extra-solar system planet formation and detection, star formation and stellar evolution, stellar nucleosynthesis, and the properties of atoms and molecules of relevance to stellar astronomy; Galactic Astronomy-studies on the composition, structure and evolution of the Milky Way galaxy and nearby galaxies (research may focus on the stellar populations in these galaxies, the characteristics of star clusters, the interstellar medium, and the properties of atomic and molecular constituents of the interstellar medium); and Extragalactic Astronomy and Cosmology-studies of the more distant Universe, with research topics including galaxy formation, evolution and interaction, active galaxies, quasars, large-scale structure, and all areas of cosmology.
Requirements: Proposals submitted to the AAG Program do not require categorization into one of the study areas identified above. Proposals may span multiple disciplines and/or areas of study and may utilize multiple techniques. Principal Investigators are encouraged to contact one of the Program Officers listed in this announcement prior to submitting a proposal to the AAG Program, particularly if the proposal will include investigators at multiple institutions. Standard NSF eligibility requirements apply.
Geographic Focus: All States
Date(s) Application is Due: Nov 15
Contact: Nigel A. Sharp; (703) 292-4905; fax (703) 292-9034; nsharp@nsf.gov
Internet: http://nsf.gov/funding/pgm_summ.jsp?pims_id=13630
Sponsor: National Science Foundation
4201 Wilson Boulevard
Arlington, VA 22230

NSF Atmospheric Chemistry Grants 3496
This program supports research to measure and model the concentration and distribution of gases and aerosols in the lower and middle atmosphere. Also supports research on the chemical reactions among atmospheric species; the sources and sinks of important trace gases and aerosols; the aqueous-phase atmospheric chemistry; the transport of gases and aerosols throughout the atmosphere; and the improved methods for measuring the concentrations of trace species and their fluxes into and out of the atmosphere. PD 98-1524
Requirements: Individuals and organizations in the following categories may submit proposals: (1) Universities and Colleges-Universities and two- and four-year colleges (including community colleges) located and accredited in the U.S., acting on behalf of their faculty members. Such organizations also are referred to as academic institutions; (2) Non-profit, non-academic organizations-Independent museums, observatories, research laboratories, professional societies and similar organizations in the U.S. that are directly associated with educational or research activities; (3) For-profit organizations-U.S. commercial organizations, especially small businesses with strong capabilities in scientific or engineering research or education. An unsolicited proposal from a commercial organization may be funded when the project is of special concern from a national point of view, special resources are available for the work, or the proposed project is especially meritorious. NSF is interested in supporting projects that couple industrial research resources and perspectives with those of universities; therefore, it especially welcomes proposals for cooperative projects involving both universities and the private commercial sector; (4) State and Local Governments-State educational offices or organizations and local school districts may submit proposals intended to broaden the impact, accelerate the pace, and increase the effectiveness of improvements in science, mathematics and engineering education in both K-12 and post-secondary levels; (5) Unaffiliated Individuals-Scientists, engineers or educators in the U.S. and U.S. citizens may be eligible for support, provided that the individual is not employed by, or affiliated with, an organization, and: (a) the proposer has demonstrated the capability and has access to any necessary facilities to carry out the project; and, (b) the proposer agrees to fiscal arrangements that, in the opinion of the NSF Division of Grants & Agreements, ensure responsible management of Federal funds. Unaffiliated individuals should contact the appropriate program before preparing a proposal for submission.
Restrictions: NSF rarely provides support to foreign organizations. NSF will consider proposals for cooperative projects involving U.S. and foreign organizations, provided support is requested only for the U.S. portion of the collaborative effort. Also, NSF does not normally support research or education activities by scientists, engineers or educators employed by Federal agencies or Federally Funded Research and Development Centers (FFRDCs). Under unusual circumstances, other Federal agencies and FFRDCs may submit proposals directly to NSF.
Geographic Focus: All States
Amount of Grant: 30,000-475,000 USD
Contact: John E. Mak, (703) 292-8524; jmak@nsf.gov
Internet: http://www.nsf.gov/funding/pgm_summ.jsp?pims_id=11692&org=ATM
Sponsor: National Science Foundation
4201 Wilson Boulevard
Arlington, VA 22230

NSF Atmospheric Sciences Mid-Size Infrastructure Opportunity Grants 3497
The Division of Atmospheric Sciences (ATM) at the National Science Foundation recognizes the need for new experimental infrastructure that directly enables areas of research traditionally supported by the Division. For the purpose of this program, mid-size infrastructure is defined as research equipment costing between $4M and $25M. The development of mid-size infrastructure is intended to be a long-term effort in ATM to phase in high priority science-enabling tools consistent with community needs, ATM goals, and the NSF strategic vision. The potential for broad community usage and support is essential. Projects to be funded as mid-size infrastructure must be multi-user facilities that have demonstrable broad community support and directly enable research activities traditionally supported by ATM. Broad community advocacy for proposed projects is essential, as evidenced by the results of decadal surveys, focused workshops, and relevant steering or advisory committee recommendations. Selected projects will be funded for one to five years. NSF 07-602
Requirements: Proposals may be submitted by: (a) Academic Institutions located in the U.S.-U.S. universities and colleges located in the U.S; for-profit organizations-U.S. commercial organizations, especially small businesses with strong capabilities in scientific or engineering research or education; non-profit, non-academic organizations-Independent museums, observatories, research labs, professional societies and similar organizations in the U.S. associated with educational or research activities. Selected projects will be funded for one or more years, not to exceed five years. The development of mid-size infrastructure is intended to be a long-term effort in ATM to phase in high priority science-enabling tools consistent with community needs, ATM goals, and the NSF strategic vision. Design, engineering, and prototyping of mid-size infrastructure will be considered for funding only if there is valid justification for why this activity cannot be funded at the program level.
Restrictions: FFRDC's (Federally Funded Research and Development Centers) affiliated with agencies other than the National Science Foundation are NOT eligible to submit a proposal to this competition.
Geographic Focus: All States
Date(s) Application is Due: Jan 11; Jun 2
Amount of Grant: 4,000,000-25,000,000 USD
Contact: Robert M. Robinson, (703) 292-8529; rmrobins@nsf.gov
Internet: http://www.nsf.gov/publications/pub_summ.jsp?ods_key=nsf07602
Sponsor: National Science Foundation
4201 Wilson Boulevard
Arlington, VA 22230

NSF Atomic Molecular and Optical Physics Grants 3498

The Atomic Molecular and Optical Physics program encompasses four sub-areas of this broad discipline: Precision Measurements, Atomic and Molecular Dynamics, Atomic and Molecular Structure, and Optical Physics. Research supported in the first three sub-areas includes activities in quantum control, cooling and trapping of atoms and ions, low-temperature collision dynamics, the collective behavior of atoms in weakly interacting gases (Bose-Einstein Condensates and dilute Fermi degenerate systems), precision measurements of fundamental constants, and the effects of electron correlation on structure and dynamics. In Optical Physics, support is provided in areas such as nonlinear response of isolated atoms to intense, ultra-short electromagnetic fields, the atom-cavity interaction at high fields, and quantum properties of the electromagnetic field.
Geographic Focus: All States
Date(s) Application is Due: Sep 30
Contact: Robert Dunford; (703) 292-7173; fax (703) 292-9078; rdunford@nsf.gov
Internet: http://www.nsf.gov/funding/pgm_summ.jsp?pims_id=13622
Sponsor: National Science Foundation
4201 Wilson Boulevard
Arlington, VA 22230

NSF Biological Databases and Informatics (BDI) Grants 3499

This program seeks to encourage new approaches to the management, analysis, and dissemination of biological knowledge for the benefit of both the scientific community and the broader public. The BD&I program is especially interested in the development of informatics tools and resources that have the potential to advance all fields of biology under the purview of the Directorate for Biological Sciences at the NSF. Annual deadline date is the second Monday in July. Refer to brochure NSF 05-577.
Requirements: Consortia of eligible individuals or organizations may apply, but a single individual or organization must accept overall management responsibility. International collaboration is encouraged; however, financial support for any non-U.S. participant organization must be provided from within the participant's country or other non-U.S. sources.
Geographic Focus: All States
Date(s) Application is Due: Jul 9
Amount of Grant: 50,000-500,000 USD
Contact: Manfred Zorn; (703) 292-8470; dbidba@nsf.gov
Internet: http://www.nsf.gov/pubs/2005/nsf05577/nsf05577.txt
Sponsor: National Science Foundation
4201 Wilson Boulevard
Arlington, VA 22230

NSF Biological Field Stations and Marine Laboratories (FSML) Grants 3500

This program supports refurbishment, rehabilitation, enhancement, and construction, including laboratory renovations, at these research resources. Sites are required to have a plan for archiving (in electronic form) and making shareable scientific databases available on the unique data they obtain. Support can be provided for necessary computer equipment, connectivity, and development of these scientific database. Deadline is the first Friday in March each year. NSF 01-59
Requirements: U.S. colleges and universities, free-standing research and education institutions, and U.S. chartered corporations with formally constituted research and education programs at field stations or marine laboratories are eligible to apply.
Geographic Focus: All States
Amount of Grant: Up to 500,000 USD
Contact: Dr. Gerald Selzer, Program Director; (703) 292-8470; gselzer@nsf.gov
Internet: http://www.nsf.gov/pubs/2001/nsf0159/nsf0159.htm
Sponsor: National Science Foundation
4201 Wilson Boulevard
Arlington, VA 22230

NSF Biological Information Technology and Systems 3501

This program supports high risk/high return research at the interface of biology and information technology. Biological systems have enormous capabilities as powerful and agile control systems for robotic and regulatory systems, and for pattern recognition, adaptability, information storage, retrieval and processing, sensor fusion and other information-handling tasks. Biology often performs orders of magnitude better than systems based on today's silicon device technologies. Determining what needs to and what can be learned about information processing in biological systems should lead to important new information systems (algorithms, software and systems) and technologies (computer platforms, sensors, robotic devices, etc.). The initial phase of this program focuses on developing computational models and theories for the information processing mechanisms encountered in biological systems that will lead to new information technology systems and hardware platforms. While these new information technology systems will not necessarily or exclusively be implemented in biological matter, the program emphasizes hybrid (bio-silical) systems, particularly as a means for experimenting with and validating new theories of biological information technologies and systems. Annual deadline is the second Friday in February; optional letter of intent is requested 4 weeks in advance of the due date. NSF 01-102
Geographic Focus: All States
Amount of Grant: 100,000-500,000 USD
Contact: James Hickman, Advisor; (703) 292-8980; jhickman@nsf.gov
Internet: http://www.nsf.gov/pubs/2001/nsf01102/nsf01102.htm
Sponsor: National Science Foundation
4201 Wilson Boulevard
Arlington, VA 22230

NSF Biological Oceanography Research Grants 3502

This program supports research in marine ecology which encompasses relationships among marine organisms and their interactions with the environment of the sea. Funded projects focus on understanding variability within marine ecosystems, including ocean margins, continental shelves, central gyres, basins, as well as the Great Lakes. Major studies aim to determine the roles of organisms in global-scale processes. Areas of research include ecosystem and biogeochemical processes, community and population ecology, behavioral, reproductive and life-history ecology, physiological and chemical ecology, molecular, cellular and biochemical studies, evolutionary ecology, and sometimes systematic biology and paleoecology. Submit proposals at any time; allow approximately six months for review and processing. Target dates indicate proximity of review panel meetings.
Geographic Focus: All States
Date(s) Application is Due: Feb 15; Aug 15
Contact: David L. Garrison, (703) 292-7588; fax (703) 292-9085; dgarriso@nsf.gov
Internet: http://www.nsf.gov/funding/pgm_summ.jsp?pims_id=11696&org=OCE&from=home
Sponsor: National Science Foundation
4201 Wilson Boulevard
Arlington, VA 22230

NSF Biological Physics (BP) 3503

The program supports projects in which the analytical and experimental tools of physics are applied to the study of problems originating in the living world. Both experimental and theoretical projects will be considered, although the main focus of the program is in the experimental area. Of particular interest are projects in which new experimental approaches are brought to bear on a well-identified problem. These approaches should at the same time have the potential for broad applicability to a set of similar problems, thereby adding to the set of tools the scientist has for addressing biological problems in general. While the problems under study must be important to advancing understanding of the living world in a meaningful way, particular emphasis will be placed on those projects in which the lessons learned from the application serve to foster new concepts and ideas that expand the intellectual basis of physics. The program funds individual investigators, although collaborative proposals between physicists and biologists are welcome.
Requirements: Unrestricted (i.e., open to any type of entity)
Geographic Focus: All States
Date(s) Application is Due: Jul 31
Amount of Grant: 135,000-500,000 USD
Contact: Ramona Winkelbauer, Program Contact; (703) 292-7390; winkelb@nsf.gov
Internet: http://nsf.gov/funding/pgm_summ.jsp?pims_id=6673&org=NSF&sel_org=NSF&from=fund
Sponsor: National Science Foundation
4201 Wilson Boulevard
Arlington, VA 22230

NSF Biological Research Collections (BRC) 3504

The program provides support for biological collection enhancement, computerization of specimen-related data, research to develop better methods for specimen curation and collection management, and activities such as symposia and workshops to investigate support and management of biological collections. Biological collections supported include those housing natural history specimens and jointly curated collections such as frozen tissues and other physical samples, e.g. DNA libraries and digital images. Such collections provide the materials necessary for research in a broad area of biological sciences. NSF 06-569
Requirements: Proposals are accepted from U.S. organizations, including colleges and universities that maintain research collections, natural history museums including herbaria, and other collections administered by independent organizations or by state, county, or local governments; non-federal and non-profit research organizations that maintain collections; and field stations, marine laboratories, botanical gardens, zoological parks, and aquaria that maintain research collections that document biological diversity. The size of an organization is not a factor in determining eligibility. The maximum that may be requested in a proposal is $500,000. Numbers of awards and average award size/duration are subject to the availability of funds.
Restrictions: In any single round of the BRC competition, only one proposal may be submitted from any individual collection within an organization.
Geographic Focus: All States
Date(s) Application is Due: Jul 28
Amount of Grant: Up to 500,000 USD
Contact: W. Carl Taylor; (703) 292-8470; fax (703) 292-9063; dbibrc@nsf.gov
Internet: http://www.nsf.gov/publications/pub_summ.jsp?ods_key=nsf06569
Sponsor: National Science Foundation
4201 Wilson Boulevard
Arlington, VA 22230

NSF Biomedical Engineering (BME) Grants 3505

The mission of the Biomedical Engineering (BME) Program is to: provide opportunities to develop novel ideas into discovery-level and transformative projects that integrate engineering and life science principles in solving biomedical problems that serve humanity in the long-term; and advance both engineering and life sciences with biomedical engineering projects that are at the interface of engineering and biomedical sciences. The BME program supports fundamental, transformative, and discovery research applied to biological systems.
Geographic Focus: All States
Date(s) Application is Due: Mar 1; Sep 15

Contact: Semahat S. Demir; (703) 292-7950; fax (703) 292-9098; sdemir@nsf.gov
Internet: http://www.nsf.gov/funding/pgm_summ.jsp?pims_id=501023
Sponsor: National Science Foundation
4201 Wilson Boulevard
Arlington, VA 22230

NSF Biomedical Engineering and Engineering Healthcare Grants **3506**
The mission of the Biomedical Engineering and Engineering Healthcare cluster is to provide opportunities to develop novel ideas into projects that integrate engineering and life science principles in solving biomedical problems that serve humanity. The cluster focuses on high impact transforming technologies for deriving information from cells, tissues, organs, and organ systems, extraction of useful information from complex biomedical signals, new approaches to the design of structures and materials for eventual medical use, biophotonics, and new methods of controlling living systems. This cluster is also directed toward the characterization, restoration, and/or substitution of normal functions in humans.
Geographic Focus: All States
Contact: Marcia Rawlings, Information Technology Specialist; (703) 292-7956; fax (703) 292-9013; mrawling@nsf.gov
Internet: http://www.nsf.gov/funding/pgm_summ.jsp?pims_id=501032
Sponsor: National Science Foundation
4201 Wilson Boulevard
Arlington, VA 22230

NSF Biomolecular Systems Cluster Grants **3507**
The Biomolecular Systems Cluster emphasizes the structure, function, dynamics, interactions, and interconversions of biological molecules. The context for such studies can range from investigations of individual macromolecules to the large-scale integration of metabolic and energetic processes. Research supported by this cluster includes development of cutting-edge technologies integrating theoretical, computational, and experimental approaches to the study of biological molecules and their functional complexes; mechanistic studies of the regulation and catalysis of enzymes and RNA, and higher-order characterization of the biochemical processes by which all organisms acquire, transform, and utilize energy from substrates. This cluster emphasizes the importance of multi-disciplinary research carried out at the interfaces of biology, physics, chemistry, mathematics and computer science, and engineering.
Requirements: Standard NSF Grant Proposal Guidelines apply (see http://www.nsf.gov/publications/pub_summ.jsp?ods_key=gpg).
Geographic Focus: All States
Date(s) Application is Due: Jan 12; Jul 12
Contact: Kamal Shukla; (703) 292-7131; fax (703) 292-9061; kshukla@nsf.gov
Internet: http://www.nsf.gov/funding/pgm_summ.jsp?pims_id=12771
Sponsor: National Science Foundation
4201 Wilson Boulevard
Arlington, VA 22230

NSF Biophotonics, Advanced Imaging, and Sensing for Human Health Grants **3508**
Innovative basic research in photonics, imaging, and sensing that is very fundamental in science and engineering is needed to lay the foundation for new technologies beyond those that are mature and ready for application in medical diagnostics and therapies. Developing molecularly specific sensing (molecular photonics), imaging, and monitoring systems with high sensitivity and resolution would be an enormous accomplishment with powerful applications to both biology and medicine. Low cost diagnostics will require novel integration of photonics, molecular biology, and material science. Complex biosensors capable of detecting and discriminating among large classes of biomolecules could be important not only to biology and medicine, but also to environmental sensing and homeland security.
Geographic Focus: All States
Date(s) Application is Due: Mar 1; Sep 15
Contact: Leon Esterowitz; (703) 292-7942; fax (703) 292-9098; lesterow@nsf.gov
Internet: http://www.nsf.gov/funding/pgm_summ.jsp?pims_id=501025
Sponsor: National Science Foundation
4201 Wilson Boulevard
Arlington, VA 22230

NSF Biotechnology, Biochemical, and Biomass Engineering (BBBE) Grants **3509**
The program deals with fundamental problems involved in the processing and manufacturing of products of economic importance by effectively utilizing renewable resources of biological origin and bioinformatics originating from genomic and proteomic information. The BBBE program emphasizes basic engineering and biological research that advances the fundamental knowledge base that contributes to a better understanding of cellular and biomolecular processes (in vivo, in vitro, and/or ex vivo) and eventually to the development of generic enabling technology and practical application. Quantitative assessments of bioprocesses and their rates at the levels of gene regulation and expression, signal transduction pathways, posttranslational protein processing, enzymes in reaction systems, metabolic pathways, cells and tissues in cultivation, and biological systems including animal, plant, microbial and insect cells, etc. are considered vital to the successful research projects in the BBBE program. Research projects supported through the BBBE program include, but are not limited to: fermentation technology; enzyme technology; recombinant DNA technology; cell culture technology; ex vivo and therapeutic stem cell culture technology; metabolic engineering; tissue engineering; nanobiotechnology; quantitative systems biotechnology; biosensor development; and food processing with special focus on the safety of the nation's food supply.
Geographic Focus: All States
Date(s) Application is Due: Mar 1; Sep 15
Contact: Frederick G. Heineken; (703) 292-7944; fax (703) 292-9098; fheineke@nsf.gov
Internet: http://www.nsf.gov/funding/pgm_summ.jsp?pims_id=501024
Sponsor: National Science Foundation
4201 Wilson Boulevard
Arlington, VA 22230

NSF Biotic Surveys and Inventories Grants **3510**
The program invites proposals to document levels of diversity among living and fossil species throughout the world, to include viruses, prokaryotes, and eukaryotes from all marine and terrestrial habitats. The program supports research to record and document the diversity of life on earth, both past and present, as a prologue to investigations of patterns and processes and the development of plans for conservation of that diversity. Well-curated, permanent collections and computerized databases are strongly encouraged as products of such support. Proposals should involve collecting specimens of extant and extinct organisms as samples of taxa of geographic regions and/or geologic horizons; conducting inventories of existing collections; developing and disseminating electronic databases of the collected or inventoried specimens and taxa; and producing biotic treatments, authenticated species lists, catalogs, manuals, keys, expert identification systems, and/or other documents in electronic and printed form. Applications are due the first Friday in November each year. All proposals must be submitted by Fastlane. NSF 98-158
Restrictions: The program does not provide support for phylogenetic, monographic, biogeographic, genetic, or ecological studies. The program does not accept proposals to fund assays of genetic diversity within single species, projects to monitor diversity over time, ecological comparison of the diversity in two or more sites, or studies of the functional significance of diversity at the ecosystem level.
Geographic Focus: All States
Amount of Grant: 30,000-15,000 USD
Contact: Program Director; (703) 306-1481; fax (703) 306-0817; bsirev@nsf.gov
Internet: http://www.nsf.gov/pubs/1998/nsf98158/nsf98158.htm
Sponsor: National Science Foundation
4201 Wilson Boulevard
Arlington, VA 22230

NSF Catalysis and Biocatalysis Grants **3511**
The Catalysis and Biocatalysis program primarily supports fundamental and applied research on: kinetics and mechanisms of important catalyzed chemical reactions as they relate to the production of chemicals, fuels, and specialized materials; characterization of chemical and biochemical phenomena occurring at or near solid surfaces and interfaces; electrocatalytic processes having engineering significance or commercial potential; sustainability, environmental catalysis, and basic research related to green chemistry or utilization of biorenewable resources; kinetic modeling and theory of heterogeneous, homogeneous, and biocatalysis; fundamental aspects of reactive deposition and processing for thin film materials; interactions between chemical reactions and transport processes in reactive systems, and the use of this information in the design or control of complex chemical reactors. This program promotes multidisciplinary research in: chemical kinetics of heterogeneous, homogeneous, and biocatalysis; synthesis and processing of chemically functional materials; interfacial, electrochemical, and photochemical reaction processes; theory and modeling of reaction processes at surfaces; synthesis and characterization of catalysts that function at the nanoscale.
Geographic Focus: All States
Date(s) Application is Due: Mar 1; Sep 15
Contact: John Regalbuto, (703) 292-8371; jregalbu@nsf.gov
Internet: http://www.nsf.gov/funding/pgm_summ.jsp?pims_id=13360
Sponsor: National Science Foundation
4201 Wilson Boulevard
Arlington, VA 22230

NSF Cell Systems Cluster Grants **3512**
The Cellular Systems Cluster, one of three thematic areas within the Division of Molecular and Cellular Biosciences, supports research, across all taxa, into the structure and organization of cells and the dynamics of cellular processes. The Cellular Systems cluster is interested not only in traditional areas of cell biology (such as the organization, function, and dynamics of membranes, organelles and other subcellular compartments, and intracellular and transmembrane signal transduction mechanisms and cell-cell signaling processes) but also in the development of quantitative, theory-driven approaches to cell biology that integrate experimental studies at the molecular genetic, biochemical, biophysical, transcriptomic and proteomic levels. Network theory (e.g., as applied to signal transduction) and molecular dynamic modeling (e.g., as applied to the structure/function relationships of cellular structures) are also of particular interest.
Requirements: While proposals using approaches and model systems traditional in the field of cell biology are welcome, studies focused on novel, unique approaches and on non-traditional model organisms are encouraged.
Geographic Focus: All States
Date(s) Application is Due: Jan 12; Jul 12
Contact: LaJoyce Debro, Program Director; (703) 292-7135; ldebro@nsf.gov
Internet: http://www.nsf.gov/funding/pgm_summ.jsp?pims_id=12772
Sponsor: National Science Foundation
4201 Wilson Boulevard
Arlington, VA 22230

NSF Center for Synthesis in Biological Evolution **3513**

NSF requests proposals to establish a Center for Synthesis in Biological Evolution. This center will serve the needs of the evolutionary biology community by providing mechanisms to foster synthetic, collaborative, cross-disciplinary studies. It will play a pivotal role in the further unification of the biological sciences as it draws together knowledge from disparate biological fields to increase our general understanding of biological design and function. Finally, the center will play a critical role in organizing and synthesizing evolutionary knowledge that will be useful to policy makers, government agencies, educators and society. A preliminary proposal is due by September 17; full proposal, by request only, is due February 1. Annual deadline dates change; contact program staff for exact dates. NSF 03-570
Geographic Focus: All States
Date(s) Application is Due: Feb 1; Sep 17
Contact: Samuel Scheiner; (703) 292-8481; fax (703) 292-9064; sscheine@nsf.gov
Internet: http://www.nsf.gov/pubs/2003/nsf03570/nsf03570.htm
Sponsor: National Science Foundation
4201 Wilson Boulevard
Arlington, VA 22230

NSF Centers for Ocean Education Excellence (COSEE) Grants 3514
Centers for Ocean Education Excellence (COSEE), supports partnerships between ocean science researchers, educators, and informal science organizations, providing the public a deeper understanding of the ocean and its influence on each person's quality of life and our national prosperity. Efforts to support recruitment of minorities to ocean science are also supported through OCE Education and Geo Diversity (OEDG).
Geographic Focus: All States
Contact: Elizabeth L. Rom; (703) 292-7709; fax (703) 292-9085; elrom@nsf.gov
Internet: http://www.nsf.gov/funding/pgm_summ.jsp?pims_id=13548&org=OCE&from=home
Sponsor: National Science Foundation
4201 Wilson Boulevard
Arlington, VA 22230

NSF Centers of Research Excellence in Science and Technology (CREST) and HBCU Research Infrastructure for Science and Engineering (RISE) Grants 3515
The Centers of Research Excellence in Science and Technology (CREST) program makes resources available to enhance the research capabilities of minority-serving institutions through the establishment of centers that effectively integrate education and research. CREST promotes the development of new knowledge, enhancements of the research productivity of individual faculty, and an expanded presence of students historically underrepresented in STEM disciplines. This solicitation requests proposals for: CREST centers; supplements for partnerships applied to existing CREST awards; HBCU Research Infrastructure for Science & Engineering (HBCU-RISE) proposals; supplements for diversity collaboration for projects co-funded with NSF's Small Business Innovation Research and Small Business Technology Transfer (SBIR/STTR) programs, which are administered by NSF's Directorate for Engineering; and new projects in the Directorate for Education and Human Resources track: Innovation through Institutional Integration.
Geographic Focus: All States
Date(s) Application is Due: Feb 24; Feb 27; Mar 13; Aug 25
Contact: Toni Edquist; (703) 292-4649; fax (703) 292-4649; tedquist@nsf.gov
Internet: http://www.nsf.gov/funding/pgm_summ.jsp?pims_id=6668&org=ENG&from=home
Sponsor: National Science Foundation
4201 Wilson Boulevard
Arlington, VA 22230

NSF Chemical and Biological Separations (CBS) Grants 3516
The Chemical and Biological Separations (CBS) program supports fundamental research on novel methods and materials for separation processes. These processes are central to the chemical, biochemical, materials, energy, and pharmaceutical industries. A fundamental understanding of the interfacial, transport, and thermodynamic behavior of multiphase chemical systems as well as quantitative descriptions of processing characteristics in the process-oriented industries is critical for efficient resource management and effective environmental protection. The program encourages proposals that address emerging research areas and technologies, have a high degree of interdisciplinary thought coupled with knowledge creation, and integrate education and research. Areas of emphasis include separation of biological molecules and separations that lead to environmentally benign processing.
Geographic Focus: All States
Date(s) Application is Due: Mar 1; Sep 15
Contact: Rosemarie D. Wesson; (703) 292-7070; fax (703) 292-9098; rwesson@nsf.gov
Internet: http://www.nsf.gov/funding/pgm_summ.jsp?pims_id=13363
Sponsor: National Science Foundation
4201 Wilson Boulevard
Arlington, VA 22230

NSF Chemical Oceanography Research Grants 3517
The Chemical Oceanography Program supports research into the chemical components, reaction mechanisms, and geochemical pathways within the ocean and at its interfaces with the solid earth and the atmosphere. Major emphases include: studies of material inputs to and outputs from marine waters; orthochemical and biological production and transformation of chemical compounds and phases within the marine system; and the determination of reaction rates and study of equilibria. The Program encourages research into the chemistry, distribution, and fate of inorganic and organic substances introduced into or produced within marine environments including those from estuarine waters to the deep sea.
Geographic Focus: All States
Date(s) Application is Due: Feb 15; Aug 15
Contact: Donald Rice; (703) 292-8582; fax (703) 292-9085; drice@nsf.gov
Internet: http://www.nsf.gov/funding/pgm_summ.jsp?pims_id=11698&org=OCE&from=home
Sponsor: National Science Foundation
4201 Wilson Boulevard
Arlington, VA 22230

NSF Chemistry Research Experiences for Undergraduates (REU) 3518
The program supports active research participation by undergraduate students in any of the areas of research funded by the National Science Foundation. The NSF Division of Chemistry funds about 20-25 REU Sites a year (depending on the availability of funds) as part of the NSF-wide REU activity. Projects involve students in meaningful ways in ongoing research programs or in research projects specifically designed for the program. Two mechanisms are featured for support of student research: (1) REU Sites are based on independent proposals to initiate and conduct projects that engage a number of students in research. REU Sites may be based in a single discipline or academic department, or on interdisciplinary or multi-department research opportunities with a coherent intellectual theme. Proposals with an international dimension are welcome. A partnership with the Department of Defense supports REU Sites in DoD-relevant research areas. (2) REU Supplements may be requested for ongoing NSF-funded research projects or may be included as a component of proposals for new or renewal NSF grants or cooperative agreements. NSF 07-569.
Requirements: For REU Site proposals, a single individual should be designated as the Principal Investigator. This individual will be responsible for overseeing all aspects of the award. However, one additional person may be designated as Co-Principal Investigator, if developing and operating the REU Site would involve such shared responsibility. Other anticipated research supervisors should be listed as Non-Co-PI Senior Personnel. Undergraduate student participants in either Sites or Supplements must be citizens or permanent residents of the United States or its possessions. An undergraduate student is a student who is enrolled in a degree program (part-time or full-time) leading to a baccalaureate or associate degree. Students who are transferring from one college or university to another and are enrolled at neither institution during the intervening summer may participate. High school graduates who have been accepted at an undergraduate institution but who have not yet started their undergraduate study are also eligible to participate. Students who have received their bachelor's degrees and are no longer enrolled as undergraduates are generally not eligible to participate. For REU Sites, a significant fraction of the student participants should come from outside the host institution or organization. Some NSF directorates encourage inclusion in the REU program of K-12 teachers of science, technology, engineering, and mathematics. Please contact the appropriate disciplinary program officer for guidance.
Restrictions: Students may not apply to NSF to participate in REU activities. Students apply directly to REU Sites and should consult the directory of active REU Sites.
Geographic Focus: All States
Contact: Wilfredo Colon; (703) 292-8171; fax (703) 292-9061; wcolon@nsf.gov
Internet: http://www.nsf.gov/funding/pgm_summ.jsp?pims_id=503210
Sponsor: National Science Foundation
4201 Wilson Boulevard
Arlington, VA 22230

NSF Chemistry Research Instrumentation and Facilities: Departmental Multi-User Instrumentation (CRIF:MU) 3519
This program is structured to enable the NSF's Division of Chemistry to respond to a variety of needs for infrastructure that promotes research and education in areas traditionally supported by the Division. The Departmental Multi-User Instrumentation component of CRIF (CRIF:MU) provides funds to universities, colleges, and consortia thereof for the purchase of multi-user instruments. The maximum request is $500,000 for instrumentation. Additional funds may be requested for personnel who are needed to support cyber-enhanced projects. The deadline is the fourth Monday of June for full proposals. Refer to NSF brochure NSF 05-578.
Geographic Focus: All States
Amount of Grant: Up to 500,000 USD
Contact: Wade Sisk; (703) 292-8840; fax (703) 292-9037; wsisk@nsf.gov
Internet: http://www.nsf.gov/pubs/2005/nsf05578/nsf05578.htm
Sponsor: National Science Foundation
4201 Wilson Boulevard
Arlington, VA 22230

NSF Chemistry Research Instrumentation and Facilities Grants 3520
This program provides funds to research institutions and consortia thereof for the purchase of multi-user instruments, for major instrumentation development and construction, and for the establishment and support of multi-user research facilities in the chemical sciences. This program is structured to enable the NSF, through its Division of Chemistry, to respond to a variety of needs for infrastructure to undergird advanced research and education in chemistry. Proposal deadline is the fourth Monday in June annually.
Requirements: Proposals may be submitted by U.S. institutions of higher education and independent nonprofit research institutions.
Geographic Focus: All States
Amount of Grant: Up to 6,000,000 USD
Contact: Wade Sisk; (703) 292-4454; fax (703) 292-9037; wsisk@nsf.gov
Internet: http://www.nsf.gov/staff/staff_bio.jsp?lan=wsisk&org=NSF
Sponsor: National Science Foundation
4201 Wilson Boulevard
Arlington, VA 22230

NSF CISE Accelerating Discovery in Science and Engineering Through 3521

Petascale Simulations and Analysis (PetaApps) Grants

This solicitation seeks proposals to develop the future simulation and analysis tools that can use petascale computing to advance the frontiers of scientific and engineering research. Proposals are sought from researchers aiming to capitalize on emerging petascale computing architectures, catalyzing potentially transformative research. NSF's emphasis is on implementation and exploitation of forefront techniques. Proposers must be prepared to demonstrate that they have a research problem that requires and can exploit petascale computing capabilities. Proposals from or including junior researchers are encouraged as one of the goals of this solicitation is to build a community capable of using petascale computing. The focus of the proposal may be in any of the following areas: Atmospheric Sciences; Astronomy and Astrophysics; Chemistry; Computer Science; Earth Sciences; Engineering; Materials Science; Mathematics and Statistics; Ocean Sciences; Physics; Social, Behavioral and Economic Sciences.
Requirements: An individual may be the PI or co-PI in no more than two proposal that responds to this solicitation. There is no limit on the number of proposals with which an individual may be associated as senior personnel to provide specific expertise.
Geographic Focus: All States
Date(s) Application is Due: Oct 30
Amount of Grant: 1,125,000-2,000,000 USD
Contact: Almadena Chtchelkanova; (703) 292-8910; achtchel@nsf.gov
Internet: http://www.nsf.gov/pubs/2008/nsf08592/nsf08592.htm
Sponsor: National Science Foundation
4201 Wilson Boulevard
Arlington, VA 22230

NSF CISE Advanced Learning Technologies (ALT) Grants 3522

Through the Advanced Learning Technologies (ALT) program, CISE and EHR support learning technologies research that enables radical improvements in learning through innovative computer and information technologies, and advances research in computer science, information technology, learning, and cognitive science through the unique challenges posed by learning environments and learning technology platforms. Integrative research approaches that build across disciplines and establish tight linkages among theory, experiment, and design are strongly encouraged. Under this program, CISE and EHR will support research and prototyping work that is built around well defined learning and technology goals, and leads to generalizable advances in knowledge and technology. Large-scale studies, infrastructure, large-scale system development, or educational materials development will not be supported under this program. Both learning and technology goals are expected to be well articulated and significant: educational applications of off-the-shelf technologies or projects that neither model nor measure learning will not be supported under this program.
Requirements: Proposals that do not address both learning and technology goals may be returned without review by the program.
Geographic Focus: All States
Date(s) Application is Due: Apr 25
Amount of Grant: 100,000-200,000 USD
Contact: Amy L. Baylor; (703) 292-8930; fax (703) 292-9073; abaylor@nsf.gov
Internet: http://www.nsf.gov/funding/pgm_summ.jsp?pims_id=12834&org=CISE&sel_org=CISE&from=fund
Sponsor: National Science Foundation
4201 Wilson Boulevard
Arlington, VA 22230

NSF CISE Assembling the Tree of Life (ATOL) Grants 3523

Projects for Assembling the Tree of Life are expected to be ambitious, large scale, and when appropriate to involve multiple investigators from multiple disciplines, likely from multiple organizations, and to include training, outreach, and dissemination components. Tree of Life projects that are taxon-oriented will focus on phylogenetic resolution of large lineages or clades; this taxon focus is not intended to deflect interest in and attention to theoretical or analytical issues, particularly when the clade under study raises critical questions about the suitability or power of current phylogenetic methods of analysis, such as complexities caused by reticulate evolution and lateral gene transfer. In addition to hypothesis-driven work, Tree of Life projects may also be method or theory-oriented, in which case they will address major analytical or computational problems in phylogenetic research and phyloinformatics. The taxon-focus and method-focus approaches described here are intended for guidance only, and not as constraints on innovative projects for Assembling the Tree of Life.
Requirements: Individuals and organizations in the following categories may submit proposals: universities and colleges; non-profit, non-academic organizations; for-profit organizations; state and local governments; unaffiliated individuals; foreign organizations; and other federal agencies.
Geographic Focus: All States
Date(s) Application is Due: Mar 14
Amount of Grant: Up to 3,000,000 USD
Contact: Patrick Herendeen, (703) 292-7184; pherende@nsf.gov
Internet: http://www.nsf.gov/pubs/2008/nsf08515/nsf08515.htm
Sponsor: National Science Foundation
4201 Wilson Boulevard
Arlington, VA 22230

NSF CISE Broadening Participation in Computing (BPC) Grants 3524

The Broadening Participation in Computing (BPC) program is intended to significantly increase the number of domestic students receiving post secondary degrees in the computing disciplines, with an emphasis on those communities that have a longstanding underrepresentation in computing: persons with disabilities, women, and minorities. Included minorities are African Americans, Hispanics, American Indians, Alaska Natives, Native Hawaiians, and Pacific Islanders. Members of these groups participate in computing at rates well below their proportionate representation in the population. Specifically, the BPC program aims to develop and implement innovative models for recruiting, mentoring, and retaining students in post secondary programs in the computing disciplines, with particular attention to students from underrepresented groups. Since the lack of role models in the professoriate can be a barrier to participation, the program also aims to develop effective strategies for encouraging students who want to pursue academic careers in computing and can serve as these role models. While the focus is on implementation, a project may include complementary research that informs the design of its activities. PIs are encouraged to include social scientists in any research, evaluation, and assessment activities.
Requirements: There is no organizational limit. Alliance proposals must include at least one partner that is a degree-granting, academic institution of higher learning located in the U.S., its territories or possessions, or a consortium of such institutions. A person can be a PI or CoPI on at most one Alliance or Alliance Extension proposal (although he or she may serve in other capacities on additional proposals). The PI on an Extension proposal must be a PI or CoPI on the original BPC Alliance award.
Geographic Focus: All States
Date(s) Application is Due: May 21
Contact: Janice Cuny, (703) 292-8489; fax (703) 292-9010; jcuny@nsf.gov
Internet: http://www.nsf.gov/pubs/2007/nsf07548/nsf07548.htm
Sponsor: National Science Foundation
4201 Wilson Boulevard
Arlington, VA 22230

NSF CISE Cluster Exploratory (CluE) Grants 3525

The CluE program invites researchers to submit proposals that describe new, innovative use of the cluster, and/or that probe the possibilities and fundamental limits of the computing paradigm it enables. Since this program provides researchers with access to the Google-IBM resource, proposals must clearly describe why access to the cluster will significantly advance the line of research inquiry.
Requirements: Proposals may only be submitted by the following: non-profit, non-academic organizations—independent museums, observatories, research labs, professional societies and similar organizations in the U.S. associated with educational or research activities; and universities and colleges—universities and two- and four-year colleges (including community colleges) located and accredited in the U.S., acting on behalf of their faculty members. Such organizations also are referred to as academic institutions.
Geographic Focus: All States
Date(s) Application is Due: Jul 17
Contact: Jim French, (703) 292-8930; jfrench@nsf.gov
Internet: http://www.nsf.gov/pubs/2008/nsf08560/nsf08560.htm
Sponsor: National Science Foundation
4201 Wilson Boulevard
Arlington, VA 22230

NSF CISE Collaborative Research in Computational Neuroscience Grants 3526

Computational neuroscience provides a theoretical foundation and a rich set of technical approaches for understanding the principles and dynamics of the nervous system. Building on the theory, methods, and findings of computer science, neuroscience, biology, psychology, the mathematical and physical sciences, engineering, and other fields, computational neuroscience employs a broad spectrum of analytical and experimental approaches to study function, organization, and computational strategies across all levels of the nervous system. Two classes of proposals will be considered in response to this solicitation: research proposals describing new collaborative research projects, and data sharing proposals to enable sharing of data and other resources. In general, appropriate scientific areas of investigations may be related to any of the participating funding organizations.
Restrictions: In response to this solicitation, an investigator may participate as PI or Co-PI in no more than one research proposal per review cycle, and in no more than one data sharing proposal per review cycle.
Geographic Focus: All States
Date(s) Application is Due: Feb 26; Oct 30
Amount of Grant: 300,000-1,250,000 USD
Contact: Kenneth Whang, CRCNS Program Coordinator; (703) 292-5149; fax (703) 292-9073; kwhang@nsf.gov
Internet: http://www.nsf.gov/pubs/2008/nsf08514/nsf08514.htm
Sponsor: National Science Foundation
4201 Wilson Boulevard
Arlington, VA 22230

NSF CISE Communicating Research to Public Audiences Grants 3527

The purpose of Communicating Research to Public Audiences grants is to promote the discovery, integration, dissemination, and employment of new knowledge in service to society and to achieve excellence in U.S. science, technology, engineering, and mathematics (STEM) education at all levels. These grants will provide an opportunity for Principal Investigators (PIs) to explain in non-technical terms the methods and/or results of their research to a broad and diverse audience. Projects with funding levels up to $75,000 will be supported to communicate the processes and results of current research awards from any NSF directorate or the Office of Polar Programs to public audiences in order to assist in the broader dissemination of research results and to promote science and technological literacy for the general public in an out-of-school setting. The grant can be used for any activity that falls within the definition of an informal science education activity such as media presentations, exhibits, or youth-based activities, in order to disseminate research results, research in progress, or research methods.

Requirements: Individuals and organizations in the following categories may submit proposals: U.S. universities and two-and four-year colleges (including community colleges) acting on behalf of their faculty members; independent museums, observatories, research laboratories, professional societies and similar organizations in the U.S. that are directly associated with educational or research activities; U.S. commercial organizations (especially small businesses with strong capabilities in scientific or engineering research or education); state and local governments; scientists, engineers or educators in the U.S. and U.S. citizens; foreign organizations; and a limited number of other types of Federal agencies.
Restrictions: NSF rarely provides support to foreign organizations. However, NSF will consider proposals for cooperative projects involving U.S. and foreign organizations, provided support is requested only for the U.S. portion of the collaborative effort.
Geographic Focus: All States
Amount of Grant: Up to 75,000 USD
Contact: Sylvia M. James; (703) 292-5333; fax (703) 292-9044; sjames@nsf.gov
Internet: http://www.nsf.gov/pubs/2003/nsf03509/nsf03509.html
Sponsor: National Science Foundation
4201 Wilson Boulevard
Arlington, VA 22230

NSF CISE Community-Based Data Interoperability Networks Grants 3528
This NSF crosscutting program supports community efforts to provide for broad interoperability with the goal of enhancing interaction and information sharing across all of the areas of science and engineering research and education represented at the National Science Foundation. The program supports the formation of community-based Data Interoperability Networks groups that enable communities to work together in the development of effective data interoperability strategies. Each Network holds dual responsibilities for: enabling broad community engagement in the development of consensus and agreement on strategies, priorities, and best approaches for achieving broad interoperability; and providing the technical expertise necessary to turn consensus and agreement into robust interoperability frameworks along with the appropriate tools and resources for their broad use and implementation.
Requirements: Individuals and organizations in the following categories may submit proposals: U.S. universities and two-and four-year colleges (including community colleges) acting on behalf of their faculty members; independent museums, observatories, research laboratories, professional societies and similar organizations in the U.S. that are directly associated with educational or research activities; U.S. commercial organizations (especially small businesses with strong capabilities in scientific or engineering research or education); state and local governments; scientists, engineers or educators in the U.S. and U.S. citizens; foreign organizations; and a limited number of other types of Federal agencies.
Restrictions: NSF rarely provides support to foreign organizations. However, NSF will consider proposals for cooperative projects involving U.S. and foreign organizations, provided support is requested only for the U.S. portion of the collaborative effort.
Geographic Focus: All States
Date(s) Application is Due: Jul 23
Amount of Grant: 750,000-1,250,000 USD
Contact: Maria Burka, (703) 292-7030; mburka@nsf.gov
Internet: http://www.nsf.gov/pubs/2007/nsf07565/nsf07565.htm
Sponsor: National Science Foundation
4201 Wilson Boulevard
Arlington, VA 22230

NSF CISE Computer and Network Systems (CNS): Core Grants 3529
CISE's Division of Computer and Network Systems (CNS) supports research and education projects that develop new knowledge in two core programs: the Computer Systems Research (CSR) program; and the Networking Technology and Systems (NeTS) program. Proposers are invited to submit proposals in three project classes, which are defined as follows: Small Projects—up to $500,000 total budget with durations up to three years; Medium Projects—$500,001 to $1,200,000 total budget with durations up to four years; and Large Projects—$1,200,001 to $3,000,000 total budget with durations up to five years. Small Project proposals are due by December 17, Medium Project proposals are due by August 31, and Large Project proposals are due by November 28.
Requirements: Proposals may only be submitted by the following: non-profit, non-academic organizations—independent museums, observatories, research labs, professional societies and similar organizations in the U.S. associated with educational or research activities; and universities and two- and four-year colleges (including community colleges) located and accredited in the U.S., acting on behalf of their faculty members.
Geographic Focus: All States
Date(s) Application is Due: Aug 31; Nov 28; Dec 17
Amount of Grant: 500,000 USD
Contact: Krishna Kant, (703) 292-4776; kkant@nsf.gov
Internet: http://www.nsf.gov/pubs/2008/nsf08576/nsf08576.htm
Sponsor: National Science Foundation
4201 Wilson Boulevard
Arlington, VA 22230

NSF CISE Computing and Communication Foundations Core Grants 3530
CISE's Division of Computing and Communication Foundations (CCF) supports research and education projects that develop new knowledge in three core programs: the Algorithmic Foundations program; the Communications and Information Foundations program; and the Software and Hardware Foundations program. Proposers are invited to submit proposals in three project classes, which are defined as follows: Small Projects—up to $500,000 total budget with durations up to three years; Medium Projects—$500,001 to $1,200,000 total budget with durations up to four years; and Large Projects—$1,200,001 to $3,000,000 total budget with durations up to five years. Small Project proposals are due by December 17, Medium Project proposals are due by August 31 (except for October 31 in 2008), and Large Project proposals are due by November 28.
Requirements: Proposals may only be submitted by the following: non-profit, non-academic organizations—independent museums, observatories, research labs, professional societies and similar organizations in the U.S. associated with educational or research activities; and universities and two- and four-year colleges (including community colleges) located and accredited in the U.S., acting on behalf of their faculty members.
Geographic Focus: All States
Date(s) Application is Due: Aug 30; Nov 28; Dec 17
Amount of Grant: 500,000 USD
Contact: Richard Beigel, (703) 292-8910; rbeigel@nsf.gov
Internet: http://www.nsf.gov/pubs/2008/nsf08577/nsf08577.htm
Sponsor: National Science Foundation
4201 Wilson Boulevard
Arlington, VA 22230

NSF CISE Computing Research Infrastructure (CRI) Grants 3531
The CISE Computing Research Infrastructure (CRI) program drives discovery and learning in the computing disciplines by supporting the creation, enhancement and operation of world-class computing research infrastructure. Further, through the CRI program CISE seeks to ensure that individuals from a diverse range of academic institutions, including minority-serving and predominantly undergraduate institutions, have access to such infrastructure. The CRI program supports two classes of awards: (1) Institutional Infrastructure awards support either the creation of new computing research infrastructure or the enhancement of existing computing research infrastructure to enable world-class research and education opportunities at the awardee and collaborating institutions. (2) Community Infrastructure awards support the planning for computing research infrastructure, or the creation of new computing infrastructure, or the enhancement of existing computing research infrastructure to enable world-class research and education opportunities for broadly-based communities of researchers and educators that extend well beyond the awardee institutions. Furthermore, CI awards support the operation of such infrastructure, ensuring that awardee institutions are well-positioned to provide a high quality of service to community researchers and educators expected to use the infrastructure to realize their research and education goals. NSF 08-570
Requirements: Individuals and organizations in the following categories may submit proposals: U.S. universities and two-and four-year colleges (including community colleges) acting on behalf of their faculty members; independent museums, observatories, research laboratories, professional societies and similar organizations in the U.S. that are directly associated with educational or research activities; U.S. commercial organizations (especially small businesses with strong capabilities in scientific or engineering research or education); state and local governments; scientists, engineers or educators in the U.S. and U.S. citizens; foreign organizations; and a limited number of other types of Federal agencies.
Restrictions: NSF rarely provides support to foreign organizations. NSF will consider proposals for cooperative projects involving U.S. and foreign organizations, provided support is requested only for the U.S. portion of the collaborative effort. NSF does not normally support research or education activities by scientists, engineers or educators employed by Federal agencies or Federally Funded Research and Development Centers (FFRDCs).
Geographic Focus: All States
Date(s) Application is Due: Aug 5; Sep 22
Amount of Grant: 200,000-750,000 USD
Contact: Tanya Korelsky, Program Director; (703) 292-8930; tkorelsk@nsf.gov
Internet: http://www.nsf.gov/funding/pgm_summ.jsp?pims_id=12810&govDel=U.S.NSF_39
Sponsor: National Science Foundation
4201 Wilson Boulevard
Arlington, VA 22230

NSF CISE Cross-Cutting Grants 3532
This solicitation seeks proposals in three programs of interest to all three CISE divisions, and which pose important and timely research and education challenges optimally addressed by researchers from a range of computing fields or sub-fields and beyond, including: data-intensive computing; network science and engineering; and trustworthy computing. Proposals submitted to each of the cross-cutting programs will be reviewed collaboratively by the three CISE divisions, with the goal of creating a comprehensive portfolio of awards with high-impact potential. CISE expects that, over time, these cross-cutting programs will evolve or be absorbed in our core programs, and that new cross-cutting programs will be introduced. Proposers are invited to submit proposals in three project classes, which are defined as follows: small projects—up to $500,000 total budget with durations up to three years; medium projects—$500,001 to $1,200,000 total budget with durations up to four years; and large projects—$1,200,001 to $3,000,000 total budget with durations up to five years.
Requirements: Proposals may only be submitted by the following: non-profit, non-academic organizations—independent museums, observatories, research labs, professional societies and similar organizations in the U.S. associated with educational or research activities; and universities and two- and four-year colleges (including community colleges) located and accredited in the U.S., acting on behalf of their faculty members. Such organizations also are referred to as academic institutions.
Geographic Focus: All States
Date(s) Application is Due: Aug 30; Nov 28; Dec 17
Contact: Jim French, Data-intensive Computing Program Contact; (703) 292-8930; fax (703) 292-9073; jfrench@nsf.gov

Internet: http://www.nsf.gov/pubs/2008/nsf08578/nsf08578.htm
Sponsor: National Science Foundation
4201 Wilson Boulevard
Arlington, VA 22230

NSF CISE Educational Innovation Grants 3533

Grants support innovative educational activities at the undergraduate level in computer and information science and engineering that transfer research results into the undergraduate curriculum. Projects supported are expected to act as national models of excellence by being prototypes of educational experiences for use by a broader segment of the CISE community. Proposals may address a variety of educational activities, including, but not limited to, the development of courses, instructional technologies, software, and other educational materials. Institutions must match at least one-third of the total cost. Refer to brochure NSF 00-33 for program announcement. Deadline is generally the second Tuesday in March each year.
Requirements: Proposals may be submitted from any U.S. college or university that currently offers baccalaureate degree programs in any discipline supported by the CISE Directorate.
Geographic Focus: All States
Date(s) Application is Due: Apr 25
Contact: Program Contact; (703) 306-1980; fax (703) 292-9030; eipd@nsf.gov
Internet: http://www.nsf.gov/pubs/2000/nsf0033/nsf0033.htm
Sponsor: National Science Foundation
4201 Wilson Boulevard
Arlington, VA 22230

NSF CISE Increasing the Participation and Advancement of Women in Academic Science and Engineering Careers (ADVANCE) Grants 3534

The goal of the ADVANCE Program is to increase the participation of women in the scientific and engineering workforce through the increased representation and advancement of women in academic science and engineering careers. Through these awards, NSF seeks to support new and systemic approaches to improving the climate for women in U.S. academic institutions and to facilitate women's retention and advancement to the highest ranks of academic leadership. Creative approaches to realize the goal of this program are sought from women and men. Proposals may include international activities that directly support the goals of ADVANCE. This program will support the following types of ADVANCE Projects: Partnerships for Adaptation, Implementation, and Dissemination (PAID) Awards (deadline January 17); Institutional Transformation (IT) Awards (deadline December 6); and Institutional Transformation Planning Grants (deadline December 6).
Requirements: Institutional Transformation proposals may be submitted by academic institutions of higher learning in the U.S., its territories or possessions, or the Commonwealth of Puerto Rico, that award degrees in a field supported by NSF. Partnerships for Adaptation, Implementation and Dissemination proposals may be submitted by: academic institutions of higher learning that award degrees in a field supported by NSF; professional societies; or other not-for-profit organizations.
Geographic Focus: All States
Date(s) Application is Due: Jan 17; Dec 6
Contact: Jessie A. DeAro; (703) 292-5350; fax (703) 292-9018; jdearo@nsf.gov
Internet: http://www.nsf.gov/funding/pgm_summ.jsp?pims_id=5383&org=CISE&sel_org=CISE&from=fund
Sponsor: National Science Foundation
4201 Wilson Boulevard
Arlington, VA 22230

NSF Climate and Large-Scale Dynamics (CLD) Grants 3535

The goals of the Program are to: advance knowledge about the processes that force and regulate the atmosphere's synoptic and planetary circulation, weather and climate; and sustain the pool of human resources required for excellence in synoptic and global atmospheric dynamics and climate research. Research topics include: theoretical, observational and modeling studies of the general circulation of the stratosphere and troposphere; synoptic scale weather phenomena; processes that govern climate; the causes of climate variability and change; methods to predict climate variations; extended weather and climate predictability; development and testing of parameterization of physical processes; numerical methods for use in large-scale weather and climate models; the assembly and analysis of instrumental and/or modeled weather and climate data; data assimilation studies; development and use of climate models to diagnose and simulate climate and its variations and change. PD 06-5740
Requirements: Individuals and organizations in the following categories may submit proposals: universities and colleges; non-profit, non-academic organizations; and for-profit organizations (an unsolicited proposal from a commercial organization may be funded when the project is of special concern from a national point of view, special resources are available for the work, or the proposed project is especially meritorious. NSF is interested in supporting projects that couple industrial research resources and perspectives with those of universities; therefore, it especially welcomes proposals for cooperative projects involving both universities and the private commercial sector; state and local governments—state educational offices or organizations and local school districts may submit proposals intended to broaden the impact, accelerate the pace, and increase the effectiveness of improvements in science, mathematics and engineering education in both K-12 and post-secondary levels; unaffiliated individuals—scientists, engineers or educators in the U.S. and U.S. citizens may be eligible for support, provided that the individual is not employed by, or affiliated with, an organization, and: (a) the proposer has demonstrated the capability and has access to any necessary facilities to carry out the project; and, (b) the proposer agrees to fiscal arrangements that, in the opinion of the NSF Division of Grants and Agreements, ensure responsible management of Federal funds. Unaffiliated individuals should contact the appropriate program before preparing a proposal for submission.
Restrictions: NSF rarely provides support to foreign organizations. NSF will consider proposals for cooperative projects involving U.S. and foreign organizations, provided support is requested only for the U.S. portion of the collaborative effort. NSF does not normally support research or education activities by scientists, engineers or educators employed by Federal agencies or Federally Funded Research and Development Centers (FFRDCs). Under unusual circumstances, other Federal agencies and FFRDCs may submit proposals directly to NSF.
Geographic Focus: All States
Contact: Jay S. Fein; (703) 292-8527; fax (703) 292-9022; jfein@nsf.gov
Internet: http://nsf.gov/funding/pgm_summ.jsp?pims_id=11699&org=ATM
Sponsor: National Science Foundation
4201 Wilson Boulevard
Arlington, VA 22230

NSF Collaborations in Mathematical Geosciences (CMG) 3536

The primary purposes of the CMG activity are to promote cutting-edge research in areas that require the collaboration of experts in both the geosciences and the mathematical sciences and to promote the training of researchers with skills in both the mathematical sciences and geosciences. We are particularly interested in supporting collaborations in areas where such interdisciplinary collaboration is presently uncommon. We are also particularly interested in supporting collaborations between young investigators. It is not the intent of this activity to provide general support for infrastructure. NSF 05-535
Geographic Focus: All States
Date(s) Application is Due: Feb 1
Amount of Grant: Up to 83,000,000 USD
Contact: Xuming He; (703) 292-4876; fax (703) 292-9032; xhe@nsf.gov
Internet: http://www.nsf.gov/pubs/2005/nsf05535/nsf05535.htm
Sponsor: National Science Foundation
4201 Wilson Boulevard
Arlington, VA 22230

NSF Collaborative Research in Chemistry 3537

This program is designed to promote interdisciplinary collaborative research in a coherent, defined project at the forefront of the chemical sciences. CRC proposals will involve three or more investigators with complementary expertise. Co-investigators may include researchers with backgrounds in diverse areas of chemistry and other science and engineering disciplines appropriate to the proposed research. The use of cyber-infrastructure to enable and enhance collaborations is encouraged. Projects should be scientifically focused in areas supported by the NSF Division of Chemistry, limited in duration, and substantial in their scope and impact. Listed deadlines are for letters of internet; full proposals are by invitation only. Annual deadline dates may vary; contact program staff for exact dates. NSF 05-606
Requirements: Only U.S. academic institutions and nonprofit, nonacademic organizations may submit proposals. The PI must be affiliated with a U.S. academic institution or nonprofit, non-academic organization. Other investigators may be affiliated with U.S. academic institutions, nonprofit, nonacademic organizations, industry, government laboratories, or international institutions.
Restrictions: No CRC award funds may go directly to industry, government laboratories or international institutions.
Geographic Focus: All States
Date(s) Application is Due: Aug 11; Nov 14
Amount of Grant: Up to 25,000,000 USD
Contact: Katharine Covert; (703) 292-4950; fax (703) 292-9037; kcovert@nsf.gov
Internet: http://www.nsf.gov/pubs/2005/nsf05606/nsf05606.htm
Sponsor: National Science Foundation
4201 Wilson Boulevard
Arlington, VA 22230

NSF Communicating Research to Public Audiences 3538

Communicating Research to Public Audiences is a component of the Informal Science Education program (ISE) in the Division of Elementary, Secondary, and Informal Education. ISE projects provide rich and stimulating contexts and experiences for individuals of all ages, interests, and backgrounds to increase their appreciation for, and understanding of, science, technology, engineering, and mathematics (STEM) in out-of-school settings. Requests for up to $75,000 will be considered to support projects that communicate to public audiences the process and results of current research that is being supported by any NSF directorate through informal science education activities, such as media presentations, exhibits, or youth-based activities. The purpose of these efforts is to disseminate research results, research in progress, or research methods. NSF 03-509
Requirements: The Principal Investigator (PI) must have an active NSF research award; a letter of support from the cognizant Program Officer for the research award is required. (NSF research awards do not include Small Grants for Exploratory Research Awards; Conference, Symposia, and Workshops grants; Dissertation Improvement Awards; or Post-doc Fellowships.) There is no fixed deadline. Proposals may be submitted at anytime, but at least six months prior to anticipated start date. The maximum award amount is $75,000. The maximum award period is 24 months. It is expected that the product (e.g. exhibit, film, program) will have a life beyond the expiration of the award. Estimated program budget, number of awards, and average award size/duration are subject to the availability of funds.
Restrictions: One proposal per active research award.
Geographic Focus: All States
Contact: Sylvia M. James; (703) 292-5333; fax (703) 292-9044; sjames@nsf.gov

Internet: http://www.nsf.gov/funding/pgm_summ.jsp?pims_id=5362&org=DGE&sel_org=DGE&from=fund
Sponsor: National Science Foundation
4201 Wilson Boulevard
Arlington, VA 22230

NSF Computational Mathematics Research Grants 3539

This program supports mathematical research in areas of science where computing plays a central and essential role, emphasizing algorithms design, numerical methods and their analysis, and symbolic methods. The prominence of computation in the research is a hallmark of the program. Proposals ranging from single-investigator projects that develop and analyze innovative computational methods to interdisciplinary team projects that not only create new mathematical and computational techniques but use them to model, study, and solve important application problems are encouraged.
Geographic Focus: All States
Contact: Junping Wang; (703) 292-4488; fax (703) 292-9032; msteuerw@nsf.gov
Internet: http://www.nsf.gov/funding/pgm_summ.jsp?pims_id=5390&org=DMS&from=home
Sponsor: National Science Foundation
4201 Wilson Boulevard
Arlington, VA 22230

NSF Condensed Matter Physics Research Grants 3540

This program supports fundamental, experimental, and combined experiment and theory projects on the physics of solid, liquid, and amorphous systems. Phenomena of interest include phase transitions; localization; electronic, magnetic, and lattice structure; superconductivity; elementary excitations, including electronic, magnetic, plasma, and lattice; transport, magnetic, and optical properties; and nonlinear dynamics. Low-temperature physics is represented by research on quantum fluids and solids as well as two-dimensional electron systems. Soft condensed matter research includes partially ordered fluids, colloid physics. Characterization and analysis of new materials by novel methods and research on condensed matter under extreme conditions-such as low temperatures, high pressures, and high magnetic fields-are of interest. Development of new experimental techniques to carry out proposed projects is encouraged. Apply to PD 03-1710
Geographic Focus: All States
Date(s) Application is Due: Nov 2
Contact: Wendy Fuller-Mora; (703) 292-4931; fax (703) 292-9035; wfullerm@nsf.gov
Internet: http://www.nsf.gov/funding/pgm_summ.jsp?pims_id=5666
Sponsor: National Science Foundation
4201 Wilson Boulevard
Arlington, VA 22230

NSF Conferences, Workshops, & Special Meetings in the Mathematical Sciences 3541

NSF supports research, education, and infrastructure for the mathematical sciences in a variety of modes, including support for symposia, workshops, regional conferences, and special research years. These provide exposure for scholarly work, encourage graduate students and junior scientists early in their careers, and they influence the sense of participation, the concept of a role-model, and the perception of current opportunities in a given scientific field. proposals normally request funding in the range of $5,000 to $25,000, although awards of up to $50,000 have occasionally been made. Their duration is normally for one year. NSF 05-540
Requirements: Proposals may be submitted by universities, colleges, and nonprofit, nonacademic organizations in support of individual investigators or small groups.
Geographic Focus: All States
Date(s) Application is Due: Aug 23
Amount of Grant: 5,000-25,000 USD
Internet: http://www.nsf.gov/publications/pub_summ.jsp?ods_key=nsf05540
Sponsor: National Science Foundation
4201 Wilson Boulevard
Arlington, VA 22230

NSF Conferences, Workshops, and Special Meetings in the Mathematical Sciences Grants 3542

For conferences, workshops, and special meetings, the Division of Mathematical Sciences (DMS) invites proposals of two types: regular conference, symposia, and workshop proposals; and proposals for special meetings, which comprise longer-term or larger-scale activities that more widely engage and connect the mathematical sciences community, such as special research years or semesters, multi-institutional regional meetings, and summer schools. Regular proposals are submitted to the cognizant DMS programs according to those programs' usual deadlines or target dates. These proposals normally request funding in the range of $5,000 to $25,000, although awards of up to $50,000 have occasionally been made. Their duration is normally for one year. Proposals for special meetings are submitted to the cognizant DMS programs but at the common deadline stated in this solicitation. Special meetings proposals may request funding of any amount and for durations of up to three years, but most awards are expected to be in the range of $50,000 to $150,000 per year. This is not a change in the sort of proposals that DMS is willing to accept or to fund; it is simply a reminder to the mathematical sciences research community that this opportunity is available.
Geographic Focus: All States
Date(s) Application is Due: Aug 27
Contact: Director; (202) 293-1170; fax (202) 293-3412; rosier@math.georgetown.edu
Internet: http://www.nsf.gov/funding/pgm_summ.jsp?pims_id=11701
Sponsor: National Science Foundation
4201 Wilson Boulevard
Arlington, VA 22230

NSF Continental Dynamics Research Grants 3543

Grants support multidisciplinary research that focuses on an improved understanding of the processes governing the origin, structure, composition, and dynamical evolution of the continents and continental building blocks. The program is particularly oriented toward projects whose scope and complexity require a cooperative or multi-institutional approach and multiyear planning and execution. It is intended that the program would fund only relatively large projects that do not fit easily within earth sciences project support and that have broad support of major sections of the earth science community. Preliminary proposals are required by April 1; full proposals must be received by November 15. NSF 04-512
Requirements: Proposals may be submitted by academic institutions and nonprofit research organizations.
Geographic Focus: All States
Date(s) Application is Due: Apr 1; Nov 15
Contact: Leonard Johnson, Program Director, Directorate for Geosciences, Division of Earth Sciences; (703) 292-8559; fax (703) 292-9025; lejohnso@nsf.gov
Internet: http://www.nsf.gov/pubs/2004/nsf04512/nsf04512.htm
Sponsor: National Science Foundation
4201 Wilson Boulevard
Arlington, VA 22230

NSF Cooperative Studies Of The Earth's Deep Interior (CSEDI) Grants 3544

The Division of Earth Sciences (EAR) invites the submission of proposals for collaborative, interdisciplinary studies of the Earth's interior within the framework of the community-based initiative known as Cooperative Studies of the Earth's Deep Interior (CSEDI). Funding will support basic research on the character and dynamics of the Earth's mantle and core, their influence on the evolution of the Earth as a whole, and on processes operating within the deep interior that affect or are expressed on the Earth's surface. Projects may employ any combination of field, laboratory, and computational studies with observational, theoretical, or experimental approaches. Support is available for research and research infrastructure through grants and cooperative agreements awarded in response to investigator-initiated proposals from U.S. universities and other eligible institutions. Multidisciplinary work is required. EAR will consider co-funding of projects with other agencies and supports international work and collaborations. NSF 06-578
Requirements: Individuals and organizations in the following categories may submit proposals: (1) Universities and Colleges-Universities and two- and four-year colleges (including community colleges) located and accredited in the U.S., acting on behalf of their faculty members. (2) Non-profit, non-academic organizations-Independent museums, observatories, research laboratories, professional societies and similar organizations in the U.S. that are directly associated with educational or research activities (3.) For-profit organizations-U.S. commercial organizations, especially small businesses with strong capabilities in scientific or engineering research or education. An unsolicited proposal from a commercial organization may be funded when the project is of special concern from a national point of view, special resources are available for the work, or the proposed project is especially meritorious. NSF is interested in supporting projects that couple industrial research resources and perspectives with those of universities; therefore, it especially welcomes proposals for cooperative projects involving both universities and the private commercial sector. (4) State and Local Governments-State educational offices or organizations and local school districts may submit proposals intended to broaden the impact, accelerate the pace, and increase the effectiveness of improvements in science, mathematics and engineering education in both K-12 and post-secondary levels. (5) Unaffiliated Individuals-Scientists, engineers or educators in the U.S. and U.S. citizens may be eligible for support, provided that the individual is not employed by, or affiliated with, an organization, and: (a) the proposer has demonstrated the capability and has access to any necessary facilities to carry out the project; and, (b) the proposer agrees to fiscal arrangements that, in the opinion of the NSF Division of Grants & Agreements, ensure responsible management of Federal funds. Unaffiliated individuals should contact the appropriate program before preparing a proposal for submission.
Restrictions: NSF rarely provides support to foreign organizations. NSF will consider proposals for cooperative projects involving U.S. and foreign organizations, provided support is requested only for the U.S. portion of the collaborative effort. NSF does not normally support research or education activities by scientists, engineers or educators employed by Federal agencies or Federally Funded Research and Development Centers (FFRDCs).
Geographic Focus: All States
Date(s) Application is Due: Sep 25
Contact: Leonard Johnson, Program Director, Directorate for Geosciences, Division of Earth Sciences; (703) 292-8559; fax (703) 292-9025; lejohnso@nsf.gov
Internet: http://www.nsf.gov/funding/pgm_summ.jsp?pims_id=11702&govDel=U.S.NSF_39
Sponsor: National Science Foundation
4201 Wilson Boulevard
Arlington, VA 22230

NSF Coupling, Energetics, and Dynamics of Atmospheric Regions Research Grants 3545

CEDAR is a broad-based, community-initiated, upper atmospheric research program. The goal is to understand the behavior of atmospheric regions from the middle atmosphere upward through the thermosphere and ionosphere into the exosphere in terms of coupling, energetics, chemistry, and dynamics on regional and global scales. These processes are related to the sources of perturbations that propagate upward from the lower atmosphere as well as to solar radiation and particle inputs from above. The activities within this program combine observations, theory and modeling. Estimated program budget, number of awards and average award size/duration are subject to the

availability of funds. Normally, CEDAR awards are made for a duration of three years, but proposers may request from one to five years of funding provided the requested duration is adequately justified. NSF estimates making 10 to 15 continuing grant awards. The typical award size will be about $85,000 per year. The maximum award size will be about $150,000 per year. NSF 06-561 (replaces Document NSF 05-554)
Requirements: Individuals and organizations in the following categories may submit proposals: (1) Universities and Colleges-Universities and two- and four-year colleges (including community colleges) located and accredited in the U.S., acting on behalf of their faculty members. Such organizations also are referred to as academic institutions; (2) Non-profit, non-academic organizations-Independent museums, observatories, research laboratories, professional societies and similar organizations in the U.S. that are directly associated with educational or research activities; (3) For-profit organizations-U.S. commercial organizations, especially small businesses with strong capabilities in scientific or engineering research or education. An unsolicited proposal from a commercial organization may be funded when the project is of special concern from a national point of view, special resources are available for the work, or the proposed project is especially meritorious. NSF is interested in supporting projects that couple industrial research resources and perspectives with those of universities; therefore, it especially welcomes proposals for cooperative projects involving both universities and the private commercial sector; (4) State and Local Governments-State educational offices or organizations and local school districts may submit proposals intended to broaden the impact, accelerate the pace, and increase the effectiveness of improvements in science, mathematics and engineering education in both K-12 and post-secondary levels; (5) Unaffiliated Individuals-Scientists, engineers or educators in the U.S. and U.S. citizens may be eligible for support, provided that the individual is not employed by, or affiliated with, an organization, and: (a) the proposer has demonstrated the capability and has access to any necessary facilities to carry out the project; and, (b) the proposer agrees to fiscal arrangements that, in the opinion of the NSF Division of Grants & Agreements, ensure responsible management of Federal funds. Unaffiliated individuals should contact the appropriate program before preparing a proposal for submission.
Restrictions: NSF rarely provides support to foreign organizations. NSF will consider proposals for cooperative projects involving U.S. and foreign organizations, provided support is requested only for the U.S. portion of the collaborative effort. Also, NSF does not normally support research or education activities by scientists, engineers or educators employed by Federal agencies or Federally Funded Research and Development Centers (FFRDCs). Under unusual circumstances, other Federal agencies and FFRDCs may submit proposals directly to NSF.
Geographic Focus: All States
Date(s) Application is Due: May 10
Amount of Grant: 85,000 USD
Contact: Shaun Young, Program Assistant; (703) 292-8521; slyoung@nsf.gov
Internet: http://www.nsf.gov/publications/pub_summ.jsp?ods_key=nsf06561
Sponsor: National Science Foundation
4201 Wilson Boulevard
Arlington, VA 22230

NSF Course, Curriculum, and Laboratory Improvement Research Grants **3546**
The goal of this program is improvement in the quality of science, technology, engineering, and mathematics (STEM) education for all undergraduate students. The program supports efforts to create new learning materials and teaching strategies, develop faculty expertise, implement educational innovations, assess learning and evaluate innovations, and conduct research on STEM teaching and learning. The program supports three types of projects representing three different phases of development, ranging from small, exploratory investigations to large comprehensive projects. NSF 06-536
Geographic Focus: All States
Date(s) Application is Due: Jan 10
Contact: Program Contact; (703) 292-8666; undergrad@nsf.gov
Internet: http://www.nsf.gov/funding/pgm_summ.jsp?pims_id=5741
Sponsor: National Science Foundation
4201 Wilson Boulevard
Arlington, VA 22230

NSF CreativeIT Grants **3547**
The goal of the CreativeIT Program is to fund research that focuses on creativity to produce simultaneous advances in both computer science and creative cognition, creativity support tools, engineering design or science. The CreativeIT Program solicits proposals for projects that explore the synergies of cross disciplinary research in creativity and computer science and information technology. Information technology is playing an increasing role in extending the capability of human creative thinking and problem solving. The study of creativity as a way to advance computer science and information technology can lead to new models of creative computational processes, innovative approaches to education that encourage creativity, innovative modes of research that include creative professionals, and new technology to support human creativity. A better understanding of creativity and it's role in computer science research, encouraging creativity in education, and supporting creativity with new information technology will improve American competitiveness and innovation.
Requirements: All collaborative proposals submitted as separate submissions from multiple organizations must be submitted via the NSF FastLane system. Individuals and organizations in the following categories may submit proposals: (1) Universities and Colleges-Universities and two- and four-year colleges (including community colleges) located and accredited in the U.S., acting on behalf of their faculty members. (2) Non-profit, non-academic organizations-Independent museums, observatories, research laboratories, professional societies and similar organizations in the U.S. that are directly associated with educational or research activities (3.) For-profit organizations-U.S. commercial organizations, especially small businesses with strong capabilities in scientific or engineering research or education. An unsolicited proposal from a commercial organization may be funded when the project is of special concern from a national point of view, special resources are available for the work, or the proposed project is especially meritorious. NSF is interested in supporting projects that couple industrial research resources and perspectives with those of universities; therefore, it especially welcomes proposals for cooperative projects involving both universities and the private commercial sector. (4) State and Local Governments-State educational offices or organizations and local school districts may submit proposals intended to broaden the impact, accelerate the pace, and increase the effectiveness of improvements in science, mathematics and engineering education in both K-12 and post-secondary levels. (5) Unaffiliated Individuals-Scientists, engineers or educators in the U.S. and U.S. citizens may be eligible for support, provided that the individual is not employed by, or affiliated with, an organization, and: (a) the proposer has demonstrated the capability and has access to any necessary facilities to carry out the project; and, (b) the proposer agrees to fiscal arrangements that, in the opinion of the NSF Division of Grants & Agreements, ensure responsible management of Federal funds. Unaffiliated individuals should contact the appropriate program before preparing a proposal for submission.
Restrictions: NSF rarely provides support to foreign organizations. NSF will consider proposals for cooperative projects involving U.S. and foreign organizations, provided support is requested only for the U.S. portion of the collaborative effort. NSF does not normally support research or education activities by scientists, engineers or educators employed by Federal agencies or Federally Funded Research and Development Centers (FFRDCs).
Geographic Focus: All States
Date(s) Application is Due: Sep 26
Amount of Grant: 200,000-800,000 USD
Contact: Mary Lou Maher; (703) 292-7242; fax (703) 292-9073; mmaher@nsf.gov
Internet: http://www.nsf.gov/funding/pgm_summ.jsp?pims_id=501096&govDel=U.S.NSF_39
Sponsor: National Science Foundation
4201 Wilson Boulevard
Arlington, VA 22230

NSF Cultural Anthropology Research Grants **3548**
The program promotes basic scientific research on the causes and consequences of human social and cultural variation. The program solicits research proposals of theoretical importance in all substantive and theoretical subfields within the discipline of cultural anthropology. Types of support include dissertation improvement grants, research experiences for graduates, supplements; and research experiences for undergraduates, supplements. Proposals submitted through Fastlane should use PD 98-1390 Cultural Anthropology when selecting a program announcement/solicitation number. Submission and program guidelines are available online.
Geographic Focus: All States
Date(s) Application is Due: Jan 15; Aug 15
Contact: Deborah Winslow, Program Director; (703) 292-7315; fax (703) 292-9068; dwinslow@nsf.gov
Internet: http://www.nsf.gov/funding/pgm_summ.jsp?pims_id=5388
Sponsor: National Science Foundation
4201 Wilson Boulevard
Arlington, VA 22230

NSF Cyber-Enabled Discovery and Innovation (CDI) **3549**
Cyber-Enabled Discovery and Innovation (CDI) is NSF's bold five-year initiative to create revolutionary science and engineering research outcomes made possible by innovations and advances in computational thinking. Computational thinking is defined comprehensively to encompass computational concepts, methods, models, algorithms, and tools. Applied in challenging science and engineering research and education contexts, computational thinking promises a profound impact on the Nation's ability to generate and apply new knowledge. Collectively, CDI research outcomes are expected to produce paradigm shifts in our understanding of a wide range of science and engineering phenomena and socio-technical innovations that create new wealth and enhance the national quality of life. NSF 08-604
Requirements: Proposals may only be submitted by the following: (1) Non-profit, non-academic organizations: Independent museums, observatories, research labs, professional societies and similar organizations in the U.S. associated with educational or research activities; or (2) Universities and Colleges: Universities and two- and four-year colleges (including community colleges) located and accredited in the U.S., acting on behalf of their faculty members. Such organizations also are referred to as academic institutions. Preliminary proposals are required and must be submitted via the NSF FastLane system, even if full proposals will be submitted via Grants.gov.
Restrictions: Limit of 2 proposals per Principal Investigator (PI). Projects that make use of existing computational concepts, methods, models, algorithms, and tools to significantly advance only one field of science or engineering should be submitted to an appropriate NSF program in that field instead of to CDI.
Geographic Focus: All States
Date(s) Application is Due: Dec 8
Contact: Mary Lou Maher; (703) 292-8930; fax (703) 292-9073; mmaher@nsf.gov
Internet: http://www.nsf.gov/funding/pgm_summ.jsp?pims_id=503163&org=DGE&sel_org=DGE&from=fund
Sponsor: National Science Foundation
4201 Wilson Boulevard
Arlington, VA 22230

NSF Decision, Risk, and Management Science Research Grants **3550**
This program supports scientific research directed at increasing the understanding and effectiveness of decision making by individuals, groups, organizations, and society.

Disciplinary and interdisciplinary research, doctoral dissertation research, and workshops are funded in the areas of judgment and decision making; decision analysis and decision aids; risk analysis; perception, and communication; societal and public policy decision making; management science and organizational design. The program also supports small grants for exploratory research of a time-critical or high-risk, potentially transformative nature. Funded research must be relevant to an operational or applied context, grounded in theory, based on empirical observation, or subject to empirical validation, and generalizable. Proposals are accepted at any time from investigators in all fields of science. The target dates indicate the proximity of panel review meetings. Refer to NSF brochure PD 98-1321 for program announcement. Full Proposal Target dates are January 18 and August 18.
Geographic Focus: All States
Date(s) Application is Due: Jan 11; Aug 11
Contact: Robert O'Connor; (703) 292-7263; fax (703) 292-9068; roconnor@nsf.gov
Internet: http://www.nsf.gov/funding/pgm_summ.jsp?pims_id=5423
Sponsor: National Science Foundation
4201 Wilson Boulevard
Arlington, VA 22230

NSF Deep Underground Science and Engineering Laboratory Grants 3551
The particle physics, cosmology/astrophysics and nuclear physics communities have identified the need for a deep underground laboratory infrastructure in order to address some of the most compelling, transformational questions at the frontier of their disciplines. While such an underground infrastructure is envisioned to specifically address forefront physics research, it may also provide cost effective opportunities for other communities to address important new research areas. The funds awarded for the proposals selected in S4 will allow the proposing team to complete the design phase(s) through Preliminary Design or beyond, as defined in the LFM, within the three year solicitation funding period.
Geographic Focus: All States
Date(s) Application is Due: Jan 9
Contact: Richard J. Fragaszy, Program Director; (703) 292-7011; rfragasz@nsf.gov
Internet: http://www.nsf.gov/funding/pgm_summ.jsp?pims_id=503136
Sponsor: National Science Foundation
4201 Wilson Boulevard
Arlington, VA 22230

NSF Developing Global Scientists and Engineers International Research Experiences for Students 3552
The United States needs to educate a globally-engaged science and engineering workforce capable of performing in an international research environment in order to remain at the forefront of world science and technology. To support this aim, the Developing Global Scientists and Engineers program provides highest quality international research experiences for U.S. students. Whereas the International Research Experiences for Students (IRES) component of the program supports groups of U.S. undergraduate or graduate students conducting research abroad in collaboration with foreign investigators, the Doctoral Dissertation Enhancement Projects (DDEP) component supports the dissertation research abroad of one doctoral student in collaboration with a foreign investigator. In addition to the activities described in this solicitation, the Office of International Science and Engineering (OISE) supports other targeted international research and education experiences for early-career scientists and engineers via the Research Experience for Undergraduates program, the East Asia and Pacific Summer Institutes for U.S. Graduate Students, the Pan-American Advanced Studies Institutes (for advanced graduate students and post-doctoral fellows), and the International Research Fellowship Program (for post-doctoral fellows or new faculty). Anticipated annual funding is estimated as: International Research Experiences for Students (IRES): $2,000,000 to fund up to 14 awards; Doctoral Dissertation Enhancement Projects (DDEP): $100,000 to fund up to 10 awards. Estimated program budget, number of awards and average award size/duration are subject to the availability of funds. NSF 04-036
Requirements: This program provides educational opportunities for Undergraduate Students and Graduate Students, however proposals must be submitted by a U.S. institution, organization, or professional society. Doctoral Dissertation Enhancement Project (DDEP) proposals must be submitted by the faculty advisor of the graduate student whose dissertation project is the subject of the proposal. On International Research Experiences for Students (IRES) proposals, indirect costs are allowable, consistent with NSF's general policy. Student stipends, travel, and subsistence costs for this training experience are typically listed as participant support costs and as such are not subject to indirect costs. Off-campus indirect rates are applicable to activities at foreign sites.
Restrictions: No indirect costs are allowable on Doctoral Dissertation Enhancement Projects (DDEP) proposals.
Geographic Focus: All States
Date(s) Application is Due: Feb 15; Sep 15
Amount of Grant: 87,500 USD
Contact: Program Officer; (703) 292-5111
Internet: http://www.nsf.gov/funding/pgm_summ.jsp?pims_id=12831
Sponsor: National Science Foundation
4201 Wilson Boulevard
Arlington, VA 22230

NSF Director's Award for Distinguished Teaching Scholars (DTS) 3553
The award recognizes and rewards individuals who have contributed significantly to the scholarship of their discipline and to the education of students in science, technology, engineering, and mathematics, and exemplify the ability to engage productively in both research and education. The program is part of NSF's efforts to foster an academic culture that values a scholarly approach to both research and education. The award is the highest honor bestowed by the NSF for excellence in both teaching and research in science, technology, education, and mathematics fields, or in educational research related to these disciplines. The awards will be conferred at a special ceremony recognizing the awardees. Preliminary proposals (required) are due in October; full proposals are due in February. Contact a program officer for updated deadline dates. Guidelines are available online. NSF 04-594
Requirements: NSF is most interested in supporting projects that promote scholarly work in STEM education; enhance its value in the higher education community; have an impact beyond the Awardees' institutions; and that maintain the prestige of the award. Scientists, engineers, and educators usually initiate proposals that are officially submitted by their employing organization.
Restrictions: Although there is no limit on the number of candidates nominated from an institution, no more than one award will be made to a single institution in a fiscal year's competition.
Geographic Focus: All States
Date(s) Application is Due: Feb 25; Nov 19
Amount of Grant: Up to 300,000 USD
Contact: Myles G. Boylan; (703) 292-4617; fax (703) 292-9048; mboylan@nsf.gov
Internet: http://www.nsf.gov/funding/pgm_summ.jsp?pims_id=8170
Sponsor: National Science Foundation
4201 Wilson Boulevard
Arlington, VA 22230

NSF Doctoral Dissertation Improvement Grants in the Directorate for Biological Sciences (DDIG) 3554
This program awards grants in selected areas of the biological sciences. These grants provide partial support of doctoral dissertation research to improve the overall quality of research. Allowed are costs for doctoral candidates to participate in scientific meetings, to conduct research in specialized facilities or field settings, and to expand an existing body of dissertation research. NSF 08-564.
Requirements: U.S. institutions and organizations that are eligible for awards from the National Science Foundation, including colleges, universities, and other nonprofit research organizations such as botanical gardens, marine and freshwater institutes, and natural history museums may submit proposals. See Chapter I, Section E of the NSF Grant Proposal Guide (GPG) for specific definitions of these categories of proposers. The NSF encourages collaborations with scientists at foreign organizations; however, primary support for any foreign participants' activities must be secured through their own national sources. A student must have advanced to candidacy for a Ph.D. degree before the submission deadline to be eligible to submit a proposal. A statement that the student has advanced to candidacy for a Ph.D., signed and dated by the department chairperson, graduate dean, or similar administrative official is required. The proposal must be submitted through regular organizational channels by the dissertation advisor(s) on behalf of a graduate student who is at the point of initiating or is already conducting dissertation research. The student must be enrolled at a U.S. institution, but need not be a U.S. citizen. Organizations should limit applications to outstanding dissertation proposals with unusual financial requirements that cannot be met otherwise. Preference may be given to projects that are underway and for which feasibility is demonstrated.
Restrictions: Indirect costs not allowed.
Geographic Focus: All States
Date(s) Application is Due: Nov 20
Amount of Grant: Up to 15,000 USD
Contact: IOS Program Officer; (703) 292-8423; ddig-ios@nsf.gov
Internet: http://www.nsf.gov/funding/pgm_summ.jsp?pims_id=5234&org=BIO&sel_org=BIO&from=fund
Sponsor: National Science Foundation
4201 Wilson Boulevard
Arlington, VA 22230

NSF East Asia and Pacific Summer Institutes for U.S. Graduate Students 3555
The East Asia and Pacific Summer Institutes provide U.S. graduate students in science and engineering: 1) first-hand research experience in Australia, China, Japan, Korea, New Zealand, Singapore or Taiwan; 2) an introduction to the science and science policy infrastructure of the respective location; and 3) orientation to the society, culture and language. The primary goals of EAPSI are to introduce students to East Asia and Pacific science and engineering in the context of a research setting, and to help students initiate scientific relationships that will better enable future collaboration with foreign counterparts. The institutes last approximately eight weeks from June to August. Awardees will receive a $5,000 stipend, an allowance of up to $2,500 for international travel, and will be supported to attend a pre-departure orientation in the Washington, D.C. area. Foreign co-sponsoring organizations will provide additional support to cover EAPSI students' living expenses abroad during the period of the summer institutes, and will provide an orientation to the science environment and culture(s) of each location. NSF Solicitation 08-603.
Requirements: Applicants: (1) Must be a citizen or permanent resident of the United States by the deadline; (2) Must be enrolled in a research-oriented master's or Ph.D. degree program at a U.S. institution located in the United States; (3) Pursuing studies in fields of science and engineering research and education supported by the National Science Foundation; and, (4) Pursuing studies in fields that are supported by the foreign co-sponsoring organization. Applications are submitted directly by the individual graduate student, unlike standard NSF proposals that are submitted through the principal investigator's U.S. authorized organizational representative. In the EAPSI FastLane application process (Section V.D.), the applicant acts as the Authorized Organizational Representative (AOR).

Restrictions: One application per student per year. Previous EAPSI awardees may not receive funding to the same location (Australia, China, Japan, Korea, New Zealand, Singapore or Taiwan).
Geographic Focus: All States
Date(s) Application is Due: Dec 8
Amount of Grant: 5,000 USD
Contact: Program Manager; (703) 292-8704; fax (703) 292-9175; eapinfo@nsf.gov
Internet: http://www.nsf.gov/publications/pub_summ.jsp?ods_key=nsf07584
Sponsor: National Science Foundation
4201 Wilson Boulevard
Arlington, VA 22230

NSF Ecosystem Science Grants 3556
This program supports mechanistic or empirical investigations of whole-system ecological processes and relationships in the following areas: biogeochemistry, such as studies of decomposition, global and regional elemental budgets, and biotic versus abiotic controls of nutrient cycles; primary productivity, particularly ecophysiology within an ecosystem framework; and landscape dynamics with an emphasis on quantitative models of disturbances, ecosystem resilience, and successional patterns. Research on natural, managed and disturbed ecosystems is supported, including terrestrial, freshwater, wetland, coastal (including salt marsh and mangrove), and human-dominated environments.
Requirements: Proposals will be considered that focus on advancing ecosystem science through either the pursuit of new theoretical paradigms or novel modeling efforts. Proposals that, in whole or part, strive to develop new techniques can be supported when a compelling argument exists that there is the potential for a major advance in ecosystem research. Inter- and multi-disciplinary proposals that fall across traditional programmatic boundaries are welcomed and encouraged. Proposals may focus on the cycling of non-nutrient elements, but proposals that are ecotoxicological in orientation will not be considered. Apply to PD 04-7381.
Geographic Focus: All States
Date(s) Application is Due: Jan 9; Jul 9
Contact: Todd A. Crowl; (703) 292-7870; fax (703) 292-9064; tcrowl@nsf.gov
Internet: http://www.nsf.gov/funding/pgm_summ.jsp?pims_id=12822&org=DEB&sel_org=DEB&from=fund
Sponsor: National Science Foundation
4201 Wilson Boulevard
Arlington, VA 22230

NSF Education and Interdisciplinary Research (EIR) Grants 3557
The program supports activities in conjunction with NSF-wide programs such as Faculty Early Career Development (CAREER), Research Experiences for Undergraduates (REU), and programs aimed at women, minorities, and persons with disabilities. Further information about all of these programs and activities is available in the Crosscutting Investment Strategies section of the NSF Guide to Programs. The program also supports activities that seek to improve the education and training of physics students (both undergraduate and graduate), such as curriculum development for upper-level physics courses, and activities that are not included in specific programs elsewhere within NSF. The program supports research at the interface between physics and other disciplines and extending to emerging areas. Broadening activities related to research at the interface with other fields, possibly not normally associated with physics, also may be considered.
Geographic Focus: All States
Date(s) Application is Due: Sep 30
Contact: Marvin Goldberg, Program Director; (703) 292-7374; mgoldber@nsf.gov
Internet: http://www.nsf.gov/funding/pgm_summ.jsp?pims_id=5610
Sponsor: National Science Foundation
4201 Wilson Boulevard
Arlington, VA 22230

NSF Electronic and Photonic Materials (EPM) Grants 3558
The goal of this program is to advance the field of electronics and photonics through basic, potentially transformative materials science research. The scope of the program encompasses the discovery and understanding of materials and material combinations with potential for major technological advantages. Program focus is on identification and understanding of fundamental atomic and molecular level mechanisms and phenomena associated with synthesis and processing of electronic and photonic materials. High risk, high payoff research is encouraged. Research topics include, but are not limited to, nucleation and growth of thin films and nanostructures; self-assembly; nanostructure definition and etching processes; interface bonding and structure; crystal and interface defects; doping; bulk crystal growth; and interrelationships between synthesis/processing, structure, and properties.
Requirements: NSF welcomes proposals on behalf of all qualified scientists, engineers and educators. The Foundation strongly encourages women, minorities and persons with disabilities to participate fully in its programs. Scientists, engineers and educators usually initiate proposals that are officially submitted by their employing organization. Before formal submission, the proposal may be discussed with appropriate NSF program staff. Graduate students are not encouraged to submit research proposals, but should arrange to serve as research assistants to faculty members. NOTE: If the closing date for the submission window falls on a weekend, the closing date moves to the following Monday. The last date of the submission window is an absolute deadline date and proposals must be received by NSF by 5:00 p.m. submitter's local time on that date.
Restrictions: The sponsor discourages the submission of more than one proposal from the same Principal Investigator during the proposal-submission window.
Geographic Focus: All States
Date(s) Application is Due: Oct 31
Contact: Z. Charles Ying, EPM Program Director; (703) 292-8428; cying@nsf.gov
Internet: http://www.nsf.gov/funding/pgm_summ.jsp?pims_id=5353&org=MPS&sel_org=MPS&from=fund
Sponsor: National Science Foundation
4201 Wilson Boulevard
Arlington, VA 22230

NSF Elementary Particle Physics (EPP) Grants 3559
The program supports research on the properties and interactions of elementary particles, the most fundamental building blocks of matter, at the frontiers of energy and sensitivity. Research includes the exploration of quarks and leptons and the interactions among these elementary constituents. The program supports university groups working at major accelerator laboratories, including those operated by the Department of Energy, and university groups involved in the construction of detectors for the Large Hadron Collider at the European Organization for Nuclear Research (CERN). The program also supports the Cornell Electron Storage Ring (CESR), which produces electron and positron colliding beams that allow detailed studies by university groups of b-meson physics and upsilon physics, and facilitates an aggressive program of synchrotron radiation research at the Cornell High-Energy Synchrotron Source, which is operated by the Division of Materials Research. CESR is among the highest luminosity electron-positron colliders in the world in this energy range. CESR also maintains a vigorous program of accelerator research and development. The annual deadline is the last Wednesday in September.
Requirements: It is recommended that proposals come in near the general Physics Division target date. Individuals and organizations in the following categories may submit proposals: (1) Universities and Colleges-Universities and two- and four-year colleges (including community colleges) located and accredited in the U.S., acting on behalf of their faculty members. Such organizations also are referred to as academic institutions. Institutions located outside the U.S. fall under (6) below. (2) Non-profit, non-academic organizations-Independent museums, observatories, research laboratories, professional societies and similar organizations in the U.S. that are directly associated with educational or research activities. (3) For-profit organizations-U.S. commercial organizations, especially small businesses with strong capabilities in scientific or engineering research or education. An unsolicited proposal from a commercial organization may be funded when the project is of special concern from a national point of view, special resources are available for the work, or the proposed project is especially meritorious. NSF is interested in supporting projects that couple industrial research resources and perspectives with those of universities; therefore, it especially welcomes proposals for cooperative projects involving both universities and the private commercial sector. (4) State and Local Governments-State educational offices or organizations and local school districts may submit proposals intended to broaden the impact, accelerate the pace, and increase the effectiveness of improvements in science, mathematics and engineering education in both K-12 and post-secondary levels. (5) Unaffiliated Individuals-Scientists, engineers or educators in the U.S. and U.S. citizens may be eligible for support, provided that the individual is not employed by, or affiliated with, an organization, and (a) the proposer has demonstrated the capability and has access to any necessary facilities to carry out the project; and (b) the proposer agrees to fiscal arrangements that, in the opinion of the NSF Division of Grants & Agreements, ensure responsible management of Federal funds. Unaffiliated individuals should contact the appropriate program before preparing a proposal for submission. (6) Foreign organizations-NSF rarely provides support to foreign organizations. NSF will consider proposals for cooperative projects involving U.S. and foreign organizations, provided support is requested only for the U.S. portion of the collaborative effort.
Restrictions: Other Federal agencies-NSF does not normally support research or education activities by scientists, engineers or educators employed by Federal agencies or Federally Funded Research and Development Centers (FFRDCs). Under unusual circumstances, other Federal agencies and FFRDCs may submit proposals directly to NSF. A proposed project is only eligible for support if it meets one or more of the following exceptions, as determined by a cognizant NSF Program Officer: (a) Special Projects. Under exceptional circumstances, research or education projects at other Federal agencies or FFRDCs that can make unique contributions to the needs of researchers elsewhere or to other specific NSF objectives may receive NSF support. (b) National and International Programs. The Foundation may fund research and logistic support activities of other Government agencies or FFRDCs directed at meeting the goals of special national and international research programs for which the Foundation bears special responsibility, such as the U.S. Antarctic Research Program. (c) International Travel Awards. In order to ensure appropriate representation or availability of a particular expertise at an international conference, staff researchers of other Federal agencies may receive NSF international travel awards. Proposers who think their project may meet one of the exceptions listed above should contact a cognizant NSF Program Officer before preparing a proposal for submission. In addition, a scientist, engineer or educator who has a joint appointment with a university and a Federal agency (such as a Veterans Administration Hospital, or with a university and a FFRDC) may submit proposals through the university and may receive support if he/she is a bona fide faculty member of the university, although part of his/her salary may be provided by the Federal agency. Preliminary inquiry should be made to the appropriate program before preparing a proposal for submission.
Geographic Focus: All States
Date(s) Application is Due: Sep 30
Contact: Marvin Goldberg, Program Director; (703) 292-7374; mgoldber@nsf.gov
Internet: http://www.nsf.gov/funding/pgm_summ.jsp?pims_id=5624&org=PHY&from=home
Sponsor: National Science Foundation
4201 Wilson Boulevard
Arlington, VA 22230

NSF Emerging Frontiers in Research and Innovation (EFRI) Grants 3560

The Directorate for Engineering at the National Science Foundation has established the Office of Emerging Frontiers in Research and Innovation (EFRI) to serve a critical role in focusing on important emerging areas in a timely manner. The EFRI Office is launching a new funding opportunity for interdisciplinary teams of researchers to embark on rapidly advancing frontiers of fundamental engineering research. For this program, they will consider proposals that aim to investigate emerging frontiers in the following two specific research areas: (1) BioSensing & BioActuation: Interface of Living and Engineered Systems (BSBA), and (2) Hydrocarbons from Biomass (HyBi). EFRI seeks proposals with transformative ideas that represent an opportunity for a significant shift in fundamental engineering knowledge with a strong potential for long term impact on national needs or a grand challenge. It is anticipated that 11 or more standard grants will be made (4-year awards).
Requirements: Proposals may be submitted by a single organization or a group of organizations consisting of a lead organization in partnership with one or more partner organizations. Only U.S. academic institutions with significant research and degree-granting education programs in disciplines normally supported by NSF are eligible to be the lead organization. Principal investigators are encouraged to form synergistic collaborations among researchers and with private and public sector organizations, government laboratories, and scientists and engineers at foreign organizations where appropriate, though no NSF funds will be provided to those organizations. Submission of Letters of Intent is required, due by October 14. Submission of Preliminary Proposals is required, due by December 2. Full proposals will be due by April 30. Each project team may receive support of up to a total of $500,000 per year for up to four years, pending the availability of funds. It is not expected that all awards will receive the maximum amount; the size of awards will depend upon the type of research program proposed. See the website for the detailed solicitation, which includes guidelines, requirements, and application instructions.
Restrictions: The principal investigator and co-principal investigators may participate in one proposal submitted to this program.
Geographic Focus: All States
Date(s) Application is Due: Apr 30; Oct 14; Dec 2
Amount of Grant: Up to 500,000 USD
Contact: Sohi Rastegar; (703) 292-8305; srastega@nsf.gov
Internet: http://www.nsf.gov/funding/pgm_summ.jsp?pims_id=13708
Sponsor: National Science Foundation
4201 Wilson Boulevard
Arlington, VA 22230

NSF Engineering Research Centers (ERC) Grants 3561

The goal of the Generation Three (Gen-3) Engineering Research Centers (ERC) Program is to create a culture of innovation in engineering research and education that links scientific discovery to technological innovation through transformational engineered systems research in order to advance technology and produce engineering graduates who will be creative innovators in a global economy. These ERCs will be at the forefront as the U.S. competes in the 21st century global economy where R&D resources and engineering talent are internationally and domestically distributed. Recognizing that optimizing efficiency and product quality is no longer sufficient for U.S. industry to remain competitive, these ERCs will optimize academic engineering research and education to stimulate increased innovation. They will develop this culture of discovery and innovation through a symbiotic relationship between academic researchers, small innovative firms, and larger industrial and practitioner partners. These ERCs will build bridges from science-based discovery to technological innovation by focusing on research needed to realize transforming engineered systems. They will have the opportunity to partner with foreign universities and provide unique opportunities for research and learning collaboration that will prepare U.S. engineering graduates for leadership in innovation in a global economy. Their faculty will be diverse and talented individuals who will prepare diverse and talented domestic and international graduates who can function in a global world where design and production efforts cross national borders. Their transforming engineering education programs will strategically impart the capacity to create and exploit knowledge for technological innovation. Awards will be made as cooperative agreements between NSF and the lead university, with subawards to the partner institutions as appropriate for their role in the ERC. ERCs are expected to compensate pre-college teachers for time spent carrying out research in the ERCs laboratories and the materials needed to develop the courseware. 07-521
Requirements: Only U.S. academic institutions with undergraduate, masters, and doctoral engineering programs of sufficient depth and breadth appropriate to support the vision may submit proposals as the lead institution. Proposals must be submitted by the lead institution with subaward budgets for partner domestic academic institutions. As a multi-university effort, the lead and partner institutions share the responsibility for the ERC. The other university partner institutions must have graduate or undergraduate engineering programs or both. At least one of these academic partners will be a university or college that serves students predominantly underrepresented in engineering. These partners also must include a few middle and high schools commited to long-term sustained partnerships to bring engineering concepts into the classroom and engage talented older pre-college students in the ERC's research programs as Young Scholars in order to raise the enrollment and diversity of domestic students in engineering and science degree program. The partners must also include an organization devoted to speeding entrepreneurship or innovation as discussed above. Full proposals must include letters from firms/agencies indicating that they are committed to the ERC's industrial/practitioner partnership program. Proposals may include a foreign university as a long-term partner in research and education. Only universities able to provide headquarters space for the proposed ERC may submit preliminary and invited full proposals as the lead institution. This headquarters should provide space for the administrative staff, the director, the deputy director, and physical and virtual conferencing. It may also provide space for facilities to support the ERC's research needs. The center director must be a tenured faculty member in an engineering department at the lead institution. The director's doctoral degree must be in engineering or an associated field of science; if the latter, she/he must have substantial career experience in engineering and a joint appointment in an engineering department. A one to two-page letter of intent is required to facilitate the NSF review process. The letter should be submitted via FastLane no later than the date specified in this solicitation.
Restrictions: Separately submitted collaborative proposals are not acceptable and will be returned without review.
Geographic Focus: All States
Date(s) Application is Due: Feb 2; May 3
Contact: Lynn Preston; (703) 292-5358; fax (703) 292-9051; lpreston@nsf.gov
Internet: http://www.nsf.gov/funding/pgm_summ.jsp?pims_id=5502&org=ENG&sel_org=ENG&from=fund
Sponsor: National Science Foundation
4201 Wilson Boulevard
Arlington, VA 22230

NSF Engineering Research Grants 3562

The Directorate for Engineering makes catalytic investments in knowledge creation and intellectual growth that can in turn spur technological innovation and economic growth in an increasingly swift, complex, and interconnected world. Although the research supported by the directorate is fundamental in nature, much of it focuses on societal needs. Over the long term, the directorate's investments contribute to an innovation system that enables the creation of valuable new products, new services, and new and more productive enterprises that enhance future economic strength, security, and quality of life. The directorate supports the following research activity programs: bioengineering and environmental systems (BES), chemical and transport systems (CTS), civil and mechanical systems (CMS), design, manufacture, and industrial innovation (DMII), electrical and communications systems (ECS), and engineering education and centers (EEC).
Requirements: Public and private colleges and universities; nonprofit institutions; and profit-making institutions including small businesses, federal, state, and local government agencies are eligible to apply.
Geographic Focus: All States
Amount of Grant: 5,000-4,000,000 USD
Contact: Joanne Culbertson, (703) 292-4602; fax (703) 292-9013; jculbert@nsf.gov
Internet: http://www.eng.nsf.gov
Sponsor: National Science Foundation
4201 Wilson Boulevard
Arlington, VA 22230

NSF Evolution of Developmental Systems Grants 3563

The Developmental Systems Cluster supports research aimed at understanding how interacting developmental processes give rise to the emergent properties of organisms. A systems level approach to understanding these processes, at the molecular, cellular, and organismal levels of organization, requires the use of molecular, genetic, biochemical, and physiological techniques as well as techniques from outside biology. The Evolution of Developmental Systems programmatic area supports research to discover the developmental processes shared by all organisms and those singular ones that produce diversity (phenotypic variation within a species and/or between species).
Requirements: The most frequent recipients of support are academic institutions and nonprofit research groups. In special circumstances, grants also are awarded to other types of institutions and to individuals.
Geographic Focus: All States
Date(s) Application is Due: Jan 12; Jul 12
Contact: Thomas Jack; (703) 292-8417; fax (703) 292-9153; tjack@nsf.gov
Internet: http://www.nsf.gov/funding/pgm_summ.jsp?pims_id=501087
Sponsor: National Science Foundation
4201 Wilson Boulevard
Arlington, VA 22230

NSF Extragalactic Astronomy and Cosmology Research Grants 3564

The program supports studies of the more distant Universe. Research topics include galaxy formation, evolution and interaction; active galaxies; quasars; large-scale structure; and all areas of cosmology. Proposals may span multiple disciplines and/or areas of study and may utilize multiple techniques. Principal Investigators are encouraged to contact one of the Program Officers listed in this announcement prior to submitting a proposal to the AAG Program, particularly if the proposal will include investigators at multiple institutions.
Requirements: Scientists initiate research proposals, which are usually submitted on their behalf by their employing institutions.
Geographic Focus: All States
Date(s) Application is Due: Nov 15
Contact: Nigel A. Sharp; (703) 292-4905; fax (703) 292-9034; nsharp@nsf.gov
Internet: http://www.nsf.gov/funding/pgm_summ.jsp?pims_id=13630
Sponsor: National Science Foundation
4201 Wilson Boulevard
Arlington, VA 22230

NSF Faculty Early Career Development (CAREER) Grants 3565

The Faculty Early Career Development (CAREER) Program is a Foundation-wide activity that offers the National Science Foundation's most prestigious awards in support of junior faculty who exemplify the role of teacher-scholars through outstanding research, excellent education and the integration of education and research within the context of the mission of their organizations. Such activities should build a firm foundation

for a lifetime of leadership in integrating education and research. NSF encourages submission of CAREER proposals from junior faculty members at all CAREER-eligible organizations and especially encourages women, members of underrepresented minority groups, and persons with disabilities to apply. This program also includes the NSF component of the Presidential Early Career Awards for Scientists and Engineers (PECASE). Deadlines vary according to the field of study. NSF 08-557
Requirements: Proposals may be submitted by (a) Academic institutions in the U.S., its territories or possessions, and the Commonwealth of Puerto Rico that award degrees in fields supported by NSF; or, (b) Non-profit, non-degree-granting organizations such as museums, observatories or research labs may also be eligible to submit proposals, if the eligibility requirements of the PI's position are satisfied. There is no limit on the number of proposals that an organization may submit. Each Principal Investigator may only submit one CAREER proposal per annual competition.
Restrictions: A Principal Investigator may not participate in more than three competitions.
Geographic Focus: All States
Date(s) Application is Due: Jul 21; Jul 22; Jul 23
Amount of Grant: 150,000-200,000 USD
Contact: Elizabeth L. Rom; (703) 292-7709; fax (703) 292-9085; elrom@nsf.gov
Internet: http://www.nsf.gov/funding/pgm_summ.jsp?pims_id=503214&from=fund
Sponsor: National Science Foundation
4201 Wilson Boulevard
Arlington, VA 22230

NSF Focused Research Groups in the Mathematical Sciences Grants 3566
The purpose of this program is to allow groups of researchers to respond to recognized scientific needs of pressing importance, to take advantage of current scientific opportunities, or to prepare the ground for anticipated significant scientific developments in the mathematical sciences. It is expected that some of these groups will include, in addition to mathematical scientists, researchers from other science and engineering disciplines appropriate to the proposed research. The activity will support projects for which the collective effort by a group of researchers is necessary to reach the scientific goals. Projects should also be timely, limited in duration to up to three years, and substantial in their scope and impact. Fastlane use is required. A letter of intent is requested by August 21; full application is due by September 18.
Requirements: Proposals may be submitted by U.S. academic institutions and other U.S. nonprofit organizations in support of small groups.
Geographic Focus: All States
Date(s) Application is Due: Aug 21; Sep 18
Amount of Grant: 800,000 USD
Contact: Henry A. Warchall; (703) 292-4861; fax (703) 292-9032; hwarchal@nsf.gov
Internet: http://www.nsf.gov/funding/pgm_summ.jsp?pims_id=5671
Sponsor: National Science Foundation
4201 Wilson Boulevard
Arlington, VA 22230

NSF Foundations Grants 3567
The program in Foundations supports research in mathematical logic and the foundations of mathematics, including proof theory, recursion theory, model theory, set theory, and infinitary combinatorics. For full proposals submitted via FastLane, standard Grant Proposal Guidelines apply. The full proposal target date is October 6.
Geographic Focus: All States
Date(s) Application is Due: Oct 6
Contact: Tomek Bartoszynski, (703) 292-4885; fax (703) 292-9032; tbartosz@nsf.gov
Internet: http://www.nsf.gov/funding/pgm_summ.jsp?pims_id=5548&org=DMS&from=home
Sponsor: National Science Foundation
4201 Wilson Boulevard
Arlington, VA 22230

NSF Galactic Astronomy Research Grants 3568
This program supports studies on the composition, structure and evolution of the Milky Way galaxy and nearby galaxies. Research may focus on the stellar populations in these galaxies; the characteristics of star clusters; the interstellar medium; and the properties of atomic and molecular constituents of the interstellar medium.
Requirements: Scientists initiate research proposals, which are usually submitted on their behalf by their employing institutions.
Geographic Focus: All States
Date(s) Application is Due: Sep 30
Contact: Nigel A. Sharp; (703) 292-4905; fax (703) 292-9034; nsharp@nsf.gov
Internet: http://www.nsf.gov/funding/pgm_summ.jsp?pims_id=13630
Sponsor: National Science Foundation
4201 Wilson Boulevard
Arlington, VA 22230

NSF Genes and Genome Systems Cluster Grants 3569
Grants support studies on genomes and genetic mechanisms in all organisms, whether prokaryote, eukaryote, phage, or virus. Proposals on the structure, maintenance, expression, transfer, and stability of genetic information in DNA, RNA, and proteins and how those processes are regulated are appropriate. Areas of interest include genome organization, molecular and cellular evolution, replication, recombination, repair, and vertical and lateral transmission of heritable information. Of equal interest are the processes that mediate and regulate gene expression, such as chromatin structure, epigenetic phenomena, transcription, RNA processing, editing and degradation, and translation. The use of innovative in vivo and/or in vitro approaches, including biochemical, physiological, genetic, genomic, and/or computational methods, is encouraged, as is research at the interfaces of biology, physics, chemistry, mathematics and computer science, and engineering.
Requirements: Standard NSF Grant Proposal Guidelines apply (see http://www.nsf.gov/publications/pub_summ.jsp?ods_key=gpg).
Geographic Focus: All States
Date(s) Application is Due: Jan 12; Jul 12
Contact: Patrick P. Dennis; (703) 292-7145; fax (703) 292-9061; pdennis@nsf.gov
Internet: http://www.nsf.gov/funding/pgm_summ.jsp?pims_id=12780
Sponsor: National Science Foundation
4201 Wilson Boulevard
Arlington, VA 22230

NSF Geometric Analysis Research Grants 3570
The program supports research on differential geometry and its relation to partial differential equations and variational principles; aspects of global analysis, including the differential geometry of complex manifolds and geometric Lie group theory; geometric methods in modern mathematical physics; and geometry of convex sets, integral geometry, and related geometric topics.
Requirements: Apply to PD 07-1265 as follows: For full proposals submitted via FastLane: standard Grant Proposal Guidelines apply. For full proposals submitted via Grants.gov: NSF Grants.gov Application Guide; A Guide for the Preparation and Submission of NSF Applications via Grants.gov Guidelines apply (Note: The NSF Grants.gov Application Guide is available on the Grants.gov website and on the NSF website at: http://www.nsf.gov/bfa/dias/policy/docs/grantsgovguide.pdf)
Restrictions: All collaborative proposals submitted as separate submissions from multiple organizations must be submitted via the NSF FastLane system.
Geographic Focus: All States
Contact: Joanna Kania-Bartoszynska; (703) 292-4881; fax (703) 292-9032; jkaniaba@nsf.gov
Internet: http://www.nsf.gov/funding/pgm_summ.jsp?pims_id=5549
Sponsor: National Science Foundation
4201 Wilson Boulevard
Arlington, VA 22230

NSF Geophysics Grants 3571
The Geophysics Program supports basic research in the physics of the solid earth to explore its composition, structure, and processes. Laboratory, field, theoretical, and computational studies are supported. Topics include seismicity, seismic wave propagation, and the nature and occurrence of earthquakes; the earth's magnetic, gravity, and electrical fields; the earth's thermal structure; and geodynamics. Supported research also includes geophysical studies of active deformation, including geodesy, and studies of the properties and behavior of earth materials in support of geophysical observation and theory.
Requirements: Individuals and organizations in the following categories may submit proposals: (1) Universities and Colleges-Universities and two- and four-year colleges (including community colleges) located and accredited in the U.S., acting on behalf of their faculty members. Such organizations also are referred to as academic institutions. Institutions located outside the U.S. fall under (6) below. (2) Non-profit, non-academic organizations-Independent museums, observatories, research laboratories, professional societies and similar organizations in the U.S. that are directly associated with educational or research activities. (3) For-profit organizations-U.S. commercial organizations, especially small businesses with strong capabilities in scientific or engineering research or education. An unsolicited proposal from a commercial organization may be funded when the project is of special concern from a national point of view, special resources are available for the work, or the proposed project is especially meritorious. NSF is interested in supporting projects that couple industrial research resources and perspectives with those of universities; therefore, it especially welcomes proposals for cooperative projects involving both universities and the private commercial sector. (4) State and Local Governments-State educational offices or organizations and local school districts may submit proposals intended to broaden the impact, accelerate the pace, and increase the effectiveness of improvements in science, mathematics and engineering education in both K-12 and post-secondary levels. (5) Unaffiliated Individuals-Scientists, engineers or educators in the U.S. and U.S. citizens may be eligible for support, provided that the individual is not employed by, or affiliated with, an organization, and (a) the proposer has demonstrated the capability and has access to any necessary facilities to carry out the project; and (b) the proposer agrees to fiscal arrangements that, in the opinion of the NSF Division of Grants & Agreements, ensure responsible management of Federal funds. Unaffiliated individuals should contact the appropriate program before preparing a proposal for submission. (6) Foreign organizations-NSF rarely provides support to foreign organizations. NSF will consider proposals for cooperative projects involving U.S. and foreign organizations, provided support is requested only for the U.S. portion of the collaborative effort.
Restrictions: Other Federal agencies-NSF does not normally support research or education activities by scientists, engineers or educators employed by Federal agencies or Federally Funded Research and Development Centers (FFRDCs). Under unusual circumstances, other Federal agencies and FFRDCs may submit proposals directly to NSF. A proposed project is only eligible for support if it meets one or more of the following exceptions, as determined by a cognizant NSF Program Officer: (a) Special Projects. Under exceptional circumstances, research or education projects at other Federal agencies or FFRDCs that can make unique contributions to the needs of researchers elsewhere or to other specific NSF objectives may receive NSF support. (b) National and International Programs. The Foundation may fund research and logistic support activities of other Government agencies or FFRDCs directed at meeting the goals of special national and international research

programs for which the Foundation bears special responsibility, such as the U.S. Antarctic Research Program. (c) International Travel Awards. In order to ensure appropriate representation or availability of a particular expertise at an international conference, staff researchers of other Federal agencies may receive NSF international travel awards. Proposers who think their project may meet one of the exceptions listed above should contact a cognizant NSF Program Officer before preparing a proposal for submission. In addition, a scientist, engineer or educator who has a joint appointment with a university and a Federal agency (such as a Veterans Administration Hospital, or with a university and a FFRDC) may submit proposals through the university and may receive support if he/she is a bona fide faculty member of the university, although part of his/her salary may be provided by the Federal agency. Preliminary inquiry should be made to the appropriate program before preparing a proposal for submission. Proposals that have been declined are not eligible for resubmission for one year from the original date of submission and must be substantially revised to be considered. A proposal that has not been substantially revised will be returned without review as per the NSF Grant Proposal Guide.
Geographic Focus: All States
Date(s) Application is Due: Dec 1
Contact: Robin Reichlin; (703) 292-8556; fax (703) 292-9025; rreichli@nsf.gov
Internet: http://www.nsf.gov/funding/pgm_summ.jsp?pims_id=13682
Sponsor: National Science Foundation
4201 Wilson Boulevard
Arlington, VA 22230

NSF George E. Brown Jr. Network for Earthquake Engineering Simulation Research (NEESR) 3572
The Division of Civil, Mechanical and Manufacturing Innovation (CMMI) in the Directorate for Engineering (ENG) of the National Science Foundation (NSF) invites proposals for research that uses the George E. Brown, Jr. Network for Earthquake Engineering Simulation (NEES) to advance knowledge discovery and innovation for: earthquake and tsunami loss reduction of our nation's civil infrastructure; and new experimental simulation techniques and instrumentation for NEES. NEES comprises a network of earthquake engineering experimental equipment sites available for experimentation on-site or in the field and through telepresence. NEES equipment sites include shake tables, geotechnical centrifuges, a tsunami wave basin, unique large-scale testing laboratory facilities, and mobile and permanently installed field equipment. The NEES networking cyberinfrastructure connects, via Internet2, the equipment sites as well as provides telepresence, a curated central data repository, simulation tools, and collaborative tools for facilitating on-line planning, execution, and post-processing of experiments. Projects proposed and supported under this solicitation must use one or more of the NEES equipment sites.
Requirements: U.S. universities and colleges and U.S. nonprofit, nonacademic organizations may submit NEESR proposals as the lead organization.
Geographic Focus: All States
Date(s) Application is Due: Mar 27
Amount of Grant: Up to 9,000,000 USD
Contact: Joy M. Pauschke, (703) 292-7024; fax (703) 292-9053; jpauschk@nsf.gov
Internet: http://www.nsf.gov/funding/pgm_summ.jsp?pims_id=6192
Sponsor: National Science Foundation
4201 Wilson Boulevard
Arlington, VA 22230

NSF Geospace Environment Modeling (GEM) 3573
GEM is a broad-based, community-initiated research program on the physics of the Earth's magnetosphere and the coupling of the magnetosphere to the atmosphere and to the solar wind. The purpose of the GEM program is to support basic research into the dynamical and structural properties of geospace, leading to the construction of a global Geospace General Circulation Model (GGCM) with predictive capability. This GGCM model will be modularized and will complement parallel developments of magnetohydrodynamic models. The strategy for achieving GEM goals is to undertake a series of campaigns, in both theory and observational modes, each focusing on particular aspects of the geospace environment. Estimated program budget, number of awards and average award size/duration are subject to the availability of funds. The typical award size is approximately $85,000 per year with a duration of three years. Proposers may request from one to five years of funding provided the requested duration is adequately justified. The maximum award size is $175,000 per year. The estimated number of awards is 8 to 12 new awards in each year. NSF 04-576 (replaces Document 02-122)
Requirements: Individuals and organizations in the following categories may submit proposals: (1) Universities and Colleges-Universities and two- and four-year colleges (including community colleges) located and accredited in the U.S., acting on behalf of their faculty members. Such organizations also are referred to as academic institutions; (2) Non-profit, non-academic organizations-Independent museums, observatories, research laboratories, professional societies and similar organizations in the U.S. that are directly associated with educational or research activities; (3) For-profit organizations-U.S. commercial organizations, especially small businesses with strong capabilities in scientific or engineering research or education. An unsolicited proposal from a commercial organization may be funded when the project is of special concern from a national point of view, special resources are available for the work, or the proposed project is especially meritorious. NSF is interested in supporting projects that couple industrial research resources and perspectives with those of universities; therefore, it especially welcomes proposals for cooperative projects involving both universities and the private commercial sector; (4) State and Local Governments-State educational offices or organizations and local school districts may submit proposals intended to broaden the impact, accelerate the pace, and increase the effectiveness of improvements in science, mathematics and engineering education in both K-12 and post-secondary levels; (5) Unaffiliated Individuals-Scientists, engineers or educators in the U.S. and U.S. citizens may be eligible for support, provided that the individual is not employed by, or affiliated with, an organization, and: (a) the proposer has demonstrated the capability and has access to any necessary facilities to carry out the project; and, (b) the proposer agrees to fiscal arrangements that, in the opinion of the NSF Division of Grants & Agreements, ensure responsible management of Federal funds. Unaffiliated individuals should contact the appropriate program before preparing a proposal for submission.
Restrictions: NSF rarely provides support to foreign organizations. NSF will consider proposals for cooperative projects involving U.S. and foreign organizations, provided support is requested only for the U.S. portion of the collaborative effort. Also, NSF does not normally support research or education activities by scientists, engineers or educators employed by Federal agencies or Federally Funded Research and Development Centers (FFRDCs). Under unusual circumstances, other Federal agencies and FFRDCs may submit proposals directly to NSF.
Geographic Focus: All States
Date(s) Application is Due: Oct 15
Amount of Grant: 80,000 USD
Contact: Robert M. Robinson; (703) 292-8529; fax (703) 292-9022; rmrobins@nsf.gov
Internet: http://www.nsf.gov/publications/pub_summ.jsp?ods_key=nsf04576
Sponsor: National Science Foundation
4201 Wilson Boulevard
Arlington, VA 22230

NSF Grant Opportunities for Academic Liaison with Industry (GOALI) 3574
Grant Opportunities for Academic Liaison with Industry (GOALI) promotes university-industry partnerships by making project funds or fellowships/traineeships available to support an eclectic mix of industry-university linkages. Special interest is focused on affording the opportunity for: faculty, postdoctoral fellows, and students to conduct research and gain experience in an industrial setting; industrial scientists and engineers to bring industry's perspective and integrative skills to academe; and interdisciplinary university-industry teams to conduct research projects. This solicitation targets high-risk/high-gain research with a focus on fundamental research, new approaches to solving generic problems, development of innovative collaborative industry-university educational programs, and direct transfer of new knowledge between academe and industry. GOALI seeks to fund transformative research that lies beyond that which industry would normally fund.
Geographic Focus: All States
Contact: Donald Senich; (703) 292-7082; fax (703) 292-9056; dsenich@nsf.gov
Internet: http://www.nsf.gov/funding/pgm_summ.jsp?pims_id=13706
Sponsor: National Science Foundation
4201 Wilson Boulevard
Arlington, VA 22230

NSF Grants for the Analysis of Science and Technology (S&T) Resources 3575
NSF welcomes proposals for research, workshops, and studies leading to improved approaches to indicator development and presentation, new S&T indicator development, and better understanding of the S&T enterprise in the United States and globally under this program. The development of new approaches to the analysis and presentation of data as indicators, with a particular focus on effective methods of presenting complex information as indicators is a priority. Although it is not required, PIs are encouraged to submit a letter of intent one month before submitting a proposal.
Requirements: Proposals may be submitted by colleges, universities, and nonprofit or commercial organizations, or combinations of such organizations.
Geographic Focus: All States
Date(s) Application is Due: Sep 18
Amount of Grant: Up to 400,000 USD
Contact: Derek Hill, Science Resources Statistics; (703) 292-7805; dhill@nsf.gov
Internet: http://www.nsf.gov/pubs/2000/nsf00111/nsf00111.htm
Sponsor: National Science Foundation
4201 Wilson Boulevard
Arlington, VA 22230

NSF Gravitational Physics Research Grants 3576
This funding program emphasizes the theory of strong gravitational fields and their application to astrophysics and cosmology, computer simulations of strong and gravitational fields, and gravitational radiation; and construction of a quantum theory of gravity. The program oversees the management of the construction, commissioning, and operation of the Laser Interferometer Gravity Wave Observatory (LIGO), and provides support for LIGO users and other experimental investigations in gravitational physics and related areas.
Requirements: Individuals and organizations in the following categories may submit proposals: (1) Universities and Colleges-Universities and two- and four-year colleges (including community colleges) located and accredited in the U.S., acting on behalf of their faculty members. Such organizations also are referred to as academic institutions. Institutions located outside the U.S. fall under (6) below. (2) Non-profit, non-academic organizations-Independent museums, observatories, research laboratories, professional societies and similar organizations in the U.S. that are directly associated with educational or research activities. (3) For-profit organizations-U.S. commercial organizations, especially small businesses with strong capabilities in scientific or engineering research or education. An unsolicited proposal from a commercial organization may be funded when the project is of special concern from a national point of view, special resources are available for the work, or the proposed project is especially meritorious. NSF is interested in supporting projects that couple industrial research resources and perspectives with those of universities; therefore, it especially welcomes proposals for cooperative projects involving both universities and the private commercial sector. (4) State and Local Governments-State educational offices or organizations and local school districts may submit proposals

intended to broaden the impact, accelerate the pace, and increase the effectiveness of improvements in science, mathematics and engineering education in both K-12 and post-secondary levels. (5) Unaffiliated Individuals-Scientists, engineers or educators in the U.S. and U.S. citizens may be eligible for support, provided that the individual is not employed by, or affiliated with, an organization, and (a) the proposer has demonstrated the capability and has access to any necessary facilities to carry out the project; and (b) the proposer agrees to fiscal arrangements that, in the opinion of the NSF Division of Grants & Agreements, ensure responsible management of Federal funds. Unaffiliated individuals should contact the appropriate program before preparing a proposal for submission. (6) Foreign organizations-NSF rarely provides support to foreign organizations. NSF will consider proposals for cooperative projects involving U.S. and foreign organizations, provided support is requested only for the U.S. portion of the collaborative effort.
Restrictions: Other Federal agencies-NSF does not normally support research or education activities by scientists, engineers or educators employed by Federal agencies or Federally Funded Research and Development Centers (FFRDCs). Under unusual circumstances, other Federal agencies and FFRDCs may submit proposals directly to NSF. A proposed project is only eligible for support if it meets one or more of the following exceptions, as determined by a cognizant NSF Program Officer: (a) Special Projects. Under exceptional circumstances, research or education projects at other Federal agencies or FFRDCs that can make unique contributions to the needs of researchers elsewhere or to other specific NSF objectives may receive NSF support. (b) National and International Programs. The Foundation may fund research and logistic support activities of other Government agencies or FFRDCs directed at meeting the goals of special national and international research programs for which the Foundation bears special responsibility, such as the U.S. Antarctic Research Program. (c) International Travel Awards. In order to ensure appropriate representation or availability of a particular expertise at an international conference, staff researchers of other Federal agencies may receive NSF international travel awards. Proposers who think their project may meet one of the exceptions listed above should contact a cognizant NSF Program Officer before preparing a proposal for submission. In addition, a scientist, engineer or educator who has a joint appointment with a university and a Federal agency (such as a Veterans Administration Hospital, or with a university and a FFRDC) may submit proposals through the university and may receive support if he/she is a bona fide faculty member of the university, although part of his/her salary may be provided by the Federal agency. Preliminary inquiry should be made to the appropriate program before preparing a proposal for submission.
Geographic Focus: All States
Date(s) Application is Due: Sep 30
Contact: Beverly Berger; (703) 292-7372; fax (703) 292-9078; bberger@nsf.gov
Internet: http://www.nsf.gov/funding/pgm_summ.jsp?pims_id=5628
Sponsor: National Science Foundation
4201 Wilson Boulevard
Arlington, VA 22230

NSF Historically Black Colleges and Universities (HBCU) Grants 3577
This program is committed to enhancing the quality of undergraduate science, technology, engineering, and mathematics (STEM) education and research at Historically Black Colleges and Universities (HBCUs). Support is available for Implementation Projects, Planning Grants, Education Research Projects, and Targeted Infusion Projects. Additional information is available on the internet. This program provides educational opportunities for Undergraduate Students. This program provides indirect funding for students at this level or focuses on educational developments for this group such as curricula development, training or retention. Fastlane submission is required for collaborative proposals. NSF 06-606 provides additional information.
Requirements: The organizational eligibility has been broadened to include accredited two-year HBCUs that do not currently have undergraduate science, technology, engineering and mathematics (STEM) degree programs. These HBCUs are encouraged to aply for funding to enhance the quality of their general STEM courses and programs and/or to establish new undergraduate programs.
Geographic Focus: All States
Date(s) Application is Due: Nov 14; Dec 15
Amount of Grant: Up to 7,000,000 USD
Contact: Jessie A. DeAro; (703) 292-5350; fax (703) 292-9018; jdearo@nsf.gov
Internet: http://www.nsf.gov/funding/pgm_summ.jsp?pims_id=5481
Sponsor: National Science Foundation
4201 Wilson Boulevard
Arlington, VA 22230

NSF Hydrologic Sciences Research Grants 3578
Hydrologic Sciences focuses on the flow of water and transport processes within streams, soils, and aquifers. Particular attention is given to spatial and temporal heterogeneity of fluxes and storages of water and chemicals over a wide range of scales, to geolimnology and to interfaces with the landscape, microbial communities, and coastal areas. Studies may also deal with processes in aqueous geochemistry and with the physical, chemical, and biological processes within water bodies. Study of these processes requires expertise from many basic sciences and mathematics, and proposals often require joint review with related programs.
Geographic Focus: All States
Contact: L. Douglas James; (703) 292-8549; fax (703) 292-9025; ldjames@nsf.gov
Internet: http://www.nsf.gov/funding/pgm_summ.jsp?pims_id=13684
Sponsor: National Science Foundation
4201 Wilson Boulevard
Arlington, VA 22230

NSF Industry/University Cooperative Research Centers (I/UCRC) Grants 3579
The Industry/University Cooperative Research Centers (I/UCRCs) program develops long-term partnerships among industry, academe, and government. The centers are catalyzed by a small investment from the National Science Foundation (NSF) and are primarily supported by industry center members, with NSF taking a supporting role in their development and evolution. Each center is established to conduct research that is of interest to both the industry and the center. An I/UCRC contributes to the Nation's research infrastructure base and enhances the intellectual capacity of the engineering and science workforce through the integration of research and education. 2 to 8 full center awards and 4-12 planning grants are awarded annually. NSF 07-537
Requirements: U.S. academic institutions with graduate research programs may apply. The center director must be the PI and come from the lead institution. A PI must be a tenured faculty member (or equivalent) at an institution eligible to submit an I/UCRC proposal. A Letter of Intent must be approved by the NSF Program Director before the institution may submit a planning grant proposal. Institutions must have been awarded a planning grant before they are eligible to submit a full center proposal. Letters of Intent are due by the first Friday in January or the fourth Friday in June. Planning Grant and Full Center Proposals are due by the fourth Friday in March or by the fourth Friday in September.
Restrictions: Each institution is limited to a maximum of two single institution I/UCRC awards. Institutions which currently have two single institution awards are not eligible to submit another single institution proposal. There is no limit on the number of proposals that may be submitted when collaborating with another institution. It is permissible for an institution to have two single institution center awards and also submit a proposal to be a site in a multi-institutional center.
Geographic Focus: All States
Contact: Alex Schwarzkopf; (703) 292-5359; fax (703) 292-9051; aschwarz@nsf.gov
Internet: http://www.nsf.gov/funding/pgm_summ.jsp?pims_id=5501
Sponsor: National Science Foundation
4201 Wilson Boulevard
Arlington, VA 22230

NSF Information and Intelligent Systems: Core Grants 3580
CISE's Division of Information and Intelligent Systems (IIS) supports research and education projects that develop new knowledge in three core programs: the Human-Centered Computing program; the Information Integration and Informatics program; and the Robust Intelligence program. The Division of Information and Intelligent Systems (IIS) studies the inter-related roles of people, computers, and information. IIS supports research and education activities that: develop new knowledge about the role of people in the design and use of information technology; increase our capability to create, manage, and understand data and information in circumstances ranging from personal computers to globally-distributed systems; and advance our understanding of how computational systems can exhibit the hallmarks of intelligence. Proposers are invited to submit proposals in three project classes, which are defined as follows: Small Projects—up to $500,000 total budget with durations up to three years; Medium Projects—$500,001 to $1,200,000 total budget with durations up to four years; and Large Projects—$1,200,001 to $3,000,000 total budget with durations up to five years.
Geographic Focus: All States
Date(s) Application is Due: Aug 30; Nov 28; Dec 17
Contact: William S. Bainbridge; (703) 292-8930; fax (703) 292-9073; wbainbri@nsf.gov
Internet: http://www.nsf.gov/funding/pgm_summ.jsp?pims_id=13707
Sponsor: National Science Foundation
4201 Wilson Boulevard
Arlington, VA 22230

NSF Innovation and Organizational Change (IOC) 3581
This program program supports scientific research directed at advancing understanding of innovation and organizational phenomena. Levels of analysis may include (but are not limited to) individuals, groups and/or institutional arrangements. Disciplinary perspectives may include (but are not limited to) organization theory, organizational behavior, organizational sociology, social and industrial psychology, public administration, computer and information sciences, complexity sciences, decision and management sciences. Research methods may span a broad variety of qualitative and quantitative methods, including (but not limited to) archival analyses, surveys, simulation studies, experiments, comparative case studies, and network analyses. Research may involve industrial, educational, service, government, not-for-profits, voluntary organizations or interorganizational arrangements.
Requirements: The most frequent recipients of support are academic institutions, although awards are occasionally made to profit-making organizations, individuals, and government agencies at all levels. Research must be grounded in theory and generalizable. It must advance our scientific understanding of innovation and organizations. Scientific inquiries that are relevant to real problems and organizations in generalizable ways are encouraged.
Restrictions: Proposals that aim to implement or evaluate innovations or particular organizational changes rather than to advance fundamental, generalizable knowledge about innovation and organizations are not appropriate for IOC.
Geographic Focus: All States
Date(s) Application is Due: Feb 2
Amount of Grant: Up to 22,000,000 USD
Contact: Robbie Brown, (703) 292-7264; rbrown@nsf.gov
Internet: http://nsf.gov/funding/pgm_summ.jsp?pims_id=5378
Sponsor: National Science Foundation
4201 Wilson Boulevard
Arlington, VA 22230

NSF Innovative Technology Experiences for Students and Teachers (ITEST) 3582

ITEST supports the research and development of innovative models for engaging K-12 students in authentic experiences that build their capacity to participate in the science, technology, engineering, and mathematics (STEM) and information and communications technology (ICT) workforce of the future. ITEST projects must include students and may include teachers. ITEST invests in four types of projects: Strategies, Research, Scale-up, and Resource Center projects. 15 to 20 grants will be awarded depending on the availability of funds.
Requirements: This program provides educational opportunities for K-12 Educators. This program provides indirect funding for students at this level or focuses on educational developments for this group such as curricula development, training or retention. NSF welcomes proposals on behalf of all qualified scientists, engineers and educators. Before formal submission, the proposal may be discussed with appropriate NSF program staff. The Foundation strongly encourages women, minorities and persons with disabilities to participate fully in its programs. Submission of Letters of Intent is required and are due on September 20 (due by 5 p.m. proposer's local time). Full proposals should be submitted via FastLane or Grants.gov on or before November 13 (due by 5 p.m. proposer's local time).
Geographic Focus: All States
Date(s) Application is Due: Nov 13
Contact: Marie Mapes, Program Manager; (703) 292-2768; mmapes@nsf.gov
Internet: http://www.nsf.gov/funding/pgm_summ.jsp?pims_id=5467&org=EHR&sel_org=EHR&from=fund
Sponsor: National Science Foundation
4201 Wilson Boulevard
Arlington, VA 22230

NSF Instrument Development for Biological Research (IDBR) 3583
The IDBR program provides support for development of the following: (1) New instrumentation that addresses emerging biological research needs with the capacity to transform biological research; (2) Novel instrument configurations (hybrids, etc.) that demonstrably address a fundamental biological research challenge; (3) Instrumentation that leverages advances in micro- and nano-fabrication and device design to provide new capabilities in measurement and observation of biological phenomena; (4) Concept and proof-of-concept of novel instruments for biological research; (5) New instruments that provide new capabilities for detection, measurement, and/or observation of biological phenomena, or that greatly extend currently achievable sensitivity, accuracy or resolution; (6) Sensors that meet emerging biological research needs and that have the potential to transform biological research at any level of organization (7) Data acquisition and analysis tools that, through the development of novel devices/instruments, meet biological research needs. The program is not limited to the above mentioned activities. Any proposal that is designed to meet the goals of the Program will be considered. Proposals must, in all cases, demonstrate a strong connection to biological research needs through, for example, collaboration with biological researchers as well as demonstrate that the need for the instrumentation is firmly based on biological research drivers. NSF Program Solicitation 08-566.
Requirements: The program encourages proposals which conduct collaborative and planning activities such as workshops and the development of virtual organization frameworks. Those activities which promote interactions between the instrument development community and biological researchers as well as innovative networking strategies that foster research collaborations or enable new instrument development directions are especially encouraged. Activities which increase participation of colleagues at smaller institutions, minority-serving institutions, community colleges, and K-12 students and teachers are also recommended. Recognizing that the development and use of novel instrumentation have become increasingly integral to activities supported by all BIO programs, the IDBR program will place a higher priority on proposals that create novel instrumentation that enables biological research in a new and potentially transformative way and that serves a broad user community. Check the NSF Grant Proposal Guide (GPG) available electronically on the NSF website at: http://www.nsf.gov/publications/pub_summ.jsp?ods_key=gpg for other general eligibility guidelines.
Restrictions: The program does not support research or technique development activities, except to the extent these are required as part of the development of the new or improved instrument, or for the testing of its utility. Projects emphasizing the development of new research techniques should be addressed to an appropriate research program. The anticipated uses of the instrumentation to be developed or improved should include areas of research that fall within the scope of the Directorate for Biological Sciences (see BIO Home Page at http://www.nsf.gov/bio). Projects aimed at instrumentation whose primary use will be in studies of the etiology, diagnosis or treatment of physical or mental disease, abnormality, or malfunction in human beings or animals, is not supported by IDBR. Similarly, the development or testing of drugs or of instruments whose primary application is in pharmaceutical chemistry are not eligible for support. Such projects should be addressed to an appropriate program in another NSF Directorate or to another agency. Projects in which the main portion of the instrument development activity will be subcontracted to a Federally Funded Research and Development Center (FFRDC) or a commercial (for profit) organization are also not eligible for support by IDBR, and should be addressed to an appropriate NSF program or to another agency.
Geographic Focus: All States
Date(s) Application is Due: Aug 28
Contact: Robyn Hannigan; (703) 292-8470; fax (703) 292-9063; dbiiid@nsf.gov
Internet: http://www.nsf.gov/publications/pub_summ.jsp?ods_key=nsf08566&org=NSF
Sponsor: National Science Foundation
4201 Wilson Boulevard
Arlington, VA 22230

NSF Integrative Graduate Education and Research Traineeship (IGERT) 3584
The Integrative Graduate Education and Research Traineeship (IGERT) program has been developed to meet the challenges of educating U.S. Ph.D. scientists and engineers who will pursue careers in research and education, with the interdisciplinary backgrounds, deep knowledge in chosen disciplines, and technical, professional, and personal skills to become, in their own careers, leaders and creative agents for change. The program is intended to catalyze a cultural change in graduate education, for students, faculty, and institutions, by establishing innovative new models for graduate education and training in a fertile environment for collaborative research that transcends traditional disciplinary boundaries. It is also intended to facilitate diversity in student participation and preparation, and to contribute to a world-class, broadly inclusive, and globally engaged science and engineering workforce.
Requirements: Academic institutions in the United States and its territories that grant the PhD degree and have research and training programs in the sciences and engineering are invited to submit preproposals. Projects involving more than one U.S. institution are eligible, but a single institution must accept overall management responsibility. Collaborating institutions need not be academic.
Geographic Focus: All States
Date(s) Application is Due: Mar 13; Sep 14
Amount of Grant: 800,000 USD
Contact: Carol Van Hartesveldt; (703) 292-8696; fax (703) 292-9048; cvanhart@nsf.gov
Internet: http://www.nsf.gov/funding/pgm_summ.jsp?pims_id=12759
Sponsor: National Science Foundation
4201 Wilson Boulevard
Arlington, VA 22230

NSF Interagency Opportunities in Metabolic Engineering Research Grants 3585
This solicitation describes a collaborative effort among the Department of Agriculture, Department of Commerce, Department of Defense, Department of Energy, Environmental Protection Agency, National Aeronautics and Space Administration, National Institute of General Medical Sciences (National Institutes of Health), and the National Science Foundation. The intent of this interagency solicitation is to provide an opportunity for an interagency granting activity in the area of metabolic engineering (ME). The eight participating agencies or departments are providing research funding and agency in-kind support such as equipment, laboratory space, personnel time, and materials in support of this solicitation. Upon conclusion of the review process, meritorious proposals may be recommended for funding by a participating agency or department. Each participating agency will make its own awards and the subsequent grant administration procedures will be in accordance with the individual policies of the awarding agency or department.
Requirements: Academic and nonprofit institutions, industrial organizations, and government (federal, state, and local) laboratories, are eligible. Investigators are strongly encouraged to discuss their idea for a preproposal with agency staff member(s).
Geographic Focus: All States
Contact: Frederick G. Heineken, Program Director; (703) 292-7944; fax (703) 292-9098; fheineke@nsf.gov
Internet: http://www.nsf.gov/funding/pgm_summ.jsp?pims_id=5494
Sponsor: National Science Foundation
4201 Wilson Boulevard
Arlington, VA 22230

NSF Interfacial Processes and Thermodynamics Research Grants 3586
The Interfacial Processes and Thermodynamics program supports fundamental research in engineering areas related to: interfacial phenomena; mass transport phenomena; and solution phase equilibrium thermodynamics. Currently, emphasis is placed on molecular engineering approaches at interfaces, especially as applied to the processing of soft materials. Molecules at interfaces with functional interfacial properties are of special interest. These interfacial molecules may have biomolecular functions at the micro and nanoscale. Interfacial materials are generally formed through molecular self-directed, -templated, and/or -assembly, and they are driven primarily by thermodynamic intermolecular forces. In some cases, these interfacial processes may also be supplemented by weak chemical reactions.
Geographic Focus: All States
Date(s) Application is Due: Mar 1; Sep 15
Contact: Robert M. Wellek; (703) 292-8370; fax (703) 292-9054; rwellek@nsf.gov
Internet: http://www.nsf.gov/funding/pgm_summ.jsp?pims_id=13362
Sponsor: National Science Foundation
4201 Wilson Boulevard
Arlington, VA 22230

NSF Linguistics Grants 3587
The Linguistics Program supports basic science in the domain of human language, encompassing investigations of the grammatical properties of individual human languages, and of natural language in general. Research areas include syntax, semantics, morphology, phonetics, and phonology. The program encourages projects that are interdisciplinary in methodological or theoretical perspective, and that address questions that cross disciplinary boundaries, such as (but not limited to): What are the psychological processes involved in the production, perception, and comprehension of language? What are the computational properties of language and/or the language processor that make fluent production, incremental comprehension or rapid learning possible? How do the acoustic and physiological properties of speech inform our theories of language and/or language processing? What role does human neurobiology play in shaping the various components of our linguistic capacities? How does language develop in children? What social and cultural factors underlie language variation and change?
Requirements: The Linguistics Program accepts proposals for a variety of project types: research proposals from scholars with PhDs or equivalent degrees, proposals for Doctoral Dissertation Research Improvement Grants, and CAREER proposals. The program

will also consider proposals for workshops, conferences, and training activities. NSF welcomes proposals on behalf of all qualified scientists, engineers and educators. The Foundation strongly encourages women, minorities and persons with disabilities to participate fully in its programs. Scientists, engineers and educators usually initiate proposals that are officially submitted by their employing organization. Before formal submission, the proposal may be discussed with appropriate NSF program staff. Graduate students are not encouraged to submit research proposals, but should arrange to serve as research assistants to faculty members. Some NSF divisions accept proposals for Doctoral Dissertation Research Grants when submitted by a faculty member on behalf of the graduate student. Proposal deadlines are January 15 and July 15 annually, or the next federal business day if the target date falls on a weekend or a federal holiday.
Geographic Focus: All States
Date(s) Application is Due: Jan 15; Jul 15
Contact: Dr. Joan Maling; (703) 292-8046; fax (703) 292-9068; jmaling@nsf.gov
Internet: http://nsf.gov/funding/pgm_summ.jsp?pims_id=5408
Sponsor: National Science Foundation
4201 Wilson Boulevard
Arlington, VA 22230

NSF Living Stock Collections (LSC) Grants 3588
The program supports operation of and improvements in outstanding collections of living organisms used in basic biological research. In addition to the collected organisms, biological reagents relevant to their study (such as isolated DNA, DNA libraries, antibodies, etc.) may be included in the collection when appropriate. To facilitate public access to collections supported on an ongoing basis by the LSC Program, all such collections must have an up-to-date web site that provides information about the contents of the collection and about procedures for ordering strains. The program provides support for two types of projects. Short-term projects are one-time awards (up to 36 months) leading to innovative handling of living stocks or to well-defined improvements in existing collections, including those not otherwise supported by LSC. Long-term projects (up to 60 months) support ongoing operation of significant collections. Collections receiving long-term support are expected to receive support from user fees. NSF 06-574
Requirements: The LSC Program accepts proposals from U.S. colleges, universities and non-profit organizations. Submission of proposals that are similar or identical to proposals submitted to another federal agency is allowed with the prior approval of NSF and of the other agency. Proposers planning such dual submission must contact the LSC Program Director in advance. Full proposals must be submitted via Grants.gov.
Restrictions: The program does not support the establishment of new collections.
Geographic Focus: All States
Date(s) Application is Due: Jul 2
Contact: Gerald B. Selzer; (703) 292-8470; fax (703) 292-9063; gselzer@nsf.gov
Internet: http://www.nsf.gov/funding/pgm_summ.jsp?pims_id=9189
Sponsor: National Science Foundation
4201 Wilson Boulevard
Arlington, VA 22230

NSF Louis Stokes Alliances for Minority Participation (LSAMP) Grants 3589
This program is aimed at increasing the quality and quantity of students successfully completing science, technology, engineering, and mathematics (STEM) baccalaureate degree programs, and increasing the number of students interested in, academically qualified for and matriculated into programs of graduate study. LSAMP supports sustained and comprehensive approaches that facilitate achievement of the long-term goal of increasing the number of students who earn doctorates in SMET fields, particularly those from populations underrepresented in SMET fields. The program goals are accomplished through the formation of alliances. Phase I awards place emphasis on aggregate baccalaureate production. Phase II awards augment the Phase I emphasis with attention to individual student retention and progression to baccaluareate degrees. Phase III awards augment the Phase I and Phase II emphases with attention to aggregate student progression to graduate school entry. NSF 03-520
Requirements: The program is limited to alliances of academic institutions which have exemplary records over several years of enrolling and retaining significant numbers of undergraduate students underrepresented in SMET disciplines. The PI must be either the Chief Executive Officer or the Chief Academic Officer of the lead institution. Proposers are required to prepare and submit all proposals for this announcement/ solicitation through the FastLane system. Victoria Smoot, Fastlane, (703) 292-4677; email: vsmoot@nsf.com
Geographic Focus: All States
Date(s) Application is Due: Oct 15
Amount of Grant: Up to 6,000,000 USD
Contact: A. James Hicks; (703) 292-4668; fax (703) 292-9018; ahicks@nsf.gov
Internet: http://www.nsf.gov/funding/pgm_summ.jsp?pims_id=13646
Sponsor: National Science Foundation
4201 Wilson Boulevard
Arlington, VA 22230

NSF Magnetospheric Physics Research Grants 3590
Supports research on the magnetized plasma envelope of the outer atmosphere, including energization by the solar wind; the origin of geomagnetic storms and substorms; the population by solar and ionospheric sources; the origin of electric fields; the coupling among the magnetosphere, ionosphere, and atmosphere; and waves and instabilities in the natural plasma. Also supported are ground-based observational programs at high latitudes and laboratory experiments applicable to the geospace environment. Theoretical research programs may include numerical simulations using a variety of MHD, hybrid and particle codes. The analysis of data from all sources, whether ground-based or from spacecraft, is also supported.
Requirements: Proposals may be submitted by academic institutions, nonacademic and nonprofit research organizations, profit-making and private research organizations, and individuals. Occasionally, NSF sponsors efforts by other government agencies, particularly for field programs. Apply to PD 98-5750 as follows: (a) For full proposals submitted via FastLane: standard Grant Proposal Guidelines apply. (b) For full proposals submitted via Grants.gov: NSF Grants.gov Application Guide; A Guide for the Preparation and Submission of NSF Applications via Grants.gov Guidelines apply.
Geographic Focus: All States
Contact: Kile Baker; (703) 292-8519; fax (703) 292-9023; kbaker@nsf.gov
Internet: http://www.nsf.gov/funding/pgm_summ.jsp?pims_id=11725
Sponsor: National Science Foundation
4201 Wilson Boulevard
Arlington, VA 22230

NSF Major Research Instrumentation Program (MRI) Grants 3591
The MRI program assists in the acquisition or development of major research instrumentation that is, in general, too costly for support through other NSF programs. This program seeks to improve the quality and expand the scope of research and research training in science and engineering, and to foster the integration of research and education by providing instrumentation for research-intensive learning environments. For proposals over $2 million, requests must be for the acquisition of a single instrument. For proposals requesting $2 million or less, investigators may seek support for instrument development or for acquisition of a single instrument, a large system of instruments, or multiple instruments that share a common or specific research focus. Submission of Letters of Intent is required (only for acquisition requests between $2 million and $4 million). NSF 08-503
Requirements: Proposals may only be submitted by the following: (1) U.S. colleges, universities and organizations of higher education located in the U.S., its territories and possessions; (2) U.S. independent research museums located in the U.S., its territories and possessions; (3) U.S. independent nonprofit research organizations located in the U.S., its territories and possessions, including consortia whose members consist only of organizations described in items (1) and (2). (Requests for instrumentation that will be located at a Federally Funded Research and Development Center (FFRDC) must be submitted as consortium proposals. This is the only mechanism by which instrumentation, whether purchased or developed, can be placed at an FFRDC.); (4) U.S. small businesses located in the U.S., its territories and possessions are eligible for instrument development support as private sector partners with submitting organizations; they may not submit proposals as a lead organization. The deadline is the fourth Thursday in January. Proposers are required to prepare and submit all proposals for this program solicitation through use of the NSF FastLane system.
Restrictions: The MRI program will NOT support proposal requests for: Computer networks used for general-purposes; General laboratory equipment or assorted instruments that do not share a common or specific research or research training focus; Instrumentation used primarily for standard science and engineering courses; Renovation or modernization of research facilities, fixed equipment, or facilities such as research vessels, airplanes, large telescopes, and supercomputing centers. The term 'research facilities' refers to the bricks-and-mortar physical plant in which sponsored or unsponsored research activities (including research training) take place, including related infrastructure, systems (e.g., HVAC and power systems, toxic waste removal systems), and fixed equipment. The term 'fixed equipment' refers to the permanent components of a research facility that are integral (i.e., built in, rather than affixed) to the facility (e.g., clean rooms, fume hoods, elevators, laboratory casework); their removal would affect the integrity or basic operation of the facility.
Geographic Focus: All States
Date(s) Application is Due: Jan 22
Amount of Grant: Up to 4,000,000 USD
Contact: Randy Phelps, (703) 292-8040; rphelps@nsf.gov
Internet: http://nsf.gov/funding/pgm_summ.jsp?pims_id=5260
Sponsor: National Science Foundation
4201 Wilson Boulevard
Arlington, VA 22230

NSF Marine Geology and Geophysics Research Grants 3592
This program supports research on all aspects of geology and geophysics of the ocean basins and margins, as well as the Great Lakes. The program includes: (a) Structure, tectonic evolution and volcanic activity of the ocean basins, the continental margins, the mid-ocean ridges, and island arc systems; (b) Processes controlling exchange of heat and chemical species between seawater and ocean rocks; (c) Genesis, chemistry, and mineralogic evolution of marine sediments; (d) Processes controlling deposition, erosion and transport of marine sediments; (e) Past ocean circulation patterns and climates; (f) and, Interactions of continental and marine geologic processes.
Requirements: Individuals and organizations in the following categories may submit proposals: (1) Universities and Colleges-Universities and two- and four-year colleges (including community colleges) located and accredited in the U.S., acting on behalf of their faculty members. Such organizations also are referred to as academic institutions. Institutions located outside the U.S. fall under (6) below. (2) Non-profit, non-academic organizations-Independent museums, observatories, research laboratories, professional societies and similar organizations in the U.S. that are directly associated with educational or research activities. (3) For-profit organizations-U.S. commercial organizations, especially small businesses with strong capabilities in scientific or engineering research or education. An unsolicited proposal from a commercial organization may be funded when the project is of special concern from a national point of view, special resources are available for the work, or the proposed project is especially meritorious. NSF is interested in

supporting projects that couple industrial research resources and perspectives with those of universities; therefore, it especially welcomes proposals for cooperative projects involving both universities and the private commercial sector. (4) State and Local Governments-State educational offices or organizations and local school districts may submit proposals intended to broaden the impact, accelerate the pace, and increase the effectiveness of improvements in science, mathematics and engineering education in both K-12 and post-secondary levels. (5) Unaffiliated Individuals-Scientists, engineers or educators in the U.S. and U.S. citizens may be eligible for support, provided that the individual is not employed by, or affiliated with, an organization, and (a) the proposer has demonstrated the capability and has access to any necessary facilities to carry out the project; and (b) the proposer agrees to fiscal arrangements that, in the opinion of the NSF Division of Grants & Agreements, ensure responsible management of Federal funds. Unaffiliated individuals should contact the appropriate program before preparing a proposal for submission. (6) Foreign organizations-NSF rarely provides support to foreign organizations. NSF will consider proposals for cooperative projects involving U.S. and foreign organizations, provided support is requested only for the U.S. portion of the collaborative effort.
Restrictions: Other Federal agencies-NSF does not normally support research or education activities by scientists, engineers or educators employed by Federal agencies or Federally Funded Research and Development Centers (FFRDCs). Under unusual circumstances, other Federal agencies and FFRDCs may submit proposals directly to NSF. A proposed project is only eligible for support if it meets one or more of the following exceptions, as determined by a cognizant NSF Program Officer: (a) Special Projects. Under exceptional circumstances, research or education projects at other Federal agencies or FFRDCs that can make unique contributions to the needs of researchers elsewhere or to other specific NSF objectives may receive NSF support. (b) National and International Programs. The Foundation may fund research and logistic support activities of other Government agencies or FFRDCs directed at meeting the goals of special national and international research programs for which the Foundation bears special responsibility, such as the U.S. Antarctic Research Program. (c) International Travel Awards. In order to ensure appropriate representation or availability of a particular expertise at an international conference, staff researchers of other Federal agencies may receive NSF international travel awards. Proposers who think their project may meet one of the exceptions listed above should contact a cognizant NSF Program Officer before preparing a proposal for submission. In addition, a scientist, engineer or educator who has a joint appointment with a university and a Federal agency (such as a Veterans Administration Hospital, or with a university and a FFRDC) may submit proposals through the university and may receive support if he/she is a bonafide faculty member of the university, although part of his/her salary may be provided by the Federal agency. Preliminary inquiry should be made to the appropriate program before preparing a proposal for submission.
Geographic Focus: All States
Date(s) Application is Due: Feb 15; Aug 15
Contact: Bilal Haq; (703) 292-8581; fax (703) 292-9085; bhaq@nsf.gov
Internet: http://www.nsf.gov/funding/pgm_summ.jsp?pims_id=11726&org=OCE&from=home
Sponsor: National Science Foundation
4201 Wilson Boulevard
Arlington, VA 22230

NSF Materials Research (DMR) Grants **3593**

The Division of Materials Research (DMR) supports a wide range of programs that address fundamental phenomena in materials, materials synthesis and processing, structure and composition, properties and performance, and materials education. Research program areas include condensed matter physics; solid-state chemistry and polymers; metals, ceramics, and electronic materials; materials theory; and materials research science and engineering centers. DMR funding modes include support for individual investigators, groups, centers, national facilities, and instrumentation.
Requirements: NSF welcomes proposals on behalf of all qualified scientists, engineers and educators. The Foundation strongly encourages women, minorities and persons with disabilities to participate fully in its programs. Scientists, engineers and educators usually initiate proposals that are officially submitted by their employing organization. Before formal submission, the proposal may be discussed with appropriate NSF program staff. Graduate students are not encouraged to submit research proposals, but should arrange to serve as research assistants to faculty members. The window for submitting unsolicited proposals to DMR begins on September 1 and ends on October 31, annually. If the closing date for the submission window falls on a weekend, the closing date moves to the following Monday. The last date of the submission window is an absolute deadline date and proposals must be received by NSF by 5:00 p.m. submitter's local time on that date.
Geographic Focus: All States
Date(s) Application is Due: Oct 31
Contact: Ian Robertson, Division Director; (703) 292-4937; iroberts@nsf.gov
Internet: http://www.nsf.gov/div/index.jsp?div=dmr
Sponsor: National Science Foundation
4201 Wilson Boulevard
Arlington, VA 22230

NSF Materials Research Science and Engineering Centers (MRSEC) **3594**

Grants support interdisciplinary materials research and education while addressing fundamental problems in science and engineering. MRSECs require outstanding research quality and intellectual breadth, provide support for research infrastructure and flexibility in responding to new opportunities, and strongly emphasize the integration of research and education. These centers foster active collaboration between universities and other sectors, including industry, and they constitute a national network of university-based centers in materials research. MRSECs address problems of a scope or complexity requiring the advantages of scale and interdisciplinary interaction provided by a center. A MRSEC may be located at a single institution, or may involve two or more institutions in partnership. Preliminary proposals are due September 5; full proposals are due January 18. Annual deadline dates may vary; contact program staff for exact dates. NSF 01-109
Requirements: Proposals may be submitted by universities in support of center-type activities. FastLane Submission is available.
Restrictions: Only two pre-proposals may be submitted by any one institution. Any one individual may be the Principal Investigator for only one pre-proposal. The same limitations apply to full proposals.
Geographic Focus: All States
Date(s) Application is Due: Jan 18
Amount of Grant: Up to 24,000,000 USD
Contact: Thomas P. Rieker, Program Director; (703) 292-4914; trieker@nsf.gov
Internet: http://www.nsf.gov/funding/pgm_summ.jsp?pims_id=5295&org=NSF&sel_org=NSF&from=fund
Sponsor: National Science Foundation
4201 Wilson Boulevard
Arlington, VA 22230

NSF Materials World Network: Cooperative Activity in Materials Research Between U.S. Investigators and their Counterparts Abroad (MWN) **3595**

The program is intended to foster collaboration in materials and condensed matter research between investigators in the U.S. and their counterparts abroad. It includes joint activities between NSF and funding organizations in Africa, through the U.S.-Africa Materials Collaboration; Asia, Europe and other countries or regions. The estimated number of awards is 20 to 40, depending on quality of proposals and availability of funds. The total anticipated funding amount is $2,500,000 to $4,500,000. Estimated total funding, number of awards and average award size/duration are subject to quality of proposals and availability of funds. Awards may be standard or continuing grants. Proposals submitted must have clear relevance to fundamental materials and condensed matter research supported by the NSF Division of Materials Research (DMR). Projects not having this focus will not be considered for funding. NSF 07-574
Requirements: NSF will accept proposals from U.S. universities and two- and four-year colleges (including community colleges) addressing collaborations between researchers from the U.S. and participating countries or regions. Concurrently, investigators at non-U.S. research institutions should submit to the counterpart funding organization in their country or region a request for support of their side of the collaboration. NSF will consider support of all appropriate research costs for the U.S. side of such collaborations, with the expectation that funding or research organizations from the appropriate countries or regions will consider supporting the costs of the non-U.S. participants. In addition to proposals for collaborative materials research and education, proposals to interface networks of U.S. investigators and counterpart networks abroad will also be considered. Proposals from networks of U.S. researchers to NSF in support of activities such as workshops, meetings, brief personnel exchanges, etc., may be linked to counterpart networks abroad supported, for example, by the European Science Foundation. Proposed projects are expected to offer students and junior researchers the opportunity to participate in an international research and education experience and, more generally, for integrating research and training in an international environment. Proposals that include exchange of students and post-doctoral research associates between the U.S. and abroad are strongly encouraged, as well as proposals from junior faculty and members of underrepresented groups in science and engineering. Of special interest to NSF are proposals including activities that build upon and expand current cyber infrastructure capabilities, such as remote use of instrumentation, database creation and use, visualization and virtual experimentation, virtual networking, etc, to enhance and advance the international collaboration.
Restrictions: Proposals from investigators at non-U.S. institutions will not be accepted. An investigator may be Principal Investigator in only one proposal submitted to this program.
Geographic Focus: All States
Date(s) Application is Due: Nov 13
Contact: Dr. Carmen Huber; (703) 292-4939; fax 703-292-9036; chuber@nsf.gov
Internet: http://www.nsf.gov/funding/pgm_summ.jsp?pims_id=12820
Sponsor: National Science Foundation
4201 Wilson Boulevard
Arlington, VA 22230

NSF Mathematical Biology Grants **3596**

The Mathematical Biology Program supports research in areas of applied and computational mathematics with relevance to the biological sciences. Successful proposals are mathematically innovative and address challenging problems of interest to members of the biological community. Projects may include development of mathematical concepts and tools traditionally seen in other disciplinary programs within the Division of Mathematical Sciences, e.g., topology, probability, statistics, and computation, etc. To receive appropriate and timely review, such proposals should be submitted directly to the relevant disciplinary program, but will be considered for co-review by the Mathematical Biology program which may be selected as a secondary program.
Geographic Focus: All States
Date(s) Application is Due: Jan 13
Contact: Mary Ann Horn; (703) 292-4879; fax (703) 292-9032; mhorn@nsf.gov
Internet: http://www.nsf.gov/funding/pgm_summ.jsp?pims_id=5690
Sponsor: National Science Foundation
4201 Wilson Boulevard
Arlington, VA 22230

NSF Microbial Observatories (MO) and Microbial Interactions and Processes **3597**

Program activities will support research to discover and characterize novel microorganisms, microbial consortia, communities, activities and other novel properties, and to study their

roles in diverse environments. The long-term goal of this activity is to develop a network of sites or microbial observatories in different habitats to study and understand microbial diversity over time and across environmental gradients. Projects supported are expected to establish or participate in an established, Internet-accessible knowledge network to disseminate information resulting from these activities. In addition, educational and outreach activities such as formal or informal training in microbial biology, and activities that will broaden the participation of underrepresented groups in microbial research and education are expected. Fastlane submission is required. NSF 03-571
Requirements: Proposals may be submitted by universities or other NSF eligible institutions in support of individual investigators or small groups.
Restrictions: A PI may submit only one proposal and he/she may only collaborate in one other proposal as a co-PI.
Geographic Focus: All States
Date(s) Application is Due: Jul 23
Amount of Grant: Up to 5,000,000 USD
Contact: Ronald Weiner , Program Director; (703) 292-8442; rweiner@nsf.gov
Internet: http://www.nsf.gov/pubs/2003/nsf03571/nsf03571.pdf
Sponsor: National Science Foundation
4201 Wilson Boulevard
Arlington, VA 22230

NSF National Astronomy and Ionosphere Center (NAIC) Grants 3598
NSF supports the National Astronomy and Ionosphere Center (NAIC), a visitor-oriented national research center devoted to scientific investigations in radio and radar astronomy and atmospheric sciences. NAIC provides telescope users with a wide range of research and observing instrumentation including receivers, transmitters, movable line feeds, and digital data acquisition and processing equipment. The center has a permanent staff of scientists, engineers, and technicians who are available to help visiting investigators with their observing programs. The NAIC facilities and instrumentation are available on a competitive basis to qualified scientists from all over the world. NAIC is operated and managed for NSF by Cornell University.
Geographic Focus: All States
Contact: Program Director; (703) 292-8820; fax (703) 292-9034
Internet: http://www.nsf.gov/funding/pgm_summ.jsp?pims_id=5652&org=AST&from=home
Sponsor: National Science Foundation
4201 Wilson Boulevard
Arlington, VA 22230

NSF National Space Weather Program 3599
Proposals should benefit and advance the National Space Weather Program's goal to mitigate the adverse effects of space weather by providing timely, accurate, and reliable space environment observations, specifications, and forecasts. An effective program requires a strong commitment to basic research in many areas of space-related science. Proposals may be submitted for basic research in solar, heliospheric, magnetospheric, ionospheric, and thermospheric physics. NSF 07-520
Geographic Focus: All States
Date(s) Application is Due: Jan 16
Amount of Grant: Up to 100,000 USD
Contact: Richard Behnke; (703) 292-8518; fax (703) 292-9023; rbehnke@nsf.gov
Internet: http://www.nsf.gov/funding/pgm_summ.jsp?pims_id=5575
Sponsor: National Science Foundation
4201 Wilson Boulevard
Arlington, VA 22230

NSF Natural, Social, and System Sciences and Research Support and Logistics 3600
The Program supports research in glaciology and in the atmospheric, biological, earth, contaminants, and ocean sciences. This program provides core support for disciplinary research in the Arctic and coordinates its support of arctic research with the Directorates for Geosciences, Mathematical and Physical Sciences, and Biological Sciences. Areas of special interest include marine and terrestrial ecosystems, arctic adaptations, atmospheric chemistry, exploration of the Arctic Ocean, as well as arctic geological and glaciological processes. The goal of this program is to gain a better understanding of the Earth's biological, geological, chemical, and social processes, and the interactions of ocean, land, atmosphere, biological, and human systems. Approximately 130-160 awards are expected to be made, with a combination of standard and continuing awards. NSF 00-96
Geographic Focus: All States
Date(s) Application is Due: Feb 15; Aug 8
Contact: Dr. Neil R. Swanberg, Program Officer; (703) 306-1029; nswanber@nsf.gov
Internet: http://www.nsf.gov/pubs/2000/nsf0096/nsf0096.htm
Sponsor: National Science Foundation
4201 Wilson Boulevard
Arlington, VA 22230

NSF Nuclear Physics Research Grants 3601
The program supports research at the frontiers of nuclear science, including: properties and behavior of nuclei and nuclear matter under extreme conditions, and/or as they relate to astrophysical phenomena; the quark-gluon basis for the structure and dynamics of nuclear matter; phase transitions of nuclear matter from normal nuclear density and temperature to the predicted high-temperature quark-gluon plasma; and basic interactions and fundamental symmetries. This research involves many venues, including low-energy to multi-GeV electrons and photons; intermediate-energy light ions; low-energy to relativistic heavy ions, including radioactive beams; cold and ultra-cold neutrons; as well as non-accelerator-based experiments. The program supports university user groups executing experiments at a large number of laboratories in the United States and abroad, and a national user facility: the National Superconducting Cyclotron Laboratory, a superconducting, heavy-ion cyclotron facility at Michigan State University. The program also supports smaller accelerator facilities, such as those at Florida State University and the University of Notre Dame.
Requirements: Individuals and organizations in the following categories may submit proposals: (1) Universities and Colleges-Universities and two- and four-year colleges (including community colleges) located and accredited in the U.S., acting on behalf of their faculty members. Such organizations also are referred to as academic institutions. Institutions located outside the U.S. fall under (6) below. (2) Non-profit, non-academic organizations-Independent museums, observatories, research laboratories, professional societies and similar organizations in the U.S. that are directly associated with educational or research activities. (3) For-profit organizations-U.S. commercial organizations, especially small businesses with strong capabilities in scientific or engineering research or education. An unsolicited proposal from a commercial organization may be funded when the project is of special concern from a national point of view, special resources are available for the work, or the proposed project is especially meritorious. NSF is interested in supporting projects that couple industrial research resources and perspectives with those of universities; therefore, it especially welcomes proposals for cooperative projects involving both universities and the private commercial sector. (4) State and Local Governments-State educational offices or organizations and local school districts may submit proposals intended to broaden the impact, accelerate the pace, and increase the effectiveness of improvements in science, mathematics and engineering education in both K-12 and post-secondary levels. (5) Unaffiliated Individuals-Scientists, engineers or educators in the U.S. and U.S. citizens may be eligible for support, provided that the individual is not employed by, or affiliated with, an organization, and (a) the proposer has demonstrated the capability and has access to any necessary facilities to carry out the project; and (b) the proposer agrees to fiscal arrangements that, in the opinion of the NSF Division of Grants & Agreements, ensure responsible management of Federal funds. Unaffiliated individuals should contact the appropriate program before preparing a proposal for submission. (6) Foreign organizations-NSF rarely provides support to foreign organizations. NSF will consider proposals for cooperative projects involving U.S. and foreign organizations, provided support is requested only for the U.S. portion of the collaborative effort.
Restrictions: Other Federal agencies-NSF does not normally support research or education activities by scientists, engineers or educators employed by Federal agencies or Federally Funded Research and Development Centers (FFRDCs). Under unusual circumstances, other Federal agencies and FFRDCs may submit proposals directly to NSF. A proposed project is only eligible for support if it meets one or more of the following exceptions, as determined by a cognizant NSF Program Officer: (a) Special Projects. Under exceptional circumstances, research or education projects at other Federal agencies or FFRDCs that can make unique contributions to the needs of researchers elsewhere or to other specific NSF objectives may receive NSF support. (b) National and International Programs. The Foundation may fund research and logistic support activities of other Government agencies or FFRDCs directed at meeting the goals of special national and international research programs for which the Foundation bears special responsibility, such as the U.S. Antarctic Research Program. (c) International Travel Awards. In order to ensure appropriate representation or availability of a particular expertise at an international conference, staff researchers of other Federal agencies may receive NSF international travel awards. Proposers who think their project may meet one of the exceptions listed above should contact a cognizant NSF Program Officer before preparing a proposal for submission. In addition, a scientist, engineer or educator who has a joint appointment with a university and a Federal agency (such as a Veterans Administration Hospital, or with a university and a FFRDC) may submit proposals through the university and may receive support if he/she is a bonafide faculty member of the university, although part of his/her salary may be provided by the Federal agency. Preliminary inquiry should be made to the appropriate program before preparing a proposal for submission.
Geographic Focus: All States
Date(s) Application is Due: Sep 30
Contact: Bradley D. Keister; (703) 292-7377; fax (703) 292-9078; bkeister@nsf.gov
Internet: http://www.nsf.gov/funding/pgm_summ.jsp?pims_id=5627&org=PHY&from=home
Sponsor: National Science Foundation
4201 Wilson Boulevard
Arlington, VA 22230

NSF Ocean Drilling (OD) Grants 3602
ODP activities are a unique, international effort to explore the earth's crust beneath oceans for the purpose of revealing the composition, structure, and history of the submerged portion of the earth's surface. Ocean drilling involves logging and collecting geologic samples from the floor of the deep ocean basins through rotary coring and hydraulic piston coring. The logs and samples of the cores are made available to qualified scientists throughout the world for individual research projects. Drilling programs have sampled sites in the North Atlantic, Norwegian Sea, Mediterranean Sea, southern and equatorial Atlantic, Pacific Ocean off the west coast of South America, Weddell Sea off Antarctica, Indian Ocean and western and equitorial Pacific Ocean. Drilling operations are managed by Texas A&M University; logging is managed by Lamont-Doherty Geological Observatory. The general contractor for the overall management and operation of ODP is Joint Oceanographic Institutions. Support for participation and drilling-related research performed by U.S. scientists is provided by NSF. This support focuses on investigations of potential drilling regions, especially by means of regional geophysical field studies; feasibility and initial development of downhole instruments and techniques; and downhole geophysical and geochemical experiments. In addition, NSF will consider proposals for studies leading to long-range definition of future drilling objectives. To be considered for support, proposed projects should be clearly relevant to the drilling plans of the international drilling community and focus on predrilling or concurrent drilling activities. Additional support for U.S. scientists may be obtained via the Joint U.S. Science Advisory

Committee (U.S.SAC), which includes support for U.S. scientists participating on the drillship and for necessary follow-up studies related to initial publication of drilling results; and planning activities, such as workshops, to define concepts and develop problem-related drilling programs, including U.S. participation in Joint Oceanographic Institutions for Deep Earth Sampling (JOIDES). Also, requests for proposals may be issued for other surveys, regional and topical synthesis of existing data, and development of downhole tools and instrumentation as these tasks are identified. Contact the NSF for further information and list of addresses to which specific proposals should be sent.
Geographic Focus: All States
Date(s) Application is Due: Feb 15; Aug 15
Contact: James F. Allan; (703) 292-8144; fax (703) 292-9085; jallan@nsf.gov
Internet: http://www.nsf.gov/funding/pgm_summ.jsp?pims_id=13524&org=NSF&sel_org=NSF&from=fund
Sponsor: National Science Foundation
4201 Wilson Boulevard
Arlington, VA 22230

NSF Oceanographic Centers and Facilities Grants **3603**
NSF supports construction, conversion, acquisition, and operation of major shared-use oceanographic facilities. The University-National Oceanographic Laboratory System (UNOLS) schedules these facilities and expeditionary programs. This program supports expensive facilities that are necessary for NSF-funded research and training of oceanographers. Examples of these facilities are ships, submersibles, large shipboard equipment, and shared-use instruments to collect and analyze data. NSF encourages local contributions from nonfederal funds; however, there is no fixed requirement for institutional contributions.
Geographic Focus: All States
Contact: Linda Goad; (703) 292-7706; fax (703) 292-9085; lgoad@nsf.gov
Internet: http://www.nsf.gov/funding/pgm_summ.jsp?pims_id=12725&org=OCE&from=home
Sponsor: National Science Foundation
4201 Wilson Boulevard
Arlington, VA 22230

NSF Ocean Technology and Interdisciplinary Coordination Grants **3604**
The Oceanographic Technology and Interdisciplinary Coordination (OTIC) Program supports a broad range of research and technology development activities. Unsolicited proposals are accepted for instrumentation development that has broad applicability to ocean science research projects and that enhance observational, experimental or analytical capabilities of the ocean science research community. Specific announcements for funding opportunities are made for additional projects involving Improvements in Facilities, Communications, and Equipment at Biological Field Stations and Marine Laboratories (FSML) and the National Ocean Partnership Program.
Geographic Focus: All States
Date(s) Application is Due: Feb 15; Aug 15
Contact: Kandace S. Binkley, (703) 292-7577; fax (703) 292-9085; kbinkley@nsf.gov
Internet: http://www.nsf.gov/funding/pgm_summ.jsp?pims_id=12724&org=OCE&from=home
Sponsor: National Science Foundation
4201 Wilson Boulevard
Arlington, VA 22230

NSF OCE Education Grants **3605**
OCE Education supports efforts to integrate ocean research and education via three main program areas, which are: Research Experiences for Undergraduates (REU) Site Program—this program provides funding to Universities and Marine Laboratories that allows them to offer summer internships to undergraduate students who would like to participate in ocean-related research efforts. Proposals may be submitted annually (August deadline); Faculty Early Career Development (CAREER)—this program supports pre-tenure researchers who would like to combine their research efforts with excellent educational programs. Proposals may be submitted annually (July deadline); and Centers for Ocean Education Excellence (COSEE)—this program supports partnerships between ocean science researchers, educators, and informal science organizations, providing the public a deeper understanding of the ocean and its influence on each person's quality of life and our national prosperity.
Geographic Focus: All States
Contact: Elizabeth L. Rom; (703) 292-7709; fax (703) 292-9085; elrom@nsf.gov
Internet: http://www.nsf.gov/funding/pgm_summ.jsp?pims_id=13548&org=OCE&from=home
Sponsor: National Science Foundation
4201 Wilson Boulevard
Arlington, VA 22230

NSF Paleoclimate Grants **3606**
This program supports research on the natural evolution of Earth's climate with the goal of providing a baseline for present variability and future trends through improved understanding of the physical, chemical, and biological processes that influence climate over the long-term. The Geosciences Directorate and the Office of Polar Programs have joined in creating the annual Paleo Perspectives on Climate Change (P2C2) competition in paleoclimate global change research. Researchers are encouraged to consider the P2C2 competition as a possible source of support for their global change research. Proposals are welcome at any time during the year for the Paleoclimate Program but, investigators are encouraged to submit proposals early in the fiscal year. PD 98-1530 Since proposals eligible for funding in the P2C2 competition are not eligible for funding in the Paleoclimate Program, researchers are strongly advised to contact the Director of the Paleoclimate Program for guidance as to the suitability of their proposed research for either program.
Requirements: Individuals and organizations in the following categories may submit proposals: (1) Universities and Colleges-Universities and two- and four-year colleges (including community colleges) located and accredited in the U.S., acting on behalf of their faculty members. Such organizations also are referred to as academic institutions; (2) Non-profit, non-academic organizations-Independent museums, observatories, research laboratories, professional societies and similar organizations in the U.S. that are directly associated with educational or research activities; (3) For-profit organizations-U.S. commercial organizations, especially small businesses with strong capabilities in scientific or engineering research or education. An unsolicited proposal from a commercial organization may be funded when the project is of special concern from a national point of view, special resources are available for the work, or the proposed project is especially meritorious. NSF is interested in supporting projects that couple industrial research resources and perspectives with those of universities; therefore, it especially welcomes proposals for cooperative projects involving both universities and the private commercial sector; (4) State and Local Governments-State educational offices or organizations and local school districts may submit proposals intended to broaden the impact, accelerate the pace, and increase the effectiveness of improvements in science, mathematics and engineering education in both K-12 and post-secondary levels; (5) Unaffiliated Individuals-Scientists, engineers or educators in the U.S. and U.S. citizens may be eligible for support, provided that the individual is not employed by, or affiliated with, an organization, and: (a) the proposer has demonstrated the capability and has access to any necessary facilities to carry out the project; and, (b) the proposer agrees to fiscal arrangements that, in the opinion of the NSF Division of Grants & Agreements, ensure responsible management of Federal funds. Unaffiliated individuals should contact the appropriate program before preparing a proposal for submission.
Restrictions: NSF rarely provides support to foreign organizations. NSF will consider proposals for cooperative projects involving U.S. and foreign organizations, provided support is requested only for the U.S. portion of the collaborative effort. Also, NSF does not normally support research or education activities by scientists, engineers or educators employed by Federal agencies or Federally Funded Research and Development Centers (FFRDCs). Under unusual circumstances, other Federal agencies and FFRDCs may submit proposals directly to NSF.
Geographic Focus: All States
Contact: David J. Verardo, (703) 292-8527; dverardo@nsf.gov
Internet: http://www.nsf.gov/funding/pgm_summ.jsp?pims_id=12727&org=ATM
Sponsor: National Science Foundation
4201 Wilson Boulevard
Arlington, VA 22230

NSF Paleo Perspectives on Climate Change (P2C2) Grants **3607**
The goal of research funded under the interdisciplinary P2C2 solicitation is to utilize key geological, chemical, and biological records of climate system variability to provide insights into the mechanisms and rate of change that characterized Earth's past climate variability, the sensitivity of Earth's climate system to changes in forcing, and the response of key components of the Earth system to these changes. Important scientific objectives of P2C2 are to: 1) provide comprehensive paleoclimate data sets that can serve as model test data sets analogous to instrumental observations; and 2) enable transformative syntheses of paleoclimate data and modeling outcomes to understand the response of the longer-term and higher magnitude variability of the climate system that is observed in the geological record. Collaborative Proposals. All collaborative proposals submitted as separate submissions from multiple organizations must be submitted via the NSF FastLane system. Solicitation: 08-505
Geographic Focus: All States
Date(s) Application is Due: Feb 4; Oct 15
Amount of Grant: 101,000 USD
Contact: Paul E. Filmer; (703) 292-8551; fax (703) 292-9025; pfilmer@nsf.gov
Internet: http://www.nsf.gov/funding/pgm_summ.jsp?pims_id=5750
Sponsor: National Science Foundation
4201 Wilson Boulevard
Arlington, VA 22230

NSF Particle and Nuclear Astrophysics (PNA) Grants **3608**
The program supports university research in particle and nuclear astrophysics, including experiments or research and design projects in underground facilities. Current supported activities are high-energy cosmic ray studies, solar and high-energy neutrino astrophysics, the study of gamma ray bursts, and searches for dark matter. Funding is also provided for accelerator-based nuclear astrophysics studies of stellar processes, nucleosynthesis, and processes related to cosmology and the early universe. Currently, the Deep Underground Science and Engineering Laboratory (DU.S.EL) program is funding Research and Development studies for experiments that need to be located in an underground laboratory to reduce backgrounds. NSF PD 06-1643
Requirements: It should be noted that proposals that are submitted to the PNA program and are requesting in excess of $1,000,000 may, at the discretion of the Program Officer, be subjected to an additional level of scrutiny in the form of a cost review which would take place before the annual PNA panel that meets to discuss all of the submitted proposals. Individuals and organizations in the following categories may submit proposals: (1) Universities and Colleges-Universities and two- and four-year colleges (including community colleges) located and accredited in the U.S., acting on behalf of their faculty members. (2) Non-profit, non-academic organizations-Independent museums, observatories, research laboratories, professional societies and similar organizations in the U.S. that are directly associated with educational or research activities (3.) For-profit

organizations-U.S. commercial organizations, especially small businesses with strong capabilities in scientific or engineering research or education. An unsolicited proposal from a commercial organization may be funded when the project is of special concern from a national point of view, special resources are available for the work, or the proposed project is especially meritorious. NSF is interested in supporting projects that couple industrial research resources and perspectives with those of universities; therefore, it especially welcomes proposals for cooperative projects involving both universities and the private commercial sector. (4) State and Local Governments-State educational offices or organizations and local school districts may submit proposals intended to broaden the impact, accelerate the pace, and increase the effectiveness of improvements in science, mathematics and engineering education in both K-12 and post-secondary levels. (5) Unaffiliated Individuals-Scientists, engineers or educators in the U.S. and U.S. citizens may be eligible for support, provided that the individual is not employed by, or affiliated with, an organization, and: (a) the proposer has demonstrated the capability and has access to any necessary facilities to carry out the project; and, (b) the proposer agrees to fiscal arrangements that, in the opinion of the NSF Division of Grants & Agreements, ensure responsible management of Federal funds. Unaffiliated individuals should contact the appropriate program before preparing a proposal for submission.
Restrictions: NSF rarely provides support to foreign organizations. NSF will consider proposals for cooperative projects involving U.S. and foreign organizations, provided support is requested only for the U.S. portion of the collaborative effort. NSF does not normally support research or education activities by scientists, engineers or educators employed by Federal agencies or Federally Funded Research and Development Centers (FFRDCs).
Geographic Focus: All States
Date(s) Application is Due: Sep 24
Contact: Marvin Goldberg, Program Director; (703) 292-7374; mgoldber@nsf.gov
Internet: http://www.nsf.gov/funding/pgm_summ.jsp?pims_id=5633&govDel=U.S.NSF_39
Sponsor: National Science Foundation
4201 Wilson Boulevard
Arlington, VA 22230

NSF Partnership for Advancing Technologies in Housing Grants **3609**
The objective of this program is to support the objectives of the PATH program, which are to: improve the quality, affordability, durability, environmental performance, and energy efficiency of today's new and existing homes; strengthen the technology infrastructure of the United States; and help create the next generation of American housing. In addition to proposals in the above areas of technology and technology management, NSF and PATH are also interested in proposals involving partnerships between research institutions, industrial enterprises, local government, and/or other broadly-defined research and development participants relevant to the home building industry. NSF 01-45
Requirements: Proposals may be submitted by U.S. academic institutions in support of individual investigators or small groups.
Geographic Focus: All States
Date(s) Application is Due: Apr 17
Amount of Grant: Up to 150,000 USD
Contact: Robert Wellek, Deputy Division Director; (703) 292-8370; rwellek@nsf.gov
Internet: http://www.nsf.gov/pubs/2001/nsf0145/nsf0145.htm
Sponsor: National Science Foundation
4201 Wilson Boulevard
Arlington, VA 22230

NSF Partnerships for Innovation (PFI) Grants **3610**
The program is intended to forge connections between new knowledge created in the discovery process to learning and innovation, while broadening the participation of people and institutions in NSF activities. For the purposes of this program, innovation is defined as the transformation of knowledge into the products, processes, systems and services that fuel economic development, create wealth and generate improvements in the national standard of living. The goals of this program are to stimulate the transformation of knowledge created by the national research and education enterprise into innovations that create new wealth, build strong local, regional and national economies, and improve the national well-being; broaden the participation of all types of academic institutions and all citizens in NSF activities to more fully meet the broad workforce needs of the national innovation enterprise; and create or enhance enabling infrastructure necessary to foster and sustain innovation in the long-term. Cost sharing is required at a level of 10 percent of the requested NSF budget. An optional letter of intent is due March 10; full proposal is due May 17. Annual deadline dates may vary; contact program staff for exact dates. NSF 04-556
Requirements: Proposals must be submitted by U.S. universities and colleges located in the U.S., its territories and possessions, including universities, two-and four-year colleges and community colleges, hereafter referred to as universities and colleges, or academic institutions. A Partnership proposal must be submitted as a single administrative package by the lead institution. The submitting institution must serve as the lead institution or partner in the proposed partnership, that is, as the institution responsible for overall management of the partnership activity with administrative responsibility for the proposed effort. Partners may include, but are not limited to, other academic institutions, not-for-profit organizations, private sector firms, entrepreneurs, venture capitalists, state and local government entities, trade and professional associations, and Federal laboratories. At a minimum, proposed partnerships must include academic institutions and private sector organizations. Partnerships that include state/local government entities are strongly encouraged.
Geographic Focus: All States
Date(s) Application is Due: Jun 3
Amount of Grant: 36,000-600,000 USD
Contact: John Hurt, Program Director; (703) 292-5332; fax (703) 292-9051; jhurt@nsf.gov
Internet: http://www.nsf.gov/pubsys/ods/getpub.cfm?ods_key=nsf04556
Sponsor: National Science Foundation
4201 Wilson Boulevard
Arlington, VA 22230

NSF Partnerships for International Research and Education (PIRE) **3611**
The Partnerships for International Research and Education (PIRE) program seeks to catalyze a higher level of international engagement in the U.S. science and engineering community by supporting innovative, international research and education collaborations. The program will enable U.S. scientists and engineers to establish collaborative relationships with international colleagues in order to advance new knowledge and discoveries at the frontiers of science and engineering and to promote the development of a diverse, globally-engaged U.S. scientific and engineering workforce. The PIRE program will support bold, forward-looking research whose successful outcome results from all partners-U.S. and foreign-providing unique contributions to the research endeavor. It is also intended to facilitate greater student preparation for and participation in international research collaboration, and to contribute to the development of a diverse, globally-engaged U.S. science and engineering workforce. The program aims to support partnerships that will strengthen the capacity of institutions, multi-institutional consortia, and networks to engage in and benefit from international research and education collaborations. This program provides educational opportunities for undergraduate students, graduate students, and postdoctoral fellows. This program supports institutions which may provide support to individuals at those institutions.
Requirements: Submission of Preliminary Proposals is required. Full proposals will be by invitation only. Proposals may only be submitted by universities that granted one (1) or more PhD in a science or engineering field since January 1, 2006, accredited in, and having a campus located in the United States. Data on science and engineering PhD production is available from The Survey of Earned Doctorates for 2006. Institutions that are not on this list but that have granted one (1) or more PhD in a science or engineering field since 2006 should contact PIRE Program staff regarding eligibility. Proposers are strongly encouraged to develop synergistic collaborations with a wide range of U.S. academic institutions, including four-year and two-year institutions, in order to take advantage of complementary strengths in international research and/or education and broaden the impact of the proposed PIRE program. They are also encouraged to develop synergistic linkages with other domestic and international partners that will enhance the PIRE project, including research laboratories, professional societies, corporations, institutions of informal learning (e.g., museums, zoos), etc.
Restrictions: Eligible institutions may submit up to three (3) preliminary proposals maximum.
Geographic Focus: All States
Date(s) Application is Due: Feb 26
Contact: Elizabeth Lyons; (703) 292-7256; fax (703) 292-9067; PIRE-info@nsf.gov
Internet: http://www.nsf.gov/funding/pgm_summ.jsp?pims_id=12819&org=OISE&from=home
Sponsor: National Science Foundation
4201 Wilson Boulevard
Arlington, VA 22230

NSF Partnerships in Astronomy & Astrophysics Research and Education **3612**
The objective of the program is to enhance diversity in astronomy and astrophysics research and education by stimulating the development of formal, potentially long-term, collaborative research and education relationships between minority-serving colleges and universities and the NSF-supported facilities, projects or faculty members at research institutions (including private observatories). As retention of traditionally underrepresented students is key to the success of this program, a solid mentoring scheme is vital. PAARE awards are expected to help build a foundation for students at an early point in their careers and help these students make the next step in their careers. The partnerships are intended to provide education and mentoring as a form of professional development to help these students succeed long after their work on the PAARE research. Funded activities might include, but are not limited to, the development of collaborative and mutually beneficial astronomy-related research and education projects, support for graduate and undergraduate students and exchanges of faculty and students. Well-prepared high-school students may also participate. Of special interest are activities based on research and education connections between the participants, and designed to increase recruitment, retention and degree attainment by members of groups underrepresented in astronomy and astrophysics research. Proposers are encouraged to contact NSF program staff to discuss the appropriateness of the planned activities.
Requirements: Proposals may only be submitted by minority-serving colleges and universities, including Historically Black Colleges and Universities (HBCUs), Hispanic-serving institutions (HSIs), and Tribally Controlled Colleges (TCCs). Each proposal must be submitted in collaboration with (and may include a subaward to) one or more NSF-supported facilities, projects or partnering faculty members at research institutions. Colleges and universities eligible to participate in this activity must offer baccalaureate degrees in physics or astronomy and meet at least one of the following criteria: 1) Be designated by the Department of Higher Education in FY 1994 as a Hispanic-serving Institution (HSI) under Title III of the Higher Education Act of 1965, as amended [See 20 U.S.C 1059 (copyright symbol); Public Law 102-325, Section 316, July 22, 1992]. 2) Be designated by the Department of Education as a Historically Black College or University (HBCU) under Title III of the Higher Education Act of 1965, as amended (see 34 CFR 608.2). 3) Be cited as a Tribal College or University in Section 532 of the Equity in Education Land-Grant Status October of 1964; Tribally Controlled Community College Assistance Act of 1978, Public Law 95-471. 4) Be a college or university whose undergraduate enrollment consists of at least 50% one or more ethnic minorities underrepresented in science and engineering in the U.S., and not a Doctoral/Research University-Extensive institution in the Carnegie classification (for the Carnegie

classification see http://www.carnegiefoundation.org/classifications/index.asp?key=785). African American, Hispanic Americans, Native Americans or Native Alaskans and Hawaiian/Pacific islanders are ethnic minorities underrepresented in science and engineering in the U.S.. Partnering institutions or organizations may be any institution or organization eligible for NSF funding that supports a research program in astronomy and astrophysics including AST-funded facilities and also private observatories. Research groups and individual scientists at research institutions are also eligible partners, provided their institution shows evidence of commitment to the proposed program as demonstrated in a letter from the institution describing its support of the partnership. Proposals may include subawards to the partnering institution, provided it has demonstrated a commitment to the activity at least for the duration of the PAARE grant.
Geographic Focus: All States
Date(s) Application is Due: Aug 3
Contact: Tammy Bosler; (703) 292-8248; fax (703) 292-9034; tbosler@nsf.gov
Internet: http://nsf.gov/funding/pgm_summ.jsp?pims_id=501046&org=NSF&sel_org=NSF&from=fund
Sponsor: National Science Foundation
4201 Wilson Boulevard
Arlington, VA 22230

NSF Perception, Action & Cognition Research Grants 3613
This program supports research on perception, action and cognition including the development of these capacities. Emphasis is on research strongly grounded in theory. Research topics include vision, audition, haptics, attention, memory, reasoning, written and spoken discourse, motor control, and developmental issues in all topic areas. The program encompasses a wide range of theoretical perspectives, such as symbolic computation, connectionism, ecological, nonlinear dynamics, and complex systems, and a variety of methodologies including both experimental studies and modeling. Research involving acquired or developmental deficits is appropriate if the results speak to basic issues of perception, action, and cognition.
Geographic Focus: All States
Date(s) Application is Due: Feb 1; Aug 1
Amount of Grant: 1,100-49,000,000 USD
Contact: Vincent R. Brown, Program Director; (703) 292-7305; vrbrown@nsf.gov
Internet: http://www.nsf.gov/funding/pgm_summ.jsp?pims_id=5686
Sponsor: National Science Foundation
4201 Wilson Boulevard
Arlington, VA 22230

NSF Physical and Dynamic Meteorology (PDM) Grants 3614
Physical and Dynamic Meteorology supports research involving studies of cloud physics; atmospheric electricity; radiation; boundary layer and turbulence; the initiation, growth, and propagation of gravity waves; all aspects of mesoscale meteorological phenomena, including their morphological, thermodynamic, and kinematic structure; development of mesoscale systems and precipitation processes; and transfer of energy between scales. The program also sponsors the development of new techniques and devices for atmospheric measurements. PD 98-1522
Requirements: Individuals and organizations in the following categories may submit proposals: (1) Universities and Colleges-Universities and two- and four-year colleges (including community colleges) located and accredited in the U.S., acting on behalf of their faculty members. Such organizations also are referred to as academic institutions; (2) Non-profit, non-academic organizations-Independent museums, observatories, research laboratories, professional societies and similar organizations in the U.S. that are directly associated with educational or research activities; (3) For-profit organizations-U.S. commercial organizations, especially small businesses with strong capabilities in scientific or engineering research or education. An unsolicited proposal from a commercial organization may be funded when the project is of special concern from a national point of view, special resources are available for the work, or the proposed project is especially meritorious. NSF is interested in supporting projects that couple industrial research resources and perspectives with those of universities; therefore, it especially welcomes proposals for cooperative projects involving both universities and the private commercial sector; (4) State and Local Governments-State educational offices or organizations and local school districts may submit proposals intended to broaden the impact, accelerate the pace, and increase the effectiveness of improvements in science, mathematics and engineering education in both K-12 and post-secondary levels; (5) Unaffiliated Individuals-Scientists, engineers or educators in the U.S. and U.S. citizens may be eligible for support, provided that the individual is not employed by, or affiliated with, an organization, and: (a) the proposer has demonstrated the capability and has access to any necessary facilities to carry out the project; and, (b) the proposer agrees to fiscal arrangements that, in the opinion of the NSF Division of Grants & Agreements, ensure responsible management of Federal funds. Unaffiliated individuals should contact the appropriate program before preparing a proposal for submission.
Restrictions: NSF rarely provides support to foreign organizations. NSF will consider proposals for cooperative projects involving U.S. and foreign organizations, provided support is requested only for the U.S. portion of the collaborative effort. Also, NSF does not normally support research or education activities by scientists, engineers or educators employed by Federal agencies or Federally Funded Research and Development Centers (FFRDCs). Under unusual circumstances, other Federal agencies and FFRDCs may submit proposals directly to NSF.
Geographic Focus: All States
Amount of Grant: 50,000-2,000,000 USD
Contact: Andrew G. Detwiler; (703) 292-8524; fax (703) 292-9022; adetwile@nsf.gov
Internet: http://www.nsf.gov/funding/pgm_summ.jsp?pims_id=12728&org=ATM
Sponsor: National Science Foundation
4201 Wilson Boulevard
Arlington, VA 22230

NSF Physical Anthropology Research Grants 3615
The program supports basic research in areas related to human evolution and contemporary human biological variation. Research areas supported by the program include, but are not limited to, human genetic variation, human adaptation, human osteology and bone biology, human and nonhuman primate paleontology, functional anatomy, and primate socioecology. Grants supported in these areas are united by an underlying evolutionary framework, and often a consideration of adaptation as a central theoretical theme. Many proposals also have a biocultural orientation. The program frequently serves as a bridge within NSF between the social and behavioral sciences and the natural and physical sciences, and proposals are commonly jointly reviewed and funded with other programs. Refer to NSF brochure PD 98-1392.
Geographic Focus: All States
Date(s) Application is Due: Jul 1; Dec 3
Contact: Dr. Mark Weiss; (703) 292-7321; fax (703) 292-9068; mweiss@nsf.gov
Internet: http://www.nsf.gov/funding/pgm_summ.jsp?pims_id=5407
Sponsor: National Science Foundation
4201 Wilson Boulevard
Arlington, VA 22230

NSF Physical Chemistry (PCHEM) Grants 3616
This program supports 2 broad areas: (1) Experimental Physical Chemistry-Supports experimental research directed at the molecular level of understanding of the physical properties of chemical systems. Experimental methodologies employed include frequency domain and time domain spectroscopic techniques covering the entire range of the electromagnetic spectrum, time-resolved dynamical studies of state-selected and mass-selected systems, and reactive scattering in molecular beams. Chemical systems studied range from single isolated molecules or ions to clusters, liquids, and solids. Chemical properties of interest include molecular structure and the shape of the ground and excited electronic-state potential energy surfaces, chemical dynamics of unimolecular and bimolecular chemical processes, time-resolved internal energy redistribution and state-to-state dynamics in molecular systems, and solute/solvent interactions in clusters and liquids. (2) Theoretical and Computational Chemistry-Supports the development of theoretical and computational methods for chemical problems. Methods of special interest include, but are not limited to, electronic structure techniques, statistical mechanics, molecular dynamics, and simulation methods for molecular or supramolecular systems. Areas of application span the full range of chemical systems from small molecules to macromolecules and degrees of aggregation from single molecules or small clusters to macroscopic systems. While application areas may involve any chemical system, including biological systems or materials, the goal of the program is to support the development of new theoretical and computational methodologies that will be broadly applicable to a range of chemical problems.
Requirements: Proposals to the Physical Chemistry Program should be submitted during the Division of Chemistry's submission windows that are that are open between July 1 and July 31 and then between November 1 and November 30. Proposals may be submitted at any time the windows are open.
Geographic Focus: All States
Date(s) Application is Due: Jul 31; Nov 30
Contact: Tanja Pietrass, (703) 292-2170; tpietras@nsf.gov
Internet: http://www.nsf.gov/funding/pgm_summ.jsp?pims_id=5674&org=NSF&more=Y
Sponsor: National Science Foundation
4201 Wilson Boulevard
Arlington, VA 22230

NSF Physical Oceanography Research Grants 3617
The Physical Oceanography Program supports research on a wide range of topics associated with the structure and movement of the ocean, with the way in which it transports various quantities, with the way the ocean's physical structure interacts with the biological and chemical processes within it, and with interactions between the ocean and the atmosphere, solid earth and ice that surround it.
Geographic Focus: All States
Date(s) Application is Due: Feb 15; Aug 15
Contact: Eric Itsweire; (703) 292-8582; fax (703) 292-9085; eitsweir@nsf.gov
Internet: http://www.nsf.gov/funding/pgm_summ.jsp?pims_id=12729&org=OCE&from=home
Sponsor: National Science Foundation
4201 Wilson Boulevard
Arlington, VA 22230

NSF Physics at the Information Frontier (PIF) Grants 3618
This program (PIF) is intended to provide support for physics proposals in three subareas: computational physics; information intensive physics; and quantum information science and revolutionary computing. The computational physics subarea focuses on computational problems in physics requiring significant long-term code development, and/or medium to large collaboratories involving physicists or physicists interacting with applied mathematicians and computer scientists. Priority will be given to proposals which: advance computational capabilities crossing intellectual boundaries in physics; result in dramatic new computational advances within a specific sub-area of physics; or provide generally useful software tools for the physics community. Information intensive physics seeks proposals to develop rapid, secure and efficient access to physics data stores rising from Petabytes (today) to Exabytes (in 10 years) via heterogeneous and distributed computing resources and networks of varying capability and reliability and to develop

internally consistent approaches to the usage of common resources required in the multiple collaborations and serving virtual science organizations on a global scale. Priority will be given to proposals which: develop tools which can serve a broad community within physics or reach out to other communities in need of rapid, secure access to large data stores; or which bring dramatic new capabilities to a specific sub-area of physics.
Geographic Focus: All States
Date(s) Application is Due: Oct 28
Contact: Barry I. Schneider; (703) 292-7383; fax (703) 292-9060; bschneid@nsf.gov
Internet: http://www.nsf.gov/funding/pgm_summ.jsp?pims_id=6681
Sponsor: National Science Foundation
4201 Wilson Boulevard
Arlington, VA 22230

NSF Physics Frontiers Centers (PFC) Grants **3619**
This program supports university-based centers and institutes where the collective efforts of a larger group of individuals can enable transformational advances in the most promising research areas. The program is designed to foster major breakthroughs at the intellectual frontiers of physics by providing needed resources such as combinations of talents, skills, disciplines, and/or specialized infrastructure, not usually available to individual investigators or small groups, in an environment in which the collective efforts of the larger group can be shown to be seminal to promoting significant progress in the science and the education of students. Activities supported are in all sub-fields of physics within the purview of the Division of Physics: atomic, molecular, optical, plasma, elementary particle, nuclear, astro-, gravitational, and biological physics. Interdisciplinary projects at the interface between these physics areas and other disciplines and physics sub-fields, e.g. biology, quantum information science, mathematical physics, condensed matter physics, and emerging areas of physics are also included. The successful PFC activity will demonstrate: the potential for a profound advance in physics; creative, substantive activities aimed at enhancing education, diversity, and public outreach; potential for broader impacts, e.g., impacts on other field(s) and benefits to society; and a synergy or value-added rationale that justifies a center- or institute-like approach.
Requirements: Proposals may be submitted by universities and colleges in the United States. In cases involving multi-institutional consortia, a single entity must accept overall management responsibility.
Restrictions: No more than two pre-proposals may be submitted by any one institution.
Geographic Focus: All States
Date(s) Application is Due: Jan 30
Amount of Grant: 500,000-4,000,000 USD
Contact: Ramona Winkelbauer; (703) 292-7390; fax (703) 292-9078; rwinkelb@nsf.gov
Internet: http://www.nsf.gov/funding/pgm_summ.jsp?pims_id=5305
Sponsor: National Science Foundation
4201 Wilson Boulevard
Arlington, VA 22230

NSF Physics of Living Systems (PoLS) Grants **3620**
PoLS will encourage research that will emphasize the physical principles of organization and function of living systems, including the exploration of artificial life forms. While the problems under study must be important to advancing our understanding of the living world in a quantitative way, particular emphasis will be placed on those projects in which lessons learned from the biological application also expand the intellectual range of physics. Awards will cover a broad spectrum of physics approaches in biology, ranging from the physical principles and mechanisms at the single cell level such as cellular organization (e.g. cytoskeleton), energy metabolism, gene regulation and intracellular and intercellular communication, to collective behavior and evolution of complexity in life forms and living populations of organisms. This systems approach in physics has been very successful in understanding inanimate systems, and has the potential to bring deep understanding of the world of animated, replicating systems, through falsifiable phenomenological theories. In vitro systems, if necessary for understanding of the physical principles in a living organism, will be supported as well. The program funds individual investigators, although collaborative proposals between physicists and biological researchers are welcome.
Geographic Focus: All States
Date(s) Application is Due: Jul 31
Contact: Krastan B. Blagoev; (703) 292-4666; fax (703) 292-9078; kblagoev@nsf.gov
Internet: http://www.nsf.gov/funding/pgm_summ.jsp?pims_id=6673
Sponsor: National Science Foundation
4201 Wilson Boulevard
Arlington, VA 22230

NSF Physics Research Grants **3621**
The Physics Division of the Mathematical and Physical Sciences Directorate (MPS) supports research projects in the areas of atomic, molecular, optical, and plasma physics, elementary particle physics, nuclear physics, and gravitational physics. The division also supports a significant level of activity in nuclear and particle astrophysics, which involves substantial linkages between the NSF nuclear and particle physics programs and with the astronomy division. Much of the research supported by the division is heavily involved in computational physics and nonlinear science. Research in theoretical physics is supported in a separate program within the division, and the theoretical physics program has strong linkages to programs in all of the other divisions in the MPS Directorate. The research activities of the division are linked to education. Research programs support and involve students at the postdoctoral, graduate, and undergraduate levels.
Geographic Focus: All States
Contact: Joseph L. Dehmer; (703) 292-7370; fax (703) 292-9078; jdehmer@nsf.gov
Internet: http://www.nsf.gov/mps/divisions/phy/about/c_programs.htm
Sponsor: National Science Foundation
4201 Wilson Boulevard
Arlington, VA 22230

NSF Physiology and Ethology Research Grants **3622**
The Physiology and Ethology Cluster of thematic areas is located within the Division of Integrative Biology and Neuroscience (IBN) and supports integrative studies of physiological functions at the genomic, cellular, systemic, and organismal levels, and animal behavior in both field and laboratory settings. Also considered are Long-Term Research in Environmental Biology (LTREB) proposals. The cluster supports research on: the mechanism, development, function, and evolution of all animal behavior, including behavioral ecology and evolution; nonhuman learning and cognition; behavioral genetics; development of behavior; and behavioral physiology and motivation, including behavioral endocrinology, animal communication, and animal orientation. Also included are studies that address ecological or evolutionary questions in the areas of morphology, comparative physiology, physiological ecology, and biomechanics of plants, animals, protists, fungi, and bacteria, with emphasis on the study of whole organisms, living or extinct.
Requirements: Scientists affiliated with U.S. institutions are eligible to apply.
Restrictions: Clinical research is not considered for support.
Geographic Focus: All States
Contact: William E. Zamer; (703) 292-7894; fax (703) 292-9154; wzamer@nsf.gov
Internet: http://www.nsf.gov/funding/pgm_summ.jsp?pims_id=12730
Sponsor: National Science Foundation
4201 Wilson Boulevard
Arlington, VA 22230

NSF Planetary Astronomy (PLA) Research Grants **3623**
The grants award research support for theoretical and observational studies of the detailed structure and composition of planetary surfaces, interiors, atmospheres, and satellites; the nature of small bodies (asteroids and comets); and the origin and development of the solar system. All applications must be submitted electronically via FastLane.
Requirements: Scientists initiate research proposals, which are usually submitted on their behalf by their employing institutions.
Geographic Focus: All States
Date(s) Application is Due: Nov 15
Contact: Nigel A. Sharp; (703) 292-4905; fax (703) 292-9034; nsharp@nsf.gov
Internet: http://www.nsf.gov/funding/pgm_summ.jsp?pims_id=13630
Sponsor: National Science Foundation
4201 Wilson Boulevard
Arlington, VA 22230

NSF Plant, Fungal and Microbial Developmental Systems Grants **3624**
The Developmental Systems Cluster supports research aimed at understanding how interacting developmental processes give rise to the emergent properties of organisms. A systems level approach to understanding these processes, at the molecular, cellular, and organismal levels of organization, requires the use of molecular, genetic, biochemical, and physiological techniques as well as techniques from outside biology. The Plant, Fungal and Microbial Development programmatic area supports research that addresses developmental processes in plants from algae to angiosperms, microbes and fungi.
Requirements: The most frequent recipients of support are academic institutions and nonprofit research groups. In special circumstances, grants also are awarded to other types of institutions and to individuals.
Geographic Focus: All States
Date(s) Application is Due: Jan 12; Jul 12
Contact: Thomas Jack; (703) 292-8417; fax (703) 292-9153; tjack@nsf.gov
Internet: http://www.nsf.gov/funding/pgm_summ.jsp?pims_id=501087
Sponsor: National Science Foundation
4201 Wilson Boulevard
Arlington, VA 22230

NSF Plant Genome Research Program (PGRP) **3625**
The goals of this program are to support basic research in plant genomics, and to accelerate the acquisition and utilization of new knowledge and innovative approaches to elucidate fundamental biological processes in plants. The focus is on plants of economic importance and plant processes of potential economic value. NSF 06-581
Requirements: Proposals are invited from U.S. academic institutions, U.S. non-profit research organizations including museums, research laboratories, professional societies and similar organizations in the U.S. that are directly associated with educational or research activities, and consortia of only the eligible organizations listed here. When a consortium of eligible organizations submits a proposal, it must be submitted as a single proposal with one organization serving as the lead and all other organizations as subawardees. Separately submitted collaborative proposals will not be accepted and will be returned without review.
Restrictions: Organizations ineligible to submit to this program solicitation may not receive subawards. There continues to be a limit on the number of proposals in which an investigator can be included. An investigator may only submit one proposal per fiscal year as a principal investigator or a co-principal investigator in response to this Program Solicitation. Proposals received in excess of this limit may be returned without review.
Geographic Focus: All States
Date(s) Application is Due: Jan 21; Oct 26
Contact: Jane Silverthorne, Program Director; (703) 292-8470; dbipgr@nsf.gov

Internet: http://www.nsf.gov/pubs/2006/nsf06581/nsf06581.htm
Sponsor: National Science Foundation
4201 Wilson Boulevard
Arlington, VA 22230

NSF Plasma Physics Grants 3626

The plasma physics program jointly with DoE funds research in the fundamental physics of plasmas. Research areas include plasma turbulence and shocks, turbulent and nonlocal, collisional transport with and without strong magnetic fields, non-neutral plasmas, cold plasmas, strongly-coupled and dusty plasmas, laser-plasma interactions, ultra-short pulse and/or ultra-intense laser plasma interactions, high-energy-density plasmas, and low temperature plasmas. Both theoretical and experimental research is included. Note, that there is often coordination between the Physics Division and other organizations within the Mathematical and Physical Science Directorate as well as in the Engineering and Geophysical Sciences Directorates to co-review proposals that are plasma related but cross disciplinary boundaries.
Geographic Focus: All States
Contact: Richard L. Berger, (703) 292-2183; fax (703) 292-9078; rlberger@nsf.gov
Internet: http://www.nsf.gov/funding/pgm_summ.jsp?pims_id=503252
Sponsor: National Science Foundation
4201 Wilson Boulevard
Arlington, VA 22230

NSF Polar Programs Division Research Grants 3627

The program seeks to strengthen and enhance the national scientific enterprise through the expansion of fundamental knowledge and increased understanding of the polar regions; to encourage and support basic research in the arctic and antarctic focused on the solid earth, glacial and sea ice, terrestrial ecosystems, the oceans, the atmosphere, and beyond. Major objectives include understanding of the natural phenomena and processes in the Antarctic and Arctic regions. Support is also provided for science and technology centers, undergraduate student research, facility enhancement, instrumentation, and laboratory equipment.
Requirements: Eligible applicants include public and private colleges and universities; nonacademic research institutions; private profit organizations; and, under special circumstances, unaffiliated scientists.
Geographic Focus: All States
Contact: Karl A. Erb; (703) 292-8030; fax (703) 292-9081; kerb@nsf.gov
Internet: http://www.nsf.gov/funding/pgm_list.jsp?org=OPP
Sponsor: National Science Foundation
4201 Wilson Boulevard
Arlington, VA 22230

NSF Polymers Research Grants 3628

The program supports basic research and education on polymeric materials. The program portfolio is mainly experimental and highly diverse with components of materials science, chemistry, physics, and related disciplines. Polymers are studied across the nano-to-macro continuum through fundamental materials-focused scientific approaches. Areas addressed include synthesis, molecular assembly, characterization, phase behavior, structure, morphology, and properties. Particular focus is on new materials or materials with novel and/or superior properties, as well as on fundamental understanding of structure-property relationships. Proposals of high potential impact for research and education in the above areas are welcome.
Geographic Focus: All States
Contact: Andrew Lovinger, Program Director; (703) 292-4933; alovinge@nsf.gov
Internet: http://www.nsf.gov/funding/pgm_summ.jsp?pims_id=5357
Sponsor: National Science Foundation
4201 Wilson Boulevard
Arlington, VA 22230

NSF Population and Evolutionary Processes Grants 3629

The Population and Evolutionary Processes Cluster supports research on population properties that lead to variation within and among populations and among species. Approaches include empirical and theoretical studies of microevolution, organismal adaptation, geographical differentiation, natural hybridization and speciation, as well as processes that lead to macroevolutionary patterns of trait evolution. Please note that studies focusing on interactions among species should be directed to the Ecological Biology Cluster. The Population and Evolutionary Processes Cluster funds projects within the Population and Evolutionary Processes Program as well as CAREER, OPU.S. and LTREB.
Geographic Focus: All States
Date(s) Application is Due: Jan 9; Jul 9
Contact: Nancy J. Huntly, Program Director; (703) 292-8980; nhuntly@nsf.gov
Internet: http://www.nsf.gov/funding/pgm_summ.jsp?pims_id=12824
Sponsor: National Science Foundation
4201 Wilson Boulevard
Arlington, VA 22230

NSF Postdoctoral Research Fellowships in Biology (PRFB) 3630

The Directorate for Biological Sciences (BIO) awards Postdoctoral Research Fellowships in Biology to recent recipients of the doctoral degree for research and training in selected areas of biology supported by BIO to encourage independence early in their research careers and to permit them to pursue their research and training goals in the most appropriate research locations regardless of the availability of funding for the Fellows at that site. The fellowships are also designed to provide active mentoring of the Fellows by the sponsoring scientists who will benefit from having additional members in their research groups. The research and training plan of each fellowship must address important scientific questions in contemporary biology within the scope of the BIO Directorate and the specific guidelines in this fellowship program solicitation. Because the fellowships are offered only to postdoctoral scientists early in their careers, doctoral advisors are encouraged to discuss the availability of BIO fellowships with their graduate students early in their doctoral programs. Fellowships are awards to individuals, not institutions, and are administered by the Fellows. Fellowship stipends are being adjusted on a yearly basis. NSF 07-580.
Requirements: Only individuals may apply. NSF postdoctoral fellowships are awards to individuals, and applications are submitted directly by applicants to NSF. However, applications must include sponsoring scientists' statements and the applicants must affiliate with institutions (e.g., colleges and universities, and privately-sponsored nonprofit institutes and museums, government agencies and laboratories, and, under special conditions, for-profit organizations) anywhere in the world. Applicants must: (1) be U.S. citizens (or nationals) or legally admitted permanent residents of the United States (i.e., have a 'green card') at the time of application; (2) earn or plan to earn the doctoral degree in a scientific or engineering field prior to the requested start date of the fellowship; (3) either currently be a graduate student or, at the deadline date, have served in a position requiring the doctoral degree for no more than 12 full time months since earning the degree. There is a one time exception to this criterion for Biological Informatics in November, 2007 only, as described below; (4) must present a research and training plan that falls within the purview of BIO and includes the required information for the specific competition as described below; (5) select a host institution and sponsoring scientist different from the doctoral degree and current position or provide compelling justification for why such a change is not being proposed; (6) not have received Federal funding of more than $20,000 as PI or co-PI (except graduate fellowships and doctoral dissertation improvement grants); (7) not have submitted concurrently the same project to another NSF program; (8) not be a named participant on any other proposal submitted to NSF, including regular research proposals, concurrent with the fellowship application, regardless of who is the named principal investigator; AND (9) meet other limitations on eligibility imposed within the specific competitive areas, if any. There is no limit on the number of applicants that an institution may host. For now, submission can only be made using NSF's FastLane system, given that Grants.gov does not support all required forms. Applications are due the first Monday in November by 5 p.m. proposer's local time.
Geographic Focus: All States
Date(s) Application is Due: Nov 0
Amount of Grant: 60,000-66,000 USD
Contact: Peter McCartney, Program Director; (703) 292-8470; pmccartn@nsf.gov
Internet: http://www.nsf.gov/publications/pub_summ.jsp?ods_key=nsf07580&org=BIO
Sponsor: National Science Foundation
4201 Wilson Boulevard
Arlington, VA 22230

NSF Presidential Awards for Excellence in Mathematics and Science Teaching 3631

This program provides national recognition of outstanding elementary, middle, and secondary school teachers. Every year, two mathematics teachers and two science teachers are selected for the honor from each state and each of the four jurisdictions listed. The program seeks to enhance the status and visibility of the teaching profession by identifying and rewarding exemplary teachers. Awardees are brought to Washington, DC, for a week of activities and professional interaction. The continuation of this interaction is encouraged, and opportunities for individual and collective leadership are facilitated. The deadline for nomination within each state is usually mid-March. Twelve state/jurisdiction finalists are forwarded to a national section committee that selects the four presidential awardees. In addition to the award, a grant from NSF is included to be used at the teacher's discretion to improve science and/or mathematics education at the awardee's home institution. Annual deadline dates may vary; contact program staff for exact dates. NSF 04-578
Requirements: Eligible applicants are teachers of K-12 science and/or mathematics in a public or private school who plan to teach for the following year. Teachers must also be full-time employees and have at least five years of K-12 teaching experience in science and/or mathematics.
Geographic Focus: All States
Date(s) Application is Due: May 1
Amount of Grant: 10,000 USD
Contact: Mark Saul; (703) 292-5092; fax (703) 292-9044; msaul@nsf.gov
Internet: http://www.nsf.gov/funding/pgm_summ.jsp?pims_id=12802
Sponsor: National Science Foundation
4201 Wilson Boulevard
Arlington, VA 22230

NSF Presidential Awards for Excellence in Science, Math, and Engineering Mentoring 3632

The PAESMEM Program seeks to identify outstanding mentoring efforts that enhance the participation of groups (i.e., women, minorities, and persons with disabilities) that are underrepresented in science, technology, engineering, and mathematics. This program provides educational opportunities for Undergraduate Students, Graduate Students, Postdoctoral Fellows, K-12 Educators. The awardees serve as leaders in the national effort to develop fully the Nation's human resources in science, technology, engineering, and mathematics.
Requirements: Individuals and institutions must be nominated by a colleague, student, administrator, or another organization.
Geographic Focus: All States
Date(s) Application is Due: May 2
Amount of Grant: 10,000 USD
Contact: David L. Temple, (703) 292-5111; fax (703) 292-9018; dtemple@nsf.gov

Internet: http://www.nsf.gov/funding/pgm_summ.jsp?pims_id=5473&org=EHR&from=fund
Sponsor: National Science Foundation
4201 Wilson Boulevard
Arlington, VA 22230

NSF Presidential Early Career Awards for Scientists and Engineers Grants 3633
Each year NSF selects nominees for the Presidential Early Career Awards for Scientists and Engineers (PECASE) from among the most meritorious new CAREER awardees. Selection for this award is based on two important criteria: innovative research at the frontiers of science and technology that is relevant to the mission of the sponsoring organization or agency, and community service demonstrated through scientific leadership, education or community outreach. These awards foster innovative developments in science and technology, increase awareness of careers in science and engineering, give recognition to the scientific missions of the participating agencies, enhance connections between fundamental research and national goals, and highlight the importance of science and technology for the Nation's future. Individuals cannot apply for PECASE. These awards are initiated by the participating federal agencies. At NSF, up to twenty nominees for this award are selected each year from among the PECASE-eligible CAREER awardees who are most likely to become the leaders of academic research and education in the twenty-first century. The White House Office of Science and Technology Policy makes the final selection and announcement of the awardees.
Requirements: Applicant must be employed at an institution in the United States, its territories or possessions, or Puerto Rico, that awards a baccalaureate or advanced degree in a field supported by NSF; be in their first or second full-time tenure-track academic appointment and have begun the first tenure-track or appointment (at any institution) on or after October 1, 1996, and before October 1, 2001; not be tenured or have held tenure on or before October 1, 2001; and not be a previous recipient of a CAREER or PECASE award. Contact program officer for exact requirements.
Restrictions: No salary support for other Senior Personnel is permitted, in either the primary budget or in any sub awards.
Geographic Focus: All States
Date(s) Application is Due: Jul 21; Jul 22; Jul 23
Amount of Grant: 400,000-500,000 USD
Contact: Elizabeth L. Rom; (703) 292-7709; fax (703) 292-9085; elrom@nsf.gov
Internet: http://www.nsf.gov/funding/pgm_summ.jsp?pims_id=5262
Sponsor: National Science Foundation
4201 Wilson Boulevard
Arlington, VA 22230

NSF Probability Research Grants 3634
The Probability Program supports research on the theory and applications of probability. Subfields include discrete probability, stochastic processes, limit theory, interacting particle systems, stochastic differential and partial differential equations, and Markov processes. Research in probability which involves applications to other areas of science and engineering is especially encouraged.
Geographic Focus: All States
Date(s) Application is Due: Nov 7
Contact: Tomek Bartoszynski, (703) 292-4885; fax (703) 292-9032; tbartosz@nsf.gov
Internet: http://www.nsf.gov/funding/pgm_summ.jsp?pims_id=5555&org=DMS&from=home
Sponsor: National Science Foundation
4201 Wilson Boulevard
Arlington, VA 22230

NSF Process and Reaction Engineering Grants 3635
The Process and Reaction Engineering program supports fundamental and applied research on: rates and mechanisms of important classes of catalyzed and uncatalyzed chemical reactions as they relate to the design, production, and application of catalysts, chemical processes, biochemical processes, and specialized materials; chemical and biochemical phenomena occurring at or near solid surfaces and interfaces; electrochemical and photochemical processes of engineering significance or with commercial potential; design and optimization of complex chemical and biochemical processes; dynamic modeling and control of process systems and individual process units; reactive processing of polymers, ceramics, and thin films; and interactions between chemical reactions and transport processes in reactive systems, and the use of this information in the design of complex chemical and biochemical reactors.
Geographic Focus: All States
Date(s) Application is Due: Mar 1; Sep 15
Contact: Maria Burka, (703) 292-7030; mburka@nsf.gov
Internet: http://www.nsf.gov/funding/pgm_summ.jsp?pims_id=13361
Sponsor: National Science Foundation
4201 Wilson Boulevard
Arlington, VA 22230

NSF Project 2010 Grants 3636
The Directorate for Biological Sciences (BIO) announces its intention to continue support of research to determine the functions of all genes in the model plant Arabidopsis thaliana by the year 2010. Individual investigators or groups of investigators will be supported to conduct creative and innovative, genome-wide or systems-level research designed to determine, using all available means, the functions of Arabidopsis genes. This year, as in the past, the Program will focus on: research on exemplary networks using high throughput methods and integrating modeling with experimental data to understand the gene circuitry underlying basic plant processes; projects that will develop experimental and computational methods, tools, and resources for enabling a broad community of scientists to conduct functional genomics research on Arabidopsis; and projects to perform genome-wide analyses of the gene function. Approximately 12 grants totaling $1,000,000 will be awarded. Full proposals should be submitted via FastLane or Grants.gov.
Requirements: Proposals are accepted only from U.S. academic institutions, U.S. non-profit research organizations including museums, research laboratories, professional societies and similar organizations in the U.S. that are directly associated with educational or research activities, and consortia of such organizations with appropriate research and educational facilities. The eligibility criteria apply to both the main and sub-awardees.
Geographic Focus: All States
Date(s) Application is Due: Jan 21
Contact: Parag R. Chitnis; (703) 292-8443; fax (703) 292-9061; bio-2010@nsf.gov
Internet: http://www.nsf.gov/funding/pgm_summ.jsp?pims_id=5337
Sponsor: National Science Foundation
4201 Wilson Boulevard
Arlington, VA 22230

NSF Research Coordination Networks in Biological Sciences Grants 3637
The goal of this program is to encourage and foster communications and collaborations among scientists with common goals and interests. Groups of investigators are supported to communicate and coordinate their research efforts across disciplinary, organizational, institutional, and geographical boundaries. The proposed networking activities should have a theme as a focus of its collaboration. The focus could be on a broad research question, a specific group of organisms, or particular technologies or approaches. Innovative ideas for implementing novel networking strategies to promote research coordination and collaboration that enable new research directions or advancement of a field are especially encouraged. Although the research coordination networks are expected to be multi-institutional, a single institution must serve as the submitting institution.
Requirements: U.S. academic institutions and U.S. nonprofit research institutions are eligible.
Geographic Focus: All States
Amount of Grant: 50,000-100,000 USD
Contact: Peter H. McCartney, Program Director; (703) 292-8470; pmccartn@nsf.gov
Internet: http://www.nsf.gov/funding/pgm_summ.jsp?pims_id=11691
Sponsor: National Science Foundation
4201 Wilson Boulevard
Arlington, VA 22230

NSF Research Experiences for Undergraduates (REU) Grants 3638
The Research Experiences for Undergraduates (REU) program supports active research participation by undergraduate students in any of the areas of research funded by the National Science Foundation. REU projects involve students in meaningful ways in ongoing research programs or in research projects specifically designed for the REU program. This solicitation features two mechanisms for support of student research: (1) REU Sites are based on independent proposals to initiate and conduct projects that engage a number of students in research. REU Sites may be based in a single discipline or academic department, or on interdisciplinary or multi-department research opportunities with a coherent intellectual theme. Proposals with an international dimension are welcome. A partnership with the Department of Defense supports REU Sites in DoD-relevant research areas. (2) REU Supplements may be requested for ongoing NSF-funded research projects or may be included as a component of proposals for new or renewal NSF grants or cooperative agreements. NSF is particularly interested in increasing the numbers of women, underrepresented minorities, and persons with disabilities in research. REU projects are strongly encouraged to involve students who are members of these groups. (Underrepresented minorities are African Americans, Hispanics, American Indians, Alaska Natives, and Native Hawaiians or Other Pacific Islanders.) Historically, the vast majority of REU participants have been junior- or senior-level undergraduates—students who have typically already committed to a major in science or engineering. So that the REU program can succeed in attracting students into science and engineering who might not otherwise consider those majors and careers, projects are also encouraged, when appropriate, to involve students at earlier stages in their college experience. NSF anticipates investing approximately $57 million (pending availability of funds) in approximately 150 new Site awards and 1,600 new Supplement awards. REU Site awards have durations of one to five years, with a three-year duration being typical in most NSF directorates. The term of REU Supplements may not exceed that of the underlying research project. NSF 05-592
Requirements: NSF welcomes proposals on behalf of all qualified scientists, engineers and educators. The Foundation strongly encourages women, minorities and persons with disabilities to participate fully in its programs. Undergraduate student participants in either Sites or Supplements must be citizens or permanent residents of the United States or its possessions. An undergraduate student is a student who is enrolled in a degree program (part-time or full-time) leading to a baccalaureate or associate degree. Students who are transferring from one college or university to another and are enrolled at neither institution during the intervening summer may participate. High school graduates who have been accepted at an undergraduate institution but who have not yet started their undergraduate study are also eligible to participate. For REU Sites, a significant fraction of the student participants should come from outside the host institution or organization. Some NSF directorates encourage inclusion in the REU program of K-12 teachers of science, technology, engineering, and mathematics.
Restrictions: Students who have received their bachelor's degrees and are no longer enrolled as undergraduates are generally not eligible to participate.
Geographic Focus: All States
Date(s) Application is Due: Jun 5; Aug 18; Sep 13
Contact: Elizabeth L. Rom; (703) 292-7709; fax (703) 292-9085; elrom@nsf.gov
Internet: http://www.nsf.gov/funding/pgm_summ.jsp?pims_id=5517
Sponsor: National Science Foundation

4201 Wilson Boulevard
Arlington, VA 22230

NSF Research on Gender in Science and Engineering (GSE) Grants **3639**

The program seeks to broaden the participation of girls and women in all fields of science, technology, engineering, and mathematics (STEM) education by supporting research, the diffusion of research-based innovations, and extension services in education that will lead to a larger and more diverse domestic science and engineering workforce. Typical projects will contribute to the knowledge base addressing gender-related differences in learning and in the educational experiences that affect student interest, performance, and choice of careers; how pedagogical approaches and teaching styles, curriculum, student services, and institutional culture contribute to causing or closing gender gaps that persist in certain fields. Projects will communicate and apply findings, evaluation results, and proven good practices and products to a wider community. It is estimated that 15 to 22 grants will be awarded per year; a mix of Research Awards, Diffusion of Research-based Innovations Awards, and Extension Services Awards. Research grants will be up to 3 years. Diffusion of Research-based Innovations grants will be up to 3 years. Extension Services grants are for five years, with years 4 and 5 depending on performance. The total anticipated funding amount is $5,000,000 for new grants in all tracks, pending availability of funds. NSF 07-578

Requirements: The categories of proposers eligible to submit proposals to the National Science Foundation are identified in the Grant Proposal Guide, Chapter I, Section E. A Preliminary Proposal is required prior to submitting a Full Proposal. (Deadlines for Preliminary Proposals: Research Proposals-Nov. 5, Extension Services Proposals-Jan. 7.) The program seeks to build resources-developing the nation's knowledge capital, social capital, and human capital — toward the goal of broadening the participation of girls and young women in STEM education from kindergarten through undergraduate education. The following types of projects will be considered: (1) Research projects: investigate factors behind the underrepresentation of girls and women in STEM education; societal, formal and informal educational systems' interaction with individuals that encourage or discourage interest and persistence in study or careers in certain STEM fields. (2) Diffusion of Research-Based Innovation projects: provide a mechanism for informing a wider audience (e.g., teachers, faculty, guidance counselors, parents, etc.) about issues, research findings, and strategies for changing educational practice. (3) Extension Services: provide training and consulting services to educators and institutions, to enable them to adopt and embed proven gender-inclusive policies and practices in pedagogy, the design of curriculum materials, student support programs, educator and faculty development. Extension services employ a 'train-the-trainer' model and are based on a 'unified program of change.' The program does not currently fund intervention or education projects that directly serve students as their primary purpose. Research projects may involve students as subjects only if the intervention is an integral part of gathering data and if the findings from the intervention would substantially answer the research questions posed within the context of theory and hypotheses. There should be meaningful control groups also included in the design. Those wishing to undertake direct intervention or education service projects are encouraged to search the NSF web site and other publications for appropriate funding programs. All proposals submitted to GSE are already intended to broaden participation in STEM. Proposals that, in addition, include a partnership with institutions serving underrepresented populations may request up to $100,000 more (for a total of $350,000) for Diffusion of Research-Based Innovation activities. The partnership must be documented with, at a minimum, strong letters of support from the partnering institutions.

Geographic Focus: All States
Date(s) Application is Due: Jan 7; Feb 25; Apr 7; Nov 5
Contact: Tayana Casseus, Program Assistant; (703) 292-4684; tcasseus@nsf.gov
Internet: http://www.nsf.gov/funding/pgm_summ.jsp?pims_id=5475
Sponsor: National Science Foundation
4201 Wilson Boulevard
Arlington, VA 22230

NSF Research to Aid Persons with Disabilities (RAPD) Grants **3640**

The Research to Aid Persons with Disabilities (RAPD) program supports research that will lead to the development of new technologies, devices, or software for persons with disabilities. Research may be supported that is directed to the characterization, restoration, and/or substitution of human functional ability or cognition, or to the interaction of persons with disabilities and their environment. Areas of particular recent interest are disability-related research in neuroscience/neuroengineering and rehabilitation robotics. Emphasis is placed on significant advancement of fundamental engineering and scientific knowledge and not on incremental improvements. Proposals should advance discovery or innovation beyond the frontiers of current knowledge in disability-related research. Applicants are encouraged to contact a program director prior to submitting a proposal.

Geographic Focus: All States
Date(s) Application is Due: Mar 1; Sep 15
Contact: Ted A. Conway, (703) 292-7091; tconway@nsf.gov
Internet: http://www.nsf.gov/funding/pgm_summ.jsp?pims_id=501021
Sponsor: National Science Foundation
4201 Wilson Boulevard
Arlington, VA 22230

NSF SBE Doctoral Dissertation Research Improvement Grants **3641**

This program award grants to doctoral students to improve the quality of dissertation research. These grants provide funds for items not normally available through the student's university. Additionally, these grants allow doctoral students to undertake significant data-gathering projects and to conduct field research in settings away from their campus that would not otherwise be possible. Proposals are judged on the basis of their scientific merit, including the theoretical importance of the research question and the appropriateness of the proposed data and methodology to be used in addressing the question. NSF 06-605

Requirements: The student applying must be enrolled at a U.S. academic institution, but need not be a U.S. citizen. The proposal must be submitted by the dissertation advisor(s) on behalf of the graduate student who is at the point of initiating or already conducting dissertation research. Proposals from women, minorities, and persons with disabilities are strongly encouraged. Items such as budget limitations, target dates and/or deadlines, page length restrictions, and review procedures vary widely across programs. Please consult the relevant program's webpage for specific information and contact the program director if necessary.

Restrictions: This program does not support research with disease-related goals, including research on the etiology, diagnosis, or treatment of physical or mental disease, abnormality, or malfunction of human beings, animals or plants. Funds may not be used as a stipend for the student, for tuition, textbooks, journals, or for the typing, reproduction, or publication costs of the student's dissertation.

Geographic Focus: All States
Amount of Grant: Up to 12,500 USD
Contact: Program Coordinator; (703) 292-5111
Internet: http://nsf.gov/funding/pgm_summ.jsp?pims_id=13453&org=NSF&sel_org=NSF&from=fund
Sponsor: National Science Foundation
4201 Wilson Boulevard
Arlington, VA 22230

NSF Scholarships in Science, Technology, Engineering, and Mathematics **3642**

This program supports scholarships for academically talented but financially needy students, enabling them to pursue associate, baccalaureate, or graduate-level degrees in science, technology, engineering, or mathematics. Academic institutions apply for awards to support scholarship activities, and are responsible for selecting scholarship recipients, reporting demographic information about student scholars, and managing the S-STEM project at the institution. An optional letter of intent is requested by January 15, October 10 or July 10, depending on the competition period; full proposal is due January 28. Annual deadline dates may vary; contact program staff for exact dates. NSF 07-524

Requirements: Proposals may only be submitted by Institutions of higher education (as defined in section 101 (a) of the Higher Education Act of 1965) in the United States and its territories that grant associate, baccalaureate, or graduate degrees in the following disciplines: biological sciences (except medicine and other clinical fields); physical sciences, including physics, chemistry, astronomy, and materials science; mathematical sciences; computer and information sciences; geosciences; engineering; technology areas associated with the preceding fields (for example, biotechnology, chemical technology, engineering technology, information technology, etc.). The Principal Investigator must be a faculty member currently teaching within one of the S-STEM disciplines who can provide the leadership required to ensure the success of the project. Projects involving more than one department within an institution are eligible, but a single Principal Investigator must accept overall management responsibility. Other members of the S-STEM project management team may be listed as Co-Principal investigators.

Restrictions: An institution may submit no more than one proposal per competition.

Geographic Focus: All States
Date(s) Application is Due: Jan 15; Jul 10; Oct 10
Amount of Grant: Up to 600,000 USD
Contact: Pratibha Varma-Nelson, (703) 292-4653; pvarmane@nsf.gov
Internet: http://www.nsf.gov/funding/pgm_summ.jsp?pims_id=5257
Sponsor: National Science Foundation
4201 Wilson Boulevard
Arlington, VA 22230

NSF Science and Society (S&S) **3643**

The program seeks proposals that examine questions that arise in the interactions of engineering, science, technology, and society. There are four components: Ethics and Values in Science, Engineering and Technology (EVS); History and Philosophy of Science, Engineering and Technology (HPS); Social Studies of Science, Engineering and Technology (SSS); and Studies of Policy, Science, Engineering and Technology (SPS). The components overlap, but are distinguished by the different scientific and scholarly orientations they take to the subject matter, as well as by different focuses within the subject area. This program offers the following modes of support: 1) S&S Scholars Awards; 2) Standard Research Grants and Grants for Collaborative Research; 3) S&S Postdoctoral Fellowships; 4) S&S Professional Development Fellowships; 5) Doctoral Dissertation Research Improvement Grants; 6) Small Grants for Training and Research; 7) Conference and Workshop Awards; 8) Other Funding Opportunities.

Requirements: 1) Scholar Awards: Normally made to U.S. academic institutions, although an individual who is not affiliated with an appropriate U.S. academic institution may submit a proposal as an independent scholar. In that case, the scholar must be a U.S. citizen or national, or have permanent resident status. Awards may provide support for full-time academic year (nine months) research, including salary, fringe benefits, and other direct costs, up to a ceiling that is ordinarily $70,000 for total direct costs. 2) Standard Research Grants and Grants for Collaborative Research: These awards are made to U.S. academic institutions. Awards can include proposals for research, infrastructure or education projects. These proposals ordinarily do not require full-time investigator support like that for Scholars' Awards. These grants can also support projects that require several investigators, advisors, or collaboration among Principal Investigators, including investigators at different institutions. They may also involve postdoctoral researchers, or graduate or undergraduate student assistants. 3) Postdoctoral Fellowships: Available for S&S researchers within 5 years of receipt of the Ph.D. degree. Fellows must be U.S. citizens or nationals, or have permanent resident status. Proposals must describe both a

training and a research component, and the site for the Fellowship must be different from the institution where the Fellow received the Ph.D. degree. 4) Professional Development Fellowships: All Fellows must be U.S. citizens or nationals, or have permanent resident status. These fellowships are available for researchers trained in all areas of Science and Society who wish to improve and expand their skills in the areas of science or engineering, and conversely for physical and natural scientists and engineers who desire training in S&S disciplines. 5) Doctoral Dissertation Research Improvement Grants: Doctoral students who are enrolled in U.S. graduate programs are eligible to apply. The dissertation advisor is the principal investigator. Doctoral students must have passed the qualifying exams, completed all course work required for the degree, and had the dissertation topic approved prior to receiving the award. These awards provide funds for dissertation research expenses not normally available through the student's university. The dissertation advisor is the principal investigator on these proposals; the doctoral student should be listed as co-principal investigator. 6) Small Grants for Training and Research (SGTR): Available for postdoctoral researchers within 5 years of receipt of the Ph.D. and for graduate students who are regularly admitted students in S&S graduate programs. All Postdoctoral Fellows must be U.S. citizens or nationals, or have permanent resident status. 7) Conference and Workshop Support: Can help to support national and international conferences, symposia, and research workshops that enable scientists, engineers, researchers in S&S areas of support, policy makers, and representatives of interested groups to develop, evaluate, and share new research findings. Also supports projects on the interactions of engineering, science, technology and society that emphasize capacity building. Such activities can include national summer workshops for graduate students or faculty, or projects by professional societies to develop concentrations in the ethical, philosophical, historical and social context of science and engineering for undergraduate or graduate level science and engineering students. 8) Other Funding: may provide supplemental funding to existing awards in order to create research experiences for undergraduates (REU). See the REU Announcement in the listings of NSF funding opportunities.
Geographic Focus: All States
Date(s) Application is Due: Feb 1; Aug 1
Contact: Stephen Zehr, (703) 292-7318; szehr@nsf.gov
Internet: http://nsf.gov/funding/pgm_summ.jsp?pims_id=5324&org=NSF&sel_org=NSF&from=fund
Sponsor: National Science Foundation
4201 Wilson Boulevard
Arlington, VA 22230

NSF Science and Technology Studies (STS) Grants **3644**
This program supports research and related activities that contribute to systematic understanding of the character and development of science and technology, including their cultural, intellectual, material,and social dimensions. The program supports research on the nature and development of technology and science, both in the past and present; on differences in the nature of theory and evidence in various fields of science and engineering; and on the interactions among science, technology and society. Proposals are welcomed from various disciplinary perspectives, including history of science, history of technology, philosophy of science, and various social sciences, including sociology, anthropology, and political science. Fastlane submission is required. NSF-01-159
Requirements: Scientists, engineers and educators usually initiate proposals that are officially submitted by their employing organization. Before formal submission, the proposal may be discussed with appropriate NSF program staff. Graduate students are not encouraged to submit research proposals, but should arrange to serve as research assistants to faculty members. Some NSF divisions accept proposals for Doctoral Dissertation Research Grants when submitted by a faculty member on behalf of the graduate student. The Foundation also provides support specifically for women and minority scientists and engineers, scientists and engineers with disabilities, and faculty at primarily undergraduate academic institutions.
Geographic Focus: All States
Date(s) Application is Due: Feb 1; Aug 1
Contact: John Perhonis, Program Director; (703) 292-7279; jperhoni@nsf.gov
Internet: http://www.nsf.gov/pubs/2001/nsf01159/nsf01159.html
Sponsor: National Science Foundation
4201 Wilson Boulevard
Arlington, VA 22230

NSF Scientific Computing Research Environments for the Mathematical Sciences (SCREMS) Grants **3645**
The Division of Mathematical Sciences (DMS) of the National Science Foundation plans a limited number of awards for the support of computing environments for research in the mathematical sciences. Scientific Computing Research Environments for the Mathematical Sciences (SCREMS) proposals are for computing environments dedicated to research in the mathematical sciences. Proposals may request support for the purchase of computing equipment and limited support for professional systems administrators or programmer personnel for research computing needs. These grants are intended to support research projects of high quality that require access to advanced computing resources. Requests for routine upgrades of standard desk-environment workstations or laptop computers are not appropriate for this program. Awards are made to provide support for specific research projects rather than to provide general computing capacity. Proposers are encouraged to include projects involving symbolic and algebraic computations, numerical computations and simulations, and graphical representations (visualization) in aid of the research. The total number of awards is estimated between 8 and 12, and the total funding amount is approximately $1,000,000. Estimated program budget, number of awards and average award size/duration are subject to the availability of funds. NSF 07-502
Requirements: Proposals may be submitted by U.S. educational institutions with ongoing research programs in mathematics, applied mathematics, or statistics. Proposals involving inter-institutional or inter-departmental sharing arrangements are welcome. SCREMS proposals are primarily for funding for equipment. If appropriate justification is made, a limited amount of funding may be requested for certain personnel costs. When appropriate and cost-effective, requests for group or departmental servers may be suitable for the SCREMS program. The net costs (after discounts) of the equipment portion should be at least $50,000. Requests to SCREMS may be as high as $200,000, provided a case is made for substantial impact and cost-effectiveness.
Restrictions: Identical proposals may not be submitted to both the SCREMS and MRI (Major Research Instrumentation) solicitations.
Geographic Focus: All States
Date(s) Application is Due: Jan 24
Contact: Florence Rabanal, MPS FastLane Contact; (703) 292-8808
Internet: http://www.nsf.gov/funding/pgm_summ.jsp?pims_id=5616
Sponsor: National Science Foundation
4201 Wilson Boulevard
Arlington, VA 22230

NSF Sedimentary Geology and Paleobiology Grants **3646**
Sedimentary Geology and Paleobiology supports studies of: (1) the changing aspects of life, ecology, environments, and biogeography in past geologic time based on fossil plants, animals, and microbes; (2) all aspects of the Earth's sedimentary carapace insights into geological processes recorded in its historical records and rich organic and inorganic resources locked in rock sequences; (3) the science of dating and measuring the time sequence of events and rates of geological processes of the Earth's past sedimentary and biological record; (4) the geologic record of the production, transportation, and deposition of physical and chemical sediments; and (5) understanding the complexities of Earth's deep time climate systems. The Sedimentary Geology and Paleobiology Program especially encourages integrative studies at the national and international levels that seek to link subdisciplines, such as paleoclimatology, paleogeography, and paleoenvironmental and paleoecologic reconstructions. The program is committed to supporting the most meritorious research in any relevant area, including interdisciplinary and multidisciplinary research, as well as research involving international collaboration. The program is especially interested in proposals in emerging fields.
Requirements: Individuals and organizations in the following categories may submit proposals: (1) Universities and Colleges-Universities and two- and four-year colleges (including community colleges) located and accredited in the U.S., acting on behalf of their faculty members. Such organizations also are referred to as academic institutions. Institutions located outside the U.S. fall under (6) below. (2) Non-profit, non-academic organizations-Independent museums, observatories, research laboratories, professional societies and similar organizations in the U.S. that are directly associated with educational or research activities. (3) For-profit organizations-U.S. commercial organizations, especially small businesses with strong capabilities in scientific or engineering research or education. An unsolicited proposal from a commercial organization may be funded when the project is of special concern from a national point of view, special resources are available for the work, or the proposed project is especially meritorious. NSF is interested in supporting projects that couple industrial research resources and perspectives with those of universities; therefore, it especially welcomes proposals for cooperative projects involving both universities and the private commercial sector. (4) State and Local Governments-State educational offices or organizations and local school districts may submit proposals intended to broaden the impact, accelerate the pace, and increase the effectiveness of improvements in science, mathematics and engineering education in both K-12 and post-secondary levels. (5) Unaffiliated Individuals-Scientists, engineers or educators in the U.S. and U.S. citizens may be eligible for support, provided that the individual is not employed by, or affiliated with, an organization, and (a) the proposer has demonstrated the capability and has access to any necessary facilities to carry out the project; and (b) the proposer agrees to fiscal arrangements that, in the opinion of the NSF Division of Grants & Agreements, ensure responsible management of Federal funds. Unaffiliated individuals should contact the appropriate program before preparing a proposal for submission. (6) Foreign organizations-NSF rarely provides support to foreign organizations. NSF will consider proposals for cooperative projects involving U.S. and foreign organizations, provided support is requested only for the U.S. portion of the collaborative effort.
Restrictions: Proposals that have been declined are not eligible for resubmission for one year from the original date of submission and must be substantially revised to be considered. A proposal that has not been substantially revised will be returned without review as per the NSF Grant Proposal Guide.
Geographic Focus: All States
Date(s) Application is Due: Jan 16; Jul 16
Contact: Paul E. Filmer, Program Director; (703) 292-7858; pfilmer@nsf.gov
Internet: http://www.nsf.gov/funding/pgm_summ.jsp?pims_id=13691
Sponsor: National Science Foundation
4201 Wilson Boulevard
Arlington, VA 22230

NSF Small Business Innovation Research (SBIR) Grants **3647**
The Small Business Innovation Research (SBIR) Program stimulates technological innovation in the private sector by strengthening the role of small business concerns in meeting Federal research and development needs, increasing the commercial application of federally supported research results, and fostering and encouraging participation by socially and economically disadvantaged and women-owned small businesses.
Requirements: Eligible to apply are small business firms—those organized for profit, individually owned or operated, not dominant in the field in which they are bidding, and with an average of no more than 500 employees in all affiliated firms. In addition,

the primary employment of the principal investigator must be with the small business firm at the time of the award.
Geographic Focus: All States
Amount of Grant: 5,000-4,000,000 USD
Contact: Gregory T. Baxter; (703) 292-7795; fax (703) 292-9057; gbaxter@nsf.gov
Internet: http://www.nsf.gov/funding/pgm_summ.jsp?pims_id=503361
Sponsor: National Science Foundation
4201 Wilson Boulevard
Arlington, VA 22230

NSF Small Business Technology Transfer Programs Phase I Grants **3648**
The small business programs stimulate technological innovation in the private sector by strengthening the role of small business concerns in meeting Federal research and development needs, increasing the commercial application of federally supported research results, and fostering and encouraging participation by socially and economically disadvantaged and women-owned small businesses. The Small Business Technology Transfer Program (STTR) requires researchers at universities and other research institutions to play a significant intellectual role in the conduct of each STTR project. These university-based researchers, by joining forces with a small company, can spin-off their commercially promising ideas while they remain primarily employed at the research institution. Letters of Intent are due January 14, with the full proposal deadline February 25.
Requirements: Primary employment must be with the small business concern at the time of an award.
Geographic Focus: All States
Date(s) Application is Due: Jan 14; Feb 25
Amount of Grant: Up to 100,000 USD
Contact: Cheryl F. Albus; (703) 292-7051; fax (703) 292-9057; calbus@nsf.gov
Internet: http://www.nsf.gov/funding/pgm_summ.jsp?pims_id=5527
Sponsor: National Science Foundation
4201 Wilson Boulevard
Arlington, VA 22230

NSF Social Psychology Research Grants **3649**
The Social Psychology Program at NSF supports basic research on human social behavior, including cultural differences and development over the life span. Among the many research topics supported are: attitude formation and change, social cognition, personality processes, interpersonal relations and group processes, the self, emotion, social comparison and social influence, and the psychophysiological and neurophysiological bases of social behavior. The scientific merit of a proposal depends on four important factors: the problems investigated must be theoretically grounded; the research should be based on empirical observation or be subject to empirical validation; the research design must be appropriate to the questions asked; and the proposed research must advance basic understanding of social behavior. Annual full proposal target dates are July 15 and January 15.
Requirements: Except where a program solicitation establishes more restrictive eligibility criteria, individuals and organizations in the following categories may submit proposals: universities and colleges; non-profit, non-academic organizations; for-profit organizations; state and local governments; unaffiliated individuals; foreign organizations; and other Federal agencies.
Geographic Focus: All States, All Countries
Date(s) Application is Due: Jan 15; Jul 15
Contact: Sally Dickerson; (703) 292-7277; fax (703) 292-9068; sdickers@nsf.gov
Internet: http://www.nsf.gov/funding/pgm_summ.jsp?pims_id=5712
Sponsor: National Science Foundation
4201 Wilson Boulevard
Arlington, VA 22230

NSF Sociology Grants **3650**
The Sociology Program supports basic research on all forms of human social organization — societies, institutions, groups and demography — and processes of individual and institutional change. The Program encourages theoretically focused empirical investigations aimed at improving the explanation of fundamental social processes. Included is research on organizations and organizational behavior, population dynamics, social movements, social groups, labor force participation, stratification and mobility, family, social networks, socialization, gender roles, and the sociology of science and technology. The Program supports both original data collections and secondary data analysis that use the full range of quantitative and qualitative methodological tools. The Sociology Program also funds doctoral dissertation research to defray direct costs associated with conducting research, for example, dataset acquisition, additional statistical or methodological training, meeting with scholars associated with original datasets, and fieldwork away from the student's home campus.
Requirements: NSF welcomes proposals on behalf of all qualified scientists, engineers and educators. The Foundation strongly encourages women, minorities and persons with disabilities to participate fully in its programs. Before formal submission, the proposal may be discussed with appropriate NSF program staff. Graduate students are not encouraged to submit research proposals, but should arrange to serve as research assistants to faculty members.
Geographic Focus: All States
Date(s) Application is Due: Feb 15
Contact: Dr. Patricia E. White; (703) 292-8762; fax (703) 292-9195; pwhite@nsf.gov
Internet: http://www.nsf.gov/funding/pgm_summ.jsp?pims_id=5369&org=SES&from=home
Sponsor: National Science Foundation
4201 Wilson Boulevard
Arlington, VA 22230

NSF Solar, Heliospheric, and INterplanetary Environment (SHINE) Grants **3651**
The solar physics program in the Upper Atmospheric Research Section of the Division of Atmospheric Sciences, National Science Foundation (NSF), solicits proposals to investigate the connections between eruptive events and magnetic phenomena on the Sun and the corresponding solar wind structures in the inner heliosphere. SHINE research focuses upon the connections between eruptive events and magnetic phenomena on the Sun and the corresponding solar wind structures in the inner heliosphere. SHINE fosters research on those processes by which magnetic fields and particles produced by the Sun permeate interplanetary space and on the mechanisms by which these fields and particles are transported to geospace through the inner heliosphere. Proposals may be submitted for any funding amount up to $200,000 per year. A maximum duration of five years will be considered in those exceptional cases where the justification and promise are compelling. Typically, SHINE awards are for three years, but proposers may request from one to five years of funding, provided the requested duration is justified. NSF expects to fund approximately 6 to 8 standard or continuing grants annually, depending on the quality of submissions and the availability of funds. NSF 04-585 (replaces Document NSF 02-189)
Requirements: Individuals and organizations in the following categories may submit proposals: (1) Universities and Colleges-Universities and two- and four-year colleges (including community colleges) located and accredited in the U.S., acting on behalf of their faculty members. Such organizations also are referred to as academic institutions; (2) Non-profit, non-academic organizations-Independent museums, observatories, research laboratories, professional societies and similar organizations in the U.S. that are directly associated with educational or research activities; (3) For-profit organizations-U.S. commercial organizations, especially small businesses with strong capabilities in scientific or engineering research or education. An unsolicited proposal from a commercial organization may be funded when the project is of special concern from a national point of view, special resources are available for the work, or the proposed project is especially meritorious. NSF is interested in supporting projects that couple industrial research resources and perspectives with those of universities; therefore, it especially welcomes proposals for cooperative projects involving both universities and the private commercial sector; (4) State and Local Governments-State educational offices or organizations and local school districts may submit proposals intended to broaden the impact, accelerate the pace, and increase the effectiveness of improvements in science, mathematics and engineering education in both K-12 and post-secondary levels; (5) Unaffiliated Individuals-Scientists, engineers or educators in the U.S. and U.S. citizens may be eligible for support, provided that the individual is not employed by, or affiliated with, an organization, and: (a) the proposer has demonstrated the capability and has access to any necessary facilities to carry out the project; and, (b) the proposer agrees to fiscal arrangements that, in the opinion of the NSF Division of Grants & Agreements, ensure responsible management of Federal funds. Unaffiliated individuals should contact the appropriate program before preparing a proposal for submission.
Restrictions: NSF rarely provides support to foreign organizations. NSF will consider proposals for cooperative projects involving U.S. and foreign organizations, provided support is requested only for the U.S. portion of the collaborative effort. Also, NSF does not normally support research or education activities by scientists, engineers or educators employed by Federal agencies or Federally Funded Research and Development Centers (FFRDCs). Under unusual circumstances, other Federal agencies and FFRDCs may submit proposals directly to NSF.
Geographic Focus: All States
Date(s) Application is Due: Aug 27
Amount of Grant: Up to 200,000 USD
Contact: Paul Bellaire; (703) 292-8529; fax (703) 292-9023; pbellair@nsf.gov
Internet: http://www.nsf.gov/publications/pub_summ.jsp?ods_key=nsf04585
Sponsor: National Science Foundation
4201 Wilson Boulevard
Arlington, VA 22230

NSF Solar Terrestrial Research Grants **3652**
This program supports research into processes by which energy in diverse forms is generated by the sun, transported to earth, and ultimately deposited in the terrestrial environment. Major topics include space weather impacts, helioseismology, the solar dynamo, the solar activity cycle, magnetic flux emergence, solar flares and eruptive activity, coronal mass ejections, solar wind heating, solar energetic particles, interactions with cosmic rays, and solar wind/magnetosphere boundary problems. Studies on terrestrial influences include solar spectral irradiance changes, solar constant changes and climactic impacts, C14 and sun/climate connections, solar activity and its effects on the terrestrial environment, and sun/weather variations on various time scales. Proposals may be submitted at any time during the year for all programs except those involving the allocation of observational and computing facilities.
Geographic Focus: All States
Contact: Paul Bellaire; (703) 292-8519; fax (703) 292-9023; pbellair@nsf.gov
Internet: http://www.nsf.gov/funding/pgm_summ.jsp?pims_id=12741&org=NSF&sel_org=NSF&from=fund
Sponsor: National Science Foundation
4201 Wilson Boulevard
Arlington, VA 22230

NSF Starter Grants for Postdoctoral Fellows in Microbial Biology **3653**
The purpose of this program is to support training and research in a host institution on the basic biology of protozoan, microalgal, fungal, archaeal, bacterial, and viral species that are not generally considered to be model organisms (e.g., E. coli, Saccharomyces cerevesiae, and TMV). The use of model organisms in comparative studies with nonmodel organisms is not excluded. Deadline is the first Monday in October. NSF 99-142

Requirements: Applicants must be: U.S. citizens, nationals, or legally admitted permanent resident aliens; may not have earned the doctoral degree more than 18 months prior to the deadline; must propose a research and training plan in microbial biology; may not have previously received a federal research grant; may not submit a research and training plan duplicated in another NSF proposal; and may not have been involved in the research before June 1 of the deadline year. Fastlane submission is required. NSF 99-142
Geographic Focus: All States
Date(s) Application is Due: Oct 1
Amount of Grant: 50,000 USD
Contact: Ms. Carter Kimsey; (703) 292-8470; ckimsey@nsf.gov
Internet: http://www.nsf.gov/funding/pgm_summ.jsp?pims_id=13441
Sponsor: National Science Foundation
4201 Wilson Boulevard
Arlington, VA 22230

NSF Statistics Research Grants **3654**
The Statistics Program supports research in statistical theory and methods, including research in statistical methods for applications to any domain of science and engineering. The theory forms the base for statistical science. The methods are used for stochastic modeling, and the collection, analysis and interpretation of data. The methods characterize uncertainty in the data and facilitate advancement in science and engineering. The Program encourages proposals ranging from single-investigator projects to interdisciplinary team projects.
Geographic Focus: All States
Date(s) Application is Due: Nov 7
Contact: Gabor Szekely; (703) 292-8869; fax (703) 292-9032; wrundell@nsf.gov
Internet: http://www.nsf.gov/funding/pgm_summ.jsp?pims_id=5556&org=DMS&from=home
Sponsor: National Science Foundation
4201 Wilson Boulevard
Arlington, VA 22230

NSF Stellar Astronomy and Astrophysics Research Grants **3655**
The Astronomy and Astrophysics Research Grants (AAG) Program provides individual investigator and collaborative research grants for observational, theoretical, laboratory and archival data studies in all areas of astronomy and astrophysics, including but not limited to the following areas of study: Planetary Astronomy; Galactic Astronomy; and Extragalactic Astronomy and Cosmology. Proposals may span multiple disciplines and/or areas of study and may utilize multiple techniques. Principal Investigators are encouraged to contact one of the Program Officers listed in this announcement prior to submitting a proposal to the AAG Program, particularly if the proposal will include investigators at multiple institutions.
Requirements: Scientists initiate research proposals, which are usually submitted on their behalf by their employing institutions.
Geographic Focus: All States
Date(s) Application is Due: Nov 15
Contact: Nigel A. Sharp; (703) 292-4905; fax (703) 292-9034; nsharp@nsf.gov
Internet: http://www.nsf.gov/funding/pgm_summ.jsp?pims_id=13630
Sponsor: National Science Foundation
4201 Wilson Boulevard
Arlington, VA 22230

NSF Synthesis and Modeling Project of the U.S. Joint Global Ocean Flux Study **3656**
The purpose of this announcement is to invite innovative research proposals to augment the ongoing Synthesis and Modeling Project (SMP) of the U.S. Joint Global Ocean Flux Study (U.S.JGOFS). As part of a multinational oceanographic effort, U.S.JGOFS has as its primary objectives the determination and understanding of processes controlling the biogeochemical cycling of carbon in the ocean and the prediction of response of marine biogeochemical processes to climate change. Now in its fourth year, the SMP is focused on the resolution of major unanswered questions during the course of U.S.JGOFS and related activities and on the modeling of the global ocean carbon cycle. NSF 01-103
Requirements: Scientists, engineers, and educators usually initiate proposals that are officially submitted by their employing organization.
Restrictions: Graduate students are not encouraged to submit research proposals, but should arrange to serve as research assistants to faculty members.
Geographic Focus: All States
Date(s) Application is Due: Sep 13
Amount of Grant: 300,000-430,000 USD
Contact: Phillip Taylor, Program Officer; (703) 306-1587; prtaylor@nsf.gov
Internet: http://www.nsf.gov/pubs/2000/nsf00103/nsf00103.htm
Sponsor: National Science Foundation
4201 Wilson Boulevard
Arlington, VA 22230

NSF Systematic Biology and Biodiversity Inventories Grants **3657**
The Systematic Biology and Biodiversity Inventories Cluster supports research in taxonomy and systematics that contributes to: 1) using phylogenetic methods to understand the evolution of life in time and space, 2) discovery, description, and cataloguing global species diversity, and 3) organizing information from the above in efficiently retrievable forms that best meet the needs of science and society. The Systematic Biology and Biodiversity Inventories Cluster funds projects within the two Programs, Systematic Biology and Biodiversity Inventories, in addition to the PEET and PBI solicitations listed below. In addition, the cluster participates in AToL and other related funding opportunities. The SBBI Cluster continues to encourage and support studies that seek to synthesize available and new species-level taxonomic information in the context of providing revisionary treatments and predictive classifications for particular groups of organisms. NSF PD 047374.
Requirements: The most frequent recipients are academic institutions and nonprofit research groups. In special circumstances, grants also are awarded to other types of institutions and to individuals. Standard NSF Grant Proposal Guidelines apply (see http://www.nsf.gov/publications/pub_summ.jsp?ods_key=gpg).
Geographic Focus: All States
Date(s) Application is Due: Jan 9; Jul 9
Contact: Maureen Kearney, Program Director; (703) 292-7187; mkearney@nsf.gov
Internet: http://www.nsf.gov/publications/pub_summ.jsp?ods_key=pd047374&org=NSF
Sponsor: National Science Foundation
4201 Wilson Boulevard
Arlington, VA 22230

NSF Tectonics Research Grants **3658**
The Tectonics Program supports a broad range of field, laboratory, computational, and theoretical investigations aimed at understanding the evolution and deformation of continental lithosphere through time. Proposals to elucidate the processes that act on the lithosphere at various time-scales and length-scales, either at depth or the surface, are encouraged. Because understanding such large-scale phenomena commonly requires a variety of expertise and methods, the Tectonics Program supports integrated research involving the disciplines of structural geology, petrology, geochronology, sedimentology, stratigraphy, geomorphology, rock mechanics, paleomagnetics, geodesy, and other geophysical techniques.
Requirements: Proposals will be accepted from colleges, universities, and other U.S. institutions with formal research programs in the earth sciences.
Geographic Focus: All States
Date(s) Application is Due: Jun 1; Dec 1
Contact: David M. Fountain; (703) 292-8552; fax (703) 292-9025; dfountai@nsf.gov
Internet: http://www.nsf.gov/funding/pgm_summ.jsp?pims_id=13673&org=NSF&sel_org=NSF&from=fund
Sponsor: National Science Foundation
4201 Wilson Boulevard
Arlington, VA 22230

NSF Theoretical Atomic, Molecular, and Optical Physics (TAMOP) Grants **3659**
The Theoretical Atomic, Molecular, and Optical Physics (TAMOP) program supports theoretical and computational research in all areas of atomic structure, the molecular structure of small molecules, electron, and atomic collisions, photoionization and photodetachment of electrons from atoms and small molecules, time-dependent interactions with atoms and small molecules, quantum optics, ultracold phenomena in Bose and Fermi gases, and quantum information. Investigations primarily directed toward a theoretical understanding of larger molecules or condensed matter systems should be directed toward the appropriate programs in the Division of Chemistry or the Division of Materials Research.
Geographic Focus: All States
Date(s) Application is Due: Sep 30
Contact: Barry I. Schneider; (703) 292-7383; fax (703) 292-9060; bschneid@nsf.gov
Internet: http://www.nsf.gov/funding/pgm_summ.jsp?pims_id=503273
Sponsor: National Science Foundation
4201 Wilson Boulevard
Arlington, VA 22230

NSF Thermal Transport Processes Research Grants **3660**
The Thermal Transport Processes program supports engineering research aimed at gaining a basic understanding of the microscopic and macroscopic levels of thermal transport phenomena (heat and mass transfer) underlying energy conversion and conservation, the synthesis and processing of materials, cooling and heating of infrastructure and equipment, the interaction of industrial processes with the environment, the propulsion of air and land-based vehicles, and thermal phenomena in biological and environmental systems. The program supports fundamental research and engineering education in transport processes that are driven by thermal gradients, and their manipulation to achieve engineering goals. This engineering science forms an important part of the intellectual infrastructure of a wide array of modern technologies.
Geographic Focus: All States
Date(s) Application is Due: Mar 1; Sep 15
Contact: Theodore L. Bergman, (703) 292-7494; tbergman@nsf.gov
Internet: http://www.nsf.gov/funding/pgm_summ.jsp?pims_id=13367
Sponsor: National Science Foundation
4201 Wilson Boulevard
Arlington, VA 22230

NSF Topology Research Grants **3661**
The program supports research on algebraic topology, including homotopy theory, ordinary and extraordinary homology and cohomology, cobordism theory, and K-theory; topological manifolds and cell complexes, fiberings, knots, and links; differential topology and actions of groups of transformations; geometric group theory; and general topology and continua theory.
Geographic Focus: All States
Date(s) Application is Due: Nov 3
Contact: Mark F. Feshbach; (703) 292-5371; fax (703) 292-9032; mfeshbac@nsf.gov
Internet: http://www.nsf.gov/funding/pgm_summ.jsp?pims_id=5551&org=DMS&from=home
Sponsor: National Science Foundation

4201 Wilson Boulevard
Arlington, VA 22230

NSF Tribal Colleges and Universities Program (TCUP) **3662**
This program provides awards to enhance the quality of science, technology, engineering and mathematics(STEM) instructional and outreach programs, with an emphasis on the leveraged use of information technologies at tribal colleges and universities, Alaskan native-serving institutions and native Hawaiian-serving institutions. Support is available for the implementation of comprehensive institutional approaches to strengthen STEM teaching and learning in ways that improve access to, retention within, and graduation from STEM programs, particularly those that have a strong technological foundation. Typical project implementation strategies include curriculum enhancement, faculty professional development, undergraduate research and community service, academic enrichment, infusion of technology to enhance STEM instruction, collaborations, and other activities that meet institutional and community needs. Annual deadline dates may vary; contact program staff for exact dates. Approximately $6 million available funds-9 awards total-approximately 6 implementation awards resulting from Full Proposals will be awarded as cooperative agreements annually, and approximately 3 planning grants resulting from Planning Proposals will be awarded annually. NSF 04-602
Requirements: Tribal colleges and universities, Alaskan Native-serving institutions, and Native Hawaiian-serving institutions are eligible to apply. The Principal Investigator is expected to be the chief academic officer of the institution, or other senior academic officer responsible for oversight and management of curriculum and instructional policies for the institution.
Restrictions: Only one proposal (for either a planning grant or an implementation award) per institution in any one year.
Geographic Focus: All States
Date(s) Application is Due: Feb 5; Nov 20
Contact: Lura Chase; (703) 292-8682; fax (703) 292-9018; lchase@nsf.gov
Internet: http://www.nsf.gov/funding/pgm_summ.jsp?pims_id=5483
Sponsor: National Science Foundation
4201 Wilson Boulevard
Arlington, VA 22230

NSF Undergraduate Research and Mentoring in the Biological Sciences **3663**
This program will fund projects that have strong research and mentoring activities designed to prepare students for successful entry into graduate programs. URM will support projects involving the recruitment, retention and development of undergraduate students, especially those from underrepresented groups, for the purpose of preparing them for graduate study in the biological sciences. Proposed projects are expected to create a URM program that will actively engage students in interesting and exciting research ideas, provide hands-on research experience, and develop their academic skills. URM will enable institutions to create innovative programs that will increase the number and diversity of students who enter graduate research programs in the biological sciences. URM will support well-defined research and mentoring activities that will enable students, especially underrepresented minorities, to become independent thinkers and effective communicators, as well as professional development activities that will expose students to exciting careers in biology. Students are expected to be recruited early in their academic career. The project should frame all student activities around an effective mentoring strategy. Factors that are known to increase the likelihood of success for students to enter, and complete, graduate research programs should be emphasized. Approximately $4,000,000 will be available for this program, subject to the availability of funds. Under the solicitation, requests may be submitted for funding amounts up to a total of $1,000,000 for up to 5 years. However, funding beyond the first year is contingent upon satisfactory progress in the program. The program expects to make at least eight awards, depending upon the quality of submissions and the availability of funds. NSF 06-591
Requirements: Preliminary Proposals (deadline: Sep. 18) are required prior to submitting a Full Proposal (deadline: Mar. 4). Proposals may only be submitted by academic institutions located in the United States. Collaborative Proposals from two or more institutions are allowed. The proposing institution is expected to foster an enriched and intellectually stimulating research and educational environment. URM proposals should involve year-round mentoring and include a major emphasis on direct student participation in research leading to publishable data. URM projects may be based in a single discipline or academic department, or on interdisciplinary or multi-department research opportunities. A proposal should reflect the unique combination of the proposing organization's interests and capabilities and those of any partnering organizations. Partnerships between minority-serving institutions, especially Historically Black Colleges and Universities (HBCU), Tribal Colleges and Universities (TCU), and Hispanic-Serving Institutions (HSI), community colleges, and research intensive institutions are encouraged, if such partnerships will result in a diverse pool of qualified participants as well as an enriched research environment for the students. Undergraduate student participants supported with NSF funds must be citizens or permanent residents of the United States or its possessions. An undergraduate student is one who is enrolled in a degree program (part-time or full-time) leading to a baccalaureate or associates degree. High school graduates who have not yet enrolled, and students who have received their bachelor's degrees and are no longer enrolled as undergraduates, are not eligible.
Restrictions: All collaborative proposals submitted as separate submissions from multiple organizations must be submitted via the NSF FastLane system.
Geographic Focus: All States
Date(s) Application is Due: Mar 4; Sep 18
Amount of Grant: Up to 1,000,000 USD
Contact: URM Program Director; (703) 292-8470; biourm@nsf.gov
Internet: http://www.nsf.gov/funding/pgm_summ.jsp?pims_id=500036
Sponsor: National Science Foundation
4201 Wilson Boulevard
Arlington, VA 22230

NSF University Radio Observatories Program (AST-URO) **3664**
The URO program provides support for operations, technical enhancements, scientific investigations, and student training at university-based radio facilities. A primary function of funded University Radio Observatories is to provide focused research environments that foster hands-on student access to state-of-the-art radio astronomy instrumentation. UROs also serve the general astronomical community by providing open access to 30%-50% of available telescope time. NSF Solicitation 08-543
Requirements: URO awards are for three-year intervals, and are renewable under the proposal review procedures described below. UROs fully funded by NSF are expected to provide the general astronomical community with access to between one-third and one-half of all telescope observing time. UROs should also develop and maintain a strong program of student education and training in radio astronomy, and should conduct public outreach activities appropriate to the location and nature of the facility. Proposals may only be submitted by universities and two- and four-year colleges (including community colleges) located and accredited in the U.S., acting on behalf of their faculty members. New applicants are strongly encouraged to contact the Program Director prior to preparation of a full proposal for an evaluation of the suitability of the proposed project to the URO program.
Geographic Focus: All States
Date(s) Application is Due: May 30
Contact: Richard E. Barvainis; (703) 292-4891; fax (703) 292-9034; rbarvai@nsf.gov
Internet: http://www.nsf.gov/funding/pgm_summ.jsp?pims_id=5292
Sponsor: National Science Foundation
4201 Wilson Boulevard
Arlington, VA 22230

NSF Upper Atmospheric Facilities (UAF) Grants **3665**
The major goal of the Upper Atmospheric Facilities (UAF) is to promote basic research on the structure and dynamics of the Earth's upper atmosphere. Research efforts utilizing these facilities have strong links to the Aeronomy Program and the Magnetospheric Physics Program. The National Science Foundation supports four large incoherent-scatter radar facilities and the SuperDARN coherent scatter radar system. The incoherent-scatter radars are located along a longitudinal chain from Greenland to Peru. Each of the incoherent-scatter facilities is also equipped with powerful optical diagnostic instruments. The SuperDARN consists of a number of coherent-scatter HF radars in both the northern and southern hemispheres.
Requirements: Standard NSF Grant Proposal Guidelines apply (see http://www.nsf.gov/publications/pub_summ.jsp?ods_key=gpg). There are no deadlines or target dates for proposals sent in to any of the Upper Atmospheric Research Section (UARS) base programs. However, it is recommended that PIs try to send in proposals early in the fiscal year. Certain special programs within UARS do have deadlines for proposal submissions.
Geographic Focus: All States
Contact: Robert Robinson; (703) 292-8529; fax (703) 292-9023; rmrobins@nsf.gov
Internet: http://www.nsf.gov/funding/pgm_summ.jsp?pims_id=12808
Sponsor: National Science Foundation
4201 Wilson Boulevard
Arlington, VA 22230

Nuffield Foundation Children and Families Grants **3666**
The foundation was founded for the advancement of health through medical research and teaching and for the care of the elderly through scientific research and education. The Children and Families program supports work to help ensure that the legal and institutional framework is best adapted to meet the needs of children and families. At present, particular interests include (but are not limited to): work that links education and child development, either in the case of adolescent mental health or younger children; work that considers policies relevant to child welfare in a broader institutional context: parents' paid working patterns; childcare and early years provision; work that considers especially the well-being of children growing up in adverse conditions, and what institutional responses may be appropriate; work in family law, including cohabitation, child contact, child support; and work in child protection and placement (adoption and fostering) but only when it raises significant issues. Where a proposal is for a research study, Trustees are interested in the dispassionate examination of evidence. It notes that evidence is likely to be different in different cases, for different types of children and families, and are more likely to support work that takes this approach.
Requirements: Usually grants are made only to United Kingdom organizations, and support work that will be mainly based in the UK, although the Trustees welcome proposals for collaborative projects involving partners in European or Commonwealth countries.
Restrictions: The Foundation does not make grants for the running costs of voluntary bodies but will consider making a contribution to voluntary sector overheads on funded projects.
Geographic Focus: All States, United Kingdom
Date(s) Application is Due: Jan 8; May 7; Sep 3
Amount of Grant: 5,000-150,000 GBP
Contact: Sharon Witherspoon, 44-171-631-0566; fax 44-171-323-4877; sfwpa@nuffieldfoundation.org or info@nuffieldfoundation.org
Internet: http://www.nuffieldfoundation.org/go/grants/cplfj/page_30.html
Sponsor: Nuffield Foundation
28 Bedford Square
London, WC1B 3JS England

Nuffield Foundation Law and Society Grants **3667**

The foundation was founded for the advancement of health through medical research and teaching and for the care of the elderly through scientific research and education. The Foundation has long had an interest in the area of law. Its current objectives in this area are: to promote developments in the legal system that will improve its accessibility to all people; to promote wider access to legal services and advice and a better understanding of the obstacles to access to justice; to fund research and promote developments in alternative dispute resolution; to help promote a greater knowledge of the rights and duties of the individual, including those of the European citizen; to examine the implications of new human rights' obligations on civil (not criminal) justice; and to help promote a greater public understanding of the role of law in society and of the legal system.
Requirements: Usually grants are made only to United Kingdom organizations, and support work that will be mainly based in the UK, although the Trustees welcome proposals for collaborative projects involving partners in European or Commonwealth countries.
Geographic Focus: All States, United Kingdom
Date(s) Application is Due: Jan 8; May 7; Sep 3
Amount of Grant: 5,000-150,000 GBP
Contact: Sharon Witherspoon, 44-171-631-0566; fax 44-171-323-4877; sfwpa@nuffieldfoundation.org or info@nuffieldfoundation.org
Internet: http://www.nuffieldfoundation.org/go/grants/accesstojustice/page_57.html
Sponsor: Nuffield Foundation
28 Bedford Square
London, WC1B 3JS England

Nuffield Foundation Oliver Bird Rheumatism Grants **3668**
The Fund will offer grants to up to five United Kingdom Higher Education Institutions where students of outstanding potential will be recruited into doctoral training. The institutions will provide multidisciplinary environments for researching problems in rheumatic disease, and be able to develop both the students' awareness of the broader context of these diseases and the professional skills required of future scientists. For each studentship the grant will provide: a stipend of not less than 13,500GBP per annum plus United Kingdom fees; research expenses at a rate of 8,000GBP per annum; and a meeting and travel allowance of 1,000GBP per annum. In addition the Fund will provide a grant to each institution of 25,000GBP over the five years of the program to support such activities as: meetings within the Collaborative Center to promote the field and assist the recruitment and training of doctoral students; the hosting of a program meeting to be attended by all program participants from the other Collaborative Centers; and occasional bursaries for interested undergraduates to work alongside the doctoral students and potential supervisors or to attend meetings in the Collaborative Centers.
Geographic Focus: All States, United Kingdom
Date(s) Application is Due: Feb 28
Amount of Grant: 27,500 GBP
Contact: Vicki Hughes, 44-171-631-0566; oliverbird@nuffieldfoundation.org
Internet: http://www.nuffieldfoundation.org/go/grants/oliverbird/page_86.html
Sponsor: Nuffield Foundation
28 Bedford Square
London, WC1B 3JS England

Nuffield Foundation Open Door Grants **3669**
The Foundation keeps an open door to proposals of exceptional merit for research projects or practical innovations that lie outside our main program areas, but that meet Trustees' wider interests. These must have some bearing on the Foundation's widest charitable object. Particular interest is given to projects which identify change or interventions which will have a practical implications for policy or practice, or that will improve the quality of research evidence in areas of public debate. Through the Open Door, the Foundation may also identify emerging areas that justify more sustained attention.
Requirements: Usually grants are made only to United Kingdom organizations, and support work that will be mainly based in the UK, although the Trustees welcome proposals for collaborative projects involving partners in European or Commonwealth countries.
Geographic Focus: All States, United Kingdom
Date(s) Application is Due: Jan 8; May 7; Sep 3
Amount of Grant: 5,000-150,000 GBP
Contact: Sharon Witherspoon, 44-171-631-0566; fax 44-171-323-4877; sfwpa@nuffieldfoundation.org or info@nuffieldfoundation.org
Internet: http://www.nuffieldfoundation.org/go/grants/opendoor/page_115.html
Sponsor: Nuffield Foundation
28 Bedford Square
London, WC1B 3JS England

NWHF Community-Based Participatory Research Grants **3670**
The Northwest Health Foundation supports community-based participatory research as a tool to generate meaningful information about community health and build the capacity of groups united by common challenges to organize and advocate for change. The RFP will seek to support community-based participatory research (CBPR) projects that: build community capacity for policy advocacy; and apply research findings to support policy change that improves community health.
Requirements: 501(c)3 educational institutions and governmental entities are eligible. Generally speaking, foundation programs serve the entire state of Oregon, and the following Washington counties: Clark, Cowlitz, Pacific, Skamania, and Wahkiakum. Specific programs and funds may have additional geographic restrictions. Applying organizations must sign a statement of nondiscrimination in leadership, staffing, and services.
Restrictions: Support will not be considered for basic operations, or for salaries or other expenses associated with ongoing programs.
Geographic Focus: Oregon, Washington
Date(s) Application is Due: Sep 8
Contact: Judith Woodruff; (503) 220-1955; fax (503) 220-1355; judith@nwhf.org
Internet: http://www.nwhf.org/index.php?/apply/cbpr
Sponsor: Northwest Health Foundation
221 NW Second Avenue, Suite 300
Portland, OR 97209

NWHF Mark O. Hatfield Research Fellowship **3671**
The Mark O. Hatfield Research Fellowship Program was established in 2005 to honor the legacy of Senator Mark O. Hatfield, founding Board Chairman of the Northwest Health Foundation (NWHF). The purpose of the Fellowship program is to advance the research career of health practitioners who serve communities in Oregon and southwest Washington. The fellowship is open to health professionals from a variety of disciplines who wish to launch, renew, or expand their research activities in biomedical, applied, or translational research. A secondary purpose is to increase the intellectual resources of research-based institutions. One two-year fellowship of up to $150,000 will be awarded. Funds will be distributed to the fellow's sponsoring institution in two annual disbursements of half of the amount awarded.
Requirements: Eligible candidates must: be a clinical, allied health, public health, or social service provider (early-career researchers specializing in translational research are also encouraged to apply); possess (or be in the process of completing) a doctoral degree or equivalent in an appropriate field related to medicine, nursing, health care, public health, or biomedical sciences; obtain sponsorship from a research-based institution that can act as fiscal manager of the award; and demonstrate a commitment to communities in Oregon and southwest Washington.
Geographic Focus: All States
Date(s) Application is Due: Mar 16
Amount of Grant: Up to 150,000 USD
Contact: R. David Rebanal, Program Officer; (503) 220-1955; drebanal@nwhf.org
Internet: http://www.nwhf.org/index.php?/apply/hatfield
Sponsor: Northwest Health Foundation
221 NW Second Avenue, Suite 300
Portland, OR 97209

NWO Added Value Plus for the Social Sciences Grants **3672**
Advancing knowledge utilization is one of the Netherlands Organization for Scientific Research's (NWO) targets. The policy for encouraging knowledge utilization has various elements. NWO facilitates the awareness of researchers so that third parties can possibly use their research results and can pay due attention to knowledge utilization. Furthermore, NWO is introducing grants that must facilitate the interaction between researchers and potential users of knowledge. Added Value grants, a joint initiative between the Divisions for the Earth and Life Sciences (ALW), Humanities (GW) and Social Sciences (MaGW), has been set up with the aim of supporting researchers who wish to make their research results from current or recently concluded projects (more) suitable and available for third parties. Added Value is a one-off initiative that each of the participating divisions will realize in its own manner appropriate to the field it covers. The aim of Added Value Plus for the Social Sciences is to encourage researchers in the domain of the social and behavioural sciences to make their research results usable in collaboration with their partners and/or users and to increase the involvement of partners and users. Added Value Plus is aimed at research within the so-called Topsectors. Proposals can be submitted if they address the research questions raised in one of the following topsectors: Agro Food, Creative Industry, Energy, Logistics or the Social Infrastructure Agenda. The deadline for application is October 15, and the funding range is 10,000 to 50,000 Euro.
Requirements: Applications can be submitted by teams that consist of researchers from NWO-funded research from the open types of support, namely the Open Competition (also TOP and Onderzoekstalent), Innovational Research Incentives Scheme and partners/users. One of the researchers (lecturer, senior lecturer, professor employed by faculties or institutes of a Dutch university or an NWO or KNAW institute) acts as the main applicant. Applications preferably are in Dutch.
Restrictions: The project grant is for a maximum period of 12 months.
Geographic Focus: All States, Netherlands
Date(s) Application is Due: Oct 15
Amount of Grant: 10,000-50,000 EUR
Contact: C.E.A.M. Raat, +31 (0) 70 344 09 08; c.raat@nwo.nl
Internet: http://www.nwo.nl/nwohome.nsf/pages/NWOP_89PH6X_Eng
Sponsor: Netherlands Organization for Scientific Research
P.O. Box 93138
The Hague, NL2509 Netherlands

NWO Aspasia Grants **3673**
Aspasia is designed to alleviate the under-representation of women in the upper echelons of academia. The aim of Aspasia is to encourage the promotion of female academics to senior lecturer or professorial level. The award available for the promotion of each grant recipient is 100,000 Euro. The award offered for the promotion of an unsuccessful, but judged very good or excellent candidate, is 200,000 Euro. This means that the latter group receive no grant under the Innovative Research Incentives Scheme but are awarded a larger subsidy under Aspasia. The award may be used in a number of alternative ways. A condition of grant is that the Board must use at least 50,000 Euro of the award to fund more generic diversity policy measures by the university or faculty to increase the upward movement of female staff within the institution.
Requirements: Aspasia is linked to NWO's competitive Vidi and Vici grant schemes. To qualify as an Aspasia candidate, a female applicant for a Vidi or Vici grant must either have been awarded such a grant, or have had her application judged very good or excellent

following the interview procedure but been unsuccessful in obtaining one: Executive Boards that promote such applicants to senior lecturer or professor within a year of the Vidi or Vici grant award decision may, subject to certain conditions, qualify for a premium.
Geographic Focus: Netherlands
Amount of Grant: 100,000-200,000 EUR
Contact: Aspasia Coordinator; +31 (0) 70 349 40 11; aspasia@nwo.nl
Internet: http://www.nwo.nl/nwohome.nsf/pages/NWOA_4YYAJV_Eng
Sponsor: Netherlands Organization for Scientific Research
P.O. Box 93138
The Hague, NL2509 Netherlands

NWO Athena Grants **3674**
In the Netherlands the percentage of women who work in the natural and technical sciences decreases the higher the position. For example, women are underrepresented in assistant, associate and full professorships. With NWO Athena Grants, the Division for the Chemical Sciences (CW) hopes to increase the flow of women in the large group of post-doctorates with a temporary appointment to a tenured scientific position. This program will encourage the appointment of talented female researchers within chemistry, who have acquired a Veni subsidy, as tenured assistant professors (or a similar permanent position at a research institute). The Athena award is worth 100,000 Euro and is given to the Veni laureate for a maximum period of three years. The grant can be used for the salary of the candidate and for the appointment of additional personnel (PhDs, post-doctorate candidates, and technicians), costs for the acquisition of materials and equipment and/or databases, travel costs and visits to conferences and research institutes.
Requirements: The award will be given to female researchers who have been awarded a Veni subsidy from the Division for Chemical Sciences and who, during the course of the Veni project, receive a tenured appointment as assistant professor at a university or a comparable position at a research institute.
Geographic Focus: Netherlands
Amount of Grant: 100,000 EUR
Contact: Dr. Varsha Kapoerchan, +31 (0) 70 349 44 42; v.kapoerchan@nwo.nl
Internet: http://www.nwo.nl/nwohome.nsf/pages/NWOA_6ZXCX3_Eng
Sponsor: Netherlands Organization for Scientific Research
P.O. Box 93138
The Hague, NL2509 Netherlands

NWSA Abafazi-Africana Women's Studies Essay Award **3675**
The purpose of the writing awards is to discover, encourage, and promote the intellectual development of emerging black women scholars who engage in critical theoretical discussions and/or analyses about feminist/womanist issues among women of color. Scholarly essays may be on any subject relevant to black girl children, women's issues, and/or experiences in the United States or throughout the diaspora.
Requirements: NWSA members will be given preference over nonmembers.
Geographic Focus: All States
Date(s) Application is Due: Feb 15
Contact: Program Contact; (301) 403-0524; fax (301) 403-4137; nwsaoffice@nwsa.org
Internet: http://www.nwsa.org/wocscol.html
Sponsor: National Women's Studies Association
7100 Baltimore Avenue, Suite 502
College Park, MD 20740

NWSA Graduate Scholarship in Jewish Women's Studies **3676**
The scholarship will be awarded to a student who, in the fall, will be engaged in the research or writing stages of a master's thesis or PhD dissertation in the interdisciplinary field of women's studies. The research project must be on women and must enhance the NWSA mission.
Requirements: Applicants must be NWSA members at the time of application.
Geographic Focus: All States
Date(s) Application is Due: Mar 1
Amount of Grant: 1,000 USD
Contact: Administrator; (301) 403-0525; fax (301) 403-4137; nwsaoffice@nwsa.org
Internet: http://www.nwsa.org
Sponsor: National Women's Studies Association
7100 Baltimore Avenue, Suite 502
College Park, MD 20740

NWSA Graduate Scholarships **3677**
NWSA will award the scholarship to a student who, in the fall of the year of the award, will be engaged in the research or writing stages of a master's Thesis or PhD dissertation in the interdisciplinary field of women's studies. The research project must be on women and must enhance the NWSA mission.
Requirements: NWSA members are eligible.
Geographic Focus: All States
Date(s) Application is Due: Feb 15
Amount of Grant: 1,000 USD
Contact: Administrator; (301) 403-0525; fax (301) 403-4137; nwsaoffice@nwsa.org
Internet: http://www.nwsa.org/gradscol.html
Sponsor: National Women's Studies Association
7100 Baltimore Avenue, Suite 502
College Park, MD 20740

NWSA Graduate Scholarships in Lesbian Studies **3678**
One scholarship is awarded per year to support a graduate student doing research for, or writing a master's thesis or PhD dissertation in, lesbian studies. Applicants need not be enrolled in a women's studies program. Preference is given to candidates who are NWSA members.
Geographic Focus: All States
Date(s) Application is Due: Feb 15
Amount of Grant: 500 USD
Contact: Administrator; (301) 403-0525; fax (301) 403-4137; nwsaoffice@nwsa.org
Internet: http://www.nwsa.org/sslesbian.html
Sponsor: National Women's Studies Association
7100 Baltimore Avenue, Suite 502
College Park, MD 20740

NYAM Audrey and William H. Helfand Fellowship **3679**
The Fellowship supports work in the humanities as they apply to medicine and health, including works of non-fiction, visual or performing arts, biography and memoir, as well as scholarly research in a humanistic discipline other than the history of medicine. Although residence is not obligatory, preference will be given applicants whose projects require use of the resources of the Academy Library and who plan to spend time at the Academy. Besides completing a research or creative project, each fellow will be expected to make a public presentation at the Academy and submit a final report.
Requirements: Applications are invited from anyone, regardless of citizenship, academic discipline, or academic status. Preference will be given to (1) those whose research will take advantage of resources that are uniquely available at the Academy, and (2) scholars or creative artists in the early stages of their careers.
Geographic Focus: All States
Date(s) Application is Due: Mar 3
Amount of Grant: Up to 5,000 USD
Contact: Administrator; (212) 822-7314; fax (212) 822-7338; history@nyam.org
Internet: http://www.nyam.org/grants/history.shtml
Sponsor: New York Academy of Medicine
1216 Fifth Avenue
New York, NY 10029-5202

NYAM Charles A. Elsberg Fellowship **3680**
The Fellowship in Neurological Surgery was established to support research training in the specialty of neurological surgery for individuals who have completed, or will shortly complete, accredited residency training in neurological surgery, and who intend to use research training for continued development of academic careers in this field. The Elsberg Fellowship program seeks to support individuals in supervised programs that will develop the candidate's capacity to perform independent clinical or laboratory research.
Requirements: Candidates who have, or will have completed an accredited neurological surgery residency program by June 30, and whose career direction suggests eligibility for subsequent research support are invited to apply. In general, applicants are expected to be at the start of their academic careers, having completed their residency training within the past year or two. Candidates must be board certified or eligible for certification by the beginning of the grant period. The proposed research must be conducted at an institution located in the United States under the mentorship of a senior researcher.
Geographic Focus: All States
Amount of Grant: 50,000 USD
Contact: Administrator; (212) 822-7204; fax (212) 822-7338; elsberg@nyam.org
Internet: http://www.nyam.org/grants/elsberg.shtml
Sponsor: New York Academy of Medicine
1216 Fifth Avenue
New York, NY 10029-5202

NYAM David E. Rogers Fellowships **3681**
The Fellowship is meant to enrich the educational experiences of medical and dental students through projects that bear on medicine and dentistry as social enterprises, such as enterprises devoted to the capacity of these professions in any and all of their expressions to serve human needs. Particularly, the focus is on the needs of underserved or disadvantaged patients or populations. The content of the Fellowship might include clinical investigation, health policy analysis, activities linking biomedicine, the social infrastructure and human need, or community activities.
Requirements: Available to first-year medical and dental students for support of projects to be executed during the summer between the first and second years of medical or dental school.
Geographic Focus: All States
Amount of Grant: 3,500 USD
Contact: Office of Medical and Academic Affairs; (212) 822-7204; fax (212) 996-7338; rogers@nyam.org
Internet: http://www.nyam.org/grants/rogers.shtml
Sponsor: New York Academy of Medicine
1216 Fifth Avenue
New York, NY 10029-5202

NYAM Edward N. Gibbs Memorial Lecture and Award in Nephrology **3682**
The New York Academy of Medicine seeks nominations of physician scientists who have dedicated their careers to advances in nephrology or are presently making cutting-edge discoveries in the field. The distinguished recipient of this honor will present his or her research at a lecture before a broad audience of scientists and clinicians, to be held at The New York Academy of Medicine in the fall. The award recipient will receive the Edward N. Gibbs Memorial Medal and an honorarium of $7,500; all travel expenses associated with the lecture will be paid by the Edward N. Gibbs Memorial Endowment. It is expected that the manuscript from this lecture will be submitted to a scholarly journal for publication.
Requirements: Candidates must hold an MD degree and be citizens of the United States to be considered for this award.

Geographic Focus: All States
Date(s) Application is Due: Mar 1
Amount of Grant: 7,500 USD
Contact: Program Coordinator; (212) 822-7204; fax (212) 822-7338; gibbs@nyam.org
Internet: https://www.nyam.org/grants/gibbs.shtml
Sponsor: New York Academy of Medicine
1216 Fifth Avenue
New York, NY 10029-5202

NYAM Edwin Beer Research Fellowship **3683**

The Edwin Beer Research Fellowship seeks to increase the number of investigators in urology and urology-related fields by providing transitional funding in support of research conducted by qualified candidates. The program provides two-year grants of up to $80,000 for research projects commencing July of the award year. The grants are made to the awardees' institutions for the direct support of their salary and research activities. Fellows are required to devote at least fifty percent of their time to the research funded by the program.
Requirements: Candidates must hold an MD, PhD, or equivalent degree and are expected to conduct research in a supervised program in the United States. Applicants should have completed residency and fellowship or postdoctoral training, having demonstrated their ability to perform and direct research, and should currently be under the mentorship of a senior researcher. Fellowships are awarded to qualified candidates who are at a stage of transition prior to receiving major independent funding. Because the program seeks to increase the number of investigators in urology and urology-related fields, applicants must have a direct association with the department of urology at their sponsoring institution. Candidates are not required to be United States citizens; however, non-citizen applicants are required to provide assurances of employment eligibility to perform the proposed research.
Geographic Focus: All States
Date(s) Application is Due: Feb 2
Amount of Grant: Up to 80,000 USD
Contact: Office of Medical and Academic Affairs; (212) 822-7204; fax (212) 822-7338; beer@nyam.org
Internet: http://www.nyam.org/grants/beer.shtml
Sponsor: New York Academy of Medicine
1216 Fifth Avenue
New York, NY 10029-5202

NYAM Ferdinand C. Valentine Fellowship **3684**

The Fellowship offers one-year, $50,000 fellowships in support of research by individuals who have completed, or will shortly complete, residency training acceptable to the American Board of Urology, and who intend to pursue academic research careers in urology.
Requirements: Fellows must commit no less than fifty percent of their time to the supported research and preference is given to investigators conducting their research at institutions in the greater New York area.
Restrictions: Indirect costs and fringe benefits will not be paid.
Geographic Focus: All States
Date(s) Application is Due: Feb 2
Amount of Grant: 50,000 USD
Contact: Administrator; (212) 822-7204; fax (212) 822-7338; valentine@nyam.org
Internet: http://www.nyam.org/grants/valentine.shtml
Sponsor: New York Academy of Medicine
1216 Fifth Avenue
New York, NY 10029-5202

NYAM Ferdinand C. Valentine Student Research Grants **3685**

Up to four $3,500 grants for students pursuing mentored research projects in urology will be awarded to MD candidates who are either enrolled in a New York-area medical school or plan to conduct research at a New York-area institution. The grants typically support 8-12 week urology research projects initiated during the summer between the first and second years of medical school, but medical school students in any year of their education may apply. Students are encouraged to pursue research projects that extend beyond this startup period.
Requirements: Applicants must provide a one-page research proposal, as well as a letter of support from the faculty mentor supervising the research.
Geographic Focus: All States
Date(s) Application is Due: Feb 2
Amount of Grant: 3,500 USD
Contact: Office of Medical and Academic Affairs; (212) 822-7204; fax (212) 996-7338; valentine@nyam.org
Internet: http://www.nyam.org/grants/valentine_student.shtml
Sponsor: New York Academy of Medicine
1216 Fifth Avenue
New York, NY 10029-5202

NYAM Gladys Brooks Internship **3686**

The Internship at the New York Academy of Medicine is designed to provide an opportunity for training in book and paper conservation. Paid internships will last for 2-3 months, depending on the number of days spent in the lab. The internship will include: examining and applying book and paper conservation principles; understanding the steps in determining conservation needs; learning which materials and structures are appropriate for use in repairing, binding and storing rare or fragile materials; and executing the agreed-upon conservation treatments. Interns will become familiar with conservation activities through extensive hands-on experience and observations in a working book and paper conservation laboratory.
Requirements: Applicants for this internship must: have some experience in the field of bookbinding and/or book conservation, and possess a desire to gain expertise through hands-on work in a functioning conservation laboratory; have aesthetic appreciation and the manual dexterity to work with delicate and fragile materials; submit a detailed letter of interest, a current resume, and three professional recommendations-all to the address below; have student or working visas if they are not U.S. citizens.
Geographic Focus: All States
Date(s) Application is Due: Dec 15
Amount of Grant: 5,000 USD
Contact: Anne Hillam, (212) 822-7365
Internet: http://www.nyam.org/initiatives/im-histc-i.shtml
Sponsor: New York Academy of Medicine
1216 Fifth Avenue
New York, NY 10029-5202

NYAM Glorney-Raisbeck Fellowships **3687**

The Fellowships in Cardiovascular Diseases are one-year grants funded by the Corlette Glorney Foundation that are awarded for basic science, clinical, or epidemiological research in cardiovascular diseases. The program seeks to enhance the potential of exceptional candidates to become physician scientists by supporting individuals in supervised programs that develop the candidate's ability to perform independent research.
Requirements: Candidates must hold an MD or equivalent degree, and are expected to have completed residency by the start of the grant period. The applicant pool includes physicians in training throughout the United States, but in order to be considered for this program candidates must pursue the funded research in a supervised program at an institution located in New York City, Long Island, or Westchester County. Candidates must be United States citizens or permanent visa holders. Renewal of awards to previous recipients will be considered on a competitive basis.
Geographic Focus: All States
Date(s) Application is Due: Feb 2
Amount of Grant: Up to 60,000 USD
Contact: Administrator; (212) 822-7204; fax (212) 822-7338; glorney-raisbeck@nyam.org
Internet: http://www.nyam.org/grants/glorneyraisbeck.shtml
Sponsor: New York Academy of Medicine
1216 Fifth Avenue
New York, NY 10029-5202

NYAM Glorney-Raisbeck Medical Student Grants **3688**

Up to four $4,000 grants, consisting of a $3,500 student stipend plus $500 in support of the laboratory or department at which the student will conduct research, will be awarded to MD candidates who are either enrolled in a New York-area medical school or plan to conduct research at a New York-area institution. The grants typically support 8-12 week mentored cardiovascular research projects initiated during the summer between the first and second years of medical school, but medical school students in any year of their education may apply. Students are encouraged to pursue research projects that extend beyond this startup period.
Requirements: Applicants must provide a one-page research proposal, as well as a letter of support from the faculty mentor supervising the research.
Geographic Focus: All States
Date(s) Application is Due: Feb 2
Amount of Grant: 4,000 USD
Samples: Daniel Belkin; Olakunle Ogunrinade; Rushita Patel (2009; Ujas Shah.
Contact: Administrator; (212) 822-7204; fax (212) 822-7338; glorney_raisbeck@nyam.org
Internet: http://www.nyam.org/grants/raisbeckstudent.shtml
Sponsor: New York Academy of Medicine
1216 Fifth Avenue
New York, NY 10029-5202

NYAM Lewis Rudin Glaucoma Grants **3689**

This $50,000 prize is awarded for the most outstanding scholarly article on glaucoma published in a peer-reviewed journal during the previous calendar year. Nominations are solicited from leaders of the ophthalmologic community and reviewed in the spring of the award year. One prize winner is recommended for ratification by the Academy's Board of Trustees by the Lewis Rudin Glaucoma Prize Selection Committee, a group of nationally recognized experts in glaucoma research chaired by David H. Abramson, MD of Memorial Sloan-Kettering Cancer Center and Weill Medical College of Cornell University. The winner is announced in July of the award year.
Requirements: Nominees for this award must be the first or last author of the published work and must hold primary responsibility for the research. Authors may nominate themselves but may not submit multiple articles and must choose only one article to submit for consideration by the committee. All authors of the article receive recognition, however the monetary prize is granted solely to the primary researcher specified in the nomination. Copies of the published article must accompany the completed nomination form.
Geographic Focus: All States
Date(s) Application is Due: May 1
Amount of Grant: 50,000 USD
Contact: Coordinator; (212) 822-7204; fax (212) 822-7338; rudinglaucoma@nyam.org
Internet: https://www.nyam.org/grants/rudin.shtml
Sponsor: New York Academy of Medicine
1216 Fifth Avenue
New York, NY 10029-5202

NYAM Louis L. Seaman Medical Student Research Grants **3690**

Up to four $3,500 grants for students pursuing mentored research projects in microbiology will be awarded to MD candidates who are either enrolled in a New York-area medical school or plan to conduct research at a New York-area institution. The grants typically support 8-12 week microbiology research projects initiated during the summer between the first and second years of medical school, but medical school students in any year of their education may apply. Students are encouraged to pursue research projects that extend beyond the start up period.
Requirements: Applicants must provide a one-page research proposal, as well as a letter of support from the faculty mentor supervising the research.
Geographic Focus: All States
Contact: Administrator; (212) 822-7204; fax (212) 996-7338; seaman@nyam.org
Internet: http://www.nyam.org/grants/studentseamansform.shtml
Sponsor: New York Academy of Medicine
1216 Fifth Avenue
New York, NY 10029-5202

NYAM Mary and David Hoar Fellowship **3691**
The Mary and David Hoar Fund was established to promote research in the prevention and treatment of hip fractures. The Fund provides a two-year grant of $100,000 to support pilot programs in clinical, epidemiologic and health services research in the following areas related to hip fractures: clinical care and treatment; prevention; rehabilitation; and model program development.
Requirements: Candidates must hold an MD, PhD, or equivalent degree and are expected to conduct research in a supervised program in the greater New York area. Preferential consideration will be given to new investigators for whom this award is likely to provide interim support pending future funding, or to more senior investigators who are making a change in career or research direction for whom this award will facilitate the transition. Candidates are not required to be United States citizens; however, non-citizen applicants are required to provide assurances, such as permanent resident status, of legal eligibility for employment to perform the proposed research throughout the award period, regardless of the disposition of this or any other pending grants.
Geographic Focus: All States
Contact: Program Administrator; (212) 822-7204; fax (212) 822-7338; hoar@nyam.org
Internet: http://www.nyam.org/grants/hoar.shtml
Sponsor: New York Academy of Medicine
1216 Fifth Avenue
New York, NY 10029-5202

NYAM Paul Klemperer Fellowship **3692**
The Fellowship supports research using the Academy Library's resources for scholarly study of the history of medicine. It is intended specifically for a scholar in residence at the Academy Library. Although residence is not obligatory, preference will be given applicants whose projects require use of the resources of the Academy Library and who plan to spend time at the Academy. Besides completing a research or creative project, each fellow will be expected to make a public presentation at the Academy and submit a final report.
Requirements: Applications are invited from anyone, regardless of citizenship, academic discipline, or academic status. Preference will be given to (1) those whose research will take advantage of resources that are uniquely available at the Academy, and (2) scholars or creative artists in the early stages of their careers.
Geographic Focus: All States
Date(s) Application is Due: Mar 3
Contact: Office of the Academy Historian; (212) 822-7314; history@nyam.org
Internet: http://www.nyam.org/grants/history.shtml
Sponsor: New York Academy of Medicine
1216 Fifth Avenue
New York, NY 10029-5202

NYAM Student Essay Grants **3693**
The New York Academy of Medicine invites entries for the New York Academy of Medicine Student Essay Prize, awarded to the best unpublished essay by a graduate student in a medical, public health, pharmacy, or nursing program in the United States. Essays should address topics in the history of public health or medicine as they relate to urban health issues; social or environmental factors in the health of urban populations, institutional histories, or specific diseases may be considered. The winner will receive $500, and the winning essay will receive expedited review for possible publication in the Journal of Urban Health. Honorable Mention prizes may also be awarded at the discretion of the Prize Committee.
Requirements: The contest is open to students in accredited professional degree programs in medicine, nursing, pharmacy, and public health. The writer must have been a student at the time the essay was written.
Geographic Focus: All States
Amount of Grant: 500 USD
Contact: Coordinator; (212) 822-7314; fax (212) 822-7338; historyessay@nyam.org
Internet: https://www.nyam.org/grants/studentessay.shtml
Sponsor: New York Academy of Medicine
1216 Fifth Avenue
New York, NY 10029-5202

NYCT Biomedical Research Grants **3694**
The Trust supports projects that strengthen preventive health care, improve access to services, promote the efficient use of health resources, and develop the skills and independence of people with special needs. The Trust provides funding for biomedical research projects in four areas: blood diseases, cancer, leprosy and tuberculosis. Grants for these projects are awarded in response to requests for proposals.
Requirements: Grants are made primarily to colleges, universities, and research organizations located in the five boroughs of New York City. Grants for programs outside the area are generally from funds designated for specific charities or have been made at the suggestion of donors.
Restrictions: The trust does not make grants to individuals, offer general or capital funding, endowments, building construction/renovation, deficit financing, films, or religion.
Geographic Focus: New York
Amount of Grant: 5,000-100,000 USD
Contact: Joyce M. Bove, Senior Vice President; (212) 686-0010, ext. 552; fax (212) 534-8528; info@nycommunitytrust.org or grants@nycommunitytrust.org
Internet: http://www.nycommunitytrust.org/ForGrantSeekers/GrantmakingGuidelines/HealthandPeoplewithSpecialNeeds/tabid/207/Default.aspx
Sponsor: New York Community Trust
909 Third Avenue, 22nd Floor
New York, NY 10022

NYCT Blindness and Visual Disabilities Grants **3695**
The Trust supports projects that strengthen preventive health care, improve access to services, promote the efficient use of health resources, and develop the skills and independence of people with special needs. The goals of this program are to support program innovation and reform and eliminate service gaps. Specifically, grants are made to projects that: improve services to those people with visual disabilities who are presently underserved, such as the elderly, minorities, people with multiple disabilities, people in institutions, and youth in transition from school to work (projects need not be limited to direct service, and may include advocacy and organizational improvement); involve people with visual disabilities more fully in community activities; expand programs that identify people with vision problems at an early stage and link them with appropriate resources; empower persons with visual disabilities by enabling them to participate in planning programs that affect them; and support research in the prevention and treatment of blindness (this may include clinical, epidemiological, and applied studies). The grant review process takes from two to six months. Grants range from $5,000 to $100,000; an average grant is around $60,000.
Requirements: Grants are made primarily to nonprofit organizations located in the five boroughs of New York City. Grants for programs outside the area are generally from funds designated for specific charities or have been made at the suggestion of donors.
Restrictions: The trust does not make grants to individuals, offer general or capital funding, endowments, building construction/renovation, deficit financing, films, or religion.
Geographic Focus: New York
Amount of Grant: 5,000-100,000 USD
Contact: Joyce M. Bove, Senior Vice President; (212) 686-0010, ext. 552; fax (212) 534-8528; info@nycommunitytrust.org or grants@nycommunitytrust.org
Internet: http://www.nycommunitytrust.org/ForGrantSeekers/GrantmakingGuidelines/HealthandPeoplewithSpecialNeeds/tabid/207/Default.aspx
Sponsor: New York Community Trust
909 Third Avenue, 22nd Floor
New York, NY 10022

NYCT Blood Disease Research Grants **3696**
The New York Community Trust is a charitable foundation concerned with the needs of the Greater New York area. The NYCT has a special fund to be used in support of research in the field of blood diseases. A panel of recognized hematologists will convene to review proposals solicited from the major medical research institutions on a competitive basis. It is estimated that about $700,000 will be available this year to support such research projects. A total of no more than two projects may be submitted from any one organization. Basic laboratory research. Candidates may apply for support totaling $100,000 for up to 2 years, but funds for any one project year may not exceed $50,000. Second-year funding is contingent upon submission of a satisfactory progress report at the end of the first year. Research that may have application to New York City residents is of particular interest to The Trust. Young investigators will be given special consideration.
Requirements: Proposals are due no later than 5:00 p.m. on September 25th. Additional guidelines and forms are available online at the NYCT website.
Geographic Focus: New York
Date(s) Application is Due: Sep 25
Amount of Grant: 100,000 USD
Contact: Joyce M. Bove, Senior Vice President; (212) 686-0010, ext. 552; fax (212) 534-8528; info@nycommunitytrust.org or grants@nycommunitytrust.org
Internet: http://www.nycommunitytrust.org/BloodDiseaseResearch/tabid/504/Default.aspx
Sponsor: New York Community Trust
909 Third Avenue, 22nd Floor
New York, NY 10022

NYCT New York City Environment Grants **3697**
The New York City program focuses on four urban environmental issues: promoting a more effective and sustainable solid waste management program through consumer education, research, and advocacy, with an emphasis on waste reduction and equitable distribution of solid waste facilities; expanding open space and parks through advocacy, constituency building, and support of local planning and greening activities; reclaiming the waterfront and brownfields through technical assistance, advocacy, and partnerships between community development and environmental groups; reducing air pollutants and other environmental toxins through cleaner fuels and engines, mass transit improvements, and elimination of indoor pollutants and the concentration of polluting facilities in poor

communities. In the New York metropolitan region, the NYCT aims are, to preserve and restore habitat, promote a more sustainable transportation system, protect open space and drinking water, and redevelop brownfields.
Requirements: Grants are made primarily to nonprofit organizations located in the five boroughs of New York City. Grants for programs outside the area are generally from funds designated for specific charities or have been made at the suggestion of donors.
Restrictions: The trust does not make grants to individuals, offer general or capital funding, endowments, building construction/renovation, deficit financing, films, or religion.
Geographic Focus: New York
Contact: Joyce M. Bove, Senior Vice President; (212) 686-0010, ext. 552; fax (212) 534-8528; info@nycommunitytrust.org or grants@nycommunitytrust.org
Internet: http://www.nycommunitytrust.org/ForGrantSeekers/GrantmakingGuidelines/CommunityDevelopmentandtheEnvironment/tabid/204/Default.aspx
Sponsor: New York Community Trust
909 Third Avenue, 22nd Floor
New York, NY 10022

NYFA Artist Fellowships **3698**

The New York Foundation for the Arts (NYFA) was founded in 1971 to empower artists at critical stages in their creative lives. Artists' Fellowships are $7,000 cash awards made to individual originating artists living and working in the state of New York for unrestricted use. Grants are awarded in 15 artistic disciplines, with applications accepted in five categories each year in rotating fashion. In the 1212 and 1215 cycles, fellowships will be awarded in painting, folk/traditional arts, interdisciplinary work, fiction, and video/film. In the 1213 and 1216 cycles, fellowships will be awarded in choreography, music and sound, architecture/environmental structures and design, play writing and screenwriting, and photography. In the 1214 and 1217 cycles, fellowships will be awarded in digital/electronic arts, crafts and sculpture, nonfiction literature, poetry, and printmaking/drawing/book arts.
Requirements: Applicants must have resided in New York State for two years prior to application. Each recipient will be required to provide public service within the state of New York.
Geographic Focus: New York
Date(s) Application is Due: Oct 3
Amount of Grant: 7,000 USD
Contact: Susan Ball, Interim Director of Programs; (212) 366-6900, ext. 217; fax (212) 366-1778; sball@nyfa.org or fellowships@nyfa.org
Internet: http://www.nyfa.org/level2.asp?id=1&fid=1
Sponsor: New York Foundation for the Arts
20 Jay Street, 7th Floor
Brooklyn, NY 11201

NYFA Shelly Pinz Fellowship **3699**

New York Foundation for the Arts Fellowships are administered by NYFA with leadership support from the New York State Council on the Arts. The Shelly Pinz Fellowship will support one Fellowship in Poetry annually. Shelly Pinz (1944-2004) was a New York songwriter, poet and psychotherapist and is best known for co-writing the Lemon Pipers' 1968 hit "Green Tambourine" as well as her volume of poetry and lyrics titled "Courage to Think." The Shelly Pinz Fellowship was developed in Ms. Pinz's memory to support the careers of poets and composers. The announcement of the Shelly Pinz Fellowship in Poetry will take place at a poetry club, hosted by NYFA, where a reading of selections from "Courage to Think" will occur.
Requirements: The Award is open to poets of all nationalities, provided they are residents of New York State. All those who apply to NYFA's Artist Fellowship Program in Poetry will automatically be considered for the Shelly Pinz Fellowship. Applicants must have resided in New York State for two years prior to application. Each recipient will be required to provide public service within the state of New York.
Geographic Focus: New York
Date(s) Application is Due: Nov 5
Contact: Susan Ball, Interim Director of Programs; (212) 366-6900, ext. 321; fax (212) 366-1778; sball@nyfa.org
Internet: http://www.nyfa.org/level3.asp?id=46&fid=1&sid=1
Sponsor: New York Foundation for the Arts
20 Jay Street, 7th Floor
Brooklyn, NY 11201

NYPL Martin Duberman Visiting Scholars **3700**

Annually the New York Public Library provides stipends for up to three Martin Duberman Visiting Scholars. The stipends support travel to New York City and related expenses to do research in the Library's premier LGBT (Lesbian, Gay, Bisexual, and Transgender) history collections. The Martin Duberman Visiting Scholars are funded by the generous support of Martin Duberman and Eli Zal.
Requirements: The program is limited to emerging scholars, those without permanent academic appointments, or those who are unaffiliated with an academic institution. Recipients must supply a written summary of their findings upon completion of their work. Interested applicants should submit a 3–5 page research proposal specifying the relevant collections at the Library for their project, a draft budget and itinerary for their planned trip, a curriculum vitae, and an appropriate letter of recommendation.
Geographic Focus: All States
Date(s) Application is Due: May 15
Amount of Grant: 1,000-8,500 USD
Contact: Jason Baumann, Scholars Manager; (212) 930-0887; JBaumann@nypl.org
Internet: http://www.nypl.org/node/113221
Sponsor: New York Public Library
Stephen A. Schwarzman Building, 476 Fifth Avenue
New York, NY 10018

NYPL Short-Term Research Fellowships **3701**

The New York Public Library provides short-term fellowships to support visiting scholars conducting research in the Library's unique research and special collections. Fellowships stipends are available to scholars from outside the New York metropolitan area engaged in graduate-level, post-doctoral, or independent research. Scholars researching in the humanities including but not limited to art history, cultural studies, history, literature, performing arts and photography are welcome to apply. Application guidelines are available on the website.
Requirements: Applicants must be United States citizens or permanent residents with the legal right to work in the United States.
Geographic Focus: All States
Date(s) Application is Due: Mar 18
Amount of Grant: 4,000 USD
Contact: Fellowship Coordinator; (212) 930-0653; short.term@nypl.org
Internet: http://www.nypl.org/help/about-nypl/fellowships-institutes/short-term-research-fellowships
Sponsor: New York Public Library
Stephen A. Schwarzman Building, 476 Fifth Avenue
New York, NY 10018

NYSCA Dance: Services to the Field Grants **3702**

Services to the Field grants provides support to service organizations that offer professional services for the advancement of dance groups and individuals. Services may include technical assistance in administrative and/or new technological areas, cooperative management, booking and promotional services, research facilities, archival documentation, projects that utilize technology, and performance spaces.
Requirements: Applicants must register with the Council before February 22 to be considered for funding. In addition, applicants should demonstrate how Council support would meet the needs of the professional dance field. A paid administrator on staff is required. Applications can be obtained at the website and should be submitted online. First-time applicants are required to call staff before the registration deadline.
Restrictions: The request amount should not exceed 50% of the total project budget.
Geographic Focus: New York
Date(s) Application is Due: Apr 1
Contact: Beverly D'Anne, Senior Program Officer; (212) 741-3232 or (212) 741-3331; bdanne@nysca.org or dlim@nysca.org
Internet: http://www.nysca.org/public/guidelines/dance/services.htm
Sponsor: New York State Council on the Arts
175 Varick Street
New York, NY 10014-4604

NYSCA Folk Arts: Exhibitions Grants **3703**

Exhibition grants are offered for an exhibition of the work of living folk artists, to explore issues in the folk arts, or to interpret the traditional culture of a specific community through its folk arts. The proposed exhibition should enhance understanding and appreciation of the folk arts for general audiences as well as communities represented in the exhibition. The exhibition may include objects of folk art, photographic representations of folk culture, and/or multimedia kiosks or other installations in the exhibition utilizing new technologies. Requests may include field research expenses for the development of an exhibition and expenses for an exhibition catalog.
Requirements: Applicants must register with the Council before February 22 to be considered for funding. The proposed exhibition should focus on traditions currently practiced or within living memory, and may incorporate relevant historical works of folk art. First-time applicants must call program staff before the registration deadline.
Restrictions: Funding awarded for exhibitions cannot exceed 50% of project expenses. Support is not available for programming involving artists who appropriate, interpret, or revive the traditions of other communities. Folk Arts emphasizes support for presentations grounded in the traditional modes of practicing folk art, and not programming involving choreography, theatricality, or stylization that significantly alters traditions.
Geographic Focus: New York
Date(s) Application is Due: Apr 1
Contact: Robert Baron, Senior Program Officer; (212) 741-7755 or (212) 741-7143; rbaron@nysca.org or rputnam@nysca.org
Internet: http://www.nysca.org/public/guidelines/folk_arts/exhibitions.htm
Sponsor: New York State Council on the Arts
175 Varick Street
New York, NY 10014-4604

NYSCA Folk Arts: Presentation Grants **3704**

The Presentation grants support activities that present folk arts and artists to the public, including concerts, festivals, lecture/demonstrations, residencies, and programs presenting oral narratives. Applications may involve one type of activity or a series of thematically related activities that may involve different modes of presentation. Folk arts events and series are also supported.
Requirements: Applicants must register with the Council by February 22 to be considered for funding. Applications should demonstrate a clearly articulated thematic focus for the event or series requested. Also, artists must receive appropriate fees for their presentations, and when travel is involved, adequate per diem. Presentations should also involve personnel and consultants with appropriate cultural expertise and technical skills. These may include, but are not limited to: folklorists, ethnomusicologists, graphic designers, publicists and

specialists in lighting, sound reinforcement, stage management, promotion and marketing. In addition, the folk art activity being presented must have its own interpretive materials, and promotion for the program should substantially publicize the folk arts event and the artists involved in the activity. First-applicants must contact program staff before the registration deadline. Applications can be found at the website and should be submitted online.
Restrictions: This grant will not fund more than 50% of the project expenses. Applicants may submit no more than one application in this funding category. Folk Arts emphasizes support for presentations grounded in the traditional modes of practicing folk art, and not programming involving choreography, theatricality, or stylization that significantly alters traditions. Support is also not available for programming involving artists who appropriate, interpret, or revive the traditions of other communities.
Geographic Focus: New York
Date(s) Application is Due: Apr 1
Contact: Robert Baron; (212) 741-7755 or (212) 741-7847; rbaron@nysca.org
Internet: http://www.nysca.org/public/guidelines/folk_arts/presenting.htm
Sponsor: New York State Council on the Arts
175 Varick Street
New York, NY 10014-4604

NYSCA Folk Arts: Regional and County Folk Arts Grants **3705**
Folk Arts, in partnership with regional and local cultural organizations, supports folk arts programs which are regional and county-wide in scope. Support for these programs is intended to generate field research and programming for diverse audiences on an ongoing basis, directed by qualified individuals with experience and expertise in the public folklore field. Grants may be used towards the salaries of staff folklorists employed year-round and expenses associated with their production of public programs, field research on local traditions, and assistance to individual folk artists and other organizations interested in developing folk arts activities.
Requirements: Applicants must register with the Council by February 22 to be considered for funding. In addition, an applicant must have: received support from Folk Arts for the past three years; a minimum of four significant public programs that are organized each year; programs to be based to a significant extent upon field research, and work plans that include sufficient time for field research. Also, field research materials collected must be maintained in a locally accessible archive. Staff folklorists supported through this category must have experience organizing folk arts public programs and academic training in folklore or ethnomusicology. The staff folklorist must be employed for a minimum of three days each week. Applicant organizations are also expected to have a long-term commitment to folk arts services and programming. Program income should come from a variety of sources. First-time applicants are required to call staff before the registration deadline.
Restrictions: The request amount should not exceed 50% of the total project budget. Folk Arts emphasizes support for presentations grounded in the traditional modes of practicing folk art, and not programming involving choreography, theatricality, or stylization that significantly alters traditions. Support is also not available for programming involving artists who appropriate, interpret, or revive the traditions of other communities.
Geographic Focus: New York
Date(s) Application is Due: Apr 1
Contact: Robert Baron; (212) 741-7755 or (212) 741-7143; rbaron@nysca.org
Internet: http://www.nysca.org/public/guidelines/folk_arts/regional.htm
Sponsor: New York State Council on the Arts
175 Varick Street
New York, NY 10014-4604

NYSCA Individual Artists: Film, Media & New Technology Production Grants **3706**
Production support is offered annually to an individual artist project for film and video, radio, sound art, or computer-generated art. The project can be a combination of media, including but not limited to, work for tape, installation, single-channel work, or interactive disc formats. Individual Artists supports projects utilizing production methods and tools ranging from the simple to state-of-the-art technologies. The Council extends equal consideration to a wide range of artists and genres. Electronic media artists or filmmakers who wish to collaborate with artists in other disciplines may apply for production support. Priority is given to projects that emphasize the collaborative process and whose goal is to produce a distinctive work resulting from the meeting of disciplines.
Requirements: Applicants must register with the Council by February 22 in order to be considered for funding. Also, artists: who receive support for a specific phase of production of a project must complete that phase before re-applying for additional funds for the same project; applying for a documentary project are only eligible for post-production support. A work-in-progress of a minimum of 15 minutes is required. In addition, a sample of a previously completed work is required. First time directors may still apply, but it is recommended that the work sample should represent a project approaching the rough-cut phase. First-time applicants must call program staff before the registration deadline.
Restrictions: Production support is not provided for projects that take another discipline as their major focus or aim to provide straightforward documentation of another art form. In addition, organizations may not impose a fee on artists for submitting their applications. Artists may be involved as principal personnel in connection with only one project request. Artists receiving support in this category cannot apply for support in the following two years. (e.g., if funded in FY11, you must wait until FY14 to apply again). This allows the Council to reach a greater number of artists with its limited funds.
Geographic Focus: New York
Date(s) Application is Due: Apr 1
Amount of Grant: 5,000-25,000 USD
Contact: Helen Cash Jackson; (212) 741-7148; hcash@nysca.org
Internet: http://www.nysca.org/public/guidelines/individual_artists/film_media.htm
Sponsor: New York State Council on the Arts
175 Varick Street
New York, NY 10014-4604

NYSCA Museum: Project Support Grants **3707**
Project Support grants offer an opportunity for museums or related service organizations to seek support for projects or programs which are essential to maintain and/or increase service to their audience/communities. Such arts and cultural projects may include, but are not limited to: exhibitions, education programs, public programs, salary support for a curator or educator, interpretation, collections research, catalogs, audience development, and services to the field. Salary Support prioritizes museums which are poised to move forward in their development by adding an additional staff member for a professional collections-based or education position.
Requirements: Applicants must register with the Council by February 22 to be considered for funding. Museums must have been open to the public for a minimum of one year prior to the registration deadline and must provide services at least 120 days per year. Applications can be found at the website and should be submitted online. First-time applicants must contact the Museum staff before the registration deadline.
Restrictions: The requested amount shall not exceed 50% of the total project budget. In addition, awards cannot be made for more than three years, and salary support can only partially cover salary and benefits. Organizations applying for or receiving Museum General Operating Support or General Program Support are not eligible for funding in this category. Also, applicants may only submit one Museum request.
Geographic Focus: New York
Date(s) Application is Due: Apr 1
Contact: Kristin Herron, Senior Program Officer; (212) 741-7848; kherron@nysca.org
Internet: http://www.nysca.org/public/guidelines/museums/projects.htm
Sponsor: New York State Council on the Arts
175 Varick Street
New York, NY 10014-4604

NYSCA Music: Services to the Field Grants **3708**
Services to the Field grants offer support to organizations that provide professional services to music organizations, artists, and research facilities. Such services may include, but are not limited to, technical assistance in administrative areas, projects that utilize technology, booking and promotional services, and concert and artist management. Support is also available for publications that specifically benefit the professional music field or its audience.
Requirements: Applicants must register with the Council by February 22 to be eligible for funding. Also, organizations must have a paid administrator to be eligible for an award. Applications are available at the website and should be submitted online. First-time applicants must contact Music staff before the registration deadline.
Restrictions: Request amounts should not exceed 50% of the total project budget. NYSCA also does not fund any projects that are not directed toward programs in New York State, competitions or contests, or out-of-state travel expenses.
Geographic Focus: New York
Date(s) Application is Due: Apr 1
Contact: Beverly D'Anne, Senior Program Officer; (212) 741-3232; bdanne@nysca.org
Internet: http://www.nysca.org/public/guidelines/music/services.htm
Sponsor: New York State Council on the Arts
175 Varick Street
New York, NY 10014-4604

NYZS/Wildlife Conservation Society Research Grants **3709**
The society operates an international program of research and action that has a direct bearing on the conservation of threatened wild animals and habitats. The research development program is designed to support individual research projects that lead to concrete advances in the understanding and conservation of wildlife. Each project must incorporate clearly identifiable, specific conservation components. To obtain an application, send a one- or two-page cover letter containing a summary of the proposed research. The RFP will not limit any individual from applying, however, most of our grantees are either professional conservationists from the country of research, and/or postgraduates pursuing a higher degree.
Requirements: Any individual is eligible to apply. Applications will be accepted for field research in Africa, Asia, and Latin America (including Mexico.)
Restrictions: Organizations are not eligible for funding. Previous research fellows are not eligible for funding for the same project. Faculty and/or research advisors should not be listed as principal investigators unless they plan to conduct and achieve the majority of the field work. The principal researcher must write the proposal. Applications written on behalf of another individual are not allowed. Translation by a third party is permitted, however, applications are accepted in English, French and Spanish. Support will not be provided for conferences, airfare to scientific meetings, legal actions, building of permanent field stations, salaries at institutions, overhead costs, or expensive laboratory analyses. Stipends will be considered only when the investigator has no other source of support. Research will not be supported in North America (except Mexico), Europe, or Australia.
Geographic Focus: All States
Date(s) Application is Due: Mar 1; Sep 1
Amount of Grant: 1,000-25,000 USD
Contact: Coordinator; (718) 220-6828; fax (718) 364-4275; fellowship@wcs.org
Internet: http://www.wcs.org/international/rfp
Sponsor: New York Zoological Society / the Wildlife Conservation Society
2300 Southern Boulevard
Bronx, NY 10460

OAH/IEHS John Higham Travel Grants **3710**

The grant is given in memory of John Higham (1920-2003), past president of both organizations and a towering figure in immigration, ethnic, and intellectual history. Travel grants of $500 are awarded to three (3) graduate students each year. Funds are to be used by graduate students toward costs of attending the OAH/IEHS annual meeting. The successful candidates will have a preferred area of concentration in American Immigration and/or American Ethnic and/or American Intellectual history. The committee will seek some balance by gender, region of country, and type of university (e.g., major research university, and second tier).
Geographic Focus: All States
Date(s) Application is Due: Dec 3
Amount of Grant: 500 USD
Contact: Kara Hamm; (812) 855-7311; fax (812) 855-0696; khamm@oah.org
Internet: http://www.oah.org/awards/awards.higham.index.html
Sponsor: Organization of American Historians
112 N Bryan Avenue, P.O. Box 5457
Bloomington, IN 47408-4141

OAH Avery O. Craven Award 3711
The award is given each year for the most original book on the coming of the Civil War, the Civil War years, or the era of Reconstruction, with the exception of works of purely military history. The exception recognizes and reflects Craven's Quaker convictions. Avery O. Craven was president of the OAH 1963-1964. One copy of each entry, clearly labeled Avery O. Craven Award Entry, must be mailed directly to committee members. Guidelines are available online.
Requirements: Each entry must be published during the period January 1 through December 31 of the current year.
Geographic Focus: All States
Date(s) Application is Due: Oct 1
Amount of Grant: 500 USD
Contact: Kara Hamm; (812) 855-7311; fax (812) 855-0696; khamm@oah.org
Internet: http://www.oah.org/awards/awards.craven.index.html
Sponsor: Organization of American Historians
112 N Bryan Avenue, P.O. Box 5457
Bloomington, IN 47408-4141

OAH Binkley-Stephenson Award 3712
The Binkley-Stephenson Award has honored scholarly work since 1967. The award recognizes the best article published in the Journal of American History during the preceding calendar year. The journal is published quarterly in March, June, September, and December. It is widely recognized as an important source of research on all aspects of American history. William C. Binkley was president of the OAH 1944-1946 and editor of the Journal of American History 1956-1963.
Geographic Focus: All States
Amount of Grant: 500 USD
Contact: Kara Hamm; (812) 855-7311; fax (812) 855-0696; khamm@oah.org
Internet: http://www.oah.org/awards/awards.binkleystephenson.index.html
Sponsor: Organization of American Historians
112 N Bryan Avenue, P.O. Box 5457
Bloomington, IN 47408-4141

OAH Darlene Clark Hine Award 3713
The Darlene Clark Hine Award is given annually by the Organization of American Historians to the author of the best book in African American women's and gender history. The award is named for Darlene Clark Hine, a pioneer in African American women's and gender history and past president of the OAH (2001-2002).
Requirements: Individuals as well as editors are encouraged to submit nominations. Each entry must be published during the period January 1 through December 31 of the award year.
Geographic Focus: All States
Date(s) Application is Due: Oct 1
Amount of Grant: 500 USD
Contact: Kara Hamm; (812) 855-7311; fax (812) 855-0696; khamm@oah.org
Internet: http://www.oah.org/awards/awards.hine.index.html
Sponsor: Organization of American Historians
112 N Bryan Avenue, P.O. Box 5457
Bloomington, IN 47408-4141

OAH David Thelen Award 3714
The OAH sponsors this biennial award (formerly the Foreign Language Article Prize through 1998) for the best article on American history published in a foreign language. The winning article will be published in the Journal of American History. An article should be concerned with the past (recent or distant) or with issues of continuity and change. It should also be concerned with events or processes that began, developed, or ended in what is now the United States. It should make a significant and original contribution to the understanding of U.S. history. Comparative and international studies that fall within these guidelines are welcome.
Requirements: Entries must have been published during the preceding two calendar years.
Geographic Focus: All States
Date(s) Application is Due: May 1
Amount of Grant: 500 USD
Contact: Kara Hamm; (812) 855-7311; fax (812) 855-0696; khamm@oah.org
Internet: http://www.oah.org/awards/awards.thelen.index.html
Sponsor: Organization of American Historians
112 N Bryan Avenue, P.O. Box 5457
Bloomington, IN 47408-4141

OAH Ellis W. Hawley Prize 3715
The prize is awarded annually for the best book-length historical study of the political economy, politics, or institutions of the United States, in its domestic or international affairs, from the Civil War to the present. Eligible works include book-length historical studies, written in English, published during a given calendar year. Dissertations completed by October 1 also are eligible. One copy of each entry, clearly labeled Ellis W. Hawley Prize Entry, must be mailed directly to committee members. Guidelines are available online.
Requirements: Each entry must be published during the period January 1 through December 31 of the current year,
Geographic Focus: All States
Date(s) Application is Due: Oct 1
Amount of Grant: 500 USD
Contact: Kara Hamm; (812) 855-7311; fax (812) 855-0696; khamm@oah.org
Internet: http://www.oah.org/awards/awards.hawley.index.html
Sponsor: Organization of American Historians
112 N Bryan Avenue, P.O. Box 5457
Bloomington, IN 47408-4141

OAH Erik Barnouw Award 3716
One or two awards are given annually in recognition of outstanding reporting or programming on network or cable television, or in documentary film, concerned with American history, the study of American history, and/or the promotion of history. The winning film(s) or video program(s) will be screened, and the award(s) will be presented at the annual meeting in March. Films and videos selected as honorable mentions will be shown at the annual meeting, along with the Barnouw winner, as part of the OAH Screening History film series. One copy of each entry, clearly labeled Erik Barnouw Award Entry, must be mailed directly to committee members.
Requirements: Films and video programs released between January 1 and December 31 of the current year are eligible for entry.
Geographic Focus: All States
Date(s) Application is Due: Jan 10
Amount of Grant: 1,000 USD
Contact: Kara Hamm; (812) 855-7311; fax (812) 855-0696; khamm@oah.org
Internet: http://www.oah.org/awards/awards.barnouw.index.html
Sponsor: Organization of American Historians
112 N Bryan Avenue, P.O. Box 5457
Bloomington, IN 47408-4141

OAH Frederick Jackson Turner Award 3717
The award is given annually for an author's first book dealing with some significant phase of American history. The work must be the first book-length study of history published by the author and must have been published or scheduled for publication in the calendar year before the award is given. A copy of each entry should be submitted. Final page proofs may be used for books to be published after September 1 and before January 1 of the following year. The publisher of the winning book will receive a complimentary ad for the book in the Journal of American History and a certificate and the author will receive a medal.
Requirements: If the author is a PhD, she/he must have received the degree no earlier than seven years prior to submitting the manuscript for publication. Each entry must be published during the period January 1 through December 31 of the current year.
Restrictions: Each press may submit one entry per year.
Geographic Focus: All States
Date(s) Application is Due: Oct 1
Amount of Grant: 1,000 USD
Contact: Kara Hamm; (812) 855-7311; fax (812) 855-0696; khamm@oah.org
Internet: http://www.oah.org/awards/awards.turner.index.html
Sponsor: Organization of American Historians
112 N Bryan Avenue, P.O. Box 5457
Bloomington, IN 47408-4141

OAH Friend of History Award 3718
First given in 2005, the Friend of History Award recognizes an institution or organization, or an individual working primarily outside college or university settings, for outstanding support for historical research, the public presentation of American history, or the work of the OAH. Recipients will be selected by the OAH Executive Committee from nominations that should include a statement of no more than five pages describing the nominee's distinguished contributions and, if for an individual, a CV of no more than five pages.
Geographic Focus: All States
Amount of Grant: 500 USD
Contact: Kara Hamm; (812) 855-7311; fax (812) 855-0696; khamm@oah.org
Internet: http://www.oah.org/awards/awards.friendsofhistory.index.html
Sponsor: Organization of American Historians
112 N Bryan Avenue, P.O. Box 5457
Bloomington, IN 47408-4141

OAH Germany Residency Grants 3719
The committee seeks applications from OAH members who are established scholars affiliated with an American or Canadian University interested in leading an advanced undergraduate/graduate student seminar focusing on one aspect of U.S. History. The

seminar will be based on a topic in U.S. History of an applicant's design, and all fields and methodologies are welcome. The language of instruction for the seminar is English. The Residency Program will provide round-trip airfare, housing for thirty days, a modest honorarium (1,500 Euros), a graduate assistant, and office space. Tubingen is located in Southwestern Germany, 25 miles south of Stuttgart. It is one of Germany's oldest and most prestigious institutions of higher education and has a vibrant community engaged in the study of the United States.
Requirements: Applicants must be members of the OAH, have a Ph.D., and be scholars of American history.
Geographic Focus: All States
Date(s) Application is Due: Oct 1
Contact: Kara Hamm; (812) 855-7311; fax (812) 855-0696; khamm@oah.org
Internet: http://www.oah.org/programs/germany/index.html
Sponsor: Organization of American Historians
112 N Bryan Avenue, P.O. Box 5457
Bloomington, IN 47408-4141

OAH Huggins-Quarles Award **3720**
The award is given annually to one or two minority graduate students at the dissertation research stage of their PhD program. To apply, a student should submit a brief two-page abstract of his/her dissertation project, along with a one-page budget explaining travel and research plans. Each application must be accompanied by a letter from the dissertation adviser attesting to the student's status and the ways in which the award will facilitate the completion of the dissertation project. Include email addresses for both the applicant and the adviser, if available. One complete copy of each application (including cover letter, abstract, budget, and reference letter) should be clearly labeled Huggins-Quarles Award Entry and received, by the listed deadline date, by each member of the committee. Guidelines are available online.
Geographic Focus: All States
Date(s) Application is Due: Dec 3
Amount of Grant: 1,000 USD
Contact: Kara Hamm; (812) 855-7311; fax (812) 855-0696; khamm@oah.org
Internet: http://www.oah.org/awards/awards.hugginsquarles.index.html
Sponsor: Organization of American Historians
112 N Bryan Avenue, P.O. Box 5457
Bloomington, IN 47408-4141

OAH James A. Rawley Prize **3721**
This prize is given each year for a book dealing with the history of race relations in the United States. Each entry must be published during the period January 1 through December 31 in the year of deadline date. One copy of each entry, clearly labeled James A. Rawley Prize Entry, must be mailed directly to each committee member. The award will be presented at the OAH annual meeting. The publisher will receive a certificate of merit.
Geographic Focus: All States
Date(s) Application is Due: Oct 1
Amount of Grant: 1,000 USD
Contact: Kara Hamm; (812) 855-7311; fax (812) 855-0696; khamm@oah.org
Internet: http://www.oah.org/awards/awards.rawley.index.html
Sponsor: Organization of American Historians
112 N Bryan Avenue, P.O. Box 5457
Bloomington, IN 47408-4141

OAH Japanese Residencies Grants **3722**
The OAH sends two American scholars to Japanese universities annually for two-week residencies. Round-trip airfare to Japan, housing, and modest daily expenses are covered by the award. There, in English, the American historians give lectures and seminars in their specialty and provide individual consultation to Japanese scholars, graduate students and sometimes undergraduates studying American history and culture. Visitors also participate in the collegial life of their host institutions and help expand personal scholarly networks between Japan and the U.S. The aim of the program is to contribute to the expansion of personal scholarly networks between the two countries. Round-trip airfare to Japan, housing, and modest daily expenses are covered. Awardees are also encouraged to explore Japan before or after their two-week residency at their own expense.
Requirements: Applicants must be members of the OAH, have a Ph.D., and be scholars of American history.
Geographic Focus: All States
Date(s) Application is Due: Oct 15
Contact: Kara Hamm; (812) 855-7311; fax (812) 855-0696; khamm@oah.org
Internet: http://www.oah.org/programs/residencies/index.html
Sponsor: Organization of American Historians
112 N Bryan Avenue, P.O. Box 5457
Bloomington, IN 47408-4141

OAH Lerner-Scott Dissertation Prize **3723**
This prize is given annually to the best doctoral dissertation in women's history. Each application must contain a letter of support from a faculty member at the degree-granting institution, along with an abstract, table of contents and sample chapter from the dissertation. Also include email addresses for both the applicant and the adviser, if available. One complete copy of each entry must be received by each member of the prize committee by the listed deadline. Finalists will be asked to submit a complete copy of the dissertation at a later date. One copy of each entry, clearly labeled Lerner-Scott Dissertation Prize Entry, must be mailed directly to committee members. Guidelines are available online.
Requirements: A dissertation must be completed during the period July 1 of last year through June 30 of the current year.
Geographic Focus: All States
Date(s) Application is Due: Oct 1
Amount of Grant: 1,000 USD
Contact: Kara Hamm; (812) 855-7311; fax (812) 855-0696; khamm@oah.org
Internet: http://www.oah.org/awards/awards.lernerscott.index.html
Sponsor: Organization of American Historians
112 N Bryan Avenue, P.O. Box 5457
Bloomington, IN 47408-4141

OAH Liberty Legacy Foundation Award **3724**
The award will be given for the best book on any historical aspect of the struggle for civil rights in the United States from the nation's founding to the present. One copy of each entry must be received by each committee member by the listed application deadline date. Final page proofs may be used for books to be published after the deadline and before 1 January. If a final page proof is submitted, a bound copy of the entry must be received no later than 7 January. The author of the winning book will receive a monetary prize and a certificate. The publisher of the book will be honored with a certificate of merit. One copy of each entry, clearly labeled 2006 Liberty Legacy Foundation Award, must be mailed directly to committee members. Guidelines are avaialable onlin.
Requirements: Each entry must be published during the period 1 January through 31 December.
Geographic Focus: All States
Date(s) Application is Due: Oct 1
Amount of Grant: 2,000 USD
Contact: Kara Hamm; (812) 855-7311; fax (812) 855-0696; khamm@oah.org
Internet: http://www.oah.org/awards/awards.liberty.index.html
Sponsor: Organization of American Historians
112 N Bryan Avenue, P.O. Box 5457
Bloomington, IN 47408-4141

OAH Louis Pelzer Memorial Award **3725**
Candidates for graduate degrees are invited to submit essays for the Louis Pelzer Memorial Award competition. Essays may deal with any period or topic in the history of the United States. The winning essay will be published in the Journal of American History. Significance of the subject matter, literary craftsmanship and competence in the handling of evidence are some of the factors that will be considered in judging the essays. Manuscripts should be clearly labeled (list current year) Louis Pelzer Memorial Award Entry and and addressed to the committee chair). The winner will be notified by the organization and furnished with details of the annual meeting and the awards presentation. Guidelines are available online.
Requirements: Essays should be submitted in quintuplicate and should not exceed 7000 words (including endnotes) in length. The endnotes should be triple-spaced. Because manuscripts are judged anonymously, the author's name and graduate program should appear only on a separate cover page.
Geographic Focus: All States
Date(s) Application is Due: Nov 30
Amount of Grant: 500 USD
Contact: Kara Hamm; (812) 855-7311; fax (812) 855-0696; khamm@oah.org
Internet: http://www.oah.org/awards/awards.pelzer.index.html
Sponsor: Organization of American Historians
112 N Bryan Avenue, P.O. Box 5457
Bloomington, IN 47408-4141

OAH Merle Curti Award **3726**
The award is given annually to the best work published in the last two years in the field of American social history (even-numbered years) and intellectual history (odd-numbered years). The author of the winning book will receive a monetary prize, a certificate, and a medal. The publisher(s) of the book(s) will be honored with a certificate(s) of merit. The award will be presented at the annual meeting of the OAH. One copy of each entry, clearly labeled Merle Curti Award Entry, must be mailed directly to the members of the chair. Guidelines are available online.
Geographic Focus: All States
Date(s) Application is Due: Oct 1
Amount of Grant: 2,000 USD
Contact: Kara Hamm; (812) 855-7311; fax (812) 855-0696; khamm@oah.org
Internet: http://www.oah.org/awards/awards.curti.index.html
Sponsor: Organization of American Historians
112 N Bryan Avenue, P.O. Box 5457
Bloomington, IN 47408-4141

OAH Ray A. Billington Prize **3727**
The award is given biennially for the best book on American frontier history published during the last two years. American frontier history is defined broadly so as to include the pioneer periods of all geographical areas and comparisons between American frontiers and others. The deadline is for even-numbered years. One copy of each entry, clearly labeled Ray Allen Billington Prize Entry, must be mailed directly to each committee member. Guidelines are available online.
Geographic Focus: All States
Date(s) Application is Due: Oct 1
Amount of Grant: 1,000 USD
Contact: Kara Hamm; (812) 855-7311; fax (812) 855-0696; khamm@oah.org
Internet: http://www.oah.org/awards/awards.billington.index.html

Sponsor: Organization of American Historians
112 N Bryan Avenue, P.O. Box 5457
Bloomington, IN 47408-4141

OAH Richard W. Leopold Prize **3728**

This award is designed to improve contacts and interrelationships within the historical profession where an increasing number of history-trained scholars hold distinguished positions in governmental agencies. This prize recognizes the significant historical work being done by historians outside academe. The award is given every two years for the best book written by a historian connected with federal, state, or municipal government, in the areas of foreign policy, military affairs broadly construed, the historical activities of the federal government, or biography in one of the foregoing areas. One copy of each entry, clearly labeled Richard W. Leopold Prize Entry, must be mailed directly to committee members.
Requirements: Each entry must be published during the two-year period between January 1 of the last year and December 31 of the current year. The winner must have been employed in a government position for at least five years.
Geographic Focus: All States
Date(s) Application is Due: Oct 1
Amount of Grant: 1,500 USD
Contact: Kara Hamm; (812) 855-7311; fax (812) 855-0696; khamm@oah.org
Internet: http://www.oah.org/awards/awards.leopold.index.html
Sponsor: Organization of American Historians
112 N Bryan Avenue, P.O. Box 5457
Bloomington, IN 47408-4141

OAH Roy Rosenzweig Distinguished Service Award **3729**

The OAH Roy Rosenzweig Distinguished Service Award is presented each year to an individual or individuals whose contributions have significantly enriched our understanding and appreciation of American history. Beginning in 2008, the award is named in memory of Roy Rosenzweig, who was the Mark and Barbara Fried Chair and founding director of the Center for History and New Media at George Mason University. Rosenzweig, a pioneer in the field of digital humanities, was actively involved in OAH for more than fifteen years, serving on its nominating board, program committees, executive board, and was an OAH Distinguished Lecturer. In 2011, George Mason University renamed the Center for History and New Media to honor his memory. Rosenzweig passed away in 2007 at the age of 57. Those eligible for this award include scholars, teachers, writers, public historians and officeholders who have shaped or deepened our sensitivity to the past.
Geographic Focus: All States
Date(s) Application is Due: Sep 10
Contact: Kara Hamm; (812) 855-7311; fax (812) 855-0696; khamm@oah.org
Internet: http://www.oah.org/awards/awards.distinguished.index.html
Sponsor: Organization of American Historians
112 N Bryan Avenue, P.O. Box 5457
Bloomington, IN 47408-4141

OAH Willi Paul Adams Award **3730**

The OAH sponsors a biennial award for the best book on American history published in a foreign language. The award was formerly known as the Foreign Language Book Prize. The OAH defines both history and American broadly. To be eligible, a book should be concerned with the past (recent or distant) or with issues of continuity and change. It should also be concerned with events or processes that began, developed, or ended in what is now the United States. Comparative and international studies that fall within these guidelines are welcomed. Authors of eligible books are invited to nominate their work. Under unusual circumstances unpublished manuscripts will be considered.
Requirements: Each entry must have been published during the two-year period July 1 through June 30 of the current year.
Geographic Focus: All States
Date(s) Application is Due: May 1
Contact: Kara Hamm; (812) 855-7311; fax (812) 855-0696; khamm@oah.org
Internet: http://www.oah.org/awards/awards.adams.index.html
Sponsor: Organization of American Historians
112 N Bryan Avenue, P.O. Box 5457
Bloomington, IN 47408-4141

Oak Foundation Environmental Grants **3731**

The foundation awards grants worldwide to address international social and environmental issues, particularly those that have a major impact on the lives of the disadvantaged. In the climate change and marine conservation programs, the Foundation typically invites organizations to submit letters of enquiry based on annual funding priorities, internal guidelines, budget cycle and timelines. If you have not been contacted by a Program Officer directly to submit a letter of enquiry, the chances are relatively low that an organization will be considered for funding. If you decide to submit an unsolicited letter of inquiry, the Foundation will respond within three months.
Requirements: Within this program, Oak funds organizations throughout Europe, the United States, Canada, and Belize.
Restrictions: Oak does not provide support to individuals, and does not provide funding for scholarships or tuition assistance for undergraduate or postgraduate studies. The Foundation also does not fund religious organizations for religious purposes, election campaigns, or general fund-raising drives.
Geographic Focus: All States, Albania, Andorra, Armenia, Austria, Azerbaijan, Belarus, Belgium, Belize, Bosnia & Herzegovina, Bulgaria, Canada, Croatia, Cyprus, Czech Republic, Denmark, Estonia, Finland, France, Germany, Greece, Hungary, Iceland, Ireland, Italy, Kosovo, Latvia, Liechtenstein, Lithuania, Luxembourg, Macedonia, Malta, Moldova, Monaco, Montenegro, Netherlands, Norway, Poland, Portugal, Romania, Russia, San Marino, Serbia, Slovakia, Slovenia, Spain, Sweden, Switzerland, Turkey, Ukraine, United Kingdom, Vatican City
Amount of Grant: 25,000-2,000,000 USD
Contact: Leonardo Lacerda; naep@oakfnd.org or eep@oakfnd.ch
Internet: http://www.oakfnd.org/node/22
Sponsor: Oak Foundation
Case Postale 115 58, Avenue Louis Casai
Geneva, Cointrin 1216 Switzerland

Obesity Society Grants **3732**

The Obesity Society's Grants Program was founded to promote, reward, and encourage research in the field of obesity. Pilot grants of up to $25,000 for a 1-year period will be offered to foster and stimulate new research ideas in any area of investigation related to obesity.
Requirements: Members of The Obesity Society (domestic and international) are eligible to apply. Individuals must be qualified by training and experience to conduct his or her responsibilities on the research project.
Geographic Focus: All States
Amount of Grant: 25,000 USD
Contact: Sadie Campbell, (301) 563-6526; fax (301) 563-6595; scampbell@obesity.org
Internet: http://www.obesity.org/about-us/obesity-society-grants.htm
Sponsor: Obesity Society
8757 Georgia Avenue, Suite 1320
Silver Spring, MD 20910

OBSSR Behavioral and Social Science Research on Understanding & Reducing Health Disparities Grants **3733**

The purpose of this program is to encourage behavioral and social science research on the causes and solutions to health and disabilities disparities in the U. S. population. Health disparities between, on the one hand, racial/ethnic populations, lower socioeconomic classes, and rural residents and, on the other hand, the overall U.S. population are major public health concerns. Emphasis is placed on research in and among three broad areas of action: public policy; health care; and disease/disability prevention. Particular attention is given to reducing health gaps among groups. Proposals that utilize an interdisciplinary approach, investigate multiple levels of analysis, incorporate a life-course perspective, and/or employ innovative methods such as system science or community-based participatory research are particularly encouraged. The NIH anticipates supporting 20 to 30 awards. Because the nature and scope of the proposed research will vary from application to application, it is anticipated that the size and duration of each award will also vary.
Requirements: Application can be made by the following institutions and organizations: public/state controlled institutions of higher education; private institutions of higher education; nonprofits with 501(c)3 IRS status (other than institutions of higher education); nonprofits without 501(c)3 IRS status (other than institutions of higher education); small businesses; for-profit organizations (other than small businesses); state governments; U.S. territories or possessions; Indian/Native American tribal governments (federally recognized); Indian/Native American tribal governments (other than Federally recognized); Indian/Native American tribally designated organizations; non-domestic (non-U.S.) entities (foreign organizations); Hispanic-serving institutions; historically Black colleges and universities (HBCUs); tribally controlled colleges and universities (TCCUs); Alaska Native and Native Hawaiian serving institutions; regional organization; eligible agencies of the Federal government; and faith-based or community based organizations.
Geographic Focus: All States
Date(s) Application is Due: Feb 5; Jun 5; Oct 5
Contact: Ronald P. Abeles, (301) 496-7859; fax (301) 435-8779; abeles@nih.gov
Internet: http://grants.nih.gov/grants/guide/pa-files/PAR-07-379.html
Sponsor: Office of Behavioral and Social Sciences Research
31 Center Drive, Building 31, Room B1C19
Bethesda, MD 20892-2027

Ohio Learning Network Grants **3734**

The network funds programs to help Ohioans increase work-related knowledge and skills through college credits; assist faculty in teaching using emerging technologies; and better understand the role of technology in learning. Grants support emerging needs, learning communities, distance learning research, partnership grants, and technology initiatives/innovations.
Geographic Focus: Ohio
Amount of Grant: Up to 100,000 USD
Contact: Judith M. Leach; (614) 485-6738 or (614) 995-3240; jleach@oln.org
Internet: http://www.oln.org/about_oln/grants.php
Sponsor: Ohio Learning Network
35 E Chestnut Street, 8th Floor
Columbus, OH 43215-2541

OISE Muriel and Danny Fung Graduate Scholarships **3735**

The purpose of this award is to give support to students enrolled in graduate studies in the Department of Sociology and Equity Studies in Education at OISE, University of Toronto. Priority consideration is given to students who are not eligible to receive the University of Toronto's Graduate Funding Package. Specifically, the Muriel and Danny Fung Graduate Award will be awarded annually to a student who is registered full time in the second year or later year of their program and who's academic and community work promote racial equality and social justice.
Restrictions: Student must be registered full-time.
Geographic Focus: Canada
Date(s) Application is Due: Mar 15

Contact: Kathleen Gallagher, (416) 978-0160; kathleen.gallagher@utoronto.ca
Internet: http://www.oise.utoronto.ca/sese/Students/Financial_Scholarships/
Sponsor: Ontario Institute for Studies in Education
252 Bloor Street West
Toronto, ON M5S 1V6 Canada

OISE William Waters Scholarships in Urban Education **3736**
Established in 2005, the Centre for Urban Schooling (CU.S.) is dedicated to improving the quality of the teaching and learning experiences available to children and youth living in under-served urban communities. The Centre for Urban Schooling conducts research, offers educational programs and professional development opportunities, and supports students, teachers and their communities to develop school reform initiatives with a focus on equity for all students. With more than twenty affiliated professors and researchers from OISE's departments, CU.S. is quickly gaining recognition as Canada's leading research and programming center in issues of urban education. OISE is offering four $30,000 scholarships to promote excellence in teaching in the urban classroom. The successful candidates will be experienced teachers entering a full-time Master's program with an interest in questions of social justice and school success for students from economically disadvantaged neighborhoods attending under-performing schools. In addition to the requirements set out by the graduate program, the successful scholarship recipients will be invited to join the Centre for Urban Schooling and participate in the wide variety of activities related to research, professional development, advocacy and teacher education promoted by the Center over the course of the year.
Requirements: Students who are enrolled or will be enrolled full-time in a master's program will be eligible to be considered for assistantships for the first two years of their program. Students who are enrolled or will be enrolled full-time in a doctoral program will be eligible to be considered for assistantships for the first five years of their program.
Geographic Focus: Canada
Date(s) Application is Due: Jan 16
Amount of Grant: 30,000 CAD
Contact: Kathleen Gallagher, (416) 978-0160; kathleen.gallagher@utoronto.ca
Internet: http://www.oise.utoronto.ca/sese/Students/Financial_Scholarships/
Sponsor: Ontario Institute for Studies in Education
252 Bloor Street West
Toronto, ON M5S 1V6 Canada

Oklahoma City Community Programs and Grants **3737**
The community foundation functions to enhance the quality of life in Oklahoma by awarding grants to nonprofits in its areas of interest, including arts and cultural programs, education, health care, health associations, and human services. Types of support include general operating support, continuing support, annual campaigns, building/renovation, equipment, emergency funds, program development, seed grants, fellowships, scholarship funds, research, and matching funds. The grants assist organizations with activities that provide new or expanded services to the community in response to new opportunities and changing circumstances. The community foundation views its role as a partner with other organizations and regards contributions and participation by other individuals and groups as an essential element in grant activities. Especially encouraged are programs and services that are oriented toward the entire population and are not restricted only to those people who live in a particular geographic location or who are members of a particular organization. The foundation encourages nonprofit organizations to cooperate and pool their resources to meet community needs. The use of volunteers as a significant program component is considered favorably as is active participation of the agency's board of directors and a broad base of financial support in the community. Grants are generally made for a specified time period and are nonrenewable. Applications and guidelines for each program are available on the Web site.
Requirements: 501(c)3 nonprofit organizations serving the greater Oklahoma City area, and programs that benefit persons living in the greater Oklahoma City metropolitan area, are eligible.
Restrictions: Grants are not awarded to individuals or to benefit specific individuals. No faxed or emailed proposals are accepted.
Geographic Focus: Oklahoma
Contact: Grants Manager; (405) 235-5603; fax (405) 235-5612; info@occf.org
Internet: http://www.occf.org/occf/community_programs/index.php
Sponsor: Oklahoma City Community Foundation
P.O. Box 1146
Oklahoma City, OK 73103-1146

Olfactory Research Fund Grants **3738**
The fund sponsors innovative scientific research in the olfactory arts and sciences. Grants promote the study of olfaction and also support educational and outreach programs about the sense of smell. The fund is especially interested in supporting research that integrates the study of olfaction with current issues in developmental, perceptual, social and cognitive psychology, and related disciplines. The fund is currently seeking proposals for research that focuses on the influence of olfaction in the mental and physical aspects of performance. Contact organization for submission deadlines.
Requirements: Applicants must have a doctoral degree.
Restrictions: Generally, the following types of research are ineligible for support: studies not directly related to olfaction and/or the human reaction to odors; studies exclusively on animal olfaction without direct relevance to humans; or invasive human studies.
Geographic Focus: All States
Contact: Theresa Molnar, Executive Administrator; (212) 725-2755; fax (212) 779-9072; tmolnar@senseofsmell.org or info@senseofsmell.org
Internet: http://www.senseofsmell.org/resources/research.php
Sponsor: Olfactory Research Fund
145 E 32nd Street
New York, NY 10016

Olin Corporation Charitable Trust Grants **3739**
The charitable trust supports nonprofits in company-operating areas. Grants focus on three priority categories: education, with a science and technology emphasis; programs that include Olin employees as volunteers and that impact communities in which Olin employees work and live; and conservation, environmental education, and environmental research. Types of support include general operating budgets, continuing support, annual campaigns, seed money, emergency funds, building funds, equipment, land acquisition, special projects, research, publications, conferences and seminars, internships, scholarship funds, employee-related scholarships, fellowships, and employee matching gifts. U.S.-based nonprofits with an international focus are supported primarily through matching gifts. There are no application deadlines. Applicants should submit a letter of request. The board meets in January to consider requests.
Requirements: Support is directed primarily to communities where Olin has a major employee presence.
Restrictions: Grants are not awarded to individuals or for endowments.
Geographic Focus: All States
Amount of Grant: 50-50,000 USD
Contact: Program Contact; (314) 480-1400
Internet: http://www.olin.com/about/charitable.asp
Sponsor: Olin Corporation Charitable Trust
190 Carondelet Plaza, Suite 1530
Clayton, MO 63105-3443

Olin Fellowships **3740**
The fellowships are offered to North American residents seeking to improve their knowledge or skills in advanced salmon-related fields while also seeking solutions to current problems in Atlantic salmon biology, management, or conservation. The fellowships may be applied toward a wide range of endeavors, including salmon management, graduate study, and research. Applicants need not be enrolled in a degree program. The awards are tenable at any accredited university or research laboratory or in an active management program. Application forms may be obtained from the Atlantic Salmon Federation, P.O. Box 807, Calais, ME, 04619, U.S.A; however, the completed form must be received at the federation's Saint Andrews office by the deadline date. Applicants will be advised of awards by May 15.
Requirements: Applicants must be legal residents of the United States or Canada and must complete the formal application form to which should be attached a statement of qualifications, transcript of grades if applicable, description of the program/project for which the fellowship is sought, and all supporting documentation required.
Geographic Focus: All States
Date(s) Application is Due: Mar 15
Amount of Grant: 1,000-3,000 CAD
Contact: Muriel Ferguson, Program Contact; (506) 529-1033
Internet: http://www.asf.ca/Awards/Awards.html
Sponsor: Atlantic Salmon Federation
P.O. Box 5200
Saint Andrews, NB E0G 2X0 Canada

OM Association Outstanding Dissertation Awards Program **3741**
The association honors doctoral candidates whose dissertations significantly advance the knowledge and understanding of education theories, concepts, and practices related to OM's mission to develop creative problem-solving skills important to a changing world. OM's mission includes the encouragement and development of cooperation, self-respect, and the appreciation and understanding of others. Winners will be notified by September 1.
Requirements: Any doctoral candidate may enter if his/her dissertation was completed and approved over the last three academic years from an institution of higher education that is sanctioned by a state, regional, or provincial accrediting agency.
Geographic Focus: All States
Date(s) Application is Due: Jun 1
Contact: Awards Coordinator; (609) 881-3596
Internet: http://www.isop.ucla.edu/eas/fellowships/om-assoc.htm
Sponsor: OM Association
P.O. Box 547
Glassboro, NJ 08028

OneSight Research Foundation Block Grants **3742**
Formerly known as the Pearle Vision Foundation, the OneSight Research Foundation is dedicated to sight preservation through research and education. A significant portion of their grant dollars will be awarded to fund research and treatment for diabetic eye diseases, hoping to find a cure for diabetic retinopathy, the number one cause of adult blindness in the United States.
Requirements: U.S. 501(c)3 nonprofits are eligible for grant support. Requests for funding should cover a period of time not to exceed one year. Exceptions to this policy will be considered on an individual basis.
Restrictions: Grants will not be considered for endowments or operating expenses.
Geographic Focus: All States
Date(s) Application is Due: Jun 30; Dec 31
Amount of Grant: 5,000-50,000 USD
Contact: Trina Parasiliti; (972) 277-6191; fax (972) 277-6414; tparasil@onesight.org
Internet: http://www.onesight.org/northamerica/na/
Sponsor: OneSight Research Foundation
2465 Joe Field Road
Dallas, TX 75229

ONF Small Grants 3743

These grant awards are designed to support oncology nursing research conducted by beginning and experienced researchers. The majority of the small grants are for general investigator-initiated oncology topics. Funding preference is given to research that addresses the ONS Research Priorities and Research Agenda. A letter of intent is requested by October 1; full proposal is due November 1. Annual deadline dates may vary; contact program staff for exact dates.
Requirements: The principal investigator must be actively involved in some aspect of cancer patient care, education, or research. Funding preference is given to projects that involve nurses in the design and conduct of the research activity and that promote oncology nursing theory and a scientific basis for practice.
Geographic Focus: All States
Date(s) Application is Due: Oct 1; Nov 1
Amount of Grant: 10,000 USD
Contact: Grants Administrator; research@ons.org
Internet: http://www.ons.org/awards/foundawards/smallGrants.shtml
Sponsor: Oncology Nursing Foundation
125 Enterprise Drive, RIDC Park W
Pittsburgh, PA 15275-1214

ONR Biological & Biomedical Division Grants 3744

The Biomedical and Biological Division's mission is to conduct research and technology demonstration programs directed at maintaining the survival, health and performance of Navy and Marine Corps personnel during training, routine operations, special operations, and in time of war. The Division's program is organized into five research programs: Casualty Care & Management, Casualty Prevention, Undersea Medicine, Stress Physiology and Naval Biosciences and Biocentric Technology. Each research program supports a focused initiative, the Force Health Protection Future Naval Capability, that will deliver specific technology products in the FY08-FY12 timeframe. Submission of the Research Proposal should always be preceded by submission of a Preproposal and a discussion of that Preproposal with the assigned Program Officer.
Requirements: Grants are awarded to academic, industrial, and nonprofit organizations, with strong emphasis on university research.
Restrictions: New grants or contracts are typically started near the beginning of the fiscal year (October 1). New proposals for the next fiscal year should be submitted no later than June 1, unless otherwise directed by the BAA. Competitive renewals should be submitted no later than six months prior to the award termination date after contacting the Program Officer. Conference proposals should be submitted at least four months prior to the conference.
Geographic Focus: All States
Contact: Dr. Igor Vodyanoy, (703) 696-4109 or (703) 588-2427; fax (703) 696-5940; igor.vodyanoy@navy.mil
Internet: http://www.onr.navy.mil/sci_tech/34/342/
Sponsor: Oncology Nursing Foundation
125 Enterprise Drive, RIDC Park W
Pittsburgh, PA 15275-1214

ONR Bisson Prize for Naval Technology Achievement 3745

This award is presented to recognize and reward an individual's successes in technology achievement. It is granted annually to a current or former navy scientist, engineer, or S&T program manager that has a significant, direct transitional impact from S&T to naval operations or systems. Such impact shall be so outstanding as to be widely recognized in the navy and the relevant technical community nationally. While the impact must have been realized in the year prior to nomination, it is recognized that the efforts leading to transition may have been ongoing for many years. The award consists of an individual plaque, a certificate, and a cash award. Additionally, the recipient's name is placed on the master plaque located on the Honorary Wall of Fame at ONR Headquarters.
Requirements: High-quality science and/or technology accomplishment, well-conducted transition of that accomplishment from S&T to naval acquisition or operations, and a significant impact upon the war fighting and/or peacekeeping capabilities of naval forces are criteria for the award. At the time of the achievement cited as the basis for the prize, nominees must have been military or civilian employees of the navy in positions managing or performing S&T.
Geographic Focus: All States
Date(s) Application is Due: Feb 15
Amount of Grant: 3,000 USD
Contact: Administrator; (703) 696-5031; fax (703) 696-5940; onrpao@onr.navy.mil
Internet: http://www.onr.navy.mil/sci_tech/34/341/training/transitions/bisson.htm
Sponsor: U.S. Department of Defense
One Liberty Center, 875 N. Randolph Street, Suite 1425
Arlington, VA 22203-1995

ONR Casualty Care and Management Research Grants 3746

This program addresses applied and advanced research and technology efforts in trauma medicine for use in austere environments with a focus on control of internatl bleeding without surgical access. Program interests include treatment of hemorrhagic shock, agents that control acute pain without impairing cognitive performance, therapeutics or other strategies that effect resuscitation without the use of fluids, and automated care (closed loop-control) of casualties during transport. Products of this research must ultimately be stable to ambient temperatures, small in size, light in weight, function in small volumes, be affordable, be patentable, and have the ability to be administered by a minimally trained individual. Submission of the Research Proposal should always be preceded by submission of a Preproposal and a discussion of that Preproposal with the assigned Program Officer.
Requirements: Grants are awarded to academic, industrial, and nonprofit organizations, with strong emphasis on university research.
Restrictions: New grants or contracts are typically started near the beginning of the fiscal year (October 1). New proposals for the next fiscal year should be submitted no later than June 1, unless otherwise directed by the BAA. Competitive renewals should be submitted no later than six months prior to the award termination date after contacting the Program Officer. Conference proposals should be submitted at least four months prior to the conference.
Geographic Focus: All States
Contact: Dr. Michael Given, (703) 696-4055 or (703) 588-2427; fax (703) 696-5940; michael.given@navy.mil
Internet: http://www.onr.navy.mil/sci_tech/34/342/cas_care.asp
Sponsor: U.S. Department of Defense
One Liberty Center, 875 N. Randolph Street, Suite 1425
Arlington, VA 22203-1995

ONR Casualty Prevention Grants 3747

This program addresses research and technology efforts focused on reducing casualties by enhancing an individual's situational awareness and countering threats from disease, battle and non-battle injuries. Solutions can be targeted by diverse disciplines such as biological, engineering and behavioral sciences, with the objective of developing improved methods, models, and techniques for understanding, preventing or mitigating sources of military casualties. Program interests include, but are not limited to: developing non-invasive or minimally invasive procedures for identifying infection by pathogenic organisms; characterizing blunt trauma effects of body armor; mitigating thermal effects during diving; and protecting personnel from lethal and non-lethal effects from lasers. Products of this program must function with minimal negative impact on operational capability. Submission of the Research Proposal should always be preceded by submission of a Preproposal and a discussion of that Preproposal with the assigned Program Officer.
Requirements: Grants are awarded to academic, industrial, and nonprofit organizations, with strong emphasis on university research.
Restrictions: New grants or contracts are typically started near the beginning of the fiscal year (October 1). New proposals for the next fiscal year should be submitted no later than June 1, unless otherwise directed by the BAA. Competitive renewals should be submitted no later than six months prior to the award termination date after contacting the Program Officer. Conference proposals should be submitted at least four months prior to the conference.
Geographic Focus: All States
Contact: Elizabeth A. Montcalm-Smith, (703) 696-2580 or (703) 588-2427; fax (703) 696-5940; elizabeth.montcalmsm@navy.mil
Internet: http://www.onr.navy.mil/sci_tech/34/342/cas_prevent.asp
Sponsor: U.S. Department of Defense
One Liberty Center, 875 N. Randolph Street, Suite 1425
Arlington, VA 22203-1995

ONR Circuits/Architectures Grants 3748

This program seeks innovation in mixed signal circuits, architectures and related all-encompassing aspects with emphasis on higher operating frequency, greater signal dynamic range. Innovations are expected to eventually lead to improvements in complex circuits such as analog-to-digital converters (ADCs), digital-to-analog converters (DACs), and direct high power conversion DACs at frequencies of microwave and millimeter wave. Examples are distributed sigma-delta and switched-mode amplifier based circuits and architectures which can be advantageously exploited for generation of multiple simultaneous RF/Microwave beams in electronically scanned arrays. Techniques that show potential to operate efficiently at 25 GHz and beyond are highly sought after. Also of interest are methods for rigorous electromagnetic modeling of complex mixed signal circuits with a potential to provide 30 dB improvement in dynamic range over current state of art. Submission of the Research Proposal should always be preceded by submission of a Preproposal and a discussion of that Preproposal with the assigned Program Officer. Proposals are preferred by April 1 of each year.
Requirements: Grants are awarded to academic, industrial, and nonprofit organizations, with strong emphasis on university research.
Geographic Focus: All States
Date(s) Application is Due: Apr 1
Amount of Grant: 115,000 USD
Contact: Dr. Daniel Purdy, (703) 588-1920 or (703) 696-5031; fax (703) 696-5940; dan.purdy@navy.mil or 312_GHz@onr.navy.mil
Internet: http://www.onr.navy.mil/sci_tech/31/312/circuits_architechtures.asp
Sponsor: U.S. Department of Defense
One Liberty Center, 875 N. Randolph Street, Suite 1425
Arlington, VA 22203-1995

ONR Cognitive Science Basic Research Grants 3749

The overall technical objectives of this program are to provide formal, computational theories of the fundamental information processing characteristics of the human cognitive architecture. Within this context, four distinct areas are attractive loci for future research efforts. They include: theoretical foundations for socio-cognitive architectures; perception, metacognition and cognitive control; representing and reasoning about uncertainty; and skill acquisition and learning. The application of much of the basic research generated through investigations of cognitive architectures is toward the development of artificially intelligent tutoring technology for training. Submission of the Research Proposal should always be preceded by submission of a Preproposal and a discussion of that Preproposal with the assigned Program Officer.

Requirements: Grants are awarded to academic, industrial, and nonprofit organizations, with strong emphasis on university research. Investigators are encouraged to contact the staff prior to formally submitting a proposal.
Restrictions: New grants or contracts are typically started near the beginning of the fiscal year (1 October). New proposals for the next fiscal year should be submitted no later than June 1, unless otherwise directed. Competitive renewals should be submitted no later than six months prior to the award termination date after contacting the Program Officer. Conference proposals should be submitted at least four months prior to the conference.
Geographic Focus: All States
Contact: Dr. Paul Bello, (703) 696-4318 or (703) 696-5031; paul.bello@navy.mil
Internet: http://www.onr.navy.mil/sci_tech/34/341/training_cog.asp
Sponsor: U.S. Department of Defense
One Liberty Center, 875 N. Randolph Street, Suite 1425
Arlington, VA 22203-1995

ONR Communications Systems Grants 3750

This division supports programs in surface and aerospace surveillance, communications, and electronic combat. Research areas in surface and aerospace surveillance include sensors, primarily radar and electro- optic/infrared, and associated signal and image processing methods. Areas of interest in communications include military radio communications with an emphasis on anti-jam and low-probability-of-intercept techniques, radio networks, and dynamic internetworking. Electronic combat research areas include threat warning systems, electronic support measures, decoys, and electronic countermeasures. Navigation research includes GPS anti-jam and associated inertial navigation techniques. Submission of the Research Proposal should always be preceded by submission of a Preproposal and a discussion of that Preproposal with the assigned Program Officer.
Requirements: Grants are awarded to academic, industrial, and nonprofit organizations, with strong emphasis on university research.
Restrictions: New grants or contracts are typically started near the beginning of the fiscal year (October 1). New proposals for the next fiscal year should be submitted no later than June 1, unless otherwise directed by the BAA. Competitive renewals should be submitted no later than six months prior to the award termination date after contacting the Program Officer. Conference proposals should be submitted at least four months prior to the conference.
Geographic Focus: All States
Contact: Dr. Santanu Das, (703) 588-1036 or (703) 588-2427; santanu.das@navy.mil
Internet: http://www.onr.navy.mil/sci_tech/31/313/default.asp
Sponsor: U.S. Department of Defense
One Liberty Center, 875 N. Randolph Street, Suite 1425
Arlington, VA 22203-1995

ONR Computational Analysis Grants 3751

The Program develops and exploits methods in modern and classical mathematical analysis with emphasis on mathematical and computational models of physical phenomena. Naval interest in waves, flows, materials, structures, and information processing motivates research in the areas of multi-phase, multi-physics, multi-material problems; fluid transport and mixing; predictability of models for nonlinear dynamics; motion and still image analysis, processing, and enhancement; and inverse problems arising from acoustic and electromagnetic wave propagation and scattering. Submission of the Research Proposal should always be preceded by submission of a Preproposal and a discussion of that Preproposal with the assigned Program Officer.
Requirements: Grants are awarded to academic, industrial, and nonprofit organizations, with strong emphasis on university research.
Restrictions: New grants or contracts are typically started near the beginning of the fiscal year (October 1). New proposals for the next fiscal year should be submitted no later than June 1, unless otherwise directed by the BAA. Competitive renewals should be submitted no later than six months prior to the award termination date after contacting the Program Officer. Conference proposals should be submitted at least four months prior to the conference.
Geographic Focus: All States
Contact: Dr. Reza Malek-Madani, (703) 696-4314 or (703) 588-2427; fax (703) 696-5940; reza.malekmadani@navy.mil
Internet: http://www.onr.navy.mil/sci_tech/31/311/computational_analysis.asp
Sponsor: U.S. Department of Defense
One Liberty Center, 875 N. Randolph Street, Suite 1425
Arlington, VA 22203-1995

ONR Computational Neuroscience & Biorobotics Grants 3752

This program aims to discover the principles of control, sensorimotor integration and biomechanics that underlie the agility, quietness, dexterity, and intelligence of swimming and flying biological systems. We seek to understand the robust dynamically stable locomotion of legged animals in unknown terrestrial and aquatic environments, and to formalize these principles to facilitate engineering invention and design. The program seeks to build a bridge between the science of biology and the art of Naval engineering by laying emphasis on the distillation of science in biology and the implementation of the kernel of it on existing Navy products to make significant gains in performance via novel inventions. Research proposals are expected to meld biology and engineering. Of particular interest are the following areas: interdisciplinary research on biology-based high-lift mechanisms and their incorporation on to existing Navy products like propulsors; principles of neural control and hybrid systems for biological and biology-based high-lift devices and system conceptualization, analysis and integration of biology-inspired high-lift mechanisms, artificial-muscle based actuators and their neural control for the development of advanced underwater vehicles and microflying vehicles; research on the neurobiological mechanisms and principles of control of fish pectoral fins, including closed loop control from biosensors; and visual guidance and sensorimotor integration for legged (especially bipedal robots) or wheeled robots that support human robot interaction. Submission of the Research Proposal should always be preceded by submission of a Preproposal and a discussion of that Preproposal with the assigned Program Officer.
Requirements: Grants are awarded to academic, industrial, and nonprofit organizations, with strong emphasis on university research. Investigators are encouraged to contact the staff prior to formally submitting a proposal.
Restrictions: New grants or contracts are typically started near the beginning of the fiscal year (1 October). New proposals for the next fiscal year should be submitted no later than June 1, unless otherwise directed. Competitive renewals should be submitted no later than six months prior to the award termination date after contacting the Program Officer. Conference proposals should be submitted at least four months prior to the conference.
Geographic Focus: All States
Contact: Dr. Thomas McKenna, (703) 696-4503 or (703) 696-5031; fax (703) 696-5940; tom.mckenna@navy.mil
Internet: http://www.onr.navy.mil/sci_tech/34/341/ne_biorobotics.asp
Sponsor: U.S. Department of Defense
One Liberty Center, 875 N. Randolph Street, Suite 1425
Arlington, VA 22203-1995

ONR Defense University Research Instrumentation Program (DURIP) Grants 3753

DURIP is a multi-agency DoD program within the University Research Initiative designed to improve the capabilities of U.S. institutions of higher education to conduct research and to educate scientists and engineers in areas important to national defense by providing funds for the acquisition of research equipment. Each year that DURIP is in effect, a complete announcement will provide detailed program information, including the deadline date for submitting proposals. DURIP proposals submitted to ONR should facilitate research in an area of interest to ONR, as described in the Science and Technology section of the ONR Website. Potential proposers may contact the appropriate program managers, listed with each area of interest, to explore possible mutual interests before submitting proposals. Proposals are evaluated competitively. The evaluation criteria are listed in the current program announcement.
Requirements: Grants are awarded to academic, industrial, and nonprofit organizations, with strong emphasis on university research.
Geographic Focus: All States
Contact: Dr. William Lukens, (703) 696-4668 or (703) 588-2427; fax (703) 696-5940; william.luckens1@navy.mil
Internet: http://www.onr.navy.mil/sci_tech/3t/corporate/durip.asp
Sponsor: U.S. Department of Defense
One Liberty Center, 875 N. Randolph Street, Suite 1425
Arlington, VA 22203-1995

ONR Device and Circuit Cooling Grants 3754

This program investigates fundamental properties associated with heat transport and cooling of semiconductor and superconductor materials, devices, and circuits. Of particular attention are interfaces between dissimilar materials exhibiting differing characteristic acoustic impedances. Submission of the Research Proposal should always be preceded by submission of a Preproposal and a discussion of that Preproposal with the assigned Program Officer.
Requirements: Grants are awarded to academic, industrial, and nonprofit organizations, with strong emphasis on university research.
Restrictions: New grants or contracts are typically started near the beginning of the fiscal year (October 1). New proposals for the next fiscal year should be submitted no later than June 1, unless otherwise directed by the BAA. Competitive renewals should be submitted no later than six months prior to the award termination date after contacting the Program Officer. Conference proposals should be submitted at least four months prior to the conference.
Geographic Focus: All States
Contact: Dr. Deborah Van Vechten, (703) 696-4219 or (703) 588-2427; fax (703) 696-5940; deborah.vanvechten@navy.mil
Internet: http://www.onr.navy.mil/sci_tech/31/312/device_circuit_cooling.asp
Sponsor: U.S. Department of Defense
One Liberty Center, 875 N. Randolph Street, Suite 1425
Arlington, VA 22203-1995

ONR DOD National Defense Science & Engineering Graduate Fellowships 3755

The NDSEG Fellowship Program is a joint program of the United States Army, Navy and Air Force within the University Research Initiative (URI), designed to increase the number of U.S. citizens trained in disciplines of science and engineering important to defense goals. DoD awards approximately 100-150 new three-year graduate fellowships each year to individuals for study and research leading to doctoral degrees in, or closely related to, the disciplines of aeronautical and astronautical engineering; biosciences, chemical engineering; chemistry; cognitive, neural, and behavioral sciences; electrical engineering; geosciences; civil engineering; computer and computational sciences; materials science and engineering; mathematics; mechanical engineering; naval architecture and ocean engineering; oceanography; and physics. The Fellowships are intended for students at or near the beginning of graduate study. The tenure of an NDSEG Fellowship is 36 months. Consult the current program announcement for stipend levels. In addition to stipends, the DoD pays the Fellow's full tuition and required fees. The deadline for application is in January.

Requirements: Applicants must be citizens or nationals of the United States who intend to pursue a Ph.D. in one of the designated fields. Applications are encouraged from women, underrepresented minorities and persons with disabilities. Recipients of new three-year fellowships must begin tenure not later than the fall following award.
Geographic Focus: All States
Contact: Dr. Anthony V. Junior, (703) 696-0453 or (703) 588-2427; fax (703) 696-5940; anthony.junior@navy.mil
Internet: http://www.onr.navy.mil/sci_tech/3t/corporate/gradfe.asp
Sponsor: U.S. Department of Defense
One Liberty Center, 875 N. Randolph Street, Suite 1425
Arlington, VA 22203-1995

ONR Electronic and Optical Ceramics Research Grants **3756**
The goal of the program is to further the understanding of piezoelectric crystals and ceramics. Lying at the heart of most SONAR transducers, these materials and structures perform the essential roles of producing strong interrogating sound pulse and also detecting very weak incoming sound waves. As looking through the sea for navigation, threat detection and weapon guidance is a critical Navy need, ONR has designated this area as a National Naval Responsibility. Submission of the Research Proposal should always be preceded by submission of a Preproposal and a discussion of that Preproposal with the assigned Program Officer.
Requirements: Grants are awarded to academic, industrial, and nonprofit organizations, with strong emphasis on university research.
Restrictions: New grants or contracts are typically started near the beginning of the fiscal year (October 1). New proposals for the next fiscal year should be submitted no later than June 1, unless otherwise directed by the BAA. Competitive renewals should be submitted no later than six months prior to the award termination date after contacting the Program Officer. Conference proposals should be submitted at least four months prior to the conference.
Geographic Focus: All States
Contact: Dr. Wallace Smith, (703) 696-0284 or (703) 588-2427; fax (703) 696-5940; wallace.smith1@navy.mil
Internet: http://www.onr.navy.mil/sci_tech/ocean/322_processes/prog_op.htm
Sponsor: U.S. Department of Defense
One Liberty Center, 875 N. Randolph Street, Suite 1425
Arlington, VA 22203-1995

ONR Electronic Combat Technology Grants **3757**
The goal of electronic combat is to disrupt, deny, and exploit enemy use of the electromagnetic (EM) spectrum while ensuring its secure friendly use. To that end, ONR's Electronic Combat Technology Program provides improved threat warning systems, electronic support measures, decoys, and electronic countermeasures (radio frequency and EO/IR) as well as new and improved communication and navigation countermeasures. Submission of the Research Proposal should always be preceded by submission of a Preproposal and a discussion of that Preproposal with the assigned Program Officer.
Requirements: Grants are awarded to academic, industrial, and nonprofit organizations, with strong emphasis on university research.
Restrictions: New grants or contracts are typically started near the beginning of the fiscal year (October 1). New proposals for the next fiscal year should be submitted no later than June 1, unless otherwise directed by the BAA. Competitive renewals should be submitted no later than six months prior to the award termination date after contacting the Program Officer. Conference proposals should be submitted at least four months prior to the conference.
Geographic Focus: All States
Contact: Dr. Peter E. Craig, (703) 696-0114 or (703) 588-2427; fax (703) 696-5940; peter.craig@navy.mil
Internet: http://www.onr.navy.mil/sci_tech/31/312/electronic_combat.asp
Sponsor: U.S. Department of Defense
One Liberty Center, 875 N. Randolph Street, Suite 1425
Arlington, VA 22203-1995

ONR Electronics, Sensors, and Network Research Grants **3758**
This division conducts an integrated program of 6.1, 6.2, and 6.3 research into the uses of the electromagnetic spectrum in support on Navy and Marine Corps needs. Specific applied programs include research in the areas of surface and aerospace surveillance, communications, electronic combat, and navigation. All these areas are supported by a broad research program in electronics, focused on the reducing the cost weight and size of transmit and receive systems. Two over arching foci of this division are: the development of technologies and techniques to support adaptive persistent surveillance; and the development of digital/RF technologies and techniques to support active aperture phased arrays capable of performing multiple functions simultaneously. Submission of the Research Proposal should always be preceded by submission of a Preproposal and a discussion of that Preproposal with the assigned Program Officer.
Requirements: Grants are awarded to academic, industrial, and nonprofit organizations, with strong emphasis on university research.
Restrictions: New grants or contracts are typically started near the beginning of the fiscal year (October 1). New proposals for the next fiscal year should be submitted no later than June 1, unless otherwise directed by the BAA. Competitive renewals should be submitted no later than six months prior to the award termination date after contacting the Program Officer. Conference proposals should be submitted at least four months prior to the conference.
Geographic Focus: All States
Contact: Dr. Harry Dietrich, (703) 696-0240 or (703) 588-2427; fax (703) 696-5940; harry.dietrich@navy.mil
Internet: http://www.onr.navy.mil/sci_tech/31/312/
Sponsor: U.S. Department of Defense
One Liberty Center, 875 N. Randolph Street, Suite 1425
Arlington, VA 22203-1995

ONR Electronics Division Research Grants **3759**
Projects within this division cover all areas of electronics, including the growth of electronic materials, understanding of surfaces and interfaces, semiconductor and magnetic material theory, microelectronics, nanoscale electronic devices, nanolithography, novel electronic device concepts, memory, high-power RF and switching components, vacuum electronics, high-temperature superconductor electronics, optoelectronics, solid-state lasers and light emitters, multiband IR and UV sensors, ionospheric effects, and innovative electronic circuits. Submission of the Research Proposal should always be preceded by submission of a Preproposal and a discussion of that Preproposal with the assigned Program Officer.
Requirements: Grants are awarded to academic, industrial, and nonprofit organizations, with strong emphasis on university research.
Restrictions: New grants or contracts are typically started near the beginning of the fiscal year (October 1). New proposals for the next fiscal year should be submitted no later than June 1, unless otherwise directed by the BAA. Competitive renewals should be submitted no later than six months prior to the award termination date after contacting the Program Officer. Conference proposals should be submitted at least four months prior to the conference.
Geographic Focus: All States
Contact: Christopher Rouin, (703) 696-0107 or (703) 588-2427; fax (703) 696-5940; chris.rouin@navy.mil
Internet: http://www.onr.navy.mil/sci_tech/31/default.asp
Sponsor: U.S. Department of Defense
One Liberty Center, 875 N. Randolph Street, Suite 1425
Arlington, VA 22203-1995

ONR Environmental Quality Research Grants **3760**
The program seeks to develop technology that leads to affordable environmental compliance and pollution prevention afloat and ashore. The program focuses on technologies that are directed toward environmentally sound ships, shore-related facilities, and sediment issues. Specific research interests includes sensors and improved cleaning methods. The program supports basic research, applied research, and advanced technology development. Currently, there are two primary focus areas: Environmentally Benign Marine Antifouling Coatings; and Waste Treatment/Reduction. Submission of the Research Proposal should always be preceded by submission of a Preproposal and a discussion of that Preproposal with the assigned Program Officer.
Requirements: Grants are awarded to academic, industrial, and nonprofit organizations, with strong emphasis on university research.
Restrictions: New grants or contracts are typically started near the beginning of the fiscal year (October 1). New proposals for the next fiscal year should be submitted no later than June 1, unless otherwise directed by the BAA. Competitive renewals should be submitted no later than six months prior to the award termination date after contacting the Program Officer. Conference proposals should be submitted at least four months prior to the conference.
Geographic Focus: All States
Contact: Stephen McElvany, (703) 696-1449 or (703) 588-2427; steve.mcelvany.navy.mil
Internet: http://www.onr.navy.mil/sci_tech/33/332/
Sponsor: U.S. Department of Defense
One Liberty Center, 875 N. Randolph Street, Suite 1425
Arlington, VA 22203-1995

ONR Expeditionary Maneuver Warfare & Combating Terrorism Department Grants **3761**
The primary focus of this Program is to lead the Department of the Navy's Science and Technology effort that develops future combat capabilities for Naval Expeditionary Maneuver Warfare and the Department's role in Combating Terrorism through the exploitation and subsequent application of Science and Technology in order to enhance the ability of the Navy-Marine Corps team to achieve assured access and conduct decisive operations as the naval portion of a Joint campaign. Submission of the Research Proposal should always be preceded by submission of a Preproposal and a discussion of that Preproposal with the assigned Program Officer.
Requirements: Grants are awarded to academic, industrial, and nonprofit organizations, with strong emphasis on university research.
Restrictions: New grants or contracts are typically started near the beginning of the fiscal year (October 1). New proposals for the next fiscal year should be submitted no later than June 1, unless otherwise directed by the BAA. Competitive renewals should be submitted no later than six months prior to the award termination date after contacting the Program Officer. Conference proposals should be submitted at least four months prior to the conference.
Geographic Focus: All States
Contact: John Moniz, (703) 696-2492 or (703) 588-2427; fax (703) 696-5940; john.moniz@navy.mil
Internet: http://www.onr.navy.mil/sci_tech/30/
Sponsor: U.S. Department of Defense
One Liberty Center, 875 N. Randolph Street, Suite 1425
Arlington, VA 22203-1995

ONR Faculty Sabbatical Leave Grants **3762**
The Faculty Sabbatical Leave Program provides science and engineering faculty the opportunity to conduct research at Navy laboratories while on sabbatical leave. Participants in the Sabbatical Leave Program will receive a monthly stipend making up the difference between salary and sabbatical leave pay from their home institution. In addition, participants will receive reimbursement for travel to and from the laboratory site and a relocation allowance for those who must relocate their residence during their sabbatical

leave tenure. Sabbatical leave appointments are contingent on the availability of sabbatical leave positions. Appointments are for a minimum of one semester and a maximum of one year. This is a residential sabbatical — participants must conduct research on site.
Requirements: Faculty should submit applications at least 6 months prior to their proposed sabbatical leave starting date.
Geographic Focus: All States
Contact: Dr. Anthony Junior, (703) 696-0453 or (703) 588-2427; fax (703) 696-5940; anthony.junior@navy.mil
Internet: http://www.onr.navy.mil/sci_tech/3t/corporate/sabbat.asp
Sponsor: U.S. Department of Defense
One Liberty Center, 875 N. Randolph Street, Suite 1425
Arlington, VA 22203-1995

ONR Historically Black Colleges and Universities Future Engineering Faculty Fellowships 3763
This fellowship program is designed to develop and attract qualified engineering faculty to Historically Black Colleges and Universities (HBCUs) with engineering programs. Each year, three recipients who have agreed to join the engineering faculty of an HBCU after receiving their degrees are competitively selected for study and research support leading to doctoral degrees in engineering. Disciplines include the following: aerospace engineering; electrical engineering; mechanical engineering; manufacturing science and engineering; civil engineering; ocean engineering; and chemical engineering. The tenure of an ONR-HBCU Future Engineering Faculty Fellowship is generally 36 months. In addition to stipends, the program pays the Fellow's full tuition and fees (not to include room and board).
Requirements: Applicants must be citizens or nationals of the United States who intend to pursue a Ph.D. in the designated fields and, in return for the support, agree to join the faculty of an HBCU. Applicants must be nominated by one of the participating HBCUs. Applications are especially encouraged from women, underrepresented minorities and persons with disabilities.
Geographic Focus: All States
Contact: Dr. Anthony Junior, (703) 696-0453 or (703) 588-2427; fax (703) 696-5940; anthony.junior@navy.mil
Internet: http://www.onr.navy.mil/sci_tech/3t/corporate/hbec.asp
Sponsor: U.S. Department of Defense
One Liberty Center, 875 N. Randolph Street, Suite 1425
Arlington, VA 22203-1995

ONR Independent Research & Development Grants 3764
Independent Research & Development (IR & D) is a DOD sponsored program that encourages contractors to pursue independent research and development projects that are of potential interest to DOD. Through this program, DOD seeks to encourage maintenance of a strong national technology base. The IR & D program allows contractors to recover some of the costs incurred in doing independent research and development as a necessary cost of doing business in a high technology environment. In addition, the Navy benefits from this program by being able to avoid duplication of effort and being able to tailor its R & D efforts to meet future needs that industry R & D efforts are not directed towards. Section 824 of P.L. 101-510 defines potential interest to include activities intended to accomplish any of the following: enabling superior performance of future United States' weapon systems and components; reducing acquisition costs and life-cycle costs of military systems; strengthening the defense industrial base and the technology base of the United States; enhancing the industrial competitiveness of the United States; promoting the development of technologies identified as critical under section 2522 of Title 10, the United States Code; increasing the development and promotion of efficient and effective applications of dual-use technologies; and providing efficient and effective technologies for achieving such environmental benefits as improved environmental data gathering, environmental cleanup and restoration, pollution reduction in manufacturing, environmental conservation, and environmentally safe management of facilities. Submission of the Research Proposal should always be preceded by submission of a Preproposal and a discussion of that Preproposal with the assigned Program Officer.
Requirements: Grants are awarded to academic, industrial, and nonprofit organizations, with strong emphasis on university research.
Restrictions: New grants or contracts are typically started near the beginning of the fiscal year (October 1). New proposals for the next fiscal year should be submitted no later than June 1, unless otherwise directed by the BAA. Competitive renewals should be submitted no later than six months prior to the award termination date after contacting the Program Officer. Conference proposals should be submitted at least four months prior to the conference.
Geographic Focus: All States
Contact: Dorothy Vincent, (703) 696-4792 or (703) 588-2427; fax (703) 696-5940; dorothy.vincent@navy.mil
Internet: http://www.onr.navy.mil/sci_tech/3t/transition/irad.asp
Sponsor: U.S. Department of Defense
One Liberty Center, 875 N. Randolph Street, Suite 1425
Arlington, VA 22203-1995

ONR Ionosphere Interactions (Space) Research Grants 3765
This program investigates physical and chemical processes that govern active and passive electromagnetic spectrum scattering from the Earth's surface, and propagation through the upper atmosphere and the near space environment. Of particular interest for surface effects is short water wave roughness modulation mechanisms; surfactant effects; intermittency in wave breaking and non-linear water waves. Research is directed toward improving the knowledge base for development of mechanistic EO/EM clutter models and automatic target recognition, and to investigate techniques that invert sensor information for the development of algorithms for assimilation into environmental models. Additional interests include electromagnetic scattering theory, microwave properties, scattering surface characterization, and wave and flux modulation mechanisms. Submission of the Research Proposal should always be preceded by submission of a Preproposal and a discussion of that Preproposal with the assigned Program Officer.
Requirements: Grants are awarded to academic, industrial, and nonprofit organizations, with strong emphasis on university research.
Restrictions: New grants or contracts are typically started near the beginning of the fiscal year (October 1). New proposals for the next fiscal year should be submitted no later than June 1, unless otherwise directed by the BAA. Competitive renewals should be submitted no later than six months prior to the award termination date after contacting the Program Officer. Conference proposals should be submitted at least four months prior to the conference.
Geographic Focus: All States
Contact: Dr. Robert McCoy, (703) 696-8699 or (703) 588-2427; fax (703) 696-5940; bob.mccoy@navy.mil
Internet: http://www.onr.navy.mil/sci_tech/32/321/space.asp
Sponsor: U.S. Department of Defense
One Liberty Center, 875 N. Randolph Street, Suite 1425
Arlington, VA 22203-1995

ONR Joint Capability Technology Demonstration (JCTD) Grants 3766
The Joint Capability Technology Demonstration (JCTD) Program is dedicated to: supporting the combatant commands (CoComs) by addressing their joint and coalition challenges that are specific to their theater or mission; bridging the gaps in joint and coalition warfighting capabilities that have been identified by the Joint Capabilities Integration and Development System (JCIDS); and bringing new technology-based capabilities to the joint and coalition warfighter. Proposals competing for project starts should submit all documentation by January 15 each year.
Requirements: Grants are awarded to academic, industrial, and nonprofit organizations, with strong emphasis on university research. Investigators are encouraged to contact the staff prior to formally submitting a proposal.
Geographic Focus: All States
Date(s) Application is Due: Jan 15
Contact: John Barkley, (703) 696-3158 or (703) 696-5031; fax (703) 696-5940; john.barkley@navy.mil or jctdcoord@onr.navy.mil
Internet: http://www.onr.navy.mil/sci_tech/3t/jctd/
Sponsor: U.S. Department of Defense
One Liberty Center, 875 N. Randolph Street, Suite 1425
Arlington, VA 22203-1995

ONR Magnetics Grants 3767
This program investigates magnetic films and electronic devices made therefrom. The scope of this program encompasses materials synthesis and device/circuit design ranging from non-volatile memory through microwave circulators and isolators. Submission of the Research Proposal should always be preceded by submission of a Preproposal and a discussion of that Preproposal with the assigned Program Officer.
Requirements: Grants are awarded to academic, industrial, and nonprofit organizations, with strong emphasis on university research.
Restrictions: New grants or contracts are typically started near the beginning of the fiscal year (October 1). New proposals for the next fiscal year should be submitted no later than June 1, unless otherwise directed by the BAA. Competitive renewals should be submitted no later than six months prior to the award termination date after contacting the Program Officer. Conference proposals should be submitted at least four months prior to the conference.
Geographic Focus: All States
Contact: Dr. Chagaan Baatar, (703) 696-0483 or (703) 588-2427; fax (703) 696-5940; chagaan.baatar@navy.mil
Internet: http://www.onr.navy.mil/sci_tech/31/312/magnetics.asp
Sponsor: U.S. Department of Defense
One Liberty Center, 875 N. Randolph Street, Suite 1425
Arlington, VA 22203-1995

ONR Manufacturing Technology Grants 3768
Managed within the Office of Naval Research (ONR), the Navy ManTech Program provides for the development of enabling manufacturing technology and the transition of this technology for the production and sustainment of Navy weapon systems to support the Fleet. Navy ManTech is currently focused on shipbuilding affordability. Reducing the acquisition cost of current and future platforms is a critical goal of the Navy, and ManTech aids in achieving this goal by developing and transitioning key manufacturing technology. Transition of technology is key to Navy ManTech success. Advances in manufacturing technology are useful to the Navy only if they result in implementation in the production of weapon systems. Transition of manufacturing processes to private and government industrial entities that manufacture and repair systems and components for the Fleet is the goal of every ManTech project. Submission of the Research Proposal should always be preceded by submission of a Preproposal and a discussion of that Preproposal with the assigned Program Officer.
Requirements: Grants are awarded to academic, industrial, and nonprofit organizations, with strong emphasis on university research.
Restrictions: New grants or contracts are typically started near the beginning of the fiscal year (October 1). New proposals for the next fiscal year should be submitted no later than June 1, unless otherwise directed by the BAA. Competitive renewals should be submitted no later than six months prior to the award termination date after contacting the Program Officer. Conference proposals should be submitted at least four months prior to the conference.
Geographic Focus: All States

Contact: John Carney, (703) 696-0352 or (703) 588-2427; fax (703) 696-5940; john.u.carney@navy.mil
Internet: http://www.onr.navy.mil/sci_tech/3t/mantech/default.asp
Sponsor: U.S. Department of Defense
One Liberty Center, 875 N. Randolph Street, Suite 1425
Arlington, VA 22203-1995

ONR Marine Meteorology and Atmospheric Effects Research Grants **3769**
This program sponsors integrated basic, applied, and developmental research with emphasis on improving the modeling and prediction of environmental parameters critical to Navy and Marine Corps platform, sensor, and weapons performance. The program includes research and development, leading to enhanced environmental support for operations, training, mission planning, and systems development, and focuses on addressing Navy and Marine Corps real-time, high-resolution environmental parameter requirements to support tactical sensors and operations in littoral zones worldwide. Topics of interest include meso-scale coastal phenomena, such as coastally trapped disturbances; data assimilation incorporating high data rate, asynchronous sensors (radar, lidar, etc.); global, mesoscale, and on-scene modeling focusing on the marine atmosphere and/or coastal zone; simulation and visualization of weather phenomena; marine boundary layer process, including aerosols and marine convective and nonconvective clouds; environmental effects on electromagnetic and electrooptic propagation; and western and southern Pacific tropical cyclone behavior and evolution in motion and structure. The encouraged submission date is June 15, though unsolicited programs are accepted at any time. Submission of the Research Proposal should always be preceded by submission of a Preproposal and a discussion of that Preproposal with the assigned Program Officer.
Requirements: Grants are awarded to academic, industrial, and nonprofit organizations, with strong emphasis on university research.
Geographic Focus: All States
Date(s) Application is Due: Jun 15
Contact: Dr. Ronald Ferek, (703) 696-0518 or (703) 588-2427; ron.ferek@navy.mil
Internet: http://www.onr.navy.mil/sci_tech/32/322/meteorology_atmospheric.asp
Sponsor: U.S. Department of Defense
One Liberty Center, 875 N. Randolph Street, Suite 1425
Arlington, VA 22203-1995

ONR Maritime Sensing Grants **3770**
The Maritime Sensing program conducts multidisciplinary science and technology development of acoustic and non-acoustic sensors systems as a means to automatically develop and maintain comprehensive tactical and strategic awareness in the maritime undersea battle space. These systems may be carried as on-board equipment for Navy surface ships, submarines, aircraft, and unmanned vehicles or may deployed from those platforms as mobile, drifting, or fixed ocean surveillance systems operating under autonomous or remote control. The intent of this program is to provide next-generation acoustic and non-acoustic sensing technology for the tactical, surveillance, and undersea warfare missions of the Navy. Project scopes may range from component-level research and development to system-level technology demonstrations. Submission of the Research Proposal should always be preceded by submission of a Preproposal and a discussion of that Preproposal with the assigned Program Officer. A planning letter must be received by close of business on May 23rd in order to receive full consideration.
Requirements: Grants are awarded to academic, industrial, and nonprofit organizations, with strong emphasis on university research.
Restrictions: New grants or contracts are typically started near the beginning of the fiscal year. Note that the goal is to issue awards early in fiscal year. Planning letters should be received by close of business June 28. New proposals for the next fiscal year should be submitted no later than September 30, unless otherwise directed by the BAA. Competitive renewals should be submitted no later than six months prior to the award termination date after contacting the Program Officer. Conference proposals should be submitted at least four months prior to the conference.
Geographic Focus: All States
Date(s) Application is Due: Sep 30
Contact: Mike Wardlaw, Team Leader; (703) 588-2427 or (703) 588-0615; fax (540) 653-5234; mike.wardlaw@navy.mil
Internet: http://www.onr.navy.mil/Science-Technology/Departments/Code-32/All-Programs/Ocean-Systems-321/Maritime-Sensing.aspx
Sponsor: U.S. Department of Defense
One Liberty Center, 875 N. Randolph Street, Suite 1425
Arlington, VA 22203-1995

ONR Mathematical, Computer, and Information Sciences Research Grants **3771**
Sponsored programs in this area focus on experimental and theoretical research, as well as technology in these areas with applications across near-, mid-, and far-term applications. The research programs range from fundamental investigations into mathematical foundations for models, computability, and algorithms; solid state materials and electronic devices; and image and signal analysis in the far-term to C2 applications and infrastructure, sensor technology, and robust navigation technology in the near-term. The technology programs seek to exploit the results of research opportunities to enhance surveillance capabilities, communications, command and control, new concepts for electronic devices, application of information sciences to complex problems including human-computer interaction, navigation, and electronic warfare. Submission of the Research Proposal should always be preceded by submission of a Preproposal and a discussion of that Preproposal with the assigned Program Officer.
Requirements: Grants are awarded to academic, industrial, and nonprofit organizations, with strong emphasis on university research.
Restrictions: New grants or contracts are typically started near the beginning of the fiscal year (October 1). New proposals for the next fiscal year should be submitted no later than June 1, unless otherwise directed by the BAA. Competitive renewals should be submitted no later than six months prior to the award termination date after contacting the Program Officer. Conference proposals should be submitted at least four months prior to the conference.
Geographic Focus: All States
Contact: Christopher Rouin, (703) 696-0107 or (703) 588-2427; fax (703) 696-5940; chris.rouin@navy.mil
Internet: http://www.onr.navy.mil/sci_tech/31/default.asp
Sponsor: U.S. Department of Defense
One Liberty Center, 875 N. Randolph Street, Suite 1425
Arlington, VA 22203-1995

ONR Medical Science and Technology Division Grants **3772**
The Biomedical & Biological Division's mission is to conduct research and technology demonstration programs directed at maintaining the survival, health and performance of Navy and Marine Corps personnel during training, routine operations, special operations, and in time of war. Its primary goals include: increasing the survival of casualties through immediate, life-saving treatment and stabilization; preventing personnel injury caused by the stresses of demanding Naval occupations and environments; and ensuring the fitness and physical readiness of Navy and Marine Corps personnel. Submission of the Research Proposal should always be preceded by submission of a Preproposal and a discussion of that Preproposal with the assigned Program Officer.
Requirements: Grants are awarded to academic, industrial, and nonprofit organizations, with strong emphasis on university research.
Restrictions: New grants or contracts are typically started near the beginning of the fiscal year (October 1). New proposals for the next fiscal year should be submitted no later than June 1, unless otherwise directed by the BAA. Competitive renewals should be submitted no later than six months prior to the award termination date after contacting the Program Officer. Conference proposals should be submitted at least four months prior to the conference.
Geographic Focus: All States
Contact: Dr. Michael Given, (703) 696-4055 or (703) 588-2427; fax (703) 696-5940; michael.given@navy.mil
Internet: http://www.onr.navy.mil/sci_tech/34/342/
Sponsor: U.S. Department of Defense
One Liberty Center, 875 N. Randolph Street, Suite 1425
Arlington, VA 22203-1995

ONR Microelectronics Applied Research Grants **3773**
This program addresses analog and digital VLSI, radiation hardened electronics, high speed A/D converters, advanced microelectronic and nanoelectronic circuits, as well as special purpose architectures including non-volatile memory, non-destructive read-out memory, and chip-to-chip optical connections. Also included are smart switching systems integrating efficient high power switches and their associated control components. Semiconductor material. Proposals are preferred by April 1 of each year. Submission of the Research Proposal should always be preceded by submission of a Preproposal and a discussion of that Preproposal with the assigned Program Officer.
Requirements: Grants are awarded to academic, industrial, and nonprofit organizations, with strong emphasis on university research.
Restrictions: New grants or contracts are typically started near the beginning of the fiscal year (October 1). New proposals for the next fiscal year should be submitted no later than June 1, unless otherwise directed by the BAA. Competitive renewals should be submitted no later than six months prior to the award termination date after contacting the Program Officer. Conference proposals should be submitted at least four months prior to the conference.
Geographic Focus: All States
Contact: Dr. Ingham Mack, (703) 696-4825 or (703) 588-2427; fax (703) 696-5940; ingham.mack@navy.mil
Internet: http://www.onr.navy.mil/sci_tech/31/312/microelectronics.asp
Sponsor: U.S. Department of Defense
One Liberty Center, 875 N. Randolph Street, Suite 1425
Arlington, VA 22203-1995

ONR Multidisciplinary Research Program of the URI (MURI) Grants **3774**
The Multidisciplinary Research Program of the University Research Initiative (MURI) is a multi-agency DoD program that supports research teams whose efforts intersect more than one traditional science and engineering discipline. Multidisciplinary team effort can accelerate research progress in areas particularly suited to this approach. Multidisciplinary research also can help to hasten the transition of research findings to practical application. MURI awards are made in research topics specified by the participating defense agencies each year that the program is in force. Specified topics change each year. Awards are typically for a period of three years (funded incrementally or as options) with two additional years possible as options to bring the total award to five-years, and at a funding level ranging from half a million to about a million dollars per year, with the size of the award dependent upon the topic, technical goals, and availability of appropriations. White papers are due on October 31, 4:00 p.m. Eastern Daylight Time, and full proposals are due on January 9, 4:00 p.m. Eastern Standard Time. Submission of the Research Proposal should always be preceded by submission of a Preproposal and a discussion of that Preproposal with the assigned Program Officer.
Requirements: Grants are awarded to academic, industrial, and nonprofit organizations, with strong emphasis on university research.
Geographic Focus: All States
Date(s) Application is Due: Jan 9; Oct 31

Contact: Dr. William Lukens, (703) 696-4668 or (703) 588-2427; fax (703) 696-5940; william.luckens1@navy.mil
Internet: http://www.onr.navy.mil/sci_tech/3t/corporate/muri.asp
Sponsor: U.S. Department of Defense
One Liberty Center, 875 N. Randolph Street, Suite 1425
Arlington, VA 22203-1995

ONR Naval Research Enterprise Intern Program (NREIP) Grants 3775
This ten-week intern program is designed to provide opportunities for undergraduate and graduate students to participate in research, under the guidance of an appropriate research mentor, at a participating Navy laboratory. Interns will be selected based upon academic achievement, personal statements, recommendation, and career and research interests. NREIP stipends are $5,500 for undergraduates and $6,500 for graduate students.
Requirements: Applicants must be: U.S. citizens (Permanent Resident Alien status is considered at some labs); junior, senior, or graduate students; and enrolled at a participating university.
Geographic Focus: All States
Amount of Grant: 5,500-6,500 USD
Contact: Dr. Anthony Junior, (703) 696-0453 or (703) 588-2427; anthony.junior@navy.mil
Internet: http://www.onr.navy.mil/sci_tech/3t/corporate/nreip.asp
Sponsor: U.S. Department of Defense
One Liberty Center, 875 N. Randolph Street, Suite 1425
Arlington, VA 22203-1995

ONR Navy Postdoctoral Fellowships 3776
The Navy, through the Naval Research Laboratory (NRL) , sponsors a Postdoctoral Fellowship Program at NRL and a number of Naval R&D centers and laboratories. The objective of this program is to encourage the involvement of creative, capable and highly trained scientists and engineers who have received a Ph.D. or equivalent within the prior seven years in research areas of interest and relevance to the Navy. The program is administered for NRL by the American Society for Engineering Education. Information about the program is also available at the ASEE website.
Geographic Focus: All States
Contact: Dr. Anthony Junior, (703) 696-0453 or (703) 588-2427; anthony.junior@navy.mil
Internet: http://www.onr.navy.mil/sci_tech/3t/corporate/postfe.asp
Sponsor: U.S. Department of Defense
One Liberty Center, 875 N. Randolph Street, Suite 1425
Arlington, VA 22203-1995

ONR Optical and Biological Oceanography Research Grants 3777
Funded projects focus on elucidating the optical and biological processes in the world ocean. Investments in optics support basic and applied long-term research leading to a more complete understanding of how light interacts with the ocean environment, including the surface, volume, and sea floor. Biological oceanographic investments are directed primarily at understanding the distribution, growth, and abundance of biota in the coastal ocean and shallow-water sediments, focusing especially on those organisms that affect or control the bulk optical and acoustical properties of the ocean. Overarching, long-term investments within the program are directed at developing better observational capabilities, in situ sensors, ocean color algorithms, and autonomous systems, and distilling the accumulated knowledge into multidisciplinary, predictive models. Planning letters are requested by March 30, prior to submitting a proposal. The department encourages telephone inquiries to discuss ideas.
Requirements: Grants are awarded to academic, industrial, and nonprofit organizations, with strong emphasis on university research.
Geographic Focus: All States
Date(s) Application is Due: Mar 30
Contact: Dr. Steve Ackleson, (703) 696-4732 or (703) 588-2427; steve.ackleson@navy.mil
Internet: http://www.onr.navy.mil/sci_tech/32/322/ocean_optics_biology.asp
Sponsor: U.S. Department of Defense
One Liberty Center, 875 N. Randolph Street, Suite 1425
Arlington, VA 22203-1995

ONR Radio Frequency (RF) Sensor Processing Research Grants 3778
This program investigates physical and chemical processes that govern active and passive electromagnetic spectrum scattering from the earth's surface, and propagation through the upper atmosphere and the near space environment. Algorithms and hardware (including photonic) for storing, relaying, processing, extracting and exploiting target information from radio frequency (RF) sensors and arrays in real-time are of interest. Submission of the Research Proposal should always be preceded by submission of a Preproposal and a discussion of that Preproposal with the assigned Program Officer.
Requirements: Grants are awarded to academic, industrial, and nonprofit organizations, with strong emphasis on university research.
Restrictions: New grants or contracts are typically started near the beginning of the fiscal year (October 1). New proposals for the next fiscal year should be submitted no later than June 1, unless otherwise directed by the BAA. Competitive renewals should be submitted no later than six months prior to the award termination date after contacting the Program Officer. Conference proposals should be submitted at least four months prior to the conference.
Geographic Focus: All States
Contact: George McNeal, (703) 696-4845 or (703) 588-2427; george.mcneal@navy.mil
Internet: http://www.onr.navy.mil/sci_tech/31/312/processing_rf.asp
Sponsor: U.S. Department of Defense
One Liberty Center, 875 N. Randolph Street, Suite 1425
Arlington, VA 22203-1995

ONR Radio Frequency Solid State Devices Applied Research Grants 3779
The program objective is to develop solid state (e.g., wide bandgap semiconductor) low noise and power devices and circuits for use in power supplies, communications, fire control, surveillance, EW, and multifunctional RF systems. Emphasis is on decade wide instantaneous bandwidth, excellent linearity, high (e.g., 60%) power added efficiency, and power output of at least 5 times that of GaAs-based devices. Control components such as limiters, circulators, and filters are also included. Submission of the Research Proposal should always be preceded by submission of a Preproposal and a discussion of that Preproposal with the assigned Program Officer. Proposals are preferred by April 1 of each year.
Requirements: Grants are awarded to academic, industrial, and nonprofit organizations, with strong emphasis on university research.
Geographic Focus: All States
Date(s) Application is Due: Apr 1
Contact: Dr. Ingham Mack, (703) 696-4825 or (703) 588-2427; ingham.mack@navy.mil
Internet: http://www.onr.navy.mil/sci_tech/31/312/solid_state_devices.asp
Sponsor: U.S. Department of Defense
One Liberty Center, 875 N. Randolph Street, Suite 1425
Arlington, VA 22203-1995

ONR Sea Platforms and Weapons Division Grants 3780
The division is focused on coordinating the transition of technologically superior systems and equipment that will enhance warfighting capabilities at reduced total ownership costs to ship acquisition programs. Research topic areas of interest that contribute to these focus areas include: naval power systems; platform/hull forms; and ship systems and systems integration. Submission of the Research Proposal should always be preceded by submission of a Preproposal and a discussion of that Preproposal with the assigned Program Officer.
Requirements: Grants are awarded to academic, industrial, and nonprofit organizations, with strong emphasis on university research.
Restrictions: New grants or contracts are typically started near the beginning of the fiscal year (October 1). New proposals for the next fiscal year should be submitted no later than June 1, unless otherwise directed by the BAA. Competitive renewals should be submitted no later than six months prior to the award termination date after contacting the Program Officer. Conference proposals should be submitted at least four months prior to the conference.
Geographic Focus: All States
Contact: Dr. Daniel Tam, (703) 696-4204 or (703) 588-2427; daniel.tam1@navy.mil
Internet: http://www.onr.navy.mil/sci_tech/33/333/
Sponsor: U.S. Department of Defense
One Liberty Center, 875 N. Randolph Street, Suite 1425
Arlington, VA 22203-1995

ONR Semiconductor Devices Grants 3781
This program investigates electronic properties of wide bandgap and silicon-based semiconductors and novel devices. The program is focussed on novel high power microwave and switching devices for broadband radar, EW, communications, and multifunctional systems applications and for high power control applications. Submission of the Research Proposal should always be preceded by submission of a Preproposal and a discussion of that Preproposal with the assigned Program Officer.
Requirements: Grants are awarded to academic, industrial, and nonprofit organizations, with strong emphasis on university research.
Restrictions: New grants or contracts are typically started near the beginning of the fiscal year (October 1). New proposals for the next fiscal year should be submitted no later than June 1, unless otherwise directed by the BAA. Competitive renewals should be submitted no later than six months prior to the award termination date after contacting the Program Officer. Conference proposals should be submitted at least four months prior to the conference.
Geographic Focus: All States
Contact: Dr. Harry Dietrich, (703) 696-0240 or (703) 588-2427; harry.dietrich@navy.mil
Internet: http://www.onr.navy.mil/sci_tech/31/312/semiconductor_devices.asp
Sponsor: U.S. Department of Defense
One Liberty Center, 875 N. Randolph Street, Suite 1425
Arlington, VA 22203-1995

ONR Small Business Innovation Research (SBIR/STTR) Grants 3782
The ONR SBIR program is an integral part of the Naval Science and Technology program. ONR SBIR topics address naval needs and should ultimately transition to the fleet through enabling technologies for future naval capabilities or directly to naval acquisition programs. Topics are derived from ONR Program Officers via periodic calls which are issued by the ONR SBIR Program Manager to the ONR Department Heads. ONR SBIR topics are published and solicited with those of other Naval components in the Department of Defense SBIR Program Solicitations. Qualified small businesses may submit proposals for these topics which are then competitively evaluated by ONR Program Officers. Awards are made to the best proposing companies within allotted funding. Funding for this program is not appropriated, but provided annually under the statutory 2-1/2% set-aside from ONR extramural research, development, testing and evaluation (RDT&E) funds. DoD goals are for all Phase Is to be under contract within 4 months of receipt and for all Phase IIs to be under contract within 6 months following the end of the Phase I contract. Timeline/calendar of events is enclosed for planning purposes and to help meet these goals.
Requirements: A business may apply that is independently owned and operated; is not dominant in the field of operation in which it is proposing; has its principal place of business located in the United States; is organized for profit; is at least 51 percent owned, or in the case of a publicly owned business, at least 51 percent of its voting stock is owned by United States citizens or lawfully admitted permanent resident aliens; and including its affiliates, has a number of employees amounting to 500 or less.

Geographic Focus: All States
Contact: John Williams, (703) 696-0342 or (703) 588-2427; john.williams6@navy.mil
Internet: http://www.onr.navy.mil/sci_tech/3t/sbir_sttr/sbir_onr.asp
Sponsor: U.S. Department of Defense
One Liberty Center, 875 N. Randolph Street, Suite 1425
Arlington, VA 22203-1995

ONR Small Business Technology Transfer (STTR) Grants **3783**
This mission-oriented science and technology program seeks to produce commercializable products through scientific research areas. The program requires that small businesses and non-profit research institutions submit a joint proposal responsive to broad topics defined in the Navy section of the annual DOD STTR solicitation. Unlike the Navy's SBIR program, the Navy STTR program focuses on a different science or technology area each year. You may register to receive the DoD Solicitation on-line via the DoD SBIR Program.
Requirements: A business may apply that is independently owned and operated; is not dominant in the field of operation in which it is proposing; has its principal place of business located in the United States; is organized for profit; is at least 51 percent owned, or in the case of a publicly owned business, at least 51 percent of its voting stock is owned by United States citizens or lawfully admitted permanent resident aliens; and including its affiliates, has a number of employees amounting to 500 or less.
Geographic Focus: All States
Contact: John Williams, (703) 696-0342 or (703) 588-2427; john.williams6@navy.mil
Internet: http://www.onr.navy.mil/sci_tech/3t/sbir_sttr/sttr_pilot.asp
Sponsor: U.S. Department of Defense
One Liberty Center, 875 N. Randolph Street, Suite 1425
Arlington, VA 22203-1995

ONR Summer Faculty Research Grants **3784**
The Summer Faculty Research Program provides science and engineering faculty members from institutions of higher education the opportunity to participate in research at Navy laboratories for a ten week period during the summer. Participants work with professional peers in the Navy laboratories on research tasks of mutual interest. Information about the research interests of the participating laboratories is available via the ASEE website. The research for the summer is defined in advance through correspondence and an optional pre-program visit to the research site, for which funds are provided. Three levels of appointment are available: Summer Faculty Fellow, Senior Summer Faculty Fellow, and Distinguished Summer Faculty Fellow. Stipends vary by level of appointment and are set each year. In addition, each Fellow will be reimbursed for his/her personal travel for an optional pre-program visit to the research site.
Requirements: The ONR Summer Faculty Research Program is open to United States citizens or nationals and to legal permanent residents holding a teaching or research position at a U.S. college or university. Some restrictions may apply to permanent residents regarding eligibility for a DoD security clearance.
Geographic Focus: All States
Contact: Dr. Anthony Junior, (703) 696-0453 or (703) 588-2427; anthony.junior@navy.mil
Internet: http://www.onr.navy.mil/sci_tech/3t/corporate/summfac.asp
Sponsor: U.S. Department of Defense
One Liberty Center, 875 N. Randolph Street, Suite 1425
Arlington, VA 22203-1995

ONR Superconducting Electronics Grants **3785**
This program's objective is to develop new and improved electronic devices and circuits using superconducting materials that are relevant to RF systems applications. Circuits clocking above 20 GHz are the focus. Improvements in packaging and memory technologies are also sought. Submission of the Research Proposal should always be preceded by submission of a Preproposal and a discussion of that Preproposal with the assigned Program Officer. Proposals are preferred by June 1 of each year.
Requirements: Grants are awarded to academic, industrial, and nonprofit organizations, with strong emphasis on university research.
Restrictions: New grants or contracts are typically started near the beginning of the fiscal year (October 1). New proposals for the next fiscal year should be submitted no later than June 1, unless otherwise directed by the BAA. Competitive renewals should be submitted no later than six months prior to the award termination date after contacting the Program Officer. Conference proposals should be submitted at least four months prior to the conference.
Geographic Focus: All States
Contact: Dr. Deborah Van Vechten, (703) 696-4219 or (703) 588-2427; fax (703) 696-5940; deborah.vanvechten@navy.mil
Internet: http://www.onr.navy.mil/sci_tech/31/312/superconducting_electronics.asp
Sponsor: U.S. Department of Defense
One Liberty Center, 875 N. Randolph Street, Suite 1425
Arlington, VA 22203-1995

ONR Surface/Aerospace Surveillance Grants **3786**
This program is devoted to the development of enhanced sensor technology, primarily radar and electro-optic/infrared (EO/IR) and associated signal, image and array processing methods. With respect to radar, areas of emphasis include wideband, frequency-agile, multifunction radar; phased-array affordability; the enhancement of radar functionality through the integration of new operating modes such as synthetic aperture radar (SAR), inverse synthetic aperture radar (ISAR), moving target indication (MTI) and moving target imaging (MTIm); and automatic target recognition (ATR). With respect to EO/IR, areas of emphasis include EO/IR shared apertures; multispectral sensing; clutter, background and target modeling; dim target detection; and active/passive classification/discrimination techniques. Additionally, there are efforts underway that seek to develop techniques for target tracking, sensor/data fusion, sensor resource management, and multi-sensor integration. Submission of the Research Proposal should always be preceded by submission of a Preproposal and a discussion of that Preproposal with the assigned Program Officer.
Requirements: Grants are awarded to academic, industrial, and nonprofit organizations, with strong emphasis on university research.
Restrictions: New grants or contracts are typically started near the beginning of the fiscal year (October 1). New proposals for the next fiscal year should be submitted no later than June 1, unless otherwise directed by the BAA. Competitive renewals should be submitted no later than six months prior to the award termination date after contacting the Program Officer. Conference proposals should be submitted at least four months prior to the conference.
Geographic Focus: All States
Contact: Dr. Michael Pollock, (703) 696-4202 or (703) 696-1438; fax (703) 696-5940; michael.a.pollock@navy.mil or 312_APEI@onr.navy.mil
Internet: http://www.onr.navy.mil/sci_tech/31/312/surveillance.asp
Sponsor: U.S. Department of Defense
One Liberty Center, 875 N. Randolph Street, Suite 1425
Arlington, VA 22203-1995

ONR Target Tracking and Sensor Fusion Grants **3787**
This program's primary purpose is to support research programs in the development of target tracking, sensor/data fusion, sensor resource management, and multi-sensor integration. Submission of the Research Proposal should always be preceded by submission of a Preproposal and a discussion of that Preproposal with the assigned Program Officer.
Requirements: Grants are awarded to academic, industrial, and nonprofit organizations, with strong emphasis on university research.
Restrictions: New grants or contracts are typically started near the beginning of the fiscal year (October 1). New proposals for the next fiscal year should be submitted no later than June 1, unless otherwise directed by the BAA. Competitive renewals should be submitted no later than six months prior to the award termination date after contacting the Program Officer. Conference proposals should be submitted at least four months prior to the conference.
Geographic Focus: All States
Contact: Dr. Rabinder Madan, (703) 696-4217 or (703) 588-2427; fax (703) 696-5940; rabinder.madan@navy.mil
Internet: http://www.onr.navy.mil/sci_tech/31/311/tracking_fusion.asp
Sponsor: U.S. Department of Defense
One Liberty Center, 875 N. Randolph Street, Suite 1425
Arlington, VA 22203-1995

ONR Young Investigator Grants **3788**
The objectives of the Young Investigator Program (YIP) are to attract to naval research outstanding new faculty members at institutions of higher education, to support their research, and to encourage their teaching and research careers. ONR anticipates making about 24 new awards annually. Awards of up to $170,000 per year for three years, with the possibility of additional support for capital equipment or collaborative research with a Navy laboratory, are made, based on research proposals and supporting materials. Special attention will be given to proposals in naval priority research areas, listed in the announcement.
Requirements: This program is open to U.S. citizens, nationals, and permanent residents holding tenure track or permanent faculty positions at U.S. institutions of higher education who have received graduate degrees (Ph.D. or equivalent) on or after the date published in the full announcement.
Geographic Focus: All States
Date(s) Application is Due: Jan 12
Amount of Grant: 170,000 USD
Contact: Dr. William Lukens, (703) 696-4668 or (703) 588-2427; fax (703) 696-5940; william.luckensl@navy.mil
Internet: http://www.onr.navy.mil/sci_tech/3t/corporate/yip.asp
Sponsor: U.S. Department of Defense
One Liberty Center, 875 N. Randolph Street, Suite 1425
Arlington, VA 22203-1995

ONS Major Research Grants **3789**
Applications are invited for research proposals which focus on increasing understanding of and management of understudied symptoms and/or symptom clusters in diverse populations and across settings. Priority will be given to proposals that address management of symptoms and/or symptom clusters that include dermatologic problems in cancer patients. Examples of symptom clusters include dermatitis, pain, and sleep/wake disturbances or fatigue, depression, and sleep/wake disturbances. Other symptom clusters and symptoms will be considered. Justification must be provided for selecting the cluster and/or the rationale for the research needed for the symptom. Membership in ONS is not required.
Requirements: The principal investigator must be actively involved in some aspect of cancer patient care, education, or research.
Geographic Focus: All States
Amount of Grant: 50,000 USD
Contact: Research Team; (412) 859-6298; fax (877) 369-5497; research@ons.org
Internet: http://www.ons.org/awards/foundawards
Sponsor: Oncology Nursing Society
125 Enterprise Drive
Pittsburgh, PA 15275-1214

Ontario Arts Council Integrated Arts Grants **3790**
The Program supports a wide range of integrated arts practice and interdisciplinary work through the research, development, creation, production and presentation of

original work. This includes new practices, hybrid forms and performance. Integrated arts are creative art forms, artistic processes or practices that use two or more artistic disciplines including visual arts, music, theater, literature, film and dance. Applicants may also choose to combine these above-mentioned disciplines with design, architecture or disciplines outside the arts. Each of these disciplines has expanded over the years to include aspects of the others. The program seeks to fund innovative integration of disciplines where there is a contribution to the development of each discipline and the artist(s). This program funds: artistic research; exploration and experimentation (including workshops/residencies where the artist(s) are integrating disciplines); collaborations fusing film making, visual/installation art, literature, dance, theater, music, architecture, design or any other disciplines; creation/presentation/production of new artistic work integrating two or more art forms; interdisciplinary performance pieces; and projects that combine artificial and natural processes. Aboriginal, culturally diverse, francophone and regional organizations receiving operating grant support are eligible to apply for additional project funding in any program outside their granting offices. Grants in research and development are generally awarded in the $3,000 to $8,000 range depending on the stage of development. Production or Presentation grants are generally awarded in the $5,000 to $15,000 range.
Requirements: This program is open to professional Ontario-based artists, collectives, teams of artists and not-for-profit organizations engaged in integrated arts development, production and presentation. Music theater and opera projects are no longer eligible in this program.
Restrictions: The following are not eligible for project funding: events/activities that take place before the deadline date; events/activities that have been completed before the results of the competition have been announced, which will be approximately 4 months after the deadline date; capital expenses;
Geographic Focus: All States, Canada
Date(s) Application is Due: Mar 17; Nov 1
Amount of Grant: Up to 20,000 USD
Contact: Georgina Braoudakis, French Services Director; (416) 969-7417 or (800) 387-0058, ext. 7417; gbraoudakis@arts.on.ca
Internet: http://www.arts.on.ca/Page120.aspx
Sponsor: Ontario Arts Council
151 Bloor Street West, 5th Floor
Toronto, ON M5S 1T6 Canada

Open Spaces Sacred Places National Awards **3791**
The initiative offers planning grant support for cross-disciplinary teams to develop proposals for integrated design and research studies to understand the impact of contemplative greenspace on urban dwellers. The awards will directly support the design and implementation of significant new Open Spaces Sacred Places sites designed to address critical human factors related to the physical, mental, and spiritual well-being of individuals living in challenging urban environments. The awards will simultaneously sponsor research on the human impact of encountering and experiencing these open, sacred places in nature in the urban setting.
Requirements: Eligible grantees include any 501(c)3 tax-exempt charitable organizations that are classified as public charities and not as private foundations. Such public charities include: a. Institutions that serve a large number of constituents, such as universities, schools, hospitals, medical research organizations, religious institutions, and governmental units and agencies as more specifically described in Section 170 (b)(1)(A)(i-vi) of the Internal Revenue Code (IRC); b. Publicly supported organizations that receive more than one-third of their revenues from modest donors and/or fees and sales of services and goods performed in conducting their charitable activities for a large number of persons as more specifically described in Section 509 (a)(2) of the IRC; c. Organizations created to support and benefit one or more other public charities as more specifically described in Section 509 (a)3 of the IRC. Eligible teams must be cross-disciplinary in makeup and must include an identified Firesoul or Firesouls. Eligible projects must have both an institutional and a community-based partner.
Geographic Focus: All States
Date(s) Application is Due: Jun 30; Sep 1
Amount of Grant: Up to 50,000 USD
Contact: Christine Tanabe, Director of Communications & Program Development; (410) 268-1376; fax (410) 268-1379; info@tkffdn.org
Internet: http://www.opensacred.org/grants/request-for-proposal
Sponsor: TKF Foundation
410 Severn Avenue, Suite 216
Annapolis, MD 21403

Oppenstein Brothers Foundation Grants **3792**
Grants are awarded in the metropolitan area of Kansas, City, Missouri, primarily for social services and early childhood, elementary, secondary, adult-basic, vocational, and higher education; family planning and services; social services and welfare agencies; Jewish welfare organizations; and programs for youth, the handicapped, disadvantaged, mentally ill, homeless, minorities, and elderly. Additional areas of interest include arts/cultural programs, the performing arts, museums, health care and health organizations, and AIDS research. The foundation considers requests for building and renovation, capital campaigns, curriculum development, emergency funds, equipment, general operating support, program development, seed money, technical support, conferences and seminars, consulting services, and matching funds. Application guidelines are available on request. The board meets every other month. Deadlines are generally three weeks prior to board meetings. Notification of award will take place within two to four months.
Requirements: Nonprofit organizations serving the metropolitan area of Kansas City, Missouri are eligible to apply.
Restrictions: The foundation primarily supports 501(c)3 nonprofit organizations serving the metropolitan area of Kansas City, Missouri. Grants are not awarded to support individuals or for annual campaigns.
Geographic Focus: Missouri
Amount of Grant: 5,000-15,000 USD
Contact: Beth Radtke, Program Officer; (816) 234-2577
Sponsor: Oppenstein Brothers Foundation
922 Walnut Street, Suite 200
Kansas City, MO 64106-1809

Orchard Foundation Grants **3793**
The foundation makes grants to eligible New York and New England nonprofit organizations in its areas of interest, including environment (air quality, biodiversity, fresh and coastal waters, forests, toxic substances, and pollution prevention); children, youth, and families (child welfare systems, literacy, and pregnancy prevention); and campaign finance reform. The foundation will occasionally provide seed money, operating support, and start-up capital to smaller home-grown organizations provided that strong local support has been obtained for a project in the form of membership and contributions, and that the entire mission of the organization coincides with a field of interest to the foundation. The listed deadlines are for concept letters.
Requirements: The foundation focuses on Massachusetts, Maine, New Hampshire, New York, Vermont, Connecticut, and Rhode Island. The foundation accepts letters of inquiry from groups in those seven states as well as from national groups with regional offices or projects in the area.
Restrictions: Grants are not made to individuals or for endowments, annual or capital campaigns, religious programs, any religion-affiliated organization, conference participation/travel unrelated to current foundation grant, research efforts unrelated to advocacy interests of the foundation, scholarships, fellowships, building projects, equipment needs, film and video projects, land acquisition, animal hospitals/rehabilitation centers, or groups that focus on specific diseases or conditions. Loans are not made.
Geographic Focus: Connecticut, Maine, Massachusetts, New Hampshire, New York, Rhode Island, Vermont
Date(s) Application is Due: Mar 1; Sep 1
Amount of Grant: 3,000-15,000 USD
Contact: Executive Director; (207) 799-0686; orchard@maine.rr.com
Internet: http://home.maine.rr.com/orchard
Sponsor: Orchard Foundation
P.O. Box 2587
Portland, ME 04116

Orchid Research Fellowships **3794**
The purpose of this fellowship program is to encourage doctoral candidates in their pursuit of fundamental and applied research on orchids; to attract future generations of scientists to orchidology; and to address critical research needs in the biology of orchids. Fellows spend up to three years working on orchid-related dissertation projects that lead to the PhD degree from accredited institutions in the United States. The candidate's dissertation project must deal with any aspect of orchid biology in the disciplines of physiology, molecular biology, structure, systematics, cytology, ecology, and/or evolution. The fellowship is limited to three consecutive years. Contact the American Orchid Society for application deadline.
Requirements: A candidate must be enrolled full time in a doctoral program of an accredited academic institution in the United States and should submit a current curriculum vita, transcripts of all college coursework, synopsis of the proposed research, and three letters of recommendation. Women, minorities, and persons with disabilities are especially encouraged to apply.
Restrictions: Fellowships will be awarded to the institution sponsoring the candidate. Institutional indirect costs are not to be paid from the awards.
Geographic Focus: All States
Contact: Program Contact; (561) 404-2050; fax (561) 404-2100; pgiust@aos.org
Internet: http://orchidweb.org/aos/research/page07.aspx
Sponsor: American Orchid Society
16700 AOS Lane
Delray Beach, FL 33446-4351

Organic Farming Research Foundation Grants **3795**
The mission of the foundation is to sponsor research related to organic farming practices. Grants support research into organic farming methods, dissemination of research results to organic farmers and to growers interested in making the transition to organic production systems, and education of the general public about organic farming issues.
Requirements: Farmers, researchers, organic advocacy organizations, and other individuals or entities may apply. Projects should involve farmers in both design and execution and take place on working organic farms whenever possible and appropriate.
Geographic Focus: All States
Date(s) Application is Due: Jul 15; Dec 15
Amount of Grant: Up to 15,000 USD
Contact: Jane Sooby, (831) 426-6606; fax (831) 426-6670; jane@ofrf.org
Internet: http://www.ofrf.org/research/index.html
Sponsor: Organic Farming Research Foundation
303 Potrero Street, Suite 29-203, P.O. Box 440
Santa Cruz, CA 95061

Organization for Tropical Studies (OTS) Fellowships **3796**
Organization of Tropical Studies (OTS) fellowships are intended to assist with thesis-related research in tropical biology and similar fields and are available in amounts up

to a maximum of $3,650. Among the available fellowships are the following named awards: the William L. Brown Fellowship; the David and Deborah Clark Fellowship; the Rexford Daubenmire Fellowship; the Dole Food Fellowship; the Emily P. Foster Fellowship; the Lillian and Murray Slatkin Fellowship; the Christiane and Christopher Tyson Fellowship (for U.S. and Latin-American graduate students); the GlaxoWellcome Endowment (for Latin-American graduate students only); the Peace Frogs Fellowships (for herpetological research); the Rowe Family Fellowships (for U.S. and Latin-American graduate students); and Water Quality Environmental Outreach. More information about these awards is available from the OTS fellowships webpage. Awards are intended primarily to cover up to fifty days of station fees (food, lodging, and station access) at OTS field stations; current rates are $63 for overnight visits and $31.50 for day visits. Other expenses that may be included in the budget, to a maximum of $500, are research- and export-permit fees, local travel, and reimbursement in whole or in part of one round-trip airline ticket. OTS has a single application process for most of its fellowships; generally applicants do not need to specify the award for which they are applying (with a few exceptions mentioned at the website). Application guidelines and a downloadable cover sheet are available from the OTS Fellowships webpage. The cover sheet and supporting documents must be submitted via email by the deadline date to the Education Director. Notification of decisions will take place about two months after each deadline. OTS is a nonprofit consortium of 64 universities and research institutions in the United States, Costa Rica, Peru, Canada, Mexico, South Africa, and Australia. Founded in 1963, OTS has a mission to provide leadership in education, research and the responsible use of natural resources in the tropics. To this end, graduate and undergraduate field courses are offered in the pure and applied areas of the natural sciences in Costa Rica, Peru, Mexico, the Brazilian Amazon, and South Africa. In addition, OTS operates three research stations in varied ecosystems in Costa Rica: La Selva in the Caribbean lowland rain forest, Palo Verde in the Pacific lowland dry forest, and Las Cruces in a mid-elevation rain forest on Costa Rica's southern Pacific slope. OTS also coordinates and facilitates research, provides environmental-education programs, and works with local agencies on issues of conservation, natural-resource management and land-use practices.
Requirements: OTS fellowships are open to graduate students enrolled in degree programs at OTS member institutions and to OTS course alumni. (A list of member institutions is available at the OTS website from the About-Us Quick Link.) Applicants interested in the Water Quality Environmental Outreach program must contact Dr. Catherine Pringle, University of Georgia, cpringle@uga.edu, before submitting an application.
Restrictions: Awards will be restricted to OTS field stations (La Selva, Palo Verde, and Las Cruces); proposals for research at other sites will not be considered. Only one research award may be held at a time. Awards do not cover capital equipment; minor amounts of materials and supplies may be considered, but are best covered by other sources of funds. Salaries or stipends for the investigator are not allowed, but salaries for field assistants will be considered if the work cannot be carried out safely or efficiently by one person.
Geographic Focus: All States, All Countries
Date(s) Application is Due: Mar 1; Oct 1
Amount of Grant: Up to 3,650 USD
Contact: Education Director; (506) 2524-0607; fax (506) 2524-0608; academic@ots.ac.cr
Internet: http://ots.ac.cr/index.php?option=com_content&task=view&id=170&Itemid=344
Sponsor: Organization for Tropical Studies
Duke University, 410 Swift Avenue
Durham, NC 27705

Organization for Tropical Studies Pilot Awards for Exploratory Research **3797**
Organization for Tropical Studies (OTS) Pilot Awards for Exploratory Research are intended to assist with thesis-related research in tropical biology and similar fields and are available in amounts up to a maximum of $2,075. Awards are intended primarily to cover up to 25 days of station fees (food, lodging and station access) at OTS field stations; current rates are $63 for overnight visits and $31.50 for day visits. Other expenses that may be included in the budget, to a maximum of $500, are research- and export-permit fees, local travel, and reimbursement in whole or in part of one round-trip airline ticket. Application guidelines and a downloadable cover sheet are available from the OTS Fellowships webpage. The cover sheet and supporting documents must be submitted via email by the deadline date to the Education Director. Notification of decisions will take place about two months after each deadline. The Organization for Tropical Studies (OTS) is a nonprofit consortium of 64 universities and research institutions in the United States, Costa Rica, Peru, Canada, Mexico, South Africa, and Australia. Founded in 1963, OTS has a mission to provide leadership in education, research and the responsible use of natural resources in the tropics. To this end, graduate and undergraduate field courses are offered in the pure and applied areas of the natural sciences in Costa Rica, Peru, Mexico, the Brazilian Amazon, and South Africa. In addition, OTS operates three research stations in varied ecosystems in Costa Rica: La Selva in the Caribbean lowland rain forest, Palo Verde in the Pacific lowland dry forest, and Las Cruces in a mid-elevation rain forest on Costa Rica's southern Pacific slope. OTS also coordinates and facilitates research, provides environmental-education programs, and works with local agencies on issues of conservation, natural-resource management and land-use practices.
Requirements: Organization for Tropical Studies (OTS) fellowships are open to graduate students enrolled in degree programs at OTS member institutions and to OTS course alumni. (A list of member institutions is available at the OTS website from the About-Us Quick Link.)
Restrictions: Only one research award may be held at a time. Awards do not cover capital equipment; minor amounts of materials and supplies may be considered, but are best covered by other sources of funds. Salaries or stipends for the investigator are not allowed, but salaries for field assistants will be considered if the work cannot be carried out safely or efficiently by one person.
Geographic Focus: All States, All Countries
Date(s) Application is Due: Mar 1; Oct 1
Amount of Grant: Up to 2,075 USD
Contact: Education Director; (506) 2524-0607; fax (506) 2524-0608; academic@ots.ac.cr
Internet: http://ots.ac.cr/index.php?option=com_content&task=view&id=170&Itemid=344
Sponsor: Organization for Tropical Studies
Duke University, 410 Swift Avenue
Durham, NC 27705

Orgone Biophysical Research Lab Lou Hochberg Awards **3798**
Hochberg Awards provide cash awards and recognition to high school and university students in several categories, and to published authors and journalists, for research and open-theme essays that are topically focused upon Wilhelm Reich's sociological discoveries. The awards are open to students and journalists worldwide, but must be submitted in the English language. Awards are made in the following categories: thesis and dissertation awards—for the best accepted and approved university level thesis or dissertation addressing Wilhelm Reich's sociological discoveries; university/college essay awards— given for the best university/college-level student research paper (maximum 50 pages) addressing Wilhelm Reich's sociological discoveries; university thesis/dissertation improvement and implementation grants—for undertaking original graduate-level research addressing Wilhelm Reich's sociological discoveries; and outstanding research and journalism appearing in popular or academic publications— for research and writing on Wilhelm Reich's discoveries, with subsequent publication of accurate articles in the popular press and/or academic journals. The journalist award is open to any article focused upon Reich's sociological and natural scientific discoveries that is published in either a regional and/or national publication (newspaper, magazine, journal). Listed deadlines are for university/college essay awards and high school essay awards.
Requirements: Competitions are open to students worldwide. Submissions must be in English, typed, and presented in triplicate no later than May 15 for the spring/summer competition or December 15 for the fall/winter competition.
Geographic Focus: All States
Date(s) Application is Due: May 15; Dec 15
Amount of Grant: 500-1,500 USD
Contact: Award Administrator; (541) 552-0118; info@orgonelab.org
Internet: http://www.orgonelab.org/hochberg.htm
Sponsor: Orgone Biophysical Research Lab
P.O. Box 1148
Ashland, OR 97520

ORISE Civilian Radioactive Waste Management Fellowship Program **3799**
These one-year, renewable fellowships are offered for graduate work in fields related to the management of spent nuclear fuel and high-level radioactive wastes. Application is usually made before or during the first year of graduate school. Benefits include payment of tuition and fees, monthly stipend, and one required summer practicum at a U.S. DOE national laboratory. Applications are due the last Monday in January.
Requirements: Graduate students at participating universities in the earth sciences, engineering, materials science, or radiation sciences are eligible.
Geographic Focus: All States
Amount of Grant: 34,300 USD
Contact: M. Carl Wheeler, Program Manager; (865) 241-6704; Wheelerc@orau.gov
Internet: http://www.orau.gov/orise/edu/DOE/gi-gCRWM.htm
Sponsor: Oak Ridge Institute for Science and Education
P.O. Box 117
Oak Ridge, TN 37831-0117

ORISE Department of Commerce Internship for Postsecondary Students **3800**
This program offers opportunities to participate in hands-on education and training related to the Department of Commerce mission. The program provides funding of: $550 for graduate students; $450 weekly stipend; $125 weekly dislocation allowance. Deadlines are the third Tuesday in January for summer; July 30 for fall semester; and December 31 for spring semester.
Requirements: Undergraduate and graduate students who are U.S. citizens are eligible.
Geographic Focus: All States
Date(s) Application is Due: Jul 30; Dec 31
Contact: Pat Pressley, Program Specialist; (865) 576-3409; presslep@orau.gov
Internet: http://www.orau.gov/orise/edu/DOC/gi-gugIPS.htm
Sponsor: Oak Ridge Institute for Science and Education
P.O. Box 117
Oak Ridge, TN 37831-0117

ORISE Environmental Management Participation for the U.S. Army Environmental Center **3801**
An interagency agreement between DOE and the U.S. Army Environmental Center makes it possible for ORISE to offer research grants in environmental programs involving restoration, compliance, conservation, pollution prevention, validation, demonstration, transfer, quality assurance and quality control, training, information management and reporting, and related resource management and planning programming. The duration of the program is 10 weeks to three months; some sabbatical appointments are available, as are full-time and part-time appointments. A negotiable stipend is based on regular salary; reimbursement for travel is included. Applications are accepted at any time.
Requirements: Applicants must be college or university full-time faculty, K-12 educators, and U.S. citizens.
Geographic Focus: All States
Contact: Kim Myers , Recruiter; (410) 436-7258; kim.myers3@us.army.mil
Internet: http://www.orau.gov/orise/edu/U.S.ARMY/gi-fEMP.htm

Sponsor: Oak Ridge Institute for Science and Education
P.O. Box 117
Oak Ridge, TN 37831-0117

ORISE Faculty Internship at the U.S. Army Center for Health Promotion and Preventive Medicine 3802

The program is available through an interagency agreement between DOE and the U.S. Army Center for Health Promotion and Preventive Medicine. It provides opportunities and support for applied clinical research and training activities in such areas as environmental health engineering, entomology, ionizing and nonionizing radiation, occupational and environmental health, industrial hygiene and worksite hazards, environmental sanitation and hygiene, and laboratory services. The duration of the program is 10 weeks to one year; some sabbatical appointments are available, as are full-time and part-time appointments. The stipend is negotiable based on regular salary, and reimbursement for travel is included. Applications are accepted at any time.
Requirements: Applicants must be college or university full-time faculty, K-12 educators, and U.S. citizens.
Geographic Focus: All States
Contact: Kim Myers , Recruiter; (410) 436-7258; kim.myers3@us.army.mil
Internet: http://www.orau.gov/orise/edu/U.S.ARMY/gi-fHPPM.htm
Sponsor: Oak Ridge Institute for Science and Education
P.O. Box 117
Oak Ridge, TN 37831-0117

ORISE Faculty Research Participation at the Centers for Disease Control and Prevention 3803

This program provides opportunities and support to conduct research on infectious diseases, environmental health, and occupational safety and health. One program provides support to conduct projects and studies in areas supportive of epidemiological programs. The duration of the programs is 10 weeks to three months; some sabbatical appointments are offered up to one year, and full-time and part-time appointments are available. The stipends are negotiable based on regular salary. Applications are accepted at any time.
Requirements: Applicants must be college or university full-time faculty members.
Geographic Focus: All States
Contact: Tom Richmond, Program Manager; (865) 576-2194; richmont@orau.gov
Internet: http://www.orau.gov/orise/edu/CDC/gi-fRP.htm
Sponsor: Oak Ridge Institute for Science and Education
P.O. Box 117
Oak Ridge, TN 37831-0117

ORISE Faculty Research Participation at the National Center for Toxicological Research 3804

The program provides opportunities and support for research in biological effects of potentially toxic chemicals and deriving solutions to problems in toxicology that have a major impact on human health and the environment. Benefits include a negotiable stipend based on regular salary and reimbursement for travel and moving. Applications are accepted at any time.
Requirements: U.S. college and university full-time faculty members in the areas of toxicology, pharmacology, chemistry, biological sciences, mathematics, computer science, medicine, or other related scientific disciplines are eligible to apply.
Geographic Focus: All States
Contact: Sherry Edwards, Program Specialist; (865) 576-7385; warrens@orau.gov
Internet: http://www.orau.gov/orise/edu/FDA/gi-fNCTR.htm
Sponsor: Oak Ridge Institute for Science and Education
P.O. Box 117
Oak Ridge, TN 37831-0117

ORISE Faculty Research Participation at the U.S. Army Joint P.O.W/MIA Accounting Command, Central Identification Laboratory 3805

This project offers opportunities to participate in research to locate, exhume, and identify the remains of individuals killed during military service. The program can last from 10 weeks to one year. Applications are accepted on a year-round basis.
Requirements: College or university full-time faculty are eligible to apply.
Geographic Focus: All States
Contact: Norma Ward, Program Assistant; (865) 241-2890; wardn@orau.gov
Internet: http://www.orau.gov/orise/edu/U.S.ARMY/gi-fCIL.htm
Sponsor: Oak Ridge Institute for Science and Education
P.O. Box 117
Oak Ridge, TN 37831-0117

ORISE Fusion Energy Postdoctoral Research 3806

Magnetic fusion energy collaborative research opportunities are available on fusion energy research and development projects at DOE Office of Fusion Energy-supported laboratories and contractor sites for appropriate science and engineering disciplines. Awards cover one year of support and are renewable. Limited reimbursement for health insurance costs and inbound travel and moving are included.
Requirements: Applicants must have completed the doctoral degree within the last three years.
Geographic Focus: All States
Date(s) Application is Due: Mar 1
Amount of Grant: 50,400 USD
Contact: Tom Richmond, Program Manager; (865) 576-2194; richmont@orau.gov
Internet: http://www.orau.gov/orise/edu/DOE/gi-pdFEPR.htm
Sponsor: Oak Ridge Institute for Science and Education
P.O. Box 117
Oak Ridge, TN 37831-0117

ORISE Fusion Energy Sciences Fellowship Program 3807

DOE and ORISE offer fellowships for graduate work in magnetic fusion science and engineering in the following disciplines: physical sciences; engineering; and mathematics. Application is usually made before or during the first year of graduate school. The deadline is the last Monday in January, and the program duration is one year. Tuition and fees are provided.
Requirements: Doctoral students are eligible.
Geographic Focus: All States
Date(s) Application is Due: Jan 31
Amount of Grant: 25,500-31,200 USD
Contact: Tom Richmond, Program Manager; (865) 576-2194; richmont@orau.gov
Internet: http://www.orau.gov/orise/edu/DOE/gi-gFES.htm
Sponsor: Oak Ridge Institute for Science and Education
P.O. Box 117
Oak Ridge, TN 37831-0117

ORISE Higher Education Research Experiences at Oak Ridge National Laboratory for Faculty 3808

Opportunities exist to participate in ongoing energy research in the following areas: computer sciences; engineering; mathematics; and physical and natural sciences. The stipend is based on regular salary; limited travel reimbursement is provided. The duration of the program is 10 weeks to three months; some sabbatical appointments up to 12 months are offered. Proposals are due the first Monday in February.
Requirements: College or university full-time faculty who are U.S. citizens or permanent residents are eligible to apply.
Geographic Focus: All States
Contact: Ruth Keller, Project Manager; (865) 574-7798; kellerr@orau.gov
Internet: http://www.orau.gov/orise/edu/ornl/gi-fHERE.htm
Sponsor: Oak Ridge Institute for Science and Education
P.O. Box 117
Oak Ridge, TN 37831-0117

ORISE Laboratory Technology Program 3809

opportunities exist to receive hands-on experience in technical areas. Current opportunities are in mechanical engineering technology. Stipends are based on academic level or degree. Applications are accepted on a year-round basis.
Requirements: Undergraduate and graduate students who are U.S. citizens or legal permanent residents are eligible.
Geographic Focus: All States
Contact: Sam Held, Project Manager; (865) 576-8223; helds@orau.gov
Internet: http://www.orau.gov/orise/edu/ornl/gi-ugIPLT.htm
Sponsor: Oak Ridge Institute for Science and Education
P.O. Box 117
Oak Ridge, TN 37831-0117

ORISE National Library of Medicine, Center for Biotechnology Information Scientific Visitors Program 3810

This program offers opportunities to participate in informatics research and training programs under the discipline(s) of biochemistry, computer science, mathematics, medical research, molecular biology, and structural biology.
Requirements: Faculty, postdoctorals, and research scientists are eligible to apply.
Geographic Focus: All States
Contact: Elizabeth Kittrell, Project Manager; (865) 241-3319; kittrell@orau.gov
Internet: http://www.orau.gov/orise/edu/nlm/gi-fpdgNCBI.htm
Sponsor: Oak Ridge Institute for Science and Education
P.O. Box 117
Oak Ridge, TN 37831-0117

ORISE National Oceanic and Atmospheric Administration Faculty Research Participation 3811

Opportunities exist to participate in research and development relating to science, math, and engineering in the following disciplines: life sciences and physical sciences. Limited travel reimbursement is provided. The duration of the program is 10 weeks in the summer; some on-campus support to continue research begun during the summer is provided. Applications are accepted at any time.
Requirements: College or university full-time faculty who are U.S. citizens or permanent residents are eligible.
Geographic Focus: All States
Contact: Jennifer Garren, Program Specialist; (865) 576-5300; garrenj@orau.gov
Internet: http://www.orau.gov/orise/edu/NOAA/gi-fRP.htm
Sponsor: Oak Ridge Institute for Science and Education
P.O. Box 117
Oak Ridge, TN 37831-0117

ORISE National Oceanic and Atmospheric Administration Student Research Participation 3812

Opportunities exist to participate in research and development relating to science, math, and engineering in the following disciplines: computer science, engineering, life sciences, and physical sciences. Applications are accepted at any time. The duration of the program is 10 weeks in the summer.
Requirements: Undergraduate and graduate students who are U.S. citizens or permanent residents are eligible.
Geographic Focus: All States

Amount of Grant: 420 USD
Contact: Jennifer Garren, Program Specialist; (865) 576-5300; garrenj@orau.gov
Internet: http://www.orau.gov/orise/edu/NOAA/gi-gugSRP.htm
Sponsor: Oak Ridge Institute for Science and Education
P.O. Box 117
Oak Ridge, TN 37831-0117

ORISE Nuclear Regulatory Commission Historically Black Colleges and Universities Faculty Research Participation 3813
The U.S. Nuclear Regulatory Commission affords faculty members from historically black colleges and universities the opportunity to participate in research at various federal facilities across the United States. The duration of the program varies: 10 weeks to three months for summer appointments, and one year for sabbatical appointments and on-campus appointments during the academic year. The deadline for applications is the third Tuesday in January. Travel reimbursement is provided.
Requirements: Applicants must be full-time faculty members from historically black institutions and U.S. citizens or permanent residents.
Geographic Focus: All States
Amount of Grant: 9,500-45,600 USD
Contact: Pat Pressley, Program Specialist; (865) 576-3409; presslep@orau.gov
Internet: http://www.orau.gov/orise/edu/NRC/gi-fHBCU.htm
Sponsor: Oak Ridge Institute for Science and Education
P.O. Box 117
Oak Ridge, TN 37831-0117

ORISE Nuclear Regulatory Commission Historically Black Colleges and Universities Student Research Participation 3814
Opportunities exist to participate in research relating to science and engineering in laboratories where NRC research is being conducted in the following disciplines: computer sciences, engineering, geosciences, health physics, human factors, materials science, mathematics, molecular/radiation biology, physical sciences, and statistics-related nuclear material control and accounting. The duration of the program is 10 to 12 weeks in the summer; some one-year appointments are available. Annual deadline dates may vary; contact program staff for exact dates.
Requirements: Undergraduate and graduate students from historically black colleges and universities, who are U.S. citizens or permanent residents, are eligible.
Geographic Focus: All States
Contact: Pat Pressley, Program Specialist; (865) 576-3409; presslep@orau.gov
Internet: http://www.orau.gov/orise/edu/NRC/gi-gugHBCU.htm
Sponsor: Oak Ridge Institute for Science and Education
P.O. Box 117
Oak Ridge, TN 37831-0117

ORISE Oak Ridge National Laboratory Postdoctoral Research Associates 3815
This program provides opportunities and support for research and training in a broad range of science and engineering activities related to basic sciences, energy, and the environment, as well as access to modern and extensive research facilities. Appointments are for one year and are renewable. Applications are accepted at any time. Stipends are based on research area and degree; reimbursement for inbound travel and moving is available. In addition, full-time temporary employees of Oak Ridge Associated Universities receive benefits.
Requirements: Applicants must have completed the doctoral degree within the last five years; others are considered on a case-by-case basis.
Geographic Focus: All States
Contact: Bonnie DeJarnette, Program Specialist; (865) 241-4832; dejarneb@orau.gov
Internet: http://www.orau.gov/orise/edu/ornl/gi-pdRA.htm
Sponsor: Oak Ridge Institute for Science and Education
P.O. Box 117
Oak Ridge, TN 37831-0117

ORISE Postgraduate Environmental Management Research Participation at the U.S. Army Environmental Center 3816
An interagency agreement between DOE and the U.S. Army Environmental Center makes the program possible. It provides research in environmental programs involving restoration, compliance, conservation, pollution prevention, validation, demonstration, transfer, quality assurance and quality control, training, information management and reporting, and related resource management and planning programming. The duration of the program is one year, and is renewable. Stipends are based on research area and degree, and limited reimbursement for inbound moving and travel is included. Applications are accepted at any time.
Requirements: Applicants must be U.S. citizens and must have completed a bachelor's or undergraduate degree within the last three years; others are considered on a case-by-case basis.
Geographic Focus: All States
Contact: Kim Myers , Recruiter; (410) 436-7258; kim.myers3@us.army.mil
Internet: http://www.orau.gov/orise/edu/U.S.ARMY/gi-pdEMP.htm
Sponsor: Oak Ridge Institute for Science and Education
P.O. Box 117
Oak Ridge, TN 37831-0117

ORISE Postgraduate Internship at the Office of Water 3817
These one-year renewable grants provide opportunities and support for studies related to development and implementation of drinking water regulations. Internships are located at the Office of Ground Water and Drinking Water in Cincinnati, OH and Washington, DC. Applications are accepted at any time. Awards are granted for one year and are renewable. Stipends are based on research area and degree; limited reimbursement for inbound travel and moving is available.
Requirements: Applicants must have completed a bachelor's, master's, or doctoral degree within the last three years. Others are considered on a case-by-case basis.
Geographic Focus: All States
Contact: Sharon Kern, (865) 576-9361; (865) 576-9361; kerns@orau.gov
Internet: http://www.orau.gov/orise/edu/EPA/gi-rgpdIOW.htm
Sponsor: Oak Ridge Institute for Science and Education
P.O. Box 117
Oak Ridge, TN 37831-0117

ORISE Postgraduate Internship at the U.S. Army Center for Health Promotion and Preventive Medicine 3818
An interagency agreement between DOE and the U.S. Army Center for Health Promotion and Preventive Medicine provides opportunities and support for applied clinical research and training activities in such areas as environmental health engineering, entomology, ionizing and nonionizing radiation, occupational and environmental health, industrial hygiene and worksite hazards, environmental sanitation and hygiene, and laboratory services. The duration of the program is one year, and is renewable. Stipends are based on research area and degree, and limited reimbursement for inbound travel and moving is included. Applications are accepted at any time.
Requirements: Applicants must be U.S. citizens and must have completed a bachelor's, master's, or doctoral degree within the last three years; others are considered on a case-by-case basis.
Geographic Focus: All States
Contact: Kim Myers , Recruiter; (410) 436-7258; kim.myers3@us.army.mil
Internet: http://www.orau.gov/orise/edu/U.S.ARMY/gi-pdCHPPM.htm
Sponsor: Oak Ridge Institute for Science and Education
P.O. Box 117
Oak Ridge, TN 37831-0117

ORISE Postgraduate Research Participation at the Center for Biologics Evaluation and Research 3819
An interagency agreement between DOE and FDA provides opportunities and support for research concerning biologics. The duration of the program is one year, and is renewable. Stipends are based on research area and degree, and limited reimbursement for inbound travel and moving is included. Applications are accepted at any time.
Requirements: Applicants must have completed a master's or doctoral degree within the last three years; others are considered on a case-by-case basis.
Geographic Focus: All States
Contact: Sherry Edwards, Program Specialist; (865) 576-7385; warrens@orau.gov
Internet: http://www.orau.gov/orise/edu/FDA/gi-rgpdCBER.htm
Sponsor: Oak Ridge Institute for Science and Education
P.O. Box 117
Oak Ridge, TN 37831-0117

ORISE Postgraduate Research Participation at the Centers for Disease Control and Prevention 3820
An interagency agreement between DOE and CDC provides opportunities and support to conduct research on infectious diseases, environmental health, and occupational safety and health; and for projects and studies in areas supportive of epidemiological programs. Programs are offered by the Division of AIDS, Sexually Transmitted Diseases, and Tuberculosis Laboratory Research; Division of Environmental Health Laboratory Sciences; Division of Mycotic and Bacterial Diseases; Division of Prevention Research and Analytic Methods; Division of Viral and Rickettsial Control and Prevention; and NIOSH. The programs last up to one year, and are renewable. Stipends are based on research area and degree. Applications are accepted at any time.
Requirements: Applicants must have completed a bachelor's, master's, or doctoral degree within the last three years; others are considered on a case-by-case basis.
Geographic Focus: All States
Contact: Tom Richmond, Program Manager; (865) 576-2194; richmont@orau.gov
Internet: http://www.orau.gov/orise/edu/CDC/gi-rgpdCDC.htm
Sponsor: Oak Ridge Institute for Science and Education
P.O. Box 117
Oak Ridge, TN 37831-0117

ORISE Postgraduate Research Participation at the National Exposure Research Laboratory 3821
An interagency agreement between DOE and EPA provides research opportunities in environmental areas in the following disciplines: physical sciences, environmental sciences, life sciences, engineering, and related sciences. Laboratories are located in Athens, GA, and Cincinnati, OH. The duration of the program is one year, and is renewable. Stipends are based on research area and degree, and limited reimbursement for inbound moving and travel is included. Applications are accepted at any time.
Requirements: Applicants must have completed a bachelor's or graduate degree within the last three years; others are considered on a case-by-case basis.
Geographic Focus: All States
Contact: Sharon Kern, Project Manager; (865) 576-9361; kerns@orau.gov
Internet: http://www.orau.gov/orise/edu/EPA/gi-rgpdNERL.htm
Sponsor: Oak Ridge Institute for Science and Education
P.O. Box 117

Oak Ridge, TN 37831-0117

ORISE Postgraduate Research Participation at the U.S. Army Central Identification Laboratory, Hawaii 3822

This program offers opportunities to participate in research to locate, exhume, and identify the remains of individuals killed during military service.
Requirements: Eligible applicants must have completed a doctoral degree degree within the last three years; others considered on a case-by-case basis. U.S. citizens preferred.
Geographic Focus: All States
Contact: Norma Ward, Program Assistant; (865) 241-2890; wardn@orau.gov
Internet: http://www.orau.gov/orise/edu/U.S.ARMY/gi-pdCIL.htm
Sponsor: Oak Ridge Institute for Science and Education
P.O. Box 117
Oak Ridge, TN 37831-0117

ORISE Savannah River Site Faculty Research Participation Program 3823

Opportunities exist to participate in ongoing energy research in the following areas: computer sciences; engineering; mathematics; and physical and natural sciences. The stipend is based on regular salary; limited travel reimbursement is provided. The duration of the program is 10 weeks to three months; some sabbatical appointments of up to 12 months are offered. Proposals due the first Monday in February for summer appointments; other applications are accepted on a year-round basis.
Requirements: College or university full-time faculty who are U.S. citizens are eligible.
Geographic Focus: All States
Contact: Ruth Keller, Project Manager; (865) 574-7798; kellerr@orau.gov
Internet: http://www.orau.gov/orise/edu/SRS/gi-fRP.htm
Sponsor: Oak Ridge Institute for Science and Education
P.O. Box 117
Oak Ridge, TN 37831-0117

ORISE Savannah River Site Postgraduate Research Program 3824

DOE and the Savannah River Site provide opportunities and support to conduct collaborative research that complements ongoing Savannah River research projects. The duration of the program is one year, and is renewable. Stipends are based on research area and degree, and limited reimbursement for inbound travel and moving is included. Applications are accepted at any time.
Requirements: Students must have completed a graduate degree within the last three years and be U.S. citizens.
Geographic Focus: All States
Contact: Ruth Keller, Project Manager; (865) 574-7798; kellerr@orau.gov
Internet: http://www.orau.gov/orise/edu/SRS/gi-rgpdRPP.htm
Sponsor: Oak Ridge Institute for Science and Education
P.O. Box 117
Oak Ridge, TN 37831-0117

ORISE Savannah River Site Professional Internship Program 3825

The Savannah River Site provides full-time and part-time opportunities for students enrolled in advanced degree programs to participate in research projects for their theses or dissertations. The duration of the program is three to 12 consecutive months. Travel (one round trip) to center/campus and off-campus tuition fees are required by the home institution.
Requirements: Applicants must be U.S. citizens and be undergraduate students, postbaccalaureates, or graduate students.
Geographic Focus: All States
Date(s) Application is Due: Feb 15; Jun 1; Oct 1
Contact: Ruth Keller, Project Manager; (865) 574-7798; kellerr@orau.gov
Internet: http://www.orau.gov/orise/edu/SRS/gi-gugrgPIP.htm
Sponsor: Oak Ridge Institute for Science and Education
P.O. Box 117
Oak Ridge, TN 37831-0117

ORISE Student Environmental Management Participation Program for the U.S. Army Environmental Center 3826

An interagency agreement between DOE and the U.S. Army Environmental Center provides research in environmental programs involving restoration, compliance, conservation, pollution prevention, validation, demonstration, transfer, quality assurance and quality control, training, information management and reporting, and related resource management and planning programming. The duration of the program is three months to one year, and full-time or part-time appointments are available. Stipends are based on research areas and academic classification.
Requirements: Students currently enrolled in undergraduate or graduate degree programs and who are U.S. citizens are eligible.
Geographic Focus: All States
Contact: Kim Myers , Recruiter; (410) 436-7258; kim.myers3@us.army.mil
Internet: http://www.orau.gov/orise/edu/U.S.ARMY/gi-gugEMP.htm
Sponsor: Oak Ridge Institute for Science and Education
P.O. Box 117
Oak Ridge, TN 37831-0117

ORISE Student Internship Program at the U.S. Army Center for Health Promotion and Preventive Medicine 3827

An interagency agreement between DOE and the U.S. Army Center for Health Promotion and Preventive Medicine provides opportunities and support for applied research and development in the areas of environmental health engineering programs, projects, and activities. The duration of the program is three months to one year, and full-time or part-time appointments are available. Applications are accepted at any time.
Requirements: Students who are U.S. citizens and are currently enrolled in undergraduate or graduate degree programs are eligible.
Geographic Focus: All States
Contact: Kim Myers , Recruiter; (410) 436-7258; kim.myers3@us.army.mil
Internet: http://www.orau.gov/orise/edu/U.S.ARMY/gi-gugHPPM.htm
Sponsor: Oak Ridge Institute for Science and Education
P.O. Box 117
Oak Ridge, TN 37831-0117

ORISE Student Research Participation at the Agency for Toxic Substances and Disease Registry 3828

An interagency agreement between DOE and the Agency for Toxic Substances and Disease Registry (Public Health Service) provides opportunities and support for research in exposure and disease registries, health investigations, public health assessments, toxicological profiles, emergency response, and health education. The duration of the program is 10 weeks to 12 months, and part-time appointments are available. A stipend based on academic status and research area is provided, plus limited travel reimbursement. Off-campus tuition and fees are required by the home institution. Applications are accepted at any time.
Requirements: Undergraduate and graduate students are eligible.
Geographic Focus: All States
Date(s) Application is Due: Feb 1
Contact: Tasha Powell, Program Specialist; (865) 574-3911; powellt@orau.gov
Internet: http://www.orau.gov/orise/edu/ATSDR/gi-gugsrp.htm
Sponsor: Oak Ridge Institute for Science and Education
P.O. Box 117
Oak Ridge, TN 37831-0117

ORISE Student Research Participation at the Centers for Disease Control and Prevention 3829

An interagency agreement between DOE and CDC provides opportunities and support to conduct research on infectious diseases, environmental health, and occupational safety and health; and to conduct projects and studies in areas supportive of epidemiological programs. The program is offered by the Division of AIDS, Sexually Transmitted Diseases, and Tuberculosis Laboratory Research; Division of Environmental Health Laboratory Sciences; Division of Mycotic and Bacterial Diseases; Division of Prevention Research and Analytic Methods; and NIOSH. The duration of the programs is one month to one year, and full-time or part-time appointments are available. Stipends are based on research area and academic classification. Applications are accepted at any time.
Requirements: Undergraduate and graduate students are eligible.
Geographic Focus: All States
Contact: Tom Richmond, Program Manager; (865) 576-2194; richmont@orau.gov
Internet: http://www.orau.gov/orise/edu/CDC/gi-gugRP.htm
Sponsor: Oak Ridge Institute for Science and Education
P.O. Box 117
Oak Ridge, TN 37831-0117

ORISE Student Research Participation at the National Center for Toxicological Research 3830

Full-time or part-time appointments, made for ten weeks to one year, provide opportunities and support for research participation in biological effects of potentially toxic chemicals and deriving solutions to problems in toxicology that have a major impact on human health and the environment. Stipends are based on research areas and academic classification. This is an interagency agreement between the U.S. Department of Energy and the National Center for Toxicological Research, with positions located in Jefferson, AR, at the research center. Applications are accepted at any time for academic-year appointments, March 1 for summer appointments.
Requirements: Applicants must be students currently enrolled in undergraduate or graduate degree programs.
Geographic Focus: All States
Date(s) Application is Due: Mar 1
Contact: Tasha Powell, Program Specialist; (865) 574-3911; powellt@orau.gov
Internet: http://www.orau.gov/orise/edu/FDA/gi-gugNCTR.htm
Sponsor: Oak Ridge Institute for Science and Education
P.O. Box 117
Oak Ridge, TN 37831-0117

ORISE University Radioactive Ion Beam (UNIRIB) Consortium 3831

This program provides opportunities for basic research on rare, short-lived atomic nuclei at the Holifield Radioactive Ion Beam Facility. For best results, apply by February 1 for summer appointments.
Requirements: Faculty members, graduate and postgraduate students, and research fellows are eligible to apply.
Geographic Focus: All States
Contact: H. Kennon Carter, Program Manager; (865) 576-2642; carterk@orau.gov
Internet: http://www.orau.gov/orise/edu/DOE/gi-fgrgpdURIB.htm
Sponsor: Oak Ridge Institute for Science and Education
P.O. Box 117
Oak Ridge, TN 37831-0117

Oscar Rennebohm Foundation Grants 3832

The foundation awards grants to nonprofit organizations in the Madison, WI area. Areas of interest, include: higher education, the arts, environmental conservation, health

and social service agencies. Types of support include building construction/renovation, equipment acquisition, and research. There are no application forms or deadlines.
Geographic Focus: Wisconsin
Samples: University of Wisconsin Foundation, Madison, WI, $2,275,000; Second Harvest Foodbank of Southern Wisconsin, Madison, WI, $140,000; Bethel Community Services Corporation, Madison, WI, $40,000.
Contact: Grants Administrator; (608) 274-5991
Sponsor: Oscar Rennebohm Foundation
P.O. Box 5187
Madison, WI 53705

OSF-Baltimore Community Fellowships 3833
The Baltimore Community Fellowships program works to meet the challenges facing Baltimore's most needy communities by identifying and supporting social innovators of unusual promise and providing them with the ingredients to ensure their ideas have a stable foundation and long-term viability. The Fellowships program seeks dynamic activists and social entrepreneurs interested in implementing projects that address problems in underserved communities in Baltimore City. Funding of $60,000 is available over 18 months. The application is available at the OSF-Baltimore website, and candidates may be placed on an application mailing list.
Requirements: Fellowships are for initiatives that take place in Baltimore City. Applicants may come from any field, including—but not limited to—business, management, the arts, law, medicine, education, architecture, and engineering. Individuals from underserved communities and people of color are strongly encouraged to apply. Applicants need not be from Baltimore City, but should be knowledgeable about social and economic justice issues affecting Baltimore's communities. Fellows must be willing to participate fully in meetings scheduled for the Community Fellows.
Restrictions: Individuals already receiving wages or a salary for the proposed project from a host organization are not eligible. Fellows may not use stipends to supplant funding for activities or projects that the host organization is already implementing. The program does not award fellowships to conduct research or to implement lobbying initiatives.
Geographic Focus: All States
Amount of Grant: 60,000 USD
Contact: Pamela King, Director, Baltimore Fellowships and Initiatives; (410) 234-1091, ext. 210; fax (410) 234-2816; pamela.king@opensocietyfoundations.org
Internet: http://www.opensocietyfoundations.org/grants/baltimore-community-fellowships
Sponsor: Open Society Foundations
201 North Charles Street, Suite 1300
Baltimore, MD 21201

OSF Access to Medicines Fellowships in Eastern Europe and Central Asia 3834
This newly created and competitive Fellowship program aims to build a cadre of motivated leaders, activists, lawyers, and health care professionals in Eastern Europe and Central Asia who have the knowledge and leadership abilities to build a vocal access to medicines movement, and propel and mobilize others for increased access to medicines advocacy. Up to five Fellows will be selected for a Fellowship period of 18 months. During the Fellowship, candidates will remain in their home organizations, but be expected to devote a minimum of five hours per week to the Fellowship, including work on an individual project. The Fellowship will include extensive virtual learning opportunities and three in-person meetings, all focused on developing expertise on technical and political access to medicines issues, as well as advocacy and leadership skills. A small honorarium will be offered to the fellows. The application and additional guidelines are available at the website.
Requirements: Applicants must meet the following *Requirements:* university diploma with at least two to three years work experience (if no university diploma, seven years work experience in a related field); experience working in the NGO sector; advocacy experience; background in health, access to medicines, law, economics, accountability and/or social justice issues, including appreciation of access to medicines issues with an understanding of current access challenges at a national/regional/global level; work in disease fields, which have acute access issues such an HIV, hepatitis C, cancer and diabetes; written support from organization, including letter from Executive Director stating support, with acknowledgment of expected time commitment of the candidate in the Fellowship; the ability to speak, write and read in English and working knowledge of Russian; willingness to travel; and based or working in East European and Central Asian region.
Geographic Focus: Albania, Belarus, Bosnia & Herzegovina, Bulgaria, Croatia, Czech Republic, Estonia, Hungary, Kazakhstan, Kosovo, Kyrgyzstan, Latvia, Lithuania, Macedonia, Moldova, Montenegro, Poland, Romania, Russia, Serbia, Slovakia, Slovenia, Tajikistan, Turkmenistan, Ukraine, Uzbekistan
Date(s) Application is Due: Jun 4
Contact: Anna Segelman, Administrative Assistant; (212) 548-0600; fax (212) 548-4600; anna.segelman@opensocietyfoundations.org
Internet: http://www.opensocietyfoundations.org/grants/access-medicines-eastern-europe-and-central-asia
Sponsor: Open Society Foundations
400 West 57th Street, 3rd Floor
New York, NY 10019

OSF Alumni Grants 3835
The Alumni Grant program offers support to Open Society Scholarship alumni to further expand knowledge gained and make a positive contribution to their home country. The competition is open only to Open Society Scholarship Programs alumni of the following programs: Afghan Communications Scholarship Program; American University in Bulgaria; Asia Pacific Scholarship Consortium; Cambridge and Oxford Hospitality Schemes; Civil Service Awards-Moldova; CNOU.S.-OSI Program; DAAD-OSI Program; Environmental Law/Management Program; Faculty Development Program; Georgian Program for Education Professionals; Global Supplementary Grant Program; Mongolian Professional Fellowship Program; Muskie/FSA Graduate Fellowship Program; OSF Scholarship in European Studies (Maastricht and Aarhus Universities); Palestinian Faculty Development Program; Palestinian Rule of Law Program; Social Work Fellowship Program; Soros Supplementary Grant Program; Supplementary Grant Program Asia (formerly the Supplementary Grant Program–Burma); Supplementary Grant Program – Former Yugoslavia; Undergraduate Exchange Program/Virtual University; and the United Kingdom Scholarship Programs (formerly the Chevening –OSI Scholarships). The application and additional guidelines are available at the website.
Requirements: Alumni must be in compliance with the conditions of any past scholarship or Open Society Foundation grant; reside in their home country or region during the project year; and present a project related to or applicable to the home country or region.
Restrictions: The program will not support: individual travel grants for participation in conferences, workshops, classes; travel to or from countries not served by the Foundations; expenses taking place outside countries served by the Foundations; additional degrees, courses, or certificates for individual alumnus/a; projects which replicate existing efforts within the country (i.e. local Soros Foundation projects); educational advising centers, or resource centers, for study abroad; or purchasing of office equipment (computers, photocopiers, fax machines).
Geographic Focus: All States, All Countries
Date(s) Application is Due: May 31
Contact: Zarina Usmanova, Program Coordinator; +44-207-031-0200; fax +44-207-031-0201; zarina.usmanova@opensocietyfoundations.org
Internet: http://www.opensocietyfoundations.org/grants/alumni-grant-program
Sponsor: Open Society Foundations
400 West 57th Street, 3rd Floor
New York, NY 10019

OSF Central Asia and Caucasus Research and Training Initiative-Junior Fellowships 3836
The Central Asia and Caucasus Research and Training Initiative (CARTI) aims to contribute to the internationalization of scholarship and development of indigenous capacities for independent and critical academic work in the social sciences and humanities in the region of post-Soviet countries of South Caucasus and Central Asia, Afghanistan and Mongolia. The program targets scholars in the early stages of their postgraduate training and recent graduates of doctoral programs, and provides support and training for original research by university-based academics in their home academic environment. CARTI helps motivated academics during crucial early stages of their careers to gain critical competence with the foundational and current scholarship, and develop active involvement in international intellectual exchange and debate in their subjects. With the support of CARTI, the Fellows engage state-of-the-art resources, theories, methodologies and approaches to their research and teaching in the subjects and areas of importance to the social transformations underway in the region. Junior Fellowships support individuals in early stages of their formal doctoral studies (such as aspirantura), and focuses on developing ideas and skills for original academic research in the social sciences and humanities. Junior Fellowships provide a medium for testing and challenging the theories, methods, and assumptions in the core of individual research agendas, while guiding and assisting the Fellows in their first independent research efforts. CARTI Fellowships are awarded for a period of 24 months, with ongoing review and evaluation procedures in place. The Fellowships typically offer: funding for research activities, communication, access to necessary equipment, materials and resources; expenses for exchange visits between the Fellows and the international scholars; and a Fellowship stipend. The application and additional guidelines are available at the website.
Requirements: The program is open to individuals who are nationals and residents of one of the following countries: Afghanistan, Armenia, Azerbaijan, Georgia, Kazakhstan, Kyrgyzstan, Mongolia, Tajikistan, Turkmenistan, and Uzbekistan. Candidates for the Junior Fellowships must hold a diploma of higher education (of at least a Specialist or Masters level or the equivalent) and be formally registered at a Doctoral studies program (a Ph.D. or equivalent program, such as aspirantura, leading towards a first Doctoral, such as a Candidate of Science degree) in the humanities and social sciences in the region. The PhD enrollment requirement may be waived for applicants from Afghanistan and Turkmenistan. Interested individuals should inquire with the program administration prior to submitting the application.
Geographic Focus: Afghanistan, Armenia, Azerbaijan, Kazakhstan, Kyrgyzstan, Mongolia, Tajikistan, Turkmenistan, Uzbekistan
Date(s) Application is Due: Nov 1
Contact: Oleksandr Shtokvych, Senior Program Manager; +36 1 882 6153; fax +36 1 882 3112; oshtokvych@osi.hu or carti@osi.hu
Internet: http://www.opensocietyfoundations.org/grants/central-asia-and-caucasus-research-and-training-initiative
Sponsor: Open Society Foundations
400 West 57th Street, 3rd Floor
New York, NY 10019

OSF Central Asia and Caucasus Research and Training Initiative-Senior Fellowships 3837
The Central Asia and Caucasus Research and Training Initiative (CARTI) aims to contribute to the internationalization of scholarship and development of indigenous capacities for independent and critical academic work in the social sciences and humanities in the region of post-Soviet countries of South Caucasus and Central Asia, Afghanistan and Mongolia. The program targets scholars in the early stages of their postgraduate training and recent graduates of doctoral programs and provides support

and training for original research by university-based academics in their home academic environment. CARTI helps motivated academics during crucial early stages of their careers to gain critical competence with the foundational and current scholarship and develop active involvement in international intellectual exchange and debate in their subjects. With the support of CARTI, the Fellows engage state-of-the-art resources, theories, methodologies and approaches to their research and teaching in the subjects and areas of importance to the social transformations underway in the region. Senior Fellowships are open to young scholars who have recently received their first Doctoral (e.g. Candidate of Science) degrees in the social sciences and humanities from the universities in their home countries/region. The Fellowships support innovative and critical research agendas and development of international research and teaching partnerships with special emphasis on publication of scholarly work, active contribution to public debate and integrating the new scholarly thinking and outcomes and experience of research into teaching. CARTI Fellowships are awarded for a period of 24 months, with ongoing review and evaluation procedures in place. The Fellowships typically offer: funding for research activities, communication, access to necessary equipment, materials and resources; expenses for exchange visits between the fellows and the international scholars; support for teaching and curriculum development activities; and a Fellowship stipend. The application and additional guidelines are available at the website.
Requirements: The program is open to individuals who are nationals and residents of one of the following countries: Afghanistan, Armenia, Azerbaijan, Georgia, Kazakhstan, Kyrgyzstan, Mongolia, Tajikistan, Turkmenistan, and Uzbekistan. Candidates for the Senior Fellowships must possess a first Doctoral or equivalent (such as a Candidate of Science) degree in the humanities or social sciences, and hold full-time teaching positions at departments of social sciences or humanities at university level institutions in the program region.
Geographic Focus: Afghanistan, Armenia, Azerbaijan, Kazakhstan, Kyrgyzstan, Mongolia, Tajikistan, Turkmenistan, Uzbekistan
Date(s) Application is Due: Nov 1
Contact: Oleksandr Shtokvych, Senior Program Manager; +36 1 882 6153; fax +36 1 882 3112; oshtokvych@osi.hu or carti@osi.hu
Internet: http://www.opensocietyfoundations.org/grants/central-asia-and-caucasus-research-and-training-initiative
Sponsor: Open Society Foundations
400 West 57th Street, 3rd Floor
New York, NY 10019

OSF Civil Service Awards **3838**
The Civil Service Awards provide fellowships for master's degree study, offering professional training and development to public sector employees engaged in policy analysis and implementation. The program is offered in Georgia and Moldova. The program is designed to meet the professional development needs of mid- to top-level staff of governmental ministries and agencies. The fellowship targets career civil servants who demonstrate both academic and professional excellence and the potential to become leaders, decision-makers, and agents of change in their home countries. Upon completing the degrees, grantees return to employment at a position determined by the home government for a commitment of three years. Applications and additional guidelines for each country are available at the website.
Requirements: For Georgia, the awards are open to employees of the two cooperating federal ministries and their associated agencies: The Ministry of Labor, Health and Social Affairs of Georgia; and The Ministry of Environmental Protection and Natural Resources of Georgia. For Moldova, the awards are open to public sector employees and are intended to benefit the following federal entities and their associated agencies: Ministry of Education; Ministry of Labor, Social Protection, and Family; Ministry of the Economy; Ministry of Transportation; Ministry of Agriculture; Ministry of Foreign Affairs and European Integration; and The State Chancellery.
Geographic Focus: Georgia, Moldova
Contact: Nino Chinchaladze, Program Coordinator; +995 32 252615; cie@osgf.ge
Internet: http://www.opensocietyfoundations.org/grants/civil-service-awards
Sponsor: Open Society Foundations
400 West 57th Street, 3rd Floor
New York, NY 10019

OSF Doctoral Fellowships-Georgia and Kyrgyzstan **3839**
The Doctoral Fellows Program is designed to provide the highest research and teaching qualifications to individuals from Georgia and Kyrgyzstan who are positioned to become leading scholars in select fields of social work, public health, education, and psychology. The awards will support flexible doctoral programs that enable fellows to study and conduct research both abroad and at home, thereby maintaining professional engagement with their home institutions while completing the degree. The application and additional guidelines are available at the website.
Geographic Focus: Kyrgyzstan
Contact: Nino Chinchaladze; +995 32 250463; fax +995 32 252615; nino@osgt.ge
Internet: http://www.opensocietyfoundations.org/grants/doctoral-fellows-program
Sponsor: Open Society Foundations
400 West 57th Street, 3rd Floor
New York, NY 10019

OSF Fellowships **3840**
The Open Society Fellowship supports outstanding individuals from around the world. The Fellowship enables innovative professionals-including journalists, activists, academics, and practitioners-to work on projects that inspire meaningful public debate, shape public policy, and generate intellectual ferment within the Open Society Foundations. Funding is open to considering a broad range of work products, including: publications (e.g. books, articles, blogs); creative public education projects including online media, pre-production research and post-production outreach for documentary films; and efforts to seed new campaigns or organizations. Stipends range from $80,000 to $100,000. The application is available at the website.
Requirements: The Open Society Fellowship accepts proposals from anywhere in the world. Applicants should possess a deep understanding of their chosen subject and a track record of professional accomplishment. Past and current fellows have included journalists, activists, academics, and practitioners in a variety of fields. Successful applicants will be eager to exploit the many resources offered by the Open Society Foundations and be prepared to engage constructively with our global network. Ideal fellows are specialists who can see beyond the basics of their field, and possess the tenacity to complete a project of exceptional merit. Proficiency in spoken English is required. Applicants who are uncertain whether their topic fits within the foundations' focus areas are invited to submit a brief letter of inquiry, accompanied by a CV, before proceeding with the online application process.
Restrictions: The fellowship does not fund enrollment for degree or non-degree study at academic institutions, including dissertation research.
Geographic Focus: All States, All Countries
Date(s) Application is Due: Feb 1
Amount of Grant: 80,000-100,000 USD
Contact: Christine Seisun; (212) 548-0600; christine.seisun@opensocietyfoundations.org
Internet: http://www.opensocietyfoundations.org/grants/open-society-fellowship
Sponsor: Open Society Foundations
400 West 57th Street, 3rd Floor
New York, NY 10019

OSF Global Faculty Grants-Track 1 **3841**
The Global Faculty Grants offer professional development and research support to faculty in the social sciences and humanities in select countries. The program aims to: encourage the pursuit of and continued engagement with academic careers; generate new approaches to curricular and pedagogical practice; support the creation of internationally peer-reviewed scholarship; boost the quality of teaching and research in the social science and humanities; and build and sustain local and international academic networks. Track 1 is offered to academics in the following countries: Nepal; Armenia; Azerbaijan; Georgia; Kazakhstan; Kyrgyzstan; Mongolia; and Tajikistan. Track I is devoted primarily to developing the teaching and research expertise of younger faculty with an eye towards encouraging improved course content via improved research and pedagogy. These Fellowships last up to 10 months, are not renewable, and are tenable at universities arranged by program administrators. The application and additional guidelines are available at the website. Individual country contact information is also available at the website.
Geographic Focus: Armenia, Azerbaijan, Kazakhstan, Kyrgyzstan, Mongolia, Nepal, Tajikistan
Contact: Richa Shivakoti, Track 1 Contact; +977-1-4472807; scholarship@asd.org.np
Internet: http://www.opensocietyfoundations.org/grants/global-faculty-grants-program
Sponsor: Open Society Foundations
400 West 57th Street, 3rd Floor
New York, NY 10019

OSF Global Faculty Grants-Track 2 **3842**
The Global Faculty Grants Program offers professional development and research support to faculty in the social sciences and humanities in select countries. The program aims to: encourage the pursuit of and continued engagement with academic careers; generate new approaches to curricular and pedagogical practice; support the creation of internationally peer-reviewed scholarship; boost the quality of teaching and research in the social science and humanities; and build and sustain local and international academic networks. Track II is offered to academics in Nepal; Afghanistan; Armenia; Azerbaijan; Georgia; Kazakhstan; Kyrgyzstan; Mongolia; Tajikistan; Egypt; Jordan; and Palestine. Track II is devoted to helping well-established faculty to pursue research leading to publication at universities outside of their home country. Fellowships last up to 5 months, and are renewable once, after residence in the home country for a minimum one year. The application and additional guidelines are available at the website, along with individual country contacts.
Geographic Focus: Afghanistan, Armenia, Azerbaijan, Egypt, Jordan, Kazakhstan, Kyrgyzstan, Mongolia, Nepal, Palestinian Authority, Tajikistan
Contact: Mohammad Yusuf Yahyapoor, Youth and Civil Society Officer; +93 (0) 778 816 380; yusufm@osafg.org or york.cambridge.s@gmail.com
Internet: http://www.opensocietyfoundations.org/grants/global-faculty-grants-program
Sponsor: Open Society Foundations
400 West 57th Street, 3rd Floor
New York, NY 10019

OSF Global Supplementary Grants **3843**
The Open Society Foundations offer supplementary grants to students from select countries of Southeastern Europe, the former Soviet Union, Mongolia, the Middle East, and South Asia. The purpose of the program is to enable qualified students to pursue doctoral studies in the humanities and social sciences at accredited universities in Asia, Australia, Europe, the Middle East, and North America. Applicants must complete the appropriate application based on the continent of study. Those applying to universities in North America and Europe will need to submit two applications. Both applications are available at the OSF website.
Requirements: The Global Supplementary Grant Program (GSGP) is available to nationals of the following countries: Afghanistan, Albania, Armenia, Azerbaijan, Belarus, Bosnia, Croatia, Egypt, Georgia, Iraq, Jordan, Kazakhstan, Kosovo, Kyrgyzstan, Lebanon, Macedonia, Moldova, Mongolia, Montenegro, Nepal, Palestine, Russia, Serbia, Syria, Tajikistan, Turkmenistan, Ukraine, and Uzbekistan. Applications according to region are available at the OSF website.
Restrictions: Funding is supplemental only. Applicants must be able to demonstrate additional support from other sources. Students pursuing doctorates in the medical, physical,

chemical, technical or natural sciences as well as fine or performing arts are not eligible for this grant. Ineligible fields of study include: business administration/management training; computer science; finance/banking/marketing; engineering; hard and natural sciences (physics, biology, chemistry, etc.); mathematics; and medical sciences.
Geographic Focus: Afghanistan, Albania, Armenia, Azerbaijan, Belarus, Bosnia & Herzegovina, Croatia, Egypt, Iraq, Jordan, Kazakhstan, Kosovo, Kyrgyzstan, Lebanon, Macedonia, Moldova, Mongolia, Montenegro, Nepal, Palestinian Territory, Russia, Serbia, Syria, Tajikistan, Turkmenistan, Ukraine, Uzbekistan
Contact: Russell Koff; (212) 548-0600; fax (212) 548-4679; rkoff@sorosny.org
Margareta Mamaliga, Program Manager; +44-207-031-0200; fax +44-207-031-0201; margareta.mamaliga@opensocietyfoundations.org
Internet: http://www.opensocietyfoundations.org/grants/global-supplementary-grant
Sponsor: Open Society Foundations
400 West 57th Street, 3rd Floor
New York, NY 10019

OSF HESP Mobility Grants **3844**

The Higher Education Support Program (HESP) Mobility Grants are designed to support student visits to and from HESP Network Institutions. Up to ten grants are awarded every year. Humanities, social sciences, business, and management disciplines are eligible to apply. Eligible expenses include travel-related expenses (travel, visa, insurance); living expenses; accommodations; and book allowance. Applications may be submitted at any time, but must be submitted at least two months prior to the proposed visit. The application and additional guidelines are available at the website.
Requirements: All application materials must be in English or accompanied by an English translation. Applicants must be able to demonstrate that the grant will be used to enhance the educational experience at the home institution for purposes related directly to the studies undertaken there. Potential applicants should note the level of studies (undergraduate, master's, doctoral) offered at the respective institution, and consider applying only to a program that corresponds to the level of study at the home institution. Students enrolled at higher education institutions in Central, Eastern, and South Eastern Europe; the former Soviet Union; and Mongolia who intend to spend a visiting semester in one of those institutions are eligible to apply for a visit of up to one semester in length.
Restrictions: The following activities are not eligible for funding: travel to and from Central European University; travel to an institution in the home country; travel from an EU country (except from HESP Network Institutions); study at summer schools; short course and seminar attendance; and participation in any kind of conference and congress. Individuals may receive a mobility grant only once. Simultaneous application to different institutions is not allowed. Applications for study during the last academic year of undergraduate study will be not considered.
Geographic Focus: Albania, Austria, Belarus, Bosnia & Herzegovina, Bulgaria, Czech Republic, Germany, Greece, Hungary, Kosovo, Liechtenstein, Macedonia, Moldova, Montenegro, Poland, Romania, Russia, Slovakia, Slovenia, Switzerland, Ukraine
Contact: Elena Naumkina, Program Officer; +36-1-882-3100; fax +36-1-882-3112; elena.naumkina@opensocietyfoundations.org or hespmobility@osi.hu
Internet: http://www.opensocietyfoundations.org/grants/hesp-mobility-program
Sponsor: Open Society Foundations
400 West 57th Street, 3rd Floor
New York, NY 10019

OSF Human Rights and Governance Internships **3845**

The Open Society Human Rights Internships are intended to attract recent graduates of masters and LL.M. programs in the social sciences, law, and humanities to the human rights sector, giving them the opportunity to gain first-hand research experience in the field of human rights, government accountability, or penal reform. Successful candidates receive opportunity to work in one of the leading human rights and accountability advocacy organizations in Central and Eastern Europe and the former Soviet Union. Internships will last six months. Additional guidelines, along with the application, are located at the Foundations website.
Requirements: Eligible candidates must fulfill all three of the following *Requirements:* graduate of an Open Society Fellowship from the Scholarship Programs, or a graduate of the Central European University's Human Rights Program, or Comparative Constitutional Law Program; graduated in the past three years from LL.M. or MA studies in the social sciences, law or humanities; and be a citizen of a country from Central and Eastern Europe, the former Soviet Union, or Mongolia.
Geographic Focus: Albania, Armenia, Austria, Belarus, Belgium, Bulgaria, Czech Republic, Estonia, Hungary, Kazakhstan, Kosovo, Latvia, Liechtenstein, Lithuania, Macedonia, Moldova, Mongolia, Montenegro, Poland, Romania, Russia, Serbia, Slovakia, Slovenia, Switzerland, Ukraine
Contact: Elizabeth Eagen, Program Manager; (212) 548-0600; fax (212) 548-4600; elizabeth.eagen.opensocietyfoundation.org
Internet: http://www.opensocietyfoundations.org/grants/human-rights-internship
Sponsor: Open Society Foundations
400 West 57th Street, 3rd Floor
New York, NY 10019

OSF Intellectual Property Reform Initiative Grants **3846**

The aim of the Intellectual Property Reform Initiative is to bring global intellectual property rules back into balance. The Initiative supports efforts that will: advocate for the adoption of the World Intellectual Property Organization (WIP.O.) Treaty for the Visually Impaired; explore the need for international standards for fair use rights in other areas such as distance education, research, and library services; and advocate for strong fair use rights in influential and regional areas. The Information Program will also support efforts that will conduct research and educate policy makers about the threat of widespread "collateral damage" posed by copyright enforcement proposals; and advocate for strong human rights and due process protections in copyright enforcement. Concept letters are accepted at any time. Additional information is available at the OSF website.
Requirements: Applicants should send a one-page concept paper to the Foundations contact with the following information: a brief description of the project goals and planned activities; information about the applicant organization and project partners; and how much the project will cost. Applicants should expect a response within two months.
Geographic Focus: All States, All Countries
Contact: Darius Cuplinskas, Director, Information Program; +44-207-031-0200; fax +44-207-031-0201; darius.cuplinskas@opensocietyfoundations.org
Internet: http://www.opensocietyfoundations.org/grants/intellectual-property-reform-initiative
Sponsor: Open Society Foundations
400 West 57th Street, 3rd Floor
New York, NY 10019

OSF Justice Initiative Fellowships **3847**

The Open Society Justice Initiative, together with Central European University (CEU), calls for applications for its Justice Initiative Fellowship Program. Up to eight Fellowships will be awarded. The Fellowships are a two-year program of study and practical work experience, whose objective is to support and further develop a network of lawyers and activists working internationally on human rights issues. During the first year of the Fellowship, Fellows will undertake a human rights degree program at CEU's Legal Studies Department (either an M.A. or LL.M., depending on their undergraduate degree). This first year of study also includes a three-month internship with a leading European human rights organization. Fellows will receive full funding for their CEU studies. During the Fellowship's second year, they are employed by their nominating NGO to conduct human rights advocacy on a non-profit basis. Their work can range from providing legal services, to undertaking human rights litigation, to providing training and human rights education. The Fellows' salary during this second year of practical work will be paid by the Justice Initiative in the form of a grant to the nominating organization. Compensation will be based on the nominating NGO's practices as well as local pay scales. Additional information and guidelines are available at the website. Applications must be submitted through the online system at the website.
Requirements: Applicants from the following regions are eligible: Central and Eastern Europe, the former Soviet Union, Africa, Asia, the Middle East, and Central/South America. Applicants must be nominated by a national non-governmental organization (NGO) concerned with human rights. Criteria for selection will include the applicant's experience, his/her potential to contribute to the protection and promotion of human rights, and the suitability of the applicant's proposed role in the nominating NGO. The applicant must demonstrate a strong commitment to human rights, with a university degree and a high degree of proficiency in English.
Geographic Focus: All Countries
Date(s) Application is Due: Jan 24
Contact: Andrea Jenei, Department Coordinator; +361 327-3205; jenei@ceu.hu
Internet: http://www.opensocietyfoundations.org/grants/justice-initiative-fellows-program
Sponsor: Open Society Foundations
400 West 57th Street, 3rd Floor
New York, NY 10019

OSF Palestinian Rule of Law Program Fellowships **3848**

The Palestinian Rule of Law (PROL) Program provides for graduate-level studies in law leading to an LLM (Masters of Laws) degree. The PROL program is predicated upon a belief that the institutionalization of the rule of law will be a key component of political, economic, and social development in the Palestinian territories and underlie any lasting, peaceful resolution of the Palestinian-Israeli conflict. It is a goal of the PROL Program that Fellows, upon their return home, will contribute to the economic and democratic growth of the Palestinian territories and serve as leaders in their fields of specialization. The Program places participants at host law schools in the U.S. and the Central European University in Budapest, an OSF-affiliated institution providing American-style teaching and U.S.-accredited qualifications. Before beginning their degree programs, Fellows attend the intensive OSF Summer School in Istanbul which includes legal writing instruction. During the academic year, Fellow conferences are held to increase the group's cohesiveness as a networked cohort. Fellows who successfully complete the one-year LLM degree can use program funding to undertake an internship/work placement at a nonprofit or nongovernmental institution that furthers their practical training in specialized fields. The Fellowship can be in the U.S., Occupied Palestinian Territories, or elsewhere. However, a primary goal of the Program is that Fellows use their newly acquired skills towards institutionalizing the rule of law and the values of an open society upon their return home. The application and additional guidelines are available at the website.
Requirements: Applicants must be lawyers or law graduates and residents in the Gaza Strip or West Bank, including East Jerusalem.
Geographic Focus: Egypt, Israel, Jordan, Palestinian Territory
Contact: Joseph Glicksberg, Senior Program Manager; (212) 548-0600; fax (212) 548-4679; joseph.glicksberg@opensocietyfoundations.org
Internet: http://www.opensocietyfoundations.org/grants/palestinian-rule-law-program
Sponsor: Open Society Foundations
400 West 57th Street, 3rd Floor
New York, NY 10019

OSF Social Work Fellowships **3849**

The Social Work Fellowship Program for Jordan allows chosen fellows to study at Columbia University in the United States for a two-year graduate program in social work.

Upon the conclusion of the Fellowship, Fellows return home to apply their new knowledge in practice. Approximately five awards are granted. The program is designed to provide training in social work to implement reform, create policy, and foster the development of social work. The application and additional guidelines are available at the website.
Requirements: The competition is merit-based and open to those meeting the following criteria: resident of Jordan; hold an undergraduate degree (4 or 5 year) at application deadline with an excellent academic record; spoken and written proficiency in English at the time of application; able to demonstrate professional aptitude and leadership potential in social work; and committed to returning to home country after completion of the program.
Geographic Focus: Jordan
Date(s) Application is Due: Nov 26
Contact: Haya Kasht; 962-6-5929994, ext. 135; fax 962-6-5929996; hkasht@amideast.org
Internet: http://www.opensocietyfoundations.org/grants/social-work-fellowship-program
Sponsor: Open Society Foundations
400 West 57th Street, 3rd Floor
New York, NY 10019

OSF Soros Justice Advocacy Fellowships 3850

The Soros Justice Fellowships fund outstanding individuals to implement innovative projects that advance reform and spur debate on a range of issues facing the U.S. criminal justice system. The Fellowships Program is part of a larger effort within the Open Society Foundations' Criminal Justice Fund to reduce the destructive impact of current criminal justice policies on the lives of individuals, families, and communities in the United State by challenging the overreliance on incarceration and harsh punishment, and ensuring a fair and equitable system of justice. Advocacy Fellowships fund outstanding individuals—including lawyers, advocates, grassroots organizers, activist academics, and others with important perspectives—to initiate innovative policy advocacy projects at the local, state, and national levels that will have a measurable impact on one or more of the Open Society Foundations' U.S. criminal justice priorities. Projects may range from litigation to public education to coalition-building to grassroots mobilization to policy-driven research. Advocacy Fellowships are 18 months in duration, may be implemented in conjunction with a host organization, and Fellows are expected to make their projects their full-time work during the term of the Fellowship. The Institute strongly encourages applications for projects that demonstrate a clear understanding of the intersection of criminal justice issues with the particular needs of low-income communities, communities of color, immigrants, gay, lesbian, bisexual and transgender people, women and children, and those disproportionately affected by harsh criminal justice policies; as well as applications for projects that cut across various criminal justice fields and related sectors, such as education, health and mental health, housing and employment. Those with personal experience with these issues are also encouraged to apply. Additional guidelines, along with the application, are available at the website.
Requirements: Fellowship projects should address one or more of the following Criminal Justice Fund priorities: expose the excessive and economically destructive costs of incarceration; promote just and effective sentencing practices; foster policies and practices that reduce prison and corrections populations; abolish the privatization of prisons, detention centers, and correctional supervision; combat the criminalization of marginalized populations; eliminate unreasonable barriers to the reintegration of people returning from prison, as well as challenge the stigmatization of people with criminal records; end the prosecution, sentencing and incarceration of children as adults; abolish the death penalty; promote the civic engagement and leadership of people with criminal records; eliminate the use of long-term solitary confinement in prison; challenge harsh immigration enforcement and detention policies and practices; promote new approaches to drug policy; reform discriminatory and abusive police and prosecution practices; improve indigent defense services and systems; and reduce unnecessary pretrial detention.
Restrictions: The Fellowships Program does not fund: enrollment for degree or non-degree study at academic institutions, including dissertation research; projects that address criminal justice issues outside the U.S. (applicants themselves, however, can be based outside the U.S., provided their work directly pertains to a U.S. issue); past recipients of a Soros Justice Fellowship; or lobbying activities.
Geographic Focus: All States
Date(s) Application is Due: Jul 30
Amount of Grant: 58,700-110,250 USD
Contact: Zsolt Bobis; (212) 548-0600; zsolt.bobis@opensocietyfoundation
Internet: http://www.opensocietyfoundations.org/grants/soros-justice-fellowships
Sponsor: Open Society Foundations
400 West 57th Street, 3rd Floor
New York, NY 10019

OSF Soros Justice Media Fellowships 3851

The Soros Justice Fellowships fund outstanding individuals to implement innovative projects that advance reform and spur debate on a range of issues facing the U.S. criminal justice system. The Fellowships are part of a larger effort within the Open Society Foundations' Criminal Justice Fund to reduce the destructive impact of current criminal justice policies on the lives of individuals, families, and communities in the U.S. by challenging the overreliance on incarceration and harsh punishment, and ensuring a fair and equitable system of justice. The Soros Media Fellowships support writers, print and broadcast journalists, bloggers, filmmakers, and other individuals with distinct voices proposing to complete media projects for local, regional and national markets that engage the public and spur debate on one or more of the Open Society Foundations' U.S. criminal justice priorities. The Fellowships aim to mitigate the time, space, and market constraints that often discourage individuals from pursuing important but marginalized, controversial or unpopular issues in a comprehensive manner. Media Fellowships are 12 months in duration, and Fellows are expected to make their projects their full-time work during the term of the Fellowship. Open Society strongly encourages applications for projects that demonstrate a clear understanding of the intersection of criminal justice issues with the particular needs of low-income communities, communities of color, immigrants, gay, lesbian, bisexual and transgender people, women and children, and those disproportionately affected by harsh criminal justice policies; as well as applications for projects that cut across various criminal justice fields and related sectors, such as education, health and mental health, housing and employment. The Institute also encourages applications from individuals directly affected by, or with significant direct personal experience with, the policies, practices and systems their projects seek to address. Additional guidelines and the application are available at the website.
Requirements: Applicants who are uncertain whether their proposed project fits with the Fellowship guidelines may submit an email inquiry before submitting the full application. All Fellowship projects should address one or more of the following Criminal Justice Fund priorities: expose the excessive and economically destructive costs of incarceration; promote just and effective sentencing practices; foster policies and practices that reduce prison and corrections populations; abolish the privatization of prisons, detention centers, and correctional supervision; combat the criminalization of marginalized populations; eliminate unreasonable barriers to the reintegration of people returning from prison, as well as challenge the stigmatization of people with criminal records; end the prosecution, sentencing and incarceration of children as adults; abolish the death penalty; promote the civic engagement and leadership of people with criminal records; eliminate the use of long-term solitary confinement in prison; challenge harsh immigration enforcement and detention policies and practices; promote new approaches to drug policy; reform discriminatory and abusive police and prosecution practices; improve indigent defense services and systems; and reduce unnecessary pretrial detention.
Restrictions: Funding is not available for enrollment for degree or non-degree study at academic institutions, including dissertation research. Projects that address criminal justice issues outside the U.S. are not eligible, but applicants can be based outside the U.S., if their work pertains to a U.S. issue. Funding is also not available for past recipients of the Social Justice Fellowship, or lobbying activities.
Geographic Focus: All States
Date(s) Application is Due: Jul 30
Amount of Grant: 58,000-110,250 USD
Contact: Zsolt Bobis; (212) 548-0600; zsolt.bobis@opensocietyfoundation
Internet: http://www.opensocietyfoundations.org/grants/soros-justice-fellowships
Sponsor: Open Society Foundations
400 West 57th Street, 3rd Floor
New York, NY 10019

OSF Think Tank Fellowships 3852

The Think Tank Fund of the Open Society Foundations invites talented and dedicated young researchers to apply for a fellowship that would allow them to spend up to three months in one of the leading think tanks in Central and Eastern Europe. In accordance with the Think Tank Fund's overall mission, this fellowship program strives to: to equip junior researchers with tools necessary to effectively influence policy-making processes and develop innovative and quality policy products; to contribute to institutional development of think tanks by broadening their professional networks, improving the junior employee skills and increasing their motivation for better participation and generation of new ideas; and to enrich portfolios and networks of think tanks by bringing in researchers with different background. Fellowships of up to $7,000 are given and cover travel expenses, along with reasonable accommodations and living costs. Regular monthly salary is maintained by a sponsoring organization during the full period of the fellowship. Detailed guidelines are available at the website. Applications are accepted at any time.
Requirements: Applicants should be permanent residents of one of the following countries: Armenia, Azerbaijan, Bulgaria, the Czech Republic, Estonia, Hungary, Georgia, Latvia, Lithuania, Moldova, Poland, Romania, Slovakia, Slovenia, and Ukraine. Full time employees at a think tank, policy center, or NGO with a strong research component in one of the Think Tank Fund target countries for at least one year at the time of the application and plan to stay for at least one more year may also apply. Applicants must be fluent in written and spoken English and between the ages of 24 and 35 years of age at the time of the application.
Geographic Focus: Armenia, Azerbaijan, Bulgaria, Czech Republic, Estonia, Hungary, Latvia, Lithuania, Moldova, Poland, Romania, Slovakia, Slovenia, Ukraine
Contact: Goran Buldioski, Program Director; +36 1 882 3121 or +36 1 882 3100; fax +36 1 882 3101; ttffellowships@osi.hu or goran.buldioski@opensocietyfoundations.og
Internet: http://www.opensocietyfoundations.org/grants/think-tank-fund-fellowships
Sponsor: Open Society Foundations
400 West 57th Street, 3rd Floor
New York, NY 10019

OSF Thomas J. Alexander Fellowships 3853

Fellows will have the opportunity to join one of the world's leading teams in this field at the OECD Headquarters to conduct research and analysis of the extensive OECD educational dataset (from PISA, TALIS, PIAAC, etc.), which is as renowned as it is unique in scale, scope and value. The unique emphasis of the Fellowships on quality and equity in education, particularly in emerging economies, makes them an opportunity to strengthen capacity for quantitative research and analysis on ideas that are crucial for open society. The research will enable an informed response to the growing need for evidence in educational policy and practice. Fellows will work at OECD headquarters in Paris with the team of analysts. They will be expected to spend at least three months and up to a maximum of 12 months in Paris. Fellows are expected to produce a publishable research paper or report at the end of their fellowship. The application and additional guidelines are available at the OSF website.
Requirements: Alexander Fellows will typically come from research institutions, governmental agencies, research centers, ministries, or professional organizations. Ideally they will a hold a Ph.D., have experience with quantitative research methods and

large-scale surveys, and have had articles published in scientific journals. The program encourages applicants from a variety of disciplines. Proposals focusing on improving equity in education are particularly encouraged. Applicants should explain how their work would benefit the broad public and promote quality and equity in their countries' education systems. A call for proposals will be issued twice a year. Only online applications will be accepted. Applications will be considered by a group of experts in the field of international education assessment. Candidates will be selected on the basis of the originality, potential, and technical quality of the proposals as well as on the qualifications and experience of the applicant and the value added by their time working at OECD headquarters in Paris.
Geographic Focus: All States, All Countries
Date(s) Application is Due: Dec 21
Contact: Diana Tramontano, Administrative Assistant; +33 1 45 24 79 04 or +33 1 45 24 82 00; fax +33 1 45 24 85 00; diana.tramontano@oecd.org
Internet: http://www.oecd.org/edu/thomasjalexanderfellowship.htm
Sponsor: Open Society Foundations
400 West 57th Street, 3rd Floor
New York, NY 10019

OSRI Biological Monitoring: Intertidal Resources at Risk to Oil Spills 3854
OSRI is seeking proposals to conduct surveys of unique, dominant, and/or previously undescribed, intertidal resources at risk to oil spills in Prince William Sound (PWS) and the Copper River Delta (CRD). OSRI anticipates surveys staging out of Cordova, Alaska. Professional services sought through this RFP include the survey prelogistics (design, equipment preparation, etc.), data acquisition, data analysis (including preparing a GIS of the data), and reporting of results. All sampling shall be conducted synoptic with GPS measurements, ensuring the acquisition of data that is both temporally and geographically encoded. The contractor is responsible for providing an electronic copy of the second survey in ARCINFO file format with the final report.
Requirements: Qualified scientists, engineers, and educators are eligible to apply.
Restrictions: Proposals will not be accepted electronically or by fax.
Geographic Focus: All States
Contact: Penny Oswalt; (907) 424-5800, ext. 224; fax (907) 424-5820; poswalt@pwssc.org
Internet: http://www.pws-osri.org
Sponsor: Oil Spill Recovery Institute
P.O. Box 705
Cordova, AK 99574

OTS Francis J. Bossuyt Fellowship 3855
Organization of Tropical Studies (OTS) fellowships in general assist with thesis-related research in tropical biology and similar fields and are available in amounts up to a maximum of $3,650. Awards are intended primarily to cover fees (food, lodging, and station access) at OTS field stations. Other expenses that may be included in the budget, to a maximum of $500, are research- and export-permit fees, local travel, and reimbursement in whole or in part of one round-trip airline ticket. OTS's Bossuyt fellowship was created as a memorial to the life of Francis Bossuyt (1969-2000). At the time of his untimely death, Francis had been collecting his final year of data for his dissertation on Titi monkeys at the Cocha Cashu Biological Station in Manu National Park in Peru. In recognizing Francis' love for the Titi Monkeys and for field-oriented research, the Bossuyt family established the endowed fellowship for the following priority areas: research on animal behavior (preference for Callicebus) in Cocha Cashu and the surrounding areas; research on other subjects in Cocha Cashu; research by Peruvian-students on tropical ecosystems in Peru; and research at other locations in the tropics. Application guidelines and a downloadable cover sheet are available from the OTS Fellowship webpage. The cover sheet and supporting documents must be submitted via email by the deadline date to the Education Director. Notification of decisions will take place about two months after each deadline. The Organization for Tropical Studies (OTS) is a nonprofit consortium of 64 universities and research institutions in the United States, Costa Rica, Peru, Canada, Mexico, South Africa, and Australia. Founded in 1963, OTS has a mission to provide leadership in education, research and the responsible use of natural resources in the tropics. To this end, graduate and undergraduate field courses are offered in the pure and applied areas of the natural sciences in Costa Rica, Peru, Mexico, the Brazilian Amazon, and South Africa. In addition, OTS operates three research stations in varied ecosystems in Costa Rica: La Selva in the Caribbean lowland rain forest, Palo Verde in the Pacific lowland dry forest, and Las Cruces in a mid-elevation rain forest on Costa Rica's southern Pacific slope. OTS also coordinates and facilitates research, provides environmental-education programs, and works with local agencies on issues of conservation, natural-resource management and land-use practices.
Requirements: In general, Organization for Tropical Studies (OTS) fellowships are open to graduate students enrolled in degree programs at OTS member institutions and to OTS course alumni. (A list of member institutions is available at the OTS website from the About-Us Quick Link.)
Restrictions: Only one OTS research award may be held at a time. Awards do not cover capital equipment; minor amounts of materials and supplies may be considered, but are best covered by other sources of funds. Salaries or stipends for the investigator are not allowed, but salaries for field assistants will be considered if the work cannot be carried out safely or efficiently by one person.
Geographic Focus: All States, All Countries, Peru
Date(s) Application is Due: Mar 1; Oct 1
Amount of Grant: Up to 3,650 USD
Contact: Education Director; (506) 2524-0607; fax (506) 2524-0608; academic@ots.ac.cr
Internet: http://ots.ac.cr/index.php?option=com_content&task=view&id=170&Itemid=344#proposal
Sponsor: Organization for Tropical Studies
Duke University, 410 Swift Avenue
Durham, NC 27705

OTS Hovore-Horn Fellowships 3856
Organization of Tropical Studies (OTS) fellowships in general assist with thesis-related research in tropical biology and similar fields. Funded through an endowment established in 2005 by biologist Frank T. Hovore in honor of his parents, Alice and Andre Horn, OTS's Hovore fellowship also honors Frank himself, who succumbed to a heart attack September 2006 while doing what he loved, collecting beetles in Ecuador. Interested candidates should propose projects with a strong field component and should emphasize the exploration, discovery and description of biological diversity (e.g., field surveys, faunal inventories, or ecological studies) using insects and other terrestrial arthropods as study organisms. Researchers may work in their own or another Latin-American country. Awards are available in the range of $500 to $1500 and are intended primarily to cover fees (food, lodging, and station access) at Organization for Tropical Studies (OTS) field stations; Other expenses that may be included in the budget, to a maximum of $500, are research- and export-permit fees, local travel, and reimbursement in whole or in part of one round-trip airline ticket. Application guidelines and a downloadable cover sheet are available from the OTS Fellowship webpage. Interested applicants should submit their proposals in Spanish via email along with the other required supporting materials by the deadline date to the contact information given. Notification of decisions will take place about two months after each deadline. OTS is a nonprofit consortium of 64 universities and research institutions in the United States, Costa Rica, Peru, Canada, Mexico, South Africa, and Australia. Founded in 1963, OTS has a mission to provide leadership in education, research and the responsible use of natural resources in the tropics. To this end, graduate and undergraduate field courses are offered in the pure and applied areas of the natural sciences in Costa Rica, Peru, Mexico, the Brazilian Amazon, and South Africa. In addition, OTS operates three research stations in varied ecosystems in Costa Rica: La Selva in the Caribbean lowland rain forest, Palo Verde in the Pacific lowland dry forest, and Las Cruces in a mid-elevation rain forest on Costa Rica's southern Pacific slope. OTS also coordinates and facilitates research, provides environmental-education programs, and works with local agencies on issues of conservation, natural-resource management and land-use practices.
Requirements: The Hovore-Horn research fellowships in entomology are available to all Latin-American graduate and advanced-undergraduate students interested in Neotropical entomology. In general applicants should be enrolled in degree programs at OTS member institutions. (A list of member institutions is available at the OTS website from the About-Us Quick Link.)
Restrictions: Only one research award may be held at a time. Awards do not cover capital equipment; minor amounts of materials and supplies may be considered, but are best covered by other sources of funds. Salaries or stipends for the investigator are not allowed, but salaries for field assistants will be considered if the work cannot be carried out safely or efficiently by one person.
Geographic Focus: Argentina, Belize, Bolivia, Brazil, Caribbean, Chile, Colombia, Costa Rica, Cuba, Dominican Republic, Ecuador, El Salvador, Guatemala, Haiti, Honduras, Mexico, Nicaragua, Panama, Paraguay, Peru, Uruguay, Venezuela
Date(s) Application is Due: Mar 1; Oct 1
Amount of Grant: 500-1,500 USD
Contact: Education Director; (506) 2524-0607; fax (506) 2524-0608; academic@ots.ac.cr
Internet: http://ots.ac.cr/index.php?option=com_content&task=view&id=170&Itemid=344#proposal
Sponsor: Organization for Tropical Studies
Duke University, 410 Swift Avenue
Durham, NC 27705

OTS Stanley Smith Horticultural Fellowships 3857
Organization for Tropical Studies (OTS) fellowships in general assist with thesis-related research in tropical biology and similar fields and are available in amounts up to a maximum of $3,650. Awards are intended primarily to cover up to fifty days of station fees (food, lodging, and station access) at OTS field stations; current rates are $63 for overnight visits and $31.50 for day visits. Other expenses that may be included in the budget, to a maximum of $500, are research and export permit fees, local travel, and reimbursement in whole or in part of one round-trip airline ticket. OTS's Stanley Smith fellowship is funded by the Stanley Smith Horticultural Trust in memory of Mr. Smith who was a British businessman, horticulturist, and frequent visitor to Las Cruces' Wilson Botanical Garden. The fellowship provides funding for horticultural work and systematics research on tropical plants of interest to the Wilson Botanical Garden. OTS has a single application process for most of its fellowships; generally applicants do not need to specify the award for which they are applying (with a few exceptions which are mentioned at the website). Application guidelines and a downloadable cover sheet are available from the OTS Fellowships webpage. The cover sheet and supporting documents must be submitted via email by the deadline date to the Education Director. Notification of decisions will take place about two months after each deadline. The Organization for Tropical Studies (OTS) is a nonprofit consortium of 64 universities and research institutions in the United States, Costa Rica, Peru, Canada, Mexico, South Africa, and Australia. Founded in 1963, OTS has a mission to provide leadership in education, research and the responsible use of natural resources in the tropics. To this end, graduate and undergraduate field courses are offered in the pure and applied areas of the natural sciences in Costa Rica, Peru, Mexico, the Brazilian Amazon, and South Africa. In addition, OTS operates three research stations in varied ecosystems in Costa Rica: La Selva in the Caribbean lowland rain forest, Palo Verde in the Pacific lowland dry forest, and Las Cruces in a mid-elevation rain forest on Costa Rica's southern Pacific slope. OTS also coordinates and facilitates research, provides environmental-education programs, and works with local agencies on issues of conservation, natural-resource management and land-use practices.

Requirements: In general OTS fellowships are open to graduate students enrolled in degree programs at OTS member institutions and to OTS course alumni. (A list of member institutions is available at the OTS website from the About-Us Quick Link.) Curatorial expertise in tropical plant groups is essential to obtain approval for a Stanley Smith Horticultural fellowship.
Restrictions: Only one research award may be held at a time. Awards do not cover capital equipment; minor amounts of materials and supplies may be considered, but are best covered by other sources of funds. Salaries or stipends for the investigator are not allowed, but salaries for field assistants will be considered if the work cannot be carried out safely or efficiently by one person.
Geographic Focus: All States, All Countries
Date(s) Application is Due: Mar 1; Oct 1
Contact: Zak Zahawi; (506) 2773-4004 ext. 3100; zahawi@ots.ac.cr
Internet: http://ots.ac.cr/index.php?option=com_content&task=view&id=170&Itemid=344#proposal
Sponsor: Organization for Tropical Studies
Duke University, 410 Swift Avenue
Durham, NC 27705

OTS Undergraduate NAPIRE Summer Internships **3858**
With funding from the National Science Foundation (NSF), the Organization for Tropical Studies (OTS) introduces Native-American and Pacific-Islander undergraduate students to ecological research in the tropics. As part of the eight-week program, students will apply scientific methods to ecological inquiry and discovery in a collaborative and team-oriented environment. Students will complete an individual field-research project including experimental design, data collection, analysis and presentation of results. In addition to their original research projects, students will participate in lectures, seminars, and other activities that focus on biodiversity, tropical ecology, conservation, and human ecology as well as the long-term research at the OTS biological stations. Participants will have the opportunity to interact with several Central-American indigenous groups and to understand their efforts towards sustainable livelihoods and tropical-forest conservation. Each student will be assigned a research mentor who will guide their independent project and provide support during the program. The program will be held in Costa Rica in June and July. Students will be in residence mostly at OTS's Las Cruces Biological Station during the eight-week research experience, with some travel to other sites. The award covers the cost of room, board, and travel to and from Costa Rica. Students also receive a $4000 stipend and funds to help cover costs of field equipment. Application for the internship requires submission of a completed application form (via email), two letters of recommendation, and official transcripts from all universities attended (via standard mail to the address specified at the website). Detailed guidelines and forms are available from the OTS NAPIRE-Application-Instructions webpage. Deadlines may vary from year to year. The Organization for Tropical Studies (OTS) is a nonprofit consortium of 64 universities and research institutions in the United States, Costa Rica, Peru, Canada, Mexico, South Africa, and Australia. Founded in 1963, OTS has a mission to provide leadership in education, research and the responsible use of natural resources in the tropics. To this end, graduate and undergraduate field courses are offered in the pure and applied areas of the natural sciences in Costa Rica, Peru, Mexico, the Brazilian Amazon, and South Africa. In addition, OTS operates three research stations in varied ecosystems in Costa Rica: La Selva in the Caribbean lowland rain forest, Palo Verde in the Pacific lowland dry forest, and Las Cruces in a mid-elevation rain forest on Costa Rica's southern Pacific slope. OTS also coordinates and facilitates research, provides environmental-education programs, and works with local agencies on issues of conservation, natural-resource management and land-use practices.
Requirements: U.S. citizens and permanent residents who are undergraduate students enrolled in accredited institutions in the United States (and Pacific Island Territories) may apply to the program.
Restrictions: Incoming freshmen and graduating seniors are not eligible. Students must be AMP students at LSAMP institutions. SLAMP is the Louis Stokes Alliance for Minority Participation, an NSF program that was designed to foster achievement by minority students seeking careers in science, technology, engineering and mathematics (STEM).
Geographic Focus: All States, Guam, Marshall Islands, Northern Mariana Islands, American Samoa, Micronesia, Palau
Date(s) Application is Due: Feb 28
Contact: Kattia Mendez, Undergraduate Program Assistant; (506) 2524-0607; fax (506) 2524-0608; kattia.mendez @ ots.ac.cr or napire.program@ots.ac.cr
Internet: http://ots.ac.cr/index.php?option=com_content&task=view&id=320&Itemid=450
Sponsor: Organization for Tropical Studies
Duke University, 410 Swift Avenue
Durham, NC 27705

Outstanding Junior Investigator Program in High Energy Physics **3859**
The purpose of this program is to support the development of individual research programs of outstanding scientists early in their careers. Awards made under this program will help to maintain the vitality of university research and assure continued excellence in the teaching of physics.
Requirements: Applications should be from tenure-track faculty investigators who are currently involved in experimental or theoretical high energy physics or accelerator physics research, and should be submitted through a U.S. academic institution.
Geographic Focus: All States
Date(s) Application is Due: Nov 15
Amount of Grant: 70,000 USD
Contact: Dr. Jeffrey Mandula; (301) 903-4829; jeffrey.mandula@science.doe.gov
Internet: http://www.science.doe.gov/grants/FAPN05-26.html
Sponsor: U.S. Department of Energy
19901 Germantown Road, MS FE-3
Germantown, MD 20874-1290

Overbrook Foundation Grants **3860**
The foundation strives to improve the lives of people by supporting projects that protect human and civil rights, advance the self-sufficiency and well being of individuals and their communities, and conserve the natural environment. The foundation's environment program supports organizations working to develop better consumption and production habits in the United States and in Latin America (currently Brazil, Mexico and Ecuador only). In Latin America the primary objective is to conserve the planet's dwindling biodiversity. The foundation's human and civil rights program is multifaceted and provides support to the following: reproductive health rights; domestic criminal system reform efforts; lesbian, gay, bisexual, and transgender rights; gun violence prevention and international human rights. An emphasis is placed on work that protects the rights of children and women. The foundation supports projects both domestically and internationally (with a particular focus in Latin America and South Africa). Generally speaking, the foundation's domestic programs support policy development, advocacy, coalition building, research, legal and other strategies likely to impact large classes of people or shape issues important to the foundation's mission. In its international funding, the foundation also supports organizations providing direct services. The first step in the application process is to Submit a letter of inquiry; full application is by invitation. Guidelines are available online.
Requirements: 501(c)3 tax-exempt organizations are eligible.
Restrictions: The foundation does not make grants to individuals and rarely for endowments, building campaigns, deficit financing, or religious purposes. Grant funds may not be used to participate in or intervene in any political campaign on behalf of or in opposition to any candidate for public office or to conduct, directly or indirectly, any voter registration drive (within 4945(d)2 of the IRS code). Grants are not made for debt reduction. Funding for conferences, publications, and media is limited to projects directly related to priorities in the foundation's program areas.
Geographic Focus: All States
Contact: Administrator; (212) 661-8710; apply@overbrookfoundation.org
Internet: http://www.overbrook.org
Sponsor: Overbrook Foundation
122 E 42nd Street, Suite 2500
New York, NY 10168

Pacific Islanders in Communications Grants **3861**
The Media Fund is the prime vehicle to spark the development of Pacific Islander programming for national public television. The Media Fund is comprised of the following programs: Open Call, an annual call that invites proposals for television projects in all stages of production; the Completion Fund, that is designed to quickly respond to urgent post-production needs; the Short Film Initiative, that encourages and builds narrative storytelling; Travel Awards to support producers' professional development; Acquisitions; and Executive Productions. All stages of production, from research and development through completion, are supported. Projects selected under the Media Fund must present the Pacific Islander experience in an authentic and educational manner, as well as meet PBS' high standards of journalistic balance, storytelling, and technical excellence. A training component is an essential part of each project to help develop future Pacific Islander filmmakers.
Requirements: Applicants must be U.S. citizens, have primary artistic and editorial control, and have a nonprofit fiscal sponsor to receive and administer PIC funds.
Restrictions: Industrial films and videos, instructional videos, student productions, and similar non-broadcast categories are not eligible. An applicant may not apply for both a research and development grant and a production grant for the same project in the same grant cycle.
Geographic Focus: All States
Amount of Grant: 15,000-50,000 USD
Contact: Grants Administrator; (808) 591-0059; fax (808) 591-1114; info@piccom.org
Internet: http://www.piccom.org/services.php
Sponsor: Pacific Islanders in Communications
1221 Kapiolani Boulevard, Suite 6A-4
Honolulu, HI 96814

Packard Foundation Fellowships for Science and Engineering **3862**
The Foundation established Fellowships for Science and Engineering to allow the nation's most promising professors to pursue science and engineering research early in their careers with few funding restrictions and limited reporting requirements. Each year the Foundation invites the presidents of 50 universities to nominate two early-career professors each from their institutions. An advisory panel of distinguished scientists and engineers carefully reviews the nominations and selects 16 Fellows to receive individual grants of $875,000, distributed over five years. Requests for nominations will be sent to university presidents by January 31. Nominations are due to the Foundation by March 15. Application materials must be submitted by April 20, with awards announced by October 15.
Requirements: Fellows must be faculty members who are eligible to serve as principal investigators engaged in research in the natural and physical sciences or engineering. They must also be within the first three years of their faculty careers. Disciplines that are considered include physics, chemistry, mathematics, biology, astronomy, computer science, earth science, ocean science, and all branches of engineering. Interested applicants should review the Fellowship Guidelines posted on the Foundation's website. If their university is on the "Invited Institutions" list and if they meet the eligibility requirements, the university's Office of Sponsored Research or department chair may be contacted to discuss the selection process.
Restrictions: Candidates engaged in research in the social sciences are not considered. Nominees will not be able to access the online application system until they have been identified to the Foundation by the university president.

Geographic Focus: All States
Amount of Grant: 175,000 USD
Contact: Xiao-Wei Wang, Program Specialist; (650) 948-7658 or (650) 917-7275; fax (650) 948-2957; fellows@packard.org or xwang@packard.org
Internet: http://www.packard.org/what-we-fund/conservation-and-science/packard-fellowships-for-science-and-engineering/
Sponsor: David and Lucile Packard Foundation
343 Second Street
Los Altos, CA 94022

Padilla Bay Foundation Grants 3863
The Padilla Bay Foundation is a membership-based non-profit organization formed in 1987. Its mission is to help preserve the Padilla Bay estuary through support of the Padilla Bay National Estuarine Research Reserve. Foundation programs include: support for education; support for research; and special projects involving the environment. Interested applicants should contact the Foundation directly.
Geographic Focus: Washington
Contact: Kay Reinhardt; (360) 757-3234; foundation@padillabay.gov
Internet: http://padillabay.gov/npfriends.asp
Sponsor: Padilla Bay Foundation
P.O. Box 1305
Mount Vernon, WA 98273

Padilla Bay National Estuarine Research Reserve Graduate Research Fellowships 3864
Grants are available on a competitive basis to students admitted to or enrolled in a full-time masters or doctoral program at U.S. accredited colleges and universities. Fellowships may be funded for up to three years to support thesis research. The Graduate Research Fellowship provides students the opportunity for hands-on training in ecological monitoring. The award amount is $20,000 per year and may be used for research, tuition, and other expenses. The application deadline is November 1.
Geographic Focus: All States
Amount of Grant: 20,000 USD
Contact: Susan White, Research Coordinator; (360) 428-1558; susan.white@noaa.gov
Internet: http://padillabay.gov/researchfellowships.asp#RA
Sponsor: National Oceanic and Atmospheric Administration
10441 Bayview-Edison Road
Mt. Vernon, WA 98273-9668

Padilla Bay National Estuarine Research Reserve Research Assistantships 3865
Research assistantships will be awarded to graduate students seeking to fund thesis work in Padilla Bay or work comparing Padilla Bay with other Washington bays or estuaries or coastal zone management issues anywhere in Washington. These assistantships require the sponsorship of a college/university faculty member who will supervise the student's thesis in coordination with the Reserve's research staff. A contract with the student totaling $5,000 will be provided over the period July 1 to December 31 of the following year (18 months). The Padilla Bay NERR is particularly interested in proposals addressing: priorities of the Puget Sound Partnership; either historical connections of Padilla Bay to the Skagit and Samish flats, or restoration of estuarine habitats and processes adjacent to Padilla Bay; the potential impacts on eelgrasses and on other plants and animals in Padilla Bay that may result from diversion of Skagit River floodwater to Swinomish Channel; the interaction and ecology of native eelgrass, Zostera marina, and Japanese eelgrass, Z. japonica; or research and resource management issues on Hat Island in Padilla Bay. First consideration will be given to applications received by May 16 each year.
Requirements: Applicants should usually be accepted into a Master's or Doctoral program in marine science, biology, botany, zoology, or one of the related natural or physical sciences.
Geographic Focus: All States
Contact: Dr. Douglas Bulthuis; (360) 428-1558; bulthuis@padillabay.gov
Internet: http://padillabay.gov/researchfellowships.asp#RA
Sponsor: National Oceanic and Atmospheric Administration
10441 Bayview-Edison Road
Mt. Vernon, WA 98273-9668

Pancreatic Cancer Action Network-AACR Fellowship 3866
The Pancreatic Cancer Action Network-AACR Fellowship represents a joint effort to attract young scientist to careers in pancreatic cancer research. The Fellowship is open to Postdoctoral Fellows and Clinical Research Fellows working at an academic, medical, or research institution who will be in the 1st, 2nd, or 3rd year of their postdoctoral training at the start of the grant term on July 1. The Fellowship provides a one-year grant of $45,000 to support the salary and benefits of the Fellow while working on a mentored pancreatic cancer research project. A partial amount of funds, up to 25% of the total grant, may be designated for direct research support. The research proposed for funding may be basic, translational, clinical or epidemiological in nature and must have direct application and relevance to pancreatic cancer. It is anticipated that one Fellowship will be funded. AACR requires applicants submit both an online and a paper application. The online application must be submitted by 12:00 noon (United Sates Eastern Time) on October 28, using the proposalCENTRAL website at http://proposalcentral.altum.com. The paper copy with original signatures and all required documents must be postmarked and sent no later then October 30.
Requirements: Applicants must have acquired a doctoral degree (including Ph.D., M.D., D.O., D.C., N.D., D.D.S., D.V.M., Sc.D., D.N.S., Pharm.D., or equivalent doctoral degree, or a combined clinical and research doctoral degree) in a related field and may not currently be candidate for another degree. Applicants must be in the 1st, 2nd, or 3rd year of Postdoctoral or Clinical Research Fellowship at an academic, medical, or research institution within the United States at the start of the Fellowship term on July 1.
Restrictions: Scientific investigators or health professionals who are funded by the tobacco industry for any project, or whose named mentors in the case of mentored grants are funded by the tobacco industry for any project, may not apply and will not be eligible for AACR grants.
Geographic Focus: All States
Date(s) Application is Due: Oct 28; Oct 30
Amount of Grant: 45,000 USD
Contact: Hanna Hopfinger; (267) 646-0665; fax (215) 440-9372; hanna.hopfinger@aacr.org
Internet: http://www.aacr.org/home/scientists/research-funding—fellowships/postdoctoral-fellowships.aspx
Sponsor: American Association for Cancer Research
615 Chestnut Street, 17th Floor
Philadelphia, PA 19106-4404

Pancreatic Cancer Action Network-AACR Pathway to Leadership Grants 3867
The Pancreatic Cancer Action Network-AACR Pathway to Leadership Grant represents a joint effort to ensure the future leadership of pancreatic cancer research by supporting outstanding early career investigators beginning in their postdoctoral research positions and continuing through their successful transition to independence. Applicants must be in the first five years of their postdoctoral or clinical research fellowships at the time of application and not already have a full-time tenure track assistant professor position (or equivalent). The Pathway to Leadership Grant will provide up to five years of support, for a total of $600,000, consisting of two phases. The initial Mentored Phase lasts up up two years, during which time recipients will receive $75,000 per year and are expected to work closely with mentors to develop their research. During the subsequent three years, recipients are expected to be independent research positions and will be funded at $150,000 per year. AACR requires applicants to submit both an online and a paper application. Online applications must be completed by 12:00 noon (United States Eastern Daylight Time) on October 28, using the proposalCENTRAL website. The paper copy with original signatures and all supporting documents must be postmarked and sent no later than October 30.
Requirements: Applicants must be in mentored training positions at an academic, medical, or research institution within the United States. There are no citizenship requirements. The applicant institution will be the mentored phase institution. Employees or subcontractors of a government entity or for-profit private industry are not eligible. Exceptions may apply if an applicant holds a full-time position at a Veterans' Hospital or national laboratory in the U.S. Contact AACR before submitting an application to determine your eligibility. Applicants must be Postdoctoral or Clinical Research Fellows who completed all doctoral degree requirements no more then five years prior to the start of the grant term.
Geographic Focus: All States
Date(s) Application is Due: Oct 28; Oct 30
Amount of Grant: 600,000 USD
Contact: Elizabeth Martin; (267) 646-0664; fax (215) 440-9372; grants@aacr.org
Internet: http://www.aacr.org/home/scientists/research-funding—fellowships/postdoctoral-fellowships.aspx
Sponsor: American Association for Cancer Research
615 Chestnut Street, 17th Floor
Philadelphia, PA 19106-4404

Parapsychology Foundation Grants 3868
The grant program is designed to advance the understanding of parapsychology. Grants are given to scientists, universities, and laboratories for original study, research, and experiments in extrasensory perception, psychokinesis, and precognition.
Restrictions: The foundation does not consider proposals for travel or for general graduate and undergraduate studies.
Geographic Focus: All States
Amount of Grant: 3,000-5,000 USD
Contact: Contact; (212) 628-1550; fax (212) 628-1559; info@parapsychology.org
Internet: http://www.parapsychology.org/dynamic/040100.html
Sponsor: Parapsychology Foundation
P.O. Box 1562
New York, NY 10021

Parkersburg Area Community Foundation Arts in Education Special Grants 3869
Because high quality arts programs aid in the development of competent, creative, and critically thinking adults, the West Virginia Department of Education recommends schools provide more time for the delivery of the art curriculum. The arts in education initiative will make grants for school-level projects and programs that address West Virginia educational content standards and link arts content to science, technology, engineering and/or math (STEM) concepts, as well as help local schools to develop partnerships with arts institutions and higher education. Priority funding consideration will be given to projects that: bring arts programming into schools, using the arts to enhance or introduce science, technology, engineering and/or math (STEM) concepts and that advance state standards and curricular goals; promote collaboration and partnerships to help sustain interest in science, technology, engineering and/or math education and the arts past the period of this grant; offer evidence of a 10% local match for the grant (in-kind or direct support); and projects that did not receive an arts in education award the previous year. Examples of possible projects and additional proposal criteria and instructions are available at the website. Organizations may apply through the online application process.
Requirements: To receive consideration, proposals must include all of the following elements: evidence of potential for increased collaboration among educators/schools with arts groups and/or community artists directly in the school environment during the standard school year for K-12 public school students for the current school year (accredited private and parochial schools in the service region with at least 100 students may apply

for grant consideration); show potential for creative application/s of all forms of the arts in traditional education, ranging from the visual, to dance, to music, to performance arts; show increased integration of the arts in core disciplines, emphasizing science, technology, engineering and math, to support student advancement and success with clear evidence of the tie between the arts activities to West Virginia content standards that will be pursued; and show evidence that students are engaged in the project and its implementation.
Restrictions: A 10% match for the amount of the grant request should be shown in the budget narrative.
Geographic Focus: West Virginia
Date(s) Application is Due: Oct 1
Amount of Grant: Up to 5,000 USD
Contact: Jane Winters; (304) 428-4438 or (866) 428-4438; jane.winters@pacfwv.com
Internet: http://www.pacfwv.com/grant-guidelines-and-applications
Sponsor: Parkersburg Area Community Foundation
501 Avenuery Street, P.O. Box 1762
Parkersburg, WV 26102-1762

Park Foundation Grants **3870**
The foundation awards grants to nonprofit organizations in central New York and areas in the U.S. Southeast. Areas of interest include education and higher education, animal welfare, environment, television, and general charitable giving. Types of support include general operating support, program development, professorships, seed money, fellowships, scholarship funds, research grants, and matching funds. There are no application deadlines. The board meets in March, June, August, October, and December. Obtain application forms.
Requirements: 501(c)3 nonprofits in he eastern U.S., primarily in central NY, Washington, DC, and North Carolina are eligible.
Restrictions: For-profit organizations and individuals are not eligible.
Geographic Focus: All States
Amount of Grant: 5,000-100,000 USD
Contact: Linda Madeo, Executive Director; (607) 272-9124; fax (607) 272-6057
Sponsor: Park Foundation
P.O. Box 550
Ithaca, NY 14851

Partnership for Cures Charles E. Culpeper Scholarships in Medical Science **3871**
The program awards scholarships in medical science to support the career development of academic physicians. In direct costs, $100,000 per year to the sponsoring institution will be committed for up to three consecutive years. It is expected that a significant portion of the award will be used as a contribution to the Scholar's salary. Up to $50,000 (plus fringe benefits) of this sum may be used for this purpose; the rest for support of research including personnel and some travel costs. Eight percent will be added to each award for indirect costs. All scientific research relevant to human health is eligible for consideration.
Requirements: Candidates must: be U.S. citizens or aliens who have been granted permanent U.S. residence (proof required); hold the M.D. degree from a U.S. medical school or the equivalent of an M.D. degree from an educational institution equivalent to a United States medical school; and be judged worthy of support by virtue of the quality of their research proposals and their potential for successful careers in academic medicine. Applicants must have at least one year of post-doctoral clinical training.
Restrictions: Individuals who have achieved a rank above assistant professor are ineligible for this program, and an awardee who is promoted above the rank of assistant professor before the initiation of the scholarship will be ineligible to receive the award.
Geographic Focus: All States
Date(s) Application is Due: Aug 16
Amount of Grant: 108,000 USD
Contact: Dr. Bruce E. Bloom; (312) 696-1366; fax (312) 263-0939; Bruce@4cures.org
Internet: http://www.4cures.org/home/culpeper_scholar_program_research
Sponsor: Partnership for Cures
70 West Madison Street, Suite 1500
Chicago, IL 60602

Partnership for Cures Two Years To Cures Grants **3872**
The Initiative is intended to bring researchers, funders and institutions together to produce better treatments and cures for patients with disease right now. The Program works to quickly bring together: breakthrough scientific discoveries about diseases and treatments; FDA approved drugs and other medical therapies (Eastern and Western) with new potential uses; research teams at top institutions ready to proceed with urgency and focus; and independent validation, compressed time lines, and financial leverage. Projects must be able to create a direct patient impact in 24 months from the date of initiation of the project. Budgets can range from $35,000 to $500,000. The Program is accepting research projects in any catastrophic disease, but has special interest in the following diseases: lupus, Crohns disease, MS, ALS, protein intolerances in children, ischemic colitis, autism, type I diabetes, lung cancer, osteosarcoma, Fragile X syndrome, Jewish genetic diseases, and Parkinsons disease. There is no deadline for submissions, but funders are looking for projects right now.
Requirements: Each project must come with a minimum of 30% matching funds. Those funds can come from the institution itself, from government or other public funders, or from individuals, foundations or other private funders.
Restrictions: No more than 10% of the budget can be allocated to overhead, travel and publications.
Geographic Focus: All States
Amount of Grant: 35,000-500,000 USD
Contact: Dr. Bruce E. Bloom; (312) 696-1366; fax (312) 263-0939; Bruce@4cures.org
Internet: http://www.4cures.org/home/funding_opportunities
Sponsor: Partnership for Cures
70 West Madison Street, Suite 1500
Chicago, IL 60602

Patrick and Anna M. Cudahy Fund Grants **3873**
The Patrick and Anna Cudahy Fund is a general purpose foundation which primarily supports organizations in Wisconsin and the metropolitan Chicago area. The principal areas of interest are social service, youth, and education with some giving for the arts, and other areas. Some support is also given for local and national programs concerned with public interest and environmental issues. A few grants are given for international programs but only to those which are represented by a United States based organization.
Requirements: Nonprofits organizations in Wisconsin and the metropolitan Chicago area are eligible. See Foundations website for additional guidelines and application form: http://www.cudahyfund.org/Guideliines.htm
Restrictions: Organizations may submit only one proposal during any calendar year. Requests are not considered for: organizations and projects primarily serving a local constituency outside of Wisconsin and the Chicago metropolitan area; organizations outside the United States who are not represented by a United States based 501 [c]3 organization; grants to individuals; loans or endowments.
Geographic Focus: Illinois, Wisconsin
Date(s) Application is Due: Jan 5; Apr 5; Jul 5; Oct 5
Amount of Grant: 5,000-25,000 USD
Contact: Janet S. Cudahy, President; (847) 866-0760; fax (847) 475-0679
Internet: http://www.cudahyfund.org/Guideliines.htm
Sponsor: 1609 Sherman Avenue, #207
Evanston, IL 60201

Patron Saints Foundation Grants **3874**
The Patron Saints Foundation is a private foundation that provides grants to public charities that improve the health of individuals residing in the West San Gabriel Valley through health care programs which are consistent with the moral and religious teachings of the Roman Catholic Church. Grants will be made to qualified public charities to sponsor charitable, scientific or educational health care programs which are not inconsistent with the moral and religious teachings of the Roman Catholic Church, in the following health care categories: community health services (direct health care services); community health care education; capital expenditures; equipment and supplies; medical research. Grants typically range from $5,000 to $15,000. Larger or smaller grants are at the discretion of The Patron Saints Foundation Board of Directors based on the merits of the project.
Requirements: California 501(c)3 tax-exempt organizations serving the West San Gabriel Valley are eligible, including: the cities of Alhambra, Arcadia, Duarte, El Monte, La Canada Flintridge, Monrovia, Monterey Park, Pasadena, Rosemead, San Gabriel, San Marino, Sierra Madre, South El Monte, South Pasadena, Temple City and the unincorporated areas of Los Angeles County known as Altadena and South San Gabriel along with the unincorporated portions of Los Angeles County within the West San Gabriel Valley.
Restrictions: Grants are not awarded for endowment funds, political activities, travel, surveys or fund raising activities.
Geographic Focus: California
Date(s) Application is Due: Mar 1; May 1
Amount of Grant: 5,000-15,000 USD
Contact: Kathleen T. Shannon; (626) 564-0444; patronsaintsfdn@sbcglobal.net
Internet: http://www.patronsaintsfoundation.org/index.php?nav=Grant%20Guidelines
Sponsor: Patron Saints Foundation
260 S Los Robles Avenue, Suite 201
Pasadena, CA 91101-3614

Paul and Daisy Soros Fellowships for New Americans **3875**
The program offers support for graduate study by New Americans, individuals who hold Green Cards or who have been naturalized as U.S. citizens or are children of two parents who are both naturalized citizens. The program offers maintenance and tuition grants for up to two years of graduate study in the United States. Fellows must have shown potential in the fields for which they seek further education; the capacity for creativity, initiative, originality, and work; and the commitment to the values of the U.S. Constitution and Bill of Rights. Applications are available from college academic advisors, the program's Web site, or by contacting the program directly.
Requirements: Applicants must have either the bachelor's degree or be in the final year of undergraduate study and must not be older than 30 years of age as of the listed application deadline.
Geographic Focus: All States
Date(s) Application is Due: Nov 1
Amount of Grant: 20,000 USD
Contact: Administrator; (212) 547-6926; pdsoros_fellows@sorosny.org
Internet: http://www.pdsoros.org
Sponsor: Paul and Daisy Soros Fellowships for New Americans
400 W 59th Street
New York, NY 10019

Paul G. Allen Family Foundation Grants **3876**
The single foundation, created through the consolidation of Allen's six previous foundations (The Allen Foundation for the Arts, The Paul G. Allen Charitable Foundation, The Paul G. Allen Foundation for Medical Research, The Paul G. Allen Forest Protection Foundation, The Allen Foundation for Music, and The Paul G. Allen Virtual Education Foundation), will continue to focus on the Allen family's philanthropic interests in the areas of arts and

culture, youth engagement, community development and social change, and scientific and technological innovation. The Arts and Culture Program fosters creativity and promotes critical thinking by helping strong arts organizations become sustainable and supporting projects that feature innovative and diverse artistic forms. The Youth Engagement Program improves the way young people learn by supporting organizations that use innovative teaching strategies and provide opportunities for children to address issues relevant to their lives. The Community Development and Social Change Program promotes individual and community development by supporting initiatives and organizations that provide access to resources and opportunities. The Scientific and Technological Innovation Program advances promising scientific and technology research that has the potential to enhance understanding and stewardship of the world in which we live. Organizations may only receive one grant per year. Organizations must not have any delinquent final reports due to any of the Paul G. Allen Foundations for previous grants. Grantseekers are encouraged to apply through the online application process, where basic organizational and project information will be requested. Guidelines are available online.
Requirements: 501(c)3 tax-exempt organizations, status from the Internal Revenue government entities, and IRS-recognized tribes are eligible. Eligible organizations must be located in, or serving populations of, the Pacific Northwest, which includes Alaska, Idaho, Montana, Oregon, and Washington.
Restrictions: In general, the foundation will not consider requests for general fund drives, annual appeals, or federated campaigns; special events or sponsorships; direct grants, scholarships, or loans for the benefit of specific individuals; projects of organizations whose policies or practices discriminate on the basis of race, ethnic origin, sex, creed, or sexual orientation; contributions to sectarian or religious organizations whose principle activity is for the benefit of their own members or adherents; loans or debt retirement; projects that will benefit the students of a single school; general operating support for ongoing activities; or projects not aligned with the foundation's specified program areas. 509(a) private foundations are ineligible.
Geographic Focus: Alaska, Idaho, Montana, Oregon, Washington
Date(s) Application is Due: Mar 31; Sep 30
Contact: Administrator; (206) 342-2030; info@pgafamilyfoundation.org
Internet: http://www.pgafamilyfoundation.org
Sponsor: Paul G. Allen Family Foundation
505 Fifth Avenue S, Suite 900
Seattle, WA 98104

Paul Marks Prizes for Cancer Research **3877**

The prizes recognize outstanding young investigators who have made significant contributions to increase the understanding of cancer or improve the treatment of the disease through basic or clinical research. The winners will present their work at the center, be honored at a dinner, and share a cash award of $150,000. Nomination packets must include a letter from the nominator outlining the significance of the accomplishments for which the candidate should be recognized; a one-page scientific biography of the candidate; a list of up to eight of the candidate's significant published papers with a brief (fewer than 100 words) explanation of the importance of each one; and the candidate's curriculum vita. Up to three supporting letters may also be submitted. Annual deadline date may vary; contact program staff for exact dates.
Requirements: Nominees are required to be age 45 or younger at the time of the submission deadline.
Geographic Focus: All States
Date(s) Application is Due: Apr 29
Amount of Grant: 150,000 USD
Contact: Paizhe Pressley, Program Contact; (212) 639-6561; presslep@mskcc.org
Internet: http://www.mskcc.org/mskcc/html/53983.cfm
Sponsor: Memorial Sloan-Kettering Cancer Center
1275 York Avenue
New York, NY 10021

Paul P. Fidler Research Grants **3878**

The grant is designed to encourage the development and dissemination of knowledge that has potential to improve the experiences of college students in transition. The award includes a cash stipend, travel to two national conferences, a presentation at a national conference, and priority consideration for publication. The competition is open to faculty, staff, and graduate students who plan to conduct research on issues of college student transitions. Cross-institutional research teams are encouraged to apply.
Requirements: Applicants for the grant must complete a research proposal as outlined by the National Resource Center. Only complete proposals will be considered.
Restrictions: The following individuals are not eligible: National Resource Center staff and fellows; current members of the Center's National Advisory Board; and students, staff, and faculty of the University of South Carolina system.
Geographic Focus: All States
Date(s) Application is Due: Jul 16
Amount of Grant: 5,000 USD
Contact: Angela Griffin; (803) 777-2134; AngelaG@mailbox.sc.edu
Internet: http://www.sc.edu/fye/research/grant/index.html
Sponsor: National Resource Center
University of South Carolina, 1728 College Street
Columbia, SC 29208

Paul Stock Foundation Grants **3879**

The foundation awards grants in Park County, WY, for higher education, university support, recreation, and human services. Types of support include annual campaigns, building construction/renovation, research grants, and scholarships to individuals. The board meets in July and December. There are no application forms.
Requirements: Giving primarily in Park County, WY; student aid is limited to those who have resided in WY for one year or more.
Geographic Focus: Wyoming
Contact: Charles Kepler, President; (307) 587-5275
Sponsor: Paul Stock Foundation
P.O. Box 2020
Cody, WY 82414-2020

PC Fred H. Bixby Fellowships **3880**

The Population Council offers a Fellowship program to expand training opportunities for social scientists and biomedical researchers in the health and population fields. This Fellowship is geared toward developing-country nationals in the early stages of their careers, and to those with a demonstrated commitment to remaining in their home countries to build capacity in local institutions or returning home after working/studying abroad. The Council's Fred H. Bixby Fellowship is a ten-year program which offers a limited number of Fellowships each year. The Bixby Fellowship allows candidates to work with experienced mentors in the Council's network of offices all over the world. Fellows work on projects in the areas of HIV and AIDS, poverty, gender, and youth, and reproductive health. The term of appointment is two years, with the second year contingent upon a successful evaluation and review of performance in the first year. The Fellowship is a full-time commitment, meaning no additional research, studies, or work may be pursued for the duration of the Fellowship. Awards consist of a monthly stipend, health insurance, an allowance for relocation expenses, and attendance and travel to one international professional meeting per year. The Fellow is responsible for all local accommodations and living costs. Fellows will be provided with office space in the Council's office, plus access to office support facilities and a notebook computer.
Requirements: Candidates must have recently completed a Ph.D. or equivalent degree within the last five years in the social sciences, public health, or biomedical sciences. All applicants should have previous direct experience with either biomedical research, program research, or policy-relevant social science research. Applicants must be legal citizens of a developing country and proficient in English. Candidates interested in applying for a Fred H. Bixby Fellowship must first email the Fellowship Coordinator with a letter indicating professional objectives and goals, a short statement about research interests and plans for the proposed Fellowship period, as well as the name(s) of the mentor(s) (up to three) with whom they would like to work. They should also include a copy of their curriculum vitae. The Fellowship Coordinator will then initiate contact with the appropriate mentor(s) and instruct each candidate on how to proceed. Additional information about the application process is available at the Council website.
Restrictions: Candidates with master's-level degrees in public health or a related field may be eligible only if they have considerable skills, background, experience, and peer-reviewed publications, including 2–3 years of relevant job experience post-master's.
Geographic Focus: All Countries
Date(s) Application is Due: Jan 31
Contact: Coordinator; (212) 339-0500; bixbyfellowship@popcouncil.org
Internet: http://www.popcouncil.org/what/bixby.asp
Sponsor: Population Council
One Dag Hammarskjold Plaza
New York, NY 10017

PDF-AANF Clinician-Scientist Development Awards **3881**

The Parkinson's Disease Foundation (PDF) and the American Academy of Neurology Foundation (AANF) collaborated in 2007 to create the Clinician Scientist Development Award to support our mutual interest in training outstanding residency-level clinicians in the field of Parkinson's disease. Successful applicants will receive a three-year award of $85,000. This award supports three years of research training in an environment where talented young clinicians address problems in Parkinson's disease with the most current scientific tools. It is expected that upon completion of the program, participants will be committed to a research or combined clinical/research career in PD and will be in line to direct robust research programs relevant to PD. This award is given every three years. It will be offered again until 2011.
Requirements: Eligibility & Guidelines: applicants must hold an M.D., D.O., or equivalent clinical degree from an accredited institution; applicants must have completed residency training but be less than seven years from completion of residency when funding begins; there is no citizenship requirement; however, the individual applying for the award must be licensed to practice medicine in the United States at the time of application; may only submit one application per department. The department is considered to be the one at which the applicant's training and research program will occur; each applicant must agree to a phone interview with a member of the PDF-AANF Advisory Committee, if necessary, during the application review period.
Geographic Focus: All States
Amount of Grant: 85,000 USD
Contact: Terry Heinz, Grants Administrator; (651) 695-2746; theinz@aan.com
Internet: http://www.pdf.org/en/grant_funding_fellow#pdflabres
Sponsor: Parkinson's Disease Foundation
1359 Broadway, Suite 1509
New York, NY 10018

PDF International Research Grants **3882**

The International Research Grants Program (IRGP) is designed to promote innovative research that has a high potential to significantly advance the knowledge of Parkinson's disease (PD) but little likelihood of securing funding through more traditional sources. By supporting novel, high risk/high reward research, the goal of the IRGP is to enable

investigators to demonstrate the feasibility of their ideas while generating preliminary data necessary for the support of future funding. At its core, research supported by the IRGP must be directly relevant to the treatment and/or understanding of Parkinson's disease. Successful IRGP projects are: novel research hypotheses of Parkinson's disease; inventive in terms of methodology or approach; and clinical, preclinical, or basic research proposals that will directly impact Parkinson's disease or its treatment. IRGP awards will now reach a maximum of $75,000 per year for two years (compared to the previous maximum of $50,000 per year for one year of research), subject to review of first year progress. Research grants are now also eligible for an additional ten percent in indirect costs. All applications must be submitted online through the PDF website; no paper copies will be accepted. Visit grants.pdf.org, after Monday, November 2, to create an account, see more detailed instructions and submit your application. The deadline for submission of grant applications will be 11:59 PM EST on Tuesday, February 2.
Requirements: The IRGP is open to applicants both nationally and internationally and basic, translational and clinical research proposals are eligible for support. All applicants must possess a Ph.D. or a M.D. (or equivalent). Preference will be given to scientists who are at an early stage in their professional careers.
Restrictions: Investigators who receive funding from PDF may not receive funding for other projects supported by PDF during the same year, nor receive funding from other foundations or institutions for the same project during that year.
Geographic Focus: All States
Date(s) Application is Due: Feb 2
Amount of Grant: 75,000 USD
Contact: Robin Anthony Elliott; (212) 923-4700; fax (212) 923-4778; info@pdf.org
Internet: http://www.pdf.org/en/grant_funding_irg#focus
Sponsor: Parkinson's Disease Foundation
1359 Broadway, Suite 1509
New York, NY 10018

PDF Postdoctoral Fellowships for Basic Scientists 3883
The Postdoctoral Fellowships for Basic Scientists are one-year fellowships for young scientists, fresh from their Ph.D. training, to study at major research institutions. The programs grants funding in the amount of $40,000 for one year as well as a research allowance of $5,000 to be used at the discretion of the Fellow, with approval of the sponsor/mentor, to pay for such items as books, training courses, travel costs (up to $2,000) or a computer. There is no provision for the deduction of postdoctoral taxes, institutional overhead or fees. However, if the Fellow submits a written request, $3,000 of the research allowance may be used to defray the cost of premiums for health insurance. Fellows may competitively reapply for continued funding after their first year for a total of three years of support. In these instances, a progress report of the current research would be due along with the new application. All applications must be submitted online through the PDF website; no paper copies will be accepted. The deadline for submission of grant applications will be 11:59 PM ET on Tuesday, February 2. Visit grants.pdf.org after Monday, November 2, to create an account, see more detailed instructions and submit your application. Final grant decisions are expected by mid-May.
Requirements: The applicant must have completed a Ph.D. or M.D. and must identify a research mentor who will serve as mentor for the fellow and supervisor of the research.
Geographic Focus: All States
Date(s) Application is Due: Feb 2
Amount of Grant: 45,000 USD
Contact: Robin Anthony Elliott; (212) 923-4700; fax (212) 923-4778; info@pdf.org
Internet: http://www.pdf.org/en/grant_funding_fellow#pdflabres
Sponsor: Parkinson's Disease Foundation
1359 Broadway, Suite 1509
New York, NY 10018

PDF Postdoctoral Fellowships for Neurologists 3884
For young clinicians who have completed their neurology residency and are seeking clinical research experience, PDF offers the Postdoctoral Fellowships for Neurologists, one-year awards in the amount of $55,000. A research allowance of $5,000 may be used at the discretion of the Fellow, with approval of the mentor, to pay for such items as books, training courses, travel costs (up to $2,000) or a computer. There is no provision for the deduction of postdoctoral taxes, institutional overhead or fees. However, if the Fellow submits a written request, $3,000 of the research allowance may be used to defray the cost of premiums for health insurance. Fellows may competitively reapply for continued funding after their first year for a total of three years of support. In these instances, a progress report of the current research would be due along with the new application. All applications must be submitted online through the PDF website; no paper copies will be accepted. The deadline for submission of grant applications will be 11:59 PM EST on Tuesday, February 2. Visit grants.pdf.org, after Monday, November 2, to create an account, see more detailed instructions and submit your application. Final decisions are expected by mid-May.
Requirements: Applicants seeking a Postdoctoral Fellowship for Neurologists must have an M.D. and be completing or have completed a neurology residency.
Restrictions: Applicants may not have their own lab.
Geographic Focus: All States
Date(s) Application is Due: Feb 2
Amount of Grant: 55,000 USD
Contact: Robin Anthony Elliott; (212) 923-4700; fax (212) 923-4778; info@pdf.org
Internet: http://www.pdf.org/en/grant_funding_fellow#pdflabres
Sponsor: Parkinson's Disease Foundation
1359 Broadway, Suite 1509
New York, NY 10018

PDF Summer Fellowships 3885
PDF's Summer Fellowship Program is used to support students from advanced undergraduates to graduate and medical students in their pursuit of Parkinson's-related summer research projects. The goal of the Summer Fellowship is to cultivate an early interest in Fellows into the cause(s) and possible treatments for Parkinson's disease. Fellows work under the close supervision of a sponsor who is an expert in the Parkinson's community and oversees the project. Typically, summer fellowships are offered for 10 weeks of laboratory work with an award of $3,000. The application period for Summer Student Fellowships will open on Monday, November 2, apply by 11:59 PM ET on Friday, January 15.
Requirements: Undergraduate students, graduate students and medical students are all eligible for summer fellowships. Each applicant must identify a mentor, with whom he or she will conduct the proposed project.
Geographic Focus: All States
Date(s) Application is Due: Jan 15
Amount of Grant: 3,000 USD
Contact: Robin Anthony Elliott; (212) 923-4700; fax (212) 923-4778; info@pdf.org
Internet: http://www.pdf.org/en/grant_funding_fellow#pdflabres
Sponsor: Parkinson's Disease Foundation
1359 Broadway, Suite 1509
New York, NY 10018

Peabody Foundation Grants 3886
The Peabody Foundation was established in 1894 in Massachusetts. Grants provide care, treatment, rehabilitation, education, and assistance to children with physical disabilities as well as encouraging and supporting medical research in the causes of crippling disease, particularly in children. The majority of recipient organizations will be located in the Boston, Massachusetts area. Proposals may be submitted at any time and are reviewed in the spring. Funding ranges from $20,000 to $200,000. Interested organizations should contact the Administrative Director for application procedures and eligibility information.
Restrictions: Grants are not made to individuals. Giving is limited to Massachusetts, with emphasis on the Boston area.
Geographic Focus: Massachusetts
Date(s) Application is Due: Feb 1
Amount of Grant: 20,000-200,000 USD
Contact: Judi Mullen, Director; (508) 728-8780; jemullen12@comcast.net
Jonathan Bashein, Office Administrator; (617) 345-1000 or (617) 310-4100; fax (617) 345-1300; jbashein@nixonpeabody.com
Sponsor: Peabody Foundation
100 Summer Street
Boston, MA 02110-2131

Peacock Foundation Grants 3887
Established by Henry B. Peacock, Jr. in 1947, the mission of Peacock Foundation, Inc. is to enhance and promote the good health and well being of children, families, and underprivileged persons in Southeast Florida, through contributions, gifts, and grants to eligible nonprofit organizations. The priorities of Peacock Foundation, Inc. include: making grants to human services providers that promote youth development, assist abused or neglected children, women, and the elderly, and seek to reduce abuse, prevent homelessness, and end hunger in our community; supporting educational programs in the arts and the environment, as well as special education for disabled persons; contributing to medical research, health care organizations, and hospitals.
Requirements: All applicants must be IRS recognized 501(c)3 public charities classified as not a private foundation, registered with the Department of Agriculture to solicit funds in Florida, when applicable, and located in and/or of significant benefit to residents to the Southeast Florida counties of Miami-Dade, Broward, or Monroe. In order for a proposal to be considered for funding, the applicant first must send a brief letter of inquiry. See Foundations website for letter of inquiry guidelines: http://www.peacockfoundationinc.org/review_progress.html.
Restrictions: Peacock Foundation, Inc. does not fund: capital campaigns, construction, or renovation projects; deficit financing or debt reduction; conferences or festivals; fundraising events or advertising; special events or athletic events; individuals; lobbying to influence legislation; religious organizations, unless engaged in a significant project benefiting the entire community.
Geographic Focus: Florida
Contact: Joelle Allen, Executive Director; (305) 373-1386
Internet: http://www.peacockfoundationinc.org/eligibility.html
Sponsor: Peacock Foundation
100 SE Second Street, Suite 2370
Miami, FL 33131

Pediatric Brain Tumor Foundation Research Grants 3888
The Pediatric Brain Tumor Foundation (PBTF) is a 501(c)3 nonprofit charitable organization whose mission is as follows: to find the cause of and cure for childhood brain tumors by supporting medical research; to increase public awareness about the severity and prevalence of childhood brain tumors; to aid in the early detection and treatment of childhood brain tumors; to support a national database on all primary brain tumors; and to provide educational and emotional support for children and families affected by this life-threatening disease. PBTF grants are meant to expand scientists' knowledge of the causes of pediatric brain tumors and should demonstrate the potential to advance discoveries into new treatment methodologies. Each year PBTF issues a Request for Application for Basic or Translational Research support. The foundation accepts requests for seed grants (start-up research programs) and for fellowships. Guidelines and application are available at the foundation website.
Requirements: Research projects must be hypothesis-driven basic or translational scientific research on pediatric brain tumors. Applicants with MDs must have no more than five

years of post-residency training as of September of the current year of application. Doctoral applicants must have no more than five years postdoctoral laboratory experience as of current year of application. Applicant organizations must be 501(c)3 institutions located in the U.S., Canada, and Australia.
Restrictions: Grant requests must not exceed $50,000 over the period of one year.
Geographic Focus: All States, Australia, Canada
Date(s) Application is Due: Sep 7
Amount of Grant: Up to 50,000 USD
Contact: Dianne Traynor, President and Chairman of the Board; (800) 253-6530 or (828) 665-6891; fax (828) 665-6894; dtraynor@pbtfus.org or pbtfus@pbtfus.org
Internet: http://www.pbtfus.org/medcomm/research/requests-for-applications.html
Sponsor: Pediatric Brain Tumor Foundation of the United States
302 Ridgefield Court
Asheville, NC 28806

Pediatric Cancer Research Foundation Grants 3889
Since its establishment in 1982 as a grass-roots organization, the Pediatric Cancer Research Foundation (PCRF) has focused its efforts upon improving the care, quality of life, and survival rate of children with malignant diseases. With these ends in mind, PCRF concentrates on laboratory research that will translate into immediate treatment for children with cancer. PCRF primarily funds research that brings innovative new drugs and treatment regimens to children. Proposals should be based on molecular, cellular, or integrated systems; be conceptually innovative; and have a clear plan for study and future clinical use of the factual data obtained. Current areas of foundation funding include stem-cell transplantation, stem-cell biology, molecular oncology, and molecular and cellular genetics. Awards will generally be made for a one- to three-year period at the Board's discretion. Renewals for additional years may be made for promising projects as recognized by the Scientific Review Committee. Prospective applicants can download the application form at the PCRF website. Completed applications must be received at the PCRF office by 5:00 PM on May 1st. If a deadline falls on a Saturday or Sunday, the applications are due the next business day.
Requirements: Applications may be submitted by individuals holding a M.D., Ph.D., or equivalent degree and working in domestic non-profit organizations such as universities, colleges, hospitals, or laboratories (the "sponsoring institution"). Applications may be multi-institutional in nature. Applicants currently holding a grant may re-apply for extension of similar research by submitting a Renewal Application.
Restrictions: No overhead or indirect costs to institutions will be funded.
Geographic Focus: All States
Date(s) Application is Due: May 1
Contact: Joseph M. Galosic; (949) 859-6312; fax (949) 859-6323; admin@pcrf-kids.org
Internet: http://www.pcrf-kids.org/
Sponsor: Pediatric Cancer Research Foundation
9272 Jeronimo Road, Suite 122
Irvine, CA 92618

Pell Grants 3890
This program provides grants to students attending eligible institutions of higher education to help them pay for their educational costs.
Requirements: Undergraduate students enrolled as regular students in an eligible program at an eligible institution of higher education and making satisfactory academic progress are eligible to apply. The applicants must be U.S. citizens or eligible noncitizens and have a high school diploma, a GED, or demonstrate the ability to benefit from the program offered.
Geographic Focus: All States
Amount of Grant: 400-4,050 USD
Contact: Program Specialist; (800) 433-3243
Internet: http://www.ifap.ed.gov/IFAPWebApp/index.jsp
Sponsor: U.S. Department of Education
400 Maryland Avenue SW
Washington, DC 20202-5960

Penn Graduate School of Education Dean's Fellowships 3891
The Dean's Fellowship provides funding to doctoral level students and includes, but is not limited to, a tuition scholarship, a living stipend, and a research and teaching apprenticeship. Fellowships are taxable and include a service requirement in the form of the apprenticeship, which promotes the development of the student's research and teaching skills. There is no application for Dean's Fellowships; award decisions are made by the academic division based on the application and communicated at the time of the admissions decision. All incoming Ph.D. students receive a fellowship package, which includes tuition and fee support, health insurance, a living stipend, and a research and teaching apprenticeship. The fellowship covers four years of study and requires that students maintain full-time enrollment.
Requirements: A limited number of Dean's Fellowships are awarded to incoming full-time, traditional Ed.D. students who show high academic potential.
Geographic Focus: All States
Contact: Dr. Andrew Porter; (215) 898-7014; fax (215) 746-6884; andyp@gse.upenn.edu
Internet: http://www.gse.upenn.edu/admissions_financial/scholarships#fellowships
Sponsor: Penn Graduate School of Education
3700 Walnut Street
Philadelphia, PA 19104-6216

Penn Graduate School of Education Foreign Language Area Studies Fellowship 3892
The Foreign Language and Area Studies (FLAS) Fellowship is a tuition scholarship and a living stipend awarded to meritorious undergraduate students and graduate students undergoing training in modern foreign languages and related area or international studies, or with the international aspects of professional or fields of study. FLAS fellowships are funded by the U.S. Department of Education and administered by the University of Pennsylvania's Title VI National Resource Centers. FLAS awards are available only for specific languages, and are contingent on federal funding. Awards are made for the Academic Year or Summer. Academic Year Fellows must be admitted to or enrolled in undergraduate, graduate, or professional programs at the University of Pennsylvania. FLAS awards may be used in some cases for students participating in official overseas language programs and in very limited cases for dissertators.
Requirements: Applicants must be U.S. citizens or permanent residents of the United States. Students receiving FLAS Fellowships must be enrolled in full-time study for the duration of the FLAS award and must take one language course and one related area or international studies course each semester.
Geographic Focus: All States
Contact: Dr. Andrew Porter; (215) 898-7014; fax (215) 746-6884; andyp@gse.upenn.edu
Internet: http://www.gse.upenn.edu/admissions_financial/scholarships#fellowships
Sponsor: Penn Graduate School of Education
3700 Walnut Street
Philadelphia, PA 19104-6216

Penn Graduate School of Education Pre-Doctoral Training in Interdisciplinary Methods for Field-Based Research in Education 3893
Penn GSE offers a Pre-Doctoral Training Program designed to support and enhance the Ph.D. training of students in different disciplines who are committed to careers in applied, field-based education research. The initiative involves six University of Pennsylvania schools and multiple school districts in Pennsylvania and New Jersey and is supported by the Institute of Education Sciences (IES) in the U.S. Department of Education. Fellows receive financial support, access to specialized resources and experts in the field of study, and become part of an emerging national network of students, faculty and practitioners conducting interdisciplinary research in the education sciences.
Geographic Focus: All States
Contact: Dr. Andrew Porter; (215) 898-7014; fax (215) 746-6884; andyp@gse.upenn.edu
Internet: http://www.gse.upenn.edu/admissions_financial/scholarships#fellowships
Sponsor: Penn Graduate School of Education
3700 Walnut Street
Philadelphia, PA 19104-6216

Penn Graduate School of Education Research Assistantships 3894
Research Assistantships provide a partial or full living stipend and/or tuition support for doctoral students in exchange for assisting faculty with eligible research projects. Research Assistantships provide valuable mentored research experiences for doctoral students and are often offered in conjunction with other forms of aid to create a financial assistance package.
Geographic Focus: All States
Contact: Dr. Andrew Porter; (215) 898-7014; fax (215) 746-6884; andyp@gse.upenn.edu
Internet: http://www.gse.upenn.edu/admissions_financial/scholarships#assistantships
Sponsor: Penn Graduate School of Education
3700 Walnut Street
Philadelphia, PA 19104-6216

Pennsylvania State University Education Graduate Assistantships 3895
The program prepares graduates for education careers in research, administration, and policy analysis. Students work closely with established scholars in research or teaching assistantships. Graduate assistantships include full tuition remission and a monthly stipend. Application and guidelines are available online.
Geographic Focus: All States
Date(s) Application is Due: Feb 1
Contact: Department of Education Policy Studies; (814) 865-1480; Highered@psu.edu
Internet: http://www.ed.psu.edu/hied/financial2.asp
Sponsor: Pennsylvania State University
400 Rackley Building
University Park, PA 16802

Penny Severns Breast, Cervical and Ovarian Cancer Research Grants 3896
The Penny Severns Breast, Cervical and Ovarian Cancer Research Fund is a special fund within the state treasury that is used for breast, cervical and ovarian cancer research grants. Revenue sources include general revenue funds, income tax contributions and gifts, as well as grants and awards from private foundations, nonprofit organizations and other governmental entities or persons. Grants support research in areas related to breast, cervical and ovarian cancer prevention, etiology, pathogenesis, early detection, treatment and behavioral sciences. Research also may include clinical trials. One-year grants are available with the possibility of two subsequent 12-month renewals. Standard research grants are intended to develop and advance the understanding, techniques and modalities effective in early detection, prevention, cure, screening and treatment of breast, cervical and ovarian cancers and may include clinical trials. The maximum award for standard research grants will not exceed $75,000. Fellowship research grants support supervised post-doctoral research training. These grants are intended to further develop the skills necessary for a career in breast, cervical and ovarian cancer research. Fellowship research grants will receive maximum annual funding of $35,000. Special consideration is given to single-year applications and to applications for pilot projects that have the potential for subsequent funding from other sources. Continuation grants, with a maximum of two additional years, are available for those grantees that upon peer review demonstrate progress toward stated goals.
Requirements: Although a majority of the applications submitted are bio-medical in nature, researchers in the fields of behavioral and social sciences also are encouraged to

apply. Funding is granted to institutions, not individuals. Research is conducted by an individual(s) under the authority of an institution.
Geographic Focus: Illinois
Contact: Coordinator; 217-524-6088; fax 217-557-3326; dph.owhline@illinois.gov
Internet: http://www.idph.state.il.us/about/womenshealth/grants/penny.htm
Sponsor: Illinois Department of Public Health
535 W. Jefferson Street, First Floor
Springfield, IL 62761-0001

Percy B. Ferebee Endowment Grants **3897**
Grants from the Percy B. Ferebee Endowment are awarded to support charitable, scientific and literary projects, and in particular, governmental and civic projects designed to further the cultural, social, economic and physical well-being of residents of Cherokee, Clay, Graham, Jackson, Macon and Swain Counties of North Carolina and the Cherokee Indian Reservation. Grants are also awarded in the form of scholarships to assist worthy and talented young men and women who reside in said counties in pursuing their college/university degree education within the state of North Carolina. The deadline for the submission of a grant application is September 30. The deadline for scholarship applications is January 31. Application forms are available online. Applicants will receive notice acknowledging receipt of the grant request, and subsequently be notified of the grant declination or approval.
Requirements: 501 (c)3 non-profits in the Cherokee, Clay, Graham, Jackson, Macon and Swain Counties of North Carolina are eligible to apply for grants. Proposals should be submitted in the following format: completed Common Grant Application Form; an original Proposal Statement; an audited financial report and a current year operating budget; a copy of your official IRS Letter with your tax determination; a listing of your Board of Directors. Proposal Statements (second item in the above Format) should answer these questions: what are the objectives and expected outcomes of this program/project/request; what strategies will be used to accomplish your objective; what is the timeline for completion; if this is part of an on-going program, how long has it been in operation; what criteria will you use to measure success; if the request is not fully funded, what other sources can you engage; an Itemized budget should be included; please describe any collaborative ventures. Prior to the distribution of funds, all approved grantees must sign and return a Grant Agreement Form, stating that the funds will be used for the purpose intended. Progress reports and Completion reports must also be filed as required for your specific grant. All current grantees must be in good standing with required documentation prior to submitting new proposals to any foundation. Scholarship recipients must be a resident of these areas and must attend a college or university in the state of North Carolina. Contact the Foundation for additional application requirements.
Restrictions: Grants are not made for political purposes, nor to organizations which discriminate on the basis of race, ethnic origin, sexual or religious preference, age or gender.
Geographic Focus: North Carolina
Date(s) Application is Due: Jan 31; Sep 30
Contact: Wachovia Bank, N.A., Trustee; grantinquiries6@wachovia.com
Internet: https://www.wachovia.com/foundation/v/index.jsp?vgnextoid=5d6852199c0aa110VgnVCM1000004b0d1872RCRD&vgnextfmt=default
Sponsor: Percy B. Ferebee Endowment
Wachovia Bank, NC6732, 100 North Main Street
Winston Salem, NC 27150

Perkin Fund Grants **3898**
The Perkin Fund supports projects and programs in the fields of astronomy, medicine, and scientific research, as well as limited giving to leading organizations in the arts, education, and social services. Most grants are awarded to well-established institutions. However, small and medium-sized institutions doing significant work in the fields of interest also are encouraged to apply. Deadlines are March 15 and September 15, with board meetings in May and November. Final notification of funding is usually two to four weeks.
Requirements: U.S. nonprofit institutions and organizations are eligible. Applicants should submit a letter with a detailed description of their project, the amount of funding requested, and a copy of their IRS determination letter.
Restrictions: The Fund does not grant to individuals or institutions outside the United States. Funding is limited to Connecticut, Massachusetts, and New York.
Geographic Focus: Connecticut, Massachusetts, New York
Date(s) Application is Due: Mar 15; Sep 15
Amount of Grant: 20,000-100,000 USD
Contact: Winifred Gray, Treasurer; (978) 468-2266; theperkinfund@verizon.net
Sponsor: Perkin Fund
176 Bay Road, P.O. Box 2220
South Hamilton, MA 01982-2232

Perl 6 Microgrants **3899**
The program was established by a donation from Best Practical Solutions to help support Perl 6 development. Grants of $500 will be awarded to a range of Perl 6-related projects over the life of the grant program. Proposals will be accepted on a rolling schedule.
Requirements: Accepted grants might be for coding, documentation, testing or even writing articles about Perl 6. The program isn't tied to any one implementation of Perl 6 — they are interested in seeing proposals related to Pugs, Perl 6 on Parrot, Perl 6 on Perl 5 or any other Perl 6 implementation, and in general, projects that can be completed in 4-6 calendar weeks. To submit a grant proposal, send an email to perl6- microgrants@perl.org with the following information: a) A two to three paragraph summary of the work you intend to do; b) A quick bio; c) A brief description of what 'success' will mean for your project (How will we know you're done?); d) Where (if anywhere) you've discussed your project in the past; e) Where you'll be blogging about your progress. (Twice-weekly blog posts are a requirement for getting your grant money.)
Geographic Focus: All States
Amount of Grant: 500 USD
Contact: Jesse Vincent, Grant Manager; jesse@bestpractical.com
Internet: http://www.perlfoundation.org/patrick_michaud_awarded_perl_6_development_grant_as_a_joint_initiative_by_the_perl_foundation_and_mozilla_foundation
Sponsor: Perl Foundation
6832 Mulderstraat
Grand Ledge, MI 48837

Perpetual Trust for Charitable Giving Grants **3900**
The Perpetual Trust for Charitable Giving was established in 1957 to support and promote quality educational, human-services, and health-care programming for underserved populations. Special consideration is given to medical aid and medical research organizations, as well as to institutions of higher learning. Grant requests for general operating support are strongly encouraged. Program support will also be considered. Small, program-related capital expenses may be included in general operating or program requests. To better support the capacity of nonprofit organizations, multi-year funding requests are strongly encouraged. Applicants must apply online at the grant website. Applicants are strongly encouraged to do the following before applying: review the downloadable state application procedures for additional helpful information and clarifications; review the downloadable online-application guidelines at the grant website; review the trust's funding history (link is available from the grant website); review the online application questions in advance; and review the list of required attachments. These will generally include: a list of board members, financial statements (audited, reviewed, or compiled by independent auditor); an organization summary; a list of other funding sources; an IRS Determination letter; and other required documents. All attachments must be uploaded in the online application as PDF, Word, or Excel files. The application deadline for the Perpetual Trust is 11:59 p.m. on September 1. Applicants will be notified of grant decisions before November 30.
Restrictions: The trust does not support requests from individuals, organizations attempting to influence policy through direct lobbying, or any political campaigns.
Geographic Focus: Massachusetts
Date(s) Application is Due: Sep 1
Contact: Miki C. Akimoto, Vice President; (866) 778-6859; miki.akimoto@baml.com
Internet: https://www.bankofamerica.com/philanthropic/fn_search.action
Sponsor: Perpetual Trust for Charitable Giving
225 Franklin Street, 4th Floor, MA1-225-04-02
Boston, MA 02110

Peter Debye Award in Physical Chemistry **3901**
This award is given annually to encourage and reward outstanding research in physical chemistry. A nominee must have accomplished outstanding research of a theoretical or experimental nature in the field. Some consideration should be given to the nominee's success as a mentor and colleague. The award will be granted without regard to age or nationality. The award consists of $5,000 and a certificate.
Requirements: Any individual, except a member of the award committee, may submit one nomination or seconding letter for the award in any given year. The nominating documents consist of a letter of not more than 1000 words containing an evaluation of the nominee's accomplishments and a specific identification of the work to be recognized, a biographical sketch including date of birth, and a list of publications and patents authored by the nominee.
Restrictions: Self-nominations are not accepted.
Geographic Focus: All States
Date(s) Application is Due: Nov 1
Amount of Grant: 5,000 USD
Contact: Felicia Dixon, Awards Administrator; (800) 227-5558 or (202) 872-4408; fax (202) 776-8008; f_dixon@acs.org or awards@acs.org
Internet: http://portal.acs.org/portal/acs/corg/content?_nfpb=true&_pageLabel=PP_ARTICLEMAIN&node_id=1319&content_id=CTP_004543&use_sec=true&sec_url_var=region1
Sponsor: American Chemical Society
1155 Sixteenth Street, NW
Washington, DC 20036-4801

Peter F. Drucker Award for Nonprofit Innovation **3902**
The award is given each November to three nonprofit organizations in recognition of an innovative, existing program that has made a difference in the lives of the people it serves. The annual awards (one first place, two runners up) are accompanied by unrestricted cash prizes designed to further the work of the recipient organizations. Application and guidelines are available online.
Requirements: Organization must be a registered 501(c)3 nonprofit organization and must have been launched no earlier than January 1, 2000, to be eligible.
Geographic Focus: All States
Date(s) Application is Due: Jun 8
Amount of Grant: 2,500-25,000 USD
Contact: Coordinator; (212) 224-1174; fax (212) 224-2508; info@leadertoleader.org
Internet: http://www.leadertoleader.org/events/index.html
Sponsor: Leader to Leader Institute
320 Park Avenue, 3rd Floor
New York, NY 10022-6839

Peter F. McManus Charitable Trust Grants **3903**

The Trust supports alcohol and mental health research on the causes of alcoholism and other substance abuses. Proposed projects may examine basic, clinical, social, and environmental causes of substance abuse.
Requirements: The Trust gives primarily to 501(c)3 organizations in Illinois, New York, Pennsylvania, and Texas. Applicants should submit no more than a three page proposal that includes the following: the qualifications of the key personnel, including the principal investigator's bio; a copy of the IRS determination letter; a detailed description of the project and the amount requested; and a copy of the most current organizational budget and/or project budget. Applicants are notified within three months of submission.
Restrictions: No funding is available for treatment of alcohol or substance abuse. No more than 10% of the grant amount may be used for indirect costs.
Geographic Focus: Illinois, New York, Pennsylvania, Texas
Date(s) Application is Due: Aug 31
Amount of Grant: Up to 50,000 USD
Contact: Katharine Lidz, Trust Administrator; (610) 647-4974
Sponsor: Peter F. McManus Charitable Trust
31 Independence Court
Wayne, PA 19087-5824

Pew Charitable Trusts Biomedical Research Grants **3904**
Pew invests in an array of programs related to science that are aimed at improving the quality of scientific research as well as making data widely available. These projects are working toward solutions to environmental, health and safety dilemmas. Pew also has a decades-long commitment to support groundbreaking research by biomedical researchers early in their careers. All of these initiatives are part of our broader effort to study and promote nonpartisan policy solutions for pressing and emerging problems affecting the American public and the global community. Pew's experts partner with leading authorities in science, the environment and associated fields to conduct research and advance fact-based solutions to compelling problems. Pew targets the frontiers of scientific knowledge where it can make a difference.
Requirements: Grants are made only to colleges and universities, as well as 501(c)3 tax-exempt organizations that are not private foundations.
Restrictions: Grants are not made to individuals or for endowments, capital campaigns, unsolicited construction requests, debt reduction, or scholarships or fellowships that are not part of a program initiated by the Trusts.
Geographic Focus: All States
Contact: Susan A. Magill, Managing Director; (202) 552-2129 or (202) 552-2000; fax (202) 552-2299; smagill@pewtrusts.org or info@pewtrusts.com
Internet: http://www.pewtrusts.org/our_work_category.aspx?id=336
Sponsor: Pew Charitable Trusts
2005 Market Street, Suite 1700
Philadelphia, PA 19103-7077

Pew Fellowships in the Arts **3905**
Fellowships support artists working in a wide variety of performing, visual, and literary disciplines. Awards are made to artists working in 12 different discipline categories, which rotate on a four-year cycle. Applications are accepted annually for the three discipline categories under consideration in that year. Fields of application for the current year are performance art; poetry; and sculpture and installation. The grants provide financial support directly to the artists so that they may have the opportunity to dedicate themselves to creative pursuits exclusively. The program aims to provide such support at moments in artists' careers when a concentration on artistic growth and exploration is most likely to have the greatest impact on an artist's long-term professional development. Guidelines and application are available online. Annual deadline dates may vary; contact program staff for exact dates.
Requirements: All applicants must be practicing artists of demonstrated commitment and professional accomplishment, and be able to provide evidence of such. Refer to specific eligibility requirements under appropriate field of application. Applicants must be verifiable Pennsylvania residents of Bucks, Chester, Delaware, Montgomery, or Philadelphia County for two years or longer immediately prior to the application deadline; and maintain permanent residence in the five-county Philadelphia area through the selection process. Artists who are granted fellowships must maintain permanent residence in the five-county Philadelphia area during their fellowship period.
Restrictions: Artists are not eligible if they are: not able to provide evidence of their professional achievements within their field of application; a matriculated student (full-time or part-time) in a degree-granting program or equivalent at the time of application or during the selection process; a current employee of PFA or The Pew Charitable Trusts, or immediate family member of such an individual; or a previous recipient of a Pew Fellowship in the Arts.
Geographic Focus: Pennsylvania
Date(s) Application is Due: Dec 2
Amount of Grant: 50,000 USD
Contact: Administrator; (215) 875-2285; fax (215) 875-2276; pewarts@mindspring.com
Internet: http://www.pewarts.org/apply.html
Sponsor: Pew Charitable Trusts
2005 Market Street, Suite 1700
Philadelphia, PA 19103-7077

Pew Scholars in the Biomedical Sciences **3906**
This program is designed to support young investigators of outstanding promise in the basic and clinical sciences relevant to the advancement of human health. The funding of the awards is provided by The Pew Charitable Trusts. The award is intended to provide assured support, during their earlier years, for junior members of the faculty as they establish their laboratories. It is hoped that the assurance provided through the program will encourage successful applicants to be more venturesome in their research and future applications for support than would otherwise be likely. ALL applicants must be nominated by their institution and complete the online application.
Requirements: The program is open to individuals with a doctorate in medicine, other health related professions, or biomedical sciences. Candidates must hold full- time appointments at the rank of assistant professor or independent researcher on the faculty of the sponsoring institution.
Restrictions: A biomedical scholar cannot apply on his/her own for this scholarship but must apply through a selected institution.
Geographic Focus: All States
Date(s) Application is Due: Nov 1
Amount of Grant: 60,000 USD
Contact: Leslie Tucker; (215) 575-4850; shans@pewtrusts.com or info@pewtrust.com
Internet: http://futurehealth.ucsf.edu/biomed/scholdes.html
Sponsor: Pew Charitable Trusts
2005 Market Street, Suite 1700
Philadelphia, PA 19103-7077

Pezcoller Foundation-AACR International Award for Cancer Research **3907**
The Award for Cancer Research was established in 1997 to annually recognize a scientist: who has made a major scientific discovery in basic cancer research OR who has made significant contributions to translational cancer research; who continues to be active in cancer research and has a record of recent, noteworthy publications; and whose ongoing work holds promise for continued substantive contributions to progress in the field of cancer. More than one scientist may be co-nominated and selected to share the Award when their investigations are closely related in subject matter and have resulted in work that is worthy of the Award. The Award consists of an unrestricted grant of Euro 75,000 and a commemorative plaque.
Requirements: Eligible candidates are cancer researchers affiliated with institutions in academia, industry, or government that are involved in cancer research, cancer medicine, or cancer-related biomedical science anywhere in the world.
Restrictions: Institutions or organizations are not eligible for the Award.
Geographic Focus: All States
Date(s) Application is Due: Sep 15
Amount of Grant: 75,000 EUR
Contact: Monique P. Eversley, Staff Associate; (267) 646-0576 or (215) 440-9300, ext. 1400; fax (215) 440-9372; monique.eversley@aacr.org or awards@aacr.org
Internet: http://www.aacr.org/home/scientists/scientific-achievement-awards/pezcoller-aacr-international-award.aspx
Sponsor: Pezcoller Foundation / American Association for Cancer Research
615 Chestnut Street, 17th Floor
Philadelphia, PA 19106-4404

Pfizer Advancing Research in Transplantation Science Research Awards **3908**
The mission of the Pfizer ARTS award program is to support basic and clinical science through a competitive, independent, peer-reviewed program that advances the understanding and medical knowledge of mTOR pathway inhibition in transplantation and related areas. Pfizer will support clinical and basic science research exploring mTOR pathway inhibition in the following areas: mechanisms of alloimmunity and allograft injury; mechanisms of cellular and molecular responses; pharmacogenomics and pharmacogenetics in transplantation; mechanisms, incidence, and treatment of adverse events; patient outcomes in transplantation; and novel applications in areas of unmet medical need. Projects are expected to yield results that will advance the medical knowledge in the research areas outlined above and that will be presented at scientific meetings and published in peer-reviewed journals. Proposed projects should last no longer than two years in duration. Award funding is limited to $250,000 per award for the entirety of the project. Funds awarded may be used to cover the awardees' salary and fringe benefits, as well as other direct expenses and indirect expenses incurred during the research project. Direct expenses may include the salary of technical associates; the purchase of relevant laboratory supplies; direct expenses necessary for proper conduct of research (e.g., subject-related costs, study-related personnel costs, diagnostic fees/services, data management expenses). Indirect expenses may include travel expenses to scientific meetings, software licensing fees, publication costs, etc. The selection of research proposals will be performed by an independent, external review committee comprised of academic clinicians/scientists who are experts in the field. Applicants must submit their proposals through the Integrated System for Pfizer Investigator Initiated Research (INSPIIRE) web portal, the link to which is available from the competitive grant announcement web page. Applications must be received by 11:59 p.m. eastern daylight time on the deadline date. Due dates may vary from year to year. Interested applicants are advised to verify current deadlines at the website or by contacting Pfizer. Pfizer, Inc. is an American multinational research-based pharmaceutical company that was founded in 1849 by two cousins Charles Pfizer and Charles Erhart. It topped the 2009 list of America's most generous companies, giving $2.3 billion in products and cash. Pfizer's grants and contributions include the following types of support: support for medical, scientific, and patient organizations; research; lobbying and political contributions; medical education grants; medical and academic partnerships; healthcare charitables; sponsorships; and special events.
Requirements: Applicants must have a professional degree (M.D., Ph.D., Pharm.D., or equivalent) and reside in the United States. Proposals will be reviewed on the following criteria: relevance to clinical transplantation; scientific excellence; originality; and realistic expectations for near term (3-5 years) clinical application.
Restrictions: Other support (funding and/or drug) provided by governmental agencies, non-governmental entities, or other pharmaceutical companies may be used to support the project;

however, there should not be any overlap between the components of the research project supported by the Pfizer ARTS award program and those supported by another source.
Geographic Focus: All States
Date(s) Application is Due: Apr 30
Amount of Grant: Up to 250,000 USD
Contact: Mikael Dolssten, President; (212) 733-2323
Pfizer Medical Representative; (212) 733-2323; Questions.PfizerARTS@pfizer.comQuestions.PfizerART
Internet: http://www.pfizerarts.com/
Sponsor: Pfizer
235 East 42nd Street
New York, NY 10017-5755

Pfizer Anti Infective Research EU Grants **3909**

Pfizer's Anti-Infectives Research Foundation seeks high-quality and innovative anti-infectives research which may result in excellence in patient care. The foundation anticipates making non-renewable awards of approximately £50,000-£100,000 each for research to be undertaken within a twelve-month period. Downloadable application guidelines are available from the grant web page. Applications must be submitted to Pfizer through the company's global investigator-initiated research (INSPIIRE) web portal. First time applicants must "Create an Account". Applications must be received by the deadline date. Due dates and focus areas may vary from year to year; applicants are encouraged to verify current submission requirements at the website or by contacting Pfizer. Pfizer, Inc. is an American multinational research-based pharmaceutical company that was founded in 1849 by two cousins Charles Pfizer and Charles Erhart. It topped the 2009 list of America's most generous companies, giving $2.3 billion in products and cash. Pfizer's grants and contributions include the following types of support: support for medical, scientific, and patient organizations; research; lobbying and political contributions; medical education grants; medical and academic partnerships; healthcare charitables; sponsorships; and special events.
Requirements: Applications for funding are open to all clinical practitioners and scientists working in the UK who hold medical or pharmacy degrees or Ph.D.s and who do not work for Pfizer. Applications should focus on translational (clinically-oriented) research that will impact patient outcomes. Applications regarding research in the late-laboratory phase will also be considered. Proposed projects should be in the following areas of Pfizer's therapeutic focus (however these may change from year to year): clinical diagnostics for invasive aspergillosis; epidemiology and management of serious fungal infections (Aspergillus and Candida) in the UK; cost-effective approaches to management of Complicated Skin and Soft Tissue Infections (cSSTI)s involving Multi-Drug-Resistant (MDR) organisms; and antibiotic stewardship programmes looking at cost and outcomes where the aim is to drive down resistant pathogens in the hospital environment. In exceptional circumstances, applications outside of the stated criteria, such as proposals of high scientific merit that will advance current clinical knowledge in the field of anti-infectives, may also be considered.
Restrictions: Pfizer reserves the right to not support projects that fall outside of Pfizer's internal compliance procedures or those that replicate existing research.
Geographic Focus: United Kingdom
Date(s) Application is Due: Mar 31
Amount of Grant: 50,000-100,000 GBP
Contact: Mikael Dolssten, President; (212) 733-2323
Coordinator; (212) 733-2323 or (203) 316-9059; Air.Foundation@Pfizer.com
Internet: http://www.pfizer.com/research/investigator/competitive_grant_programs.jsp
Sponsor: Pfizer
235 East 42nd Street
New York, NY 10017-5755

Pfizer ASPIRE EU Antifungal Research Awards **3910**

Pfizer invites investigators to submit research proposals that evaluate early diagnosis of invasive mould infections. Pfizer will fund 5 awards up to € 50,000 each for a one-year period. The grants will be awarded to the host institutions on behalf of the awardees. Each award amount includes direct costs (labor and study costs), institutional overhead costs, and indirect costs (additional expenses such as publication). Prior to submitting an application, applicants are requested to have a "medical to medical conversation" with their Pfizer representative. Applicants can email iir@pfizer.com to obtain the appropriate Pfizer contact information for their region. Downloadable application guidelines and online submission instructions are available at the grant website. Applications must be submitted to Pfizer through an online submission website. Applicants should visit http://www.europeaspire.org and click on "apply". From there they will be routed to Pfizer's global investigator-initiated research (INSPIIRE) web portal where they should click on "Submit an IIR Request" and follow the online instructions. First-time applicants must "Create an Account". Applications must be received by the deadline date. (Due dates may vary from year to year; applicants are encouraged to verify current deadlines at the website or by contacting Pfizer.) Pfizer, Inc. is an American multinational research-based pharmaceutical company that was founded in 1849 by two cousins Charles Pfizer and Charles Erhart. It topped the 2009 list of America's most generous companies, giving $2.3 billion in products and cash. Pfizer's grants and contributions include the following types of support: support for medical, scientific, and patient organizations; lobbying and political contributions; medical education grants; medical and academic partnerships; healthcare charitables; sponsorships; and special events. Particular to the grant opportunity listed on this page, Pfizer offers a competitive research grants program ASPIRE (Advancing Science Through Pfizer-Investigator Research Exchange). ASPIRE grant opportunities are available by geographic area, specifically: worldwide, North America, Europe, Asia, and Australia. Notices of opportunities are posted on Pfizer's competitive grants and regional ASPIRE websites. Prospective applicants may fill out a web-based form at the main ASPIRE website (http://www.aspireresearch.org/) to be notified of future competitive grant opportunities via email.
Requirements: Applicants must have a professional degree (M.D., Ph.D., Pharm.D., or equivalent) and reside in one of the EU 16 countries: Austria, Belgium, Denmark, Finland, France, Germany, Greece, Ireland, Italy, Netherlands, Norway, Portugal, Spain, Sweden, Switzerland, or the United Kingdom. Applicants must have a strong academic career interest in the research focus area. Investigators are expected to generate epidemiological, microbiological, and clinical data to better understand how non-culture-based diagnostic methods (e.g., imaging or detection of galactomannan, ß-D-glucan, or nucleic acid) can contribute to early diagnosis of invasive mould disease. Proposals should be clinical in design. Prospective evaluations are preferred however, well-designed retrospective analyses may also be considered for inclusion into the overall proposal. An important criterion in protocol assessment will be how the research contributes to improvement in the quality of care. It is expected that results will be presented at scientific meetings and published in peer reviewed journals. All applications will be formally reviewed by an independent committee of European medical experts in relevant fields. Committee members will evaluate proposals using the following criteria: scientific merit of the research proposal; qualifications of the applicant; relevance of proposed research to the program's mission; and evidence of a suitable research environment.
Restrictions: No other government, non-governmental, or industry-sponsored projects may cover the same work scope as the grant application to the EUROPE ASPIRE Program. However, a EUROPE ASPIRE grant may be related to other funding from foundations or government agencies, as long as there is no direct overlap. The institutional overhead costs cannot exceed 28% of the direct costs. Pfizer does not pay overhead on indirect costs.
Geographic Focus: Austria, Belgium, Denmark, Finland, France, Germany, Greece, Ireland, Italy, Netherlands, Norway, Portugal, Spain, Sweden, Switzerland, United Kingdom
Date(s) Application is Due: Mar 11
Amount of Grant: Up to 50,000 USD
Contact: Mikael Dolssten, President; (212) 733-2323
Internet: http://www.europeaspire.com/antifungals/index.html
Sponsor: Pfizer
235 East 42nd Street
New York, NY 10017-5755

Pfizer ASPIRE EU Dupuytren's Contracture Research Awards **3911**

Current gold standard treatment for the management of Dupuytren's contracture is surgical excision involving removal of the affected palmar fascial tissue. There is however growing interest in nonsurgical, outpatient-based treatments that could provide reduced morbidity, show a decreased rate of recurrence, and give patients with Dupuytren's contracture an improved quality of life when compared with traditional surgical management. Pfizer is funding three awards up to € 60,000 each, for one year, to conduct clinical research into surgical alternatives to treating Dupuytren's contracture. Prior to submitting an application, applicants are requested to have a "medical to medical conversation" with their Pfizer representative. Applicants can email iir@pfizer.com to obtain the appropriate contact information for their region. Downloadable application guidelines and online submission instructions are available at the grant website. Applications must be submitted to Pfizer through an online submission website. Applicants should visit http://www.europeaspire.org and click on "apply". From there they will be routed to Pfizer's global investigator-initiated research (INSPIIRE) web portal where they should click on "Submit an IIR Request" and follow the online instructions. First-time applicants must "Create an Account". Applications must be received by the deadline date. (Due dates may vary from year to year; applicants are encouraged to verify current deadlines at the website or by contacting Pfizer.) Pfizer, Inc. is an American multinational research-based pharmaceutical company that was founded in 1849 by two cousins Charles Pfizer and Charles Erhart. It topped the 2009 list of America's most generous companies, giving $2.3 billion in products and cash. Pfizer's grants and contributions include the following types of support: support for medical, scientific, and patient organizations; lobbying and political contributions; medical education grants; medical and academic partnerships; healthcare charitables; sponsorships; and special events. Particular to the grant opportunity listed on this page, Pfizer offers a competitive research grants program ASPIRE (Advancing Science Through Pfizer-Investigator Research Exchange). ASPIRE grant opportunities are available by geographic area, specifically: worldwide, North America, Europe, Asia, and Australia. Notices of opportunities are posted on Pfizer's competitive grants and regional ASPIRE websites. Prospective applicants may fill out a web-based form at the main ASPIRE website (http://www.aspireresearch.org/) to be notified of future competitive grant opportunities via email.
Requirements: Applicants must have a professional degree (M.D., Ph.D., Pharm.D., or equivalent) and reside in one of the EU 16 countries: Austria, Belgium, Denmark, Finland, France, Germany, Greece, Ireland, Italy, Netherlands, Norway, Portugal, Spain, Sweden, Switzerland, or the United Kingdom. Applicants must have a strong interest in use of Collagenase Clostridium Histcolytium (CCH) in relation to the following areas of interest: various injection techniques; treatment of various cords; timing of finger extension; splinting techniques and their outcomes after treatment; Pre- and Post-treatment pain management; studies looking at Mode of Action using imaging techniques; studies splitting the volume of the reconstituted product into multiple injection sites in the same cord (0.58 mg dose stays the same); and studies assessing treatments aiming at decreasing the extent of local inflammatory reaction (e.g. use of ice packs after injection). Non-drug studies can include the development of hand functionality instruments. It is expected that the results will be presented at scientific meetings and published in peer reviewed journals. An independent committee of European medical experts in the field of hand surgery will review the research proposals and select the grant recipients. In their evaluation, committee members will consider the following criteria: scientific merit of the

research proposal; qualifications of the applicant; relevance of proposed research to the program's mission; and evidence of a suitable research environment.
Restrictions: No other government, non-governmental, or industry-sponsored projects may cover the same work scope as the grant application to the EUROPE ASPIRE Program. However, a EUROPE ASPIRE Program grant may be related to other funding from foundations or government agencies, as long as there is no direct overlap. Due to the competitive nature of these awards, Pfizer cannot provide any additional funding and/or drug support beyond what has been requested and approved by the external review panel.
Geographic Focus: Austria, Belgium, Denmark, Finland, France, Germany, Greece, Ireland, Italy, Netherlands, Norway, Portugal, Spain, Sweden, Switzerland, United Kingdom
Date(s) Application is Due: Jun 15
Amount of Grant: Up to 60,000 EUR
Contact: Mikael Dolssten, President; (212) 733-2323
Internet: http://www.europeaspire.org/Dupuytrens/index.html
Sponsor: Pfizer
235 East 42nd Street
New York, NY 10017-5755

Pfizer ASPIRE EU Emerging Mechanisms of Resistance Antibacterial Research Awards 3912

Pfizer invites investigators to submit research proposals with the primary objective of further understanding the emerging mechanisms of resistance of selected Gram-negative bacteria and the susceptibility of these pathogens to antimicrobial agents. Pfizer will fund five proposals up to 50,000 Euros each for a one-year period. Funds can be used to cover the awardees' salary and fringe benefits, as well as other direct expenses incurred during the research project. Direct expenses may include the salary of technical associates, the purchase of relevant laboratory supplies, and direct expenses necessary for proper conduct of research (e.g. subject-related costs, study-related personnel costs, diagnostic fees/services, data management expenses). Indirect costs may include travel expenses to scientific meetings, software licensing fees, publication costs, etc. Prior to submitting an application, applicants are requested to have a "medical to medical conversation" with their Pfizer representative. Applicants can email iir@pfizer.com to obtain the appropriate Pfizer contact information for their region. Downloadable application guidelines and online submission instructions are available at the grant website. Applications must be submitted to Pfizer through an online submission website. Applicants should visit http://www.europeaspire.org and click on "apply". From there they will be routed to Pfizer's global investigator-initiated research (INSPIIRE) web portal where they should click on "Submit an IIR Request" and follow the online instructions. First-time applicants must "Create an Account". Applications must be received by the deadline date. (Due dates may vary from year to year; applicants are encouraged to verify current deadlines at the website or by contacting Pfizer.) Pfizer, Inc. is an American multinational research-based pharmaceutical company that was founded in 1849 by two cousins Charles Pfizer and Charles Erhart. It topped the 2009 list of America's most generous companies, giving $2.3 billion in products and cash. Pfizer's grants and contributions include the following types of support: support for medical, scientific, and patient organizations; lobbying and political contributions; medical education grants; medical and academic partnerships; healthcare charitables; sponsorships; and special events. Particular to the grant opportunity listed on this page, Pfizer offers a competitive research grants program ASPIRE (Advancing Science Through Pfizer-Investigator Research Exchange). ASPIRE grant opportunities are available by geographic area, specifically: worldwide, North America, Europe, Asia, and Australia. Notices of opportunities are posted on Pfizer's competitive grants and regional ASPIRE websites. Prospective applicants may fill out a web-based form at the main ASPIRE website (http://www.aspireresearch.org/) to be notified of future competitive grant opportunities via email.
Requirements: Applicants must have a professional degree (M.D., Ph.D., Pharm.D., or equivalent) and reside in one of the EU 16 countries: Austria, Belgium, Denmark, Finland, France, Germany, Greece, Ireland, Italy, Netherlands, Norway, Portugal, Spain, Sweden, Switzerland, or the United Kingdom. Applicants must have a strong academic career interest in the research focus area. Research proposals should focus on the activity of tigecycline alone or in combination with other antibiotics on Enterobacteriaceae (mainly Escherichia coli and Klebsiella pneumoniae) showing multi-drug or extreme-drug resistance to other antibiotics. This includes epidemiological studies, microbiological in-vitro susceptibility studies; In-vivo studies in animal model; and studies for determination of antimicrobial mechanisms of resistance. It is expected that results will be presented at scientific meetings and published in peer reviewed journals. All applications will be formally reviewed by an independent committee of European medical experts in relevant fields. Committee members will evaluate proposals using the following criteria: scientific merit of the research proposal; qualifications of the applicant; relevance of proposed research to the program's mission; and evidence of a suitable research environment.
Restrictions: No other government, non-governmental, or industry-sponsored projects may cover the same work scope as the grant application to the EUROPE ASPIRE Program. However, a EUROPE ASPIRE grant may be related to other funding from foundations or government agencies, as long as there is no direct overlap. No more than 28% of the funds may be used for institutional overhead costs. The following topics fall outside of the scope of the EUROPE ASPIRE Program: general education and / or training; public health; and support for ongoing clinical programs that are part of an organization's routine operations.
Geographic Focus: Austria, Belgium, Denmark, Finland, France, Germany, Greece, Ireland, Italy, Netherlands, Norway, Portugal, Spain, Sweden, Switzerland, United Kingdom
Date(s) Application is Due: Mar 11
Amount of Grant: Up to 50,000 USD
Samples: Dr. Pos, Germany-Tigecycline resistance mediated by efflux pumps: Molecular mechanism of tigecycline binding and transport by tripartite RND efflux systems (2012); Dr. Jonas, Germany-Detection of Tigecycline Resistance Determinants by Comparative Transcriptome Analysis of Enterobacter cloacae strains (2012); Dr. Bartolini, Italy-Infections caused by Klebsiella pneumoniae producing KPC-type carbapenemases in Florence University Hospital: molecular epidemiology, clinical impact, treatment outcome and risk factors (2012).
Contact: Mikael Dolssten, President; (212) 733-2323
Internet: http://www.europeaspire.com/Tigecycline/index.html
Sponsor: Pfizer
235 East 42nd Street
New York, NY 10017-5755

Pfizer ASPIRE EU MRSA Nosocomial Pneumonia & MRSA Complicated Skin & Soft Tissue Infections Antibacterial Research Awards 3913

Pfizer invites investigators in the EU 16 to submit proposals to conduct clinical research generating data that will lead to a better understanding of the safety and efficacy of linezolid to enhance the clinical care of patients with MRSA nosocomial pneumonia and MRSA complicated skin and soft tissue infections. Pfizer will fund five proposals up to 50,000 Euros each for a one-year period. Funds can be used to cover the awardees' salary and fringe benefits, as well as other direct expenses incurred during the research project. Direct expenses may include the salary of technical associates, the purchase of relevant laboratory supplies, and direct expenses necessary for proper conduct of research (e.g. subject-related costs, study-related personnel costs, diagnostic fees/services, data management expenses). Indirect costs may include travel expenses to scientific meetings, software licensing fees, and publication costs, etc. Prior to submitting an application, applicants are requested to have a "medical to medical conversation" with their Pfizer representative. Applicants can email iir@pfizer.com to obtain the appropriate Pfizer contact information for their region. Downloadable application guidelines and online submission instructions are available at the grant website. Applications must be submitted to Pfizer through an online submission website. Applicants should visit http://www.europeaspire.org and click on "apply". From there they will be routed to Pfizer's global investigator-initiated research (INSPIIRE) web portal where they should click on "Submit an IIR Request" and follow the online instructions. First-time applicants must "Create an Account". Applications must be received by the deadline date. (Due dates may vary from year to year; applicants are encouraged to verify current due dates at the website or by contacting Pfizer.) Pfizer, Inc. is an American multinational research-based pharmaceutical company that was founded in 1849 by two cousins Charles Pfizer and Charles Erhart. It topped the 2009 list of America's most generous companies, giving $2.3 billion in products and cash. Pfizer's grants and contributions include the following types of support: support for medical, scientific, and patient organizations; lobbying and political contributions; medical education grants; medical and academic partnerships; healthcare charitables; sponsorships; and special events. Particular to the grant opportunity listed on this page, Pfizer offers a competitive research grants program ASPIRE (Advancing Science Through Pfizer-Investigator Research Exchange). ASPIRE grant opportunities are available by geographic area, specifically: worldwide, North America, Europe, Asia, and Australia. Notices of opportunities are posted on Pfizer's competitive grants and regional ASPIRE websites. Prospective applicants may fill out a web-based form at the main ASPIRE website (http://www.aspireresearch.org/) to be notified of future competitive grant opportunities via email.
Requirements: Applicants must have a professional degree (M.D., Ph.D., Pharm.D., or equivalent) and reside in one of the EU 16 countries: Austria, Belgium, Denmark, Finland, France, Germany, Greece, Ireland, Italy, Netherlands, Norway, Portugal, Spain, Sweden, Switzerland, or the United Kingdom. Applicants must have a strong academic career interest in the research focus area and propose innovative clinical studies which clinically evaluate linezolid in treatment of new or recurrent infections due to MRSA in nosocomial pneumonia or in complicated skin and soft tissue infections in patient populations with diabetes, renal insufficiency, morbid obesity, vascular disease, and pulmonary comorbidities. Proposals should be clinical in design. Prospective evaluations are preferred however well-designed retrospective analyses may also be considered for inclusion into the overall proposal to generate and test hypotheses. It is expected that results will be presented at scientific meetings and published in peer reviewed journals. All applications will be formally reviewed by an independent committee of European medical experts in relevant fields. Committee members will evaluate proposals using the following criteria: scientific merit of the research proposal; qualifications of the applicant; relevance of proposed research to the program's mission; and evidence of a suitable research environment.
Restrictions: No other government, non-governmental, or industry-sponsored projects may cover the same work scope as the grant application to the EUROPE ASPIRE Program. However, a EUROPE ASPIRE grant may be related to other funding from foundations or government agencies, as long as there is no direct overlap. Due to the competitive nature of these awards, Pfizer cannot provide any additional funding and/ or drug support beyond what has been requested and approved by the external review panel. No more than 28% of the funds may be used for institutional overhead costs. The following topics fall outside of the scope of the EUROPE ASPIRE Program: general education and / or training; public health; and support for ongoing clinical programs that are part of an organization's routine operations.
Geographic Focus: Austria, Belgium, Denmark, Finland, France, Germany, Greece, Ireland, Italy, Netherlands, Norway, Portugal, Spain, Sweden, Switzerland, United Kingdom
Date(s) Application is Due: Mar 11
Amount of Grant: Up to 50,000 EUR
Contact: Mikael Dolssten, President; (212) 733-2323
Coordinator; (212) 733-2323; questions.aspire.mrsa@pfizer.com or IIR@pfizer.com
Internet: http://www.europeaspire.com/Linezolid2012/eligibility.html
Sponsor: Pfizer
235 East 42nd Street
New York, NY 10017-5755

Pfizer ASPIRE North America Broad Spectrum Antibiotics for the Treatment of Gram-Negative or Polymicrobial Infections Research Awards 3914

While considerable clinical trial data exists for tygecycline in a variety of patients with known infections, most of this information is from controlled clinical trials which leave patients with certain disease complexities underrepresented. Pfizer invites investigators to submit innovative clinical research proposals that evaluate tigecycline's role in the treatment of infections within approved indications. Studies with well-designed retrospective analyses are preferred, however small pilot studies that attempt to test hypotheses will be considered as long as the study can be completed within 1 year. Funds may be used to cover the awardees' salary and fringe benefits, as well as other direct expenses incurred during the research project. Direct expenses may include the salary of technical associates; the purchase of relevant laboratory supplies; direct expenses necessary for proper conduct of research (e.g. subject-related costs, study-related personnel costs, diagnostic fees/services, and data management expenses). Indirect costs may include travel expenses to scientific meetings, software licensing fees, publication costs, etc. Downloadable application guidelines are available at the grant website. Applicants must submit their proposals through the Integrated System for Pfizer Investigator Initiated Research (INSPIIRE) web portal, the link to which is available from the competitive grant announcement on the ASPIRE website. Applications must be received by 11:59 p.m. eastern daylight time on the deadline date. Due dates may vary from year to year. Applicants are encouraged to verify current deadlines at the website or by contacting Pfizer. Pfizer, Inc. is an American multinational research-based pharmaceutical company that was founded in 1849 by two cousins Charles Pfizer and Charles Erhart. It topped the 2009 list of America's most generous companies, giving $2.3 billion in products and cash. Pfizer's grants and contributions include the following types of support: support for medical, scientific, and patient organizations; lobbying and political contributions; medical education grants; medical and academic partnerships; healthcare charitables; sponsorships; and special events. Particular to the grant opportunity listed on this page, Pfizer offers a competitive research grants program ASPIRE (Advancing Science Through Pfizer-Investigator Research Exchange). ASPIRE grant opportunities are available by geographic area, specifically: worldwide, North America, Europe, Asia, and Australia. Notices of opportunities are posted on Pfizer's competitive grants and regional ASPIRE websites. Prospective applicants may fill out a web-based form at the ASPIRE website to by notified of future competitive grant opportunities via email.

Requirements: Applicants must have a professional degree (M.D., Ph.D., Pharm.D., or equivalent) and reside in the United States. Applications submitted by junior investigators (within 5 years of terminal training) will be afforded preferential consideration. Applicants must have a strong academic career interest in clinical evaluations in the following focus areas and patient populations: the disease states of complicated intra-abdominal infections, complicated skin and skin-structure infection, and community-acquired bacterial pneumonia; immunocompromised patient populations; patient populations older than fifty-five years of age; patient populations suffering from diabetes, renal insufficiency, pulmonary comorbidities, vascular disease, obesity, or thermal injuries; and patient populations that have transplants or that are critically ill. Within the specified patient populations with infection types of interest, investigators are expected to generate exploratory clinical data in patients receiving tigecycline to better understand disease pathogenesis, risk factors, host response and treatment outcomes. It is expected that results will be presented at scientific meetings and published in peer-reviewed journals. ASPIRE applications will be reviewed by an independent, external review committee comprised of medical and scientific experts. Review committee members will consider the following criteria in evaluating proposals: scientific merit of the research proposal; qualifications of the applicant; relevance of the proposal to the ASPIRE program's mission; evidence of the applicant's commitment to an academic research career, and evidence of a suitable research environment.

Restrictions: No other government, non-governmental, or industry-sponsored funding may cover the same work scope as the grant application to the ASPIRE program; however the proposal may be related to other funding as long as there is no direct overlap. The following topics fall outside of the scope of the ASPIRE Antibacterial Research Program: general education and/or training; public health; support for ongoing clinical programs that are part of an organization's routine operations; and proposals that do not evaluate tigecycline. Investigators may consider evaluating additional clinical measures (e.g. length of stay, readmission, ventilator days, etc.). However, proposals seeking to financially quantify outcomes are outside the scope of the program. Studies designed to address how patients / payers use evidence about Pfizer products to make coverage and reimbursement decisions are prohibited.

Geographic Focus: All States

Date(s) Application is Due: Apr 24

Contact: Mikael Dolssten, President; (212) 733-2323

Internet: https://www.aspireresearch.org/ASPIRE-AR_U.S./ASPIRE%20AR%20Website%20Tab%20Send/tigecycline/index.html

Sponsor: Pfizer
235 East 42nd Street
New York, NY 10017-5755

Pfizer ASPIRE North America Hemophilia Research Awards 3915

Pfizer's ASPIRE Hemophilia Research awards are intended to support basic-science, translational, and clinical research that advances medical knowledge, pathogenesis, and treatment of hemophilia and to support academic research as well as the career development of promising young and established scientists. Projects are expected to yield results that will advance the medical knowledge of Hemophilia A or Hemophilia B. It is expected that results will be presented at scientific meetings and published in peer reviewed journals. Proposed projects should last 1-2 years and cost approximately $125,000/year, inclusive of overhead costs. Funds may be used to cover the awardees' salary and fringe benefits, as well as other direct expenses incurred during the research project. Direct expenses may include the salary of technical associates; the purchase of relevant laboratory supplies; direct expenses necessary for proper conduct of research (e.g. subject-related costs, study-related personnel costs, diagnostic fees/services, data management expenses). Indirect costs may include travel expenses to scientific meetings, software licensing fees, publication costs, etc. Research proposals will be evaluated and selected by an independent, external expert panel comprised of nationally known academic clinicians. Applicants must submit their proposals through the Integrated System for Pfizer Investigator Initiated Research (INSPIIRE) web portal, the link to which is available from the competitive grant announcement on the ASPIRE website (http://www.aspireresearch.org/). Applications must be received by 11:59 p.m. eastern daylight time on the deadline date. Deadlines may vary from year to year. Interested applicants are advised to verify the current deadline date at the website or by contacting Pfizer. Pfizer, Inc. is an American multinational research-based pharmaceutical company that was founded in 1849 by two cousins Charles Pfizer and Charles Erhart. It topped the 2009 list of America's most generous companies, giving $2.3 billion in products and cash. Pfizers grants and contributions include the following types of support: support for medical, scientific, and patient organizations; lobbying and political contributions; medical education grants; medical and academic partnerships; healthcare charitables; sponsorships; and special events. Particular to the grant opportunity listed on this page, Pfizer offers a competitive research grants program ASPIRE (Advancing Science Through Pfizer-Investigator Research Exchange). ASPIRE grant opportunities are available by geographic area, specifically: worldwide, North America, Europe, Asia, and Australia. Notices of opportunities are posted on Pfizer's competitive grants and regional ASPIRE websites. Prospective applicants may fill out a web-based form at the ASPIRE website to by notified of future competitive grant opportunities via email.

Requirements: Applicants must have a professional degree (M.D., Ph.D., Pharm.D., or equivalent) and reside in the United States. Applicants should consider the following components when developing their research proposals: in vitro, in vivo, and clinical investigations; evaluation of disease pathogenesis and host response; and comparison of host response and clinical impact. Proposals will be evaluated on the following criteria: scientific merit of the research proposal; qualifications of the applicant; relevance of the proposed research to Pfizer's hematology program mission; evidence of the applicant's commitment to an academic research career; and evidence of a suitable research environment.

Restrictions: ASPIRE Hemophilia award winners may apply for awards in the future; awards are based upon scientific merit commensurate with funding that is available. No other government, non-governmental, or industry-sponsored funding may cover the same work scope as the grant application to the ASPIRE program; however the proposal may be related to other funding as long as there is no direct overlap.

Geographic Focus: All States

Date(s) Application is Due: May 14

Amount of Grant: 100,000-200,000 USD

Contact: Mikael Dolssten, President; (212) 733-2323

Internet: https://www.aspireresearch.org/ASPIRE-Hemophilia/index.html

Sponsor: Pfizer
235 East 42nd Street
New York, NY 10017-5755

Pfizer ASPIRE North America Narrow Spectrum Antibiotics for the Treatment of MRSA Research Awards 3916

While considerable clinical trial data exists for linezolid in a variety of patients with known infections, most of this information is from controlled clinical trials which leave patients with certain disease complexities underrepresented. Pfizer invites investigators to submit innovative clinical research proposals that evaluate linezolid's role in the treatment of infections in complex adult patients within approved indications. Studies with well-designed retrospective analyses are preferred, however small pilot studies that attempt to test hypotheses will be considered as long as the study can be completed within 1 year. Funds may be used to cover the awardees' salary and fringe benefits, as well as other direct expenses incurred during the research project. Direct expenses may include the salary of technical associates; the purchase of relevant laboratory supplies; direct expenses necessary for proper conduct of research (e.g. subject-related costs, study-related personnel costs, diagnostic fees/services, and data management expenses). Indirect costs may include travel expenses to scientific meetings, software licensing fees, publication costs, etc. Downloadable application guidelines are available at the grant website. Applicants must submit their proposals through the Integrated System for Pfizer Investigator Initiated Research (INSPIIRE) web portal, the link to which is available from the competitive grant announcement on the ASPIRE website. Applications must be received by 11:59 p.m. eastern daylight time on the deadline date. Due dates may vary from year to year. Applicants are encouraged to verify current deadlines at the website or by contacting Pfizer. Pfizer, Inc. is an American multinational research-based pharmaceutical company that was founded in 1849 by two cousins Charles Pfizer and Charles Erhart. It topped the 2009 list of America's most generous companies, giving $2.3 billion in products and cash. Pfizer's grants and contributions include the following types of support: support for medical, scientific, and patient organizations; lobbying and political contributions; medical education grants; medical and academic partnerships; healthcare charitables; sponsorships; and special events. Particular to the grant opportunity listed on this page, Pfizer offers a competitive research grants program ASPIRE (Advancing Science Through Pfizer-Investigator Research Exchange). ASPIRE grant opportunities are available by geographic area, specifically: worldwide, North America, Europe, Asia, and Australia. Notices of opportunities are posted on Pfizer's competitive grants and regional ASPIRE websites. Prospective applicants may fill out a web-based form at the ASPIRE website to by notified of future competitive grant opportunities via email.

Requirements: Applicants must have a professional degree (M.D., Ph.D., Pharm.D., or equivalent) and reside in the United States. Applications submitted by junior investigators (within 5 years of terminal training) will be afforded preferential consideration. Applicants

must have a strong academic career interest in clinical evaluations of linezolid in the following focus areas and populations: the disease states of nosocomial pneumonia (e.g. HAP, VAP, HCAP), skin and soft tissue infection (e.g. surgical site infection, wound, and complicated cellulitis), and diabetic foot infection without osteomyelitis; patient populations with diabetes, renal insufficiency, pulmonary dysfunction (e.g. COPD), peripheral vascular disease, morbid obesity, and immunosuppression; older adults stratified by age (e.g. 60-70 years, 79-80 years, >80 years); adults stratified by age (e.g. 18-35 years, 36-55 years, 56-72 years, >72 years; the critically ill; pharmakinetics, host immune response; and rate of resolution of signs/symptoms of infection. Within the specified patient populations with infection types of interest, investigators are expected to generate exploratory clinical data to better understand the disease pathogenesis, risk factors, host response, and treatment outcomes. It is expected that results will be presented at scientific meetings and published in peer reviewed journals. ASPIRE applications will be reviewed by an independent, external review committee comprised of medical and scientific experts. Review committee members will consider the following criteria in evaluating proposals: scientific merit of the research proposal; qualifications of the applicant; relevance of the proposal to the ASPIRE program's mission; evidence of the applicant's commitment to an academic research career, and evidence of a suitable research environment.
Restrictions: No other government, non-governmental, or industry-sponsored funding may cover the same work scope as the grant application to the ASPIRE program; however the proposal may be related to other funding as long as there is no direct overlap. The following topics fall outside of the scope of the ASPIRE Antibacterial Research Program: general education and/or training; public health; support for ongoing clinical programs that are part of an organization's routine operations; and proposals that do not evaluate linezolid. Investigators may consider evaluating additional clinical measures (e.g. length of stay, readmission, ventilator days, etc.). However, proposals seeking to financially quantify outcomes are outside the scope of the program. Studies designed to address how patients / payers use evidence about Pfizer products to make coverage and reimbursement decisions are prohibited.
Geographic Focus: All States
Date(s) Application is Due: May 1
Amount of Grant: 50,000-100,000 USD
Contact: Mikael Dolssten, President; (212) 733-2323
Internet: https://www.aspireresearch.org/ASPIRE-AR_U.S./ASPIRE%20AR%20Website%20Tab%20Send/linezolid/index.html
Sponsor: Pfizer
235 East 42nd Street
New York, NY 10017-5755

Pfizer ASPIRE North America Rheumatology Research Awards **3917**
Pfizer's ASPIRE Rheumatology Research awards are designed to increase the understanding of health disparities as they relate to rheumatoid arthritis in Hispanic and other underrepresented minority populations. Specifically, proposals should address the following areas: epidemiology and current trends in diagnosis and management; clinical phenotypic differences; and genomic research into the heterogeneity of and within the targeted population. Pfizer will provide $100,000 to fund research conducted over a two-year period. Applicants must submit their proposals through the Integrated System for Pfizer Investigator Initiated Research (INSPIIRE) web portal, the link to which is available from the competitive grant announcement on the ASPIRE website (http://www.aspireresearch.org/). Applications must be received by 11:59 p.m. eastern daylight time on July 10. Pfizer, Inc. is an American multinational research-based pharmaceutical company that was founded in 1849 by two cousins Charles Pfizer and Charles Erhart. It topped the 2009 list of America's most generous companies, giving $2.3 billion in products and cash. Pfizer's grants and contributions include the following types of support: support for medical, scientific, and patient organizations; lobbying and political contributions; medical education grants; medical and academic partnerships; healthcare charitables; sponsorships; and special events. Particular to the grant opportunity listed on this page, Pfizer offers a competitive research grants program ASPIRE (Advancing Science Through Pfizer-Investigator Research Exchange). ASPIRE grant opportunities are available by geographic area, specifically: worldwide, North America, Europe, Asia, and Australia. Notices of opportunities are posted on Pfizer's competitive grants and regional ASPIRE websites. Prospective applicants may fill out a web-based form at the ASPIRE website to by notified of future competitive grant opportunities via email.
Requirements: Applicants must have a professional degree (M.D., Ph.D., Pharm.D., or equivalent) and reside in the United States. Investigators are expected to generate exploratory data to better understand the variance in clinical presentation, severity of illness, response to therapy, and predictors of efficacy and tolerability in the targeted population. ASPIRE review committee members will consider the following criteria in evaluating proposals: scientific merit of the research proposal; qualifications of the applicant; relevance of the proposal to the ASPIRE program's mission; evidence of the applicant's commitment to an academic research career, evidence of a suitable research environment; and evidence of a mentorship program.
Restrictions: ASPIRE Rheumatology Research awards are limited to one award per investigator per lifetime. No other government, non-governmental, or industry-sponsored funding may cover the same work scope as the grant application to the ASPIRE program; however the proposal may be related to other funding as long as there is no direct overlap.
Geographic Focus: All States
Date(s) Application is Due: Jul 10
Amount of Grant: Up to 100,000 USD
Contact: Mikael Dolsssten, President; (212) 733-2323
Internet: http://www.pfizer.com/research/investigator/competitive_grant_programs.jsp
Sponsor: Pfizer
235 East 42nd Street
New York, NY 10017-5755

Pfizer ASPIRE Worldwide Endocrine Young Investigator Grants **3918**
The mission of the ASPIRE Young Investigator Awards in Endocrine Research is to advance the medical knowledge of the fundamental mechanisms of endocrine disease and the mechanisms of action of current and potential future treatments to improve the care of patients with endocrine disorders. Pfizer invites basic-science, innovative-translational, and clinical-research proposals that evaluate endocrine disease, disease mechanisms, and outcomes, primarily in the growth-hormone insulin-like growth-factor field. Funding will be provided to research proposals that address understanding disease mechanisms, genetic factors involved in disease development, factors affecting response to therapeutic agents, identification of novel biomarkers, and identification of novel therapeutic targets or therapeutic approaches. Pfizer will fund up to three awards of $50,000 each for one year. Applicants must submit their proposals through the Integrated System for Pfizer Investigator Initiated Research (INSPIIRE) web portal, the link to which is available from the competitive grant announcement on the ASPIRE website (http://www.aspireresearch.org/). Applications must be received by 11:59 p.m. eastern daylight time on June 15. Interested applicants can download an application guide from this location for more detailed information about application requirements. Pfizer, Inc. is an American multinational research-based pharmaceutical company that was founded in 1849 by two cousins Charles Pfizer and Charles Erhart. It topped the 2009 list of America's most generous companies, giving $2.3 billion in products and cash. Pfizer's grants and contributions include the following types of support: support for medical, scientific, and patient organizations; lobbying and political contributions; medical education grants; medical and academic partnerships; healthcare charitables; sponsorships; and special events. Particular to the grant opportunity listed on this page, Pfizer offers a competitive research grants program ASPIRE (Advancing Science Through Pfizer-Investigator Research Exchange). ASPIRE grant opportunities are available by geographic area, specifically: worldwide, North America, Europe, Asia, and Australia. Notices of opportunities are posted on Pfizer's competitive grants and regional ASPIRE websites. Prospective applicants may fill out a web-based form at the ASPIRE website to by notified of future competitive grant opportunities via email.
Requirements: Applicants must hold a professional degree (M.D., Ph.D., Pharm.D., or equivalent), be qualified to carry out the proposed investigation, and still be in the early stages of their career (preferably within ten years of having completed a post-graduate degree). ASPIRE review committee members will consider the following criteria in evaluating proposals: scientific merit of the research proposal; qualifications of the applicant; relevance of the proposal to the ASPIRE program's mission; evidence of the applicant's commitment to an academic research career, evidence of a suitable research environment; and evidence of suitable mentors.
Restrictions: No other government, non-governmental, or industry-sponsored funding may cover the same work scope as the grant application to the ASPIRE program; however the proposal may be related to other funding as long as there is no direct overlap. Due to the Foreign Corrupt Practices Act (FCPA), submissions selected for awards must go through an internal review at Pfizer to ensure provision of the grant is appropriate and that relevant information regarding potentially influencing government official status, as well as beneficiaries and controllers information, is captured, when applicable.
Geographic Focus: All States, All Countries
Date(s) Application is Due: Jun 15
Amount of Grant: 50,000 USD
Contact: Mikael Dolsssten, President; (212) 733-2323
Internet: http://www.pfizer.com/research/investigator/competitive_grant_programs.jsp
Sponsor: Pfizer
235 East 42nd Street
New York, NY 10017-5755

Pfizer Australia Cancer Research Grants **3919**
Pfizer Australia Cancer Research Grants were established in 2007 to support and advance clinical research in the field of oncology. Grants are available up to AUD $55,000 (including the Australian Goods and Services Tax) and can be used to cover part salaries, equipment, research materials, and expenses. Proposals should involve clinical or translational research. Translation applications are required to describe how the outcomes from the proposed research project will potentially impact clinical practice, policy, or further research in cancer control. Applications must be submitted via an online application form which can be accessed at the website from January to May. Submission deadlines may vary from year to year. Interested applicants are encouraged to verify current due dates at the website or by contacting Pfizer Australia. All applications are reviewed by two expert referees and Pfizer (for proposals involving Pfizer compounds); final decisions are then made by the Australian Physicians Independent Committee (APIC), composed of eight leading specialists. Pfizer Australia awards up to $1.5 million in grants to support medical graduates conducting research in the areas of neuroscience, oncology, paediatric endocrinology, and related areas. Grants are awarded for 12 months and assist researchers to establish a basis for further research funding from granting bodies such as the National Health and Medical Research Council.
Requirements: Eligible applicants will have entered the field of research, (or returned after an appropriate break) within the last five years of having attained post-graduate research or clinical qualifications (e.g., Ph.D./M.D. or College Fellowship). The principal applicant must be a citizen or permanent resident of Australia. Grant recipients must conduct the majority of the research within Australia. Pfizer Australia Cancer Research Grants are awarded with the expectation that timely submissions for presentation will be made at appropriate medical meetings or research forums and that papers will be submitted for publication in appropriate peer-reviewed medical journals. The grant recipient will be required to provide Pfizer with an opportunity to review any proposed publications.

Restrictions: Previous Cancer Research Grant recipients are ineligible. Only one Pfizer Australia Cancer Research Grant will be allocated to a single applicant in a given year although applications will be accepted and reviewed for more than one project. Applicants are not eligible to receive funding from more than one Pfizer Australia Research Grant program in the same financial year.
Geographic Focus: Australia
Date(s) Application is Due: May 11
Amount of Grant: Up to 55,000 AUD
Contact: Trudy Snape, Grants Coordinator; +61 2 9850 3973
Internet: https://www.pfizergrants.com.au/Grants/CancerResearch.aspx
Sponsor: Pfizer
235 East 42nd Street
New York, NY 10017-5755

Pfizer Australia Neuroscience Research Grants **3920**
The Pfizer Australia Neuroscience Research Grants were established in 2002 to support and advance basic and clinical research and to encourage young researchers to enter or continue in the field of neuroscience. Grants are available up to AUD $44,000 (including the Goods and Services Tax) and can be used to cover part salaries, equipment, research materials, and expenses. Applications must be submitted via an online application form which can be accessed on the website between January 1 and May 11. Submission deadlines may vary from year to year; interested applicants are encouraged to verify current due dates by visiting the website or by contacting Pfizer Australia. Applications are reviewed by two expert referees and by Pfizer (for proposals involving Pfizer compounds); final decisions are then made by the Australian Physicians Independent Committee (APIC), composed of eight leading specialists. Pfizer Australia awards up to $1.5 million in grants to support medical graduates conducting research in the areas of neuroscience, oncology, paediatric endocrinology, and related areas. Grants are awarded for 12 months and assist researchers to establish a basis for further research funding from granting bodies such as the National Health and Medical Research Council.
Requirements: The interested applicant must meet the following eligibility criteria: be an Australian citizen or permanent resident; have entered or returned to the field of research within the last five years; and be a medical graduate who has obtained or is in process of obtaining specialist qualifications or a doctorate or who is currently in advanced training. Applications are welcome for research into any aspect of cognitive decline, mood disorders or pain. Proposals should involve clinical research or basic research involving humans. The majority of the research must be conducted within Australia. Pfizer grants are awarded with the expectation that timely submissions for presentation will be made at appropriate medical meetings or research forums and that papers will be submitted for publication in appropriate peer-reviewed medical journals. The grant recipient will be required to provide Pfizer with an opportunity to review any proposed publications.
Restrictions: Applications will not be accepted from established researchers including professors, associate professors, or CIAs with any competitive grant of $50,000 or above, or in receipt of a fellowship of $100,000 or more. Previous successful applicants are not eligible. Only one Neuroscience Research Grant will be allocated to a single applicant in a given year although applications will be accepted and reviewed for more than one project. Applicants will not be eligible to receive funding from more than one Pfizer Australia Research Program in the same financial year.
Geographic Focus: Australia
Date(s) Application is Due: May 11
Amount of Grant: Up to 44,000 AUD
Contact: Trudy Snape, Grants Coordinator; +61 2 9850 3973; trudy.snape@pfizer.com
Internet: https://www.pfizergrants.com.au/Grants/NeuroscienceResearch.aspx
Sponsor: Pfizer
235 East 42nd Street
New York, NY 10017-5755

Pfizer Australia Paediatric Endocrine Care Research Grants **3921**
The Australia Paediatric Endocrine Care (APEC) research grants were established in 2007 to support and advance clinical research in the field of Paediatric Endocrinology Care and to encourage young researchers to enter or continue in the field. Grants are available up to AUD $55,000 (including the Australian Goods and Services Tax) and can be used to cover part salaries, equipment, research materials, and expenses. Applications must be submitted via an online application form which can be accessed on the website from January 1-March 23. Submission deadlines may vary from year to year. Interested applicants are encouraged to verify current due dates by visiting the website or by contacting Pfizer Australia. Applications are reviewed by two expert referees and by Pfizer (for proposals involving Pfizer compounds); final decisions are then made by the Australian Physicians Independent Committee (APIC), composed of eight leading specialists. Pfizer Australia awards up to $1.5 million in grants to support medical graduates conducting research in the areas of neuroscience, oncology, paediatric endocrinology, and related areas. Grants are awarded for 12 months and assist researchers to establish a basis for further research funding from granting bodies such as the National Health and Medical Research Council.
Requirements: The interested applicant must meet the following criteria to be eligible to apply: be a medical graduate currently residing, working, and conducting research within Australasia; be registered to practice medicine in Australasia; be under the age of 40 at time of application; and be a current trainee in Paediatric Endocrinology in Australia or within five years of conferring of FRACP (Fellow of the Royal Australasian College of Physicians) or within five years of accreditation of Paediatric Endocrinology training from the Endocrinology SAC (Specialist Advisory Committee) of the Royal Australian College of Physicians. Pfizer grants are awarded with the expectation that timely submissions for presentation will be made at appropriate medical meetings or research forums and that papers will be submitted for publication in appropriate peer-reviewed medical journals. The grant recipient will be required to provide Pfizer with an opportunity to review any proposed publications.
Restrictions: Awardees must remain in Australasia for the duration of the study.
Geographic Focus: All Countries
Date(s) Application is Due: Mar 23
Amount of Grant: Up to 55,000 AUD
Contact: Trudy Snape, Grants Coordinator; +61 2 9850 3973
Internet: https://www.pfizergrants.com.au/grants/PaediatricEndocrine.aspx
Sponsor: Pfizer
235 East 42nd Street
New York, NY 10017-5755

PFizer Compound Transfer Agreements **3922**
Pfizer has a pure-powder-transfer program encompassing all of the company's publicly-known compounds and routinely enters Compound Transfer Agreements (CTAs) that provide compounds free of charge to third parties to conduct agreed-upon studies. To request a Pfizer pure-substance compound, applicants should complete the submission form located on the Integrated System for Pfizer Investigator Initiated Research (INSPIIRE) web portal, the link to which is available on the Compound-Transfer-Program web page. Processing time from initial submission to shipment of the approved compound(s) takes approximately 12 weeks. Applicants will receive an auto-reply email confirming their submission has been completed and can track the status of their submissions on the INSPIIRE portal. Pfizer, Inc. is an American multinational research-based pharmaceutical company that was founded in 1849 by two cousins Charles Pfizer and Charles Erhart. It topped the 2009 list of America's most generous companies, giving $2.3 billion in products and cash. Most of Pfizer's philanthropic activities are managed by its Corporate Responsibility Department through two foundations, the Pfizer Foundation and the Pfizer Patient Assistance Foundation. Pfizer's philanthropic activities are generally divided among its global health programs, its various grants and contributions programs, and its work with health care professionals. Pfizer's grants and contributions include the following types of support: support for medical, scientific, and patient organizations; lobbying and political contributions; medical education grants; medical and academic partnerships; healthcare charitables; sponsorships; and special events.
Requirements: An attachment with a full description of the specific research study proposed should be included with the applicant's on-line submission. Pfizer reviews each request for content, value, and consistency with Pfizer's direction for the requested compound and reserves the right to deny any request the reviewers feel does not advance the compound(s) or add value to human or animal health. Pfizer requires a manuscript or report of the conclusions to bring closure to the study.
Restrictions: Studies involving humans or requesting funding support should be submitted through Pfizer's Investigator Initiated Research program.
Geographic Focus: All States, All Countries
Contact: Mikael Dolssten, President; (212) 733-2323
Pfizer Country Medical Representative or Regional Research Specialist; (212) 733-2323; iir@pfizer.com
Internet: http://www.pfizer.com/research/rd_works/compound_transfer_program.jsp
Sponsor: Pfizer
235 East 42nd Street
New York, NY 10017-5755

Pfizer Foundation Epilepsy Scholarship Award **3923**
The scholarship honors 16 outstanding students who have overcome the challenges of epilepsy, been successful in school, done well in activities out of the classroom, and shown the ability to make the most out of college or graduate school. Experts in education and medicine will choose the winners. Information and application are available online.
Requirements: An applicant must be under a doctor's care for epilepsy; a high school senior who has applied to college; a freshman, sophomore, or junior in college; or a college senior who has applied to graduate school.
Geographic Focus: All States
Date(s) Application is Due: Mar 1
Amount of Grant: 3,000 USD
Contact: Scholarship Coordinator; (800) 292-7373; Scholarship@edencomgroup.com
Internet: http://www.epilepsy-scholarship.com
Sponsor: Eden Communications Group
515 Valley Street, Suite 200
Maplewood, NJ 07040

Pfizer Global Investigator-Initiated Research Grants **3924**
The mission and purpose of Pfizer's Investigator-Initiated Research (IIR) program is to provide support for investigator-initiated research that advances medical and scientific knowledge about Pfizer products and that generates promising medical interventions. This global program is open to all researchers who are interested in conducting their own research. Pfizer support is typically provided in the form of funding and/or drugs (either formulated or pure compound). Pfizer accepts concept submissions for IIR grant requests. If a concept submission is of interest, Pfizer will issue the applicant a follow-up request for a full submission. Requests are submitted to Pfizer through the Integrated System for Pfizer Investigator Initiated Research (INSPIIRE) web portal, the link to which is available on the grant web page. Prior to submitting a grant request, applicants should contact a Pfizer Regional Medical Research Specialist (RMRS), a Pfizer Country Medical Representative, or other Pfizer Medical personnel for more information. Applicants may use the email address provided in the contact section of this page to request the appropriate Pfizer contact information. Interested applicants can download a program brochure from the grant web page to obtain detailed information about application requirements. Pfizer,

Inc. is an American multinational research-based pharmaceutical company that was founded in 1849 by two cousins Charles Pfizer and Charles Erhart. It topped the 2009 list of America's most generous companies, giving $2.3 billion in products and cash. Most of Pfizer's philanthropic activities are managed by its Corporate Responsibility Department through two foundations, the Pfizer Foundation and the Pfizer Patient Assistance Foundation. Pfizer's philanthropic activities are generally divided among its global health programs, its various grants and contributions programs, and its work with health care professionals. Pfizer's grants and contributions include the following types of support: support for medical, scientific, and patient organizations; lobbying and political contributions; medical education grants; medical and academic partnerships; healthcare charitables; sponsorships; and special events.

Requirements: Following are types of research eligible for support: clinical studies of approved and unapproved uses involving approved or unapproved Pfizer drugs; observational studies, such as epidemiology studies and certain outcomes research studies where the primary focus is the scientific understanding of disease; other types of independent research on disease states, including novel diagnostic screening tools and surveys where Pfizer has no direct commercial interest; and in vitro or animal studies which include a request for funding. Applicants should note that Pfizer requires the following documents before it can initiate support (either drug and/or funding) of an IIR request: internal review board (IRB)/ ethics committee approval; regulatory response documentation-Investigational New Drug (IND) or Compound Transfer Agreement (CTA) documentation if applicable, a fully executed IIR agreement, and a final study protocol.

Restrictions: Pfizer will not support the following types of requests: requests for ongoing or new research without an associated study protocol or synopsis; general education and training activities; support for ongoing clinical programs that are part of an organization's routine operations; start-up funds to establish new clinical or research programs or to expand existing programs; purchases of capital equipment unrelated to the study or that would generate revenue; construction funds to build new facilities; and hiring of staff that are not dedicated to the study.

Geographic Focus: All States, All Countries

Samples: Interested applicants may download alphabetical listings of grant recipients per year from Pfizer's Transparency In Grants webpage at http://www.pfizer.com.

Contact: Mikael Dolssten, President; (212) 733-2323

Internet: http://www.pfizer.com/research/investigator/investigator_initiated_research.jsp

Sponsor: Pfizer

235 East 42nd Street

New York, NY 10017-5755

Pfizer Global Research Awards for Nicotine Independence **3925**

The mission of Pfizer's Global Research Awards for Nicotene Independence (GRAND) Program is to advance the understanding of the mechanisms of tobacco and nicotine dependence and its treatment. Projects should yield results that merit first submission as abstracts to scientific meetings and, subsequently, publication in peer-reviewed journals. Pfizer intends to fund at least five grants in amounts between $50,000 and $200,000. Funds from the GRAND program may be used to support the applicant's salary and fringe benefits, technical salaries, and supplies. Applicants must submit their proposals through the Integrated System for Pfizer Investigator Initiated Research (INSPIIRE) web portal, the link to which is available from the competitive-grant-announcement web page. Applications must be received by 11:59 p.m. eastern daylight time on June 15. Interested applicants can also download an application guide from the announcement web page for more detailed information about application requirements. Pfizer, Inc. is an American multinational research-based pharmaceutical company that was founded in 1849 by two cousins Charles Pfizer and Charles Erhart. It topped the 2009 list of America's most generous companies, giving $2.3 billion in products and cash. Pfizer's grants and contributions include the following types of support: support for medical, scientific, and patient organizations; research; lobbying and political contributions; medical education grants; medical and academic partnerships; healthcare charitables; sponsorships; and special events.

Requirements: Applicants holding an M.D., a Ph.D., or equivalent degree are eligible to apply. Research projects should aim to provide information that could directly improve the use of pharmacotherapy in clinical practice. The GRAND review committee will consider the following criteria in evaluating proposals: scientific merit of the research proposal; qualifications of the applicant; relevance of the proposed research to the advancement of the treatment of nicotine dependence; evidence of a suitable research environment; and feasibility of the project. Prior to submitting an application via the INSPIIRE portal, applicants should notify the GRAND coordinator via the current email address given in the guidelines (at this writing, applications@grand2012.org) so that the GRAND coordinator can expedite and track the processing of the application.

Restrictions: The following topics fall outside of the scope of the GRAND Program in the current year: animal studies; genetics; epidemiology (unrelated to pharmacotherapy, e.g. harm and prevalence of dependence); tobacco policy; public health; community interventions; economic and social factors; and educational programs. No other government, non-governmental, or industry-sponsored funding may cover the same work scope as the grant application to the GRAND program; however the proposal may be related to other funding as long as there is no direct overlap. If the conduct of the research results in any invention or discovery by the awardee that relates to a Pfizer product, the awardee will grant to Pfizer a perpetual, royalty-free worldwide, non-exclusive license to each such invention.

Geographic Focus: All States, All Countries

Date(s) Application is Due: Jun 15

Amount of Grant: 50,000-200,000 USD

Contact: Mikael Dolssten, President; (212) 733-2323

Internet: http://www.pfizer.com/research/investigator/competitive_grant_programs.jsp

Sponsor: Pfizer

235 East 42nd Street

New York, NY 10017-5755

Pfizer Inflammation Competitive Research Awards (UK) **3926**

The Pfizer Inflammation Competitive Research Programme (I-CRP) seeks innovative research proposals in the field of inflammation which may further the understanding of certain rheumatology and dermatology conditions and contribute to excellence in patient care. Pfizer anticipates making non-renewable awards of approximately £50,000-£100,000 each for research to be undertaken within a time period of 12-18 months. Prior to submitting an application, applicants are requested to have a "medical to medical conversation" with their Pfizer representative. Applicants can email iir@pfizer.com to obtain the appropriate Pfizer contact information for their region. Downloadable application guidelines are available at the grant website. Applications must be submitted to Pfizer through the company's global investigator-initiated research (INSPIIRE) web portal. Applicants should click on "Submit an IIR Request" at that site and follow the online instructions. First-time applicants must "Create an Account". Applications must be received by the deadline date. (Due dates may vary from year to year; applicants are encouraged to verify current due dates at the website or by contacting Pfizer.) Pfizer, Inc. is an American multinational research-based pharmaceutical company that was founded in 1849 by two cousins Charles Pfizer and Charles Erhart. It topped the 2009 list of America's most generous companies, giving $2.3 billion in products and cash. Pfizer's grants and contributions include the following types of support: support for medical, scientific, and patient organizations; research; lobbying and political contributions; medical education grants; medical and academic partnerships; healthcare charitables; sponsorships; and special events.

Requirements: Clinical practitioners and scientists working in the United Kingdom who hold a medical or pharmacy degree or a Ph.D., do not directly work for Pfizer or any other pharmaceutical company, and are not members of the current I-CRP Grant Awarding Committee are eligible to apply. Application should focus on translational (clinically-oriented) research that will impact patient outcomes and relate to an area aligned to Pfizer's areas of therapeutic focus: Rheumatoid Arthritis, Psoriatic Arthritis, Anklylosing Spondylitis, Psoriasis, or Juvenile Idiopathic Arthritis. (Proposals outside these high-interest areas will also be considered but only supported if deemed of exceptional scientific merit.) Studies are not necessarily expected to involve treatment with a Pfizer medication. It is expected that results will be presented at scientific meetings and published in peer-reviewed journals. All applications will be formally reviewed by an independent committee of European medical experts in rheumatology and dermatology. Projects that are awarded funding through I-CRP may also be eligible for National Institute for Health Research (NIHR)'s Clincial Research Network (CRN) support.

Restrictions: Pfizer reserves the right not to support projects that fall outside of Pfizer's internal compliance policies and procedures or those that replicate existing research. Pfizer is unable to fund direct comparative-treatment studies primarily involving competitor products, studies where the risk/benefit ratio to patients is not obviously positive, or studies that involve use of compounds in development. Due to the competitive nature of the awards, Pfizer cannot provide any additional funding and/or drug support beyond what has been requested and approved by the external review panel. Current Grant Awarding Committee members cannot be an applicant or co-applicant on a proposal submitted for the programme, nor be an author on any publication ensuing from the award. Committee members will not participate in the review of any proposal that poses a conflict of interest. Proposals received from current Grant Awarding Committee members' institutions will still be allowed but the scoring and review decisions of these proposals will be made by the committee excluding the potentially conflicted member.

Geographic Focus: United Kingdom

Date(s) Application is Due: Apr 12

Amount of Grant: 50,000-100,000 GBP

Contact: Mikael Dolssten, President; (212) 733-2323

Coordinator; (212) 733-2323; ICRP@pfizer.com or iir@pfizer.com

Internet: http://www.pfizer.com/research/investigator/competitive_grant_programs.jsp

Sponsor: Pfizer

235 East 42nd Street

New York, NY 10017-5755

Pfizer Medical Education Track Two Grants **3927**

The mission of the Pfizer medical-education-grants program is to cooperate with health-care delivery organizations and professional associations to narrow professional practice gaps through support of continuing education, learning, and continuous-improvement strategies that result in measurable improvement in competence, performance, or patient outcomes. Applicants may apply for two types of medical-education grants: healthcare quality improvement (track one) grants; and annual meetings for purposes of emerging science/knowledge exchange (track two) grants. The purpose of track-two grants is to support the live presentation of original research and the exchange of emerging clinical information. Eligible providers can request funding to support live (fact-to-face) activities at major national and regional conferences and congresses in the following clinical areas: amyloidosis (familial amyloid polyneuropathy); cardiovascular metabolic risk (lipids, diabetes); chronic obstructive pulmonary disease; cough and cold; diabetic peripheral neuropathy; dietary supplements; fibromyalgia; Gaucher disease; growth disorders; hemophilia; infectious disease; multiple sclerosis; oncology (breast cancer, gastrointestinal stromal tumors, leukemia, lung cancer, non-Hodgkin's lymphoma, pancreatic neuroendocrine tumors, and renal cell carcinoma); pain; pulmonary arterial hypertension; rheumatoid arthritis; smoking cessation; transplantation (kidney); urology (overactive bladder); women's health (menopause); and support for the Continuing Medical Education (CME)/Continuing Education (CE)/Continuing Professional Development (CPD) profession. Track two grants have five submission cycles annually: January 1 through February 15 for activities starting in April; March 1 through April 15 for activities starting in June; May 1 through June 15 for activities starting in August; July

1 through August 15 for activities starting in October; and September 1 through October 15 for activities starting in December. Regional meeting requests are capped at $25,000 and national meeting requests have a $50,000 cap. Application must be made through the Medical Education Group's online Grant Management System (GMS) which is also available at the website. Applicants must first register themselves in the GMS to create an account. Pfizer, Inc. is an American multinational research-based pharmaceutical company that was founded in 1849 by two cousins Charles Pfizer and Charles Erhart. It topped the 2009 list of America's most generous companies, giving $2.3 billion in products and cash. Most of Pfizer's philanthropic activities are managed by its Corporate Responsibility Department through two foundations, the Pfizer Foundation and the Pfizer Patient Assistance Foundation. Pfizer's philanthropic activities are generally divided among its global health programs, its various grants and contributions programs, and its work with health care professionals. Pfizer's grants and contributions include the following types of support: support for medical, scientific, and patient organizations; lobbying and political contributions; medical education grants; medical and academic partnerships; healthcare charitables; sponsorships; and special events.
Requirements: Eligibility is based on the following criteria: whether or not the activity aligns with Pfizer's posted areas of interest (see Description); whether the activity is a live (face-to-face) annual activity which serves as a platform for the exchange of new clinical and scientific information and reaches a national or regional audience; whether the activity was developed by the applicant organization specifically for its professional members or constituents; whether the activity is an established part of the applicant organization's ongoing educational program; and whether the activity (if new) is clearly based on an assessment of the educational needs of the applicant's target audience.
Restrictions: Only one request per annual meeting will be accepted.
Geographic Focus: All States, All Countries
Date(s) Application is Due: Feb 15; Apr 15; Jun 15; Aug 15; Oct 15
Amount of Grant: 25,000-50,000 USD
Contact: Caroline Roan; (866) 634-4647 or (212) 209-8997; healthcharitables@pfizer.com
Internet: www.pfizermededgrants.com
Sponsor: Pfizer
235 East 42nd Street
New York, NY 10017-5755

Pfizer Research Initiative Dermatology Grants (Germany) 3928
Pfizer GmbH seeks proposals to study the effectiveness and safety of treating dermatologic diseases with TNF-blockade. Downloadable application guidelines (in German) are available from the grant web page. Interested applicants are encouraged to contact the Medical Science Liason for more information. Pfizer, Inc. is an American multinational research-based pharmaceutical company that was founded in 1849 by two cousins Charles Pfizer and Charles Erhart. It topped the 2009 list of America's most generous companies, giving $2.3 billion in products and cash. Pfizer's grants and contributions include the following types of support: support for medical, scientific, and patient organizations; research; lobbying and political contributions; medical education grants; medical and academic partnerships; healthcare charitables; sponsorships; and special events. Pfizer, Inc. is an American multinational research-based pharmaceutical company that was founded in 1849 by two cousins Charles Pfizer and Charles Erhart. It topped the 2009 list of America's most generous companies, giving $2.3 billion in products and cash. Pfizer's grants and contributions include the following types of support: support for medical, scientific, and patient organizations; research; lobbying and political contributions; medical education grants; medical and academic partnerships; healthcare charitables; sponsorships; and special events.
Geographic Focus: Germany
Date(s) Application is Due: Aug 1
Amount of Grant: Up to 60,000 EUR
Contact: Dr. Ekkehard Lange; +49 (0) 30 550055-52975; ekkehard.lange@pfizer.com
Internet: http://www.pfizer.com/research/investigator/competitive_grant_programs.jsp
Sponsor: Pfizer
235 East 42nd Street
New York, NY 10017-5755

Pfizer Research Initiative Rheumatology Grants (Germany) 3929
Pfizer GmbH seeks proposals to study significant or pathophysiological issues in treating rheumatic diseases with TNF-alpha blockade. Downloadable application guidelines (in German) are available from the grant web page. Interested applicants are encouraged to contact the Medical Science Liason for more information. Pfizer, Inc. is an American multinational research-based pharmaceutical company that was founded in 1849 by two cousins Charles Pfizer and Charles Erhart. It topped the 2009 list of America's most generous companies, giving $2.3 billion in products and cash. Pfizer's grants and contributions include the following types of support: support for medical, scientific, and patient organizations; research; lobbying and political contributions; medical education grants; medical and academic partnerships; healthcare charitables; sponsorships; and special events. Pfizer, Inc. is an American multinational research-based pharmaceutical company that was founded in 1849 by two cousins Charles Pfizer and Charles Erhart. It topped the 2009 list of America's most generous companies, giving $2.3 billion in products and cash. Pfizer's grants and contributions include the following types of support: support for medical, scientific, and patient organizations; research; lobbying and political contributions; medical education grants; medical and academic partnerships; healthcare charitables; sponsorships; and special events.
Geographic Focus: Germany
Date(s) Application is Due: Aug 1
Amount of Grant: Up to 60,000 EUR
Contact: Dr. Ekkehard Lange; +49 (0) 30 550055-52975; ekkehard.lange@pfizer.com
Internet: http://www.pfizer.com/research/investigator/competitive_grant_programs.jsp
Sponsor: Pfizer
235 East 42nd Street
New York, NY 10017-5755

Pharmacia-ASPET Award for Experimental Therapeutics 3930
The Pharmacia-ASPET Award in Experimental Therapeutics is given annually to recognize and stimulate outstanding research in pharmacology and experimental therapeutics-basic laboratory or clinical research that has had, or potentially will have, a major impact on the pharmacological treatment of disease. The winner will receive a $2,500 honorarium, a plaque, hotel and economy airfare for the winner and spouse to the award ceremony at the ASPET annual meeting.
Requirements: Nominations must be made by ASPET members; however, the nominee need not be a member. No restrictions exist as to age or institutional affiliation of nominee.
Geographic Focus: All States
Date(s) Application is Due: Sep 15
Amount of Grant: 2,500 USD
Contact: Christine Carrico, Executive Officer; (301) 634-7060; fax (301) 634-7061; ccarrico@aspet.org or info@aspet.org
Internet: http://www.aspet.org/public/awards/exp_ther_award.html
Sponsor: Pfizer / American Society for Pharmacology and Experimental Therapeutics
9650 Rockville Pike
Bethesda, MD 20814-3995

PHF Andrew W. Mellon Postdoctoral Fellowships in the Humanities 3931
Five one-year residential postdoctoral fellowships are available for untenured junior scholars who are no more than eight years out of their doctorate. The programs of the Penn Humanities Forum are conceived through yearly topics that invite broad interdisciplinary collaboration. Research proposals are invited on the annual topic in all areas of humanistic study, except educational curriculum-building and the performing arts. Candidates from all humanistic disciplines are eligible, as well as those in allied areas such as Anthropology and History of Science. Fellows teach one undergraduate course each of two terms in addition to conducting their research. The fellowship is open to all scholars, national and international. Guidelines and application are available online.
Requirements: The program is open to scholars who are not yet tenured (may not be tenured during the fellowship year) and who are no more than eight years out of their doctorate. The PhD is the only terminal degree eligible.
Geographic Focus: All States
Date(s) Application is Due: Oct 15
Amount of Grant: 42,000 USD
Contact: Jennifer Conway, (215) 898-8220; conwayj@sas.upenn.edu
Internet: http://humanities.sas.upenn.edu/mellon_description.shtml
Sponsor: Penn Humanities Forum, University of Pennsylvania
3619 Locust Walk
Philadelphia, PA 19104-6213

PHF Penn Faculty Research Fellowships 3932
Applications are solicited from standing faculty members in the humanities and allied departments at Penn for the Penn Humanities Forum's Mellon Faculty Research Fellowships. Available are junior faculty awards of $5000 disbursed as salary, as well as senior faculty awards of $10,000 paid directly to departments to offset the cost of a course release. Fellows must be engaged in research directly related to the Forum's theme for the year in which the award is granted, and are expected to participate in the Forum's weekly Mellon Research Seminars.
Requirements: Applicants must be Penn State University faculty in the humanities or allied departments.
Geographic Focus: All States
Date(s) Application is Due: Feb 2
Amount of Grant: 15,000 USD
Contact: Jennifer Conway, (215) 898-8220; conwayj@sas.upenn.edu
Internet: http://humanities.sas.upenn.edu/faculty_description.shtml
Sponsor: Penn Humanities Forum, University of Pennsylvania
3619 Locust Walk
Philadelphia, PA 19104-6213

PHF Regional Mellon Faculty Research and Cultural Fellowships 3933
The PHF research stipends of $3,000 are intended for faculty in the humanities from Philadelphia-area colleges and universities and for department directors and curators of Philadelphia-area arts and cultural organizations. The stipends are designed to encourage interdisciplinary exchange among the region's academies and cultural institutions. Fellows must be engaged in research directly related to the Forum's theme for the year in which the award is granted, and are expected to participate in the Forum's weekly Mellon Research Seminars.
Requirements: Applicants must be faculty members of Philadelphia-area colleges or universities, or department directors and curators of Philadelphia-area arts and cultural organizations.
Geographic Focus: Pennsylvania
Date(s) Application is Due: Feb 2
Amount of Grant: 3,000 USD
Contact: Jennifer Conway, (215) 898-8220; conwayj@sas.upenn.edu
Internet: http://humanities.sas.upenn.edu/regional_description.shtml
Sponsor: Penn Humanities Forum, University of Pennsylvania
3619 Locust Walk
Philadelphia, PA 19104-6213

PHF Undergraduate Humanities Forum Research Fellowships 3934

Each year, a number of Fellowships are offered to promote undergraduate research in the humanities and cultivate the importance of humanistic thought across all disciplines. Interested candidates should be conducting work related to the Forum's theme for the year in which the award is granted. Calls for applications are solicited in February of the year preceding the fellowship. One $1500 PHF Coordinating Research Fellowship and approximately eighteen PHF Research Fellowships at $600 each are available each year. Applicants are drawn from students across Penn, as well as from the university's twelve undergraduate College Houses.
Geographic Focus: All States
Date(s) Application is Due: Mar 15
Amount of Grant: 600-1,500 USD
Contact: Jennifer Conway, (215) 898-8220; conwayj@sas.upenn.edu
Internet: http://humanities.sas.upenn.edu/undergraduate_description.shtml
Sponsor: Penn Humanities Forum, University of Pennsylvania
3619 Locust Walk
Philadelphia, PA 19104-6213

Phi Beta Kappa Award in Science 3935

The award is offered for outstanding contributions by scientists to the literature of science. The intent of the award is to encourage literate and scholarly interpretations of the physical and biological sciences and mathematics. Entries must be submitted by the publisher who should first obtain a copy of conditions of eligibility from Phi Beta Kappa. By March 31, applicants should ship one copy of the book to the appropriate awards committee. By April 10, if the submission is eligible, applicants will receive the shipping list for judges, and should ship one book to each judge by April 30.
Requirements: To be eligible, biographies of scientists must have a substantial critical emphasis on their scientific research. Entries must be original works published in the United States during the 12-month period between May 1 (of the past year) and the April 30 (of the current year) deadline date. Entries ordinarily will be the work of a single writer. Exceptions may be made for books written by small teams of scholars working in close collaboration.
Restrictions: Entries should not be of a technical or highly specialized character. Monographs and compendiums are not eligible.
Geographic Focus: All States
Date(s) Application is Due: Apr 30
Amount of Grant: 2,500 USD
Contact: Sandra Beasley; (202) 265-3808; SBeasley@pbk.org or info@pbk.org
Internet: http://staging.pbk.org/AM/Template.cfm?Section=Book_Awards&Template=/CM/HTMLDisplay.cfm&ContentID=2646
Sponsor: Phi Beta Kappa Society
1606 New Hampshire Avenue NW
Washington, DC 20009

Phi Delta Kappa Outstanding Doctoral Dissertation Awards 3936

These biennial awards recognize scholars who show promise for noteworthy contributions to educational research. One award is given in each of Phi Delta Kappa's regions in the fall of even-numbered years. Winners receive a monetary award and a certificate of recognition. Each entry must include a summary of the dissertation and a completed application form. Application forms, nominating guidelines, and criteria for judging are available online.
Requirements: Competition is open to both Phi Delta Kappa members and nonmembers, but all applicants must be nominated by a Phi Delta Kappa chapter. The nominee must be a recent graduate of an accredited college or university. The nominee's major field must be in education, or the dissertation must focus on an educational topic. The nominee's dissertation must have been completed (approved by the doctoral committee and accepted by the graduate school or school of education) between September 1, 2002, and August of the current year.
Geographic Focus: All States
Date(s) Application is Due: Aug 31
Amount of Grant: 500 USD
Contact: Carol Langdon, (800) 766-1156 or (812) 339-1156; fax (812) 339-0018; clangdon@pdkintl.org
Internet: http://pdkintl.org/edres/cedr_p2.htm
Sponsor: Phi Delta Kappa International
408 N Union Street, P.O. Box 789
Bloomington, IN 47402-0789

Phi Kappa Phi Foundation Love of Learning Awards 3937

The Phi Kappa Phi Love of Learning Awards help fund post-baccalaureate studies and/or career development for active Phi Kappa Phi members to include (but not be limited to) graduate or professional studies, doctoral dissertations, continuing education, career development, travel related to teaching/studies, etc. Submission guidelines and a current application are available at the website.
Requirements: The awards are open to all active Phi Kappa Phi members who have completed their baccalaureate studies in any major by the application submission deadline, and have not received a Phi Kappa Phi Fellowship or Award of Excellence grant.
Restrictions: Recipients of the Fellowship award are not eligible to apply.
Geographic Focus: All States
Date(s) Application is Due: Jun 30
Contact: Maria Davis; (800) 804-9880; LoveofLearning@PhiKappaPhi.org
Internet: http://www.phikappaphi.org/Web/Awards/Love_of_Learning.html
Sponsor: Phi Kappa Phi Foundation
7576 Goodwood Boulevard
Baton Rouge, LA 70806

Phi Kappa Phi Graduate Fellowships 3938

The Phi Kappa Phi Fellowships help fund graduate work for students entering graduate school within one year of their baccalaureate degree. However, persons of any age having delayed graduate study may apply through the chapter of their initiation or chapter of current membership for consideration as the chapter nominee. Any major is eligible as long as the applicant is a PKP member in good standing. Applications must be filed with the secretary of the local chapter by the established deadline on special forms available at the PKP website. Each chapter selects one applicant to compete for the award. Five fellowships of $15,000 each, and 50 fellowships of $5,000 each are available. Specific guidelines, a current application, advice for applicants, and personal statement samples are available at the PKP website.
Requirements: Recipients must be active members of Phi Kappa Phi and have initiated plans to enroll for an advanced degree in a recognized graduate or professional school. Applications will be accepted from individuals selected for membership but not yet initiated. Applications from international students and schools are eligible as long as the student's PKP membership is in good standing.
Restrictions: Applicants who have earned ten or more graduate credits before the current academic year are not eligible. The deadline for the application is usually spring, but dates vary for individual chapters.
Geographic Focus: All States, All Countries
Contact: Local Chapter; (800) 804-9880; fax (225) 388-4900; awards@phikappaphi.org
Internet: http://www.phikappaphi.org/Web/Awards/Fellowship.html
Sponsor: Phi Kappa Phi Foundation
7576 Goodwood Boulevard
Baton Rouge, LA 70806

Phi Kappa Phi Scholar Award 3939

The Phi Kappa Phi Scholar Award honors those who demonstrate the ideals of the Society through their activities, achievements, and scholarship. The Award recognizes excellence in teaching, research, and public service. The Award is given every three years. Chapters and active members are encouraged to nominate an individual for this award; self-nominations are encouraged. Although nominees are often faculty members at colleges and universities, a chapter or any active member may nominate an active Society member who is not institution-affiliated provided all criteria are met. The Award recipient will be presented with a check for $1,000, a one year active membership in Phi Kappa Phi, a recognition certificate, and an invitation to the annual Convention. Travel, lodging and meal expenses incurred by the recipient are reimbursed. Specific criteria and a nomination form are located at the website.
Requirements: Applicants must be Society members in good standing; hold a distinguished record of past scholarly achievements; possess documented evidence of national or international prominence and visibility; demonstrated consistency of excellence in teaching and/or service; and a distinguished record of honors, publications, patents, inventions or other types of recognition for excellence and scholarly achievement.
Geographic Focus: All States, All Countries
Date(s) Application is Due: Feb 15
Contact: Scholar Award Contact; (800) 804-9880; Scholar@PhiKappaPhi.org
Internet: http://www.phikappaphi.org/Web/Awards/PKP_Scholar_and_Artist.html
Sponsor: Phi Kappa Phi Foundation
7576 Goodwood Boulevard
Baton Rouge, LA 70806

Phi Upsilon Omicron Alumni Research Grant 3940

Phi Upsilon Omicron National Honor Society was founded in 1910 at the University of Minnesota in the area of Family and Consumer Sciences. The purpose of the society's Alumni Research Grant is to support post-graduate research in family and consumer sciences. Proposed research should contribute to the knowledge base of family and consumer sciences or further the purposes of Phi Upsilon Omicron, especially in regard to academic excellence, leadership development, and/or professional service. The grant is awarded on even-numbered years. Application forms may be obtained from the Society's National Office (see contact section on this page). The application packet must be postmarked no later than November 1. Applicants will be notified of the disposition of their application following the meeting of the Alumni Research Grant Advisory Committee.
Requirements: One or more of the following entities are eligible to apply: individual society alumni; society alumni committees or task forces; and alumni chapters.
Geographic Focus: All States
Date(s) Application is Due: Nov 1
Amount of Grant: 2,500 USD
Contact: Susan Rickards; (304) 368-0612; rickards@phiu.org or info@phiu.org
Internet: http://phiu.org/awards.htm
Sponsor: Phi Upsilon Omicron, Inc.
P.O. Box 329
Fairmont, WV 26555

Phi Upsilon Omicron Candle Fellowships 3941

Phi Upsilon Omicron is a national honor society in Family and Consumer Sciences that was founded in 1910 at the University of Minnesota. The society awards two Candle fellowships annually for study leading to a master's degree in family and consumer sciences. The application guidelines can be downloaded from the fellowship website. The application packet must be postmarked no later than February 1. The society offers numerous graduate fellowship opportunities to its members. Applicants may apply for more than one fellowship opportunity with the same application packet and should list the name(s) of the fellowships for which application is being made.
Requirements: Applicant must be a member of Phi Upsilon Omicron who is enrolled in or planning to enroll in graduate school. Selection will be based upon the following criteria:

applicant's scholastic record; honors and recognitions; participation in honor society as well as professional, community, and other organizations; scholarly work; statement of professional goals; and recommendations.
Geographic Focus: All States
Date(s) Application is Due: Feb 1
Amount of Grant: 750-1,500 USD
Contact: Susan Rickards; (304) 368-0612; rickards@phiu.org or info@phiu.org
Internet: http://phiu.org/fellowships.htm#
Sponsor: Phi Upsilon Omicron, Inc.
P.O. Box 329
Fairmont, WV 26555

Phi Upsilon Omicron Diamond Anniversary Fellowships 3942
Phi Upsilon Omicron is a national honor society in Family and Consumer Sciences that was founded in 1910 at the University of Minnesota. Two Diamond Anniversary fellowships will be awarded per year for study at the master's or doctoral level in family and consumer sciences or a related area. The application guidelines can be downloaded from the fellowship website. The application packet must be postmarked no later than February 1. The society offers numerous graduate fellowship opportunities to its members. Applicants may apply for more than one fellowship opportunity with the same application packet and should list the name(s) of the fellowships for which application is being made.
Requirements: Applicant must be a member of Phi Upsilon Omicron and have been accepted into (or be currently enrolled in) a graduate program in consumer and family sciences. Selection will be based upon the following criteria: applicant's scholastic record; honors and recognitions; participation in honor society as well as professional, community, and other organizations; scholarly work; statement of professional goals; and recommendations.
Geographic Focus: All States
Date(s) Application is Due: Feb 1
Amount of Grant: 750-1,500 USD
Contact: Susan Rickards, Executive Director; (304) 368-0612; rickards@phiu.org or info@phiu.org
Internet: http://phiu.org/fellowships.htm#
Sponsor: Phi Upsilon Omicron, Inc.
P.O. Box 329
Fairmont, WV 26555

Phi Upsilon Omicron Florence Fallgatter Distinguished Service Award 3943
Phi Upsilon Omicron National Honor Society was founded in 1910 at the University of Minnesota in the area of Family and Consumer Sciences. The society's Florence Fallgatter Distinguished Service Award is granted biennially to a qualified alumni member of the society for outstanding achievement in family and consumer sciences. The recipient's expenses to attend the society's biennial meeting will be paid, and the member will be recognized with a plaque at the meeting. The nomination deadline is November 1. Interested applicants should contact the society for information on nomination procedures.
Requirements: A nominee must exhibit the following characteristics: be an alumni member of the society; have finished his or her bachelor's degree no less than ten years ago; have demonstrated excellence in business, dietetics/nutrition, education, cooperative extension; journalism; research; social services, or volunteer services; and have had a significant positive impact on the family and consumer sciences profession.
Geographic Focus: All States
Date(s) Application is Due: Nov 1
Contact: Susan Rickards; (304) 368-0612; rickards@phiu.org or info@phiu.org
Internet: http://phiu.org/awards.htm
Sponsor: Phi Upsilon Omicron, Inc.
P.O. Box 329
Fairmont, WV 26555

Phi Upsilon Omicron Founders Fellowship 3944
Phi Upsilon Omicron is a national honor society in Family and Consumer Sciences that was founded in 1910 at the University of Minnesota. The society's Founders Fellowship supports advanced graduate study in the field of consumer and family sciences. The application guidelines can be downloaded from the fellowship website. The application packet must be postmarked no later than February 1. The society offers numerous graduate fellowship opportunities to its members. Applicants may apply for more than one fellowship opportunity with the same application packet and should list the name(s) of the fellowships for which application is being made.
Requirements: Applicant must be a member of Phi Upsilon Omicron, have completed at least half the credit-hour requirements toward the doctorate in some area of family and consumer sciences, and have had several years of successful employment in the profession. Selection will be based upon the following criteria: applicant's scholastic record; honors and recognitions; participation in honor society as well as professional, community, and other organizations; scholarly work; statement of professional goals; and recommendations.
Geographic Focus: All States
Date(s) Application is Due: Feb 1
Amount of Grant: 750-1,500 USD
Contact: Susan Rickards; (304) 368-0612; rickards@phiu.org or info@phiu.org
Internet: http://phiu.org/fellowships.htm#
Sponsor: Phi Upsilon Omicron, Inc.
P.O. Box 329
Fairmont, WV 26555

Phi Upsilon Omicron Geraldine Clewell Doctoral Fellowship 3945
Phi Upsilon Omicron is a national honor society in Family and Consumer Sciences that was founded in 1910 at the University of Minnesota. The society's Geraldine Clewell Doctoral Fellowship is awarded annually to a doctoral student in family and consumer sciences or a related area. The application guidelines can be downloaded from the fellowship website. The application packet must be postmarked no later than February 1. Applicants should note that one of the three required recommendations must come from their major advisor. The society offers numerous graduate fellowship opportunities to its members. Applicants may apply for more than one fellowship opportunity with the same application packet and should list the name(s) of the fellowships for which application is being made.
Requirements: Applicants must be members of Phi Upsilon Omicron. Preference is given to a student who desires to teach at the college/university level. Selection will be based upon the following criteria: applicant's scholastic record; honors and recognitions; participation in honor society as well as professional, community, and other organizations; scholarly work; statement of professional goals; and recommendations.
Geographic Focus: All States
Date(s) Application is Due: Feb 1
Amount of Grant: 750-1,500 USD
Contact: Susan Rickards; (304) 368-0612; rickards@phiu.org or info@phiu.org
Internet: http://phiu.org/fellowships.htm#
Sponsor: Phi Upsilon Omicron, Inc.
P.O. Box 329
Fairmont, WV 26555

Phi Upsilon Omicron Geraldine Clewell Masters Fellowship 3946
Phi Upsilon Omicron is a national honor society in Family and Consumer Sciences that was founded in 1910 at the University of Minnesota. The society offers numerous graduate fellowship opportunities to its members. The society's Geraldine Clewell Masters Fellowship is awarded annually to a student pursuing a master's degree in family and consumer sciences or related area. The application guidelines can be downloaded from the fellowship website. The application packet must be postmarked no later than February 1. Applicants should note that one of the three required recommendations must come from their major advisor. The society offers numerous graduate fellowship opportunities to its members. Applicants may apply for more than one fellowship opportunity with the same application packet and should list the name(s) of the fellowships for which application is being made.
Requirements: Applicants must be members of Phi Upsilon Omicron. Preference is given to a student majoring in family and consumer-sciences education and who desires to teach at the elementary/secondary level. Selection will be based upon the following criteria: applicant's scholastic record; honors and recognitions; participation in honor society as well as professional, community, and other organizations; scholarly work; statement of professional goals; and recommendations.
Geographic Focus: All States
Date(s) Application is Due: Feb 1
Amount of Grant: 750-1,500 USD
Contact: Susan Rickards; (304) 368-0612; rickards@phiu.org or info@phiu.org
Internet: http://phiu.org/fellowships.htm#
Sponsor: Phi Upsilon Omicron, Inc.
P.O. Box 329
Fairmont, WV 26555

Phi Upsilon Omicron Jean Dearth Dickerscheid Fellowship 3947
Phi Upsilon Omicron is a national honor society in Family and Consumer Sciences that was founded in 1910 at the University of Minnesota. The society's Jean Dearth Dickerscheid Fellowship is awarded annually to a doctoral student who is pursuing his or her degree in family and consumer sciences or a related area and has earned at least one other degree in consumer and family sciences. The application guidelines can be downloaded from the fellowship website. The application packet must be postmarked no later than February 1. The society offers numerous graduate fellowship opportunities to its members. Applicants may apply for more than one fellowship opportunity with the same application packet and should list the name(s) of the fellowships for which application is being made.
Requirements: Applicants must be members of Phi Upsilon Omicron and intend to pursue a career in academia. Selection will be based upon the following criteria: applicant's scholastic record; honors and recognitions; participation in honor society as well as professional, community, and other organizations; scholarly work; statement of professional goals; and recommendations.
Geographic Focus: All States
Date(s) Application is Due: Feb 1
Amount of Grant: 750-1,500 USD
Contact: Susan Rickards; (304) 368-0612; rickards@phiu.org or info@phiu.org
Internet: http://phiu.org/fellowships.htm#
Sponsor: Phi Upsilon Omicron, Inc.
P.O. Box 329
Fairmont, WV 26555

Phi Upsilon Omicron Margaret Drew Alpha Fellowship 3948
Phi Upsilon Omicron is a national honor society in Family and Consumer Sciences that was founded in 1910 at the University of Minnesota. The society's Margaret Drew Alpha Fellowship supports graduate study in a field of family and consumer sciences with preference given to dietetics or food and nutrition. The application guidelines can be downloaded from the fellowship website. The application packet must be postmarked no later than February 1. The society offers numerous graduate fellowship opportunities to its members. Applicants may apply for more than one fellowship opportunity with the same application packet and should list the name(s) of the fellowships for which application is being made.

Requirements: Applicant must be a member of Phi Upsilon Omicron. Selection will be based upon the following criteria: applicant's scholastic record; honors and recognitions; participation in honor society as well as professional, community, and other organizations; scholarly work; statement of professional goals; and recommendations.
Geographic Focus: All States
Date(s) Application is Due: Feb 1
Amount of Grant: 750-1,700 USD
Contact: Susan Rickards; (304) 368-0612; rickards@phiu.org or info@phiu.org
Internet: http://phiu.org/fellowships.htm#
Sponsor: Phi Upsilon Omicron, Inc.
P.O. Box 329
Fairmont, WV 26555

Phi Upsilon Omicron Presidents Research Fellowship 3949
Phi Upsilon Omicron is a national honor society in Family and Consumer Sciences that was founded in 1910 at the University of Minnesota. The society's President Research Fellowship supports graduate research in family and consumer sciences at the master's, doctoral, or postdoctoral level. The application guidelines can be downloaded from the fellowship website. The application packet must be postmarked no later than February 1. The society offers numerous graduate fellowship opportunities to its members. Applicants may apply for more than one fellowship opportunity with the same application packet and should list the name(s) of the fellowships for which application is being made.
Requirements: Applicant must be a member of Phi Upsilon Omicron. Selection will be based upon the following criteria: the organization and need for the research as stated in the applicant's research prospectus; applicant's scholastic record; honors and recognitions; participation in honor society as well as professional, community, and other organizations; scholarly work; statement of professional goals; and recommendations.
Geographic Focus: All States
Date(s) Application is Due: Feb 1
Amount of Grant: 750-1,500 USD
Contact: Susan Rickards; (304) 368-0612; rickards@phiu.org or info@phiu.org
Internet: http://phiu.org/fellowships.htm#
Sponsor: Phi Upsilon Omicron, Inc.
P.O. Box 329
Fairmont, WV 26555

PhRMA Foundation Health Outcomes Post Doctoral Fellowships 3950
The program provides stipend support for individuals engaged in a research training program that will create or extend their credentials in health outcomes. The intent of this program is to support post doctoral career development activities of individuals prepared (or preparing) to engage in research that will strengthen representation of health outcomes in schools of pharmacy, medicine, nursing and public health. To accomplish these goals, support will be provided for a two-year period to selected individuals who are beginning careers in health outcomes research and who give promise of outstanding development as researchers. The award, consisting of a $40,000 annual stipend, is made to the institution on behalf of the fellow. The award is intended solely as a stipend and may not be used otherwise.
Requirements: Eligible applicants include well-trained graduates from Pharm. D., M.D., and Ph.D. programs who seek to further develop and refine their research skills through formal postdoctoral training. They must also have a firm commitment from an accredited U.S. university and be a U.S. citizen or permanent resident. The application must include a research plan written by the applicant, the mentor's research record, and a description of how the mentored experience will enhance the applicant's career development in health outcomes research. The sponsor (mentor) of the post doctoral program must describe how the goals of the research training program will be accomplished and provide assurance that key collaborating mentors endorse and are willing to support the training plan.
Restrictions: Applications, however good, which do not meet the aims of the program will be disapproved.
Geographic Focus: All States
Amount of Grant: 40,000 USD
Contact: Eileen M. Cannon, (202) 572-7756; fax (202) 572-7799; info@phrmafoundation.org
Internet: http://www.phrmafoundation.org/awards/outcomes/postdoc.php
Sponsor: Pharmaceutical Research and Manufacturers of America Foundation
950 F Street NW, Suite 300
Washington, DC 20004

PhRMA Foundation Health Outcomes Research Starter Grants 3951
The purpose of the program is to offer financial support to individuals beginning their independent research careers at the faculty level. The program provides a research grant of $30,000 per year for up to two years. This program supports individuals beginning independent research careers in academia who do not have other substantial sources of research.
Requirements: Those holding the academic rank of instructor or assistant professor and investigators at the doctoral level with equivalent positions are eligible to apply, providing their proposed research is neither directly nor indirectly subsidized to any significant degree by an extramural support mechanism. The program is not intended for those in postdoctoral training programs. However, individuals in postdoctoral training scheduled to conclude and who will hold an academic appointment by January 1 may apply. Applicants must be sponsored by the department or unit within which the proposed research is to be undertaken. The grant is made to the university on behalf of the applicant and with the understanding that the university will administer the funds. Schools of medicine, pharmacy, public health, nursing, dentistry and schools of other areas where appropriate are eligible for this award. Applications must be submitted by an accredited school in the U.S., and all applicants must be U.S. citizens or permanent residents. Before an individual is eligible to apply for a PhRMA Foundation award, the applicant must first have a firm commitment from a U.S. university.
Restrictions: This program is not offered as a means to augment substantially funded research efforts. It is intended to offer support for researchers who are starting their independent research efforts. This award is granted in part based on need. If an individual currently has or is guaranteed substantial funding, they should not apply.
Geographic Focus: All States
Amount of Grant: 30,000 USD
Contact: Eileen M. Cannon, (202) 572-7756; fax (202) 572-7799; info@phrmafoundation.org
Internet: http://www.phrmafoundation.org/awards/outcomes/starter.php
Sponsor: Pharmaceutical Research and Manufacturers of America Foundation
950 F Street NW, Suite 300
Washington, DC 20004

PhRMA Foundation Health Outcomes Sabbatical Fellowships 3952
The fellowship provides stipend support for individuals engaged in a multidisciplinary research training program that will create or extend their credentials in health outcomes. The program enables faculty with active research programs to work outside of their home institution for periods of six months to one year to learn new skills or develop new collaborations that will enhance their research and research training capabilities in health outcomes. The applicant and mentor of the program must describe how the multidisciplinary goals of the research experiential program will be accomplished and provide assurance that key collaborating mentors endorse and are willing to support the plan. Matching funds must be provided by the home institution. Awards may be activated beginning January 1 or on the first day of any month thereafter, up to and including December 1. Guidelines and application are available online.
Requirements: Eligible applicants must (1) hold a PhD, Pharm.D., M.D. or Sc.D. degree in a field of study logically or functionally related to the proposed post doctoral activities, (2) hold a faculty appointment that imparts eligibility for a sabbatical leave from their home institution, (3) have institutional approval of a sabbatical plan that includes partial salary that matches the PhRMA stipend, (4) hold an endorsement from a mentor who agrees to sponsor the applicant's visiting scientist activity, and (5) be a U.S. citizen or permanent resident. Matching funds must be provided by the home institution.
Restrictions: The program provides no other subsidies (travel, tuition, fringe benefit costs, etc.), and indirect costs to the institution are not provided.
Geographic Focus: All States
Amount of Grant: 40,000 USD
Contact: Eileen M. Cannon, (202) 572-7756; fax (202) 572-7799; info@phrmafoundation.org
Internet: http://www.phrmafoundation.org/awards/outcomes/sabbatical.php
Sponsor: Pharmaceutical Research and Manufacturers of America Foundation
950 F Street NW, Suite 300
Washington, DC 20004

PhRMA Foundation Informatics Post Doctoral Fellowships 3953
The postdoctoral program provides stipend support for individuals engaged in a multidisciplinary research training program that will create or extend their credentials in informatics. The intent of this program is to support postdoctoral career development activities of individuals preparing to engage in research that will bridge the gap between experimental and computational approaches in genomic and biomedical studies. It is anticipated that this research training will be accomplished in academic and/or industrial laboratory settings where multidisciplinary teams are organized to address problems that span the range of biological complexity rather than focus on the application of single technologies. The award, consisting of a $40,000 annual stipend, is made to the institution on behalf of the fellow. The award is intended solely as a stipend and may not be used otherwise.
Requirements: Eligible applicants must either hold a PhD degree in a field of study logically or functionally related to the proposed postdoctoral activities, or expect to receive the PhD before activating the award. They must also have a firm commitment from an accredited U.S. university and be a U.S. citizen or permanent resident. The sponsor (mentor) of the postdoctoral program must describe how the multidisciplinary goals of the research training program will be accomplished and provide assurance that key collaborating mentors endorse and are willing to support the training plan.
Restrictions: The program provides no other subsidies (travel, tuition, fringe benefit costs, etc.) and indirect costs to the institution are not provided. The second year of this award is contingent upon a progress report approved by the Foundation and submission of a financial report.
Geographic Focus: All States
Amount of Grant: 40,000 USD
Contact: Eileen M. Cannon, (202) 572-7756; fax (202) 572-7799; info@phrmafoundation.org
Internet: http://www.phrmafoundation.org/awards/informatics/postdoc.php
Sponsor: Pharmaceutical Research and Manufacturers of America Foundation
950 F Street NW, Suite 300
Washington, DC 20004

PhRMA Foundation Informatics Pre Doctoral Fellowships 3954
The program is designed for candidates who expect to complete the requirements for the Ph.D. in the fields of informatics in two years or less from the time the fellowship begins. It aims at supporting promising students during their thesis research. The award is made to the university on behalf of the fellow. The fellowship provides a stipend of $20,000 a year payable monthly for a minimum of one year and a maximum of two years. Of the $20,000 awarded annually, up to $500 a year may be used for incidentals directly associated with the thesis research preparation (e.g., secretarial help, artwork, books, travel, etc.). Applications

will be accepted for a minimum of one year and a maximum of two years. The program assumes that the fellows will devote full-time (including summers) to their research.
Requirements: The fellowship program of pre doctoral support is designed to assist full-time, in-residence Ph.D. candidates in the fields of informatics who are enrolled in U.S.A schools of medicine, pharmacy, dentistry or veterinary medicine. The program supports full-time advanced students who will have completed the bulk of their pre-thesis requirements (at least two years of study) and are engaged in thesis research as Ph.D. Candidates by the time the award is activated. Students just starting in graduate school should not apply. Highest priority will be given to applications who support the goal of the Informatics Programs. The goal of the program is to promote the use of informatics in an integrative approach to the understanding of biological and disease processes. Informatics awards will support career development of scientists engaged in computational and experimental research to integrate cutting-edge information technology with advanced biological, chemical, and pharmacological sciences. Applications must be submitted by an accredited U.S. school, and all applicants must be U.S. citizens or permanent residents.
Restrictions: The fellowship may not be used to supplement funds from other fellowships, traineeships or assistantships, unless necessary to make stipend levels compliant with institutional policy. The focus of the program is to assist in the predoctoral training of the candidate. It should not be viewed as a program to fund a research project. Indirect costs are not provided to the institution and PhRMA Foundation grant funds may not be used for this purpose.
Geographic Focus: All States
Amount of Grant: 20,000 USD
Contact: Eileen M. Cannon, (202) 572-7756; fax (202) 572-7799; info@phrmafoundation.org
Internet: http://www.phrmafoundation.org/awards/informatics/predoc.php
Sponsor: Pharmaceutical Research and Manufacturers of America Foundation
950 F Street NW, Suite 300
Washington, DC 20004

PhRMA Foundation Informatics Research Starter Grants 3955

This program supports individuals beginning independent research careers in academia who do not have other substantial sources of research. The areas of interest within this program consists of research that supports career development of scientists engaged in computational and experimental research to integrate cutting edge-edge information technology with advanced biological, chemical, and pharmacological sciences in: Genetics (Molecular, Medical-human, Pharmaco, Population); Genomics (Function, Structural, Toxico, Pharmaco, Comparative); Proteomics; Biological pathways. The program provides a research grant of $30,000 per year for up to two years. The 'starter' aspect of the program strives to assist those individuals who are establishing careers as independent investigators in the field of informatics. The program is not offered as a means to augment an ongoing research effort or is the grant intended to be used for any direct effort to obtain further extramural funding. The funds are to be used to conduct the proposed research.
Requirements: Those holding the academic rank of instructor or assistant professor and investigators at the doctoral level with equivalent positions are eligible to apply for these research starter grants, providing their proposed research is neither directly nor indirectly subsidized to any significant degree by an extramural support mechanism. The program is not intended for those in postdoctoral training programs. However, individuals in postdoctoral training scheduled to conclude and who will hold an academic appointment by January 1 may apply. Applicants must be sponsored by the department or unit in which the proposed research is to be undertaken. The grant is made to the university on behalf of the applicant and with the understanding that the university will administer the funds. Applications must be submitted by an accredited U.S. school, and all applicants must be U.S. citizens or permanent residents.
Restrictions: Research projects that extend or develop the proprietary value of specific drug products are not acceptable in this program. This exclusion does not preclude research in which specific drug products are used to test hypotheses that have a general applicability. The funds may not be used for salary support of the grantee, but may be used to support technical assistance. No more than $500 a year may be used for travel to professional meetings by the grantee. Indirect costs are not provided to the institution, and grant funds may not be used for this purpose.
Geographic Focus: All States
Amount of Grant: 30,000 USD
Contact: Eileen M. Cannon, (202) 572-7756; fax (202) 572-7799; info@phrmafoundation.org
Internet: http://www.phrmafoundation.org/awards/informatics/starter.php
Sponsor: Pharmaceutical Research and Manufacturers of America Foundation
950 F Street NW, Suite 300
Washington, DC 20004

PhRMA Foundation Informatics Sabbatical Fellowships 3956

The fellowship supports individuals engaged in a multidisciplinary research training program that will create or extend their credentials in informatics. The program enables faculty with active research programs to work outside of their home institution for periods of six months to one year to learn new skills or develop new collaborations that will enhance their research and research training capabilities in informatics. Applicants are expected to engage in research that supports career development of scientists engaged in computational and experimental research to integrate cutting edge information technology with advanced biological, chemical, and pharmacological sciences in the fields of genetics, genomics, and proteomics biological pathways. Emphasis will be placed on the development of new informatics technologies that demonstrate the translation of genomic data into an elucidation and understanding of biological and disease processes. The applicant and mentor of the program must describe how the multidisciplinary goals of the research experiential program will be accomplished and provide assurance that key collaborating mentors endorse and are willing to support the plan. The award provides up to $40,000 stipend support for mid-career scientists to engage in an academic year or calendar year experiential program intended to redirect their core research focus to an area of emerging importance to pharmaceutical research and development.
Requirements: Eligible applicants must hold a PhD or MD degree in a field of study logically or functionally related to the proposed postdoctoral activities; hold a faculty appointment that imparts eligibility for a sabbatical leave from their home institution; have institutional approval of a sabbatical plan that includes partial salary that matches the PhRMA stipend; and be a U.S. citizen or permanent resident. Preference will be given to those individuals whose research combines novel computational methods with experimental validation. Emphasis will be placed on the development of new informatics technologies that demonstrate the translation of genomic data into an elucidation and understanding of biological and disease processes. Matching funds must be provided by the home institution.
Restrictions: The program provides no other subsidies (travel, tuition, fringe benefit costs, etc.), and indirect costs to the institution are not provided.
Geographic Focus: All States
Amount of Grant: 40,000 USD
Contact: Eileen M. Cannon, (202) 572-7756; fax (202) 572-7799; info@phrmafoundation.org
Internet: http://www.phrmafoundation.org/awards/informatics/sabbatical.php
Sponsor: Pharmaceutical Research and Manufacturers of America Foundation
950 F Street NW, Suite 300
Washington, DC 20004

PhRMA Foundation Paul Calabresi Medical Student Research Fellowship 3957

The fellowship is offered to medical or dental students who have substantial interests in research and teaching careers in pharmacology-clinical pharmacology and who are willing to spend full-time in a specific research effort within a pharmacology or clinical pharmacology unit. Fellowships are available for a minimum period of three months or any period of time up to 24 months. The commitment must be full-time. The student may undertake this investigative effort at his/her own school or at another institution. Requests for a fellowship are to be submitted online by the appropriate representative of the school or university. The fellowship is made to the institution on behalf of the student and the institution must agree to administer the funds. The program offers funding of up to $1,500 per month. It is expected that the funds will be distributed to the fellow in regular payments over the period of the fellowship by the recipient school. The funds may be used towards the stipend, tuition or affiliated fees for the fellow. Fellowships are available for a minimum period of six months or any period of time up to 24 months. However, if the student's award period is 24 months, the maximum stipend is still $18,000.
Requirements: A candidate must be enrolled in a U.S. medical/dental school and have finished at least one year of the school curriculum. Priority consideration will be given to those candidates who project strong commitments to careers in the field of clinical pharmacology. The applicant must be sponsored by the pharmacology or clinical pharmacology program in which the investigative project is to be undertaken. The sponsoring unit must have a demonstrated commitment to training in the fields of pharmacology or clinical pharmacology. Applications must be submitted by an accredited U.S. school, and all applicants must be U.S. citizens or permanent residents.
Restrictions: There are no indirect costs provided to the institution nor may any of the funds be used for this purpose. Once an individual has received a Medical Student Fellowship, that individual may not reapply.
Geographic Focus: All States
Amount of Grant: Up to 18,000 USD
Contact: Eileen M. Cannon, (202) 572-7756; fax (202) 572-7799; info@phrmafoundation.org
Internet: http://www.phrmafoundation.org/awards/clinical/medstudent.php
Sponsor: Pharmaceutical Research and Manufacturers of America Foundation
950 F Street NW, Suite 300
Washington, DC 20004

PhRMA Foundation Pharmaceutics Postdoctoral Fellowships 3958

The purpose of the fellowships is to encourage graduates from PhD programs in pharmaceutics to continue to refine their research skills through formal postdoctoral training. Applicants should indicate a strong determination to continue their research careers in pharmaceutics following completion of the fellowship. Applications will be accepted for postdoctoral research training extending over a minimum of one year and a maximum of two years. Each fellowship carries an annual stipend of $40,000 which can be supplemented by the sponsor and/or his/her institution.
Requirements: To be eligible, applicants must either hold a PhD degree in pharmaceutics from a school of pharmacy accredited by the American Council on Pharmaceutical Education or expect to receive such a degree before activating the fellowship. The mentor of the fellow must be in an accredited school of pharmacy. Alternatively, a student with a Ph.D. from a school of pharmacy may use the fellowship to study with a mentor in any science or engineering department. An equally important aspect of eligibility is that suitable facilities for the necessary training and research must be available to the applicant. The program with which the applicant is, or will be, associated must provide an environment where the applicant's potential can be developed to the fullest extent possible.
Restrictions: The award is intended solely as a stipend and may not be used otherwise. If necessary, the institution may supplement the award to a level that is consistent with other postdoctoral fellowships it currently offers. Research projects that extend or develop the proprietary value of specific drug products are not acceptable in this program. The program provides no other subsidies (travel, tuition, fringe benefit costs, etc.) and indirect costs to the institution are not provided.
Geographic Focus: All States
Contact: Eileen M. Cannon, (202) 572-7756; fax (202) 572-7799; info@phrmafoundation.org
Internet: http://www.phrmafoundation.org/awards/pharmaceutics/postdoc.php
Sponsor: Pharmaceutical Research and Manufacturers of America Foundation

950 F Street NW, Suite 300
Washington, DC 20004

PhRMA Foundation Pharmaceutics Pre Doctoral Fellowships 3959

To encourage the entry of more qualified individuals into this area, the program aims at supporting promising students during their thesis research by providing assistance in the form of stipend and funds to cover costs incidental to the training. Applications will be accepted for a minimum of one year and a maximum of two years. The program is designed for applicants who expect to complete the requirements for the Ph.D. in pharmaceutics in two years or less from the time the fellowship begins. For the purposes of this program, pharmaceutics includes basic pharmaceutics, biopharmaceutics and pharmaceutical technology. The fundamental aspects of pharmacokinetics are not included since these are covered by PhRMA Foundation awards in pharmacology and toxicology.
Requirements: To be eligible, applicants must hold a B.S., M.S., or Pharm.D. degree in pharmacy or a related area such as chemistry or biology from an accredited school in the U.S. A candidate enrolled in a Pharm.D./Ph.D. program should not be taking required clinical course work or clinical clerkships during the tenure of the fellowship. Also, before an individual is eligible to apply for a PhRMA Foundation award, the applicant must first have a firm commitment from a U.S. university, and must be a U.S. citizen or permanent resident. The program assumes that the fellows will devote full-time (including summers) to their research. The fellowship may not be used as a supplement to funds from other fellowships or assistantships. However, this does not preclude institutions supplementing a fellowship with their own funds in order to bring the stipend within the normal range of the institution.
Restrictions: Research projects which extend or develop the proprietary value of specific drug products are not acceptable in this program. This exclusion does not preclude research in which specific drug products are used to test hypotheses which have a general applicability. Applications, however good, which do not meet the aims of the program will be disapproved.
Geographic Focus: All States
Contact: Eileen M. Cannon, (202) 572-7756; fax (202) 572-7799; info@phrmafoundation.org
Internet: http://www.phrmafoundation.org/awards/pharmaceutics/predoc.php
Sponsor: Pharmaceutical Research and Manufacturers of America Foundation
950 F Street NW, Suite 300
Washington, DC 20004

PhRMA Foundation Pharmaceutics Research Starter Grants 3960

The purpose of the starter grants is to offer financial support to individuals beginning their independent research careers at the faculty level. The areas of interest within this program are research efforts in the fields of pharmaceutics. The program provides a research grant of $30,000 per year for up to two years. For the purposes of this program, pharmaceutics includes basic pharmaceutics, biopharmaceutics, pharmaceutical technology and pharmaceutical biotechnology. The fundamental aspects of pharmacokinetics are not included since they are covered by PhRMA Foundation programs in pharmacology and toxicology. The 'starter' aspect of the program strives to assist those individuals who are establishing careers as independent investigators in the fields listed above.
Requirements: Those holding the academic, tenure track, rank of instructor or assistant professor and investigators at the doctoral level with equivalent positions are eligible to apply for these research starter grants, providing their proposed research is neither directly nor indirectly subsidized to any significant degree by an extramural support mechanism. It is anticipated that this research experience will occur in an academic and/or industrial laboratory setting. Applications must be submitted by a school of pharmacy accredited by the American Council on Pharmaceutical Education, and all applicants must be U.S. citizens or permanent residents. The grant is made to the university on behalf of the applicant and with the understanding that the university will administer the funds. Only schools of pharmacy are eligible for this award.
Restrictions: The program is not offered as a means to augment an ongoing research effort. Nor is the grant intended to be used for any direct effort to obtain further extramural funding. The funds are to be used to conduct the proposed research. The funds may not be used for salary support of the grantee, but may be used to support technical assistance. No more than $500 a year may be used for travel to professional meetings by the grantee. Indirect costs are not provided to the institution, and grant funds may not be used for this purpose.
Geographic Focus: All States
Contact: Eileen M. Cannon, (202) 572-7756; fax (202) 572-7799; info@phrmafoundation.org
Internet: http://www.phrmafoundation.org/awards/pharmaceutics/starter.php
Sponsor: Pharmaceutical Research and Manufacturers of America Foundation
950 F Street NW, Suite 300
Washington, DC 20004

PhRMA Foundation Pharmaceutics Sabbatical Fellowships 3961

The purpose of this program is to enable faculty with active research programs to work outside of their home institution for periods of 6 months to one year to learn new skills or develop new collaborations that will enhance their research and research training capabilities in pharmaceutics. For the purposes of this program, pharmaceutics includes basic pharmaceutics, biopharmaceutics, pharmaceutical technology and pharmaceutical biotechnology. The fundamental aspects of pharmacokinetics are not included since they are covered by PhRMA Foundation programs in pharmacology and toxicology. It is anticipated that this research experience will occur in an academic and/or industrial laboratory setting. The award provides up to $40,000 stipend support for mid-career scientists to engage in an academic year or calendar year experiential program intended to redirect their core research focus to an area of emerging importance to pharmaceutical research and development.
Requirements: Eligible applicants must hold a PhD degree in pharmaceutics from a School of Pharmacy accredited by the American Council on Pharmaceutical Education; hold a faculty appointment that imparts eligibility for a sabbatical leave from their home institution; have institutional approval of a sabbatical plan that includes partial salary that matches the PhRMA stipend; hold an endorsement from a mentor who agrees to sponsor the applicants visiting scientist activity; and be a U.S. citizen or permanent resident. The applicant and mentor of the program must describe how the multidisciplinary goals of the research experiential program will be accomplished and provide assurance that key collaborating mentors endorse and are willing to support the plan. Matching funds must be provided by the home institution.
Restrictions: The program provides no other subsidies (travel, tuition, fringe benefit costs, etc.) and indirect costs to the institution are not provided.
Geographic Focus: All States
Amount of Grant: 40,000 USD
Contact: Eileen M. Cannon, (202) 572-7756; fax (202) 572-7799; info@phrmafoundation.org
Internet: http://www.phrmafoundation.org/awards/pharmaceutics/sabbatical.php
Sponsor: Pharmaceutical Research and Manufacturers of America Foundation
950 F Street NW, Suite 300
Washington, DC 20004

PhRMA Foundation Pharmacology/Toxicology Post Doctoral Fellowships 3962

The postdoctoral program provides stipend support for individuals engaged in a multidisciplinary research training program that will create or extend their credentials in pharmacology or toxicology. The intent of this program is to support postdoctoral career development activities of individuals prepared (or preparing) to engage in research that integrates information on molecular or cellular mechanisms of action with information on the effect of an agent in the intact organism. It is anticipated that this research training will be accomplished in academic and/or industrial laboratory settings where multidisciplinary teams are organized to integrate informatics, molecular, cell, and systems biology with pharmacology/toxicology research. Funding is for up to 2 years but not beyond the 3rd year of the existing Post Doc.
Requirements: Eligible applicants must either hold a PhD degree or appropriate terminal research doctorate in a field of study logically or functionally related to the proposed post doctoral activities; or expect to receive the PhD before activating the award. They must also have a firm commitment from an accredited U.S. university and be a U.S. citizen or permanent resident. The sponsor (mentor) must describe how the multidisciplinary goals of the research training program will be accomplished and provide assurance that key collaborating mentors endorse and are willing to support the training plan. The award, consisting of a $40,000 annual stipend, is made to the institution on behalf of the fellow. The award is intended solely as a stipend and may not be used otherwise. If necessary, the institution may supplement the award to a level that is consistent with other postdoctoral fellowships it currently offers.
Restrictions: The program provides no other subsidies (travel, tuition, fringe benefit costs, etc.) and indirect costs to the institution are not provided. The second year of this award is contingent upon a progress report approved by the foundation and submission of a financial report.
Geographic Focus: All States
Amount of Grant: 40,000 USD
Contact: Eileen M. Cannon, (202) 572-7756; fax (202) 572-7799; info@phrmafoundation.org
Internet: http://www.phrmafoundation.org/awards/pharmacology/postdoc.php
Sponsor: Pharmaceutical Research and Manufacturers of America Foundation
950 F Street NW, Suite 300
Washington, DC 20004

PhRMA Foundation Pharmacology/Toxicology Pre Doctoral Fellowships 3963

The program is designed for candidates who expect to complete the requirements for the Ph.D. in the fields of pharmacology or toxicology in two years or less from the time the fellowship begins. The award is made to the university on behalf of the fellow. The fellowship provides a stipend of $20,000 a year payable monthly for a minimum of one year and a maximum of two years. Of the $20,000 awarded annually, up to $500 a year may be used for incidentals directly associated with the thesis research preparation (e.g., secretarial help, artwork, books, travel, etc.). Awards may be activated beginning January 1 or on the first day of any month thereafter, up to and including August 1. The second year of a fellowship is contingent upon certification by the thesis advisor that satisfactory progress is being achieved.
Requirements: The fellowship program of pre doctoral support is designed to assist full-time, in-residence Ph.D. candidates in the fields of pharmacology or toxicology who are enrolled in U.S.A schools of medicine, pharmacy, dentistry or veterinary medicine. The program supports full-time advanced students who will have completed the bulk of their pre-thesis requirements (at least two years of study) and are engaged in thesis research as Ph.D. Candidates by the time the award is activated. Students just starting in graduate school should not apply. In the field of pharmacology or toxicology, highest priority will be given to applications for research that attempts to integrate information on the mechanism of action of a drug or chemical at a molecular or cellular level with the drug's effects in the intact laboratory animal or human, encompassing the potential influences of biochemical, physiological or behavioral systems. The pharmacology/toxicology research and teaching mission of the unit awarding the degree must be apparent as a core mission component on their web site. Applications must be submitted by an accredited U.S. school, and all applicants must be U.S. citizens or permanent residents. The program assumes that the fellows will devote full-time (including summers) to their research. The PhRMA Foundation fellowship may not be used to supplement funds from other fellowships, traineeships or assistantships, unless necessary to make stipend levels compliant with institutional policy. The focus of the program is to assist in the predoctoral training of the candidate. It should not be viewed as a program to fund a research project.
Restrictions: A candidate enrolled in an M.D./Ph.D. program should not be taking required clinical course work or clerkships during the tenure of the fellowship. The foundation will accept only one application per academic institution. Exceptions may be

considered for institutions with multiple campuses. Indirect costs are not provided to the institution and PhRMA Foundation grant funds may not be used for this purpose.
Geographic Focus: All States
Amount of Grant: 20,000 USD
Contact: Eileen M. Cannon, (202) 572-7756; fax (202) 572-7799; info@phrmafoundation.org
Internet: http://www.phrmafoundation.org/awards/pharmacology/predoc.php
Sponsor: Pharmaceutical Research and Manufacturers of America Foundation
950 F Street NW, Suite 300
Washington, DC 20004

PhRMA Foundation Pharmacology/Toxicology Research Starter Grants **3964**

The purpose of the program is to offer financial support to individuals beginning their independent research careers at the faculty level. The areas of interest within this program are research efforts in the fields of pharmacology and drug toxicology. The program provides a research grant of $30,000 per year for up to two years. The 'starter' aspect of the program strives to assist those individuals who are establishing careers as independent investigators in the fields listed above. The program is not offered as a means to augment an ongoing research effort. The funds are to be used to conduct the proposed research. This program supports individuals beginning independent research careers in academia who do not have other substantial sources of research. Applicants will be judged on the scientific merit of the proposed research, and on the degree of financial need.
Requirements: Those holding the academic rank of instructor or assistant professor and investigators at the doctoral level with equivalent positions are eligible to apply for these research starter grants, providing their proposed research is neither directly nor indirectly subsidized to any significant degree by a competitive extramural grant. The program is not intended for those in postdoctoral training programs. However, individuals in postdoctoral training scheduled to conclude and who will hold an academic appointment by January 1 may apply. Applicants must be sponsored by the department or unit in which the proposed research is to be undertaken. The sponsoring unit must have responsibility for pharmacology/toxicology teaching and research as part of its core mission. If the pharmacology/toxicology mission is not apparent on the sponsoring unit's web site, the sponsor letter must describe how the academic appointment will support the applicant's career development in pharmacology/toxicology. The grant is made to the university on behalf of the applicant, with the understanding that the university will administer the funds. Schools of medicine (human or veterinary), pharmacy and dentistry are eligible for this award. In the field of pharmacology or toxicology, highest priority will be given to applications for research that attempts to integrate information on the mechanism of action of a drug or chemical at a molecular or cellular level with the drug's effects in the intact laboratory animal or human, encompassing the potential influences of biochemical, physiological or behavioral systems. Applications must be submitted by an accredited U.S. school, and all applicants must be U.S. citizens or permanent residents.
Restrictions: Research projects that extend or develop the proprietary value of specific drug products are not acceptable in this program. This exclusion does not preclude research in which specific drug products are used to test hypotheses that have a general applicability. The funds may not be used for salary support of the grantee, but may be used to support technical assistance. No more than $500 a year may be used for travel to professional meetings by the grantee. Indirect costs are not provided to the institution, and grant funds may not be used for this purpose. Applicants holding multi-year national competitive awards or grants supporting their independent research will not be considered. If a national competitive grant is obtained after the application has been submitted, the Foundation must be so informed.
Geographic Focus: All States
Amount of Grant: 30,000 USD
Contact: Eileen M. Cannon, (202) 572-7756; fax (202) 572-7799; info@phrmafoundation.org
Internet: http://www.phrmafoundation.org/awards/pharmacology/starter.php
Sponsor: Pharmaceutical Research and Manufacturers of America Foundation
950 F Street NW, Suite 300
Washington, DC 20004

PhRMA Foundation Pharmacology/Toxicology Sabbatical Fellowships **3965**

This program provides stipend funding to enable faculty members at all levels with active research programs an opportunity to work at other institutions for periods of six months to one year to learn new skills or develop new collaborations that will enhance their research and research training activities in pharmacology/toxicology. Applicants are expected to engage in multidisciplinary research that integrates information on molecular or cellular mechanisms of action with information on the effect of an agent in the intact organism. It is anticipated that this research experience will occur in academic and/or industrial laboratory settings as part of a multidisciplinary team organized to integrate informatics, molecular, cell, and systems biology with pharmacology/toxicology research. The award provides up to $40,000 stipend support for mid-career scientists to engage in an academic year or calendar year experiential program intended to redirect their core research focus to an area of emerging importance to pharmaceutical research and development.
Requirements: Eligible applicants must hold a PhD degree or appropriate terminal doctorate and record of research accomplishment in a field of study logically or functionally related to the proposed post doctoral activities; hold a faculty appointment that imparts eligibility for a sabbatical leave from their home institution; have institutional approval of a sabbatical plan that includes partial salary that matches the PhRMA stipend; hold an endorsement from a mentor who agrees to sponsor the applicants visiting scientist activity; and be a U.S. citizen or permanent resident. The applicant and mentor of the program must describe how the multidisciplinary goals of the research experiential program will be accomplished and provide assurance that key collaborating mentors endorse and are willing to support the plan. Matching funds must be provided through the university.
Restrictions: The program provides no other subsidies (travel, tuition, fringe benefit costs, etc.) and indirect costs to the institution are not provided.
Geographic Focus: All States
Amount of Grant: 40,000 USD
Contact: Eileen M. Cannon, (202) 572-7756; fax (202) 572-7799; info@phrmafoundation.org
Internet: http://www.phrmafoundation.org/awards/pharmacology/sabbatical.php
Sponsor: Pharmaceutical Research and Manufacturers of America Foundation
950 F Street NW, Suite 300
Washington, DC 20004

Pi Lambda Theta Research Grants **3966**

Research Grants are awarded in amounts up to $2,500 for clearly conceptualized research proposals concerning education. A research grant proposal can be for an independent project or a part of a larger project, such as a dissertation. Research grants are funded by the Pi Lambda Theta Educational Endowment and the Maxine Northrop Stephanian Fund.
Requirements: Grant available to Pi Lambda Theta members only.
Geographic Focus: All States
Date(s) Application is Due: Apr 1; Oct 1
Amount of Grant: Up to 2,500 USD
Contact: Pam Todd; (812) 339-3411; fax (812) 339-3462; pam@pilambda.org
Internet: http://www.pilambda.org/benefits/awards/ScholarshipsAwards.html
Sponsor: Pi Lambda Theta
4101 E Third Street, P.O. Box 6626
Bloomington, IN 47407-6626

Pinkerton Foundation Grants **3967**

The Pinkerton Foundation is an independent grantmaking foundation established in 1966 by Robert Allan Pinkerton with the broad directive to reduce the incidence of crime and to prevent juvenille delinquency. The Foundation's principal program interests are focused on economically disadvantaged children, youth and families, and severely learning disabled children and adults of borderline intelligence. The Foundation supports efforts to strengthen and expand community-based programs for children, youth and families in New York City. The Foundation also occasionally funds research, demonstration and evaluation projects in its principal program areas. While grants for direct service projects are usually limited to New York City, those with potential for national impact or replication may go beyond this geographic limitation. The Foundation's Board of Directors have two grantmaking meetings per year, in May and in December. Letters of inquiry are welcome throughout the year. Additional guidelines are available at the Foundation's website.
Requirements: Grants are awarded primarily to New York City 501(c)3 nonprofit public charitable organizations.
Restrictions: The foundation does not grant requests for emergencies, medical research, direct provision of health care, religious education, conferences, publications, or capital projects.
Geographic Focus: All States
Date(s) Application is Due: Feb 1; Sep 1
Amount of Grant: 25,000-250,000 USD
Samples: Reel Works Teen Filmmaking, Brooklyn, NY, $35,000—operating support for filmmaking program that serves up to 180 high school students; Madison Square Boys & Girls Club (New York, NY) $250,000—for after school and summer program operations at Madison's clubhouses; New York Cares, Inc., New York, NY, $20,000—in support of their SAT Prep Program.
Contact: Joan Colello, (212) 332-3385; fax (212) 332-3399; pinkfdn@pinkertonfdn.org
Internet: http://fdncenter.org/grantmaker/pinkerton
Sponsor: Pinkerton Foundation
610 Fifth Avenue, Suite 316
New York, NY 10020

Pioneer Hi-Bred Honorary Plant Breeding Fellowships **3968**

The Pioneer Hi-Bred Honorary Plant Breeding Fellowships honor the contributions of 10 outstanding Pioneer plant breeders to the progress of science and agriculture. These fellowships are intended to promote graduate education in plant breeding and genetics, and are expected to attract high-quality students whose future careers will continue to advance the science of plant breeding. A Fellowship Committee established by Pioneer will review applications and select recipients based on their qualifications, including academic excellence; proposed study program area, with consideration given to agronomic crops such as corn, soybeans, canola, sorghum, sunflower, wheat and alfalfa; future professional plans; and leadership and other experiences predicting future success. The funding limit is $25,000 per year for the duration of the fellowship.
Requirements: Scholarship candidates must be nominated and sponsored by a Pioneer Hi-Bred employee.
Geographic Focus: All States
Contact: Tabare Abadie, (515) 270-3200; fax (515) 334-4415; Tabare.Abadie@pioneer.com
Internet: http://www.pioneer.com/web/site/portal/menuitem.07196fe7964ae318bc0c0a03d10093a0/
Sponsor: Pioneer Hi-Bred International
6900 NW 62nd Avenue, P.O. Box 246
Johnston, IA 50131

Pioneer Hi-Bred International Scholarships **3969**

The Pioneer Hi-Bred International Scholarship is designed to contribute to agricultural research and education by providing financial support for outstanding students studying for graduate level degrees in plant breeding overseas. Funding is considered to be an unrestricted gift. Funding decisions will be reviewed annually and any changes will be communicated by September 30 of each year. For specific details about the program, contact the listed individual by email.

Geographic Focus: All States
Contact: Peter Freymark, (515) 270-3200 or (800) 247-6803, ext. 3915; fax (515) 334-4415; Peter.Freymark@pioneer.com
Internet: http://www.pioneer.com/web/site/portal/menuitem.07196fe7964ae318bc0c0a03d10093a0/
Sponsor: Pioneer Hi-Bred International
6900 NW 62nd Avenue, P.O. Box 246
Johnston, IA 50131

Pioneer Hi-Bred Society Fellowships 3970
The Pioneer Hi-Bred Society Fellowship is designed to contribute to agricultural research and education by providing financial support for outstanding students studying for graduate level degrees in plant breeding. Leading societies that support plant improvement graduate programs in crops of interest to Pioneer have been selected by the Fellowship Sub-committee. The institutions are responsible for student selection and administering the program. Funded societies include: Agronomic Science Foundation; American Phytopathological Society; American Society of Plant Biologists; the Consultative Group on International Agricultural Research (CGIAR); and the Entomological Foundation.
Requirements: Pioneer prefers this grant to be used solely to fund a student(s) progressing towards a graduate degree in plant breeding from the funded institution. No university overhead will be paid from these funds.
Geographic Focus: All States
Contact: David Harwood, (515) 270-3200 or (800) 247-6803, ext. 3915; fax (515) 334-4415; David.Harwood@pioneer.com
Internet: http://www.pioneer.com/web/site/portal/menuitem.07196fe7964ae318bc0c0a03d10093a0/
Sponsor: Pioneer Hi-Bred International
6900 NW 62nd Avenue, P.O. Box 246
Johnston, IA 50131

Pioneer Hi-Bred University Fellowships 3971
The Pioneer Hi-Bred University Plant Breeding Fellowships are designed to contribute to agricultural research and education by providing financial support for outstanding students working toward graduate level degrees in plant breeding. Leading universities that support plant improvement graduate programs in crops of interest to Pioneer have been selected by the Fellowship Subcommittee. The institutions are responsible for student selection and administering the program. Funded institutions currently include: Cornell University; North Carolina State University; University of Illinois; University of Minnesota; and University of Wisconsin-Madison.
Geographic Focus: All States
Contact: Stephanie Raber, (515) 270-3200 or (800) 247-6803, ext. 3915; fax (515) 334-4415; Stephanie.Raber@pioneer.com
Internet: http://www.pioneer.com/web/site/portal/menuitem.07196fe7964ae318bc0c0a03d10093a0/
Sponsor: Pioneer Hi-Bred International
6900 NW 62nd Avenue, P.O. Box 246
Johnston, IA 50131

Piper Trust Healthcare and Medical Research Grants 3972
The Piper Trust's grantmaking focuses on Virginia Galvin Piper's commitment to improving the quality of life for residents of Maricopa County. Piper Trust's particular interest lies with projects that benefit young children, adolescents and older adults in Maricopa County. The Trust makes grants to faith-based organizations that serve these target populations in a manner consistent with program guidelines. For Healthcare and Medical Research grants, the trust is most interested in proposals that address improved facilities for children, adolescents and older adults; better trained healthcare workforce; increased access to basic healthcare; and, centers for advancement in personalized medicine.
Requirements: Piper Trust makes grants to actively operating Section 501(c)3 organizations in Maricopa County. These organizations must have been in operation for at least three years from the effective date of their IRS ruling. Special rules apply to private foundations and 509(a)3 (Type III) organizations. There are no deadlines on initial proposals, and letters of inquiry throughout the year are reviewed throughout the year. If the Trust asks for a full proposal, its disposition depends on its completeness and the meeting schedule of the Piper trustees.
Geographic Focus: Arizona
Contact: Terri Leon, Program Officer; (480) 556-7121; tleon@pipertrust.org
Internet: http://pipertrust.org/our-grants/healthcare-medical-research/
Sponsor: Virginia G. Piper Charitable Trust
1202 East Missouri Avenue
Phoenix, AZ 85014

Pittsburgh Foundation Medical Research Grants 3973
Through the generosity of individuals and families who have established medical research funds with The Pittsburgh Foundation, the Foundation provides support for this cutting edge research. Based on the donors' fields of interest, the Foundation provides grants for the following: Richard S. Caliguiri Fund-for research in the field of amyloidosis; The Walter L. Copeland Fund-for cranial research (disease of the brain/neurological disease), to be conducted within the Department of Neurological Surgery at the University of Pittsburgh; John F. and Nancy A. Emmerling Fund-for research in the field of neuropsychobiology (with emphasis on biochemistry) of mental disorders, with the goal of discovering the causes and cure, of amelioration, of mental disorders. The principal investigator must have full time status at the University of Pittsburgh; Albert B. Ferguson, M.D. Orthopaedic Fund-for orthopaedic research; GLBT Health and Wellness Fund-to assess the incidence, cause or cure of certain diseases (i.e., cancer, heart disease, diabetes, arthritis, AIDS and Alzheimer's disease) in the Gay, Lesbian, Bisexual and Transgender (GLBT) community; Medical Research Funding-for research in the field of cancer, heart disease, arthritis, diabetes, multiple sclerosis and geriatric diseases (with focuses particularly on Alzheimer's disease and other related diseases of the elderly).
Requirements: Pennsylvania not-for-profit institutions with 501(c)3 tax-exempt status are invited to apply for funding. Support is given to investigators in Allegheny County with a proven record of performance in the area under investigation. Guidelines and applications vary, see: http://www.pittsburghfoundation.org/page32428.cfm for a complete listing of each grant.
Geographic Focus: Pennsylvania
Contact: Judy Powell, Administrative Support; (412) 394-4294; powellj@pghfdn.org
Internet: http://www.pittsburghfoundation.org/page32428.cfm
Sponsor: Pittsburgh Foundation
5 PPG Place, Suite 250
Pittsburgh, PA 15222-5414

PKD Foundation Lillian Kaplan International Prize for the Advancement in the Understanding of Polycystic Kidney Disease 3974
The prize is a joint venture between the foundation and the International Society of Nephrology. The winning nominee of the initial award received the prize in Berlin; it has been awarded every other year since. Each award consists of a monetary award, a framed certificate with an appropriate citation, and a designated sculpture to clinicians and scientists who have made outstanding discoveries in basic science or who have made outstanding clinical contributions to the care of individuals. This is the largest cash prize for a clinical medical specialty in the world signaling the prestige and importance it carries. PKD experts from both organizations judge nominations.
Requirements: Awards are made to residents of any country without restriction to gender, race, religion, creed, or nationality. Nomination letters should describe the key role of the nominee(s) in a biomedical advance relevant to PKD and should be accompanied by a full curriculum vita with a complete mailing address and at least two supporting letters that detail a major discovery or sustained advance. If two individuals are nominated to share the prize, the reasons for why the prize would be shared must be clearly delineated.
Geographic Focus: All States
Date(s) Application is Due: Feb 1
Amount of Grant: 50,000 USD
Samples: Dr. Corinne Antignac, Necker Hospital, Paris, France, $50,000; Dr. Lisa Guay-Woodford, University of Alabama at Birmingham, $50,000; Dr. Friedhelm Hildebrandt, University of Michigan, $50,000.
Contact: Dan Lara, Government Relations Manager; (816) 268-8480; fax (816) 931-8655; larad@pkdcure.org or pkdcure@pkdcure.org
Internet: http://www.pkdcure.org
Sponsor: PKD Foundation
9221 Ward Parkway, Suite 400
Kansas City, MO 64114-3367

PKD Foundation Research Grants 3975
The PKD Foundation, established in 1982 by Joseph H. Bruening and Jared J. Grantham, leads the fight against polycistic kidney disease through vital research funding and patient education. Annually, the PKD Foundation funds more than $4.2 million in vital PKD research. Most recently, clinical trials in cardiovascular health, Rapamycin, Somatostatin, Tolvaptan, and more have offered help and hope to the 12.5 million worldwide with the disease.
Geographic Focus: All States
Amount of Grant: 25,000-1,000,000 USD
Contact: Dan Lara, Government Relations Manager; (816) 268-8480; fax (816) 931-8655; larad@pkdcure.org or pkdcure@pkdcure.org
Internet: http://www.pkdcure.org/Research/tabid/82/Default.aspx
Sponsor: PKD Foundation
9221 Ward Parkway, Suite 400
Kansas City, MO 64114-3367

Plan for Social Excellence Grants 3976
Projects supported by the foundation take place at various levels including early childhood education, technology, after-school mentoring, and higher education. Additionally, the foundation supports programs that are fluid and responsive to the needs of individual schools and communities rather than programs that attempt to address these needs through a system-wide process of reform. Fields of interest include: Grants to schools or other forms of educational institutions for innovative projects; Publication of monographs and videotapes related to the work of the foundation; Advocacy on behalf of children in selected issues; and, Support of research in specific education areas.
Requirements: Projects desiring consideration for funding must be for either the replication of Plan's projects or new initiatives. If not previously invited to apply via an RFP, submit a proposal for the funding of a project or program that is in line with the vision and the mission of the foundation. Review the guidelines posted on the website for clarification. NOTE: The deadline for proposals to be funded during the following school year is by 5:00 P.M. EST on the first Friday in May.
Restrictions: The foundation does not make operating grants.
Geographic Focus: All States
Date(s) Application is Due: May 4
Contact: Dr. Mario J. Pena; (813) 282-1966; fax (813) 282-1679; mariopena@pfse.org
Sponsor: Plan for Social Excellence
2502 Rocky Point Drive, Suite 880
Tampa, FL 33607

Plastic Surgery Educational Foundation/AAO-HNSF Combined Grant 3977
The PSEF/AAO-HNSF Combined Grant is designed to promote interdisciplinary research by encouraging a collaborative research effort between the specialties of Plastic Surgery and Otolaryngology. Recipients of this grant must demonstrate a potential for excellence in research and training and a commitment to an academic research career in either Plastic Surgery or Otolaryngology.
Requirements: Applicants must have demonstrated a potential for excellence in research and teaching and serious commitment to an academic research career in otolaryngology or plastic surgery. Priority will be given to senior residents, fellows or faculty who have completed residencies or fellowships within four years of the application receipt date.
Geographic Focus: All States
Date(s) Application is Due: Dec 15
Amount of Grant: Up to 20,000 USD
Contact: Carol V. Wargo, Director of Corporate Relations; (847) 228-3358 or (847) 228-9900; fax (847) 228-9131; cvw@plasticsurgery.org or cschmieden@plasticsurgery.org
Internet: http://www.plasticsurgery.org/medical_professionals/research/Grant-Applications.cfm
Sponsor: Plastic Surgery Educational Foundation
444 E Algonquin Road
Arlington Heights, IL 60005

Plastic Surgery Educational Foundation Basic Research Grants 3978
The Plastic Surgery Educational Foundation and the American Society for Aesthetic Plastic Surgery offer annual research grants to provide seed money to U.S. and Canadian surgeons for research projects in the field of plastic and reconstructive surgery. Nonmembers of PSEF must submit a letter of sponsorship from a foundation member.
Requirements: Applicant must be a plastic surgeon, MD, or PhD working in plastic surgery.
Geographic Focus: All States
Date(s) Application is Due: Jan 17
Amount of Grant: 5,000 USD
Samples: Dr. James Anthony, U of California (San Francisco, CA)—for research on wound cytokine levels, $4600. Dr. Steven Buchman, U of Michigan (MI)—for research on cranial suture closure, $4998. Dr. James Doyle, U of Massachusetts Medical Ctr (MA)—for research questioning IL-10 as an immunosuppressive factor in chronic wound fluid, $5000.
Contact: Carol V. Wargo, Director of Corporate Relations; (847) 228-3358 or (847) 228-9900; fax (847) 228-9131; cvw@plasticsurgery.org or cschmieden@plasticsurgery.org
Internet: http://www.plasticsurgery.org/medical_professionals/research/Grant-Applications.cfm
Sponsor: Plastic Surgery Educational Foundation
444 E Algonquin Road
Arlington Heights, IL 60005

Plastic Surgery Educational Foundation Research Fellowship Grants 3979
The Research Fellowship Grants are awarded for the purpose of encouraging research and academic career development in Plastic Surgery. The PSEF Research Fellowship is for training in any area of Plastic Surgery and is designed for those who wish to supplement one year of clinical training for a research experience. The PSEF/KCI Wound Care Research Fellowship is open to plastic surgery residents and recent plastic surgery resident graduates focusing on a project in wound care management.
Geographic Focus: All States
Samples: Dr. Emmanuella Joseph (Boston, MA)—senior award winner for clinical research, $3000. Dr. Jayesh Panchal (Oklahoma City, OK)—junior award winner for clinical research, $3000. Dr. W.P. Andrew Leww (Boston, MA)—senior award winner for basic science, $3000. Dr. Vaishali Doolabh (Saint Louis, MO)—junior award winner for basic science, $3000.
Contact: Carol V. Wargo, Director of Corporate Relations; (847) 228-3358 or (847) 228-9900; fax (847) 228-9131; cvw@plasticsurgery.org or cschmieden@plasticsurgery.org
Internet: http://www.plasticsurgery.org/medical_professionals/research/Grant-Applications.cfm
Sponsor: Plastic Surgery Educational Foundation
444 E Algonquin Road
Arlington Heights, IL 60005

Playboy Foundation Grants 3980
The Playboy Foundation seeks to foster social change by confining its grants and other support to projects of national impact and scope involved in fostering open communication about, and research into, human sexuality, reproductive health and rights; protecting and fostering civil rights and civil liberties in the United States for all people, including women, people affected and impacted by HIV/AIDS, gays and lesbians, racial minorities, the poor and the disadvantaged; and eliminating censorship and protecting freedom of expression and First Amendment rights. The Foundation does not accept unsolicited proposals, but welcomes letters of inquiry via post for the areas of interest noted above. Grants awarded by the Foundation are typically up to $10,000.
Restrictions: The foundation will not consider religious programs, individual needs, capital campaigns, endowments, scholarships, or fellowships; social services, including residential care, clinics, treatment, or recreation programs; national health, welfare, educational, or cultural organizations, or their state affiliates; or government agencies or projects.
Geographic Focus: All States
Amount of Grant: 5,000-10,000 USD
Contact: Executive Director; (312) 373-2437 or (312) 751-8000; giving@playboy.com
Internet: http://www.playboyenterprises.com/foundation
Sponsor: Playboy Foundation
680 North Lake Shore Drive
Chicago, IL 60611

Pollock-Krasner Foundation Grants 3981
The foundation seeks to find artists of recognized merit throughout the world who have current financial need, whether professional or personal, for whom grants will assist in pursuing their careers. The amount of the grant is determined by the individual circumstances of the artist. The foundation's mission is to aid individuals who have worked as professional artists over a significant period of time. The foundation cannot respond to application requests by telephone or in person. Write to the office for guidelines. Applications are accepted anytime.
Requirements: The foundation welcomes applications from painters, sculptors, printmakers, and artists who work on paper.
Restrictions: The foundation will not accept applications from commercial artists, photographers, video artists, filmmakers, craftmakers, or any artist whose work primarily falls into these categories. Grants are not made to students nor to support academic The Foundation does not make grants to pay for past debts, legal fees, the purchase of real estate, moves to other cities, or to pay for the costs of installations, commissions, or projects ordered by others. With very few exceptions, the foundation will not fund travel expenses. study.
Geographic Focus: All States
Contact: Program Officer; (212) 517-5400; fax (212) 288-2836; grants@pkf.org
Internet: http://www.arts.state.va.us/pkfgrants.htm
Sponsor: Pollock-Krasner Foundation
863 Park Avenue
New York, NY 10021

Polytechnic University International Research Studentships 3982
Research Studentships support full-time PhD or MPhil students. The period of the studentship is decided by the department according to the student's ability, academic background, or experience. Recipients spend up to six hours of training each week within their department. The training involves assisting teaching and scholarly activities and is relevant to students' programs of study.
Requirements: MPhil and PhD students admitted to the university are eligible.
Restrictions: Part-time students are not eligible.
Geographic Focus: All States
Contact: Research; (852) 2766-5025; fax (852) 2355-7651; roro@inet.polyu.edu.hk
Internet: http://www.me.polyu.edu.hk/virtual/Acad_prog/mphil.htm#Studentships
Sponsor: Polytechnic University
Chung Sze Yuen Building, Room AG328
Hung Hom, Kowloon Hong Kong

Postdoctoral Research Fellowships in Gerontological Public Health 3983
The Center for Research on Health and Aging (CRHA) at the university offers postdoctoral fellowships in gerontological public health. These are two- to three-year appointments sponsored by the National Institute on Aging. This program focuses on developing research skills through mentoring, a comprehensive curriculum, hands-on research experience, presentations and publications, and opportunities for grantsmanship. The goal of the program is to foster and hone the research skills of future faculty and academic researchers in health and aging. This is an integrated multidisciplinary program with academic faculty and researchers in public health, gerontology, geriatrics, epidemiology, biostatistics, psychology, sociology, disability and human development, medicine and occupational therapy. Each postdoc fellow will be matched with a mentor from the Training Program Core Faculty. This primary mentor will work with the fellow to guide and collaborate on research projects. The fellowship includes a salary, with reimbursement for health insurance and yearly travel support. Additional financial support is also available. All fellows will have office space, a computer, phone, and access to university resources through the CRHA. Applications are accepted at any time for open positions. Guidelines are available online.
Requirements: Candidates should have received (or be very close to receiving) a doctoral degree in public health, gerontology, sociology, medicine, disability studies, or a related discipline.
Geographic Focus: All States
Contact: Dr. Thomas Prohaska, (312) 413-9830; fax (312) 413-9835; prohaska@uic.edu
Internet: http://www.uic.edu/depts/ovcr/hrpc/centers/fellow_gphpost.html
Sponsor: University of Illinois at Chicago
1747 W Roosevelt Road
Chicago, IL 60608

Pott Foundation Grants 3984
Established in 1963 by Herman T. Pott, The Pott Foundation is a non-profit organization supporting children, education and health and human services. Most of the grant recipients are located in Herman Pott's adopted city of St. Louis, Missouri. The Foundation distributes approximately $1.5 million annually to over 100 charities.
Requirements: Non-profit 501(c)3 organization are eligible to apply in Missouri. Mail grant applications to: The Pott Foundation, C/O U.S. Bank, N.A., The Private Client Reserve, Attn: Carol Eaves, Mail Loc: SL-MO-CTCS, 10 North Hanley, Clayton, MO 63105.
Geographic Focus: Missouri
Date(s) Application is Due: Apr 1
Contact: Carol Eaves, (314) 418-8317
Sponsor: Pott Foundation
10 North Hanley
Clayton, MO 63105

Potts Memorial Foundation Grants 3985

The Potts Memorial Foundation is a private foundation, incorporated in 1922 in New York. The Foundation gives on a national and international basis. The foundation was established to provide for the care, treatment, and rehabilitation of persons afflicted with tuberculosis; support for tuberculosis eradication, including fellowship programs for physicians. Grants given for tuberculosis and related disorders.
Requirements: Application form not required, Applicants should submit a detailed description of project and amount of funding requested. Proposal must be submitted via mail.
Restrictions: No grants to individuals, or for endowment funds or matching gifts; no loans.
Geographic Focus: All States
Date(s) Application is Due: Apr 15; Sep 15
Amount of Grant: 15,000-45,000 USD
Samples: Johns Hopkins University, Baltimore, Maryland, $41,800; Columbia Memorial Hospital, Hudson, New York, $20,000;
Contact: Charles Inman, Secretary; (518) 828-3365
Sponsor: Potts Memorial Foundation
444 Warren Street
Hudson, NY 12534-2415

Praxair Foundation Grants 3986
The foundation supports nonprofit organizations primarily in the communities where Praxair operates. Preference is given to projects in the following areas: higher education—science, engineering, and environmental studies and programs that encourage minority students to pursue studies in these areas; community service—United Way (matching employee donations), public Libraries (technology upgrading, acquiring new materials, or initiating innovative programs), employee volunteer grants (grants to nonprofit organizations where employees volunteer); public policy and the environment; and international organizations operating in countries where Praxair is expanding its presence, primarily in Asia. All proposals should describe projects that serve the needs of the community at large. Special attention will be given to innovative programs that can be started and completed within one year of the award. Goals should be clearly defined and measurable. The level of support varies depending upon the nature of the project and how closely it fits with foundation guidelines. To apply for a grant, complete the Praxair Foundation Funding Request form.
Requirements: 501(c)3 tax-exempt organizations are eligible.
Geographic Focus: All States
Amount of Grant: 1,000-50,000 USD
Contact: Grants Administrator; (800) 772-9247; fax (716) 879-2040
Internet: http://www.praxair.com/praxair.nsf/AllContent/1E607B9456292F1185 256 BF800779733?OpenDocument
Sponsor: Praxair Foundation
39 Old Ridgebury Road
Danbury, CT 06810

Price Family Charitable Fund Grants 3987
Established in 1983, the Price Family Charitable Fund serves the San Diego, Carlsbad and, San Marcos, California region. The foundation is interested in supporting the economically disadvantaged, giving primarily for education and philanthropy purposes. The types of support include: annual campaigns; fellowships; program evaluation; scholarship funds; scholarships to individuals and; general operating support. There's no formal application to submit. Potential grantees much first submit a letter of Inquiry including: project goals, objectives and expected results (maximum of one page); project narrative (maximum of five pages); project budget (maximum of one page); a list of grants, if any, received by organization in the last 12 months for this program/project (sources and amounts); a list of the organization's current board of directors, including each member's name, profession, and office help on the board, if any.
Requirements: Non-profit organizations with 501(c)3 status or governmental units such as public schools or city departments in are available for funding. The Foundation gives primarily in the following region of: San Diego; Carlsbad; San Marcos, California.
Restrictions: Grants are not made: to organizations whose primary purpose is religious, or for propagandizing, influencing legislation and/or elections, promoting voting registration, for political candidates, political campaigns or organizations engaged in political activities; or to federal appeals or to organizations the collect funds for redistribution to other non-profit groups. Unsolicited requests for funds are not accepted.
Geographic Focus: California
Contact: Terry Malavenda, (858) 551-2330
Sponsor: Price Family Charitable Fund
7979 Ivanhoe Avenue, Suite 520
La Jolla, CA 92037-4513

Princess Grace Foundation Theater, Dance, and Film Grants 3988
The Princess Grace Foundation Grants for Theater, Dance, and Film are awarded to individuals, through their affiliation with nonprofit organizations, in the areas of dance, theater, and film. The grants are made on an annual basis by application for scholarships, fellowships, and apprenticeships. This unique collaborative process fills voids in the artistic community. Scholarships, apprenticeships, and fellowships give emerging artists the financial assistance and moral encouragement to focus on artistic excellence, while monetary support for the nominating organizations eases fund raising challenges, directing resources toward the creative process. The March 31 application deadline is for theater and playwriting, April 30 for dance, and June 1 for film. The application and specific guidelines for each category of funding are available at the Foundation website.
Requirements: Applicants must be U.S. citizens or have permanent resident status. They must also be studying with a company in the United States. All applicants except playwrights must be nominated by a school department chair or dean or company artistic director. The nominating organization must be a registered 501(c)3 company.
Restrictions: Short training programs, intensives, and film festivals are not eligible for funding.
Geographic Focus: All States
Date(s) Application is Due: Mar 31; Apr 30; May 15; Jun 1
Amount of Grant: 5,000-25,000 USD
Contact: Toby Boshak; (212) 317-1470; fax (212) 317-1473; pgfusa@pgfusa.com
Internet: http://www.pgfusa.com/awards/grants/index.html
Sponsor: Princess Grace Foundation-U.S.A
150 East 58th Street, 25th Floor
New York, NY 10155

Princeton University Center for Human Values Graduate Prize Fellowships 3989
The Graduate Prize Fellowship program is intended to recognize and support post-generals graduate students with distinguished academic records in any discipline whose dissertation research centrally involves the critical study of human values. In addition to providing tangible support, a main purpose is to enable people working on topics in human values to engage with each other both informally and in a faculty-led fellows' seminar, where they present their work in progress. The principal considerations in the selection of fellows will be the promise of excellence in an applicant's dissertation work, the applicant's capacity to contribute constructively to the fellows' group and the likelihood that participation in the program will help to shape and improve the applicant's dissertation. Accordingly, preference will be given to applicants who are in the relatively early stages of work on their dissertations. Graduate Prize Fellowships provide a stipend of $32,000 for twelve months.
Requirements: The fellowships are for Princeton University graduate students only. Fellows must be enrolled for the duration of the fellowship and may not take on any teaching or other employment. Except in extraordinary cases, we do not expect to support students who would be beyond their fifth year of enrollment during the fellowship year.
Geographic Focus: All States
Date(s) Application is Due: Mar 6
Amount of Grant: 32,000 USD
Contact: Erin Graham, Program Coordinator; (609) 258-1460 or (609) 258-4798; fax (609) 258-2729; egraham@princeton.edu
Internet: http://uchv.princeton.edu/fellowships_awards/graduate_prize_fellowships.php
Sponsor: Princeton University
5 Ivy Lane, 304 Louis Marx Hall
Princeton, NJ 08544

Princeton University Center for Human Values Graduate Student Top-Up Prizes 3990
The University Center for Human Values offers a limited number of top-up prizes to admitted graduate students who show promise in the area of ethics, political theory/philosophy, and human values. The prize is a one-time only award that is added to the stipend during the first year of study. Each department may nominate several students for this award. All applicants who have been offered admission are eligible.
Geographic Focus: All States
Contact: Erin Graham, Program Coordinator; (609) 258-1460 or (609) 258-4798; fax (609) 258-2729; egraham@princeton.edu
Internet: http://uchv.princeton.edu/fellowships_awards/topup.php
Sponsor: Princeton University
5 Ivy Lane, 304 Louis Marx Hall
Princeton, NJ 08544

Princeton University Center for Human Values Harold T. Shapiro Postdoctoral Research Associate in Bioethics 3991
The Harold T. Shapiro Postdoctoral Research Associate position supports outstanding scholars studying ethical issues arising from developments in medicine or the biological sciences. We particularly encourage proposals focusing on problems of practical importance that have broader theoretical interest. The Shapiro Research Associate spends from one to three years at Princeton conducting research and teaching the equivalent of one course each year. He or she participates in the Ira W. DeCamp Seminar in Bioethics and is invited to participate in other activities of the University Center for Human Values, including a year-long research seminar for visiting fellows and Center faculty. The term of the position is one year with the possibility of extension for up to two further years. Princeton University offers competitive salary and employee benefits.
Requirements: Applicants must have completed all the requirements for the Ph.D., M.D., or other equivalent doctoral degree by the start of the appointment and must not have held the degree for more than three years. An applicant may have a background in any area of ethical studies, not necessarily in bioethics. However, the proposed research plan should be related to the field of bioethics.
Geographic Focus: All States
Contact: Erin Graham, Program Coordinator; (609) 258-1460 or (609) 258-4798; fax (609) 258-2729; egraham@princeton.edu
Internet: http://uchv.princeton.edu/fellowships_awards/hts_bioethics.php
Sponsor: Princeton University
5 Ivy Lane, 304 Louis Marx Hall
Princeton, NJ 08544

Princeton University Center for Human Values Laurance S. Rockefeller Visiting Fellowships 3992
The University Center for Human Values invites applications for Laurance S. Rockefeller Visiting Faculty Fellowships for the academic year. Fellows devote an academic year in residence at Princeton to do research and writing about topics involving human values in public and private life. The program is open to scholars in all disciplines provided their research plans qualify.

In recent years fellows have been drawn from fields including philosophy, political theory, literature, history, classics, economics and law, but this list is not meant to be exhaustive.
Requirements: Fellows are expected to reside in or around Princeton and to be active contributors to the intellectual life of the Center. This includes participating in a weekly seminar attended by fellows and Center faculty to discuss work in progress and in various other seminars, colloquia and lectures sponsored by the Center. Fellows enjoy access to Firestone Library and to a wide range of activities throughout the University. Candidates should have a doctorate or equivalent professional degree and a strong record of research publications appropriate to their career stage.
Geographic Focus: All States
Date(s) Application is Due: Nov 5
Contact: Erin Graham, Program Coordinator; (609) 258-1460 or (609) 258-4798; fax (609) 258-2729; egraham@princeton.edu
Internet: http://uchv.princeton.edu/fellowships_awards/lsr_visiting_fellowships.php
Sponsor: Princeton University
5 Ivy Lane, 304 Louis Marx Hall
Princeton, NJ 08544

Princeton University Center for Human Values Political Philosophy Travel Grants 3993
The University Center for Human Values, along with the Program in Political Philosophy, offers Political Philosophy Research and Travel grants, supported by a fund established by Amy Gutmann, former provost of the University and founding director of the University Center for Human Values. Individual grants of up to $2,000 are awarded to graduate students working on topics in political philosophy and enrolled in the departments that participate in the program (Classics, History, Philosophy, Politics, Religion) for summer travel.
Requirements: The graduate student must be enrolled at the university during the time of travel and grants are to be used in support of research and travel related to research. The intent is to increase the overall support provided by the University for excellent work in the field of political philosophy. The terms of the fund specify that grants must be used to augment resources already available at Princeton from other sources.
Geographic Focus: All States
Amount of Grant: Up to 2,000 USD
Contact: Erin Graham, Program Coordinator; (609) 258-1460 or (609) 258-4798; fax (609) 258-2729; egraham@princeton.edu
Internet: http://uchv.princeton.edu/fellowships_awards/graduate_student_travel_grants.php
Sponsor: Princeton University
5 Ivy Lane, 304 Louis Marx Hall
Princeton, NJ 08544

Princeton University Center for Human Values Postdoctoral Research Associate in Values and Public Policy 3994
The University Center for Human Values and the Woodrow Wilson School of Public and International Affairs invite applications for postdoctoral positions in Values and Public Policy. The program aims to support highly promising scholars trained in moral and political philosophy, political theory, normative economics and related areas to develop a research agenda in the ethical dimensions of public policy. Candidates selected will undertake a research project exploring a normative problem arising in an area of public policy in which the Woodrow Wilson School sponsors research. They will be affiliated with a unit of the School that can inform and support their work. They will also be invited to participate in programs of the University Center for Human Values. Candidates will be expected to contribute one course each year to the School's undergraduate program on a topic related to ethical issues in public policy, subject to sufficient enrollment and approval by the Dean of the Faculty. The term of appointment is one year, normally beginning September 1, with the expectation of renewal for one further year assuming good performance. Applicants are expected to be in residence for the duration of the appointment. Princeton offers competitive salary and employee benefits. Candidates should submit an online application.
Requirements: Applicants must have completed the requirements for the Ph.D. by the start of the appointment and must not have held the degree for more than three years by that date. They may not be employed by another institution during the term of their Princeton appointment.
Geographic Focus: All States
Date(s) Application is Due: Dec 3
Contact: Erin Graham, Program Coordinator; (609) 258-1460 or (609) 258-4798; fax (609) 258-2729; egraham@princeton.edu
Internet: http://uchv.princeton.edu/fellowships_awards/vpp_postdoc.php
Sponsor: Princeton University
5 Ivy Lane, 304 Louis Marx Hall
Princeton, NJ 08544

Princeton University Center for Human Values Short Movie Prize 3995
The UCHVFest is to be an annual gathering of the Center community to celebrate the achievements and participation of the Center's undergraduate students in their engagement with human values. The Center's undergraduate activities include the Undergraduate Certificate Program in Values and Public Life, the Film Forum, the Environmental Affairs Forum, and the Human Values Forum. The UCHVFest adds another way for undergraduates to engage with human values through the Center, namely, by making a short movie addressing the current year's theme. Selected movies will be screened and a $1,000 Short Movie Prize will be awarded. Undergraduate students enrolled at Princeton during academic year are invited to submit entries no later than February 15.
Requirements: The movie should be no more than seven minutes and should address the current year's theme.
Geographic Focus: All States
Date(s) Application is Due: Feb 15
Amount of Grant: 1,000 USD
Contact: Erum Syed; (609) 258-2739 or (609) 258-4798; erum@princeton.edu
Internet: http://uchv.princeton.edu/fellowships_awards/short_movie_prize.php
Sponsor: Princeton University
5 Ivy Lane, 304 Louis Marx Hall
Princeton, NJ 08544

Princeton University Lewis Center for the Arts Hodder Fellowships 3996
The Hodder Fellowship will be given to writers and other artists of exceptional promise to pursue independent projects at Princeton University during the academic year. Hodder Fellows may be writers, composers, choreographers, visual artists, performance artists, or other kinds of artists or humanists who have much more than ordinary intellectual and literary gifts; they are selected more for promise than for performance. Given the strength of the applicant pool, most successful Fellows have published a first book or have similar achievements in their own fields; the Hodder is designed to provide Fellows with the studious leisure to undertake significant new work. Hodder Fellows spend an academic year at Princeton, but no formal teaching is involved. A $68,000 stipend is provided. Apply by November 1 through the Princeton Human Resources website.
Requirements: One need not be a U.S. citizen to apply.
Restrictions: Fellowships cannot fund work leading to the Ph.D. degree.
Geographic Focus: All States
Date(s) Application is Due: Nov 1
Amount of Grant: 68,000 USD
Contact: Michael Cadden, Acting Chair; (609) 258-1500; fax (609) 258-2230; mcadden@princeton.edu or oconnorm@princeton.edu
Internet: http://www.princeton.edu/arts/lewis_center/society_of_fellows/
Sponsor: Princeton University
185 Nassau Street
Princeton, NJ 08544

Princeton University Library Research Grants 3997
The Friends of the Princeton University Library offers short-term research grants to promote scholarly use of the research collections of the library. The Program in Hellenic Studies supports a limited number of library fellowships in Hellenic Studies, and the Cotsen Children's Library supports research in its collection on aspects of children's books. These Library Research Grants are meant to help defray expenses incurred in traveling to and residing in Princeton during the tenure of the grant. The length of the grant will depend on the applicant's research proposal, but is usually up to one month. Application information and guidelines are available at the website.
Requirements: Applicants are asked to read the FAQ at the website, then complete an online application form, and submit a Word or PDF file (preferred) containing a budget form, a curriculum vitae or resume, and a research proposal not exceeding one thousand words in length. Applicants must also arrange for two confidential letters of recommendation to be sent directly to the Library Research Grants Committee. The proposal should address specifically the relevance to the proposed research of unique resources found in the Princeton University Library collections. Applications will be considered for scholarly use of archives, manuscripts, rare books, and other rare and unique holdings of the Department of Rare Books and Special Collections, the Mudd Library, the Marquand Library of Art and Archaeology, and the East Asian Library (Gest Collection). Prospective grantees are urged to consult the Library's home page for detailed descriptions of the collections, especially those in the Department of Rare Books and Special Collections. Applicants should have specific Princeton resources in mind as they prepare their proposals. A committee consisting of members of the faculty, the library staff, and the Friends of the Library will award the grants on the basis of the relevance of the proposal to unique holdings of the library, the significance of the project, and the applicant's scholarly qualifications.
Restrictions: Recommendation letters must be received from an institutional email address, and not a commercial provider such as hotmail. The general circulating collections and electronic resources of the Princeton University Library are not relevant for purposes of these grants.
Geographic Focus: All States
Date(s) Application is Due: Jan 15
Amount of Grant: Up to 3,500 USD
Contact: Research Grants Contact; (609) 258-3184; fax (609) 258-2324; rbsc@princeton.edu or pulgrant@princeton.edu
Internet: http://www.princeton.edu/rbsc/fellowships/f_ships.html
Sponsor: Princeton University Library
One Washington Road
Princeton, NJ 08544

Princeton University Whiting Postdoctoral Fellowships 3998
The program awards two postdoctoral fellowships to humanities scholars early in their careers to undertake and/or complete a major project, interact with a group of diverse postdoctoral fellows and regular faculty, and contribute to the undergraduate teaching program at the university. Fellows are appointed to a home department during the one-year tenure and will be expected to be in residence spending most of their time on their own research and to present a seminar on their work at least once during the year in their department of appointment. In addition, fellows are expected to teach one undergraduate course during the academic year. The fellowship may be renewable for one additional year. Interested applicants should send, in addition to a letter of application, a current curriculum vita, including educational background, professional experience, and prior publications; an abstract of their completed PhD dissertation, a two-page statement of their proposed research and writing

project for the fellowship; and two confidential letters of recommendation, one from their dissertation supervisor and one from another established scholar in the field.
Requirements: Scholars who have been awarded a PhD from a U.S. institution of higher education between January 1, 2001, and January 1 2003, in the fields of architecture (history, theory, and criticism), art history, language and literature, history (including history of science), East Asian Studies, Near Eastern Studies, philosophy (including political philosophy and classical philosophy), and religious studies are eligible.
Geographic Focus: All States
Date(s) Application is Due: Feb 15
Amount of Grant: 37,000 USD
Contact: Sandy Sussman, (609) 258-3035; ssussman@princeton.edu
Internet: http://gradschool.princeton.edu/financial/fellowships/competitive/honorifics/#comp000044bf59cc0000001c223c1d
Sponsor: Princeton University
The Graduate School, Clio Hall
Princeton, NJ 08544

Project AWARE Foundation Grants **3999**
The foundation's grant program supports worthwhile aquatic conservation projects. Grants are awarded to projects that have a direct benefit to the aquatic environment such as: public education (formal and informal); grassroots conservation and enhancement projects; environmentally focused research that leads to conservation measures; public awareness initiatives; environmental assessment and monitoring projects; and volunteer-supported community activism. Micro grants provide funding for localized grassroots efforts, including but not limited to, beach and underwater cleanups, mooring buoy programs, research projects, local education, and public awareness programs. Macro grants support large-scale regional, national, or international efforts, including but not limited to, broad-based education and public awareness programs and high profile conservation projects. Quarterly board meetings are typically scheduled during the third week of the month. The program operates in the United States, Australia, and the United Kingdom.
Restrictions: Grants do not support legislative advocacy to influence policy; political campaigns; projects whose methods are not environmentally accepted; overhead expenses including salaries, stipends, benefits or tuition; capital expenses including computer hardware/software or office furnishings; travel and living expenses; dive equipment or instruction; or products designed for resale.
Geographic Focus: All States
Date(s) Application is Due: Feb 15; May 15; Aug 15; Nov 15
Amount of Grant: Up to 1,000 USD
Contact: Grants Administrator; information@projectaware.org
Internet: http://www.projectaware.org/americas/english/grants.asp
Sponsor: Project AWARE Foundation
30151 Tomas Street
Rancho Santa Margarita, CA 92688

Prostate Cancer Foundation Research Grants **4000**
The heart of the Prostate Cancer Foundation's effort to to find better treatments and a cure for recurrent prostate cancer is its research awards programs. Each year, the PCF invests millions of dollars in research projects that have the greatest potential to advance the fight against prostate cancer. The PCF's research awards program has three components: Competitive Awards, Special Awards and Donor-Directed Awards. The foundation favors research applications focused on translational and clinical programs with potential for near term benefit for patients with recurrent prostate cancer. The listed application deadline is for competitive awards. Application guidelines and form are available online.
Requirements: Applications from individuals with or without institutional affiliations will be considered.
Restrictions: For-profits or government agencies are not eligible for consideration.
Geographic Focus: All States
Date(s) Application is Due: Dec 2
Amount of Grant: 25,000-100,000 USD
Contact: Grants Administrator; (310) 570-4700; fax (310) 570-4701
Internet: http://www.prostatecancerfoundation.org/site/pp.asp?c=itIWK2OSG&b=68790
Sponsor: Prostate Cancer Foundation
1250 Fourth Street, Suite 360
Santa Monica, CA 90401

Province of Alberta Queen Elizabeth II Graduate Scholarships **4001**
Administered by Alberta Scholarship Programs, this scholarship commemorates the Queen's visit to Alberta in May of 2005. Applicants are selected by their Alberta institution. Contact the faculty of graduate studies at your institution for more information about the nomination process.
Requirements: Candidates for the scholarships or fellowships must be enrolled full-time in a faculty of graduate studies in Alberta. Candidates must be either Canadian citizens or landed immigrants at the time of taking up the award.
Geographic Focus: All States, Canada
Amount of Grant: 10,800-15,000 USD
Contact: Scholarships and Bursaries; (780) 427-8640 or (780) 310-0000; fax (780) 427-1288; scholarships@gov.ab.ca
Internet: http://www.alis.gov.ab.ca/scholarships/info.asp?EK=39
Sponsor: Alberta Heritage Scholarship Endowment Fund
Capital Health Centre, South Tower 12th Floor, 10030-107 Street
Edmonton, AB T5J 3E4 Canada

Prudential Foundation Education Grants **4002**
The Prudential Foundation's areas of interest are ready-to-learn programs, ready-to-work programs, and ready-to-live programs. In order to promote sustainable communities and improve social outcomes for community residents, the Foundation focuses its strategy in the following educational areas: education leadership to support reform in public education by increasing the capacity of educators, parents, and community residents to implement public school reform; and youth development to build skills and competencies needed for young people to be productive citizens (this includes expanding arts education opportunities and supporting effective out-of-school-time programs for young people). Finally, the Foundation also funds organizations whose efforts influence policy that adapts promising practices and evidence-based approaches to instruction and learning in schools. Types of support include operating support, continuing support, annual campaigns, seed money, matching funds and employee matching gifts, consulting services, technical assistance, employee-related scholarships, research, capital campaigns, conferences and seminars, and projects/programs. The Foundation is especially interested in proposals that anticipate and address potential major problems. Funds are targeted to areas where Prudential has a strong presence. Applicant should make initial contact with a brief letter to determine whether a more detailed proposal would be acceptable.
Requirements: The Prudential Foundation supports nonprofit, charitable organizations, and programs whose mission and operations are broad and non-discriminatory. The Foundation focuses its resources to support organizations whose activities address social needs or benefit underserved groups and communities. Priority in order of preference goes to programs in Newark, New Jersey, and surrounding communities; Los Angeles, California; Hartford, Connecticut; New York, New York; Chicago, Illinois; Jacksonville, Florida; Atlanta, Georgia; Minneapolis, Minnesota; Philadelphia and Scranton, Pennsylvania; Houston and Dallas, Texas; Dubuque, Iowa; Phoenix, Arizona; and New Orleans, Louisiana.
Restrictions: The Foundation does not fund: organizations that are not tax-exempt under paragraph 501(c)3 of the U.S. Internal Revenue Code; labor, religious, political, lobbying, or fraternal groups—except when these groups provide needed services to the community at large; direct grants or scholarships to individuals; support for single-disease health groups; or good will advertising.
Geographic Focus: Arizona, California, Connecticut, Florida, Georgia, Illinois, Iowa, Louisiana, Minnesota, New Jersey, New York, Pennsylvania, Texas
Amount of Grant: Up to 1,000,000 USD
Contact: Lata Reddy, Director of Programs and Operations; (973) 802-4791; community.resources@prudential.com
Internet: http://www.prudential.com/view/page/public/12373
Sponsor: Prudential Foundation
751 Broad Street, 15th Floor
Newark, NJ 07102-3777

PSG Mentored Clinical Research Awards **4003**
The Mentored Clinical Research Award (MCRA) for new investigators is funded by a grant from the Parkinson's Disease Foundation (PDF) to the Parkinson Study Group (PSG). The PDF has partnered with the PSG to encourage the professional and scientific development of young investigators on their path to independence. To this end, this grant will support a new investigator for a one year project in patient oriented research in Parkinson's disease or other parkinsonian disorders under the mentorship of an experienced investigator. The training should lead a junior investigator to gain skills in clinical research. The research plan should address unmet needs of people living with PD, have the potential for broad application among the PD community, and lead to advances in clinically relevant treatment options. An award of $75,000 will be available this year. An electronic copy of the proposal in Microsoft Word or pdf-format must be received by the PSG on or before Friday, March 26. Applications should be sent by email to: Roseanna.Battista@ctcc.rochester.edu with a cover note that includes the candidate's name and the title of proposal. The candidate's proposal should be formatted according to the guidelines indicated in the toolkit available on the PSG website http://www.parkinson-study-group.org/PSGToolkit.asp. Applicants will be notified of the results of these reviews by May 21.
Requirements: Appropriate applicants for the MCRA are clinicians and scientists who are within 5 years of having completed formal training (this includes PhD professionals working in the field of Parkinson's disease). Fellows may apply. See website for additional guidelines.
Geographic Focus: All States
Date(s) Application is Due: Mar 26
Amount of Grant: 75,000 USD
Contact: Roseanna Battista; roseanna.battista@ctcc.rochester.edu
Internet: http://www.parkinson-study-group.org/Resources/PSG%20SRC%20Mentored%20Clinical%20Research%20Award%20RFP%202010-2011%20FINAL.pdf
Sponsor: Parkinson Study Group
University of Rochester, 1351 Mt. Hope Avenue, Suite 223
New York, NY 14620

PVA Research Foundation Grants **4004**
The foundation, established by the Paralyzed Veterans of America, supports innovative research and fellowships that improve the lives of those with spinal cord injury and disease (SCI/D). The Foundation funds five research categories: laboratory research in the basic sciences to find a cure for SCI/D; clinical and functional studies of the medical, psychosocial and economic effects of SCI/D, and interventions to alleviate these effects; design and development of assistive technology for people with SCI/D, which includes improving the identification, selection and utilization of these devices; fellowships for postdoctoral scientists, clinicians and engineers to encourage training and specialization in the field of spinal cord research; and conferences and symposia that provide opportunities for collaboration and interaction among scientists, health care providers and others involved in the SCI/D research community. The PVA Research Foundation online application site will open on May 1 to receive new applications. The annual deadline is September 1.

Requirements: Eligible grantee institutions must be located in the United States or Canada. However, investigators and fellows are not required to be U.S. or Canadian citizens. All grant applicants must have a professional degree: Ph.D. or M.D. preferred, master's degree acceptable under some circumstances. Senior fellows are encouraged to apply as principal investigators. Post-doctoral individuals are eligible to apply for fellowship support within four years of receiving a Ph.D. or completing M.D. residency. Graduate students can participate in Foundation-related research and be paid from a Foundation award.
Restrictions: Graduate students cannot apply for a Foundation grant either as a fellow or as a principal investigator.
Geographic Focus: All States, Canada
Date(s) Application is Due: Sep 1
Amount of Grant: 50,000-75,000 USD
Contact: Barbara Zupnik, Grants Portfolio Manager; (202) 416-7652 or (800) 424-8200; fax (202) 416-7641; barbaraz@pva.org or foundations@pva.org
Internet: http://www.pva.org/site/c.ajIRK9NJLcJ2E/b.6305827/k.7268/PVA_Research_Foundation.htm
Sponsor: Paralyzed Veterans of America Research Foundation
801 Eighteenth Street, NW
Washington, DC 20006-3517

R.B. Terry Charitable Foundation Grants **4005**
The foundation supports higher education institutions primarily in Raleigh, NC, and Woodberry Forest, VA. There are no application deadlines of forms. Submit a letter of inquiry to begin the application process.
Requirements: Higher education institutions in Raleigh, NC, and Woodberry Forest, VA, are eligible.
Restrictions: Individuals are not eligible.
Geographic Focus: All States
Samples: North Carolina State U (Raleigh, NC)—to establish a new veterinary hospital at the College of Veterinary Medicine, $20 million over 10 years.
Contact: Grants Administrator
Sponsor: R.B. Terry Charitable Foundation
P.O. Box 1009
High Point, NC 27261

R.C. Baker Foundation Grants **4006**
The foundation makes grants to U.S. nonprofit organizations for projects and programs that support social services for youth and the elderly, crime prevention, education, religion (Christian, Episcopal, Friends, Jewish, Methodist, and Presbyterian), scientific research, culture, and health. Support will be provided for fellowships and scholarships, general operating grants, challenge/matching grants, emergency funds, building funds, equipment, continuing support, annual campaigns, capital campaigns, renovation projects, and special projects. Submit cover letter with proposal. The board meets in June and November to consider requests.
Requirements: Nonprofit organizations are eligible. $25,000 to
Restrictions: Anaheim Memorial Medical Center, Anaheim, CA-$40,000; Harvey Mudd College, Claremont, CA-$25,000;
Date(s) Application is Due: May 1; Oct 1
Amount of Grant: 1,000-280,000 USD
Samples: Anaheim Memorial Medical Center, Anaheim, CA-$40,000; Harvey Mudd College, Claremont, CA-$25,000;
Contact: Frank Scott, Chairman; (714) 750-8987
Sponsor: R.C. Baker Foundation
P.O. Box 6150
Orange, CA 92863-6150

R.H. Wilhelm Award in Chemical Reaction Engineering **4007**
Sponsored by ExxonMobil Research and Engineering Company and administered by AIChE, the award is given to recognize significant and new contributions in chemical reaction engineering. Recipients shall have been judged as advancing the frontiers of chemical reaction engineering with an emphasis on originality, creativity, and novelty of concept application. The recipient need not be a member of the institute.
Requirements: All members of AIChE, as well as other interested persons, are urged to nominate deserving candidates. All information and supporting documents are to be included with the completed form; nomination forms and instructions may be obtained from the institute.
Geographic Focus: All States
Date(s) Application is Due: Feb 15
Amount of Grant: 3,000-3,500 USD
Contact: Gordon Ellis, Awards Administrator; (646) 495-1348; fax (646) 495-1503; gorde@aiche.org or awards@aiche.org
Internet: http://www.aiche.org/About/Awards/RHWilhelmAward.aspx
Sponsor: American Institute of Chemical Engineers
3 Park Avenue
New York, NY 10016

RACGP/APHCRI Indigenous Health Award **4008**
The College supports research which makes a difference to health outcomes, policy and practice, and the capacity of general practice to engage Aboriginal and Torres Strait Islanders' health issues. It particularly encourages a systematic approach to strategic and policy-driven research focusing on primary care, developing collaborative approaches and the building of research capacity within Indigenous populations and communities. The Award is valued at $2500 and is available for a 12 month period to provide support for: indigenous medical students; medical students to gain experience in indigenous health; indigenous medical graduates undertaking GP training; registrars and GPs undertaking research and educational projects in indigenous health, or; other activities supporting the provision of high quality health care to indigenous people.
Requirements: Applicants for the award must be either a medical student or a GP or GP registrar who is a member of the RACGP.
Restrictions: Unless otherwise indicated, the Award must be used in the year for which it is awarded. The Award will be paid only to incorporated bodies that have an ABN and ACN and not to individuals. By applying for the Award each applicant will be taken to have consented to permit the RACGP and APHCRI to publicise their name, institution, and title of their research project.
Geographic Focus: All States, Australia
Date(s) Application is Due: May 8
Amount of Grant: 2,500 USD
Contact: Samantha Fernandes; 03-8699-0496; fax 03-8699-0400; research@racgp.org.au
Internet: http://www.racgp.org.au/researchfoundation/awards/indigenous
Sponsor: Royal Australian College of General Practitioners
College House, 1 Palmerston Crescent
South Melbourne, VIC 3205 Australia

RACGP/CONROD Research Fellowship **4009**
Fellowships are available to support research into general practice patient populations who have experienced motor vehicle related trauma. Fellowships encompass all types of motor-vehicle related trauma including (but not limited to) a particular interest in whiplash. Research grants of up $20,000 per year may be awarded to applicants who conduct general practice research in any of the following areas: social research on risk taking behavior in young males, the effects of the compensation process on patients with motor vehicle related injuries, and prevention of the effects of motor vehicle trauma at a community level; clinical research on short and/or long term management of patients who have had motor vehicle accidents, GP management of whiplash, and the influence of psychotherapeutic agents on driving safety; medical education and medical informatics research relating to implementing guidelines for dealing with impaired drivers, or; other areas of research that could involve the role of the GP in the rehabilitation needs of persons who have sustained motor vehicle related injuries, the assessment of the elderly driver, supporting carers of patients with motor vehicle related trauma, and assessing young people at high risk of motor vehicle accidents.
Requirements: Applicants for the RACGP / CONROD Research Fellowship must be a GP or GP registrar and a member of the RACGP.
Restrictions: Unless otherwise indicated, the Fellowship must be used in the year for which it is awarded. The Fellowship will be paid only to incorporated bodies that have an ABN and ACN and not to individuals. By applying for the Fellowship each applicant will be taken to have consented to permit the RACGP to publicise their name, institution, and title of their research project.
Geographic Focus: All States, Australia
Date(s) Application is Due: May 8
Amount of Grant: Up to 20,000 USD
Contact: Samantha Fernandes; 03-8699-0496; fax 03-8699-0400; research@racgp.org.au
Internet: http://www.racgp.org.au/researchfoundation/awards/conrod
Sponsor: Royal Australian College of General Practitioners
College House, 1 Palmerston Crescent
South Melbourne, VIC 3205 Australia

RACGP Alan Chancellor Award **4010**
The Alan Chancellor Award is presented each year to the GP considered to be the best first time presenter of a research paper at the RACGP Annual Scientific Convention. The prize is $500 and an engraved medallion and is given in memory of Dr Chancellor.
Requirements: To be eligible for the Award, an individual must be: a GP or GP registrar; a member of the RACGP; in the early stages of their research career; a first-time presenter of a research paper at the RACGP Annual Scientific Conference, and; a major contributor to the research study being presented.
Restrictions: By applying for the award each applicant will be taken to have consented to permit RACGP to publicise their name, their institution, and the title of their research project.
Geographic Focus: All States, Australia
Date(s) Application is Due: Sep 1
Contact: Samantha Fernandes; 03-8699-0496; fax 03-8699-0400; research@racgp.org.au
Internet: http://www.racgp.org.au/researchfoundation/awards/chancellor
Sponsor: Royal Australian College of General Practitioners
College House, 1 Palmerston Crescent
South Melbourne, VIC 3205 Australia

RACGP Cardiovascular Research Grants in General Practice **4011**
Individual grants of up to $25,000 will be offered for a period of one year to fund projects investigating various aspects of cardiovascular disease including hypertension, stroke, lipid disorder, health promotion and lifestyle changes. The Cardiovascular Research Grants (CVRGs) in General Practice are a joint initiative of the RACGP and the Australian Association of Academic General Practice (AAAGP).
Requirements: To be eligible for the Cardiovascular Research Grants, the chief investigator must be a GP or GP registrar who is a member of the RACGP or the AAAGP. New and emerging researchers are particularly encouraged to apply.
Restrictions: Unless otherwise indicated, Grants must be used in the year for which they are awarded. Grants will be paid only to incorporated bodies that have an ABN and ACN and

not to individuals. By applying for a Grant each applicant will be taken to have consented to permit the RACGP to publicise their name, institution, and title of their research project.
Geographic Focus: All States, Australia
Date(s) Application is Due: May 8
Amount of Grant: Up to 25,000 USD
Contact: Samantha Fernandes; 03-8699-0496; fax 03-8699-0400; research@racgp.org.au
Internet: http://www.racgp.org.au/researchfoundation/awards/cvrg
Sponsor: Royal Australian College of General Practitioners
College House, 1 Palmerston Crescent
South Melbourne, VIC 3205 Australia

RACGP Chris Silagy Research Scholarship **4012**
The Chris Silagy Research Scholarship is designed to support a twelve month research project into an aspect of evidence-based primary health care by a general practitioner researcher. The scholarship is to the value of $20,000 per annum which includes a return economy airfare (booked by the College) to either attend a meeting of the Cochrane Collaboration or visit a Cochrane Research Centre during the year of their scholarship. Scholarships are awarded by the Chris Silagy Research Scholarship Committee, a subcommittee of the RACGP National Research Awards Committee.
Requirements: Applicants for the Chris Silagy Research Scholarship must be: a General Practitioner or General Practice Registrar; a member of the RACGP, and; currently undertaking a higher degree (Masters, PHD, MD) or have completed such a course in the past 5 years.
Restrictions: Unless otherwise indicated, the Scholarship must be used in the year for which it is awarded. The Scholarship will be paid only to incorporated bodies that have an ABN and ACN and not to individuals. By applying for the Scholarship each applicant will be taken to have consented to permit the RACGP to publicise their name, institution, and title of their research project.
Geographic Focus: All States, Australia
Date(s) Application is Due: May 8
Amount of Grant: 20,000 USD
Contact: Samantha Fernandes; 03-8699-0496; fax 03-8699-0400; research@racgp.org.au
Internet: http://www.racgp.org.au/researchfoundation/awards/silagy
Sponsor: Royal Australian College of General Practitioners
College House, 1 Palmerston Crescent
South Melbourne, VIC 3205 Australia

RACGP Family Medical Care, Education and Research Grants **4013**
The aim of the grant is to encourage and support general practitioners in the early stages of their research career to conduct research into an aspect of primary health care. The objective of the grant is to advance education in, and research into, medical knowledge and science. The grants are available for a 12 month period to provide new and emerging GP researchers with an opportunity to conduct research.
Requirements: To be eligible for the FMCER grants, the chief investigator must be a GP or GP registrar who is a member of the RACGP and must have a project that relates clearly to general practice.
Restrictions: Unless otherwise indicated, the Grant must be used in the year for which it is awarded. The Grant will be paid only to incorporated bodies that have an ABN and ACN and not to individuals. By applying for the Grant each applicant will be taken to have consented to permit the RACGP to publicise their name, institution, and title of their research project.
Geographic Focus: All States, Australia
Date(s) Application is Due: May 8
Amount of Grant: Up to 10,000 USD
Contact: Samantha Fernandes; 03-8699-0496; fax 03-8699-0400; research@racgp.org.au
Internet: http://www.racgp.org.au/researchfoundation/awards/fmcer
Sponsor: Royal Australian College of General Practitioners
College House, 1 Palmerston Crescent
South Melbourne, VIC 3205 Australia

RACGP Grieve Memorial Award **4014**
The award, which is to the value of $2,500, is designed to support a plan of study in medical education for the general practitioner or for research in general practice and must be used with 12 months of being awarded. The award is administered by the RACGP National Research Awards Committee and funded by the RACGP Council.
Requirements: Applicants for the award must be a GP or GP registrar who is a member of the RACGP.
Restrictions: Unless otherwise indicated, the Award must be used in the year for which it is awarded. The Award will be paid only to incorporated bodies that have an ABN and ACN and not to individuals. By applying for the Award each applicant will be taken to have consented to permit the RACGP to publicise their name, institution, and title of their research project.
Geographic Focus: All States, Australia
Date(s) Application is Due: May 8
Amount of Grant: 2,500 USD
Contact: Samantha Fernandes; 03-8699-0496; fax 03-8699-0400; research@racgp.org.au
Internet: http://www.racgp.org.au/researchfoundation/awards/pwhgrieve
Sponsor: Royal Australian College of General Practitioners
College House, 1 Palmerston Crescent
South Melbourne, VIC 3205 Australia

RACGP National Asthma Council Research Award **4015**
The Award is awarded annually by the National Asthma Council to the best presenter of an asthma related abstract or poster at the RACGP Annual Scientific Convention. The award is to the value of $500 and is presented at the closing ceremony.
Requirements: To be eligible for the award, an individual must be: a GP or GP registrar; a member of the RACGP; in the early stages of their research career; presenting an asthma-related paper or poster at the RACGP Annual Scientific Conference, and; a major contributor to the research study being presented.
Restrictions: By applying for the award each applicant will be taken to have consented to permit RACGP to publicise their name, their institution, and the title of their research project.
Geographic Focus: All States, Australia
Date(s) Application is Due: Sep 1
Amount of Grant: 500 USD
Contact: Samantha Fernandes; 03-8699-0496; fax 03-8699-0400; research@racgp.org.au
Internet: http://www.racgp.org.au/researchfoundation/awards/nac
Sponsor: Royal Australian College of General Practitioners
College House, 1 Palmerston Crescent
South Melbourne, VIC 3205 Australia

RACGP Rex Walpole Travelling Fellowship **4016**
The fellowship, which is to the value of $2500, is designed to support a plan of travel for a general practitioner and must be used within 12 months of being awarded. The fellowship is administered by the RACGP National Research Awards Committee and funded by the RACGP Victorian Faculty. Applications will be judged on benefits gained from travel to the individual, the profession, and the College and value for money.
Requirements: Applicants for the Fellowship must be a GP or GP registrar who is a member of the RACGP.
Restrictions: Unless otherwise indicated, the Fellowship must be used in the year for which it is awarded. The Fellowship will be paid only to incorporated bodies that have an ABN and ACN and not to individuals. By applying for the Fellowship each applicant will be taken to have consented to permit the RACGP to publicise their name, institution, and title of their research project.
Geographic Focus: All States, Australia
Date(s) Application is Due: May 8
Contact: Samantha Fernandes; 03-8699-0496; fax 03-8699-0400; research@racgp.org.au
Internet: http://www.racgp.org.au/researchfoundation/awards/rexwalpole
Sponsor: Royal Australian College of General Practitioners
College House, 1 Palmerston Crescent
South Melbourne, VIC 3205 Australia

RACGP Vicki Kotsirilos Integrative Medicine Grants **4017**
The Vicki Kotsirilos Integrative Medicine Grant is to encourage research that assists the integration of safe and ethical forms of natural and complementary medicines or therapies into mainstream general practice. The Grant is designed to encourage research that assists the integration of safe and ethical forms of natural and complementary medicines or therapies into mainstream general practice. A Grant of up $5,000 is offered to applicants conducting general practice research in the following areas: clinical research that contributes to the knowledge base concerning the efficacy of natural and complementary medicines or therapies; any project that supports research in behavioural and lifestyle factors impacting on health eg. diet, exercise, stress management, and; research that contributes to the efficient dissemination of the evidence regarding integrative medicine.
Requirements: Applicants must be: a General Practitioner or General Practice Registrar, and; a member of the RACGP.
Restrictions: Unless otherwise indicated, the Grant must be used in the year for which it is awarded. The Grant will be paid only to incorporated bodies that have an ABN and ACN and not to individuals. By applying for the Grant each applicant will be taken to have consented to permit the RACGP to publicise their name, institution, and title of their research project.
Geographic Focus: All States, Australia
Date(s) Application is Due: May 8
Contact: Samantha Fernandes; 03-8699-0496; fax 03-8699-0400; research@racgp.org.au
Internet: http://www.racgp.org.au/researchfoundation/awards/kotsirilos
Sponsor: Royal Australian College of General Practitioners
College House, 1 Palmerston Crescent
South Melbourne, VIC 3205 Australia

Radcliffe Institute Carol K. Pforzheimer Student Fellowships **4018**
The Arthur and Elizabeth Schlesinger Library on the History of Women in America invites Harvard undergraduates to make use of the library's collections with competitive awards of amounts up to $2,500 for relevant research projects. Preference will be given to applicants pursuing research in the history of work and the family, community service and volunteerism, the culinary arts, or women's health. The research may be, but is not required to be, in connection with a project for academic credit. Applications will be evaluated on the significance of the research and the project's potential contribution to the advancement of knowledge as well as its creativity in drawing on the library's holdings. The awards may be used to cover photocopying, other incidental research expenses, or living expenses in lieu of term time or summer employment. See website for current deadlines and application.
Requirements: Each application must include: the Schlesinger Library Grants and Fellowships cover page; a project description (no longer than four double-spaced pages in twelve-point font) indicating the purpose of the research, the Schlesinger Library holdings to be consulted, and the significance of these holdings to the project overall; a one-page bibliography of principal secondary sources consulted in designing the research; a résumé no longer than two pages; the name of one reference who has agreed to send a supporting letter; and a proposed budget indicating how the funds requested will be spent.
Restrictions: Only Harvard undergraduates are eligible. Funding does not cover the purchase of durable equipment.
Geographic Focus: All States
Amount of Grant: Up to 2,500 USD

Contact: (617) 496-3048; science@radcliffe.edu
Internet: http://www.radcliffe.edu/schles/pforzheimer_grant.aspx
Sponsor: Radcliffe Institute for Advanced Study
10 Garden Street
Cambridge, MA 02138

Radcliffe Institute Dissertation Grants **4019**
The Arthur and Elizabeth Schlesinger Library on the History of Women in America invites scholars whose dissertation research requires use of the library's collections to apply for research support. Applications are evaluated on the significance of the research and the project's potential contribution to the advancement of knowledge as well as its creativity in drawing on the library's holdings. Grants cover travel expenses, living expenses, photocopying, and other incidental research expenses.
Requirements: Applicants must be a doctoral candidate in a relevant field and have an approved dissertation topic. Priority is given to those whose projects require use of materials available only in the Schlesinger Library. A brief report about work accomplished under the grant will be due no later than one year from the date of the award.
Restrictions: Non-U.S. citizens are eligible but should contact the library regarding visas and other required paperwork prior to applying.The grant does not cover the purchase of durable equipment.
Geographic Focus: All States, All Countries
Amount of Grant: Up to 3,000 USD
Contact: Grants Administrator; (617) 495-8647; slgrants@radcliffe.edu
Internet: http://www.radcliffe.edu/schles/dissertation_grant.aspx
Sponsor: Radcliffe Institute for Advanced Study
10 Garden Street
Cambridge, MA 02138

Radcliffe Institute Fellowships for Creative Artists **4020**
The Radcliffe Institute is unique among centers for advanced study across the country in its commitment to encouraging and developing the creative arts along with all other disciplines. Through fellowships, public lectures and exhibits, the Institute offers artists not only a quiet place to create or compose, but also an opportunity to showcase their new work.
Requirements: Scholars in any field with a doctorate or appropriate terminal degree at least two years prior to appointment (by December of the prior year) in the area of the proposed project are eligible to apply. Only scholars who have published at least two articles or monographs are eligible to apply. Artists and writers need not have a PHD or an MFA to apply; however, they must meet other specific eligibility requirements. Fellows are expected to be free of their regular commitments so they may devote themselves full time to the work outlined in their proposal. Since this is a residential fellowship, fellows are expected to reside in the Boston area during that period and to have their primary office at the Institute so that they can participate fully in the life of the community. Applicants are strongly encouraged to use the online application system.
Restrictions: Former fellows of the Radcliffe Institute Fellowship Program (1999 to present) are not eligible to apply.
Geographic Focus: All States
Date(s) Application is Due: Oct 1
Amount of Grant: 70,000 USD
Contact: Science Office; (617) 496-3048; science@radcliffe.edu
Internet: http://www.radcliffe.edu/research/arts.aspx
Sponsor: Radcliffe Institute for Advanced Study
10 Garden Street
Cambridge, MA 02138

Radcliffe Institute Individual Residential Fellowships **4021**
The Radcliffe Institute Fellowship Program is a scholarly community where individuals pursue advanced work across a wide range of academic disciplines, professions, and creative arts. Radcliffe Institute fellowships are designed to support scholars, scientists, artists, and writers of exceptional promise and demonstrated accomplishment who wish to pursue work in academic and professional fields and in the creative arts. In recognition of Radcliffe's historic contributions to the education of women and to the study of issues related to women, the Radcliffe Institute sustains a continuing commitment to the study of women, gender, and society. Applicants need not focus on gender, however. Women and men from across the United States and throughout the world are encouraged to apply. email fellowships@radcliffe.edu for questions regarding humanities, social sciences, or creative arts fellowships. email sciencefellowships@radcliffe.edu for questions regarding natural science and mathematics fellowships.
Requirements: Scholars in any field with a doctorate or appropriate terminal degree at least two years prior to appointment (by December of the prior year) in the area of the proposed project are eligible to apply. Only scholars who have published at least two articles or monographs are eligible to apply. Artists and writers need not have a PHD or an MFA to apply; however, they must meet other specific eligibility requirements. Fellows are expected to be free of their regular commitments so they may devote themselves full time to the work. Applicants must reside in the Boston area, with their primary office at the Institute. Applications are available online, but applicants should refer to website for current deadlines and application requirements for each academic discipline.
Restrictions: Former fellows of the Radcliffe Institute Fellowship Program (1999 to present) are not eligible to apply.
Geographic Focus: All States, All Countries
Amount of Grant: 35,000-70,000 USD
Contact: (617) 496-1324
Internet: http://www.radcliffe.edu/fellowships/apply.asp
Sponsor: Radcliffe Institute for Advanced Study
10 Garden Street
Cambridge, MA 02138

Radcliffe Institute Oral History Grants **4022**
The Arthur and Elizabeth Schlesinger Library invites scholars who are conducting oral history interviews relevant to the history of women or gender in the United States to apply for funding support. This grant stipulates that the interviews take place in accordance with guidelines of the Oral History Association, that consent is obtained from interviewees for their words to be viewed by researchers worldwide, and that true copies or transcripts of the original recording of the oral interviews, as well as copies of the consent forms, be deposited in the Schlesinger Library. Applicants may request support to cover travel expenses, living expenses, transcription services, and incidental research materials. Applications will be evaluated on the significance of the research and documentation planned and the project's potential contribution to the advancement of knowledge.
Requirements: Each application must include: the Schlesinger Library Grants and Fellowships cover page; a project description (no longer than four double-spaced pages) indicating the purpose of the research, the persons to be interviewed, and the interview arrangements; a curriculum vitae of no more than two pages; the name of one reference who has agreed to send a supporting letter; and a proposed budget indicating how the requested funds will be spent. A brief report about work accomplished under the grant will be due no later than one year from the date of the award.
Restrictions: The grant does not cover funding for durable equipment. Non-U.S. citizens are eligible but should contact the library regarding visas and other required paperwork prior to applying.
Geographic Focus: All States, All Countries
Amount of Grant: Up to 3,000 USD
Contact: Administrator; (617) 495-8647; fax (617) 496-4640; slgrants@radcliffe.edu
Internet: http://www.radcliffe.edu/schles/oral_history_grant.aspx
Sponsor: Radcliffe Institute for Advanced Study
10 Garden Street
Cambridge, MA 02138

Radcliffe Institute Research Support Grants **4023**
The Arthur and Elizabeth Schlesinger Library on the History of Women in America invites scholars to apply for research support. Applications are evaluated on the significance of the research and the project's potential contribution to the advancement of knowledge as well as its creativity in drawing on the library's holdings. Grants cover travel and living expenses, photocopying, and other incidental research expenses.
Requirements: Each application must include: the Schlesinger grants and fellowships cover page; a detailed project description; a curriculum vitae; name of one reference; and a detailed proposed budget. Applicants must have a doctoral degree or equivalent research and writing experience. Priority is given to those who demonstrate research productivity and whose projects require use of materials available only at the Schlesinger Library. Applicants should refer to the grant website for requirement details and current deadlines.
Restrictions: Non-U.S. citizens are eligible but should contact the library regarding visas and other required paperwork prior to applying. The grant does not cover the purchase of durable equipment such as laptop computers.
Geographic Focus: All States, All Countries
Amount of Grant: Up to 3,000 USD
Contact: Grants Administrator; (617) 495-8647; slgrants@radcliffe.edu
Internet: http://www.radcliffe.edu/schles/research_grant.aspx
Sponsor: Radcliffe Institute for Advanced Study
10 Garden Street
Cambridge, MA 02138

Radcliffe Institute Residential Group Fellowships **4024**
The Radcliffe Institute Fellowship Program is a scholarly community where individuals pursue advanced work across a wide range of academic disciplines, professions, and creative arts. Radcliffe Institute fellowships are designed to support scholars, scientists, artists, and writers of exceptional promise and demonstrated accomplishment who wish to pursue work in academic and professional fields and in the creative arts. In recognition of Radcliffe's historic contributions to the education of women and to the study of issues related to women, the Radcliffe Institute sustains a continuing commitment to the study of women, gender, and society. Applicants' projects need not focus on gender, however. Women and men from across the United States and throughout the world are encouraged to apply.
Requirements: Scholars in any field with a doctorate or appropriate terminal degree at least two years prior to appointment (by December of the prior year) in the area of the proposed project are eligible to apply. All Humanists, Mathematicians, and Social and Natural Scientists applicants must have earned their doctorate at least one year prior to application. Creative Artists applicants need not have a PHD or an MFA to apply; however, they must meet other specific eligibility requirements. Applicants should refer to the website's eligibility guidelines for further details and current application deadlines.
Restrictions: Former fellows of the Radcliffe Institute Fellowship Program (1999 to present) are not eligible to apply. Only scholars who have published at least two articles or monographs are eligible to apply.
Geographic Focus: All States, All Countries
Amount of Grant: 35,000-70,000 USD
Contact: Judith Vichniac, Associate Dean, Fellowship Program; (617) 495-8212 or (617) 496-1324; fax (617) 495-8136
Internet: http://radcliffe.onlineapplicationportal.com/misc/guidelines/
Sponsor: Radcliffe Institute for Advanced Study
10 Garden Street
Cambridge, MA 02138

Rajiv Gandhi Foundation Grants 4025

The foundation conducts activities in India in the areas of the arts and humanities, conservation and the environment, economic affairs, education, international affairs, law and human rights, medicine and health, science and technology, and social welfare. Main projects are concerned with literacy and primary education, empowering women and children, aid for the disabled, health care services in rural and poor urban areas, and decentralized government. Types of support include research grants to institutions and scholarships, fellowships, and prizes to individuals.

Geographic Focus: All States

Contact: Jawahar Bhawan, Director; (091-11) 23755117 or (091-11) 23312456; fax (091-11) 23755119; info@rgfindia.com

Internet: http://www.rgfindia.com

Sponsor: Rajiv Gandhi Foundation
Dr. Rajendra Prasad Road
New Delhi, 110 001 India

RA Kleinhans Fellowship for Research in Tropical Nontimber Forest Products 4026

These fellowships are offered to further sustainable use of tropical forest resources for the benefit of indigenous peoples, concurrent with conservation of biodiversity of forest ecosystems. The program promotes research into practical means for managing and using tropical forest resources without destroying the integrity of the ecosystem. The geographical focus of research is currently restricted to Latin America. However, projects with potential for other parts of the world are encouraged. Research should merge forest conservation and business prospects and must lead to the development of a product or marketing technique that will provide income for community groups living within or near a tropical forest. Research should culminate in an economically sound strategy that encourages sustainable extraction of tropical forest resources for food, fiber, medicine, and other products for which there is an existing market. Research on products encouraging the reforestation of degraded areas is encouraged. Annual deadline dates may vary; contact the alliance or visit the Web site for exact dates. Please do not fax applications. Please do not send videos. Photos, maps and other visual materials are encouraged but cannot be returned.

Requirements: Citizenship is unrestricted. Applicants with master's degrees in forestry, ecology, environmental science, or a related discipline may apply. However, doctoral candidates and postdoctoral researchers are preferred. Applicants may substitute relevant experience for degrees. Applicants are required to submit their completed application in digital (electronic) form by email to SRodriguez@RA.org. Check Web site for application requirements.

Geographic Focus: All States

Date(s) Application is Due: Dec 31

Amount of Grant: 15,000 USD

Contact: Deanna Newsom, (212) 677-1900 or (888) 693-2784; dnewsom@ra.org

Internet: http://www.rainforest-alliance.org/kleinhans.cfm?id=main

Sponsor: Rainforest Alliance
665 Broadway, Suite 500
New York, NY 10012

Ralph F. Hirschmann Award in Peptide Chemistry 4027

This award, supported by Merck Research Laboratories, is given annually to recognize and encourage outstanding achievements in the chemistry, biochemistry, and biophysics of peptides. The award is granted without regard to age or nationality of the recipient. The award consists of $5,000 and a certificate. Up to $2,500 for travel expenses to the meeting at which the award will be presented will be reimbursed.

Requirements: Any individual, except a member of the award committee, may submit one nomination or seconding letter for each award in any given year. Nominating documents consist of a letter of not more than 1000 words containing an evaluation of the nominee's accomplishments and a specific identification of the work to be recognized, a biographical sketch including date of birth, and a list of publications and patents authored by the nominee.

Restrictions: Self-nominations are not accepted.

Geographic Focus: All States

Date(s) Application is Due: Nov 1

Amount of Grant: 5,000 USD

Contact: Felicia Dixon, Awards Administrator; (800) 227-5558 or (202) 872-4408; fax (202) 776-8008; f_dixon@acs.org or awards@acs.org

Internet: http://portal.acs.org/portal/acs/corg/content?_nfpb=true&_pageLabel=PP_ARTICLEMAIN&node_id=1319&content_id=CTP_004547&use_sec=true&sec_url_var=region1

Sponsor: American Chemical Society
1155 Sixteenth Street, NW
Washington, DC 20036-4801

Ralph Waldo Emerson Award 4028

This award is offered annually for studies of the intellectual and cultural condition of the human race. The purpose of the award is to recognize comprehensive studies that contribute significantly to historical, philosophical, or religious interpretations of the human condition. History, philosophy, and religion are conceived in sufficiently broad terms to permit the inclusion of appropriate work in related fields. Biographies of public figures may be eligible if their critical emphasis is primarily on the intellectual and cultural condition of humanity. Entries must be submitted by publishers who have first obtained the conditions of eligibility from Phi Beta Kappa. By March 31, applicants should ship one copy of the book to the appropriate awards committee. By April 10, if the submission is eligible, applicants will receive the shipping list for judges, and should ship one book to each judge by April 30.

Requirements: Entries must be original works published in the United States during the 12-month period between May 1 (of the past year) and the April 30 (of the current year) deadline date. Entries will ordinarily be the work of a single writer. Exceptions may be made for books written by small teams of scholars working in close collaboration. Authors must be U.S. citizens or residents.

Restrictions: Narrowly limited or purely technical studies will not be considered. In general, reports on research and monographs as such are not eligible. Works of fiction and unpublished manuscripts are not eligible.

Geographic Focus: All States

Date(s) Application is Due: Apr 30

Contact: Sandra Beasley, Awards Coordinator; (202) 265-3808; fax (202) 986-1601; SBeasley@pbk.org or info@pbk.org

Internet: http://staging.pbk.org/AM/Template.cfm?Section=Book_Awards&Template=/CM/HTMLDisplay.cfm&ContentID=2646

Sponsor: Phi Beta Kappa Society
1606 New Hampshire Avenue NW
Washington, DC 20009

RAND Corporation Fellowships in Population Studies and the Study of Aging 4029

The RAND Postdoctoral Training Program in the Study of Aging enables outstanding junior scholars in demographic and aging research to sharpen their analytic skills, learn to communicate research results effectively, and advance their research agenda. Housed within the Labor and Population Program, the program blends formal and informal training and extensive collaboration with distinguished researchers in a variety of disciplines. Fellowships are for one year, renewable for a second. Each fellow receives a competitive annual stipend, travel stipend and health insurance. Each fellow, in regular consultation with a staff adviser, will design a program that reflects his or her research needs and interests, and stage in their research career.

Requirements: Fellows must be U.S. citizens or permanent residents, and must have completed a Ph.D. (or its equivalent) in a relevant discipline before they begin the program. The program is open to new scholars in the field of demographic and aging research, as well as individuals who already have research experience on these topics. For example, in some cases, fellows are on leave from an academic position so they can engage in research and writing full time, and extend their research agenda and funding base. Applicants should submit a completed application form, curriculum vitae, two or more letters of recommendation, a sample of written work, and a statement of their research goals for the Fellows program. Application forms are available online (in PDF or in interactive format) or by writing to the address below. Members of the program's steering committee will rank applications on the basis of academic performance, the match between the applicant's research interests and those of the RAND staff, and the applicant's ability to benefit from the program. Members of minorities are encouraged to apply.

Geographic Focus: All States

Date(s) Application is Due: Feb 9

Contact: Diana Malouf, (310) 393-0411, ext. 6462; diana_malouf@rand.org

Internet: http://www.rand.org/labor/fellows/aging.html

Sponsor: RAND Corporation
1776 Main Street, P.O. Box 2138
Santa Monica, CA 90407-2138

RAND Corporation Graduate Student Summer Associateships 4030

The program introduces outstanding graduate students to RAND, an institution that conducts research on a wide range of national security problems and domestic and international social policy issues. Students receive a stipend and are given the opportunity to conduct research that can be completed during the three months they are at RAND. The program receives about 500 applications each year for the 20+ positions. The associates who end up finding matches with a research project do so because their skill sets match the needs of the project. The program runs in the summer months only. Summer associates work at RAND full time for a 12-week period. Positions are available in RAND's three major U.S. offices-Santa Monica, Arlington, and Pittsburgh. All summer associates are collocated with project mentors. The location of the project mentor determines the location of the summer associate. The summer stipend in 2008 was about $12,000 (before taxes) for the 12 weeks of full time research.

Requirements: The program is designed for full time students who have completed at least two years of graduate work leading to a doctorate or professional degree (e.g., law or medical degree, professional engineer certificate). Students must be enrolled full time in a graduate degree program during the spring and fall of the current school year to be considered for the program. U.S. citizenship is not necessary except for positions that require security clearances. Most of RAND's professional hiring is at the PhD level, so the summer program is oriented toward individuals who are generally within a year or two of completing their doctorates. However, sometimes there are projects that match with the skills of graduate students in other stages of their education who also have significant work experience. To apply, go to the website at http://www.rand.org/jobs/. Click Find a Job, click Advanced Search, select Research-Summer Associate Prog under Job Families, and click the Search button. Click on the position and click the Apply Now button. Follow the prompts for application submission.

Restrictions: The program does not hire undergraduates for the summer. Also, those students graduating prior to the summer of employment are not eligible to apply. The program typically does not consider Master's degree students without significant work or research experience. Postdoctoral fellows are not eligible for the summer program.

Geographic Focus: All States

Date(s) Application is Due: Jan 31

Contact: Summer Director; (310) 393-0411; Summer_Director@rand.org

Internet: http://www.rand.org/about/edu_op/fellowships/gsap/

Sponsor: RAND Corporation
1776 Main Street, P.O. Box 2138
Santa Monica, CA 90407-2138

RAND Corporation Military Fellowships 4031

RAND acts as a host to active-duty service members from all branches of our nation's Armed Forces through the Military Fellows Program. Typically, fellows are stationed at RAND's Santa Monica location for one year. Fellows are given access to all aspects of RAND's research community and are encouraged to participate actively in RAND research projects, seminars, and discussion groups.
Requirements: Requirements and procedures for the Military Fellows Program differ by branch of service: Army, Navy, Air Force, Marines, Coast Guard. Additionally, RAND conducts a fellowship with the Department of Homeland Security. Guidelines and contact information for each branch are available at the website.
Geographic Focus: All States
Contact: Gene Gritton; (310) 393-0411, ext. 6933; Gene_Gritton@rand.org
Internet: http://www.rand.org/natsec_area/fellowships.html
Sponsor: RAND Corporation
1776 Main Street, P.O. Box 2138
Santa Monica, CA 90407-2138

RAND Corporation Transatlantic Postdoctoral Fellowships for International Relations and Security (TAPIR) 4032

The program is open to candidates who have recently received their doctorate in social and political sciences or economics and whose research focuses on topics of international relations and security. Fellowships are granted for a duration of 24 months to prepare Fellows for a career in policy-oriented and international research at renowned think tanks and political consulting research institutes. Fellows spend three eight-month stays at research institutions or think tanks participating in the program-at least one on the Eastern, and one on the Western side of the Atlantic. Participating institutes include: Carnegie Endowment for International Peace, Institut fran?§ais des relations internationales (IFRI), European Union Institute for Security Studies (EUISS), RAND Corporation, Chatham House, SAIS Center for Transatlantic Relations, the Institute for Strategic and International Studies (IEEI), the Center for International Relations (CIR), the Center for Security Studies (CSS) and the German Institute for International and Security Affairs/Stiftung Wissenschaft und Politik (SWP). Fellows will be supervised by a researcher at each host institute, while being integrated into the relevant research units or projects. Fellows will have the opportunity to work on issues of major and immediate political significance, and to publish their work in journals or paper series that will be read by decision-makers at national and international levels. The fellowship aims to involve participants actively in the work and research of leading international think tanks. It is not intended to primarily fund the completion of major individual publications or of academic peer-reviewed journal articles. The fellowship consists of a monthly stipend of Euro 2,000 and a one-time travel allowance of Euro 3,500 for travel to and from the host institutions. Health insurance and fees for residence permits are not included.
Requirements: To be eligible, candidates must have successfully defended their Ph.D. by the start of the program (October). An exception can be made for candidates who can prove that their thesis has been accepted by the jury and requires no further work. Candidates must have a strong interest in working in a policy-oriented international environment and must demonstrate their ability to do so by previous international academic experience. U.S., Canadian and European nationals can apply to the program. Applications should contain: (1) five-page research statement outlining the candidate's current as well as longer-term academic interests. This does not have to be limited to transatlantic issues; (2) Curriculum Vitae, including a list of publications; (3) Two letters of recommendation. All files, including letters of recommendation, should be sent in PDF format to transatlanticfellowship@swp-berlin.org by the published deadline date.
Geographic Focus: All States, Albania, Andorra, Armenia, Austria, Azerbaijan, Belarus, Belgium, Bosnia & Herzegovina, Bulgaria, Canada, Croatia, Cyprus, Czech Republic, Denmark, Estonia, Finland, France, Georgia, Germany, Greece, Hungary, Iceland, Ireland, Italy, Kosovo, Latvia, Liechtenstein, Lithuania, Luxembourg, Macedonia, Malta, Moldova, Monaco, Montenegro, Norway, Poland, Portugal, Romania, Russia, San Marino, Serbia, Slovakia, Slovenia, Spain, Sweden, Switzerland, The Netherlands, Turkey, Ukraine, United Kingdom, Vatican City
Date(s) Application is Due: Feb 16
Contact: Tobias von Lossow, Stiftung Wissenschaft und Politik; 49 (0)30 88007-106; fax 49 (0)30 88007- 5321, ext. 100; tobias.vonlossow@swp-berlin.org
Internet: http://www.rand.org/about/edu_op/fellowships/transatlantic/
Sponsor: RAND Corporation
1200 S Hayes Street
Arlington, VA 22202

Rathmann Family Foundation Grants 4033

The foundation's main funding areas are education, with priority given to science and math; the arts; children and youth health organizations; and preservation of the environment. Types of support include general operating support, continuing support, capital campaigns, equipment acquisition, endowment funds, program development, conferences and seminars, seed grants, curriculum development, fellowships, internships, scholarship funds, research, and matching funds. There are no specific deadlines or application forms, and interested parties should begin by contacting the Foundation directly.
Requirements: Grants are awarded to organizations in the San Francisco Bay, California area; the Annapolis, Maryland area; the Seattle, Washington area; the Philadelphia, Pennsylvania area: and metropolitan Minneapolis/Saint Paul, Minnesota.
Restrictions: Grants are not awarded to/for private foundations; religious organizations for religious activities; civil rights; social action; advocacy organizations; fraternal groups; political purposes; mental health counseling; individuals; or fundraisers, media events, public relations, annual appeals, or propaganda.
Geographic Focus: California, Maryland, Minnesota, Pennsylvania, Washington
Amount of Grant: 1,000-100,000 USD
Contact: Rick Rathmann, (410) 349-2376; fax (410) 349-2377
Sponsor: Rathmann Family Foundation
1290 Bay Dale Drive, P.O. Box 352
Arnold, MD 21012

Rayonier Foundation Grants 4034

The foundation awards grants to eligible nonprofit organizations in its areas of interest, including children and youth, community development, disadvantaged (economically), education, engineering, environmental programs, families, health care and hospitals, libraries, minorities, performing arts, recreation science, social services, technology, and volunteerism. Types of support include annual campaigns, building construction/renovation, continuing support, employee matching gifts, employee-related scholarships, endowments, equipment acquisition, general operating support, matching/challenge grants, performing arts, program development, recreation, science, research grants, scholarship funds, and seed grants.
Requirements: Nonprofits in Florida, Georgia, and Washington are eligible.
Geographic Focus: Florida, Georgia, Washington
Amount of Grant: 250-25,000 USD
Contact: Charles H. Hood, President; (904) 357-9100; fax (904) 357-9101; info@rayonier.com
Sponsor: Rayonier Foundation
50 North Laura Street, Stuite 1900
Jacksonville, FL 32202

RCPSC/AMS CanMEDs Research and Development Grants 4035

The primary objective of the grant is to support research, development, and/or implementation of projects to enhance specialty education in the promotion of the key roles for specialists as defined by CanMEDS. Applications may be individual or joint, involving one or more departments, disciplines, or faculties of medicine. It is not necessary to be a Fellow of the RCPSC or a physician; however, a Fellow of the Royal College must contribute to the project. Eligible projects may include faculty development, course development and/or medical education research that both apply and promote the CanMEDS roles. Budgets may include requests to defray the costs for travel, accommodation and meals for investigators to disseminate the project findings. One or more grants of up to $25,000 will be awarded each year.
Requirements: Applicants must be involved in teaching and/or medical education research in Canada. Applications that include research must meet contemporary standards of educational research, which include the relevance of the research question, the quality of the research design, ethical review of the submission, accountability for funds received, and the ability to complete the project during the funding period. Applications that include project development must include an evaluative component assessing the impact of the project.
Restrictions: Budgets should not include salary or honoraria for investigators. Ineligible costs include tuition fees, capital equipment costs (e.g. computer hardware, office equipment), overhead costs, university or other organization administrative fees.
Geographic Focus: All States, Canada
Date(s) Application is Due: Mar 2
Amount of Grant: Up to 25,000 USD
Contact: Allison Shaw, Awards and Grants Section; (800) 668-3740, ext. 355 or (613) 730-8177; fax (613) 730-2410; awards@rcpsc.edu
Internet: http://rcpsc.medical.org/awards/ams_e.php
Sponsor: Royal College of Physicians and Surgeons of Canada
774 Echo Drive
Ottawa, ON K1S 5N8 Canada

RCPSC Detweiler Traveling Fellowships 4036

Fellowships are made to improve the quality of medical and surgical practice in Canada. The fellowships will enable the recipients to visit medical centers in Canada or abroad to study or gain experience in the use or application of new knowledge or techniques in their fields or to further the pursuit of a fundamental or clinical research problem. Five or more fellowships offered annually to a maximum of $21,000 will be reserved for junior applicants. Two additional fellowships offered annually will be reserved for senior applicants: one will be granted to a Fellow who has held Royal College certification for five years or more and has been in private practice for at least three years, and one will be granted to a Fellow who has held Royal College certification for five years or more and has been in academic practice for at least three years. Each award will be to a maximum of $10,500.
Requirements: Applicants must be Fellows of the Royal College in good standing, or residents in the final year of their training who will be admitted to Fellowship in the Royal College prior to December 31st.
Restrictions: Travel fellowships will not be awarded retroactively.
Geographic Focus: All States, Canada
Date(s) Application is Due: Sep 30
Amount of Grant: 10,500-21,000 USD
Contact: Allison Shaw, Awards and Grants Section; (800) 668-3740, ext. 355 or (613) 730-8177; fax (613) 730-2410; awards@rcpsc.edu
Internet: http://rcpsc.medical.org/awards/detweiler_e.php
Sponsor: Royal College of Physicians and Surgeons of Canada
774 Echo Drive

Ottawa, ON K1S 5N8 Canada

RCPSC International Traveling Fellowship 4037

The objective of the fellowship is to enable fellows residing outside of Canada to study in a Canadian medical center in order to acquire a new knowledge and skills applicable to clinical practice, undertake clinical research, or to enable fellows residing in Canada to practice and teach in a less developed country. The fellowship is applicable for visits of six to 12 months duration. Visits may be broken into separate time blocks but must be completed within two years. An amount equivalent to 75% of the award will be given at the beginning of the fellowship and the remainder upon receipt of the final report at the end of the fellowship. The fellowship may be used to supplement financial assistance granted from other sources.
Requirements: Applicants must be Fellows of the Royal College in good standing, or residents in the final year of their training who will be admitted to Fellowship in the Royal College prior to December 31.
Geographic Focus: All States, Canada
Date(s) Application is Due: Sep 30
Amount of Grant: 10,500-2,100 USD
Contact: Allison Shaw, Awards and Grants Section; (800) 668-3740, ext. 355 or (613) 730-8177; fax (613) 730-2410; awards@rcpsc.edu
Internet: http://rcpsc.medical.org/awards/international_e.php
Sponsor: Royal College of Physicians and Surgeons of Canada
774 Echo Drive
Ottawa, ON K1S 5N8 Canada

RCPSC Medical Education Research Grants 4038

The primary objective of the grant is to support quality research in Canada relevant to the field of postgraduate medical education or continuing professional development. Applications may be individual or joint, involving one or more departments, disciplines, or faculties of medicine. It is not necessary to be a Fellow of the RCPSC or a physician; however, a Fellow of the Royal College must contribute to the research project. The following areas of research are encouraged: curriculum design and implementation; instructional methods; assessment; program evaluation; continuing education; professional development; and faculty development. S
Requirements: Applicants must be involved in teaching and/or medical education research in Canada. Applications must meet contemporary standards of educational research, which include the relevance of the research question, the quality of the research design, ethical review of the submission, accountability for funds received, and the ability to complete the project during the funding period. One or more grants, of up to $25,000 per year of study, will be awarded annually.
Restrictions: Ineligible costs include tuition fees, capital equipment costs (e.g. computer hardware, office equipment), overhead costs, university or other organization administrative fees.
Geographic Focus: All States, Canada
Date(s) Application is Due: Mar 2
Contact: Allison Shaw, Awards and Grants Section; (800) 668-3740, ext. 355 or (613) 730-8177; fax (613) 730-2410; awards@rcpsc.edu
Internet: http://rcpsc.medical.org/awards/MERG_e.php
Sponsor: Royal College of Physicians and Surgeons of Canada
774 Echo Drive
Ottawa, ON K1S 5N8 Canada

RCPSC Medical Education Traveling Fellowships 4039

The objective of the fellowship is to enable the recipient to acquire knowledge and expertise in the field of medical education. Upon completion, the recipient will be able to use education principles to develop educational programs and proceed to the evaluation of learning experiences at the undergraduate, postgraduate, or continuing education level. Fellowships are applicable for visits of six to 12 months duration. Up to 75 percent of the fellowship stipend will be given at the beginning of the fellowship and the remainder upon receipt of the final report at the end of the fellowship. The fellowship may be used to supplement financial assistance granted from other sources.
Requirements: Applicants must be Fellows of the Royal College in good standing, or residents in the final year of their training who will be admitted to Fellowship in the Royal College prior to December 31st.
Geographic Focus: All States, Canada
Date(s) Application is Due: Sep 30
Amount of Grant: 15,000-30,000 USD
Contact: Allison Shaw, Awards and Grants Section; (800) 668-3740, ext. 355 or (613) 730-8177; fax (613) 730-2410; awards@rcpsc.edu
Internet: http://rcpsc.medical.org/awards/medical_e.php
Sponsor: Royal College of Physicians and Surgeons of Canada
774 Echo Drive
Ottawa, ON K1S 5N8 Canada

RCPSC Royal College Fellowship for Studies in Medical Education 4040

The goal of this fellowship is to increase the number and quality of professionally trained medical educators in Canada by providing training in the science of medical education to selected promising candidates. The program is to help specialists acquire knowledge and skills in the field of medical education in order to develop educational programs and evaluation methods that apply to any phase of the continuum of medical education. This fellowship funding is for one year. Applicants may reapply for one additional year of funding; however, repeat or renewal applications will be considered on a competitive basis with all other applications received in the same year, and will require an outline of the progress of the previous year's work.
Requirements: Applicants must register in a Masters or PhD level university program leading to a degree in Education, or a subject field/program of study closely related to medical education. The candidate must be registered and attending university as a full time or part time student, either within Canada or abroad. Applicants must be Canadian citizens or permanent residents who, are Fellows of the Royal College in good standing or residents who, at the time of the application, are registered in a program accredited by the Royal College.
Geographic Focus: All States, Canada
Date(s) Application is Due: Sep 30
Amount of Grant: 45,000 USD
Contact: Allison Shaw, Awards and Grants Section; (800) 668-3740, ext. 355 or (613) 730-8177; fax (613) 730-2410; awards@rcpsc.edu
Internet: http://rcpsc.medical.org/awards/fellow_e.php
Sponsor: Royal College of Physicians and Surgeons of Canada
774 Echo Drive
Ottawa, ON K1S 5N8 Canada

RCPSC Royal College Visiting Professorships in Medical Research 4041

The intent of this grant is to provide medical students and residents with the opportunity to meet established researchers from other centers in order to promote medical research as a career choice. It is intended that the visiting professor have the opportunity to interact with both undergraduate medical students and residents through rounds and other teaching sessions, and that the visit not be department-specific. Such interaction will provide the visiting professor with opportunities to expound on the challenges and rewards of a career in medical research. A $3,000 grant is available for each medical faculty.
Geographic Focus: All States, Canada
Date(s) Application is Due: Sep 30
Amount of Grant: 3,000 USD
Contact: Ines Martin, (800) 668-3740, ext. 212 or (613) 730-6212; fax (613) 730-2410; imartin@rcpsc.edu
Internet: http://rcpsc.medical.org/awards/visiting-medresearch_e.php
Sponsor: Royal College of Physicians and Surgeons of Canada
774 Echo Drive
Ottawa, ON K1S 5N8 Canada

RCPSC Walter C. MacKenzie Fellowship in Surgery 4042

The fellowship will enable the recipient to visit one or more centers to acquire surgical research or clinical surgical training. Knowledge gained should have a direct application to the recipient's clinical surgical practice. One or more fellowship are offered annually, depending on budget. Tenable for visits of two weeks to 12 months duration; visits may be broken into separate time blocks but must be completed within two years. An amount equivalent to 75% of the award will be given at the beginning of the fellowship, and the remainder upon receipt of a final report at the end of the fellowship. The fellowship may be used to supplement financial assistance granted from other sources.
Requirements: Applications are accepted from practicing surgeons who are fellows of the Royal College in good standing at the time of application and who reside in Canada. Practicing surgeons are those who have spent at least 50 percent of their professional time in active clinical surgery over the last five years.
Geographic Focus: All States, Canada
Date(s) Application is Due: Sep 30
Amount of Grant: 7,500-180,000 USD
Contact: Allison Shaw, Awards and Grants Section; (800) 668-3740, ext. 355 or (613) 730-8177; fax (613) 730-2410; awards@rcpsc.edu
Internet: http://rcpsc.medical.org/awards/mackenzie_e.php
Sponsor: Royal College of Physicians and Surgeons of Canada
774 Echo Drive
Ottawa, ON K1S 5N8 Canada

RCY-NSF Ocean Sciences Division Research Grants 4043

This division supports research to improve understanding of the ocean and ocean basins. Basic research programs support individual scientists, small groups of cooperating scientists, and some large coordinated projects. Ocean sciences also supports efforts to develop, acquire, and operate the instrumentation and facilities needed to carry out these research programs. Programs include ocean sciences instrumentation development, oceanographic instrumentation, shipboard scientific support equipment, and ocean sciences research support and its subprograms of physical oceanography, chemical oceanography, marine geology and geophysics, and biological oceanography; oceanographic centers and facilities; and the ocean drilling program.
Geographic Focus: All States
Date(s) Application is Due: Feb 15; Aug 15
Contact: David L. Garrison, (703) 292-7588; fax (703) 292-9085; dgarriso@nsf.gov
Internet: http://www.geo.nsf.gov/oce/ocefund.htm
Sponsor: National Science Foundation
4201 Wilson Boulevard
Arlington, VA 22230

RealNetworks Helix Community Grants 4044

The grant program will ensure the Helix DNA platform incorporates cutting edge research advances and has the widest extensibility. Ground-breaking research proposals, inventive implementations, and creative project ideas from independent developers, the academic and research communities, or any nonprofit or commercial enterprise engaged in digital media research and development are welcome. Proposals are encouraged to develop solutions for IP transmission of streaming data over terrestrial and wireless networks. Proposal guidelines are available online. Submit proposals by email.

Requirements: Independent open source developers, faculty members and the students under their purview from U.S.-based academic institutions recognized by the Council of the Higher Education Accreditation and the U.S. Department of Education, industry-recognized research organizations, 501(c)3 nonprofit organizations, and publicly or privately held companies are eligible.
Geographic Focus: All States
Date(s) Application is Due: Oct 3
Amount of Grant: Up to 75,000 USD
Contact: Helix Community Manager; (206) 674-2700; fax 206-674-2698; grants@helixcommunity.org
Internet: https://www.helixcommunity.org
Sponsor: RealNetworks
P.O. Box 91123
Seattle, WA 98111-9223

Regenstrief General Internal Medicine General Pediatrics Research Fellowship 4045
The Regenstrief Institute and the Division of General Internal Medicine offers a 2-year research oriented fellowship (with an optional third year) to prepare physicians for academic careers in general internal medicine and general pediatrics. Fellows receive formal didactic training through classes and seminars taught by faculty from both the Institute as well as Indiana University.
Requirements: Candidates for the Fellowship must hold the M.D. or equivalent degree, and must be U.S. citizens or permanent residents. Minority applicants are encouraged to apply.
Geographic Focus: All States
Contact: Kurt Kroenke, Program Director; (317) 630-7447; fax (317) 630-6611; KKroenke@regenstrief.org
Internet: http://www.regenstrief.org/training/general-internal-medicine-pediatric-research-fellowship-program
Sponsor: Regenstrief Institute
410 West 10th Street, Suite 2000
Indianapolis, IN 46202-3012

Regenstrief Geriatrics Fellowship 4046
The Department of Medicine and the Division of General Internal Medicine and Geriatrics offer a 2-year fellowship in Geriatric Medicine (with an optional third year). Residents who have completed three years of training in internal medicine or family practice and who are board eligible or board certified may apply for the fellowship. Fellows will be trained in clinical geriatrics, research, teaching, and/or administration in preparation for a successful career in academic geriatric medicine. Formal instruction will be provided by a structured curriculum with goals and objectives that will be met by clinical experiences in a variety of settings, didactic seminars to augment the clinical experiences, and seminars in research, teaching and administration. For further information on the Geriatric Fellowship contact the Program Director.
Geographic Focus: All States
Contact: Glenda Westmorland, Program Director; (317) 630-6906; fax (317) 630-2667; GWestmorland@regenstrief.org
Internet: http://www.regenstrief.org/training/geriatrics-fellowship-program
Sponsor: Regenstrief Institute
410 West 10th Street, Suite 2000
Indianapolis, IN 46202-3012

Regenstrief Master Of Science In Clinical Research/CITE Program 4047
Indiana University has received funding from the National Institute of Health through a K-30 grant to develop and implement the Clinical Investigation and Translational Education (CITE) program. The purpose of this program is to prepare health care professionals for a career in clinical research. Following completion of the program, graduates can embark on a career in clinical research with the skills necessary to successfully compete for grant funding, conduct and analyze research findings, and publish their work in scientific journals. By participating in the program, CITE trainees will complete a two-year formal clinical research curriculum, at the end of which they will receive a Master of Science in Clinical Research degree. The program is designed for participants that already have or are in training for a doctoral degree (e.g., MD, PhD, DNS, DDS, DPT).
Requirements: Most participants will already have a doctoral degree (e.g., MD, PhD, DNS, DDS, or DPT) or be in training for a doctoral degree. Potential applicants include, but are not limited to, the following: fellows or other health care professional trainees who have substantial protected time for clinical research; junior faculty who have career awards or support from their division or department head for participation in the CITE program; other faculty who previously have focused on basic research or other academic activities but now are committed to pursuing a career in clinical research CITE constitutes the formal didactic requirements for certain types of federal training grants (such as K-23's) and other career awards. Individuals in these programs would be expected to participate in CITE unless they were enrolled in some other type of formal degree program.
Geographic Focus: All States
Contact: Kurt Kroenke, Program Director; (317) 630-7447; fax (317) 630-6611; KKroenke@regenstrief.org
Internet: https://www.indianactsi.org/cite/citeapplication
Sponsor: Regenstrief Institute
1050 Wishard Boulevard, RG-6
Indianapolis, IN 46202-2872

Regenstrief Medical Informatics NLM Research Fellowships 4048
The Regenstrief Institute, an internationally recognized leader in medical informatics and health services research, offers National Library of Medicine supported fellowships in the use of computer technology to improve the quality and delivery of healthcare. Interacting closely with faculty mentors, fellows participate in a two-year (with optional third year) comprehensive training program to prepare for roles in developing, implementing and studying clinical information systems. The medical informatics fellowship is designed for physicians, other clinicians holding doctorates, and PhDs in relevant fields, with previous computer experience who desire to become part of a community of scholars and clinicians at the forefront of medical and technological innovations. The primary intent of this fellowship is to train fellows for careers in medical informatics research positions in academic or state and federal government programs, although graduates are also sought after by the health care and the medical systems development industry.
Requirements: Applicants must either hold an M.D. or equivalent degree and we prefer that M.D. applicants have completed their medical residency program or have a Ph.D. in a field related to medicine or technology. Applicants must have significant training or experience with computers. Due to limitations imposed by our NLM funded training programs, applicants must be U.S. citizens or permanent residents. Applications from women and minorities are encouraged.
Geographic Focus: All States
Contact: Medical Research Fellowship Office; (317) 423-5670; fax (317) 423-5695
Internet: http://www.regenstrief.org/medinformatics/fellowship
Sponsor: Regenstrief Institute
410 West 10th Street, Suite 2000
Indianapolis, IN 46202-3012

Regents Professional Opportunity Scholarships 4049
Students pursuing a career in one of several professions may be eligible to apply for the scholarships. Degree programs include: associate degree, available at two-year community colleges and some four-year colleges—dental hygiene, dietetics/nutrition, massage therapy assistant, ophthalmic dispensing, physical therapy assistant, registered physician assistant, and veterinary technician; baccalaureate degree, available at four-year colleges only—certified public accountancy, architecture, athletic trainer, dietetics/nutrition, professional engineering, interior design, landscape architecture, registered nurse, occupational therapy, pharmacy, physical therapy, registered physician assistant, and veterinary technician; master's degree, graduate programs—acupuncture, architecture, audiology, landscape architecture, midwifery, occupational therapy, physical therapy, registered physician assistant, social work, speech language, and pathology; doctoral degree (Drive, PhD, PharmD)—chiropractic, optometry, pharmacy, podiatry, psychology, and veterinary medicine; and doctor of law (JD). Priority is given to individuals who are economically disadvantaged and/or who are members of a minority group that is historically underrepresented in the chosen profession; and enrolled in or graduate from SEEK, College Discovery, EOP, or HEOP opportunity programs.
Requirements: Students must study full time and be matriculated in an approved program of study in New York State, be New York State residents, and be U.S. citizens or qualifying noncitizens. Upon completion of study, the student must work as a licensed professional one year for each annual payment received. Employment must be in the studied profession and must be in New York State.
Geographic Focus: New York
Amount of Grant: 1,000-5,000 USD
Contact: Program Coordinator; (518) 486-1319 or (518) 473-1574
Internet: http://www.hesc.com/Content.nsf/SFC/2/NYS_Regents_Professional_Opportunity_Scholarships
Sponsor: New York State Higher Education Services Corporation
Education Building Addition, Room 1071
Albany, NY 12255

Rehabilitation Nursing Foundation Research Grants 4050
The Foundation makes available research grants totaling $44,500. RNF offers funding of up to $30,000 for rehabilitation nursing research which is awarded in the form of multiple grants. A New Investigator Award, offered to encourage nurses who are novice researchers, will be awarded for up to $10,000. Up to two grants from the remaining amount will be awarded to recipients, who will be designated RNF Research Fellows. Research proposals that address clinical practice or educational or administrative dimensions of rehabilitation nursing are accepted. Quantitative and qualitative research projects will be reviewed. Funding may be awarded in the form of multiple grants. The Research Grant Committee reviews proposals once a year. Proposals must be postmarked by February 1. Grant recipients will be notified in September, and funding begins January of the following year for the selected proposals.
Requirements: The principal investigator for the research project must be a registered nurse who is active in rehabilitation or who demonstrates interest in and significant contributions to rehabilitation nursing. Membership in the Association of Rehabilitation Nurses is not required. Graduate student researchers are eligible for funding.
Geographic Focus: All States
Date(s) Application is Due: Feb 1
Amount of Grant: Up to 44,500 USD
Contact: Program Director; (847) 375-4710 or (800) 229-7530; info@rehabnurse.org
Internet: http://www.rehabnurse.org/research/researchgrants.html
Sponsor: Rehabilitation Nursing Foundation
4700 W Lake Avenue
Glenview, IL 60025-1485

Reinberger Foundation Grants 4051
Clarence T. Reinberger was born in 1894 on Cleveland's west side, and began his business career in the 1920's as a pioneer in the automobile replacement parts field. Starting as

a clerk in the Cleveland retail store of the National Automotive Parts Association's (NAPA) Automotive Parts Company, he became president of that company in 1948. In the 1960's the Automotive Parts Company merged with the Genuine Parts Company of Atlanta. Mr. Reinberger held the position of Chairman of the Board of Genuine Parts Company until his death in 1968. Louise Fischer Reinberger was born in Germany. After graduation from high school in the United States, she was employed by the Halle Brothers Company, a large Cleveland department store. The Reinberger Foundation was established by Mr. and Mrs. Reinberger in 1966. Mr. Reinberger left a substantial bequest to the Foundation at his death in 1968. Upon Mrs. Reinberger's death in 1984, the major portion of her estate was also bequeathed to the Foundation. Although the Reinbergers had no children, the foundation continues to be managed by several generations of Mr. Reinberger's family. Since its inception, the foundation has distributed over $91,000,000 to the non-profit community. The foundation divides its support among the following program areas: Arts, Culture, and Humanities; Education; Human Service-Health; and Human Service-Other. Categories supported under Arts, Culture and Humanities include museums, visual arts, performing arts, media and communication, arts education, zoos, and public recreation. Categories supported under Education include K-12 schools, early childhood education, adult education/literacy, libraries, and workforce development. Categories supported under Human Service-Health include hospitals and clinics, substance abuse prevention/treatment, medical and disease research, disease prevention, and speech and hearing. Categories supported under Human Service-Other include children and youth services, residential and home care, emergency food programs, youth development, domestic violence, and temporary housing/homeless shelters. Letters of Inquiry for grants of any type may be submitted according to the following program-area schedule: Education-March 1; Human Service (Health)-June 1; Human Service (Other)-September 1; and Arts, Culture, and Humanities-December 1. Letters of Inquiry may either be sent by U.S. Mail or emailed and are due in the foundation office by the deadline date. Full proposals are accepted only at the request of the foundation.
Requirements: Applicants must be 501(c)3 organizations. Preferential consideration is given to organizations serving Northeast Ohio, or the greater-Columbus area.
Restrictions: No loans are made, nor are grants given to individuals. The Reinberger Foundation does not make more than one grant to a particular organization during a given calendar year, nor will new proposals be considered until existing multi-year commitments have been paid.
Geographic Focus: Ohio
Date(s) Application is Due: Mar 1; Jun 1; Sep 1; Dec 1
Samples: Cleveland Zoological Society, Cleveland, Ohio, $100,000; Cuyahoga Community College Foundation, Cleveland, Ohio, $37,500; The Free Medical Clinic of Greater Cleveland, Cleveland, Ohio, $50,000; Community Kitchen, Inc., Columbus, Ohio, $5,000.
Contact: Karen R. Hooser; (216) 292-2790; info@reinbergerfoundation.org
Internet: http://reinbergerfoundation.org/apply.html
Sponsor: Reinberger Foundation
30000 Chagrin Boulevard #300
Cleveland, OH 44122

Religion and Health: Effects, Mechanisms, and Interpretation RFP **4052**
The Center seeks proposals to elucidate how religious involvement (religious attendance, worship, altruistic and caring activities) influences individual and community health. The goals of this research program are to: quantitatively document effects on health (where health is broadly defined); clarify the biological, social, psychological mechanisms involved; and interpret what the findings mean for individual, congregational, and community health through trans-disciplinary collaboration and deliberation. This RFP focuses on research that examines the contribution of context and caring as part of the mechanism that accounts for the effects of religion on health, in particular the effects of membership and practices within a faith community. Letters of Intent are due July 15. Thirty finalists will be invited to submit full proposals; seven $200,000 grants will be awarded to outstanding research proposals. Finalists will be selected by Aug 5, with full proposals due by November 1, and awards announced on January 1.
Geographic Focus: All States
Date(s) Application is Due: Jul 15
Amount of Grant: 200,000 USD
Contact: Catherine Craver; (919) 660-7578; fax (919) 684-8569; catherine.craver@duke.edu
Internet: http://www.dukespiritualityandhealth.org/proposals/
Sponsor: Center for Spirituality, Theology and Health at Duke Medical Center
Box 3825, Duke University Medical Center
Durham, NC 27710

Research Corporation Cottrell Scholars Awards **4053**
Cottrell Scholar Awards (CSA) originate from the Foundation's concern with the apparent separation of teaching and research in Ph.D. institutions. The main goal is to promote and support the university scholar model. University scholars are faculty members who have both excellent research programs and excellent approaches to student learning at the undergraduate level. Desired outcomes include: a cultural shift in Ph.D.-granting institutions toward valuing the university scholar; increased attraction and retention of undergraduates in science; and increased undergraduates from Ph.D.-granting institutions pursuing graduate degrees. A key objective is to build a community of outstanding scholar-educators who are dedicated to becoming leaders in both research and teaching and who collectively have the potential to change the way science is taught nationally.
Requirements: Eligible applicants must be tenure-track faculty members at United States institutions whose primary appointment is in a Bachelor's and Ph.D.-granting department of astronomy, biochemistry, biophysics, chemistry, or physics, but not in a school of medicine or engineering. Applicants must demonstrate a strong commitment to lower division undergraduate education and high prospects for becoming a national leader in their field of research. Potential applicants begin the online submission process by completing an eligibility quiz. If eligible, applicants gain access to the application packet and instructions for electronic submission.
Geographic Focus: All States
Date(s) Application is Due: Aug 1
Amount of Grant: 75,000-75,000 USD
Contact: Silvia Ronco, Ph.D., Program Officer; (520) 571-1111; fax (520) 571-1119; sronco@rescorp.org or csa@rescorp.org
Internet: http://www.rescorp.org/grants-and-awards/cottrell-scholar-awards/about-the-program
Sponsor: Research Corporation for Science Advancement
4703 East Camp Lowell Drive, Suite 201
Tucson, AZ 85712

Research Corporation Multi-Investigator Cottrell College Science Awards **4054**
Multi-Investigator Cottrell College Science Awards (CCSA) are a new initiative aimed at helping start sustainable, collaborative programs of research by cross-disciplinary teams of faculty from science departments in primarily undergraduate institutions. Grants are focused on early career faculty. The research project should be one that could not be effectively attacked by an individual researcher or a group of researchers within the same discipline. Rather it is aimed at projects that, by their complexity and interdisciplinary nature, require a cross-disciplinary team approach to achieve significant progress and sustainability. Funding provides $75,000 for a two-investigator team and $100,000 for a three-investigator team. The award duration is two years with a single, one-year, extension available on request. An institutional match of $25,000 is required.
Requirements: An eligible team will consist of at least two, and no more than three, faculty from at least two distinct science departments, all from within the same institution. One of the team members must be from a department of chemistry, physics or astronomy. Only team members from non-Ph.D.-granting departments are eligible. While one team member may be a tenured faculty, the remaining team members must be tenure-track faculty within six years of their first faculty appointment at the time of an award. An on-line eligibility quiz for both institutional and team eligibility is available. Eligible teams will be linked to a pre-proposal application form to be submitted by September 15. Pre-proposals most closely meeting the review criteria and considered to have the highest prospects for establishing a sustainable program of cross-disciplinary collaborative research will be invited to submit the full application with an application target date of November 15.
Restrictions: Team members may have previously held a Single Investigator CCSA, but cannot have an active award at the time of application. Team members may not have both a Single-Investigator and Multi-Investigator CCSA application pending at the same time. Applications containing members from a Department or School of Engineering or Medicine will not be accepted.
Geographic Focus: All States
Contact: Silvia Ronco, Ph.D.; (520) 571-1111; fax (520) 571-1119; sronco@rescorp.org
Internet: http://www.rescorp.org/grants-and-awards/cottrell-college-science-awards/multi-investigator
Sponsor: Research Corporation for Science Advancement
4703 East Camp Lowell Drive, Suite 201
Tucson, AZ 85712

Research Corporation Scialog Grants **4055**
Scialog® Grants seek to accelerate the work of 21st-century transformational science through funding research, intensive dialog and community building. Each multi-year initiative will promote scientific innovation in the face of a complex research challenge that serves as a driver in contemporary science. Successful grantees will be asked to address a few narrowly focused issues on a particular research initiative and to communicate with one another in an annual closed conference for the purpose of sharing insights and building further collaborations. Recent funding was for Scialog: Solar Energy Conversion. This funding cycle has ended. Please check the website for the announcement of a new round of funding. Funding is $100,000 for an individual or $250,000 for a team.
Requirements: The application process begins with completion of the online eligibility quiz. Applicants who pass the eligibility quiz will be given access to necessary forms to submit a proposal. However, final determination of eligibility will be made by a Program Officer.
Restrictions: Multiple proposals from the same investigator are not accepted. There is, however, no limit on the number of proposals submitted from the same institution.
Geographic Focus: All States
Contact: Silvia Ronco, Ph.D.; (520) 571-1111; fax (520) 571-1119; sronco@rescorp.org
Internet: http://www.rescorp.org/grants-and-awards/scialog/solar-energy-conversion
Sponsor: Research Corporation for Science Advancement
4703 East Camp Lowell Drive, Suite 201
Tucson, AZ 85712

Research Corporation Single-Investigator Cottrell College Science Awards **4056**
Single-Investigator Cottrell College Science Awards (CCSA) support research in astronomy, chemistry, physics and closely related fields that significantly overlap these three disciplines at public and private, predominantly undergraduate institutions. Proposed projects are judged on the basis of scientific merit, originality, significance, feasibility, overlap with the three core disciplines and the ability of the institutional environment to sustain the activity. The involvement of undergraduate students in the research is expected, and is an important factor in most awards. The overarching goal is providing "seed" funding for significant research that will enhance the professional and scholarly development of beginning faculty at predominantly undergraduate institutions, along with their students. Funding is expected to help early career faculty establish sustainable programs of research. Faculty are expected to explore new areas of science, to make new discoveries that contribute to their discipline, and to initiate new research programs that can be sustained by other

extramural funding sources, as well as with institutional support. Successful programs are not summer research projects, but rather those undertaken with continuity throughout the academic year. A key element is the involvement of undergraduate students in meaningful ways and thereby raising the quality of undergraduate education. Two-year awards are $35,000, and an institutional match of $10,000 is required.
Requirements: The principal investigator must have a faculty appointment at a United States college or university in a department of astronomy, chemistry or physics; or, if from another department, propose research that significantly overlaps with research in astronomy, chemistry or physics. The applicant's home department must offer at least baccalaureate, but not doctoral, degrees. At the time of application the applicant must be within the first three years of her/his first tenure-track appointment. The potential of a proposed research project to add to fundamental scientific knowledge is a prime criterion in its evaluation, as is its potential for developing into a long-term viable program capable of attracting future support from other agencies. Other factors considered are the environment for research at the institution, student participation, and the contribution the research will make to the college's science programs. An on-line eligibility quiz is available. Applicants meeting the basic eligibility requirements will be directed to submit a brief pre-proposal by September 15.
Restrictions: Faculty with a previous CCSA award or whose appointment is in a department or school of engineering or medicine are not eligible. Expenses are limited to five budget categories: equipment, supplies, student stipends, faculty stipends and travel away from the home institution to conduct research. No indirect costs or fringe benefits (other than FICA at 7.65 percent) are allowed.
Geographic Focus: All States
Date(s) Application is Due: Nov 15
Contact: Silvia Ronco, Ph.D., (520) 571-1111; fax (520) 571-1119; sronco@rescorp.org
Internet: http://www.rescorp.org/grants-and-awards/cottrell-college-science-awards/single-investigator
Sponsor: Research Corporation for Science Advancement
4703 East Camp Lowell Drive, Suite 201
Tucson, AZ 85712

Research Program at Earthwatch Grants **4057**
The program considers proposals for field research in any discipline that can gainfully employ nonspecialists in the implementation of a carefully constructed pure or applied research project. The volunteers, who are recruited and screened to meet scientists' needs, are highly educated citizens dedicated to improving environmental understanding. The center encourages proposals that are interdisciplinary and/or transnational. Specific requests for proposals include the life sciences, earth sciences, social sciences, and interdisciplinary. A typical project would employ five to 10 volunteers each on three to six sequential teams. Teams normally spend one to three weeks in the field. Shorter and longer teams are encouraged where appropriate, as are larger or smaller teams. Preliminary proposals should be submitted 12 to 14 months in advance of anticipated field dates. Full proposals are invited upon review of preliminary materials and will be peer reviewed. Proposals are accepted and reviewed year-round.
Requirements: Earthwatch primarily supports postdoctoral or equivalent scholarship, including principal investigators with commensurate life experience. Proposals are welcomed from advanced scholars of any nationality, covering any geographic region. Applicants intending to conduct research in foreign countries must include host-country nationals in their research staffs.
Geographic Focus: All States
Contact: Center for Field Research; (800) 776-0188 or (978) 461-0081; fax (978) 461-2332; research@earthwatch.org
Internet: http://www.earthwatch.org/research/grantslists.html
Sponsor: Earthwatch Institute
3 Clock Tower Place, Suite 100, Box 75
Maynard, MA 01754-0075

Rett Syndrome Research Foundation Grants **4058**
The foundation (RSRF) awards research grants and postdoctoral fellowships for international biomedical research for the treatment and cure of Rett Syndrome. Relevant disciplines include but are not limited to developmental biology, molecular biology, genetics, biochemistry, cellular physiology, neurology, immunology, pharmacology, gastroenterology, orthopedics, and experimental neurotherapeutics. Treatment-related research is especially encouraged. RSRF seeks to encourage research that encompasses innovative approaches and techniques that will lead to NIH and other funding sources. The foundation is committed to funding senior investigators as well as junior, provided they have a supportive environment. Teams working together from different disciplines, institutions as well as industry are encouraged to apply. Grants are made for technical support, supplies, equipment, and relevant travel. No salary support is provided for the Principal Investigator. Indirect costs of up to 10 percent of direct costs may be included. Letters of intent must be received by the listed application deadline; full proposals are by invitation. Guidelines are available online.
Geographic Focus: All States
Date(s) Application is Due: Jan 31
Amount of Grant: Up to 50,000 USD
Contact: Monica Coenraads; (203) 445-9233; fax (203) 445-9234; monica@rsrf.org
Internet: http://www.rsrf.org/researchers/4.7.1.html
Sponsor: Rett Syndrome Research Foundation
4600 Devitt Drive
Cincinnati, OH 45246

Reuter Foundation Grants **4059**
The Reuter Foundation's areas of interest lie in Social Services to help the poor, needy, homeless and disadvantaged in the Cleveland OH and Collin County TX areas.
Requirements: There is no special form for grant applications. However, each proposal must include a one-page cover letter describing the project, the amount of funds requested, a statement that you are not a 509(a)(3) agency, the contact person, and email address. The preparation of this cover letter is important because if it is not worded convincingly, the remainder of the proposal may not be read. The cover letter and proposal should be sent to: Proposal@ReuterFdn.org and signed by the organizations Executive Director.
Restrictions: Except in unusual circumstances grants do not support endowment programs; construction or renovation of buildings; equipment purchases; artistic or civic projects; debt reductions; benefit functions; conferences, travel, or media projects; graduate studies, scholarships; or to influence legislation or participate in any political campaigns.
Geographic Focus: Ohio, Texas
Samples: The Wilkerson Center, Dallas, TX $5,000-for general operating expenses related to their new Pleasant Grove facility in support of poor and needy area families,; New Life Community, Cleveland, OH $15,000-to help support homeless, jobless single mothers and their children with shelter, job training, job retention, money management, parenting, budgeting, and preparation for GED; Suicide Prevention Alliance, Beachwood, OH $5,000-to support instructor salaries in presenting teenage suicide prevention and depression identification in local high schools;
Contact: Administrator; (216) 961-1141; fax (216) 651-1777; mail@ReuterFdn.org
Internet: http://www.reuterfdn.org/apply.htm
Sponsor: Reuter Foundation
14430 Indian Creek Drive
Cleveland, OH 44130

RGK Foundation Grants **4060**
RGK Foundation awards grants in the broad areas of Education, Community, and Medicine/Health. The foundation's primary interests within Education include programs that focus on formal K-12 education (particularly mathematics, science and reading), teacher development, literacy, and higher education. Within Community, the foundation supports a broad range of human services, community improvement, abuse prevention, and youth development programs. Human service programs of particular interest to the foundation include children and family services, early childhood development, and parenting education. The foundation supports a variety of Community Improvement programs including those that enhance non-profit management and promote philanthropy and voluntarism. Youth development programs supported by the foundation typically include after-school educational enrichment programs that supplement and enhance formal education systems to increase the chances for successful outcomes in school and life. The foundation is also interested in programs that attract female and minority students into the fields of mathematics, science, and technology. The foundation's current interests in the area of Medicine/Health include programs that promote the health and well-being of children, programs that promote access to health services, and Foundation-initiated programs focusing on ALS.
Requirements: Although there are no geographic restrictions to the foundation's grantmaking program, the foundation no longer accepts unsolicited requests for international agencies or programs. All applicants must complete an electronic Letter of Inquiry from the Web site as the first step. RGK Foundation will entertain one electronic Letter of Inquiry (LOI) per organization in a twelve-month period. While the foundation occasionally awards grants for operating expenses, capital campaigns, endowments, and international projects, such grants are infrequent and usually initiated by the foundation. Grants are made only to nonprofit organizations certified as tax exempt under Sections 501(c)3 or 170(c) of the IRS code. Hospitals, educational institutions, and governmental institutions meeting these requirements are eligible to apply.
Restrictions: The foundation refrains from funding annual funds, galas or other special-event fundraising activities; debt reduction; emergency or disaster relief efforts; dissertations or student research projects; indirect/administrative costs; sectarian religious activities, political lobbying, or legislative activities; institutions that discriminate on the basis of race, creed, gender, or sexual orientation in policy or in practice; or loans, scholarships, fellowships, or grants to individuals; unsolicited requests for international organizations or programs or for ALS research projects.
Geographic Focus: All States
Amount of Grant: 1,000-25,000 USD
Contact: Suzanne Haffey, Grants Associate; (512) 474-9298; fax (512) 474-7281; shaffey@rgkfoundation.org
Internet: http://www.rgkfoundation.org/public/guidelines
Sponsor: RGK Foundation
1301 West 25th Street, Suite 300
Austin, TX 78705-4236

Rhode Island Foundation Grants **4061**
The foundation seeks to promote philanthropic activities that will improve the living conditions and well-being of the inhabitants of Rhode Island. Grants for capital and operating purposes principally to agencies working in the fields of education, health care, the arts and cultural affairs, youth, the aged, social services, urban affairs, historic preservation, and the environment.
Requirements: Rhode Island nonprofits organizations may apply. See Foundations website for additional guidelines.
Restrictions: The foundation does not make grants for endowments, research, religious groups for religious purposes, hospital equipment, capital needs of health organizations, or to educational institutions for general operating expenses.
Geographic Focus: Rhode Island
Samples: Ocean State Action Fund, Cranston, RI, $50,000-for continued funding for grassroots organizing, coalition work, and public communications in support of affordable, quality health care for all Rhode Islanders; Community Housing Land

Trust of Rhode Island, Providence, RI, $50,000-for new program targeting foreclosed properties or those close to being foreclosed upon; Sakonnet Preservation Association, Little Compton, RI, $25,000-for three-community partnership to preserve drinking water quality and open space in Newport County;
Contact: Owen Heleen, Vice President for Grant Programs; (401) 427-4009 or (401) 274-4564; fax (401) 331-8085; oheleen@rifoundation.org
Internet: http://www.rifoundation.org/Nonprofits/GrantOpportunities/tabid/175/Default.aspx
Sponsor: Rhode Island Foundation
1 Union Station
Providence, RI 02903

Rhode Island State Government Internships **4062**
The Rhode Island State Government Internships give undergraduate and graduate students the opportunity to participate in state government operations by interning in executive, legislative, and judicial departments. A spring internship and a summer internship are offered. Under the spring program, students receive academic credit toward their major field of study. Students are required to complete papers, work eight to 10 hours per week within a field placement, attend weekly seminars, and complete a take-home midterm and final exam. During the eight-week summer internship program, July and August, interns work 35 hours per week within their chosen field placement, are responsible for time sheets, and are responsible for mandatory student progress reports and final assessment. The summer program has a Rhode Island residency clause. A stipend is awarded on a monthly basis for the two-month period. Work study students are welcome to apply. Applicants may apply to the contact address or access information at the website.
Requirements: Juniors and seniors in college with at least a 2.5 grade point average, along with an interest in state government, are eligible to apply. Applicants for the summer program must be residents of Rhode Island attending an out-of-state college or university.
Restrictions: Financial aid and grants are not awarded.
Geographic Focus: Rhode Island
Contact: Robert Gemma; (401) 222-6782; fax (401) 222-4447; intern@rilin.state.ri.us
Internet: http://www.rilin.state.ri.us/intern/
Sponsor: Rhode Island State Internship Program
State House, Room 8AA
Providence, RI 02903

Richard and Rhoda Goldman Fund Grants **4063**
The fund is interested in supporting programs that will have a significant positive impact in an array of fields, including culture, the environment, population, Jewish affairs, violence prevention, children and youth, the elderly, and social and human services. Types of support include general operating support, continuing support, capital campaigns, program development, and seed money. Funds will be allocated primarily in the San Francisco Bay area of California and to organizations benefiting this area. There are no application deadlines.
Requirements: Nonprofit organizations serving primarily the San Francisco Bay area of California are eligible.
Restrictions: The fund does not accept applications for research or award grants or scholarships for individuals, conferences, documentary films, or fund-raisers. Unsolicited proposals for support of arts organizations or institutions of primary, secondary, or higher education will not be accepted.
Geographic Focus: California
Amount of Grant: 1,000-500,000 USD
Contact: Sam Salkin; (415) 345-6300; fax (415) 345-9686; info@goldmanfund.org
Internet: http://www.goldmanfund.org
Sponsor: Richard and Rhoda Goldman Fund
211 Lincoln Boulevard, P.O. Box 29924
San Francisco, CA 94129

Richard and Susan Smith Family Foundation Grants **4064**
The foundation awards grants to Massachusetts nonprofit organizations in its areas of interest, including arts and culture, children and youth, disadvantaged (economically), education (early childhood, elementary, and higher), minorities, biomedical research, health care and social services delivery. Types of support include annual campaigns, building construction/renovation, capital campaigns, curriculum development, general operating grants, program/project support, seed grants, and service delivery programs. There are no application forms.
Requirements: Massachusetts nonprofits serving the greater Boston area are eligible.
Restrictions: Grants do not support political activities; religious activities; individuals; or requests for deficit financing or endowment funds.
Geographic Focus: Massachusetts
Date(s) Application is Due: Mar 15; Aug 15
Samples: Compassionate Care ALS, West Falmouth, MA $25,000-for the purchase of a cargo van to transport adaptive equipment to and from patient homes; United Teen Equality Center, Lowell, MA $944,780-funding provides violence prevention, youth development, education and community organizing programs for the youth center, the grant is payable over 5 years;
Contact: David Ford; (617) 278-5200; fax (617) 278-5250; dford@smithfamilyfoundation.net
Internet: http://www.smithfamilyfoundation.net
Sponsor: Richard and Susan Smith Family Foundation
1280 Boylston Street, Suite 100
Chestnut Hill, MA 02467

Richard Carley Hunt Memorial Postdoctoral Fellowships **4065**
A limited number of nonrenewable postdoctoral fellowships are awarded each year to scholars within five years after receipt of the doctorate in anthropology to aid in the write-up of research results for publication. Funds may be used for any items necessary to the completion and/or write-up of research. Applicants who require funds during the first half of the following year should meet the May 1 deadline. Projects scheduled to begin in July or later may be submitted by the November 1 deadline. Decisions for each application cycle will be announced six to eight months after the deadline date. Only one application may be submitted during any 12-month period. Eligible individuals may request application forms.
Requirements: Doctoral-level students and researchers and postdoctoral scholars in anthropology and closely related disciplines are eligible to apply. Qualified scholars are eligible without regard to nationality or institutional affiliation. Applicants must hold the PhD at the time of application.
Restrictions: Funds do not provide for salary and/or fringe benefits of applicant; nonproject personnel; travel to meetings; or institutional overhead. Expenses incurred prior to the effective date of an award also will not be covered.
Geographic Focus: All States
Date(s) Application is Due: May 1; Nov 1
Amount of Grant: Up to 40,000 USD
Contact: Program Contact; (212) 683-5000; info@wennergren.org
Internet: http://www.wennergren.org/programsirg.html
Sponsor: Wenner-Gren Foundation for Anthropological Research
470 Park Avenue S, 8th Floor
New York, NY 10016-6819

Richard Davoud Donchian Foundation Grants **4066**
The Richard Davoud Donchian Foundation was founded after Mr. Donchian's death in April of 1993. Extending Mr. Donchian's lifetime passion, its mission is to help others meet their potential and achieve high degrees of personal and professional success. The Foundation's Board of Directors is continuous in its efforts to preserve the Donchian legacy of leadership and integrity, as well as its commitment to learning and personal growth. Consequently, the Foundation concentrates its primary giving activities in the areas of ethical leadership in business and community affairs; education, personal development and literacy; and moral, ethical and spiritual advancement in all areas of life. Applications are accepted throughout the year and are reviewed in the order they are received. The Foundation encourages pilot initiatives that test new program models.
Requirements: The majority of the Foundation's grantmaking is focused in the northeastern United States, although, occasionally, grants may be made in other regions of the country and/or abroad. All applicants must have tax-exempt 501(c)3 status as a non-profit organization as defined by the Internal Revenue Service. The applicant must have an active board of directors with policy-making authority. Grants may range from a few thousand dollars up to $50,000. In unique circumstances, the Foundation does consider a more significant grant for a program having a major impact in one or more of our areas of interest. Of particular interest to the Foundation are organizations that promote partnerships and collaborative efforts among multiple groups and organizations. Priority will be given to requests that show specific plans for funding beyond the present and program innovation and tangible outcomes, with an emphasis on opportunities for significant and lasting social improvement.
Restrictions: The Foundation generally will not provide grants for the following: organizations not determined to be tax-exempt under section 501(c)3 of the Internal Revenue Code; individuals; general fundraising drives; endowments; government agencies; or organizations that subsist mainly on third party funding and have demonstrated no ability or expended little effort to attract private funding.
Geographic Focus: Connecticut, Delaware, District of Columbia, Florida, Maine, Maryland, New Jersey, New York, North Carolina, South Carolina, Vermont, Virginia
Amount of Grant: Up to 50,000 USD
Contact: Donchian Administrator, (203) 629-8552; fax (203) 547-6112; rdd@fsllc.net
Internet: http://www.foundationservices.cc/RDD2/grantrequests.htm#Guidelines
Sponsor: Richard Davoud Donchian Foundation
640 W. Putnam Avenue, 3rd Floor
Greenwich, CT 06830

Richard P. Covert Scholarship **4067**
Two scholarships are awarded annually, one each at the undergraduate and graduate levels, in the field of health care information and management systems. Funding is provided by HIMSS. Completed application form, available from the society, must be accompanied by relevant transcripts, a technical paper and/or an essay, and a faculty recommendation. Annual deadline dates may vary; contact the program office for exact dates.
Requirements: Applicant must be at least a junior when the scholarship is awarded and must be a student member of HIMSS. Previous winners are not eligible, although an undergraduate winner is eligible if pursuing a graduate degree.
Geographic Focus: All States
Date(s) Application is Due: Oct 31
Contact: John Page; (312) 644-4467; fax (312) 664-6143; jpage@himss.org
Internet: http://www.himss.org/ASP/scholarship_ms.asp
Sponsor: Healthcare Information and Management Systems Society
230 E Ohio Street, Suite 500
Chicago, IL 60611-3269

Rich Foundation Grants **4068**
The foundation awards grants to nonprofit organizations in Atlanta, GA, to support community funds, performing arts, cultural programs, higher education, social services, homeless, youth, hospitals, heart disease, theater, and AIDS. Types of support include annual campaigns, building funds, continuing support, equipment acquisition, operating budgets, technical assistance, and research. Contact the office for application forms. The board meets quarterly to review requests.
Requirements: Nonprofits in the Atlanta, GA, area are eligible for grant support.

Geographic Focus: Georgia
Date(s) Application is Due: Mar 15; Jun 15; Sep 15; Dec 15
Amount of Grant: 5,000-100,000 USD
Contact: Anne Poland Berg, Grant Consultant; (404) 262-2266
Sponsor: Rich Foundation
11 Piedmont Center, Suite 204
Atlanta, GA 30305

Righteous Persons Foundation Grants **4069**
The foundation, established by Steven Spielberg, makes grants nationally for projects to revitalize Jewish life, help young people learn about Judaism, and promote tolerance among people of all faiths and ethnicities. Many of the awards are for projects that explore new ideas and opportunities with a special emphasis on young people and that include many projects that may have otherwise had a difficult time obtaining funding. Projects that are national in scope or can serve as models for other communities will receive preference.
Requirements: U.S. 501(c)3 nonprofits are eligible.
Restrictions: Grants are not awarded to support endowments, capital campaigns, building funds, university faculty chairs, individual synagogues or day schools, research, or the publication of books or magazines.
Geographic Focus: All States
Amount of Grant: 5,000-400,000 USD
Contact: Rachel Levin, Program Officer; (310) 314-8393; fax (310) 314-8396
Sponsor: Righteous Persons Foundation
2800 28th Street, Suite 105
Santa Monica, CA 90405

Robert A. Levy Fellowship in Law and Liberty **4070**
The School of Law at the university invites applications from PhD and ABD economists and political scientists who wish to pursue a JD at George Mason University. Qualified applicants will be evaluated for their promise as law and economics scholars capable of making significant contributions to understanding the institutions of a free society.
Requirements: To be eligible for a Levy Fellowship, an applicant must either have earned a PhD in economics, philosophy, political science, finance, or a related field from an accredited university or have successfully completed all course work in one of the above fields and have passed a general exam for a PhD in one of those fields from an accredited university. Candidates must take the LSAT no later than February 2006, register with the Law School Data Assembly Service (LSDAS), and apply for admission to the George Mason University School of Law.
Geographic Focus: All States
Date(s) Application is Due: Apr 1
Amount of Grant: 22,000 USD
Contact: Director; (703) 993-8066; fax (703) 993-8088; djohnsen@gmu.edu
Internet: http://www.gmu.edu/departments/law/admission/levy.html
Sponsor: George Mason University
301 N Fairfax Drive
Arlington, VA 22201-4498

Robert and Patricia Switzer Foundation Fellowships **4071**
The foundation awards 20 environmental fellowships to graduate students in New England and California each year. The foundation seeks to identify and support individuals with the potential to contribute significantly to improving the natural environment. All applicants for fellowships must be nominated in writing by a faculty member, advisor, or an environmental professional before application materials are made available to the applicant. The foundation has a partnership with The San Francisco Foundation to manage the nomination, application and selection process for the program in California (225 Bush Street, Suite 500; San Francisco, CA 94104-4224). Nominations are due on the listed application deadline.
Requirements: To be eligible for a fellowship in California or New England, a candidate must be a U.S. citizen; have completed either one semester of course work at the master's level or one year of doctoral course work; be enrolled in an accredited institution in California or New England (Connecticut, Maine, Massachusetts, New Hampshire, Rhode Island, and Vermont); and have superior academic qualifications.
Geographic Focus: All States
Date(s) Application is Due: Jan 6
Contact: Erin Lloyd, (207) 338-5654; erin@switzernetwork.org
Internet: http://www.switzernetwork.org/fellowship-overview.html
Sponsor: Robert and Patricia Switzer Foundation
P.O. Box 293
Belfast, ME 04915-0293

Robert B McMillen Foundation Grants **4072**
The Robert B McMillen Foundation is a non-profit charitable foundation. The Foundation offers funding in the states of Washington and Alaska. The two areas of intent are: medical and social enhancement. Fifty percent of the annual funding is earmarked for medical research. The Foundation will consider making grants to non-profit organizations involved in researching cardiology, lipid and organ transplants. Twenty-five percent of annual giving is earmarked for social areas including, but not limited to, Goodwill, Salvation Army & United Way. Preference is given to organizations and/or programs that use art as the vehicle to impact communities and change individual lives.
Requirements: The Foundation does not accept uninvited medical or social proposals. Contact the Foundations office for information on, how to be invited to make a medical or social proposal to the McMillen Foundation.
Restrictions: The Foundation do not make grants to: religious, fraternal or other organizations primarily benefiting their own members; other private foundations nor political organizations; for-profit entities; other than scholarships, the foundation will not make multi-year grants.
Geographic Focus: Alaska, Washington
Contact: Cassandra Town; (425) 313-5711; cassandra@mcmillenfoundation.org
Internet: http://www.mcmillenfoundation.org/mission.htm
Sponsor: Robert B McMillen Foundation
55 1st Place NW, Suite 2, P.O. Box 1523
Issaquah, WA 98027

Robert Bowne Foundaion Edmund A. Stanley, Jr. Research Grants **4073**
The foundation provides research grants as part of the Afterschool Matters Initiative to foster high quality cutting-edge research that has a lasting impact on the field. Research should support either original empirical research in or about community-based youth programs during the non-school hours or research syntheses or policy analyses of community-based youth programs.
Requirements: This is a national grant competition, and research may be conducted in any state. Contact foundation for application details.
Geographic Focus: All States
Amount of Grant: 10,000 USD
Contact: Anne Lawrence; (212) 792-6250; anne.lawrence@bownefoundation.org
Internet: http://www.bownefoundation.org/index.php?option=com_content&view=category&layout=blog&id=74&Itemid=96
Sponsor: Robert Bowne Foundation
6 East 39th Street, 10th Floor
New York, NY 10016

Robert E. and Evelyn McKee Foundation Grants **4074**
The foundation awards grants in the following categories: civic, cultural, and religion; education and scholarships; hospitals; medical and medical research; welfare, rehabilitation, and mental health; United Way; and youth activities. Grants awarded for scholarships are made to a few selected high schools and universities, who select scholarship recipients. Religious contributions are basically limited to the local Episcopal churches. The majority of funding is made to local charitable organizations or through local affiliates of national or statewide organizations.
Requirements: 501(c)3 tax-exempt organizations are eligible.
Restrictions: Funding is not available to organizations outside the United States or those limited by race or ethnic origin. The foundation refrains from awarding grants for endowments or deficit financing. Grants are not made to other foundations, except for a local community foundation.
Geographic Focus: Texas
Date(s) Application is Due: Dec 15
Amount of Grant: 1,000-10,000 USD
Contact: Administrator; (915) 581-4025; mckeefoundation@directway.com
Internet: http://www.mckeefoundation.org
Sponsor: Robert E. and Evelyn McKee Foundation
5835 Cromo Drive, Suite 1
El Paso, TX 79912

Robert G. Cabell III and Maude Morgan Cabell Foundation Grants **4075**
The foundation supports capital projects of Virginia nonprofit organizations in the areas of arts and culture, community development, higher education, historic preservation, the arts and culture, religion (Christian, Episcopal, Protestant, and Roman Catholic), social services, and community welfare. Preference will be given to the Richmond, Virginia, area. There are no application forms. Applicants may submit a written proposal that includes a brief description of the overall mission of the organization, a brief description of how the specific project supports the overall mission, a project budget, the level and type of support for the project from the local area, applicant's current operating budget and most recent budget, list of governing board and its officers, evidence of 501(c)3 status, and cover letter. Proposals are considered at grant review meetings each spring and fall.
Requirements: Virginia 501(c)3 tax-exempt organizations that are not classified as private foundations or private operating foundations are eligible.
Restrictions: The foundation does not make grants to individuals, primary or secondary schools, or state-supported organizations; nor will grants be made for debt reduction, endowments (except for faculty development), general operating expenses, research, or scholarships.
Geographic Focus: District of Columbia, Maryland, Virginia
Date(s) Application is Due: Mar 1; Sep 1
Amount of Grant: 25,000-500,000 USD
Contact: Jill A. McCormick; (804) 780-2050; cabell.foundation@gmail.com
Sponsor: Robert G. Cabell III and Maude Morgan Cabell Foundation
901 E Cary Street, Suite 1402
Richmond, VA 23219-4037

Robert Leet and Clara Guthrie Patterson Trust Grants **4076**
The Robert E. Leet and Clara Guthrie Patterson Trust was established in 1980 to support and promote the investigation of human disease and the alleviation of human suffering through improved treatment. While the funding history shows that the trust has recently awarded grants, the grant website states that the trust is not currently accepting applications. Individuals desiring more information about funding opportunities from the Robert E. Leet and Clara Guthrie Patterson Trust are advised to contact the Senior Foundation Officer for further information.
Geographic Focus: Connecticut, New Jersey, New York
Contact: Carmen Britt; (860) 657-7019; carmen.britt@baml.com
Internet: https://www.bankofamerica.com/philanthropic/fn_search.action

Sponsor: Robert Leet and Clara Guthrie Patterson Trust
200 Glastonbury Boulevard, Suite # 200, CT2-545-02-05
Glastonbury, CT 06033-4056

Robert M. Hearin Foundation Grants **4077**

The foundation awards grants to four-year colleges and universities in Mississippi. Areas of interest include program development and improvement, research, scholarships, and faculty development. There are no application deadlines or forms.
Geographic Focus: Mississippi
Amount of Grant: 200,000-650,000 USD
Contact: Daisy Blackwell, (601) 366-8363; fax (601) 961-6876
Sponsor: Robert M. Hearin Foundation
P.O. Box 16505
Jackson, MS 39207

Roberto Longhi Foundation Fellowships **4078**

Fellowships are available for nine months at the institute for those who are seriously dedicated to research in the history of art, particularly Italian painting from the 13th to the 18th centuries. Eligible are Italian citizens who possess degrees from Italian universities with theses in the history of art, and non-Italian citizens who have fulfilled the preliminary requirements for doctoral degrees in the history of art at an accredited university or institution of equal standing. Fellows begin their research in October; they must collaborate on a specific group research project selected by the scientific committee while doing research on their own projects. Fellows are also required to attend seminars, lectures, and other activities arranged by the institute.
Requirements: Application must contain the candidate's biographical data, transcript of undergraduate and graduate records, copy of degree thesis (if any) and other original works, a curriculum vita indicating knowledge of foreign languages spoken and written, two letters of reference from persons acquainted with the candidate's work, the subject of the research that the candidate is interested in pursuing, and two passport photographs. Fellows must live in Florence for the duration of the fellowship.
Restrictions: Applicants must be under thirty-two before the deadline date. Fellows may not make arrangements with any other institution during the period of the fellowship.
Geographic Focus: All States
Date(s) Application is Due: May 15
Contact: Phillip Grosshans; (919) 681-5132; fax (919) 684-2418; phillipg@duke.edu
Internet: http://researchfunding.duke.edu/detail.asp?OppID=4988
Sponsor: Duke University
P.O. Box 104010
Durham, NC 27708

Robert Schalkenbach Foundation Grants **4079**

The foundation seeks to support projects that advocate ideas and programs compatible with Henry George's philosophy of economic justice. George's central thesis was that policies should avoid the perpetuation of unearned privilege, particularly those forms that allow economic rent, or socially-produced surplus, to be privatized. Applicants who have an understanding of these issues and who recognize that Georgist philosophy represents a 'third way' that is distinct from capitalism and socialism or from progressive and libertarian ideals. The foundation will consider two types of projects: 1) Research-projects that deepen understanding of how Georgist principles are connected with issues and debates in the social science or humanities disciplines. 2) Policy-policy-oriented projects that demonstrate how Georgist philosophy yields practical results in a variety of policy settings. Traditionally, the Foundation has primarily supported projects which advocate a gradual tax shift from buildings to land within the context of local property tax reform. At present, the foundation wishes to expand the areas in which it will consider policy-oriented applications. Proposals of this type should demonstrate the capacity not only to evaluate the likely consequences of a policy but also to identify the businesses, advocacy groups, and other organizations that might be mobilized on its behalf, and the arguments that might be used to persuade them to join a coalition.
Requirements: Applicants who are are not prior recipients of a Schalkenbach funding should first submit a one-page letter of inquiry by email. The letter of inquiry should indicate the general nature of the proposed project and the amount of funding that will be requested. Above all, it should indicate specifically how the project will advance the aim of the foundation in promoting the ideas of Henry George either in policy-oriented activities or intellectual research. The foundation also encourages applicants to pursue a variety of topics and issues rather than focusing on the two-rate property tax as the sole practical consequence of Georgist philosophy. Proposals that promote that idea will continue to be considered, but they will be given lower priority than other policy-oriented proposals.
Restrictions: The foundation is not primarily interested in projects that deal specifically with Henry George as a person. Also, it not generally receptive to applications for projects that are primarily libertarian in orientation.
Geographic Focus: All States
Amount of Grant: 2,000-6,000 USD
Contact: Mark A. Sullivan; (212) 683-6424; msullivan@schalkenbach.org
Internet: http://www.schalkenbach.org/funding-program.html
Sponsor: Robert Schalkenbach Foundation
90 John Street, Suite 501
New York, NY 10038

Robert Sterling Clark Foundation Government Accountability Grants **4080**

The underlying premise of the Foundation's interest in government performance is that government agencies and employees will deliver better services to the public in a more cost-effective manner if their activities are examined, evaluated and held up to public view. In supporting advocacy organizations, the first of the Foundation's objectives is to safeguard the well-being of those who are most dependent on government programs. Thus, many of its grantees seek to improve policies that affect low-income New Yorkers and other vulnerable individuals served by publicly funded agencies (e.g. children in foster care, the elderly, and families attempting to move from welfare to work). With regard to the last of these, the Foundation is particularly interested in promoting the implementation of programs to help the adults in such families obtain education, training, and job placement as well as access to child care. A second objective of this grant-making program is to help ensure that all children in the State receive an adequate education. Finally, the Foundation's grant-making also supports some efforts to improve city and state policies in areas such as solid waste management, land use planning, and protection of parks and open space.
Geographic Focus: New York
Amount of Grant: 1,000-200,000 USD
Contact: Margaret C. Ayers, Executive Director; (212) 288-8900; fax (212) 288-1033; rcsf@rsclark.org
Internet: http://www.rsclark.org/index.php?page=government-accountability
Sponsor: Robert Suiterling Clark Foundation
135 East 64th Street
New York, NY 10065-7045

Robert Sterling Clark Foundation Reproductive Rights and Health Grants **4081**

The objective of the Foundation's work in this field is to protect and expand women's access to reproductive health services in the belief that the ability to control one's fertility and prevent unintended pregnancy is fundamental to the advancement of women's opportunities. The strategy employed by the Foundation to achieve this objective is to support policy analysis, advocacy, litigation, research, message development, and/or organizing aimed at: promoting the implementation of laws, policies and practices that will enable all women to have access to comprehensive reproductive health information and services, including emergency contraception and abortion; challenging laws and legal decisions that undermine reproductive rights and developing legal theory in support of these rights; mobilizing an informed grassroots constituency; developing new pro-choice messages, based on current opinion research, that resonate with a broad public; informing the development of judicial selection criteria that will produce non-ideological courts; and promoting the use of comprehensive sexuality education curricula in the schools.
Requirements: The Foundation makes support available to national and regional organizations working on these issues.
Geographic Focus: All States
Amount of Grant: 1,000-200,000 USD
Contact: Margaret C. Ayers; (212) 288-8900; fax (212) 288-1033; rcsf@rsclark.org
Internet: http://www.rsclark.org/index.php?page=reproductive-right-and-health
Sponsor: Robert Suiterling Clark Foundation
135 East 64th Street
New York, NY 10065-7045

Robert W. Deutsch Foundation Grants **4082**

The foundation provides support to technology programs in institutions of higher education in Maryland. Types of support include: continuing support, curriculum development, fellowships, general operating support, internship funds, organizational development, program development, research, and seed grants. There are no application deadlines or forms; submit initial inquiries by email or postal mail.
Restrictions: No support for religious or political organizations.
Geographic Focus: Maryland
Samples: University of Maryland-College Park, College Park, MD. $250,000-For nanotechnology research, payable over 1 year.
Contact: Jane Brown; (410) 252-9244; fax (410) 252-0997; rdeutsch@rwd.com
Sponsor: Robert W. Deutsch Foundation
5521 Research Park Drive
Baltimore, MD 21228-4664

Roche Organ Transplant Research Foundation Grants **4083**

The Roche Organ Transplantation Research Foundation (ROTRF) is committed to funding a broad range of transplantation-related research, with the ultimate aim of improving long-term outcomes for people with organ transplants. The ROTRF supports important and innovative projects in clinical research applicable to the understanding and care of human organ transplant recipients and related tissue transplants, such as islet transplantation. The ROTRF gives priority to clinically oriented research projects, such as observational clinical studies, or studies that use human transplant samples for laboratory examinations, and focus on understanding the pathogenesis and prevention of human transplantation diseases, including organ injury and rejection, viral infection, and late graft deterioration. Furthermore, the ROTRF may consider funding research that addresses related clinical issues in non-transplant patients, including hepatitis C infection and other viral infections, and inflammatory processes relevant to human organ transplantation. These related clinical issues should demonstrate their relevance to disease phenotypes in organ transplant patients. The Trustees will also consider funding studies that investigate transplant populations, ethics, organ preservation and allocation, and healthcare delivery. Moreover, the ROTRF will welcome research in new emerging technologies that examine the pathogenesis of human disease states in organ transplantation. If animal models are used to complement the clinical research, a higher priority will be given to those that simulate pathological states and mechanisms operating in human organ transplantation. Application is made through the online letter of intent.

Applicants are given a submission ID number which allows them to access and edit the letter of intent an any time until the next submission deadline.
Requirements: To be considered for a grant award, applications should have demonstrable applicability to human organ transplantation. Investigators working in areas such as antibody-mediated rejection and antibody formation, graft pathology during rejection events, histocompatibility, infectious agents, and disease phenotypes in transplant patients are encouraged to apply. The Trustees will also consider funding studies that investigate transplant populations, ethics, organ preservation and allocation, and health care delivery. Furthermore, the ROTRF will welcome research in new emerging technologies that examine the pathogenesis of human disease states in organ transplantation.
Restrictions: Applications must be submitted in English. Only established, faculty member of academic staff at universities, transplant centers and research institutes are eligible as Principal Applicants. Every applicant can submit only one Letter of Intent for each funding cycle. Every ROTRF grantee (including co-investigator) can only hold one ROTRF grant at a time.
Geographic Focus: All States, All Countries
Date(s) Application is Due: Oct 1
Contact: Administrator; (41) 41 377 53 53; fax (41) 41 377 53 34; admin@rotrf.org
Internet: http://www.rotrf.org
Sponsor: Roche Organ Transplant Research Foundation
Postfach 222
Meggen, CH-6045 Switzerland

Rockefeller Archive Center Ehrlich Grants **4084**
Through a generous gift, the Rockefeller Archive Center has available funds to support short-term research in the Paul Ehrlich Collection at the Archive Center. The Ehrlich Collection is the largest and most important body of original and facsimile materials documenting the life and scientific-medical research program of Nobel Laureate Paul Ehrlich (1854-1915). Application to the Ehrlich grant program may be made at any time and, if judged worthy, awards will be made within a month of the receipt of the applicant's two letters of recommendation. Reimbursement of up to $2,500 will be made upon receipt of original expense receipts after the completion of the grant recipient's research visit.
Requirements: Application to this program is open to U.S. citizens and citizens of foreign countries. Prospective applicants are required to contact the Archive Center with an initial description of the applicant's research.
Restrictions: Grant recipients must commence their research at the Center within one year of being notified of the award. The grant recipients are also, requested to submit a short report on their research at the Center. Edited versions of or excerpts from these reports may be used in the Center's publications. Recipients also are asked to provide the Center with a copy of any publication resulting from research conducted as a result of the grant.
Geographic Focus: All States
Contact: Camilla Harris, Grants Administrator; (914) 366-6311 or (914) 631-4505; fax (914) 631-6017; charris@rockarch.org or archive@rockarch.org
Internet: http://www.rockarch.org/grants/ehrlichgrant.php
Sponsor: Rockefeller Archive Center
15 Dayton Avenue
Sleepy Hollow, NY 10591-1598

Rockefeller Archive Center Grants in Aid Program **4085**
Rockefeller Archive Center's competitive grant-in-aid program provides awards for reimbursement of up to $5000 for certain receipted expenses related to research conducted at the RAC. The research grants are designed to foster, promote, and support research by serious scholars in the collections located at the Rockefeller Archive Center, which include the records of the Rockefeller Family, The Rockefeller University, The Rockefeller Foundation, and other philanthropies and associated individuals. The deadline for applications is November 15th each year, and awards are announced at the end of the following February. Grant recipients must commence their research at the Center within one year of being notified of the award. Grant recipients are requested to submit a short report on their research at the Center. Edited versions of or excerpts from these reports may be used in the Center's publications. Recipients also are asked to provide the Center with a copy of any publication resulting from research conducted as a result of the grant.
Requirements: Application to the grant-in-aid program is open to U.S. citizens and citizens of foreign countries. Certain U.S. Government requirements will apply to non-U.S. citizens.
Restrictions: Awards are made to individuals. Institutions are not eligible to apply. The grant does not support research at other institutions, and it does not provide general tuition support.
Geographic Focus: All States
Date(s) Application is Due: Nov 15
Contact: Camilla Harris, Grants Administrator; (914) 366-6311 or (914) 631-4505; fax (914) 631-6017; charris@rockarch.org or archive@rockarch.org
Internet: http://www.rockarch.org/grants/generalgia.php
Sponsor: Rockefeller Archive Center
15 Dayton Avenue
Sleepy Hollow, NY 10591-1598

Rockefeller Archive Center Research Grants **4086**
The foundation awards grants for travel and research at the Rockefeller Archive Center, which contains records of the Rockefeller Foundation, Rockefeller University, and other affiliated philanthropic organizations and individuals. The center also will award grants to support research on the history of basic medical research, the life of the Nobel laureate Paul Ehrlich, and on postcolonial studies.
Requirements: Scholarly reseearchers, particularly students preparing dissertations, are eligible.
Restrictions: The Archive Center's programs do not support research at other institutions, and they do not provide general tuition support.
Geographic Focus: All States
Contact: Camilla Harris, Grants Administrator; (914) 366-6311 or (914) 631-4505; fax (914) 631-6017; charris@rockarch.org or archive@rockarch.org
Internet: http://www.rockarch.org/grants/
Sponsor: Rockefeller Foundation
15 Dayton Avenue
Sleepy Hollow, NY 10591-1598

Rockefeller Archive Center Visiting Archivist Fellowship **4087**
The Rockefeller Archive Center invites applications for a Visiting Archivist Fellowship for a professional archivist from: South and Central America, including Mexico; the Caribbean; Africa; south and central Asia; China; Korea; Taiwan; the Philippines; Indonesia; and the Pacific Ocean states. The Rockefeller Archive Center will host a visiting archivist for at least two weeks, but no longer than one month for the purposes of enhancing professional development, and extending knowledge of the Center's collections. Applications must be postmarked by January 31st each year, and awards will be announced at the end of April. Upon completion of the fellowship, the visiting archivist can receive reimbursement for up to $8000 in receipted expenses for round-trip Economy Class air fare; housing, food and local transportation.
Requirements: Applicants must be, or recently have been, employed as archivists, and demonstrably able (1) to improve their archival skills through a residency at the Center, and (2) to contribute to a better understanding of the Center's documentation of the applicant's nation or region. The successful applicant will have a B.A. or equivalent, professional archival training, and 2-5 years experience as an archivist. He/she should be seeking to expand his/her experience in order to develop new services or technologies in his/her home institution. The applicant must have a strong command of spoken and written English, and must provide evidence to that effect. Potential applicants are required to contact an archivist at the Archive Center by email at archive@rockarch.org or by fax at (914) 631-6017 well in advance of completing their applications with a description of their fellowship project.
Geographic Focus: All States
Date(s) Application is Due: Jan 31
Amount of Grant: 8,000 USD
Contact: Camilla Harris; (914) 366-6311 or (914) 631-4505; charris@rockarch.org
Internet: http://www.rockarch.org/grants/visitingarchivist.php
Sponsor: Rockefeller Archive Center
15 Dayton Avenue
Sleepy Hollow, NY 10591-1598

Rockefeller Brothers Fund Fellowship in Nonprofit Law **4088**
The fellowship permits one fellow each year to spend one year in residence at the Vera Institute of Justice in New York City, working closely with Vera's general counsel and special counsel on legal issues. The fellowship will provide a law school graduate with exposure to a wide variety of legal and organizational issues encountered by nonprofit organizations. It also will provide the opportunity to identify an emerging or changing area of law with particular significance for nonprofits. Fellows will be selected without regard to race, color, religion, gender, political beliefs, national origin, disability, age, or sexual orientation.
Requirements: Graduates of accredited U.S. law schools are eligible.
Geographic Focus: All States
Date(s) Application is Due: Dec 5
Amount of Grant: 47,000 USD
Contact: Erin Bainbridge Ortiz; (212) 998-6168; fax (212) 995-3149; ncpl.info@nyu.edu
Internet: http://www.rbf.org/grantsdatabase/grantsdatabase_show.htm?doc_id=616213
Sponsor: National Center on Philanthropy and the Law
110 W Third Street, Room 205
New York, NY 10012

Rockefeller Foundation Humanities Fellowships **4089**
Fellowships in the humanities and the study of culture will be offered at 21 host institutions that were selected for their potential to promote new work in the humanities. The fellowships are meant to serve scholars who are testing disciplinary boundaries or moving into newer fields of inquiry within the humanities. Although the majority of the fellow's time will be spent pursuing his or her own research toward publication, the residency may involve participation in seminars, conferences, or other collaborative activities within the host program. Complete information about eligibility, stipends, and procedures for application is available directly from current host institutions (listed in application). Individuals apply for resident fellowships at the host institutions.
Requirements: Residential fellowships are meant to serve scholars who are testing disciplinary boundaries or moving into newer fields of inquiry within the humanities. Applications are invited from programs in colleges and universities and other institutions, such as libraries, museums, and community cultural spaces, with engaged communities of scholars and other resources of value to emerging fields of the humanities.
Restrictions: Awards may not be used for the completion of graduate studies, advanced training, curriculum development projects, or for writing poetry or fiction.
Geographic Focus: All States
Contact: Grants Administrator; (212) 869-8500; fax (212) 764-3468
Internet: http://www.rockfound.org/bellagio/bellagio.shtml
Sponsor: Rockefeller Foundation
420 Fifth Avenue
New York, NY 10018-2702

Rockefeller Foundation MAP Fund Grants **4090**

MAP works to build a thriving, risk-welcoming contemporary performance field by providing project-specific funding to playwrights, choreographers, directors, composers and performers experimenting in any performance tradition or discipline. MAP believes that exploration drives human progress, in art as in science and medicine. Sometimes the path of progress is lighted by an act of the imagination on the part of the artist. MAP exists to facilitate these acts and help bring them to their fullest realization. Up to 40 grants per annual cycle, ranging from $10,000 to $45,000. The average award amount is $25,000.
Requirements: Applications must come from organizations based in the United States that have current nonprofit federal tax status (501(c)3). Unincorporated artists or ensembles may apply to MAP through a fiscal sponsor. Organizations and artists must demonstrate at least 2 years professional experience. MAP supports only projects that contain a live performance. Eligible projects must not have premiered anywhere in the world before September 1.
Restrictions: Artists will not receive support for more than one project per funding cycle. Artists supported through the fund for the past two consecutive years will not be eligible for support in the current year but may reapply the following year.
Geographic Focus: All States
Date(s) Application is Due: Jan 11
Amount of Grant: Up to 50,000 USD
Contact: Moira Brennan, Program Director; (212) 226-1677; fax (212) 226-7665; mapinfo@mapfund.org
Internet: http://www.mapfund.org/apply.html
Sponsor: Rockefeller Foundation
73 Spring Street, Suite 401
New York, NY 10012

Rockefeller Foundation Scholarly Residencies **4091**
A Bellagio residency provides time for critical thinking, disciplined work, individual reflection, and collegial engagement, uninterrupted by the usual professional and personal demands. The Center typically offers one-month residencies for no more than 12 scholars and scientists at a time. Individuals in any discipline and from any part of the world are welcome to apply. Space is reserved for both academic projects, as well as projects that align with the Foundation's mission to expand opportunities for poor or vulnerable people and to help ensure that globalization's benefits are shared more widely. Residency cohorts include artists, writers, non-governmental organization practitioners, policymakers, scholars and scientists from around the world. Meals and informal presentations of work provide an opportunity for residents to engage with each other. During special dinners, residents also have the chance to interact with the participants of international conferences that are hosted in various buildings on the Center's property. The combination of private time for work and reflection and informal gatherings is an important part of life at the Center.
Restrictions: Accommodations are not available for children and family members other than spouses/life partners, nor for friends or pets.
Geographic Focus: All States
Amount of Grant: 37,000 USD
Contact: Residencies Administrator; (212) 869-8500; fax (212) 764-3468
Internet: http://www.rockfound.org/bellagio/bel_scholar.shtml
Sponsor: Rockefeller Foundation
420 Fifth Avenue
New York, NY 10018-2702

Rockford College Faculty Fellowships **4092**
Faculty Fellows use funds to explore how active citizenship can be applied to their curriculum and/or research interests. In some cases, this is a continuation of work already being done. For other faculty members, the fellowship affords an opportunity to explore a new dimension of their discipline. The overall goals of the program include the following: supporting the practice of the scholarship of engagement; developing as faculty leaders in modeling best practices, recruiting other faculty, and institutionalizing socially responsive knowledge to complement each student's education; encouraging Faculty Fellows to build a strong and productive social and intellectual community of scholars—coming together across disciplinary and community boundaries, reading and sharing work, discussing ideas, and grounding that work in new theoretical and conceptual contexts; and providing Fellows opportunities to conduct research that builds knowledge about civic engagement or is "civically engaged" and developing curricular approaches that prepare students to be effective citizens.
Requirements: All full-time and adjunct faculty members of Rockford College are eligible to apply for these fellowships.
Geographic Focus: Illinois
Contact: Ryan Kessler, Student Advocate; (815) 394-4262 or (815) 226-4062; fax (815) 394-5174; SAS@rockford.edu
Internet: http://www.rockford.edu/?page=JACCEScholarshipProg
Sponsor: Rockford College
5050 E State Street
Rockford, IL 61108-2393

Rockwell International Corporate Trust Grants **4093**
Grants are awarded in the areas of education and youth development, with emphasis in math, science, and engineering; and culture and the arts, with emphasis on youth educational programs. Rockwell also contributes to health, human services, and civic organizations and also gives special consideration to qualifying organizations in which employees are involved as volunteers. Types of support include capital grants, challenge/matching grants, conferences and seminars, development grants, endowments, research grants, scholarship funds, and general operating grants. Awards include single-year, multiple-year, and provisional continuing support of specific programs. Applicants should submit full, detailed proposals. Proposals are accepted at any time. At the Web site, click on About Us, then Corporate Citizenship, and then Rockwell Automation Corporation Trust.
Restrictions: Organizations are not eligible if they have not received a permanent, tax-exempt ruling determination from the federal government; if they cannot provide current full, certified, audited financial statements; or if they are private foundations. Funding will not be considered for the following purposes: general endowments, deficit reduction, grants to individuals, federated campaigns, organizations or projects outside the United States, religious organizations for religious purposes, or fraternal or social organizations.
Geographic Focus: All States
Amount of Grant: 1,000-50,000 USD
Contact: Rockwell International Corporation General Information; (562) 797-3311
Internet: http://www.rockwell.com
Sponsor: Rockwell International Corporate Trust
1201 S Second Street
Milwaukee, WI 53204

Rocky Mountain Mineral Law Foundation Grants **4094**
Grants are awarded to persons or organizations for meritorious projects in natural resources law such as the preparation of teaching materials; specialized faculty or student research; law school seminars, lectures, short courses, and symposia; and publication of substantive articles. Preference will be given to Member Law Schools of the Rocky Mountain Mineral Law Foundation (see the website for a full listing).
Requirements: No special application form is required; submit a letter and supporting data including: your qualifications to undertake the project; a detailed explanation of your proposal; a budget of your total anticipated expenses, exclusive of project salary/remuneration for application; and, the amount of support you are requesting from the foundation. Applications should be submitted electronically to csherwood@rmmlf.org.
Restrictions: Personal remuneration or honoraria are not within the scope of these grants.
Geographic Focus: Arizona, California, Colorado, Idaho, Kansas, Louisiana, Montana, Nebraska, Nevada, New Mexico, North Dakota, Oklahoma, Oregon, South Dakota, Texas, Utah, Washington, Wyoming, Canada
Contact: Carla Sherwood, (303) 321-8100, ext. 107; csherwood@rmmlf.org
Internet: http://www.rmmlf.org/geninfo/schgrant.htm
Sponsor: Rocky Mountain Mineral Law Foundation
9191 Sheridan Boulevard, Suite 203
Westminster, CO 80031

Roeher Institute Research Grants **4095**
The institute, through the support of the Scottish Rite Charitable Foundation of Canada, offers research grants to associates, associations, and agencies in a broad range of fields relating to human services and intellectual disability. The renewable grants are offered for one year. Research projects must be under the direction of the Roeher Institute academic associates or other university faculty or under researchers approved by the Roeher Institute. Areas constituting research priorities for the Roeher Institute are issues affecting people with intellectual disabilities; integration of people who have an intellectual disability into society; prevention, early identification, and minimization of disabling conditions; and strategies for social change that improve the quality of life of persons with intellectual disabilities. Applicants are asked to submit a summary of their research project, budget, and financial requirements.
Requirements: Applications are invited from university faculty, department heads, supervisors, Roeher Institute associates, and/or consultants.
Geographic Focus: All States
Contact: Cameron Crawford, President; (416) 661-9611; fax (416) 661-5701; cameronc@roeher.ca or cameronc@worldchat.com
Internet: http://www.roeher.ca/english/about/about.htm
Sponsor: Roeher Institute
4700 Keele Street, York University, Kinsmen Building.
North York, ON M3J 1P3 Canada

Roger Adams Award in Organic Chemistry **4096**
This award, sponsored by Organic Reactions, Inc. and Organic Syntheses, Inc., and administered by the American Chemical Society, is given biennially in odd-numbered years to recognize and encourage outstanding contributions to research in organic chemistry defined in its broadest sense. The award will be presented biennially in odd-numbered years at the ACS National Meeting. The recipient will deliver a lecture at the Biennial National Organic Chemistry Symposium of the American Chemical Society National Meeting and Symposium. The award is granted to an individual without regard to nationality. Nominations will be accepted by the indicated deadline date in even-numbered years.
Requirements: Any individual, except a member of the award committee, may submit one nomination or seconding letter for the award in any given year. The nominating documents consist of a letter of not more than 1000 words containing an evaluation of the nominee's accomplishments and a specific identification of the work to be recognized, a biographical sketch including date of birth, and a list of publications and patents authored by the nominee.
Restrictions: Self-nominations are not accepted.
Geographic Focus: All States
Date(s) Application is Due: Nov 1
Amount of Grant: Up to 25,000 USD
Contact: Felicia Dixon, Awards Administrator; (800) 227-5558 or (202) 872-4408; fax (202) 776-8008; f_dixon@acs.org or awards@acs.org
Internet: http://portal.acs.org/portal/acs/corg/content?_nfpb=true&_pageLabel=PP_ARTICLEMAIN&node_id=1319&content_id=CTP_004550&use_sec=true&sec_url_var=region1
Sponsor: American Chemical Society

1155 Sixteenth Street, NW
Washington, DC 20036-4801

Roger K. Summit Scholarship **4097**

The three awards are made in North America; Europe, the Middle East, and Africa; and Asia-Pacific to graduate students in library and information science who have demonstrated outstanding interest or performance in electronic information services. Scholarship recipients are selected based on academic achievement; demonstrated interest in electronic information services, based on course work, research and experience; proficiency on Dialog's online services and products; and faculty recommendations. Application and guidelines are available online.
Requirements: Graduate students in library and information science are eligible.
Geographic Focus: All States
Amount of Grant: 5,000 USD
Contact: Betty Jo Hibberd, Scholarship Administrator; Betty.Jo.Hibberd@dialog.com
Internet: http://gep.dialog.com/scholarship
Sponsor: Dialog Corporation
11000 Regency Parkway, Suite 10
Cary, NC 27511

Rohm and Haas Company Grants **4098**

The company awards grants in five key philanthropic categories: education, environment, civic and community, health and human services, and arts and culture. Education grants support programs linking education in science, technology, and math to workplace and career opportunities; and educational enrichment for students, particularly in the areas of math, science, technology, and the environment either during or after school hours (including weekends, summer, and other times when school is not in session). Environment grants support after-school programs that build environmental awareness and understanding; sustainability programs that educate and promote development that meets the needs of the present society without compromising the ability of future generations to meet their own needs; pollution prevention and waste management, environmental conservation and biodiversity, and energy and water conservation; and community beautification. Civic and community grants support regional efforts that build local competitiveness and strengthen the economic and social base in Rohm and Haas host communities; programs that cut across multiple giving categories and/or that impact key stakeholders in Rohm and Haas communities; and programs that focus on volunteerism. Health and human services grants support public safety awareness; disaster preparedness; structured youth development such as leadership training; and structured youth activities and recreation. Arts and culture grants support organizational capacity building in the form of plan implementation—technology implementation and/or enhancement, communications and marketing, outreach and audience development, financial management and fundraising, board development, strategic development, professional development, and educational program development. Application and guidelines are available online.
Requirements: Eligible national and international organizations include performing and visual arts, humanities and historical societies, and museums. Preference is given to requests from Chicago, IL, and the greater Philadelphia area.
Restrictions: Grants are not awarded to individuals or for fundraising, advertising, or testimonials.
Geographic Focus: All States
Amount of Grant: 250-50,000 USD
Contact: Alexandra Samuels, (215) 592-3644; alexandra_samuels@rohmhaas.com
Internet: http://www.rohmhaas.com/community/giving/guidelines/guidelines.html
Sponsor: Rohm and Haas Company
100 Independence Mall W
Philadelphia, PA 19106-2399

Romanell-Phi Beta Kappa Professorship **4099**

The Professorship is awarded annually to scholars in the field of philosophy, without restriction to any one school of philosophical thought. The professorship is intended to recognize not only distinguished achievement but also the recipient's contribution or potential contribution to public understanding of philosophy. The professorship carries with it a stipend of $7,500, which will be awarded in addition to the recipient's regular salary.
Requirements: Although the awardee need not be a member of Phi Beta Kappa, he or she must be on the faculty of an institution sheltering a chapter of Phi Beta Kappa.
Geographic Focus: All States
Amount of Grant: 7,500 USD
Contact: Sandra Beasley, Awards Coordinator; (202) 265-3808; fax (202) 986-1601; SBeasley@pbk.org or info@pbk.org
Internet: http://staging.pbk.org/AM/Template.cfm?Section=Romanell_Professorship&Template=/CM/HTMLDisplay.cfm&ContentID=2295
Sponsor: Phi Beta Kappa Society
1606 New Hampshire Avenue NW
Washington, DC 20009

Ronald Breslow Award for Achievement in Biomimetic Chemistry **4100**

The purpose of the Award is to recognize outstanding contributions to the field of biomimetic chemistry. Special consideration will be given to individuals who are scientifically active. The award consists of $5,000 and a certificate. Reasonable travel expenses to the meeting at which the award will be presented will be reimbursed.
Requirements: Any individual, except a member of the award committee, may submit one nomination or seconding letter for the award in any given year. The nominating documents consist of a letter of not more than 1,000 words containing an evaluation of the nominee's accomplishments and a specific identification of the work to be recognized, a biographical sketch including date of birth, and a list of publications and patents authored by the nominee.
Restrictions: Self-nominations are not accepted.
Geographic Focus: All States
Date(s) Application is Due: Nov 1
Amount of Grant: 5,000 USD
Contact: Felicia Dixon, Awards Administrator; (800) 227-5558 or (202) 872-4408; fax (202) 776-8008; f_dixon@acs.org or awards@acs.org
Internet: http://portal.acs.org/portal/acs/corg/content?_nfpb=true&_pageLabel=PP_ARTICLEMAIN&node_id=1319&content_id=CTP_004551&use_sec=true&sec_url_var=region1
Sponsor: American Chemical Society
1155 Sixteenth Street, NW
Washington, DC 20036-4801

Ronald McDonald House Charities Grants **4101**

Ronald McDonald House Charities provides grants to nonprofit, tax-exempt organizations whose national or global programs help improve the health and well being of children under 21. Types of support include capital grants, challenge/matching grants, formula grants, professorships, program grants, research grants, scholarships, seed money grants, and visiting scholars. The board meets quarterly to review, select, and award grants to organizations that have demonstrated an ability to respond to the needs of specific groups of children in a definitive, hands-on manner that yields measurable results.
Requirements: To be considered for funding, an applicant must be designated a not-for-profit, tax-exempt charitable organization. U.S. charities must have a current 501(c)3 tax-exempt status letter on file with the Internal Revenue Service. Organizations seeking funding should have: a broad base of funding support; management and staff capacity to effectively execute the project; a longer-term organizational strategic plan (evidence to ensure program and organization sustainability); and a clear, concise plan for project evaluation with outcome measurement that shows potential for meaningful change. Organizations seeking funding should have a specific program that: directly improves the health and well being of children; addresses a significant funding gap or critical opportunity; has long-term impact in terms of replication or reach; produces measurable results; and is sustainable without relying on RMHC funding. Prior to submitting a full proposal, organizations must submit a letter of inquiry which will be reviewed by the Board. If interested, the Board will invite organizations to submit a formal grant proposal.
Restrictions: The charity does not fund to projects/programs that are local in scope and that do not benefit children under 21; to reduce debt; for annual fund appeals; for medical research; for for-profit organizations; for fundraising sponsorships; for programs/projects administered by activist groups; for projects that have already been completed; for sectarian or religious purposes; for scholarships and fellowships to individuals; to propagandize or influence elections or legislation; for advertising or fundraising drives; to intermediary funding agencies; for endowment campaigns; for individuals.
Geographic Focus: All States
Date(s) Application is Due: Mar 5; Sep 7
Contact: Michael Singer; (630) 623-7048; fax (630) 623-7488; grants@rmhc.org
Internet: http://rmhc.org/what-we-do/grants/how-to-apply/
Sponsor: Ronald McDonald House Charities
1 Kroc Drive
Oak Brook, IL 60523

Rosalinde and Arthur Gilbert Foundation/AFAR New Investigator Awards in Alzheimer's Disease **4102**

The major goal of this program is to support important research in areas in which more scientific investigation is needed to improve the prevention, diagnosis, and treatment of Alzheimer's Disease. The program will also serve to encourage junior investigators in the United States and Israel to pursue research and academic careers in the neurosciences, and Alzheimer's disease in particular. Projects in basic and translational research related to Alzheimer's disease (AD) that are clinically relevant, will be considered. Examples of promising areas of research include, but are not limited to: basic mechanisms of aging in the central nervous system; genetics of AD; neuroimaging and precursors of AD; cellular and Molecular pathways of AD; biological Markers of AD; exercise and dietary factors; neurogenesis and AD; impact of environmental agents in CNS aging and early AD; inflammation; cardiovascular and cerebrovascular factors; drug discovery. It is anticipated that 5 grants of up to $75,000 each will be awarded. Applicants may propose to use the award over the course of one or two years as justified by the proposed research. Funds may not be requested for overhead or indirect costs. Funding will begin July 1.
Requirements: The applicant must be an independent investigator with assigned independent space and must have received a junior faculty appointment (instructor, assistant professor or equivalent) no sooner then July 1st. The proposed research must be conducted at any type of not-for-profit setting in the United States or Israel. Five criteria are used to determine the merit of an application: qualifications of the applicant; quality of the proposed research; relevance of the proposal to how mechanisms of aging may lead to AD; excellence of the research environment; likelihood that the project will advance the applicant's career in basic research on the mechanisms of aging and AD. All candidates must submit applications endorsed by their institution. The deadline for receipt of applications and supporting materials from candidates based in the U.S. is December 16th, at 5:00 p.m. EST. The deadline for receipt of applications and supporting materials from candidates based in Israel is December 30th. See website for instruction sheet and application for complete application procedures. Incomplete applications cannot be considered.
Restrictions: The Gilbert/AFAR Research Grant Program does not provide support for: Postdoctoral fellows in the laboratory of a senior investigator; investigators who have already received major extramural funding for research on Alzheimer's disease and/or

aging (such as an R01 grant); Senior faculty, i.e. at the rank of Associate Professor level or higher; NIH Intramural program employees. Applicants cannot apply for the AFAR Research Grant Program or Glenn/AFAR BIG Program.
Geographic Focus: All States, Israel
Date(s) Application is Due: Dec 16; Dec 30
Amount of Grant: 75,000 USD
Contact: Grants Manager; (212) 703-9977 or (888) 582-2327; fax (212) 997-0330; grants@afar.org or info@afar.org
Internet: http://afar.org/Gilbert.html
Sponsor: American Federation for Aging Research
55 West 39th Street, 16th Floor
New York, NY 10018

Rosalynn Carter Institute John and Betty Pope Fellowships **4103**
The John and Betty Pope Fellowship in Caregiving is a unique program of the Rosalynn Carter Institute for Caregiving at Georgia Southwestern State University. The fellowship provides financial support for outstanding individuals pursuing training and careers in fields related to caregiving, and provides a structure and course of study intended to prepare them for leadership in the field. The certificate program of study includes courses in psychology, nursing, sociology and special education taken over up to four years of study at GSW. The amount of the award is determined annually and is currently set at $3750 per academic year.
Requirements: Award of Pope Fellowships is currently limited to students pursuing a degree program at Georgia Southwestern State University and who are or will be enrolled in the Caregiver Specialist Certificate at GSW. Applicants must have a grade point average of 3.0 as well as an established interest in academic study related to caregiving or a record of outstanding professional or informal caregiving. In addition, Pope Fellows must: enroll for minimum of 12 credit hours each semester (undergraduate) or 9 credit hours (graduate) for a full award; attend Pope Fellowship orientation and other required meetings; and make steady progress in fulfilling requirements of the Caregiver Specialist Certificate.
Geographic Focus: All States
Date(s) Application is Due: Feb 15
Amount of Grant: 6,000-24,000 USD
Contact: Leisa Easom, Pope Fellowship Director; (229) 928-1234; fax (229) 931-2663; leasom@gsw.edu or rci@canes.gsw.edu
Internet: http://www.rosalynncarter.org/PopeFellowship/
Sponsor: Georgia Southwestern State University
800 Wheatley Street
Americus, GA 31709-4379

Rose Community Foundation Health Grants **4104**
The Rose Community Foundation recognizes that improving access to quality care requires well-informed, visionary leaders. For this reason, the Foundation promotes initiatives that develop health-policy and public-health leadership. While the Foundation invests in health leadership that can improve access to quality care over the long term, it also works to effect more immediate change. Committed to improving access to care for low-income children, youth and families, the Foundation continues to support efforts to enroll them in programs such as Child Health Plan Plus (CHP+), a publicly supported health insurance program. Because at least half of all premature deaths are the result of lifestyle choices that individuals may be able to change, the Foundation also supports health promotion and disease prevention programs that encourage healthy choices and discourage behaviors that lead to illness and injury. In addition, the Foundation funds community organizations that provide information about the steps individuals can take to prevent such conditions as diabetes, teen pregnancy, HIV, heart disease and cancer.
Requirements: Colorado 501(c)3 tax-exempt organizations serving the residents of Adams, Arapahoe, Boulder, Denver, Douglas, and Jefferson Counties are eligible.
Geographic Focus: Colorado
Contact: Whitney G. Connor, Program Officer; (303) 398-7410 or (303) 398-7400; fax (303) 398-7430; wconnor@rcfdenver.org
Internet: http://www.rcfdenver.org/programs_health.htm
Sponsor: Rose Community Foundation
600 South Cherry Street, Suite 1200
Denver, CO 80246-1712

Rosenthal Writer-in-Residence Program **4105**
The Rosenthal Writer-in-Residence Program brings professional nonfiction writers to the Shorenstein Center to provide an opportunity for them to conduct research and work on a specific project, as well as interact with a community of scholars and students. In addition to a writing project, the Writer will teach student workshops and participate in Shorenstein Center events. Candidates for the program should be writers with an established project and a solid history of published books or essays in the area of press, policy or public affairs. Writers who wish to be considered for the Rosenthal Writer-in-Residence Program should send a one-page letter describing their writing project, along with their curriculum vitae.
Geographic Focus: All States
Contact: Alex S. Jones, (617) 495-8269; fax (617) 495-8696; alex_jones@harvard.edu
Internet: http://www.hks.harvard.edu/presspol/fellowships/fellowship_rosenthal.html
Sponsor: John F. Kennedy School of Government, Harvard University
79 John F Kennedy Street
Cambridge, MA 02138

Ross Foundation Grants **4106**
The Ross Foundation, established in 1966, supports organizations involved with education, arts and cultural enrichment, community beautification and improvement, historical preservation, mental health and people who are developmentally disabled, and forestry research and conservation management. Giving is limited to the Arkadelphia and Clark County, Arkansas region. Contact the Foundation for an application form to begin your proposal.
Requirements: Non-profits operating in the Arkadelphia and Clark County, Arkansas region are eligible to apply for funding. Applicants submit the following: copy of IRS Determination Letter; brief history of organization and description of its mission; copy of most recent annual report/audited financial statement/990; listing of board of directors, trustees, officers and other key people and their affiliations; detailed description of project; amount of funding requested, and 6 copies of the proposal.
Geographic Focus: Arkansas
Date(s) Application is Due: Jan 1; Mar 31; May 1; Jul 31; Sep 1; Nov 30
Amount of Grant: 3,000-20,000 USD
Contact: Mary Elizabeth Eldridge, Director of Programs; (870) 246-9881; fax (870) 246-9674; meeldridge@rossfoundation.us
Sponsor: Ross Foundation
P.O. Box 335
Arkadelphia, AR 71923

Royal Belgian Numismatic Society Research Grants **4107**
Each year the society awards a number of small research grants, available to fellows and nonfellows, for any field of numismatic research. There is no standard application form, but applicants should set out the purpose of their project; what they plan to achieve; detailed costings; and what other sources of finance they have available, have applied for, or intend to apply for. Applications in writing, with the names of two referees, should be submitted; the awards should be announced within two months of the closing date.
Requirements: Contenders must be no more than 35 years old as of the deadline date; participation is not restricted to holders of diplomas, titles, nor to nationality.
Geographic Focus: All States, All Countries
Date(s) Application is Due: Dec 31
Amount of Grant: 635 EUR
Contact: Department of Coins and Medals; jm@bvdmc.com
Internet: http://www.numisbel.be/
Sponsor: Royal Belgian Numismatic Society
Boulevard de l'Empereur 4 à 1000 Bruxelles
Brussels, Belgium

Royal Caribbean Cruises Ocean Fund **4108**
The Royal Caribbean Cruises Ocean Fund supports efforts to restore and maintain a healthy marine environment, minimize the impact of human activity on the environment, and promote awareness of ocean and coastal issues and respect for marine life. Ocean Fund grants are made to a variety of nonprofit groups and institutions whose activities are directly related to marine conservation, including initiatives in research, education and innovative technologies. Previous funding for organizations around the world include include projects that relate to ocean science, climate change, key marine species, education and innovative technologies. Previously funded projects are available at the Ocean Fund website.
Requirements: Organizations seeking grant funding from the Ocean Fund are welcome to email throughout the year to introduce their organization, mission, programs and potential projects relevant to the mission of the Ocean Fund within the body of the email. Due to the volume of email inquiries, only those organizations who are requested to submit a letter of interest will receive a personalized response to their email. The Ocean Fund grant selection process is by invitation only.
Restrictions: Any attachments to emails will not be reviewed.
Geographic Focus: All States
Amount of Grant: 20,000-40,000 USD
Contact: Ocean Fund Administrator; (305) 539-6573; oceanfund@rccl.com
Internet: http://www.theoceanfund.com/about
Sponsor: Royal Caribbean Cruises
1050 Caribbean Way
Miami, FL 33132-2096

Royal Norwegian Embassy Kavli Prizes **4109**
The Kavli Prize is awarded every two years and includes three separate prizes awarded for outstanding scientific research in the fields of astrophysics, neuroscience and nanoscience. The Prize, which consists of a cash award, a scroll, and a medal, is intended not only as an accolade for the individual scientists concerned, but also as a recognition of important research efforts. It aims to promote international cooperation between researchers and to increase awareness among the general public of the importance of research. The Norwegian Academy of Science and Letters has full responsibility for selecting the prize winners to ensure that this process is independent of the funding organization. The Academy has set up three selection committees made up of top international scientists within the fields for which the prize is awarded. Four of the committee members are Nobel Laureates. The committee members have been chosen in close cooperation with the corresponding institutions in the U.S.A, China, France, the UK and Germany.
Requirements: The nomination process is open to all who wish to nominate candidates. The prize can be awarded to a single person for profound scientific achievements or shared for closely related fundamental contributions. Specific guidelines and a current nomination form, which must be completed online in one session, are available at the website.
Restrictions: Self nominations are not accepted, and deceased persons may not be nominated.
Geographic Focus: All States, All Countries
Date(s) Application is Due: Dec 1

Amount of Grant: 1,000,000 USD
Contact: Eirik Lislerud, Co-Coordinator; +47 221236640; eirik.lislerud@dnva.no
Internet: http://www.norway.org/studywork/research/prizes/Kavli/
Sponsor: Royal Norwegian Embassy in Washington
2720 34th Street NW
Washington, DC 20008

Royal Scottish Geographical Society Expedition Grants **4110**
The RSGS is an educational charity the aims of which are to advance the science of Geography and create a greater understanding of the wider world, in the belief that geographical knowledge makes a significant difference to the way in which we view and shape the world we live in. The Society wishes to encourage expeditions having a geographical research and exploration component, where the research projects are integrated in an overall objective. The expedition grants committee will give preference to projects with a substantial geographical content. If the expedition is not in the fields of Human or Physical Geography, there should be a geographical aspect of a related discipline with the emphasis on spatial distributions, ecology, environmental science, ethnology, surveying or exploration.
Requirements: The closing date is January 31 for expeditions planned to take place in the following 15 months. Essential criteria include: Expedition composed of at least three persons; University expeditions to spend a minimum of three weeks in the field, and school expeditions a minimum of two weeks; Expedition should originate in Scotland or have at least one Scottish member; Geographical / scientific research program planned; Relevant training / expertise among party; Realistic logistical and financial planning underway. The person nominated as your contact person should also be readily accessible, and should respond promptly to any correspondence from the Society. Download the required application form from the sponsor's website.
Restrictions: Grants are not available for purely sporting or adventure expeditions. Expeditions composed of teams whose members are each pursuing research on individual unrelated topics are not favoured, and no support is available for post-graduate field-work for a higher degree, nor for the support of post-doctoral research.
Geographic Focus: All States
Date(s) Application is Due: Jan 31
Amount of Grant: 900 GBP
Contact: Bruce Maclachlan, Marketing and Publicity Officer; 0[0 44]1361 882 444; bruce.maclachlan@virgin.net
Internet: http://www.geo.ed.ac.uk/rsgs/expedits/
Sponsor: Royal Scottish Geographical Society
40 George Street
Glasgow, G1 1QE Scotland

Royal Scottish Geographical Society Small Research Grants **4111**
The program is designed to provide individual members of the Society in higher education institutions with a grant not exceeding 1000 GBP in support of geographical research. Typical grants can be in the region of 250-500 GBP. This award is given specifically to help with the cost of data collection and fieldwork. Applications are particularly welcomed from postgraduates and those early in their research careers.
Requirements: Research grants are available to individuals only, and applicants must be members of the Royal Scottish Geographical Society. (Membership is open to all.) The required application form can be downloaded from the RSGS website.
Restrictions: No more than one application can be accepted from an individual per year, and re-applications for the same project are not acceptable. Awards cannot be made retrospectively.
Geographic Focus: All States
Date(s) Application is Due: Jan 31
Amount of Grant: Up to 1,000 GBP
Contact: Small Research Grants Administrative Officer; 0141 552 3330; fax 0141 552 3331; rsgs@strath.ac.uk
Internet: http://www.geo.ed.ac.uk/rsgs/grants1.htm
Sponsor: Royal Scottish Geographical Society
40 George Street
Glasgow, G1 1QE Scotland

Royal Scottish Geographical Society Small Travel Grants **4112**
The program is designed to provide individual members of the Society in higher education institutions with grants up to a maximum of 500 GBP in support of geographical research. Awards are given specifically to help with the cost of travel and subsistence associated with attendance at conferences and symposia where the applicant intends to present a research paper. Applications for travel grants are welcomed particularly from postgraduates and those early in their research careers. Typical grants are in the region of 250-500 GBP.
Requirements: Grants are available to individuals only, and applicants must be members of the Royal Scottish Geographical Society. (Membership is open to all.) The required application form can be downloaded from the RSGS website.
Restrictions: No more than one application can be accepted from an individual per year, and re-applications for the same project are not acceptable. Awards cannot be made retrospectively.
Geographic Focus: All States
Date(s) Application is Due: Jan 31
Amount of Grant: Up to 500 GBP
Contact: Travel Grants Administrative Officer
Internet: http://www.geo.ed.ac.uk/rsgs/grants2.htm
Sponsor: Royal Scottish Geographical Society
40 George Street
Glasgow, G1 1QE Scotland

Roy J. Carver Charitable Trust Higher Education Grants **4113**
The trust supports Iowa and western Illinois colleges and universities. The Trust's interests in this area extend to those projects that strengthen institutions and offer innovative learning opportunities to students enrolled in baccalaureate and graduate programs throughout the grant-making region of Iowa and parts of Illinois. Typically, funding is awarded to capital or programmatic initiatives that seek to substantively improve instruction or increase access to educational resources, with the goal of better preparing students for future academic and professional challenges. Projects incorporating interdisciplinary approaches and demonstrating the potential for lasting impact among a diverse range of students are especially encouraged. A majority of awards have centered on teaching and research in the natural sciences.
Requirements: Grants are made to 501(c)3 tax-exempt colleges and universities located in Iowa and western Illinois.
Restrictions: The trust does not support annual campaigns or ongoing operations, direct grants to individuals, religious activities, organizations without 501(c)3 status, fund-raising benefits or program advertising, or political parties or candidates.
Geographic Focus: Illinois, Iowa
Amount of Grant: 10,000-150,000 USD
Contact: Dr. Troy Ross; (563) 263-4010; fax (563) 263-1547; info@carvertrust.org
Internet: https://www.carvertrust.org/index.php?page=35
Sponsor: Roy J. Carver Charitable Trust
202 Iowa Avenue
Muscatine, IA 52761-3733

Roy J. Carver Charitable Trust Medical and Science Research Grants **4114**
The trust supports Iowa and western Illinois nonprofit organizations and projects in the area of medical and scientific research. The goal of the program in medical and scientific research is to provide support for innovative investigation that may hold great promise for advancing scientific knowledge and improving human health. In this context, the Trust has recognized a variety of university-based scientific endeavors, most notably in the fields of medicine, engineering and the natural sciences. Although grants are occasionally awarded to address institutional capital needs, the focus of this program area is typically on the work of specific laboratories, often emphasizing multidisciplinary approaches and the development of untried, inherently risky research directions. Applicants are encouraged to contact staff prior to the development of a proposal. The board considers requests at quarterly meetings in January, April, July, and October. Approved grant payments are made semiannually in April and October.
Requirements: Grants are made to 501(c)3 tax-exempt scientific organizations in Iowa and western Illinois.
Restrictions: The trust does not support annual campaigns or ongoing operations, direct grants to individuals, religious activities, organizations without 501(c)3 status, fund-raising benefits or program advertising, or political parties or candidates.
Geographic Focus: Illinois, Iowa
Date(s) Application is Due: Feb 15; May 15; Aug 15; Nov 15
Contact: Dr. Troy Ross; (563) 263-4010; fax (563) 263-1547; info@carvertrust.org
Internet: https://www.carvertrust.org/index.php?page=56
Sponsor: Roy J. Carver Charitable Trust
202 Iowa Avenue
Muscatine, IA 52761-3733

RRF General Grants **4115**
The program funds service, education, research, and advocacy projects. The foundation is particularly interested in innovative projects that have the potential to change practice, policy, or delivery systems. The foundation's programs seek to improve the availability and quality of community-based and institutional long-term care programs; expand opportunities for older persons to play meaningful roles in society; support selected basic, applied, and policy research into the causes and solutions of significant problems of the aged; and increase the number of professionals and paraprofessionals adequately prepared to serve the elderly. Proposals fall into three general categories: research, model projects and service, and education and training. Guidelines are available online.
Requirements: Direct service projects that seek to improve the availability and quality of community-based and institutional long-term care programs; expand opportunities for older persons to play meaningful roles in society; support selected applied and policy research into the causes and solutions of significant problems of the aged; increase the number of professionals and paraprofessionals adequately prepared to serve the elderly.
Restrictions: The foundation will only consider proposals from applicants in Illinois, Indiana, Iowa, Kentucky, Missouri, Wisconsin, and Florida. Priority is given to nonprofit organizations serving the Chicago metropolitan area.
Geographic Focus: Florida, Illinois, Indiana, Iowa, Kentucky, Missouri, Wisconsin
Date(s) Application is Due: Feb 1; May 1; Aug 1
Amount of Grant: Up to 15,000,000 USD
Contact: Marilyn Hennessy; (773) 714-8080; fax (773) 714-8089; info@rrf.org
Internet: http://www.rrf.org/generalProgram.htm
Sponsor: Retirement Research Foundation
8765 W Higgins Road, Suite 430
Chicago, IL 60631-4170

RSC Abbyann D. Lynch Medal in Bioethics **4116**
The award was created to honour bioethicist and Past-President of AMS, Professor Abbyann D. Lynch, C.M., Ph.D., for her substantial achievements and support of biomedical ethics in Canada. Its purpose is to raise the profile of ethics in medicine and science by honouring leading Canadian exponents in these fields and bring recognition to their work. The bronze medal and a cash amount of $2,000 are offered every year if there is a suitable candidate.

Requirements: The medal is awarded for a major contribution in bioethics by a Canadian. The contribution may be a book, a report, a scholarly article, a monograph or series of articles, which will normally be published in the two years preceding the nomination. Candidates shall be Canadian citizens or persons having permanent resident status, and who have lived in Canada during the three years preceding the date of nomination. Nominations shall be made by three persons with the initiator being a Fellow of the Society. The form for nomination can be downloaded from the sponsor's website.
Restrictions: Self-nomination for an award is not permitted.
Geographic Focus: All States, Canada
Date(s) Application is Due: Dec 1
Amount of Grant: 2,000 CAD
Contact: Erin Hearty; (613) 991-6990; fax (613) 991-6996; fellowship@rsc-src.ca
Internet: http://www.rsc.ca/index.php?page_id=61&lang_id=1&award_id=16
Sponsor: Royal Society of Canada
170 Waller Street
Ottawa, ON K1N 9B9 Canada

RSC Alice Wilson Award 4117
At the behest of the Committee for Advancement of Women in Scholarship, the award is given yearly to a woman of outstanding academic qualifications who is entering a career in scholarship or research at the postdoctoral level. The recipient is chosen from the current year's female winners of postdoctoral fellowships from the Medical Research Council, the Social Sciences and Humanities Research Council, and the Natural Sciences and Engineering Research Council, each council in turn by year. The award consists of a diploma and cash award.
Requirements: Society fellows may nominate Canadian citizens and persons who have been Canadian residents for the five years preceding the date of the attribution of the award.
Geographic Focus: All States, Canada
Date(s) Application is Due: Dec 1
Amount of Grant: 1,000 CAD
Contact: Jeanne Salo, Program Supervisor; (613) 991-9007; jsalo@rsc.ca
Internet: http://www.rsc.ca/index.php?page_id=61&lang_id=1&award_id=11
Sponsor: Royal Society of Canada
170 Waller Street
Ottawa, ON K1N 9B9 Canada

RSC Bancroft Award 4118
Awarded every two years if there is a suitable candidate, this award is given for publication, instruction, and research in the geological and geophysical sciences that have noticeably contributed to public understanding and appreciation of the subject. Nominations must be received by the listed application deadline. A diploma is included with the award.
Requirements: Candidates are to be to Canadian citizens or persons who have been Canadian residents for the five years prior to award date. Nominations must be made by three persons with the initiator being a fellow of the society and consist of citation of not more than 250 words stating concisely why the nominee should be considered, curriculum vita, list of publications, and detailed appraisal of nominee's work. The nomination dossier will remain valid for five years from the year of first nomination.
Restrictions: Self-nomination for an award is not permitted.
Geographic Focus: All States, Canada
Date(s) Application is Due: Dec 1
Amount of Grant: 2,500 CAD
Contact: Jeanne Salo, Program Supervisor; (613) 991-9007; jsalo@rsc.ca
Internet: http://www.rsc.ca/index.php?page_id=61&lang_id=1&award_id=11
Sponsor: Royal Society of Canada
170 Waller Street
Ottawa, ON K1N 9B9 Canada

RSC Jason A. Hannah Medal 4119
The Jason A. Hannah Medal was established in 1976 by RSC with the assistance of Associated Medical Services Incorporated, through its Hannah Institute for the History of Medicine. The medal was established to honor the late President and Managing Director of AMS Inc., Dr. Jason A. Hannah, and to bring recognition to the work of Canadian research in the history of medicine. The award is normally made for a book published in the two years preceding its nomination. The bronze medal and a cash amount are offered every year if there is a suitable nomination.
Requirements: The medal is awarded for an important Canadian publication in the history of medicine. In this context, the word 'Canadian' refers to the citizenship or residence of the author, or to a content clearly relevant to Canadian medicine and health care. Nominations shall be made by three persons with the initiator being a Fellow of the Society.
Geographic Focus: All States, Canada
Date(s) Application is Due: Mar 1
Contact: Erin Hearty; (613) 991-6990; fax (613) 991-6996; fellowship@rsc-src.ca
Internet: http://www.rsc.ca/index.php?page_id=61&lang_id=1&award_id=14
Sponsor: Royal Society of Canada
170 Waller Street
Ottawa, ON K1N 9B9 Canada

RSC John L. Synge Award 4120
The award, established to honor John Lighton Synge, one of the first mathematicians working in Canada to obtain international recognition by research in mathematics, is given for outstanding research in any of the branches of the mathematical sciences. The award consists of a diploma and a cash amount, and is awarded at irregular intervals if there is a suitable candidate.
Requirements: The award is given for outstanding research in any of the branches of the mathematical sciences. Some preference is given to candidates whose age is not over 40 in the year of the award. Society fellows may nominate Canadian citizens or persons having permanent resident status, and who have lived in Canada during the three years preceding the date of nomination.
Restrictions: Self-nomination for an award is not permitted.
Geographic Focus: All States, Canada
Date(s) Application is Due: Dec 1
Amount of Grant: 2,500 CAD
Contact: Jeanne Salo, Program Supervisor; (613) 991-9007; jsalo@rsc.ca
Internet: http://www.rsc.ca/english/awards_synge.html
Sponsor: Royal Society of Canada
170 Waller Street
Ottawa, ON K1N 9B9 Canada

RSC McLaughlin Medal 4121
This medal and cash award are given annually for important research of sustained excellence in any branch of medical sciences. Nominations must be received by the listed application deadline.
Requirements: Candidates are to be Canadian citizens or persons who have been Canadian residents for the five years prior to award date. Nominations must be made by three persons, with the initiator being a fellow of the society, and consist of citation of not more than 250 words stating concisely why the nominee should be considered, curriculum vita, list of publications, and detailed appraisal of nominee's work. The nomination dossier will remain valid for three years from the year of first nomination.
Restrictions: Self-nomination for an award is not permitted.
Geographic Focus: All States, Canada
Date(s) Application is Due: Dec 1
Amount of Grant: 2,500 CAD
Contact: Erin Hearty; (613) 991-6990; fax (613) 991-6996; fellowship@rsc-src.ca
Internet: http://www.rsc.ca/index.php?page_id=61&lang_id=1&award_id=17
Sponsor: Royal Society of Canada
170 Waller Street
Ottawa, ON K1N 9B9 Canada

RSC Miroslaw Romanowski Medal 4122
The medal is awarded for significant contributions to the resolution of scientific aspects of environmental problems or for important improvements to the quality of an ecosystem in all aspects, terrestrial, atmospheric, and aqueous, brought about by scientific means. The award, which consists of a medal and cash award, is made every year if there is a suitable candidate. Nominations must be received by the listed application deadline. An annual lecture series for the award recipient is also associated to the medal.
Requirements: Society fellows may nominate Canadian citizens and persons who have been Canadian residents for the five years preceding the date of the attribution of the award.
Restrictions: Self-nomination for an award is not permitted.
Geographic Focus: All States, Canada
Date(s) Application is Due: Dec 1
Contact: Jeanne Salo, Program Supervisor; (613) 991-9007; jsalo@rsc.ca
Internet: http://www.rsc.ca/index.php?page_id=61&lang_id=1&award_id=6
Sponsor: Royal Society of Canada
170 Waller Street
Ottawa, ON K1N 9B9 Canada

RSC Pierre Chauveau Medal 4123
The Pierre Chauveau Medal was established in 1951 to honour the memory of Pierre J.O. Chauveau (1820-1890), FRSC, writer, orator, educator, Canadian statesman and the second President of RSC (1883-1884). He was the first Premier of Quebec (1867-1872) and Speaker of the Senate (1874). The medal is awarded for a distinguished contribution to knowledge in the humanities other than Canadian literature and Canadian history. The silver medal is offered every two years if there is a suitable candidate.
Requirements: Candidates shall be Canadian citizens or persons having permanent resident status, and who have lived in Canada during the three years preceding the date of nomination. Nominations shall be made by three persons with the initiator being a Fellow of the Society. The form for nomination can be downloaded from the sponsor's website.
Restrictions: Self-nomination for an award is not permitted.
Geographic Focus: All States, Canada
Date(s) Application is Due: Dec 1
Contact: Jeanne Salo, Program Supervisor; (613) 991-9007; jsalo@rsc.ca
Internet: http://www.rsc.ca/index.php?page_id=61&lang_id=1&award_id=3
Sponsor: Royal Society of Canada
170 Waller Street
Ottawa, ON K1N 9B9 Canada

RSC Rutherford Memorial Medals 4124
Two awards are given annually, one in physics and one in chemistry, if there are suitable candidates for outstanding research in any branch of physics and chemistry. In recognition of Lord Rutherford's own research carried out in Canada at a relatively young age, some preference will be given to candidates whose age is not over 40 in the year of the award. Whenever possible, the society subsidizes the travel expenses of the award winners to attend the annual meeting at which they are to receive the awards. Nominations must be received by the listed application deadline. Each gold plated silver medal is accompanied by a cash amount.

Requirements: Candidates are to be Canadian citizens or persons having permanent resident status, and who have lived in Canada during the three years preceding the date of nomination. Nominations must be made by three persons, with the initiator being a fellow of the society, and consist of a citation of not more than 250 words stating concisely why the nominee should be considered, curriculum vita, list of publications, and detailed appraisal of nominee's work. The nomination dossier will remain valid for three years from the year of first nomination.
Restrictions: Self-nomination for an award is not permitted.
Geographic Focus: All States, Canada
Date(s) Application is Due: Dec 1
Contact: Jeanne Salo, Program Supervisor; (613) 991-9007; jsalo@rsc.ca
Internet: http://www.rsc.ca/index.php?page_id=61&lang_id=1&award_id=7
Sponsor: Royal Society of Canada
170 Waller Street
Ottawa, ON K1N 9B9 Canada

RSC Sir John William Dawson Medal **4125**
The award honors Sir John William Dawson, the first president of the society and the foremost Canadian scientist and educator of his day: geologist, naturalist, teacher, author, man of religion, educational administrator, and promoter of learning. The award is made every two years, if there is a suitable candidate, and honors important and sustained contributions by one individual in at least two different domains in the general areas of interest of the society or in interdisciplinary research. Nominations must be received by the listed application deadline. The award includes a silver medal and a cash amount
Requirements: Candidates shall be Canadian citizens or persons having permanent resident status, and who have lived in Canada during the three years preceding the date of nomination. Nominations shall be made by three persons with the initiator being a Fellow of the Society. The form for nomination can be downloaded from the sponsor's website.
Restrictions: Self-nomination for an award is not permitted.
Geographic Focus: All States, Canada
Date(s) Application is Due: Dec 1
Amount of Grant: 2,500 CAD
Contact: Jeanne Salo, Program Supervisor; (613) 991-9007; jsalo@rsc.ca
Internet: http://www.rsc.ca/index.php?page_id=61&lang_id=1&award_id=4
Sponsor: Royal Society of Canada
170 Waller Street
Ottawa, ON K1N 9B9 Canada

RSC Thomas W. Eadie Medal **4126**
The medal is awarded annually in recognition of major contributions to any field through engineering or applied science, with preference given to those having an impact on communications, in particular the development of the Internet. Nominations must be received by the listed application deadline. The award includes a bronze medal and a cash amount.
Requirements: Candidates are to be Canadian citizens or persons who have been Canadian residents for the five years prior to award date. Nominations must be made by three persons, with the initiator being a fellow of the society, and consist of citation of not more than 250 words stating concisely why the nominee should be considered, curriculum vita, list of publications, and detailed appraisal of nominee's work. The nomination dossier will remain valid for five years from the year of first nomination.
Restrictions: Self-nomination for an award is not permitted.
Geographic Focus: All States, Canada
Date(s) Application is Due: Dec 1
Amount of Grant: 3,000 CAD
Contact: Erin Hearty; (613) 991-6990; fax (613) 991-6996; fellowship@rsc-src.ca
Internet: http://www.rsc.ca/index.php?page_id=61&lang_id=1&award_id=5
Sponsor: Royal Society of Canada
170 Waller Street
Ottawa, ON K1N 9B9 Canada

Rumsfeld Foundation Graduate Fellowships **4127**
The Rumsfeld Foundation Graduate Fellowships provide funding for promising students with an interest in serving the nation through policy-relevant fields or by pursuing a career in public service. Depending on the circumstances of each selected student, he or she can receive a stipend for the academic year and tuition assistance for a combined total of up to $25,000 per student. Fellowships are awarded on the basis of recommendations from fourteen advisers at the following institutions: Georgetown University; Princeton University; University of Pennsylvania; Harvard University; Stanford University; University of Virginia; Indiana University; University of Chicago; Yale University; Johns Hopkins University; University of Notre Dame; George Washington University; and Missouri State University. These fellows include students pursuing a Master's, Ph.D., or Juris Doctorate with concentrations including political and theological ethics, public management, economics, international affairs, philosophy, strategic studies, Middle East Politics, and East Asian civilizations. Once nominations are submitted, the Foundation trustees meet to review and determine the fellowships to be granted.
Requirements: Fellowship recipients must be nominated and attend one of the 14 universities listed.
Geographic Focus: All States
Amount of Grant: Up to 25,000 USD
Contact: Lawrence Di Rita, Trustee; contact@rumseldfoundation.org
Internet: http://www.rumsfeldfoundation.org/focus/our/fellowships
Sponsor: Rumsfeld Foundation
1718 M Street NW, #366
Washington, DC 20036

Russell Family Foundation Environmental Sustainability Grants **4128**
The foundation is committed to improving protection of the environment in western Washington, with an emphasis on the waters of Puget Sound. The foundation focuses its grantmaking in this program on Puget Sound, Environmental Education and Green Business and supports systemic and community-centered strategies that encourage collaboration across sectors and, where possible, leverage other support, build partnerships and reach new audiences. Though not a requirement for funding, they welcome projects that effectively address and connect several of the above-mentioned areas of interest.
Requirements: A Letter of Inquiry is required prior to any formal application. If the foundation is interested in considering a grant request, they will ask for a more detailed proposal. Full proposals are accepted by invitation only.
Restrictions: Requests are accepted from 501(c)3 organizations. Nonprofit entities such as public schools and school districts also may apply. Organizations must be located in and/or provide services within the Puget Sound region. The region includes 12 counties: Clallam, Jefferson, Kitsap, Mason, Thurston, Pierce, King, Snohomish, Skagit, Whatcom, Island, and San Juan.
Geographic Focus: Washington
Date(s) Application is Due: Jan 8; Jul 16
Amount of Grant: 1,000-35,000 USD
Contact: Grant Administrator; info@trff.org
Internet: http://www.trff.org/grant_programs/env_education.asp
Sponsor: Russell Family Foundation
P.O. Box 2567
Gig Harbor, WA 98335

Russell Sage Foundation Grants **4129**
The foundation pursues four programs: a program of research on the future of work concerned with the causes and consequences of the decline in demand for low-skill workers in advanced economies; a program of research on current U.S. immigration that focuses on the adaptation of the second generation to American society; the social psychology of cultural contact that focuses on improving relations between racial and ethnic groups in schools, workplaces, and neighborhood settings; and the examination of social inequality on a number of dimensions, including family well-being, educational quality and opportunity, health care and coverage, legal services and criminal justice, political participation and representation, banking and credit, neighborhoods and housing, pension provision, environmental quality, and even access to computers and the Internet. The board meets in June and November. Unless otherwise specified in the request for proposals in a particular research competition, the deadline for submissions to the November funding round is mid-August, and to the June round is mid-March.
Requirements: Higher education institutions, research groups, and researchers with PhD degrees may submit grant applications. All funded research is U.S.-focused.
Restrictions: Support is not granted for predoctoral research.
Geographic Focus: All States
Amount of Grant: 35,000-500,000 USD
Contact: Grants Administrator; (212) 750-6000; fax (212) 371-4761; info@rsage.org
Internet: http://www.russellsage.org/programs/index.shtml
Sponsor: Russell Sage Foundation
112 East 64th Street
New York, NY 10021

Ruth and Vernon Taylor Foundation Grants **4130**
The foundation awards grants to nonprofit organizations in the areas of arts and humanities, civic and public affairs, secondary schools, higher education, environment, hospitals, human services, health, youth services, and social services. Types of support include general operating support, building construction and renovation, endowment funds, and research. The foundation suggests that initial contact be made in writing, since unsolicited requests for funds are not accepted. The Board meets in May and September.
Requirements: Organizations located in Colorado, Illinois, Montana, New Jersey, New York, Pennsylvania, Texas, or Wyoming are eligible.
Restrictions: Grants are not awarded to individuals.
Geographic Focus: Colorado, Illinois, Montana, New Jersey, New York, Pennsylvania, Texas, Wyoming
Amount of Grant: 1,000-20,000 USD
Contact: Douglas Taylor, Trustee; (303) 893-5284; fax (303)893-8263
Sponsor: Ruth and Vernon Taylor Foundation
518 17th Street, Suite 1670
Denver, CO 80202

Ruth Estrin Goldberg Memorial for Cancer Research Grants **4131**
The Ruth Estrin Goldberg Memorial for Cancer Reearch Grants are intended to support cancer research projects being conducted by an investigator at a recognized hospital, university, or other research institution. Grants are awarded for one year with possible renewal. Application forms and guidelines are available upon request.
Requirements: The principal investigator must be affiliated with an accredited institution.
Restrictions: Focus is limited to New Jersey, New York, and Pennsylvania.
Geographic Focus: New Jersey, New York, Pennsylvania
Amount of Grant: 10,000-15,000 USD
Contact: Rhoda Goodman, Grants Contact; (908) 686-5508 or (973) 467-3838
Sponsor: Ruth Estrin Goldberg Memorial for Cancer Research
c/o Greenberg and Company
Springfield, NJ 07081

RWJF Active Living Research Grants 4132

Active Living Research supports research examining how environments and policies impact physical activity, especially among ethnic minorities and children living in low-income communities. Findings are expected to inform environmental and policy changes that will promote active living among children and families. the Foundation encourages experts in fields such as public health, public administration, law, economics, transportation, medicine, recreation, urban planning, architecture, the behavioral sciences, child care and education to work together to identify promising approaches for increasing physical activity and preventing obesity among children and families.
Requirements: Requirements vary from one Call for Proposals to the next.
Restrictions: The program does not provide support for international activities, or for institutions in other countries.
Geographic Focus: All States
Contact: James F. Sallis, Program Director; (619) 260-5534; fax (619) 260-1510; ALR@projects.sdsu.edu
Internet: http://activelivingresearch.org/grantsearch/granteeresources
Sponsor: Robert Wood Johnson Foundation
3900 Fifth Avenue, Suite 310
San Diego, CA 92103

RWJF Changes in Health Care Financing and Organization Grants 4133

The Robert Wood Johnson Foundation, through its Changes in Health Care Financing and Organization (HCFO) initiative, supports research, policy analysis and evaluation projects that provide policy leaders timely information on health care policy, financing and organization issues. Supported projects include: examining significant issues and interventions related to health care financing and organization and their effects on health care costs, quality and access; and exploring or testing major new ways to finance and organize health care that have the potential to improve access to more affordable and higher quality health services. Grants will be awarded in two categories: small grants for projects requiring $100,000 or less and projected to take up to twelve months or less; and large grants for projects requiring more than $100,000 and/or projected to take longer than twelve months.
Requirements: Researchers, as well as practitioners and public and private policy-makers working with researchers, are eligible to submit proposals through their organizations. Projects may be initiated from within many disciplines, including health services research, economics, sociology, political science, public policy, public health, public administration, law and business administration. RWJF encourages proposals from organizations on behalf of researchers who are just beginning their careers, who can serve either individually as principal investigators or as part of a project team comprising researchers or other collaborators with more experience. Only organizations and government entities are eligible to receive funding under this program. Preference will be given to applicants that are either public entities or nonprofit organizations that are tax-exempt under Section 501(c)3 of the Internal Revenue Code and are not private foundations as defined under Section 509(a). Proposals for this solicitation must be submitted via the RWJF online system. Applicants should direct inquiries to the national program office.
Restrictions: Grants are not available for ongoing general operating expenses or existing deficits; endowment or capital costs, including construction, renovation, or equipment purchases; basic biomedical research; conferences, symposia, publications, or media projects, unless they are clearly related to the foundation's goals; research on unapproved drug therapies or devices; international programs and institutions; or direct support to individuals.
Geographic Focus: All States
Amount of Grant: Up to 500,000 USD
Contact: Bonnie J. Austin, HCFO Deputy Director; (202) 292-6700 or (202) 292-6756; fax (202) 292-6800; bonnie.austin@academyhealth.org
Internet: http://www.hcfo.org/funding
Sponsor: Robert Wood Johnson Foundation
Route 1 and College Road East, P.O. Box 2316
Princeton, NJ 08543-2316

RWJF Changes in Health Care Financing and Organization Small Grants 4134

The RWJF Changes in Health Care Financing and Organization Small Grant program supports policy analysis, research, evaluation and demonstration projects that provide policy leaders timely information on health care policy and financing issues. Supported projects include: examining significant issues and interventions related to health care financing and organization and their effects on health care costs, quality and access; and exploring or testing major new ways to finance and organize health care that have the potential to improve access to more affordable and higher quality health services. The grant period is twelve or fewer months for small grants of up to $100,000. There are no application deadlines. Program information and the online application process are available online.
Requirements: Researchers, as well as practitioners and public and private policy-makers working with researchers, are eligible to apply. Projects may be initiated from within many disciplines, including health services research, economics, sociology, political science, public policy, public health, public administration, law and business administration. Multidisciplinary teams and researchers who are just beginning their careers-perhaps teaming with a more senior researcher to develop the analytic approach-are especially encouraged to apply. Preference will be given to applicants that are public agencies or are tax-exempt under Section 501(c)3 of the Internal Revenue Code and are not private foundations as defined under Section 509(a). Proposals for this program must be submitted through the RWJF Grantmaking Online system.
Restrictions: Grant funds may not be used to subsidize individuals for the cost of health care, to construct or renovate facilities, for lobbying, or as a substitute for funds currently being used to support similar activities.
Geographic Focus: All States
Amount of Grant: Up to 100,000 USD
Contact: Bonnie Austin; (202) 292-6700; fax (202) 292-6800; hcfo@academyhealth.org
Internet: http://www.rwjf.org/content/rwjf/en/grants/calls-for-proposals/2011/changes-in-health-care-financing-and-organization—small-grants.html
Sponsor: Robert Wood Johnson Foundation
Route 1 and College Road East, P.O. Box 2316
Princeton, NJ 08543-2316

RWJF Childhood Obesity Grants 4135

Through its Childhood Obesity Grant program, RWJF has developed three integrated strategies to reverse the childhood obesity epidemic: funding local organizations to make positive changes at the community level, advocating for healthier policies in the public and private sectors, and providing grants to researchers and evaluators to strengthen the evidence about what works. The Foundation's grant making is centered on advancing these strategies in ways that make it easier for all children to lead healthy lives. The program is intended to improve understanding of school, community, state and national policies and environmental factors affecting youth diet, physical activity, obesity, and tobacco, alcohol and drug use, and to evaluate the effectiveness of interventions to prevent youth obesity.
Requirements: Eligible applicants for research grants are scholars associated with educational institutions, research organizations, health care providers or other public or nonprofit organizations. Preference will be given to public entities or nonprofit organizations tax exempt under Section 501(c)3 of the Internal Revenue Code. Applicants may be independent of or associated with organizations currently receiving grants.
Geographic Focus: All States
Contact: C. Tracy Orleans, Senior Program Officer/Senior Scientist; (609) 627-5962 or (877) 843-7953; cto@rwjf.org
Internet: http://www.rwjf.org/en/about-rwjf/program-areas/childhood-obesity/programs-and-grants.html
Sponsor: Robert Wood Johnson Foundation
Route 1 and College Road East, P.O. Box 2316
Princeton, NJ 08543-2316

RWJF Clinical Scholars Grants 4136

RWJF Clinical Scholars Grants foster the development of leaders who will transform U.S. health and health care. These future leaders will be equipped to work with communities and to conduct innovative research as a result of the didactic and experiential learning they receive at each of four chosen training sites (University of California, Los Angeles; University of Michigan; University of Pennsylvania; and Yale University). The program's focus emphasizes community-based research and leadership training. The goal of the program is the integration of scholars' clinical training with skills in program development and research methodology to help them find solutions for issues in public policy, community health, and health services research. The program offers graduate-level study and research as part of a university-based post-residency training program. Up to 29 scholars will be selected for appointments beginning the following year. Under the program, the universities provide Scholars with their stipends and health insurance. Stipends currently range from $67,000 to $75,500 per year, depending on the training site. Limited financial support is also available at each university for expenses associated with Scholars' research, including travel.
Requirements: To be eligible, physicians must be: U.S. citizens; be committed to a career consonant with the program's purposes and priorities; be highly regarded by those responsible for their clinical training; and plan to complete the clinical requirements of their residency/fellowship training by the date of entry into the program (except for surgeons). Scholars may not hold appointments as sub-specialty fellows during their tenure in the program.
Geographic Focus: All States
Date(s) Application is Due: Feb 28
Amount of Grant: 67,000-75,500 USD
Contact: Desmond Kimo Runyan, National Program Director; (919) 843-8261 or (919) 843-1351; fax (919) 843-2666; Desmond.Runyan@childrenscolorado.org or rwjcsp_admin@med.unc.edu
Internet: http://rwjcsp.unc.edu/
Sponsor: Robert Wood Johnson Foundation
Route 1 and College Road East, P.O. Box 2316
Princeton, NJ 08543-2316

RWJF Disparities Research for Change Grants 4137

The RWJF Disparities Research for Change program was created to support rigorous evaluations of practical and replicable solutions designed to reduce and eliminate disease specific racial and ethnic health care disparities. It seeks to improve the quality of health care provided to patients affected by racial and ethnic disparities by: Granting funds to discover and evaluate practical and replicable solutions designed to reduce and eliminate disease specific racial and ethnic health care disparities; Focusing on interventions aimed at health care delivery for one or more of the following health concerns: cardiovascular disease, depression, or diabetes; Conducting systematic reviews of the literature regarding racial and ethnic health care disparities interventions; and Disseminating results from these research efforts and systematic reviews to encourage health care systems to address racial and ethnic gaps in care.
Requirements: For-profit organizations are eligible to apply, however preference will be given to applicants that are public agencies or are tax-exempt under Section 501(c)3 of the Internal Revenue Code and are not private foundations as defined under Section 509(a). Proposals must focus on innovative interventions that reduce racial and ethnic disparities in the treatment of cardiovascular disease, depression, and diabetes. There are no age restrictions for the target populations of the interventions proposed for evaluation using program funds. Although there are no restrictions on the age of target populations, proposals targeting pediatric and geriatric populations are encouraged. Applications are

accepted in two phases. The first phase is a brief proposal submission open to all interested research teams. The second phase is a full proposal submission and will be accepted by invitation only. All applications, both the brief and full proposals, must be submitted through the Robert Wood Johnson Foundation's Grantmaking Online system.
Restrictions: Individuals are not eligible to apply. Stand-alone prevention interventions such as tobacco cessation or exercise programs are not eligible unless they are one part of a larger, more comprehensive, intervention. The program will only directly fund health care delivery institutions. However, eligible institutions may partner/sub-contract with academic or research groups.
Geographic Focus: All States
Amount of Grant: 100,000-275,000 USD
Contact: Marshall H. Chin, Director; (866) 344-9800 or (773) 702-4769; fax (773) 702-4620; mchin@medicine.bsd.uchicago.edu or info@solvingdisparities.org
Internet: http://www.solvingdisparities.org/
Sponsor: Robert Wood Johnson Foundation
Route 1 and College Road East, P.O. Box 2316
Princeton, NJ 08543-2316

RWJF Evaluating Innovations in Nursing Education Grants **4138**
The RWJF Evaluating Innovations in Nursing Education (EIN) will support evaluations of interventions that expand teaching capacity or promote faculty recruitment and retention in nursing schools. The program aims to increase the number of nursing school graduates by evaluating strategies that address the nurse faculty shortage. An important activity of the program is disseminating successful strategies so they can be replicated in other nursing education settings. The Robert Wood Johnson Foundation (RWJF) awards most grants through calls for proposals (CFPs) issued for national programs. Typically, RWJF's national programs are administered by staff affiliated with an outside institution selected for their substantive expertise relevant to the program. Established in 2008, the EIN NP.O. is located at the Center for State Health Policy at Rutgers University, in New Brunswick, New Jersey. The NP.O. provides overall research direction, management, technical assistance, and monitoring of the EIN grants.
Requirements: Nursing schools or health care organizations who will act as the applicant institution and subcontract to the evaluation team are eligible. Preference will be given to applicant organizations that are tax-exempt under Section 501(c)3 of the Internal Revenue Code and are not private foundations as defined under Section 509(a). The Robert Wood Johnson Foundation makes grants to organizations in the United States and its territories only.
Geographic Focus: All States
Amount of Grant: Up to 300,000 USD
Contact: Joanne Fuccello, Deputy Director; (848) 932-4670; fax (732) 932-0069; info@evaluatinginnovationsinnursing.org
Internet: http://www.evaluatinginnovationsinnursing.org/
Sponsor: Robert Wood Johnson Foundation
Route 1 and College Road East, P.O. Box 2316
Princeton, NJ 08543-2316

RWJF Harold Amos Medical Faculty Development Research Grants **4139**
The Harold Amos Medical Faculty Development program was established to increase the number of faculty from historically disadvantaged backgrounds who can achieve senior rank in academic medicine and dentistry and who will encourage and foster the development of succeeding classes of such physicians and dentists. Four-year postdoctoral research awards are offered to universities, schools of medicine and dentistry and research institutions to support the research and career development of physicians and dentists from historically disadvantaged backgrounds who are committed to developing careers in academic medicine and dentistry and to serving as role models for students and faculty of similar background. The program defines the term historically disadvantaged to mean the challenges facing individuals because of their race, ethnicity, socioeconomic status or similar factors. The program will fund up to nine four-year awards, with scholars to receive an annual stipend of up to $75,000 each, complemented by a $30,000 annual grant toward support of research activities.
Requirements: Eligible to apply are minority physicians who are U.S. citizens at the time of application, have excelled in their post-secondary education within the United States, are now completing or will have completed formal clinical training, and are committed to academic careers.
Geographic Focus: All States
Date(s) Application is Due: Mar 15
Amount of Grant: 30,000-105,000 USD
Contact: National Program Officer; (317) 278-0500; amfdp@starpower.net
Internet: http://www.amfdp.org/for-applicants
Sponsor: Robert Wood Johnson Foundation
Route 1 and College Road East, P.O. Box 2316
Princeton, NJ 08543-2316

RWJF Harold Amos Medical Faculty Development Scholars **4140**
The Harold Amos Medical Faculty Development Program supports basic/biomedical, clinical, and health services/epidemiology research. The purpose of the AMFDP award is to facilitate the transition of the newly-trained clinician who wishes to develop into an independent investigator. A completed application consists of two parts: the completion of the online proposal process, and submission of other (paper) supporting documents. The maximum amount of support is $420,000 over a four-year period. This includes an annual $75,000 stipend, plus $30,000 per year for research support.
Requirements: To be eligible, applicants must be physicians from historically disadvantaged backgrounds (ethnic, financial, or educational) who are U.S. citizens or permanent residents at the time of application deadline; have excelled in their education; are now completing or have completed their formal clinical training; are prepared to devote four consecutive years to research; are committed to pursuing academic careers; and are committed to improving the health status of the under-served, decreasing health disparities, or serving as role models for students and faculty from historically disadvantaged backgrounds Preference will be given to physicians who have recently completed their formal clinical training. In order to pursue the advanced research training required by this program, applicants must identify faculty mentors with whom a research plan will be developed. The mentor's experience in the supervision of trainees and the adequacy of the mentor's research environment and support will be strongly considered.
Geographic Focus: All States
Date(s) Application is Due: Mar 15
Amount of Grant: Up to 105,000 USD
Contact: Nina Ardery, Deputy Director; (301) 565-4080 or (317) 278-0500; fax (301) 565-4088; nardery@erols.com or amfdp@starpower.net
Internet: http://www.amfdp.org/for-applicants
Sponsor: Robert Wood Johnson Foundation
Route 1 and College Road East, P.O. Box 2316
Princeton, NJ 08543-2316

RWJF Health and Society Scholars **4141**
The RWJF Health and Society Scholars program is a two-year fellowship designed to build the nation's capacity for research, leadership, and action to address the broad range of factors affecting health. The goal of the program is to improve health by training scholars to rigorously investigate the connections between biological, behavioral, environmental, economic, and social determinants of health; and develop, evaluate, and disseminate knowledge and interventions based on integration of these determinants. Each year the program will enable up to twelve outstanding individuals who have completed doctoral training to engage in an intensive two-year program at one of four nationally prominent universities. Scholars will have access to a full range of university resources and will receive an annual stipend of $80,000, plus health insurance from their university site. Scholars will also have access at each university to financial support for research-related expenses, training workshops and travel to professional meetings.
Requirements: Individuals must apply only through the foundation's online system. To be eligible, scholars must: have completed doctoral training by the time of entry into the program (August or September) in one of a variety of fields including, but not limited, to the behavioral and social sciences, the biological and natural sciences, health professions, public policy, public health, history and ethics; have significant research experience; clearly connect their research interests to substantive population health concerns; and be citizens or permanent residents of the United States or its territories. Applications from candidates with diverse backgrounds are encouraged.
Restrictions: Individuals placed in the program must be willing to relocate to the institution designated during the match process for the duration of the two-year program.
Geographic Focus: All States
Date(s) Application is Due: Sep 20
Amount of Grant: 89,000 USD
Contact: Gerard P. Lebeda, (212) 419-3566; fax (212) 419-3569; hss@nyam.org
Internet: http://www.healthandsocietyscholars.org
Sponsor: Robert Wood Johnson Foundation
Route 1 and College Road East, P.O. Box 2316
Princeton, NJ 08543-2316

RWJF Health Policy Fellowships **4142**
The purpose of RWJF Health Policy Fellowships is to give outstanding mid-career health professionals in academic and community-based settings an opportunity to learn more about the legislative process through direct participation. Fellows actively contribute to the formulation of national health policies and accelerate their careers as leaders in health policy. Up to six grants of up to $165,000 each will be made. Each fellow will receive up to $94,000 for the Washington stay in salary plus fringe benefits or fellowship stipend. Fellows will receive an additional allowance for relocation subject to limitations provided in detail on the program's Web site. November 13 is the deadline for receipt of applications submitted online.
Requirements: Applicants may have backgrounds in the following disciplines: allied health professions; biomedical sciences; dentistry; economics or other social sciences; health services organization and administration; medicine; nursing; public health; or social and behavioral health. Applicants must be U.S. citizens or have permanent residency status at the time of application. Potential applicants must obtain documented sponsorship from an eligible institution to support their application. Nonprofit health care organizations and academic centers are eligible sponsors. Eligible institutions may sponsor more than one candidate, but no more than one individual will be selected from any one sponsoring organization.
Restrictions: No additional direct costs (such as laptops or PDAs) or indirect costs are paid for during the first year of this program.
Geographic Focus: All States
Date(s) Application is Due: Nov 13
Amount of Grant: Up to 165,000 USD
Contact: Marie Michnich, Director; (202) 334-1506; fax (202) 334-3862; mmichnich@nas.edu or info@healthpolicyfellows.org
Internet: http://www.rwjf.org/content/rwjf/en/grants/calls-for-proposals/2013/health-policy-fellows-.html
Sponsor: Robert Wood Johnson Foundation
Route 1 and College Road East, P.O. Box 2316
Princeton, NJ 08543-2316

RWJF Healthy Eating Research Grants 4143

Healthy Eating Research is a national program that supports research to identify, analyze, and evaluate environmental and policy strategies that can promote healthy eating among children and prevent childhood obesity. Special emphasis is given to research projects that benefit children in the low-income and racial/ethnic populations at highest risk for obesity. The program funds two types of research grants: studies to identify and/or evaluate promising school food environment and policy changes (12- to 18-month awards up to $100,000 each, and 18- to 36-month awards up to $400,000 each); and analyses of macro-level policy or system determinants of school food environments and policies (12- to 18-month awards up to $75,000 each). Applications must be submitted through the online process in response to future calls for proposals. Calls for Proposals and their guidelines are available online.

Requirements: Preference will be given to those applicants that may be either public entities or nonprofit organizations that are tax-exempt under Section 501(c)3 of the Internal Revenue Code. Applicant organizations must be based in the United States or U.S. Territories. The focus of this program is the United States; studies of policies in other countries will be considered only to the extent that they may directly inform U.S. policy. The experience and qualifications of the research team is one of the primary criteria for proposal review. A doctorate or terminal degree is preferred for the principal investigator.

Restrictions: This program does not fund demonstration projects. The foundation does not award grants to private individuals or applicants from outside the U.S. or U.S. Territories. No faxed, emailed or mailed proposals will be accepted.

Geographic Focus: All States

Amount of Grant: Up to 400,000 USD

Contact: Kathy Kosiak, Research Coordinator; (800) 578-8636; fax (612) 624-9328; healthyeating@umn.edu

Internet: http://www.healthyeatingresearch.org/

Sponsor: Robert Wood Johnson Foundation
Route 1 and College Road East, P.O. Box 2316
Princeton, NJ 08543-2316

RWJF Interdisciplinary Nursing Quality Research Initiative 4144

The Interdisciplinary Nursing Quality Research Initiative has most recently awarded two-year grants to support the generation, dissemination, and translation of research to improve the contributions of nursing to the quality of hospital care. It will continue to support interdisciplinary teams of nurse scholars and scholars from other disciplines to address gaps in knowledge about the relationship between nursing and health care quality. Topic areas may include: identifying, developing, and testing new measures to advance quality nursing research; and testing the effects of implementing the nursing-sensitive performance measures endorsed by the National Quality Forum. Investigator-initiated ideas that are consistent with the broader set of program goals also will be considered. The two-phase application process begins with online submittal of a brief proposal; full proposals are by invitation, and are due approximately four months after invitations to submit are given. All current and future calls for proposals are detailed at the web site.

Requirements: Teams of researchers who are U.S. citizens or permanent residents are eligible. Each team must have at least one nursing researcher and at least one researcher from another discipline. Preference will be given to 501(c)3 applicants that are not 509(a) private foundations.

Geographic Focus: All States

Amount of Grant: Up to 300,000 USD

Contact: Lori A. Melichar, Senior Program Officer; (301) 571-8161 or (800) 978-8309; INQRIhelpdesk@constellagroup.com

Internet: http://www.inqri.org

Sponsor: Robert Wood Johnson Foundation
Route 1 and College Road East, P.O. Box 2316
Princeton, NJ 08543-2316

RWJF Investigator Awards in Health Policy Research 4145

The Investigator Awards in Health Policy Research program supports researchers whose cross-cutting and innovative ideas promise to contribute meaningfully to improving health and health policy in America. Each year, the program awards 24- to 36-month grants of up to $335,000 to highly qualified individuals to fund approximately ten research projects that have national policy relevance. The program funds projects with a broader scope than traditional NIH or NSF grants. Data-driven health services research focusing on a narrowly defined research question is not appropriate for this program, nor are pilot projects that are designed to evaluate clinical care in a specific geographic area or to demonstrate service delivery programs or financing strategies.

Requirements: Individuals in the United States who are affiliated either with a U.S.-based educational institution or a 501(c)3 tax-exempt organization are eligible. Applications are welcomed from investigators in fields such as anthropology, business, demography, economics, engineering, ethics, genetics, health and social policy, history, journalism, law, medicine, nursing, political science, psychology, public health, science policy, social work, sociology and others. The program seeks a diverse group of applicants including minorities, researchers early in their careers, and individuals in nonacademic settings.

Restrictions: Past awardees and teams of more than two researchers are not eligible.

Geographic Focus: All States

Amount of Grant: Up to 335,000 USD

Contact: Melissa Manolis; (617) 353-9220, ext 1; fax (617) 353-9227; rwjfihp@bu.edu

Internet: http://www.investigatorawards.org

Sponsor: Robert Wood Johnson Foundation
Route 1 and College Road East, P.O. Box 2316
Princeton, NJ 08543-2316

RWJF New Connections Junior Investigator Grants 4146

New Connections provides funding for research that aims to expand the diversity of perspectives that inform Robert Wood Johnson Foundation (RWJF) programming and introduce new researchers and scholars to RWJF, while simultaneously helping to meet staff needs for data analysis. The program invites Junior Investigators—scholars from historically disadvantaged and underrepresented communities—to submit proposals that address programming priorities for one of RWJF's program areas. The program offers up to ten awards that will extend for a period of 24-months for up to $100,000 to conduct projects based on RWJF team research and evaluation priorities. Junior Investigators are expected to commit between .25-.75 FTE during the grant period. In addition to providing support for the researcher's time, the award can be used to cover research expenses including support for mentors and methodological training. August 29 is the annual deadline for proposals. Applicants will be invited to submit full proposals by October 17, with November 28 being the final deadline for online application submission.

Requirements: This program seeks scholars who have historically been underrepresented in research activities supported by RWJF. This includes people from ethnic or racial minority groups, first-generation college graduates and people from low-income communities. The initiative will support research activities that address specific questions identified by RWJF programming teams and portfolios. Contact the sponsor for the programming focus. There are two stages in the application process.

Restrictions: Full proposal submission is by invitation only.

Geographic Focus: All States

Date(s) Application is Due: Nov 28

Amount of Grant: Up to 100,000 USD

Contact: Debra Joy Perez, Senior Program Officer; (215) 732-2200; fax (215) 732-8123; djperez@fas.harvard.edu or info@rwjf-newconnections.org

Internet: http://www.rwjf-newconnections.org/apply-for-grant

Sponsor: Robert Wood Johnson Foundation
Route 1 and College Road East, P.O. Box 2316
Princeton, NJ 08543-2316

RWJF New Connections Mid-Career Consultants Grants 4147

New Connections provides funding for research that aims to expand the diversity of perspectives that inform Robert Wood Johnson Foundation (RWJF) programming and introduce new researchers and scholars to RWJF, while simultaneously helping to meet staff needs for data analysis. The program invites Mid-career Consultants — scholars from historically disadvantaged and underrepresented communities who have between 10 to 15 years of research and/or evaluation experience — to address program priorities for RWJF's teams and portfolios. These three funding opportunities for Mid-Career Consultants consist of grants or contracts of up to $100,000. The grant period for consultants is twelve months. Past Calls for Proposals (CFPs) for Consultants have focused on evaluation, syntheses, and qualitative work although consultant projects may span an array of topics. August 29 is the annual deadline for proposals. Applicants will be invited to submit full proposals by October 17, with November 28 being the final deadline for online application submission.

Requirements: This program seeks scholars who have historically been underrepresented in research activities supported by RWJF. This includes people from ethnic or racial minority groups, first-generation college graduates and people from low-income communities. The initiative will support research activities that address specific questions identified by RWJF programming teams and portfolios. Contact the sponsor for the programming focus. There are two stages in the application process.

Geographic Focus: All States

Date(s) Application is Due: Nov 28

Amount of Grant: Up to 100,000 USD

Contact: Debra Joy Perez, Senior Program Officer; (215) 732-2200; fax (215) 732-8123; djperez@fas.harvard.edu or info@rwjf-newconnections.org

Internet: http://www.rwjf-newconnections.org/apply-for-grant

Sponsor: Robert Wood Johnson Foundation
Route 1 and College Road East, P.O. Box 2316
Princeton, NJ 08543-2316

RWJF Nurse Faculty Scholars 4148

The goal of the Robert Wood Johnson Foundation Nurse Faculty Scholars (NFS) program is to develop the next generation of national leaders in academic nursing through career development awards for outstanding junior nursing faculty. The program aims to strengthen the academic productivity and overall excellence of nursing schools by providing mentorship, leadership training, and salary and research support to junior faculty. The Nurse Faculty Scholars program is part of the Foundation's Human Capital Portfolio, which aims to ensure that the nation has a diverse, well-trained workforce to meet the health care needs of all Americans. Up to 12 awards of up to $350,000 each over three years will be available. February 12 is the deadline for receipt of applications.

Requirements: Accredited colleges and universities within the United States are eligible to apply. To be eligible, candidates must meet the following criteria: be a registered nurse with a research doctorate in nursing or a related discipline; be a junior faculty member in an accredited school of nursing with an academic position that could lead to tenure; after completing the doctorate, be in an academic position that could lead to tenure for at least two years and no more than five years at the start of the program; identify at least one senior leader in the school of nursing to serve as a primary (nursing) mentor for academic career planning, and to provide access to organizations, programs and colleagues helpful to the candidate's work as a scholar; identify at least one senior researcher in the university with similar or complementary research interests to serve as a research mentor; and be a U.S citizen or permanent resident at the time of application.

Geographic Focus: All States

Date(s) Application is Due: Feb 12
Contact: Maryjoan D. Ladden; (877) 738-0737; fax (410) 614-8285; rwjfnfs@jhu.edu
Internet: http://www.rwjf.org/applications/solicited/npo.jsp?FUND_ID=55121
Sponsor: Robert Wood Johnson Foundation
Route 1 and College Road East, P.O. Box 2316
Princeton, NJ 08543-2316

RWJF Partners Investing in Nursing's Future Grants **4149**
The Partners Investing in Nursing's Future Grants program is a collaborative initiative of the Robert Wood Johnson Foundation (RWJF) and the Northwest Health Foundation (NWHF), with the intention of addressing nursing issues at the community level through funding partnerships with local and regional foundations. Up to ten awards of up to $250,000 each will be made for projects lasting up to 24 months. Local foundations, with other partners (such as state workforce investment boards, hospital and long-term-care associations, foundations and other funding sources) will match these awards with at least $1 for every $2 provided by the program. Funding will be commensurate with the size and scope of the proposed activity and the foundation's experience with nursing issues. Funding is not intended to support the continuation of ongoing projects, but it may support new programs led by foundations that are already working in the field of nursing.
Requirements: Eligible institutions include local or regional private, family or community foundations and public charities. Eligible foundations are those classified as tax-exempt under Section 501(c)3 as a public charity or private foundation, a nonexempt charitable trust treated as a private foundation under Section 4947(a)(1), or organizations that claim status as private operating foundations under Section 4942(j)3 or 4942(j)5 of the Internal Revenue Code. Government entities, corporate grant-makers and others may participate as part of funding collaboratives; however, they may not serve as the applicant institution. All proposals to this program must address one or more of these topics: Diversity; Educational Infrastructure and Faculty Development; Public Health; Long-Term Care; and/or Mid-Level Management. Proposals for planning as well as implementation projects will be considered.
Geographic Focus: All States
Contact: Renee' Jensen Reinhardt, Program Officer; (971) 230-0093; fax (503) 220-1335; renee@nwhf.org or info@PartnersInNursing.org
Internet: http://www.partnersinnursing.org/
Sponsor: Robert Wood Johnson Foundation
Route 1 and College Road East, P.O. Box 2316
Princeton, NJ 08543-2316

RWJF Pioneer Portfolio Grants **4150**
The Pioneer Portfolio Grant program aims to support novel, high-return ideas that may have far-reaching impact on people's health, the quality of care they receive and the systems through which that care is provided. It seeks ideas not only from the mainstream of health and health care but also looks to sources outside of these fields for innovations that might have transformative impact. Several projects apply approaches from diverse sectors such as finance, design and entertainment to forge new solutions in health and health care arenas.
Requirements: Preference is given to public agencies, public charities or are tax-exempt under section 501(c)3 of the Internal Revenue Code. Organizations must be located in the United States or one of its territories.
Restrictions: The organization does not fund any of the following: Ongoing general operating expenses or existing deficits; Endowment or capital costs; Basic biomedical research; Research on drug therapies or devices; International programs; Direct support of individuals; Lobbying of any kind. Additionally, RWJF no longer funds projects in the following areas: End-of-life care; Long-term care; Specific chronic conditions (that are not part of the strategies of other Program Areas); Physical activity for adults age 50 or older.
Geographic Focus: All States
Contact: Deborah H. Bae; (609) 627-5812 or (877) 843-7953; dbae@rwjf.org
Internet: http://www.rwjf.org/en/grants/what-we-fund.html
Sponsor: Robert Wood Johnson Foundation
Route 1 and College Road East, P.O. Box 2316
Princeton, NJ 08543-2316

Ryan Gibson Foundation Grants **4151**
The Foundation welcomes any and all research grant requests from around the country. The main focus is on blood related cancers. Applicants should submit a one to two page proposal with the following information: description of the procedure and or drug that is to be tested/ developed, and what the desired/expected result will be; some background on the researcher and the facility that will be used (past projects success and or failures should be included); and a budget of anticipated costs. Mail all requests to the following address: The Ryan Gibson Foundation, Attn: Research Proposal, 18111 Preston Road, Suite 650, Dallas, TX 75252
Geographic Focus: All States
Contact: National Director; (214) 987-0008; fax (214) 365-9293; mail@trgf.org
Internet: http://www.trgf.org/grants.shtml
Sponsor: Ryan Gibson Foundation
5910 North Central Expressway, Suite 770
Dallas, TX 75206

S. D. Bechtel, Jr. Foundation / Stephen Bechtel Fund Preventive Healthcare and Selected Research Grants **4152**
The Foundation?fund believe that access to quality care, preventive health care and research programs can help individuals of all ages increase life expectancy and improve their quality of life. The directors have a particular interest in promoting nutritional health, fitness, and wellness in order to increase positive health outcomes. The Foundation/Fund support organizations working to achieve the following objectives: enabling San Francisco Bay Area community clinics and school-based health centers to improve the quality and delivery of medical care; enhancing and expand school-based efforts in the San Francisco Bay Area that improve the nutritional health and fitness of children; and advancing research related to Alzheimer's, aging, nutrition or other Board-determined specific research initiatives.
Requirements: The primary geographic focus of the Foundation/Fund is the San Francisco Bay area. The Foundation/Fund: support non-profit organizations providing quality programs in science, technology, engineering and math (STEM) education, environment, environmental education, character and citizenship development and preventive healthcare and selected research; provide capital support; provide operational support; and provide project support.
Restrictions: The Foundation/Fund do not provide endowment funding, international grants, or grants for individuals.
Geographic Focus: California
Date(s) Application is Due: Oct 1
Amount of Grant: 5,000-25,000 USD
Contact: Coordinator; (415) 284-8675; fax (415) 284-8571; sdbjr@sdbjrfoundation.org
Internet: http://www.sdbjrfoundation.org/program_areas.htm#health
Sponsor: S. D. Bechtel, Jr. Foundation / Stephen Bechtel Fund
P.O. Box 193809
San Francisco, CA 94119-3809

S.E.VEN Fund Annual Grants **4153**
The sponsor is a virtual non-profit entity run by entrepreneurs whose strategy is to markedly increase the rate of diffusion of enterprise-based solutions to poverty. This is accomplished by targeted investment that fosters thought leadership through books, films and websites; supporting role models-whether they are entrepreneurs or innovative firms-in developing nations; and shaping a new discourse in government, the press and the academy around private-sector innovation, prosperity and progressive human values. This program is intended to catalyze, support and disseminate research on questions of economic development, prosperity and entrepreneurship. Funding is particularly targeted toward new frontiers in enterprise-based solutions to poverty and innovative ideas that are unlikely to be supported by conventional funding sources.
Requirements: S.E.VEN welcomes the participation of entrepreneurs, business people, investors, academics, laypeople, and philanthropists. To begin the process, add your name to the database online. Further information will be emailed to you.
Geographic Focus: All States
Amount of Grant: Up to 100,000 USD
Contact: Elizabeth Hooper, Executive Director; info@sevenfund.org
Internet: http://www.sevenfund.org/research-funding-areas.html
Sponsor: Social Equity Venture Fund (S.E.VEN)
1770 Massachusetts Avenue, 247
Cambridge, MA 02140

S.E.VEN Fund Mini-Grants **4154**
The sponsor is a virtual non-profit entity run by entrepreneurs whose strategy is to markedly increase the rate of diffusion of enterprise-based solutions to poverty. This is accomplished by targeted investment that fosters thought leadership through books, films and websites; supporting role models-whether they are entrepreneurs or innovative firms-in developing nations; and shaping a new discourse in government, the press and the academy around private-sector innovation, prosperity and progressive human values. Mini-grant requests should align with the S.E.VEN Fund's funding priorities and do not exceed $10,000. The sponsor considers funding requests for mini-grants on an ongoing basis; the competition runs three times annually.
Requirements: This program's funding is intended to catalyze, support and disseminate research on questions of economic development, prosperity and entrepreneurship. Funding is particularly targeted toward new frontiers in enterprise-based solutions to poverty and innovative ideas that are unlikely to be supported by conventional funding sources. Funded projects are typically research-based or involve the creative use of the media-books, film, the Web, organizing conferences, etc. You can apply for funding by filling out the online mini-grant request form.
Restrictions: The sponsor is unable to provide funding for direct services.
Geographic Focus: All States
Amount of Grant: Up to 10,000 USD
Contact: Elizabeth Hooper, Executive Director; info@sevenfund.org
Internet: http://www.sevenfund.org/submit-mini-grants.php
Sponsor: Social Equity Venture Fund (S.E.VEN)
1770 Massachusetts Avenue, 247
Cambridge, MA 02140

S.E.VEN Fund Open Enterprise Solutions to Poverty RFP Grants **4155**
The RFP competition focuses on academic and scientific research in the field of Enterprise-based Solutions to Poverty. The competition will award up to two (2) research grants of no more than $100,000 each for a duration of no more than twelve months. There is no geographic or other limitation on who may compete for these funds-anyone may compete for these funds.
Requirements: Anyone may submit an Initial Proposal in order to compete for a SEVEN RFP grant as long as the proposal reflects the scope and guidelines stated in the call for RFPs. The winning parties will be required to enter into a contract with SEVEN prior to final award. Since this is a program run by both practitioners and academics, SEVEN may be a bit more flexible than traditional funding agencies in the definition of a Principal Investigator. Thus, they may accept applications from non-academics as the PI or a co-investigator if his/her experience and background warrant it. Acceptable use of grant funds include: student or postdoctoral salary and benefits for part of the academic year; summer salary and teaching buyout for academics; support for specific projects during sabbaticals; assistance

in writing or publishing books; modest allowance for justifiable equipment, computers, publication charges, and other supplies; modest travel allowance; development of workshops, conferences, or lecture series for professionals; or, justified overhead of at most 20%.
Geographic Focus: All States
Date(s) Application is Due: Oct 15
Amount of Grant: 100,000 USD
Contact: Elizabeth Hooper, Executive Director; info@sevenfund.org
Internet: http://www.sevenfund.org/enterprise-solutions-poverty/index.php
Sponsor: Social Equity Venture Fund (S.E.VEN)
1770 Massachusetts Avenue, 247
Cambridge, MA 02140

S.E.VEN Fund WHY Prize **4156**
The sponsor's objective with The WHY Prize is to stimulate innovation in the specific domain of Enterprise Solutions to Poverty by encouraging collaboration; specifically, they will support collaboration between macroeconomists and anthropologists. The WHY Prize intends to catalyze cross-fertilization across disciplines and spur progress in finding sustainable solutions to poverty. The prize is a one-time award of $50,000.
Requirements: Anyone may submit a nomination in order to compete for The WHY Prize as long as the proposal reflects the scope and guidelines stated in the call for nominations. Individuals may nominate themselves or others for consideration. Nominated teams must include one individual who holds a masters or doctoral degree in economics and anthropology respectively. Other disciplines may join the collaboration to broaden the perspective beyond these two disciplines. The WHY prize will reward a team that has published in either a major venue like the Atlantic Monthly, New York Times Magazine, Financial Times, Businessweek, Forbes, Foreign Policy, Foreign Affairs etc. or in a peer reviewed journal, with a view toward producing further work such as a book or additional articles in a peer reviewed format. The published piece must add to the dialogue around enterprise solutions to poverty by proposing solutions that represent the best integration of actionable insights and thought leadership. Nominees may submit an article, paper, or book for consideration by applying on the sponsor's website.
Restrictions: Only applications submitted through the form on the sponsor's website are accepted.
Geographic Focus: All States
Date(s) Application is Due: Oct 7
Amount of Grant: 50,000 USD
Contact: Elizabeth Hooper, Executive Director; info@sevenfund.org
Internet: http://www.sevenfund.org/press-release-mar-2008-WHY-prize.html#qanda
Sponsor: Social Equity Venture Fund (S.E.VEN)
1770 Massachusetts Avenue, 247
Cambridge, MA 02140

Sachs Foundation Scholarships **4157**
The foundation uses its funds to improve the well-being of African American residents of Colorado. Funds are provided for undergraduate studies. Graduate grants are available to individuals who received undergraduate grants. The foundation will finance each awardee for four years. Applications are released and accepted from January 1 until March 1 each calendar year; applicants are informed in May regarding their eligibility for an interview in June.
Requirements: All applicants must be African American U.S. citizens and residents of Colorado for at least five years. Funds are only available for young students graduating from accredited high schools with at least a 3.4 grade point average.
Geographic Focus: Colorado
Date(s) Application is Due: Mar 1
Amount of Grant: Up to 4,000 USD
Contact: Grants Administrator; (719) 633-2353; sachs@frii.com
Internet: http://www.sachsfoundation.org/scholarships.htm
Sponsor: Sachs Foundation
90 South Cascade Avenue, Suite 1410
Colorado Springs, CO 80903

Sage Foundation Grants **4158**
The foundation awards grants nationwide, with emphasis on the Midwest, to nonprofit organizations in support of higher education, legal education, and secondary education; seniors; hospitals; disabled; Catholic giving and welfare; youth; child welfare; social services; and cultural programs. Types of support include annual campaigns, building construction/renovation, capital campaigns, challenge grants, endowments, equipment acquisition, general operating grants, multi-year support, project development, research, and scholarship funds. There are no application deadlines or forms; submit a letter of inquiry. The board meets quarterly.
Geographic Focus: All States
Amount of Grant: 1,000-250,000 USD
Contact: Melissa Sage Fadim; (810) 227-7660 or (212) 737-7311
Sponsor: Sage Foundation
P.O. Box 1919
Brighton, MI 48116

Saint Mary's College Fellowships **4159**
The Center for Women's InterCultural Leadership (CWIL) seeks to increase awareness of the importance of intercultural dialogue in today's world, to highlight the leadership roles women have adopted in such dialogue, and to foster the next generation of women leaders. In pursuit of these goals, CWIL offers visiting and faculty Fellowships. All appointments are for one year with the possibility of a renewal for a second year. CWIL Fellows have three general types of responsibilities: research; teaching courses relating to cultural or women's leadership issues; and assisting the CWIL and/or the department in its programmatic and community activities. These activities will vary depending on the type of Fellowship and the needs of the sponsoring department.
Geographic Focus: All States
Contact: Linda Biggins; (574) 284-4051; fax (574) 284-4141; cwil@saintmarys.edu
Internet: http://www.centerforwomeninleadership.org/research-and-scholarship/fellows
Sponsor: Saint Mary's College
Center for Women's Intercultural Leadership
Notre Dame, IN 46556-5001

Salem Foundation Grants **4160**
The primary purpose of the Foundation is support of higher education, cultural activities, health services, hospital building funds, and Christian giving. Its fields of interest include: arts; Christian agencies and churches; the disabled; education (secondary and higher); the environment; family services; federated giving programs; health care (particularly heart and circulatory disease research); hospitals; human services; museums; and Protestant federated giving programs. Types of support include: annual campaigns, building/renovation projects, capital campaigns, general operating support, and program development. There is no formal application process and no deadlines. Please contact via letter.
Restrictions: Giving primarily in Florida, Maine and Minnesota. No grants to individuals.
Geographic Focus: Florida, Maine, Minnesota
Amount of Grant: Up to 10,000 USD
Contact: Robert S. Parish, President; (952) 476-6292
Sponsor: Salem Foundation
2181 Springwood Road
Wayzata, MN 55391-2254

Samueli Foundation Health Grants **4161**
The Foundation considers grants to agencies, primarily in Orange County, California, that serve the community, and whose programs meet the guidelines listed. Grants are usually approved for a defined period of time, but may be paid over a multi-year period. In the area of Health, its goals are to: support programs promoting alternative and integrative medicine; and utilize creative and innovative approaches to education, prevention, treatment and research of disease and illness.
Requirements: 501(c)3 tax-exempt organizations are eligible.
Restrictions: The foundation does not fund umbrella fund raising organizations, political campaigns, or grants to individuals.
Geographic Focus: All States
Date(s) Application is Due: Nov 24
Amount of Grant: 1,000-50,000 USD
Contact: Maggie Sherer; (949) 760-4400; fax (949) 759-5707; Info@samueli.org
Internet: http://www.samueli.org/health.aspx
Sponsor: Samueli Foundation
2101 East Coast Highway, 3rd Floor
Corona del Mar, CA 92625

Samueli Institute Scientific Research Grants **4162**
The Institute operates in conjunction with the Uniformed Services University of Health Sciences and Universitatsklinikum Freiburg Institute. Its mission is is to transform health care through the scientific exploration of healing. The Institute is one of an elite group of organizations in the nation with a track record in complementary and integrative medicine, healing relationships and military medical research. The major purpose of funding is to investigate healing processes and their application in promoting health and wellness, preventing illness, and treating disease. Researchers should respond to Requests for Applications or Program Announcements.
Geographic Focus: All States
Amount of Grant: 2,000-600,000 USD
Contact: Matt Fritts, (703) 299-4800; fax (703) 299-4800; mfritts@SamueliInstitute.org
Internet: http://www.siib.org/funding/investigator-home/218-SIIB.html
Sponsor: Samueli Institute
1737 King Street, Suite 600
Alexandria, VA 22314

San Antonio Area Foundation Grants **4163**
Grants are usually awarded in the areas of health care and biomedical research, community and social services, arts and culture, education (early childhood education, higher education, medical schools, nursing schools, adult basic education and literacy), and animal services. Types of support include operating budgets, continuing support, annual campaigns, seed grants, emergency funds, equipment acquisition, matching funds, scholarship funds, research, lectureships, professorships, and building renovations. Applicants must submit a letter of intent to apply; if the letter is approved, the foundation will send a request for a proposal to the applicant. The foundation only reviews full proposals from applicants whose letters of intent have been approved. If asked to submit a proposal, the applicant will be sent an application package with a letter of notification in early February.
Requirements: Grants are made to organizations in the San Antonio, Texas, area.
Geographic Focus: Texas
Date(s) Application is Due: Nov 15
Amount of Grant: 250-250,000 USD
Contact: Lydia R. Saldana, Program Officer; (210) 228-3753 or (210) 225-2243; fax (210) 225-1980; lsaldana@saafdn.org or info@saafdn.org
Internet: http://www.saafdn.org/NetCommunity/Page.aspx?pid=254

Sponsor: San Antonio Area Foundation
110 Broadway Street, Suite 230
San Antonio, TX 78205-1948

San Diego Women's Foundation Grants 4164
The San Diego Women's Foundation (SDWF) strengthens and improves women's capacities to engage in significant philanthropy in the San Diego region. The foundation chooses one focus area of grant making each year. Its primary focus areas include: education, arts and culture, environment, employment and economic development, civil society, and health and human services. Check the web site each September for that year's focus guidelines.
Requirements: Only 501(c)3 organization within San Diego County should apply.
Geographic Focus: California
Date(s) Application is Due: Dec 6
Amount of Grant: 25,000 USD
Contact: Tracy Johnson, Director; (619) 814-1374 or (619) 235-2300; fax (619) 239-1710; tracy@sdfoundation.org
Internet: http://www.sdwomensfoundation.org/
Sponsor: San Diego Women's Foundation
2508 Historic Decatur Road, Suite 200
San Diego, CA 92106

Sandler Program for Asthma Research Grants 4165
The mission of the foundation is to develop important new pathways of investigation in basic research regarding asthma. The program, which is open to investigators from all fields, particularly encourages applications from investigators not currently studying asthma. Innovation and risk are strongly encouraged. The program especially (but not exclusively) seeks applications in the following areas: development of new animal models for asthma; studies in humans or of human materials (excluding clinical trials); genetics of asthma; and technology development relevant to asthma. Three-year senior investigator awards and three-year junior investigator awards will be made. Awards to junior investigators are intended to support investigators who have demonstrated exceptional early accomplishment as independent investigators, permitting the expansion of their work into the field of asthma; they are not start-up funds. The proposed work should be directed toward uncovering basic mechanisms in the pathogenesis of asthma.
Requirements: Senior investigators will have well-established research programs and an international reputation for their research. They will usually hold a full-time academic appointment as professor, tenured associate professor, or the equivalent. Junior investigators will have already established an independent research program. They will usually hold a full-time academic appointment as an assistant professor, early associate professor, or the equivalent. Investigators may apply from any U.S. or Canadian nonprofit organization that can and will provide the necessary facilities and infrastructure for the research, and that will assure compliance with guidelines for animal and human studies established by the U.S. Department of Health and Human Services. Applicants from Canadian institutions must also provide evidence that the institution has been designated as a public charity by the U.S. Internal Revenue Service.
Geographic Focus: All States
Date(s) Application is Due: Feb 8
Amount of Grant: 150,000-250,000 USD
Contact: William E. Seaman; (415) 514-0730; seamanb@sandlerresearch.org
Internet: http://www.sandlerresearch.org
Sponsor: Sandler Foundation
Box 0509, UCSF
San Francisco, CA 94143-0509

Sara Elizabeth O'Brien Trust Grants 4166
The Sara Elizabeth O'Brien Trust was established in 1981 to support charitable organizations that provide treatment or care to those who are blind or have cancer. The O'Brien Trust also supports medical research in the areas of blindness or cancer. The majority of grants from the O'Brien Trust are one year in duration; on occasion, multi-year support is awarded. Applicants must apply online at the grant website. Applicants are strongly encouraged to do the following before applying: review the downloadable state application procedures for additional helpful information and clarifications; review the downloadable online-application guidelines at the grant website; review the trust's funding history (link is available from the grant website); review the online application questions in advance; and review the list of required attachments. These will generally include: a list of board members, financial statements (audited, reviewed, or compiled by independent auditor); an organization summary; a list of other funding sources; an IRS Determination letter; and other required documents. All attachments must be uploaded in the online application as PDF, Word, or Excel files. The application deadline for the Sara Elizabeth O'Brien Trust is 11:59 p.m. on October 15. Applicants will be notified of grant decisions before December 31.
Requirements: Applicants must have 501(c)3 tax-exempt status.
Restrictions: The trust does not support requests from individuals, organizations attempting to influence policy through direct lobbying, or any political campaigns.
Geographic Focus: Massachusetts
Date(s) Application is Due: Oct 15
Contact: Michealle Larkins; (866) 778-6859; michealle.larkins@baml.com
Internet: https://www.bankofamerica.com/philanthropic/fn_search.action
Sponsor: Sara Elizabeth O'Brien Trust
225 Franklin Street, 4th Floor, MA1-225-04-02
Boston, MA 02110

Sarasota Memorial Healthcare Foundation Grants 4167
The Foundation is an organization with the mission to improve the delivery of health care in the Sarasota area through the acquisition and utilization of philanthropic funds. Although the Foundation primarily supports Sarasota Memorial Health Care System, it also awards grants to other area nonprofit organizations to meet the community's changing health care needs. Grants totaling more than $32,000,000 have been provided since the Foundation's establishment in 1976. Current fields of interest include: health care; medical care; and community health systems.
Requirements: Health care programs and organization in the Sarasota, Florida, region are eligible to apply.
Geographic Focus: Florida
Contact: Alexandra Quarles, President; (941) 917-1286; alex-quarles@smh.com
Internet: http://www.smhf.org
Sponsor: Sarasota Memorial Healthcare Foundation
1515 S Osprey Avenue, Suite B4
Sarasota, FL 34239

SARE Graduate Student Grants-North Central Region 4168
The North Central Region SARE (NCR-SARE) Graduate Student Grants support projects by graduate students that address sustainable agriculture issues and are part of the student's degree program. At least 15 projects of up to $10,00 each are funded each year. Projects may last up to 36 months. Funded proposals have contributed to farmer or rancher profitability, environmental quality, and the enhancement of the quality of life of farmers or ranchers, rural communities, and society as a whole. NCR-SARE strongly encourages students to involve farmers and ranchers in their grant projects. NCR-SARE usually issues a call for proposals in the fall, and graduate student grant proposals are due in January. Usually, the Council announces the awards in the spring. Funds are typically available in the fall. Specific guidelines and information, including a video and list of previously funded projects, are available at the website.
Restrictions: Graduate students may only receive one NCR-SARE Graduate Student award during their graduate student career. Projects must benefit agriculture in the North Central Region.
Geographic Focus: Illinois, Indiana, Iowa, Kansas, Michigan, Minnesota, Missouri, Nebraska, North Dakota, Ohio, South Dakota, Wisconsin
Date(s) Application is Due: Jan 1
Contact: Beth Nelson; (612) 626-4436; fax (612) 626-3132; Schre002@umn.edu
Internet: http://www.northcentralsare.org/Grants/Our-Grant-Programs/Graduate-Student-Grant-Program
Sponsor: Sustainable Agriculture Research and Education
University of Maryland
College Park, MD 20742-6715

SARE Graduate Student Grants-Northeast Region 4169
The SARE Graduate Student Grants are available for graduate students who are researching topics in sustainable agriculture that will serve the interests of farmers and agricultural service providers, such as extension staff, NRCD personnel, and others in the agricultural community. Successful Graduate Student Grant proposals will address inquiries in sustainable agriculture that are important to farmers, Cooperative Extension staff, and agricultural researchers. Awards may encompass a wide range of topics including, but not limited to cropping systems, pest management, livestock health, farm energy production, farm labor, soil quality, and the marketing of local food. SARE funds can be used for supplies, labor, equipment specific to the project (which must then become the property of the college or university), farm equipment rental or operating charges specific to the project, and travel expenses specific to the project and consistent with institutional mileage and per-diem rates. The grant can also be used for indirect costs up to 10% of your project's direct costs. Specific guidelines, examples of previously funded projects, and application submission instructions are available at the website.
Requirements: Grants are open to graduate students enrolled at an accredited college, university, or veterinary school in the Northeast region. Applicants must have a faculty advisor who will endorse the proposal, oversee the research, and act as the official principal investigator. Applications must come from one qualified individual. Groups or two-person teams may not participate. Collaboration is encouraged, but one person must be in charge of the project, ensuring that the work is done and reports are filed.
Restrictions: Grant funds cannot be used for any of the following: graduate student tuition remission; tuition payment; curriculum fees; travel to scholarly meetings (unless essential to the project, such as for a presentation of project results); travel outside the country; costs associated with preparing your thesis; the purchase of classroom texts; or capital costs such as the purchase of land, buildings, livestock, orchards, greenhouses, or other major farm improvements.
Geographic Focus: Connecticut, Delaware, District of Columbia, Maine, Maryland, Massachusetts, New Hampshire, New Jersey, New York, Pennsylvania, Rhode Island, Vermont, West Virginia
Date(s) Application is Due: May 25
Contact: Judy Gifford; (410) 648-5753; fax (443) 480-1853; gifford@baybroadband.net
Internet: http://www.nesare.org/Grants/Get-a-Grant/Graduate-Student-Grant
Sponsor: Sustainable Agriculture Research and Education
University of Maryland
College Park, MD 20742-6715

SARE Graduate Student Grants-Southern Region 4170
Sustainable Agriculture Research and Education Graduate Student Grants support sustainable agriculture research funding for Master's and Ph.D. students. Funds may be used for the following purposes: supplies, including software; equipment (any equipment obtained becomes the property of the university that the funded graduate student

attends); labor; special texts not readily available; farm equipment rental or operating charges; and travel and per diem necessary for the project. The application and additional guidelines are available at the website.
Requirements: Full time Master's or Ph.D. students at accredited institutions in the southern region, with research projects that address issues of sustainable agriculture of current and potential importance to the southern region, may apply for funding. The grant will fund a project maximum of $11,000 for up to three years work on a student's project.
Restrictions: Only graduate students enrolled in academic programs in the southern region may apply for funding. Funds may not be used for the following purposes: graduate student stipends; travel to scholarly meetings or any international travel; preparation of thesis copy; publication costs; purchase of classroom books or payment of tuition; and permanent capital improvements. Applicants may receive only one SARE Graduate Student Grant during their graduate student career.
Geographic Focus: Arkansas, Florida, Georgia, Kentucky, Louisiana, Mississippi, North Carolina, Oklahoma, Puerto Rico, South Carolina, Tennessee, Texas, U.S. Virgin Islands, Virginia
Date(s) Application is Due: Jun 1
Amount of Grant: Up to 11,000 USD
Contact: Paige Patton; (770) 412-4787; fax (770) 412-4789; ppatton@uga.edu
Internet: http://www.southernsare.org/Grants/Types-of-Grants/Graduate-Student-Grants
Sponsor: Sustainable Agriculture Research and Education
University of Maryland
College Park, MD 20742-6715

SARE Graduate Student Grants in Sustainable Agriculture-Western Region 4171
The Western SARE Graduate Student Grants in Sustainable Agriculture offer grants of up to $25,000 that may last up to two years. Grants are awarded to the applicant's university, with the funds to be assigned to the graduate student's project. An initial 90% of the award will be paid by reimbursement of allowable project expenses. The final 10% will be reimbursed after Western SARE has received and approved a final project report. Funds may be used for: outreach expenses such as holding a field day, tours, making a brochure, or creating a webpage; supplies, including software; small tools and equipment (this becomes property of the graduate student's university) not to exceed $5,000 per piece of equipment; hourly labor for documented work on the Western SARE project; farm equipment rental or operating charges; out-of-town travel and per diem necessary for the project (as prescribed by your university rules); local travel (where personal auto is used and reimbursed) needed for project; or publication costs (not to exceed $1,000). Specific guidelines and additional information about how to apply and proposals will be judged are available at the website.
Requirements: Applicants may receive only one grant during his or her graduate program. Western SARE will consider proposals from master's and Ph.D. students enrolled at accredited colleges or universities in the Western region. The student must be considered full-time, according to his or her institution's requirements, at the time the proposal is submitted. Proposals must address issues in sustainable agriculture of current and potential importance to the Western Region. Proposals are requested within the following guidelines: plant health and production and plant products; animal health and production and animal products; food safety, nutrition, and health; renewable energy, natural resources and environment; agriculture systems and technology; and agriculture economics and rural communities. Western SARE requires annual progress reports, a final report and one copy each of all theses, dissertations, and all other publications that are developed through the funded project. Project reports are compiled each year and shared with producers, interested policymakers and leaders at community, state, regional and national levels.
Restrictions: Funds may not be used for: preparation of thesis copy; purchase of classroom books or payment of tuition; permanent improvements to a farm or ranch, such as constructing or remodeling a building; graduate student stipends; or any single piece of equipment costing more than $5,000.
Geographic Focus: Alaska, Arizona, California, Colorado, Guam, Hawaii, Idaho, Montana, Nevada, New Mexico, Northern Mariana Islands, Oregon, Utah, Washington, Wyoming, American Samoa, Micronesia
Date(s) Application is Due: May 27
Amount of Grant: Up to 25,000 USD
Contact: Robert Newhall, Deputy Director; (435) 797-2257; fax (435) 797-3344; bob.newhall@usu.edu or wsare@usu.edu
Internet: http://www.westernsare.org/Grants/Types-of-Grants
Sponsor: Sustainable Agriculture Research and Education
University of Maryland
College Park, MD 20742-6715

SARE Large Systems Research Grants-Southern Region 4172
The SARE Large Systems Research Grants are designed to assure research capacity in addition to gaining a near-term result to a specific research question. They are also designed to improve the opportunities for teams to engage in interdisciplinary activities. This ongoing process of research design followed by research activities, analysis, and redesign does not fit standard short-term modes of funding common in many grant programs. The Southern Region SARE Program hopes to address that funding gap. Project maximum is $100,000 per year for three years with opportunity for additional three-year renewals. Additional guidelines and the application is available at the website.
Requirements: Large Systems Grants are for teams who already have successful agricultural systems research in place but need support to accomplish additional long-term research goals. Research activities that not only keep the research team and infrastructure together over time but that also allow the teams to analyze and assess results over time, integrating new questions that arise into the long-term research protocol.
Geographic Focus: Alabama, Arkansas, Georgia, Kentucky, Louisiana, Mississippi, North Carolina, Oklahoma, Puerto Rico, South Carolina, Tennessee, Texas, U.S. Virgin Islands, Virginia
Amount of Grant: Up to 100,000 USD
Contact: Sandra Young; (770) 229-3212; fax (770) 412-4790; syoung@griffin.uga.edu
Internet: http://www.southernsare.org/Grants/Types-of-Grants/Large-Systems-Research-Grants
Sponsor: Sustainable Agriculture Research and Education
University of Maryland
College Park, MD 20742-6715

SARE On-Farm Research Grants-Southern Region 4173
The Southern Region SARE program recognizes the value and importance of on-farm research in developing sustainable solutions to agricultural production problems by offering funding opportunities to Extension, Natural Resources Conservation Service personnel, university researchers, and non-governmental organizations who currently work with farmers and conduct on-farm research with at least one farmer/rancher cooperator. The On-Farm Research Grant program assists these individuals and organizations in conducting on-farm research or marketing projects related to sustainable agriculture to help the farmers they work with find solutions to various agricultural production issues. Successful proposals include: cooperators — on-farm research projects must include at least one farmer and/or rancher project cooperator; an outreach plan — to provide producers, researchers, and Extension personnel with the ability to learn from project results (outreach plans may include workshops, field days, fact sheets, or brochure); and clear goals — clear and simple goals help ensure a successful project. On-Farm Research Grant funds may be used for the following purposes: costs of sampling and sample analysis; materials and supplies needed for the project; outreach expenses such as holding a field day; travel (maximum 51 cents/mile) needed for the project; hired labor, including farmer/rancher labor beyond their normal farming duties; and refreshments at field days. Specific guidelines and the application are available at the website.
Requirements: Cooperative Extension, NRCS, university personnel, governmental and non-governmental organizations who work with farmers and ranchers, and are interested in conducting on-farm research or marketing projects related to sustainable agriculture are eligible to apply for the grant. On-farm research and marketing projects that promote sustainable agriculture are eligible to apply. Projects should be innovative, generate results that are useful beyond one year, and produce information that many farmers can use.
Restrictions: Funding may not be used for the following purposes: starting or expanding a farm, farming operation or NGO; equipment; permanent improvements to a farm or ranch, (e.g. planting an orchard, buying a herd, permanent fencing, permanent irrigation, permanent greenhouses or buildings); lunches or other full meals at field days or large gatherings; testing of commercial products.
Geographic Focus: Alabama, Arkansas, Florida, Georgia, Kentucky, Mississippi, North Carolina, Oklahoma, Puerto Rico, South Carolina, Tennessee, Texas, U.S. Virgin Islands, Virginia
Date(s) Application is Due: Nov 1
Amount of Grant: Up to 15,000 USD
Contact: Paige Patton; (770) 412-4787; fax (770) 412-4789; ppatton@uga.edu
Internet: http://www.southernsare.org/Grants/Types-of-Grants/On-Farm-Research-Grants
Sponsor: Sustainable Agriculture Research and Education
University of Maryland
College Park, MD 20742-6715

SARE Producer Research and Education Grants-Western Region 4174
The Producer Grants are conducted by agricultural producers with support and guidance from a technical advisor. Individual farmers or ranchers may apply for up to $15,000, and a group of three or more producers may apply for up to $25,000. Producers typically use their grants to conduct on-site experiments that can improve their operations and the environment, and can be shared with other producers. Grant recipients may also focus on marketing and organic production. Funding may be used for the following purposes: mini-grants to producers to compensate for field research and educational activities; cost of field sampling, crop analysis, and educational surveys; materials and supplies need for the project; small tools and equipment; outreach expenses such as a field day, tours, making a brochure, or creating a webpage; local or out of town travel need for the project; hired labor for heavy equipment; and producer labor for project activities. Specific guidelines, the application, and instructions about contacting state officers are available at the website.
Requirements: All proposals must have at least one producer. Applicants are producers if their primary occupation is farming or ranching; they have a farm tax number, or; they are a part-time producer with at least $1,000 documented annual income from their operation. The producer is fiscally responsible for the project and, if funded, provides his or her Social Security number. All proposals must include the following: yearly plan for outreach to other producers; a list of the educational materials they plan to produce and use in that outreach; and expected changes in these producers.
Geographic Focus: Alaska, Arizona, California, Colorado, Guam, Hawaii, Idaho, Montana, Nevada, New Mexico, Northern Mariana Islands, Oregon, Utah, Washington, Wyoming, American Samoa, Micronesia
Date(s) Application is Due: Dec 2
Amount of Grant: 15,000-25,000 USD
Contact: Robert Newhall; (435) 797-2183; fax (435) 797-3344; bob.newhall@usu.edu
Internet: http://www.westernsare.org/Grants/Types-of-Grants
Sponsor: Sustainable Agriculture Research and Education
University of Maryland
College Park, MD 20742-6715

SARE Professional + Producer Grants-Western Region 4175

With a Professional + Producer Research & Education Grant, an agricultural professional, working with at least five producers, develops a proposal to conduct research and education on a sustainable agricultural topic and incorporate such items as: on-farm demonstrations, farmer-to-farmer educational outreach, and other items to assist in producer adoption in an area of sustainable agriculture. The goal is to achieve results that can be communicated to producers and professionals – information that can improve income, the environment, communities, and quality of life for all citizens. Producers will also have the opportunity to expand their knowledge and expertise in sustainable agriculture. Sustainable agriculture is defined as an integrated system of plant and animal production practices having a site-specific application that will, over the long term: satisfy human food and fiber needs; enhance environmental quality and the natural resource base upon which the agriculture economy depends; make the most efficient use of nonrenewable resources and on-farm resources and integrate, where appropriate, natural biological cycles and controls; sustain the economic viability of farm operations; and enhance the quality of life for farmers and society as a whole. On-farm education and experience of FFA, 4-H and other similar aged youth groups dealing with sustainable agricultural techniques and practices are encouraged. The SARE Council typically selects proposals diverse in subject matter and geography and that demonstrate outcomes that farmers and ranchers in the region can successfully adopt. Specific guidelines and instructions for application are available at the website.

Requirements: A minimum of five producers (separate operations) are required to participate in the development of on-farm research and demonstration with educational outreach. Each producer must be an independent and separate operation. Any agriculture/natural resource professional in the Western region may apply. These professionals are cooperative extension educators/agents and specialists, USDA-NRCS field staff, agricultural consultants, nonprofit staff members or other agricultural professionals assisting producers at the local level. Congress mandates that the SARE grant program depart from "business as usual." To that end, the Administrative Council requires that producers are involved from start to finish in the planning, design, implementation, and educational outreach of any SARE-funded project. Projects should: relate ideas to the definition of sustainable agriculture; address the changes that could come from the adoption of the results by other producers; and detail creative educational outreach plans that deliver this knowledge to other producers in the Western Region.

Geographic Focus: Alaska, Arizona, California, Colorado, Guam, Hawaii, Idaho, Montana, Nevada, New Mexico, Northern Mariana Islands, Oregon, Utah, Washington, Wyoming, American Samoa, Micronesia

Date(s) Application is Due: Dec 2

Amount of Grant: Up to 50,000 USD

Contact: Robert Newhall; (435) 797-2183; fax (435) 797-3344; bob.newhall@usu.edu

Internet: http://www.westernsare.org/Grants/Types-of-Grants

Sponsor: Sustainable Agriculture Research and Education
University of Maryland
College Park, MD 20742-6715

SARE Professional Development Grants-Northeast Region 4176

The Professional Development (PD) Grants fund outcome-based projects that train Cooperative Extension educators and other agricultural service providers in sustainable techniques and concepts. Projects must be directed toward increasing the skill and understanding of these service providers and consistent with SARE's larger goal of broad farmer adoption of sustainable practices. The program emphasizes training the trainers, disseminating sustainable practices and technologies, advancing new content, and seeking measurable, verifiable change that leads to the improved and widespread practice of agricultural sustainability. Professional Development projects can cover a wide range of topics, such as ecological weed management, labor, whole-farm nutrient planning, tillage, season extension, carbon trading, and the needs of new farmers. Grants focus on training extension and NRCS staff, agricultural consultants, and other service providers who work directly with farmers and advise them on their farm and business practices. Projects should seek a cascade effect so that the people who took the training are positioned to teach others. SARE funds have also been used to produce books, videos, webcasts, and other media tools to improve participant understanding of sustainable techniques. Specific guidelines and application instructions are available at the website.

Requirements: Anyone can apply for a PD grant, but applicants should have direct and substantial experience delivering agricultural education to extension and other agricultural service providers. All applicants should demonstrate a firm grasp of current barriers and issues in sustainable agriculture. Proposals typically come from extension staff and the agricultural nonprofit sector, as well as college and university faculty and staff, and people who work on research farms and experiment stations. Northeast SARE uses outcome funding to select and follow through on its PD projects. All PD proposals should address the Northeast SARE outcome statement ("agriculture in the Northeast will be diversified and profitable, providing healthful products to its customers; it will be conducted by farmers who manage resources wisely, who are satisfied with their lifestyles, and have a positive influence on their communities and the environment.). Applicants not familiar with outcome funding should read the instruction guides and handouts on the website. The outcome approach permeates the entire application process, uses a distinct vocabulary, and operates on specific assumptions. Applicants must be conversant in outcome funding to successfully apply for a PD grant.

Restrictions: Applicants must be located in the Northeast SARE region.

Geographic Focus: Connecticut, Delaware, District of Columbia, Maine, Maryland, Massachusetts, New Hampshire, New Jersey, New York, Pennsylvania, Rhode Island, Vermont, West Virginia

Date(s) Application is Due: Oct 20

Amount of Grant: 60,000-200,000 USD

Contact: Tom Morris; (860) 486-0637; fax (860) 486-0682; thomas.morris@uconn.edu

Internet: http://www.nesare.org/Grants/Get-a-Grant/Professional-Development-Grant

Sponsor: Sustainable Agriculture Research and Education
University of Maryland
College Park, MD 20742-6715

SARE Research and Education Grants-North Central Region 4177

The North Central Region SARE (NCR-SARE) Research and Education Grants award funding to researchers and educators involved in projects that explore and promote environmentally sound, profitable, and socially responsible food and/or fiber systems. Research and Education projects include a strong outreach component and significant farmer/rancher or other end user involvement from inception of the idea through implementation of the project. Many projects are interdisciplinary and/or multi-institutional, involving a broad range of agricultural interests. Project coordinators in the past have explored sustainable agriculture under the following topics: biocontrol, crop production, education and extension, networking, livestock production, marketing, quality of life, soil quality, value-added marketing, waste management, water quality, and weed control. Awards range from $10,000 to $200,000, with projects lasting up to 36 months. Coordinators are encouraged to explore social, environmental, and economic aspects of whole systems, keeping in mind that projects must have a strong outreach component. Specific guidelines, previously funded projects, and application instructions are available at the website.

Geographic Focus: Illinois, Indiana, Iowa, Kansas, Michigan, Minnesota, Missouri, Nebraska, North Dakota, Ohio, South Dakota, Wisconsin

Date(s) Application is Due: Nov 1

Amount of Grant: Up to 175,000 USD

Contact: Beth Nelson; (612) 626-4436; fax (612) 626-3132; Schre002@umn.edu

Internet: http://www.northcentralsare.org/Grants/Our-Grant-Programs/Research-and-Education

Sponsor: Sustainable Agriculture Research and Education
University of Maryland
College Park, MD 20742-6715

SARE Research and Education Grants-Northeast Region 4178

The Research and Education Grants fund outcome-based projects offering research, education, and demonstration projects that benefit farmers and explore new sustainable farm practices. Projects should be directed toward research results that will translate quickly into farmer benefit. Farmers, extension staff, and other stakeholders should be involved in planning project content. Research and Education projects can cover a wide range of topics. Northeast SARE has recently funded cover crop trials, farm-based energy, new approaches to livestock health, nutrient management and recycling, agritourism, and new directions in integrated crop and pest management. Whatever the topic, the project outcome must focus on a measurable change in farm practices and management that improves sustainability. SARE funds have also been used to produce books, videos, webcasts, and other media tools to improve the understanding of sustainable techniques. Specific guidelines, instructions for the application process, and a list of previously funded projects are available at the website.

Requirements: Anyone can apply for the grant, but applicants should have experience doing agricultural research, preferably research in cooperation with commercial farmers and applied research that will translate quickly into improved practices on the ground. All applicants must demonstrate a firm grasp of current barriers and issues in sustainable agriculture. Proposals typically come from university and extension staff, the agricultural nonprofit sector, and from those working on research farms and experiment stations. There is no requirement that applicants have a specific affiliation, but there is a clear expectation that applicants are credible and experienced in the science of agriculture and agricultural sustainability and understand the importance of profitability, good stewardship, and sustainable communities.

Geographic Focus: Connecticut, Delaware, District of Columbia, Maine, Maryland, Massachusetts, New Hampshire, New Jersey, New York, Pennsylvania, Rhode Island, Vermont, West Virginia

Date(s) Application is Due: Oct 20

Contact: Judy Gifford; (410) 648-5753 or (443) 480-1853 cell; gifford@baybroadband.net

Internet: http://www.nesare.org/Grants/Get-a-Grant/Research-and-Education-Grant

Sponsor: Sustainable Agriculture Research and Education
University of Maryland
College Park, MD 20742-6715

SARE Research and Education Grants-Southern Region 4179

Research and Education projects are conducted by teams of interdisciplinary researchers specializing in sustainable agricultural systems. At the core of sustainable agriculture research is the application of a systems approach to research that aims to understand how a complex system functions as a whole, often beginning with a conceptual model. This approach recognizes that agricultural systems are complex and that interaction among components determines characteristics of the system. Research and Education Grants encourage that systems research approach in sustainable agriculture. Successful proposals include systems orientation and farmer or end-user involvement. Specific guidelines, application procedures, and samples of previous funded projects are available at the website.

Requirements: Research-based activities with an educational/outreach component that focus on developing sustainable agricultural systems or moving existing systems toward sustainable agriculture are eligible to apply. The following entities may apply for funding: teams of researchers from public and private institutions such as 1862 and 1890 land-grant universities, environmental agencies, historically African American universities, USDA-ARS labs, non-governmental organizations, and community based organizations. All Research and Education Grant applicants must propose systems oriented research that fits one of three grant categories: production research — focused on actual production

methods, this kind of research has made up the bulk of SARE's project portfolio and has developed techniques that have become common tools for farmers; postharvest/food systems research — these projects examine what happens past the farm gate such as in the markets, distribution systems and policy making. This category can serve as a funding path for social science researchers to also make a difference in our farm and food systems; and a combination of production and postharvest/food systems research — the ultimate in systems research connects what goes on in the ground with everything that happens after a crop is harvested, including adding value, marketing, infrastructure for processing and transportation, as well as policy making.
Geographic Focus: Alabama, Arkansas, Florida, Georgia, Kentucky, Louisiana, Mississippi, North Carolina, Oklahoma, Puerto Rico, South Carolina, Tennessee, Texas, Virginia, Virgin Islands (U.S.)
Amount of Grant: Up to 300,000 USD
Contact: Sandra Young; (770) 229-3212; fax (770) 412-4790; syoung@griffin.uga.edu
Internet: http://www.southernsare.org/Grants/Types-of-Grants/Research-Education-Grants
Sponsor: Sustainable Agriculture Research and Education
University of Maryland
College Park, MD 20742-6715

SARE Research and Education Grants-Western Region **4180**
The SARE Research and Education Grants involve scientists, producers, and others using interdisciplinary approaches to address issues related to sustaining agriculture. The program encourages projects that: recognize the importance of whole-farm or whole-ranch systems approaches; utilize multiple disciplines (project cooperators) to assist in the project; produce impacts and outcomes that can be identified, measured and disseminated; address weak links or information gaps in a whole-farm or whole-ranch system; include educational outreach to producers and educational products to state extension and the Natural Resource Conservation Service (NRCS) field offices; incorporate research and education efforts with on-farm/ranch projects; use both research methods and educational outreach components to accomplish the SARE goal of producer adoption on the ground. The project's primary force should always: conduct both research and extension/education outreach on the sustainable agriculture project with producers; produce scholarly and educational materials; conduct outreach and disseminate the findings to producers and agricultural professionals (extension and USDA-NRCS, sustainable and conventional agricultural organizations, nonprofits, and NGOs); and assist with producer adoption of the findings. Specific guidelines and information about state offices and contacts is available at the website.
Requirements: Applicants must show relevance to the national definition of sustainable agriculture; include a minimum of five separate producers for on-farm testing; provide educational outreach plans to producers and agricultural producers; and include quantifiable scholarly and educational products with distribution techniques to producers and extension/NRCS field office staffs.
Geographic Focus: Alaska, Arizona, California, Colorado, Guam, Hawaii, Idaho, Montana, Nevada, New Mexico, Northern Mariana Islands, Oregon, Utah, Washington, Wyoming, American Samoa, Micronesia
Contact: Robert Newhall; (435) 797-2183 or (435) 797-2257; bob.newhall@usu.edu
Internet: http://www.westernsare.org/Grants/Types-of-Grants
Sponsor: Sustainable Agriculture Research and Education
University of Maryland
College Park, MD 20742-6715

Sarkeys Foundation Grants **4181**
Governed by a dedicated Board of Trustees, the foundation that bears SJ Sarkeys' name is deeply committed to furthering his vision to improve the quality of life in Oklahoma. The Foundation provides grants to a diverse group of nonprofit organizations and institutions, almost all of which are located in Oklahoma. Major areas of foundation support include education, arts and cultural endeavors, scientific research, animal welfare, social service and human service needs, and cultural and humanitarian programs of regional significance. Grant proposals are considered at the April and October meetings of the board of trustees.
Requirements: Most organizations classified by the IRS as being a 501(c)3 that is not a private foundation and is headquartered and offering services in Oklahoma may apply. Preference is given to organizations that have been in operation at least 3 years. Organizations are required to submit a Letter of Inquiry to determine whether they meet the criteria and priorities for funding. Representatives are encouraged to speak with a program officer for more information and to ask any questions about the process. An organization may submit one request in a twelve month period.
Restrictions: The Sarkeys Foundation will not fund: local programs appropriately financed within the community; direct mail solicitations and annual campaigns; out of state institutions; hospitals; operating expenses; purchase of vehicles; grants to individuals; responsibility for permanent financing of a program; programs whose ultimate intent is to be profit making; start-up funding for new organizations; feasibility studies; grants which trigger expenditure responsibility by Sarkeys Foundation; direct support to government agencies; individual public or private elementary or secondary schools, unless they are serving the needs of a special population which are not being met elsewhere; and religious institutions and their subsidiaries.
Geographic Focus: Oklahoma
Date(s) Application is Due: Feb 3; Aug 1
Amount of Grant: Up to 50,000 USD
Contact: Susan Frantz; (405) 364-3703; susan@sarkeys.org or sarkeys@sarkeys.org
Internet: http://www.sarkeys.org/grant_guidelines.html
Sponsor: Sarkeys Foundation
530 East Main Street
Norman, OK 73071

SASS Birgit Baldwin Fellowship **4182**
The Society for the Advancement of Scandinavian Study (SASS) offers the Birgit Baldwin Fellowship in Scandinavian Studies, a grant to encourage the research and writing of dissertations for the Ph.D. in topics concerned with Scandinavian (Nordic) literature or film that can most effectively be pursued in the archives and libraries of the Nordic countries. The stipend is $20,000 payable in two installments for travel to and living expenses in the Nordic region during one academic year. Specific guidelines and a current application are available online. Applicants submit materials to sass.baldwin.fellowship@gmail.com.
Requirements: Applicants must be members in good standing of the Society for the Advancement of Scandinavian Study, both during the year in which they submit applications and in the fellowship year. After completion of the fellowship year, the Baldwin Fellow is invited to give a short report at the next annual meeting of SASS. Applicants should submit a dissertation proposal that includes research plans and evidence of competency (including linguistic competency) to carry them out, supported by letters from two scholars. The successful applicant will demonstrate the significance of the subject, the need to work in Scandinavia (the Nordic region), and the feasibility of accomplishing the objectives of the research plans.
Restrictions: Applicants must be earning their degrees in North American universities. The Fellowship recipient is required to enroll in an international medical insurance policy for the duration of their stay overseas. SASS will issue the first fellowship check only after receiving a copy of the medical insurance policy and approving it.
Geographic Focus: All States
Date(s) Application is Due: Nov 1
Amount of Grant: 20,000 USD
Contact: Mabel Williams; (801) 422-8326; mabelwilliams@hotmail.com
Internet: http://www.scandinavianstudy.org/site/index.php?option=com_content&view=article&id=57&Itemid=62
Sponsor: Society for the Advancement of Scandinavian Study
Brigham Young University
Provo, UT 84602-6702

SASS Einar and Eva Lund Haugen Memorial Scholarship **4183**
The Society for the Advancement of Scandinavian Study (SASS) offers the Einar and Eva Lund Haugen Memorial Scholarship. The scholarship supports outstanding graduate students whose doctoral dissertations examine a Scandinavian or Scandinavian-American topic in History and the Social Sciences. Scandinavia is defined as the five countries of the Nordic region: Norway, Sweden, Finland, Denmark, and Iceland. The grant can be used for all aspects of dissertation work. Preference will be given to those who need to conduct research in archives and libraries or to pursue fieldwork or excavation in places remote from their home institutions. The current application and specific guidelines, including a list of all materials required, are available at the website.
Requirements: Applicants must be enrolled in a doctoral program in North America. They must be members in good standing of the Society for the Advancement of Scandinavian Study, both during the year in which they submit applications and in the year of funding. A successful applicant must prove completion of course work and preliminary requirements for a doctoral degree before receiving the grant. After completion of the period funded by the scholarship, the recipient is invited to give a short report at the annual meeting of SASS.
Geographic Focus: All States, Denmark, Finland, Iceland, Norway, Sweden
Date(s) Application is Due: Jan 15
Amount of Grant: 5,000 USD
Contact: Mabel Williams; (801) 422-8326; mabelwilliams@hotmail.com
Internet: http://www.scandinavianstudy.org/site/index.php?option=com_content&view=article&id=88&Itemid=70
Sponsor: Society for the Advancement of Scandinavian Study
Brigham Young University
Provo, UT 84602-6702

Savoy Foundation Post-Doctoral and Clinical Research Fellowships **4184**
Fellowships of the Savoy Foundation are available to Canadian researchers and to foreign nationals or for projects conducted in Canada. Post-doctoral and clinical research fellowships of $30,000 will be awarded to scientists or medical specialists (Ph.D. or M.D.) wishing to carry out a full time research project in the field of epilepsy. The grant will be for one (1) year, but will be renewable, exceptionally, once upon request. As a rule, the modalities of attribution and amounts awarded will be similar to those of governmental funding agencies (MRC or FRSQ). The fellow with the highest mark will received the prestigious Steriade-Savoy award of an additional $1,500.
Restrictions: Fellows funded by the Savoy Foundation must not cumulate two major awards.
Geographic Focus: All States, Canada
Date(s) Application is Due: Jan 15
Amount of Grant: 30,000 USD
Contact: Coordinator; (450) 358-9779; fax (450) 346-1045; epilepsy@savoy-foundation.ca
Internet: http://www.savoy-foundation.ca/eng/mission/defaut.htm
Sponsor: Savoy Foundation
230 Foch Street
St Jean Sur Richelieu, QC J3B 6B2 Canada

Savoy Foundation Research Grants **4185**
A small number of research grants are available each year to clinicians and/or established scientists working on epilepsy or related subjects. These grants will be attributed for the following purposes: launching of a new research project, preliminary studies in preparation of a more substantial request to be addressed to another agency, continuation or completion of a project, sharing of expenses, or a limited project not requiring funding in excess of the limit. As a rule, these grants are not renewable. Contact the foundation for application forms.

Requirements: Research grants are available to Canadian citizens or for projects conducted in Canada.
Geographic Focus: All States, Canada
Date(s) Application is Due: Jan 15
Amount of Grant: 25,000 USD
Contact: Administrator; (450) 358-9779; fax (450) 346-1045; epilepsy@savoy-foundation.ca
Internet: http://www.savoy-foundation.ca/eng/mission/defaut.htm
Sponsor: Savoy Foundation
230 Foch Street
St Jean Sur Richelieu, QC J3B 6B2 Canada

Savoy Foundation Studentships **4186**
Studentships of the Savoy Foundation are available to Canadian researchers and to foreign nationals or for projects conducted in Canada. Studentships will be awarded to meritorious applicants wishing to acquire training and pursue research in a biomedical discipline, the health sciences or social sciences related to epilepsy. These studentships last one year, but will be renewable three times, upon request, for a maximum duration of four years. The stipend will be $15,000 per year. The student with the highest mark will receive the prestigious Van Gelder-Savoy award of an additional $1,500. An annual sum of $1,000 will be allocated to the laboratory or institution as additional support for the research project.
Requirements: To be eligible, the candidate must have a good university record (B. Sc., M.D. or equivalent diploma) and have ensured that a qualified researcher affiliated to a university and/or hospital, will supervise his/her work. Concomitant registration in a graduate program (M. Sc. or Ph. D.) is encouraged.
Restrictions: Students funded by Savoy must not cumulate two studentship awards.
Geographic Focus: All States, Canada
Date(s) Application is Due: Jan 15
Contact: Caroline Savoy, (450) 358-9779; fax (450) 346-1045; epileps@savoy-foundation.ca
Internet: http://www.savoy-foundation.ca/eng/mission/defaut.htm
Sponsor: Savoy Foundation
230 Foch Street
St Jean Sur Richelieu, QC J3B 6B2 Canada

Schomburg Center Scholars-in-Residence Fellowships **4187**
Schomburg Center Scholars-in-Residence Fellowships assist those scholars and professionals whose research in the black experience can benefit from extended access to the Center's resources. Fellowships allow recipients to spend six months or a year in residence with access to resources at the Schomburg Center and other centers of The New York Public Library. The program encourages research and writing on black history and culture, facilitates interaction among participating scholars, and provides widespread dissemination of findings through lectures, publications, and colloquia and seminars. It encompasses projects in African, Afro-American, and Afro-Caribbean history and culture. Funding support for the program has been provided by the Ford Foundation, the Rockefeller Foundation, the Aaron Diamond Foundation, the Irene Diamond Foundation, the Samuel I. Newhouse Foundation, the Andrew W. Mellon Foundation and the National Endowment for the Humanities. Fellowships are awarded for continuous periods of six or twelve months at the Schomburg Center with maximum stipends of $30,000 for six months and $60,000 for twelve months. Applications are available on the website.
Requirements: Fellowships are open to scholars studying the history, literature, and culture of peoples of African descent from a humanistic perspective and to professionals in fields related to the Schomburg Center's collections and program activities. Projects in the social sciences, science and technology, psychology, education, and religion are eligible if they utilize a humanistic approach and contribute to humanistic knowledge.
Restrictions: Creative writing (works of poetry and fiction), projects that result in a performance, and dissertation research are not eligible.
Geographic Focus: All States
Date(s) Application is Due: Dec 1
Contact: Fellowship Coordinator; (212) 491-2200
Internet: http://www.nypl.org/locations/tid/64/node/131
Sponsor: Schomburg Center for Research in Black Culture
515 Malcolm X Boulevard
New York, NY 10037-1801

Schramm Foundation Grants **4188**
The foundation awards grants to Colorado nonprofit organizations in its areas of interest, including arts and culture, civic affairs, community development, education (elementary, secondary, and higher), health care, housing, humanities, medical research, science, social services delivery, technology, women's issues, and youth. Types of support include building construction/renovation, continuing support, equipment acquisition, general operating support, matching/challenge grants, program development, and scholarship funds. Applications are accepted from July 1 through August 31 (postmarked).
Requirements: Colorado 501(c)3 nonprofit organizations are eligible. Preference is given to requests from the Denver area. Applications must clearly express the reason(s) for the request, attach financial statements and copy of exemption letter.
Restrictions: Grants do not support advertising, advocacy organizations, individuals, international organizations, political organizations, religious organizations, school districts, special events, or veterans organizations.
Geographic Focus: Colorado
Date(s) Application is Due: Aug 31
Contact: Gary Kring, President; (303) 861-8291
Sponsor: Schramm Foundation
800 Grant Street, Suite 330
Denver, CO 80203-2944

Scientific Communities Initiative Grants **4189**
The coalition is awarding grants to digital ventures offering information in the areas of science, technology, and medicine (STM). The purpose of the grants is to stimulate and accelerate the creation of new university-based information communities serving users in key STM fields. Proposed projects should offer strategic alternatives to inefficiencies in the current, traditional scientific communication process. Projects should provide increased responsiveness to the needs of scientists—as authors and information users—and to libraries by offering an effective new electronic venue for publication of research; aggregating content/links needed by targeted users in sufficient breadth and mass; addressing important objectives of universities and colleges; delivering convenient, efficient access to this information in a Web environment optimized to the needs of scientists in the target discipline; and fostering a more open and competitive marketplace in which academe exercises an increased role in assuring broad dissemination of research. Partnerships and collaborations among universities and professional societies are encouraged. Letters of intent must meet the listed application deadline; full applications will be requested.
Geographic Focus: All States
Amount of Grant: 100,000-500,000 USD
Contact: Heather Joseph, (202) 296-2296; sparc@arl.org
Internet: http://www.arl.org/sparc/core/index.asp?page=f17
Sponsor: Scholarly Publishing and Academic Resources Coalition
21 Dupont Circle, Suite 800
Washington, DC 20036

Scleroderma Foundation Established Investigator Grants **4190**
The Scleroderma Foundation is seeking applications from promising established investigators both inside and outside the field of SSc who wish to propose pilot studies to obtain preliminary data dealing with a highly innovative and/or highly relevant theme related to SSc. Applications along with the instructions are available at the SF website.
Requirements: Applications may be submitted by domestic and foreign nonprofit organizations, public and private, such as universities, colleges, hospitals, and laboratories. Applicants must have a doctorate degree in medicine, osteopathy, veterinary medicine, or one of the sciences; must have completed a postdoctoral fellowship by the grant award date; and have been principal investigator on grants from the foundation or other national, private, or government agencies. Applicants may request up to $75,000. Up to 8% of the direct costs per year of award may be set aside as indirect costs. The indirect cost amount will be subtracted from the yearly total [up to $75,000) of the award) per year for up to two years (total for 2 years not to exceed $150,000).
Restrictions: Funding are not renewable. Before completion of this grant, investigators are encouraged to seek more substantial continuing support for research through other grant mechanisms through private or government agencies. Replacement of the principal investigator on this award is not permitted. There will be no routine escalation for future years. Awards are contingent on the availability of funds and the receipt of sufficiently meritorious applications meeting the stated eligibility requirements.
Geographic Focus: All States, All Countries
Date(s) Application is Due: Sep 15
Amount of Grant: Up to 75,000 USD
Contact: Tracey Sperry;)978) 463-5843; fax (978) 463-5809; tsperry@scleroderma.org
Internet: http://www.scleroderma.org/site/PageServer?pagename=prof_research_types#.UCz0vXAZzbg
Sponsor: Scleroderma Foundation
300 Rosewood Drive, Suite 105
Danvers, MA 01923

Scleroderma Foundation New Investigator Grants **4191**
The Scleroderma Foundation is seeking applications from promising new investigators who hold faculty or equivalent positions and who wish to pursue a career in research related to systemic sclerosis (scleroderma). This grant will support promising research that is likely to lead to individual research project grants. Awards are contingent on the availability of funds and the receipt of sufficiently meritorious applications meeting the state eligibility requirements. The application is available at the SF website.
Requirements: Applications may be submitted by domestic non-profit organizations, public and private such as universities, colleges, hospitals and laboratories. Applicants must have a doctoral degree in Medicine, Osteopathy, Veterinary Medicine or one of the sciences and must have completed a postdoctoral fellowship by the grant award date. Applicants may request up to $50,000. Up to 8% of the direct costs per year of award may be set aside as indirect costs. The indirect costs will be subtracted from the yearly total (up to $50,000 of the award) per year for up to three years (total for three years not to exceed $150,000).
Restrictions: Foreign organizations and institutions are not eligible. These new investigator grants may not be used to support thesis or dissertation research or fellowship training. Applicants who have been a Principal Investigator on grants from the Scleroderma Foundation or other national, private or government agencies other than fellowship grants are not eligible for this award. Awards are not renewable. Before completion of this grant, investigators are encouraged to seek continuing support for research through other grant mechanisms through private or government agencies. Replacement of the Principal Investigator on this award is not permitted. There will be no routine escalation for future years.
Geographic Focus: All States
Date(s) Application is Due: Sep 15
Contact: Tracey Sperry;)978) 463-5843; fax (978) 463-5809; tsperry@scleroderma.org
Internet: http://www.scleroderma.org/site/PageServer?pagename=prof_research_types#.UCzz-3AZzbg
Sponsor: Scleroderma Foundation
300 Rosewood Drive, Suite 105
Danvers, MA 01923

Searle Freedom Trust Grants 4192

The Searle Freedom Trust was founded by Daniel C. Searle in 1998 to foster research and education on public policy issues that affect individual freedom and economic liberty. Through its grantmaking, the foundation seeks to develop solutions to the country's most important and challenging domestic policy issues. Recognizing that good policymaking relies on the availability of high-quality research, the foundation invests primarily in scholarship that results in the publication of books, journal articles, and policy papers. Funding is typically provided in the form of research grants, fellowships, and other types of targeted project support. The Searle Freedom Trust also provides funding for public interest litigation and supports outreach to the public through a variety of forums, including sponsorship of research conferences and seminars, film and journalism projects, and new media initiatives. With funding help from the Trust, university and think tank scholars investigate a wide range of issues such as tax and budget policy; cost-benefit analysis of regulatory practices and proposals; the workings of the legal system; environmental policy; social welfare reform; and K-12 and higher education policy. The Board meets three times a year. Proposals should be submitted through the Trust's online system.

Requirements: In addition to the application, organizations must submit a project description; the project budget; the organization's annual budget; any supporting materials; a list of key supporters; and evidence of the organization's non-profit status.

Restrictions: The Trust will typically not support building projects; individuals; for profit organizations; organizations primarily depend on government funds; capital campaigns; or endowments. Funding is also not available to support unspecified overhead costs associated with university research projects.

Geographic Focus: All States

Contact: Kimberly Dennis; (202) 375-7820; kdennis@searlefreedomtrust.org

Internet: http://www.searlefreedomtrust.org/applicationguidelines.html

Sponsor: Searle Freedom Trust

1055 Thomas Jefferson Street, NW, Suite L26

Washington, DC 20007

Searle Scholars Grants 4193

The aim of the Searle Scholars Program (administered by the Kinship Foundation) is to identify junior faculty members who have demonstrated a potential for doing innovative research in medicine, chemistry, and the biological sciences. Approximately 15 scholars are selected each year to receive the award. Each scholar receives $100,000 per year to support their research for three years. Submissions are restricted to selected candidates from invited institutions only and are not open to individuals who have not been selected by their institution to participate in the competition. The list of invited institutions can be found on the Participating Institutions page on the Program's website. Individuals from an invited institution who are considering submitting an application should consult with the individual at their institution who handles outside awards to determine how many applications their institution has been allowed to submit, as well as how selections will be made within their institution.

Requirements: Applicants will be expected to be pursuing independent research careers in biochemistry, cell biology, genetics, immunology, neuroscience, pharmacology, related areas in chemistry, medicine, and the biological sciences. Candidates should be in their first or second year at the assistant professor level. This appointment must be a tenure-track position and must be in an academic department of an invited, degree-granting institution. Potential applicants whose institutions do not have tenure-track appointments should consult with the scientific director of the program prior to preparing an application.

Restrictions: Applications from any institution not invited to participate will not be considered. No more than two applications will be accepted from any one institution.

Geographic Focus: All States

Date(s) Application is Due: Sep 28

Contact: Jennifer Nesler, Administrative and Grants Coordinator; (312) 803-6200; fax (312) 803-6201; jennifer.nesler@kinshipfoundation.org

Internet: http://www.searlescholars.net/go.php?id=5

Sponsor: Searle Scholars Program

303 West Madison Street, Suite 1800

Chicago, IL 60606

Seattle Foundation Medical Funds Grants 4194

The Seattle Foundation administers the Medical Funds program to support medical research of potential benefit to the community and to address specific healthcare needs. In the area of medical research, grants are available in the fields of cancer, cardio-pulmonary disease, multiple sclerosis and diabetes. Special consideration will be given to research projects related to immunology, oncology, neurology, molecular biology and genetics. Grants will also be given to requests for specific equipment. Grants are also available to organizations that are administering projects addressing the healthcare needs of low-income children. In this area, grants will be given to requests for specific equipment and support for capital campaigns or facility renovation projects. A total amount of $200,000 is available for grants annually in this entire program. Typically no more than $50,000 is disbursed to any one organization.

Requirements: Preference will be given to organizations/institutions with the capacity to disseminate research findings and who receive regular support from other recognized funding sources (including the federal government).

Restrictions: These funds are not to be used for patient care. Research activities can include patient care, but the primary purpose of these grants is to support the purchase of equipment used in medical research.

Geographic Focus: Washington

Contact: Ceil Erickson; (206) 515-2131 or (206) 515-2109; fax (206) 622-7673; grantmaking@seattlefoundation.org or c.erickson@seattlefoundation.org

Internet: http://www.seattlefoundation.org/nonprofits/medicalfunds/Pages/MedicalFunds.aspx

Sponsor: Seattle Foundation

1200 Fifth Avenue, Suite 1300

Seattle, WA 98101-3151

Seaver Institute Grants 4195

The Institute's focuses its giving in four primary areas: scientific and medical research, education, public affairs, and the cultural arts. Seed grants are awarded to nonprofit organizations for projects that offer the potential for a significant advancement in their fields. The institute seeks proposals that incorporate state-of-the-art scientific research, exceptional educational development, unique approaches to the creative arts, and the ramifications of shifting societal realities. There are no application deadlines.

Requirements: U.S. 501(c)3 nonprofits are eligible. Interested applicants should submit an inquiry letter of two or three pages and include the goals of the project, a preliminary budget, and the qualifications of the organization and principal investigator. If interested, the institute will invite a full proposal.

Restrictions: Grants are not made for construction, endowments, ongoing projects, operating budgets, deficits, or individual scholarships.

Geographic Focus: All States

Amount of Grant: 1,000 USD

Contact: Victoria Seaver Dean, President; (310) 979-0298; vsd@theseaverinstitute.org

Sponsor: Seaver Institute

12400 Wilshire Boulevard

Los Angeles, CA 90025

SeaWorld and Busch Gardens Conservation Fund Grants 4196

The fund's grants support wildlife conservation projects conducted by recognized charitable organizations and noted scientists throughout the world. Grants support projects in the areas of species research, terrestrial; species research, aquatic; habitat protection; animal rescue and rehabilitation; and conservation education. Grants are awarded on a semiannual basis. Both guidelines and application are available online.

Requirements: Charitable organizations and scientists worldwide are eligible.

Geographic Focus: All States

Date(s) Application is Due: Jun 1; Dec 1

Amount of Grant: 5,000-25,000 USD

Contact: Fred Jacobs, Senior Director; (314) 613-6077 or (877) 792-4332; fax (314) 613-6079; fred.jacobs@anheuser-busch.com or mailbox@swbgfund.org

Internet: http://www.seaworld.org/conservation-matters/conservation-fund/index.htm

Sponsor: SeaWorld and Busch Gardens Conservation Fund

423 Lynch Street, Building 260-2

Saint Louis, MO 63118

Self Foundation Grants 4197

The Self Family Foundation primarily serves the Greenwood, South Carolina and surrounding counties. The Self Family Foundation will consider requests from other regions of South Carolina, if they have statewide impact. The Foundation supports non-profit organizations in the following areas of interest: Education-enhancing children's school readiness and ability to achieve educational goals with success; working together with schools, families, youth-serving organizations and other community organizations to support the intellectual and social development of young people; Arts, Culture & History-increasing awareness of and access to activities that broaden the horizons and knowledge of all residents, within Greenwood, South Carolina; Civic & Community-improving neighborhood resources and community attitudes and services, emphasizing the utilization of existing facilities, skills and talents to create a stronger, more unified community through collaboration with organizations, businesses, public institutions and residents; Health and Human Services-promoting individual and community wellness, prevention and literacy.

Requirements: IRS 501(c)3 tax-exempt or 509(a) nonprofit organizations in South Carolina are eligible. Letters of Inquiry or requests for information may be emailed to: application@selffoundation.org or mailed to Foundation. Contact Foundation Program Officer by email or phone prior to submitting proposal. All applicants must have the financial potential to sustain the project on a continuing basis after any funding is approved.

Restrictions: No grants to individuals and no loans available.

Geographic Focus: South Carolina

Date(s) Application is Due: Feb 15; May 15; Aug 15; Nov 15

Contact: Grants Administrator; (864) 953-2441 or (864) 941-4036; fax (864) 941- 4091; info@selffoundation.org

Internet: http://www.selffoundation.org/grantmaking4.htm

Sponsor: Self Family Foundation

P.O. Box 1017

Greenwood, SC 29648-1017

Seneca Nation Higher Education Grants 4198

This program seeks research information that can be used to revitalize and stabilize the Seneca Language. This research will examine the following influences on second-language learning: curriculum, instruction, standards, assessment, school and classroom settings, and educational leadership and teacher professional development, as well as, family and community collaboration with schools.

Requirements: Enrolled tribal members of the Seneca Nation of Indians may apply.

Geographic Focus: All States

Contact: Carrie Haynes, Director; (716) 532-3341; fax (716) 532-3269

Internet: http://www.sni.org/eae.html

Sponsor: Seneca Nation of Indians

12861 Route 438

Irving, NY 14081

Sensient Technologies Foundation Grants 4199

The foundation supports organizations involved with arts and culture; children/youth, services; community/economic development; education; education, fund raising/fund distribution; education, research; family services; food services; general charitable giving; health care; higher education; homeless, human services; hospitals (general); human services; medical research, institute; mental health/crisis services; nutrition; performing arts; residential/custodial care, hospices; United Ways and Federated Giving Programs; Urban/community development; and voluntarism promotion. The foundation's types of support include: annual campaigns; capital campaigns; emergency funds; endowments; general operating support; matching/challenge support; program development; research; and scholarship funds.
Requirements: An application form is not required. Submit a letter of inquiry as an initial approach. The advisory board meets in January and June/July or as needed, with deadline for proposal review one month prior to board meetings.
Restrictions: The foundation gives primarily to its areas of interest in Indianapolis, Indiana; St. Louis, Missouri; and Milwaukee, Wisconsin. It does not support sectarian religious, fraternal, or partisan political organizations. It does not give grants to individuals.
Geographic Focus: Indiana, Missouri, Wisconsin
Contact: Douglas L. Arnold, (414) 271-6755; fax (414) 347-4783
Sponsor: Sensient Technologies Foundation
777 E Wisconsin Avenue
Milwaukee, WI 53202-5304

Sexually Transmitted Diseases Postdoctoral Fellowships 4200

The ASHA Research Fund fellowships program is intended to expand the number of trained scientists working in the field of sexually transmissible diseases research by providing postdoctoral candidates with the opportunity to spend two years in a challenging, stimulating, and instructive STD research environment under the sponsorship of an established researcher. Applications for research training relative to any aspect of the sexually transmissible diseases are welcome. ASHA has supported projects in chancroid, human papilloma virus (HPV), group B streptococcal infection, herpes, gonorrhea, chlamydia, HIV, pelvic inflammatory disease (PID), and syphilis. In funding HIV-related research, ASHA gives priority to researchers exploring the relationship between HIV and other STDs. Although applications covering fellowship training in basic investigations are not discouraged, it is the intent of this program to focus on research of short- or medium-term duration. Requests for applications are usually issued in May of odd-numbered years, with an application deadline of October 15, and the award being given the following spring for a fellowship period commencing on July 1.
Requirements: Applications are requested from MDs, PhDs, DSc who will be conducting two-year research projects under the sponsorship of established STD investigators. Applicants are required to study at established U.S. institutions.
Restrictions: ASHA fellowships may not be used in support of faculty positions unless approved by ASHA's management. Current ASHA fellows are not eligible to apply for other ASHA fellowships.
Geographic Focus: All States
Date(s) Application is Due: Nov 15
Amount of Grant: 35,500-36,000 USD
Contact: Grant Administrator; (919) 361-8400; fax (919) 361-8425
Internet: http://www.ashastd.org/programs/arf.html
Sponsor: American Social Health Association
P.O. Box 13827
Research Triangle Park, NC 27709

SFI Principal Investigator Program Grant 4201

Science Foundation Ireland advances national scientific progress by awarding grants for research on a competitive basis in those fields of science, engineering, software engineering, computer science, and mathematics that underpin biotechnology and information and communications technology. Grants support the research program costs of the principal investigator and research group. Eligible costs include relevant research expenses, including personnel, equipment, consumables, collaboration, and travel; contributions to salaries for staff hired specifically to carry out the research program; and access to necessary special facilities, such as synchrotron beam lines or high-speed computing. Grants may be held for five years. Application and guidelines are available online.
Requirements: All applicants will be expected to have a proven record of internationally recognized independent research accomplishments and to have at least three years of independent research experience beyond the PhD or equivalent.
Geographic Focus: All States
Contact: Program Contact; 353 (0)1 607 3285; fax 353 (0)1 607 3201; info-@-sfi.ie
Internet: http://www.sfi.ie/content/content.asp?section_id=510&language_id=1
Sponsor: Science Foundation Ireland
Wilton Park House, Wilton Place
Dublin 2, Ireland

SfN Chapters Graduate Student Travel Awards 4202

SfN's Chapters Graduate Student Travel Award honors outstanding graduate students nominated by their local SfN chapter. The Award provides a stipend to help defray travel expenses and provide complimentary meeting registration fees. The application is available on the SfN website, along with a detailed list of attachments required.
Requirements: Each SfN chapter may submit one nomination. Nominees must be the first author of an abstract to be presented at the annual meeting. The nominee and his or her advisor should be members of the SfN. Candidates must be graduate students who have advanced to candidacy for a Ph.D. Only one applicant from a given laboratory may be considered by a chapter, and a candidate may not be nominated two years in a row. Candidates may not be recipients of another SfN travel award in the same year.
Geographic Focus: All States, Puerto Rico, Argentina, Australia, Brazil, Canada, China, Denmark, India, Israel, Malaysia, Mexico, New Zealand, Nigeria, Norway, Singapore, South Korea, Spain, Switzerland, Turkey, Ukraine, United Kingdom
Date(s) Application is Due: Jun 1
Contact: James W. Geddes; (859) 323-5135; fax (859) 257-5737; jgeddes@uky.edu
Internet: http://www.sfn.org/index.aspx?pagename=TravelAwards_gradtravel
Sponsor: Society for Neuroscience
1121 14th Street, NW, Suite 1010
Washington, DC 20005

SfN Chapters Postdoctoral Trainee Travel Awards 4203

The Society for Neuroscience (SfN) Postdoctoral Trainee Travel Award honors outstanding postdoctoral trainees nominated by local SfN chapters. Awardees are chosen based on the scientific merit of their abstract, evidence of outreach activities, and letters of nomination from the principal investigator or mentor and the local nominating chapter. Twenty-five awardees are chosen. Recipients attend a special reception held in their honor at the annual meeting where they will also receive their award checks.
Requirements: Each chapter may submit one nomination to the SfN Membership and Chapters Committee. The nominee must meet the following criteria: the nominee must be the first author of an abstract to be presented at the annual meeting; the nominee and his or her advisor/PI must be SfN members and show proof of their membership; and the nominee must have a Ph.D. Only one applicant from a given laboratory may be considered by a chapter in any given calendar year. A nominee may not be nominated for an award two years in a row, or a recipient of another SfN travel award in the same year. Nomination packets must include the SfN online application form; a copy of the abstract submitted for the SfN annual meeting; a one page recommendation letter from the nominee's mentor/advisor that discusses the nominee's contributions to neuroscience research; and the nominee's CV. The chapter's representative must submit a one page letter of nomination which describes the chapter's selection process and the nominee's involvement in and contributions to their chapter.
Geographic Focus: All States, Puerto Rico, Argentina, Australia, Brazil, Canada, China, Denmark, India, Israel, Malaysia, Mexico, New Zealand, Nigeria, Norway, Singapore, South Korea, Spain, Switzerland, Turkey, Ukraine, United Kingdom
Date(s) Application is Due: Jun 1
Contact: James W. Geddes; (859) 323-5135; fax (859) 257-5737; jgeddes@uky.edu
Internet: http://www.sfn.org/index.aspx?pagename=TravelAwards_postdoctravel
Sponsor: Society for Neuroscience
1121 14th Street, NW, Suite 1010
Washington, DC 20005

SfN Donald B. Lindsley Prize in Behavioral Neuroscience 4204

The Donald B. Lindsley Prize in Behavioral Neuroscience recognizes a young neuroscientist's outstanding Ph.D. thesis in the general area of behavioral neuroscience (defined as neuroscience research involving behavioral variables or oriented toward the solution of behavioral problems). The recipient receives a $2,500 prize; and air or ground transportation, two nights hotel, and complimentary registration for the Society's annual meeting. The Society's President presents the prize at a lecture at the annual meeting. All nomination materials must be submitted through the Society's award nomination link at the Society's website.
Requirements: Nominees must have a thesis submitted and approved between January 1 and December 31 of the preceding calendar year. Nominations must be made or endorsed by SFN members, although nominees do not need to be SFN members.
Restrictions: Self nomination is not permitted. No person may nominate more than one candidate. Current Society officers and councilors are ineligible for nomination.
Geographic Focus: All States, Puerto Rico, Argentina, Australia, Brazil, Canada, China, Denmark, India, Israel, Malaysia, Mexico, New Zealand, Nigeria, Norway, Singapore, South Korea, Spain, Switzerland, Turkey, Ukraine, United Kingdom
Date(s) Application is Due: Jun 1
Contact: Robert Desimone; (617) 324-0141; fax (627) 452-4119; desimone@mit.edu
Internet: http://www.sfn.org/index.aspx?pagename=FellowshipAndAwards_lindsley
Sponsor: Society for Neuroscience
1121 14th Street, NW, Suite 1010
Washington, DC 20005

SfN Graduate Student Travel Awards 4205

The Society for Neuroscience offers Graduate Student Travel Awards, administered by the Professional Development Committee (PDC), to honor outstanding graduate researchers. The Award includes complimentary registration and a stipend to offset travel expenses for the SfN annual meeting. Awardees are chosen on the basis of the scientific merit of their abstracts, a letter of nomination from a principal investigator or advisor, research and career goals, and essay. Award checks are presented at the SfN's annual meeting. The application is located at the SfN website.
Requirements: Applicants must be the first author of an abstract to be presented at the annual meeting; members of the SfN (along with their advisor); a graduate student who has advanced to doctoral candidacy; and studying with a laboratory that has not been a recipient of the travel award the previous year. Application packages must include: the completed application form; the applicant's abstract submission confirmation form for the SfN annual meeting; curriculum vitae including education, honors/awards, publications (abstracts and manuscripts) in the required format; one-page letter of recommendation from the applicant's advisor; essay by the applicant about a woman scientist who has inspired him/her (no more than 500 words); and a description of the applicant's research and career goals (no more than 250 words).
Geographic Focus: All States, Puerto Rico, Argentina, Australia, Brazil, Canada, China, Denmark, India, Israel, Malaysia, Mexico, New Zealand, Nigeria, Norway, Singapore, South Korea, Spain, Switzerland, Turkey, Ukraine, United Kingdom

Contact: Michael Neal Lehman; (601) 984-1607; fax (601) 984-1655; mlehman@umc.edu
Internet: http://www.sfn.org/index.aspx?pagename=WomeninNeuroscience_cwintravelgrad
Sponsor: Society for Neuroscience
1121 14th Street, NW, Suite 1010
Washington, DC 20005

SfN Jacob P. Waletzky Award **4206**
Each year, the Society for Neuroscience (SfN) honors some of the best research and achievements by neuroscientists around the world with more than $500,000 in prizes and other compensation, such as complimentary travel and registration for SfN's annual meeting. These awards recognize scientists at all stages of their careers for a variety of activities, including research that expands knowledge of the brain and its functions, outreach programs that educate the public about neuroscience, and mentoring efforts that cultivate achievement for the next generation of neuroscientists. The Waletzky Award is presented to a scientist who has conducted research or plans to conduct research in the area of substance abuse and the brain and nervous system. Applicants may contact the Society for current information about the Award.
Requirements: Nominations must be made or endorsed by a Society member, although nominees do not need to be Society members. All materials are submitted using the Society's award nomination process located at the Society's website.
Geographic Focus: All States, Puerto Rico, Argentina, Australia, Brazil, Canada, Denmark, India, Israel, Malaysia, Mexico, New Zealand, Nigeria, Norway, Singapore, South Korea, Spain, Switzerland, Turkey, Ukraine, United Kingdom
Date(s) Application is Due: Jun 22
Amount of Grant: 25,000 USD
Contact: Coordinator; (202) 962-4000; fax (202) 962-4941; awards@sfn.org
Internet: http://www.sfn.org/index.aspx?pagename=awards_all
Sponsor: Society for Neuroscience
1121 14th Street, NW, Suite 1010
Washington, DC 20005

SfN Janett Rosenberg Trubatch Career Development Award **4207**
The Trubatch Award recognizes individuals who have demonstrated originality and creativity in research. The Award also promotes success during academic transitions prior to tenure. The recipient receives a $2,000 award and complimentary registration to the Society's annual meeting. Two recipients are selected each year. The Awards are presented during the Society's annual meeting. All materials must be submitted using the Society's online nomination site.
Requirements: Nominees with a Ph.D. or M.D. (or equivalent) are eligible until they have completed ten years of full-time work. Candidates working in non-academic environments are eligible if their work is published and meets academic standards. Nominees must not have received university tenure by the nomination deadline. Nominations must be made or endorsed by a Society member, although nominees do not need to be Society members.
Restrictions: Self-nomination is not permitted. No person may nominate more than one candidate. Current Society officers and councilors are ineligible for nomination.
Geographic Focus: All States, Puerto Rico, Argentina, Australia, Brazil, Canada, China, Denmark, India, Israel, Malaysia, Mexico, New Zealand, Nigeria, Norway, Singapore, South Korea, Spain, Switzerland, Turkey, Ukraine, United Kingdom
Date(s) Application is Due: Jun 1
Amount of Grant: 2,000 USD
Contact: Leslie P. Tolbert; (520) 621-3513; fax (520) 621-7507; tolbert@email.arizona.edu
Internet: http://www.sfn.org/index.aspx?pagename=WomeninNeuroscience_cwincda
Sponsor: Society for Neuroscience
1121 14th Street, NW, Suite 1010
Washington, DC 20005

SfN Louise Hanson Marshall Specific Recognition Award **4208**
The Louise Hanson Marshall Special Recognition Award recognizes an individual who has significantly promoted the professional development of women in neuroscience through teaching, organizational leadership, public advocacy, or other efforts that are not necessarily related to research. The recipient receives the Marshall Award, along with complimentary transportation, hotel, and registration for the SfN annual meeting. The Award is presented at the Celebration of Women in Neuroscience luncheon held during the SfN annual meeting. All nominating materials should be submitted using SfN's award nomination site.
Requirements: Nominees must have served the professional through SfN and/or related organizations. They must also have attained recognition at national or international levels as a scientist, educator, businessperson, or administrator in neuroscience. In addition, nominees must have demonstrated a high degree of imagination, innovation, and initiative in the pursuit of neuroscience, and facilitated mentoring or the advancement of young women in neuroscience. Nominees may only be made by SfN members, but the nominees do not need to be SfN members.
Restrictions: Self nomination is not permitted. No one may nominate more than one candidate.
Geographic Focus: All States, Puerto Rico, Argentina, Australia, Brazil, Canada, China, Denmark, India, Israel, Malaysia, Mexico, New Zealand, Nigeria, Norway, Singapore, South Korea, Spain, Switzerland, Turkey, Ukraine, United Kingdom
Date(s) Application is Due: Jun 8
Contact: Leslie P. Tolbert; (520) 621-3513; fax (520) 621-7507; tolbert@email.arizona.edu
Internet: http://www.sfn.org/index.aspx?pagename=WomeninNeuroscience_cwinsra
Sponsor: Society for Neuroscience
1121 14th Street, NW, Suite 1010
Washington, DC 20005

SfN Mika Salpeter Lifetime Achievement Award **4209**
The Mika Salpeter Lifetime Achievement Award recognizes an individual with outstanding career achievements in neuroscience who has also significantly promoted the professional advancement of women in neuroscience. The recipient receives a $5,000 award, in addition to complimentary air or ground transportation, hotel, and registration to the SfN annual meeting. The nominee must have exhibited a dedication to facilitating the mentoring and entry of young women into neuroscience or to the advancement of women in neuroscience; sustained exceptional achievements in neuroscience as evidenced by publications, inventions, and/or awards; served the profession through SfN and/or related organizations; recognized at national or international levels as a scientist, educator, businessperson, or administrator in neuroscience; and demonstrated a high degree of imagination, innovation, and initiative in the pursuit of neuroscience.
Requirements: Nominations must be made or endorsed by a Society member, although nominees do not need to be Society members. All nomination materials are submitted using SfN's award nomination site.
Restrictions: Self nomination is not permitted, and no person may nominate more than one candidate. Current SfN officers and councilors are ineligible for nomination.
Geographic Focus: All States, Puerto Rico, Argentina, Australia, Brazil, Canada, China, Denmark, India, Israel, Malaysia, Mexico, New Zealand, Nigeria, Norway, Singapore, South Korea, Spain, Switzerland, Turkey, Ukraine, United Kingdom
Date(s) Application is Due: Jun 8
Amount of Grant: 5,000 USD
Contact: Leslie P. Tolbert; (520) 621-3513; fax (520) 621-7507; tolbert@email.arizona.edu
Internet: http://www.sfn.org/index.aspx?pagename=WomeninNeuroscience_cwinlaa
Sponsor: Society for Neuroscience
1121 14th Street, NW, Suite 1010
Washington, DC 20005

SfN Postdoctoral Fellow Travel Awards **4210**
The Society for Neuroscience offers Postdoctoral Fellow Travel Awards, administered by the Professional Development Committee (PDC), to honor outstanding postdoctoral researchers. The Award includes complimentary registration and a stipend to offset travel expenses for the SfN annual meeting. Awardees are chosen on the basis of the scientific merit of their abstracts, a letter of nomination from a principal investigator or advisor, research and career goals, and essay. Award checks will be presented at the SfN's annual meeting.
Requirements: Applicants must be the first author of an abstract to be presented at the annual meeting, and a postdoctoral fellow. The applicant and his or her advisor must be members of SfN. The applicant should also be studying at a laboratory that has not been the recipient of the travel award in the previous year. Application packages must include: completed application form; the applicant's abstract submission confirmation form for the SfN annual meeting; a one-page letter of recommendation from the applicant's advisor; curriculum vitae, including education, date at which nominee received PhD, honors/awards, and publications (abstracts and manuscripts) in the required format; essay by the applicant about a woman scientist who has inspired him/her (no more than 500 words). All essays must be the applicant's own words and ideas. Using someone else's work without citing the source is not permissible-this includes copying, rephrasing, and borrowing facts, statistics, or other material; and description of the applicant's research and career goals (no more than 250 words).
Geographic Focus: All States, Puerto Rico, Argentina, Australia, Brazil, Canada, China, Denmark, India, Israel, Malaysia, Mexico, New Zealand, Nigeria, Norway, Singapore, South Korea, Spain, Switzerland, Turkey, Ukraine, United Kingdom
Contact: Michael Neal Lehman; (601) 984-1607; fax (601) 984-1655; mlehman@umc.edu
Internet: http://www.sfn.org/index.aspx?pagename=WomeninNeuroscience_cwintravelpostdoc
Sponsor: Society for Neuroscience
1121 14th Street, NW, Suite 1010
Washington, DC 20005

SfN Ralph W. Gerard Prize in Neuroscience **4211**
The Ralph W. Gerard Prize in Neuroscience honors an outstanding scientist who has made significant contributions to neuroscience throughout his or her career. The Gerard Prize is the highest recognition conferred by the Society, and open to all scientists with contributions to the field of neuroscience. The recipient receives $25,000, in addition to complimentary air or ground transportation, hotel, and registration for the SfN annual meeting.
Requirements: Nominations must be made or endorsed by Society member, although nominees do not need to be Society members. All nomination materials need to be submitted using SfN's award nomination site.
Restrictions: Self nominations are not permitted, and no person may nominate more than one candidate. Current SfN officers and councilors are ineligible for nomination.
Geographic Focus: All States, Puerto Rico, Argentina, Australia, Brazil, Canada, China, Denmark, India, Israel, Malaysia, Mexico, New Zealand, Nigeria, Norway, Singapore, Spain, Switzerland, Turkey, Ukraine, United Kingdom
Date(s) Application is Due: Jun 22
Amount of Grant: 25,000 USD
Contact: William T. Newsome III, Ph.D., Chairperson; (650) 725-5814; fax (650) 725-3958; bill@monkeybiz.stanford.edu
Internet: http://www.sfn.org/index.aspx?pagename=FellowshipAndAwards_gerard
Sponsor: Society for Neuroscience
1121 14th Street, NW, Suite 1010
Washington, DC 20005

SfN Ricardo Miledi Neuroscience Training Program 4212

The Ricardo Miledi Neuroscience Training Program (RMNTP) strives to attract, support, and retain the best students and young professionals in the field of neuroscience within Latin America and the Caribbean (LAC). The overall goal of the program has been to enhance the training and career development of highly promising neuroscience trainees from the LAC countries by providing an annual neuroscience course that emphasizes hands-on training. This four week course is held the entire month of March and will provide a comprehensive view of relevant concepts and current techniques in neuroscience. The course has been divided into two modules. The first module will cover neural development, ion channels and receptors, synaptic transmission, neurogenesis, and synaptogenesis. The second module will explore neuronal circuits, integration, sensory physiology, learning, memory and circadian rhythms. Students will join the laboratory of one of the participating institutions for hands-on training. Students selected for the training program also receive the Hugo Arechiga Fellowship, which includes an SfN membership, as well as registration and travel expenses to the SfN annual meeting. The application and additional background information about the program are available at the SfN website.

Requirements: The Program is open to LAC neuroscience graduate students. Course locations rotate annually among colleges and universities in Latin America and the Caribbean. Students must be able to communicate in English as most lectures and discussions will take place in English. Students will be required to present and discuss the results of their findings.

Geographic Focus: Puerto Rico, U.S. Virgin Islands, Antigua & Barbuda, Argentina, Aruba, Bahamas, Barbados, Belize, Bolivia, Brazil, Cayman Islands, Chile, Colombia, Costa Rica, Cuba, Dominica, Dominican Republic, Ecuador, El Salvador, French Guiana, Grenada, Guadeloupe, Guatemala, Guyana, Haiti, Honduras, Jamaica, Martinique, Mexico, Nicaragua, Panama, Paraguay, Peru, Saint Kitts And Nevis, St. Lucia, St. Vincent and the Grenadines, Suriname, Trinidad and Tobago, Turks and Caicos Islands, Uruguay, Venezuela

Date(s) Application is Due: Nov 1

Contact: Kelly McCabe; (202) 962-4000; fax (202) 962-4941; kmccabe@sfn.org

Internet: http://www.sfn.org/index.aspx?pagename=FellowshipAndAwards_rmntp

Sponsor: Society for Neuroscience
1121 14th Street, NW, Suite 1010
Washington, DC 20005

SfN Swartz Prize for Theoretical and Computational Neuroscience 4213

The Swartz Prize is presented annually to an individual whose activities have produced a significant cumulative contribution to theoretical models or computational methods in neuroscience, or who has made a noteworthy advance in theoretical or computational neuroscience. The recipient receives $25,000, in addition to complimentary transportation, hotel, and registration to the SfN annual meeting, where the Prize is presented. The recipient also receives a lecture slot at the Dynamical Neuroscience Satellite Event, sponsored by the NIMH. All nomination materials must be submitted using SfN's award nomination site.

Requirements: Nominations are open to all individuals who have contributed to theoretical or computational neuroscience. Nominations should be made or endorsed by SfN members, although nominees do not need to be SfN members.

Restrictions: Self nomination is not permitted, and no person may nominate or write a letter of recommendation for more than one candidate. Current SfN officers and councilors are not eligible for nomination.

Geographic Focus: All States, Puerto Rico, Argentina, Australia, Brazil, Canada, China, Denmark, India, Israel, Malaysia, Mexico, New Zealand, Nigeria, Norway, Singapore, South Korea, Spain, Switzerland, Turkey, Ukraine, United Kingdom

Date(s) Application is Due: Jun 22

Amount of Grant: 25,000 USD

Contact: J. Anthony Movshon; (212) 998-7880; fax (212) 995-4183; movshon@nyu.edu

Internet: http://www.sfn.org/index.aspx?pagename=FellowshipAndAwards_Swartz

Sponsor: Society for Neuroscience
1121 14th Street, NW, Suite 1010
Washington, DC 20005

SfN Travel Awards for the Federation of European Neuroscience Societies Forum 4214

The SfN offers up to 15 travel awards of $2,000 each to support the participation of graduate students from the U.S., Canada, and Mexico at the Federation of European Neuroscience Societies (FENS) Forum, held every two years. Recipients of the travel awards, selected by the SfN International Affairs Committee, are based on the scientific merit of their abstract submitted to the Forum, and the strength of related application materials. Applicants will be emailed about the status of their application. Recipients must attend the Forum to receive their award checks.

Requirements: The applicant must be a U.S., Canadian, or Mexican citizen or permanent resident and graduate student who has advanced to candidacy for a Ph.D. The nominee must be the first author of an abstract to be presented at the Forum. Both the applicant and his or her advisor must be SfN members, and the applicant may not be a recipient of another SfN travel award in the same year. In addition to the online application form, applicants must submit a paragraph describing what they hope to achieve by attending the Forum; a copy of the abstract that would be submitted for the Forum; a one page letter of recommendation from their advisor; and a one page CV which includes their education, honors/awards, and publications (abstracts and manuscripts).

Geographic Focus: All States, Canada, Mexico

Date(s) Application is Due: Dec 9

Amount of Grant: 2,000 USD

Contact: Kelly McCabe; (202) 962-4000; fax (202) 962-4941; kmccabe@sfn.org

Internet: http://www.sfn.org/index.aspx?pagename=TravelAwards_FENS

Sponsor: Society for Neuroscience
1121 14th Street, NW, Suite 1010
Washington, DC 20005

SfN Travel Awards for the International Brain Research Organization World Congress 4215

The Society for Neuroscience (SfN) offers up to 15 travel awards of $2,000 each to support the participation of U.S., Canadian, and Mexican graduate students at the IBRO World Congress held every four years. Recipients of the travel awards are selected by the SfN International Affairs Committee, based on the scientific merit of the abstract submitted to the Congress and the strength of related application materials. Award checks are presented at the conference. Applicants are notified by email of the status of their application. Deadlines vary according to the date of the World Congress.

Requirements: Applicants must be U.S., Canadian, or Mexican citizens or permanent residents, and graduate students who have advanced to candidacy for a Ph.D. Nominees must be first authors of an abstract to be presented at the conference. Applicants and their advisors must be SfN members. Applicants may not be recipients of another SfN travel award in the same award year. In addition to the online application form, applicants must submit: a paragraph describing what they hope to achieve by attending the conference; a copy of the abstract submitted for the conference; a one page letter of recommendation from their advisor; and a one page CV which includes their education, honors/awards, and publications (abstracts and manuscripts).

Geographic Focus: All States, Canada, Mexico

Amount of Grant: 2,000 USD

Contact: Kelly McCabe, Global Programs Assistant; (202) 962-4000; fax (202) 962-4941; kmccabe@sfn.org or globalaffairs@sfn.org

Internet: http://www.sfn.org/index.aspx?pagename=TravelAwards_IBRO

Sponsor: Society for Neuroscience
1121 14th Street, NW, Suite 1010
Washington, DC 20005

SfN Travel Awards for the Japan Neuroscience Society Meeting 4216

The Society for Neuroscience (SfN) and the Japan Neuroscience Society (JNS) support a joint travel award exchange program for graduate students or postdoctoral investigators to their respective annual meetings. The Society sponsors five eligible members from Canada, Mexico, and the U.S. who intend on participating in the JNS meeting in Japan. The JNS will also sponsor five members to attend the SfN annual meeting in the U.S. (locations vary). Recipients are selected on the merit of their abstract, CV, and letters of recommendation. Students are selected by the International Affairs Committee. Award checks are distributed to selectees following the meeting. Awardees will need to provide a certificate of attendance and airline boarding passes to receive their award.

Requirements: Applicants must be Ph.D. candidates or postdoctoral scientists studying or working in Canada, Mexico, or the U.S. They must also be citizens or permanent residents of those locations. Candidates must be SfN members at the time of application and first author on an abstract to be presented at the JNS meeting. Candidates may not have been selected for another SfN travel award within the last calendar year. Preference is given to those who have not previously had extensive experience living or working in Japan. In addition to the online application, candidates must submit a paragraph describing what the applicant hopes to achieve by attending the JNS meeting; a copy of the abstract to be submitted; a one page recommendation from their advisor; and the applicant's CV including information on education, date of Ph.D. candidacy, honors/awards, and publications.

Geographic Focus: All States, Canada, Japan, Mexico

Date(s) Application is Due: Mar 5

Contact: Kelly McCabe, Global Programs Assistant; (202) 962-4000; fax (202) 962-4941; kmccabe@sfn.org or globalaffairs@sfn.org

Internet: http://www.sfn.org/index.aspx?pagename=TravelAwards_JNS

Sponsor: Society for Neuroscience
1121 14th Street, NW, Suite 1010
Washington, DC 20005

SfN Undergraduate Student Travel Awards 4217

The Society for Neuroscience offers Undergraduate Student Travel Awards, administered by the Committee on Neuroscience Departments and Programs (CNDP), to honor outstanding undergraduate student researchers who are nominated by Institutional Program (IP) Member Programs. The award includes complimentary registration and a stipend to offset travel expenses for the SfN annual meeting. Awardees are chosen on the basis of the scientific merit of their abstracts, their participation in the presented research, a letter of nomination from their research advisor, and a personal statement regarding how attendance at the SfN meeting will enhance their professional goals. The application is available at the SfN website, along with detailed information about the application attachments.

Requirements: Applicants must meet the following criteria: full-time students at a four-year accredited university or college that is affiliated with an IP Member Program in good standing; member of SfN; first author of an abstract to be presented at the annual meeting; and the research project must have been conducted while the applicant was enrolled as an undergraduate.

Geographic Focus: All States, Puerto Rico, Argentina, Australia, Brazil, Canada, China, Denmark, India, Israel, Malaysia, Mexico, New Zealand, Nigeria, Norway, Singapore, South Korea, Switzerland, Turkey, Ukraine, United Kingdom

Date(s) Application is Due: Jun 1

Contact: David R. Riddle; (336) 716-4677; fax (336) 716-4534; driddle@wakehealth.edu

Internet: http://www.sfn.org/index.aspx?pagename=TravelAwards_CNDP

Sponsor: Society for Neuroscience
1121 14th Street, NW, Suite 1010
Washington, DC 20005

SfN Young Investigator Award 4218

The Young Investigator Award recognizes the outstanding achievements and contributions by a young neuroscientist who has recently received his or her advanced professional degree. The recipient receives a $15,000 award, along with complimentary ground or air transportation, hotel, and registration for the SfN annual meeting, where the awards are presented. Nominations must be submitted using SfN's award nomination site.
Requirements: Nominees must be active in research at the time the award is given, and have received a Ph.D. or M.D. (or equivalent) within the past ten years. Nominations must be made or endorsed by an SfN members, but nominees do not need to be SfN members.
Restrictions: Self-nomination is not permitted. No person may nominate more than one candidate. Current SfN officers and councilors are ineligible for nomination.
Geographic Focus: All States, Puerto Rico, Argentina, Australia, Brazil, Canada, China, Denmark, India, Israel, Malaysia, Mexico, New Zealand, Nigeria, Norway, Singapore, South Korea, Spain, Switzerland, Turkey, Ukraine, United Kingdom
Date(s) Application is Due: Jun 1
Amount of Grant: 15,000 USD
Contact: Tamas F. Freund, Chairperson; 3636-1-21; freund@koki.hu
Internet: http://www.naccho.org/project30.cfm
Sponsor: Society for Neuroscience
1121 14th Street, NW, Suite 1010
Washington, DC 20005

Shastri Indo-Canadian Institute Arts Fellowships 4219

Arts fellowships are for visual and performing artists who are recognized and emerging practitioners of an Indian art form and wish to undergo training or engage in related activity in India to improve their skills or expand their repertoire; or are not practitioners of an Indian art form but wish to undergo training in India in an Indian art form to develop their own work in new, interesting, and creative ways. Arts Fellowships are for a minimum of four months and a maximum of twelve months. Artists at all stages of their career are welcome to apply. All fellowships include: one round-trip excursion airfare from the place of affiliation in Canada to the place of affiliation in India by the most direct and least expensive route; a living allowance of Rs 21,600 per month; an accountable research allowance of Rs 5,400 per month (receipts and expenditure statements required); and a reimbursement of affiliation fee up to a maximum of Rs 8,000 (to be reimbursed by the Shastri India Office on presentation of receipts).
Restrictions: Eligibility is restricted to citizens and permanent residents of Canada who are a minimum of 18 years of age.
Geographic Focus: Canada
Date(s) Application is Due: Jun 30
Amount of Grant: 81,000-324,000 INR
Contact: Mahmuda Aldeen; (403) 220-7467; fax (403) 289-0100; sici@ucalgary.ca
Internet: http://www.sici.org/programmes/details/arts-fellowship-2012-2013/
Sponsor: Shastri Indo-Canadian Institute
1402 Education Tower, 2500 University Drive NW
Calgary, AB T2N 1N4 Canada

Shastri Indo-Canadian Institute Collaborative Research Project Grants 4220

The Shastri Indo-Canadian Institute with funding support from the Department of Education and Ministry of Human Resource Development, invites proposals for Collaborative Research Projects. Applications from Indian and Canadian Shastri member institutions are welcome. The purpose of these projects is to fill information gaps for policy makers and others who influence policy-making processes. Grant applicants should demonstrate their ability to bring forward innovative ideas, as well as how they plan to share their findings with stake holders and at the grass roots level. The intention behind the projects should be to promote awareness among students and the public in the areas of: environment management; economic development and competitiveness; trade; democracy; rule of law; human rights; managing diversity; energy; public health; poverty alleviation; sustainable development; India-Canada relations; or gender equality.
Requirements: Indian and Canadian Shastri member institutions are eligible to apply. Applicants are required to name a partner institution in the other country. The partner institution may or may not be a Shastri member.
Geographic Focus: Canada, India
Date(s) Application is Due: Nov 1
Contact: Mahmuda Aldeen; (403) 220-7467; fax (403) 289-0100; sici@ucalgary.ca
Internet: http://www.sici.org/programmes/details/collaborative-research-projects-2011-2013/
Sponsor: Shastri Indo-Canadian Institute
1402 Education Tower, 2500 University Drive NW
Calgary, AB T2N 1N4 Canada

Shastri Indo-Canadian Institute Doctoral Student Research Fellowships 4221

Doctoral Student Research Fellowships are awarded to students enrolled in Doctoral programs outside India with a significant India Studies focus, to enable them to conduct research in India towards their degree. Students must be affiliated with an Indian institution. All fellowships include: one round-trip excursion airfare from the place of affiliation in Canada to the place of affiliation in India by the most direct and least expensive route; a living allowance of Rs 21,600 per month; an accountable research allowance of Rs 5,400 per month (receipts and expenditure statements required); and a reimbursement of affiliation fee up to a maximum of Rs 8,000 (to be reimbursed by the Shastri India Office on presentation. The Institute may offer support for more than one year provided satisfactory standing and progress are maintained. Student Graduate Degree Fellowships are for a minimum of three months to a maximum of twelve months. The awards are made for the April 1st through March 31st period. Fellowships for successful applicants are subject to approval of their research topics/areas by the Indian Consulate or High Commission closest to their city of residence.
Requirements: Students must be registered in a graduate program. Canadian citizens enrolled in graduate programs outside Canada are eligible for student fellowships.
Restrictions: Eligibility is restricted to citizens and permanent residents of Canada who are a minimum of 18 years of age. Applicants who have accepted a grant or award from the Shastri Institute within the last 12 months are not eligible to apply. Individuals who have received previous grants or awards from the Shastri Institute and have yet to submit their final reports are not eligible to apply.
Geographic Focus: Canada
Date(s) Application is Due: Jun 30
Amount of Grant: 81,000-324,000 INR
Contact: Mahmuda Aldeen; (403) 220-7467; fax (403) 289-0100; sici@ucalgary.ca
Internet: http://www.sici.org/programmes/details/doctoral-student-research-fellowship-2012-2013/
Sponsor: Shastri Indo-Canadian Institute
1402 Education Tower, 2500 University Drive NW
Calgary, AB T2N 1N4 Canada

Shastri Indo-Canadian Institute Faculty Research Fellowships 4222

Faculty fellowships support research and training in India by Canadian scholars with full- or part-time appointments in a Canadian institution of higher education. Retired faculty who maintain affiliation with a Canadian institution of higher education are eligible to apply if, upon return to Canada, they will continue to contribute to India studies through either teaching or research. Fellowships are awarded on the basis of detailed proposals, which are peer reviewed. Faculty research fellowships are for scholars who have demonstrated excellence in a specific field of India studies. Faculty training fellowships are for scholars who have had little or no previous involvement with India studies and wish to undertake an intensive program of study at an Indian institution; wish to change their disciplinary focus within India studies; wish to develop or enhance their knowledge of an Indian language; or wish to carry out comparative research in India. All fellowships include: one round-trip excursion airfare from the place of affiliation in Canada to the place of affiliation in India by the most direct and least expensive route; a living allowance of Rs 21,600 per month; an accountable research allowance of Rs 5,400 per month (receipts and expenditure statements required); and a reimbursement of affiliation fee up to a maximum of Rs 8,000 (to be reimbursed by the Shastri India Office on presentation of receipts).
Requirements: Eligibility is restricted to citizens and permanent residents of Canada who are a minimum of 18 years of age. Eligible Scholars hold full or part-time appointments in an institution of higher education. Retired faculty who maintain affiliation with an institution of higher education are eligible to apply if upon return to Canada they will continue to contribute to India Studies through either teaching or research.
Restrictions: Applicants who have accepted a grant or award from the Shastri Institute within the last 12 months are not eligible to apply. Individuals who have received previous grants or awards from the Shastri Institute and have yet to submit their final reports are not eligible to apply.
Geographic Focus: Canada
Date(s) Application is Due: Jun 30
Amount of Grant: 81,000-324,000 INR
Contact: Mahmuda Aldeen; (403) 220-7467; fax (403) 289-0100; sici@ucalgary.ca
Internet: http://www.sici.org/programmes/details/faculty-research-fellowship-2012-2013/
Sponsor: Shastri Indo-Canadian Institute
1402 Education Tower, 2500 University Drive NW
Calgary, AB T2N 1N4 Canada

Shastri Indo-Canadian Institute Faculty Training Fellowships 4223

Faculty Training Fellowships are for scholars: who have had little or no previous involvement with India Studies and wish to undertake an intensive program of study at an Indian institution; who wish to change their disciplinary focus within India Studies; who wish to develop or enhance their knowledge of an Indian Language; and who wish to carry out comparative research in India. Academic Fellowships are for a minimum of three to a maximum of twelve months. All fellowships include: one round-trip excursion airfare from the place of affiliation in Canada to the place of affiliation in India by the most direct and least expensive route; a living allowance of Rs 21,600 per month; an accountable research allowance of Rs 5,400 per month (receipts and expenditure statements required); and a reimbursement of affiliation fee up to a maximum of Rs 8,000 (to be reimbursed by the Shastri India Office on presentation of receipts). Faculty Training Fellowships are for a minimum of three months to a maximum of twelve months. The awards are made for the April 1st through March 31st period. Fellowships for successful applicants are subject to approval of their research topics/areas by the Indian Consulate or High Commission closest to their city of residence.
Requirements: Faculty Fellowships support research and training in India by Canadian scholars with full- or part-time appointments in a Canadian institution of higher education. Retired faculty who maintain affiliation with a Canadian institution of higher education are eligible to apply if upon return to Canada they will continue to contribute to India studies through either teaching or research. They are awarded on the basis of detailed proposals, which are peer reviewed.
Restrictions: Eligibility is restricted to citizens and permanent residents of Canada who are a minimum of 18 years of age. Applicants who have accepted a grant or award from the Shastri Institute within the last 12 months are not eligible to apply. Individuals who have received previous grants or awards from the Shastri Institute and have yet to submit their final reports are not eligible to apply.
Geographic Focus: Canada
Date(s) Application is Due: Jun 30

Amount of Grant: 81,000-324,000 INR
Contact: Mahmuda Aldeen; (403) 220-7467; fax (403) 289-0100; sici@ucalgary.ca
Internet: http://www.sici.org/programmes/details/faculty-training-fellowship-2012-2013/
Sponsor: Shastri Indo-Canadian Institute
1402 Education Tower, 2500 University Drive NW
Calgary, AB T2N 1N4 Canada

Shastri Indo-Canadian Institute Graduate Degree Fellowships from an Indian Institution 4224
Student Graduate Degree Fellowships are given to students who wish to enroll in an Indian institution of higher learning to obtain a graduate degree. Applicants should seek confirmation from the India Office that the Government of India approves the proposed institution as a degree-granting body and that their current Canadian university supports their proposed Indian program and will allow transfer of credits. The applicant must have the support of the current chairperson of the department of study in Canada. Students should note that a Canadian university might not accept an Indian degree and courses as equivalent. All fellowships include: one round-trip excursion airfare from the place of affiliation in Canada to the place of affiliation in India by the most direct and least expensive route; a living allowance of Rs 21,600 per month; an accountable research allowance of Rs 5,400 per month (receipts and expenditure statements required); and a reimbursement of affiliation fee up to a maximum of Rs 8,000 (to be reimbursed by the Shastri India Office on presentation. The Institute may offer support for more than one year provided satisfactory standing and progress are maintained. Student Graduate Degree Fellowships are for a minimum of six months to a maximum of twelve months (extension could be offered under special circumstances). The awards are made for the April 1st through March 31st period. Fellowships for successful applicants are subject to approval of their research topics/areas by the Indian Consulate or High Commission closest to their city of residence.
Requirements: Eligibility is restricted to citizens and permanent residents of Canada who are a minimum of 18 years of age. Students must be registered in a graduate program (Canadian citizens enrolled in graduate programs outside Canada are eligible for student fellowships).
Restrictions: Applicants who have accepted a grant or award from the Shastri Institute within the last 12 months are not eligible to apply. Individuals who have received previous grants or awards from the Shastri Institute and have yet to submit their final reports are not eligible to apply.
Geographic Focus: Canada
Date(s) Application is Due: Jun 30
Amount of Grant: 81,000-324,000 INR
Contact: Mahmuda Aldeen; (403) 220-7467; fax (403) 289-0100; sici@ucalgary.ca
Internet: http://www.sici.org/graduate-degree-fellowships-degree-from-an-indian-institution-2012-2013/
Sponsor: Shastri Indo-Canadian Institute
1402 Education Tower, 2500 University Drive NW
Calgary, AB T2N 1N4 Canada

Shastri Indo-Canadian Institute Librarian Fellowships 4225
Librarian Fellowships are awarded for Canadian university librarians to undertake: an intensive program of study at an Indian Institution; enhancement of knowledge of and Indian Language; or enhancement of research support for south Asian studies. Academic Fellowships are for a minimum of three to a maximum of twelve months. Fellowships for successful applicants are subject to approval of their research topics/areas by the Indian Consulate or High Commission closest to their city of residence. All fellowships include: one round-trip excursion airfare from the place of affiliation in Canada to the place of affiliation in India by the most direct and least expensive route; a living allowance of Rs 21,600 per month; an accountable research allowance of Rs 5,400 per month (receipts and expenditure statements required); and a reimbursement of affiliation fee up to a maximum of Rs 8,000 (to be reimbursed by the Shastri India Office on presentation of receipts). Librarian Fellowships are for a minimum of three months to a maximum of twelve months. The awards are made for the April 1st to March 31st period. Fellowships for successful applicants are subject to approval of their research topics/areas by the Indian Consulate or High Commission closest to their city of residence.
Requirements: Faculty Fellowships support research and training in India by Canadian scholars with full- or part-time appointments in a Canadian institution of higher education. Retired faculty who maintain affiliation with a Canadian institution of higher education are eligible to apply if upon return to Canada they will continue to contribute to India studies through either teaching or research. They are awarded on the basis of detailed proposals, which are peer reviewed.
Restrictions: Eligibility is restricted to citizens and permanent residents of Canada who are a minimum of 18 years of age. Applicants who have accepted a grant or award from the Shastri Institute within the last 12 months are not eligible to apply. Individuals who have received previous grants or awards from the Shastri Institute and have yet to submit their final reports are not eligible to apply.
Geographic Focus: Canada
Date(s) Application is Due: Jun 30
Amount of Grant: 81,000-324,000 INR
Contact: Mahmuda Aldeen; (403) 220-7467; fax (403) 289-0100; sici@ucalgary.ca
Internet: http://www.sici.org/programmes/details/librarian-fellowship-2012-2013/
Sponsor: Shastri Indo-Canadian Institute
1402 Education Tower, 2500 University Drive NW
Calgary, AB T2N 1N4 Canada

Shastri Indo-Canadian Institute Master's Student Research Fellowships 4226
Master's Student Research Fellowships are awarded to students enrolled in Master's programs outside India with a significant India Studies focus, to enable them to conduct research in India towards their degree. Students should be affiliated with an Indian institution.
Requirements: Eligible are Canadian citizens and landed immigrants with full- or part-time appointments in Canadian institutions who plan to return to Canada following the tenure of the fellowship. All fellowships include: one round-trip excursion airfare from the place of affiliation in Canada to the place of affiliation in India by the most direct and least expensive route; a living allowance of Rs 21,600 per month; an accountable research allowance of Rs 5,400 per month (receipts and expenditure statements required); and a reimbursement of affiliation fee up to a maximum of Rs 8,000 (to be reimbursed by the Shastri India Office on presentation of receipts).
Restrictions: Eligibility is restricted to citizens and permanent residents of Canada who are a minimum of 18 years of age.
Geographic Focus: Canada
Date(s) Application is Due: Jun 30
Amount of Grant: 81,000-324,000 INR
Contact: Mahmuda Aldeen; (403) 220-7467; fax (403) 289-0100; sici@ucalgary.ca
Internet: http://www.sici.org/programmes/details/masters-student-research-fellowship-2012-2013/
Sponsor: Shastri Indo-Canadian Institute
1402 Education Tower, 2500 University Drive NW
Calgary, AB T2N 1N4 Canada

Shastri Indo-Canadian Institute Millennium Development Research Grants 4227
Funds will be provided to a Canadian or Indian member university on behalf of a binational research teams to undertake public policy relevant research toward one of the eight United Nations Millennium Development Goals. Grant recipients will be required to disseminate their findings, both to the academic and to the broader community. Acceptable expenses include: travel and accommodation for participants, meals, communications costs (audio, video conferencing, promotional materials), organizational support, translation, copying and printing, dissemination of research, research assistant allowance, and reasonable out of pocket expenses. Travel expenses may either be for Canadians going to India or for Indians coming from India or domestic travel for research purposes. Total awards are $80,000 maximum.
Requirements: The lead applicant may be an individual researcher/faculty member or research team affiliated with a Canadian or Indian member institution of the Shastri Indo-Canadian Institute. Applicant must have a partner in the other country. Grant recipients will be required to integrate consideration of gender equality and environmental sustainability in their research model. (Preferences will be given to: applicants who partner with business, and or government; and applicants who also partner with business, civil society organizations and/or government.
Restrictions: The Shastri Institute will not pay institutional overhead.
Geographic Focus: Canada
Date(s) Application is Due: Aug 31
Amount of Grant: Up to 80,000 CAD
Contact: Mahmuda Aldeen; (403) 220-7467; fax (403) 289-0100; sici@ucalgary.ca
Internet: http://www.sici.org/programmes/details/shastri-millennium-development-research-grants/
Sponsor: Shastri Indo-Canadian Institute
1402 Education Tower, 2500 University Drive NW
Calgary, AB T2N 1N4 Canada

Shastri Indo-Canadian Institute Partnership Development Seed Grants 4228
The purpose of the grant is to be a catalyst for scholarly activities which will foster the creation of new knowledge and contribute to formal institutional research agreements or other long-term partnerships. Funds may be used to support a binational seminar, conference, meeting, or other activities that: build awareness of potential collaboration opportunities with Indian institutions; promote sharing of knowledge, information, resources and expertise; showcase existing Canada-India research collaborations and their impact on society; and assist researchers to take advantage of Canada-India collaborative opportunities. The Shastri Institute will provide five grants to Canadian or Indian institutions of higher learning to support catalyst activities such as workshops, seminars, conferences, scholarly discussions and planning meetings that are expected to lead to significant research collaborations with universities or research institutions from the other country. Acceptable expenses include: travel and accommodation for speakers and participants, meals, communication costs (audio, video conferencing, promotional materials), organizational support, translation, copying and printing. Travel expenses may either be for Canadians going to India or for Indians coming to Canada. Applications which promote partnerships with business, civil society organizations and/or government are especially welcome.
Requirements: Individual researcher/faculty member or research teams affiliated with a Canadian or Indian university are eligible. The lead applicant must be employed or affiliated with a Canadian or Indian member institution.
Restrictions: The Shastri Institute will not pay institutional overhead.
Geographic Focus: Canada
Amount of Grant: 10,000 USD
Contact: Mahmuda Aldeen; (403) 220-7467; fax (403) 289-0100; sici@ucalgary.ca
Internet: http://www.sici.org/programmes/details/partnership-development-seed-grant-2011-2012/
Sponsor: Shastri Indo-Canadian Institute
1402 Education Tower, 2500 University Drive NW
Calgary, AB T2N 1N4 Canada

Shastri Indo-Canadian Institute Post-Doctoral Research Fellowships 4229
Post-doctoral Research Fellowships are open to scholars who have completed a Ph.D. and wish to do research in India. Applicants need not have an academic position. These awards will not normally be made if it has been greater than five years since completion of a doctoral degree. All fellowships include: one round-trip excursion airfare from the place of affiliation in Canada to the place of affiliation in India by the most direct and least expensive route; a living allowance of Rs 21,600 per month; an accountable research allowance of Rs 5,400 per month (receipts and expenditure statements required); and a reimbursement of affiliation fee up to a maximum of Rs 8,000 (to be reimbursed by the Shastri India Office on presentation. The Institute may offer support for more than one year provided satisfactory standing and progress are maintained. Student Graduate Degree Fellowships are for a minimum of three months to a maximum of twelve months (extension could be offered under special circumstances). The awards are made for the April 1st through March 31st period. Fellowships for successful applicants are subject to approval of their research topics/areas by the Indian Consulate or High Commission closest to their city of residence.
Requirements: Eligibility is restricted to citizens and permanent residents of Canada who are a minimum of 18 years of age.
Restrictions: Applicants who have accepted a grant or award from the Shastri Institute within the last 12 months are not eligible to apply. Individuals who have received previous grants or awards from the Shastri Institute and have yet to submit their final reports are not eligible to apply.
Geographic Focus: Canada
Date(s) Application is Due: Jun 30
Amount of Grant: 81,000-324,000 INR
Contact: Mahmuda Aldeen; (403) 220-7467; fax (403) 289-0100; sici@ucalgary.ca
Internet: http://www.sici.org/programmes/details/post-doctoral-fellowship-2012-2013/
Sponsor: Shastri Indo-Canadian Institute
1402 Education Tower, 2500 University Drive NW
Calgary, AB T2N 1N4 Canada

Shastri Indo-Canadian Institute Research Collaboration Development Grants 4230
The Shastri Indo Canadian Institute announces a new institutional grant program for its Canadian and Indian Member Institutions, the Institutional Research Collaboration Development Grant (IRCDG). The objective of IRCDG is to support the Institute's research collaboration programs in the area of socio-economic development with focus on sustainable development. Themes of particular interests to the Institute are food security, child and youth education and employment, environment and energy, gender and health, and good governance. The program, made possible by the funding that the Institute receives from the Canadian International development Agency (CIDA), is to encourage new institutional investments in these areas of research. Small matching grants of up to CAD $10,000 are available to selected member institutions with commitment of equivalent amount of funds to initiate and develop research projects or programs that address CIDA's development priorities, are based on bi-national collaboration, and will lead to future research in these broad areas.
Requirements: The initial phase of the collaboration will have to be completed within three months. Canadian and Indian member institutions are eligible to apply for this grant. They may submit only one application (whether they are the lead institution or not). Other academic or development partners may be involved in a project but do not qualify as co-applicants. Either applicant, in India or in Canada, can be the lead applicant.
Geographic Focus: Canada
Date(s) Application is Due: Sep 30
Amount of Grant: Up to 10,000 USD
Contact: Mahmuda Aldeen; (403) 220-7467; fax (403) 289-0100; sici@ucalgary.ca
Internet: http://www.sici.org/programmes/details/shastri-institutional-research-collaboration-development-grant/
Sponsor: Shastri Indo-Canadian Institute
1402 Education Tower, 2500 University Drive NW
Calgary, AB T2N 1N4 Canada

Shastri Indo-Canadian Institute Student Language Training Fellowships 4231
Student Language Training Fellowships are awarded to students enrolled in graduate degree programs outside India, who have taken training in an Indian language in Canada and wish to pursue advanced studies in that language in India. The award is tenable only at recognized educational institutions that award appropriate diplomas or certificates. Student Language Training Fellowships are for a minimum of three months to a maximum of twelve months. The awards are made for the April 1st to March 31st period. Fellowships for successful applicants are subject to approval of their research topics/areas by the Indian Consulate or High Commission closest to their city of residence. All fellowships include: one round-trip excursion airfare from the place of affiliation in Canada to the place of affiliation in India by the most direct and least expensive route; a living allowance of Rs 21,600 per month; an accountable research allowance of Rs 5,400 per month (receipts and expenditure statements required); and a reimbursement of affiliation fee up to a maximum of Rs 8,000 (to be reimbursed by the Shastri India Office on presentation of receipts).
Requirements: Eligibility is restricted to citizens and permanent residents of Canada who are a minimum of 18 years of age. Students must be registered in a graduate program (Canadian citizens enrolled in graduate programs outside Canada are eligible for student fellowships).
Restrictions: Applicants who have accepted a grant or award from the Shastri Institute within the last 12 months are not eligible to apply. Individuals who have received previous grants or awards from the Shastri Institute and have yet to submit their final reports are not eligible to apply.
Geographic Focus: Canada
Date(s) Application is Due: Jun 30
Amount of Grant: 8,100-324,000 INR
Contact: Mahmuda Aldeen; (403) 220-7467; fax (403) 289-0100; sici@ucalgary.ca
Internet: http://www.sici.org/programmes/details/student-language-training-2012-2013/
Sponsor: Shastri Indo-Canadian Institute
1402 Education Tower, 2500 University Drive NW
Calgary, AB T2N 1N4 Canada

Shastri Indo-Canadian Institute University of Ottawa Overseas India Summer Course Grants 4232
Shastri Indo-Canadian Institute and The University of Ottawa in affiliation with the University of Mumbai announce the University of Ottawa Overseas India Summer Course, with a focus on the religious and cultural tradition of Jainism. This unique program will provide students from all over Canada an opportunity to participate in a cultural and educational immersion in India. In India, the program will be based at the University of Mumbai with students residing on the U of M campus. Students will also spend several days in Varanasi, as well as visit the Ajanta-Ellora caves. The dates for the program are: May 25-June 17. Students in the program are eligible to receive 3 credits (graduate or undergraduate credits, depending on their degree programs) through the accrediting host institution, the University of Ottawa. The reduced cost rate of the program is the cost of registration for the University of Ottawa 3-credit course (approx. $650), plus $2500 for the India portion which includes application fees, airfare to and from India, health insurance, accommodation, three vegetarian meals a day, flights/trains/buses within India, course materials, field trip costs.
Requirements: Graduate Students with an interest in India Studies and advanced Undergraduate Students who have completed 3 years of undergraduate study, including some component of India Studies coursework (in Religious Studies, History, or Philosophy etc.) or have a particular academic interest in India Studies are eligible.
Geographic Focus: Canada
Date(s) Application is Due: Dec 15
Contact: Mahmuda Aldeen; (403) 220-7467; fax (403) 289-0100; sici@ucalgary.ca
Internet: http://www.sici.org/programmes/details/shastri-institute-university-of-ottawa-study-jainism-in-india-summer-co/
Sponsor: Shastri Indo-Canadian Institute
1402 Education Tower, 2500 University Drive NW
Calgary, AB T2N 1N4 Canada

Shell Oil Company Foundation Grants 4233
Shell makes monetary grants to nonprofit organizations within the U.S. Priority consideration will be given to organizations in or near communities where Shell Oil Company or its affiliates in the United States have a major presence. Shell Oil Company and the Shell Oil Company Foundation will consider charitable contributions to eligible nonprofit organizations for select programs that fall within our focused areas of contributions: health and welfare; community outreach; workforce development / education and; environment. There are no deadline dates for submitting a proposal, proposals are reviewed on a regular basis.
Requirements: Nonprofit organizations in company operating locations are eligible.
Restrictions: Shell Oil Company and the Shell Oil Company Foundation will not consider contributions for the following purposes: individuals; private Foundations; non-profit organizations without a current 501(c)3 exempt status; conferences or symposia; endowment funds; fraternal and labor organizations; capital campaigns; conferences, workshops, or seminars not directly related to Shell business interests; religious organizations that do not serve the general public on a non-denominational basis; organizations located in or benefiting nations other than U.S. and its territories; organizational operating expenses.
Geographic Focus: All States
Amount of Grant: 10,000-50,000 USD
Contact: A.H. Myres; (713) 241-3616; fax (713) 241-3329; socfoundation@shellus.com
Internet: http://www.shell.us/home/content/usa/responsible_energy/shell_in_the_society/giving_back/grant/grant_request_060208.html
Sponsor: Shell Oil Company Foundation
One Shell Plaza, 910 Louisiana, Suite 4478A
Houston, TX 77002

SHGAPE Vincent P. DeSantis Prize 4234
The prize will be awarded in odd-numbered years for the best book treating any aspect of United States history in the period 1865-1920 published during the two years preceding the year of the award. It must be the author's first book. Books written in English and published between January 1, 2007 and December 31, 2008 are eligible for the 2009 prize. Authors or their publishers should send a copy of the book to each committee member by October 1, 2008. The prize will be awarded at the SHGAPE luncheon held at the Organization of American Historians annual meeting.
Geographic Focus: All States
Date(s) Application is Due: Dec 31
Contact: Thomas R. Pegram, (410) 617-2856; tpegram@loyola.edu
Internet: http://www.h-net.msu.edu/announce/show.cgi?ID=156867
Sponsor: Society for Historians of the Gilded Age and Progressive Era
310 Auditorium Building, Michigan State University
East Lansing, MI 48824-1120

Shorenstein Center Fellowships 4235
The Joan Shorenstein Center offers a one-semester fellowship at Harvard's Kennedy School of Government. Fellows are domestic or international journalists, scholars, and/or policymakers who are interested in the influence of the press on public policy and politics. The purpose of the Fellowship Program is to advance existing research in

press/politics and to provide an opportunity for distinguished experts to reflect on their discipline. The goal is to foster a collegial and intellectual environment that will enrich and complement one's knowledge of the field. The primary focus for a fellow in residence is a paper (approximately 15 pages in length) on a topic examining and analyzing the influence of the press on politics and public policy in the domestic or international arena. A stipend of $30,000 disbursed in four installments over the semester. Office space, computer, printer and telephone are provided.
Requirements: Domestic and international applicants are eligible.
Restrictions: Travel and living expenses are not covered by the Center.
Geographic Focus: All States
Date(s) Application is Due: Feb 1
Amount of Grant: 30,000 USD
Contact: Edith Holway; (617) 495-8269; fax (617) 495-8696; edith_holway@harvard.edu
Internet: http://www.hks.harvard.edu/presspol/fellowships/index.html
Sponsor: John F. Kennedy School of Government, Harvard University
79 John F Kennedy Street
Cambridge, MA 02138

Short-Term Travel to Japan Grants 4236
Grants of a maximum of $3,000 are available to cover expenses while in Japan conducting a specific project explicitly related to Japan which can be accomplished in the period of time requested. These grants are intended for short-term research trips by scholars who are already familiar with Japan and with their topic, but who need time in Japan in order to complete their work.
Requirements: Grants are made only to people with a Ph.D. or comparable professional qualification. Ph.D. candidates are not eligible for this program.
Restrictions: Grantees must use American carriers for any international transportation to be reimbursed under this program. Applicants must not have received funds in this category within the past five years.
Geographic Focus: All States
Date(s) Application is Due: Feb 1; Oct 1
Amount of Grant: 3,000 USD
Contact: Teresa Spence; (734) 665-2490; fax (734) 665-3801; tspence@asian-studies.org
Internet: http://www.asian-studies.org/grants/main.htm#NEAC-JAPAN
Sponsor: Northeast Asia Council/Association for Asian Studies
1021 E Huron Street
Ann Arbor, MI 48104

SHOT Abbott Payson Usher Prize 4237
The SHOT Abbott Payson Usher Prize was established to encourage the publication of original research of the highest standard. It is awarded annually to the author of the best scholarly work published during the preceding three years under the auspices of the Society for the History of Technology. A certificate is included with the $400 prize.
Geographic Focus: All States
Amount of Grant: 400 USD
Contact: Bernie Carlson; (434) 987-6230; fax (434) 975-2190; shot@virginia.edu
Internet: http://www.historyoftechnology.org/awards/usher.html
Sponsor: Society for the History of Technology
P.O. Box 400744
Charlottesville, VA 22904-4744

SHOT Brooke Hindle Postdoctoral Fellowships 4238
THIS Fellowship in the History of Technology honors the contribution of Brooke Hindle to the work of the Society for the History of Technology. The fellowship is for $10,000 and may be used for any purpose connected with research or writing in the history of technology for a period of not less than four months during the year following the award. Applicants should be intending either to prepare a dissertation for publication as articles or as a monograph, whether or not this involves fresh primary research, or to develop a new project based on primary research.
Requirements: Applicants must hold a doctorate in the history of technology or a related field, normally awarded within the preceding four years, or expect to have graduated by the time of the award. Applications must be in English.
Geographic Focus: All States
Date(s) Application is Due: Apr 15
Amount of Grant: 10,000 USD
Contact: Bernie Carlson; (434) 987-6230; fax (434) 975-2190; shot@virginia.edu
Internet: http://www.historyoftechnology.org/awards/hindle.html
Sponsor: Society for the History of Technology
P.O. Box 400744
Charlottesville, VA 22904-4744

SHOT Dibner Award for Excellence in Museum Exhibits 4239
The Dibner Award for Excellence in Museum Exhibits was established in 1985 to recognize excellence in museums and museum exhibits that interpret the history of technology, industry, and engineering to the general public. Winning exhibits, in addition to being well designed and produced, should raise pertinent historical issues. Artifacts and images should be used in a manner that interests, teaches, and stimulates both the general public and historians. The award consists of a plaque and up to $1,000 to cover expenses for a member of the design team to accept the award at the SHOT awards banquet. The Society especially encourages nominations from local and regional historical societies. Virtual exhibits are not eligible for the award.
Requirements: Anyone, including the institution or individual responsible for its creation, may nominate an exhibit for the Dibner Award.
Restrictions: Exhibits are eligible for this award if they have been open to the public for no more than 24 months before the deadline for nominations.
Geographic Focus: All States
Date(s) Application is Due: May 1
Amount of Grant: 1,000 USD
Contact: Bernie Carlson; (434) 987-6230; fax (434) 975-2190; shot@virginia.edu
Internet: http://www.historyoftechnology.org/awards/dibner.html
Sponsor: Society for the History of Technology
P.O. Box 400744
Charlottesville, VA 22904-4744

SHOT Eugene S. Ferguson Prize 4240
The Eugene S. Ferguson Prize is awarded by SHOT for outstanding and original reference work that will support future scholarship in the history of technology. The Ferguson Prize recognizes work that is in the tradition of scholarly excellence established by Eugene S. Ferguson (1916-2004), SHOT's pioneering bibliographer, a founding member of the Society (President, 1977-78; da Vinci Medalist, 1977), museum curator and exhibit catalog author, editor, annotator, university professor, and scholar of the history of engineering and technology. The prize consists of a plaque and a cash award of $2,500.
Requirements: Works published in the four calendar years prior to the year of the award are eligible for consideration. Works that extend beyond the printed text, including other media, such as CDs, World Wide Web sites, and electronic data bases and tools, keeping in mind that the principal criterion of support [of] future scholarship for judging excellence requires that such works exhibit a degree of permanence similar to that of books.
Geographic Focus: All States
Date(s) Application is Due: Apr 1
Amount of Grant: 2,500 USD
Contact: Bernie Carlson; (434) 987-6230; fax (434) 975-2190; shot@virginia.edu
Internet: http://www.historyoftechnology.org/awards/ferguson.html
Sponsor: Society for the History of Technology
P.O. Box 400744
Charlottesville, VA 22904-4744

SHOT IEEE Life Members Prize in Electrical History 4241
This prize is awarded annually to the best paper in electrical history published during the previous year. The prize is sponsored by the life members of the Institute of Electrical and Electronics Engineers (IEEE). The competition is open to any historical paper published in a learned journal or magazine if it treats the art or engineering aspects of electrotechnology and its practitioners. Electrotechnology encompasses power, electronics, telecommunications, and computer science. The cash portion of the prize is shared among joint authors; individual certificates are presented to each author. The prize consists of a cash award of $500 and a certificate.
Requirements: The article must be written in English.
Geographic Focus: All States
Date(s) Application is Due: Apr 15
Contact: Bernie Carlson; (434) 987-6230; fax (434) 975-2190; shot@virginia.edu
Internet: http://www.historyoftechnology.org/awards/ieee.html
Sponsor: Society for the History of Technology
P.O. Box 400744
Charlottesville, VA 22904-4744

SHOT International Scholars 4242
The SHOT International Scholars program is intended to foster a stronger international community for the study of the history of technology by identifying and helping non-U.S. historians of technology to participate in the meetings and governance of the society and receive recognition from their respective governments and national academic communities. Appointments will be for two-year terms beginning each January. International scholars will be made full members of the society and receive complimentary subscriptions to the society's journal. They will be strongly encouraged to attend the annual meetings during their terms. In addition, they will be asked to prepare at least one report or essay review on the history of technology scene in their country for possible presentation to, or publication by, the society. Up to seven scholars are appointed each year.
Requirements: Scholars must reside and work in underrepresented countries outside the United States; they may be both junior and more advanced scholars. An individual may be nominated, or may nominate her/himself, by sending a letter containing a brief statement about the individual (current status, employment, field of interest and research, etc.), a brief curriculum vita, a list of publications, and a fax number or email address.
Geographic Focus: All Countries
Date(s) Application is Due: May 30
Contact: Bernie Carlson; (434) 987-6230; fax (434) 975-2190; shot@virginia.edu
Internet: http://www.historyoftechnology.org/international_scholars.html
Sponsor: Society for the History of Technology
P.O. Box 400744
Charlottesville, VA 22904-4744

SHOT Joan Cahalin Robinson Prize 4243
This prize is awarded annually for the best paper presented at the SHOT annual meeting by a historian under 30 years of age or by a person past 30 who is an accredited graduate student or candidate for a higher degree and who is presenting his or her first paper at a SHOT annual meeting. Candidates for the award are judged not only on the quality of the historical research and scholarship demonstrated in the written papers submitted to the Robinson Prize Committee before the meeting, but also on the effectiveness of their oral presentations. A certificate is included with the prize.

Geographic Focus: All States
Date(s) Application is Due: Jul 1
Contact: Bernie Carlson; (434) 987-6230; fax (434) 975-2190; shot@virginia.edu
Internet: http://www.historyoftechnology.org/awards/robinson.html
Sponsor: Society for the History of Technology
P.O. Box 400744
Charlottesville, VA 22904-4744

SHOT Leonardo da Vinci Medal 4244
The highest recognition from the Society for the History of Technology is the Leonardo da Vinci Medal, presented to an individual who has made an outstanding contribution to the history of technology, through research, teaching, publications, and other activities. Andras Beck (formerly of the Hungarian Academy of Arts) designed the medal, the face of which shows Leonardo's head modeled after the artist's self-portrait. The reverse design shows (in the words of the sculptor) "the basic sources of energy: water, wind, and fire." The prize consists of a medal and a certificate. SHOT invites members to nominate candidates for the DaVinci Medal. To put forward a person's name for consideration, send a letter of nomination, describing the person's outstanding contributions to the history of technology.
Geographic Focus: All States
Contact: Bernie Carlson; (434) 987-6230; fax (434) 975-2190; shot@virginia.edu
Internet: http://www.historyoftechnology.org/awards/davinci.html
Sponsor: Society for the History of Technology
P.O. Box 400744
Charlottesville, VA 22904-4744

SHOT Melvin Kranzberg Dissertation Fellowship 4245
The Melvin Kranzberg Dissertation Fellowship is presented annually to a doctoral student engaged in the preparation of a dissertation on the history of technology, broadly defined. This award is in memory of the co-founder of the Society and honors Melvin Kranzberg's many contributions to developing the history of technology as a field of scholarly endeavor and SHOT as a professional organization. The $4,000 award is unrestricted and may be used in any way that the winner chooses to advance the research and writing of his or her dissertation. Possible uses include underwriting the costs of travel to archival collections; photocopying or microfilming; translation of documents; and so on.
Requirements: Students from institutions of higher learning anywhere in the world who are working on projects in the history of technology are eligible to apply; doctoral candidates from outside the United States are especially encouraged to submit application materials. Applicants must have completed all requirements for their doctorate except for the dissertation by September 1 of the year the award is made.
Restrictions: The award may not be used for university tuition or fees.
Geographic Focus: All States, All Countries
Date(s) Application is Due: Apr 15
Amount of Grant: 4,000 USD
Contact: Bernie Carlson; (434) 987-6230; fax (434) 975-2190; shot@virginia.edu
Internet: http://www.historyoftechnology.org/awards/kranzberg.html
Sponsor: Society for the History of Technology
P.O. Box 400744
Charlottesville, VA 22904-4744

SHOT NASA Fellowship 4246
The NASA Fellowship in the History of Space Technology, offered by SHOT and supported by the National Aeronautics and Space Administration (NASA) History Division, will fund one predoctoral or postdoctoral fellow for up to one academic year to undertake a research project related to the history of space technology. The fellowship may support advanced research related to all aspects of space history, leading to publications on the history of space technology broadly considered, including cultural and intellectual history, institutional history, economic history, history of law and public policy, and history of engineering and management. The fellowship carries a stipend of $17,000, paid quarterly. Fellows will carry out their research projects using personal office space, equipment, and supplies.
Requirements: Applicants must possess a doctorate in history of technology or in a closely related field, or be enrolled as a student in a doctoral degree program and have completed all requirements for the Ph.D., except the dissertation, in history of technology or a related field. Eligibility is not limited to U.S. citizens or residents.
Restrictions: Funds may not be used to support tuition or fees.
Geographic Focus: All States, All Countries
Date(s) Application is Due: Apr 1
Contact: Bernie Carlson; (434) 987-6230; fax (434) 975-2190; shot@virginia.edu
Internet: http://www.historyoftechnology.org/awards/nasa.html
Sponsor: Society for the History of Technology
P.O. Box 400744
Charlottesville, VA 22904-4744

SHOT Samuel Eleazar and Rose Tartakow Levinson Prize 4247
This Samuel Eleazar and Rose Tartakow Levinson Prize is awarded annually by the Society for the History of Technology for an original essay in the history of technology that is the author's first work intended for publication. The essay must explicitly examine in some detail a technology or technological device within the framework of social or intellectual history. To nominate an essay, all materials should be forwarded by email in MS Word or PDF format.
Requirements: Manuscripts already published or accepted for publication are not eligible. In order to be considered, manuscripts must be in English and of a length suitable for publication as an article in Technology and Culture-approximately 7,500 words.
Geographic Focus: All States
Date(s) Application is Due: Apr 15
Amount of Grant: 450 USD
Contact: Bernie Carlson; (434) 987-6230; fax (434) 975-2190; shot@virginia.edu
Internet: http://www.historyoftechnology.org/awards/levinson.html
Sponsor: Society for the History of Technology
P.O. Box 400744
Charlottesville, VA 22904-4744

SHRM Dissertation Awards 4248
These awards are given to four promising researchers in the field of human resource management at the dissertation stage of their doctoral education. This annual competition is funded by the SHRM Foundation, in partnership with the HR Division of the Academy of Management and is intended to support the dissertation research of doctoral candidates. The winners of these dissertation grants each receive a plaque, $5,000 from the SHRM Foundation, and are honored at the Academy of Management's annual meeting.
Requirements: The applicant's dissertation must address a phenomenon that is of importance to the field of human resources and have potential applicability to HR practice. Applicants must be in good standing in a doctoral program in human resource management or a closely related field. Both the applicant and their dissertation advisor (or faculty sponsor in cases where there is not a formal advisor) must be members of the HR Division of the Academy of Management. Applicants must have had their dissertation proposals approved by their dissertation committees prior to application (in schools where there is not a formal proposal defense, documentation (an email from a faculty advisor/sponsor) is needed showing that the student's proposed research design has been deemed acceptable.
Restrictions: Applicants must be in the dissertation stage of their doctoral degree. Applicants who will have defended their dissertation by the submission deadline or who will do so within 6 months after the deadline are ineligible.
Geographic Focus: All States
Date(s) Application is Due: Feb 1
Amount of Grant: 5,000 USD
Contact: Lynn M. Shore, (619) 594-4309; lshore@mail.sdsu.edu
Internet: http://www.shrm.org/foundation/EducationGrants.asp
Sponsor: Society for Human Resource Management Foundation
5500 Campanile Drive
San Diego, CA 92128

SHRM Foundation Research Grants 4249
The Society for Human Resource Management (SHRM) Foundation seeks to fund high-impact research on all aspects of human resource management with a focus on addressing current challenges or understanding emerging trends. Any topic may be funded however the research must be aimed at an academic audience while also having direct, actionable implications for HR practice. It is crucial that there is continuity between the research questions proposed and the methods used. The research should also be able to reasonably generalize across people and settings. Investigators should include a statement in the proposal regarding the individuals, groups, industry sectors or countries for which their findings are expected to generalize, given the sample and study design.
Requirements: The Foundation welcomes proposals from any country. Submissions must be written in English and applications will be read by reviewers whose first language is English. There is no requirement that researchers be based in the United States. In recent years, the funding rate has been similar for researchers both inside and outside the U.S. The SHRM Foundation encourages submitting a formal letter of inquiry to find out whether or not your proposal will be a good fit with SHRM Foundation funding criteria. Overhead up to 15% of direct costs may be requested when required by the PI's institution. However the maximum overall grant budget must not exceed $200,000 USD.
Restrictions: Employees or national board members of SHRM, HRCI or the SHRM Foundation and members of their immediate family are not eligible to apply for SHRM Foundation grants, awards or scholarships.
Geographic Focus: All States, All Countries
Date(s) Application is Due: Apr 1; Oct 1
Amount of Grant: Up to 200,000 USD
Contact: Lynn McFarland; (803) 767-6992; lynnmcfarland@bellsouth.net
Internet: http://www.shrm.org/about/foundation/research/pages/applyforgrant.aspx
Sponsor: Society for Human Resource Management Foundation
1800 Duke Street
Alexandria, VA 22314

SHRM Michael R. Losey Human Resource Research Award 4250
The Award recognizes outstanding contributions by a human resources researcher or professional to the human resource field. The award acknowledges significant past research accomplishments and facilitates continuing contributions by an individual to the human resource management field.
Restrictions: The Award is open to everyone, excluding current SHRM, SHRM Foundation and Human Resource Certification Institute Board memebers and members of SHRM volunteer panels.
Geographic Focus: All States
Date(s) Application is Due: Jan 22
Amount of Grant: 50,000 USD
Samples: Gary P. Latham, Ph.D; Frank Schmidt; Dr. Lee Dyer.
Contact: Nancy Woolever, (703) 535-6457; loseyaward@shrm.org
Internet: http://www.shrm.org/loseyaward
Sponsor: Society for Human Resource Management
1800 Duke Street
Alexandria, VA 22314

SIAM/ACM Award in Computational Science and Engineering 4251

Established in 2002 and first awarded in 2003, the prize is awarded every other year by the Society for Industrial and Applied Mathematics (SIAM) and the Association for Computing Machinery (ACM) in the area of computational science in recognition of outstanding contributions to the development and use of mathematical and computational tools and methods for the solution of science and engineering problems. The award will include a certificate containing the citation and a cash prize, the amount of which will be determined by the Executive Committee of the SIAM Board. Normally the amount of the prize will be approximately the available proceeds from the endowment minus travel expenses from the endowment as noted below. Travel expenses for the award recipient(s) to the meeting will normally be paid by the prize fund.
Requirements: The prize is intended to recognize either one individual or a group of individuals for outstanding research contributions to the field of computational science and engineering. The contribution(s) for which the award is made must be publicly available and may belong to any aspect of computational science in its broadest sense.
Geographic Focus: All States
Date(s) Application is Due: Sep 30
Amount of Grant: 5,000 USD
Contact: Dr. John B. Bell; (215) 382-9800; fax (215) 386-7999; littleton@siam.org
Internet: http://www.acm.org/awards/siam_call.html
Sponsor: Society for Industrial and Applied Mathematics and the Association for Computing Machinery
3600 University City Science Center
Philadelphia, PA 19104-2688

Sidgmore Family Foundation Grants 4252

The Sidgmore Family Foundation honors the legacy of John W. Sidgmore by taking a proactive approach to helping others succeed. The Foundation desires to use its resources to find creative and innovative solutions so that people may achieve their full potential and become responsible, healthy and productive members of society. In recognition that an impoverished environment limits the possibilities for people to develop and thrive, the Sidgmore Family Foundation is particularly interested in funding organizations that: improve the quality of education and teacher training; further the advancement of knowledge in the field of medicine with a special emphasis on hearing and cardiology; utilize entrepreneurial skills to explore and develop creative, scalable, and sustainable solutions to critical social problems; and, provide support and services to those in need in the Washington DC area. Grants are awarded to organizations that have a clear, replicable plan for success, measured sustainable results, and high approval ratings from charity evaluator organizations, such as Charity Navigator. The Foundation also awards multi-year grants that can sustain a program or project.
Requirements: Nonprofit 501(c)3 organizations are eligible. Preference is given to organizations that serve residents in the Washington, D.C. metropolitan area, Maryland, and Virginia. Applicants must begin the process with an initial Letter of Inquiry (LOI). LOIs will receive a response if the Foundation wishes to receive a proposal from your organization.
Restrictions: The Sidgmore Family Foundation does not make grants to individuals, national health organizations, government agencies, or political and public policy advocacy groups.
Geographic Focus: All States
Contact: M. Gelbwaks, Director; (516) 541-2713; SidgmoreFound@aol.com
Internet: http://www.sidgmorefoundation.com/#application_process
Sponsor: Sidgmore Family Foundation
71 Leewater Avenue
Massapequa, NY 11758

Sid W. Richardson Foundation Grants 4253

Grants are provided to tax-exempt organizations in Texas in the areas of education (museums, learning centers, day schools, K-12 schools, higher education institutions, and business and economic education), health (medical schools, organ donor registries, hospitals, disease prevention, health science centers, and nursing associations), arts (arts councils, visual and performing arts festivals, museums, ballet, symphony orchestra, and arts education programs), and human services (boys' and girls' clubs, united funds, the elderly, crime prevention, the disabled, housing opportunities, food programs, and drug and alcohol abuse prevention); specific goals within these larger areas change periodically. Types of support include operating budgets, seed grants, building construction funds, equipment acquisition, endowment funds, research, publications, conferences and seminars, matching funds, continuing support, and projects/programs. Award amounts vary depending on proposed projects.
Requirements: Proposals must cover programs and projects within Texas.
Restrictions: Grants are not made to individuals.
Geographic Focus: Texas
Date(s) Application is Due: Jan 15
Amount of Grant: 10,000-100,000 USD
Contact: Valleau Wilkie Jr., Executive Director; (817) 336-0494; fax (817) 332-2176
Internet: http://www.sidrichardson.org/GrantInfo.html
Sponsor: Sid W. Richardson Foundation
309 Main Street
Fort Worth, TX 76102

Sigma Theta Tau International /American Nurses Foundation Grant 4254

Sigma Theta Tau International (STTI), the only international nursing honor society worldwide, is a global community of nurse leaders dedicated to using knowledge, scholarship, service, and learning to improve the health of the world's people. STTI members belong to 475 chapters in 90 countries. STTI and the American Nurses' Foundation (ANF) have joined forces to provide funding for the STTI/ANF research grant. The purpose of the grant is to encourage the research career development of nurses through support of research conducted by beginning nurse researchers or experienced nurse researchers who are entering a new field of study. Applications must be submitted directly to STTI via their online submission system by the May 1 deadline date. Funding will begin in October. Links to an FAQ, application-review criteria, and help with grant-writing are available at the website.
Requirements: Eligible applicants must meet the following criteria: be a registered nurse with a current license; have a master's or doctoral degree or be enrolled in a doctoral program; have a clinical nursing research topic; submit a completed research application packagbe and signed research agreement; and be ready to implement the research project when funding is received. Preference will be given to Sigma Theta Tau members, other qualifications being equal. Allocation of funds is based on the quality of the proposed research, the future promise of the applicant, and the applicant's research budget.
Restrictions: Funds for this grant do not cover expenses incurred prior to the funding date.
Geographic Focus: All States, All Countries
Date(s) Application is Due: May 1
Amount of Grant: Up to 7,500 USD
Contact: Research Services; (888) 634-7575 or (317) 634-8171; fax (317) 634-8188; research@stti.iupui.edu
Internet: http://www.nursingsociety.org/Research/Grants/Pages/grant_anf.aspx
Sponsor: Sigma Theta Tau International
550 W. North Street
Indianapolis, IN 46202-2156

Sigma Theta Tau International / Council for the Advancement of Nursing Science Grants 4255

Sigma Theta Tau International (STTI), the only international nursing honor society worldwide, is a global community of nurse leaders dedicated to using knowledge, scholarship, service, and learning to improve the health of the world's people. STTI members belong to 475 chapters in 90 countries. STTI and the Council for the Advancement of Nursing Science (CANS) have joined forces to provide funding for the STTI/CANS research grant. The purpose of the grant is to encourage qualified nurses to improve the health of the world's people through research. Proposals for clinical, educational or historical research, including plans for broadly disseminating the research findings, may be submitted for the grant. Applications must be submitted directly to STTI via their online submission system by the July 1 deadline date. Funding will begin in November. Links to an FAQ, application-review criteria, and help with grant-writing are available at the website.
Requirements: Eligible applicants must meet the following criteria: be a member in good standing of both STTI and CANS; hold at minimum a Master's degree or equivalent; submit a completed research application via STTI's on-line submission system; be ready to implement research project when funding is received; submit a final report; submit a completed abstract to Virginia Henderson International Nursing Library; submit the completed project abstract for presentation at the Sigma Theta Tau or Council for Advancement of Nursing Science research meeting; and credit research-grant partners in all publications and presentations of the research.
Restrictions: Funds for this grant do not cover expenses incurred prior to the funding date.
Geographic Focus: All States, All Countries
Date(s) Application is Due: Jul 1
Amount of Grant: Up to 5,000 USD
Contact: Research Services; (888) 634-7575 or (317) 634-8171; fax (317) 634-8188; research@stti.iupui.edu
Internet: http://www.nursingsociety.org/Research/SmallGrants/Pages/CANS.aspx
Sponsor: Sigma Theta Tau International
550 W. North Street
Indianapolis, IN 46202-2156

Sigma Theta Tau International / Oncology Nursing Society Grant 4256

Sigma Theta Tau International (STTI), the only international nursing honor society worldwide, is a global community of nurse leaders dedicated to using knowledge, scholarship, service, and learning to improve the health of the world's people. STTI members belong to 475 chapters in 90 countries. STTI and the Oncology Nurses Society (ONS) have joined forces to provide funding through ONS's Small Research Grants program for this grant. From the perspectives of both organizations, the purpose of the grant is to stimulate clinically-related oncology-nursing research, increase the knowledge base for oncology-nursing practice, and to prepare future oncology-nurse researchers. Application guidelines are available through ONS; applications must be submitted directly to ONS. Contact information and a link to the ONS website is available at the grant website.
Requirements: Eligible applicants must meet the following criteria: be a registered nurse actively involved in some aspect of cancer-patient care, education or research and have earned a master's degree. Other qualifications being equal, preference will be given to Sigma Theta Tau International members.
Geographic Focus: All States, All Countries
Date(s) Application is Due: Oct 1
Amount of Grant: 10,000 USD
Contact: Debbie Kubiak, Grants Coordinator; (412) 859-6224 or (866) 357-4667; fax (877) 369-5497; dkubiak@ons.org
Internet: http://www.nursingsociety.org/Research/Grants/Pages/grant_ons.aspx
Sponsor: Sigma Theta Tau International
550 W. North Street
Indianapolis, IN 46202-2156

Sigma Theta Tau International Doris Bloch Research Award 4257

Sigma Theta Tau International (STTI), the only international nursing honor society worldwide, is a global community of nurse leaders dedicated to using knowledge, scholarship, service, and learning to improve the health of the world's people. STTI members belong to 475 chapters in 90 countries. The purpose of STTI's Doris Bloch Research Award is to encourage nurses to contribute to the advancement of nursing through research. Allocation of funds is based on the quality of the proposed research, the future promise of the applicant, and the applicant's research budget. Applications from novice researchers who have received no other national research funds are encouraged and will receive preference for funding, other aspects being equal. Preference will be given to Sigma Theta Tau International members, other qualifications being equal. Applications must be submitted via STTI's online submission system. A link to the submission system will be available in July. Funding will begin in June. Links to an FAQ, application-review criteria, and help with grant-writing are available at the website. Funding is provided by a gift to the honor society's Research Endowment from the Doris Bloch estate.
Requirements: Eligible applicants must meet the following criteria: be a registered nurse with a current license; hold a master's or doctoral degree or be enrolled in a doctoral program; submit a completed research application package and signed research agreement; be ready to implement the research project when funding is received; complete the project within one (1) year of funding; submit to STTI a final report; and submit a completed abstract to STTI's Virginia Henderson International Nursing Research Library.
Geographic Focus: All States, All Countries
Date(s) Application is Due: Dec 1
Amount of Grant: Up to 5,000 USD
Contact: Research Services; (888) 634-7575 or (317) 634-8171; research@stti.iupui.edu
Internet: http://www.nursingsociety.org/Research/Grants/Pages/grant_bloch.aspx
Sponsor: Sigma Theta Tau International
550 W. North Street
Indianapolis, IN 46202-2156

Sigma Theta Tau Small Grants 4258

Sigma Theta Tau International (STTI), the only international nursing honor society worldwide, is a global community of nurse leaders dedicated to using knowledge, scholarship, service, and learning to improve the health of the world's people. STTI members belong to 475 chapters in 90 countries. The purpose of STTI's Small Grants Program is to encourage nurses to contribute to the advancement of nursing through research. Ten to fifteen grants are awarded annually. Allocation of funds is based on the quality of the proposed research, the future promise of the applicant, and the applicant's research budget. Applications from novice researchers who have received no other national research funds are encouraged and will receive preference for funding, other aspects being equal. Preference will be given to Sigma Theta Tau members, other qualifications being equal. Applications must be submitted via STTI's online submission system. A link to the submission system will be available in July. Funding will begin in June. Links to an FAQ, application-review criteria, and help with grant-writing are available at the website.
Requirements: Eligible applicants must meet the following criteria: be a registered nurse with a current license; have a master's or doctoral degree or be enrolled in a doctoral program; have a clinical-nursing research topic; submit a completed research application package and signed research agreement; be ready to implement the research project when funding is received; complete the project within one (1) year of funding; submit a final report; and submit a completed abstract to STTI's Virginia Henderson International Nursing Research Library.
Restrictions: Funds for this grant do not cover expenses incurred prior to funding.
Geographic Focus: All States, All Countries
Date(s) Application is Due: Dec 1
Amount of Grant: Up to 5,000 USD
Contact: Research Services; (888) 634-7575 or (317) 634-8171; research@stti.iupui.edu
Internet: http://www.nursingsociety.org/Research/Grants/Pages/small_grants.aspx
Sponsor: Sigma Theta Tau International
550 W. North Street
Indianapolis, IN 46202-2156

Sigma Xi Grants in Aid of Research 4259

Research awards are made to support specific scientific projects in any field of science or engineering. Awards are limited to the graduate and undergraduate levels of study. Awards are made payable to the individual recipients. No part of the award may be used for payment of indirect costs to recipient's institution. Applications are accepted online only.
Requirements: Eligible to apply are nationals of any country who are students in undergraduate- or graduate-degree programs with a clearly outlined project of scientific investigation. Applicants need not be members of the society. See website for complete listing of requirements and restrictions.
Geographic Focus: All States
Date(s) Application is Due: Mar 15; Oct 15
Contact: Coordinator; (800) 243-6534 or (919) 549-4691; giar@sigmaxi.org
Internet: http://www.sigmaxi.org/programs/giar/index.shtml
Sponsor: Sigma Xi, the Scientific Research Society
P.O. Box 13975
Research Triangle Park, NC 27709

Sigrid Juselius Foundation Grants 4260

The Sigrid Juselius Foundation, offers funding for domestic medical research. To enable established foreign medical researchers to work in a Finnish institution for 1-12 months, the Foundation also offers grants to nationals of other countries. Resources are allocated to senior and post doc researchers. The grants cover all disciplines of medical science. The grants can cover personal expenses (living costs) and travel.
Requirements: The Foundation does not award direct grants to foreign medical research. A foreign scientist interested in coming over to Finland has to get in touch with a Finnish scientist of his particular field, and the application should be made on the part of the Finnish host together with the proposed visiting researcher. A electronic application is available on the Sigrid Juselius Foundation website.
Restrictions: The funds are not available for studies or doctoral theses.
Geographic Focus: All States
Date(s) Application is Due: Apr 15; Sep 15
Contact: Funding Officer; 358 20 710 9082; fax 358 20 710 9089
Internet: http://www.sigridjuselius.fi/index.php?page_id=48
Sponsor: Sigrid Juselius Foundation
Aleksanterinkatu 48 B
Helsinki, FI-00100 Finland

Simone and Cino del Duca Grand Prix Awards 4261

The Simone and Cino del Duca Foundation of the Institut de France awards each year a scientific Grand Prix based on the recommendations of a jury of members of the Institut de France. The amount of the Grand Prix will be 364,000 euros: 250,000 euros to finance the work of the award-winning group, 50,000 euros awarded to the director of the research group, and an additional 64,000 euros towards hiring one or more French or foreign postdoctoral researchers to be selected by the group director.
Requirements: The Grand Prix will reward an internationally recognized French or European mathematician proposing an ambitious research project based on the collective work of a group under his/her direction. The proposal should describe the scientific objectives and the means that will be used to implement the project (duration of the project, invitations, meetings, seminars, etc.).
Geographic Focus: All States, Albania, Andorra, Armenia, Austria, Azerbaijan, Belarus, Belgium, Bosnia & Herzegovina, Bulgaria, Croatia, Cyprus, Czech Republic, Denmark, Estonia, Finland, France, Georgia, Germany, Greece, Hungary, Iceland, Ireland, Italy, Kosovo, Latvia, Liechtenstein, Lithuania, Luxembourg, Macedonia, Malta, Moldova, Monaco, Montenegro, Norway, Poland, Portugal, Romania, Russia, San Marino, Serbia, Slovakia, Slovenia, Spain, Sweden, Switzerland, The Netherlands, Turkey, Ukraine, United Kingdom, Vatican City
Date(s) Application is Due: Jan 25
Amount of Grant: Up to 364,000 EUR
Contact: Grants Administrator; 01 47 66 01 21; fax 01 46 22 45 02; fondation-del-duca@institut-de-france.fr
Internet: http://www.enseignementsup-recherche.gouv.fr/cid23192/call-for-applications-for-the-grand-prix-scientifique-simone-and-cino-del-duca-2009.html
Sponsor: Institut de France
23, quai Conti
Paris, 75270 France

Sioux Falls Area Community Foundation Community Fund Grants 4262

The purpose of the Sioux Falls Area Foundation (SFACF)'s unrestricted grantmaking program is to provide support across a wide spectrum of charitable needs and interests. Grantmaking categories include Arts and Humanities (e.g., theatre, music, arts, dance, cultural development, historic preservation, library programs, and museums); Community Affairs and Development (e.g., citizen participation, public use of parks and recreation, administration of justice, economic development, employment, and training); Education (e.g., lifelong-learning activities in formal educational settings, support of educational facilities and systems, and scholarships); Environment (e.g., protection of natural areas, conservation of energy, prevention and elimination of pollution or hazardous waste, wildlife protection, and water quality); Health (e.g., improvement of healthcare, prevention of substance abuse; support of mental-health needs, and medical research); Human Services (e.g., assistance to families, youth, the elderly, disabled, special groups, social service providers, and those who stand in need); and Religion (e.g., support for churches, religious institutions, and religion programs). SFACF offers two grant programs from its unrestricted funds: Spot Grants for projects up to $3,000 and Community Fund Grants for projects over $3,000. The majority of Community Fund Grants are made in the range of $5,000-$10,000. Proposals for Community Fund Grants must be submitted using a standard application form (available from the SFACF office or downloadable from the website). Applications are accepted anytime and will be reviewed by the Grants Committee at their next scheduled meeting. Meetings take place six times a year: January, March, May, July, September, and November (a schedule is posted at the SFACF website).
Requirements: Nonprofit organizations serving residents in the Sioux Falls, South Dakota area (Minnehaha, Lincoln, McCook, and Turner counties) are eligible to apply. SFACF considers grant requests for programs that require start-up funds to address important community needs or opportunities, expansion of programs that meet important community needs or opportunities, assistance to organizations weathering unforeseen or unusual financial crises; programs that increase an organization's capacity to advance its mission more efficiently or effectively; and programs or studies that inform the community's understanding of needs or opportunities. Requests are evaluated by the following criteria: comparative benefit to the community; the organization's capacity to achieve the stated objectives; the amount of support requested versus the number of people benefited; a well-planned approach to achieving stated objectives; a reasonable expectation that the program can be sustained over time (where applicable); the organization's history of working collaboratively to address community needs and opportunities; and when applicable, the organization's past SFACF grant performance.
Restrictions: SFACF does not consider grant requests for individuals, national fundraising efforts, political advocacy, and sectarian religious programs. The following types of requests are discouraged: large capital improvements or construction drives; ongoing operational support; reduction or elimination of organizational deficits; reimbursement

of expenses undertaken prior to submission of a grant application; computer hardware and software, unless these are the focus of a new or enhanced program; public art for which approval and placement has not yet been secured; and multi-year requests.
Geographic Focus: South Dakota
Date(s) Application is Due: Jan 1; Mar 1; May 1; Jul 1; Sep 1; Nov 1
Contact: Candy Hanson; (605) 336-7055 ext. 12; chanson@sfacf.org
Internet: http://www.sfacf.org/AboutGrants.aspx
Sponsor: Sioux Falls Area Community Foundation
300 N Philips Avenue, Suite 102
Sioux Falls, SD 57104-6035

Sioux Falls Area Community Foundation Spot Grants (Unrestricted) **4263**
The purpose of the Sioux Falls Area Foundation (SFACF)'s unrestricted grantmaking program is to provide support across a wide spectrum of charitable needs and interests. Grantmaking categories include Arts and Humanities (e.g., theatre, music, arts, dance, cultural development, historic preservation, library programs, and museums); Community Affairs and Development (e.g., citizen participation, public use of parks and recreation, administration of justice, economic development, employment, and training); Education (e.g., lifelong-learning activities in formal educational settings, support of educational facilities and systems, and scholarships); Environment (e.g., protection of natural areas, conservation of energy, prevention and elimination of pollution or hazardous waste, wildlife protection, and water quality); Health (e.g., improvement of healthcare, prevention of substance abuse; support of mental-health needs, and medical research); Human Services (e.g., assistance to families, youth, the elderly, disabled, special groups, social service providers, and those who stand in need); and Religion (e.g., support for churches, religious institutions, and religion programs). SFACF offers two grant programs from its unrestricted funds: Community Fund Grants for projects over $3,000 and Spot Grants for projects up to $3,000. Spot Grant proposals may be submitted at any time and do not require SFACF's standard application form. Applicants should include the following components in their requests: a typed summary of their program in two pages or fewer; signatures of the organization's executive director and board chair; a board of directors roster; a copy of the organization's IRS tax determination letter; and a project budget. In most cases, SFACF will review and respond to Spot Grant requests within two weeks of receipt. SFACF has provided complete guidelines and an informative FAQ at their website.
Requirements: Nonprofit organizations serving residents in the Sioux Falls, South Dakota area (Minnehaha, Lincoln, McCook, and Turner counties) are eligible to apply. SFACF considers grant requests for programs that require start-up funds to address important community needs or opportunities, expansion of programs that meet important community needs or opportunities, assistance to organizations weathering unforeseen or unusual financial crises; programs that increase an organization's capacity to advance its mission more efficiently or effectively; and programs or studies that inform the community's understanding of needs or opportunities. Requests are evaluated by the following criteria: comparative benefit to the community; the organization's capacity to achieve the stated objectives; the amount of support requested versus the number of people benefited; a well-planned approach to achieving stated objectives; a reasonable expectation that the program can be sustained over time (where applicable); the organization's history of working collaboratively to address community needs and opportunities; and when applicable, the organization's past SFACF grant performance.
Restrictions: SFACF does not consider grant requests for individuals, national fundraising efforts, political advocacy, and sectarian religious programs. The following types of requests are discouraged: large capital improvements or construction drives; ongoing operational support; reduction or elimination of organizational deficits; reimbursement of expenses undertaken prior to submission of a grant application; computer hardware and software, unless these are the focus of a new or enhanced program; public art for which approval and placement has not yet been secured; and multi-year requests.
Geographic Focus: South Dakota
Amount of Grant: Up to 3,000 USD
Contact: Candy Hanson; (605) 336-7055 ext. 12; fax (605) 336-0038; chanson@sfacf.org
Internet: http://www.sfacf.org/sfacf/aboutgrants.aspx
Sponsor: Sioux Falls Area Community Foundation
300 N Philips Avenue, Suite 102
Sioux Falls, SD 57104-6035

Siragusa Foundation Health Services & Medical Research Grants **4264**
The Siragusa Foundation, established in 1950 by Ross D. Siragusa, is a private family foundation that is committed to honoring its founder by sustaining and developing Chicago's extraordinary nonprofit resources. The Foundation's Health Services and Medical Research Program is balanced between medical research organizations, which are concerned with scientific inquiry and the exploration of treatment methods, and health service organizations, which provide direct social and clinical care. The Foundation primarily funds fellowships, single-disease research programs and clinical treatment services. The Foundation believes it is critical to fund research efforts, health support services, patient care services and treatment, and advocacy programs for both adults and youth, all of which are highly complementary to each other.
Requirements: Nonprofit 501(c)3 organization in Chicago are eligible to apply. Before submitting a formal proposal to the Foundation, prospective applicants should submit a two-page letter of inquiry outlining the background of the organization and the proposed project. The letter of inquiry should describe succinctly the project for which funding is requested, including how it relates to the Foundation's interests, the target audience, the estimated budget and request amount. After reviewing the letter of inquiry, the Foundation may or may not request a formal proposal.
Restrictions: The Foundation limits potential grantees from applying for funds to once per calendar year. The Foundation does not accept unsolicited proposals from outside the Chicago area and does not support individuals or political advocacy. The foundation typically does not support capital expenditures or endowments.
Geographic Focus: Illinois
Contact: Kristen M. Buerster; (312) 755-0064; fax (312) 755-0069
Internet: http://www.siragusa.org/pages/program_areas/13.php
Sponsor: Siragusa Foundation
1 East Wacker Drive, Suite 2910
Chicago, IL 60601

Sir Dorabji Tata Trust Individual Education and Travel Grants **4265**
The trust offers several scholarships/education grants for higher education: General Education Grants, Sir Dorab Tata Merit Scholarships, Dr SR Moolgaokar Scholarship, M. Phil scholarships, and a scholarship of Rs 6,000 per annum is also given to a trainee at the Training Ship Rajendra Chanakya. The Trust also offers Travel grants for short-term studies abroad as well as for attending conferences, presentation of papers for research and sports related activities. The grants range from Rs 20,000 to Rs 50,000. Applications are received throughout the year and the grants are processed based on an internal appraisal process. A panel of educationists interview candidates recommended by their respective colleges. These scholarships are provided till the student completes his or her first degree.
Requirements: The application may be made by way of a letter, by email, or in person. The following documents as may be applicable should accompany the form: Marksheets from SSC onwards (college studies) / school report (school studies); Proof of family income-salary certificate or tax returns or similar proof; Proof of residence (ration card); Student's Identity Card; Copy of the fee receipts; Proof of the college being approved by Association of Indian Universities and the course being approved by Accreditation Board of AICTE. Each scholarship/award also has additional specific guidelines-see the website for more details.
Restrictions: Forms received late will not be accepted. Limited to India.
Geographic Focus: All States, Abkhazia, Afghanistan, Armenia, Azerbaijan, Bahrain, Bangladesh, Bhutan, British Indian Ocean Territory, Brunei, Burma (Myanmar), Cambodia, China, Christmas Island, Cocos, Cyprus, Georgia, Hong Kong, India, Indonesia, Iran, Iraq, Israel, Japan, Jordan, Kazakhstan, Kuwait, Kyrgyzstan, Laos, Lebanon, Macau, Malaysia, Maldives, Mongolia, Nagorno-Karabakh, Nepal, North Korea, Northern Cyprus, Oman, Pakistan, Palestinian Authority, Philippines, Qatar, Russia, Saudi Arabia, Singapore, South Korea, South Ossetia, Sri Lanka, Syria, Taiwan, Tajikistan, Thailand, Timor-Lester, Turkey, Turkmenistan, United Arab Emirates, Uzbekistan, Vietnam, Yemen
Date(s) Application is Due: Sep 30
Contact: R.M. Lala; 91 22 6665 8282; fax 91 22 2204 5427; sdtt@sdtatatrust.com
Internet: http://www.dorabjitatatrust.org/ind_grants/ind_ietg.asp
Sponsor: Sir Dorabji Tata Trust
Bombay House, 24 Homi Mody Street
Mumbai, 400 001 India

Skaggs Foundation Grants **4266**
The Skaggs Foundation was established in 1962 in California, and currently provides funding in Alaska. The foundations area of interest include: animals/wildlife; preservation/protection; children services; special education; environmental education; environment; natural resources; marine science. Types of support available with these funds include: continuing support; endowments; equipment; general/operating support; internship funds; land acquisition; matching/challenge support; program evaluation; research; technical assistance.
Requirements: Nonprofit organizations are eligible, preference is given to requests from Alaska. The initial approach should be a one or two page letter of inquiry. There is no formal application form. Applicants should submit the following: copy of IRS Determination Letter; copy of current year's organizational budget and/or project budget; one copy of the proposal.
Restrictions: No grants to individuals.
Geographic Focus: Alaska
Date(s) Application is Due: May 1; Oct 1
Contact: Samuel D. Skaggs Jr., President; (907) 463-4843
Sponsor: Skaggs Foundation
P.O. Box 20510
Juneau, AK 99802-0510

Skin Cancer Foundation Research Grants **4267**
Funding is provided annually for basic research, clinical studies, and educational programs related to skin cancer. The foundation's grants program is intended for pilot projects, which, if successful, can be further developed in order to be eligible for larger grants from other sources. Grants include: The Dr. Patricia Wexler Research Grant Awards (one-year, $10,000 grant for a research project); The Theodore S. Tromovitch Memorial Award (one-year, $5,000 grant for a research project); The Melissa K. Bambino Memorial Award (one-year, $8,000 grant for a research project); The Robert N. Cooke Memorial Award (one-year, $3,000 grant for a research project); Pevonia Botantica Research Award (one-year, $10,000 grant for a research project)
Requirements: Candidates must comply with all directions on the application form (available at the website). All projects are selected on the basis of their applicability to the foundation's mission of reducing the incidence, morbidity and mortality of skin cancer. Research must be conducted at institutions within the U.S. Preference is given to projects that address, at the basic science and clinical level, improved methods of prevention, detection, and treatment of skin cancers.
Restrictions: Overhead or indirect costs are not eligible.
Geographic Focus: All States
Date(s) Application is Due: Oct 1

Amount of Grant: 3,000-10,000 USD
Contact: Grants Administrator; (800) 754-6490; info@skincancer.org
Internet: http://www.skincancer.org/research-grants-2010.html
Sponsor: Skin Cancer Foundation
149 Madison Avenue, Suite 901
New York, NY 10016

Smith Family Awards for Excellence in Biomedical Research 4268

For over 20 years, the Smith Family Foundation has been supporting groundbreaking medical research through the Smith Family Awards Program for Excellence in Biomedical Research. Its mission is to launch the careers of newly independent biomedical researchers with the ultimate goal of achieving medical breakthroughs. Applications focus on all fields of basic biomedical science and may also be submitted by investigators in physics, chemistry and engineering. Three year awards in the amount of $300,000 ($100,000 per year inclusive of 5% indirect costs) target new faculty who are within two years of their first independent faculty appointment as of July 1st of the application year.
Requirements: Applicants, nominated by their institutions, must be full-time faculty at nonprofit academic, medical or research institutions in Massachusetts, at Brown University or at Yale University. Applications from women scientists are encouraged. Applications must be received by 12:00 Noon, U.S. Eastern time of the published deadline date.
Geographic Focus: Massachusetts
Date(s) Application is Due: Jul 16
Amount of Grant: 300,000 USD
Contact: Gay Lockwood, Senior Program Officer; (617) 279-2240 ext. 702; fax (617) 423-4619; glockwood@hria.org
Internet: http://hria.org/tmfservices/tmfgrants/smith.html
Sponsor: Health Resources in Action
95 Berkeley Street
Boston, MA 02116

Smith Richardson Foundation Domestic Public Research Fellowship 4269

The Domestic Public Policy program supports projects that will help the public and policy makers understand and address critical challenges facing the United States. An overarching goal of the Foundation's grant making is to support projects that help stimulate and inform important public policy debates. To that end, the Foundation supports research on and evaluation of existing public policies and programs, as well as projects that inject new ideas into public debates. The Foundation is interested in a wide range of topics. Education and school reform have been central to the Foundation's grant making in recent years, including efforts to assess the effectiveness of various school reform initiatives, such as charter schools. The Foundation has expanded on its interest in public finance issues by beginning to explore the future of the Medicare program and the larger challenge of coping with expected increases in the costs of federal entitlement programs. The Foundation also provides support for projects that assess the impact of immigration on American society, examine the effectiveness of our regulatory policies, and explore the interaction between the policy making process and the political process through support for projects on congressional redistricting and the conduct and financing of political campaigns. The Foundation continues to support projects that inform the debate over policies designed to assist disadvantaged families.
Requirements: Fellowships will only be awarded to individual researchers and not to research teams. He or she must hold a position as a full-time faculty member at a college or university or as a full-time fellow at a public policy think tank or research organization in the United States. An applicant should explain how he or she meets these requirements in the cover letter to his or her research proposal.
Geographic Focus: All States
Amount of Grant: 60,000 USD
Contact: Mark Steinmeyer; (203) 222-6222; fax (203) 222-6282; msteinmeyer@srf.org
Internet: http://www.srf.org/grants/domestic.php
Sponsor: Smith Richardson Foundation
60 Jesup Road
Westport, CT 06880

Smith Richardson Foundation Strategy and Policy Fellowship 4270

The Smith Richardson Foundation's International Security and Foreign Policy program announces the Strategy and Policy Fellows Program, its annual grant competition to support young scholars and policy thinkers on American foreign policy, international relations, international security, military policy, and diplomatic and military history. The purpose of the program is to strengthen the U.S. community of scholars and researchers conducting policy analysis in these fields. The Foundation will award at least three research grants of $60,000 each to enable the recipients to research and write a book. Within the academic community, this program supports junior or adjunct faculty, research associates, and post-docs who are engaged in policy-relevant research and writing. Within the think tank community, the program supports members of the rising generation of policy thinkers who are focused on U.S. strategic and foreign policy issues. The Foundation must receive all Strategy and Policy and Fellows proposals by 5:00 p.m., EDT, June 14.
Requirements: An applicant must have a Ph.D. by the time of the deadline, preferably in Political Science, Public Policy, Policy Analysis, International Political Economy, or History. The proposal must be submitted by the institution the applicant has an affiliation and be accompanied by a letter from the institution's representative who is authorized to accept the grant on behalf of the institution. The applicant should also explain how he or she meets all of these requirements in a cover letter to the proposal. The recipients are likely to be under the age of 40. There is, however, no formal age requirement.
Geographic Focus: All States
Date(s) Application is Due: Jun 14
Amount of Grant: 60,000 USD
Contact: Dr. Nadia Schadlow; (203) 222-6222; fax (203) 222-6282; Schadlow@srf.org
Internet: http://www.srf.org/grants/international.php
Sponsor: Smith Richardson Foundation
60 Jesup Road
Westport, CT 06880

Smith Richardson Foundation World Politics and Statecraft Fellowship 4271

The Smith Richardson Foundation announces the World Politics and Statecraft Fellowship program, its annual grant competition to support Ph.D. dissertation research on American foreign policy, international relations, international security, strategic studies, area studies, and diplomatic and military history. The fellowship's objective is to support the research and writing of policy-relevant dissertations through funding of fieldwork, archival research, and language training. In evaluating applications, the Foundation will accord preference to those projects that could directly inform U.S. policy debates and thinking, rather than dissertations that are principally focused on abstract theory or debates within a scholarly discipline. The Foundation will award up to twenty grants of $7,500 each. The Foundation must receive all World Politics and Statecraft Fellowship proposals by 5:00 p.m., EDT, October 15.
Geographic Focus: All States
Amount of Grant: 7,500 USD
Contact: Dr. Nadia Schadlow; (203) 222-6222; fax (203) 222-6282; Schadlow@srf.org
Internet: http://www.srf.org/grants/world_politics.php
Sponsor: Smith Richardson Foundation
60 Jesup Road
Westport, CT 06880

Smithsonian Biodiversity Genomics and Bioinformatics Postdoctoral Fellowships 4272

The Smithsonian Institution (SI) Postdoctoral Fellowships in Biodiversity Genomics and Bioinformatics promote collaborative research in these fields (60%), plus a well-defined outreach component (40%) oriented toward building genomics expertise in the greater SI research community. Research should involve comparative genomic approaches such as phylogenomics, population genomics, metagenomics or transcriptomics, and have a component that involves significant bioinformatics analysis. Your proposal should also detail in at least one page your bioinformatics outreach plan, which can include collaborative work with other SI projects and teams, training workshops or development of software, pipelines or tutorials. We plan to fill three or more Fellowships in the area of Biodiversity Genomics and are especially interested in recruiting a team of Fellows who will work together to advance bioinformatics at the Institution. Coordinated applications dealing with multiple genomic approaches are strongly encouraged. Proposals are due April 1 each year.
Requirements: Applicants must propose to conduct research in-residence for a period of 12 to 24 months. Applicants must have completed or be near completion of the Ph.D. Recipients who have not completed the Ph.D. at the time of application must provide proof of completion of the degree before the fellowship begins.
Geographic Focus: All States
Date(s) Application is Due: Apr 1
Contact: Bryan T. Fair; (202) 275-0655 or (202) 633-7070; siofg@si.edu
Internet: http://www.si.edu/ofg/fell.htm
Sponsor: Smithsonian Institution
470 L'Enfant Plaza SW, Suite 7102, MRC 902 P.O. Box 37012
Washington, DC 20013-7012

Smithsonian Environmental Research Center Graduate Student Fellowships 4273

Graduate Student Fellowships are offered to students formally enrolled in a graduate program of study, who have completed at least one semester, and not yet been advanced to candidacy if in a PhD program. Applicants must submit a proposal for research in a discipline which is pursued at the Smithsonian Institution. The term is 10 weeks with a stipend of $6,500. Fellows are selected in a single, annual competition. The annual application deadline is January 15 of the proposed year of sponsorship.
Geographic Focus: All States
Date(s) Application is Due: Jan 15
Amount of Grant: 6,500 USD
Contact: Dan Gustafson, Internship, Fellowship, and Volunteer Coordinator; (443) 482-2217 or (443) 482-2200; fax (443) 482-2380; gustafsond@si.edu
Internet: http://www.serc.si.edu/pro_training/fellowships/fellowships.aspx
Sponsor: Smithsonian Institution
647 Contees Wharf Road, P.O. Box 28
Edgewater, MD 21037

Smithsonian Environmental Research Center Internships 4274

The Smithsonian Environmental Research Center's (SERC) Internship program offers undergraduate and beginning graduate students a unique opportunity to gain hands-on experience in the fields of environmental research and education. This program enables students to work on specific projects under the direction of SERC's professional staff and is tailored to provide the maximum educational benefit to each participant. Projects include terrestrial, atmospheric and estuarine environmental research within the disciplines of ecology, biology, chemistry, microbiology, botany, zoology, mathematics and physics. Projects are also offered in environmental education and environmental information management. Although students will become familiar with all the research being conducted by a particular staff member and with much of the research of SERC in general, individuals will devote most of their time to a project and content area of particular interest to them. Students will have the opportunity to expand their knowledge

in a selected field of study and to learn a variety of research techniques through firsthand experience. At the conclusion of the internship, student participants will be expected to present the findings of their independent projects in a formal seminar to the SERC community. As an additional educational benefit, SERC maintains an ongoing seminar program that is designed to bring in noted outside speakers on a variety of topics in the environmental sciences. During the program's spring and summer seasons, field trips to research facilities and Smithsonian museums conducting similar programs are also coordinated. Selected candidates will receive a stipend of $475.00 per week. Limited on-site dormitory space may be available for $95.00 per week.
Requirements: The Smithsonian Environmental Research Center will consider applications from currently enrolled undergraduate and beginning graduate students, or students who have recently graduated from an undergraduate or Masters program. Applicants must be in a position to commit fully to the completion of a project. U.S. citizenship is not a requirement to participate in this program. Selection is based upon the student's academic credentials, extent of relevant training or experience, letters of recommendation, and the congruence of the student's expressed goals with those of the internship program.
Restrictions: SERC does not supply board, although cooking facilities and utensils are available.
Geographic Focus: All States
Date(s) Application is Due: Feb 1; Jun 1; Nov 15
Amount of Grant: 475-5,700 USD
Contact: Dan Gustafson, Internship, Fellowship, and Volunteer Coordinator; (443) 482-2217 or (443) 482-2200; fax (443) 482-2380; gustafsond@si.edu
Internet: http://www.serc.si.edu/pro_training/internships/internships.aspx
Sponsor: Smithsonian Institution
647 Contees Wharf Road, P.O. Box 28
Edgewater, MD 21037

Smithsonian Environmental Research Center Postdoctoral/Senior Fellowships 4275
Smithsonian Environmental Research Center Postdoctoral and Senior Fellowships are offered to scholars who have held their degree or equivalent for less than seven years. Senior fellowships are offered to scholars who held their degree or equivalent for seven years or more. Applicants must submit a detailed proposal including a justification for conducting research in-residence at the Smithsonian. The term is 12-24 months at SERC with the second year of funding contigent upon satisfactory progress and availability of funds. Both fellowships have annual stipend of $45,000 plus allowances. Fellows are selected in a single, annual competition. The annual application deadline is January 15 of the proposed year of sponsorship.
Geographic Focus: All States
Date(s) Application is Due: Jan 15
Amount of Grant: 45,000-90,000 USD
Contact: Dan Gustafson, Internship, Fellowship, and Volunteer Coordinator; (443) 482-2217 or (443) 482-2200; fax (443) 482-2380; gustafsond@si.edu
Internet: http://www.serc.si.edu/pro_training/fellowships/fellowships.aspx
Sponsor: Smithsonian Institution
647 Contees Wharf Road, P.O. Box 28
Edgewater, MD 21037

Smithsonian Environmental Research Center Predoctoral Fellowships 4276
Predoctoral Fellowships are offered to doctoral candidates who have completed preliminary course work and examinations. The applicant must submit a detailed proposal including a justification for conducting their research in-residence at the Institution. Candidates must have the approval of their universities to conduct doctoral research at the Smithsonian. The term is 3 to 12 months with annual stipend of $30,000 plus allowances. Fellows are selected in a single, annual competition. The annual application deadline is January 15 of the proposed year of sponsorship.
Geographic Focus: All States
Date(s) Application is Due: Jan 15
Amount of Grant: 7,500-30,000 USD
Contact: Dan Gustafson, Internship, Fellowship, and Volunteer Coordinator; (443) 482-2217 or (443) 482-2200; fax (443) 482-2380; gustafsond@si.edu
Internet: http://www.serc.si.edu/pro_training/fellowships/fellowships.aspx
Sponsor: Smithsonian Institution
647 Contees Wharf Road, P.O. Box 28
Edgewater, MD 21037

Smithsonian Graduate Student Fellowships 4277
Students actively engaged in graduate study at any level may apply to study and conduct research for 10-week periods under the guidance of Smithsonian staff members. Applicants must propose to conduct research in disciplines that are pursued at the Smithsonian (American history, art, and culture; history of science and technology; history of art, design, crafts, and decorative arts; anthropology, archaeology, linguistics, and ethnic studies; evolutionary, systematic, behavioral, and environmental biology and conservation; earth, mineral, and planetary sciences; and materials characterization and conservation) and must plan to spend their entire tenure in residence. These fellowships allow students to conduct research for periods of three to twelve months. Applications are posted to the Web site in September.
Requirements: Applicants must have completed coursework and preliminary examinations for the doctoral degree, and must be engaged in dissertation research. In addition, candidates must have the approval of their universities to conduct their doctoral research at the Smithsonian.
Geographic Focus: All States
Date(s) Application is Due: Jan 15
Amount of Grant: 6,500 USD
Contact: Bryan T. Fair; (202) 275-0655 or (202) 633-7070; siofg@si.edu
Internet: http://www.si.edu/ofg/fell.htm
Sponsor: Smithsonian Institution
470 L'Enfant Plaza SW, Suite 7102, MRC 902 P.O. Box 37012
Washington, DC 20013-7012

Smithsonian Latino Studies Fellowships 4278
Smithsonian Latino Studies Fellowships provide opportunities to U.S. Hispanic predoctoral students and postdoctoral and senior scholars to pursue research topics that relate to Latino art, culture, and history. Interdisciplinary subjects are encouraged and can be undertaken at more than one of the Smithsonian museums and/or research units, and advised by one or more of the Smithsonian research staff members. These types of fellowships are available at both the senior and postdoctoral level. A research and travel allowance is made available to cover additional costs of spending up to one-third of the fellowship tenure away from the Smithsonian, if appropriate and necessary, but not at the fellow's home institution. Applications are posted to the Web site in September.
Requirements: Postdoctoral fellowships are offered to scholars who have held a PhD or equivalent for less than seven years. Senior fellowships are offered to scholars who have held a PhD or equivalent for seven years or more. Predoctoral fellowships are offered to doctoral candidates who have completed preliminary course work and examinations, and have been advanced to candidacy.
Geographic Focus: All States
Date(s) Application is Due: Jan 15
Amount of Grant: 30,000-45,000 USD
Contact: Bryan T. Fair; (202) 275-0655 or (202) 633-7070; siofg@si.edu
Internet: http://www.si.edu/ofg/fell.htm
Sponsor: Smithsonian Institution
470 L'Enfant Plaza SW, Suite 7102, MRC 902 P.O. Box 37012
Washington, DC 20013-7012

Smithsonian Minority Visiting Student Awards 4279
Through the Minority Student Awards Program the Office of Fellowships offers internships and visiting student awards to increase participation of U.S. minority groups who are underrepresented in Smithsonian scholarly programs, in the disciplines of research conducted at the Institution, and in the museum field. Visiting Student Awards are available for currently enrolled advanced graduate students. Visiting Students pursue independently designed research projects in association with Smithsonian staff. Students should contact the Office of Fellowships for application information. The stipend amount is $550 per week for up to ten weeks.
Requirements: U.S. minority undergraduate and graduate students are invited to apply.
Geographic Focus: All States
Date(s) Application is Due: Feb 1; Oct 15
Contact: Bryan T. Fair; (202) 275-0655 or (202) 633-7070; siofg@si.edu
Internet: http://www.si.edu/ofg/fell.htm
Sponsor: Smithsonian Institution
470 L'Enfant Plaza SW, Suite 7102, MRC 902 P.O. Box 37012
Washington, DC 20013-7012

Smithsonian Molecular Evolution Fellowships 4280
Postdoctoral Fellowships in Molecular Evolution are available to support research that uses the resources and research opportunities offered at the National Zoological Park (Washington, D.C.), the National Museum of Natural History (Washington, D.C.) or the Smithsonian Tropical Research Institute (Panama). The stipend being offered is currently $45,000 for one year (12 months) to $90,000 for two years (24 months). The annual deadline is January 15 for awards to begin on or after June 1.
Geographic Focus: All States
Date(s) Application is Due: Jan 15
Contact: Bryan T. Fair; (202) 275-0655 or (202) 633-7070; siofg@si.edu
Internet: http://www.si.edu/ofg/fell.htm
Sponsor: Smithsonian Institution
470 L'Enfant Plaza SW, Suite 7102, MRC 902 P.O. Box 37012
Washington, DC 20013-7012

Smithsonian Museum Conservation Institute Analytical Studies Group Internships 4281
The Smithsonian Museum Conservation Institute in conjunction with the NSF REU Sites Program is offering short-term opportunities for undergraduates to work on research projects with staff members of its technical studies and research group. The Smithsonian's Museum Conservation Institute (MCI) is the center for specialized technical collections research and conservation for all Smithsonian museums, collections, and research centers. MCI combines state-of-the-art instrumentation and scientific techniques to provide technical research and characterization of objects. Internships will take place during the summer. Acceptable majors include, but are not limited to: museum studies, art history, archaeology, paleontology, chemistry, biochemistry, biology, ecology, earth science, chemical engineering, and materials science engineering. A stipend will be offered for a period of up to 10 weeks.
Requirements: Applicants must be currently enrolled as an undergraduate in a university program or scheduled to matriculate in the fall of the current year. Basic chemistry and mathematical coursework is preferred; laboratory experience is not required. Applicants must be U.S. citizens or permanent residents.
Geographic Focus: All States
Date(s) Application is Due: Mar 30
Contact: Paula T. DePriest, Deputy Director; (301) 238-1206 or (301) 238-1240; fax (301) 238-3709; DepriestP@si.edu or MCIweb@si.edu

Internet: http://si.edu/MCI/english/professional_development/2012TechInternships.html
Sponsor: Smithsonian Institution
4210 Silver Hill Road
Suitland, MD 20746

Smithsonian Museum Conservation Institute Internships for Research Projects 4282

The Smithsonian's Museum Conservation Institute is offering short-term opportunities to work on research projects with staff members. The internships are intended mainly for the summer, but consideration will be given to projects at other times during the year. Preference will be given to applicants currently enrolled in conservation training programs or graduate students in related subjects. A stipend may be offered for a period of up to 10 weeks. Projects will be supervised by a number of staff members (listed at the website), whom applicants are urged to contact in advance of application. Applications should be made online prior to the annual March 30 deadline, and selection notifications will be sent by April 15.
Geographic Focus: All States
Date(s) Application is Due: Mar 30
Contact: Paula T. DePriest, Deputy Director; (301) 238-1206 or (301) 238-1240; fax (301) 238-3709; DepriestP@si.edu or MCIweb@si.edu
Internet: http://si.edu/MCI/english/professional_development/2012ResearchInternships.html
Sponsor: Smithsonian Institution
4210 Silver Hill Road
Suitland, MD 20746

Smithsonian Museum Conservation Institute Research Post-Doctorate/Post Graduate Fellowships 4283

Fellowships at the Smithsonian Museum Conservation Institute offers fellowship opportunities involving a variety of artifact analysis, preservation, and conservation treatment specialties. MCI's commitment is to enhance the experience of the fellow or intern, providing specialized technical and scientific training. Such training contributes significantly to the qualifications of the students in their subsequent professional employment. Applications are currently invited for a post-doctoral or a post-graduate fellowship position at the mass spectrom'try and proteomics laboratory at the Smithsonian's Museum Conservation Institute. The aim of the laboratory is to develop mass spectrometry and proteomics technologies relevant to museums' specimens. Projects include but are not limited to: proteomics, biological dating using various MS techniques such as amino acid racemization; analysis of insoluble proteinaceous or polymeric materials; analysis of paints, inks, etc. using surface ionization techniques; and development of portable separation-mass spectrometry devices for onsite chemical/biological analysis. Qualified applicants should send a letter of interest and resume, including lists of publications and references to Dr. Mehdi Moini (moinim@si.edu).
Requirements: The successful applicant will have a Ph.D. or Master's degree in the area of separation, microfabrication, mass spectrometry, or proteomics. Working experience with Thermo LTQ Orbitrap Veloes, ABI Qstar and 4700 TOF/TOF, as well as with capillary electrophoresis and nano-LC is highly desirable.
Geographic Focus: All States
Contact: Paula T. DePriest , Deputy Director; (301) 238-1206 or (301) 238-1240; fax (301) 238-3709; DepriestP@si.edu or MCIweb@si.edu
Internet: http://si.edu/MCI/english/professional_development/2012ResearchPostDoctorate.html
Sponsor: Smithsonian Institution
4210 Silver Hill Road
Suitland, MD 20746

Smithsonian National Air and Space Museum Aviation Space Writers Foundation Grant 4284

The National Air and Space Museum is offering a grant of $5,000, to be awarded in even-numbered years (2012, 2014, 2016 etc.), to support research on aerospace topics. The product created as a result of the grant must be in any form suitable for potential public dissemination in print, electronic, or broadcast or other visual medium, including, but not limited to a book manuscript, a video or film script, or monograph. Potential topics might be contemporary aviation or space events that are of interest to the general public; significant persons, historical events or trends that illuminate the history of human flight in air and space; records or compendia of aerospace source material. Funds may be used to support research travel and expenses, or to produce a product of the research as outlined.
Requirements: Recipients of the award need not be in residence at the National Air and Space Museum. In addition, award winners are required to provide a summary report in the form of a memorandum.
Geographic Focus: All States
Date(s) Application is Due: Jan 15
Contact: Collette Williams; (202) 633-2648; fax (202) 786-2447; williamsce@si.edu
Internet: http://airandspace.si.edu/getinvolved/fellow/writer_grant.cfm
Sponsor: Smithsonian Institution
Sixth and Independence Avenue SW, Room 3313, MRC 312
Washington, DC 20560

Smithsonian National Air and Space Museum Charles A. Lindbergh Chair in Aerospace History 4285

Charles A. Lindbergh (1902-1974) was one of the most significant figures in the history of aviation. On May 20-21, 1927, he made a historic nonstop solo flight in the Ryan NY-P Spirit of St. Louis from New York to Paris that changed the face of aviation for all time. As a result, he was given international honors and acclaimed a hero. The Charles A. Lindbergh Chair in Aerospace History is a competitive twelve-month fellowship open to senior scholars with distinguished records of publication who are at work on, or anticipate being at work on, books in aerospace history. Support is available for replacement of salary and benefits up to a maximum of $100,000 a year. Tenure is typically for an academic year (September through August) and applications are due January 15 of the year preceding the award of the fellowship.
Requirements: The Lindbergh Chair is open to established and recognized senior scholars with distinguished records of publication who are at work on, or anticipate being at work on, books in aerospace history. Residency in the Washington, DC area is a requirement. Interaction with Museum professional staff is highly desired and expected, including attendance at selected professional seminars and colloquia.
Geographic Focus: All States
Date(s) Application is Due: Jan 31
Contact: Collette Williams; (202) 633-2648; fax (202) 786-2447; williamsce@si.edu
Internet: http://airandspace.si.edu/getinvolved/fellow/lindfellow.cfm
Sponsor: Smithsonian Institution
Sixth and Independence Avenue SW, Room 3313, MRC 312
Washington, DC 20560

Smithsonian National Air and Space Museum Guggenheim Fellowships 4286

The Daniel and Florence Guggenheim Foundation was established in 1924 to advance the "well-being of mankind throughout the world." The Foundation supported Robert Goddard's work on the development of liquid-propelled rockets and established many research centers and scholarly activities concerned with the aerospace sciences and exploration. The National Air and Space Museum Guggenheim Fellowships are designed to promote research into, and writing about, the history of aviation and space flight. Candidates are encouraged to pursue programs of research and writing that support publication of works that are scholarly in tone and substance and intended for publication as articles in peer-reviewed journals or in book form from a reputable publisher (in the case of post postdoctoral applicants) or in a doctoral dissertation (in the case of pre-docs). Each fellow will work closely with staff members who have similar interests. Staff members of the Aeronautics Division currently conduct research into aviation history including such themes as the growth and impact of aeronautics on society, the evolution of aircraft technology, and the development of air transport and military aviation. The Division emphasizes both U.S. and international aspects of aviation history. Members of the Space History Division conduct research in the history of post-war science and technology and the history of space flight, and have written major works on the history of rocketry and the origins of space science.
Requirements: The Guggenheim Fellowships are a competitive three- to twelve-month in-residence fellowship for pre- or postdoctoral research in aviation and space history. Predoctoral applicants should have completed preliminary course work and examinations and be engaged in dissertation research. Postdoctoral applicants should have received their Ph.D. within the past seven years. An annual stipend of $30,000 for predoctoral candidates and $45,000 for postdoctoral candidates will be awarded, with limited additional funds for travel and miscellaneous.
Geographic Focus: All States
Date(s) Application is Due: Jan 15
Amount of Grant: 30,000-45,000 USD
Contact: Collette Williams; (202) 633-2648; fax (202) 786-2447; williamsce@si.edu
Internet: http://airandspace.si.edu/getinvolved/fellow/gfellow.cfm
Sponsor: Smithsonian Institution
Sixth and Independence Avenue SW, Room 3313, MRC 312
Washington, DC 20560

Smithsonian National Air and Space Museum Postdoctoral Earth and Planetary Sciences Fellowship 4287

The National Air and Space Museum has established the Postdoctoral Earth and Planetary Sciences Fellowship to support scientific research in this area. Scientists in the Center for Earth and Planetary Studies concentrate on geologic and geophysical research of the Earth and other terrestrial planets, using remote sensing data obtained from Earth-orbiting and interplanetary spacecraft. Research also focuses on global environmental change. Appointments can be made for one or more years. Stipends are compatible with NRC postdoctoral fellowships in the applicant's field. Applications for fellowships at the National Air and Space Museum are typically posted in November. The application deadline is January 15.
Geographic Focus: All States
Date(s) Application is Due: Jan 15
Contact: Collette Williams; (202) 633-2648; fax (202) 786-2447; williamsce@si.edu
Internet: http://airandspace.si.edu/getinvolved/fellow/cepsfell.cfm
Sponsor: Smithsonian Institution
Sixth and Independence Avenue SW, Room 3313, MRC 312
Washington, DC 20560

Smithsonian National Air and Space Museum Verville Fellowships 4288

Alfred V. Verville (1890-1970), an innovative designer, made numerous contributions to civilian and military aviation during his 47 years in the aerospace industry. The Verville Fellowship is designed to promote research into, and writing about, the history of aviation and space flight. Candidates are encouraged to pursue programs of research and writing professional in tone and substance, but addressed to an audience with broad interests. Each fellow will work closely with staff members who have similar interests. Staff members of the Aeronautics Division currently conduct research into aviation history including such themes as the growth and impact of aeronautics on society, the evolution of aircraft technology, and the development of air transport and military aviation. The Division emphasizes both U.S. and international aspects of aviation history.

Members of the Space History Division conduct research in the history of post-war science and technology and the history of space flight, and have written major works on the history of rocketry and the origins of space science.
Requirements: The Verville Fellowship is a competitive nine- to twelve-month in-residence fellowship intended for the analysis of major trends, developments, and accomplishments in the history of aviation or space studies. The fellowship is open to all interested candidates with demonstrated skills in research and writing. An advanced degree in history, engineering, or related fields is not a requirement. An annual stipend of $55,000 will be awarded for a 12-month fellowship, with limited additional funds for travel and miscellaneous expenses.
Geographic Focus: All States
Date(s) Application is Due: Jan 15
Amount of Grant: 55,000 USD
Contact: Collette Williams; (202) 633-2648; fax (202) 786-2447; williamsce@si.edu
Internet: http://airandspace.si.edu/getinvolved/fellow/vfellow.cfm
Sponsor: Smithsonian Institution
Sixth and Independence Avenue SW, Room 3313, MRC 312
Washington, DC 20560

Smithsonian National Museum of American History Lemelson Center Fellowships 4289
The Lemelson Center Fellows Program supports projects that present creative approaches to the study of invention and innovation in American society. These include, but are not limited to, historical research and documentation projects, resulting in publications, exhibitions, educational initiatives, and multimedia products. The Lemelson Center invites application covering a broad spectrum of research topics that resonate with its mission to foster a greater understanding of invention and innovation, broadly defined. However, the Center especially encourages project proposals that will illuminate the role of women inventors; inventors with disabilities; inventors from diverse backgrounds; or any inventions and technologies associated with groups that are traditionally under-represented in the historical record. Fellowship tenure is based upon the applicants' stated needs (and available funding) up to a maximum of ten weeks. The term is up to ten weeks in length, with pre-doctoral fellows earning a $575 stipend per week, and post-doctoral and professionals earning an $870 stipend per week.
Requirements: The Center offers fellowships to pre-doctoral graduate students, postdoctoral scholars, and other professionals who have completed advanced training. Fellows are expected to reside in the Washington, D.C. area, to participate in the Center's activities, and to make presentations on their work to colleagues at the museum.
Geographic Focus: All States
Date(s) Application is Due: Jan 15
Amount of Grant: 575-8,700 USD
Contact: Omar Eaton-Martínez; (202) 633-3556; fax (202) 786-2851
Eric Hintz, Historian; (202) 633-3734; fax (202) 357-4517; hintze@si.edu
Internet: http://americanhistory.si.edu/getinvolved/fellowindex.cfm
Sponsor: Smithsonian Institution
14th Street and Constitution Avenue, NW
Washington, DC 20013

Smithsonian National Museum of National History Lemelson Center Travel to Collections Awards 4290
The Travel to Collections Awards Program supports research on the history of invention and innovation based on the holdings of the Museum's Archives Center and curatorial divisions. The Archives Center holds more than 20,000 feet of archival materials. The collections are particularly strong in personal papers and business records documenting the history of American enterprise and technology. To encourage use of its invention-related collections, the Lemelson Center offers five short-term travel awards per year. Awards are $150 per day for ten business days maximum. The travel award may be used to cover transportation and living expenses, and copying pertinent archival resources. Applications are accepted from June 1 to July 1 each year.
Requirements: Scholars, graduate students, and independent researchers not residing or attending school within commuting distance of the National Museum of American History may apply for this program. Recipients must commence their research at the National Museum of American History within one year of being notified of the award.
Restrictions: Awards may not be used to extend other Smithsonian appointments. Only one award can be offered to a visitor within a twelve-month period.
Geographic Focus: All States
Date(s) Application is Due: Jul 1
Amount of Grant: 150-1,500 USD
Contact: Omar Eaton-Martínez; (202) 633-3556; fax (202) 786-2851
Internet: http://invention.smithsonian.org/resources/research_travel.aspx
Sponsor: Smithsonian Institution
14th Street and Constitution Avenue, NW
Washington, DC 20013

Smithsonian Native American Community Scholars Awards 4291
The Office of Fellowships offers Community Scholars Awards to Native Americans who are formally or informally related to a Native American community to undertake individually designed research projects related to Native American topics and using Native American resources at the Smithsonian. The stipend is equal to $150 per day for up to 21 days, with a travel allowance and small research allowance also being offered. Deadlines are February 1 (to start after June 1), June 1 (to start after October 1), and October 1 (to start after January 1).
Requirements: Native American undergraduate and graduate students are eligible.
Geographic Focus: All States
Date(s) Application is Due: Feb 1; Jun 1; Oct 1
Amount of Grant: Up to 3,500 USD
Contact: Bryan T. Fair; (202) 275-0655 or (202) 633-7070; siofg@si.edu
Internet: http://www.si.edu/ofg/fell.htm
Sponsor: Smithsonian Institution
470 L'Enfant Plaza SW, Suite 7102, MRC 902 P.O. Box 37012
Washington, DC 20013-7012

Smithsonian Native American Visiting Student Awards 4292
Appointments are available for currently enrolled advanced Native American graduate students. Visiting students pursue independent research in association with Smithsonian staff. Stipends are per day or per week and include a travel allowance and small research allowance. This award is granted three times a year, depending upon when the funds are requested. The stipend is $150 per day for up to a maximum of 21 days, and $550 per week for three to ten weeks. A travel allowance and a small research allowance is also offered. Deadlines are February 1 (to start after June 1), June 1 (to start after October 1), and October 1 (to start after January 1).
Requirements: Native American graduate students are eligible.
Geographic Focus: All States
Date(s) Application is Due: Feb 1; Jun 1; Oct 1
Amount of Grant: 150-5,500 USD
Contact: Bryan T. Fair; (202) 275-0655 or (202) 633-7070; siofg@si.edu
Internet: http://www.si.edu/ofg/fell.htm
Sponsor: Smithsonian Institution
470 L'Enfant Plaza SW, Suite 7102, MRC 902 P.O. Box 37012
Washington, DC 20013-7012

Smithsonian Postdoctoral Fellowships 4293
Postdoctoral Fellowships of three to twelve months are available for scholars who have held the doctoral degree or equivalent for fewer than seven years as of the application deadline. Proposals for research may be offered in fields of study conducted at the institute (American history, art, and culture; history of science and technology; history of art, design, crafts, and decorative arts; anthropology, archaeology, linguistics, and ethnic studies; evolutionary, systematic, behavioral, and environmental biology and conservation; earth, mineral, and planetary sciences; and materials characterization and conservation). The fellowship term is three to 12 months. Applications are posted on the Web site each September.
Requirements: Applicants must have a PhD or equivalent degree for less than seven years.
Geographic Focus: All States
Date(s) Application is Due: Jan 15
Amount of Grant: 45,000 USD
Contact: Bryan T. Fair; (202) 275-0655 or (202) 633-7070; siofg@si.edu
Internet: http://www.si.edu/ofg/fell.htm
Sponsor: Smithsonian Institution
470 L'Enfant Plaza SW, Suite 7102, MRC 902 P.O. Box 37012
Washington, DC 20013-7012

Smithsonian Postgraduate Fellowships in Conservation of Museum Collections 4294
Smithsonian Postgraduate Fellowships in Conservation of Museum Collections are offered to recent graduates of masters programs in art conservation or the equivalent or conservation scientists, including those at the postdoctoral level, who wish to conduct research and gain further training in Smithsonian conservation laboratories for a period of one year. Additional facilities may be available to museum or archives fellows for analytical work at the Museum Conservation Institute (MCI). Fellowship duration in this category is three to twelve months. A stipend of $35,000 is being offered, plus allowances. The deadline is January 15 for awards to begin on or after June 1. Applicants are urged to apply concurrently to all other Smithsonian programs for which they may be eligible.
Geographic Focus: All States
Date(s) Application is Due: Jan 15
Amount of Grant: 35,000 USD
Contact: Bryan T. Fair; (202) 275-0655 or (202) 633-7070; siofg@si.edu
Internet: http://www.si.edu/ofg/fell.htm
Sponsor: Smithsonian Institution
470 L'Enfant Plaza SW, Suite 7102, MRC 902 P.O. Box 37012
Washington, DC 20013-7012

Smithsonian Predoctoral Fellowships 4295
Fellowships support independently conceived and designed research study by predoctoral students to work with Smithsonian staff members and use Smithsonian collections, facilities, and other special resources. Proposals are accepted for research in the areas of study conducted at the institute (American history, art, and culture; history of science and technology; history of art, design, crafts, and decorative arts; anthropology, archaeology, linguistics, and ethnic studies; evolutionary, systematic, behavioral, and environmental biology and conservation; earth, mineral, and planetary sciences; and materials characterization and conservation). The fellowship term is three to 12 months. Applications are posted to the Web site in September.
Requirements: Candidates must have completed all coursework and examinations and be ready to engage in dissertation research.
Geographic Focus: All States
Date(s) Application is Due: Jan 15
Amount of Grant: 30,000 USD
Contact: Bryan T. Fair; (202) 275-0655 or (202) 633-7070; siofg@si.edu

Internet: http://www.si.edu/ofg/fell.htm
Sponsor: Smithsonian Institution
470 L'Enfant Plaza SW, Suite 7102, MRC 902 P.O. Box 37012
Washington, DC 20013-7012

Smithsonian Senior Fellowships 4296

Senior Fellowships of three to twelve months are available for scholars who have held the doctoral degree or equivalent for more than seven years as of the application deadline. Applications for these fellowships may be made up to eighteen months in advance. Stipends for senior fellowships are the same as for the postdoctoral program, but the Smithsonian's stipend may be matched by other sources of funding such as a sabbatical salary. Each year, appointments are made to distinguished senior scientists for research in the fields of anthropology, natural history, environmental science and ecological processes, tropical biology, African art and culture, oriental art, astrophysics and related topics, history of science and technology, American cultural and social history, or art history. Prior to submitting applications, interested senior scientists should consult with Smithsonian staff in their chosen fields. Applications are posted to the Web site in September.
Requirements: Applicants must have held the PhD or equivalent degree for at least seven years.
Geographic Focus: All States
Date(s) Application is Due: Jan 15
Amount of Grant: 45,000 USD
Contact: Bryan T. Fair; (202) 275-0655 or (202) 633-7070; siofg@si.edu
Internet: http://www.si.edu/ofg/fell.htm
Sponsor: Smithsonian Institution
470 L'Enfant Plaza SW, Suite 7102, MRC 902 P.O. Box 37012
Washington, DC 20013-7012

Smithsonian Tropical Research Institute Short Term Fellowships 4297

The Smithsonian Tropical Research Institute (STRI), is a division of the Smithsonian Institution in Washington DC and maintains research facilities for marine and terrestrial research at various locations on the Isthmus of Panama. Disciplines include ecology, anthropology, paleontology, paleoecology, evolutionary biology, molecular phylogenetics, biogeography, animal behavior, neurobiology, soils sciences, and physiology of tropical plants and animals. Support is provided to carry out short-term research projects in the tropics. Fellowships may provide a modest stipend to cover living expenses while at STRI (currently $750/month); a modest research allowance that rarely exceeds $2000 and is usually less; and roundtrip coach airfare. Applicants are encouraged to seek additional sources of funding, and partial awards are often given.
Requirements: Before submitting a formal application all applicants should consult with STRI research staff, who will serve as potential advisors, to confirm that they are willing to supervise and support the proposed project. Proposals without the support of named advisors will not be funded. When corresponding with advisors applicants are strongly encouraged to ask for advice on the scientific merit of the proposed research, methodology and logistics, and for written comments on preliminary drafts of the proposals. Proposals may be submitted in English or Spanish. Download the complete guidelines from the sponsor's website.
Restrictions: Projects usually are complete in themselves and their maximum duration is 3 months; extensions are awarded only in exceptional circumstances.
Geographic Focus: All States
Date(s) Application is Due: Feb 15; May 15; Aug 15; Nov 15
Amount of Grant: 750-9,000 USD
Contact: Adriana Bilgray, (507) 212-8031; fax (507) 212-8150; fellows@si.edu
Internet: http://www.stri.org/english/education_fellowships/fellowships/index.php
Sponsor: Smithsonian Tropical Research Institute
Office of Academic Programs, Unit 0948
AP.O., AA 34002-0948

SNM/ Covidien Seed Grant in Molecular Imaging/Nuclear Medicine Research 4298

The Society of Nuclear Medicine is pleased to offer the SNM/Covidien Seed Grant in Nuclear Medicine Research. This grant is designed to assist researchers in conducting new and innovative pilot projects that have potential for future support from foundations, corporations or government agencies.
Requirements: Applicant Eligibility: basic or clinical scientists employed by academic and research-oriented organizations; applicants from any country may apply; applicants must hold a full-time faculty position in an educational institution when the award period begins; applicants must have completed all advanced training and be no more than five years post training; applicants must not have served as a principal investigator on peer-reviewed grants totaling $50,000 or more in a single calendar year; preference will be given to individuals who have demonstrated great potential for a research career in the field of nuclear medicine/molecular imaging.
Restrictions: No grantee may receive more than one SNM research grant in any one year. The grant will not pay institutional overhead costs or indirect costs. No salary support for principal investigator or co-principal investigators. Grants will be limited to a time span of one year; no extensions are allowed
Geographic Focus: All States
Date(s) Application is Due: Feb 20
Amount of Grant: 25,000 USD
Contact: Nicole Kern; (703) 652-6795, ext. 1255; fax (703) 708-9015; nkern@snm.org
Internet: http://interactive.snm.org/index.cfm?PageID=2255
Sponsor: Society of Nuclear Medicine
1850 Samuel Morse Drive
Reston, VA 20190

SNM Molecular Imaging Research Grant For Junior Medical Faculty 4299

The objective of this program is to provide salary support for one junior faculty member in an academic/research setting to enable them to engage in Molecular Imaging research related to diagnostic or therapeutic applications. The grant will support new, innovative pilot projects in any area of molecular imaging research. The grant may supplement other funds, but not duplicate other support.
Requirements: Applicant Eligibility: applicants must have MD degree (or equivalent) and have completed a residency and be certified in their specialty; applicants must be practicing in an academic/research setting as a faculty member in the United States; applicants must be within 5 years of their initial faculty appointment with an academic rank of instructor or assistant professor (or equivalent); applicants must provide a detailed proposal of the proposed research activity; applicants must not have received grant/contract amounts totaling $50,000.00 or more in a single calendar year as the principle investigator; preference will be given to individuals who have demonstrated great potential for a research career in the field of molecular imaging; multidisciplinary projects are encouraged; only one application per faculty member is accepted per grant; no grantee may receive more than one SNM research grant in any one year. Likewise no other research grant request or continuation will be considered until a satisfactory summary of an earlier grant is received; applicants are required to be a SNM member at the time of application.
Restrictions: The grant will not pay institutional overhead costs or indirect costs
Geographic Focus: All States
Date(s) Application is Due: Jan 20
Amount of Grant: 100,000 USD
Contact: Nicole Kern; (703) 652-6795, ext. 1255; fax (703) 708-9015; nkern@snm.org
Internet: http://interactive.snm.org/index.cfm?PageID=7252
Sponsor: Society of Nuclear Medicine
1850 Samuel Morse Drive
Reston, VA 20190

SNM Pilot Research Grants 4300

The pilot research grants are designed to help basic or clinical scientists in the early stages of their career conduct research that may lead to further funding. The Mitzi and William Blahd, MD Pilot Research Grant is awarded to the applicant with the highest-ranked proposal. Preference will be given to individuals who have demonstrated great potential for a research career in the field of nuclear medicine/molecular imaging.
Requirements: Applicant Eligibility: basic or clinical scientists employed by academic and research-oriented organizations in North America are eligible; applicants must not be no more than five years post training.
Geographic Focus: All States
Date(s) Application is Due: Feb 20
Amount of Grant: 25,000 USD
Contact: Nicole Mitchell, Program Manager; (703) 652-6795; nkern@snm.org
Internet: http://interactive.snm.org/index.cfm?PageID=2255
Sponsor: Society of Nuclear Medicine
1850 Samuel Morse Drive
Reston, VA 20190

SNM Pilot Research Grants in Nuclear Medicine/Molecular Imaging 4301

The pilot research grants are designed to help basic or clinical scientists in the early stages of their career conduct research that may lead to further funding.
Requirements: Applicant Eligibility: basic or clinical scientists employed by academic and research-oriented organizations in North America are eligible; applicants must be no more than five years post training; applicant must not have served as the principal investigator of a peer-reviewed grant for more than $50,000 in a single calendar year; preference will be given to individuals who have demonstrated great potential for a research career in the field of nuclear medicine/molecular imaging.
Restrictions: The grant will not pay salaries, institutional overhead costs or indirect costs. Funds may supplement other funds, but not duplicate other support. Grants will be limited to a time span of one year. No grantee may receive more than one grant in any one year.
Geographic Focus: All States
Date(s) Application is Due: Feb 20
Amount of Grant: 25,000 USD
Contact: Nicole Kern, SNM Program Manager; (703) 652-6795, ext. 1255; fax (703) 708-9015; nkern@snm.org
Internet: http://interactive.snm.org/index.cfm?PageID=2255
Sponsor: Society of Nuclear Medicine
1850 Samuel Morse Drive
Reston, VA 20190

SNM Postdoctoral Molecular Imaging Scholar Grants 4302

The postdoctoral program will support a two-year research endeavor that promotes integration of molecular imaging into the career of the trainee. The funds maybe used for new, innovative pilot projects in any area of molecular imaging research. The grant may supplement other funds, but not duplicate other support. A summary report and an accounting of funds will be requested at the halfway point and upon completion of the research project. Grantees are expected to submit results of the research as an abstract for presentation at the SNM Annual Meeting and/or as a scientific manuscript to the Journal of Nuclear Medicine. A final report is due within 60 days after completion of the research project.
Requirements: Applicant Eligibility: MD, MD/PhD, and PhD applicants are eligible; applicants from any country may apply; applicants must hold a full-time position in an educational institution when the award starts; applicants must have completed all basic training and be no more than 2 years post training; applicants must not have served as a principal investigator on peer-reviewed grants totaling $50,000 or more in a single

calendar year; preference will be given to individuals who have demonstrated great potential for a research career in the field of molecular imaging; only one application per trainee is accepted; no grantee may receive more than one SNM research grant in any one year. Likewise no other research grant request or continuation will be considered until a satisfactory summary of an earlier grant is received.
Restrictions: Grant extensions are not often allowed, but will be reviewed and decided upon by the SNM Grants and Awards Committee if requested. The grant will not pay institutional overhead costs or indirect costs.
Geographic Focus: All States
Date(s) Application is Due: Jan 20
Amount of Grant: 60,000 USD
Contact: Nicole Kern; (703) 652-6795, ext. 1255; fax (703) 708-9015; nkern@snm.org
Internet: http://interactive.snm.org/index.cfm?PageID=7252
Sponsor: Society of Nuclear Medicine
1850 Samuel Morse Drive
Reston, VA 20190

SNM Student Fellowship Awards 4303
The SNM Student Fellowships are designed to stimulate interest in molecular imaging/nuclear medicine by supporting their full-time participation in clinical and basic research activities for three months (or less). These Fellowships are made possible by support of the Education and Research Foundation for SNM.
Requirements: Candidates from any country may apply. Students must be enrolled in a medical, pharmacy or graduate school, or be an undergraduate student demonstrating outstanding competence in nuclear medicine and/or molecular imaging research. Fellowship must be carried out during a three-month that starts no earlier then March and completed no later then December 31st.
Geographic Focus: All States
Date(s) Application is Due: Jan 23
Amount of Grant: 3,000 USD
Contact: Nicole Kern; (703) 652-6795, ext. 1255; fax (703) 708-9015; nkern@snm.org
Internet: http://interactive.snm.org/index.cfm?PageID=2254
Sponsor: Society of Nuclear Medicine
1850 Samuel Morse Drive
Reston, VA 20190

SNM Tetalman Memorial Young Investigator Awards 4304
Established by the family and friends of Dr. Tetalman, this award honors the work of a young investigator who is pursuing a career in nuclear medicine. It is named in memory of a highly respected and productive clinician and researcher, Mark Tetalman, M.D., whose career was cut tragically short. The award is based on a submitted paper supporting current research efforts and accomplishments, teaching, clinical service and administration. Service to the Society and a commitment to academic nuclear medicine will also be considered.
Requirements: Applicants must be 36 years old or younger as of July 1 of the application year and must have obtained certification in Nuclear Medicine or Nuclear Radiology or have completed a PhD program within the last seven years. Complete applications must be must be received, not simply post-marked by November 14th. One original unstapled copy of the completed application must be sent to: SNM Development Office, Attention: SNM Committee on Awards
Geographic Focus: All States
Date(s) Application is Due: Nov 14
Amount of Grant: 2,500 USD
Contact: Nicole Kern; (703) 652-6795, ext. 1255; fax (703) 708-9015; nkern@snm.org
Internet: http://interactive.snm.org/index.cfm?PageID=2253
Sponsor: Society of Nuclear Medicine
1850 Samuel Morse Drive
Reston, VA 20190

Society for Imaging Informatics in Medicine (SIIM) Emeritus Mentor Grants 4305
The mission of SIIM, The Society for Imaging Informatics in Medicine, is to advance computer applications and information technology in medical imaging through education and research. In support of this mission, the SIIM Research and Development Committee offers a $20,000 grant for a retired, or semi-retired, expert in imaging informatics to act as a mentor for a trainee (student, resident, fellow) or junior faculty member for hypothesis-driven research. Any area of research involving medical imaging informatics is eligible for support, as well as development of new hardware, software, or processes to support electronic imaging practice, education, or research.
Requirements: A senior, experienced investigator jointly applies with a resident, fellow, graduate student, or junior faculty for a SIIM emeritus mentor grant. Preference will be given to projects at SIIM institutions and to projects that contribute to the development of careers of new or prospective researchers. This grant may not be applied for in conjunction with another SIIM grant covering the same subject material.
Geographic Focus: All States
Date(s) Application is Due: Aug 3
Amount of Grant: 20,000 USD
Contact: Grant Manager; (703) 723-0432; fax (703) 723-0415; grants@siimweb.org or info@siimweb.org
Internet: http://www.scarnet.org/index.cfm?id=63
Sponsor: Society for Imaging Informatics in Medicine
19440 Golf Vista Plaza, Suite 330
Leesburg, VA 20176-8264

Society for Imaging Informatics in Medicine (SIIM) Micro Grants 4306
The mission of SIIM, The Society for Imaging Informatics in Medicine, is to advance computer applications and information technology in medical imaging through education and research. In support of this mission, the SIIM Research and Development Committee offers a $5,000 grant for hypothesis-driven experiments of short duration and high risk. Prior preliminary data is not required. An imaging informatics project that needs resources falls between what a researcher is willing to fund personally or is more than his/her department provides in seed money and what is typically requested in grant applications. Examples include trying unconventional hardware or software in an imaging informatics experiment, trying a particular piece of hardware for a user-interface design, paying for short duration consulting assistance, or paying for an educational course that the researcher needs to enhance a particular study.
Requirements: There are no restrictions on who may apply for the grant. Trainees (residents, fellows, and graduate students) are required to include a letter from a faculty member, acting as an advisor, to show that the applicant's department supports the project. Preference will be given to projects at SIIM institutions and to projects that may yield high value results or provide enough preliminary data to justify additional work and application for traditional grant funding.
Geographic Focus: All States
Date(s) Application is Due: Aug 3
Amount of Grant: 5,000 USD
Contact: Grant Manager; (703) 723-0432; fax (703) 723-0415; grants@siimweb.org
Internet: http://www.scarnet.org/index.cfm?id=63
Sponsor: Society for Imaging Informatics in Medicine
19440 Golf Vista Plaza, Suite 330
Leesburg, VA 20176-8264

Society for Imaging Informatics in Medicine (SIIM) Research Grants 4307
The mission of SIIM, The Society for Imaging Informatics in Medicine, is to advance computer applications and information technology in medical imaging through education and research. In support of this mission, the SIIM Research and Development Committee offers two $50,000 grants for original research. Any area of research involving medical imaging informatics is eligible for support, as well as development of new hardware, software, or processes to support electronic imaging practice, education, or research.
Requirements: Residents, fellows, graduate students, or faculty may apply for SIIM research grants. Preference will be given to projects at SIIM institutions and to projects that contribute to the development of careers of new or prospective researchers. Residents, fellows, and students must conduct their projects under the guidance of experienced investigators. Grant recipients will not be eligible for concurrent support through other SIIM grants.
Geographic Focus: All States
Date(s) Application is Due: Aug 3
Amount of Grant: 50,000 USD
Contact: Grant Manager; (703) 723-0432; fax (703) 723-0415; grants@siimweb.org
Internet: http://www.scarnet.org/index.cfm?id=63
Sponsor: Society for Imaging Informatics in Medicine
19440 Golf Vista Plaza, Suite 330
Leesburg, VA 20176-8264

Society for Imaging Informatics in Medicine (SIIM) Small Grant for Product Support Development 4308
The mission of SIIM, The Society for Imaging Informatics in Medicine, is to advance computer applications and information technology in medical imaging through education and research. In support of this mission, the SIIM Research and Development Committee offers a $20,000 grant to support the development of a product or tool for use in the field of imaging informatics. Any area of product or tool development for use in the field of medical imaging informatics is eligible for support (e.g., open source application development, work flow measurement tools, replacement PACS change management tools).
Requirements: Residents, fellows, graduate students, or faculty may apply for product support development grants. Preference will be given to projects at SIIM institutions and to projects that contribute to the development of careers of new or prospective researchers. Residents, fellows, and students must conduct their projects under the guidance of experienced investigators.
Geographic Focus: All States
Date(s) Application is Due: Aug 3
Amount of Grant: 20,000 USD
Contact: Grant Manager; (703) 723-0432; fax (703) 723-0415; grants@siimweb.org
Internet: http://www.scarnet.org/index.cfm?id=63
Sponsor: Society for Imaging Informatics in Medicine
19440 Golf Vista Plaza, Suite 330
Leesburg, VA 20176-8264

Society for Imaging Informatics in Medicine (SIIM) Small Training Grants 4309
The mission of SIIM, The Society for Imaging Informatics in Medicine, is to advance computer applications and information technology in medical imaging through education and research. In support of this mission, the SIIM Research and Development Committee offers a $20,000 grant targeted at providing training opportunities in the field of imaging informatics for students and imaging informatics professionals. Any area of training within medical imaging informatics is eligible for support. The grant is designed to support hands-on training or experience with a designated mentor or sponsor in the imaging informatics arena. It is not intended to support tuition or other traditional education-related expenses. The period of training must last a minimum of three months and may extend no longer than twelve months.
Requirements: Residents, fellows, graduate students, or faculty members at recognized institutions may apply for SIIM training grants. Preference will be given to training

experiences at SIIM institutions and to those that contribute to the careers of new or prospective imaging informatics personnel.
Geographic Focus: All States
Date(s) Application is Due: Aug 3
Amount of Grant: 20,000 USD
Contact: Grant Manager; (703) 723-0432; fax (703) 723-0415; grants@siimweb.org
Internet: http://www.scarnet.org/index.cfm?id=63
Sponsor: Society for Imaging Informatics in Medicine
19440 Golf Vista Plaza, Suite 330
Leesburg, VA 20176-8264

Society for Research in Child Development Executive Branch Policy Fellowships 4310
The fellowships are designed to provide greater interaction between the developmental research community and federal research programs and policies. Fellows spend a year in Washington working as resident scholars in a federal agency that sponsors or uses developmental research, providing advice and guidance to programs, and learning about federal science policy processes.
Requirements: Applicants must have doctoral-level degrees and exceptional competence in an area of child development research. PhDs, MDs, EdDs, and DSWs are encouraged to apply.
Geographic Focus: All States
Date(s) Application is Due: Dec 15
Amount of Grant: 45,000-65,000 USD
Contact: Director; (202) 336-5926; fax (202) 336-5953; SRCD@apa.org
Internet: http://www.srcd.org/policyfellowships.html
Sponsor: Society for Research in Child Development
750 First Street NE
Washington, DC 20002-4242

Society for the Arts in Healthcare Environmental Arts Research Grant 4311
The American Art Resources and Society for the Arts in Healthcare Environmental Arts Research Grant program offers a cash award sponsored by American Art Resources in Houston, Texas. The program is provided to a Principal Investigator to conduct or supplement Arts in Healthcare research. Selection is made on the merit, innovation and feasibility of the research proposal.
Requirements: Applicant organizations must be based in the United States.
Geographic Focus: All States
Contact: Anita Boles; (202) 299-9770; fax (202) 299-9887; anita@thesah.org
Internet: http://www.thesah.org/template/page.cfm?page_id=570
Sponsor: Society for the Arts in Healthcare
2437 15th Street NW, B South
Washington, DC 20009

Society of Cosmetic Chemists Allan B. Black Award Sponsored by Presperse 4312
Dedicated to the advancement of cosmetic science, the Society strives to increase and disseminate scientific information through meetings and publications. By promoting research in cosmetic science and industry, and by setting high ethical, professional and educational standards, it strives to improve the qualifications of cosmetic scientists. Its mission is to further the interests and recognition of cosmetic scientists while maintaining the confidence of the public in the cosmetic and toiletries industry. With the Allan B. Black Award, a scroll and honorarium are awarded annually for the Best Paper on make-up technology either presented at an annual meeting, a seminar, or published in the Journal of Cosmetic Science, the Official Journal of the Society of Cosmetic Chemists.
Geographic Focus: All States
Contact: Theresa Cesario, (212) 668-1500; fax (212) 668-1504; TCesario@SCConline.org
Internet: http://www.scconline.org/website/about_scc/awards_fellowships.shtml
Sponsor: Society of Cosmetic Chemists
120 Wall Street, Suite 2400
New York, NY 10005-4088

Society of Cosmetic Chemists Award Sponsored by McIntyre Group, Ltd. 4313
Dedicated to the advancement of cosmetic science, the Society strives to increase and disseminate scientific information through meetings and publications. By promoting research in cosmetic science and industry, and by setting high ethical, professional and educational standards, it strives to improve the qualifications of cosmetic scientists. Its mission is to further the interests and recognition of cosmetic scientists while maintaining the confidence of the public in the cosmetic and toiletries industry. This award, given in the form of a scroll and honorarium, are awarded for the best paper presented at the Society's Annual Scientific Meeting.
Geographic Focus: All States
Contact: Theresa Cesario, (212) 668-1500; fax (212) 668-1504; TCesario@SCConline.org
Internet: http://www.scconline.org/website/about_scc/awards_fellowships.shtml
Sponsor: Society of Cosmetic Chemists
120 Wall Street, Suite 2400
New York, NY 10005-4088

Society of Cosmetic Chemists Award Sponsored by The HallStar Company 4314
Dedicated to the advancement of cosmetic science, the Society strives to increase and disseminate scientific information through meetings and publications. By promoting research in cosmetic science and industry, and by setting high ethical, professional and educational standards, it strives to improve the qualifications of cosmetic scientists. Its mission is to further the interests and recognition of cosmetic scientists while maintaining the confidence of the public in the cosmetic and toiletries industry. Offered in the form of a scroll and honorarium, this honor is awarded for the Best Paper that makes the greatest scientific contribution to knowledge in the field of protecting against or ameliorating damage to human skin caused by exposure to UV radiation presented at the Society's Annual Scientific Seminar.
Geographic Focus: All States
Contact: Theresa Cesario, (212) 668-1500; fax (212) 668-1504; TCesario@SCConline.org
Internet: http://www.scconline.org/website/about_scc/awards_fellowships.shtml
Sponsor: Society of Cosmetic Chemists
120 Wall Street, Suite 2400
New York, NY 10005-4088

Society of Cosmetic Chemists Chapter Best Speaker Award 4315
Dedicated to the advancement of cosmetic science, the Society strives to increase and disseminate scientific information through meetings and publications. By promoting research in cosmetic science and industry, and by setting high ethical, professional and educational standards, it strives to improve the qualifications of cosmetic scientists. Its mission is to further the interests and recognition of cosmetic scientists while maintaining the confidence of the public in the cosmetic and toiletries industry. The Chapter Best Speaker Award is given for the best scientific paper presented before each Chapter during a given calendar year. Recommendations for this award originates with the Chapters and are submitted to the National Office. A scroll and honorarium ($250 for a manuscript submission and $150 for an abstract submission) is provided by National for presentation by the Chapter.
Geographic Focus: All States
Contact: Theresa Cesario, (212) 668-1500; fax (212) 668-1504; TCesario@SCConline.org
Internet: http://www.scconline.org/website/about_scc/awards_fellowships.shtml
Sponsor: Society of Cosmetic Chemists
120 Wall Street, Suite 2400
New York, NY 10005-4088

Society of Cosmetic Chemists Chapter Merit Award 4316
Dedicated to the advancement of cosmetic science, the Society strives to increase and disseminate scientific information through meetings and publications. By promoting research in cosmetic science and industry, and by setting high ethical, professional and educational standards, it strives to improve the qualifications of cosmetic scientists. Its mission is to further the interests and recognition of cosmetic scientists while maintaining the confidence of the public in the cosmetic and toiletries industry. The Chapter Merit Award is awarded to a Chapter Member for outstanding service, dedication and voluntary services over a period of time. A scroll is provided by National for presentation by the Chapter. Recommendations for this award originates with each Chapter and are submitted to the National Office.
Geographic Focus: All States
Contact: Theresa Cesario, (212) 668-1500; fax (212) 668-1504; TCesario@SCConline.org
Internet: http://www.scconline.org/website/about_scc/awards_fellowships.shtml
Sponsor: Society of Cosmetic Chemists
120 Wall Street, Suite 2400
New York, NY 10005-4088

Society of Cosmetic Chemists Frontier of Science Award 4317
Dedicated to the advancement of cosmetic science, the Society strives to increase and disseminate scientific information through meetings and publications. By promoting research in cosmetic science and industry, and by setting high ethical, professional and educational standards, it strives to improve the qualifications of cosmetic scientists. Its mission is to further the interests and recognition of cosmetic scientists while maintaining the confidence of the public in the cosmetic and toiletries industry. The Frontier of Science Award is an honorarium given to a speaker who has achieved exceptional national or international stature in the scientific community for delivering a lecture at the Annual Scientific Meeting.
Geographic Focus: All States
Contact: Theresa Cesario, (212) 668-1500; fax (212) 668-1504; TCesario@SCConline.org
Internet: http://www.scconline.org/website/about_scc/awards_fellowships.shtml
Sponsor: Society of Cosmetic Chemists
120 Wall Street, Suite 2400
New York, NY 10005-4088

Society of Cosmetic Chemists Hans Schaeffer Award 4318
Dedicated to the advancement of cosmetic science, the Society strives to increase and disseminate scientific information through meetings and publications. By promoting research in cosmetic science and industry, and by setting high ethical, professional and educational standards, it strives to improve the qualifications of cosmetic scientists. Its mission is to further the interests and recognition of cosmetic scientists while maintaining the confidence of the public in the cosmetic and toiletries industry. The Hans Schaeffer Award, offered in the form of a scroll and honorarium, are awarded annually for the Most Innovative Paper presented at either an annual meeting or a seminar.
Geographic Focus: All States
Contact: Theresa Cesario, (212) 668-1500; fax (212) 668-1504; TCesario@SCConline.org
Internet: http://www.scconline.org/website/about_scc/awards_fellowships.shtml
Sponsor: Society of Cosmetic Chemists
120 Wall Street, Suite 2400
New York, NY 10005-4088

Society of Cosmetic Chemists Joseph P. Ciaudelli Award 4319
Dedicated to the advancement of cosmetic science, the Society strives to increase and disseminate scientific information through meetings and publications. By promoting research in cosmetic science and industry, and by setting high ethical, professional and educational standards, it strives to improve the qualifications of cosmetic scientists. Its mission is to further the interests and recognition of cosmetic scientists while maintaining the confidence of the public in the cosmetic and toiletries industry. The Joseph P. Ciaudelli Award, offered

in the form of a scroll and honorarium, are bestowed annually for the best article appearing in the Journal of Cosmetic Science on the subject of hair care technology.
Geographic Focus: All States
Contact: Theresa Cesario, (212) 668-1500; fax (212) 668-1504; TCesario@SCConline.org
Internet: http://www.scconline.org/website/about_scc/awards_fellowships.shtml
Sponsor: Society of Cosmetic Chemists
120 Wall Street, Suite 2400
New York, NY 10005-4088

Society of Cosmetic Chemists Keynote Award Lecture Sponsored by Ruger Chemical Corporation **4320**
Dedicated to the advancement of cosmetic science, the Society strives to increase and disseminate scientific information through meetings and publications. By promoting research in cosmetic science and industry, and by setting high ethical, professional and educational standards, it strives to improve the qualifications of cosmetic scientists. Its mission is to further the interests and recognition of cosmetic scientists while maintaining the confidence of the public in the cosmetic and toiletries industry. The Keynote Award is an honorarium given to a speaker who has achieved exceptional national or international stature in the scientific community for delivering a lecture at the Annual Scientific Meeting.
Geographic Focus: All States
Contact: Theresa Cesario, (212) 668-1500; fax (212) 668-1504; TCesario@SCConline.org
Internet: http://www.scconline.org/website/about_scc/awards_fellowships.shtml
Sponsor: Society of Cosmetic Chemists
120 Wall Street, Suite 2400
New York, NY 10005-4088

Society of Cosmetic Chemists Literature Award **4321**
Dedicated to the advancement of cosmetic science, the Society strives to increase and disseminate scientific information through meetings and publications. By promoting research in cosmetic science and industry, and by setting high ethical, professional and educational standards, it strives to improve the qualifications of cosmetic scientists. Its mission is to further the interests and recognition of cosmetic scientists while maintaining the confidence of the public in the cosmetic and toiletries industry. Through the Literature Award, a scroll and honorarium are presented annually to the author or authors of scientific papers in basic research judged to be an outstanding contribution to cosmetic science and technology.
Geographic Focus: All States
Contact: Theresa Cesario, (212) 668-1500; fax (212) 668-1504; TCesario@SCConline.org
Internet: http://www.scconline.org/website/about_scc/awards_fellowships.shtml
Sponsor: Society of Cosmetic Chemists
120 Wall Street, Suite 2400
New York, NY 10005-4088

Society of Cosmetic Chemists Maison G. de Navarre Medal **4322**
Dedicated to the advancement of cosmetic science, the Society strives to increase and disseminate scientific information through meetings and publications. By promoting research in cosmetic science and industry, and by setting high ethical, professional and educational standards, it strives to improve the qualifications of cosmetic scientists. Its mission is to further the interests and recognition of cosmetic scientists while maintaining the confidence of the public in the cosmetic and toiletries industry. The Society's highest honor, the Maison G. de Navarre Medal, recognizes individuals for their technical contributions to cosmetic science. The award is presented annually.
Geographic Focus: All States
Contact: Theresa Cesario, (212) 668-1500; fax (212) 668-1504; TCesario@SCConline.org
Internet: http://www.scconline.org/website/about_scc/awards_fellowships.shtml
Sponsor: Society of Cosmetic Chemists
120 Wall Street, Suite 2400
New York, NY 10005-4088

Society of Cosmetic Chemists Merit Award **4323**
Dedicated to the advancement of cosmetic science, the Society strives to increase and disseminate scientific information through meetings and publications. By promoting research in cosmetic science and industry, and by setting high ethical, professional and educational standards, it strives to improve the qualifications of cosmetic scientists. Its mission is to further the interests and recognition of cosmetic scientists while maintaining the confidence of the public in the cosmetic and toiletries industry. The Merit Award is presented annually for outstanding service and distinguished leadership in Society activities.
Geographic Focus: All States
Contact: Theresa Cesario, (212) 668-1500; fax (212) 668-1504; TCesario@SCConline.org
Internet: http://www.scconline.org/website/about_scc/awards_fellowships.shtml
Sponsor: Society of Cosmetic Chemists
120 Wall Street, Suite 2400
New York, NY 10005-4088

Society of Cosmetic Chemists Robert A. Kramer Lifetime Service Award **4324**
Dedicated to the advancement of cosmetic science, the Society strives to increase and disseminate scientific information through meetings and publications. By promoting research in cosmetic science and industry, and by setting high ethical, professional and educational standards, it strives to improve the qualifications of cosmetic scientists. Its mission is to further the interests and recognition of cosmetic scientists while maintaining the confidence of the public in the cosmetic and toiletries industry. The Robert A. Kramer Lifetime Service Award, given at the discretion of the Board of Directors, goes to an individual who has given extraordinary service over the course of his or her membership.
Geographic Focus: All States
Contact: Theresa Cesario, (212) 668-1500; fax (212) 668-1504; TCesario@SCConline.org
Internet: http://www.scconline.org/website/about_scc/awards_fellowships.shtml
Sponsor: Society of Cosmetic Chemists
120 Wall Street, Suite 2400
New York, NY 10005-4088

Society of Cosmetic Chemists Shaw Mudge Award **4325**
Dedicated to the advancement of cosmetic science, the Society strives to increase and disseminate scientific information through meetings and publications. By promoting research in cosmetic science and industry, and by setting high ethical, professional and educational standards, it strives to improve the qualifications of cosmetic scientists. Its mission is to further the interests and recognition of cosmetic scientists while maintaining the confidence of the public in the cosmetic and toiletries industry. The Shaw Mudge Award, given in the form of a scroll and honorarium, are awarded for the best paper presented at the Society's Annual Scientific Seminar.
Geographic Focus: All States
Contact: Theresa Cesario, (212) 668-1500; fax (212) 668-1504; TCesario@SCConline.org
Internet: http://www.scconline.org/website/about_scc/awards_fellowships.shtml
Sponsor: Society of Cosmetic Chemists
120 Wall Street, Suite 2400
New York, NY 10005-4088

Society of Cosmetic Chemists Student Poster Awards **4326**
Dedicated to the advancement of cosmetic science, the Society strives to increase and disseminate scientific information through meetings and publications. By promoting research in cosmetic science and industry, and by setting high ethical, professional and educational standards, it strives to improve the qualifications of cosmetic scientists. Its mission is to further the interests and recognition of cosmetic scientists while maintaining the confidence of the public in the cosmetic and toiletries industry. The Student Poster Award, sponsored by DD-Chemco, Inc., is given annually in conjunction with the Annual Scientific Seminar. Honorariums are given for first, second, third and fourth place Best Student Poster.
Geographic Focus: All States
Contact: Theresa Cesario, (212) 668-1500; fax (212) 668-1504; TCesario@SCConline.org
Internet: http://www.scconline.org/website/about_scc/awards_fellowships.shtml
Sponsor: Society of Cosmetic Chemists
120 Wall Street, Suite 2400
New York, NY 10005-4088

Society of Cosmetic Chemists Young Scientist Awards **4327**
The Young Scientist Awards are presented each year to recognize the outstanding achievements of new cosmetic scientists throughout the industry. Each year every manufacturer, supplier or other organization in the personal care industry may select an individual in their employ as a recipient of this award. The criteria for selection is made by the individual company. Information is sent in March of each year and is posted on the SCC Website. The submission deadline is July 1st each year. The awards may be presented at an SCC Chapter Meeting or through an internal corporate presentation. Recipient names are listed in the SCC Winter Newsletter.
Requirements: Recipients be an active member of the Society for at least one year and working in the industry less than 10 years.
Geographic Focus: All States
Date(s) Application is Due: Jul 1
Contact: Theresa Cesario, (212) 668-1500; fax (212) 668-1504; TCesario@SCConline.org
Internet: http://www.scconline.org/website/about_scc/awards_fellowships.shtml
Sponsor: Society of Cosmetic Chemists
120 Wall Street, Suite 2400
New York, NY 10005-4088

Society of Naval Architects and Marine Engineers Scholarships **4328**
The society annually awards both graduate and undergraduate scholarships to encourage study in naval architecture, marine engineering, ocean engineering or marine industry related fields. All applicants must plan to study in the United States or Canada. The listed application deadline is for graduate scholarships; undergraduate scholarships are available at and administered through select universities. Guidelines are available online.
Requirements: Applicants are required to be U.S. or Canadian citizens with character, leadership, and general promise for a future in the marine industry, as well as scholastic ability and desire to pursue advanced study.
Geographic Focus: All States
Date(s) Application is Due: Feb 1
Amount of Grant: 10,000 USD
Contact: Philip Kimball; (201) 798-4800, ext. 3024; efaustino@sname.org
Internet: http://www.sname.org/membership/Scholar.html
Sponsor: Society of Naval Architects and Marine Engineers
601 Pavonia Avenue
Jersey City, NJ 07306

Sony Corporation of America Corporate Philanthropy Grants **4329**
Sony Corporation of America consists of three operating companies as well as the corporate headquarters, which is based in New York City: Sony Electronics, Inc. (headquarters in San Diego, California); Sony Pictures Entertainment, Inc. (headquarters in Culver City, California); and Sony Music Entertainment, Inc. (headquarters in New York City). Each company, as well as the overall corporation, has its own philanthropic priorities and resources (e.g. grants, product donations, and recordings and screenings) that benefit a multitude of causes. Taken together the corporation's areas of interest

cover arts and culture; health and human services; civic and community outreach, education; the environment; disaster response; and volunteerism. The core of Sony's various corporate philanthropy programs are their contributions to the communities in which Sony employees work and live; however, the corporation and subsidiaries contribute to national nonprofits as well. In the past, types of support have included general operating budgets, continuing support, annual campaigns, seed grants, building construction, equipment acquisition, endowment funds, employee matching gifts, internships, and employee-related scholarships. Sony Corporation of America and its subsidiaries welcome requests for support throughout the year. There is no grant application form. Requests must be submitted in writing to the corporation or its operating companies. Contact information is given on this page and at the website. Guidelines for what to include in the application as well as more information on types of programs the corporation and/or its subsidiaries have supported are available at the grant website. Notification of grant-request approval or rejection will be made in writing within one month of receipt of all proposed materials.
Requirements: U.S. nonprofits, including schools and school districts, are eligible.
Restrictions: The corporation does not consider multi-year requests for support. The following types of organizations will not be funded: organizations that discriminate on the basis of race, color, creed, gender, religion, age, national origin, or sexual orientation; partisan political organizations, committees, or candidates and public office holders; religious organizations in support of their sacramental or theological functions; labor unions; endowment or capital campaigns of national origin; organizations whose prime purpose is to influence legislation; testimonial dinners in general; for-profit publications or organizations seeking advertisements or promotional support; individuals seeking self-advancement; foreign or non-U.S.-based organizations; and organizations whose mission is outside of the U.S.
Geographic Focus: All States
Contact: Janice Pober; (310) 244-7737
Internet: http://www.sony.com/SCA/philanthropy/guidelines.shtml
Sponsor: Sony Corporation of America
550 Madison Avenue, 33rd Floor
New York, NY 10022-3211

South Florida Student Shark Program (SFSSP) 4330

The South Florida Student Shark Program (SFSSP) is a collaborative, multi-disciplinary research and education program supporting the career development for graduate, undergraduate as well as high school students. Founded in 2006, the SFSSP is a partnership among the University of Miami Rosenstiel School, The Explorers Club, the NOAA LMRCSC & the Herbert W. Hoover Foundation. Focusing on the study and conservation of Floridian shark species, mangrove fish habitat and the Florida watershed, the program encourages students to take an active role in modern scientific education & research. The SFSSP's full-immersion approach allows students to actively grow as future scientists, while supporting the important ongoing research crucial to shark conservation. Primary objectives are to: provide practical, hands-on marine science education and self initiated research project opportunities for high school, undergraduate and graduate college students in the marine science field. Educational and training opportunities accruing from the proposed work will also especially be made available to underrepresented (gender, ethnicity, disability) students; conduct numerous studies that will aid in the monitoring of biological and environmental conditions of Floridian marine ecosystems; eliminate general misconceptions about sharks and the Florida watershed by disseminating the results of this study and creating public awareness of the perils faced by sharks and the current threats to the Florida watershed; and help students (high school, undergraduate and graduate college) gain and develop useful skills in marine science and play an active role in creating, conducting and disseminating the program through service learning interactions. Contact the office for further details.
Geographic Focus: All States
Contact: Neil Hammerschlag; (305) 421-4356 or (954) 815-0920; nhammerschlag@rsmas.miami.edu
Internet: http://cufer.rsmas.miami.edu/wordpress/?page_id=105
Sponsor: Rosenstiel School, University of Miami
4600 Rickenbacker Causeway
Miami, FL 33149

Special Olympics Eunice Kennedy Shriver (EKS) Fellowships 4331

The Eunice Kennedy Shriver (EKS) Program aims to develop leadership on the field of intellectual disability through development of cross-sector bonds among professionals working in the field of intellectual disability in developing countries. Fellowships will be granted to candidates with a commitment to NGO or governmental efforts that support people with intellectual disabilities. Participants will receive an intense, hands-on learning experience that will benefit them both professionally and personally. Fellows will receive Mentors in their Host Programs that will support their learning. Likewise, the Special Olympics Program in their home country will provide aid in their six-month volunteer project upon their return. Fellows will also receive programmatic support from a dedicated team of professionals based in the Washington, D.C Special Olympics headquarters office. All housing, food, transportation to and from the host country and other travel as directed by Special Olympics, Inc. will be covered at the U.S. government per diem rate. In addition, a stipend for other living expenses will be awarded to each Fellow in the Program.
Requirements: Fellows will be financially responsible for obtaining their passports and visas.
Geographic Focus: All States
Contact: Lisa Dietz; (202) 824-0228; fax (202) 824-0200; ldietz@specialolympics.org
Internet: http://www.specialolympics.org/EKS_fellowship.aspx
Sponsor: Special Olympics
1133 19th Street, NW
Washington, DC 20036-3604

Special Olympics Health Professions Student Grants 4332

The purpose of this program is to engage health professions students to work with persons with intellectual disabilities as a way of filling in a gap that exists in most health program curricula. The program promotes short-term projects exploring issues that impact the health and well-being of all persons with intellectual disabilities, including, but not limited to, Special Olympics athletes. Projects may include: data collection and analysis on issues impacting persons with intellectual disabilities; measurement of attitudes, opinions and behaviors of health professionals, coaches, family/caregivers and athletes; follow-up assessments of existing programs; or health promotion projects. Projects that involve collaborations with Special Olympics Programs or other CDC grant recipients (e.g., state and local health departments) are encouraged. Special Olympics is not currently prioritizing specific research areas for funding, as the program's primary goal is to engage health professions students to work with the population of persons with intellectual disabilities. All awards will be made to the institution on behalf of the student grantee. Funding must include all expenses and no indirect costs may be covered by this grant. Applications are accepted year-round. Application and guidelines are available online.
Requirements: Health professions students at the bachelor, master, doctorate and post-doctorate levels from accredited institutions or programs in the United States or other countries are eligible. Fields or disciplines may include, audiology, dentistry, pre-med, medicine, nursing, optometry, physical therapy, podiatry, public health, social work, law, public administration, or other related health areas. All students??? projects will require a faculty advisor.
Geographic Focus: All States
Amount of Grant: Up to 5,000 USD
Contact: Darcie Mersereau, Manager, Research and Evaluation; (202) 628-3600 or (800) 700-8585; fax (202) 824-0200; 2006hastudentgrant@specialolympics.org
Internet: http://www.specialolympics.org/content.aspx?id=6336&terms=student
Sponsor: Special Olympics
1133 19th Street, NW
Washington, DC 20036-3604

Special Olympics Youth Fan Grants 4333

Special Olympics is the leader in cutting-edge research and evaluation to better understand the many challenges faced by people with intellectual disabilities and the significant impact of Special Olympics on their lives. The program's independent research is also a driving force for realizing improved policies, laws and rights for people with intellectual disabilities around the world. Projects may be research-based or programmatic (conferences, health promotion programs, etc), but if they are programmatic they must still contain an evaluation component.
Geographic Focus: All States
Contact: Kathleen Palermo, Grants Coordinator – Healthy Athletes; (202) 628-3630 or (202) 824-0238; fax (202) 628-3926; kpalermo@specialolympics.org
Internet: http://www.specialolympics.org/research_studies.aspx
Sponsor: Special Olympics
1133 19th Street, NW
Washington, DC 20036-3604

Spencer Foundation Education Dissertation Fellowships 4334

The Dissertation Fellowship program seeks to encourage a new generation of scholars from a wide range of disciplines and professional fields to undertake research relevant to the improvement of education. These $25,000 fellowships support individuals whose dissertations show potential for bringing fresh and constructive perspectives to the history, theory, or practice of formal or informal education anywhere in the world. This highly competitive program aims to identify the most talented researchers conducting dissertation research related to education. Like all Spencer Foundation programs, the Dissertation Fellowship program receives many more applications than it can fund. This year, up to 600 applications are anticipated and up to 25 fellowships will be awarded.
Requirements: Grants are made only to institutions, not directly to individuals. Applicants nominated by their institutions must be candidates for the doctoral degree in any field of study at a graduate school in the United States. Dissertation topics must concern education. All predissertation requirements must be completed by June 1 of the upcoming year.
Restrictions: Awards are not designed for individuals in the fields of guidance or curriculum development, nor for those planning to be school administrators.
Geographic Focus: All States
Date(s) Application is Due: Nov 1
Amount of Grant: 20,000 USD
Contact: Annie Brinkman, Grants Manager; (312) 274-6526 or (312) 274-6511; fax (312) 337-0282; fellows@spencer.org or abrinkman@spencer.org
Internet: http://www.spencer.org/content.cfm/naed_spencer-dissertation-fellowships-in-education
Sponsor: Spencer Foundation
625 N Michigan Avenue, Suite 1600
Chicago, IL 60611

Spencer Foundation Major Research Grants 4335

Funded research projects vary widely, ranging from medium-sized studies that can be completed within a year by an individual researcher to more extensive collaborative studies that last several years. There are no funding priorities for subjects of research; projects originate from research ideas initiated in the field by scholars and other researchers. Investigators of projects with budgets over $40,000 are asked to submit a preliminary proposal. If the Foundation is interested in the project, the Principal Investigator will be invited to submit a full proposal. Normally, applicants will be informed about the status of the preliminary proposal within two to four weeks of receipt.

Requirements: Ordinarily, principal investigators applying for a Major Research Grant must be affiliated with a school district, a college or university, a research facility, or a cultural institution. Researchers must also have an earned doctorate in an academic discipline or professional field or appropriate experience in the teaching profession.
Restrictions: Grantees may not receive two research grants simultaneously from the foundation. The foundation does not fund direct interventions or evaluations of programs. The foundation does not pay government-approved overhead rates on research grants; overhead requests of more than $75,000 may not exceed 15 percent of the requested direct costs. The foundation does not pay indirect costs on research grants of $75,000 or less.
Geographic Focus: All States
Amount of Grant: 40,000-350,000 USD
Contact: Annie Brinkman, Grants Manager; (312) 274-6511 or (312) 337-7000; fax (312) 337-0282; abrinkman@spencer.org or majgrant@spencer.org
Internet: http://www.spencer.org/content.cfm/budgets-over-40000
Sponsor: Spencer Foundation
625 N Michigan Avenue, Suite 1600
Chicago, IL 60611

Spencer Foundation Postdoctoral Fellowships **4336**
Administered by the National Academy of Education, the postdoctoral fellowships are designed to promote scholarship in the United States and abroad on matters relevant to the improvement of education in all its forms. Scholars anywhere in the world who have completed their doctorates within the last five years and who wish to conduct research related to education may apply.
Geographic Focus: All States
Contact: Annie Brinkman, Grants Manager; (312) 274-6526 or (312) 274-6511; fax (312) 337-0282; fellows@spencer.org or abrinkman@spencer.org
Internet: http://www.spencer.org/content.cfm/naed-spencer-postdoctoral-fellowships
Sponsor: Spencer Foundation
625 N Michigan Avenue, Suite 1600
Chicago, IL 60611

Spencer Foundation Small Research Grants **4337**
The Spencer Foundation was established in 1962 by Lyle M. Spencer. The Foundation received its major endowment upon Spencer's death in 1968 and began formal grant making in 1971. The Foundation is intended, by Spencer's direction, to investigate ways in which education, broadly conceived, can be improved around the world. From the first, the Foundation has been dedicated to the belief that research is necessary to the improvement in education. The Foundation is thus committed to supporting high-quality investigation of education through its research programs and to strengthening and renewing the educational research community through its fellowship and training programs and related activities. The Foundation has identified the following five areas of inquiry: education and social opportunity; organizational learning; purposes and values of education; teaching, learning, and instructional resources; and field-initiated proposals. Small grants are those lass than $40,000.
Requirements: Initial proposals are accepted at any time, regardless of the level of funding requested. Applicants must possess an earned doctorate and be affiliated with a school, college, university, or cultural institution. Educators currently employed in K-12 settings who are not pursuing a graduate degree are also eligible.
Restrictions: Principal investigators are not permitted to receive two research grants simultaneously. Projects may not extend beyond a year in duration, and no extensions will be granted. Grants do not support indirect costs.
Geographic Focus: All States
Amount of Grant: 1,000-40,000 USD
Contact: Annie Brinkman; (312) 274-6511 or (312) 337-7000; abrinkman@spencer.org
Internet: http://www.spencer.org/programs/grants/research_grants.htm
Sponsor: Spencer Foundation
625 N Michigan Avenue, Suite 1600
Chicago, IL 60611

SPIE A.E. Conrady Award **4338**
The A. E. Conrady Award is and sponsored Optical Research Associates. The awards given annually in recognition of exceptional contributions in design, construction, and testing of optical systems and instrumentation. The recognition of this award is based on developments of new equipment, techniques, and applications for designing, testing, analyzing, and/or evaluating optical systems, components, and theories.
Requirements: Nominations must be accompanied by the following: completed award nomination form; brief citation stating the accomplishments that make the nominee deserving of this particular award; a one-page narrative description of the significant aspects of the nominee's career. Curriculum vitae listing educational background, positions held, publications, awards, honors, activities and offices held within the Society; a minimum of two (2) letters of reference from other than the nominator. Diversity of support is preferred; not all letters of recommendation should come from nominee's organization.
Geographic Focus: All States
Date(s) Application is Due: Oct 1
Amount of Grant: 2,000 USD
Contact: Coordinator; (360) 676-3290 or (888) 504-8171; awards@spie.org
Internet: http://spie.org/app/awards/nominate.aspx?redir=app/awards/nominate.aspx&ID=x2894
Sponsor: International Society Advancing Light-Based Research
1000 20th Street, P.O. Box 10
Bellingham, WA 98225

SPIE Chandra S. Vikram Award in Optical Metrology **4339**
The Chandra S. Vikram Award in Optical Metrology is given annually in recognition of exceptional contributions to the field of optical metrology. The award may be presented for a specific achievement, development, or invention of significant importance to optical metrology, or may be given for lifetime achievement.
Requirements: Nominations must be accompanied by the following: completed award nomination form; brief citation stating the accomplishments that make the nominee deserving of this particular award; a one-page narrative description of the significant aspects of the nominee's career. Curriculum vitae listing educational background, positions held, publications, awards, honors, activities and offices held within the Society; a minimum of two (2) letters of reference from other than the nominator. Diversity of support is preferred; not all letters of recommendation should come from nominee's organization.
Geographic Focus: All States
Date(s) Application is Due: Oct 1
Amount of Grant: 2,000 USD
Contact: Coordinator; (360) 676-3290 or (888) 504-8171; awards@spie.org
Internet: http://spie.org/app/awards/nominate.aspx?redir=app/awards/nominate.aspx&ID=x2894
Sponsor: International Society Advancing Light-Based Research
1000 20th Street, P.O. Box 10
Bellingham, WA 98225

SPIE Dennis Gabor Award **4340**
The Dennis Gabor Award is presented annually in recognition of outstanding inventive accomplishments in electro-optical systems, especially those which further the development of holographic imaging, and metrology applications.
Requirements: Nominations must be accompanied by the following: completed award nomination form; brief citation stating the accomplishments that make the nominee deserving of this particular award; a one-page narrative description of the significant aspects of the nominee's career. Curriculum vitae listing educational background, positions held, publications, awards, honors, activities and offices held within the Society; a minimum of two (2) letters of reference from other than the nominator. Diversity of support is preferred; not all letters of recommendation should come from nominee's organization.
Geographic Focus: All States
Date(s) Application is Due: Oct 1
Amount of Grant: 2,000 USD
Contact: Coordinator; (360) 676-3290 or (888) 504-8171; awards@spie.org
Internet: http://spie.org/app/awards/nominate.aspx?redir=app/awards/nominate.aspx&ID=x2894
Sponsor: International Society Advancing Light-Based Research
1000 20th Street, P.O. Box 10
Bellingham, WA 98225

SPIE Frits Zernike Award **4341**
The Frits Zernike Award for Microlithography is given annually for outstanding accomplishments in microlithographic technology, especially those furthering the development of semiconductor lithographic imaging solutions.
Requirements: Nominations must be accompanied by the following: completed award nomination form; brief citation stating the accomplishments that make the nominee deserving of this particular award; a one-page narrative description of the significant aspects of the nominee's career. Curriculum vitae listing educational background, positions held, publications, awards, honors, activities and offices held within the Society; a minimum of two (2) letters of reference from other than the nominator. Diversity of support is preferred; not all letters of recommendation should come from nominee's organization.
Geographic Focus: All States
Date(s) Application is Due: Oct 1
Amount of Grant: 2,000 USD
Contact: Coordinator; (360) 676-3290 or (888) 504-8171; awards@spie.org
Internet: http://spie.org/app/awards/nominate.aspx?redir=app/awards/nominate.aspx&ID=x2894
Sponsor: International Society Advancing Light-Based Research
1000 20th Street, P.O. Box 10
Bellingham, WA 98225

SPIE George W. Goddard Award **4342**
The George W. Goddard Award is given annually in recognition of exceptional achievement in optical or photonic instrumentation for aerospace applications.
Requirements: Nominations must be accompanied by the following: completed award nomination form; brief citation stating the accomplishments that make the nominee deserving of this particular award; a one-page narrative description of the significant aspects of the nominee's career. Curriculum vitae listing educational background, positions held, publications, awards, honors, activities and offices held within the Society; a minimum of two (2) letters of reference from other than the nominator. Diversity of support is preferred; not all letters of recommendation should come from nominee's organization.
Geographic Focus: All States
Date(s) Application is Due: Oct 1
Amount of Grant: 2,000 USD
Contact: Coordinator; (360) 676-3290 or (888) 504-8171; awards@spie.org
Internet: http://spie.org/app/awards/nominate.aspx?redir=app/awards/nominate.aspx&ID=x2894
Sponsor: International Society Advancing Light-Based Research
1000 20th Street, P.O. Box 10
Bellingham, WA 98225

SPIE Gold Medal of the Society Award 4343

The Gold Medal of the Society is the highest honor the Society bestows. Beginning in 1977, it has been awarded annually in recognition of outstanding engineering or scientific accomplishments in optics, electro-optics, or photographic technologies or applications. The recipient(s) shall have made an exceptional contribution to the advancement of relevant technology.
Requirements: Nominations must be accompanied by the following: completed award nomination form; brief citation stating the accomplishments that make the nominee deserving of this particular award; a one-page narrative description of the significant aspects of the nominee's career. Curriculum vitae listing educational background, positions held, publications, awards, honors, activities and offices held within the Society; a minimum of two (2) letters of reference from other than the nominator. Diversity of support is preferred; not all letters of recommendation should come from nominee's organization.
Geographic Focus: All States
Date(s) Application is Due: Oct 1
Amount of Grant: 10,000 USD
Contact: Coordinator; (360) 676-3290 or (888) 504-8171; awards@spie.org
Internet: http://spie.org/app/awards/nominate.aspx?redir=app/awards/nominate.aspx&ID=x2894
Sponsor: International Society Advancing Light-Based Research
1000 20th Street, P.O. Box 10
Bellingham, WA 98225

SPIE Technology Achievement Award 4344

The SPIE Technology Achievement Award is given in recognition of exceptional contributions in design, construction, and testing of optical systems and instrumentation.
Requirements: Nominations must be accompanied by the following: completed award nomination form; brief citation stating the accomplishments that make the nominee deserving of this particular award; a one-page narrative description of the significant aspects of the nominee's career. Curriculum vitae listing educational background, positions held, publications, awards, honors, activities and offices held within the Society; a minimum of two (2) letters of reference from other than the nominator. Diversity of support is preferred; not all letters of recommendation should come from nominee's organization.
Geographic Focus: All States
Date(s) Application is Due: Oct 1
Amount of Grant: 2,000 USD
Contact: Coordinator; (360) 676-3290 or (888) 504-8171; awards@spie.org
Internet: http://spie.org/app/awards/nominate.aspx?redir=app/awards/nominate.aspx&ID=x2894
Sponsor: International Society Advancing Light-Based Research
1000 20th Street, P.O. Box 10
Bellingham, WA 98225

SPIRE Postdoctoral Fellowships 4345

The Seeding Postdoctoral Innovators in Research and Education (SPIRE) Postdoctoral Fellowship Program was one of the first grantees to receive funding in 1999 from the special Institutional Research and Academic Career Development Award (IRACDA) program, which is now part of the division of Training, Workforce Development, & Diversity under the National Institute for General Medical Sciences (NIGMS). The goal of SPIRE was to address the growing crisis in science education by investing in tomorrow's science scholars and educators and providing them with a holistic approach to postgraduate professional development. SPIRE strives to blend an outstanding research environment with broad professional development activities and hands-on classroom experiences for those scientists who want to combine research with undergraduate and graduate education. Fellowship stipends will be competitive with federal guidelines as set by the National Institutes of Health (NIH), with annual increases as allowed by these guidelines. Health insurance will also be provided. In addition, each fellowship includes an allowance for research and teaching activities, travel, support for mentoring an undergraduate student, workshops on workplace skills, and participation in annual SPIRE events. Each fellowship is three years.
Requirements: Applicants can be either new PhDs or postdocs with up to two years postgraduate training at the time of appointment. Candidates from groups underrepresented in the biological, biomedical, chemical, and related STEM fields in sciences are strongly urged to apply and applicants from diverse backgrounds are welcome. According to the NIH regulations for postdoctoral training grant eligibility: Applicants must be U.S. citizens or non-citizen nationals, or have been lawfully admitted for permanent residence and possess an Alien Registration Receipt Card or some other verification of legal status as a permanent resident. Non-citizen nationals, although not U.S. citizens, owe permanent allegiance to the U.S.. They are generally born in lands, which are not states but are under U.S. sovereignty, jurisdiction, or administration. Individuals on temporary or student visas are not eligible for SPIRE fellowships.
Geographic Focus: All States
Contact: Brian Rybarczyk, Ph.D., (919) 962-2505; brybar@unc.edu
Internet: http://spire.unc.edu/apply/
Sponsor: University of North Carolina at Chapel Hill
CB#4010
Chapel Hill, NC 27599

SPJ Eugene C. Pulliam Fellowship for Editorial Writing 4346

The Pulliam Fellowship is awarded to an outstanding editorial writer to help broaden his or her journalistic horizons and knowledge of the world. The award can be used to cover the cost of study, research, and/or travel in any field. In some cases, the fellowship results in editorials and other writings, including books. The Fellowship has no formal application form. Each applicant should submit thc following materials: a cover letter stating the purpose and nature of the proposed study, a time line for accomplishing the work, and a plan for how the stipend will be used. The cover letter should contain complete contact information for the applicant including mailing address, telephone number and email address; the editor's endorsement; a one-page professional biography and summary of professional experience; and five samples of editorials or columns.
Requirements: To be eligible for a Pulliam fellowship, a candidate must: hold a position as a part-time or full-time editorial writer or columnist at a news publication located in the U.S; have at least three years experience as an editorial writer or columnist; demonstrate outstanding writing and analytical abilities; secure assurances by the editor or publisher that the applicant will be allowed sufficient time to pursue the fellowship without jeopardizing employment. (fellows do not have to leave their jobs.); demonstrate ability and intent to publish work within 18 months of selection. (If selected, work must be published within 18 months of receiving the fellowship). The selected applicant must provide a post-fellowship written report on how funds were used. Each Fellowship recipient will become a mentor to the following year's recipient.
Geographic Focus: All States
Date(s) Application is Due: Jun 22
Amount of Grant: 75,000 USD
Contact: Sue Kopen Katcef, Vice Chair, Awards Committee; (301) 405-7526 or (301) 405-2399; fax (301) 314-9166; skatcef@spg.org
Internet: http://www.spj.org/a-pulliamfellow.asp
Sponsor: Society of Professional Journalists
3909 North Meridian Street
Indianapolis, IN 46208

SPJ Pulliam/Kilgore Freedom of Information Internships 4347

The Society of Professional Journalists (SPJ) annually awards two Pulliam/Kilgore Freedom of Information Internships. One intern works in the offices of the Society's First Amendment legal counsel in Washington, D.C. The other intern works at the Society's National Headquarters in Indianapolis. The interns research and write about freedom of information issues while assisting the Society's Freedom of Information Committee and First Amendment legal counsel in preparing SPJ's annual Freedom of Information Report. Other responsibilities may include working on the SPJ website and assisting with SPJ programs and initiatives. Interns receive a stipend of $400 per week and work during a mutually agreeable ten week period during the summer. Recipients are notified in February, and should arrange to start their internships in late May or early June. The application and specific guidelines for each internship are available at the SPJ website.
Requirements: Candidates must be senior or graduate journalism students, or law students with journalism backgrounds. To be considered, applicants must include the following: a completed application (located at the website); resume; a one page essay illustrating their understanding of Freedom of Information issues and what they expect to gain from the internship; and no more than five writing samples.
Geographic Focus: All States
Date(s) Application is Due: Jan 16
Amount of Grant: 4,000 USD
Contact: Joe Skeel, Executive Director; (317) 927-8000, ext. 216; fax (317) 920-4789; jskeel@spj.org
Internet: http://www.spj.org/i-pullkilgore.asp
Sponsor: Society of Professional Journalists
3909 North Meridian Street
Indianapolis, IN 46208

SRC AMD/Mahboob Khan Fellowship 4348

The Mahboob Khan Fellowship was created in memory of Dr. Khan and in recognition of his support of SRC students, especially the Fellows. Advanced Micro Devices (AMD) partially funds this Fellowship. Succeeding Fellowships will be awarded as the current recipient completes doctoral studies. AMD selects this Fellow from the applications received during the regular Graduate Fellowship Program (GFP) application process. In addition to the academic and citizenship criteria, AMD seeks to identify a student with technical interest specifically aligned with that of AMD. As part of the GFP, this Fellowship provides full tuition and fees and a competitive stipend for up to five years of doctoral study. Fellowships also include an unrestricted annual gift of $2,000 for use by the student's faculty advisor in support of the student. The Fellowships will continue to completion of the degree program, but for no more than five years. Continuation will be contingent only upon satisfactory progress in the proposed course of study and research as determined by the faculty advisor and GRC. The Graduate Fellowships are offered through a national competition and are awarded based on outstanding academic achievement in the field of microelectronics and the relevance of the proposed research to SRC/semiconductor industry goals. Application materials are typically available in November, and completed applications are due mid-February of the following year with awards made in April. All named Fellowships will be awarded consecutively with the next award being made at the completion of the current Fellowship. Notice of Fellowship competition is through universities having GRC-sponsored research contracts (see the website for a complete list of universities and contacts), and other sources as appropriate.
Requirements: This Fellowship is open to students who are U.S. citizens or have permanent resident, refugee, or political asylum status in the U.S; are pursuing or planning to pursue a Ph.D. degree with research relevant to microelectronics under the guidance of a GRC-approved faculty member and with at least two years to completion of the doctoral degree; will be performing research under a sponsored research program approved by GRC; and are willing to provide a copy of their Ph.D. dissertation to GRC for publication to the SRC website.
Geographic Focus: All States
Date(s) Application is Due: Feb 15
Contact: Fellowship Coordinator; (919) 941-9400; fax (919) 941-9450; apply@src.org

Internet: http://www.src.org/student-center/fellowship/#gfp
Sponsor: Semiconductor Research Corporation
P.O. Box 12053
Research Triangle Park, NC 27709-2053

SRC Aristotle Award **4349**
The Aristotle Award was created by the SRC Board of Directors to recognize SRC-supported faculty whose deep commitment to the educational experience of SRC students has had a profound and continuing impact on their professional performance and a significant impact for members over a long period of time. The Award is intended to acknowledge outstanding teaching in its broadest sense, emphasizing student advising and teaching during the research project, thereby contributing to the maturation of the student. The following equally-weighted selection criteria will be considered by the Awards Committee: innovation in student advising; provision of learning experiences for students in sound execution of semiconductor technology (device physics, process, design, etc.), and deployment of statistical methodology such as design of experiments; creation of an environment that encourages development in leadership skills; and student exposure to and interaction with industry.
Requirements: Nominations are accepted by email from the SRC community including the SRC Board of Directors, Advisory Boards, Industrial Liaisons, university researchers (self-nominations will not be accepted), government agencies that participate in the SRC research program, and alumni of the SRC student programs. Nominations must include: the nomination form (located at the website); a nomination narrative covering each of the four selection criteria; and supporting documentation.
Geographic Focus: All States
Date(s) Application is Due: Apr 18
Contact: Lee Ann Clewell; (919) 941-9444; fax (919) 941-9450; clewell@src.org
Internet: http://www.src.org/award/aristotle/
Sponsor: Semiconductor Research Corporation
P.O. Box 12053
Research Triangle Park, NC 27709-2053

SRC Company-Named Fellowships **4350**
The Company-Named Fellowships were created to recognize outstanding academic and research achievement and to increase the number of Graduate Fellows in research areas of special interest to the co-sponsoring companies. These Fellowships are named for the companies providing matching funding. Fellows are chosen from the applicants for the Graduate Fellowship Program (GFP) based on the eligibility requirements used to select the GFP Fellows, and on the match between the student's proposed research and the co-sponsoring company's research interests. The recipient will be mentored for the term of the Fellowship by an industry advisor from the co-sponsoring company, and will be awarded internships and other support from the co-sponsoring company as appropriate. As a part of the Graduate Fellowship Program, this Fellowship provides full tuition and fees and a competitive stipend for up to five years of doctoral study. Fellowships also include an unrestricted annual gift of $2,000 for use by the student's faculty advisor in support of the student. The Fellowships will continue to completion of the degree program, but for no more than five years. The Graduate Fellowships are offered through a national competition and are awarded based on outstanding academic achievement in the field of microelectronics and the relevance of the proposed research to SRC/semiconductor industry goals. Application materials are typically available in November, and completed applications are due mid-February of the following year with awards made in April. Additional information is available at the SRC website.
Requirements: Fellowships are open to students who: are U.S. citizens or have permanent resident, refugee, or political asylum status in the U.S; are pursuing or planning to pursue a Ph.D. degree with research relevant to microelectronics under the guidance of a GRC-approved faculty member and with at least two years to completion of the doctoral degree; will be performing research under a sponsored research program approved by GRC; and are willing to provide a copy of his/her Ph.D. dissertation to GRC for publication to the SRC website.
Geographic Focus: All States
Date(s) Application is Due: Feb 15
Contact: Coordinator; (919) 941-9400; fax (919) 941-9450; apply@src.org
Internet: http://www.src.org/student-center/fellowship/
Sponsor: Semiconductor Research Corporation
P.O. Box 12053
Research Triangle Park, NC 27709-2053

SRC Company-Named Scholarships **4351**
Company-Named scholarships were created to increase the number of Master's Scholarships in research areas of special interest to the co-sponsoring companies. These scholarships are named for the companies providing matching funding. The recipient of a company-named scholarship will be mentored for the term of the scholarship by an industry advisor from the co-sponsoring company, and will be awarded internships and other support from the co-sponsoring company as appropriate. As part of the Master's Scholarship program, these scholarships provide full tuition and fees and a competitive stipend for up to two years of master's level study. Scholarships also include an unrestricted gift of $2,000 for use by the student's faculty advisor in support of the student. The Scholarships will continue to completion of the degree program, but for no more than two years. Continuation will be contingent only upon satisfactory progress in the proposed course of study and research as determined by the faculty advisor and Global Research Collaboration (GRC). Scholarships are offered through a national competition and are awarded based on minority or female status and outstanding academic achievement. Students in BS/MS degree programs are encouraged to apply for the Master's Scholarships at the beginning of their senior year. Application materials are typically available in November, and completed applications are due mid-February of the following year with awards made in April. Notice of scholarship competition is through universities having GRC-sponsored research contracts, and other sources as appropriate. Candidates may email or write directly to SRC for an application.
Requirements: Scholarships are open to students who are women or are members of an underrepresented minority category (African-American, Hispanic, Native American); meet admission requirements for graduate school at a GRC-participating university (see website for a complete list of universities and contacts); are U.S. citizens or have permanent resident, refugee, or political asylum status in the U.S; are planning to pursue a master's degree with research relevant to microelectronics under the guidance of a GRC-approved faculty member and under a GRC-approved research program; and are willing to provide a copy of their master's thesis where appropriate for publication to the SRC secure website.
Geographic Focus: All States
Date(s) Application is Due: Feb 15
Contact: Coordinator; (919) 941-9400; fax (919) 941-9450; apply@src.org
Internet: http://www.src.org/student-center/fellowship/
Sponsor: Semiconductor Research Corporation
P.O. Box 12053
Research Triangle Park, NC 27709-2053

SRC Core Research Grants **4352**
Global Research Collaboration (GRC) supports two kinds of research: core and custom. Core research is selected by representatives of the entire GRC membership. Currently, core research is about 80% of the total research budget. GRC, as a mission-oriented research organization, funds research directly responsive to its defined agenda. The funding terms for contracts vary; however, the GRC trend is toward multi-year research efforts (typically three years) with an initial 12-month funding term. At the end of the multi-year period, researchers may re-compete for new support.
Requirements: Proposals should provide sufficient information to permit an evaluation of the intended research, its technical merits, innovative approach, and relevance to the GRC research agenda. The author should presume general technical knowledge in the field by reviewers, but should not assume that the merits of the proposal are obvious. The capability of the institution to adequately support the proposed research and the technical expertise of the researchers to conduct this research should be justified. Refer to the Proposal Submission Guide, located at the website. for specific authoring and submission instructions.
Geographic Focus: All States
Contact: Core Research Coordinator; (919) 941-9400; fax (919) 941-9450
Internet: http://www.src.org/program/grc/about/core-research/
Sponsor: Semiconductor Research Corporation
P.O. Box 12053
Research Triangle Park, NC 27709-2053

SRC Cross-Disciplinary Research (CSR) Grants **4353**
The goal of the Cross-Disciplinary Semiconductor Research Program is to foster exploratory, multi-disciplinary, high-risk university research leading to novel high-payoff solutions for the technology challenges faced by the semiconductor industry at and beyond the time horizons of the SIA International Technology Roadmap for Semiconductors (ITRS). Successful projects will involve synergistic research across normal disciplinary boundaries that offers innovative/disruptive solutions supporting the continuation of the Moore's Law cost/performance improvement cadence. Possible targets for the research can address semiconductor technologies ranging from factory systems through device technologies to packaged systems. Multi-disciplinary projects leading to divergent applications of semiconductor technologies will also be considered. The role of the Program is to stimulate non-traditional thinking about the issues facing the industry. The Program is also intended to complement the traditional GRC science area research programs. Cross-Disciplinary Semiconductor Research projects are exploratory in nature and should lead to research programs that address challenges identified by ITRS. Funding for a project is provided on a non-renewable grant basis, with maximum funding not to exceed $40,000.
Requirements: International university faculty are eligible. Participants in the program are expected to participate in an annual workshop to describe the results of their study, and to develop a proposal for follow-on funding for expanded programs by the GRC Core Research Program. Follow-on program funding will depend on the availability of science area funds and alignment with the science area strategic plans.
Geographic Focus: All Countries
Amount of Grant: Up to 40,000 USD
Contact: Victor Zhirnov; (919) 941-9400; fax (919) 941-9450; victor.zhirnov@src.org
Internet: http://www.src.org/program/grc/csr/
Sponsor: Semiconductor Research Corporation
P.O. Box 12053
Research Triangle Park, NC 27709-2053

SRC Educational Science Conference Grants **4354**
The intent of the Program is to foster links between the research field of Educational Science and international research, and to encourage mobility in Swedish higher education. The grant caters for Swedish arrangers of international conferences relating to the research field of Educational Science. Grants for invitations to foreign lecturers of international repute are (with respect to travel and accommodation costs) given priority. To receive a grant, it is important that the congress, conference or symposium is considered to be of manifest interest for the research area of education science.
Requirements: The applicant with scientific responsibility must have a doctorate or corresponding skills.
Restrictions: Support for conferences is awarded only to cover costs (travel and lodging) for a limited number of invited foreign lecturers. Grants are not provided for other participants, for administration or for social-security contributions.

Geographic Focus: All States
Date(s) Application is Due: Feb 15
Amount of Grant: Up to 100,000 SEK
Contact: Helene Sundewall; 46 8 546 44 159; fax 46 8 546 44 180; helene.sundewall@vr.se
Internet: http://www.vr.se/mainmenu/applyforgrants/callforproposals/grantsview.4.64a4ae3e1092b3475b6800063.html?resourceId=-1418&languageId=2
Sponsor: Swedish Research Council
Klarabersgviadukten 82
Stockholm, SE-103 78 Sweden

SRC Educational Science Doctoral Grants 4355

The Program's function is to promote research of high scientific quality, relevant to teacher training and professional teaching activity. A number of national doctoral programs have already been established (in mathematics, specializing in subject teaching; adult learning; aesthetic learning processes; and Swedish language specializing in subject teaching). The Research Council's Committee for Educational Science now seeks, in particular, to encourage applications for admission to national, multidisciplinary doctoral programs with the emphasis on Educational Science. Funds are granted for a maximum of four years.
Requirements: The program must: include several cooperating HEIs, which should include at least one with teacher training; have a formalized organization with a program director or coordinator in charge of implementation; offer a coherent study program for doctoral students on a well-developed theme, or a well-defined research program; be responsible for its own admission, curricula and monitoring; and must provide a relevant range of courses for doctoral students. The HEIs taking part must jointly offer a good research environment for the specific theme or program concerned. The number of doctoral students funded with grants from the Research Council's CES must not exceed five, and the total amount applied for should not exceed SEK 2 million.
Geographic Focus: All States
Contact: Helene Sundewall; 46 8 546 44 159; fax 46 8 546 44 180; helene.sundewall@vr.se
Internet: http://www.vr.se/responsibilities/researchfunding/educationalscience.4.69f66a93108e85f68d48000693.html
Sponsor: Swedish Research Council
Klarabersgviadukten 82
Stockholm, SE-103 78 Sweden

SRC Educational Science Grants 4356

The Swedish Government set up the Committee for Educational Science (CES) as a special expert body in the Swedish Research Council. The Committee's function is to promote research of high scientific quality, relevant to teacher training and professional teaching activity. Accordingly, this means research on learning, knowledge formation, education and teaching. During the CES's existence, the pressure of applications has increased by almost 300 percent. For the current year, CES had a budget of approximately 128 million SEK. It also supports the development of research closely associated with practice, i.e. closely connected with teaching activity. Research support in Educational Science is subject to a requirement of co-funding from, and cooperation between, participating higher education institutions (HEIs).
Geographic Focus: All States
Contact: Helene Sundewall; 46 8 546 44 159; fax 46 8 546 44 180; helene.sundewall@vr.se
Internet: http://www.vr.se/responsibilities/researchfunding/educationalscience.4.69f66a93108e85f68d48000693.html
Sponsor: Swedish Research Council
Klarabersgviadukten 82
Stockholm, SE-103 78 Sweden

SRC Educational Science Grants for Invitation of Visiting Scientists 4357

The grant is intended to promote the internationalization of Swedish research in educational sciences. Costs for guest researcher's travel expenses, board and lodging, and in exceptional cases a stipend payment are allowable. Grants are normally administered by a Swedish university/institution or another Swedish public organization with research as its purpose. Applications should be submitted electronically.
Requirements: Individual researchers at Swedish university departments may apply. The applicant with scientific responsibility must have obtained a doctorate or have an equivalent qualification. The application should be written in Swedish or English.
Geographic Focus: All States
Date(s) Application is Due: Mar 3
Contact: Helene Sundewall; 46 8 546 44 159; fax 46 8 546 44 180; helene.sundewall@vr.se
Internet: http://www.vr.se/mainmenu/applyforgrants/callforproposals/grantsview.4.64a4ae3e1092b3475b6800063.html?resourceId=-1382&languageId=2
Sponsor: Swedish Research Council
Klarabersgviadukten 82
Stockholm, SE-103 78 Sweden

SRC Educational Science Grants for Travels to Foreign Institutions as a Guest Scientist 4358

The intent is to foster links between the research field of Educational Science and international research, and to encourage mobility in Swedish higher education. Costs for guest researcher's travel expenses, board and lodging, and in exceptional cases a stipend payment are allowable. The grant is intended for longer periods of time (up to six months). The abstract in the online form should include a description of: the foreign seat of learning which the researcher wishes to visit; the period envisaged; brief reasons for the visit; and the program planned. Applications should be submitted electronically.
Requirements: Individual researchers at Swedish university departments may apply. The application should be written in Swedish or English.
Geographic Focus: All States
Date(s) Application is Due: Mar 3
Contact: Helene Sundewall; 46 8 546 44 159; fax 46 8 546 44 180; helene.sundewall@vr.se
Internet: http://www.vr.se/mainmenu/applyforgrants/callforproposals/grantsview.4.64a4ae3e1092b3475b6800063.html?resourceId=-1385&languageId=2#kontakt
Sponsor: Swedish Research Council
Klarabersgviadukten 82
Stockholm, SE-103 78 Sweden

SRC Educational Science Junior Research Position Grants 4359

The Program's function is to promote research of high scientific quality, relevant to teacher training and professional teaching activity. This grant relates mainly to salary costs and usually covers four years. An applicant for a junior research position must, in the application, propose the subject area for and location of the position sought. If an application is successful, the applicant is appointed at the HEI or research institute concerned.
Geographic Focus: All States
Contact: Helene Sundewall; 46 8 546 44 159; fax 46 8 546 44 180; helene.sundewall@vr.se
Internet: http://www.vr.se/responsibilities/researchfunding/educationalscience.4.69f66a93108e85f68d48000693.html
Sponsor: Swedish Research Council
Klarabersgviadukten 82
Stockholm, SE-103 78 Sweden

SRC Educational Science Language Checking Grants 4360

The Council supports basic research in educational science (research into training, education, teaching and learning). Basic research/curiosity-driven research refers to research that provides new knowledge and contributes explanations and understanding. Scientific quality is the deciding factor in the assessment of applications. Educational science research may deal with, for example, different kinds of learning, the link between the nature of the content, teaching and learning or about learning in different social, cultural and institutional contexts including working life. It may concern diversity aspects of learning and memory such as gender, ethnicity, social background, talent, disabilities, etc. or learning at different stages of life, such as lifelong learning. Other examples are research into the effects of educational innovations, educational reforms or different educational arrangements, and the way the education system and informal learning function under the influence of different social, economic and political conditions. Important aspects are the possibilities for research to shed light on topical and fundamental issues in educational practice and the teaching profession. Applications must be made entirely electronically.
Requirements: The applicant with scientific responsibility must have obtained a doctorate or have an equivalent qualification. The application must be written in Swedish or English. Grants are normally administered by a university/institution (host university/institution). A full application comprises the filled-in online form and appendices. Note that Appendix S, signed, must be submitted to the Swedish Research Council in paper form.
Restrictions: Grants are not awarded to studies of a more developmental nature, development of an activity or follow-up.
Geographic Focus: All States
Date(s) Application is Due: Oct 1
Amount of Grant: 15,000 SEK
Contact: Helene Sundewall; 46 8 546 44 159; fax 46 8 546 44 180; helene.sundewall@vr.se
Internet: http://www.vr.se/mainmenu/applyforgrants/callforproposals/grantsview.4.64a4ae3e1092b3475b6800063.html?resourceId=-1480&languageId=2#kontakt
Sponsor: Swedish Research Council
Klarabersgviadukten 82
Stockholm, SE-103 78 Sweden

SRC Educational Science Medium-Cost Scientific Equipment Grants 4361

The Program's function is to promote research of high scientific quality, relevant to teacher training and professional teaching activity. This grant is for purchase of equipment costing between 170,000 and 2,000,000 SEK, and is disbursed as a lump sum. One of these applicants receives project funds from the Research Council for the activity to which the equipment relates.
Geographic Focus: All States
Amount of Grant: 170,000-2,000,000 SEK
Contact: Helene Sundewall; 46 8 546 44 159; fax 46 8 546 44 180; helene.sundewall@vr.se
Internet: http://www.vr.se/responsibilities/researchfunding/educationalscience.4.69f66a93108e85f68d48000693.html
Sponsor: Swedish Research Council
Klarabersgviadukten 82
Stockholm, SE-103 78 Sweden

SRC Educational Science Network Grants 4362

The Program's function is to promote research of high scientific quality, relevant to teacher training and professional teaching activity. This grant is for a network of researchers in specific subject areas within Educational Science.
Geographic Focus: All States
Contact: Helene Sundewall; 46 8 546 44 159; fax 46 8 546 44 180; helene.sundewall@vr.se
Internet: http://www.vr.se/responsibilities/researchfunding/educationalscience.4.69f66a93108e85f68d48000693.html
Sponsor: Swedish Research Council
Klarabersgviadukten 82
Stockholm, SE-103 78 Sweden

SRC Educational Science Postdoctoral Fellowships 4363

These fellowships are designed to enable educational science researchers with Swedish doctorates or exams from EUI (European University Institute) to stay at foreign universities or research establishments. The Committee wants to contribute to strengthening current research as well as supporting research within less researched fields, preferably interdisciplinary or multidisciplinary. The Committee especially welcomes applications from young researchers who wish to qualify in the field of educational science. The maximum fellowship period of 24 months is reduced by the length of time for which the applicant, since obtaining a doctorate, has held a postdoctoral position, a similar grant or a fellowship, at a higher education institution or faculty or company other than the one where the doctorate, or equivalent, was awarded. The proposed research project must be essentially different from the applicant's previous research. Essentially different means a new direction, new ways of thinking and/or new orientation of the research. The amount of the fellowship will be the standard grant rate, calculated according to costs in the host country. Applications to the Swedish Research Council must be made entirely electronically, and the application may be written in either Swedish or English.

Requirements: A Swedish doctorate or exam from EUI (European University Institute) is required. The doctorate must have been awarded within three years before and latest at the same day as the deadline for applications. Sick leave, service in the Armed Forces, doctor's employment as a preregistration house officer, work for trade unions and student organizations, parental leave and other circumstances of this kind are not counted in the three years. There is no age limit for applicants.

Geographic Focus: All States

Date(s) Application is Due: Feb 20

Contact: Helene Sundewall; 46 8 546 44 159; fax 46 8 546 44 180; helene.sundewall@vr.se

Internet: http://www.vr.se/mainmenu/applyforgrants/callforproposals/grantsview.4.64a4ae3e1092b3475b6800063.html?resourceId=-1445&languageId=2

Sponsor: Swedish Research Council
Klarabersgviadukten 82
Stockholm, SE-103 78 Sweden

SRC Educational Science Project Research Grants 4364

The Program's function is to promote research of high scientific quality, relevant to teacher training and professional teaching activity. This grant is for research during a specific period of time, to carry out a defined research task. Normally, the grant is awarded for up to three years. A doctoral student's position may be included. Up to SEK 170,000 in equipment costs may be applied for.

Restrictions: Salary costs for a person employed as a junior researcher or assistant lecturer is not funded.

Geographic Focus: All States

Contact: Helene Sundewall; 46 8 546 44 159; fax 46 8 546 44 180; helene.sundewall@vr.se

Internet: http://www.vr.se/responsibilities/researchfunding/educationalscience.4.69f66a93108e85f68d48000693.html

Sponsor: Swedish Research Council
Klarabersgviadukten 82
Stockholm, SE-103 78 Sweden

SRC Educational Science Travel Grants 4365

The intent is to foster links between the research field of Educational Science and international research, and to encourage mobility in Swedish higher education. To accomplish this, the Committee for Educational Science (CES) has allocated funds for conference and travel grants. Travel grants are intended first and foremost to enable post-doc researchers at Swedish seats of learning who have graduated not more than five years before the proposed travel to establish personal contacts with foreign researchers, and enjoy a short stay within a foreign research environment. CES also aims to promote opportunities for researchers, especially those with newly awarded PhDs, to stay briefly in other research environments in Sweden. Grants are provided primarily for researchers with recent doctorates at Swedish higher education institutions (HEIs), for the following purposes: attending congresses, conferences or symposia-provided that the researcher has received a personal invitation to offer an academic contribution; and spending brief periods (not exceeding one month) at other Swedish HEIs or universities abroad. Types of cost include travel costs, board and lodgings and specified conference expenses.

Requirements: The applicant with scientific responsibility must have a doctorate or corresponding skills.

Geographic Focus: All States

Contact: Helene Sundewall; 46 8 546 44 159; fax 46 8 546 44 180; helene.sundewall@vr.se

Internet: http://www.vr.se/mainmenu/applyforgrants/callforproposals/grantsview.4.64a4ae3e1092b3475b6800063.html?resourceId=-1461&languageId=2#kontakt

Sponsor: Swedish Research Council
Klarabersgviadukten 82
Stockholm, SE-103 78 Sweden

SRC Education Science Grants for Postdoctoral Positions in Sweden 4366

The grant will enable education science researchers with Swedish or non-Swedish doctorates (PhDs or equivalent) to work at Swedish higher education institutions or research establishments. The maximum fellowship period of 24 months is reduced by the length of time for which the applicant, since obtaining a doctorate, has held a postdoctoral position, a similar grant or a fellowship, at a higher education institution or faculty or company other than the one where the doctorate, or equivalent, was awarded. The host university/institution or equivalent is the employer and determines which form of employment the position will have.

Requirements: Individual education science researchers may apply. Swedish doctorate or foreign degree equivalent to a doctorate is required. The doctorate must have been awarded within three years before and latest at the same day as the deadline for applications. Sick leave, service in the Armed Forces, doctor's employment as a preregistration house officer, work for trade unions and student organizations, parental leave and other circumstances of this kind are not included in the three years. There is no age limit for applicants.

Geographic Focus: All States

Date(s) Application is Due: Feb 20

Contact: Helene Sundewall; 46 8 546 44 159; fax 46 8 546 44 180; helene.sundewall@vr.se

Internet: http://www.vr.se/mainmenu/applyforgrants/callforproposals/grantsview.4.64a4ae3e1092b3475b6800063.html?resourceId=-1392&languageId=2

Sponsor: Swedish Research Council
Klarabersgviadukten 82
Stockholm, SE-103 78 Sweden

SRC Franco Cerrina Fellowship 4367

The Cerrina Fellowship encourages, sponsors, and supports young scientists and technologists whose work integrates leading-edge research into new application areas, crossing boundaries between traditional or established disciplines, with original ideas and solutions. The Franco Cerrina Fellowship will be awarded to students who best represent the spirit of Cerrina by taking research in new directions in areas such as silicon/biological interfaces and applications, beyond-CMOS computational devices, nanotechnology and nanofabrication, and opening new frontiers in electrical and computer engineering.

Geographic Focus: All States

Contact: Fellowship Administrator; (919) 941-9400; fax (919) 941-9450

Internet: http://www.src.org/program/srcea/cerrina/

Sponsor: Semiconductor Research Corporation
P.O. Box 12053
Research Triangle Park, NC 27709-2053

SRC Graduate Fellowships 4368

The Graduate Fellowship Program (GFP) addresses the issues of improving educational opportunities at the doctoral level and supplying a relevantly educated workforce for the semiconductor industry. The objectives of the program are to encourage academically gifted U.S./permanent resident students to pursue doctoral degrees in research areas consistent with SRC program goals; and to develop high quality doctoral graduates for member companies and U.S. universities. The GFP was created to attract exceptionally talented students with U.S. citizenship to academic areas of interest to SRC members. The program has since been opened to students holding permanent resident, refugee, or political asylum status in the U.S. While the Fellows are not required to take employment within the SRC community upon graduation, they are strongly encouraged to do so, and assistance is provided in finding appropriate employment in an SRC member company, U.S. government agency, or U.S. university. Graduating Fellows entering member companies facilitate the transfer of new science and technology from the participating universities to the supporting organizations. Fellows joining faculties at universities worldwide carry with them the expertise to stimulate new research activities and to encourage additional student interest in semiconductor-related fields. Students who are supported by the GFP are actively involved in GRC-approved sponsored research; however, the scope of the student's work may be beyond the scope of the research defined by the research sponsor. The Fellows are encouraged to conduct research leading to novel, high-payoff solutions for the technology challenges faced by the semiconductor industry at and beyond the time horizons of the International Technology Roadmap for Semiconductors. The Graduate Fellowships are offered through a national competition and are awarded based on outstanding academic achievement in the field of microelectronics and the relevance of the proposed research to SRC/semiconductor industry goals. Application materials are typically available in November, and completed applications are due mid-February of the following year with awards made in April. Notice of Fellowship competition is through universities having GRC-sponsored research contracts, and other sources as appropriate. A list of those universities and their contacts is posted on the SRC website.

Requirements: The Fellowships are open to students who are U.S. citizens or have permanent resident, refugee, or political asylum status in the U.S; are pursuing or planning to pursue a Ph.D. degree with research relevant to microelectronics under the guidance of a GRC-approved faculty member and with at least two years to completion of the doctoral degree; will be performing research under a sponsored research program approved by GRC; and are willing to provide a copy of his/her Ph.D. dissertation to GRC for publication to the SRC website. It is expected that all Fellows will complete their Ph.D. degrees in an area relevant to microelectronics and that upon completion of the program, secure employment with an SRC member or U.S. government agency or a faculty position in an accredited four-year U.S. college or university.

Geographic Focus: All States, All Countries

Date(s) Application is Due: Feb 15

Contact: Fellowship Coordinator; (919) 941-9400; fax (919) 941-9450; apply@src.org

Internet: http://www.src.org/student-center/fellowship/#gfp

Sponsor: Semiconductor Research Corporation
P.O. Box 12053
Research Triangle Park, NC 27709-2053

SRC GRC and SEMATECH Joint Projects Grants 4369

GRC and SEMATECH, in order to expedite the transition of research results into applications with even more value benefiting members, have formed a partnership to conduct applied research at jointly funded centers. Through the collaborative process GRC funds research concepts, develops research options, and narrows the research options. GRC and SEMATECH jointly fund a transition period to develop the feasibility of the research, and transition of the research concepts is made to technology hardening which is funded by SEMATECH. Additional information about currently funded projects is available at the SRC website.

Geographic Focus: All States
Contact: Robert Havemann; (919) 941-9400; fax (919) 941-9450; bob.havemann@src.org
Internet: http://www.src.org/program/grc/about/sematech-joint-projects/
Sponsor: Semiconductor Research Corporation
P.O. Box 12053
Research Triangle Park, NC 27709-2053

SRC Guest Professorship Program 4370

The Swedish Research Council provides funding for three Guest Professorships, as well as a professorship in the Swedish language at the University of Tartu, Estonia. The Olof Palme Guest Professorship is intended for research of importance for efforts to bring about peace. The Kerstin Hesselgren Guest Professorship, instituted in memory of Sweden's first female member of Parliament (the Riksdag), is awarded to an eminent woman researcher from outside of Sweden. The Guest Professorship for Artistic Development is intended to support arts and humanities programs. Deadlines vary, and applicants should contact the SRC for further details.
Geographic Focus: All States, Sweden
Contact: Berit Angusti, Senior Research Administrator; +46 (0)8 546 44 315; fax +46 8 546 44 180; berit.angusti@vr.se
Internet: http://www.vr.se/mainmenu/organisation/aboutus/scientificcouncilforhumanitiesandsocialsciences.4.69f66a93108e85f68d4800011579.html;jsessionid=C3A13BBA053853B83ABC978C5F432BA5?contactPage=4.69f66a93108e85f68d4800011579&state=showCreateMessage&sv.url=12.20a
Sponsor: Swedish Research Council
Klarabersgviadukten 82
Stockholm, SE-103 78 Sweden

SRC Hans J. Coufal Fellowship 4371

The SRC/NRI Hans J. Coufal Fellowship was created in memory of Hans J. Coufal, the founding director of the Nanoelectronics Research Initiative (NRI), for the study of nanoelectronics-related disciplines. The Coufal Fellowship encourages non-traditional thinking and exploratory, high-risk research leading to novel, high-payoff solutions for challenges faced by the semiconductor industry at and beyond the time horizons of the International Technology Roadmap for Semiconductors. The recipient is expected to work on an NRI-funded research program, but the student's work is encouraged to go beyond the scope of the research defined in the research contract. The Coufal Fellowship is administered by NRI and provides full tuition and fees, a competitive stipend, and a $2,000 annual unrestricted gift to the recipient's faculty advisor. The Fellowship will continue to completion of the degree program, but for no more than five years. Continuation will be contingent only upon satisfactory progress in the proposed course of study and research as determined by the faculty advisor and NRI.
Requirements: The Fellowship will be awarded to a graduate student who has completed a bachelor's degree and is pursuing, or is planning to pursue, a doctoral program in an academic area appropriate to the Fellowship. Although this Fellowship is not restricted to any particular area of nanoelectronics, preference will be given to proposed work in Dr. Coufal's primary areas of interest (nanoelectronic device technologies, physics, and materials). The recipient will hold citizenship or permanent resident status in the U.S.
Geographic Focus: All States
Date(s) Application is Due: Feb 1
Contact: Fellowship Administrator; (919) 941-9400; fax (919) 941-9450
Internet: http://www.src.org/student-center/fellowship/#coufal
Sponsor: Semiconductor Research Corporation
P.O. Box 12053
Research Triangle Park, NC 27709-2053

SRC Humanities and Social Science Postdoctoral Fellowships 4372

These fellowships are designed to enable humanities and social science researchers with Swedish doctorates or exams from EUI (European University Institute) to stay at foreign universities or research establishments. The maximum fellowship period of 24 months is reduced by the length of time for which the applicant, since obtaining a doctorate, has held a postdoctoral position, a similar grant or a fellowship, at a higher education institution or faculty or company other than the one where the doctorate, or equivalent, was awarded. The proposed research project must be essentially different from the applicant's previous research. Essentially different means a new direction, new ways of thinking and/or new orientation of the research. The amount of the fellowship will be the standard grant rate, calculated according to costs in the host country. Applications to the Swedish Research Council must be made entirely electronically, and the application may be written in either Swedish or English.
Requirements: A Swedish doctorate or exam from EUI (European University Institute) is required. The doctorate must have been awarded within three years before and latest at the same day as the deadline for applications. Sick leave, service in the Armed Forces, doctor's employment as a preregistration house officer, work for trade unions and student organizations, parental leave and other circumstances of this kind are not counted in the three years. There is no age limit for applicants.
Geographic Focus: All States
Date(s) Application is Due: Feb 20
Contact: Berit Angusti; +46 (0)8 546 44 315; fax +46 8 546 44 180; berit.angusti@vr.se
Internet: http://www.vr.se/mainmenu/applyforgrants/callforproposals/grantsview.4.64a4ae3e1092b3475b6800063.html?resourceId=-1439&languageId=2
Sponsor: Swedish Research Council
Klarabersgviadukten 82
Stockholm, SE-103 78 Sweden

SRC Humanities and Social Sciences Grants 4373

Researchers in this Program study the workings of society and the creation and dissemination of cultural values. A researcher can obtain a project research grant to perform a research task, as a rule for three years, to cover salary, materials, travel, publication and other expenses. Long-term support for up to six years is also given to leading researchers who are expected to increase, develop and convey knowledge in their research fields. Professors can apply for funds for sabbatical terms, i.e. remuneration for six months full-time research. Grants may also be provided for temporary research, junior research and doctoral student positions in fields assigned priority by the Scientific Council. Examples of the Research Council's international involvement are its collaboration within the EU Sixth Framework Programme, in the thematic area of Citizens and governance in a knowledge-based society.
Geographic Focus: All States
Contact: Berit Angusti; +46 (0)8 546 44 315; fax +46 8 546 44 180; berit.angusti@vr.se
Internet: http://www.vr.se/responsibilities/researchfunding/humanitiesandsocialsciences.4.69f66a93108e85f68d480001122.html
Sponsor: Swedish Research Council
Klarabersgviadukten 82
Stockholm, SE-103 78 Sweden

SRC Humanities and Social Sciences Grants for Postdoctoral Positions In Sweden 4374

The grant will enable researchers with Swedish or non-Swedish doctorates (PhDs or equivalent) to work at Swedish higher education institutions or research establishments. The maximum fellowship period of 24 months is reduced by the length of time for which the applicant, since obtaining a doctorate, has held a postdoctoral position, a similar grant or a fellowship, at a higher education institution or faculty or company other than the one where the doctorate, or equivalent, was awarded. The host university/institution or equivalent is the employer and determines which form of employment the position will have.
Requirements: Individual researchers may apply. Swedish doctorate or foreign degree equivalent to a doctorate is required. The doctorate must have been awarded within three years before and latest at the same day as the deadline for applications. Sick leave, service in the Armed Forces, doctor's employment as a pre-registration house officer, work for trade unions and student organizations, parental leave and other circumstances of this kind are not included in the three years. There is no age limit for applicants to the Swedish Research Council.
Geographic Focus: All States
Date(s) Application is Due: Feb 20
Contact: Berit Angusti; +46 (0)8 546 44 315; fax +46 8 546 44 180; berit.angusti@vr.se
Internet: http://www.vr.se/mainmenu/applyforgrants/callforproposals/grantsview.4.64a4ae3e1092b3475b6800063.html?resourceId=-1387&languageId=2
Sponsor: Swedish Research Council
Klarabersgviadukten 82
Stockholm, SE-103 78 Sweden

SRC International Fellowship 4375

The International Fellowship was first awarded in recognition of GRC's international status. As part of the Graduate Fellowship Program, this fellowship provides full tuition and fees and a competitive stipend for up to five years of doctoral study. Fellowships also include an unrestricted annual gift of $2,000 for use by the student's faculty advisor in support of the student. Applicants may contact the SRC about the status of this Fellowship, which is offered as funding becomes available.
Geographic Focus: All Countries
Date(s) Application is Due: Feb 15
Contact: Fellowship Coordinator; (919) 941-9400; fax (919) 941-9450; apply@src.org
Internet: http://www.src.org/student-center/fellowship/
Sponsor: Semiconductor Research Corporation
P.O. Box 12053
Research Triangle Park, NC 27709-2053

SRC Landmark Innovation Award 4376

The SRC Intellectural Assets Study Group established the Landmark Innovation Award to be given to a SRC university researcher/researchers when an SRC-supported invention proves to have a long-term, significant benefit for the industry. Given the nature of this award, the selection criteria is intentionally general in nature. The impact of the innovation may be technological or economic in nature, and preferably should have created a broad or globally recognized benefit for the industry. Examples that could support an award are significant cost savings or revenue generation for SRC member companies, compelling reasons for SRC membership, proven commercialization and implementation in industry, direct/indirect beneficial impact on the industry, and long term benefits to the industry and/or society. The cash award is shared equally among key contributors who develop innovative patented technology that significantly enhances the productivity and competitiveness of the semiconductor industry. There intentionally has been no specific dollar amount of this award, thus the impact of the landmark innovation can be correspondingly rewarded. Nominations may be submitted to the SRC Intellectual Property Office for consideration by the award committee and the SRC Board of Directors. The nomination form is available at the SRC website.
Requirements: Nominations may be submitted at any time, directed to SRC for the consideration of the SRC Board. Nominations are accepted from selected members of the SRC community, including the SRC Board of Directors, SRC Technical Advisory Boards, mentors, member company employees, university principal investigators, and research task leaders. Nominations may be sent via email or directly to the SRC address, and will be confirmed by receipt. No particular form or format is required, and any

additional supporting documentation (letters, publication bibliographical information, etc.) that helps distinguish a nomination is welcome.
Geographic Focus: All States
Contact: Director; (919) 941-9400; fax (919) 941-9450; legal@src.org
Internet: http://www.src.org/award/landmark/
Sponsor: Semiconductor Research Corporation
P.O. Box 12053
Research Triangle Park, NC 27709-2053

SRC Mahboob Khan Outstanding Industry Liaison Award 4377
The SRC Industry Liaison Program is one of the most effective methods developed by SRC to provide its members with early access to key technologies, substantial leverage in the explicit direction of technology development, and ultimate transfer of semiconductor technology from the research base to the supporting industry. Each year SRC recognizes individuals who, in serving as SRC Industry Liaisons, have had significant impact on the technical contributions of the research, are deeply committed to the education of graduate students, and have been instrumental in transferring the research results to industry. The following selection criteria is used by the committee: the liaison's role in developing the workfoce; the liaison's effectiveness and involvement in aligning research to industry needs, as well as guidance and material help during all phases of the research process; and the liaison's role ion technology transfer from the university community to the industrial community. The nomination form is available at the website.
Requirements: Nominations for individuals serving as SRC Industry Liaisons for a period of at least one year are accepted from members of the SRC community. Past winners may only receive the award once for an individual task.
Geographic Focus: All States
Date(s) Application is Due: Apr 18
Contact: Lee Ann Clewell; (919) 941-9444; fax (919) 941-9450; leeann.clewell@src.org
Internet: http://www.src.org/award/khan-liaison/call/
Sponsor: Semiconductor Research Corporation
P.O. Box 12053
Research Triangle Park, NC 27709-2053

SRC Master's Scholarships 4378
The Master's Scholarship Program (MSP) addresses issues of improving educational opportunities at the master's level for students in underrepresented minority categories. The objectives of the program are to encourage academically gifted U.S./permanent resident students in these populations to pursue graduate research in areas consistent with Global Research Collaboration (GRC) goals; and to develop a cadre of highest quality minority candidates for doctoral study and hire by GRC companies. The MSP Scholarships provide full tuition and fees and a competitive stipend for up to two years of master's level study. Scholarships also include an unrestricted gift of $2,000 for use by the student's faculty advisor in support of the student. The Scholarships will continue to completion of the degree program, but for no more than two years. Continuation will be contingent only upon satisfactory progress in the proposed course of study and research as determined by the faculty advisor and GRC. The Master's Scholarships are offered through a national competition and are awarded based on minority or female status and outstanding academic achievement. Students in BS/MS degree programs are encouraged to apply for the Master's Scholarships at the beginning of their senior year. Application materials are typically available in November, and completed applications are due mid-February of the following year with awards made in April. Notice of scholarship competition is through universities having GRC-sponsored research contracts, and other sources as appropriate. Applications are available by email request or by writing directly to the SRC.
Requirements: The MSP is open to students who are women or are members of an underrepresented minority category (African-American, Hispanic, Native American); meet admission requirements for graduate school at a GRC-participating university (see website for a complete list of universities and contacts); are U.S. citizens or have permanent resident, refugee, or political asylum status in the U.S; are planning to pursue a master's degree with research relevant to microelectronics under the guidance of a GRC-approved faculty member and under a GRC-approved research program; and are willing to provide a copy of his/her master's thesis where appropriate for publication to the SRC secure website.
Geographic Focus: All States
Date(s) Application is Due: Feb 15
Contact: Scholarship Coordinator; (919) 941-9400; fax (919) 941-9450; apply@src.org
Internet: http://www.src.org/student-center/fellowship/
Sponsor: Semiconductor Research Corporation
P.O. Box 12053
Research Triangle Park, NC 27709-2053

SRC Medical Research Grants 4379
The Scientific Council for Medicine supports research throughout the discipline of Medical Science, i.e. Medicine, Odontology, Pharmaceutics and Health Sciences. Medical research supported by the Scientific Council ranges from studies at molecular and cell level, and research involving laboratory animal science and patients, to epidemiological studies on groups of human subjects. The Program offers grants for research and junior research positions; grants for research periods for clinical researchers; grants for research stays (sabbaticals) abroad; project research grants (distributed, as a rule, for three years for salary, materials, travel, publication, etc); international postdoctoral fellowships and grants for postdoctoral positions in Sweden; networks; symposium; research equipment; collaboration grant and grant for inviting a visiting scientist.
Geographic Focus: All States
Contact: Paola Norlin; +46 (0)8 546 44 311; fax 46 8 546 44 180; paola.norlin@vr.se
Internet: http://www.vr.se/responsibilities/researchfunding/medicine.4.69f66a93108e85f68d48000837.html
Sponsor: Swedish Research Council
Klarabersgviadukten 82
Stockholm, SE-103 78 Sweden

SRC Medical Science Grants for Large International Conferences in Sweden 4380
The Council has decided to call for grant applications to arrange occasional, large international conferences addressing topics in medical science in Sweden. Applications must be received no later than May 30. Large international conferences play an important role in the research process since they provide a forum for researchers from different countries to disseminate research results and exchange research findings. For applicants to be granted economic support, the scientific goals and conference program must be judged to have high scientific priority. Furthermore, the scientific content of the conference should contribute to the advancement of an area supported and prioritized by the Council. The grants will be framework grants that may be used to cover expenses, e.g. for travel and per diem costs for participants, venue rental, meeting publications, and other costs relevant to the scientific program. The number of conference participants is unrestricted.
Requirements: The applicant should be the individual responsible for the scientific content of the conference and shall have a doctorate or corresponding qualifications. Other members of the organizing committee or initiators of the conference may be listed as associates. The application shall be written in English and submitted electronically via VR-Direct.
Geographic Focus: All States
Date(s) Application is Due: May 30
Amount of Grant: Up to 1,000,000 SEK
Contact: Agneta Axelsson, Executive Secretary, Research Administrator; 46 0 8 546 44 275; fax 46 0 8 546 44 180; agneta.axelsson@vr.se
Internet: http://www.vr.se/mainmenu/applyforgrants/callforproposals/grantsview.4.64a4ae3e1092b3475b6800063.html?resourceId=-1477&languageId=2
Sponsor: Swedish Research Council
Klarabersgviadukten 82
Stockholm, SE-103 78 Sweden

SRC Medical Science Grants for Postdoctoral Positions in Sweden 4381
The grant will enable Medical Science researchers with Swedish or non-Swedish doctorates (PhDs or equivalent) to work at Swedish higher education institutions or research establishments. The maximum fellowship period of 24 months is reduced by the length of time for which the applicant, since obtaining a doctorate, has held a postdoctoral position, a similar grant or a fellowship, at a higher education institution or faculty or company other than the one where the doctorate, or equivalent, was awarded. The host university/institution or equivalent is the employer and determines which form of employment the position will have.
Requirements: Individual researchers may apply. Swedish doctorate or foreign degree equivalent to a doctorate is required. The doctorate must have been awarded within three years before and latest at the same day as the deadline for applications. Sick leave, service in the Armed Forces, doctor's employment as a pre-registration house officer, work for trade unions and student organizations, parental leave and other circumstances of this kind are not included in the three years. There is no age limit for applicants.
Geographic Focus: All States
Date(s) Application is Due: Feb 20
Contact: Paola Norlin; +46 (0)8 546 44 311; fax 46 8 546 44 180; paola.norlin@vr.se
Internet: http://www.vr.se/mainmenu/applyforgrants/callforproposals/grantsview.4.64a4ae3e1092b3475b6800063.html?resourceId=-1389&languageId=2
Sponsor: Swedish Research Council
Klarabersgviadukten 82
Stockholm, SE-103 78 Sweden

SRC Medical Science Postdoctoral Fellowships 4382
These fellowships are designed to enable medical science researchers with Swedish doctorates or exams from EUI (European University Institute) to stay at foreign universities or research establishments. The maximum fellowship period of 24 months is reduced by the length of time for which the applicant, since obtaining a doctorate, has held a postdoctoral position, a similar grant or a fellowship, at a higher education institution or faculty or company other than the one where the doctorate, or equivalent, was awarded. The proposed research project must be essentially different from the applicant's previous research. Essentially different means a new direction, new ways of thinking and/or new orientation of the research. The amount of the fellowship will be the standard grant rate, calculated according to costs in the host country. Applications to the Swedish Research Council must be made entirely electronically, and the application may be written in either Swedish or English.
Requirements: A Swedish doctorate or exam from EUI (European University Institute) is required. The doctorate must have been awarded within three years before and latest at the same day as the deadline for applications. Sick leave, service in the Armed Forces, doctor's employment as a preregistration house officer, work for trade unions and student organizations, parental leave and other circumstances of this kind are not counted in the three years. There is no age limit for applicants.
Geographic Focus: All States
Date(s) Application is Due: Feb 20
Contact: Paola Norlin; +46 (0)8 546 44 311; fax 46 8 546 44 180; paola.norlin@vr.se
Internet: http://www.vr.se/mainmenu/applyforgrants/callforproposals/grantsview.4.64a4ae3e1092b3475b6800063.html?resourceId=-1442&languageId=2
Sponsor: Swedish Research Council
Klarabersgviadukten 82
Stockholm, SE-103 78 Sweden

SRC Natural and Engineering Science Conference Grants 4383

The main purpose of the program is to enable eminent foreign lecturers to be invited to and attend international conferences. The conference program must be of high quality, and the lecturer invited must be a researcher of international repute. An application for a conference grant may be submitted at any time of the year, though applications are not dealt with between 15 May and 15 August. There are no special forms for applications for conference grants. The maximum conference grant is SEK 100,000.

Requirements: Applicants must have obtained their PhDs by the date of the application. The conference must be international and to be organized in Sweden, another Nordic country or one of the Baltic countries by Swedish arrangers, and it must cover a research field that is supported, or could be supported, by the Swedish Research Council. An application for a conference grant must be received not later than two months before the conference is to take place.

Restrictions: Support for conferences is awarded only to cover costs (travel and lodging) for a limited number of invited foreign lecturers. Grants are not provided for other participants, for administration or for social-security contributions.

Geographic Focus: All States

Amount of Grant: 100,000 SEK

Contact: Mona Berggren; 46 8 546 44 246; fax 46 8 546 44 180; mona.berggren@vr.se

Internet: http://www.vr.se/mainmenu/applyforgrants/callforproposals/grantsview.4.64a4ae3e1092b3475b6800063.html?resourceId=-357&languageId=2#kontakt

Sponsor: Swedish Research Council
Klarabersgviadukten 82
Stockholm, SE-103 78 Sweden

SRC Natural and Engineering Science Grants for Postdoctoral Positions in Sweden 4384

The grant will enable natural and engineering science researchers with Swedish or non-Swedish doctorates (PhDs or equivalent) to work at Swedish higher education institutions or research establishments. The maximum fellowship period of 24 months is reduced by the length of time for which the applicant, since obtaining a doctorate, has held a postdoctoral position, a similar grant or a fellowship, at a higher education institution or faculty or company other than the one where the doctorate, or equivalent, was awarded. The host university/institution or equivalent is the employer and determines which form of employment the position will have.

Requirements: Individual researchers may apply. Swedish doctorate or foreign degree equivalent to a doctorate is required. The doctorate must have been awarded within three years before and latest at the same day as the deadline for applications. Sick leave, service in the Armed Forces, doctor's employment as a pre-registration house officer, work for trade unions and student organizations, parental leave and other circumstances of this kind are not included in the three years. There is no age limit for applicants.

Geographic Focus: All States

Date(s) Application is Due: Feb 20

Contact: Annika Johansen; +46 (0)8 546 44 251; fax +46 8 546 44 180; annika.johansen@vr.se

Internet: http://www.vr.se/mainmenu/applyforgrants/callforproposals/grantsview.4.64a4ae3e1092b3475b6800063.html?resourceId=-1390&languageId=2#kontakt

Sponsor: Swedish Research Council
Klarabersgviadukten 82
Stockholm, SE-103 78 Sweden

SRC Natural and Engineering Science Postdoctoral Fellowships 4385

These fellowships are designed to enable natural and engineering science researchers with Swedish doctorates or exams from EUI (European University Institute) to stay at foreign universities or research establishments. The maximum fellowship period of 24 months is reduced by the length of time for which the applicant, since obtaining a doctorate, has held a postdoctoral position, a similar grant or a fellowship, at a higher education institution or faculty or company other than the one where the doctorate, or equivalent, was awarded. The proposed research project must be essentially different from the applicant's previous research. Essentially different means a new direction, new ways of thinking and/or new orientation of the research. The amount of the fellowship will be the standard grant rate, calculated according to costs in the host country. Applications to the Swedish Research Council must be made entirely electronically, and the application may be written in either Swedish or English.

Requirements: A Swedish doctorate or exam from EUI (European University Institute) is required. The doctorate must have been awarded within three years before and latest at the same day as the deadline for applications. Sick leave, service in the Armed Forces, doctor's employment as a preregistration house officer, work for trade unions and student organizations, parental leave and other circumstances of this kind are not counted in the three years. There is no age limit for applicants.

Geographic Focus: All States

Date(s) Application is Due: Feb 20

Contact: Annika Johansen; +46 (0)8 546 44 251; fax +46 8 546 44 180; annika.johansen@vr.se

Internet: http://www.vr.se/mainmenu/applyforgrants/callforproposals/grantsview.4.64a4ae3e1092b3475b6800063.html?resourceId=-1443&languageId=2

Sponsor: Swedish Research Council
Klarabersgviadukten 82
Stockholm, SE-103 78 Sweden

SRC Natural and Engineering Science Senior Research Positions in Sweden 4386

With its Senior Research Positions in Sweden, the Council seeks to recruit senior researchers and renew Swedish research. This kind of position is intended as a step on the career ladder for an independent, established researcher at senior lecturer (Associate Professor/Docent) level, who is in an active research phase. In the Swedish system this phase is, as a rule, the period between a junior research position/forskarassistentanstallning and permanent employment as a senior university lecturer or professor. The primary obligation of appointees is to conduct research. In addition, these researchers must supervise doctoral students, assess the doctoral students' performance and teach, mainly in postgraduate studies. It is also possible to carry out research outside Sweden for brief spells during the grant period, provided this is done in accordance with the host higher education institution (HEI) and that the research is of benefit to Swedish research in the Natural and Engineering Sciences. Research periods abroad lasting longer than six months in any given three-year period must be approved by the Swedish Research Council. One researcher will be appointed for each position listed.

Requirements: A person who has obtained a doctorate (PhD) or has a non-Swedish degree deemed equivalent to a doctorate, for example a Candidate of Sciences in Russia, is eligible for a Senior Research Position. The doctorate must have been awarded within the ten years prior to the end of the application period. A person whose doctorate was awarded more than ten years previously may also be considered if special reasons apply. Special reasons are: leave of absence due to ill-health; service in the Armed Forces; functions in trade unions and student organizations; parental leave; and other, similar circumstances.

Geographic Focus: All States

Date(s) Application is Due: Feb 20

Contact: Annika Johansen; +46 (0)8 546 44 251; fax +46 8 546 44 180; annika.johansen@vr.se

Internet: http://www.vr.se/mainmenu/applyforgrants/callforproposals/grantsview.4.64a4ae3e1092b3475b6800063.html?resourceId=-1378&languageId=2

Sponsor: Swedish Research Council
Klarabersgviadukten 82
Stockholm, SE-103 78 Sweden

SRC Natural and Engineering Sciences Grants 4387

The Scientific Council for Natural and Engineering Sciences is an expert body that takes decisions on financial support for Swedish basic research. Members of the Council are appointed by the research community and the Government. Their priority ranking of applications is based on assessment of the quality of the research project and the applicant's skill. Research for which support is provided may be conducted throughout the spectrum of basic research in the natural and engineering sciences. Examples are: the Earth and our climate; biology and biotechnology; biodiversity; chemistry and chemical engineering; physics and applied physics; electronics; system engineering; astronomy; materials science; computer science; and mechanical engineering.

Geographic Focus: All States

Amount of Grant: 600,000 SEK

Contact: Mona Berggren; 46 8 546 44 246; fax 46 8 546 44 180; mona.berggren@vr.se

Internet: http://www.vr.se/responsibilities/researchfunding/naturalandengineeringsciences.4.69f66a93108e85f68d48000652.html

Sponsor: Swedish Research Council
Klarabersgviadukten 82
Stockholm, SE-103 78 Sweden

SRC Natural and Engineering Science Travel Grants 4388

Travel grants are available for natural science and engineering science researchers both domestic and abroad. Grants for researchers' attendance at congresses, conferences or symposia are allowable, provided that the researcher concerned has received a personal invitation by the arrangers to hold a lecturer, and if the meeting is deemed to be of obvious interest to Swedish Natural and/or Engineering Sciences. An application for a travel grant must be received not later than two months before the date of departure. Applications for grants below 6,000 SEK will not be considered, and no applications are dealt with between May 15 and August 15.

Requirements: Doctoral students can obtain grants only in exceptional cases. This does not apply to travel relating to observations or experiments at facilities with which the Swedish Research Council has special agreements.

Restrictions: Poster presentations are not approved. Grants are not provided for accompanying family members.

Geographic Focus: All States

Amount of Grant: 6,000 SEK

Contact: Mona Berggren; 46 8 546 44 246; fax 46 8 546 44 180; mona.berggren@vr.se

Internet: http://www.vr.se/mainmenu/applyforgrants/callforproposals/grantsview.4.64a4ae3e1092b3475b6800063.html?resourceId=-857&languageId=2#kontakt

Sponsor: Swedish Research Council
Klarabersgviadukten 82
Stockholm, SE-103 78 Sweden

SRC Peter Verhofstadt Fellowship 4389

The Verhofstadt Fellowship was created by SRC in memory of Peter Verhofstadt, SRC Chief Scientist and Co-Executive Director of MARCO, an SRC subsidiary that manages the Focus Center Research Program. The Verhofstadt Fellowship funds study in the physical sciences, specifically chemistry, physics, mathematics, or biophysics. This Fellowship strives to stimulate non-traditional thinking and encourage exploratory, high-risk research leading to novel, high-payoff solutions for challenges faced by the semiconductor industry at and beyond the time horizons of the International Technology Roadmap for Semiconductors. The recipient is expected to work on a GRC-funded research program, but the student's work is encouraged to be beyond the scope of the research defined in the research program. As part of the Graduate Fellowship Program, the Verhofstadt Fellowship provides full tuition and fees and a competitive stipend for up to five years of doctoral study. Fellowships also include an unrestricted annual gift of $2,000 for use by the student's faculty advisor in support of the student. The Graduate Fellowships are offered through a national competition and are awarded based on outstanding academic achievement in the field of microelectronics and the relevance

of the proposed research to SRC semiconductor industry goals. Application materials are typically available in November, and completed applications are due mid-February of the following year with awards made in April.
Requirements: The Fellowship is open to students who: are U.S. citizens or have permanent resident, refugee, or political asylum status in the U.S; are pursuing or planning to pursue a Ph.D. degree with research relevant to microelectronics under the guidance of a GRC-approved faculty member and with at least two years to completion of the doctoral degree; will be performing research under a sponsored research program approved by GRC; and are willing to provide a copy of his/her Ph.D. dissertation to GRC for publication to the SRC website.
Geographic Focus: All States, All Countries
Date(s) Application is Due: Feb 15
Contact: Fellowship Coordinator; (919) 941-9400; fax (919) 941-9450; apply@src.org
Internet: http://www.src.org/student-center/fellowship/
Sponsor: Semiconductor Research Corporation
P.O. Box 12053
Research Triangle Park, NC 27709-2053

SRC Research Customization Program (RCP) Grants **4390**
The Research Customization Program (RCP) was launched to increase member benefits by providing the flexibility for eligible members to directly select a few university projects that are of particular interest to that member. The projects that a member selects must be consistent with the Global Research Collaboration (GRC) mission, responsive to the needs articulated by the International Technology Roadmap for Semiconductors (ITRS), and suitable for university research. Unless a unique situation exists, the nominal commitment is for three years of at least $50,000 per year, the minimum for a project with a Ph.D. candidate to achieve meaningful results. Additional information about the selection process is available at the SRC website.
Requirements: Eligibility requirements are specific and must be followed. A member company ascertains the amount of funding available for RCP. The Finance Department provides quarterly reports outlining the available funding to the Executive TAB (ETAB) representative and Board of Director of eligible members. A member company makes a preliminary selection after reviewing a short written description of the planned research. The ETAB representative for the member company makes the final selection of the project after receiving inputs from within the company. The ETAB representative then verifies the nature of the proposed research with the Principal Investigator (PI), provides the PI with the name of the Industry Liaison(s), obtains a Proposal ID from SRC, and requests the PI to submit a formal proposal in the standard format to GRC. The ETAB representative provides an official authorization for funding to GRC. When the GRC has received both the proposal and the member company authorization, the project is reviewed by the Research Management Committee (RMC). Projects meeting the RCP project guidelines will be approved for funding. A standard GRC contract is made with the university (a list of supporting universities is posted on the website). The appropriate Science Area implements and manages the project in a manner similar to the other projects in the research portfolio. Additional information is available at the website.
Geographic Focus: All States
Contact: Coordinator; (919) 941-9400; fax (919) 941-9450; legal@src.org
Internet: http://www.src.org/program/grc/about/custom-research/
Sponsor: Semiconductor Research Corporation
P.O. Box 12053
Research Triangle Park, NC 27709-2053

SRC Research Infrastructures Grants **4391**
The Program supports the establishment and use of infrastructure for Swedish research of the highest scientific quality. Research infrastructures include, for example, research installations and facilities for register research or extensive data networks, used nationally or internationally by several research groups, that are required for high-quality research in several research fields. The Program processes grant applications for: expensive scientific equipment; large databases; planning; and operations. Four review groups deal with applications within the Program's sphere of responsibility. These are the groups for: infrastructure for astronomy and subatomic research; infrastructure for molecular, cell and material research; infrastructure for research on the Earth and its near surroundings; and infrastructure for e-Science.
Geographic Focus: All States
Amount of Grant: Up to 100,000,000 SEK
Contact: Finn Karlsson; +46 (0)8 546 44 247; fax 46 (0) 8 546 44 247 180; finn.karlsson@vr.se
Internet: http://www.vr.se/responsibilities/researchfunding/researchinfrastructures.4.69f66a93108e85f68d480001169.html
Sponsor: Swedish Research Council
Klarabersgviadukten 82
Stockholm, SE-103 78 Sweden

SRC Robert M. Burger Fellowship **4392**
The Robert M. Burger Fellowship was created by SRC in recognition of Dr. Burger's ten years of service to SRC as Vice President and Chief Scientist. This Fellowship supports doctoral study in non-technology topics related to microelectronics and the semiconductor industry. Relevant topics include public policy (e.g. regulatory, tax/fiscal, immigration, or federal investment in research and education), management, technology policy, fiscal, innovation, workforce education, or economic issues. As part of the Graduate Fellowship Program, the Burger Fellowship provides full tuition and fees and a competitive stipend for up to five years of doctoral study. Fellowships also include an unrestricted annual gift of $2,000 for use by the student's faculty advisor in support of the student. The Graduate Fellowships are offered through a national competition and are awarded based on outstanding academic achievement in the field of microelectronics and the relevance of the proposed research to SRC/semiconductor industry goals. Application materials are typically available in November, and completed applications are due mid-February of the following year with awards made in April.
Requirements: The Burger Fellowship is awarded to a graduate student who is a U.S. citizen, has completed a master's degree, and who is pursuing or planning to pursue a doctoral program.
Geographic Focus: All States
Date(s) Application is Due: Feb 15
Contact: Coordinator, (919) 941-9400; fax (919) 941-9450; apply@src.org
Internet: http://www.src.org/student-center/fellowship/
Sponsor: Semiconductor Research Corporation
P.O. Box 12053
Research Triangle Park, NC 27709-2053

SRC Simon Karecki Fellowship **4393**
The Simon Karecki Fellowship Fund was created in Simon's memory to continue his research and to provide other outstanding young people with the opportunity to make their contributions to environmental research. The Fund will make an annual cash award or fellowship for outstanding contributions to environmental research. The Karecki Fellowship is awarded to a doctoral student who is performing or is planning to perform research through the Center. The Fellowship will be awarded competitively on the basis of quality, relevance, and timeliness of the proposed research; demonstrated academic and leadership ability, and the appropriateness of the institutional environment for the research. The Awards are based on academic excellence, demonstrated concern for the environment, including significant contribution to research in environmentally benign semiconductor manufacturing, demonstrated ability for technical communication, and demonstrated ability to work with others and foster cooperation. The level of annual funding is determined by the Education Alliance and the Advisory Board.
Geographic Focus: All States
Contact: Fellowship Administrator; (919) 941-9400; fax (919) 941-9450; srcea@src.org
Internet: http://www.src.org/program/srcea/karecki/
Sponsor: Semiconductor Research Corporation
P.O. Box 12053
Research Triangle Park, NC 27709-2053

SRC Technical Excellence Award **4394**
The Technical Excellence Award was established by the Board of Directors as an incentive and recognition program for research of exceptional value to SRC members. The $5,000 award is shared equally among key contributors for innovative technology that significantly enhances the productivity/competitiveness of the semiconductor industry. Nominations are accepted from all members of the SRC community for work performed during previous years. Most nominations are from industry, indicating the extent to which results from SRC-funded programs are being applied or likely will be applied by the industry. The award is presented during the fall of each year. Nomination forms and a description of the narrative required with the form are posted at the website.
Requirements: In reviewing each nomination for the award, the Awards Review Committee considers creativity, relevance to SRC/industry research objectives, value to industry, and the researcher's role in technology transfer. The Awards Committee presents its recommendations to the Board of Directors; the Board approves and confirms all award recipients.
Geographic Focus: All States
Date(s) Application is Due: Apr 18
Amount of Grant: 5,000 USD
Contact: Lee Ann Clewell, Awards Coordinator; (919) 941-9400 or (919) 941-9444; fax (919) 941-9450; clewell@src.org
Internet: http://www.src.org/award/tech-excellence/
Sponsor: Semiconductor Research Corporation
P.O. Box 12053
Research Triangle Park, NC 27709-2053

SRC Undergraduate Research Opportunities **4395**
Undergraduate Research Opportunities (URO) at the Semiconductor Research Corporation offer an innovative program co-sponsored by SRC's Education Alliance and the Intel Foundation that provides undergraduates with valuable research experience and mentoring. Participating undergraduates not only gain confidence in their ability to perform hands-on research, but they also come to appreciate the doors that are opened to those with an advanced degree. The objectives of the SRC URO program are to increase academic retention among undergraduates who express interest in physical science and engineering disciplines, and encourage them to continue their education beyond their four-year degree. The SRC URO program seeks to attract a diverse student population, including women and under-represented minorities. The program also offers workshops and resources to assist students with the graduate school application process. In addition to working on research projects that address the needs of SRC member companies, students have meaningful interactions with industry scientists and engineers, as well as with researchers from a broad academic community.
Requirements: Applicants should contact the URO Coordinator at their college for further information about applying. The following universities host SRC URO programs: Carnegie Mellon University; Cornell University; Georgia Institute of Technology; Howard University; North Carolina A&T; Oregon State University; Portland State University; Purdue University; University of California at Berkeley; University of California at Los Angeles; University of Illinois at Urbana-Champaign; University of Michigan; University of Texas at Austin; and the University of Washington.
Geographic Focus: All States
Contact: Coordinator; (919) 941-9400; fax (919) 941-9450; educationalliance@src.org

Internet: http://www.src.org/program/srcea/uro/
Sponsor: Semiconductor Research Corporation
P.O. Box 12053
Research Triangle Park, NC 27709-2053

SSHRC-APC Automotive Partnership Canada 4396
Automotive Partnership Canada (APC) provides $145 million in research funding over five years to support significant, collaborative research and development activities to benefit the entire Canadian automotive industry. This initiative is a partnership between five federal research and granting agencies under the Industry Canada umbrella, which has identified collaborative, industry-driven automotive research as a priority. APC's mission is to support research and development that will move the Canadian automotive sector to a greater level of innovation. APC's emphasis will be on transformative, integrated projects or programs that will give Canadian industry and academia the resources required to further their automotive research and development ambitions. Smaller projects are also eligible for support through APC. However, the urgent requirement to boost automotive research will require big ideas and wide-ranging resources. All research to be supported must fall under at least one of the ten priority areas: improving the automobile's environmental performance and impact-lighter weight and/or sustainable materials; advanced powertrain; energy storage; and application of alternative fuels; the cognitive car-vehicular software; electronics and mechatronics for safety and performance enhancement; wireline and wireless communications for vehicular application; next generation manufacturing-manufacturing processes for mass reduction, cost reduction, and quality improvement; and improving manufacturing flexibility and efficiency. APC will consider proposals in social sciences and humanities that fall outside the priority areas if they follow these two *Requirements:* they must maintain and enhance the competitiveness of the Canadian automotive sector; and the research must be carried out with a non-academic partner(s) that can apply and use the research results. APC proposals are accepted on an ongoing basis. This allows proposals to come forward as industry needs arise and also allows for proposals to be uniquely crafted according to the requirements of the research and industry challenges at hand. There are no minimum or maximum dollar amounts that can be requested. Projects can range from six months to five years in duration. However, all projects should first be defined by the scope of research and development that needs to be carried out, after which an appropriate project budget and duration can be determined. Additional information and guidelines, including possible research partners, are posted on the APC website.
Requirements: Researchers must first contact the APC Project Office to discuss their project aims, scope, industrial involvement, and other parameters. If their idea fits the mandate of the APC, they are asked to prepare a written overview of the proposed research. If the project still fits within the APC parameters, a full proposal is requested. The Project Office will provide advice and feedback, review drafts, and work with the applicant until a satisfactory full proposal is completed.
Restrictions: Unsolicited full proposals are not accepted. The APC definition of "automotive" includes vehicles licensed for on-road driving (cars, light trucks, heavy trucks, buses and recreational vehicles). The definition does not include agricultural vehicles, military vehicles, tracked vehicles, off-road and rail vehicles.
Geographic Focus: All States, Canada
Contact: John Wood; (905) 403-0600; john.wood@nserc-crsng.gc.ca
Internet: http://www.apc-pac.ca/About-Renseignements/Background-Contexte_eng.asp
Sponsor: Social Sciences and Humanities Research Council of Canada
350 Albert Street, P.O. Box 1610
Ottawa, ON K1P 6G4 Canada

SSHRC-CIHR-SC Research Data Centres Joint Initiative 4397
Decision-makers need an up-to-date and in-depth understanding of Canadian society to help them respond not only to today's needs, but to anticipate tomorrow's as well. This need is underlined by a growing demand for analytical output from the rich sources of data collected by Statistics Canada. The Canadian initiative on social statistics access to research data centres (RDC) is a strategic joint initiative of SSHRC, the Canadian Institutes of Health Research and Statistics Canada. This initiative strives to promote research and training in the application of social statistics. RDCs provide researchers with access, in a secure university setting, to microdata from population and household surveys. The centres are staffed by Statistics Canada employees. They are operated under the provisions of the Statistics Act in accordance with all the confidentiality rules and are accessible only to researchers with approved projects who have been sworn in under the Statistics Act as 'deemed employees. These RDCs are located throughout the country, so researchers do not need to travel to Ottawa to access Statistics Canada microdata. Applications may be submitted at any time. Results will be announced approximately eight weeks after submission. Additional information is posted on the Statistics Canada website.
Requirements: The application process and guidelines are based on the affiliation of the Principal Investigator and the type of research being conducted at a Research Data Centre (RDC). Proposals depend on the type of research academics involved. Complete descriptions of the proposal process by category of researcher (faculty, masters or doctoral student, by academic course, those working for academic institutions who are not members of the Canadian Research Data Centre Network, or type of government research funded) are posted on the Statistics Canada website.
Geographic Focus: All States, Canada
Contact: Mika Oehling, Program Officer, Partnerships Portfolio; (613) 992-4227; fax (613) 947-0223; mika.oehling@sshrc-crsh.gc.ca
Internet: http://www.statcan.gc.ca/rdc-cdr/index-eng.htm
Sponsor: Social Sciences and Humanities Research Council of Canada
350 Albert Street, P.O. Box 1610
Ottawa, ON K1P 6G4 Canada

SSHRC-FHSS Awards to Scholarly Publications 4398
The Awards to Scholarly Publications (ASPP) provides funding to help authors and academic presses publish scholarly books that make a significant original contribution to knowledge in the humanities and social sciences in Canada, but would not be published for commercial purposes. Up to 180 subsidies of $8,000 are available each year. Five subsidies of $12,000 are also available for the translation of scholarly works. Detailed eligibility criteria for authors and publishers and a list of frequently asked questions are posted on the FHSS website.
Restrictions: Funding is not available for the publication of textbooks, technical reports, original works of poetry, fiction and drama, scholarly journals and articles, and conference proceedings. If the manuscript is a "trade" or popular publication (self-help book, volume of personal reflections, etc.), it is not eligible for funding.
Geographic Focus: Canada
Amount of Grant: 8,000 CAD
Contact: Jessica Clark; (613) 238-6112, ext. 352; fax (613) 238-6114; ideas-idees.ca
Internet: http://www.ideas-idees.ca/aspp
Sponsor: Social Sciences and Humanities Research Council of Canada
350 Albert Street, P.O. Box 1610
Ottawa, ON K1P 6G4 Canada

SSHRC-IDRC International Partnerships for Sustainable Societies 4399
The SSHRC-IDRC joint initiative supports international partnerships to produce high-quality research to inform academic, public, and policy debates in ways that can help create just, inclusive, and sustainable social and economic development. Thematic areas include information and networks; inclusive growth; governance, security, and justice; and green growth. Partnerships must be led by a project director based in an eligible low- or middle-income country (LMIC), according to the World Bank listing, and a project director based in Canada. At the letter of intent stage, up to 10 applicants will be awarded a $20,000 grant to prepare a formal application. However, SSHRC may consider requests of up to $30,000 if well justified. Requests for support at the formal application stage would typically be in the range of $500,000 to $2,500,000 over four to seven years. The project director based in an eligible LMIC may request up to $1,250,000 from IDRC and the project director based in Canada may request up to $1,250,000 from SSHRC.
Requirements: The Canadian project director must submit a proposal on behalf of the Canadian and LMIC partnership to SSHRC's Partnership Grants – Letter of Intent funding opportunity. This proposal will describe in its entirety the research program, the partnership, and the budget request from SSHRC and IDRC. In addition, the LMIC project director must submit the IDRC Supplementary Information Form to IDRC. SSHRC and IDRC will combine the application forms to conduct the merit review process. Applicants who are successful at the LOI stage will be invited to submit, again jointly, a formal application. They will be given instructions on how to complete the application at that time.
Geographic Focus: Canada
Date(s) Application is Due: Feb 15
Amount of Grant: 20,000-1,250,000 CAD
Contact: David O'Brien; (613) 236-6163, ext 2052; fax (613) 238-7230; dobrien@idrc.ca
Internet: http://www.idrc.ca/EN/Funding/Competitions/Pages/CompetitionDetails.aspx?CompetitionID=48
Sponsor: Social Sciences and Humanities Research Council of Canada
350 Albert Street, P.O. Box 1610
Ottawa, ON K1P 6G4 Canada

SSHRC-NSERC College and Community Innovation Enhancement Grants 4400
The objective of the Innovation Enhancement Grants is to increase innovation at the community and/or regional level by enabling Canadian colleges to increase their capacity to work with local small and medium-sized companies. They support applied research and collaborations that facilitate commercialization, as well as technology transfer, adaptation and adoption of new technologies. Grants will stimulate applied research that brings together expertise from diverse fields such as natural sciences and engineering, social sciences and humanities, and/or health sciences to address business-driven challenges and opportunities. Grants provide funding to colleges on a competitive basis to enhance their applied research capacity, and carry out applied research and technology transfer activities in collaboration with, and to the benefit of small to medium size companies in particular. These grants will focus on an area where the college has recognized expertise, that meet local or regional needs, and where there is the potential to increase economic development of the community. Grants are awarded for either a two-year or a five-year period. For the two-year grant, $100,000 per year for each of two years is provided. For the five-year grant, a base grant of up to $500,000 per year is provided for the first three years. During the course of the grants, colleges are expected to increase their effective collaboration with local or regional companies, and other existing community resources. To ensure increased and sustained commitment and involvement from college partners in years four and five of the five-year grants, grants will provide four-fifths of the average annual base funding received in the first three years. It is expected that other sources of support will be available from business and other partners (either cash or in-kind contributions). Additional guidelines, including entry-level and five year grants requirements and descriptions, are available at the NSERC website.
Requirements: Canadian colleges and universities are eligible to apply. These institutions must offer programs in the natural sciences, engineering, social sciences, humanities and/or health sciences. Faculty members involved in the grants must be involved in these disciplines. Institutions must also provide the space, facilities and services to enable its natural sciences, engineering, social sciences, humanities and/or health sciences faculty members to conduct research. Additional requirements and guidelines for businesses are available at the NSERC website.
Geographic Focus: Canada

Contact: Eric Bergeron, Program Officer, Partnerships Portfolio; (613) 996-1422 or (613) 944-5802; fax (613) 947-0223; eric.bergeron@sshrc-crsh.gc.ca
Internet: http://www.nserc-crsng.gc.ca/Professors-Professeurs/RPP-PP/CCI-ICC_eng.asp
Sponsor: Social Sciences and Humanities Research Council of Canada
350 Albert Street, P.O. Box 1610
Ottawa, ON K1P 6G4 Canada

SSHRC-Public Safety Canada Kanishka Project Initiative **4401**
The Kanishka Project Initiative funds research on pressing questions for Canada on terrorism and preventing violent extremism. The Project strives to understand what terrorism means in the Canadian context, how it is changing over time, and what can be done to support effective policies and programs to counter terrorism and violent extremism in Canada. Funding is available to researchers from a variety of disciplines to support a range of activities, including research, workshops, and publications. Research themes include ideological extremism and violence; perception and emotion; collective dynamics and resilience; and organization and effectiveness. An extensive list of guidelines and funding opportunities is posted on the Public Safety Canada website. Descriptions of ongoing projects are also included.
Geographic Focus: Canada
Amount of Grant: Up to 2,500,000 CAD
Contact: Brett Kubicek, Manager of Research and Academic Relations; (800) 830-3118 or (613) 944-4875; fax (613) 943-5186; Kanishka@ps-sp.gc.ca
Kanishka Project Coordinator; (613) 943-1007; partnershipgrants@sshrc-crsh.gc.ca
Internet: http://www.publicsafety.gc.ca/prg/ns/ai182/kpcp/knsh-eng.aspx
Sponsor: Social Sciences and Humanities Research Council of Canada
350 Albert Street, P.O. Box 1610
Ottawa, ON K1P 6G4 Canada

SSHRC Aid to Scholarly Journals **4402**
The objective of the Aid to Scholarly Journals is to increase access to, and readership for, original research results in the social sciences and humanities through Canadian scholarly journals. Through this funding opportunity, SSHRC will contribute to the broad dissemination and impact of original research results in social science and humanities scholarship through Canadian scholarly journals. Funding will also assist journals as they seek to take advantage of advances in communication technologies. Funds will be awarded to help defray the costs of publishing scholarly articles, to assist with distribution costs, and to support journal organizations in their transitions to digital media. A list of eligible expenses is available at the SSHRC website. Funding is available for three years, with up to $30,000 per year. Up to $5,000 a year may also be requested for: distribution costs, or those costs directly related to the circulation of the journal, be they either print or electronic (printing and mailing costs, fees paid to electronic-service providers, or support of existing websites); and transition costs, or those costs directly related to the transition to an electronic medium (the purchase of new computers, including software, costs related to the launch of a website, and staff training). Detailed estimates of cost must be provided for these items. For this competition, a scholarly journal is defined as a peer-reviewed academic publication that disseminates the results of original scholarship. Subscription-based journals, journals using an open-access business model, and journals using a "moving wall" arrangement to provide delayed open access are eligible for support. Also, for this competition, an article is defined as a record of previously unpublished, high-quality, original research or scholarship that has been accepted for publication after a process of peer review. Eligible article types include: research or theoretical articles and/or essays, and literature reviews, provided they meet the other criteria of being original, peer-reviewed and high-quality. Applicants will be required in the application to justify their choices of peer-reviewed articles. The application and additional guidelines are available at the website.
Requirements: To be eligible for support under this program, a journal must meet the definition of a scholarly journal; submit each article to be published to a rigorous, independent peer-review process; founded at least two years prior to the application deadline and published within that timeframe at least four issues, each having at least three peer-reviewed articles, or, for electronic journals not publishing in issue format, at least 12 peer-reviewed articles; a minimum of 250 regular readers, as demonstrated through a detailed web-usage report or other verifiable documentation, such as a list of subscribers; an editor-in-chief who is a faculty member of a Canadian postsecondary institution; an editorial board responsible for the intellectual direction and content of the journal of which at least one-third of the board's members are scholars affiliated with a Canadian postsecondary institution; and published primarily in English and/or French.
Restrictions: Article types that are not eligible for funding include: introductions; individual book, article or event reviews; letters to the editor; or editorials. A journal is not eligible if it disseminates previously published research results, or mainly publishes material assembled by non-specialists from secondary sources; is a professional or trade magazine, bulletin, newsletter, or advocacy publication, or is instructional in nature; is sponsored or published by, or on behalf of, a government department, agency, or commission; is edited and published solely by students; any member of its editorial board is under SSHRC sanction for financial or research misconduct; has an operating surplus equal to or greater than one year's operating expenses; or a non-Canadian publisher owns the exclusive copyright to its contents.
Geographic Focus: Canada
Date(s) Application is Due: Jun 30
Contact: Julie Lafontaine; (613) 947-9654; fax (613) 992-7635; julie.lafontaine@sshrc.ca
Internet: http://www.sshrc-crsh.gc.ca/funding-financement/programs-programmes/scholarly_journals-revues_savantes-eng.aspx
Sponsor: Social Sciences and Humanities Research Council of Canada
350 Albert Street, P.O. Box 1610
Ottawa, ON K1P 6G4 Canada

SSHRC Aid to Small Universities **4403**
The objective of this program is to enable small universities to develop and strengthen focused research capacity in the social sciences and humanities. Examples of activities that the ASU program supports: start-up costs/partial funding of research centers; stipends to doctoral students, provided the program of studies is related to the ASU grant; agenda-setting seminars; visiting scholars (travel and stipend); organization of colloquia or symposia; and seed funding for collaborative research or the development of partnerships.
Requirements: Eligible institutions must have active degree-granting status in the SSHRC fields at the undergraduate level or beyond, hold membership in the Association of Universities and Colleges of Canada, have fewer than 250 full-time faculty in the social sciences and humanities, and be independent of the federal government for the purpose of faculty employment status.
Geographic Focus: Canada
Date(s) Application is Due: Dec 1
Amount of Grant: Up to 30,000 CAD
Contact: Lorraine Anderson, Team Leader, Research Portfolio; (613) 947-4233; fax (613) 992-7635; lorraine.anderson@sshrc-crsh.gc.ca
Internet: http://www.sshrc-crsh.gc.ca/funding-financement/programs-programmes/aid_small_universities-aide_petites_universites-eng.aspx
Sponsor: Social Sciences and Humanities Research Council of Canada
350 Albert Street, P.O. Box 1610
Ottawa, ON K1P 6G4 Canada

SSHRC Banting Postdoctoral Fellowships **4404**
The Banting Postdoctoral Fellowships strive to attract and retain top-tier postdoctoral talent, nationally and internationally; develop their leadership potential; and position them for success as research leaders, contributing to Canada's economic, social, and research based growth through a research-intensive career. Selection criteria is based on research excellence and leadership in the research domain; quality of applicant's proposed research program; and institutional commitment and demonstrated synergy between applicant and institutional strategic priorities. The Fellowships are distributed equally across Canada's three federal granting agencies: the Canadian Institutes of Health Research (CIHR), the Natural Sciences and Engineering Research Council (NSERC) and the Social Sciences and Humanities Research Council (SSHRC). Up to 70 new two-year awards are given annually, with up to 140 awards active at any given time. Candidates may apply through the ResearchNet application process at the Banting website. Additional information and frequently asked questions are also posted on the website.
Requirements: Canadian citizens, permanent residents, and international citizens may apply. Applicants must fulfill or have fulfilled all degree requirements for a Ph.D., its equivalent, or health professional degree (a complete list of eligible health professionals is posted on the website). Applicants must contact the hot research institution whose endorsement they wish to receive, and continue contact to discuss benefits and their rights and responsibilities.
Restrictions: Applicants outside Canada may only hold their Banting Fellowship at a Canadian institution. Canadian or permanent resident applicants who receive their Ph.D. from an international institution may only hold the Banting Fellowship at a Canadian institution. Awards must be taken up no earlier than April 1 and no later than October 1 following the year of application.
Geographic Focus: All States, All Countries, Canada
Date(s) Application is Due: Nov 1
Amount of Grant: 70,000 USD
Contact: Banting Fellowship Coordinator; (613) 943-7777; banting@researchnet-recherchenet.ca or fellowships@sshrc-crsh.gc.ca
Internet: http://www.sshrc-crsh.gc.ca/funding-financement/programs-programmes/fellowships/banting-eng.aspx
Sponsor: Social Sciences and Humanities Research Council of Canada
350 Albert Street, P.O. Box 1610
Ottawa, ON K1P 6G4 Canada

SSHRC Bora Laskin National Fellowship in Human Rights Research **4405**
The purpose of this fellowship is to support research, preferably of a multidisciplinary or interdisciplinary nature, and to develop Canadian expertise in the field of human rights, with emphasis on themes and issues relevant to Canadian human rights. The Fellowship supports research in all areas of humanities and social sciences, including journalism and law. Preference is given to multidisciplinary or interdisciplinary research projects. One non-renewable fellowship, tenable for 12 months in Canada, is awarded every year. The award consists of a $45,000 stipend plus $10,000 for research and research-related travel expenses.
Requirements: Applicants must be affiliated with a Canadian postsecondary institution. Preference will be given to applicants with at least five years of proven research experience in their field. Awards must be used between April 1 and October 1 of the year for which they are awarded.
Restrictions: The Fellowship is not intended for scholars just beginning their research careers, and may not be used to complete the requirements for a degree program.
Geographic Focus: All States, Canada
Amount of Grant: 55,000 CAD
Contact: Fellowships and Institutional Grants Contact; (613) 943 7777; fax (613) 943 1329; fellowships@sshrc-crsh.gc.ca
Internet: http://www.sshrc-crsh.gc.ca/funding-financement/programs-programmes/fellowships/bora_laskin-eng.aspx
Sponsor: Social Sciences and Humanities Research Council of Canada
350 Albert Street, P.O. Box 1610
Ottawa, ON K1P 6G4 Canada

SSHRC Digging into Data Challenge Grants 4406

The Social Sciences and Humanities Research Council of Canada and several other organizations launch the next round of the Digging into Data Challenge, a competition for humanities and social sciences research on cyberscholarship. As the world becomes increasingly digital, new techniques will be needed to search, analyze, and understand these everyday materials. Digging into Data challenges the research community to help create the new research infrastructure for 21st century scholarship. In recognition of the international nature of cyberinfrastructure/e-science, the purpose of the program is to bring together international research teams to advance research and to share their results openly, so that others may learn from them. Along with SSHRC, the following research funders are also involved: the Arts and Humanities Research Council (United Kingdom), the Economic and Social Research Council (United Kingdom), the Institute of Museum and Library Services (United States), the Joint Information Systems Committee (United Kingdom), the National Endowment for the Humanities (United States), the National Science Foundation (United States), and the Netherlands Organisation for Scientific Research (Netherlands). Specific objectives are: to promote the development and deployment of innovative research techniques in large-scale data analysis; to foster interdisciplinary collaboration among scholars in the humanities, social sciences, computer sciences, information sciences, and other fields, around questions of text and data analysis; to promote international collaboration; and to work with data repositories that hold large digital collections to ensure efficient access to these materials for research. Applicants will form international teams from at least two of the participating countries. Winning teams will receive grants from two or more of the funding agencies and, two years later, will be invited to show off their work at a special conference sponsored by the eight funders. Total project funding is approximately U.S. $4.8 million.
Geographic Focus: All States, Canada, Netherlands, United Kingdom
Amount of Grant: Up to 4,800,000 CAD
Contact: Gail Zboch; (613) 943-1148; gail.zboch@sshrc-crsh.gc.ca
Internet: http://www.sshrc.ca/web/apply/program_descriptions/digging_into_data_e.asp
Sponsor: Social Sciences and Humanities Research Council of Canada
350 Albert Street, P.O. Box 1610
Ottawa, ON K1P 6G4 Canada

SSHRC Doctoral Fellowships 4407

Doctoral fellowships are offered to assist in developing research and scholarly resources in the social sciences and humanities. Fellowships are for one, two, three, or four years for universities in Canada or abroad. Direct applications are due November 7, while university applications are due January 9. Applicants are also eligible for the Bombardier Doctoral Scholarship worth $35,000 per year, although the Bombardier Scholarship may only be used at Canadian universities. Applicants apply for both awards through one application form, located at the website. Doctoral Fellowship holders are also eligible for department-awarded Queen's Fellowship of $15,000, William Taylor Fellowship of $5,000, and Aileen Ross Fellowship of $10,000.
Requirements: Candidates must be Canadian citizens or have permanent resident (landed immigrant) status at the time of application and must have completed one year of doctoral study or a master's degree by the date of taking the award. Applicants must be pursuing doctoral studies in the social sciences or humanities.
Restrictions: Applicants must not have already received a scholarship or fellowship from SSHRC, NSERC, or CIHR, and not be applying to NSERC or CIHR.
Geographic Focus: Canada
Date(s) Application is Due: Jan 9; Nov 7
Amount of Grant: 20,000 CAD
Contact: Division of Fellowships and Institutional Grants; (613) 943-7777; fax (613) 943-1329; fellowships@sshrc.ca
Internet: http://www.sshrc-crsh.gc.ca/funding-financement/programs-programmes/fellowships/doctoral-doctorat-eng.aspx#a5
Sponsor: Social Sciences and Humanities Research Council of Canada
350 Albert Street, P.O. Box 1610
Ottawa, ON K1P 6G4 Canada

SSHRC Graduate Scholarships-Michael Smith Foreign Study Supplements 4408

The Canada Graduate Scholarships-Michael Smith Foreign Study Supplements (CGS-FSS) program will support high-calibre Canadian graduate students in building global linkages and international networks through the pursuit of exceptional research experiences in Canada. By accessing international scientific research and training, CGS-FSS recipients will contribute to strengthening the potential for collaboration between Canadian universities and affiliated research institutions and universities, or other research institutions outside of Canada. The defined research study period abroad to be covered must be no less than three months and no more than six months. Supplements of up to $6,000 will be awarded to Canadian citizens or permanent residents who hold an active Joseph-Armand Bombardier, Alexander Graham Bell or Frederick Banting and Charles Best Canada Graduate Scholarship (CGS) at the master's or doctoral level, or a Vanier CGS at the doctoral level. A total of 250 CGS-MSFSS awards are available each year, and are divided among the three granting agencies to reflect the distribution of CGS awards across the social sciences and humanities, natural sciences and engineering, and health-related disciplines: 45 for the Canadian Institutes of Health Research (CIHR); 80 for the Natural Sciences and Engineering Research Council (NSERC); and 125 for SSHRC.
Requirements: To be eligible to apply for this program, candidates must be a Canadian citizen or permanent resident of Canada, and be registered at a Canadian institution at the time of application. Eligible institutions abroad include foreign universities, international or national research institutes, or other institutions that actively carry out research.
Geographic Focus: Canada
Date(s) Application is Due: Jun 10; Oct 10
Amount of Grant: Up to 6,000 CAD
Contact: Training Portfolio Coordinator; (613) 995-2694; scholfin@sshrc-crsh.gc.ca
Internet: http://www.sshrc-crsh.gc.ca/funding-financement/programs-programmes/fellowships/cgs_foreign_study-besc_etudes_etranger-eng.aspx
Sponsor: Social Sciences and Humanities Research Council of Canada
350 Albert Street, P.O. Box 1610
Ottawa, ON K1P 6G4 Canada

SSHRC Insight Development Grants 4409

Insight Development Grants support research in its initial stages. The grants enable the development of new research questions, as well as experimentation with new methods, theoretical approaches and/or ideas. Funding is provided for short-term research development projects of up to two years, proposed by individuals or teams. The objectives of the Insight program are to build knowledge and understanding from disciplinary, interdisciplinary and/or cross-sector perspectives through support for the best researchers; support new approaches to research on complex and important topics; provide a high-quality research training experience for students; fund research expertise that relates to societal challenges and opportunities; and mobilize research knowledge, to and from academic and non-academic audiences, with the potential to lead to intellectual, cultural, social and economic influence, benefit and impact. Proposed projects may involve case studies, pilot initiatives, and critical analyses of existing research. Projects may also involve national and international research collaboration, and the exploration of new ways of producing, structuring and mobilizing knowledge within and across disciplines and sectors. Funding is available for emerging and established scholars. Emerging scholar funding is used to develop new research questions and/or approaches. Funding for established scholars provides support to explore new research questions and/or approaches that are distinct from the applicant's previous research. The SSHRC also welcomes applications involving research-creation. Insight Development Grants are adjudicated by academic and/or other experts drawn from, but not limited to, the following research fields: Group 1-history; medieval studies; classics; literature; fine arts; philosophy; religious studies; and related fields: Group 2-anthropology; archaeology; linguistics; translation; political science; public administration; law; criminology; geography; urban planning and environmental studies; and related fields: Group 3-business and management; economics; and related fields: Group 4-sociology; demography; communication studies; journalism; media studies; gender studies; cultural studies; library and information science; and related fields: Group 5-education; psychology; social work; career guidance; and related fields. In the application form, applicants will be asked to indicate which of the five groups they consider most appropriate for their proposal, and to select a primary discipline and area of research from a detailed list on the application form. The application form and Canadian Common CV are available at the SSHRC website.
Requirements: Applications may be submitted by an individual researcher or team of researchers. Applicants must be affiliated with a Canadian university or college at the time of application.
Restrictions: Insight Development Grants are not intended to support large-scale initiatives. Long-term support for research is offered through SSHRC's Insight Grants. Research projects should be clearly defined and in the early stages of the research process. Insight Development Grant funding is not intended to support ongoing research for established scholars.
Geographic Focus: All States, Canada
Date(s) Application is Due: Feb 1
Amount of Grant: 7,000-75,000 CAD
Contact: Fatima Bidas; (613) 943-1560; fax (613) 992-7635; fatima.bidas@sshrc-crsh.gc.ca
Internet: http://www.sshrc-crsh.gc.ca/funding-financement/programs-programmes/insight_development_grants-subventions_de_developpement_savoir-eng.aspx
Sponsor: Social Sciences and Humanities Research Council of Canada
350 Albert Street, P.O. Box 1610
Ottawa, ON K1P 6G4 Canada

SSHRC Insight Grants 4410

Insight Grants support research excellence in the social sciences and humanities. Funding is available to both emerging and established scholars for long-term research initiatives of three to five years. The Insight Grants funding opportunity supports research proposed by scholars and judged worthy of funding by their peers and/or other experts. Insight Grants research initiatives may be undertaken by an individual researcher or a team of researchers working in collaboration. Insight Grants are adjudicated by committees of academic and/or other experts drawn from, but not limited to, the following research fields: Group 1-history; medieval studies; classics; literature; fine arts; philosophy; religious studies; and related fields: Group 2-anthropology; archaeology; linguistics; translation; political science; public administration; law; criminology; geography; urban planning and environmental studies; and related fields: Group 3-business and management; economics; and related fields: Group 4-sociology; demography; communication studies; journalism; media studies; gender studies; cultural studies; library and information science; and related fields: and Group 5-education, psychology; social work; and related fields. In the application form, applicants will be asked to indicate which of the five groups they consider most appropriate for their proposal, and to select a primary discipline and area of research from a detailed list available on the application form. The maximum value of a grant is $500,000, over three to five years. A minimum request of $7,000 is required in at least one of the years. A maximum of $200,000 is available in a single year. Additional guidelines are available at the website.
Requirements: Applications may be submitted by an individual researcher or a team of researchers. All applicants must be affiliated with a Canadian university, college, or research institute. Before the full application, applicants must first submit a notice of intent to the SSHRC.

Geographic Focus: All States, Canada
Date(s) Application is Due: Oct 15
Amount of Grant: 7,000-500,000 CAD
Contact: Fatima Bidas; (613) 943-1560; fax (613) 992-7635; fatima.bidas@sshrc-crsh.gc.ca
Internet: http://www.sshrc-crsh.gc.ca/funding-financement/programs-programmes/insight_grants-subventions_savoir-eng.aspx
Sponsor: Social Sciences and Humanities Research Council of Canada
350 Albert Street, P.O. Box 1610
Ottawa, ON K1P 6G4 Canada

SSHRC Institutional Grants **4411**

Under this program, SSHRC provides annual block grants for eligible Canadian postsecondary institutions in order to fund small-scale research activities by their faculty in the social sciences and humanities. The program assists Canadian universities in developing, increasing, or strengthening research excellence in social sciences and humanities; assisting researchers beginning their research career to become competitive in grant competitions at the national level; helping established researchers with modest funding requirements or those wishing to reorient and strengthen their research programs; and supporting national and international dissemination and collaboration. The application and additional guidelines, including the terms of the funding, are posted on the SSHRC website.
Requirements: To be eligible for funding, a postsecondary institution must be: eligible to manage SSHRC funds (see SSHRC's list of eligible institutions); an institutional member of the Association of Universities and Colleges of Canada (AUCC) and/or the Association of Canadian Community Colleges (ACCC); and have active degree-granting status in at least one of the social sciences and humanities disciplines; or the institution must be eligible to manage SSHRC funds (see SSHRC's list of eligible institutions); be an institutional member of the AUCC and/or the ACCC; affiliated or federated with an eligible institution, but grant their own degrees, receive their own operating budget directly from the provincial government, and have their own board of directors; and offer programs in at least one of the social sciences and humanities disciplines.
Restrictions: Institutions must adhere to the following *Restrictions:* applicants must maintain an affiliation with the postsecondary institution for the duration of the grant period; visiting fellows and scholars are not eligible, nor are the costs incurred for their travel and residency (although postdoctoral fellows affiliated with the institution are eligible to receive support provided they meet the other eligibility criteria); students are not eligible to hold their own award (but SSHRC encourages the employment of students in SIG-funded research); and support must be for scholarly activity in the humanities and the social sciences.
Geographic Focus: Canada
Date(s) Application is Due: Dec 1
Amount of Grant: 5,000 CAD
Contact: Lorraine Anderson, Team Leader, Research Portfolio; (613) 947-4233; fax (613) 992-7635; lorraine.anderson@sshrc-crsh.gc.ca
Internet: http://www.sshrc-crsh.gc.ca/funding-financement/programs-programmes/institutional_grants-subventions_institutionnelles-eng.aspx
Sponsor: Social Sciences and Humanities Research Council of Canada
350 Albert Street, P.O. Box 1610
Ottawa, ON K1P 6G4 Canada

SSHRC Joseph-Armand Bombardier CGS Doctoral Scholarships **4412**

The J.-Armand Bombardier CGS Doctoral Scholarships programs aim to develop research skills and assist in the training of highly-qualified personnel by supporting students who demonstrate a high standard of scholarly achievement in undergraduate and graduate studies in the social sciences and humanities. Scholarships are valued at $35,000 per annum for 36 months. Scholarship applicants are also eligible for the SSHRC Doctoral Fellowships at $20,000 per annum for 12, 24, 36, or 48 months. Holders of the Bombardier Scholarship are also eligible for consideration of the department-awarded Queen's Fellowship of $15,000, William Taylor Fellowship of $5,000, and the Aileen Ross Fellowship of $10,000. Direct applications are due November 7, while university applications are due January 9.
Requirements: To qualify to apply for the J.-Armand Bombardier CGS Doctoral Awards programs, applicants must: be a citizen or permanent resident of Canada; be applying for support to pursue a first PhD (or equivalent); be pursuing doctoral studies (or a combined MA/PhD) in the social sciences or humanities; not have already received a scholarship or fellowship from SSHRC, NSERC or CIHR to undertake or complete a doctoral degree or combined MA/PhD; not be applying in the upcoming academic year to NSERC or CIHR; or not be pursuing a degree program that combines undergraduate and graduate degrees.
Restrictions: The Scholarship may only be used at an eligible university in Canada. A detailed list of those universities is located at the website.
Geographic Focus: Canada
Date(s) Application is Due: Jan 9; Nov 7
Amount of Grant: 35,000 CAD
Contact: Doctoral Awards Coordinator; (613) 943-7777 or (613) 995-4273; fax (613) 943-1329; fellowships@sshrc.ca
Internet: http://www.sshrc-crsh.gc.ca/funding-financement/programs-programmes/fellowships/doctoral-doctorat-eng.aspx#a5
Sponsor: Social Sciences and Humanities Research Council of Canada
350 Albert Street, P.O. Box 1610
Ottawa, ON K1P 6G4 Canada

SSHRC Jules and Gabrielle Leger Fellowship **4413**

This fellowship is offered in odd-numbered years to assist in developing research and scholarly resources in the social sciences and humanities. The fellowship is awarded for research and writing on the historical and contemporary contribution of the Crown and its representatives, federal and provincial, in addition to the political, constitutional, cultural, intellectual, and social life of the country. One non-renewable fellowship is offered every second year and is tenable for 12 months. There are no restrictions on the place of tenure. The award consists of a $40,000 stipend plus $10,000 for research and research-related travel expenses. This award may not be used to complete the requirements for a degree program. Awards must be taken up between April 1 and October 1 of the year for which they are awarded. Additional guidelines are available at the SSHRC website.
Requirements: Applicants must be affiliated with a Canadian postsecondary institution. This fellowship is not intended for scholars at the beginning of their research careers. Preference will be given to applicants with at least five years of proven research experience in their field. Please note that all applicants, as well as anyone else participating in any capacity in the proposed project, must not be under SSHRC sanction for financial or research misconduct.
Geographic Focus: All States, Canada
Date(s) Application is Due: Oct 1
Amount of Grant: 50,000 CAD
Contact: Program Officer; (613) 943-7777; fellowships@sshrc-crsh.gc.ca
Internet: http://www.sshrc-crsh.gc.ca/funding-financement/programs-programmes/fellowships/leger-eng.aspx
Sponsor: Social Sciences and Humanities Research Council of Canada
350 Albert Street, P.O. Box 1610
Ottawa, ON K1P 6G4 Canada

SSHRC Metropolis Project **4414**

The purpose of the annual Metropolis National Research Competition is to select and finance a major national-level policy research project in the field of immigration and diversity. The findings of the research will be of direct relevance to policy-makers within the Government of Canada and, more specifically, to the main funding partners of the Metropolis Project. At a broader level, the objectives of the competition are to produce a pan-Canadian research study that informs federal policy on issues surrounding immigration and diversity; produce knowledge that affects federal policy and programs related to immigration and diversity; widely disseminate, in collaboration with the Metropolis secretariat, the results of this research through a range of knowledge transfer/mobilization activities targeting federal policy-makers and the research community; and create knowledge that informs and expands the broader academic discourse on immigration and diversity. The maximum value of the grant is $125,000 for 12 to 18 months. Extensions are not available. The application and detailed guidelines are posted at the website.
Requirements: Researchers affiliated with Canadian postsecondary educational institutions are invited to submit a proposal. Successful applicants must seek affiliation with one of the five Metropolis centres (an up-to-date list is found at the website). Individuals who wish to become affiliated with a Metropolis centre can apply by completing the appropriate application forms, available on the Metropolis website, or by contacting one of the centres directly.
Geographic Focus: Canada
Date(s) Application is Due: Oct 11
Amount of Grant: Up to 125,000 CAD
Contact: Mika Oehling; (613) 992-4227; mika.oehling@sshrc-crsh.gc.ca
Internet: http://www.sshrc-crsh.gc.ca/funding-financement/programs-programmes/metropolis-eng.aspx
Sponsor: Social Sciences and Humanities Research Council of Canada
350 Albert Street, P.O. Box 1610
Ottawa, ON K1P 6G4 Canada

SSHRC Postdoctoral Fellowships **4415**

The Postdoctoral Fellowships support the most promising new scholars in the social sciences and humanities and assists them in establishing a research base at an important time in their research careers. Postdoctoral Fellowships provide stipendiary support to recent PhD graduates who are: undertaking original research; publishing research findings; developing and expanding personal research networks; broadening teaching experience; and preparing to become competitive in national research grants competitions. Posdoctoral Fellowships are valued at approximately $38,000 per year, plus a research allowance of up to $5,000 per award. These are non-renewable fellowships, tenable for a minimum of 12 months and a maximum of 24 months. Fellowships will normally be awarded to candidates affiliated with a university other than that which awarded the PhD. Recipients of this Fellowship are also eligible for department consideration of the Postdoctoral Prize of $10,000. Additional guidelines and the application are available at the SSHRC website.
Requirements: To apply to this program, applicants must be a citizen or permanent resident of Canada; have earned a first doctorate from a recognized university within three years of the application date or have completed a first doctorate no earlier than five years before the application date, but have had their career interrupted or delayed for the purpose of child rearing, illness, or health-related family responsibilities; not hold a tenure or tenure-track faculty position; have finalized arrangements for affiliation with a recognized university or research institution; and have applied not more than twice before for a SSHRC Postdoctoral Fellowship.
Restrictions: Applicants who obtain their PhD from a foreign university are eligible to apply only if they wish to hold their award at a Canadian university. Applicants are eligible to apply to hold their award at a foreign university only if their PhD was earned at a Canadian university.
Geographic Focus: All States, Canada
Date(s) Application is Due: Oct 4
Contact: Fatima Bidas, Research Training Portfolio Contact; (613) 943-7777; fax (613) 943-1329; fatima.bidas@sshrc-crsh.gc.ca or fellowships@sshrc-crsh.gc.ca
Internet: http://www.sshrc-crsh.gc.ca/funding-financement/programs-programmes/fellowships/postdoctoral-doctorat-eng.aspx
Sponsor: Social Sciences and Humanities Research Council of Canada
350 Albert Street, P.O. Box 1610
Ottawa, ON K1P 6G4 Canada

SSHRC Sport Participation Research Initiative: Doctoral Award Supplements 4416

Applicants for a SSHRC doctoral award who indicate their interest on the application form and who are working in one or more of the target areas will be considered for a Sport Canada Doctoral Award Supplement. Those target areas are the following: identification and overcoming of barriers to participation in sport; training of participants, volunteers, coaches and administrators in sport; development, monitoring and evaluation of policies designed to enhance participation in sport; development of capacity and infrastructure that will enhance participation in sport; and identification and assessment of the benefits and outcomes of participation in sport. Those who already hold a SSHRC doctoral award and are working in one or more of the target areas are also eligible to apply for this supplement. For applicants registered at a Canadian postsecondary institution, the postsecondary institution sets the deadline; for all others, November 7 is the deadline.

Requirements: Applicants for, and holders of, a SSHRC doctoral award who are working in one or more of the target areas of interest, and who are affiliated with an eligible Canadian university are eligible to apply.

Geographic Focus: Canada

Amount of Grant: Up to 10,000 CAD

Contact: Fatima Bidas, Program Officer, Research Portfolio; (613) 943-1560; fax (613) 992-7635; fatima.bidas@sshrc-crsh.gc.ca

Joanne Kay, Senior Research and Policy Analyst; (613) 294-3036; fax (514) 939-1918; joanne.kay@pch.gc.ca

Internet: http://www.sshrc-crsh.gc.ca/funding-financement/programs-programmes/sport_can-eng.aspx#atarget

Sponsor: Social Sciences and Humanities Research Council of Canada
350 Albert Street, P.O. Box 1610
Ottawa, ON K1P 6G4 Canada

SSHRC Sport Participation Research Initiative: Postdoctoral Fellowship Supplements 4417

Applicants for a SSHRC Postdoctoral Fellowship who indicate their interest on the application form and who are working in one or more of the following target areas will be considered for a Sport Canada Postdoctoral Fellowship Supplement. Those areas of research are: identification and overcoming of barriers to participation in sport; training of participants, volunteers, coaches and administrators in sport; development, monitoring and evaluation of policies designed to enhance participation in sport; development of capacity and infrastructure that will enhance participation in sport; and identification and assessment of the benefits and outcomes of participation in sport. Sport Participation Research Initiative Postdoctoral Fellowship Supplements are tenable for 12 months and worth $10,000, in addition to the annual value of the SSHRC Postdoctoral Fellowship. Additional guidelines are posted on the SSHRC website.

Requirements: Applicants for a SSHRC Postdoctoral Fellowship who are working in one or more of the target areas of interest, and who are affiliated with an eligible Canadian postsecondary institution, are eligible to apply.

Restrictions: Only applications that are recommended for funding at SSHRC are referred to Sport Canada.

Geographic Focus: All States, Canada

Date(s) Application is Due: Oct 4

Amount of Grant: 10,000 CAD

Contact: Fatima Bidas, Program Officer, Research Portfolio; (613) 943-1560; fax (613) 992-7635; fatima.bidas@sshrc-crsh.gc.ca

Internet: http://www.sshrc-crsh.gc.ca/funding-financement/programs-programmes/sport_can-eng.aspx

Sponsor: Social Sciences and Humanities Research Council of Canada
350 Albert Street, P.O. Box 1610
Ottawa, ON K1P 6G4 Canada

SSHRC Therese F. Casgrain Fellowship 4418

The Therese F. Casgrain Fellowship supports research on women and social change in Canada. This Fellowship supports research in the field of social justice, particularly in defense of individual rights and the promotion of the economic and social interests of Canadian women. The Foundation is particularly interested in research that includes a discussion of public policy options and makes concrete recommendations for change. The non-renewable Fellowship, tenable for 12 months and only in Canada, is available every fourth year. The award consists of a $40,000 stipend, paid in three installments, of which up to $10,000 may be used for travel and research expenses. The successful applicant must take up the award between April 1 and October 1 following the announcement of awards.

Requirements: Applicants must be Canadian citizens or permanent residents of Canada, hold a doctorate degree or equivalent advanced professional degree, have at least five years of proven research experience, and not be under SSHRC sanction for financial or research misconduct. Affiliation with a university or an appropriate research institute or similar organization is desirable, but is not a condition of the award.

Geographic Focus: Canada

Amount of Grant: Up to 40,000 CAD

Contact: Fellowships and Institutional Grants Contact; (613) 943-7777; fax (613) 943-1329; fellowships@sshrc-crsh.gc.ca

Internet: http://www.sshrc-crsh.gc.ca/funding-financement/programs-programmes/fellowships/casgrain-eng.aspx

Sponsor: Social Sciences and Humanities Research Council of Canada
350 Albert Street, P.O. Box 1610
Ottawa, ON K1P 6G4 Canada

SSRC-Van Alen Fellowships 4419

The SSRC-Van Alen New York Prize Fellowships support work on the intersection between social dynamics and the built environment. Since its inception, two rounds of Fellowships have been offered in collaboration with the Van Alen Institute, an independent nonprofit architectural organization, on issues of urban sustainability and the 'spatialization' of information policy. Grants are awarded to partnerships between social scientists and architectural/design specialists. Investigations may take a range of formats, including installations, demonstrations, performances, and workshops, but are by their involvement of the general public. Additional information is available at the SSRC website.

Geographic Focus: All States

Contact: Joe Karaganis, Program Director; (718) 517-3673 or (212) 377-2700; fax (212) 377-2727; karaganis@ssrc.org

Internet: http://www.ssrc.org/fellowships/ssrc-van-alen-fellows-program/

Sponsor: Social Science Research Council
One Pierrepont Plaza, 15th Floor
Brooklyn, NY 11201

SSRC Abe Fellowship for Journalists 4420

The Abe Fellowship for Journalists is designed to encourage in-depth coverage of topics of pressing concern to the U.S. and Japan through individual short-term policy-related projects. Applicants are invited to submit proposals on one of the following three themes: traditional and non-traditional approaches to security and diplomacy (transnational terrorism, internal ethnic and religious strife, infectious diseases, food safety, climate change, and non-proliferation, as well as the role of cultural initiatives in peace building); global and regional economic issues (regional and bilateral trade arrangements, international financial stability, globalization and the mitigation of its adverse consequences, sustainable urbanization, and environmental degradation); and social and cultural issues (demographic change, immigration, the role of civil society and media as champions of the public interest, social enterprise, corporate social responsibility, and revitalization of multi-cultural urban areas). Fellows are expected to produce an analytical article or feature story that will inform public debate or a policy community. The application and additional requirements are available at the SSRC website.

Requirements: The AFJ competition is open to citizens of the U.S. and Japan with at least five years of professional journalistic experience with newspapers, newsmagazines, wire services, and on-line news organizations. Freelancers are also eligible. Nationals of other countries must be permanent residents of the U.S. or Japan, or have a long-term affiliation with the American or Japanese journalistic communities. U.S.-based applicants with no previous journalistic employment in Japan have priority. Japan-based applicants with no previous journalistic employment in the U.S. have priority.

Restrictions: Fellows may receive salary from their employers, but cannot carry out assignments while on the award.

Geographic Focus: All States, Japan

Date(s) Application is Due: Sep 15

Amount of Grant: Up to 26,100 USD

Contact: Fernando Rojas, U.S; (202) 377-2700; fax (212) 377-2727; abe@ssrc.org
fax +81-3-5369-6042; ssrcABE@gol.com or tozaki@gol.com

Internet: http://www.ssrc.org/fellowships/abe-fellowship-for-journalists/

Sponsor: Social Science Research Council
One Pierrepont Plaza, 15th Floor
Brooklyn, NY 11201

SSRC Abe Fellowships 4421

The Abe Fellowship is designed to encourage international multidisciplinary research on topics of pressing global concern. The program seeks to foster the development of a new generation of researchers who are interested in policy-relevant topics of long-range importance and who are willing to become key members of a bilateral and global research network built around such topics. Rather than seeking to promote greater understanding of a single country, the Abe Fellowship encourages research on the experiences and future challenges of the U.S. and Japan in a comparative or global perspective. The program committee seeks applications for research focusing explicitly on policy-relevant and contemporary issues that have a comparative or transnational perspective and that draw the study of the U.S. and Japan into wider disciplinary or theoretical debates. Applicants are invited to submit proposals for research in the social sciences or the humanities relevant to any one or combination of the following three themes: global issues; problems common to industrial and industrializing societies; or issues that pertain to U.S.-Japan relations. Applications are welcome from scholars and non-academic research professionals. Funding for the program is provided by the Japan Foundation Center for Global Partnership. The application and additional information are available at the SSRC website.

Requirements: Applications are welcome from scholars and non-academic research professionals. Competition is open to citizens of the U.S. and Japan as well as to nationals of other countries who can demonstrate strong and serious long-term affiliations with research communities in Japan or the U.S. Applicants must hold a Ph.D. or the terminal degree in their field, or have attained an equivalent level of professional experience.

Geographic Focus: All States

Date(s) Application is Due: Sep 1

Contact: Fernando Rojas; (212) 377-2000; fax (212) 377-2727; abe@ssrc.org

Internet: http://www.ssrc.org/fellowships/abe-fellowship/

Sponsor: Social Science Research Council
One Pierrepont Plaza, 15th Floor
Brooklyn, NY 11201

SSRC African Peacebuilding Network Research Grants **4422**
The African Peacebuilding Network (APN) of the Social Science Research Council (SSRC) invites research grant applications from African researchers, policy analysts, and practitioners working on conflict and peacebuilding at universities and research institutions or regional governmental and non-governmental organizations in Africa. A core component of the APN, the Research Grants program is a vehicle for enhancing the quality and visibility of African peacebuilding research regionally and globally and for making such knowledge accessible to key policymakers and research centers of excellence in Africa and around the world. Substantial support is available for research and analysis on issues such as: conflict prevention, mediation, management, resolution, and transformation; environmental change and conflict; post-conflict democratization, governance, and reconstruction; the relationship between peacebuilding and statebuilding, including state-society relations and state reconstruction; transitional justice, reconciliation, social and economic justice; peacekeeping and peace support operations; and disarmament, demobilization, and reinsertion. The APN is interested in innovative projects that demonstrate strong potential for producing high-quality research and analysis that can lead to practical action on peacebuilding and/or facilitate inter-regional collaboration and networking among African researchers and practitioners. Additional proposal guidelines are available at the SSRC website. Grants are awarded on a competitive, peer-reviewed basis and are intended to support six to seven months of research. Up to fifteen grants of a maximum of $15,000 will be awarded. The application is available through the SSRC's application portal at its website.
Requirements: Applicants must be African citizens currently residing in a Sub-Saharan African country. Academics must hold a faculty or research position at an African university or research organization and have either a Ph.D. or a master's, with at least two years of research experience. Policy analysts and practitioners must be based in Africa at a regional or sub-regional institution, a government agency, or a non-governmental, media, or civil society organization and have a master's with at least two years of work experience.
Geographic Focus: Angola, Benin, Botswana, Burkina Faso, Burundi, Cameroon, Cape Verde, Central African Republic, Chad, Comoros, Congo, Democratic Republic of, Cote d' Ivoire (Ivory Coast), Djibouti, Equatorial Guinea, Eritrea, Ethiopia, Gabon, Gambia, Ghana, Guinea, Guinea-Bissau, Kenya, Lesotho, Liberia, Madagascar, Malawi, Mali, Mauritania, Mauritius, Mozambique, Namibia, Niger, Nigeria, Rwanda, Sao Tome & Principe, Senegal, Seychelles, Sierra Leone, South Africa, Sudan, Swaziland, Tanzania, Togo, Uganda, Zambia, Zimbabwe
Date(s) Application is Due: Jun 15
Amount of Grant: Up to 15,000 USD
Contact: Kirin Bains, Research Coordinator; (718) 517-3679 or (212) 377-2700; fax (212) 377-2727; kbains@ssrc.org
Internet: http://www.ssrc.org/fellowships/apn-research-grants/
Sponsor: Social Science Research Council
One Pierrepont Plaza, 15th Floor
Brooklyn, NY 11201

SSRC Collaborative Research Grants on Environment and Health in China **4423**
The collaborative grants program was begun to stimulate new research on environment and health in China that is directly linked to policy and practice, to encourage collaboration across disciplines, and to help develop a network of scholars working in this field. Each year seven to eight grants of $10,000 to 15,000 each are given to Chinese institutes for projects that: integrate environment or health perspectives into research currently focused on a single domain by sharing existing data or conducting limited additional data collection; develop a new focus on environment and health within the context of ongoing research on development; broaden the scope of analysis through collaboration with partners who bring additional disciplinary perspectives or international experience; or work with NGO or government partners to make existing research more accessible and useful. Additional information is available at the SSRC website.
Requirements: Each project must address an environmental health problem that affects a substantial number of people and it must be clearly useful in generating or disseminating knowledge that will inform policy or the work of government or civil society actors. In order to ensure feasibility, applicants must demonstrate the necessary access to data, or relationships with local communities or government agencies, to conduct the project successfully. If the project is being carried out at the local level, researchers must either live and work in the same province, or otherwise demonstrate how the project will build local capacity to work on environment and health issues.
Geographic Focus: China
Amount of Grant: 10,000-15,000 USD
Contact: Chao Cai; (212) 377-2700; fax (212) 377-2727; chao@ssrc.org
Internet: http://www.ssrc.org/fellowships/grants-for-collaborative-research-on-environment-and-health-in-china/
Sponsor: Social Science Research Council
One Pierrepont Plaza, 15th Floor
Brooklyn, NY 11201

SSRC Dissertation Proposal Development Faculty Field Fellowships **4424**
The Dissertation Proposal Development Fellowship (DPDF) Faculty Field Competition is open to tenured humanities and social sciences faculty interested in creating or reinvigorating interdisciplinary fields of study through the training of the next generation of researchers. Selected research directors guide the development of effective doctoral dissertation proposals within innovative fields by helping fellows sharpen the focus of their research and identify appropriate methods of investigation and analysis. Annually, the DPDF program selects up to five research fields, each proposed by two senior faculty with different institutional affiliations and, as relevant, different disciplinary specializations. Faculty selected as field research directors design and lead two workshops for student fellows, in coordination with DPDF staff, and serve as mentors to the fellows during the course of their summer pre-dissertation research. Proposed research fields should be domains of inquiry centered on a core set of issues and questions that can be valuably addressed by diverse scholarship. Fields should address topics of broad public concern that can be approached from multiple intellectual, societal, and geographic orientations. Fields should also address topics that encompass broad geographic areas rather than focus on one specific country. Research fields proposed by pairs of faculty who bring distinct perspectives from any different disciplines are welcome, but fields stemming jointly from the humanities and social sciences are particularly encouraged. Additional information, frequently asked questions, and a sample application are available at the SSRC website.
Requirements: Applicants must apply in pairs, and be based at different university, with at least one applicant based in the U.S. Both applicants must be available to attend the spring and fall workshops in their entirety (workshop dates found at website). Both applicants must be tenured at the time of the application and trained in different disciplines or bring different methodological frameworks to their proposed research field. At least one applicant must be based at a doctoral-degree granting university. However, both applicants must demonstrate records of carrying out research activities and mentoring graduate students.
Restrictions: Research fields proposed to be led by more than two people will not be accepted.
Geographic Focus: All States
Date(s) Application is Due: Sep 30
Amount of Grant: Up to 13,000 USD
Contact: Marika Dunn, Program Officer; (718) 517-3645 or (212) 377-2700; fax (212) 377-2727; dunn@ssrc.org
Internet: http://www.ssrc.org/fellowships/dpdf-faculty-fields-competition/
Sponsor: Social Science Research Council
One Pierrepont Plaza, 15th Floor
Brooklyn, NY 11201

SSRC Dissertation Proposal Development Student Fellowships **4425**
The Dissertation Proposal Development Fellowship (DPDF) is organized to help early-stage graduate students in the humanities and social sciences formulate effective doctoral dissertation proposals. Each year, the program offers training within different interdisciplinary fields of study under the leadership of pairs of tenured senior faculty, who define the fields and serve as research directors for groups of 12 graduate students. The students participate in two workshops: one in the spring, to prepare students to undertake summer research that will inform the design of their dissertation research; the other in the fall, to help students apply their summer research experiences to writing dissertation and funding proposals. Students may apply for up to $5,000 to cover summer research costs. Working together, research directors and graduate students design research that will help to shape evolving fields in the humanities and social science. Current research fields include: Black Atlantic Studies; The Political Economy of Redistribution; Visual Culture; Critical Studies of Science and Technology Policy; Animal Studies; Human Dimensions of Global Environmental Change; Muslim Modernities; Science/Art Studies; Global Indigenous Politics; and Migration and Gender Studies. Research fields change every year. The application and additional information are available at the SSRC website.
Requirements: The DPDF program is open to doctoral students in social sciences or humanities disciplines who have completed their major course requirements, and are beginning to design research proposals. Students who have completed their comprehensive, general, or qualifying exams are eligible to apply as long as their dissertation proposals will be formally approved by their department after the fall DPDF workshop. Typically such students will be second and third year graduate students, but first and fourth year student may, under exceptional circumstances, be eligible. Applicants must be enrolled full-time in a Ph.D. program within an accredited university in the U.S., unless indicated otherwise for a particular field.
Geographic Focus: All States
Date(s) Application is Due: Sep 30
Amount of Grant: Up to 5,000 USD
Contact: Schuyler Marquez, Program Assistant; (718) 517-3704 or (212) 377-2700; fax (212) 377-2727; marquez@ssrc.org
Internet: http://www.ssrc.org/fellowships/dpdf-fellowship/
Sponsor: Social Science Research Council
One Pierrepont Plaza, 15th Floor
Brooklyn, NY 11201

SSRC Drugs, Security and Democracy Fellowships **4426**
The Drugs, Security and Democracy (DSD) Fellowships support research on organized crime, drug policy, issues of governance, and associated topics across the social sciences and related disciplines. The fellowship seeks to develop a concentration of researchers who are interested in policy-relevant outcomes and membership in a global interdisciplinary network. Funded research must address the relationship among at least two of the following three themes: drugs — potential topics include, but are not limited to, international and regional drug policy, drug trafficking, organized crime, drug production, and impact on communities including youth delinquency and gangs; security — potential topics include, but are not limited to, issues of traditional and non-traditional security, public safety, the pluralization of security actors, the role of law enforcement, the accountability of police forces, formal and informal strategies to increase security, violence, instability, immigration and border security; and/or democracy — potential topics include, but are not limited to, issues of governance, state responses to organized violence, civil society networks and how they mobilize against organized crime and drugs, the framing of incentives to develop appropriate policies, freedom of the press, impunity, corruption, and the relationship between state and non-state actors. Examples of possible research topics that address the relationship among two of the three themes include: substance control in border regions, violence and electoral campaigns, human rights and security policy,

and media and violence and global drug flows. The program encourages interdisciplinary projects and those that address transnational and transregional issues. Applications from researchers in professions outside of academia are welcome. Additional information and requirements are available at the SSRC website.
Requirements: The DSD Program features two competitions: Dissertation Fellowship — open to PHD and JSD candidates worldwide who have an approved dissertation prospectus by July 1, but have not completed writing for final submission; and Research Fellowship — applications are accepted from PHD recipients worldwide who have completed their PHD within seven years of the application deadline and researchers in Latin America or the Caribbean without a PHD but with a master's or the terminal degree in their field or equivalent professional experience.
Geographic Focus: All States, Puerto Rico, Antigua & Barbuda, Argentina, Aruba, Bahamas, Barbados, Bolivia, Brazil, British Virgin Islands, Cayman Islands, Chile, Colombia, Costa Rica, Cuba, Dominican Republic, Ecuador, El Salvador, Grenada, Guadeloupe, Guatemala, Haiti, Honduras, Jamaica, Martinique, Mexico, Nicaragua, Panama, Paraguay, Peru, Saint Kitts And Nevis, St. Lucia, Trinidad and Tobago, Turks and Caicos Islands, Uruguay, Venezuela, Virgin Islands (U.S.)
Date(s) Application is Due: Jan 20
Contact: Jessica Mack; (212) 377-2700 or (718) 517-3656; mack@ssrc.org
Internet: http://www.ssrc.org/fellowships/dsd-fellowship/
Sponsor: Social Science Research Council
One Pierrepont Plaza, 15th Floor
Brooklyn, NY 11201

SSRC Eurasia Program Dissertation Development Awards **4427**
The Eurasia Program offers financial and academic support to graduate students in the early stages of dissertation development, Ph.D. candidates near completion of their doctoral programs in the social sciences and related humanities, and young scholars within five years of the completion of their Ph.D. Dissertation Development Awards (DDAs) are intended to provide one academic year of support to enable the prompt completion of a Ph.D dissertation. Fellows will participate in professionalization activities and a spring conference, and contribute to the Eurasia Program's new working paper and policy brief series. The application is available through the Council's online application system. Frequently asked questions are also available at the website.
Requirements: Applicants must be citizens or permanent residents of the U.S., and have completed all requirements for the Ph.D. except for the dissertation at the time the fellowship begins. Applicants must have obtained ABD status (must have completed all of the requirements for the Ph.D degree except for the dissertation) by the application submission deadline. Applicants should demonstrate the ability to complete their dissertation by the end of the fellowship support period, and demonstrate evidence of not less than 10% University cost-sharing. This may be an in-kind contribution, or may come in the form of a tuition remission, a waiver of required fees, or health coverage.
Restrictions: Only the following countries are eligible for funding under the Eurasian Program: Armenia, Azerbaijan, Belarus, Georgia, Kazakhstan, Kyrgyzstan, Moldova, Russia, Tajikistan, Turkmenistan, Ukraine and Uzbekistan.
Geographic Focus: All States
Contact: Denise Mishiwiec; (212) 377-2700; fax (212) 377-2727; mishiwiec@ssrc.org
Internet: http://www.ssrc.org/fellowships/eurasia-fellowship/
Sponsor: Social Science Research Council
One Pierrepont Plaza, 15th Floor
Brooklyn, NY 11201

SSRC Eurasia Program Pre-Dissertation Awards **4428**
The Eurasia Program provides financial and academic support to graduate students in the early stages of dissertation development; Ph.D. candidates near completion of their doctoral programs in the social sciences and related humanities; and young scholars within five years of the completion of their Ph.D. Pre-Dissertation Awards (PDAs) enable early stage graduate students to perform initial field assessments of up to 4 weeks for archival exploration, preliminary interviews, and other forms of feasibility studies related to their dissertations. Proposals should reflect a clear plan for initial field assessment, require a budget of less than $4,000, and clearly articulate the policy relevance of the proposed project. The Council anticipates that six scholars will be able to gain firsthand knowledge of their proposed field sites, establish contacts within local communities, meet with local scholars, and gain insight into how their dissertation topic resonates with regional intellectual, political and social currents. Frequently asked questions are available at the Council website, in additional to the online application process.
Requirements: Applicants must be citizens or permanent residents of the United States, and have completed all requirements for the Ph.D. except for the dissertation at the time the fellowship commences. Applicants must have obtained ABD status (must have completed all of the requirements for the Ph.D degree except for the dissertation) by the application submission deadline. Applicants should demonstrate the ability to complete their dissertation by the end of the fellowship support period, and demonstrate evidence of not less than 10% University cost-sharing.
Restrictions: The following countries are eligible for funding under the SSRC Eurasia Fellowship Program: Armenia, Azerbaijan, Belarus, Georgia, Kazakhstan, Kyrgyzstan, Moldova, Russia, Tajikistan, Turkmenistan, Ukraine and Uzbekistan.
Geographic Focus: All States
Date(s) Application is Due: Dec 1
Contact: Denise Mishiwiec; (212) 377-2700; fax (212) 377-2727; mishiwiec@ssrc.org
Internet: http://www.ssrc.org/fellowships/eurasia-fellowship/
Sponsor: Social Science Research Council
One Pierrepont Plaza, 15th Floor
Brooklyn, NY 11201

SSRC International Dissertation Research Fellowships **4429**
The International Dissertation Research Fellowship (IDRF) offers nine to twelve months of support for up to 80 graduate students in the humanities and social sciences who are enrolled in doctoral programs in the U.S. and conducting dissertation research outside of the U.S. IDRF promotes research that is situated in a specific discipline and geographical region but is also informed by interdisciplinary and cross-regional perspectives. Research topics may address all periods in history, but applicants should be alert to the broader implications of their research as it relates to contemporary issues and debates. The program invites proposals for empirical and site-specific dissertation research outside the U.S. It will consider applications for dissertation research grounded in a single site, informed by broader cross-regional and interdisciplinary perspectives, as well as applications for multi-sited, comparative, and transregional research. Fellowship amounts vary depending on the research plan, with a per-fellowship average of $20,000. The fellowship includes participation in an SSRC-funded interdisciplinary workshop upon the completion of IDRF-funded research. The application and additional information are available at the SSRC website.
Requirements: The program is open to graduate students in the humanities and social sciences-regardless of citizenship-enrolled in doctoral programs in the U.S. Applicants must complete all Ph.D. requirements except on-site research by the time the fellowship begins or by December, whichever comes first.
Restrictions: Proposals that identify the U.S. as a case for comparative inquiry are welcome; however, proposals that require no field research outside the U.S. are not eligible. The IDRF program expects fellows to remain at their research site(s) for the full nine- to twelve-month funding period. The IDRF program will not support study at foreign universities, conference participation, or dissertation write-up. The program does not accept applications from Ph.D. programs in law, business, medicine, nursing or journalism, nor does it accept applications in doctoral programs that do not lead to a Ph.D.
Geographic Focus: All States
Date(s) Application is Due: Nov 3
Amount of Grant: Up to 19,000 USD
Contact: Elsa Ransom, Program Manager; (718) 517-3642 or (212) 377-2700; fax (212) 377-2727; ransom@ssrc.org
Internet: http://www.ssrc.org/fellowships/idrf-fellowship/
Sponsor: Social Science Research Council
One Pierrepont Plaza, 15th Floor
Brooklyn, NY 11201

SSRC Islamic Traditions and Muslim Societies in World Contexts Grants **4430**
The Islamic Traditions and Muslim Societies in World Contexts Grants support public engagement by academic experts on Islamic traditions and Muslim societies. The grants program seeks to encourage projects permitting the dissemination of academic research on Islamic traditions and Muslim societies to targeted constituencies, with particular emphasis on media and policymakers. With this program the SSRC looks to nurture the public relevancy of area studies research by promoting strategic partnerships that will break down barriers and build connections between academic researchers, journalists, policymakers, and practitioners. The SSRC also seeks to assist faculty in developing the necessary skills they require to be able to engage specialized nonacademic audiences on the topic of Islam in the world. Funds are available to innovative projects that will effect change by creating robust and sustainable capacities for public and policy engagement on university campuses. Projects which include strong faculty participation and leadership will be given funding priority as will projects that disseminate expert knowledge born of context specific research. Successful applications will demonstrate usage of adequate distribution channels for all deliverables intended to reach targeted constituencies. The application and additional information are available at the SSRC website.
Requirements: Grants are available to interdisciplinary research centers on U.S. university campuses, including but not limited to Title VI National Resource Centers. Research centers are eligible to apply individually for grants up to $50,000 to be used over a period of fifteen months. Extraordinary projects involving multiple collaborating centers or universities may be considered for grants up to $100,000.
Geographic Focus: All States
Date(s) Application is Due: Dec 3
Amount of Grant: Up to 100,000 USD
Contact: Thomas Asher, Program Director; (718) 517-3661 or (212) 377-2700; fax (212) 377-2727; asher@ssrc.org
Internet: http://www.ssrc.org/fellowships/islam-and-muslims-in-world-contexts/
Sponsor: Social Science Research Council
One Pierrepont Plaza, 15th Floor
Brooklyn, NY 11201

SSRC Japan Society for the Promotion of Science (JSPS) Fellowships **4431**
The Japan Society for the Promotion of Science (JSPS) Fellowship Program provides recent PhD recipients and ABDs (please see program eligibility requirements) with opportunities to conduct research in Japan under the leadership of a host researcher. Fellows are encouraged to advance their own research and at the same time closely collaborate with young Japanese researchers and contribute to Japanese research communities. Fellows are selected based on nominations made by the SSRC Japan Advisory Board. The SSRC Japan Advisory Board recommends to JSPS up to 20 candidates annually from the social sciences and humanities. Award offers are made by JSPS in early-summer for fellowships that must commence between April 1 and November 30 (long-term) or April 1 and March 31 (short-term). Applications are welcome from all social science and humanities disciplines and need not be explicitly related to the study of Japan.

Requirements: Applicants must be U.S. citizens or permanent residents at the time of application and submit proof of affiliation with an eligible host research institution in Japan as part of the application packet. Permanent residents must provide a copy of a permanent resident card (green card). Citizens of other countries are eligible for short-term fellowships if they have completed a master's or PhD course at an institution of higher education in the U.S. and, upon completing the course, have for at least three continuous years conducted high-level research at a university in the U.S. Priority will be given to researchers who are not Japanese citizens.
Restrictions: JSPS does not extend fellowships to scholars employed in the institutions under jurisdiction of the U.S. Department of Defense. Scholars who have previously received funding from JSPS for 12 months or longer are not eligible to apply for JSPS fellowships.
Geographic Focus: All States
Date(s) Application is Due: Dec 1
Contact: Fernando Rojas, Fellowships Associate; (212) 377-2700 or (718) 517-3639; fax (212) 377-2727; japan@ssrc.org or rojas@ssrc.org
Internet: http://www.ssrc.org/fellowships/jsps-fellowship/
Sponsor: Social Science Research Council
One Pierrepont Plaza, 15th Floor
Brooklyn, NY 11201

SSRC Mellon Mays Predoctoral Research Grants 4432

The SSRC-Mellon Mays Graduate Initiatives Program offers three grants (Graduate Studies Enhancement, Predoctoral Research Development and Dissertation Completion) designed to meet specific needs during the doctoral degree process. The requirements of each category are described in detail at the website. The application period for each is from May 1 through November 1.
Requirements: Applicants must have been selected as Mellon Mays Fellows as undergraduates. Fellows may apply for one grant per year and must be enrolled in a doctoral program in a Mellon-designated field of study, or have filed a petition for inclusion of another field.
Geographic Focus: All States
Date(s) Application is Due: Nov 1
Amount of Grant: Up to 5,000 USD
Contact: Adam Radwan, Program Associate; (212) 377-2700 or (718) 517-3617; fax (212) 377-2727; radwan@ssrc.org or mellonmays@ssrc.org
Internet: http://www.ssrc.org/fellowships/mellon-mays-predoctoral-research-grants/
Sponsor: Social Science Research Council
One Pierrepont Plaza, 15th Floor
Brooklyn, NY 11201

SSRC New Directions in the Study of Prayer Research Grants 4433

The New Directions in the Study of Prayer program invites proposals from scholars in all disciplines for studies that will enhance knowledge of the social, cultural, psychological, and cognitive dimensions of prayer, and of its origins, variations, and correlations in human life. Supported with funding from the John Templeton Foundation, and developed in conjunction with the SSRC's program on Religion and the Public Sphere, the project is led by a multi-disciplinary SSRC advisory committee. Approximately 25 research grants, ranging from $50,000 to $200,000, are awarded. Both individual and collaborative projects will be considered, and a small number of journalism grants of up to $50,000 each will also be awarded. All grantees will be asked to participate in a series of interdisciplinary workshops, conferences, and online initiatives organized in conjunction with the project.
Requirements: Candidates must submit a letter of intent before pursuing a project.
Geographic Focus: All States
Amount of Grant: 50,000-200,000 USD
Contact: Taline Cox, Program Assistant; (718) 517-3682 or (212) 377-2700; fax (212) 377-2727; cox@ssrc.org
Internet: http://www.ssrc.org/programs/new-directions-in-the-study-of-prayer/
Sponsor: Social Science Research Council
One Pierrepont Plaza, 15th Floor
Brooklyn, NY 11201

SSRC Next Generation Social Sciences in Africa: Doctoral Dissertation Completion Fellowships 4434

The Social Science Research Council offers Fellowships to support the advancement of social science faculty toward completion of doctoral degrees and to promote next generation social science research in Ghana, Nigeria, South Africa, Tanzania, and Uganda. The Fellowships support dissertations and research on peace, security, and development topics. The Doctoral Dissertation Completion Fellowship supports a one-year leave from teaching responsibilities and a stipend up to $15,000 to permit the completion of a dissertation that advances research on peace, security, and development topics.
Requirements: All applicants must be citizens of and reside in a sub-Saharan African country while holding a current faculty position at an accredited college or university in Ghana, Nigeria, South Africa, Tanzania, or Uganda. Applicants for any of the funding opportunities offered through this program must have a master's degree, and be working toward completion of the doctoral degree. The program seeks to promote diversity and encourages women to apply. To be eligible for the dissertation completion fellowship, applicants must be in the final year of writing and have completed at least one dissertation chapter.
Geographic Focus: Ghana, Nigeria, South Africa, Tanzania, Uganda
Date(s) Application is Due: Dec 1
Amount of Grant: Up to 15,000 USD
Contact: Kimberly Greenberg, Program Associate; (718) 517-3663 or (212) 377-2700; fax (212) 377-2727; greenberg@ssrc.org or nextgenafrica@ssrc.org
Internet: http://www.ssrc.org/fellowships/next-generation-social-sciences-in-africa-doctoral-dissertation-completion-fellowship/
Sponsor: Social Science Research Council
One Pierrepont Plaza, 15th Floor
Brooklyn, NY 11201

SSRC Next Generation Social Sciences in Africa: Doctoral Dissertation Proposal Fellowships 4435

The Social Science Research Council offers approximately 40 Fellowships each year to support the advancement of social science faculty toward completion of doctoral degrees and to promote next generation social science research in Ghana, Nigeria, South Africa, Tanzania, and Uganda. The Fellowships support dissertations and research on peace, security, and development topics. The Doctoral Dissertation Proposal Fellowship supports short-term research costs of up to $3,000 to develop a doctoral dissertation proposal. The application and additional information are available at the SSRC website.
Requirements: All applicants must be citizens of and reside in a sub-Saharan African country while holding a current faculty position at an accredited college or university in Ghana, Nigeria, South Africa, Tanzania, or Uganda. Applicants for any of the funding opportunities offered through this program must have a master's degree, and be working toward completion of the doctoral degree. The program seeks to promote diversity and encourages women to apply. To be eligible for the Dissertation Proposal Fellowship, applicants must be admitted to a graduate program but have yet to undertake dissertation research.
Geographic Focus: Ghana, Nigeria, South Africa, Tanzania, Uganda
Date(s) Application is Due: Dec 1
Amount of Grant: Up to 3,000 USD
Contact: Kimberly Greenberg, Program Associate; (718) 517-3663 or (212) 377-2700; fax (212) 377-2727; greenberg@ssrc.org or nextgenafrica@ssrc.org
Internet: http://www.ssrc.org/fellowships/next-generation-social-sciences-in-africa-proposal-fellowship/
Sponsor: Social Science Research Council
One Pierrepont Plaza, 15th Floor
Brooklyn, NY 11201

SSRC Next Generation Social Sciences in Africa: Doctoral Dissertation Research Fellowships 4436

The Social Science Research Council offers approximately 40 Fellowships each year to support the advancement of social science faculty toward completion of doctoral degrees and to promote next generation social science research in Ghana, Nigeria, South Africa, Tanzania, and Uganda. The Fellowships support dissertations and research on peace, security, and development topics. The doctoral dissertation research Fellowship supports 9-12 months of dissertation research costs of up to $15,000. The application and additional information are available at the SSRC website.
Requirements: All applicants must be citizens of and reside in a sub-Saharan African country while holding a current faculty position at an accredited college or university in Ghana, Nigeria, South Africa, Tanzania, or Uganda. Applicants for any of the funding opportunities offered through this program must have a master's degree, and be working toward completion of the doctoral degree. The program seeks to promote diversity and encourages women to apply. To be eligible for the $15,000 dissertation research fellowship, applicants must have an approved dissertation proposal, but will not yet have undertaken research of nine months or more.
Geographic Focus: Ghana, Nigeria, South Africa, Tanzania, Uganda
Date(s) Application is Due: Dec 1
Amount of Grant: Up to 15,000 USD
Contact: Kimberly Greenberg, Program Associate; (718) 517-3663 or (212) 377-2700; fax (212) 377-2727; greenberg@ssrc.orc or nextgenafrica@ssrc.org
Internet: http://www.ssrc.org/fellowships/next-generation-social-sciences-in-africa-research-fellowship/
Sponsor: Social Science Research Council
One Pierrepont Plaza, 15th Floor
Brooklyn, NY 11201

SSRC Our Shared Past Grants 4437

Our Shared Past grants promote the development of international scholarly communities committed to analyzing history curriculum, and reframing the teaching of world history through the identification of new scholarship and the development of new curricular content that illustrate shared cultural, economic, military, religious, social, and scientific networks and practices as well as shared global norms and values that inform world history and society. The project will encourage both the synthesis of existing scholarship on these topics and the exploration of concrete ways that this reframing can be successfully introduced into teaching curriculum in European, Middle Eastern, North African, or North American contexts. Our Shared Past grants may support workshops that convene scholars and advisory boards as well as partnerships between regional educational organizations and advocates. Additionally, projects may use a grant to support public-facing events that allow for engagement with curriculum reform around world history. Funds are available to projects that will develop robust networks of faculty who possess the expertise to study existing world history curriculum (K–14, i.e., at pre-collegiate and collegiate levels) in a specific national or regional setting and who are positioned to propose alternative curriculum design based on that scholarship. Successful projects will make use of relevant distribution channels for all project deliverables. Five grants are expected to be awarded. Organizations must apply through the application portal at the SSRC website. Hard copy applications will not be accepted.
Requirements: Grants will be made to research teams or research centers on U.S. university campuses, including, but not limited to, Title VI National Resource Centers. Research

teams/consortia may include partner institutions from the U.S. and/or Middle East/ North Africa region, but the principal investigator must be based at a U.S. university.
Geographic Focus: All States
Date(s) Application is Due: May 31
Amount of Grant: Up to 75,000 USD
Contact: Kimberly Greenberg, Program Associate; (212) 377-2700; fax (212) 377-2727; greenburg@ssrc.org or osp@ssrc.org
Internet: http://www.ssrc.org/fellowships/our-shared-past/
Sponsor: Social Science Research Council
One Pierrepont Plaza, 15th Floor
Brooklyn, NY 11201

SSRC Postdoctoral Fellowship for Transregional Research: Inter-Asian Contexts and Connections 4438

The Postdoctoral Fellowship for Transregional Research: Inter-Asian Contexts and Connections program is a pilot postdoctoral fellowship program to support transregional research under the rubric Inter-Asian Contexts and Connections. Its purpose is to strengthen the understanding of issues and geographies that may not fit neatly into existing divisions of academia or the world and to develop new approaches, practices, and opportunities in international, regional, and area studies in the U.S. Funded by the Andrew W. Mellon Foundation, these fellowships will help junior scholars (those at the postdoctoral stage, one to seven years out of the Ph.D.) complete first books and/or undertake second projects. In addition to funding research, the program creates networks and shared resources to support Fellows well beyond the grant period. The Postdoctoral Fellowship for Transregional Research will provide promising scholars important support at critical junctures in their careers. Fifteen fellowships are awarded over the two-year course of the pilot program.
Geographic Focus: All States
Contact: Holly Danzeisen, Associate Director, Eurasia and Inter-Asia; (718) 517-3662 or (212) 377-2700; fax (212) 377-2727; danzeisen@ssrc.org or transregional@ssrc.org
Internet: http://www.ssrc.org/fellowships/transregional-research-fellowship/
Sponsor: Social Science Research Council
One Pierrepont Plaza, 15th Floor
Brooklyn, NY 11201

SSSP C. Wright Mills Award 4439

The Society for the Study of Social Problems established the C. Wright Mills Award in 1964. The SSSP annually gives its prestigious C. Wright Mills Award to the author of what the committee considers to be the most outstanding book written in the tradition of C. Wright Mills and his dedication to a search for a sophisticated understanding of the individual and society. Past winners have included Patricia Hill Collins, William Julius Wilson, Pun Ngai, Pierrette Hondagneu-Sotelo, Charles Tilly, Theda Skocpol, Rosabeth Moss Kanter, Harry Braverman, Frances Fox Piven-Richard Cloward, Jacqueline P. Wiseman, Laud Humphreys, and Jerome H. Skolnick. The books nominated must have been published in the year preceding the year of the award.
Geographic Focus: All States
Date(s) Application is Due: Dec 15
Amount of Grant: 500 USD
Contact: Michele Smith Koontz, Administrative Officer; (865) 689-1531; fax (865) 689-1534; mkoontz3@utk.edu or sssp@utk.edu
Internet: http://www.sssp1.org/index.cfm/m/259
Sponsor: Society for the Study of Social Problems
901 McClung Tower, University of Tennessee
Knoxville, TN 37996-0490

SSSR Distinguished Article Award 4440

The annual award was established to recognize the most outstanding article published by a SSSR member or members within the past year (Scholars and graduate students from all fields are invited to join). The award will be presented at the annual meeting. It includes a plaque and $1,000.
Requirements: An article may be nominated by the editor of the journal in which it appeared, by its author, or by any other SSSR member deeming it to be of outstanding quality. Articles must have been published in refereed journals. Nominated articles must be sent by May 15 to the committee chair. The committee will judge each article's quality of scholarship, importance to the field, and relevance to the larger society. Any article involving some aspect of the scientific study of religion is eligible.
Restrictions: Submissions published solely in edited volumes will not be considered.
Geographic Focus: All States
Date(s) Application is Due: May 15
Amount of Grant: 1,000 USD
Contact: Peter Beyer, Chair; pbeyer@uottawa.ca
Internet: http://www.sssrweb.org/Awards.cfm#8
Sponsor: Society for the Scientific Study of Religion
IUPUI Center for the Study of Religion and American Culture, 425 University Boulevard
Indianapolis, IN 46202-5140

SSSR Student Paper Award 4441

The award recognizes the best paper scheduled for presentation by a student at the annual meeting. (Scholars and graduate students from all fields are invited to join.) SSSR wishes to (1) encourage scholarly work and participation by students in SSSR meetings, and (2) give students the opportunity to receive constructive feedback and criticism on their work by professional colleagues and mentors. The award will be presented at the annual meeting. It includes a plaque and $1,000.
Requirements: Authors of papers submitted for this competition must be graduate or undergraduate students at institutions of higher learning anywhere in the world, but papers must be submitted in English. Any co-authors must also be students. To be eligible, a paper must not have been published or accepted for publication. At least one author for the paper must be registered for the SSSR conference. Submission of the paper implies a commitment for an author to appear and present the paper at the conference. Student authors who intend to enter this competition may wish to notify the Student Paper Award Committee Chair via email at the time of their abstract submission, but they are not required to do so.
Restrictions: Neither faculty nor postdoctoral co-authors are permitted.
Geographic Focus: All States
Date(s) Application is Due: Apr 15; Aug 1
Amount of Grant: 1,000 USD
Contact: Carolyn Warner, Chair; cwarner@asu.edu
Internet: http://www.sssrweb.org/Awards.cfm#9
Sponsor: Society for the Scientific Study of Religion
IUPUI Center for the Study of Religion and American Culture, 425 University Boulevard
Indianapolis, IN 46202-5140

Stackpole-Hall Foundation Grants 4442

The Stackpole-Hall Foundation awards grants to nonprofit organizations and institutions designed to enhance the social welfare of the Pennsylvania area. The Foundation concentrates on the areas of education, health care, cultural, youth development, social welfare, environmental, and community development. Priority is given to Elk County. Grants are awarded in support of matching, seed, partnership, and under certain circumstances, operational grants. The Foundation trustees meet four times a year to award grants. Proposals are due a month prior to the meeting for which they will be considered. Application forms are not required. Detailed instructions about information to include with the letter of inquiry are located at the Foundation website.
Requirements: The Foundation considers requests from organizations which qualify as churches; governmental organizations; or 501(c)3 organizations; or are classified as not being a private foundation under section 509(a) of the federal tax code.
Restrictions: Unless usual circumstances exist, requests for operating or endowment grants are given low priority. Funding is not available for individuals.
Geographic Focus: Pennsylvania
Amount of Grant: Up to 50,000 USD
Contact: William Conrad, Executive Director; (814) 834-1845; fax (814) 834-1869; stackpolehall@windstream.net
Internet: http://www.stackpolehall.org/application.html
Sponsor: Stackpole-Hall Foundation
44 South St. Marys Street
St. Marys, PA 15857

Stanford Center on Adolescence Youth Purpose Research Grants 4443

Small grants for dissertation and postdoctoral research will be awarded in March for research to be completed by September. The Stanford Center on Adolescence aims to spread the recognition of youth purpose as an important area of scientific research by encouraging all inquiries that shed light on topics related to purpose among young people, such as adolescent motivation, personal meaning, or positive youth development, in an array of settings found in contemporary society. The Center operationally defines purpose as a stable and generalized intention to accomplish something that is meaningful to the self and that leads to productive engagement with some aspect of the world beyond the self. Through its own research, the Center seeks to address questions about how purpose develops, the antecedents of purpose, the difference that purpose makes in the lives of young people, and how to best cultivate purpose in youth. The Center encourages researchers from all disciplines to submit proposals for projects that will contribute to the scientific study of this research topic. Applicants may request up to $10,000 in funding for research-related expenses.
Requirements: Applicants must be affiliated with an accredited college or university and should be U.S citizens or permanent residents. Postdoctoral applicants must have received their Ph.D. within the past five years.
Geographic Focus: All States
Contact: Lisa Staton; (650) 725-8205; fax (650) 725-8207; stanfordcoa@stanford.edu
Internet: http://www.stanford.edu/group/adolescent.ctr/researchawards
Sponsor: Stanford Center on Adolescence
505 Lasuen Mall, Stanford University
Stanford, CA 94305-3083

Stanford Humanities Center Arts Writer/Practitioner Fellowship 4444

The Stanford Humanities Center and the Stanford Institute for Creativity and the Arts (SiCa) are offering one year-long fellowship to an arts practitioner who is also a writer, scholar, or critic pursuing a research or critical project in the arts. Fellows are expected to remain in residence, living in a ten-mile radius of the university, during the regular academic year (mid-September through mid-June). The arts practitioner/writer fellow will be in residence with other fellows at the Humanities Center and is expected to make an intellectual contribution to Stanford University by participating in the scholarly life of the Humanities Center, e.g. regularly attending lunch and participating in weekly research presentations. Applicants should be interested in engaging with the intellectual life at Stanford and exploring connections with departments and other centers on campus. Fellows are awarded a stipend of up to $60,000 and a housing and moving allowance of up to $15,000. Medical insurance is not provided. Applications are submitted via the center's online application system which becomes accessible at the center's website each August. All applications must be in English. Annual deadline dates may vary. The center notifies applicants of the outcome in late March. Application guidelines are available at the website.

Requirements: Applicants may practice in any field of art (visual, performance, composition, mixed- and inter-media, architecture and design) but must demonstrate significant professional accomplishment as arts practitioners and as critics or scholars. Fellowships will be awarded on the basis of a scholarly or critical project in the arts and not on the basis of art production. Applicants may, but need not, be affiliated with an academic institution; independent practitioners are welcome to apply. Applications will be judged on the promise of the specific research project being proposed, the research project's potential interest to scholars in different fields of the humanities, the originality and intellectual distinction of the candidate's previous work, and the applicant's perceived ability to engage in collegial interaction and to contribute to the discussion of presentations.
Geographic Focus: All States, All Countries
Amount of Grant: Up to 75,000 USD
Contact: Robert Barrick, Fellowship Program Administrator; (650) 723-3052 ext. 3-3054; fax (650) 723-1895; rbarrick@stanford.edu or humanities-center@stanford.edu
Internet: http://shc.stanford.edu/fellowships/arts/#arts_answer_6
Sponsor: Stanford University
424 Santa Teresa Street
Stanford, CA 94305-4015

Stanford Humanities Center External Faculty Fellowships 4445

Founded in 1980, the Stanford Humanities Center is a multidisciplinary research institute at Stanford University dedicated to advancing knowledge about culture, philosophy, history, and the arts. The center's external-faculty fellowships are offered to non-Stanford scholars (both to members of humanities departments as traditionally defined and to other scholars seriously interested in humanistic issues). Fellowship duration is one full academic year. Fellows are expected to remain in residence, living in a ten-mile radius of the university, during the regular academic year (mid-September through mid-June). Additionally fellows are expected to make contributions to the intellectual life of the Stanford community by taking part in a Geballe Research Workshop, regularly attending lunches, and participating in weekly research presentations. Fellows are awarded a stipend of up to $60,000 and a housing and moving allowance of up to $15,000. Medical insurance is not provided. Applications are submitted via the center's online application system which becomes accessible at the center's website each August. All applications must be in English. Annual deadline dates may vary. The center notifies applicants of the fellowship competition outcome in late March. Application guidelines are available at the website.
Requirements: External fellowships are intended primarily for non-Stanford scholars who are currently teaching in or affiliated with an academic institution; however, independent non-Stanford scholars may also apply. All applicants must have a Ph.D. External fellowships fall into two categories: senior fellowships for well-established scholars, and junior fellowships for scholars who during the fellowship year will be at least three but no more than 10 years beyond receipt of their Ph.D. Research projects must be in the humanities. The center is open to projects employing information technology in humanities research. Applications will be judged on the promise of the specific research project being proposed, the research project's potential interest to scholars in different fields of the humanities, the originality and intellectual distinction of the candidate's previous work, and the applicant's perceived ability to engage in collegial interaction and to contribute to the discussion of presentations.
Restrictions: Creative arts projects are not eligible for funding. Former external faculty fellows are not eligible to apply. Scholars who have received major research grants within the last three years generally will not be as competitive as other applicants.
Geographic Focus: All States, All Countries
Date(s) Application is Due: Oct 15
Amount of Grant: Up to 75,000 USD
Contact: Robert Barrick, Fellowship Program Administrator; (650) 723-3052 ext. 3-3054; fax (650) 723-1895; rbarrick@stanford.edu or humanities-center@stanford.edu
Internet: http://shc.stanford.edu/fellowships/non-stanford-faculty/
Sponsor: Stanford University
424 Santa Teresa Street
Stanford, CA 94305-4015

Stanley Smith Horticultural Trust Grants 4446

The Stanley Smith Horticultural Trust was created in 1970 by May Smith, in honor of her late husband. The trust typically makes grants to botanical gardens, arboreta, and universities primarily in North and South America to support education and research in ornamental horticulture. Specifically, the trust is interested in funding organizations pursuing the following activities: the advancement of research in ornamental horticulture and the publication of the results of such research; assisting in the creation, development, preservation, and maintenance of gardens accessible to the public for educational purposes; promotion of the environmentally responsible introduction, cultivation, and distribution of plants which have ornamental horticultural value; assisting in the publication of books or other works relating to the science of horticulture; and informal and/or formal educational activities which further ornamental horticulture. The trust typical awards grants in the amount of $20,000. Organizations seeking funding should first confirm that their projects meet the trust's funding interests and eligibility requirements by taking the eligibility quiz which is accessible from the grantseeker's webpage at the trust's website. Grant seekers meeting the qualifications should send an email to the Grants Director with a brief description of their project. Organizations with projects that appear to be qualified and that are of interest to the trust will be sent current application guidelines along with an invitation to submit a full proposal. Proposals are accepted until August 15 and are reviewed by the trustees once a year in November. Funding of approved grants occurs in December for use in the following calendar year.
Requirements: Organizations that are recognized as 501(c)3 public charities by the United States Internal Revenue Service and non-U.S. organizations that can demonstrate that they would meet the requirements for such status are eligible to apply. The trust provides support for projects in North and South America in the following areas: development of programs and projects; salaries; physical improvements; signage; access; equipment; publications; and, under some circumstances, general operations.
Restrictions: The trust does not fund endowments, indirect costs, or projects concerned primarily with agriculture, environmental issues, science education, or horticultural therapy.
Geographic Focus: All States, All Countries
Date(s) Application is Due: Aug 15
Amount of Grant: Up to 20,000 USD
Contact: Dr. Thomas F. Daniel; (415) 379-5350; tdaniel@calacademy,org
Internet: http://www.adminitrustllc.com/stanley-smith-horticultural-trust/
Sponsor: Stanley Smith Horticultural Trust
c/o Adminitrust LLC
Sausalito, CA 94965

Starr Foundation Grants 4447

The foundation awards grants to nonprofit organizations in the areas of education, medicine and health care, human needs, public policy, and culture. Education grants support scholarships for deserving students, groups that offer need-based financial aid, and international-exchange programs. Human needs grants support food programs for the poor, job training, literacy, adequate housing, and programs for the disabled. Medicine and health care grants include capital grants to hospitals, significant research grants, and grants to assist in the provision of health care to underserved communities. Public policy grants support international relations and democratic institutions around the world. Culture grants go to exchange programs and provide services to the elderly and/or disabled. The foundation rarely funds local charities outside of New York City but may fund national organizations that serve communities outside of New York. Types of support include multiyear grants, general operating grants, and capacity-building grants. There are no application deadlines. The board meets throughout the year.
Geographic Focus: All States
Amount of Grant: Up to 70,000,000 USD
Contact: Florence Davis; (212) 770-6882; florence.davis@starrfdn.org
Internet: http://fdncenter.org/grantmaker/starr
Sponsor: Starr Foundation
70 Pine Street, 14th Floor
New York, NY 10270

State Farm Companies Doctoral Dissertation Award 4448

Doctoral dissertation awards are designed to stimulate research and development of knowledge of use to the insurance industry and to increase the number of qualified teachers of insurance in the colleges and universities of the country. Applications are sent to deans and directors of doctoral programs at select colleges and universities in December. Completed applications, nomination forms, and required attachments must be received in the Foundation office by March 31. A grant is given to any college that submits eligible nominations.
Requirements: Awards are available to doctoral candidates who have completed a major portion of their doctoral programs and who are at the dissertation stage and are majoring in insurance or a related field of study. Applicants must be U.S. citizens.
Restrictions: Only applicants nominated by the director of the school's doctoral program qualify as candidates. Direct applications will not be considered.
Geographic Focus: All States
Date(s) Application is Due: Mar 31
Amount of Grant: 10,000-13,000 USD
Contact: Assistant Secretary; (309) 766-2039 or (309) 766-2161; info@statefarm.com
Internet: http://www.statefarm.com/about/part_spos/grants/doctoral.asp
Sponsor: State Farm Companies Foundation
1 State Farm Plaza, B-4
Bloomington, IL 61710-0001

State Scholarships Foundation Grants 4449

The foundation awards scholarships for postgraduate/postdoctoral studies in Greece to nationals from Western Europe, United States, Canada, Australia, and Japan. The scholarships support postgraduate/doctoral studies (MA/Msc, up to three years); PhD studies (up to four years); postdoctoral research studies (six months to one year); further education in the Greek language, literature, philosophy, history, and art (for professors of Greek in foreign universities, for six months up to one year); and specialization in fine arts (for up to two years).
Requirements: Application is open to nationals of the Balkan countries, countries of Western Europe, United States, Canada, Australia, and Japan. In most cases, the applicant must hold a graduate degree from a foreign university (outside of Greece) and a first postgraduate degree or equivalent. Medical doctors should have either completed their specialization or hold a postgraduate degree. For postdoctoral research studies, applicants must hold a PhD degree. Their first university degree must be from a foreign university. For further education in the Greek language, literature, philosophy, history, or art, applicants must be professors in a department of Greek studies of a foreign university; and a validated certificate of the relevant university department abroad attesting the profession and employment of the applicant must also be submitted. Applicants must have an excellent knowledge of English or French.
Geographic Focus: All States, Albania, Algeria, Andorra, Angola, Armenia, Australia, Austria, Azerbaijan, Belarus, Belgium, Benin, Bosnia & Herzegovina, Botswana, Bulgaria, Burkina Faso, Burundi, Cameroon, Canada, Cape Verde, Central African Republic, Chad, Comoros, Congo, Congo, Democratic Republic of, Cote d' Ivoire (Ivory Coast), Croatia, Cyprus, Czech Republic, Denmark, Djibouti, Egypt, Equatorial Guinea, Eritrea, Estonia, Ethiopia, Finland, France, Gabon, Gambia, Georgia,

Germany, Ghana, Greece, Guinea, Guinea-Bissau, Hungary, Iceland, Ireland, Italy, Japan, Kenya, Kosovo, Latvia, Lesotho, Liberia, Libya, Liechtenstein, Lithuania, Luxembourg, Macedonia, Madagascar, Malawi, Mali, Malta, Mauritania, Mauritius, Moldova, Monaco, Montenegro, Morocco, Mozambique, Namibia, Niger, Nigeria, Norway, Poland, Portugal, Romania, Russia, Rwanda, San Marino, Sao Tome & Principe, Senegal, Serbia, Seychelles, Sierra Leone, Slovakia, Slovenia, Somalia, South Africa, Spain, Sudan, Swaziland, Sweden, Switzerland, The Netherlands, Turkey, Ukraine, United Kingdom, Vatican City
Date(s) Application is Due: Feb 28
Amount of Grant: 60,000 EUR
Contact: M. Mikedaki-Drakaki, Chief of Section; 3010-3726300; fax 3010-3312759
Internet: http://www.iky.gr
Sponsor: State Scholarships Foundation
14 Lyssicratous Street
Athens, 105 58 Greece

STC Research Grants 4450

For many years, the Society for Technical Communication has offered research grants to talented individuals doing research on topics that are of concern and interest to the Society's mainstream membership. To date nearly $300,000 has been awarded for research grants. Along with other STC initiatives to increase member satisfaction and educational opportunities, the STC Research Grants program is currently being re-envisioned and reformulated. STC remains committed to funding quality research of value to academic and industry practitioners and teachers.
Requirements: Proposals may be submitted by student members, regular members, and nonmembers of the STC.
Geographic Focus: All States
Date(s) Application is Due: Jun 30; Dec 1
Amount of Grant: Up to 10,000 USD
Contact: Rachel Spilka, Manager; (414) 962-7254; spilka@uwm.edu
Internet: http://www.stc.org/edu/grantsResearch01.asp
Sponsor: Society for Technical Communication
4520 N Sheffield Avenue
Shorewood, WI 53211

STC Scholarships in Technical Communication 4451

Each year the society awards two scholarships to undergraduates and two scholarships to graduates to assist students who are pursuing established degree programs for careers in some area of technical communication. The completed application form must be accompanied by a certified copy of the latest transcript from the school currently attended and two letters of recommendation from faculty members in the technical communication field.
Requirements: Full-time undergraduate students having upper-class standing in a two- or four-year degree program in technical communication and full-time graduate students working toward a master's or doctoral degree in technical communication may apply.
Geographic Focus: All States
Date(s) Application is Due: Feb 15
Amount of Grant: 1,500 USD
Contact: Elreatha Matthews, Education Manager; (571) 366-1903 or (703) 522-4114; fax (703) 522-2075; elreatha.matthews@stc.org or stc@stc.org
Internet: http://www.stc.org/edu/scholarshipInfo01.asp
Sponsor: Society for Technical Communication
901 North Stuart Street, Suite 904
Arlington, VA 22203-1822

Strake Foundation Grants 4452

The Foundation supports hospitals, schools, colleges, and Catholic charities, as well as projects focusing in adult basic education and literacy, museums, and arts and culture. Support is considered for operating budgets, capital campaigns, special projects, research, matching funds and general purposes.
Requirements: Awards are made to 501(c)3 organizations located only in the United States, primarily in Texas. Organizations may submit only one request per calendar year. There are no set amounts for requests, however awards generally range between $2,000 and $20,000 with a few exceptions as high as $50,000.
Restrictions: Awards are not made to support individuals, nor for deficit financing, consulting services, technical assistance, publications, or loans.
Geographic Focus: All States
Contact: George Strake, Jr., President; (713) 216-2400; foundation@strake.org
Sponsor: Strake Foundation
712 Main Street, Suite 3300
Houston, TX 77002

Streilein Foundation for Ocular Immunology Visiting Professorships 4453

The SFOI Visiting Professorship is intended to stimulate the spread of knowledge concerning the immunological causes and approaches for managing ocular diseases. The SFOI Visiting Professorship should provide a clear scientific and educational benefit for the host institution. The Professorship will support a short visit of a senior scientist or clinician with a track record of ocular immunology research to a host institution. An important feature of the Visiting Professorship is to encourage and mentor young investigators. The visiting professor shall contribute one full day of teaching and interacting with students, residents, postdoctoral fellows, and faculty. The host institution will receive a $5,000 award to pay for coach airfare, and/or standard ground transportation, hotel (up to two nights), meals, honorarium, and expenses associated with this visit (e.g., refreshments, advertising). Transportation by personal car will be reimbursed at $0.51/mile. Meal allowance will be set by the host institution. Also included is an honorarium of $1,500 paid at the time expense reimbursement is submitted.
Requirements: Any institution of higher education or research institute is eligible to apply. The application for a SFOI Visiting Professorship must be submitted by a departmental Chairman of an educational institution, the Director of a research institute, or an official of similar stature from the host organization. The application must be received by the SFOI by July 1st of the calendar year in which the professorship is intended.
Geographic Focus: All States
Date(s) Application is Due: Jul 1
Amount of Grant: 6,500 USD
Contact: Joan Stein-Streilein, Ph.D., Treasurer; (617) 912-7494; fax (617) 912-0105; jstein@vision.eri.harvard.edu
Internet: http://www.streilein-foundation.org/fellowships.html
Sponsor: Streilein Foundation for Ocular Immunology
P.O. Box 6104
Boston, MA 02114

Stroke Association Allied Health Professional Research Bursaries 4454

The Research Bursary Awards are primarily intended for the purpose of training nurses and therapists, but consideration will be given to other health professionals. These bursaries will be awarded to UK departments that can demonstrate a track record and current participation in stroke research. The host department will be required to provide appropriate supervision and a research training program suitable for equipping the candidate for a career in stroke research, and to enable them to obtain a post graduate qualification. Bursaries may be taken up on a part-time, pro-rata basis.
Requirements: Research must take place in the United Kingdom.
Geographic Focus: All States, United Kingdom
Amount of Grant: 30,000-60,000 GBP
Contact: Research Department; (020) 7566-0345 or (020) 7490-2686; fax (020) 7251-9096; research@stroke.org.uk
Internet: http://www.stroke.org.uk/research/research_we_fund/allied_health_professional_bursaries/index.html
Sponsor: Stroke Association
Stroke House, 240 City Road
London, EC1V 2PR England

Stroke Association Clinical Fellowships 4455

Fellowships are awarded to individuals for training in stroke medicine, and are intended to enable a specialist registrar to gain the appropriate clinical experience and accreditation required for a career in the prevention, treatment or rehabilitation of stroke.
Requirements: Applicants are required to nominate a supervisor and training program in a host institution that has been approved for stroke medicine training by the Stroke Medicine Sub-specialty Advisory Committee (SSAC). The funding is not to be taken as part of a one year taught course or for those who have completed the SpR training program and have certification.
Geographic Focus: All States, United Kingdom
Date(s) Application is Due: Jan 6
Amount of Grant: 40,000 GBP
Contact: Research Department; (020) 7566-0345 or (020) 7490-2686; fax (020) 7251-9096; research@stroke.org.uk
Internet: http://www.stroke.org.uk/research/research_we_fund/fellowships/index.html
Sponsor: Stroke Association
Stroke House, 240 City Road
London, EC1V 2PR England

Stroke Association Research Project Grants 4456

Research project grants are awarded to doctors, professors, research fellows, therapists, psychologists and nurses at universities and hospitals across the UK. Project grants can last for up to three years. The Stroke Association's priorities for funding include research into the prevention, treatment, rehabilitation and long-term care of stroke patients. The majority of funded projects are patient orientated. The Stroke Association has recently undertaken a review of its activities and has developed a new Research and Development Strategy. As a result, applications for Research Project Grants have been revised and new application forms have been developed. All applications for funding by The Stroke Association will now undergo review by a panel of service users in addition to expert peer review. Therefore all applicants will now be required to complete a Project Grant Lay Summary Form, as well as the main Research Project Grant Application Form. Details of the new procedures and forms can be found on online.
Requirements: Research must take place in the United Kingdom.
Restrictions: Applications that are not submitted on the new forms will not be accepted for this application round.
Geographic Focus: All States, United Kingdom
Date(s) Application is Due: Feb 5
Amount of Grant: 70,000-210,000 GBP
Contact: Research Department; (020) 7566-0345 or (020) 7490-2686; fax (020) 7251-9096; research@stroke.org.uk
Internet: http://www.stroke.org.uk/research/apply_for_funding/index.html
Sponsor: Stroke Association
Stroke House, 240 City Road
London, EC1V 2PR England

STTI/ATI Educational Assessment Nursing Research Grants 4457

The purpose of this grant, co-sponsored by Assessment Technologies Institute (ATI), with the Honor Society of Nursing, Sigma Theta Tau International will support research that encourages the appropriate use of a standardized assessment program in a nursing education curriculum, including, but not limited to: use of formative assessment to drive remediation; use of summative assessment to guide curricular evaluation and revision; timing of assessment administration within a program of study; identification of key markers of at-risk student status.
Requirements: Eligibility Criteria: registered nurse with current license; have a master's or doctoral degree or be enrolled in a doctoral program; submit a completed research application package and signed research agreement via the online submission system; ready to implement research project when funding is received; submit a final report; submit completed abstract to the Virginia Henderson Research Library and credit research grant partners in all publications and presentations of the research.
Geographic Focus: All States
Date(s) Application is Due: Jul 1
Amount of Grant: 4,500 USD
Contact: Research Services; (888) 634-7575 or (317) 634-8171; fax (317) 634-8188; research@stti.iupui.edu
Internet: http://www.nursingsociety.org/Research/SmallGrants/Pages/STTIATI.aspx
Sponsor: Sigma Theta Tau International
550 W. North Street
Indianapolis, IN 46202-2156

STTI Alpha Eta Collaborative Research Grants 4458

The purpose of the STTI Alpha Eta Collaborative Research Grant is to provide up to U.S.$2000 research grant to Tau Lambda-at-large chapter member(s) for a new research project in Africa. Funds for this grant are provided by Alpha Eta Chapter.
Requirements: Eligibility Criteria: member of Tau Lambda-at-Large residing in Africa; submit online a completed research application package and signed research agreement via the online submission system; ready to implement research project when funding is received; submit a final report; submit completed abstract to Virginia Henderson International Nursing Research Library and credit research grant partners in all publications and presentations of the research.
Geographic Focus: All States, Algeria, Angola, Benin, Botswana, Burkina Faso, Burundi, Cameroon, Cape Verde, Central African Republic, Chad, Comoros, Congo, Congo, Democratic Republic of, Cote d' Ivoire (Ivory Coast), Djibouti, Egypt, Equatorial Guinea, Eritrea, Ethiopia, Gabon, Gambia, Ghana, Guinea, Guinea-Bissau, Kenya, Lesotho, Liberia, Libya, Madagascar, Malawi, Mali, Mauritania, Mauritius, Morocco, Mozambique, Namibia, Niger, Nigeria, Rwanda, Sao Tome & Principe, Senegal, Seychelles, Sierra Leone, Somalia, South Africa, Sudan, Swaziland
Date(s) Application is Due: Dec 1
Amount of Grant: 2,000 USD
Contact: Research Services; (888) 634-7575 or (317) 634-8171; fax (317) 634-8188; research@stti.iupui.edu
Internet: http://www.nursingsociety.org/Research/SmallGrants/Pages/AlphaEta.aspx
Sponsor: Sigma Theta Tau International
550 W. North Street
Indianapolis, IN 46202-2156

STTI Emergency Nurses Association Foundation Grant 4459

The ENA Foundation and Sigma Theta Tau International, Inc. (STTI) have combined resources to offer this annual research grant. The ENA Foundation/STTI research grant provides funding for research that will advance the specialized practice of emergency nursing. All relevant research topics will be considered. Priority will be given to research studies relating to the ENA/ENA Foundation research initiatives.
Requirements: Eligibility Criteria: principal investigator must be a registered nurse; team members may be from other disciplines; principal investigator must have a master's degree; submit a completed research application package and signed research agreement; be ready to start, or have already started, the research project. An electronic Research Grant Application is available on the ENA Foundation website. All applications MU.S.T be returned to ENA Foundation.
Restrictions: Application packets submitted to Sigma Theta Tau International will not be reviewed.
Geographic Focus: All States
Date(s) Application is Due: Mar 1
Amount of Grant: 6,000 USD
Contact: Department of Research; (847) 460-4119; fax (847) 460-4005; res@ena.org
Internet: http://www.nursingsociety.org/Research/SmallGrants/Pages/grant_ena.aspx
Sponsor: Emergency Nurses Association Foundation
915 Lee Street
Des Plaines, IL 60016 -6569

STTI Environment of Elder Care Nursing Research Grants 4460

The Environment of Elder Care Nursing Research Grant co-sponsored by Hill-Rom Company will advance the science of nursing through research focused on critical aspects of elder care including clear lungs, no falls, safe skin, patient comfort and ease-of-use. Applications from novice researchers who have received no other national research funds are encouraged. Preference will be given to Sigma Theta Tau International members, other qualifications being equal. Allocation of funds is based on the quality of the proposed research, the future promise of the applicant, and the applicant's research
Requirements: Eligibility Criteria: registered nurse with a current license; hold at minimum a Master's degree or its equivalent, and/or be enrolled in a Doctoral program; submission of completed research application packet and a signed research agreement via our on-line submission system; ready to implement research project when funding is received; submit to STTI a final report; submit completed abstract to STTI's Virginia Henderson International Nursing Research Library; and submit an abstract for publication to Hill-Rom regarding the research done as a product of the grant, and credit research grant partners in all publications and presentations of the research.
Geographic Focus: All States
Date(s) Application is Due: Jul 1
Amount of Grant: Up to 9,000 USD
Contact: Research Services; (888) 634-7575 or (317) 634-8171; fax (317) 634-8188; research@stti.iupui.edu
Internet: http://www.nursingsociety.org/Research/SmallGrants/Pages/Environmentof Care.aspx
Sponsor: Sigma Theta Tau International
550 W. North Street
Indianapolis, IN 46202-2156

STTI Joan K. Stout RN Research Grants 4461

The purpose of the Sigma Theta Tau International/Joan K. Stout, RN, Research Grant is to advance ongoing evidence-based study by nurse researchers on the impact of the practice of simulation education in schools of nursing and clinical care settings. The allocation of funds is based upon a research project that is ready for implementation. The proposed research project should be designed to ensure the ongoing practice of nurse-led simulation in improving quality of care in clinical and/or academic settings with the potential for further funding and ongoing research. Funds provided by interest from a research endowment established at Sigma Theta Tau International Foundation for Nursing, with a donation by the Hugoton Foundation.
Requirements: Eligibility Criteria: registered nurse with current license; have a master's degree, and/or be enrolled in a Doctoral program; submit a completed research application via online submission system; ready to implement research project when funding is received; complete project within one (1) year of funding; submit to STTI a final report; submit completed abstract to STTI's Virginia Henderson International Nursing Research Library.
Geographic Focus: All States
Date(s) Application is Due: Jul 1
Amount of Grant: 5,000 USD
Contact: Research Services; (888) 634-7575 or (317) 634-8171; fax (317) 634-8188; research@stti.iupui.edu
Internet: http://www.nursingsociety.org/Research/SmallGrants/Pages/JoanKStout.aspx
Sponsor: Sigma Theta Tau International
550 W. North Street
Indianapolis, IN 46202-2156

STTI National League for Nursing Grants 4462

Funds for this grant are made possible from a gift from Dr. Diane Billings to Sigma Theta Tau International Foundation for Nursing's research endowment. The purpose of this grant is to support research that advances the science of nursing education and learning through the use of technology in dissemination of knowledge.
Requirements: Eligibility Criteria: preference given to Sigma Theta Tau International and/or National League for Nursing members; grant proposals must advance the science of nursing education through the use of technology in the dissemination of knowledge; registered nurse with current license; hold a master's or doctoral degree or be enrolled in a doctoral program; submit a completed research application package and signed research agreement; ready to implement research project when funding is received. All applications must be submitted via the online submission system (see STTI website)
Geographic Focus: All States
Date(s) Application is Due: Jun 1
Amount of Grant: 5,000 USD
Contact: Research Services; (888) 634-7575 or (317) 634-8171; fax (317) 634-8188; research@stti.iupui.edu
Internet: http://www.nursingsociety.org/Research/SmallGrants/Pages/grant_nln.aspx
Sponsor: Sigma Theta Tau International
550 W. North Street
Indianapolis, IN 46202-2156

STTI Rosemary Berkel Crisp Research Awards 4463

The purpose of the STTI Rosemary Berkel Crisp Research Awards is to support nursing research in the critical areas of women's health, oncology, and pediatrics. The allocation of funds is based upon a research project that is ready for implementation in the area of women's health, oncology, or pediatrics; the quality of the proposed research, future potential of the application, appropriateness of the research budget, and feasibility of time frame. Funds provided by interest from a gift from The Harry L. Crisp, II and Rosemary Berkel Crisp Foundation to the honor society's Research Endowment.
Requirements: Eligibility Criteria: registered nurse with current license; hold a master's or doctoral degree or be enrolled in a doctoral program; submit a completed research application package and signed research agreement via the online submission system; ready to implement research project when funding is received; member of Sigma Theta Tau International; Note: Some preference will be given to applicants residing in Illinois, Missouri, Arkansas, Kentucky, and Tennessee.
Geographic Focus: All States
Date(s) Application is Due: Dec 1
Amount of Grant: Up to 5,000 USD
Contact: Research Services; (888) 634-7575 or (317) 634-8171; fax (317) 634-8188; research@stti.iupui.edu

Internet: http://www.nursingsociety.org/Research/SmallGrants/Pages/grant_crisp.aspx
Sponsor: Sigma Theta Tau International
550 W. North Street
Indianapolis, IN 46202-2156

STTI Virginia Henderson Clinical Research Grants **4464**

The purpose of the Sigma Theta Tau International Virginia Henderson Clinical Research Grant is to encourage the research career development of clinically based nurses through support of clinically oriented research. Allocation of funds is based on a research project ready for implementation, the quality of the proposed research, the future potential of the applicant, appropriateness of the research budget and feasibility of the time frame. Funds for this biennial grant are provided by interest from the Virginia Henderson Clinical Research Endowment Fund.
Requirements: Eligibility Criteria: registered nurse actively involved in some aspect of health care delivery, education or research in a clinical setting; hold a master's or doctoral degree or be enrolled in a doctoral program; submit a completed research application package and signed research agreement via the online submission system; ready to implement research project when funding is received; member of Sigma Theta Tau International.
Geographic Focus: All States
Date(s) Application is Due: Dec 1
Amount of Grant: Up to 5,000 USD
Contact: Research Services; (888) 634-7575 or (317) 634-8171; fax (317) 634-8188; research@stti.iupui.edu
Internet: http://www.nursingsociety.org/Research/SmallGrants/Pages/grant_VHL.aspx
Sponsor: Sigma Theta Tau International
550 W. North Street
Indianapolis, IN 46202-2156

Summerfield G. Roberts Award **4465**

A cash award is presented at the annual membership meeting in the spring to the author of a work of creative writing on the Republic of Texas. The purpose of this award is to encourage literary effort and research about historical events and personalities during the days of the Republic of Texas, 1836-1846, and to stimulate interest in this period. The judges will determine which manuscript best portrays the spirit, character, strength, and deeds of men and women who lived in the Republic of Texas. The manuscripts must be written or published during the calendar year for which the award is given. There is no word limit on the material submitted for the award. The manuscripts may be either fiction or nonfiction, poems, essays, short stories, novels, or biographies. Judges for a given year are the winners of the award in the last three years. Five copies of each entry must be mailed to the office shown for the use of the judges and to add to the archives; no copies are returned.
Requirements: The competition is open to all writers everywhere; they need not reside in Texas, nor must the publishers be in Texas.
Geographic Focus: All States
Date(s) Application is Due: Jan 15
Amount of Grant: 2,500 USD
Samples: New Orleans and the Texas Revolution, By Edward L. Millier, $2500; Frontier Blood, the Saga of the Parker Family by Jo Ella Powell Exley.
Contact: Secretary; (979) 245-6644; fax (979) 244-3819; srttexas@srttexas.org
Internet: http://www.srttexas.org/sumfield.html
Sponsor: Sons of the Republic of Texas
1717 Eighth Street
Bay City, TX 77414

Sun Academic Excellence Grants **4466**

The Sun Academic Excellence Grant program has helped numerous schools kick-start their projects by providing equipment grants and access to free training and certifications. Funding for grants focus on areas such as: curriculum integration—the integration of Sun Technology in undergraduate and graduate level courses through the development and use of model curricula; community development—the sponsorship of a Sun or Open Source community development projects that align with the academic and research interests of your students, faculty and researchers and result in put-backs to the code; localization—the localization of the code or development tools and resources to languages other than English; community outreach—to promote broad awareness of Sun technologies on campus and participation of faculty, students and other members of the academic community in Sun communities through events, workshops and end-user communications; Java software development—NetBeans IDE plug-ins that result in put backs to the Net Beans community; application development—development of an application, technology or solution using the Sun platform which can be deployed in the Education, Life Science, Healthcare or Government industry; and thin client—deployment of Sun Ray thin clients; ideally 100 or more seats. Grants are made on a published quarterly calendar and are for hardware donations only.
Requirements: In the United States, higher education eligible institutions must have: 3304(f) or a 501(c)3 or Section 115 tax exempt status; and be a degree granting institution or be involved in basic research. Outside of the United States, higher education institutions must either be degree-granting or involved in basic research and be recognized as a non-commercial enterprise by their national government. To be eligible as a primary or secondary educational institution, the following guidelines must be met: public or private formal full-time educational institution offering classes in the range or equivalence of grades Primary and Secondary (K12); or licensed, certified or accredited to operate as a school or school administrative unit in the state or country of operation.
Restrictions: Grants do not include maintenance, service or technical support. Non-profit institutions with 501(c)3 status which are neither degree-granting nor directly involved with basic research may be eligible if they directly support the academic and research communities in their mission to create and disseminate knowledge, and if they can demonstrate financial need.
Geographic Focus: All States
Date(s) Application is Due: Feb 26; May 31; Aug 31; Nov 30
Contact: Linda Rogers, Senior Director of Strategy; (650) 336-0487 or (650) 960-1300; fax (650) 856-2114; corpaffrs@corp.sun.com
Internet: http://www.sun.com/solutions/landing/industry/education/aeg.xml
Sponsor: Sun Microsystems Corporation
4150 Network Circle
Santa Clara, CA 95054

Sun Grid Research Grants **4467**

Sun fosters close relationships with researchers in academia, national labs and nonprofit organizations to create the future of computer and information technology. Sun Grid Research Grants award hours of central processing units (CPUs) on the Sun Grid Compute Utility. Sun Grid helps educators and researchers derive immediate benefits from an open, grid-based computing infrastructure on a utility basis providing choice and control over how IT is purchased. The Sun External Research Office collaborates with the academic community to define and create the future of computer and information technology. Sun and the university work to transform key research outcomes into future products.
Requirements: In the United States, higher education eligible institutions must have: 3304(f) or a 501(c)3 or Section 115 tax exempt status; and be a degree granting institution or be involved in basic research. Outside of the United States, higher education institutions must either be degree-granting or involved in basic research and be recognized as a non-commercial enterprise by their national government.
Geographic Focus: All States
Date(s) Application is Due: Feb 26; May 31; Aug 31; Nov 30
Contact: Linda Rogers, Senior Director of Strategy; (650) 336-0487 or (650) 960-1300; fax (650) 856-2114; corpaffrs@corp.sun.com
Internet: http://be.sun.com/teleweb/grantprog/pdf/STAR_DS_new_brandV5.pdf
Sponsor: Sun Microsystems Corporation
4150 Network Circle
Santa Clara, CA 95054

Sun Microsystems Foundation Summer Academic Enrichment Grants **4468**

The Sun Foundation donates dollars to provide access to academic enrichment programs for low-income and under represented students. Through the Sun Foundation's Summer Academic Enrichment Program, cash grants have been awarded to non-profit organizations in the Greater Boston area, Denver Metro Area, California Bay Area, and U.K. (Scotland, England). The grant recipients' innovative programs help youth develop skills through hands-on collaborative learning. Additionally, the programs encourage students to pursue further studies in math, science, technology and engineering. The Sun Foundation's educational grants support the early development of talent necessary to meet the high-tech workforce needs of the future.
Requirements: U.S. 3304(f) or 501(c)3 tax-exempt organizations that serve students in the metropolitan regions of: Boston, Massachusetts; Denver, Colorado; and California Bay are eligible to apply, as well as the international areas of the United Kingdom and Scotland.
Geographic Focus: California, Colorado, Massachusetts, United Kingdom
Samples: Girls For A Change, San Jose, California—to help over one thousand girls become critical thinkers, innovators and leaders; InternalDrive Foundation, Campbell, California—to fund the Hands-On Tech Outreach Program, a summer computer camp in Santa Clara; UMass Lowell Summer DESIGNCAMP, Lowell, Massachusetts——to improve critical thinking and problem solving skills of 360 students.
Contact: Linda Rogers, Senior Director of Strategy; (650) 336-0487 or (650) 960-1300; fax (650) 856-2114; corpaffrs@corp.sun.com
Internet: http://www.sun.com/aboutsun/comm_invest/grants.html
Sponsor: Sun Microsystems Foundation
901 San Antonio Road, MS MPK15-204
Palo Alto, CA 94303

Sunny and Abe Rosenberg Foundation Grants **4469**

The Rosenberg Foundation awards grants to New York City area nonprofit organizations. Funding priorities include programs that promote excellence in public education; artistic and cultural education for public school students; and activities that promote learning for disadvantaged children and youth. The Foundation also awards support to supplementary services for children currently in foster care and young adults moving out of the foster system. Typically, an awarded grant ranges from $10,000 to $50,000. The Foundation does not accept unsolicited proposals.
Requirements: New York City-area nonprofit organizations are eligible. Applicants should contact the Foundation to see if their project may be eligible for funding.
Restrictions: The Foundation does not support advocacy, litigation, or political causes. Funding is also not available for individuals or other private foundations.
Geographic Focus: New York
Amount of Grant: 10,000-50,000 USD
Contact: Maryanne Passafiume; (212) 993-6230; mpassafiume@rosenbergfoundation.org
Internet: http://www.rosenbergfoundation.org/
Sponsor: Sunny and Abe Rosenberg Foundation
Ansonia Station
New York, NY 10023

Surdna Foundation Arts Teachers Fellowships 4470

The Surdna Foundation announces the annual national initiative to support the artistic revitalization of outstanding arts teachers in public arts high schools. Surdna recognizes that arts teachers often lack the time and resources to reconnect with other arts professionals and with the artistic processes they teach. Through the Surdna Arts Teachers Fellowship Program (SATF), Fellows design individualized courses of study that provide both immersion in their own creative work and the opportunity to interact with other professional artists in their fields. Surdna believes that this approach to professional development will enhance the effectiveness of arts teachers and will directly benefit the young people they teach. All permanently assigned full- and part-time arts faculty in public arts high schools are eligible. Twenty awards of $5,500 each, with a complementary grant of $1,500 to the Fellow's school to support post-fellowship activities, will be made. The fellowship award may be used to defray the costs of tuition and other fees, room and board, travel, purchase of materials and/or equipment for personal art-making, childcare, and other relevant expenses.

Requirements: All permanently assigned full- and part-time arts faculty in public arts high schools are eligible. Applicants must be minimally in their fifth year of teaching arts in high school, and plan to continue as an arts teacher in their public arts high school during the upcoming school year.

Geographic Focus: All States

Date(s) Application is Due: Feb 10

Amount of Grant: Up to 6,500 USD

Contact: Kimberly Bartosik, SATF Program Director; (212) 557-0010, ext. 256; fax (212) 557-0003; artsfellowship@surdna.org

Internet: http://www.surdna.org/what-we-fund/thriving-cultures/109.html

Sponsor: Surdna Foundation
330 Madison Avenue, 30th Floor
New York, NY 10017-5001

Surdna Foundation Sustainable Environments Grants 4471

The Surdna Foundation seeks to create just and sustainable communities where consumption and conservation are balanced and innovative solutions to environmental problems improve people's lives. It works from a sustainable development perspective to demonstrate that a healthy environment is the backbone of a healthy economy and a democratic society. The Foundation funds three key related priority areas-Climate Change, Green Economy, and Transportation and Smart Growth-that aim to transform how Americans work, consume, and move. Together these will help make the theory of a carbon free society into a practical and achievable reality for communities across the United States. The Surdna Foundation accepts applications on an ongoing basis. However, grants are approved three times per year: in February, May and September. Applicants should use the online letter of inquiry to apply for funding.

Requirements: 501(c)3 organizations may apply. The foundation urges applicants to send two- to three-page letters of intent before sending proposals. IRS nonprofit status certification, recent audited financial statements, and project budget should also be included.

Restrictions: Generally, the Surdna Foundation does not support: programs addressing toxics, hazardous waste, land and habitat conservation; animal welfare, biodiversity and ocean management; individuals; or academic fellowships.

Geographic Focus: All States

Amount of Grant: 15,000-1,000,000 USD

Contact: Edward Skloot, Executive Director; (212) 557-0010; fax (212) 557-0003; grants@surdna.org or executivedirector@surdna.org

Internet: http://www.surdna.org/what-we-fund/sustainable-environments.html

Sponsor: Surdna Foundation
330 Madison Avenue, 30th Floor
New York, NY 10017-5001

Susan G. Komen Breast Cancer Foundation Brinker Awards for Scientific Distinction 4472

Established through a partnership between Susan G. Komen for the Cure and Brinker International in 1992, the Brinker Award for Scientific Distinction is a marquee award that recognizes pioneering work in breast cancer research and treatment. The Brinker Award honors leading scientists for their significant achievements and contributions to basic and translational science and clinical practice that have advanced the fight to save lives and realize our vision of a world without breast cancer. Since 1992, the roster of Komen Brinker Award laureates has grown to include names of researchers who have made the most significant advances in breast cancer research and medicine. The Brinker Award for Scientific Distinction is given annually in two categories. The Basic Science Award is presented to a researcher whose scientific discoveries or novel technologies have added substantively to our understanding of breast cancer and the intrinsic molecular processes that drive the disease, and/or whose work has bridged the gap between basic research and patient care. The Clinical Research Award is presented to a clinical or translational researcher who has advanced the identification of new prevention, detection or treatment approaches for breast cancer and promoted their incorporation into clinical care.

Requirements: Breast cancer researchers who are affiliated with nonprofit institutions (academic, industry, government, or other) in the United States and abroad are eligible.

Geographic Focus: All States

Date(s) Application is Due: Jun 30

Amount of Grant: 20,000 USD

Contact: Award Administrator; (972) 855-1600; fax (972) 855-1605

Internet: http://ww5.komen.org/ResearchGrants/BrinkerAward.html

Sponsor: Susan G. Komen Breast Cancer Foundation
5005 LBJ Freeway, Suite 250
Dallas, TX 75244

Susan G. Komen Breast Cancer Foundation Career Catalyst Research Grants 4473

Career Catalyst Research (CCR) Grants are intended to foster promising breast cancer researchers who are in the early stages of their faculty careers by providing support for up to three years of "protected time" for research career development under the guidance of a mentor committee. These Grants provide funding for research projects that have significant potential to advance our understanding of breast cancer and lead to reductions in breast cancer incidence and/or mortality within the next decade. It is expected that awardees will launch independent research careers and successfully compete for subsequent research project funding in breast cancer following the successful completion of a Career Catalyst Research Grant. CCRs provide support for hypothesis-driven research projects that have significant potential to advance our understanding of breast cancer and lead to reductions in breast cancer incidence and/or mortality within the next decade. Studies focusing on quality of life or survivorship issues are not appropriate for this mechanism. Applicants may request funding of up to $150,000 per year (combined direct and indirect costs) for up to three years.

Requirements: Applicants and Institutions must conform to the eligibility criteria to apply for a CCR Grant (eligibility requirements must be met at the time of full application submission). Applicants must: have a doctoral degree, including MD, PhD, DrPH, DO, or equivalent; currently hold a faculty appointment or have a formal offer letter at the time of application, verified by the Letter of Institutional Support; not have held any faculty appointment, including non-tenure and tenure track appointments combined, for more than a total of 6 years at the time of application, verified by the Letter of Institutional Support; not currently hold or simultaneously apply for a Komen Investigator-Initiated Research Grant; not currently be a Principal Investigator on an existing R01 research grant; have adequate space to conduct proposed research and protected time for research, verified by the Letter of Institutional Support; and ensure that all past and current Komen-funded grants are up to date and in compliance with all Komen requirements; e.g. Applicants are not required to be U.S. citizens or residents. Institutions must be a nonprofit institution or organization anywhere in the world.

Geographic Focus: All States, All Countries

Date(s) Application is Due: Dec 17

Amount of Grant: Up to 450,000 USD

Contact: Director, Grants and Sponsored Programs; (972) 855-1600 or (866) 921-9678; fax (972) 855-1605; helpdesk@komengrantsaccess.org

Internet: http://ww5.komen.org/uploadedFiles/Content/ResearchGrants/GrantPrograms/new-CareerCatalystRFA2012-2.pdf

Sponsor: Susan G. Komen Breast Cancer Foundation
5005 LBJ Freeway, Suite 250
Dallas, TX 75244

Susan G. Komen Breast Cancer Foundation Challege Grants: Breast Cancer and the Environment 4474

Major advances have been made in understanding the biology and diversity of breast cancer, but much more remains to be discovered about the many causes of breast cancer – particularly what contributions a diverse array of environmental factors may be making – and how to prevent it. The challenges are many: The scientific community has been presented with conflicting and inconclusive results from past studies. With increased knowledge of the complexity of breast cancer biology, it has become apparent that future research into environmental influences will need to focus on early-life exposures, associations with specific tumor types, and gene-environment interactions. Adding to the complexity of this task is the fact that for a wide array of exposures the assessment methodologies, tools and resources are limited. Through its Challenge Grants: Breast Cancer and the Environment program, the Foundation seeks to address key research needs in the field of environmental contributions to breast cancer risk. Susan G. Komen for the Cure and its Scientific Advisory Board have requested that the Institute of Medicine (IOM) review the current evidence on environmental risk factors for breast cancer, consider gene-environment interactions in breast cancer, explore evidence-based actions that might reduce the risk of breast cancer, and recommend research needed in these areas. At the conclusion of their report, the IOM issued 13 recommendations for further research, of which three have been chosen by the Komen Scientific Advisory Board as the subject of Challenge Grants: Studies of Occupational Cohorts and Other Highly Exposed Populations; New Exposure Assessment Tools; and Minimizing Exposure to Ionizing Radiation.

Requirements: U.S. nonprofit organizations and international institutions will be eligible.

Geographic Focus: All States

Amount of Grant: Up to 250,000 USD

Contact: Director; (972) 855-1600; fax (972) 855-1605; grants@komen.org

Internet: http://ww5.komen.org/ResearchGrants/FundingOpportunities.html

Sponsor: Susan G. Komen Breast Cancer Foundation
5005 LBJ Freeway, Suite 250
Dallas, TX 75244

Susan G. Komen Breast Cancer Foundation Challege Grants: Investigator Initiated Research 4475

Susan G. Komen Breast Cancer Foundation Challenge Grants seek to stimulate exploration of new ideas and novel approaches to research on breast cancer and the environment, through a life course approach, that have significant potential to lead to reductions in breast cancer incidence and/or mortality within the next decade. In the area of Studies of Occupational Cohorts and Other Highly-Exposed Populations, the research seeking to study populations exposed to potentially cancer-causing agents or external toxins, chemicals, hormones or radiation, such as occupational cohorts, persons with event-related high exposures, or patient groups receiving high-dose or long-term medical treatments, including medical radiation. In the area of New Exposure Assessment Tools, the research is seeking to improve methodologies for measuring, across the life course, personal exposure to and biologically effective doses of environmental

factors that may alter risk for or susceptibility to breast cancer. In the area of Minimizing Exposure to Ionizing Radiation, comparative effectiveness research is seeking to assess the relative benefits and harms of medical radiation, including imaging procedures and diagnostic/follow-up radiologic tests using algorithms in common practice. Applicants may request either three or four years of funding as follows: up to a total of $750,000 over three years (combined direct and indirect costs); or up to a total of $1,000,000 over four years (combined direct and indirect costs). Reasonable compensation of advocates is allowed when advocates perform services that would otherwise be a contracted expense. Compensation may be in the form of salary, per-hour compensation, or honoraria. Additionally, grant funds can be used for advocate participation in scientific conferences that would enhance their knowledge and skills related to the research project.
Requirements: A wide range of nonprofits, including research groups and health and human service agencies, are eligible to apply. Budgets are not required to be equivalent across each year of the Grant, but rather should reflect the costs appropriate to support the research project each year.
Restrictions: Personnel on the project are limited to a base salary at or below $250,000 per year, and PIs must provide a 10% minimal level of effort. Equipment costs are limited to no more than 30% of total direct costs. Indirect costs cannot exceed 25% of total direct costs (including any indirect costs paid through subcontracts or consortia). Costs that are not allowed include: travel costs; publication costs and meeting-related poster printing costs are allowed; graduate and postdoctoral fellow tuition; visa costs; or professional membership dues.
Geographic Focus: All States
Date(s) Application is Due: Dec 20
Amount of Grant: Up to 1,000,000 USD
Contact: Director, Grants and Sponsored Programs; (972) 855-1600 or (866) 921-9678; fax (972) 855-1605; helpdesk@komengrantsaccess.org
Internet: http://ww5.komen.org/uploadedFiles/Content/ResearchGrants/GrantPrograms/new-Challenge-InvestInitiatedRFA2012-2.pdf
Sponsor: Susan G. Komen Breast Cancer Foundation
5005 LBJ Freeway, Suite 250
Dallas, TX 75244

Susan G. Komen Breast Cancer Foundation Challenge Grants: Career Catalyst Research 4476
Challenge Grants-Career Catalyst Research are intended to foster promising breast cancer researchers who are in the early stages of their faculty careers by providing support for up to three years of protected time for research career development under the guidance of a mentor committee. It is expected that awardees will launch independent research careers and successfully compete for subsequent research project funding in breast cancer following the successful completion of the Grant. These Grants provide funding for research projects that have significant potential to advance our understanding of breast cancer and the environment, through a life course approach, that have significant potential to lead to reductions in breast cancer incidence and/or mortality within the next decade. Applications will be accepted within the following three focus areas: studies of occupational cohorts and other highly-exposed populations; new exposure assessment tools; and minimizing exposure to ionizing radiation. The pre-application deadline is September 10 annually, and there is a limit of $450,000 combined direct and indirect costs ($150,000 per year for up to 3 years).
Requirements: CCR grants are open to scientists who have held faculty positions for no more than six years at the time of full application to achieve research independence. Applicants and Institutions must conform to the eligibility criteria to apply for a CCR Grant (eligibility requirements must be met at the time of full application submission). Applicants must: have a doctoral degree, including MD, PhD, DrPH, DO, or equivalent; currently hold a faculty appointment or have a formal offer letter at the time of application, verified by the Letter of Institutional Support; not have held any faculty appointment, including non-tenure and tenure track appointments combined, for more than a total of 6 years at the time of application, verified by the Letter of Institutional Support; not currently hold or simultaneously apply for a Komen Investigator-Initiated Research Grant; not currently be a Principal Investigator on an existing R01 research grant; have adequate space to conduct proposed research and protected time for research, verified by the Letter of Institutional Support; and ensure that all past and current Komen-funded grants are up to date and in compliance with all Komen requirements; e.g. Applicants are not required to be U.S. citizens or residents. Institutions must be a nonprofit institution or organization anywhere in the world.
Geographic Focus: All States, All Countries
Date(s) Application is Due: Dec 17
Amount of Grant: Up to 450,000 USD
Contact: Director, Grants and Sponsored Programs; (972) 855-1600 or (866) 921-9678; fax (972) 855-1605; helpdesk@komengrantsaccess.org
Internet: http://ww5.komen.org/uploadedFiles/Content/ResearchGrants/GrantPrograms/Challenge-CareerCatalystRFA2012-2.pdf
Sponsor: Susan G. Komen Breast Cancer Foundation
5005 LBJ Freeway, Suite 250
Dallas, TX 75244

Susan G. Komen Breast Cancer Foundation Investigator Initiated Research Grants 4477
Susan G. Komen Breast Cancer Foundation Investigator Initiated Research Grants seek to stimulate exploration of new ideas and novel approaches in breast cancer research and clinical practice that have significant potential to lead to reductions in breast cancer incidence and/or mortality within the next decade. Applications addressing topics other than those described below will be administratively withdrawn from consideration, and will not be reviewed or scored: Novel Therapeutics and/or Resistance-Therapeutic Implications of Tumor Genomics; Biology of Breast Cancer-Implications of the Immune System in Breast Cancer Biology; Prevention/Early Detection-New Strategies for Early Detection; and Disparities in Breast Cancer Outcomes-Outcomes of Specific Populations after Diagnosis. The pre-application submission deadline is September 10, with the full application due December 17. $1,000,000 in combined direct and indirect costs are allowable, which equals $250,000 per year for up to 4 years.
Requirements: Grants will be awarded to a single Principal Investigator (PI) or one PI and one Co-Principal Investigator (Co-PI). PIs and Co-PIs: must have a doctoral degree, including MD, PhD, DrPH, DO, or equivalent; must have a full time faculty appointment at the time of application; may only apply in the final year of their funding and may not hold both grants simultaneously; must ensure that all past and current Komen-funded Grants are up to date and in compliance with all Komen requirements; and are not required to be U.S. citizens or residents. Primary Institution must be a nonprofit institution or organization anywhere in the world.
Geographic Focus: All States, All Countries
Date(s) Application is Due: Dec 17
Amount of Grant: Up to 1,000,000 USD
Contact: Director, Grants and Sponsored Programs; (972) 855-1600 or (866) 921-9678; fax (972) 855-1605; helpdesk@komengrantsaccess.org
Internet: http://ww5.komen.org/uploadedFiles/Content/ResearchGrants/GrantPrograms/InvestInitiatedRFA2012-2-new2a.pdf
Sponsor: Susan G. Komen Breast Cancer Foundation
5005 LBJ Freeway, Suite 250
Dallas, TX 75244

Swann Foundation Fellowships 4478
The fellowship is awarded annually to assist ongoing scholarly research and writing projects in the field of caricature and cartoon. Although research must be in the field of caricature and cartoon, there is no limitation regarding the place or time period covered. There is also no restriction on which university department might oversee a project proposed for the fellowships, provided the subject pertains to caricature or cartoon art. This program is administered by the Library of Congress Prints and Photographs Division. Application forms are available on the Web site.
Requirements: To be eligible, an applicant must be a candidate for an MA or PhD degree in a university of the United States, Canada, or Mexico working toward the completion of a dissertation or thesis for that degree or engaged in postgraduate research within three years of receiving his or her MA or PhD.
Geographic Focus: All States, Canada, Mexico
Date(s) Application is Due: Feb 15
Amount of Grant: 15,000 USD
Contact: Martha Kennedy, (202) 707-9115; fax (202) 707-6647; swann@loc.gov
Internet: http://lcweb.loc.gov/rr/print/swann/swannhome.html
Sponsor: Swann Foundation Fund
9140 E Hampton Drive
Capitol Heights, MD 20743-3809

Swedish Natural Science Research Council Grants 4479
The Swedish Natural Science Research Council offers grants for senior research positions, postgraduate fellowships, junior researchers, lectureships, adjunct professorships, and council professorships. Applications for grants from NFR for financing employment at the universities are due in mid-May each year. For more information, contact the senior scientific secretaries of the committees: scientific council for humanities and social sciences, scientific council for medicine, scientific council for natural and engineering sciences, and committee for educational science.
Geographic Focus: All States
Contact: Administrator; +46-8-546 44 000; fax +46-8-546 44 180; vetenskapsradet@vr.se
Internet: http://www.vr.se/mainmenu/applyforgrants/callforproposals.4.64a4ae3e1092b3475b6800055.html
Sponsor: Swedish Natural Science Research Council
Regeringsgatan 56
Stockholm, SE-103 78 Sweden

Swedish Women's Educational Association International Scholarships 4480
SWEA is a global non-profit organization for Swedish speaking women who live or have lived abroad. Its primary objective objective is to protect the Swedish language, support and inform about culture and traditions. Each year a scholarship is awarded for studies at the graduate level in Swedish literature, language, or linguistics. The award is directed specifically toward candidates who are well into the process of writing their dissertation.
Requirements: Students of any nationality may apply. A good knowledge of the Swedish language, written or spoken, is required. Applicant must reside permanently outside Sweden and be studying at a non-Swedish university.
Geographic Focus: All States
Date(s) Application is Due: Mar 1
Amount of Grant: 7,000 USD
Contact: Scholarship Committee; (702) 873-7399; fax (702) 873-0492; office@swea.org
Internet: http://www.swea.org/inenglish___3175.aspx
Sponsor: Swedish Women's Educational Association International
P.O. Box 30190
Las Vegas, NV 89173-0190

SWE Scholarships 4481
The SWE Scholarship Program provides financial assistance to women admitted to accredited baccalaureate or graduate programs, in preparation for careers in engineering, engineering technology and computer science. Currently, SWE disbursed more than 180 new and renewed scholarships valued at $470,000. Grants are announced in the summer (sophomore, junior, senior and graduate students) and late summer/early fall (freshmen and reentry students) for

use during the following academic year. Grant payments are made in the fall for corporate-sponsored scholarships and in both the fall and spring for endowed scholarships, upon proof of registration. Scholarships and fellowships range from $1,000 to $10,000 each.
Requirements: All SWE-administered scholarships are open only to women majoring in engineering in a school, college, or university with an accredited engineering program, and who will be in a specified year of study during the academic year following presentation of the grant. Reentry scholarships are limited to women who have been out of the engineering job market for a minimum of two years and who will return to school for an engineering program, undergraduate or graduate.
Geographic Focus: All States
Date(s) Application is Due: Feb 15
Amount of Grant: 1,000-10,000 USD
Contact: Reena Singhall Lee, (312) 596-5223; reenasinghallee@gmail.com
Internet: http://societyofwomenengineers.swe.org/index.php?option=com_content&task=view&id=222&Itemid=229
Sponsor: Society of Women Engineers
120 S La Salle Street, Suite 1515
Chicago, IL 60603

Swiss National Science Foundation DORE (DO REsearch) Conference Grants 4482
DORE (DO REsearch) is the Foundation's promotional tool for practical research at Universities of Applied Sciences and Universities of Teacher Education. DORE supports practical research in the fields of social work, health, music and theater, art, education, applied psychology and applied linguistics. Awards are granted for research projects, publications, courses for junior scientists, and scientific conferences. A maximum amount of 10,000 Swiss francs is awarded for one scientific conference to primarily pay for invited speakers' travel and accommodation fees. Applications are to be submitted no later than three months prior to the planned event.
Geographic Focus: All States, Austria, Czech Republic, Germany, Hungary, Liechtenstein, Poland, Slovakia, Slovenia, Switzerland
Amount of Grant: 10,000 CHF
Contact: Secretariat; +41 (0) 31 308 22 22; fax +41 (0) 31 301 30 09; dore@snf.ch
Internet: http://www.snf.ch/e/funding/conferences/doreconferences/seiten/default.aspx
Sponsor: Swiss National Science Foundation
Wildhainweg 3, P.O. Box 8232
Bern, CH-3001 Switzerland

Swiss National Science Foundation Fellowships for Advanced Researchers 4483
Fellowships for advanced researchers enable young scientists planning to follow an academic career to benefit from a stay abroad in order to increase their knowledge and scientific reputation. These fellowships for overseas visits in any discipline are supported by the Foundation, except experimental and clinical medicine. Funding includes personal maintenance, a fixed sum for travel expenses and may include a contribution towards research and conference expenses. The amount of the fellowship is based on personal details, family obligations and cost of living in the host country and runs for a period of 12 to 36 months.
Requirements: Applicant requirements include; a doctorate degree; independent research activity, with at least one year of research experience at post-doc level at the time the application is submitted; Swiss nationality, residence or settlement permit, cross-border commuter permit; and connections with Swiss research.
Geographic Focus: All States, Austria, Czech Republic, Germany, Hungary, Liechtenstein, Poland, Slovakia, Slovenia, Switzerland
Date(s) Application is Due: Feb 1; Aug 1
Contact: Secretariat; +41 (0) 31 308 22 22; fax +41 (0) 31 301 30 09; dore@snf.ch
Internet: http://www.snf.ch/e/funding/individuals/advancedresearchers/seiten/default.aspx
Sponsor: Swiss National Science Foundation
Wildhainweg 3, P.O. Box 8232
Bern, CH-3001 Switzerland

Swiss National Science Foundation Fellowships for Prospective Researchers 4484
Fellowships for prospective researchers offer young scientists, who are at the point of starting their research career, the possibility to stay at a research institution abroad. These fellowships for a stay oversea are available in all disciplines supported by the Swiss National Science Foundation (SNSF). Funding includes personal maintenance, a fixed sum for travel expenses and may include a contribution towards research and conference expenses. The amount of the fellowship is based on personal details, family obligations and cost of living in the host country. For informations about the fellowships duration, an applicant should contact the local Research Commission.
Requirements: Applicants must have: a doctorate (lic., diploma or master's degree may be acceptable in some cases at the end of a doctoral thesis); for a postdoctoral fellowship, at the latest 3 years after obtaining the doctorate; for a doctoral fellowship at the end of the thesis, at the earliest 2 years after obtaining the licentiate, diploma or master's degree; Swiss nationality, residence or settlement permit, or cross-border commuter permit; and connections with Swiss research.
Geographic Focus: All States, Austria, Czech Republic, Germany, Hungary, Liechtenstein, Poland, Slovakia, Slovenia, Switzerland
Contact: Secretariat; +41 (0) 31 308 22 22; fax +41 (0) 31 301 30 09; dore@snf.ch
Internet: http://www.snf.ch/e/funding/individuals/prospective/seiten/default.aspx
Sponsor: Swiss National Science Foundation
Wildhainweg 3, P.O. Box 8232
Bern, CH-3001 Switzerland

Swiss National Science Foundation Marie Heim-Vogtlin (MHV) Grants 4485
To increase the number of women in research, the Foundation annually awards a limited number of Marie Heim-Vogtlin (MHV)-subsidies. These subsidies are awarded after a two-phase selection procedure. The subsidies aim to facilitate the recipients research and integration in a host institution at a Swiss University. The program provides financial support where no other means of such integration exists or where female researchers cannot qualify for other funding programs. The subsidies are aimed for doctoral and postdoctoral candidates within all disciplines funded by the SNSF. The objective is to support well-qualified female scientists who are or were forced to accept an interruption or reduction of their research activities due to family obligations or a change of residence as a result of their partner's career development. Applications for these subsidies may only be submitted in response to a specific call.
Requirements: A candidate's occupation degree for the project must be at least 50%. In order to achieve sustained integration, the host institution must furnish not only a written declaration of acceptance and job confirmation but also a continuing commitment following expiry of the MHV-subsidy.
Geographic Focus: All States, Austria, Czech Republic, Germany, Hungary, Liechtenstein, Poland, Slovakia, Slovenia, Switzerland
Contact: Secretariat; +41 (0) 31 308 22 22; fax +41 (0) 31 301 30 09; dore@snf.ch
Internet: http://www.snf.ch/e/funding/individuals/mhv/seiten/default.aspx
Sponsor: Swiss National Science Foundation
Wildhainweg 3, P.O. Box 8232
Bern, CH-3001 Switzerland

Swiss National Science Foundation Scientific Conferences Grants 4486
The Swiss National Science Foundation (SNSF) provides funding for scientific conferences (especially colloquiums and symposiums) if these are of obvious interest to the Swiss research community. Funding is mainly intended to cover travel and accommodation expenses for guest speakers from abroad. Applications for conference grants are not subject to a deadline. However, organizers must submit their applications at least six months before the date of the planned event.
Requirements: The conference venues must be in Switzerland.
Geographic Focus: All States, Austria, Czech Republic, Germany, Hungary, Liechtenstein, Poland, Slovakia, Slovenia, Switzerland
Amount of Grant: Up to 10,000 CHF
Contact: Secretariat; +41 (0) 31 308 22 22; fax +41 (0) 31 301 30 09; dore@snf.ch
Internet: http://www.snf.ch/e/funding/conferences/scientificconferences/seiten/default.aspx
Sponsor: Swiss National Science Foundation
Wildhainweg 3, P.O. Box 8232
Bern, CH-3001 Switzerland

Sybil G. Jacobs Award for Outstanding Use of Tobacco Industry Documents 4487
The award recognizes a person who has made a significant and well-recognized contribution to the health of the public in the recent past through use of tobacco documents. The award also honors innovation in the use and application of tobacco industry documents to improve the public's health and, where applicable, to further the goals of tobacco prevention and control in order to help build a world where young people reject tobacco and anyone can quit. Those nominated should be individuals who have made a notable impact through innovative use of tobacco industry documents as applied to research, policy, or advocacy.
Requirements: At least one of the following two criteria must be met by the nominated individual: Nominees must have made a remarkable research, policy, or advocacy contribution with the use of tobacco industry documents, and/or; Nominees must have employed innovative, creative approaches to the employment of tobacco industry documents that result in an improvement in the health or public awareness of a community or nation. Nominations may be made by individuals such as colleagues, peers, coworkers, instructors/professors, governments, or CBOs that are qualified to make such a nomination because of their familiarity with the nominee's work and contribution.
Restrictions: Nominees must not have any affiliation with the tobacco industry. Members of the awards committee may not nominate potential recipients, although they may provide written support as part of a nomination package. Employees and directors of the American Legacy Foundation and their relatives are not eligible.
Geographic Focus: All States
Date(s) Application is Due: Feb 6
Amount of Grant: 7,500 USD
Contact: Virginia Lockmuller; (202) 454-5555; awards@americanlegacy.org
Internet: http://www.legacyforhealth.org/awards/
Sponsor: American Legacy Foundation
1724 Massachusetts Avenue NW
Washington, DC 20036

Syracuse University Graduate Fellowships 4488
Syracuse University makes available 100 graduate fellowships annually, created to support graduate training for students with outstanding qualifications and the potential to make significant contributions in their future professions. The competition is open to applicants in all fields of graduate study at Syracuse University. Full tuition for an academic year and a stipend are provided.
Requirements: Full-time students must be nominated by an academic department.
Geographic Focus: All States
Contact: Fellowships Administrator; (315) 443-4492; grad@syr.edu
Internet: http://gradsch.syr.edu/student/FinResor.htm
Sponsor: Syracuse University
207 Bowne Hall
Syracuse, NY 13244

T. James Kavanagh Foundation Grants 4489

A large percentage of funding supports the Catholic church and religious associations, along with support for U.S. Roman Catholic schools, with emphasis on Pennsylvania and southern New Jersey. The foundation also makes music awards. Types of support include annual campaigns, continuing support, equipment, general operating support, program development, research, and scholarship funds. Proposals should be submitted preferably by the end of February, July, or October. The board meets in March, August, and November to consider requests.

Requirements: Giving is strictly limited to the U.S., with emphasis on southern NJ and PA. Any Roman Catholic affiliate, church, school, college, or hospital will be considered.

Restrictions: Grants will not be awarded outside of the United States, not even to U.S. missions or to help their agencies abroad. No funding will be given for individuals; endowment funds; seed money; deficit financing; land acquisition; publications; conferences; scholarships or fellowships; matching gifts; or loans.

Geographic Focus: All States

Amount of Grant: 2,000-10,000 USD

Contact: Thomas Kavanagh, Trustee; (610) 356-4606

Sponsor: T. James Kavanagh Foundation

234 E State Street

Sharon, PA 16146

T.R. Higgins Lectureship Award 4490

The award recognizes an outstanding lecturer and author whose technical research paper or papers may be identified as an outstanding contribution to engineering literature on fabricated structural steel published during the eligibility period. The award will be made to a nominated individual on the basis of his/her reputation as a lecturer combined with an evaluation of the originality, the contribution to engineering knowledge, the projected future significance, the clarity of presentation, and the value to the fabricated structural steel industry, as found by the jury in the paper or papers named in the nomination. Nominations must be received by the listed deadline. Contact the institute to verify deadline date.

Requirements: The author must be a permanent resident of the United States. The paper or papers must have been published in a professional journal that provides for a discussion of papers by the engineering profession within the recent five-year period.

Geographic Focus: All States

Date(s) Application is Due: Aug 1

Amount of Grant: 10,000 USD

Contact: Award Administrator; (312) 670-2400; fax (312) 670-5403

Internet: http://www.aisc.org/Content/NavigationMenu/About_AISC/Competition_and_Awards/T_R__Higgins_Lectureship/T_R__Higgins_Lectureship.htm

Sponsor: American Institute of Suiteel Construction

One E Wacker Drive, Suite 3100

Chicago, IL 60601-2001

Takemi Program in International Health Research and Advanced Training Fellowships 4491

This midcareer program offers a limited number of fellowship positions for research and advanced training on critical issues of international health, especially those relating to developing countries. This interdisciplinary program has four major objectives: to create better methods for mobilizing and using health resources in both rich and poor countries; to promote cooperative research and comparative analysis of health policies and programs in different countries; to study transnational causes of health conditions such as population migration, disease transmission, and air and water pollution; and to bring together leading health professionals and scholars from many nations for research and training. Fellowship positions cover estimated costs for the program, which includes 10 months in residence at the Harvard School of Public Health. Contact program staff for deadlines.

Requirements: Applicants should have completed graduate degrees and demonstrated potential leadership capacity in their home countries. They are expected to show strong promise and appropriate preparation to enable them to benefit from a period at Harvard. Applicants may come from the disciplines of medicine, law, public health, economics, management, and social sciences. Applicants should submit a curriculum vita and publications list, a proposal of research and writing to be undertaken during the fellowship, and three letters of reference. The proposal should not exceed five typed pages and should include a clear statement of objectives, significance or relevance, data sources, and methodology.

Restrictions: Clinical research will not be funded. Individuals enrolled in degree-granting programs are ineligible.

Geographic Focus: All States

Contact: Coordinator; (617) 432-0686; fax (617) 432-1251; takemi@hsph.harvard.edu

Internet: http://www.hsph.harvard.edu/takemi

Sponsor: Harvard University

665 Huntington Avenue, Building 1-1207

Boston, MA 02115-6021

Talbert and Leota Abrams Foundation 4492

The Talbert and Leota Abrams Foundation awards grants to organizations in Michigan in its fields of interest, including higher education, adult basic education and literacy programs, libraries and reading, and community funds. Grants support research and program development. The board meets in May to consider requests.

Requirements: Nonprofit organizations in Michigan, primarily central Michigan, are eligible. Applicants submit a two page letter of inquiry, along with a copy of their 501(c)3 exemption. Churches may apply for nonsectarian, nonreligious grants.

Restrictions: Grants do not support athletic activities, churches for sectarian use, individuals, operating expense, loans, or traveling expenses. The Foundation does not fund the maintenance of existing programs or services.

Geographic Focus: Michigan

Date(s) Application is Due: May 31

Contact: Joe Foster Jr., Secretary; (517) 706-5791

Sponsor: Talbert and Leota Abrams Foundation

400 Lake Lansing Road, Suite F

Lansing, MI 48912-3674

Tandem Launch Technologies Grants 4493

Tandem Launch Technologies is currently accepting proposals for their latest round of multimedia invention development funding. This unique acceleration program provides entrepreneurial university inventors with up to $1 million worth of funding, business and technical development resources, and connections in the multimedia industry. Multimedia projects appropriate for the program are scalable to the consumer electronics market, fit with a licensing model, and require a budget of less than $1 million to develop an initial prototype. Projects in the following fields can be submitted: video processing and encoding; display technologies; camera and image capture; audio processing and encoding; natural human computer interaction; and natural speech synthesis and recognition.

Requirements: The TandemLaunch model is only suitable for technologies with the following characteristics: protectable intellectual property, ideally through patents; scalable to the broader consumer electronics and not limited to niche markets; amenable to a licensing business model; development of an initial demonstration prototype with an overall budget of less than $1M requiring no restricted capital equipment; and support from dedicated and highly skilled inventors with a keen interest in commercializing the technology.

Geographic Focus: All States, All Countries, Canada

Date(s) Application is Due: Aug 15

Amount of Grant: Up to 1,000,000 USD

Contact: Katie Young Morris, Project Manager; (438) 380-5435, ext. 209; Catherine.young-morris@tandemlaunchtech.com or projects@tandemlaunchtech.com

David Zangwill, (438) 380-5435, ext. 204; david.zangwill@tandemlaunchtech.com

Internet: http://www.tandemlaunchtech.com/proposals.php

Sponsor: TandemLaunch Technologies

245 Victoria Avenue, Suite 200

Westmount, QC H3Z 2M6 Canada

Tanner Humanities Center Research Fellowships 4494

The Tanner Humanities Center awards fellowships to encourage, support, and disseminate important humanistic research. Each fellow will spend one year or one semester conducting research full-time, in residence at the center. In addition, fellows will present work-in-progress talks throughout the year. The deadline for faculty applicants is October 1 to their department chair and October 15 to the center. Visiting faculty must apply by December 1. The deadline for graduate students to apply for doctoral fellowships is March 1. Guidelines and application forms for each fellowship are available online.

Requirements: Applicants must have received the PhD by August two years prior to the year of application. Eligible fields include anthropology, communication, history (including art history and criticism, film history, musicology, and theater history), comparative religion, ethnic studies, jurisprudence, languages and linguistics, literature, philosophy, women's studies, and historical or philosophical applications of the social and natural sciences.

Restrictions: The center does not consider applications for projects leading to completion of advanced degrees.

Geographic Focus: All States

Date(s) Application is Due: Mar 1; Oct 1; Oct 15; Dec 1

Contact: Manager; (801) 581-7989; Tanner.Humanities@hum.utah.edu

Internet: http://vegeta.hum.utah.edu/humcntr/applications.html

Sponsor: University of Utah

380 S 1400 E, Room 201

Salt Lake City, UT 84112-0312

Tau Beta Pi Association Fellowships 4495

The Fellowship Program is Tau Beta Pi's single most important project for the advancement of engineering education and the profession. It is a time-proven idea which was pioneered by Tau Beta Pi in the honor-society field. The plan was adopted by the 1928 Convention as a result of the recommendation and labors of President A. D. Moore. Since the beginning, the purpose has been to finance, for a select group of members chosen for merit and need, a year of graduate study at the colleges of their choice. A distinguishing feature of Tau Beta Pi fellowships has always been that they are as free of binding rules as possible. Tau Beta Pi fellows are free to do graduate work in any field which will enable them to contribute to the engineering profession. The only specific duty of awardees is to write a report at the completion of the fellowship year summarizing their work. The annual application deadline is February 1.

Requirements: Tau Beta Pi Constitution and Bylaws specifically restrict these fellowships to initiated members of the Tau Beta Pi Association on the date the fellowship is awarded.

Geographic Focus: All States

Date(s) Application is Due: Feb 1

Contact: Steve Pierre; (865) 546-4578; fax (865) 546-4579; fellowship@tbp.org

Internet: http://www.tbp.org/fellowships.cfm

Sponsor: Tau Beta Pi Association

P.O. Box 2697

Knoxville, TN 37901-2697

Taub Foundation Grants for MDS Research 4496

The Henry and Marilyn Taub Foundation works with The Medical Foundation, a division of Health Resources in Action (HRiA), to select the most qualified applicants. HRiA is a nonprofit organization in Boston that advances public health and medical research. The Taub Foundation Grants Program for Myelodysplastic Syndromes (MDS) Research

was created to support high-impact, innovative translational research to understand the underlying causes of MDS and to advance its treatment and prevention. The Program specifically focuses on MDS research, exclusive of AML and MPN. Studies focusing on molecular genetics, epigenetics, splicing factors, stem cells, the microenvironment and novel therapeutic targets relevant to MDS are encouraged. Three-year awards of $600,000 ($200,000 per year, inclusive of 10% indirect costs) will be made to independent investigators working in non-profit, non-governmental academic, medical, or research institutions within the United States. Collaborative efforts are encouraged.
Requirements: All applicants must hold a faculty appointment at a nonprofit, governmental academic, medical, or research institution in the United States. Applicants do not need to be nominated by their institutions. United States citizenship is not required; visa documentation is not required. Awards are not restricted to investigators currently working in MDS. Applications from investigators in other fields and collaborative efforts are encouraged. The complete initial online application must be received by The Medical Foundation by March 25, 12:00 Noon, U.S. Eastern Standard time.
Restrictions: Only one application may be submitted per applicant.
Geographic Focus: All States
Date(s) Application is Due: Mar 25
Amount of Grant: 600,000 USD
Contact: Erin Johnstone; (617) 279-2240 ext. 710; ejohnstone@hria.org
Internet: http://hria.org/tmfservices/tmfgrants/taub.html
Sponsor: Henry and Marilyn Taub Foundation
300 Frank West Burr Boulevard, 7th Floor
Teaneck, NJ 07666

TD4HIM Foundation Grants **4497**
The San Francisco 49ers Foundation is the non-profit community funding extension of the San Francisco 49ers. Now in its 16th year, the 49ers Foundation supports development programs for underserved youth that keep them safe, on track and in school. A significant portion of its funding goes toward family violence prevention programs and activities that teach youth leadership and respect.
Requirements: 501(c)3 organizations and programs in the greater Bay Area that are in alignment with the mission of the 49ers Foundation: keeping kids safe, on track and in school, are eligible to apply.
Restrictions: Please be advised that the 49ers Foundation does not fund the following areas: land, building or vehicle purchase; endowments; political campaigns; religious organizations (projects to promote religion); loans of any type; personal scholarships; travel; drives; specific individuals; organizations based outside the greater San Francisco Bay Area.
Geographic Focus: All States
Contact: TD4HIM Program Coordinator; (408) 829-3381
Internet: http://www.49ers.com/community/foundation.html
Sponsor: TD4HIM Foundation
60 Shoreline Circle
Incline Village, NV 89451-9529

Teachers College Minority Postdoctoral Fellowships **4498**
The program aims to increase the number of minority faculty members at graduate schools of education. It provides recent doctorate recipients the opportunity to develop a program of research and participate in a mentored program at a graduate research university for one year. Fellows conduct research as outlined in the application proposals, provide written reports of the research programs, and participate in an academic department with the option of teaching a course or providing a series of special lectures, to be decided. Fellowships may provide campus housing and limited relocation costs.
Requirements: Candidates must have earned the doctorate degree in any area within the past three years and be committed to research and teaching in an education-related area.
Geographic Focus: All States
Date(s) Application is Due: Jan 6
Amount of Grant: 33,000 USD
Contact: Young Jin Jo; (212) 678-3998; jo@exchange.tc.columbia.edu
Internet: http://www.tc.edu/dean/mpost.htm
Sponsor: Columbia University Teachers College
204 Main Hall, 525 W 120th Street, P.O. Box 54
New York, NY 10027-6696

Temple University George D. McDowell Fellowship **4499**
The fellowship provides a stipend and access to photographs and newspaper clippings in the archives of the Philadelphia Evening Bulletin newspaper, a resource for research on the Philadelphia metropolitan area. The program provides support for a six-week fellowship to do research and study that results in a scholarly work. Application and guidelines are available online.
Requirements: Undergraduate and graduate students, independent scholars, educators, writers, filmmakers, and anyone interested in the history of Philadelphia are invited to apply.
Restrictions: Current employees of Temple University are ineligible.
Geographic Focus: All States
Date(s) Application is Due: Feb 15
Amount of Grant: 3,000 USD
Contact: Margaret Jerrido; (215) 204-5750; fax (212) 204-3681; urban@library.temple.edu
Internet: http://library.temple.edu/collections/urbana
Sponsor: Temple University
1210 W Berks Street
Philadelphia, PA 19122

Tessie Agan Award Competition **4500**
This is an annual award competition open to undergraduate and graduate students to promote and encourage careers in the field of housing education. Students submit a research or position paper on some aspect or current issue in housing. The awards committee of the AAHE selects both graduate and undergraduate winners. Award winners must present their papers at the AAHE national conference. The call for papers is in March; deadlines for submission are posted in the In House newsletter. Specific guidelines are available through the AAHE headquarters. Contact program staff for availability.
Geographic Focus: All States
Date(s) Application is Due: Apr 30
Amount of Grant: 250-750 USD
Contact: JoAnn Emmell; (540) 231-9259; fax (540) 231-1697; jemmel@vt.edu.
Internet: http://www.housingeducators.org/MEMAREA/HeraAwards.shtml
Sponsor: American Association of Housing Educators
Illinois State University, 5060 FCS Dept.
Normal, IL 61790-5060

Textron Corporate Contributions Grants **4501**
Textron Inc. was founded in 1923 as a small textile company and has since become one of the world's best known multi-industry companies. Textron focuses philanthropic giving in the following areas: workforce development and education; healthy families/vibrant communities; and sponsorships. In the area of workforce development and education, the company focuses on job-training and employment development (eg., school-to-work, welfare-to-work, job-training for underserved-audiences, literacy, and English-as a-Second-Language programs). In the area of healthy families/vibrant communities, the company focuses on arts and culture (with emphasis on outreach programs that enhance learning and target low- and moderate-income individuals), community revitalization (eg., affordable housing and economic development in low-income areas), and health and human-service organizations (eg., food pantries, homeless shelters, and services for low-income residents). In the area of sponsorships, the company encourages volunteerism and sponsors worthwhile events that benefit the communities where employees live and work. Textron's grant history has included funding of general-operating costs, capital campaigns, building construction/renovation, equipment acquisition, program development, conferences and seminars, publication, seed money, fellowships, scholarship funds, research, technical assistance, consulting services, and matching funds. Downloadable guidelines (PDF) and application (Word document) are available online. The completed application and required accompanying documentation must be received via mail by the deadline date.
Requirements: Textron targets its giving to nonprofit agencies located in its headquarters state of Rhode Island and those locations where the company has divisional operations. Organizations outside of Rhode Island should contact the Textron company in their area; a listing of Textron businesses along with their contact information can be accessed by clicking the "Contact Us" link at the Textron website.
Restrictions: Textron will review only one request per organization during a 12-month period. Contributions will not be made to the following types of organizations: organizations without 501(c)3 status as defined by the Internal Revenue Service; individuals; political, fraternal or veterans organizations; religious institutions when the grant would support sectarian activities; and organizations that discriminate by race, creed, gender, ethnicity, sexual orientation, disability, age or any other basis prohibited by law.
Geographic Focus: Georgia, Illinois, Kansas, Louisiana, Maryland, Massachusetts, New York, North Carolina, Pennsylvania, Rhode Island, Texas, Germany, Great Britain
Date(s) Application is Due: Mar 1; Sep 1
Contact: Karen Warfield; (401) 421-2800; fax (401) 457-2225
Internet: http://www.textron.com/about/commitment/corp-giving/
Sponsor: Textron Charitable Trust
40 Westminster Street
Providence, RI 02903

The Ray Charles Foundation Grants **4502**
Although musician Ray Charles was blind since the age of seven, he always maintained that blindness was not a handicap, but rather that the inability to hear music constituted the truer loss. Because of this philosophy, he began making contributions to the field of hearing impairment and often anonymously contributed to cochlear implants for individuals who couldn't afford such surgeries. In 1986, Mr. Charles founded "The Robinson Foundation for Hearing Disorders" which was later renamed "The Ray Charles Foundation." Convinced of the tremendous need of education for youth, Mr. Charles also directed the foundation to make donations and support institutions and organizations for educational purposes. Over the past 23 years, the Ray Charles Foundation has provided financial donations to various institutions involved in the areas of hearing disorders as well as education. The foundation prefers to fund programs and projects that show promise of strengthening the community beyond the grant period and that offer maximum community impact to achieve long-term results. Types of funding the foundation provides are for capital, program, and core support. Capital funding is available for land, facility, and equipment purchases and renovations or new construction. Renovation and construction projects will be invited only for organizations that have already raised a substantial amount of their fundraising goal. The project must either be in construction or have a firm construction start date. Program funding is available for both new and expansion projects. While these must evidence a viable fundraising and sustainability plan, requests for program development or enhancement activiies will still be considered. Core support funding is available for established and well-managed organizations and programs. Communication with Foundation Directors is discouraged. Applicants to the Ray Charles Foundation should simply fill out and submit the foundation's Grant Application Packet via a mail carrier to the address given. Guidelines, instructions, and links to form are given at the website. Applications are accepted continually throughout the year.

Requirements: 501(c)3 organizations that have been in existence for at least three years and that are incorporated and delivering services to the United States are eligible to apply. Programs must advance the mission of the Foundation in the following program categories: Education, Hearing Disorders, and Culture and the Arts. Educational programs eligible for funding include, but are not limited to, after-school programs and academic activities aimed at college preparation for underprivileged youth, and academic/therapy services for youth who are blind and/or deaf. Hearing Disorder programs eligible for funding include research and treatment in the area of hearing disorders and educational programs and resources for youth with hearing disorders. Culture and Arts programs eligible for funding include institutions and museums whose mission it is to provide musical and cultural education and access to underprivileged and disadvantaged youth.
Restrictions: The Foundation does not fund organizations and programs outside of the United States. Most grants will be limited to one year in duration. Multi-year funding will be considered on an exceptional case by case basis. The Foundation generally does not approve grants to organizations on a continuing annual basis. If an agency applies for a 'one year grant' more than three years in a row, they will be declined and asked to wait at least two years before applying again. More obligations and responsibilities of grant recipients are listed at the website.
Geographic Focus: All States
Contact: Coordinator; (323) 737-8000; fax (323) 737-0148; info@theraycharlesfoundation.org
Internet: http://www.theraycharlesfoundation.org/GrantQualifications.html
Sponsor: Ray Charles Foundation
2107 W. Washington Boulevard
Los Angeles, CA 90018

The World Food Prize 4503
The international prize recognizes — without regard to race, religion, nationality, or political beliefs — the achievements of individuals who have advanced human development by improving the quality, quantity, or availability of food in the world. The award recognizes contributions in any field involved in the world food supply—food and agriculture science and technology, manufacturing, marketing, nutrition, economics, poverty alleviation, political leadership, and the social sciences. The prize emphasizes the importance of a nutritious and sustainable food supply for all people. Nomination form and complete details are available online. Annual deadline dates may vary; contact program staff for exact dates.
Geographic Focus: All States
Date(s) Application is Due: Mar 1
Amount of Grant: 250,000 USD
Samples: Dr. Per Pinstrup-Andersen, Director General of the International Food Policy Research Institute—for research that enabled several governments to reform their food subsidy programs, dramatically increasing food availability to the most poor in each country, and for the establishment of programs in which families receive food subsidies when children stay in school, $250,000.
Contact: Judith Pim, Director of Secretariat Operations; (515) 245-3796; fax (515) 245-3785; jpim@worldfoodprize.org
Internet: http://worldfoodprize.org/Home/tonominate.htm
Sponsor: World Food Prize Foundation
666 Grand Avenue, Suite 1700
Des Moines, IA 50309

Third Millennium Foundation Fellowships 4504
Through the Third Millennium Foundation's Fellowship programs, college students, graduate students and activists receive financial and professional support that enables them to build their capacity across a full range of human rights issues. In doing so, the Foundation aims to develop a cadre of young activists from different countries whose understanding of, commitment to, and effectiveness in human rights work has been enhanced. These Fellowships provide a unique opportunity for in-depth exposure to and reflection on a broad range of perspectives, situations and experiences related to human rights work.
Geographic Focus: All States
Contact: Manager; (212) 421-5244; fax (212) 421-5243; tmf.usa@verizon.net
Internet: http://www.seedsoftolerance.org/initiative-fellowships.html
Sponsor: Third Millennium Foundation
340 West 12th Street
New York, NY 10014

Third Millennium Foundation Grants 4505
The Third Millennium Foundation's principal goal is to support programs and initiatives designed to promote tolerance, particularly among the young. Its work is focused on developing young children's understanding of and respect for the differences that exist among themselves, especially those related to culture, ethnicity, gender, race, and socioeconomic status. The Foundation emphasizes art in its grant making and continues to explore creative new avenues for unlearning intolerance. Foundation programs build on research findings demonstrating that gaining self-control and the ability to relate positively to others is developed in early childhood and that family, educators, and child caregivers play significant roles in shaping lifelong attitudes. The Foundation makes grants in the United States and internationally, and supports arts and human rights programs. Its programmatic emphasis is on education, children's rights and capacity building for grassroots organizations.
Geographic Focus: All States
Contact: Grants Manager; (212) 421-5244; fax (212) 421-5243; tmf.usa@verizon.net
Internet: http://www.seedsoftolerance.org/initiative-grantmaking.html
Sponsor: Third Millennium Foundation
340 West 12th Street
New York, NY 10014

Thomas and Dorothy Leavey Foundation Grants 4506
Thomas Leavey and partner John C. Tyler began the Farmer's Insurance Company in 1928. The foundation was begun in 1952 by Leavey and his wife, Dorothy. Thomas Leavey died in 1980 and Dorothy actively led the foundation until her death in 1998. Giving primarily in southern California, the Foundation offers support for: hospitals, medical research, higher and secondary education, Catholic church groups; and provides scholarships to children of employees of Farmers Group.
Requirements: The foundation gives primarily in southern California.
Geographic Focus: California
Amount of Grant: 10,000-2,000,000 USD
Contact: Kathleen Leavey McCarthy, Chair; (310) 551-9936
Sponsor: Thomas and Dorothy Leavey Foundation
10100 Santa Monica Boulevard, Suite 610
Los Angeles, CA 90067-4110

Thomas Austin Finch, Sr. Foundation Grants 4507
The purpose of the Finch Foundation is the improvement of the mental, moral and physical well-being of the inhabitants of Thomasville, North Carolina, with emphasis on improving education, improving health related facilities and attracting new business to the community. However, grants are made for a wide variety of charitable causes throughout the greater Thomasville area. In general, grants are awarded for seed money, matching funds and general purposes. Primary consideration is given to projects of a non-recurring nature or to start up funding of limited duration. The foundation meets twice a year to review grant requests. Requests must be submitted by February 15th for the Spring Meeting and October 15th for the Fall Meeting. Application forms are available online. Applicants will receive notice acknowledging receipt of the grant request, and subsequently be notified of the grant declination or approval.
Requirements: Eligible applicants are IRS 501 (c)3 non-profit organizations located within the corporate limits of the City of Thomasville, North Carolina or, organizations where the benefits of the grant will inure primarily to the residents of Thomasville. Proposals should be submitted in the following format: completed Common Grant Application Form; an original Proposal Statement; an audited financial report and a current year operating budget; a copy of your official IRS Letter with your tax determination; a listing of your Board of Directors. Proposal Statements (second item in the above Format) should answer these questions: what are the objectives and expected outcomes of this program/project/request; what strategies will be used to accomplish your objective; what is the timeline for completion; if this is part of an on-going program, how long has it been in operation; what criteria will you use to measure success; if the request is not fully funded, what other sources can you engage; an Itemized budget should be included; please describe any collaborative ventures. Prior to the distribution of funds, all approved grantees must sign and return a Grant Agreement Form, stating that the funds will be used for the purpose intended. Progress reports and Completion reports must also be filed as required for your specific grant. All current grantees must be in good standing with required documentation prior to submitting new proposals to any foundation.
Restrictions: Grants are generally not made for typical operational or maintenance-oriented purposes, political purposes, nor to organizations which discriminate on the basis of race, ethnic origin, sexual or religious preference, age or gender.
Geographic Focus: North Carolina
Date(s) Application is Due: Feb 15; Oct 15
Amount of Grant: 1,500-100,000 USD
Contact: Wachovia Bank, N.A., Trustee; grantinquiries6@wachovia.com
Sponsor: Thomas Austin Finch, Sr. Foundation
Wachovia Bank, NC6732, 100 North Main Street
Winston Salem, NC 27150

Thomas C. Ackerman Foundation Grants 4508
The Thomas C. Ackerman Foundation was founded on August 6, 1991. By Spring, 1992, substantial assets had been received from the Thomas C. Ackerman Trust, funded as a result of Mr. Ackerman's death on February 13, 1991; the Board of Directors was fully constituted; and the guidelines and objectives for grant making were adopted. The Foundation's areas of interest include: arts and culture; education; and health and human services. To be considered for a grant from the Thomas C. Ackerman Foundation, an organization must initially submit a Letter of Intent online. Directions for completing the grant application will be provided to an organization, if it is invited to submit one. Awards generally range from $1,000 to $40,000.
Requirements: California nonprofits, primarily in San Diego County, are eligible.
Restrictions: It is a policy of the Foundation not to provide continuous support for any project to the extent that the project becomes dependent upon the Foundation for its continued existence. The Foundation may make multi-year commitments on occasion. Grants are not made to individuals. Generally, the Foundation does not support conferences or symposia. While occasional grant support is given to religious organizations, those grants are made for direct support of nonsectarian educational or service projects and not for projects which are of primary benefit to members of a particular religion or belief or which primarily promote a particular religion or belief. The Foundation will not consider grants relating to human medical or biomedical research.
Geographic Focus: California
Amount of Grant: 1,000-40,000 USD
Contact: Lynne Newman; (619) 741-0113; info@AckermanFoundation.org
Internet: http://www.ackermanfoundation.org/
Sponsor: Thomas C. Ackerman Foundation
600 W Broadway, Suite 2600
San Diego, CA 92101-3391

Thome Foundation Awards in Age-Related Macular Degeneration Research 4509

The Edward N. & Della L. Thome Memorial Foundation was created in 2002 to advance the health of older adults through the support of direct service projects and medical research on diseases and disorders affecting older adults. In keeping with the Foundation's mission, the goal of the Awards Program is to support translational research that will lead to improved therapies for individuals suffering from age-related macular degeneration (AMD). As steward of the Thome Memorial Foundation, Bank of America N.A., works with The Medical Foundation division's Scientific Review Committees to select the most qualified candidates. Two-year awards of $500,000 ($250,000 per year) will be made to investigators who hold a faculty appointment at a non-profit academic, medical, or research instituion in the United States.

Requirements: Successful research proposals will extend recent basic research findings regarding the underlying mechanisms of AMD. Examples of funding areas include, but are not limited to, genetically engineered animal models, the discovery and testing of small molecule therapies directed at promising targets including immune- or inflammation-related pathways associated with AMD, local drug delivery systems, and neuroprotective strategies. Basic research and clinical trials are outside the scope of this Program. However, clinical studies that involve the evaluation of new imaging modalities with no potential risks to human subjects are eligible. Eligibility is not limited to those investigators currently working in age-related macular degeneration research; scientists who have conducted research exploring the biologic causes of related disorders and/or similar translational research programs are encouraged to apply. All applicants must hold a faculty appointment at a nonprofit academic, medical, or research institution in the United States. Applicants do not need to be nominated by their institutions. United States citizenship is not required; visa documentation is not required. Applications must be submitted by 12:00 Noon, U.S. Eastern time of the published deadline date.

Restrictions: Only one application may be submitted per applicant.

Geographic Focus: All States

Date(s) Application is Due: Jun 13

Amount of Grant: Up to 500,000 USD

Contact: Erin Johnstone; (617) 279-2240 ext. 710; ejohnstone@hria.org

Internet: http://hria.org/tmfservices/tmfgrants.html

Sponsor: Health Resources in Action
95 Berkeley Street
Boston, MA 02116

Thome Foundation Awards in Alzheimer's Disease Drug Discovery Research 4510

The Edward N. & Della L. Thome Memorial Foundation was created in 2002 to advance the health of older adults through the support of direct service projects and medical research on diseases and disorders affecting older adults. In keeping with the foundation's mission, the goal of the research program is to support pilot studies towards innovative drug discovery research that will lead to improved therapies for individuals suffering from Alzheimer's disease. Examples of funding areas include the design, synthesis, and development of target compounds or the modification of existing compounds to improve drug effectiveness and safety as well as other approaches within the field of medicinal chemistry. Researchers dedicated to the validation and testing of target compounds, small molecule therapies, nanotechnologies, or similar techniques are also encouraged to apply. Two-year awards of up to $200,000 ($100,000 per year) will be made (all awards inclusive of 10% indirect costs). Applications must be submitted through the online application system and must be received by 12:00 Noon, U.S. Eastern time of the published deadline date.

Requirements: All applicants must hold a faculty appointment at a non-profit academic, medical, non-governmental or research institution in the United States. U.S. citizenship is not required. Applicants do not need to be nominated by their institutions. Preference will be given to originality of ideas, regardless of faculty seniority.

Restrictions: Genetic studies, biomarker research, neuroimaging, clinical studies, basic research and new target discovery are currently outside the scope of this program.

Geographic Focus: All States

Date(s) Application is Due: Jun 15

Contact: Erin Johnstone; (617) 279-2240, ext. 710; ejohnstone@hria.org

Internet: http://hria.org/tmfservices/tmfgrants/thomead.html

Sponsor: Health Resources in Action
95 Berkeley Street
Boston, MA 02116

Thompson Charitable Foundation Grants 4511

Established in 1987, the Foundation supports organizations involved with education, Christian organizations, Autism research, health, human services, youth, including funding for capital and building improvements for human service organizations and educational institutions. Giving is limited to Bell, Clay, Laurel, and Leslie counties, Kentucky; Anderson, Blount, Knox, and Scott counties, Tennessee; and Buchanan and Tazewell counties, Virginia.

Requirements: Nonprofits operating in Bell, Clay, Laurel, and Leslie counties, Kentucky; Anderson, Blount, Knox, and Scott counties, Tennessee; and Buchanan and Tazewell counties, Virginia are eligible to apply. Initial contact should be a letter, no more then 2 pages long. The letter should include: a statement of the problem, the project will address; a detailed description of project, and amount of funding requested.

Restrictions: No support for religious or political organizations, budget deficits or endowments.

Geographic Focus: Kentucky, Tennessee, Virginia

Date(s) Application is Due: Mar 31; Sep 1

Amount of Grant: 2,000-500,000 USD

Contact: Debbie Black; (865) 588-0491; fax (865) 588-4496; debbie@cf.org

Sponsor: Thompson Charitable Foundation
800 South Gay Street, Suite 2021
Knoxville, TN 37929-9710

Thomson Reuters / MLA Doctoral Fellowships 4512

The purpose of the Thomson Reuters/MLA Doctoral Fellowship is to foster and encourage superior students to conduct doctoral work in health-sciences librarianship or information science by providing support to individuals who have been admitted to candidacy. The award was originally established in 1986 by the Institute of Scientific Information (ISI, which is now a part of Thomson Reuters) and is administered by the Medical Library Association (MLA). The award is granted bi-annually in even-numbered years. The fellowship pays $2,000 in one payment toward the following types of expenses: project expenses; related travel expenses; or augmentation of another larger, separately-funded project relevant to health-science librarianship and which is part of the requirements of a doctoral degree. Guidelines and application materials are available at the website. All materials and letters of support must be received by MLA via email, fax, or mail by December 1. email applications and documents will be accepted as PDF or MS Word files only. Founded in 1898, the Medical Library Association (MLA) is a nonprofit, educational organization of more than 1,100 institutions and 3,600 individual members in the health sciences information field committed to educating health-information professionals, supporting health-information research, promoting access to the world's health-sciences information, and working to ensure that the best health information is available to all.

Requirements: Applicants must satisfy the following eligibility criteria: be graduates of an American Library Association (ALA)-accredited school of library science; be candidates in a Ph.D. program with emphsis on biomedical and health-related information science; and be citizens of or have permanent residence status in the United States or Canada. Preference will be given to applicants who have at least 75% of their course work completed and dissertation prospectus either approved or in the approval process.

Restrictions: Past recipients of the Thomson Reuters / MLA Doctoral Fellowship are ineligible. The following types of expenses are not supported: tuition fees; tuition-related expenses; and living expenses.

Geographic Focus: All States, Canada

Date(s) Application is Due: Dec 1

Amount of Grant: Up to 2,000 USD

Contact: Carla Funk, Program Contact; (312) 419-9094, ext. 14; fax (312) 419-8950; mlapd2@mlahq.org or grants@mlahq.org

Internet: http://www.mlanet.org/awards/grants/index.html

Sponsor: Medical Library Association
65 East Wacker Place, Suite 1900
Chicago, IL 60601-7246

Thoracic Foundation Grants 4513

The major emphasis of the Thoracic Foundation of Boston is to offer support for medical research in thoracic diseases. Contributions are also made to hospitals providing service to thoracic and cardiac cases. Fields of interest include: asthma research; cancer research; children and youth services; heart and circulatory diseases; specialty hospitals; medical school education; and smoking cessation. In addition, funds are awarded to assist with publications, conferences, and seminars in the same field. Grants are awarded thoughout Massachusetts, but primarily in the Boston metropolitan area. A letter of application may be submitted at any time, offering a detailed description of the project and amount of funding requested. Awards range from $5,000 to $75,000.

Geographic Focus: District of Columbia, Massachusetts

Contact: Bradley R. Cook, Trustee, c/o Taylor, Ganson, and Perrin, LLP; (617) 951-2777; fax (617) 951-0989

Sponsor: Thoracic Foundation
160 Federal Street, 20th Floor
Boston, MA 02110-1700

Thornton Foundation Grants 4514

The foundation awards grants to Calilfornia nonprofit organizations in its areas of interest, including arts/ and culture and higher/secondary education. Types of support include annual campaigns, building construction/renovation, continuing support, endowment funds, and research. There are no application deadlines.

Requirements: California nonprofit organizations are eligible.

Geographic Focus: California

Amount of Grant: 1,000-200,000 USD

Contact: Charles Thornton Jr., President; (213) 629-3867

Sponsor: Thornton Foundation
523 W Sixth Street, Suite 636
Los Angeles, CA 90014

Thrasher Research Fund Early Career Awards 4515

The Fund recognizes that young investigators may find it difficult to remain in pediatric research because of a lack of funding. Therefore, the purpose of this program is to encourage the development of medical research in child health by awarding small grants to new researchers, helping them gain a foothold in this important area. The Fund is open to a variety of research topics important to children's health. Both incidence and severity are considered when determining the significance of a problem being studied. The program is particularly interested in applicants that show great potential to impact that field of children's health through medical research. Both an applicant's aptitude and inclination toward research are considered. The quality of the mentor and the mentoring relationship are also considered to be important predictors of success. There are three funding cycles per year for Early Career Award Program grants.

Requirements: Eligible applicants are: physicians who are in a residency/fellowship training program, or who completed that program no more than one year before the date of submission of the concept paper; post-doctoral researchers who received the doctoral level degree no more than three years prior to the date of submission of the concept paper. Each project needs to

be under the guidance of a mentor. The qualifications and experience of the mentor will be considered in the evaluation of the application. A mentor may have only one Thrasher Research Fund Early Career Awardee at a time. There are no restrictions with regard to citizenship. The Fund is open to applications from institutions both inside and outside the United States.
Restrictions: A new investigator who holds a National Institutes of Health (NIH) K award or a Clinical and Translational Science Award (CTSA) is not eligible to apply for the Thrasher Early Career Award.
Geographic Focus: All States, All Countries
Date(s) Application is Due: Mar 30; Jul 27; Nov 30
Contact: Megan Duncan; (801) 240-4720; DuncanME@thrasherresearch.org
Internet: http://www.thrasherresearch.org/sites/www_thrasherresearch_org/Default.aspx?page=231
Sponsor: Thrasher Research Fund
68 S. Main Street, Suite 400
Salt Lake City, UT 84101

Thrasher Research Fund Grants **4516**
The Thrasher Research Fund provides grants for pediatric medical research. Because significant solutions for many children's health problems remain undiscovered, the Fund invites a broad array of applications. The Thrasher Research Fund seeks to foster an environment of creativity and discovery aimed at finding solutions to children's health problems. The Fund awards grants for research that offers substantial promise for meaningful advances in prevention and treatment of children's diseases, particularly research that offers broad-based applications.
Requirements: Principal Investigators must be qualified in terms of education and experience to conduct research. A doctoral-level degree is required. There are no citizenship or residency requirements. The Fund is open to applications from institutions both inside and outside the United States.
Restrictions: The fund does not award grants for educational programs; general operating expenses; general bridge funding for incomplete projects; construction or renovation of buildings or facilities; loans, student aid, scholarships, tuition; or, support of other funding organizations. Grants generally are not awarded for conferences, workshops, or symposia. Proposals in the areas of research on human fetal tissue or behavioral science research will not be considered.
Geographic Focus: All States, All Countries
Amount of Grant: 100,000-350,000 USD
Contact: Megan Duncan; (801) 240-4720; DuncanME@thrasherresearch.org
Internet: http://www.thrasherresearch.org/sites/www_thrasherresearch_org/Default.aspx?page=28
Sponsor: Thrasher Research Fund
68 S. Main Street, Suite 400
Salt Lake City, UT 84101

TIAA-CREF Institute Research Grants **4517**
The program was established to encourage scholars to pursue research in areas of interest to TIAA-CREF, such as retirement income security issues and strategic issues in higher education. esearch findings from many grant projects have been published in journals such as the American Economic Review, Journal of Finance, the Journal of Economic Theory, and the Journal of Financial Services Research.
Requirements: Researchers who are interested in applying for a grant are strongly advised to submit a brief letter of inquiry to determine whether the Institute's present interests and budget allow consideration of the request. The letter should be submitted via email. The letter should include the following four items : 1) a short description of the project, 2) estimated overall budget, 3) period of time for which funds are requested, and 4) names of researchers who will be conducting the research. The Institute will respond to the inquiry and may ask the grant seeker to submit a formal proposal.
Geographic Focus: All States
Contact: Paul Yakoboski; (800) 842-2733, ext. 3663; pyakoboski@tiaa-cref.org
Internet: http://www.tiaa-crefinstitute.org/research/grants/index.html
Sponsor: TIAA-CREF Institute
730 Third Avenue
New York, NY 10017-3206

TIAA-CREF Paul A. Samuelson Award **4518**
The Samuelson Award recognizes outstanding scholarly writing on issues related to lifelong financial security. It is given each year in recognition of an outstanding research publication containing ideas that the public and private sectors can use to maintain and improve Americans financial well being. The work must be theoretical or empirical in nature and must cover a subject directly relevant to lifelong financial security, such as: retiree healthcare issues; the design of public and private programs providing income or health benefits in retirement; the functioning of insurance markets; the design of investment products and strategies for pension funds and other long-term savings plans; the impact of tax and regulatory policies on savings, retirement, and insurance products and programs; and the changing nature of consumption needs and labor market activities over the life cycle. The winner(s) will announced late in the year and is honored at a reception during the annual meeting of the Allied Social Science Association. The application is available at the TIAA-CREF website.
Requirements: Individuals who have published research, in either book or article form, between January 1 of the previous year and June 20 of the current year are eligible.
Restrictions: Publications previously submitted will not be considered. Publications authors by current employees and trustees of TIAA-CREF, judges of the award, or their close relatives will not be considered. Collected paper in an edited volume are not eligible as a submission.
Geographic Focus: All States
Date(s) Application is Due: Sep 10
Amount of Grant: Up to 10,000 USD
Contact: David P. Richardson, Ph.D., Award Coordinator; (800) 842-2252
Internet: http://www.tiaa-crefinstitute.org/awards/samuelson/index.html
Sponsor: Teachers Insurance and Annuity Association College Retirement Equities Fund
730 Third Avenue
New York, NY 10017-3206

Timoshenko Medal **4519**
This medal is bestowed in recognition of distinguished contributions to applied mechanics without restriction to nationality or profession. A certificate is included with the award. Listed deadline is for nomination.
Requirements: Any individual member, group of members, or committee may nominate candidates for this award.
Geographic Focus: All States
Date(s) Application is Due: Oct 1
Amount of Grant: 1,000 USD
Contact: Program Contact; (212) 591-7736; fax (212) 705-7739; ditullio@asme.org or infocentral@asme.org
Internet: http://www.asme.org/honors/ms71/saa/timoshenko.html
Sponsor: American Society of Mechanical Engineers
3 Park Avenue
New York, NY 10016

TIRF Doctoral Dissertation Grants **4520**
Since 2002, The International Research Foundation for English Language Education (TIRF) has supported students completing their doctoral research on topics related to the foundation's priorities. Each year applicants who have been advanced to candidacy in legitimate PhD or EdD programs are invited to submit proposals for Doctoral Dissertation Grants (DDGs). Grants will be awarded in amounts up to $5,000 USD per awardee. Each year the highest rated Doctoral Dissertation Grant proposal is awarded in the name of the late Russell N. Campbell. Dr. Campbell, a Past President of TESOL, was one of the founding fathers of the Foundation.
Requirements: Applicants must be enrolled in a legitimate doctoral program; be advanced to candidacy and have had a research plan approved by a faculty committee at their university; write a proposal which is clearly related to TIRF's research priorities; follow specific instructions located in the call for proposals; and have their research supervisor submit an official letter of support attesting to the applicant's readiness to complete the doctoral dissertation. Please note that DDG applicants do not need to be U.S. citizens. Each year the research priority for competition is announced. Check the website for this year's priority.
Geographic Focus: All States
Date(s) Application is Due: May 14
Amount of Grant: Up to 5,000 USD
Contact: Ryan Damerow, fax 831-647-6650; info@tirfonline.org
Internet: http://www.tirfonline.org/research-grants/doctoral-dissertation-grants/
Sponsor: International Research Foundation for English Language Education
177 Webster Street, #220
Monterey, CA 93940

Tobias Landau Fellowships in Marine Biology **4521**
The foundation awards field study support for marine science research in the United States and Israel. Americans will work at the Heinz Steinitz Marine Biology Laboratory in Eilat, Israel, and Israelis at marine stations in the United States. Cooperative support is welcomed. Research proposals are required; application forms giving details are available from the foundation. All applications are reviewed by an international panel on marine biology. Awards are made on the basis of scientific merit, novelty of the research ideas, and their suitability to the resources and location of the Eilat Station or the appropriate American station if the applicant is an Israeli.
Geographic Focus: All States
Amount of Grant: Up to 5,000 USD
Contact: Dr. Helen Hayes, Director, Tobias Landau Fellowships; (828) 692-5170
Sponsor: Tobias Landau Foundation of the United Jewish Endowment Fund
105 Echo Drive
Hendersonville, NC 28739

Todd Brock Family Foundation Grants **4522**
Established in Texas in 2007, the Todd Brock Family Foundation offers support for educational programs, children, athletics, research, and community projects, as well Protestant agencies and churches. A formal application is required, and applicants should forward the entire proposal to the office. The Foundation rarely offers funding outside of Texas. There are no deadlines for submitting a completed proposal, and grants have most recently ranged from $300 to $250,000.
Geographic Focus: Texas
Samples: Memorial High School Boosters, Houston, Texas. $1,000-funding for annual gold tournament; Legacy Christian Academy, Beaumont, Texas, $250,000-for general operations; Yellowstone Academy, Houston, Texas, $25,000-support for low-income children and families.
Contact: Todd O. Brock, President; (409) 833-6226; fax (409) 832-3019
Sponsor: Todd Brock Family Foundation
1670 E Cardinal Drive, P.O. Box 306
Beaumont, TX 77704-0306

Tourette Syndrome Association Post-Doctoral Fellowships 4523

The Tourette Syndrome Association is interested in supporting any research that has the potential to contribute to the understanding of Tourette Syndrome, including genetics, pathogenesis, pathophysiology, clinical treatments and animal model development. Post-doctoral research fellowships of up to $40,000 for one year are awarded. Applications are evaluated on the following information: candidate's qualifications and objectives; candidate's experience relevant to the project; methodology; significance and/or relevance to the TS field; percentage of time to be devoted to project; adequacy and availability of research facilities and other project support; and the ability to complete the project in a timely manner. The application and deadline information are available at the website.
Requirements: Candidates must have a M.D., Ph.D. or equivalent. Previous experience in the field of movement disorders is desirable, but not essential. International applicants are welcome.
Geographic Focus: All States, All Countries
Contact: Heather Cowley, Fellowship Administrator; (718) 224-2999, ext. 247 or (718) 224-2999, ext. 222; fax (718) 279-9596; heather.cowley@tsa-usa.org
Internet: http://tsa-usa.org/research.html
Sponsor: Tourette Syndrome Association
42-40 Bell Boulevard
Bayside, NY 11361-2820

Tourette Syndrome Association Research Grants 4524

The Tourette Syndrome Association (TSA) is interested in receiving applications from any research study that may contribute to the understanding of Tourette Syndrome, including genetics, pathogenesis, pathophysiology, clinical treatments and animal model development. Research grants of up to $75,000 for one year or up to $150,000 for two years are awarded. The TSA will consider reasonable requests for the following: salaries (NIH salary caps apply); laboratory supplies and other research-related expenses; travel expenses for investigators (up to $1,500 per year); research equipment up to $5,000; and indirect costs of up to 10% may be included. Applicants are evaluated on the following information: candidate's qualifications and objectives; their relevant experience to the project; methodology; significance/relevance to the TS field; percentage of time to be devoted to the project; adequacy and availability of research facilities and other project support; and the ability to complete the project in a timely manner. For preliminary screening, a pre-proposal briefly describing the scientific basis and relevance of the proposed project is required. Preproposal applications are available at the website. Applicants for grants involving clinical treatment are encourage to read the posted clinical studies documents at the website. Payment schedules will vary according to one year or two year awards. Applicants invited to submit a full research proposal will be contacted with instructions.
Requirements: Investigators are required to have an advanced degree, such as a Ph.D., M.D., or equivalent. International applicants are welcome to apply. Investigators for nonprofit and for-profit organizations may apply. Previous experience in the field of movement disorders is desirable, but not essential.
Geographic Focus: All States, All Countries
Date(s) Application is Due: Nov 1
Contact: Heather Cowley, TSA Research Grant Coordinator; (718) 224-2999, ext. 247 or (718) 224-2999, ext. 222; fax (718) 279-9596; heather.cowley@tsa-usa.org
Internet: http://tsa-usa.org/research.html
Sponsor: Tourette Syndrome Association
42-40 Bell Boulevard
Bayside, NY 11361-2820

TRDRP California Research Awards 4525

This program is focused on research projects that address questions specific to tobacco-related disease or tobacco control issues in California. The awards will provide funding for average annual direct costs of $140,000; $170,000 per year for research involving human subjects. Full indirect costs are paid to eligible institutions. Maximum duration is 3 years.
Requirements: Principal Investigators from California not-for-profit organizations are eligible for funding. Research undertaken with TRDRP funds must be conducted primarily in California. The Principal Investigator must conduct and supervise the research project directly and in person. Underrepresented racial and ethnic investigators are encouraged to submit applications. To be competitive, proposals must present a compelling justification, including evidence if available, supporting the California-specific criterion. The proposal should be fully developed, scientifically rigorous, and include sound background information, hypothesis, and promising preliminary studies or preliminary data.
Restrictions: All submitted applications will be screened for their direct relevance to tobacco use or tobacco-related disease. Proposed research that is not relevant will not undergo peer review for scientific merit. It is incumbent on the applicant to make a compelling case that the proposed work is directly relevant to the mission of TRDRP.
Geographic Focus: California
Date(s) Application is Due: Jan 15
Contact: M.F. Bowen, (510) 987-9811; mf.bowen@ucop.edu
Internet: http://www.trdrp.org/fundingopps/callawardmechs.asp#calrchawd
Sponsor: Tobacco-Related Disease Research Program
Office of the President, 300 Lakeside Drive, 6th Floor
Oakland, CA 94612-3550

TRDRP Dissertation Research Awards 4526

This award is intended to support the dissertation research of a doctoral candidate pursuing tobacco-related research. The award is designed for students advanced to candidacy no later than July 1, and initiating their dissertation research. Maximum of $20,000 annual direct costs averaged over the duration of the award (maximum 2 years) for stipend, supplies, and domestic travel. This award also provides an additional maximum of $10,000 per year for tuition/enrollment fee remission, fringe benefits, and health insurance.
Requirements: Applications in all relevant research areas are welcomed, but applications in the social/behavioral sciences and in public policy are encouraged. The applicant and principal mentor must be affiliated with an academic research institution. U.S. citizenship is not a requirement. The candidate must commit a minimum of 80 percent time to the research project.
Restrictions: No equipment purchases are allowed. No indirect costs are allowed.
Geographic Focus: California
Date(s) Application is Due: Jan 15
Amount of Grant: 1-30,000 USD
Contact: M.F. Bowen, (510) 987-9811; mf.bowen@ucop.edu
Internet: http://www.trdrp.org/fundingopps/callawardmechs.asp
Sponsor: Tobacco-Related Disease Research Program
Office of the President, 300 Lakeside Drive, 6th Floor
Oakland, CA 94612-3550

TRDRP Exploratory/Developmental Research Awards 4527

The purpose of these grants is to gather preliminary data or demonstrate proof-of-principle (i.e., pilot projects), or to conduct a research project within the specified limits of money and time. The ultimate goal of these awards is to provide the foundation for proposals for fully-developed research project awards from other funding programs. Total direct costs of $250,000 for the entire project duration (maximum of 2 years). Indirect costs are paid to eligible institutions.
Requirements: Principal Investigators from California not-for-profit organizations are eligible for funding. Research undertaken with TRDRP funds must be conducted primarily in California. The Principal Investigator must conduct and supervise the research project directly and in person. Underrepresented racial and ethnic investigators are encouraged to submit applications. Although Principal Investigators may submit as many proposals for these awards as they wish for separate projects, no more than one Exploratory/Developmental Research Award will be made to any individual PI in the same year. Only proposals that address one of TRDRP's Primary research area will be eligible and will undergo peer review. Exploratory/Developmental Research Awards are for new projects only; that is, competing renewal applications will not be accepted.
Restrictions: All submitted applications will be screened for their direct relevance to tobacco use or tobacco-related disease. Proposed research that is not relevant will not undergo peer review for scientific merit. It is incumbent on the applicant to make a compelling case that the proposed work is directly relevant to the mission of TRDRP.
Geographic Focus: All States
Date(s) Application is Due: Jan 15
Amount of Grant: 250,000 USD
Contact: M.F. Bowen, (510) 987-9811; mf.bowen@ucop.edu
Internet: http://www.trdrp.org/fundingopps/callawardmechs.asp#explordev
Sponsor: Tobacco-Related Disease Research Program
Office of the President, 300 Lakeside Drive, 6th Floor
Oakland, CA 94612-3550

TRDRP New Investigator Awards 4528

This award mechanism is specifically designed to enable new investigators to initiate an independent research program in a tobacco-related field. These are awards for investigators who have a doctoral degree, have not yet served as a principal investigator of a research project grant (e.g., an NIH R01-type grant), and have had no more than five years experience as an independent investigator, i.e., no more than five years since completing formal postdoctoral training, or since the doctoral degree if no postdoctoral training. Direct costs of $90,000 per year can be averaged over the duration of the award. The indirect cost rate is capped at 8% for eligible institutions.
Requirements: New investigators must commit to a minimum of 50 percent time to the research project for the award duration.
Restrictions: Current or past new investigator awardees from TRDRP, other UC Special Research Programs, or other agencies (for example, current or past recipients of NIH K-series career development awards) are not eligible.
Geographic Focus: California
Date(s) Application is Due: Jan 15
Amount of Grant: 90,000 USD
Contact: M.F. Bowen, (510) 987-9811; mf.bowen@ucop.edu
Internet: http://www.trdrp.org/fundingopps/callawardmechs.asp
Sponsor: Tobacco-Related Disease Research Program
Office of the President, 300 Lakeside Drive, 6th Floor
Oakland, CA 94612-3550

TRDRP Participatory Research Awards: CARA/SARA 4529

The purpose of the Community Academic Research Awards (CARA) and the School Academic Research Awards (SARA) is to stimulate and support collaborations between community-based organizations/ schools with academic investigators. These awards are for a collaborative partnership to perform scientifically rigorous research into tobacco control issues that are identified as important and meaningful to specific communities/schools in the state.
Requirements: The applicant partners must demonstrate the use of methods that are relevant, culturally sensitive, and appropriate in terms defined and accepted by the participating communities/schools. Applicants are encouraged to consider TRDRP's Primary research areas and and the TEROC recommended research strategies that lend themselves to a participatory research approach.
Restrictions: All submitted applications will be screened for their direct relevance to tobacco use or tobacco-related disease. Proposed research that is not relevant will not undergo peer review for scientific merit. It is incumbent on the applicant to make a compelling case that the proposed work is directly relevant to the mission of TRDRP.

Geographic Focus: California
Date(s) Application is Due: Jan 15
Amount of Grant: 100,000-300,000 USD
Contact: M.F. Bowen, (510) 987-9811; mf.bowen@ucop.edu
Internet: http://www.trdrp.org/fundingopps/callawardmechs.asp
Sponsor: Tobacco-Related Disease Research Program
Office of the President, 300 Lakeside Drive, 6th Floor
Oakland, CA 94612-3550

TRDRP Postdoctoral Fellowship Awards 4530
These are awards for individuals to obtain postdoctoral research training under a designated mentor. The application must be prepared and submitted exclusively by the fellow and must outline an original research project (separate from the project of a mentor). Letters of support addressing the candidate's training, potential, and the commitment of the mentor and the department to the candidate's career development are essential.
Requirements: Principal Investigators from California not-for-profit organizations are eligible for funding. Research undertaken with TRDRP funds must be conducted primarily in California. The Principal Investigator must conduct and supervise the research project directly and in person. Underrepresented racial and ethnic investigators are encouraged to submit applications. To be eligible, the candidate must be recognized by the applicant institution as a postdoctoral fellow no later than July 1 of the year intended for the fellowship. U.S. citizenship is not a requirement. The fellow must commit a minimum of 75 percent time to the research project. Direct costs of $45,000 per year averaged over the duration of the award. No indirect costs are allowed. Maximum duration is 3 years.
Restrictions: All submitted applications will be screened for their direct relevance to tobacco use or tobacco-related disease. Proposed research that is not relevant will not undergo peer review for scientific merit. It is incumbent on the applicant to make a compelling case that the proposed work is directly relevant to the mission of TRDRP.
Geographic Focus: California
Date(s) Application is Due: Jan 15
Amount of Grant: 45,000 USD
Contact: M.F. Bowen, (510) 987-9811; mf.bowen@ucop.edu
Internet: http://www.trdrp.org/fundingopps/callawardmechs.asp#postdoc
Sponsor: Tobacco-Related Disease Research Program
Office of the President, 300 Lakeside Drive, 6th Floor
Oakland, CA 94612-3550

TRDRP Research Grants 4531
TRDRP will consider unsolicited requests to support research-related activities that build research infrastructure, disseminate research findings, or stimulate new research directions. Support can be requested for scientific conferences to assess tobacco's impact on California's population; or to allow tobacco investigators to evaluate, in a timely manner, new and breaking trends in tobacco control or tobacco-related disease research. The TRDRP Scientific Advisory Committee will make recommendations regarding funding. These opportunities will be limited in number, scope, cost, and duration. Contact a TRDRP research administrator regarding the appropriateness of your proposal prior to submission.
Requirements: In order to qualify for funding, the planned activities must be directly related to one or more of TRDRP's Priority research areas. The activity must primarily take place in California, involve California investigators, and include, where applicable, discussants and speakers funded by TRDRP. Unsolicited requests may be submitted at any time. Requests will be evaluated expeditiously by ad-hoc peer review when appropriate.
Restrictions: All submitted applications will be screened for their direct relevance to tobacco use or tobacco-related disease. Proposed research that is not relevant will not undergo peer review for scientific merit. It is incumbent on the applicant to make a compelling case that the proposed work is directly relevant to the mission of TRDRP.
Geographic Focus: California
Contact: M.F. Bowen, (510) 987-9811; mf.bowen@ucop.edu
Internet: http://www.trdrp.org/fundingopps/callawardmechs.asp
Sponsor: Tobacco-Related Disease Research Program
Office of the President, 300 Lakeside Drive, 6th Floor
Oakland, CA 94612-3550

Trevithick Book Prize Award 4532
This award is given each year for the encouragement of innovative excellence in research in the Committee on Resources and Environment program. The application should consist of a concise 1000-word description of the candidate's research. Clearly drafted diagrams (maximum of four) may be used. The supervisor should append his/her comments on the innovation and quality of the work and its importance to the field. A rigid format is not defined, but the following topics should be covered: aim, methods, results, interpretation of results, and implications for and relations to the general field of study.
Requirements: The competition is open to all graduate students currently registered in the RESR program who are completing their theses prior to April 1 of the current academic year. Those completing their theses after this date are eligible for the competition the following year. A student may enter the competition only once.
Geographic Focus: All States
Date(s) Application is Due: Jun 30
Amount of Grant: Up to 200 CAD
Contact: Connie Baines; (403) 220-5690; cbaines@ucalgary.ca
Internet: http://www.grad.ucalgary.ca/funding/internal_scholarships/lev_4/trevithick.htm
Sponsor: University of Calgary
2500 University Drive NW, Earth Sciences Building, Room 720
Calgary, AB T2N 1N4 Canada

Trinity College Ann Plato Fellowship 4533
The fellowship, named for a 19th-century African-American poet, essayist, and teacher, supports a doctoral student who is an American citizen and is engaged in writing his or her dissertation. Applicants from historically underrepresented groups, including African Americans, Asian Americans, Native Americans, Pacific Islanders, and U.S. Latinos, are especially encouraged to apply. Preference will be given to candidates working in fields currently supported by the college. The fellow enjoys faculty status, delivers a formal, public lecture in the fall semester, and teaches one course in the spring semester. The fellow is expected to be in residence during the fellowship tenure and to become engaged in the Trinity College community. The fellowship provides a stipend; a campus apartment; an office; use of a computer; library privileges at Trinity, including the Watkinson Library, and its consortial colleges; and ready access to Hartford-area archives, including the Connecticut Historical Society, the Wadsworth Atheneum, the state library, the Cities Data Center, and the Harriet Beecher Stowe Center. Appointment is for one academic year. Applicants should send a letter of interest, curriculum vita, a copy of their dissertation proposal, a 10 to 20 page writing sample, and three letters of recommendation to the Ann Plato Search Committee.
Geographic Focus: All States
Date(s) Application is Due: Dec 15
Amount of Grant: 32,000 USD
Contact: Janet Marotto, gradgrnt@indiana.edu
Internet: http://www.indiana.edu/~gradgrnt/v14n2/plato.html
Sponsor: Trinity College
Williams 232
Hartford, CT 06106

Triological Society Research Career Development Awards 4534
The purpose of this award is to provide support for the research career development of otolaryngologists (head and neck surgeons) who have made a commitment to focus their research endeavors on patient-oriented research, e.g., clinical trials, translational research. Research training supported by this award may be related to any research questions relevant to the specialty of otolaryngology, as long as it is demonstrated that the training will have a direct impact on the applicant's ability to pursue his/her long-range research objectives. Grants can be for one to two years, are non-renewable, and have a $40,000 maximum award limitation. There are approximately six grants awarded annually. Letters of Intent are due by December 15, with full proposals received by January 15 annually.
Requirements: Otolaryngologists-head and neck surgeons who hold full-time, part-time or contributed service medical school faculty appointments and who have made a commitment to focus their research endeavors on patient-oriented research may apply. Applicants must be sponsored by the Chair of his/her division or department and by an official representative of the institution which would administer the award and in whose name the application is formally submitted.
Geographic Focus: All States
Date(s) Application is Due: Jan 15; Dec 15
Amount of Grant: Up to 40,000 USD
Contact: Stephanie L. Jones; (703) 519-1586 or (703) 836-4444; sljones@entnet.org
Internet: http://www.triological.org/researchgrants.htm
Sponsor: Triological Society
555 North 30th Street
Omaha, NE 68131

Turtle Conservation Fund Grants 4535
The mission of Turtle Conservation Fund (TCF) is to ensure that no species of tortoise or freshwater turtle becomes extinct and that sustainable populations of all species persist and thrive in the wild. TCF provides the critical funding necessary for facilitating and coordinating global turtle conservation initiatives. As one of its main functions, TCF administers a turtle conservation and research grants program. Awards are granted to organizations or individuals for specific conservation or research projects dealing with highly endangered tortoises or freshwater turtles, but not marine turtles, with either partial or full support as funding allows. TCF Awards at the present time are approximately in the $2,000 to $5,000 range per project, with two funding cycles annually. Total disbursement has averaged between $30,000 to $50,000 per year to date, and is growing as the TCF grows.
Requirements: Priority for funding is given to projects that focus on tortoises or freshwater turtles that are already highly threatened according to the IUCN Red List criteria, either Critically Endangered or Endangered, or which may be highly threatened but not yet officially classified as such on the Red List. Award recipients enter into contractual agreement with either Conservation International or Chelonian Research Foundation to produce the proposed work (these two non-profit organizations serve as TCF's financial umbrella organizations). Award recipients are also encouraged to publish at least partial results of the supported research in the international scientific turtle journal, Chelonian Conservation and Biology, published by Chelonian Research Foundation. Applications must be submitted electronically and must include the required TCF Proposal Cover Sheet
Restrictions: TCF does not provide support for indirect organizational overhead expenses associated with administration of grants, and will not fund requests for such indirect costs.
Geographic Focus: All States
Date(s) Application is Due: May 1; Nov 1
Amount of Grant: 2,000-5,000 USD
Contact: Hugh R. Quinn, Co-Chair; 605-342-0583; DoubleHQ@aol.com
Internet: http://www.turtleconservationfund.org/grants/
Sponsor: Turtle Conservation Fund
2011 Crystal Drive, Suite 500
Arlington, VA 22202

Twentieth-Century Japan Research Awards 4536
Two $1,500 grants are available to support research in the library's Prange Collection and East Asia Collection on topics related to the period of the Allied Occupation of Japan and its aftermath, 1945-1960. Holders of the Ph.D. or an equivalent degree are eligible to apply, as are graduate students who have completed all requirements for the doctorate except the dissertation. The competition is open to scholars in all parts of the world and from any discipline, but historical topics are preferred. Applicants should send a curriculum vitae and a two- to three-page description of their chosen research project.
Restrictions: University of Maryland faculty, staff, and students may not apply.
Geographic Focus: All States
Date(s) Application is Due: Nov 9
Contact: Director; (301) 405-8739; fax (301) 314-9399; historycenter@umd.edu
Internet: http://www.history.umd.edu/HistoryCenter/awards.htm
Sponsor: Nathan and Jeanette Miller Center for Historical Studies
Department of History, 2115 Francis Scott Key Hall, University of Maryland
College Park, MD 20742-7315

U.S. Department of Education American Overseas Research Centers 4537
This program provides grants to establish or operate overseas research centers that promote postgraduate research, exchanges, and area studies. Grants may be used to pay for all or a portion of the cost of establishing or operating a center or program. Costs may include faculty and staff stipends and salaries; faculty, staff, and student travel; operation and maintenance of overseas facilities; teaching and research materials; the acquisition, maintenance, and preservation of library collections; travel for visiting scholars and faculty members who are teaching or conducting research; preparation for and management of conferences; and the publication and dissemination of material for the scholars and general public.
Requirements: Eligible Applicants are Consortia of institutions of higher education that: receive more than 50 percent of their funding from public or private U.S. sources; have a permanent presence in the country in which the center is located; and, are tax-exempt nonprofit organizations described in Section 501(c)3 of the Internal Revenue Code. Note: Competitions for funding are held every four years. The most recent competition: FY 2007.
Geographic Focus: All States
Date(s) Application is Due: Mar 15
Contact: Cheryl Gibbs, (202) 502-7634; fax (202) 502-7859; cheryl.gibbs@ed.gov
Internet: http://www.ed.gov/programs/iegpsaorc/index.html
Sponsor: U.S. Department of Education
1990 K Street NW, 6th Floor
Washington, DC 20006-8521

U.S. Department of Education Disability and Rehabilitation Research and Related Projects 4538
The purpose of the program is to plan and conduct research, demonstration projects, training, and related activities to improve the lives of individuals with disabilities. The projects are quite varied, though all are aimed at fulfilling NIDRR's (National Institute on Disability and Rehabilitation Research) overarching goals of inclusion, integration, employment, and self-sufficiency for people with disabilities.
Requirements: Institutions of Higher Education, States, Nonprofit Organizations are eligible to apply. Public or private agencies—including for-profit agencies, Indian tribes, and tribal organizations may also apply.
Geographic Focus: All States
Contact: Donna Nangle, (202) 245-7462; fax (202) 245-7323; donna.nangle@ed.gov
Internet: http://www.ed.gov/programs/drrp/index.html
Sponsor: U.S. Department of Education
400 Maryland Avenue SW, Room 6030, PCP
Washington, DC 20202-2700

U.S. Department of Education Foreign Languages & Area Studies Fellowships 4539
The mission of the International Education Programs Service (IEPS) is to meet the national needs for expertise and competence in foreign languages and area or international studies. This program provides grants to a select number of public and private institutions of higher education to enable them to offer competitive academic year and summer fellowships to graduate students annually. The disciplinary structure of a FLAS program may be interdisciplinary or multi-disciplinary, including the humanities, social sciences, or other professional studies, and must include the study of the languages of the geographical area of specialization. Institutions compete for fellowship allocations every four years.
Requirements: Eligible applicants include Institutions of higher education that offer performance-based language instruction. A student is eligible to receive a fellowship if he or she: is a graduate student; is a citizen, national, or permanent resident of the United States; is accepted for enrollment or is enrolled in a program that combines modern foreign language training with area or international studies; and shows potential for high academic achievement based on grade point average, class ranking, or similar measures determined by the institution. Institutions should email questions to ope_nrc-flas@ed.gov or visit the IEPS Web site at www.ed.gov/programs/iegpsflasf for more information, including: application deadline, application package, technical assistance, workshop and meeting schedules, and program statute and regulations. Students should visit the IEPS Web site at http://www.ed.gov/programs/iegpsflasf/awards.html for a list of currently funded FLAS institutions and contact the schools directly regarding selection procedures and competition schedules.
Geographic Focus: All States
Contact: Rod Schwartz, (202) 502-7805; rod.schwartz@ed.gov
Internet: http://www.ed.gov/programs/iegpsflasf/index.html
Sponsor: U.S. Department of Education
1990 K Street NW, 6th Floor
Washington, DC 20006-8521

U.S. Department of Education Graduate Assistance in Areas of National Need 4540
This program provides fellowships, through academic departments and programs of IHEs, to assist graduate students with excellent records who demonstrate financial need and plan to pursue the highest degree available in their course study at the institution in a field designated as an area of national need. Grants are awarded to programs and institutions to sustain and enhance the capacity for teaching and research in areas of national need.
Requirements: Institutions of Higher Education (IHEs) may apply. Academic departments and programs of IHEs that provide courses of study leading to a graduate degree may apply. Nondegree granting institutions may submit joint proposals with degree-granting IHEs.
Restrictions: Students are ineligible to apply for funding.
Geographic Focus: All States
Date(s) Application is Due: Mar 16
Amount of Grant: 130,656-261,312 USD
Contact: Gary Thomas, (202) 502-7767 or (202) 502-7779; fax (202) 502-7859; ope_gaann_program@ed.gov
Internet: http://www.ed.gov/programs/gaann/index.html
Sponsor: U.S. Department of Education
1990 K Street NW, 6th Floor
Washington, DC 20006-8521

U.S. Department of Education Jacob K. Javits Fellowships 4541
This program provides fellowships to students of superior academic ability-selected on the basis of demonstrated achievement, financial need, and exceptional promise-to undertake study at the doctoral and Master of Fine Arts level in selected fields of arts, humanities, and social sciences. Panels of experts appointed by the Javits Fellowship Board (Board) select fellows according to criteria established by the Board. Students must demonstrate financial need by filing the Free Application for Federal Student Aid (FAFSA). The selected fields of study supported by the Javits Fellowship Program are available at http://www.ed.gov/programs/jacobjavits/javits-fieldsofstudy.html. For fiscal year 2008, the maximum stipend was $30,000, and the institutional payment was $12,891.
Requirements: Eligibility is limited to U.S. citizens or nationals, permanent residents of the U.S., or citizens of any one of the Freely Associated States. Applicants must be eligible to be accepted to or currently attending a graduate program leading to a doctorate or a terminal master's degree in an eligible field of study at an institution of higher education approved by an accrediting agency recognized by the Secretary of the Department of Education. Eligibility is limited to individuals who, as Javits fellows, will be eligible to receive any grant, loan, or work assistance pursuant to Section 484 of the Higher Education Act of 1965, as amended.
Restrictions: Applicants attending a foreign institution are ineligible. Students who have already received a Javits fellowship in previous years are not eligible.
Geographic Focus: All States
Date(s) Application is Due: Oct 3
Contact: Carmen Gordon, (202) 502-7542; OPE_Javits_Program@ed.gov
Internet: http://www.ed.gov/programs/jacobjavits/index.html
Sponsor: U.S. Department of Education
1990 K Street NW, 6th Floor
Washington, DC 20006-8521

U.S. Department of Education National Resource Centers (NRC) 4542
The program provides grants to establish, strengthen, and operate language and area or international studies centers that will be national resources for teaching any modern foreign language. Grants support: instruction in fields needed to provide full understanding of areas, regions or countries; research and training in international studies; work in the language aspects of professional and other fields of study; and instruction and research on issues in world affairs.
Requirements: Institutions of higher education or consortia of institutions of higher education may apply. A comprehensive or undergraduate National Resource Center (NRC): 1) teaches at least one modern foreign language; 2) provides (a) instruction in fields necessary to provide a full understanding of the areas, regions, or countries in which the languages are commonly used; (b) resources for training and research in international and foreign language aspects of professional and other fields of study; or(c) opportunities for training and research on issues in world affairs that concern one or more countries; 3) provides outreach and consultative services on a national, regional and local basis; 4) maintains linkages with overseas institutions of higher education and other organizations that may contribute to the teaching and research of the Center. Funds for undergraduate travel are allowable only in conjunction with a formal program of supervised study in the subject area on which the center focuses.
Restrictions: Equipment costs exceeding ten percent of the grant are not allowable. Grant funds may not be used to supplant.
Geographic Focus: All States
Date(s) Application is Due: Nov 14
Amount of Grant: 90,000-500,000 USD
Contact: Rod Schwartz, (202) 502-7805; rod.schwartz@ed.gov
Internet: http://www.ed.gov/programs/iegpsnrc/index.html
Sponsor: U.S. Department of Education
1990 K Street NW, 6th Floor
Washington, DC 20006-8521

U.S. Department of Education Native Hawaiian Education Grants 4543
The purpose of this program is to develop innovative education programs to assist native Hawaiians and to supplement and expand programs and authorities in the area of education. Authorized activities include, among others: early education and care programs; family-based education centers; beginning reading and literacy programs; activities to address the needs of gifted and talented native Hawaiian students; special

education programs; professional development for educators; and activities to enable native Hawaiian students to enter and complete postsecondary education programs.
Requirements: Nonprofit Organizations, Native Hawaiian education organizations; Native Hawaiian community-based organizations; public and private nonprofit organizations, agencies, and institutions (including State Education Agencies [SEAs], Local Education Agencies [LEA]s, and Institutions of Higher Education [IHEs]) with experience in developing or operating Native Hawaiian programs or programs of instruction in the Native Hawaiian language; and consortia thereof may apply.
Geographic Focus: All States
Date(s) Application is Due: Mar 24
Contact: Stephen Balkcom, (202) 260-2737; Stephen.Balkcom@ed.gov
Internet: http://www.ed.gov/programs/nathawaiian/index.html
Sponsor: U.S. Department of Education
400 Maryland Avenue SW, Room 3E214
Washington, DC 20202-6200

U.S. Department of Education Rehabilitation Engineering Research Centers Grants 4544

RERCs support activities that: (1) lead to the development of methods, procedures, and devices that will benefit individuals with disabilities, especially those with the most severe disabilities; or (2) involve technology for the purposes of enhancing opportunities for meeting the needs of and addressing the barriers confronted by individuals with disabilities in all aspects of their lives. Types of activities supported include: the development of technological systems for persons with disabilities; stimulation of the production and distribution of equipment in the private sector; and clinical evaluations of equipment. Awards are for five years, except that grants to new recipients or to support new or innovative research may be made for fewer than five years.
Requirements: States, Institutions of Higher Education (IHEs), Nonprofit Organizations, public or private agencies, including for-profit agencies, Indian tribes, and tribal organizations may apply. RERCs must be operated by or in collaboration with one or more institutions of higher education or nonprofit organizations.
Geographic Focus: All States
Date(s) Application is Due: Apr 1; Aug 1
Amount of Grant: 850,000-1,000,000 USD
Contact: Donna Nangle, (202) 245-7462; fax (202) 245-7323; donna.nangle@ed.gov
Internet: http://www.ed.gov/programs/rerc/index.html
Sponsor: U.S. Department of Education
400 Maryland Avenue SW, Room 5051, PCP
Washington, DC 20202-2700

U.S. Department of Education Rehabilitation Research Training Centers 4545

The RRTCs conduct coordinated and advanced programs of research, training, and information dissemination. Each RRTC has a major program of research in a particular area, such as mental illness, vocational rehabilitation (VR), or independent living, which is specified by the National Institute on Disability and Rehabilitation Research (NIDRR). The RRTCs must serve as centers of national excellence and national or regional resources for providers and individuals with disabilities and their representatives. RRTC awards are for five years, except that grants to new recipients or to support new or innovative research may be made for fewer than five years.
Requirements: Note: Each year, competitions are held in specific areas that determine the types of projects. Eligible applicants include Institutions of Higher Education and Nonprofit Organizations. States, public or private agencies-including for-profit agencies, Indian tribes, and tribal organizations may also apply. Rehabilitation research and training centers must be operated by or in collaboration with: (1) one or more institutions of higher education or (2) one or more providers of rehabilitation or other appropriate services.
Geographic Focus: All States
Date(s) Application is Due: Aug 21; Aug 22
Amount of Grant: Up to 675,000 USD
Contact: Donna Nangle, (202) 245-7462; fax (202) 245-7323; donna.nangle@ed.gov
Internet: http://www.ed.gov/programs/rrtc/index.html
Sponsor: U.S. Department of Education
400 Maryland Avenue SW, Room 5051, PCP
Washington, DC 20202-2700

U.S. Department of Education Ronald E. McNair Postbaccalaureate Achievement Grants 4546

This program prepares participants for doctoral studies through involvement in research and other scholarly activities. Participants are from disadvantaged backgrounds and have demonstrated strong academic potential. Institutions work closely with participants as they complete their undergraduate requirements. Institutions encourage participants to enroll in graduate programs and then track their progress through to the successful completion of advanced degrees. The goal is to increase the attainment of Ph.D. degrees by students from underrepresented segments of society. Projects provide: academic counseling, financial aid assistance, mentoring, research opportunities, seminars, summer internships, and tutoring. Guidance for students seeking admission and financial aid for graduate programs also is supported.
Requirements: Institutions of higher education and combinations of institutions of higher education may apply. Students who qualify for McNair must be enrolled in a degree-granting program at an eligible institution. In all projects, at least two-thirds of the participants must be low-income, potential first-generation college students. The remaining participants may be from groups that are underrepresented in graduate education.
Geographic Focus: All States
Amount of Grant: 200,000-300,000 USD
Contact: Eileen Bland, Team Leader (CU.S.T); (202) 502-7730; eileen.bland@ed.gov
Internet: http://www.ed.gov/programs/triomcnair/index.html
Sponsor: U.S. Department of Education
1990 K Street NW, Suite 7000
Washington, DC 20006-8510

U.S. Department of Education United States-Russia Program: Improving Research and Educational Activities in Higher Education 4547

The program provides grants that demonstrate partnerships between Russian and American institutions of higher education that contribute to the development and promotion of educational opportunities between the two nations, particularly in the areas of mutual foreign language learning and the cooperative study of mathematics and science. Collaborative projects comprised of partnerships of Russian and American universities and colleges will be supported.
Requirements: Institutions of Higher Education (IHEs) or combinations of IHEs and other public and private nonprofit institutions and agencies are eligible to apply. Applicants must select one of three academic discipline as the subject area for their grant proposal: (A) Environmental Science Studies; (B) Biotechnology; or, (C) Any discipline, other than (A) and (B). Applications are invited from institutions of higher education with the capacity to contribute to a collaborative project in the areas listed with a Russian institution. The consortium partners, through promoting the study of and communication in foreign languages, are expected to increase awareness and understanding of the two cultures, and to strengthen the professional and scholarly ties between the two countries. Russian institutions will apply to The Russian Ministry of Education and Science for funding.
Restrictions: Any application that proposes a budget exceeding $150,000 for a single budget period of 12 months will be rejected. The Assistant Secretary for Postsecondary Education may change the maximum amount through a notice published in the Federal Register.
Geographic Focus: All States, Russia
Date(s) Application is Due: Jul 7
Amount of Grant: 189,000 USD
Samples: 2008: Stanford University, CA-$188,718
Contact: Krish Mathur, Coordinator, U.S.-Russia Program; (202) 502-7512; fax (202) 502-7877; Krish.Mathur@ed.gov
Internet: http://www.ed.gov/programs/fipserussia/index.html
Sponsor: U.S. Department of Education
1990 K Street NW, Room 6155
Washington, DC 20006-8544

UCLA / RAND Corporation Post-Doctoral Fellowships 4548

The post-doctoral training program is jointly housed in the RAND Health Sciences program and the UCLA School of Public Health. Post-doctoral fellows have the option of pursuing a Masters degree in the UCLA Department of Health Services. The grant covers all tuition and fees, as well as a stipend, which currently ranges from $35,568 to $51,036, depending on the number of years of post-doctoral experience. Travel money is also available for fellows to attend professional meetings.
Requirements: To be eligible, candidates must have obtained a Ph.D., medical degree, or other professional doctoral degree. Awards will last two years. Fellows will be required to attend at least 8 courses at UCLA or RAND, including three courses in health services research methods, two in statistics, and three in health services issue areas. Those pursuing a Masters degree must complete a total of 12 courses. Upon arrival, fellows will align with an ongoing research project at RAND or UCLA. RAND and UCLA have numerous ongoing projects in the health services research area. These fall into several areas, including: access/utilization among the underserved; assessing quality, effectiveness, outcomes; health behaviors; alternative delivery systems; hospital and physician payment; health care cost containment; technology assessment; private health insurance; international comparisons. By the end of the first year, the fellow will identify a research idea and write a proposal. The proposal may be addressed to the principal investigator of an ongoing research project, or to an outside funding agency. By the end of the program, the fellow will prepare an article on research conducted during the fellowship, and submit it to a peer reviewed journal. This can either be done alone, or in collaboration with the research director or other member of the fellow's research team.
Geographic Focus: All States
Amount of Grant: 35,568-51,036 USD
Contact: Communications Office; (310) 393-0411, ext. 7775; RAND_Health@rand.org
Internet: http://www.ph.ucla.edu/hs/post_doctoral.html
Sponsor: RAND Corporation
1776 Main Street, P.O. Box 2138
Santa Monica, CA 90407-2138

UCLA Research Grant Program in Ethnic Studies 4549

The institute, in conjunction with the the American Indian Studies Center, the Asian American Studies Center, Bunche Center for African American Studies, and the Chicano Studies Research Center, invites applications for support of research on African Americans, American Indians, Asian Americans, and Chicanas/os. The IAC also is working with the University of California Committee on Latino Research and the Chicano Studies Research Center to offer research grants in Latino Policy Studies. Prior to submission of the application, applicants should discuss their proposal with the assistant director or director of the appropriate center, or in the case of interethnic proposals, with the assistant directors or directors of the centers involved. The program is on a reimbursement basis only. Guidelines and application are available online.
Requirements: UCLA faculty, staff, graduate students, and IAC postdoctoral fellows are eligible.

Restrictions: Funds for the purchase of permanent equipment will be provided only under exceptional circumstances. Conference travel, whether the applicant is presenting or attending, is ineligible.
Geographic Focus: All States
Date(s) Application is Due: Apr 28
Amount of Grant: Up to 7,000 USD
Contact: Ethnic Studies Contact; (310) 825-1233; IACcoordinator@gdnet.ucla.edu
Internet: http://www.gdnet.ucla.edu/iacweb/rgweber.htm
Sponsor: Institute of American Cultures
1237 Murphy Hall, Box 951419
Los Angeles, CA 90095-1419

UCR Graduate Student in the Humanities Fellowship **4550**
The College of Humanities, Arts and Social Sciences and the Center for Ideas and Society are pleased to announce a new one-year fellowship for exceptional graduate students in the humanities who have advanced to candidacy and are in the final stages of completing their dissertations. This program is designed to support advanced doctoral students and to encourage the collaboration, interdisciplinary dialogue, and innovation that are fundamental to research in the arts and humanities in the University of California. Along with Faculty Research Fellows, Graduate Fellows will be part of a new UC Society of Fellows in the Humanities. During the fellowship year, UC Graduate Fellows will be affiliated with the Center for Ideas and Society, and are expected to give a lecture sometime during their fellowship year. They will also be asked to take part in the activities of the Center, and will be eligible for workspace within it. Graduate Fellows will be invited to participate in an annual meeting of UC Faculty Research Fellows and UC Graduate Fellows and to submit a brief essay for inclusion in an e-publication disseminated by the UC Humanities Research Institute. Information about the application is available at the website.
Requirements: Graduate students need not apply separately for the UC Graduate Fellowship in the Humanities. The Center will select Fellowship holders from among those who apply for the Graduate Division's Dissertation Year Program Awards (http://graduate.ucr.edu/fin_aid.html).
Restrictions: Applicants must be humanities graduate students at the UCR.
Geographic Focus: All States
Contact: Laura Lozon, Associate Director; (951) 827-1555; laura.lozon@ucr.edu
Internet: http://www.ideasandsociety.ucr.edu/fellowships_grants/UC_grad_fellows/index.html
Sponsor: University of California, Riverside
900 University Avenue
Riverside, CA 92521

UCR Humanities Graduate Student Research Grants **4551**
The Humanities Graduate Student Research Grants assist graduate students involved in original research or creative projects in the humanities, as broadly defined by the National Endowment for the Humanities. The bill creating the National Endowment for the Humanities indicates "...the term humanities includes, but is not limited to, the study of the following: language; linguistics; literature; history; jurisprudence; philosophy; archaeology; comparative religion; ethics; the history, criticism, theory and practice of the arts; those aspects of the social sciences which have humanistic content and employ humanistic methods; and the study and application of the humanities to the human environment with particular attention to the relevance of the humanities to the current condition of national life..." The Endowment supports projects involving additional humanities-based disciplines not explicitly named in the law, such as the history of science and cultural anthropology. These awards are primarily intended for research proposals that relate to a student's dissertation work. However, if funds permit, other proposals will be considered. The application and specific guidelines regarding research protocol and expenditures are available at the website.
Requirements: Applicants must be graduate students enrolled in the humanities at UCR.
Restrictions: Expenditures must be within the categories budgeted in the application. Written permission is required for budget adjustments prior to spending funds. Major purchases of non-expendable equipment will not be funded. Reusable supplies and equipment purchased from these funds remain the property of UCR. Research grants are neither for stipends nor for the actual preparation of thesis copy such as typing, charts, or photographs.
Geographic Focus: All States
Date(s) Application is Due: Mar 15
Amount of Grant: Up to 1,500 USD
Contact: Laura Lozon, Associate Director; (951) 827-1555; laura.lozon@ucr.edu
Internet: http://www.ideasandsociety.ucr.edu/fellowships_grants/grad_research_grants/index.html
Sponsor: University of California, Riverside
900 University Avenue
Riverside, CA 92521

UCR Resident Fellowships **4552**
Each quarter four UCR faculty are awarded Resident Fellowships to support collaborative research. The purpose of this program is to promote multi-disciplinary and cross-disciplinary humanistic scholarship by assembling groups of scholars from different disciplines to examine areas of common interest. Normally fellowships are awarded to groups, not to individuals, and should include at least some members from the College of Humanities, Arts, and Social Sciences. Preference will ordinarily be given to those who have not been Resident Fellows. Home departments are provided with $5,000 funding for release time for the Faculty Fellows. In exchange, the fellow recipients are relieved of departmental and teaching obligations during their residence, which is normally during the quarter with the lightest teaching load. Fellows are provided with work space at the Center for Ideas and Society and meet in seminar at least once a week. The format of the meeting as well as the duration is at the discretion of the Fellows. Resident Fellows are expected to take part in the Center's activities, such as attending Center sponsored talks. Faculty who receive Fellowships become eligible again in four years. The application package is available at the website.
Requirements: Applicants must be faculty members from the College of Humanities, Arts, and Social Sciences.
Restrictions: Preference will be given to those who have not been Resident Fellows.
Geographic Focus: All States
Date(s) Application is Due: Dec 5
Amount of Grant: 5,000 USD
Contact: Laura Lozon, Assistant Director; (951) 827-1555; fax (951) 827-6377; laura.lozon@ucr.edu or ideassoc@citrus.ucr.edu
Internet: http://www.ideasandsociety.ucr.edu/fellowships_grants/resident_fellows/index.html
Sponsor: University of California, Riverside
900 University Avenue
Riverside, CA 92521

UE Explore Undergraduate Research Grants **4553**
This an undergraduate research program designed to provide students and faculty the opportunity to work together on research projects. The ideas for the research projects may be generated by the students or the faculty. There are both summer projects and semester projects, which are performed during the academic year. For the on-campus summer projects, students receive a stipend for 10 weeks, free residence housing during that period, and grant money to pay for literature searches, books, and equipment. Students receive grant money for the semester projects. A wide range of projects have been funded that cover the humanities, arts, sciences, and professional schools. Learn more about who is eligible and how to obtain funds through the web site.
Requirements: Students who satisfy and one of the following criteria are welcome to apply: students in good standing in the Honors Program; students who have a 3.5 GPA or above (based on at least 60 semester hours of which 30 must be earned at UE) (these hour requirements must be satisfied before the beginning of the project); academic scholarship winners (trustee, president or faculty); students recommended by consensus of as academic department or school (the consensus must be verified in writing from the department chair or dean).
Geographic Focus: All States
Amount of Grant: Up to 5,500 USD
Contact: Program Contact; (800) 423-8633; admission@evansville.edu
Internet: http://www2.evansville.edu/undergraduateresearch
Sponsor: University of Evansville
1800 Lincoln Avenue
Evansville, IN 47722

UEF Advancement of Engineering Grants **4554**
The purpose of these grants is to provide either start-up funding for promising programs, or incremental funding for significant opportunities in established programs. Major impact must be possible within the level of UEF's, possibly leveraged, funding. National programs are preferable to regional ones. Interengineering programs are preferable to single discipline programs. Proposals in the following areas are encouraged but all proposals are welcome: precollegiate technical programs; midcareer assistance for engineers; opportunities for retired engineers to volunteer their skills; public awareness of engineers and engineering. Listed deadline is for concept letters; full proposals are by invitation.
Geographic Focus: All States
Date(s) Application is Due: Jun 1
Amount of Grant: 20,000-50,000 USD
Contact: Dr. David Belden; (973) 244-2328; fax (973) 882-5155; engfnd@aol.com
Internet: http://www.uefoundation.org
Sponsor: United Engineering Foundation
P.O. Box 70
Mount Vernon, VA 22121-0070

UICC American Cancer Society International Fellowships for Beginning Investigators **4555**
The international fellowships fund 12-month basic or clinical research projects in the areas of epidemiology, prevention, detection, diagnosis, treatment, and psycho-oncology that foster a bidirectional flow of knowledge, experience, expertise, and innovation to and from the United States. Candidates should be in the early stages of their careers. Applications that are geared to the development of specific cancer control measures in developing and central and east European countries are particularly encouraged. Six to eight fellowships, with tenures beginning in April, will be awarded each year.
Requirements: Eligible candidates should hold assistant professorships or similar positions at their home institutes and have, ordinarily, a minimum of two and a maximum of ten years of postdoctoral experience after obtaining their MD or PhD degrees or equivalents.
Restrictions: Candidates who are physically present at the proposed host institute while their applications are under consideration are not eligible for UICC fellowships.
Geographic Focus: All States
Date(s) Application is Due: Dec 1
Amount of Grant: 45,000 USD
Contact: Beate Vought; +41 22 809 1843; fax +41 22 809 1810; fellows@uicc.ch
Internet: http://www.uicc.org/index.php?option=com_content&task=view&id=14259&Itemid=182
Sponsor: International Union Against Cancer
62, route de Frontenex
Geneva, CH-1207 Switzerland

UICC International Cancer Technology Transfer (ICRETT) Fellowships 4556
The program helps to facilitate rapid international transfer of cancer research techiques and technology and clinical management skills. It is hoped that funding will support the exchange of knowledge and enhance skills in basic, clinical, behavioral, and epidemiological areas of cancer research, in cancer control and prevention (including tobacco control). Participants should acquire appropriate clinical management, diagnostic and therapeutic skills for effective application and use in the home organization upon their return. There are 120 to 150 awards available annually, with applications accepted at any time and notification generally within 60 days of registration.
Requirements: Eligible applicants include appropriately qualified investigators, pathologists, epidemiologists, cancer registrars, behavioral scientists, tobacco control activists, laboratory technicians, clinicians and general practitioners, generally in the early stages of their careers.
Geographic Focus: All States
Amount of Grant: 3,400 USD
Contact: Beate Vought; +41 22 809 1843; fax +41 22 809 1810; fellows@uicc.ch
Internet: http://www.uicc.org/index.php?option=com_content&task=view&id=14262&Itemid=185
Sponsor: International Union Against Cancer
62, route de Frontenex
Geneva, CH-1207 Switzerland

UICC Raisa Gorbachev Memorial International Cancer Fellowships 4557
This award is funded by UICC International Cancer Foundation and enables qualified beginning pathologists to carry out specific research or training project relating to the many fields of oncology, and should be centred on an appropriate training period in modern diagnostic tools in the field of pathology. The Fellowship is conditional on Fellows returning to the home institutes/country at the end of the fellowship period. The original Fellowships cannot be prolonged, or run concurrently with other awards, even those funded by other agencies. They can however, be extended subject to the written approval of home and host supervisors and at no additional cost to UICC i.e. up to a total duration of 3 months. Should return of the Fellow to their home country be delayed past these 3 months, 50 percent of the travel award must be reimbursed to UICC.
Requirements: The candidate must possess the appropriate professional qualifications and experience according to the specifications described on the website and must currently be engaged in cancer research, clinical oncology practice, or cancer society work. To permit effective communication at the host institute, the candidate must have adequate fluency in a common language. The candidate must also be on the staff payroll of a university, research laboratory or institute, hospital, oncology unit to where they will return at the end of a fellowship.
Restrictions: Candidates attached to commercial entities are not eligible. Fellowships cannot be granted to candidates who are already physically present at the proposed host institute whilst their applications are under consideration.
Geographic Focus: All States
Contact: Beate Vought; +41 22 809 1843; fax +41 22 809 1810; fellows@uicc.ch
Internet: http://www.uicc.org/index.php?option=com_content&task=view&id=15956&Itemid=301
Sponsor: International Union Against Cancer
62, route de Frontenex
Geneva, CH-1207 Switzerland

UICC Yamagiwa-Yoshida Memorial International Cancer Study Grants 4558
These study grants enable cancer investigators from any country to establish bilateral research projects abroad that exploit complementary materials or skills, including advanced training in experimental methods or special techniques. There are 14 to 16 three-to six-month grants awarded each year. Selection takes place twice annually, with recipients notified in mid-April and mid-October. Grants should be taken up within six months following notification. Travel awards contribute toward the least expensive international two-way airfare or other appropriate form of transport. Financial support is not provided for dependents. Application materials are available on the Web site.
Requirements: Candidates should have appropriate scientific or medical qualifications and be actively engaged in cancer research.
Geographic Focus: All States
Date(s) Application is Due: Jan 15; Jul 1
Amount of Grant: 30,000-60,000 USD
Contact: Beate Vought; +41 22 809 1843; fax +41 22 809 1810; fellows@uicc.ch
Internet: http://www.uicc.org/index.php?option=com_content&task=view&id=15958&Itemid=302
Sponsor: International Union Against Cancer
62, route de Frontenex
Geneva, CH-1207 Switzerland

UIC Rockefeller Foundation Resident Fellowships in the Humanities 4559
The Latin American and Latino Studies Program at the university will host a Rockefeller Humanities Residency site. The program seeks one junior and one senior fellow whose research interests explore the diversification and internal hybridity of Latino communities in Chicago. Themes explored may include interlatino relations, interracial dynamics, and interactions and transculturations between native-born Latinas/os and more recent Latin American immigrants. The fellowships include stipend, fringe benefits, and travel and research expenses. Guidelines are available online.
Geographic Focus: All States
Date(s) Application is Due: Feb 1
Amount of Grant: 37,100-47,700 USD
Contact: Lorena Garca, (312) 996-8749; fax (312) 996-1796; lorena@uic.edu
Internet: http://www.uic.edu/las/latamst/rockefeller1.htm
Sponsor: University of Illinois at Chicago
1527 University Hall, 601 S Morgan Street
Chicago, IL 60607-7115

UNC Carolina Postdoctoral Program for Faculty Diversity 4560
As part of a continuing commitment to building a culturally diverse intellectual community and advancing scholars from underrepresented groups in higher education, The University of North Carolina at Chapel Hill (UNC) Carolina Postdoctoral Program for Faculty Diversity makes available postdoctoral research appointments for a period of two years. The purpose of the program is to develop scholars from underrepresented groups for possible tenure track appointments at UNC and other research universities. Postdoctoral scholars will be engaged full-time in research and may teach only one course per fiscal year. The program supports 10 postdoctoral scholars, engaged full-time in research and writing for a two-year term starting July 1. Fellows are provided in the program with research funds, program laptop, as well as department mentor, office, computer, health benefits and $39,874 annual salary. Meetings and workshops are structured for the purposes of networking, social interaction and supporting professional development of diversity postdoctoral scholars within the program.
Requirements: Applicants who will have completed their doctoral degree no later than July 1 of the current academic year and no earlier than July 1 of four years prior to the current academic year are eligible to apply. The primary criterion for selection is evidence of scholarship potentially competitive for tenure track appointments at UNC and other research universities. An important secondary criterion is the support of prospective departments. Preference will be given to U.S. citizens and permanent residents. UNC strongly encourages applications from African American, Native American and Hispanic scholars.
Geographic Focus: All States
Date(s) Application is Due: Jan 4
Amount of Grant: 39,874 USD
Contact: Judson H. Fraley; (919) 962-9982; jhfraley@email.unc.edu
Internet: http://research.unc.edu/carolina-postdocs/
Sponsor: University of North Carolina at Chapel Hill
301 Bynum Hall, CB #4100
Chapel Hill, NC 27599-4100

UNCFSP NAI-MIRS Summer Sabbatical Awards 4561
The NASA Astrobiology Institute Minority Institution Research Support (NAI-MIRS) Summer Sabbatical Awards provide opportunities for researchers working in minority institutions to collaborate with NASA Astrobiology Institute (NAI) Research Teams. In 2005, NAI partnered with Tennessee State University (TSU) to broaden the program by organizing summer sabbaticals for faculty from all minority serving institutions at selected NAI Team sites. The goals of NAI-MIRS are to give faculty at minority serving institutions an opportunity to expand their research, to increase diversity at all levels of the NAI, and to strengthen the participation of minority institutions in the field of astrobiology through research opportunities and ongoing collaborations. See website for application information.
Requirements: Applicants must be: faculty members at a U.S. college or university designated by the U.S. Department of Education as a Minority Serving Institution; able to provide a commitment from their institution to include astrobiology in university programs; qualified to conduct astrobiology competitive research in the host and their home institution; U.S. Citizen or U.S. Permanent Resident.
Geographic Focus: All States
Date(s) Application is Due: Mar 16
Amount of Grant: 26,500 USD
Contact: Jacqueline Howard-Matthews, Director; (703) 205-8145 or (703) 677-3400; fax (703) 205-7645; jhowardmatthews@uncfsp.org
Internet: http://www.uncfsp.org/spknowledge/default.aspx?page=program.view&areaid=5&contentid=570&type=ANMT&nview=y
Sponsor: United Negro College Fund Special Programs Corporation
2750 Prosperity Avenue, Suite 600
Fairfax, VA 22031

UNCFSP NASA Administrator's Fellowship Program 4562
As one of UNCFSP's premier fellowships, the NASA Administrator's Fellowship Program (NAFP) enhances the professional development for NASA employees as well as STEM faculty from Minority Institutions (MIs) while building and strengthening the STEM infrastructure at MIs. NASA employee fellows teach and/or conduct research and STEM faculty fellows conduct research at NASA Centers. All NAFP fellows engage in a comprehensive, integrated professional development curriculum to equip them with the skills to become progressive change agents in their organizations and institutions.
Requirements: Open to U.S. citizens, including residents of Puerto Rico, the Virgin Islands, and other U.S. territories. Applicants must hold a Ph.D., Sc.D, or equivalent (less than 20 years) in a STEM discipline OR have expertise in a NASA-related discipline.
Geographic Focus: All States
Date(s) Application is Due: Jan 21
Contact: Melissa C. Green, Director; (703) 205-7636 or (703) 677-3400; fax (703) 205-7645; Melissa.Green@uncfsp.org
Internet: http://www.uncfsp.org/nasa/nafp
Sponsor: United Negro College Fund Special Programs Corporation
2750 Prosperity Avenue, Suite 600
Fairfax, VA 22031

UNCFSP NASA Harriett G. Jenkins Predoctoral Fellowships 4563

The mission of the program is to increase the numbers of women, minorities, and persons with disabilities participating in the math, science, engineering, and technology (MSET) disciplines. Up to 20 fellows will be selected annually to receive three-year fellowships to support graduate education leading to doctoral degrees in NASA-related disciplines, including aeronautics/aerospace, environnemental sciences, astronomy, life sciences, bioengineering, materials sciences, biology, mathematics, chemistry, meteorology, computers/computer science, physical sciences, earth sciences, physics/health physics, engineering, and science education. Application and guidelines are available online.

Requirements: Applicants must meet the following: must be a member of an underrepresented group, which includes women, ethnic minorities and persons with disabilities majoring in a Science, Technology, Engineering or Math (STEM) discipline; must either be within the first 3 years of a graduate program at an accredited, U.S. university OR be in the process of applying to an accredited STEM graduate program; cannot participate in any other federally-funded fellowship or scholarship program; fellows are required to be enrolled as a full-time student. Although the fellowship is not renewable during the three-year period, the availability of support for subsequent years is 1) subject to availability of funds and 2) contingent upon certification to NASA by the fellow's institution that the fellow is making satisfactory academic progress toward an advanced degree

Geographic Focus: All States

Date(s) Application is Due: Feb 15

Amount of Grant: 16,000-22,000 USD

Contact: Melissa C. Green, Director; (703) 205-7636 or (703) 677-3400; fax (703) 205-7645; Melissa.Green@uncfsp.org

Internet: http://www.uncfsp.org/spknowledge/default.aspx?page=program.view&areaid=1&contentid=177&typeid=jpfp

Sponsor: United Negro College Fund Special Programs Corporation

2750 Prosperity Avenue, Suite 600

Fairfax, VA 22031

UNCFSP NASA Science and Technology Institute Summer Scholars 4564

The NASA Science and Technology Institute (NSTI) Summer Scholars Program, a component of the Institute, is a ten-week summer research experience for undergraduates in science, technology, engineering, and mathematics (STEM) disciplines at a NASA Center. NSTI Scholars will be matched with a mentor to conduct NASA-relevant research. Scholars will be placed at one of three NASA centers including: Ames Research Center in Moffett Field, CA, Johnson Space Center in Houston, TX or Glenn Research Center in Cleveland, OH.

Requirements: Applicants must meet the following: must be a member of an underrepresented group, which includes women, ethnic minorities and persons with disabilities majoring in a Science, Technology, Engineering or Math (STEM) discipline; be a current college freshmen, sophomore or junior; maintain a minimum 3.0 on a 4.0 grading scale;

Geographic Focus: All States

Date(s) Application is Due: Feb 20

Contact: Melissa C. Green, Director; (703) 205-7636 or (703) 677-3400; fax (703) 205-7645; Melissa.Green@uncfsp.org

Internet: NASA Science and Technology Institute (NSTI) Summer Scholars Program

Sponsor: United Negro College Fund Special Programs Corporation

2750 Prosperity Avenue, Suite 600

Fairfax, VA 22031

UNC School of Information and Library Science Doctoral Fellowships 4565

The School of Information and Library Science (SILS) at the University of North Carolina seeks applicants interested in earning a doctoral degree in digital curation. The fellowship offers the following: a 20 hour a week position as a Research Fellow in digital curation; an annual two year stipend; health coverage with in-state tuition; and extensive opportunities to meet leaders in the digital curation research and practice areas with workshops and symposia.

Requirements: Applicants should follow the regular application procedures found on the SILS Ph.D. admissions page. Students are encouraged to apply early, as this ensures consideration for the greatest amount of university funding. In addition to the required written statement of the intended research focus, applicants will write a separate essay elaborating on their educational goals and how they are related to the goals of DigCCurr II. See the DigCCurr II website and promotional flyer for more details, including current deadlines and application.

Restrictions: Applicants must be U.S. citizens.

Geographic Focus: All States

Amount of Grant: 38,000 USD

Contact: Dr. Helen Tibbo, Fellowship Contact; (919) 962-8063; tibbo@email.unc.edu

Dr. Cal Lee, Fellowship Contact; (919) 962-7024; callee@ils.unc.edu

Internet: http://ils.unc.edu/digccurr/aboutII.html

Sponsor: University of North Carolina-Chapel Hill

Campus Box 3360, Manning Hall

Chapel Hill, NC 27599-3360

UNESCO Co-Sponsored Fellowships 4566

UNESCO has adopted a proactive policy towards the encouragement of partnerships with governments, institutions of higher education and foundations in order to increase training opportunities for other Member States. Under this policy, Member States, foundations, institutions and private donors may offer fellowships (contributions-in-kind), or may finance fellowships (cash contributions, or either through cost-sharing arrangements and through total funding) in fields within UNESCO's competence. The terms for each program under the Co-sponsored Fellowships Scheme are negotiated and agreed upon with each donor. UNESCO cooperates with the donor country or institution in administering the training scheme jointly. The Organization announces the scheme to the targeted Member States, solicits and collects applications, and prepares a short-list of candidates, which is then submitted to the donor to make the final selection. The Fellowships Section then grants the fellowships to the selected beneficiaries. Deadlines and Fellowship amounts vary by country and particular Fellowship. Additional information is available at the website.

Geographic Focus: All Countries

Contact: Irina Bokova; +33 (0) 1 45 68 10 00; fax +33 (0) 1 45 66 94 29

Internet: http://portal.unesco.org/en/ev.php-URL_ID=14133&URL_DO=DO_TOPIC&URL_SECTION=201.html

Sponsor: United Nations Educational, Scientific, and Cultural Organization

7 place de Fontenoy

Paris 07 SP, 75352

UNESCO Fellowships 4567

The UNESCO Fellowships are designed to provide an opportunity for study and training to qualified persons who are seeking to pursue advance studies in the fields of education, natural sciences, social and human sciences, culture, communication, information and informatics. Deadlines and placement of Fellows vary by country and fields of discipline. The application and detailed guidelines are available at the website.

Requirements: All applications must be channeled through the National Commissions of the candidate's country and must be addressed to the Director-General of UNESCO. Applicants must have exceptional merit, and demonstrate that they possess outstanding potential to contribute to the social, cultural and/or economic development of their country to be considered for a UNESCO fellowships. Candidates are usually persons who have already begun their careers, are at an advanced stage of their university education, and are pursuing short term post graduate studies of six months or less. Priority is given to women, youth, the disadvantaged, underdeveloped countries, and member states of Africa. Candidates should be in good health and no more than 45 years of age.

Restrictions: The UNESCO Secretariat cannot accept requests directly from individuals. Undergraduate students who wish to pursue studies in their own country cannot be beneficiaries of UNESCO fellowships.

Geographic Focus: All States, All Countries

Contact: Irina Bokova; +33 (0) 1 45 68 10 00; fax +33 (0) 1 45 66 94 29

Internet: http://portal.unesco.org/en/ev.php-URL_ID=17616&URL_DO=DO_TOPIC&URL_SECTION=201.html

Sponsor: United Nations Educational, Scientific, and Cultural Organization

7 place de Fontenoy

Paris 07 SP, 75352

Union Pacific Foundation Community and Civic Grants 4568

The Foundation has a strong interest in promoting organizational effectiveness among nonprofits. To that end, this Foundation dedicates the majority of their grants to help nonprofit organizations build their capacity, increase their impact, and operate more efficiently and effectively. Grants are made primarily to proposals in the areas of community and civic, and health and human services. The community and civic grants category focuses on assisting community-based organizations and related activities that improve and enrich the general quality of life in the community. This category includes organizations such as aquariums, botanical gardens, children's museums, history/science museums, public libraries, public television and radio, and zoos.

Requirements: The Foundation will accept only online applications; printed copies of the application are not available and will not be accepted. Grants are made to institutions located in communities served by Union Pacific Corporation and its operating company Union Pacific Railroad Company.

Restrictions: The Foundation will not consider a request from or for: individuals; organizations/projects/programs that do not fit within the Foundation's funding priorities; organizations without a Section 501(c)3 public charity determination letter from the Internal Revenue Service; organizations that channel grant funds to third parties; organizations whose dominant purpose is to influence legislation or participate/intervene in political campaigns on behalf of or against any candidate for public office; organization/projects/programs for which the Foundation is asked to serve as the sole funder; organizations that already have an active multi-year Union Pacific Foundation grant; religious organizations for non-secular programs (i.e. programs which promote religious doctrine); organizational deficits; local affiliates of national health/disease-specific organizations; non-U.S.-based charities; organizations whose program activities are mainly international; elementary or secondary schools; athletic programs or events; donations of railroad equipment; conventions, conferences or seminars; fellowships or research; loans; labor organizations; or organizations whose programs have a national scope.

Geographic Focus: Arizona, Arkansas, California, Colorado, Idaho, Illinois, Iowa, Kansas, Louisiana, Minnesota, Missouri, Montana, Nebraska, Nevada, New Mexico, Oklahoma, Oregon, Tennessee, Texas, Utah, Washington, Wisconsin, Wyoming

Date(s) Application is Due: Aug 15

Contact: Darlynn Myers, Director; (402) 271-5600; fax (402) 501-2291; upf@up.com

Internet: http://www.up.com/found/grants.shtml

Sponsor: Union Pacific Foundation

1400 Douglas Street, Stop 1560

Omaha, NE 68179

Union Pacific Foundation Health and Human Services Grants 4569

The Foundation has a strong interest in promoting organizational effectiveness among nonprofits. To that end, this foundation dedicates the majority of their grants to help nonprofit organizations build their capacity, increase their impact, and operate more

efficiently and effectively. Grants are made primarily to proposals in the areas of community and civic, and health and human services. The health and human services category assists organizations dedicated to improving the level of health care or providing human services in the community. Types of support include general operating support, continuing support, capital campaigns, building construction and renovation, curriculum development, equipment acquisition, program development, and matching funds.
Requirements: The Foundation will accept only online applications; printed copies of the application are not available and will not be accepted. Grants are made to institutions located in communities served by Union Pacific Corporation and its operating company Union Pacific Railroad Company.
Restrictions: The Foundation generally will not consider a request from or for: individuals; organizations/projects/programs that do not fit within the Foundation's funding priorities; organizations without a Section 501(c)3 public charity determination letter from the Internal Revenue Service; organizations that channel grant funds to third parties; organizations whose dominant purpose is to influence legislation or participate/intervene in political campaigns on behalf of or against any candidate for public office; organization/projects/programs for which the Foundation is asked to serve as the sole funder; organizations that already have an active multi-year Union Pacific Foundation grant; religious organizations for non-secular programs; organizational deficits; local affiliates of national health/disease-specific organizations; non U.S.-based charities; organizations whose program activities are mainly international; elementary or secondary schools; athletic programs or events; donations of railroad equipment; conventions, conferences or seminars; fellowships or research; loans; labor organizations; or organizations whose programs have a national scope.
Geographic Focus: Arizona, Arkansas, California, Colorado, Idaho, Iowa, Kansas, Louisiana, Minnesota, Missouri, Montana, Nebraska, Nevada, New Mexico, Oklahoma, Oregon, Tennessee, Texas, Utah, Washington, Wisconsin, Wyoming
Date(s) Application is Due: Aug 15
Contact: Darlynn Myers, Director; (402) 271-5600; fax (402) 501-2291; upf@up.com
Internet: http://www.up.com/found/grants.shtml
Sponsor: Union Pacific Foundation
1400 Douglas Street, Stop 1560
Omaha, NE 68179

United Methodist Church Asian, Black, Hispanic and Native American History Research Grants **4570**

The purpose of the Asian, Black, Hispanic and Native American History Research award is to promote excellence in research and writing in the history of these four groups in the United Methodist Church or its antecedents. Grants totaling $3,000 will be available with each grant being at least $1,000 awarded at the discretion of the committee of judges. Selection is made by a committee consisting of three persons: one historian who is a member of The United Methodist Church, one member of the General Commission on Archives and History and one member chosen at large by the General Secretary of the General Commission. Submitted proposals should be postmarked no later than December 31. Grant recipients will be announced by May 1 of the following year.
Requirements: The candidate must submit an application in English and include biographical information; a description of the project, including its significance, what resources will be used, and how it will be made available to the public; the expected date of completion; and an itemized budget. The candidate must submit three letters of recommendation in English from persons who can provide evidence of his/her capability. A reference copy of the final product will be donated to the General Commission on Archives and History of the United Methodist Church. Submit materials to the office.
Geographic Focus: All States
Date(s) Application is Due: Dec 31
Amount of Grant: 1,000-3,000 USD
Contact: Robert J. Williams, General Secretary; (973) 408-3189; fax (973) 408-3909; rwilliams@gcah.org or research@gcah.org
Internet: http://www.gcah.org/AwardsGrants/RacialEthnic.htm
Sponsor: United Methodist Church
36 Madison Avenue, P.O. Box 127
Madison, NJ 07940

United Methodist Church Center for the Evangelical UBH Audrie E. Reber Memorial Award **4571**

The Center for the Evangelical United Brethren Heritage is offering an annual award to an undergraduate, seminary, or graduate student enrolled in an institution of higher education who submits the best paper on some aspect of Evangelical United Brethren Church history, theology or church life. The Evangelical United Brethren heritage includes the history of its predecessor bodies, the Evangelical Church and the Church of the United Brethren in Christ, as well as the wider Pietist heritage from which these bodies derive. The award is in memory of Audrie E. Reber, former missionary, author and active church-woman who was a Founding Member of the Center and a Member of its Advisory Board. The winning paper may be published in Methodist History, and a cash award of $700 will be given.
Requirements: Three clean copies of each paper must be submitted accompanied by a letter of nomination from the instructor under whose supervision it was written. The paper must not be longer than 20 double-spaced typewritten or computer processed pages. It must be properly footnoted and bear evidence of thorough and reliable handling of sources using the annotation standards accepted at the school. A bibliography also must be included. By April 1st each year, all competing papers must be in the hands of the Director of The Center for the Evangelical United Brethren Heritage, United Theological Seminary, 4501 Denlinger Road, Dayton, OH 45426.
Restrictions: The decision of the judges is final. No award will be made if the judges determine that no paper submitted merits the award.
Geographic Focus: All States
Date(s) Application is Due: Apr 1
Amount of Grant: 700 USD
Contact: Robert J. Williams, General Secretary; (973) 408-3189; fax (973) 408-3909; rwilliams@gcah.org or research@gcah.org
Internet: http://www.gcah.org/site/c.ghKJI0PHIoE/b.4619069/k.552C/EUB_Audrie_E_Reber_Award.htm
Sponsor: United Methodist Church
36 Madison Avenue, P.O. Box 127
Madison, NJ 07940

United Methodist Church John Harrison Ness Memorial Award **4572**

The General Commission on Archives and History of The United Methodist Church offers an annual award of $500 (first prize) and $300 (second prize) to the students enrolled in the M. Div. program (or its equivalent) in United Methodist or other seminaries accredited by the Association of Theological Schools who submit the best papers on some aspect of United Methodist history broadly conceived. Guidelines are available online.
Requirements: The paper must have been judged by its author's professor as one of quality. Once the paper has passed his or her critical judgment, the professor will submit it to the General Commission on Archives and History. The paper must be properly footnoted and bear evidence of thorough and reliable handling of sources using the annotation standards accepted at the seminary. A bibliography also must be included. The paper should normally range in length from 4,000 to 8,000 words. Three copies of the paper are to be submitted for judging. The winning paper(s) may be published in 'Methodist History', at the discretion of the editor.
Restrictions: The decision of the judges will be final. No prize will be awarded if the judges decide that no paper submitted has the required quality of scholarship for an award.
Geographic Focus: All States
Date(s) Application is Due: Feb 1
Amount of Grant: 300-500 USD
Contact: Robert J. Williams; (973) 408-3189; fax (973) 408-3909; rwilliams@gcah.org
Internet: http://www.gcah.org/site/c.ghKJI0PHIoE/b.3526379/k.8DA6/John_Ness_Award.htm
Sponsor: United Methodist Church
36 Madison Avenue, P.O. Box 127
Madison, NJ 07940

United Methodist Church Josephine Forman Scholarship **4573**

The scholarship is named for Josephine Forman, archivist for eighteen years of the Southwest Texas Conference of The United Methodist Church and author of We Finish to Begin: A History of Travis Park United Methodist Church, 1846–1991. As Conference Archivist, Ms. Forman collecteded records from closed churches along with administrative records for the Conference. For many years she was also the Church Historian at Travis Park United Methodist Church in San Antonio. The purpose of the Josephine Forman Scholarship is to provide financial support to minority students pursuing graduate education in archival science, to encourage students to pursue a career as an archivist, and to promote the diversification of the American archives profession. The scholarship is given to applicants who demonstrate excellent potential for scholastic and personal achievement and who manifest a commitment both to the archives profession and to advancing diversity concerns within it. The selection process is managed by the General Commission on Archives and History in partnership with the Society of American Archivists.
Requirements: The applicant must be: a citizen or permanent resident of the United States; of American Indian/Alaska Native, Asian, Black/African American, Hispanic/Latino, or Native Hawaiian/Other Pacific Islander descent; currently enrolled in a graduate program or a multi-course program in archival administration, or have applied to such a program for the next academic year; and enrolled in a graduate program and begin school no later than September 1 or the fall semester/quarter immediately following the award. The applicant shall have completed no more than half of the credit requirements toward her/his graduate degree at the time of the award
Geographic Focus: All States
Amount of Grant: 10,000 USD
Contact: Robert J. Williams; (973) 408-3189; fax (973) 408-3909; rwilliams@gcah.org
Internet: http://www.gcah.org/site/c.ghKJI0PHIoE/b.6364989/k.C50B/Josephine_Forman_Scholarship.htm
Sponsor: United Methodist Church
36 Madison Avenue, P.O. Box 127
Madison, NJ 07940

United Methodist Church Women in History Research Grant **4574**

The purpose of the Women in United Methodist History Research Grant is to provides seed money (travel, secretarial services, etc.) for research projects relating specifically to the history of women in the United Methodist Church or its antecedents. Grants totaling $2,500 will be available with each grant being at least $1,000 awarded at the discretion of the committee of judges. A copy of the final product will be deposited with the General Commission on Archives and History of The United Methodist Church.
Requirements: Selection is made by a committee consisting of three persons who are historians of women in United Methodism. Proposals on women of color and on history at the grass roots level are especially encouraged. Application guidelines are available online. The candidate must submit an application which includes the following: vita or resume; description of the project, including the significance of the research; description of the final product-applicants are encouraged to consider a variety of formats (written, audiovisual, oral history, bibliographies, archival guides, etc.); timetable for the project, including the expected date of completion; budget; indication of how the research results will be disseminated. The candidate must submit three letters of recommendation from persons who can provide evidence of his/her scholarly capability.

Restrictions: The grant money is not to be used for equipment, publications costs, or researcher's salary.
Geographic Focus: All States
Date(s) Application is Due: Dec 31
Amount of Grant: 1,000-2,500 USD
Contact: Robert J. Williams; (973) 408-3189; fax (973) 408-3909; rwilliams@gcah.org
Internet: http://www.gcah.org/site/c.ghKJI0PHIoE/b.3526385/k.D947/Women_in_UM_History_Research.htm
Sponsor: United Methodist Church
36 Madison Avenue, P.O. Box 127
Madison, NJ 07940

United Methodist Church Women in History Writing Award 4575

One Women in History Writing Award is given each year for an original manuscript on the history of women in the United Methodist Church or its antecedents. Consideration will be given to completed, original manuscripts of article length. The winning manuscript is considered for publication in Methodist History. A copy of the winning entry will be deposited with the General Commission on Archives and History of the United Methodist Church. Guidelines are available online.
Requirements: Manuscripts should not be longer than twenty double-spaced, typewritten pages. Footnotes and bibliography should follow the Kate Turabian Manual of Style. Selection is made by a committee consisting of three persons who are historians of women in United Methodism.
Geographic Focus: All States
Date(s) Application is Due: Dec 31
Amount of Grant: 500 USD
Contact: Robert J. Williams; (973) 408-3189; fax (973) 408-3909; rwilliams@gcah.org
Internet: http://www.gcah.org/site/c.ghKJI0PHIoE/b.3526381/k.7DAA/Women_in_UM_History_Award.htm
Sponsor: United Methodist Church
36 Madison Avenue, P.O. Box 127
Madison, NJ 07940

United Methodist Church World is My Parish United Methodist History Research Grant 4576

The World is My Parish United Methodist History Research Grant was created to enable the collection of written and oral history and encourage research regarding the development of The United Methodist Church and its antecedents in any of the conferences not in the United States. Grants totaling $3,000 will be available with each grant being at least $1,000 awarded at the discretion of the committee of judges. A reference copy of the final product will be donated to the General Commission on Archives and History of the United Methodist Church. The Commission will send copies to other United Methodist general agencies as appropriate.
Requirements: The applicant must submit an application which includes the following: (if at all possible the application should be in English but the applicant's language will be accepted) Biographical information; description of the project, including its significance, what resources will be used, and how it will be made available to the public; expected date of completion; and, an itemized budget. The applicant must submit three letters of recommendation from persons who can provide evidence of his/her capability.
Geographic Focus: All States
Date(s) Application is Due: Dec 31
Contact: Robert J. Williams; (973) 408-3189; fax (973) 408-3909; rwilliams@gcah.org
Internet: http://www.gcah.org/site/c.ghKJI0PHIoE/b.4619067/k.B158/World_is_My_Parish_Research.htm
Sponsor: United Methodist Church
36 Madison Avenue, P.O. Box 127
Madison, NJ 07940

United Methodist Committee on Relief Global Health Grants 4577

UMCOR Global Health programs work internationally with more than 300 United Methodist hospitals and clinics, using education as well as preventative and curative measures to confront major health issues such as malnutrition, maternal and child mortality, HIV and AIDS, and malaria. Our programs also seek to increase access to clean water and better sanitation. In the United States, UMCOR Global Health programs connect people to resources and support networks and advocate for people with disabilities. UMCOR's holistic approach to health empowers people to take charge of their own lives. This strategy emphasizes education and the development of local resources so that improvements in public health are sustained over time. UMCOR's programs reach entire communities, especially the most vulnerable, and value local cultural practices as sources of preventative and curative health care. The goals are to: end poverty and hunger; attain universal education; achieve gender equality; promote child health; promote maternal health; combat HIV/AIDS and other disease of poverty; attain environmental sustainability; and build a global partnership.
Restrictions: This program is only for United Methodist and Methodist-related projects outside the U.S.
Geographic Focus: All States, All Countries
Contact: Shannon Trilli, Director of Global Health Initiatives; (212) 870-3870 or (212) 870-3951; strilli@umcor.org or umcor@umcor.org
Internet: http://new.gbgm-umc.org/umcor/http://www.umcor.org/UMCOR/Programs/Global-Health/Global-Health-and-Developmentwork/health/malaria/
Sponsor: United Methodist Committee on Relief
475 Riverside Drive, Room 1520
New York, NY 10115

United Methodist Communications Stoody-West Fellowship 4578

United Methodist Communications gives two $6,000 Stoody-West Fellowships per year; one of these goes to a United Methodist student. The fellowships are given to students who intend to pursue careers in religious journalism in their post-graduate studies at an accredited U.S. college or university. The purpose of the fellowship is to enhance the recipient's professional competence and thereby help to perpetuate the standards exemplified by Ralph Stoody and Arthur West. Religious journalism is broadly interpreted to include audiovisual, electronic, and print journalism. Half of the grant is paid in August or September after the recipient is enrolled in the graduate program; the second half is paid in December. Applications are available on the Web site.
Requirements: Applicant should submit three letters of recommendation, an official academic transcript mailed directly by the school, a statement about interest in religious journalism and future plans, four examples of journalistic work, a list of published articles, and a recent photograph.
Restrictions: Funds are not paid for summer sessions.
Geographic Focus: All States
Date(s) Application is Due: Mar 15
Amount of Grant: 6,000 USD
Contact: Scholarship Administrator; (615) 742-5769 or (615) 742-5400; fax (615) 742-5485; scholarships@umcom.org or foundation@umcom.org
Internet: http://archives.umc.org/interior.asp?ptid=1&mid=2664
Sponsor: Foundation for United Methodist Communications
P.O. Box 320
Nashville, TN 37202-0320

United States Institute of Peace-Jennings Randolph Peace Dissertation Scholarships 4579

The Jennings Randolph Program for International Peace awards nonresidential Peace Dissertation Scholarships to students at U.S. universities who are writing doctoral dissertations on topics related to peace, conflict and international security. Each year the program awards approximately ten scholarships. Funding lasts for ten months starting in September. Doctoral students from any county may apply, and projects in all disciplines are welcome. Scholars receive $20,000 for 10 months. Candidates may apply through the online process system at the U.S.IP website.
Requirements: Citizens of any country may apply. Applicants must be enrolled in recognized doctoral programs in accredited universities in the United States. Successful candidates must have completed all course work and examinations towards their doctoral degrees by the time their scholarship funding begin. Proposals should be consistent with the Institute's mandate and present a research agenda with clear relevance to policy issues. Historical topics are appropriate if they promise to shed light on contemporary issues. Area studies projects and single case studies will be competitive if they focus on conflict and its resolution, apply to other regions and cases, or both.
Restrictions: Peace Scholar awards may not be made for projects that constitute policymaking for a government agency or private organization, focus to any substantial degree on conflicts within U.S. domestic society, or adopt a partisan, advocacy or activist stance. Peace Scholar awards may not be deferred. They generally may not be combined with any other major award or fellowship except in special circumstances and with the written approval of the Institute.
Geographic Focus: All States, All Countries
Date(s) Application is Due: Dec 14
Amount of Grant: 20,000 USD
Contact: Steven Riskin, Senior Program Officer, Fellowships; (202) 429-3869 or (202) 457-1700; fax (202) 429-6063; interviews@usip.org
Internet: http://www.usip.org/grants-fellowships/jennings-randolph-peace-scholarship-dissertation-program
Sponsor: United States Institute of Peace
2301 Constitution Avenue, NW
Washington, DC 20037

United States Institute of Peace-Jennings Randolph Senior Fellowships 4580

The Jennings Randolph Senior Fellowships provide scholars, policy analysts, policy makers, journalists, and other experts with opportunities to spend time in residence at the Institute, reflecting and writing on pressing international peace and security challenges. Applications are invited from all disciplines and professions. Senior Fellowships usually last for ten months, starting in October, but shorter-term fellowships are also available. Fellowships are open to citizens of any country. The Institute awards between 10 and 12 fellowships per year. The program attempts to match the recipient's earned income during the year preceding the fellowships, up to a maximum of $100,000 for 10 months. The Institute will provide coverage of 80% of health premiums for the Fellow and his/her eligible dependents, with a cap of $500 per month. The Institute will also cover travel to and from Washington, D.C., for Fellows and their dependents. Each Fellow is provided with a part-time research assistant during his/her fellowship. The Institute does not provide housing in Washington DC, but it provides information on housing, schools and daycare.
Requirements: Citizens of any country may apply. Non-U.S. citizens without permanent resident status must obtain a J-1 exchange visitor visa to participate in the Fellowship Program. This status requires recipients to reside in their home country for two years following the fellowship before applying for the H or L visa, or for permanent residency in the United States. There is no specific educational degree requirement for candidates. Fellows come from a variety of professional backgrounds and from early, middle and late stages of their careers. Priority is given to proposals deemed likely to make timely and significant contributions to the understanding and resolution of ongoing and emerging conflicts and other challenges to international peace and security. Historical topics are appropriate if they promise to shed light on contemporary issues. Area studies projects and

single-case studies will be competitive if they focus on conflict and its resolution, apply to other regions and cases, or both. Fellows are expected to be at the Institute and participate in its daily life. Fellows are expected to devote full attention to their Fellowship work in order to complete their projects within the period of residency. The Institute requires first right of review for manuscripts produced as a result of Fellowship support.
Restrictions: Senior Fellow awards may not be granted for projects that constitute policymaking for a government agency or private organization, focus to any substantial degree on conflicts within U.S. domestic society, or adopt a partisan, advocacy, or activist stance. Joint applications (two or more applicants for a single project) will not be accepted. An Institute Fellowship may not be deferred or combined with any other major award.
Geographic Focus: All States, All Countries
Date(s) Application is Due: Sep 10
Amount of Grant: Up to 100,000 USD
Contact: Elizabeth Cole, Senior Program Officer; (202) 429-3869 or (202) 457-1700; fax (202) 429-6063; interviews@usip.org
Internet: http://www.usip.org/grants-fellowships/jennings-randolph-senior-fellowship-program
Sponsor: United States Institute of Peace
2301 Constitution Avenue, NW
Washington, DC 20037

United States Institute of Peace-Transatlantic Post-Doc Fellowship for International Relations and Security 4581
The Transatlantic Post-Doc Fellowship for International Relations and Security (TAPIR) is open to candidates who have recently received their doctorate in social and political sciences or economics, and whose research focuses on topics of international relations and/or international peace and security issues. Up to five Fellowships are awarded each year. Fellowships are granted for a duration of 24 months to prepare Fellows for a career in policy-oriented and international research at renowned think tanks and political consulting research institutes. Fellows spend three eight-month stays at institutions participating in the program-at least one on the Eastern and one on the Western side of the Atlantic. During their stay, Fellows are supervised by a specialist at the Institute, while being integrated into the relevant research units or projects. Fellows will have the opportunity to work on issues of major and immediate political significance, and to publish their work in journals or paper series that will be read by decision-makers at national and international levels. The Fellowship aims to involve participants actively in the work and research of leading international think tanks. It is not intended to primarily fund the completion of major individual publications or of academic peer-reviewed journal articles. The Fellowship consists of a monthly stipend; a monthly allowance for health insurance; and a one-time travel allowance for travel to and from the host institutions. The TAPIR Fellowship is administered by the Stiftung Wissenschaft und Politik in Germany. A list of participating institutes is posted on their website.
Requirements: To be eligible for the Transatlantic Post-Doc Fellowship, candidates must successfully defend their Ph.D. by the beginning of their Fellowship. Candidates must have a strong interest in working in a policy-oriented international environment and must demonstrate their ability to do so by previous international academic experience. Applicants must either be citizens of the U.S., Canada or Europe (in a broader geographical definition) OR must have resided in the U.S., Canada or Europe at least for two-and-a-half years in the five years before the application deadline.
Restrictions: Fees for residence permits are not included.
Geographic Focus: All States, Albania, Andorra, Armenia, Austria, Azerbaijan, Belarus, Belgium, Bosnia & Herzegovina, Bulgaria, Canada, Croatia, Cyprus, Czech Republic, Denmark, Estonia, Finland, France, Germany, Greece, Hungary, Iceland, Ireland, Italy, Kazakhstan, Latvia, Liechtenstein, Lithuania, Luxembourg, Macedonia, Malta, Moldova, Monaco, Montenegro, Netherlands, Norway, Poland, Portugal, Romania, Russia, San Marino, Serbia, Slovakia, Spain, Sweden, Switzerland, Turkey, Ukraine, United Kingdom, Vatican City
Date(s) Application is Due: Feb 1
Amount of Grant: Up to 51,500 EUR
Contact: Tobias von Lossow, TAPIR Coordinator; +49 (0)30 88007-106; fax +49 (0)30 88007-5106 / -100; tobias.vonlossow@swp-berlin.org
Internet: http://www.usip.org/grants-fellowships
Sponsor: United States Institute of Peace
2301 Constitution Avenue, NW
Washington, DC 20037

United States Institute of Peace Annual Grants 4582
The Institute's Annual Grants support innovative peacebuilding projects involving research, the identification of promising models and effective practices, the development of practitioner resources and tools, the development and delivery of education, training and dialogue programs, and the production of films, radio programs, and other media. It funds projects focused on preventing, managing, and resolving violent conflict and promoting post-conflict peacebuilding in settings outside the borders of the United States. Awards support activities that apply across a broad range of relevant disciplines, skills, and approaches. The Institute welcomes proposals of an interdisciplinary or multidisciplinary nature. Topic areas of interest include, but are not limited to: conflict analysis and prevention; mediation and conflict resolution; postconflict peace and stability operations; religion and peacemaking; women and girls in conflict and peacebuilding; rule of law and transitional justice; economies and conflict; social, psychological, and physical impacts of war and conflict; and media and conflict. Candidates may access the online application process at the Institute website. Additional guidelines and a detailed list of previously funded projects are also available at the website.
Requirements: The Institute may provide grant support to domestic and international nonprofit organizations and individuals including the following: institutions of post-secondary, community, and secondary education; public and private education, training or research institutions, and libraries. Individuals whose proposals are funded will be required to identify a nonprofit organization to receive and manage the grant monies. When applicants are employed by an eligible institution, such as a college or university, the Institute requires that grants be made to the institution rather than to the individual.
Restrictions: Support for degree work is not eligible in this program. Only one application per competition may be submitted by the same project director. Unsuccessful applicants of previous grant program competitions may not submit the same application unless it has been substantially revised. The Institute will not accept applications that list Institute participants, consultants, or project personnel members of the Institute's Board of Directors or staff. In addition, any application that lists the Institute as a collaborator in the project will not be accepted. Individuals who are currently working as Institute contractors may not be eligible to apply for grant support.
Geographic Focus: All States, All Countries
Date(s) Application is Due: Oct 1
Amount of Grant: 50,000-120,000 USD
Contact: Steven Riskin; (202) 429-3869 or (202) 457-1700; fax (202) 429-6063
Internet: http://www.usip.org/grants-fellowships/annual-grant-competition
Sponsor: United States Institute of Peace
2301 Constitution Avenue, NW
Washington, DC 20037

United Theological Seminary Merit-Based Scholarships 4583
Merit-based scholarships are awarded to students who demonstrate exceptional academic ability. This aid covers up to 60 percent of tuition and can be awarded together with need-based aid to provide up to 100 percent of tuition. Scholarships include: the George D. Dayton Scholarship—granted up to four years to Masters of Divinity and Master of Arts in Religious Leadership students preparing for positions of professional church leadership; and the Presidential Scholarship—granted for up to three years to students in Master of Arts in Religion and Theology, Master of Arts in Theology and the Arts, or Master of Arts in Women Studies.
Requirements: Successful candidates must hold a baccalaureate degree from a regionally accredited college or university; be admitted in good standing to United Theological Seminary as a degree candidate; take the Graduate Record Examination (GRE) test; and demonstrate general excellence in academic and intellectual achievement, personal characteristics and qualities, leadership capacity in both the interpersonal and group contexts, service orientation, and faith commitment.
Geographic Focus: Minnesota
Date(s) Application is Due: Jan 31
Contact: Michelle Turnau, Director of Financial Aid and Housing; (651) 255-6117 or (651) 633-4311; fax (651) 633-4315; mturnau@unitedseminary.edu
Internet: http://www.unitedseminary.edu/admissions/financialaid.asp
Sponsor: United Theological Seminary of the Twin Cities
300 Fifth Street NW
New Brighton, MN 55112-2598

University of British Columbia Graduate Student Fellowships 4584
The Faculty of Graduate Studies offers merit-based fellowship awards to full-time graduate students registered at the University of British Columbia. Fellowship awards are open to all disciplines. Application forms are available on the Web site. All UGF application materials must be received by the deadline set by the department.
Requirements: Applicants must be graduate students with a first class standing in each of the last two years of full-time study. Students from any country and from all fields are eligible.
Geographic Focus: All States
Contact: Faculty of Graduate Studies; (604) 822-8501 or (604) 822-0976; fax (604) 822-5802; grad-award@mercury.ubc.ca
Internet: http://www.grad.ubc.ca/awards/index.asp?menu=004,000,000,000
Sponsor: University of British Columbia
180-6371 Crescent Road
Vancouver, BC V6T 1Z2 Canada

University of Calgary Nursing Alumni Scholarship 4585
One scholarship per year is given to a Canadian citizen or landed immigrant of high academic standing who is registered in or admitted to a full-time master's program, who intends to make a career in nursing, and who is a registered member of the University of Calgary Faculty of Nursing Alumni Association. The scholarship may be used for any area or discipline at the master's level related to nursing. Annual deadline dates may vary; contact program staff for exact dates.
Requirements: To be eligible, applicants must be Canadian citizens or landed immigrants; graduate students in a master's program who intend to make a career in nursing; and registered members of the University of Calgary Faculty of Nursing Alumni Association.
Geographic Focus: All States, Canada
Date(s) Application is Due: Dec 2
Amount of Grant: 1,000 CAD
Contact: Maria Lavoie, Graduate Scholarship Officer; (403) 220-8483; lavoiem@ucalgary.ca or ucawards@ucalgary.ca
Internet: http://www.grad.ucalgary.ca/funding/internal_scholarships/lev_4/nurs_alum_burs.htm
Sponsor: University of Calgary
2500 University Drive NW, Earth Sciences Building, Room 720
Calgary, AB T2N 1N4 Canada

University of Colorado Doctoral Dissertation Fellowship in Media, Religion, and Culture 4586

Grants are designed to coincide with the development of dissertation research in the early stages of candidacy for students engaged in research projects involving media, religion, and culture. Fellows will be selected from various universities and disciplines on the basis of their fellowship proposals and will be expected to reside in their current location (the fellowship does not require relocation). Fellows will be brought together for master seminars at which leading scholars in media, religion, and culture will interact with fellows, their advisors, and the research team at the University of Colorado's School of Journalism and Mass Communication. The fellowships stipend may be used to cover research and other expenses as desired by the recipient.
Geographic Focus: All States
Date(s) Application is Due: Mar 31
Amount of Grant: 12,000-32,000 USD
Contact: Monica Emerich, Fellowship Coordinator; monica.emerich@colorado.edu
Internet: http://www.colorado.edu/journalism/mcm/mrc/mrc-fellowships.htm
Sponsor: University of Colorado
1171 University Avenue, Box 478
Boulder, CO 80309-0478

University of Georgia Counseling and Psychological Services Postdoctoral Fellowship 4587

The program is designed to develop advanced skills in the practice of clinical and counseling psychology. The postdoctoral program provides fellows an opportunity to refine and expand experience in the following areas: clinical practice with individuals, couples and groups, assessment, crisis intervention, workshops, and educational programming. Postdoctoral fellows will work in close collaboration with senior staff and will function as professionals with many of the same responsibilities. Fellows attend all case conferences, professional development workshops, and staff meetings. In addition to the stipend, fellows are also eligible for staff benefits, including health and dental insurance, retirement, vacation, sick leave, professional development, exercise facility, library privileges, and licensing preparation time. Guidelines are available online. Annual deadline dates may vary; contact program staff for exact dates.
Requirements: Admission requirements include completion of all professional degree requirements from a regionally accredited institution of higher education or an APA/CPA–accredited program and completion of an APA approved predoctoral internship. Fellow must have received diploma before first day of fellowship or have a letter from Director of Graduate Studies verifying that all degree requirements have been met. Previous counseling center experience is preferred.
Geographic Focus: All States
Date(s) Application is Due: Mar 11
Amount of Grant: 25,000 USD
Contact: Dr. Al Clarke; (706) 542-2273; aclarke@uhs.uga.edu
Internet: http://www.uhs.uga.edu/CAPS/postdoc/index.html
Sponsor: University of Georgia
University of Georgia Health Center
Athens, GA 30602-1755

University of London Visiting Fellowships 4588

Visiting fellowships support work in any one or more of the areas covered by the university's research institutes and programs, including advanced legal studies, classical studies, Commonwealth studies, English studies, Germanic and romance studies, historical research, study of the Americas, the Warburg Institute, and the philosophy program. The fellowships, tenable for up to six consecutive months between September and June, offer travel, accommodation, and research expenses. Application forms are available from the office.
Requirements: Fellowships are available to applicants of professorial or equivalent status for the purpose of supporting research in London.
Geographic Focus: All States
Amount of Grant: 16,000 GBP
Contact: Program Contact, School of Advanced Study; (020) 7862-8659; fax (020) 7862-8657; Deans.Office@sas.ac.uk
Internet: http://www.sas.ac.uk/School_Fellowships
Sponsor: University of London
Senate House, Malet Street, Room 310
London, WC1E 7HU England

University of Southern Mississippi H.A. & Margret Rey Translation Fellowships 4589

This fellowship program, established by the Rey Foundation and the de Grummond Children's Literature Collection at the University of Southern Mississippi, awards grants to afford the hiring of translators (i.e. German, Portuguese, French, Russian) for the purpose of translating the H. A. and Margret Rey documents into English to further facilitate use in research. Grants totaling up to $1,664 may be made to applicants who demonstrate experience in translating.
Requirements: Recipients will be required to produce completed translations of papers from the H.A. and Margret Rey collection in the de Grummond Children's Literature Collection.
Geographic Focus: All States
Date(s) Application is Due: Oct 31
Amount of Grant: Up to 1,664 USD
Contact: Ellen Ruffin; (601) 266-4349; fax (601) 266-6269; ellen.ruffin@usm.edu
Internet: http://www.lib.usm.edu/legacy/degrum/public_html/html/research/re-fellowships.shtml
Sponsor: University of Southern Mississippi
118 College Drive, #5148
Hattiesburg, MS 39406-5148

University of Southern Mississippi Keats/Domanska/de Grummond Children's Literature Research Fellowships 4590

This program, established by the Ezra Jack Keats Foundation, the Janina Domanska Literary Estate, and the de Grummond Children's Literature Research Collection of the University of Southern Mississippi, awards grants to scholars engaged in projects based substantially on the holdings of the de Grummond collection. The purpose of the program is to encourage imaginative and productive research in children's literature. The de Grummond Collection focuses on American and British children's literature, historical as well as contemporary. Original manuscripts and illustrations are a special emphasis of the collection. Grants of up to $1,582 may be made to applicants of any discipline who engage in projects which require substantial use of the de Grummond Collection. Awards are based upon the applicant's scholarly qualifications, the merits of the proposal submitted, and the appropriateness of the proposal to the resources of the de Grummond Collection. Funding can be used for travel, lodging, and meals. Applications are available at the website.
Requirements: Applicants may be researchers in any discipline who engage in projects that require substantial use of the de Grummond collection. Preference is given to projects that have a good likelihood of being published.
Geographic Focus: All States
Date(s) Application is Due: Oct 31
Amount of Grant: Up to 1,582 USD
Contact: Ellen Ruffin; (601) 266-4349; fax (601) 266-6269; ellen.ruffin@usm.edu
Internet: http://www.lib.usm.edu/legacy/degrum/public_html/html/research/re-fellowships.shtml
Sponsor: University of Southern Mississippi
118 College Drive, #5148
Hattiesburg, MS 39406-5148

Upjohn Institute Dissertation Award 4591

The Upjohn Institute supports and conducts policy-relevant research on employment and unemployment-related issues. This award, for the best PhD dissertation on employment-related issues, may come from any discipline. Dissertations are evaluated by a panel of economists using the following criteria: policy relevance, technical quality of the research, potential impact on real-world problems, and presentation. Applicants are advised that they will need to supply a copy of the entire dissertation if they are selected as a finalist. A first prize and up to two honorable mention awards will be made.
Requirements: Any individual whose dissertation has been accepted during the 24-month period of July 1 to June 30, 2006, is eligible.
Geographic Focus: All States
Date(s) Application is Due: Jul 7
Amount of Grant: 750-2,000 USD
Contact: Grant Coordinator; (616) 343-5541; fax (616) 343-3308; webmaster@upjohninstitute.org
Internet: http://www.upjohninst.org/dissert.html
Sponsor: W.E. Upjohn Institute for Employment Research
300 S Westnedge Avenue
Kalamazoo, MI 49007-4686

Ursula Thrush Peace Seed Grants 4592

Ursula Thrush founded the Maria Montessori School of the Golden Gate and Teacher Training Center in San Francisco and helped to establish several other Montessori programs, including The Science of Peace Task Force and the Montessori Peace Academy. Her dedication to fulfilling Maria Montessori's vision for peace through children opened doors to many Montessori educators, inspiring them to include peace education in their classrooms and communities. Since 2004, the AMS Peace Committee has awarded annual Peace Seed Grants to encourage educators to promote peace through their teaching. Lesley Nan Haberman, headmistress of The Family Schools in New York City, initiated the grant in memory of her dear friend Ursula Thrush.
Requirements: Montessorians who have formulated a project fostering education for peace are invited to apply. Applicants must have a Montessori background. The project must further education for peace and must reach a significant number of children and/or educators. The recipient of the grant must share the results of his/her project with the AMS community. The project goal must be accomplished in the year following receipt of the grant. The recipient must show how he/she will be accountable for the use of the funds.
Geographic Focus: All States
Date(s) Application is Due: Feb 1
Amount of Grant: 1,100 USD
Contact: Judi Bauerlein, (212) 358-1250; fax (212) 358-1256; Judi4trees@sbcglobal.net
Internet: http://www.amshq.org/peaceseedconnection.htm#thrushGrant
Sponsor: American Montessori Society
281 Park Avenue South
New York, NY 10010-6102

U.S.A Funds Access to Education Scholarships 4593

U.S.A Funds is a nonprofit organization that helps students from groups underrepresented in higher education pursue a college degree. U.S.A awards scholarships to low income students, with special consideration for students who are members of an ethnic minority group or who have a documented physical disability. The scholarship recipients must either plan to enroll or be enrolled in full- or half-time undergraduate or full-time graduate- or professional-degree coursework at an accredited two- or four-year college, university or vocational/technical school. See website for current guidelines and application. The recipients were selected based on additional factors, including the student's past academic performance and future potential, leadership and participation in school and community activities, work experience, career and educational aspirations and goals, and any unusual personal or family circumstances.

Requirements: Eligibility includes high-school seniors, currently enrolled college students, or incoming college students who plan to enroll or are enrolled in full- or half-time undergraduate or full-time graduate coursework at an accredited two- or four-year college, university, or vocational/technical school; have an annual adjusted gross family income of $35,000 or less, as documented on a federal income-tax return on which the applicant is claimed as a federal tax exemption; and are U.S. citizens or eligible non-citizens (permanent U.S. residents with a Green Card, or refugees with a Departure Record (Form I-94) from the U.S. Immigration and Naturalization Service.
Restrictions: Applicants may not be in default on a previous federal education loan.
Geographic Focus: All States
Amount of Grant: 1,500 USD
Contact: Scholarship Management Services; (800) 537-4180; contact@usafunds.org
Internet: http://www.usafunds.org/about_usa_funds/index.html
Sponsor: U.S.A Funds
P.O. Box 6028
Indianapolis, IN 46206-6028

USAID Accelerating Progress Against Tuberculosis in Kenya Grants **4594**
The United States Agency for International Development (USAID) is seeking applications for a Cooperative Agreement to provide funding in support of a program entitled Accelerating Progress Against TB (APA-K) in Kenya. The main objective of this program is to extend access to quality-assured TB services in all districts and for all forms of TB, through the identification and implementation of evidence-based activities that support and/or complement the activities of the Government of Kenya's Division of Leprosy, TB and Lung Diseases. It is expected that a single grant with be funded in the amount of $40,000,000. The deadline for application is April 15.
Requirements: In response to this RFA, only local Kenyan organizations are eligible to apply.
Geographic Focus: Kenya
Date(s) Application is Due: Apr 15
Amount of Grant: 40,000,000 USD
Contact: Jennifer Kiiru, Acquisitions and Assistance Specialist; +254-20-862-2848 or +254-20-862-2000; fax +254-20-862-2680; jkiiru@usaid.gov or usaidke@usaid.gov
Internet: http://www.grants.gov/search/search.do;jsessionid=Ty6nRyTfyLlh9Bp501FpjptxyhJL02HR2D01sPWcc1s9JVyBsr3x!1962156644?oppId=223813&mode=VIEW
Sponsor: U.S. Agency for International Development
Ronald Reagan Building
Washington, DC 20523-1000

USAID Comprehensive District-Based Support for Better HIV/TB Patient Outcomes Grants **4595**
The United States Government, as represented by the United States Agency for International Development (USAID) Mission to Southern Africa, is seeking applications from organizations interested in implementing a five-year comprehensive district-based HIV-related services support program for better patient outcomes, as fully described in this Request for Applications (RFA). The purpose of this program is to strengthen SAG systems in order to improve patient outcomes and prevent HIV by supporting comprehensive clinic-based HIV-related services in select districts. All support will be in line with SAG national, provincial and district policies, standards, guidelines, and implementation plans as well as the Partnership Framework between the U.S.G and SAG. The program will build and capitalize on the accomplishments and lessons learned since the national ARV rollout began in 2004. It will focus on institutionalization of routine and consistent use of systems designed to improve efficiencies in district and sub-district management that ultimately improve patient and public health outcomes.
Requirements: Any local South African non-governmental organizations (NGOs), private voluntary organizations (PVOs), Faith-based and community organizations, and for-profit companies willing to forego profit will be eligible for award under this RFA. In support of the Agency's interest in fostering a larger assistance base and expanding the number and sustainability of development partners, USAID encourages applications from potential new partners.
Geographic Focus: South Africa
Amount of Grant: Up to 20,000,000 USD
Contact: Beatrice Lumanade, Acquisitions and Assistance Specialist; +27 12 452 2377 or +27 12 452 2000; blumande@usaid.gov or applications4@usaid.gov
Internet: http://www07.grants.gov/search/search.do;jsessionid=1v8kR3HpjBFsc7Z47pRgjfSFjbcHqr2Jf6p2BXMYZTRdnTrpWtT4!1321711693?oppId=169154&mode=VIEW
Sponsor: U.S. Agency for International Development
Ronald Reagan Building
Washington, DC 20523-1000

USAID Development Innovation Accelerator Broad Agency Announcement Grants **4596**
This Broad Agency Announcement (BAA) seeks opportunities to co-create, co-design, co-invest, and collaborate in basic and applied research and development for Science, Technology, Innovation, and Partnership (STIP). The United States Agency for International Development (USAID) invites organizations and companies to participate with USAID in response to a Critical Development Challenge Addendum issued under this BAA, as described, to provide innovations and technologies that further USAID's development goal of dramatically improving or saving the lives of over 200 million people over the next five years. Whether cost share, match, and/or leverage are required will be determined by the individual BAA Addendum, final award type, and/or the mutual agreement of the parties. Nothing in the BAA precludes reasonable cost sharing, matching, leveraging, or other exchange of resource arrangements, and proposers are encouraged to suggest creative approaches to resourcing projects.
Requirements: Public, private, for-profit, and non-profit organizations, as well as institutions of higher education, public international organizations, non-governmental organizations, U.S. and non-U.S. government organizations, and international donor organizations are eligible under this BAA, unless otherwise stated in the individual BAA Addendum.
Geographic Focus: All States
Contact: Agreement Specialist; DIAglobalscale@usaid.gov
Internet: https://www.fbo.gov/index?s=opportunity&mode=form&id=3f0db7bc1098a3704e4823db9395ce38&tab=core&_cview=1
Sponsor: U.S. Agency for International Development
Ronald Reagan Building
Washington, DC 20523-1000

USAID Development Innovation Ventures Grants **4597**
As part of its commitment to increase investments and engagement in cost-efficient innovations, USAID launched Development Innovation Ventures (DIV) as a way of producing development outcomes more effectively and cost-efficiently while managing risk and obtaining leverage. Through DIV, USAID seeks to identify and rigorously test promising projects with the potential to significantly (rather than incrementally) improve development outcomes, and help replicate and scale projects that are proven successful. USAID/DIV expects its most successful of investments will have an accelerated growth path to reach tens of millions of beneficiaries worldwide within 10 years. DIV Grants will support projects at three stages: Stage 1, the development needed to support proof of concept and feasibility; Stage 2, implementing the project at large scale with rigorous impact testing; and Stage 3, transitioning innovations to widespread adoption throughout one country and/or additional adoption in other countries. For Stage 1 applicants, a Concept Paper and its attachments are the only application documents required. Stage 2 and 3 applicants may submit a Concept Paper and wait to be invited (or not) to submit a Full Application. Alternatively Stage 2 and 3 applicants may elect to forego the Concept Paper and submit a Full Application. Interested applicants must first submit an Letters of Interest using the LOI form posted. Only invited applicants will receive the Full Application form.
Requirements: USAID/DIV welcomes applications from many types of organizations, including: U.S. non-governmental organizations (NGOs); non-U.S. NGOs; faith-based organizations; U.S. private businesses; non-U.S private businesses; business and trade associations; international organizations; U.S. colleges and universities (public and private); non-U.S. colleges and universities (public and private); civic groups; regional organizations; and foundations.
Restrictions: All applicants must be legally recognized organizational entities under applicable law. Individuals and governments are not eligible to apply directly at this time. The participation of foreign government organizations under this APS is possible only through an approved subaward agreement with a prime recipient.
Geographic Focus: All States, All Countries
Date(s) Application is Due: Apr 15
Amount of Grant: 100,000-15,000,000 USD
Contact: Tameka Laws, Grantor; (202) 567-5066; divapplication@usaid.gov
Internet: http://idea.usaid.gov/organization/div
Sponsor: U.S. Agency for International Development
Ronald Reagan Building
Washington, DC 20523-1000

USAID Family Planning and Reproductive Health Methods Grants **4598**
The United States Agency for International Development (USAID) is seeking concept papers from qualified U.S. nonprofit non-Governmental Organizations (NGOs) for a program titled Family Planning and Reproductive Health Methods to Address Unmet Need, for funding of Cooperative Agreements. The purpose of this APS is to support the research, development, and introduction of technologies and approaches that better meet the needs of women and girls as their sexual and reproductive health concerns change over time. The General Objectives of the APS are to: refine existing FP methods to address method-related reasons for non-use; respond to product-related issues about currently available FP methods that arise at purchase and/or from the field, and that may affect provider/user perceptions and/or the supply chain; develop new FP methods that address method-related reasons for non-use, and/or fill gaps in the existing method mix; conduct research to foster the introduction and uptake of new and/or underutilized woman-initiated methods, particularly non-hormonal barriers, contraceptive vaginal rings, and fertility awareness methods based on knowledge and monitoring of the menstrual cycle; and develop multipurpose prevention technologies (MPTs) that address the simultaneous risks of unintended pregnancy, HIV, and other sexually transmitted infections (STIs) – particularly Herpes Simplex Virus (HSV) and Human Papillomavirus (HPV). The funding range of awards is $500,000 to $80,000,000.
Requirements: Eligible organizations include registered U.S. and non-U.S. private non-governmental organizations, non-profit organizations, and for-profit organizations willing to forego profit.
Geographic Focus: All States
Date(s) Application is Due: Jan 14
Amount of Grant: 500,000-80,000,000 USD
Contact: Franklin Goode, Contract Specialist; (202) 567-5316; fgoode@usaid.gov
Internet: http://www07.grants.gov/search/search.do;jsessionid=kGnXRdGLdv3SbzZvRgnNLtS1jvT1Z8qs4Tcf1kvwg8y5TtT8hcN1!1884755759?oppId=215873&mode=VIEW
Sponsor: U.S. Agency for International Development
Ronald Reagan Building
Washington, DC 20523-1000

USAID Microbicides Research, Development, and Introduction Grants **4599**
Clinical trials of vaginal microbicide have been tested recently in South Africa to determine efficacy of reduction of HIV infection. Microbicide research, development, and introduction activities which could potentially be supported through this APS mechanism are expected to contribute substantially to the Principles of the Global Health Initiative (GHI), particularly in promoting research and innovation; implementing a woman- and girl-centered approach; strengthening and leveraging inputs from multilateral organizations, global health partnerships, and the private sector; and encouraging country ownership and leadership. Potential activities could also contribute significantly to strengthening the health system building blocks related to commodities and procurement, service delivery, and leadership and governance. It is expected that gender analyses and other gender considerations will continue to inform the design and implementation of activities implemented as part of the Agency's microbicide program. All awards will also be subject to USAID environmental requirements, including completion of an initial environmental examination.
Requirements: To qualify for funding, applicants must: be either U.S. or non-U.S. organizations registered and working in USAID priority countries; and have substantial experience (5 or more years) working directly with relevant HIV/AIDS-related activities in multiple countries in sub-Saharan Africa. Relevancy of experience will be determined by USAID.
Restrictions: Non-eligible entities include other U.S. Government agencies and departments. Other U.S. Government agencies and departments may not apply for USAID funding under this APS.
Geographic Focus: All States, All Countries
Amount of Grant: 100,000-2,250,000 USD
Contact: Kathleen Bumpass; (202) 567-5008; kbumpass@usaid.gov
Internet: http://www.grants.gov/search/search.do;jsessionid=JLnfQ2dXjGsLJMk6qcndy nccd2Y8QYb2vGLdchNvr5WhGqZ80RGb!99807221?oppId=187913&mode=VIEW
Sponsor: U.S. Agency for International Development
Ronald Reagan Building
Washington, DC 20523-1000

USAID Palestinian Community Infrastructure Development Grants **4600**
The Palestinian Community Infrastructure Development (PCID) Program aims to increase Palestinian access to water, sanitation and other small and medium scale community infrastructure in the West Bank. This focus will be complemented by a minimum stand-by capability to respond rapidly to needs in Gaza should political and security conditions require. Any interventions in Gaza will be identified and discussed by USAID and the implementing partner on a case by case basis after award. No interventions in Gaza are contemplated at this time. Accordingly, PCID will meet the following objectives: increase access to water and sanitation systems, reduce water losses and enhance water and sanitation system sustainability; increase the impact of USAID-funded programs in health, democracy and governance, education and the private sector by addressing underlying infrastructure needs which may be limiting the impact of those programs; and respond rapidly to urgent water, sanitation and other public infrastructure needs. The estimated ceiling for this RFA is $100 million.
Requirements: U.S.-based non-governmental organizations are eligible to submit applications. Applicants must have established financial management, monitoring and evaluation, internal control systems, and policies and procedures that comply with established U.S. Government standards, laws, and regulations.
Restrictions: Applications from non-U.S. organizations as well as Public International Organizations will not be considered.
Geographic Focus: All States
Date(s) Application is Due: Mar 9
Amount of Grant: Up to 100,000,000 USD
Contact: Claudia Koziol, Agreements Officer; 972-3-519-8507 or 972-3-511-4868; fax ckoziol@usaid.gov; WBG-OCM-packages@usaid.gov
Internet: http://www.federalgrants.com/Palestinian-Community-Infrastructure-Development-34136.html
Sponsor: U.S. Agency for International Development
Ronald Reagan Building
Washington, DC 20523-1000

USAID Research and Innovation for Health Supply Chain Systems and Commodity Security Grants **4601**
The purpose of this APS is to publicize the United States Government's (U.S.G) plan to fund a limited number of awards through the U.S. Agency for International Development's (USAID) Washington Global Health Bureau (GH) to support research on a focused set of health supply chain systems and related commodity security issues in low and middle income countries. Funds may support: formative, analytical research on challenges to improving supply chain systems and the enabling environments for commodity security; and/or development and testing of interventions in the field based on formative research conducted under this APS or from other sources. The awards under this APS will contribute to advancing and supporting health programs worldwide, contributing to ending preventable child and maternal deaths, creating an AIDS-free generation, and protecting communities from infectious diseases. In addition, this APS will address several key focus areas of the Global Health Initiative (GHI), including building sustainability through health systems strengthening and accelerating results through research and innovation. The objectives of the APS are listed below: improve health supply chain systems through transformative changes that use industry best practice; improve the quantity, reliability, and efficiency of financing for health commodities and supply chain systems; and improve governance and accountability for commodity security.
Requirements: To qualify for funding, organizations must be: U.S. organizations or non-U.S. organizations registered and working in USAID-assisted countries; and academic organizations (universities, professional schools, or two- or four-year colleges, including community colleges), or research institutes, professional societies or similar organizations that are directly associated with research activities.
Geographic Focus: All States
Date(s) Application is Due: May 29
Amount of Grant: 500,000-2,000,000 USD
Contact: Acquisitions Specialist; support@grants.gov
Internet: http://www.federalgrants.com/Research-and-Innovation-for-Health-Supply-Chain-Systems-and-Commodity-Security-46262.html
Sponsor: U.S. Agency for International Development
Ronald Reagan Building
Washington, DC 20523-1000

USAID Strengthening RMNCH Through Indian Health Professional Associations and Scaling Up Uptake Grants **4602**
USAID Strengthening RMNCH Through Indian Health Professional Associations and Scaling Up Uptake Grants generally targets India's vulnerable, marginalized, and underserved populations spread across rural areas, urban slums, and tribal pockets – including those currently outside the realm of effective service provision for reasons related to economics, gender, and social inclusion (notably females across the life cycle). The expected number of awards is six, and each will range from $3,650,000 to $8,000,000. The annual deadline date is April 9.
Requirements: USAID seeks applications from local India organizations.
Geographic Focus: All States, India
Date(s) Application is Due: Apr 9
Amount of Grant: 3,650,000-8,000,000 USD
Contact: Vandana Vats, Acquisitions and Assisstance Specialist; 91-11-24198568 or 91-11-24198000; fax 91-11-24198612; vvats@usaid.gov or indiarco@usaid.gov
Internet: http://www.grants.gov/search/search.do;jsessionid=6wkXRm4XlPrL2gnTX wynW1rnnBXH1gDJQDZdGVySjBTTpwmYrsQT!1762381823?oppId=229933&mode=VIEW
Sponsor: U.S. Agency for International Development
Ronald Reagan Building
Washington, DC 20523-1000

U.S.C Annenberg/Getty Arts Journalism Fellowships **4603**
Funded by a grant from the J. Paul Getty Trust, the U.S.C Annenberg/Getty Arts Fellowship seeks to establish a new standard of excellence in arts and culture coverage. The program's philosophy is guided by a core belief in first-hand encounters with artists, arts administrators, and journalism colleagues. The three-week residencies in Los Angeles familiarize the fellows with the city's vibrant cultural scene as a means to strengthen their intellectual and analytical journalism skills, fill them with new story ideas, and make them consider their hometowns in new lights. The program period is April 22 through May 13. Application and guidelines are available online.
Requirements: Requirements for Eligibility: candidates must be mid-career critics, reporters, writers, columnists, editors, broadcasters or producers who specialize in arts, culture or entertainment for or a newspaper, magazine, broadcast or online media organization; must have at least three years of steady professional arts journalism in your background; the program is NOT intended for entry-level journalists, students or journalism educators. Freelancers and/or self-employed individuals who regularly publish online, in print or over the airwaves ARE eligible, as long as their application is accompanied by a letter of recommendation by a newsroom editor/manager at an organization that regularly publishes or broadcasts their work OR by an experienced journalist who can speak to their work in convincing detail; work examples for the application must have been published or produced between May and June of previous year; this is a national and international program, applicants from anywhere MAY apply; advanced degrees or specialized training in the arts are NOT required; we do not discriminate on the basis of race, religion, national or international origin, color, sex, age, veteran status, or disability. It is our intention that all qualified applications be given equal opportunity and that our selection is based upon merit alone.
Geographic Focus: All States
Date(s) Application is Due: Jan 27
Contact: Sasha Anawalt, (213) 740-3914; fax (212) 754-4073; uscgetty@usc.edu
Internet: http://annenberg.usc.edu/Home/CentersandPrograms/ProfessionalEducation/GettyArtsJourn.aspx
Sponsor: University of Southern California
300 S Grand Avenue, Suite 3950
Los Angeles, CA 90071

U.S.CD/NOAA Illinois-Indiana Sea Grant Core Proposals **4604**
The Illinois-Indiana Sea Grant (IISG) program funds two to five projects during each two-year cycle. Requests for proposals are usually issued in December of even years for funding beginning 14 months later, in February of the following even year. IISG begins the proposal selection process by requesting preproposals. Preproposals are screened by an internal committee appointed by the IISG director for their applicability to IISG topic areas, potential to impact management decisions, appearance of scientific rigor, and funding level. Preproposal selection is the only opportunity the IISG committee has to independently select proposal topics of most interest. An IISG internal committee meets to make final funding recommendations based on the expert panel ranking and the research funds available. Recommendations are submitted to the National Sea Grant Office for approval. Upon approval, researchers are asked to submit any necessary final paperwork. This information is included in an Omnibus Proposal submitted to the National Sea Grant Office in November of each year. Projects start March 1 of the following year.

Geographic Focus: Illinois, Indiana
Contact: Tomas Hook, Associate Director of Research; (765) 496-6799 or (217) 333-6444; fax (217) 333-8046; thook@purdue.edu or iisg@illinois.edu
Internet: http://www.iisgcp.org/funding/core.html
Sponsor: U.S. Department of Commerce
1401 Constitution Avenue, NW, Room 5128
Washington, DC 20230

U.S.C Pentecostal and Charismatic Research Initiative Individual Grants 4605
The Pentecostal and Charismatic Research Initiative (PCRI) request for proposals (RFP) for Individual Grants is designed to provide a scholarly framework based in the social sciences and financial support for research into Pentecostalism and charismatic Christianity in four specific geographic regions: Africa, Asia, Latin America, and the former Soviet Union. PCRI will award up to 15 grants in this area. $1,500,000 in grant funding is available for the individual grants RFP.
Requirements: Doctoral students and scholars affiliated with universities, colleges, and research centers may apply. Applicants may request up to $100,000 over two years (24 months) to support scholarly research, although not all projects can be supported at this level. Research projects should engage topics of interest not only to Pentecostal specialists, but to a broader group of scholars interested in the dynamics of social change and the potentially dynamic role of religion in this process. Proposals must be focused on one of four regions (Africa, Asia, Latin America, or the former Soviet Union). Comparative research projects are also eligible, including projects that explore the connections between Western countries and one or more of the four regions mentioned above. PCRI is particularly interested in the economic implications of this fast growing movement. However, scholars from all social science disciplines are encouraged to apply (e.g., anthropologists, sociologists, political scientists, etc.). Grants must be administered through a scholarly institution such as a university, seminary, or research center.
Restrictions: Projects involving only theological questions are not eligible under this RFP, but collaborative research involving theologians may be appropriate. No more than 10% of a grant may be devoted to a combination of indirect costs, overhead or administration.
Geographic Focus: All States
Date(s) Application is Due: Aug 1
Amount of Grant: Up to 100,000 USD
Contact: Ms. Brie Loskota; (213) 740-8562; fax (213) 740-5810; pcri@usc.edu
Internet: http://www.usc.edu/schools/college/crcc/pcri/individuals/
Sponsor: Center for Religion and Civic Culture, University of Southern California
825 Bloom Walk, ACB 439
Los Angeles, CA 90089-1481

U.S.C Pentecostal and Charismatic Research Initiative Regional Center Grants 4606
The Pentecostal and Charismatic Research Initiative (PCRI) request for proposals (RFP) for regional centers is designed to provide a scholarly framework based in the social sciences and financial support for research into Pentecostal and charismatic Christianity in four specific geographic regions: Africa, Asia, Latin America, and the former Soviet Union. PCRI will award up to seven center grants. $2,000,000 in grant funding is available for the regional center RFP. Grants will be awarded to a single institution, but a project may include scholars from surrounding cities or even countries in the region.
Requirements: Grants are intended to strengthen or establish research hubs in a region and create new clusters of scholars working together. These centers should develop consultations and conferences, and provide logistical support for the associated scholars, graduate students, and research staff. Research projects should engage topics of interest not only to Pentecostal specialists, but to a broader group of scholars interested in the dynamics of social change and the potentially dynamic role of religion in this process. Funding will be limited to research that has the potential to significantly impact the 'big picture' understanding of Pentecostal and charismatic Christianity and that addresses the role of religion in human society. Proposals must be focused on one of four regions (Africa, Asia, Latin America, or the former U.S.SR). Comparative research projects are also eligible, including projects that explore the connections between Western countries and one or more of the four regions mentioned above. While proposals may include scholars from North America and Europe, the core component of research staff must be located in the region of focus. Also, at least one of the Principal Investigators must reside in Africa, Asia, Latin America, or the former Soviet Union. Grants will be awarded based on academic merit and not on the affiliation (secular, denomination, etc.) of the research center.
Restrictions: No more than 10% of a grant may be devoted to a combination of indirect costs, overhead or administration.
Geographic Focus: All States
Date(s) Application is Due: Aug 1
Amount of Grant: Up to 500,000 USD
Contact: Ms. Brie Loskota; (213) 740-8562; fax (213) 740-5810; pcri@usc.edu
Internet: http://www.usc.edu/schools/college/crcc/pcri/centers/
Sponsor: Center for Religion and Civic Culture, University of Southern California
825 Bloom Walk, ACB 439
Los Angeles, CA 90089-1481

U.S. CRDF CGP-Climate Change and Energy Competition For Joint U.S. Ukrainian Research Teams 4607
The U.S. Civilian Research and Development Foundation (CRDF) is a nonprofit organization authorized by the U.S. Congress and established in 1995 by the National Science Foundation. This unique organization promotes international scientific and technical collaboration through grants, technical resources, and training. CRDF is currently accepting proposals for the 2009 Cooperative Grants Program (CGP) Climate Change and Energy Competition. This competition is open only to joint teams of scientists from the United States and Ukraine. This program provides awards of up to $90,000 U.S. Dollars (USD) for up to two years of support to joint research teams of U.S. and Ukrainian scientists.
Requirements: CRDF will accept applications from all scientific disciplines as long as the research proposed is basic in nature. The program will consider proposals from any of the following disciplines: Biology, Chemistry, Geology, Mathematics, Physics, or Electronic, Material, Mechanical, and Manufacturing Engineering. CRDF strongly encourages applicants to contact a CGP Program Manager at cgp@crdf.org if they are unsure whether their research topic is eligible for this competition. Eligible research projects should be submitted to one of the two Focus Areas of this competition: Climate Change Impacts and Climate Change Solutions. Proposals eligible for Focus Area I: Climate Change Impacts should i) address the impacts of climate change on human and biological systems ii) measure, monitor, and model the processes that will provide accurate future projections of climatic and environmental changes or iii) study regionally specific feedbacks associated with the climate change. Proposals eligible for Focus Area II: Climate Change Solutions should address solutions to climate change such as developing technologies in energy, land and water management, and materials science that may mitigate the impact of climate change. Each proposal submitted to CGP must have one Ukrainian Principal Investigator and one U.S. Principal Investigator, who share overall responsibility for the project in their respective countries, coordinating all project participants and institutions. Each Principal Investigator must: possess the degree of Ph.D., M.D., or the equivalent research experience; have at least five (5) scientific publications in peer-reviewed scientific literature (if the applicant has received his or her Ph.D., M.D., or the equivalent in the past six (6) years, three (3) publications are required), and, work full-time in a civilian research environment. The Ukrainian Principal Investigator must be a citizen and permanent resident of Ukraine. All Ukrainian participants must reside in Ukraine for the majority of the grant period. The U.S. Principal Investigator and participants on the U.S. team must be U.S. citizens or permanent residents of the U.S. and must reside in the U.S. during the entire course of the project. Graduate students on the U.S. team may be foreign nationals, but they must be enrolled in an accredited degree program at a U.S. institution. All projects must be oriented toward non-military objectives and must be carried out in a civilian research environment. (U.S. Government Laboratory employees working on civilian-oriented research and development projects are eligible to apply to this program.) For Ukrainian applicants only: Ukrainian-Language Text of Project Narrative, Expert Conclusion, and Author's Statement submitted to CRDF counterpart agency. For U.S. For-Profit companies only: Letter of Commitment to contribute 10% of Ukrainian expenses and coverage of all U.S. expenses. For Ukrainian For-Profit companies only: Letter of Commitment to contribute up to 5-10% of total Ukrainian expenses. For proposals with in-kind contributions: Letter of Commitment from contributing U.S. and/ or Ukrainian institution is required. For proposals involving Human Subjects research: Human Subjects Protocol Summary Form and supporting information, such as Institutional Review Board/Internal Ethical Committee Approval, Human Subjects Governmental Approval, and Confirmation of Federal-Wide Assurance. No more than 84% ($76,000) of the funds awarded to each project must be used for the Ukrainian team's expenses, including institutional support. No more than 16% ($14,000) may be used for the U.S. team's expenses.
Geographic Focus: All States, Ukraine
Date(s) Application is Due: Dec 1
Amount of Grant: 90,000 USD
Contact: Mr. Stuart Politi; (703) 526-9720; fax (703) 526-9721; cgp@crdf.org
Internet: http://www.crdf.org/funding/funding_show.htm?doc_id=1014809
Sponsor: U.S. Civilian Research and Development Foundation
1530 Wilson Boulevard, 3rd Floor
Arlington, VA 22209

U.S. CRDF CGP-RFBR Climate Change and Energy Competition For Joint U.S.-Russian Research Teams 4608
The U.S. Civilian Research and Development Foundation (CRDF) is a nonprofit organization authorized by the U.S. Congress and established in 1995 by the National Science Foundation. This unique organization promotes international scientific and technical collaboration through grants, technical resources, and training. CRDF and the Russian Foundation for Basic Research (RFBR) are currently accepting proposals for the 2009 Cooperative Grants Program (CGP)-RFBR Climate Change and Energy Competition. This competition is open only to joint teams of scientists from the United States and the Russian Federation. The total maximum grant amount is $55,000 U.S. Dollars (USD) from CRDF and 1,200,000 Russian Rubles (RUB) from RFBR (a combined total of approximately $93,000 USD) for up to two years of research and development.
Requirements: CRDF and RFBR will accept applications from all scientific disciplines as long as the research proposed is basic in nature. All proposals must be submitted to both CRDF and RFBR. The program will consider proposals from any of the following disciplines: Biology, Chemistry, Geology, Mathematics, Physics, or Engineering. CRDF strongly encourages applicants to contact a CGP Program Manager at cgp@crdf.org if they are unsure whether their research topic is eligible for this competition. Eligible research projects should be submitted to one of the two Focus Areas of this competition: Climate Change Impacts and Climate Change Solutions. Proposals eligible for Focus Area I: Climate Change Impacts should i) address the impacts of climate change on human and biological systems ii) measure, monitor, and model the processes that will provide accurate future projections of climatic and environmental changes or iii) study regionally specific feedbacks associated with the climate change. Proposals eligible for Focus Area II: Climate Change Solutions should address solutions to climate change such as developing technologies in energy, land and water management, and materials science that may mitigate the impact of climate change. Each proposal submitted to the CGP-RFBR Climate Change and Energy competition must have one Russian Principal Investigator and one U.S. Principal Investigator, who share overall responsibility for the project in their respective countries, coordinating all project

participants and institutions. Each Principal Investigator must: possess the degree of Ph.D., M.D., or the equivalent research experience; have at least five (5) scientific publications in peer-reviewed scientific literature (if the applicant has received his or her Ph.D., M.D., or the equivalent in the past six (6) years, three (3) publications are required); and, work full-time in a civilian research environment. (U.S. Government Laboratory employees working on civilian-oriented research and development projects are eligible to apply to this program.) The Russian Principal Investigator and participants of the Russian team must be citizens or permanent residents of the Russian Federation. All Russian participants must reside in the Russian Federation for the majority of the grant period. The U.S. Principal Investigator and participants on the U.S. team must be U.S. citizens or permanent residents and must reside in the U.S. during the course of the CGP-RFBR project. Graduate students on the U.S. team may be foreign nationals, but they must be enrolled full-time in an accredited degree program at a U.S. institution. All projects must be oriented toward non-military objectives and must be carried out in a civilian research environment. Special *Requirements:* Clearance for Former Weapons Scientists; Proposal Overview Form submitted to CRDF Moscow office; U.S. For-Profit companies only: Letter of Commitment to contribute 10% of Russian expenses and coverage of all U.S. expenses; Russian For-Profit companies only: Letter of Commitment to contribute up to 5-10% of total Russian expenses; Proposals with In-Kind contributions: Letter of Commitment from contributing U.S. and/ or Russian institution. For proposals involving Human Subjects research: Protocol Summary Form and supporting information, such as Institutional Review Board/Internal Ethical Committee Approval, Human Subjects Governmental Approval, and Confirmation of Federal-Wide Assurance.
Restrictions: Proposals oriented towards applied research will not be accepted.
Geographic Focus: All States, Russia
Date(s) Application is Due: Dec 1
Amount of Grant: 93,000 USD
Contact: Mr. Stuart Politi; (703) 526-9720; fax (703) 526-9721; cgp@crdf.org
Internet: http://www.crdf.org/funding/funding_show.htm?doc_id=1014806
Sponsor: U.S. Civilian Research and Development Foundation
1530 Wilson Boulevard, 3rd Floor
Arlington, VA 22209

U.S. CRDF CGP ETF 2009 Energy Research Competition For Joint U.S. Estonian Research Teams 4609

The U.S. Civilian Research and Development Foundation (CRDF) is a nonprofit organization authorized by the U.S. Congress and established in 1995 by the National Science Foundation. This unique organization promotes international scientific and technical collaboration through grants, technical resources, and training. In cooperation with the Estonia Science Foundation (ETF), CRDF is announcing a joint 2009 Energy Research Competition which provides awards of approximately $50,000 for up to two years of support to joint research teams of U.S. and Estonian scientists and engineers. Applications must fall under one of the following two (2) areas of energy research and development: Clean Fossil Fuels Energy and Renewable Energy.
Requirements: All proposals must be related to renewable energy or clean fossil fuel energy. The program will consider proposals from all natural sciences, specifically from any of the following disciplines: Biology, Chemistry, Geology, Mathematics, Physics, or Electronic, Material, Mechanical, and Manufacturing Engineering. CRDF strongly encourages applicants to contact a CGP Program Manager at cgp@crdf.org if they are unsure whether their research topic is eligible for this competition. The total maximum grant size is $50,000 for up to two years of research and development. The Estonian team is eligible to receive 78% of the funds for their expenses, including institutional support. The U.S. team is eligible for 22% of the award total. These funds will be allocated separately and administered directly to the individuals and institutions involved. Proposals submitted to the competition must have one Estonian Principal Investigator and one U.S. Principal Investigator, who share the overall responsibility for the project in their respective countries and coordinate all project participants and institutions. Each Principal Investigator must: possess the degree of Ph.D., M.D., or the equivalent research experience; have at least five (5) scientific publications in peer-reviewed scientific literature; and, work full-time in a civilian research environment. The Estonian Principal Investigator and participants on the Estonian team must be citizens or permanent residents of Estonia. All Estonian participants must reside in Estonia during the project period. The U.S. Principal Investigator and participants of the U.S. team must be U.S. citizens or permanent residents of the U.S. and must reside in the U.S. during the project period. U.S. graduate students may be foreign nationals, but they must be enrolled in an accredited degree program at a U.S. institution. All projects must be oriented toward non-military objectives and must be carried out in a civilian research environment. U.S. project participants are encouraged to travel to the partner labs to ensure an effective exchange of ideas and research findings. For U.S. For-Profit companies only: Letter of Commitment to contribute 10% of Estonian expenses and coverage of all U.S. expenses is required. For Estonian For-Profit companies only: Letter of Commitment to contribute up to 5-10% of total Estonian expenses is required. For Proposals with in-kind contributions: Letter of Commitment from contributing U.S. and/ or Estonian institution is required.
Restrictions: CRDF reserves the right to restrict the participation of any individual or institution from its programs. CRDF complies with all U.S. laws and regulations pertaining to export control and the participation of foreign nationals or institutions in its activities. It is the policy of CRDF not to conduct any transactions with U.S. restricted entities without appropriate authorization from the U.S. Government.
Geographic Focus: All States, Estonia
Date(s) Application is Due: Dec 1
Amount of Grant: 50,000 USD
Contact: Mr. Stuart Politi; (703) 526-9720; fax (703) 526-9721; cgp@crdf.org
Internet: http://www.crdf.org/funding/funding_show.htm?doc_id=1014817
Sponsor: U.S. Civilian Research and Development Foundation
1530 Wilson Boulevard, 3rd Floor
Arlington, VA 22209

U.S. CRDF Leishmaniasis: Collaborative Research Opportunities in North Africa and the Middle East 4610

The U.S. Civilian Research and Development Foundation (CRDF) is a nonprofit organization authorized by the U.S. Congress and established in 1995 by the National Science Foundation. This unique organization promotes international scientific and technical collaboration through grants, technical resources, and training. CRDF, with funding from the National Institute for Allergy and Infectious Diseases (NIAID), announces a targeted, collaborative grant competition on Leishmaniasis in the North Africa and the Middle East (MENA) Region. This grant competition is a follow-on activity to the June 2009 interdisciplinary research conference in Tunisia sponsored by NIAID that brought together regional and international scientists to discuss research opportunities on sand fly biology, leishmania parasites, immunology, leishmaniasis epidemiology, animal reservoirs, clinical manifestations, genetics, diagnostics, treatment, and prevention of leishmaniasis in North Africa and the Middle East. These follow-on grants will provide up to one- year's support to joint research teams of U.S. and MENA scientists, with the possibility of up to a one year no-cost extension, if justified. CRDF and NIAID anticipate that the research implemented through these awards will lay the foundation for future grant proposals to NIAID and other funding agencies. The average anticipated grant will provide up to $30,000 U.S. dollars for a period of one year. At least 80% of the funds awarded to each project will be used for project-related expenses of the MENA component of the research team, including institutional support. No more than 20% may be used for U.S. team expenses. As part of the grant, the CRDF strongly encourages a member of the MENA team to visit the U.S. team's laboratory, and members of the U.S. team are encouraged to visit the MENA laboratory.
Requirements: Each proposal submitted to the CRDF must have at least one MENA Principal Investigator and at least one U.S. Principal Investigator, who share overall responsibility for the project in their respective countries, coordinating all project participants and institutions. Each proposal must meet each of the following eligibility criteria: 1) Only workshop attendees and their institutions are eligible to apply. 2) Each Principal Investigator must work full-time in a civilian research environment. (U.S. Government Laboratories working on civilian-oriented R&D projects are eligible to apply to this program.) 3) The U.S. Principal Investigator and participants of the U.S. team may be foreign nationals, but must reside in the U.S during the course of the grant award. Graduate students on the U.S. team may be foreign nationals, but they must be enrolled in an accredited degree program at a U.S. institution. A PI from Navy Medical Research Unit 3 (NAMRU3) may serve as either a U.S. or MENA researcher. 4) All projects must be in one of the following subject areas: Sand fly biology; Leishmania Parasites; Immunology; Leishmaniasis Epidemiology; Animal reservoirs; Clinical manifestations, treatment, and prevention of leishmaniasis in North Africa and the Middle East (but not clinical trials involving interventions); Genetics; Diagnostics. 5) Only one proposal will be accepted from a U.S. or MENA PI. 6) All projects must be oriented toward non-military objectives and must be carried out in a civilian research environment. Use of human subjects requires appropriate approvals from all institutions involved including Local Ethics Committee (LEC) and U.S. IBR. Middle East countries: Afghanistan, Behrain, Iran, Iraq, Israel, Jordan, Kyrgyzstan, Kuwait, Lebanon, Oman, Pakistan, Palestine, Qatar, Saudi Arabia, Syria, Tajikistan, Turkey, Turkmenistan, United Arab Emirates, Uzbekistan, Yemen. North African countries: Algeria, Egypt, Libya, Morocco, Sudan, Tunisia, Western Sahara.
Restrictions: No clinical trials involving interventions will be accepted for this competition.
Geographic Focus: All States
Date(s) Application is Due: Sep 30
Amount of Grant: Up to 30,000 USD
Contact: Mr. Stuart Politi; (703) 526-9720; fax (703) 526-9721; cgp@crdf.org
Internet: http://www.crdf.org/funding/funding_show.htm?doc_id=949719
Sponsor: U.S. Civilian Research and Development Foundation
1530 Wilson Boulevard, 3rd Floor
Arlington, VA 22209

U.S. CRDF Science and Technology Entrepreneurship Program Business Partnership Grant Competition Armenia 4611

The U.S. Civilian Research and Development Foundation (CRDF) is a nonprofit organization authorized by the U.S. Congress and established in 1995 by the National Science Foundation. This unique organization promotes international scientific and technical collaboration through grants, technical resources, and training. The STEP Business Partnership Grants Competition, conducted by CRDF, the Ministry of Economy of RA and Enterprise Incubator Foundation (EIF), promotes research and development (R&D) partnerships between companies (Company) and teams of researchers/scientists (Technology Team) to develop new commercial opportunities of economic benefit to both parties.
Requirements: Projects must have R&D as their core task and must include preliminary market assessment, customer needs analysis and business development components. Projects that demonstrate their ability to enable revenue creation and/or attract non-STEP Partner organization sources of funding will be given priority consideration in the selection process. All goals and outcomes of the projects must be defined and measurable. STEP BPG projects are limited to a maximum of 12 months. Funds for projects are provided by the CRDF and the Ministry of Economy of RA. Funds are limited to $15,000. At least 10% of company cash contribution is mandatory. New Science Team/Company Partnership is preferable. All funds must be used for Science Team expenses. At first, written proposals should be submitted to EIF for review on eligibility and completeness of information provided. Complete proposals must be submitted to CRDF via electronic proposal submission (EPS) and emailed to the EIF by (18:00) local time on the following submission deadlines to be eligible. Proposals for EIF should be both in English and Armenian. Each proposal submitted to the STEP BPG is required to have a single Principal Investigator (PI) from a Company and a single Principal Investigator (PI) from the Science Team who will be

responsible for coordinating all participating personnel, organizations, and work. Scientists and engineers may participate concurrently on more than one EIF/CRDF grant, but cannot be the PI on more than one active EIF/CRDF project or exceed full-time work status on the combined EIF/CRDF projects. Projects are required to be in research and development fields that are oriented toward commercial, non-military objectives and must be carried out in a civilian research environment. The participants of the Science Team are required to be scientists or engineers at institutes, universities, companies, or other organizations involved in research and development. The PIs and all participants of the Science Team are required to be citizens of and must currently reside and work in Armenia. The primary partner of the Company Team is required to be a for-profit company based in Armenia. In order to be considered Armenia-based, a company must have registered offices in Armenia and its facilities and personnel associated with the project must be located in this country. Joint U.S.-Armenia for-profit ventures are eligible to participate. The PI of the Company Team is required to be a citizen of Armenia and must currently reside and work in Armenia.
Restrictions: The Company and the organization employing the team of scientists may not be subsidiaries of or owned, either wholly or partially, by one another. Organizations and individuals of the Science Team working for organizations that are barred by the U.S. Government from receiving U.S. financial support, materials, or services are not eligible to participate in BPG program.
Geographic Focus: All States, Ukraine
Date(s) Application is Due: Oct 5
Amount of Grant: 15,000 USD
Contact: Natalia Pipia; (703) 526-9720; fax (703) 526-9721; step@crdf.org
Internet: http://www.crdf.org/funding/funding_show.htm?doc_id=990740
Sponsor: U.S. Civilian Research and Development Foundation
1530 Wilson Boulevard, 3rd Floor
Arlington, VA 22209

U.S. CRDF STEP Business Partnership Grant Competition Ukraine **4612**

The U.S. Civilian Research and Development Foundation (CRDF) is a nonprofit organization authorized by the U.S. Congress and established in 1995 by the National Science Foundation. This unique organization promotes international scientific and technical collaboration through grants, technical resources, and training. The Science and Technology Entrepreneurship Program Business Partnership Grants competition (STEP BPG), conducted by CRDF and the Ministry of Education and Science (MES) promotes research and development (R&D) partnerships between companies (Company) and teams of scientists (Science Team) to develop new commercial opportunities of economic benefit to both parties. At the first stage of the STEP Program, two workshops will be carried out in Lviv and Sevastopol, Ukraine which will provide basic training on technology commercialization, and intellectual property protection. Workshops will also further clarify STEP program goals and objectives and will introduce the BPG competition terms and conditions.
Requirements: STEP BPG projects must have research and development (R&D) as their core task and must include a preliminary market assessment; a customer needs analysis, and business development components. Projects that demonstrate their ability to enable revenue creation and/or attract non-STEP Partner organization sources of funding will be given priority consideration in the selection process. All goals and outcomes of the projects must be well defined and measurable. STEP BPG projects are limited to a maximum of 12 months. Funds for projects are provided by the CRDF and MES. Funds are limited to $24,000. All funds must be used for Science Team expenses. In support of the goals of the Business Partnership Grant competition to increase the number of linkages between the science and business sectors, preference will be given to newly-formed science- business partnerships. Preference will also be given to projects with at least 10% company cash contribution. Each proposal submitted to the STEP BPG competition is required to have a single Principal Investigator (PI) from a Company and a single Principal Investigator (PI) from the Science Team who will be responsible for coordinating all participating personnel, organizations, and work. Scientists and engineers may participate concurrently on more than one CRDF grant, but cannot be the PI on more than one active CRDF project or exceed full-time work status on the combined CRDF projects. The STEP BPG projects are required to be in research and development fields that are oriented toward commercial, non-military objectives and must be carried out in a civilian research environment. All proposals are required to be submitted to CRDF in English. Two copies of all proposals should be submitted in printed form to the MES in Ukrainian. The participants of the Science Team are required to be scientists or engineers at institutes, universities, companies, or other organizations involved in research and development. The PIs and all participants are required to be citizens of and must currently reside and work in Ukraine. The primary partner is required to be a for-profit company based in Ukraine or in the U.S; thus, a company must have registered offices in Ukraine or in the U.S. and its facilities and personnel associated with the project must be located in these countries. Joint U.S.-Ukrainian for-profit ventures are eligible to participate. The PI of the Company Team is required to be a citizen of and must currently reside and work in Ukraine. Note: The MES and CRDF discourage the inclusion of an 'administrative' superior in a proposal in any role for which individual financial support is requested. An 'Administrative Superior' is defined as an individual who has direct administrative authority over the Principal Investigator, and who works within the same division, laboratory, or unit as the Principal Investigator. Institutional leadership including university rectors and vice rectors and institute directors and deputy directors are also defined as 'administrative superiors.' This definition does not apply to other individuals of higher 'rank' within the institution but who work outside the division, laboratory or unit of the Principal Investigator.
Restrictions: The Company and the organization employing the team of scientists may not be subsidiaries of or owned, either wholly or partially, by one another. Organizations and individuals working for organizations that are barred by the U.S. Government from receiving U.S. financial support, materials, or services are not eligible to participate in CRDF programs. No grant funds can be used for company expenses.
Geographic Focus: All States, Ukraine
Date(s) Application is Due: Dec 20
Amount of Grant: 24,000 USD
Contact: Natalia Pipia; (703) 526-9720; fax (703) 526-9721; step@crdf.org
Internet: http://www.crdf.org/funding/funding_show.htm?doc_id=1034127
Sponsor: U.S. Civilian Research and Development Foundation
1530 Wilson Boulevard, 3rd Floor
Arlington, VA 22209

USDA 1890 Institution Teaching, Research, and Extension Capacity Building Grants **4613**

The 1890 Institution Teaching, Research, and Extension Capacity Building Grants (CBG) are intended to strengthen teaching, research and extension programs in the food and agricultural sciences by building the institutional capacities of the 1890 land-grant institutions, and Tuskegee University. The CBG program supports projects that strengthen teaching programs in the food and agricultural sciences in the need areas of curriculum design, materials development, and faculty development. CBG also supports projects that strengthen research and extension programs in need areas of studies and experimentation, and extension program development support systems. The CBG also support integrated project grants. Applications submitted to CBG must address at least one of the following NIFA strategic goals: sustainable bioenergy; food security; childhood obesity prevention; climate change; or food safety. The electronic applications, abstracts of funded projects, and specific guidelines are available at the website.
Requirements: Applicants must be 1890 land-grant institutions to be eligible for funding.
Geographic Focus: Alabama, Arkansas, Delaware, Georgia, Kentucky, Louisiana, Maryland, Mississippi, Nebraska, North Carolina, Oklahoma, South Carolina, Tennessee, Texas, Virginia, West Virginia
Date(s) Application is Due: Mar 27
Amount of Grant: 150,000-600,000 USD
Contact: Edwin Lewis; (202) 690-4565; fax (202) 720-2030; elewis@nifa.usda.gov
Internet: http://www.nifa.usda.gov/fo/1890capacity.cfm
Sponsor: U.S. Department of Agriculture
1400 Independence Avenue, SW, Stop 2201
Washington, DC 20250-2201

USDA Agriculture and Food Research Initiative Competitive Grants-Food Security **4614**

The USDA Agricultural and Food Research Initiative Competitive Grants focuses on the societal challenge to keep American agriculture competitive while ending world hunger. In the Food Security Challenge Area, specific program areas are designed to achieve the long-term outcome of increasing food availability and decreasing the number of food insecure individuals. Project types supported include single-function extension projects, multi-function integrated research, education and/or extension projects, and Food and Agricultural Science Enhancement (FASE) Grants. Applicants should refer to the online request for application to find the specific contact person for their project. Information about the application, criteria, and particular guidelines for each grant area are located at the website. Funding is based on grant categories.
Requirements: All proposals submitted to this challenge area must be multidisciplinary in which two or more disciplines collaborate to address a common problem. These collaborations, where appropriate, may involve an integration of the biological, physical, chemical, or social sciences.
Geographic Focus: All States
Date(s) Application is Due: Mar 1
Contact: Margo Holland; (202) 401-5044; mholland@nifa.usda.gov
Internet: http://www.csrees.usda.gov/funding/rfas/afri.html
Sponsor: U.S. Department of Agriculture
1400 Independence Avenue, SW, Stop 2201
Washington, DC 20250-2201

USDA Alaska Native-Serving and Native Hawaiian-Serving Institutions Education Competitive Grants **4615**

The purpose of the program is to promote and strengthen the ability of Alaska Native-Serving Institutions and Native Hawaiian-Serving Institutions to carry out education, applied research, and related community development programs. NIFA intends this program to address educational needs, as determined by each institution, within a broadly defined arena of food and agricultural sciences-related disciplines. Priority will be given to those projects that enhance educational equity for under represented students; strengthen institutional educational capacities; prepare students for careers related to the food, agricultural, and natural resource systems of the United States; and maximize the development and use of resources to improve food and agricultural sciences teaching programs. Samples of previous projects are available at the website. Applications must be filed electronically through Grants.gov.
Requirements: Private and state controlled institutions of higher education are eligible to apply. Individual public or private, nonprofit Alaska Native-serving and Native Hawaiian-serving institutions of higher education that meet the definitions of Alaska Native-Serving Institution or Native Hawaiian-Serving Institution established in Title III, Part A of the Higher Education Act of 1965, are also eligible to apply.
Geographic Focus: Alaska, Hawaii
Date(s) Application is Due: Mar 1
Contact: Saleia Afele-Faamuli; (202) 720-0384; sfaamuli@nifa.usda.gov
Internet: http://www.nifa.usda.gov/funding/rfas/alaska.html
Sponsor: U.S. Department of Agriculture
1400 Independence Avenue, SW, Stop 2201
Washington, DC 20250-2201

USDA Animal Health and Disease Research Grants **4616**

The purpose of this funding is to increase animal health and disease research activities at accredited State veterinary schools or colleges or agricultural experiment stations that conduct animal health and disease research. To support animal health and disease research at eligible Schools and Colleges of Veterinary Medicine, and State Agricultural Experiment Stations whose purpose is to improve the health and productivity of food animals and horses through effective prevention, control or treatment of disease, reduction of losses from transportation and other hazards, and protect human health through control of animal diseases transmissible to people.

Requirements: Eligibility is restricted to schools and colleges of veterinary medicine and state agricultural experiment stations.

Geographic Focus: All States

Amount of Grant: Up to 100,000 USD

Contact: Peter Johnson; (202) 401-1896; fax (202) 205-3641; pjohnson@nifa.usda.gov

Internet: https://www.cfda.gov/?s=program&mode=form&tab=step1&id=1ddd2def6f16faee9d008a7ed30a8a7b

Sponsor: U.S. Department of Agriculture
1400 Independence Avenue, SW, Stop 2201
Washington, DC 20250-2201

USDA Basic and Applied Research Grants **4617**

USDA Basic and Applied Research Grant program has as its primary mission to make agricultural research discoveries, evaluate alternative ways of attaining research goals, and provide scientific technical information. In conducted these project grants, research is conducted that is in cooperation with and is correlated with the Agricultural Research Service's in-house research programs and projects. Limited discretionary research funds are periodically made available. A peer review panel considers each proposal, evaluates the qualifications of applicants in line with research to be undertaken and determines priority for final negotiations of the grant. The grant range is $1,000 to $25,000, with an average award being $3,000. There are no identified deadlines.

Requirements: Nonprofit institutions of higher education or other nonprofit research organizations, whose primary purpose is conducting scientific research, are eligible to apply. Preapplication coordination is required. Environmental impact information is not required for this program.

Geographic Focus: All States

Amount of Grant: 1,000-25,000 USD

Contact: Kathleen S. Townson; (301) 504-1702; kathleen.townson@ars.usda.gov

Internet: http://www.ars.usda.gov/Research/Research.htm

Sponsor: U.S. Department of Agriculture
Jamie L. Whitten Building
Washington, DC 20250

USDA Biotechnology Risk Assessment Grants **4618**

The purpose of the USDA Biotechnology Risk Assessment Grants (BRAG) program is to support the generation of new information that will assist Federal regulatory agencies in making science-based decisions about the effects of introducing into the environment genetically engineered organisms (GE), including plants, microorganisms (including fungi, bacteria, and viruses), arthropods, fish, birds, mammals and other animals excluding humans. The BRAG program accomplishes its purpose by providing Federal regulatory agencies with scientific information relevant to regulatory issues. Although investigators are not required to perform actual risk assessments in the research they propose, they should design studies that will provide information useful to regulators for making science-based decisions in their assessments of genetically-modified organisms. Organizations are required to submit the Grants.gov online application available on its website. Specific guidelines for the individual proposals and Grants.gov are located at the NIFA website and within the request for application.

Requirements: Eligible applicants include any public or private research or educational institution or organization.

Restrictions: Submission of a full proposal with a corresponding letter of intent will not be considered for review. Awards will not be made for food safety risk assessment or risk management, health risk assessment or risk management of humans or domestic food animals exposed to GE organisms, social or economic research, methods for seed storage, clinical trials, commercial product development, product marketing strategies, product stewardship, or other research deemed inappropriate to risk assessment or risk management relative to the environment.

Geographic Focus: All States

Date(s) Application is Due: Feb 1

Amount of Grant: 25,000-1,000,000 USD

Contact: Shing Kwok; (202) 401-6060; skwok@nifa.usda.gov
Jack Okamuro, Program Leader for Crop Production and Protection; (301) 504-5912 or (202) 285-9520 cell; jack.okamuro@ars.usda.gov

Internet: http://www.nifa.usda.gov/funding/rfas/biotech_risk.html

Sponsor: U.S. Department of Agriculture
1400 Independence Avenue, SW, Stop 2201
Washington, DC 20250-2201

USDA Cochran Fellowships **4619**

The Cochran Fellowship Program (CFP) provides U.S.-based agricultural training opportunities for senior and mid-level specialists and administrators from public and private sectors who are concerned with agricultural trade, agribusiness development, management, policy, and marketing. The program offers short-term training opportunities, most ranging from 1-3 weeks depending on the objectives of the program. Participants meet with professionals in their fields, participate in field observations and industry visits, experience on-the-job training, attend university courses and seminars, or participate in a combination of these experiences. CFP works closely with U.S. agricultural trade and market development associations to identify potential areas of training for countries participating in the Program. Recruitment typically occurs between December and March each year. Most training activities are then conducted between March and October. However, most Eurasian recruitment takes place from June to August, and most Eurasian training activities are conducted January to June. Applicants should refer to the website for guidelines and contact information for their specific country.

Requirements: Program participation is open to the staff of agribusinesses, government departments, universities, and other agricultural organizations. In their own countries, applicants may be managers, technicians, scientists, professors, administrators, or policy makers. All participants must be in good health. Applicants are required to have a thorough physical exam prior to traveling to the U.S., as well as a medical letter stating he/she is fit to travel. Participants who are proficient in the oral and written usage of the English language are preferred. Applications are required to be completed in English (preferably typed). The officer conducts an initial screening of applications to determine whether the proposed study program is appropriate and is consistent with the country strategy. The application must include a detailed description of the specific training requested by the applicant.

Restrictions: The Cochran Program does not fund international travel to and from the participating country, except in the following countries: Armenia, Azerbaijan, Bulgaria, Georgia, Kazakhstan, Kyrgyzstan, Moldova, Tajikistan, Turkmenistan, and Uzbekistan.

Geographic Focus: All Countries

Contact: Scott Lewis, Chief, Cochran Fellowship Branch; (202) 720-9033 or (202) 690-1734; fax (202) 690-0349; CochranFellowships@fas.usda.gov

Internet: http://www.fas.usda.gov/icd/cochran/cochran.asp

Sponsor: U.S. Department of Agriculture
1400 Independence Avenue SW
Washington, DC 20250

USDA Cooperative Extension System Education Grants **4620**

The program seeks to help people improve their lives and communities through an educational process that uses scientific knowledge focused on issues critical to the economic, agricultural, societal, health/safety, and environmental progress of all Americans. In addition, the program seeks to identify and solve farm, home, and community problems through the practical application of research findings of the USDA and land-grant colleges and universities. The cooperative extension system is a future-oriented, self-renewing, national educational network providing excellence in programs that focus on contemporary issues and needs of people.

Requirements: Grants are available to designated 1862, 1890, and 1994 land-grant institutions in the 50 states and Puerto Rico, Guam, the Virgin Islands, American Samoa, Micronesia, Northern Marianas, and the District of Columbia.

Geographic Focus: All States

Amount of Grant: 890,000-20,000,000 USD

Contact: Andrea L. Brandon, Assistant Director; (202) 720-9181 or (202) 720-4423; abrandon@nifa.usda.gov

Internet: http://www.csrees.usda.gov/nea/education/education.cfm

Sponsor: U.S. Department of Agriculture
1400 Independence Avenue, SW, Stop 2201
Washington, DC 20250-2201

USDA Decadal and Regional Climate Prediction Using Earth System Models Grants **4621**

The USDA Decadal and Regional Climate Prediction Using Earth System Models (EaSM) Grants seek to improve and expand current modeling capabilities in order to substantively contribute to the advancement of reliable regional and decadal climate predictions. This solicitation will capitalize on the synergy between development of climate models, their use in both the assessment and attribution of climate variability and impacts, and the development of approaches to effectively inform adaptation policy. This interdisciplinary scientific challenge calls for the development and application of next-generation Earth System Models that include coupled and interactive representations of such things as ocean and atmospheric currents, human activities, agricultural working lands and forests, urban environments, biogeochemistry, atmospheric chemistry, the water cycle and land ice. This solicitation seeks to attract scientists from the disciplines of geosciences, social sciences, agricultural and biological sciences, mathematics and statistics, physics, and chemistry. Successful proposals will develop intellectual excitement in the participating disciplinary communities and engage diverse interdisciplinary teams with sufficient breadth to achieve the scientific objectives. The project encourages proposals that have strong broader impacts, including public access to data and other research products of general interest, as well as educational, diversity, or societal impacts. The following are specific areas of interest to the funding agencies for EaSM: (i) research that has the potential to dramatically improve predictive capabilities; (ii) prediction and attribution studies; (iii) research that addresses critical issues linking relevant earth system processes over a variety of spatial and temporal scales; (iv) research that examines the relationships between climate variability and change to human and natural environments from the human perspective; (v) development and applications of metrics, methods, and tools for testing and evaluating climate and climate impact predictions and their uncertainty characterization. The electronic application and specific guidelines are available at the website. The submission of the application is through the National Science Foundation.

Requirements: Proposals may only be submitted by the following: universities and colleges, including community colleges; non-profit and non-academic organizations; for-profit organizations; and federal agencies and federally funded research and development centers.

Geographic Focus: All States

Date(s) Application is Due: May 11

Amount of Grant: 300,000-1,000,000 USD
Contact: Nancy Cavallaro, National Program Leader; (202) 401-5176 or (202) 401-6070; ncavallaro@nifa.usda.gov or easm2@nsf.gov
Internet: http://www.nifa.usda.gov/fo/earthsystemmodeling.cfm
Sponsor: U.S. Department of Agriculture
1400 Independence Avenue, SW, Stop 2201
Washington, DC 20250-2201

USDA Distance Education Grants for Higher Education Institutions in Insular Areas 4622

The Distance Education Grants Program for Higher Education Institutions in Insular Areas (DEG) is a NIFA-administered competitive grants program focused on improving formal, postsecondary agricultural sciences education. The USDA program strives to strengthen resident instruction, curriculum, and teaching programs in the food and agricultural sciences in these insular areas through distance education technology. The purpose of the DEG program is to promote and strengthen the ability of insular area institutions to: (1) acquire the equipment, instrumentation, networking capability, hardware and software, digital network technology, and infrastructure necessary to teach technology in the classroom; (2) develop and provide educational services to prepare students or faculty seeking a degree or certificate; (3) provide teacher education, library and media specialist training, and preschool and teacher aid certification to individuals who seek to acquire or enhance technology skills; (4) implement a joint project to provide education regarding technology in the classroom; or (5) provide leadership development to administrators, board members, and faculty of eligible institutions with institutional responsibility for technology education. Funded projects improve the economic health and viability of rural communities through the development of degree programs emphasizing new and emerging employment opportunities in agriscience and agribusiness. Projects should also address the national challenge to increase the number and diversity of students entering other agriculture related science, technology, engineering, and mathematics (STEM) disciplines (i.e., having an agricultural sciences workforce representative of the nation's population). Funding varies according to project goals and objectives. The electronic application and specific guidelines are available at the website.
Requirements: Applications must state how the funded project will address the two distance education program goals: to increase the number and diversity of students who will pursue the complete a 2- or 4-year postsecondary degree in the food and agricultural sciences, or other STEM fields closely related to the food and agricultural sciences, and encourage study in areas that contribute to any of the five NIFA priority areas (global food security and hunger, climate change, sustainable bioenergy, childhood obesity, and food safety); and to enhance the quality of postsecondary instruction in order to help meet current and future national food and agricultural sciences workplace needs. Applications must also give a compelling educational challenge, indicating how the methodology is both unique and with merit; offer significant promise of adoption by others; and the expectation that impacts will continue once grant funds end. Individual land-grant colleges and universities and other institutions that have secured land-grant status, and are located in insular areas are eligible for awards under this program, either as direct applicants or as parties to a consortium agreement. Eligible insular areas include the Commonwealth of Puerto Rico, Guam, American Samoa, the Commonwealth of the Northern Mariana Islands, the Federated States of Micronesia, the Republic of the Marshall Islands, the Republic of Palau, and the Virgin Islands of the United States.
Geographic Focus: Guam, Puerto Rico, U.S. Virgin Islands, American Samoa
Date(s) Application is Due: Mar 1
Amount of Grant: 70,000-750,000 USD
Contact: Saleia Afele-Faamuli; (202) 720-0384; sfaamuli@nifa.usda.gov
Internet: http://www.nifa.usda.gov/fo/distanceeducationgrant.cfm
Sponsor: U.S. Department of Agriculture
1400 Independence Avenue, SW, Stop 2201
Washington, DC 20250-2201

USDA Emerging Markets Grants 4623

The purpose of the program is to assist U.S. organizations, public and private, to improve the market access and to develop and promote U.S. agricultural products and/or processes in low- to middle-income countries that offer promise of emerging market opportunities in the near to medium term. This is to be accomplished by providing U.S. technical assistance through projects and activities in those emerging markets consistent with U.S. foreign policy interests. The program funds technical assistance activities to leverage the export and marketing of U.S. agricultural products to emerging markets. The emphasis is on marketing opportunities where there are risks that the private sector would not normally undertake alone, with funding provided on a project-by-project basis. The program is intended to support primarily small- to medium-sized U.S. firms that may need federal assistance in realizing or maintaining access in overseas markets. Application deadline is announced in the Federal Register.
Requirements: Any U.S. agricultural and/or agribusiness organization, university, or state department of agriculture is eligible to participate in the program. Priority will be given to those proposals that include significant support and involvement by private industry. Proposals from research and consulting organizations will be considered if they provide evidence of substantial participation by U.S. industry. U.S. market development cooperators may seek funding to address priority, market-specific issues and to undertake activities not already serviced by or unsuitable for funding under other FAS marketing programs.
Restrictions: No proposal will be considered without the element of cost-sharing.
Geographic Focus: All States
Date(s) Application is Due: May 20
Amount of Grant: Up to 500,000 USD
Contact: Maureen Quinn; (202) 720-7115; fax (202) 720-1727; maureen.quinn@usda.gov
Internet: http://www.fas.usda.gov/mos/em-markets/em-markets.asp
Sponsor: U.S. Department of Agriculture
1400 Independence Avenue SW
Washington, DC 20250

USDA Food and Agricultural Sciences National Needs Graduate Fellowship Grants 4624

The Food and Agricultural Sciences National Needs Graduate Fellowship Grants support training students for master's and doctoral degrees in food, agricultural, and natural resource sciences. It also supports special international study or thesis/dissertation research travel allowances for eligible USDA NNF beneficiaries. The objectives of the NNF program, working collaboratively with eligible institutions, are: supporting the training of students at the master's and doctoral levels in order to fill the nationally identified expertise shortage areas in the food and agricultural sciences; encouraging and recruiting outstanding students, including those from traditionally underrepresented groups, to enter and complete graduate training in the food and agricultural sciences; increasing the quality and diversity of students completing graduate degrees in the food and agricultural sciences; increasing the number of students with graduate degrees in the food, agricultural and related sciences awarded by at least 50 per annum; and supporting at least 10 eligible National Needs Fellows to build science competency and have international exposure within the Targeted Expertise Shortage Area (TESA) in food and agricultural sciences through the IRTA awards. Applications are accepted from institutions that offer a graduate degree in at least one of the following targeted expertise shortage areas: animal and plant production; forest resources; agricultural educators and communicators; agricultural management and economics; food science and human nutrition; sciences for agricultural biosecurity; and training in integrative biosciences for sustainable food and agricultural systems. Specific guidelines and additional information, including abstracts for previously funded projects, are located at the website. The application is available at grants.gov.
Requirements: Applications may be submitted by institutions having the capacity to carry out the teaching of food and agricultural sciences, including land grant institutions, public and private colleges and institutions, and colleges and universities with significant minority enrollments. Applications also may be submitted by research foundations maintained by eligible colleges or universities. NNF stipend funds for fellowships must be used to: (1) support the same Fellow at the master's degree level who maintains satisfactory progress and is eligible to receive grant support for a maximum of 24 months during a 36 period, and/or (2) support the same Fellow at the doctoral degree level who maintains satisfactory progress and is eligible to receive grant support for a maximum of 36 months within a 42 month period.
Restrictions: Indirect costs are not eligible for funding. NIFA provides a per Fellow cost-of-education allowance each year in lieu of indirect costs.
Geographic Focus: All States
Date(s) Application is Due: Jun 1
Amount of Grant: 4,500-262,000 USD
Contact: Ray A. Ali; (202) 720-2727; fax (202) 690-0062; rali@nifa.usda.gov
Internet: http://www.nifa.usda.gov/fo/graduateandpostgraduatefellowshipsserd.cfm
Sponsor: U.S. Department of Agriculture
1400 Independence Avenue, SW, Stop 2201
Washington, DC 20250-2201

USDA Foreign Market Development Grants 4625

The Foreign Market Development Program (FMD), also known as the cooperator program, is administered by the Foreign Agricultural Service (FAS) of the U.S. Department of Agriculture (USDA). The goal of the program is to develop, maintain, and expand long-term export markets for U.S. agricultural products. The Cooperator program is designed to create, expand, and maintain foreign markets for U.S. agricultural commodities and products through cost-share assistance. Financial assistance under the Cooperator program will be made available on a competitive basis and applications will be reviewed against the evaluation criteria contained herein and in the Cooperator program regulations. All U.S. agricultural commodities, except tobacco, are eligible for consideration. Under the Cooperator program, the FAS enters into agreements with eligible nonprofit U.S. trade organizations to share the cost of certain overseas marketing and promotion activities.
Requirements: To participate in the Cooperator program, an applicant must: be a nonprofit U.S. agricultural trade organization; and agree to contribute resources to its proposed promotional activities.
Geographic Focus: All States
Contact: Maureen Quinn, Director; (202) 720-7115 or (202) 720-4327; fax (202) 720-1727; maureen.quinn@usda.gov or podadmin@fas.usda.gov
Sally Klusaritz, Deputy Director; (202) 720-3448 or (202) 720-4327; fax (202) 720-1727; sally.klusaritz@usda.gov or podadmin@fas.usda.gov
Internet: http://www.fas.usda.gov/mos/programs/fmdprogram.asp
Sponsor: U.S. Department of Agriculture
1400 Independence Avenue SW
Washington, DC 20250

USDA Hazardous Fuels Woody Biomass Utilization Grants 4626

The U.S. Forest Service, State and Private Forestry, Technology Marketing Unit, located at the Forest Products Laboratory, requests proposals for wood energy projects that require engineering services. These projects will use woody biomass, such as material removed from forest restoration activities, wildfire hazardous fuel treatments, insect and disease mitigation, forest management due to catastrophic weather events, and/or thinning overstocked stands. The woody biomass shall be used in a bioenergy facility that uses commercially proven technologies to produce thermal, electrical or liquid/gaseous bioenergy. Grant funding must be used to further the planning of such facilities by

funding the engineering services necessary for final design and cost analysis. Examples of projects might include the following: engineering design of a woody biomass boiler for steam at a sawmill, hospital or school; non-pressurized hot water system for various applications; and biomass power generation facility. Specific guidelines and current application information are available at the website.
Requirements: The following entities are eligible for funding: state and county governments; city or township governments; special district governments; independent school districts; public, private, or state controlled institutions of higher education; federally recognized tribal governments; public housing authorities; nonprofits with or without 501(c)3 status; for profit organizations; and small businesses.
Geographic Focus: All States
Date(s) Application is Due: Mar 1
Amount of Grant: 50,000-250,000 USD
Contact: Susan LeVan-Green, Program Manager; (608) 231-9518; slevan@fs.fed.us
Internet: http://www.forestsandrangelands.gov/Woody_Biomass/opportunities.shtml
Sponsor: U.S. Department of Agriculture
1400 Independence Avenue SW
Washington, DC 20250

USDA Higher Education Challenge Grants 4627
The USDA National Institute of Food and Agriculture requests applications for the Higher Education Challenge Grants Program (HEC) to stimulate and enable colleges and universities to provide the quality of education necessary to produce baccalaureate or higher degree level graduates capable of strengthening the Nation's food and agricultural, scientific, and professional workforce. Projects supported by the USDA Higher Education Challenge Grants will: address a state, regional, national, or international educational need; involve a creative or non-traditional approach toward addressing the need that can serve as a model to others; encourage and facilitate better working relationships in the university science and education community, as well as between universities and the private sector; and result in benefits that will likely transcend the project duration and USDA support. Electronic application information is available at the website.
Requirements: Applications may be submitted by: U.S. public or private nonprofit colleges and universities offering a baccalaureate or first professional degree in at least one discipline or area of the food and agricultural sciences; land-grant colleges and universities; colleges and universities having significant minority enrollments and the ability to carry out the teaching of food and agricultural sciences; and other colleges and universities having a demonstrable capacity to carry out the teaching of food and agricultural sciences.
Restrictions: Organizations are required to match 25% of the USDA funds awarded from non-federal sources. Funds may not be used for the renovation or reburbishment of research, education, or extension space; the purchase or installation of fixed equipment in such space, or to plan, repair, rehabilitate, acquire, construct building or facilities (i.e., greenhouses, laboratories, barns, or other structures). Promotional items such as t-shirts and other give-a-ways, and food functions such as cookouts or other social meal gatherings are considered entertainment, and are also not allowable expenses. Federal Cost Principles governing this grant program state that items not eligible for federal funding may also not be used as an applicant's contribution toward meeting matching requirements.
Geographic Focus: All States, All Countries
Date(s) Application is Due: Mar 30
Amount of Grant: 30,000-750,000 USD
Contact: Gregory Smith, National Program Leader; (202) 720-1973; gsmith@csrees.usda.gov or HEC@nifa.usda.gov
Internet: http://www.csrees.usda.gov/fo/fundview.cfm?fonum=1082
Sponsor: U.S. Department of Agriculture
1400 Independence Avenue, SW, Stop 2201
Washington, DC 20250-2201

USDA Hispanic-Serving Institutions Education Grants 4628
The Hispanic-Serving Institutions Education Grants are intended to promote and strengthen the ability of Hispanic-Serving Institutions to carry out higher education programs in the food and agricultural sciences. The programs aim is to attract outstanding students and produce graduates capable of enhancing the nation's food and agricultural scientific and professional work force. Award recipients may subcontract to organizations not eligible to apply provided such organizations are necessary for the conduct of the project. The budget for subcontractors must be included with the corresponding budget narrative. Applicants are encouraged to collaborate with a USDA agency. Regular project applications must be received by February 9, while collaborative applications are due March 8. Application information, including samples of application packages and frequently asked questions, are available at the website.
Requirements: Only public or other non-profit Hispanic-Serving Institutions are eligible to apply for this program. To qualify as an Hispanic-Serving Institution, applicants must at the time of application, have an enrollment of undergraduate full-time equivalent students that is at least 25 percent. By submitting an application, the organization certifies that it is eligible to receive funding under this program.
Geographic Focus: All States
Amount of Grant: Up to 1,100,000 USD
Contact: Irma Lawrence, National Program Leader; (202) 720-2082 or (202) 720-1793; fax (202) 720-2030; ilawrence@nifa.usda.gov
Internet: http://www.csrees.usda.gov/fo/fundview.cfm?fonum=1094
Sponsor: U.S. Department of Agriculture
1400 Independence Avenue, SW, Stop 2201
Washington, DC 20250-2201

USDA Integrated Pest Management Competitive Grants-Western Region 4629
The USDA Regional Integrated Pest Management Competitive Grants (RIPM) support the continuum of research and extension efforts needed to increase the implementation of IPM methods. The RIPM program supports projects that develop individual pest control tactics, integrate individual tactics into an IPM system, and develop and implement extension and education programs. The program is administered by the land-grant university system's four regional IPM Centers (North Central, Northeastern, Southern, Western) in partnership with NIFA. Funding will support projects that promote cooperative efforts across appropriate disciplines, with linkages between research and extension efforts, and components of existing or emerging pest management systems. The proposal should describe the role of each member of the multi-disciplinary team and their responsibilities to the project. The electronic application, specific guidelines, and abstracts of previously funded projects are available at the website.
Requirements: Organizations eligible to receive funding include the following: state agricultural experiment stations; 1862 and 1890 land-grant colleges and universities, including Tuskegee University and West Virginia State University; 1994 land-grant colleges and universities; research foundations established by land-grant colleges and universities; colleges and universities receiving funds under the Act of October 10, 1962 (16 U.S.C 582a et seq.); accredited schools or colleges of veterinary medicine; and the University of the District of Columbia. Organizations eligible to receive Extension awards are 1862 and 1890 land-grant colleges and universities, including Tuskegee University and West Virginia State University, and the University of the District of Columbia.
Restrictions: Funding may not be used for renovation or refurbishment of research, education, or extension space; the purchase or installation of fixed equipment in such space; or the planning, repair, rehabilitation, acquisition, or construction of buildings or facilities.
Geographic Focus: Alaska, Arizona, California, Colorado, Guam, Hawaii, Idaho, Montana, Nevada, New Mexico, Northern Mariana Islands, Oregon, Utah, Washington, Wyoming, American Samoa, Micronesia, Palau
Date(s) Application is Due: Feb 28
Amount of Grant: 180,000 USD
Contact: Robert Nowierski, National Program Leader; (202) 401-4900; fax (202) 401-4888; rnowierski@nifa.usda.gov
Internet: http://www.nifa.usda.gov/funding/rfas/ipm_western.html
Sponsor: U.S. Department of Agriculture
1400 Independence Avenue, SW, Stop 2201
Washington, DC 20250-2201

USDA Integrated Research Education and Extension Grants-Organic Transitions 4630
The Integrated Research Education and Extension Grants-Organic Transitions are soliciting proposals to assess the impact of organic systems on environmental outcomes in soil conservation, land stewardship, soil carbon sequestration, and greenhouse gas emissions. Fieldwork to set up treatments or collect data on organic practices must be done on certified organic land, or land in transition to organic certification. Practices and systems to be addressed include those associated with organic crops, organic animal production (including dairy), and organic systems integrating plant and animal production. Applicants are encouraged to develop content suitable for delivery through eXtension. Funding may be requested to: establish an eXtension Community of Practice (CoP); or exhance an existing CoP, such as eOrganic, to endure that information and technology transfer reaches potential adopters as quickly as possible. Projects are expected to use field and modeling information to demonstrate the benefits or challenges to environmental services and climate change litigation posed by implementing certified organic practices and/or combinations of practices. Specific guidelines, abstracts of funded projects, workshops, proposal examples, and grant writing tips are available at the website. The application is available at grants.gov.
Requirements: The following entities are eligible to apply: 1862, 1890, and 1994 land grant institutions; Hispanic serving institutions; and private and public institutions of higher education. Applicants are expected to describe stakeholder involvement in problem identification, planning, implementation and evaluation. Applications describing multi-state, multi-institutional, multidisciplinary, multifunctional activities and combinations are encouraged. S single university demonstrating significant collaboration with various agencies or organizations within the host state, as appropriate to project goals, may also be competitive. Applicants are strongly encouraged to assemble project teams that include those with expertise in research, education, extension, and evaluation. Projects should plan to deliver applied production information to producers and/or students. Applicants are also encouraged to describe how results at the field and farm scale might be extrapolated beyond the experimental boundaries of the proposed project.
Restrictions: If a grant provides a particular benefit to a specific agricultural commodity, the grant recipient is required to provide funds awarded on a dollar-for-dollar basis from non-federal sources with cash and/or in-kind contributions. NIFA may waive the matching funds requirement for a grant if it is determined that: the results of the project, while of particular benefit to a specific agricultural commodity, are likely to be applicable to agricultural commodities generally; or the project involves a minor commodity, the project deals with scientifically important research, and the grant recipient is unable to satisfy the matching funds requirement. Projects in organic agriculture research typically qualify for a waiver of the matching requirement.
Geographic Focus: All States
Date(s) Application is Due: Apr 25
Contact: Steven Smith; (202) 401-6134; fax (202) 401-1782; sismith@nifa.usda.gov
Internet: http://www.csrees.usda.gov/fo/organictransitionsprogram.cfm
Sponsor: U.S. Department of Agriculture
1400 Independence Avenue, SW, Stop 2201
Washington, DC 20250-2201

USDA International Science and Education Competitive Grants 4631

The International Science and Education Competitive Grants (ISE) supports activities that will improve the capabilities of American colleges and universities to conduct international collaborative research, extension and teaching. The grants are expected to enhance the international content of curricula; ensure that faculty work beyond the U.S. and bring lessons learned back home; promote international research partnerships; enhance the use and application of foreign technologies in the U.S; and strengthen the role that colleges and universities play in maintaining U.S. competitiveness. Multi-year proposals of up to four years may be submitted. Partnerships with minority serving institutions and/or community college are encouraged. Current applications, specific guidelines, and abstracts of previously funded projects are available at the website.

Requirements: Public and private colleges and universities are eligible to apply, in addition to 1862, 1890, and 1994 land-grant institutions.

Restrictions: Internationalization of U.S. campus-based programs is the primary focus of the ISE program. Proposals that do not show benefit to the U.S. will not be reviewed. Since the ISE program is an integrated program, proposals should focus on at least two of the three main components (agricultural research, extension and teaching). Single function proposals will not be reviewed.

Geographic Focus: All States

Date(s) Application is Due: Jan 26

Contact: Patricia Fulton, National Program Leader; (202) 690-3852 or (202) 720-3801; fax (202) 690-2355; pfulton@nifa.usda.gov

Internet: http://www.nifa.usda.gov/fo/internationalscienceandeducation.cfm

Sponsor: U.S. Department of Agriculture
1400 Independence Avenue, SW, Stop 2201
Washington, DC 20250-2201

USDA Interregional Research Project Minor Crop Pest Management Grants 4632

The primary goal of the IR-4 program is to provide safe, effective, and economical pest management solutions for growers of minor/specialty crops. IR-4 facilitates crop protection by providing expert assistance with product development and registration. IR-4 positively impacts the ability of research and extension personnel, producers, and grower organizations to provide a continuous supply of affordable minor/specialty crops to the public. Successful applicants will demonstrate the capacity and commitment required to accelerate the registration of newer, reduced risk pest management tools; expedite access to these pest management tools for minor/specialty crop growers; and conduct efficacy research to identify new and more effective minor/specialty crop pest management solutions. The FY 2012 IR-4 Program will fund four geographically-based IR-4 Regional Centers and an IR-4 National Headquarters. NIFA does not require matching support for this program, and matching resources will not be factored into the review process. The electronic application, specific guidelines, and abstracts of previously funded projects are available at the website.

Requirements: Land-grant colleges and universities (1862, 1890, and 1994) and state agricultural experiment stations are eligible to apply. Applications may be submitted by state agricultural experiment stations, research foundations established by land grant colleges and universities, colleges and universities receiving funds under the McIntire-Stennis Act, and accredited schools or colleges of veterinary medicine for the purpose of facilitating or expanding ongoing state-federal food and agricultural research programs. Applicants should clearly describe a plan for achieving three core objectives: to maintain regulatory clearances of effective crop protection agents for high-value, minor/specialty food crops and for minor uses (special circumstance needs) on major food crops with special emphasis on lower risk chemicals and uses that are compatible with integrated pest management (IPM) programs; to support research to enhance the development and registration of bio-pesticides for use in food and non-food pest management programs; and to support research on crop protection products that will expand their uses on ornamental crops (nursery, floral, turf, and other non-food crop systems) to allow management of new and important pest species.

Restrictions: Indirect costs and tuition remissions are not allowable costs. Costs that are a part of an institution's indirect cost pool (e.g., administrative or clerical salaries) may not be reclassified as direct costs for the purpose of making them allowable. Funding may not be used for the renovation or refurbishment of research, education, or extension space; the purchase or installation of fixed equipment in such space; or the planning, repair, rehabilitation, acquisition, or construction of buildings or facilities.

Geographic Focus: All States

Date(s) Application is Due: Mar 20

Amount of Grant: 1,000,000-4,800,000 USD

Contact: Monte Johnson; (202) 401-1108; fax (202) 401-1782; mpjohnson@nifa.usda.gov

Internet: http://www.nifa.usda.gov/funding/rfas/minor_crop.html

Sponsor: U.S. Department of Agriculture
1400 Independence Avenue, SW, Stop 2201
Washington, DC 20250-2201

USDA Market Access Grants 4633

The Market Access Program (MAP), formerly the Market Promotion Program, uses funds from the U.S. Department of Agriculture's (USDA) Commodity Credit Corporation (CCC) to help U.S. producers, exporters, private companies, and other trade organizations finance promotional activities for U.S. agricultural products. The MAP encourages the development, maintenance, and expansion of commercial export markets for agricultural commodities. Activities financed include consumer promotions, market research, technical assistance, and trade servicing.

Requirements: To participate in the MAP, an applicant must be a nonprofit U.S. agricultural trade organization, a nonprofit SRTG, a nonprofit U.S. agricultural cooperative, or a State government agency. A small-sized U.S. commercial entity may participate through a MAP participant.

Geographic Focus: All States

Contact: Maureen Quinn, Director; (202) 720-7115 or (202) 720-4327; fax (202) 720-1727; maureen.quinn@usda.gov or podadmin@fas.usda.gov
Sally Klusaritz, Deputy Director; (202) 720-3448 or (202) 720-4327; fax (202) 720-1727; sally.klusaritz@usda.gov or podadmin@fas.usda.gov

Internet: http://www.fas.usda.gov/mos/programs/map.asp

Sponsor: U.S. Department of Agriculture
1400 Independence Avenue SW
Washington, DC 20250

USDA McIntire-Stennis Cooperative Forestry Research Grants 4634

The USDA McIntire-Stennis Grants assist states in continuing programs of research at forestry schools and in developing a trained pool of scientists to conduct forestry research. Categories of forestry research supported include reforestation and management of land for the production of crops of timber and other related products of the forest; management of forest and related watershed lands to improve conditions of waterflow and to protect resources against flood and erosion; management of forest and related rangeland for production of forage, domestic livestock and game, and improvement of food and habitat for wildlife; management of forest lands for outdoor recreation; protection of forest resources against fire, insects, diseases, and other destructive agents; utilization of wood and other forest products; development of sound policies for the management of forest lands, and the harvesting and marketing of forest products; and other studies necessary to obtain the fullest and most effective use of forest resources. The application package must be obtained from grants.gov, with the appropriate funding number entered. Current specific instructions are available at the website.

Requirements: Applicants must be state or territorial institutions.

Geographic Focus: All States

Date(s) Application is Due: Aug 29

Amount of Grant: 33,000-1,200,000 USD

Contact: Linda Toppin; 202-690-2563; ltoppin@nifa.usda.gov

Internet: http://www.csrees.usda.gov/business/awards/formula/mcintire_stennis.html

Sponsor: U.S. Department of Agriculture
1400 Independence Avenue, SW, Stop 2201
Washington, DC 20250-2201

USDA National Integrated Food Safety Initiative Grants 4635

The National Integrated Food Safety Initiative Grants support food safety projects that demonstrate an integrated approach to solving problems in applied food safety research, education, or extension. Special emphasis is given to research describing multi-functional activities, such as research and extension, research and education, or extension and education. Applications describing multi-state, multi-institutional, multidisciplinary, and multifunctional activities are encouraged. If international activities (e.g. partnerships, exchanges, travel, etc) are proposed, then applicants should describe indicators that will be used to assess those activities. Project teams should include university, government, and industry collaborators. Specific criteria (standard vs. special emphasis projects, conferences, and bridge funding), samples of previous proposals, the electronic application, and workshop presentations are available at the website.

Requirements: The following entities are eligible to apply: 1862, 1890, and 1994 land grant institutions; Hispanic serving institutions (four year and two year, if the institution offers an associate degree in an agriculture-related field); and private and public institutions of higher education, particularly institutions that serve high-risk, under-served, or hard-to-reach populations. If a grant provides a particular benefit to a specific agricultural commodity, the grant recipient is required to match the USDA funds awarded on dollar-for-dollar basis from non-federal sources with cash and in-kind contributions.

Geographic Focus: All States

Date(s) Application is Due: Jun 27

Contact: Jan Singleton; (202) 401-1954; fax (202) 401-4888; jsingleton@nifa.usda.gov

Internet: http://www.nifa.usda.gov/fo/nationalintegratedfoodsafetyinitiative.cfm

Sponsor: U.S. Department of Agriculture
1400 Independence Avenue, SW, Stop 2201
Washington, DC 20250-2201

USDA Norman E. Borlaug International Agricultural Science and Technology Fellowships 4636

The Norman E. Borlaug International Agricultural Science and Technology Fellowships promote food security and economic growth by increasing scientific knowledge and collaborative research to improve agricultural productivity. It accomplishes this by: providing training opportunities for early and mid-career researchers and policymakers from developing and middle-income countries; fostering collaborative research to improve agricultural productivity; facilitating the transfer of new science and agricultural technologies to strengthen agricultural practices; and addressing obstacles to the adoption of technology, such as inadequate policies and regulations. Fellowships range from six to 12 weeks depending on the research topic and funding availability. Areas of training and research can be in any agriculture-related field including plant pathology, entomology, veterinary sciences, microbiology, agricultural economics, food safety, sanitary and phytosanitary topics, natural resources management, agricultural biotechnology, global climate change, and agricultural policy. Allowable training topics vary by country and year and can be found in the country-specific application announcements posted on the website.

Requirements: Candidates must be female citizens of the county in which the Fellowship is offered, hold a minimum of a Master's degree, and be in the early to middle stage of their professional career, with a good working knowledge of the English language. They must also have at least two years of practical work experience, with current employment in a university, government, non-profit, or private agricultural research entity in their county, with the intention to continue working in their home country after the Fellowship is completed.

Restrictions: Passports must be valid for at least six months from the date the applicant enters the U.S.
Geographic Focus: Bangladesh, Burundi, Comoros, Djibouti, Eritrea, Ethiopia, Ghana, Guatemala, Haiti, Honduras, Kenya, Liberia, Madagascar, Malawi, Mali, Mauritius, Mayotte, Mozambique, Nigeria, Reunion, Rwanda, Senegal, Seychelles, Somalia, Tanzania, Uganda, Zambia, Zimbabwe
Contact: Karina Ramos-Hides; (202) 720-3736; karina.ramos@fas.usda.gov
Internet: http://www.fas.usda.gov/icd/borlaug/Borlaug.asp
Sponsor: U.S. Department of Agriculture
1400 Independence Avenue SW
Washington, DC 20250

USDA Organic Agriculture Research and Extension Initiative **4637**
The USDA Organic Agriculture Research and Extension Initiative (OREI) seeks to solve critical organic agriculture issues, priorities, or problems through the integration of research and extension activities. The purpose of this program is to fund projects that will enhance the ability of producers and processors who have already adopted organic standards to grow and market high quality organic agricultural products. Priority concerns include biological, physical, and social sciences, including economics. The OREI is particularly interested in projects that emphasize research and outreach that assist farmers and ranchers with whole farm planning. Projects should plan to deliver applied production information to producers. Fieldwork must be done on certified organic land or on land in transition to organic certification, as appropriate to project goals and objectives. Funding amounts are based on specific project categories.
Requirements: The following are eligible to apply for the initiative: public and state controlled institutions of higher education; nonprofits having a 501(c)3 status with the IRS, other than institutions of higher education; nonprofits that do not have a 501(c)3 status with the IRS, other than institutions of higher education; private institutions of higher education; individuals; for profit organizations other than small businesses; and small businesses. The following entities are also eligible to apply: state agricultural experiment stations; colleges and universities; university research foundations; other research institutions and organizations; federal agencies; national laboratories; private organizations or corporations; individuals who are United States citizens or nationals; or a group consisting of two or more of the above entities.
Restrictions: Recipients are required to provide funds or in-kind support from non-federal sources in an amount at least equal to the amount provided by the funding agency.
Geographic Focus: All States
Date(s) Application is Due: Mar 9
Amount of Grant: 50,000-2,000,000 USD
Contact: Steven Smith; (202 401-6134; fax (202) 401-1782; sismith@nifa.usda.gov
Internet: http://www.nifa.usda.gov/funding/rfas/OREI.html
Sponsor: U.S. Department of Agriculture
1400 Independence Avenue, SW, Stop 2201
Washington, DC 20250-2201

USDA Plant and Animal Disease, Pest Control, and Animal Care Grants **4638**
The Plant and Animal Disease, Pest Control, and Animal Care Grants help protect U.S. agriculture from economically injurious plant and animal diseases and pests, guarantee the safety and potency of veterinary biologics, and ensure the humane treatment of animals. Project and training grants are awarded to eligible organizations to carry out these objectives. Possible projects include conducting surveys, inspections to detect and appraise infestations, eradication and control activities, and carrying out regulatory actions to prevent interstate spread of infestations and diseases. Additional guidelines and application instructions are available at the website for the Catalog of Federal Domestic Assistance (project number 10.025). Applicants may contact their state Animal and Plant Health Inspection Service office.
Requirements: Eligible to apply are foreign, state, local, and U.S. territorial government agencies; nonprofit institutions of higher education; and nonprofit associations or organizations requiring federal support to eradicate, control, or assess the status of injurious plant and animal diseases and pests that are a threat to regional or national agriculture and to conduct related demonstration projects. A letter from applicants seeking support or cooperation with the Animal and Plant Health Inspection Service is required. Applicants must indicate the nature and severity of the specific problem and explain how they will address the problem. Applications are approved upon determination that the project will contribute toward accomplishment of the Agency's overall mission and meet any established project evaluation and/or selection criteria.
Restrictions: Grant recipients share some project or program costs. The cost-sharing arrangements are developed between the USDA and the recipients in advance. Individuals are not eligible for funding.
Geographic Focus: All States
Contact: Eileen M. Berke; (301) 734-8330; eileen.m.berke@aphis.usda.gov
Internet: https://www.cfda.gov/index?s=program&mode=form&tab=step1&id=4eef76daa8c24580301ca56617a3462e
Sponsor: U.S. Department of Agriculture
4700 River Road
Riverdale, MD 20737

USDA Plant Feedstock Genomics for Bioenergy: A Joint Research Grant **4639**
The National Institute of Food and Agriculture's (NIFA) Competitive Programs and the Department of Energy's Office of Science, Office of Biological and Environmental Research (OBER) announce the inter-agency program to support genomics-based research that will lead to the improved use of biomass and plant feedstocks for the production of fuels such as ethanol or renewable chemical feedstocks. Specifically, applications are sought for fundamental research on plants that will improve biomass characteristics, biomass yield, or sustainability. Systems biology approaches to identify genetic indicators enabling plants to be efficiently bred or manipulated, or research to predict phenotype from underlying genotype leading to improved feedstock characterization and sustainability are also encouraged. Applicants may request up to three years of funding. Please note that applicants will submit applications through DOE and a pre-application is requested rather than a letter of intent. The electronic application, specific guidelines, and abstracts for previously funded projects are available at the website.
Requirements: The following are eligible to apply for funding: 1862, 1890, and 1994 land grant institutions; public and private institutions; small businesses and all other for profit businesses; Hispanic serving institutions; nonprofits with or without 501(c)3; and state agricultural experiment stations. Federal agencies, national laboratories, private organizations or corporations, and individuals are also eligible to apply.
Restrictions: Research that seeks to increase starch content for improved nutrient qualities or to facilitate the digestion and fermentation of starch to produce sugars and other bio-based products or bio-fuels is not eligible for funding. Research that seeks to increase grain yield or seed-oil productions is also not eligible.
Geographic Focus: All States
Date(s) Application is Due: Feb 24
Amount of Grant: 200,000-500,000 USD
Contact: Dr. Ed Kaleikau; (202) 401-1931; fax (202) 401-1782; ekaleikau@nifa.usda.gov
Internet: http://www.nifa.usda.gov/fo/plantfeedstockgenomicsforbioenergy.cfm
Sponsor: U.S. Department of Agriculture
1400 Independence Avenue, SW, Stop 2201
Washington, DC 20250-2201

USDA Policy Research Center Grants **4640**
The USDA Policy Research Center (PRC) Grants are available for centers to conduct research and education programs that are objective, operationally independent, and external to the federal government. They should also study the effect of public policies and trade agreements in the following areas: the farm and agricultural sectors (including commodities, livestock, dairy, and specialty crops, such as fruits and vegetables, tree nuts, and plants used for medicinal purposes); the environment; rural families, households, and economies; and consumers, food, and nutrition. Research may be either disciplinary or interdisciplinary and include the following activities: quantify implications of public policies and regulations; develop theoretical and research methods; collect, analyze, and disseminate data for policy makers, analysts, and individuals; and develop programs to train future analysts. A PRC is defined as having the analytical capacity to either perform policy analysis across multiple sets of public policy issues or have a specialized capacity in a single policy area. A center may reside in a single institution or be a collaborative effort across multiple institutions. Matching support is not required. Instructions for the Grants.gov application, specific qualifications guidelines, and evaluation criteria are available at the website.
Requirements: Applications may be submitted by state agricultural experiment stations, colleges and universities, other research institutions and organizations, private organizations, corporations, and individuals. Award recipients may subcontract to organizations not eligible to apply provided such organizations are necessary for the conduct of the projec.
Geographic Focus: All States
Date(s) Application is Due: Feb 13
Contact: Robbin Shoemaker, National Program Leader; (202) 720-5468; fax (202) 401-5179; rshoemaker@nifa.usda.gov
Internet: http://www.nifa.usda.gov/funding/rfas/prc.html
Sponsor: U.S. Department of Agriculture
1400 Independence Avenue, SW, Stop 2201
Washington, DC 20250-2201

USDA Rangeland Research Grants **4641**
The primary purpose of the Rangeland Research Grants is to provide U.S. agricultural producers, rural landowners, and land managers with integrated science strategies to make informed land management decisions with an emphasis on enhancing the restoration and sustainable integrity of U.S. rangelands. Rangeland research should employ a methodology that integrates approaches to discover, understand, and address phenomena that are too broad to be explained by standard disciplinary science. Priority consideration will be given to the following areas: rangeland restoration, management of rotational grazing, and management of co-grazing. Successful applicants will: document the experience level of key personnel working within a collaborative, integrated rangeland research project; demonstrate the ability to participate in a comprehensive nationally coordinated effort that shares data and its analysis, and derived conclusions; involve novel or non-traditional partners (e.g., non-governmental organizations, public or private foundations, social scientists or psychologists) to move programs toward improved rangeland management and behavior change in stakeholders; demonstrate capacity to conduct efficient and successful integrated rangeland research projects; demonstrate ability to develop new scientific methodologies and innovations that transcend standard disciplinary science; develop multidisciplinary research projects which result in an enhanced land and resource management knowledge base; and conceptualize and implement research projects that support the purpose and goals of the research grants. Specific guidelines, including abstracts of previously funded projects, are available at the website. The application is available at grants.gov.
Requirements: Land-grant colleges and universities, state agricultural experiment stations, in addition to colleges, universities, and federal laboratories having a demonstrable capacity in rangeland research are eligible to apply for funding. Award recipients may subcontract to organizations not eligible to apply provided such organizations are necessary for the conduct of the project. Applicants are required to provide 50% matching funds from non-federal sources for all proposed federal funds being sought. Non-federal matching contributions, such as cash and third party in kind, are accepted under this program.

Restrictions: Grant funds may not be used for the following: renovation or refurbishment of research, education, or extension space; the purchase or installation of fixed equipment in such space; or the planning, repair, rehabilitation, acquisition, or construction of buildings or facilities. Funding is also not available for indirect costs, tuition remission, or time extensions beyond five years.
Geographic Focus: All States
Date(s) Application is Due: Jul 15
Amount of Grant: 350,000-500,000 USD
Contact: James Dobrowolski; (202) 401-5016; jdobrowolski@nifa.usda.gov
Internet: http://www.nifa.usda.gov/fo/rangelandresearchprogram.cfm
Sponsor: U.S. Department of Agriculture
1400 Independence Avenue, SW, Stop 2201
Washington, DC 20250-2201

USDA Regional Integrated Pest Management Grants-North Central Region 4642
The USDA Regional Integrated Pest Management Competitive Grants (RIPM) support the ongoing research and extension efforts needed to increase the implementation of IPM methods. The RIPM program supports projects that develop individual pest control tactics, integrate individual tactics into an IPM system, and develop and implement extension and education programs. The program is administered by the land-grant university system's four regional IPM Centers (North Central, Northeastern, Southern, Western) in partnership with NIFA. Funding will support projects that promote cooperative efforts across appropriate disciplines, with linkages between research and extension efforts, and components of existing or emerging pest management systems. The proposal should describe the role of each member of the multi-disciplinary team and their responsibilities to the project. The electronic application, specific guidelines, and abstracts of previously funded projects are available at the website.
Requirements: The following organizations are eligible for funding: state agricultural experiment stations; 1862 and 1890 land-grant colleges and universities, including Tuskegee University and West Virginia State University; 1994 land-grant colleges and universities; research foundations established by land-grant colleges and universities; colleges and universities receiving funds under the Act of October 10, 1962 (16 U.S.C 582a et seq.); accredited schools or colleges of veterinary medicine; and the University of the District of Columbia. Organizations eligible to receive extension awards are 1862 and 1890 land-grant colleges and universities, including Tuskegee University and West Virginia State University, and the University of the District of Columbia.
Restrictions: Funding may not be used for renovation or refurbishment of research, education, or extension space; the purchase or installation of fixed equipment in such space; or the planning, repair, rehabilitation, acquisition, or construction of buildings or facilities.
Geographic Focus: Illinois, Indiana, Iowa, Kansas, Michigan, Minnesota, Missouri, Nebraska, North Dakota, Ohio, South Dakota, Wisconsin
Date(s) Application is Due: Feb 28
Amount of Grant: Up to 175,000 USD
Contact: Robert Nowierski; (202 401-4900; fax (202) 401-4888; rnowierski@nifa.usda.gov
Internet: http://www.nifa.usda.gov/fo/regionalintegratedpestmgtnorthcentral.cfm
Sponsor: U.S. Department of Agriculture
1400 Independence Avenue, SW, Stop 2201
Washington, DC 20250-2201

USDA Regional Integrated Pest Management Grants-Northeastern Region 4643
The USDA Regional Integrated Pest Management Competitive Grants (RIPM) support the continuum of research and extension efforts needed to increase the implementation of IPM methods. The RIPM program supports projects that develop individual pest control tactics, integrate individual tactics into an IPM system, and develop and implement extension and education programs. The program is administered by the land-grant university system's four regional IPM Centers (North Central, Northeastern, Southern, Western) in partnership with NIFA. Funding will support projects that promote cooperative efforts across appropriate disciplines, with linkages between research and extension efforts, and components of existing or emerging pest management systems. The proposal should describe the role of each member of the multi-disciplinary team and their responsibilities to the project. The electronic application, specific guidelines, and abstracts of previously funded projects are available at the website.
Requirements: The following organizations are eligible to receive funding: state agricultural experiment stations; 1862 and 1890 land-grant colleges and universities, including Tuskegee University and West Virginia State University; 1994 land-grant colleges and universities; research foundations established by land-grant colleges and universities; colleges and universities receiving funds under the Act of October 10, 1962 (16 U.S.C 582a et seq.), accredited schools or colleges of veterinary medicine; and the University of the District of Columbia. Organizations eligible to receive extension awards are: 1862 and 1890 land-grant colleges and universities, including Tuskegee University and West Virginia State University, and the University of the District of Columbia.
Restrictions: Funding may not be used for renovation or refurbishment of research, education, or extension space; the purchase or installation of fixed equipment in such space; or the planning, repair, rehabilitation, acquisition, or construction of buildings or facilities.
Geographic Focus: Connecticut, Delaware, District of Columbia, Maine, Maryland, Massachusetts, New Hampshire, New Jersey, New York, Pennsylvania, Rhode Island, Vermont, West Virginia
Date(s) Application is Due: Feb 28
Contact: Robert Nowierski; (202) 401-4900; rnowierski@nifa.usda.gov
Internet: http://www.nifa.usda.gov/fo/regionalintegratedpestmgtnortheastern.cfm
Sponsor: U.S. Department of Agriculture
1400 Independence Avenue, SW, Stop 2201
Washington, DC 20250-2201

USDA Regional Integrated Pest Management Grants-Southern Region 4644
The USDA Regional Integrated Pest Management Competitive Grants (RIPM) support the ongoing research and extension efforts needed to increase the implementation of IPM methods. The RIPM program supports projects that develop individual pest control tactics, integrate individual tactics into an IPM system, and develop and implement extension and education programs. The program is administered by the land-grant university system's four regional IPM Centers (North Central, Northeastern, Southern, Western) in partnership with NIFA. Funding will support projects that promote cooperative efforts across appropriate disciplines, with linkages between research and extension efforts, and components of existing or emerging pest management systems. The proposal should describe the role of each member of the multi-disciplinary team and their responsibilities to the project. The electronic application, specific guidelines, and abstracts of previously funded projects are available at the website.
Requirements: Organizations eligible to receive funding include the following: state agricultural experiment stations; 1862 and 1890 land-grant colleges and universities, including Tuskegee University and West Virginia State University; 1994 land-grant colleges and universities; research foundations established by land-grant colleges and universities; colleges and universities receiving funds under the Act of October 10, 1962 (16 U.S.C 582a et seq.); accredited schools or colleges of veterinary medicine; and the University of the District of Columbia. Organizations eligible to receive Extension awards are 1862 and 1890 land-grant colleges and universities, including Tuskegee University and West Virginia State University, and the University of the District of Columbia.
Restrictions: Funding may not be used for renovation or refurbishment of research, education, or extension space; the purchase or installation of fixed equipment in such space; or the planning, repair, rehabilitation, acquisition, or construction of buildings or facilities.
Geographic Focus: Alabama, Arkansas, Florida, Georgia, Kentucky, Louisiana, Mississippi, North Carolina, Oklahoma, Puerto Rico, South Carolina, Tennessee, Texas, U.S. Virgin Islands, Virginia
Date(s) Application is Due: Feb 28
Amount of Grant: Up to 171,000 USD
Contact: Robert Nowierski; (202) 401-4900; rnowierski@nifa.usda.gov
Internet: http://www.nifa.usda.gov/fo/regionalintegratedpestmgtsouthern.cfm
Sponsor: U.S. Department of Agriculture
1400 Independence Avenue, SW, Stop 2201
Washington, DC 20250-2201

USDA Resident Instruction Grants for Higher Education in Insular Areas 4645
The purpose of this program is to promote and strengthen the ability of insular area institutions to carry out teaching and education programs within a broadly defined arena of food and agricultural sciences-related disciplines. Insular areas include the Commonwealth of Puerto Rico, Guam, American Samoa, the Commonwealth of the Northern Mariana Islands, the Federated States of Micronesia, the Republic of the Marshall Islands, the Republic of Palau, and the Virgin Islands of the United States. By strengthening institutional educational capacities in instruction and curriculum, and enhancing the quality of teaching and learning, NIFA intends that this program will help insular area institutions meet their unique needs. This program also will assist insular area institutions to make more efficient use of existing educational funds by providing resources for partnerships between faculties at insular area and mainland institutions. The electronic application and specific guidelines are available at the website.
Requirements: Eligible applicants include higher education institutions located in an insular area with a capacity to carry out teaching and extension programs in the food and agricultural sciences. Individual land-grant colleges and universities, and other institutions that have secured land-grant status through federal legislation, and located in insular areas are automatically eligible for awards under this program, either as direct applicants or as parties to a consortium agreement.
Geographic Focus: Guam, Marshall Islands, Northern Mariana Islands, Puerto Rico, U.S. Virgin Islands, American Samoa, Micronesia, Palau
Date(s) Application is Due: Mar 1
Amount of Grant: 70,000-850,000 USD
Contact: Saleia Afele-Faamuli, National Program Leader; (202) 720-0384; fax (202) 720-4924; sfaamuli@nifa.usda.gov
Internet: http://www.nifa.usda.gov/fo/residentinstructionforinsularareas.cfm
Sponsor: U.S. Department of Agriculture
1400 Independence Avenue, SW, Stop 2201
Washington, DC 20250-2201

USDA Scientific Cooperation Exchange Program 4646
The Scientific Cooperation Exchange Program (SCEP) with the People's Republic of China supports international exchanges to promote bilateral scientific exchanges designed to promote agricultural cooperation, development, and trade. The program offers opportunities for U.S. teams of up to five members to initiate linkages with potential long-term collaborators at Chinese institutions throughout the People's Republic of China (PRC). Under a special USDA Scientific Technological Exchange Agreement with the PRC, the Chinese Ministry of Agriculture offers two-week visits to China designed to meet the scientific objectives of each team. Priority will be given to innovative proposals that address food security, climate change, food safety, animal and plant health, biotechnology and other emerging technologies, food processing and product development, non-food product development, and sustainable plant and animal management. Specific guidelines, including funding and cost share requirements, program purpose and priorities, focus areas and recommended topics, application and submission information, and frequently asked questions are available at the website.
Requirements: Proposals may be received from universities and colleges, including community colleges, accredited in and having a campus located in the United States, acting on behalf of

their institutions. U.S. non-profit research organizations, including research laboratories, professional societies, or similar organizations in the U.S. that are directly associated with education, extension, research, or capacity building activities; private organizations; federal and state agencies; or consortia led by the eligible organizations listed here are also eligible. A proposal from a consortium of organizations must be submitted as a single proposal with one U.S. organization serving as the lead and all other organizations as team members and sub-awardees, when applicable. The Principal Investigator (PI) must hold a position at an eligible U.S. institution. Especially encouraged are proposals from smaller academic institutions, as well as minority serving institutions, researchers early in their career, extension agents, and women that hold positions at an eligible institution.
Restrictions: The U.S. team is responsible for the international airfare between the U.S. and China. MOA will cover all in-country program related expenses, including lodging, transportation, meals, translations, and cultural activities. USDA does not provide any funding to support U.S. teams.
Geographic Focus: All States
Date(s) Application is Due: Jul 8
Contact: Khaliaka Meardry; (202) 690-1866; Khaliaka.Meardry@fas.usda.gov
Nora Banks, Program Contact; (202) 720-3904; Nora.Banks@fas.usda.gov
Internet: http://www.fas.usda.gov/icd/grants/scep/scep.asp
Sponsor: U.S. Department of Agriculture
1400 Independence Avenue SW
Washington, DC 20250

USDA Scientific Cooperation Research Program **4647**
The Scientific Cooperation Research Program (SCRP) supports joint research, extension, and education projects among domestic and international agricultural professionals. These projects address issues including agricultural trade and market access, animal and plant health, biotechnology, food safety and security, and sustainable natural resource management in the U.S. and internationally. The objective of SCRP is to reduce global poverty and hunger by supporting applied scientific research, extension, or education projects that aim to address challenges faced by smallholder farmers in emerging economies. For the purposes of this announcement, smallholders, as defined by the World Bank, are farmers that own or lease less than 124 acres of land (less than 50 hectares). Since 1980, the program has supported more than 400 projects with approximately 95 partnering countries, enhancing the technical skills of more than 1,000 agricultural professionals. Specific guidelines, along with application and submission information, countries defined by the World Bank as eligible emerging economies, and frequently asked questions are available at the website.
Requirements: Proposals may be received from academic institutions such as universities and colleges, including community colleges, accredited in and having a campus located in the U.S., acting on behalf of their institution. Non-profit research organizations in the U.S., including research laboratories, professional societies, or similar organizations that are directly associated with education, extension, research, or capacity building activities; private organizations; federal and state agencies; or consortia led by these eligible organizations. A proposal from a consortium of organizations must be submitted as a single proposal with one U.S. organization serving as the lead and all other organizations as team members and sub-awardees when applicable. All proposals must include foreign collaborations with institutions in emerging economies that have diplomatic relations with the U.S. All proposals should address one of the three focus areas: improving agricultural productivity, creating sustainable agricultural systems, or building regional or global trade capacities. Priority will be given to: minority serving institutions; countries defined by the World Bank as emerging economies; and projects that address challenges to women. Proposals should also utilize the knowledge and technologies of the scientific communities to help aid in developing practical solutions to these challenges. Projects must include foreign collaborations and may not exceed two years. Funding may be allocated to foreign collaborators through sub-awards. All proposals must match at least 25% of the requested funds; third party cost sharing is allowed.
Restrictions: Projects may not exceed two years. Funding will be on a cost reimbursable basis. FAS reserves the right to allocate partial funding or terminate funding at any time. Scholarships, stipends, sabbaticals, tuition, tuition remission, general purpose equipment (no particular scientific, technical, or programmatic purpose), entertainment, capital improvements, thank you gifts, and other expenses not directly related to the project are not eligible for funding.
Geographic Focus: All States
Date(s) Application is Due: Jul 5
Amount of Grant: Up to 45,000 USD
Contact: Khaliaka Meardry, Program Contact; (202) 690-1866; Khaliaka.Meardry@fas.usda.gov or SCRPProposals@fas.usda.gov
Nora Banks, Program Contact; (202) 720-3904; Nora.Banks@fas.usda.gov
Internet: http://www.fas.usda.gov/icd/SCRP/default.asp
Sponsor: U.S. Department of Agriculture
1400 Independence Avenue SW
Washington, DC 20250

USDA Small Business Innovation Research Grants-Phase I **4648**
The Small Business Innovation Research Grant (SBIR)-Phase I provides funding to qualified small businesses to support high quality, advanced concepts research related to important scientific problems and opportunities in agriculture that could lead to significant public benefit if successful. Objectives of the SBIR program include the following: stimulating technological innovation in the private sector; strengthening the role of small businesses in meeting federal research and development needs; increasing private sector commercialization of innovations derived from USDA-supported research and development efforts; and fostering and encouraging participation by women-owned and socially and economically disadvantaged small business firms in technological innovation. Selected areas for research are forests and related resources; plant production and protection; animal production and protection; air, water, and soils; food science and nutrition; rural and community development; aquaculture; and industrial applications. Phase I awards will be made for periods normally not to exceed six months to determine the scientific or technical feasibility of ideas in the selected research areas. Phase II awards will be made to firms with approaches that appear sufficiently promising as a result of Phase I studies for a period not to exceed 24 months. The electronic application, specific guidelines, and abstracts from previously funded projects are available on the website.
Requirements: Small, for-profit businesses are eligible to apply. Each organization submitting a proposal must qualify as a small business for research or research and development purposes. Participation by university faculty or government scientists in SBIR projects is strongly encouraged.
Restrictions: Phase I winners who have previously been rejected for Phase II funding are not eligible to apply. Funding is not available for loans or establishing a business.
Geographic Focus: All States
Date(s) Application is Due: Jul 13
Amount of Grant: 70,000-100,000 USD
Contact: Dr. Charles Cleland, National Program Leader; (202) 401-4002 or (202) 401-5852; fax (202) 401-6070; ccleland@nifa.usda.gov
Internet: http://www.csrees.usda.gov/fo/fundview.cfm?fonum=1220
Sponsor: U.S. Department of Agriculture
1400 Independence Avenue, SW, Stop 2201
Washington, DC 20250-2201

USDA Small Business Innovation Research Grants-Phase II **4649**
The USDA Small Business Innovation Research Grants (SBIR)-Phase II provide funding to qualified small businesses to support high quality, advanced concepts research related to important scientific problems and opportunities in agriculture that could lead to significant public benefit if successful. Objectives of the SBIR program include the following: stimulating technological innovation in the private sector; strengthening the role of small businesses in meeting Federal research and development needs; increasing private sector commercialization of innovations derived from USDA-supported research and development efforts; and fostering and encouraging participation by women-owned and socially and economically disadvantaged small business firms in technological innovation. The electronic application, specific guidelines, and abstracts from previously funded projects are available on the website.
Requirements: To be eligible for Phase II, applicants must be Phase I winners as described in the Request for Application. Small businesses and small proprietorships that are in business for profit are eligible to submit applications to this program. Each organization submitting a proposal must qualify as a small business concern for research or research and development purposes. Participation by university faculty or government scientists in SBIR projects is strongly encouraged.
Restrictions: Phase I winners who have previously been rejected for Phase II funding are not eligible to apply. Funding is not available for loans or establishing a business.
Geographic Focus: All States
Date(s) Application is Due: Mar 1
Amount of Grant: Up to 450,000 USD
Contact: Charles Cleland, National Program Leader; (202) 401-5852 or (202) 401-4002; fax (202) 401-6070; ccleland@nifa.usda.gov
Internet: http://www.nifa.usda.gov/fo/sbir.cfm
Sponsor: U.S. Department of Agriculture
1400 Independence Avenue, SW, Stop 2201
Washington, DC 20250-2201

USDA Special Research Grants **4650**
The program supports research to facilitate or expand promising breakthroughs in areas of the food and agricultural sciences of importance to the nation and to facilitate or expand ongoing state-federal food and agricultural research programs. Areas of basic and applied research are generally limited to high-priority problems of a regional or national scope. Areas currently considered are water quality, integrated pest management, potato breeding, and range land research. Application deadlines are announced each fiscal year in the Federal Register and may also be obtained by calling the Office of Extramural Programs.
Requirements: State agricultural experiment stations, all colleges and universities, other research institutions and organizations, federal agencies, private organizations or corporations, and individuals having a demonstrable capacity to conduct research to facilitate or expand promising breakthroughs in areas of the food and agricultural sciences of importance to the United States are eligible to apply.
Geographic Focus: All States
Amount of Grant: Up to 95,000,000 USD
Contact: Andrea L. Brandon; (202) 720-9181; abrandon@nifa.usda.gov
Internet: http://www.csrees.usda.gov/funding/rfas/pmap.html
Sponsor: U.S. Department of Agriculture
1400 Independence Avenue, SW, Stop 2201
Washington, DC 20250-2201

USDA Technical Agricultural Assistance Research Grants **4651**
The Technical Agricultural Assistance Research Grants identify and apply the most appropriate solutions to international agricultural problems, and increase the capabilities of U.S. educational institutions and nonprofit agencies in agricultural technical assistance and research. Technical assistance provided through these reimbursable agreements benefit foreign governments and related agricultural institutions in their countries. Examples of previous projects include the following: development of effective management concepts and methodologies for implementation of technical agricultural and environmental protection

projects; hurricane rehabilitation activities in Central America and the Caribbean, emphasizing watershed rehabilitation, food safety, and food security; sound natural resources management, agricultural and environmental policy reform; micro-finance for economic and rural development in developing countries; and strengthening of host government institutions in post-conflict environments. Programs currently funded include provision of technical assistance for expanding the use of effective program management concepts and technologies in developing countries to promote food security; guidance on natural resources management policy, procedures and techniques; trade capacity building; post-disaster and post-conflict reconstruction; and integrated approaches to international agricultural development and environmental protection. Cooperative agreements are generally funded for one to two years. Specific guidelines, including how to submit requests for proposal packages, are available at the USDA/FAS website.
Requirements: U.S. institutions of higher learning, and public and private/nonprofit organizations whose primary purpose is agriculture, natural resources management, and/or rural development (including those located in U.S. territories).
Geographic Focus: All States
Amount of Grant: 30,000-600,000 USD
Contact: Grant Pettrie, Program Contact; (202) 690-1924; grant.pettrie@usda.gov
Internet: https://www.cfda.gov/index?s=program&mode=form&tab=core&id=3e3bdec d95f214083e79af9ad9a83de3
Sponsor: U.S. Department of Agriculture
1400 Independence Avenue SW
Washington, DC 20250

USDA Tribal College Research Grants **4652**
The USDA Tribal College Research Grants are designed to assist 1994 Land-Grant Institutions (Tribal Colleges) in building institutional research capacity through applied projects that address student educational needs and solve community, reservation or regional problems. Awards are available to conduct applied research that enhances educational or career prospects of American Indians through student-developed research. Collaboration with 1862 or 1890 Land Grant Institutions is a requirement. Eligible institutions may propose projects in any discipline of the food and agricultural sciences. Examples include but are not limited to: developing new techniques for increased/more efficient bio-energy and biomass conversion through implementing sustainable use of natural resources; research on sustainable agriculture, sustainable forestry, biotechnology, agribusiness management and marketing, or aquaculture; research on human nutrition, including health and obesity prevention; research on natural resource issues including phyto-remediation of soils; water quality and quantity; alternative and sustainable energy sources; waste management; and ecology of grasslands, forest and animal systems; and research on the interventions and approaches that promote the participation of American Indian students in the food and agricultural sciences. The electronic application, specific guidelines (including tribal colleges eligible for funding), evaluation criteria, and abstracts from previously funded projects are available at the website.
Requirements: Applicants must be 1994 land grant institutions. Grant applications submitted under this program must certify that the research will be performed in collaboration with at least one 1862 or 1890 Land-Grant college or university.
Restrictions: Colleges must collaborate with an 1862 or 1890 land grant institution, but as the lead institution, the 1994 land grant institution must retain at least 60% of the award funds. The use of grant funds to plan, acquire, or construct a building or facility is not allowed under this program. With prior approval, some grant funds may be used for minor alterations, renovations, or repairs deemed necessary to retrofit existing teaching or research spaces in order to carry out a funded project. However, requests to use grant funds for such purposes must demonstrate that the alterations, renovations, or repairs are essential to achieving the major purpose of the project. Grant funds may not be used for endowment investing.
Geographic Focus: Arizona, California, Kansas, Michigan, Minnesota, Montana, Nebraska, New Mexico, North Dakota, South Dakota, Washington, Wisconsin
Date(s) Application is Due: Mar 14
Contact: Tim Grosser; (202) 690-0402; fax (202) 720-3945; tgrosser@nifa.usda.gov
Internet: http://www.nifa.usda.gov/fo/tribalcollegesresearchtcrgp.cfm
Sponsor: U.S. Department of Agriculture
1400 Independence Avenue, SW, Stop 2201
Washington, DC 20250-2201

USDA Wildlife Services Grants **4653**
The objectives of the Wildlife Services Grants are to reduce damage caused by mammals and birds and those mammal and bird species that are reservoirs for zoonotic diseases. Wherever feasible, humane methods will be employed. Recipients are to work closely with state departments of fish and game, agriculture, health, and counties to alleviate wild animal damage. Those who receive funding are expected to conduct surveys and campaigns to reduce wild animal damage, develop methods to control wild animal damage, and provide technical advice and assistance. There are no deadlines and funding is available year round. Specific guidelines and additional information is available at the CFDA website (project 10.028). For direct technical assistance, state fish and game departments should be contacted.
Requirements: State and local governments, federally recognized tribal governments, public/private nonprofit organizations, nonprofit institutions of higher education, and individuals are eligible to apply.
Geographic Focus: All States
Contact: Eileen M. Berke; (301) 734-8330; eileen.m.berke@aphis.usda.gov
Internet: https://www.cfda.gov/index?s=program&mode=form&tab=step1&id=12c239 6f3a0d6463d2d9132339523995
Sponsor: U.S. Department of Agriculture
4700 River Road
Riverdale, MD 20737

USDC/NOAA Anadromous Fish Conservation Act Grants **4654**
These projects are conducted for the conservation, development, and enhancement of anadromous fishery resources (those that migrate from salt to fresh water for spawning) including similar species in the Great Lakes and Lake Champlain. All projects must be cleared with the State fishery agency of the State that the work is carried out in. Many projects funded under this program are critical elements of larger multi-funded programs to manage, restore or enhance United States anadromous species such as Pacific and Atlantic salmon, Atlantic and shortnose sturgeon, and American Shad and river herrings. Funds may be used for spawning area improvement, installment of fishways, data collection, construction of fish protection devices and hatcheries, and research to improve management and increase anadromous fish resources. This program is also supported by the Fish and Wildlife Service of the DOI. Applications are accepted at any time for one- to three-year projects. The project application should be submitted 90 days in advance of desired effective date.
Requirements: Any interested person or organization may propose a cooperative undertaking; however, all proposals must be coordinated with the state fishery agency having responsibility for the resource to be affected by the proposal.
Restrictions: Funds may not be used for law enforcement, public relations, or construction of facilities and vessels, whose primary purpose is to commercially harvest, handle, and process fishery products and may not be used for projects in the Columbia River Basin with the exception of the state of Idaho. The federal share of a project cost shall usually not exceed 50 percent. Federal funds for projects supporting an interstate fishery management plan may be up to 90 percent.
Geographic Focus: All States
Amount of Grant: 2,000-400,000 USD
Contact: Michelle Winowitch, (301) 713-3155; fax (301) 713-0596; Michelle.Winowitch@noaa.gov or grants.information@noaa.gov
Internet: http://www.nmfs.noaa.gov/sfa/state_federal/State-Federal-WEB/afca.htm
Sponsor: U.S. Department of Commerce
1401 Constitution Avenue, NW, Room 5128
Washington, DC 20230

USDC/NOAA Coastal and Marine Fish Habitat Protection Grants **4655**
NOAA Fisheries and the National Fish and Wildlife Foundation are soliciting projects to support the protection of coastal and marine fish habitats through NFHP Fish Habitat Partnerships. Successful proposals will: provide protection of coastal and marine fish habitats through cooperative, non-regulatory approaches as a primary benefit of the project; demonstrate how proposed actions will result in quantifiable fish habitat protection; address a strategic habitat protection priority of one or more of the eight coastally-focused Fish Habitat Partnerships, as identified in a Partnership's strategic plan; support knowledge transfer and lessons learned to other entities protecting fish habitat; and be supported by the relevant NFHP Fish Habitat Partnership, local constituencies and affected landowners. The total funding available for this program is $135,000. Awards considered for funding under this initiative are likely to fall within the $15,000 to $50,000 range and should meet project selection criteria above while demonstrating a strong, on-the ground benefit to coastal or marine habitat(s). Applicants are encouraged to provide non-federal match of $2 for every $1 of grant funds requested. Eligible non-federal matching sources can include cash, in-kind donations, and/or volunteer labor.
Requirements: All persons, organizations, and agencies (excluding U.S. Federal Government) working on projects to protect coastal and marine fish habitats are eligible to apply for funding.
Restrictions: Federal agencies may not receive funds, but may be listed as partners on an application. No part the grant funds or the non-federal match may be used to: support litigation expenses or lobbying activities; cover permanent federal employee salary expenses; supplement shortfalls in government agency budgets; support basic research; support basic planning, outreach, or education projects without an on-the-ground component.
Geographic Focus: All States
Date(s) Application is Due: Jun 29
Amount of Grant: 15,000-50,000 USD
Contact: Cara Rose; (503) 417-8700, ext. 6008; cara.rose@nfwf.org
Internet: http://www.nfwf.org/AM/Template.cfm?Section=Search&TEMPLATE=/CM/ContentDisplay.cfm&CONTENTID=24847
Sponsor: U.S. Department of Commerce
1401 Constitution Avenue, NW, Room 5128
Washington, DC 20230

USDC/NOAA Community-Based Habitat Restoration Grants **4656**
The Nature Conservancy (TNC) and the National Oceanic and Atmospheric Administration (NOAA) Community-based Restoration Program (CRP) is intended to bring together interested groups, public, private, tribal and non-profit organizations to implement habitat restoration projects to benefit NOAA trust resources (coastal and marine species and their habitats). This innovative program recognizes the significant role that partnerships can play in making habitat restoration happen within communities, and acknowledges that habitat restoration is often best implemented through technical and monetary support provided at a community level. The program funds grass-roots marine and coastal habitat restoration projects that will benefit andromous fish species, commercial and recreational resources, and endangered and threatened species. The program invites the public to submit proposals for available funding to implement grass-roots habitat restoration projects that will benefit living marine resources, including andromous fish. The Federal Funding Opportunity describes the conditions under which applications (project proposals) will be accepted under the CRP and describes criteria under which applications will be evaluated for funding consideration. Projects funded through the CRP will be expected to have strong on-the-ground habitat restoration components that provide educational and social benefits for people and the communities in addition to long-term ecological habitat improvements

for NOAA trust resources. Proposals selected for funding through this solicitation will be implemented through a project grant or cooperative agreement mechanism. Preference will be given to projects at priority sites identified through Marine Ecoregional Assessments and other TNC priority setting approaches at the state and territory level.
Requirements: Eligible applicants are institutions of higher education, hospitals, other non-profits, commercial (for profit) organizations, organizations under the jurisdiction of foreign governments, international organizations, and state, local and Indian tribal governments whose projects have the potential to benefit NOAA trust resources. Non-federal match is required at the rate of 1:1. The 1:1 match can come from a variety of (non-federal) public and private sources and can include in-kind goods and services from project partners.
Restrictions: Applications from federal agencies or employees of Federal agencies will not be considered.
Geographic Focus: All States
Date(s) Application is Due: Apr 13
Amount of Grant: 25,000-250,000 USD
Contact: Boze Hancock; (401) 874-6121; bhancock@tnc.org
Internet: http://www.habitat.noaa.gov/funding/northeast.html
Sponsor: U.S. Department of Commerce
1401 Constitution Avenue, NW, Room 5128
Washington, DC 20230

USDC/NOAA Community-Based Marine Debris Removal Grants **4657**
The Program provides funding to catalyze the implementation of locally driven, community-based marine debris prevention and removal projects that will benefit coastal habitat, waterways, and NOAA trust resources including diadromous fish. Projects funded through the MDP have strong on-the-ground habitat components involving the removal of marine debris and derelict fishing gear that will provide educational and social benefits for people and their communities in addition to long-term ecological habitat improvements for NOAA trust resources. The principal objective of the Program is to provide federal financial and technical assistance to grass-roots, community-based activities that improve living marine resource habitats through the removal of marine debris and the prevention of debris entering the marine environment, and that promote stewardship and a conservation ethic for NOAA trust resources. Generally, costs that are allowable include salaries and equipment and supplies (as long as these are necessary and reasonable specifically for the purpose of the award). Typical project awards will range from $15,000 to $150,000. NOAA will not accept proposals for less than $15,000 or proposals for more than $250,000 under this solicitation.
Requirements: Eligible applicants are institutions of higher education, other non-profits, commercial (for profit) organizations, organizations under the jurisdiction of foreign governments, international organizations, and state, local and Indian tribal governments whose projects have the potential to benefit NOAA trust resources. Applicants are encouraged to demonstrate a minimum 1:1 non-Federal match for MDP funds requested to conduct the proposed project.
Restrictions: Applications from federal agencies or employees of Federal agencies will not be considered.
Geographic Focus: All States
Date(s) Application is Due: Nov 1
Amount of Grant: 15,000-250,000 USD
Contact: Tom Barry, (301) 427-8653; tom.barry@noaa.gov
Internet: http://www.habitat.noaa.gov/funding/marinedebris.html
Sponsor: U.S. Department of Commerce
1401 Constitution Avenue, NW, Room 5128
Washington, DC 20230

USDC/NOAA Marine Fisheries Initiative (MARFIN) Grants **4658**
The Marine Fisheries Initiative (MARFIN) is a competitive Federal assistance program that funds projects seeking to optimize research and development benefits from U.S. marine fishery resources through cooperative efforts involving the best research and management talents to accomplish priority activities. Projects funded under MARFIN provide answers for fishery needs covered by the NMFS Strategic Plan, available from National Marine Fisheries Service (NMFS), particularly those goals relating to: rebuilding over-fished marine fisheries, maintaining currently productive fisheries, and integrating conservation of protected species and fisheries management. Funding priorities for MARFIN are formulated from recommendations received from non-federal scientific and technical experts and from NMFS research and operations officials. The NMFS Southeast Regional Office anticipates awarding approximately ten projects that will range from $25,000 to $175,000 per year for each project (not to exceed $175,000 per year). The total federal amount that may be requested shall not exceed $175,000 for a one-year project, $350,000 for a two-year project, and $525,000 for a three-year project.
Requirements: Eligible applicants may be institutions of higher education, nonprofits, commercial organizations, individuals, state, local and Indian tribal governments.
Restrictions: Federal agencies or institutions are not eligible. Foreign governments, organizations under the jurisdiction of foreign governments, and international organizations are excluded for purposes of this solicitation
Geographic Focus: All States
Date(s) Application is Due: Aug 9
Amount of Grant: 25,000-525,000 USD
Contact: Ellie Francisco Roche; (727) 824-5324; Ellie.Roche@noaa.gov
Internet: http://sero.nmfs.noaa.gov/grants/marfin.htm
Sponsor: U.S. Department of Commerce
1401 Constitution Avenue, NW, Room 5128
Washington, DC 20230

USDC/NOAA National Marine Aquaculture Initiative Grants **4659**
NOAA launched the National Marine Aquaculture Initiative (NMAI) in 1998 to encourage scientific research and development to support a robust, sustainable domestic marine aquaculture industry. The competition fosters dynamic partnerships that channel resources toward the development of sustainable aquaculture technologies. Projects often involve partnerships among commercial companies, research institutions, universities, state governments, and coastal communities. Currently, NOAA Sea Grant's National Marine Aquaculture Initiative is making $3.2 million available through a Sea Grant Aquaculture Research competitive grants program. Funds will support marine aquaculture research projects as part of the overall plan to support the development of environmentally and economically sustainable ocean, coastal, and Great Lakes aquaculture. Priorities for this competition include: research to inform specific regulatory decisions; research that supports multi-use spatial planning; and socio-economic research targeted to understand aquaculture in a larger context. The deadline for receipt of preproposals via electronic mail at the National Sea Grant Office is 4:00 p.m. Eastern Time on February 7.
Requirements: Institutions of higher education, nonprofit organizations, commercial organizations, State, local and Indian tribal governments and individuals are eligible.
Restrictions: Federal agencies and their personnel are not permitted to receive federal funding under this competition; however, federal scientists can serve as uncompensated partners or co-Principal Investigators on research proposals. Directors of the state Sea Grant Programs are not eligible to compete for funds under this announcement, although for administrative purposes, they will be considered to be the Principal Investigator for all awards made to their state programs.
Geographic Focus: All States
Date(s) Application is Due: Feb 7
Amount of Grant: 50,000-500,000 USD
Contact: Dr. Gene Kim, (301) 734-1281; oar.hq.sg.aquaculture@noaa.gov
Internet: http://www.grants.gov/search/search.do;jsessionid=MynvQR9LnVNshhngjY2THgjbvFLty914kZ1Fwnq4PVT5shWzCxgy!1460891328?oppId=135993&mode=VIEW
Sponsor: U.S. Department of Commerce
1401 Constitution Avenue, NW, Room 5128
Washington, DC 20230

USDC/NOAA National Marine Fisheries Service Sea Grant Joint Graduate Fellowships **4660**
NOAA National Sea Grant Office and NOAA Fisheries established a Graduate Fellowship Program in two specialized areas: population dynamics and marine resource economics. Population dynamics is the study of fish populations as affected by fishing mortality, growth, recruitment and natural mortality. Ph.D. candidates interested in the population dynamics of living marine resources and the development and implementation of quantitative methods for assessing their status can receive up to three years of funding. Ph.D. students in marine resource economics, concentrating on the conservation and management of living marine resources, can receive two years of funding. The four main goals of the NOAA Fisheries/Sea Grant Fellowship Program are: to encourage qualified applicants to pursue careers in either population dynamics and stock assessment or in marine resource economics; to increase available expertise related to these fields; to foster closer relationships between academic scientist and NOAA Fisheries; and to provide real-world experience to graduate students and accelerate their career development.
Requirements: The Fisheries Fellowship is available to U.S. citizens who are graduate students enrolled in PhD degree programs in academic institutions in the United States and its territories.
Geographic Focus: All States
Date(s) Application is Due: Jan 20
Contact: Terry Smith; (301) 734-1077, ext. 1075; terry.smith@noaa.gov
Internet: http://www.seagrant.noaa.gov/funding/rfp.html#nmfs
Sponsor: U.S. Department of Commerce
1401 Constitution Avenue, NW, Room 5128
Washington, DC 20230

USDC/NOAA Nature Conservancy Community-based Matching Grants **4661**
The objectives of TNC and NOAA's Community-based Restoration Program (CRP) are to bring together interested groups, public, private, tribal and non-profit organizations to implement habitat restoration projects to benefit NOAA trust resources (coastal and marine species and their habitats). This innovative program recognizes the significant role that partnerships can play in making habitat restoration happen within communities, and acknowledges that habitat restoration is often best implemented through technical and monetary support provided at a community level. NOAA and TNC are looking to support community-based restoration of a diversity of habitat types and no habitats are excluded. The office will consider any innovative restoration project nationwide that supports NOAA trust resources, and particularly those projects that have a multi-species benefit or emphasize Ecosystem-Based Management. A focal area of particular interest, though not exclusive or limiting, is native shellfish (bivalve) restoration projects. Typical grants will be in the range of $25,000 to $85,000.
Geographic Focus: All States
Date(s) Application is Due: Apr 13
Amount of Grant: 25,000-85,000 USD
Contact: Boze Hancock; (401) 874-6121; bhancock@tnc.org
Internet: http://www.habitat.noaa.gov/funding/northeast.html
Sponsor: U.S. Department of Commerce
1401 Constitution Avenue, NW, Room 5128
Washington, DC 20230

USDC/NOAA Restoration Center National and Regional Partnership Grants 4662
For more than 20 years, the NOAA Restoration Center has worked to maintain strong relationships with a number of non-profit organizations focused on restoring coastal habitat nationwide. These partnerships help the Restoration Center leverage greater funds and effort for restoration through match and in-kind contributions such as funding, land easements, staff expertise, volunteer support, and other services. In return, the NOAA Restoration Center offers technical and permitting assistance in restoring habitats from coast to coast. The office offers funding for multi-year national and regional habitat restoration Partnerships that will result in implementation of a wide-range of habitat restoration projects-from locally-driven, hands-on projects that emphasize stewardship, to mid-scale, watershed-scale projects that yield significant ecological and socioeconomic benefits. Grants range from $500,000 to $1,000,000.
Requirements: Eligible applicants are institutions of higher education, non-profits, commercial (for profit) organizations, U.S. Territories, and state, local and Indian tribal governments.
Restrictions: Applications from Federal agencies or employees of Federal agencies will not be considered. Federal agencies are strongly encouraged to work with states, non-governmental organizations, municipal and county governments, conservation corps organizations and others that are eligible to apply.
Geographic Focus: All States
Date(s) Application is Due: Sep 30
Amount of Grant: 500,000-1,000,000 USD
Contact: Robin Bruckner, (301) 713-0174; Robin.Bruckner@noaa.gov
Melanie Gange, (301) 713-0174; Melanie.Gange@noaa.gov
Internet: http://www.habitat.noaa.gov/funding/partnerships.html
Sponsor: U.S. Department of Commerce
1401 Constitution Avenue, NW, Room 5128
Washington, DC 20230

USDC/NOAA Saltonstall-Kennedy Grants 4663
The Saltonstall-Kennedy Act established a fund (known as the S-K fund) that the Secretary of Commerce uses to provide grants or cooperative agreements for fisheries research and development projects addressing aspects of U.S. fisheries, including, but not limited to, harvesting, processing, marketing, and associated infrastructures. U.S. fisheries include any fishery, commercial or recreational, that is, or may be, engaged in by citizens or nationals of the United States, or citizens of the Northern Mariana Islands (NMI), the Republic of the Marshall Islands, Republic of Palau, and the Federated States of Micronesia.
Requirements: Any U.S. citizen or national, or any citizen of the Northern Mariana Islands, the Republic of the Marshall Islands, Republic of Palau, or the Federated States of Micronesia; or group including state and local governments may apply.
Restrictions: Employees of federal agencies and regional fishery management councils and their employees are not eligible.
Geographic Focus: All States
Date(s) Application is Due: Oct 1
Amount of Grant: 100,000-300,000 USD
Contact: Daniel Namur, Program Manager; (301) 713-2259; Dan.Namur@noaa.gov
Internet: http://www.nmfs.noaa.gov/mb/financial_services/skhome.htm
Sponsor: U.S. Department of Commerce
1401 Constitution Avenue, NW, Room 5128
Washington, DC 20230

USDC Afghanistan and Iraq Investment and Reconstruction Task Forces Internships 4664
The Afghanistan and Iraq Investment and Reconstruction Task Forces of the U.S. Department of Commerce serve as clearinghouses of information for U.S. companies interested in doing business in Iraq and Afghanistan. The Task Force works closely with other U.S. government agencies, Afghan and Iraqi government agencies, and international organizations to provide U.S. companies counseling on the latest developments and business opportunities in Iraq and Afghanistan. The Task Forces seek unpaid international trade interns for one or two competitive positions each semester (Spring, Summer, Fall). This is an excellent opportunity for students to learn about how the U.S. government operates and contributes to Afghanistan and Iraq reconstruction efforts. Interns work on both Task Forces.
Requirements: Qualifications include: concurrently enrolled undergraduate or graduate students interested in learning how the U.S. government operates and contributes to Afghanistan and Iraq reconstruction efforts; majoring in international affairs, international trade, economics, or business; excellent research, writing, and computer skills; available at least 20 hours per week; and U.S. citizenship required.
Geographic Focus: All States
Contact: Tracy M. Rollins; (202) 482-0073; SABITApply@mail.doc.gov
Internet: http://www.trade.gov/afghanistan/tg_aftf_003400.asp
Sponsor: U.S. Department of Commerce
1401 Constitution Avenue, NW
Washington, DC 20230

USDC Dissertation Fellowships 4665
The Census Bureau Dissertation Fellowship program seeks to award a limited number of contracts each year to Ph.D. -granting accredited institutions of higher education in the United States to fund the dissertation research of doctoral candidates who propose and investigate solutions to problems on the list of research topics of primary interest to the Census Bureau. When fully implemented, up to ten dissertation fellowships will be awarded annually for up to $50,000 each. Dissertation Fellowships will be awarded to institutions of higher education for doctoral candidates of unusual promise and ability whose general fields of research relate to statistics, economics, and demography. Direct connections to the cited listing of research topics is essential. Exceptionally strong candidates writing master's theses that address these problems will also be considered.
Requirements: The application materials and supporting documents for each Dissertation Fellow must be submitted by his or her university. A Dissertation Fellow must: be a doctoral candidate or exceptionally strong master's candidate; be funded by a contract awarded to institution for tuition, research materials, travel, and salary; devote full-time effort, be enrolled full-time, and in residence at Fellow's Institution; report progress biannually; collaborate on-site with Census Bureau staff for one to two months; and present research results in a Census Bureau seminar. United States Citizenship is required.
Geographic Focus: All States
Date(s) Application is Due: Feb 28
Amount of Grant: Up to 50,000 USD
Contact: Dr. Tommy Wright; (301) 763-1702; tommy.wright@census.gov
Carol A. Gentry, carol.a.gentry@census.gov
Internet: http://www.census.gov/research/opportunities/
Sponsor: U.S. Department of Commerce
4600 Silver Hill Road
Washington, DC 20233

USDC Low-Power Television and Translator Upgrade Grants 4666
The Low-Power Television and Translator Upgrade program reimburses the eligible costs to upgrade analog low power television broadcast stations, Class A television stations, television translator stations, and television booster stations in rural communities to digital transmission. Applications are received the first business day of each month, and the final closing date for receipt of Upgrade program applications will be July 2.
Requirements: Eligible stations must be broadcasting a licensed digital signal before applying for a grant.
Geographic Focus: All States
Date(s) Application is Due: Jul 2
Contact: Dr. Bernadette McGuire-Rivera, Associate Administrator; (202) 482-5802; fax (202) 501-8009; bmcguire-rivera@ntia.doc.gov
Internet: http://www.ntia.doc.gov/category/low-power-television-and-translator-program
Sponsor: U.S. Department of Commerce
1401 Constitution Avenue, NW
Washington, DC 20230

USDC Market Development Cooperator Program (MDCP) Grants 4667
The goal of the Market Development Cooperator Program is to develop, maintain, and expand foreign markets for non-agricultural goods and services produced in the United States. The intended beneficiaries of the program are U.S. producers of non-agricultural goods or services that seek to export such goods or services. The International Trade Administration (ITA) encourages applicants to propose activities that would be most appropriate to the market development needs of their U.S. industry or industries. Examples of activities might include: opening an overseas office or offices to perform a variety of market development services for companies joining a consortium to avail themselves of such services; detailing a private-sector representative to a U.S.&FCS post in accordance with 15 U.S.C 4723(c); commissioning overseas market research, participating in overseas trade exhibitions, and trade missions to promote U.S. exports, and/or hosting reverse trade missions; and conducting U.S. product demonstrations abroad. The applicant must contribute at least two dollars for each federal dollar provided. Full guidelines are available on the website or may be obtained by contacting the office listed. No award will exceed $300,000. ITA anticipates concluding seven (7) to ten (10) awards.
Requirements: U.S. trade associations; nonprofit industry organizations; state trade departments and their regional associations, including centers for international trade development; and private industry firms or groups of firms in cases where no entity described above represents that industry are eligible.
Geographic Focus: All States
Date(s) Application is Due: Feb 21
Amount of Grant: Up to 300,000 USD
Contact: Brad Hess; (202) 482-2969; fax (202) 482-5828; Brad.Hess@trade.gov
Internet: http://ita.doc.gov/td/mdcp/
Sponsor: U.S. Department of Commerce-1
14th Street and Constitution Avenue NW, Herbert C Hoover Building
Washington, DC 20230

USDC Planning and Local Technical Assistance Grants 4668
The Economic Development Administration announces general policies and application procedures for grant-based investments under the Planning and Local Technical Assistance Programs. These programs will help communities develop the planning and technical expertise to support communities and regions in their comprehensive, entrepreneurial, and innovation-based economic development efforts. Resulting in increased private investment and higher-skill, higher-wage jobs in areas experiencing substantial and persistent economic distress, these programs are designed to enhance the competitiveness of regions. EDA's Planning and Local Technical Assistance Programs are two of the six economic development assistance programs the agency operates. Under the Planning Program, EDA provides assistance to eligible recipients to create regional economic development plans in order to stimulate and guide the economic development efforts of a community or region. EDA's Local Technical Assistance Program helps eligible recipients fill the knowledge and information gaps that may prevent leaders in the public and nonprofit sectors in economically distressed regions from making optimal decisions on local economic development issues.
Requirements: Eligible applicants include: district organizations; Indian tribes or a consortium of Indian tribes; states, cities or other political subdivisions of a state,

including special purpose units of a state or local government engaged in economic or infrastructure development activities, or a consortium of political subdivisions; institutions of higher education or a consortium of institutions of higher education; or public or private non-profit organizations or associations acting in cooperation with officials of a political subdivision of a state.
Restrictions: EDA is not authorized to provide grants directly to individuals or to for-profit entities seeking to start or expand a private business.
Geographic Focus: All States
Contact: Lindsey Pangretic, Director, Research and National Technical Assistance Division; (202) 482-4085 or (202) 482-2900; grants@eda.doc.gov
Internet: http://www.grants.gov/search/search.do;jsessionid=kDW2PsLT1zdv3HLW1Bpwx3yQvQbpJPt1XnmTfyM1yGJpBP99tt2g!-757993493?oppId=58876&mode=VIEW
Sponsor: U.S. Department of Commerce
1401 Constitution Avenue, NW, Suite 7800
Washington, DC 20230

USDC Postdoctoral Research Grants 4669
The objectives of the Census Bureau Postdoctoral Research Program are to provide postdoctoral candidates of unusual promise and ability opportunities for research on problems, largely of their own choice, that are compatible with the interests of the Census Bureau, and to contribute to the overall efforts of federal statistical agencies. A Postdoctoral Researcher is a resident researcher and a temporary employee of the Census Bureau. Postdoctoral Researchers are analogous to fellows or similar temporary researchers at the postdoctoral level in universities and other organizations. Postdoctoral Research positions are not intended to be, or to compete with, permanent professional career positions. Awardees are offered two-year appointments, but extensions are possible with special considerations. An evaluation is conducted after one year to ensure that the Postdoctoral Researcher is making suitable progress. An applicant's training and research experience may be in any appropriate discipline or combination of disciplines required for the proposed research. The annual base salary for a Census Bureau Postdoctoral Researcher starts from $74,872 to $97,333 with adjustments depending on qualifications and experience. For fullest consideration, submit application by January 31.
Requirements: No commitment on the part of a Postdoctoral Researcher or the Census Bureau with regard to later employment is implied or should be inferred by the offer or acceptance of an award as a Postdoctoral Researcher. Postdoctoral Researchers must devote their full-time effort to the research program and must be in residence at the Census Bureau during the program. No period of tenure may be spent in residence at another agency or institution. No additional monetary aid or other remuneration may be accepted from another appointment, fellowship, or similar grant during the period of the program. Postdoctoral Research positions are awarded only to persons who have held the doctorate less than six years or are in the process of receiving a doctorate degree at the time of application. If a doctorate has been held for more than six years, refer to the ASA/NSF/Census Bureau Fellowship Program for another research opportunity. Awardees must hold the Ph.D. or other earned research degree recognized in the U.S. as equivalent to the Ph.D. or must present acceptable evidence of having completed all the formal academic requirements for the degree before appointment. Applicants must have demonstrated ability for creative research.
Geographic Focus: All States
Amount of Grant: 74,872-97,333 USD
Contact: Chris A. Cowan; (301) 480-3538; chris.cowan@hhs.gov
Internet: http://www.census.gov/hrd/www/jobs/prp.html
Sponsor: U.S. Department of Commerce
4600 Silver Hill Road
Washington, DC 20233

USDC Presidential Management Fellowships 4670
For the past 27 years, the Presidential Management Fellows (PMF) Program and its predecessor, the Presidential Management Intern (PMI) Program, have been attracting outstanding masters, law, and doctoral-level students to the Federal service. The PMF Program seeks candidates who display excellent communication, interpersonal, planning and analytical skills, decision-making ability, and the potential to perform in leadership positions. This rigorous two-year paid program includes 80-hours of training each year, rotational opportunities, challenging assignments, and opportunities to network between agencies. The Census Bureau is supportive of the PMF program and has a strong commitment to training and professional development. Current and former PMFs work in numerous positions throughout the Bureau and engage in a variety of diverse tasks. As a PMF at Census, you could have the opportunity to work with current data on topics such as unemployment, business productivity, poverty, and health insurance. This data has a profound impact on policy formation, evaluation, and decision making in both the public and private sectors.
Geographic Focus: All States
Contact: Chris A. Cowan; (301) 480-3538; chris.cowan@hhs.gov
Internet: http://www.census.gov/hrd/www/jobs/pmip.html
Sponsor: U.S. Department of Commerce
4600 Silver Hill Road
Washington, DC 20233

USDC State Broadband Initiative Grants 4671
The National Telecommunications and Information Administration's State Broadband Initiative implements the purposes of both the American Recovery and Reinvestment Act and the Broadband Data Improvement Act, which envisioned a comprehensive program — led by state entities or non-profit organizations working at their direction — to help integrate broadband and information technology into state and local economies. Through the State Broadband Initiative (formerly called the State Broadband Data and Development grant program), NTIA awarded a total of $293 million to 56 grantees, one from each of the 50 states, five territories, and the District of Columbia, or their designees. Grantees are using this funding to support the efficient and creative use of broadband technology to better compete in the digital economy. These state-created efforts vary depending on local needs but include programs to assist small businesses and community institutions in using technology more effectively, conduct research on barriers to broadband adoption, implement innovative applications that increase access to government services and information, and support state and local task forces to expand broadband access and adoption. Since accurate data is critical for broadband planning, another purpose of the State Broadband Initiative is to assist states in gathering data twice a year on the availability, speed, and location of broadband services, as well as the broadband services that community institutions use. The data will be used by NTIA to update its publicly searchable, interactive National Broadband Map.
Geographic Focus: All States
Contact: Dr. Bernadette McGuire-Rivera, Associate Administrator; (202) 482-5802; fax (202) 501-8009; bmcguire-rivera@ntia.doc.gov
Internet: http://www.ntia.doc.gov/category/state-broadband-initiative
Sponsor: U.S. Department of Commerce
1401 Constitution Avenue, NW
Washington, DC 20230

USDD Acoustic Transduction Materials and Devices Grants 4672
Research on innovative materials and device structures is sought that enhances the performance of the electro-mechanical transducers used by the Department of Navy to generate, detect and suppress undersea sound waves. The present Office of Naval Research program has two thrusts. The first aims to devise and validate first-principles quantum-mechanical methods to evaluate the properties of materials with structural phase transitions; this will enable the exploration of the properties of new materials systems in advance of their synthesis. The second focuses on the new high-coupling, high-strain relaxor piezoelectric single crystals; efforts underway aim to produce and characterize these piezocrystals and to design/fabricate innovative transducers from them for Navy SONAR systems. Applications include: Mine Identification SONAR; UUV Acoustic Modem; Vector Sensor Towed Arrays; Echocardiography, Medical Ultra-Sound; and 4D Realtime Volumetric Harmonic Imaging. Submission of the Research Proposal should always be preceded by submission of a Preproposal and a discussion of that Preproposal with the assigned Program Officer. Applications for the upcoming fiscal year should be received by September 30.
Requirements: Grants are awarded to academic, industrial, and nonprofit organizations, with strong emphasis on university research. Investigators are encouraged to contact the staff prior to formally submitting a proposal.
Restrictions: New grants or contracts are typically started near the beginning of the fiscal year. New proposals for the next fiscal year should be submitted no later than September 30, unless otherwise directed. Competitive renewals should be submitted no later than six months prior to the award termination date after contacting the Program Officer. Conference proposals should be submitted at least four months prior to the conference.
Geographic Focus: All States
Date(s) Application is Due: Sep 30
Contact: Dr. Wallace Smith; (703) 696-0284; wallace.smith1@navy.mil
Internet: http://www.onr.navy.mil/Science-Technology/Departments/Code-33/All-Programs/332-naval-materials/Acoustic-Transduction-Materials-Devices.aspx
Sponsor: U.S. Department of Defense
One Liberty Center, 875 N. Randolph Street, Suite 1425
Arlington, VA 22203-1995

USDD Active Aperture Array Grants 4673
The Office of Naval Research (ONR) Active Aperture Array Program seeks to develop innovative electronics technologies that will enable the revolutionary architectures and capabilities of future electronically scanned array (ESA) apertures. The emphasis is on enabling affordable solutions without sacrificing advanced performance at RF frequencies from microwave through millimeter wave. Research efforts are currently focused on: improvement of solid state device parameters including: switching speed,bandwidth, noise, dynamic range, linearity and robustness to dense signal environments; increasing the performance and efficiency of component technologies via advanced semiconductor III-V materials such as InP, GaN, SiC and others such as evolving nanotechnology materials; demonstrating size reduction of components needed to fit within an array lattice; and mixed analog and digital conversion circuits and signal processing architectures. Grants for the upcoming fiscal year must be received by September 30.
Requirements: Grants are awarded to academic, industrial, and nonprofit organizations, with strong emphasis on university research.
Geographic Focus: All States
Date(s) Application is Due: Sep 30
Contact: Dr. Stephen Pappert; (703) 696-5268; stephen.pappert@navy.mil
Internet: http://www.onr.navy.mil/Science-Technology/Departments/Code-31/All-Programs/312-Electronics-Sensors/Active-Aperture-Array.aspx
Sponsor: U.S. Department of Defense
One Liberty Center, 875 N. Randolph Street, Suite 1425
Arlington, VA 22203-1995

USDD Active Authentication Grants 4674
DARPA is soliciting innovative research proposals in support of the development of new software-based biometric modalities. Proposed research should investigate innovative approaches that enable revolutionary advances in science, devices, or systems. The Active Authentication Grants program seeks to change the current focus from user proxies (e.g., passwords and CACs) when validating identity on the Department of Defense IT systems to a focus on the individual. Within this program, the intention is to focus on the unique

factors that make up the individual, also known as their biometrics, without requiring the deployment of additional hardware sensors. Research resulting from these Grants will support that overall program intent by investigating novel software-based biometric modalities that can be used to provide meaningful and continual authentication when later integrated into a cyber security system. Application information is available on the website. Multiple awards not exceeding $500,000 per effort are expected.
Requirements: All responsible sources capable of satisfying Government requirements are eligible. Historically black colleges and universities, small businesses, small disadvantaged businesses and minority institutions are encouraged to apply.
Restrictions: Specifically excluded is research that primarily results in evolutionary improvements to the existing state of practice.
Geographic Focus: All States
Date(s) Application is Due: Mar 6
Contact: Richard Guidorizzi; (703) 526-6630; ActiveAuthentication@darpa.mil
Internet: http://www.grants.gov/search/search.do;jsessionid=Kk1NP54Tt0jpnMLC6J7vB0Sh53T9jH1YXXSfy5LnVN1hCmPD7TKp!-1993000306?oppId=138033&mode=VIEW
Sponsor: U.S. Department of Defense
3701 North Fairfax Drive
Arlington, VA 21702-1714

USDD Advances in Bioscience for Airmen Performance Grants 4675
The USDD Air Force Research Laboratory Advances in Bioscience for Airmen Performance Grants are intended for innovative science and technology projects to support advanced bioscience research. Specifically, unique and innovative research concepts are sought in the following technical areas: applied biotechnology; biobehavioral performance; biomechanics; and counterproliferation. The purpose is to augment core research activities and facilities to provide radically new military capabilities that improve warfighter performance and combat effectiveness. The research will provide for a broad range of revolutionary and evolutionary technologies to meet the critical needs of the warfighter for full dimensional protection, survivability, and performance enhancement. Grants will be offered for multiple years, and applicants may submit white papers at any time. Applicants whose white papers are determined to meet the needs of the Air Force will be invited to submit a full proposal. Total annual funding is expected to be more than $8,000,000 with approximately ten awards being made per year.
Requirements: This is an unrestricted solicitation. Small businesses are encouraged to propose on all or any part of this solicitation.
Geographic Focus: All States
Date(s) Application is Due: Sep 30
Contact: Angela G. Lauofo; (937) 255-5937; Angela.Lauofo@wpafb.af.mil
Internet: http://www.grants.gov/search-grants.html?agencies%3DDOD|Department%20of%20Defense
Sponsor: U.S. Department of Defense
1400 Defense Pentagon
Washington, DC 20301-1400

USDD Aerospace Medicine, Clinical Research, Human Performance Research, and Expeditionary Medicine Grants 4676
The USDD Air Force Research Laboratory is offering Aerospace Medicine, Clinical Research, Human Performance Research, and Expeditionary Medicine Grants. The Grants will support technology research primarily focused in the following research and technology areas. For aerospace and expeditionary medicine and clinical research the areas are: advanced diagnostics; therapeutics; occupational toxicology; expeditionary ground medicine and aeromedical evacuation; product developmental requirements; general acceleration and hypobaric test and evaluation; aircrew and battlefield airman health and performance; and clinical and translational research. For human performance and systems integration, research is needed in order to optimize warfighter effectiveness in all environments including, but not limited to: hot and humid, cold and dry, and high altitude. Interested applicants may submit white papers at any time. Applicants submitting white papers assessed as meeting Air Force needs will be asked to submit a technical and cost proposal. This is a five-year, open-ended solicitation. Five or six awards annually are anticipated.
Requirements: This is an unrestricted solicitation. Small businesses are encouraged to propose on all or any part of this solicitation.
Geographic Focus: All States
Date(s) Application is Due: Dec 15
Contact: Nicholas J. Voiles; 937-656-9808; nicholas.voiles@wpafb.af.mil
Internet: http://www.grants.gov/search-grants.html?agencies%3DDOD|Department%20of%20Defense
Sponsor: U.S. Department of Defense
1400 Defense Pentagon
Washington, DC 20301-1400

USDD Air Force Defense Research and Development Rapid Innovation Grants 4677
Air Force Research Laboratory Contracting Directorate (AFRL/PK) is soliciting white papers for the Air Force Defense Research and Development Rapid Innovation Fund (RIF) Program to facilitate the rapid insertion of innovative technologies into military systems or programs meeting critical national security needs. This BAA is open to any technologies, products, or processes at the component, subsystem, or system level demonstrating near-term potential to address Air Force requirements or improve existing Air Force acquisition programs addressing specific areas outlined above. Proposal invitation will be determined based on evaluation of white papers using the criteria published in this BAA. Total program value is $50 million; however, no individual award is anticipated to exceed $3 million. The Government may provide additional funds to be used for new awards. The Air Force anticipates selecting 16 or more awards from this announcement. White papers are due no later than 3:00 pm EST, November 27.
Requirements: This is an unrestricted solicitation. Small businesses are encouraged to propose on all or any part of this solicitation. Cost share is encouraged but not required.
Geographic Focus: All States
Date(s) Application is Due: Nov 27
Amount of Grant: Up to 3,000,000 USD
Contact: Kristina J. Croake; (937) 255-4758; kristina.croake@wpafb.af.mil
Internet: http://www07.grants.gov/search/search.do;jsessionid=8Fy2QFPFZ6dnVjThhgpY2M1P1w8z2MSLs39J9s3FDRPZQyJ1tn9Q!2071562142?oppId=202933&mode=VIEW
Sponsor: U.S. Department of Defense
1400 Defense Pentagon
Washington, DC 20301-1400

USDD Air Force Medical Support Agency/Modernization Directorate Research/Development and Innovations Grants 4678
The USDD Air Force Research Lab, on behalf of the Air Force Medical Support Agency (AFMSA)/Modernization Directorate is offering Research/Development and Innovations Grants. The Grants will ensure the AFMSA identifies and develops new knowledge and technologies to fulfill critical mission requirements. The intent of the Grants is to fund research targeted towards Air Force specific needs. AFMSA is soliciting white papers (and later technical/cost proposals) on the following research effort: (1) force health protection; (2) enroute care; (3) occupational medicine/in garrison; (4) expeditionary medicine; and (5) human performance. Interested applicants may submit white papers by the published due date. Selected applicants will be invited to submit a full proposal. Ten to twenty awards are anticipated per year with an estimated value of $100,000 to $1,000,000 annually, with total contract values of up to $3,000 over a three-year period. This is a five-year open ended solicitation. It is anticipated that updated requests for white papers will be posted on a yearly basis. Updated requests will likely occur in the first quarter of each calendar year. The continuation of requests for white papers and the submission dates should be verified from the website. Instructions for white paper submission can be found on the website.
Requirements: This is an unrestricted solicitation. Small businesses are encouraged.
Geographic Focus: All States
Contact: Kimberly Rhoads; (937) 255-0908; Kimberly.Rhoads@wpafb.af.mil
Internet: http://www07.grants.gov/search/search.do;jsessionid=91nSQFGd2zP2KG6lvQ_mQ3ftJX0xlTV251SLm6RJvgtpJHyHlJsTy!2071562142?oppId=106033&mode=VIEW
Sponsor: U.S. Department of Defense
1400 Defense Pentagon
Washington, DC 20301-1400

USDD Air Warfare and Naval Weapons Applications Grants 4679
Naval unique or essential projects in this division are involved with applied research and advanced technology aligned with current and future naval capability gaps and innovative naval prototypes. The portfolio of current future naval capability programs includes: Discriminate and Provide Terminal Guidance, Enhanced Weapon Systems Technologies, Helicopter Low-Level Operations, Laser-Based Landing Aids, and Increased Capabilities Against Moving and Stationary Targets. The Innovative Naval Prototype portfolio consists of the Electromagnetic Railgun and the Free Electron Laser programs. Submission of the Research Proposal should always be preceded by submission of a Pre-proposal and a discussion of that Pre-proposal with the assigned Program Officer.
Requirements: Grants are awarded to academic, industrial, and nonprofit organizations, with strong emphasis on university research.
Restrictions: New grants or contracts are typically started near the beginning of the fiscal year (October 1). New proposals for the next fiscal year should be submitted no later than June 1, unless otherwise directed by the BAA. Competitive renewals should be submitted no later than six months prior to the award termination date after contacting the Program Officer. Conference proposals should be submitted at least four months prior to the conference.
Geographic Focus: All States
Contact: Roger Ellis, (703) 696-9504 or (703) 588-2427; roger.ellis@navy.mil
Internet: http://www.onr.navy.mil/Science-Technology/Departments/Code-35/All-Programs/air-warfare-352.aspx
Sponsor: U.S. Department of Defense
One Liberty Center, 875 N. Randolph Street, Suite 1425
Arlington, VA 22203-1995

USDD Alloys and Joining Grants 4680
Structural metals are the basis of the modern Navy. Ship construction depends critically upon the assembly of metal plates bonded to each other and to an underlying metal support framework. The physical and mechanical properties of these metals limit the designs of ships, aircraft and ground vehicles. The chemistry and solid-state physics of alloy formation certainly plays an important role in determining the properties of any particular alloy. This Office of Naval Research program aims to design an alloy to possess optimal mechanical properties for a particular application, and the highest corrosion resistance, with the lightest weight. The joining and welding of metals also pervades the military. The integrity of a welded joint, under all possible environmental conditions, is critically important in ships and aircraft. Research supported in this program focuses closely on the physics, chemistry and materials science of welds to understand the origins of bonding strength and adhesion between different joined substances. Submission of the Research Proposal should always be preceded by submission of a Preproposal and a discussion of that Preproposal with the assigned Program Officer. Applications for the upcoming fiscal year should be received by September 30.

Requirements: Grants are awarded to academic, industrial, and nonprofit organizations, with strong emphasis on university research. Investigators are encouraged to contact the staff prior to formally submitting a proposal.
Restrictions: New grants or contracts are typically started near the beginning of the fiscal year. New proposals for the next fiscal year should be submitted no later than September 30, unless otherwise directed. Competitive renewals should be submitted no later than six months prior to the award termination date after contacting the Program Officer. Conference proposals should be submitted at least four months prior to the conference.
Geographic Focus: All States
Date(s) Application is Due: Sep 30
Contact: Dr. William Mullins, Program Officer; (703) 696-0707; fax (703) 696-0934; william.m.mullins@navy.mil
Internet: http://www.onr.navy.mil/Science-Technology/Departments/Code-33/All-Programs/332-naval-materials/Alloys-Joining.aspx
Sponsor: U.S. Department of Defense
One Liberty Center, 875 N. Randolph Street, Suite 1425
Arlington, VA 22203-1995

USDD Antifouling/Fouling Release Coatings Grants 4681
A significant limit to the performance of any marine platform arises from surface fouling by marine organisms. The response to this is to develop nontoxic approaches to antifouling and fouling release coatings. The state-of-the-art antifouling coatings are copper rosin or copper self polishing coatings, while fouling-release coatings are silicone-based. The overall program is broad with contributions from surface and adhesion science, analytical chemistry, and marine biology. Portions of this program are allied with the polymer science programs and are focused on the synthesis of new coatings. Current approaches include improved silicones, low surface energy coatings, amphiphilic, zwitterionic-containing and textured/patterned coatings with complex topographies. The long-term goals of this Office of Naval Research (ONR) program are to develop new approaches to antifouling or foul-release coatings with the following attributes: non-toxic or environmentally benign; durable in order to allow 12 -year dry-dock intervals; improved anti-fouling performance over existing copper-based paints. As marine fouling involves living organisms, there is a close cooperation between this program and programs maintained by Dr. Linda Chrisey of the ONR Human and Bioengineered Systems Division. ONR welcomes research ideas and suggestions from a variety of scientific and technical disciplines. Submission of the Research Proposal should always be preceded by submission of a Preproposal and a discussion of that Preproposal with the assigned Program Officer. Applications for the upcoming fiscal year should be received by September 30.
Requirements: Grants are awarded to academic, industrial, and nonprofit organizations, with strong emphasis on university research. Investigators are encouraged to contact the staff prior to formally submitting a proposal.
Restrictions: New grants or contracts are typically started near the beginning of the fiscal year. New proposals for the next fiscal year should be submitted no later than September 30, unless otherwise directed. Competitive renewals should be submitted no later than six months prior to the award termination date after contacting the Program Officer. Conference proposals should be submitted at least four months prior to the conference.
Geographic Focus: All States
Date(s) Application is Due: Sep 30
Contact: Dr. Stephen McElvany, Program Officer; (703) 696-1449; fax (703) 696-6887; steve.mcelvaney@navy.mil
Internet: http://www.onr.navy.mil/Science-Technology/Departments/Code-33/All-Programs/332-naval-materials/Antifouling-Fouling-Release-Coating.aspx
Sponsor: U.S. Department of Defense
One Liberty Center, 875 N. Randolph Street, Suite 1425
Arlington, VA 22203-1995

USDD Applications and Transitions Grants 4682
The Office of Naval Research (ONR) Applications and Transitions Division supports programs in surface and aerospace surveillance, communications, and electronic combat. Research areas in surface and aerospace surveillance include sensors, primarily radar and electro-optic/infrared, and associated signal and image processing methods. Areas of interest in communications include military radio communications with an emphasis on anti-jam and low-probability-of-intercept techniques, radio networks, and dynamic internetworking. Electronic combat research areas include threat warning systems, electronic support measures, decoys and electronic countermeasures. Navigation research includes GPS anti-jam and associated inertial navigation techniques. Applications for the upcoming fiscal year should be received by September 30.
Requirements: Grants are awarded to academic, industrial, and nonprofit organizations, with strong emphasis on university research.
Restrictions: New grants or contracts are typically started near the beginning of the fiscal year. New proposals for the next fiscal year should be submitted no later than September 30, unless otherwise directed by the BAA. Competitive renewals should be submitted no later than six months prior to the award termination date after contacting the Program Officer. Conference proposals should be submitted at least four months prior to the conference.
Geographic Focus: All States
Date(s) Application is Due: Sep 30
Contact: Michael Pollock, Division Director; (703)696-4202 or (703) 696-4217; fax (703) 696-3098; michael.a.pollock@navy.mil
Internet: http://www.onr.navy.mil/Science-Technology/Departments/Code-31/All-Programs/313-applications-transitions.aspx
Sponsor: U.S. Department of Defense
One Liberty Center, 875 N. Randolph Street, Suite 1425
Arlington, VA 22203-1995

USDD Applied and Computational Analysis Grants 4683
The Office of Naval Research (ONR) Applied Computational Analysis Program develops and exploits methods in modern and classical mathematical analysis with emphasis on mathematical and computational models of physical phenomena. Naval interest in waves, flows, materials, structures, and information processing motivates research in the areas of multi-phase, multi-physics, multi-material problems; fluid transport and mixing; predictability of models for nonlinear dynamics; and inverse problems arising from acoustic and electromagnetic wave propagation and scattering. The Applied Computational Analysis Program currently has three thrusts: Multi-Scale/Multi-Physics Modeling; Dynamical Systems and Oceanic and Atmospheric Modeling; and Inverse Methods. Proposers are encouraged to contact the program officer to discuss their research interest prior to the submission of formal proposals.
Requirements: Grants are awarded to academic, industrial, and nonprofit organizations, with strong emphasis on university research.
Restrictions: New grants or contracts are typically started near the beginning of the fiscal year (1 October). New proposals for the next fiscal year are ongoing. Competitive renewals should be submitted no later than six months prior to the award termination date after contacting the Program Officer. Conference proposals should be submitted at least four months prior to the conference.
Geographic Focus: All States
Contact: Dr. Reza Malek-Madani; (703) 696-0195; reza.malekmadani@navy.mil
Internet: http://www.onr.navy.mil/Science-Technology/Departments/Code-31/All-Programs/311-Mathematics-Computers-Research/Applied-Computational-Analysis.aspx
Sponsor: U.S. Department of Defense
One Liberty Center, 875 N. Randolph Street, Suite 1425
Arlington, VA 22203-1995

USDD Applied Computational Analysis Grants 4684
The Office of Naval Research (ONR) Applied Computational Analysis Program develops and exploits methods in modern and classical mathematical analysis with emphasis on mathematical and computational models of physical phenomena. Naval interest in waves, flows, materials, structures, and information processing motivates research in the areas of multi-phase, multi-physics, multi-material problems; fluid transport and mixing; predictability of models for nonlinear dynamics; and inverse problems arising from acoustic and electromagnetic wave propagation and scattering. The Applied Computational Analysis Program currently has three thrusts: Multi-Scale/Multi-Physics Modeling; Dynamical Systems and Oceanic and Atmospheric Modeling; and Inverse Methods. Proposers are encouraged to contact the program officer to discuss their research interest prior to the submission of formal proposals. The annual deadline is September 30.
Requirements: Grants are awarded to academic, industrial, and nonprofit organizations, with strong emphasis on university research.
Restrictions: New grants or contracts are typically started near the beginning of the fiscal year (October 1). New proposals for the next fiscal year should be submitted no later than June 1, unless otherwise directed by the BAA. Competitive renewals should be submitted no later than six months prior to the award termination date after contacting the Program Officer. Conference proposals should be submitted at least four months prior to the conference.
Geographic Focus: All States
Contact: Dr. Reza Malek-Madani; (703) 696 4314; reza.malekmadani@navy.mil
Internet: http://www.onr.navy.mil/Science-Technology/Departments/Code-31/All-Programs/311-Mathematics-Computers-Research/Applied-Computational-Analysis.aspx
Sponsor: U.S. Department of Defense
One Liberty Center, 875 N. Randolph Street, Suite 1425
Arlington, VA 22203-1995

USDD Arctic and Global Prediction Grants 4685
The Office of Naval Research (ONR) Arctic and Global Prediction Program is motivated by the rapid decline in summer ice extent that has occurred in recent years. This has prompted renewed Department of Navy interest in understanding and predicting the Arctic physical environment at a variety of time and space scales. The ability to predict the location of the ice edge, the space-time evolution of the ice cover, and the ice thickness will be particularly critical to safe naval operations in the Arctic. To achieve this, models integrating the ocean, waves, ice and atmosphere must be able to represent the physical processes, interactions and feedbacks involved in the seasonal evolution of ice extent, area, thickness and volume. Consequently, the program has three focus areas: developing integrated ocean-ice-wave-atmosphere Earth System models for improved prediction at a variety of time scales; improving understanding of the physical environment and processes in the Arctic Ocean; and investigating new technologies, e.g., sensors, platforms and communications, for sustained operation and observation in the challenging Arctic environment. This announcement will remain open until September 30.
Requirements: The Arctic and Global Prediction Program is committed to contributing to Navy STEM (Science, Technology and Mathematics) education efforts, particularly the education of the next generation of scientists specializing in the physics of the Arctic Ocean system. Investigators are encouraged to include students (undergraduate and graduate) and post-doctoral researchers in their research plans. As part of this commitment, the program will also consider requests for Early Student Support (ESS) to assist beginning graduate students working with ONR-funded investigators.
Geographic Focus: All States
Contact: Dr. Martin Jeffries; (703) 696-7825; martin.jeffries@navy.mil
Internet: http://www.onr.navy.mil/Science-Technology/Departments/Code-32/All-Programs/Atmosphere-Research-322/Arctic-Global-Prediction.aspx
Sponsor: U.S. Department of Defense
One Liberty Center, 875 N. Randolph Street, Suite 1425
Arlington, VA 22203-1995

USDD ARO Broad Agency Announcement Grants 4686

The United States Army Research Office (ARO) solicits proposals for basic and scientific research in mechanical sciences, environmental sciences, mathematical sciences, computer sciences, electronics, computational and information sciences, physics, chemistry, life sciences, and materials science. Interested applicants are strongly encouraged to make preliminary inquiries as to the general need for the type of research effort contemplated before preparing a proposal. Each technical area has a point of contact identified in the full announcement at the website. This is a multiple year continuously open Broad Agency Announcement (BAA). Amendments may be posted. Applicants are encouraged monitor FedBizOpps.gov (https://www.fbo.gov/) and/or Grants.gov (http://grants.gov/applicants/find_grant_opportunities.jsp) for updates. The ARO prefers proposals for a 3-year period which include a brief summary of work contemplated for each 12-month period so that awards may be negotiated for an entire 3-year program or for individual 1-year increments of the total.

Requirements: Proposals may be submitted by degree-granting universities, nonprofit organizations, or industrial concerns. Proposals are encouraged from historically black colleges and universities and from minority institutions.

Geographic Focus: All States

Date(s) Application is Due: Mar 31

Contact: Kevin Bassler; (919) 549-4295; kevin.j.bassler.civ@mail.mil

Internet: http://www.grants.gov/search/search.do;jsessionid=T23pP9VJLR1BL3prQrvr7Rq3JwPh12ckQPsJ17XClKWQcZh8ScJj!-2099600874?oppId=41952&mode=VIEW

Sponsor: U.S. Department of Defense
4400 Martin Road, Redstone Arsenal
Huntsville, AL 35898

USDD Atomic, Molecular and Quantum Physics Grants 4687

The research supported under this program seeks primarily to understand and exploit the interactions of light, atoms, ions, and molecules. Areas supported may include: laser cooling and trapping of both ions and neutral atoms; Bose-Einstein condensation; atom laser concepts; coherent control, e.g., of chemical reactions; femtosecond pulse shaping; quantum computing; and quantum state teleportation. Also supported are novel theoretical methods for computational electronic structure. Challenges addressed through the atomic and molecular physics program include: obtaining ultra-high precision spectroscopy; overcoming limitations in atomic time and frequency standards, leading, for example, to the pursuit of the physics of ultra-cold collisions; the ability to control chemical dynamics and branching ratios; and the ability to predict properties of materials, and devise novel new materials (e.g., through a better understanding of strong electron correlations). Many of these subprograms and goals are interrelated and interdependent. Submission of the research proposal should always be preceded by submission of a preproposal and a discussion of that preproposal with the assigned Program Officer. Proposals are preferred by April 1 of each year.

Requirements: Grants are awarded to academic, industrial, and nonprofit organizations, with strong emphasis on university research.

Restrictions: New grants or contracts are typically started near the beginning of the fiscal year (1 October). New proposals for the next fiscal year should be submitted no later than April 1, unless otherwise directed. Competitive renewals should be submitted no later than six months prior to the award termination date after contacting the Program Officer. Conference proposals should be submitted at least four months prior to the conference.

Geographic Focus: All States

Date(s) Application is Due: Apr 1

Contact: Dr. Charles W. Clark, (703) 696-5267 or (703) 696-5031; fax (703) 696-5940; charles.clark3@navy.mil or 312_AM@onr.navy.mil

Internet: http://www.onr.navy.mil/sci_tech/31/312/atomic_molecular_physics.asp

Sponsor: U.S. Department of Defense
One Liberty Center, 875 N. Randolph Street, Suite 1425
Arlington, VA 22203-1995

USDD Basic Research Challenge: Carbon Molecular Electronics Grants 4688

The Office of Naval Research is interested in receiving proposals on carbon-based molecular electronics. The objective are to encourage research and innovation in bottom-up chemical synthesis and assembly of carbon, particularly graphene, based electronic devices and circuits with atomic precision and Angstrom resolution. The program will address the challenges and limitations in top-down fabricated graphene nanoelectronic devices where structures are too large (limited by lithography) and intrinsic graphene properties are often masked by edge roughness and hard to measure. The long term vision of the program is to enable a new kind of molecular electronics, a single giant molecule made of carbon or other related materials that functions as electronic, magnetic or optical devices or circuits. White paper submission is not required but strongly encouraged. White papers are accepted electronically. Full proposals must be submitted through Grants.gov. Two to six awards are anticipated. The target budget for single investigator awards is $150,000 or less. The target budget for teams of no more than four is $600,000 or less. The award period will be three years plus two additional years of option period.

Requirements: All responsible sources from academia and industry are eligible. Historically black colleges and universities and minority institutions are encouraged to apply and to join others in submitting proposals.

Geographic Focus: All States

Date(s) Application is Due: May 1

Contact: Jennifer Brown; (703) 588-2432; jennifer.brown4@navy.mil

Internet: http://www.onr.navy.mil/Contracts-Grants/Funding-Opportunities/Special-Notices.aspx

Sponsor: U.S. Department of Defense
One Liberty Center, 875 N. Randolph Street, Suite 1425
Arlington, VA 22203-1995

USDD Basic Research Challenge: Decentralized Online Optimization Grants 4689

The Office of Naval Research is offering a Basic Research Challenge: Decentralized Online Optimization. The objective is to develop scientific principles and algorithms for solving decentralized, online optimization problems. Proposals should scientifically address efficient, robust computational techniques and have the following characteristics: (1) The techniques can be applied to problems possessing a high degree of decentralization, and in which there is limited communication between system components; (2) The techniques can be applied to problems in which not all relevant information of the environment is known a priori, and is revealed incrementally to individual system components; information spreads through the system as communications become available and with potentially substantial time delays; and (3) The techniques are capable of producing high-quality solutions in a reasonable amount of time. Solution quality is measured against optimality, and solution time is measured against the time scale of changes in the environment. Analysis of these measures is expected to be mathematically rigorous. White papers are not required but are strongly encouraged for all applicants. White papers may be submitted electronically, and proposals may be submitted through Grants.gov. Eight to ten awards are anticipated, and the project period may be from one to five years. Questions regarding this announcement may be submitted electronically no later than two weeks prior to the due date. Answers will be addressed in the form of an amendment and will be posted to the following web pages: Federal Business Opportunities (https://www.fbo.gov/); Grants.gov (http://www.grants.gov/); and ONR Broad Agency Announcement (http://www.onr.navy.mil/Contracts-Grants/Funding-Opportunities/Special-Notices.aspx).

Geographic Focus: All States

Date(s) Application is Due: May 15

Amount of Grant: 100,000-150,000 USD

Contact: Gordon Jaquith; 703) 696-7831; gordon.jaquith@navy.mil

Internet: http://www.onr.navy.mil/Contracts-Grants/Funding-Opportunities/Special-Notices.aspx

Sponsor: U.S. Department of Defense
One Liberty Center, 875 N. Randolph Street, Suite 1425
Arlington, VA 22203-1995

USDD Basic Research for Combating Weapons of Mass Destruction Broad Agency Announcement Grants 4690

The Defense Threat Reduction Agency (DTRA) seeks Basic Research for Combating Weapons of Mass Destruction (C-WMD) Grants. Grants are focused on soliciting basic research projects that support the DTRA mission to safeguard America and its allies from WMD (e.g., chemical, biological, radiological, nuclear, and high-yield explosives) by providing capabilities to reduce, eliminate, and counter the threat and mitigate its effects. To this end DTRA seeks research across broad strategic thrust areas: (1) science of WMD sensing and recognition; (2) cognitive and information science; (3) science for protection; (4) science to defeat WMD; and (5) science to secure WMD. Pre-application white papers are due January 13, and invitation-only proposals are accepted from April 2 through 20 only. Single scope awards will average $150,000 per year, and multidisciplinary awards will average $350,000 per year. This is a Broad Agency Announcement (BAA) which will have multiple, consecutive opportunities/periods not yet determined. It is anticipated that there will be at least one opportunity per year announced in the December timeframe. There may be additional opportunities/periods in any year. Subsequent submission opportunities/periods with corresponding topics, application packages, and applicable deadlines, will be published via amendments to this BAA. Applicants are encouraged to monitor Grants.gov Grant Opportunities (http://grants.gov/applicants/find_grant_opportunities.jsp) and the DTRA Basic and Fundamental Research Portal (www.dtrasubmission.net/portal) for announcements and updates.

Requirements: Applicants eligible to submit pre-application white papers and proposals include the following United States and foreign-based equivalents: accredited degree-granting colleges and universities; industrial/commercial basic research centers, including small businesses with a portfolio predominantly in basic research proposing a project with significant participation (minimum 30% of total effort value on the proposed project) by one or more accredited degree-granting colleges and/or universities; and not-for-profit organizations/basic research centers with a portfolio predominantly in basic research proposing a project with significant participation (minimum 30% of total effort value on the proposed project) by one or more accredited degree-granting colleges and/or universities. Proof of 501(c)3 status may be required for not-for-profits. For foreign-based establishments entirely based outside the United States and/or its territories, proof of not-for-profit status may be required.

Geographic Focus: All States, All Countries

Contact: Gena Leake; (703) 767-5870; HDTRA1-BRCWMD-A@dtrasubmission.net

Internet: http://www.dtra.mil/Research.aspx

Sponsor: U.S. Department of Defense
6200 Meade Road
Fort Belvoir, VA 22060-5264

USDD Basic Research Initiative Grants 4691

The Air Force Office of Scientific Research (AFOSR) manages the basic research investment for the United States Air Force (U.S.AF). As a part of the Air Force Research Laboratory (AFRL), AFOSR's technical experts foster and fund research within the AFRL, universities, and industry laboratories to ensure the transition of research results to support U.S.AF needs. The Basic Research Initiative Grants are intended to initiate a basic research program that begins to build the foundational understanding of the factors that influence human reliance on autonomous systems and offer both practical guidance for the Air Force and future research avenues for applied research. The research encompassed will focus on a set of studies for targeting the psychological, neurological, or contextual elements of human reliance on autonomous systems. The intent is to identify general principles of human trust, and the results should not be bound to a particular system.

Suggested research areas may include: (1) empirical studies to reveal the psychological, organizational, technological, and group-level factors that shape reliance on automation; (2) neuroscience and psychophysiology research to establish the markers for said reliance; (3) case studies of Department of Defense users with experience using autonomous vehicles to identify practical lessons learned and realistic perspectives for use or disuse of such systems; (4) anthropological/cross-cultural science to understand the barriers to optimal use of automation in other countries; (5) cross-service studies to reveal potential differences in perceptions/use of automated systems; (6) harnessing and understanding relevant and practical metrics for establishing trustworthiness of autonomous systems (e.g., verification and validation); and (7) understanding the impact of different generations (i.e., age groups) on reliance. Three to four awards of $100,000 to $250,000 for three years are expected. White papers are strongly encouraged prior to submitting a full proposal. There is no deadline for submission. This announcement remains open until superseded.
Requirements: All responsible potential applicants from academia and industry are eligible.
Geographic Focus: All States
Date(s) Application is Due: Mar 30
Amount of Grant: Up to 750,000 USD
Contact: Dr. Joseph Lyons; (703) 696-6207; Joseph.Lyons@afosr.af.mil
Internet: http://www07.grants.gov/search/search.do;jsessionid=cTNwP5vW3XRK6sZQhKynnCW1wJYNK5n2LXd28SkRTWH312MZ1Q9D!-1993000306?oppId=129053&mode=VIEW
Sponsor: U.S. Department of Defense
875 North Randolph Street, Suite 325, Room 3112
Arlington, VA 22203-1768

USDD Battlespace Environment Division Research Interests Grant **4692**
The USDD Air Force Research Laboratory Space Vehicles Research Directorate, Battlespace Environment Division, is interested in receiving proposals to advance the state-of-the-art and scientific knowledge in space and missile technology. Proposals that either enable future space systems to meet performance, reliability, maintainability, supportability and affordability goals, or which enhance performance or significantly reduce cost, schedule, or risk of existing technologies, are of particular interest. This is a multiple year open-ended Broad Agency Announcement (BAA). Periodically proposal call announcements (CALLS) may be issued to request proposals for specific research efforts under technical topic areas. These subsequent CALLS will contain specific descriptions of the research effort to be addressed, anticipated period of performance, information peculiar to the specific research technical topic area, and the expected dollar range for proposals received under the CALL. No proposals should be submitted against this baseline BAA. Proposal due dates and times will be specified in each CALL. A minimum of one award per specific CALL is anticipated. Anticipated funding for this multiple-year program (not per award) is approximately $9 million to $10 million per fiscal year.
Geographic Focus: All States
Date(s) Application is Due: Nov 10
Amount of Grant: Up to 58,000,000 USD
Contact: Jean Barnes; (505) 846-4695; jean.barnes@kirtland.af.mil.
Dr. Joel B. Mozer, Technical Specialist; (505) 853-0059; Joel.Mozer@kirtland.af.mi
Internet: http://www.grants.gov/search-grants.html?agencies%3DDOD|Department%20of%20Defense
Sponsor: U.S. Department of Defense
1400 Defense Pentagon
Washington, DC 20301-1400

USDD Breast Cancer Clinical Translational Research Grants **4693**
Applications for the Breast Cancer Research Program (BCRP) are being solicited by the Assistant Secretary of Defense for Health Affairs, Defense Health Program. The BCRP was established in 1992 to support innovative research focused on ending breast cancer. The BCRP challenges the scientific community to design research that will address the urgency of the vision to end breast cancer. Specifically, the BCRP seeks to accelerate high-impact research, encourage innovation and stimulate creativity, bring new investigators into the breast cancer field, and facilitate multidisciplinary collaborations. The BCRP Clinical Translational Research (CTR) Grants are intended to promote significant improvements over current approaches to breast cancer prevention and therapy, such as studies that will prevent primary or recurrent breast cancer and studies that may result in a new treatment to prevent breast cancer progression to metastasis. CTR Grants support research projects that are likely to have a major impact on breast cancer by applying promising research findings to patients with, or populations at risk for, breast cancer. Principal Investigators (PIs) and their collaborators may have originated projects in their laboratories that will form the basis for 1-2 years of advanced translational research leading to a prospective clinical trial to be conducted during this award period. Alternatively, PIs may leverage partnerships with industry. The ability to conduct the required translational research and early phase clinical trial during the award period must be demonstrated. Investigators must demonstrate availability of and access to an appropriate patient population that will support meaningful clinical outcomes. Submission is a multi-step process requiring both pre-application submission through the CDMRP eReceipt System (https://cdmrp.org/) and application submission through Grants.gov (http://www.grants.gov/). The maximum period of performance is 5 years. The maximum allowable direct costs for the entire period is $12M plus indirect costs.
Requirements: PIs must be at or above the level of assistant professor (or equivalent).
Restrictions: Applicants are encouraged to sign up on Grants.gov for "send me change notification emails" by following the link on the Synopsis page for the Program Announcement/Funding Opportunity. If the application package is updated or changed, the original version of the application package may not be accepted by Grants.gov.
Geographic Focus: All States
Date(s) Application is Due: Aug 15
Contact: Grants Officer; (301) 682-5507; help@cdmrp.org
Internet: http://www.grants.gov/search/search.do;jsessionid=G6z2PM6MT4vFzPGvmYl7TPJTc9lpLwG0CHlsTPC1jw2TJXJRC87n!488296553?oppId=146134&mode=VIEW
Sponsor: U.S. Department of Defense
1400 Defense Pentagon
Washington, DC 20301-1400

USDD Breast Cancer Era of Hope Scholar Grants **4694**
The Era of Hope Scholar Grants support individuals who are early in their careers and have high potential for innovation in breast cancer research. Since the intent is to recognize creative and innovative individuals rather than projects, the central feature of the award is the innovative contribution that the Principal Investigator (PI) can make toward ending breast cancer. The PI should articulate a vision that challenges current dogma and demonstrates an ability to look beyond tradition and convention. Experience in breast cancer research is not required; however, the application must focus on breast cancer and the PI must commit at least 50% of his/her full-time professional effort during the award period to breast cancer research. Five awards are anticipated. The maximum period of performance is 5 years, and the maximum direct costs for the entire period is $2.5M plus indirect costs.
Requirements: Individuals should be exceptionally talented scientists who have demonstrated that they are the "best and brightest" in their field(s) through extraordinary creativity, vision, and productivity. They should have demonstrated experience in forming effective partnerships and collaborations and should exhibit strong potential for leadership in the breast cancer research community. The PI must be an independent investigator within 6 years of his/her last mentored position as of the deadline.
Restrictions: Postdoctoral fellows, clinical fellows (including residents and interns), and other researchers in training positions are not eligible for this award.
Geographic Focus: All States
Date(s) Application is Due: Apr 19
Contact: Grants Officer; (301) 682-5507; help@cdmrp.org
Internet: http://www07.grants.gov/search/search.do;jsessionid=kvB0PNRhp6q1GBx0mTyPhVY6Lp0lyn05T431P0d9jvdVvysv7tSH!488296553?oppId=146114&mode=VIEW
Sponsor: U.S. Department of Defense
1400 Defense Pentagon
Washington, DC 20301-1400

USDD Breast Cancer Idea Grants **4695**
The Idea Award is designed to promote new ideas that are still in the early stages of development and have the potential to yield highly impactful data and new avenues of investigation. This mechanism supports conceptually innovative, high-risk/high-reward research that could lead to critical discoveries or major advancements that will accelerate progress toward ending breast cancer. Applications should include a well-formulated, testable hypothesis based on strong scientific rationale. Innovation and impact are the most important aspects of the Idea Award. Applications that demonstrate exceptional scientific merit but lack innovation and high potential impact do not meet the intent of the Idea Award. Presentation of preliminary data is not consistent with the intent of the Idea Award mechanism. While the inclusion of preliminary data is not prohibited, the strength of the application should not rely on preliminary data, but on the innovative idea and the potential impact it will provide. Submission is a multi-step process requiring both pre-application submission through the CDMRP eReceipt System (https://cdmrp.org/) and application submission through Grants.gov (http://www.grants.gov/). Approximately fifty awards are anticipated. The maximum period of performance is 3 years. The maximum allowable direct costs for the entire period is $375,000 plus indirect costs.
Requirements: Investigators at all academic levels (or equivalent) are eligible.
Restrictions: Applicants are encouraged to sign up on Grants.gov for "send me change notification emails" by following the link on the Synopsis page for the Program Announcement/Funding Opportunity. If the application package is updated or changed, the original version of the application package may not be accepted by Grants.gov.
Geographic Focus: All States
Date(s) Application is Due: Aug 15
Contact: Grants Officer; (301) 682-5507; help@cdmrp.org
Internet: http://www07.grants.gov/search/search.do;jsessionid=xrZyPNZPBMTJ6V24yPlrLtPxyWHP8L5d3WzzXHTGqVwCSZdY5VyK!488296553?oppId=146113&mode=VIEW
Sponsor: U.S. Department of Defense
1400 Defense Pentagon
Washington, DC 20301-1400

USDD Breast Cancer Impact Grants **4696**
The Breast Cancer Research Program (BCRP) Impact Grants support research projects (from small to largescale) that specifically focus on scientific and clinical breast cancer issues, which, if successfully addressed, will revolutionize the understanding, prevention, and/or treatment of breast cancer and make major advances toward the goal of ending the disease. The BCRP particularly encourages applications that focus on poorly understood issues that are of critical significance in breast cancer, such as those related to prevention, susceptibility, recurrence, or metastasis. The proposed work must be based on sound overall research and fully supported by preliminary data and/or published reports. The most important aspect is the potential of the proposed research to have an unprecedented impact on reducing breast cancer incidence, recurrence, and/or mortality. The research project may be from any discipline or combination of disciplines, including basic, translational, clinical (clinical trials are allowed), behavioral, and/or epidemiological research. The potential impact may be near-term or long-term, but it must be significant and non-incremental. Applications are encouraged that include meaningful and productive collaborations between investigators. The Partnering PI Option is structured to accommodate two

PIs, who will each receive a separate award. One partner is identified as the Initiating PI and the other partner is identified as the Partnering PI. The Initiating and Partnering PIs have different submission requirements; however, both PIs should contribute to the preparation of a single application. The collaborative partners may have expertise in similar or disparate scientific disciplines, but each partner is expected to bring different strengths to the application. New collaborations are encouraged but not required. It is the responsibility of the collaborating investigators to describe how their combined expertise in the collaboration will better address the research question and explain why the work should be done together rather than through separate efforts. Submission is a multi-step process requiring both pre-application submission through the CDMRP eReceipt System (https://cdmrp.org/) and application submission through Grants.gov (http://www.grants.gov/). Five awards are anticipated. The maximum period of performance is 5 years. The maximum allowable direct costs for the entire period is $2M plus indirect costs.
Requirements: PIs must be at or above the level of Assistant Professor (or equivalent).
Restrictions: Applicants are encouraged to sign up on Grants.gov for "send me change notification emails" by following the link on the Synopsis page for the Program Announcement/Funding Opportunity. If the application package is updated or changed, the original version of the application package may not be accepted by Grants.gov.
Geographic Focus: All States
Date(s) Application is Due: Aug 15
Contact: Grants Officer; (301) 682-5507; help@cdmrp.org
Internet: http://www07.grants.gov/search/search.do;jsessionid=3JX5PNlGM67F6dpRvnM4TBpcZtQzdPD1pnJYwTGzv4FVp28xw3xg!488296553?oppId=146133&mode=VIEW
Sponsor: U.S. Department of Defense
1400 Defense Pentagon
Washington, DC 20301-1400

USDD Breast Cancer Innovation Grants **4697**
The Breast Cancer Research Program (BCRP) Innovator Grants support visionary individuals who have demonstrated creativity, innovative work, and leadership in any field including, but not limited to, breast cancer. The Innovator Grants will provide these individuals with the funding and freedom to pursue their most novel, visionary, high-risk ideas that could ultimately lead to ending breast cancer. Since the intent of the Innovator Grant mechanism is to recognize creative and innovative individuals rather than projects, the central feature of the award is the innovative contribution that the Principal Investigator (PI) can make toward ending breast cancer. Experience in breast cancer research is not required; however, the application must focus on breast cancer, and the PI must commit at least 50% of his/her full-time professional effort during the award period to breast cancer research. Submission is a multi-step process requiring both pre-application submission through the CDMRP eReceipt System (https://cdmrp.org/) and application submission through Grants.gov (http://www.grants.gov/). Three awards are anticipated. The maximum period of performance is 5 years. The maximum allowable direct costs for the entire period is $5M plus indirect costs.
Requirements: PIs must be at or above the level of associate professor (or equivalent). The PI should have a past record of creativity, promise for continued innovation in future work, and a vision that challenges current dogma and demonstrates an ability to look beyond tradition and convention. The PI is also expected to have demonstrated success at forming and leading effective partnerships and collaborations.
Restrictions: Previous recipients of the BCRP Innovator Grants and BCRP Integration Panel members are ineligible. Applicants are encouraged to sign up on Grants.gov for "send me change notification emails" by following the link on the Synopsis page for the Program Announcement/Funding Opportunity. If the application package is updated or changed, the original version of the application package may not be accepted by Grants.gov.
Geographic Focus: All States
Date(s) Application is Due: Aug 2
Contact: Grants Officer; (301) 682-5507; help@cdmrp.org
Internet: http://www07.grants.gov/search/search.do;jsessionid=TR2GPNndhVCGQGnL6n9z32GJB9hkxn2NDRVlLLqDhVQGNF5Fhc2B!488296553?oppId=146115&mode=VIEW
Sponsor: U.S. Department of Defense
1400 Defense Pentagon
Washington, DC 20301-1400

USDD Breast Cancer Research Postdoctoral Fellowships **4698**
The Postdoctoral Fellowships support exceptionally talented recent doctoral or medical graduates in their pursuit of innovative, high-impact breast cancer research during their postdoctoral training and allow them to obtain the necessary experience for an independent career at the forefront of breast cancer research. These individuals should be exceptionally talented scientists who have demonstrated that they are the "best and brightest" of their peers. Applicants must exhibit a strong desire to pursue a career in breast cancer research. Under this award mechanism, the postdoctoral fellow is considered the Principal Investigator (PI) and, as such, should write the project narrative with appropriate direction from the mentor. While the PI is not required to have previous experience in breast cancer research, the application must focus on breast cancer research. Submission is a multi-step process requiring both pre-application submission through the CDMRP eReceipt System (https://cdmrp.org/) and application submission through Grants.gov (http://www.grants.gov/). Approximately 22 awards are anticipated. The maximum period of performance is 3 years. The maximum allowable direct costs for the entire period of performance are $300,000 plus indirect costs.
Requirements: At the time of application, all eligible PIs must have: successfully completed the requirements for a doctoral or medical degree; been in the laboratory or clinical research setting in which the proposed research is to be performed for no more than 2 years as a postdoctoral researcher; and a total of less than 4 years of postdoctoral research experience (excluding clinical residency or fellowship training).
Restrictions: Only postdoctoral fellows are eligible for this award. Faculty members, instructors, and all other non-postdoctoral positions are not eligible to apply. Applicants are encouraged to sign up on Grants.gov for "send me change notification emails" by following the link on the Synopsis page for the Program Announcement/Funding Opportunity. If the application package is updated or changed, the original version of the application package may not be accepted by Grants.gov.
Geographic Focus: All States
Date(s) Application is Due: Apr 19
Contact: Fellowship Coordinator; (301) 682-5507; help@cdmrp.org
Internet: http://www.grants.gov/search/search.do;jsessionid=3n5KPNzLnlcmMXkBykjzS0JKc293H3XwKHBBpn5ZqpB0HL2y36XN!488296553?oppId=146095&mode=VIEW
Sponsor: U.S. Department of Defense
1400 Defense Pentagon
Washington, DC 20301-1400

USDD Breast Cancer Transformative Vision Grant **4699**
The Breast Cancer Research Program (BCRP) Transformative Vision Grant supports research projects that will fulfill an extraordinary vision for dramatically affecting the prevention or treatment of breast cancer. The Grant requires a plan that will test and achieve the vision as quickly as possible through the translation of the research ideas to individuals with, and/or those at risk for, breast cancer. The scope of the effort may include a broad spectrum of research spanning from basic to clinical studies. There are two critical components. The first is vision and impact. A vision for a new approach that will have a revolutionary impact on the prevention or treatment of breast cancer must be articulated. The time to the final impact may vary, but the success of the vision must be transformative and significantly advance the goal of ending breast cancer. The second is implementation. The vision must be supported by a detailed plan that identifies critical milestones, outlines the innovations and technical solutions that will be implemented to accomplish the milestones, and explains how these solutions will be translated to individuals with, and/or those at risk for, breast cancer. It is expected that the proposed plan will present an exceptional level of innovation and creativity and that the Principal Investigator (PI) will assemble the team necessary to realize the vision. Submission is a multi-step process requiring both pre-application submission through the CDMRP eReceipt System (https://cdmrp.org/) and application submission through Grants.gov (http://www.grants.gov/). One award is anticipated. The maximum period of performance is 5 years. The maximum allowable direct costs for the entire period is $12M plus indirect costs.
Requirements: PIs must be at or above the level of Assistant Professor (or equivalent).
Restrictions: Applicants are encouraged to sign up on Grants.gov for "send me change notification emails" by following the link on the Synopsis page for the Program Announcement/Funding Opportunity. If the application package is updated or changed, the original version of the application package may not be accepted by Grants.gov.
Geographic Focus: All States
Date(s) Application is Due: Aug 15
Contact: Grants Officer; (301) 682-5507; help@cdmrp.org
Internet: http://www.grants.gov/search/search.do;jsessionid=vQJqPN4Q9PjYKC9JVyhJgwGfQj3dL3Sn40jyCLgQs42pN0X6wdsb!488296553?oppId=146116&mode=VIEW
Sponsor: U.S. Department of Defense
1400 Defense Pentagon
Washington, DC 20301-1400

USDD Broad Agency Announcement for the Naval Medical Research and Development Center-Fredrick Grants **4700**
The Naval Medical Research and Development Center-Frederick has a broad role in competitive funding of meritorious research across a spectrum of science and engineering disciplines. This Broad Agency Announcement (BAA) is to solicit proposals for innovative approaches in infectious disease research, non-communicable disease research, and combat casualty care research. Amendments will be posted to one or more of the following web pages: FedBizOpps (https://www.fbo.gov/); Grants.gov (http://www.grants.gov/), or Naval Medical Logistics Command Research and Grants (http://www.nmlc.med.navy.mil/DBU-RnG.asp). Applicants are encouraged to monitor these sites. Applicants are strongly encouraged to contact the Grants Officer prior to submitting a proposal. Funding and the number of awards is indeterminate. Selection of research projects is based on the evaluation of the proposal, programmatic review and the availability of funds.
Requirements: All responsible sources from academia and industry are eligible. Historically black colleges and universities and minority institutions are encouraged to apply.
Geographic Focus: All States
Date(s) Application is Due: Dec 31
Contact: Christopher Sherman; (301) 619-2157; NMLC-Research@med.navy.mil
Internet: http://www.grants.gov/search/search.do;jsessionid=vKBrPDydsD287nDBQ7lpKh0LBwgGkYc3QTGFG2XGDQMHCF5L2fG1!-757993493?oppId=138693&mode=VIEW
Sponsor: U.S. Department of Defense
693 Neiman Street
Fort Detrick, MD 21702-9203

USDD Broad Agency Announcement Grants **4701**
The Department of Defense HIV/AIDS Prevention Program (DHAPP) has successfully engaged over 80 countries in efforts to combat HIV/AIDS among their respective military services. DHAPP is a partner collaborating with the United States State Department, Health and Human Services, United States Agency for International Development, and Centers for Disease Control and Prevention, in the President's Emergency Plan for AIDS Relief (PEPFAR). This BAA is intended to solicit existing partners and establish new partners in order to expand the DHAPP program. Proposals should focus on rapidly extending HIV/AIDS services. Applicants are encouraged to target specific needs with a

practical business plan, using small grass-roots organizations to provide community-based services as a way to enhance organic capabilities and sustainability. Concept Papers may be submitted at any time during this period. Early submission is strongly encouraged due to the possible lack of funding for the later evaluation periods. The period of performance is generally one year, with up to two additional phases subject to availability of funding. Applicants are encouraged to propose additional phases if appropriate. This is a multiple year BAA which may be revised as needed. Changes will be posted at Grants.gov (http://grants.gov/applicants/find_grant_opportunities.jsp). Applicants are encouraged to sign up on Grants.gov for "send me change notification emails" by following the link on the synopsis page for the program announcement/funding opportunity.
Requirements: All responsible sources from academia, industry, and non-governmental organizations are eligible. Applicants respondents must demonstrate the active support of the in-country military and the Department of Defense representative in the corresponding United States Embassy in the planning and execution of proposals.
Restrictions: Funding is available for Angola, Botswana, Caribbean Regional, Central America Regional, Cote d' Ivoire, Democratic Republic of the Congo, Dominican Republic, Ethiopia, Ghana, Guyana, India, Indonesia, Lesotho, Malawi, Mozambique, Namibia, Rwanda, South Africa, Swaziland, Uganda, Ukraine, Vietnam and Zambia. A description of the work needed, along with the program areas and estimated funding is available in the BAA found at the website. All questions should be submitted to the points of contact by email.
Geographic Focus: All States
Date(s) Application is Due: Sep 30
Amount of Grant: 300,000-700,000 USD
Contact: Latrice Rubenstein; (619) 532-4357; Latrice.Rubenstein@navy.mil
Internet: http://www.grants.gov/search/search.do;jsessionid=1GhLPGQSVtvhVkKQYJgRnyFGWgBT8SQvFXjYv31mW6Ky7klFQ1Vx!-757993493?oppId=67733&mode=VIEW
Sponsor: U.S. Department of Defense
5450 Carlistle Park, P.O. Box 2050
Mechanicsburg, PA 17055-0791

USDD Broad Agency Announcement HIV/AIDS Prevention Grants **4702**
The Department of Defense HIV/AIDS Prevention Program (DHAPP) has successfully engaged over 80 countries in efforts to combat HIV/AIDS among their respective military services. DHAPP is a partner collaborating with the United States State Department, Health and Human Services, United States Agency for International Development, and Centers for Disease Control and Prevention, in the President's Emergency Plan for AIDS Relief (PEPFAR). This Broad Agency Announcement (BAA) allows DHAPP to assist countries where militaries are not funded under PEPFAR by soliciting existing partners and establishing new partners in furtherance of DHAPP and partner military program goals. Proposals should focus on rapidly extending HIV/AIDS services. Applicants are encouraged to target specific needs with a practical business plan, using small grass-roots organizations to provide community-based services as a way to enhance organic capabilities and sustainability. Concept papers are required and full proposals are by invitation only. Early submission is strongly encouraged. The period of performance is generally one year, with up to two additional phases subject to availability of funding. Applicants are encouraged to propose additional phases if appropriate. This is a multiple year BAA which may be revised as needed. Based on the availability of funds, up to three evaluation cycles will be conducted. Changes will be posted at Grants.gov (http://grants.gov/applicants/find_grant_opportunities.jsp). Applicants are encouraged to sign up on Grants.gov for "send me change notification emails" by following the link on the synopsis page for the program announcement/funding opportunity.
Requirements: All responsible sources from academia, industry, and non-governmental organizations are eligible. Applicants must demonstrate the active support of the in-country military and the USDD representative in the corresponding United States Embassy in the planning and execution of proposals.
Restrictions: Funding is available for Ghana and El Salvador. A description of the work needed, along with the program areas and estimated funding is available in the BAA found at the website. All questions should be submitted to the points of contact by email.
Geographic Focus: All States
Amount of Grant: 60,000-500,000 USD
Contact: Latrice Rubenstein; (619) 556-5276; Latrice.Rubenstein@navy.mil
Internet: http://www07.grants.gov/search/search.do;jsessionid=DBT6PFBSpD4JnP9CyYnwCqZjv7ZnwjK79s2tl1KPTPCTvZRwpQJn!1471941753?oppId=141233&mode=VIEW xLjJrqQ0vdnQM3Wt08p5Y171y8Gj4qPxb!748281696?oppId=143433&mode=VIEW
Sponsor: U.S. Department of Defense
5450 Carlistle Park, P.O. Box 2050
Mechanicsburg, PA 17055-0791

USDD Broad Agency Announcement Research and Development Grants **4703**
The United States Army Engineer Research and Development Center (ERDC) has issued a Broad Agency Announcement (BAA) for various research and development topic areas. This BAA solicits proposals for basic and applied research and that part of development not related to the development of a specific system or hardware procurement. To be eligible the technology or methodology must be either basic research, applied research, advanced technology development not for a specific system/hardware, or demonstration and validation. This BAA will only be used when meaningful proposals with varying technical/scientific approaches can be reasonably anticipated. The ERDC consists of the Coastal and Hydraulics Lab, the Geotechnical and Structures Lab, the Environmental Lab, and the Information Technology Lab in Vicksburg, Mississippi; the Cold Regions Research and Engineering Lab in Hanover, New Hampshire; the Construction Engineering Research Lab in Champaign, Illinois; and the Topographic Engineering Center in Alexandria, Virginia. The ERDC is responsible for conducting research in the broad fields of hydraulics, dredging, coastal engineering, instrumentation, oceanography, remote sensing, geotechnical engineering, earthquake engineering, soil effects, vehicle mobility, self-contained munitions, military engineering, geophysics, pavements, protective structures, aquatic plants, water quality, dredged material, treatment of hazardous waste, wetlands, physical/mechanical/ chemical properties of snow and other frozen precipitation, infrastructure and environmental issues for installations, computer science, telecommunications management, energy, facilities maintenance, materials and structures, engineering processes, environmental processes, land and heritage conservation, and ecological processes. Applicants are encouraged to make preliminary inquiries about the research considered before preparing a proposal. Points of contact are listed for each research areas for each laboratory. The proposal should present the proposed research effort in sufficient detail to allow ERDC to evaluate the scientific merit and relevance of the proposed research and to determine funding availability. The ERDC also supports conferences and symposia in special areas of science that bring experts together to discuss recent research or educational findings or to expose other researchers or advanced graduate students to new research and educational techniques. The ERDC encourages the convening, in the United States, of major international conferences, symposia, and assemblies of international alliances. Conference support proposals should be submitted a minimum of six months prior to the date of the conference.
Requirements: Eligible applicants are education institutions, non-profit organizations, and private industries.
Geographic Focus: All States
Date(s) Application is Due: Jan 31
Contact: Lisa Luckett; (601) 634-2633; Lisa.K.Luckett@usace.army.mil
Internet: http://www07.grants.gov/search/search.do;jsessionid=bZGQPD0CJrznhzqf4llwszNtgyn7k5pDLGcj7LW1TGcLRGGJYQpd!-757993493?oppId=139493&mode=VIEW
Sponsor: U.S. Department of Defense
3909 Halls Ferry Road
Vicksburg, MS 39180-6199

USDD Bulk Nanostructured Materials Grants **4704**
This program takes advantage of the unusual properties of nanomaterials to build hybrid structural materials as combinations or composites of both nano and regular materials in order to obtain new materials with enhanced properties compared to the original constituents. For example, there is interest in developing materials with greatly increased strength over that of the components without, at the same time, increasing the composite's brittleness or fragility. In addition, work is underway to develop thermal sprayed nanostructured coatings. Such coatings may offer greatly increased resistance to wear and corrosion especially in moving/rotating parts. Submission of the Research Proposal should always be preceded by submission of a Preproposal and a discussion of that Preproposal with the assigned Program Officer. Applications for the upcoming fiscal year should be received by September 30.
Requirements: Grants are awarded to academic, industrial, and nonprofit organizations, with strong emphasis on university research. Investigators are encouraged to contact the staff prior to formally submitting a proposal.
Restrictions: New grants or contracts are typically started near the beginning of the fiscal year. New proposals for the next fiscal year should be submitted no later than September 30, unless otherwise directed. Competitive renewals should be submitted no later than six months prior to the award termination date after contacting the Program Officer. Conference proposals should be submitted at least four months prior to the conference.
Geographic Focus: All States
Date(s) Application is Due: Sep 30
Contact: Lawrence Kabacoff, Program Officer; (703) 696-0283 or (703) 696-3073; fax (703) 796-0934; lawrence.kabacoff@navy.mil
Internet: http://www.onr.navy.mil/Science-Technology/Departments/Code-33/All-Programs/332-naval-materials/Bulk-Nanostructured-Materials.aspx
Sponsor: U.S. Department of Defense
One Liberty Center, 875 N. Randolph Street, Suite 1425
Arlington, VA 22203-1995

USDD Capacitors for Pulsed Power Applications Grants **4705**
The Capacitors for Pulsed Power Applications program's goal is to increase the energy density of large dielectric capacitors suitable for handling large amounts of power in conditioners, converters and pulsed power application. The current state of the art capacitors are based on metallized biaxially oriented polypropylene and feature high-breakdown strength, low dielectric losses and graceful failure. Approaches currently in the program include improved synthesis and characterization of olefinic polymers, PVDF and other polar polymers, and composite and nanocomposite approaches. Within the program, there is collaboration between synthetic, physical, and theoretical chemists, materials scientists, and engineers. Submission of the Research Proposal should always be preceded by submission of a Preproposal and a discussion of that Preproposal with the assigned Program Officer. Applications for the upcoming fiscal year should be received by September 30.
Requirements: Grants are awarded to academic, industrial, and nonprofit organizations, with strong emphasis on university research. Investigators are encouraged to contact the staff prior to formally submitting a proposal.
Restrictions: New grants or contracts are typically started near the beginning of the fiscal year. New proposals for the next fiscal year should be submitted no later than September 30, unless otherwise directed. Competitive renewals should be submitted no later than six months prior to the award termination date after contacting the Program Officer. Conference proposals should be submitted at least four months prior to the conference.
Geographic Focus: All States
Date(s) Application is Due: Sep 30
Contact: Dr. Paul Armistead; (703) 696-4315; paul.armistead@navy.mil
Internet: http://www.onr.navy.mil/Science-Technology/Departments/Code-33/All-Programs/332-naval-materials/Capacitors-Pulsed-Power-Applications.aspx

Sponsor: U.S. Department of Defense
One Liberty Center, 875 N. Randolph Street, Suite 1425
Arlington, VA 22203-1995

USDD Care for the Critically Injured Burn Patient Grants 4706

The Assistant Secretary of Defense for Health Affairs, Defense Health Program Medical Research and Development Office is soliciting proposals for the Defense Medical Research and Development Program (DMRDP). The Telemedicine and Advanced Technology Research Center (TATRC) is administering the application process, and the United States Army Medical Research Acquisition Activity (U.S.AMRAA) is issuing this program announcement and will be negotiating all resulting awards. The goal is to advance the state of medical science in those areas of most pressing need and relevance to today's battlefield experience, in this case burn injuries. The objectives of the DMRDP are to discover and explore innovative approaches to protect, support, and advance the health and welfare of military personnel, families, and communities; to accelerate the transition of medical technologies into deployable products; and to accelerate the translation of advances in knowledge into new standards of care for injury prevention, treatment of casualties, rehabilitation, and training systems that can be applied in theater or in the clinical facilities of the Military Health System. The research results are expected to increase the body of knowledge available to professionals and practitioners in health, medical science and related fields. The research impact is expected to benefit the military community and civilian community, particularly as it relates to the following priority areas: (1) multicenter development and assessment of the impact of checklists on standardization of burn care and improvement of outcomes in burn patients; (2) multicenter studies of ICU-based rehabilitation outcomes, i.e., evaluation of physical and rehabilitation technologies/therapies directed at short-term ICU outcomes, understanding the effect of rehabilitation on pathophysiology in the ICU burn patient, and documenting the efficacy of these immediate ICU interventions on long-term quality-of-life outcomes such as the return-to-work/duty rate; (3) device/drug development and validation for inhalation lung injury; and (4) research into the formation of hypertrophic scarring and the development of therapies to manage the condition. Pre-applications are due by December 2, and applications are by invitation only. Three to six awards are anticipated.
Requirements: Principal investigators must be independent investigators at any academic level from academia, research institutions, industry, local, state or federal governments, and private foundations that possess the skills, knowledge, and resources necessary to carry out the proposed research. Applications from investigators within the military services and other federal agencies are encouraged, as are applications involving multidisciplinary collaborations among academia, industry, the military services, the Department of Veterans Affairs, and other federal government agencies.
Geographic Focus: All States
Date(s) Application is Due: Mar 9
Contact: Mary Rico, Grants Officer; (703) 518-4726; mary.rico@amedd.army.mil
Internet: http://www07.grants.gov/search/search.do;jsessionid=YYJpP9pQHVhR7BFjdtNMBm1grFHnpNT4H8rRHtKrSJtHcGVFsQnJ!712472910?oppId=129414&mode=VIEW
Sponsor: U.S. Department of Defense
1400 Defense Pentagon
Washington, DC 20301-1400

USDD Cellular Materials Grants 4707

Structural cellular materials, either as stochastic foams or integrated into various periodic configurations, are highly porous structures with 20 percent or less of their interior volume occupied by material. These materials, therefore, are useful in a number of naval applications. To enhance our scientific understanding of these materials, the Office of Naval Research (ONR) supports basic research into the behavior of cellular materials under the influence of dynamic deformation processes. An objective is to integrate different materials with different structures and orientations in order to get superior thermal conductance in complex structures. In addition, ONR is very interested in improved protection from fragments or blast under static and dynamic load conditions. ONR seeks to explore further the use of different materials and materials combinations to enhance overall protection to the warfighter. Experimental characterization of material structures under blast and/or fragmentation threats should yield physics-based models and validated algorithms that, in turn, should enable optimal design of blast and fragmentation resistant structures. Finally, the use of light-weight structural cellular materials for thermal management in non-electronic systems is of interest. Submission of the Research Proposal should always be preceded by submission of a Preproposal and a discussion of that Preproposal with the assigned Program Officer. Applications for the upcoming fiscal year should be received by September 30.
Requirements: Grants are awarded to academic, industrial, and nonprofit organizations, with strong emphasis on university research. Investigators are encouraged to contact the staff prior to formally submitting a proposal.
Restrictions: New grants or contracts are typically started near the beginning of the fiscal year. New proposals for the next fiscal year should be submitted no later than September 30, unless otherwise directed. Competitive renewals should be submitted no later than six months prior to the award termination date after contacting the Program Officer. Conference proposals should be submitted at least four months prior to the conference.
Geographic Focus: All States
Date(s) Application is Due: Sep 30
Contact: Dr. David Shifler; (703) 696-0285 or (703) 696-0934; david.shifler@navy.mil
Internet: http://www.onr.navy.mil/Science-Technology/Departments/Code-33/All-Programs/332-naval-materials/Cellular-Materials.aspx
Sponsor: U.S. Department of Defense
One Liberty Center, 875 N. Randolph Street, Suite 1425
Arlington, VA 22203-1995

USDD Chemical and Biological Defense Innovations and Technologies Grants 4708

The Space and Naval Warfare Systems Center (SPAWAR) is soliciting proposals for basic research, applied research and advanced research in areas relating to the advancement of technology to meet USDD requirements for chemical, biological, and pandemic disease defense. This Broad Agency Announcement seeks white papers and full proposals for proof-of-concept research to develop an in vitro platform of human organ constructs that could recapitulate the human response to threat agents and accurately predict human safety, efficacy, and pharmacokinetics of candidate Medical Countermeasures. Proposed research should investigate unique and innovative approaches for: rapid detection and diagnosis of chemical and biological threats (to include chemical, viral, bacterial, and protozoan); dissemination of information to local, state, and federal authorities for action; and defining and developing next generation capabilities in chemical, biological and pandemic disease defense.
Geographic Focus: All States
Date(s) Application is Due: Dec 11
Contact: David Hill, Contract Specialist; (619) 553-4349; david.l.hill3@navy.mil
Internet: http://www07.grants.gov/search/search.do;jsessionid=hjvmPLnJ6qX61np2LnHvy1yhzb8Tr8TKLGL2n31L1LJ6tvgZgYPD!748281696?oppId=133833&mode=VIEW
Sponsor: U.S. Department of Defense
P.O. Box 190022
North Charleston, SC 29419-9022

USDD Clinical Functional Assessment Investigator-Initiated Grants 4709

The United States Army Medical Research and Materiel Command (U.S.AMRMC) is seeking health information technology research that may lead to outcomes and products that may qualify as advanced development activities within the Department of Defense (USDD). The results of this research will support the Military Health System in the adoption and implementation of clinical functional disability assessments and standards, by leveraging the skills and talents of United States based businesses and teaching/research academic institutions for success. The intent is to create enduring initiatives and collaborations that will develop and support sustained business and enterprise solutions for the USDD as well as organizations such as the Veterans Benefits Administration, Centers for Medicare and Medicaid Services, and the general public. Proposals are invited that develop products and systems and generate new knowledge that supports medical protection and treatment for USDD personnel and beneficiaries. Military relevance will be a primary criterion for evaluation in all areas. Pre-proposals are due by November 30, and full proposals are by invitation only. One or two awards are anticipated.
Requirements: Eligible applicants include United States based businesses, teaching research academic institutions, and multi-center collaborations, to include international partners, that develop products and systems and generate new knowledge that supports medical protection and treatment for USDD personnel and beneficiaries. Principal investigators must be independent investigators at any academic level from academia, research institutions, government, industry, and private foundations that possess the skills, knowledge, and resources necessary to carry out the proposed research.
Geographic Focus: All States
Date(s) Application is Due: Mar 15
Amount of Grant: 1,500,000 USD
Contact: Mary Rico, Grants Officer; (703) 518-4726; mary.rico@amedd.army.mil
Internet: http://www07.grants.gov/search/search.do;jsessionid=YYJpP9pQHVhR7BFjdtNMBm1grFHnpNT4H8rRHtKrSJtHcGVFsQnJ!712472910?oppId=127813&mode=VIEW
Sponsor: U.S. Department of Defense
1400 Defense Pentagon
Washington, DC 20301-1400

USDD Collaborative Research in Structural Sciences 4710

The focus of this program will be on the following three technical thrust areas: adaptive, spatial and temporal, multi-scaling strategies for long duration path-dependent combined-environment structural simulation; damage-integrated, structural-thermal, reduced-order-models for realistic structural components; and high-fidelity, continuum-level, constitutive and material damage evolution models. A major goal within AFRL is to establish cooperative relationships with external researchers in order to increase the agility and responsiveness of our research efforts. We wish to pursue a collaborative effort in Structural Science. The objective of the collaboration is to provide research in aerospace structural design through a broad and agile base of expertise. The collaborative research is expected to further new inter-disciplinary research areas of unique interest to the Air Force; for example, faculty members at universities can hire top graduate students and formulate research projects of long-term interest to AFRL. By conducting this collaborative research, AFRL can tap into the required expertise to help accelerate innovative research in structures in a variety of areas. The Air Force anticipates awarding one (1) award for this announcement, but reserves the right to award all, part, or none of the proposals received and reserves the right to award grants or cooperative agreements. The base value is estimated to be **$2,500,000**, with a **$1,000,000** priced Option, for a total program value of **$3,500,000.**
Requirements: This is an unrestricted solicitation. Small businesses are encouraged to propose on all or any part of this solicitation.
Geographic Focus: All States
Date(s) Application is Due: Oct 31
Amount of Grant: Up to 3,500,000 USD
Contact: Noreen Bennett; (937) 656-9671; noreen.bennett@wpafb.af.mil
Internet: http://www07.grants.gov/search/search.do;jsessionid=8Fy2QFPFZ6dnVjThhgpY2M1P1w8z2MSLs39J9s3FDRPZQyJ1tn9Q!2071562142?oppId=202057&mode=VIEW
Sponsor: U.S. Department of Defense
1400 Defense Pentagon
Washington, DC 20301-1400

USDD Command and Control Grants 4711

The Office of Naval Research (ONR) Command and Control program focuses on developing new approaches to support tactical decision makers and information superiority. This is performed by producing distributed situation assessment of a commander's environment through development of a common tactical picture; developing collaborative, distributed mission plans; monitoring and dynamically replanning mission execution as required; understanding their environment by being able to access distributed data bases intelligently through use of software agents; effectively using sensor information and making resource allocation decisions and information operations. Specific research involves real-time computing, decision aids and collaborative workspaces; secure distributed architecture; and information warfare techniques to protect secure information. The Command and Control program has several focus areas that build upon the central theme of providing distributed fusion, information processing and information management within a service-oriented architecture, including: Management of Coordinated Electronic Surveillance; Management of Coordinated Electronic Surveillance Higher Level Situation and Knowledge-Based Services; Data Workflow Management Across Globally Netted Maritime Headquarters With Maritime Operations Center; and Dynamic Management Across Wide Variety of Networks. The annual deadline is ongoing.

Requirements: Grants are awarded to academic, industrial, and nonprofit organizations, with strong emphasis on university research.

Geographic Focus: All States

Contact: Dr. Gary Toth; (703) 696-4961; fax (703) 696-2611; gary.toth@navy.mil

Internet: http://www.onr.navy.mil/Science-Technology/Departments/Code-31/All-Programs/311-Mathematics-Computers-Research/Command-Control.aspx

Sponsor: U.S. Department of Defense

One Liberty Center, 875 N. Randolph Street, Suite 1425

Arlington, VA 22203-1995

USDD Communications and Networking Discovery and Invention Grants 4712

The goal of the Communications and Networking Program within the Office of Naval Research is to support the FORCEnet vision by developing measurable advances in technology that can directly enable and enhance end-to-end connectivity and quality-of-service for mission-critical information exchange among such widely dispersed naval, joint, and coalition forces. The vision is to provide high throughput robust communications and networking to ensure all warfighters, from the operational command to the tactical edge, have access to information, knowledge, and decision-making necessary to perform their assigned tasks. Proposals are sought under the following focus areas: (1) Back-end electronics devices/sub-systems/architectures/algorithms for ultra-wideband apertures with simultaneous low noise, low distortion, and very high instantaneous dynamic range; (2) Innovative approaches for spectrum co-existence underlay/overlay, interference cancellation, etc.) of military waveforms with commercial communications; (3) High pulse power (variable pulse rep frequency) fiber laser emission in the blue-green (450-520 nm) using all-fiber power-scaling and combining techniques. Fully-passive (or minimal electrical/thermal control) filter designs with high transmissivity, narrow optical bandpass (<0.5 nm), and large field-of-view in the blue-green region. High speed (Gbps), single-photon counting, blue-green detectors and high performance receiver waveforms/protocols with relaxed sync/timing accuracies; (4) Machine learning techniques and other novel approaches for autonomous network management; and (5) Robust tools/approaches for de-confliction of traffic prioritization policies, according to Commander's Intent, over a distributed, tactical network. The Office of Naval Research is also receptive to highly innovative ideas in other general communications and networking areas that are not within the designated focus areas specified above but nonetheless are important to the Navy/Marine Corps. White papers are accepted by mail with delivery confirmation or by commercial carrier. An expanded oral presentation will be requested from applicants whose proposals are identified as being of "particular value" to the Navy. Following oral presentations, applicants whose technology is still considered as having "particular value" to the Navy will be encouraged to submit detailed technical and cost proposals. Average awards are $500,000 per year. The award period may be from one to three years.

Requirements: All responsible sources from academia and industry are eligible. Historically black colleges and universities and minority institutions are encouraged to apply. Teams are encouraged and may submit proposals in any and all areas.

Geographic Focus: All States

Date(s) Application is Due: Jun 28

Contact: Sheila Richardson; (703) 696-1575; sheila .richardson@navy.mil

Dr. Santanu Das, Contract Specialist; (703) 696-1575; Santanu.Das@navy.mil

Internet: http://www.onr.navy.mil/Contracts-Grants/Funding-Opportunities/Broad-Agency-Announcements.aspx

Sponsor: U.S. Department of Defense

One Liberty Center, 875 N. Randolph Street, Suite 1425

Arlington, VA 22203-1995

USDD Communications and Networking Grants 4713

The Office of Naval Research (ONR) Communications and Networking Program supports basic and applied research and development of antennas, radio communications and wireless networking relevant to naval applications. Interest areas in antenna technology include electrically small antennas, wideband multifunction antennas, compatibility of phased array antennas with naval platforms and marine environments, directional beam forming/steering techniques, and special-purpose submarine communication antenna systems. Areas of interest in radio communications include anti-jam and low-probability-of-intercept techniques, SATCOM performance enhancements, interference mitigation, adaptive equalization, bandwidth efficient modulation, cognitive radio for dynamic spectrum management, and high data rate tactical communications techniques including communications at speed and depth (for submarines). Areas of interest in wireless networks include mobile ad-hoc wireless networking algorithms/protocols, end-to-end Quality-of-Service, joint/coalition interoperability, service oriented tactical networking, mission-based policy and network controls and management. Submission of the Research Proposal should always be preceded by submission of a Pre-proposal and a discussion of that Pre-proposal with the assigned Program Officer.

Requirements: Grants are awarded to academic, industrial, and nonprofit organizations, with strong emphasis on university research.

Restrictions: New grants or contracts are typically started near the beginning of the fiscal year (October 1). New proposals for the next fiscal year should be submitted no later than June 1, unless otherwise directed by the BAA. Competitive renewals should be submitted no later than six months prior to the award termination date after contacting the Program Officer. Conference proposals should be submitted at least four months prior to the conference.

Geographic Focus: All States

Contact: Santanu Das, Program Officer; (703) 588-1036 or (703) 588-2427; fax (703) 696-5940; santanu.das@navy.mil

Internet: http://www.onr.navy.mil/Science-Technology/Departments/Code-31/All-Programs/312-Electronics-Sensors/Communications-Networking.aspx

Sponsor: U.S. Department of Defense

One Liberty Center, 875 N. Randolph Street, Suite 1425

Arlington, VA 22203-1995

USDD Computer-Aided Materials Design Grants 4714

Advances in theory (e.g., classical, quantum and statistical mechanics), along with computing hardware and software, now make it possible to suggest molecules, their structures and assemblies (viz., materials) and to compute expected properties, in some cases, even before actual synthesis and processing have been carried out. With computer-aided design, the Office of Naval Research (ONR) believes that fundamental science can be explored, the cost of research and development can be substantially lowered, and, in the process, optimally designed materials can be derived. This Computer-Aided Materials Design (CAMD) program currently focuses on alloys, superconducting materials, new polymer composites, high-temperature materials, enhanced capacitors and batteries, and anti-fouling materials. Several topic areas reflect cooperation with other programs in the division and in ONR. General computer-aided organic or inorganic chemical synthesis is not an element of this program. Materials and systems of emphasis are alloys, polymer composites and materials relevant to power and energy. Submission of the Research Proposal should always be preceded by submission of a Preproposal and a discussion of that Preproposal with the assigned Program Officer. Applications for the upcoming fiscal year should be received by September 30.

Requirements: Grants are awarded to academic, industrial, and nonprofit organizations, with strong emphasis on university research. Investigators are encouraged to contact the staff prior to formally submitting a proposal.

Restrictions: New grants or contracts are typically started near the beginning of the fiscal year. New proposals for the next fiscal year should be submitted no later than September 30, unless otherwise directed. Competitive renewals should be submitted no later than six months prior to the award termination date after contacting the Program Officer. Conference proposals should be submitted at least four months prior to the conference.

Geographic Focus: All States

Date(s) Application is Due: Sep 30

Contact: Dr. Kenny Lipkowitz, Program Officer; (703) 696-0487 or (703) 696-0707; fax (703) 696-0934; kenny.lipkowitz@navy.mil

Internet: http://www.onr.navy.mil/Science-Technology/Departments/Code-33/All-Programs/332-naval-materials/Computer-Aided-Materials-Design.aspx

Sponsor: U.S. Department of Defense

One Liberty Center, 875 N. Randolph Street, Suite 1425

Arlington, VA 22203-1995

USDD Crowd Sourced Formal Verification Grants 4715

DARPA-Information Innovations Office is soliciting innovative research proposals in the area of crowd sourced formal verification of software. Proposed research should investigate innovative approaches that enable revolutionary advances in science, devices, or systems. The goal of the Crowd Sourced Formal Verification (CSFV) program is to make formal verification of software more cost-effective by enabling non-specialists to participate productively in the formal verification process. The approach is to transform the verification task into a more accessible task by creating a game that reflects a relevant abstraction of the program for verification purposes, is intuitively understandable by ordinary people, and is fun to play. Completion of the game effectively helps the formal verification tool complete the corresponding formal program verification proof. The scope of the game-based formal verification technical area is limited to automatic game construction, bidirectional interfaces to formal verification tools, and necessary modifications to those tools. As the CSFV program goals involve certifying the absence of certain security vulnerabilities in code, logically sound verification techniques are required. The CSFV program seeks improved compiler optimization by reusing annotations useful for verification to provide more precise control and data flow information. Finally, a research integrator role will bring the verification tools and games together in a coherent system and website. Multiple awards are anticipated. Annual funding is expected to range from $4.7 to $11.05 million. The level of funding for individual awards has not been determined. Application instructions can be found on the website.

Requirements: All responsible sources capable of satisfying the government's needs are eligible. Historically black colleges and universities, small businesses, small disadvantaged businesses and minority institutions are encouraged to submit proposals and team with others to submit proposals.

Restrictions: Specifically excluded from CSFV is the construction of new formal verification tools. Also excluded is research that primarily results in evolutionary improvements to the existing state of practice.

Geographic Focus: All States
Date(s) Application is Due: Mar 20
Contact: Dean Drew, Program Manager; (703) 526-6630; CSFV@darpa.mil
Internet: http://www.grants.gov/search/search.do;jsessionid=Kk1NP54Tt0jpnMLC6J7vB0Sh53T9jH1YXXSfy5LnVN1hCmPD7TKp!-1993000306?oppId=135453&mode=VIEW
Sponsor: U.S. Department of Defense
3701 North Fairfax Drive
Arlington, VA 21702-1714

USDD Defense Sciences Research and Technology Grants 4716

The mission of the Defense Advanced Research Projects Agency (DARPA) Defense Sciences Office (DSO) is to pursue and exploit fundamental science and innovation for national defense. The Defense Sciences Reach and Tecnology Grants solicit proposal abstracts and full proposals for advanced research and development in a variety of enabling technical areas. Proposal abstracts and full proposals may be either basic or applied research. Applicants should demonstrate that their proposed effort is aimed at high-risk/high-payoff technologies that have the potential for making, in the five- to ten-year timeframe, revolutionary rather than incremental improvements to national security, including emerging threats and operational challenges. While there is no specific requirement for cost and duration of the proposed effort, it is recommended that proposers include a timeline that plots the most critical issues on the path to success. Topic areas include: physical sciences; material science; biology; neuroscience; mathematics; and testing and demonstration. Each area is described in detail at the website and should be reviewed by applicants. Abstracts/proposals that are not within the topic areas described may be considered out of scope and may not be evaluated. Proposal abstracts and full proposals may be submitted at any time until the deadline.
Requirements: All responsible sources capable of satisfying the government's needs are eligible. Historically black colleges and universities, small businesses, small disadvantaged businesses and minority institutions are encouraged to apply.
Restrictions: Abstracts/proposals for the integration of existing technologies or development of systems will be considered out of scope and will not be evaluated. The preferred method of communication is email.
Geographic Focus: All States
Contact: Dr. Jon Mogford, Technical Specialist; DARPA-BAA-11-65@darpa.mil
Internet: http://www.grants.gov/search/search.do;jsessionid=ph8qP51B1TnR0qyJGvGvmlVF3KKh1lLjbHlxpj3GQQ7FxlpPTB0h!-1993000306?oppId=112994&mode=VIEW
Sponsor: U.S. Department of Defense
3701 North Fairfax Drive
Arlington, VA 22203-6630

USDD Developer Tools for Medical Education Public Physiology Research Platform Grant 4717

The Assistant Secretary of Defense for Health Affairs, Defense Health Program is soliciting proposals for a Developer Tools for Medical Education Public Physiology Research Platform (DTME-PRP) Grant. The DTME-PRP Grant will be issued and awarded by the United States Army Medical Research Acquisition Activity (U.S.AMRAA) and managed by the Army's Telemedicine and Advanced Technology Research Center. This initiative is primarily focused on the creation of publicly accessible, free, open-license, and/or open-source tools for creation of medical training and education content and technology. Principally, the DTME initiative seeks to support the strategic goals of the Joint Program Committee 1 – Medical Training and Health Information Sciences Research Program through the leveraging of technologies created by this effort, finding cost efficiencies through utilization of free tools and allowing more commonality for military medical training technology systems. It is envisioned that expanding a research effort to create this capability to an open public trial and experimentation basis will allow for feedback from a larger part of the medical training community than would otherwise be possible from a single group. Feedback regarding novel uses of the evolving research platform and the data created by this research will lead to improvements in the science regarding simulated physiology for use in public and private sector medical education. Proposals should address research that will do the following: (1) create a public physiology research platform for use by the medical simulation development and education community; (2) enable accurate and consistent simulated physiology across multiple training applications and platforms; (3) allow for community contributions to this effort; (4) encourage greater use of scientifically-based physiological and pharmacological predictive technologies; (5) lower barriers to creating medical training content through availability of this resource; (6) serve the training needs of the military; and (7) expand the corpus of knowledge regarding the use of simulated physiology for medical education. Approximately $7 million is available, and one award is anticipated. The maximum period of performance is five years. Pre-proposals are due by January 10, and full proposal submissions are by invitation only.
Requirements: Eligible organizations must be extramural and include nonprofit, public, and private organizations, such as universities, colleges, hospitals, laboratories, and commercial firms. Historically black colleges and universities/minority institutions are encouraged to apply.
Geographic Focus: All States
Date(s) Application is Due: Apr 26
Amount of Grant: 7,000,000 USD
Contact: Mary Rico, Grants Officer; (703) 518-4726; mary.rico@amedd.army.mil
Internet: http://www.grants.gov/search/search.do;jsessionid=Q21ZP9lV4mvG5TgvVqhZ2YcTpzV1NQypcLBss0zn38199gQ1WYBM!-2099600874?oppId=130394&mode=VIEW
Sponsor: U.S. Department of Defense
1400 Defense Pentagon
Washington, DC 20301-1400

USDD Development of Quantum Computing Algorithms Grants 4718

The United States Army Research Office (ARO) together with the National Security Agency is soliciting proposals to develop new quantum computing algorithms for hard computational problems, develop insights into the power of quantum computation, and consider issues of quantum complexity and computability. Proposals for research in quantum algorithms should primarily be to devise novel quantum algorithms for solving mathematically and computationally hard problems from such diverse fields as algebra, number theory, geometry, analysis, optimization, graph theory, differential equations, combinatorics, topology, logic, and simulation. Quantum algorithms that are developed should focus on constructive solutions for specific tasks and on general methodologies for expressing and analyzing algorithms tailored to specific problems. Complexity analysis such as upper and lower bounds on algorithms, including developing new methodologies for deriving such bounds, is encouraged. Analysis of the robustness of an algorithm against uncertainty (e.g. small perturbations in the problem's defining parameters or input states) is also encouraged. Investigators should presuppose the existence of a fully functional quantum computer and consider what algorithmic tasks are particularly well suited to such a machine. A necessary component of this research will be to compare the efficiency of the quantum algorithm to the best existing classical algorithm for the same problem. Although quantum algorithm proposals may consider general architectural constraints (e.g. nearest neighbor only gates) for implementing algorithms, they should otherwise concentrate on developing the algorithm. Quantum algorithm proposals may consider computational models other than the circuit model (e.g. the adiabatic model), however proposals should assiduously avoid tying algorithms to specific physical models. Applicants are strongly encouraged to submit white papers. Full proposals are required to apply. This is a multiple year Broad Agency Announcement (BAA). Applicants are encouraged to periodically check the following websites for updates and/or amendments: FedBizOpps.gov (https://www.fbo.gov/), Grants.gov (http://grants.gov/applicants/find_grant_opportunities.jsp), and the ARO website (http://www.arl.army.mil/www/default.cfm?Action=6&Page=8).
Requirements: Degree-granting universities (foreign and domestic), nonprofit organizations, and industrial concerns (large and small businesses) are eligible. Proposals are encouraged from historically black colleges and universities and from minority institutions.
Geographic Focus: All States, All Countries
Date(s) Application is Due: May 8
Amount of Grant: 1,000,000 USD
Contact: Jessica Leach, Grants Officer; (919) 549-4387; jessica.e.leach.civ@mail.mil
Internet: http://www.grants.gov/search/search.do;jsessionid=G6z2PM6MT4vFzPGvmYl7TPJTc9lpLwG0CHlsTPC1jw2TJXJRC87n!488296553?oppId=146413&mode=VIEW
Sponsor: U.S. Department of Defense
4400 Martin Road, Redstone Arsenal
Huntsville, AL 35898

USDD Dielectric Films for Capacitors Grants 4719

Dielectric films for capacitors are critical for the development of energy- and power-dense systems for shipboard use impulse power, power conditioning and back-up power applications. The Office of Naval Research program focuses on the materials and structures that have the capacity to increase energy density, reducing weight and volume allocation for capacitor banks while extending service life in extreme conditions. Current efforts are focused on multi-scale models of dielectric charge storage and transfer including failure mechanisms, synthesis of novel polymer and composite dielectric materials, optimization for rapid, controlled discharge rates and development of high temperature (150 – 200 C) polymer dielectrics. Applications include: conditioners; capacitors; rail guns; auxiliary power; and power electronic inverters and converters. Submission of the Research Proposal should always be preceded by submission of a Preproposal and a discussion of that Preproposal with the assigned Program Officer. Applications for the upcoming fiscal year should be received by September 30.
Requirements: Grants are awarded to academic, industrial, and nonprofit organizations, with strong emphasis on university research. Investigators are encouraged to contact the staff prior to formally submitting a proposal.
Restrictions: New grants or contracts are typically started near the beginning of the fiscal year. New proposals for the next fiscal year should be submitted no later than September 30, unless otherwise directed. Competitive renewals should be submitted no later than six months prior to the award termination date after contacting the Program Officer. Conference proposals should be submitted at least four months prior to the conference.
Geographic Focus: All States
Date(s) Application is Due: Sep 30
Contact: Dr. Paul Armistead; (703) 696-4315; paul.armistead@navy.mil
Sponsor: U.S. Department of Defense
One Liberty Center, 875 N. Randolph Street, Suite 1425
Arlington, VA 22203-1995

USDD Electro-Optical (EO)/Infrared (IR) Sensor Processing Grants 4720

The Office of Naval Research (ONR) Electro-Optical (EO) / Infrared (IR) Sensors and Sensor Processing Program objective is to detect, classify/identify, and localize/geolocate air, sea-surface, and ground targets by developing innovative sensor technology and by improving the performance of the signal and image processing techniques associated with electro-optical (EO) and infrared (IR) sensors (passive and active) that operate from the visible through longwave infrared bands. Proposers are encouraged to contact the program officer to discuss their research interest prior to the submission of formal proposals.
Requirements: Grants are awarded to academic, industrial, and nonprofit organizations, with strong emphasis on university research.
Restrictions: New grants or contracts are typically started near the beginning of the fiscal year (October 1). New proposals for the next fiscal year should be submitted no later than September 30, unless otherwise directed by the BAA. Competitive renewals

should be submitted no later than six months prior to the award termination date after contacting the Program Officer. Conference proposals should be submitted at least four months prior to the conference.
Geographic Focus: All States
Date(s) Application is Due: Sep 30
Contact: Ravindra Athale; (703) 696-2237; fax (703) 696-2206; ravindra.athale@navy.mil
Internet: http://www.onr.navy.mil/Science-Technology/Departments/Code-31/All-Programs/312-Electronics-Sensors/EO-IR-Sensor-Processing.aspx
Sponsor: U.S. Department of Defense
One Liberty Center, 875 N. Randolph Street, Suite 1425
Arlington, VA 22203-1995

USDD Electrochemical Materials Grants **4721**
The Office of Naval Research Electrochemical Materials Program is focused on developing a fundamental understanding of charge (electron and ion) storage, transport and transfer mechanisms and applying that knowledge to develop materials, materials architectures and devices that address unique Navy and Marine Corps application needs, such as current and future logistics fueled devices and air-independent propulsion for unmanned undersea vehicles (UUVs). Current foci in this program include in-situ probes for determining reaction mechanisms in high temperature electrochemical systems, multi-scale modeling to meaningfully span atomistic to cell level behavior, challenges unique to three-dimensional battery architectures and fundamental understanding of lithium-based battery catastrophic failure modes. Submission of the Research Proposal should always be preceded by submission of a Preproposal and a discussion of that Preproposal with the assigned Program Officer. Applications for the upcoming fiscal year should be received by September 30.
Requirements: Grants are awarded to academic, industrial, and nonprofit organizations, with strong emphasis on university research. Investigators are encouraged to contact the staff prior to formally submitting a proposal.
Restrictions: New grants or contracts are typically started near the beginning of the fiscal year. New proposals for the next fiscal year should be submitted no later than September 30, unless otherwise directed. Competitive renewals should be submitted no later than six months prior to the award termination date after contacting the Program Officer. Conference proposals should be submitted at least four months prior to the conference.
Geographic Focus: All States
Date(s) Application is Due: Sep 30
Contact: Dr. Michele Anderson; (703) 696-1938; michele.anderson1@navy.mil
Internet: http://www.onr.navy.mil/Science-Technology/Departments/Code-33/All-Programs/332-naval-materials/Electrochemical-Materials.aspx
Sponsor: U.S. Department of Defense
One Liberty Center, 875 N. Randolph Street, Suite 1425
Arlington, VA 22203-1995

USDD Electromagnetic Materials Grants **4722**
The Office of Naval Research (ONR) Electromagnetic Materials Program develops advanced electronic and magnetic materials critical to the advancement of electronic and magnetic devices. Included are III-V semiconductors, oxides, multi-ferroic and multi-functional materials, magnetic films and materials that address micro-scale thermal management, integration and/or novel functionality. Focused experimental and theoretical work is supported. This program currently has three major thrust areas: growth and integration of multi-ferroic materials and their hetero-structures to achieve novel functional devices including, but not limited to, uncooled magnetic sensors, microwave circulators, isolators and tunable filters and phase shifters; III-V device technologies to maximize RF performance metrics including power, linearity, bandwidth, efficiency and noise by improving materials to overcome fundamental limitations such as thermal resistance or capacitance; and emerging materials research that offers the potential for unprecedented functionality such as bandwidth, power handling, linearity, tunability or other new capabilities. Investigators are encouraged to contact the staff prior to formally submitting a proposal. Submissions for the upcoming fiscal year must be received by September 30.
Requirements: Grants are awarded to academic, industrial, and nonprofit organizations, with strong emphasis on university research.
Geographic Focus: All States
Date(s) Application is Due: Sep 30
Contact: Dr. Daniel Green; (703) 696-4220; daniel.s.green1@navy.mil
Internet: http://www.onr.navy.mil/Science-Technology/Departments/Code-31/All-Programs/312-Electronics-Sensors/Electronic-Materials.aspx
Sponsor: U.S. Department of Defense
One Liberty Center, 875 N. Randolph Street, Suite 1425
Arlington, VA 22203-1995

USDD Electronic Warfare Grants **4723**
The Office of Naval Research (ONR) Electronic Warfare Program goal is to control the electromagnetic spectrum (EMS) by exploiting, deceiving, or denying enemy use of the spectrum while ensuring its use by friendly forces. To that end, ONR's EW Technology Program provides naval forces (including Navy and Marine Corps) with improved threat warning systems; electronic warfare support (ES); decoys and countermeasures against weapon tracking and guidance systems; electronic attack (EA) against adversary command, control, communications, computers, intelligence, surveillance, and reconnaissance (C4ISR); and electronic protection (EP) of our own weapons and C4ISR from intentional and unintentional interference.
Requirements: All responsible sources from academia and industry are eligible. Historically black colleges and universities and minority institutions are encouraged to apply. Teams are encouraged and may submit proposals in any and all areas.
Geographic Focus: All States
Date(s) Application is Due: Jul 11
Amount of Grant: 350,000-500,000 USD
Contact: Peter Craig, Program Officer; (703) 696-1575; peter.craig@navy.mil
Internet: http://www.onr.navy.mil/Science-Technology/Departments/Code-31/All-Programs/312-Electronics-Sensors/Electronic-Warfare.aspx
Sponsor: U.S. Department of Defense
One Liberty Center, 875 N. Randolph Street, Suite 1425
Arlington, VA 22203-1995

USDD Entry-Level Faculty Awards in Ocean Acoustics **4724**
The purpose of the USDD Entry-Level Faculty Awards in Ocean Acoustics is to encourage research careers in ocean acoustics or the related disciplines of undersea signal processing, marine structural acoustics and transducer materials science by supporting the research of outstanding young university faculty members. Applicants must be U.S. citizens and must apply within the first four years following the start of his or her first tenure-track appointment. Awardees will be selected based on past performance and quality of proposal. These awards are 36 months in duration and support young faculty scientists at $100,000 per year. ONR expects to grant up to two new entry-level faculty awards per year.
Requirements: A detailed technical proposal in the format described in Planning Letter/Pre-Proposal and Proposal Process is required.
Geographic Focus: All States
Date(s) Application is Due: Aug 1
Contact: Dr. Ellen Livingston, Program Officer; (703) 696-4203 or (703) 696-5031; fax (703) 696-5940; ellen.livingston@navy.mil
Internet: http://www.onr.navy.mil/Science-Technology/Departments/Code-32/All-Programs/Atmosphere-Research-322/Ocean-Acoustics/Awards-Ocean-Acoustics.aspx
Sponsor: U.S. Department of Defense
One Liberty Center, 875 N. Randolph Street, Suite 1425
Arlington, VA 22203-1995

USDD Extended Solids Grants **4725**
The Defense Advanced Research Projects Agency (DARPA) is soliciting innovative multidisciplinary research proposals to develop and demonstrate synthesis methods of materials with superior properties (>1.5X over SOA) that currently can only be made using ultrahigh pressure techniques, such that processing under conditions amenable to production scales (<2000 K and 0.5 GPa) and stability under ambient conditions is achieved. Extended solids are polymorphs/phases of simple molecules that are currently formed under ultrahigh pressure conditions where strong intermolecular bonding and tight crystal packing can be induced, leading to dramatic changes in physical, mechanical, and functional properties. Abstracts are encouraged in advance of full proposals in order to provide potential applicants with a rapid response to minimize unnecessary effort. Multiple awards are anticipated.
Requirements: All responsible sources capable of satisfying the government's needs are eligible. Historically black colleges and universities, small businesses, small disadvantaged businesses and minority institutions are encouraged to apply and to join others in submitting proposals.
Geographic Focus: All States
Date(s) Application is Due: May 22
Contact: Dr. Judah Goldwasser; DARPA-BAA-12-20@darpa.mil
Internet: http://www07.grants.gov/search/search.do;jsessionid=YntGPNHNj6Hytpk2g4rJnQhr3MnThpc36qvhLJLLRnPSFKpbhWdL!488296553?oppId=145014&mode=VIEW
Sponsor: U.S. Department of Defense
3701 North Fairfax Drive
Arlington, VA 22203-6630

USDD Fires Thrust Grants **4726**
The Office of Naval Research (ONR) Fires thrust seeks to enable warfighters employed in small, distributed units with tools to locate and decisively destroy larger enemy forces by applying timely, reliable, precise and accurate fires from a myriad of platforms. Products will equip warfighters with integrated, light-weight optics and sensors to see through all battlefield conditions (day, night, low-light, obscuration) and light-weight, organic, advanced weapons for the rapid, accurate, effective application of firepower. Productions would also equip tactical units frequently operating well beyond conventional parameters of direct fire mutual support. The increased intelligence capabilities delivered by company intelligence cells will generate more potential targets in the future. Technology investment areas include: developing targeting technologies (detection, locating, identification, designation, and tracking) for faster, more precise engagements, while simplifying tasks for the warfighter (specific interests include: precision location of targets, reduction of ballistic flight error, and eye-safe, low-power, designation technology); providing more capable, lighter weight ammunition across the spectrum of lethality (to include contained collateral damage situations); with increased reliability, range and precision; for direct and indirect fire weapons, small arms through major caliber; with low costs and logistics burden, and increased operating, transportation and storage safety; and increased shelf life with reduced storage maintenance requirements (specific interests include affordable guided flight control, improved first round Probability of Kill (Pk) and novel seeker technologies); increasing the capabilities and reducing the weight of the Marine Ground Combat Element (GCE) through Marine Rifle Company organic weapons systems and Navy Expeditionary Combat Command organic weapons systems (specific interests include: reducing weight and cost, increasing range and safety, and extending the service life); and investigating the fundamental science of energetic materials that will lead to the development of ammunition propellants with improved performance and minimized safety hazards. Specific interests include: improving timing and initiation behavior; reducing shock, impact, storage and transportation hazards; furthering the understanding of material properties at the nanoscale.

The Expeditionary Maneuver and Combating Terrorism Department welcomes innovative ideas and white papers in response to the ONR long-range broad agency announcement.
Requirements: Grants are awarded to academic, industrial, and nonprofit organizations, with strong emphasis on university research.
Restrictions: New grants or contracts are typically started near the beginning of the fiscal year (October 1). New proposals for the next fiscal year should be submitted no later than September 30, unless otherwise directed by the BAA. Competitive renewals should be submitted no later than six months prior to the award termination date after contacting the Program Officer. Conference proposals should be submitted at least four months prior to the conference.
Geographic Focus: All States
Date(s) Application is Due: Sep 30
Contact: Dan Simons, Thrust Manager; (540) 653-4984; dan.simons@navy.mil
Internet: http://www.onr.navy.mil/Science-Technology/Departments/Code-30/All-Programs/Fires.aspx
Sponsor: U.S. Department of Defense
One Liberty Center, 875 N. Randolph Street, Suite 1425
Arlington, VA 22203-1995

USDD Force Protection Thrust Grants 4727
The Office of Naval Research (ONR) Force Protection thrust seeks to develop and mature technologies that provide protection from myriad modes of enemy attack throughout the spectrum of warfare, including concepts such as asymmetric and irregular warfare and distributed operations which concentrate on the small unit (battalion and below) and individual warfighter levels. End products will include protective systems expeditionary in nature, lightweight and capable of providing a far greater degree of performance than any comparable system currently available. The functional areas of investigation are explosive hazard defeat through detection, breaching and neutralization of all explosive hazards, counter sniper, counter rocket, artillery, and mortar (CRAM), counter-bomber and personal protective equipment. Technology investment areas include detection, neutralization and mitigation. The Expeditionary Maneuver and Combating Terrorism Department welcomes innovative ideas and white papers in response to the ONR long-range broad agency announcement.
Requirements: Grants are awarded to academic, industrial, and nonprofit organizations, with strong emphasis on university research.
Restrictions: New grants or contracts are typically started near the beginning of the fiscal year. New proposals for the next fiscal year should be submitted no later than September 30, unless otherwise directed by the BAA. Competitive renewals should be submitted no later than six months prior to the award termination date after contacting the Program Officer. Conference proposals should be submitted at least four months prior to the conference.
Geographic Focus: All States
Date(s) Application is Due: Sep 30
Contact: Lee Mastroianni; (703) 696-3073; lee.mastroianni@navy.mil
Internet: http://www.onr.navy.mil/Science-Technology/Departments/Code-30/All-Programs/Force-Protection.aspx
Sponsor: U.S. Department of Defense
One Liberty Center, 875 N. Randolph Street, Suite 1425
Arlington, VA 22203-1995

USDD Fundamental Research to Counter Weapons of Mass Destruction Grants 4728
This is a Broad Agency Announcement (BAA) by the Defense Threat Reduction Agency (DTRA). It is an extramural endeavor that combines the fundamental research, educational program, or other effort needs of DTRA and other Department of Defense interests. DTRA safeguards America and its allies from weapons of mass destruction (WMD) and provides capabilities to reduce, eliminate and counter the threat and effects from chemical, biological, radiological, nuclear, and high yield explosives. DTRA seeks to identify, adopt, and adapt emerging, existing and revolutionary sciences that may demonstrate high payoff potential to Counter-WMD (C-WMD) threats. This BAA solicits ideas and topic-based pre-application white papers for long-term challenges that offer a significant contribution in the following: the current body of knowledge; the understanding of phenomena and observable facts; advancement in revolutionary technology; and new concepts for technology application or that may have impact on future C-WMD threat reduction or capabilities. A portion of this effort is expected to be devoted to awards for science, technology, engineering, and mathematics education programs with a C-WMD focus; such as, but not limited to, postdoctoral fellowships, stipends, degrees, visiting scientist programs, student exchange programs, and development of accredited C-WMD curricula. There are 6 thrust areas: (1) science of WMD sensing and recognition; (2) cognitive, information and network science; (3) science for protection; (4) science to defeat WMD; (5) science to secure WMD; and (6) cooperative counter WMD research with global partners. All non-topic based proposed efforts must be coordinated with the relevant technical point of contact prior to submission. Contact information is listed in the BAA on the website. On average 15 to 30 individual awards are expected per year. There are two categories of awards. Single scope awards (averaging $150K per year) are research projects or efforts that focus on exploratory aspects of a unique problem, a high risk approach, or innovative research in subjects with potential for high impact to C-WMD science. Multidisciplinary awards (averaging $350K per year) are research projects or efforts that involve a comprehensive program of innovative research in an interdisciplinary area with potential for high impact. Application packages and instructions vary by topic and thrust area. Applicants should carefully review the instructions found in the BAA at the website. Generally there are no deadlines for pre-application white papers, and the deadline for invited proposals will be provided in the letter of invitation. Proposal invitations are limited to available funds. Timeline details are available at the DTRA Basic and Fundamental Research Community Portal (www.dtrasubmission.net/portal). Applicants are responsible for checking the site for changes and/or updates.
Requirements: Applicants eligible to submit pre-application white papers and proposals include the following United States and foreign-based equivalents: accredited degree-granting colleges and universities; industrial/commercial basic research centers, including small businesses with a portfolio predominantly in basic research proposing a project with significant participation (minimum 30% of total effort value on the proposed project) by one or more accredited degree-granting colleges and/or universities; and not-for-profit organizations/basic research centers with a portfolio predominantly in basic research proposing a project with significant participation (minimum 30% of total effort value on the proposed project) by one or more accredited degree-granting colleges and/or universities. Proof of 501(c)3 status may be required for not-for-profits. For foreign-based establishments entirely based outside the United States and/or its territories, proof of not-for-profit status may be required.
Geographic Focus: All States, All Countries
Contact: Gena Leake; (703) 767-5870; HDTRA1-09-14-FRCWMD@dtra.mil
Internet: http://www.dtra.mil/Research.aspx
Sponsor: U.S. Department of Defense
6200 Meade Road
Fort Belvoir, VA 22060-5264

USDD Future Naval Fuels Grants 4729
During the 21st Century, the availability of petroleum-derived fuels will decline, causing a global shift towards alternative fuels. Additionally, because the United States currently imports approximately 60 percent of our current oil needs, our energy security is potentially subject to uncertainties that could impact the operations of the Navy and Marine Corps assets. The Navy uses two specified fuels for aircraft and shipboard propulsion: JP-5 and NATO F-76. They are compatible with seawater compensation and have a higher flash point that meets international requirements for shipboard usage. A number of key system parameters must be investigated in determining the impact of alternative fuel changes on the combustions systems (gas turbines and diesel engines) that power Navy and Marine Corps assets. Alternative fuels aboard Navy ships must be compatible with current Navy fuels, tolerant to seawater compensation, have flash point characteristics equivalent to current Navy fuels, have long-term storage capabilities, resistance to biocontamination, and will not negatively impact the current Navy fuels logistics. Research needs to address a methodology for fuel characterization and combustion qualification for acceptable operation of current and proposed gas turbines and diesel engines that may be deployed by the Navy and Marine Corps. We need to understand the impact of varying fuel stock and/or fuel processing on fuel properties and corresponding effects on engine operations, reliability and life. Research must also explore the long-term impact, if any, of alternative fuels on the engine materials and components. Investigations will include validated models that can predict the combustion dynamics, flame stability and emissions of varying fuel constituents. By understanding petroleum-based fuel characteristics and combustion dynamics, opportunities exist to design synthetic fuels with higher heat specific heat content that could extend operational range of assets. Submission of the Research Proposal should always be preceded by submission of a Preproposal and a discussion of that Preproposal with the assigned Program Officer. Applications for the upcoming fiscal year should be received by September 30.
Requirements: Grants are awarded to academic, industrial, and nonprofit organizations, with strong emphasis on university research. Investigators are encouraged to contact the staff prior to formally submitting a proposal.
Restrictions: New grants or contracts are typically started near the beginning of the fiscal year. New proposals for the next fiscal year should be submitted no later than September 30, unless otherwise directed. Competitive renewals should be submitted no later than six months prior to the award termination date after contacting the Program Officer. Conference proposals should be submitted at least four months prior to the conference.
Geographic Focus: All States
Date(s) Application is Due: Sep 30
Contact: Sharon Beermann-Curtin, Program Officer; (703) 588-2358; fax (703) 696-0934; sharon.beermanncurti@navy.mil
Internet: http://www.onr.navy.mil/Science-Technology/Departments/Code-33/All-Programs/332-naval-materials/Future-Naval-Fuels.aspx
Sponsor: U.S. Department of Defense
One Liberty Center, 875 N. Randolph Street, Suite 1425
Arlington, VA 22203-1995

USDD Graduate Traineeship Awards in Ocean Acoustics 4730
USDD Graduate Traineeship Awards in Ocean Acoustics, for study and research leading to a doctoral degree, will be given to individuals who have demonstrated a special aptitude and desire for advanced training in ocean acoustics or the related disciplines of undersea signal processing, marine structural acoustics and transducer materials science. Applicants must have received their baccalaureate or master degrees within three years of the year in which they apply and must plan to attend an institution with doctoral granting programs in the ocean sciences or related fields. Special consideration will be given to students who have passed doctoral candidate qualifying exams. These awards are 36 months in duration and cover the student's full tuition, salary and required fees. Applications should include a letter of intent (describing applicant's research interests and academic institution), three reference letters, and official transcripts. ONR expects to grant up to five new graduate traineeships per year.
Geographic Focus: All States
Date(s) Application is Due: Aug 1
Contact: Dr. Ellen Livingston, Program Officer; (703) 696-4203 or (703) 696-5031; fax (703) 696-5940; ellen.livingston@navy.mil
Internet: http://www.onr.navy.mil/Science-Technology/Departments/Code-32/All-Programs/Atmosphere-Research-322/Ocean-Acoustics/Awards-Ocean-Acoustics.aspx
Sponsor: U.S. Department of Defense
One Liberty Center, 875 N. Randolph Street, Suite 1425
Arlington, VA 22203-1995

USDD High Assurance Cyber Military Systems Grants 4731

The Defense Advanced Research Projects Agency (DARPA) is soliciting innovative research proposals in the area of the clean-slate development of software for high-assurance cyber-physical systems. Proposed research should investigate innovative approaches that enable revolutionary advances in science or systems. The goal of the High-Assurance Cyber Military Systems (HACMS) Grants program is to create technology for the construction of high-assurance, cyber-physical systems, where high assurance is defined to mean functionally correct and satisfying appropriate safety and security properties. Achieving this goal requires a fundamentally different approach from what the software community has taken to date. Consequently, HACMS will adopt a clean-slate, formal methods–based approach that enables semi-automated code synthesis from executable, formal specifications. In addition to generating code, such a synthesizer will produce a machine-checkable proof that the generated code satisfies the functional specification as well as security and safety policies. A key technical challenge is the development of techniques to ensure that such proofs are composable, allowing the construction of high-assurance systems out of high-assurance components. The HACMS program is structured as a 4.5-year effort, split into three 18-month phases. The five technical areas are: military vehicle experts; operating systems; control systems; research integration and red team. Multiple awards are anticipated.

Requirements: All responsible sources capable of satisfying the government's needs are eligible. Historically black colleges and universities, small businesses, small disadvantaged businesses and minority institutions are encouraged to apply and to team with others to submit proposals.

Restrictions: Specifically excluded is research that primarily results in evolutionary improvements to the existing state of practice.

Geographic Focus: All States

Date(s) Application is Due: Jul 10

Contact: Kathleen Fisher, Program Manager; (703) 526-6630; HACMS@darpa.mil

Internet: http://www.grants.gov/search/search.do;jsessionid=G6z2PM6MT4vFzPGvmYl7TPJTc9lpLwG0CHlsTPC1jw2TJXJRC87n!488296553?oppId=146074&mode=VIEW

Sponsor: U.S. Department of Defense
3701 North Fairfax Drive
Arlington, VA 21702-1714

USDD High School Apprenticeship Program and Undergraduate Research Apprenticeship Grants 4732

The Army Research Office (ARO) intends to conduct the High School Apprenticeship Program (HSAP) and Undergraduate Research Apprenticeship Program (URAP) for another year. HSAP/URAP will fund the science, technology, engineering, or math (STEM) apprenticeship of promising high school juniors, seniors, and undergraduate college students to work in a university structured research environment under the direction of existing ARO-sponsored principal investigators (PI) serving as mentors. The overall objectives include: ensuring a high quality and high quantity STEM workforce in the United States; fostering student interest in STEM subject areas; facilitating follow-on student opportunities including university enrollment and Department of Defense scholarships/internships; and ultimately exposing the apprentices to the attractive, meaningful, and challenging military and civilian career STEM opportunities. The PI/mentor is principally responsible for identifying meaningful and significant duties and responsibilities under the auspices of the current grant and will provide the day-to-day guidance and oversight for the apprentice. Awards will not exceed $3,000 per apprentice. Each modification can have no more than two apprentices for grants and no more than six apprentices for UARCs. Application instructions are on the website.

Requirements: Awards will be made as add-ons to existing single-investigator grants, Multidisciplinary University Research Initiative grants, and University Affiliated Research Centers that have at least 12 months' period of performance remaining from the date of proposal submission. An apprentice must be at least a junior in high school, i.e., started the fall semester of their junior year or be an undergraduate college student.

Geographic Focus: All States

Date(s) Application is Due: Feb 29

Contact: Ashley Wade; (919) 549-4205; ashley.e.wade.civ@mail.mil

Internet: http://www07.grants.gov/search/search.do;jsessionid=mSS0P8dKJX8QFKRfJpvnv9203KK2snNrHqMGLhnWRnhrnjMMCcks!733686662?oppId=140735&mode=VIEW

Sponsor: U.S. Department of Defense
4400 Martin Road, Redstone Arsenal
Huntsville, AL 35898

USDD HIV/AIDS Prevention: Military Specific HIV/AIDS Prevention, Care, and Treatment for Non-PEPFAR Funded Countries Grants 4733

The Department of Defense HIV/AIDS Prevention Program (DHAPP) has successfully engaged over 80 countries in efforts to combat HIV/AIDS among their respective military services. DHAPP is a partner collaborating with the United States State Department, Health and Human Services, United States Agency for International Development, and Centers for Disease Control and Prevention, in the President's Emergency Plan for AIDS Relief (PEPFAR). This Broad Agency Announcement (BAA) allows DHAPP to assist countries where militaries are not funded under PEPFAR and to solicit existing partners and establish new partners in furtherance of DHAPP and partner military program goals. Proposals should focus on rapidly extending HIV/AIDS services. Applicants are encouraged to target specific needs with a practical business plan, using small grass-roots organizations to provide community-based services as a way to enhance organic capabilities and sustainability. Concept papers are required and full proposals are by invitation only. Submission within the first cycle, ending February 21, is strongly encouraged. The period of performance is generally one year, with up to two additional phases subject to availability of funding. Applicants are encouraged to propose additional phases if appropriate. This is a multiple year BAA which may be revised as needed. Based on the availability of funds, up to three evaluation cycles will be conducted. Changes will be posted at Grants.gov (http://grants.gov/applicants/find_grant_opportunities.jsp). Applicants are encouraged to sign up on Grants.gov for "send me change notification emails" by following the link on the synopsis page for the program announcement/funding opportunity.

Requirements: All responsible sources from academia, industry, and non-governmental organizations are eligible. Applicants respondents must demonstrate the active support of the in-country military and the Department of Defense representative in the corresponding United States Embassy in the planning and execution of proposals.

Restrictions: Funding is available for Chad, Niger, Mali, and Burkina Faso. A description of the work needed, along with the program areas and estimated funding is available in the BAA found at the website. All questions should be submitted to the points of contact by email.

Geographic Focus: All States

Date(s) Application is Due: Sep 30

Amount of Grant: 30,000-290,000 USD

Contact: Latrice Rubenstein; (619) 556-5276; Latrice.Rubenstein@navy.mil

Internet: http://www.grants.gov/search/search.do;jsessionid=2zGTPFTZnGnR2n0hFVQxLjJrqQ0vdnQM3Wt08p5Y171y8Gj4qPxb!748281696?oppId=141233&mode=VIEW

Sponsor: U.S. Department of Defense
5450 Carlistle Park, P.O. Box 2050
Mechanicsburg, PA 17055-0791

USDD HIV/AIDS Prevention Program Information Systems Development Grants 4734

The Department of Defense HIV/AIDS Prevention Program (DHAPP) has successfully engaged over 80 countries in efforts to combat HIV/AIDS among their respective military services. DHAPP is a partner collaborating with the United States State Department, Health and Human Services, United States Agency for International Development, and Centers for Disease Control and Prevention, in the President's Emergency Plan for AIDS Relief. This BAA is intended to solicit existing partners and establish new partners to support the development of a DHAPP.org web site. This site will serve as a source for program updates, with both a public side and a password protected intranet side. A secure online database is needed and needs to aggregate all individual partners' submissions into a single file for download by the DHAPP operations staff in a common format, such as XL and CSV and TXT. Other site capacities should include a file structure to upload and download files, internet and email messaging, and database functions to support information gathering and analysis. Proposed work should include an infrastructure plan and a plan for support and/or turnover of the site. The period of performance is generally one year, with additional phases at the discretion of the government. Applicants are encouraged to propose additional phases if appropriate. This is a multiple year BAA which may be revised as needed. Based on the availability of funds, up to three evaluation cycles will be conducted. Changes will be posted at Grants.gov (http://grants.gov/applicants/find_grant_opportunities.jsp). Applicants are encouraged to sign up on Grants.gov for "send me change notification emails" by following the link on the synopsis page for the program announcement/funding opportunity.

Requirements: All responsible sources from academia, industry, and non-governmental organizations are eligible. Applicants must demonstrate the active support of the in-country military and the USDD representative in the corresponding United States Embassy in the planning and execution of proposals. Applicants should have prior experience with development and hosting of websites preferably for funding multiple partners, from many countries and cultures and languages, and experience in training staff to use and maintain web based databases. Applicants should also have experience with development and hosting of USDD websites and be familiar with the most recent Information Assurance requirements.

Geographic Focus: All States

Date(s) Application is Due: Sep 30

Contact: Latrice Rubenstein; (619) 556-5276; Latrice.Rubenstein@navy.mil

Internet: http://www.afosr.af.mil/pdfs/BAA2005-1.pdf

Sponsor: U.S. Department of Defense
5450 Carlistle Park, P.O. Box 2050
Mechanicsburg, PA 17055-0791

USDD Hull Grooming/Advanced Hull Cleaning Technologies Grants 4735

While all marine operations hope eventually to be able to prevent bio-fouling at low cost and no maintenance, at present it is periodically necessary to mechanically clean a ship's hull. This requires periodic maintenance for the removal of marine biofouling. Because this is a labor-intensive operation, the Navy is developing autonomous robotic devices that can continuously "groom" the hull surface. Some of the challenges, apart from building an autonomous device, are associated with the development of sensor systems (detectors and analysis-and-control software) that can provide both navigation and path planning and discrimination between clean and fouled surfaces. While large diver-operated aggressive cleaning methods currently exist, a more proactive continuous grooming of the hull that occurs at the onset of biofouling settlement/colonization by autonomous vehicles should reduce maintenance costs and maintain a clean hull that will minimize fuel consumption and greenhouse gas emissions. Approaches to grooming methods that minimize surface damage or are non-contact in nature are of extreme interest to the development of this technology. Submission of the Research Proposal should always be preceded by submission of a Preproposal and a discussion of that Preproposal with the assigned Program Officer. Applications for the upcoming fiscal year should be received by September 30.

Requirements: Grants are awarded to academic, industrial, and nonprofit organizations, with strong emphasis on university research. Investigators are encouraged to contact the staff prior to formally submitting a proposal.

Restrictions: New grants or contracts are typically started near the beginning of the fiscal year. New proposals for the next fiscal year should be submitted no later than September

30, unless otherwise directed. Competitive renewals should be submitted no later than six months prior to the award termination date after contacting the Program Officer. Conference proposals should be submitted at least four months prior to the conference.
Geographic Focus: All States
Date(s) Application is Due: Sep 30
Contact: Dr. Stephen McElvany, Program Officer; steve.mcelvaney@navy.mil
Internet: http://www.onr.navy.mil/Science-Technology/Departments/Code-33/All-Programs/332-naval-materials/Hull-Grooming-Cleaning-Technologies.aspx
Sponsor: U.S. Department of Defense
One Liberty Center, 875 N. Randolph Street, Suite 1425
Arlington, VA 22203-1995

USDD Human Performance Training and Education Grants **4736**
The Office of Naval Research (ONR) Human Performance Training and Education (HPT&E) thrust seeks to develop, evaluate and deliver scientifically proven methodologies and technologies that enable the cognitive and physical superiority of Marines. This thrust seeks to develop technologies, architectures and systems that accelerate the training at the individual and team level by enhancing physical and cognitive preparedness. Efforts are focused on providing tools, technologies and methodologies to support the overall goal of either the physical or cognitive technology areas of investment by advancing the delivery of their science into systems for use by the warfighter. In the area of physical performance, ONR's HPT&E thrust is focused on technologies that provide protection against disease and injury as well as greater resilience and optimized physical and physiological performance across the spectrum of conflict at both individual and team levels. This technology area of investment includes: Warfighter Physiology-conducting assessments of the physiological impact of extended full-spectrum conflicts in order to develop methods for mitigating the physiological decline inherent to these types of operations; and Physical Preparedness-pursuing technologies and methodologies for improving and sustaining the warfighter's physical readiness. In the area of cognitive performance, HPT&E's thrust is concentrated on developing human performance training technologies that can train warriors faster and to a higher standard, and resulting in increased retention. It also pursues technologies that improve the ability of Marines to perform cognitively demanding mission-specific tasks, including: Adaptive (Dynamic) Learning-assessing the development of the trainee's cognition and behavior in real time in order to adapt the training environment to the individual and team; Advanced Language Learning-expanding ONR's understanding of how the brain develops during language learning in order to leverage this knowledge to create better training methods, enhanced skill retention and to incorporate socioeconomic knowledge; Experiential Learning-developing first-person immersive training that can provide the visual, auditory, tactile and olfactory stimuli of a complex battlefield within all spectrum of warfare, including culturally realistic, reactive, dynamic and synthetic entities that allow realistic interaction with civilians and opposing forces; and Perception and Perceptuomotor-developing tactically relevant perceptual skills and control interfaces in order to enhance situational awareness.
Requirements: Grants are awarded to academic, industrial, and nonprofit organizations, with strong emphasis on university research.
Restrictions: New grants or contracts are typically started near the beginning of the fiscal year (October 1). New proposals for the next fiscal year should be submitted no later than June 1, unless otherwise directed by the BAA. Competitive renewals should be submitted no later than six months prior to the award termination date after contacting the Program Officer. Conference proposals should be submitted at least four months prior to the conference.
Geographic Focus: All States
Date(s) Application is Due: Jun 1
Contact: Peter Squire, Thrust Manager; (703) 696-2580 or (703) 588-2547; fax (540) 653-3607; peter.squire@navy.mil
Internet: http://www.onr.navy.mil/Science-Technology/Departments/Code-30/All-Programs/Human-Performance-Training.aspx
Sponsor: U.S. Department of Defense
One Liberty Center, 875 N. Randolph Street, Suite 1425
Arlington, VA 22203-1995

USDD Human Social, Cultural and Behavior Modeling Thrust Grants **4737**
The Office of Naval Research (ONR) Human Social, Culture and Behavior Modeling Program invests in research on building capability through the development of a knowledge base, building models and creating training capacity in order to understand, predict and shape human behavior cross-culturally. The program seeks to: understand the human, social, cultural and behavioral factors that influence human behavior; improve our ability to model these influences and understand their impact on human behavior at the individual-, group- and society-level of analysis; improve computational modeling and simulation capabilities, visualization software toolsets, and training/mission rehearsal systems that provide forecasting capabilities for socio-cultural responses; and develop and demonstrate an integrated set of model description data (metadata), information systems and procedures that will facilitate assessment of the software engineering quality of sociocultural behavior models, their theoretical foundation and the translation of theory into model constructs.
Requirements: Grants are awarded to academic, industrial, and nonprofit organizations, with strong emphasis on university research.
Geographic Focus: All States
Contact: Dr. Ivy Estabrooke; (703) 588-2396; fax (703) 696-1211; ivy.estabrooke@navy.mil
Internet: http://www.onr.navy.mil/Science-Technology/Departments/Code-30/All-Programs/Human-Behavioral-Sciences.aspx
Sponsor: U.S. Department of Defense
One Liberty Center, 875 N. Randolph Street, Suite 1425
Arlington, VA 22203-1995

USDD Image Analysis and Understanding Grants **4738**
The Office of Naval Research (ONR) Intelligent and Autonomous Systems program is concerned with the environments in which warfighters operate; these environments are uncontrolled, complex and dynamically changing. Autonomous agents (cyber or physical) do not yet have the level of intelligence needed to operate in such open, uncertain and unpredictable environments either independently or alongside warfighters. The program's main objectives are to develop principles of machine intelligence, efficient computational methods, algorithms and tools for building versatile smart agents that can perform missions autonomously with minimal human supervision, and collaborate seamlessly with teams of warfighters and other agents. The following are the focus areas of this program: Building Blocks of Machine Intelligence; Autonomy for Teams of Unmanned Vehicles; and Human-Agent Collaboration. Proposers are encouraged to contact the program officer to discuss their research interest prior to the submission of formal proposals. The annual deadline for each upcoming fiscal year is September 30.
Requirements: Grants are awarded to academic, industrial, and nonprofit organizations, with strong emphasis on university research.
Restrictions: New grants or contracts are typically started near the beginning of the fiscal year. New proposals for the next fiscal year should be submitted no later than September 30, unless otherwise directed by the BAA. Competitive renewals should be submitted no later than six months prior to the award termination date after contacting the Program Officer. Conference proposals should be submitted at least four months prior to the conference.
Geographic Focus: All States
Contact: Dr. Behzad Kamgar-Parsi, Program Officer; (703) 696-5754; fax (703 696-2611; behzad.kamgarparsi@navy.mil
Internet: http://www.onr.navy.mil/Science-Technology/Departments/Code-31/All-Programs/311-Mathematics-Computers-Research/Image-Analysis-Understanding.aspx
Sponsor: U.S. Department of Defense
One Liberty Center, 875 N. Randolph Street, Suite 1425
Arlington, VA 22203-1995

USDD Image Optimization for Small Focal Plane Arrays Grants **4739**
While the Office of Naval Research (ONR) continues to develop larger format Focal Plane Arrays (FPAs) to provide wide area coverage sensors, it has been recognized that this approach will lead to ever increasing FPA costs. A more cost effective approach could be defined by an investigation of how to achieve sensor performance using smaller, more affordable FPAs by optimizing the imagery from these systems to meet detection, tracking and identification requirements. This Broad Agency Announcement (BAA) seeks to develop techniques to optimize imagery from passive and active Electro-Optics Infrared (EO-IR) imaging sensors based on smaller, less expensive FPAs. Some past and on-going ONR efforts include coded sensor processing, compressive sensing, and "Lucky Imaging" algorithms. There may also be techniques for adding processing to read out integrated circuits (ROICs) that will enable better utilization of smaller FPAs. Although this effort is not intended to cover the development of new EO-IR sensor hardware, including FPAs and ROICs, the impact of this investigation may lead to the proposal and implementation of new sensor hardware architectures. The overall intent of the BAA is the development of image enhancement techniques and methodologies that can be incorporated in processing of imagery from a variety of different small-format EO-IR sensor architectures. This approach can greatly improve the Navy's ability to achieve Information Dominance with a broader spectrum of lower-cost sensors. Work funded may include basic research, applied research and some advanced technology development. All applicants must submit white papers which are accepted electronically only. An expanded oral presentation will be requested from applicants whose proposals are identified as being of "particular value" to the Navy. Following oral presentations, applicants whose technology is still considered as having "particular value" to the Navy will be encouraged to submit full proposals. Two or three awards are anticipated. The award period may be from one to three years.
Requirements: All responsible sources from academia and industry are eligible. Historically black colleges and universities and minority institutions are encouraged to apply.
Geographic Focus: All States
Date(s) Application is Due: May 1
Contact: Rebecca Foster; (703) 696-2972; rebecca.d.foster@navy.mil
Internet: http://www.onr.navy.mil/Contracts-Grants/Funding-Opportunities.aspx
Sponsor: U.S. Department of Defense
One Liberty Center, 875 N. Randolph Street, Suite 1425
Arlington, VA 22203-1995

USDD Information Innovation Office Grants **4740**
The Defense Advanced Research Projects Agency (DARPA) is soliciting innovative research proposals of interest to the Information Innovation Office (I2O). Proposed research should investigate innovative approaches that enable revolutionary advances in science, devices, or systems. I2O seeks unconventional approaches that are outside the mainstream, undertaking directions that challenge assumptions and have the potential to radically change established practice. I2O is interested in submissions related to the focus areas of understanding, empowering, and connecting. I2O seeks to enable understanding through the development of computational and analytic techniques that can process and assimilate the huge volumes of data generated by modern sensor, information, and communication systems, and by incorporating these techniques and new enabling sensor technologies in intelligence, surveillance, and reconnaissance exploitation and language processing systems that can classify behaviors, correlate events, identify trends, detect anomalies, and generate alerts. I2O works to empower warfighters at all echelons to meet challenges by developing advanced computing technologies, including modeling and simulation, visualization, knowledge management, geographic information systems, semantic web techniques, social networks, crowd-sourced, and Web 2.0/3.0 techniques, and implementing these in command and control systems that automate and improve

the application of information to military decision-making, planning, training, mission rehearsal, and operations support. I2O aims to ensure connection by developing more resilient computing and communication technologies and by creating new approaches to ensure the security of defense networks and computing systems. These technologies will enable more rapid detection, analysis, and mitigation of cyber threats; more efficient utilization of network resources; and more effective monitoring and control of computing and communications infrastructure. Multiple awards are anticipated. The level of funding for individual awards has not been determined and will depend on the quality of the proposals received and the availability of funds. Applicants may submit an abstract in advance of a proposal to reduce potential effort in proposal preparation. DARPA will respond to abstracts indicating whether there is interest for the proposed work. Proposals may be submitted and will be reviewed irrespective of comments or feedback received in response to the abstract. Abstracts and proposals may be submitted until the deadline and will be reviewed on a rolling basis through the submission due date.
Requirements: All responsible sources capable of satisfying the government's needs are eligible. Historically black colleges and universities, small businesses, small disadvantaged businesses and minority institutions are encouraged to submit proposals and team with others to submit proposals.
Restrictions: Specifically excluded is research that results primarily in evolutionary improvements to the existing state of the art.
Geographic Focus: All States
Date(s) Application is Due: May 30
Contact: I2O BAA Coordinator; (703) 526-6630; DARPA-BAA-11-34@darpa.mil
Internet: http://www07.grants.gov/search/search.do;jsessionid=hZQpP58bNPTzkMhJ98DNlppVHLHxTBY69kms2gTlh8TRHTbGSz7g!-1993000306?oppId=97194&mode=VIEW
Sponsor: U.S. Department of Defense
3701 North Fairfax Drive
Arlington, VA 21702-1714

USDD Information Integration Grants **4741**
The Office of Naval Research (ONR) Information Integration program is concerned with basic research in probability (a mathematical framework to represent the likelihood of phenomena with uncertain outcomes) and statistics (the process of collecting data and extracting and visualizing the information contained in the data). Recently, connection with several areas of mathematics such as analysis, geometry, algebra, combinatorics and theoretical computer science has generated a number of new scientific ideas with important applications. Such synergism will likely be required to develop innovative techniques for solving the challenging problems discussed below. Several focus areas build around the central themes of information representation, discovery, processing and integration, including: Information Representation and Discovery; Information Integration and Data Fusion; Decision Making Under Uncertainty; and Image Analysis for Intelligence, Surveillance, Reconnaissance, Targeting. The annual deadline is September 30.
Geographic Focus: All States
Date(s) Application is Due: Sep 30
Contact: Dr. Wen Masters; (703) 696-3191 or (703) 696-3193; wen.masters@navy.mil
Internet: http://www.onr.navy.mil/Science-Technology/Departments/Code-31/All-Programs/311-Mathematics-Computers-Research/Information-Integration.aspx
Sponsor: U.S. Department of Defense
One Liberty Center, 875 N. Randolph Street, Suite 1425
Arlington, VA 22203-1995

USDD Innovative Systems for Military Missions-Tactical Technology Office Grants **4742**
The Tactical Technology Office (TTO) of the Defense Advanced Research Projects Agency (DARPA) is soliciting executive summaries, white papers and proposals for advanced research and development of Innovative Systems for Military Missions. Innovative Systems are integrated systems or critical systems components, which often incorporate emerging advanced technologies, and which enable revolutionary improvements to the capability, efficiency and effectiveness of the military. TTO seeks responses relating to three mission thrust areas: (1) advanced weapons systems, (2) advanced platforms, and (3) advanced space systems. TTO's solicitation focuses on the high risk/high payoff development, integration, demonstration and evaluation of innovative systems or critical systems components enabled by, and incorporating, new or emerging technologies. Proposed efforts must also show significant promise to provide the United States military with revolutionary new mission capabilities, and/or enable significant increases in mission effectiveness. Innovative system concepts of interest to TTO typically address emerging technical opportunities, advanced systems concepts, emergent threats, and/or new technology-enabled concepts of operation. TTO strongly encourages respondents to adopt a complete systems engineering approach to the problems. Executive summaries, white papers, and proposals are accepted. Applicants are strongly encouraged to submit a one page executive summary describing the technology area and new ideas of interest prior to submitting a white paper or proposal. Executive summaries will be reviewed as they are received. Based on a favorable review of an executive summary, applicants will be invited to submit a six page white paper describing the proposer's proposed concept prior to submitting a proposal. A DARPA Program Manager may contact applicants to discuss the executive summary or white paper. This procedure is intended to minimize unnecessary effort in proposal preparation and review. Submission of an executive summary and/or white paper is not required prior to submitting a full proposal. Executive summaries are due by December 16. White papers are due by January 6. Full proposals are due by February 17. Multiple awards are anticipated. The number of awards made will depend on the quality of proposals and the availability of funds. In general awards are anticipated to be for less than $1 million and less than 18 months duration.
Requirements: All responsible sources capable of satisfying the government's needs are eligible. Historically black colleges and universities, small businesses, small disadvantaged businesses and minority institutions are encouraged to submit proposals and join others in submitting proposals.
Geographic Focus: All States
Date(s) Application is Due: Feb 17
Contact: David Neyland, Director; (703) 526-6630; DARPA-BAA 11-13@darpa.mil
Internet: http://www.grants.gov/search/search.do;jsessionid=qFTNP7SNp7qTyygkG1CvWMc011NgYX1NyBCQv3QlQvSpTCtbLZhb!733686662?oppId=75453&mode=VIEW
Sponsor: U.S. Department of Defense
3701 North Fairfax Drive
Arlington, VA 22203-6630

USDD Intelligence, Surveillance and Reconnaissance Thrust Grants **4743**
The Office of Naval Research (ONR) Intelligence, Surveillance and Reconnaissance (ISR) thrust seeks to develop and leverage advanced technologies for applications in future intelligence, surveillance and reconnaissance systems. It also seeks to enhance situational awareness to enable real-time tactical decision making for distributed operations and provide proactive and predictive capabilities for asymmetric and irregular warfare. ISR's technology investment areas include: Sensor Fields; Relevant and Situational Information on Demand; and Actionable Intelligence for Expeditionary and Irregular Warfare. The Expeditionary Maneuver and Combating Terrorism Department welcomes innovative ideas and white papers in response to the ONR long-range broad agency announcement.
Requirements: Grants are awarded to academic, industrial, and nonprofit organizations, with strong emphasis on university research.
Restrictions: New grants or contracts are typically started near the beginning of the fiscal year (October 1). New proposals for the next fiscal year should be submitted no later than September 30, unless otherwise directed by the BAA. Competitive renewals should be submitted no later than six months prior to the award termination date after contacting the Program Officer. Conference proposals should be submitted at least four months prior to the conference.
Geographic Focus: All States
Date(s) Application is Due: Sep 30
Contact: Martin Kruger; (703)696-5349; fax (703) 696-4274; martin.krugerl@navy.mil
Internet: http://www.onr.navy.mil/Science-Technology/Departments/Code-30/All-Programs/Intelligence-Surveillance-ISR.aspx
Sponsor: U.S. Department of Defense
One Liberty Center, 875 N. Randolph Street, Suite 1425
Arlington, VA 22203-1995

USDD Intelligent and Autonomous Systems Grants **4744**
The Office of Naval Research (ONR) Intelligent and Autonomous Systems program is concerned with the environments in which warfighters operate; which are uncontrolled, complex and dynamically changing. Autonomous agents (cyber or physical) do not yet have the level of intelligence needed to operate in such open, uncertain and unpredictable environments either independently or alongside warfighters. The program's main objectives are to develop principles of machine intelligence, efficient computational methods, algorithms and tools for building versatile smart agents that can perform missions autonomously with minimal human supervision and collaborate seamlessly with teams of warfighters and other agents. Program focus areas include: Building Blocks of Machine Intelligence; Autonomy for Teams of Unmanned Vehicles; and Human-Agent Collaboration. Proposers are encouraged to contact the Program Officer to discuss their research interest prior to the submission of formal proposals. The annual deadline for each upcoming fiscal year is September 30.
Requirements: Grants are awarded to academic, industrial, and nonprofit organizations, with strong emphasis on university research.
Restrictions: New grants or contracts are typically started near the beginning of the fiscal year (October 1). New proposals for the next fiscal year should be submitted no later than September 30, unless otherwise directed by the BAA. Competitive renewals should be submitted no later than six months prior to the award termination date after contacting the Program Officer. Conference proposals should be submitted at least four months prior to the conference.
Geographic Focus: All States
Date(s) Application is Due: Sep 30
Contact: Dr. Behzad Kamgar-Parsi, Program Officer; (703) 696-5754 or (703) 588-2427; fax (703) 696-5940; behzad.kamgar-parsi@.navy.mil or 311_IS@onr.navy.mil
Internet: http://www.onr.navy.mil/Science-Technology/Departments/Code-31/All-Programs/311-Mathematics-Computers-Research/Intelligent-Autonomous-Systems.aspx
Sponsor: U.S. Department of Defense
One Liberty Center, 875 N. Randolph Street, Suite 1425
Arlington, VA 22203-1995

USDD Ionospheric Modification Grants **4745**
The Office of Naval Research, Ionospheric Modifications program investigates the interaction of high power, high frequency electromagnetic radiation with the ionosphere for the purpose of generating low frequency communication signals. The approach is to observe, diagnose, and interpret: naturally occurring, highly variable phenomena associated with the high latitude ionosphere; and the relatively small but useful perturbations that can be induced for communication applications. Emphasis is given to research to be conducted at the High-Frequency Active Auroral Research Program Observatory in Gakona, Alaska, but research at other facilities may also be considered, if supportive of the program goals. Average awards range at $50,000. Applications for the upcoming fiscal year should be received by September 30.
Requirements: Grants are awarded to academic, industrial, and nonprofit organizations, with strong emphasis on university research.

Restrictions: New grants or contracts are typically started near the beginning of the fiscal year. New proposals for the next fiscal year should be submitted no later than September 30, unless otherwise directed by the BAA. Competitive renewals should be submitted no later than six months prior to the award termination date after contacting the Program Officer. Conference proposals should be submitted at least four months prior to the conference.
Geographic Focus: All States
Date(s) Application is Due: Sep 30
Amount of Grant: 25,000-75,000 USD
Contact: Craig Selcher, Program Officer; Craig_Selcher@onr.navy.mil
Internet: http://www.onr.navy.mil/Science-Technology/Departments/Code-31/All-Programs/312-Electronics-Sensors/Ionospheric-Modification.aspx
Sponsor: U.S. Department of Defense
One Liberty Center, 875 N. Randolph Street, Suite 1425
Arlington, VA 22203-1995

USDD Littoral Geosciences and Optics Grants **4746**
The Office of Naval Research Littoral Geosciences and Optics program combines the former Coastal Geosciences and Environmental Optics programs and will leverage the combined expertise of our investigators to better address the multidisciplinary nature of many new basic research areas. We will continue to support basic and applied research efforts directed at solving nearshore, geoscience, optical and remote sensing problems in addition to new collaborative projects. Main areas of interest include predicting the 4D coastal environment and predicting seafloor characteristics in the global ocean margins. Areas of interest are motivated by on-going and future Navy and Marine Corps needs, including littoral operations in nearshore, estuarine, and riverine environments; prediction of sediment transport phenomena across a broad spectrum of coastal environments; and assessment of sources of sonar and optical clutter, among many others. Basic research interests include: nearshore, estuarine and riverine processes; remote sensing of coastal and riverine environments; and sediment transport processes. Studies by individual investigators and collaborative teams are often undertaken in conjunction with other programs at ONR, including the Physical Oceanography, Ocean Engineering and Marine Systems, and Ocean Acoustics programs. Focused, field-oriented research programs are undertaken as Departmental Research Initiatives (DRIs) or as Multidisciplinary University Research Initiatives (MURIs), each of which typically has a five-year timeframe and substantial funding for field studies. Submission of the Research Proposal should always be preceded by submission of a Preproposal and a discussion of that Preproposal with the assigned Program Officer.
Requirements: Grants are awarded to academic, industrial, and nonprofit organizations, with strong emphasis on university research.
Restrictions: New grants or contracts are typically started near the beginning of the fiscal year. New proposals for the next fiscal year should be submitted no later than September 30, unless otherwise directed by the BAA. Competitive renewals should be submitted no later than six months prior to the award termination date after contacting the Program Officer. Conference proposals should be submitted at least four months prior to the conference.
Geographic Focus: All States
Date(s) Application is Due: Sep 30
Contact: Dr. Thomas Drake, Division Director; (703) 696-1206 or (703) 588-2427; fax (703) 696-5940; tom.drake@navy.mil or 321_CG@onr.navy.mil
Internet: http://www.onr.navy.mil/Science-Technology/Departments/Code-32/All-Programs/Atmosphere-Research-322/Littoral-Geosciences-Optics.aspx
Sponsor: U.S. Department of Defense
One Liberty Center, 875 N. Randolph Street, Suite 1425
Arlington, VA 22203-1995

USDD Logistics Thrust Grants **4747**
The Office of Naval Research (ONR) Logistics thrust seeks to provide Marines of the future with a precisely tailored level of sustained logistic support from sea-based platforms to rapidly maneuver forces ashore. Logistics delivery systems of the future will be more responsive and flexible, enabling Marines to out-pace rapidly changing operational scenarios. Likewise, delivered logistic commodities will provide more operational value per unit weight, enhancing combat unit self-sufficiency and maneuverability. Finally, operational units will benefit from technologies that maximize equipment readiness by minimizing both down-time and maintenance requirements. Technology areas of investment include: Asset Visibility; Logistic Transport; Operational Sustainment; and Maintenance Reduction. Proposers are encouraged to contact the program officer to discuss their research interest prior to the submission of formal proposals.
Requirements: Grants are awarded to academic, industrial, and nonprofit organizations, with strong emphasis on university research.
Restrictions: New grants or contracts are typically started near the beginning of the fiscal year (October 1). New proposals for the next fiscal year should be submitted no later than September 30, unless otherwise directed by the BAA. Competitive renewals should be submitted no later than six months prior to the award termination date after contacting the Program Officer. Conference proposals should be submitted at least four months prior to the conference.
Geographic Focus: All States
Contact: Cody Reese, Program Officer; reesec@onr.navy.mil
Internet: http://www.onr.navy.mil/Science-Technology/Departments/Code-30/All-Programs/Logistics.aspx
Sponsor: U.S. Department of Defense
One Liberty Center, 875 N. Randolph Street, Suite 1425
Arlington, VA 22203-1995

USDD Long Range Broad Agency Announcement for Navy and Marine Corps Science and Technology Grants **4748**
The Office of Naval Research is interested in receiving proposals for long-range science and technology projects which offer potential for advancement and improvement of Navy and Marine Corps operations. This is a Broad Agency Announcement intended for proposals related to basic research, applied research, or advanced technology development. White papers and full proposals are expected to be unclassified; however, classified proposals are permitted. Proposals may be submitted at any time until the closing of this BAA, which is September 30.
Requirements: All responsible sources from academia and industry are eligible to apply. Historically black colleges and universities and minority institutions are encouraged to apply and to join others in submitting proposals.
Geographic Focus: All States
Date(s) Application is Due: Sep 30
Contact: Ganesh Krish; (703) 588-2784 or (703) 696-4046; Ganesh.Krish@navy.mil
Internet: http://www.onr.navy.mil/Contracts-Grants/Funding-Opportunities/Broad-Agency-Announcements.aspx
Sponsor: U.S. Department of Defense
One Liberty Center, 875 N. Randolph Street, Suite 1425
Arlington, VA 22203-1995

USDD Maneuver Thrust Grants **4749**
The Office of Naval Research (ONR) Maneuver thrust explores technologies to increase the warfighting capabilities and effectiveness of the Marine Corps Air Ground Task Force with emphasis on improving survivability and enhancing maneuver and maneuver-enabling systems in both decentralized and asymmetric warfare. Marine forces of the future will be significantly more agile, lethal, mobile and survivable. This thrust's technology investment areas include: Survivability-providing survivability within the battlefield in order to support expeditionary maneuver warfare; Advanced Mobility-increasing operational tempo and gaining temporal dominance in order to achieve positional advantage; and Maneuver Enablers-providing maneuver-enabling technologies to separate warfighters from potentially hazardous missions and to increase efficiency and economy of force. Proposers are encouraged to contact the program officer to discuss their research interest prior to the submission of formal proposals.
Requirements: Grants are awarded to academic, industrial, and nonprofit organizations, with strong emphasis on university research.
Restrictions: New grants or contracts are typically started near the beginning of the fiscal year (October 1). New proposals for the next fiscal year should be submitted no later than September 30, unless otherwise directed by the BAA. Competitive renewals should be submitted no later than six months prior to the award termination date after contacting the Program Officer. Conference proposals should be submitted at least four months prior to the conference.
Geographic Focus: All States
Contact: Jeff Bradel, Thrust Manager; (703) 588-2552; jeff.bradel@navy.mil
Internet: http://www.onr.navy.mil/Science-Technology/Departments/Code-30/All-Programs/Maneuver.aspx
Sponsor: U.S. Department of Defense
One Liberty Center, 875 N. Randolph Street, Suite 1425
Arlington, VA 22203-1995

USDD Marine Mammals and Biological Oceanography Grants **4750**
ONR's Marine Mammals and Biological Oceanography program was created late in 2007 and replaced two previous ONR programs (the Marine Mammals and Biology Program in Code 34 and the biological component of the Ocean Optics & Biology Program in Code 32). The Marine Mammals and Biological Oceanography program supports basic and applied research and technology development related to understanding the effects of sound on marine mammals, including physiological, behavioral, ecological effects and population-level effects. Our current program thrusts include, but are not limited to: Monitoring and Detection; Integrated Ecosystem Research; Effects of Sound on Marine Life; and Models and Databases for Environmental Compliance. Pre-proposals are required, and are due April 15. An invitation for full proposals will be sent out in either June or July, with final drafts due by the end of August.
Requirements: Grants are awarded to academic, industrial, and nonprofit organizations, with strong emphasis on university research.
Geographic Focus: All States
Date(s) Application is Due: Aug 30
Contact: Michael Weise; (703) 696-4533; fax (703) 696-2007; michael.j.weise@navy.mil
Internet: http://www.onr.navy.mil/sci_tech/32/322/marine_mammals_biology.asp
Sponsor: U.S. Department of Defense
One Liberty Center, 875 N. Randolph Street, Suite 1425
Arlington, VA 22203-1995

USDD Mathematical Optimization and Operations Research Grants **4751**
The Office of Naval Research (ONR) Mathematical Optimization and Operations Research program supports basic and applied research in mathematical optimization. The primary focus of the program's basic-research portion is linear, nonlinear, integer and combinatorial optimization. Theoretical development, algorithm design and analysis, computational techniques, and software prototypes for solving large-scale problems are of interest. Techniques that produce provably optimal or near-optimal solutions, as well as techniques applicable to large problem domains are favored. This includes, but is not limited to, cutting-plane and polyhedral techniques for mixed-integer programming and interior-point and first-order algorithms for conic/convex optimization. Innovative techniques for dealing with uncertainty from stochastic optimization, robust optimization and simulation-based optimization are of growing

interest. The primary focus of the applied-research portion of the program is on the development of innovative models and solution techniques for Navy-relevant problems in resource allocation, such as those arising in mission planning, command and control, and logistics. Current application areas of interest are maritime mission planning and unmanned sensor management and allocation. New proposals for the next fiscal year should be submitted no later than September 30.
Requirements: Grants are awarded to academic, industrial, and nonprofit organizations, with strong emphasis on university research.
Restrictions: New grants or contracts are typically started near the beginning of the fiscal year. New proposals for the next fiscal year should be submitted no later than September 30, unless otherwise directed by the BAA. Competitive renewals should be submitted no later than six months prior to the award termination date after contacting the Program Officer. Conference proposals should be submitted at least four months prior to the conference.
Geographic Focus: All States
Contact: Dr. Donald Wagner, Program Officer; (703) 696-4313; don.wagner@navy.mil
Internet: http://www.onr.navy.mil/Science-Technology/Departments/Code-31/All-Programs/311-Mathematics-Computers-Research/Mathematical-Optimization.aspx
Sponsor: U.S. Department of Defense
One Liberty Center, 875 N. Randolph Street, Suite 1425
Arlington, VA 22203-1995

USDD Medical Practice Initiative Breadth of Medical Practice and Disease Frequency Exposure Grants 4752

USDD Medical Practice Initiative Breadth of Medical Practice and Disease Frequency Exposure (MPI-BMP) Grants are being solicited by the Assistant Secretary of Defense for Health Affairs, Defense Health Program (DHP). The MPI is primarily focused on the development of medical training systems and competency assessment tools for the sustainment of military medical readiness. The preliminary focus will be on determining the viability of extracting relevant metrics from existing data for use in the determination of skills degradation and using this data to assist in identifying gaps. The MPI-BMP does not only concern itself with degradation from being removed from practice, but also looks at the breadth of clinical exposures and looks across each physician's expected clinical knowledge domain. The data collected in the course of this research will aid both public and private sector organizations in their efforts to improve the quality of care provided in the outpatient setting. The primary, but not exclusive, purpose is research that will: (1) identify when and why degradation of cognitive clinical skills occurs; (2) create validated analytical tools that can predict the probable onset of cognitive skills degradation and determine, with specificity, when skill or knowledge areas have degraded or will be likely to degrade; and (3) propose methods and tools which will enable physicians to preemptively refresh knowledge and maintain familiarity and fluency across the expected competencies for their specialty. Pre-proposals are due December 22, and proposals are by invitation only. Two awards are anticipated. The maximum period of performance is three years.
Requirements: Eligible organizations must be extramural and include nonprofit, public, and private organizations, such as universities, colleges, hospitals, laboratories, and commercial firms. Historically black colleges and universities/minority institutions are encouraged to apply.
Geographic Focus: All States
Date(s) Application is Due: Apr 19
Amount of Grant: 4,000,000 USD
Contact: Mary Rico, Grants Officer; (703) 518-4726; mary.rico@amedd.army.mil
Internet: http://www07.grants.gov/search/search.do;jsessionid=YYJpP9pQHVhR7BFjdtNMBm1grFHnpNT4H8rRHtKrSJtHcGVFsQnJ!712472910?oppId=127853&mode=VIEW
Sponsor: U.S. Department of Defense
1400 Defense Pentagon
Washington, DC 20301-1400

USDD Medical Practice Initiative Procedural Skill Decay and Maintenance Grants 4753

USDD Medical Practice Initiative Procedural Skill Decay and Maintenance (MPI-PSD) Grants are awarded by the United States Army Medical Research Acquisition Activity (U.S.AMRAA) and managed by the Army's Telemedicine and Advanced Technology Research Center. The MPI is primarily focused on the development of medical training systems and competency assessment tools for the sustainment of military medical readiness. The preliminary focus will be on determining the viability of extracting relevant metrics from existing data for use in the determination of skills degradation and using this data to assist in identifying gaps. The data collected in the course of this research will aid both public and private sector organizations in their efforts to improve the quality of care provided in the outpatient setting. The primary, but not exclusive, purpose is research that will: (1) identify when and why degradation of psychomotor and procedural clinical skills occurs; (2) create validated analytical tools that can predict the probable onset of skills2 degradation and determine, with specificity, when skills have degraded or will be likely to degrade; and (3) propose methods and tools which will enable physicians or surgeons to preemptively refresh the expected competencies for their specialty. The study must focus on one of more of the following physician practices: (1) surgical; (2) intensive care medicine; and (3) outpatient invasive procedures. Pre-proposals are due December 22, and proposals are by invitation only. Two awards are anticipated. The maximum period of performance is three years. The maximum period of performance is three years.
Requirements: Eligible organizations must be extramural and include nonprofit, public, and private organizations, such as universities, colleges, hospitals, laboratories, and commercial firms. Historically black colleges and universities/minority institutions are encouraged to apply.
Geographic Focus: All States
Date(s) Application is Due: Apr 19
Amount of Grant: 2,000,000 USD
Contact: Mary Rico, Grants Officer; (703) 518-4726; mary.rico@amedd.army.mil
Internet: http://www07.grants.gov/search/search.do;jsessionid=YYJpP9pQHVhR7BFjdtNMBm1grFHnpNT4H8rRHtKrSJtHcGVFsQnJ!712472910?oppId=127873&mode=VIEW
Sponsor: U.S. Department of Defense
1400 Defense Pentagon
Washington, DC 20301-1400

USDD Metamaterials for Radion Frequency and Optical Applications Grants 4754

This is a multiple year Broad Agency Announcement (BAA). The objective is to encourage a flow of supplementary and/or complementary technologies for electromagnetic metamaterials. Such technologies may include active electronic matching networks, frequency selective structures, and periodic structure with engineered dispersion, including photonic band-gap materials, acoustic metamaterials. Innovative advancements in using metamaterials and/or complementary technologies that leverage commercial economies of scale are highly desired. The areas of focus are radio frequency, optical, acoustic and bio-related metamaterials. White papers only are being accepted. White papers will be reviewed to determine if full proposals will be requested. The award period is expected to be between 9 and 24 months with individual awards ranging from $50K to $1M. Ten awards are anticipated.
Requirements: This is an unrestricted solicitation. Small businesses are encouraged to apply.
Geographic Focus: All States
Contact: Dawn M. Dalhamer; (937) 255-5372; dawn.dalhamer@wpafb.af.mil
Internet: http://www07.grants.gov/search/search.do;jsessionid=91nSQFGd2zP2KG6lvQmQ3ftJX0xlTV251SLm6RJvgtpJHyHlJsTy!2071562142?oppId=145253&mode=VIEW
Sponsor: U.S. Department of Defense
1400 Defense Pentagon
Washington, DC 20301-1400

USDD Mixed Signal, Signal Processing Devices, Circuits and Architecture Grants 4755

The Office of Naval Research Electronic Circuits Program seeks new approaches to achieving 100+ GHz logic and circuits, which includes investigations in devices, architecture, power dissipation, materials and related all-encompassing aspects. This work is expected to lead to the ability to efficiently digitize and synthesize RF signals at frequencies up to and beyond 50 GHz as well as to amplitude, phase and frequency modulate these signals. Additional objectives of this work are expected to lead to digital beam-forming and steering of electronic antenna arrays and the use of analog and digital signal processing and noise shaping techniques to improve the dynamic range of RF transmit/receive systems. Techniques to control amplitude and phase jitter throughout RF systems are also relevant. Proposal submissions are ongoing, but proposals are preferred by April 1 each year.
Requirements: Grants are awarded to academic, industrial, and nonprofit organizations, with strong emphasis on university research.
Restrictions: New grants or contracts are typically started near the beginning of the fiscal year (October 1). New proposals for the next fiscal year should be submitted no later than April 1, unless otherwise directed by the BAA. Competitive renewals should be submitted no later than six months prior to the award termination date after contacting the Program Officer. Conference proposals should be submitted at least four months prior to the conference.
Geographic Focus: All States
Date(s) Application is Due: Apr 1
Contact: Dr. Stephen Pappert, Program Officer; (703) 696-5268 or (619) 553-5704; stephen.pappert@navy.mil
Internet: http://www.onr.navy.mil/Science-Technology/Departments/Code-31/All-Programs/312-Electronics-Sensors/RF-Signal-Circuits-Architecture.aspx
Sponsor: U.S. Department of Defense
One Liberty Center, 875 N. Randolph Street, Suite 1425
Arlington, VA 22203-1995

USDD Multidisciplinary Research Program of the University Researc Initiative Grant 4756

The MURI program supports basic research in science and engineering at U.S. institutions of higher education (hereafter referred to as "universities") that is of potential interest to DoD. The program is focused on multidisciplinary research efforts where more than one traditional discipline interact to provide rapid advances in scientific areas of interest to the DoD. As defined by the DoD, basic research is systematic study directed toward greater knowledge or understanding of the fundamental aspects of phenomena and of observable facts without specific applications towards processes or products in mind. It includes all scientific study and experimentation directed toward increasing fundamental knowledge and understanding in those fields of the physical, engineering, environmental, and life sciences related to long-term national security needs. It is farsighted high payoff research that provides the basis for technological progress. The DoD's basic research program invests broadly in many specific fields to ensure that it has early cognizance of new scientific knowledge.
Requirements: Proposals from a team of university investigators are warranted when the necessary expertise in addressing the multiple facets of the topics may reside in different universities, or in different departments in the same university. By supporting multidisciplinary teams, the program is complementary to other DoD basic research programs that support university research through single-investigator awards. Proposals shall name one Principal Investigator (PI) as the responsible technical point of contact.
Geographic Focus: All States
Date(s) Application is Due: Dec 10
Contact: Dr. Bill Lukens; (703) 696-4668; william.lukens1@navy.mil
Internet: http://www07.grants.gov/search/search.do;jsessionid=8Fy2QFPFZ6dnVjThhgpY2M1P1w8z2MSLs39J9s3FDRPZQyJ1tn9Q!2071562142?oppId=196533&mode=VIEW

Sponsor: U.S. Department of Defense
875 North Randolph Street, Suite 325, Room 3112
Arlington, VA 22203-1768

USDD Nano-Bio Manufacturing Institute Grants **4757**

The objective of the NBMI is to accelerate risk assessment and the development of robust mitigation approaches for critical nano-bio materials and fabrication innovations by developing common technology and manufacturing platforms through public-private partnerships between federal laboratories, DoD acquisition centers, academia, U.S. manufacturing enterprises, and commercial industries. To achieve this objective, the institute is envisioned to create a community of large and small companies, academia, federal laboratories, and the states to accelerate innovation by investing in industrially-relevant nano-bio manufacturing technologies. Ultimately, a fully functioning NBMI will have both virtual and physical entities that will enable partners to bridge the gap between basic research and product transition by developing verified and validated open architecture processing-design-performance, forecasting, and risk assessment tools for integration of new nano-bio materials and process technologies. Additionally, it will provide shared assets to help companies' access cutting-edge capabilities and equipment, and create an unparalleled environment to educate and train students and workers in advanced manufacturing skills. A successful institute will serve as a technical center of excellence, providing the infrastructure to support manufacturing enterprises at a prototype level that are relevant and scalable; and ensuring that the key capabilities for DoD needs is available.
Requirements: This is a restricted solicitation limited to a non-profit 501(c) designated] organization.
Geographic Focus: All States
Date(s) Application is Due: Nov 13
Amount of Grant: Up to 4,400,000 USD
Contact: Randall Wallace; (937) 255-0816; randall.wallace@wpafb.af.mil
Internet: http://www07.grants.gov/search/search.do;jsessionid=8Fy2QFPFZ6dnVjThhgpY2M1P1w8z2MSLs39J9s3FDRPZQyJ1tn9Q!2071562142?oppId=203973&mode=VIEW
Sponsor: U.S. Department of Defense
1400 Defense Pentagon
Washington, DC 20301-1400

USDD Nanoscale Electronics Grants **4758**

The Office of Naval Research (ONR) Nanoscale Electronics Program objective is to study the electronic, optical and magnetic properties of, and potential device functionalities in, nanometer scale materials and structures. In order to accomplish this objective, the program will address science and technology challenges at multiple levels: material level-synthesis, characterization and understanding of novel nanomaterials with desired functionalities is a challenging task; device and circuit level-need to invent and demonstrate entirely new classes of devices and circuits that fully utilize the properties of nanomaterials; architecture level-conventional von Neumann architecture is not scalable; and fabrication and manufacturing-as we approach sub-10nm scale, current fabrication techniques have difficulty with resolution and/or throughput Potential proposers are encouraged to communicate with the program officer prior to white paper submission. White papers are preferred by April 30 of each year. Typically awards range from $50,000 to $400,000.
Requirements: Grants are awarded to academic, industrial, and nonprofit organizations, with strong emphasis on university research.
Restrictions: Competitive renewals should be submitted no later than six months prior to the award termination date after contacting the Program Officer.
Geographic Focus: All States
Date(s) Application is Due: Apr 30
Amount of Grant: 50,000-400,000 USD
Contact: Dr. Chagaan Baatar, Program Officer; (703) 696-0483 or (703) 588-2427; fax (703) 696-5940; chagaan.baatar@navy.mil
Internet: http://www.onr.navy.mil/sci_tech/31/312/nanoscale_electronics.asp
Sponsor: U.S. Department of Defense
One Liberty Center, 875 N. Randolph Street, Suite 1425
Arlington, VA 22203-1995

USDD National Consortium for Measurement and Signature Intelligence Research Grants **4759**

Measurement and Signature Intelligence (MASINT) is an intelligence discipline that employs a broad range of scientific developments to gather foreign intelligence. MASINT oversight and direction is provided by the National MASINT Office. The National Consortium for MASINT Research as the focal point for MASINT innovation provides cutting-edge MASINT research for the Intelligence Community. A formal consortium of laboratories, academia, and industry forms the core of the advanced MASINT research and development program and creates a dynamic forum for diverse research partners to the benefit of MASINT. Applicants are invited to present related work, on-going research activities and proposed future activities associated with the following areas: (1) remote assessment of missile performance characteristics such as location, thrust, throw weight, warhead accuracy, and defensive capabilities; (2) remote assessment and detection of weapons of mass destruction such as nuclear, biological, chemical and radiological weapons (excluding improvised explosive devices utilizing standard explosives such as dynamite, TNT, C4, and ANFO); and (3) remote assessment and detection of directed energy weapons (which include all lasers primarily designed as weapons as well as high-powered microwave and electromagnetic pulse weapons). Applicants are encouraged to first submit white papers for evaluation. Feedback will be provided. White papers are due February 1. Although proposals will be accepted through the published deadline, to ensure timely consideration they must be received by April 4. Proposals submitted after April 4 will be reviewed and considered if additional funds become available.
Requirements: All responsible sources from academia (colleges and universities) are eligible, including academia from the British Commonwealth nations (Canada, Great Britain, Australia and New Zealand). Historically black colleges and universities and minority institutions are encouraged to apply.
Geographic Focus: All States, Australia, Canada, Great Britain, New Zealand
Date(s) Application is Due: Dec 31
Contact: Janet Norton; (808) 473-7669; janet.norton@navy.mil
Internet: http://www.grants.gov/search/search.do;jsessionid=1GhLPGQSVtvhVkKQYJgRnyFGWgBT8SQvFXjYv31mW6Ky7klFQ1Vx!-757993493?oppId=137613&mode=VIEW
Sponsor: U.S. Department of Defense
5450 Carlistle Park, P.O. Box 2050
Mechanicsburg, PA 17055-0791

USDD National Geospatial-Intelligency Agency Academic Research Grants **4760**

The National Geospatial-Intelligence Agency (NGA) welcomes all innovative ideas for path-breaking research that may advance the geospatial intelligence (GEOINT) mission. The NGA mission is to provide timely, relevant, and accurate GEOINT in support of national security objectives. GEOINT is the exploitation and analysis of imagery and geospatial information to describe, assess and visually depict physical features and geographically referenced activities on the earth. GEOINT consists of imagery, imagery intelligence, and geospatial information. NGA offers a variety of critical GEOINT products in support of United States national security objectives and natural disasters, including aeronautical, geodesy, hydrographic, imagery, targeting information and topographical. The NGA Academic Research Program (NARP) is focused on innovative, far-reaching basic and applied research in science, technology, engineering and mathematics that has the potential to advance the GEOINT mission. The objective of the NARP is to support innovative, high-payoff research that provides the basis for revolutionary progress in areas of science and technology affecting the needs and mission of NGA. This research also supports the National System for Geospatial Intelligence, which is the combination of technology, systems and organizations that gather, produce, distribute and consume geospatial data and information. The end result is aimed at advancing GEOINT capabilities by improving analytical methods, enhancing and expanding systems capabilities, and leveraging resources for common NSG goals. Amendments will be posted to Grants.gov (http://www.grants.gov/) and to the Intelligence Community ARC (https://arc.westfields.net). Applicants are encouraged to periodically check these websites for updates and amendments and to set up alerts. White papers may be submitted at any time, but no later than April 30, 2013. Proposals may be submitted at any time, but no later than September 30, 2013. Applicants are highly encouraged to submit white papers prior to submitting proposals in order for NGA to determine the level of interest in the proposed research. Awards will be made on a rolling basis subject to availability of funds.
Requirements: United States institutions of higher education are eligible. Proposals may be submitted in any of the following four categories for which the applicant is eligible: (1) university research initiatives; (2) historically black colleges and universities and minority institutions; (3) outstanding new science and technical innovative researcher; and (4) research collaboration forum.
Geographic Focus: All States
Date(s) Application is Due: Sep 13
Amount of Grant: 300,000-4,800,000 USD
Contact: Eric Rauch, Grants Officer; (571) 577-2462; Eric.Rauch@nga.mil
Internet: http://www.grants.gov/search/search.do;jsessionid=zmNmPD4RGXRJS8JcMcT2THyjRxWSX2LclYL005wssgPQdCq513Y7!-757993493?oppId=141713&mode=VIEW
Sponsor: U.S. Department of Defense
7500 GEOINT Drive
Springfield, VA 22150-7500

USDD Naval Research Facilities Grants **4761**

The Research Facilities program manages, under a charter party agreement with selected universities, the operation and maintenance of all oceanographic research vessels owned by ONR for Department of the Navy. As part of the management of the ONR fleet, the program actively supports fleet modernization to maintain the level of excellence that is required to support cutting edge oceanographic research. Other oceanographic platforms such as unmanned underwater vehicles, aerial unmanned vehicles may be used to collect field data through the Research Facilities Program. Submission of the Research Proposal should always be preceded by submission of a Preproposal and a discussion of that Preproposal with the assigned Program Officer. Applications for the upcoming fiscal year should be received by September 30.
Requirements: Grants are awarded to academic, industrial, and nonprofit organizations, with strong emphasis on university research. Investigators are encouraged to contact the staff prior to formally submitting a proposal.
Restrictions: New grants or contracts are typically started near the beginning of the fiscal year. New proposals for the next fiscal year should be submitted no later than September 30, unless otherwise directed by the BAA. Competitive renewals should be submitted no later than six months prior to the award termination date after contacting the Program Officer. Conference proposals should be submitted at least four months prior to the conference.
Geographic Focus: All States
Date(s) Application is Due: Sep 30
Contact: Robert Schnoor, Team Leader; tim.schnoor@navy.mil
Internet: http://www.onr.navy.mil/Science-Technology/Departments/Code-32/All-Programs/Ocean-Systems-321/Navy-Research-Facilities.aspx
Sponsor: U.S. Department of Defense
One Liberty Center, 875 N. Randolph Street, Suite 1425
Arlington, VA 22203-1995

USDD Navigation and Timekeeping Grants 4762

The Office of Naval Research (ONR) Navigation and Timekeeping Program seeks new and innovative navigation technologies that will provide more accurate, reliable, maintainable and affordable systems for Naval air, surface, subsurface and ground platforms and forces. Precision navigation and timekeeping are essential for many modem naval and maritime systems, and it is essential that navigation and timekeeping services be made available to platforms and weapons at the highest level of accuracy and with the highest possible confidence at reasonable cost. Lack of precise navigation and timekeeping technologies may jeopardize the success of military operations. In the past six years, ONR has been conducting a series of Science and Technology Projects In the following three technology areas; GPS Anti-Jam Technology, Precision Time and Time Transfer Technology, and Non-GPS Navigation Technology. Proposers are encouraged to contact the program officer to discuss their research interest prior to the submission of formal proposals. Applications for the upcoming fiscal year should be received by September 30.
Requirements: Grants are awarded to academic, industrial, and nonprofit organizations, with strong emphasis on university research.
Restrictions: New grants or contracts are typically started near the beginning of the fiscal year. New proposals for the next fiscal year should be submitted no later than September 30, unless otherwise directed by the BAA. Competitive renewals should be submitted no later than six months prior to the award termination date after contacting the Program Officer. Conference proposals should be submitted at least four months prior to the conference.
Geographic Focus: All States
Date(s) Application is Due: Sep 30
Contact: John C. Kim, Program Officer; (703) 696-4214; john.c.kim1@navy.mil
Internet: http://www.onr.navy.mil/Science-Technology/Departments/Code-31/All-Programs/312-Electronics-Sensors/Navigation-Timekeeping.aspx
Sponsor: U.S. Department of Defense
One Liberty Center, 875 N. Randolph Street, Suite 1425
Arlington, VA 22203-1995

USDD Non-Destructive Evaluation and Prognostics: Advanced Sensors and Technologies Grants 4763

Integrated system health management (SHM) for platforms systems and subcomponents is now an integral part of planning the design and construction of future assets (e.g., ships, aircraft, vehicles, buildings). Health management of critical assets depends on two equally important processes. The first is health diagnostics; the second is health and lifetime prediction, or prognostics. The diagnostic step certainly feeds vital information to the prognostic arm. A variety of sensors have been developed, and continue to be developed or enhanced, to detect and analyze vibrations, current surges, stress and strain, as well as to see and identify defects such as cracks and corrosion pits. An example: spectroscopic oil analysis can be a valuable tool to identify potential stressors to engines and engine lifetime. Sensor development focuses on efficiency improvements through miniaturization, sensitivity enhancement and energy harvesting to reduce power requirements. The prognostic part of the program is perhaps the more difficult at this time. Prognosis involves prediction, and prediction can only be reliably made if one knows the physics, chemistry and allied engineering of the underlying failure mechanisms. This Office of Naval Research program supports work in a variety of fields to address these issues. Submission of the Research Proposal should always be preceded by submission of a Preproposal and a discussion of that Preproposal with the assigned Program Officer. Applications for the upcoming fiscal year should be received by September 30.
Requirements: Grants are awarded to academic, industrial, and nonprofit organizations, with strong emphasis on university research. Investigators are encouraged to contact the staff prior to formally submitting a proposal.
Restrictions: New grants or contracts are typically started near the beginning of the fiscal year. New proposals for the next fiscal year should be submitted no later than September 30, unless otherwise directed. Competitive renewals should be submitted no later than six months prior to the award termination date after contacting the Program Officer. Conference proposals should be submitted at least four months prior to the conference.
Geographic Focus: All States
Date(s) Application is Due: Sep 30
Contact: Dr. Ignacio Perez; (703) 696-0688; fax (703) 696-0934; ignacio.perez1@navy.mil
Internet: http://www.onr.navy.mil/Science-Technology/Departments/Code-33/All-Programs/332-naval-materials/NDE-Sensors-Prognostics.aspx
Sponsor: U.S. Department of Defense
One Liberty Center, 875 N. Randolph Street, Suite 1425
Arlington, VA 22203-1995

USDD Non-Destructive Evaluation and Prognostics: Fatigue and Fracture Damage Protection Grants 4764

The Non-Destructive Evaluation and Prognostics: Fatigue and Fracture Damage Protection program focuses on fatigue and damage accumulation of structural metals and alloys, including nanostructured materials. Fatigue focuses on quantifying the damage parameters in terms of ?K (cyclic damage) and Kmax (static damage), which are key measures of damage in materials. Because damage frequently arises from a combination of mechanical and corrosion effects, these mechanisms are the primary areas of focus in the program, with the goal of a reliable fatigue damage life prediction capability. Because they are important metals to the Navy, there is a focus on aluminum, steel and titanium alloys. In nanostructured alloys, research involves bi-modal and tri-modal microstructures of 5xxx aluminum alloys and their cumulative effects, in order to characterize fracture and fatigue as a function of processing variables. Submission of the Research Proposal should always be preceded by submission of a Preproposal and a discussion of that Preproposal with the assigned Program Officer. Applications for the upcoming fiscal year should be received by September 30.
Requirements: Grants are awarded to academic, industrial, and nonprofit organizations, with strong emphasis on university research. Investigators are encouraged to contact the staff prior to formally submitting a proposal.
Restrictions: New grants or contracts are typically started near the beginning of the fiscal year. New proposals for the next fiscal year should be submitted no later than September 30, unless otherwise directed. Competitive renewals should be submitted no later than six months prior to the award termination date after contacting the Program Officer. Conference proposals should be submitted at least four months prior to the conference.
Geographic Focus: All States
Date(s) Application is Due: Sep 30
Contact: Dr. A.K. Vasudevan; (703) 696-8181; vasu.vasudevan@navy.mil
Internet: http://www.onr.navy.mil/Science-Technology/Departments/Code-33/All-Programs/332-naval-materials/NDE-Fatigue-Fracture.aspx
Sponsor: U.S. Department of Defense
One Liberty Center, 875 N. Randolph Street, Suite 1425
Arlington, VA 22203-1995

USDD NRL Broad Agency Announcement Grants 4765

The Naval Research Laboratory (NRL) is the Navy's coporate laboratory. NRL conducts basic and applied research for the Navy in a variety of scientific and technical disciplines. The basic research program is driven by perceptions about future requirements of the Navy. NRL conducts most of its research program at its own facilities but also funds some related research such as anticipated by this announcement. The Navy's operational effectiveness depends on its ability to keep pace with rapidly developing technologies. Proposals may range from theoretical studies to proof-of-concept to include fabrication and delivery of a prototype. This is a multiple year Broad Agency Announcement (BAA). Amendments may be published at Grants.gov (http://grants.gov/applicants/find_grant_opportunities.jsp) and/or at FedBizOpps (https://www.fbo.gov/). Applicants are encouraged to monitor these sites or to sign up for notification of changes. White papers may be submitted and will be reviewed within a reasonable period of time, generally 30 days. Instructions for formal proposals, as well as classified proposals, can be found on the website.
Requirements: Eligible applicants include any non-governmental entity, including commercial firms, institutions of higher education with degree-granting programs in science or engineering (universities), or by consortia led by such concern. NRL encourages participation by small businesses, veteran-owned small businesses, service-disabled veteran-owned small businesses, small disadvantaged businesses, HUBZone small businesses, woman-owned small businesses, and historically black colleges and universities and minority institutions.
Geographic Focus: All States
Contact: Mary Johnson; (202) 767-2021 or (301) 342-3744; nrlproposals@nrl.navy.mil
Internet: http://www07.grants.gov/search/search.do;jsessionid=GY1VPGFLJ5bRMmthrHlhcPGQlVWZ6kYqwnpZ12n2TCMlzNywLMnh!-757993493?oppId=100133&mode=VIEW
Sponsor: U.S. Department of Defense
4555 Overlook Avenue Southwest
Washington, DC 20375

USDD Ocean Acoustics Research Grants 4766

The Ocean Acoustics Research program supports research that addresses an understanding of the physics of the generation, propagation, and scattering of narrowband and broadband acoustic (and elastic) waves in the temporally and spatially varying ocean environment. Though research in acoustical signal processing is no longer a significant thrust of this program, support is granted to research in collaboration with the Undersea Signal Processing Program, which also manages development and demonstration efforts. The Ocean Acoustics program contains three primary thrusts and a miscellaneous thrust. The Shallow-Water Acoustics thrust seeks to understand the propagation and scattering of low frequency (10Hz to a few kHz) acoustic energy within the shallow-water ocean environment. Areas of research include investigations of the dominant shallow-water scattering mechanisms, the conversion of seafloor-incident acoustic energy into elastic body waves and interface waves, and acoustic propagation through linear and nonlinear internal waves. The High-Frequency Acoustics thrust seeks to understand the interaction of high frequency (few kHz to thousand kHz) sound with the ocean environment. Components of this thrust include the propagation of sound through an intervening turbulent or stochastic medium; scattering from rough surfaces, biologics, and bubbles; and penetration/propagation within the porous seafloor. The Long-Range Propagation thrust seeks to understand the behavior of sound as it propagates over very long ranges (several hundred kilometers to several thousand kilometers) in the ocean. The main area of interest is understanding the effect of ocean internal waves and mesoscale on transmitted broadband acoustic signals. Submission of the Research Proposal should always be preceded by submission of a Preproposal and a discussion of that Preproposal with the assigned Program Officer. Applications for the upcoming fiscal year should be received by September 30.
Requirements: Grants are awarded to academic, industrial, and nonprofit organizations, with strong emphasis on university research. Investigators are encouraged to contact the staff prior to formally submitting a proposal.
Restrictions: New grants or contracts are typically started near the beginning of the fiscal year. New proposals for the next fiscal year should be submitted no later than September 30, unless otherwise directed. Competitive renewals should be submitted no later than six months prior to the award termination date after contacting the Program Officer. Conference proposals should be submitted at least four months prior to the conference.
Date(s) Application is Due: Sep 30
Contact: Dr. Robert Headrick; (703)696-4135; bob.headrick@navy.mil
Internet: http://www.onr.navy.mil/Science-Technology/Departments/Code-32/All-Programs/Atmosphere-Research-322/Ocean-Acoustics.aspx
Sponsor: U.S. Department of Defense
One Liberty Center, 875 N. Randolph Street, Suite 1425
Arlington, VA 22203-1995

USDD Ocean Engineering and Marine Systems Grants 4767
The goals of the Ocean Engineering and Marine Systems program are to improve the knowledge base of fluid-structure interactions for engineering designs and to develop and demonstrate new technologies for expeditionary operations and special warfare. The investigators conduct multidisciplinary science and technology efforts in the behavior of ocean systems in general as well as specifically for the Marine Corps, Navy, Special Forces, and Navy Explosive Ordnance Disposal. The basic research program focuses on fluid forcing and structural response mechanisms and the coupled nonlinear interactions of fluids with generic structural components and unmanned platform systems. Technology development efforts are primarily in the areas of detection and imaging technologies applicable to underwater and buried objects, technologies for neutralization of explosive devices, underwater life support technologies, surface and subsurface transport systems for Navy Special Forces, mobile autonomous platform system technologies for very shallow water, surf zone and beach zone reconnaisance, and technologies leading to capability for rapid wide-area clearance of mines and obstacles from the surf zone. Submission of the Research Proposal should always be preceded by submission of a Preproposal and a discussion of that Preproposal with the assigned Program Officer. Applications for the upcoming fiscal year should be received by September 30.
Requirements: Grants are awarded to academic, industrial, and nonprofit organizations, with strong emphasis on university research. Investigators are encouraged to contact the staff prior to formally submitting a proposal.
Restrictions: New grants or contracts are typically started near the beginning of the fiscal year. New proposals for the next fiscal year should be submitted no later than September 30, unless otherwise directed. Competitive renewals should be submitted no later than six months prior to the award termination date after contacting the Program Officer. Conference proposals should be submitted at least four months prior to the conference.
Geographic Focus: All States
Date(s) Application is Due: Sep 30
Contact: Dr. Thomas Swean, Team Leader; (703) 696-4025 or (703) 696-5031; fax (703) 696-5940; tom.swean@navy.mil
Internet: http://www.onr.navy.mil/Science-Technology/Departments/Code-32/All-Programs/Ocean-Systems-321/Ocean-Engineering-Marine-Systems.aspx
Sponsor: U.S. Department of Defense
One Liberty Center, 875 N. Randolph Street, Suite 1425
Arlington, VA 22203-1995

USDD Office-Wide Broad Agency Announcement Microsystems Technology Office Grant 4768
The mission of the Defense Advanced Research Projects Agency's (DARPA) Microsystems Technology Office (MTO) is to exploit breakthroughs in materials, devices, circuits, and mathematics to develop beyond leading edge microsystems components with revolutionary performance and functionality to enable new platform capability for the Department of Defense. To execute this mission, MTO supports revolutionary research in electronics, photonics, MEMS, algorithms, and combined microsystems technology to deliver new capabilities to sense, communicate, energize, actuate, and process data and information for the war fighter. These Grants seek revolutionary research ideas for topics not being addressed by ongoing MTO programs or other published solicitations. These Grants are primarily, but not solely, intended for early stage research (studies) that may lead to larger, focused, MTO programs in the future. Studies are defined as single phase efforts of short duration (less than 12 months) costing less than $1,000,000. Whereas proposers are strongly encouraged to submit studies, multi-phase efforts will also be considered. multi-phase efforts are efforts with more than one phase that may span longer than 12 months and that may cost more than $1,000,000. Multi-phase efforts must also cover discernable research topic areas that lead to clearly defined, quantitative technical metrics. Multiple awards are anticipated. Funding awarded will depend on the quality of the proposals received and the availability of funds. Submission of abstracts in advance of a full proposal is strongly encouraged. This procedure is intended to minimize unnecessary effort in proposal preparation and review. Early submission of abstracts and proposals is strongly encouraged because selections may be made at any time during the period of solicitation. Abstracts and proposals may be submitted until the deadline.
Requirements: All responsible sources capable of satisfying the government's needs are eligible. Historically black colleges and universities, small businesses, small disadvantaged businesses and minority institutions are encouraged to submit proposals and join others in submitting proposals.
Geographic Focus: All States
Date(s) Application is Due: Sep 1
Contact: Dr. Thomas Lee; (703) 526-6630; DARPA-BAA-10-35@darpa.mil
Internet: http://www.grants.gov/search/search.do;jsessionid=qFTNP7SNp7qTyygkG1CvWMc011NgYX1NyBCQv3QJQvSpTCtbLZhb!733686662?oppId=52311&mode=VIEW
Sponsor: U.S. Department of Defense
3701 North Fairfax Drive
Arlington, VA 22203-6630

USDD Office of Naval Research (ONR) Command, Control, Computers and Communication (C4) Thrust Grants 4769
The Office of Naval Research (ONR) Command, Control, Computers and Communication (C4) Thrust seeks to improve C4 capabilities for naval war fighters with an emphasis on mall units, asymmetric and irregular warfare, information analysis and communication, and istributed operations. Its technology investment areas include network-centric warfare and interoperability, over-the-horizon communications gateways, and small-unit technologies. In the area of network-centric warfare and interoperability, ONR seeks to nearly ubiquitous communications and availability of information for naval warfighters, particularly in austere environments. This technology area of investment includes: mobility management; and information assurance application tolerance. In the area of over-the-horizon communications, ONR seeks to provide long-reach communications enabling technologies that allow lower-echelon warfighters to exploit the global network. This would include beyond-line-of-sight (BLoS) communications, which provide reach-back to higher echelons by all users, and networked RF distribution systems, which would provide access to the Global Information Grid services by appropriate echelons. In the area of small-unit technologies, ONR seeks to close technology gaps to allow small-unit warfighters to gain timely, accurate, and nearly complete situational awareness.
Requirements: Grants are awarded to academic, industrial, and nonprofit organizations, with strong emphasis on university research.
Restrictions: New grants or contracts are typically started near the beginning of the fiscal year (October 1). New proposals for the next fiscal year should be submitted no later than June 1, unless otherwise directed by the BAA. Competitive renewals should be submitted no later than six months prior to the award termination date after contacting the Program Officer. Conference proposals should be submitted at least four months prior to the conference.
Geographic Focus: All States
Date(s) Application is Due: Jun 1
Contact: John Moniz; (703) 696-2492 or (703)696-2611; john.moniz@navy.mil
Internet: http://www.onr.navy.mil/Science-Technology/Departments/Code-30/All-Programs/C4.aspx
Sponsor: U.S. Department of Defense
One Liberty Center, 875 N. Randolph Street, Suite 1425
Arlington, VA 22203-1995

USDD Organic Photovoltaics Grants 4770
The Office of Naval Research Organic Photovoltaics program's purpose is to develop low-cost, light-weight, rugged thin-film organic photovoltaics with overall power conversion efficiencies of 15 percent or greater. This technology would enable the in-field implementation of flexible, disposable photovoltaic devices such as sensors, solar tent canopies and battery rechargers – greatly reducing the logistical burden of maintaining power generation capabilities in extensible positions. Current focus areas include the development of advanced materials for separate functionalities (hole/electron transport, sensitizing), investigating the effects of interfaces and constituent morphology on cell performance, developing tools to optimize device efficiency and develop reproducible material processing capabilities for affordability. Applications include: man-portable sunlight-powered devices; disposable sensors; and battery rechargers.
Requirements: Grants are awarded to academic, industrial, and nonprofit organizations, with strong emphasis on university research. Investigators are encouraged to contact the staff prior to formally submitting a proposal.
Restrictions: New grants or contracts are typically started near the beginning of the fiscal year. New proposals for the next fiscal year should be submitted no later than September 30, unless otherwise directed. Competitive renewals should be submitted no later than six months prior to the award termination date after contacting the Program Officer. Conference proposals should be submitted at least four months prior to the conference.
Geographic Focus: All States
Date(s) Application is Due: Sep 30
Contact: Dr. Paul Armistead; (703) 696-4315; paul.armistead@navy.mil
Internet: http://www.onr.navy.mil/Science-Technology/Departments/Code-33/All-Programs/332-naval-materials/Organic-Photovoltaics.aspx
Sponsor: U.S. Department of Defense
One Liberty Center, 875 N. Randolph Street, Suite 1425
Arlington, VA 22203-1995

USDD Physical Oceanography Grants 4771
The Physical Oceanography program supports process-oriented and hypothesis-driven science and technology in the area of physical oceanography and in the applications of oceanic research tools and techniques for fleet usage. In response to post-Cold War naval strategy and tactics, increased emphasis is given to the littoral, defined here as the oceanographic region encompassing the continental shelf and slope and the adjacent deep water. The approach is principally on field observations with theory and modeling expected to be closely integrated. The following thrusts are supported: air-sea interaction: impact on mixed layer dynamics for improved parameterizations in oceanic and atmospheric models; internal waves/turbulence: development of a global littoral model and a low level of open ocean work; marginal seas/straits: identify key physics, develop archetypes, generalize results; shelf/slope dynamics: shelf/slope coupling, parameterizations for improved slope/shelf modeling; boundary current/slope interaction: quantify key physics for enhancement of improved regional modeling; fleet meteorological and oceanographic support: develop tools, techniques, and observations of direct impact to activities at sea. An overarching objective of the program is to foster transition of research products such as numerical and theoretical models, analysis algorithms, in situ data, sea-going instrumentation, and platforms into operational naval systems. In addition, support is also available for critical evaluation of the impact of the environment on fleet exercises that test and evaluate operational naval systems. Regular planning letters are dure by June 30, and applications for the upcoming fiscal year should be received by September 30.
Requirements: Grants are awarded to academic, industrial, and nonprofit organizations, with strong emphasis on university research.
Restrictions: New grants or contracts are typically started near the beginning of the fiscal year. New proposals for the next fiscal year should be submitted no later than September 30, unless otherwise directed by the BAA. Competitive renewals should be submitted no later than six months prior to the award termination date after contacting the Program Officer. Conference proposals should be submitted at least four months prior to the conference.
Geographic Focus: All States

Date(s) Application is Due: Sep 30
Contact: Dr. Terri Paluszkiewicz; (703) 696-6680; terri.paluszkiewicz@navy.mil
Internet: http://www.onr.navy.mil/Science-Technology/Departments/Code-32/All-Programs/Atmosphere-Research-322/Physical-Oceanography.aspx
Sponsor: U.S. Department of Defense
One Liberty Center, 875 N. Randolph Street, Suite 1425
Arlington, VA 22203-1995

USDD Polymeric Organic Materials: Conducting/Semiconducting Polymers Grants 4772

The Office of Naval Research Functional Polymeric/Organic Materials program supported the research that led to the discovery of high-electrical conductivity in doped conjugated polymers. In the past, the program supported research towards improving the stability of conducting polymers and developing conducting polymers for several anti-static applications. As semiconducting properties emerged, the research was focused on organic light-emitting diodes and non-linear optical applications. At this time, the primary focus is on photovoltaic devices and the design, synthesis, assembly, fabrication and characterization of the active materials in these devices. Research is primarily conducted on organic and hybrid thin-film approaches including bulk heterojunction and planar device architectures. Developing low-cost, potentially disposable, light-weight flexible organic and/or polymeric thin-film photovoltaics is critical for military use in distributed operations. Possible applications include the ability to recharge batteries in-field as well as helping to power sensors and unmanned undersea vehicles. Closely related to this program is work on thin-film dielectric capacitors for energy storage. Applications include: flexible organic and/or polymeric thin-film photovoltaics; nonlinear optical films; data processing; light-emitting diodes; and piezoelectric and electrostrictive materials for acoustic sensing or transduction. Submission of the Research Proposal should always be preceded by submission of a Preproposal and a discussion of that Preproposal with the assigned Program Officer. Applications for the upcoming fiscal year should be received by September 30.
Requirements: Grants are awarded to academic, industrial, and nonprofit organizations, with strong emphasis on university research. Investigators are encouraged to contact the staff prior to formally submitting a proposal.
Restrictions: New grants or contracts are typically started near the beginning of the fiscal year. New proposals for the next fiscal year should be submitted no later than September 30, unless otherwise directed. Competitive renewals should be submitted no later than six months prior to the award termination date after contacting the Program Officer. Conference proposals should be submitted at least four months prior to the conference.
Geographic Focus: All States
Date(s) Application is Due: Sep 30
Contact: Dr. Paul Armistead; (703) 696-4315; paul.armistead@navy.mil
Internet: http://www.onr.navy.mil/Science-Technology/Departments/Code-33/All-Programs/332-naval-materials/Functional-Polymeric-Organic-Materials.aspx
Sponsor: U.S. Department of Defense
One Liberty Center, 875 N. Randolph Street, Suite 1425
Arlington, VA 22203-1995

USDD Polymer Matrix Composites Grants 4773

The Office of Naval Research Polymer Matrix Composites program focuses on advancing understanding to exploit research on carbon nanotube fibers. These fibers have been found to possess remarkable tensile strength compared to ordinary carbon and other fibers. Fiber re-enforcement of polymer matrices has been used for decades to increase the strength of polymeric materials. This program supports efforts to increase the yield and decrease the cost of producing sufficiently large quantities of carbon nanotube fibers to build and test polymer matrix components. In addition, research is encouraged to find optimum blends of polymeric materials to use with the fibers. The concomitant testing of materials is supported. Submission of the Research Proposal should always be preceded by submission of a Preproposal and a discussion of that Preproposal with the assigned Program Officer. Applications for the upcoming fiscal year should be received by September 30.
Requirements: Grants are awarded to academic, industrial, and nonprofit organizations, with strong emphasis on university research. Investigators are encouraged to contact the staff prior to formally submitting a proposal.
Restrictions: New grants or contracts are typically started near the beginning of the fiscal year. New proposals for the next fiscal year should be submitted no later than September 30, unless otherwise directed. Competitive renewals should be submitted no later than six months prior to the award termination date after contacting the Program Officer. Conference proposals should be submitted at least four months prior to the conference.
Geographic Focus: All States
Date(s) Application is Due: Sep 30
Contact: Dr. Ignacio Perez; (703) 696-0688; ignacio.perez1@navy.mil
Internet: http://www.onr.navy.mil/Science-Technology/Departments/Code-33/All-Programs/332-naval-materials/Polymer-Matrix-Composites.aspx
Sponsor: U.S. Department of Defense
One Liberty Center, 875 N. Randolph Street, Suite 1425
Arlington, VA 22203-1995

USDD Postdoctoral Fellowship Awards in Ocean Acoustics 4774

USDD Postdoctoral Fellowship Awards in Ocean Acoustics will be given to outstanding doctoral degree holders who wish to continue research in ocean acoustics or the related disciplines of undersea signal processing, marine structural acoustics and transducer materials science. Applicants must identify and provide a letter of recommendation from an ocean scientist or acoustician at an academic institution or naval laboratory with whom they propose to work. Special consideration will be given to applications that include collaboration with a scientist at a naval research laboratory or naval warfare center division. These postdoctoral awards are 24 months in duration and cover salary and benefits. A detailed planning letter is required. ONR expects to grant up to three such awards per year.
Geographic Focus: All States
Date(s) Application is Due: Aug 1
Contact: Dr. Ellen Livingston, Program Officer; (703) 696-4203 or (703) 696-5031; fax (703) 696-5940; ellen.livingston@navy.mil
Internet: http://www.onr.navy.mil/Science-Technology/Departments/Code-32/All-Programs/Atmosphere-Research-322/Ocean-Acoustics/Awards-Ocean-Acoustics.aspx
Sponsor: U.S. Department of Defense
One Liberty Center, 875 N. Randolph Street, Suite 1425
Arlington, VA 22203-1995

USDD Power and Thermal Management Technology Development Grants 4775

The objective of the Power and Thermal Management Technology Development Grants is to obtain cutting edge technology from industry and academia that the Air Force Research Laboratory, Propulsion Directorate, Wright Research Site, has not previously considered for funding. Contracts will be awarded that provide a broad range of highly unique evolutionary and revolutionary technology advances in three application areas: integrated power and thermal subsystems; special purpose power and thermal; and battlespace fuels. White papers may be submitted at any time. Review is anticipated to take approximately 45 working days. Applicants whose white papers meet Air Force needs will be asked to submit a technical and cost proposal within 30 working days. Instructions and the format for white paper and proposal submission can be found on the website. Awards will normally be made for up to three years. This is an initial Broad Agency Announcement. Please see the website for updates.
Requirements: This is an unrestricted solicitation. Small businesses are encouraged to propose on all or any part of this solicitation.
Geographic Focus: All States
Amount of Grant: 200,000-3,000,000 USD
Contact: Lori Shirdon; (937) 255-8077; lori.shirdon@wpafb.af.mil
Sarah E.P. Chaffee, Contract Specialist; (937) 255-5499; Sarah.Chaffee@wrafb.af.mil
Internet: http://www07.grants.gov/search/search.do;jsessionid=91nSQFGd2zP2KG6lvQ_mQ3ftJX0xlTV251SLm6RJvgtpJHyHlJsTy!2071562142?oppId=134153&mode=VIEW
Sponsor: U.S. Department of Defense
1400 Defense Pentagon
Washington, DC 20301-1400

USDD Power Efficiency Revolution for Embedded Computing Technologies Grants 4776

The Defense Advanced Research Projects Agency (DARPA) is soliciting proposals for innovative research and development in the area of embedded power efficient computing. The goal of the Power Efficiency Revolution for Embedded Computing Technologies (PERFECT) Grants program is to provide the technologies and techniques to overcome the power efficiency barriers that currently constrain embedded computing systems capabilities and limit the potential of future embedded systems. This goal will be achieved by taking a revolutionary approach to processing power efficiency. This approach includes near threshold voltage operation and massive heterogeneous processing concurrency, combined with techniques to effectively utilize the resulting concurrency and tolerate the resulting increased rate of soft errors. Multiple awards are anticipated. Funding awarded will depend on the quality of the proposals received and the availability of funds. Submission of abstracts in advance of a full proposal is strongly encouraged. This procedure is intended to minimize unnecessary effort in proposal preparation and review. Early submission of abstracts is encouraged. Abstracts will be reviewed in the order received. Detailed instructions for submitting abstracts and proposals are provided on the website. Abstracts are due on or before 4:00 pm eastern time February 23, and proposals are due on or before 4:00 pm eastern time April 16.
Requirements: All responsible sources capable of satisfying the government's needs are eligible. Historically black colleges and universities, small businesses, small disadvantaged businesses and minority institutions are encouraged to submit proposals and join others in submitting proposals.
Geographic Focus: All States
Contact: Dr. Charlie Holland; (703) 526-6630; DARPA-BAA-12-24@darpa.mil
Internet: http://www.grants.gov/search/search.do;jsessionid=qFTNP7SNp7qTyygkG1CvWMc011NgYX1NyBCQv3QlQvSpTCtbLZhb!733686662?oppId=142433&mode=VIEW
Sponsor: U.S. Department of Defense
3701 North Fairfax Drive
Arlington, VA 22203-6630

USDD President's Emergency Plan for AIDS Relief Military Specific HIV Seroprevalence and Behavioral Epidemiology Risk Suvey Grants 4777

The Department of Defense HIV/AIDS Prevention Program (DHAPP) has successfully engaged over 80 countries in efforts to combat HIV/AIDS among their respective military services. DHAPP is a partner collaborating with the United States State Department, Health and Human Services, United States Agency for International Development, and Centers for Disease Control and Prevention, in the President's Emergency Plan for AIDS Relief (PEPFAR). This Broad Agency Announcement (BAA) is to solicit existing partners and establish new partners to assist select military troops with performing an independent HIV Seroprevalence and Behavioral Epidemiology Risk Survey. Proposals should focus on rapidly extending HIV/AIDS services. Applicants are encouraged to target specific needs with a practical business plan, using small grass-roots organizations to provide community-based services as a way to enhance organic capabilities and sustainability. Concept papers are required and full proposals are by invitation only. Concept papers may be submitted at any time. Early submission is strongly encouraged due to the possible lack

of funding for the later evaluation periods. The period of performance is generally one year, with up to two additional phases subject to availability of funding. Applicants are encouraged to propose additional phases if appropriate. This is a multiple year BAA which may be revised as needed. Based on the availability of funds, up to three evaluation cycles will be conducted. Changes will be posted at Grants.gov (http://grants.gov/applicants/find_grant_opportunities.jsp). Applicants are encouraged to sign up on Grants.gov for "send me change notification emails" by following the link on the synopsis page for the program announcement/funding opportunity.
Requirements: All responsible sources from academia, industry, and non-governmental organizations are eligible. Applicants must demonstrate the active support of the in-country military and the USDD representative in the corresponding United States Embassy in the planning and execution of proposals.
Restrictions: Countries covered by this BAA are: Democratic Republic of Congo, Liberia, Malawi, Mali, Sierra Leone, South Africa, and Vietnam.
Geographic Focus: All States
Date(s) Application is Due: Sep 30
Contact: Latrice Rubenstein; (619) 556-5276; Latrice.Rubenstein@navy.mil
Internet: http://www.grants.gov/search/search.do;jsessionid=1GhLPGQSVtvhVkKQYJgRnyFGWgBT8SQvFXjYv31mW6Ky7klFQ1Vx!-757993493?oppId=65295&mode=VIEW
Sponsor: U.S. Department of Defense
5450 Carlistle Park, P.O. Box 2050
Mechanicsburg, PA 17055-0791

USDD Propulsion Materials Grants 4778

Reliable high-temperature propulsion materials are crucial to improved engine efficiency, reduced fuel costs and decreased maintenance/total life cycle costs. The useful life of a material in any application depends on diverse factors such as the marine environment, temperature and cyclic activities, and mechanical stress. The materials also need to resist oxidation, corrosion or alternating cycles of oxidation and corrosion. The research in this Office of Naval Research (ONR) topic area involves, in part, the thermodynamics and kinetics of materials interactions and materials stability under marine operating environments and temperatures. In order to eventually provide optimal materials for application, research is needed to develop models to assist the development of new materials and to help predict the life of both existing and new materials. Basic research in this area should help ONR understand how to reduce and manage materials instabilities and degradation. The research to investigate the mechanisms that lead to materials degradation should also explore how the mechanisms fundamentally relate to or depend on mechanics, interdiffusion, coatings and materials chemistry, as well as the environmental parameters of temperature, pressure and humidity. Computational approaches to creating new materials and material processes or improving the performance of current materials are encouraged. Current materials include Ni-base single-crystal superalloys, ceramic matrix composites, ceramics, Mo-based superalloys, thermal barrier coatings, environmental coatings and other coatings resistant to oxidation and corrosion environments. Submission of the Research Proposal should always be preceded by submission of a Preproposal and a discussion of that Preproposal with the assigned Program Officer. Applications for the upcoming fiscal year should be received by September 30.
Requirements: Grants are awarded to academic, industrial, and nonprofit organizations, with strong emphasis on university research. Investigators are encouraged to contact the staff prior to formally submitting a proposal.
Restrictions: New grants or contracts are typically started near the beginning of the fiscal year. New proposals for the next fiscal year should be submitted no later than September 30, unless otherwise directed. Competitive renewals should be submitted no later than six months prior to the award termination date after contacting the Program Officer. Conference proposals should be submitted at least four months prior to the conference.
Geographic Focus: All States
Date(s) Application is Due: Sep 30
Contact: Dr. David Shifler; (703) 696-0285; fax (703) 696-0934; david.shifler@navy.mil
Internet: http://www.onr.navy.mil/Science-Technology/Departments/Code-33/All-Programs/332-naval-materials/Structural-Propulsion-Materials.aspx
Sponsor: U.S. Department of Defense
One Liberty Center, 875 N. Randolph Street, Suite 1425
Arlington, VA 22203-1995

USDD Radar and Signal Processing Grants 4779

RADAR, an acronym for radio detection and ranging, is a method of locating distant targets by sending bursts of electromagnetic radiation and measuring their refl ections. It can be used for determining both the location and speed of an object. RADAR senses the environment by detecting and characterizing electromagnetic radiation that is reflected from objects within the environment. The Office of Naval Research (ONR) RADAR and Signal Processing Program has the objective of developing an affordable, fully automated network of time-coordinated mono-static, bi-static and passive surveillance sensors providing real-time tracking, identification, and engagement information with persistent wide area awareness. Electromagnetic propagation and scattering with emphasis on antennas; target, clutter and background modeling; and sensor performance prediction. Applications for the upcoming fiscal year must be received by September 30.
Requirements: Grants are awarded to academic, industrial, and nonprofit organizations, with strong emphasis on university research.
Restrictions: New grants or contracts are typically started near the beginning of the fiscal year. New proposals for the next fiscal year should be submitted no later than September 30, unless otherwise directed by the BAA. Competitive renewals should be submitted no later than six months prior to the award termination date after contacting the Program Officer. Conference proposals should be submitted at least four months prior to the conference.
Geographic Focus: All States
Date(s) Application is Due: Sep 30
Contact: Michael Pollock, Division Director; (703) 696-4202 or (703)696-4217; fax (703) 696-3098; michael.a.pollock@navy.mil
Internet: http://www.onr.navy.mil/Science-Technology/Departments/Code-31/All-Programs/312-Electronics-Sensors/Radar-Systems-Signal-Processing.aspx
Sponsor: U.S. Department of Defense
One Liberty Center, 875 N. Randolph Street, Suite 1425
Arlington, VA 22203-1995

USDD Radio Frequency-Field Programmable Gate Arrays Grants 4780

The Defense Advanced Research Projects Agency (DARPA) is soliciting innovative research and development proposals in the area of Radio Frequency Field Programmable Gate Arrays (RF-FPGAs), a thrust within the DARPA Microsystems Technology Office's Adaptive RF Technology program. The goal of RF-FPGA is to enable a common hardware architecture that facilitates reutilization of the same set of RF front-end components across disparate applications through programmability of the transceiver chain. RF-FPGAs will impact the areas of communications, electronic warfare, and signal intelligence by eliminating redundant and costly hardware development required for the adoption or recognition of a new wireless function or waveform. Multiple awards are anticipated. The funding amount of awards will depend on the quality of the proposals received and the availability of funds. Applicants are strongly encouraged to submit an abstract in advance of a full proposal. This procedure is intended to minimize unnecessary effort in proposal preparation and review. Instructions for submission of abstracts and proposals can be found at the website. Abstracts are due on or before 4:00 pm eastern time December 7. Proposal are due on or before 4:00 pm eastern time February 16.
Requirements: All responsible sources capable of satisfying the government's needs are eligible. Historically black colleges and universities, small businesses, small disadvantaged businesses and minority institutions are encouraged to submit proposals and join others in submitting proposals.
Geographic Focus: All States
Contact: Dr. William J. Chappel, Program Manager; (703) 526-6630 or (571) 218-4507; William.Chappel@darpa.mil or DARPA-BAA-12-13@darpa.mil
Internet: http://www.grants.gov/search/search.do;jsessionid=qFTNP7SNp7qTyygkG1CvWMc011NgYX1NyBCQv3QlQvSpTCtbLZhb!733686662?oppId=130353&mode=VIEW
Sponsor: U.S. Department of Defense
3701 North Fairfax Drive
Arlington, VA 22203-6630

USDD Research and Analytical Support for the 711th Human Performance Wing's Human Effectiveness Directorate Grants 4781

The goal of the UDDD's Air Force Research Laboratory Grants is to obtain technical and analytical support for the conduct of research at the 711th Human Performance Wing's (HPW) Human Effectiveness Directorate. This includes all locations (Wright-Patterson Air Force Base, Ohio; Brooks City-Base, Texas; and Mesa Research Site, Arizona). Support may take the following forms: provide appropriately training and experienced undergraduate and graduate students, post-doctoral fellows and faculty for direct work on 711 HPW's research, analyses, and studies; provide faculty to supervise students to perform advanced research analysis and studies; match faculty expertise to designated research and study needs with 711 HPW; and provide a pool of subjects for research studies. This is a multi-year offering and applicants may submit white papers at any time. Applicants whose white paper are considered to meet the needs of the Air Force will be invited to submit a full proposal. Early white paper submission is encouraged. Individual Grants will normally be for 12 months or less.
Requirements: This is an unrestricted solicitation. Small businesses are encouraged to propose on all or any part of this solicitation.
Geographic Focus: All States
Date(s) Application is Due: Sep 30
Contact: Pamela J. Covey; (937) 255-8958; pamela.covey@wpafb.af.mil
Linda Lange, Technical Specialist; (937) 255-0406; Helen.Williams@wpafb.af.mil
Internet: http://www.grants.gov/search-grants.html?agencies%3DDOD|Department%20of%20Defense
Sponsor: U.S. Department of Defense
1400 Defense Pentagon
Washington, DC 20301-1400

USDD Research and Studies for the Office of Net Assessment Grants 4782

The Department of Defense's Office of Net Assessment conducts and sponsors analyses that compare the standing, trends, and future prospects of United States and foreign military capability and military potential. Priority is given to assessing aspects of the security environment and parts of the world that are likely to change in the future and that may present threats or opportunities to the United States. Assessments may focus on specific theaters, regions, functions, mission areas, major weapons categories, doctrine, etc., as well as on demographic, economic, and political developments that may affect the power and strategies of nations. Proposals will be considered that undertake comprehensive approaches to one or more of the following areas of interest or that investigate one or more particular aspect(s) or case(s) that belong to the broader areas listed. There are five specific areas of interest and one area for additional research topics. They include: (1) future naval warfare; (2) proliferated world; (3) space, (4) sustaining current areas of United States advantage; (5) history (military); and (6) additional research topics (other germane topics may be of interest). The content and format for applications can be found on the website. Proposals should be submitted as early as possible and are valid for a period of one year after receipt. Awards will be made for a period of five years on a bi-monthly basis. The period of performance typically ranges from one to five years.

Requirements: All profit and non-profit organizations as well as institutions of higher education are eligible.
Geographic Focus: All States
Date(s) Application is Due: Jun 15
Contact: Sherry Goodman; 703-588-1138; Sherry.Goodman.ctr@whs.mil
Internet: http://www07.grants.gov/search/search.do;jsessionid=d9p2Py1N5cBy6yvd10QnSrTq1ZLMMBCfXw6ShFKyzlmJ2LHnQVxt!-992301837?oppId=46090&mode=VIEW
Sponsor: U.S. Department of Defense
1400 Defense Pentagon
Washington, DC 20301-1400

USDD Research Initiatives at the Naval Postgraduate School Grants 4783

The Naval Postgraduate School (NPS) is interested in receiving proposals for research initiatives which offer potential for advancement and improvement in the NPS core mission of graduate education and research. This includes research initiatives across a spectrum of science and engineering, business and policy, operational and informational sciences, and interdisciplinary disciplines that support the NPS' graduate education and research mission. Work funded may include basic research, applied research and some advanced research. Applicants are strongly encouraged to contact an NPS point of contact whose program best matches their field of interest. See the website to locate the appropriate point of contact. Applicants are encouraged to submit a white paper prior to submitting a full proposal. The funding amount and award period will vary depending on the research area and the technical approach pursued by the applicant.
Requirements: All responsible sources from academia and industry are eligible. Historically black colleges and universities and minority institutions are encouraged to apply.
Geographic Focus: All States
Date(s) Application is Due: Mar 31
Contact: Janet Norton; (808) 473-7669; janet.norton@navy.mil
Internet: http://www.grants.gov/search/search.do;jsessionid=1GhLPGQSVtvhVkKQYJgRnyFGWgBT8SQvFXjYv31mW6Ky7klFQ1Vx!-757993493?oppId=81013&mode=VIEW
Sponsor: U.S. Department of Defense
5450 Carlistle Park, P.O. Box 2050
Mechanicsburg, PA 17055-0791

USDD Research Interest of the Air Force Office of Scientific Research Grants 4784

The Air Force Office of Scientific Research (AFOSR) plans, coordinates, and executes the Air Force Research Laboratory's (AFRL) basic research program in response to technical guidance from AFRL and requirements of the Air Force; fosters, supports, and conducts research within Air Force, university, and industry laboratories; and ensures transition of research results to support U.S.AF needs. The focus of AFOSR is on research areas that offer significant and comprehensive benefits to our national warfighting and peacekeeping capabilities. These areas are organized and managed in three scientific directorates. First is Aerospace, Chemical and Material Sciences (RSA), which focuses on meeting the basic research challenges related to future air and space power by leading the discovery and development of fundamental science and engineering across three integrated research focus areas of: aero-structure interaction and control; complex materials and structures; and energy, power and propulsion. Second is Physics and Electronics, which generates the fundamental knowledge needed to advance Air Force operational capabilities and is categorized into the following three broad areas of: complex electronics and fundamental quantum processes; plasma physics and high energy density nonequilibrium processes; and optics, electromagnetics, communication, and signal processing. Third is Mathematics, Information and Life Sciences, which supports a wide range of fundamental mathematics, information and computer science, biological sciences, and human behavioral/performance research to provide novel options that increase operational flexibility and performance of systems and environments of relevance to the Air Force. Research is primarily encompassed in the following four focus areas: information and complex networks research; decision-making research; dynamical systems, optimization, and control research; and natural materials and systems research. This announcement remains open until superseded. Proposals may be submitted at any time, and various contacts are listed in the proposal.
Geographic Focus: All States
Contact: Contracting Officer; afosr.baa@afosr.af.mil
Internet: http://www07.grants.gov/search/search.do;jsessionid=PFWqP5wPQ5Nv21lbxsnTTFmj1q6hm7L3TBt6kKJBk30JGMPvHcG8!-1993000306?oppId=88213&mode=VIEW
Sponsor: U.S. Department of Defense
875 North Randolph Street, Suite 325, Room 3112
Arlington, VA 22203-1768

USDD Research Interests of the U.S. Air Force Academy Grants 4785

The United States Air Force Academy (U.S.AFA) invests in an active research program for three main reasons. First and foremost, research significantly enhances the cadet learning experience. Research is done by, for and with cadets who work alongside fellow cadets and faculty mentors. Secondly, research programs provide opportunities for essential faculty development. Research broadens and deepens the experience base of the faculty and infuses current, relevant, state-of-the-art and cutting-edge applications and examples into the curriculum. Third, U.S.AFA strives to conduct research to enhance the ability of the Air Force to perform its mission. There are currently ongoing research projects spanning topics as diverse as super cooled cesium atoms, cyber security, spatial disorientation and homeland defense. This is a Broad Agency Announcement (BAA) by U.S.AFA to solicit proposals for basic and applied research from business and academia. This strategy is an acquisition tool for U.S.AFA with the flexibility to solicit proposals and make awards to develop technologies to meet present and future Air Force research needs as technology issues are identified. Proposals are invited for research in many broad areas which are described in detail in the BAA on the website. This BAA will remain open until superseded. White papers may be submitted at any time and are reviewed as received. Proposal call announcements (Calls) may be issued against this BAA. Calls will contain specific instruction including further technical details, application instructions and due dates.
Requirements: All responsible, potential applicants from academia and industry are eligible. Proposals are particularly encouraged from small businesses, historically black colleges and universities, minority institutions and minority researchers.
Restrictions: Questions about a technical research area should be directed to the technical program manager for each specific research topic listed in the BAA.
Geographic Focus: All States
Contact: Jessica Niebuhr; (719) 333-3907; Jessica.Niebuhr@us.af.mil
Internet: http://www07.grants.gov/search/search.do;jsessionid=8Fy2QFPFZ6dnVjThhgpY2M1P1w8z2MSLs39J9s3FDRPZQyJ1tn9Q!2071562142?oppId=49713&mode=VIEW
Sponsor: U.S. Department of Defense
4102 Pinion Drive
Colorado Springs, CO 80840

USDD RF Semiconductors, RF Solid State Amplifiers and Wide Bandgap Materials Grants 4786

The Office of Naval Research (ONR) RF Semiconductor Devices, RF Solid State Amplifiers and Wide Bandgap Materials Program's objective is to develop solid state (e.g., wide bandgap semiconductor) low noise, high speed, and power devices and circuits for use in surveillance, electronic warfare, multifunctional RF systems communications, and power applications. Emphasis is on device advancements that enable a significant improvement in broadband performance, high linearity, and/or high (e.g., 60%) power added efficiency. Power output of at least five times that of GaAs-based devices. Emerging applications for wide bandgap materials and devices in the areas of transmitter and receiver components, mixed signal and digital circuits are also included. The program investigates electronic properties of wide bandgap, III-V and silicon-based semiconductors and novel devices; focusing on novel high power microwave and switching devices for broadband radar, electronic warfare, communications, and multifunctional systems applications and for power control applications. Proposers are encouraged to contact the program officer to discuss their research interest prior to the submission of formal proposals.
Requirements: Grants are awarded to academic, industrial, and nonprofit organizations, with strong emphasis on university research.
Restrictions: New grants or contracts are typically started near the beginning of the fiscal year (October 1). New proposals for the next fiscal year should be submitted no later than September 30, unless otherwise directed by the BAA. Competitive renewals should be submitted no later than six months prior to the award termination date after contacting the Program Officer. Conference proposals should be submitted at least four months prior to the conference.
Geographic Focus: All States
Date(s) Application is Due: Sep 30
Contact: Dr. Paul Maki, Program Officer; (703) 696-3060 or (703) 588-2427; fax (703) 696-8496; paul.maki@navy.mil
Internet: http://www.onr.navy.mil/Science-Technology/Departments/Code-31/All-Programs/312-Electronics-Sensors/Semiconductor-Devices.aspx
Sponsor: U.S. Department of Defense
One Liberty Center, 875 N. Randolph Street, Suite 1425
Arlington, VA 22203-1995

USDD RF Superconducting Technologies Grants 4787

The Office of Naval Research (ONR) RF Superconducting Technologies Program's objective is to develop new and improved electronic devices and circuits using superconducting materials that are relevant to RF systems applications. Digital and mixed signal circuits clocking above 20 GHz are the focus. ADC and interference cancellation are specific topics of interest. Improvements in packaging and memory technologies are also sought. Basic research on the fundamental physics of phonons as they interact with light and of heat transport and cooling are of interest. Proposers are encouraged to contact the program officer to discuss their research interest prior to the submission of formal proposals. Basic research proposals are preferred by June 1 of each year, while applied research proposals should be submitted in approximately March in response to an annual specific broad agency announcement from the ONR electronics program.
Requirements: Grants are awarded to academic, industrial, and nonprofit organizations, with strong emphasis on university research.
Geographic Focus: All States
Date(s) Application is Due: Jun 1
Contact: Deborah Van Vechten; (703) 696-4219; deborah.vanvechten@navy.mil
Internet: http://www.onr.navy.mil/Science-Technology/Departments/Code-31/All-Programs/312-Electronics-Sensors/RF-Superconducting-Technologies.aspx
Sponsor: U.S. Department of Defense
One Liberty Center, 875 N. Randolph Street, Suite 1425
Arlington, VA 22203-1995

USDD Sabbatical Leave Fellowships 4788

The Sabbatical Leave program provides fellowship appointments for a minimum of one semester to a maximum of one year in length. The purpose of sabbatical leave is to provide an opportunity for faculty members to engage in scholarly, creative, professional, research, or other academic activities that will enhance the faculty member's further contributions to their institution. There are three levels of appointment: summer faculty fellow, senior summer faculty fellow and distinguished summer faculty fellow. Participants in the Sabbatical Leave program receive stipends ranging from $1,400 to $1,900 per week. Each fellow will be reimbursed for expenses incurred on an optional pre-program visit to the sponsoring laboratory and one round-trip encompassing travel to the sponsoring

laboratory at the beginning of the program and travel back to their home residence at the end of the program. Relocation assistance is provided to qualifying participants.
Requirements: Sabbatical leave appointments are contingent on the availability of sabbatical leave positions. Participants must conduct research on-site. Prospective applicants should contact the research facility at which they are interested in working to jointly develop a suitable research proposal. Also required is an estimate of base salary and percentage of salary provided by the home institution while on sabbatical leave.
Restrictions: Faculty members from Historically Black Colleges and Universities, Minority Institutions, American Indian Tribal Colleges and Universities, and Hispanic Serving Institutions (HBCU/MI/TCU/HIS), as designated by the U.S. Department of Education, are especially encouraged to apply.
Geographic Focus: All States
Amount of Grant: 14,000-19,000 USD
Contact: V. Nadine Daniel, Point of Contact; (202) 289-9806, ext. 126; daniel.vontilla.ctr@navy.mil or ndaniel@aghe.org
Internet: http://onr.asee.org/about_the_sabbatical_leave_program
Sponsor: U.S. Department of Defense
One Liberty Center, 875 N. Randolph Street, Suite 1425
Arlington, VA 22203-1995

USDD Science, Technology, Engineering and Mathematics (STEM) Grants 4789
The United States Army Engineer Research and Development Center (ERDC) seeks proposals to stimulate young pupils in the sciences, technology, engineering and mathematics (STEM). Depending upon the availability of appropriated funds, ERDC may: (1) make multiple awards; and (2) consider options exercisable for multi-year performance. Funding is limited and proposals are primarily sought in the not-to-exceed $30,000 range; however, larger awards may be considered when appropriate.
Requirements: All responsible sources from educational institutions and non-profit organizations are eligible. Historically black colleges and universities and minority institutions are encouraged to submit proposals and join others in submitting proposals. The target students attend primary, middle, and high schools in the following locations; Warren County, Mississippi; Champaign County, Illinois; Grafton County, New Hampshire; Orange County, Vermont; Windsor County, Vermont; Sullivan County, New Hampshire; Arlington County, Virginia; and Alexandria, Virginia. The target students may attend public or private schools.
Geographic Focus: All States
Date(s) Application is Due: May 15
Amount of Grant: 30,000 USD
Contact: Kimberly Dulaney, Contract Specialist; (601) 634-3954 or (601) 634-3188; kimberly.l.dulaney@usace.army.mil
Internet: http://www.grants.gov/search/search.do;jsessionid=3KlVPDgTC2SJq9ZMk9Kvn22nkSfKbwL3b2s0bWQRJBd1rxNjKVnF!-757993493?oppId=141513&mode=VIEW
Sponsor: U.S. Department of Defense
3909 Halls Ferry Road
Vicksburg, MS 39180-6199

USDD Science and Tecnology for Warfighter Training and Aiding Grants 4790
The 711th Human Performance Wing (HPW), Warfighter Readiness Research Division, conducts pioneering basic, applied, and advanced research and development in support of a variety of research and operational needs. The objectives of the Grants are to research, demonstrate, evaluate, and transition human performance methods and technology to enable warfighters to have the right skills, knowledge, experience and capabilities at the right time to make the right decisions. The objective will be met through the following three science and technology areas: cognitive modeling; immersive environments; and continuous learning. This is a multi-year offering and applicants may submit white papers at any time. Applicants whose white paper are considered to meet the needs of the Air Force will be invited to submit a full proposal. Early white paper submission is encouraged. Estimated funding over five years is $49 million. Three to five awards per year are anticipated.
Requirements: This is an unrestricted solicitation. Small businesses are encouraged to propose on all or any part of this solicitation.
Geographic Focus: All States
Date(s) Application is Due: Sep 30
Contact: Angela G. Lauofo; (937) 255-5091; Angela.Lauofo@wpafb.af.mil
Internet: http://www.grants.gov/search-grants.html?agencies%3DDOD|Department%20of%20Defense
Sponsor: U.S. Department of Defense
1400 Defense Pentagon
Washington, DC 20301-1400

USDD Sensor Innovative Research Grants 4791
The Electromagnetics Technology Division (RYH), Sensors Directorate (RY), Air Force Research Laboratory (AFRL), is looking to conduct research in the following areas: Antenna Technology, Opto-Electronic Technology, Electromagnetic Scattering, and Infrared Sensor Technology. The Air Force anticipates awarding up to 25 awards for this announcement. Potential applicants should see the full proposal for specific details concerning the description of efforts and primary objectives. Typical grant awards will range from $50,000 to $1,000,000.
Requirements: This is an unrestricted solicitation. Small businesses are encouraged to propose on all or any part of this solicitation.
Geographic Focus: All States
Amount of Grant: 50,000-1,000,000 USD
Contact: Brian Lowe, Contract Officer; (937) 255-3585; brian.lowe@wpafb.af.mil
Internet: http://www07.grants.gov/search/search.do;jsessionid=8Fy2QFPFZ6dnVjThhgpY2M1P1w8z2MSLs39J9s3FDRPZQyJ1tn9Q!2071562142?oppId=162173&mode=VIEW
Sponsor: U.S. Department of Defense
1400 Defense Pentagon
Washington, DC 20301-1400

USDD Ship Systems and Engineering Research Grants 4792
The division is focused on providing technologically superior warfighting capabilities at reduced total ownership costs for surface and subsurface platforms through investments in basic and applied research, and advanced technology development of programs in the following areas: Advanced Sea Platforms; Sea Platform Survivability; and Surface Ship Hydrodynamics. Research topic areas of interest that contribute to these focus areas include: electrical power generation systems and equipment; energy conversion, distribution and storage; electro-mechanics; thermoelectrics; phototonics; thermal energy management; stability and control of electrical networks; ship design and analysis; systems engineering; ship systems integration; hydromechanics; maneuvering and seakeeping; hydroacoustics; structural mechanics; solid mechanics; thermo-mechanics; structural acoustics; structural dynamics; computational mechanics; advanced structural systems; protection systems and design methodology; automation; control of dynamic systems; damage control and firefighting; control of acoustic and non-acoustic signatures; high-speed platforms; sense and respond logistics; and inter- and intra-ship material movement and tracking. Submission of the Research Proposal should always be preceded by submission of a Preproposal and a discussion of that Preproposal with the assigned Program Officer. Applications for the upcoming fiscal year should be received by September 30.
Requirements: Grants are awarded to academic, industrial, and nonprofit organizations, with strong emphasis on university research.
Restrictions: New grants or contracts are typically started near the beginning of the fiscal year. New proposals for the next fiscal year should be submitted no later than September 30, unless otherwise directed. Competitive renewals should be submitted no later than six months prior to the award termination date after contacting the Program Officer. Conference proposals should be submitted at least four months prior to the conference.
Geographic Focus: All States
Date(s) Application is Due: Sep 30
Contact: Dr. John Pazik, Division Director; (703) 696-4404; john.pazik@navy.mil
Internet: http://www.onr.navy.mil/Science-Technology/Departments/Code-33/All-Programs/331-ship-systems-engineering.aspx
Sponsor: U.S. Department of Defense
One Liberty Center, 875 N. Randolph Street, Suite 1425
Arlington, VA 22203-1995

USDD Signal Processing for Networked Sensing Grants 4793
A majority of performance improvements in sensors, networking and communication connectivity are expected to come from conceptualization of new systems and innovations in signal processing. The Office of Naval Research Signal Processing for Networked Sensing program's goal is to leverage developments in signal processing techniques to bring about orders of magnitude improvements in sensing, target resolution, small-target detectability, multi-target tracking and to address hard problems in sensor data fusion, track fusion and communications. The program addresses hard problems such as: secure error free data communication in networked sensors; processing tracks for highly maneuvering and dense target scenarios; tracking and identifying clustered, grouped, disintegrating targets; achieving a reliable conflict free track per target view in a multi-target scenario across networked sensors; and maintaining tracks and a consistent target picture in the presence of clutter, out-of-order measurements, electronic interference and miscellaneous disrupting challenges. The submission deadline is ongoing.
Requirements: Grants are awarded to academic, industrial, and nonprofit organizations, with strong emphasis on university research.
Geographic Focus: All States
Contact: Dr. Rabinder Madan; (703) 696-4217; rabinder@madan@navy.mil
Internet: http://www.onr.navy.mil/Science-Technology/Departments/Code-31/All-Programs/311-Mathematics-Computers-Research/Signal-Processing-Networked-Sensing.aspx
Sponsor: U.S. Department of Defense
One Liberty Center, 875 N. Randolph Street, Suite 1425
Arlington, VA 22203-1995

USDD Small University Grants 4794
The USDD's Air Force Research Laboratory Space Vehicles Directorate is offering Small University Grants in space related basic, applied, and advanced research projects. The scope of the research includes the entire spectrum of space that is applicable to the Air Force including all tangentially space-related research. Research includes, but is not limited to, fundamental research in: space system components and technologies; space environment and surveillance; and space system integration. Proposals may be submitted at any time. However, applicants must call the technology points of contact to make sure funding is available for the technology area of consideration before submitting a proposal. Each technology area has a specific contact and the names and telephone numbers can be found in the full announcement at the website. Proposals are reviewed and evaluated as they are received. Multiple Grants of various values are anticipated with a period of performance from one to two years. At minimum one Grant per fiscal year is planned. Anticipated annual funding in total is expected to be $790,000 to $990,000. It is anticipated that Grants will remain open for a period of five years. Any updates or possible cancellation will be issued as an amendment and can be found on the website.
Requirements: Only United States and United States territory educational institutions are eligible.
Geographic Focus: All States

Date(s) Application is Due: Aug 26
Amount of Grant: Up to 4,950,000 USD
Contact: LaDonna Nivens; (505) 853-2279; amanda.qualls@kirtland.af.mil
Internet: http://www.grants.gov/search-grants.html?agencies%3DDOD|Department%20of%20Defense
Sponsor: U.S. Department of Defense
1400 Defense Pentagon
Washington, DC 20301-1400

USDD Software and Computer Systems Grants **4795**

The Office of Naval Research Software and Computer Systems program focuses on developing the fundamental principles and models for the design and construction of complex software systems that meet required assurances for security, safety, reliability and performance. It builds upon foundations of mathematical logics and computational complexity but takes into account the uncertainties and constraints of the operational environment. The overall goal is to measurably improve the ability of complex software-intensive systems to meet the information-processing challenges of future naval systems. The program has several thrust areas that build around the theme of information as a computational resource to be managed, secured and shared in different contexts: Principles for Correctness and Security Properties; Trusted Network Computing; Secure Information Management, Sharing and Interaction; Information Assurance of Cyber-Physical Systems; and Quantum Information Sciences for Secure Computing and Communication. The annual deadline for the next fiscal year is September 30.
Requirements: Grants are awarded to academic, industrial, and nonprofit organizations, with strong emphasis on university research.
Restrictions: New grants or contracts are typically started near the beginning of the fiscal year. New proposals for the next fiscal year should be submitted no later than September 30, unless otherwise directed by the BAA. Competitive renewals should be submitted no later than six months prior to the award termination date after contacting the Program Officer. Conference proposals should be submitted at least four months prior to the conference.
Geographic Focus: All States
Date(s) Application is Due: Sep 30
Contact: Dr. J. Sukarno Mertoguno; (703) 696-0107; sukarno@mertoguno@navy.mil
Internet: http://www.onr.navy.mil/Science-Technology/Departments/Code-31/All-Programs/311-Mathematics-Computers-Research/Software-Computing-Systems.aspx
Sponsor: U.S. Department of Defense
One Liberty Center, 875 N. Randolph Street, Suite 1425
Arlington, VA 22203-1995

USDD Solid and Liquid Waste Treatment Grants **4796**

The Office of Naval Research Solid and Liquid Waste Treatment program seeks to enhance the science and technology base in order to assist the development of sensors, materials, methods and, ultimately, systems for the treatment or destruction of solid and liquid wastes both on board ships and at shore-side support activities. Areas of interest include: novel approaches to on-board treatment of both oily and non-oily wastewaters to meet discharge standards; reliable sensor and control systems to automate shipboard waste treatment and reduce the manpower burden; novel approaches to reduce the threat of the transport of non-indigenous species in ballast operations; and alternative approaches to treat or remediate solid and hazardous wastes. Submission of the Research Proposal should always be preceded by submission of a Preproposal and a discussion of that Preproposal with the assigned Program Officer. Applications for the upcoming fiscal year should be received by September 30.
Requirements: Grants are awarded to academic, industrial, and nonprofit organizations, with strong emphasis on university research. Investigators are encouraged to contact the staff prior to formally submitting a proposal.
Restrictions: New grants or contracts are typically started near the beginning of the fiscal year. New proposals for the next fiscal year should be submitted no later than September 30, unless otherwise directed. Competitive renewals should be submitted no later than six months prior to the award termination date after contacting the Program Officer. Conference proposals should be submitted at least four months prior to the conference.
Geographic Focus: All States
Date(s) Application is Due: Sep 30
Contact: Dr. Stephen McElvany; (703) 696-1449; steve.mcelvaney@navy.mil
Internet: http://www.onr.navy.mil/Science-Technology/Departments/Code-33/All-Programs/332-naval-materials/Solid-Liquid-Waste-Treatment.aspx
Sponsor: U.S. Department of Defense
One Liberty Center, 875 N. Randolph Street, Suite 1425
Arlington, VA 22203-1995

USDD Solid Mechanics Grants **4797**

The field of solid mechanics bridges research in materials science and structural mechanics and design. The Office of Naval Research Solid Mechanics program supports research in mechanics of deformation, damage evolution and failure of composite hull structural materials in the naval environment, when subjected to complex three-dimensional static, cyclic and dynamic loading conditions. The program aims to establish physically based models for material behavior, with predictive capabilities. The current focus is on marine composites and sandwich structures. Areas of interest include: 3D failure modes, mode coupling and failure criteria; effects of moisture, sea water, hydrostatic pressure and temperature; durability and life prediction; effects of dynamic loading (e.g. impact, shock, blast implosion); mechanics of nanocomposites and multifunctional composites; concepts for mitigation of damage and failure, and life extension; concepts for the enhancement of thermo-mechanical properties and structural performance; and materials systems of interest include composites and sandwich structures for naval ship hulls. Submission of the Research Proposal should always be preceded by submission of a Preproposal and a discussion of that Preproposal with the assigned Program Officer. Applications for the upcoming fiscal year should be received by September 30.
Requirements: Grants are awarded to academic, industrial, and nonprofit organizations, with strong emphasis on university research. Investigators are encouraged to contact the staff prior to formally submitting a proposal.
Restrictions: New grants or contracts are typically started near the beginning of the fiscal year. New proposals for the next fiscal year should be submitted no later than September 30, unless otherwise directed. Competitive renewals should be submitted no later than six months prior to the award termination date after contacting the Program Officer. Conference proposals should be submitted at least four months prior to the conference.
Geographic Focus: All States
Date(s) Application is Due: Sep 30
Contact: Dr. Yapa Rajapakse; (703) 696-4405; yapa.rajapakse@navy.mil
Internet: http://www.onr.navy.mil/Science-Technology/Departments/Code-33/All-Programs/332-naval-materials/Solid-Mechanics.aspx
Sponsor: U.S. Department of Defense
One Liberty Center, 875 N. Randolph Street, Suite 1425
Arlington, VA 22203-1995

USDD Strategic Technologies Grants **4798**

The Defense Advanced Research Projects Agency's (DARPA) Strategic Technology Office (STO) is soliciting innovative proposals for the performance of research, development, design, and testing that directly supports STO. This includes finding difficult targets; communications, networks and electronic warfare; shaping the environment; and foundational technologies that support multiple STO focus areas. The recommended order of submissions is executive summary, abstract, and proposal; however, it is not mandatory to submit an executive summary and/or abstract before submitting a full proposal. Applicants are strongly encouraged to submit an executive summary prior to abstracts or proposals. This procedure is intended to minimize unnecessary effort in proposal preparation and review. Instructions for submitting executive summaries, abstracts, and proposals can be found on the website. Multiple awards are possible. The funding awarded will depend on the quality of the proposals received and the availability of funds.
Requirements: All responsible sources capable of satisfying the government's needs are eligible. Historically black colleges and universities, small businesses, small disadvantaged businesses and minority institutions are encouraged to submit proposals and join others in submitting proposals.
Geographic Focus: All States
Date(s) Application is Due: Jan 12
Contact: BAA Coordinator; (703) 526-6630; fax DARPA-BAA-12-09@DARPA.mil
Internet: http://www.grants.gov/search/search.do;jsessionid=qFTNP7SNp7qTyygkG1CvWMc011NgYX1NyBCQy3QlQvSpTCtbLZhb!733686662?oppId=137993&mode=VIEW
Sponsor: U.S. Department of Defense
3701 North Fairfax Drive
Arlington, VA 22203-6630

USDD U.S. Army Medical Research and Materiel Command Broad Agency Announcement for Extramural Medical Research Grants **4799**

The United States Army Medical Research and Materiel Command's mission is to provide solutions to medical problems of importance to the American warfighter at home and abroad. The scope of this effort and the priorities attached to specific projects are influenced by changes in military and civilian medical science and technology, operational requirements, military threat assessments, and national defense strategies. Research areas of interest for the Extramural Medical Research Grants include the following: military infectious diseases; combat casualty care; military operational medicine; clinical and rehabilitative medicine; medical biological defense; medical chemical defense; telemedicine and advanced technology; and special programs (including health care delivery; detection, diagnosis, control or eradication of specified diseases, conditions, or syndromies; and the advancement of military medical interests). This Broad Agency Announcement (BAA) is intended to solicit extramural research and development ideas. Funding has not been set aside specifically for this announcement and the number of awards has not been determined. Applicants are encouraged to submit proposals that span their entire research project up to five years. Pre-proposals and full proposals may be submitted at any time until the deadline.
Requirements: Eligible applicants include the following: private/public/state controlled institutions of higher education; Hispanic-serving institutions; historically black colleges and universities/minority institutions; tribally controlled colleges and universities; Alaska Native and Native Hawaiian serving institutions; nonprofits with 501(c)3 status (other than institutions of higher education; small businesses; for-profit organizations (other than small businesses); Indian/Native American tribal governments; Indian/Native American tribally designated organizations; non-domestic (non-United States) entities (foreign organizations); and federal, state, and local government agencies.
Geographic Focus: All States
Date(s) Application is Due: Sep 30
Contact: Grants Officer; (703) 518-4726; (QA.BAA@amedd.army.mil),
Internet: http://www07.grants.gov/search/search.do;jsessionid=24YPPC4SdRy7mYT61phb5n46fQty11L2LySnpQGjn4qRpnSwgvG0!712472910?oppId=125873&mode=VIEW
Sponsor: U.S. Department of Defense
1400 Defense Pentagon
Washington, DC 20301-1400

USDD Ultra-High Temperature Materials Grants 4800

This Office of Naval Research program seeks to develop materials, structures and computational tools that will enable the development of expendable systems that operate above 2500°C with no ablation or erosion. The primary application areas are naval hypersonics technologies, including advanced missile systems and Electromagnetic Railgun projectile. This program is also developing high-temperature structural insulator materials that mechanically support and thermally isolate internal systems (e.g., sensors, guidance mechanisms, weapons payload) from high-heat loads generated by burning fuel and/or aerothermal heating during launch/firing/re-entry of advanced naval weapons systems. Understanding the fundamental relationships between composition, microstructure and physical response (thermal and mechanical) at the micro- and meso-scale is of high importance, as is the potential for low-cost manufacturing techniques. Primary applications include: high-temperature structural insulators; electromagnetic railgun projectiles; missile systems; leading edges; and rocket nozzles. Submission of the Research Proposal should always be preceded by submission of a Preproposal and a discussion of that Preproposal with the assigned Program Officer. Applications for the upcoming fiscal year should be received by September 30.
Requirements: Grants are awarded to academic, industrial, and nonprofit organizations, with strong emphasis on university research. Investigators are encouraged to contact the staff prior to formally submitting a proposal.
Restrictions: New grants or contracts are typically started near the beginning of the fiscal year. New proposals for the next fiscal year should be submitted no later than September 30, unless otherwise directed. Competitive renewals should be submitted no later than six months prior to the award termination date after contacting the Program Officer. Conference proposals should be submitted at least four months prior to the conference.
Geographic Focus: All States
Date(s) Application is Due: Sep 30
Contact: Dr. Eric Wuchina; (301) 227-3949; fax (301) 227-4732; eric.wuchina@navy.mil
Internet: http://www.onr.navy.mil/Science-Technology/Departments/Code-33/All-Programs/332-naval-materials/Ultra-High-Temperature-Materials.aspx
Sponsor: U.S. Department of Defense
One Liberty Center, 875 N. Randolph Street, Suite 1425
Arlington, VA 22203-1995

USDD Undersea Signal Processing Grants 4801

The USDD Undersea Signal Processing program's goal is to develop signal processing algorithms that improve the Navy's ability to detect, identify and locate submarines in shallow and deep ocean environments. To this end, we fund basic and applied research projects at universities, federal laboratories and industry. The program is organized into the following technical thrusts: Active Sonar Signal Processing-projects focus on active sonar systems for wide area surveillance, distributed search and battle group defense; Passive Sonar Signal Processing-projects focus on signal processing algorithms that detect, classify and locate modern diesel-electric submarines at tactically useful ranges; Data Fusion Technology-the program's objective is to develop new ways to fuse sonar data from multiple sensors; and Fundamental Research Initiatives-this program supports basic research in statistical and ocean acoustic signal processing projects conducted at universities throughout North America. Submissions for fiscal year planning letters are due by the close of business, June 14. Applications for the upcoming fiscal year should be received by September 30.
Requirements: Grants are awarded to academic, industrial, and nonprofit organizations, with strong emphasis on university research.
Restrictions: New grants or contracts are typically started near the beginning of the fiscal year. Fiscal year planning letters are due by the close of business, June 14. New proposals for the next fiscal year should be submitted no later than September 30, unless otherwise directed by the BAA. Competitive renewals should be submitted no later than six months prior to the award termination date after contacting the Program Officer. Conference proposals should be submitted at least four months prior to the conference.
Geographic Focus: All States
Contact: John Tague, Team Leader; (703) 696-4399; john.a.tague@navy.mil
Internet: http://www.onr.navy.mil/Science-Technology/Departments/Code-32/All-Programs/Ocean-Systems-321/Undersea-Signal-Processing.aspx
Sponsor: U.S. Department of Defense
One Liberty Center, 875 N. Randolph Street, Suite 1425
Arlington, VA 22203-1995

USDD Vacuum Devices Grants 4802

The Office of Naval Research (ONR) Vacuum Devices Program develops techniques and technologies for compact high-power, high-frequency vacuum electronic amplifiers exhibiting high gain, bandwidth, and efficiency. Of particular interest is the development of compact high power amplifiers that can be applied to the millimeter and sub-millimeter wave frequency range. Geometry and physics-based computationally efficient algorithms, models and simulation based design are included. Applications for the upcoming fiscal year should be received by September 30.
Requirements: Grants are awarded to academic, industrial, and nonprofit organizations, with strong emphasis on university research.
Restrictions: New grants or contracts are typically started near the beginning of the fiscal year. New proposals for the next fiscal year should be submitted no later than September 30, unless otherwise directed by the BAA. Competitive renewals should be submitted no later than six months prior to the award termination date after contacting the Program Officer. Conference proposals should be submitted at least four months prior to the conference.
Geographic Focus: All States
Date(s) Application is Due: Sep 30
Contact: Ingham Mack, Program Officer; (703) 696-4825; ingham.mack@navy.mil
Internet: http://www.onr.navy.mil/Science-Technology/Departments/Code-31/All-Programs/312-Electronics-Sensors/Vacuum-Devices.aspx
Sponsor: U.S. Department of Defense
One Liberty Center, 875 N. Randolph Street, Suite 1425
Arlington, VA 22203-1995

USDD Warfighter Interface Technologies Advanced Research Grants 4803

The 711th Human Performance Wing, Human Effectiveness Directorate is soliciting white papers for innovative technologies to support the development of Warfighter Interface Technologies Advanced Research Programs. The objective is to address highly unique and innovative research and development concept areas in Warfighter Interface Technologies. Proposed work should address the innovative and strategic thought of the 21st century adversaries, and develop new concepts to counter these adversaries with innovative warfighter interface capabilities. The technology research will be primarily focused in the following four areas: battlespace acoustics; battlespace visualizations; human role in autonomous systems; and distributed decision making. This is a multi-year offering and applicants may submit white papers at any time. Applicants whose white paper are considered to meet the needs of the Air Force will be invited to submit a full proposal. Early white paper submission is encouraged. Anticipated annual funding ranges from $5,900,000 to $9,900,000. Ten awards are expected per year.
Requirements: This is an unrestricted solicitation. Small businesses are encouraged to propose on all or any part of this solicitation.
Geographic Focus: All States
Date(s) Application is Due: Sep 30
Contact: Angela G. Lauofo; (937) 656-5937; Angela.Lauofo@wpafb.af.mil
Internet: http://www.grants.gov/search-grants.html?agencies%3DDOD|Department%20of%20Defense
Sponsor: U.S. Department of Defense
1400 Defense Pentagon
Washington, DC 20301-1400

USDD Water Desalination and Purification Grants 4804

The Office of Naval Research (ONR) Water Desalination/Purification program is currently an applied research and development program focusing on maturing promising technologies for incorporation into prototype desalination systems. The goal is to improve the Navy desalination capability in terms of increased operational availability in littoral waters, reduced maintenance, decreased size and weight, and reduced energy usage. The systems must be designed for the operational constraints of Navy ships as detailed in ONR BAA 09-013: Component Development for Advanced Shipboard Desalination. The fiscal year will consist of two-year efforts to mature promising component level technologies followed by development of complete shipboard desalination systems. Though the program is quite applied now, ONR involvement derives from prior basic research on the development of novel membranes for shipboard wastewater treatment (solid and liquid waste treatment). Submission of the Research Proposal should always be preceded by submission of a Preproposal and a discussion of that Preproposal with the assigned Program Officer. Applications for the upcoming fiscal year should be received by September 30.
Requirements: Grants are awarded to academic, industrial, and nonprofit organizations, with strong emphasis on university research. Investigators are encouraged to contact the staff prior to formally submitting a proposal.
Restrictions: New grants or contracts are typically started near the beginning of the fiscal year. New proposals for the next fiscal year should be submitted no later than September 30, unless otherwise directed. Competitive renewals should be submitted no later than six months prior to the award termination date after contacting the Program Officer. Conference proposals should be submitted at least four months prior to the conference.
Geographic Focus: All States
Date(s) Application is Due: Sep 30
Contact: Dr. Paul Armistead, Program Officer; (703) 696-4315; fax (703) 696-0001; paul.armistead@navy.mil
Internet: http://www.onr.navy.mil/Science-Technology/Departments/Code-33/All-Programs/332-naval-materials/Water-Desalination-Purification.aspx
Sponsor: U.S. Department of Defense
One Liberty Center, 875 N. Randolph Street, Suite 1425
Arlington, VA 22203-1995

USDD Workplace Violence in the Military Grants 4805

The Defense Health Program is seeking applications for the Psychological Health and Traumatic Brain Injury (PH/TBI) Research Program for Workplace Violence in the Military. The PH/TBI Research Program was established to complement ongoing Department of Defense (USDD) efforts towards promoting optimal care for PH (including workplace violence across the USDD service areas) and TBI in the areas of prevention, detection and intervention. United States warfighters endure increasingly demanding and high tempo conditions both in garrison and in the combat field of operations in order to keep pace with ongoing wartime mission requirements. In addition to the loss of lives overseas, warfighters have lost their lives on an Army post stemming from acts of violence in the workplace. The tragic shooting of military personnel at Fort Hood in November 2009 highlighted the need for the USDD to thoroughly review its approach to force health protection and to broaden their historical focus on hostile external threats. The USDD force protection policies, programs, and procedures need to be improved and integrated into a cohesive and comprehensive approach. This Grant program stems directly from recommendations yielded from the Independent Review Related to Fort Hood, which determined that USDD programs, policies, process and procedures that address identification of indicators for violence and radicalization are outdated, incomplete, and fail to include key indicators of potentially violent behaviors.

It was also determined that currently the USDD has no risk assessment system available to supervisors and commanders to assist them in the identification, mitigation, and tracking of internal threats. The Independent Review outlined actions needed to identify behavioral indicators of violence. Applications are sought to support a retrospective study and a prospective study across USDD service areas to develop a set of empirical behavioral-based predictors of potential violence. Applications may include a combined, two-tiered approach that would first involve the retrospective study with those results informing the design of the prospective study. Submitters may also choose to propose only a retrospective or a prospective study. Approximately $8M is available, and five awards are anticipated. Pe-applications should be submitted by February 27.
Requirements: Applications from investigators within the military services and other federal agencies are encouraged, as are applications involving multidisciplinary collaborations among academia, industry, the military services, the Department of Veterans Affairs, and other federal government agencies. A principal investigator must be an independent investigator at any academic level from government, academia, research institutions, industry, and private foundations that possess the skills, knowledge, and resources necessary to carry out the proposed research.
Geographic Focus: All States
Date(s) Application is Due: Apr 26
Amount of Grant: 3,000,000 USD
Contact: Grants Manager; (703) 674-2500; programannouncements@tatrc.org.
Internet: http://www.grants.gov/search/search.do;jsessionid=Q2lZP9lV4mvG5TgvVqhZ2YcTpzV1NQypcLBss0zn38199gQ1WYBM!-2099600874?oppId=140173&mode=VIEW
Sponsor: U.S. Department of Defense
1400 Defense Pentagon
Washington, DC 20301-1400

Van Alen Institute New York Prize Fellowship 4806
The fellowship program supports emerging research and experimental practice in public architecture. Fellows are given an opportunity to pursue advanced independent study and to generate projects that are accessible to the general public. Projects may be an ongoing component or an outcome of fellows' research. They can range across media, temporalities, dimensions and scales in the form of exhibitions, installations, performances, symposia and other formats. The institute supports projects that provoke new ways of thinking about and making public space, that critically interrogate how and for whom spaces operate as 'public', and that engage public audiences in imaginative and productive ways.
Requirements: he fellowship program is open to emerging practitioners and scholars from the design and planning disciplines, and other fields in the arts, humanities and sciences that foster critical and creative inquiry into contemporary forms of public space. The fellowship is international in scope-the institute welcomes applications from around the world, across the country, and around the region. Candidates may be early to mid-career individuals in the design and planning disciplines, arts, humanities or sciences. Design professionals, artists and other practitioners must hold at least a Bachelors' degree. Scholars should hold a terminal degree as defined by their field of specialty. Candidates must be able to spend a minimum of three months in continuous residence in New York City; must be able to develop and present projects within the fellowship period; and, must possess English proficiency. Joint applications are welcome, provided all parties meet eligibility requirements.
Restrictions: Students are not eligible. Work for the completion of a degree is not eligible.
Geographic Focus: All States
Date(s) Application is Due: May 9
Amount of Grant: 5,000-15,000 USD
Contact: Jessica Blaustein; (212) 924-7000, ext. 5; jblaustein@vanalen.org
Internet: http://www.vanalen.org/nyprize/fellowship_overview.html
Sponsor: Van Alen Institute
30 W 22nd Street, 6th Floor
New York, NY 10010

Vancouver Foundation Grants and Community Initiatives Program 4807
Vancouver Foundation's Grants and Community Initiatives program provides Community Impact Grants in nine fields of interest: Animal Welfare; Arts and Culture; Children, Youth and Families; Education; Environment; Health and Social Development; Health and Medical Research; Youth Homelessness; Youth Philanthropy. The Vancouver Foundation has a two-stage application process. The first stage is a letter of intent. This consists of a brief proposal to determine basic suitability. The second stage is submission of a full grant application by a specific date. The letter of intent may be submitted online. You will be notified, if you are invited to submit a full grant application. The following is a list of deadline dates for the letters of intent and applications. Animal Welfare, Arts and Culture, Children, Youth and Families, Education, Environment, and Health and Social Development grants have a deadline date of: Letter of Intent, June 18, Application, August 6. Youth Homelessness grants: Letter of Intent, August 6, Application, September 24. Youth Philanthropy grants: Letter of Intent, September 17, Application, October 1. Health and Medical Research grants deadline dates are undetermined, at this time.
Requirements: Under our Grants and Community Initiatives program, Vancouver Foundation funds eligible applicants, which include: registered charities and qualified donees under the Income Tax Act; First Nations that may be considered a public body performing a function of government.
Restrictions: The Foundation does not fund the following: 100% of a proposal's costs; an organization's ongoing operational or core expenses; retroactive funding, or for any expenses to be incurred prior to the Foundation's decision date; debt retirement or reserves; mortgage pay-downs; conferences, competitions, symposia, annual events, or travel to/attendance at such events; office equipment and furniture; activities of religious organizations that serve primarily their membership and/or their direct religious purposes, unless the community at large will benefit significantly; sabbatical leaves, student exchanges; medical facilities or equipment; league-based sports and recreation programs; library acquisitions and construction; school construction; publication of books; academic or dissertation research; research with human subjects; endowment matching grants; capital requests, with the exceptions noted above; activities previously supported through government funding; bursaries, scholarships and awards; school trips, annual events or equipment; grants to individuals or businesses; funding requests if reports on any previous grants are overdue.
Geographic Focus: All States, Canada
Contact: Mark Gifford, Director; (604) 629-5362; markg@vancouverfoundation.ca
Internet: http://www.vancouverfoundation.ca/grants/communityinitiatives.htm
Sponsor: Vancouver Foundation
555 West Hastings Street, Suite 1200
Vancouver, BC V6B 4N6 Canada

Vanguard Public Foundation Grant Funds 4808
Vanguard provides funding to groups in northern California under four categories; the Social Justice Fund, the Community Institution Building Program, the Technical Assistance and Capacity Building Program, and the Social Justice Sabbatical Fund. Types of support include seed money, technical assistance, and operating expenses. Priority is given to projects without access to traditional funding sources because they may be thought to be risky, controversial, or of low priority. Vanguard strongly encourages coalitions that emphasize joint strategies and projects. Visit the web site for application guidelines. The next deadline to submit a proposal for the Social Justice Fund is October 14 and for a Social Justice Sabbatical is December 15. Annual deadline dates may vary; contact program staff for exact dates.
Requirements: Vanguard funds new or existing organizations involved in direct organizing or advocacy that are based in northern California (i.e., all counties north of Monterey). At the time a grant is made, an organization must either be tax-exempt or have a fiscal sponsor.
Restrictions: Vanguard does not fund capital campaigns or equipment purchase; organizations involved in direct services, research, or education, unless the proposed project clearly has an organizing component and cannot be supported within the general program budget; production costs for film; out-of-state travel; one-time conferences or events not integrally related to ongoing organizing; organizations with access to traditional funding sources and budgets of over $200,000 (unless the project for which funds are requested is unlikely to attract support because of its risky or controversial character); and costs already incurred.
Geographic Focus: California
Date(s) Application is Due: Oct 14; Dec 15
Contact: Grants Director; (415) 487-2111; fax (415) 487-2124; danielle@vanguardsf.org or vpf@vanguardsf.org
Internet: http://www.vanguardsf.org
Sponsor: Vanguard Public Foundation
383 Rhode Island Street, Suite 301
San Francisco, CA 94103

Vernon K. Krieble Foundation Grants 4809
The Foundation was established in 1984 by the family of Professor Vernon K. Krieble, scientist, educator, inventor, and entrepreneur. Recognizing that the Foundation's assets are the product of a free and democratic society, the founders considered it fitting that those assets be used "to further democratic capitalism and the preserve and promote a society of free, educated, healthy and creative individuals." The Foundation offers support to non-profit charitable and educational organizations that demonstrate leadership in furthering the original objectives, so that future generations can aspire to and achieve their full potential in a free society. Awards range from $2,500 to $50,000. There are no deadlines.
Requirements: Nonprofit 501(c)3 organizations are eligible. Funding is provided only for those organizations and projects which involve public policy research and education on issues supporting the preservation, and in some cases the restoration, or freedom and democracy in the United States, according to the principles of the Founding Fathers. Written proposals should include a summary of the project, the project budget, the amount requested, the qualifications of individuals involved, and a copy of the organization's Internal Revenue Service determination letter.
Geographic Focus: All States
Contact: Helen E. Krieble, President; (303) 758-3956; fax (303) 488-0068
Internet: http://www.krieble.org/grants
Sponsor: Vernon K. Krieble Foundation
1777 S Harrison Street, Suite 807
Denver, CO 80210

Victor E. Speas Foundation Grants 4810
The Victor E. Speas Foundation was established to provide medical care for the needy, to further medical research, and to support and promote quality educational, cultural, human-services, and health-care programming. In the area of arts, culture, and humanities, the foundation supports programming that: fosters the enjoyment and appreciation of the visual and performing arts; strengthens humanities and arts-related education programs; provides affordable access; enhances artistic elements in communities; and nurtures a new generation of artists. In the area of education, the foundation supports programming that: promotes effective teaching; improves the academic achievement of, or expands educational opportunities for, disadvantaged students; improves governance and management; strengthens nonprofit organizations, school leadership, and teaching; and bolsters strategic initiatives of area colleges and universities. In the area of health, the foundation supports programming that improves the delivery of health care to the indigent, uninsured, and other vulnerable populations and addresses health and health-care problems that intersect with social factors. In the area of human services, the foundation funds programming that: strengthens agencies that deliver critical human services and maintains the community's safety net and helps agencies respond to federal, state, and local public policy changes.

In the area of community improvement, the foundation funds capacity-building and infrastructure-development projects including: assessments, planning, and implementation of technology for management and programmatic functions within an organization; technical assistance on wide-ranging topics, including grant writing, strategic planning, financial management services, business development, board and volunteer management, and marketing; and mergers, affiliations, or other restructuring efforts. Grant requests for general operating support and program support will be considered. Grants from the foundation are one year in duration. Application materials are available for download at the grant website. There are no application deadlines for the Victor E. Speas Foundation. Proposals are reviewed on an ongoing basis.
Requirements: Applicants must have 501(c)3 tax-exempt status and serve the residents of Kansas City, Missouri. Two copies of the completed application must be mailed.
Restrictions: Grant requests for capital support will not be considered. The trust does not support requests from individuals, organizations attempting to influence policy through direct lobbying, or any political campaigns.
Geographic Focus: Missouri
Contact: Spence Heddens; (816) 292-4301; Spence.heddens@baml.com
Internet: https://www.bankofamerica.com/philanthropic/fn_search.action
Sponsor: Victor E. Speas Foundation
1200 Main Street, 14th Floor, P.O. Box 219119
Kansas City, MO 64121-9119

Viking Children's Fund Grants **4811**
The fund supports the needs of children in Minnesota and the upper Midwest. Grants support research at the University of Minnesota Department of Pediatrics into the major diseases and disorders of childhood; and nonprofits that focus efforts in the areas of health education and family services supporting the well-being of children in the upper Midwest.
Requirements: Nonprofit organizations in Minnesota and the upper Midwest, including Iowa, North Dakota, South Dakota, and Wisconsin, are eligible.
Geographic Focus: Iowa, Minnesota, North Dakota, South Dakota, Wisconsin
Date(s) Application is Due: Apr 1
Amount of Grant: 500-5,000 USD
Contact: Patrick Leopold, Fund Coordinator; (952) 828-6500; info@vikings.com
Internet: http://www.vikings.com/commchildfund.html
Sponsor: Minnesota Vikings Football Club
9520 Viking Drive
Eden Prairie, MN 55344

Virginia A. Hodgkinson Research Prize **4812**
In partnership with the Association for Research on Nonprofit Organizations and Voluntary Action, the research prize is awarded annually in recognition of a book that informs nonprofit policy and practice. Recipients are researchers and practitioners who work in or with the voluntary sector, and through their work, have achieved local, regional, national, or international impact and recognition. Submissions from researchers and practitioners representing any discipline, nonprofit field, or organization are welcome. Multidisciplinary approaches, submissions from young scholars and practitioners working in collaboration with others, and from researchers and practitioners at the start of their careers are encouraged. Two research prizes will be awarded.
Requirements: Each submission will be judged on the contribution the research makes to policy and practice. The research article, report, or book must have been published within three calendar years prior to the year of the award. Books can be considered more than once. Entries must be published by a university press or commercial publishing house. Submissions from scholars representing any discipline are welcome. In addition, submissions are encouraged from young scholars, practitioners working in collaboration with university-based researchers, and scholars working on multi-disciplinary approaches. Self-nominations will be accepted.
Restrictions: Prior prize recipients are ineligible for a five-year period after their award is received.
Geographic Focus: All States
Date(s) Application is Due: Jun 30
Amount of Grant: 1,000-2,000 USD
Contact: Thomas Jeavons, Executive Director; (317) 684-2120; tjeavons@arnova.org
Internet: http://www.independentsector.org/programs/research/vah_research_prize.html
Sponsor: Independent Sector
1200 18th Street NW, Suite 200
Washington, DC 20036

Virginia Foundation for the Humanities Discretionary Grants **4813**
The Virginia Foundation for the Humanities (VFH) was established in 1974 as a nonprofit organization dedicated to promoting the humanities, and to using the humanities to address issues of broad public concern. Since 1974, the VFH has awarded more than 3,000 grants, supporting tens of thousands of separate activities, and serving audiences in every city and county in Virginia. Goals of the VFH grant program are as follows: to encourage the development of high-quality educational programs in the humanities; to support accessible programs that reach the broadest possible audience in Virginia; to support the work of humanities institutions-museums, libraries, historical societies, colleges, and universities-as well as other non-profit organizations working within the humanities; to explore the stories that define Virginia and its people; and to address the issues that are most important to communities in Virginia. Projects supported include exhibits, public forums and discussions, media programs (film, video, radio, and digital media), publications, research, teachers' institutes and seminars, oral history projects, lectures and conferences, and other kinds of programs that draw on the resources of the humanities, address important issues, and enrich the cultural life of the state. The program names six key-priority areas of interest: Books, Reading, and Literacy; Rights and Responsibilities; Media and Culture; Violence and Community; Science, Technology, and Society; Virginia History. Other identified areas of interest are teacher education programs (especially those related to Virginia's Standards of Learning); African American history and culture; Native American history and culture; history and culture of minority communities in Virginia; Virginia's folklife and traditional culture(s); and the future of rural Virginia. The foundation's interest is not confined to only these topics. Discretionary Grants allow VFH to make smaller grants (up to $3,000) at any time of the year. These grants may be used to plan larger projects or to implement programs which require only a modest amount of funding. Decisions are normally made within four weeks following receipt of the application. Applications should be submitted via the VFH online system (available at the foundation's website) which requires as a first step, creation of an account and password. Grant guidelines are available at the website in downloadable Word format. Applicants should contact VFH staff in advance before submitting a Discretionary Grant proposal. Draft proposals are strongly encouraged and are recommended to be submitted at least three weeks prior to the deadline to allow time for a thorough staff review.
Requirements: Any incorporated non-profit organization in Virginia is eligible to apply. Incorporated non-profit organizations based outside of Virginia are also eligible if their project deals with a subject directly related to Virgina and a significant audience within the state is anticipated. All VFH grants must be matched with at least an equal amount of cost share, which can be in the form of cash or in-kind contributions from non-Federal sources. Sources and amounts of anticipated cost share should be indicated at the time of the proposal.
Restrictions: VFH Grants may not be used to support the following: advocacy or political action programs that promote a particular solution or point of view; creative or performing arts, unless they are used in a supporting role to enhance discussion of issues or interpretation; research or writing unless these are integral to programs having a direct public audience; subvention of publications; courses for credit, except those designed especially for teachers; acquisition of equipment; building construction, restoration, or preservation; meals, other than necessary travel expenses for program personnel; international travel; indirect costs; and projects whose primary audience is children or youth.
Geographic Focus: All States
Amount of Grant: Up to 3,000 USD
Contact: David Bearinger; (434) 924-3296; fax (434) 296-4714; dab@virginia.edu
Internet: http://virginiahumanities.org/grants/
Sponsor: Virginia Foundation for the Humanities
145 Ednam Drive
Charlottesville, VA 22903-4629

Virginia Foundation for the Humanities Open Grants **4814**
The Virginia Foundation for the Humanities (VFH) was established in 1974 as a nonprofit organization dedicated to promoting the humanities, and to using the humanities to address issues of broad public concern. Since 1974, the VFH has awarded more than 3,000 grants, supporting tens of thousands of separate activities, and serving audiences in every city and county in Virginia. Goals of the VFH grant program are as follows: to encourage the development of high-quality educational programs in the humanities; to support accessible programs that reach the broadest possible audience in Virginia; to support the work of humanities institutions-museums, libraries, historical societies, colleges, and universities-as well as other non-profit organizations working within the humanities; to explore the stories that define Virginia and its people; and to address the issues that are most important to communities in Virginia. Projects supported include exhibits, public forums and discussions, media programs (film, video, radio, and digital media), publications, research, teachers' institutes and seminars, oral history projects, lectures and conferences, and other kinds of programs that draw on the resources of the humanities, address important issues, and enrich the cultural life of the state. The program names six key-priority areas of interest: Books, Reading, and Literacy; Rights and Responsibilities; Media and Culture; Violence and Community; Science, Technology, and Society; and Virginia History. Other identified areas of interest are teacher-education programs (especially those related to Virginia's Standards of Learning); African-American history and culture; Native-American history and culture; history and culture of minority communities in Virginia; Virginia's folklife and traditional culture(s); and the future of rural Virginia. The foundation's interest is not confined to only these topics. Open Grant applications are usually considered in three annual grant cycles with deadlines of October 15, February 1, and May 1. Applications should be submitted via the VFH online system (available at the foundation's website) which requires as a first step, creation of an account and password. Grant guidelines are available at the website in downloadable Word format. All applicants are urged to contact the staff prior to submission of their proposals. Draft proposals are strongly encouraged and are recommended to be submitted at least three weeks prior to the deadline to allow time for a thorough staff review.
Requirements: Any incorporated non-profit organization in Virginia is eligible to apply. Incorporated non-profit organizations based outside of Virginia are also eligible if their project deals with a subject directly related to Virgina and a significant audience within the state is anticipated. All VFH grants must be matched with at least an equal amount of cost share, which can be in the form of cash or in-kind contributions from non-Federal sources. Sources and amounts of anticipated cost share should be indicated at the time of the proposal.
Restrictions: VFH Grants may not be used to support the following: advocacy or political action programs that promote a particular solution or point of view; creative or performing arts, unless they are used in a supporting role to enhance discussion of issues or interpretation; research or writing unless these are integral to programs having a direct public audience; subvention of publications; courses for credit, except those designed especially for teachers; acquisition of equipment; building construction, restoration, or preservation; meals, other than necessary travel expenses for program personnel; international travel; indirect costs; and projects whose primary audience is children or youth.
Geographic Focus: All States, Canada
Date(s) Application is Due: Feb 1; May 1; Oct 15

Contact: David Bearinger; (434) 924-3296; fax (434) 296-4714; dab@virginia.edu
Internet: http://virginiahumanities.org/grants/
Sponsor: Virginia Foundation for the Humanities
145 Ednam Drive
Charlottesville, VA 22903-4629

Virginia Foundation for the Humanities Residential Fellowships **4815**

The Virginia Foundation for the Humanities (VFH) supports scholarly work on the humanities and in the public interest. The foundation's residential fellowship program is located at its facility in Charlottesville, Virginia, in the foothills of the Blue Ridge Mountains. The fellowship offers time, space, and resources to scholars and writers bringing the humanities to visibility-applying the tools of history, philosophy, ethics, cultural studies, and literary criticism to matters of public concern. Fellowships are awarded for one semester or for a full year. Stipends range up to $15,000 per semester. Fellows have visiting-faculty status at the nearby University of Virginia as well as the opportunity to get to know VFH staff members and become involved, formally or informally, in a variety of public humanities projects. Fellows also appear on With Good Reason, a VFH Radio program. Candidates must apply through an online application system at the foundation website. Applicants will need to create a login account to use this system. Letters of recommendation must be mailed separately to the contact information given.
Requirements: The foundation seeks applications that are intellectually stimulating, imaginative, and accessible to the public. There are no restrictions on topic, and applications are invited from across the broad spectrum of the humanities. Candidates may be faculty members in the humanities, independent scholars, or others working on projects in the humanities and need not have advanced degrees. However, the foundation generally does not support dissertation work and strongly encourages postdoctoral applicants to apply for projects other than dissertation revisions.
Restrictions: The foundation does not support work toward a degree. Former VFH Fellows must wait three years before applying for another fellowship. Former Fellows are welcome to apply for space (without stipend) at any time.
Geographic Focus: All States, All Countries
Date(s) Application is Due: Dec 1
Contact: Ann Spencer, Program Associate, Research and Education; (434) 924-3296; fax (434) 296-4714; aspencer@virginia.edu or VFHfellowships@virginia.edu
Internet: http://virginiahumanities.org/fellowships/
Sponsor: Virginia Foundation for the Humanities
145 Ednam Drive
Charlottesville, VA 22903-4629

Virginia Historical Society Research Fellowships **4816**

The society offers fellowships of up to four weeks a year to help defray travel expenses for commuting researchers who wish to conduct research using collections of the society. Awards include the Andrew W. Mellon Research Fellowships, the Betty Sams Christian Fellowships in business history, the Frances Lewis Fellowships in women's studies, and the Reese Fellowships in American Bibliography and the History of the Book in the Americas. Recipients are expected to work on a regular basis in the society's reading room. Applicants whose research promises to result in significant publication, including the Virginia Magazine of History and Biography, will receive primary consideration. Applicants should send an original and three copies of the following: cover letter, resume, two letters of recommendation (which may be sent separately), and a description of their research project not longer than two double-spaced pages that also states the length of the award requested.
Requirements: Applications are welcome from doctoral candidates who reside outside of the Richmond metropolitan area.
Restrictions: Undergraduates, master's students, and graduate students not yet admitted to PhD candidacy are not eligible.
Geographic Focus: Virginia
Date(s) Application is Due: Feb 1
Contact: Nelson Lankford, (804) 342-9672; fax (804) 355-2399; nlankford@vahistorical.org
Internet: http://www.vahistorical.org/research/fellowships.htm
Sponsor: Virginia Historical Society
428 N Boulevard
Richmond, VA 23220

Volkswagen of America Corporate Contributions Grants **4817**

The corporation accepts proposals broadly from programs servicing needs in healthcare, wellness, children, medical research, education and diversity.
Requirements: There are no specific deadlines, however the review committee does meet quarterly to go over proposals received. The committee considers the type of organization, its mission and the intended use of the funds.
Restrictions: Grants are available to 501(c)3 organizations only.
Geographic Focus: All States
Contact: Tara Jones, (248) 754-4693; tara.jones@vw.com
Sponsor: Volkswagen of America
3800 Hamlin Road
Auburn Hills, MI 48326

W. C. Griffith Foundation Grants **4818**

The W. C. Griffith Foundation was established in 1959 by a donation of William C. Griffith, and Ruth Perry Griffith. The Foundation gives primarily in Indianapolis-Carmel, Indiana area. The Foundation provides support primarily for hospitals, health associations, medical and cancer research, the arts, including music and museums, community funds and development, higher, secondary, and other education, family planning services, child welfare, the homeless, the environment, libraries, and Christian religious organizations.
Requirements: Applicants should submit the following: detailed description of project and amount of funding requested; 1 Copy of the proposal.
Geographic Focus: Indiana
Amount of Grant: 1,000-45,000 USD
Samples: Eiteljorg Museum of American Indians and Western Art, Indianapolis, IN, $25,000; Providence Cristo Rey High School, Indianapolis, IN, $10,000;
Contact: Trust Coordinator; fax (317) 693-2504
Sponsor: W.C. Griffith Foundation
101 West Ohio Street, Suite 1450
Indianapolis, IN 46204-1998

W.C. Griffith Foundation Grants **4819**

The foundation awards grants, primarily in Indiana, to nonprofit organizations in its areas of interest, including arts and culture, cancer and medical research, children and youth services, Christian churches and organizations, community development, education, environmental programs, health organizations and hospitals, higher education, homeless services, human reproduction and fertility, library science and libraries, minorities, museums, music and performing arts, secondary education, and social services delivery. Types of support include building construction/renovation, capital campaigns, and continuing support. There are no application forms or deadlines. The board meets in June and November; proposals should be received in May and October.
Requirements: Nonprofit organizations are eligible. Preference is given to requests from Indianapolis, Indiana.
Geographic Focus: All States
Amount of Grant: 500-50,000 USD
Contact: Curt Farran, c/o National City Bank of Indiana; (317) 267-7262
Sponsor: W.C. Griffith Foundation
101 West Washington Street, Suite 600 East
Indianapolis, IN 46255

W. David Smith, Jr. Graduate Student Paper Award **4820**

This award, sponsored by the DuPont Chemical Company Division of AIChE, is given annually to an individual for published work in the application of computing and systems technology to chemical engineering. The work must have been done by the individual while pursuing graduate or undergraduate studies in chemical engineering. A plaque is included with the award. Submit nominations to Dr. Karen High, School of Chemical Engineering-423 EN, Oklahoma State University, Stillwater, OK 74078; high@okstate.edu; phone: (405) 744-9112; fax: (405)744-6338
Requirements: All members as well as other interested persons are urged to nominate deserving candidates. All information and supporting documentation are to be included with completed forms; nominating forms and instructions may be obtained from the institute.
Geographic Focus: All States
Date(s) Application is Due: Apr 15
Amount of Grant: 1,500 USD
Contact: Programs Director; (212) 591-7096; fax (212) 591-8898; awards@aiche.org
Internet: http://students.aiche.org/honors/awarddetail.asp?AwardID=30
Sponsor: American Institute of Chemical Engineers
3 Park Avenue
New York, NY 10016

W. Frank Blair Award **4821**

This award, to encourage excellence in written data presentation, is given annually to a student for the best paper, published or unpublished, pertaining to the natural sciences in the Chihuahuan Desert Region. Papers must be based on original research performed by the author and must be in English, 7500 words or less.
Requirements: Undergraduate and graduate students are eligible for the award. Entries must be accompanied by a letter from a department chairperson that certifies the author's status as a student. Entries must be accompanied by a letter from a department chairperson stating that the entrant is enrolled as a full-time student at the time the entry is submitted.
Geographic Focus: All States
Date(s) Application is Due: Nov 1
Amount of Grant: 500 USD
Contact: Award Coordinator; (432) 364-2499; fax (432) 364-2686; manager@cdri.org
Internet: http://www.cdri.org/Blair%20Award/Blair.html
Sponsor: Chihuahuan Desert Research Institute
P.O. Box 905
Fort Davis, TX 79734

W.K. Kellogg Foundation Healthy Kids Grants **4822**

The Foundation believes that children need nutrition, stimulation, healthy living conditions and access to quality health care. It helps many of them get all four, by funding organizations that improve birth outcomes and first food experiences, create access to healthy foods, improve health services, and educate families and communities about the inter-related factors that determine well-being. The Foundation especially focuses on children who are disadvantaged by multiple societal factors, a disproportionate percentage of whom are children of color.
Requirements: To be eligible for support, your organization or institution, as well as the purpose of the proposed project, must qualify under regulations of the United States Internal Revenue Service. The Kellogg Foundation does not have any submission deadlines. Grant applications are accepted throughout the year and are reviewed at its headquarters in Battle Creek, Michigan, or in its regional office in Mexico.
Restrictions: In general, the Foundation does not provide funding for: operational phases of established programs; capital requests (which includes the construction, purchase,

renovation, and/or furnishing of facilities); equipment; conferences and workshops; films, television and/or radio programs; endowments; development campaigns; or research/studies (unless they are an integral part of a larger program budget being considered for funding).
Geographic Focus: All States
Contact: Deborah A. Rey; (269) 969-2133 or (269) 968-1611; fax (269) 968-0413
Internet: http://www.wkkf.org/what-we-support/healthy-kids.aspx
Sponsor: W.K. Kellogg Foundation
1 Michigan Avenue East
Battle Creek, MI 49017-4012

W.M. Keck Foundation Medical Research Grants **4823**
The W.M. Keck Foundation concentrates on strengthening studies and programs in accredited universities and colleges and major, independent medical research institutions. Medical Research grants seek to advance the frontiers of the life sciences by supporting basic research that is high-risk and has the potential to transform its field. Successful projects are distinctive and novel in their approach to problems, push the edge of their field, or question the prevailing paradigm. Past grants have been awarded to major universities, medical schools, and independent research institutions to support pioneering biological research, including the development of new technologies, instrumentation, or methodologies. Historically, grants range from $500,000 to $5 million, and are typically $2 million or less. Applicants are strongly urged to contact Foundation staff well in advance of submitting a Phase I Application. The best times for these contacts are between January 1 and February 15 leading up to a May 1 submittal, or between July 1 and August 15 leading up to a November 1 submittal.
Requirements: Research universities, medical colleges and major private independent scientific and medical research institutes are eligible. Organizations must be exempt from federal taxation as defined by Section 501(c)3 of the Internal Revenue Code and be designated as a public charity (and not a private foundation) as defined by Section 509(a)1 or 509(a)2 or 170(b) of the Internal Revenue Code. If the institution is located in the State of California, the organization must also be exempt from California State Franchise or Income Tax under Section 23701(d) of the Revenue and Taxation Code.
Restrictions: Grants are not considered for: general operating expenses, endowments or deficit reduction; general and federated campaigns, including fundraising events, dinners or mass mailings; individuals; conference or seminar sponsorship; book publication and film or theater productions; public policy research; institutions that are located outside the United States; institutions that do not have at least three consecutive full, certified, audited financial statements; conduit organizations, unified funds or organizations that use grant funds from donors to support other organizations or individuals; or institutions that are subsidiaries or affiliates of larger entities that do not have a separate board of directors and independent audited financial statements.
Geographic Focus: All States
Date(s) Application is Due: May 1; Nov 1
Amount of Grant: 500,000-5,000,000 USD
Contact: Margie Antonetti, Grants and Database Specialist; (213) 680-3833; fax (213) 614-0934; mantonetti@wmkeck.org or MedRsch@wmkeck.org
Internet: http://www.wmkeck.org/grant-programs/medicalresearch.html
Sponsor: W.M. Keck Foundation
550 S Hope Street, Suite 2500
Los Angeles, CA 90071

W.M. Keck Foundation Science and Engineering Research Grants **4824**
The W.M. Keck Foundation concentrates on strengthening studies and programs in accredited universities and colleges and major, independent medical research institutions. Science and Engineering Research grants seek to benefit humanity by supporting projects that are distinctive and novel in their approach, question the prevailing paradigm, or have the potential to break open new territory in their field. Past grants have been awarded to major universities and independent research institutions to support pioneering science and engineering research, the development of new technologies, and to facilitate the purchase of advanced instruments where such instruments would further specific research ventures. Historically, grants range from $500,000 to $5 million, and are typically $2 million or less. Applicants are strongly urged to contact Foundation staff well in advance of submitting a Phase I Application. The best times for these contacts are between January 1 and February 15 leading up to a May 1 submittal, or between July 1 and August 15 leading up to a November 1 submittal.
Requirements: Research universities, medical colleges and major private independent scientific and medical research institutes are eligible. Organizations must be exempt from federal taxation as defined by Section 501(c)3 of the Internal Revenue Code and be designated as a public charity (and not a private foundation) as defined by Section 509(a)1 or 509(a)2 or 170(b) of the Internal Revenue Code. If the institution is located in the State of California, the organization must also be exempt from California State Franchise or Income Tax under Section 23701(d) of the Revenue and Taxation Code.
Restrictions: Grants are not considered for: general operating expenses, endowments or deficit reduction; general and federated campaigns, including fundraising events, dinners or mass mailings; individuals; conference or seminar sponsorship; book publication and film or theater productions; public policy research; institutions that are located outside the United States; institutions that do not have at least three consecutive full, certified, audited financial statements; conduit organizations, unified funds or organizations that use grant funds from donors to support other organizations or individuals; or institutions that are subsidiaries or affiliates of larger entities that do not have a separate board of directors and independent audited financial statements.
Geographic Focus: All States
Date(s) Application is Due: May 1; Nov 1
Amount of Grant: 500,000-5,000,000 USD
Contact: Margie Antonetti, Grants and Database Specialist; (213) 680-3833; fax (213) 614-0934; mantonetti@wmkeck.org or SciEng@wmkeck.org
Internet: http://www.wmkeck.org/grant-programs/science-engineering.html
Sponsor: W.M. Keck Foundation
550 S Hope Street, Suite 2500
Los Angeles, CA 90071

W.M. Keck Foundation Undergraduate Education Grants **4825**
The W.M. Keck Foundation concentrates on strengthening studies and programs in accredited universities and colleges and major, independent medical research institutions. The Foundation's Undergraduate Education program promotes distinctive learning and research experiences in science, engineering, and the liberal arts at four-year undergraduate colleges only in Foundation designated states, or through national organizations that address undergraduate needs. Historically, grants range from $200,000 to $1 million, and are typically under $500,000. On a case-by-case basis, the Foundation may consider the award of smaller seed grants for planning and piloting. Applicants are strongly urged to contact Foundation staff well in advance of submitting a Phase I Application. The best times for these contacts are between January 1 and February 15 leading up to a May 1 submittal, or between July 1 and August 15 leading up to a November 1 submittal.
Restrictions: Grants are not considered for: general operating expenses, endowments or deficit reduction; general and federated campaigns, including fundraising events, dinners or mass mailings; individuals; conference or seminar sponsorship; book publication and film or theater productions; public policy research; institutions that are located outside the United States; institutions that do not have at least three consecutive full, certified, audited financial statements; conduit organizations, unified funds or organizations that use grant funds from donors to support other organizations or individuals; or institutions that are subsidiaries or affiliates of larger entities that do not have a separate board of directors and independent audited financial statements.
Geographic Focus: Alaska, Arizona, California, Colorado, Hawaii, Idaho, Kansas, Louisiana, Mississippi, Montana, Nevada, New Mexico, Oklahoma, Oregon, Texas, Utah, Washington, Wyoming
Date(s) Application is Due: May 1; Nov 1
Amount of Grant: 200,000-1,000,000 USD
Contact: Margie Antonetti, Grants and Database Specialist; (213) 680-3833; fax (213) 614-0934; mantonetti@wmkeck.org or UGProgram@wmkeck.org
Internet: http://www.wmkeck.org/grant-programs/undergraduate-program.html
Sponsor: W.M. Keck Foundation
550 S Hope Street, Suite 2500
Los Angeles, CA 90071

Wabash Center Project Grants **4826**
The Center provides grants for activities that enhance teaching and learning in the fields of religion and theology. Proposals need to demonstrate that the project improves practical application of teaching and learning methods, creates a supportive environment for teachers, promotes a sustained conversation about pedagogy, or supports research and study of pedagogical issues. Projects might involve meetings of scholars across schools, regions, denominations, or subject areas to think abut syllabi, teaching strategies, student learning styles, or teaching as a vocation. Projects might involve a single department experimenting with creative ways to enhance teaching or a doctoral institution helping their students prepare for teaching careers. Project Grants require a full application process and are awarded at three different times during the year for up to $20,000.
Requirements: Full-time faculty members in religion departments of colleges and universities in North America (United States and Canada) and in theological schools accredited by the Association of Theological Schools are eligible to apply for these grants.
Restrictions: Activities not funded include routine curriculum revision or course development; international travel for research, attendance at disciplinary conferences, or the publication of conference papers; or the preparation of textbooks.
Geographic Focus: All States
Date(s) Application is Due: Jan 1; May 1; Sep 1
Contact: Rev. Dr. Paul O. Myhre, Associate Director; (765) 361-6436 or (800) 655-7117; fax (765) 361-6051; myhrep@wabash.edu
Internet: http://www.wabashcenter.wabash.edu/grants/types.aspx
Sponsor: Wabash Center for Teaching and Learning in Theology and Religion
301 W Wabash Avenue, P.O. Box 352
Crawfordsville, IN 47933

Wabash Center Small Project Grants **4827**
The Center provides grants for activities that enhance teaching and learning in the fields of religion and theology. Proposals need to demonstrate that the project improves practical application of teaching and learning methods, creates a supportive environment for teachers, promotes a sustained conversation about pedagogy, or supports research and study of pedagogical issues. Projects might involve meetings of scholars across schools, regions, denominations, or subject areas to think abut syllabi, teaching strategies, student learning styles, or teaching as a vocation. Projects might involve a single department experimenting with creative ways to enhance teaching or a doctoral institution helping their students prepare for teaching careers. Small Project Grants have a short application procedure and can be approved anytime throughout the year for: up to $2,500; and group or individual projects.
Requirements: Full-time faculty members in religion departments of colleges and universities in North America (United States and Canada) and in theological schools accredited by the Association of Theological Schools are eligible to apply for these grants.
Restrictions: Activities not funded include routine curriculum revision or course development; international travel for research, attendance at disciplinary conferences, or the publication of conference papers; or the preparation of textbooks.

Geographic Focus: All States
Amount of Grant: Up to 2,500 USD
Contact: Rev. Dr. Paul O. Myhre, Associate Director; (765) 361-6436 or (800) 655-7117; fax (765) 361-6051; myhrep@wabash.edu
Internet: http://www.wabashcenter.wabash.edu/grants/index.html
Sponsor: Wabash Center for Teaching and Learning in Theology and Religion
301 W Wabash Avenue, P.O. Box 352
Crawfordsville, IN 47933

Wabash Center Waorkshop and Colloquy Participant Grants 4828
The Center provides funds for projects that enhance teaching and learning in the fields of religion or theology. It seeks to fund: projects that improve the practical application of teaching and learning methods; projects that create supportive environments for teachers; and projects that promote a sustained conversation about pedagogy. In this category, the Center offers particular grant programs to workshop and colloquy participants, such as: the Summer Workshop Fellowship and Grant Program; Colloquy Follow-up Grants; and Workshop on Teaching College Introductory Courses Departmental Grants. Only participants in these workshop or colloquy programs are eligible for these grants.
Requirements: Full-time faculty members in religion departments of colleges and universities in North America (United States and Canada) and in theological schools accredited by the Association of Theological Schools are eligible to apply for these grants.
Restrictions: Activities not funded include routine curriculum revision or course development; international travel for research, attendance at disciplinary conferences, or the publication of conference papers; or the preparation of textbooks.
Geographic Focus: All States
Contact: Rev. Dr. Paul O. Myhre, Associate Director; (765) 361-6436 or (800) 655-7117; fax (765) 361-6051; myhrep@wabash.edu
Internet: http://www.wabashcenter.wabash.edu/grants/index.html
Sponsor: Wabash Center for Teaching and Learning in Theology and Religion
301 W Wabash Avenue, P.O. Box 352
Crawfordsville, IN 47933

Wallace Foundation Grants 4829
The foundation has three current objectives: strengthen education leadership to improve student achievement; improve after-school learning opportunities; and expand participation in arts and culture. Specifically, the fund seeks to develop effective school leaders to improve student learning, create new standards for community participation in the arts, and provide high-quality informal learning opportunities in communities for children and families. Most grants are awarded as part of funds-initiated programs, and unsolicited proposals are rarely funded. Nevertheless, organizations may submit a one- to two-page letter of inquiry describing the project, the organization, the estimated total for the project, and the portion requiring funding.
Requirements: U.S. nonprofit organizations are eligible. In most cases, the foundation identifies and evaluates prospective grantees through the issuance of Requests for Proposals or other careful screening processes. This also means that unsolicited proposals are rarely funded. Nevertheless, you may submit an inquiry by email briefly describing the project, your organization, the estimated total for the project and the portion requiring funding to: The Wallace Foundation, grantrequest@wallacefoundation.org.
Restrictions: Grants do not support religious or fraternal organizations; international programs; conferences; historical restoration; health, medical, or social service programs; environmental/conservation programs; capital campaigns; emergency funds or deficit financing; private foundations; or individuals.
Geographic Focus: All States
Amount of Grant: 75,000-4,000,000 USD
Contact: Holly Dodge, Grants Administrator; (212) 251-9700; fax (212) 679-6990
Internet: http://www.wallacefoundation.org/GrantsPrograms/GrantApproach/Pages/FundingGuidelines.aspx
Sponsor: Wallace Foundation
5 Penn Plaza, 7th Floor
New York, NY 10001

Walter J. Jensen Fellowship for French Language, Literature, and Culture 4830
The Fellowship is awarded for at least six months of study in France. The purpose of the award is to help educators and researchers improve education in standard French language, literature and culture and in the study of standard French in the United States. The fellowship is awarded annually and has a stipend of at least $10,000. Phi Beta Kappa will also cover a single round-trip, economy-class ticket for the recipient to travel to France; some additional support may be available to those with dependents.
Requirements: Candidates must be U.S. citizens under the age of 40 who can demonstrate their career does or will involve active use of the French language. They must have earned a bachelor's degree from an accredited four-year institution with a 3.0 minimum GPA in French language and literature as a major. They must demonstrate superior competence in French, according to the standards established by the American Association of Teachers of French.
Geographic Focus: All States
Amount of Grant: 10,000 USD
Contact: Sandra Beasley, Awards Coordinator; (202) 265-3808; fax (202) 986-1601; SBeasley@pbk.org or info@pbk.org
Internet: http://staging.pbk.org/AM/Template.cfm?Section=Jensen_Fellowship&Template=/CM/HTMLDisplay.cfm&ContentID=2648
Sponsor: Phi Beta Kappa Society
1606 New Hampshire Avenue NW
Washington, DC 20009

Washington Area Women's Foundation Strategic Opportunity and Partnership Fund Grants 4831
In response to the findings in "A Portrait of Women & Girls in the Washington Metropolitan Area," the Women's Foundation created Stepping Stones to help build the economic security and financial independence of low-income, women-headed families. Stepping Stones is the first multi-year initiative of its kind for the region and will be the major focus of grants and program work. One area of funding under the Stepping Stones umbrella is known as Strategic Opportunity and Partnership Fund Grants. This Fund provides grants to nonprofit organizations engaged in research, policy advocacy, grassroots efforts, capacity building or communications and media activities that support the overall goals of Stepping Stones.
Requirements: Nonprofits in the Washington metropolitan area are eligible to apply.
Geographic Focus: District of Columbia
Contact: Nicole Cozier; (202) 347-7737, ext. 203; fax (202) 347-7739; ncozier@wawf.org
Internet: http://thewomensfoundation.org/our-work/stepping-stones/strategic-opportunity-and-partnership-fund/
Sponsor: Washington Area Women's Foundation
1331 H Street, NW, Suite 1000
Washington, DC 20005

WAWH Barbara Penny Kanner Award 4832
The Western Association of Women Historians will award $500 to the best scholarly bibliographical and historical guide to research focused on women or gender history. The Kanner Award is intended to promote the practice of bibliomethodology or autobiography in historical context. The bibliomethodology award should reflect the critical tools of the historian's craft as they have been developed to provide research guides rather than library catalogs. The autobiography in historical context award should reflect the craft of history as developed and interpreted in individual lives. Book-length submissions are preferred but substantial guides in other forms (articles or book chapters) may also be considered. The bibliographical and the autobiography awards will be given in alternate years, with submissions for each award considered for two-year intervals from the date of publication.
Requirements: Recipient must be a member of the association.
Geographic Focus: All States
Date(s) Application is Due: Jan 15
Amount of Grant: 500 USD
Contact: Regina Lark; (310) 206-5898; rlark@women.ucla.edu
Internet: http://www.wawh.org/2005awardinfo.html#Kanner
Sponsor: Western Association of Women Historians
P.O. Box 758
Knights Landing, CA 95645

WAWH Founders Dissertation Fellowship 4833
The association promotes the professional achievement of female historians working in any field of history. Usually one fellowship is offered per year to members of the association.
Requirements: Applicants must be members of the WAWH, advanced to candidacy, writing the dissertation at the time of the application, and expecting to receive the PhD.
Geographic Focus: All States
Date(s) Application is Due: Jan 15
Amount of Grant: 1,000 USD
Contact: Alexandra Nickliss, Dept of Social Science, City College of San Francisco; (415) 239-3283; anicklis@ccsf.org
Internet: http://www.wawh.org/awardsandprizes.html#FDF
Sponsor: Western Association of Women Historians
50 Phelan Avenue
San Francisco, CA 94112

WAWH Sierra Award 4834
The award is given each year for the best monograph published by a Western Association of Women Historians member during the previous year. The monograph must be based on original research (not anthology or edited work) in any field of history.
Requirements: Individuals submitting works for the award should either be members residing in the western states (Alaska, Arizona, California, Colorado, Hawaii, Idaho, Kansas, Montana, Nebraska, Nevada, New Mexico, North Dakota, Oklahoma, Oregon, South Dakota, Texas, Utah, Washington, Wyoming, and Canadian western provinces [Alberta, British Columbia]); or, if resident elsewhere, individuals should have been paid WAWH members for three years (though not necessarily consecutive years) prior to submission.
Geographic Focus: All States
Date(s) Application is Due: Jan 15
Amount of Grant: 250 USD
Contact: Jessica Weiss, (510) 885-3239; jessica.weiss@csueastbay.edu
Internet: http://www.wawh.org/awardsandprizes.html#Sierra
Sponsor: Western Association of Women Historians
25800 Carlos Bee Boulevard
Hayward, CA 94542-3014

Wayne and Gladys Valley Foundation Grants 4835
The foundation awards grants to California nonprofits in its primary areas of interest, including education at all levels, health care and medical research, youth and families, human services, sciences and engineering, and religion. Types of support include general operating support, building construction/renovation, program development, professorships, research, and matching funds. There are no application forms or deadlines. The board meets in February, May, September, and November to consider requests.
Requirements: California nonprofits in Alameda, Contra Costa, and Santa Clara Counties are eligible.

Geographic Focus: California
Contact: Michael Desler; (510) 466-6060; fax (510) 466-6067; info@wgvalley.org
Sponsor: Wayne and Gladys Valley Foundation
1939 Harrison Street, Suite 510
Oakland, CA 94612-3532

Web We Want Small Grants 4836

The Web We Want campaign exists to support global, national and regional efforts to make the Web universal, free and open. The campaign is focused on creating a world where everyone, everywhere is free to collaborate and create without fear of censorship or persecution. A limited number of small grants are available to support national advocacy and campaigning efforts to make the web free, open, socially relevant and accessible to all. Applications will be accepted in either the Rapid Response or Surveillance Action Research categories. The Rapid Response has a rolling deadline and are micro-grants designed to help when you face an unexpected crisis or strategic opportunity that demands more resources fast. Surveillance Action Research grants will enable national groups or coalitions to investigate their own country's surveillance policies, laws and practices. Funds are available for action research to provide solid evidence to underpin your own local advocacy, media work and campaigning. The results of the country research will also be published under your group's name in a special report on global surveillance. Applications will be accepted until September 30.
Requirements: Applicants must join the Web We Want mailing list (signup at the website) prior to applying. Applications are encouraged from a broad and diverse range of groups and organizations, around the world. Priority will go to non-profit organisations in the global South and to those who show that they have the partnerships and follow-up mechanisms needed to reach, influence and mobilise others in their country. Rapid Response applicants may apply for any amount between $1,000 and $2,000 and can submit an application at any time. Decisions will be made within three (3) working days. If the application is successful funds will be administered within five (5) working days of approval decision. Surveillance Action Research applications can apply for any amount between $1,000 and $3,000 towards the costs of research, report writing and follow-up media work and advocacy based on your findings. All successful grant applicants will be listed publicly, with a brief description of project the grant is supporting on webwewant.org, unless the grant applicant shows that this would put the project at risk.
Restrictions: Grants cannot be used to fund the development or implementation of applications, software, computer equipment, computer literacy training, technology start-ups, hubs etc. These grants are only for advocacy, campaigning and influencing activities.
Geographic Focus: All States, All Countries
Date(s) Application is Due: Sep 30
Internet: http://webwewant.org/about/
Sponsor: World Wide Web Foundation
1889 F Street NW, 2nd Floor
Washington, DC 20006

Weizmann Institute of Science Postdoctoral Fellowships 4837

The institute operates in Israel and internationally in the fields of education and science (biology, biophysics-biochemistry, chemistry, mathematical sciences, and physics). Postdoctoral fellowships and visiting professorships are available. Applications and information are posted on the Web site.
Requirements: The postdoctoral fellowships are open to citizens of all countries. Candidates must have received a PhD degree or equivalent within three years of the submission deadline.
Geographic Focus: All States
Date(s) Application is Due: Jan 1; May 15
Contact: Fellowship Office; 972-8-9344553; fax 972-8-9344114; postdoc@weizmann.ac.il
Internet: http://www.weizmann.ac.il/feinberg/postdoc_fell_weiz.shtml
Sponsor: Weizmann Institute of Science
P.O. Box 26
Rehovot, 76100 Israel

Welch Award in Chemistry 4838

The purpose of the award is to foster and encourage basic chemical research and to recognize, in a substantial manner, the value of chemical research contributions for the benefit of mankind. Any person can be considered for the award. The award is intended to recognize contributions that have not previously been rewarded in a similar manner. All nominations must include a brief biographical sketch of the proposed nominee, a list of 25 of the nominee's most important scientific publications, a statement of the research areas in chemistry on which the nomination is based, and not more than three seconding nomination letters. Nominations with supporting documents should be sent to the office.
Requirements: Nominations will be solicited from selected scientific organizations and individuals. Nominations received from other organizations and individuals will also be considered, though no person may nominate oneself.
Restrictions: Directors and officers of the foundation and members of the scientific advisory board of the foundation are not eligible.
Geographic Focus: All States
Date(s) Application is Due: Feb 1
Contact: Carla Atmar, Grant Programs Coordinator; (713) 961-9963, ext. 13 or (713) 961-9884; fax (713) 961-5168
Internet: http://www.welch1.org/Awards/WelchAwardinChemist0943/index.asp
Sponsor: Welch Foundation
5555 San Felipe, Suite 1900
Houston, TX 77056-2732

Welch Foundation Departmental Grants 4839

The purpose of the departmental grant is to support chemical research by members of the chemistry department faculty, and to provide an opportunity for students to study chemistry in a less structured way. The grant is also designed to provide students with the tools they need to gain and develop an interest in pursuing a career in chemistry. This funding also ensures that students will have scholarships, lab equipment and chemicals and can travel to industry meetings and conferences. These grants are awarded to smaller and medium-sized educational institutions in Texas.
Restrictions: Applications are restricted to smaller and medium-sized educational institutions in Texas.
Geographic Focus: Texas
Date(s) Application is Due: Feb 1
Amount of Grant: 150,000 USD
Contact: Carla Atmar; (713) 961-9963, ext. 13 or (713) 961-9884; fax (713) 961-5168
Internet: http://www.welch1.org/GrantsPrograms/DepartmentalGrants/index.asp
Sponsor: Welch Foundation
5555 San Felipe, Suite 1900
Houston, TX 77056-2732

Welch Foundation Endowed Chairs 4840

The purpose of an endowed chair is to provide a faculty position for an eminent scientist, thus further increasing the basic scientific research in chemistry and other sciences in Texas. Historically, the institution must be willing to match the endowment funds dollar for dollar. The Welch Foundation accepts grant requests throughout the grant year, and the Foundation has no required application form for an endowed chair proposal.
Requirements: To be considered for an endowed chair, a Texas educational institution must have a Ph.D. program in chemistry.
Geographic Focus: Texas
Contact: Carla Atmar, Grant Programs Coordinator; (713) 961-9963, ext. 13 or (713) 961-9884; fax (713) 961-5168
Internet: http://www.welch1.org/GrantsPrograms/EndowedChairs/index.asp
Sponsor: Welch Foundation
5555 San Felipe, Suite 1900
Houston, TX 77056-2732

Welch Foundation Norman Hackeman Award in Chemical Research 4841

The purpose of the award is to encourage and recognize young chemical scientists in Texas for their past research endeavors. The award is meant not only to recognize the work of a scientist, but to encourage those who are embarking on careers dedicated to increasing fundamental understanding of chemistry. The award is not intended to finance any further research or activities of the recipient, nor is the recipient required to render any future services as a condition of receiving the award.
Requirements: A candidate for the award must be 40 years of age or younger on June 30 for the year in which nominated. Any Texas college or university may nominate two candidates for any given year. The nomination should normally be by letter from the institution in which the individual is a faculty member. To be eligible, a nominee should have spent the most recent three years ending June 30 doing chemical research in Texas.
Restrictions: Directors and officers of the foundation and members of the scientific advisory board of the foundation are not eligible. Self-nominations are not accepted.
Geographic Focus: Texas
Date(s) Application is Due: Sep 15
Contact: Carla Atmar, Grant Programs Coordinator; (713) 961-9963, ext. 13 or (713) 961-9884; fax (713) 961-5168
Internet: http://www.welch1.org/Awards/NormanHackermanAwar11BD/index.asp
Sponsor: Welch Foundation
5555 San Felipe, Suite 1900
Houston, TX 77056-2732

Welch Foundation Research Grants 4842

The foundation supports fundamental chemical research at educational institutions within the state of Texas. Each research grant provides a minimum of $150,000 in funding over a three year period to support research in chemistry by a full-time tenured or tenure-track faculty member who serves as principal investigator. Each research grant may utilize its funds on stipends, fellowships, scholarships, permanent scientific equipment, expendable scientific items, services, equipment maintenance, publication expenses, and travel.
Restrictions: Applications are restricted to universities, colleges, or other educational institutions located in the state of Texas. Only one proposal may be submitted naming a particular principal investigator.
Geographic Focus: Texas
Date(s) Application is Due: Feb 1
Contact: Carla Atmar, Grant Programs Coordinator; (713) 961-9963, ext. 13 or (713) 961-9884; fax (713) 961-5168
Internet: http://www.welch1.org/GrantsPrograms/ResearchGrants/index.asp
Sponsor: Welch Foundation
5555 San Felipe, Suite 1900
Houston, TX 77056-2732

Wellcome Trust Biomedical Science Grants 4843

Wellcome Trust supports research into all aspects of science: from molecules and cells vital to life, through the spread of diseases or the vectors of disease across the globe, to clinical and public health research to improve the quality of healthcare. Research can be based in the laboratory, the clinic or the field, and may involve experimental or theoretical approaches. The Trust not only funds scientists but also supports undergraduate and

doctoral students, clinicians, dentists and veterinarians. It funds the best researchers with the most innovative and exciting ideas in the form of personal and research support, collaborations and support for symposia, conferences and workshops. A significant proportion of the funding is devoted to support those working overseas.
Requirements: Overseas applicants should fulfill the eligibility criteria for principal applicants. Principal applicants must be established researchers who are expected to apply from an eligible organization, able to sign up to the Wellcome Trust's Grant Conditions, and should normally: hold an academic or research post (or equivalent); have at least five years' postdoctoral (or equivalent) research experience (ndividuals who hold an established lectureship or a well-recognized fellowship, but who have less than five years postdoctoral experience are also eligible; other researchers with less than five years' experience should contact the Grants Information Desk for further advice); be in receipt of salary funding for the duration of the grant requested or, if this is not in place, have a position that is underwritten by their employing institution for the duration of the grant.
Geographic Focus: All States, All Countries
Contact: Kevin Moses, Director of Science; +44 (0)20 7611 5757 or +44 (0)20 7611 8888; fax +44 (0)20 7611 8258; sciencegrants@wellcome.ac.uk
Internet: http://www.wellcome.ac.uk/Funding/Biomedical-science/index.htm
Sponsor: Wellcome Trust
Gibbs Building, 215 Euston Road
London, GLONDON NW1 2BE United Kingdom

Wellcome Trust New Investigator Awards **4844**
The Wellcome Trust is particularly interested in receiving applications from university lecturers within the first five years of their independent research careers. New Investigator Awards are intended to support strong researchers who are in the early stages of their independent research careers and have already shown that they can innovate and drive advances in their field of study. Candidates should be no more than five years from appointment to their first academic position. These Awards provide flexible support at a level and length appropriate to enable the best researchers to address the most important questions about health and disease. Awards may equally be small or large, but candidates should be able to articulate a compelling vision for their research, while ensuring that their proposal and requested funding is appropriate to their research experience to date. The research should contribute to the Trust's vision to achieve extraordinary improvements in human and animal health, and the proposal should describe how it will address one or more of our five major challenges. Prospective applicants are encouraged to contact the Trust if they wish to discuss their proposal or their suitability for this scheme.
Requirements: To be eligible for a New Investigator Award you should be based in the UK, Republic of Ireland or a low- or middle-income country, and you should: have an established academic post at an eligible higher education or research institution (you are employed on a permanent, open-ended or long-term rolling contract, salaried by your host institution); and be no more than five years from appointment to your first established academic post on the date you submit your application. More experienced researchers who have obtained significant, independent research support prior to their first academic appointment will be advised to apply for a Senior Investigator Award. Applicants are also eligible to apply if you have a written guarantee of an
Geographic Focus: Afghanistan, Albania, Algeria, American Samoa, Angola, Argentina, Armenia, Azerbaijan, Bangladesh, Belarus, Belize, Benin, Bhutan, Bolivia, Bosnia & Herzegovina, Botswana, Brazil, Bulgaria, Burkina Faso, Burundi, Cambodia, Cameroon, Cape Verde, Central African Republic, Chad, China, Colombia, Comoros, Congo, Congo, Democratic Republic of, Costa Rica, Cote d' Ivoire (Ivory Coast), Cuba, Djibouti, Dominica, Dominican Republic, Ecuador, Egypt, El Salvador, Eritrea, Ethiopia, Fiji, Gabon, Gambia, Ghana, Grenada, Guatemala, Guinea, Guinea-Bissau, Guyana, Haiti, Honduras, India, Indonesia, Iran, Iraq, Ireland, Jamaica, Jordan, Kazakhstan, Kenya, Kiribati, Korea, Kosovo, Kyrgyzstan, Laos, Lebanon, Lesotho, Liberia, Libya, Macedonia, Madagascar, Malawi, Malaysia, Maldives, Mali, Mauritania, Mauritius, Mexico, Micronesia, Moldova, Mongolia, Montenegro, Morocco, Mozambique, Myanmar (Burma), Namibia, Nepal, Nicaragua, Niger, Nigeria, Pakistan, Palau, Panama, Papua New Guinea, Paraguay, Peru, Philippines, Romania, Rwanda, Sao Tome & Principe, Senegal, Serbia, Seychelles, Sierra Leone, Solomon Islands, Somalia, South Africa, South Sudan, Sri Lanka, St. Lucia, St. Vincent and the Grenadines, Suriname, Swaziland, Syrian Arab Republic, Tajikistan, Tanzania, Thailand, Timor-Lester, Togo, Tonga, Tunisia, Turkey, Turkmenistan, Tuvalu, Uganda, Ukraine, United Kingdom, Uzbekistan, Vanuatu, Venezuela, Vietnam, Yemen, Zambia, Zimbabwe
Date(s) Application is Due: Mar 14; Jul 14; Nov 14
Contact: Kevin Moses, Director of Science; +44 (0)20 7611 5757 or +44 (0)20 7611 8888; fax +44 (0)20 7611 8258; sciencegrants@wellcome.ac.uk
Internet: http://www.wellcome.ac.uk/Funding/Biomedical-science/Funding-schemes/Investigator-Awards/WTX059284.htm
Sponsor: Wellcome Trust
Gibbs Building, 215 Euston Road
London, GLONDON NW1 2BE United Kingdom

Wellesley Centers for Women Postdoctoral Fellowships **4845**
The program provides systematic research training for candidates who want to develop expertise in research on childhood and adolescence and investigate variations in race and ethnicity, gender, and social class and how these interact with risk and resiliency factors in human development. The training program consists of three components: training and career development; collaborating with the preceptor on the preceptor's externally funded research projects; and independent research of the fellow's choosing conducted under the guidance of his/her preceptor. One of the program's goals is to identify and develop scholars from underrepresented groups who will launch independent research careers through external funding. Guidelines are available online.
Requirements: Candidates must be U.S. citizens or permanent residents. Researchers with PhDs, MDs, or comparable doctoral degrees from accredited foreign institutions also are eligible.
Geographic Focus: All States
Date(s) Application is Due: Jan 1
Amount of Grant: 40,000 USD
Contact: Sumru Erkut, (781) 283-2533; serkut@wellesley.edu
Internet: http://www.wcwonline.org/postdoc/index.html
Sponsor: Wellesley College
106 Central Street
Wellesley, MA 02181-8200

Wenner-Gren Foundation African Fellowships **4846**
The Fellowships are intended to provide support for African students undertaking study leading to a Ph.D. at a South African university that can provide them with international-level training in anthropology (including biological anthropology and archaeology). The annual fellowship is $17,500 and can be used towards travel, living expenses, tuition, student fees, insurance, books, research expenses, and any other relevant categories of expenditure while studying at the Host institution. The fellowship is renewable for up to two additional years upon successful completion of each preceding year's study. A separate application can be made for an additional year of funding to support dissertation write-up. The Program has a two-stage application process. The first stage is a preliminary inquiry. The deadline for the preliminary inquiry is December 1. If a candidate is considered eligible based on the review of the preliminary inquiry he/she will be invited to submit a full application. The deadline for the full application is December 15. Final decisions are made by January 15.
Requirements: The Fellowship is intended for young African scholars who may not otherwise be able to pursue a doctoral degree in anthropology. Normally, applicants will be under 35 years of age at the time they begin their fellowship. They must be citizens and residents of an African country at the time of application. They must also be members of an underrepresented group in academic anthropology/archaeology.
Geographic Focus: All States, Algeria, Angola, Benin, Botswana, Burkina Faso, Burundi, Cameroon, Cape Verde, Central African Republic, Chad, Comoros, Congo, Congo, Democratic Republic of, Cote d' Ivoire (Ivory Coast), Djibouti, Egypt, Equatorial Guinea, Eritrea, Ethiopia, Gabon, Gambia, Ghana, Guinea, Guinea-Bissau, Kenya, Lesotho, Liberia, Libya, Madagascar, Malawi, Mali, Mauritania, Mauritius, Morocco, Mozambique, Namibia, Niger, Nigeria, Rwanda, Sao Tome & Principe, Senegal, Seychelles, Sierra Leone, Somalia, South Africa, Sudan, Swaziland
Date(s) Application is Due: Dec 1; Dec 15
Amount of Grant: 17,500 USD
Contact: Leslie C. Aiello; (212) 683-5000; fax (212) 683-9151; laiello@wennergren.org
Internet: http://www.wennergren.org/programs/programs_show.htm?doc_id=377114&attrib_id=13232
Sponsor: Wenner-Gren Foundation for Anthropological Research
470 Park Avenue S, 8th Floor
New York, NY 10016-6819

Wenner-Gren Foundation Conference and Workshop Grants **4847**
In accordance with the mission of the Foundation, priority is given to events that foster the creation of an international community of research scholars in anthropology and advance significant and innovative anthropological research. Conferences are defined as public events that are comprised primarily of oral and poster presentations to a larger audience of anthropologists. Priority is given to major conferences sponsored by large international anthropological organizations. Workshops are defined as working meetings that focus on developing and debating topical issues in theoretical anthropology. Priority is given to those workshops that devote the majority of time to discussion and debate rather than to the presentation of papers. Application deadlines for Conference and Workshop Grants are June 1 and December 1.
Requirements: Applications must be made by a professional anthropologist who is the primary organizer of the proposed conference or workshop. A co-applicant can be specified on the application.
Restrictions: Applications are not accepted from students without a co-applicant who has a doctorate and holds an established academic position. Individual requests for travel support to attend meetings are not accepted. Symposia within larger professional meetings are generally not supported. Limited funds are available to aid the participation of Third World scholars in international conferences, which are provided in the form of grants to the organizer(s).
Geographic Focus: All States
Date(s) Application is Due: Jun 1; Dec 1
Amount of Grant: Up to 15,000 USD
Contact: Leslie C. Aiello; (212) 683-5000; fax (212) 683-9151; laiello@wennergren.org
Internet: http://www.wennergren.org/programs/programs_show.htm?doc_id=370402&attrib_id=13235
Sponsor: Wenner-Gren Foundation for Anthropological Research
470 Park Avenue S, 8th Floor
New York, NY 10016-6819

Wenner-Gren Foundation Dissertation Fieldwork Grants **4848**
These grants are awarded to individuals to aid doctoral dissertation or thesis research. The program contributes to the Foundation's overall mission to support basic research in anthropology and to ensure that the discipline continues to be a source of vibrant and significant work that furthers understanding of humanity's cultural and biological origins, development, and variation. The Foundation supports research that demonstrates a clear link to anthropological theory and debates, and promises to make a solid contribution

to advancing these ideas. There is no preference for any methodology, research location, or subfield. The Foundation particularly welcomes proposals that employ a comparative perspective, can generate innovative approaches or ideas, and/or integrate two or more subfields. Grants are non-renewable and provide a maximum of U.S. $25,000. Applicants should submit application materials using the Foundation's online application submission procedure as well as send printed copies to the Foundation by regular mail.
Requirements: Students must be enrolled in a doctoral program (or equivalent, if applying from outside the United States) at the time of application. Students of all nationalities are eligible to apply. There is no time limit on the duration of the grant, and funding may be requested to cover distinct research phases.
Restrictions: Funding is not provided for salary and/or fringe benefits of applicant, tuition, non-project personnel, travel to meetings, institutional overhead, or other support for writing, publication assistance, and film- or video making (unless inherent to the research). Expenses incurred prior to the effective date of an award will not be covered.
Geographic Focus: All States
Date(s) Application is Due: May 1; Nov 1
Amount of Grant: Up to 25,000 USD
Contact: Leslie C. Aiello; (212) 683-5000; fax (212) 683-9151; laiello@wennergren.org
Internet: http://www.wennergren.org/programs/programs_show.htm?doc_id=367834&attrib_id=13232
Sponsor: Wenner-Gren Foundation for Anthropological Research
470 Park Avenue S, 8th Floor
New York, NY 10016-6819

Wenner-Gren Foundation Historical Archives Grants **4849**
The objective of this program is to encourage the preservation of unpublished records and other materials of value for research on the history of anthropology. Grants are offered for two purposes: to assist senior scholars (or their heirs) holding unpublished research materials and personal papers of significant value for the study of the history of Anthropology, with the expenses of preparing and transferring them for archival deposit; and to aid oral-history interviews with senior anthropologists. Applicants must first submit a Preliminary Inquiry to determine their eligibility. If the project is deemed eligible for consideration, an application form will be sent. There is no fixed deadline. Inquiries and applications are reviewed as they are received.
Restrictions: Funds are not provided to establish archival facilities or develop collections already held in archival deposit.
Geographic Focus: All States
Amount of Grant: Up to 15,000 USD
Contact: Leslie C. Aiello; (212) 683-5000; fax (212) 683-9151; laiello@wennergren.org
Internet: http://www.wennergren.org/programs/programs_show.htm?doc_id=369096&attrib_id=13236
Sponsor: Wenner-Gren Foundation for Anthropological Research
470 Park Avenue S, 8th Floor
New York, NY 10016-6819

Wenner-Gren Foundation Hunt Postdoctoral Fellowships **4850**
The fellowships support the writing-up of already completed research. The fellowship is awarded to scholars in the earlier stages of their careers, when they frequently lack the time and resources to develop their research for publication. Scholars with a Ph.D. in hand for no more than ten years (from the application deadline) are eligible to apply. A maximum of eight Hunt Postdoctoral Fellowships are awarded annually.
Requirements: Applicants must have a Ph.D. or equivalent at the time of application. Applicants must have received a Ph.D. or equivalent within ten years of the application deadline.
Restrictions: Aid is not provided for salary and/or fringe benefits of applicant, family expenses, or institutional overhead. Support is not normally provided for master's studies.
Geographic Focus: All States
Date(s) Application is Due: May 1; Nov 1
Amount of Grant: 40,000 USD
Contact: Leslie C. Aiello; (212) 683-5000; fax (212) 683-9151; laiello@wennergren.org
Internet: http://www.wennergren.org/programs/programs_show.htm?doc_id=368531&attrib_id=13233
Sponsor: Wenner-Gren Foundation for Anthropological Research
470 Park Avenue S, 8th Floor
New York, NY 10016-6819

Wenner-Gren Foundation International Collaborative Research Grants **4851**
Grants are available to assist anthropological research projects undertaken jointly by two (or more) investigators from different countries. These grants are renewable for a second period of research. The purpose of the program is to encourage collaborations in which the principal investigators bring different and complementary perspectives, knowledge, and/or skills. Priority is given to new collaborations, but an independent project within a larger, long-term collaboration can be considered.
Requirements: Projects must involve at least one principal investigator from outside the United States, Canada, Western Europe, Australia, or New Zealand. Both investigators must meet the qualification for Post-PhD Grants (i.e., holding the doctorate or equivalent in anthropology or a related discipline).
Restrictions: Projects primarily for other purposes, such as training, education, or writing, are not eligible under this program.
Geographic Focus: All States
Date(s) Application is Due: Jun 1; Dec 1
Amount of Grant: Up to 30,000 USD
Contact: Leslie C. Aiello; (212) 683-5000; fax (212) 683-9151; laiello@wennergren.org
Internet: http://www.wennergren.org/programs/programs_show.htm?doc_id=368683&attrib_id=13233
Sponsor: Wenner-Gren Foundation for Anthropological Research
470 Park Avenue S, 8th Floor
New York, NY 10016-6819

Wenner-Gren Foundation Post-PhD Research Grants **4852**
Grants are awarded to individuals holding a Ph.D. or equivalent degree to support individual research projects. The program contributes to the Foundation's overall mission to support basic research in anthropology and to ensure that the discipline continues to be a source of vibrant and significant work that furthers our understanding of humanity's cultural and biological origins, development, and variation. The Foundation supports research that demonstrates a clear link to anthropological theory and debates, and promises to make a solid contribution to advancing these ideas. There is no preference for any methodology, research location, or subfield. The Foundation particularly welcomes proposals that employ a comparative perspective, can generate innovative approaches or ideas, and/or integrate two or more subfields. Grants are for a maximum of U.S. $25,000 towards research expenses.
Requirements: Applications are accepted from scholars who are close to completing their doctorate or equivalent degree; however, the grant cannot be awarded until the degree is completed.
Geographic Focus: All States
Date(s) Application is Due: May 1; Nov 1
Amount of Grant: Up to 25,000 USD
Contact: Leslie C. Aiello; (212) 683-5000; fax (212) 683-9151; laiello@wennergren.org
Internet: http://www.wennergren.org/programs/programs_show.htm?doc_id=368389&attrib_id=13233
Sponsor: Wenner-Gren Foundation for Anthropological Research
470 Park Avenue S, 8th Floor
New York, NY 10016-6819

Wenner-Gren Foundation Short-Term Fellowships **4853**
The Fellowships are intended to provide support for doctoral students or post-doctoral scholars who are within ten years of receiving their doctorate to enhance their research capacity in the field of anthropology. The fellowship may be used to: pursue research in a library outside the applicant's home country; or take advantage of a training opportunity outside the applicant's home country. The fellowships provide funding of up to $5000 and are not renewable. The length of the library or training visit should normally be no more than 3 months in duration. The award can be used towards relevant expenses for travel, subsistence, fees, insurance, books, etc. There is no deadline for the Wadsworth Short-Term Fellowships; however, there is a two-stage application process. The first stage is the preliminary inquiry. If an applicant is considered eligible based on the review of the preliminary inquiry, he/she will be invited to submit a full application.
Requirements: Applicants must be from countries where anthropology is underrepresented and where there are limited resources to send scholars overseas for training or library research. Student applicants must be registered for a doctoral degree in anthropology or a related sub-discipline at a university in a qualifying country. Post-doctoral applicants must be within ten years of completing their doctoral degree and hold an academic position in a university in a qualifying country.
Geographic Focus: All States
Amount of Grant: 5,000 USD
Contact: Leslie C. Aiello; (212) 683-5000; fax (212) 683-9151; laiello@wennergren.org
Internet: http://www.wennergren.org/programs/programs_show.htm?doc_id=376103&attrib_id=13232
Sponsor: Wenner-Gren Foundation for Anthropological Research
470 Park Avenue S, 8th Floor
New York, NY 10016-6819

Wenner-Gren Foundation Wadsworth International Fellowships **4854**
The Fellowships are intended to provide support for students undertaking study leading to a Ph.D. or equivalent doctoral degree at universities where they can receive international-level training in anthropology. These fellowships are available only to students from countries where anthropology is underrepresented and where there are limited resources to send students overseas for training. Priority is given to applicants who have not already begun graduate training outside of their home countries. Applications from individuals who are more advanced in their doctoral studies will, however, be considered. The annual fellowship is $17,500 and can be used towards travel, living expenses, tuition, income taxes, student fees, insurance, books, research expenses and any other relevant categories of expenditure while studying at the Host Institution. The fellowship is renewable for up to two additional years upon successful completion of each preceding year's study or research. A separate application can be made for an additional year of funding to support dissertation write-up. The Program has a two-stage application process. The first stage is a preliminary inquiry. The deadline for the preliminary inquiry is February 1. If a candidate is considered eligible based on the review of the preliminary inquiry he/she will be invited to submit a full application. The deadline for the full application is March 1. Final decisions are made by May 1.
Requirements: Applicants must be from countries where anthropology is underrepresented and where there are limited resources to send students overseas for training. Normally, applicants will be under 35 years of age at the time they begin their fellowship.
Restrictions: Applicants wishing to study anthropology at universities in their home country are not eligible to apply.
Geographic Focus: All States
Date(s) Application is Due: Feb 1; Mar 1
Amount of Grant: 17,500 USD
Contact: Leslie C. Aiello; (212) 683-5000; fax (212) 683-9151; laiello@wennergren.org

Internet: http://www.wennergren.org/programs/programs_show.htm?doc_id=368103&attrib_id=13232
Sponsor: Wenner-Gren Foundation for Anthropological Research
470 Park Avenue S, 8th Floor
New York, NY 10016-6819

Western Regional Graduate Program **4855**
Residents of 14 western states are eligible to attend distinguished graduate programs at reduced tuition rates through the Western Regional Graduate Program, a program coordinated by WICHE. Qualified residents of Alaska, Arizona, Colorado, Hawaii, Idaho, Montana, Nevada, New Mexico, North Dakota, Oregon, South Dakota, Utah, Washington, and Wyoming may enroll in any of the qualified graduate programs at institutions in those states (list available on the Web site). Students pay resident tuition at public institutions or substantially reduced tuition at private institutions. Application deadlines are set by the institution and department. Students should apply directly to the programs, not WICHE; however, WICHE will provide an information brochure upon request.
Geographic Focus: Alaska, Arizona, Colorado, Hawaii, Idaho, Montana, Nevada, New Mexico, North Dakota, Oregon, South Dakota, Utah, Washington, Wyoming
Date(s) Application is Due: Nov 18
Contact: Program Coordinator; (303) 541-0200; info-sep@wiche.edu
Internet: http://wiche.edu/SEP/WRGP
Sponsor: Western Interstate Commission for Higher Education
P.O. Box 9752
Boulder, CO 80301-9752

Westinghouse Charitable Giving Grants **4856**
The foundation makes charitable contributions to community priorities primarily where Westinghouse has a presence. Areas of emphasis include education—elementary, secondary, and high school programs that emphasize math and science (some non-fine arts programs); health and welfare—to improve quality of life in the community; and civic and social affairs—to support economic development, environmental quality, and the preservation of public safety. Within each area, the foundation actively encourages programs that meet the needs of special populations, such as the disadvantaged, the young, the elderly, the gifted, people with disabilities, minorities, and women. Because of the foundation's approach to funding in Westinghouse plant locations, all organizations, except colleges and universities, should submit their requests for support to the nearest Westinghouse location. Applications are accepted on an ongoing basis.
Requirements: 501(c)3 tax-exempt organizations are eligible.
Restrictions: The following organizations and activities are ineligible: individuals; political contributions; religious organizations; highly specialized health, medical, or welfare programs; fine arts organizations/programs; organizations that discriminate by race, color, creed, gender, or national origin; hospitals; colleges, universities, and two-year institutions; capital improvement or building projects; medical education; chairs or professorships; liberal arts, fine arts, or similar educational programs; educational research programs, regardless of academic discipline; graduate education; general endowment; equipment purchases at universities; United Way-affiliated agencies; tickets; or memberships.
Geographic Focus: All States
Amount of Grant: Up to 5,000 USD
Contact: Program Officer; (412) 374-6824; fax (412) 642-4874
Internet: http://www.westinghousenuclear.com/E2.asp
Sponsor: Westinghouse Electric Corporation
P.O. Box 355, ECE 575C
Pittsburgh, PA 15230-0355

Weyerhaeuser Company Foundation Grants **4857**
The foundation's mission is to improve the quality of life in communities where Weyerhaeuser has a major presence and to provide leadership that increases public understanding of issues where society's needs intersect with the interests of the forest products industry. Community service grants include awards in the fields of education and youth, health and welfare, civic and community improvement, and culture and the arts. In addition to community grants, industry-related awards are made to educational institutions, environmental groups, and professional organizations that promote further understanding of how the forest products industry responds to a changing society. Types of support include seed money, building construction/renovation, equipment acquisition, employee-related scholarships, publication, conferences and seminars, fellowships, lectureships, operating budgets, research, program development, employee matching gifts, and technical assistance. Applicants should send a short letter that introduces the project and sponsoring organization and provide tax-exempt status evidence. If further consideration is warranted, the foundation may ask for additional information or a formal proposal. Proposals may be submitted at any time.
Requirements: Applying organizations must have nonprofit, tax-exempt status. To be considered for funding, an organization must: serve a community within a 50-mile radius of a major Weyerhaeuser facility; support a state-wide issue of interest to the Foundation and Weyerhaeuser in the key states of Alabama, Arkansas, Louisiana, Mississippi, North Carolina, Oklahoma, Oregon or Washington; or support a selected, high-priority national or international initiative directly related to the sustainability and importance of working forests. A limited number of smaller awards are also made to other locales where fewer employees are based.
Restrictions: Grants are not awarded to individuals or for political campaigns, activities that influence legislation, religious organizations seeking funds for theological purposes, or funds to purchase tickets or tables at fundraising benefits.
Geographic Focus: Alabama, Arizona, Arkansas, California, Colorado, District of Columbia, Georgia, Idaho, Illinois, Kentucky, Louisiana, Maryland, Michigan, Minnesota, Mississippi, Missouri, New Hampshire, New Jersey, New Mexico, North Carolina, Ohio, Oklahoma, Oregon, Pennsylvania, South Carolina, Texas, Utah, Virginia, Washington, West Virginia, Wisconsin
Date(s) Application is Due: Aug 31
Contact: Anne Levya, Team Coordinator; (253) 924-3159; fax (253) 924-3658; foundation@weyerhaeuser.com or anne.leyva@weyerhaeuser.com
Internet: http://www.weyerhaeuser.com/Sustainability/Foundation
Sponsor: Weyerhaeuser Company Foundation
P.O. Box 9777
Federal Way, WA 98063-9777

Whitley Awards **4858**
The scheme supports pragmatic, long-lasting field-based nature conservation projects. A major award from the scheme should form the bulk of a project's required funding. The program is interested in wide nature conservation issues, not simply individual wildlife species preservation projects.
Requirements: Applications from leaders of teams undertaking appropriate projects anywhere in the world are encouraged. Close involvement of host country researchers and local institutions is essential. Conservationists from any country working anywhere in the world are welcome to apply. Projects involving a single visit to a country are unlikely to be funded.
Restrictions: PhD project and undergraduate expeditions are ineligible.
Geographic Focus: All States
Date(s) Application is Due: Nov 18
Amount of Grant: 30,000 GBP
Contact: Whitley Award Scheme Contact; +44 (0)20 7602 3443; fax +44 (0)20 7603 3935; info@whitleyaward.org
Internet: http://www.whitley-award.org
Sponsor: Whitley Laing Foundation
50 Queensdale Road
London, W11 4SA

WHOI Geophysical Fluid Dynamics Fellowships **4859**
The Geophysical Fluid Dynamics Fellowship involve an intense ten-week summer research experience and vigorous discussions of concepts that span different disciplines. Participants include graduate students and researchers from a variety of fields who share a common interest in the nonlinear dynamics of rotating, stratified fluids. Under the supervision of staff, Fellows pursue a research project and present a lecture and a written report for a proceedings volume. Participants receive a stipend of $5,600 and an allowance for travel expenses within the U.S. Up to ten Fellowships are available. Applications and a list of required materials are available at the website.
Requirements: Applicants should be graduate students in any field sharing a common interest in the nonlinear dynamics of rotating, stratified fluids. These fields include classical fluid dynamics, physical oceanography, meteorology, astrophysics, planetary atmospheres, geological fluid dynamics, hydromagnetics, physics and applied mathematics. Fellows will be expected to attend the full ten weeks of the Program, complete a research project under the supervision of the staff, and present a lecture and a written report for a proceedings volume at the end of the program.
Geographic Focus: All States
Date(s) Application is Due: Feb 15
Amount of Grant: 5,600 USD
Contact: Janet Fields, GFD Program Coordinator; (508) 289-2950; fax (508) 457-2115; jfields@whoi.edu or gfd@whoi.edu
Internet: http://www.whoi.edu/main/gfd/fellowships
Sponsor: Woods Hole Oceanographic Institution
266 Woods Hole Road
Woods Hole, MA 02543

WHOI Guest Student Internships **4860**
The Woods Hole Oceanographic Institution offers Guest Student Internships for a limited number of graduate and undergraduate students. Guest Student positions are extremely limited, and the ability to place a Guest Student depends on whether the student's interests and academic background coincide with available openings at the time the application is received. Scientists may invite their own Guest Students, so it is strongly recommended that a prospective student contact the scientist(s) with whom he or she is interested in working. Applications are accepted year-round and evaluated on a rolling basis, and are available at the admissions website.
Requirements: Applicants must be current undergraduate or graduate students. Advanced high school students enrolled in Upper Cape and Island schools may be eligible.
Restrictions: Guest Students are responsible for arranging their own room and board, and are expected to arrive with their own health insurance. Woods Hole does not issue academic credit for Guest Student work, but a description and evaluation of a student's work is available upon request.
Geographic Focus: All States
Contact: Michelle McCafferty, Undergraduate Programs Coordinator; (508) 289-2219; fax (508) 457-2188; mmccafferty@whoi.edu
Internet: http://www.whoi.edu/gueststudent/
Sponsor: Woods Hole Oceanographic Institution
266 Woods Hole Road
Woods Hole, MA 02543

WHOI Marine Policy Fellowships 4861

Qualified individuals in the social sciences are invited to apply their training and expertise to the economic, legal and policy issues that arise from use of the world's oceans. At the Institution's Marine Policy Center, emphasis is placed on multidisciplinary research to advance the conservation and management of coastal and marine resources. The work of Marine Policy Fellowship scholars integrates law, policy analysis, and statistics with WHOI's basic strengths in ocean sciences. Current areas of research concentration include fisheries and aquaculture; the economics of coastal communities; marine spatial planning and ecosystem-based management; sea level rise and shoreline change; and water resources management, with a particular focus on coastal aquifers and the effects of climate change. The main objectives of the program are the following: provide support and experience to Research Fellows interested in marine policy issues; offer opportunities for interdisciplinary application of social sciences and natural sciences to marine policy problems; and to conduct research and convey information necessary for the development of effective local, national, and international ocean policy. The Fellowship Program also provides successful applicants the opportunity to participate in Woods Hole seminars and study groups and to develop collaborative research with other members of the scientific and technical staff at the Institution. Workshops and conferences involving senior policymakers and scholars are occasionally held as part of the Center's program. Recipients will receive a stipend of $57,000 for a period of one year and are eligible for group health and dental insurance. In addition, modest research and travel funds will be made available. Recipients are appointed for one year; however, depending upon the nature and scope of the recipient's effort, opportunities may be available to continue a professional affiliation with the Center. The application and information about additional materials needed are located at the website.
Requirements: Applicants must have completed a doctoral level degree or possess equivalent professional qualifications through career experience. The Center also welcomes experienced professionals who can arrange a leave or sabbatical.
Geographic Focus: All States
Date(s) Application is Due: Jan 5
Amount of Grant: 57,000 USD
Contact: Janet Fields; (508) 289-2950; jfields@whoi.edu or postdoc@whoi.edu
Internet: http://www.whoi.edu/page.do?pid=8027
Sponsor: Woods Hole Oceanographic Institution
266 Woods Hole Road
Woods Hole, MA 02543

WHOI Ocean Science Journalism Fellowship 4862

The Ocean Science Journalism Fellowship is designed to introduce science journalists to the interdisciplinary and wide-ranging fields of oceanography and ocean engineering. Through seminars, laboratory visits, and brief field expeditions, Fellows gain access to new research findings. They also gain fundamental background information in engineering, marine biology, geology and geophysics, marine chemistry and geochemistry, and physical oceanography. Topics include the following: harmful algal blooms to deep-sea hydrothermal vents; seafloor earthquakes to ice-sheet dynamics; the ocean's role in climate change to the human impact on fisheries and coastline change; and ocean instruments and observatories to underwater robots. The one-week residential experience offers travel allowance and room and board. International journalists are welcome to apply. The application and a list of required materials are available at the website. Applications are due by May 17, but acceptance begins in early January.
Requirements: The Fellowship is open to professional writers, producers, and editors working for print, broadcast, radio, and Internet media. Ideal candidates will have at least two years of writing, producing, or editing experience for a general-interest audience.
Restrictions: Freelancers, book authors, and writers from non-traditional media are also considered, but should be prepared to show a substantial portfolio of work. The application may be emailed, but required materials should be mailed.
Geographic Focus: All States, All Countries
Date(s) Application is Due: May 17
Contact: Erin Koenig; (508) 289-2270; ekoenig@whoi.edu or media@whoi.edu
Internet: http://www.whoi.edu/page.do?pid=8887
Sponsor: Woods Hole Oceanographic Institution
266 Woods Hole Road
Woods Hole, MA 02543

WHOI Postdoctoral Fellowships 4863

Postdoctoral Fellows with national or international competitively awarded fellowships may be appointed at WHOI, provided they are able to locate a scientist on the staff willing to act as a sponsor. Once a sponsor has been located, and the appointment has been approved by the sponsor's Department chair and the Dean, the recipient will be appointed the appropriately named Postdoctoral Fellow at the Institution. An official letter of notification of the award from the national or international awarding organization must also be received. There is no deadline. The appointment time period will coincide with the fellowship award time period.
Requirements: Fellows must find a scientist on staff with the Institution who is willing to act as their sponsor.
Geographic Focus: All States
Contact: Janet Fields, Postdoctoral Coordinator; (508) 289-2950; fax (508) 457-2115; jfields@whoi.edu or postdoc@whoi.edu
Internet: http://www.whoi.edu/page.do?pid=8029
Sponsor: Woods Hole Oceanographic Institution
266 Woods Hole Road
Woods Hole, MA 02543

WHOI Postdoctoral Scholar Awards 4864

Postdoctoral Scholar awards are offered annually for 18 months to recipients of new or recent doctorates in the fields of chemistry, engineering, geology, geophysics, mathematics, meteorology, physics, and biology as well as oceanography. The competitive awards are designed to further the education and training of the applicant with primary emphasis placed on the individual's research promise. The 18 month appointment includes $57,000 per year, plus $28,500 for the additional six months, plus a relocation allowance. All appointees are eligible for group health and dental insurance. Limited support is available for travel expenses, equipment supplies, and special services. Additional guidelines are posted on the website. The application and a description of additional required materials are available at the website.
Requirements: In order to be eligible for one of these awards, applicants must have received their doctoral degree within the three years. As a condition of the award, successful applicants must be able to present a certified copy of their diploma, or a certified letter from the appropriate college or university official stating that they have completed all of the requirements for their degree including the completion of their final thesis, before they begin their scholar appointment. However, applicants do not need to have already received their degree prior to applying.
Restrictions: MIT/WHOI Joint Program Students, and those holding any type of WHOI appointment at the post-Ph.D. level during the 12 months prior to April 1st of the Scholar award year are not eligible for the Scholar competition.
Geographic Focus: All States
Date(s) Application is Due: Jan 5
Amount of Grant: 85,500 USD
Contact: Janet Fields, Postdoctoral Coordinator; (508) 289-2950; fax (508) 457-2115; jfields@whoi.edu or postdoc@whoi.edu
Internet: http://www.whoi.edu/apo/postdoctoral/scholarship/
Sponsor: Woods Hole Oceanographic Institution
266 Woods Hole Road
Woods Hole, MA 02543

WHOI Summer Student Fellowships 4865

The Summer Student Fellowships bring bright, talented students to Woods Hole to learn more about ocean science, attend lectures by Woods Hole scientists, and conduct independent research working alongside world-class scientists. All Fellows are expected to work on a project selected in collaboration with their sponsor(s) that will provide meaningful results in one summer's work. Project topics span the vast spectrum of research in ocean sciences and engineering conducted in WHOI's science departments and the Woods Hole Field Station of the U. S. Geological Survey (U.S.GS). Fellowship recipients have the opportunity to attend and participate in a busy schedule of talks, seminars and a hands-on, one-day, at-sea practicum cruise focusing on data collection and sampling methods with advanced oceanographic technology and instruments. Fellows also have opportunities to interact with current MIT/WHOI Joint Program graduate students with a question and answer session designed specifically for that purpose with the Ethics in Science Workshop. Fellows receive $520 per week for the ten to twelve week program. A stipend for round-trip travel to Woods Hole is also available. International students are welcome to apply. Additional guidelines and program requirements are posted on the website. The application is available at the Admissions page.
Requirements: Students who have completed their junior year at colleges or universities by the start of the Fellowship period are eligible to apply. Preference is given to students studying in any of the fields of science or engineering including but not limited to the fields of biology, chemistry, engineering, geology, geophysics, mathematics, meteorology, physics, oceanography, and marine policy. Students must have at least a tentative interest in the ocean sciences, oceanographic engineering, mathematics, or marine policy.
Geographic Focus: All States, All Countries
Date(s) Application is Due: Feb 15
Amount of Grant: 5,200 USD
Contact: Michelle McCafferty, Undergraduate Programs Coordinator; (508) 289-2219; fax (508) 457-2188; mmccafferty@whoi.edu or education@whoi.edu
Internet: http://www.whoi.edu/main/summer-student-fellowship
Sponsor: Woods Hole Oceanographic Institution
266 Woods Hole Road
Woods Hole, MA 02543

WHOI Undergraduate Fellowships for Minorities 4866

The Undergraduate Fellowships for Minorities are awarded to minority undergraduates enrolled at U.S. colleges or universities who have completed at least one year of undergraduate study, and who have academic interests in physical or natural sciences, mathematics, engineering, or marine policy. Through the Minority Fellowship program, WHOI's aim is to provide promising students from underrepresented groups with a meaningful first-hand introduction to research in oceanography, oceanographic engineering, or marine policy. Fellows are selected based on the applicant's previous academic and scientific achievements and promise as future ocean scientists or ocean engineers. Minority Fellowships may be awarded for a ten- to twelve-week period in the summer or for a semester during the academic year, and may be renewed the following year. Fellowships include a stipend of $500 per week, in addition to housing and travel expenses. Important consideration is given to matching each Fellow with an appropriate advisor on the Assistant, Associate or Senior Scientific Staff or a Senior Technical Staff member. Applicants are encouraged to contact a WHOI assistant, associate, or senior scientist or a senior technical staff member who may be willing to provide an educational opportunity in their lab. Applicants are also encouraged to visit the U.S.G.S. Woods Hole Science Center website for potential sponsors from among their team of research scientists. The application and additional requirements are posted on the

website. Applications are accepted from November 5 through February 15, with award decisions made in mid-March.
Requirements: Applicants must be U.S. citizens or permanent residents of the U.S. Those eligible to apply include the following minority groups: African American or Black; Asian American; Chicano, Mexican American, Puerto Rican and/or other Hispanic; Native American, Native Alaskans, and Native Hawaiians. Applicants must have completed at least one year of undergraduate study, and have an academic interest in physical or natural sciences, mathematics, engineering or marine policy.
Geographic Focus: All States
Date(s) Application is Due: Feb 15
Amount of Grant: 5,000 USD
Contact: Michelle McCafferty; (508) 289-2219; mmccafferty@whoi.edu
Internet: http://www.whoi.edu/minorityfellow/
Sponsor: Woods Hole Oceanographic Institution
266 Woods Hole Road
Woods Hole, MA 02543

WILD Foundation Grants **4867**
Founded in 1974, WILD is the only international organization dedicated entirely and explicitly to wilderness protection around the world. WILD works to protect the planet's last wild places and the wildlife and people who depend upon them, because wilderness areas provide essential social, spiritual, biological and economic benefits. The WILD Foundation focuses on four main program areas: World Wilderness Congress, wilderness policy and research, communications and field projects. Refer to the Foundations website for more detailed information on each program of interest. There are no specific deadlines with which to adhere. Contact the Foundation for further application information and guidelines.
Geographic Focus: All States
Contact: Charlotte Baron; (303) 442-8811; fax (303) 442-8877; info@wild.org
Internet: http://www.wild.org/main/about/
Sponsor: WILD Foundation
717 Poplar Avenue
Boulder, CO 80304

Wilhelm Sander Foundation Grants **4868**
The purpose of the Foundations grants are to support research in the field of human medicine with a focus on cancer research.
Restrictions: The grant is limited to Germany and Switzerland.
Geographic Focus: All States, Germany, Sweden
Contact: Dr. J. Koppenhofer, (089) 544-1870; fax (089) 544-18720; info@sanst.de
Internet: http://www.wilhelm-sander-stiftung.de
Sponsor: Wilhelm Sander Foundation
Goethe route 74
Munich, 80336 Germany

Willamette University Minority Graduate Fellowship **4869**
The fellowship supports a year-long residence at the university for a minority scholar to complete his or her dissertation. The PhD may be in any discipline in Williamette's undergraduate liberal arts curriculum. The program is intended to prepare minority scholars for careers in American higher education and also exposes such students to doctoral studies. The fellowship extends from approximately September 1 through August 31, during which time the dissertation is to be completed. It also may include teaching, either individually or as part of a team. Applications are available on the Web site and will be accepted until the position is filled.
Requirements: Applicants must be U.S. citizens and have completed all PhD requirements.
Geographic Focus: All States
Amount of Grant: 27,500 USD
Contact: Dr. Lawrence Cress, Dean; (503) 370-6285 or (503) 370-6300
Internet: http://www.willamette.edu/mgf
Sponsor: Willamette University
900 State Street
Salem, OR 97301

Willary Foundation Grants **4870**
The Willary Foundation supports projects that are interesting, creative, and imaginative and that benefit communities in northeastern Pennsylvania. The Foundation is particularly interested in projects that support leadership and the development of leadership in business, the economy, education, human services, government, the arts, media, and research. The Foundation gives preference to efforts that have a ripple effect in the community and those that are conducted in conjunction with other sources of funding. Willary seeks to foster both individuals and groups with unique, innovative, or unusual ideas and efforts. The application and samples of previously funded projects are available at the Foundation website.
Requirements: Individuals and nonprofits in Lackawanna and Luzerne counties, Pennsylvania, are eligible to apply.
Restrictions: Willary will not consider applications for capital campaigns and annual drives.
Geographic Focus: Pennsylvania
Date(s) Application is Due: Mar 10; Aug 27
Amount of Grant: 1,200-50,000 USD
Contact: Linda Donovan; (570) 961-6952; fax (570) 961-7269; info@willary.org
Internet: http://www.willary.org
Sponsor: Willary Foundation
P.O. Box 283
Scranton, PA 18501-0937

William A. Klinger Research Award **4871**
The AGC Education and Research Foundation is proud to support construction industry-related research through its William A. Klinger Research Award. Research topics are developed by the Foundation Board of Directors in cooperation with the Associated General Contractors of America's Education Committee, and RFPs are sent to universities across the country for consideration.
Requirements: RFPs for upcoming Klinger Awards are made available in August and sent to universities across the country for consideration. To be placed on the mailing list for future RFPs, please send an email to Floretta Slade, Director of Programs (see contact info for email address).
Geographic Focus: All States
Date(s) Application is Due: Nov 1
Amount of Grant: 25,000 USD
Contact: Floretta Slade; (703) 837-5342; fax (703) 837-5402; sladef@agc.org
Internet: http://www.agc.org/page.ww?section=AGC+Foundation&name=Industry-Related+Research
Sponsor: AGC Education & Research Foundation
2300 Wilson Boulevard, Suite 400
Arlington, VA 22201

William and Charlotte Parks Foundation for Animal Welfare Grants **4872**
The foundation awards grants to support animal protection to both animal protection organizations and to individual scholars who are pursuing a course of study that might advance the goals of the trust founders. Organizations may receive funds for capital projects, for general operating expenses, or for specific projects. Doctoral or post-doctoral candidates may receive funds to support a specific research project. These funds must be distributed through an organization with 501(c)3 status. Deadlines are May 1 for project, operating, and capital grants; and December 1 for PhD/veterinary fellowships. Application and guidelines are available online.
Requirements: 501(c)3 tax-exempt organizations are eligible. A non-U.S. applicant must be registered as a charitable organization in the home country.
Restrictions: Grants will not normally be awarded to improve animal health, for local spay/neuter assistance, to save endangered species, to rehabilitate wildlife, or to support political candidates. Grants focused on the conservation or protection of wild animal populations will not be funded. Grants will not ordinarily be made to organizations with an annual income of more than $1 million or a large asset base, with the exception of institutions of higher learning.
Geographic Focus: All States
Date(s) Application is Due: May 1; Dec 1
Contact: Donna Pease, (301) 548-7726; info@parksfoundation.org
Internet: http://www.parksfoundation.org
Sponsor: William and Charlotte Parks Foundation
700 Professional Drive
Gaithersburg, MD 20879

William and Flora Hewlett Foundation Education Grants **4873**
The William and Flora Hewlett Foundation has been making grants since 1967 to solve social and environmental problems at home and around the world. Two types of Education Grants are funded. The first is deeper learning which delivers the skills and knowledge students need to succeed in a world that is changing at an unprecedented pace. Deeper learning prepares students to: master core academic content; think critically and solve complex problems; work collaboratively; communicate effectively; and learn how to learn (e.g., self-directed learning). The second is open educational resources which are high-quality, openly licensed, online educational materials that offer an extraordinary opportunity for people everywhere to share, use, and reuse knowledge. Open educational resources demonstrate great potential as a mechanism for instructional innovation as networks of teachers and learners share best practices.
Requirements: Applicants whose project or organization aligns closely with the Foundation's goals and strategies, may submit a brief on-line letter of inquiry. The Foundation will provide Education Grants supporting deeper learning to organizations that: promote policies or strategies that create incentives for K-12 schools and community colleges to focus on deeper learning; build the capacity of the education system and teaching practices both online and in the classroom to reach large numbers of students with deeper learning; support proof points, including model K-12 schools and community colleges, and fund research that promotes deeper learning as an attainable and necessary goal for all students; and develop new, innovative models to increase access for all students and to improve deeper learning. The Foundation will provide Education Grants supporting open educational resources primarily by supporting the infrastructure needed to sustain a well-functioning open educational ecosystem and demonstration projects that strengthen student access to deeper learning in K-12 and community colleges.
Restrictions: The Foundation does not consider requests to fund student aid, individual scholarships, construction, equipment and computer purchases, health research, or health education programs. Few, if any, letters of inquiry are funded. A full proposal should not be submitted unless expressly requested. The majority of awards are made to organizations by invitation. Due to the volume of submissions the Foundation does not respond to phone calls or emails regarding letters of inquiry. Submissions via the online letter of inquiry form receive a response within thirty days or as soon thereafter as possible.
Geographic Focus: All States
Contact: Barbara Chow, (650) 234-4500; fax (650) 234-4501; info@hewlett.org
Internet: http://www.hewlett.org/Grantseekers
Sponsor: William and Flora Hewlett Foundation
2121 Sand Hill Road
Menlo Park, CA 94025

William and Flora Hewlett Foundation Environmental Grants 4874

The William and Flora Hewlett Foundation has been making grants since 1967 to solve social and environmental problems at home and around the world. The Foundation's Environmental Grants goals are to: conserve the Western United States and Canada for wildlife and people; slow global climate change by reducing greenhouse gas emissions; ensure that the United States energy supply is clean and consumption is efficient; and address environmental problems that disproportionately affect disadvantaged communities in the San Francisco Bay Area. Support for western conservation support four key strategies: land (protecting large open spaces); water (restoring river flows and conserve areas near rivers and streams); energy (reducing fossil fuel development and increase energy efficiency and renewable energy sources; and broad-based support (building broad-based support for land, water, and energy goals among key stakeholders). Support for energy and climate pursue strategies in three areas: global climate policy; national energy policy; and sustainable transportation. Support for the San Francisco Bay Area is provided to regional organizations partnering with residents of these communities to: improve and expand urban parks; support outdoor recreation programs for youth; and improve public transportation available to these communities.

Requirements: Applicants whose project or organization aligns closely with the Foundation's goals and strategies, may submit a brief on-line letter of inquiry. Western conservation is supported in twelve states and three Canadian provinces: Alaska, Arizona, California, Colorado, Idaho, Montana, Nevada, New Mexico, Oregon, Washington, Wyoming, British Columbia, Alberta, and Yukon.

Restrictions: The Foundation does not consider requests to fund student aid, individual scholarships, construction, equipment and computer purchases, health research, or health education programs. Few, if any, letters of inquiry are funded. A full proposal should not be submitted unless expressly requested. The majority of awards are made to organizations by invitation. Due to the volume of submissions the Foundation does not respond to phone calls or emails regarding letters of inquiry. Submissions via the online letter of inquiry form receive a response within thirty days or as soon thereafter as possible.

Geographic Focus: All States, Canada

Contact: Tom Steinbach; (650) 234-4500; fax (650) 234-4501; info@hewlett.org

Internet: http://www.hewlett.org/programs/environment-program

Sponsor: William and Flora Hewlett Foundation
2121 Sand Hill Road
Menlo Park, CA 94025

William Blair and Company Foundation Grants 4875

Contributing to the community is an important part of the culture of William Blair and Company. The Foundation was officially established in Illinois in 1980, with giving primarily centered around metropolitan Chicago. All partners of the firm contribute part of their individual share of profits to the Foundation. Donation requests are made to the Foundation by partners and employees. The Foundation supports a broad range of causes including: civic affairs; public safety; arts and culture; higher education; youth-oriented activities; healthcare research; cultural affairs; and civic charities. Types of support include: annual campaigns; building and renovation; capital campaigns; general operating support; endowments; fellowships; internship programs; and scholarship funding. There are no specific deadlines or applications forms required. Funding typically ranges from $500 to $25,000.

Requirements: Requests can be made by sending a letter with a general description of the organization and its special purpose. Activities should have a significant impact on the Chicago metropolitan area.

Geographic Focus: Illinois

Contact: E. David Coolidge III, Vice President; (312) 236-1600

Sponsor: William Blair and Company Foundation
222 W Adams Street, 28th Floor
Chicago, IL 60606

William G. and Helen C. Hoffman Foundation Grants 4876

Helen C. Hoffman resided in the Village of South Orange, New Jersey. Her foundation was established in 1998 in memory of herself and her husband after the death of their daughter Corinne Blair. Her testamentary wish was to establish this foundation for charitable, religious, scientific, literary, and educational purposes. Her preference was to support blindness and its cure. Approximately 90% of the grants will provide support to the blind and, to medical research for the prevention of blindness. The remaining 10% will fund annual grants in the following areas of interest: education, the arts, environment and, social/civic causes. Requests must be received by January 15 for the March meeting and, August 22 for the October meeting. Application forms are available online. Applicants will receive notice acknowledging receipt of the grant request, and subsequently be notified of the grant declination or approval.

Requirements: Any U.S. 501(c)3 non-profit organizations may apply. Proposals should be submitted in the following format: completed Common Grant Application Form; an original Proposal Statement; an audited financial report and a current year operating budget; a copy of your official IRS Letter with your tax determination; a listing of your Board of Directors. Proposal Statements (second item in the above Format) should answer these questions: what are the objectives and expected outcomes of this program/project/request; what strategies will be used to accomplish your objective; what is the timeline for completion; if this is part of an on-going program, how long has it been in operation; what criteria will you use to measure success; if the request is not fully funded, what other sources can you engage; an Itemized budget should be included; please describe any collaborative ventures. Prior to the distribution of funds, all approved grantees must sign and return a Grant Agreement Form, stating that the funds will be used for the purpose intended. Progress reports and Completion reports must also be filed as required for your specific grant. All current grantees must be in good standing with required documentation prior to submitting new proposals to any foundation.

Restrictions: Grants are not made for political purposes, nor to organizations which discriminate on the basis of race, ethnic origin, sexual or religious preference, age or gender.

Geographic Focus: All States

Date(s) Application is Due: Jan 15; Aug 22

Amount of Grant: 5,000-15,000 USD

Contact: Wachovia Bank, N.A., Trustee; grantinquiries2@wachovia.com

Internet: https://www.wachovia.com/foundation/v/index.jsp?vgnextoid=522852199c0aa110VgnVCM1000004b0d1872RCRD&vgnextfmt=default

Sponsor: William G. and Helen C. Hoffman Foundation
190 River Road, NJ3132
Summit, NJ 07901

William G. McGowan Charitable Fund Grants 4877

The fund awards grants to eligible nonprofit organizations that are devoted to three specific philanthropic goals: developing the talents and gifts of the very young, especially those who have been disenfranchised by virtue of low income status, inner-city conditions or family situations; funding for selected areas of medical research directed toward finding cures for, or relieving the pain and suffering of, those afflicted with debilitating conditions or diseases; and the McGowan Scholars program, which provides financial assistance to selected college and university students interested in pursuing an undergraduate or graduate degree at a college or university business school accredited by the International Association for Management Education or Association of Collegiate Business Schools and Programs.

Requirements: Nonprofit organizations in the following states and cities are eligible: Chicago, IL; District of Columbia and its Virginia suburbs; metropolitan Kansas City, western New York; and northeast Pennsylvania.

Restrictions: The fund does not support multiyear grants; accept requests online; or fund art or theatrical activities, building funds, church renovation campaigns, or endowments.

Geographic Focus: California, District of Columbia, Illinois, Missouri, New York, Pennsylvania, Virginia

Date(s) Application is Due: Jan 2; May 1; Sep 1

Amount of Grant: 10,000-100,000 USD

Contact: Bernard Goodrich, Executive Director; (301) 320-8570; fax (301) 320-8627; goodric@aol.com or info@mcgowanfund.org

Internet: http://www.mcgowanfund.org/guidelines.html

Sponsor: William G. McGowan Charitable Fund
P.O. Box 40515
Washington, DC 20016-0515

William H. Adams Foundation for ALS Grants 4878

Established in 1999, the William H. Adams Foundation researches and funds programs specializing in the treatment of Amyotrophic Lateral Sclerosis (ALS) or Lou Gehrig's disease. Since there are no formal applications required, interested parties should contact the Foundation directly. Most recent grant amounts have ranged from $10,000 to $250,000, and are either in the form of general donations or research specific. There are no specific annual deadlines for making application.

Geographic Focus: All States

Amount of Grant: 10,000-250,000 USD

Contact: Ellen Adams, President; (415) 592-8151

Sponsor: William H. Adams Foundation for ALS
1918 43rd Avenue
San Francisco, CA 94116-1025

William N. and Myriam Pennington Foundation Grants 4879

The foundation supports education, health, and medical research in Nevada. Funding priority is given to organizations serving children and youth, the elderly, and the economically disadvantaged. Types of support include general operating support, building construction/renovation, equipment acquisition, and scholarships to individuals.

Requirements: Nevada nonprofits may apply for grant support.

Geographic Focus: Nevada

Amount of Grant: 2,500-200,000 USD

Contact: Kent Green, Foundation Manager; (775) 333-9100

Sponsor: William N. and Myriam Pennington Foundation
441 W Plumb Lane
Reno, NV 89509

William T. Grant Foundation Distinguished Fellows Awards 4880

The goal of the program is to increase the supply of, demand for, and use of high-quality research in the service of improved youth outcomes and well-being. To accomplish this goal, the program gives researchers the opportunity to immerse themselves in practice or policy settings and allows influential mid-career practitioners and policymakers to work and be mentored in research settings. Each Fellow will receive up to $175,000 for the total duration of the fellowship, which may range between six months and two years, depending on the proposed design. The time structure of the fellowship is deliberately flexible. Fellowships are permitted to extend up to two years provided that the fellowship activity amounts to a minimum of half of a person-year spent at the fellowship site(s) over the duration of the project. (Thus, the minimum duration is full-time for six months at the fellowship site.) The foundation will also provide small grants up to $25,000 to the fellowship site(s) to defray the costs associated with hosting a Distinguished Fellow.

Requirements: Influential mid-career practitioners, policymakers, and researchers are encouraged to apply. Support is restricted to researchers seeking to work in policy or practice settings, or policymakers and practitioners seeking to work in research settings. The deadline for receipt of Letters of Inquiry is January 10. All fellowships must support activities that help the Distinguished Fellow advance the foundation's Current Research Priorities

(download the most current listing at the website). Grant awards will be made to the applicant's employer, which must be a non-profit private or governmental organization.
Restrictions: The foundation reserves the right to terminate the award if the Fellow's project changes substantially from that originally proposed by the Fellow. Additionally, approval will be required from the foundation if the Fellow wishes to transfer the award to another institution.
Geographic Focus: All States
Date(s) Application is Due: Jan 8
Amount of Grant: Up to 175,000 USD
Contact: Sharon Brewster; (212) 752-0071; fax (212) 752-1398; sbrewster@wtgrantfdn.org
Internet: http://www.wtgrantfoundation.org/funding_opportunities/fellowships/william_t__grant_distinguished_fellows
Sponsor: William T. Grant Foundation
570 Lexington Avenue, 18th Floor
New York, NY 10022-6837

William T. Grant Foundation Research Grants 4881
The foundation supports research to improve the lives of youth ages 8 to 25 in the United States. This is done primarily by investing in high quality empirical studies. The foundation's Current Research Interests are understanding and improving social settings (i.e., families, schools, peer groups, organizations, programs, etc.), their effects on youth, and the use and influence of scientific evidence. Within these interests, the foundation is interested in descriptive and intervention studies. There are no geographical boundaries for its support of research projects.
Requirements: The grants are limited, without exception, to established tax-exempt, private organizations and institutions. To apply for a major grant, the Principal Investigator must submit a Letter of Inquiry to the grants coordinator briefly describing the project or program and the financial needs. Letters of Inquiry are due on the deadline dates listed. If it is determined that the project falls within the current program interests and priorities, a full proposal will be requested. Full proposal deadlines are four to six months prior to the quarterly board meetings in March and October.
Restrictions: The foundation does not support or make contributions to building funds, fundraising drives, endowment funds, general operating budgets, or scholarships. Grants are made to organizations or institutions, not individuals.
Geographic Focus: All States
Date(s) Application is Due: Jan 3; Apr 1; Nov 1
Amount of Grant: 200,000-500,000 USD
Contact: Nancy Rivera-Torres, Research Grants Coordinator; (212) 752-0071; fax (212) 752-1398; nrivera@wtgrantfdn.org
Internet: http://www.wtgrantfoundation.org/funding_opportunities/research_grants
Sponsor: William T. Grant Foundation
570 Lexington Avenue, 18th Floor
New York, NY 10022-6837

William T. Grant Foundation Scholars Program 4882
The William T. Grant Scholars Program supports promising early-career researchers from diverse disciplines, who have demonstrated success in conducting high-quality research and are seeking to further develop and broaden their expertise. Candidates are nominated by a supporting institution and must submit five-year research plans that demonstrate creativity, intellectual rigor, and a commitment to continued professional development. Every year, four to six William T. Grant Scholars are selected and each receives $350,000 distributed over a five-year period.
Requirements: To be eligible for consideration, applicants must: be employed at a nonprofit institution, either in the United States or abroad; submit a project that is consistent with the Foundation's Current Research Interests; address issues that have compelling relevance for theory, and policies or practices, affecting the settings of youth ages 8 to 25 in the United States or a vulnerable subpopulation of those youth; and have received their terminal degree within seven years of submitting their application.
Restrictions: The Foundation rarely funds international work. The William T. Grant Scholars award must not replace the institution's current support of the applicant's research nor should it duplicate any other funding, received or pending.
Geographic Focus: All States
Date(s) Application is Due: Jul 6
Amount of Grant: 350,000 USD
Contact: Irene Williams, Grants Coordinator; (212) 752-0071; fax (212) 752-1398; iwilliams@wtgrantfdn.org or info@wtgrantfdn.org
Internet: http://www.wtgrantfoundation.org/funding_opportunities/fellowships/william_t__grant_scholars/william_t_grant_scholars
Sponsor: William T. Grant Foundation
570 Lexington Avenue, 18th Floor
New York, NY 10022-6837

William T. Grant Foundation Understanding the Acquisition, Interpretation, and Use of Research Evidence in Policy and Practice Grants 4883
The William T. Grant Foundation has a longstanding interest in supporting research that can inform policy and practice. The Foundation's particular focus is on policies and practices that affect youth ages 8 to 25 in the United States. In this area, there are significant gaps between research and policy, and between research and practice. Researchers express frustration that policymakers and practitioners do not use or misuse research findings. Policymakers and practitioners suggest that research is often not relevant to their work or is not easily accessible or understood. Many researchers, research funders, and intermediary organizations have sought to address these gaps by encouraging the production of more rigorous research evidence, better research syntheses, and improved approaches to disseminating research evidence. Policymakers have also tried to improve the connection between research and practice by mandating the use of research findings through law or regulation. Relatively little research attention has been devoted to understanding the user sideâ€"that is, studying what affects policymakers' and practitioners' acquisition, interpretation, and use of research evidence. At the Foundation, we believe studies of this topic will increase our understanding of how to improve the production and subsequent use of research for and in policy and practice. For the next several years, the Foundation anticipates supporting a group of research projects, with award amounts ranging from $100,000 to $600,000, covering direct and indirect costs for two to three years of work. Applicants should submit letters of inquiry by May 12. Selected applicants will be nvited to submit full proposals, which will be due by October 6. Funding decisions will be made at the Board of Trustees meeting in June and awards will be made available shortly thereafter.
Requirements: To be eligible for consideration, applicants must: be employed at a nonprofit institution, either in the United States or abroad. In most instances, these organizations will have 501(c)3 tax exempt status, set up to receive and process grant awards. However, in rare cases, colleges or universities (such as Arizona State University) will not have 501(c)3 tax exemption. If your institution fits into this category, the Foundation can instead accept an IRS determination letter which states that it is classified as a school under sections 509a1 and 170b1Aii of the IRS Code. Submit a project that is consistent with the stated goals of the RFP; and reflect high standards of evidence and rigorous methods, commensurate with the RFP's goals.
Geographic Focus: All States
Date(s) Application is Due: May 12
Amount of Grant: 100,000-600,000 USD
Contact: Nancy Rivera-Torres; (212) 752-0071; fax (212) 752-1398; nrivera@wtgrantfdn.org
Internet: http://www.wtgrantfoundation.org/funding_opportunities/research_grants
Sponsor: William T. Grant Foundation
570 Lexington Avenue, 18th Floor
New York, NY 10022-6837

Winn Feline Foundation/AVMF Excellence In Feline Research Award 4884
The Winn Foundation has partnered with the American Veterinary Medical Foundation (AVMF) to recognize the veterinary research scientist whose work mirrors Winn's efforts to improve the lives of "Every Cat, Every Day". The Winn/AVMF Excellence in Feline Research Award is accompanied by a $2,500 cash award, a crystal cat sculpture, plus up to $1,000 to offset round trip airfare and other travel expenses. The Research Award is paired with a matching scholarship award by the AVMF for a veterinary student interested in feline medicine. For nomination forms and additional information related to supporting materials, award criteria, and the selection process, visit www.avma.org/awards.
Requirements: Applicants should include a cover letter, a list of significant research achievements, and a concise curriculum vita (CV) including relevant publications. The cover letter must include the name of the award; the name, mailing address, telephone number(s), email address, and college and year of graduation of the nominee; the nature of the nominee's professional activity (type of practice or type of salaried work); organizational memberships (professional and scientific) of the nominee; a narrative sketch of the nominee's professional background; and a statement pertaining to the nominee's qualifications for the award. To be considered, nominees must have demonstrated Winn's mission to advance feline health and welfare through research; have a pattern of following good research study protocols; complete progress reports in a timely manner; and have their research published and acknowledge Winn as a donor of grants when appropriate.
Restrictions: No self-nominations will be considered.
Geographic Focus: All States
Date(s) Application is Due: Feb 1
Contact: Janet Wolf, Executive Director; (856) 447-9798; winnfeline@aol.com
Eileen Hoblit, Award Contact; (800) 248-2862, ext. 6778; fax (847) 925-1329
Internet: http://www.winnfelinehealth.org/Pages/Researchers.html
Sponsor: Winn Feline Foundation
390 Amwell Road, Suite 402
Hillsborough, NJ 08844

Winn Feline Foundation Grants 4885
The foundation supports research studies into medical problems affecting cats with the Winn Feline Foundation Grant and the Miller Trust Grant. The Foundation supports research on feline asthma, anemia, inflammatory bowel disease, hyperthyroidism, FIP, megacolon, polycystic kidney disease, transdermal drug delivery, liver disease, and vaccine-associated fibrosarcoma. Studies general in scope are primarily encouraged; however, the foundation is also interested in projects that address problems unique to individual breeds. Continuation of grants awarded will be considered. Application guidelines and current deadlines are available online.
Requirements: Applicants may be faculty veterinarians, postdoctoral fellows, practicing veterinarians, or veterinary students.
Restrictions: The foundation does not fund salaries of principal investigators, major equipment expenditures, travel or indirect costs.
Geographic Focus: All States
Amount of Grant: Up to 35,000 USD
Contact: Janet Wolf; (908) 359-1184; fax (908) 359-7619; winn@winnfelinehealth.org
Internet: http://www.winnfelinehealth.org
Sponsor: Winn Feline Foundation
390 Amwell Road, Suite 402
Hillsborough, NJ 08844

Winston Churchill Foundation Scholarships 4886

The program offers American citizens of exceptional ability and outstanding achievement the opportunity to pursue graduate studies in engineering, mathematics, or the sciences at the University of Cambridge. The foundation awards at least thirteen scholarships and hopes to increase that number to fifteen soon. The one-year awards lead to the Masters of Philosophy (MPhil), the Certificate of Post-Graduate Study (CPGS in different fields), the Certificate of Advanced Study (CAS in theoretical or applied mathematics), and a Diploma (in Computer Science). On occasion the foundation offers the award to an individual who has made outstanding contributions exemplifying Churchill's attributes and ideals. is worth between $44,000 and $50,000. It covers all University and College tuition and fees (currently about $25,000). In addition, students receive a living allowance of 10,000 GBP if enrolled in a nine-month academic program and 12,000 GBP if enrolled in a full-year academic program. They also receive an allowance of up to $1,000 for travel to and from the United Kingdom. The foundation also offers the possibility of a Special Research Grant of up to $2,000; this grant may cover travel for presentations at international conferences, short stays at another university or institute for special research, and other activities. Married students should consult with the Foundation about the possibility of additional support.
Requirements: Applicants for a Churchill Scholarship must be citizens of the United States and must be a senior who is enrolled in one of the institutions participating in the program competition or a student who has recently graduated from one of those institutions. (List available at www.winstonchurchillfoundation.org/Institutions.html.) Upon taking up a scholarship, a Churchill Scholar must be between the ages of 19 and 26, hold a bachelor's degree or its equivalent, and may not have attained a doctorate. Instructions, forms and applications are available at the sponsor's website.
Geographic Focus: All States
Date(s) Application is Due: Nov 12
Amount of Grant: 44,000-50,000 USD
Contact: Peter C. Patrikis; (212) 752-3200; fax (212) 246-8330; info@winstonchurchillfoundation.org
Internet: http://www.winstonchurchillfoundation.org/Scholarships.html
Sponsor: Winston Churchill Foundation of the United States
600 Madison Avenue, 16th Floor
New York, NY 10022-1615

Winterthur Museum Dissertation Fellowships 4887

Founded almost 60 years ago by collector and horticulturist Henry Francis du Pont, Winterthur consists of a prestigious museum of American decorative arts, a research library that serves scholars around the world, and a 1,000 acre preserve located on the grounds of du Pont's childhood home. The organization offers various opportunities to both the public and the academic community. Winterthur's Dissertation Fellowships support doctoral candidates in conducting research and writing their dissertations in areas of social and cultural history, including material culture, architecture, decorative arts, design, consumer culture, garden and landscape studies, Shaker studies, travel and tourism, the Atlantic World, and objects in literature. Dissertation fellowships last four to nine months. Fellows receive a stipend of up to $7,000 per semester. Housing is available on the Winterthur grounds and is within easy walking distance of the library and museum where fellows enjoy many staff privileges, including 24-hour access to the library. Detailed guidelines and a downloadable application form are available at the website. The preferred method of submission is email; however, applicants may also mail the completed application and required supporting documentation (essay, bibliography, vita, etc.) to the address given. All applicants are strongly encouraged to search Wintercat (the library's online catalog), visit Winterthur, and contact staff members to discuss potential research projects. The suitability of a project to Winterthur's collections is the primary consideration of the fellowship award committee.
Requirements: Applicants must be doctoral candidates researching or writing their dissertation on a topic that fits well with Winterthur's collections.
Restrictions: Candidates applying for a Dissertation Fellowship or a National Endowment for the Humanities Fellowship may ask to also be considered for a short-term research fellowship in the event that the primary application is not accepted.
Geographic Focus: All States, All Countries
Date(s) Application is Due: Jan 15
Amount of Grant: Up to 7,000 USD
Contact: Rosemary Krill; (302) 888-4637; fax (302) 888-4870; rkrill@winterthur.org
Internet: http://www.winterthur.org/?p=535
Sponsor: Winterthur Museum, Garden, and Library
5105 Kennett Pike
Wilmington, DE 19735

Winterthur Research Fellowships 4888

Founded almost 60 years ago by collector and horticulturist Henry Francis du Pont, Winterthur consists of a prestigious museum of American decorative arts, a research library that serves scholars around the world, and a 1,000 acre preserve located on the grounds of du Pont's childhood home. The organization offers various opportunities to both the public and the academic community. Winterthur offers one- to three-month research fellowships to encourage research in America's artistic, cultural, intellectual, and social history. The stipend is $1,500 per month. While applicants need not apply for a specific named fellowship, Winterthur will designate certain awards, for instance: the Faith Andrews Fellowships (for the study of Shaker life and material culture); the Robert Lee Gill Fellowships (for research on American decorative arts, painting, architecture, or historic preservation); and the Neville McD. Thompson Fellowships (for the study of domestic life, late 19th- and early 20th-century design and material culture). Grants are available to academic, museum, and independent scholars, and to support dissertation research. Resources include a library of approximately one-half million imprints, manuscripts, visual materials, and printed ephemera supporting interdisciplinary study of American life into the early 20th century. Housing is available on the grounds. Fellows are provided with a library carrel and 24-hour access to the library's circulating collections. Detailed application guidelines and a downloadable form are available at the website. The preferred method of submission is email; however, applicants may also mail the completed application and required supporting documentation (essay, bibliography, vita, etc.) to the address given. All applicants are strongly encouraged to search Wintercat (the library's online catalog), visit Winterthur, and contact staff members to discuss potential research projects. The suitability of a project to Winterthur's collections is the primary consideration of the fellowship award committee.
Requirements: Applications are welcome from academic, museum, and independent scholars, including graduate students.
Geographic Focus: All States, All Countries
Date(s) Application is Due: Jan 15
Amount of Grant: 1,500-5,500 USD
Contact: Rosemary Krill, Senior Lecturer; (302) 888-4637; rkrill@winterthur.org
Internet: http://www.winterthur.org/?p=535
Sponsor: Winterthur Museum, Garden, and Library
5105 Kennett Pike
Wilmington, DE 19735

WMU Department of Spanish Travel Study Abroad Award 4889

The Department of Spanish Travel Study Abroad Award is intended for Spanish majors or minors. The purpose of these awards is to encourage students to further develop their Spanish language skills and to broaden their understanding of other cultures and peoples. These awards may be used for travel and/or study abroad during any part of the academic year or summer. Preference will be given to those who enroll in WMU sponsored programs.
Requirements: Applicants must have: a declared major or minor in the Department of Spanish; completed a minimum of two courses in the Spanish major/minor by the end of the semester of application; and a minimum overall GPA of 3.25, plus a GPA of 3.5 in the Spanish major or minor.
Restrictions: No student may receive a Travel/Study Abroad Award more than once. Recipients of this award are not eligible for the President's Award or Scholarship nor are the recipients of the President's Award/Scholarships eligible for a Travel/Study Abroad Award.
Geographic Focus: All States
Contact: Irma Lopez; (269) 387-3023; fax (269) 387-3103; spanish@wmich.edu
Internet: http://www.wmich.edu/spanish/scholarships-awards/travstud.htm
Sponsor: Western Michigan University
420 Sprau Tower
Kalamazoo, MI 49008-5338

Wolfsonian-FIU Fellowships 4890

The Wolfsonian is a museum, library, and research center donated by Mitchell Wolfson, collector of decorative and propaganda arts, to Florida International University in 1997. The museum explores, through its exhibitions and special programs, the critically important role of design at the height of the industrial age (1885-1945) in the context of social, political, and technological issues. The collections comprise approximately 120,000 objects from the period in a variety of media including furniture; industrial-design objects; works in glass, ceramics, and metal; rare books; periodicals; ephemera; works on paper; paintings; textiles; and medals. The Wolfsonian-FIU Fellowship program promotes scholarly research on The Wolfsonian's collections. Since its inception in 1995, the program has supported projects from a wide range of academic fields. Fellowships are awarded for full-time research at The Wolfsonian, generally for periods of three to four weeks. Fellowships include a stipend, accommodations, and round-trip travel. The timing of dates will be negotiated with individual awardees. The annual submission deadline is December 31 for a residency beginning no earlier than July 1 of the following year. Applicants are strongly encouraged to discuss their project with the Fellowship Coordinator prior to submitting their application. Applications are submitted via an online form accessible from the fellowship's webpage at the Wolfsonian website.
Requirements: The program is open to holders of master's or doctoral degrees, Ph.D. candidates, and to others who have a significant record of professional achievement in relevant fields. Scholars from outside of the United States are eligible. Fellowships are awarded without regard to race, color, gender, religion, national origin, sexual orientation, age, or disability.
Geographic Focus: All States, All Countries
Date(s) Application is Due: Dec 31
Contact: Coordinator; (305) 535-2623 or 305-531-1001; research@thewolf.fiu.edu
Internet: http://www.wolfsonian.org/research-library/fellowships
Sponsor: Florida International University
1001 Washington Avenue
Miami Beach, FL 33139

Woodrow Wilson Doctoral Dissertation Fellowships in Women's Studies 4891

The Woodrow Wilson Dissertation Fellowship in Women's Studies encourages original and significant research about women that crosses disciplinary, regional, or cultural boundaries. The Women's Studies Fellowships are provided to Ph.D. candidates at institutions in the United States who will complete their dissertations during the fellowship year. Fellows receive $3,000 to be used for expenses connected with the dissertation. These may include, but are not limited to, travel, books, microfilming, taping, and computer services.
Requirements: Applications will be judged on originality and significance to women's studies, scholarly validity, the applicant's academic preparation and ability to accomplish the work, and whether the dissertation will be completed within a reasonable time period.

Evidence of commitment to women's studies is also considered. Collaborative work is not eligible. Applications must be filed using the online form provided on the Woodrow Wilson National Fellowship Foundation website. A completed application comprises the online application form, two letters of recommendation, and a current transcript.
Geographic Focus: All States
Date(s) Application is Due: Oct 13
Amount of Grant: 3,000 USD
Contact: Dana DiLullo; (609) 452-7007; WWTeachingFellowships@woodrow.org
Internet: http://www.woodrow.org/fellowships/women_gender/index.php
Sponsor: Woodrow Wilson National Fellowship Foundation
5 Vaughn Drive, Suite 300
Princeton, NJ 08540-6313

Woodrow Wilson MMUF Dissertation Grants 4892
The MMUF Dissertation Grants are available to graduate students who participated in the Mellon Mays Undergraduate Fellowship Program. The grants provide graduate students at the critical juncture of completing their graduate degrees with support to spend a year finishing the writing of the dissertation. The dissertation grant offers up to $20,000 for a 12-month period. Fellows are encouraged to apply for dissertation grants from their home institutions. Winners who do not receive dissertation support from their own institutions are eligible to receive the full amount. The grant will be reduced by funding from other sources to ensure that each awardee receives a maximum award from all sources of $20,000. Woodrow Wilson will ask graduate schools to waive tuition for winners of the MMUF Dissertation Grant.
Requirements: Applicants must be Mellon Mays Undergraduate Fellows and candidates for the Ph.D. degree in fields recognized under the terms of the Mellon Mays Undergraduate Fellowship Program. Candidates must have completed all pre-dissertation requirements preceding the application deadline. Specifically, each applicant must have passed all comprehensive examination, completed all coursework for the degree, received approval of the dissertation topic, and completed most or all of the fieldwork and/or research necessary to begin writing. Applications must include the following: a three-page prospectus; a draft dissertation chapter; a three-page personal statement; an official graduate transcript; two letters of recommendation-one from the dissertation director and one from another academic knowledgeable about the Fellow's academic performance and/or familiar with the Fellow's dissertation project and its contribution to the field of scholarship; the official budget form; and a declaration of previous fellowship awards. Applicants who have held a similar (non-institutional) dissertation award for the final year of dissertation writing are not eligible. At present, applications for the MMUF Dissertation Grants may be submitted in hard copy only; there is no online application as yet. If you are interested in applying to the MMUF Dissertation Grants Program or need more information, please email Sylvia Sheridan, Assistant Director.
Geographic Focus: All States
Date(s) Application is Due: Dec 19
Amount of Grant: 20,000 USD
Contact: Sylvia Sheridan; (609) 452-7007; fax (609) 452-0066; mmuftr@woodrow.org
Internet: http://www.woodrow.org/fellowships/opportunity/dissertation/index.php
Sponsor: Woodrow Wilson National Fellowship Foundation
5 Vaughn Drive, Suite 300
Princeton, NJ 08540-6313

Woodrow Wilson MMUF Travel & Research Grants 4893
The MMUF Travel & Research Grants are available to graduate students who participated in the Mellon Mays Undergraduate Fellowship Program. The grants provide eligible graduate students with the financial means to complete their research prior to the start of dissertation writing. Funding is for one summer or one semester of travel and/or research support. The MMUF Travel & Research grants provide up to $5,000. Grants may be used to cover dissertation research, travel to/from research sites, and photocopying of documents and/or the purchase of access to databases for research purposes.
Requirements: Eligibility: applicants must be Mellon Mays Undergraduate Fellows and candidates for the Ph.D. degree in fields recognized under the terms of the Mellon Mays Undergraduate Fellowship Program; must have passed all comprehensive examinations, completed all course work for the degree, and selected a dissertation topic that has been approved by the dissertation advisor. Applications must include the following: the official application form; a resume; a personal statement as outlined in the application; an official budget form. Funding may be applied to travel to research sites, cost of meals and lodging at research sites, photocopying or microfilming of documents, purchase of access to research databases, fees for use of research facilities, and other research-related expenses; an official graduate transcript; and a brief letter of recommendation (two-page maximum) from the dissertation advisor endorsing the request for funding.
Restrictions: The award may not be used to purchase computer software or hardware, and may not be used for general living expenses.
Geographic Focus: All States
Date(s) Application is Due: Mar 20
Amount of Grant: 5,000 USD
Contact: Sylvia Sheridan; (609) 452-7007; fax (609) 452-0066; pickeringfaf@woodrow.org
Internet: http://www.woodrow.org/fellowships/opportunity/research/index.php
Sponsor: Woodrow Wilson National Fellowship Foundation
5 Vaughn Drive, Suite 300
Princeton, NJ 08540-6313

Woodrow Wilson National Fellowship Foundation Travel and Research Grants 4894
The grants provide eligible graduate students with the financial means to complete their research prior to the start of dissertation writing. Funding is for one summer or one semester of travel and/or research support. Awards are based on the scholarly merit of individual applications received. If you are interested in the MMUF Travel & Research Grants Program or need more information, email Sylvia Sheridan, Assistant Director.
Requirements: Applicants must be Mellon Mays Undergraduate Fellows and candidates for the Ph.D. degree in fields recognized under the terms of the Mellon Mays Undergraduate Fellowship Program. All candidates must have passed all comprehensive examinations, completed all course work for the degree, and selected a dissertation topic that has been approved by the dissertation advisor. Applications must include the following: the official application form; a resume; a personal statement as outlined in the application; an official budget form. Funding may be applied to travel to research sites, cost of meals and lodging at research sites, photocopying or microfilming of documents, purchase of access to research databases, fees for use of research facilities, and other research-related expenses; an official graduate transcript; and a brief letter of recommendation (two-page maximum) from the dissertation advisor endorsing the request for funding.
Restrictions: The award may not be used to purchase computer software or hardware, and may not be used for general living expenses.
Geographic Focus: All States
Date(s) Application is Due: Mar 20
Amount of Grant: 5,000 USD
Contact: Sylvia Sheridan; (609) 452-7007; fax (609) 452-0066; pickeringfaf@woodrow.org
Internet: http://www.woodrow.org/fellowships/opportunity/research/index.php
Sponsor: Woodrow Wilson National Fellowship Foundation
5 Vaughn Drive, Suite 300
Princeton, NJ 08540-6313

Woodward Fund Grants 4895
Grants support nonprofit institutions, corporations, and associations that are located in Georgia or one of its neighboring states and are organized and operated exclusively for religious, educational, and charitable and scientific purposes. Grantmaking focuses on capital projects. The distribution committee makes its decisions based on materials provided in the grant request. The trust does not require a formal application form, but requests a proposal outlining the project or program and containing the following information: project/program description; complete itemized project budget, including project schedule; need for the project; other funding sources, including the amount received from each source; names and qualifications of those conducting the project; objectives and how they will be achieved; a brief description of the applicant organization; method and criteria for project evaluation; list of officers, board of directors, or trustees; copy of the 501(c)3 designation letter; and plans for recognition of the trust in the project. Applicants should not contact committee members personally concerning a proposal or anticipated approach to the fund.
Requirements: 501(c)3 nonprofit organizations in Georgia and its neighboring states are eligible. Government agencies to which contributions by individuals are made deductible from income by IRS laws also are eligible.
Restrictions: Grants are not awarded to individuals or for scholarships or student loans.
Geographic Focus: Alabama, Florida, Georgia, North Carolina, South Carolina, Tennessee
Date(s) Application is Due: May 1; Nov 1
Contact: Alice Sheets; grantinquiries8@wachovia.com
Internet: https://www.wachovia.com/charitable_services/woodward_overview.asp
Sponsor: David, Helen, and Marian Woodward Fund
3280 Peachtree Road NE, Suite 400, MC G0141-041
Atlanta, GA 30305

World Bank Graduate Scholarship Program 4896
With grant funding from the government of Japan, the World Bank has established scholarships for graduate studies leading to a higher degree in a development-related social science. The program is aimed at increasing the number of high-quality professionals in economic and social development. Scholarships are available to applicants from developing countries to pursue a career in a development field. It is anticipated that each recipient will return to his/her home country at the conclusion of studies and apply the knowledge and skills acquired to help accelerate the pace of economic and social development. The scholarship provides full tuition, the cost of medical and accident insurance obtained through the university, a monthly subsistence allowance for one person, and paid economy-class travel. Contact the organization for application packet. Application forms for the Regular Program in English, French, Spanish, and Portuguese are available from the JJ/WBGSP Secretariat from September through January. Forms are available at World Bank offices in many countries, as well as online.
Requirements: Qualified applicants must be nationals of a World Bank member country currently eligible to borrow, younger than age 45 (and preferably under age 35), of good character, and in good health. They must have attained the bachelor's degree or equivalent in law, business or public administration, or any other development-related technical discipline. Candidates must have a superior academic record with at least three, preferably four to five, years of recent experience in a public agency or public service organization engaged in development-related activities. Admittance to two universities renowned for development research and teaching where the recipient will pursue a master's or doctorate degree is also required.
Restrictions: The program does not sponsor undergraduate studies, distance learning programs, short-term training, conferences, seminars, thesis writing, research projects, and fields of studies not related to development. Executive directors, their alternates, and staff of the World Bank Group (the World Bank, the International Finance Corporation, and the International Development Association), including consultants, as well as close relatives, are ineligible. MBA applicants are ineligible. Those with refugee or asylum status are ineligible.

Geographic Focus: All States
Date(s) Application is Due: Mar 31
Contact: Secretariat; (202) 473-1000; fax (202) 477-6391; opportunities@worldbank.org
Internet: http://go.worldbank.org/NV4RFT1BP0
Sponsor: World Bank
1818 H Street NW
Washington, DC 20433

WWF Russell E. Train Education for Nature Alumni Grants 4897

Education for Nature (EFN) is committed to ensuring that our alumni continue to advance in their careers and become a voice for conservation. By providing these ongoing education opportunities, EFN Alumni can gain up-to-date training and knowledge needed to advance conservation efforts in their home countries and regions. EFN Alumni Grants support ongoing education and training for former Train Fellows or Scholars who have earned a degree through EFN. Many Alumni use these opportunities to present a paper at a conference, attend a workshop or short-term training courses offered by universities and other institutions, or to conduct research within a WWF priority place.
Requirements: To be eligible for an EFN Alumni Grant, applicants must: be a former Russell E. Train Fellow or Scholar; have submitted all reports required by their grant contracts; have completed their two-year work requirement in their home country; have a bank account in their own name that can receive U.S. dollars; not have received an EFN Alumni Grant from EFN in the past three years.
Restrictions: Grant requests may not exceed U.S. $3,500 and alumni must observe a three-year waiting period before requesting another grant.
Geographic Focus: All States, All Countries
Amount of Grant: Up to 3,500 USD
Contact: Mercedes Gonzalez, Senior Program Officer; (202) 778-9612; efn@wwfus.org or mercedes.gonzalez@wwfus.org
Internet: http://worldwildlife.org/initiatives/efn-alumni-grants
Sponsor: World Wildlife Fund
P.O. Box 97180
Washington, DC 20090-7180

WWF Russell E. Train Fellowships 4898

Russell E. Train Fellowships support individuals pursuing a master's or doctoral degree in conservation. Each year, WWF supports a small number of committed conservationists from target countries to receive financial support for their studies and field research. Applicants can apply to attend any university around the world and must return to their home countries to work in conservation for at least two years after completing their degree. Applicants from select WWF-U.S. priority countries must meet all of the eligibility criteria to be considered for a grant. Fellows will receive financial support for tuition, books, travel to and from the training institution, and room and board for an academic program for a maximum of $20,000 USD (per year) for up to two years; limited costs for field research required by the academic institution to obtain the degree; long-term professional development support through the EFN Alumni Program.
Requirements: To be eligible for a Train Fellowship, applicants must: be a citizen and legal resident of a participating country; have a minimum of two year's work experience in conservation; be enrolled, admitted, or have applied to an institution of higher education; and, begin their studies within one year from the application deadline for their respective countries. Participating countries: Cameroon, Central African Republic, Democratic Republic of Congo, Gabon, Kenya, Malawi, Mozambique, Nepal, Republic of Congo, Tanzania, Thailand, and Vietnam. Eligibility criteria may vary slightly among countries.
Restrictions: Individuals who have received a fellowship from EFN in the past and who wish to pursue a second degree are not eligible.
Geographic Focus: Cameroon, Central African Republic, Congo, Congo, Democratic Republic of, Gabon, Kenya, Malawi, Mozambique, Nepal, Tanzania, Thailand, Vietnam
Date(s) Application is Due: Feb 28
Amount of Grant: Up to 20,000 USD
Contact: Mercedes Gonzalez, Senior Program Officer; (202) 778-9612; efn@wwfus.org or mercedes.gonzalez@wwfus.org
Internet: http://worldwildlife.org/initiatives/russell-e-train-fellowships
Sponsor: World Wildlife Fund
P.O. Box 97180
Washington, DC 20090-7180

Xerox Foundation Grants 4899

Xerox Foundation Grants assist a variety of social, civic and cultural organizations that provide broad-based programs and services in cities where our employees work and live. The Foundation also remains committed to a program of grants to colleges and universities to prepare qualified men and women for careers in business, science, government, and general education. The Foundation seeks to further advance knowledge in science and technology, and to enhance learning opportunities for minorities and the disadvantaged. Worldwide, Xerox philanthropy tries to engage national leadership in addressing major social problems and to support programs in education, employability and cultural affairs. Other areas of particular focus include programs responsive to the national concern for the environment and the application of information technology. Large grants may be approved for more than one year (multi-year grants). All organizations that have previously received support on an annual basis from the Foundation, must re-submit a request each year to be evaluated for continued support.
Requirements: Grants are made only to 501(c)3 and 509(a) organizations. No specific application form is used. Requests for grants/funding should be submitted in letter form describing the project or program. This request should contain the legal name of the organization, the official contact person, its tax- exempt status, a brief description of its activities and programs, the purpose for which the grant is being requested, the benefits expected, the plans for evaluation, the projected budget, and the expected sources and amount of needed funds.
Restrictions: The foundation declines requests to support individuals; capital grants (new construction or renovation); endowments or endowed chairs; organizations supported by United Way, unless permission has been granted by United Way to a member agency to conduct a capital fund drive or a special benefit; political organizations or candidates; religious or sectarian groups; or municipal, county, state, federal, or quasi-government agencies.
Geographic Focus: All States
Amount of Grant: 500-50,000 USD
Contact: Joseph M. Cahalan, President; (203) 968-2453 or (203) 968-3000
Internet: http://www.xerox.com/about-xerox/citizenship/xerox-foundation/enus.html
Sponsor: Xerox Foundation
45 Glover Avenue, P.O. Box 4505
Norwalk, CT 06856-4505

Xoran Technologies Resident Research Grant 4900

The purpose of this grant is to stimulate original resident research in otolaryngology projects that are well-conceived and scientifically valid, with potential to clarify the role of head and neck computed tomography in diagnosis or treatment. Awards are for one year, are non renewable, and have a $10,000 maximum limit. Generally, a single award is made available annually. Letters of Intent are due on December 15, with full proposals due by January 15.
Requirements: Application is open to residents of an accredited otolaryngolgoy-head and neck surgery training program in the U.S. and Canada for projects that are well-conceived and scientifically valid, with potential to clarify the role of the head and neck computed tomography in diagnosis or treatment.
Geographic Focus: All States
Date(s) Application is Due: Jan 15; Dec 15
Amount of Grant: Up to 10,000 USD
Contact: Stephanie L. Jones; (703) 519-1586 or (703) 836-4444; sljones@entnet.org
Internet: http://www.entnet.org/EducationAndResearch/coreGrants.cfm
Sponsor: Xoran Technologies
One Prince Street
Alexandria, VA 22314-3357

Yale University Beinecke Library Fellowships for Visiting Scholars 4901

The Yale University Beinecke Rare Book and Manuscript Library offers short-term Fellowships to support visiting scholars pursuing post-doctoral or equivalent research in its collections. The Fellowships are designed to provide access to the library for scholars living outside the greater New Haven area. Fellowships pay for travel to and from New Haven, and include a living allowance of $4,000 per month. Granted for one month, Fellowships must be taken between September 1 and May 31. Recipients are expected to be in residence during the period of their award, and are encouraged to participate in the activities of Yale University. The application is located at the Beinecke website, along with a list of frequently asked questions about the application process.
Requirements: Applicants must explain in specific detail the relationship of the Beinecke collections to their project and its significance within the larger field of scholarly work. Along with the online application form, candidates must submit the following items to the selection committee: their curriculum vitae; a brief research proposal; a detailed list of specific research materials to be consulted at the library during their Fellowship; and two letters of recommendation, either sealed, or sent directly from the recommenders.
Restrictions: Students enrolled in degree programs are not eligible.
Geographic Focus: All States
Date(s) Application is Due: Dec 7
Amount of Grant: 4,000 USD
Contact: Jennifer Castellon; (203) 432-2977 or (203) 432-7325; fax (203) 432-4047; jennifer.castellon@yale.edu or beinecke.library@yale.edu
Internet: http://www.library.yale.edu/beinecke/brbleduc/brblapplyvisiting.html
Sponsor: Yale University Library
Sterling Memorial Library
New Haven, CT 06520-8240

Yale University Beinecke Library Research Fellowships for Yale Graduate and Professional School Students 4902

The Beinecke Library offers Research Fellowships for Yale Graduate and Professional School students who wish to use Beinecke collections as a primary resource for their dissertations or culminating projects. All awarded Fellows are expected to dedicate a minimum of 25 hours of research in the Beinecke collections per week throughout the Fellowship period (attendance is monitored and stipends are adjusted to reflect shortened research periods). At the end of the research period, Fellows are also asked to provide a brief statement indicating how this Fellowship support was used to accomplish research goals. Graduate students seeking these Fellowships must have completed their course work, passed their qualifying examinations, and be prepared to pursue research based upon an approved prospectus. Professional school students seeking these Fellowships must be engaged in research for an approved culminating project required for their degree, with the expectation that their project will be undertaken in the final year of their program. Short-term and long-term Fellowships are available, depending on the needs of the applicant. Short-term Fellowships for $3,000 per month are available for one, two, or three months, and may be held at any time. Long-term Fellowships for $13,650 are for one full semester. The application is available at the Beineke website.
Requirements: Along with the application, the following materials must be submitted: a curriculum vitae; a proposal explaining in detail the specific relationship between the Beinecke Collectons and the applicant's research; a detailed list of specific research materials

to be consulted at the Beineke; and two sealed letters of recommendation. Graduate students must submit an approved or pending prospectus. Professional School students must submit a statement explaining how their project fulfills the culminating requirement in their school.
Restrictions: Only students whose proposals demonstrate a need for extensive use of Beinecke materials will be considered for these awards. Awarded Fellows may not hold concurrent teaching positions or any concurrent Fellowships (other than, when appropriate, the standard Graduate School support). Applications from previous recipients of these Fellowships will be considered only in exceptional circumstances.
Geographic Focus: All States
Date(s) Application is Due: Mar 8; Oct 19
Contact: Jennifer Castellon; (203) 432-7325; jennifer.castellon@yale.edu
Internet: http://www.library.yale.edu/beinecke/brbleduc/brblapplystudent_research.html
Sponsor: Yale University Library
Sterling Memorial Library
New Haven, CT 06520-8240

Young Scientists Summer Program **4903**
IIASA's annual 3-month Young Scientists Summer Program (YSSP) offers research opportunities to talented young researchers whose interests correspond with IIASA's ongoing research on issues of global environmental, economic and social change. Many of IIASA's National Member Organizations (NMOs) provide special grants to enable young scientists from their countries, or foreign students studying in their country, to participate in the program. From June through August accepted participants work within the Institute's Research Programs under the guidance of IIASA scientific staff. The YSSP provides a unique opportunity for participants to: advance their research under the direct supervision of an experienced IIASA scientist, and at the same time contribute to IIASA's ongoing scientific agenda; broaden their research interests by working in IIASA's interdisciplinary and international research environment; build contacts with IIASA's worldwide network of collaborators and with other YSSP fellows. The summer program will run from 1 June until 31 August.
Requirements: Interested candidates should have research experience corresponding to a level typical of a researcher about two years prior to receiving a PhD or equivalent degree.
Geographic Focus: All States
Date(s) Application is Due: Jan 15
Contact: Barbara Hauser; 43 2236 807 0; fax 43 2236 71 313; hauser@iiasa.ac.at
Internet: http://www.iiasa.ac.at/Admin/YSP/IIASA_YSSP.html
Sponsor: International Institute for Applied Systems Analysis
Norton's Woods, 136 Irving Street
Cambridge, MA 02138-1996

Yves Rocher Foundation Terra Ficaria Prize for Scientific Research **4904**
Since its inception, the Foundation has upheld and promoted Yves Rocher's strongest conviction that the future of humanity depends upon its ability to respect and live in harmony with nature. To that end, the Foundation spearheads international programs to protect endangered species. In 2004, the Foundation further demonstrated its commitment to the environment by creating the Terra Ficaria Scientific Award to recognize research in plant studies for the betterment of humanity. Since then, scientists have been honored for their contribution with a monetary prize of 20,000 Euros to continue the development of their research.
Geographic Focus: All States
Amount of Grant: 20,000 EUR
Contact: Christina Hane, (001) 450-442-9555, ext. 2158; rpr@yrnet.com
Internet: http://www.yves-rocher-fondation.org/ca/en/preserving_nature/terra_ficaria/
Sponsor: Yves Rocher Foundation
2199 Fernand Lafontaine
Longueuil, QC J4G 2V7 Canada

Zonta International Foundation Amelia Earhart Fellowships **4905**
The Zonta International Foundation awards 35 Fellowships of $10,000 each annually to assist women in advanced study and research in a suitable area of aerospace science or engineering. These Amelia Earhart Fellowships are open to women of any nationality and may be used at any institution of higher learning offering accredited graduate courses and degrees in aerospace studies. The Fellowship may be used for tuition, books and fees, or living expenses. Requests for a second award by current fellows undergo the same evaluation procedures as the initial application; a third award may be made in exceptional cases. The successful candidate identifies the relevance of her field of study to aerospace-related science or aerospace-related engineering, demonstrates outstanding academic credentials, and specifically outlines her professional or career goals as well as detailing an original focus in her research project. Applications for the Fellowships must be received or post-marked by November 15.
Requirements: Women of any nationality pursuing a Ph.D./doctoral degree who demonstrate a superior academic record in the field of aerospace-related sciences and aerospace-related engineering are eligible.
Geographic Focus: All States, All Countries
Date(s) Application is Due: Nov 15
Amount of Grant: 10,000 USD
Contact: Martina Gamboa, Programs Coordinator; (630) 928-1400; fax (630) 928-1559; zontafdtn@zonta.org or programs@zonta.org
Internet: http://www.zonta.org/WhatWeDo/InternationalPrograms/AmeliaEarhart Fellowship.aspx
Sponsor: Zonta International Foundation
1211 W 22nd Street, Suite 900
Oak Brook, IL 60523

Subject Index

AIDS
25th Anniversary Foundation Grants, 6
Abbott Fund Global AIDS Care Grants, 219
Abbott Fund Science Education Grants, 220
AMA Foundation Seed Grants for Research, 601
amfAR Fellowships, 809
amfAR Mathilde Krim Fellowships in Basic Biomedical Research, 810
amfAR Research Grants, 811
Ann & Robert H. Lurie Family Foundation Grants, 859
ASM/CDC Fellowships in Infectious Disease and Public Health Microbiology, 1100
ASM Merck Irving S. Sigal Memorial Awards, 1177
Bernard Van Leer Foundation Grants, 1336
Blowitz-Ridgeway Foundation Grants, 1349
Bodenwein Public Benevolent Foundation Grants, 1355
Boston Foundation Grants, 1361
Bristol-Myers Squibb Foundation Fellowships, 1375
Bristol-Myers Squibb Foundation Global HIV/AIDS Initiative Grants, 1376
Bristol-Myers Squibb Foundation Health Disparities Grants, 1377
Chestnut Hill Charitable Foundation, Inc Grants, 1538
Community Found AIDS Endowment Awards, 1614
Community Foundation for S Arizona Grants, 1618
Coors Brewing Corporate Contributions Grants, 1644
Cowles Charitable Trust Grants, 1652
Dade Community Foundation Grants, 1712
Doris Duke Charitable Foundation Clinical Scientist Development Award, 1836
Doris Duke Charitable Foundation Operations Research on AIDS Care and Treatment in Africa Grants, 1839
Duchossois Family Foundation Grants, 1867
Elisabeth Severance Prentiss Foundation Grants Program, 1905
Elizabeth Glaser Int'l Leadership Awards, 1906
Elizabeth Glaser Scientist Award, 1907
Frances L. and Edwin L. Cummings Memorial Fund Grants, 2046
Ittleson Foundation AIDS Grants, 2436
Johnson & Johnson Corporate Contributions, 2508
Judith Graham Pool Postdoctoral Research Fellowship in Hemophilia, 2529
M. Bastian Family Foundation Grants, 2636
McCarthy Family Foundation Grants, 2717
NFID New Investigator Matching Grants, 2989
NIA AIDS and Aging: Behavioral Sciences Prevention Research Grants, 3083
NIA Medical Management of Older Patients with HIV/AIDS Grants, 3107
NIDDK Transmission of Human Immunodeficiency Virus In Semen, 3187
NIMH AIDS and Aging: Behavioral Sciences Prevention Research Grants, 3249
NINR AIDS and Aging: Behavioral Sciences Prevention Research Grants, 3260
Oppenstein Brothers Foundation Grants, 3792
ORISE Faculty Research Participation at the Centers for Disease Control and Prevention, 3803
ORISE Postgraduate Research Participation at the Centers for Disease Control and Prevention, 3820
ORISE Student Research Participation at the Centers for Disease Control and Prevention, 3829
Pittsburgh Foundation Medical Research Grants, 3973
Playboy Foundation Grants, 3980
Rhode Island Foundation Grants, 4061
Rich Foundation Grants, 4068
USAID Comprehensive District-Based Support for Better HIV/TB Patient Outcomes Grants, 4595
USAID Microbicides Research, Development, and Introduction Grants, 4599
USAID Strengthening RMNCH Through Indian Health Professional Associations and Scaling Up Uptake Grants, 4602
USDD Broad Agency Announcement Grants, 4701
USDD Broad Agency Announcement HIV/AIDS Prevention Grants, 4702
USDD HIV/AIDS Prevention: Military Specific HIV/AIDS Prevention, Care, and Treatment for Non-PEPFAR Funded Countries Grants, 4733
USDD HIV/AIDS Prevention Program Information Systems Development Grants, 4734
USDD President's Emergency Plan for AIDS Relief Military Specific HIV Seroprevalence and Behavioral Epidemiology Risk Suvey Grants, 4777
Victor E. Speas Foundation Grants, 4810

AIDS Counseling
Ann & Robert H. Lurie Family Foundation Grants, 859
Community Found AIDS Endowment Awards, 1614
NIA Medical Management of Older Patients with HIV/AIDS Grants, 3107
USAID Microbicides Research, Development, and Introduction Grants, 4599
USAID Strengthening RMNCH Through Indian Health Professional Associations and Scaling Up Uptake Grants, 4602

AIDS Education
Abbott Fund Science Education Grants, 220
amfAR Fellowships, 809
amfAR Mathilde Krim Fellowships in Basic Biomedical Research, 810
amfAR Research Grants, 811
Community Found AIDS Endowment Awards, 1614
Fairlawn Foundation Grants, 1974
Frances L. and Edwin L. Cummings Memorial Fund Grants, 2046
Ittleson Foundation AIDS Grants, 2436
McCarthy Family Foundation Grants, 2717
PC Fred H. Bixby Fellowships, 3880
USAID Microbicides Research, Development, and Introduction Grants, 4599
USAID Strengthening RMNCH Through Indian Health Professional Associations and Scaling Up Uptake Grants, 4602

AIDS Prevention
Abbott Fund Global AIDS Care Grants, 219
Abbott Fund Science Education Grants, 220
Community Found AIDS Endowment Awards, 1614
Elizabeth Glaser Int'l Leadership Awards, 1906
Frances L. and Edwin L. Cummings Memorial Fund Grants, 2046
Ittleson Foundation AIDS Grants, 2436
NIA AIDS and Aging: Behavioral Sciences Prevention Research Grants, 3083
NIDDK Transmission of Human Immunodeficiency Virus In Semen, 3187
NIMH AIDS and Aging: Behavioral Sciences Prevention Research Grants, 3249
NINR AIDS and Aging: Behavioral Sciences Prevention Research Grants, 3260
USAID Microbicides Research, Development, and Introduction Grants, 4599
USAID Strengthening RMNCH Through Indian Health Professional Associations and Scaling Up Uptake Grants, 4602
USDD Broad Agency Announcement Grants, 4701
USDD Broad Agency Announcement HIV/AIDS Prevention Grants, 4702
USDD HIV/AIDS Prevention: Military Specific HIV/AIDS Prevention, Care, and Treatment for Non-PEPFAR Funded Countries Grants, 4733
USDD HIV/AIDS Prevention Program Information Systems Development Grants, 4734
USDD President's Emergency Plan for AIDS Relief Military Specific HIV Seroprevalence and Behavioral Epidemiology Risk Suvey Grants, 4777

Aboriginal Studies
AFUW Padnendadlu Graduate Bursary, 364
AFUW Padnendadlu Postgraduate Bursary, 365
AFUW Padnendadlu Undergraduate Bursary, 366
AOCS Analytical Division Student Award, 874
ATA Educational Trust Grants - Curriculum Resource or Research Project, 1253
Australasian Institute of Judicial Administration Seed Funding Grants, 1268
CTCRI Knowledge Synthesis Grants, 1666
RACGP/APHCRI Indigenous Health Award, 4008
Vancouver Foundation Grants and Community Initiatives, 4807

Abortion
Lalor Foundation Postdoctoral Fellowships, 2580
Playboy Foundation Grants, 3980

Academic Achievement
3M Foundation Community Giving Grants, 5
AASF Lydia Donaldson Tutt-Jones Research Grant, 203
AIR Research Grants/Dissertation Fellowships, 470
Alice Tweed Tuohy Foundation Grants Program, 569
American Wine Society Educational Foundation Scholarship Program, 808
ASTA Academic Scholarships, 1224
AWWA Academic Achievement Awards, 1283
Carnegie Corporation of New York Grants, 1432
Clarence E. Heller Charitable Foundation Grants, 1575
Fluor Foundation Grants, 2028
FTE North American Doctoral Fellowships, 2061
Gilbreth Memorial Fellowships, 2141
GNOF IMPACT Grants for Education, 2154
Hispanic Scholarships, 2270
Lumina Foundation for Education Grants, 2627
Lynde & Harry Bradley Foundation Fellowships, 2634
Lynde and Harry Bradley Foundation Grants, 2635
Mattel Children's Foundation Grants, 2685
Mattel International Grants Program, 2686
Nellie Mae Education Foundation District-Level Change Grants, 2968
Nellie Mae Education Foundation State Level Systems Change Grants, 2971
NMF National Medical Association Awards for Medical Journalism, 3323
NMF National Medical Association Emerging Scholar Awards, 3324
NMF National Medical Association Patti LaBelle Award, 3325
NMF Ralph W. Ellison Prize, 3327
NSERC Howard Alper Postdoctoral Prize, 3441
NSF Activities in Science, Engineering, and Mathematics for Persons with Disabilities, 3460
RGK Foundation Grants, 4060
SfN Janett Rosenberg Trubatch Career Development Award, 4207
SfN Louise Hanson Marshall Specific Recognition Award, 4208
Victor E. Speas Foundation Grants, 4810

Accidents
Jacob and Valeria Langeloth Foundation Grants, 2455

Accounting
EFWA Accounting Research Grants, 1901
Regents Professional Opportunity Scholarships, 4049

Accounting Education
Deloitte Foundation Grants, 1735
EFWA Accounting Research Grants, 1901

Acoustics
ASHFound Research Grant in Speech Science, 1083
ASME J.P. Hartog Award, 1134
ASME Per Bruel Gold Medal for Noise Control and Acoustics, 1153
ONR Maritime Sensing Grants, 3770
ONR Optical and Biological Oceanography Research Grants, 3777
ONR Young Investigator Program Grants, 3788
USDD Acoustic Transduction Materials and Devices Grants, 4672
USDD Ocean Acoustics Research Grants, 4766

Acupuncture
Regents Professional Opportunity Scholarships, 4049

Addictions
AACAP-NIDA K12 Career Development Awards, 54
AACAP Educational Outreach Program for Residents in Alcohol Research, 58
American Psychiatric Association Minority Medical Student Summer Externship in Addiction Psychiatry, 785
CTCRI Idea Grants, 1665
CTCRI Knowledge Synthesis Grants, 1666
CTCRI Researcher Travel Grants, 1668
CTCRI Research Planning Grants, 1669
CTCRI Student Research Grants, 1670
CTCRI Workshop and Learning Opportunities Grants, 1671
Johnson & Johnson Corporate Contributions, 2508
MGM Resorts Foundation Community Grants, 2753
National Center for Responsible Gaming Centers of Excellence in Gambling Research Grants, 2850
National Center for Responsible Gaming Conference Scholarships, 2851
National Center for Responsible Gaming Early Stage Investigator Grants, 2852
National Center for Responsible Gaming Exploration Grants, 2853
National Center for Responsible Gaming Independent Scientist Grants, 2854
National Center for Resp Gaming Large Grants, 2855
National Center for Responsible Gaming Postdoctoral Fellowships, 2856
National Center for Resp Gaming Seed Grants, 2857
National Center for Responsible Gaming Travel Grants, 2858
NCI/NIDA Transdisciplinary Tobacco Use Research Centers Grants, 2912
NIDA Pilot and Feasibility Studies in Preparation for Drug Abuse Prevention Trials, 3130
Pfizer Global Research Awards for Nicotine Independence, 3925
TRDRP California Research Awards, 4525
TRDRP New Investigator Awards, 4528

Adhesion
American Chemical Society Award in Applied Polymer Science, 663

Adolescent Health
Appalachian Regional Commission Health Care Grants, 915
Blowitz-Ridgeway Foundation Grants, 1349
Delta Air Lines Foundation Health and Wellness Grants, 1736
DHHS Emergency Medical Services for Children Program, 1755
DHHS Oral Health Promotion Research Across the Lifespan, 1759
DTE Energy Foundation Health and Human Services Grants, 1865
E.W. "Al" Thrasher Awards, 1887
Fairlawn Foundation Grants, 1974
Flinn Foundation Grants Programs, 2024
Foundation for Seacoast Health Grants, 2041
General Service Foundation Human Rights and Economic Justice Grants, 2108
Lloyd A. Fry Foundation Health Grants, 2620
NINR Chronic Illness Self-Management in Children and Adolescents Grants, 3261
Partnership for Cures Two Years To Cures Grants, 3872

Adolescent Psychiatry
AACAP Educational Outreach Program for Child and Adolescent Psychiatry Residents, 56
AACAP Educational Outreach Program for General Psychiatry Residents, 57
AACAP Elaine Schlosser Lewis Award for Research on Attention-Deficit Disorder, 59
AACAP George Tarjan Award for Contributions in Developmental Disabilities, 60
AACAP Irving Philips Award for Prevention, 61
AACAP Jeanne Spurlock Lecture and Award on Diversity and Culture, 62
AACAP Jeanne Spurlock Minority Medical Student Clinical Fellowship in Child and Adolescent Psychiatry, 63
AACAP Klingenstein Third Generation Foundation Award for Research in Depression or Suicide, 66
AACAP Mary Crosby Congressional Fellowships, 68
AACAP Norbert and Charlotte Rieger Award for Scientific Achievement, 69
AACAP Pilot Research Awards, Supported by Eli Lilly and Company, 72
AACAP Rieger Psychodynamic Psychothe Award, 74
AACAP Rieger Service Award for Excellence, 75
AACAP Cancro Academic Leadership Award, 76
AACAP Robinson-Cunningham Award, 77
AACAP Sidney Berman Award for the School-Based Study and Intervention for Learning Disorders and Mental Ilness, 78
AACAP Simon Wile Leadership in Consultation Award, 79
AACAP Systems of Care Special Program Scholarships, 81
American Psychiatric Foundation Proposals, 791
American Psychiatric Foundation James H. Scully Jr., M.D., Educational Fund Grants, 792
Blowitz-Ridgeway Foundation Grants, 1349
NARSAD Ruane Prize for Child and Adolescent Psychiatric Research, 2835

Adolescent Psychology
Blowitz-Ridgeway Foundation Grants, 1349
NARSAD Ruane Prize for Child and Adolescent Psychiatric Research, 2835
NIMH Early Identification and Treatment of Mental Disorders in Children and Adolescents Grants, 3252
William T. Grant Foundation Research Grants, 4881

Adolescents
Connecticut Community Foundation Grants, 1627
Deborah Munroe Noonan Mem Research Grants, 1729
Dell Foundation Open Grants, 1734
DHHS Comprehensive Community Mental Health Services Grants for Children with Serious Emotional Disturbances, 1754
FCD Child Development Grants, 1984
Robert Bowne Foundaion Edmund A. Stanley, Jr. Research Grants, 4073
Stanford Center on Adolescence Youth Purpose Research Grants, 4443
USAID Comprehensive District-Based Support for Better HIV/TB Patient Outcomes Grants, 4595

Adoption
ACF Adoption Opportunities Project Grants, 270
Dave Thomas Foundation for Adoption Grants, 1723

Adult Basic Education
Bill and Melinda Gates Foundation Financial Services for the Poor Grants, 1346
IRA Jeanne S. Chall Research Fellowship, 2410
IRA Pearson Foundation-IRA-Rotary Awards, 2412
Joyce Foundation Employment Grants, 2520
McCarthy Family Foundation Grants, 2717
McCune Charitable Foundation Grants, 2719
Reinberger Foundation Grants, 4051
Xerox Foundation Grants, 4899

Adult Development
NSF Social Psychology Research Grants, 3649

Adult and Continuing Education
Allyn Foundation Grants, 574
Auburn Foundation Grants, 1263
Ball Brothers Foundation General Grants, 1291
Bodenwein Public Benevolent Foundation Grants, 1355
Booth-Bricker Fund Grants, 1359
Boston Foundation Grants, 1361
Carl and Lily Pforzheimer Foundation Grants, 1427
CIGNA Foundation Grants, 1568
Coors Brewing Corporate Contributions Grants, 1644
Cowles Charitable Trust Grants, 1652
Credit Union Found of British Columbia Grants, 1654
Cruise Industry Charitable Foundation Grants, 1660
Dept of Ed Adult Education--National Leadership Activities Grants, 1744
Emma B. Howe Memorial Foundation Grants, 1920
Evjue Foundation Grants, 1963
Frances L. and Edwin L. Cummings Memorial Fund Grants, 2046
George Foundation Grants, 2115
Harold Simmons Foundation Grants, 2203
HRSA Nurse Education, Practice, Quality and Retention Grants, 2290
HSBC Contributions Community Grants, 2291
Lumina Foundation for Education Grants, 2627
McCune Charitable Foundation Grants, 2719
McCune Foundation Education Grants, 2721
NBME Stemmler Medical Education Research Fund Grants, 2899
Norcliffe Foundation Grants, 3415
Oppenstein Brothers Foundation Grants, 3792
Pfizer Medical Education Track Two Grants, 3927
Richard and Rhoda Goldman Fund Grants, 4063
San Antonio Area Foundation Grants, 4163
Sioux Falls Area Community Foundation Community Fund Grants (Unrestricted), 4262
Sioux Falls Area Comm Foundation Spot Grants, 4263
Strake Foundation Grants, 4452
Talbert and Leota Abrams Foundation, 4492
Xerox Foundation Grants, 4899

Adults
Mabel H. Flory Charitable Trust Grants, 2646
NHLBI Targeted Approaches to Weight Control for Young Adults Grants, 3065

Advanced Imaging
Grass Foundation Marine Biological Laboratory Advanced Imaging Fellowships, 2162
NSF Biophotonics, Advanced Imaging, and Sensing for Human Health Grants, 3508
USDD Image Analysis & Understanding Grants, 4738

Advertising
American Academy of Advertising Distinguished Service Award, 607
American Academy of Advertising Doctoral Dissertation Competition Awards, 608
American Academy of Advertising Ivan Preston Outstanding Contribution to Research on Advertising Award, 609
American Academy of Advertising Kim Rotzoll Award for Advertising Ethics and Social Responsibility, 610
American Academy of Advertising Research Fellowships, 611
Charles Stewart Mott Foundation Anti-Poverty Program, 1528
Duke University Libraries Hartman Center Travel Grants, 1875
Duke University Libraries J. Walter Thompson Research Fellowships, 1876
Duke Univ Libraries Mary Lily Research Grants, 1877
NCI/NIDA Transdisciplinary Tobacco Use Research Centers Grants, 2912
NYSCA Music: Services to the Field Grants, 3708
Wolfsonian-FIU Fellowships, 4890

Aeronautical/Astronautical Engineering
American Society of Naval Engineers Scholarship, 800
ASME Spirit of St. Louis Medal, 1165
Emil Buehler Perpetual Trust Grants, 1919
Nanyang Technological University Lee Kuan Yew Postdoctoral Fellowship, 2826
Zonta International Foundation Amelia Earhart Fellowships, 4905

Aeronautics

ASME Spirit of St. Louis Medal, 1165
DSO Defense Sciences Research and Technology Grants, 1863
NASA Cooperative Agreement Notice for Technology Development Grants, 2837
NASA Graduate Student Researchers Program Fellowships, 2839
NASA Marslink Initiative, 2841
NASA Undergraduate Student Researchers Project Grants, 2843
ONR Surface/Aerospace Surveillance Grants, 3786
Smithsonian National Air and Space Museum Charles A. Lindbergh Chair in Aerospace History, 4285
Smithsonian National Air and Space Museum Guggenheim Fellowships, 4286
Smithsonian National Air and Space Museum Verville Fellowships, 4288
UNCFSP NASA Harriett G. Jenkins Predoctoral Fellowships, 4563
USDD Air Warfare and Naval Weapons Applications Grants, 4679
USDD Research and Analytical Support for the 711th Human Performance Wing's Human Effectiveness Directorate Grants, 4781
USDD Warfighter Interface Technologies Advanced Research Programs Grants, 4803
Zonta International Foundation Amelia Earhart Fellowships, 4905

Aeronomy
Abdus Salam ICTP Group Associates Grants, 228
Abdus Salam ICTP Junior Associates Grants, 231
Abdus Salam ICTP Postgraduate Diploma Grants, 235
Abdus Salam ICTP Regular Associates Grants, 237
Abdus Salam ICTP Senior Associates Grants, 239
Abdus Salam ICTP Training and Research in Italian Laboratories Grants, 240
NCAR Postdoctoral Fellowships in the Atmospheric Sciences, 2902
NSF Arctic Natural Science Grants in Atmospheric Sciences, 3483

Aerospace
ACSM NASA Space Physiology Research Grants, 319
ASME Spirit of St. Louis Medal, 1165
ISGC Graduate Fellowships, 2430
ISGC Informal Education Grants, 2431
ISGC Multiple Undergraduate Research Grants, 2432
ISGC Research Seed Grants, 2433
NASA Cooperative Agreement Notice for Technology Development Grants, 2837
NASA Graduate Student Researchers Program Fellowships, 2839
NASA New Aeronautics Competition for University Students, 2842
NASA Undergraduate Student Researchers Project Grants, 2843
Smithsonian National Air and Space Museum Aviation Space Writers Foundation Grant, 4284
Smithsonian National Air and Space Museum Charles A. Lindbergh Chair in Aerospace History, 4285
Smithsonian National Air and Space Museum Guggenheim Fellowships, 4286
Smithsonian National Air and Space Museum Verville Fellowships, 4288
USDD Applications and Transitions Grants, 4682
USDD Research Interest of the Air Force Office of Scientific Research Grants, 4784
USDD Small University Grants, 4794

Africa
American Historical Association Bernadotte E. Schmitt Research Grants in European, African, or Asian History, 715
American Historical Association James Henry Breasted Prize, 729
Annenberg Foundation Grants, 862
ASM-UNESCO Leadership Grant for International Educators, 1099
DOI African Elephant Conservation Grants, 1820
Doris Duke Charitable Foundation Operations Research on AIDS Care and Treatment in Africa Grants, 1839
Douglass Institute Postdoctoral Fellowships, 1851
Elizabeth Glaser Int'l Leadership Awards, 1906
ExxonMobil Foundation Malaria Grants, 1968
Fulbright Traditional Scholar Program in Sub-Saharan Africa, 2081
Kellogg Institute for International Studies Experiencing The World Fellowships, 2539
Kellogg Institute for International Studies Grants for Conferences and Workshops, 2543
Kellogg Institute for International Studies Grants for Distinguished Lectureships, 2544
Kellogg Institute for International Studies Grants for Events, 2545
Kellogg Institute for International Studies Supplemental Awards for PhD Students, 2550
NSF Materials World Network: Cooperative Activity in Materials Research between US Investigators and their Counterparts Abroad, 3595
Schomburg Center Scholars-in-Residence Fellows, 4187
Smithsonian Senior Fellowships, 4296
SSRC Next Generation Social Sciences in Africa: Dissertation Completion Fellowships, 4434
SSRC Next Generation Social Sciences in Africa: Doctoral Dissertation Proposal Fellowships, 4435
SSRC Next Generation Social Sciences in Africa: Doctoral Dissertation Research Fellowships, 4436
STTI Alpha Eta Collaborative Research Grants, 4458
Wenner-Gren Foundation African Fellowships, 4846

Africa, Central
SSRC Next Generation Social Sciences in Africa: Dissertation Completion Fellowships, 4434

Africa, Eastern
SSRC Next Generation Social Sciences in Africa: Dissertation Completion Fellowships, 4434

Africa, Northern
Fulbright Critical Lang Enhancement Awards, 2068
Fulbright Traditional Scholar Program in the Near East and North Africa Region, 2083
IREX MENA Media Emerging Leaders Fellows, 2421
SSRC Next Generation Social Sciences in Africa: Dissertation Completion Fellowships, 4434
SSRC Our Shared Past Grants, 4437

Africa, Sub-Saharan
ACLS African Humanities Program in Ghana, Nigeria, South Africa, Tanzania, and Uganda, 279
Fulbright Traditional Scholar Program in Sub-Saharan Africa, 2081
SSRC African Peacebuilding Network Research Grants, 4422
SSRC Next Generation Social Sciences in Africa: Dissertation Completion Fellowships, 4434

Africa, Western
NMF General Electric Med Scholars Fellowships, 3322
SSRC Next Generation Social Sciences in Africa: Dissertation Completion Fellowships, 4434

African American Students
AAA Minority Dissertation Fellowship Program, 25
Auburn University President's Graduate Opportunities Program, 1264
Consortium for Graduate Study in Management Minority Fellowships, 1642
KFF Barbara Jordan Health Policy Scholars, 2556
Miller Brewing Corporate Contributions Program Grants, 2774
MLA Grad Scholarship for Minority Students, 2787
NAACP Agnes Jones Jackson Grad Scholarships, 2809
NACC David Stevenson Fellowships, 2812
NACC William Diaz Fellowships, 2813
NMF National Medical Association Patti LaBelle Award, 3325
Sachs Foundation Scholarships, 4157
UNC Carolina Postdoctoral Program for Faculty Diversity, 4560

African American Studies
Alabama Humanities Foundation Mini Grants, 480
Alabama Humanities Foundation Planning/Consultant Grants, 481
American Historical Association Martin A. Klein Prize in African History, 737
ASU Graduate College Dean's Fellowships, 1236
Douglass Institute Postdoctoral Fellowships, 1851
Douglass Institute Predoctoral Fellowships, 1852
EU Robert W. Woodruff Library Fellowships, 1959
Frederick Douglass Summer Fellowships, 2052
Frederick Douglass Summer Scholars Fellowship, 2053
Institute of American Cultures Graduate and Predoctoral Fellowships, 2383
Institute of American Cultures Postdoctoral and Visiting Scholar Fellowships, 2384
Virginia Foundation for the Humanities Discretionary Grants, 4813
Virginia Foundation for Humanities Open Grants, 4814

African Americans
AASF Lydia Donaldson Tutt-Jones Memorial Research Grant, 203
African American Heritage Grants, 357
American Historical Association Martin A. Klein Prize in African History, 737
American Sociological Association Minority Fellowships, 805
AMS Howard Mayer Brown Fellowships, 827
ASM Robert D. Watkins Graduate Research Fellowship, 1183
Boston Foundation Grants, 1361
CAPAA Mini Grant for Research Related to African Americans in Education, 1424
CMA Foundation Grants, 1582
CMS Historically Black Colleges and Universities Health Services Research Grants, 1585
Coca-Cola Foundation Grants, 1593
Connecticut Community Foundation Grants, 1627
FDHN Student Research Fellowships, 2001
Helen Carr Minority Fellowships, 2229
Minority Faculty Fellowship Program, 2781
NEH Awards for Faculty at Historically Black Colleges and Universities, 2949
NWSA Abafazi-Africana Women's Studies Essay Award, 3675
OAH Avery O. Craven Award, 3711
OAH Darlene Clark Hine Award, 3713
OAH Frederick Jackson Turner Award, 3717
RWJF Harold Amos Medical Faculty Development Research Grants, 4139
SRC AMD/Mahboob Khan Fellowship, 4348
SRC Company-Named Scholarships, 4351
SRC Master's Scholarships, 4378
UCLA Research Grant in Ethnic Studies, 4549
UNC Carolina Postdoctoral Program for Faculty Diversity, 4560
United Methodist Church Asian, Black, Hispanic and Native American History Research Grants, 4570
Virginia Foundation for the Humanities Discretionary Grants, 4813
Virginia Foundation for Humanities Open Grants, 4814

African Art
Christensen Fund Regional Grants, 1554
Smithsonian Senior Fellowships, 4296

After-School Programs
Arizona Commission on the Arts Education Projects Grants, 996
Carnegie Corporation of New York Grants, 1432
Charles Stewart Mott Foundation Anti-Poverty Program, 1528
CIRCLE Civic Education at the High School Level Research Grants, 1571
Edward W. and Stella C. Van Houten Memorial Fund Grants, 1899

Emma B. Howe Memorial Foundation Grants, 1920
Helen Bader Foundation Grants, 2228
Luther I. Replogle Foundation Grants, 2631
Mattel Children's Foundation Grants, 2685
Mattel International Grants Program, 2686
Motorola Foundation Innovation Generation Collaborative Grants, 2797
Oklahoma City Community Grants, 3737
Paul G. Allen Family Foundation Grants, 3876
Pinkerton Foundation Grants, 3967
Plan for Social Excellence Grants, 3976
Robert Bowne Foundaion Edmund A. Stanley, Jr. Research Grants, 4073
Rohm and Haas Company Grants, 4098
Wallace Foundation Grants, 4829
William Blair and Company Foundation Grants, 4875
William G. and Helen C. Hoffman Found Grants, 4876

Age Discrimination
Allstate Foundation Safe and Vital Communities Grants, 573
Dept of Ed Women's Educational Equity Program Grants, 1752

Aging/Gerontology
AAFCS International Graduate Fellowships, 154
AAFCS National Graduate Fellowships, 155
Abington Foundation Grants, 243
AFAR CART Fund Grants, 351
AFAR Medical Student Training in Aging Research Program, 352
AFAR Paul Beeson Career Development Awards in Aging Research for the Island of Ireland, 353
AFAR Research Grants, 354
AGHE Graduate Scholarships and Fellowships, 369
Ahn Family Foundation Grants, 379
Alfred P. Sloan Found Working Longer Grants, 566
American Academy of Nursing Building Academic Geriatric Nursing Capacity Scholarships, 614
American Academy of Nursing Claire M. Fagin Fellowships, 615
American Academy of Nursing Mayday Grants, 616
American Society on Aging Graduate Student Research Award, 801
APSA Congressional Health and Aging Policy Fellowships, 931
ATS Henry Luce III Fellowships, 1258
Brookdale Found Leadership in Aging Fellowships, 1381
Bupa Foundation Medical Research Grants, 1386
Caesars Foundation Grants, 1407
Century Foundation Grants, 1489
Claude Pepper Foundation Grants, 1578
Clay Foundation Grants, 1579
DHHS Special Programs for the Aging Training, Research, and Discretionary Projects and Programs Grants, 1761
Donald W. Reynolds Foundation Aging and Quality of Life Grants, 1832
Duke University Postdoctoral Research Fellowships in Aging, 1880
Ellison Medical Foundation/AFAR Julie Martin Mid-Career Award in Aging Research, 1910
Ellison Medical Foundation/AFAR Postdoctoral Fellows in Aging Research Program, 1911
Glenn/AFAR Breakthroughs in Geron Awards, 2147
Glenn/AFAR Research Grants for Postdoc Fellows, 2148
Greater Worcester Community Foundation Jeppson Memorial Fund for Brookfield Grants, 2165
Hawaii Community Foundation Health Education and Research Grants, 2221
Helen Bader Foundation Grants, 2228
James H. Cummings Foundation Grants, 2460
Kavli Foundation Research Grants, 2533
Leonard and Helen R. Stulman Charitable Foundation Grants, 2602
M.D. Anderson Foundation Grants, 2637
McLean Contributionship Grants, 2727
Mericos Foundation Grants, 2742
NHLBI Independent Scientist Award, 3028
NIA Aging, Oxidative & Cell Death Grants, 3081
NIA Aging Research Dissertation Awards to Increase Diversity, 3082
NIA AIDS and Aging: Behavioral Sciences Prevention Research Grants, 3083
NIA Alzheimer's Disease Core Centers Grants, 3084
NIA Archiving and Development of Socialbehavioral Datasets in Aging Related Studies Grants, 3088
NIA Awards to Support Research on the Biology of Aging in Invertebrates Grants, 3089
NIA Bioenergetics, Fatigability, and Activity Limitations in Aging, 3091
NIA Diversity in Medication Use and Outcomes in Aging Populations Grants, 3092
NIA Effects of Gene-Social Environment Interplay on Health and Behavior in Later Life Grants, 3093
NIA Harmonization of Longitudinal Cross-National Surveys of Aging Grants, 3097
NIA Health Behaviors and Aging Grants, 3098
NIA Healthy Aging through Behavioral Economic Analyses of Situations Grants, 3099
NIA Higher-Order Cognitive Functioning and Aging Grants, 3100
NIA Human Biospecimen Resources for Aging Research Grants, 3101
NIA Independent Scientist Award, 3103
NIA Malnutrition in Older Persons Grants, 3104
NIA Mechanisms, Measurement, and Management of Pain in Aging: from Molecular to Clinical, 3105
NIA Medical Management of Older Patients with HIV/AIDS Grants, 3107
NIA Mentored Research Scientist Devel Award, 3109
NIA Minority Dissertation Grants in Aging, 3110
NIA Multicenter Study on Exceptional Survival in Families: The Long Life Family Study Grants, 3112
NIA Nathan Shock Centers of Excellence in Basic Biology of Aging Grants, 3113
NIA Network Infrastructure for Emerging Behavioral and Social Research Areas in Aging, 3114
NIA Pilot Research Grant Program, 3115
NIA Postdoctoral Training in Research on Aging in Canada, 3116
NIA Promoting Careers in Aging and Health Disparities Research Grants, 3117
NIA Renal Function and Chronic Kidney Disease in Aging Grants, 3118
NIA Role of Apolipoprotein E, Lipoprotein Receptors and CNS Lipid Homeostasis in Brain Aging and Alzheimer's Disease, 3119
NIA Role of Nuclear Receptors in Tissue and Organismal Aging Grants, 3120
NIA Support of Scientific Meetings as Cooperative Agreements, 3121
NIA Thyroid in Aging Grants, 3122
NIA Transdisciplinary Research on Fatigue and Fatigability in Aging Grants, 3123
NIA Translational Research at the Aging/Cancer Interface Grants, 3124
NIA Vulnerable Dendrites and Synapses in Aging and Alzheimer's Disease Grants, 3125
NIH Research on Sleep and Sleep Disorders, 3236
NIMH AIDS and Aging: Behavioral Sciences Prevention Research Grants, 3249
NINR AIDS and Aging: Behavioral Sciences Prevention Research Grants, 3260
NSF Biomedical Engineering and Engineering Healthcare Grants, 3506
Piper Trust Healthcare & Med Research Grants, 3972
RAND Corporation Fellowships in Population Studies and the Study of Aging, 4029
Rhode Island Foundation Grants, 4061
Rosalinde and Arthur Gilbert Foundation/AFAR New Investigator Awards in Alzheimer's Disease, 4102
STTI Environment of Elder Care Nursing Research Grants, 4460
Thome Foundation Awards in Age-Related Macular Degeneration Research, 4509
Thome Foundation Awards in Alzheimer's Disease Drug Discovery Research, 4510

Agribusiness
McCune Charitable Foundation Grants, 2719
MDARD Specialty Crop Block Grant Program-Farm Bill, 2730
SARE Research and Education Grants - North Central Region, 4177
SARE Research and Education Grants - Northeast Region, 4178
USDA Cochran Fellowships, 4619
USDA Small Business Innovation Research Grants - Phase I, 4648
USDA Small Business Innovation Research Grants - Phase II, 4649

Agricultural Collectives/Cooperatives
National Council of Farmer Cooperatives Graduate Awards, 2860

Agricultural Commodities
AOCS Timothy L. Mounts Award, 900
MDARD Specialty Crop Block Grant Program-Farm Bill, 2730
NWO Added Value Plus for Soc Sciences Grants, 3672
USDA Interregional Research Project Minor Crop Pest Management Grants, 4632
USDA Policy Research Center Grants, 4640

Agricultural Economics
BARD Research Grants, 1299
Honeybee Health Improvement Project Grants, 2276
MDARD Specialty Crop Block Grant Program-Farm Bill, 2730
National Council of Farmer Cooperatives Graduate Awards, 2860
SARE Graduate Student Grants - NE Region, 4169
SARE Graduate Student Grants in Sustainable Agriculture - Western Region, 4171
SARE Professional + Producer Grants - Western Region, 4175
USDA 1890 Institution Teaching, Research, and Extension Capacity Building Grants, 4613
USDA Distance Education Grants for Higher Education Institutions in Insular Areas, 4622
USDA Food and Agricultural Sciences National Needs Graduate Fellowship Grants, 4624
USDA Norman E. Borlaug International Agricultural Science and Technology Fellowships, 4636
USDA Resident Instruction Grants for Higher Education in Insular Areas, 4645
USDA Tribal College Research Grants, 4652

Agricultural Engineering
ASME Ralph Coats Roe Medal, 1158
BARD Research Grants, 1299
USDA 1890 Institution Teaching, Research, and Extension Capacity Building Grants, 4613
USDA Distance Education Grants for Higher Education Institutions in Insular Areas, 4622
USDA Resident Instruction Grants for Higher Education in Insular Areas, 4645
USDA Tribal College Research Grants, 4652

Agricultural Extension
Phi Upsilon Omicron Florence Fallgatter Distinguished Service Award, 3943
SARE Graduate Student Grants - NE Region, 4169
SARE Graduate Student Grants - So Region, 4170
SARE Large Systems Research Grants - Southern Region, 4172
SARE On-Farm Research Grants - So Region, 4173
SARE Producer Research and Education Grants-Western Region, 4174
SARE Professional + Producer Grants - Western Region, 4175
SARE Professional Development Grants - Northeast Region, 4176
SARE Research and Education Grants - North Central Region, 4177
SARE Research and Education Grants - Northeast Region, 4178
SARE Research and Education Grants - Western Region, 4180

USDA 1890 Institution Teaching, Research, and Extension Capacity Building Grants, 4613
USDA Cochran Fellowships, 4619
USDA Distance Education Grants for Higher Education Institutions in Insular Areas, 4622
USDA Food and Agricultural Sciences National Needs Graduate Fellowship Grants, 4624
USDA Integrated Pest Management Competitive Grants - Western Region, 4629
USDA International Science and Education Competitive Grants, 4631
USDA Interregional Research Project Minor Crop Pest Management Grants, 4632
USDA National Integrated Food Safety Initiative Grants, 4635
USDA Norman E. Borlaug International Agricultural Science and Technology Fellowships, 4636
USDA Regional Integrated Pest Management Grants - North Central Region, 4642
USDA Regional Integrated Pest Management Grants - Northeastern Region, 4643
USDA Regional Integrated Pest Management Grants - Southern Region, 4644
USDA Scientific Cooperation Research Program, 4647
USDA Technical Agricultural Assistance Research Grants, 4651

Agricultural Finance
National Council of Farmer Cooperatives Graduate Awards, 2860

Agricultural Geography
USDA 1890 Institution Teaching, Research, and Extension Capacity Building Grants, 4613
USDA Distance Education Grants for Higher Education Institutions in Insular Areas, 4622
USDA Resident Instruction Grants for Higher Education in Insular Areas, 4645
USDA Tribal College Research Grants, 4652

Agricultural Management
SARE Large Systems Research Grants - Southern Region, 4172
SARE Professional + Producer Grants - Western Region, 4175
USDA Food and Agricultural Sciences National Needs Graduate Fellowship Grants, 4624
USDA Integrated Pest Management Competitive Grants - Western Region, 4629
USDA Regional Integrated Pest Management Grants - North Central Region, 4642
USDA Regional Integrated Pest Management Grants - Northeastern Region, 4643
USDA Regional Integrated Pest Management Grants - Southern Region, 4644
USDA Scientific Cooperation Exchange Program, 4646

Agricultural Planning/Policy
Clarence E. Heller Charitable Foundation Grants, 1575
Great Lakes Protection Fund Grants, 2170
Institute for Agriculture and Trade Policy Food and Society Fellowships, 2381
MDARD Specialty Crop Block Grant Program-Farm Bill, 2730
SARE Graduate Student Grants - NE Region, 4169
SARE Graduate Student Grants in Sustainable Agriculture - Western Region, 4171
SARE Large Systems Research Grants - Southern Region, 4172
SARE Professional + Producer Grants - Western Region, 4175
SARE Research and Education Grants - Southern Region, 4179
The World Food Prize, 4503
USDA 1890 Institution Teaching, Research, and Extension Capacity Building Grants, 4613
USDA Biotechnology Risk Assessment Grants, 4618
USDA Cochran Fellowships, 4619
USDA Cooperative Extension System Education Grants, 4620
USDA Distance Education Grants for Higher Education Institutions in Insular Areas, 4622
USDA Emerging Markets Grants, 4623
USDA Foreign Market Development Grants, 4625
USDA Market Access Program Grants, 4633
USDA Norman E. Borlaug International Agricultural Science and Technology Fellowships, 4636
USDA Resident Instruction Grants for Higher Education in Insular Areas, 4645

Agriculture
Abdus Salam ICTP/IAEA Sandwich Training Education Grants, 226
ALSAM Foundation Grants, 579
American-Scandinavian Foundation Short-Term Training Program Grants, 606
American Chemical Society AGFD Advancement of the Application of Agricultural and Food Chemistry Award, 626
American Chemical Society AGFD Teranishi Graduate Fellowship Award, 628
American Chemical Society AGFD Undergraduate Research Symposium Travel Grants, 629
American Chemical Society AGFD Young Scientist Award, 630
American Chemical Society AGRO Award for Innovation in Chemistry of Agriculture, 631
American Chemical Society AGRO Educ Awards, 632
American Chemical Society AGRO Fellows, 633
American Chemical Society AGRO International Award for Research in Agrochemicals, 634
American Chemical Society AGRO Young Scientist Research Awards, 636
American Chemical Society Herman Frasch Foundation Grants, 680
Anderson Foundation Grants, 842
ASA and USDA/NASS Research Fellow and Associate Program, 1033
ASPRS William A. Fischer Mem Scholarship, 1216
BARD Postdoctoral Fellowships, 1298
BARD Research Grants, 1299
C. Alma Baker Trust Grants, 1403
Canada-U.S. Fulbright International Science and Technology Award, 1417
Canada-U.S. Fulbright Senior Specialists Program Grants, 1418
Canada-U.S. Fulbright Traditional Scholars, 1419
Canada Graduate Scholarships and NSERC Postgraduate Scholarships, 1420
CCFF Distinguished Agriscience Scie Award, 1450
Ekhagastiftelsen Foundation Grants, 1903
Fulbright Alumni Initiatives Awards Grants, 2064
Fulbright Distinguished Chairs Program Grants, 2069
Fulbright European Union Scholar-in-Residence Program Grants, 2071
Fulbright New Century Scholars Program Grants, 2075
Fulbright Occasional Lecturer Program Grants, 2076
Fulbright Specialists Program Grants, 2079
Fulbright Traditional Scholar Program in Europe and Eurasia, 2080
Fulbright Traditional Scholar Program in Sub-Saharan Africa, 2081
Fulbright Traditional Scholar Program in the East Asia/Pacific Region, 2082
Fulbright Traditional Scholar Program in the Near East and North Africa Region, 2083
Fulbright Traditional Scholar Program in the South and Central Asia Region, 2084
Fulbright Traditional Scholar Program in the Western Hemisphere, 2085
Fulbright Visiting Specialists Program: Direct Access to the Muslim World, 2087
GCA Katharine M. Grosscup Scholarships, 2099
Gruber Foundation Weizmann Institute Awards, 2184
Honeybee Health Improvement Project Grants, 2276
Institute for Agriculture and Trade Policy Food and Society Fellowships, 2381
John Deere Foundation Grants, 2500
Library Company of Philadelphia Barra Foundation International Research Fellowships, 2606
Library Company of Philadelphia Prizes, 2608
Library Company of Philadelphia Program in Early American Economy and Society Dissertation-Level Fellowships, 2609
Library Company of Philadelphia Program in Early American Economy and Society One-Month Fellowships, 2610
Library Company of Philadelphia Program in Early American Economy and Society Post-Doctoral Research Fellowships, 2611
McCune Charitable Foundation Grants, 2719
MDARD Specialty Crop Block Grant Program-Farm Bill, 2730
Mellon Post-Doctoral Fellowship in Sustainable Agriculture, 2740
Monfort Family Foundation Grants, 2794
RSC Thomas W. Eadie Medal, 4126
USDA Basic and Applied Research Grants, 4617
USDA Emerging Markets Grants, 4623
USDA Food and Agricultural Sciences National Needs Graduate Fellowship Grants, 4624
USDA Foreign Market Development Grants, 4625
USDA Higher Education Challenge Grants, 4627
USDA Hispanic-Serving Institutions Education Grants, 4628
USDA Market Access Program Grants, 4633
USDA Small Business Innovation Research Grants - Phase I, 4648
USDA Special Research Grants, 4650
USDA Wildlife Services Grants, 4653

Agriculture Education
C. Alma Baker Trust Grants, 1403
Conservation, Food, and Health Foundation Grants for Developing Countries, 1641
McCune Charitable Foundation Grants, 2719
SARE Graduate Student Grants - NE Region, 4169
SARE Research and Education Grants - North Central Region, 4177
SARE Research and Education Grants - Northeast Region, 4178
USDA 1890 Institution Teaching, Research, and Extension Capacity Building Grants, 4613
USDA Agriculture and Food Research Initiative Competitive Grants - Food Security, 4614
USDA Alaska Native-Serving and Native Hawaiian-Serving Institutions Education Grants, 4615
USDA Basic and Applied Research Grants, 4617
USDA Distance Education Grants for Higher Education Institutions in Insular Areas, 4622
USDA Integrated Pest Management Competitive Grants - Western Region, 4629
USDA Interregional Research Project Minor Crop Pest Management Grants, 4632
USDA National Integrated Food Safety Initiative Grants, 4635
USDA Norman E. Borlaug International Agricultural Science and Technology Fellowships, 4636
USDA Organic Agriculture Research and Extension Initiative, 4637
USDA Policy Research Center Grants, 4640
USDA Regional Integrated Pest Management Grants - North Central Region, 4642
USDA Regional Integrated Pest Management Grants - Northeastern Region, 4643
USDA Regional Integrated Pest Management Grants - Southern Region, 4644
USDA Resident Instruction Grants for Higher Education in Insular Areas, 4645
USDA Tribal College Research Grants, 4652

Agriscience
American Chemical Society AGRO Award for Innovation in Chemistry of Agriculture, 631
American Chemical Society AGRO Educ Awards, 632
American Chemical Society AGRO Fellows, 633
American Chemical Society AGRO Young Scientist Research Awards, 636
CCFF Distinguished Agriscience Scie Award, 1450
Honeybee Health Improvement Project Grants, 2276

USDA Decadal and Regional Climate Prediction Using Earth System Models Grants, 4621

Agronomy
American Chemical Society AGRO Award for Innovation in Chemistry of Agriculture, 631
American Chemical Society AGRO Educ Awards, 632
American Chemical Society AGRO Fellows, 633
American Chemical Society AGRO International Award for Research in Agrochemicals, 634
American Chemical Society AGRO Sterling B. Hendricks Memorial Lectureship, 635
American Chemical Society AGRO Young Scientist Research Awards, 636
GCSAA Essay Contest, 2102
NWO Added Value Plus for Soc Sciences Grants, 3672
USDA 1890 Institution Teaching, Research, and Extension Capacity Building Grants, 4613
USDA Distance Education Grants for Higher Education Institutions in Insular Areas, 4622
USDA Food and Agricultural Sciences National Needs Graduate Fellowship Grants, 4624
USDA Resident Instruction Grants for Higher Education in Insular Areas, 4645
USDA Tribal College Research Grants, 4652

Alaskan Natives
ASM Robert D. Watkins Graduate Research Fellowship, 1183
DHHS Native American Programs, 1758
FDHN Student Research Fellowships, 2001
MLA Grad Scholarship for Minority Students, 2787
NAGPRA Museum Consultation/Documentation Grants, 2822
NAGPRA Museum Repatriation Grants, 2823
NEH Awards for Faculty at Tribal Colleges and Universities, 2951
NHLBI Training Program Grants for Institutions that Promote Diversity, 3067
NSF Tribal Colleges and Universities Program, 3662
UCLA Research Grant in Ethnic Studies, 4549
USDA Alaska Native-Serving and Native Hawaiian-Serving Institutions Education Grants, 4615

Alcohol Education
AACAP Educational Outreach Program for Residents in Alcohol Research, 58
Coors Brewing Corporate Contributions Grants, 1644
Fairlawn Foundation Grants, 1974

Alcohol/Alcoholism
AAA Foundation for Traffic Safety Grants, 24
AACAP Educational Outreach Program for Residents in Alcohol Research, 58
Achelis Foundation Grants, 274
Adaptec Foundation Grants, 340
Alcohol Misuse and Alcoholism Research Grants, 512
Alex Stern Family Foundation Grants, 542
amfAR Fellowships, 809
amfAR Mathilde Krim Fellowships in Basic Biomedical Research, 810
amfAR Research Grants, 811
Berks County Community Foundation Grants, 1333
Bodman Foundation Grants, 1356
Bupa Foundation Multi-Country Grant, 1387
Carpenter Foundation Grants, 1439
Community Foundation for S Arizona Grants, 1618
ConocoPhillips Grants Program, 1629
Coors Brewing Corporate Contributions Grants, 1644
D.F. Halton Foundation Grants, 1692
Farmers Insurance Corporate Giving Grants, 1981
Jackson Foundation Grants, 2453
Johnson & Johnson Corporate Contributions, 2508
Joint Center for Poverty Research Grants, 2511
Musgrave Foundation Grants, 2807
NIAAA Independent Scientist Award, 3078
NIAAA Mechanisms of Alcohol and Drug-Induced Pancreatitis Grants, 3079
NIH Research on Sleep and Sleep Disorders, 3236
NIMH Early Identification and Treatment of Mental Disorders in Children and Adolescents Grants, 3252
Peter F. McManus Charitable Trust Grants, 3903
Rayonier Foundation Grants, 4034
SfN Jacob P. Waletzky Award, 4206
Sid W. Richardson Foundation Grants, 4253
Sioux Falls Area Comm Foundation Spot Grants, 4263
Victor E. Speas Foundation Grants, 4810

Algebra
ARO Discrete Mathematics and Computer Science Research Contracts, 1009
NSF Algebra, Number Theory and Combinatorics Grants, 3468
NSF Analysis Grants, 3469
NSF Scientific Computing Research Environments for the Mathematical Sciences Grants, 3645
NSF Topology Research Grants, 3661

Allergy
AAAAI ARTrust Mini Grants for Allied Health, 14
AAAAI ARTrust Mini Grants for Clinical Res, 15
AAAAI ARTrust Mini Grants in Faculty Devel, 16
AAAAI Distinguished Clinician Award, 17
AAAAI Distinguished Layperson Award, 18
AAAAI Distinguished Scientist Award, 19
AAAAI Distinguished Service Award, 20
AAAAI Fellows-in-Training Abstract Award, 21
AAAAI Fellows-in-Training Grants, 22
AAAAI Fellows-in-Training Travel Grants, 23
Coca-Cola Foundation Grants, 1593
NIAID Independent Scientist Award, 3102

Allied Health
MLA David A. Kronick Traveling Fellowship, 2784
MLA Murray Gottlieb Prize Essay Award, 2788

Allied Health Education
AFUW Winifred E. Preedy Postgraduate Bursary, 368
APSA Robert Wood Johnson Foundation Health Policy Fellowships, 944
DHHS Oral Health Promotion Research Across the Lifespan, 1759
Hospital Libraries Section / MLA Professional Development Grants, 2283
Josiah Macy Jr. Foundation Grants, 2516
MLA David A. Kronick Traveling Fellowship, 2784
MLA Donald Lindberg Research Fellowship, 2785

Alternative Education
Joyce Foundation Education Grants, 2519

Alternative Fuels
Alfred P. Sloan Foundation Research Fellowships, 560
American Chemical Society Green Chem Grants, 679
BMW of N America Charitable Contributions, 1353
Colorado Bioscience Discovery Evaluation Grants, 1598
DOE Engineering Research Grants, 1789
DOE Initial H-Prize Competition for Breakthrough Advances in Materials for Hydrogen Storage, 1801
DOE State Energy Program Special Projects, 1815
Hydrogen Student Design Contest, 2311
SSHRC-APC Automotive Partnership Canada, 4396
USDD Future Naval Fuels Grants, 4729

Alternative Medicine
AIHS Alberta/Pfizer Translational Res Grants, 435
Blowitz-Ridgeway Foundation Grants, 1349
Ekhagastiftelsen Foundation Grants, 1903
Fannie E. Rippel Foundation Grants, 1978
GPP Charles E. Culpeper Foundation Biomedical Pilot Projects Initiative Grants, 2159
Jenifer Altman Foundation Grants, 2484
NCCAM Exploratory Developmental Grants for Complementary and Alternative Medicine Studies of Humans, 2903
NCCAM Ruth L. Kirschstein National Research Service Awards for Postdoctoral Training in Complementary and Alternative Medicine, 2904
NINR Quality of Life for Individuals at the End of Life Grants, 3266
RACGP Vicki Kotsirilos Integrative Med Grants, 4017
Susan G. Komen Breast Cancer Foundation Challege Grants: Investigator Initiated Research, 4475

Alternative Modes of Education
ASME Frank von Flue Award, 1121
Joyce Foundation Education Grants, 2519
Nellie Mae Education Foundation Public Understanding Grants, 2969
Nellie Mae Education Foundation Research and Development Grants, 2970
Parkersburg Area Community Foundation Arts in Education Special Grants, 3869

Alzheimer's Disease
AHAF Alzheimer's Disease Research Grants, 370
AIHS Alberta/Pfizer Translational Res Grants, 435
Alzheimer's Association Conference Grants, 589
Alzheimer's Association Development of New Cognitive and Functional Instruments Grants, 590
Alzheimer's Association Everyday Technologies for Alzheimer Care Grants, 591
Alzheimer's Association Investigator-Initiated Research Grants, 592
Alzheimer's Association Mentored New Investigator Research Grants to Promote Diversity, 593
Alzheimer's Association Neuronal Hyper Excitability and Seizures in Alzheimer's Disease Grants, 594
Alzheimer's Association New Investigator Research Grants, 595
Alzheimer's Association New Investigator Research Grants to Promote Diversity, 596
Alzheimer's Association U.S.-U.K. Young Investigator Exchange Fellowships, 597
Alzheimer's Association Zenith Fellows Awards, 598
Brookdale Foundation Leadership in Aging Fellowships, 1381
Caesars Foundation Grants, 1407
Fairlawn Foundation Grants, 1974
Hawaii Community Foundation Health Education and Research Grants, 2221
Helen Bader Foundation Grants, 2228
NIA Alzheimer's Disease Core Centers Grants, 3084
NIA Alzheimer's Disease Drug Devel Grants, 3085
NIA Alzheimer's Disease Pilot Clinical Trials, 3086
NIA Alzheimer's Disease Res Centers Grants, 3087
NIA Grants for Alzheimer's Disease Drug Disc, 3096
NIA Role of Apolipoprotein E, Lipoprotein Receptors and CNS Lipid Homeostasis in Brain Aging and Alzheimer's Disease, 3119
NIA Vulnerable Dendrites and Synapses in Aging and Alzheimer's Disease Grants, 3125
Northwestern Mutual Foundation Grants, 3430
Pfizer Australia Neuroscience Research Grants, 3920
Pittsburgh Foundation Medical Research Grants, 3973
Rosalinde and Arthur Gilbert Foundation/AFAR New Investigator Awards in Alzheimer's Disease, 4102
Thome Foundation Awards in Alzheimer's Disease Drug Discovery Research, 4510

American History
ALA Gale Cengage History Research and Innovation Award, 488
American Historical Association Martin A. Klein Prize in African History, 737
American Historical Association Raymond J. Cunningham Prize, 742
BMI Woody Guthrie Fellowships, 1352
CAA Wyeth Publication Grant, 1405
IES Faculty Fellowships, 2341
Irving and Rose Crown Scholarships and Fellowships in American History, 2427
NL American Society for Eighteenth-Century Studies Fellowships, 3279
NL Arthur and Lila Weinberg Fellowships for Independent Scholars and Researchers, 3280
NL Associated Colleges of the Midwest/Great Lakes College Association Faculty Fellowships, 3281

NL British Academy Fellowship for Study in the United Kingdom, 3283
NL Frances C. Allen Fellowships, 3284
NL Grad Scholar-in-Residence Opportunities, 3285
NL Herzog August Bibliothek Wolfenbuttel Fellowship, 3286
NL Lawrence Lipking Fellowship at Newberry, 3287
NL Lester J. Cappon Fellowship in Doc Editing, 3288
NL Lloyd Lewis Fellowships in American Hist, 3289
NL Mellon Postdoctoral Research Fellowships, 3294
NL Midwest Modern Lang Assoc Fellowship, 3299
NL Monticello College Foundation Fellowship for Women, 3304
NL National Endowment for the Humanities Fellowships, 3310
NL Newberry Consortium on American Indian Studies Faculty Fellowships, 3311
NL Northeast Modern Language Assocation Fellowship, 3314
NL Scholar-in-Residence Opportunity, 3315
NL Short-Term Resident Fellowships for Individual Research, 3317
NL South Central Modern Languages Association Fellowship, 3318
NL Susan Kelly Power and Helen Hornbeck Tanner Fellowship, 3319
Radcliffe Institute Carol K. Pforzheimer Student Fellowships, 4018
Smithsonian Graduate Student Fellowships, 4277
Smithsonian Minority Visiting Student Awards, 4279
Smithsonian National Museum of American History Lemelson Center Fellowships, 4289
Smithsonian National Museum of History Lemelson Center Travel to Collections Awards, 4290
Smithsonian Native American Community Scholars Awards, 4291
Smithsonian Native American Visiting Student Awards, 4292
Smithsonian Postdoctoral Fellowships, 4293
Smithsonian Predoctoral Fellowships, 4295
Wolfsonian-FIU Fellowships, 4890

American Indian Culture
American Indian Studies Research Inst Fellowship, 746
IES Faculty Fellowships, 2341
McCune Charitable Foundation Grants, 2719
NL American Society for Eighteenth-Century Studies Fellowships, 3279
NL Arthur and Lila Weinberg Fellowships for Independent Scholars and Researchers, 3280
NL Associated Colleges of the Midwest/Great Lakes College Association Faculty Fellowships, 3281
NL British Academy Fellowship for Study in the United Kingdom, 3283
NL Frances C. Allen Fellowships, 3284
NL Grad Scholar-in-Residence Opportunities, 3285
NL Herzog August Bibliothek Wolfenbuttel Fellowship, 3286
NL Lawrence Lipking Fellowship at Newberry, 3287
NL Lester J. Cappon Fellowship in Documentary Editing, 3288
NL Mellon Postdoctoral Research Fellowships, 3294
NL Midwest Modern Lang Assoc Fellowship, 3299
NL Monticello College Foundation Fellowship for Women, 3304
NL National Endowment for the Humanities Fellowships, 3310
NL Newberry Consortium on American Indian Studies Faculty Fellowships, 3311
NL Newberry Consortium on American Indian Studies Graduate Student Fellowships, 3312
NL Northeast Modern Language Assocation Fellowship, 3314
NL Scholar-in-Residence Opportunity, 3315
NL Short-Term Resident Fellowships for Individual Research, 3317
NL South Central Modern Languages Association Fellowship, 3318
NL Susan Kelly Power and Helen Hornbeck Tanner Fellowship, 3319

American Indian History
American Indian Studies Research Inst Fellowship, 746
IES Faculty Fellowships, 2341
McCune Charitable Foundation Grants, 2719
NL American Society for Eighteenth-Century Studies Fellowships, 3279
NL Arthur and Lila Weinberg Fellowships for Independent Scholars and Researchers, 3280
NL Associated Colleges of the Midwest/Great Lakes College Association Faculty Fellowships, 3281
NL British Academy Fellowship for Study in the United Kingdom, 3283
NL Frances C. Allen Fellowships, 3284
NL Grad Scholar-in-Residence Opportunities, 3285
NL Herzog August Bibliothek Wolfenbuttel Fellowship, 3286
NL Lawrence Lipking Fellowship at Newberry, 3287
NL Lester J. Cappon Fellowship in Documentary Editing, 3288
NL Lloyd Lewis Fellowships in American Hist, 3289
NL Mellon Postdoctoral Research Fellowships, 3294
NL Midwest Modern Lang Assoc Fellowship, 3299
NL Monticello College Foundation Fellowship for Women, 3304
NL National Endowment for the Humanities Fellowships, 3310
NL Newberry Consortium on American Indian Studies Faculty Fellowships, 3311
NL Newberry Consortium on American Indian Studies Graduate Student Fellowships, 3312
NL Northeast Modern Language Assocation Fellowship, 3314
NL Scholar-in-Residence Opportunity, 3315
NL Short-Term Resident Fellowships for Individual Research, 3317
NL South Central Modern Languages Association Fellowship, 3318
NL Susan Kelly Power and Helen Hornbeck Tanner Fellowship, 3319

American Indian Language
American Indian Studies Research Inst Fellowship, 746
IES Faculty Fellowships, 2341
McCune Charitable Foundation Grants, 2719
NL American Society for Eighteenth-Century Studies Fellowships, 3279
NL Arthur and Lila Weinberg Fellowships for Independent Scholars and Researchers, 3280
NL Associated Colleges of the Midwest/Great Lakes College Association Faculty Fellowships, 3281
NL British Academy Fellowship for Study in the United Kingdom, 3283
NL Frances C. Allen Fellowships, 3284
NL Grad Scholar-in-Residence Opportunities, 3285
NL Herzog August Bibliothek Wolfenbuttel Fellowship, 3286
NL Lawrence Lipking Fellowship at Newberry, 3287
NL Lester J. Cappon Fellowship in Documentary Editing, 3288
NL Mellon Postdoctoral Research Fellowships, 3294
NL Midwest Modern Lang Assoc Fellowship, 3299
NL Monticello College Foundation Fellowship for Women, 3304
NL National Endowment for the Humanities Fellowships, 3310
NL Newberry Consortium on American Indian Studies Faculty Fellowships, 3311
NL Newberry Consortium on American Indian Studies Graduate Student Fellowships, 3312
NL Northeast Modern Language Assocation Fellowship, 3314
NL Scholar-in-Residence Opportunity, 3315
NL Short-Term Resident Fellowships for Individual Research, 3317
NL South Central Modern Languages Association Fellowship, 3318
NL Susan Kelly Power and Helen Hornbeck Tanner Fellowship, 3319

American Recovery and Reinvestment Act
NIH Recovery Act Limited Competition: Academic Research Enhancement Awards, 3230
NIH Recovery Act Limited Competition: Biomedical Research, Development, and Growth to Spur the Acceleration of New Technologies Pilot Program Grants, 3231
NIH Recovery Act Limited Competition: Small Business Catalyst Awards for Accelerating Innovative Research Grants, 3233
NIH Recovery Act Limited Competition: Supporting New Faculty Recruitment to Enhance Research Resources through Biomedical Research Core Centers Grants, 3234

American Studies
AAS Kate B. and Hall J. Peterson Fellowships, 204
AAS Legacy Fellowships, 205
Ahmanson and Getty Postdoctoral Fellowships, 377
Alabama Humanities Foundation Mini Grants, 480
Alabama Humanities Foundation Planning/Consultant Grants, 481
Albert M. Greenfield Foundation Dissertation Fellowship, 507
American Historical Association Albert B. Corey Prize in Canadian-American Relations, 711
American Historical Association Albert J. Beveridge Grants for Research in the History of the Western Hemisphere, 713
American Historical Association J. Franklin Jameson Fellowship, 725
American Historical Association John H. Dunning Prize, 732
American Philosophical Society Library Resident Research Fellowships, 769
APSA E.E. Schattschneider Award, 932
Canada-U.S. Fulbright International Science and Technology Award, 1417
Canada-U.S. Fulbright Senior Specialists Program Grants, 1418
Canada-U.S. Fulbright Traditional Scholars, 1419
Center for the Study of Religion Fellowships, 1485
Cushwa Center Research Travel Grants, 1673
DOS Study of the United States Institutes, 1846
Fulbright Alumni Initiatives Awards Grants, 2064
Fulbright Distinguished Chairs Program Grants, 2069
Fulbright European Union Scholar-in-Residence Program Grants, 2071
Fulbright New Century Scholars Program Grants, 2075
Fulbright Occasional Lecturer Program Grants, 2076
Fulbright Specialists Program Grants, 2079
Fulbright Traditional Scholar Program in Europe and Eurasia, 2080
Fulbright Traditional Scholar Program in Sub-Saharan Africa, 2081
Fulbright Traditional Scholar Program in the East Asia/Pacific Region, 2082
Fulbright Traditional Scholar Program in the Near East and North Africa Region, 2083
Fulbright Traditional Scholar Program in the South and Central Asia Region, 2084
Fulbright Traditional Scholar Program in the Western Hemisphere, 2085
Fulbright Visiting Specialists Program: Direct Access to the Muslim World, 2087
Herbert Hoover Presidential Library Association Travel Grants, 2242
Library Company of Philadelphia Barra Foundation International Research Fellowships, 2606
Library Company of Philadelphia Prizes, 2608
Library Company of Philadelphia Program in Early American Economy and Society Dissertation-Level Fellowships, 2609
Library Company of Philadelphia Program in Early American Economy and Society One-Month Fellowships, 2610
Library Company of Philadelphia Program in Early American Economy and Society Post-Doctoral Research Fellowships, 2611

Maine Community Foundation Keepers Preservation Education Grants, 2651
McCune Charitable Foundation Grants, 2719
MCEAS Dissertation Fellowships, 2723
National Endowment for the Humanities Fellowships at Winterthur, 2867
NPS American Battlefield Protection Grants, 3435
OAH Avery O. Craven Award, 3711
OAH Darlene Clark Hine Award, 3713
OAH Frederick Jackson Turner Award, 3717
OAH Merle Curti Award, 3726
OAH Ray A. Billington Prize, 3727
Smithsonian National Museum of History Lemelson Center Travel to Collections Awards, 4290
Smithsonian Senior Fellowships, 4296
University of London Visiting Fellowships, 4588
Winterthur Museum Dissertation Fellowships, 4887
Winterthur Research Fellowships, 4888

Amyotrophic Lateral Sclerosis
Judith and Jean Pape Adams Charitable Foundation ALS Grants, 2528
William H. Adams Foundation for ALS Grants, 4878

Analytical Techniques
AACC International Edith A. Christensen Award, 87
American Chemical Society ANYL Award in Electrochemistry, 642
AOCS George Schroepfer Medal, 879
AOCS Herbert J. Dutton Award, 883

Anatomy
M.J. Murdock Charitable Trust College Research Program for Life Sciences, 2638
NSF Physical Anthropology Research Grants, 3615
PhRMA Foundation Pharmacology/Toxicology Post Doctoral Fellowships, 3962

Ancient History
AAR Rome Prize Fellowships, 202
American Historical Association James Henry Breasted Prize, 729

Anemia
ACCP Pharmacotherapy Investigator Development Research Awards, 266
Taub Foundation Grants for MDS Research, 4496

Anesthesiology
Duke University Adult Cardiothoracic Anesthesia and Critical Care Medicine Fellowships, 1869
Duke University Ambulatory and Regional Anesthesia Fellowships, 1870
Duke Univ Obstetric Anesthesia Fellowships, 1878
NIGMS Mentored Clinical Scientist Research Career Development Awards, 3197

Animal Behavior/Ethology
Animal Behavior Society C.H. Turner Award, 849
Animal Behavior Society Developing Nations Research Grants, 850
Animal Behavior Society Diversity Grant, 851
Animal Behavior Society Latin American Travel Awards, 852
Animal Behavior Society Student Research Grants, 853
McCune Charitable Foundation Grants, 2719
NSF Physiology and Ethology Research Grants, 3622

Animal Care
Adam Richter Charitable Trust Grants, 339
Banfield Charitable Trust Grants, 1295
Batchelor Foundation Grants, 1305
Beim Foundation Grants, 1331
Dade Community Foundation Grants, 1712
Florida Sea Turtle Grants Program, 2027
Margaret T. Morris Foundation Grants, 2659
Matilda R. Wilson Fund Grants, 2684
McCune Charitable Foundation Grants, 2719
NFWF Bird Conservation Initiative Grants, 2997
NFWF National Whale Conservation Grants, 3004
NFWF Wildlife and Habitat Conservation Initiative Grants, 3009
San Antonio Area Foundation Grants, 4163
Vancouver Foundation Grants and Community Initiatives, 4807
William and Charlotte Parks Foundation for Animal Welfare Grants, 4872
Winn Feline Foundation/AVMF Excellence In Feline Research Award, 4884
Winn Feline Foundation Grants, 4885

Animal Development
McCune Charitable Foundation Grants, 2719

Animal Diseases/Pathology
ACVIM Foundation Clinical Investigation Grants, 334
National Turkey Federation Research Award, 2887
USDA Animal Health & Disease Grants, 4616
USDA Plant and Animal Disease, Pest Control, and Animal Care Grants, 4638
USDA Wildlife Services Grants, 4653
Winn Feline Foundation/AVMF Excellence In Feline Research Award, 4884

Animal Ecology
NFWF/Exxon Save the Tiger Fund Grants, 2995
Turtle Conservation Fund Grants, 4535
USDA Scientific Cooperation Exchange Program, 4646

Animal Feeds/Nutrition
ASNS Bio-Serv Award in Experimental Animal Nutrition, 1188
USDA 1890 Institution Teaching, Research, and Extension Capacity Building Grants, 4613
USDA Distance Education Grants for Higher Education Institutions in Insular Areas, 4622
USDA Norman E. Borlaug International Agricultural Science and Technology Fellowships, 4636
USDA Resident Instruction Grants for Higher Education in Insular Areas, 4645
USDA Tribal College Research Grants, 4652

Animal Genetics/Breeding
AALAS Foundation Grants, 182
BARD Research Grants, 1299
International Crane Foundation Internships, 2396
NAAB Coordinated Research Grants Program, 2808
National Turkey Federation Research Award, 2887
NSF Living Stock Collections Grants, 3588
NSF Physiology and Ethology Research Grants, 3622

Animal Physiology/Morphology
NSF Physiology and Ethology Research Grants, 3622

Animal Rescue
Turtle Conservation Fund Grants, 4535

Animal Research Policy
Alternatives Research and Development Foundation Alternatives in Education Grants, 581
Alternatives Research and Development Foundation Grants, 582
NCRR Novel Approaches to Enhance Animal Stem Cell Research, 2935
Vancouver Foundation Grants and Community Initiatives, 4807
William and Charlotte Parks Foundation for Animal Welfare Grants, 4872

Animal Rights
Adam Richter Charitable Trust Grants, 339
Bodenwein Public Benevolent Foundation Grants, 1355
Cleveland-Cliffs Foundation Grants, 1580
David Bohnett Foundation Grants, 1724
Frank Stanley Beveridge Foundation Grants, 2051
Helen V. Brach Foundation Grants, 2231
Park Foundation Grants, 3870
Rhode Island Foundation Grants, 4061
USDA Plant and Animal Disease, Pest Control, and Animal Care Grants, 4638
Vancouver Foundation Grants and Community Initiatives, 4807
William and Charlotte Parks Foundation for Animal Welfare Grants, 4872

Animal Science
ACVIM Foundation Clinical Investigation Grants, 334
BARD Research Grants, 1299
IMLS Conservation Project Support Grants, 2363
National Turkey Federation Research Award, 2887
NFWF National Whale Conservation Grants, 3004
NSF Animal Developmental Mechanisms Grants, 3471
NSF Cell Systems Cluster Grants, 3512
NSF Hydrologic Sciences Research Grants, 3578
USDA 1890 Institution Teaching, Research, and Extension Capacity Building Grants, 4613
USDA Distance Education Grants for Higher Education Institutions in Insular Areas, 4622
USDA Food and Agricultural Sciences National Needs Graduate Fellowship Grants, 4624
USDA Plant and Animal Disease, Pest Control, and Animal Care Grants, 4638
USDA Resident Instruction Grants for Higher Education in Insular Areas, 4645
USDA Scientific Cooperation Exchange Program, 4646
USDA Small Business Innovation Research Grants - Phase I, 4648
USDA Tribal College Research Grants, 4652

Animal Welfare
Annenberg Foundation Grants, 862
Herbert & Adrian Woods Foundation Grants, 2239
Matilda R. Wilson Fund Grants, 2684
McCune Charitable Foundation Grants, 2719
Sarkeys Foundation Grants, 4181
SeaWorld and Busch Gardens Conservation Fund Grants, 4196
William and Charlotte Parks Foundation for Animal Welfare Grants, 4872

Animals as Pets
Banfield Charitable Trust Grants, 1295
Mericos Foundation Grants, 2742

Animals for Assistance/Therapy
Blum-Kovler Foundation Grants, 1351
Lumpkin Family Found Healthy People Grants, 2629

Anorexia Nervosa
Davis Foundation Postdoctoral Fellowships in Eating Disorders Research, 1727
Klarman Family Foundation Grants in Eating Disorders Research, 2558
NIMH Early Identification and Treatment of Mental Disorders in Children and Adolescents Grants, 3252

Antarctica
NSF Antarctic Aeronomy and Astrophysics Program Grants, 3472
NSF Antarctic Earth Sciences Program Grants, 3474
NSF Antarctic Glaciology Program Grants, 3475
NSF Antarctic Organisms & Ecosystems Grants, 3478
NSF Integrative Graduate Education and Research Traineeship, 3584
NSF Polar Programs Division Research Grants, 3627

Antennae
ARO Mobile, Wireless Communications and Networks Research Contracts, 1012
USDD Sensor Innovative Research Grants, 4791

Anthropology
AAA Minority Dissertation Fellowship Program, 25
AAB Berlin Prize Fellowships, 48
ACLS Charles A. Ryskamp Research Fellowships, 281
ACLS Frederick Burkhardt Residential Fellowships for Recently Tenured Scholars, 288
Alfred P. Sloan Foundation Energy and Environment Grants, 552
American Philosophical Soc Judson Daland Prize, 766

American Philosophical Society Lewis and Clark Fund for Exploration and Field Research, 767
American Philosophical Society Phillips Fund Grants for Native American Research, 770
Andrew W. Mellon/ACLS Dissertation Completion Fellowships, 844
Anthropological Center for Training in Global and Environmental Change Grad Assistantships, 870
Australasian Institute of Judicial Administration Seed Funding Grants, 1268
Canada-U.S. Fulbright International Science and Technology Award, 1417
Canada-U.S. Fulbright Senior Specialists Program Grants, 1418
Canada-U.S. Fulbright Traditional Scholars, 1419
Center for East Asian Studies Fellowships, 1474
Christine Wilson Award, 1558
Douglass Institute Postdoctoral Fellowships, 1851
Field Museum Visiting Scholarships, 2011
Fulbright Alumni Initiatives Awards Grants, 2064
Fulbright Distinguished Chairs Program Grants, 2069
Fulbright European Union Scholar-in-Residence Program Grants, 2071
Fulbright New Century Scholars Program Grants, 2075
Fulbright Occasional Lecturer Program Grants, 2076
Fulbright Specialists Program Grants, 2079
Fulbright Traditional Scholar Program in Europe and Eurasia, 2080
Fulbright Traditional Scholar Program in Sub-Saharan Africa, 2081
Fulbright Traditional Scholar Program in the East Asia/Pacific Region, 2082
Fulbright Traditional Scholar Program in the Near East and North Africa Region, 2083
Fulbright Traditional Scholar Program in the South and Central Asia Region, 2084
Fulbright Traditional Scholar Program in the Western Hemisphere, 2085
Fulbright Visiting Specialists Program: Direct Access to the Muslim World, 2087
Harry Frank Guggenheim Foundation Dissertation Fellowships, 2207
Horowitz Foundation for Social Policy Grants, 2280
Horowitz Found for Social Policy Special Awards, 2281
IMLS 21st Century Museum Prof Grants, 2362
IMLS National Leadership Grants, 2366
Jacobs Research Funds Small Grants Program, 2456
Jeffrey Campbell Grad Fellow in Anthropology, 2482
Maine Community Foundation Keepers Preservation Education Grants, 2651
Mellon Post-Doctoral Fellowship in Sustainable Agriculture, 2740
MPI for Social Anthropology Department II Postdoctoral Grants, 2798
MPI for Social Anthropology Max Planck Fellow Research Group Law, Organizations, Science, and Technology, 2799
National Geographic Society Expeditions Council Grants, 2872
National Geographic Society Field Research and Exploration Grants, 2873
NCFR Innovation Grants, 2908
North Carolina Arts Council Folklife Internship, 3418
NSF Archaeology & Archeaometry Res Grants, 3482
NSF Arctic Social Sciences Grants, 3492
NSF SBE Doctoral Dissertation Research Improvement Grants, 3641
PHF Penn Faculty Research Fellowships, 3932
PHF Regional Mellon Faculty Research and Cultural Fellowships, 3933
PHF Undergraduate Humanities Forum Research Fellowships, 3934
Richard Carley Hunt Memorial Postdoctoral Fellowships, 4065
S.E.VEN Fund WHY Prize, 4156
Smithsonian Graduate Student Fellowships, 4277
Smithsonian Minority Visiting Student Awards, 4279
Smithsonian Native American Community Scholars Awards, 4291
Smithsonian Native American Visiting Student Awards, 4292
Smithsonian Postdoctoral Fellowships, 4293
Smithsonian Predoctoral Fellowships, 4295
Smithsonian Senior Fellowships, 4296
SSHRC Insight Development Grants, 4409
SSHRC Insight Grants, 4410
SSRC Mellon Mays Predoctoral Research Grants, 4432
Wenner-Gren Foundation African Fellowships, 4846
Wenner-Gren Foundation Conference and Workshop Grants, 4847
Wenner-Gren Foundation Dissertation Fieldwork Grants, 4848
Wenner-Gren Found Historical Archives Grants, 4849
Wenner-Gren Foundation Hunt Postdoctoral Fellowships, 4850
Wenner-Gren Foundation International Collaborative Research Grants, 4851
Wenner-Gren Found Post-PhD Research Grants, 4852
Wenner-Gren Found Short-Term Fellowships, 4853
Wenner-Gren Foundation Wadsworth International Fellowships, 4854

Anthropology, Cultural
Anthropological Center for Training in Global and Environmental Change Grad Assistantships, 870
Australasian Institute of Judicial Administration Seed Funding Grants, 1268
Christine Wilson Award, 1558
Field Museum Visiting Scholarships, 2011
Foundation for the Advancement of Mesoamerican Studies Grants, 2042
Leakey Foundation Paleoanthropology Award, 2600
MPI for Social Anthropology Department II Postdoctoral Grants, 2798
NSF/NEH Documenting Endangered Languages Project Grants and Fellowships: data, infrastructure and computational methods, 3455
NSF Cultural Anthropology Research Grants, 3548
PHF Penn Faculty Research Fellowships, 3932
PHF Regional Mellon Faculty Research and Cultural Fellowships, 3933
PHF Undergraduate Humanities Forum Research Fellowships, 3934
Smithsonian Graduate Student Fellowships, 4277
Smithsonian Minority Visiting Student Awards, 4279
Smithsonian Native American Community Scholars Awards, 4291
Smithsonian Native American Visiting Student Awards, 4292
Smithsonian Predoctoral Fellowships, 4295
Wenner-Gren Foundation Dissertation Fieldwork Grants, 4848

Anthropology, Physical
Anthropological Center for Training in Global and Environmental Change Grad Assistantships, 870
Christine Wilson Award, 1558
Field Museum Visiting Scholarships, 2011
Jeffrey Campbell Grad Fellow in Anthropology, 2482
Leakey Foundation Paleoanthropology Award, 2600
NSF Physical Anthropology Research Grants, 3615
Wenner-Gren Foundation Dissertation Fieldwork Grants, 4848
Wenner-Gren Foundation Hunt Postdoctoral Fellowships, 4850

Anthropology, Social
Anthropological Center for Training in Global and Environmental Change Grad Assistantships, 870
Australasian Institute of Judicial Administration Seed Funding Grants, 1268
Christine Wilson Award, 1558
Field Museum Visiting Scholarships, 2011
MPI for Social Anthropology Department II Postdoctoral Grants, 2798
PHF Penn Faculty Research Fellowships, 3932
PHF Regional Mellon Faculty Research and Cultural Fellowships, 3933
PHF Undergraduate Humanities Forum Research Fellowships, 3934
USDD Basic Research Initiative Grants, 4691
Wenner-Gren Foundation Dissertation Fieldwork Grants, 4848
Wenner-Gren Foundation Hunt Postdoctoral Fellowships, 4850

Antibiotics
NFID New Investigator Matching Grants, 2989
Pfizer Anti Infective Research EU Grants, 3909
Pfizer ASPIRE EU Emerging Mechanisms of Resistance Antibacterial Research Awards, 3912
Pfizer ASPIRE EU MRSA Nosocomial Pneumonia & MRSA Complicated Skin & Soft Tissue Infections Antibacterial Research Awards, 3913
Pfizer ASPIRE North America Broad Spectrum Antibiotics for the Treatment of Gram-Negative or Polymicrobial Infections Research Awards, 3914
Pfizer ASPIRE North America Narrow Spectrum Antibiotics for the Treatment of MRSA Research Awards, 3916

Antivirals
Doris Duke Charitable Foundation Operations Research on AIDS Care and Treatment in Africa Grants, 1839
Pfizer Anti Infective Research EU Grants, 3909

Appalachia
Berea College Appalachian Music Fellowships, 1332

Aquaculture/Hydroponics
BARD Research Grants, 1299
SARE Producer Research and Education Grants-Western Region, 4174
SARE Research and Education Grants - Western Region, 4180
USDA 1890 Institution Teaching, Research, and Extension Capacity Building Grants, 4613
USDA Distance Education Grants for Higher Education Institutions in Insular Areas, 4622
USDA Resident Instruction Grants for Higher Education in Insular Areas, 4645
USDA Small Business Innovation Research Grants - Phase I, 4648
USDA Tribal College Research Grants, 4652
USDC/NOAA Marine Fisheries Grants, 4658
USDC/NOAA National Marine Aquaculture Initiative Grants, 4659
USDC/NOAA Nature Conservancy Community-based Matching Grants, 4661

Aquariums
IMLS 21st Century Museum Prof Grants, 2362
IMLS National Leadership Grants, 2366

Aquatic Ecology
HRF Mark B. Bain Graduate Fellowships, 2287
SeaWorld and Busch Gardens Conservation Fund Grants, 4196
USDC/NOAA National Marine Aquaculture Initiative Grants, 4659

Arabic Language/Literature
Fulbright Critical Lang Enhancement Awards, 2068

Arboreta
IMLS 21st Century Museum Prof Grants, 2362
IMLS National Leadership Grants, 2366
Stanley Smith Horticultural Trust Grants, 4446

Archaeology
AAR Postdoctoral Fellowships, 201
ACC Ford Foundation Fellowships, 253
ACC Humanities Fellowship Program, 255
ACC Residency Program in Asia, 267
ACLS Charles A. Ryskamp Research Fellowships, 281
ACLS Frederick Burkhardt Residential Fellowships for Recently Tenured Scholars, 288

AIAR Andrew W. Mellon Foundation Fellowships, 392
AIAR Annual Professorships, 393
AIAR Ernest S. Frerichs Fellow and Program Coordinator Grant, 394
AIAR George A. Barton Fellowships, 395
Alabama Humanities Foundation Mini Grants, 480
Alabama Humanities Foundation Planning/Consultant Grants, 481
American Numismatic Society Eric P. Newman Graduate Seminar Grants, 761
American Philosophical Soc Judson Daland Prize, 766
American Philosophical Society Lewis and Clark Fund for Exploration and Field Research, 767
ARIT/NEH Fellowships for Research in Turkey, 990
ARIT Fellowships, 991
ARIT Kenan T. Erim Fellowship for Research at Aphrodisias, 993
ASCSA Fellowships, 1052
ASCSA Jacob Hirsch Fellowship, 1054
ASCSA Weiner Laboratory Faunal Studies Research Fellowship, 1056
Australasian Institute of Judicial Administration Seed Funding Grants, 1268
Bogliasco Foundation Liguria Study Center Residencies, 1358
Canada-U.S. Fulbright International Science and Technology Award, 1417
Canada-U.S. Fulbright Senior Specialists Program Grants, 1418
Canada-U.S. Fulbright Traditional Scholars, 1419
Cen Cal Diving and Aquatic Studies Scholarship, 1471
Colonel Stanley R. McNeil Foundation Grants, 1597
Earhart Foundation Institutional Grants, 1888
Foundation for the Advancement of Mesoamerican Studies Grants, 2042
Fulbright Alumni Initiatives Awards Grants, 2064
Fulbright Distinguished Chairs Program Grants, 2069
Fulbright European Union Scholar-in-Residence Program Grants, 2071
Fulbright New Century Scholars Program Grants, 2075
Fulbright Occasional Lecturer Program Grants, 2076
Fulbright Specialists Program Grants, 2079
Fulbright Traditional Scholar Program in Europe and Eurasia, 2080
Fulbright Traditional Scholar Program in Sub-Saharan Africa, 2081
Fulbright Traditional Scholar Program in the East Asia/Pacific Region, 2082
Fulbright Traditional Scholar Program in the Near East and North Africa Region, 2083
Fulbright Traditional Scholar Program in the South and Central Asia Region, 2084
Fulbright Traditional Scholar Program in the Western Hemisphere, 2085
Fulbright Visiting Specialists Program: Direct Access to the Muslim World, 2087
Gerda Henkel Foundation Grants, 2127
Gladys Krieble Delmas Foundation Grants for Venetian Research, 2144
Kluge Center Kislak Fellows in American Studies, 2562
Kluge Center Kislak Short-Term Fellowships in American Studies, 2563
Luther I. Replogle Foundation Grants, 2631
Maine Community Foundation Keepers Preservation Education Grants, 2651
Mary Isabel Sibley Fellowship, 2672
NAGPRA Museum Consultation/Documentation Grants, 2822
NAGPRA Museum Repatriation Grants, 2823
National Geographic Society Field Research and Exploration Grants, 2873
NOAA Ocean Exploration Omnibus: Marine Archaeology Grants, 3397
NSF Archaeology & Archeaometry Res Grants, 3482
NSF Arctic Social Sciences Grants, 3492
PHF Penn Faculty Research Fellowships, 3932
PHF Regional Mellon Faculty Research and Cultural Fellowships, 3933
PHF Undergraduate Humanities Forum Research Fellowships, 3934
Smithsonian Graduate Student Fellowships, 4277
Smithsonian Minority Visiting Student Awards, 4279
Smithsonian Native American Community Scholars Awards, 4291
Smithsonian Native American Visiting Student Awards, 4292
Smithsonian Postdoctoral Fellowships, 4293
Smithsonian Predoctoral Fellowships, 4295
SSHRC Insight Development Grants, 4409
SSHRC Insight Grants, 4410
Wenner-Gren Foundation African Fellowships, 4846
Wenner-Gren Foundation Conference and Workshop Grants, 4847
Wenner-Gren Foundation Dissertation Fieldwork Grants, 4848
Wenner-Gren Foundation International Collaborative Research Grants, 4851
Wenner-Gren Found Short-Term Fellowships, 4853
Wenner-Gren Foundation Wadsworth International Fellowships, 4854

Archaeology Education

Andrew W. Mellon/ACLS Dissertation Completion Fellowships, 844
Bogliasco Foundation Liguria Study Center Residencies, 1358
Luther I. Replogle Foundation Grants, 2631

Archaeometry

NSF Archaeology & Archeaometry Res Grants, 3482

Architectural Engineering

AIA Latrobe Prize, 390

Architecture

AAB Berlin Prize Fellowships, 48
AAF Richard M Hunt Architecture Fellowship, 169
AAR Rome Prize Fellowships, 202
AAUW Selected Professions Fellowships, 217
ACC Humanities Fellowship Program, 255
ACC Residency Program in Asia, 267
AIA Latrobe Prize, 390
American Society of Naval Engineers Scholarship, 800
ARIT Kenan T. Erim Fellowship for Research at Aphrodisias, 993
Athenaeum of Philadelphia Senior Fellowships, 1256
Beverly Willis Architecture Foundation Grants and Fellowships, 1340
Bogliasco Foundation Liguria Study Center Residencies, 1358
Cintas Foundation Fellowships, 1570
DAAD Study & Internship Program in Germany, 1708
Energy Foundation Buildings Grants, 1922
Energy Foundation Power Grants, 1924
Enterprise Community Partners Rose Architectural Fellowships, 1929
Enterprise Community Partners Terwilliger Fellowship, 1930
Furthermore Grants in Publishing, 2091
Getty Research Grants for Institutions - Archival Grants, 2131
Getty Research Grants for Institutions - Art History Research Grants, 2132
Getty Research Grants for Institutions - Cataloguing of Museum Collections Grants, 2133
Gladys Krieble Delmas Foundation Grants for Venetian Research, 2144
Kate Neal Kinley Memorial Fellowship, 2532
Maine Community Foundation Keepers Preservation Education Grants, 2651
National Endowment for the Humanities Fellowships at Winterthur, 2867
Northwood University/Alden B. Dow Creativity Center Summer Residency Program, 3432
NYFA Artist Fellowships, 3698
OSF-Baltimore Community Fellowships, 3833
Regents Professional Opportunity Scholarships, 4049
SSRC-Van Alen Fellowships, 4419
Stanford Humanities Center Arts Writer/Practitioner Fellowship, 4444
Van Alen Institute New York Prize Fellowship, 4806
Winterthur Museum Dissertation Fellowships, 4887
Winterthur Research Fellowships, 4888

Architecture Education

Athenaeum of Philadelphia Senior Fellowships, 1256
Bogliasco Foundation Liguria Study Center Residencies, 1358
Enterprise Community Partners Rose Architectural Fellowships, 1929
Enterprise Community Partners Terwilliger Fellowship, 1930
Getty Research Grants for Institutions - Cataloguing of Museum Collections Grants, 2133
Yale University Beinecke Library Research Fellowships for Yale Grad & Professional School Students, 4902

Architecture History

ACC Humanities Fellowship Program, 255
ACC Residency Program in Asia, 267
ACLS Frederick Burkhardt Residential Fellowships for Recently Tenured Scholars, 288
Andrew W. Mellon/ACLS Dissertation Completion Fellowships, 844
Getty Research Grants for Institutions - Archival Grants, 2131
Getty Research Grants for Institutions - Art History Research Grants, 2132
Maine Community Foundation Keepers Preservation Education Grants, 2651

Archives

AAB Berlin Prize Fellowships, 48
ALA Jan Merrill-Oldham Professional Development Grant, 491
American Academy of Religion Individual Research Grants, 618
Cushwa Center Research Travel Grants, 1673
Getty Research Grants for Institutions - Archival Grants, 2131
Getty Research Grants for Institutions - Art History Research Grants, 2132
Grammy Foundation Grants, 2161
Marine Corps Historical Center Beeler-Raider Fellowship, 2662
NHPRC Archives - Basic Projects Grants, 3070
NHPRC Archives - Detailed Processing Projects Grants, 3071
NHPRC Digitizing Historical Records Grants, 3072
NHPRC Electronic Records Projects Grants, 3073
NHPRC Historical Editing Fellowships, 3074
NHPRC Publishing Historical Records Grants, 3075
NHPRC State and National Archival Partnership Grants, 3076
NL Newberry Library/Ecole Nationale des Chartes Exchange Fellowship, 3313
NSF/NEH Documenting Endangered Languages Project Grants and Fellowships: data, infrastructure and computational methods, 3455
Rockefeller Archive Center Ehrlich Grants, 4084
Rockefeller Archive Center Grants in Aid, 4085
Rockefeller Archive Center Visiting Archivist Fellowship, 4087
Temple Univ George D. McDowell Fellowship, 4499
United Methodist Church Josephine Forman Scholarship, 4573

Arctic

NOAA Arctic Research Program Grants, 3330
NSF Arctic Natural Science Grants in Atmospheric Sciences, 3483
NSF Arctic Nat Science Grants in Earth Sciences, 3485
NSF Arctic Natural Science Grants in Glaciology, 3486
NSF Arctic Nat Science Grants in Ocean Science, 3487
NSF Arctic Natural Sciences Program Grants, 3488
NSF Arctic Research and Education Grants, 3489
NSF Arctic Research Opportunities, 3490
NSF Arctic Research Support and Logistics Program Grants, 3491
NSF Arctic Social Sciences Grants, 3492

NSF Arctic System Science Program Grants, 3493
NSF Integrative Graduate Education and Research Traineeship, 3584
NSF Natural, Social, and System Sciences and Research Support and Logistics Program, 3600
NSF Polar Programs Division Research Grants, 3627
USDD Arctic and Global Prediction Grants, 4685

Arctic Ocean
NOAA Arctic Research Program Grants, 3330
NSF Arctic Research Support and Logistics Program Grants, 3491
USDD Arctic and Global Prediction Grants, 4685

Area Studies
ACLS International and Area Studies Fellowships, 289
Dept of Ed International Research and Studies Project Grants, 1745
Horowitz Foundation for Social Policy Grants, 2280
Horowitz Found for Social Policy Special Awards, 2281
National Security Education Program David L. Boren Graduate Fellowships, 2884
SSRC Eurasia Program Pre-Dissertation Awards, 4428
SSRC Mellon Mays Predoctoral Research Grants, 4432
U.S. Department of Education American Overseas Research Centers, 4537
U.S. Department of Education Foreign Languages and Area Studies Fellowships, 4539

Arrhythmia
NHLBI New Approaches to Arrhythmia Detection and Treatment Grants for SBIR, 3042
NHLBI New Approaches to Arrhythmia Detection and Treatment Grants for STTR, 3043
NHLBI Targeting Calcium Regulatory Molecules for Arrhythmia Prevention Grants, 3066

Art Appreciation
Arthur F. and Alice E. Adams Charitable Foundation Grants, 1024
Carrier Corporation Contributions Grants, 1440
Helen Bader Foundation Grants, 2228
North Carolina Arts Council Artist Fellowships, 3417
NYSCA Folk Arts: Exhibitions Grants, 3703
NYSCA Folk Arts: Presentation Grants, 3704
NYSCA Folk Arts: Regional and County Folk Arts Programs Grants, 3705
NYSCA Museum: Project Support Grants, 3707
San Diego Women's Foundation Grants, 4164

Art Conservation
AAR Rome Prize Fellowships, 202
Christensen Fund Regional Grants, 1554
IMLS Conservation Project Support Grants, 2363
McCune Charitable Foundation Grants, 2719
NYSCA Folk Arts: Presentation Grants, 3704
NYSCA Folk Arts: Regional and County Folk Arts Programs Grants, 3705
NYSCA Museum: Project Support Grants, 3707
Smithsonian Postgraduate Fellowships in Conservation of Museum Collections, 4294

Art Criticism
ACC Humanities Fellowship Program, 255
ACC Starr Foundation Fellowship Program, 268
Alabama Humanities Foundation Mini Grants, 480
Alabama Humanities Foundation Planning/Consultant Grants, 481
PHF Penn Faculty Research Fellowships, 3932
PHF Regional Mellon Faculty Research and Cultural Fellowships, 3933
PHF Undergraduate Humanities Forum Research Fellowships, 3934

Art Education
3M Foundation Community Giving Grants, 5
ACC Fellowships and Projects, 252
ACC Mekong Region Fellowship, 258
Ahn Family Foundation Grants, 379
Alces Foundation Grants, 510
ALZA Corporate Contributions Grants, 588
Anderson Center Residency Program, 841
Andrew W. Mellon/ACLS Dissertation Completion Fellowships, 844
Andy Warhol Foundation for the Visual Arts Curatorial Research Fellowships, 847
Angels Baseball Foundation Grants, 848
Annenberg Foundation Grants, 862
Arizona Commission on the Arts Education Projects Grants, 996
Arizona Commission on the Arts Folklorist Residencies, 997
Arthur F. and Alice E. Adams Charitable Foundation Grants, 1024
Ashland Corporate Contributions Grants, 1087
Bayer Foundation Grants, 1309
CAA Professional Development Fellowships for Graduate Students, 1404
Carl and Lily Pforzheimer Foundation Grants, 1427
Christensen Fund Regional Grants, 1554
CMU Arts in Society Fellowship, 1587
Connecticut Community Foundation Grants, 1627
Coughlin-Saunders Foundation Grants Program, 1648
DAAD Research Grants for Doctoral Candidates and Young Academics and Scientists, 1703
DAAD Research Stays for University Academics and Scientists, 1705
DAAD Scholarships for Artists: Study Visits for Academics, 1707
Ford Motor Company Fund Grants Program, 2033
George A Ohl Jr. Foundation Grants, 2112
George Gund Foundation Grants, 2117
Helen Bader Foundation Grants, 2228
Holberg Memorial Fund School Project, 2274
Jacob and Hilda Blaustein Foundation Grants, 2454
Jovid Foundation Grants, 2518
Kate Neal Kinley Memorial Fellowship, 2532
Kress Foundation Art History Pre-Doctoral Fellowships, 2576
Luther I. Replogle Foundation Grants, 2631
Massachusetts Cultural Cou Artist Fellowships, 2677
McCune Charitable Foundation Grants, 2719
McGregor Fund Grants, 2725
NAEA Research Grants, 2814
NAEF Teacher Incentive Grants, 2815
North Carolina Arts Council Artist Fellowships, 3417
NYSCA Folk Arts: Regional and County Folk Arts Programs Grants, 3705
NYSCA Museum: Project Support Grants, 3707
Parkersburg Area Community Foundation Arts in Education Special Grants, 3869
Peacock Foundation Grants, 3887
Radcliffe Inst Fellowships for Creative Artists, 4020
Rockwell International Corporate Trust Grants, 4093
San Diego Women's Foundation Grants, 4164
Sid W. Richardson Foundation Grants, 4253
Sunny and Abe Rosenberg Foundation Grants, 4469
Surdna Foundation Arts Teachers Fellowships, 4470
Third Millennium Foundation Grants, 4505
Vancouver Foundation Grants and Community Initiatives, 4807
Westinghouse Charitable Giving Grants, 4856

Art History
AAB Berlin Prize Fellowships, 48
AAR Postdoctoral Fellowships, 201
ACC Fellowships and Projects, 252
ACC Ford Foundation Fellowships, 253
ACC Humanities Fellowship Program, 255
ACC Mekong Region Fellowship, 258
ACC Residency Program in Asia, 267
ACLS Charles A. Ryskamp Research Fellowships, 281
ACLS Frederick Burkhardt Residential Fellowships for Recently Tenured Scholars, 288
AIAR Andrew W. Mellon Foundation Fellowships, 392
American Numismatic Society Eric P. Newman Graduate Seminar Grants, 761
ARIT Kenan T. Erim Fellowship for Research at Aphrodisias, 993
ASCSA Fellowships, 1052
CAA Professional Development Fellowships for Graduate Students, 1404
CAA Wyeth Publication Grant, 1405
Center for East Asian Studies Fellowships, 1474
Foundation for the Advancement of Mesoamerican Studies Grants, 2042
Gerda Henkel Foundation Grants, 2127
Germanistic Society of America Fellowships, 2128
Getty Research Grants for Institutions - Archival Grants, 2131
Getty Research Grants for Institutions - Art History Research Grants, 2132
Getty Research Grants for Institutions - Cataloguing of Museum Collections Grants, 2133
Getty Research Grants for Institutions - Critical Reference Resources Grants, 2134
Getty Research Institute Nonresidential Grants - Collaborative Research Grant, 2135
Getty Research Institute Nonresidential Grants - Curatorial Research Fellowships, 2136
Getty Research Institute Nonresidential Grants - Postdoctoral Fellowships, 2137
Henry Luce Foundation/ACLS Dissertation Fellowships in American Art, 2235
Huntington Library Dana and David Dornsife Fellowships, 2303
Huntington Library National Endowment for the Humanities Fellowships, 2307
Huntington Library Travel Grants and Exchange Fellowships for Study in Great Britain, 2309
Kress Foundation Art History Pre-Doctoral Fellowships, 2576
Maine Community Foundation Keepers Preservation Education Grants, 2651
McCune Charitable Foundation Grants, 2719
National Endowment for the Humanities Fellowships at Winterthur, 2867
North Carolina Arts Council Artist Fellowships, 3417
NYPL Short-Term Research Fellowships, 3701
NYSCA Folk Arts: Presentation Grants, 3704
NYSCA Folk Arts: Regional and County Folk Arts Programs Grants, 3705
NYSCA Museum: Project Support Grants, 3707
PHF Penn Faculty Research Fellowships, 3932
PHF Regional Mellon Faculty Research and Cultural Fellowships, 3933
PHF Undergraduate Humanities Forum Research Fellowships, 3934
Roberto Longhi Foundation Fellowships, 4078
Smithsonian Senior Fellowships, 4296
SSRC Mellon Mays Predoctoral Research Grants, 4432
Winterthur Museum Dissertation Fellowships, 4887
Winterthur Research Fellowships, 4888

Art Museums
Blum-Kovler Foundation Grants, 1351
McCune Charitable Foundation Grants, 2719
Mericos Foundation Grants, 2742
Meyer and Stephanie Eglin Foundation Grants, 2747
NAEA Research Grants, 2814
NYSCA Museum: Project Support Grants, 3707
Smithsonian Postgraduate Fellowships in Conservation of Museum Collections, 4294

Art Works/Artifacts
Arthur F. and Alice E. Adams Charitable Foundation Grants, 1024
McCune Charitable Foundation Grants, 2719
National Endowment for the Humanities Fellowships at Winterthur, 2867
NYSCA Folk Arts: Exhibitions Grants, 3703
NYSCA Museum: Project Support Grants, 3707
Winterthur Museum Dissertation Fellowships, 4887
Winterthur Research Fellowships, 4888
Wolfsonian-FIU Fellowships, 4890

Art in Public Places
Alces Foundation Grants, 510
Arthur F. and Alice E. Adams Charitable Foundation Grants, 1024

NYSCA Folk Arts: Exhibitions Grants, 3703
NYSCA Folk Arts: Regional and County Folk Arts Programs Grants, 3705

Art, Experimental
Connecticut Community Foundation Grants, 1627
Florida Division of Cultural Affairs Individual Artist Fellowships, 2026
Manitoba Arts Council Aboriginal Arts Creative Development Pilot Projects Grants, 2652
NYFA Artist Fellowships, 3698

Arthritis
Arthritis Foundation of Australia Grants, 1022
Dorr Inst for Arthritis Research & Education, 1843
Fairlawn Foundation Grants, 1974
Mabel H. Flory Charitable Trust Grants, 2646
NIAMS Pilot & Feasibility Clinical Grants in Arthritis & Musculoskeletal & Skin Diseases, 3111
Pfizer ASPIRE North America Rheumatology Research Awards, 3917
Pfizer Inflammation Competitive Res Awards, 3926
Pittsburgh Foundation Medical Research Grants, 3973

Artificial Intelligence
DOE Engineering Research Grants, 1789
NSF Info & Intelligent Systems Core Grants, 3580
ONR Computational Neuroscience & Biorobotics Grants, 3752

Artificial Reproduction
NAAB Coordinated Research Grants Program, 2808
USAID Family Planning and Reproductive Health Methods Grants, 4598

Artist Studios
Manitoba Arts Council Dance Creation/Production Grants, 2654
NYSCA Individual Artists: Film, Media and New Technology Production Grants, 3706

Artists in Residence
Andy Warhol Foundation for the Visual Arts Curatorial Research Fellowships, 847
Arizona Commission on the Arts Education Projects Grants, 996
Arizona Commission on the Arts Folklorist Residencies, 997
Conseil des arts de Montreal Residency of the Americas, 1640
North Carolina Arts Council Residency Grants, 3420
NYSCA Folk Arts: Presentation Grants, 3704
NYSCA Music: Services to the Field Grants, 3708
Rockefeller Foundation MAP Fund Grants, 4090
Rockefeller Foundation Scholarly Residencies, 4091

Artists' Fellowships
AAB Berlin Prize Fellowships, 48
ACC Hong Kong Arts Program Fellowships, 254
ACC Japan-US Arts Program Fellowships, 256
ACC Philippines Fellowship Program, 261
ACC Starr Foundation Fellowship Program, 268
ACC Taiwan Fellowship Program, 269
Andy Warhol Foundation for the Visual Arts Curatorial Research Fellowships, 847
Arkansas Arts Council Individual Fellowships, 999
CICF Efroymson Contemporary Arts Fellows, 1563
Cintas Foundation Fellowships, 1570
Delaware Division of the Arts Individual Artist Fellowships, 1732
Henry Luce Foundation/ACLS Dissertation Fellowships in American Art, 2235
John Simon Guggenheim Memorial Foundation Fellowships, 2507
Manitoba Arts Council Dance Creation/Production Grants, 2654
Massachusetts Cultural Cou Artist Fellowships, 2677
National Endowment for the Arts Literature Fellowships: Translation Projects, 2866
National Gallery of Art Kress Interpretive Fellow, 2869
NEA National Heritage Fellowships in the Folk and Traditional Arts, 2944
Nevada Arts Council Artists Fellowships, 2984
North Dakota Council on the Arts Individual Artist Fellowships, 3426
NYFA Shelly Pinz Fellowship, 3699
NYSCA Individual Artists: Film, Media and New Technology Production Grants, 3706
Pew Fellowships in the Arts, 3905
Pollock-Krasner Foundation Grants, 3981
Princess Grace Foundation Theater, Dance, and Film Grants, 3988

Arts Administration
Alcatel-Lucent Technologies Foundation Grants, 509
Andy Warhol Foundation for the Visual Arts Curatorial Research Fellowships, 847
CCA Alberta Creative Development Initiative: Grants to Individuals, 1446
CDECD Elizabeth L. Mahaffey Arts Administration Fellowship, 1468
Conseil des arts de Montreal Residency of the Americas, 1640
Kennedy Center Fellowship at the DeVos Institute of Arts Management, 2552
Kennedy Center Summer Fellowships for International Arts Managers, 2553
North Carolina Arts Council Folklife Public Program Grants, 3419
Ontario Arts Council Integrated Arts Grants, 3790
Sid W. Richardson Foundation Grants, 4253

Arts Festivals
Angels Baseball Foundation Grants, 848
Macquarie Bank Foundation Grants, 2650
North Carolina Arts Council Folklife Public Program Grants, 3419
NYSCA Folk Arts: Presentation Grants, 3704
NYSCA Folk Arts: Regional and County Folk Arts Programs Grants, 3705
Ontario Arts Council Integrated Arts Grants, 3790

Arts and Culture
3M Foundation Community Giving Grants, 5
Annenberg Foundation Grants, 862
Assisi Foundation of Memphis General Grants, 1221
Bank of America Charitable Foundation Matching Gifts, 1297
Boeing Company Contributions Grants, 1357
CAA Wyeth Publication Grant, 1405
CCA Alberta Creative Development Initiative: Grants to Individuals, 1446
CIC Smithsonian Institution Fellowships, 1564
CMU Arts in Society Fellowship, 1587
Davis Family Foundation Grants, 1726
Eaton Charitable Fund Grants, 1893
George Kress Foundation Grants, 2119
Harvey Randall Wickes Foundation Grants, 2218
Helen Bader Foundation Grants, 2228
Herbert & Adrian Woods Foundation Grants, 2239
Maine Community Foundation Keepers Preservation Education Grants, 2651
McCune Charitable Foundation Grants, 2719
McGraw-Hill Companies Community Grants, 2724
McGregor Fund Grants, 2725
Meyer Memorial Trust Responsive Grants, 2748
Norcliffe Foundation Grants, 3415
NYSCA Dance: Services to the Field Grants, 3702
NYSCA Folk Arts: Exhibitions Grants, 3703
NYSCA Folk Arts: Presentation Grants, 3704
NYSCA Folk Arts: Regional and County Folk Arts Programs Grants, 3705
NYSCA Individual Artists: Film, Media and New Technology Production Grants, 3706
NYSCA Museum: Project Support Grants, 3707
OSF-Baltimore Community Fellowships, 3833
Paul G. Allen Family Foundation Grants, 3876
Perkin Fund Grants, 3898
Rohm and Haas Company Grants, 4098
Sarkeys Foundation Grants, 4181
Schomburg Center Scholars-in-Residence Fellows, 4187
Sioux Falls Area Community Foundation Community Fund Grants (Unrestricted), 4262
Sioux Falls Area Comm Foundation Spot Grants, 4263
Smithsonian Graduate Student Fellowships, 4277
Smithsonian Minority Visiting Student Awards, 4279
Smithsonian Native American Community Scholars Awards, 4291
Smithsonian Native American Visiting Student Awards, 4292
Smithsonian Postdoctoral Fellowships, 4293
Smithsonian Predoctoral Fellowships, 4295
Sony Corporation of America Corporate Philanthropy Grants, 4329
Stanford Humanities Center Arts Writer/Practitioner Fellowship, 4444
Stanford Humanities Center External Faculty Fellowships, 4445
Sunny and Abe Rosenberg Foundation Grants, 4469
Textron Corporate Contributions Grants, 4501
The Ray Charles Foundation Grants, 4502
Victor E. Speas Foundation Grants, 4810
Virginia Foundation for the Humanities Discretionary Grants, 4813
Virginia Foundation for Humanities Open Grants, 4814
Virginia Foundation for the Humanities Residential Fellowships, 4815
Willary Foundation Grants, 4870

Arts, Fine
Anderson Foundation Grants, 842
ASU Graduate College Completion Fellowships, 1235
Helen Bader Foundation Grants, 2228
Jewish Federation of Metropolitan Chicago Scholarships, 2486
Stanford Humanities Center Arts Writer/Practitioner Fellowship, 4444

Arts, General
AAB Berlin Prize Fellowships, 48
AAUW American Dissertation Fellowships, 209
AAUW American Postdoctoral Research Leave Fellowships, 210
AAUW American Summer/Short-Term Research Publication Grants, 211
Achelis Foundation Grants, 274
Acuity Charitable Foundation Grants, 333
Adam Richter Charitable Trust Grants, 339
Ahn Family Foundation Grants, 379
Albuquerque Community Foundation Grants, 508
Alcatel-Lucent Technologies Foundation Grants, 509
Alces Foundation Grants, 510
Alcoa Foundation Grants, 511
Alexander von Humboldt Feodor-Lynen Research Fellowships for Experienced Researchers, 518
Alexander von Humboldt Foundation German Chancellor Fellowships for Prospective Leaders from the People's Republic of China, 526
Alexander von Humboldt Foundation German Chancellor Fellowships for Prospective Leaders from the Russian Federation, 527
Alexander von Humboldt Foundation Research Awards, 537
Alexander von Humboldt Foundation Research Awards for Top Researchers in Germany, 538
Alex Stern Family Foundation Grants, 542
Alfred P. Sloan Foundation Public Understanding of Science and Technology-Theater Grants, 559
Alice Tweed Tuohy Foundation Grants Program, 569
Alvin and Lucy Owsley Foundation Grants, 584
American Schlafhorst Foundation Grants, 796
Anderson Foundation Grants, 842
Angels Baseball Foundation Grants, 848
Annenberg Foundation Grants, 862
Applied Biosystems Grants, 916
ARIT/NEH Fellowships for Research in Turkey, 990
ARIT Fellowships, 991
Arizona Cardinals Grants, 995
Arizona Public Service Corporate Giving Program Grants, 998

Arthur F. and Alice E. Adams Charitable Foundation Grants, 1024
ASU Graduate College Completion Fellowships, 1235
ATS Henry Luce III Fellowships, 1258
Banfi Vintners Foundation Grants, 1296
Bank of America Charitable Foundation Matching Gifts, 1297
Barra Foundation Project Grants, 1301
Batchelor Foundation Grants, 1305
Bayer Foundation Grants, 1309
Beim Foundation Grants, 1331
Berks County Community Foundation Grants, 1333
Bodenwein Public Benevolent Foundation Grants, 1355
Bosque Foundation Grants, 1360
Boston Foundation Grants, 1361
Burlington Northern Santa Fe Foundation Grants, 1389
Bush Foundation Artist Fellowships, 1391
Bydale Foundation Grants, 1400
Byron W. and Alice L. Lockwood Grants, 1401
Cabot Corporation Foundation Grants, 1406
Campbell Soup Foundation Grants, 1415
Canada-U.S. Fulbright International Science and Technology Award, 1417
Canada-U.S. Fulbright Senior Specialists Program Grants, 1418
Canada-U.S. Fulbright Traditional Scholars, 1419
Carl and Eloise Pohlad Family Grants, 1426
Carl C. Icahn Foundation Grants, 1428
Carpenter Foundation Grants, 1439
Charles Lafitte Foundation Grants, 1525
Chautauqua Region Community Found Grants, 1533
Chestnut Hill Charitable Foundation, Inc Grants, 1538
Chiles Foundation Grants, 1552
CIGNA Foundation Grants, 1568
Claneil Foundation Grants, 1574
Cleveland-Cliffs Foundation Grants, 1580
Coca-Cola Foundation Grants, 1593
Coeta and Donald Barker Foundation Grants, 1594
Communities Foundation of Texas Grants, 1613
Community Found for Greater Buffalo Grants, 1615
Community Found for Muskegon County Grants, 1617
Community Foundation for S Arizona Grants, 1618
Community Found of E Central Illinois Grants, 1619
Compton Foundation Grants Program, 1623
Connecticut Community Foundation Grants, 1627
ConocoPhillips Grants Program, 1629
Cord Foundation Grants, 1645
Coughlin-Saunders Foundation Grants Program, 1648
Cowles Charitable Trust Grants, 1652
Cudd Foundation Grants, 1672
DAAD Study & Internship Program in Germany, 1708
Dade Community Foundation Grants, 1712
DaimlerChrysler Corporation Fund Grants, 1714
Dell Foundation Open Grants, 1734
DeMatteis Family Foundation Grants, 1740
Donald and Sylvia Robinson Family Foundation Grants, 1830
Dorothy Rider Pool Health Care Grants, 1841
Drs. Bruce and Lee Foundation Grants, 1859
E.L. Wiegand Foundation Grants, 1885
Edward W. and Stella C. Van Houten Memorial Fund Grants, 1899
Eugene M. Lang Foundation Grants, 1957
Evanston Community Foundation Grants, 1962
Ezra Jack Keats/Kerlan Collection Memorial Fellowship, 1970
Farmers Insurance Corporate Giving Grants, 1981
Fidelity Foundation Grants, 2007
Fishman Family Foundation Grants, 2020
Ford Motor Company Fund Grants Program, 2033
Forrest C. Lattner Foundation Grants, 2035
Foundation for Enhancing Communities Grants, 2038
Frank Stanley Beveridge Foundation Grants, 2051
Fulbright Alumni Initiatives Awards Grants, 2064
Fulbright Distinguished Chairs Program Grants, 2069
Fulbright European Union Scholar-in-Residence Program Grants, 2071
Fulbright New Century Scholars Program Grants, 2075
Fulbright Occasional Lecturer Program Grants, 2076
Fulbright Specialists Program Grants, 2079
Fulbright Traditional Scholar Program in Europe and Eurasia, 2080
Fulbright Traditional Scholar Program in Sub-Saharan Africa, 2081
Fulbright Traditional Scholar Program in the East Asia/Pacific Region, 2082
Fulbright Traditional Scholar Program in the Near East and North Africa Region, 2083
Fulbright Traditional Scholar Program in the South and Central Asia Region, 2084
Fulbright Traditional Scholar Program in the Western Hemisphere, 2085
Fulbright Visiting Specialists Program: Direct Access to the Muslim World, 2087
Fund for Folk Culture Artists Support Grants, 2088
General Motors Foundation Grants, 2107
George A Ohl Jr. Foundation Grants, 2112
George Frederick Jewett Foundation Grants, 2116
George Gund Foundation Grants, 2117
George S. and Dolores Dore Eccles Grants, 2120
George W. Codrington Charitable Grants, 2122
Getty Scholars and Visiting Scholars Grants, 2138
Gilroy and Lillian P. Roberts Charitable Foundation Grants, 2142
Greater Saint Louis Community Found Grants, 2164
Greater Worcester Community Foundation Jeppson Memorial Fund for Brookfield Grants, 2165
Green Diamond Charitable Contributions, 2171
H.A. and Mary K. Chapman Charitable Grants, 2193
H.J. Heinz Company Foundation Grants, 2195
Harold Alfond Foundation Grants, 2200
Harold Simmons Foundation Grants, 2203
Harry W. Bass, Jr. Foundation Grants, 2210
Helen Bader Foundation Grants, 2228
Henry W. Bull Foundation Grants, 2237
Herbert H. and Grace A. Dow Found Grants, 2241
Herb Society of America Research and Education Grants, 2243
Hill Crest Foundation Grants, 2267
Hoglund Foundation Grants, 2271
Hutchinson Community Foundation Grants, 2310
IIE Rockefeller Foundation Bellagio Center Residencies, 2355
J.B. Reynolds Foundation Grants, 2449
Jackson Foundation Grants, 2453
Jacob and Hilda Blaustein Foundation Grants, 2454
James M. Collins Foundation Grants, 2461
John Ben Snow Memorial Trust Grants, 2497
John Deere Foundation Grants, 2500
John G. Duncan Trust Grants, 2501
John W. Alden Trust Grants, 2510
Joseph Alexander Foundation Grants, 2513
Josiah W. and Bessie H. Kline Foundation Grants, 2517
Kenneth T. and Eileen L. Norris Found Grants, 2554
Kovler Family Foundation Grants, 2575
Lillian S. Wells Foundation Grants, 2614
Lowe Foundation Grants, 2622
Lucile Horton Howe and Mitchell B. Howe Foundation Grants, 2624
Ludwick Family Foundation Grants, 2625
Luther I. Replogle Foundation Grants, 2631
M. Bastian Family Foundation Grants, 2636
M.J. Murdock Charitable Trust General Grants, 2641
Macquarie Bank Foundation Grants, 2650
Manitoba Arts Council Aboriginal Arts Creative Development Pilot Projects Grants, 2652
Marcia and Otto Koehler Foundation Grants, 2656
Margaret T. Morris Foundation Grants, 2659
Mary K. Chapman Foundation Grants, 2673
Mary S. and David C. Corbin Foundation Grants, 2675
Matilda R. Wilson Fund Grants, 2684
Max and Victoria Dreyfus Foundation Grants, 2687
Maximilian E. and Marion O. Hoffman Foundation Grants, 2688
McCombs Foundation Grants, 2718
McKnight Doctoral Fellowships, 2726
Mericos Foundation Grants, 2742
Meyer and Pepa Gold Family Foundation Grants, 2746
MGN Family Foundation Grants, 2754
Monfort Family Foundation Grants, 2794
Motorola Foundation Grants, 2796
National Gallery of Art Kress Interpretive Fellow, 2869
NEH Collaborative Research Grants, 2952
NEH Digital Humanities Implementation Grants, 2953
NEH Digital Humanities Start-Up Grants, 2954
Nelda C. and H.J. Lutcher Stark Found Grants, 2967
New Hampshire Charitable Foundation Grants, 2985
Nordson Corporation Foundation Grants, 3416
Northwestern Mutual Foundation Grants, 3430
Northwood University/Alden B. Dow Creativity Center Summer Residency Program, 3432
NSF Antarctic Artists and Writers Grants, 3473
Oklahoma City Community Grants, 3737
Oppenstein Brothers Foundation Grants, 3792
Oscar Rennebohm Foundation Grants, 3832
Patrick and Anna M. Cudahy Fund Grants, 3873
Peacock Foundation Grants, 3887
Perpetual Trust for Charitable Giving Grants, 3900
Pinkerton Foundation Grants, 3967
Price Family Charitable Fund Grants, 3987
Rajiv Gandhi Foundation Grants, 4025
Rathmann Family Foundation Grants, 4033
Rhode Island Foundation Grants, 4061
Richard and Susan Smith Family Found Grants, 4064
Righteous Persons Foundation Grants, 4069
Robert B McMillen Foundation Grants, 4072
Robert G. Cabell III and Maude Morgan Cabell Foundation Grants, 4075
Rockwell International Corporate Trust Grants, 4093
Ronald McDonald House Charities Grants, 4101
Ross Foundation Grants, 4106
Ruth and Vernon Taylor Foundation Grants, 4130
Sage Foundation Grants, 4158
Salem Foundation Grants, 4160
San Antonio Area Foundation Grants, 4163
Schramm Foundation Grants, 4188
Seaver Institute Grants, 4195
Self Foundation Grants, 4197
Sensient Technologies Foundation Grants, 4199
Sioux Falls Area Community Foundation Community Fund Grants (Unrestricted), 4262
Sioux Falls Area Comm Foundation Spot Grants, 4263
Smithsonian Latino Studies Fellowships, 4278
Society for the Arts in Healthcare Environmental Arts Research Grant, 4311
Sony Corporation of America Corporate Philanthropy Grants, 4329
SSHRC Doctoral Fellowships, 4407
SSHRC Graduate Scholarships-Michael Smith Foreign Study Supplements, 4408
SSHRC Joseph-Armand Bombardier CGS Doctoral Scholarships, 4412
SSRC Dissertation Proposal Development Faculty Field Fellowships, 4424
SSRC Dissertation Proposal Development Student Fellowships, 4425
Stackpole-Hall Foundation Grants, 4442
Stanford Humanities Center Arts Writer/Practitioner Fellowship, 4444
Strake Foundation Grants, 4452
Thomas C. Ackerman Foundation Grants, 4508
Thornton Foundation Grants, 4514
United Theological Seminary Merit-Based Scholarships, 4583
University of Southern Mississippi H.A. & Margret Rey Translation Fellowships, 4589
Vancouver Foundation Grants and Community Initiatives, 4807
W.C. Griffith Foundation Grants, 4819
W. C. Griffith Foundation Grants, 4818
Wallace Foundation Grants, 4829
Weyerhaeuser Company Foundation Grants, 4857
WILD Foundation Grants, 4867
William G. and Helen C. Hoffman Found Grants, 4876
Wolfsonian-FIU Fellowships, 4890
Xerox Foundation Grants, 4899

Asia

ACC Asian Art and Religion Fellowships, 251
ACC Ford Foundation Fellowships, 253

ACC Humanities Fellowship Program, 255
ACC John D. Rockefeller III Award, 257
ACC Residency Program in Asia, 267
ACC Starr Foundation Fellowship Program, 268
Ahmanson and Getty Postdoctoral Fellowships, 377
American Historical Association Bernadotte E. Schmitt Research Grants in European, African, or Asian History, 715
American Historical Association James Henry Breasted Prize, 729
American Historical Association John K. Fairbank Prize in East Asian History, 733
Asia Foundation Exchange Programs, 1090
ASM-UNESCO Leadership Grant for International Educators, 1099
Florence Tan Moeson Fellowships, 2025
Fulbright Traditional Scholar Program in the East Asia/Pacific Region, 2082
IREX Individual Advanced Research Opportunities Grants, 2419
IREX International Leadership in Education Program Grants, 2420
IREX Policy-Connect Collaborative Grants, 2422
IREX Short-Term Travel Grants, 2424
Kellogg Institute for International Studies Experiencing The World Fellowships, 2539
Kellogg Institute for International Studies Grants for Conferences and Workshops, 2543
Kellogg Institute for International Studies Grants for Distinguished Lectureships, 2544
Kellogg Institute for International Studies Grants for Events, 2545
Kellogg Institute for International Studies Supplemental Awards for PhD Students, 2550
MPI for Social Anthropology Department II Postdoctoral Grants, 2798
SSRC Postdoc Fellowship for Transregional Research: Inter-Asian Contexts and Connections, 4438

Asia, Central
ACIE Title VIII Research Scholar Fellowships, 276
Florence Tan Moeson Fellowships, 2025

Asia, East
Florence Tan Moeson Fellowships, 2025
Fulbright Traditional Scholar Program in the East Asia/Pacific Region, 2082
Rumsfeld Foundation Graduate Fellowships, 4127

Asia, Southeast
American Historical Association John F. Richards Prize in South Asian History, 731
Florence Tan Moeson Fellowships, 2025
Fulbright Traditional Scholar Program in the East Asia/Pacific Region, 2082
JSPS Bilateral Exchange Programs with Foreign Academic Institutions, 2524

Asia, Southern
American Historical Association John F. Richards Prize in South Asian History, 731
Florence Tan Moeson Fellowships, 2025
NSF Materials World Network: Cooperative Activity in Materials Research between US Investigators and their Counterparts Abroad, 3595
Shastri Indo-Canadian Inst Librarian Fellows, 4225

Asia, Western
Florence Tan Moeson Fellowships, 2025

Asian American Studies
AAJA Fellowships, 181
Asian Pacific American Institute for Congressional Studies Fellowships, 1091
Florence Tan Moeson Fellowships, 2025
Institute of American Cultures Graduate and Predoctoral Fellowships, 2383
Institute of American Cultures Postdoctoral and Visiting Scholar Fellowships, 2384

Asian Americans
AAJA Fellowships, 181
American Sociological Association Minority Fellowships, 805
AMS Howard Mayer Brown Fellowships, 827
Asian Pacific American Institute for Congressional Studies Fellowships, 1091
John R. Oishei Foundation Grants, 2505
MLA Grad Scholarship for Minority Students, 2787
UCLA Research Grant in Ethnic Studies, 4549
United Methodist Church Asian, Black, Hispanic and Native American History Research Grants, 4570

Asian Arts
ACC Asian Art and Religion Fellowships, 251
ACC Ford Foundation Fellowships, 253
ACC Humanities Fellowship Program, 255
ACC John D. Rockefeller III Award, 257
ACC Starr Foundation Fellowship Program, 268
ACC Taiwan Fellowship Program, 269
Christensen Fund Regional Grants, 1554
Florence Tan Moeson Fellowships, 2025
Smithsonian Senior Fellowships, 4296

Asian Languages/Literature
ACIE Title VIII Combined Research and Language Training Fellowships, 275
Shastri Indo-Canadian Inst Librarian Fellows, 4225

Assistantships, General
NOAA Pacific Islands Assistantships, 3399
Penn Graduate School of Education Research Assistantships, 3894

Assisted-Living Programs
McLean Contributionship Grants, 2727
Reinberger Foundation Grants, 4051

Assistive Technology
Deborah Munroe Noonan Mem Research Grants, 1729
NYSCA Dance: Services to the Field Grants, 3702
U.S. Department of Education Rehabilitation Engineering Research Centers Grants, 4544

Asthma
AAAAI ARTrust Mini Grants for Allied Health, 14
AAAAI ARTrust Mini Grants for Clinical Res, 15
AAAAI Distinguished Service Award, 20
AAAAI Fellows-in-Training Abstract Award, 21
AAAAI Fellows-in-Training Grants, 22
AAFA Investigator Research Grants, 153
ACAAI Foundation Research Grants, 244
Fairlawn Foundation Grants, 1974
NHLBI Airway Smooth Muscle Function and Targeted Therapeutics in Human Asthma Grants, 3014
NHLBI Clinical Centers for the NHLBI Asthma Network Grants, 3024
RACGP National Asthma Council Res Award, 4015
Sandler Program for Asthma Research Grants, 4165

Astrobiology
American Philosophical Soc Judson Daland Prize, 766
American Philosophical Society Lewis and Clark Fund for Exploration and Field Research, 767
American Philosophical Society Lewis and Clark Fund for Exploration & Research in Astrobiology, 768

Astronautics
ASME Spirit of St. Louis Medal, 1165
Zonta International Foundation Amelia Earhart Fellowships, 4905

Astronomy
Alfred P. Sloan Foundation History of Science and Technology Grants, 553
Anna-Greta and Holger Crafoords Grants, 858
Arecibo Observatory Grants, 988
Astronomical Society of the Pacific Amateur Achievement Award, 1228
Astronomical Society of the Pacific Dorothea Klumpke-Roberts Award, 1229
Astronomical Society of the Pacific Las Cumbres Amateur Outreach Award, 1230
Astronomical Society of the Pacific Maria and Eric Muhlmann Award, 1231
Astronomical Society of the Pacific Priscilla and Bart Bok Awards, 1232
Astronomical Society of the Pacific Richard H. Emmons Award for Excellence in College Astronomy Teaching, 1233
Astronomical Society of the Pacific Robert J. Trumpler Award, 1234
Bower Award for Achievement in Science, 1364
Canada Graduate Scholarships and NSERC Postgraduate Scholarships, 1420
Carnegie Postdoctoral Research Fellowships in Astronomy, 1437
Education Through Astronomy and Space Science Initiative, 1896
Gruber Foundation Cosmology Prize, 2177
Gruber Foundation International Astronomical Union Fellowships, 2179
ISGC Graduate Fellowships, 2430
ISGC Informal Education Grants, 2431
ISGC Multiple Undergraduate Research Grants, 2432
ISGC Research Seed Grants, 2433
M.J. Murdock Charitable Trust College Science Research Program, 2639
National Geographic Society Expeditions Council Grants, 2872
National Geographic Society Field Research and Exploration Grants, 2873
NPSC Grad Fellowships in the Physical Sciences, 3436
NSF Antarctic Aeronomy and Astrophysics Program Grants, 3472
NSF Astronomy and Astrophysics Postdoctoral Fellowships, 3494
NSF Astronomy and Astrophysics Res Grants, 3495
NSF CISE Accelerating Discovery in Science and Engineering Through Petascale Simulations and Analysis Grants, 3521
NSF CISE Cross-Cutting Program Grants, 3532
NSF CISE Increasing the Participation and Advancement of Women in Academic Science and Engineering Careers Grants, 3534
NSF Extragalactic Astronomy and Cosmology Research Grants, 3564
NSF Galactic Astronomy Research Grants, 3568
NSF National Astronomy and Ionosphere Center Grants, 3598
NSF Partnerships in Astronomy & Astrophysics Research and Education, 3612
NSF Physics Research Grants, 3621
NSF Planetary Astronomy Research Grants, 3623
NSF Stellar Astronomy and Astrophysics Research Grants, 3655
NSF University Radio Observatories Program, 3664
Perkin Fund Grants, 3898
Research Corporation Cottrell Scholars Awards, 4053
Research Corporation Multi-Investigator Cottrell College Science Awards, 4054
Research Corporation Scialog Grants, 4055
Research Corporation Single-Investigator Cottrell College Science Awards, 4056
Roy J. Carver Charitable Trust Higher Education Grants, 4113
Roy J. Carver Charitable Trust Medical and Science Research Grants, 4114
Sigma Xi Grants in Aid of Research, 4259
SPIE George W. Goddard Award, 4342
SRC Natural and Engineering Sciences Grants, 4387
UNCFSP NASA Harriett G. Jenkins Predoctoral Fellowships, 4563

Astrophysics
Abdus Salam ICTP Laurea Magistralis Grants, 229
Abdus Salam ICTP Postdoctoral Positions in Theoretical High Energy Physics, 234

Abdus Salam Joint ICTP/SISSA PhD Programme in Physics and Mathematics Grants, 242
Alfred P. Sloan Foundation Sloan Digital Sky Survey Grants, 564
AMNH Graduate Student Fellowships, 815
APS Hans A. Bethe Prize, 960
Bower Award for Achievement in Science, 1364
Brinson Foundation Grants, 1372
Fermilab Postdoctoral Fellowship Programs, 2004
ISGC Graduate Fellowships, 2430
JILA Visiting Fellowships and Postdoctoral Research Associateships, 2489
Kavli Foundation Research Prizes, 2534
NSF Antarctic Aeronomy and Astrophysics Program Grants, 3472
NSF Antarctic Research Grants, 3479
NSF Astronomy and Astrophysics Postdoctoral Fellowships, 3494
NSF CISE Accelerating Discovery in Science and Engineering Through Petascale Simulations and Analysis Grants, 3521
NSF CISE Cross-Cutting Program Grants, 3532
NSF CISE Increasing the Participation and Advancement of Women in Academic Science and Engineering Careers Grants, 3534
NSF Extragalactic Astronomy and Cosmology Research Grants, 3564
NSF Gravitational Physics Research Grants, 3576
NSF Particle and Nuclear Astrophysics Grants, 3608
NSF Partnerships in Astronomy & Astrophysics Research and Education, 3612
NSF Physics Frontiers Centers Grants, 3619
NSF Physics Research Grants, 3621
NSF Stellar Astronomy and Astrophysics Research Grants, 3655
Royal Norwegian Embassy Kavli Prizes, 4109
Roy J. Carver Charitable Trust Higher Education Grants, 4113
Smithsonian Senior Fellowships, 4296
WHOI Geophysical Fluid Dynamics Fellowships, 4859

At-Risk Students
Hilda and Preston Davis Foundation Grants, 2266
Sidgmore Family Foundation Grants, 4252
The Ray Charles Foundation Grants, 4502
Victor E. Speas Foundation Grants, 4810

At-Risk Youth
ACT Awards, 330
Boston Foundation Grants, 1361
Charles Lafitte Foundation Grants, 1525
CMA Foundation Grants, 1582
Commonweal Foundation Community Assistance Grants, 1602
ConocoPhillips Grants Program, 1629
Cruise Industry Charitable Foundation Grants, 1660
EPA Children's Health Protection Grants, 1933
Herbert & Adrian Woods Foundation Grants, 2239
Hilda and Preston Davis Foundation Grants, 2266
Jenifer Altman Foundation Grants, 2484
Jovid Foundation Grants, 2518
Lillian S. Wells Foundation Grants, 2614
Luther I. Replogle Foundation Grants, 2631
Mattel Children's Foundation Grants, 2685
Mattel International Grants Program, 2686
MGM Resorts Foundation Community Grants, 2753
OSF Soros Justice Advocacy Fellowships, 3850
OSF Soros Justice Media Fellowships, 3851
Paul G. Allen Family Foundation Grants, 3876
Pinkerton Foundation Grants, 3967
Richard Davoud Donchian Foundation Grants, 4066
Ronald McDonald House Charities Grants, 4101
Sidgmore Family Foundation Grants, 4252
Sunny and Abe Rosenberg Foundation Grants, 4469
The Ray Charles Foundation Grants, 4502
Victor E. Speas Foundation Grants, 4810
W.K. Kellogg Foundation Healthy Kids Grants, 4822

Athletics
ACSM Raymond and Rosalee Weiss Research Endowment Grant, 322
Regents Professional Opportunity Scholarships, 4049
Textron Corporate Contributions Grants, 4501
Todd Brock Family Foundation Grants, 4522

Atmosphere: Composition, Disturbance, Physics
DOE Program for Ecosystem Research, 1810
NSF Aeronomy Grants, 3466
NSF Natural, Social, and System Sciences and Research Support and Logistics Program, 3600
NSF Solar, Heliospheric, and INterplanetary Environment Grants, 3651
NSF Solar Terrestrial Research Grants, 3652

Atmospheric Radiation
NCAR Postdoctoral Fellowships in the Atmospheric Sciences, 2902
NSF Aeronomy Grants, 3466
NSF Particle and Nuclear Astrophysics Grants, 3608

Atmospheric Sciences
Abdus Salam ICTP Joint PhD Programme in Environmental Fluid Mechanics Grants, 230
Abdus Salam ICTP Postgraduate Diploma Grants, 235
Abdus Salam ICTP Schools, Conferences, Workshops and Extended Research Workshops Grants, 238
Alfred P. Sloan Foundation Sloan Digital Sky Survey Grants, 564
AMS/Industry Government Graduate Fellowships, 817
AMS/Industry Minority Scholarships, 818
AMS Grad Fellowships in History of Science, 824
AMS Undergraduate Scholarships, 837
Arecibo Observatory Grants, 988
ASEE Army Research Laboratory Postdoctoral Fellowships, 1059
Cooperative Institute for Research in Environmental Sciences Visiting Fellows Program, 1643
DOE Environmental Processes Research Grants, 1790
DOE Program for Ecosystem Research, 1810
Dreyfus Postdoctoral Program in Environmental Chemistry Grant, 1855
Father James B. Macelwane Annual Awards, 1983
Global Change Education Program, 2151
M.J. Murdock Charitable Trust College Science Research Program, 2639
NASA Goddard Visiting Fellowships, 2838
NCAR Graduate Research Fellowships, 2901
NCAR Postdoctoral Fellowships in the Atmospheric Sciences, 2902
NOAA Abrupt Climate Change Studies Grants, 3329
NOAA Arctic Research Program Grants, 3330
NOAA Assessing Atlantic Meridional Overturning Circulation Program Grants, 3331
NOAA Atmospheric Composition and Climate Program Grants, 3333
NOAA Climate Change Data and Detection Program Grants, 3344
NOAA Climate Dynamics and Experimental Prediction Program Grants, 3345
NOAA Climate Prediction Program for the Americas Grants, 3346
NOAA Climate Variability and Predictability Program Grants, 3348
NOAA Dr. Nancy Foster Scholarships, 3367
NOAA Ernest F. Hollings Scholarships, 3370
NOAA Global Carbon Cycle Program Grants, 3374
NOAA Graduate Sciences Program Grants, 3375
NOAA Regional Integrated Sciences and Assessments Program Grants, 3402
NOAA Scientific Data Stewardship Grants, 3403
NOAA Sector Applications Research Grants, 3405
NOAA Transition of Research Applications to Climate Services Grants, 3408
NSF Aeronomy Grants, 3466
NSF Antarctic Ocean and Atmospheric Sciences Grants, 3477
NSF Antarctic Research Grants, 3479
NSF Arctic Natural Science Grants in Atmospheric Sciences, 3483
NSF Arctic Natural Sciences Program Grants, 3488
NSF Atmospheric Chemistry Grants, 3496
NSF Atmospheric Sciences Mid-Size Infrastructure Opportunity Grants, 3497
NSF CISE Accelerating Discovery in Science and Engineering Through Petascale Simulations and Analysis Grants, 3521
NSF CISE Cross-Cutting Program Grants, 3532
NSF CISE Increasing the Participation and Advancement of Women in Academic Science and Engineering Careers Grants, 3534
NSF Collaborations in Math Geosciences, 3536
NSF Coupling, Energetics, and Dynamics of Atmospheric Regions Research Grants, 3545
NSF Geospace Environment Modeling, 3573
NSF Hydrologic Sciences Research Grants, 3578
NSF Magnetospheric Physics Research Grants, 3590
NSF National Astronomy and Ionosphere Center Grants, 3598
NSF National Space Weather Program, 3599
NSF Physical and Dynamic Meteorology Grants, 3614
NSF Polar Programs Division Research Grants, 3627
NSF Solar, Heliospheric, and INterplanetary Environment Grants, 3651
NSF Solar Terrestrial Research Grants, 3652
NSF Upper Atmospheric Facilities Grants, 3665
ONR Electronic and Optical Ceramics Research Grants, 3756
ONR Marine Meteorology and Atmospheric Effects Research Grants, 3769
ONR Radio Frequency Sensor Processing Research Grants, 3778
ONR Radio Frequency Solid State Devices Applied Research Grants, 3779
ORISE National Oceanic and Atmospheric Admin Student Research Participation, 3812
SPIE George W. Goddard Award, 4342
Swiss National Science Foundation Fellowships for Advanced Researchers, 4483
Swiss National Science Foundation Fellowships for Prospective Researchers, 4484
Swiss National Science Foundation Marie Heim-Vogtlin Grants, 4485
W.M. Keck Foundation Science and Engineering Research Grants, 4824
Wellcome Trust Biomedical Science Grants, 4843

Atomic and Molecular Structure
NSF Atomic Molecular & Optical Physics Grants, 3498
NSF Electronic and Photonic Materials Grants, 3558
USDD Research Interests of the US Air Force Academy Grants, 4785

Attention Deficit Hyperactivity Disorder
AACAP Elaine Schlosser Lewis Award for Research on Attention-Deficit Disorder, 59
AACAP Pilot Research Award for Attention-Deficit Disorder, 70

Audience Development
ALZA Corporate Contributions Grants, 588
Andy Warhol Foundation for the Visual Arts Curatorial Research Fellowships, 847
Connecticut Community Foundation Grants, 1627
NYSCA Museum: Project Support Grants, 3707
Wallace Foundation Grants, 4829

Audio Production
Audio Engineering Society Grad Scholarships, 1265
Florida Division of Cultural Affairs Individual Artist Fellowships, 2026
Manitoba Arts Council Dance Creation/Production Grants, 2654
North Carolina Arts Council Folklife Pub Grants, 3419
NYSCA Individual Artists: Film, Media and New Technology Production Grants, 3706
Tandem Launch Technologies Grants, 4493
WHOI Ocean Science Journalism Fellowship, 4862

Audiology
ASHA Students Preparing for Academic & Research Careers Award, 1075
ASHFoundation Graduate Student Scholarships for Minority Students, 1078
ASHFoundation Graduate Student with a Disability Scholarship, 1079
ASHFoundation New Century Scholars Program Doctoral Scholarships, 1080
ASHFoundation New Century Scholars Grant, 1081
ASHFoundation Res Grant for New Investigators, 1082
ASHFoundation Student Research Grants in Audiology, 1084
Regents Professional Opportunity Scholarships, 4049
Special Olympics Health Professions Stu Grants, 4332

Audiovisual Materials
AASL Research Grant, 207
Dept of Ed Star Schools Program Grants, 1749
Manitoba Arts Council Dance Creation/Production Grants, 2654

Australia
AFUW Diamond Jubilee Bursary, 359
AFUW Doreen McCarthy, Barbara Crase, Cathy Candler & Brenda Nettle Bursaries, 360
AFUW Fellowships, 361
AFUW Georgina Sweet Fellowships, 362
AFUW Heather Latz Bursary, 363
AFUW Padnendadlu Graduate Bursary, 364
AFUW Padnendadlu Postgraduate Bursary, 365
AFUW Padnendadlu Undergraduate Bursary, 366
AFUW South Australia Jean Gilmore, Thenie Baddams & Daphne Elliott Bursaries, 367
AFUW Winifred E. Preedy Postgraduate Bursary, 368
American Australian Association Australia To USA Fellowships, 619
American Australian Association USA to Australia Fellowships, 620
Australasian Institute of Judicial Administration Seed Funding Grants, 1268
Australian Postdoctoral Research Fellowships, 1270
Australian Postgraduate Research Fellowships, 1271
Australian Research Fellowships/Queen Elizabeth II Fellowships, 1272
Beryl Henderson Foundation Grants, 1337

Autism
Deborah Munroe Noonan Mem Research Grants, 1729
Ireland Family Foundation Grants, 2415
NWHF Mark O. Hatfield Research Fellowship, 3671
Partnership for Cures Two Years To Cures Grants, 3872
Thompson Charitable Foundation Grants, 4511

Autoimmunity
AIHS Alberta/Pfizer Translational Res Grants, 435
Glaucoma Research Pilot Project Grants, 2146
Scleroderma Foundation Established Investigator Gra[illegible]90

Automotive Engineering
ASCE Jack E. Leisch Memorial National Graduate Fellowship, 1041
BMW of N America Charitable Contributions, 1353
SSHRC-APC Automotive Partnership Canada, 4396

Automotive Industry
SSHRC-APC Automotive Partnership Canada, 4396

Aviation
ASEE Army Laboratory Postdoc Fellowships, 1059
Aviation Research Grants, 1273
Center of Excellence in General Aviation Grants, 1486
Emil Buehler Perpetual Trust Grants, 1919
NOAA Collaborative Science, Technology, and Applied Research Program Grants, 3354
Smithsonian National Air and Space Museum Aviation Space Writers Foundation Grant, 4284
Smithsonian National Air and Space Museum Charles A. Lindbergh Chair in Aerospace History, 4285

Aviation History
Smithsonian National Air and Space Museum Charles A. Lindbergh Chair in Aerospace History, 4285
Smithsonian National Air and Space Museum Guggenheim Fellowships, 4286
Smithsonian National Air and Space Museum Verville Fellowships, 4288

Azerbaijan
Fulbright Critical Lang Enhancement Awards, 2068
IREX Individual Advanced Research Opportunities Grants, 2419
IREX Policy-Connect Collaborative Grants, 2422

Bacteriology
ASM/CDC Fellowships in Infectious Disease and Public Health Microbiology, 1100
NSF Physiology and Ethology Research Grants, 3622
ORISE Faculty Research Participation at the Centers for Disease Control and Prevention, 3803
ORISE Postgraduate Research Participation at the Centers for Disease Control and Prevention, 3820
ORISE Student Research Participation at the Centers for Disease Control and Prevention, 3829
Pfizer ASPIRE EU Emerging Mechanisms of Resistance Antibacterial Research Awards, 3912
Pfizer ASPIRE EU MRSA Nosocomial Pneumonia & MRSA Complicated Skin & Soft Tissue Infections Antibacterial Research Awards, 3913
Pfizer ASPIRE North America Narrow Spectrum Antibiotics for the Treatment of MRSA Research Awards, 3916

Baking
AACC International Milling & Baking Division's John C. Halverson Memorial Lectureship, 92

Band Music
NYSCA Music: Services to the Field Grants, 3708

Bangladesh
Fulbright Traditional Scholar Program in the South and Central Asia Region, 2084

Banking
Bill and Melinda Gates Foundation Financial Services for the Poor Grants, 1346
Equipment Leasing and Finance Research Grants, 1943
Germanistic Society of America Fellowships, 2128

Bankruptcy
American College of Bankruptcy Grants, 699

Baptist Church
Booth-Bricker Fund Grants, 1359
Bosque Foundation Grants, 1360
MacDonald-Peterson Foundation Grants, 2649

Basic Living Expenses
IBRO Asia Reg APRC Exchange Fellowships, 2323
IBRO InEurope Short Stay Grants, 2325
JSA JLab Sabbatical Research Leave Grants, 2522
MBL Burr & Susie Steinbach Summer Fellow, 2696

Basic Skills Education
Allyn Foundation Grants, 574
Auburn Foundation Grants, 1263
Ball Brothers Foundation General Grants, 1291
Bodenwein Public Benevolent Foundation Grants, 1355
Booth-Bricker Fund Grants, 1359
Boston Foundation Grants, 1361
Carl and Lily Pforzheimer Foundation Grants, 1427
CIGNA Foundation Grants, 1568
Cowles Charitable Trust Grants, 1652
Cruise Industry Charitable Foundation Grants, 1660
Dept of Ed Adult Education--National Leadership Activities Grants, 1744
Evjue Foundation Grants, 1963
Frances L. and Edwin L. Cummings Memorial Fund Grants, 2046
George Foundation Grants, 2115
Harold Simmons Foundation Grants, 2203
Joyce Foundation Employment Grants, 2520
McCune Foundation Education Grants, 2721
Norcliffe Foundation Grants, 3415
Oppenstein Brothers Foundation Grants, 3792
Richard and Rhoda Goldman Fund Grants, 4063
Robert Bowne Foundaion Edmund A. Stanley, Jr. Research Grants, 4073
San Antonio Area Foundation Grants, 4163
Sony Corporation of America Corporate Philanthropy Grants, 4329
Strake Foundation Grants, 4452
Talbert and Leota Abrams Foundation, 4492

Behavioral Medicine
ACSM Coca-Cola Company Doctoral Student Grant on Behavior Research, 313
Cystic Fibrosis Canada Fellowships, 1677
James S. McDonnell Found Research Grants, 2466
James S. McDonnell Foundation Scholar Awards, 2467
NCIC Canadian Cancer Society Research Grants, 2919
NCI Stages of Breast Development: Normal to Metastatic Disease Grants, 2932
NEI Ruth L. Kirschstein National Research Service Award Short-Term Institutional Research Training Grants, 2963
NHLBI Ruth L. Kirschstein National Research Service Awards for Individual Postdoctoral Fellows, 3057
NHLBI Ruth L. Kirschstein National Research Service Awards for Individual Predoctoral Fellowships to Promote Diversity in Health-Related Research, 3058
NIA AIDS and Aging: Behavioral Sciences Prevention Research Grants, 3083
NIH Research on Sleep and Sleep Disorders, 3236
NIH Ruth L. Kirschstein National Research Service Awards for Individual Postdoctoral Fellows, 3239
NIMH AIDS and Aging: Behavioral Sciences Prevention Research Grants, 3249
NINR AIDS and Aging: Behavioral Sciences Prevention Research Grants, 3260

Behavioral Sciences
ACS Research Scholar Grants in Psychosocial and Behavioral and Cancer Control Research, 329
Alzheimer's Association Investigator-Initiated Research Grants, 592
ASID Foundation Joel Polsky Academic Achievement Award, 1095
Cystic Fibrosis Canada Fellowships, 1677
Cystic Fibrosis Canada Studentships, 1683
Cystic Fibrosis Canada Summer Studentships, 1684
Duke University Postdoctoral Research Fellowships in Aging, 1880
James S. McDonnell Foundation Scholar Awards, 2467
James S. McDonnell Foundation Understanding Human Cognition Awards, 2468
Marine Corps Historical Center Dissertation Fellowship, 2663
National Academy of Education Spencer Postdoctoral Fellowships, 2845
National Center for Responsible Gaming Centers of Excellence in Gambling Research Grants, 2850
National Center for Responsible Gaming Early Stage Investigator Grants, 2852
National Center for Responsible Gaming Exploration Grants, 2853
National Center for Responsible Gaming Independent Scientist Grants, 2854
National Center for Resp Gaming Large Grants, 2855
National Center for Responsible Gaming Postdoctoral Fellowships, 2856
National Center for Resp Gaming Seed Grants, 2857
National Center for Responsible Gaming Travel Grants, 2858
NCI/NIDA Transdisciplinary Tobacco Use Research Centers Grants, 2912
NCI Cancer Education and Career Development, 2917
NCI Exploratory Grants for Behavioral Research in Cancer Control, 2928

USDA Plant Feedstock Genomics for Bioenergy: A Joint Research Grant, 4639
USDD Power and Thermal Management Technology Development Grants, 4775

Biomaterials
NSF Condensed Matter Physics Research Grants, 3540

Biomechanics
ASME H.R. Lissner Award, 1127
NSF Physiology and Ethology Research Grants, 3622
USDD Advances in Bioscience for Airmen Performance Grants, 4675

Biomedical Education
AAAS Award for Public Engagement with Science, 26
AIHS Full-Time M.D./Ph.D. Studentships, 447
AIHS Knowledge Exchange Visiting Professors, 451
AIHS Knowledge Exchange - Visiting Scientists, 452
AIHS Media Summer Fellowship, 453
AIHS Part-Time Studentships, 455
Bristol-Myers Squibb Clinical Outcomes and Research Grants, 1374
HHMI Biomedical Research Grants for Int'l Scientists Infectious Diseases and Parasitology, 2251
HHMI Biomedical Research Grants for International Scientists in Canada and Latin America, 2252
HHMI Biomedical Research Grants for International Scientists in the Baltics, Central and Eastern Europe, Russia, and Ukraine, 2253
HHMI Gilliam Fellowships for Advanced Study, 2254
HHMI Initiative for Colleges Grants, 2256
HHMI Initiative for Universities Grants, 2257
HHMI Physician-Scientist Early Career Award, 2260
HHMI Undergraduate Science Education Grants for Professors, 2263
Mt. Sinai Health Care Foundation Academic Medicine and Bioscience Grants, 2803
NCI Ruth L. Kirschstein National Research Service Award Institutional Research Training Grants, 2931
NHLBI Biomedical Research Training Program for Individuals from Underrepresented Groups, 3019
NHLBI Summer Internship Program in Biomedical Research, 3064
NIH Summer Internship Program in Biomedical Research, 3248
NYCT Biomedical Research Grants, 3694
PC Fred H. Bixby Fellowships, 3880
Piper Trust Healthcare & Med Research Grants, 3972

Biomedical Engineering
ASME Ralph Coats Roe Medal, 1158
IIE Whitaker International Fellowships and Scholarships, 2356
MBL Albert & Ellen Grass Summer Fellowships, 2692
MBL Associates Summer Fellowships, 2694
MBL Burr & Susie Steinbach Summer Fellow, 2696
MBL E.E. Just Summer Fellowship for Minority Scientists, 2697
MBL Erik B. Fries Summer Fellowships, 2698
MBL Eugene and Millicent Bell Fellowships, 2699
MBL Evelyn & Mel Spiegal Summer Fellowship, 2700
MBL Frank R. Lillie Summer Fellowship, 2701
MBL Fred Karush Library Readership, 2703
MBL Gruss Lipper Family Foundation Summer Fellowship, 2704
MBL H. Keffer Hartline and Edward F. MacNichol, Jr. Fellowships, 2705
MBL Herbert W. Rand Summer Fellowship, 2706
MBL James E. and Faith Miller Memorial Summer Fellowship, 2707
MBL John M. Arnold Award, 2708
MBL Lucy B. Lemann Mem Summer Fellowship, 2710
MBL M.G.F. Fuortes Summer Fellowships, 2711
MBL Nikon Summer Fellowship, 2712
MBL Stephen W. Kuffler Summer Fellowships, 2715
MBL William Townsend Porter Summer Fellowships for Minority Investigators, 2716
NHLBI Intramural Research Training Awards, 3029
NSF Biomedical Engineering Grants, 3505
NSF Biomedical Engineering and Engineering Healthcare Grants, 3506
NSF Emerging Frontiers in Research and Innovation Grants, 3560
ONR Biological & Biomedical Division Grants, 3744
Smith Family Awards for Excellence in Biomedical Research, 4268
W.M. Keck Foundation Science and Engineering Research Grants, 4824

Biomedical Research
25th Anniversary Foundation Grants, 6
A.O. Smith Community Grants, 12
AAAS Award for Public Engagement with Science, 26
AACR Award for Outstanding Achievement in Cancer Research, 128
AACR Team Science Award, 148
AAMC Award for Distinguished Research, 183
Achelis Foundation Grants, 274
Adaptec Foundation Grants, 340
AFAR CART Fund Grants, 351
Affymetrix Corporate Contributions Grants, 355
AHRQ Individual Awards for Postdoctoral Fellows Ruth L. Kirschstein National Research Service Awards, 389
AIChE Professional Progress Award for Outstanding Progress in Chemical Engineering, 423
AIHS Alberta/Pfizer Translational Res Grants, 435
AIHS Clinical Fellowships, 436
AIHS Fast-Track Fellowships, 440
AIHS ForeFront Internships, 442
AIHS Full-Time Fellowships, 445
AIHS Full-Time Health Research Studentships, 446
AIHS Full-Time M.D./Ph.D. Studentships, 447
AIHS Full-Time Studentships, 448
AIHS Heritage Youth Researcher Summer Science Grants, 449
AIHS Interdisciplinary Team Grants, 450
AIHS Knowledge Exchange Visiting Professors, 451
AIHS Knowledge Exchange - Visiting Scientists, 452
AIHS Part-Time Fellowships, 454
AIHS Part-Time Studentships, 455
AIHS Proposals for Special Initiatives, 456
AIHS Research Prize, 457
AIHS Summer Studentships, 458
Albany Medical Center Prize in Medicine and Biomedical Research, 502
Albert and Mary Lasker Foundation Awards, 505
Albert and Mary Lasker Foundation Clinical Research Scholars, 506
Alcoa Foundation Grants, 511
Allyn Foundation Grants, 574
ALSAM Foundation Grants, 579
American Chemical Society Award in Separations Science and Technology, 671
American Foodservice Charitable Trust Grants, 710
American Philosophical Society Library Resident Research Fellowships, 769
Amgen Foundation Grants, 812
ANA/Grass Foundation Award in Neuroscience, 839
ANA Wolfe Neuropathy Research Prize, 840
Annenberg Foundation Grants, 862
Anthony R. Abraham Foundation Grants, 869
AOCS George Schroepfer Medal, 879
AOCS Industrial Oil Products Div Student Award, 885
APHL Emerging Infectious Diseases Fellowships, 914
Arizona Cardinals Grants, 995
Arthur F. and Alice E. Adams Charitable Foundation Grants, 1024
ASCO/UICC International Cancer Technology Transfer Fellowships, 1044
ASCO Advanced Clinical Research Award in Colorectal Cancer, 1045
ASCO Advanced Clinical Research Awards in Breast Cancer, 1046
ASCO Advanced Clinical Research Awards in Sarcoma, 1047
ASCO Long-Term International Fellowships, 1048
ASM Abbott Laboratories Award in Clinical and Diagnostic Immunology, 1101
Australian Academy of Science Grants, 1269
Aviation Research Grants, 1273
Barth Syndrome Foundation Research Grants, 1302
Batchelor Foundation Grants, 1305
BCBSNC Foundation Grants, 1318
Beckman Coulter Community Relations Grants, 1325
Beim Foundation Grants, 1331
Blue Shield of California Grants, 1350
Blum-Kovler Foundation Grants, 1351
Bodman Foundation Grants, 1356
Booth-Bricker Fund Grants, 1359
Bosque Foundation Grants, 1360
Brinson Foundation Grants, 1372
Bristol-Myers Squibb Clinical Outcomes and Research Grants, 1374
Bristol-Myers Squibb Foundation Fellowships, 1375
Bristol-Myers Squibb Foundation Global HIV/AIDS Initiative Grants, 1376
Bristol-Myers Squibb Foundation Health Disparities Grants, 1377
Brookdale Foundation Leadership in Aging Fellowships, 1381
Burden Trust Grants, 1388
BWF Institutional Program Unifying Population and Laboratory Based Sciences Grants, 1397
BWF Preterm Birth Initiative Grants, 1399
Byron W. and Alice L. Lockwood Grants, 1401
Callaway Golf Company Foundation Grants, 1410
Campbell Soup Foundation Grants, 1415
Carl C. Icahn Foundation Grants, 1428
Catherine Holmes Wilkins Foundation Grants, 1442
CCFF Chairman's Dist Life Sciences Award, 1449
Chapin Hall International Fellowships in Children's Policy Research, 1517
Charles A. King Trust Postdoctoral Research Fellowships, 1518
Charles Edison Fund Grants, 1520
Charles H. Revson Foundation Grants, 1524
Charles Lafitte Foundation Grants, 1525
Chest Foundation Geriatric Development Research Awards, 1536
Chiles Foundation Grants, 1552
Chiron Foundation Community Grants, 1553
CJ Foundation for SIDS Research Grants, 1573
Claude Pepper Foundation Grants, 1578
CMA Foundation Grants, 1582
CMS Research and Demonstration Grants, 1586
Connecticut Health Foundation Health Grants, 1628
Conquer Cancer Foundation of ASCO Career Development Award, 1630
Conquer Cancer Foundation of ASCO Comparative Effectiveness Research Professorship in Breast Cancer, 1631
Conquer Cancer Foundation of ASCO Improving Cancer Care Grants, 1633
Conquer Cancer Foundation of ASCO International Innovation Grant, 1636
Conquer Cancer Foundation of ASCO Medical Student Rotation Grants, 1637
Conquer Cancer Foundation of ASCO Translational Research Professorships, 1639
Croucher Foundation Grants, 1659
CSTE CDC/CSTE Applied Epidemiology Fellowships, 1664
Cystic Fibrosis Canada Clinical Fellowships, 1674
Cystic Fibrosis Canada Fellowships, 1677
Cystic Fibrosis Canada Research Grants, 1678
Cystic Fibrosis Canada Senior Scientist Research Training Awards, 1680
Cystic Fibrosis Canada Studentships, 1683
Cystic Fibrosis Canada Summer Studentships, 1684
Cystic Fibrosis Canada Visiting Allied Health Professional Awards, 1686
Cystic Fibrosis Canada Visiting Scientist Awards, 1688
Cystic Fibrosis Research Eliz Nash Fellowships, 1689
Cystic Fibrosis Research New Horizons Campaign Grants, 1690
Cystic Fibrosis Trust Research Grants, 1691
DAAD Re-invitation Program for Former Scholarship Holders, 1701

DAAD Research Grants for Doctoral Candidates and Young Academics and Scientists, 1703
DAAD Research Stays for University Academics and Scientists, 1705
Dana Foundation Science and Health Grants, 1716
Del E. Webb Foundation Grants, 1733
DeMatteis Family Foundation Grants, 1740
Denton A. Cooley Foundation Grants, 1743
Doris Duke Charitable Foundation Clinical Interfaces Award Program, 1833
Doris Duke Charitable Foundation Clinical Research Fellowships for Medical Students, 1834
Doris Duke Charitable Foundation Clinical Scientist Development Award Bridge Grants, 1837
Doris Duke Charitable Foundation Distinguished Clinical Scientist Award Program, 1838
Dorr Inst for Arthritis Research & Education, 1843
Duke University Postdoctoral Research Fellowships in Aging, 1880
Edwin S. Webster Foundation Grants, 1900
Elisabeth Severance Prentiss Foundation Grants Program, 1905
Ellison Medical Foundation/AFAR Postdoctoral Fellows in Aging Research Program, 1911
Emerson Charitable Trust Grants, 1918
Emma B. Howe Memorial Foundation Grants, 1920
Eugene M. Lang Foundation Grants, 1957
Eugene McDermott Foundation Grants, 1958
ExxonMobil Foundation Malaria Grants, 1968
Eye-Bank for Sight Restoration and Fight for Sight Summer Student Research Fellowship, 1969
Fairlawn Foundation Grants, 1974
FAMRI Clinical Innovator Awards, 1975
FAMRI Young Clinical Scientist Awards, 1976
FIC Research Collaboration Award, 2006
Fidelity Foundation Grants, 2007
Fight for Sight-Streilein Foundation for Ocular Immunology Research Award, 2012
Fight for Sight Grants-in-Aid, 2013
Fight for Sight Post-Doctoral Awards, 2014
Fight for Sight Summer Student Fellowships, 2015
Fishman Family Foundation Grants, 2020
Flinn Foundation Grants Programs, 2024
Forrest C. Lattner Foundation Grants, 2035
Foundation of the American Thoracic Society Research Grants, 2044
Francis C. Wood Scholar in Residence Program, 2047
Frank Stanley Beveridge Foundation Grants, 2051
Frederick Gardner Cottrell Foundation Grants, 2054
Fritz B. Burns Foundation Grants, 2057
Fund for Scientific Research--Flanders Grants, 2089
George Foundation Grants, 2115
Gerber Foundation Grants, 2126
Gil and Dody Weaver Foundation Grants, 2139
GPP Charles E. Culpeper Foundation Biomedical Pilot Projects Initiative Grants, 2159
Greenspun Family Foundation Grants, 2173
Gruber Foundation Int'l Research Award, 2180
H.A. and Mary K. Chapman Charitable Grants, 2193
Harold R. Bechtel Testamentary Charitable Trust Grants, 2202
Harry A. and Margaret D. Towsley Found Grants, 2205
Healthcare Foundation of New Jersey Grants, 2222
Hearst Foundations Health Grants, 2225
HHMI-NIBIB Interfaces Initiative Grants, 2248
HHMI-NIH Cloister Research Scholars, 2249
HHMI Biomedical Research Grants for Int'l Scientists Infectious Diseases and Parasitology, 2251
HHMI Biomedical Research Grants for International Scientists in Canada and Latin America, 2252
HHMI Biomedical Research Grants for International Scientists in the Baltics, Central and Eastern Europe, Russia, and Ukraine, 2253
HHMI Gilliam Fellowships for Advanced Study, 2254
HHMI Grants and Fellowships Programs, 2255
HHMI International Research Scholars Program, 2258
HHMI Physician-Scientist Early Career Award, 2260
HHMI Research Training Fellowships, 2262
HHMI Undergraduate Science Education Grants for Professors, 2263
Hilda and Preston Davis Foundation Grants, 2266
HomeBanc Foundation Grants, 2275
IBRO Europe Regional Grants for Western European Workshops and Symposia, 2324
J.B. Reynolds Foundation Grants, 2449
J.W. Kieckhefer Foundation Grants, 2452
James H. Cummings Foundation Grants, 2460
Jeffress Awards in Interdisciplinary Research, 2481
John R. Oishei Foundation Grants, 2505
John S. Dunn Research Foundation Grants, 2506
Johnson & Johnson Corporate Contributions, 2508
Joseph Alexander Foundation Grants, 2513
Josiah W. and Bessie H. Kline Foundation Grants, 2517
Kavli Foundation Research Grants, 2533
Kenneth T. and Eileen L. Norris Found Grants, 2554
Kettering Fund Grants, 2555
Klarman Family Foundation Grants in Eating Disorders Research, 2558
Komen Greater NYC Clinical Research Enrollment Grants, 2568
Komen Greater NYC Community Breast Health Grants, 2569
Lance Armstrong Foundation Research Grants, 2585
Leon Lowenstein Foundation Grants, 2603
Lillian S. Wells Foundation Grants, 2614
Lucile Horton Howe and Mitchell B. Howe Foundation Grants, 2624
Lumpkin Family Found Healthy People Grants, 2629
Lymphatic Education & Research Network Additional Support Grants for NIH-funded F32 Postdoctoral Fellows, 2632
Lymphatic Education & Research Network Postdoctoral Fellowships, 2633
M.J. Murdock Charitable Trust General Grants, 2641
Margaret T. Morris Foundation Grants, 2659
Mary Gates Research Scholarships, 2670
Mary K. Chapman Foundation Grants, 2673
Mary Kay Foundation Cancer Research Grants, 2674
Mary S. and David C. Corbin Foundation Grants, 2675
MBL Associates Summer Fellowships, 2694
MBL Baxter Postdoctoral Summer Fellowship, 2695
MBL Burr & Susie Steinbach Summer Fellow, 2696
MBL Eugene and Millicent Bell Fellowships, 2699
MBL Evelyn & Mel Spiegal Summer Fellowship, 2700
MBL Frank R. Lillie Summer Fellowship, 2701
MBL Frederik B. and Betsy G. Bang Summer Fellowships, 2702
MBL Gruss Lipper Family Foundation Summer Fellowship, 2704
MBL Herbert W. Rand Summer Fellowship, 2706
MBL James E. and Faith Miller Memorial Summer Fellowship, 2707
MBL John M. Arnold Award, 2708
MBL Lucy B. Lemann Mem Summer Fellowship, 2710
MBL William Townsend Porter Summer Fellowships for Minority Investigators, 2716
McCarthy Family Foundation Grants, 2717
McCombs Foundation Grants, 2718
McLean Contributionship Grants, 2727
Mericos Foundation Grants, 2742
Meyer and Pepa Gold Family Foundation Grants, 2746
MGN Family Foundation Grants, 2754
MLA Donald Lindberg Research Fellowship, 2785
Monfort Family Foundation Grants, 2794
Mt. Sinai Health Care Foundation Academic Medicine and Bioscience Grants, 2803
Murphy Institute Center for Ethics and Public Affairs Faculty Fellowships, 2805
National MPS Society Grants, 2880
NCI Ruth L. Kirschstein National Research Service Award Institutional Research Training Grants, 2931
NEI Ruth L. Kirschstein National Research Service Award Short-Term Institutional Research Training Grants, 2963
NFL Charities Medical Grants, 2994
NHLBI Ancillary Studies in Clinical Trials, 3015
NHLBI Biomedical Research Training Program for Individuals from Underrepresented Groups, 3019
NHLBI Independent Scientist Award, 3028
NHLBI Intramural Research Training Awards, 3029
NHLBI Investigator Initiated Multi-Site Clinical Trials, 3030
NHLBI Lymphatics in Health and Disease in the Digestive, Urinary, Cardiovascular and Pulmonary Systems, 3032
NHLBI Mentored Career Development Award to Promote Faculty Diversity/Re-Entry in Biomedical Research, 3035
NHLBI Mentored Clinical Scientist Research Career Development Awards, 3036
NHLBI Mentored Patient-Oriented Research Career Development Awards, 3037
NHLBI Mentored Quantitative Research Career Development Awards, 3038
NHLBI Midcareer Investigator Award in Patient-Oriented Research, 3040
NHLBI Pathway to Independence Awards, 3045
NHLBI Research Demonstration and Dissemination Grants, 3051
NHLBI Ruth L. Kirschstein National Research Service Award Institutional Research Training Grants, 3056
NHLBI Ruth L. Kirschstein National Research Service Awards for Individual Postdoctoral Fellows, 3057
NHLBI Ruth L. Kirschstein National Research Service Awards for Individual Predoctoral Fellowships to Promote Diversity in Health-Related Research, 3058
NHLBI Ruth L. Kirschstein National Research Service Awards for Individual Predoctoral MD/PhD Fellows and Other Dual Degree Fellows, 3059
NHLBI Ruth L. Kirschstein National Research Service Awards for Individual Senior Fellows, 3060
NHLBI Ruth L. Kirschstein National Research Service Award Short-Term Institutional Research Training Grants, 3061
NHLBI Short-Term Research Education Program to Increase Diversity in Health-Related Research, 3062
NHLBI Summer Internship Program in Biomedical Research, 3064
NIAAA Independent Scientist Award, 3078
NIAAA Mentored Clinical Scientist Research Career Development Awards, 3080
NIA Alzheimer's Disease Core Centers Grants, 3084
NIAID Independent Scientist Award, 3102
NIA Independent Scientist Award, 3103
NIA Mentored Clinical Scientist Research Career Development Awards, 3108
NIDCD Mentored Clinical Scientist Research Career Development Award, 3131
NIDDK Ancillary Studies to Major Ongoing NIDDK and NHLBI Clinical Research Studies, 3136
NIDDK Erythropoiesis: Components and Mechanisms Grants, 3153
NIEHS Hazardous Materials Worker Health and Safety Training Grants, 3190
NIEHS Hazmat Training at Doe Nuclear Weapons Complex Grants, 3191
NIEHS Independent Scientist Award, 3192
NIGMS Ruth L. Kirschstein National Research Service Awards for Individual Predoctoral Fellows in PharmD/PhD Grants, 3200
NIH Academic Research Enhancement Awards, 3204
NIH Biomedical Research on the International Space Station, 3207
NIH Earth-Based Research Relevant to the Space Environment, 3213
NIH Exceptional, Unconventional Research Enabling Knowledge Acceleration Grants, 3215
NIH Exploratory/Developmental Research Grant, 3216
NIH Exploratory Innovations in Biomedical Computational Science & Tech Grants, 3217
NIH Independent Scientist Award, 3221
NIH Innovations in Biomedical Computational Science and Technology Grants, 3222
NIH Innovations in Biomedical Computational Science and Technology Initiative Grants for SBIR, 3223
NIH Innovations in Biomedical Computational Science and Technology Initiative Grants for STTR, 3224
NIH Mentored Clinical Scientist Devel Award, 3225
NIH Mentored Research Scientist Development Awards, 3227

Jackson Foundation Grants, 2453
Johnson & Johnson Corporate Contributions, 2508
Joint Center for Poverty Research Grants, 2511
Lillian S. Wells Foundation Grants, 2614
Lucile Horton Howe and Mitchell B. Howe Foundation Grants, 2624
Musgrave Foundation Grants, 2807
NIAAA Mechanisms of Alcohol and Drug-Induced Pancreatitis Grants, 3079
NIDA Mentored Clinical Scientist Research Career Development Awards, 3129
NIDA Pilot and Feasibility Studies in Preparation for Drug Abuse Prevention Trials, 3130
NIH Research on Sleep and Sleep Disorders, 3236
NIMH Early Identification and Treatment of Mental Disorders in Children and Adolescents Grants, 3252
Olin Corporation Charitable Trust Grants, 3739
OSF Soros Justice Advocacy Fellowships, 3850
OSF Soros Justice Media Fellowships, 3851
Patrick and Anna M. Cudahy Fund Grants, 3873
Peter F. McManus Charitable Trust Grants, 3903
Rayonier Foundation Grants, 4034
Sid W. Richardson Foundation Grants, 4253
Sioux Falls Area Community Foundation Community Fund Grants (Unrestricted), 4262
Sioux Falls Area Comm Foundation Spot Grants, 4263
SSRC Drugs, Security & Democracy Fellows, 4426
Stackpole-Hall Foundation Grants, 4442
Victor E. Speas Foundation Grants, 4810

Dyslexia
Talbert and Leota Abrams Foundation, 4492

Early Childhood Development
Better Way Foundation Grants, 1338
NCFR Innovation Grants, 2908
Piper Trust Healthcare & Med Research Grants, 3972

Early Childhood Education
AASF Lydia Donaldson Tutt-Jones Memorial Research Grant, 203
ATA Educational Trust Grants - Curriculum Resource or Research Project, 1253
Better Way Foundation Grants, 1338
BMW of N America Charitable Contributions, 1353
Harry A. and Margaret D. Towsley Found Grants, 2205
IRA Jeanne S. Chall Research Fellowship, 2410
Joyce Foundation Education Grants, 2519
McCune Charitable Foundation Grants, 2719
McGregor Fund Grants, 2725
Paul G. Allen Family Foundation Grants, 3876
Phi Upsilon Omicron Alumni Research Grant, 3940
Phi Upsilon Omicron Florence Fallgatter Distinguished Service Award, 3943
Reinberger Foundation Grants, 4051
Third Millennium Foundation Fellowships, 4504

Earth Science Education
Alfred P. Sloan Foundation Deep Carbon Observatory Grants, 547
DAAD Research Internships in Science and Engineering, 1704
DAAD RISE Professional Internships, 1706
NOAA Spherical Display Systems for Earth System Science - Installations and Content Development Grants, 3407
Northrop Grumman Foundation Grants, 3429
W.M. Keck Foundation Science and Engineering Research Grants, 4824

Earth Sciences
Abdus Salam ICTP Schools, Conferences, Workshops and Extended Research Workshops Grants, 238
Alfred P. Sloan Foundation Deep Carbon Observatory Grants, 547
American Chemical Society Petroleum Research Fund Undergraduate Research Grants, 695
AMNH Graduate Student Fellowships, 815
ASCE Trent R. Dames and William W. Moore Engineering Fellowship, 1043
Carnegie Postdoctoral Research Fellowships, 1436
Chrysalis Scholarships, 1559
Cooperative Institute for Research in Environmental Sciences Visiting Fellows Program, 1643
DOE Basic Energy Sciences Grants, 1772
DOI Earthquake Hazards Reduction Res Grants, 1821
DOI Geological Research and Data Acquisition Grants, 1822
Earthwatch Institute Student Fellowships, 1890
GEM PhD Science Fellowships, 2106
GSA Research Grants, 2185
IIE KAUST Graduate Fellowships, 2353
ISGC Graduate Fellowships, 2430
NASA Goddard Visiting Fellowships, 2838
NATO Advanced Research Workshops Program, 2894
NATO Advanced Study Institutes Program, 2895
NATO Collaborative Linkage Grants, 2896
NATO Expert Visits Program Grants, 2897
NOAA Spherical Display Systems for Earth System Science - Installations and Content Development Grants, 3407
Northrop Grumman Foundation Grants, 3429
NSF Antarctic Research Grants, 3479
NSF Arctic Nat Science Grants in Earth Sciences, 3485
NSF Arctic Natural Sciences Program Grants, 3488
NSF CISE Accelerating Discovery in Science and Engineering Through Petascale Simulations and Analysis Grants, 3521
NSF CISE Cross-Cutting Program Grants, 3532
NSF CISE Increasing the Participation and Advancement of Women in Academic Science and Engineering Careers Grants, 3534
NSF Continental Dynamics Research Grants, 3543
NSF Cooperative Studies Of The Earth's Deep Interior Grants, 3544
NSF Geophysics Grants, 3571
NSF Hydrologic Sciences Research Grants, 3578
NSF Paleoclimate Grants, 3606
NSF Paleo Perspectives on Climate Grants, 3607
NSF Polar Programs Division Research Grants, 3627
NSF Sediment Geology & Paleobiology Grants, 3646
NSF Tectonics Research Grants, 3658
Research Program at Earthwatch Grants, 4057
Roy J. Carver Charitable Trust Higher Education Grants, 4113
Smithsonian Graduate Student Fellowships, 4277
Smithsonian Minority Visiting Student Awards, 4279
Smithsonian National Air and Space Museum Postdoctoral Earth and Planetary Sciences Fellowship, 4287
Smithsonian Native American Visiting Student Awards, 4292
Smithsonian Postdoctoral Fellowships, 4293
Smithsonian Predoctoral Fellowships, 4295
SRC Natural and Engineering Sciences Grants, 4387
SSHRC Digging into Data Challenge Grants, 4406
SSRC Mellon Mays Predoctoral Research Grants, 4432
Swedish Natural Science Res Council Grants, 4479
Swiss National Science Foundation Fellowships for Advanced Researchers, 4483
Swiss National Science Foundation Fellowships for Prospective Researchers, 4484
Swiss National Science Foundation Marie Heim-Vogtlin Grants, 4485
UNCFSP NASA Harriett G. Jenkins Predoctoral Fellowships, 4563
USDA Decadal and Regional Climate Prediction Using Earth System Models Grants, 4621

Earthquake Engineering
DOI Earthquake Hazards Reduction Res Grants, 1821
NSF George E. Brown Jr. Network for Earthquake Engineering Simulation Research, 3572

Eating Disorders
Davis Foundation Postdoctoral Fellowships in Eating Disorders Research, 1727
Hilda and Preston Davis Foundation Grants, 2266
Klarman Family Foundation Grants in Eating Disorders Research, 2558
NEDA/AED Charron Family Research Grant, 2945
NEDA/AED Joan Wismer Research Grant, 2946
NEDA/AED Tampa Bay Eating Disorders Task Force Award, 2947

Ecology
AAAS Roger Revelle Fellowship in Global Stewardship, 38
Alaska Conservation Foundation Watchable Wildlife Conservation Trust Grants, 498
Alfred P. Sloan Foundation Energy and Environment Grants, 552
American Philosophical Soc Judson Daland Prize, 766
American Philosophical Society Lewis and Clark Fund for Exploration and Field Research, 767
AMNH Lerner-Gray Grants for Marine Research, 816
AT&T Industrial Ecology Faculty Fellowships, 1251
Brainerd Foundation Grants, 1368
Clarence E. Heller Charitable Foundation Grants, 1575
Conservation, Food, and Health Foundation Grants for Developing Countries, 1641
Cooperative Institute for Research in Environmental Sciences Visiting Fellows Program, 1643
DOE Environmental Processes Research Grants, 1790
DOI Rhinoceros and Tiger Conservation Grants, 1824
DOI Upper Mississippi River System Long-Term Resource Monitoring Program Grants, 1825
Edmund Niles Huyck Preserve Graduate and Postgraduate Research Grants, 1894
EPA Temporally Integrated Monitoring of Ecosystems and Long-Term Monitoring Grants, 1942
Franklin Mosher Baldwin Fellowships, 2050
G. Unger Vetlesen Foundation Grants, 2093
Global Change Education Program, 2151
HRF Hudson River Improvement Grants, 2286
HRF New York City Environmental Grants for Newton Creek, 2288
HSBC Contributions Community Grants, 2291
Illinois DNR Youth Recreation Corps Grants, 2361
International Crane Foundation Internships, 2396
IREX Short-Term Travel Grants, 2424
Jenifer Altman Foundation Grants, 2484
John D. and Catherine T. MacArthur Foundation Global Challenges Program Grants, 2498
M.J. Murdock Charitable Trust College Research Program for Life Sciences, 2638
MacArthur Foundation Conservation and Sustainable Development Grants, 2648
MBL Albert & Ellen Grass Summer Fellowships, 2692
MBL Associates Summer Fellowships, 2694
MBL Burr & Susie Steinbach Summer Fellow, 2696
MBL E.E. Just Summer Fellowship for Minority Scientists, 2697
MBL Evelyn & Mel Spiegal Summer Fellowship, 2700
MBL Frank R. Lillie Summer Fellowship, 2701
MBL Gruss Lipper Family Foundation Summer Fellowship, 2704
MBL H. Keffer Hartline and Edward F. MacNichol, Jr. Fellowships, 2705
MBL Herbert W. Rand Summer Fellowship, 2706
MBL James E. and Faith Miller Memorial Summer Fellowship, 2707
MBL Lucy B. Lemann Mem Summer Fellowship, 2710
MBL M.G.F. Fuortes Summer Fellowships, 2711
MBL William Townsend Porter Summer Fellowships for Minority Investigators, 2716
NCAR Postdoctoral Fellowships in the Atmospheric Sciences, 2902
NFWF Pulling Together Initiative Grants, 3007
NOAA Coastal Ocean Program Grants, 3353
NOAA Cooperative Research Program Grants, 3359
NOAA Great Lakes Habitat Restoration Partnership Grants, 3376
NOAA National Estuarine Research Reserve System Graduate Research Fellowships, 3390
Northwest Fund for the Environment Grants, 3431
NSF-CONACyT Collaborative Research Opportunities, 3452
NSF Arctic Nat Science Grants in Earth Sciences, 3485
NSF Biological Oceanography Research Grants, 3502

NSF Doctoral Dissertation Improvement Grants in the Directorate for Biological Sciences, 3554
NSF Ecosystem Science Grants, 3556
NSF Physical Anthropology Research Grants, 3615
NSF Physiology and Ethology Research Grants, 3622
NSF Sediment Geology & Paleobiology Grants, 3646
Orchid Research Fellowships, 3794
Organization for Tropical Studies Fellowships, 3796
Organization for Tropical Studies Pilot Awards for Exploratory Research, 3797
OTS Francis J. Bossuyt Fellowship, 3855
OTS Hovore-Horn Fellowships, 3856
OTS Stanley Smith Horticultural Fellowships, 3857
OTS Undergrad NAPIRE Summer Interns, 3858
Robert and Patricia Switzer Found Fellowships, 4071
Smithsonian Environmental Research Center Graduate Student Fellowships, 4273
Smithsonian Environmental Research Center Internships, 4274
Smithsonian Environmental Research Center Postdoctoral/Senior Fellowships, 4275
Smithsonian Environmental Research Center Predoctoral Fellowships, 4276
Smithsonian Senior Fellowships, 4296
SSRC Mellon Mays Predoctoral Research Grants, 4432

Ecology, Aquatic
Alaska Conservation Foundation Watchable Wildlife Conservation Trust Grants, 498
Great Lakes Fishery Ecological & Biological Fisheries Research to Inform Management Grants, 2169
Great Lakes Protection Fund Grants, 2170
HRF Graduate Fellowships, 2285
HRF New York City Environmental Grants for Newton Creek, 2288
NFWF Shell Marine Habitat Program Grants, 3008
NOAA Ballast Water Management Demonstration Program (RDTE Facility) Grants, 3334
NOAA Ballast Water Management Demonstration Program (Technologies and Practices) Grants, 3335
NOAA Ballast Water Technology Demonstration Program (Research, Development, Testing and Evaluation Facility) Grants, 3336
NOAA Ballast Water Technology Demonstration Program (Treatment Technology Demonstration Project) Grants, 3337
NOAA Ecological Forecasting Grants, 3368
NOAA Financial Assistance for National Centers for Central Coastal Ocean Science, 3372
NOAA General Coral Reef Conservation Grants, 3373
NOAA Great Lakes Habitat Restoration Partnership Grants, 3376
NSF Antarctic Organisms & Ecosystems Grants, 3478
NSF Biotic Surveys and Inventories Grants, 3510
RSC Miroslaw Romanowski Medal, 4122

Ecology, Atmospheric
NSF Arctic System Science Program Grants, 3493
RSC Miroslaw Romanowski Medal, 4122
Smithsonian Environmental Research Center Graduate Student Fellowships, 4273
Smithsonian Environmental Research Center Internships, 4274
Smithsonian Environmental Research Center Postdoctoral/Senior Fellowships, 4275
Smithsonian Environmental Research Center Predoctoral Fellowships, 4276

Ecology, Environmental Education
EPA Southeastern Multi-Media and Geographic Priority Project Grants, 1939
EPA Temporally Integrated Monitoring of Ecosystems and Long-Term Monitoring Grants, 1942
USAID Palestinian Community Infrastructure Development Grants, 4600

Ecology, Terrestrial
Alaska Conservation Foundation Watchable Wildlife Conservation Trust Grants, 498
Global Change Education Program, 2151
NSF Biotic Surveys and Inventories Grants, 3510
NSF Polar Programs Division Research Grants, 3627
RSC Miroslaw Romanowski Medal, 4122
Smithsonian Environmental Research Center Graduate Student Fellowships, 4273
Smithsonian Environmental Research Center Internships, 4274
Smithsonian Environmental Research Center Postdoctoral/Senior Fellowships, 4275
Smithsonian Environmental Research Center Predoctoral Fellowships, 4276

Econometrics/Forecasting
Abdus Salam ICTP Schools, Conferences, Workshops and Extended Research Workshops Grants, 238
Alfred P. Sloan Foundation Economic Implications of the Great Recession Grants, 549
Alfred P. Sloan Foundation Empirical Economic Research Enablers Grants, 550

Economic Development
Achelis Foundation Grants, 274
Alcatel-Lucent Technologies Foundation Grants, 509
Alfred P. Sloan Foundation Behavioral Economics and Household Finance Grants, 544
Alfred P. Sloan Foundation Economic Implications of the Great Recession Grants, 549
Allstate Foundation Safe and Vital Communities Grants, 573
AMI Semiconductors Corporate Grants, 813
Anderson Foundation Grants, 842
Bay and Paul Foundations, Inc Grants, 1307
Bayer Foundation Grants, 1309
Berks County Community Foundation Grants, 1333
Bikes Belong Foundation Research Grants, 1343
Bodman Foundation Grants, 1356
Boeing Company Contributions Grants, 1357
Brico Fund Grants, 1371
Campbell Soup Foundation Grants, 1415
Carl and Eloise Pohlad Family Grants, 1426
Carnegie Corporation of New York Grants, 1432
Charles M & Mary D Grant Foundation Grants, 1527
Charles Stewart Mott Anti-Poverty Grants, 1528
Charles Stewart Mott Foundation Grants, 1529
Dade Community Foundation Grants, 1712
DHHS Independence Demonstration Program, 1757
DOT Transportation Statistics Research Grants, 1850
George Gund Foundation Grants, 2117
Green Diamond Charitable Contributions, 2171
Harold R. Bechtel Testamentary Charitable Trust Grants, 2202
Harry B. and Jane H. Brock Foundation Grants, 2206
Helen Bader Foundation Grants, 2228
HUD Community Outreach Partnership Centers Grants, 2292
Hutchinson Community Foundation Grants, 2310
IIE David L. Boren Fellowships, 2350
Illinois DCEO Coal Revival Grants, 2360
Illinois DNR Youth Recreation Corps Grants, 2361
IREX International Leadership in Education Program Grants, 2420
Jackson Foundation Grants, 2453
John D. and Catherine T. MacArthur Foundation Global Challenges Program Grants, 2498
John Deere Foundation Grants, 2500
Kellogg Institute for International Studies Collaborative Faculty Grants, 2536
Lynde & Harry Bradley Foundation Fellowships, 2634
Lynde and Harry Bradley Foundation Grants, 2635
M.J. Murdock Charitable Trust General Grants, 2641
MacArthur Foundation Conservation and Sustainable Development Grants, 2648
MacDonald-Peterson Foundation Grants, 2649
McCune Charitable Foundation Grants, 2719
Melville Charitable Trust Grants, 2741
MGM Resorts Foundation Community Grants, 2753
NSF Engineering Research Grants, 3562
Percy B. Ferebee Endowment Grants, 3897
RA Kleinhans Fellowship for Research in Tropical Nontimber Forest Products, 4026
Rhode Island Foundation Grants, 4061
Ross Foundation Grants, 4106
Shastri Indo-Canadian Institute Collaborative Research Project Grants, 4220
Shastri Indo-Canadian Institute Partnership Development Seed Grants, 4228
Sioux Falls Area Community Foundation Community Fund Grants (Unrestricted), 4262
Sioux Falls Area Comm Foundation Spot Grants, 4263
SSHRC-IDRC International Partnerships for Sustainable Societies, 4399
SSRC Next Generation Social Sciences in Africa: Dissertation Completion Fellowships, 4434
SSRC Next Generation Social Sciences in Africa: Doctoral Dissertation Proposal Fellowships, 4435
SSRC Next Generation Social Sciences in Africa: Doctoral Dissertation Research Fellowships, 4436
Textron Corporate Contributions Grants, 4501
Thomas and Dorothy Leavey Foundation Grants, 4506
US CRDF Science and Technology Entrepreneurship Program Business Partnership Grant Competition Armenia, 4611
US CRDF STEP Business Partnership Grant Competition Ukraine, 4612
USDA Cooperative Extension System Education Grants, 4620
USDC/NOAA Nature Conservancy Community-based Matching Grants, 4661
USDC Planning and Local Technical Assistance Grants, 4668
W.K. Kellogg Foundation Healthy Kids Grants, 4822
Westinghouse Charitable Giving Grants, 4856
World Bank Graduate Scholarship Program, 4896

Economic History
Alfred P. Sloan Foundation Behavioral Economics and Household Finance Grants, 544
Alfred P. Sloan Foundation Economic Implications of the Great Recession Grants, 549
American Historical Association Fellowship in Aerospace History, 717
Library Company of Philadelphia Prizes, 2608
SSRC Our Shared Past Grants, 4437

Economic Justice
Alfred P. Sloan Foundation Behavioral Economics and Household Finance Grants, 544
Gruber Foundation Prize for Global Justice and Women's Rights, 2182
OSF-Baltimore Community Fellowships, 3833
Shastri Indo-Canadian Institute Collaborative Research Project Grants, 4220

Economic Opportunities
Alfred P. Sloan Foundation Behavioral Economics and Household Finance Grants, 544
Alfred P. Sloan Foundation Economic Analysis of Science and Technology Grants, 548
Alfred P. Sloan Foundation Economic Implications of the Great Recession Grants, 549
Alfred P. Sloan Foundation Empirical Economic Research Enablers Grants, 550
Washington Area Women's Foundation Strategic Opportunity and Partnership Fund Grants, 4831

Economic Science
Alfred P. Sloan Foundation Economic Analysis of Science and Technology Grants, 548
Alfred P. Sloan Foundation Empirical Economic Research Enablers Grants, 550
NSF CISE Accelerating Discovery in Science and Engineering Through Petascale Simulations and Analysis Grants, 3521
NSF CISE Cross-Cutting Program Grants, 3532
NSF CISE Increasing the Participation and Advancement of Women in Academic Science and Engineering Careers Grants, 3534
W.M. Keck Foundation Science and Engineering Research Grants, 4824

Economic Self-Sufficiency
Abington Foundation Grants, 243
Boeing Company Contributions Grants, 1357
Charles Stewart Mott Anti-Poverty Grants, 1528
Eaton Charitable Fund Grants, 1893
Emma B. Howe Memorial Foundation Grants, 1920
FCD Child Development Grants, 1984
Jovid Foundation Grants, 2518
Surdna Found Sustainable Environments Grants, 4471
TIAA-CREF Paul A. Samuelson Award, 4518
U.S. Department of Education Rehabilitation Research Training Centers, 4545
USDA Norman E. Borlaug International Agricultural Science and Technology Fellowships, 4636
Washington Area Women's Foundation Strategic Opportunity and Partnership Fund Grants, 4831

Economic Stimulus
Alfred P. Sloan Foundation Behavioral Economics and Household Finance Grants, 544
Alfred P. Sloan Foundation Economic Implications of the Great Recession Grants, 549
NIH Recovery Act Limited Competition: Academic Research Enhancement Awards, 3230
NIH Recovery Act Limited Competition: Biomedical Research, Development, and Growth to Spur the Acceleration of New Technologies Pilot Program Grants, 3231
NIH Recovery Act Limited Competition: Small Business Catalyst Awards for Accelerating Innovative Research Grants, 3233
NIH Recovery Act Limited Competition: Supporting New Faculty Recruitment to Enhance Research Resources through Biomedical Research Core Centers Grants, 3234

Economic Theory
Alfred P. Sloan Foundation Behavioral Economics and Household Finance Grants, 544
Alfred P. Sloan Foundation Economic Implications of the Great Recession Grants, 549
Alfred P. Sloan Foundation Empirical Economic Research Enablers Grants, 550
Century Foundation Grants, 1489

Economics
AAB Berlin Prize Fellowships, 48
AAR Postdoctoral Fellowships, 201
ACLS Charles A. Ryskamp Research Fellowships, 281
ACLS Frederick Burkhardt Residential Fellowships for Recently Tenured Scholars, 288
Alex C. Walker Educational and Charitable Foundation Grants, 541
Alfred P. Sloan Foundation Behavioral Economics and Household Finance Grants, 544
Alfred P. Sloan Foundation Civic Grants, 545
Alfred P. Sloan Foundation Economic Analysis of Science and Technology Grants, 548
Alfred P. Sloan Foundation Economic Implications of the Great Recession Grants, 549
Alfred P. Sloan Foundation Empirical Economic Research Enablers Grants, 550
Alfred P. Sloan Energy and Environment Grants, 552
Alfred P. Sloan Foundation Research Fellowships, 560
American Chemical Society Award in Analytical Chemistry, 662
Appraisal Institute Education Trust Scholarships, 917
ASA/NSF/BLS Fellowships, 1030
ASA/NSF/Census Research Fellowships, 1031
ASCE Trent R. Dames and William W. Moore Engineering Fellowship, 1043
AT&T Industrial Ecology Faculty Fellowships, 1251
Canada-U.S. Fulbright International Science and Technology Award, 1417
Canada-U.S. Fulbright Senior Specialists Program Grants, 1418
Canada-U.S. Fulbright Traditional Scholars, 1419
Carnegie Corporation of New York Grants, 1432
CDC Steven M. Teutsch Prevention Effectiveness Fellowships, 1467
Century Foundation Grants, 1489
Chapin Hall International Fellowships in Children's Policy Research, 1517
Consortium for Graduate Study in Management Minority Fellowships, 1642
DAAD Study & Internship Program in Germany, 1708
Dean Witter Foundation Grants, 1728
DHHS Native American Programs, 1758
DOE Fusion Energy Science Technology Division Grants, 1796
DOT Transit Planning and Research Grants, 1849
Douglass Institute Postdoctoral Fellowships, 1851
Duke Univ Libraries Mary Lily Research Grants, 1877
Earhart Foundation Institutional Grants, 1888
ESRI Conservation Program Grants, 1952
Ford Motor Company Fund Grants Program, 2033
Fulbright Alumni Initiatives Awards Grants, 2064
Fulbright Distinguished Chairs Program Grants, 2069
Fulbright European Union Scholar-in-Residence Program Grants, 2071
Fulbright New Century Scholars Program Grants, 2075
Fulbright Occasional Lecturer Program Grants, 2076
Fulbright Specialists Program Grants, 2079
Fulbright Traditional Scholar Program in Europe and Eurasia, 2080
Fulbright Traditional Scholar Program in Sub-Saharan Africa, 2081
Fulbright Traditional Scholar Program in the East Asia/Pacific Region, 2082
Fulbright Traditional Scholar Program in the Near East and North Africa Region, 2083
Fulbright Traditional Scholar Program in the South and Central Asia Region, 2084
Fulbright Traditional Scholar Program in the Western Hemisphere, 2085
Fulbright Visiting Specialists Program: Direct Access to the Muslim World, 2087
George S. and Dolores Dore Eccles Grants, 2120
Germanistic Society of America Fellowships, 2128
Gladys Krieble Delmas Foundation Grants for Venetian Research, 2144
Harry Frank Guggenheim Foundation Dissertation Fellowships, 2207
Harvard University John M. Olin Prizes, 2214
Herb Society of America Research and Education Grants, 2243
Herman Kahn Fellowship, 2247
Horowitz Foundation for Social Policy Grants, 2280
Horowitz Found for Social Policy Special Awards, 2281
Horton Research Grants, 2282
Ian N. McKinnon Memorial Fellowship, 2314
IIE Leonora Lindsley Memorial Fellowships, 2354
IREX Edmund S. Muskie Graduate Fellowship, 2417
J.S. Mill Fellowships in Economics for Exceptional Students, 2451
James S. McDonnell Foundation Complex Systems Collaborative Activity Awards, 2465
Japan Foundation Japanese Studies Fellowships, 2476
Japan Foundation New York Japanese Studies Fellowships, 2478
JMO Law and Legal System Grants, 2491
JMO Public Policy Research Grants, 2492
John B. Bene Fellowships in Social Forestry, 2496
Joseph L. Fisher Dissertation Awards, 2515
Jovid Foundation Grants, 2518
Library Company of Philadelphia Barra Foundation International Research Fellowships, 2606
Library Company of Philadelphia Program in Early American Economy and Society Dissertation-Level Fellowships, 2609
Library Company of Philadelphia Program in Early American Economy and Society One-Month Fellowships, 2610
Library Company of Philadelphia Program in Early American Economy and Society Post-Doctoral Research Fellowships, 2611
Maine Community Foundation Keepers Preservation Education Grants, 2651
Marine Corps Historical Center Dissertation Fellowship, 2663
Mellon Post-Doctoral Fellowship in Sustainable Agriculture, 2740
MSI Alden G. Clayton Dissertation Proposal Competition, 2800
NAPM Doctoral Dissertation Grant, 2827
National Council of Farmer Cooperatives Graduate Awards, 2860
NATO Science for Peace Grants, 2898
NCAR Postdoctoral Fellowships in the Atmospheric Sciences, 2902
NCFR Innovation Grants, 2908
NCHS National Center for Health Statistics Postdoctoral Research Appointments, 2910
NCI/NIDA Transdisciplinary Tobacco Use Research Centers Grants, 2912
NEFE Grants, 2948
NFID Adv Vaccinology Course Travel Grant, 2986
NIA Healthy Aging through Behavioral Economic Analyses of Situations Grants, 3099
NOAA Communicating Hurricane Information Grants, 3355
NSF Arctic Social Sciences Grants, 3492
NSF Integrative Graduate Education and Research Traineeship, 3584
NSF Partnerships for Innovation Grants, 3610
NSF SBE Doctoral Dissertation Research Improvement Grants, 3641
PhRMA Foundation Health Outcomes Post Doctoral Fellowships, 3950
PhRMA Foundation Informatics Post Doctoral Fellowships, 3953
Rajiv Gandhi Foundation Grants, 4025
RAND Corporation Graduate Student Summer Associateships, 4030
RAND Corporation Transatlantic Postdoctoral Fellowships for Int'l Relations & Security, 4032
Rayonier Foundation Grants, 4034
Robert A. Levy Fellowship in Law and Liberty, 4070
Robert Schalkenbach Foundation Grants, 4079
RSC Thomas W. Eadie Medal, 4126
Rumsfeld Foundation Graduate Fellowships, 4127
Russell Sage Foundation Grants, 4129
RWJF Investigator Awards in Health Policy Res, 4145
SRC Robert M. Burger Fellowship, 4392
SSHRC Insight Development Grants, 4409
SSHRC Insight Grants, 4410
SSRC Int'l Dissertation Research Fellowships, 4429
Takemi Program in International Health Research and Advanced Training Fellowships, 4491
United States Institute of Peace - Transatlantic Post-Doc Fellowship for International Relations and Security, 4581
Upjohn Institute Dissertation Award, 4591
USDC Market Development Cooperator Program Grants, 4667
WHOI Marine Policy Fellowships, 4861
Young Scientists Summer Program, 4903

Economics Education
3M Foundation Community Giving Grants, 5
Alcoa Foundation Grants, 511
Alex C. Walker Educational and Charitable Foundation Grants, 541
Alfred P. Sloan Foundation Behavioral Economics and Household Finance Grants, 544
Alfred P. Sloan Foundation Economic Implications of the Great Recession Grants, 549
Allstate Foundation Safe and Vital Communities Grants, 573
Andrew W. Mellon/ACLS Dissertation Completion Fellowships, 844
Coca-Cola Foundation Grants, 1593
Dean Witter Foundation Grants, 1728
FMC Foundation Grants, 2029
George W. Codrington Charitable Grants, 2122
McCune Charitable Foundation Grants, 2719
MSI Alden G. Clayton Dissertation Proposal Competition, 2800
NCSS/FASSE/CUFA Inquiry Grant, 2936
NCSS Exemplary Res in Social Studies Award, 2937

NCSS Jean Dresden Grambs Distinguished Career Research in Social Studies Award, 2938
NCSS L Metcalf Exemplary Dissertation Award, 2939
Searle Freedom Trust Grants, 4192
Sid W. Richardson Foundation Grants, 4253

Ecosystems
Brainerd Foundation Grants, 1368
Conservation, Food, and Health Foundation Grants for Developing Countries, 1641
EPA Temporally Integrated Monitoring of Ecosystems and Long-Term Monitoring Grants, 1942
GNOF Gulf Coast Oil Spill Grants, 2153
HRF Mark B. Bain Graduate Fellowships, 2287
Illinois DNR Youth Recreation Corps Grants, 2361
MacArthur Foundation Conservation and Sustainable Development Grants, 2648
Organization for Tropical Studies Fellowships, 3796
Organization for Tropical Studies Pilot Awards for Exploratory Research, 3797
OTS Francis J. Bossuyt Fellowship, 3855
OTS Hovore-Horn Fellowships, 3856
OTS Undergrad NAPIRE Summer Interns, 3858
SARE Large Systems Research Grants - Southern Region, 4172

Editing
American Historical Association J. Franklin Jameson Award In Outstanding Editorial Achievement, 724
DAAD Intensive Language Courses Grants, 1699
NYSCA Individual Artists: Film, Media and New Technology Production Grants, 3706
WHOI Ocean Science Journalism Fellowship, 4862

Education
1976 Foundation Grants, 7
A.L. Mailman Family Foundation Grants, 11
AAAS Science and Technology Policy Fellowships: Health, Education and Human Services, 43
AAG E. Willard and Ruby S. Miller Award in Geography, 172
Aaron & Cecile Goldman Family Found Grants, 200
AAUW Community Action Grants, 213
Abbott Fund Global AIDS Care Grants, 219
Abington Foundation Grants, 243
A Charitable Foundation Grants, 273
Achelis Foundation Grants, 274
ACT Awards, 330
Acuity Charitable Foundation Grants, 333
AERA Minority Fellowships, 343
Ahn Family Foundation Grants, 379
AIChE Charles M.A. Stine Award, 399
AIChE David Himmelblau Award for Innovations in Computer-Based Chem Engineering Education, 405
Akonadi Foundation Anti-Racism Grants, 477
Albert and Margaret Alkek Foundation Grants, 504
Albuquerque Community Foundation Grants, 508
Alces Foundation Grants, 510
Alcoa Foundation Grants, 511
Alcon Foundation Grants Program, 513
Alex Stern Family Foundation Grants, 542
Alice Tweed Tuohy Foundation Grants Program, 569
All Saints Educational Trust Grants, 572
Allstate Foundation Safe and Vital Communities Grants, 573
ALZA Corporate Contributions Grants, 588
American Academy of Arts and Sciences Hellman Fellowships, 612
American Academy of Arts and Sciences Visiting Scholars, 613
American Chemical Society Corporate Associates Seed Grants, 676
American Family Insurance Corporate Giving Program Grants, 709
American Foodservice Charitable Trust Grants, 710
American Historical Association Award for Scholarly Distinction, 714
American Montessori Society Thesis and Dissertation Awards, 758
American Psychiatric Foundation Proposals, 791
American Schlafhorst Foundation Grants, 796
Amgen Foundation Grants, 812
Anderson Foundation Grants, 842
Andrew Family Foundation Opportunity Knocks Grants, 843
Angels Baseball Foundation Grants, 848
Ann Arbor Area Community Foundation Grants, 860
Annenberg Foundation Grants, 862
Ann Peppers Foundation Grants, 863
Anthony R. Abraham Foundation Grants, 869
Applied Biosystems Grants, 916
Arcadia Foundation Grants, 983
Arie and Ida Crown Memorial Grants, 989
Arizona Cardinals Grants, 995
Arizona Public Service Corporate Giving Program Grants, 998
ARO Conference and Symposia Grants, 1007
Arthritis Foundation of Australia Grants, 1022
Arthur and Rochelle Belfer Foundation Grants, 1023
ASGH Award for Excellence in Human Genetics Education, 1066
Ashland Corporate Contributions Grants, 1087
ASID Foundation Grants, 1093
ASID Foundation Joel Polsky Academic Achievement Award, 1095
ASM Undergraduate Research Fellowship, 1186
ASR Fichter Research Grant, 1217
Assisi Foundation of Memphis General Grants, 1221
ASU Graduate College Completion Fellowships, 1235
ATA Doctoral Fellowships in Education, 1252
ATA Education Research Award, 1254
ATA Nadene M. Thomas Grad Research Bursary, 1255
Australasian Institute of Judicial Administration Seed Funding Grants, 1268
Australian Academy of Science Grants, 1269
Axe-Houghton Foundation Grants, 1287
Bailey Family Foundation Grants, 1289
Bank of America Charitable Foundation Matching Gifts, 1297
Barra Foundation Project Grants, 1301
Batchelor Foundation Grants, 1305
Bayer Foundation Grants, 1309
BCBSM Foundation Student Award Program, 1316
Beckman Coulter Community Relations Grants, 1325
Beim Foundation Grants, 1331
Berks County Community Foundation Grants, 1333
Blowitz-Ridgeway Foundation Early Childhood Development Research Award, 1348
Blowitz-Ridgeway Foundation Grants, 1349
BMW of N America Charitable Contributions, 1353
Booth-Bricker Fund Grants, 1359
BPW Career Advancement Scholarships, 1366
Brinson Foundation Grants, 1372
Bristol-Myers Squibb Foundation Health Education Grants, 1378
Burden Trust Grants, 1388
Burton G. Bettingen Grants, 1390
Byron W. and Alice L. Lockwood Grants, 1401
C. Alma Baker Trust Grants, 1403
Cabot Corporation Foundation Grants, 1406
Callaway Golf Company Foundation Grants, 1410
Campbell Soup Foundation Grants, 1415
Canada-U.S. Fulbright International Science and Technology Award, 1417
Canada-U.S. Fulbright Senior Specialists Program Grants, 1418
Canada-U.S. Fulbright Traditional Scholars, 1419
Cape Branch Foundation Grants, 1425
Carl and Eloise Pohlad Family Grants, 1426
Carl and Lily Pforzheimer Foundation Grants, 1427
Carl C. Icahn Foundation Grants, 1428
Carl Gellert and Celia Berta Gellert Foundation Grants, 1429
Carpenter Foundation Grants, 1439
Carrier Corporation Contributions Grants, 1440
Catholic University Life Cycle Institute Research Fellowship, 1444
CCCF Dora M Brown Seminary Scholarships, 1448
CDC Grants for Violence-Related Injury Prevention Research, 1463
Center for the Study of Philanthropy Senior International Fellowships, 1483
Chapin Hall International Fellowships in Children's Policy Research, 1517
Charles H. Price II Family Foundation Grants, 1523
Charles H. Revson Foundation Grants, 1524
Charles Lafitte Foundation Grants, 1525
Charles Stewart Mott Anti-Poverty Grants, 1528
Charles Stewart Mott Foundation Grants, 1529
Chautauqua Region Community Found Grants, 1533
Chestnut Hill Charitable Foundation, Inc Grants, 1538
Chiron Foundation Community Grants, 1553
Christensen Fund Regional Grants, 1554
CIA Undergraduate Internship - Analysis, 1562
Claneil Foundation Grants, 1574
Clarence E. Heller Charitable Foundation Grants, 1575
Clay Foundation Grants, 1579
Cleveland-Cliffs Foundation Grants, 1580
Coca-Cola Foundation Grants, 1593
Colonel Stanley R. McNeil Foundation Grants, 1597
Commonweal Foundation Grants, 1603
Communities Foundation of Texas Grants, 1613
Community Found for Greater Buffalo Grants, 1615
Community Found for Muskegon County Grants, 1617
Community Foundation for S Arizona Grants, 1618
Community Found of E Central Illinois Grants, 1619
Community Foundation of the Virgin Islands Judith A. Towle Environmental Studies Grants, 1621
Compton Foundation International Fellowships, 1624
Compton Foundation Mentor Fellowships, 1625
Connecticut Community Foundation Grants, 1627
Conquer Cancer Foundation of ASCO Comparative Effectiveness Research Professorship in Breast Cancer, 1631
Conquer Cancer Foundation of ASCO Translational Research Professorships, 1639
Cord Foundation Grants, 1645
Corning Incorporated Found Educational Grants, 1646
Coughlin-Saunders Foundation Grants Program, 1648
Cowles Charitable Trust Grants, 1652
Croucher Foundation Grants, 1659
Cruise Industry Charitable Foundation Grants, 1660
Cudd Foundation Grants, 1672
D.F. Halton Foundation Grants, 1692
DAAD Re-invitation Program for Former Scholarship Holders, 1701
DAAD Research Grants, 1702
DAAD Research Grants for Doctoral Candidates and Young Academics and Scientists, 1703
DAAD Research Stays for University Academics and Scientists, 1705
DAAD Scholarships for Artists: Study Visits for Academics, 1707
DAAD Study & Internship Program in Germany, 1708
Dade Community Foundation Grants, 1712
DaimlerChrysler Corporation Fund Grants, 1714
Dana Foundation Science and Health Grants, 1716
Davis Family Foundation Grants, 1726
DEED Student Research Grants and Internships, 1731
Del E. Webb Foundation Grants, 1733
Dell Foundation Open Grants, 1734
Delta Dental Master's Thesis Award, 1737
DeMatteis Family Foundation Grants, 1740
Dept of Ed Safe and Drug-Free Schools and Communities State Grants, 1748
Detroit Lions Charities Grants, 1753
DOE Faculty and Student Teams Grants, 1792
Drs. Bruce and Lee Foundation Grants, 1859
E.L. Wiegand Foundation Grants, 1885
Eaton Charitable Fund Grants, 1893
Edward Bangs Kelley and Elza Kelley Foundation Grants, 1898
Edward W. and Stella C. Van Houten Memorial Fund Grants, 1899
Edwin S. Webster Foundation Grants, 1900
Emma B. Howe Memorial Foundation Grants, 1920
Ensign-Bickford Foundation Grants, 1928
Eugene M. Lang Foundation Grants, 1957
Eugene McDermott Foundation Grants, 1958
Evan and Susan Bayh Foundation Grants, 1960

Evanston Community Foundation Grants, 1962
Ewing Halsell Foundation Grants, 1964
Farmers Insurance Corporate Giving Grants, 1981
FASSE Demonstration Project Grant, 1982
FCD Child Development Grants, 1984
Fetzer Institute Research Grants, 2005
Fidelity Foundation Grants, 2007
Fishman Family Foundation Grants, 2020
Fluor Foundation Grants, 2028
FMC Foundation Grants, 2029
Fordham Prizes, 2032
Ford Motor Company Fund Grants Program, 2033
Forrest C. Lattner Foundation Grants, 2035
Foundation for Enhancing Communities Grants, 2038
Frances C. and William P. Smallwood Foundation Grants, 2045
Frank Stanley Beveridge Foundation Grants, 2051
Frederick Douglass Summer Scholars Fellowship, 2053
Fritz B. Burns Foundation Grants, 2057
Fulbright Alumni Initiatives Awards Grants, 2064
Fulbright Distinguished Chairs Program Grants, 2069
Fulbright English Teaching Assistantships, 2070
Fulbright European Union Scholar-in-Residence Program Grants, 2071
Fulbright International Education Administrators Seminar Program Grants, 2073
Fulbright New Century Scholars Program Grants, 2075
Fulbright Occasional Lecturer Program Grants, 2076
Fulbright Specialists Program Grants, 2079
Fulbright Traditional Scholar Program in Europe and Eurasia, 2080
Fulbright Traditional Scholar Program in Sub-Saharan Africa, 2081
Fulbright Traditional Scholar Program in the East Asia/Pacific Region, 2082
Fulbright Traditional Scholar Program in the Near East and North Africa Region, 2083
Fulbright Traditional Scholar Program in the South and Central Asia Region, 2084
Fulbright Traditional Scholar Program in the Western Hemisphere, 2085
Fulbright Visiting Specialists Program: Direct Access to the Muslim World, 2087
G. Unger Vetlesen Foundation Grants, 2093
General Motors Foundation Grants, 2107
Genuardi Family Foundation Grants, 2109
George A Ohl Jr. Foundation Grants, 2112
George B. Storer Foundation Grants, 2113
George Frederick Jewett Foundation Grants, 2116
George Gund Foundation Grants, 2117
George Kress Foundation Grants, 2119
George W. Codrington Charitable Grants, 2122
Georgia Power Foundation Grants, 2125
Gerber Foundation Grants, 2126
Getty Research Institute Nonresidential Grants - Collaborative Research Grant, 2135
Getty Research Institute Nonresidential Grants - Curatorial Research Fellowships, 2136
Gil and Dody Weaver Foundation Grants, 2139
Gilroy and Lillian P. Roberts Charitable Foundation Grants, 2142
Greater Saint Louis Community Found Grants, 2164
Greater Worcester Community Foundation Jeppson Memorial Fund for Brookfield Grants, 2165
Green Diamond Charitable Contributions, 2171
GUITS Post-Doctoral Summer Travel-Research Grants, 2189
H.A. and Mary K. Chapman Charitable Grants, 2193
H.J. Heinz Company Foundation Grants, 2195
Harold Alfond Foundation Grants, 2200
Harold R. Bechtel Charitable Remainder Uni-Trust Grants, 2201
Harry A. and Margaret D. Towsley Found Grants, 2205
Harry B. and Jane H. Brock Foundation Grants, 2206
Harry Frank Guggenheim Foundation Dissertation Fellowships, 2207
Harry W. Bass, Jr. Foundation Grants, 2210
Harvard Postdoctoral Fellowships on Education, 2211
Harvard University Graduate School of Education Fellowships and Awards, 2213
Helen Bader Foundation Grants, 2228
Helen V. Brach Foundation Grants, 2231
Henry Luce Foundation Special Projects in Higher Education Grants, 2236
Herbert H. and Grace A. Dow Found Grants, 2241
HHMI Biomedical Research Grants for Int'l Scientists Infectious Diseases and Parasitology, 2251
HHMI Biomedical Research Grants for International Scientists in Canada and Latin America, 2252
HHMI Biomedical Research Grants for International Scientists in the Baltics, Central and Eastern Europe, Russia, and Ukraine, 2253
HHMI Initiative for Colleges Grants, 2256
HHMI Initiative for Universities Grants, 2257
HHMI Precollege Outreach Initiative for Biomedical Research Institutions Grants, 2261
Hilda and Preston Davis Foundation Grants, 2266
Hilton Hotels Corporate Giving Program Grants, 2268
Hirtzel Memorial Foundation Grants, 2269
Hoglund Foundation Grants, 2271
HomeBanc Foundation Grants, 2275
HSBC Contributions Community Grants, 2291
HUD Community Outreach Partnership Centers Grants, 2292
Human Source Foundation Grants, 2297
Hutchinson Community Foundation Grants, 2310
IIE David L. Boren Scholarships, 2351
IRA Teacher as Researcher Grants, 2414
Ireland Family Foundation Grants, 2415
IREX International Leadership in Education Program Grants, 2420
IREX University Administration Support, 2425
ISGC Informal Education Grants, 2431
ITEA Maley/FTE Scholarship, 2435
Ittleson Foundation Mental Health Grants, 2438
J.W. Kieckhefer Foundation Grants, 2452
Jackson Foundation Grants, 2453
Jacob and Hilda Blaustein Foundation Grants, 2454
Jacob and Valeria Langeloth Foundation Grants, 2455
James Graham Brown Foundation Fellowships, 2459
Jewish Federation of Metropolitan Chicago Scholarships, 2486
JM Foundation Grants, 2490
John Deere Foundation Grants, 2500
John G. Duncan Trust Grants, 2501
John P. McGovern Foundation Grants, 2504
John R. Oishei Foundation Grants, 2505
Johnson & Johnson Corporate Contributions, 2508
John Stauffer Charitable Trust Grants, 2509
John W. Alden Trust Grants, 2510
Josiah W. and Bessie H. Kline Foundation Grants, 2517
Joyce Foundation Education Grants, 2519
Kavli Foundation Research Grants, 2533
Kosciuszko Foundation Tuition Scholarships, 2574
Kovler Family Foundation Grants, 2575
Lady Davis Fellowships, 2578
Leahi Fund, 2599
Leon Lowenstein Foundation Grants, 2603
Lucile Horton Howe and Mitchell B. Howe Foundation Grants, 2624
Ludwick Family Foundation Grants, 2625
Luther I. Replogle Foundation Grants, 2631
Lynde & Harry Bradley Foundation Fellowships, 2634
Lynde and Harry Bradley Foundation Grants, 2635
M.J. Murdock Charitable Trust General Grants, 2641
Mabel Y. Hughes Charitable Trust Grants, 2647
Macquarie Bank Foundation Grants, 2650
Marcia and Otto Koehler Foundation Grants, 2656
Marion Gardner Jackson Charitable Trust Grants, 2664
Martin Luther King Jr. Graduate Memorial Scholarships, 2668
Mary K. Chapman Foundation Grants, 2673
Mattel Children's Foundation Grants, 2685
Mattel International Grants Program, 2686
Max and Victoria Dreyfus Foundation Grants, 2687
Maximilian E. and Marion O. Hoffman Foundation Grants, 2688
McCarthy Family Foundation Grants, 2717
McCombs Foundation Grants, 2718
McCune Charitable Foundation Grants, 2719
McGregor Fund Grants, 2725
McLean Contributionship Grants, 2727
Mead Johnson Nutritionals Med Educ Grants, 2734
Meadows Foundation Grants, 2736
Melville Charitable Trust Grants, 2741
Mericos Foundation Grants, 2742
MGM Resorts Foundation Community Grants, 2753
MGN Family Foundation Grants, 2754
Miller Brewing Corporate Contributions Program Grants, 2774
Millstone Foundation Grants, 2776
Monfort Family Foundation Grants, 2794
Musgrave Foundation Grants, 2807
NAGC Masters and Specialists Award, 2821
NASA Marslink Initiative, 2841
NASA New Aeronautics Competition for University Students, 2842
National Academy of Education Spencer Postdoctoral Fellowships, 2845
National Education Reporting Fellowships, 2861
National Housing Endowment Challenge/Build/Grow Grant, 2877
Nelda C. and H.J. Lutcher Stark Found Grants, 2967
NFL Charities Medical Grants, 2994
NFWF National Whale Conservation Grants, 3004
NFWF Nature Restoration Trust Grants, 3005
NFWF Puget Sound Marine Conservation Fund Grants, 3006
NIH School Interventions to Prevent Obesity, 3244
NIMH Short Courses in Neuroinformatics, 3254
NOAA Chesapeake Bay Office B-WET (Bay Watershed Education and Training) Grants, 3339
NOAA Chesapeake Bay Office Cooperative Research Program Grants, 3340
NOAA Chesapeake Bay Office Fisheries Science Program Grants, 3341
NOAA Chesapeake Bay Office Non-Native Oyster Research Grants, 3342
NOAA Chesapeake Bay Office Submerged Aquatic Vegetation Culture and Restoration Grants, 3343
NOAA Cooperative Unit for Fisheries Education and Research Internships, 3360
NOAA Faculty and Student Intern Research, 3371
NOAA Miami Lab Summer Internships, 3382
NOAA National Climatic Data Center Interns, 3387
Norcliffe Foundation Grants, 3415
Nordson Corporation Foundation Grants, 3416
Northwestern Mutual Foundation Grants, 3430
Northwood University/Alden B. Dow Creativity Center Summer Residency Program, 3432
NRA Foundation Grants, 3437
NSF Antarctic Artists and Writers Grants, 3473
NSF Arctic Research and Education Grants, 3489
NSF CreativeIT Grants, 3547
NSF Education & Interdisciplinary Res Grants, 3557
NSF Innovation and Organizational Change, 3581
NSF Integrative Graduate Education and Research Traineeship, 3584
NSF Partnerships for Innovation Grants, 3610
NYSCA Museum: Project Support Grants, 3707
OAH Friend of History Award, 3718
Ohio Learning Network Grants, 3734
OISE Muriel & Danny Fung Grad Scholarships, 3735
OISE William Waters Scholarships in Urban Education, 3736
Oklahoma City Community Grants, 3737
OM Association Outstanding Dissertation Awards Program, 3741
Organic Farming Research Foundation Grants, 3795
OSF-Baltimore Community Fellowships, 3833
OSF Doctoral Fellows Georgia & Kyrgyzstan, 3839
OSF Palestinian Rule of Law Fellowships, 3848
Park Foundation Grants, 3870
Patrick and Anna M. Cudahy Fund Grants, 3873
Paul G. Allen Family Foundation Grants, 3876
Paul P. Fidler Research Grants, 3878
Peabody Foundation Grants, 3886
Peacock Foundation Grants, 3887
Penn Graduate School of Education Dean's Fellowships, 3891

Penn Graduate School of Education Foreign Language Area Studies Fellowship, 3892
Penn Graduate School of Education Pre-Doctoral Training in Interdisciplinary Methods for Field-Based Research in Education, 3893
Penn Graduate School of Education Research Assistantships, 3894
Percy B. Ferebee Endowment Grants, 3897
Perkin Fund Grants, 3898
Perpetual Trust for Charitable Giving Grants, 3900
Phi Delta Kappa Outstanding Doctoral Dissertation Awards, 3936
Pi Lambda Theta Research Grants, 3966
Pinkerton Foundation Grants, 3967
Plan for Social Excellence Grants, 3976
Playboy Foundation Grants, 3980
Praxair Foundation Grants, 3986
Price Family Charitable Fund Grants, 3987
Prudential Foundation Education Grants, 4002
R.C. Baker Foundation Grants, 4006
RACGP Grieve Memorial Award, 4014
RACGP Rex Walpole Travelling Fellowship, 4016
Radcliffe Institute Carol K. Pforzheimer Student Fellowships, 4018
Rajiv Gandhi Foundation Grants, 4025
Rathmann Family Foundation Grants, 4033
Rayonier Foundation Grants, 4034
Regenstrief General Internal Medicine - General Pediatrics Research Fellowship Program, 4045
Rehabilitation Nursing Found Research Grants, 4050
Research Corporation Cottrell Scholars Awards, 4053
RGK Foundation Grants, 4060
Rhode Island Foundation Grants, 4061
Richard and Rhoda Goldman Fund Grants, 4063
Richard and Susan Smith Family Found Grants, 4064
Richard Davoud Donchian Foundation Grants, 4066
Robert Bowne Foundaion Edmund A. Stanley, Jr. Research Grants, 4073
Rockwell International Corporate Trust Grants, 4093
Rohm and Haas Company Grants, 4098
Ronald McDonald House Charities Grants, 4101
Rosalynn Carter Institute John and Betty Pope Fellowships, 4103
Ross Foundation Grants, 4106
Royal Caribbean Cruises Ocean Fund, 4108
Roy J. Carver Charitable Trust Medical and Science Research Grants, 4114
RRF General Program Grants, 4115
RSC Sir John William Dawson Medal, 4125
Russell Sage Foundation Grants, 4129
Ruth and Vernon Taylor Foundation Grants, 4130
San Diego Women's Foundation Grants, 4164
Sarkeys Foundation Grants, 4181
Schomburg Center Scholars-in-Residence Fellows, 4187
Self Foundation Grants, 4197
Sensient Technologies Foundation Grants, 4199
Shell Oil Company Foundation Grants, 4233
SHRM Dissertation Awards, 4248
Sidgmore Family Foundation Grants, 4252
South Florida Student Shark Program, 4330
Spencer Foundation Education Dissertation Fellowships, 4334
Spencer Foundation Major Research Grants, 4335
Spencer Foundation Small Research Grants, 4337
SRC Educational Science Doctoral Grants, 4355
SRC Educational Science Grants, 4356
SRC Educational Science Grants for Invitation of Visiting Scientists, 4357
SRC Educational Science Grants for Travels to Foreign Institutions as a Guest Scientist, 4358
SRC Educational Science Junior Position Grants, 4359
SRC Educational Science Lang Checking Grants, 4360
SRC Educational Science Medium-Cost Scientific Equipment Grants, 4361
SRC Educational Science Network Grants, 4362
SRC Educational Science Postdoc Fellowships, 4363
SRC Educational Science Project Res Grants, 4364
SRC Educational Science Travel Grants, 4365
SRC Education Science Grants for Postdoctoral Positions in Sweden, 4366
SRC Medical Science Grants for Postdoctoral Positions in Sweden, 4381
SSHRC Insight Development Grants, 4409
SSHRC Insight Grants, 4410
Sun Academic Excellence Grants, 4466
Sun Microsystems Foundation Summer Academic Enrichment Grants, 4468
T. James Kavanagh Foundation Grants, 4489
Third Millennium Foundation Fellowships, 4504
Thomas Austin Finch, Sr. Foundation Grants, 4507
Thomas C. Ackerman Foundation Grants, 4508
Thompson Charitable Foundation Grants, 4511
UNESCO Fellowships, 4567
Ursula Thrush Peace Seed Grants, 4592
USAID Palestinian Community Infrastructure Development Grants, 4600
Vancouver Foundation Grants and Community Initiatives, 4807
Vanguard Public Foundation Grant Funds, 4808
W.C. Griffith Foundation Grants, 4819
W. C. Griffith Foundation Grants, 4818
Wallace Foundation Grants, 4829
Wayne and Gladys Valley Foundation Grants, 4835
Weizmann Institute of Science Postdoctoral Fellowships, 4837
Welch Foundation Endowed Chairs, 4840
Welch Foundation Research Grants, 4842
Westinghouse Charitable Giving Grants, 4856
Weyerhaeuser Company Foundation Grants, 4857
Willary Foundation Grants, 4870
William G. and Helen C. Hoffman Found Grants, 4876
William G. McGowan Charitable Fund Grants, 4877
William N. and Myriam Pennington Foundation Grants, 4879
Woodward Fund Grants, 4895
World Bank Graduate Scholarship Program, 4896
Xerox Foundation Grants, 4899
Yale University Beinecke Library Research Fellowships for Yale Grad & Professional School Students, 4902

Education Reform

Achelis Foundation Grants, 274
Bodman Foundation Grants, 1356
Boeing Company Contributions Grants, 1357
Carnegie Corporation of New York Grants, 1432
Charles Stewart Mott Foundation Grants, 1529
ETS Sylvia Taylor Johnson Minority Fellowship in Educational Measurement, 1955
Fordham Prizes, 2032
GNOF IMPACT Grants for Education, 2154
IREX University Administration Support, 2425
Joyce Foundation Education Grants, 2519
NFIE Found Learning & Leadership Grants, 2992
Richard Davoud Donchian Foundation Grants, 4066
Spencer Foundation Major Research Grants, 4335
William and Flora Hewlett Foundation Education Grants, 4873

Education and Work

Radcliffe Institute Carol K. Pforzheimer Student Fellowships, 4018
Xerox Foundation Grants, 4899

Educational Administration

AERA Research Grants, 344
Harvard University Administrative Fellowships, 2212
Harvard University Graduate School of Education Fellowships and Awards, 2213
IREX University Administration Support, 2425
Lilly Endowment Teacher Creativity Fellowships, 2616
Lumina Foundation for Education Grants, 2627
Meadows Foundation Grants, 2736
Nellie Mae Education Foundation District-Level Change Grants, 2968
Nellie Mae Education Foundation State Level Systems Change Grants, 2971
Pennsylvania State University Education Graduate Assistantships, 3895
RSC Sir John William Dawson Medal, 4125
Wallace Foundation Grants, 4829

Educational Evaluation/Assessment

A.O. Smith Community Grants, 12
AERA Minority Fellowships, 343
AIR/IES-NCES Postdoctoral Policy Fellowships in Washington, 466
AIR National Summer Data Policy Institute Fellowships, 469
ETS Sylvia Taylor Johnson Minority Fellowship in Educational Measurement, 1955
IBRO Latin America Regional Funding for Neuroscience Schools, 2327
IRA Albert J. Harris Awards, 2405
NAEF Teacher Incentive Grants, 2815
Nellie Mae Education Foundation Public Understanding Grants, 2969
Nellie Mae Education Foundation Research and Development Grants, 2970
Nellie Mae Education Foundation State Level Systems Change Grants, 2971
OSF Thomas J. Alexander Fellowships, 3853
RCPSC Royal College Fellowship for Studies in Medical Education, 4040
TIRF Doctoral Dissertation Grants, 4520

Educational Finance

AERA Research Grants, 344
Allstate Foundation Safe and Vital Communities Grants, 573
TIAA-CREF Paul A. Samuelson Award, 4518

Educational Humanities

National Academy of Education Spencer Postdoctoral Fellowships, 2845

Educational Instruction

ACLS Fellowships, 287
American Montessori Society Thesis and Dissertation Awards, 758
ASM International Fellowship for Latin America and the Caribbean, 1175
ASM Int'l Fellowships for Asia and Africa, 1176
Bayer Clinical Scholarship Award, 1308
BCBSM Foundation Student Award Program, 1316
ITEA Maley/FTE Scholarship, 2435
Motorola Foundation Innovation Generation Collaborative Grants, 2797
Nellie Mae Education Foundation Public Understanding Grants, 2969
Nellie Mae Education Foundation State Level Systems Change Grants, 2971
Parkersburg Area Community Foundation Arts in Education Special Grants, 3869
San Diego Women's Foundation Grants, 4164
Ursula Thrush Peace Seed Grants, 4592
USDA Distance Education Grants for Higher Education Institutions in Insular Areas, 4622

Educational Planning/Policy

Achelis Foundation Grants, 274
AERA Dissertation Grants, 342
AERA Minority Fellowships, 343
AERA Research Grants, 344
AIR/IES-NCES Postdoctoral Policy Fellowships in Washington, 466
Bodman Foundation Grants, 1356
Bristol-Myers Squibb Foundation Health Education Grants, 1378
Center for Education of Women Visiting Scholar Research Grants, 1475
DOE Albert Einstein Distinguished Educator Fellowship Program, 1770
ETS Predoctoral Summer Program, 1954
GNOF IMPACT Grants for Education, 2154
Harvard Postdoctoral Fellowships on Education, 2211
Herman Kahn Fellowship, 2247
IBRO Latin America Regional Funding for Neuroscience Schools, 2327
IREX Regional Policy Symposium Grants, 2423
IREX University Administration Support, 2425
IREX US Embassy Policy Specialist Grants, 2426

Japan Foundation Center for Global Partnership Institutional Grants, 2473
Joyce Foundation Education Grants, 2519
Lumina Foundation for Education Grants, 2627
National Assessment of Educational Progress Visiting Scholar Program, 2846
Nellie Mae Education Foundation District-Level Change Grants, 2968
Nellie Mae Education Foundation Research and Development Grants, 2970
Nellie Mae Education Foundation State Level Systems Change Grants, 2971
Pennsylvania State University Education Graduate Assistantships, 3895
Searle Freedom Trust Grants, 4192
SRC Educational Science Postdoc Fellowships, 4363
Wallace Foundation Grants, 4829
William T. Grant Foundation Distinguished Fellows Awards, 4880

Educational Psychology
ETS Predoctoral Summer Program, 1954
Institute of Noetic Sciences Research Internships, 2386
Ittleson Foundation Mental Health Grants, 2438
RCPSC Royal College Fellowship for Studies in Medical Education, 4040

Educational Technology
Alcatel-Lucent Technologies Foundation Grants, 509
Arizona Commission on the Arts Education Projects Grants, 996
Bay and Paul Foundations, Inc Grants, 1307
Corning Incorporated Found Educational Grants, 1646
Dept of Ed Star Schools Program Grants, 1749
Educational Communications & Tech Awards, 1895
Frederick Douglass Summer Scholars Fellowship, 2053
Hilda and Preston Davis Foundation Grants, 2266
ITEA Maley/FTE Scholarship, 2435
Nellie Mae Education Foundation State Level Systems Change Grants, 2971
NSF CISE Educational Innovation Grants, 3533
RCPSC Royal College Fellowship for Studies in Medical Education, 4040
Sun Microsystems Foundation Summer Academic Enrichment Grants, 4468

Educational Testing/Measurement
AERA Research Grants, 344
ETS Predoctoral Summer Program, 1954
National Assessment of Educational Progress Visiting Scholar Program, 2846
NBME Stemmler Medical Education Research Fund Grants, 2899
SRC Educational Science Lang Checking Grants, 4360
TIRF Doctoral Dissertation Grants, 4520

Educational Theory
AERA Minority Fellowships, 343
Harvard Postdoctoral Fellowships on Education, 2211
IREX University Administration Support, 2425
Nellie Mae Education Foundation Research and Development Grants, 2970
Pennsylvania State University Education Graduate Assistantships, 3895
Spencer Foundation Education Dissertation Fellowships, 4334
Spencer Foundation Small Research Grants, 4337
SRC Educational Science Lang Checking Grants, 4360
SRC Educational Science Postdoc Fellowships, 4363

Elder Abuse
Peacock Foundation Grants, 3887

Elderly
AAA Foundation for Traffic Safety Grants, 24
Alex Stern Family Foundation Grants, 542
American Schlafhorst Foundation Grants, 796
AMI Semiconductors Corporate Grants, 813
Ann Arbor Area Community Foundation Grants, 860
Ann Peppers Foundation Grants, 863
Arthur and Rochelle Belfer Foundation Grants, 1023
Avon Products Foundation Grants, 1275
BCBSNC Foundation Grants, 1318
Bodenwein Public Benevolent Foundation Grants, 1355
Boston Foundation Grants, 1361
Burden Trust Grants, 1388
Callaway Golf Company Foundation Grants, 1410
Charles H. Farnsworth Trust Grants, 1521
Chatlos Foundation Grants Program, 1532
Commonwealth Fund Harkness Fellowships in Health Care Policy and Practice, 1606
Commonwealth Fund Quality of Care for Frail Elders Grants, 1610
Comprehensive Health Education Found Grants, 1622
ConocoPhillips Grants Program, 1629
DHHS Special Programs for the Aging Training, Research, and Discretionary Projects and Programs Grants, 1761
Donald W. Reynolds Foundation Aging and Quality of Life Grants, 1832
Edward W. and Stella C. Van Houten Memorial Fund Grants, 1899
Elisabeth Severance Prentiss Foundation Grants, 1905
Fannie E. Rippel Foundation Grants, 1978
George A Ohl Jr. Foundation Grants, 2112
George Foundation Grants, 2115
Hearst Foundations Health Grants, 2225
Herbert H. and Grace A. Dow Found Grants, 2241
Hilda and Preston Davis Foundation Grants, 2266
Hoglund Foundation Grants, 2271
Mericos Foundation Grants, 2742
MGM Resorts Foundation Community Grants, 2753
Musgrave Foundation Grants, 2807
NAMM Foundation Music Research Grants, 2825
NCI Stages of Breast Development: Normal to Metastatic Disease Grants, 2932
Norcliffe Foundation Grants, 3415
Nuffield Foundation Law and Society Grants, 3667
Nuffield Foundation Open Door Grants, 3669
Oppenstein Brothers Foundation Grants, 3792
R.C. Baker Foundation Grants, 4006
RAND Corporation Fellowships in Population Studies and the Study of Aging, 4029
Reuter Foundation Grants, 4059
Richard and Rhoda Goldman Fund Grants, 4063
Rosalynn Carter Institute John and Betty Pope Fellowships, 4103
RRF General Program Grants, 4115
Schramm Foundation Grants, 4188
Sid W. Richardson Foundation Grants, 4253
STTI Environment of Elder Care Nursing Research Grants, 4460
Thomas Austin Finch, Sr. Foundation Grants, 4507
TIAA-CREF Institute Research Grant Program, 4517
Victor E. Speas Foundation Grants, 4810
Westinghouse Charitable Giving Grants, 4856
William N. and Myriam Pennington Foundation Grants, 4879

Electoral Systems
APSA Warren E. Miller Fellowship in Electoral Politics, 949
Carnegie Corporation of New York Grants, 1432
Ford Foundation Peace and Social Justice Grants, 2031

Electric Power
CIA Undergraduate Co-Op Program - Engineer/ Program Management Engineer, 1561
DHHS Native American Programs, 1758
DOE Combustion Systems Research Grants, 1778

Electric Power Generation
ASME James N. Landis Medal, 1136
ASME Prime Movers Committee Award, 1156
DEED Grants, 1730
DEED Student Research Grants and Internships, 1731
DOE Electric Energy Systems Research Grants, 1782
DOE Reliable Electric Power Delivery Grants, 1812
DOE Vehicle and Engine Research and Development Grants, 1819

Electric Powered Systems
ASME Prime Movers Committee Award, 1156

Electrochemistry
American Chemical Society ANYL Award in Electrochemistry, 642
USDD Electrochemical Materials Grants, 4721

Electromagnetics
DOE Reliable Electric Power Delivery Research Grants, 1812
NSF Geophysics Grants, 3571
ONR Electronics Division Research Grants, 3759
ONR Ionosphere Interactions Research Grants, 3765
ONR Marine Meteorology and Atmospheric Effects Research Grants, 3769
ONR Medical Science and Technology Division Grants, 3772
ONR Radio Frequency Sensor Processing Research Grants, 3778
ONR Radio Frequency Solid State Devices Applied Research Grants, 3779
ONR Young Investigator Program Grants, 3788
USDD Electromagnetic Materials Grants, 4722
USDD National Consortium for Measurement and Signature Intelligence Research Grants, 4759
USDD Naval Research Facilities Grants, 4761
USDD Sensor Innovative Research Grants, 4791

Electronic Media
Reinberger Foundation Grants, 4051
SSHRC Digging into Data Challenge Grants, 4406
USDD Electronic Warefare Grants, 4723
WHOI Ocean Science Journalism Fellowship, 4862

Electronic Warfare
USDD Electronic Warefare Grants, 4723
USDD Sensor Innovative Research Grants, 4791

Electronics
ASME Allan Kraus Thermal Management Medal, 1104
SRC Natural and Engineering Sciences Grants, 4387
USDD ARO Broad Agency Announcement, 4686
USDD Basic Research Challenge: Carbon Molecular Electronics Grants, 4688
USDD Office-Wide Broad Agency Announcement Microsystems Technology Office Grant, 4768
USDD Research Interest of the Air Force Office of Scientific Research Grants, 4784

Electronics, Solid State
ONR Electronic Combat Technology Grants, 3757
ONR Electronics, Sensors, and Network Research Grants, 3758
ONR Superconducting Electronics Grants, 3785
USDD Electro-Optical/Infrared Sensor Processing Grants, 4720
USDD Nanoscale Electronics Grants, 4758
USDD RF Semiconductors, RF Solid State Amplifiers and Wide Bandgap Materials Grants, 4786

Electronics/Electrical Engineering
American Society of Naval Engineers Scholarship, 800
ARO Minimum Energy Low Power Electronics Research Contracts, 1011
ARO Theoretical Physics and Nonlinear Phenomena Research Contracts, 1017
ASME Calvin W. Rice Lecture Award, 1112
ASME Kate Gleason Award, 1137
ASME Nancy DeLoye Fitzroy and Roland V. Fitzroy Medal, 1149
ASME Old Guard Early Career Award, 1150
ASME Patrick J. Higgins Award, 1152
ASME Prime Movers Committee Award, 1156
ASME Ralph Coats Roe Medal, 1158
ASME Robert Henry Thurston Lecture Award, 1160
ASME Student Section Advisor Award, 1166
ASME Y.C. Fung Young Investigator Award, 1173
ASU Graduate College Science Foundation Arizona Fellowships in Science and Engineering, 1245

AT&T Bell Labs Graduate Research Program, 1250
IEEE Electrical History Fellowship, 2339
IIE General Electric Foundation TECH Awards, 2352
IIE KAUST Graduate Fellowships, 2353
Motorola Foundation Grants, 2796
Nanyang Technological University Lee Kuan Yew Postdoctoral Fellowship, 2826
NSF-NIST Interaction in Chemistry, Materials Research, Molecular Biosciences, Bioengineering, and Chemical Engineering, 3453
NSF Engineering Research Grants, 3562
NSF Materials Research Grants, 3593
ONR Electronic Combat Technology Grants, 3757
ONR Electronics, Sensors, and Network Research Grants, 3758
ONR Electronics Division Research Grants, 3759
SHOT Brooke Hindle Postdoctoral Fellowships, 4238
SHOT Eugene S. Ferguson Prize, 4240
SHOT IEEE Life Members Prize in Elec Hist, 4241
SHOT NASA Fellowship, 4246
SPIE Dennis Gabor Award, 4340
SPIE Gold Medal of the Society Award, 4343
SRC Franco Cerrina Fellowship, 4367
US CRDF CGP-Climate Change and Energy Competition For Joint U.S.-Ukrainian Research Teams, 4607
USDD RF Semiconductors, RF Solid State Amplifiers and Wide Bandgap Materials Grants, 4786

Electrophysiology
Duke University Clinical Cardiac Electrophysiology Fellowships, 1871

Elementary Education
3M Foundation Community Giving Grants, 5
A.L. Mailman Family Foundation Grants, 11
A.O. Smith Community Grants, 12
AASF Lydia Donaldson Tutt-Jones Memorial Research Grant, 203
Acuity Charitable Foundation Grants, 333
Affymetrix Corporate Contributions Grants, 355
Alcatel-Lucent Technologies Foundation Grants, 509
Andrew Family Foundation Opportunity Knocks Grants, 843
APS/Phoenix Suns Energy and Environmental Minigrants, 919
Archer Daniels Midland Foundation Grants, 985
Astronomical Society of the Pacific Las Cumbres Amateur Outreach Award, 1230
Ball Brothers Foundation General Grants, 1291
Bay and Paul Foundations, Inc Grants, 1307
BMW of N America Charitable Contributions, 1353
Boeing Company Contributions Grants, 1357
Booth-Bricker Fund Grants, 1359
Boston Foundation Grants, 1361
Bristol-Myers Squibb Foundation Health Education Grants, 1378
Carnegie Corporation of New York Grants, 1432
Charles H. Price II Family Foundation Grants, 1523
Charles Lafitte Foundation Grants, 1525
Christensen Fund Regional Grants, 1554
Clarence E. Heller Charitable Foundation Grants, 1575
Coca-Cola Foundation Grants, 1593
Colonel Stanley R. McNeil Foundation Grants, 1597
Commonweal Foundation Community Assistance Grants, 1602
Commonweal Foundation Grants, 1603
ConocoPhillips Grants Program, 1629
DaimlerChrysler Corporation Fund Grants, 1714
Dept of Ed Jacob K. Javits Gifted and Talented Students Education Program-National Research and Development Center, 1746
Dept of Ed Star Schools Program Grants, 1749
DOE Albert Einstein Distinguished Educator Fellowship Program, 1770
Dorr Foundation Grants, 1842
Earthwatch Institute Student Fellowships, 1890
Eaton Charitable Fund Grants, 1893
Elizabeth McGraw Foundation Grants, 1908
Eugene McDermott Foundation Grants, 1958
Frances L. and Edwin L. Cummings Memorial Fund Grants, 2046
George Foundation Grants, 2115
George Frederick Jewett Foundation Grants, 2116
GNOF IMPACT Grants for Education, 2154
Greenspun Family Foundation Grants, 2173
Harold Alfond Foundation Grants, 2200
Helen Bader Foundation Grants, 2228
HSBC Contributions Community Grants, 2291
Hutchinson Community Foundation Grants, 2310
IRA Esther Zolt Teacher Research Award, 2408
IRA Helen M. Robinson Dissertation Res Award, 2409
IRA Jeanne S. Chall Research Fellowship, 2410
IRA Steven A. Stahl Research Grants, 2413
Joyce Foundation Education Grants, 2519
Leon Lowenstein Foundation Grants, 2603
Lilly Endowment Teacher Creativity Fellowships, 2616
Luther I. Replogle Foundation Grants, 2631
Margaret & James A. Elkins Jr. Found Grants, 2658
McGraw-Hill Companies Community Grants, 2724
Meadows Foundation Grants, 2736
Mericos Foundation Grants, 2742
Meyer Memorial Trust Responsive Grants, 2748
MGM Resorts Foundation Community Grants, 2753
Motorola Foundation Grants, 2796
Musgrave Foundation Grants, 2807
NEH Fellowships, 2956
Nellie Mae Education Foundation District-Level Change Grants, 2968
Nellie Mae Education Foundation Public Understanding Grants, 2969
Nellie Mae Education Foundation State Level Systems Change Grants, 2971
Nordson Corporation Foundation Grants, 3416
NSF Presidential Awards for Excellence in Mathematics and Science Teaching, 3631
NSF Presidential Awards for Excellence in Science, Math, and Engineering Mentoring, 3632
Oppenstein Brothers Foundation Grants, 3792
Phi Upsilon Omicron Geraldine Clewell Masters Fellowship, 3946
Rajiv Gandhi Foundation Grants, 4025
Reinberger Foundation Grants, 4051
Richard and Susan Smith Family Found Grants, 4064
Ronald McDonald House Charities Grants, 4101
Sage Foundation Grants, 4158
Schramm Foundation Grants, 4188
Sid W. Richardson Foundation Grants, 4253
Sioux Falls Area Community Foundation Community Fund Grants (Unrestricted), 4262
Sioux Falls Area Comm Foundation Spot Grants, 4263
Sony Corporation of America Corporate Philanthropy Grants, 4329
Sun Academic Excellence Grants, 4466
Sun Microsystems Foundation Summer Academic Enrichment Grants, 4468
Sunny and Abe Rosenberg Foundation Grants, 4469
T. James Kavanagh Foundation Grants, 4489
The Ray Charles Foundation Grants, 4502
Third Millennium Foundation Fellowships, 4504
Victor E. Speas Foundation Grants, 4810
Westinghouse Charitable Giving Grants, 4856
William and Flora Hewlett Foundation Education Grants, 4873
William Blair and Company Foundation Grants, 4875

Embryology
Carnegie Developmental Bio Research Fellows, 1433
MBL E.E. Just Summer Fellowship for Minority Scientists, 2697

Emergency Preparedness
CDC Experience Appl Epidemiology Fellowships, 1460
CSTE CDC/CSTE Applied Epidemiology Fellowships, 1664
DOI Earthquake Hazards Reduction Res Grants, 1821
GNOF Gulf Coast Oil Spill Grants, 2153
ONR Casualty Prevention Grants, 3747
ORISE Student Research Participation at the Agency for Toxic Substances and Disease Registry, 3828

Emergency Programs
Alcatel-Lucent Technologies Foundation Grants, 509
Blowitz-Ridgeway Foundation Grants, 1349
Coughlin-Saunders Foundation Grants Program, 1648
DHHS Emergency Medical Services for Children Program, 1755
DOI Earthquake Hazards Reduction Res Grants, 1821
GNOF Gulf Coast Oil Spill Grants, 2153
Helen V. Brach Foundation Grants, 2231
McGregor Fund Grants, 2725
Oklahoma City Community Grants, 3737
ONR Casualty Care and Management Research Grants, 3746
Reinberger Foundation Grants, 4051

Emergency Services
Campbell Soup Foundation Grants, 1415
Carpenter Foundation Grants, 1439
Carrier Corporation Contributions Grants, 1440
DHHS Emergency Medical Services for Children Program, 1755
GNOF Gulf Coast Oil Spill Grants, 2153
McCune Foundation Human Services Grants, 2722
Piper Trust Healthcare & Med Research Grants, 3972
Reinberger Foundation Grants, 4051
Rhode Island Foundation Grants, 4061
Union Pacific Foundation Health and Human Services Grants, 4569

Emission Control
Energy Foundation Power Grants, 1924
US CRDF CGP ETF 2009 Energy Research Competition For Joint U.S.-Estonian Research Teams, 4609
USDA Integrated Research Education and Extension Grants - Organic Transitions, 4630

Emotion
NIMH Basic and Translational Research in Emotion Grants, 3250

Emotional/Mental Health
Delta Air Lines Foundation Health and Wellness Grants, 1736
DTE Energy Foundation Health and Human Services Grants, 1865
Gerber Foundation Grants, 2126
GNOF Gulf Coast Oil Spill Grants, 2153
Herbert & Adrian Woods Foundation Grants, 2239
Nick Traina Foundation Grants, 3128
Peter F. McManus Charitable Trust Grants, 3903
USDD Workpl Violence in the Military Grants, 4805

Emotionally Disturbed
DHHS Comprehensive Community Mental Health Services Grants for Children with Serious Emotional Disturbances, 1754
Herbert & Adrian Woods Foundation Grants, 2239

Employment Opportunity Programs
Achelis Foundation Grants, 274
Bodman Foundation Grants, 1356
Boston Foundation Grants, 1361
Chiron Foundation Community Grants, 1553
Connecticut Community Foundation Grants, 1627
Cruise Industry Charitable Foundation Grants, 1660
Frank Stanley Beveridge Foundation Grants, 2051
Helen Bader Foundation Grants, 2228
Illinois DNR Youth Recreation Corps Grants, 2361
Johnson & Johnson Corporate Contributions, 2508
Jovid Foundation Grants, 2518
Joyce Foundation Employment Grants, 2520
M.D. Anderson Foundation Grants, 2637
NIDRR Field-Initiated Projects, 3189
NOAA Graduate Sciences Program Grants, 3375
Pinkerton Foundation Grants, 3967
Reuter Foundation Grants, 4059
Sioux Falls Area Community Foundation Community Fund Grants (Unrestricted), 4262
Sioux Falls Area Comm Foundation Spot Grants, 4263

Starr Foundation Grants, 4447
Textron Corporate Contributions Grants, 4501
USDA Alaska Native-Serving and Native Hawaiian-Serving Institutions Education Grants, 4615
USDA Organic Agriculture Research and Extension Initiative, 4637
USDC Planning and Local Technical Assistance Grants, 4668

Employment/Unemployment Studies
ASA/NSF/BLS Fellowships, 1030
DHHS Welfare Reform Research, Evaluations, and National Studies Grants, 1762
Joint Center for Poverty Research Grants, 2511
Joyce Foundation Employment Grants, 2520
Russell Sage Foundation Grants, 4129
Upjohn Institute Dissertation Award, 4591

Endangered Language
Endangered Language Fund Grants, 1921
NSF/NEH Documenting Endangered Languages Project Grants and Fellowships: data, infrastructure and computational methods, 3455

Endangered Species
Turtle Conservation Fund Grants, 4535

Endocrine Disorders
NIDDK Devel of Disease Biomarkers Grants, 3145
Pfizer ASPIRE Worldwide Endocrine Young Investigator Grants, 3918
Pfizer Australia Paediatric Endocrine Care Research Grants, 3921

Endocrinology
AHAF National Glaucoma Research Grants, 372
AMA Foundation Seed Grants for Research, 601
Jenifer Altman Foundation Grants, 2484
NIDDK Co-Activators and Co-Repressors in Gene Expression Grants, 3141
NIDDK Collaborative Interdisciplinary Team Science in Diabetes, Endocrinology and Metabolic Diseases Grants, 3142
NIDDK Diabetes, Endocrinology, and Metabolic Diseases NRSAs--Individual, 3146
NIDDK Endoscopic Clinical Research Grants In Pancreatic And Biliary Diseases, 3150
NIDDK Pilot and Feasibility Clinical Research Grants in Diabetes, Endocrine & Metabolic Diseases, 3170
NIDDK Proteomics: Diabetes, Obesity, And Endocrine, Digestive, Kidney, Urologic, And Hematologic Diseases, 3174
NIDDK Seeding Collaborative Interdisciplinary Team Science in Diabetes, Endocrinology and Metabolic Diseases Grants, 3179
NIDDK Small Grants for Underrepresented Minority Scientists in Diabetes and Digestive and Kidney Diseases, 3183
NIDDK Translational Research for the Prevention and Control of Diabetes and Obesity, 3186
NSF Physiology and Ethology Research Grants, 3622
Pfizer ASPIRE Worldwide Endocrine Young Investigator Grants, 3918

Endowments
Hearst Foundations Health Grants, 2225
Margaret T. Morris Foundation Grants, 2659

Energy
AIChE Transport and Energy Processes Division Award, 427
American Chemical Society Petroleum Research Fund Doctoral New Investigator Grants, 690
American Chemical Society Petroleum Research Fund Undergraduate Faculty Sabbatical Grants, 693
American Chemical Society Petroleum Research Fund Undergraduate New Investigator Grants, 694
ANL Faculty Research Participation Program, 855
ANL Graduate Thesis Parts Res Appointments, 856
ANL Laboratory-Graduate Res Appointments, 857
APS/Phoenix Suns Energy and Environmental Minigrants, 919
ASME George Westinghouse Medals, 1124
AT&T Industrial Ecology Faculty Fellowships, 1251
Aviation Research Grants, 1273
Canada Graduate Scholarships and NSERC Postgraduate Scholarships, 1420
Clarence E. Heller Charitable Foundation Grants, 1575
Colorado Clean Energy Solar Innovation Grants, 1599
Conservation, Food, and Health Foundation Grants for Developing Countries, 1641
DOE Advanced Energy Projects/Laboratory Technology Research Grants, 1768
DOE Advanced Nuclear Energy Systems Basic Research Grants, 1769
DOE Atomic, Molecular, and Optical Sciences Research Grants, 1771
DOE Basic Energy Sciences Grants, 1772
DOE Catalysis and Chemical Transformations Research Grants, 1773
DOE Chemical Energy and Chemical Engineering Research Grants, 1774
DOE Chemical Physics Research Grants, 1775
DOE Chemical Sciences Research Grants, 1776
DOE Coal Liquefaction Research Grants, 1777
DOE Energy Biosciences Research Grants, 1784
DOE Energy Conversion Tech Research Grants, 1786
DOE Energy from Municipal Waste Grants, 1787
DOE Energy Research Analyses Grants, 1788
DOE Engineering Research Grants, 1789
DOE Environmental Processes Research Grants, 1790
DOE Experimental Program to Stimulate Competitive Research, 1791
DOE Fusion Energy Science Division Grants, 1795
DOE Fusion Energy Science Technology Division Grants, 1796
DOE Geosciences Research Grants, 1797
DOE Heavy Element Chem Research Grants, 1798
DOE High-Energy Physics Research Grants, 1799
DOE Industrial Materials for the Future Grant, 1800
DOE Inventions and Innovation Grant Program, 1802
DOE Materials Sciences and Engineering Research Grants, 1803
DOE Mathematical, Information, and Computational Sciences Grants, 1804
DOE Nuclear Physics Research Grants, 1807
DOE On-Line Temperature Measurement Instrumentation for Gasification Process Control Development Grants, 1808
DOE Photochem & Radiation Research Grants, 1809
DOE Program for Ecosystem Research, 1810
DOE Programs in Fusion Energy Sciences, 1811
DOE Separations and Analysis Research Grants, 1813
DOE State Energy Program Special Projects, 1815
DOE University Coal Research Grants, 1816
DOI Geological Research and Data Acquisition Grants, 1822
Energy Foundation Climate Grants, 1923
General Motors Foundation Grants, 2107
Gilbert F. White Postdoctoral Fellowships, 2140
Gruber Foundation Weizmann Institute Awards, 2184
Honor the Earth Grants, 2279
Ian N. McKinnon Memorial Fellowship, 2314
IIE General Electric Foundation TECH Awards, 2352
Illinois Clean Energy Community Foundation Renewable Energy Grants, 2358
IREX Short-Term Travel Grants, 2424
Joseph L. Fisher Dissertation Awards, 2515
Link Foundation Energy Doctoral Fellowships, 2617
McCune Charitable Foundation Grants, 2719
NSF-NIST Interaction in Chemistry, Materials Research, Molecular Biosciences, Bioengineering, and Chemical Engineering, 3453
NWO Added Value Plus for Soc Sciences Grants, 3672
ORISE Fusion Energy Postdoctoral Research, 3806
ORISE Higher Education Research Experiences at Oak Ridge National Laboratory for Faculty, 3808
ORISE Laboratory Technology Program, 3809
ORISE Oak Ridge National Laboratory Postdoctoral Research Associates Program, 3815
ORISE Savannah River Site Faculty Research Participation Program, 3823
Shastri Indo-Canadian Institute Collaborative Research Project Grants, 4220
Surdna Found Sustainable Environments Grants, 4471
USDD Dielectric Films for Capacitors Grants, 4719
William and Flora Hewlett Foundation Environmental Grants, 4874
Young Scientists Summer Program, 4903

Energy Conservation
AIChE Transport and Energy Processes Division Award, 427
ASME Frank Kreith Energy Award, 1120
Bedding Plants Foundation Research Grants, 1330
Bullitt Foundation Grants, 1384
DEED Grants, 1730
DEED Student Research Grants and Internships, 1731
DHHS Native American Programs, 1758
DOE Competitive Financial Assistance for Renewable Energy and Energy Efficiency Programs, 1779
DOE Energy Biosciences Research Grants, 1784
DOE Energy Conservation Research and Development Grants, 1785
DOE Energy from Municipal Waste Grants, 1787
DOE Industrial Materials for the Future Grant, 1800
DOE Program for Ecosystem Research, 1810
DOE State Energy Program Special Projects, 1815
Energy Foundation Buildings Grants, 1922
Energy Foundation Climate Grants, 1923
Energy Foundation Power Grants, 1924
Energy Foundation Transportation Grants, 1925
Illinois Clean Energy Community Foundation Renewable Energy Grants, 2358
NIST Small Business Innovation Res Grants, 3273
NSF Partnership for Advancing Technologies in Housing Grants, 3609
Rohm and Haas Company Grants, 4098
Sioux Falls Area Community Foundation Community Fund Grants (Unrestricted), 4262
Sioux Falls Area Comm Foundation Spot Grants, 4263

Energy Conversion
AIChE Donald Q. Kern Award, 406
American Chemical Society Green Chem Grants, 679
DOE Energy Conversion Tech Research Grants, 1786
Donald Q. Kern Award in Heat Transfer or Energy Conversion, 1831
Illinois DCEO Coal Research Grants, 2359
Illinois DCEO Coal Revival Grants, 2360
USDD Dielectric Films for Capacitors Grants, 4719

Energy Economics
AT&T Industrial Ecology Faculty Fellowships, 1251
Colorado Clean Energy Solar Innovation Grants, 1599
Energy Foundation Power Grants, 1924
Energy Foundation Transportation Grants, 1925
McCune Charitable Foundation Grants, 2719

Energy Education
DOE Program for Ecosystem Research, 1810
Engineering Information Foundation Grants, 1927
McCune Charitable Foundation Grants, 2719

Energy Engineering
ASME Ralph Coats Roe Medal, 1158
DOE Coal Liquefaction Research Grants, 1777
DOE Industrial Materials for the Future Grant, 1800
McCune Charitable Foundation Grants, 2719

Energy Planning/Policy
AAAS Science and Technology Policy Fellowships: Energy, Environment, and Agriculture, 41
DOE Energy-Related Inventions Grants, 1783
DOE Geosciences Research Grants, 1797
DOE Industrial Materials for the Future Grant, 1800
Energy Foundation Climate Grants, 1923
Energy Foundation Power Grants, 1924
Energy Foundation Transportation Grants, 1925

Illinois Clean Energy Community Foundation Renewable Energy Grants, 2358
Japan Foundation Center for Global Partnership Institutional Grants, 2473

Energy Storage Systems
SSHRC-APC Automotive Partnership Canada, 4396

Energy Utilization
ARO Minimum Energy Low Power Electronics Research Contracts, 1011
DOE Energy Conservation Research and Development Grants, 1785
DOE Energy Conversion Tech Research Grants, 1786
HSBC Contributions Community Grants, 2291
McCune Charitable Foundation Grants, 2719

Energy, Exploration/Reserves
American Chemical Society Green Chem Grants, 679
NOAA Ocean Exploration Omnibus: Ocean Exploration Grants, 3398

Energy, Fossil
ASME James N. Landis Medal, 1136
DOE Combustion Systems Research Grants, 1778
DOE Fossil Energy Research & Devel Grants, 1794
Energy Foundation Climate Grants, 1923
Illinois DCEO Coal Research Grants, 2359
Illinois DCEO Coal Revival Grants, 2360
McCune Charitable Foundation Grants, 2719
US CRDF CGP ETF 2009 Energy Research Competition For Joint U.S.-Estonian Research Teams, 4609

Energy, Geothermal
Illinois Clean Energy Community Foundation Renewable Energy Grants, 2358
McCune Charitable Foundation Grants, 2719

Energy, Hydro
DOE Initial H-Prize Competition for Breakthrough Advances in Materials for Hydrogen Storage, 1801
Honor the Earth Grants, 2279
Hydrogen Student Design Contest, 2311

Energy, Solar
ASME Frank Kreith Energy Award, 1120
Colorado Clean Energy Solar Innovation Grants, 1599
DOE Million Solar Roofs Program, 1805
Illinois Clean Energy Community Foundation Renewable Energy Grants, 2358
McCune Charitable Foundation Grants, 2719
NSF Solar Terrestrial Research Grants, 3652
US CRDF CGP ETF 2009 Energy Research Competition For Joint U.S.-Estonian Research Teams, 4609

Energy, Wind
DOE Competitive Financial Assistance for Renewable Energy and Energy Efficiency Programs, 1779
DOE State Energy Program Special Projects, 1815
Illinois Clean Energy Community Foundation Renewable Energy Grants, 2358
McCune Charitable Foundation Grants, 2719

Engineering
AAAS Award for Public Engagement with Science, 26
AAAS Award for Science Diplomacy, 27
AAAS Award for Scientific Freedom and Responsibility, 28
AAAS Entry Point Internships, 30
AAAS Kavli Science Journalism Awards, 33
AAAS Newcomb Cleveland Prize, 36
AAAS Science and Technology Policy Fellowships: Congressional, 39
AAAS Science and Technology Policy Fellowships: Global Health and Development, 42
AAAS Science Prize for Inquiry-Based Instruction, 45
AACC International Graduate Fellowships, 91
AACR Team Science Award, 148
AAES Engineering Journalism Award, 150
AAUW American Dissertation Fellowships, 209
AAUW Selected Professions Fellowships, 217
Abdus Salam ICTP Postgraduate Diploma Grants, 235
Abdus Salam ICTP Training and Research in Italian Laboratories Grants, 240
Acuity Charitable Foundation Grants, 333
AEG Foundation Legget Fund Grants, 341
AFRL/NRC Resident Research Associateship, 358
AIChE Computational Molecular Science and Engineering Forum Graduate Student Award, 401
AIChE Research Excellence In Sustainable Engineering Award, 424
AIChE W. David Smith, Jr. Graduate Student Paper Award, 428
Alexander von Humboldt Foundation Fellowships for Experienced Researchers, 516
Alexander von Humboldt Foundation Fellowships for Postdoctoral Researchers, 517
Alexander von Humboldt Foundation Georg Forster Research Award, 525
Alexander von Humboldt Foundation Max Planck Research Award, 532
Alexander von Humboldt Foundation Philipp Franz von Siebold Award for Japanese Researchers, 534
Alfred P. Sloan Foundation Research Fellowships, 560
American-Scandinavian Foundation Short-Term Training Program Grants, 606
American Australian Association USA to Australia Fellowships, 620
American Automatic Control Council Donald P. Eckman Award, 622
American Chemical Society Award in Separations Science and Technology, 671
American Historical Association Fellowship in Aerospace History, 717
American Indian College Fund STEM Leadership Fellowships, 745
American Mathematical Society AAAS Mass Media Fellowships, 751
AMS/Industry Government Graduate Fellowships, 817
AMS/Industry Minority Scholarships, 818
AMS Grad Fellowships in History of Science, 824
AMS Undergraduate Scholarships, 837
ANL Faculty Research Leave Program, 854
ANL Faculty Research Participation Program, 855
ANL Graduate Thesis Parts Res Appointments, 856
ANL Laboratory-Graduate Res Appointments, 857
Applied Biosystems Grants, 916
ARM Faculty Fellowships, 1001
ARM Graduate Fellowships, 1002
ARM Postgraduate Fellowships, 1003
ARM Visiting Scientist Awards, 1004
ARO Short-Term Innovation Research Program, 1014
ARO Theoretical Physics and Nonlinear Phenomena Research Contracts, 1017
ARO Young Investigator Program/Presidential Early Career Awards, 1018
ASCE Arthur S. Tuttle Memorial Scholarships, 1039
ASCE Freeman Fellowships, 1040
ASCE Trent R. Dames and William W. Moore Engineering Fellowship, 1043
ASEE Army Research Laboratory Postdoctoral Fellowships, 1059
ASEE Naval Research Laboratory Postdoctoral Fellowships, 1060
ASME McDonald Mentoring Award, 1143
Astronomical Society of the Pacific Priscilla and Bart Bok Awards, 1232
ASU Graduate College Dean's Fellowships, 1236
ASU Graduate College Dissertation Fellowships, 1237
ASU Graduate College Science Foundation Arizona Bisgrove Postdoctoral Scholars, 1244
AT&T Bell Labs Graduate Research Program, 1250
Audio Engineering Society Grad Scholarships, 1265
AWIS Educational Foundation Predoc Awards, 1278
Beckman Young Investigators Program Grants, 1327
BWF Career Awards at the Scientific Interface, 1393
Canada-U.S. Fulbright International Science and Technology Award, 1417
Canada-U.S. Fulbright Senior Specialists Grants, 1418
Canada-U.S. Fulbright Traditional Scholars, 1419
Canada Graduate Scholarships and NSERC Postgraduate Scholarships, 1420
Center for Education of Women Visiting Scholar Research Grants, 1475
Center of Excellence in General Aviation Grants, 1486
CFUW Memorial Fellowship, 1515
ConocoPhillips Grants Program, 1629
Consortium for Graduate Study in Management Minority Fellowships, 1642
Corning Incorporated Found Educational Grants, 1646
DAAD Research Grants for Doctoral Candidates and Young Academics and Scientists, 1703
DAAD Research Stays for University Academics and Scientists, 1705
DAAD Study & Internship Program in Germany, 1708
DOE Basic Energy Sciences Grants, 1772
DOE Computational Science Grad Fellowships, 1780
DOE Energy-Related Inventions Grants, 1783
DOE Engineering Research Grants, 1789
DOE Experimental Program to Stimulate Competitive Research, 1791
DOE Faculty and Student Teams Grants, 1792
DOE Fossil Advanced Research and Technology Development Grants, 1793
Engineering Information Foundation Grants, 1927
Fannie and John Hertz Foundation Graduate Fellowships, 1977
Fluor Foundation Grants, 2028
Fulbright Alumni Initiatives Awards Grants, 2064
Fulbright Binational Business Grants in Mexico, 2066
Fulbright Business Grants in Spain, 2067
Fulbright Distinguished Chairs Program Grants, 2069
Fulbright European Union Scholar-in-Residence Program Grants, 2071
Fulbright New Century Scholars Program Grants, 2075
Fulbright Occasional Lecturer Program Grants, 2076
Fulbright Specialists Program Grants, 2079
Fulbright Traditional Scholar Program in Europe and Eurasia, 2080
Fulbright Traditional Scholar Program in Sub-Saharan Africa, 2081
Fulbright Traditional Scholar Program in the East Asia/Pacific Region, 2082
Fulbright Traditional Scholar Program in the Near East and North Africa Region, 2083
Fulbright Traditional Scholar Program in the South and Central Asia Region, 2084
Fulbright Traditional Scholar Program in the Western Hemisphere, 2085
Fulbright Visiting Specialists Program: Direct Access to the Muslim World, 2087
Helen Carr Minority Fellowships, 2229
HHMI-NIBIB Interfaces Initiative Grants, 2248
Ian N. McKinnon Memorial Fellowship, 2314
IIE Central Europe Summer Research Institute Summer Research Fellowship, 2349
IIE David L. Boren Fellowships, 2350
IIE Leonora Lindsley Memorial Fellowships, 2354
ISGC Informal Education Grants, 2431
L'Oreal USA Women in Science Fellowship, 2577
McKnight Doctoral Fellowships, 2726
Meadows Foundation Grants, 2736
Medtronic Foundation Fellowships, 2738
Millipore Foundation Grants, 2775
Nanyang Technological University Lee Kuan Yew Postdoctoral Fellowship, 2826
NATO Advanced Research Workshops Program, 2894
NATO Advanced Study Institutes Program, 2895
NATO Collaborative Linkage Grants, 2896
NATO Expert Visits Program Grants, 2897
NHGRI Mentored Research Scientist Development Award, 3013
NHLBI Mentored Quantitative Research Career Development Awards, 3038
NIMH Curriculum Development Award in Neuroinformatics Research and Analysis, 3251
NIST National Standard Reference Data System Research Grants, 3271

NOAA Communicating Hurricane Information Grants, 3355
Northwood University/Alden B. Dow Creativity Center Summer Residency Program, 3432
NSERC Brockhouse Canada Prize for Interdisciplinary Research in Science and Engineering Grant, 3438
NSERC Gerhard Herzberg Canada Gold Medal for Science and Engineering Grants, 3440
NSERC John C. Polanyi Awards, 3446
NSERC Research Tools and Instruments Grants, 3449
NSF-CONACyT Collaborative Research Opportunities, 3452
NSF/DOE Partnership in Basic Plasma Science and Engineering Grants, 3454
NSF Active Nano- Structures & Systems Grants, 3459
NSF Activities in Science, Engineering, and Mathematics for Persons with Disabilities, 3460
NSF ADVANCE: Increasing the Participation and Advancement of Women in Academic Science and Engineering Careers Grants, 3461
NSF Advanced Learning Technologies Grants, 3462
NSF Alan T. Waterman Award, 3467
NSF Applied Mathematics Grants, 3481
NSF Arctic Research and Education Grants, 3489
NSF Centers of Research Excellence in Science and Technology and HBCU Research Infrastructure for Science and Engineering Grants, 3515
NSF CISE Accelerating Discovery in Science and Engineering Through Petascale Simulations and Analysis Grants, 3521
NSF CISE Advanced Learning Tech Grants, 3522
NSF CISE Collaborative Research in Computational Neuroscience Grants, 3526
NSF CISE Cross-Cutting Program Grants, 3532
NSF CISE Increasing the Participation and Advancement of Women in Academic Science and Engineering Careers Grants, 3534
NSF Collaborative Research in Chemistry, 3537
NSF Deep Underground Science and Engineering Laboratory Grants, 3551
NSF Developing Global Scientists & Engineers, 3552
NSF East Asia and Pacific Summer Institutes for U.S. Graduate Students, 3555
NSF Emerging Frontiers in Research and Innovation Grants, 3560
NSF Engineering Research Centers Grants, 3561
NSF Faculty Early Career Development Grants, 3565
NSF Focused Research Groups in the Mathematical Sciences Grants, 3566
NSF Grant Opportunities for Academic Liaison with Industry, 3574
NSF Industry/University Cooperative Research Centers Grants, 3579
NSF Innovation and Organizational Change, 3581
NSF Integrative Graduate Education and Research Traineeship, 3584
NSF Interfacial Processes and Thermodynamics Research Grants, 3586
NSF Materials Research Science and Engineering Centers, 3594
NSF Partnerships for International Research and Education, 3611
NSF Presidential Early Career Awards for Scientists and Engineers Grants, 3633
NSF Process and Reaction Engineering Grants, 3635
NSF Research on Gender in Science and Engineering Grants, 3639
NSF SBE Doctoral Dissertation Research Improvement Grants, 3641
NSF Scholarships in Science, Technology, Engineering, and Mathematics, 3642
NSF Science and Society, 3643
NSF Small Business Innovation Research Grants, 3647
NSF Small Business Technology Transfer Programs Phase I Grants, 3648
NSF Tribal Colleges and Universities Program, 3662
ONR Bisson Prize for Naval Tech Achievement, 3745
ONR DOD National Defense Science & Engineering Graduate Fellowships, 3755
ONR Historically Black Colleges and Universities Future Engineering Faculty Fellowships, 3763
ONR Multidisciplinary Research Program of the URI Grants, 3774
ONR Naval Research Enterprise Intern Program Grants, 3775
ONR Summer Faculty Research Program Grants, 3784
ORISE Fusion Energy Sciences Fellowship, 3807
ORISE Higher Education Research Experiences at Oak Ridge National Laboratory for Faculty, 3808
ORISE National Oceanic and Atmospheric Administration Faculty Research Participation, 3811
ORISE National Oceanic and Atmospheric Admin Student Research Participation, 3812
ORISE Nuclear Regulatory Commission Historically Black Colleges and Universities Student Research Participation, 3814
ORISE Oak Ridge National Laboratory Postdoctoral Research Associates Program, 3815
ORISE Postgraduate Research Participation at the National Exposure Research Laboratory, 3821
ORISE Savannah River Site Faculty Research Participation Program, 3823
OSF-Baltimore Community Fellowships, 3833
Packard Foundation Fellowships for Science and Engineering, 3862
Parkersburg Area Community Foundation Arts in Education Special Grants, 3869
Rayonier Foundation Grants, 4034
Regents Professional Opportunity Scholarships, 4049
Rockwell International Corporate Trust Grants, 4093
RSC Thomas W. Eadie Medal, 4126
Sigma Xi Grants in Aid of Research, 4259
Smith Family Awards for Excellence in Biomedical Research, 4268
Society of Cosmetic Chemists Allan B. Black Award Sponsored by Presperse, 4312
Society of Cosmetic Chemists Award Sponsored by McIntyre Group, Ltd., 4313
Society of Cosmetic Chemists Award Sponsored by The HallStar Company, 4314
Society of Cosmetic Chemists Chapter Best Speaker Award, 4315
Society of Cosmetic Chemists Merit Award, 4316
Society of Cosmetic Chemists Frontier of Science Award, 4317
Society of Cosmetic Chemists Schaeffer Award, 4318
Society of Cosmetic Chemists Joseph P. Ciaudelli Award, 4319
Society of Cosmetic Chemists Keynote Award Lecture Sponsored by Ruger Chemical Corporation, 4320
Society of Cosmetic Chemists Literature Award, 4321
Society of Cosmetic Chemists Maison G. de Navarre Medal, 4322
Society of Cosmetic Chemists Merit Award, 4323
Society of Cosmetic Chemists Robert A. Kramer Lifetime Service Award, 4324
Society of Cosmetic Chemists Mudge Award, 4325
Society of Cosmetic Chemists Poster Awards, 4326
Society of Cosmetic Chemists Young Scie Awards, 4327
SPIE C.S. Vikram Award in Optical Metrology, 4339
SPIE Gold Medal of the Society Award, 4343
SPIE Technology Achievement Award, 4344
SRC Natural and Engineering Science Grants for Postdoctoral Positions in Sweden, 4384
SRC Natural and Engineering Science Postdoctoral Fellowships, 4385
SRC Natural and Engineering Science Senior Research Positions in Sweden, 4386
SRC Natural and Engineering Sciences Grants, 4387
SRC Undergraduate Research Opportunities, 4395
SSHRC Graduate Scholarships-Michael Smith Foreign Study Supplements, 4408
Sun Microsystems Foundation Summer Academic Enrichment Grants, 4468
Swedish Natural Science Res Council Grants, 4479
SWE Scholarships, 4481
Tau Beta Pi Association Fellowships, 4495
U.S. Department of Education Graduate Assistance in Areas of National Need, 4540
UEF Advancement of Engineering Grants, 4554
UNCFSP NASA Administrator's Fellowship, 4562
UNCFSP NASA Harriett G. Jenkins Predoctoral Fellowships, 4563
UNCFSP NASA Science and Technology Institute Summer Scholars Program, 4564
US CRDF CGP-RFBR Climate Change and Energy Competition For Joint U.S.-Russian Research Teams, 4608
USDC/NOAA Nature Conservancy Community-based Matching Grants, 4661
USDD Basic Research Challenge: Decentralized Online Optimization Grants, 4689
USDD Broad Agency Announcement Research and Development Grants, 4703
USDD Research Initiatives at the Naval Postgraduate School Grants, 4783
USDD Sabbatical Leave Fellowships, 4788
W.M. Keck Foundation Science and Engineering Research Grants, 4824
W.M. Keck Foundation Undergraduate Education Grants, 4825
Wayne and Gladys Valley Foundation Grants, 4835
WHOI Postdoctoral Scholar Awards, 4864
WHOI Summer Student Fellowships, 4865
WHOI Undergrad Fellowships for Minorities, 4866
Winston Churchill Foundation Scholarships, 4886
Yale University Beinecke Library Research Fellowships for Yale Grad & Professional School Students, 4902

Engineering Administration/Management

ASME Worcester Reed Warner Medal, 1172
NASA Lewis Educational Research and Collaborative Internships, 2840

Engineering Design

ASCE Jack E. Leisch Memorial National Graduate Fellowship, 1041
ASME Barnett-Uzgiris Product Safety Award, 1106
ASME J.P. Hartog Award, 1134
ASME Machine Design Award, 1140
ASME Patrick J. Higgins Award, 1152
ASME Ruth and Joel Spira Outstanding Design Educator Award, 1162
DOI Earthquake Hazards Reduction Res Grants, 1821
NSF CreativeIT Grants, 3547
NSF Engineering Research Grants, 3562
SPIE Technology Achievement Award, 4344

Engineering Education

3M Company Higher Education Grants, 4
3M Foundation Community Giving Grants, 5
AAAS Entry Point Internships, 30
ACM K Karlstrom Outstanding Educator Award, 299
Acuity Charitable Foundation Grants, 333
AEG Foundation Legget Fund Grants, 341
AERA Dissertation Grants, 342
AIChE Warren K. Lewis Award for Contributions to Chemical Engineering Education, 429
Alcatel-Lucent Technologies Foundation Grants, 509
Alcoa Foundation Grants, 511
Applied Biosystems Grants, 916
ASME Ben C. Sparks Medal, 1107
ASME Edwin F. Church Medal, 1117
ASME Frank von Flue Award, 1121
BPW Career Advancement Scholarships, 1366
Christine Mirzayan Science and Technology Policy Graduate Fellowship Program, 1556
DAAD Research Internships in Science and Engineering, 1704
DAAD RISE Professional Internships, 1706
DaimlerChrysler Corporation Fund Grants, 1714
DENSO North America Foundation Grants, 1742
DOE Computational Science Grad Fellowships, 1780
E.J. Sierleja Memorial Fellowship, 1884
Emerson Charitable Trust Grants, 1918
Engineering Information Foundation Grants, 1927
FMC Foundation Grants, 2029
Ford Motor Company Fund Grants Program, 2033
GEM MS Engineering Fellowships, 2104

GEM PhD Engineering Fellowships, 2105
HHMI Med into Grad Initiative Grants, 2259
Knoll-MacDonald Foundation Grants, 2566
Kosciuszko Foundation Tuition Scholarships, 2574
NAACP Willems Scholarships, 2810
NASA Lewis Educational Research and Collaborative Internships, 2840
Northrop Grumman Foundation Grants, 3429
NSF ADVANCE: Increasing the Participation and Advancement of Women in Academic Science and Engineering Careers Grants, 3461
NSF Developing Global Scientists & Engineers, 3552
NSF Director's Award for Distinguished Teaching Scholars, 3553
NSF Engineering Research Centers Grants, 3561
NSF Engineering Research Grants, 3562
NSF Historically Black Colleges and Universities Program Grants, 3577
NSF Innovative Technology Experiences for Students and Teachers, 3582
NSF Louis Stokes Alliances for Minority Participation Grants, 3589
NSF Presidential Awards for Excellence in Science, Math, and Engineering Mentoring, 3632
NSF Research Experiences for Undergrad Grants, 3638
NSF SBE Doctoral Dissertation Research Improvement Grants, 3641
NSF Tribal Colleges and Universities Program, 3662
ONR DOD National Defense Science & Engineering Graduate Fellowships, 3755
ONR Historically Black Colleges and Universities Future Engineering Faculty Fellowships, 3763
ONR Multidisciplinary Research Program of the URI Grants, 3774
Praxair Foundation Grants, 3986
SRC Franco Cerrina Fellowship, 4367
SRC Natural and Engineering Science Postdoctoral Fellowships, 4385
SRC Undergraduate Research Opportunities, 4395
SWE Scholarships, 4481
Tau Beta Pi Association Fellowships, 4495
U.S. Department of Education Graduate Assistance in Areas of National Need, 4540
USDD High School Apprenticeship Program and Undergraduate Research Apprenticeship Program Grants, 4732
USDD Research Initiatives at the Naval Postgraduate School Grants, 4783
USDD Science, Technology, Engineering and Mathematics Grants, 4789
W.M. Keck Foundation Undergraduate Education Grants, 4825

Engines
ASME Internal Combustion Engine Award, 1132
DOE Energy Conversion Tech Research Grants, 1786
DOE Vehicle and Engine Research and Development Grants, 1819

English
SSRC Mellon Mays Predoctoral Research Grants, 4432
TIRF Doctoral Dissertation Grants, 4520

English Education
Fulbright English Teaching Assistantships, 2070
NCTE Teacher Researcher Award, 2940
TIRF Doctoral Dissertation Grants, 4520
University of London Visiting Fellowships, 4588

English as a Second Language
National Book Scholarship Fund, 2849
Textron Corporate Contributions Grants, 4501

Enteral Nutrition
ASPEN C. Richard Fleming Grant, 1190
ASPEN Douglas Wilmore Grant, 1191
ASPEN Dudrick Research Scholar Award, 1192
ASPEN Harry M. Vars Award, 1193
ASPEN Maurice Shils Grant, 1195
ASPEN Norman Yoshimura Grant, 1196
ASPEN Nutrition Research Grants, 1197
ASPEN Scientific Abstracts Awards for Papers or Posters, 1198
ASPEN Scientific Abstracts Promising Investigator Awards, 1199

Entertainment Industry
MetLife/TCG Aha! Think it Do it Grants, 2745
NEA/TCG Career Development Grants for Theatre Directors, 2943

Entomology
AMNH Graduate Student Fellowships, 815
Honeybee Health Improvement Project Grants, 2276
ORISE Faculty Internship at the US Army Center for Health Promotion and Preventive Medicine, 3802
ORISE Postgraduate Internship at the US Army Center for Health Promotion and Preventive Med, 3818
OTS Hovore-Horn Fellowships, 3856
USDA Norman E. Borlaug International Agricultural Science and Technology Fellowships, 4636
USDA Plant and Animal Disease, Pest Control, and Animal Care Grants, 4638

Entrepreneurship
Achelis Foundation Grants, 274
Allstate Foundation Safe and Vital Communities Grants, 573
Barker Award, 1300
Bodman Foundation Grants, 1356
CHF Glenn E. and Barbara Hodsdon Ullyot Scholarship, 1540
CHF Long-Term Fellowships, 1541
CHF Robert W. Gore Fellowship in Materials Innovation, 1542
CHF Societe de Chimie Industrielle (American Section) Fellowship, 1544
CHF Travel Grants, 1545
Dean Witter Foundation Grants, 1728
Independent Institute John M. Templeton Fellowships Essay Contest, 2369
Indian Council of Social Science Research Grants, 2377
Jewish Social Entrepreneurs Fellowships, 2487
JM Foundation Grants, 2490
Monfort Family Foundation Grants, 2794
OSF-Baltimore Community Fellowships, 3833
S.E.VEN Fund Annual Grants, 4153
S.E.VEN Fund Mini-Grants, 4154
S.E.VEN Fund Open Enterprise Solutions to Poverty RFP Grants, 4155
Tandem Launch Technologies Grants, 4493

Environment
A Charitable Foundation Grants, 273
Alfred P. Sloan Foundation Energy and Environment Grants, 552
Alfred P. Sloan Foundation Science of Learning STEM Grants, 562
American Chemical Society Kenneth G. Hancock Memorial Student Award in Green Chemistry, 683
American Chemical Society Presidential Green Chemistry Challenge Awards, 696
Anderson Foundation Grants, 842
As You Sow, 1249
Caesars Foundation Grants, 1407
Charles M & Mary D Grant Foundation Grants, 1527
Energy Foundation Transportation Grants, 1925
EPA Brownfields Area-Wide Planning Grants, 1931
EPA Compliance Assistance Support for Services to the Regulated Community and Other Assistance Providers, 1934
EPA Regional Healthy Indoor Air Projects for Community Outreach and Education, 1937
EPA Research, Development, Monitoring, Public Education, Training, Demonstrations, and Studies Grants, 1938
EPA Southeastern Multi-Media and Geographic Priority Project Grants, 1939
Evan Frankel Foundation Grants, 1961
Freshwater Future Project Grants, 2056
G. Unger Vetlesen Foundation Grants, 2093
Georgia-Pacific Foundation Environment Grants, 2123
Great Lakes Fishery Commission Pilot Research Project Grants, 2166
Great Lakes Fishery Commission Sea Lamprey Research Grants, 2167
Gruber Foundation Weizmann Institute Awards, 2184
Harry A. and Margaret D. Towsley Found Grants, 2205
Honeywell Corporation Habitat and Conservation Grants, 2277
Honeywell Corporation Sustainable Opportunities Grants, 2278
HRF Graduate Fellowships, 2285
Meadows Foundation Grants, 2736
Mellon Post-Doctoral Fellowship in Sustainable Agriculture, 2740
Mericos Foundation Grants, 2742
MoEF Dr. Salim Ali Nat Wildlife Fellowship, 2792
MOEF Gandhi Paryavaran Puraskar Award, 2793
NOAA Chesapeake Bay Office B-WET (Bay Watershed Education and Training) Grants, 3339
NOAA Chesapeake Bay Office Cooperative Research Program Grants, 3340
NOAA Chesapeake Bay Office Fisheries Science Program Grants, 3341
NOAA Chesapeake Bay Office Non-Native Oyster Research Grants, 3342
NOAA Chesapeake Bay Office Submerged Aquatic Vegetation Culture and Restoration Grants, 3343
NOAA Cooperative Unit for Fisheries Education and Research Internships, 3360
NOAA Faculty and Student Intern Research, 3371
NOAA Miami Lab Summer Internships, 3382
NOAA Monkfish Research Set-Aside Grants, 3386
NOAA National Climatic Data Center Interns, 3387
NOAA Sea Scallop Research Set-Aside Program Grants, 3404
Norcliffe Foundation Grants, 3415
Northrop Grumman Corporation Grants, 3428
Oak Foundation Environmental Grants, 3731
Rohm and Haas Company Grants, 4098
Sony Corporation of America Corporate Philanthropy Grants, 4329
South Florida Student Shark Program, 4330
SSRC Collaborative Research Grants on Environment and Health in China, 4423
Susan G. Komen Breast Cancer Foundation Challege Grants: Breast Cancer and the Environment, 4474
USDC/NOAA Coastal and Marine Fish Habitat Protection Grants, 4655
USDC/NOAA Restoration Center National and Regional Partnership Grants, 4662
USDD Battlespace Environment Division Research Interests Grant, 4692
USDD Broad Agency Announcement Research and Development Grants, 4703
William and Flora Hewlett Foundation Environmental Grants, 4874
Xerox Foundation Grants, 4899

Environmental Biology
Alfred P. Sloan Foundation Barcode of Life Grants, 543
Alfred P. Sloan Foundation Microbiology of the Built Environment Grants, 554
DOE Environmental Processes Research Grants, 1790
IIE Central Europe Summer Research Institute Summer Research Fellowship, 2349
NOAA Ballast Water Management Demonstration Program (RDTE Facility) Grants, 3334
NOAA Ballast Water Management Demonstration Program (Technologies and Practices) Grants, 3335
NOAA Ballast Water Technology Demonstration Program (Research, Development, Testing and Evaluation Facility) Grants, 3336
NOAA Ballast Water Technology Demonstration Program (Treatment Technology Demonstration Project) Grants, 3337
NOAA Population Dynamics Recruiting Program Grants, 3400
Northrop Grumman Corporation Grants, 3428

NSF Biotic Surveys and Inventories Grants, 3510
NSF Doctoral Dissertation Improvement Grants in the Directorate for Biological Sciences, 3554
Smithsonian Graduate Student Fellowships, 4277
Smithsonian Minority Visiting Student Awards, 4279
Smithsonian Native American Community Scholars Awards, 4291
Smithsonian Native American Visiting Student Awards, 4292
Smithsonian Postdoctoral Fellowships, 4293
Smithsonian Predoctoral Fellowships, 4295
Susan G. Komen Breast Cancer Foundation Challege Grants: Breast Cancer and the Environment, 4474
UNCFSP NASA Harriett G. Jenkins Predoctoral Fellowships, 4563
US CRDF CGP-Climate Change and Energy Competition For Joint U.S.-Ukrainian Research Teams, 4607
US CRDF CGP-RFBR Climate Change and Energy Competition For Joint U.S.-Russian Research Teams, 4608
USDA 1890 Institution Teaching, Research, and Extension Capacity Building Grants, 4613
USDA Distance Education Grants for Higher Education Institutions in Insular Areas, 4622
USDA Norman E. Borlaug International Agricultural Science and Technology Fellowships, 4636
USDA Resident Instruction Grants for Higher Education in Insular Areas, 4645
USDA Tribal College Research Grants, 4652
USDC/NOAA Community-Based Marine Debris Removal Grants, 4657

Environmental Conservation
Animal Behavior Society C.H. Turner Award, 849
Animal Behavior Society Developing Nations Research Grants, 850
Animal Behavior Society Diversity Grant, 851
Animal Behavior Society Latin American Travel Awards, 852
Animal Behavior Society Student Research Grants, 853
ASME Dixy Lee Ray Award, 1115
As You Sow, 1249
EPA Southeastern Multi-Media and Geographic Priority Project Grants, 1939
GNOF Gulf Coast Oil Spill Grants, 2153
Honeywell Corporation Habitat and Conservation Grants, 2277
Honeywell Corporation Sustainable Opportunities Grants, 2278
HRF Hudson River Improvement Grants, 2286
HRF New York City Environmental Grants for Newton Creek, 2288
IIE Rockefeller Foundation Bellagio Center Residencies, 2355
MDEQ Local Water Quality Monitoring Grants, 2732
Meadows Foundation Grants, 2736
Meyer Memorial Trust Responsive Grants, 2748
MoEF Dr. Salim Ali Nat Wildlife Fellowship, 2792
Neotropical Grassland Conservancy Derald G. Langham Memorial Research Grant, 2973
Neotropical Grassland Conservancy Grants, 2974
Neotropical Grassland Conservancy Remembrance Fund Grants, 2975
Neotropical Grassland Conservancy Grants, 2976
Neotropical Grassland Conservancy W. J. and Virginia W. Moorhouse Research Grant, 2977
Neotropical Grassland Conservancy Wijdenes Memorial Research Grant, 2978
Northrop Grumman Corporation Grants, 3428
Rohm and Haas Company Grants, 4098
SARE Graduate Student Grants - North Central Region, 4168
SARE Professional + Producer Grants - Western Region, 4175
Smithsonian Museum Conservation Institute Analytical Studies Group Internships, 4281
Smithsonian Museum Conservation Institute Internships for Research Projects, 4282
Sony Corporation of America Corporate Philanthropy Grants, 4329
USDA Integrated Research Education and Extension Grants - Organic Transitions, 4630
USDA McIntire-Stennis Cooperative Forestry Research Grants, 4634
USDC/NOAA Saltonstall-Kennedy Grants, 4663
WWF Russell E. Train Fellowships, 4898

Environmental Design
College of Human Sciences Scholarships, 1596
GNOF Gulf Coast Oil Spill Grants, 2153
Northrop Grumman Corporation Grants, 3428
Vancouver Foundation Grants and Community Initiatives, 4807

Environmental Economics
Alfred P. Sloan Foundation Behavioral Economics and Household Finance Grants, 544
Brainerd Foundation Grants, 1368
GNOF Gulf Coast Oil Spill Grants, 2153
Jenifer Altman Foundation Grants, 2484
Russell Family Foundation Environmental Sustainability Grants, 4128
SARE Professional + Producer Grants - Western Region, 4175

Environmental Education
Ahn Family Foundation Grants, 379
Alces Foundation Grants, 510
Alfred P. Sloan Foundation Microbiology of the Built Environment Grants, 554
Allen Foundation Educational Nutrition Grants, 570
APS/Phoenix Suns Energy and Environmental Minigrants, 919
A Rocha USA Grants, 1006
As You Sow, 1249
ATA Educational Trust Grants - Curriculum Resource or Research Project, 1253
BMW of N America Charitable Contributions, 1353
Boeing Company Contributions Grants, 1357
Bullitt Foundation Grants, 1384
Cabot Corporation Foundation Grants, 1406
Carrier Corporation Contributions Grants, 1440
Clarence E. Heller Charitable Foundation Grants, 1575
Community Foundation for Greater New Haven Quinnipiac River Fund Grants, 1616
Del E. Webb Foundation Grants, 1733
DOI Rhinoceros and Tiger Conservation Grants, 1824
Dorr Foundation Grants, 1842
EPA Capacity Building Grants for Compliance Assurance and Enforcement Activities in Indian Country and Other Tribal Areas, 1932
EPA Research, Development, Monitoring, Public Education, Training, Demonstrations, and Studies Grants, 1938
EPA Southeastern Multi-Media and Geographic Priority Project Grants, 1939
EREF Doctoral Scholarship, 1945
EREF Fiessinger Doctoral Scholarship, 1946
EREF Solid Waste Research Grants, 1947
EREF Sustainability Research Grants, 1948
Explorers Club Scott Pearlman Field Awards, 1967
Georgia-Pacific Foundation Environment Grants, 2123
GNOF Gulf Coast Oil Spill Grants, 2153
Honeywell Corporation Habitat and Conservation Grants, 2277
Honeywell Corporation Sustainable Opportunities Grants, 2278
HRF Hudson River Improvement Grants, 2286
HRF New York City Environmental Grants for Newton Creek, 2288
HSBC Contributions Community Grants, 2291
Illinois DNR Youth Recreation Corps Grants, 2361
Ittleson Foundation Environment Grants, 2437
John D. and Catherine T. MacArthur Foundation Global Challenges Program Grants, 2498
MacArthur Foundation Conservation and Sustainable Development Grants, 2648
McCune Charitable Foundation Grants, 2719
Meadows Foundation Grants, 2736
National Wetlands Awards, 2889
National Wild Turkey Federation Research Grant, 2893
NFWF Acres for America Grants, 2996
NFWF Budweiser Renewable Energy and Wildlife Conservation Research Prize, 2998
NFWF Nature Restoration Trust Grants, 3005
NFWF Shell Marine Habitat Program Grants, 3008
NOAA Coastal Management Fellowships, 3352
NOAA Dr. Nancy Foster Scholarships, 3367
NOAA Ernest F. Hollings Scholarships, 3370
NOAA Graduate Sciences Program Grants, 3375
NOAA Living Marine Resources Cooperative Science Center Internships/Fellowships, 3380
NOAA Living Marine Resources Cooperative Science Center Mini Grants, 3381
NOAA Pacific Islands Assistantships, 3399
NOAA Population Dynamics Recruiting Program Grants, 3400
Northrop Grumman Corporation Grants, 3428
NYCT New York City Environment Grants, 3697
Oak Foundation Environmental Grants, 3731
Olin Corporation Charitable Trust Grants, 3739
Organization for Tropical Studies Fellowships, 3796
Organization for Tropical Studies Pilot Awards for Exploratory Research, 3797
OTS Francis J. Bossuyt Fellowship, 3855
OTS Undergrad NAPIRE Summer Interns, 3858
Padilla Bay Foundation Grants, 3863
Project AWARE Foundation Grants, 3999
Rohm and Haas Company Grants, 4098
Russell Family Foundation Environmental Sustainability Grants, 4128
SeaWorld and Busch Gardens Conservation Fund Grants, 4196
Skaggs Foundation Grants, 4266
USDA Policy Research Center Grants, 4640
USDC/NOAA Community-Based Marine Debris Removal Grants, 4657
Vancouver Foundation Grants and Community Initiatives, 4807

Environmental Effects
AIChE Research Excellence In Sustainable Engineering Award, 424
Alfred P. Sloan Foundation Microbiology of the Built Environment Grants, 554
Alfred P. Sloan Foundation Research Fellowships, 560
American Chemical Society Award for Creative Advances in Environmental Science and Technology, 652
A Rocha USA Grants, 1006
ASHRAE Graduate Student Grant-in-Aid, 1088
Brainerd Foundation Grants, 1368
Bydale Foundation Grants, 1400
Carrier Corporation Contributions Grants, 1440
Clarence E. Heller Charitable Foundation Grants, 1575
Community Foundation for Greater New Haven Quinnipiac River Fund Grants, 1616
Cooperative Institute for Research in Environmental Sciences Visiting Fellows Program, 1643
DHHS Native American Programs, 1758
DOE Atomic, Molecular, and Optical Sciences Research Grants, 1771
DOE Catalysis and Chemical Transformations Research Grants, 1773
DOE Chemical Energy and Chemical Engineering Research Grants, 1774
DOE Chemical Physics Research Grants, 1775
DOE Chemical Sciences Research Grants, 1776
DOE Fusion Energy Science Technology Division Grants, 1796
DOE Heavy Element Chem Research Grants, 1798
DOE Photochem & Radiation Research Grants, 1809
DOE Program for Ecosystem Research, 1810
DOE Separations and Analysis Research Grants, 1813
DOT Transit Planning and Research Grants, 1849
EPA Brownfields Area-Wide Planning Grants, 1931
EPA Children's Health Protection Grants, 1933

EPA Research, Development, Monitoring, Public Education, Training, Demonstrations, and Studies Grants, 1938
Georgia-Pacific Foundation Environment Grants, 2123
Gerber Foundation Grants, 2126
Global Change Education Program, 2151
Great Lakes Protection Fund Grants, 2170
IIE Rockefeller Foundation Bellagio Center Residencies, 2355
Jenifer Altman Foundation Grants, 2484
John D. and Catherine T. MacArthur Foundation Global Challenges Program Grants, 2498
NOAA Abrupt Climate Change Studies Grants, 3329
NOAA Arctic Research Program Grants, 3330
NOAA Assessing Atlantic Meridional Overturning Circulation Program Grants, 3331
NOAA Atmospheric Composition and Climate Program Grants, 3333
NOAA Climate Change Data and Detection Program Grants, 3344
NOAA Climate Dynamics and Experimental Prediction Program Grants, 3345
NOAA Climate Prediction Program for the Americas Grants, 3346
NOAA Climate Variability and Human Health Program Grants, 3347
NOAA Climate Variability and Predictability Program Grants, 3348
NOAA Global Carbon Cycle Program Grants, 3374
NOAA Regional Integrated Sciences and Assessments Program Grants, 3402
NOAA Scientific Data Stewardship Grants, 3403
NOAA Sector Applications Research Grants, 3405
NOAA Transition of Research Applications to Climate Services Grants, 3408
Northrop Grumman Corporation Grants, 3428
NSF-NIST Interaction in Chemistry, Materials Research, Molecular Biosciences, Bioengineering, and Chemical Engineering, 3453
NSF Arctic Nat Science Grants in Ocean Science, 3487
NSF Chemistry Research Experiences for Undergraduates, 3518
NSF Paleoclimate Grants, 3606
NSF Thermal Transport Processes Grants, 3660
ONR Electronics Division Research Grants, 3759
ONR Environmental Quality Research Grants, 3760
ONR Marine Meteorology and Atmospheric Effects Research Grants, 3769
ONR Medical Science and Technology Division Grants, 3772
ORISE Faculty Research Participation at the National Center for Toxicological Research, 3804
OSRI Biological Monitoring: Intertidal Resources at Risk to Oil Spills, 3854
USDA Biotechnology Risk Assessment Grants, 4618
USDA Integrated Research Education and Extension Grants - Organic Transitions, 4630
USDC/NOAA Community-Based Marine Debris Removal Grants, 4657
USDD Physical Oceanography Grants, 4771
Young Scientists Summer Program, 4903
Yves Rocher Foundation Terra Ficaria Prize for Scientific Research, 4904

Environmental Effects, Fossil Fuels

DOE Control Technology and Coal Preparation Research Grants, 1781
DOE Fossil Advanced Research and Technology Development Grants, 1793
DOE Fossil Energy Research & Devel Grants, 1794
DOE University Coal Research Grants, 1816
Energy Foundation Transportation Grants, 1925
GNOF Gulf Coast Oil Spill Grants, 2153

Environmental Engineering

ASME Dixy Lee Ray Award, 1115
ASME Ralph Coats Roe Medal, 1158
ASME Soichiro Honda Medal, 1164
Dreyfus Postdoctoral Program in Environmental Chemistry Grant, 1855
IIE KAUST Graduate Fellowships, 2353
Lawrence K. Cecil Award in Environmental Chemical Engineering, 2591
Nanyang Technological University Lee Kuan Yew Postdoctoral Fellowship, 2826
NPSC Grad Fellowships in the Physical Sciences, 3436
NSF Engineering Research Grants, 3562

Environmental Health

AAAS Science and Technology Policy Fellowships: Energy, Environment, and Agriculture, 41
AAAS Science and Technology Policy Fellowships: Health, Education and Human Services, 43
AAOHN Found Experienced Researcher Grants, 195
AAOHN Foundation New Investigator Researcher Grants, 196
Ahn Family Foundation Grants, 379
Akonadi Foundation Anti-Racism Grants, 477
Alaska Conservation Foundation Watchable Wildlife Conservation Trust Grants, 498
Alces Foundation Grants, 510
Alfred P. Sloan Foundation Microbiology of the Built Environment Grants, 554
Allen Foundation Educational Nutrition Grants, 570
American Rivers Anthony A. Lapham River Conservation Fellowship, 795
Applied Biosystems Grants, 916
A Rocha USA Grants, 1006
Ashland Corporate Contributions Grants, 1087
ASME Dixy Lee Ray Award, 1115
BMW of N America Charitable Contributions, 1353
California Wellness Foundation Work and Health Grants, 1409
Canada Graduate Scholarships and NSERC Postgraduate Scholarships, 1420
CDC Experience Appl Epidemiology Fellowships, 1460
CDC Found Tobacco Network Lab Fellowship, 1462
Community Foundation for Greater New Haven Quinnipiac River Fund Grants, 1616
CSTE CDC/CSTE Applied Epidemiology Fellowships, 1664
Donald and Sylvia Robinson Family Foundation Grants, 1830
EPA Children's Health Protection Grants, 1933
GNOF Gulf Coast Oil Spill Grants, 2153
Heineman Foundation for Research, Education, Charitable and Scientific Purposes, 2227
Honeywell Corporation Sustainable Opportunities Grants, 2278
IIE David L. Boren Fellowships, 2350
IIE Rockefeller Foundation Bellagio Center Residencies, 2355
Illinois Clean Energy Community Foundation Renewable Energy Grants, 2358
Independent Institute John M. Templeton Fellowships Essay Contest, 2369
National Geographic Society/Waitt Grants, 2870
National Geographic Society Conservation Trust Grants, 2871
National Geographic Society Expeditions Council Grants, 2872
National Geographic Society Field Research and Exploration Grants, 2873
National Geographic Soc Young Explorers Grant, 2874
NFWF Acres for America Grants, 2996
NFWF Budweiser Renewable Energy and Wildlife Conservation Research Prize, 2998
NFWF Nature Restoration Trust Grants, 3005
NIDCD Mentored Research Scientist Development Award, 3132
NIEHS Independent Scientist Award, 3192
NIEHS Mentored Clinical Scientist Research Career Development Awards, 3193
NIEHS Mentored Research Scientist Development Award, 3194
NIEHS Small Business Innovation Research Program Grants, 3195
NOAA Cooperative Institute: Eastern U.S. Continental Shelf Frontier Exploration, Research, and Technology Development Grants, 3356
NOAA National Sea Grant Oyster Disease Research Program Grants, 3395
NOAA Population Dynamics Recruiting Program Grants, 3400
Northwest Fund for the Environment Grants, 3431
NSF Climate and Large-Scale Dynamics Grants, 3535
NSF Ecosystem Science Grants, 3556
NYCT New York City Environment Grants, 3697
Oak Foundation Environmental Grants, 3731
Open Spaces Sacred Places National Awards, 3791
ORISE Faculty Internship at the US Army Center for Health Promotion and Preventive Medicine, 3802
ORISE Faculty Research Participation at the Centers for Disease Control and Prevention, 3803
ORISE Faculty Research Participation at the National Center for Toxicological Research, 3804
ORISE Postgraduate Internship at the US Army Center for Health Promotion and Preventive Med, 3818
ORISE Postgraduate Research Participation at the Centers for Disease Control and Prevention, 3820
ORISE Student Internship Program at the US Army Center for Health Promotion and Preventive Medicine, 3827
ORISE Student Research Participation at the Centers for Disease Control and Prevention, 3829
ORISE Student Research Participation at the National Center for Toxicological Research, 3830
Padilla Bay Foundation Grants, 3863
Padilla Bay National Estuarine Research Reserve Graduate Research Fellowships, 3864
Padilla Bay National Estuarine Research Reserve Research Assistantships, 3865
Peter F. McManus Charitable Trust Grants, 3903
Rayonier Foundation Grants, 4034
Robert and Patricia Switzer Found Fellowships, 4071
SeaWorld and Busch Gardens Conservation Fund Grants, 4196
SSRC Collaborative Research Grants on Environment and Health in China, 4423
UNCFSP NASA Harriett G. Jenkins Predoctoral Fellowships, 4563
USDA Organic Agriculture Research and Extension Initiative, 4637
USDC/NOAA Anadromous Fish Conservation Act Grants, 4654
USDC/NOAA Community-Based Marine Debris Removal Grants, 4657
Vancouver Foundation Grants and Community Initiatives, 4807
WILD Foundation Grants, 4867
Yves Rocher Foundation Terra Ficaria Prize for Scientific Research, 4904

Environmental Issues

GNOF Gulf Coast Oil Spill Grants, 2153
Independent Institute John M. Templeton Fellowships Essay Contest, 2369
USDC/NOAA Saltonstall-Kennedy Grants, 4663

Environmental Law

Alaska Conservation Foundation Watchable Wildlife Conservation Trust Grants, 498
Brainerd Foundation Grants, 1368
Bullitt Foundation Grants, 1384
GNOF Gulf Coast Oil Spill Grants, 2153
Honor the Earth Grants, 2279
Ittleson Foundation Environment Grants, 2437
Jessie Smith Noyes Foundation Grants, 2485
MacArthur Foundation Conservation and Sustainable Development Grants, 2648
National Wildlife Federation Internships, 2891
WILD Foundation Grants, 4867

Environmental Planning/Policy

A. V. Stout Fund Research Grants, 13
AAAS Roger Revelle Fellowship in Global Stewardship, 38
Achelis Foundation Grants, 274
Alaska Conservation Foundation Watchable Wildlife Conservation Trust Grants, 498

AMNH Graduate Student Fellowships, 815
ASME Dixy Lee Ray Award, 1115
As You Sow, 1249
Bodman Foundation Grants, 1356
Clarence E. Heller Charitable Foundation Grants, 1575
Community Foundation for Greater New Haven Quinnipiac River Fund Grants, 1616
EPA Brownfields Area-Wide Planning Grants, 1931
EPA National Estuary Program Grants, 1935
EPA Targeted Watersheds Grants, 1941
Freshwater Future Project Grants, 2056
Georgia-Pacific Foundation Environment Grants, 2123
GNOF Gulf Coast Oil Spill Grants, 2153
Great Lakes Fishery Commission Pilot Research Project Grants, 2166
Great Lakes Fishery Commission Sea Lamprey Research Grants, 2167
Honeywell Corporation Sustainable Opportunities Grants, 2278
HRF Hudson River Improvement Grants, 2286
HRF New York City Environmental Grants for Newton Creek, 2288
IIE Rockefeller Foundation Bellagio Center Residencies, 2355
Illinois Clean Energy Community Foundation Renewable Energy Grants, 2358
IREX Edmund S. Muskie Graduate Fellowship Program, 2417
Ittleson Foundation Environment Grants, 2437
Japan Foundation Center for Global Partnership Institutional Grants, 2473
John D. and Catherine T. MacArthur Foundation Global Challenges Program Grants, 2498
MacArthur Foundation Conservation and Sustainable Development Grants, 2648
National Wildlife Refuge System Centennial Scholarship Program, 2892
NFWF Shell Marine Habitat Program Grants, 3008
NOAA Cooperative Institute: Eastern U.S. Continental Shelf Frontier Exploration, Research, and Technology Development Grants, 3356
NOAA International Coral Reef Conservation Project Grants, 3377
NYCT New York City Environment Grants, 3697
Oak Foundation Environmental Grants, 3731
Praxair Foundation Grants, 3986
Shastri Indo-Canadian Institute Collaborative Research Project Grants, 4220
USDA Cooperative Extension System Education Grants, 4620
USDC/NOAA Coastal and Marine Fish Habitat Protection Grants, 4655
WILD Foundation Grants, 4867

Environmental Programs

3M Foundation Community Giving Grants, 5
A. V. Stout Fund Research Grants, 13
AAAS Science and Technology Policy Fellowships: Energy, Environment, and Agriculture, 41
Abdus Salam ICTP Laurea Magistralis Grants, 229
Adam Richter Charitable Trust Grants, 339
Ahn Family Foundation Grants, 379
Akonadi Foundation Anti-Racism Grants, 477
Alaska Conservation Foundation Watchable Wildlife Conservation Trust Grants, 498
Albuquerque Community Foundation Grants, 508
Alces Foundation Grants, 510
Alcoa Foundation Grants, 511
Alfred P. Sloan Foundation Civic Grants, 545
Allen Foundation Educational Nutrition Grants, 570
American Rivers Anthony A. Lapham River Conservation Fellowship, 795
Amgen Foundation Grants, 812
ANL Faculty Research Participation Program, 855
Ann Arbor Area Community Foundation Grants, 860
Applied Biosystems Grants, 916
Archer Daniels Midland Foundation Grants, 985
Arizona Public Service Corporate Giving Program Grants, 998
A Rocha USA Grants, 1006
ASCSA Weiner Laboratory Faunal Studies Research Fellowship, 1056
Ashland Corporate Contributions Grants, 1087
As You Sow, 1249
AT&T Industrial Ecology Faculty Fellowships, 1251
Audubon Naturalist Society Research Grants, 1266
Batchelor Foundation Grants, 1305
Beim Foundation Grants, 1331
Berks County Community Foundation Grants, 1333
BMW of N America Charitable Contributions, 1353
Bodenwein Public Benevolent Foundation Grants, 1355
Boeing Company Contributions Grants, 1357
Boston Foundation Grants, 1361
Brainerd Foundation Grants, 1368
Brian G. Dyson Foundation Grants, 1370
Brico Fund Grants, 1371
Bullitt Foundation Grants, 1384
Burton G. Bettingen Grants, 1390
Bydale Foundation Grants, 1400
Caesars Foundation Grants, 1407
Callaway Golf Company Foundation Grants, 1410
Carl and Eloise Pohlad Family Grants, 1426
Carrier Corporation Contributions Grants, 1440
Charles Stewart Mott Foundation Grants, 1529
Claneil Foundation Grants, 1574
Clarence E. Heller Charitable Foundation Grants, 1575
Cleveland-Cliffs Foundation Grants, 1580
Community Found for Greater Buffalo Grants, 1615
Community Foundation for Greater New Haven Quinnipiac River Fund Grants, 1616
Community Found for Muskegon County Grants, 1617
Community Foundation for S Arizona Grants, 1618
Community Found of E Central Illinois Grants, 1619
Community Foundation of the Virgin Islands Judith A. Towle Environmental Studies Grants, 1621
Compton Foundation Grants Program, 1623
Connecticut Community Foundation Grants, 1627
ConocoPhillips Grants Program, 1629
Coors Brewing Corporate Contributions Grants, 1644
Cowles Charitable Trust Grants, 1652
Cruise Industry Charitable Foundation Grants, 1660
Dade Community Foundation Grants, 1712
DaimlerChrysler Corporation Fund Grants, 1714
David Bohnett Foundation Grants, 1724
DOE Geosciences Research Grants, 1797
DOE Program for Ecosystem Research, 1810
Donald and Sylvia Robinson Family Foundation Grants, 1830
Drs. Bruce and Lee Foundation Grants, 1859
Edward Bangs Kelley and Elza Kelley Foundation Grants, 1898
Energy Foundation Buildings Grants, 1922
Energy Foundation Power Grants, 1924
EPA Brownfields Area-Wide Planning Grants, 1931
EPA Children's Health Protection Grants, 1933
EPA National Estuary Program Grants, 1935
EPA Regional Healthy Indoor Air Projects for Community Outreach and Education, 1937
EPA Research, Development, Monitoring, Public Education, Training, Demonstrations, and Studies Grants, 1938
EPA Southeastern Multi-Media and Geographic Priority Project Grants, 1939
EPA Targeted Watersheds Grants, 1941
EREF Solid Waste Research Grants, 1947
EREF Sustainability Research Grants, 1948
ESRI Conservation Program Grants, 1952
Evjue Foundation Grants, 1963
Ewing Halsell Foundation Grants, 1964
Florida Sea Turtle Grants Program, 2027
Ford Foundation Peace and Social Justice Grants, 2031
Forrest C. Lattner Foundation Grants, 2035
Freshwater Future Project Grants, 2056
General Motors Foundation Grants, 2107
George A Ohl Jr. Foundation Grants, 2112
George Gund Foundation Grants, 2117
Georgia-Pacific Foundation Environment Grants, 2123
Georgia Power Foundation Grants, 2125
GNOF Gulf Coast Oil Spill Grants, 2153
Greater Saint Louis Community Found Grants, 2164
Great Lakes Fishery Commission Pilot Research Project Grants, 2166
Great Lakes Fishery Commission Sea Lamprey Research Grants, 2167
Great Lakes Fishery Trust Access Grants, 2168
Green Diamond Charitable Contributions, 2171
H.A. and Mary K. Chapman Charitable Grants, 2193
Heineman Foundation for Research, Education, Charitable and Scientific Purposes, 2227
Honeywell Corporation Habitat and Conservation Grants, 2277
Honeywell Corporation Sustainable Opportunities Grants, 2278
Honor the Earth Grants, 2279
HRF Tibor T. Polgar Fellowships, 2289
HSBC Contributions Community Grants, 2291
Illinois Clean Energy Community Foundation Renewable Energy Grants, 2358
Illinois DNR Youth Recreation Corps Grants, 2361
International Peace Research Inst Oslo Grants, 2402
Ittleson Foundation Environment Grants, 2437
Jackson Foundation Grants, 2453
Jenifer Altman Foundation Grants, 2484
Jessie Smith Noyes Foundation Grants, 2485
Kavli Foundation Research Grants, 2533
Kenneth T. and Eileen L. Norris Found Grants, 2554
Kettering Fund Grants, 2555
Macquarie Bank Foundation Grants, 2650
Margaret T. Morris Foundation Grants, 2659
Mary K. Chapman Foundation Grants, 2673
McGraw-Hill Companies Community Grants, 2724
McLean Contributionship Grants, 2727
MDEQ Scrap Tire Market Development Grants, 2733
MOEF Gandhi Paryavaran Puraskar Award, 2793
Motorola Foundation Grants, 2796
National Geographic Society/Waitt Grants, 2870
National Geographic Society Conservation Trust Grants, 2871
National Geographic Society Expeditions Council Grants, 2872
National Geographic Society Field Research and Exploration Grants, 2873
National Geographic Soc Young Explorers Grant, 2874
National Wildlife Federation Internships, 2891
NATO Science for Peace Grants, 2898
Nelda C. and H.J. Lutcher Stark Found Grants, 2967
New Hampshire Charitable Foundation Grants, 2985
NFWF Acres for America Grants, 2996
NFWF Budweiser Renewable Energy and Wildlife Conservation Research Prize, 2998
NFWF Nature Restoration Trust Grants, 3005
NFWF Puget Sound Marine Conservation Fund Grants, 3006
NIST Small Business Innovation Res Grants, 3273
NMSF Ernest F. Hollings Ocean Awareness Trust Fund Grants, 3328
NOAA Ballast Water Management Demonstration Program (RDTE Facility) Grants, 3334
NOAA Ballast Water Management Demonstration Program (Technologies and Practices) Grants, 3335
NOAA Ballast Water Technology Demonstration Program (Research, Development, Testing and Evaluation Facility) Grants, 3336
NOAA Ballast Water Technology Demonstration Program (Treatment Technology Demonstration Project) Grants, 3337
NOAA California Bay Watershed Education and Training Program Grants, 3338
NOAA Coastal Ecosystem Research Project, 3350
NOAA Coastal Hypoxia Research Grants, 3351
NOAA Coastal Management Fellowships, 3352
NOAA Cooperative Institute: Eastern U.S. Continental Shelf Frontier Exploration, Research, and Technology Development Grants, 3356
NOAA Coral Reef Ecosystem Studies Grants, 3361
NOAA CRCP Coral Reef Ecosystem Monitoring Grants, 3362
NOAA CSCOR Regional Ecosystem Prediction Program: Cumulative Impacts of Stressors at the Land-Water Interface in the Mid-Atlantic, 3363

NOAA CSCOR Regional Ecosystem Prediction Program: Ecosystem Goal-Setting in Coastal Waters and Reefs of South Florida, 3364
NOAA CSCOR Regional Ecosystem Prediction Program: Invasive Species in the Great Lakes--A Regional Scale Approach, 3365
NOAA Cumulative Impacts of Multiple Stressors (MultiStress) Grants, 3366
NOAA Ecological Forecasting Grants, 3368
NOAA Ecology and Oceanography of Harmful Algal Blooms Grants, 3369
NOAA Great Lakes Habitat Restoration Partnership Grants, 3376
NOAA International Coral Reef Conservation Project Grants, 3377
NOAA Living Marine Resources Cooperative Science Center Internships/Fellowships, 3380
NOAA Living Marine Resources Cooperative Science Center Mini Grants, 3381
NOAA Mid-Atlantic Research Set-Aside Program Grants, 3384
NOAA Monitoring and Event Response for Harmful Algal Blooms Grants, 3385
NOAA National Coastal and Estuarine Research and Technology Program Grants, 3388
NOAA National Estuarine Research Reserve Land Acquisition and Construction Program Grants, 3389
NOAA National Estuarine Research Reserve System Graduate Research Fellowships, 3390
NOAA Northern Gulf of Mexico Ecosystems and Hypoxia Assessment Program Grants, 3396
NOAA Pacific Islands Assistantships, 3399
NOAA Population Dynamics Recruiting Program Grants, 3400
NOAA Regional Ecosystem Prediction Program on Coral Reef Ecosystem Studies Grants, 3401
NOAA South Florida Program Grants, 3406
Northrop Grumman Corporation Grants, 3428
Northwest Fund for the Environment Grants, 3431
NRA Foundation Grants, 3437
NSF Arctic Nat Science Grants in Earth Sciences, 3485
NSF Arctic Research Opportunities, 3490
NSF Climate and Large-Scale Dynamics Grants, 3535
NSF Ecosystem Science Grants, 3556
NSF Partnership for Advancing Technologies in Housing Grants, 3609
NYCT New York City Environment Grants, 3697
Oak Foundation Environmental Grants, 3731
Orchard Foundation Grants, 3793
ORISE Environmental Management Participation for the US Army Environmental Center, 3801
ORISE Postgraduate Environmental Management Research Participation at the US Army Environmental Center, 3816
ORISE Student Environmental Management Participation Program for the US Army Environmental Center, 3826
Oscar Rennebohm Foundation Grants, 3832
Overbrook Foundation Grants, 3860
Padilla Bay Foundation Grants, 3863
Padilla Bay National Estuarine Research Reserve Graduate Research Fellowships, 3864
Padilla Bay National Estuarine Research Reserve Research Assistantships, 3865
Park Foundation Grants, 3870
Patrick and Anna M. Cudahy Fund Grants, 3873
Paul G. Allen Family Foundation Grants, 3876
Peacock Foundation Grants, 3887
Praxair Foundation Grants, 3986
Project AWARE Foundation Grants, 3999
Rajiv Gandhi Foundation Grants, 4025
Rathmann Family Foundation Grants, 4033
Rayonier Foundation Grants, 4034
Reinberger Foundation Grants, 4051
Rhode Island Foundation Grants, 4061
Richard and Rhoda Goldman Fund Grants, 4063
Robert and Patricia Switzer Found Fellowships, 4071
Robert Schalkenbach Foundation Grants, 4079
Ross Foundation Grants, 4106
Royal Caribbean Cruises Ocean Fund, 4108
RSC Miroslaw Romanowski Medal, 4122
Russell Family Foundation Environmental Sustainability Grants, 4128
Ruth and Vernon Taylor Foundation Grants, 4130
Salem Foundation Grants, 4160
San Diego Women's Foundation Grants, 4164
SeaWorld and Busch Gardens Conservation Fund Grants, 4196
Shastri Indo-Canadian Institute Partnership Development Seed Grants, 4228
Shell Oil Company Foundation Grants, 4233
Sioux Falls Area Comm Foundation Spot Grants, 4263
Skaggs Foundation Grants, 4266
Sony Corporation of America Corporate Philanthropy Grants, 4329
Stackpole-Hall Foundation Grants, 4442
STC Research Grants, 4450
Surdna Found Sustainable Environments Grants, 4471
Thompson Charitable Foundation Grants, 4511
UNCFSP NASA Harriett G. Jenkins Predoctoral Fellowships, 4563
Union Pacific Foundation Community and Civic Grants, 4568
USDA Organic Agriculture Research and Extension Initiative, 4637
USDC/NOAA Anadromous Fish Conservation Act Grants, 4654
USDC/NOAA Coastal and Marine Fish Habitat Protection Grants, 4655
USDC/NOAA Community-Based Habitat Restoration Program Grants, 4656
USDC/NOAA Community-Based Marine Debris Removal Grants, 4657
Vancouver Foundation Grants and Community Initiatives, 4807
W. C. Griffith Foundation Grants, 4818
Westinghouse Charitable Giving Grants, 4856
Weyerhaeuser Company Foundation Grants, 4857
William G. and Helen C. Hoffman Found Grants, 4876
Yves Rocher Foundation Terra Ficaria Prize for Scientific Research, 4904

Environmental Protection

ASME Dixy Lee Ray Award, 1115
Caesars Foundation Grants, 1407
EPA Compliance Assistance Support for Services to the Regulated Community and Other Assistance Providers, 1934
EPA Regional Healthy Indoor Air Projects for Community Outreach and Education, 1937
EPA Research, Development, Monitoring, Public Education, Training, Demonstrations, and Studies Grants, 1938
Freshwater Future Project Grants, 2056
Georgia-Pacific Foundation Environment Grants, 2123
Great Lakes Fishery Commission Sea Lamprey Research Grants, 2167
Honeywell Corp Sustainable Opportunities Grants, 2278
Meadows Foundation Grants, 2736
MoEF Dr. Salim Ali Nat Wildlife Fellowship, 2792
MOEF Gandhi Paryavaran Puraskar Award, 2793
Oak Foundation Environmental Grants, 3731
Sioux Falls Area Community Foundation Community Fund Grants (Unrestricted), 4262
Sioux Falls Area Comm Foundation Spot Grants, 4263
USDA Organic Agriculture Research and Extension Initiative, 4637
USDA Technical Agricultural Assistance Research Grants, 4651
USDC/NOAA Restoration Center National and Regional Partnership Grants, 4662

Environmental Research

A Charitable Foundation Grants, 273
EPA Brownfields Area-Wide Planning Grants, 1931
EPA Regional Healthy Indoor Air Projects for Community Outreach and Education, 1937
EPA Research, Development, Monitoring, Public Education, Training, Demonstrations, and Studies Grants, 1938
EPA Southeastern Multi-Media and Geographic Priority Project Grants, 1939
EPA Surveys, Studies, Research, Investigations, Demonstrations, and Special Purpose Activities Relating to the Clean Air Act, 1940
Georgia-Pacific Foundation Environment Grants, 2123
GNOF Gulf Coast Oil Spill Grants, 2153
Great Lakes Fishery Commission Pilot Research Project Grants, 2166
Honeywell Corporation Sustainable Opportunities Grants, 2278
Illinois DCEO Coal Research Grants, 2359
Mary Gates Research Scholarships, 2670
MDEQ Scrap Tire Market Development Grants, 2733
MoEF Dr. Salim Ali Nat Wildlife Fellowship, 2792
Oak Foundation Environmental Grants, 3731
OTS Undergrad NAPIRE Summer Interns, 3858
SARE Large Systems Research Grants - Southern Region, 4172
Searle Freedom Trust Grants, 4192
SRC GRC & SEMATECH Joint Projects Grants, 4369
SRC Simon Karecki Fellowship, 4393
USDA Basic and Applied Research Grants, 4617
USDA Biotechnology Risk Assessment Grants, 4618
USDA International Science and Education Competitive Grants, 4631
USDA Interregional Research Project Minor Crop Pest Management Grants, 4632
USDA Policy Research Center Grants, 4640
USDA Wildlife Services Grants, 4653
USDD ARO Broad Agency Announcement, 4686
USDD Battlespace Environment Division Research Interests Grant, 4692
WILD Foundation Grants, 4867

Environmental Services

Carl C. Icahn Foundation Grants, 1428
EPA Compliance Assistance Support for Services to the Regulated Community and Other Assistance Providers, 1934
Georgia-Pacific Foundation Environment Grants, 2123
Oak Foundation Environmental Grants, 3731

Environmental Studies

Abdus Salam ICTP Postgraduate Diploma Grants, 235
Abdus Salam ICTP Training and Research in Italian Laboratories Grants, 240
AIChE Environmental Division Graduate Student Paper Award, 407
AIChE Environmental Division Undergraduate Student Paper Award, 408
AIChE Lawrence K. Cecil Award in Environmental Chemical Engineering, 415
AINSE Honours Scholarships, 460
AINSE John Ferris Memorial Award, 462
AINSE Postgraduate Research Awards, 463
Alexander von Humboldt Foundation International Climate Protection Fellowships, 529
Alfred P. Sloan Foundation Barcode of Life Grants, 543
Alfred P. Sloan Foundation Microbiology of the Built Environment Grants, 554
American Chemical Society Award for Affordable Green Chemistry, 650
AMS/Industry Government Graduate Fellowships, 817
AMS/Industry Minority Scholarships, 818
AMS Grad Fellowships in History of Science, 824
AMS Undergraduate Scholarships, 837
ANL Faculty Research Participation Program, 855
ANL Graduate Thesis Parts Res Appointments, 856
ANL Laboratory-Graduate Res Appointments, 857
APS/Phoenix Suns Energy and Environmental Minigrants, 919
ASCE Trent R. Dames and William W. Moore Engineering Fellowship, 1043
ASME Dixy Lee Ray Award, 1115
Aviation Research Grants, 1273
AWRA Richard A. Herbert Mem Scholarship, 1281
Brainerd Foundation Grants, 1368
Canada-U.S. Fulbright International Science and Technology Award, 1417

Canada-U.S. Fulbright Senior Specialists Program Grants, 1418
Canada-U.S. Fulbright Traditional Scholars, 1419
Cooperative Institute for Research in Environmental Sciences Visiting Fellows Program, 1643
DOE Program for Ecosystem Research, 1810
Earthwatch Institute Student Fellowships, 1890
Edmund Niles Huyck Preserve Graduate and Postgraduate Research Grants, 1894
EPA Surveys, Studies, Research, Investigations, Demonstrations, and Special Purpose Activities Relating to the Clean Air Act, 1940
EREF Sustainability Research Grants, 1948
Explorers Club Scott Pearlman Field Awards, 1967
Fulbright Alumni Initiatives Awards Grants, 2064
Fulbright Distinguished Chairs Program Grants, 2069
Fulbright European Union Scholar-in-Residence Program Grants, 2071
Fulbright New Century Scholars Program Grants, 2075
Fulbright Occasional Lecturer Program Grants, 2076
Fulbright Specialists Program Grants, 2079
Fulbright Traditional Scholar Program in Europe and Eurasia, 2080
Fulbright Traditional Scholar Program in Sub-Saharan Africa, 2081
Fulbright Traditional Scholar Program in the East Asia/Pacific Region, 2082
Fulbright Traditional Scholar Program in the Near East and North Africa Region, 2083
Fulbright Traditional Scholar Program in the South and Central Asia Region, 2084
Fulbright Traditional Scholar Program in the Western Hemisphere, 2085
Fulbright Visiting Specialists Program: Direct Access to the Muslim World, 2087
Fund for Scientific Research--Flanders Grants, 2089
GCA Summer Environmental Scholarships, 2100
Georgia-Pacific Foundation Environment Grants, 2123
Gilbert F. White Postdoctoral Fellowships, 2140
Honeywell Corporation Habitat and Conservation Grants, 2277
HRF Hudson River Improvement Grants, 2286
HRF New York City Environmental Grants for Newton Creek, 2288
IFT Foundation Grants, 2345
Ittleson Foundation Environment Grants, 2437
John B. Bene Fellowships in Social Forestry, 2496
Joseph L. Fisher Dissertation Awards, 2515
Loeb Fellowships in Advanced Environmental Design Studies, 2621
NATO Advanced Research Workshops Program, 2894
NATO Advanced Study Institutes Program, 2895
NATO Collaborative Linkage Grants, 2896
NATO Expert Visits Program Grants, 2897
NMSF Ernest F. Hollings Ocean Awareness Trust Fund Grants, 3328
NOAA Ecological Forecasting Grants, 3368
NOAA International Coral Reef Conservation Project Grants, 3377
NOAA Joint Research Institutes Grants, 3379
NOAA National Estuarine Research Reserve Land Acquisition and Construction Program Grants, 3389
NSF-NIST Interaction in Chemistry, Materials Research, Molecular Biosciences, Bioengineering, and Chemical Engineering, 3453
NSF Arctic System Science Program Grants, 3493
NSF Natural, Social, and System Sciences and Research Support and Logistics Program, 3600
NSF Paleoclimate Grants, 3606
NSF Solar, Heliospheric, and INterplanetary Environment Grants, 3651
NYCT New York City Environment Grants, 3697
Oak Foundation Environmental Grants, 3731
Olin Corporation Charitable Trust Grants, 3739
ONR Electronic and Optical Ceramics Research Grants, 3756
ONR Marine Meteorology and Atmospheric Effects Research Grants, 3769
ONR Optical and Biological Oceanography Research Grants, 3777
ORISE Oak Ridge National Laboratory Postdoctoral Research Associates Program, 3815
ORISE Postgraduate Research Participation at the National Exposure Research Laboratory, 3821
RA Kleinhans Fellowship for Research in Tropical Nontimber Forest Products, 4026
Richard and Rhoda Goldman Fund Grants, 4063
Robert Schalkenbach Foundation Grants, 4079
Smithsonian Environmental Research Center Graduate Student Fellowships, 4273
Smithsonian Environmental Research Center Internships, 4274
Smithsonian Environmental Research Center Postdoctoral/Senior Fellowships, 4275
Smithsonian Environmental Research Center Predoctoral Fellowships, 4276
Smithsonian Senior Fellowships, 4296
Society for the Arts in Healthcare Environmental Arts Research Grant, 4311
SSHRC Insight Development Grants, 4409
SSHRC Insight Grants, 4410
Starr Foundation Grants, 4447
Trevithick Book Prize Award, 4532
U.S. Department of Education United States-Russia Program: Improving Research and Educational Activities in Higher Education, 4547
UNCFSP NASA Harriett G. Jenkins Predoctoral Fellowships, 4563
USCD/NOAA Illinois-Indiana Sea Grant Core Proposals, 4604
US CRDF CGP-Climate Change & Energy Competition For U.S.-Ukrainian Res Teams, 4607
US CRDF CGP-RFBR Climate Change and Energy Competition For Joint U.S.-Russian Research Teams, 4608
USDA 1890 Institution Teaching, Research, and Extension Capacity Building Grants, 4613
USDA Distance Education Grants for Higher Education Institutions in Insular Areas, 4622
USDA Norman E. Borlaug International Agricultural Science and Technology Fellowships, 4636
USDA Rangeland Research Grants, 4641
USDA Resident Instruction Grants for Higher Education in Insular Areas, 4645
USDA Tribal College Research Grants, 4652
WILD Foundation Grants, 4867
Yves Rocher Foundation Terra Ficaria Prize for Scientific Research, 4904

Enzymes/Enzymology

NHLBI Career Transition Awards, 3021
NIDDK Development of Assays for High-Throughput Drug Screening Grants, 3144
NSF Biomolecular Systems Cluster Grants, 3507

Epidemiology

AACR-American Cancer Society Award for Research Excellence in Epidemiology & Prevention, 112
AACR Basic Cancer Research Fellowships, 130
AACR Career Development Awards for Pediatric Cancer Research, 132
AACR Judah Folkman Career Development Award for Anti-Angiogenesis Research, 140
AACR Outstanding Investigator Award for Breast Cancer Research, 145
ACS Mentored Research Scholar Grant in Applied and Clinical Research, 314
ACSM Paffenbarger-Blair Fund for Epidemiological Research on Physical Activity Grants, 321
AHAF Alzheimer's Disease Research Grants, 370
AMSSM Foundation Research Grants, 836
APHL Emerging Infectious Diseases Fellowships, 914
ASM-PAHO Infectious Diseases Epidemiology and Surveillance Fellowships, 1098
Bristol-Myers Squibb Foundation Fellowships, 1375
Bristol-Myers Squibb Foundation Global HIV/AIDS Initiative Grants, 1376
CDC Experience Appl Epidemiology Fellowships, 1460
CDC Foundation Atlanta International Health Fellowships, 1461
CDC Preventive Med Residency & Fellowship, 1465
CSTE CDC/CSTE Applied Epidemiology Fellowships, 1664
IARC Expertise Transfer Fellowship, 2315
IARC Postdoctoral Fellowships for Training in Cancer Research, 2316
IARC Visiting Scientist Award for Sen Scientists, 2317
James S. McDonnell Foundation Complex Systems Collaborative Activity Awards, 2465
Landon Foundation-AACR Innovator Award for Cancer Prevention Research, 2586
Landon Foundation-AACR Innovator Award for International Collaboration, 2587
NCI/NIDA Transdisciplinary Tobacco Use Research Centers Grants, 2912
NCI Basic Cancer Research in Cancer Health Disparities Grants, 2916
NCI Cancer Education and Career Development, 2917
NCIC Terry Fox Postdoc Research Fellowships, 2923
NCI Stages of Breast Development: Normal to Metastatic Disease Grants, 2932
NEI Clinical Study Planning Grant, 2959
NEI Clinical Vision Research Devel Award, 2960
NFID New Investigator Matching Grants, 2989
NFID Postdoctoral Fellowships in Emerging Infectious Diseases, 2991
NHLBI National Research Service Award Programs in Cardiovascular Epidemiology and Biostatistics, 3041
NIAMS Pilot & Feasibility Clinical Grants in Arthritis & Musculoskeletal & Skin Diseases, 3111
NIDA Mentored Clinical Scientist Research Career Development Awards, 3129
NIDDK Type 2 Diabetes in the Pediatric Population Research Grants, 3188
NIGMS/NIDCR Evolution of Infectious Diseases Research Grants, 3196
NYAM Mary and David Hoar Fellowship, 3691
ORISE Faculty Research Participation at the Centers for Disease Control and Prevention, 3803
ORISE Postgraduate Research Participation at the Centers for Disease Control and Prevention, 3820
ORISE Student Research Participation at the Centers for Disease Control and Prevention, 3829
Pancreatic Cancer Network-AACR Fellowship, 3866
Pancreatic Cancer Action Network-AACR Pathway to Leadership Grants, 3867
Pfizer Anti Infective Research EU Grants, 3909
Pfizer ASPIRE EU Antifungal Research Awards, 3910
Pfizer ASPIRE EU Emerging Mechanisms of Resistance Antibacterial Research Awards, 3912
Pfizer ASPIRE North America Rheumatology Research Awards, 3917
Pfizer Global Investigator-Initiated Res Grants, 3924
Pfizer Inflammation Competitive Res Awards, 3926
Simone and Cino del Duca Grand Prix Awards, 4261
Susan G. Komen Breast Cancer Foundation Challege Grants: Investigator Initiated Research, 4475
Susan G. Komen Breast Cancer Foundation Challenge Grants: Career Catalyst Research, 4476
UICC American Cancer Society International Fellowships for Beginning Investigators, 4555
US CRDF Leishmaniasis: Collaborative Research Opportunities in N Africa & Middle East, 4610

Epigraphy

AIAR Andrew W. Mellon Foundation Fellowships, 392
Kluge Center Kislak Fellows in American Studies, 2562
Kluge Center Kislak Short-Term Fellowships in American Studies, 2563

Epilepsy

AES-Grass Young Investigator Travel Awards, 345
AES Robert S. Morison Fellowship, 346
AES Susan S. Spencer Clinical Research Epilepsy Training Fellowship, 347
ASPET Epilepsy Research Award for Outstanding Contributions to the Pharmacology of Antiepileptic Drugs, 1203
Grass Foundation Marine Biological Laboratory Fellowships, 2163

NIH Research on Sleep and Sleep Disorders, 3236
Pfizer Foundation Epilepsy Scholarship Award, 3923
Savoy Foundation Post-Doctoral and Clinical Research Fellowships, 4184
Savoy Foundation Research Grants, 4185
Savoy Foundation Studentships, 4186

Episcopal Church
Barra Foundation Project Grants, 1301
Booth-Bricker Fund Grants, 1359
Herbert & Adrian Woods Foundation Grants, 2239
MacDonald-Peterson Foundation Grants, 2649
Robert E. and Evelyn McKee Foundation Grants, 4074

Equal Opportunity
Allstate Foundation Safe and Vital Communities Grants, 573
Compton Foundation Grants Program, 1623
DaimlerChrysler Corporation Fund Grants, 1714
Dept of Ed Women's Educational Equity Grants, 1752
Vanguard Public Foundation Grant Funds, 4808

Equipment/Instrumentation
American Chemical Society ANYL Award in Chemical Instrumentation, 641
ARO Theoretical Physics and Nonlinear Phenomena Research Contracts, 1017
Ball Brothers Foundation Rapid Grants, 1292
Boeing Company Contributions Grants, 1357
Campbell Soup Foundation Grants, 1415
Chatlos Foundation Grants Program, 1532
Coughlin-Saunders Foundation Grants Program, 1648
Dept of Ed Star Schools Program Grants, 1749
E.L. Wiegand Foundation Grants, 1885
Fritz B. Burns Foundation Grants, 2057
Ludwick Family Foundation Grants, 2625
NCAR Graduate Research Fellowships, 2901
Norcliffe Foundation Grants, 3415
NSERC Research Tools and Instruments Grants, 3449
NSF Arctic Research Opportunities, 3490
NSF Atmospheric Sciences Mid-Size Infrastructure Opportunity Grants, 3497
NSF Biological Field Stations and Marine Laboratories Grants, 3500
NSF Chemistry Research Instrumentation and Facilities Program Grants, 3520
NSF Doctoral Dissertation Improvement Grants in the Directorate for Biological Sciences, 3554
NSF Historically Black Colleges and Universities Program Grants, 3577
NSF Instrument Devel for Biological Research, 3583
NSF Interagency Opportunities in Metabolic Engineering Research Grants, 3585
NSF Living Stock Collections Grants, 3588
ONR Electronic and Optical Ceramics Research Grants, 3756
Potts Memorial Foundation Grants, 3985
RCY - NSF Ocean Sciences Div Research Grants, 4043
Robert Leet and Clara Guthrie Patterson Grants, 4076
RSC Thomas W. Eadie Medal, 4126
Seattle Foundation Medical Funds Grants, 4194
SRC Educational Science Medium-Cost Scientific Equipment Grants, 4361
SRC Educational Science Project Res Grants, 4364
SRC Research Infrastructures Grants, 4391
Stanley Smith Horticultural Trust Grants, 4446
Sun Microsystems Foundation Summer Academic Enrichment Grants, 4468

Ergonomics
BMW of N America Charitable Contributions, 1353

Erythropoiesis
NIDDK Erythropoiesis: Components and Mechanisms Grants, 3153

Estuarine Sciences
EPA National Estuary Program Grants, 1935
EPA Targeted Watersheds Grants, 1941
HRF Graduate Fellowships, 2285
HRF Hudson River Improvement Grants, 2286
HRF Mark B. Bain Graduate Fellowships, 2287
HRF Tibor T. Polgar Fellowships, 2289
NOAA California Bay Watershed Education and Training Program Grants, 3338
NOAA Coastal and Estuarine Land Conservation Program Grants, 3349
NOAA Coastal Ocean Program Grants, 3353
NOAA National Coastal and Estuarine Research and Technology Program Grants, 3388
NOAA National Estuarine Research Reserve Land Acquisition and Construction Program Grants, 3389
NOAA National Estuarine Research Reserve System Graduate Research Fellowships, 3390
NSF Arctic System Science Program Grants, 3493
Padilla Bay Foundation Grants, 3863
Padilla Bay National Estuarine Research Reserve Graduate Research Fellowships, 3864
Padilla Bay National Estuarine Research Reserve Research Assistantships, 3865
Smithsonian Environmental Research Center Graduate Student Fellowships, 4273
Smithsonian Environmental Research Center Internships, 4274
Smithsonian Environmental Research Center Postdoctoral/Senior Fellowships, 4275
Smithsonian Environmental Research Center Predoctoral Fellowships, 4276

Ethics
AAAS Award for Scientific Freedom and Responsibility, 28
Affymetrix Corporate Contributions Grants, 355
Alabama Humanities Foundation Mini Grants, 480
Alabama Humanities Foundation Planning/Consultant Grants, 481
Alice Tweed Tuohy Foundation Grants Program, 569
American Academy of Advertising Kim Rotzoll Award for Advertising Ethics and Social Responsibility, 610
Center for the Study of Professional Military Ethics Resident Fellowship, 1484
Charles A. King Trust Postdoctoral Research Fellowships, 1518
Charlotte W. Newcombe Doctoral Dissertation Fellowships, 1531
ConocoPhillips Grants Program, 1629
Earhart Foundation Institutional Grants, 1888
Ethics and Excellence in Journalism Foundation Grants, 1953
Gilbreth Memorial Fellowships, 2141
Goldsmith Awards Program, 2158
Greenwall Foundation Faculty Scholars Program in Bioethics, 2175
Greenwall Foundation Fellowship Program in Bioethics and Health Policy, 2176
Marion Gardner Jackson Charitable Trust Grants, 2664
Murphy Institute Center for Ethics and Public Affairs Faculty Fellowships, 2805
Murphy Institute Center for Ethics and Public Affairs Graduate Fellowships, 2806
NFID Adv Vaccinology Course Travel Grant, 2986
NIH Research On Ethical Issues In Human Subjects Research, 3235
NSF Science and Society, 3643
Princeton University Center for Human Values Graduate Student Top-Up Prizes, 3990
Princeton University Center for Human Values Laurance S. Rockefeller Visiting Fellowships, 3992
Richard Davoud Donchian Foundation Grants, 4066
RSC Abbyann D. Lynch Medal in Bioethics, 4116
Rumsfeld Foundation Graduate Fellowships, 4127
Virginia Foundation for the Humanities Residential Fellowships, 4815

Ethnicity
CSRPC Dissertation Fellowships, 1661
CSRPC Postdoctoral Fellowship for Advanced Scholars, 1662
CSRPC Postdoc Fellowship for Junior Scholars, 1663
USDC Postdoctoral Research Grants, 4669

Ethnography
Center for East Asian Studies Fellowships, 1474
Douglass Institute Postdoctoral Fellowships, 1851
Kluge Center Kislak Fellows in American Studies, 2562
Kluge Center Kislak Short-Term Fellowships in American Studies, 2563
Smithsonian Senior Fellowships, 4296

Ethnology
ACC Ford Foundation Fellowships, 253
Australasian Institute of Judicial Administration Seed Funding Grants, 1268
Wenner-Gren Foundation Dissertation Fieldwork Grants, 4848

Ethnomusicology
SSRC Mellon Mays Predoctoral Research Grants, 4432

Etiology
AACR Gertrude Elion Cancer Research Award, 137
Alzheimer's Association Conference Grants, 589
Alzheimer's Association Development of New Cognitive and Functional Instruments Grants, 590
Alzheimer's Association Everyday Technologies for Alzheimer Care Grants, 591
Alzheimer's Association Mentored New Investigator Research Grants to Promote Diversity, 593
Alzheimer's Association Neuronal Hyper Excitability and Seizures in Alzheimer's Disease Grants, 594
Alzheimer's Association New Investigator Research Grants to Promote Diversity, 596
Alzheimer's Association U.S.-U.K. Young Investigator Exchange Fellowships, 597
Alzheimer's Association Zenith Fellows Awards, 598
IARC Expertise Transfer Fellowship, 2315
IARC Postdoctoral Fellowships for Training in Cancer Research, 2316
IARC Visiting Scientist Award for Sen Scientists, 2317
NIA Alzheimer's Disease Core Centers Grants, 3084
NYAM Edward N. Gibbs Memorial Lecture and Award in Nephrology, 3682
Penny Severns Breast, Cervical and Ovarian Cancer Research Grants, 3896

Eugenics
American Philosophical Society Library Resident Research Fellowships, 769

Europe
Ahmanson and Getty Postdoctoral Fellowships, 377
ALA WESS De Gruyter European Librarianship Study Grant, 501
American Historical Association Bernadotte E. Schmitt Research Grants in European, African, or Asian History, 715
American Historical Assoc George L. Mosse Prize, 719
American Historical Assoc George L. Beer Prize, 720
American Historical Association Herbert Baxter Adams Prize, 722
American Historical Assoc James Breasted Prize, 729
American Historical Association Michael Kraus Research Grant in American Colonial History, 738
American Historical Association Paul Birdsall Prize in European Military and Strategic History, 740
American Historical Assoc Wesley-Logan Prize, 744
BA Elisabeth Barker Fund, 1288
DAAD Undergraduate Scholarships, 1710
ESF Latsis Prize, 1951
Fulbright Traditional Scholar Program in Europe and Eurasia, 2080
IREX Short-Term Travel Grants, 2424
Kress Found Art History Pre-Doc Fellowships, 2576
MPI for Social Anthropology Department II Postdoctoral Grants, 2798

Europe, Central
ARIT Mellon Foundation Postdoctoral Research Fellowships for East European Scholars, 994
ASM-UNESCO Leadership Grant for International Educators, 1099

BA Elisabeth Barker Fund, 1288
Charles Stewart Mott Foundation Grants, 1529
IREX Individual Advanced Research Opportunities Grants, 2419
IREX Short-Term Travel Grants, 2424
OSF Human Rights and Governance Internships, 3845
UICC American Cancer Society International Fellowships for Beginning Investigators, 4555

Europe, Eastern
AAB Berlin Prize Fellowships, 48
ACLS Dissertation Fellowships in East European Studies, 285
ACLS Early Career Postdoctoral Fellowships in East European Studies, 286
ARIT Mellon Foundation Postdoctoral Research Fellowships for East European Scholars, 994
ASM-UNESCO Leadership Grant for International Educators, 1099
BA Elisabeth Barker Fund, 1288
Charles Stewart Mott Foundation Grants, 1529
Eastern Europe & the Independent States of the Former Soviet Union Research & Training Grants, 1891
Fulbright Traditional Scholar Program in Europe and Eurasia, 2080
IREX Individual Advanced Research Opportunities Grants, 2419
IREX Short-Term Travel Grants, 2424
OSF Human Rights and Governance Internships, 3845
UICC American Cancer Society International Fellowships for Beginning Investigators, 4555

European History
American Historical Association George Louis Beer Prize, 720
American Historical Assoc Leo Gershoy Award, 734
IES Faculty Fellowships, 2341
LBI-L Leo Baeck/Studienstiftung Fellowships, 2592
LBI-NY/DAAD Fellowships in Germany, 2593
LBI-NY/DAAD Fellowships in New York, 2594
LBI-NY Fritz Halbers Fellowships, 2597
LBI-NY Gerald Westheimer Career Development Fellowship, 2598
Leo Baeck Essay Prize, 2601
NL American Society for Eighteenth-Century Studies Fellowships, 3279
NL Arthur and Lila Weinberg Fellowships for Independent Scholars and Researchers, 3280
NL Associated Colleges of the Midwest/Great Lakes College Association Faculty Fellowships, 3281
NL British Academy Fellowship for Study in the United Kingdom, 3283
NL Frances C. Allen Fellowships, 3284
NL Grad Scholar-in-Residence Opportunities, 3285
NL Herzog August Bibliothek Wolfenbuttel Fellowship, 3286
NL Lawrence Lipking Fellowship at Newberry, 3287
NL Lester J. Cappon Fellowship in Documentary Editing, 3288
NL Mellon Postdoctoral Research Fellowships, 3294
NL Midwest Modern Lang Assoc Fellowship, 3299
NL Monticello College Foundation Fellowship for Women, 3304
NL National Endow for Humanities Fellowships, 3310
NL Northeast Modern Lang Assoc Fellowship, 3314
NL Scholar-in-Residence Opportunity, 3315
NL Short-Term Resident Fellowships for Individual Research, 3317
NL South Central Modern Languages Association Fellowship, 3318
NL Susan Kelly Power and Helen Hornbeck Tanner Fellowship, 3319
NL Weiss/Brown Publication Subvention Award, 3320
SSRC Our Shared Past Grants, 4437

Evolutionary Biology
Alfred P. Sloan Foundation Research Fellowships, 560
American Philosophical Society Library Resident Research Fellowships, 769
AMNH Graduate Student Fellowships, 815
AMNH Lerner-Gray Grants for Marine Research, 816
Franklin Mosher Baldwin Fellowships, 2050
MBL Albert & Ellen Grass Summer Fellowships, 2692
MBL Stephen W. Kuffler Summer Fellowships, 2715
NIGMS/NIDCR Evolution of Infectious Diseases Research Grants, 3196
NSF Center for Synthesis in Biological Evolution, 3513
NSF Doctoral Dissertation Improvement Grants in the Directorate for Biological Sciences, 3554
NSF Genes and Genome Systems Cluster Grants, 3569
NSF Physiology and Ethology Research Grants, 3622
NSF Systematic Biology and Biodiversity Inventories Grants, 3657
Orchid Research Fellowships, 3794
Smithsonian Graduate Student Fellowships, 4277
Smithsonian Minority Visiting Student Awards, 4279
Smithsonian Molecular Evolution Fellowships, 4280
Smithsonian Native American Community Scholars Awards, 4291
Smithsonian Native American Visiting Student Awards, 4292
Smithsonian Postdoctoral Fellowships, 4293
Smithsonian Predoctoral Fellowships, 4295

Excavation
ARIT Kenan T. Erim Fellowship for Research at Aphrodisias, 993

Exchange Programs, Student
IBRO InEurope Short Stay Grants, 2325

Exercise
ACSM Carl V. Gisolfi Memorial Fund Grant, 312
ACSM Coca-Cola Company Doctoral Student Grant on Behavior Research, 313
ACSM Foundation Research Grants, 317
ACSM International Student Awards, 318
ACSM NASA Space Physiology Research Grants, 319
ACSM Oded Bar-Or Int'l Scholars, 320
BCBSNC Foundation Grants, 1318
NCI Improving Diet and Physical Activity Assessment Grants, 2929
ONR Medical Science and Technology Division Grants, 3772

Exhibitions, Collections, Performances
ACC Asian Art and Religion Fellowships, 251
ACC Japan-US Arts Program Fellowships, 256
ACC Philippines Fellowship Program, 261
Andy Warhol Foundation for the Visual Arts Curatorial Research Fellowships, 847
CCH Documentary Project Research and Development Grants, 1451
Christensen Fund Regional Grants, 1554
Florence Tan Moeson Fellowships, 2025
Getty Research Institute Nonresidential Grants - Curatorial Research Fellowships, 2136
Greater Worcester Community Foundation Jeppson Memorial Fund for Brookfield Grants, 2165
Massachusetts Foundation for the Humanities Project Grants, 2678
Musgrave Foundation Grants, 2807
National Endowment for the Humanities Fellowships at Winterthur, 2867
NYSCA Folk Arts: Exhibitions Grants, 3703
NYSCA Folk Arts: Regional and County Folk Arts Programs Grants, 3705
NYSCA Museum: Project Support Grants, 3707
Seaver Institute Grants, 4195
Virginia Foundation for the Humanities Discretionary Grants, 4813
Virginia Foundation for Humanities Open Grants, 4814
Winterthur Museum Dissertation Fellowships, 4887
Winterthur Research Fellowships, 4888

Expert Systems
ARO Discrete Mathematics and Computer Science Research Contracts, 1009

Explosions and Explosives
USDD Ocean Engineering and Marine Systems Grants, 4767

Eye Diseases
1st Touch Foundation Grants, 2
1976 Foundation Grants, 7
AHAF Macular Degeneration Research Grants, 371
CNIB Baker Applied Research Fund Grants, 1588
CNIB Baker New Researcher Fund Grants, 1589
CNIB Barbara Tuck MacPhee Award, 1590
CNIB Canada Glaucoma Clinical Research Council Grants, 1591
CNIB E. & Mary Hochhausen Access Technology Research Grants, 1592
Donald & Sylvia Robinson Family Found Grants, 1830
E.L. Wiegand Foundation Grants, 1885
Eye-Bank for Sight Restoration and Fight for Sight Summer Student Research Fellowship, 1969
Fairlawn Foundation Grants, 1974
Fight for Sight-Streilein Foundation for Ocular Immunology Research Award, 2012
Fight for Sight Grants-in-Aid, 2013
Fight for Sight Post-Doctoral Awards, 2014
Fight for Sight Summer Student Fellowships, 2015
George Gund Foundation Grants, 2117
Glaucoma Foundation Grants, 2145
Glaucoma Research Pilot Project Grants, 2146
MBL Plum Found John E. Dowling Fellowships, 2713
NEI Scholars Program, 2964
NEI Translational Research Program On Therapy For Visual Disorders, 2965
OneSight Research Foundation Block Grants, 3742
Sara Elizabeth O'Brien Trust Grants, 4166
Streilein Foundation for Ocular Immunology Visiting Professorships, 4453
Thome Foundation Awards in Age-Related Macular Degeneration Research, 4509

Facility Support
HRF Hudson River Improvement Grants, 2286
NYSCA Individual Artists: Film, Media and New Technology Production Grants, 3706

Faculty Development
American Chemical Society Chemical Technology Partnership Mini Grants, 674
American Chemical Society Petroleum Research Fund Doctoral New Investigator Grants, 690
American Chemical Society Petroleum Research Fund Undergraduate Faculty Sabbatical Grants, 693
ARO Young Investigator Program/Presidential Early Career Awards, 1018
ASEE Naval Research Laboratory Postdoctoral Fellowships, 1060
Beckman Young Investigators Program Grants, 1327
BWF Institutional Program Unifying Population and Laboratory Based Sciences Grants, 1397
Carpenter Foundation Grants, 1439
Deloitte Foundation Grants, 1735
FAMRI Young Clinical Scientist Awards, 1976
Graduate Institute Hodson Teacher Fellowships, 2160
HHMI/EMBO Start-up Grants for Cen Europe, 2250
Independent Institute John M. Templeton Fellowships Essay Contest, 2369
JSA JLab Sabbatical Research Leave Grants, 2522
M.J. Murdock Charitable Trust Research Start-up Grants for New Science Faculty, 2643
MBL Ann E. Kammer Mem Summer Fellowship, 2693
MBL Burr & Susie Steinbach Summer Fellow, 2696
MBL E.E. Just Summer Fellowship for Minority Scientists, 2697
MBL Eugene and Millicent Bell Fellowships, 2699
MBL Evelyn & Mel Spiegal Summer Fellowship, 2700
MBL Frank R. Lillie Summer Fellowship, 2701
MBL Gruss Lipper Family Foundation Summer Fellowship, 2704
MBL Herbert W. Rand Summer Fellowship, 2706
MBL James E. and Faith Miller Memorial Summer Fellowship, 2707

MBL Lucy B. Lemann Mem Summer Fellowship, 2710
McGregor Fund Grants, 2725
NHLBI Mentored Career Award For Faculty At Minority Institutions, 3034
NL Associated Colleges of the Midwest/Great Lakes College Association Faculty Fellowships, 3281
North Carolina Biotechnology Center Oliver Smithies Faculty Recruitment Grant, 3423
NSF Historically Black Colleges and Universities Program Grants, 3577
OSF Global Faculty Grants - Track 2, 3842
PhRMA Foundation Health Outcomes Post Doctoral Fellowships, 3950
PhRMA Foundation Pharmaceutics Research Starter Grants, 3960
Robert M. Hearin Foundation Grants, 4077
RWJF Harold Amos Medical Faculty Development Research Grants, 4139
RWJF Harold Amos Medical Faculty Development Scholars, 4140
RWJF New Connections Jun Investigator Grants, 4146
RWJF New Connections Mid-Career Consultants Grants, 4147
RWJF Nurse Faculty Scholars, 4148
Saint Mary's College Fellowships, 4159
Shastri Indo-Canadian Institute Faculty Research Fellowships, 4222
SNM Molecular Imaging Research Grant For Junior Medical Faculty, 4299
SNM Postdoctoral Molecular Imaging Scholars, 4302
UCR Resident Fellowships, 4552
UNCFSP NASA Administrator's Fellowship, 4562
USDA 1890 Institution Teaching, Research, and Extension Capacity Building Grants, 4613
USDA Higher Education Challenge Grants, 4627

Faculty Support
ALA Excellence in Academic Libraries Award, 486
CIC Summer Research Opportunities, 1565
FAMRI Young Clinical Scientist Awards, 1976
NHLBI Ruth L. Kirschstein National Research Service Awards for Individual Predoctoral Fellowships to Promote Diversity in Health-Related Research, 3058

Familial Abuse
Adaptec Foundation Grants, 340
AMWA Kathryn C. Bemmann New Investigator Grant on Violence against Women, 838
CDC Grants for Violence-Related Injury Prevention Research, 1463
Herbert & Adrian Woods Foundation Grants, 2239

Family
3M Foundation Community Giving Grants, 5
A.L. Mailman Family Foundation Grants, 11
ACF Adoption Opportunities Project Grants, 270
Achelis Foundation Grants, 274
ACS Pilot and Exploratory Projects in Palliative Care of Cancer Patients and Their Families Grants, 324
Adaptec Foundation Grants, 340
Alcatel-Lucent Technologies Foundation Grants, 509
Alex Stern Family Foundation Grants, 542
Arizona Cardinals Grants, 995
Auburn Foundation Grants, 1263
BBF Fellowships, 1311
Better Way Foundation Grants, 1338
Bodenwein Public Benevolent Foundation Grants, 1355
Bodman Foundation Grants, 1356
Booth-Bricker Fund Grants, 1359
Boston Foundation Grants, 1361
Brookdale Foundation Leadership in Aging Fellowships, 1381
Campbell Soup Foundation Grants, 1415
Carpenter Foundation Grants, 1439
Chapin Hall International Fellowships in Children's Policy Research, 1517
Charles Stewart Mott Foundation Grants, 1529
Clay Foundation Grants, 1579
College of Human Sciences Scholarships, 1596
Community Found for Greater Buffalo Grants, 1615
Connecticut Community Foundation Grants, 1627
D.F. Halton Foundation Grants, 1692
DHHS Comprehensive Community Mental Health Services Grants for Children with Serious Emotional Disturbances, 1754
DHHS Independence Demonstration Program, 1757
DOJ Juvenile Justice and Delinquency Prevention Special Emphasis Grants, 1826
Donald and Sylvia Robinson Family Foundation Grants, 1830
Emma B. Howe Memorial Foundation Grants, 1920
EPA Children's Health Protection Grants, 1933
Evanston Community Foundation Grants, 1962
FAR Fund Grants, 1980
Farmers Insurance Corporate Giving Grants, 1981
FCD Child Development Grants, 1984
Fidelity Foundation Grants, 2007
Fitzpatrick and Francis Family Business Continuity Foundation Grants, 2021
Foundation for Seacoast Health Grants, 2041
George Foundation Grants, 2115
George Gund Foundation Grants, 2117
George Kress Foundation Grants, 2119
H.J. Heinz Company Foundation Grants, 2195
Harold R. Bechtel Charitable Remainder Uni-Trust Grants, 2201
Harry Frank Guggenheim Foundation Research Grants, 2208
Helen Bader Foundation Grants, 2228
Hutchinson Community Foundation Grants, 2310
Jacob and Hilda Blaustein Foundation Grants, 2454
Jacob and Valeria Langeloth Foundation Grants, 2455
James R. Dougherty Jr. Foundation Grants, 2463
John P. McGovern Foundation Grants, 2504
Johnson & Johnson Corporate Contributions, 2508
Josiah Macy Jr. Foundation Grants, 2516
Lucile Horton Howe and Mitchell B. Howe Foundation Grants, 2624
Margaret T. Morris Foundation Grants, 2659
McCarthy Family Foundation Grants, 2717
McGregor Fund Grants, 2725
Miller Brewing Corporate Contributions Program Grants, 2774
NAPNAP Foundation Graduate Student Research Grant, 2828
National Book Scholarship Fund, 2849
NCFR Affiliate Councils Outstanding Student Research Papers, 2906
NCFR Family Therapy Section Best Research Paper Awards, 2907
NCFR Innovation Grants, 2908
NCFR Ruth Hathaway Jewson Award, 2909
NEFE Grants, 2948
NIA Multicenter Study on Exceptional Survival in Families: The Long Life Family Study Grants, 3112
NIDRR Field-Initiated Projects, 3189
Northwestern Mutual Foundation Grants, 3430
Nuffield Foundation Children & Families Grants, 3666
Orchard Foundation Grants, 3793
Perpetual Trust for Charitable Giving Grants, 3900
Pinkerton Foundation Grants, 3967
Radcliffe Institute Carol K. Pforzheimer Student Fellowships, 4018
Rayonier Foundation Grants, 4034
Reuter Foundation Grants, 4059
Rhode Island Foundation Grants, 4061
Robert Bowne Foundaion Edmund A. Stanley, Jr. Research Grants, 4073
Sarkeys Foundation Grants, 4181
Sensient Technologies Foundation Grants, 4199
Thompson Charitable Foundation Grants, 4511
Vancouver Foundation Grants and Community Initiatives, 4807
Viking Children's Fund Grants, 4811
Volkswagen of America Corporate Contributions Grants, 4817
Wallace Foundation Grants, 4829
Wayne and Gladys Valley Foundation Grants, 4835

Family Planning
Allyn Foundation Grants, 574
Conservation, Food, and Health Foundation Grants for Developing Countries, 1641
Cowles Charitable Trust Grants, 1652
Donald and Sylvia Robinson Family Foundation Grants, 1830
J.W. Kieckhefer Foundation Grants, 2452
Jacob and Hilda Blaustein Foundation Grants, 2454
Oppenstein Brothers Foundation Grants, 3792
Radcliffe Inst Individual Residential Fellowships, 4021
Radcliffe Institute Research Support Grants, 4023
Richard and Rhoda Goldman Fund Grants, 4063
Robert Sterling Clark Foundation Reproductive Rights and Health Grants, 4081
W. C. Griffith Foundation Grants, 4818

Family Practice
AAFP Foundation Joint Grants, 156
AAFP Foundation Pfizer Teacher Development Awards, 157
AAFP Foundation Wyeth Immunization Awards, 158
AAFP Research Stimulation Grants, 159
ACS Cancer Control Career Development Awards for Primary Care Physicians, 305
Dorothy Rider Pool Health Care Grants, 1841
Josiah Macy Jr. Foundation Grants, 2516
Regenstrief Geriatrics Fellowship, 4046

Family Studies
NCFR Affiliate Councils Outstanding Student Research Papers, 2906
NCFR Family Therapy Section Best Research Paper Awards, 2907
NCFR Innovation Grants, 2908
NCFR Ruth Hathaway Jewson Award, 2909

Family/Marriage Counseling
College of Human Sciences Scholarships, 1596
DHHS Promoting Safe & Stable Families Grants, 1760
NCFR Family Therapy Section Best Research Paper Awards, 2907
NCFR Ruth Hathaway Jewson Award, 2909

Farm and Ranch Management
Bill and Melinda Gates Foundation Agricultural Development Grants, 1345
Organic Farming Research Foundation Grants, 3795
SARE Graduate Student Grants - North Central Region, 4168
SARE On-Farm Research Grants - So Region, 4173
SARE Producer Research and Education Grants-Western Region, 4174
SARE Professional Development Grants - Northeast Region, 4176
SARE Research and Education Grants - Western Region, 4180

Farming
American Chemical Society AGRO International Award for Research in Agrochemicals, 634
Bill and Melinda Gates Foundation Agricultural Development Grants, 1345
MDARD Specialty Crop Block Grant Program-Farm Bill, 2730
SARE Professional + Producer Grants - Western Region, 4175
SARE Research and Education Grants - Southern Region, 4179
USDA Integrated Research Education and Extension Grants - Organic Transitions, 4630
USDA Scientific Cooperation Research Program, 4647

Fashion/Textiles Design
College of Human Sciences Scholarships, 1596
Phi Upsilon Omicron Alumni Research Grant, 3940
Phi Upsilon Omicron Florence Fallgatter Distinguished Service Award, 3943
Wolfsonian-FIU Fellowships, 4890

Fatigue/Fracture
NIA Bioenergetics, Fatigability, and Activity Limitations in Aging, 3091
NIA Transdisciplinary Research on Fatigue and Fatigability in Aging Grants, 3123
USDD Non-Destructive Eval & Prognostics: Fatigue and Fracture Damage Protection Grants, 4764

Fellowship Programs, General
A-T Children's Project Post Doctoral Fellowships, 9
AAAS Science and Technology Policy Fellowships: Diplomacy, Security, and Development, 40
AAAS Science and Technology Policy Fellowships: Energy, Environment, and Agriculture, 41
AAAS Science and Technology Policy Fellowships: Health, Education and Human Services, 43
AAAS Science and Technology Policy Overseas Fellowships at USAID Missions, 44
AACAP-NIDA K12 Career Development Awards, 54
AACAP Educational Outreach Program for Residents in Alcohol Research, 58
AACAP Jeanne Spurlock Research Fellowship in Substance Abuse and Addiction for Minority Medical Students, 64
AACAP Life Members Mentorship Grants for Medical Students, 67
AACAP Summer Medical Student Fellowships, 80
AACC International Graduate Fellowships, 91
AACC International Milling & Baking Division's M. Rella Dwyer Graduate Fellowship, 93
AACR-Colorectal Cancer Coalition Fellows Grant, 113
AACR Centennial Pre-doctoral Fellowships in Cancer Research, 134
AAF Richard M Hunt Architecture Fellowship, 169
AAUW International Doctoral Fellowships, 214
AAUW International Master's/First Professional Degree Fellowships, 215
AAUW International Postdoctoral Fellowships, 216
ACC Fellowships and Projects, 252
ACC Mekong Region Fellowship, 258
ACCP Infectious Diseases Research Fellowships, 263
ACCP Oncology Research Fellowships, 264
ACIE Title VIII Combined Research and Language Training Fellowships, 275
ACIE Title VIII Research Scholar Fellowships, 276
ACLS/New York Public Library Fellowships, 277
ACLS African Humanities Program for Dissertation Completion, 278
ACLS American Research in the Humanities in China Fellowships, 280
ACLS Collaborative Research Fellowships, 282
ACLS Digital Innovation Fellowships, 284
ACLS Dissertation Fellowships in East European Studies, 285
ACLS Early Career Postdoctoral Fellowships in East European Studies, 286
ACLS Fellowships, 287
ACLS Frederick Burkhardt Residential Fellowships for Recently Tenured Scholars, 288
ACLS International and Area Studies Fellowships, 289
ACLS New Faculty Fellows, 290
Acton Institute for the Study of Religion and Liberty Calihan Academic Fellowships, 331
ADA Foundation Dentsply International Research Fellowships, 335
ADA Foundation Scientist in Training Fellowship, 337
ADA Foundation Summer Scholars Fellowships, 338
AES-Grass Young Investigator Travel Awards, 345
AES Robert S. Morison Fellowship, 346
AES Susan S. Spencer Clinical Research Epilepsy Training Fellowship, 347
Aetna Foundation Minority Scholars Grants, 349
Aetna Foundation National Medical Fellowship in Healthcare Leadership, 350
AFAR Medical Student Training in Aging Research Program, 352
AGHE Graduate Scholarships and Fellowships, 369
AHRQ Individual Awards for Postdoctoral Fellows Ruth L. Kirschstein National Research Service Awards, 389
AIHS Media Summer Fellowship, 453
AIR/IES-NCES Postdoctoral Policy Fellowships in Washington, 466
AIR/NCES Fellowships for Graduate Study, 467
AIR/NRC Summer Institute on First-Year Assessment Undergraduate Student Fellowships, 468
AIR National Summer Data Policy Institute Fellowships, 469
ALA ACRL Doctoral Dissertation Fellowship, 478
ALA Samuel Lazerow Fellowship For Research in Collections and Technical Services in Academic and Research Libraries, 497
Albert M. Greenfield Foundation Dissertation Fellowship, 507
Alexander von Humboldt Feodor-Lynen Research Fellowships for Experienced Researchers, 518
Alexander von Humboldt Foundation Feodor-Lynen Research Fellowships for Postdoc Researchers, 519
Alexander von Humboldt Foundation Georg Forster Fellowships for Experienced Researchers, 523
Alexander von Humboldt Foundation Georg Forster Fellowships for Postdoctoral Researchers, 524
Alexander von Humboldt Foundation German Chancellor Fellowships for Prospective Leaders from the People's Republic of China, 526
Alexander von Humboldt Foundation Hezekiah Wardwell Fellowships, 528
Alexander von Humboldt Found Professorships, 535
Alpha Kappa Alpha Educational Advancement Foundation Fellowships, 575
Alston Bannerman Sabbatical Fellowships, 580
American Academy of Advertising Research Fellowships, 611
American Academy of Arts and Sciences Hellman Fellowships, 612
American Academy of Nursing Claire M. Fagin Fellowships, 615
American Australian Association Australia To USA Fellowships, 619
American Australian Association USA to Australia Fellowships, 620
American Chemical Society Joseph Breen Memorial Fellowship in Green Chemistry, 682
American College of Surgeons Traveling Fellowship to Germany, 706
American College of Surgeons Traveling Fellowship to Japan, 707
American College of Surgeons Wound Care Management Award, 708
American Indian College Fund STEM Leadership Fellowships, 745
American Mathematical Society AAAS Congressional Fellowships, 750
American Mathematical Society AAAS Mass Media Fellowships, 751
American Mathematical Society Centennial Fellowships, 752
American Rivers Anthony A. Lapham River Conservation Fellowship, 795
American Sociological Association Congressional Fellowships, 803
Andrew W. Mellon/ACLS Dissertation Completion Fellowships, 844
Andrew W. Mellon Postdoctoral Fellowships at UCLA Humanities Consortium, 846
AOCS Ralph H. Potts Memorial Fellowship, 893
AOCS Thomas H. Smouse Memorial Fellowship, 899
APHL Emerging Infectious Diseases Fellowships, 914
APSA Congressional Fellowships for Communications Scholars and Journalists, 926
APSA Congressional Fellowships for Federal Executives, 927
APSA Congressional Fellowships for International Scholars, 928
APSA Congressional Health and Aging Policy Fellowships, 931
APSA Minority Fellowships, 939
APSA Robert Wood Johnson Foundation Health Policy Fellowships, 944
ARM Faculty Fellowships, 1001
ARM Graduate Fellowships, 1002
ARM Postgraduate Fellowships, 1003
ASA/NSF Federal Statistics Fellowships, 1032
ASCE Freeman Fellowships, 1040
ASCE Jack E. Leisch Memorial National Graduate Fellowship, 1041
ASCE O.H. Ammann Research Fellowship in Structural Engineering, 1042
ASCE Trent R. Dames and William W. Moore Engineering Fellowship, 1043
ASCO Long-Term International Fellowships, 1048
ASCSA Fellowships, 1052
Asian Pacific American Institute for Congressional Studies Fellowships, 1091
ASM-PAHO Infectious Diseases Epidemiology and Surveillance Fellowships, 1098
ASM/CDC Fellowships in Infectious Disease and Public Health Microbiology, 1100
ASM International Fellowship for Latin America and the Caribbean, 1175
ASM Int'l Fellowships for Asia and Africa, 1176
ASM Microbiology Undergraduate Research Fellowship, 1178
ASM Robert D. Watkins Graduate Research Fellowship, 1183
ASM Undergraduate Research Fellowship, 1186
Aspen Institute William Randolph Hearst Endowed Fellowship for Minority Students, 1194
ASPET Summer Undergrad Research Fellows, 1208
ASPO Fellowships, 1211
ASU Graduate College Completion Fellowships, 1235
ASU Graduate College Dean's Fellowships, 1236
ASU Graduate College Dissertation Fellowships, 1237
ASU Graduate College Doctoral Enrichment Fellowships, 1238
ASU Graduate College Earl and Lenore Tripke Fellowship, 1239
ASU Graduate College Herman E. Demund Memorial Fellowships, 1240
ASU Graduate College Lattie and Elva Coor Building Great Communities Graduate Fellowship, 1241
ASU Grad College Reach for Stars Fellowship, 1243
ASU Graduate College Science Foundation Arizona Bisgrove Postdoctoral Scholars, 1244
ASU Graduate College Science Foundation Arizona Fellowships in Science and Engineering, 1245
ASU Inst for Humanities Research Fellowships, 1246
ASU Institute for Humanities Research Visiting Fellowships, 1248
ATA Doctoral Fellowships in Education, 1252
BIA Native American Grad Student Fellowships, 1342
Brookdale Foundation Leadership in Aging Fellowships, 1381
CAA Professional Development Fellowships for Graduate Students, 1404
Carrier Corporation Contributions Grants, 1440
CDC Found Tobacco Network Lab Fellowship, 1462
CDC Preventive Med Residency & Fellowship, 1465
Cedar Tree Foundation David H. Smith Conservation Research Fellowship, 1470
Center for East Asian Studies Fellowships, 1474
Center for the Study of Democratic Politics Fellowships, 1481
Central European University Fellowships, 1488
CFR Edward R. Murrow Press Fellowship, 1500
CFR International Affairs Fellowships, 1501
CFR International Affairs in Japan Fellowships, 1504
CFUW Beverley Jackson Fellowship, 1507
CFUW Canadian Home Economics Association Fellowship, 1509
CFUW Marg McWilliams Predoc Fellowship, 1513
CFUW Marion Elder Grant Fellowship, 1514
CFUW Ruth Binnie Fellowship, 1516
Charles H. Revson Foundation Grants, 1524
CIC Smithsonian Institution Fellowships, 1564
CMAP Physician Advocacy Fellowship, 1583
Compton Foundation International Fellowships, 1624
Compton Foundation Mentor Fellowships, 1625
Council on Foundations Emerging Philanthropic Leaders Fellowships, 1649

Covidien Advanced MIS Surgical Fellowships, 1650
Covidien Surgical Mini-Fellowships, 1651
CSTE CDC/CSTE Applied Epidemiology Fellowships, 1664
Cystic Fibrosis Canada Clinical Fellowships, 1674
Cystic Fibrosis Canada Fellowships, 1677
DAAD Faculty Research Visit Grants, 1697
Daimler and Benz Foundation Fellowships, 1713
Deloitte Foundation Grants, 1735
Delta Gamma Foundation Fellowships for Graduate Study, 1738
Delyte and Dorothy Morris Doctoral Fellowships, 1739
Douglass Institute Postdoctoral Fellowships, 1851
Douglass Institute Predoctoral Fellowships, 1852
Dr. Robert H. Goddard Research Fellowships, 1854
Duke University Adult Cardiothoracic Anesthesia and Critical Care Medicine Fellowships, 1869
Duke University Ambulatory and Regional Anesthesia Fellowships, 1870
Duke University Clinical Cardiac Electrophysiology Fellowships, 1871
Duke University Hyperbaric Center Fellowships, 1872
Duke Univ Interventional Cardio Fellowships, 1873
Duke Univ Obstetric Anesthesia Fellowships, 1878
Duke University Pain Management Fellowships, 1879
Duke University Postdoctoral Research Fellowships in Aging, 1880
Durfee Foundation Stanton Fellowship, 1881
Edward W. and Stella C. Van Houten Memorial Fund Grants, 1899
EMBO Long-Term Fellowships, 1916
EMBO Short-Term Fellowships, 1917
Enterprise Community Partners Terwilliger Fellowship, 1930
EPA Capacity Building Grants for Compliance Assurance and Enforcement Activities in Indian Country and Other Tribal Areas, 1932
Facebook Fellowships, 1973
FAMRI Young Clinical Scientist Awards, 1976
FCD Young Scholars Fellowships, 1985
Field Museum Graduate Student Fellowships, 2008
Field Museum Visiting Scholarships, 2011
Fleishhacker Foundation Eureka Fellowships, 2023
Florence Tan Moeson Fellowships, 2025
Ford Foundation International Fellowships, 2030
Foundation for Evangelism Harry Denman Fellowships, 2040
Frederick Douglass Summer Fellowships, 2052
Frederick Douglass Summer Scholars Fellowship, 2053
Fritz B. Burns Foundation Grants, 2057
Gates Family Foundation Fellowships, 2097
Getty Research Institute Nonresidential Grants - Postdoctoral Fellowships, 2137
Grass Foundation Marine Biological Laboratory Advanced Imaging Fellowships, 2162
Grass Foundation Marine Biological Laboratory Fellowships, 2163
Greenwall Foundation Fellowship Program in Bioethics and Health Policy, 2176
Gruber Foundation International Astronomical Union Fellowships, 2179
H.B. Earhart Fellowships, 2194
Henry Luce Foundation/ACLS Dissertation Fellowships in American Art, 2235
HRF Mark B. Bain Graduate Fellowships, 2287
Huntington Library Babara Thom Postdoctoral Fellowships, 2301
Huntington Library Clark-Huntington Joint Bibliographical Fellowship, 2302
Huntington Library Dana and David Dornsife Fellowships, 2303
Huntington Library Dibner History of Science Long-Term Fellowships, 2304
Huntington Library Mellon Fellowships, 2306
Huntington Library Short-Term Fellowships, 2308
IBRO Albert Aguayo Fellowship, 2322
IBRO Asia Reg APRC Exchange Fellowships, 2323
IBRO Regional Grants for Int'l Fellowships to U.S. Laboratory Summer Neuroscience Courses, 2330
IBRO Research Fellowships, 2331
IBRO Return Home Fellowships, 2332
IES Faculty Fellowships, 2341
IFUW Dr. A. Vibert Douglas Int'l Fellowship, 2346
IHI Quality Improvement Fellowships, 2348
IIE Whitaker International Fellowships and Scholarships, 2356
Independent Institute John M. Templeton Fellowships Essay Contest, 2369
Indiana Space Grant Consortium Fellowships, 2375
Intel Foundation PhD Fellowships, 2394
IRA Jeanne S. Chall Research Fellowship, 2410
IREX ECA Alumni Small Grants Program, 2416
IREX MENA Media Emerging Leaders Fellows, 2421
IREX University Administration Support, 2425
Irving and Rose Crown Scholarships and Fellowships in American History, 2427
ISGC Graduate Fellowships, 2430
Izaak Walton Killam Memorial Postdoc Fellowships at the Univ of British Columbia, 2447
Izaak Walton Killam Memorial Predoctoral Fellowships at the University of British Columbia, 2448
J.H. Stewart Reid Memorial Fellowship for Doctoral Studies, 2450
James and Sylvia Thayer Short -Term Research Fellowships, 2457
James Graham Brown Foundation Fellowships, 2459
Japan Foundation Center for Global Partnership NPO Fellowships, 2474
Japan Foundation Center for Global Partnership Security Studies Fellowships, 2475
Japan Foundation Japanese Studies Fellowships, 2476
Japan Foundation New York Japanese Studies Fellowships, 2478
JMO Law and Legal System Grants, 2491
JMO Public Policy Research Grants, 2492
JMO Strategic and International Studies Grants, 2493
John Dana Archbold Fellowships, 2499
JSA JLab Graduate Fellowships, 2521
Kellogg Institute for International Studies Experiencing The World Fellowships, 2539
Kellogg Institute for International Studies Faculty Fellows' Research Grants, 2540
Kellogg Institute for International Studies Quechua Summer Language Fellowships, 2548
Kennedy Center Fellowship at the DeVos Institute of Arts Management, 2552
Kennedy Center Summer Fellowships for International Arts Managers, 2553
Kobe College Corporation Graduate Fellowships, 2567
Kosciuszko Found Fellowships for Polish Citizens, 2570
Law and Public Affairs Fellowships at Princeton University, 2588
LBI-L Leo Baeck/Studienstiftung Fellowships, 2592
LBI-NY/DAAD Fellowships in Germany, 2593
LBI-NY/DAAD Fellowships in New York, 2594
LBI-NY Fred Grubel Fellowship, 2596
LBI-NY Fritz Halbers Fellowships, 2597
Library Company of Philadelphia and The Historical Society of Pennsylvania Visiting Research Fellowships, 2605
Lilly Endowment Teacher Creativity Fellowships, 2616
Loeb Fellowships in Advanced Environmental Design Studies, 2621
Lynde & Harry Bradley Foundation Fellowships, 2634
Lynde and Harry Bradley Foundation Grants, 2635
MAP International Medical Fellowships, 2655
Massachusetts Cultural Cou Artist Fellowships, 2677
Mayo Clinic Administrative Fellowship, 2689
Mayo Clinic Admin Fellowship Eau Claire, 2690
Mayo Clinic Business Consulting Fellowship, 2691
MBL Albert & Ellen Grass Summer Fellowships, 2692
MBL Ann E. Kammer Mem Summer Fellowship, 2693
MBL Associates Summer Fellowships, 2694
MBL Baxter Postdoctoral Summer Fellowship, 2695
MBL Burr & Susie Steinbach Summer Fellow, 2696
MBL E.E. Just Summer Fellowship for Minority Scientists, 2697
MBL Erik B. Fries Summer Fellowships, 2698
MBL Eugene and Millicent Bell Fellowships, 2699
MBL Evelyn & Mel Spiegal Summer Fellowship, 2700
MBL Frank R. Lillie Summer Fellowship, 2701
MBL Frederik B. and Betsy G. Bang Summer Fellowships, 2702
MBL Fred Karush Library Readership, 2703
MBL Gruss Lipper Family Foundation Summer Fellowship, 2704
MBL H. Keffer Hartline and Edward F. MacNichol, Jr. Fellowships, 2705
MBL Herbert W. Rand Summer Fellowship, 2706
MBL James E. and Faith Miller Memorial Summer Fellowship, 2707
MBL John M. Arnold Award, 2708
MBL Laura & Art Colwin Summer Fellowships, 2709
MBL Lucy B. Lemann Mem Summer Fellowship, 2710
MBL M.G.F. Fuortes Summer Fellowships, 2711
MBL Nikon Summer Fellowship, 2712
MBL Plum Found John E. Dowling Fellowships, 2713
MBL Robert Day Allen Summer Fellowship, 2714
MBL Stephen W. Kuffler Summer Fellowships, 2715
MBL William Townsend Porter Summer Fellowships for Minority Investigators, 2716
Mead Johnson Nutritionals Scholarships, 2735
MedImmune Charitable Grants, 2737
Medtronic Foundation Fellowships, 2738
Mellon Post-Doctoral Fellowship in Philosophy, 2739
Mericos Foundation Grants, 2742
MGFA Post-Doctoral Research Fellowships, 2751
MGFA Student Fellowships, 2752
MHS Teacher Fellowships, 2759
Minerva Fellowships, 2778
Minerva Short-Term Research Grants, 2779
MLA Cunningham Memorial Int'l Fellowship, 2783
MLA David A. Kronick Traveling Fellowship, 2784
Murphy Institute Center for Ethics and Public Affairs Faculty Fellowships, 2805
Murphy Institute Center for Ethics and Public Affairs Graduate Fellowships, 2806
NACC David Stevenson Fellowships, 2812
NACC William Diaz Fellowships, 2813
NASA Graduate Student Researchers Program Fellowships, 2839
National Center on Philanthropy and the Law Fellowship in Nonprofit Law, 2859
National Endowment for the Arts Jazz Masters Fellowships, 2864
National Wildlife Federation Campus Ecology Fellowships, 2890
NCCAM Ruth L. Kirschstein National Research Service Awards for Postdoctoral Training in Complementary and Alternative Medicine, 2904
NEH Fellowships for Advanced Social Science Research on Japan, 2958
NFID John P. Utz Medical Mycology Postdoctoral Fellowship, 2988
NFID Nosocomial Infection Research and Training Postdoctoral Fellowship, 2990
NHLBI Ruth L. Kirschstein National Research Service Awards for Individual Predoctoral Fellowships to Promote Diversity in Health-Related Research, 3058
NHLBI Ruth L. Kirschstein National Research Service Awards for Individual Predoctoral MD/PhD Fellows and Other Dual Degree Fellows, 3059
NHLBI Ruth L. Kirschstein National Research Service Awards for Individual Senior Fellows, 3060
NIAF Fellowship Grants, 3095
NIH Ruth L. Kirschstein National Research Service Awards for Individual Senior Fellows, 3238
NIH Ruth L. Kirschstein National Research Service Awards for Individual Predoctoral Fellowships to Promote Diversity in Health-Related Research, 3241
NIH Ruth L. Kirschstein National Research Service Awards for Individual Predoctoral MD/PhD and Other Dual Doctoral Degree Fellows, 3242
NINR Ruth L. Kirschstein National Research Service Award for Individual Predoctoral Fellows in Nursing Research, 3267
NL American Society for Eighteenth-Century Studies Fellowships, 3279
NL Arthur and Lila Weinberg Fellowships for Independent Scholars and Researchers, 3280

NL Associated Colleges of the Midwest/Great Lakes College Association Faculty Fellowships, 3281
NL British Academy Fellowship for Study in the United Kingdom, 3283
NL Frances C. Allen Fellowships, 3284
NL Herzog August Bibliothek Wolfenbuttel Fellowship, 3286
NL Lawrence Lipking Fellowship at Newberry, 3287
NL Lloyd Lewis Fellowships in American Hist, 3289
NL Mellon Postdoctoral Research Fellowships, 3294
NL Midwest Modern Lang Assoc Fellowship, 3299
NL Monticello College Foundation Fellowship for Women, 3304
NL Northeast Modern Language Assocation Fellowship, 3314
NL Short-Term Resident Fellowships for Individual Research, 3317
NL Susan Kelly Power and Helen Hornbeck Tanner Fellowship, 3319
NMF General Electric Med Scholars Fellowships, 3322
NOAA Coastal Management Fellowships, 3352
NOAA Living Marine Resources Cooperative Science Center Internships/Fellowships, 3380
NOAA National Sea Grant Industry Fellowships, 3393
Northwood University/Alden B. Dow Creativity Center Summer Residency Program, 3432
NSERC E.W.R. Steacie Memorial Fellowships, 3439
NSERC Northern Research Internships Program, 3447
NSERC Postdoctoral Fellowships Program, 3448
NSERC Undergrad Student Research Awards, 3451
NSF Astronomy and Astrophysics Postdoctoral Fellowships, 3494
NSF Postdoc Research Fellowships in Biology, 3630
NYPL Short-Term Research Fellowships, 3701
ONR DOD National Defense Science & Engineering Graduate Fellowships, 3755
ONR Historically Black Colleges and Universities Future Engineering Faculty Fellowships, 3763
ONR Navy Postdoctoral Fellowships, 3776
ONR Summer Faculty Research Program Grants, 3784
OSF Access to Medicines Fellowships in Eastern Europe and Central Asia, 3834
OSF Central Asia and Caucasus Research and Training Initiative - Junior Fellowships, 3836
OSF Central Asia and Caucasus Research and Training Initiative - Senior Fellowships, 3837
OSF Doctoral Fellows Georgia & Kyrgyzstan, 3839
OSF Fellowships, 3840
OSF Global Faculty Grants - Track 1, 3841
OSF Justice Initiative Fellowships, 3847
OSF Palestinian Rule of Law Fellowships, 3848
OSF Social Work Fellowships, 3849
OSF Soros Justice Advocacy Fellowships, 3850
OSF Think Tank Fellowships, 3852
OSF Thomas J. Alexander Fellowships, 3853
Packard Foundation Fellowships for Science and Engineering, 3862
Pancreatic Cancer Network-AACR Fellowship, 3866
Paul & Daisy Soros Fellowships for New Amer, 3875
PC Fred H. Bixby Fellowships, 3880
PDF-AANF Clinician-Scientist Devel Awards, 3881
PDF Postdoctoral Fellowships for Basic Scientists, 3883
PDF Postdoctoral Fellowships for Neurologists, 3884
PDF Summer Fellowships, 3885
Penn Graduate School of Education Dean's Fellowships, 3891
Penn Graduate School of Education Foreign Language Area Studies Fellowship, 3892
Penn Graduate School of Education Pre-Doctoral Training in Interdisciplinary Methods for Field-Based Research in Education, 3893
Phi Kappa Phi Graduate Fellowships, 3938
Phi Upsilon Omicron Candle Fellowships, 3941
Phi Upsilon Omicron Diamond Anniversary Fellowships, 3942
Phi Upsilon Omicron Founders Fellowship, 3944
Phi Upsilon Omicron Geraldine Clewell Doctoral Fellowship, 3945
Phi Upsilon Omicron Geraldine Clewell Masters Fellowship, 3946
Phi Upsilon Omicron Jean Dearth Dickerscheid Fellowship, 3947
Phi Upsilon Omicron Margaret Drew Alpha Fellowship, 3948
Phi Upsilon Omicron Presidents Res Fellowship, 3949
PhRMA Foundation Health Outcomes Sabbatical Fellowships, 3952
PhRMA Foundation Informatics Sabbatical Fellowships, 3956
PhRMA Foundation Paul Calabresi Medical Student Research Fellowship, 3957
Potts Memorial Foundation Grants, 3985
Price Family Charitable Fund Grants, 3987
Princess Grace Foundation Theater, Dance, and Film Grants, 3988
Princeton University Center for Human Values Graduate Prize Fellowships, 3989
Princeton University Center for Human Values Harold T. Shapiro Postdoctoral Research Associate in Bioethics, 3991
Princeton University Center for Human Values Postdoctoral Research Associate in Values and Public Policy, 3994
Province of Alberta Queen Elizabeth II Graduate Scholarships, 4001
PVA Research Foundation Grants, 4004
Radcliffe Institute Carol K. Pforzheimer Student Fellowships, 4018
Radcliffe Inst Individual Residential Fellowships, 4021
RAND Corporation Fellowships in Population Studies and the Study of Aging, 4029
RAND Corporation Military Fellowships, 4031
RAND Corporation Transatlantic Postdoctoral Fellowships for Int'l Relations & Security, 4032
Regenstrief General Internal Medicine - General Pediatrics Research Fellowship Program, 4045
Regenstrief Geriatrics Fellowship, 4046
Regenstrief Medical Informatics NLM Research Fellowships, 4048
Rhode Island Foundation Grants, 4061
Robert W. Deutsch Foundation Grants, 4082
Rockford College Faculty Fellowships, 4092
Rosenthal Writer-in-Residence Program, 4105
Rumsfeld Foundation Graduate Fellowships, 4127
RWJF Clinical Scholars Grants, 4136
RWJF Health and Society Scholars, 4141
RWJF Investigator Awards in Health Policy Res, 4145
RWJF New Connections Jun Investigator Grants, 4146
RWJF New Connections Mid-Career Consultants Grants, 4147
RWJF Nurse Faculty Scholars, 4148
Saint Mary's College Fellowships, 4159
SASS Birgit Baldwin Fellowship, 4182
Shastri Indo-Canadian Institute Doctoral Student Research Fellowships, 4221
Shastri Indo-Canadian Institute Faculty Training Fellowships, 4223
Shastri Indo-Canadian Institute Master's Student Research Fellowships, 4226
Shorenstein Center Fellowships, 4235
SHOT Mel Kranzberg Dissertation Fellowship, 4245
Siragusa Foundation Health Services & Medical Research Grants, 4264
Smithsonian Biodiversity Genomics and Bioinformatics Postdoctoral Fellowships, 4272
Smithsonian Graduate Student Fellowships, 4277
Smithsonian Museum Conservation Institute Research Post-Doctorate/Post-Graduate Fellowships, 4283
Smithsonian National Air and Space Museum Charles A. Lindbergh Chair in Aerospace History, 4285
Smithsonian National Air and Space Museum Postdoctoral Earth and Planetary Sciences Fellowship, 4287
Smithsonian National Museum of American History Lemelson Center Fellowships, 4289
Smithsonian Postdoctoral Fellowships, 4293
Smithsonian Predoctoral Fellowships, 4295
Special Olympics Kennedy Shriver Fellowships, 4331
Spencer Foundation Postdoctoral Fellowships, 4336
SPIRE Postdoctoral Fellowships, 4345
SPJ Eugene C. Pulliam Fellowship for Editorial Writing, 4346
SRC AMD/Mahboob Khan Fellowship, 4348
SRC Educational Science Postdoc Fellowships, 4363
SRC Education Science Grants for Postdoctoral Positions in Sweden, 4366
SRC Humanities and Social Science Postdoctoral Fellowships, 4372
SRC Medical Research Grants, 4379
SRC Medical Science Grants for Postdoctoral Positions in Sweden, 4381
SRC Medical Science Postdoctoral Fellowships, 4382
SRC Natural and Engineering Science Grants for Postdoctoral Positions in Sweden, 4384
SRC Natural and Engineering Science Postdoctoral Fellowships, 4385
SSHRC Sport Participation Research Initiative: Doctoral Award Supplements, 4416
SSHRC Sport Participation Research Initiative: Postdoctoral Fellowship Supplements, 4417
SSRC-Van Alen Fellowships, 4419
SSRC Dissertation Proposal Development Faculty Field Fellowships, 4424
SSRC Dissertation Proposal Development Student Fellowships, 4425
SSRC Japan Society for the Promotion of Science Fellowships, 4431
Stanford Humanities Center Arts Writer/Practitioner Fellowship, 4444
Stanford Humanities Center External Faculty Fellowships, 4445
Stroke Association Clinical Fellowships, 4455
Swiss National Science Foundation Fellowships for Prospective Researchers, 4484
Syracuse University Graduate Fellowships, 4488
Tau Beta Pi Association Fellowships, 4495
Thomson Reuters / MLA Doctoral Fellowships, 4512
TRDRP Postdoctoral Fellowship Awards, 4530
U.S. Department of Education Graduate Assistance in Areas of National Need, 4540
U.S. Department of Education Jacob K. Javits Fellowships, 4541
UCLA / RAND Corporation Post-Doctoral Fellowships, 4548
UCR Grad Student in Humanities Fellowship, 4550
UCR Resident Fellowships, 4552
UNC Carolina Postdoctoral Program for Faculty Diversity, 4560
UNC School of Information and Library Science Doctoral Fellowships, 4565
UNESCO Co-Sponsored Fellowships, 4566
UNESCO Fellowships, 4567
United Methodist Church Josephine Forman Scholarship, 4573
United States Institute of Peace - Jennings Randolph Senior Fellowships, 4580
United States Institute of Peace - Transatlantic Post-Doc Fellowship for International Relations and Security, 4581
University of British Columbia Graduate Student Fellowships, 4584
University of Southern Mississippi H.A. & Margret Rey Translation Fellowships, 4589
University of Southern Mississippi Keats/Domanska/de Grummond Children's Literature Research Fellowships, 4590
USDA Cochran Fellowships, 4619
USDD Postdoctoral Fellowship Awards in Ocean Acoustics, 4774
Van Alen Institute New York Prize Fellowship, 4806
Virginia Foundation for the Humanities Residential Fellowships, 4815
WHOI Geophysical Fluid Dynamics Fellowships, 4859
WHOI Marine Policy Fellowships, 4861
WHOI Postdoctoral Fellowships, 4863
WHOI Postdoctoral Scholar Awards, 4864
WHOI Summer Student Fellowships, 4865
WHOI Undergrad Fellowships for Minorities, 4866
William T. Grant Foundation Distinguished Fellows Awards, 4880

Woodrow Wilson Doctoral Dissertation Fellowships in Women's Studies, 4891
Yale University Beinecke Library Fellowships for Visiting Scholars, 4901
Yale University Beinecke Library Research Fellowships for Yale Grad & Professional School Students, 4902
Zonta International Foundation Amelia Earhart Fellowships, 4905

Feminism
American Historical Association Joan Kelly Memorial Prize in Women's History, 730
ASR Fichter Research Grant, 1217
NWSA Abafazi-Africana Women's Studies Essay Award, 3675
Radcliffe Inst Individual Residential Fellowships, 4021
Radcliffe Institute Research Support Grants, 4023

Fertilizers
American Chemical Society Herman Frasch Foundation Grants, 680

Fetus Research
BWF Preterm Birth Initiative Grants, 1399
NIH Research on Sleep and Sleep Disorders, 3236

Fiber Optics
Abdus Salam ICTP Postgraduate Diploma Grants, 235
Abdus Salam ICTP Training and Research in Italian Laboratories Grants, 240
ARO Mobile, Wireless Communications and Networks Research Contracts, 1012

Fiction
Florida Division of Cultural Affairs Individual Artist Fellowships, 2026
National Endowment for the Arts Literature Fellowships: Creative Writing, 2865
North Carolina Arts Council Residency Grants, 3420
NYFA Artist Fellowships, 3698
Princeton University Lewis Center for the Arts Hodder Fellowships, 3996
Summerfield G. Roberts Award, 4465

Film Production
Astronomical Society of the Pacific Dorothea Klumpke-Roberts Award, 1229
Carole Fielding Film/Video Production or Research Grants, 1438
Delaware Division of the Arts Individual Artist Fellowships, 1732
Florida Division of Cultural Affairs Individual Artist Fellowships, 2026
IFP Minnesota McKnight Screenwriters Fellows, 2343
IFP New York Anthony Radziwill Documentary Fund Grants, 2344
John Simon Guggenheim Memorial Foundation Fellowships, 2507
NLM Grants for Scholarly Works in Biomedicine and Health, 3298
North Carolina Arts Council Folklife Grants, 3419
NYFA Artist Fellowships, 3698
NYSCA Individual Artists: Film, Media and New Technology Production Grants, 3706
OAH Erik Barnouw Award, 3716
Playboy Foundation Grants, 3980
Princeton University Center for Human Values Short Movie Prize, 3995

Films
AAB Berlin Prize Fellowships, 48
ACC Humanities Fellowship Program, 255
ACC Residency Program in Asia, 267
ACC Starr Foundation Fellowship Program, 268
Bogliasco Foundation Liguria Study Center Residencies, 1358
Canada-U.S. Fulbright International Science and Technology Award, 1417
Canada-U.S. Fulbright Senior Specialists Program Grants, 1418
Canada-U.S. Fulbright Traditional Scholars, 1419
Carole Fielding Film/Video Production or Research Grants, 1438
CCH Documentary Project Research and Development Grants, 1451
Fulbright Alumni Initiatives Awards Grants, 2064
Fulbright Distinguished Chairs Program Grants, 2069
Fulbright European Union Scholar-in-Residence Program Grants, 2071
Fulbright New Century Scholars Program Grants, 2075
Fulbright Occasional Lecturer Program Grants, 2076
Fulbright Specialists Program Grants, 2079
Fulbright Traditional Scholar Program in Europe and Eurasia, 2080
Fulbright Traditional Scholar Program in Sub-Saharan Africa, 2081
Fulbright Traditional Scholar Program in the East Asia/Pacific Region, 2082
Fulbright Traditional Scholar Program in the Near East and North Africa Region, 2083
Fulbright Traditional Scholar Program in the South and Central Asia Region, 2084
Fulbright Traditional Scholar Program in the Western Hemisphere, 2085
Fulbright Visiting Specialists Program: Direct Access to the Muslim World, 2087
Harry Ransom Humanities Research Center Research Fellowships, 2209
IFP Minnesota McKnight Film Fellowships, 2342
Massachusetts Foundation for the Humanities Project Grants, 2678
NYSCA Individual Artists: Film, Media and New Technology Production Grants, 3706
OAH Erik Barnouw Award, 3716
OSF Soros Justice Media Fellowships, 3851
Princess Grace Foundation Theater, Dance, and Film Grants, 3988
Third Millennium Foundation Grants, 4505

Finance
Bill and Melinda Gates Foundation Financial Services for the Poor Grants, 1346
CBO Economic Policy Fellowships, 1445
College of Human Sciences Scholarships, 1596
Dean Witter Foundation Grants, 1728
Equipment Leasing and Finance Foundation Research Grants, 1943
John R. Oishei Foundation Grants, 2505
Library Company of Philadelphia Barra Foundation International Research Fellowships, 2606
Library Company of Philadelphia Prizes, 2608
Library Company of Philadelphia Program in Early American Economy and Society Dissertation-Level Fellowships, 2609
Library Company of Philadelphia Program in Early American Economy and Society One-Month Fellowships, 2610
Library Company of Philadelphia Program in Early American Economy and Society Post-Doctoral Research Fellowships, 2611
Mayo Clinic Administrative Fellowship, 2689
Mayo Clinic Admin Fellowship Eau Claire, 2690
MSI Alden G. Clayton Dissertation Proposal Competition, 2800
NEFE Grants, 2948
TIAA-CREF Paul A. Samuelson Award, 4518

Financial Aid (Scholarships and Loans)
ASM Robert D. Watkins Graduate Research Fellowship, 1183
AUPHA Corris Boyd Scholarships, 1267
Edward W. and Stella C. Van Houten Memorial Fund Grants, 1899
MLA Graduate Scholarship, 2786
MLA Grad Scholarship for Minority Students, 2787
Princeton University Center for Human Values Graduate Student Top-Up Prizes, 3990
USA Funds Access to Education Scholarships, 4593
WMU Department of Spanish Travel Study Abroad Award, 4889

Financial Education
Bill and Melinda Gates Foundation Financial Services for the Poor Grants, 1346
FINRA Investor Education Foundation General Grants, 2016
HSBC Contributions Community Grants, 2291
McCune Charitable Foundation Grants, 2719
McGraw-Hill Companies Community Grants, 2724
NEFE Grants, 2948

Financial Literacy
Bill and Melinda Gates Foundation Financial Services for the Poor Grants, 1346
Chicago Community Trust Housing Grants: Advancing Affordable Rental Housing, 1547
FINRA Investor Education Foundation General Grants, 2016
HSBC Contributions Community Grants, 2291
McGraw-Hill Companies Community Grants, 2724
TIAA-CREF Paul A. Samuelson Award, 4518

Fine Arts
Achelis Foundation Grants, 274
Berks County Community Foundation Grants, 1333
Bodenwein Public Benevolent Foundation Grants, 1355
Bodman Foundation Grants, 1356
E.L. Wiegand Foundation Grants, 1885
Frederick Douglass Summer Scholars Fellowship, 2053
Gilroy and Lillian P. Roberts Charitable Foundation Grants, 2142
IREX Short-Term Travel Grants, 2424
Kate Neal Kinley Memorial Fellowship, 2532
Ludwig Vogelstein Foundation Grants, 2626
Lumpkin Family Found Healthy People Grants, 2629
Marion Gardner Jackson Charitable Trust Grants, 2664
NYSCA Dance: Services to the Field Grants, 3702
R.C. Baker Foundation Grants, 4006
SSHRC Insight Development Grants, 4409
SSHRC Insight Grants, 4410
U.S. Department of Education Jacob K. Javits Fellowships, 4541
Yale University Beinecke Library Research Fellowships for Yale Grad & Professional School Students, 4902

Finnish Language/Literature
CIMO Scholarships for Advanced Studies of the Finnish Language at Finnish Universities, 1569

Fire Prevention
Burlington Northern Santa Fe Foundation Grants, 1389
Center of Excellence in General Aviation Grants, 1486
ConocoPhillips Grants Program, 1629
NIST Fire Research Grants, 3268
NOAA Collaborative Science, Technology, and Applied Research Program Grants, 3354
Thompson Charitable Foundation Grants, 4511

Fire/Flammability Studies
John D. and Catherine T. MacArthur Foundation Global Challenges Program Grants, 2498
NIST Fire Research Grants, 3268
NIST Materials Science and Engineering Grants, 3270
USDD RF Semiconductors, RF Solid State Amplifiers and Wide Bandgap Materials Grants, 4786

Firearms
NRA Foundation Grants, 3437

Fiscal Theory/Application
JMO Public Policy Research Grants, 2492
Robert Schalkenbach Foundation Grants, 4079
SRC Robert M. Burger Fellowship, 4392

Fish and Fisheries
American Rivers Anthony A. Lapham River Conservation Fellowship, 795
FishAmerica Foundation Marine and Anadromous Fish Habitat Restoration Grants, 2018
FishAmerica Foundation Research Grants, 2019
GNOF Gulf Coast Oil Spill Grants, 2153

Great Lakes Fishery Trust Access Grants, 2168
HRF Hudson River Improvement Grants, 2286
NFWF Freshwater Fish Conservation Initiative Grants, 2999
NFWF More Fish Partnership Fund Grants, 3003
NFWF Nature Restoration Trust Grants, 3005
NFWF Puget Sound Marine Conservation Fund Grants, 3006
NFWF Shell Marine Habitat Program Grants, 3008
NOAA Atlantic Salmon Conservation Grants, 3332
NOAA Chesapeake Bay Office B-WET (Bay Watershed Education and Training) Grants, 3339
NOAA Chesapeake Bay Office Cooperative Research Program Grants, 3340
NOAA Chesapeake Bay Office Fisheries Science Program Grants, 3341
NOAA Chesapeake Bay Office Non-Native Oyster Research Grants, 3342
NOAA Chesapeake Bay Office Submerged Aquatic Vegetation Culture and Restoration Grants, 3343
NOAA Cooperative Research Program Grants, 3359
NOAA Cooperative Unit for Fisheries Education and Research Internships, 3360
NOAA Faculty and Student Intern Research, 3371
NOAA Living Marine Resources Cooperative Science Center Mini Grants, 3381
NOAA Miami Lab Summer Internships, 3382
NOAA Mid-Atlantic Research Set-Aside Program Grants, 3384
NOAA Monkfish Research Set-Aside Grants, 3386
NOAA National Sea Grant Gulf of Mexico Oyster Industry Program Grants, 3392
NOAA National Sea Grant Oyster Disease Research Program Grants, 3395
NOAA Sea Scallop Research Set-Aside Grants, 3404
Northwest Fund for the Environment Grants, 3431
Olin Fellowships, 3740
Royal Caribbean Cruises Ocean Fund, 4108
RSC Thomas W. Eadie Medal, 4126
South Florida Student Shark Program, 4330
USDC/NOAA Anadromous Fish Conservation Act Grants, 4654
USDC/NOAA Coastal and Marine Fish Habitat Protection Grants, 4655
USDC/NOAA Community-Based Habitat Restoration Program Grants, 4656
USDC/NOAA Community-Based Marine Debris Removal Grants, 4657
USDC/NOAA Marine Fisheries Grants, 4658
USDC/NOAA National Marine Fisheries Service Sea Grant Joint Graduate Fellowships, 4660
USDC/NOAA Saltonstall-Kennedy Grants, 4663

Fitness
Anthem Blue Cross and Blue Shield School Nurse Grants, 868
BCBSNC Foundation Grants, 1318
Boeing Company Contributions Grants, 1357

Floriculture
AOS Research Grants, 903
Orchid Research Fellowships, 3794

Florida
Kluge Center Kislak Fellows in American Studies, 2562
Kluge Center Kislak Short-Term Fellowships in American Studies, 2563

Flow Phenomena
DOE Engineering Research Grants, 1789

Fluid Dynamics
Abdus Salam ICTP Group Associates Grants, 228
Abdus Salam ICTP Joint PhD Programme in Environmental Fluid Mechanics Grants, 230
Abdus Salam ICTP Junior Associates Grants, 231
Abdus Salam ICTP Regular Associates Grants, 237
Abdus Salam ICTP Senior Associates Grants, 239
APS Andreas Acrivos Dissertation Award in Fluid Dynamics, 940
APS Fluid Dynamics Prize, 956
ASME Fluids Engineering Award, 1119
ASME Robert E. Koski Medal, 1159
WHOI Geophysical Fluid Dynamics Fellowships, 4859

Fluid Mechanics
ACSE J. Waldo Smith Hydraulic Fellowship, 309
ASME Allan Kraus Thermal Management Medal, 1104
ASME Fluids Engineering Award, 1119
ASME Timoshenko Medal, 1168
DOE Energy Conversion Tech Research Grants, 1786
NSF Interfacial Processes and Thermodynamics Research Grants, 3586
NSF Physical Chemistry Grants, 3616
USDD Ocean Engineering and Marine Systems Grants, 4767
USDD Physical Oceanography Grants, 4771

Fluids Engineering
AIChE Lectureship Award in Fluidization, 416
AIChE Thomas Baron Award in Fluid-Particle Systems, 426
ASME Fluids Engineering Award, 1119
ASME Freeman Scholar Award, 1122
ASME Robert E. Koski Medal, 1159
NSF Interfacial Processes and Thermodynamics Research Grants, 3586

Folk Medicine
North Carolina Arts Council Folklife Grants, 3419
Virginia Foundation for the Humanities Discretionary Grants, 4813
Virginia Foundation for Humanities Open Grants, 4814

Folk Music
Florida Division of Cultural Affairs Individual Artist Fellowships, 2026
North Carolina Arts Council Folklife Internship, 3418
North Carolina Arts Council Folklife Public Program Grants, 3419
NYSCA Folk Arts: Presentation Grants, 3704
Virginia Foundation for the Humanities Discretionary Grants, 4813
Virginia Foundation for Humanities Open Grants, 4814

Folk/Ethnic Arts
Arizona Commission on the Arts Folklorist Residencies, 997
Bush Foundation Artist Fellowships, 1391
Florida Division of Cultural Affairs Individual Artist Fellowships, 2026
NEA National Heritage Fellowships in the Folk and Traditional Arts, 2944
North Carolina Arts Council Folklife Internship, 3418
North Carolina Arts Council Folklife Public Program Grants, 3419
NYSCA Folk Arts: Exhibitions Grants, 3703
NYSCA Folk Arts: Presentation Grants, 3704
NYSCA Folk Arts: Regional and County Folk Arts Programs Grants, 3705
Smithsonian Graduate Student Fellowships, 4277
Smithsonian Minority Visiting Student Awards, 4279
Smithsonian Native American Community Scholars Awards, 4291
Smithsonian Native American Visiting Student Awards, 4292
Smithsonian Postdoctoral Fellowships, 4293
Smithsonian Predoctoral Fellowships, 4295
Virginia Foundation for the Humanities Discretionary Grants, 4813
Virginia Foundation for Humanities Open Grants, 4814

Folklore and Mythology
Alabama Humanities Foundation Mini Grants, 480
Alabama Humanities Foundation Planning/Consultant Grants, 481
Canada-U.S. Fulbright International Science and Technology Award, 1417
Canada-U.S. Fulbright Senior Specialists Grants, 1418
Canada-U.S. Fulbright Traditional Scholars, 1419
Florida Division of Cultural Affairs Individual Artist Fellowships, 2026
Fulbright Alumni Initiatives Awards Grants, 2064
Fulbright Distinguished Chairs Program Grants, 2069
Fulbright European Union Scholar-in-Residence Program Grants, 2071
Fulbright New Century Scholars Program Grants, 2075
Fulbright Occasional Lecturer Program Grants, 2076
Fulbright Specialists Program Grants, 2079
Fulbright Traditional Scholar Program in Europe and Eurasia, 2080
Fulbright Traditional Scholar Program in Sub-Saharan Africa, 2081
Fulbright Traditional Scholar Program in the East Asia/Pacific Region, 2082
Fulbright Traditional Scholar Program in the Near East and North Africa Region, 2083
Fulbright Traditional Scholar Program in the South and Central Asia Region, 2084
Fulbright Traditional Scholar Program in the Western Hemisphere, 2085
Fulbright Visiting Specialists Program: Direct Access to the Muslim World, 2087
Virginia Foundation for the Humanities Discretionary Grants, 4813
Virginia Foundation for Humanities Open Grants, 4814
Wenner-Gren Found Historical Archives Grants, 4849

Food Analysis
USDA National Integrated Food Safety Initiative Grants, 4635

Food Banks
A Charitable Foundation Grants, 273
Anthony R. Abraham Foundation Grants, 869
Batchelor Foundation Grants, 1305
Caesars Foundation Grants, 1407
Callaway Golf Company Foundation Grants, 1410
Campbell Soup Foundation Grants, 1415
Charles H. Farnsworth Trust Grants, 1521
Frank Stanley Beveridge Foundation Grants, 2051
GNOF Gulf Coast Oil Spill Grants, 2153
H.J. Heinz Company Foundation Grants, 2195
Kenneth T. and Eileen L. Norris Found Grants, 2554
MacDonald-Peterson Foundation Grants, 2649
McCarthy Family Foundation Grants, 2717
Northwestern Mutual Foundation Grants, 3430
Pott Foundation Grants, 3984
Reinberger Foundation Grants, 4051
Textron Corporate Contributions Grants, 4501
Vancouver Foundation Grants, 4807
William Blair and Company Foundation Grants, 4875

Food Consumption
Anthem Blue Cross and Blue Shield School Nurse Grants, 868
Coca-Cola Foundation Grants, 1593

Food Distribution
AAAS Roger Revelle Fellowship in Global Stewardship, 38
Caesars Foundation Grants, 1407
Conservation, Food, and Health Foundation Grants for Developing Countries, 1641
Donald and Sylvia Robinson Family Foundation Grants, 1830
GNOF Gulf Coast Oil Spill Grants, 2153
Harris Graduate School of Public Policy Studies Research Development Grants, 2204
McGregor Fund Grants, 2725
SARE Research and Education Grants - Southern Region, 4179
Sensient Technologies Foundation Grants, 4199
Sid W. Richardson Foundation Grants, 4253
Starr Foundation Grants, 4447
The World Food Prize, 4503
Thomas C. Ackerman Foundation Grants, 4508
USDA Foreign Market Development Grants, 4625
USDA Market Access Program Grants, 4633
W.K. Kellogg Foundation Healthy Kids Grants, 4822

Food Engineering
AIChE Food, Pharmaceutical and Bioengineering Division Awards in Chemical Engineering, 410
BWF Innovation in Regulatory Science Grants, 1396
NSF-NIST Interaction in Chemistry, Materials Research, Molecular Biosciences, Bioengineering, and Chemical Engineering, 3453
USDA Distance Education Grants for Higher Education Institutions in Insular Areas, 4622
USDA Scientific Cooperation Exchange Program, 4646

Food Labeling
IFT Foundation Grants, 2345
SARE Research and Education Grants - Southern Region, 4179

Food Management
SARE Research and Education Grants - Southern Region, 4179

Food Preparation
College of Human Sciences Scholarships, 1596

Food Processing
AOCS Edible Applications Technology Division Outstanding Achievement Award, 877
AOCS Edible Applications Technology Division Student Award, 878
Institute for Agriculture and Trade Policy Food and Society Fellowships, 2381
USDA Scientific Cooperation Exchange Program, 4646

Food Production
BWF Innovation in Regulatory Science Grants, 1396
SARE Research and Education Grants - Southern Region, 4179

Food Safety
AAAS Science and Technology Policy Fellowships: Health, Education and Human Services, 43
AOCS Processing Division Student Award, 891
GNOF Gulf Coast Oil Spill Grants, 2153
IFT Foundation Grants, 2345
IIE Rockefeller Foundation Bellagio Center Residencies, 2355
The World Food Prize, 4503
USDA 1890 Institution Teaching, Research, and Extension Capacity Building Grants, 4613
USDA Agriculture and Food Research Initiative Competitive Grants - Food Security, 4614
USDA Hispanic-Serving Institutions Education Grants, 4628
USDA National Integrated Food Safety Initiative Grants, 4635
USDA Norman E. Borlaug International Agricultural Science and Technology Fellowships, 4636
USDA Organic Agriculture Research and Extension Initiative, 4637
USDA Scientific Cooperation Exchange Program, 4646
USDA Technical Agricultural Assistance Research Grants, 4651

Food Sciences
AACC International Cecil F. Pinney Travel Awards, 86
AACC International Graduate Fellowships, 91
American Chemical Society AGFD Donald A. Withycombe Graduate Fellowship Award in Food Chemistry, 627
AOCS Lipid Oxidation and Quality Division Poster Competition, 886
BWF Innovation in Regulatory Science Grants, 1396
Canada Graduate Scholarships and NSERC Postgraduate Scholarships, 1420
Coca-Cola Foundation Grants, 1593
IFT Foundation Grants, 2345
M.J. Murdock Charitable Trust College Research Program for Life Sciences, 2638
Mellon Post-Doctoral Fellowship in Sustainable Agriculture, 2740
National Turkey Federation Research Award, 2887
Phi Upsilon Omicron Alumni Research Grant, 3940
Phi Upsilon Omicron Candle Fellowships, 3941
Phi Upsilon Omicron Diamond Anniversary Fellowships, 3942
Phi Upsilon Omicron Florence Fallgatter Distinguished Service Award, 3943
Phi Upsilon Omicron Founders Fellowship, 3944
Phi Upsilon Omicron Geraldine Clewell Doctoral Fellowship, 3945
Phi Upsilon Omicron Geraldine Clewell Masters Fellowship, 3946
Phi Upsilon Omicron Jean Dearth Dickerscheid Fellowship, 3947
Phi Upsilon Omicron Margaret Drew Alpha Fellowship, 3948
Phi Upsilon Omicron Presidents Res Fellowship, 3949
The World Food Prize, 4503
USDA 1890 Institution Teaching, Research, and Extension Capacity Building Grants, 4613
USDA Distance Education Grants for Higher Education Institutions in Insular Areas, 4622
USDA Food and Agricultural Sciences National Needs Graduate Fellowship Grants, 4624
USDA Higher Education Challenge Grants, 4627
USDA Hispanic-Serving Institutions Education Grants, 4628
USDA Interregional Research Project Minor Crop Pest Management Grants, 4632
USDA National Integrated Food Safety Initiative Grants, 4635
USDA Resident Instruction Grants for Higher Education in Insular Areas, 4645
USDA Scientific Cooperation Exchange Program, 4646
USDA Small Business Innovation Research Grants - Phase I, 4648
USDA Special Research Grants, 4650
USDA Tribal College Research Grants, 4652

Food Technology
AACC International Cecil F. Pinney Travel Awards, 86
AACC International Graduate Fellowships, 91
American Chemical Society AGFD Advancement of the Application of Agricultural and Food Chemistry Award, 626
American Chemical Society AGFD Teranishi Graduate Fellowship Award, 628
American Chemical Society AGFD Young Scientist Award, 630
AOCS Edible Applications Technology Division Outstanding Achievement Award, 877
AOCS Edible Applications Technology Division Student Award, 878
AOCS Processing Distinguished Service Award, 890
AOCS Stephen S. Chang Award, 896
USDA Scientific Cooperation Exchange Program, 4646

Foods
AAAAI Distinguished Clinician Award, 17
AIChE Food, Pharmaceutical and Bioengineering Division Awards in Chemical Engineering, 410
AIChE Food, Pharmaceutical and Bioengineering Division Distinguished Service Award in Chemical Engineering, 411
American Chemical Society AGFD Advancement of the Application of Agricultural and Food Chemistry Award, 626
American Chemical Society AGFD Donald A. Withycombe Graduate Fellowship Award in Food Chemistry, 627
American Chemical Society AGFD Teranishi Graduate Fellowship Award, 628
American Chemical Society AGFD Undergraduate Research Symposium Travel Grants, 629
American Chemical Society AGFD Young Scientist Award, 630
Christine Wilson Award, 1558
Institute for Agriculture and Trade Policy Food and Society Fellowships, 2381
USDA Special Research Grants, 4650

Football
Arizona Cardinals Grants, 995
Viking Children's Fund Grants, 4811

Foreign Languages
Alabama Humanities Foundation Mini Grants, 480
Alabama Humanities Foundation Planning/Consultant Grants, 481
IIE David L. Boren Scholarships, 2351
IIE Leonora Lindsley Memorial Fellowships, 2354
Japan Foundation Six-Month Grants for Specialists in Cultural and Academic Fields, 2479
OAH David Thelen Award, 3714
OAH Willi Paul Adams Award, 3730
PHF Penn Faculty Research Fellowships, 3932
PHF Regional Mellon Faculty Research and Cultural Fellowships, 3933
PHF Undergraduate Humanities Forum Research Fellowships, 3934
SSRC Int'l Dissertation Research Fellowships, 4429
SSRC Mellon Mays Predoctoral Research Grants, 4432

Foreign Languages Education
ARIT Fellowships for Intensive Advanced Turkish Language Study, 992
DAAD Learn German in Germany Grants, 1700
Dept of Ed International Research and Studies Project Grants, 1745
Dept of Ed Star Schools Program Grants, 1749
Eastern Europe & the Independent States of the Former Soviet Union Research & Training Grants, 1891
Fulbright-Hays Seminars Abroad Bilateral Projects Grants, 2062
IIE David L. Boren Fellowships, 2350
IIE David L. Boren Scholarships, 2351
IIE Leonora Lindsley Memorial Fellowships, 2354
Institute of Turkish Studies Summer Language Study Grants in Turkey, 2391
Japan Foundation Six-Month Grants for Specialists in Cultural and Academic Fields, 2479
National Security Education Program Grants to Institutions of Higher Education, 2885
NSF East Asia and Pacific Summer Institutes for U.S. Graduate Students, 3555
Penn Graduate School of Education Foreign Language Area Studies Fellowship, 3892
SSRC Eurasia Program Pre-Dissertation Awards, 4428
U.S. Department of Education Foreign Languages and Area Studies Fellowships, 4539
U.S. Department of Education National Resource Centers, 4542

Foreign Policy
CFR Edward R. Murrow Press Fellowship, 1500
CFR International Affairs Fellowships, 1501
CFR International Affairs Fellowships in Nuclear Security, 1502
CFR Int'l Affairs Fellowships in South Korea, 1503
CFR International Affairs in Japan Fellowships, 1504
CFR Military Fellowships, 1505
CFR Stanton Nuclear Security Fellowships, 1506
Kluge Center Kissinger Chair in Foreign Policy and International Relations, 2564
Kluge Center Kissinger Lecturer Award, 2565
United States Institute of Peace - Transatlantic Post-Doc Fellowship for International Relations and Security, 4581

Forensic Law
DOI Rhinoceros and Tiger Conservation Grants, 1824
ORISE Faculty Research Participation at the U.S. Army Joint POW/MIA Accounting Command, Central Identification Laboratory, 3805
ORISE Postgraduate Research Participation at the US Army Central Identification Lab, Hawaii, 3822

Forest Ecology
Bullard Fellowship in Forest Research, 1383
GCA Tropical Botany Research Awards, 2101
ISA Hyland R. Johns Grant Program, 2428

ISA John Z. Duling Grants, 2429
Orchard Foundation Grants, 3793
Organization for Tropical Studies Fellowships, 3796
Organization for Tropical Studies Pilot Awards for Exploratory Research, 3797
OTS Francis J. Bossuyt Fellowship, 3855
OTS Undergrad NAPIRE Summer Interns, 3858
RA Kleinhans Fellowship for Research in Tropical Nontimber Forest Products, 4026
Ross Foundation Grants, 4106
USDA Hazardous Fuels Woody Biomass Utilization Grants, 4626
USDA McIntire-Stennis Cooperative Forestry Research Grants, 4634
Young Scientists Summer Program, 4903

Forest Economics
ISA Hyland R. Johns Grant Program, 2428
ISA John Z. Duling Grants, 2429
USDA McIntire-Stennis Cooperative Forestry Research Grants, 4634

Forest Products Industry
Forest Products Society Wood Award, 2034
Green Diamond Charitable Contributions, 2171
Institute of Paper Science & Tech Fellowships, 2387
ISA Hyland R. Johns Grant Program, 2428
RA Kleinhans Fellowship for Research in Tropical Nontimber Forest Products, 4026
USDA McIntire-Stennis Cooperative Forestry Research Grants, 4634
Weyerhaeuser Company Foundation Grants, 4857

Forestry Management
American-Scandinavian Foundation Short-Term Training Program Grants, 606
Great Lakes Protection Fund Grants, 2170
ISA Hyland R. Johns Grant Program, 2428
ISA John Z. Duling Grants, 2429
RSC Thomas W. Eadie Medal, 4126
USDA Hazardous Fuels Woody Biomass Utilization Grants, 4626
USDA McIntire-Stennis Cooperative Forestry Research Grants, 4634

Forestry/Forest Sciences
USDA Food and Agricultural Sciences National Needs Graduate Fellowship Grants, 4624
USDA Hazardous Fuels Woody Biomass Utilization Grants, 4626
USDA McIntire-Stennis Cooperative Forestry Research Grants, 4634

Forests and Woodlands
Canada Graduate Scholarships and NSERC Postgraduate Scholarships, 1420
Edmund Niles Huyck Preserve Graduate and Postgraduate Research Grants, 1894
Green Diamond Charitable Contributions, 2171
Jenifer Altman Foundation Grants, 2484
John B. Bene Fellowships in Social Forestry, 2496
John D. and Catherine T. MacArthur Foundation Global Challenges Program Grants, 2498
MacArthur Foundation Conservation and Sustainable Development Grants, 2648
Northwest Fund for the Environment Grants, 3431
USDA McIntire-Stennis Cooperative Forestry Research Grants, 4634
USDA Small Business Innovation Research Grants - Phase I, 4648
Weyerhaeuser Company Foundation Grants, 4857

Fossil Fuels
David Bohnett Foundation Grants, 1724
DOE Coal Liquefaction Research Grants, 1777
DOE Energy Biosciences Research Grants, 1784
DOE University Coal Research Grants, 1816
US CRDF CGP ETF 2009 Energy Competition For Joint U.S.-Estonian Research Teams, 4609

USDA Plant Feedstock Genomics for Bioenergy: A Joint Research Grant, 4639
William and Flora Hewlett Foundation Environmental Grants, 4874

Fossil Fuels, Coal
DOE Combustion Systems Research Grants, 1778
DOE Fossil Advanced Research and Technology Development Grants, 1793
DOE Fossil Energy Research & Devel Grants, 1794
Illinois DCEO Coal Revival Grants, 2360

Fossil Fuels, Natural Gas
DOE Fossil Energy Research & Devel Grants, 1794
Gas Technology Institute Basic Research, 2096
Illinois DCEO Coal Research Grants, 2359
Rocky Mountain Mineral Law Found Grants, 4094

Fossil Fuels, Petroleum
American Chemical Society Petroleum Research Fund Doctoral New Investigator Grants, 690
American Chemical Society Petroleum Research Fund Undergraduate New Investigator Grants, 694
DOE Fossil Energy Research & Devel Grants, 1794
DOE Vehicle & Engine Research & Devel Grants, 1819
Rocky Mountain Mineral Law Found Grants, 4094

Foster Care
Sunny and Abe Rosenberg Foundation Grants, 4469

French Art
Camargo Foundation Fellowships, 1412
Walter J. Jensen Fellowship for French Language, Literature, and Culture, 4830

French Language/Literature
American Historical Assoc J. Russell Major Prize, 726
ATA Educational Trust Grants - Curriculum Resource or Research Project, 1253
Camargo Foundation Fellowships, 1412
CFUW Bourse Georgette LeMoyne Award, 1508
Harry Ransom Humanities Research Center Research Fellowships, 2209
Mary Isabel Sibley Fellowship, 2672
NL Northeast Modern Language Assocation Fellowship, 3314
NL Weiss/Brown Publication Subvention Award, 3320
Walter J. Jensen Fellowship for French Language, Literature, and Culture, 4830

Friction/Wear Lubrication
ASME Burt L. Newkirk Award, 1111
ASME Marshall B. Peterson Award, 1141
ASME Mayo D. Hersey Award, 1142

Fuel Cells
Illinois Clean Energy Community Foundation Renewable Energy Grants, 2358

Fuels
USDD Future Naval Fuels Grants, 4729
USDD Power and Thermal Management Technology Development Grants, 4775

Fund-Raising
AAAAI Distinguished Layperson Award, 18
AFP Research Grants, 356
Beim Foundation Grants, 1331
Victor E. Speas Foundation Grants, 4810

Fusion, Inertial
DOE Fusion Energy Science Division Grants, 1795
DOE Fusion Energy Science Tech Div Grants, 1796
DOE Programs in Fusion Energy Sciences, 1811

Galaxies, Stellar Systems
NSF Astronomy and Astrophysics Res Grants, 3495
NSF Extragalactic Astronomy and Cosmology Research Grants, 3564
NSF Galactic Astronomy Research Grants, 3568

Gambling
National Center for Responsible Gaming Centers of Excellence in Gambling Research Grants, 2850
National Center for Responsible Gaming Conference Scholarships, 2851
National Center for Responsible Gaming Early Stage Investigator Grants, 2852
National Center for Responsible Gaming Exploration Grants, 2853
National Center for Responsible Gaming Independent Scientist Grants, 2854
National Center for Resp Gaming Large Grants, 2855
National Center for Responsible Gaming Postdoctoral Fellowships, 2856
National Center for Resp Gaming Seed Grants, 2857
National Center for Responsible Gaming Travel Grants, 2858

Gangs
Juvenile Justice Clearing House Grants, 2530

Gas Turbines
ASME Patrick J. Higgins Award, 1152
ASME R. Tom Sawyer Award, 1157

Gastroenterology
ASGE / ConMed Award for Outstanding Manuscript by a Fellow/Resident, 1061
ASGE Don Wilson Award, 1062
ASGE Endoscopic Research Awards, 1063
ASGE Endoscopic Research Career Development Awards, 1064
ASGE Given Capsule Endoscopy Res Award, 1065
CFF First- and Second-Year Clinical Fellowships, 1492
FDHN Bridging Grants, 1986
FDHN Centocor International Research Fellowship in Gastrointestinal Inflammation & Immunology, 1987
FDHN Designated Outcomes Award in Geriatric Gastroenterology, 1988
FDHN Designated Research Award in Geriatric Gastroenterology, 1989
FDHN Designated Research Award in Research Related to Pancreatitis, 1990
FDHN Fellow Abstract Prizes, 1991
FDHN Fellowship to Faculty Transition Awards, 1992
FDHN Funderburg Research Scholar Award in Gastric Biology Related to Cancer, 1993
FDHN Graduate Student Awards, 1994
FDHN Jon I. Isenberg International Research Scholar Award, 1995
FDHN June & Donald O. Castell MD, Esophageal Clinical Research Award, 1996
FDHN Moti L. & Kamla Rustgi International Travel Awards, 1997
FDHN Non-Career Research Awards, 1998
FDHN Non-Career Research Grants, 1999
FDHN Research Scholar Awards, 2000
FDHN Student Research Fellowships, 2001
FDHN TAP Endowed Designated Research Award in Acid-Related Diseases, 2002
FDHN Translational Research Awards, 2003
NFID New Investigator Matching Grants, 2989
Rett Syndrome Research Foundation Grants, 4058

Gender
Catholic University Life Cycle Institute Research Fellowship, 1444
Overbrook Foundation Grants, 3860
Radcliffe Institute Oral History Grants, 4022

Gender Equity
AAUW Community Action Grants, 213
ASR Fichter Research Grant, 1217
Harvard Women and Public Policy Postdoctoral Fellowship, 2217
NCSS Jean Dresden Grambs Distinguished Career Research in Social Studies Award, 2938
NSF ADVANCE: Increasing the Participation and Advancement of Women in Academic Science and Engineering Careers Grants, 3461

NSF Research on Gender in Science and Engineering Grants, 3639
PC Fred H. Bixby Fellowships, 3880
Rockefeller Foundation Humanities Fellowships, 4089
Rockefeller Foundation Scholarly Residencies, 4091
Saint Mary's College Fellowships, 4159
SfN Louise Hanson Marshall Specific Recognition Award, 4208
SfN Mika Salpeter Lifetime Achievement Award, 4209
Shastri Indo-Canadian Institute Collaborative Research Project Grants, 4220
Shastri Indo-Canadian Institute Partnership Development Seed Grants, 4228
United Methodist Committee on Relief Global Health Grants, 4577

Gender Studies
AAUW American Dissertation Fellowships, 209
ASU Graduate College Dean's Fellowships, 1236
Five College Women's Research Center Ford Associateships, 2022
Harry Frank Guggenheim Foundation Dissertation Fellowships, 2207
Radcliffe Institute Oral History Grants, 4022
SSHRC Insight Development Grants, 4409
SSHRC Insight Grants, 4410

Gene Regulation
NHLBI Career Transition Awards, 3021
NIDDK Co-Activators and Co-Repressors in Gene Expression Grants, 3141

Gene Therapy
ACGT Investigators Grants, 271
NHLBI Career Transition Awards, 3021

Genetic Manipulation
Gruber Foundation Rosalind Franklin Young Investigator Award, 2183
Hunter's Hope Foundation Grants, 2298

Genetics
AACC International Graduate Fellowships, 91
AFAR CART Fund Grants, 351
AFAR Research Grants, 354
Affymetrix Corporate Contributions Grants, 355
AIHS Alberta/Pfizer Translational Res Grants, 435
Alfred P. Sloan Foundation Barcode of Life Grants, 543
Alfred P. Sloan Foundation Energy and Environment Grants, 552
American Philosophical Society Library Resident Research Fellowships, 769
ASGH Award for Excellence in Human Genetics Education, 1066
ASGH C.W. Cotterman Award, 1067
ASGH Charles J. Epstein Trainee Awards for Excellence in Human Genetics Research, 1068
ASGH McKusick Leadership Award, 1069
ASGH William Allan Award, 1070
ASGH William Curt Stern Award, 1071
ASHG Public Health Genetics Fellowship, 1086
Barth Syndrome Foundation Research Grants, 1302
Bedding Plants Foundation Research Grants, 1330
Carnegie Developmental Bio Research Fellows, 1433
Charles H. Hood Foundation Child Health Research Awards, 1522
Dana Foundation Science and Health Grants, 1716
Del E. Webb Foundation Grants, 1733
Fannie E. Rippel Foundation Grants, 1978
Glaucoma Research Pilot Project Grants, 2146
Gruber Foundation Gentics Prize, 2178
Gruber Foundation Rosalind Franklin Young Investigator Award, 2183
Hawaii Community Foundation Health Education and Research Grants, 2221
Hereditary Disease Foundation John J. Wasmuth Postdoctoral Fellowships, 2244
Hereditary Disease Found Lieberman Award, 2245
Hereditary Disease Foundation Research Grants, 2246
Judith Graham Pool Postdoctoral Research Fellowship in Hemophilia, 2529
National MPS Society Grants, 2880
NCI/NIDA Transdisciplinary Tobacco Use Research Centers Grants, 2912
NCI Centers of Excellence in Cancer Communications Research, 2921
NEI Clinical Vision Research Devel Award, 2960
NFX Patient-Oriented Research Grants, 3010
NFX Summer Research Fellowships, 3011
NHGRI Mentored Research Scientist Development Award, 3013
NHLBI Career Transition Awards, 3021
NHLBI Mentored Patient-Oriented Research Career Development Awards, 3037
NHLBI Pediatric Cardiac Genomics Consortium Grants, 3046
NIA Effects of Gene-Social Environment Interplay on Health and Behavior in Later Life Grants, 3093
NIA Nathan Shock Centers of Excellence in Basic Biology of Aging Grants, 3113
NICHD Innovative Therapies and Clinical Studies for Screenable Disorders Grants, 3126
NIDCD Mentored Research Scientist Development Award, 3132
NIDDK Cell-Specific Delineation of Prostate and Genitourinary Development, 3140
NIDDK Co-Activators and Co-Repressors in Gene Expression Grants, 3141
NIDDK Erythroid Lineage Mol Toolbox Grants, 3152
NIGMS Pharmacogenomics Res Network Grants, 3198
NIGMS Predoctoral Training in Bioinformatics and Computation Biology Grants, 3199
NIGMS Short Courses on Mathematical, Statistical, and Computational Tools for Studying Biological Systems, 3201
NIH Fine Mapping and Function of Genes for Type 1 Diabetes Grants, 3218
NSF CISE Assembling the Tree of Life Grants, 3523
NSF Genes and Genome Systems Cluster Grants, 3569
NSF Physical Anthropology Research Grants, 3615
NSF Population & Evolutionary Process Grants, 3629
NSF Project 2010 Grants, 3636
Pediatric Cancer Research Foundation Grants, 3889
PhRMA Foundation Informatics Sabbatical Fellowships, 3956
Rett Syndrome Research Foundation Grants, 4058
Searle Scholars Program Grants, 4193
Sigrid Juselius Foundation Grants, 4260
Starr Foundation Grants, 4447
Susan G. Komen Breast Cancer Foundation Challege Grants: Investigator Initiated Research, 4475
Tourette Syndrome Association Post-Doctoral Fellowships, 4523
Tourette Syndrome Association Research Grants, 4524
US CRDF Leishmaniasis: Collaborative Research Opportunities in N Africa & Middle East, 4610
USDA Biotechnology Risk Assessment Grants, 4618

Genetics, Molecular
APHL Emerging Infectious Diseases Fellowships, 914
ASGH Award for Excellence in Human Genetics Education, 1066
ASGH C.W. Cotterman Award, 1067
ASGH Charles J. Epstein Trainee Awards for Excellence in Human Genetics Research, 1068
ASGH McKusick Leadership Award, 1069
ASGH William Allan Award, 1070
ASGH William Curt Stern Award, 1071
Davis Foundation Postdoctoral Fellowships in Eating Disorders Research, 1727
GPP Charles E. Culpeper Foundation Biomedical Pilot Projects Initiative Grants, 2159
Gruber Foundation Gentics Prize, 2178
Gruber Foundation Rosalind Franklin Young Investigator Award, 2183
IARC Expertise Transfer Fellowship, 2315
IARC Postdoctoral Fellowships for Training in Cancer Research, 2316
IARC Visiting Scientist Award for Sen Scientists, 2317
NIDDK Co-Activators and Co-Repressors in Gene Expression Grants, 3141
NIDDK Erythroid Lineage Mol Toolbox Grants, 3152
NIGMS/NIDCR Evolution of Infectious Diseases Research Grants, 3196
NIH Fine Mapping and Function of Genes for Type 1 Diabetes Grants, 3218
NSF Animal Developmental Mechanisms Grants, 3471
NSF Evolution of Developmental Systems Grants, 3563
NSF Genes and Genome Systems Cluster Grants, 3569
NSF Plant, Fungal and Microbial Developmental Systems Grants, 3624
Pediatric Cancer Research Foundation Grants, 3889
SfN Ralph W. Gerard Prize in Neuroscience, 4211

Genomics
ASHG Public Health Genetics Fellowship, 1086
Gruber Foundation Gentics Prize, 2178
Pfizer ASPIRE North America Rheumatology Research Awards, 3917
PhRMA Foundation Informatics Sabbatical Fellowships, 3956
Smithsonian Biodiversity Genomics and Bioinformatics Postdoctoral Fellowships, 4272

Geochemistry
American Chemical Society Green Chemistry Chapter Award for Student Affiliates, 678
American Chemical Society Joseph Breen Memorial Fellowship in Green Chemistry, 682
American Chemical Society Kenneth G. Hancock Memorial Student Award in Green Chemistry, 683
American Chemical Society Overcoming Challenges Award, 686
American Chemical Society Petroleum Research Fund New Directions Grants, 691
American Chemical Society Presidential Green Chemistry Challenge Awards, 696
American Chem Soc WCC Rising Star Award, 698
Carnegie Postdoctoral Research Fellowships, 1436
DOE Geosciences Research Grants, 1797
NSF Cooperative Studies Of The Earth's Deep Interior Grants, 3544
NSF Ecosystem Science Grants, 3556
NSF Sediment Geology & Paleobiology Grants, 3646
NSF Synthesis and Modeling Project of the US Joint Global Ocean Flux Study, 3656

Geodesy
NOAA National Height Modernization Program Grants, 3391
NSF Geophysics Grants, 3571

Geographic Quantitative Methods
NSF Arctic Social Sciences Grants, 3492

Geography
AAG Anne U. White Fund Grants, 170
AAG Dissertation Research Grants, 171
AAG E. Willard and Ruby S. Miller Award in Geography, 172
AAG International Geographic Information Fund Graduate Research Grants, 173
AAG International Geographic Information Fund Student Travel Grants, 174
AAG John Brinkerhoff Jackson Prize, 175
AAG Mel Marcus Fund Grants for Physical Geography, 176
AAG Research Grants, 177
ACLS Charles A. Ryskamp Research Fellowships, 281
ACLS Frederick Burkhardt Residential Fellowships for Recently Tenured Scholars, 288
AIAR Andrew W. Mellon Foundation Fellowships, 392
AIAR Annual Professorships, 393
AIAR Ernest S. Frerichs Fellow and Program Coordinator Grant, 394
AIAR George A. Barton Fellowships, 395
American Philosophical Soc Judson Daland Prize, 766
American Philosophical Society Lewis and Clark Fund for Exploration and Field Research, 767

Andrew W. Mellon/ACLS Dissertation Completion Fellowships, 844
Canada-U.S. Fulbright International Science and Technology Award, 1417
Canada-U.S. Fulbright Senior Specialists Program Grants, 1418
Canada-U.S. Fulbright Traditional Scholars, 1419
Canada Graduate Scholarships and NSERC Postgraduate Scholarships, 1420
Detroit Lions Charities Grants, 1753
DOI National Spatial Data Infrastructure Cooperative Agreements Program Grants, 1823
ESRI Conservation Program Grants, 1952
Fulbright Alumni Initiatives Awards Grants, 2064
Fulbright Distinguished Chairs Program Grants, 2069
Fulbright European Union Scholar-in-Residence Program Grants, 2071
Fulbright New Century Scholars Program Grants, 2075
Fulbright Occasional Lecturer Program Grants, 2076
Fulbright Specialists Program Grants, 2079
Fulbright Traditional Scholar Program in Europe and Eurasia, 2080
Fulbright Traditional Scholar Program in Sub-Saharan Africa, 2081
Fulbright Traditional Scholar Program in the East Asia/Pacific Region, 2082
Fulbright Traditional Scholar Program in the Near East and North Africa Region, 2083
Fulbright Traditional Scholar Program in the South and Central Asia Region, 2084
Fulbright Traditional Scholar Program in the Western Hemisphere, 2085
Fulbright Visiting Specialists Program: Direct Access to the Muslim World, 2087
Geographic Education Dissertation Award, 2110
Luther I. Replogle Foundation Grants, 2631
Mellon Post-Doctoral Fellowship in Sustainable Agriculture, 2740
National Geographic Society Expeditions Council Grants, 2872
National Geographic Society Field Research and Exploration Grants, 2873
NOAA National Height Modernization Program Grants, 3391
NSF Biotic Surveys and Inventories Grants, 3510
NSF SBE Doctoral Dissertation Research Improvement Grants, 3641
Royal Scottish Geographical Society Expedition Grants, 4110
Royal Scottish Geographical Society Small Research Grants, 4111
Royal Scottish Geographical Society Small Travel Grants, 4112
SSHRC Insight Development Grants, 4409
SSHRC Insight Grants, 4410

Geography Education
AAG Anne U. White Fund Grants, 170
AAG E. Willard and Ruby S. Miller Award in Geography, 172
AAG International Geographic Information Fund Graduate Research Grants, 173
AAG International Geographic Information Fund Student Travel Grants, 174
AAG Mel Marcus Fund Grants for Physical Geography, 176
Geographic Education Dissertation Award, 2110
Luther I. Replogle Foundation Grants, 2631
NCSS/FASSE/CUFA Inquiry Grant, 2936
NCSS Exemplary Res in Social Studies Award, 2937
NCSS Jean Dresden Grambs Distinguished Career Research in Social Studies Award, 2938
NCSS L Metcalf Exemplary Dissertation Award, 2939
NOAA National Height Modernization Program Grants, 3391
Royal Scottish Geographical Society Expedition Grants, 4110
Royal Scottish Geographical Society Small Research Grants, 4111

Geology
AEG Foundation Legget Fund Grants, 341
American Chemical Society AGRO Sterling B. Hendricks Memorial Lectureship, 635
American Chemical Society Award in Separations Science and Technology, 671
American Chemical Society Green Chemistry Chapter Award for Student Affiliates, 678
American Chemical Society Joseph Breen Memorial Fellowship in Green Chemistry, 682
American Chemical Society Kenneth G. Hancock Memorial Student Award in Green Chemistry, 683
American Chemical Society Overcoming Challenges Award, 686
American Chemical Society Petroleum Research Fund New Directions Grants, 691
American Chemical Society Presidential Green Chemistry Challenge Awards, 696
American Chem Soc WCC Rising Star Award, 698
American Philosophical Soc Judson Daland Prize, 766
American Philosophical Society Lewis and Clark Fund for Exploration and Field Research, 767
Anna-Greta and Holger Crafoords Grants, 858
Canada Graduate Scholarships and NSERC Postgraduate Scholarships, 1420
ConocoPhillips Grants Program, 1629
DOI Geological Research and Data Acquisition Grants, 1822
Field Museum Visiting Scholarships, 2011
GSA Research Grants, 2185
Ian N. McKinnon Memorial Fellowship, 2314
John O. Galloway Memorial Scholarship in Geology and Geophysics, 2503
Karst Research Fellowships and Grants, 2531
National Geographic Society Expeditions Council Grants, 2872
National Geographic Society Field Research and Exploration Grants, 2873
NCAR Postdoctoral Fellowships in the Atmospheric Sciences, 2902
NPSC Grad Fellowships in the Physical Sciences, 3436
NSF Antarctic Earth Sciences Program Grants, 3474
NSF Applied Mathematics Grants, 3481
NSF Arctic Natural Science Grants in Atmospheric Sciences, 3483
NSF Arctic Natural Science Grants in Glaciology, 3486
NSF Biotic Surveys and Inventories Grants, 3510
NSF Continental Dynamics Research Grants, 3543
NSF Cooperative Studies Of The Earth's Deep Interior Grants, 3544
NSF Hydrologic Sciences Research Grants, 3578
NSF Integrative Graduate Education and Research Traineeship, 3584
NSF Natural, Social, and System Sciences and Research Support and Logistics Program, 3600
NSF Paleo Perspectives on Climate Grants, 3607
NSF Sediment Geology & Paleobiology Grants, 3646
NSF Tectonics Research Grants, 3658
ORISE Nuclear Regulatory Commission Historically Black Colleges and Universities Student Research Participation, 3814
Roy J. Carver Charitable Trust Higher Education Grants, 4113
RSC Bancroft Award, 4118
RSC Sir John William Dawson Medal, 4125
SRC Natural and Engineering Science Senior Research Positions in Sweden, 4386
SSRC Mellon Mays Predoctoral Research Grants, 4432
US CRDF CGP-Climate Change and Energy Competition For Joint U.S.-Ukrainian Research Teams, 4607
US CRDF CGP-RFBR Climate Change and Energy Competition For Joint U.S.-Russian Research Teams, 4608
WHOI Postdoctoral Scholar Awards, 4864

Geology Education
AEG Foundation Legget Fund Grants, 341
NPSC Grad Fellowships in the Physical Sciences, 3436
RSC Bancroft Award, 4118

Geology, Marine
NSF Marine Geology & Geophysics Res Grants, 3592
RCY - NSF Ocean Sciences Div Research Grants, 4043
USDD Naval Research Facilities Grants, 4761
WHOI Summer Student Fellowships, 4865

Geomechanics
NSF Cooperative Studies Of The Earth's Deep Interior Grants, 3544

Geometry
Abdus Salam Joint ICTP/SISSA PhD Programme in Physics and Mathematics Grants, 242
ARO Discrete Mathematics and Computer Science Research Contracts, 1009
ASCE Jack E. Leisch Memorial National Graduate Fellowship, 1041
International Max Planck Research School Mathematics in the Sciences Fellowships, 2401
NSF Algebra, Number Theory and Combinatorics Grants, 3468
NSF Geometric Analysis Research Grants, 3570

Geomorphology
NSF Sediment Geology & Paleobiology Grants, 3646

Geophysics
Abdus Salam ICTP Group Associates Grants, 228
Abdus Salam ICTP Junior Associates Grants, 231
Abdus Salam ICTP Postgraduate Diploma Grants, 235
Abdus Salam ICTP Regular Associates Grants, 237
Abdus Salam ICTP Schools, Conferences, Workshops and Extended Research Workshops Grants, 238
Abdus Salam ICTP Senior Associates Grants, 239
Abdus Salam ICTP Training and Research in Italian Laboratories Grants, 240
American Chemical Society Green Chemistry Chapter Award for Student Affiliates, 678
American Chemical Society Joseph Breen Memorial Fellowship in Green Chemistry, 682
American Chemical Society Kenneth G. Hancock Memorial Student Award in Green Chemistry, 683
American Chemical Society Overcoming Challenges Award, 686
American Chemical Society Petroleum Research Fund New Directions Grants, 691
American Chemical Society Presidential Green Chemistry Challenge Awards, 696
American Chem Soc WCC Rising Star Award, 698
Brinson Foundation Grants, 1372
Carnegie Postdoctoral Research Fellowships, 1436
ConocoPhillips Grants Program, 1629
DOE Basic Energy Sciences Grants, 1772
DOE Geosciences Research Grants, 1797
John O. Galloway Memorial Scholarship in Geology and Geophysics, 2503
NSF Antarctic Earth Sciences Program Grants, 3474
NSF Arctic Nat Science Grants in Earth Sciences, 3485
NSF Cooperative Studies Of The Earth's Deep Interior Grants, 3544
NSF Geophysics Grants, 3571
NSF Marine Geology & Geophysics Res Grants, 3592
RCY - NSF Ocean Sciences Div Research Grants, 4043
RSC Bancroft Award, 4118
USDD Naval Research Facilities Grants, 4761
WHOI Postdoctoral Scholar Awards, 4864
WHOI Summer Student Fellowships, 4865

Geosciences
DOE Basic Energy Sciences Grants, 1772
NSF Collaborations in Math Geosciences, 3536
NSF Partnerships for International Research and Education, 3611
USDA Decadal and Regional Climate Prediction Using Earth System Models Grants, 4621
USDD Littoral Geosciences and Optics Grants, 4746

Geriatrics
AFAR Medical Student Training in Aging Research Program, 352

AHRQ Independent Scientist Award, 388
American Academy of Nursing Building Academic Geriatric Nursing Capacity Scholarships, 614
American Academy of Nursing Claire M. Fagin Fellowships, 615
American Academy of Nursing Mayday Grants, 616
Chest Foundation Geriatric Development Research Awards, 1536
Donald W. Reynolds Foundation Aging and Quality of Life Grants, 1832
Ellison Medical Foundation/AFAR Julie Martin Mid-Career Award in Aging Research, 1910
Ellison Medical Foundation/AFAR Postdoctoral Fellows in Aging Research Program, 1911
FDHN Designated Outcomes Award in Geriatric Gastroenterology, 1988
FDHN Designated Research Award in Geriatric Gastroenterology, 1989
Greenwall Foundation Bioethics Grants, 2174
Healthcare Foundation of New Jersey Grants, 2222
Leonard and Helen R. Stulman Charitable Foundation Grants, 2602
NIA Health Behaviors and Aging Grants, 3098
Regenstrief General Internal Medicine - General Pediatrics Research Fellowship Program, 4045

German Language/Literature
DAAD-Leo Baeck Institute Fellowships, 1694
DAAD German Studies Research Grants, 1698
DAAD Intensive Language Courses Grants, 1699
DAAD Learn German in Germany Grants, 1700
Germanistic Society of America Fellowships, 2128
NL Northeast Modern Language Assocation Fellowship, 3314

German Studies
AAB Berlin Prize Fellowships, 48
Alexander von Humboldt Foundation Fraunhofer-Bessel Research Awards, 521
DAAD-AICGS Research Fellowships, 1693
DAAD Emigre Memorial German Internship Program - Bundestag, 1696
DAAD German Studies Research Grants, 1698
DAAD Intensive Language Courses Grants, 1699
DAAD Study & Internship Program in Germany, 1708
DAAD Visiting Professorship Grants, 1711
Fulbright German Studies Seminar Grants, 2072
Fulbright International Education Administrators Seminar Program Grants, 2073
LBI-L Leo Baeck/Studienstiftung Fellowships, 2592
LBI-NY/DAAD Fellowships in Germany, 2593
LBI-NY/DAAD Fellowships in New York, 2594
LBI-NY Gerald Westheimer Career Development Fellowship, 2598
Leo Baeck Essay Prize, 2601
Minerva Fellowships, 2778
Minerva Short-Term Research Grants, 2779
OAH Germany Residency Program Grants, 3719
University of London Visiting Fellowships, 4588

Germany
Alexander von Humboldt Foundation Fellowships for Experienced Researchers, 516
Alexander von Humboldt Foundation Fellowships for Postdoctoral Researchers, 517
Alexander von Humboldt Foundation Georg Forster Research Award, 525
Alexander von Humboldt Foundation Konrad Adenauer Research Award, 531
Alexander von Humboldt Foundation Reimar Lust Award, 536
DAAD Emigre Memorial German Internship Program - Bundestag, 1696
DAAD Faculty Research Visit Grants, 1697
DAAD German Studies Research Grants, 1698
DAAD Intensive Language Courses Grants, 1699
DAAD Learn German in Germany Grants, 1700
DAAD Research Grants for Doctoral Candidates and Young Academics and Scientists, 1703
DAAD Research Stays for University Academics and Scientists, 1705
DAAD Scholarships for Artists: Study Visits for Academics, 1707
DAAD Study & Internship Program in Germany, 1708
DAAD Undergraduate Scholarships, 1710
DAAD Visiting Professorship Grants, 1711
Eisenhower Exchange International Fellowships, 1902
Fulbright Beginning Prof Journalism Awards, 2065
Fulbright German Studies Seminar Grants, 2072
Fulbright International Education Administrators Seminar Program Grants, 2073
Fulbright Travel Grants, 2086
Gerda Henkel Foundation Grants, 2127
Germanistic Society of America Fellowships, 2128
LBI-NY Fritz Halbers Fellowships, 2597
Minerva Fellowships, 2778
Minerva Short-Term Research Grants, 2779
NL Herzog August Bibliothek Wolfenbuttel Fellowship, 3286
OAH Germany Residency Program Grants, 3719

Gifted/Talented Education
Dept of Ed Jacob K. Javits Gifted and Talented Students Education Program-National Research and Development Center, 1746
Ford Motor Company Fund Grants Program, 2033
Lynde & Harry Bradley Foundation Fellowships, 2634
Lynde and Harry Bradley Foundation Grants, 2635
NAGC Hollingworth Award, 2820
NAGC Masters and Specialists Award, 2821
NFIE Found Learning & Leadership Grants, 2992

Girls
Mattel International Grants Program, 2686
NWSA Abafazi-Africana Women's Studies Essay Award, 3675
USAID Family Planning and Reproductive Health Methods Grants, 4598
USAID Microbicides Research, Development, and Introduction Grants, 4599

Glaciology
NSF Antarctic Glaciology Program Grants, 3475
NSF Antarctic Research Grants, 3479
NSF Arctic Natural Science Grants in Glaciology, 3486
NSF Arctic Natural Sciences Program Grants, 3488
NSF Polar Programs Division Research Grants, 3627
NSF Sediment Geology & Paleobiology Grants, 3646

Glaucoma
AHAF National Glaucoma Research Grants, 372
Columbus Foundation Ann Ellis Fund Grants, 1600
Glaucoma Foundation Grants, 2145
Glaucoma Research Pilot Project Grants, 2146
NYAM Lewis Rudin Glaucoma Grants, 3689

Global Change
Bill and Melinda Gates Foundation Agricultural Development Grants, 1345
Bill and Melinda Gates Foundation Water, Sanitation and Hygiene Grants, 1347
NASA Goddard Visiting Fellowships, 2838
SSRC African Peacebuilding Network Research Grants, 4422

Global Issues
Bill and Melinda Gates Foundation Agricultural Development Grants, 1345
Bill and Melinda Gates Foundation Water, Sanitation and Hygiene Grants, 1347
IIE David L. Boren Fellowships, 2350
IIE Rockefeller Foundation Bellagio Center Residencies, 2355
Independent Institute John M. Templeton Fellowships Essay Contest, 2369
Kellogg Institute for International Studies Collaborative Faculty Grants, 2536
National Wildlife Federation Campus Ecology Fellowships, 2890
SSRC Abe Fellowships, 4421
SSRC Eurasia Program Dissertation Development Awards, 4427
USDA Hispanic-Serving Institutions Educ Grants, 4628
USDA Scientific Cooperation Research Program, 4647

Global Warming
National Wildlife Federation Campus Ecology Fellowships, 2890

Globalization
Boeing Company Contributions Grants, 1357
Independent Institute John M. Templeton Fellowships Essay Contest, 2369
SRC Landmark Innovation Award, 4376

Glomerular Disease
NIDDK Grants for Basic Research in Glomerular Diseases, 3155

Government
AAAS Science and Technology Policy Overseas Fellowships at USAID Missions, 44
ALA NewsBank/Readex Cath Reynolds Award, 494
American Mathematical Society AAAS Congressional Fellowships, 750
Anderson Foundation Grants, 842
Arca Foundation Grants, 984
Beim Foundation Grants, 1331
Carl and Lily Pforzheimer Foundation Grants, 1427
Carnegie Corporation of New York Grants, 1432
Carnegie Endowment for International Peace Junior Fellowships, 1434
Castle Rock Foundation Grants, 1441
Center of Excellence in General Aviation Grants, 1486
Chautauqua Region Community Found Grants, 1533
Christine Mirzayan Science and Technology Policy Graduate Fellowship Program, 1556
DAAD Emigre Memorial German Internship Program - Bundestag, 1696
Eastern Europe & the Independent States of the Former Soviet Union Research & Training Grants, 1891
Ford Foundation Peace and Social Justice Grants, 2031
Gates Family Foundation Fellowships, 2097
Goldsmith Awards Program, 2158
Hatton W. Sumners Foundation for the Study and Teaching of Self Government Grants, 2219
IREX International Leadership in Educ Grants, 2420
James S. McDonnell Foundation Complex Systems Collaborative Activity Awards, 2465
Japan Foundation Center for Global Partnership Grants, 2472
Kluge Center Distinguished Chairs, 2560
NOAA Joint Research Institutes Grants, 3379
NSF Innovation and Organizational Change, 3581
NSF Integrative Graduate Education and Research Traineeship, 3584
NSF Science and Technology Studies Grants, 3644
OSF Human Rights and Governance Internships, 3845
PHF Penn Faculty Research Fellowships, 3932
PHF Regional Mellon Faculty Research and Cultural Fellowships, 3933
PHF Undergraduate Humanities Forum Research Fellowships, 3934
Playboy Foundation Grants, 3980
Rajiv Gandhi Foundation Grants, 4025
Rosenthal Writer-in-Residence Program, 4105
RWJF Childhood Obesity Grants, 4135
Shorenstein Center Fellowships, 4235
Willary Foundation Grants, 4870

Government Documents
ALA Margaret T. Lane/Virginia F. Saunders Memorial Research Award, 492
ALA ProQuest Documents to the People Award, 496

Government Regulations
JMO Public Policy Research Grants, 2492
OSF Global Faculty Grants - Track 1, 3841
OSF Thomas J. Alexander Fellowships, 3853

SSHRC Metropolis Project, 4414
USDA Norman E. Borlaug International Agricultural Science and Technology Fellowships, 4636

Government Studies
AAAS Science and Technology Policy Fellowships: Congressional, 39
AAAS Science and Technology Policy Overseas Fellowships at USAID Missions, 44
APSA Congressional Fellowships for Communications Scholars and Journalists, 926
APSA Congressional Fellowships for Federal Executives, 927
APSA Congressional Fellowships for International Scholars, 928
APSA Congressional Fellowships for Journalists, 929
APSA Congressional Fellowships for Political Scientists, 930
APSA Congressional Health and Aging Policy Fellowships, 931
APSA E.E. Schattschneider Award, 932
APSA Edward S. Corwin Award, 933
APSA Gabriel A. Almond Award, 934
APSA Harold D. Lasswell Award, 935
APSA Helen Dwight Reid Award, 936
APSA Leonard D. White Award, 937
APSA Leo Strauss Award, 938
APSA Presidency Research Fund Fellowships, 943
APSA Robert Wood Johnson Foundation Health Policy Fellowships, 944
APSA Special Fund for the Study of Women and Politics, 947
APSA Warren E. Miller Fellowship in Electoral Politics, 949
APSA William Anderson Award, 950
DAAD Emigre Memorial German Internship Program - Bundestag, 1696
Gates Family Foundation Fellowships, 2097
Helen Kellogg Institute for International Studies Residential Fellowships, 2230
Herbert Hoover Presidential Library Association Travel Grants, 2242
Rhode Island State Government Internships, 4062
Rosenthal Writer-in-Residence Program, 4105
Shorenstein Center Fellowships, 4235
Web We Want Small Grants, 4836

Government, Comparative
AAAS Science and Technology Policy Overseas Fellowships at USAID Missions, 44
APSA Gabriel A. Almond Award, 934
APSA Harold D. Lasswell Award, 935
APSA Helen Dwight Reid Award, 936
APSA Leonard D. White Award, 937
APSA Leo Strauss Award, 938
Canada-U.S. Fulbright International Science and Technology Award, 1417
Canada-U.S. Fulbright Senior Specialists Grants, 1418
Canada-U.S. Fulbright Traditional Scholars, 1419
Fulbright Alumni Initiatives Awards Grants, 2064
Fulbright Distinguished Chairs Program Grants, 2069
Fulbright European Union Scholar-in-Residence Program Grants, 2071
Fulbright New Century Scholars Program Grants, 2075
Fulbright Occasional Lecturer Program Grants, 2076
Fulbright Specialists Program Grants, 2079
Fulbright Traditional Scholar Program in Europe and Eurasia, 2080
Fulbright Traditional Scholar Program in Sub-Saharan Africa, 2081
Fulbright Traditional Scholar Program in the East Asia/Pacific Region, 2082
Fulbright Traditional Scholar Program in the Near East and North Africa Region, 2083
Fulbright Traditional Scholar Program in the South and Central Asia Region, 2084
Fulbright Traditional Scholar Program in the Western Hemisphere, 2085
Fulbright Visiting Specialists Program: Direct Access to the Muslim World, 2087

Government, Federal
AAAS Science and Technology Policy Fellowships: Congressional, 39
AAAS Science and Technology Policy Overseas Fellowships at USAID Missions, 44
AACAP Mary Crosby Congressional Fellowships, 68
APSA Congressional Fellowships for Communications Scholars and Journalists, 926
APSA Congressional Fellowships for Federal Executives, 927
APSA Congressional Fellowships for International Scholars, 928
APSA Congressional Fellowships for Journalists, 929
APSA Congressional Fellowships for Political Scientists, 930
APSA Congressional Health and Aging Policy Fellowships, 931
APSA E.E. Schattschneider Award, 932
APSA Edward S. Corwin Award, 933
APSA Gabriel A. Almond Award, 934
APSA Harold D. Lasswell Award, 935
APSA Helen Dwight Reid Award, 936
APSA Leonard D. White Award, 937
APSA Leo Strauss Award, 938
APSA Presidency Research Fund Fellowships, 943
APSA Robert Wood Johnson Foundation Health Policy Fellowships, 944
APSA Special Fund for the Study of Women and Politics, 947
APSA Warren E. Miller Fellowship in Electoral Politics, 949
APSA William Anderson Award, 950
Carnegie Endowment for International Peace Junior Fellowships, 1434
CIA Undergraduate Internship - Analysis, 1562
DHHS Emerging Leaders Program Internships, 1756
Herbert Hoover Presidential Library Association Travel Grants, 2242
IBM Center for the Business of Gov Grants, 2320
IIE David L. Boren Fellowships, 2350
IIE David L. Boren Scholarships, 2351
OAH Richard W. Leopold Prize, 3728
SSHRC-Public Safety Canada Kanishka Project Initiative, 4401
SSHRC Metropolis Project, 4414
USDC Dissertation Fellowships, 4665

Government, Local
Alfred P. Sloan Foundation Selected National Issues Grants, 563
APSA Gabriel A. Almond Award, 934
APSA William Anderson Award, 950
Ford Foundation Peace and Social Justice Grants, 2031
IBM Center for the Business of Gov Grants, 2320
Lynde & Harry Bradley Foundation Fellowships, 2634
Lynde and Harry Bradley Foundation Grants, 2635
OAH Richard W. Leopold Prize, 3728
Robert Sterling Clark Foundation Government Accountability Grants, 4080

Government, Municipal
APSA William Anderson Award, 950
IBM Center for the Business of Gov Grants, 2320
IDEM Section 319(h) Nonpoint Source Program Grants, 2338
OAH Richard W. Leopold Prize, 3728

Government, State
APSA Gabriel A. Almond Award, 934
APSA William Anderson Award, 950
Ford Foundation Peace and Social Justice Grants, 2031
IBM Center for the Business of Gov Grants, 2320
Lynde & Harry Bradley Foundation Fellowships, 2634
Lynde and Harry Bradley Foundation Grants, 2635
OAH Richard W. Leopold Prize, 3728
Rhode Island State Government Internships, 4062
Robert Sterling Clark Foundation Government Accountability Grants, 4080

Governmental Functions
Alfred P. Sloan Foundation Selected National Issues Grants, 563
APSA Warren E. Miller Fellowship in Electoral Politics, 949
CBO Economic Policy Fellowships, 1445
DAAD Emigre Memorial German Internship Program - Bundestag, 1696
Jacob and Hilda Blaustein Foundation Grants, 2454

Graduate Education
AIHS Media Summer Fellowship, 453
ALA Jan Merrill-Oldham Prof Devel Grant, 491
AOCS Analytical Division Student Award, 874
AOCS Biotechnology Division Student Awards, 876
AOCS Edible Applications Technology Division Student Award, 878
AOCS Health & Nutrition Div Student Award, 881
AOCS Health and Nutrition Poster Competition, 882
AOCS Manuchehr Eijadi Award, 887
AOCS Peter and Claire G. Kalustian Award, 888
AOCS Processing Division Student Award, 891
AOCS Ralph H. Potts Memorial Fellowship, 893
AOCS Surfactants and Detergents Division Student Award, 898
AOCS Thomas H. Smouse Memorial Fellowship, 899
ASU Graduate College Herman E. Demund Memorial Fellowships, 1240
ASU Graduate College Martha E. Bernal Memorial Award, 1242
CIC Smithsonian Institution Fellowships, 1564
CIC Traveling Scholars, 1566
Foundation for Evangelism Doc Scholarships, 2039
Grass Foundation Marine Biological Laboratory Advanced Imaging Fellowships, 2162
HRF Mark B. Bain Graduate Fellowships, 2287
ISGC Graduate Fellowships, 2430
JSA JLab Graduate Fellowships, 2521
MBL Gruss Lipper Family Foundation Summer Fellowship, 2704
MBL William Townsend Porter Summer Fellowships for Minority Investigators, 2716
MLA Rittenhouse Award, 2790
National Wildlife Federation Internships, 2891
OSF Civil Service Awards, 3838
OSF Global Supplementary Grants, 3843
OSF Palestinian Rule of Law Fellowships, 3848
Penn Graduate School of Education Dean's Fellowships, 3891
Penn Graduate School of Education Foreign Language Area Studies Fellowship, 3892
Penn Graduate School of Education Pre-Doctoral Training in Interdisciplinary Methods for Field-Based Research in Education, 3893
Penn Graduate School of Education Research Assistantships, 3894
Phi Upsilon Omicron Candle Fellowships, 3941
Phi Upsilon Omicron Diamond Anniversary Fellowships, 3942
Phi Upsilon Omicron Founders Fellowship, 3944
Phi Upsilon Omicron Presidents Res Fellowship, 3949
Princeton University Center for Human Values Graduate Student Top-Up Prizes, 3990
SfN Graduate Student Travel Awards, 4205
SfN Ricardo Miledi Neuroscience Training, 4212
SfN Travel Awards for the Federation of European Neuroscience Societies Forum, 4214
SRC AMD/Mahboob Khan Fellowship, 4348
SRC Company-Named Fellowships, 4350
SRC Company-Named Scholarships, 4351
SRC Graduate Fellowships, 4368
SRC International Fellowship, 4375
SRC Mahboob Khan Outstanding Industry Liaison Award, 4377
SRC Peter Verhofstadt Fellowship, 4389
SRC Robert M. Burger Fellowship, 4392
United States Institute of Peace - Jennings Randolph Peace Dissertation Scholarships, 4579
USDD Graduate Traineeship Awards in Ocean Acoustics, 4730

WHOI Guest Student Internships, 4860
WWF Russell E. Train Fellowships, 4898
Yale University Beinecke Library Research Fellowships for Yale Grad & Professional School Students, 4902

Grain
AACC International Cecil F. Pinney Travel Awards, 86
AACC International Engineering and Processing Division's Stanley Watson Award, 88
AACC International Graduate Fellowships, 91

Grain Processing
AACC International Milling & Baking Division's John C. Halverson Memorial Lectureship, 92

Graphic Design
Houghton Library Visiting Fellowships, 2284
NYFA Artist Fellowships, 3698
Swann Foundation Fellowships, 4478

Grassroots Leadership
Akonadi Foundation Anti-Racism Grants, 477
Brainerd Foundation Grants, 1368
Girl's Best Friend Foundation Grants, 2143
GNOF Gulf Coast Oil Spill Grants, 2153
Japan Found Center for Global Partnership Grants, 2472
OSF Soros Justice Advocacy Fellowships, 3850
Washington Area Women's Foundation Strategic Opportunity and Partnership Fund Grants, 4831

Gravitation
APS Einstein Prize, 955
NSF Arctic Nat Science Grants in Ocean Science, 3487
NSF Geophysics Grants, 3571
NSF Gravitational Physics Research Grants, 3576
NSF Physical and Dynamic Meteorology Grants, 3614
NSF Physics Frontiers Centers Grants, 3619
NSF Physics Research Grants, 3621

Great Lakes
George Gund Foundation Grants, 2117
Great Lakes Fishery Commission Pilot Research Project Grants, 2166
Great Lakes Fishery Commission Sea Lamprey Research Grants, 2167
Great Lakes Fishery Ecological & Biological Fisheries Research to Inform Management Grants, 2169
Great Lakes Protection Fund Grants, 2170
MDEQ Local Water Quality Monitoring Grants, 2732
NL Newberry Consortium on American Indian Studies Graduate Student Fellowships, 3312
NOAA Coastal Hypoxia Research Grants, 3351
NOAA Coastal Ocean Program Grants, 3353
NOAA Coral Reef Ecosystem Studies Grants, 3361
NOAA Cumulative Impacts of Multiple Stressors (MultiStress) Grants, 3366
NOAA Ecological Forecasting Grants, 3368
NOAA Ecology and Oceanography of Harmful Algal Blooms Grants, 3369
NOAA Great Lakes Habitat Restoration Partnership Grants, 3376
NOAA Monitoring and Event Response for Harmful Algal Blooms Grants, 3385
NOAA National Sea Grant John A. Knauss Marine Policy Fellowships, 3394
USDC/NOAA Anadromous Fish Conservation Act Grants, 4654
USDC/NOAA Nature Conservancy Community-based Matching Grants, 4661

Greece
ASCSA Alison M. Frantz Fellowship, 1051
ASCSA Fellowships, 1052
ASCSA Harry Bikakis Fellowship, 1053
ASCSA Jacob Hirsch Fellowship, 1054
ASCSA Summer Sessions Scholarships, 1055
ASCSA Weiner Laboratory Faunal Studies Research Fellowship, 1056
Center for Hellenic Studies Fellowships, 1476
State Scholarships Foundation Grants, 4449

Greek Language/Literature
AAR Postdoctoral Fellowships, 201
Center for Hellenic Studies Fellowships, 1476
Mary Isabel Sibley Fellowship, 2672
Princeton University Library Research Grants, 3997
State Scholarships Foundation Grants, 4449

Grief
Dell Foundation Open Grants, 1734

Growth Factors
NIH Receptors and Signaling in Bone in Health and Disease, 3229

Gulf of Mexico
NFWF Shell Marine Habitat Program Grants, 3008
NOAA Coastal Ecosystem Research Project, 3350
NOAA Coastal Hypoxia Research Grants, 3351
NOAA Cooperative Research Program Grants, 3359
NOAA Coral Reef Ecosystem Studies Grants, 3361
NOAA Cumulative Impacts of Multiple Stressors (MultiStress) Grants, 3366
NOAA Ecological Forecasting Grants, 3368
NOAA Ecology and Oceanography of Harmful Algal Blooms Grants, 3369
NOAA Monitoring and Event Response for Harmful Algal Blooms Grants, 3385
NOAA National Sea Grant Gulf of Mexico Oyster Industry Program Grants, 3392
NOAA Northern Gulf of Mexico Ecosystems and Hypoxia Assessment Program Grants, 3396
NOAA South Florida Program Grants, 3406

Gun Control
David Bohnett Foundation Grants, 1724
Overbrook Foundation Grants, 3860
Richard and Rhoda Goldman Fund Grants, 4063

HIV
25th Anniversary Foundation Grants, 6
Abbott Fund Science Education Grants, 220
Alternatives Research and Development Foundation Alternatives in Education Grants, 581
Alternatives Research and Development Foundation Grants, 582
AMA Foundation Seed Grants for Research, 601
American Psychiatric Association Minority Medical Student Fellowship in HIV Psychiatry, 784
ASM/CDC Fellowships in Infectious Disease and Public Health Microbiology, 1100
ASM ICAAC Young Investigator Awards, 1174
ASM Merck Irving S. Sigal Memorial Awards, 1177
Bernard Van Leer Foundation Grants, 1336
Blowitz-Ridgeway Foundation Grants, 1349
Bodenwein Public Benevolent Foundation Grants, 1355
Bristol-Myers Squibb Foundation Fellowships, 1375
Bristol-Myers Squibb Foundation Global HIV/AIDS Initiative Grants, 1376
Bristol-Myers Squibb Foundation Health Disparities Grants, 1377
CDC Foundation Atlanta International Health Fellowships, 1461
Community Foundation for S Arizona Grants, 1618
Elizabeth Glaser Int'l Leadership Awards, 1906
Elizabeth Glaser Scientist Award, 1907
Fairlawn Foundation Grants, 1974
Frances L. and Edwin L. Cummings Memorial Fund Grants, 2046
Jessie Smith Noyes Foundation Grants, 2485
Johnson & Johnson Corporate Contributions, 2508
Judith Graham Pool Postdoctoral Research Fellowship in Hemophilia, 2529
McCarthy Family Foundation Grants, 2717
NFID New Investigator Matching Grants, 2989
NHLBI Microbiome of the Lung and Respiratory Tract in HIV-Infected Individuals and HIV-Uninfected Controls, 3039
NIA AIDS and Aging: Behavioral Sciences Prevention Research Grants, 3083
NIA Medical Management of Older Patients with HIV/AIDS Grants, 3107
NIDA Pilot and Feasibility Studies in Preparation for Drug Abuse Prevention Trials, 3130
NIDDK Transmission of Human Immunodeficiency Virus In Semen, 3187
NIMH AIDS and Aging: Behavioral Sciences Prevention Research Grants, 3249
NINR AIDS and Aging: Behavioral Sciences Prevention Research Grants, 3260
ORISE Postgraduate Research Participation at the Centers for Disease Control and Prevention, 3820
ORISE Student Research Participation at the Centers for Disease Control and Prevention, 3829
Playboy Foundation Grants, 3980
Sexually Transmitted Diseases Postdoctoral Fellowships, 4200
USAID Comprehensive District-Based Support for Better HIV/TB Patient Outcomes Grants, 4595
USAID Microbicides Research, Development, and Introduction Grants, 4599
USAID Strengthening RMNCH Through Indian Health Professional Associations and Scaling Up Uptake Grants, 4602
USDD Broad Agency Announcement Grants, 4701
USDD Broad Agency Announcement HIV/AIDS Prevention Grants, 4702
USDD HIV/AIDS Prevention: Military Specific HIV/AIDS Prevention, Care, and Treatment for Non-PEPFAR Funded Countries Grants, 4733
USDD HIV/AIDS Prevention Program Information Systems Development Grants, 4734
USDD President's Emergency Plan for AIDS Relief Military Specific HIV Seroprevalence and Behavioral Epidemiology Risk Suvey Grants, 4777

Habitat Preservation
Green Diamond Charitable Contributions, 2171
HRF Hudson River Improvement Grants, 2286
NFWF Shell Marine Habitat Program Grants, 3008
NFWF Wildlife and Habitat Conservation Initiative Grants, 3009
Sioux Falls Area Community Foundation Community Fund Grants (Unrestricted), 4262
Sioux Falls Area Comm Foundation Spot Grants, 4263
USDC/NOAA Coastal and Marine Fish Habitat Protection Grants, 4655
USDC/NOAA Restoration Center National and Regional Partnership Grants, 4662

Habitats
American Rivers Anthony A. Lapham River Conservation Fellowship, 795
FishAmerica Foundation Marine and Anadromous Fish Habitat Restoration Grants, 2018
FishAmerica Foundation Research Grants, 2019
Florida Sea Turtle Grants Program, 2027
Illinois DNR Youth Recreation Corps Grants, 2361
NFWF Acres for America Grants, 2996
NFWF Nature Restoration Trust Grants, 3005
NFWF Puget Sound Marine Conservation Fund Grants, 3006
NFWF Wildlife and Habitat Conservation Initiative Grants, 3009
NOAA Great Lakes Habitat Restoration Partnership Grants, 3376
SeaWorld and Busch Gardens Conservation Fund Grants, 4196
USDC/NOAA Coastal and Marine Fish Habitat Protection Grants, 4655
USDC/NOAA Restoration Center National and Regional Partnership Grants, 4662

Hawaiian Natives
DHHS Native American Programs, 1758
MLA Grad Scholarship for Minority Students, 2787
NAGPRA Museum Consultation/Documentation Grants, 2822
NAGPRA Museum Repatriation Grants, 2823
NSF Tribal Colleges and Universities Program, 3662

Pacific Islanders in Communications Grants, 3861
U.S. Department of Education Native Hawaiian Education Grants, 4543
USDA Alaska Native-Serving and Native Hawaiian-Serving Institutions Education Grants, 4615

Hazardous Waste
ARO Atomic, Molecular, and Optical Physics Research Contracts, 1005
Center for Defense Information Internships, 1473
Honor the Earth Grants, 2279
HRF Hudson River Improvement Grants, 2286
Jessie Smith Noyes Foundation Grants, 2485
NIEHS Hazardous Materials Worker Health and Safety Training Grants, 3190
NIEHS Hazmat Training at Doe Nuclear Weapons Complex Grants, 3191
NSF Interfacial Processes and Thermodynamics Research Grants, 3586
OSRI Biological Monitoring: Intertidal Resources at Risk to Oil Spills, 3854
Sioux Falls Area Community Foundation Community Fund Grants (Unrestricted), 4262
Sioux Falls Area Comm Foundation Spot Grants, 4263

Hazardous Waste, Disposal/Clean-Up
DOE Control Technology and Coal Preparation Research Grants, 1781
Great Lakes Protection Fund Grants, 2170
NIEHS Hazardous Materials Worker Health and Safety Training Grants, 3190
NIEHS Hazmat Training at Doe Nuclear Weapons Complex Grants, 3191
NSF Interfacial Processes and Thermodynamics Research Grants, 3586
USDD Water Desalination & Purific Grants, 4804

Headaches
National Headache Foundation Research Grants, 2875
National Headache Foundation Seymour Diamond Clinical Fellowship in Headache Education, 2876

Health
3M Foundation Community Giving Grants, 5
Albert and Margaret Alkek Foundation Grants, 504
Appalachian Regional Commission Health Care Grants, 915
APSA Robert Wood Johnson Foundation Health Policy Fellowships, 944
ASHFoundation New Century Scholars Program Doctoral Scholarships, 1080
ASHFoundation New Century Scholars Research Grant, 1081
ASHFoundation Res Grant for New Investigators, 1082
Caesars Foundation Grants, 1407
Charles M & Mary D Grant Foundation Grants, 1527
Delta Air Lines Foundation Health and Wellness Grants, 1736
DTE Energy Foundation Health and Human Services Grants, 1865
Fidelity Foundation Grants, 2007
Gil and Dody Weaver Foundation Grants, 2139
Gruber Foundation Weizmann Institute Awards, 2184
Harry W. Bass, Jr. Foundation Grants, 2210
Herbert & Adrian Woods Foundation Grants, 2239
Hilda and Preston Davis Foundation Grants, 2266
Hill Crest Foundation Grants, 2267
Jane Bradley Pettit Foundation Health Grants, 2471
Kluge Center David B. Larson Fellowship in Health and Spirituality, 2559
Lloyd A. Fry Foundation Health Grants, 2620
Lumpkin Family Found Healthy People Grants, 2629
Mary Baker Eddy Library Summer Research Fellowships, 2669
Mattel International Grants Program, 2686
Meyer Memorial Trust Responsive Grants, 2748
MGM Resorts Foundation Community Grants, 2753
NCFR Innovation Grants, 2908
Nelda C. and H.J. Lutcher Stark Found Grants, 2967
NIH Building Interdisciplinary Research Careers in Women's Health Grants, 3209
Norcliffe Foundation Grants, 3415
Northrop Grumman Corporation Grants, 3428
Rohm and Haas Company Grants, 4098
Sigma Theta Tau International /American Nurses Foundation Grant, 4254
Sigma Theta Tau Small Grants, 4258
Sony Corporation of America Corporate Philanthropy Grants, 4329
SSHRC NSERC College and Community Innovation Enhancement Grants, 4400
SSRC Collaborative Research Grants on Environment and Health in China, 4423
Textron Corporate Contributions Grants, 4501
USAID Accelerating Progress Against Tuberculosis in Kenya Grants, 4594
USAID Development Innovation Accelerator Broad Agency Announcement Grants, 4596
USAID Research & Innov for Health Supply Chain Systems and Commodity Security Grants, 4601
USAID Strengthening RMNCH Through Indian Health Professional Associations and Scaling Up Uptake Grants, 4602
USDA Agriculture and Food Research Initiative Competitive Grants - Food Security, 4614
USDD Clinical Functional Assessment Investigator-Initiated Grants, 4709
Wellcome Trust New Investigator Awards, 4844

Health Care
1st Touch Foundation Grants, 2
AACAP Jeanne Spurlock Research Fellowship in Substance Abuse and Addiction for Minority Medical Students, 64
AACAP Summer Medical Student Fellowships, 80
AACR Scholar-in-Training Awards, 146
AACR Scholar-in-Training Awards - Special Conferences, 147
AAFA Investigator Research Grants, 153
AAMC Award for Distinguished Research, 183
Abbott Fund Global AIDS Care Grants, 219
Abbott Fund Science Education Grants, 220
Abington Foundation Grants, 243
ACAAI Foundation Research Grants, 244
AcademyHealth Alice S. Hersh New Investigator Awards, 245
AcademyHealth Nemours Child Health Services Research Awards, 247
AcademyHealth PHSR Research Article of the Year Awards, 248
ACGT Young Investigator Grants, 272
Achelis Foundation Grants, 274
ACS Pilot and Exploratory Projects in Palliative Care of Cancer Patients and Their Families Grants, 324
Acuity Charitable Foundation Grants, 333
Adam Richter Charitable Trust Grants, 339
Aetna Foundation Integrated Health Care Grants, 348
AHRF Regular Research Grants, 386
Albany Medical Center Prize in Medicine and Biomedical Research, 502
Albuquerque Community Foundation Grants, 508
Alcatel-Lucent Technologies Foundation Grants, 509
Alcoa Foundation Grants, 511
Alcohol Misuse and Alcoholism Research Grants, 512
Alice Tweed Tuohy Foundation Grants Program, 569
ALSAM Foundation Grants, 579
Alvin and Lucy Owsley Foundation Grants, 584
ALZA Corporate Contributions Grants, 588
AMA-MSS Research Poster Award, 599
AMA-WPC Joan F. Giambalvo Memorial Scholarships, 600
American-Italian Cancer Foundation Fellowships, 602
American Foodservice Charitable Trust Grants, 710
American Schlafhorst Foundation Grants, 796
Angels Baseball Foundation Grants, 848
Ann & Robert H. Lurie Family Foundation Grants, 859
Ann Peppers Foundation Grants, 863
Appalachian Regional Commission Health Care Grants, 915
Arcadia Foundation Grants, 983
Arie and Ida Crown Memorial Grants, 989
Arizona Cardinals Grants, 995
Arizona Public Service Corporate Giving Program Grants, 998
Arronson Foundation Grants, 1019
Arthritis Foundation of Australia Grants, 1022
ArvinMeritor Foundation Health Grants, 1027
Assisi Foundation of Memphis General Grants, 1221
Australasian Institute of Judicial Administration Seed Funding Grants, 1268
Ball Brothers Foundation General Grants, 1291
Banfi Vintners Foundation Grants, 1296
Barra Foundation Project Grants, 1301
Barth Syndrome Foundation Research Grants, 1302
Batchelor Foundation Grants, 1305
Baxter International Foundation William B. Graham Prize for Health Services Research, 1306
BCBSM Foundation Excellence in Research Awards for Students, 1312
BCBSM Foundation Investigator Initiated Research Grants, 1313
BCBSM Foundation Physician Investigator Research Awards, 1314
BCBSM Foundation Primary and Clinical Prevention Grants, 1315
BCBSM Foundation Student Award Program, 1316
BCBSM Frank J. McDevitt, DO, Excellence in Research Awards, 1317
BCBS of Massachusetts Foundation Grants, 1319
Becton Dickinson and Company Grants, 1328
Berks County Community Foundation Grants, 1333
Better Way Foundation Grants, 1338
Blowitz-Ridgeway Foundation Early Childhood Development Research Award, 1348
Blowitz-Ridgeway Foundation Grants, 1349
Blue Shield of California Grants, 1350
Blum-Kovler Foundation Grants, 1351
Bodenwein Public Benevolent Foundation Grants, 1355
Bodman Foundation Grants, 1356
Boeing Company Contributions Grants, 1357
Booth Bricker Fund Grants, 1359
Bosque Foundation Grants, 1360
Boston Foundation Grants, 1361
Brian G. Dyson Foundation Grants, 1370
Bristol-Myers Squibb Foundation Health Disparities Grants, 1377
Bristol-Myers Squibb Foundation Health Education Grants, 1378
Bristol-Myers Squibb Foundation Women's Health Grants, 1379
Broad Foundation IBD Research Grants, 1380
Bupa Foundation Medical Research Grants, 1386
Burlington Northern Santa Fe Foundation Grants, 1389
Byron W. and Alice L. Lockwood Grants, 1401
Caesars Foundation Grants, 1407
Callaway Golf Company Foundation Grants, 1410
Campbell Soup Foundation Grants, 1415
Carl and Eloise Pohlad Family Grants, 1426
Carl and Lily Pforzheimer Foundation Grants, 1427
Carl C. Icahn Foundation Grants, 1428
Carl Gellert and Celia Berta Gellert Foundation Grants, 1429
Carpenter Foundation Grants, 1439
Carrier Corporation Contributions Grants, 1440
CDC Cooperative Agreement for the Development, Operation, and Evaluation of an Entertainment Education Program, 1457
CDC Grants for Violence-Related Injury Prevention Research, 1463
Charles H. Farnsworth Trust Grants, 1521
Charles Lafitte Foundation Grants, 1525
Charles Stewart Mott Foundation Grants, 1529
Chatlos Foundation Grants Program, 1532
CIGNA Foundation Grants, 1568
Claneil Foundation Grants, 1574
Clarence E. Heller Charitable Foundation Grants, 1575
Clay Foundation Grants, 1579
Cleveland-Cliffs Foundation Grants, 1580
CMS Hispanic Health Services Research Grants, 1584

Coca-Cola Foundation Grants, 1593
Coeta and Donald Barker Foundation Grants, 1594
Coleman Foundation Cancer Care Grants, 1595
Colonel Stanley R. McNeil Foundation Grants, 1597
Columbus Foundation Ann Ellis Fund Grants, 1600
Commonfund Endowment Management Prize, 1601
Commonweal Foundation Community Assistance Grants, 1602
Commonweal Foundation Grants, 1603
Commonwealth Fund Affordable Health Insurance Grants, 1605
Commonwealth Fund Health Care Quality Improvement and Efficiency Grants, 1607
Commonwealth Fund Patient-Centered Coordinated Care Program Grants, 1608
Commonwealth Payment System Reform Grants, 1609
Commonwealth Fund Quality of Care for Frail Elders Grants, 1610
Commonwealth Fund Small Grants, 1611
Commonwealth Fund State High Performance Health Systems Grants, 1612
Communities Foundation of Texas Grants, 1613
Community Found for Greater Buffalo Grants, 1615
Community Found for Muskegon County Grants, 1617
Community Foundation for S Arizona Grants, 1618
Community Found of E Central Illinois Grants, 1619
Comprehensive Health Education Found Grants, 1622
Connecticut Health Foundation Health Initiative Grants, 1628
ConocoPhillips Grants Program, 1629
Conquer Cancer Foundation of ASCO Improving Cancer Care Grants, 1633
Cudd Foundation Grants, 1672
Dade Community Foundation Grants, 1712
DaimlerChrysler Corporation Fund Grants, 1714
Dana Foundation Science and Health Grants, 1716
Daviess County Community Foundation Health Grants, 1725
Davis Family Foundation Grants, 1726
Dell Foundation Open Grants, 1734
DeMatteis Family Foundation Grants, 1740
Denton A. Cooley Foundation Grants, 1743
Detroit Lions Charities Grants, 1753
DHHS Emergency Medical Services for Children Program, 1755
DHHS Oral Health Promotion Research Across the Lifespan, 1759
Doris Duke Charitable Foundation Clinical Interfaces Award Program, 1833
Dorothy Rider Pool Health Care Grants, 1841
Dr. Bob and Jean Smith Foundation Grants, 1853
DTE Energy Foundation Health and Human Services Grants, 1865
Duke University Pain Management Fellowships, 1879
Edward Bangs Kelley and Elza Kelley Foundation Grants, 1898
Edward W. and Stella C. Van Houten Memorial Fund Grants, 1899
Elisabeth Severance Prentiss Foundation Grants, 1905
Eugene M. Lang Foundation Grants, 1957
Eugene McDermott Foundation Grants, 1958
Evanston Community Foundation Grants, 1962
Ewing Halsell Foundation Grants, 1964
Fairlawn Foundation Grants, 1974
FAR Fund Grants, 1980
Farmers Insurance Corporate Giving Grants, 1981
Fidelity Foundation Grants, 2007
Flinn Foundation Grants Programs, 2024
Fluor Foundation Grants, 2028
FMC Foundation Grants, 2029
Ford Motor Company Fund Grants Program, 2033
Forrest C. Lattner Foundation Grants, 2035
Foundation for Enhancing Communities Grants, 2038
Foundation for Seacoast Health Grants, 2041
Frances L. and Edwin L. Cummings Memorial Fund Grants, 2046
Frank Stanley Beveridge Foundation Grants, 2051
General Motors Foundation Grants, 2107
General Service Foundation Human Rights and Economic Justice Grants, 2108
Genuardi Family Foundation Grants, 2109
George A Ohl Jr. Foundation Grants, 2112
George Foundation Grants, 2115
George Frederick Jewett Foundation Grants, 2116
George H. Hitchings New Investigator Award in Health Research, 2118
George Kress Foundation Grants, 2119
Georgia Power Foundation Grants, 2125
Gertrude B. Elion Mentored Medical Student Research Awards, 2130
Gilroy and Lillian P. Roberts Charitable Foundation Grants, 2142
Goddess Scholars Grants, 2155
Greater Saint Louis Community Found Grants, 2164
Green Diamond Charitable Contributions, 2171
Green Foundation Human Services Grants, 2172
H.A. and Mary K. Chapman Charitable Grants, 2193
Harold Alfond Foundation Grants, 2200
Harold R. Bechtel Charitable Remainder Uni-Trust Grants, 2201
Harold Simmons Foundation Grants, 2203
Healthcare Foundation of New Jersey Grants, 2222
Health Foundation of Southern Florida Responsive Grants, 2223
Hearst Foundations Health Grants, 2225
Heineman Foundation for Research, Education, Charitable and Scientific Purposes, 2227
Helen V. Brach Foundation Grants, 2231
Hendrick Foundation for Children Grants, 2232
Henry W. Bull Foundation Grants, 2237
HHMI Biomedical Research Grants for Int'l Scientists Infectious Diseases and Parasitology, 2251
HHMI Biomedical Research Grants for International Scientists in Canada and Latin America, 2252
HHMI Biomedical Research Grants for International Scientists in the Baltics, Central and Eastern Europe, Russia, and Ukraine, 2253
HHMI Initiative for Colleges Grants, 2256
HHMI Initiative for Universities Grants, 2257
HHMI Med into Grad Initiative Grants, 2259
HHMI Precollege Outreach Initiative for Biomedical Research Institutions Grants, 2261
Hilton Hotels Corporate Giving Program Grants, 2268
Hirtzel Memorial Foundation Grants, 2269
HUD Community Outreach Partnership Centers Grants, 2292
Hutchinson Community Foundation Grants, 2310
IHI Quality Improvement Fellowships, 2348
J.W. Kieckhefer Foundation Grants, 2452
Jackson Foundation Grants, 2453
Jacob and Hilda Blaustein Foundation Grants, 2454
Jacob and Valeria Langeloth Foundation Grants, 2455
James M. Collins Foundation Grants, 2461
James R. Dougherty Jr. Foundation Grants, 2463
Jane Bradley Pettit Foundation Health Grants, 2471
Jewish Federation of Metropolitan Chicago Scholarships, 2486
John Deere Foundation Grants, 2500
John G. Duncan Trust Grants, 2501
John P. McGovern Foundation Grants, 2504
John R. Oishei Foundation Grants, 2505
John S. Dunn Research Foundation Grants, 2506
Johnson & Johnson Corporate Contributions, 2508
John W. Alden Trust Grants, 2510
Lance Armstrong Foundation Research Grants, 2585
Lloyd A. Fry Foundation Health Grants, 2620
Lowe Foundation Grants, 2622
Ludwick Family Foundation Grants, 2625
Lumpkin Family Found Healthy People Grants, 2629
M. Bastian Family Foundation Grants, 2636
Mabel H. Flory Charitable Trust Grants, 2646
Mabel Y. Hughes Charitable Trust Grants, 2647
Macquarie Bank Foundation Grants, 2650
MAP International Medical Fellowships, 2655
Mary K. Chapman Foundation Grants, 2673
Mary S. and David C. Corbin Foundation Grants, 2675
Mattel Children's Foundation Grants, 2685
Max and Victoria Dreyfus Foundation Grants, 2687
McGregor Fund Grants, 2725
McLean Contributionship Grants, 2727
Meadows Foundation Grants, 2736
Mericos Foundation Grants, 2742
Mesothelioma Applied Research Found Grants, 2743
MGN Family Foundation Grants, 2754
Miles of Hope Breast Cancer Foundation Grants, 2773
Millipore Foundation Grants, 2775
Monfort Family Foundation Grants, 2794
NAPNAP Foundation Nursing Research Grants, 2829
NCFR Ruth Hathaway Jewson Award, 2909
Nell Warren Elkin and William Simpson Elkin Foundation, 2972
New Hampshire Charitable Foundation Grants, 2985
NIDDK Research Grants on Improving Health Care for Obese Patients, 3176
NIHCM Found Health Care Research Awards, 3210
NIHCM Foundation Health Care Television and Radio Journalism Awards, 3211
NIH Exceptional, Unconventional Research Enabling Knowledge Acceleration Grants, 3215
NIH Research Project Grants, 3237
NIH School Interventions to Prevent Obesity, 3244
NINR Mentored Research Scientist Development Award, 3265
Nordson Corporation Foundation Grants, 3416
Northrop Grumman Corporation Grants, 3428
Northwestern Mutual Foundation Grants, 3430
Northwood University/Alden B. Dow Creativity Center Summer Residency Program, 3432
NSF Innovation and Organizational Change, 3581
Nuffield Foundation Law and Society Grants, 3667
NWHF Mark O. Hatfield Research Fellowship, 3671
NYCT Blindness and Visual Disabilities Grants, 3695
Oklahoma City Community Grants, 3737
Olin Corporation Charitable Trust Grants, 3739
ONR Casualty Care and Management Research Grants, 3746
Oppenstein Brothers Foundation Grants, 3792
Oscar Rennebohm Foundation Grants, 3832
OSF Access to Medicines Fellowships in Eastern Europe and Central Asia, 3834
Patrick and Anna M. Cudahy Fund Grants, 3873
Patron Saints Foundation Grants, 3874
Peabody Foundation Grants, 3886
Peacock Foundation Grants, 3887
Pediatric Brain Tumor Found Research Grants, 3888
Percy B. Ferebee Endowment Grants, 3897
Perpetual Trust for Charitable Giving Grants, 3900
Pfizer Medical Education Track Two Grants, 3927
PhRMA Foundation Health Outcomes Research Starter Grants, 3951
PhRMA Foundation Health Outcomes Sabbatical Fellowships, 3952
Potts Memorial Foundation Grants, 3985
Price Family Charitable Fund Grants, 3987
Prostate Cancer Foundation Research Grants, 4000
R.C. Baker Foundation Grants, 4006
RACGP/APHCRI Indigenous Health Award, 4008
RACGP Cardiovascular Research Grants in General Practice, 4011
RACGP Chris Silagy Research Scholarship, 4012
RACGP Family Medical Care, Education and Research Grants, 4013
Rajiv Gandhi Foundation Grants, 4025
Rathmann Family Foundation Grants, 4033
Rayonier Foundation Grants, 4034
RCPSC Detweiler Traveling Fellowships, 4036
Regenstrief General Internal Medicine - General Pediatrics Research Fellowship Program, 4045
Rehabilitation Nursing Found Research Grants, 4050
Rhode Island Foundation Grants, 4061
Richard and Rhoda Goldman Fund Grants, 4063
Richard and Susan Smith Family Found Grants, 4064
Robert E. and Evelyn McKee Foundation Grants, 4074
Rockwell International Corporate Trust Grants, 4093
Ronald McDonald House Charities Grants, 4101
Rose Community Foundation Health Grants, 4104
RSC Thomas W. Eadie Medal, 4126
Ruth and Vernon Taylor Foundation Grants, 4130
RWJF Changes in Health Care Financing and Organization Grants, 4133

RWJF Evaluating Innovations in Nursing Education Grants, 4138
RWJF Health and Society Scholars, 4141
RWJF Pioneer Portfolio Grants, 4150
S. D. Bechtel, Jr. Foundation / Stephen Bechtel Fund Preventive Healthcare and Selected Research Grants, 4152
Salem Foundation Grants, 4160
Samueli Institute Scientific Research Grants, 4162
San Antonio Area Foundation Grants, 4163
Sandler Program for Asthma Research Grants, 4165
Sarasota Memorial Healthcare Found Grants, 4167
Sarkeys Foundation Grants, 4181
Savoy Foundation Research Grants, 4185
Schramm Foundation Grants, 4188
Self Foundation Grants, 4197
Shell Oil Company Foundation Grants, 4233
Sigma Theta Tau International / Council for the Advancement of Nursing Science Grants, 4255
Sioux Falls Area Community Foundation Community Fund Grants (Unrestricted), 4262
Sioux Falls Area Comm Foundation Spot Grants, 4263
Siragusa Foundation Health Services & Medical Research Grants, 4264
Stackpole-Hall Foundation Grants, 4442
Starr Foundation Grants, 4447
Takemi Program in International Health Research and Advanced Training Fellowships, 4491
Textron Corporate Contributions Grants, 4501
Thomas and Dorothy Leavey Foundation Grants, 4506
Thomas Austin Finch, Sr. Foundation Grants, 4507
Thomas C. Ackerman Foundation Grants, 4508
Thompson Charitable Foundation Grants, 4511
Todd Brock Family Foundation Grants, 4522
UCLA / RAND Corporation Post-Doctoral Fellowships, 4548
USAID Research & Innov for Health Supply Chain Systems and Commodity Security Grants, 4601
USDD Care for the Critically Injured Burn Patient Grants, 4706
USDD Clinical Functional Assessment Investigator-Initiated Grants, 4709
Vancouver Foundation Grants and Community Initiatives, 4807
Victor E. Speas Foundation Grants, 4810
Volkswagen of America Corporate Contributions Grants, 4817
W. C. Griffith Foundation Grants, 4818
W.C. Griffith Foundation Grants, 4819
Wayne and Gladys Valley Foundation Grants, 4835
Weyerhaeuser Company Foundation Grants, 4857
Wilhelm Sander Foundation Grants, 4868
William Blair and Company Foundation Grants, 4875
William G. and Helen C. Hoffman Found Grants, 4876
William G. McGowan Charitable Fund Grants, 4877
William N. and Myriam Pennington Foundation Grants, 4879
William T. Grant Foundation Research Grants, 4881

Health Care Access

1976 Foundation Grants, 7
Achelis Foundation Grants, 274
Adam Richter Charitable Trust Grants, 339
Angels Baseball Foundation Grants, 848
Appalachian Regional Commission Health Care Grants, 915
BCBSM Foundation Investigator Initiated Research Grants, 1313
BCBSM Foundation Physician Investigator Research Awards, 1314
BCBSNC Foundation Grants, 1318
BCBS of Massachusetts Foundation Grants, 1319
Better Way Foundation Grants, 1338
Blue Shield of California Grants, 1350
Bodman Foundation Grants, 1356
Breast Cancer Fund Grants, 1369
Brian G. Dyson Foundation Grants, 1370
Caesars Foundation Grants, 1407
California Wellness Foundation Work and Health Grants, 1409
CMS Hispanic Health Services Research Grants, 1584
Connecticut Health Foundation Health Initiative Grants, 1628
Daviess County Community Foundation Health Grants, 1725
DTE Energy Foundation Health and Human Services Grants, 1865
E.W. "Al" Thrasher Awards, 1887
FCD Child Development Grants, 1984
Flinn Foundation Grants Programs, 2024
General Service Foundation Human Rights and Economic Justice Grants, 2108
Gerber Foundation Grants, 2126
Green Foundation Human Services Grants, 2172
Healthcare Foundation of New Jersey Grants, 2222
Health Foundation of Southern Florida Responsive Grants, 2223
Hearst Foundations Health Grants, 2225
Hill Crest Foundation Grants, 2267
HRSA Nurse Education, Practice, Quality and Retention Grants, 2290
Jane Bradley Pettit Foundation Health Grants, 2471
Johnson & Johnson Corporate Contributions, 2508
Kenneth T. and Eileen L. Norris Found Grants, 2554
Lloyd A. Fry Foundation Health Grants, 2620
Lumpkin Family Found Healthy People Grants, 2629
Macquarie Bank Foundation Grants, 2650
Marion Gardner Jackson Charitable Trust Grants, 2664
Mary K. Chapman Foundation Grants, 2673
Mattel Children's Foundation Grants, 2685
Mattel International Grants Program, 2686
MGM Resorts Foundation Community Grants, 2753
MLA Donald Lindberg Research Fellowship, 2785
NYCT Blindness and Visual Disabilities Grants, 3695
Paul G. Allen Family Foundation Grants, 3876
Piper Trust Healthcare & Med Research Grants, 3972
Robert Sterling Clark Foundation Reproductive Rights and Health Grants, 4081
Rose Community Foundation Health Grants, 4104
RWJF Changes in Health Care Financing and Organization Grants, 4133
RWJF Disparities Research for Change Grants, 4137
RWJF Evaluating Innovations in Nursing Education Grants, 4138
RWJF Investigator Awards in Health Policy Res, 4145
RWJF Partners Investing in Nursing's Future, 4149
RWJF Pioneer Portfolio Grants, 4150
S. D. Bechtel, Jr. Foundation / Stephen Bechtel Fund Preventive Healthcare and Selected Research Grants, 4152
Siragusa Foundation Health Services & Medical Research Grants, 4264
UCLA / RAND Corporation Post-Doctoral Fellowships, 4548
USAID Accelerating Progress Against Tuberculosis in Kenya Grants, 4594
USAID Research & Innov for Health Supply Chain Systems and Commodity Security Grants, 4601

Health Care Administration

American Academy of Nursing Building Academic Geriatric Nursing Capacity Scholarships, 614
AUPHA Corris Boyd Scholarships, 1267
Beckman Coulter Community Relations Grants, 1325
Byron W. and Alice L. Lockwood Grants, 1401
Canada-U.S. Fulbright International Science and Technology Award, 1417
Canada-U.S. Fulbright Senior Specialists Program Grants, 1418
Canada-U.S. Fulbright Traditional Scholars, 1419
CMS Historically Black Colleges and Universities Health Services Research Grants, 1585
Commonwealth Payment System Reform Grants, 1609
Fulbright Alumni Initiatives Awards Grants, 2064
Fulbright Distinguished Chairs Program Grants, 2069
Fulbright European Union Scholar-in-Residence Program Grants, 2071
Fulbright New Century Scholars Program Grants, 2075
Fulbright Occasional Lecturer Program Grants, 2076
Fulbright Specialists Program Grants, 2079
Fulbright Traditional Scholar Program in Europe and Eurasia, 2080
Fulbright Traditional Scholar Program in Sub-Saharan Africa, 2081
Fulbright Traditional Scholar Program in the East Asia/Pacific Region, 2082
Fulbright Traditional Scholar Program in the Near East and North Africa Region, 2083
Fulbright Traditional Scholar Program in the South and Central Asia Region, 2084
Fulbright Traditional Scholar Program in the Western Hemisphere, 2085
Fulbright Visiting Specialists Program: Direct Access to the Muslim World, 2087
Health Management Scholarships and Grants for Minorities, 2224
IHI Quality Improvement Fellowships, 2348
International Fellowships in Medical Education, 2398
Mayo Clinic Administrative Fellowship, 2689
Mayo Clinic Admin Fellowship Eau Claire, 2690
Mayo Clinic Business Consulting Fellowship, 2691
NAPNAP Foundation Nursing Research Grants, 2829
National Blood Foundation Research Grants, 2848
Richard P. Covert Scholarship, 4067
RWJF Health Policy Fellowships, 4142
SfN Louise Hanson Marshall Specific Recognition Award, 4208
Takemi Program in International Health Research and Advanced Training Fellowships, 4491
UCLA / RAND Corporation Post-Doctoral Fellowships, 4548

Health Care Assessment

AACAP Jeanne Spurlock Research Fellowship in Substance Abuse and Addiction for Minority Medical Students, 64
AACAP Summer Medical Student Fellowships, 80
Commonwealth Fund Harkness Fellowships in Health Care Policy and Practice, 1606
Commonwealth Fund Health Care Quality Improvement and Efficiency Grants, 1607
RWJF Evaluating Innovations in Nursing Education Grants, 4138
UCLA / RAND Corporation Post-Doctoral Fellowships, 4548

Health Care Economics

AMI Semiconductors Corporate Grants, 813
CDC Grants for Violence-Related Injury Prevention Research, 1463
CMS Hispanic Health Services Research Grants, 1584
Foundation for Seacoast Health Grants, 2041
Percy B. Ferebee Endowment Grants, 3897
PhRMA Foundation Health Outcomes Research Starter Grants, 3951
Robert Leet and Clara Guthrie Patterson Grants, 4076
RWJF Changes in Health Care Financing and Organization Grants, 4133
UCLA / RAND Corporation Post-Doctoral Fellowships, 4548

Health Care Financing

AUPHA Corris Boyd Scholarships, 1267
BCBSM Foundation Investigator Initiated Research Grants, 1313
RWJF Changes in Health Care Financing and Organization Small Grants, 4134
Sarasota Memorial Healthcare Found Grants, 4167
UCLA / RAND Corporation Post-Doctoral Fellowships, 4548

Health Care Personnel

Connecticut Health Foundation Health Initiative Grants, 1628

Health Care Promotion

CDC Steven M. Teutsch Prevention Effectiveness Fellowships, 1467
Hill Crest Foundation Grants, 2267
Nestle Foundation Training Grant, 2983

Sigma Theta Tau International /American Nurses Foundation Grant, 4254
Sigma Theta Tau Small Grants, 4258

Health Disparities
Appalachian Regional Commission Health Care Grants, 915
Delta Air Lines Foundation Health and Wellness Grants, 1736
NHLBI Mentored Career Development Award to Promote Faculty Diversity/Re-Entry in Biomedical Research, 3035
OBSSR Behavioral and Social Science Research on Understanding and Reducing Health Disparities Grants, 3733
PC Fred H. Bixby Fellowships, 3880
Pfizer ASPIRE North America Rheumatology Research Awards, 3917
USDD Clinical Functional Assessment Investigator-Initiated Grants, 4709

Health Facilities Studies
RWJF Evaluating Innovations in Nursing Education Grants, 4138
UCLA / RAND Corporation Post-Doctoral Fellowships, 4548

Health Insurance
BCBSNC Foundation Grants, 1318
BCBS of Massachusetts Foundation Grants, 1319
Blue Shield of California Grants, 1350
California Wellness Foundation Work and Health Grants, 1409
CDC National Training Information and Exchange Program, 1464
CMS Hispanic Health Services Research Grants, 1584
CMS Research and Demonstration Grants, 1586
Commonwealth Payment System Reform Grants, 1609
Commonwealth Fund Small Grants, 1611
RWJF Changes in Health Care Financing and Organization Grants, 4133
RWJF Investigator Awards in Health Policy Res, 4145
UCLA / RAND Corporation Post-Doctoral Fellowships, 4548

Health Personnel/Professions
BCBSM Foundation Student Award Program, 1316
NHLBI Short-Term Research Education Program to Increase Diversity in Health-Related Research, 3062
RWJF Harold Amos Medical Faculty Development Scholars, 4140
RWJF Nurse Faculty Scholars, 4148
Special Olympics Health Professions Stu Grants, 4332

Health Planning/Policy
AAAS Science and Technology Policy Fellowships: Health, Education and Human Services, 43
AcademyHealth Alice S. Hersh New Investigator Awards, 245
AcademyHealth HSR Impact Awards, 246
ACS Research Scholar Grants for Health Services and Health Policy Research, 327
American Psychiatric Foundation Jeanne Spurlock Congressional Fellowship, 793
American Psychiatric Foundation Minority Fellowships, 794
Appalachian Regional Commission Health Care Grants, 915
ASHG Public Health Genetics Fellowship, 1086
AUPHA Corris Boyd Scholarships, 1267
BCBSM Foundation Excellence in Research Awards for Students, 1312
BCBSM Foundation Student Award Program, 1316
BCBSM Frank J. McDevitt, DO, Excellence in Research Awards, 1317
BCBS of Massachusetts Foundation Grants, 1319
Becton Dickinson and Company Grants, 1328
Blue Shield of California Grants, 1350
California Wellness Foundation Work and Health Grants, 1409
CDC Preventive Med Residency & Fellowship, 1465
CDC Steven M. Teutsch Prevention Effectiveness Fellowships, 1467
Chiron Foundation Community Grants, 1553
Commonwealth Fund/Harvard University Fellowship in Minority Health Policy, 1604
Commonwealth Fund Affordable Health Insurance Grants, 1605
Commonwealth Fund Harkness Fellowships in Health Care Policy and Practice, 1606
Commonwealth Fund Health Care Quality Improvement and Efficiency Grants, 1607
Commonwealth Fund Patient-Centered Coordinated Care Program Grants, 1608
Commonwealth Payment System Reform Grants, 1609
Commonwealth Fund Small Grants, 1611
Commonwealth Fund State High Performance Health Systems Grants, 1612
Doris Duke Charitable Foundation Operations Research on AIDS Care and Treatment in Africa Grants, 1839
General Service Foundation Human Rights and Economic Justice Grants, 2108
Greenwall Foundation Bioethics Grants, 2174
Greenwall Foundation Fellowship Program in Bioethics and Health Policy, 2176
Henry J. Kaiser Family Found Media Fellowships, 2234
HHMI Grants and Fellowships Programs, 2255
Jane Bradley Pettit Foundation Health Grants, 2471
Japan Foundation Center for Global Partnership Institutional Grants, 2473
KFF Barbara Jordan Health Policy Scholars, 2556
Meyer and Pepa Gold Family Foundation Grants, 2746
NCI/NIDA Transdisciplinary Tobacco Use Research Centers Grants, 2912
NOAA Climate Variability and Human Health Program Grants, 3347
RAND Corporation Graduate Student Summer Associateships, 4030
Robert Sterling Clark Foundation Reproductive Rights and Health Grants, 4081
Rose Community Foundation Health Grants, 4104
RWJF Changes in Health Care Financing and Organization Small Grants, 4134
RWJF Evaluating Innovations in Nursing Education Grants, 4138
RWJF Health Policy Fellowships, 4142
RWJF Investigator Awards in Health Policy Res, 4145
RWJF Partners Investing in Nursing's Future, 4149
Takemi Program in International Health Research and Advanced Training Fellowships, 4491
UCLA / RAND Corporation Post-Doctoral Fellowships, 4548
USAID Development Innovation Accelerator Broad Agency Announcement Grants, 4596
USAID Research & Innov for Health Supply Chain Systems and Commodity Security Grants, 4601
USAID Strengthening RMNCH Through Indian Health Professional Associations and Scaling Up Uptake Grants, 4602
William T. Grant Foundation Research Grants, 4881

Health Promotion
A-T Medical Research Foundation Grants, 10
AAFP Research Stimulation Grants, 159
Abington Foundation Grants, 243
ACGT Young Investigator Grants, 272
Allen Foundation Educational Nutrition Grants, 570
American-Italian Cancer Foundation Fellowships, 602
Ann Arbor Area Community Foundation Grants, 860
Ann Peppers Foundation Grants, 863
Anthem Blue Cross and Blue Shield School Nurse Grants, 868
Appalachian Regional Commission Health Care Grants, 915
ASHG Public Health Genetics Fellowship, 1086
ASID Foundation Irene Winifred Eno Grant, 1094
BCBSM Foundation Investigator Initiated Research Grants, 1313
BCBSNC Foundation Grants, 1318
Bikes Belong Foundation Research Grants, 1343
Bullitt Foundation Grants, 1384
California Wellness Foundation Work and Health Grants, 1409
CDC Public Health Conference Support Grant, 1466
CDC Steven M. Teutsch Prevention Effectiveness Fellowships, 1467
Charles H. Hood Foundation Child Health Research Awards, 1522
CMA Foundation Grants, 1582
Colonel Stanley R. McNeil Foundation Grants, 1597
Comprehensive Health Education Found Grants, 1622
Connecticut Community Foundation Grants, 1627
Connecticut Health Foundation Health Grants, 1628
Conservation, Food, and Health Foundation Grants for Developing Countries, 1641
Daniel Mendelsohn New Investigator Award, 1717
DHHS Oral Health Promotion Research Across the Lifespan, 1759
Dorothy Rider Pool Health Care Grants, 1841
Ewing Halsell Foundation Grants, 1964
Gerber Foundation Grants, 2126
Health Foundation of Southern Florida Responsive Grants, 2223
Hirtzel Memorial Foundation Grants, 2269
HomeBanc Foundation Grants, 2275
Institute for Agriculture and Trade Policy Food and Society Fellowships, 2381
Jacob and Hilda Blaustein Foundation Grants, 2454
Jenifer Altman Foundation Grants, 2484
John P. McGovern Foundation Grants, 2504
McCarthy Family Foundation Grants, 2717
MedImmune Charitable Grants, 2737
NAMM Foundation Music Research Grants, 2825
NAPNAP Foundation Nursing Research Grants, 2829
NCIC Terry Fox Research Studentships, 2924
Nestle Foundation Pilot Grants, 2980
NIA Health Behaviors and Aging Grants, 3098
NIH Health Promotion Among Racial and Ethnic Minority Males, 3219
NOAA Climate Variability and Human Health Program Grants, 3347
Nuffield Foundation Law and Society Grants, 3667
ONR Medical Science and Technology Division Grants, 3772
Open Spaces Sacred Places National Awards, 3791
ORISE Faculty Internship at the US Army Center for Health Promotion and Preventive Medicine, 3802
ORISE Postgraduate Internship at the US Army Center for Health Promotion and Preventive Med, 3818
ORISE Student Internship Program at the US Army Center for Health Promotion and Preventive Medicine, 3827
RGK Foundation Grants, 4060
RWJF Health and Society Scholars, 4141
RWJF Healthy Eating Research Grants, 4143
Sigma Theta Tau International /American Nurses Foundation Grant, 4254
Sigma Theta Tau Small Grants, 4258
Special Olympics Health Professions Stu Grants, 4332
Susan G. Komen Breast Cancer Foundation Career Catalyst Research Grants, 4473
Textron Corporate Contributions Grants, 4501
USAID Development Innovation Accelerator Broad Agency Announcement Grants, 4596
USAID Research & Innov for Health Supply Chain Systems and Commodity Security Grants, 4601

Health Records
Regenstrief General Internal Medicine - General Pediatrics Research Fellowship Program, 4045
USDD Medical Practice Breadth of Med Practice & Disease Frequency Exposure Grants, 4752

Health Research
AAAAI Distinguished Layperson Award, 18
AIHS/Mitacs Health Pilot Partnership Internship, 434
AIHS Alberta/Pfizer Translational Res Grants, 435
AIHS Collaborative Research and Innovation Grants - Collaborative Program, 437

AIHS Collaborative Research and Innovation Grants - Collaborative Project, 438
AIHS Collaborative Research and Innovation Grants - Collaborative Team, 439
AIHS Heritage Youth Researcher Summer Science Grants, 449
AIHS Sustainability Grants, 459
APHL Emerging Infectious Diseases Fellowships, 914
Bupa Foundation Medical Research Grants, 1386
BWF Preterm Birth Initiative Grants, 1399
CDC Found Tobacco Network Lab Fellowship, 1462
CDC Steven M. Teutsch Prevention Effectiveness Fellowships, 1467
Deborah Munroe Noonan Mem Research Grants, 1729
Delta Air Lines Foundation Health and Wellness Grants, 1736
E.W. "Al" Thrasher Awards, 1887
Lymphatic Education & Research Network Additional Support Grants for NIH-funded F32 Postdoctoral Fellows, 2632
Lymphatic Education & Research Network Postdoctoral Fellowships, 2633
Mary Gates Research Scholarships, 2670
Meyer and Pepa Gold Family Foundation Grants, 2746
NFL Charities Medical Grants, 2994
NHLBI Ruth L. Kirschstein National Research Service Awards for Individual Predoctoral MD/PhD Fellows and Other Dual Degree Fellows, 3059
Perpetual Trust for Charitable Giving Grants, 3900
Pfizer ASPIRE Worldwide Endocrine Young Investigator Grants, 3918
Seattle Foundation Medical Funds Grants, 4194
Sigma Theta Tau International Doris Bloch Research Award, 4257
Special Olympics Youth Fan Grants, 4333
USDD Clinical Functional Assessment Investigator-Initiated Grants, 4709
Wellcome Trust New Investigator Awards, 4844

Health Sciences

AAAS Science and Technology Policy Fellowships: Health, Education and Human Services, 43
AIHS ForeFront MBA Studentship Award, 443
AIHS ForeFront MBT Studentship Awards, 444
American Mathematical Society AAAS Mass Media Fellowships, 751
Charles A. King Trust Postdoctoral Research Fellowships, 1518
Duke University Adult Cardiothoracic Anesthesia and Critical Care Medicine Fellowships, 1869
Duke University Ambulatory and Regional Anesthesia Fellowships, 1870
Duke University Clinical Cardiac Electrophysiology Fellowships, 1871
Duke University Hyperbaric Center Fellowships, 1872
Duke Univ Interventional Cardio Fellowships, 1873
Frederick Douglass Summer Scholars Fellowship, 2053
Harry A. and Margaret D. Towsley Found Grants, 2205
Hoglund Foundation Grants, 2271
Hospital Libraries Section / MLA Professional Development Grants, 2283
MLA Cunningham Memorial International Fellowship, 2783
MLA David A. Kronick Traveling Fellowship, 2784
MLA Graduate Scholarship, 2786
MLA Grad Scholarship for Minority Students, 2787
MLA Murray Gottlieb Prize Essay Award, 2788
MLA Research, Development, and Demonstration Project Grant, 2789
MLA Rittenhouse Award, 2790
MLA Thomas Reuters / Frank Bradway Rogers Information Advancement Award, 2791
NCFR Innovation Grants, 2908
NHLBI Lymphatics in Health & Disease in Digestive, Cardiovascular and Pulmonary Systems, 3031
NHLBI Pathway to Independence Awards, 3045
NHLBI Research Supplements to Promote Diversity in Health-Related Research, 3054
NIA Promoting Careers in Aging and Health Disparities Research Grants, 3117
NIH Recovery Act Limited Competition: Building Sustainable Community-Linked Infrastructure to Enable Health Science Research, 3232
NIH Research Project Grants, 3237
NLM Advancing Novel Science in Women's Health Research Exploratory/Developmental Research Grants, 3291
NLM Advancing Novel Science in Women's Health Research Small Project Grants, 3292
NLM Grants for Scholarly Works in Biomedicine and Health, 3298
NLM Research Project Grants, 3306
NWHF Mark O. Hatfield Research Fellowship, 3671
NYCT Blindness and Visual Disabilities Grants, 3695
Regenstrief General Internal Medicine - General Pediatrics Research Fellowship Program, 4045
Religion and Health: Effects, Mechanisms, and Interpretation RFP, 4052
San Diego Women's Foundation Grants, 4164
Savoy Foundation Post-Doctoral and Clinical Research Fellowships, 4184
Savoy Foundation Studentships, 4186
Scleroderma Foundation Established Investigator Grants, 4190
Sigma Theta Tau International / Council for the Advancement of Nursing Science Grants, 4255
Siragusa Foundation Health Services & Medical Research Grants, 4264
SRC Medical Research Grants, 4379
SSHRC Graduate Scholarships-Michael Smith Foreign Study Supplements, 4408
STTI Emergency Nurses Assoc Found Grant, 4459
Swiss National Science Foundation Fellowships for Advanced Researchers, 4483
Swiss National Science Foundation Fellowships for Prospective Researchers, 4484
Swiss National Science Foundation Marie Heim-Vogtlin Grants, 4485
Thomson Reuters / MLA Doctoral Fellowships, 4512
US CRDF Leishmaniasis: Collaborative Research Opportunities in N Africa & Middle East, 4610

Health Services Delivery

A.O. Smith Community Grants, 12
AAHPM Hospice and Palliative Med Fellowships, 180
Abington Foundation Grants, 243
AcademyHealth Alice S. Hersh New Investigator Awards, 245
AcademyHealth HSR Impact Awards, 246
ACSM Paffenbarger-Blair Fund for Epidemiological Research on Physical Activity Grants, 321
Amgen Foundation Grants, 812
APA Young Investigator Grant Program, 912
Appalachian Regional Comm Health Care Grants, 915
Auburn Foundation Grants, 1263
AUPHA Corris Boyd Scholarships, 1267
Barth Syndrome Foundation Research Grants, 1302
Baxter International Foundation William B. Graham Prize for Health Services Research, 1306
BCBSM Foundation Investigator Initiated Research Grants, 1313
BCBSM Foundation Student Award Program, 1316
BCBSM Frank J. McDevitt, DO, Excellence in Research Awards, 1317
BCBSNC Foundation Grants, 1318
BCBS of Massachusetts Foundation Grants, 1319
Blue Shield of California Grants, 1350
California Wellness Foundation Work and Health Grants, 1409
CMA Foundation Grants, 1582
CMS Hispanic Health Services Research Grants, 1584
CMS Historically Black Colleges and Universities Health Services Research Grants, 1585
Commonwealth Fund Harkness Fellowships in Health Care Policy and Practice, 1606
Commonwealth Fund Quality of Care for Frail Elders Grants, 1610
Connecticut Health Foundation Health Grants, 1628
Daviess County Community Foundation Health Grants, 1725
Del E. Webb Foundation Grants, 1733
DeMatteis Family Foundation Grants, 1740
DHHS Oral Health Promotion Research Across the Lifespan, 1759
Doris Duke Charitable Foundation Operations Research on AIDS Care and Treatment in Africa Grants, 1839
Dorothy Rider Pool Health Care Grants, 1841
DTE Energy Foundation Health and Human Services Grants, 1865
Emma B. Howe Memorial Foundation Grants, 1920
Ewing Halsell Foundation Grants, 1964
Fairlawn Foundation Grants, 1974
Flinn Foundation Grants Programs, 2024
George Frederick Jewett Foundation Grants, 2116
Gil and Dody Weaver Foundation Grants, 2139
GPP Charles E. Culpeper Foundation Biomedical Pilot Projects Initiative Grants, 2159
Green Diamond Charitable Contributions, 2171
Hawaii Community Foundation Health Education and Research Grants, 2221
Hearst Foundations Health Grants, 2225
Hill Crest Foundation Grants, 2267
Hirtzel Memorial Foundation Grants, 2269
Hoglund Foundation Grants, 2271
HomeBanc Foundation Grants, 2275
Jane Bradley Pettit Foundation Health Grants, 2471
Johnson & Johnson Corporate Contributions, 2508
Lynde & Harry Bradley Foundation Fellowships, 2634
Lynde and Harry Bradley Foundation Grants, 2635
M.D. Anderson Foundation Grants, 2637
M.J. Murdock Charitable Trust General Grants, 2641
McCune Foundation Human Services Grants, 2722
McGraw-Hill Companies Community Grants, 2724
Miles of Hope Breast Cancer Foundation Grants, 2773
MLA Donald Lindberg Research Fellowship, 2785
NAPNAP Foundation Nursing Research Grants, 2829
NCI Exploratory Grants for Behavioral Research in Cancer Control, 2928
NIA Alzheimer's Disease Core Centers Grants, 3084
NMF General Electric Med Scholars Fellowships, 3322
Northrop Grumman Corporation Grants, 3428
NYAM Mary and David Hoar Fellowship, 3691
Pfizer Medical Education Track Two Grants, 3927
Prostate Cancer Foundation Research Grants, 4000
Reuter Foundation Grants, 4059
Rockwell International Corporate Trust Grants, 4093
Rosalynn Carter Institute John and Betty Pope Fellowships, 4103
RWJF Changes in Health Care Financing and Organization Grants, 4133
RWJF Changes in Health Care Financing and Organization Small Grants, 4134
RWJF Disparities Research for Change Grants, 4137
RWJF Evaluating Innovations in Nursing Education Grants, 4138
RWJF Harold Amos Medical Faculty Development Scholars, 4140
Salem Foundation Grants, 4160
San Diego Women's Foundation Grants, 4164
Shell Oil Company Foundation Grants, 4233
Sony Corporation of America Corporate Philanthropy Grants, 4329
Starr Foundation Grants, 4447
Textron Corporate Contributions Grants, 4501
UCLA / RAND Corporation Post-Doctoral Fellowships, 4548
Union Pacific Foundation Health and Human Services Grants, 4569
USAID Strengthening RMNCH Through Indian Health Professional Associations and Scaling Up Uptake Grants, 4602
Vancouver Foundation Grants, 4807
W.K. Kellogg Foundation Healthy Kids Grants, 4822

Health and Safety Education

Alexander and Margaret Stewart Trust Grants, 514
Avon Foundation Breast Care Fund Grants, 1274
BCBSNC Foundation Grants, 1318
Becton Dickinson and Company Grants, 1328

Blowitz-Ridgeway Foundation Early Childhood Development Research Award, 1348
Breast Cancer Fund Grants, 1369
CampusRN/AACN Nursing Scholarship Fund, 1416
Carlos and Marguerite Mason Trust Grants, 1431
CDC Public Health Conference Support Grant, 1466
Charles Lafitte Foundation Grants, 1525
Chiron Foundation Community Grants, 1553
Comprehensive Health Education Found Grants, 1622
Daviess County Community Foundation Health Grants, 1725
Dorothy Rider Pool Health Care Grants, 1841
Fairlawn Foundation Grants, 1974
Foundation for Seacoast Health Grants, 2041
Healthcare Foundation of New Jersey Grants, 2222
Health Foundation of Southern Florida Responsive Grants, 2223
HomeBanc Foundation Grants, 2275
John P. McGovern Foundation Grants, 2504
John S. Dunn Research Foundation Grants, 2506
Johnson & Johnson Corporate Contributions, 2508
Lloyd A. Fry Foundation Health Grants, 2620
McCarthy Family Foundation Grants, 2717
MedImmune Charitable Grants, 2737
NAPNAP Foundation Nursing Research Grants, 2829
National Blood Foundation Research Grants, 2848
Northrop Grumman Corporation Grants, 3428
NRA Foundation Grants, 3437
ORISE Student Research Participation at the Agency for Toxic Substances and Disease Registry, 3828
PhRMA Foundation Health Outcomes Sabbatical Fellowships, 3952
Rosalynn Carter Institute John and Betty Pope Fellowships, 4103
RWJF Active Living Research Grants, 4132
Special Olympics Health Professions Stu Grants, 4332
Susan G. Komen Breast Cancer Foundation Career Catalyst Research Grants, 4473
USDA Cooperative Extension System Education Grants, 4620
USDA Organic Agriculture Research and Extension Initiative, 4637
Viking Children's Fund Grants, 4811

Hearing
AHRF Eugene L. Derlacki Research Grants, 384
AHRF Georgia Birtman Grant, 385
AHRF Wiley H. Harrison Mem Research Award, 387
Marjorie C. Adams Charitable Trust Grants, 2665
Reinberger Foundation Grants, 4051

Hearing Impairments
AHRF Eugene L. Derlacki Research Grants, 384
AHRF Georgia Birtman Grant, 385
AHRF Wiley H. Harrison Mem Research Award, 387
ASHFoundation Graduate Student Scholarships, 1077
ASHFoundation Graduate Student Scholarships for Minority Students, 1078
ASHFoundation Graduate Student with a Disability Scholarship, 1079
Batchelor Foundation Grants, 1305
Herbert & Adrian Woods Foundation Grants, 2239
Mabel H. Flory Charitable Trust Grants, 2646
Marjorie C. Adams Charitable Trust Grants, 2665
The Ray Charles Foundation Grants, 4502

Heat Transfer
AIChE Donald Q. Kern Award, 406
AIChE Transport & Energy Processes Div Award, 427
ASME Allan Kraus Thermal Management Medal, 1104
ASME Bergles-Rohsenow Young Investigator Award in Heat Transfer, 1108
ASME Heat Transfer Memorial Award, 1128
Donald Q. Kern Award in Heat Transfer or Energy Conversion, 1831

Heating, Ventilation, and Air Conditioning
ASHRAE Graduate Student Grant-in-Aid, 1088
ASHRAE Willis H. Carrier Graduate Research Fellowship, 1089

Hematology
Bayer Clinical Scholarship Award, 1308
Chest Found Grant in Venous Thromboembolism, 1537
Chiron Foundation Community Grants, 1553
Doris Duke Charitable Foundation Clinical Scientist Development Award, 1836
Fairlawn Foundation Grants, 1974
Judith Graham Pool Postdoctoral Research Fellowship in Hemophilia, 2529
National Blood Foundation Research Grants, 2848
NHLBI Ancillary Studies in Clinical Trials, 3015
NHLBI Characterizing the Blood Stem Cell Niche Grants, 3022
NHLBI Circadian-Coupled Cellular Function in Heart, Lung, and Blood Tissue Grants, 3023
NHLBI Immunomodulatory, Inflammatory, and Vasoregulatory Properties of Transfused Red Blood Cell Units as a Function of Preparation and Storage Grants, 3027
NHLBI Mentored Career Award for Faculty at Institutions that Promote Diversity, 3033
NHLBI Mentored Career Award For Faculty At Minority Institutions, 3034
NHLBI Mentored Career Development Award to Promote Faculty Diversity/Re-Entry in Biomedical Research, 3035
NHLBI Ruth L. Kirschstein National Research Service Awards for Individual Postdoctoral Fellows, 3057
NHLBI Short-Term Research Education Program to Increase Diversity in Health-Related Research, 3062
NHLBI Training Program Grants for Institutions that Promote Diversity, 3067
NIDDK Devel of Disease Biomarkers Grants, 3145
NIDDK Education Program Grants, 3149
NIDDK Grants for Basic Research in Glomerular Diseases, 3155
NIDDK Non-Invasive Methods for Diagnosis and Progression of Diabetes, Kidney, Urological, Hematological and Digestive Diseases, 3168
NIDDK Proteomics: Diabetes, Obesity, And Endocrine, Digestive, Kidney, Urologic, And Hematologic Diseases, 3174
NIDDK Small Grants for Underrepresented Minority Scientists in Diabetes and Digestive and Kidney Diseases, 3183
NIH Bone and the Hematopoietic and Immune Systems Research Grants, 3208
NYCT Biomedical Research Grants, 3694
NYCT Blood Disease Research Grants, 3696
ONR Medical Science and Technology Division Grants, 3772
Victor E. Speas Foundation Grants, 4810

Hemophilia
Bayer Clinical Scholarship Award, 1308
Judith Graham Pool Postdoctoral Research Fellowship in Hemophilia, 2529
Pfizer ASPIRE North America Hemophilia Research Awards, 3915
Pfizer Medical Education Track Two Grants, 3927

Hepatitis
ASM/CDC Fellowships in Infectious Disease and Public Health Microbiology, 1100
Bristol-Myers Squibb Foundation Fellowships, 1375
Bristol-Myers Squibb Foundation Health Disparities Grants, 1377
NFID New Investigator Matching Grants, 2989
NIDDK Health Disparities in NIDDK Diseases, 3156
NIDDK Research Grants for Studies of Hepatitis C in the Setting of Renal Disease, 3175

Hepatology
FDHN Non-Career Research Grants, 1999
FDHN Research Scholar Awards, 2000
FDHN Translational Research Awards, 2003

Herbs
Herb Society of America Research and Education Grants, 2243

Hereditary Diseases and Disorders
AIHS Alberta/Pfizer Translational Res Grants, 435
Barth Syndrome Foundation Research Grants, 1302

Herpesviruses
Sexually Transmitted Diseases Postdoctoral Fellowships, 4200

Herpetology
Organization for Tropical Studies Fellowships, 3796

High Pressure Sciences
ASME S.Y. Zamrik PVP Medal, 1163
Carnegie Postdoctoral Research Fellowships, 1436
Ipatieff Prize, 2404

Higher Education
3M Canada National Teaching Fellowships, 3
3M Company Higher Education Grants, 4
3M Foundation Community Giving Grants, 5
Abbott Fund Global AIDS Care Grants, 219
Achelis Foundation Grants, 274
ACLS/New York Public Library Fellowships, 277
ACLS Fellowships, 287
ACLS International and Area Studies Fellowships, 289
Adam Richter Charitable Trust Grants, 339
Adaptec Foundation Grants, 340
AFUW Diamond Jubilee Bursary, 359
AFUW Doreen McCarthy, Barbara Crase, Cathy Candler & Brenda Nettle Bursaries, 360
AFUW Fellowships, 361
AFUW Georgina Sweet Fellowships, 362
AFUW Heather Latz Bursary, 363
AFUW Padnendadlu Graduate Bursary, 364
AFUW Padnendadlu Postgraduate Bursary, 365
AFUW Padnendadlu Undergraduate Bursary, 366
AFUW South Australia Jean Gilmore, Thenie Baddams & Daphne Elliott Bursaries, 367
AFUW Winifred E. Preedy Postgraduate Bursary, 368
AGHE Graduate Scholarships and Fellowships, 369
AIR/IES-NCES Postdoctoral Policy Fellowships in Washington, 466
AIR/NCES Fellowships for Graduate Study, 467
AIR/NRC Summer Institute on First-Year Assessment Undergraduate Student Fellowships, 468
AIR National Summer Data Policy Institute Fellowships, 469
Alces Foundation Grants, 510
Alfred P. Sloan Foundation Civic Grants, 545
Alfred P. Sloan Foundation Selected National Issues Grants, 563
All Saints Educational Trust Grants, 572
Allyn Foundation Grants, 574
ALSAM Foundation Grants, 579
American Chemical Society Petroleum Research Fund Doctoral New Investigator Grants, 690
American Chemical Society Petroleum Research Fund Undergraduate New Investigator Grants, 694
American Foodservice Charitable Trust Grants, 710
Anderson Foundation Grants, 842
Annenberg Foundation Grants, 862
Ann Peppers Foundation Grants, 863
Applied Biosystems Grants, 916
Archer Daniels Midland Foundation Grants, 985
Arronson Foundation Grants, 1019
Arthur and Rochelle Belfer Foundation Grants, 1023
Arthur F. and Alice E. Adams Charitable Foundation Grants, 1024
ASHA Students Preparing for Academic & Research Careers Award, 1075
Ashland Corporate Contributions Grants, 1087
ASHRAE Willis H. Carrier Graduate Research Fellowship, 1089
ASU Grad College Reach for Stars Fellowship, 1243
Athenaeum of Philadelphia Senior Fellowships, 1256
Auburn Foundation Grants, 1263
Auburn University President's Graduate Opportunities Program, 1264
Bailey Family Foundation Grants, 1289
Ball Brothers Foundation General Grants, 1291

Banfi Vintners Foundation Grants, 1296
Batchelor Foundation Grants, 1305
Becton Dickinson and Company Grants, 1328
Bernard L. Schwartz Fellows Program, 1334
Beryl Henderson Foundation Grants, 1337
Better Way Foundation Grants, 1338
Blum-Kovler Foundation Grants, 1351
BMW of N America Charitable Contributions, 1353
Bodman Foundation Grants, 1356
Boeing Company Contributions Grants, 1357
Booth-Bricker Fund Grants, 1359
Bosque Foundation Grants, 1360
Brian G. Dyson Foundation Grants, 1370
Burlington Northern Santa Fe Foundation Grants, 1389
Bydale Foundation Grants, 1400
Byron W. and Alice L. Lockwood Grants, 1401
Caesars Foundation Grants, 1407
Callaway Golf Company Foundation Grants, 1410
Campbell Soup Foundation Grants, 1415
CAPAA Mini Grant for Research Related to African Americans in Education, 1424
Carnegie Corporation of New York Grants, 1432
Center for Education of Women Visiting Scholar Research Grants, 1475
Charles H. Revson Foundation Grants, 1524
Chiles Foundation Grants, 1552
Chiron Foundation Community Grants, 1553
Christensen Fund Regional Grants, 1554
CICU James C. Ross Research Fellowship, 1567
CIGNA Foundation Grants, 1568
Clay Foundation Grants, 1579
Cleveland-Cliffs Foundation Grants, 1580
Coca-Cola Foundation Grants, 1593
Coeta and Donald Barker Foundation Grants, 1594
Colonel Stanley R. McNeil Foundation Grants, 1597
Commonfund Endowment Management Prize, 1601
Community Foundation of the Fox River Valley, 1620
ConocoPhillips Grants Program, 1629
Conquer Cancer Foundation of ASCO Drug Development Research Professorship, 1632
Coors Brewing Corporate Contributions Grants, 1644
Cord Foundation Grants, 1645
Corning Incorporated Found Educational Grants, 1646
Cowles Charitable Trust Grants, 1652
Credit Union Found of British Columbia Grants, 1654
Cystic Fibrosis Canada Fellowships, 1677
DaimlerChrysler Corporation Fund Grants, 1714
Datatel Scholars Foundation Scholarship, 1722
Davis Family Foundation Grants, 1726
Dean Witter Foundation Grants, 1728
Delta Dental Master's Thesis Award, 1737
DENSO North America Foundation Grants, 1742
Diversifying Higher Education Faculty in Illinois Program, 1764
Dorr Foundation Grants, 1842
Dr. Bob and Jean Smith Foundation Grants, 1853
DTE Energy Found Higher Learning Grants, 1866
Duke Univ Interventional Cardio Fellowships, 1873
E.L. Wiegand Foundation Grants, 1885
Educause Jane N. Ryland Fellowships, 1897
Emerson Charitable Trust Grants, 1918
Ethics and Excellence in Journalism Foundation Grants, 1953
ETS Predoctoral Summer Program, 1954
Eugene M. Lang Foundation Grants, 1957
Eugene McDermott Foundation Grants, 1958
Evjue Foundation Grants, 1963
Fluor Foundation Grants, 2028
FMC Foundation Grants, 2029
Foundation for Enhancing Communities Grants, 2038
Frances C. and William P. Smallwood Foundation Grants, 2045
Frances L. and Edwin L. Cummings Memorial Fund Grants, 2046
Frederick Gardner Cottrell Foundation Grants, 2054
Fritz B. Burns Foundation Grants, 2057
Fulbright mtvU Awards, 2074
Fulbright Scholar-in-Residence Program Grants, 2078
George B. Storer Foundation Grants, 2113
George Frederick Jewett Foundation Grants, 2116
George Kress Foundation Grants, 2119
George S. and Dolores Dore Eccles Grants, 2120
George W. Codrington Charitable Grants, 2122
Georgia Power Foundation Grants, 2125
Gilroy and Lillian P. Roberts Charitable Foundation Grants, 2142
Greenspun Family Foundation Grants, 2173
H.W. Wilson Foundation Grants, 2196
Harold Alfond Foundation Grants, 2200
Harold R. Bechtel Charitable Remainder Uni-Trust Grants, 2201
Harold R. Bechtel Testamentary Charitable Trust Grants, 2202
Harold Simmons Foundation Grants, 2203
Harry B. and Jane H. Brock Foundation Grants, 2206
Harvard University Administrative Fellowships, 2212
Harvard University Graduate School of Education Fellowships and Awards, 2213
Harvey Randall Wickes Foundation Grants, 2218
Henry W. Bull Foundation Grants, 2237
HHMI Grants and Fellowships Programs, 2255
Hill Crest Foundation Grants, 2267
Hirtzel Memorial Foundation Grants, 2269
HSBC Contributions Community Grants, 2291
HUD Community Outreach Partnership Centers Grants, 2292
Huie-Dellmon Trust Grants, 2295
IDEM Section 319(h) Nonpoint Source Program Grants, 2338
IIE David L. Boren Scholarships, 2351
Intel Bachelor's and Master's Degree Internships, 2393
Intel Foundation PhD Fellowships, 2394
J.B. Reynolds Foundation Grants, 2449
J.W. Kieckhefer Foundation Grants, 2452
Jacob and Hilda Blaustein Foundation Grants, 2454
James M. Collins Foundation Grants, 2461
James R. Dougherty Jr. Foundation Grants, 2463
Japan Foundation New York Institutional Support for Japanese Studies Grants, 2477
Japan Foundation Six-Month Grants for Specialists in Cultural and Academic Fields, 2479
JMO Public Policy Research Grants, 2492
John Ben Snow Memorial Trust Grants, 2497
John Deere Foundation Grants, 2500
John R. Oishei Foundation Grants, 2505
Johnson & Johnson Corporate Contributions, 2508
John Stauffer Charitable Trust Grants, 2509
Joseph Alexander Foundation Grants, 2513
Josiah W. and Bessie H. Kline Foundation Grants, 2517
JSPS Bilateral Exchange Programs with Foreign Academic Institutions, 2524
Kavli Foundation Research Grants, 2533
Knoll-MacDonald Foundation Grants, 2566
Kovler Family Foundation Grants, 2575
Leonard and Helen R. Stulman Charitable Foundation Grants, 2602
Lillian S. Wells Foundation Grants, 2614
Lowe Foundation Grants, 2622
Lumina Foundation for Education Grants, 2627
Lumpkin Family Found Healthy People Grants, 2629
Lynde & Harry Bradley Foundation Fellowships, 2634
Lynde and Harry Bradley Foundation Grants, 2635
M. Bastian Family Foundation Grants, 2636
M.J. Murdock Charitable Trust Exceptional Opportunity Grants, 2640
M.J. Murdock Charitable Trust General Grants, 2641
M.J. Murdock Charitable Trust Scientific Research Grants, 2644
MAA-Tensor Foundation Women and Mathematics Grants, 2645
MacDonald-Peterson Foundation Grants, 2649
Margaret & James A. Elkins Jr. Found Grants, 2658
Margaret T. Morris Foundation Grants, 2659
Marion Gardner Jackson Charitable Trust Grants, 2664
Mary S. and David C. Corbin Foundation Grants, 2675
Matilda R. Wilson Fund Grants, 2684
McCombs Foundation Grants, 2718
McCune Foundation Civic Grants, 2720
McCune Foundation Education Grants, 2721
McGraw-Hill Companies Community Grants, 2724
McGregor Fund Grants, 2725
Melville Charitable Trust Grants, 2741
Mericos Foundation Grants, 2742
Meyer and Stephanie Eglin Foundation Grants, 2747
Meyer Memorial Trust Responsive Grants, 2748
MGM Resorts Foundation Community Grants, 2753
MGN Family Foundation Grants, 2754
Millstone Foundation Grants, 2776
Minnesota War Orphans Educational Assistance Grants, 2780
Minority Faculty Fellowship Program, 2781
Motorola Foundation Grants, 2796
Motorola Foundation Innovation Generation Collaborative Grants, 2797
Musgrave Foundation Grants, 2807
National Rising Scholars Award to Advance Research on Higher Education for the Public Good, 2883
National Security Education Program Grants to Institutions of Higher Education, 2885
NEH Fellowships, 2956
Nellie Mae Education Foundation District-Level Change Grants, 2968
Nellie Mae Education Foundation Public Understanding Grants, 2969
Nellie Mae Education Foundation State Level Systems Change Grants, 2971
Nell Warren Elkin and William Simpson Elkin Foundation, 2972
NFIE Found Learning & Leadership Grants, 2992
NHPRC Digitizing Historical Records Grants, 3072
NIA Mentored Clinical Scientist Research Career Development Awards, 3108
NL Associated Colleges of the Midwest/Great Lakes College Association Faculty Fellowships, 3281
NOAA Joint Research Institutes Grants, 3379
Norcliffe Foundation Grants, 3415
Nordson Corporation Foundation Grants, 3416
North Carolina Biotechnology Center Collaborative Funding Grants, 3421
North Carolina Biotech Center Research Grants, 3424
Northwestern Mutual Foundation Grants, 3430
Notsew Orm Sands Foundation Grants, 3434
NSF Grant Opportunities for Academic Liaison with Industry, 3574
NSF Integrative Graduate Education and Research Traineeship, 3584
NSF Materials World Network: Cooperative Activity in Materials Research between US Investigators and their Counterparts Abroad, 3595
NSF Presidential Awards for Excellence in Science, Math, and Engineering Mentoring, 3632
NSF Scholarships in Science, Technology, Engineering, and Mathematics, 3642
NSF Tribal Colleges and Universities Program, 3662
NSF University Radio Observatories Program, 3664
NWO Aspasia Grants, 3673
Ohio Learning Network Grants, 3734
Olin Corporation Charitable Trust Grants, 3739
Oppenstein Brothers Foundation Grants, 3792
Oscar Rennebohm Foundation Grants, 3832
Park Foundation Grants, 3870
Patrick and Anna M. Cudahy Fund Grants, 3873
Paul & Daisy Soros Fellowships for New Amer, 3875
Paul Stock Foundation Grants, 3879
Pell Grant Program, 3890
Pennsylvania State University Education Graduate Assistantships, 3895
Percy B. Ferebee Endowment Grants, 3897
Perpetual Trust for Charitable Giving Grants, 3900
Phi Upsilon Omicron Geraldine Clewell Doctoral Fellowship, 3945
Plan for Social Excellence Grants, 3976
Pott Foundation Grants, 3984
Praxair Foundation Grants, 3986
Price Family Charitable Fund Grants, 3987
R.B. Terry Charitable Foundation Grants, 4005
R.C. Baker Foundation Grants, 4006
Radcliffe Institute Dissertation Grants, 4019
Reinberger Foundation Grants, 4051
RGK Foundation Grants, 4060

Richard and Susan Smith Family Found Grants, 4064
Rich Foundation Grants, 4068
Robert G. Cabell III and Maude Morgan Cabell Foundation Grants, 4075
Robert M. Hearin Foundation Grants, 4077
Robert W. Deutsch Foundation Grants, 4082
Rockford College Faculty Fellowships, 4092
Rockwell International Corporate Trust Grants, 4093
Ross Foundation Grants, 4106
Salem Foundation Grants, 4160
San Antonio Area Foundation Grants, 4163
Sarkeys Foundation Grants, 4181
Schramm Foundation Grants, 4188
Scientific Communities Initiative Grants, 4189
Sensient Technologies Foundation Grants, 4199
Shastri Indo-Canadian Institute Faculty Research Fellowships, 4222
Shastri Indo-Canadian Institute Research Collaboration Development Grants, 4230
Shell Oil Company Foundation Grants, 4233
SHRM Dissertation Awards, 4248
Sid W. Richardson Foundation Grants, 4253
Sioux Falls Area Community Foundation Community Fund Grants (Unrestricted), 4262
Sioux Falls Area Comm Foundation Spot Grants, 4263
Sir Dorabji Tata Trust Individual Education and Travel Grants, 4265
Sony Corporation of America Corporate Philanthropy Grants, 4329
SRC Educational Science Doctoral Grants, 4355
SRC Educational Science Grants, 4356
SRC Educational Science Junior Research Position Grants, 4359
SRC Educational Science Medium-Cost Scientific Equipment Grants, 4361
SRC Educational Science Network Grants, 4362
SRC Educational Science Project Res Grants, 4364
SSHRC Aid to Small Universities, 4403
Stackpole-Hall Foundation Grants, 4442
Starr Foundation Grants, 4447
Sun Academic Excellence Grants, 4466
Sun Microsystems Foundation Summer Academic Enrichment Grants, 4468
Talbert and Leota Abrams Foundation, 4492
Tanner Humanities Center Research Fellowships, 4494
Teachers College Minority Postdoc Fellowships, 4498
Textron Corporate Contributions Grants, 4501
The Ray Charles Foundation Grants, 4502
Thomas and Dorothy Leavey Foundation Grants, 4506
Thomas Austin Finch, Sr. Foundation Grants, 4507
Thompson Charitable Foundation Grants, 4511
Thornton Foundation Grants, 4514
TIAA-CREF Institute Research Grant Program, 4517
U.S. Department of Education American Overseas Research Centers, 4537
U.S. Department of Education Foreign Languages and Area Studies Fellowships, 4539
U.S. Department of Education Graduate Assistance in Areas of National Need, 4540
U.S. Department of Education Jacob K. Javits Fellowships, 4541
U.S. Department of Education National Resource Centers, 4542
U.S. Department of Education Native Hawaiian Education Grants, 4543
U.S. Department of Education Ronald E. McNair Postbaccalaureate Achievement Grants, 4546
U.S. Department of Education United States-Russia Program: Improving Research and Educational Activities in Higher Education, 4547
United Methodist Church Center for the Evangelical UBH Audrie E. Reber Memorial Award, 4571
USDA Higher Education Challenge Grants, 4627
USDC Dissertation Fellowships, 4665
USDC Planning and Local Technical Assistance Grants, 4668
USDD Small University Grants, 4794
Victor E. Speas Foundation Grants, 4810
Virginia Foundation for the Humanities Discretionary Grants, 4813
Virginia Foundation for Humanities Open Grants, 4814
W. C. Griffith Foundation Grants, 4818
W.C. Griffith Foundation Grants, 4819
W.K. Kellogg Foundation Healthy Kids Grants, 4822
Wabash Center Project Grants, 4826
Wabash Center Small Project Grants, 4827
Wabash Center Waorkshop and Colloquy Participant Grants, 4828
Wallace Foundation Grants, 4829
Wayne and Gladys Valley Foundation Grants, 4835
Western Regional Graduate Program, 4855
Willamette Univ Minority Grad Fellowship, 4869
William Blair and Company Foundation Grants, 4875
William G. and Helen C. Hoffman Found Grants, 4876
Woodrow Wilson MMUF Dissertation Grants, 4892
Woodrow Wilson MMUF Travel & Res Grants, 4893
Woodrow Wilson National Fellowship Foundation Travel and Research Grants, 4894
WWF Russell E. Train Education for Nature Alumni Grants, 4897
WWF Russell E. Train Fellowships, 4898

Higher Education Administration

Harvard University Graduate School of Education Fellowships and Awards, 2213
SfN Louise Hanson Marshall Specific Recognition Award, 4208

Higher Education Studies

AAUW American Postdoctoral Research Leave Fellowships, 210
ACLS African Humanities Program in Ghana, Nigeria, South Africa, Tanzania, and Uganda, 279
AERA Minority Fellowships, 343
AIR/IES-NCES Postdoctoral Policy Fellowships in Washington, 466
AIR/NCES Fellowships for Graduate Study, 467
AIR/NRC Summer Institute on First-Year Assessment Undergraduate Student Fellowships, 468
AIR National Summer Data Policy Institute Fellowships, 469
AIR Research Grants/Dissertation Fellowships, 470
Andrew W. Mellon/ACLS Dissertation Completion Fellowships, 844
ASME Graduate Teaching Fellowships, 1125
Association for the Study of Higher Education Dissertation Fellowships, 1222
Astronomical Society of the Pacific Richard H. Emmons Award for Excellence in College Astronomy Teaching, 1233
ASU Graduate College Completion Fellowships, 1235
ASU Graduate College Dean's Fellowships, 1236
ATA Doctoral Fellowships in Education, 1252
CIC Traveling Scholars, 1566
Douglass Institute Postdoctoral Fellowships, 1851
Frederick Douglass Summer Fellowships, 2052
HAF Riley Frazel Memorial Scholarship, 2197
Higher Education Graduate Scholarships, Fellowships, and Grants, 2265
IBRO Regional Grants for Int'l Fellowships to U.S. Laboratory Summer Neuroscience Courses, 2330
Maine Community Foundation Keepers Preservation Education Grants, 2651
Mayo Clinic Business Consulting Fellowship, 2691
MBL William Townsend Porter Summer Fellowships for Minority Investigators, 2716
Medtronic Foundation Fellowships, 2738
National Wildlife Federation Internships, 2891
OSF Global Supplementary Grants, 3843
OSF Justice Initiative Fellowships, 3847
Paul P. Fidler Research Grants, 3878
Rockford College Faculty Fellowships, 4092
Rumsfeld Foundation Graduate Fellowships, 4127
SARE Graduate Student Grants - North Central Region, 4168
SARE Graduate Student Grants - So Region, 4170
SARE Graduate Student Grants in Sustainable Agriculture - Western Region, 4171
Scleroderma Foundation Established Investigator Grants, 4190
SHRM Dissertation Awards, 4248
SPJ Pulliam/Kilgore Freedom of Information Internships, 4347
SRC Educational Science Doctoral Grants, 4355
SRC Educational Science Grants, 4356
SRC Educational Science Junior Research Position Grants, 4359
SRC Educational Science Medium-Cost Scientific Equipment Grants, 4361
SRC Educational Science Network Grants, 4362
SRC Educational Science Project Res Grants, 4364
SRC Guest Professorship Program, 4370
SRC Research Customization Program Grants, 4390
UCR Grad Student in Humanities Fellowship, 4550
UCR Humanities Grad Student Research Grants, 4551
UNC School of Information and Library Science Doctoral Fellowships, 4565
UNESCO Co-Sponsored Fellowships, 4566
United States Institute of Peace - Jennings Randolph Peace Dissertation Scholarships, 4579
USDA Distance Education Grants for Higher Education Institutions in Insular Areas, 4622
USDA Tribal College Research Grants, 4652

Higher Education, Private

Burlington Northern Santa Fe Foundation Grants, 1389
Coca-Cola Foundation Grants, 1593
DTE Energy Found Higher Learning Grants, 1866
F.W. Olin Foundation Construction Grant, 1972
Kenneth T. and Eileen L. Norris Found Grants, 2554
M.J. Murdock Charitable Trust College Research Program for Life Sciences, 2638
M.J. Murdock Charitable Trust College Science Research Program, 2639
M.J. Murdock Charitable Trust Research Start-up Grants for New Science Faculty, 2643
McGregor Fund Grants, 2725
Nordson Corporation Foundation Grants, 3416
Strake Foundation Grants, 4452
USDA Alaska Native-Serving and Native Hawaiian-Serving Institutions Education Grants, 4615

Higher Education, Public

DTE Energy Found Higher Learning Grants, 1866
USDA Alaska Native-Serving and Native Hawaiian-Serving Institutions Education Grants, 4615

Hispanic Education

AAA Minority Dissertation Fellowship Program, 25
APA Planning Fellowships, 910
Carl and Lily Pforzheimer Foundation Grants, 1427
Consortium for Graduate Study in Management Minority Fellowships, 1642
Diversifying Higher Education Faculty in Illinois Program, 1764
DOT Dwight David Eisenhower Transportation Fellowship Program, 1847
GEM MS Engineering Fellowships, 2104
GEM PhD Engineering Fellowships, 2105
GEM PhD Science Fellowships, 2106
General Motors Foundation Grants, 2107
Hispanic Scholarships, 2270
John R. Oishei Foundation Grants, 2505
KFF Barbara Jordan Health Policy Scholars, 2556
McCune Charitable Foundation Grants, 2719
Miller Brewing Corporate Contributions Grants, 2774
MLA Grad Scholarship for Minority Students, 2787
NACC David Stevenson Fellowships, 2812
NACC William Diaz Fellowships, 2813
National Assessment of Educational Progress Visiting Scholar Program, 2846
NEH Awards for Faculty at Institutions with High Hispanic Enrollment, 2950
NHLBI Training Program Grants for Institutions that Promote Diversity, 3067
NMF Need-Based Scholarships, 3326
UNC Carolina Postdoctoral Program for Faculty Diversity, 4560
USDA Hispanic-Serving Institutions Education Grants, 4628

Hispanic Studies
ASU Graduate College Dean's Fellowships, 1236
CMS Hispanic Health Services Research Grants, 1584
Institute of American Cultures Graduate and Predoctoral Fellowships, 2383
Institute of American Cultures Postdoctoral and Visiting Scholar Fellowships, 2384
UIC Rockefeller Foundation Resident Fellowships in the Humanities, 4559

Hispanics
American Sociological Association Minority Fellowships, 805
AMS Howard Mayer Brown Fellowships, 827
ASM Robert D. Watkins Graduate Research Fellowship, 1183
CMS Hispanic Health Services Research Grants, 1584
FDHN Student Research Fellowships, 2001
Minority Faculty Fellowship Program, 2781
NEH Awards for Faculty at Institutions with High Hispanic Enrollment, 2950
Pfizer ASPIRE North America Rheumatology Research Awards, 3917
RWJF Harold Amos Medical Faculty Development Research Grants, 4139
Smithsonian Latino Studies Fellowships, 4278
SRC AMD/Mahboob Khan Fellowship, 4348
SRC Master's Scholarships, 4378
United Methodist Church Asian, Black, Hispanic and Native American History Research Grants, 4570

Historical Preservation
AAF Richard M Hunt Architecture Fellowship, 169
Blum-Kovler Foundation Grants, 1351
Forrest C. Lattner Foundation Grants, 2035
George Kress Foundation Grants, 2119
Maine Community Foundation Keepers Preservation Education Grants, 2651
McCombs Foundation Grants, 2718
McLean Contributionship Grants, 2727
National Endowment for the Humanities Fellowships at Winterthur, 2867
NL Newberry Library/Ecole Nationale des Chartes Exchange Fellowship, 3313
Norcliffe Foundation Grants, 3415
Sioux Falls Area Community Foundation Community Fund Grants (Unrestricted), 4262
Sioux Falls Area Comm Foundation Spot Grants, 4263
Virginia Foundation for the Humanities Discretionary Grants, 4813
Virginia Foundation for Humanities Open Grants, 4814
Winterthur Museum Dissertation Fellowships, 4887
Winterthur Research Fellowships, 4888

History
AAB Berlin Prize Fellowships, 48
AAHN Grant for Historical Research, 178
AAHN Pre-Doctoral Research Grant, 179
AAR Postdoctoral Fellowships, 201
AAS Kate B. and Hall J. Peterson Fellowships, 204
AAS Legacy Fellowships, 205
Academy of American Franciscan History Dissertation Fellowships, 249
ACLS Charles A. Ryskamp Research Fellowships, 281
ACLS Frederick Burkhardt Residential Fellowships for Recently Tenured Scholars, 288
Ahmanson and Getty Postdoctoral Fellowships, 377
AIAR Annual Professorships, 393
AIAR George A. Barton Fellowships, 395
Alabama Humanities Foundation Mini Grants, 480
Alabama Humanities Foundation Planning/Consultant Grants, 481
Albert M. Greenfield Foundation Dissertation Fellowship, 507
Alice Fisher Society Fellowships, 568
Alpha Sigma Nu Book Award Competition, 578
American Historical Assoc Al J. Beveridge Award, 712
American Historical Association Albert J. Beveridge Grants for Research in the History of the Western Hemisphere, 713
American Historical Association Award for Scholarly Distinction, 714
American Historical Association Bernadotte E. Schmitt Research Grants in European, African, or Asian History, 715
American Historical Association Clarence H. Haring Prize, 716
American Historical Association Fellowship in Aerospace History, 717
American Historical Association Folger Shakespeare Library Fellowship, 718
American Historical Association George L. Mosse Prize, 719
American Historical Association George Louis Beer Prize, 720
American Historical Association Helen and Howard R. Marraro Prizes, 721
American Historical Association Herbert Baxter Adams Prize, 722
American Historical Assoc Herb Feis Award, 723
American Historical Association J. Franklin Jameson Award In Outstanding Editorial Achievement, 724
American Historical Association J. Franklin Jameson Fellowship, 725
American Historical Assoc J. Russell Major Prize, 726
American Historical Association James A. Rawley Prize in Atlantic History, 727
American Historical Association James Harvey Robinson Prize, 728
American Historical Association James Henry Breasted Prize, 729
American Historical Association Joan Kelly Memorial Prize in Women's History, 730
American Historical Association John F. Richards Prize in South Asian History, 731
American Historical Association John H. Dunning Prize, 732
American Historical Association John K. Fairbank Prize in East Asian History, 733
American Historical Assoc Leo Gershoy Award, 734
American Historical Association Michael Kraus Research Grant in American Colonial History, 738
American Historical Association Morris D. Forkosch Prize, 739
American Historical Association Paul Birdsall Prize in European Military and Strategic History, 740
American Historical Association Premio Del Rey, 741
American Historical Association Waldo J. Leland Prize, 743
American Historical Assoc Wesley-Logan Prize, 744
American Numismatic Society Eric P. Newman Graduate Seminar Grants, 761
AMS Grad Fellowships in History of Science, 824
Andrew W. Mellon/ACLS Dissertation Completion Fellowships, 844
ARIT/NEH Fellowships for Research in Turkey, 990
ARIT Fellowships, 991
ARIT Kenan T. Erim Fellowship for Research at Aphrodisias, 993
ASECS James L. Clifford Prize, 1058
ATS Henry Luce III Fellowships, 1258
Australasian Institute of Judicial Administration Seed Funding Grants, 1268
BA Elisabeth Barker Fund, 1288
Bogliasco Foundation Liguria Study Center Residencies, 1358
Canada-U.S. Fulbright International Science and Technology Award, 1417
Canada-U.S. Fulbright Senior Specialists Program Grants, 1418
Canada-U.S. Fulbright Traditional Scholars, 1419
CCWH Catherine Prelinger Award, 1454
CCWH Ida B. Wells Grad Student Fellowship, 1455
Center for East Asian Studies Fellowships, 1474
Chapin Hall International Fellowships in Children's Policy Research, 1517
Charles Edison Fund Grants, 1520
CHF Charles C. Price Fellow in Polymer History, 1539
CHF Glenn E. and Barbara Hodsdon Ullyot Scholarship, 1540
CHF Robert W. Gore Fellowship in Materials Innovation, 1542
CHF Roy G. Neville Short-Term Fellowships, 1543
CHF Societe de Chimie Industrielle (American Section) Fellowship, 1544
CHF Travel Grants, 1545
CIC Smithsonian Institution Fellowships, 1564
Colonel Stanley R. McNeil Foundation Grants, 1597
Cushwa Center Research Travel Grants, 1673
Earhart Foundation Institutional Grants, 1888
Eugene McDermott Foundation Grants, 1958
Fischelis Grants for Research in the History of American Pharmacy, 2017
Fulbright Alumni Initiatives Awards Grants, 2064
Fulbright Distinguished Chairs Program Grants, 2069
Fulbright European Union Scholar-in-Residence Program Grants, 2071
Fulbright New Century Scholars Program Grants, 2075
Fulbright Occasional Lecturer Program Grants, 2076
Fulbright Specialists Program Grants, 2079
Fulbright Traditional Scholar Program in Europe and Eurasia, 2080
Fulbright Traditional Scholar Program in Sub-Saharan Africa, 2081
Fulbright Traditional Scholar Program in the East Asia/Pacific Region, 2082
Fulbright Traditional Scholar Program in the Near East and North Africa Region, 2083
Fulbright Traditional Scholar Program in the South and Central Asia Region, 2084
Fulbright Traditional Scholar Program in the Western Hemisphere, 2085
Fulbright Visiting Specialists Program: Direct Access to the Muslim World, 2087
Gerda Henkel Foundation Grants, 2127
Germanistic Society of America Fellowships, 2128
Gladys Krieble Delmas Foundation Grants for Venetian Research, 2144
Harry Frank Guggenheim Foundation Dissertation Fellowships, 2207
Harry Ransom Humanities Research Center Research Fellowships, 2209
Herb Society of America Research and Education Grants, 2243
Houghton Library Visiting Fellowships, 2284
IAS Mellon Fellowships for Assistant Professors, 2318
IEEE Electrical History Fellowship, 2339
IMLS 21st Century Museum Prof Grants, 2362
IMLS National Leadership Grants, 2366
IREX Short-Term Travel Grants, 2424
Irving and Rose Crown Scholarships and Fellowships in American History, 2427
Kluge Center Kislak Fellows in American Studies, 2562
Kluge Center Kislak Short-Term Fellowships in American Studies, 2563
Kosciuszko Foundation Tuition Scholarships, 2574
LBI-NY Fred Grubel Fellowship, 2596
Library Company of Philadelphia NEH Long-term Post-Doctoral Fellowships, 2607
Maine Community Foundation Keepers Preservation Education Grants, 2651
Marine Corps Historical Center Dissertation Fellowship, 2663
Mary Baker Eddy Library Summer Research Fellowships, 2669
Mary Isabel Sibley Fellowship, 2672
MHS Long Term Research Fellowships, 2755
MHS Short Term Research Fellowships, 2757
MHS Suzanne and Caleb Loring Fellowship, 2758
NEH Collaborative Research Grants, 2952
NEH Digital Humanities Implementation Grants, 2953
NEH Digital Humanities Start-Up Grants, 2954
NHPRC Archives - Basic Projects Grants, 3070
NHPRC Archives - Detailed Processing Grants, 3071
NHPRC Digitizing Historical Records Grants, 3072
NHPRC Electronic Records Projects Grants, 3073
NHPRC Historical Editing Fellowships, 3074
NHPRC Publishing Historical Records Grants, 3075
NHPRC State and National Archival Partnership Grants, 3076

NL American Society for Eighteenth-Century Studies Fellowships, 3279
NL Arthur and Lila Weinberg Fellowships for Independent Scholars and Researchers, 3280
NL British Academy Fellowship for Study in the United Kingdom, 3283
NL Frances C. Allen Fellowships, 3284
NL Grad Scholar-in-Residence Opportunities, 3285
NL Mellon Postdoctoral Research Fellowships, 3294
NL Monticello College Foundation Fellowship for Women, 3304
NL National Endowment for the Humanities Fellowships, 3310
NL Scholar-in-Residence Opportunity, 3315
NL Short-Term Fellowships in the History of Cartography, 3316
NL Short-Term Resident Fellowships for Individual Research, 3317
NOAA Ocean Exploration Omnibus: Marine Archaeology Grants, 3397
North Carolina Arts Council Folklife Internship, 3418
NPS American Battlefield Protection Grants, 3435
NYAM Paul Klemperer Fellowship, 3692
NYPL Martin Duberman Visiting Scholars, 3700
NYPL Short-Term Research Fellowships, 3701
OAH/IEHS John Higham Travel Grants, 3710
OAH Binkley-Stephenson Award, 3712
OAH Darlene Clark Hine Award, 3713
OAH David Thelen Award, 3714
OAH Erik Barnouw Award, 3716
OAH Frederick Jackson Turner Award, 3717
OAH Friend of History Award, 3718
OAH Germany Residency Program Grants, 3719
OAH James A. Rawley Prize, 3721
OAH Japanese Residencies Program Grants, 3722
OAH Lerner-Scott Dissertation Prize, 3723
OAH Liberty Legacy Foundation Award, 3724
OAH Louis Pelzer Memorial Award, 3725
OAH Richard W. Leopold Prize, 3728
OAH Roy Rosenzweig Dist Service Award, 3729
OAH Willi Paul Adams Award, 3730
PHF Penn Faculty Research Fellowships, 3932
PHF Regional Mellon Faculty Research and Cultural Fellowships, 3933
PHF Undergraduate Humanities Forum Research Fellowships, 3934
Radcliffe Institute Dissertation Grants, 4019
Ralph Waldo Emerson Award, 4028
Righteous Persons Foundation Grants, 4069
Robert Schalkenbach Foundation Grants, 4079
Rockefeller Archive Center Research Grants, 4086
SASS Einar and Eva Lund Haugen Scholarship, 4183
Schomburg Center Scholars-in-Residence Fellows, 4187
Self Foundation Grants, 4197
SHGAPE Vincent P. DeSantis Prize, 4234
SHOT Mel Kranzberg Dissertation Fellowship, 4245
Sid W. Richardson Foundation Grants, 4253
Smithsonian Latino Studies Fellowships, 4278
SSHRC Insight Development Grants, 4409
SSHRC Insight Grants, 4410
SSHRC Jules and Gabrielle Leger Fellowship, 4413
SSRC Int'l Dissertation Research Fellowships, 4429
SSRC Mellon Mays Predoctoral Research Grants, 4432
Stanford Humanities Center Arts Writer/Practitioner Fellowship, 4444
Stanford Humanities Center External Faculty Fellowships, 4445
Twentieth-Century Japan Research Awards, 4536
United Methodist Church Asian, Black, Hispanic and Native American History Research Grants, 4570
United Methodist Church Center for the Evangelical UBH Audrie E. Reber Memorial Award, 4571
United Methodist Church John Harrison Ness Memorial Award, 4572
United Methodist Church Women in History Research Grant, 4574
United Methodist Church Women in History Writing Award, 4575
United Methodist Church World is My Parish United Methodist History Research Grant, 4576
University of London Visiting Fellowships, 4588
USDD Research and Studies for the Office of Net Assessment Grants, 4782
Virginia Foundation for the Humanities Discretionary Grants, 4813
Virginia Foundation for Humanities Open Grants, 4814
Virginia Foundation for the Humanities Residential Fellowships, 4815
Virginia Historical Society Research Fellowships, 4816
WAWH Barbara Penny Kanner Award, 4832
WAWH Founders Dissertation Fellowship, 4833
WAWH Sierra Award, 4834
Wenner-Gren Foundation African Fellowships, 4846
Wenner-Gren Foundation Conference and Workshop Grants, 4847
Wenner-Gren Foundation International Collaborative Research Grants, 4851
Wenner-Gren Found Short-Term Fellowships, 4853
Wenner-Gren Foundation Wadsworth International Fellowships, 4854

History Education
American Historical Association Award for Scholarly Distinction, 714
American Historical Association James Harvey Robinson Prize, 728
American Historical Association Morris D. Forkosch Prize, 739
American Society for Yad Vashem Grants, 798
Bogliasco Foundation Liguria Study Center Residencies, 1358
CCWH Ida B. Wells Grad Student Fellowship, 1455
Library Company of Philadelphia NEH Long-term Post-Doctoral Fellowships, 2607
SHGAPE Vincent P. DeSantis Prize, 4234
Virginia Historical Society Research Fellowships, 4816

History, Ancient/Classical
AAR Postdoctoral Fellowships, 201
AIAR Andrew W. Mellon Foundation Fellowships, 392
AIAR Annual Professorships, 393
American Historical Association Herbert Baxter Adams Prize, 722
American Numismatic Society Eric P. Newman Graduate Seminar Grants, 761
ARIT/NEH Fellowships for Research in Turkey, 990
ARIT Fellowships, 991
Center for Hellenic Studies Fellowships, 1476

History, Byzantine
ARIT Fellowships, 991

Holistic Medicine
AIHS Alberta/Pfizer Translational Res Grants, 435
Samueli Institute Scientific Research Grants, 4162

Home Economics
AAFCS International Graduate Fellowships, 154
AAFCS National Graduate Fellowships, 155
All Saints Educational Trust Grants, 572
CFUW Canadian Home Economics Association Fellowship, 1509
CFUW Ruth Binnie Fellowship, 1516
Phi Upsilon Omicron Alumni Research Grant, 3940
Phi Upsilon Omicron Candle Fellowships, 3941
Phi Upsilon Omicron Diamond Anniversary Fellowships, 3942
Phi Upsilon Omicron Florence Fallgatter Distinguished Service Award, 3943
Phi Upsilon Omicron Founders Fellowship, 3944
Phi Upsilon Omicron Geraldine Clewell Doctoral Fellowship, 3945
Phi Upsilon Omicron Geraldine Clewell Masters Fellowship, 3946
Phi Upsilon Omicron Jean Dearth Dickerscheid Fellowship, 3947
Phi Upsilon Omicron Marg Drew Alpha Fellowship, 3948
Phi Upsilon Omicron Presidents Res Fellowship, 3949

Home Economics Education
CFUW Ruth Binnie Fellowship, 1516
Phi Upsilon Omicron Florence Fallgatter Distinguished Service Award, 3943
Phi Upsilon Omicron Geraldine Clewell Doctoral Fellowship, 3945
Phi Upsilon Omicron Geraldine Clewell Masters Fellowship, 3946

Homeland Security
USDD Research Interests of the US Air Force Academy Grants, 4785

Homeless Shelters
Adaptec Foundation Grants, 340
Green Foundation Human Services Grants, 2172
HomeBanc Foundation Grants, 2275
McCarthy Family Foundation Grants, 2717
Melville Charitable Trust Grants, 2741
Northwestern Mutual Foundation Grants, 3430
Perpetual Trust for Charitable Giving Grants, 3900
Reinberger Foundation Grants, 4051
Textron Corporate Contributions Grants, 4501

Homelessness
Achelis Foundation Grants, 274
Bodman Foundation Grants, 1356
Boston Foundation Grants, 1361
Dade Community Foundation Grants, 1712
Eugene M. Lang Foundation Grants, 1957
George A Ohl Jr. Foundation Grants, 2112
Green Foundation Human Services Grants, 2172
Greenspun Family Foundation Grants, 2173
Helen V. Brach Foundation Grants, 2231
Hilton Hotels Corporate Giving Program Grants, 2268
Luther I. Replogle Foundation Grants, 2631
M.J. Murdock Charitable Trust General Grants, 2641
McCarthy Family Foundation Grants, 2717
Melville Charitable Trust Grants, 2741
National Book Scholarship Fund, 2849
Oppenstein Brothers Foundation Grants, 3792
OSF Soros Justice Advocacy Fellowships, 3850
OSF Soros Justice Media Fellowships, 3851
Patrick and Anna M. Cudahy Fund Grants, 3873
Peacock Foundation Grants, 3887
Perpetual Trust for Charitable Giving Grants, 3900
Reinberger Foundation Grants, 4051
Rhode Island Foundation Grants, 4061
Richard and Susan Smith Family Found Grants, 4064
Rich Foundation Grants, 4068
Sensient Technologies Foundation Grants, 4199
Thomas C. Ackerman Foundation Grants, 4508
Vancouver Foundation Grants and Community Initiatives, 4807
W.C. Griffith Foundation Grants, 4819
W. C. Griffith Foundation Grants, 4818
William G. and Helen C. Hoffman Found Grants, 4876
Xerox Foundation Grants, 4899

Homeownership
HSBC Contributions Community Grants, 2291

Homosexuals, Female
Arcus Foundation Religion and Values Grants, 987
CDC Grants for Violence-Related Injury Prevention Research, 1463
David Bohnett Foundation Grants, 1724
National LGBT Scholarship Fund, 2879
NWSA Graduate Scholarships in Lesbian Studies, 3678
NYPL Martin Duberman Visiting Scholars, 3700
Overbrook Foundation Grants, 3860
Playboy Foundation Grants, 3980
Vanguard Public Foundation Grant Funds, 4808
Xerox Foundation Grants, 4899

Homosexuals, Male
Arcus Foundation Religion and Values Grants, 987
CDC Grants for Violence-Related Injury Prevention Research, 1463
David Bohnett Foundation Grants, 1724

National LGBT Scholarship Fund, 2879
NYPL Martin Duberman Visiting Scholars, 3700
Overbrook Foundation Grants, 3860
Playboy Foundation Grants, 3980
USAID Comprehensive District-Based Support for Better HIV/TB Patient Outcomes Grants, 4595
Vanguard Public Foundation Grant Funds, 4808
Xerox Foundation Grants, 4899

Hormones
NHLBI Lymphatics in Health and Disease in the Digestive, Urinary, Cardiovascular and Pulmonary Systems, 3032
NIDDK Co-Activators and Co-Repressors in Gene Expression Grants, 3141
NIH Receptors and Signaling in Bone in Health and Disease, 3229
Sigma Xi Grants in Aid of Research, 4259

Horticulture
AACC International Graduate Fellowships, 91
Bedding Plants Found Horticulture Scholarships, 1329
Bedding Plants Foundation Research Grants, 1330
C. Alma Baker Trust Grants, 1403
CPC Catherine H. Beattie Fellowship, 1653
GCA Interchange Fellowship, 2098
GCA Katharine M. Grosscup Scholarships, 2099
Herb Society of America Research and Education Grants, 2243
International Crane Foundation Internships, 2396
Open Spaces Sacred Places National Awards, 3791
OTS Stanley Smith Horticultural Fellowships, 3857
Stanley Smith Horticultural Trust Grants, 4446
USDA 1890 Institution Teaching, Research, and Extension Capacity Building Grants, 4613
USDA Distance Education Grants for Higher Education Institutions in Insular Areas, 4622
USDA Resident Instruction Grants for Higher Education in Insular Areas, 4645
USDA Tribal College Research Grants, 4652

Hospice Care
AAHPM Hospice and Palliative Med Fellowships, 180
Alex Stern Family Foundation Grants, 542
Arizona Public Service Corporate Giving Program Grants, 998
Arronson Foundation Grants, 1019
Conquer Cancer Foundation of ASCO International Development and Education Award in Palliative Care, 1634
Dell Foundation Open Grants, 1734
Edward W. and Stella C. Van Houten Memorial Fund Grants, 1899
J.W. Kieckhefer Foundation Grants, 2452
Margaret T. Morris Foundation Grants, 2659
MGN Family Foundation Grants, 2754
Olin Corporation Charitable Trust Grants, 3739
Sensient Technologies Foundation Grants, 4199
Thompson Charitable Foundation Grants, 4511

Hospital Administration
Mayo Clinic Administrative Fellowship, 2689
Mayo Clinic Admin Fellowship Eau Claire, 2690
Mayo Clinic Business Consulting Fellowship, 2691
SfN Louise Hanson Marshall Specific Recognition Award, 4208

Hospitals
Adam Richter Charitable Trust Grants, 339
Alcoa Foundation Grants, 511
Allyn Foundation Grants, 574
American Foodservice Charitable Trust Grants, 710
Anthony R. Abraham Foundation Grants, 869
Archer Daniels Midland Foundation Grants, 985
Arthur and Rochelle Belfer Foundation Grants, 1023
Auburn Foundation Grants, 1263
Banfi Vintners Foundation Grants, 1296
Batchelor Foundation Grants, 1305
Beckman Coulter Community Relations Grants, 1325
Blum-Kovler Foundation Grants, 1351
Burden Trust Grants, 1388
Burlington Northern Santa Fe Foundation Grants, 1389
Byron W. and Alice L. Lockwood Grants, 1401
Caesars Foundation Grants, 1407
Carl and Lily Pforzheimer Foundation Grants, 1427
Carl Gellert and Celia Berta Gellert Foundation Grants, 1429
Chestnut Hill Charitable Foundation, Inc Grants, 1538
CMA Foundation Grants, 1582
Coeta and Donald Barker Foundation Grants, 1594
Commonwealth Fund Small Grants, 1611
Communities Foundation of Texas Grants, 1613
Cowles Charitable Trust Grants, 1652
Cystic Fibrosis Canada Fellowships, 1677
Davis Family Foundation Grants, 1726
Denton A. Cooley Foundation Grants, 1743
Detroit Lions Charities Grants, 1753
Edwin S. Webster Foundation Grants, 1900
Elisabeth Severance Prentiss Foundation Grants, 1905
Eugene McDermott Foundation Grants, 1958
Fannie E. Rippel Foundation Grants, 1978
FMC Foundation Grants, 2029
Ford Motor Company Fund Grants Program, 2033
Frances L. and Edwin L. Cummings Memorial Fund Grants, 2046
Fritz B. Burns Foundation Grants, 2057
George B. Storer Foundation Grants, 2113
George Kress Foundation Grants, 2119
George S. and Dolores Dore Eccles Grants, 2120
George W. Codrington Charitable Grants, 2122
Gilroy and Lillian P. Roberts Charitable Foundation Grants, 2142
Greenwall Foundation Bioethics Grants, 2174
Harold Alfond Foundation Grants, 2200
Harvey Randall Wickes Foundation Grants, 2218
Hearst Foundations Health Grants, 2225
Hilda and Preston Davis Foundation Grants, 2266
Hospital Libraries Section / MLA Professional Development Grants, 2283
Huie-Dellmon Trust Grants, 2295
Jacob and Hilda Blaustein Foundation Grants, 2454
Jacob and Valeria Langeloth Foundation Grants, 2455
James H. Cummings Foundation Grants, 2460
John S. Dunn Research Foundation Grants, 2506
Joseph Alexander Foundation Grants, 2513
Josiah W. and Bessie H. Kline Foundation Grants, 2517
Lillian S. Wells Foundation Grants, 2614
Lucile Horton Howe and Mitchell B. Howe Foundation Grants, 2624
Lumpkin Family Found Healthy People Grants, 2629
Mabel Y. Hughes Charitable Trust Grants, 2647
MacDonald-Peterson Foundation Grants, 2649
Margaret & James A. Elkins Jr. Found Grants, 2658
Mary S. and David C. Corbin Foundation Grants, 2675
Matilda R. Wilson Fund Grants, 2684
Max and Victoria Dreyfus Foundation Grants, 2687
Maximilian E. and Marion O. Hoffman Foundation Grants, 2688
McGregor Fund Grants, 2725
McLean Contributionship Grants, 2727
Meadows Foundation Grants, 2736
Mericos Foundation Grants, 2742
Meyer and Stephanie Eglin Foundation Grants, 2747
Norcliffe Foundation Grants, 3415
Notsew Orm Sands Foundation Grants, 3434
Olin Corporation Charitable Trust Grants, 3739
Oppenstein Brothers Foundation Grants, 3792
Peacock Foundation Grants, 3887
Pott Foundation Grants, 3984
R.C. Baker Foundation Grants, 4006
Rayonier Foundation Grants, 4034
Reinberger Foundation Grants, 4051
Richard and Susan Smith Family Found Grants, 4064
Rich Foundation Grants, 4068
Robert E. and Evelyn McKee Foundation Grants, 4074
Rockwell International Corporate Trust Grants, 4093
Salem Foundation Grants, 4160
Sara Elizabeth O'Brien Trust Grants, 4166
Sensient Technologies Foundation Grants, 4199
Shell Oil Company Foundation Grants, 4233
Sid W. Richardson Foundation Grants, 4253
Starr Foundation Grants, 4447
Strake Foundation Grants, 4452
Thomas and Dorothy Leavey Foundation Grants, 4506
Thomas Austin Finch, Sr. Foundation Grants, 4507
Thompson Charitable Foundation Grants, 4511
Thoracic Foundation Grants, 4513
Victor E. Speas Foundation Grants, 4810
W.C. Griffith Foundation Grants, 4819
W. C. Griffith Foundation Grants, 4818

Hotel and Restaurant Management
College of Human Sciences Scholarships, 1596
Coors Brewing Corporate Contributions Grants, 1644
Phi Upsilon Omicron Florence Fallgatter Distinguished Service Award, 3943

Housing
Adaptec Foundation Grants, 340
Ann Arbor Area Community Foundation Grants, 860
Ann Peppers Foundation Grants, 863
Arizona Cardinals Grants, 995
Batchelor Foundation Grants, 1305
Boston Foundation Grants, 1361
Byron W. and Alice L. Lockwood Grants, 1401
Campbell Soup Foundation Grants, 1415
Carl and Eloise Pohlad Family Grants, 1426
Carpenter Foundation Grants, 1439
Chautauqua Region Community Found Grants, 1533
Chicago Community Trust Housing Grants: Advancing Affordable Rental Housing, 1547
Community Found AIDS Endowment Awards, 1614
Enterprise Community Partners Rose Architectural Fellowships, 1929
Evanston Community Foundation Grants, 1962
Fannie Mae Foundation Kennedy School of Government Fellowships, 1979
Frank Stanley Beveridge Foundation Grants, 2051
George A Ohl Jr. Foundation Grants, 2112
George Gund Foundation Grants, 2117
HomeBanc Foundation Grants, 2275
HUD Community Outreach Partnership Centers Grants, 2292
HUD Urban Scholars Postdoctoral Fellowships, 2293
Hutchinson Community Foundation Grants, 2310
Jackson Foundation Grants, 2453
Luther I. Replogle Foundation Grants, 2631
Mary S. and David C. Corbin Foundation Grants, 2675
McGregor Fund Grants, 2725
Mericos Foundation Grants, 2742
Musgrave Foundation Grants, 2807
NIST Small Business Innovation Res Grants, 3273
Northwestern Mutual Foundation Grants, 3430
NSF Partnership for Advancing Technologies in Housing Grants, 3609
Olin Corporation Charitable Trust Grants, 3739
Patrick and Anna M. Cudahy Fund Grants, 3873
Paul G. Allen Family Foundation Grants, 3876
Peacock Foundation Grants, 3887
Perpetual Trust for Charitable Giving Grants, 3900
Pinkerton Foundation Grants, 3967
Reinberger Foundation Grants, 4051
Reuter Foundation Grants, 4059
Rhode Island Foundation Grants, 4061
Schramm Foundation Grants, 4188
Sid W. Richardson Foundation Grants, 4253
Starr Foundation Grants, 4447
Tessie Agan Award Competition, 4500
Textron Corporate Contributions Grants, 4501
Thompson Charitable Foundation Grants, 4511
Vancouver Foundation Grants and Community Initiatives, 4807

Human Connectome
NIH Human Connectome Project Grants, 3220

Human Development
Chapin Hall International Fellowships in Children's Policy Research, 1517
College of Human Sciences Scholarships, 1596

Grammy Foundation Grants, 2161
Kellogg Institute for International Studies Collaborative Faculty Grants, 2536
Kellogg Institute for International Studies Grants for Events, 2545
Kellogg Institute for International Studies Intramural Grants, 2546
Kellogg Institute for International Studies Kroc Undergraduate Research Grants, 2547
NIH Research on Sleep and Sleep Disorders, 3236
Spencer Foundation Education Dissertation Fellowships, 4334
USDD Human Social, Cultural and Behavior Modeling Thrust Grants, 4737
Wellesley Centers for Women Postdoc Fellows, 4845

Human Factors in Engineering
Aviation Research Grants, 1273
NLM Research Project Grants, 3306
NSF Thermal Transport Processes Grants, 3660
ORISE Nuclear Regulatory Commission Historically Black Colleges and Universities Student Research Participation, 3814

Human Genome
AACR Team Science Award, 148
ASGH Award for Excellence in Human Genetics Education, 1066
ASGH C.W. Cotterman Award, 1067
ASGH Charles J. Epstein Trainee Awards for Excellence in Human Genetics Research, 1068
ASGH McKusick Leadership Award, 1069
ASGH William Allan Award, 1070
ASGH William Curt Stern Award, 1071
Gruber Foundation Gentics Prize, 2178
NHGRI Mentored Research Scientist Development Award, 3013
NIH Human Connectome Project Grants, 3220

Human Learning and Memory
AASL Research Grant, 207
ASTD Research Award, 1226
ATA Education Research Award, 1254
Center for Judaic Studies Fellowships, 1478
Coca-Cola Foundation Grants, 1593
ETS Sylvia Taylor Johnson Minority Fellowship in Educational Measurement, 1955
Helen Bader Foundation Grants, 2228
NSF Perception, Action & Cognition Grants, 3613
RCPSC Royal College Fellowship for Studies in Medical Education, 4040
SfN Swartz Prize for Theoretical and Computational Neuroscience, 4213
Spencer Foundation Education Dissertation Fellowships, 4334
Wallace Foundation Grants, 4829

Human Population Genetics
ASGH C.W. Cotterman Award, 1067
ASGH Charles J. Epstein Trainee Awards for Excellence in Human Genetics Research, 1068
Gruber Foundation Rosalind Franklin Young Investigator Award, 2183

Human Reproduction/Fertility
Donald and Sylvia Robinson Family Foundation Grants, 1830
General Service Foundation Human Rights and Economic Justice Grants, 2108
Jessie Smith Noyes Foundation Grants, 2485
Overbrook Foundation Grants, 3860
Playboy Foundation Grants, 3980
Robert Sterling Clark Foundation Reproductive Rights and Health Grants, 4081
USAID Family Planning and Reproductive Health Methods Grants, 4598
W.C. Griffith Foundation Grants, 4819

Human Resources
Bill and Melinda Gates Foundation Financial Services for the Poor Grants, 1346
Bill and Melinda Gates Foundation Water, Sanitation and Hygiene Grants, 1347
Mayo Clinic Administrative Fellowship, 2689
Mayo Clinic Admin Fellowship Eau Claire, 2690
SHRM Dissertation Awards, 4248
SHRM Foundation Research Grants, 4249
SHRM Michael R. Losey Human Resource Research Award, 4250
USDD Human Performance Training and Education Grants, 4736

Human Services
3M Foundation Community Giving Grants, 5
Abbott Fund Global AIDS Care Grants, 219
A Charitable Foundation Grants, 273
Anderson Foundation Grants, 842
Assisi Foundation of Memphis General Grants, 1221
Batchelor Foundation Grants, 1305
Better Way Foundation Grants, 1338
Boeing Company Contributions Grants, 1357
Caesars Foundation Grants, 1407
Charles M & Mary D Grant Foundation Grants, 1527
Colonel Stanley R. McNeil Foundation Grants, 1597
DHHS Emerging Leaders Program Internships, 1756
Drs. Bruce and Lee Foundation Grants, 1859
DTE Energy Health and Human Services Grants, 1865
Frederick Douglass Summer Scholars Fellowship, 2053
Genuardi Family Foundation Grants, 2109
George Kress Foundation Grants, 2119
Gil and Dody Weaver Foundation Grants, 2139
Green Foundation Human Services Grants, 2172
Harold R. Bechtel Testamentary Trust Grants, 2202
Harry B. and Jane H. Brock Foundation Grants, 2206
Harvey Randall Wickes Foundation Grants, 2218
Herbert & Adrian Woods Foundation Grants, 2239
Hilda and Preston Davis Foundation Grants, 2266
Hill Crest Foundation Grants, 2267
Hirtzel Memorial Foundation Grants, 2269
Human Source Foundation Grants, 2297
Kovler Family Foundation Grants, 2575
M.D. Anderson Foundation Grants, 2637
MacDonald-Peterson Foundation Grants, 2649
Marion Gardner Jackson Charitable Trust Grants, 2664
McCune Foundation Human Services Grants, 2722
McGraw-Hill Companies Community Grants, 2724
Meyer Memorial Trust Responsive Grants, 2748
NCFR Affiliate Councils Outstanding Student Research Papers, 2906
Northrop Grumman Corporation Grants, 3428
Notsew Orm Sands Foundation Grants, 3434
Oppenstein Brothers Foundation Grants, 3792
Oscar Rennebohm Foundation Grants, 3832
Paul G. Allen Family Foundation Grants, 3876
Perpetual Trust for Charitable Giving Grants, 3900
Reinberger Foundation Grants, 4051
Rohm and Haas Company Grants, 4098
Sarkeys Foundation Grants, 4181
Shell Oil Company Foundation Grants, 4233
Sioux Falls Area Community Foundation Community Fund Grants (Unrestricted), 4262
Sioux Falls Area Comm Foundation Spot Grants, 4263
Sony Corporation of America Corporate Philanthropy Grants, 4329
Textron Corporate Contributions Grants, 4501
Union Pacific Foundation Health and Human Services Grants, 4569
Victor E. Speas Foundation Grants, 4810
Willary Foundation Grants, 4870

Human Subjects Policy
Alternatives Research and Development Foundation Alternatives in Education Grants, 581
Alternatives Research and Development Foundation Grants, 582
NIH Research On Ethical Issues In Human Subjects Research, 3235
ONF Small Grants Program, 3743

Humanities
AAR Rome Prize Fellowships, 202
AAS Kate B. and Hall J. Peterson Fellowships, 204
AAS Legacy Fellowships, 205
AAS Long-Term Visiting Academic Research Fellowships, 206
AAS Short-Term Visiting Academic Research Fellowships, 208
AAUW American Dissertation Fellowships, 209
AAUW American Postdoctoral Research Leave Fellowships, 210
AAUW American Summer/Short-Term Research Publication Grants, 211
ACLS/New York Public Library Fellowships, 277
ACLS African Humanities Program for Dissertation Completion, 278
ACLS American Research in the Humanities in China Fellowships, 280
ACLS Charles A. Ryskamp Research Fellowships, 281
ACLS Collaborative Research Fellowships, 282
ACLS Digital Innovation Fellowships, 284
ACLS Fellowships, 287
ACLS Frederick Burkhardt Residential Fellowships for Recently Tenured Scholars, 288
ACLS International and Area Studies Fellowships, 289
AIAR/Council of American Overseas Research Centers Fellowships for Advanced Multi-Country Research, 391
AIAR Andrew W. Mellon Foundation Fellowships, 392
AIAR Annual Professorships, 393
Aid to Scholarly Publications, 431
Alabama Humanities Foundation Mini Grants, 480
Alabama Humanities Foundation Planning/Consultant Grants, 481
Alexander von Humboldt Foundation Anneliese Maier Award, 515
Alexander von Humboldt Foundation Fellowships for Experienced Researchers, 516
Alexander von Humboldt Foundation Fellowships for Postdoctoral Researchers, 517
Alexander von Humboldt Feodor-Lynen Research Fellowships for Experienced Researchers, 518
Alexander von Humboldt Feodor Lynen Research Fellowships for Experienced Researchers, 520
Alexander von Humboldt Foundation Georg Forster Fellowships for Experienced Researchers, 523
Alexander von Humboldt Foundation Georg Forster Research Award, 525
Alexander von Humboldt Foundation JSPS Research Fellowships, 530
Alexander von Humboldt Foundation Max Planck Research Award, 532
Alexander von Humboldt Foundation NSC Taiwan Fellowships, 533
Alexander von Humboldt Foundation Philipp Franz von Siebold Award for Japanese Researchers, 534
Alexander von Humboldt Foundation Reimar Lust Award, 536
Alexander von Humboldt Foundation Research Awards, 537
Alexander von Humboldt Foundation Research Awards for Top Researchers in Germany, 538
Alexander von Humboldt Foundation Sofja Kovalevskaja Awards, 540
Alpha Sigma Nu Book Award Competition, 578
American Academy of Arts and Sciences Hellman Fellowships, 612
American Academy of Arts and Sciences Visiting Scholars, 613
American Academy of Religion Individual Research Grants, 618
American Numismatic Society Doc Fellowships, 759
American Philosophical Society Sabbatical Fellowship in the Humanities and Social Sciences, 771
amfAR Fellowships, 809
amfAR Mathilde Krim Fellowships in Basic Biomedical Research, 810
amfAR Research Grants, 811
Andrew W. Mellon/ACLS Dissertation Completion Fellowships, 844

ARIT/NEH Fellowships for Research in Turkey, 990
ARIT Fellowships, 991
ARIT Mellon Foundation Postdoctoral Research Fellowships for East European Scholars, 994
Arthur J. Ennis Fellowships, 1025
ASCSA Harry Bikakis Fellowship, 1053
ASU Institute for Humanities Res Seed Grants, 1247
Avon Products Foundation Grants, 1275
BA Larger Research Grants, 1290
Banfi Vintners Foundation Grants, 1296
BA Sir Ernest Cassel Educational Trust Grants, 1303
BA Small Research Grants, 1304
BMI Woody Guthrie Fellowships, 1352
Brookdale Foundation Leadership in Aging Fellowships, 1381
Camargo Foundation Fellowships, 1412
Canada-U.S. Fulbright International Science and Technology Award, 1417
Canada-U.S. Fulbright Senior Specialists Program Grants, 1418
Canada-U.S. Fulbright Traditional Scholars, 1419
CCH Documentary Project Research and Development Grants, 1451
CFUW Margaret Dale Philp Award, 1512
Chestnut Hill Charitable Foundation, Inc Grants, 1538
CMU Arts in Society Fellowship, 1587
Community Foundation for S Arizona Grants, 1618
Community Found of E Central Illinois Grants, 1619
Connecticut Community Foundation Grants, 1627
ConocoPhillips Grants Program, 1629
Consortium for Graduate Study in Management Minority Fellowships, 1642
DAAD Research Grants for Doctoral Candidates and Young Academics and Scientists, 1703
DAAD Research Stays for University Academics and Scientists, 1705
DAAD Scholarships for Artists: Study Visits for Academics, 1707
Dartmouth College Andrew W. Mellon Postdoctoral Fellowship, 1721
Douglass Institute Predoctoral Fellowships, 1852
Evan Frankel Foundation Grants, 1961
FASSE Demonstration Project Grant, 1982
Ford Motor Company Fund Grants Program, 2033
Forrest C. Lattner Foundation Grants, 2035
Foundation for the Advancement of Mesoamerican Studies Grants, 2042
Frederick Douglass Summer Fellowships, 2052
Frederick Douglass Summer Scholars Fellowship, 2053
Fulbright Alumni Initiatives Awards Grants, 2064
Fulbright Distinguished Chairs Program Grants, 2069
Fulbright European Union Scholar-in-Residence Program Grants, 2071
Fulbright New Century Scholars Program Grants, 2075
Fulbright Occasional Lecturer Program Grants, 2076
Fulbright Scholar-in-Residence Program Grants, 2078
Fulbright Specialists Program Grants, 2079
Fulbright Traditional Scholar Program in Europe and Eurasia, 2080
Fulbright Traditional Scholar Program in Sub-Saharan Africa, 2081
Fulbright Traditional Scholar Program in the East Asia/Pacific Region, 2082
Fulbright Traditional Scholar Program in the Near East and North Africa Region, 2083
Fulbright Traditional Scholar Program in the South and Central Asia Region, 2084
Fulbright Traditional Scholar Program in the Western Hemisphere, 2085
Fulbright Visiting Specialists Program: Direct Access to the Muslim World, 2087
Fund for Scientific Research--Flanders Grants, 2089
Gaius Charles Bolin Fellowships for Minority Graduate Students, 2095
George Frederick Jewett Foundation Grants, 2116
Gerda Henkel Foundation Grants, 2127
Getty Scholars and Visiting Scholars Grants, 2138
Greenwall Foundation Bioethics Grants, 2174
Harry Ransom Humanities Research Center Research Fellowships, 2209
Haverford College Mellon Postdoc Fellowships, 2220
Holberg Memorial Fund International Prize, 2272
Holberg Memorial Fund Nils Klim Prize, 2273
Holberg Memorial Fund School Project, 2274
Humanities Inst Visiting Scholars Fellowships, 2296
Huntington Library Babara Thom Postdoctoral Fellowships, 2301
Huntington Library Clark-Huntington Joint Bibliographical Fellowship, 2302
Huntington Library Dibner History of Science Long-Term Fellowships, 2304
Huntington Library Dibner History of Science Short-Term Fellowships, 2305
Huntington Library Mellon Fellowships, 2306
Huntington Library National Endowment for the Humanities Fellowships, 2307
Huntington Library Short-Term Fellowships, 2308
IES Faculty Fellowships, 2341
IFUW Winifred Cullis Grants, 2347
IIE Leonora Lindsley Memorial Fellowships, 2354
IIE Rockefeller Foundation Bellagio Center Residencies, 2355
Inamori Foundation Research Grants, 2368
Institute of Turkish Studies Dissertation Writing Grants, 2389
Institute of Turkish Studies Research Grants in Comparative Studies of Modern Turkey, 2390
IREX Eurasian Undergraduate Exchange Program Grants, 2418
IREX Individual Advanced Research Opportunities Grants, 2419
IREX Short-Term Travel Grants, 2424
J.B. Reynolds Foundation Grants, 2449
Japan Foundation Japanese Studies Fellowships, 2476
Japan Foundation New York Japanese Studies Fellowships, 2478
Japan Foundation Six-Month Grants for Specialists in Cultural and Academic Fields, 2479
John Simon Guggenheim Memorial Foundation Fellowships, 2507
Kellogg Institute for International Studies Andrónico Luksic Grants, 2535
Kellogg Institute for International Studies Course Development Grants, 2537
Kellogg Institute for International Studies Faculty Fellows' Research Grants, 2540
Kellogg Institute for International Studies Global Collaboration Initiative Grants, 2541
Kellogg Institute for International Studies Kroc Undergraduate Research Grants, 2547
Kluge Center Fellowships, 2561
Lady Gregory Postgraduate Fellowship, 2579
Lawrence C. Gallen Teaching Fellowship, 2590
Library Company of Philadelphia NEH Long-term Post-Doctoral Fellowships, 2607
Library of Congress Fellowships in International Studies, 2612
Massachusetts Foundation for the Humanities Project Grants, 2678
McGregor Fund Grants, 2725
National Security Education Program Grants to Institutions of Higher Education, 2885
NEH Awards for Faculty at Institutions with High Hispanic Enrollment, 2950
NEH Collaborative Research Grants, 2952
NEH Digital Humanities Implementation Grants, 2953
NEH Digital Humanities Start-Up Grants, 2954
NEH Documenting Endangered Lang Grants, 2955
NEH Fellowships, 2956
NEH Fellowships at Independent Research Institutions, 2957
NEH Fellowships for Advanced Social Science Research on Japan, 2958
Nelda C. and H.J. Lutcher Stark Found Grants, 2967
New Hampshire Charitable Foundation Grants, 2985
NHC Summer Institutes in Lit Studies Grants, 3012
NL American Society for Eighteenth-Century Studies Fellowships, 3279
NL Arthur and Lila Weinberg Fellowships for Independent Scholars and Researchers, 3280
NL Associated Colleges of the Midwest/Great Lakes College Association Faculty Fellowships, 3281
NL British Academy Fellowship for Study in the United Kingdom, 3283
NL Frances C. Allen Fellowships, 3284
NL Grad Scholar-in-Residence Opportunities, 3285
NL Herzog August Bibliothek Wolfenbuttel Fellowship, 3286
NL Lawrence Lipking Fellowship at Newberry, 3287
NL Lester J. Cappon Fellowship in Documentary Editing, 3288
NL Lloyd Lewis Fellowships in American Hist, 3289
NL Mellon Postdoctoral Research Fellowships, 3294
NL Midwest Modern Lang Assoc Fellowship, 3299
NL Monticello College Foundation Fellowship for Women, 3304
NL National Endowment for the Humanities Fellowships, 3310
NL Newberry Consortium on American Indian Studies Faculty Fellowships, 3311
NL Newberry Library/Ecole Nationale des Chartes Exchange Fellowship, 3313
NL Northeast Modern Language Assocation Fellowship, 3314
NL Scholar-in-Residence Opportunity, 3315
NL Short-Term Fellowships in the History of Cartography, 3316
NL Short-Term Resident Fellowships for Individual Research, 3317
NL South Central Modern Languages Association Fellowship, 3318
NL Susan Kelly Power and Helen Hornbeck Tanner Fellowship, 3319
NL Weiss/Brown Publication Subvention Award, 3320
Northwood University/Alden B. Dow Creativity Center Summer Residency Program, 3432
NYAM Audrey & Will Helfand Fellowship, 3679
NYPL Short-Term Research Fellowships, 3701
OSF Central Asia and Caucasus Research and Training Initiative - Junior Fellowships, 3836
OSF Central Asia and Caucasus Research and Training Initiative - Senior Fellowships, 3837
OSF Global Faculty Grants - Track 2, 3842
OSF Global Supplementary Grants, 3843
OSF HESP Mobility Grants, 3844
PHF Andrew W. Mellon Postdoctoral Fellowships in the Humanities, 3931
PHF Penn Faculty Research Fellowships, 3932
PHF Regional Mellon Faculty Research and Cultural Fellowships, 3933
PHF Undergraduate Humanities Forum Research Fellowships, 3934
Princeton University Lewis Center for the Arts Hodder Fellowships, 3996
Princeton Univ Whiting Postdoc Fellowships, 3998
Rajiv Gandhi Foundation Grants, 4025
Rockefeller Foundation Humanities Fellowships, 4089
Rockefeller Foundation Scholarly Residencies, 4091
RSC Pierre Chauveau Medal, 4123
Ruth and Vernon Taylor Foundation Grants, 4130
Schomburg Center Scholars-in-Residence Fellows, 4187
Schramm Foundation Grants, 4188
Shastri Indo-Canadian Institute Faculty Research Fellowships, 4222
Smithsonian Graduate Student Fellowships, 4277
Smithsonian Minority Visiting Student Awards, 4279
Smithsonian Native American Community Scholars Awards, 4291
Smithsonian Native American Visiting Student Awards, 4292
Smithsonian Predoctoral Fellowships, 4295
SRC Humanities and Social Science Postdoctoral Fellowships, 4372
SRC Humanities and Social Sciences Grants, 4373
SRC Humanities and Social Sciences Grants for Postdoctoral Positions In Sweden, 4374
SSHRC-APC Automotive Partnership Canada, 4396
SSHRC-CIHR-SC Research Data Centres Joint Initiative, 4397
SSHRC-FHSS Awards to Scholarly Publications, 4398

SSHRC-IDRC International Partnerships for Sustainable Societies, 4399
SSHRC-NSERC College and Community Innovation Enhancement Grants, 4400
SSHRC-Public Safety Canada Kanishka Project Initiative, 4401
SSHRC Aid to Scholarly Journals, 4402
SSHRC Aid to Small Universities, 4403
SSHRC Banting Postdoctoral Fellowships, 4404
SSHRC Bora Laskin National Fellowship in Human Rights Research, 4405
SSHRC Digging into Data Challenge Grants, 4406
SSHRC Doctoral Fellowships, 4407
SSHRC Graduate Scholarships-Michael Smith Foreign Study Supplements, 4408
SSHRC Insight Grants, 4410
SSHRC Institutional Grants, 4411
SSHRC Joseph-Armand Bombardier CGS Doctoral Scholarships, 4412
SSHRC Jules and Gabrielle Leger Fellowship, 4413
SSHRC Postdoctoral Fellowships, 4415
SSHRC Therese F. Casgrain Fellowship, 4418
SSRC Dissertation Proposal Development Faculty Field Fellowships, 4424
SSRC Dissertation Proposal Development Student Fellowships, 4425
SSRC Eurasia Program Dissertation Development Awards, 4427
SSRC Int'l Dissertation Research Fellowships, 4429
SSRC Japan Society for the Promotion of Science Fellowships, 4431
SSRC Our Shared Past Grants, 4437
SSRC Postdoc Fellowship for Transregional Research: Inter-Asian Contexts and Connections, 4438
Stanford Humanities Center Arts Writer/Practitioner Fellowship, 4444
Stanford Humanities Center External Faculty Fellowships, 4445
Swedish Natural Science Res Council Grants, 4479
Tanner Humanities Center Research Fellowships, 4494
U.S. Department of Education Jacob K. Javits Fellowships, 4541
UIC Rockefeller Foundation Resident Fellowships in the Humanities, 4559
Virginia Foundation for the Humanities Discretionary Grants, 4813
Virginia Foundation for the Humanities Residential Fellowships, 4815
Wolfsonian-FIU Fellowships, 4890
Woodrow Wilson Doctoral Dissertation Fellowships in Women's Studies, 4891

Humanities Education
AAS Kate B. and Hall J. Peterson Fellowships, 204
AAS Legacy Fellowships, 205
AAS Long-Term Visiting Academic Research Fellowships, 206
AAS Short-Term Visiting Academic Research Fellowships, 208
ACLS African Humanities Program in Ghana, Nigeria, South Africa, Tanzania, and Uganda, 279
ACLS Comparative Perspectives on Chinese Culture and Society Grants, 283
ACLS Dissertation Fellowships in East European Studies, 285
ACLS Early Career Postdoctoral Fellowships in East European Studies, 286
Andrew W. Mellon Postdoctoral Fellowships at UCLA Humanities Consortium, 846
Arthur J. Ennis Fellowships, 1025
ASU Graduate College Completion Fellowships, 1235
ASU Graduate College Dissertation Fellowships, 1237
ASU Inst for Humanities Research Fellowships, 1246
ASU Institute for Humanities Res Seed Grants, 1247
ASU Institute for Humanities Research Visiting Fellowships, 1248
Chatlos Foundation Grants Program, 1532
Erasmus Institute Fellowships, 1944
EU Center for Humanistic Inquiry Junior and Postdoctoral Fellowships, 1956
Fulbright-Hays Seminars Abroad Bilateral Projects Grants, 2062
Getty Scholars and Visiting Scholars Grants, 2138
Graduate Institute Hodson Teacher Fellowships, 2160
Harry Frank Guggenheim Foundation Research Grants, 2208
Harry Ransom Humanities Research Center Research Fellowships, 2209
Haverford College Mellon Postdoc Fellowships, 2220
Humanities Inst Visiting Scholars Fellowships, 2296
Huntington Library Babara Thom Postdoctoral Fellowships, 2301
Huntington Library Clark-Huntington Joint Bibliographical Fellowship, 2302
Huntington Library Dibner History of Science Long-Term Fellowships, 2304
Huntington Library Dibner History of Science Short-Term Fellowships, 2305
Huntington Library Mellon Fellowships, 2306
Huntington Library Short-Term Fellowships, 2308
Institute of Turkish Studies Research Grants in Comparative Studies of Modern Turkey, 2390
Kellogg Institute for International Studies Andrónico Luksic Grants, 2535
Kellogg Institute for International Studies Course Development Grants, 2537
Kellogg Institute for International Studies Global Collaboration Initiative Grants, 2541
Kellogg Institute for International Studies Kroc Undergraduate Research Grants, 2547
Kluge Center Fellowships, 2561
Lawrence C. Gallen Teaching Fellowship, 2590
Library Company of Philadelphia NEH Long-term Post-Doctoral Fellowships, 2607
Mary Gates Research Scholarships, 2670
NEH Awards for Faculty at Historically Black Colleges and Universities, 2949
NEH Awards for Faculty at Tribal Colleges and Universities, 2951
NEH Digital Humanities Implementation Grants, 2953
NEH Digital Humanities Start-Up Grants, 2954
NEH Documenting Endangered Lang Grants, 2955
NEH Fellowships, 2956
NEH Fellowships for Advanced Social Science Research on Japan, 2958
NHC Summer Institutes in Lit Studies Grants, 3012
NL Mellon Postdoctoral Research Fellowships, 3294
NL National Endowment for the Humanities Fellowships, 3310
Northwood University/Alden B. Dow Creativity Center Summer Residency Program, 3432
OSF Global Faculty Grants - Track 2, 3842
PHF Andrew W. Mellon Postdoctoral Fellowships in the Humanities, 3931
PHF Penn Faculty Research Fellowships, 3932
PHF Regional Mellon Faculty Research and Cultural Fellowships, 3933
PHF Undergraduate Humanities Forum Research Fellowships, 3934
Princeton Univ Whiting Postdoc Fellowships, 3998
Schomburg Center Scholars-in-Residence Fellows, 4187
SRC Humanities and Social Science Postdoctoral Fellowships, 4372
SRC Humanities and Social Sciences Grants for Postdoctoral Positions In Sweden, 4374
SSHRC Banting Postdoctoral Fellowships, 4404
SSHRC Digging into Data Challenge Grants, 4406
UCR Grad Student in Humanities Fellowship, 4550
UCR Humanities Grad Student Research Grants, 4551
UCR Resident Fellowships, 4552
UIC Rockefeller Foundation Resident Fellowships in the Humanities, 4559
Virginia Foundation for the Humanities Discretionary Grants, 4813
Virginia Foundation for Humanities Open Grants, 4814
Willamette Univ Minority Grad Fellowship, 4869
Wolfsonian-FIU Fellowships, 4890

Humor
Swann Foundation Fellowships, 4478

Hunger
Bill and Melinda Gates Foundation Financial Services for the Poor Grants, 1346
Campbell Soup Foundation Grants, 1415
Coors Brewing Corporate Contributions Grants, 1644
Farmers Insurance Corporate Giving Grants, 1981
Green Foundation Human Services Grants, 2172
John Deere Foundation Grants, 2500
Peacock Foundation Grants, 3887
R.C. Baker Foundation Grants, 4006
United Methodist Committee on Relief Global Health Grants, 4577
USDA Agriculture and Food Research Initiative Competitive Grants - Food Security, 4614
USDA Hispanic-Serving Institutions Education Grants, 4628
USDA Scientific Cooperation Research Program, 4647

Huntington's Disease
Hereditary Disease Foundation John J. Wasmuth Postdoctoral Fellowships, 2244
Hereditary Disease Found Lieberman Award, 2245
Hereditary Disease Foundation Research Grants, 2246
Huntington's Disease Society of America Research Fellowships, 2299
Huntington's Disease Society of America Research Grants, 2300

Hurricanes
NOAA Joint Hurricane Testbed Grants, 3378

Hydraulic Engineering
Abdus Salam ICTP Joint PhD Programme in Environmental Fluid Mechanics Grants, 230
ACSE J. Waldo Smith Hydraulic Fellowship, 309
ASCE Arthur S. Tuttle Memorial Scholarships, 1039
ASCE Freeman Fellowships, 1040

Hydrodynamics
DOE Initial H-Prize Competition for Breakthrough Advances in Materials for Hydrogen Storage, 1801
HRF Hudson River Improvement Grants, 2286
Hydrogen Student Design Contest, 2311
ONR Young Investigator Program Grants, 3788
USDD Ship Systems and Engineering Grants, 4792
WHOI Geophysical Fluid Dynamics Fellowships, 4859

Hydrogen and Synthetic Fuels
DOE Initial H-Prize Competition for Breakthrough Advances in Materials for Hydrogen Storage, 1801
Hydrogen Student Design Contest, 2311

Hydrology
AMS/Industry Government Graduate Fellowships, 817
AMS/Industry Minority Scholarships, 818
AMS Grad Fellowships in History of Science, 824
AMS Undergraduate Scholarships, 837
DOI Geological Research and Data Acquisition Grants, 1822
Horton Research Grants, 2282
International Crane Foundation Internships, 2396
NASA Goddard Visiting Fellowships, 2838
NSF Hydrologic Sciences Research Grants, 3578
NSF Sediment Geology & Paleobiology Grants, 3646

Hyperparathyroidism
NIH Receptors and Signaling in Bone in Health and Disease, 3229

Hypertension
NHLBI Research Grants on the Relationship Between Hypertension and Inflammation, 3052

Iceland
American-Scandinavian Foundation Awards for Advanced Study or Research in the U.S., 603
American-Scandinavian Foundation Fellowship for Study in Scandinavia, 604
Bilateral Scholarships, 1344

Ileitis
Crohn's Disease and Colitis Research Fellowships, 1657
Crohn's Disease and Ulcerative Colitis Senior Research Grants, 1658

Immigrants
Boston Foundation Grants, 1361
CCIS Visiting Research Fellowships, 1452
FCD Young Scholars Fellowships, 1985
HIAS Scholarship Awards, 2264
Joint Center for Poverty Research Grants, 2511
Paul & Daisy Soros Fellowships for New Amer, 3875
Rockefeller Foundation Humanities Fellowships, 4089
Rockefeller Foundation Scholarly Residencies, 4091
Russell Sage Foundation Grants, 4129
SSHRC Metropolis Project, 4414

Immigration Law
OSF Soros Justice Advocacy Fellowships, 3850
OSF Soros Justice Media Fellowships, 3851
SSHRC Metropolis Project, 4414

Immune System
ASM Abbott Laboratories Award in Clinical and Diagnostic Immunology, 1101
Lymphatic Education & Research Network Postdoctoral Fellowships, 2633
NHLBI Developmental Origins of Altered Lung Physiology and Immune Function Grants, 3025
NIH Bone and the Hematopoietic and Immune Systems Research Grants, 3208
Roche Organ Transplant Research Found Grants, 4083

Immune System Disorders
Fairlawn Foundation Grants, 1974
Lymphatic Education & Research Network Postdoctoral Fellowships, 2633
NIH Bone and the Hematopoietic and Immune Systems Research Grants, 3208
Scleroderma Established Investigator Grants, 4190

Immunization Programs
Chiron Foundation Community Grants, 1553
New Hampshire Charitable Foundation Grants, 2985

Immunogenetics
National Psoriasis Foundation Research Grants, 2882

Immunology
AAAAI ARTrust Mini Grants for Allied Health, 14
AAAAI ARTrust Mini Grants for Clinical Res, 15
AAAAI ARTrust Mini Grants in Faculty Development, 16
AAAAI Distinguished Scientist Award, 19
AAAAI Distinguished Service Award, 20
AAAAI Fellows-in-Training Abstract Award, 21
AAAAI Fellows-in-Training Grants, 22
AAAAI Fellows-in-Training Travel Grants, 23
AAO-HNSF Health Services Research Grants, 187
AAO-HNSF Maureen Hannley Research Training Awards, 188
AAO-HNSF Percy Memorial Research Award, 189
AAO-HNSF Rande H. Lazar Health Services Research Grant, 190
AAO-HNSF Resident Research Awards, 191
AAOA Foundation/AAO-HNSF Combined Research Grants, 194
AFAR Research Grants, 354
AIHS Alberta/Pfizer Translational Res Grants, 435
ASM Eli Lilly and Company Research Award, 1138
Bristol-Myers Squibb Clinical Outcomes and Research Grants, 1374
Dana Foundation Science and Health Grants, 1716
FDHN Fellowship to Faculty Transition Awards, 1992
Hawaii Community Foundation Health Education and Research Grants, 2221
M.J. Murdock Charitable Trust College Research Program for Life Sciences, 2638
MBL Frederik B. and Betsy G. Bang Summer Fellowships, 2702
National Blood Foundation Research Grants, 2848
NCI Technologies and Software to Support Integrative Cancer Biology Research Grants, 2933
NFID Adv Vaccinology Course Travel Grant, 2986
NIGMS/NIDCR Evolution of Infectious Diseases Research Grants, 3196
NIH Nonhuman Primate Immune Tolerance Cooperative Study Group Grants, 3228
NIH Ruth L. Kirschstein National Research Service Awards for Individual Postdoctoral Fellows, 3239
Rett Syndrome Research Foundation Grants, 4058
Searle Scholars Program Grants, 4193
Streilein Foundation for Ocular Immunology Visiting Professorships, 4453
US CRDF Leishmaniasis: Collaborative Research Opportunities in N Africa & Middle East, 4610

Immunopathology
AAAAI Distinguished Clinician Award, 17
National Psoriasis Foundation Research Grants, 2882

Incarceration
OSF Soros Justice Advocacy Fellowships, 3850
OSF Soros Justice Media Fellowships, 3851

Independent Living Programs
Arizona Public Service Corporate Giving Program Grants, 998
Charles H. Farnsworth Trust Grants, 1521
McLean Contributionship Grants, 2727
Mericos Foundation Grants, 2742
Reinberger Foundation Grants, 4051
Sara Elizabeth O'Brien Trust Grants, 4166
U.S. Department of Education Rehabilitation Research Training Centers, 4545

India
Elizabeth Glaser Int'l Leadership Awards, 1906
Fulbright Traditional Scholar Program in the Near East and North Africa Region, 2083
Indian Council of Social Science Research Grants, 2377
MOEF Gandhi Paryavaran Puraskar Award, 2793
Rajiv Gandhi Foundation Grants, 4025
Shastri Indo-Canadian Institute Arts Fellowships, 4219
Shastri Indo-Canadian Institute Collaborative Research Project Grants, 4220
Shastri Indo-Canadian Institute Doctoral Student Research Fellowships, 4221
Shastri Indo-Canadian Institute Faculty Research Fellowships, 4222
Shastri Indo-Canadian Institute Faculty Training Fellowships, 4223
Shastri Indo-Canadian Institute Graduate Degree Fellowships from an Indian Institution, 4224
Shastri Indo-Canadian Inst Librarian Fellows, 4225
Shastri Indo-Canadian Institute Master's Student Research Fellowships, 4226
Shastri Indo-Canadian Institute Millennium Development Research Grants, 4227
Shastri Indo-Canadian Institute Post-Doctoral Research Fellowships, 4229
Shastri Indo-Canadian Institute Research Collaboration Development Grants, 4230
Shastri Indo-Canadian Institute Student Language Training Fellowships, 4231
Shastri Indo-Canadian Institute University of Ottawa Overseas India Summer Course Grants, 4232
Sir Dorabji Tata Trust Individual Education and Travel Grants, 4265
USAID Strengthening RMNCH Through Indian Health Professional Associations and Scaling Up Uptake Grants, 4602

Indian Language/Literature
Shastri Indo-Canadian Institute Collaborative Research Project Grants, 4220
Shastri Indo-Canadian Institute Faculty Research Fellowships, 4222
Shastri Indo-Canadian Institute Faculty Training Fellowships, 4223
Shastri Indo-Canadian Institute Graduate Degree Fellowships from an Indian Institution, 4224
Shastri Indo-Canadian Institute Master's Student Research Fellowships, 4226
Shastri Indo-Canadian Institute Millennium Development Research Grants, 4227
Shastri Indo-Canadian Institute Research Collaboration Development Grants, 4230
Shastri Indo-Canadian Institute Student Language Training Fellowships, 4231

Indigenous Cultures
Bernard Van Leer Foundation Grants, 1336
Honor the Earth Grants, 2279
NL Newberry Consortium on American Indian Studies Faculty Fellowships, 3311
North Carolina Arts Council Folklife Public Program Grants, 3419
Pacific Islanders in Communications Grants, 3861

Industrial Design
ASME Barnett-Uzgiris Product Safety Award, 1106
Wolfsonian-FIU Fellowships, 4890

Industrial Engineering
AOCS Industrial Oil Products Division Student Award, 885
ASME Ralph Coats Roe Medal, 1158
ASME Worcester Reed Warner Medal, 1172
Canada Graduate Scholarships and NSERC Postgraduate Scholarships, 1420
Gilbreth Memorial Fellowships, 2141
Mayo Clinic Business Consulting Fellowship, 2691
NAPM Doctoral Dissertation Grant, 2827
NSF-NIST Interaction in Chemistry, Materials Research, Molecular Biosciences, Bioengineering, and Chemical Engineering, 3453
NSF Engineering Research Grants, 3562
NSF Industry/University Cooperative Research Centers Grants, 3579
NSF Thermal Transport Processes Grants, 3660

Industrial Hygiene
ORISE Faculty Internship at the US Army Center for Health Promotion and Preventive Medicine, 3802
ORISE Postgraduate Internship at the US Army Center for Health Promotion and Preventive Med, 3818

Industrial Relations
SSRC Abe Fellowships, 4421

Industry
ALZA Corporate Contributions Grants, 588
APS Prize for Industrial Applications of Physics, 978
AT&T Industrial Ecology Faculty Fellowships, 1251
CHF Glenn E. and Barbara Hodsdon Ullyot Scholarship, 1540
CHF Long-Term Fellowships, 1541
CHF Robert W. Gore Fellowship in Materials Innovation, 1542
CHF Societe de Chimie Industrielle (American Section) Fellowship, 1544
CHF Travel Grants, 1545
DOE Competitive Financial Assistance for Renewable Energy and Energy Efficiency Programs, 1779
DOE State Energy Program Special Projects, 1815
Institute Award for Excellence in Industrial Gases Technology, 2379
MAA-Tensor Foundation Women and Mathematics Grants, 2645
Millipore Foundation Grants, 2775
National Endowment for Democracy Discretionary Grants, 2862
NATO Science for Peace Grants, 2898
NCI/NIDA Transdisciplinary Tobacco Use Research Centers Grants, 2912
NIST Small Business Innovation Res Grants, 3273
North Carolina Biotechnology Center Collaborative Funding Grants, 3421
North Carolina Biotech Center Research Grants, 3424

NSERC Industrial R&D Fellowships, 3442
NSERC Industrial Research Chairs, 3443
NSF Grant Opportunities for Academic Liaison with Industry, 3574
NSF Industry/University Cooperative Research Centers Grants, 3579
NSF Innovation and Organizational Change, 3581
NSF Integrative Graduate Education and Research Traineeship, 3584
NWO Added Value Plus for Soc Sciences Grants, 3672
RSC Thomas W. Eadie Medal, 4126
SRC Landmark Innovation Award, 4376
USDA Small Business Innovation Res Grants, 4648
USDC/NOAA Nature Conservancy Community-based Matching Grants, 4661

Infant Psychology
Bunting Fellowship Program, 1385

Infants
AAP Nutrition Award, 198
Chautauqua Region Community Found Grants, 1533
CIGNA Foundation Grants, 1568
Foundation for Seacoast Health Grants, 2041
Gerber Foundation Grants, 2126
NIH Research on Sleep and Sleep Disorders, 3236
Sigma Xi Grants in Aid of Research, 4259
USAID Comprehensive District-Based Support for Better HIV/TB Patient Outcomes Grants, 4595

Infectious Diseases/Agents
ACCP Infectious Diseases PRN Minisabbatical Program, 262
ACCP Infectious Diseases Research Fellowships, 263
ACCP Oncology Research Fellowships, 264
APHL Emerging Infectious Diseases Fellowships, 914
ASM-PAHO Infectious Diseases Epidemiology and Surveillance Fellowships, 1098
ASM/CDC Fellowships in Infectious Disease and Public Health Microbiology, 1100
ASM ICAAC Young Investigator Awards, 1174
ASM Merck Irving S. Sigal Memorial Awards, 1177
ASM Siemens Healthcare Diagnostics Young Investigator Award, 1185
Bristol-Myers Squibb Clinical Outcomes and Research Grants, 1374
CDC Cooperative Agreement for Continuing Enhanced National Surveillance for Prion Diseases in the United States, 1456
CDC Evaluation of the Use of Rapid Testing For Influenza in Outpatient Medical Settings, 1458
CDC Experience Appl Epidemiology Fellowships, 1460
CDC Foundation Atlanta International Health Fellowships, 1461
CDC Preventive Med Residency & Fellowship, 1465
Chiron Foundation Community Grants, 1553
CSTE CDC/CSTE Applied Epidemiology Fellowships, 1664
National Blood Foundation Research Grants, 2848
NFID New Investigator Matching Grants, 2989
NFID Nosocomial Infection Research and Training Postdoctoral Fellowship, 2990
NFID Postdoctoral Fellowships in Emerging Infectious Diseases, 2991
NIAID Independent Scientist Award, 3102
NIGMS/NIDCR Evolution of Infectious Diseases Research Grants, 3196
ORISE Faculty Research Participation at the Centers for Disease Control and Prevention, 3803
ORISE Postgraduate Research Participation at the Centers for Disease Control and Prevention, 3820
ORISE Student Research Participation at the Centers for Disease Control and Prevention, 3829
Pfizer Anti Infective Research EU Grants, 3909
Pfizer ASPIRE EU Antifungal Research Awards, 3910
Pfizer ASPIRE EU Emerging Mechanisms of Resistance Antibacterial Research Awards, 3912
Pfizer ASPIRE EU MRSA Nosocomial Pneumonia & MRSA Complicated Skin & Soft Tissue Infections Antibacterial Research Awards, 3913
Pfizer ASPIRE North America Broad Spectrum Antibiotics for the Treatment of Gram-Negative or Polymicrobial Infections Research Awards, 3914
Pfizer ASPIRE North America Narrow Spectrum Antibiotics for the Treatment of MRSA Research Awards, 3916
Pfizer Medical Education Track Two Grants, 3927
USDA Animal Health & Disease Grants, 4616

Inflammation
AIHS Alberta/Pfizer Translational Res Grants, 435
Broad Foundation IBD Research Grants, 1380
NHLBI Research Grants on the Relationship Between Hypertension and Inflammation, 3052
Pfizer Inflammation Competitive Res Awards, 3926

Informatics
Abdus Salam ICTP Postgraduate Diploma Grants, 235
Abdus Salam ICTP Training and Research in Italian Laboratories Grants, 240
NHLBI Mentored Quantitative Research Career Development Awards, 3038
NLM Limited Competition for Continuation of Biomedical Informatics/Bioinformatics Resource Grants, 3301
PhRMA Foundation Informatics Sabbatical Fellowships, 3956
UNESCO Fellowships, 4567

Information Dissemination
ALA Frederick G. Kilgour Award for Research in Library and Information Technology, 487
ALA Innovation Award, 490
AT&T Industrial Ecology Faculty Fellowships, 1251
Eastern Europe & the Independent States of the Former Soviet Union Research & Training Grants, 1891
Engineering Information Foundation Grants, 1927
ESRI Conservation Program Grants, 1952
Foundation for Seacoast Health Grants, 2041
MLA Donald Lindberg Research Fellowship, 2785
MLA Thomas Reuters / Frank Bradway Rogers Information Advancement Award, 2791
National Security Education Program Grants to Institutions of Higher Education, 2885
NCI Centers of Excellence in Cancer Communications Research, 2921
Nellie Mae Education Foundation State Level Systems Change Grants, 2971
NLM Research Project Grants, 3306
NSF Activities in Science, Engineering, and Mathematics for Persons with Disabilities, 3460
NSF Biological Databases & Informatics Grants, 3499
NSF Communicating Res to Public Audiences, 3538
OSF Fellowships, 3840
Pfizer Medical Education Track Two Grants, 3927
Sigma Theta Tau International / Council for the Advancement of Nursing Science Grants, 4255
SPJ Pulliam/Kilgore Freedom of Information Internships, 4347
STC Research Grants, 4450
U.S. Department of Education Disability and Rehabilitation Research and Related Projects, 4538

Information Science Education
ALA WESS De Gruyter European Librarianship Study Grant, 501
AT&T Bell Labs Graduate Research Program, 1250
Hospital Libraries Section / MLA Professional Development Grants, 2283
MLA David A. Kronick Traveling Fellowship, 2784
MLA Donald Lindberg Research Fellowship, 2785
NSF CISE Advanced Learning Tech Grants, 3522
NSF CISE Educational Innovation Grants, 3533
NSF Innovative Technology Experiences for Students and Teachers, 3582
Roger K. Summit Scholarship, 4097
U.S. Department of Education Graduate Assistance in Areas of National Need, 4540
USDD Research Initiatives at the Naval Postgraduate School Grants, 4783

Information Science/Systems
AASL Research Grant, 207
Abdus Salam ICTP Schools, Conferences, Workshops and Extended Research Workshops Grants, 238
ALA Ingenta Research Awards, 489
ALA Phyllis Dain Library History Dissert Award, 495
ARO Minimum Energy Low Power Electronics Research Contracts, 1011
ARO Quantum Information Science Research Contracts, 1013
ARO Software and Knowledge-Based Systems Research Contracts, 1015
ARO Theoretical Physics and Nonlinear Phenomena Research Contracts, 1017
AT&T Bell Labs Graduate Research Program, 1250
Canada Graduate Scholarships and NSERC Postgraduate Scholarships, 1420
CHF Glenn E. and Barbara Hodsdon Ullyot Scholarship, 1540
CHF Long-Term Fellowships, 1541
CHF Robert W. Gore Fellowship in Materials Innovation, 1542
CHF Societe de Chimie Industrielle (American Section) Fellowship, 1544
CHF Travel Grants, 1545
DaimlerChrysler Corporation Fund Grants, 1714
EPA Pollution Prevention Grants, 1936
Frances C. and William P. Smallwood Foundation Grants, 2045
GNOF Gulf Coast Oil Spill Grants, 2153
Hospital Libraries Section / MLA Professional Development Grants, 2283
IREX Edmund S. Muskie Graduate Fellowship Program, 2417
Mayo Clinic Administrative Fellowship, 2689
McLean Contributionship Grants, 2727
MLA Cunningham Memorial International Fellowship, 2783
MLA David A. Kronick Traveling Fellowship, 2784
MLA Donald Lindberg Research Fellowship, 2785
MLA Graduate Scholarship, 2786
MLA Grad Scholarship for Minority Students, 2787
MLA Research, Development, and Demonstration Project Grant, 2789
MLA Thomas Reuters / Frank Bradway Rogers Information Advancement Award, 2791
NASA Goddard Visiting Fellowships, 2838
NIGMS Predoctoral Training in Bioinformatics and Computation Biology Grants, 3199
NLM Internet Connection for Medical Institutions Grants, 3300
NLM Research Project Grants, 3306
NSF Biological Databases and Informatics Program Grants, 3499
NSF Biological Information Tech & Systems, 3501
NSF CISE Advanced Learning Tech Grants, 3522
NSF CISE Computing Res Infrastructure Grants, 3531
NSF CreativeIT Grants, 3547
NSF Faculty Early Career Development Grants, 3565
NSF Integrative Graduate Education and Research Traineeship, 3584
NSF Physics at the Information Frontier Grants, 3618
ONR Computational Neuroscience & Biorobotics Grants, 3752
ONR Electronics Division Research Grants, 3759
ONR Mathematical, Computer, and Information Sciences Research Grants, 3771
ORISE Environmental Management Participation for the US Army Environmental Center, 3801
ORISE Postgraduate Environmental Management Research Participation at the US Army Environmental Center, 3816
ORISE Student Environmental Management Participation Program for the US Army Environmental Center, 3826
Scientific Communities Initiative Grants, 4189
SFI Principal Investigator Program Grant, 4201
SSHRC Insight Development Grants, 4409
SSHRC Insight Grants, 4410
Thomson Reuters / MLA Doctoral Fellowships, 4512

USDD Research Interest of the Air Force Office of Scientific Research Grants, 4784
Xerox Foundation Grants, 4899

Information Technology
ALA Frederick G. Kilgour Award for Research in Library and Information Technology, 487
ALA Innovation Award, 490
Alfred P. Sloan Scholarly Communication Grants, 561
ASU Graduate College Science Foundation Arizona Fellowships in Science and Engineering, 1245
Frederick Douglass Summer Scholars Fellowship, 2053
Hospital Libraries Section / MLA Professional Development Grants, 2283
MLA Donald Lindberg Research Fellowship, 2785
MLA Research, Development, and Demonstration Project Grant, 2789
Stanford Humanities Center Arts Writer/Practitioner Fellowship, 4444
Stanford Humanities Center External Faculty Fellowships, 4445
USDD Broad Agency Announcement Research and Development Grants, 4703
USDD Clinical Functional Assessment Investigator-Initiated Grants, 4709
USDD Communications and Networking Discovery and Invention Grants, 4712
USDD Crowd Sourced Formal Verif Grants, 4715
USDD Electronic Warefare Grants, 4723
USDD HIV/AIDS Prevention Program Information Systems Development Grants, 4734
USDD Information Innovation Office Grants, 4740

Information Theory
ARO Information Processing and Fusion and Circuits Research Contracts, 1010
Hospital Libraries Section / MLA Professional Development Grants, 2283
MLA David A. Kronick Traveling Fellowship, 2784
MLA Donald Lindberg Research Fellowship, 2785
NCI Centers of Excellence in Cancer Communications Research, 2921
NSF Decision, Risk, and Management Science Research Grants, 3550

Infrared Technology
NSF Advanced Technologies and Instrumentation Grants, 3463
USDD Sensor Innovative Research Grants, 4791

Infrastructure
USAID Palestinian Community Infrastructure Development Grants, 4600

Injury
CDC Experience Appl Epidemiology Fellowships, 1460
CDC Grants for Violence-Related Injury Prevention Research, 1463
CDC National Training Information and Exchange Program, 1464
CSTE CDC/CSTE Applied Epidemiology Fellowships, 1664
USDD Workpl Violence in the Military Grants, 4805

Injury, Head
AHNS/AAO-HNSF Surgeon Scientist Combined Award, 380
AHNS/AAO-HNSF Young Investigator Combined Award, 381
AHNS Alando J. Ballantyne Resident Research Pilot Grant, 382
AHNS Pilot Grant, 383
NFL Charities Medical Grants, 2994
USDD Workpl Violence in the Military Grants, 4805

Injury, Spinal Cord
Detroit Lions Charities Grants, 1753
Gruber Foundation Neuroscience Prize, 2181
NFL Charities Medical Grants, 2994
PVA Research Foundation Grants, 4004

Inner Cities
Arie and Ida Crown Memorial Grants, 989
Detroit Lions Charities Grants, 1753
Luther I. Replogle Foundation Grants, 2631
Millipore Foundation Grants, 2775
OSF-Baltimore Community Fellowships, 3833

Instruction/Curriculum Development
ASHG Public Health Genetics Fellowship, 1086
ATA Educational Trust Grants - Curriculum Resource or Research Project, 1253
Honeywell Corp Habitat & Conservation Grants, 2277
ICC Faculty Fellowships, 2333
ICC Scholarship of Engagement Faculty Grants, 2334
IRA Esther Zolt Teacher Research Award, 2408
Joyce Foundation Education Grants, 2519
Kellogg Institute for International Studies Course Development Grants, 2537
NACC William Diaz Fellowships, 2813
NAEF Teacher Incentive Grants, 2815
OSF Global Faculty Grants - Track 2, 3842
Parkersburg Area Community Foundation Arts in Education Special Grants, 3869
USDA 1890 Institution Teaching, Research, and Extension Capacity Building Grants, 4613
USDA Distance Education Grants for Higher Education Institutions in Insular Areas, 4622
USDA International Science and Education Competitive Grants, 4631

Instructional Materials and Practices
American Historical Association Morris D. Forkosch Prize, 739
ATA Education Research Award, 1254
Dept of Ed International Research and Studies Project Grants, 1745
Dept of Ed Voc Education National Programs, 1750
Dept of Ed Women's Educational Equity Grants, 1752
Hatton W. Sumners Foundation for the Study and Teaching of Self Government Grants, 2219
NSF CISE Educational Innovation Grants, 3533
NSF Course, Curriculum, and Laboratory Improvement Research Grants, 3546
USDA Higher Education Challenge Grants, 4627

Instrumentation, Scientific
American Chemical Society ANYL Award in Chemical Instrumentation, 641
Arecibo Observatory Grants, 988
Astronomical Society of the Pacific Maria and Eric Muhlmann Award, 1231
NSF Advanced Technologies and Instrumentation Grants, 3463
NSF Atmospheric Sciences Mid-Size Infrastructure Opportunity Grants, 3497
NSF Biological Field Stations and Marine Laboratories Grants, 3500
NSF Chemistry Res Instrumentation and Facilities: Departmental Multi-User Instrumentation, 3519
NSF Instrument Devel for Biological Research, 3583
NSF Major Research Instrumentation Grants, 3591
NSF National Astronomy and Ionosphere Center Grants, 3598
NSF Upper Atmospheric Facilities Grants, 3665
SPIE Gold Medal of the Society Award, 4343
USDA Higher Education Challenge Grants, 4627

Insurance/Actuarial Science
Huebner Foundation Fellowships, 2294
Starr Foundation Grants, 4447
State Farm Co Doctoral Dissertation Award, 4448

Integrated Science
NSF Antarctic Integrated System Science Grants, 3476
USDA Integrated Pest Management Competitive Grants - Western Region, 4629
USDA Regional Integrated Pest Management Grants - North Central Region, 4642
USDA Regional Integrated Pest Management Grants - Southern Region, 4644

Intellectual Freedom
AAAS Award for Scientific Freedom and Responsibility, 28
ALA Margaret T. Lane/Virginia F. Saunders Memorial Research Award, 492
MPI for Social Anthropology Max Planck Fellow Research Group Law, Organizations, Science, and Technology, 2799
OSF Intellectual Property Reform Grants, 3846
Playboy Foundation Grants, 3980
SPJ Pulliam/Kilgore Freedom of Information Internships, 4347

Intelligent Systems
USDD Chemical and Biological Defense Innovations and Technologies Grants, 4708
USDD Image Analysis & Understanding Grants, 4738
USDD Intelligence, Surveillance and Reconnaissance Thrust Grants, 4743
USDD Intelligent & Autonomous Syst Grants, 4744
USDD National Consortium for Measurement and Signature Intelligence Research Grants, 4759

Intercultural Studies
BMW of N America Charitable Contributions, 1353
Frederick Douglass Summer Fellowships, 2052
Frederick Douglass Summer Scholars Fellowship, 2053
NYPL Short-Term Research Fellowships, 3701
Saint Mary's College Fellowships, 4159

Interdisciplinary Arts
Arizona Commission on the Arts Folklorist Residencies, 997
ASU Graduate College Dissertation Fellowships, 1237
Delaware Division of the Arts Individual Artist Fellowships, 1732
Florida Division of Cultural Affairs Individual Artist Fellowships, 2026
North Carolina Arts Council Residency Grants, 3420
Stanford Humanities Center Arts Writer/Practitioner Fellowship, 4444

Interfacial Phenomena
NSF Condensed Matter Physics Research Grants, 3540
NSF Interfacial Processes and Thermodynamics Research Grants, 3586

Interior Design
ASID Foundation Grants, 1093
ASID Foundation Irene Winifred Eno Grant, 1094
ASID Foundation Joel Polsky Academic Achievement Award, 1095
ASID Foundation Joel Polsky Prize, 1096
ASID Foundation Mabelle Wilhelmina Boldt Memorial Scholarship, 1097
College of Human Sciences Scholarships, 1596
Wolfsonian-FIU Fellowships, 4890

International Affairs
AAAS Science and Technology Policy Overseas Fellowships at USAID Missions, 44
CFR Edward R. Murrow Press Fellowship, 1500
CFR International Affairs Fellowships, 1501
CFR International Affairs Fellowships in Nuclear Security, 1502
CFR Int'l Affairs Fellowships in South Korea, 1503
CFR International Affairs in Japan Fellowships, 1504
CFR Military Fellowships, 1505
CFR Stanton Nuclear Security Fellowships, 1506
GMFUS Black Sea Trust for Regional Corporation Grants, 2152
Harvard University Weatherhead Center for International Affairs Fellowships, 2215
IIPP Graduate Fellowship, 2357
Kluge Center Kissinger Chair in Foreign Policy and International Relations, 2564
Kluge Center Kissinger Lecturer Award, 2565
Rumsfeld Foundation Graduate Fellowships, 4127
Smith Richardson Foundation Strategy and Policy Fellowship, 4270

Smith Richardson Foundation World Politics and Statecraft Fellowship, 4271
United States Institute of Peace - Transatlantic Post-Doc Fellowship for International Relations and Security, 4581
USAID Devel Innovation Ventures Grants, 4597
USDC Afghanistan and Iraq Investment and Reconstruction Task Forces Internships, 4664

International Agriculture
American Chemical Society AGRO International Award for Research in Agrochemicals, 634
Bill and Melinda Gates Foundation Agricultural Development Grants, 1345
Mellon Post-Doctoral Fellowship in Sustainable Agriculture, 2740
USDA Cochran Fellowships, 4619
USDA International Science and Education Competitive Grants, 4631
USDA Norman E. Borlaug International Agricultural Science and Technology Fellowships, 4636
USDA Scientific Cooperation Research Program, 4647
USDA Technical Agricultural Assistance Research Grants, 4651

International Economics
AAAS Science and Technology Policy Overseas Fellowships at USAID Missions, 44
ConocoPhillips Grants Program, 1629
Helen Kellogg Institute for International Studies Residential Fellowships, 2230
Herman Kahn Fellowship, 2247
International Studies Graduate Fellowships, 2403
Jenifer Altman Foundation Grants, 2484
JMO Strategic and International Studies Grants, 2493
Mellon Post-Doctoral Fellowship in Sustainable Agriculture, 2740
OSF Alumni Grants, 3835
RAND Corporation Transatlantic Postdoctoral Fellowships for Int'l Relations & Security, 4032
Robert Schalkenbach Foundation Grants, 4079
United States Institute of Peace - Transatlantic Post-Doc Fellowship for International Relations and Security, 4581
USDA Norman E. Borlaug International Agricultural Science and Technology Fellowships, 4636
USDA Technical Agricultural Assistance Research Grants, 4651
World Bank Graduate Scholarship Program, 4896

International Education/Training
3M Canada National Teaching Fellowships, 3
AAAAI Fellows-in-Training Grants, 22
AIAR Andrew W. Mellon Foundation Fellowships, 392
Alexander von Humboldt Foundation Anneliese Maier Award, 515
Alexander von Humboldt Foundation Fellowships for Experienced Researchers, 516
Alexander von Humboldt Foundation Fellowships for Postdoctoral Researchers, 517
Alexander von Humboldt Feodor Lynen Research Fellowships for Experienced Researchers, 520
Alexander von Humboldt Foundation German Chancellor Fellowships for Prospective Leaders from the People's Republic of China, 526
Alexander von Humboldt Foundation German Chancellor Fellowships for Prospective Leaders from the Russian Federation, 527
Alexander von Humboldt Foundation International Climate Protection Fellowships, 529
Alexander von Humboldt Foundation Reimar Lust Award, 536
Alexander von Humboldt Foundation Research Fellowships for Postdoctoral Researchers, 539
American-Scandinavian Foundation Fellowship for Study in Scandinavia, 604
American Chemical Society Petroleum Research Fund Undergraduate Faculty Sabbatical Grants, 693
ARIT Fellowships for Intensive Advanced Turkish Language Study, 992
Asia Foundation Exchange Programs, 1090
ASM-PAHO Infectious Diseases Epidemiology and Surveillance Fellowships, 1098
ASM-UNESCO Leadership Grant for International Educators, 1099
ASM International Fellowship for Latin America and the Caribbean, 1175
ASM Int'l Fellowships for Asia and Africa, 1176
ASM Millis-Colwell Postgraduate Travel Grant, 1179
Audio Engineering Society Grad Scholarships, 1265
Australian Academy of Science Grants, 1269
Bill and Melinda Gates Foundation Agricultural Development Grants, 1345
Bogliasco Foundation Liguria Study Center Residencies, 1358
Center for the Study of Philanthropy International Fellowships, 1482
Center for the Study of Philanthropy Senior International Fellowships, 1483
Churchill Fellowships, 1560
CIMO Scholarships for Advanced Studies of the Finnish Language at Finnish Universities, 1569
CIUS Neporany Res & Teaching Fellowship, 1572
Coca-Cola Foundation Grants, 1593
Compton Foundation Grants Program, 1623
Corpus Christi College Research Scholarships, 1647
DAAD Bilateral Programs for Co-Operative Research Grants, 1695
DAAD Intensive Language Courses Grants, 1699
DAAD Re-invitation Program for Former Scholarship Holders, 1701
DAAD Research Grants, 1702
DAAD Research Grants for Doctoral Candidates and Young Academics and Scientists, 1703
DAAD Research Stays for University Academics and Scientists, 1705
DAAD Scholarships for Artists: Study Visits for Academics, 1707
DAAD Study & Internship Program in Germany, 1708
DAAD Undergraduate Scholarships, 1710
DOS Study of the United States Institutes, 1846
Elizabeth Glaser Int'l Leadership Awards, 1906
Ford Foundation International Fellowships, 2030
Fulbright Scholar-in-Residence Program Grants, 2078
GUITS Matching Seed-Money Grants for New Positions in Turkish Studies, 2188
HHMI-NIH Cloister Research Scholars, 2249
HHMI Gilliam Fellowships for Advanced Study, 2254
HHMI Physician-Scientist Early Career Award, 2260
HHMI Undergraduate Science Education Grants for Professors, 2263
IBRO Latin America Regional Funding for Neuroscience Schools, 2327
IERF Sepmeyer Research Grant Program, 2340
IIE David L. Boren Scholarships, 2351
IIE Whitaker International Fellowships and Scholarships, 2356
International Fellowships in Medical Education, 2398
IRA Steven A. Stahl Research Grants, 2413
IREX Edmund S. Muskie Graduate Fellowship Program, 2417
Kennedy Center Summer Fellowships for International Arts Managers, 2553
Kress Foundation Art History Pre-Doctoral Fellowships, 2576
Lady Gregory Postgraduate Fellowship, 2579
Marshall Scholarships, 2666
Mellon Post-Doctoral Fellowship in Sustainable Agriculture, 2740
Nanyang Technological University Lee Kuan Yew Postdoctoral Fellowship, 2826
National Security Education Program Grants to Institutions of Higher Education, 2885
NSF Conferences, Workshops, and Special Meetings in the Mathematical Sciences, 3541
NSF Developing Global Scientists & Engineers, 3552
OSF Alumni Grants, 3835
OSF Civil Service Awards, 3838
OSF Intellectual Property Reform Grants, 3846
PC Fred H. Bixby Fellowships, 3880
Pioneer Hi-Bred Hon Plant Breeding Fellowships, 3968
Polytechnic University International Research Studentships, 3982
Roger K. Summit Scholarship, 4097
SASS Birgit Baldwin Fellowship, 4182
SASS Einar and Eva Lund Haugen Memorial Scholarship, 4183
Shastri Indo-Canadian Institute Faculty Research Fellowships, 4222
Shastri Indo-Canadian Inst Librarian Fellows, 4225
Shastri Indo-Canadian Institute Post-Doctoral Research Fellowships, 4229
SHOT International Scholars, 4242
SRC Guest Professorship Program, 4370
SRC International Fellowship, 4375
SRC Simon Karecki Fellowship, 4393
State Scholarships Foundation Grants, 4449
Susan G. Komen Breast Cancer Foundation Challenge Grants: Career Catalyst Research, 4476
U.S. Department of Education American Overseas Research Centers, 4537
U.S. Department of Education United States-Russia Program: Improving Research and Educational Activities in Higher Education, 4547
UNESCO Co-Sponsored Fellowships, 4566
USC Annenberg/Getty Arts Journalism Fellows, 4603
USDA International Science and Education Competitive Grants, 4631
USDA Norman E. Borlaug International Agricultural Science and Technology Fellowships, 4636
USDA Scientific Cooperation Research Program, 4647
USDA Technical Agricultural Assistance Research Grants, 4651
USDC Afghanistan and Iraq Investment and Reconstruction Task Forces Internships, 4664
WHOI Ocean Science Journalism Fellowship, 4862

International Exchange Programs
ACC Japan-US Arts Program Fellowships, 256
ACC Philippines Fellowship Program, 261
ACC Taiwan Fellowship Program, 269
ACSM Oded Bar-Or Int'l Scholars, 320
Albuquerque Community Foundation Grants, 508
ASM Millis-Colwell Postgraduate Travel Grant, 1179
Bilateral Scholarships, 1344
Coca-Cola Foundation Grants, 1593
DAAD Bilateral Programs for Co-Operative Research Grants, 1695
Eisenhower Exchange International Fellowships, 1902
Fulbright mtvU Awards, 2074
Fulbright Occasional Lecturer Program Grants, 2076
GCA Interchange Fellowship, 2098
GMFUS Black Sea Trust for Regional Corporation Grants, 2152
IBRO Asia Reg APRC Exchange Fellowships, 2323
IBRO InEurope Short Stay Grants, 2325
IREX ECA Alumni Small Grants Program, 2416
IREX Edmund S. Muskie Graduate Fellowship, 2417
IREX Eurasian Undergraduate Exchange Program Grants, 2418
IREX Individual Advanced Research Opportunities Grants, 2419
Japan Foundation Center for Global Partnership Grants, 2472
JSPS Bilateral Exchange Programs with Foreign Academic Institutions, 2524
JSPS Fellowships for Foreign Scientists, 2525
JSPS International Joint Research Projects and Scientific Seminars, 2526
Kosciuszko Found Fellowships for Polish Citizens, 2570
Kosciuszko Foundation Teaching Fellowships for Polish Citizens, 2573
Lady Davis Fellowships, 2578
SfN Travel Awards for the Japan Neuroscience Society Meeting, 4216
SRC Educational Science Grants for Invitation of Visiting Scientists, 4357
SRC Educational Science Grants for Travels to Foreign Institutions as a Guest Scientist, 4358

SRC Humanities and Social Sciences Grants for Postdoctoral Positions In Sweden, 4374
U.S. Department of Education American Overseas Research Centers, 4537
U.S. Department of Education United States-Russia Program: Improving Research and Educational Activities in Higher Education, 4547
US CRDF CGP-Climate Change and Energy Competition For Joint U.S.-Ukrainian Research Teams, 4607
US CRDF CGP-RFBR Climate Change and Energy Competition For Joint U.S.-Russian Research Teams, 4608
USDA International Science and Education Competitive Grants, 4631
USDA Scientific Cooperation Exchange Program, 4646
USDA Scientific Cooperation Research Program, 4647

International Justice
Gruber Foundation Prize for Global Justice and Women's Rights, 2182
OSF Alumni Grants, 3835
Smith Richardson Foundation Strategy and Policy Fellowship, 4270
Smith Richardson Foundation World Politics and Statecraft Fellowship, 4271
SSRC African Peacebuilding Network Research Grants, 4422

International Organizations
OSF Civil Service Awards, 3838
UNESCO Co-Sponsored Fellowships, 4566

International Planning/Policy
CFR Edward R. Murrow Press Fellowship, 1500
CFR International Affairs Fellowships, 1501
CFR International Affairs Fellowships in Nuclear Security, 1502
CFR Int'l Affairs Fellowships in South Korea, 1503
CFR International Affairs in Japan Fellowships, 1504
CFR Military Fellowships, 1505
CFR Stanton Nuclear Security Fellowships, 1506
GMFUS Black Sea Trust for Regional Corporation Grants, 2152
International Peace Research Inst Oslo Grants, 2402
IREX Regional Policy Symposium Grants, 2423
IREX US Embassy Policy Specialist Grants, 2426
Japan Foundation Center for Global Partnership Grants, 2472
Japan Foundation Center for Global Partnership Institutional Grants, 2473
Jefferson Science Fellows Program, 2480
Jenifer Altman Foundation Grants, 2484
Kluge Center Kissinger Chair in Foreign Policy and International Relations, 2564
Kluge Center Kissinger Lecturer Award, 2565
Lynde & Harry Bradley Foundation Fellowships, 2634
Lynde and Harry Bradley Foundation Grants, 2635
Mary Ingraham Bunting Inst Peace Fellowship, 2671
NOAA International Coral Reef Conservation Project Grants, 3377
RAND Corporation Graduate Student Summer Associateships, 4030
SSRC African Peacebuilding Network Research Grants, 4422
SSRC Eurasia Program Dissertation Development Awards, 4427
United States Institute of Peace - Transatlantic Post-Doc Fellowship for International Relations and Security, 4581

International Programs
AAAS Science and Technology Policy Overseas Fellowships at USAID Missions, 44
A Charitable Foundation Grants, 273
AFAR Paul Beeson Career Development Awards in Aging Research for the Island of Ireland, 353
AIChE Kazutoshi Fujimura Award for Lifetime Achievement, 414
ALA Ingenta Research Awards, 489
Alcoa Foundation Grants, 511
Alexander von Humboldt Foundation Fellowships for Experienced Researchers, 516
Alexander von Humboldt Foundation Fellowships for Postdoctoral Researchers, 517
Alexander von Humboldt Foundation Georg Forster Fellowships for Experienced Researchers, 523
Alexander von Humboldt Foundation Georg Forster Fellowships for Postdoctoral Researchers, 524
Allen Lane Foundation Grants, 571
American-Scandinavian Foundation Awards for Advanced Study or Research in the U.S., 603
American-Scandinavian Foundation Grant for Study in Scandinavia, 605
Archer Daniels Midland Foundation Grants, 985
ARO Conference and Symposia Grants, 1007
Arronson Foundation Grants, 1019
ASCO Long-Term International Fellowships, 1048
ATS Henry Luce III Fellowships, 1258
BA Larger Research Grants, 1290
Banfi Vintners Foundation Grants, 1296
BA Small Research Grants, 1304
Beatrice Laing Trust Grants, 1324
Becton Dickinson and Company Grants, 1328
Bill and Melinda Gates Foundation Agricultural Development Grants, 1345
Bill and Melinda Gates Foundation Financial Services for the Poor Grants, 1346
Bill and Melinda Gates Foundation Water, Sanitation and Hygiene Grants, 1347
Bristol-Myers Squibb Foundation Health Disparities Grants, 1377
Broad Foundation IBD Research Grants, 1380
Burden Trust Grants, 1388
Carnegie Endowment for International Peace Junior Fellowships, 1434
CCA Diefenbaker Award to update, 1447
Center for the Study of Philanthropy International Fellowships, 1482
Center for the Study of Philanthropy Senior International Fellowships, 1483
Charles Stewart Mott Foundation Grants, 1529
Chatlos Foundation Grants Program, 1532
CIGNA Foundation Grants, 1568
Compton Foundation Grants Program, 1623
ConocoPhillips Grants Program, 1629
Conquer Cancer Foundation of ASCO International Development and Education Awards, 1635
Cystic Fibrosis Canada Visiting Allied Health Professional Awards, 1686
Cystic Fibrosis Canada Visiting Clinician Awards, 1687
Cystic Fibrosis Trust Research Grants, 1691
DAAD Re-invitation Program for Former Scholarship Holders, 1701
Duke University Clinical Cardiac Electrophysiology Fellowships, 1871
Duke Univ Interventional Cardio Fellowships, 1873
Elizabeth Glaser Scientist Award, 1907
Gruber Foundation Int'l Research Award, 2180
H.J. Heinz Company Foundation Grants, 2195
Harold Simmons Foundation Grants, 2203
Helen Bader Foundation Grants, 2228
IES Faculty Fellowships, 2341
International Human Rights Funders Grants, 2399
IRA Outstanding Dissertation of the Year Award, 2411
IREX Policy-Connect Collaborative Grants, 2422
IREX Regional Policy Symposium Grants, 2423
IREX US Embassy Policy Specialist Grants, 2426
Jacob and Hilda Blaustein Foundation Grants, 2454
James H. Cummings Foundation Grants, 2460
Jenifer Altman Foundation Grants, 2484
John Deere Foundation Grants, 2500
Kress Foundation Art History Pre-Doctoral Fellowships, 2576
Landon Foundation-AACR Innovator Award for International Collaboration, 2587
Ludwick Family Foundation Grants, 2625
Lustgarten Foundation for Pancreatic Cancer Research Grants, 2630
Mead Johnson Nutritionals Med Educ Grants, 2734
MGFA Post-Doctoral Research Fellowships, 2751
MGFA Student Fellowships, 2752
NAA Norwegian Marshall Fund Grants, 2811
NATO Advanced Research Workshops Program, 2894
NATO Advanced Study Institutes Program, 2895
NATO Collaborative Linkage Grants, 2896
NATO Expert Visits Program Grants, 2897
NATO Science for Peace Grants, 2898
NEH Fellowships at Independent Research Institutions, 2957
NFWF/Exxon Save the Tiger Fund Grants, 2995
NOAA International Coral Reef Conservation Project Grants, 3377
NSF Developing Global Scientists & Engineers, 3552
NSF Grants for the Analysis of Science and Technology Resources, 3575
NSF Materials World Network: Cooperative Activity in Materials Research between US Investigators and their Counterparts Abroad, 3595
NSF Partnerships for International Research and Education, 3611
NSF SBE Doctoral Dissertation Research Improvement Grants, 3641
NSF Synthesis and Modeling Project of the US Joint Global Ocean Flux Study, 3656
Nuffield Foundation Law and Society Grants, 3667
OSF Alumni Grants, 3835
OSF Human Rights and Governance Internships, 3845
OSF Think Tank Fellowships, 3852
OSF Thomas J. Alexander Fellowships, 3853
Patrick and Anna M. Cudahy Fund Grants, 3873
PDF International Research Grants, 3882
Potts Memorial Foundation Grants, 3985
Praxair Foundation Grants, 3986
Rett Syndrome Research Foundation Grants, 4058
Rockefeller Archive Center Research Grants, 4086
Rosalinde and Arthur Gilbert Foundation/AFAR New Investigator Awards in Alzheimer's Disease, 4102
Savoy Foundation Post-Doctoral and Clinical Research Fellowships, 4184
Savoy Foundation Research Grants, 4185
Savoy Foundation Studentships, 4186
Sigrid Juselius Foundation Grants, 4260
SNM/ Covidien Seed Grant in Molecular Imaging/ Nuclear Medicine Research, 4298
SNM Postdoctoral Molecular Imaging Scholars, 4302
SRC Cross-Disciplinary Research Grants, 4353
SRC Educational Science Conference Grants, 4354
SRC Humanities and Social Sciences Grants, 4373
SRC Medical Science Grants for Large International Conferences in Sweden, 4380
SRC Nat & Engineering Science Travel Grants, 4388
SSRC African Peacebuilding Network Research Grants, 4422
SSRC Eurasia Program Dissertation Development Awards, 4427
Swiss National Science Foundation DORE (DO REsearch) Conference Grants, 4482
Swiss National Science Foundation Scientific Conferences Grants, 4486
UNESCO Co-Sponsored Fellowships, 4566
USDA Foreign Market Development Grants, 4625
USDA Market Access Program Grants, 4633
W.K. Kellogg Foundation Healthy Kids Grants, 4822
Wenner-Gren Foundation African Fellowships, 4846
Wenner-Gren Foundation International Collaborative Research Grants, 4851
Wenner-Gren Found Short-Term Fellowships, 4853
Wenner-Gren Foundation Wadsworth International Fellowships, 4854
WILD Foundation Grants, 4867
WWF Russell E. Train Education for Nature Alumni Grants, 4897

International Relations
AAAS Award for Science Diplomacy, 27
AAAS Science and Technology Policy Fellowships: Congressional, 39
AAAS Science and Technology Policy Fellowships: Global Health and Development, 42

Alexander von Humboldt Foundation Anneliese Maier Award, 515
Alexander von Humboldt Foundation Fellowships for Experienced Researchers, 516
Alexander von Humboldt Foundation Fellowships for Postdoctoral Researchers, 517
Alexander von Humboldt Foundation German Chancellor Fellowships for Prospective Leaders from the People's Republic of China, 526
Alexander von Humboldt Foundation German Chancellor Fellowships for Prospective Leaders from the Russian Federation, 527
Alexander von Humboldt Foundation Reimar Lust Award, 536
American-Scandinavian Foundation Awards for Advanced Study or Research in the U.S., 603
American Academy of Arts and Sciences Hellman Fellowships, 612
American Academy of Arts and Sciences Visiting Scholars, 613
APSA Helen Dwight Reid Award, 936
Arca Foundation Grants, 984
Archer Daniels Midland Foundation Grants, 985
Arie and Ida Crown Memorial Grants, 989
Asia Foundation Exchange Programs, 1090
Bydale Foundation Grants, 1400
Canon Foundation Grants, 1423
Center for Defense Information Internships, 1473
Center for the Study of Professional Military Ethics Resident Fellowship, 1484
Century Foundation Grants, 1489
CFR Edward R. Murrow Press Fellowship, 1500
CFR International Affairs Fellowships, 1501
CFR International Affairs Fellowships in Nuclear Security, 1502
CFR Int'l Affairs Fellowships in South Korea, 1503
CFR International Affairs in Japan Fellowships, 1504
CFR Military Fellowships, 1505
CFR Stanton Nuclear Security Fellowships, 1506
Coca-Cola Foundation Grants, 1593
Compton Foundation Grants Program, 1623
ConocoPhillips Grants Program, 1629
DOE Nonproliferation & Nat Security Research, 1806
Donald and Sylvia Robinson Family Grants, 1830
DOS Study of the United States Institutes, 1846
Earhart Foundation Institutional Grants, 1888
Eugene McDermott Foundation Grants, 1958
Ford Foundation Peace and Social Justice Grants, 2031
Harry Frank Guggenheim Foundation Dissertation Fellowships, 2207
Harvard Women and Public Policy Postdoctoral Fellowship, 2217
Institute for Palestine Studies, Publishing, and Research Organization Grants, 2382
Institute of Global Conflict and Cooperation Research, Conference, and Teaching Grants, 2385
International Studies Graduate Fellowships, 2403
IREX Edmund S. Muskie Graduate Fellowship Program, 2417
IREX Regional Policy Symposium Grants, 2423
IREX US Embassy Policy Specialist Grants, 2426
Japan Foundation Six-Month Grants for Specialists in Cultural and Academic Fields, 2479
Jefferson Science Fellows Program, 2480
Jenifer Altman Foundation Grants, 2484
JMO Strategic and International Studies Grants, 2493
Kluge Center Kissinger Chair in Foreign Policy and International Relations, 2564
Kluge Center Kissinger Lecturer Award, 2565
Kosciuszko Foundation Tuition Scholarships, 2574
Mary Ingraham Bunting Inst Peace Fellowship, 2671
Minerva Fellowships, 2778
Minerva Short-Term Research Grants, 2779
National Endowment for Democracy Discretionary Grants, 2862
OAH Ellis W. Hawley Prize, 3715
OAH Richard W. Leopold Prize, 3728
Rajiv Gandhi Foundation Grants, 4025
RAND Corporation Transatlantic Postdoctoral Fellowships for Int'l Relations & Security, 4032
Smith Richardson Foundation Strategy and Policy Fellowship, 4270
Smith Richardson Foundation World Politics and Statecraft Fellowship, 4271
SRC Educational Science Conference Grants, 4354
SRC Natural and Engineering Science Conference Grants, 4383
SSRC Abe Fellowships, 4421
Starr Foundation Grants, 4447
U.S. Department of Education Foreign Languages and Area Studies Fellowships, 4539
U.S. Department of Education National Resource Centers, 4542
U.S. Department of Education United States-Russia Program: Improving Research and Educational Activities in Higher Education, 4547
United States Institute of Peace - Jennings Randolph Peace Dissertation Scholarships, 4579
United States Institute of Peace - Jennings Randolph Senior Fellowships, 4580
United States Institute of Peace - Transatlantic Post-Doc Fellowship for International Relations and Security, 4581
United States Institute of Peace Annual Grants, 4582
USDA Emerging Markets Grants, 4623
USDA Foreign Market Development Grants, 4625
USDA Market Access Program Grants, 4633
WHOI Marine Policy Fellowships, 4861

International Students

3M Canada National Teaching Fellowships, 3
AAUW International Doctoral Fellowships, 214
AAUW International Master's/First Professional Degree Fellowships, 215
AAUW International Postdoctoral Fellowships, 216
ACLS African Humanities Program for Dissertation Completion, 278
ACLS African Humanities Program in Ghana, Nigeria, South Africa, Tanzania, and Uganda, 279
Alexander von Humboldt Foundation German Chancellor Fellowships for Prospective Leaders from the People's Republic of China, 526
Alexander von Humboldt Foundation German Chancellor Fellowships for Prospective Leaders from the Russian Federation, 527
APSA Travel Grants, 948
DAAD Bilateral Programs for Co-Operative Research Grants, 1695
DAAD Intensive Language Courses Grants, 1699
DAAD Re-invitation Program for Former Scholarship Holders, 1701
DAAD Study Visits / Study Seminars and Practicals in Germany by Foreign Students, 1709
Grass Foundation Marine Biological Laboratory Fellowships, 2163
IBRO Latin America Regional Funding for Neuroscience Schools, 2327
IBRO Regional Grants for Int'l Fellowships to U.S. Laboratory Summer Neuroscience Courses, 2330
Independent Institute John M. Templeton Fellowships Essay Contest, 2369
Mary Gates Research Scholarships, 2670
MBL Gruss Lipper Family Foundation Summer Fellowship, 2704
MBL William Townsend Porter Summer Fellowships for Minority Investigators, 2716
OSF Alumni Grants, 3835
OSF Civil Service Awards, 3838
OSF Justice Initiative Fellowships, 3847
OSF Think Tank Fellowships, 3852
SASS Einar and Eva Lund Haugen Memorial Scholarship, 4183
SfN Travel Awards for the Japan Neuroscience Society Meeting, 4216
SRC Graduate Fellowships, 4368
SRC International Fellowship, 4375
UNESCO Co-Sponsored Fellowships, 4566
USDA Norman E. Borlaug International Agricultural Science and Technology Fellowships, 4636
WHOI Summer Student Fellowships, 4865

International Studies

3M Canada National Teaching Fellowships, 3
AAP International Travel Grants, 197
ACLS African Humanities Program in Ghana, Nigeria, South Africa, Tanzania, and Uganda, 279
ACLS American Research in the Humanities in China Fellowships, 280
ACLS Early Career Postdoctoral Fellowships in East European Studies, 286
ACLS International and Area Studies Fellowships, 289
ALA WESS De Gruyter European Librarianship Study Grant, 501
Alexander von Humboldt Feodor Lynen Research Fellowships for Experienced Researchers, 520
Alexander von Humboldt Foundation Friedrich Wilhelm Bessel Research Awards, 522
Alexander von Humboldt Foundation German Chancellor Fellowships for Prospective Leaders from the People's Republic of China, 526
Alexander von Humboldt Foundation German Chancellor Fellowships for Prospective Leaders from the Russian Federation, 527
Alexander von Humboldt Foundation JSPS Research Fellowships, 530
Coca-Cola Foundation Grants, 1593
DAAD Intensive Language Courses Grants, 1699
DAAD Research Grants for Doctoral Candidates and Young Academics and Scientists, 1703
DAAD Research Stays for University Academics and Scientists, 1705
DAAD Scholarships for Artists: Study Visits for Academics, 1707
DAAD Study & Internship Program in Germany, 1708
DAAD Study Visits / Study Seminars and Practicals in Germany by Foreign Students, 1709
Dept of Ed International Research and Studies Project Grants, 1745
Earhart Foundation Institutional Grants, 1888
Gerda Henkel Foundation Grants, 2127
H.B. Earhart Fellowships, 2194
Helen Kellogg Institute for International Studies Residential Fellowships, 2230
Henry Luce Foundation Special Projects in Higher Education Grants, 2236
International Studies Graduate Fellowships, 2403
Japan Foundation Center for Global Partnership Security Studies Fellowships, 2475
Japan Foundation New York Institutional Support for Japanese Studies Grants, 2477
JMO Strategic and International Studies Grants, 2493
Library of Congress Fellowships in Int'l Studies, 2612
Mellon Post-Doctoral Fellowship in Sustainable Agriculture, 2740
NCSS/FASSE/CUFA Inquiry Grant, 2936
NCSS Exemplary Res in Social Studies Award, 2937
NCSS Jean Dresden Grambs Distinguished Career Research in Social Studies Award, 2938
NCSS L Metcalf Exemplary Dissertation Award, 2939
NSF Developing Global Scientists & Engineers, 3552
OSF Alumni Grants, 3835
Rockefeller Foundation Humanities Fellowships, 4089
Short-Term Travel to Japan Grants, 4236
Smith Richardson Foundation World Politics and Statecraft Fellowship, 4271
SRC Educational Science Lang Checking Grants, 4360
SRC Guest Professorship Program, 4370
SRC Humanities and Social Sciences Grants for Postdoctoral Positions In Sweden, 4374
SRC Medical Research Grants, 4379
SSRC Int'l Dissertation Research Fellowships, 4429
U.S. Department of Education American Overseas Research Centers, 4537
U.S. Department of Education Foreign Languages and Area Studies Fellowships, 4539
U.S. Department of Education National Resource Centers, 4542
UNESCO Co-Sponsored Fellowships, 4566
US CRDF CGP-Climate Change and Energy Competition For Joint U.S.-Ukrainian Research Teams, 4607

US CRDF CGP-RFBR Climate Change and Energy Competition For Joint U.S.-Russian Research Teams, 4608
USDC Afghanistan and Iraq Investment and Reconstruction Task Forces Internships, 4664
World Bank Graduate Scholarship Program, 4896

International Trade and Finance
Archer Daniels Midland Foundation Grants, 985
Bedding Plants Found Horticulture Scholarships, 1329
International Studies Graduate Fellowships, 2403
IREX Regional Policy Symposium Grants, 2423
IREX US Embassy Policy Specialist Grants, 2426
Japan Foundation Center for Global Partnership Grants, 2472
Shastri Indo-Canadian Institute Collaborative Research Project Grants, 4220
Starr Foundation Grants, 4447
USDA Cochran Fellowships, 4619
USDA Emerging Markets Grants, 4623
USDA Market Access Program Grants, 4633
USDA Technical Agricultural Assistance Research Grants, 4651
USDC Market Development Cooperator Program Grants, 4667

International and Comparative Law
APSA Gabriel A. Almond Award, 934
Ford Foundation Peace and Social Justice Grants, 2031
Germanistic Society of America Fellowships, 2128
Hague Academy of International Law Summer Scholarships, 2198
Independent Institute John M. Templeton Fellowships Essay Contest, 2369
OSF Intellectual Property Reform Grants, 3846

Internet
ESRI Conservation Program Grants, 1952
IMLS 21st Century Museum Prof Grants, 2362
IMLS National Leadership Grants, 2366
IREX International Leadership in Education Program Grants, 2420
Microsoft Research Intelligent Web 3.0 Grants, 2769
MLA Rittenhouse Award, 2790
NASA Marslink Initiative, 2841
NLM Internet Connection for Medical Institutions Grants, 3300
RSC Thomas W. Eadie Medal, 4126
Scientific Communities Initiative Grants, 4189
WHOI Ocean Science Journalism Fellowship, 4862

Internship Programs
AAAS Entry Point Internships, 30
AFAR Medical Student Training in Aging Research Program, 352
AIHS/Mitacs Health Pilot Partnership Internship Grants, 434
American-Scandinavian Foundation Short-Term Training Program Grants, 606
CIA Undergraduate Co-Op Program - Engineer/ Program Management Engineer, 1561
CIA Undergraduate Internship - Analysis, 1562
Community Foundation of the Virgin Islands Judith A. Towle Environmental Studies Grants, 1621
DHHS Emerging Leaders Program Internships, 1756
Do Something TMI Intern, 1844
Do Something User Experience Research Intern, 1845
Field Museum High School Internships, 2009
Field Museum Undergraduate Internships, 2010
Kellogg Institute for International Studies Summer Internships, 2549
National Wildlife Federation Internships, 2891
NHLBI Summer Internship Program in Biomedical Research, 3064
OSF Justice Initiative Fellowships, 3847
Rhode Island State Government Internships, 4062
Robert W. Deutsch Foundation Grants, 4082
Skaggs Foundation Grants, 4266
Smithsonian Museum Conservation Institute Analytical Studies Group Internships, 4281
Smithsonian Museum Conservation Institute Internships for Research Projects, 4282
SPJ Pulliam/Kilgore Freedom of Information Internships, 4347
SRC Educational Science Doctoral Grants, 4355
SRC Educational Science Junior Research Position Grants, 4359
SRC Educational Science Network Grants, 4362
USDC Afghanistan and Iraq Investment and Reconstruction Task Forces Internships, 4664
WHOI Guest Student Internships, 4860
William Blair and Company Foundation Grants, 4875

Interstellar Studies
NSF Stellar Astronomy and Astrophysics Research Grants, 3655

Intervention Programs
BFGoodrich Collegiate Inventors Prize, 1341
Bodman Foundation Grants, 1356
CDC Steven M. Teutsch Prevention Effectiveness Fellowships, 1467
CSTE CDC/CSTE Applied Epidemiology Fellowships, 1664
DOE Energy-Related Inventions Grants, 1783
DOE Inventions and Innovation Grant Program, 1802
NIH Biobehavioral Research for Effective Sleep, 3205
NSF Partnerships for Innovation Grants, 3610
William T. Grant Foundation Research Grants, 4881

Invention and Innovation
AIHS Alberta/Pfizer Translational Res Grants, 435
AIHS Collaborative Research and Innovation Grants - Collaborative Project, 438
ALA Excellence in Academic Libraries Award, 486
ALA Innovation Award, 490
Alfred P. Sloan Foundation Public Understanding of Science and Technology-Book Grants, 555
Conquer Cancer Foundation of ASCO International Innovation Grant, 1636
NHLBI Bioengineering Approaches to Energy Balance and Obesity Grants for SBIR, 3017
OSF Fellowships, 3840
Smithsonian National Museum of American History Lemelson Center Fellowships, 4289
Smithsonian National Museum of History Lemelson Center Travel to Collections Awards, 4290
SRC Aristotle Award, 4349
SRC Cross-Disciplinary Research Grants, 4353
SRC Landmark Innovation Award, 4376
SRC Peter Verhofstadt Fellowship, 4389
SRC Robert M. Burger Fellowship, 4392

Investments and Securities
MSI Marketing Strategy Meets Wall Street Research Competition, 2801

Ionized Gases
APS Ja Clerk Maxwell Prize for Plasma Physics, 966
APS Will Allis Prize for Study of Ionized Gases, 982

Ionosphere Interactions
ONR Ionosphere Interactions Research Grants, 3765
USDD Ionospheric Modification Grants, 4745

Iran
IREX Individual Advanced Research Opportunities Grants, 2419

Iraq
USDC Afghanistan and Iraq Investment and Reconstruction Task Forces Internships, 4664

Islamic Studies
Ahmanson and Getty Postdoctoral Fellowships, 377
AIAR Andrew W. Mellon Foundation Fellowships, 392
Center for Advanced Judaic Studies Postdoctoral Fellowships, 1472
SSRC Islamic Traditions and Muslim Societies in World Contexts Grants, 4430

Isotope/Radiation Technology
Abdus Salam ICTP-ELETTRA Users Grants, 221
Abdus Salam ICTP/IAEA Sandwich Training Education Grants, 226
AINSE Postgraduate Research Awards, 463
DSO Radiation Biodosimetry Grants, 1864

Israel
American Society for Yad Vashem Grants, 798
BARD Postdoctoral Fellowships, 1298
BARD Research Grants, 1299
Donald and Sylvia Robinson Family Grants, 1830
Fishman Family Foundation Grants, 2020
Helen Bader Foundation Grants, 2228
Institute for Palestine Studies, Publishing, and Research Organization Grants, 2382
Jacob and Hilda Blaustein Foundation Grants, 2454
Lady Davis Fellowships, 2578
Millstone Foundation Grants, 2776
Minerva Fellowships, 2778
Minerva Short-Term Research Grants, 2779

Italian Art
Gladys Krieble Delmas Foundation Grants for Venetian Research, 2144
Roberto Longhi Foundation Fellowships, 4078

Italian History
American Historical Association Helen and Howard R. Marraro Prizes, 721
NIAF Fellowship Grants, 3095

Italian Language/Literature
Gladys Krieble Delmas Foundation Grants for Venetian Research, 2144
NIAF Fellowship Grants, 3095
NL Northeast Modern Language Assocation Fellowship, 3314
NL Weiss/Brown Publication Subvention Award, 3320

Italian Studies
AAR Postdoctoral Fellowships, 201
AAR Rome Prize Fellowships, 202
NIAF Fellowship Grants, 3095

Italy
AAR Postdoctoral Fellowships, 201
AAR Rome Prize Fellowships, 202
American Historical Association Helen and Howard R. Marraro Prizes, 721
Fulbright Travel Grants, 2086
Gladys Krieble Delmas Foundation Grants for Venetian Research, 2144
Roberto Longhi Foundation Fellowships, 4078

Japan
Alexander von Humboldt Foundation JSPS Research Fellowships, 530
Alexander von Humboldt Foundation Philipp Franz von Siebold Award for Japanese Researchers, 534
American College of Surgeons Traveling Fellowship to Japan, 707
American Historical Association John K. Fairbank Prize in East Asian History, 733
Center for East Asian Studies Fellowships, 1474
CFR International Affairs in Japan Fellowships, 1504
Fulbright International Education Administrators Seminar Program Grants, 2073
Japan Foundation Center for Global Partnership Grants, 2472
Japan Foundation Center for Global Partnership NPO Fellowships, 2474
Japan Foundation Japanese Studies Fellowships, 2476
Japan Foundation New York Institutional Support for Japanese Studies Grants, 2477
Japan Foundation New York Japanese Studies Fellowships, 2478
Japan Foundation Six-Month Grants for Specialists in Cultural and Academic Fields, 2479

JSPS Bilateral Exchange Programs with Foreign Academic Institutions, 2524
JSPS Fellowships for Foreign Scientists, 2525
JSPS International Joint Research Projects and Scientific Seminars, 2526
Kobe College Corporation Graduate Fellowships, 2567
NEH Fellowships for Advanced Social Science Research on Japan, 2958
OAH Japanese Residencies Program Grants, 3722
Short-Term Travel to Japan Grants, 4236
SSRC Abe Fellowship for Journalists, 4420
SSRC Abe Fellowships, 4421
SSRC Japan Society for the Promotion of Science Fellowships, 4431
Twentieth-Century Japan Research Awards, 4536

Japanese Art
ACC Japan-US Arts Program Fellowships, 256
Fulbright International Education Administrators Seminar Program Grants, 2073

Japanese Language/Literature
Japan Foundation Center for Global Partnership Grants, 2472
Japan Foundation Japanese Studies Fellowships, 2476
Japan Foundation New York Institutional Support for Japanese Studies Grants, 2477
Japan Foundation New York Japanese Studies Fellowships, 2478
Japan Foundation Six-Month Grants for Specialists in Cultural and Academic Fields, 2479
NSF East Asia and Pacific Summer Institutes for U.S. Graduate Students, 3555

Jazz
Florida Division of Cultural Affairs Individual Artist Fellowships, 2026
National Endowment for the Arts Jazz Masters Fellowships, 2864

Jewelry/Metal Arts
Florida Division of Cultural Affairs Individual Artist Fellowships, 2026

Jewish Culture
Brach Family Foundation Grants, 1367
LBI-L Leo Baeck/Studienstiftung Fellowships, 2592
LBI-NY David Baumgardt Memorial Fellowships, 2595
Leo Baeck Essay Prize, 2601
Meyer and Stephanie Eglin Foundation Grants, 2747

Jewish Services
Aaron & Cecile Goldman Family Found Grants, 200
Albuquerque Community Foundation Grants, 508
Archer Daniels Midland Foundation Grants, 985
Arie and Ida Crown Memorial Grants, 989
A Rocha USA Grants, 1006
Arronson Foundation Grants, 1019
Arthur and Rochelle Belfer Foundation Grants, 1023
Brach Family Foundation Grants, 1367
Carl C. Icahn Foundation Grants, 1428
Charles H. Revson Foundation Grants, 1524
Donald and Sylvia Robinson Family Grants, 1830
Fishman Family Foundation Grants, 2020
Gilroy and Lillian P. Roberts Charitable Foundation Grants, 2142
Girl's Best Friend Foundation Grants, 2143
Greenspun Family Foundation Grants, 2173
Healthcare Foundation of New Jersey Grants, 2222
HIAS Scholarship Awards, 2264
Jacob and Hilda Blaustein Foundation Grants, 2454
Jewish Federation of Metropolitan Chicago Scholarships, 2486
Jewish Social Entrepreneurs Fellowships, 2487
Joseph Alexander Foundation Grants, 2513
Kovler Family Foundation Grants, 2575
Meyer and Stephanie Eglin Foundation Grants, 2747
Millstone Foundation Grants, 2776
Oppenstein Brothers Foundation Grants, 3792
Righteous Persons Foundation Grants, 4069

Jewish Studies
Aaron & Cecile Goldman Family Found Grants, 200
Ahmanson and Getty Postdoctoral Fellowships, 377
AJA Ethel Marcus Memorial Fellowship in American Jewish Studies, 471
AJA Lowenstein-Wiener Summer Fellowship Awards in American Jewish Studies, 472
AJA Marguerite R. Jacobs Memorial Postdoctoral Award in American Jewish Studies, 473
AJA Rabbi Frederic A. Doppelt Memorial Fellowship in American Jewish Studies, 474
AJA Starkoff Fellowship in Amer Jewish Studies, 475
American Jewish Historical Society Awards, 747
Arthur and Rochelle Belfer Foundation Grants, 1023
Brach Family Foundation Grants, 1367
Center for Advanced Judaic Studies Postdoctoral Fellowships, 1472
Center for Judaic Studies Fellowships, 1478
DAAD-Leo Baeck Institute Fellowships, 1694
Fishman Family Foundation Grants, 2020
Hebrew University Lady Davis Trust Fellowships, 2226
Helen Bader Foundation Grants, 2228
LBI-L Leo Baeck/Studienstiftung Fellowships, 2592
LBI-NY/DAAD Fellowships in Germany, 2593
LBI-NY/DAAD Fellowships in New York, 2594
LBI-NY David Baumgardt Memorial Fellowships, 2595
LBI-NY Fred Grubel Fellowship, 2596
LBI-NY Fritz Halbers Fellowships, 2597
LBI-NY Gerald Westheimer Career Development Fellowship, 2598
Leo Baeck Essay Prize, 2601
NWSA Graduate Scholarship in Jewish Women's Studies, 3676
Oppenstein Brothers Foundation Grants, 3792
Ralph Waldo Emerson Award, 4028
Richard and Rhoda Goldman Fund Grants, 4063
Righteous Persons Foundation Grants, 4069

Job Training Programs
3M Foundation Community Giving Grants, 5
Achelis Foundation Grants, 274
Adaptec Foundation Grants, 340
Allen Lane Foundation Grants, 571
Bayer Foundation Grants, 1309
Bodman Foundation Grants, 1356
Boeing Company Contributions Grants, 1357
Chiron Foundation Community Grants, 1553
Coors Brewing Corporate Contributions Grants, 1644
DaimlerChrysler Corporation Fund Grants, 1714
Dept of Ed Vocational Rehabilitation Services Demonstration and Training Programs, 1751
DHHS Emerging Leaders Program Internships, 1756
GNOF Gulf Coast Oil Spill Grants, 2153
Heineman Foundation for Research, Education, Charitable and Scientific Purposes, 2227
Helen Bader Foundation Grants, 2228
Helen V. Brach Foundation Grants, 2231
HUD Community Outreach Partnership Centers Grants, 2292
Johnson & Johnson Corporate Contributions, 2508
Joyce Foundation Employment Grants, 2520
McGregor Fund Grants, 2725
MGM Resorts Foundation Community Grants, 2753
Miller Brewing Corporate Contributions Grants, 2774
Reuter Foundation Grants, 4059
Sioux Falls Area Community Foundation Community Fund Grants (Unrestricted), 4262
Sioux Falls Area Comm Foundation Spot Grants, 4263
Starr Foundation Grants, 4447
Textron Corporate Contributions Grants, 4501
USDA Alaska Native-Serving and Native Hawaiian-Serving Institutions Education Grants, 4615
USDA Organic Agriculture Research and Extension Initiative, 4637
WILD Foundation Grants, 4867

Journalism
AAAS Kavli Science Journalism Awards, 33
AAB Berlin Prize Fellowships, 48
AAES Engineering Journalism Award, 150
AAES National Engineering Award, 151
AAES Norm Augustine Award, 152
AAJA Fellowships, 181
AcademyHealth PHSR Research Article of the Year Awards, 248
APSA Congressional Fellowships for Communications Scholars and Journalists, 926
APSA Congressional Fellowships for International Scholars, 928
APSA Congressional Fellowships for Journalists, 929
APSA Congressional Health and Aging Policy Fellowships, 931
ASU Graduate College Herman E. Demund Memorial Fellowships, 1240
Bernard L. Schwartz Fellows Program, 1334
Canada-U.S. Fulbright International Science and Technology Award, 1417
Canada-U.S. Fulbright Senior Specialists Program Grants, 1418
Canada-U.S. Fulbright Traditional Scholars, 1419
Century Foundation Grants, 1489
CFR Edward R. Murrow Press Fellowship, 1500
DAAD Study & Internship Program in Germany, 1708
Dart Center Ochberg Fellowship, 1720
Ethics and Excellence in Journalism Grants, 1953
Fulbright Alumni Initiatives Awards Grants, 2064
Fulbright Beginning Prof Journalism Awards, 2065
Fulbright Distinguished Chairs Program Grants, 2069
Fulbright European Union Scholar-in-Residence Program Grants, 2071
Fulbright New Century Scholars Program Grants, 2075
Fulbright Occasional Lecturer Program Grants, 2076
Fulbright Specialists Program Grants, 2079
Fulbright Traditional Scholar Program in Europe and Eurasia, 2080
Fulbright Traditional Scholar Program in Sub-Saharan Africa, 2081
Fulbright Traditional Scholar Program in the East Asia/Pacific Region, 2082
Fulbright Traditional Scholar Program in the Near East and North Africa Region, 2083
Fulbright Traditional Scholar Program in the South and Central Asia Region, 2084
Fulbright Traditional Scholar Program in the Western Hemisphere, 2085
Fulbright Visiting Specialists Program: Direct Access to the Muslim World, 2087
Goldsmith Awards Program, 2158
Harry Frank Guggenheim Foundation Dissertation Fellowships, 2207
IREX Edmund S. Muskie Graduate Fellowship Program, 2417
IREX International Leadership in Education Program Grants, 2420
IWMF Public Health Fellowship, 2446
Japan Foundation Japanese Studies Fellowships, 2476
Japan Foundation New York Japanese Studies Fellowships, 2478
John Ben Snow Memorial Trust Grants, 2497
Kiplinger Program in Public Affairs Journalism Fellowships, 2557
Mary Baker Eddy Library Summer Research Fellowships, 2669
MIT Knight Science Journalism Fellowships, 2782
National Education Reporting Fellowships, 2861
National Endowment for Democracy Discretionary Grants, 2862
NIHCM Foundation Health Care Television and Radio Journalism Awards, 3211
NMF National Medical Association Awards for Medical Journalism, 3323
Orgone Biophysical Research Lab Lou Hochberg Awards, 3798
Perl 6 Microgrants, 3899
Playboy Foundation Grants, 3980
Prudential Foundation Education Grants, 4002
SPJ Eugene C. Pulliam Fellowship for Editorial Writing, 4346
SPJ Pulliam/Kilgore Freedom of Information Internships, 4347

SSHRC Bora Laskin National Fellowship in Human Rights Research, 4405
SSHRC Insight Development Grants, 4409
SSHRC Insight Grants, 4410
SSRC Abe Fellowship for Journalists, 4420
SSRC New Directions in the Study of Prayer Research Grants, 4433
United Methodist Communications Stoody-West Fellowship, 4578
USC Annenberg/Getty Arts Journalism Fellows, 4603
WHOI Ocean Science Journalism Fellowship, 4862

Journalism Education
ASU Graduate College Herman E. Demund Memorial Fellowships, 1240
Fulbright Beginning Prof Journalism Awards, 2065
Kosciuszko Foundation Tuition Scholarships, 2574
Yale University Beinecke Library Research Fellowships for Yale Grad & Professional School Students, 4902

Judaism
NFJC Gantz-Zahler Grant, 2993

Judicial/Law Administration
JMO Law and Legal System Grants, 2491
MPI for Social Anthropology Max Planck Fellow Research Group Law, Organizations, Science, and Technology, 2799
OSF Soros Justice Advocacy Fellowships, 3850

Junior High School Education
Astronomical Society of the Pacific Las Cumbres Amateur Outreach Award, 1230
Colonel Stanley R. McNeil Foundation Grants, 1597
Helen Bader Foundation Grants, 2228
Joyce Foundation Education Grants, 2519
NSF Engineering Research Centers Grants, 3561

Junior and Community Colleges
AIR Research Grants/Dissertation Fellowships, 470
ALA Excellence in Academic Libraries Award, 486
ALA WESS De Gruyter European Librarianship Study Grant, 501
Credit Union Found of British Columbia Grants, 1654
HAF Riley Frazel Memorial Scholarship, 2197
Harvey Randall Wickes Foundation Grants, 2218
Jackson Foundation Grants, 2453
Marion Gardner Jackson Charitable Trust Grants, 2664
McGraw-Hill Companies Community Grants, 2724
Miller Brewing Corporate Contributions Program Grants, 2774
Minnesota War Orphans Educational Assistance Grants, 2780
Nellie Mae Education Foundation Public Understanding Grants, 2969
NSF Activities in Science, Engineering, and Mathematics for Persons with Disabilities, 3460
NSF Louis Stokes Alliances for Minority Participation Grants, 3589
NSF Materials World Network: Cooperative Activity in Materials Research between US Investigators and their Counterparts Abroad, 3595
NSF University Radio Observatories Program, 3664
Sidgmore Family Foundation Grants, 4252
Talbert and Leota Abrams Foundation, 4492
USDA Tribal College Research Grants, 4652
Victor E. Speas Foundation Grants, 4810
Virginia Foundation for the Humanities Discretionary Grants, 4813
Virginia Foundation for Humanities Open Grants, 4814

Jurisprudence
Dennis Leslie Mahony Prize in Legal Theory, 1741

Justice
Gruber Foundation Prize for Global Justice and Women's Rights, 2182
OSF Fellowships, 3840
Princeton University Center for Human Values Short Movie Prize, 3995

Juvenile Delinquency
Boston Foundation Grants, 1361
Connecticut Community Foundation Grants, 1627
DOJ Juvenile Justice and Delinquency Prevention Special Emphasis Grants, 1826
DOJ National Institute of Corrections Training and Staff Development Grants, 1827
DOJ National Inst of Justice Visiting Fellowships, 1829
Girl's Best Friend Foundation Grants, 2143
Juvenile Justice Clearing House Grants, 2530
Oppenstein Brothers Foundation Grants, 3792

Juvenile Law
DOJ Juvenile Justice and Delinquency Prevention Special Emphasis Grants, 1826
OSF Soros Justice Advocacy Fellowships, 3850

Kazakhstan
Fulbright Traditional Scholar Program in the South and Central Asia Region, 2084
IREX Individual Advanced Research Opportunities Grants, 2419

Kidney Diseases and Disorders
ASCO/UICC International Cancer Technology Transfer Fellowships, 1044
ASCO Young Investigator Award, 1050
Conquer Cancer Foundation of ASCO Career Development Award, 1630
NIA Renal Function and Chronic Kidney Disease in Aging Grants, 3118
NIDDK Advances in Polycystic Kidney Disease Grants, 3133
NIDDK Ancillary Studies of Kidney Disease Accessing Information from Clinical Trials, Epidemiological Studies, and Databases Grants, 3135
NIDDK Calcium Oxalate Stone Diseases Grants, 3139
NIDDK Devel of Disease Biomarkers Grants, 3145
NIDDK Education Program Grants, 3149
NIDDK Enhancing Zebrafish Research with Research Tools and Techniques, 3151
NIDDK Health Disparities in NIDDK Diseases, 3156
NIDDK Mentored Research Scientist Development Award, 3161
NIDDK Multi-Center Clinical Study Cooperative Agreements, 3162
NIDDK Multi-Center Clinical Study Implementation Planning Grants, 3163
NIDDK Non-Invasive Methods for Diagnosis and Progression of Diabetes, Kidney, Urological, Hematological and Digestive Diseases, 3168
NIDDK Pilot and Feasibility Clinical Research Grants in Kidney or Urologic Diseases, 3171
NIDDK Proteomics: Diabetes, Obesity, And Endocrine, Digestive, Kidney, Urologic, And Hematologic Diseases, 3174
NIDDK Research Grants for Studies of Hepatitis C in the Setting of Renal Disease, 3175
NIDDK Secondary Analyses in Obesity, Diabetes and Digestive and Kidney Diseases Grants, 3178
NIDDK Small Grant for K08/K23 Recipients, 3182
NIDDK Small Grants for Underrepresented Minority Scientists in Diabetes and Digestive and Kidney Diseases, 3183
NIDDK Training in Clinical Investigation in Kidney and Urology Grants, 3185
NIGMS Short Courses on Mathematical, Statistical, and Computational Tools for Studying Biological Systems, 3201
NIH Nonhuman Primate Immune Tolerance Cooperative Study Group Grants, 3228
NIH Sarcoidosis: Research into the Cause of Multi-Organ Disease and Clinical Strategies for Therapy Grants, 3243
NKF Clinical Scientist Awards, 3274
NKF Franklin McDonald/Fresenius Medical Care Clinical Research Grants, 3275
NKF Professional Councils Research Grants, 3276
NKF Research Fellowships, 3277
NKF Young Investigator Grants, 3278
NYAM Edward N. Gibbs Memorial Lecture and Award in Nephrology, 3682
Pfizer ASPIRE North America Broad Spectrum Antibiotics for the Treatment of Gram-Negative or Polymicrobial Infections Research Awards, 3914
Pfizer ASPIRE Worldwide Endocrine Young Investigator Grants, 3918
Pfizer Medical Education Track Two Grants, 3927
PKD Foundation Lillian Kaplan International Prize for the Advancement in the Understanding of Polycystic Kidney Disease, 3974
PKD Foundation Research Grants, 3975

Kinesiology
Canada Graduate Scholarships and NSERC Postgraduate Scholarships, 1420

Knowledge Acceleration
NLM Exceptional, Unconventional Research Grants Enabling Knowledge Acceleration, 3295

Korea
American Historical Association John K. Fairbank Prize in East Asian History, 733
Center for East Asian Studies Fellowships, 1474
CFR Int'l Affairs Fellowships in South Korea, 1503
Fulbright International Education Administrators Seminar Program Grants, 2073

Korean Language/Literature
Fulbright Critical Lang Enhancement Awards, 2068

Labor Economics
National Housing Endowment Challenge/Build/Grow Grant, 2877

Labor Relations
National Endowment for Democracy Discretionary Grants, 2862

Land Management
Appraisal Institute Education Trust Scholarships, 917
DOI Geological Research and Data Acquisition Grants, 1822
DOI Rhinoceros and Tiger Conservation Grants, 1824
Illinois DNR Youth Recreation Corps Grants, 2361
National Housing Endowment Challenge/Build/Grow Grant, 2877
NFWF Pulling Together Initiative Grants, 3007
Northwest Fund for the Environment Grants, 3431
Organization for Tropical Studies Fellowships, 3796
Organization for Tropical Studies Pilot Awards for Exploratory Research, 3797
OTS Francis J. Bossuyt Fellowship, 3855
OTS Undergrad NAPIRE Summer Interns, 3858
Union Pacific Foundation Community and Civic Grants, 4568
USDA McIntire-Stennis Cooperative Forestry Research Grants, 4634
Weyerhaeuser Company Foundation Grants, 4857
William and Flora Hewlett Foundation Environmental Grants, 4874

Land Use Planning/Policy
DOI Earthquake Hazards Reduction Res Grants, 1821
Emma B. Howe Memorial Foundation Grants, 1920
National Housing Endowment Challenge/Build/Grow Grant, 2877
NFWF Acres for America Grants, 2996
NIEHS Mentored Clinical Scientist Research Career Development Awards, 3193
Organization for Tropical Studies Fellowships, 3796
Organization for Tropical Studies Pilot Awards for Exploratory Research, 3797
OTS Francis J. Bossuyt Fellowship, 3855
OTS Undergrad NAPIRE Summer Interns, 3858
Rocky Mountain Mineral Law Found Grants, 4094
Surdna Found Sustainable Environments Grants, 4471
Van Alen Institute New York Prize Fellowship, 4806
Young Scientists Summer Program, 4903

Landscape Architecture/Design

AAR Rome Prize Fellowships, 202
Andy Warhol Foundation for the Visual Arts Curatorial Research Fellowships, 847
Bogliasco Foundation Liguria Study Center Residencies, 1358
GCA Interchange Fellowship, 2098
Herb Society of America Research and Education Grants, 2243
Maine Community Foundation Keepers Preservation Education Grants, 2651
National Endowment for the Humanities Fellowships at Winterthur, 2867
Regents Professional Opportunity Scholarships, 4049
Winterthur Museum Dissertation Fellowships, 4887
Winterthur Research Fellowships, 4888

Language

ACLS Charles A. Ryskamp Research Fellowships, 281
ACLS Frederick Burkhardt Residential Fellowships for Recently Tenured Scholars, 288
Alabama Humanities Foundation Mini Grants, 480
Alabama Humanities Foundation Planning/Consultant Grants, 481
ARO Discrete Mathematics and Computer Science Research Contracts, 1009
Endangered Language Fund Grants, 1921
Foundation for Endangered Languages Grants, 2037
George Foundation Grants, 2115
Kellogg Institute for International Studies Quechua Summer Language Fellowships, 2548
Lady Gregory Postgraduate Fellowship, 2579
National Security Education Program David L. Boren Graduate Fellowships, 2884
NEH Documenting Endangered Lang Grants, 2955
NSF/NEH Documenting Endangered Languages Project Grants and Fellowships: data, infrastructure and computational methods, 3455
NSF SBE Doctoral Dissertation Research Improvement Grants, 3641
ONR Computational Neuroscience & Biorobotics Grants, 3752
Shastri Indo-Canadian Institute Student Language Training Fellowships, 4231
SRC Educational Science Lang Checking Grants, 4360

Language Acquisition and Development

ASHFoundation Student Research Grants in Early Childhood Language Development, 1085
IRA Jeanne S. Chall Research Fellowship, 2410
Robert Bowne Foundaion Edmund A. Stanley, Jr. Research Grants, 4073
TIRF Doctoral Dissertation Grants, 4520

Language Arts Education

Andrew W. Mellon/ACLS Dissertation Completion Fellowships, 844
Holberg Memorial Fund International Prize, 2272
IRA Jeanne S. Chall Research Fellowship, 2410
McCune Charitable Foundation Grants, 2719
NCTE Teacher Researcher Award, 2940

Laryngology

ALVRE Casselberry Award, 585
ALVRE Grant, 586
ALVRE Seymour R. Cohen Award, 587

Lasers and Masers

Abdus Salam ICTP/IAEA Sandwich Training Education Grants, 226
Abdus Salam ICTP Group Associates Grants, 228
Abdus Salam ICTP Junior Associates Grants, 231
Abdus Salam ICTP Postgraduate Diploma Grants, 235
Abdus Salam ICTP Regular Associates Grants, 237
Abdus Salam ICTP Schools, Conferences, Workshops and Extended Research Workshops Grants, 238
Abdus Salam ICTP Senior Associates Grants, 239
Abdus Salam ICTP Training and Research in Italian Laboratories Grants, 240
APS Arthur L. Schawlow Prize in Laser Science, 945
ARO Atomic, Molecular, and Optical Physics Research Contracts, 1005
USDD Atomic, Molecular and Quantum Physics Grants, 4687

Latin America

American Historical Association Albert J. Beveridge Award, 712
American Historical Association Albert J. Beveridge Grants for Research in the History of the Western Hemisphere, 713
American Historical Association Clarence H. Haring Prize, 716
American Historical Association James Henry Breasted Prize, 729
ASM-UNESCO Leadership Grant for International Educators, 1099
Bernard Lowy Fund for the Study of Tropical Botany in Latin America, 1335
Kellogg Institute for International Studies Experiencing The World Fellowships, 2539
Kellogg Institute for International Studies Grants for Conferences and Workshops, 2543
Kellogg Institute for International Studies Grants for Distinguished Lectureships, 2544
Kellogg Institute for International Studies Grants for Events, 2545
Kellogg Institute for International Studies Supplemental Awards for PhD Students, 2550
NL Audrey Lumsden-Kouvel Fellowships, 3282
Organization for Tropical Studies Fellowships, 3796
Organization for Tropical Studies Pilot Awards for Exploratory Research, 3797
OTS Francis J. Bossuyt Fellowship, 3855
OTS Hovore-Horn Fellowships, 3856
OTS Stanley Smith Horticultural Fellowships, 3857
OTS Undergrad NAPIRE Summer Interns, 3858
Overbrook Foundation Grants, 3860
RA Kleinhans Fellowship for Research in Tropical Nontimber Forest Products, 4026
Smithsonian Latino Studies Fellowships, 4278
SSRC Drugs, Security & Democracy Fellows, 4426

Latin Language/Literature

AAR Postdoctoral Fellowships, 201

Law

AAUW Selected Professions Fellowships, 217
ACLS Charles A. Ryskamp Research Fellowships, 281
ACLS Frederick Burkhardt Residential Fellowships for Recently Tenured Scholars, 288
Alabama Humanities Foundation Mini Grants, 480
Alabama Humanities Foundation Planning/Consultant Grants, 481
Alexander von Humboldt Foundation German Chancellor Fellowships for Prospective Leaders from the People's Republic of China, 526
Alexander von Humboldt Foundation German Chancellor Fellowships for Prospective Leaders from the Russian Federation, 527
American Academy of Arts and Sciences Hellman Fellowships, 612
American Academy of Arts and Sciences Visiting Scholars, 613
Andrew W. Mellon/ACLS Dissertation Completion Fellowships, 844
APSA Gabriel A. Almond Award, 934
ASCSA Harry Bikakis Fellowship, 1053
Canada-U.S. Fulbright International Science and Technology Award, 1417
Canada-U.S. Fulbright Senior Specialists Grants, 1418
Canada-U.S. Fulbright Traditional Scholars, 1419
Chapin Hall International Fellowships in Children's Policy Research, 1517
Charles Stewart Mott Foundation Anti-Poverty Program, 1528
Dave Thomas Foundation for Adoption Grants, 1723
E.L. Wiegand Foundation Grants, 1885
Fulbright Alumni Initiatives Awards Grants, 2064
Fulbright Binational Business Grants in Mexico, 2066
Fulbright Business Grants in Spain, 2067
Fulbright Distinguished Chairs Program Grants, 2069
Fulbright European Union Scholar-in-Residence Program Grants, 2071
Fulbright New Century Scholars Program Grants, 2075
Fulbright Occasional Lecturer Program Grants, 2076
Fulbright Specialists Program Grants, 2079
Fulbright Traditional Scholar Program in Europe and Eurasia, 2080
Fulbright Traditional Scholar Program in Sub-Saharan Africa, 2081
Fulbright Traditional Scholar Program in the East Asia/Pacific Region, 2082
Fulbright Traditional Scholar Program in the Near East and North Africa Region, 2083
Fulbright Traditional Scholar Program in the South and Central Asia Region, 2084
Fulbright Traditional Scholar Program in the Western Hemisphere, 2085
Fulbright Visiting Specialists Program: Direct Access to the Muslim World, 2087
Fund for Scientific Research--Flanders Grants, 2089
Gladys Krieble Delmas Foundation Grants for Venetian Research, 2144
Harry Frank Guggenheim Foundation Dissertation Fellowships, 2207
Harvard University John M. Olin Prizes, 2214
Holberg Memorial Fund International Prize, 2272
Holberg Memorial Fund Nils Klim Prize, 2273
Holberg Memorial Fund School Project, 2274
Horton Research Grants, 2282
IREX Edmund S. Muskie Graduate Fellowship Program, 2417
IREX International Leadership in Education Program Grants, 2420
Japan Foundation Japanese Studies Fellowships, 2476
Japan Foundation New York Japanese Studies Fellowships, 2478
Jewish Federation of Metropolitan Chicago Scholarships, 2486
JMO Law and Legal System Grants, 2491
JMO Public Policy Research Grants, 2492
John Ben Snow Memorial Trust Grants, 2497
Kluge Center Distinguished Chairs, 2560
Law and Public Affairs Fellowships at Princeton University, 2588
Law and Social Science Doctoral Dissertation Fellowships, 2589
LSAC Research Grants, 2623
Maine Community Foundation Keepers Preservation Education Grants, 2651
National Center on Philanthropy and the Law Fellowship in Nonprofit Law, 2859
OSF-Baltimore Community Fellowships, 3833
OSF Palestinian Rule of Law Fellowships, 3848
OSF Soros Justice Advocacy Fellowships, 3850
PHF Penn Faculty Research Fellowships, 3932
PHF Regional Mellon Faculty Research and Cultural Fellowships, 3933
PHF Undergraduate Humanities Forum Research Fellowships, 3934
Prudential Foundation Education Grants, 4002
Rajiv Gandhi Foundation Grants, 4025
Regents Professional Opportunity Scholarships, 4049
Rockefeller Brothers Fund Fellowship in Nonprofit Law, 4088
Rocky Mountain Mineral Law Found Grants, 4094
Shastri Indo-Canadian Institute Collaborative Research Project Grants, 4220
Special Olympics Health Professions Stu Grants, 4332
SSHRC Bora Laskin National Fellowship in Human Rights Research, 4405
SSHRC Insight Development Grants, 4409
SSHRC Insight Grants, 4410
Takemi Program in International Health Research and Advanced Training Fellowships, 4491
University of London Visiting Fellowships, 4588
Wallace Foundation Grants, 4829
Yale University Beinecke Library Research Fellowships for Yale Grad & Professional School Students, 4902

Law Enforcement
Alberta Law Foundation Grants, 503
ASA/BJS Statistical Methodological Research Program Grants, 1029
Burlington Northern Santa Fe Foundation Grants, 1389
DOI Rhinoceros and Tiger Conservation Grants, 1824
DOJ National Institute of Corrections Training and Staff Development Grants, 1827
DOJ National Institute of Justice Graduate Research Fellowships, 1828
DOJ National Inst of Justice Visiting Fellowships, 1829
IREX International Leadership in Education Program Grants, 2420
Kenneth T. and Eileen L. Norris Found Grants, 2554
NRA Foundation Grants, 3437

Law and Social Change
Alston Bannerman Sabbatical Fellowships, 580
American Historical Association Littleton-Griswold Prize in American Law and Society, 735
American Historical Association Littleton-Griswold Research Grants in U.S. Legal History, 736
Law and Public Affairs Fellowships at Princeton University, 2588
OSF Palestinian Rule of Law Fellowships, 3848
OSF Social Work Fellowships, 3849
OSF Soros Justice Advocacy Fellowships, 3850
OSF Soros Justice Media Fellowships, 3851
Radcliffe Institute Research Support Grants, 4023
SSRC African Peacebuilding Network Research Grants, 4422

Law and Society
AAB Berlin Prize Fellowships, 48
APSA Edward S. Corwin Award, 933
Hatton W. Sumners Foundation for the Study and Teaching of Self Government Grants, 2219
Law and Public Affairs Fellowships at Princeton University, 2588
United States Institute of Peace Annual Grants, 4582

Law, History of
American Historical Association Fellowship in Aerospace History, 717
American Historical Association Littleton-Griswold Prize in American Law and Society, 735
American Historical Association Littleton-Griswold Research Grants in U.S. Legal History, 736

Leadership
AAAAI Distinguished Scientist Award, 19
AAAS Philip Hauge Abelson Award, 37
AABB Hemphill-Jordan Leadership Award, 49
AACR-Minorities in Cancer Research Jane Cooke Wright Lectureship Awards, 119
AACR Margaret Foti Award for Leadership and Achievements in Cancer Research, 142
AAOHN Foundation New Investigator Researcher Grants, 196
AIChE Clarence G. Gerhold Award, 400
ALA Excellence in Academic Libraries Award, 486
Alexander von Humboldt Foundation German Chancellor Fellowships for Prospective Leaders from the People's Republic of China, 526
Alexander von Humboldt Foundation German Chancellor Fellowships for Prospective Leaders from the Russian Federation, 527
Arca Foundation Grants, 984
ASHRAE Willis H. Carrier Graduate Research Fellowship, 1089
ASM bioMerieux Sonnenwirth Award for Leadership in Clinical Microbiology, 1103
Bodman Foundation Grants, 1356
Boeing Company Contributions Grants, 1357
Bower Award for Achievement in Science, 1364
BP Conservation Programme Future Conservationist Awards, 1365
Bullitt Foundation Grants, 1384
Castle Rock Foundation Grants, 1441
CDC Preventive Med Residency & Fellowship, 1465
Center for Education of Women Visiting Scholar Research Grants, 1475
Chapin Hall International Fellowships in Children's Policy Research, 1517
Charles Lafitte Foundation Grants, 1525
Charles Stewart Mott Anti-Poverty Grants, 1528
Charles Stewart Mott Foundation Grants, 1529
Chicago Community Trust Fellowships, 1546
ConocoPhillips Grants Program, 1629
Cowles Charitable Trust Grants, 1652
Dean Witter Foundation Grants, 1728
DOE Albert Einstein Distinguished Educator Fellowship Program, 1770
Dorothy Rider Pool Health Care Grants, 1841
El Pomar Leadership Development and Community Engagement Postgraduate Fellowship, 1912
Energy Foundation Climate Grants, 1923
FCD Child Development Grants, 1984
FTE North American Doctoral Fellowships, 2061
Gates Family Foundation Fellowships, 2097
Gilbreth Memorial Fellowships, 2141
Heineman Foundation for Research, Education, Charitable and Scientific Purposes, 2227
Hispanic Scholarships, 2270
Ian Axford (New Zealand) Fellowships in Public Policy, 2313
IMLS 21st Century Museum Prof Grants, 2362
IMLS National Leadership Grants, 2366
Japan Foundation Center for Global Partnership Grants, 2472
John D. and Catherine T. MacArthur Foundation Global Challenges Program Grants, 2498
Johnson & Johnson Corporate Contributions, 2508
LITA/Gaylord Award for Achievement in Library and Information Technology, 2618
Lumina Foundation for Education Grants, 2627
National Urban Fellows Program, 2888
NCI Cancer Education and Career Development, 2917
NFIE Found Learning & Leadership Grants, 2992
NMF National Medical Association Awards for Medical Journalism, 3323
NMF National Medical Association Emerging Scholar Awards, 3324
NMF Ralph W. Ellison Prize, 3327
Nordson Corporation Foundation Grants, 3416
NSERC Howard Alper Postdoctoral Prize, 3441
NSF Presidential Early Career Awards for Scientists and Engineers Grants, 3633
Prudential Foundation Education Grants, 4002
Rhode Island Foundation Grants, 4061
Richard Davoud Donchian Foundation Grants, 4066
RWJF Health Policy Fellowships, 4142
Saint Mary's College Fellowships, 4159
SfN Louise Hanson Marshall Specific Recognition Award, 4208
SRC Aristotle Award, 4349
Surdna Found Sustainable Environments Grants, 4471
The World Food Prize, 4503
W.K. Kellogg Foundation Healthy Kids Grants, 4822
Wallace Foundation Grants, 4829
Willary Foundation Grants, 4870
Xerox Foundation Grants, 4899

Learning Disabilities
AACAP Pilot Research Award for Learning Disabilities, Supported by the Elaine Schlosser Lewis Fund, 71
AACAP Sidney Berman Award for the School-Based Study and Intervention for Learning Disorders and Mental Ilness, 78
IRA Albert J. Harris Awards, 2405
Pinkerton Foundation Grants, 3967

Learning Motivation
Lilly Endowment Teacher Creativity Fellowships, 2616
Nellie Mae Education Foundation District-Level Change Grants, 2968
SRC Aristotle Award, 4349
Stanford Center on Adolescence Youth Purpose Research Grants, 4443

Lectureships
AACC International Milling & Baking Division's John C. Halverson Memorial Lectureship, 92
ASME Calvin W. Rice Lecture Award, 1112
ASME Robert Henry Thurston Lecture Award, 1160
Kellogg Institute for International Studies Grants for Distinguished Lectureships, 2544
NWO Aspasia Grants, 3673

Legal Education
Alberta Law Foundation Grants, 503
Andrew W. Mellon/ACLS Dissertation Completion Fellowships, 844
BPW Career Advancement Scholarships, 1366
Christine Mirzayan Science and Technology Policy Graduate Fellowship Program, 1556
Cord Foundation Grants, 1645
Hague Academy of International Law Summer Scholarships, 2198
Jewish Federation of Metropolitan Chicago Scholarships, 2486
Kosciuszko Foundation Tuition Scholarships, 2574
Law and Public Affairs Fellowships at Princeton University, 2588
LSAC Research Grants, 2623
Massachusetts Bar Foundation Legal Intern Fellowships, 2676
Miller Brewing Corporate Contributions Program Grants, 2774
National Center on Philanthropy and the Law Fellowship in Nonprofit Law, 2859
Robert A. Levy Fellowship in Law and Liberty, 4070
Searle Freedom Trust Grants, 4192
USDC Dissertation Fellowships, 4665

Legal Procedure
Alberta Law Foundation Grants, 503
APSA Gabriel A. Almond Award, 934

Legal Profession
Alberta Law Foundation Grants, 503
LSAC Research Grants, 2623

Legal Reform
Alberta Law Foundation Grants, 503
OSF Palestinian Rule of Law Fellowships, 3848

Legal Services
Alberta Law Foundation Grants, 503
American College of Bankruptcy Grants, 699
Bodenwein Public Benevolent Foundation Grants, 1355
ConocoPhillips Grants Program, 1629
GNOF Gulf Coast Oil Spill Grants, 2153
Massachusetts Bar Foundation Legal Intern Fellowships, 2676
McCune Foundation Human Services Grants, 2722
Rhode Island Foundation Grants, 4061

Legal Systems
APSA Gabriel A. Almond Award, 934
Carrier Corporation Contributions Grants, 1440
JMO Law and Legal System Grants, 2491
Law and Public Affairs Fellowships at Princeton, 2588
Law and Social Science Doctoral Dissertation Fellowships, 2589

Leukemia
ACGT Investigators Grants, 271
AMA Foundation Seed Grants for Research, 601
NCIC Canadian Cancer Society Research Grants, 2919
NYCT Biomedical Research Grants, 3694
Pfizer Medical Education Track Two Grants, 3927
Ryan Gibson Foundation Grants, 4151

Liberal Arts Education
Holberg Memorial Fund International Prize, 2272
McCune Charitable Foundation Grants, 2719
UCR Resident Fellowships, 4552
W.M. Keck Foundation Undergraduate Education Grants, 4825

Liberty
Independent Institute John M. Templeton Fellowships Essay Contest, 2369
Searle Freedom Trust Grants, 4192

Libraries
AAB Berlin Prize Fellowships, 48
ALA ACRL Doctoral Dissertation Fellowship, 478
ALA Annual Diversity Research Grant, 479
ALA Carroll Preston Baber Research Grant, 483
ALA Frederick G. Kilgour Award for Research in Library and Information Technology, 487
ALA Ingenta Research Awards, 489
ALA NewsBank/Readex Cath Reynolds Award, 494
ALA Phyllis Dain Library History Dissert Award, 495
ALA STS Innovation in Science and Technology Librarianship Award, 499
Alcatel-Lucent Technologies Foundation Grants, 509
American Academy of Religion Individual Research Grants, 618
American Historical Association J. Franklin Jameson Fellowship, 725
Auburn Foundation Grants, 1263
Bay and Paul Foundations, Inc Grants, 1307
Bodenwein Public Benevolent Foundation Grants, 1355
Booth-Bricker Fund Grants, 1359
Brinson Foundation Grants, 1372
Burlington Northern Santa Fe Foundation Grants, 1389
Carl and Lily Pforzheimer Foundation Grants, 1427
Chautauqua Region Community Found Grants, 1533
Corning Incorporated Found Educational Grants, 1646
Cushwa Center Research Travel Grants, 1673
Frances C. and William P. Smallwood Foundation Grants, 2045
George Kress Foundation Grants, 2119
H.W. Wilson Foundation Grants, 2196
Harvey Randall Wickes Foundation Grants, 2218
Hatton W. Sumners Foundation for the Study and Teaching of Self Government Grants, 2219
Huie-Dellmon Trust Grants, 2295
IMLS 21st Century Museum Prof Grants, 2362
IMLS Grants to State Library Administrative Agencies, 2364
IMLS Laura Bush 21st Century Librarian Program Grants, 2365
IMLS National Leadership Grants, 2366
John Ben Snow Memorial Trust Grants, 2497
LITA/Gaylord Award for Achievement in Library and Information Technology, 2618
McLean Contributionship Grants, 2727
Motorola Foundation Grants, 2796
Munby Fellowship in Bibliography, 2804
NEH Fellowships at Independent Research Institutions, 2957
NHPRC Digitizing Historical Records Grants, 3072
NLM Internet Connection for Medical Institutions Grants, 3300
NSF CISE Cluster Exploratory Grants, 3525
OSF Intellectual Property Reform Grants, 3846
Rayonier Foundation Grants, 4034
Reinberger Foundation Grants, 4051
Roger K. Summit Scholarship, 4097
Scientific Communities Initiative Grants, 4189
Sioux Falls Area Community Foundation Community Fund Grants (Unrestricted), 4262
Sioux Falls Area Comm Foundation Spot Grants, 4263
Talbert and Leota Abrams Foundation, 4492
Union Pacific Foundation Community and Civic Grants, 4568
USDA Hispanic-Serving Institutions Education Grants, 4628
Virginia Foundation for the Humanities Discretionary Grants, 4813
Virginia Foundation for Humanities Open Grants, 4814
W. C. Griffith Foundation Grants, 4818
W.C. Griffith Foundation Grants, 4819

Libraries, Academic
ACRL Academic/Research Librarian of the Year Award, 303
ACRL Instruction Section Innovation Award, 304
ALA Excellence in Academic Libraries Award, 486
ALA Innovation Award, 490
ALA Marta Lange/CQ Press Award, 493
ALA ProQuest Documents to the People Award, 496
ALA Samuel Lazerow Fellowship For Research in Collections and Technical Services in Academic and Research Libraries, 497
ALA STS Innovation in Science and Technology Librarianship Award, 499
ALA WESS De Gruyter European Librarianship Study Grant, 501
CIC Traveling Scholars, 1566
EU Robert W. Woodruff Library Fellowships, 1959
F.W. Olin Foundation Construction Grant, 1972
H.W. Wilson Foundation Grants, 2196
IMLS Grants to State Library Administrative Agencies, 2364
Japan Foundation Six-Month Grants for Specialists in Cultural and Academic Fields, 2479
John Deere Foundation Grants, 2500
Minority Faculty Fellowship Program, 2781
NSF Biological Research Collections, 3504
Princeton University Library Research Grants, 3997
Roy J. Carver Charitable Trust Medical and Science Research Grants, 4114
Starr Foundation Grants, 4447
Wayne and Gladys Valley Foundation Grants, 4835

Libraries, Art
IMLS Grants to State Library Administrative Agencies, 2364

Libraries, Law
ALA Marta Lange/CQ Press Award, 493
Alberta Law Foundation Grants, 503
IMLS Grants to State Library Administrative Agencies, 2364

Libraries, Medical
Hospital Libraries Section / MLA Professional Development Grants, 2283
IMLS Grants to State Library Administrative Agencies, 2364
MLA Cunningham Memorial International Fellowship, 2783
MLA David A. Kronick Traveling Fellowship, 2784
MLA Donald Lindberg Research Fellowship, 2785
MLA Graduate Scholarship, 2786
MLA Grad Scholarship for Minority Students, 2787
MLA Murray Gottlieb Prize Essay Award, 2788
MLA Research, Development, and Demonstration Project Grant, 2789
MLA Rittenhouse Award, 2790
MLA Thomas Reuters / Frank Bradway Rogers Information Advancement Award, 2791
NLM Exploratory/Developmental Grants, 3296
NLM Grants for Scholarly Works in Biomedicine and Health, 3298
Thomson Reuters / MLA Doctoral Fellowships, 4512

Libraries, Presidential
Herbert Hoover Presidential Library Association Travel Grants, 2242

Libraries, Public
ALA ProQuest Documents to the People Award, 496
IMLS Grants to State Library Administrative Agencies, 2364
Japan Foundation Six-Month Grants for Specialists in Cultural and Academic Fields, 2479
Praxair Foundation Grants, 3986
Union Pacific Foundation Community and Civic Grants, 4568

Libraries, Research
ACRL Academic/Research Librarian of the Year Award, 303
ALA ACRL Doctoral Dissertation Fellowship, 478
ALA Excellence in Academic Libraries Award, 486
ALA Innovation Award, 490
ALA ProQuest Documents to the People Award, 496
ALA Samuel Lazerow Fellowship For Research in Collections and Technical Services in Academic and Research Libraries, 497
ALA WESS De Gruyter European Librarianship Study Grant, 501
Alberta Law Foundation Grants, 503
CIC Traveling Scholars, 1566
Clark Library Short-Term Resident Fellowships, 1577
Duke Univ Libraries Mary Lily Research Grants, 1877
Francis C. Wood Scholar in Residence Program, 2047
Harry Ransom Humanities Research Center Research Fellowships, 2209
Huntington Library National Endowment for the Humanities Fellowships, 2307
IES Faculty Fellowships, 2341
IMLS Grants to State Library Administrative Agencies, 2364
Japan Foundation Six-Month Grants for Specialists in Cultural and Academic Fields, 2479
Mary Gates Research Scholarships, 2670
MBL Fred Karush Library Readership, 2703
NL American Society for Eighteenth-Century Studies Fellowships, 3279
NL Arthur and Lila Weinberg Fellowships for Independent Scholars and Researchers, 3280
NL Associated Colleges of the Midwest/Great Lakes College Association Faculty Fellowships, 3281
NL Audrey Lumsden-Kouvel Fellowships, 3282
NL British Academy Fellowship for Study in the United Kingdom, 3283
NL Frances C. Allen Fellowships, 3284
NL Grad Scholar-in-Residence Opportunities, 3285
NL Herzog August Bibliothek Wolfenbuttel Fellowship, 3286
NL Lawrence Lipking Fellowship at Newberry, 3287
NL Lester J. Cappon Fellowship in Documentary Editing, 3288
NL Lloyd Lewis Fellowships in American Hist, 3289
NL Midwest Modern Lang Assoc Fellowship, 3299
NL Monticello College Foundation Fellowship for Women, 3304
NL National Endowment for the Humanities Fellowships, 3310
NL Newberry Library/Ecole Nationale des Chartes Exchange Fellowship, 3313
NL Northeast Modern Language Assocation Fellowship, 3314
NL Scholar-in-Residence Opportunity, 3315
NL Short-Term Fellowships in the History of Cartography, 3316
NL South Central Modern Languages Association Fellowship, 3318
NL Susan Kelly Power and Helen Hornbeck Tanner Fellowship, 3319
NSF Biological Research Collections, 3504
OSF Intellectual Property Reform Grants, 3846
Princeton University Library Research Grants, 3997
University of Southern Mississippi Keats/Domanska/de Grummond Children's Literature Research Fellowships, 4590
Wolfsonian-FIU Fellowships, 4890
Yale University Beinecke Library Fellowships for Visiting Scholars, 4901
Yale University Beinecke Library Research Fellowships for Yale Grad & Professional School Students, 4902

Libraries, School
AASL Research Grant, 207
Alabama Library Assoc CUS Research Grants, 482
ALA ProQuest Documents to the People Award, 496
H.W. Wilson Foundation Grants, 2196
IMLS Grants to State Library Administrative Agencies, 2364
Lilly Endowment Teacher Creativity Fellowships, 2616

Libraries, Special
IMLS Grants to State Library Admin Agencies, 2364
NSF Biological Research Collections, 3504

Libraries, State
ALA ProQuest Documents to the People Award, 496
IMLS Grants to State Library Administrative Agencies, 2364

Library Administration
ACRL Instruction Section Innovation Award, 304
ALA Excellence in Academic Libraries Award, 486
IMLS Grants to State Library Administrative Agencies, 2364
MLA David A. Kronick Traveling Fellowship, 2784
MLA Donald Lindberg Research Fellowship, 2785

Library Automation
ACRL Instruction Section Innovation Award, 304
ALA Excellence in Academic Libraries Award, 486
ALA Frederick G. Kilgour Award for Research in Library and Information Technology, 487
ALA Innovation Award, 490
ALA Jan Merrill-Oldham Professional Development Grant, 491
ALA ProQuest Documents to the People Award, 496
IMLS Grants to State Library Administrative Agencies, 2364
LITA/Gaylord Award for Achievement in Library and Information Technology, 2618

Library History
ALA Donald G. Davis Article Award, 484
ALA Emerald Research Grant Awards, 485
ALA Gale Cengage History Research and Innovation Award, 488
ALA Phyllis Dain Library History Dissert Award, 495
Getty Research Grants for Institutions - Art History Research Grants, 2132
IMLS Grants to State Library Admin Agencies, 2364
United Methodist Church Josephine Forman Scholarship, 4573

Library Science
ACRL Academic/Research Librarian of the Year Award, 303
ALA Annual Diversity Research Grant, 479
ALA Emerald Research Grant Awards, 485
ALA Excellence in Academic Libraries Award, 486
ALA Gale Cengage History Research and Innovation Award, 488
ALA Innovation Award, 490
ALA ProQuest Documents to the People Award, 496
ALA Ulrich's Serials Librarianship Award, 500
Canada-U.S. Fulbright International Science and Technology Award, 1417
Canada-U.S. Fulbright Senior Specialists Program Grants, 1418
Canada-U.S. Fulbright Traditional Scholars, 1419
Frances C. and William P. Smallwood Foundation Grants, 2045
Fulbright Alumni Initiatives Awards Grants, 2064
Fulbright Distinguished Chairs Program Grants, 2069
Fulbright European Union Scholar-in-Residence Program Grants, 2071
Fulbright New Century Scholars Program Grants, 2075
Fulbright Occasional Lecturer Program Grants, 2076
Fulbright Specialists Program Grants, 2079
Fulbright Traditional Scholar Program in Europe and Eurasia, 2080
Fulbright Traditional Scholar Program in Sub-Saharan Africa, 2081
Fulbright Traditional Scholar Program in the East Asia/Pacific Region, 2082
Fulbright Traditional Scholar Program in the Near East and North Africa Region, 2083
Fulbright Traditional Scholar Program in the South and Central Asia Region, 2084
Fulbright Traditional Scholar Program in the Western Hemisphere, 2085
Fulbright Visiting Specialists Program: Direct Access to the Muslim World, 2087
IREX Edmund S. Muskie Graduate Fellowship, 2417
MLA Donald Lindberg Research Fellowship, 2785
MLA Graduate Scholarship, 2786
MLA Grad Scholarship for Minority Students, 2787
MLA Research, Development, and Demonstration Project Grant, 2789
NSF Biological Research Collections, 3504
Roger K. Summit Scholarship, 4097
Shastri Indo-Canadian Inst Librarian Fellows, 4225
SSHRC Insight Development Grants, 4409
SSHRC Insight Grants, 4410
UNC School of Information and Library Science Doctoral Fellowships, 4565
W. C. Griffith Foundation Grants, 4818
W.C. Griffith Foundation Grants, 4819

Library Science Education
AASL Research Grant, 207
Alabama Library Assoc CUS Research Grants, 482
ALA Jan Merrill-Oldham Professional Development Grant, 491
ALA Ulrich's Serials Librarianship Award, 500
ALA WESS De Gruyter European Librarianship Study Grant, 501
IMLS Laura Bush 21st Century Librarian Grants, 2365
UNC School of Information and Library Science Doctoral Fellowships, 4565
Yale University Beinecke Library Research Fellowships for Yale Grad & Professional School Students, 4902

Life Sciences
AIHS ForeFront MBA Studentship Award, 443
Alexander von Humboldt Foundation Max Planck Research Award, 532
Alexander von Humboldt Foundation Philipp Franz von Siebold Award for Japanese Researchers, 534
Alfred P. Sloan Foundation Deep Carbon Observatory Grants, 547
American Australian Association USA to Australia Fellowships, 620
ANL Faculty Research Leave Program, 854
ANL Graduate Thesis Parts Res Appointments, 856
ANL Laboratory-Graduate Res Appointments, 857
Applied Biosystems Grants, 916
ASME Van C. Mow Medal, 1169
Beckman Young Investigators Program Grants, 1327
CCFF Chairman's Dist Life Sciences Award, 1449
Consortium for Graduate Study in Management Minority Fellowships, 1642
DAAD Study & Internship Program in Germany, 1708
GEM PhD Science Fellowships, 2106
HHMI Grants and Fellowships Programs, 2255
Kavli Foundation Research Grants, 2533
M.J. Murdock Charitable Trust College Research Program for Life Sciences, 2638
NATO Advanced Research Workshops Program, 2894
NATO Advanced Study Institutes Program, 2895
NATO Collaborative Linkage Grants, 2896
NATO Expert Visits Program Grants, 2897
NHLBI Pathway to Independence Awards, 3045
NSF Biomedical Engineering and Engineering Healthcare Grants, 3506
ORISE National Oceanic and Atmospheric Administration Faculty Research Participation, 3811
ORISE National Oceanic and Atmospheric Admin Student Research Participation, 3812
ORISE Postgraduate Research Participation at the National Exposure Research Laboratory, 3821
Research Program at Earthwatch Grants, 4057
RSC Thomas W. Eadie Medal, 4126
Swiss National Science Foundation Fellowships for Advanced Researchers, 4483
Swiss National Science Foundation Fellowships for Prospective Researchers, 4484
Swiss National Science Foundation Marie Heim-Vogtlin Grants, 4485
UNCFSP NASA Harriett G. Jenkins Predoctoral Fellowships, 4563
USDD ARO Broad Agency Announcement, 4686
W.M. Keck Foundation Science and Engineering Research Grants, 4824
Wellcome Trust Biomedical Science Grants, 4843

Life Skills Training
Coors Brewing Corporate Contributions Grants, 1644
Cruise Industry Charitable Foundation Grants, 1660
Pinkerton Foundation Grants, 3967
Plan for Social Excellence Grants, 3976

Light
APS Arthur L. Schawlow Prize in Laser Science, 945

Linguistic Geography
ACLS Charles A. Ryskamp Research Fellowships, 281
Foundation for Endangered Languages Grants, 2037
NEH Documenting Endangered Lang Grants, 2955

Linguistics/Philology
AAB Berlin Prize Fellowships, 48
ACLS Charles A. Ryskamp Research Fellowships, 281
AIAR Andrew W. Mellon Foundation Fellowships, 392
Alabama Humanities Foundation Mini Grants, 480
Alabama Humanities Foundation Planning/Consultant Grants, 481
Alfred P. Sloan Foundation Energy and Environment Grants, 552
American Philosophical Society Phillips Fund Grants for Native American Research, 770
Andrew W. Mellon/ACLS Dissertation Completion Fellowships, 844
ARIT/NEH Fellowships for Research in Turkey, 990
Australasian Institute of Judicial Administration Seed Funding Grants, 1268
Canada-U.S. Fulbright International Science and Technology Award, 1417
Canada-U.S. Fulbright Senior Specialists Program Grants, 1418
Canada-U.S. Fulbright Traditional Scholars, 1419
Center for East Asian Studies Fellowships, 1474
Endangered Language Fund Grants, 1921
Foundation for Endangered Languages Grants, 2037
Foundation for the Advancement of Mesoamerican Studies Grants, 2042
Fulbright Alumni Initiatives Awards Grants, 2064
Fulbright Distinguished Chairs Program Grants, 2069
Fulbright European Union Scholar-in-Residence Program Grants, 2071
Fulbright New Century Scholars Program Grants, 2075
Fulbright Occasional Lecturer Program Grants, 2076
Fulbright Specialists Program Grants, 2079
Fulbright Traditional Scholar Program in Europe and Eurasia, 2080
Fulbright Traditional Scholar Program in Sub-Saharan Africa, 2081
Fulbright Traditional Scholar Program in the East Asia/Pacific Region, 2082
Fulbright Traditional Scholar Program in the Near East and North Africa Region, 2083
Fulbright Traditional Scholar Program in the South and Central Asia Region, 2084
Fulbright Traditional Scholar Program in the Western Hemisphere, 2085
Fulbright Visiting Specialists Program: Direct Access to the Muslim World, 2087
George Foundation Grants, 2115
IES Faculty Fellowships, 2341
Jacobs Research Funds Small Grants Program, 2456
Kluge Center Kislak Fellows in American Studies, 2562
Kluge Center Kislak Short-Term Fellowships in American Studies, 2563
NEH Documenting Endangered Lang Grants, 2955
NL American Society for Eighteenth-Century Studies Fellowships, 3279
NL Arthur and Lila Weinberg Fellowships for Independent Scholars and Researchers, 3280
NL Associated Colleges of the Midwest/Great Lakes College Association Faculty Fellowships, 3281
NL British Academy Fellowship for Study in the United Kingdom, 3283
NL Frances C. Allen Fellowships, 3284
NL Grad Scholar-in-Residence Opportunities, 3285
NL Herzog August Bibliothek Wolfenbuttel Fellowship, 3286

NL Lawrence Lipking Fellowship at Newberry, 3287
NL Lester J. Cappon Fellowship in Documentary Editing, 3288
NL Mellon Postdoctoral Research Fellowships, 3294
NL Midwest Modern Lang Assoc Fellowship, 3299
NL Monticello College Foundation Fellowship for Women, 3304
NLM Research Project Grants, 3306
NL National Endowment for the Humanities Fellowships, 3310
NL Northeast Modern Language Assocation Fellowship, 3314
NL Scholar-in-Residence Opportunity, 3315
NL Short-Term Resident Fellowships for Individual Research, 3317
NL South Central Modern Languages Association Fellowship, 3318
NL Susan Kelly Power and Helen Hornbeck Tanner Fellowship, 3319
NSF/NEH Documenting Endangered Languages Project Grants and Fellowships: data, infrastructure and computational methods, 3455
NSF Arctic Social Sciences Grants, 3492
PHF Penn Faculty Research Fellowships, 3932
PHF Regional Mellon Faculty Research and Cultural Fellowships, 3933
PHF Undergraduate Humanities Forum Research Fellowships, 3934
Smithsonian Graduate Student Fellowships, 4277
Smithsonian Minority Visiting Student Awards, 4279
Smithsonian Native American Community Scholars Awards, 4291
Smithsonian Native American Visiting Student Awards, 4292
Smithsonian Postdoctoral Fellowships, 4293
Smithsonian Predoctoral Fellowships, 4295
SSHRC Insight Development Grants, 4409
SSHRC Insight Grants, 4410
Swedish Women's Educational Association International Scholarships, 4480
U.S. Department of Education Foreign Languages and Area Studies Fellowships, 4539
Wenner-Gren Foundation Dissertation Fieldwork Grants, 4848

Lipids
AOCS ACI/NBB Innovation Research Award, 872
AOCS Alton E. Bailey Award, 873
AOCS Analytical Division Student Award, 874
AOCS Australasian Award for Scientific Excellence in Lipid Research, 875
AOCS Edible Applications Technology Division Outstanding Achievement Award, 877
AOCS Edible Applications Technology Division Student Award, 878
AOCS Health & Nutrition Div Student Award, 881
AOCS Health and Nutrition Poster Competition, 882
AOCS Honored Student Award, 884
AOCS Lipid Oxidation and Quality Division Poster Competition, 886
AOCS Peter and Claire G. Kalustian Award, 888
AOCS Phospholipid Best Paper Award, 889
AOCS Processing Distinguished Service Award, 890
AOCS Ralph Holman Lifetime Award, 894
AOCS Stephen S. Chang Award, 896
AOCS Young Scientist Research Award, 902
NIH Structural Bio of Membrane Proteins Grant, 3247
Pfizer Medical Education Track Two Grants, 3927
Robert B McMillen Foundation Grants, 4072

Literacy
ALA Innovation Award, 490
Alcatel-Lucent Technologies Foundation Grants, 509
Allyn Foundation Grants, 574
Amgen Foundation Grants, 812
Andrew Family Foundation Opportunity Knocks Grants, 843
Arizona Cardinals Grants, 995
Ashland Corporate Contributions Grants, 1087
Assisi Foundation of Memphis General Grants, 1221
Auburn Foundation Grants, 1263
Ball Brothers Foundation General Grants, 1291
Bayer Foundation Grants, 1309
BBF Fellowships, 1311
Bodenwein Public Benevolent Foundation Grants, 1355
Bodman Foundation Grants, 1356
Booth-Bricker Fund Grants, 1359
Boston Foundation Grants, 1361
Brinson Foundation Grants, 1372
Cabot Corporation Foundation Grants, 1406
Carl and Lily Pforzheimer Foundation Grants, 1427
Charles Lafitte Foundation Grants, 1525
CIGNA Foundation Grants, 1568
Commonweal Foundation Grants, 1603
Coors Brewing Corporate Contributions Grants, 1644
Cowles Charitable Trust Grants, 1652
Cruise Industry Charitable Foundation Grants, 1660
Evjue Foundation Grants, 1963
Farmers Insurance Corporate Giving Grants, 1981
Frances L. and Edwin L. Cummings Memorial Fund Grants, 2046
G. Unger Vetlesen Foundation Grants, 2093
Harold Simmons Foundation Grants, 2203
Herbert & Adrian Woods Foundation Grants, 2239
Hilda and Preston Davis Foundation Grants, 2266
IRA Albert J. Harris Awards, 2405
IRA Elva Knight Research Grant, 2406
IRA Esther Zolt Academic Research Award, 2407
IRA Esther Zolt Teacher Research Award, 2408
IRA Helen M. Robinson Dissertation Res Award, 2409
IRA Jeanne S. Chall Research Fellowship, 2410
IRA Outstanding Dissertation of the Year Award, 2411
IRA Pearson Foundation-IRA-Rotary Awards, 2412
IRA Steven A. Stahl Research Grants, 2413
IRA Teacher as Researcher Grants, 2414
John P. McGovern Foundation Grants, 2504
Mattel Children's Foundation Grants, 2685
Mattel International Grants Program, 2686
McCune Foundation Education Grants, 2721
National Book Scholarship Fund, 2849
Norcliffe Foundation Grants, 3415
Nordson Corporation Foundation Grants, 3416
Orchard Foundation Grants, 3793
Percy B. Ferebee Endowment Grants, 3897
Pinkerton Foundation Grants, 3967
Rajiv Gandhi Foundation Grants, 4025
Reinberger Foundation Grants, 4051
RGK Foundation Grants, 4060
Richard and Rhoda Goldman Fund Grants, 4063
Richard Davoud Donchian Foundation Grants, 4066
Robert Bowne Foundaion Edmund A. Stanley, Jr. Research Grants, 4073
San Antonio Area Foundation Grants, 4163
Self Foundation Grants, 4197
Sony Corporation of America Corporate Philanthropy Grants, 4329
Stackpole-Hall Foundation Grants, 4442
Starr Foundation Grants, 4447
Strake Foundation Grants, 4452
Talbert and Leota Abrams Foundation, 4492
Textron Corporate Contributions Grants, 4501
Virginia Foundation for the Humanities Discretionary Grants, 4813
Virginia Foundation for Humanities Open Grants, 4814
William G. and Helen C. Hoffman Found Grants, 4876

Literary Arts
Arkansas Arts Council Individual Fellowships, 999
Marion Gardner Jackson Charitable Trust Grants, 2664

Literary Criticism
Children's Literature Assoc Research Fellowships, 1551
Christian Gauss Award, 1555
Virginia Foundation for the Humanities Residential Fellowships, 4815

Literature
AAR Postdoctoral Fellowships, 201
AAR Rome Prize Fellowships, 202
ACLS Charles A. Ryskamp Research Fellowships, 281
ACLS Frederick Burkhardt Residential Fellowships for Recently Tenured Scholars, 288
AIAR Andrew W. Mellon Foundation Fellowships, 392
AIChE Outstanding Young Researcher Award, 398
AIChE William H. Walker Award for Excellence in Contributions to Chemical Engineering Lit, 430
Alabama Humanities Foundation Mini Grants, 480
Alabama Humanities Foundation Planning/Consultant Grants, 481
Albert and Margaret Alkek Foundation Grants, 504
Alpha Sigma Nu Book Award Competition, 578
American Historical Assoc Leo Gershoy Award, 734
Andrew W. Mellon/ACLS Dissertation Completion Fellowships, 844
ARIT/NEH Fellowships for Research in Turkey, 990
ASECS/Clark Fellowships, 1057
ASID Foundation Joel Polsky Prize, 1096
Bogliasco Foundation Liguria Study Center Residencies, 1358
Canada-U.S. Fulbright International Science and Technology Award, 1417
Canada-U.S. Fulbright Senior Specialists Program Grants, 1418
Canada-U.S. Fulbright Traditional Scholars, 1419
Christian Gauss Award, 1555
Cleveland-Cliffs Foundation Grants, 1580
Delaware Division of the Arts Individual Artist Fellowships, 1732
Florida Division of Cultural Affairs Individual Artist Fellowships, 2026
Fulbright Alumni Initiatives Awards Grants, 2064
Fulbright Distinguished Chairs Program Grants, 2069
Fulbright European Union Scholar-in-Residence Program Grants, 2071
Fulbright New Century Scholars Program Grants, 2075
Fulbright Occasional Lecturer Program Grants, 2076
Fulbright Specialists Program Grants, 2079
Fulbright Traditional Scholar Program in Europe and Eurasia, 2080
Fulbright Traditional Scholar Program in Sub-Saharan Africa, 2081
Fulbright Traditional Scholar Program in the East Asia/Pacific Region, 2082
Fulbright Traditional Scholar Program in the Near East and North Africa Region, 2083
Fulbright Traditional Scholar Program in the South and Central Asia Region, 2084
Fulbright Traditional Scholar Program in the Western Hemisphere, 2085
Fulbright Visiting Specialists Program: Direct Access to the Muslim World, 2087
George Foundation Grants, 2115
Harry Frank Guggenheim Foundation Dissertation Fellowships, 2207
Harry Ransom Humanities Research Center Research Fellowships, 2209
Heineman Foundation for Research, Education, Charitable and Scientific Purposes, 2227
Helen V. Brach Foundation Grants, 2231
Herb Society of America Research and Education Grants, 2243
Houghton Library Visiting Fellowships, 2284
Huntington Library Dana and David Dornsife Fellowships, 2303
Huntington Library National Endowment for the Humanities Fellowships, 2307
Huntington Library Travel Grants and Exchange Fellowships for Study in Great Britain, 2309
IREX Short-Term Travel Grants, 2424
Josiah W. and Bessie H. Kline Foundation Grants, 2517
Kosciuszko Foundation Tuition Scholarships, 2574
Lady Gregory Postgraduate Fellowship, 2579
LBI-NY Fred Grubel Fellowship, 2596
Ludwig Vogelstein Foundation Grants, 2626
National Endowment for the Arts Literature Fellowships: Translation Projects, 2866
National Endowment for the Humanities Fellowships at Winterthur, 2867
NEH Collaborative Research Grants, 2952
NEH Digital Humanities Implementation Grants, 2953

NEH Digital Humanities Start-Up Grants, 2954
NL Audrey Lumsden-Kouvel Fellowships, 3282
NL Weiss/Brown Publication Subvention Award, 3320
Northwood University/Alden B. Dow Creativity Center Summer Residency Program, 3432
NYFA Artist Fellowships, 3698
NYPL Short-Term Research Fellowships, 3701
Ontario Arts Council Integrated Arts Grants, 3790
Paul G. Allen Family Foundation Grants, 3876
PHF Penn Faculty Research Fellowships, 3932
PHF Regional Mellon Faculty Research and Cultural Fellowships, 3933
PHF Undergraduate Humanities Forum Research Fellowships, 3934
Schomburg Center Scholars-in-Residence Fellows, 4187
SSHRC Insight Development Grants, 4409
SSHRC Insight Grants, 4410
SSRC Mellon Mays Predoctoral Research Grants, 4432
Swedish Women's Educational Association International Scholarships, 4480
Winterthur Museum Dissertation Fellowships, 4887
Winterthur Research Fellowships, 4888

Literature, Modern
DOS Study of the United States Institutes, 1846
EU Robert W. Woodruff Library Fellowships, 1959
NHC Summer Institutes in Lit Studies Grants, 3012

Liver Diseases and Disorders
FDHN Fellowship to Faculty Transition Awards, 1992
FDHN Graduate Student Awards, 1994
FDHN Moti L. & Kamla Rustgi International Travel Awards, 1997
FDHN Research Scholar Awards, 2000
NIDDK Developmental Biology and Regeneration of the Liver Grants, 3143
NIDDK Devel of Disease Biomarkers Grants, 3145
NIDDK New Technol for Liver Disease SBIR, 3166
NIDDK New Technol for Liver Disease STTR, 3167
NLM New Technol for Liver Disease Grants, 3303

Livestock
DOI Rhinoceros and Tiger Conservation Grants, 1824
USDA Animal Health & Disease Grants, 4616
USDA Integrated Research Education and Extension Grants - Organic Transitions, 4630
USDA Policy Research Center Grants, 4640

Local History
ALA Gale Cengage History Research and Innovation Award, 488
IES Faculty Fellowships, 2341
Massachusetts Foundation for the Humanities Project Grants, 2678
NHPRC Digitizing Historical Records Grants, 3072
NL American Society for Eighteenth-Century Studies Fellowships, 3279
NL Arthur and Lila Weinberg Fellowships for Independent Scholars and Researchers, 3280
NL Associated Colleges of the Midwest/Great Lakes College Association Faculty Fellowships, 3281
NL British Academy Fellowship for Study in the United Kingdom, 3283
NL Frances C. Allen Fellowships, 3284
NL Grad Scholar-in-Residence Opportunities, 3285
NL Herzog August Bibliothek Wolfenbuttel Fellowship, 3286
NL Lawrence Lipking Fellowship at Newberry, 3287
NL Lester J. Cappon Fellowship in Documentary Editing, 3288
NL Lloyd Lewis Fellowships in American Hist, 3289
NL Mellon Postdoctoral Research Fellowships, 3294
NL Midwest Modern Lang Assoc Fellowship, 3299
NL Monticello College Foundation Fellowship for Women, 3304
NL National Endowment for the Humanities Fellowships, 3310
NL Northeast Modern Language Assocation Fellowship, 3314
NL Scholar-in-Residence Opportunity, 3315
NL Short-Term Resident Fellowships for Individual Research, 3317
NL South Central Modern Languages Association Fellowship, 3318
NL Susan Kelly Power and Helen Hornbeck Tanner Fellowship, 3319
Sarkeys Foundation Grants, 4181
United Methodist Church Women in History Research Grant, 4574
Virginia Foundation for the Humanities Discretionary Grants, 4813
Virginia Foundation for Humanities Open Grants, 4814

Logic
ARO Discrete Mathematics and Computer Science Research Contracts, 1009
NSF Topology Research Grants, 3661

Logistics
Marine Corps Historical Center Dissertation Fellowship, 2663
NSF Arctic Research Support and Logistics Program Grants, 3491
NWO Added Value Plus for Soc Sciences Grants, 3672
USDD Logistics Thrust Grants, 4747
USDD Mathematical Optimization and Operations Research Grants, 4751

Long-Term Care
HRSA Nurse Education, Practice, Quality and Retention Grants, 2290
NINR Acute and Chronic Care During Mechanical Ventilation Research Grants, 3259
Sensient Technologies Foundation Grants, 4199

Lung Disease
NHLBI Ancillary Studies in Clinical Trials, 3015
NHLBI Career Transition Awards, 3021
NHLBI Microbiome of the Lung and Respiratory Tract in HIV-Infected Individuals and HIV-Uninfected Controls, 3039
NIH Sarcoidosis: Research into the Cause of Multi-Organ Disease and Clinical Strategies for Therapy Grants, 3243
Pfizer Medical Education Track Two Grants, 3927

Lupus
Partnership for Cures Two Years To Cures Grants, 3872

Lutheran Church
MacDonald-Peterson Foundation Grants, 2649

Lymphatic System
Lymphatic Education & Research Network Additional Support Grants for NIH-funded F32 Postdoctoral Fellows, 2632
Lymphatic Education & Research Network Postdoctoral Fellowships, 2633
NHLBI Lymphatics in Health & Disease in Digestive, Cardiovascular and Pulmonary Systems, 3031
NHLBI Lymphatics in Health and Disease in the Digestive, Urinary, Cardiovascular and Pulmonary Systems, 3032

Lymphoma
ACGT Investigators Grants, 271
Pfizer Medical Education Track Two Grants, 3927

Macedonia
IREX Individual Advanced Research Opportunities Grants, 2419

Macroeconomics
BC Center for Retirement Research Dissertation Fellowships, 1320
BC Steven H. Sandell Grant Program, 1323
Harris Graduate School of Public Policy Studies Research Development Grants, 2204
S.E.VEN Fund WHY Prize, 4156

Magnetics
ARO Theoretical Physics and Nonlinear Phenomena Research Contracts, 1017
Carnegie Postdoctoral Research Fellowships, 1436
DOE Programs in Fusion Energy Sciences, 1811
NSF Arctic Nat Science Grants in Earth Sciences, 3485
NSF Arctic Nat Science Grants in Ocean Science, 3487
ONR Magnetics Grants, 3767
ORISE Fusion Energy Postdoctoral Research, 3806
ORISE Fusion Energy Sciences Fellowship, 3807
WHOI Geophysical Fluid Dynamics Fellowships, 4859

Malaria
ExxonMobil Foundation Malaria Grants, 1968
United Methodist Committee on Relief Global Health Grants, 4577

Malaysia
Eisenhower Exchange International Fellowships, 1902

Mammalogy
Lalor Foundation Postdoctoral Fellowships, 2580
USDD Marine Mammals and Biological Oceanography Grants, 4750

Managed Care
CMS Research and Demonstration Grants, 1586
Mayo Clinic Administrative Fellowship, 2689

Management
American Historical Association Fellowship in Aerospace History, 717
DAAD Study & Internship Program in Germany, 1708
IREX International Leadership in Education Program Grants, 2420
Kennedy Center Fellowship at the DeVos Institute of Arts Management, 2552
NAPM Doctoral Dissertation Grant, 2827
NSF Innovation and Organizational Change, 3581
ORISE Environmental Management Participation for the US Army Environmental Center, 3801
ORISE Postgraduate Environmental Management Research Participation at the US Army Environmental Center, 3816
ORISE Student Environmental Management Participation Program for the US Army Environmental Center, 3826
OSF-Baltimore Community Fellowships, 3833
OSF HESP Mobility Grants, 3844
Robert Sterling Clark Foundation Government Accountability Grants, 4080
Rumsfeld Foundation Graduate Fellowships, 4127
SSHRC Insight Development Grants, 4409
SSHRC Insight Grants, 4410
Takemi Program in International Health Research and Advanced Training Fellowships, 4491
USDC Planning and Local Technical Assistance Grants, 4668
WHOI Marine Policy Fellowships, 4861

Management Information Systems
ALA Frederick G. Kilgour Award for Research in Library and Information Technology, 487
ALA Innovation Award, 490
Richard P. Covert Scholarship, 4067
USDD Communications and Networking Discovery and Invention Grants, 4712
USDD HIV/AIDS Prevention Program Information Systems Development Grants, 4734

Management Planning/Policy
SRC Robert M. Burger Fellowship, 4392
Washington Area Women's Foundation Strategic Opportunity and Partnership Fund Grants, 4831

Management Sciences
AAFCS International Graduate Fellowships, 154
AAFCS National Graduate Fellowships, 155
Consortium for Graduate Study in Management Minority Fellowships, 1642

DaimlerChrysler Corporation Fund Grants, 1714
General Motors Foundation Grants, 2107
Johnson & Johnson Corporate Contributions, 2508
NSF Decision, Risk, and Management Science Research Grants, 3550
NSF SBE Doctoral Dissertation Research Improvement Grants, 3641
Xerox Foundation Grants, 4899

Manufacturing
ASME M. Eugene Merchant Manuf Medal, 1139
ASME William T. Ennor Manufacturing Technology Award, 1171
CCFF Chairman's Dist Life Sciences Award, 1449
Indiana 21st Century Research and Technology Fund Awards, 2370
Indiana Waste Tire Fund Program Grants, 2376
Library Company of Philadelphia Barra Foundation International Research Fellowships, 2606
Library Company of Philadelphia Prizes, 2608
Library Company of Philadelphia Program in Early American Economy and Society Dissertation-Level Fellowships, 2609
Library Company of Philadelphia Program in Early American Economy and Society One-Month Fellowships, 2610
Library Company of Philadelphia Program in Early American Economy and Society Post-Doctoral Research Fellowships, 2611
NIST Hollings Manuf Extension Partnerships, 3269
ONR Manufacturing Technology Grants, 3768

Manufacturing Processes
ASME M. Eugene Merchant Manuf Medal, 1139
ASME Milton C. Shaw Manufacturing Research Medal, 1147
ASME William T. Ennor Manufacturing Technology Award, 1171
NSF-NIST Interaction in Chemistry, Materials Research, Molecular Biosciences, Bioengineering, and Chemical Engineering, 3453
NSF Chemistry Research Experiences for Undergraduates, 3518
ONR Manufacturing Technology Grants, 3768
SRC GRC & SEMATECH Joint Projects Grants, 4369
SSHRC-APC Automotive Partnership Canada, 4396
US CRDF CGP-Climate Change and Energy Competition For Joint U.S.-Ukrainian Research Teams, 4607

Manuscripts/Books/Music Scores
ASHA Advancing Academic-Research Award, 1072
CAA Wyeth Publication Grant, 1405
Clark Library Short-Term Resident Fellowships, 1577
Harry Ransom Humanities Research Center Research Fellowships, 2209
NYSCA Music: Services to the Field Grants, 3708

Marijuana
Marijuana Policy Project Grants, 2661

Marine Engineering
American Society of Naval Engineers Scholarship, 800
IIE KAUST Graduate Fellowships, 2353
ONR Multidisciplinary Research Program of the URI Grants, 3774
Society of Naval Architects and Marine Engineers Scholarships, 4328
USDD Ocean Engineering and Marine Systems Grants, 4767

Marine Resources
American Rivers Anthony A. Lapham River Conservation Fellowship, 795
Cen Cal Diving and Aquatic Studies Scholarship, 1471
FishAmerica Foundation Marine and Anadromous Fish Habitat Restoration Grants, 2018
FishAmerica Foundation Research Grants, 2019
NFWF Marine & Coastal Conservation Initiative Grants, 3001
NFWF Marine Debris Research and Technology Grants, 3002
NFWF More Fish Partnership Fund Grants, 3003
NFWF Nature Restoration Trust Grants, 3005
NFWF Puget Sound Marine Conservation Fund Grants, 3006
Richard and Rhoda Goldman Fund Grants, 4063
Royal Caribbean Cruises Ocean Fund, 4108
USDC/NOAA Community-Based Habitat Restoration Program Grants, 4656
USDC/NOAA Community-Based Marine Debris Removal Grants, 4657
USDC/NOAA Nature Conservancy Community-based Matching Grants, 4661
USDC/NOAA Restoration Center National and Regional Partnership Grants, 4662
USDD Human Performance Training and Education Grants, 4736
Vancouver Foundation Grants and Community Initiatives, 4807

Marine Sciences
AAAS Award for Scientific Freedom and Responsibility, 28
American Australian Association USA to Australia Fellowships, 620
AMNH Lerner-Gray Grants for Marine Research, 816
Cen Cal Diving and Aquatic Studies Scholarship, 1471
Community Foundation of the Virgin Islands Judith A. Towle Environmental Studies Grants, 1621
Dreyfus Postdoctoral Program in Environmental Chemistry Grant, 1855
Frederick Gardner Cottrell Foundation Grants, 2054
IIE KAUST Graduate Fellowships, 2353
Margaret T. Morris Foundation Grants, 2659
MBL Fred B. & Betsy Bang Summer Fellowships, 2702
NFWF Marine & Coastal Conservation Initiative Grants, 3001
NFWF Puget Sound Marine Conservation Fund Grants, 3006
NOAA Collaborative Science, Technology, and Applied Research Program Grants, 3354
NOAA Financial Assistance for National Centers for Central Coastal Ocean Science, 3372
NOAA Ocean Exploration Omnibus: Marine Archaeology Grants, 3397
NSF Antarctic Earth Sciences Program Grants, 3474
NSF Antarctic Organisms & Ecosystems Grants, 3478
NSF Arctic Nat Science Grants in Ocean Science, 3487
NSF Biological Field Stations and Marine Laboratories Grants, 3500
NSF Chemical Oceanography Research Grants, 3517
NSF Marine Geology & Geophysics Res Grants, 3592
NSF Polar Programs Division Research Grants, 3627
ONR Marine Meteorology and Atmospheric Effects Research Grants, 3769
ONR Maritime Sensing Grants, 3770
ONR Multidisciplinary Research Program of the URI Grants, 3774
ONR Optical and Biological Oceanography Research Grants, 3777
RCY - NSF Ocean Sciences Div Research Grants, 4043
Skaggs Foundation Grants, 4266
Tobias Landau Fellowships in Marine Biology, 4521
USCD/NOAA Illinois-Indiana Sea Grant Core Proposals, 4604
USDC/NOAA Coastal and Marine Fish Habitat Protection Grants, 4655
USDC/NOAA Nature Conservancy Community-based Matching Grants, 4661
USDD Antifouling/Fouling Release Coatings, 4681
USDD Human Performance Training and Education Grants, 4736
USDD Marine Mammals and Biological Oceanography Grants, 4750
USDD Naval Research Facilities Grants, 4761
USDD Ocean Engineering and Marine Systems Grants, 4767
WHOI Summer Student Fellowships, 4865
WHOI Undergrad Fellowships for Minorities, 4866

Marketing Education
American Academy of Advertising Distinguished Service Award, 607
American Academy of Advertising Doctoral Dissertation Competition Awards, 608
American Academy of Advertising Ivan Preston Outstanding Contribution to Research on Advertising Award, 609
American Academy of Advertising Kim Rotzoll Award for Advertising Ethics and Social Responsibility, 610
Kosciuszko Foundation Tuition Scholarships, 2574
MSI Alden G. Clayton Dissertation Proposal Competition, 2800
USDA Cochran Fellowships, 4619

Marketing Research
American Academy of Advertising Distinguished Service Award, 607
American Academy of Advertising Doctoral Dissertation Competition Awards, 608
American Academy of Advertising Ivan Preston Outstanding Contribution to Research on Advertising Award, 609
American Academy of Advertising Kim Rotzoll Award for Advertising Ethics and Social Responsibility, 610
American Academy of Advertising Research Fellowships, 611
AOCS Industrial Oil Products Div Student Award, 885
Bikes Belong Foundation Research Grants, 1343
Charles H. Farnsworth Trust Grants, 1521
Duke Univ Libraries Hartman Center Travel Grants, 1875
Duke University Libraries J. Walter Thompson Research Fellowships, 1876
Mary Gates Research Scholarships, 2670
MSI Alden G. Clayton Dissertation Proposal Competition, 2800
MSI Marketing Strategy Meets Wall Street Research Competition, 2801
MSI Research Support Standard Grants, 2802
USDA Foreign Market Development Grants, 4625
USDA Market Access Program Grants, 4633
USDC Market Development Cooperator Program Grants, 4667

Marketing/Public Relations
American Academy of Advertising Distinguished Service Award, 607
American Academy of Advertising Doctoral Dissertation Competition Awards, 608
American Academy of Advertising Ivan Preston Outstanding Contribution to Research on Advertising Award, 609
American Academy of Advertising Kim Rotzoll Award for Advertising Ethics and Social Responsibility, 610
American Academy of Advertising Research Fellowships, 611
Barker Award, 1300
ConocoPhillips Grants Program, 1629
DOE Energy-Related Inventions Grants, 1783
Duke Univ Libraries Mary Lily Research Grants, 1877
GNOF Gulf Coast Oil Spill Grants, 2153
Manitoba Arts Council Dance Creation/Production Grants, 2654
MDARD Specialty Crop Block Grant Program-Farm Bill, 2730
MSI Alden G. Clayton Dissertation Proposal Competition, 2800
MSI Marketing Strategy Meets Wall Street Research Competition, 2801
MSI Research Support Standard Grants, 2802
NCI/NIDA Transdisciplinary Tobacco Use Research Centers Grants, 2912
Ontario Arts Council Integrated Arts Grants, 3790
RA Kleinhans Fellowship for Research in Tropical Nontimber Forest Products, 4026
Robert Sterling Clark Foundation Government Accountability Grants, 4080
USDA Emerging Markets Grants, 4623
USDA Foreign Market Development Grants, 4625
USDA Market Access Program Grants, 4633

Mass Communication
ASU Graduate College Herman E. Demund Memorial Fellowships, 1240
IREX Edmund S. Muskie Graduate Fellowship, 2417
Virginia Foundation for the Humanities Discretionary Grants, 4813
Virginia Foundation for Humanities Open Grants, 4814

Mass Media
AAAS Kavli Science Journalism Awards, 33
Brainerd Foundation Grants, 1368
Center for Defense Information Internships, 1473
IFT Foundation Grants, 2345
IREX MENA Media Emerging Leaders Fellows, 2421
Lumina Foundation for Education Grants, 2627
Rockefeller Foundation Humanities Fellowships, 4089
SSHRC Insight Development Grants, 4409
SSHRC Insight Grants, 4410
Web We Want Small Grants, 4836
WHOI Ocean Science Journalism Fellowship, 4862

Mass Spectrometry
APHL Emerging Infectious Diseases Fellowships, 914
Frank Field & Joe Franklin Award for Outstanding Achievement in Mass Spectrometry, 2049
Smithsonian Museum Conservation Institute Research Post-Doctorate/Post-Graduate Fellowships, 4283

Mass Transfer
NIST Fire Research Grants, 3268

Massachusetts
MHS Long Term Research Fellowships, 2755
MHS Short Term Research Fellowships, 2757

Massage Therapy
Massage Therapy Foundation Practitioner Case Report Contest, 2679
Massage Therapy Foundation Research Grants, 2680
Massage Therapy Foundation Student Case Report Contest, 2681

Materials Acquisition (Books, Tapes, etc.)
Ball Brothers Foundation Rapid Grants, 1292
Fund for Folk Culture Artists Support Grants, 2088
MBL Fred Karush Library Readership, 2703
USDA 1890 Institution Teaching, Research, and Extension Capacity Building Grants, 4613
Vancouver Foundation Grants and Community Initiatives, 4807

Materials Engineering
AIChE Charles M.A. Stine Award, 399
ARO Synthesis and Processing of Materials Research Contracts, 1016
ASME Nadai Medal, 1148
ASU Graduate College Science Foundation Arizona Fellowships in Science and Engineering, 1245
AT&T Industrial Ecology Faculty Fellowships, 1251
IIE KAUST Graduate Fellowships, 2353
Nanyang Technological University Lee Kuan Yew Postdoctoral Fellowship, 2826
NSF Electronic and Photonic Materials Grants, 3558
NSF Materials Research Grants, 3593
NSF Materials World Network: Cooperative Activity in Materials Research between US Investigators and their Counterparts Abroad, 3595
US CRDF CGP-Climate Change and Energy Competition For Joint U.S.-Ukrainian Research Teams, 4607
USDD Ultra-High Temp Materials Grants, 4800

Materials Sciences
Abdus Salam ICTP Postgraduate Diploma Grants, 235
Abdus Salam ICTP Training and Research in Italian Laboratories Grants, 240
AINSE Honours Scholarships, 460
AINSE Postgraduate Research Awards, 463
American Chemical Society Petroleum Research Fund Undergraduate Research Grants, 695
ANL Faculty Research Participation Program, 855
APS James C. McGroddy Prize for New Materials, 965
ARO Degradation, Reactivity, and Protection of Materials Research Contracts, 1008
ARO Synthesis and Processing of Materials Research Contracts, 1016
ARO Theoretical Physics and Nonlinear Phenomena Research Contracts, 1017
ASEE Army Research Laboratory Postdoctoral Fellowships, 1059
AT&T Bell Labs Graduate Research Program, 1250
Canada Graduate Scholarships and NSERC Postgraduate Scholarships, 1420
Carnegie Postdoctoral Research Fellowships, 1436
CHF Glenn E. and Barbara Hodsdon Ullyot Scholarship, 1540
CHF Robert W. Gore Fellowship in Materials Innovation, 1542
CHF Societe de Chimie Industrielle (American Section) Fellowship, 1544
CHF Travel Grants, 1545
DOE Advanced Nuclear Energy Systems Basic Research Grants, 1769
DOE Basic Energy Sciences Grants, 1772
DOE Engineering Research Grants, 1789
DOE Fossil Advanced Research and Technology Development Grants, 1793
DOE Materials Sciences and Engineering Research Grants, 1803
HHMI-NIBIB Interfaces Initiative Grants, 2248
IIE KAUST Graduate Fellowships, 2353
M.J. Murdock Charitable Trust College Science Research Program, 2639
Materials Research Society Medalist Awards, 2683
Northrop Grumman Foundation Grants, 3429
NPSC Grad Fellowships in the Physical Sciences, 3436
NSF-NIST Interaction in Chemistry, Materials Research, Molecular Biosciences, Bioengineering, and Chemical Engineering, 3453
NSF CISE Accelerating Discovery in Science and Engineering Through Petascale Simulations and Analysis Grants, 3521
NSF CISE Cross-Cutting Program Grants, 3532
NSF CISE Increasing the Participation and Advancement of Women in Academic Science and Engineering Careers Grants, 3534
NSF Electronic and Photonic Materials Grants, 3558
NSF Major Research Instrumentation Grants, 3591
NSF Materials Research Grants, 3593
NSF Materials Research Science and Engineering Centers, 3594
NSF Materials World Network: Cooperative Activity in Materials Research between US Investigators and their Counterparts Abroad, 3595
NSF Polymers Research Grants, 3628
ONR Electronics Division Research Grants, 3759
ORISE Nuclear Regulatory Commission Historically Black Colleges and Universities Student Research Participation, 3814
Smithsonian Graduate Student Fellowships, 4277
Smithsonian Minority Visiting Student Awards, 4279
Smithsonian Native American Visiting Student Awards, 4292
Smithsonian Postdoctoral Fellowships, 4293
Smithsonian Predoctoral Fellowships, 4295
SRC Natural and Engineering Sciences Grants, 4387
UNCFSP NASA Harriett G. Jenkins Predoctoral Fellowships, 4563
USDD ARO Broad Agency Announcement, 4686
USDD Atomic, Molecular and Quantum Physics Grants, 4687
USDD Research Interest of the Air Force Office of Scientific Research Grants, 4784

Materials Testing
AOCS Industrial Oil Products Division Student Award, 885
ARO Degradation, Reactivity, and Protection of Materials Research Contracts, 1008
ASME Patrick J. Higgins Award, 1152
ASME Performance Test Codes Medal, 1154
AT&T Industrial Ecology Faculty Fellowships, 1251
USDD Extended Solids Grants, 4725

Materials, Metals/Alloys
ARO Synthesis and Processing of Materials Research Contracts, 1016
ARO Theoretical Physics and Nonlinear Phenomena Research Contracts, 1017
ASME Patrick J. Higgins Award, 1152
NSF Condensed Matter Physics Research Grants, 3540
NSF Small Business Innovation Research Grants, 3647
USDD Alloys and Joining Grants, 4680

Materials, Physical Properties
APS Arthur L. Schawlow Prize in Laser Science, 945
ARO Degradation, Reactivity, and Protection of Materials Research Contracts, 1008
NSF Materials Research Grants, 3593
SRC Hans J. Coufal Fellowship, 4371

Materials, Polymeric
USDD Polymeric Organic Materials: Conducting/Semiconducting Polymers Grants, 4772
USDD Polymer Matrix Composites Grants, 4773

Materials, Preparation/Fabrication
American Chemical Society Award in the Chemistry of Materials, 672
CIA Undergraduate Co-Op Program - Engineer/Program Management Engineer, 1561

Materials, Processing and Finishing
American Chemical Society Award in the Chemistry of Materials, 672
ARO Atomic, Molecular, and Optical Physics Research Contracts, 1005
ARO Degradation, Reactivity, and Protection of Materials Research Contracts, 1008
ARO Synthesis and Processing of Materials Research Contracts, 1016
NSF-NIST Interaction in Chemistry, Materials Research, Molecular Biosciences, Bioengineering, and Chemical Engineering, 3453
NSF Interfacial Processes and Thermodynamics Research Grants, 3586
NSF Materials Research Grants, 3593
NSF Thermal Transport Processes Grants, 3660

Materials, Structure and Phase
ARO Degradation, Reactivity, and Protection of Materials Research Contracts, 1008
ARO Synthesis and Processing of Materials Research Contracts, 1016
AT&T Industrial Ecology Faculty Fellowships, 1251
NSF Materials Research Grants, 3593
USDD Alloys and Joining Grants, 4680

Mathematical Physics
Abdus Salam ICTP-OEA Projects Grants, 223
Abdus Salam Scientific Meetings Grants, 224
Abdus Salam ICTP-OEA Visiting Scholars/Consultants Grants, 225
Abdus Salam ICTP/IAEA Sandwich Training Education Grants, 226
Abdus Salam ICTP Group Associates Grants, 228
Abdus Salam ICTP Junior Associates Grants, 231
Abdus Salam ICTP Mathematical Research Fellowships, 232
Abdus Salam ICTP Ramanujan Prize for Young Mathematicians, 236
Abdus Salam ICTP Schools, Conferences, Workshops and Extended Research Workshops Grants, 238
Abdus Salam ICTP Senior Associates Grants, 239
Abdus Salam Joint ICTP/SISSA PhD Programme in Physics and Mathematics Grants, 242
APS Dannie Heineman Prize for Mathematical Physics, 951
Norwegian Acad of Science & Letters Abel Prize, 3433

Smithsonian Environmental Research Center Graduate Student Fellowships, 4273
Smithsonian Environmental Research Center Internships, 4274
Smithsonian Environmental Research Center Postdoctoral/Senior Fellowships, 4275
Smithsonian Environmental Research Center Predoctoral Fellowships, 4276

Mathematics

AAAS Award for Public Engagement with Science, 26
AAAS Entry Point Internships, 30
Abdus Salam ICTP-OEA Affil Centre Grants, 222
Abdus Salam ICTP-OEA Projects Grants, 223
Abdus Salam Scientific Meetings Grants, 224
Abdus Salam ICTP-OEA Visiting Scholars/ Consultants Grants, 225
Abdus Salam ICTP/IAEA Sandwich Training Education Grants, 226
Abdus Salam ICTP Federation Arrangement Scheme Grants, 227
Abdus Salam ICTP Group Associates Grants, 228
Abdus Salam ICTP Junior Associates Grants, 231
Abdus Salam ICTP Mathematical Research Fellowships, 232
Abdus Salam ICTP Networks Grants, 233
Abdus Salam ICTP Ramanujan Prize for Young Mathematicians, 236
Abdus Salam ICTP Regular Associates Grants, 237
Abdus Salam ICTP Schools, Conferences, Workshops and Extended Research Workshops Grants, 238
Abdus Salam ICTP Senior Associates Grants, 239
Abdus Salam ITCP Postgraduate Diploma Programme in General Physics, 241
ACSE J. Waldo Smith Hydraulic Fellowship, 309
Alfred P. Sloan Foundation Energy and Environment Grants, 552
Alfred P. Sloan Foundation Research Fellowships, 560
Alfred P. Sloan Foundation Science of Learning STEM Grants, 562
American Indian College Fund STEM Leadership Fellowships, 745
American Mathematical Society AAAS Congressional Fellowships, 750
American Mathematical Society Centennial Fellowships, 752
American Mathematical Society China Exchange Travel Grants, 753
American Mathematical Society Graduate Student Travel Grants, 754
American Mathematical Society International Congress of Mathematicians Travel Grants, 755
American Math Society Simons Travel Grants, 756
American Mathematical Society Travel Grant for the Mathematical Congress of the Americas, 757
AMS/Industry Government Graduate Fellowships, 817
AMS/Industry Minority Scholarships, 818
AMS Grad Fellowships in History of Science, 824
AMS Undergraduate Scholarships, 837
ANL Faculty Research Leave Program, 854
ANL Faculty Research Participation Program, 855
ANL Graduate Thesis Parts Res Appointments, 856
ANL Laboratory-Graduate Res Appointments, 857
Anna-Greta and Holger Crafoords Grants, 858
Applied Biosystems Grants, 916
APS/Phoenix Suns Energy and Environmental Minigrants, 919
ARM Visiting Scientist Awards, 1004
ARO Discrete Mathematics and Computer Science Research Contracts, 1009
ARO Theoretical Physics and Nonlinear Phenomena Research Contracts, 1017
Astronomical Society of the Pacific Priscilla and Bart Bok Awards, 1232
AT&T Bell Labs Graduate Research Program, 1250
BWF Career Awards at the Scientific Interface, 1393
Canada-U.S. Fulbright International Science and Technology Award, 1417
Canada-U.S. Fulbright Senior Specialists Program Grants, 1418
Canada-U.S. Fulbright Traditional Scholars, 1419
Canada Graduate Scholarships and NSERC Postgraduate Scholarships, 1420
Center for Education of Women Visiting Scholar Research Grants, 1475
CFUW Memorial Fellowship, 1515
DOE Basic Energy Sciences Grants, 1772
Fannie and John Hertz Foundation Graduate Fellowships, 1977
Fluor Foundation Grants, 2028
Frederick Douglass Summer Scholars Fellowship, 2053
Fulbright Alumni Initiatives Awards Grants, 2064
Fulbright Distinguished Chairs Program Grants, 2069
Fulbright European Union Scholar-in-Residence Program Grants, 2071
Fulbright New Century Scholars Program Grants, 2075
Fulbright Occasional Lecturer Program Grants, 2076
Fulbright Specialists Program Grants, 2079
Fulbright Traditional Scholar Program in Europe and Eurasia, 2080
Fulbright Traditional Scholar Program in Sub-Saharan Africa, 2081
Fulbright Traditional Scholar Program in the East Asia/Pacific Region, 2082
Fulbright Traditional Scholar Program in the Near East and North Africa Region, 2083
Fulbright Traditional Scholar Program in the South and Central Asia Region, 2084
Fulbright Traditional Scholar Program in the Western Hemisphere, 2085
Fulbright Visiting Specialists Program: Direct Access to the Muslim World, 2087
GEM PhD Science Fellowships, 2106
Gruber Foundation Cosmology Prize, 2177
HHMI-NIBIB Interfaces Initiative Grants, 2248
IIE Central Europe Summer Research Institute Summer Research Fellowship, 2349
IIE David L. Boren Fellowships, 2350
Institute of Paper Science & Tech Fellowships, 2387
International Max Planck Research School Mathematics in the Sciences Fellowships, 2401
ISGC Informal Education Grants, 2431
L'Oreal USA Women in Science Fellowship, 2577
MBL Eugene and Millicent Bell Fellowships, 2699
McKnight Doctoral Fellowships, 2726
Motorola Foundation Innovation Generation Collaborative Grants, 2797
NCAR Graduate Research Fellowships, 2901
NCTM Edward G. Begle PreK-6 Classroom Research Grants, 2942
NHGRI Mentored Research Scientist Development Award, 3013
NHLBI Mentored Quantitative Research Career Development Awards, 3038
NIGMS/NIDCR Evolution of Infectious Diseases Research Grants, 3196
NIGMS Short Courses on Mathematical, Statistical, and Computational Tools for Studying Biological Systems, 3201
NIMH Curriculum Development Award in Neuroinformatics Research and Analysis, 3251
Northrop Grumman Foundation Grants, 3429
Norwegian Acad of Science & Letters Abel Prize, 3433
NSF Activities in Science, Engineering, and Mathematics for Persons with Disabilities, 3460
NSF Advanced Learning Technologies Grants, 3462
NSF Alan T. Waterman Award, 3467
NSF Algebra, Number Theory and Combinatorics Grants, 3468
NSF Analysis Grants, 3469
NSF Applied Mathematics Grants, 3481
NSF Arctic Research and Education Grants, 3489
NSF CISE Collaborative Research in Computational Neuroscience Grants, 3526
NSF Collaborations in Math Geosciences, 3536
NSF Computational Math Research Grants, 3539
NSF Conferences, Workshops, and Special Meetings in the Mathematical Sciences Grants, 3542
NSF Focused Research Groups in the Mathematical Sciences Grants, 3566
NSF Foundations Grants, 3567
NSF Geometric Analysis Research Grants, 3570
NSF Integrative Graduate Education and Research Traineeship, 3584
NSF Mathematical Biology Grants, 3596
NSF Physics Frontiers Centers Grants, 3619
NSF Probability Research Grants, 3634
NSF Research on Gender in Science and Engineering Grants, 3639
NSF Scholarships in Science, Technology, Engineering, and Mathematics, 3642
NSF Scientific Computing Research Environments for the Mathematical Sciences Grants, 3645
NSF Statistics Research Grants, 3654
NSF Topology Research Grants, 3661
NSF Tribal Colleges and Universities Program, 3662
ONR Computational Analysis Grants, 3751
ONR Mathematical, Computer, and Information Sciences Research Grants, 3771
ORISE Faculty Research Participation at the National Center for Toxicological Research, 3804
ORISE Fusion Energy Sciences Fellowship, 3807
ORISE Higher Education Research Experiences at Oak Ridge National Laboratory for Faculty, 3808
ORISE National Library of Medicine, Center for Biotech Info Scientific Visitors Grants, 3810
ORISE National Oceanic and Atmospheric Administration Faculty Research Participation, 3811
ORISE National Oceanic and Atmospheric Admin Student Research Participation, 3812
ORISE Nuclear Regulatory Commission Historically Black Colleges and Universities Student Research Participation, 3814
ORISE Savannah River Site Faculty Research Participation Program, 3823
Parkersburg Area Community Foundation Arts in Education Special Grants, 3869
Phi Beta Kappa Award in Science, 3935
RSC John L. Synge Award, 4120
SFI Principal Investigator Program Grant, 4201
SfN Swartz Prize for Theoretical and Computational Neuroscience, 4213
SIAM/ACM Award in Computational Science and Engineering, 4251
Simone and Cino del Duca Grand Prix Awards, 4261
Smithsonian Environmental Research Center Graduate Student Fellowships, 4273
Smithsonian Environmental Research Center Internships, 4274
Smithsonian Environmental Research Center Postdoctoral/Senior Fellowships, 4275
Smithsonian Environmental Research Center Predoctoral Fellowships, 4276
SRC Peter Verhofstadt Fellowship, 4389
SSRC Mellon Mays Predoctoral Research Grants, 4432
Swedish Natural Science Res Council Grants, 4479
U.S. Department of Education Graduate Assistance in Areas of National Need, 4540
UNCFSP NASA Administrator's Fellowship, 4562
UNCFSP NASA Harriett G. Jenkins Predoctoral Fellowships, 4563
UNCFSP NASA Science and Technology Institute Summer Scholars Program, 4564
US CRDF CGP-Climate Change and Energy Competition For Joint U.S.-Ukrainian Research Teams, 4607
US CRDF CGP-RFBR Climate Change and Energy Competition For Joint U.S.-Russian Research Teams, 4608
USDD Applied Computational Analysis Grants, 4684
USDD ARO Broad Agency Announcement, 4686
USDD Defense Sciences Research and Technology Grants, 4716
USDD Development of Quantum Computing Algorithms Grants, 4718
USDD Information Integration Grants, 4741
USDD Mathematical Optimization and Operations Research Grants, 4751
USDD Research Interest of the Air Force Office of Scientific Research Grants, 4784

Weizmann Institute of Science Postdoctoral Fellowships, 4837
WHOI Postdoctoral Scholar Awards, 4864
WHOI Summer Student Fellowships, 4865
WHOI Undergrad Fellowships for Minorities, 4866
Winston Churchill Foundation Scholarships, 4886

Mathematics Education
3M Foundation Community Giving Grants, 5
AAAS Award for Public Engagement with Science, 26
AAAS Entry Point Internships, 30
AAUW Community Action Grants, 213
Abdus Salam ICTP-OEA Affil Centre Grants, 222
Abdus Salam ICTP Networks Grants, 233
AERA Dissertation Grants, 342
Affymetrix Corporate Contributions Grants, 355
Alcoa Foundation Grants, 511
Alfred P. Sloan Foundation History of Science and Technology Grants, 553
Applied Biosystems Grants, 916
Association for Women in Mathematics Mentoring Travel Grants, 1223
ASU Graduate College Dean's Fellowships, 1236
ASU Graduate College Dissertation Fellowships, 1237
AWM Mentoring Travel Grants for Women, 1279
AWM Travel Grants for Women Researchers, 1280
Bay and Paul Foundations, Inc Grants, 1307
Chiron Foundation Community Grants, 1553
ConocoPhillips Grants Program, 1629
Dept of Ed Star Schools Program Grants, 1749
Dept of Ed Women's Educational Equity Grants, 1752
DOE Acad Creating Teacher Scientists Grants, 1766
DOE Albert Einstein Distinguished Educator Fellowship Program, 1770
Eaton Charitable Fund Grants, 1893
Education Through Astronomy and Space Science Initiative, 1896
Ford Motor Company Fund Grants Program, 2033
MAA-Tensor Foundation Women and Mathematics Grants, 2645
McCune Charitable Foundation Grants, 2719
MfA Newton Fellowships, 2749
MfA Newton Master Teacher Fellowships, 2750
Motorola Foundation Grants, 2796
Motorola Foundation Innovation Generation Collaborative Grants, 2797
NAACP Willems Scholarships, 2810
NASA Lewis Educational Research and Collaborative Internships, 2840
NCTM E. Glenadine Gibb 7-12 Classroom Research Grants, 2941
NCTM Edward G. Begle PreK-6 Classroom Research Grants, 2942
Northrop Grumman Foundation Grants, 3429
NSF Conferences, Workshops, and Special Meetings in the Mathematical Sciences, 3541
NSF Director's Award for Distinguished Teaching Scholars, 3553
NSF Historically Black Colleges and Universities Program Grants, 3577
NSF Innovative Technology Experiences for Students and Teachers, 3582
NSF Louis Stokes Alliances for Minority Participation Grants, 3589
NSF Presidential Awards for Excellence in Mathematics and Science Teaching, 3631
NSF Presidential Awards for Excellence in Science, Math, and Engineering Mentoring, 3632
NSF Research Experiences for Undergrad Grants, 3638
NSF SBE Doctoral Dissertation Research Improvement Grants, 3641
NSF Tribal Colleges and Universities Program, 3662
Rathmann Family Foundation Grants, 4033
RGK Foundation Grants, 4060
Rohm and Haas Company Grants, 4098
U.S. Department of Education Graduate Assistance in Areas of National Need, 4540
USDD Fundamental Research to Counter Weapons of Mass Destruction Grants, 4728
USDD High School Apprenticeship Program and Undergraduate Research Apprenticeship Program Grants, 4732
USDD Science, Technology, Engineering and Mathematics Grants, 4789
Westinghouse Charitable Giving Grants, 4856

Mathematics and Statistics
AAUW American Dissertation Fellowships, 209
AAUW Selected Professions Fellowships, 217
Abdus Salam Joint ICTP/SISSA PhD Programme in Physics and Mathematics Grants, 242
Norwegian Acad of Science & Letters Abel Prize, 3433
NSF CISE Accelerating Discovery in Science and Engineering Through Petascale Simulations and Analysis Grants, 3521
NSF CISE Cross-Cutting Program Grants, 3532
NSF CISE Increasing the Participation and Advancement of Women in Academic Science and Engineering Careers Grants, 3534
USDA Decadal and Regional Climate Prediction Using Earth System Models Grants, 4621
USDD Applied Computational Analysis Grants, 4684
USDD Information Integration Grants, 4741
USDD Mathematical Optimization and Operations Research Grants, 4751

Mathematics, Applied
AAAS Award for Public Engagement with Science, 26
Abdus Salam ICTP Group Associates Grants, 228
Abdus Salam ICTP Junior Associates Grants, 231
Abdus Salam ICTP Regular Associates Grants, 237
Abdus Salam ICTP Schools, Conferences, Workshops and Extended Research Workshops Grants, 238
Abdus Salam ICTP Senior Associates Grants, 239
Abdus Salam Joint ICTP/SISSA PhD Programme in Physics and Mathematics Grants, 242
Alfred P. Sloan Foundation Research Fellowships, 560
ASEE Army Research Laboratory Postdoctoral Fellowships, 1059
ASME Robert Henry Thurston Lecture Award, 1160
Canada Graduate Scholarships and NSERC Postgraduate Scholarships, 1420
DOE Mathematical, Information, and Computational Sciences Grants, 1804
IIE KAUST Graduate Fellowships, 2353
International Max Planck Research School Mathematics in the Sciences Fellowships, 2401
NIGMS Predoctoral Training in Bioinformatics and Computation Biology Grants, 3199
Northrop Grumman Foundation Grants, 3429
NSF Applied Mathematics Grants, 3481
NSF Scientific Computing Research Environments for the Mathematical Sciences Grants, 3645
RSC John L. Synge Award, 4120
USDD Development of Quantum Computing Algorithms Grants, 4718
WHOI Geophysical Fluid Dynamics Fellowships, 4859

Matrix Theory
ARO Discrete Mathematics and Computer Science Research Contracts, 1009

Mechanical Engineering
American Society of Naval Engineers Scholarship, 800
ASME Arthur L. Williston Medal, 1105
ASME Barnett-Uzgiris Product Safety Award, 1106
ASME Ben C. Sparks Medal, 1107
ASME Bergles-Rohsenow Young Investigator Award in Heat Transfer, 1108
ASME Bernard F. Langer Nuclear Codes and Standards Award, 1109
ASME Blackall Machine Tool Award, 1110
ASME Burt L. Newkirk Award, 1111
ASME Calvin W. Rice Lecture Award, 1112
ASME Charles Russ Richards Memorial Award, 1113
ASME Daniel C. Drucker Medal, 1114
ASME Edward F. Obert Award, 1116
ASME Edwin F. Church Medal, 1117
ASME Engineer-Historian Award, 1118
ASME Frank von Flue Award, 1121
ASME Freeman Scholar Award, 1122
ASME Gas Turbine Award, 1123
ASME George Westinghouse Medals, 1124
ASME Graduate Teaching Fellowships, 1125
ASME Gustus L. Larson Memorial Award, 1126
ASME Heat Transfer Memorial Award, 1128
ASME Henry Hess Award, 1129
ASME Henry R. Worthington Medal, 1130
ASME Holley Medal, 1131
ASME Internal Combustion Engine Award, 1132
ASME J. Hall Taylor Medal, 1133
ASME James Harry Potter Gold Medal, 1135
ASME James N. Landis Medal, 1136
ASME Kate Gleason Award, 1137
ASME Machine Design Award, 1140
ASME Marshall B. Peterson Award, 1141
ASME Mayo D. Hersey Award, 1142
ASME Medal, 1144
ASME Melville Medal, 1145
ASME Mel R. Green Codes & Standards Medal, 1146
ASME Nancy DeLoye Fitzroy and Roland V. Fitzroy Medal, 1149
ASME Old Guard Early Career Award, 1150
ASME Old Guard Prizes, 1151
ASME Patrick J. Higgins Award, 1152
ASME Per Bruel Gold Medal for Noise Control and Acoustics, 1153
ASME Performance Test Codes Medal, 1154
ASME Pi Tau Sigma Gold Medal, 1155
ASME R. Tom Sawyer Award, 1157
ASME Ralph Coats Roe Medal, 1158
ASME Robert Henry Thurston Lecture Award, 1160
ASME Soichiro Honda Medal, 1164
ASME Spirit of St. Louis Medal, 1165
ASME Student Section Advisor Award, 1166
ASME Thomas A. Edison Patent Award, 1167
ASME Worcester Reed Warner Medal, 1172
ASME Y.C. Fung Young Investigator Award, 1173
AT&T Bell Labs Graduate Research Program, 1250
CIA Undergraduate Co-Op Program - Engineer/ Program Management Engineer, 1561
IIE KAUST Graduate Fellowships, 2353
Institute of Paper Science & Tech Fellowships, 2387
Nanyang Technological University Lee Kuan Yew Postdoctoral Fellowship, 2826
NPSC Grad Fellowships in the Physical Sciences, 3436
NSF Engineering Research Grants, 3562
NSF George E. Brown Jr. Network for Earthquake Engineering Simulation Research, 3572
SRC Natural and Engineering Sciences Grants, 4387
Timoshenko Medal, 4519
US CRDF CGP-Climate Change and Energy Competition For Joint U.S.-Ukrainian Research Teams, 4607
USDD ARO Broad Agency Announcement, 4686

Mechanics
ASME Timoshenko Medal, 1168
ASME Warner T. Koiter Medal, 1170
BMW of N America Charitable Contributions, 1353
Timoshenko Medal, 4519

Media
Alexander von Humboldt Foundation German Chancellor Fellowships for Prospective Leaders from the People's Republic of China, 526
Alexander von Humboldt Foundation German Chancellor Fellowships for Prospective Leaders from the Russian Federation, 527
Andrew W. Mellon/ACLS Dissertation Completion Fellowships, 844
Ethics and Excellence in Journalism Foundation Grants, 1953
RealNetworks Helix Community Grants, 4044
Reinberger Foundation Grants, 4051
SSRC African Peacebuilding Network Research Grants, 4422
University of Colorado Doctoral Dissertation Fellowship in Media, Religion, and Culture, 4586

Virginia Foundation for the Humanities Discretionary Grants, 4813
Virginia Foundation for Humanities Open Grants, 4814
Web We Want Small Grants, 4836
Willary Foundation Grants, 4870

Media Arts
Carole Fielding Film/Video Production or Research Grants, 1438
CICF Efroymson Contemporary Arts Fellows, 1563
David Bohnett Foundation Grants, 1724
Delaware Division of the Arts Individual Artist Fellowships, 1732
Florida Division of Cultural Affairs Individual Artist Fellowships, 2026
Manitoba Arts Council Aboriginal Arts Creative Development Pilot Projects Grants, 2652
NEH Digital Humanities Implementation Grants, 2953
NEH Digital Humanities Start-Up Grants, 2954
NYFA Artist Fellowships, 3698
OSF Soros Justice Media Fellowships, 3851
Reinberger Foundation Grants, 4051
Stanford Humanities Center Arts Writer/Practitioner Fellowship, 4444

Medical Devices Engineering
W.M. Keck Foundation Med Research Grants, 4823

Medical Education
AACAP Jeanne Spurlock Research Fellowship in Substance Abuse and Addiction for Minority Medical Students, 64
AACAP Life Members Mentorship Grants for Medical Students, 67
AACAP Summer Medical Student Fellowships, 80
AAFP Foundation Pfizer Teacher Devel Awards, 157
AAFP Foundation Wyeth Immunization Awards, 158
AAFP Research Stimulation Grants, 159
AAO-HNS Medical Student Research Prize, 192
AAP Resident Research Grants, 199
ACS Cancer Control Career Development Awards for Primary Care Physicians, 305
ACS Doctoral Training Grants in Oncology Social Work, 308
ACS Physician Training Awards in Preventive Medicine, 323
Aetna Foundation National Medical Fellowship in Healthcare Leadership, 350
AIHS ForeFront MBA Studentship Award, 443
AIHS ForeFront MBT Studentship Awards, 444
Allyn Foundation Grants, 574
American College of Surgeons Co-Sponsored K08/K23 NIH Supplement Awards, 702
American College of Surgeons Resident Research Scholarships, 705
APA Young Investigator Grant Program, 912
Australian Academy of Science Grants, 1269
Booth-Bricker Fund Grants, 1359
BPW Career Advancement Scholarships, 1366
BWF Ad Hoc Grants, 1392
BWF Career Awards for Medical Scientists, 1394
California Wellness Foundation Work and Health Grants, 1409
Canadian Patient Safety Inst Research Grants, 1421
Catherine Holmes Wilkins Foundation Grants, 1442
Christine Mirzayan Science and Technology Policy Graduate Fellowship Program, 1556
Clowes ACS/AAST/NIGMS Jointly Sponsored Mentored Clinical Scientist Devel Award, 1581
CMAP Physician Advocacy Fellowship, 1583
Conquer Cancer Foundation of ASCO International Development and Education Awards, 1635
Cord Foundation Grants, 1645
Covidien Advanced MIS Surgical Fellowships, 1650
Cowles Charitable Trust Grants, 1652
Denton A. Cooley Foundation Grants, 1743
Donald W. Reynolds Foundation Aging and Quality of Life Grants, 1832
Doris Duke Charitable Foundation Clinical Research Fellowships for Medical Students, 1834
Dr. Bob and Jean Smith Foundation Grants, 1853
Drs. Bruce and Lee Foundation Grants, 1859
Duke University Adult Cardiothoracic Anesthesia and Critical Care Medicine Fellowships, 1869
Duke University Ambulatory and Regional Anesthesia Fellowships, 1870
Duke University Clinical Cardiac Electrophysiology Fellowships, 1871
Duke University Hyperbaric Center Fellowships, 1872
Duke Univ Interventional Cardio Fellowships, 1873
Duke Univ Obstetric Anesthesia Fellowships, 1878
Duke University Pain Management Fellowships, 1879
Elisabeth Severance Prentiss Foundation Grants Program, 1905
Flinn Foundation Grants Programs, 2024
FMC Foundation Grants, 2029
Goldman Philanthropic Partnerships Program, 2157
Harry A. and Margaret D. Towsley Found Grants, 2205
Hawaii Community Foundation Health Education and Research Grants, 2221
Healthcare Foundation of New Jersey Grants, 2222
HHMI-NIH Cloister Research Scholars, 2249
HHMI Biomedical Research Grants for Int'l Scientists Infectious Diseases and Parasitology, 2251
HHMI Biomedical Research Grants for International Scientists in Canada and Latin America, 2252
HHMI Biomedical Research Grants for International Scientists in the Baltics, Central and Eastern Europe, Russia, and Ukraine, 2253
HHMI Gilliam Fellowships for Advanced Study, 2254
HHMI Initiative for Colleges Grants, 2256
HHMI Initiative for Universities Grants, 2257
HHMI Med into Grad Initiative Grants, 2259
HHMI Physician-Scientist Early Career Award, 2260
HHMI Precollege Outreach Initiative for Biomedical Research Institutions Grants, 2261
HHMI Research Training Fellowships, 2262
HHMI Undergraduate Science Education Grants for Professors, 2263
International Fellowships in Medical Education, 2398
J.W. Kieckhefer Foundation Grants, 2452
James H. Cummings Foundation Grants, 2460
James M. Collins Foundation Grants, 2461
Jewish Federation of Metropolitan Chicago Scholarships, 2486
John S. Dunn Research Foundation Grants, 2506
Josiah Macy Jr. Foundation Grants, 2516
Kosciuszko Foundation Tuition Scholarships, 2574
Margaret & James A. Elkins Jr. Found Grants, 2658
Mead Johnson Nutritionals Med Educ Grants, 2734
Mead Johnson Nutritionals Scholarships, 2735
MLA Donald Lindberg Research Fellowship, 2785
Mt. Sinai Health Care Foundation Academic Medicine and Bioscience Grants, 2803
NBME Stemmler Medical Education Research Fund Grants, 2899
NCI Cancer Education and Career Development Program, 2917
NHLBI Biomedical Research Training Program for Individuals from Underrepresented Groups, 3019
NHLBI Ruth L. Kirschstein National Research Service Award Institutional Research Training Grants, 3056
NHLBI Ruth L. Kirschstein National Research Service Award Short-Term Institutional Research Training Grants, 3061
NHLBI Short-Term Research Education Program to Increase Diversity in Health-Related Research, 3062
NLM Internet Connection for Medical Institutions Grants, 3300
NMF California Comm Service Scholarships, 3321
NMF General Electric Med Scholars Fellowships, 3322
NMF National Medical Association Awards for Medical Journalism, 3323
NMF National Medical Association Emerging Scholar Awards, 3324
NMF National Medical Association Patti LaBelle Award, 3325
NMF Need-Based Scholarships, 3326
NMF Ralph W. Ellison Prize, 3327
Nuffield Foundation Law and Society Grants, 3667
NYAM Audrey & Will Helfand Fellowship, 3679
NYAM David E. Rogers Fellowships, 3681
Ohio Learning Network Grants, 3734
Partnership for Cures Charles E. Culpeper Scholarships in Medical Science, 3871
Pfizer Medical Education Track Two Grants, 3927
PhRMA Foundation Informatics Pre Doctoral Fellowships, 3954
Piper Trust Healthcare & Med Research Grants, 3972
Potts Memorial Foundation Grants, 3985
RACGP Grieve Memorial Award, 4014
Rathmann Family Foundation Grants, 4033
RCPSC/AMS CanMEDs Research & Dev Grants, 4035
RCPSC International Traveling Fellowship, 4037
RCPSC Medical Education Research Grants, 4038
RCPSC Medical Educ Traveling Fellowships, 4039
RCPSC Royal College Fellowship for Studies in Medical Education, 4040
RCPSC Royal College Visiting Professorships in Medical Research, 4041
Regenstrief General Internal Medicine - General Pediatrics Research Fellowship Program, 4045
Regenstrief Geriatrics Fellowship, 4046
Regenstrief Master Of Science In Clinical Research/ CITE Program, 4047
RWJF Clinical Scholars Grants, 4136
RWJF Harold Amos Medical Faculty Development Scholars, 4140
RWJF Nurse Faculty Scholars, 4148
Samueli Foundation Health Grants, 4161
San Antonio Area Foundation Grants, 4163
SNM Student Fellowship Awards, 4303
Society for Imaging Informatics in Medicine Emeritus Mentor Grants, 4305
Society for Imaging Informatics in Medicine Micro Grants, 4306
Society for Imaging Informatics in Medicine Research Grants, 4307
Society for Imaging Informatics in Medicine Small Grant for Product Support Development, 4308
Society for Imaging Informatics in Medicine Small Training Grants, 4309
Special Olympics Health Professions Stu Grants, 4332
SRC Medical Science Grants for Postdoctoral Positions in Sweden, 4381
SRC Medical Science Postdoctoral Fellowships, 4382
Starr Foundation Grants, 4447
USDD Developer Tools for Medical Education Public Physiology Research Platform Grant, 4717
Victor E. Speas Foundation Grants, 4810
Yale University Beinecke Library Research Fellowships for Yale Grad & Professional School Students, 4902

Medical Ethics
Greenwall Foundation Bioethics Grants, 2174
HHMI Grants and Fellowships Programs, 2255
MPI for Social Anthropology Max Planck Fellow Research Group Law, Organizations, Science, and Technology, 2799
NINR Acute and Chronic Care During Mechanical Ventilation Research Grants, 3259
Regenstrief Geriatrics Fellowship, 4046
RSC Abbyann D. Lynch Medal in Bioethics, 4116

Medical Informatics
AHIMA Dissertation Assistance Grants, 374
AHIMA Faculty Development Stipends, 375
AHIMA Grant-In-Aid Research Grants, 376
Fannie E. Rippel Foundation Grants, 1978
Foundation for Seacoast Health Grants, 2041
MLA Cunningham Memorial International Fellowship, 2783
MLA David A. Kronick Traveling Fellowship, 2784
MLA Graduate Scholarship, 2786
MLA Grad Scholarship for Minority Students, 2787
MLA Research, Development, and Demonstration Project Grant, 2789
MLA Rittenhouse Award, 2790
MLA Thomas Reuters / Frank Bradway Rogers Information Advancement Award, 2791

NEI Ruth L. Kirschstein National Research Service Award Short-Term Institutional Research Training Grants, 2963
NHGRI Mentored Research Scientist Development Award, 3013
NIGMS Predoctoral Training in Bioinformatics and Computation Biology Grants, 3199
NIMH Curriculum Development Award in Neuroinformatics Research and Analysis, 3251
NIMH Short Courses in Neuroinformatics, 3254
NLM Exploratory/Developmental Grants, 3296
NLM Grants for Scholarly Works in Biomedicine and Health, 3298
NLM Internet Connection for Medical Institutions Grants, 3300
NLM Research Project Grants, 3306
NSF Interagency Opportunities in Metabolic Engineering Research Grants, 3585
NSF Postdoc Research Fellowships in Biology, 3630
ORISE National Library of Medicine, Center for Biotech Info Scientific Visitors Grants, 3810
PhRMA Foundation Informatics Post Doctoral Fellowships, 3953
PhRMA Foundation Informatics Pre Doctoral Fellowships, 3954
PhRMA Foundation Informatics Research Starter Grants, 3955
PhRMA Foundation Informatics Sabbatical Fellowships, 3956
PhRMA Foundation Pharmacology/Toxicology Post Doctoral Fellowships, 3962
PhRMA Foundation Pharmacology/Toxicology Sabbatical Fellowships, 3965
Regenstrief General Internal Medicine - General Pediatrics Research Fellowship Program, 4045
Regenstrief Geriatrics Fellowship, 4046
Regenstrief Master Of Science In Clinical Research/ CITE Program, 4047
Regenstrief Medical Informatics NLM Research Fellowships, 4048
Richard P. Covert Scholarship, 4067
Society for Imaging Informatics in Medicine Emeritus Mentor Grants, 4305
Society for Imaging Informatics in Medicine Micro Grants, 4306
Society for Imaging Info in Med Research Grants, 4307
Society for Imaging Informatics in Medicine Small Grant for Product Support Development, 4308
Society for Imaging Informatics in Medicine Small Training Grants, 4309
Thomson Reuters / MLA Doctoral Fellowships, 4512
USDD Clinical Functional Assessment Investigator-Initiated Grants, 4709

Medical Physics
Abdus Salam ICTP Group Associates Grants, 228
Abdus Salam ICTP Junior Associates Grants, 231
Abdus Salam ICTP Postgraduate Diploma Grants, 235
Abdus Salam ICTP Regular Associates Grants, 237
Abdus Salam ICTP Schools, Conferences, Workshops and Extended Research Workshops Grants, 238
Abdus Salam ICTP Senior Associates Grants, 239
Abdus Salam ICTP Training and Research in Italian Laboratories Grants, 240
A Social Corporation Grants, 1189
NIGMS Support of Competitive Research Pilot Project Awards, 3202
NIGMS Support of Competitive Research Research Advancement Awards, 3203
ORISE Nuclear Regulatory Commission Historically Black Colleges and Universities Student Research Participation, 3814
Smith Family Awards for Excellence in Biomedical Research, 4268
W.M. Keck Foundation Med Research Grants, 4823

Medical Programs
ASHG Public Health Genetics Fellowship, 1086
Blowitz-Ridgeway Foundation Early Childhood Development Research Award, 1348
BWF Ad Hoc Grants, 1392
BWF Career Awards for Medical Scientists, 1394
CDC Evaluation of the Use of Rapid Testing For Influenza in Outpatient Medical Settings, 1458
CDC Preventive Med Residency & Fellowship, 1465
Community Found AIDS Endowment Awards, 1614
Conquer Cancer Foundation of ASCO Improving Cancer Care Grants, 1633
Covidien Advanced MIS Surgical Fellowships, 1650
Denton A. Cooley Foundation Grants, 1743
DHHS Emergency Medical Services for Children Program, 1755
George S. and Dolores Dore Eccles Grants, 2120
Goldman Philanthropic Partnerships Program, 2157
James M. Collins Foundation Grants, 2461
M.D. Anderson Foundation Grants, 2637
Marcia and Otto Koehler Foundation Grants, 2656
Maximilian E. and Marion O. Hoffman Foundation Grants, 2688
Mead Johnson Nutritionals Med Educ Grants, 2734
Nelda C. and H.J. Lutcher Stark Found Grants, 2967
NIGMS Support of Competitive Research Pilot Project Awards, 3202
NIGMS Support of Competitive Research Research Advancement Awards, 3203
Pittsburgh Foundation Medical Research Grants, 3973
Rayonier Foundation Grants, 4034
RCPSC/AMS CanMEDs Research and Development Grants, 4035
RCPSC Royal College Visiting Professorships in Medical Research, 4041
Regenstrief Geriatrics Fellowship, 4046
Robert B McMillen Foundation Grants, 4072
RWJF Harold Amos Medical Faculty Development Scholars, 4140
RWJF Nurse Faculty Scholars, 4148
Samueli Foundation Health Grants, 4161
Sarasota Memorial Healthcare Found Grants, 4167
SRC Medical Research Grants, 4379
SRC Medical Science Grants for Large International Conferences in Sweden, 4380
SRC Medical Science Grants for Postdoctoral Positions in Sweden, 4381
SRC Medical Science Postdoctoral Fellowships, 4382
STTI/ATI Educational Assessment Nursing Research Grants, 4457
USDD Air Force Medical Support Agency/ Modernization Directorate Research/Development and Innovations Grants, 4678
USDD Care for the Critically Injured Burn Patient Grants, 4706
USDD Medical Practice Breadth of Med Practice & Disease Frequency Exposure Grants, 4752
USDD Medical Practice Initiative Procedural Skill Decay and Maintenance Grants, 4753
USDD U.S. Army Medical Research and Materiel Command Broad Agency Announcement for Extramural Medical Research Grants, 4799
Vancouver Foundation Grants and Community Initiatives, 4807
Volkswagen of America Corporate Contributions Grants, 4817

Medical Research
AAAAI ARTrust Mini Grants for Allied Health, 14
AAAAI Distinguished Layperson Award, 18
AACAP Life Members Mentorship Grants for Medical Students, 67
ADA Foundation George C. Paffenbarger Student Research Awards, 336
AES-Grass Young Investigator Travel Awards, 345
AES Robert S. Morison Fellowship, 346
Aetna Foundation National Medical Fellowship in Healthcare Leadership, 350
ANA/Grass Foundation Award in Neuroscience, 839
ANA Wolfe Neuropathy Research Prize, 840
APHL Emerging Infectious Diseases Fellowships, 914
Arthur F. and Alice E. Adams Charitable Foundation Grants, 1024
Australian Academy of Science Grants, 1269
Blowitz-Ridgeway Foundation Early Childhood Development Research Award, 1348
Carlos and Marguerite Mason Trust Grants, 1431
Charles A. King Trust Postdoctoral Research Fellowships, 1518
CJ Foundation for SIDS Research Grants, 1573
CSTE CDC/CSTE Applied Epidemiology Fellowships, 1664
Elizabeth Glaser Int'l Leadership Awards, 1906
FAMRI Clinical Innovator Awards, 1975
FAMRI Young Clinical Scientist Awards, 1976
Grass Foundation Marine Biological Laboratory Fellowships, 2163
Gruber Foundation Int'l Research Award, 2180
Hearst Foundations Health Grants, 2225
Hirtzel Memorial Foundation Grants, 2269
IBRO Return Home Fellowships, 2332
Jeffress Awards in Interdisciplinary Research, 2481
Jeffrey Thomas Stroke Shield Foundation Research Grants, 2483
Kovler Family Foundation Grants, 2575
Lymphatic Education & Research Network Additional Support Grants for NIH-funded F32 Postdoctoral Fellows, 2632
Lymphatic Education & Research Network Postdoctoral Fellowships, 2633
M.J. Murdock Charitable Trust Scientific Research Grants, 2644
MacDonald-Peterson Foundation Grants, 2649
Margaret & James A. Elkins Jr. Found Grants, 2658
Mary Gates Research Scholarships, 2670
Mary Kay Foundation Cancer Research Grants, 2674
MBL Fred Karush Library Readership, 2703
MBL Plum Found John E. Dowling Fellowships, 2713
Mericos Foundation Grants, 2742
Meyer and Stephanie Eglin Foundation Grants, 2747
NFL Charities Medical Grants, 2994
NHLBI Ruth L. Kirschstein National Research Service Award Institutional Research Training Grants, 3056
NHLBI Ruth L. Kirschstein National Research Service Awards for Individual Predoctoral Fellowships to Promote Diversity in Health-Related Research, 3058
NHLBI Ruth L. Kirschstein National Research Service Awards for Individual Predoctoral MD/PhD Fellows and Other Dual Degree Fellows, 3059
NHLBI Ruth L. Kirschstein National Research Service Awards for Individual Senior Fellows, 3060
NHLBI Ruth L. Kirschstein National Research Service Award Short-Term Institutional Research Training Grants, 3061
Notsew Orm Sands Foundation Grants, 3434
Perpetual Trust for Charitable Giving Grants, 3900
Pfizer Advancing Research in Transplantation Science Research Awards, 3908
Pfizer ASPIRE EU Dupuytren's Contracture Research Awards, 3911
Pfizer ASPIRE North America Hemophilia Research Awards, 3915
Pfizer ASPIRE North America Narrow Spectrum Antibiotics for the Treatment of MRSA Research Awards, 3916
Pfizer Australia Cancer Research Grants, 3919
Pfizer Australia Neuroscience Research Grants, 3920
Pfizer Australia Paediatric Endocrine Care Research Grants, 3921
Pfizer Inflammation Competitive Res Awards, 3926
Pfizer Research Initiative Rheumatology Grants, 3929
Reinberger Foundation Grants, 4051
Roche Organ Transplant Research Found Grants, 4083
Ruth Estrin Goldberg Memorial for Cancer Research Grants, 4131
Sara Elizabeth O'Brien Trust Grants, 4166
Schramm Foundation Grants, 4188
Scleroderma Foundation Established Investigator Grants, 4190
SfN Jacob P. Waletzky Award, 4206
SfN Janett Rosenberg Trubatch Career Development Award, 4207
SfN Young Investigator Award, 4218

Sioux Falls Area Community Foundation Community Fund Grants (Unrestricted), 4262
Sioux Falls Area Comm Foundation Spot Grants, 4263
Smith Family Awards for Excellence in Biomedical Research, 4268
Thome Foundation Awards in Alzheimer's Disease Drug Discovery Research, 4510
USDD Air Force Medical Support Agency/ Modernization Directorate Research/Development and Innovations Grants, 4678
USDD Broad Agency Announcement for the Naval Medical Research and Development Center-Fredrick Grants, 4700
USDD Broad Agency Announcement Grants, 4701
USDD Broad Agency Announcement HIV/AIDS Prevention Grants, 4702
USDD Care for the Critically Injured Burn Patient Grants, 4706
USDD Clinical Functional Assessment Investigator-Initiated Grants, 4709
USDD Developer Tools for Medical Education Public Physiology Research Platform Grant, 4717
USDD HIV/AIDS Prevention: Military Specific HIV/ AIDS Prevention, Care, and Treatment for Non-PEPFAR Funded Countries Grants, 4733
USDD HIV/AIDS Prevention Program Information Systems Development Grants, 4734
USDD Medical Practice Breadth of Med Practice & Disease Frequency Exposure Grants, 4752
USDD Medical Practice Initiative Procedural Skill Decay and Maintenance Grants, 4753
USDD President's Emergency Plan for AIDS Relief Military Specific HIV Seroprevalence and Behavioral Epidemiology Risk Suvey Grants, 4777
USDD U.S. Army Medical Research and Materiel Command Broad Agency Announcement for Extramural Medical Research Grants, 4799
Victor E. Speas Foundation Grants, 4810
W.M. Keck Foundation Med Research Grants, 4823
Wellcome Trust New Investigator Awards, 4844

Medical Sciences
ASHFoundation New Century Scholars Program Doctoral Scholarships, 1080
ASHFoundation New Century Scholars Research Grant, 1081
ASHFoundation Res Grant for New Investigators, 1082
Conquer Cancer Foundation of ASCO Improving Cancer Care Grants, 1633
Jeffress Awards in Interdisciplinary Research, 2481
Scleroderma Foundation Established Investigator Grants, 4190
USDD Air Force Medical Support Agency/ Modernization Directorate Research/Development and Innovations Grants, 4678
USDD Clinical Functional Assessment Investigator-Initiated Grants, 4709
USDD Developer Tools for Medical Education Public Physiology Research Platform Grant, 4717
W.M. Keck Foundation Med Research Grants, 4823
W.M. Keck Foundation Science and Engineering Research Grants, 4824

Medical Technology
APHL Emerging Infectious Diseases Fellowships, 914
A Social Corporation Grants, 1189
Covidien Advanced MIS Surgical Fellowships, 1650
DHHS Emergency Medical Services for Children Program, 1755
Mead Johnson Nutritionals Med Educ Grants, 2734
NCCAM Translational Tools for Clinical Studies of CAM Interventions Grants, 2905
NCI Academic-Industrial Partnerships for Devel & Validation of In Vivo Imaging Systems and Methods for Cancer Investigations, 2913
NCI Application and Use of Transformative Emerging Technologies in Cancer Research, 2914
NIGMS Support of Competitive Research Pilot Project Awards, 3202
NIGMS Support of Competitive Research Research Advancement Awards, 3203
NLM Manufacturing Processes of Medical, Dental, and Biological Technologies Grants, 3302
ONR Medical Science and Technology Division Grants, 3772
RCPSC/AMS CanMEDs Research and Development Grants, 4035
Regenstrief General Internal Medicine - General Pediatrics Research Fellowship Program, 4045
Regenstrief Master Of Science In Clinical Research/ CITE Program, 4047
Samueli Foundation Health Grants, 4161
STTI Joan K. Stout RN Research Grants, 4461
USDD Air Force Medical Support Agency/ Modernization Directorate Research/Development and Innovations Grants, 4678
USDD Clinical Functional Assessment Investigator-Initiated Grants, 4709
W.M. Keck Foundation Med Research Grants, 4823

Medical/Diagnostics Imaging
A Social Corporation Grants, 1189
Davis Foundation Postdoctoral Fellowships in Eating Disorders Research, 1727
Grass Foundation Marine Biological Laboratory Advanced Imaging Fellowships, 2162
NCCAM Translational Tools for Clinical Studies of CAM Interventions Grants, 2905
NCI/NIDA Transdisciplinary Tobacco Use Research Centers Grants, 2912
NCI Cancer Education and Career Development Program, 2917
NCI Exploratory/Developmental Grants for Diagnostic Cancer Imaging, 2927
NHLBI Career Transition Awards, 3021
Pfizer ASPIRE EU Antifungal Research Awards, 3910
Pfizer Inflammation Competitive Res Awards, 3926
Piper Trust Healthcare & Med Research Grants, 3972
Regenstrief General Internal Medicine - General Pediatrics Research Fellowship Program, 4045
Regenstrief Medical Informatics NLM Research Fellowships, 4048
Samueli Foundation Health Grants, 4161
SNM Molecular Imaging Research Grant For Junior Medical Faculty, 4299
SNM Postdoctoral Molecular Imaging Scholars, 4302

Medicinal Chemistry
American Chemical Society Alfred Burger Award in Medicinal Chemistry, 638
Beckman Scholars Program Grants, 1326
E.B. Hershberg Award for Important Discoveries in Medicinally Active Substances, 1882
Thome Foundation Awards in Alzheimer's Disease Drug Discovery Research, 4510

Medicine
AAUW Selected Professions Fellowships, 217
American Chemical Society Award in Separations Science and Technology, 671
ASME Van C. Mow Medal, 1169
Blowitz-Ridgeway Foundation Early Childhood Development Research Award, 1348
Elisabeth Severance Prentiss Foundation Grants Program, 1905
HHMI Biomedical Research Grants for Int'l Scientists Infectious Diseases and Parasitology, 2251
HHMI Biomedical Research Grants for International Scientists in Canada and Latin America, 2252
HHMI Biomedical Research Grants for International Scientists in the Baltics, Central and Eastern Europe, Russia, and Ukraine, 2253
HHMI Initiative for Colleges Grants, 2256
HHMI Initiative for Universities Grants, 2257
HHMI Med into Grad Initiative Grants, 2259
HHMI Precollege Outreach Initiative for Biomedical Research Institutions Grants, 2261
OSF-Baltimore Community Fellowships, 3833
OSF Access to Medicines Fellowships in Eastern Europe and Central Asia, 3834
Perkin Fund Grants, 3898
Piper Trust Healthcare & Med Research Grants, 3972
Searle Scholars Program Grants, 4193
USDD Aerospace Medicine, Clinical Research, Human Performance Research, and Expeditionary Medicine Grants, 4676
USDD Air Force Medical Support Agency/ Modernization Directorate Research/Development and Innovations Grants, 4678

Medicine, History of
AIHP Sonnedecker Visiting Scholar Grants, 432
AIHP Thesis Support Grants, 433
American Philosophical Society Library Resident Research Fellowships, 769
CHF Glenn E. and Barbara Hodsdon Ullyot Scholarship, 1540
CHF Long-Term Fellowships, 1541
CHF Robert W. Gore Fellowship in Materials Innovation, 1542
CHF Societe de Chimie Industrielle (American Section) Fellowship, 1544
CHF Travel Grants, 1545
Lillian Sholtis Brunner Summer Fellowships for Historical Research in Nursing, 2615
MLA Murray Gottlieb Prize Essay Award, 2788
NLM Grants for Scholarly Works in Biomedicine and Health, 3298
NYAM Paul Klemperer Fellowship, 3692
NYAM Student Essay Grants, 3693
RSC Jason A. Hannah Medal, 4119

Medicine, Internal
AAHPM Hospice and Palliative Med Fellowships, 180
ACS Cancer Control Career Development Awards for Primary Care Physicians, 305
American Philosophical Society Daland Fellowships in Clinical Investigation, 763
CFF First- and Second-Year Clinical Fellowships, 1492
CFF Leroy Matthews Physician-Scientist Awards, 1494
CMAP Physician Advocacy Fellowship, 1583
NCCAM Translational Tools for Clinical Studies of CAM Interventions Grants, 2905
NIA Diversity in Medication Use and Outcomes in Aging Populations Grants, 3092
Regenstrief General Internal Medicine - General Pediatrics Research Fellowship Program, 4045
Regenstrief Geriatrics Fellowship, 4046
Swedish Natural Science Res Council Grants, 4479
Swiss National Science Foundation Fellowships for Advanced Researchers, 4483
Swiss National Science Foundation Fellowships for Prospective Researchers, 4484
Swiss National Science Foundation Marie Heim-Vogtlin Grants, 4485

Medicine, Palliative
Conquer Cancer Foundation of ASCO International Development and Education Award in Palliative Care, 1634

Medieval History
AAR Rome Prize Fellowships, 202
American Historical Association Herbert Baxter Adams Prize, 722
Joseph C. Wey Fellowship in Medieval Studies, 2514
NL Audrey Lumsden-Kouvel Fellowships, 3282
SSHRC Insight Development Grants, 4409
SSHRC Insight Grants, 4410

Membranes
NIH Structural Bio of Membrane Proteins Grant, 3247
NSF Cell Systems Cluster Grants, 3512

Menopause
AFAR Research Grants, 354
Pfizer Medical Education Track Two Grants, 3927

Mental Disorders
AACAP Rieger Service Award for Excellence, 75
AACAP Sidney Berman Award for the School-Based Study and Intervention for Learning Disorders and Mental Ilness, 78
Alexander and Margaret Stewart Trust Grants, 514
American Psychiatric Association/SmithKline Beecham Young Faculty Award for Research Development in Biological Psychiatry, 779
American Psychiatric Foundation Proposals, 791
American Psychiatric Foundation James H. Scully Jr., M.D., Educational Fund Grants, 792
American Psychiatric Foundation Jeanne Spurlock Congressional Fellowship, 793
American Psychiatric Foundation Minority Fellowships, 794
Auburn Foundation Grants, 1263
Catherine Holmes Wilkins Foundation Grants, 1442
Community Foundation for S Arizona Grants, 1618
Fairlawn Foundation Grants, 1974
James H. Cummings Foundation Grants, 2460
John W. Alden Trust Grants, 2510
Leonard and Helen R. Stulman Charitable Foundation Grants, 2602
NARSAD Distinguished Investigator Grants, 2831
NARSAD Independent Investigator Grants, 2833
NARSAD Ruane Prize for Child and Adolescent Psychiatric Research, 2835
NARSAD Young Investigator Grants, 2836
NFX Patient-Oriented Research Grants, 3010
NFX Summer Research Fellowships, 3011
NIEHS Small Business Innovation Research Program Grants, 3195
NIMH Early Identification and Treatment of Mental Disorders in Children and Adolescents Grants, 3252
Oppenstein Brothers Foundation Grants, 3792
OSF Soros Justice Media Fellowships, 3851
Rosalynn Carter Institute John and Betty Pope Fellowships, 4103

Mental Health
AACAP Sidney Berman Award for the School-Based Study and Intervention for Learning Disorders and Mental Ilness, 78
Abbott Fund Science Education Grants, 220
American Psychiatric Association/AACDP Research Mentorship Award, 772
American Psychiatric Assoc Early Career Award, 782
American Psychiatric Association Senior Scholar Health Services Research Award, 789
American Psychiatric Foundation Proposals, 791
American Psychiatric Foundation James H. Scully Jr., M.D., Educational Fund Grants, 792
American Psychiatric Foundation Jeanne Spurlock Congressional Fellowship, 793
American Psychiatric Foundation Minority Fellowships, 794
American Sociological Association Minority Fellowships, 805
ASU Graduate College Martha E. Bernal Memorial Award, 1242
BCBS of Massachusetts Foundation Grants, 1319
Blowitz-Ridgeway Foundation Early Childhood Development Research Award, 1348
Bodenwein Public Benevolent Foundation Grants, 1355
Burton G. Bettingen Grants, 1390
Caesars Foundation Grants, 1407
Callaway Golf Company Foundation Grants, 1410
Comprehensive Health Education Found Grants, 1622
Connecticut Community Foundation Grants, 1627
Connecticut Health Foundation Health Initiative Grants, 1628
DHHS Comprehensive Community Mental Health Services Grants for Children with Serious Emotional Disturbances, 1754
Duchossois Family Foundation Grants, 1867
Evjue Foundation Grants, 1963
Fairlawn Foundation Grants, 1974
FAR Fund Grants, 1980
Foundation for Seacoast Health Grants, 2041
Frank Stanley Beveridge Foundation Grants, 2051
Girl's Best Friend Foundation Grants, 2143
Hampton Roads Community Foundation Mental Health Research Grants, 2199
Hawaii Community Foundation Health Education and Research Grants, 2221
Ittleson Foundation Mental Health Grants, 2438
Jacob and Hilda Blaustein Foundation Grants, 2454
Leonard and Helen R. Stulman Charitable Foundation Grants, 2602
M.J. Murdock Charitable Trust General Grants, 2641
MacDonald-Peterson Foundation Grants, 2649
Margaret T. Morris Foundation Grants, 2659
Mattel Children's Foundation Grants, 2685
Mattel International Grants Program, 2686
McCarthy Family Foundation Grants, 2717
Meadows Foundation Grants, 2736
NARSAD Distinguished Investigator Grants, 2831
NARSAD Independent Investigator Grants, 2833
NARSAD Young Investigator Grants, 2836
Nick Traina Foundation Grants, 3128
NIH Research on Sleep and Sleep Disorders, 3236
Peter F. McManus Charitable Trust Grants, 3903
Rayonier Foundation Grants, 4034
Robert E. and Evelyn McKee Foundation Grants, 4074
Ross Foundation Grants, 4106
Sensient Technologies Foundation Grants, 4199
Sidgmore Family Foundation Grants, 4252
Simone and Cino del Duca Grand Prix Awards, 4261
Sioux Falls Area Community Foundation Community Fund Grants (Unrestricted), 4262
Sioux Falls Area Comm Foundation Spot Grants, 4263
Stackpole-Hall Foundation Grants, 4442
USDD Workpl Violence in the Military Grants, 4805
William T. Grant Foundation Research Grants, 4881

Mental Retardation
George Foundation Grants, 2115
John W. Alden Trust Grants, 2510
NFX Patient-Oriented Research Grants, 3010
NFX Summer Research Fellowships, 3011
Pinkerton Foundation Grants, 3967
Roeher Institute Research Grants, 4095

Mentoring Programs
AAAS Mentor Awards, 35
AACN Mentorship Grant, 110
AAJA Fellowships, 181
Aetna Foundation Minority Scholars Grants, 349
American Psychiatric Association/Kempf Fund Award for Research Development in Psychobiological Psychiatry, 775
American Psychiatric Association Minority Medical Student Summer Mentoring Program, 786
AOCS Processing Distinguished Service Award, 890
ASA Jeanne E. Griffith Mentoring Award, 1035
ASHA Advancing Academic-Research Award, 1072
ASHA Research Mentoring-Pair Travel Award, 1073
ASHA Students Preparing for Academic & Research Careers Award, 1075
ASME McDonald Mentoring Award, 1143
Association for Women in Mathematics Mentoring Travel Grants, 1223
AWM Mentoring Travel Grants for Women, 1279
Boeing Company Contributions Grants, 1357
CDC Preventive Med Residency & Fellowship, 1465
Center for the Study of Philanthropy Senior International Fellowships, 1483
CIC Summer Research Opportunities, 1565
CMA Foundation Grants, 1582
Coca-Cola Foundation Grants, 1593
Conquer Cancer Foundation of ASCO Comparative Effectiveness Research Professorship in Breast Cancer, 1631
Conquer Cancer Foundation of ASCO International Development and Education Awards, 1635
Conquer Cancer Foundation of ASCO Medical Student Rotation Grants, 1637
Conquer Cancer Foundation of ASCO Translational Research Professorships, 1639
Cruise Industry Charitable Foundation Grants, 1660
CSTE CDC/CSTE Applied Epidemiology Fellowships, 1664
Dreyfus Senior Scientist Mentor Program Grants, 1857
Elizabeth Glaser Int'l Leadership Awards, 1906
Farmers Insurance Corporate Giving Grants, 1981
FTE Doctoral Fellowships, 2059
Kennedy Center Summer Fellowships for International Arts Managers, 2553
M.J. Murdock Charitable Trust Partners in Science Grants, 2642
MAA-Tensor Foundation Women and Mathematics Grants, 2645
Mericos Foundation Grants, 2742
NAGC Masters and Specialists Award, 2821
NCI Mentored Career Development Award to Promote Diversity, 2930
NEI Mentored Clinical Scientist Development Program Award, 2961
NHLBI Mentored Career Development Award to Promote Faculty Diversity/Re-Entry in Biomedical Research, 3035
NHLBI Mentored Quantitative Research Career Development Awards, 3038
NHLBI Midcareer Investigator Award in Patient-Oriented Research, 3040
NHLBI Pathway to Independence Awards, 3045
NIA Mentored Clinical Scientist Research Career Development Awards, 3108
NIDDK Mentored Research Scientist Development Award, 3161
NIH Mentored Clinical Scientist Devel Award, 3225
NSF Presidential Awards for Excellence in Science, Math, and Engineering Mentoring, 3632
NSF Undergraduate Research and Mentoring in the Biological Sciences, 3663
OSF-Baltimore Community Fellowships, 3833
PC Fred H. Bixby Fellowships, 3880
Prudential Foundation Education Grants, 4002
SfN Louise Hanson Marshall Specific Recognition Award, 4208
Sidgmore Family Foundation Grants, 4252
Society for Imaging Informatics in Medicine Emeritus Mentor Grants, 4305
SRC Aristotle Award, 4349
Vancouver Foundation Grants and Community Initiatives, 4807
Wellesley Centers for Women Postdoc Fellows, 4845

Merchandising
College of Human Sciences Scholarships, 1596
National Turkey Federation Research Award, 2887

Mesoamerica
Kluge Center Kislak Fellows in American Studies, 2562
Kluge Center Kislak Short-Term Fellowships in American Studies, 2563

Metabolic Diseases
Astellas Foundation Research Grants, 1227
Bristol-Myers Squibb Clinical Outcomes and Research Grants, 1374
CFF Postdoctoral Research Fellowships, 1496
CFF Research Grants, 1497
MDA Neuromuscular Disease Research Grants, 2729
NIDDK Collaborative Interdisciplinary Team Science in Diabetes, Endocrinology and Metabolic Diseases Grants, 3142
NIDDK Devel of Disease Biomarkers Grants, 3145
NIDDK Diabetes, Endocrinology, and Metabolic Diseases NRSAs--Individual, 3146
NIDDK Pilot and Feasibility Clinical Research Grants in Diabetes, Endocrine & Metabolic Diseases, 3170
NIDDK Seeding Collaborative Interdisciplinary Team Science in Diabetes, Endocrinology and Metabolic Diseases Grants, 3179
NIDDK Translational Research for the Prevention and Control of Diabetes and Obesity, 3186

Metabolism
AFAR Research Grants, 354
American Chemical Society AGRO Educ Awards, 632
Astellas Foundation Research Grants, 1227
Beckman Young Investigators Program Grants, 1327
CFF Postdoctoral Research Fellowships, 1496
CFF Research Grants, 1497
NCI Application of Metabolomics for Translational and Biological Research Grants, 2915
NHLBI Career Transition Awards, 3021
NHLBI Research on the Role of Cardiomyocyte Mitochondria in Heart Disease: An Integrated Approach, 3053
NIDDK Collaborative Interdisciplinary Team Science in Diabetes, Endocrinology and Metabolic Diseases Grants, 3142
NIDDK Diet Comp & Energy Balance Grants, 3148
NIDDK Small Grants for Underrepresented Minority Scientists in Diabetes and Digestive and Kidney Diseases, 3183
NIDDK Type 2 Diabetes in the Pediatric Population Research Grants, 3188
NSF Biomolecular Systems Cluster Grants, 3507
NSF Interagency Opportunities in Metabolic Engineering Research Grants, 3585

Metallurgy
Canada Graduate Scholarships and NSERC Postgraduate Scholarships, 1420
DOE Materials Sciences and Engineering Research Grants, 1803
Dorr Foundation Grants, 1842
NSF Materials Research Grants, 3593

Meteorology
Abdus Salam ICTP Joint PhD Programme in Environmental Fluid Mechanics Grants, 230
Abdus Salam ICTP Postgraduate Diploma Grants, 235
Abdus Salam ICTP Training and Research in Italian Laboratories Grants, 240
AMS/Industry Government Graduate Fellowships, 817
AMS/Industry Minority Scholarships, 818
AMS Grad Fellowships in History of Science, 824
AMS Undergraduate Scholarships, 837
Aviation Research Grants, 1273
Canada Graduate Scholarships and NSERC Postgraduate Scholarships, 1420
DOE Environmental Processes Research Grants, 1790
Father James B. Macelwane Annual Awards, 1983
NCAR Graduate Research Fellowships, 2901
NOAA Collaborative Science, Technology, and Applied Research Program Grants, 3354
NOAA Cooperative Institute for Applied Meteorological Studies Grants, 3357
NOAA Cooperative Institute for Tropical Meteorology Research Grants, 3358
NSF Arctic Natural Science Grants in Atmospheric Sciences, 3483
NSF Magnetospheric Physics Research Grants, 3590
NSF National Space Weather Program, 3599
NSF Physical and Dynamic Meteorology Grants, 3614
NSF Solar Terrestrial Research Grants, 3652
ONR Marine Meteorology and Atmospheric Effects Research Grants, 3769
ONR Radio Frequency Sensor Processing Research Grants, 3778
ONR Radio Frequency Solid State Devices Applied Research Grants, 3779
UNCFSP NASA Harriett G. Jenkins Predoctoral Fellowships, 4563
USDD Physical Oceanography Grants, 4771
WHOI Geophysical Fluid Dynamics Fellowships, 4859
WHOI Postdoctoral Scholar Awards, 4864
WHOI Summer Student Fellowships, 4865

Methodist Church
Barra Foundation Project Grants, 1301
Foundation for Evangelism Doc Scholarships, 2039
Foundation for Evangelism Harry Denman Fellowships, 2040
MacDonald-Peterson Foundation Grants, 2649
United Methodist Church Asian, Black, Hispanic and Native American History Research Grants, 4570
United Methodist Church Center for the Evangelical UBH Audrie E. Reber Memorial Award, 4571
United Methodist Church John Harrison Ness Memorial Award, 4572
United Methodist Church Women in History Research Grant, 4574
United Methodist Church Women in History Writing Award, 4575
United Methodist Church World is My Parish United Methodist History Research Grant, 4576
United Methodist Committee on Relief Global Health Grants, 4577
United Methodist Communications Stoody-West Fellowship, 4578

Mexico
Foundation for the Advancement of Mesoamerican Studies Grants, 2042
Fulbright/Garcia Robles Grants, 2063
Fulbright Binational Business Grants in Mexico, 2066
Fulbright Public Policy Initiative Program Grants in Mexico, 2077
Fulbright Traditional Scholar Program in the Western Hemisphere, 2085
Jacobs Research Funds Small Grants Program, 2456

Microbial Degradation
NSF Starter Grants for Postdoctoral Fellows in Microbial Biology, 3653

Microbiology
AACC International Graduate Fellowships, 91
Abbott-ASM Lifetime Achievement Award, 218
APHL Emerging Infectious Diseases Fellowships, 914
ASM-PAHO Infectious Diseases Epidemiology and Surveillance Fellowships, 1098
ASM-UNESCO Leadership Grant for Int'l Educ, 1099
ASM/CDC Fellowships in Infectious Disease and Public Health Microbiology, 1100
ASM BD Award for Res in Clinical Microbiology, 1102
ASM bioMerieux Sonnenwirth Award for Leadership in Clinical Microbiology, 1103
ASM Eli Lilly and Company Research Award, 1138
ASM ICAAC Young Investigator Awards, 1174
ASM International Fellowship for Latin America and the Caribbean, 1175
ASM Int'l Fellowships for Asia and Africa, 1176
ASM Merck Irving S. Sigal Memorial Awards, 1177
ASM Microbiology Undergraduate Research Fellowship, 1178
ASM Millis-Colwell Postgraduate Travel Grant, 1179
ASM Procter & Gamble Award in Applied and Environmental Microbiology, 1180
ASM Promega Biotechnology Research Award, 1181
ASM Raymond W. Sarber Awards, 1182
ASM Robert D. Watkins Graduate Research Fellowship, 1183
ASM sanofi-aventis ICAAC Award, 1184
ASM Siemens Healthcare Diagnostics Young Investigator Award, 1185
ASM Undergraduate Research Fellowship, 1186
ASM William A Hinton Res Training Award, 1187
Bettina Bahlsen Memorial Graduate Scholarship, 1339
DOE Energy Biosciences Research Grants, 1784
Jeffress Awards in Interdisciplinary Research, 2481
M.J. Murdock Charitable Trust College Research Program for Life Sciences, 2638
MBL Albert & Ellen Grass Summer Fellowships, 2692
MBL Associates Summer Fellowships, 2694
MBL Burr & Susie Steinbach Summer Fellow, 2696
MBL E.E. Just Summer Fellowship for Minority Scientists, 2697
MBL Erik B. Fries Summer Fellowships, 2698
MBL Eugene and Millicent Bell Fellowships, 2699
MBL Evelyn & Mel Spiegal Summer Fellowship, 2700
MBL Frank R. Lillie Summer Fellowship, 2701
MBL Fred Karush Library Readership, 2703
MBL Gruss Lipper Family Summer Fellowship, 2704
MBL H. Keffer Hartline and Edward F. MacNichol, Jr. Fellowships, 2705
MBL Herbert W. Rand Summer Fellowship, 2706
MBL James E. and Faith Miller Memorial Summer Fellowship, 2707
MBL John M. Arnold Award, 2708
MBL Lucy B. Lemann Mem Summer Fellowship, 2710
MBL M.G.F. Fuortes Summer Fellowships, 2711
MBL Nikon Summer Fellowship, 2712
MBL Stephen W. Kuffler Summer Fellowships, 2715
MBL William Townsend Porter Summer Fellowships for Minority Investigators, 2716
NFID Colin L. Powell Minority Postdoctoral Fellowship in Tropical Disease Research, 2987
NFID New Investigator Matching Grants, 2989
NIGMS/NIDCR Evolution of Infectious Diseases Research Grants, 3196
NIH Ruth L. Kirschstein National Research Service Awards for Individual Postdoctoral Fellows, 3239
NSF Animal Developmental Mechanisms Grants, 3471
NSF Cell Systems Cluster Grants, 3512
NSF Evolution of Developmental Systems Grants, 3563
NSF Genes and Genome Systems Cluster Grants, 3569
NSF Microbial Observatories and Microbial Interactions and Processes, 3597
NSF Plant, Fungal and Microbial Developmental Systems Grants, 3624
NYAM Louis L. Seaman Medical Student Research Grants, 3690
Pfizer ASPIRE EU Antifungal Research Awards, 3910
Pfizer ASPIRE EU Emerging Mechanisms of Resistance Antibacterial Research Awards, 3912
Smithsonian Environmental Research Center Graduate Student Fellowships, 4273
Smithsonian Environmental Research Center Internships, 4274
Smithsonian Environmental Research Center Postdoctoral/Senior Fellowships, 4275
Smithsonian Environmental Research Center Predoctoral Fellowships, 4276
USDA Norman E. Borlaug International Agricultural Science and Technology Fellowships, 4636

Microbiome
NHLBI Microbiome of the Lung and Respiratory Tract in HIV-Infected Individuals and HIV-Uninfected Controls, 3039
Pfizer ASPIRE EU Emerging Mechanisms of Resistance Antibacterial Research Awards, 3912

Microeconomics
McCune Charitable Foundation Grants, 2719

Microelectronics
Abdus Salam ICTP Group Associates Grants, 228
Abdus Salam ICTP Junior Associates Grants, 231
Abdus Salam ICTP Regular Associates Grants, 237
Abdus Salam ICTP Senior Associates Grants, 239
ONR Microelectronics Applied Research Grants, 3773
SRC AMD/Mahboob Khan Fellowship, 4348
SRC Aristotle Award, 4349
SRC Company-Named Fellowships, 4350
SRC Company-Named Scholarships, 4351
SRC Cross-Disciplinary Research Grants, 4353
SRC Franco Cerrina Fellowship, 4367
SRC Graduate Fellowships, 4368
SRC GRC & SEMATECH Joint Projects Grants, 4369
SRC Hans J. Coufal Fellowship, 4371
SRC International Fellowship, 4375
SRC Landmark Innovation Award, 4376
SRC Mahboob Khan Outstanding Industry Liaison Award, 4377
SRC Master's Scholarships, 4378
SRC Peter Verhofstadt Fellowship, 4389
SRC Research Customization Program Grants, 4390
SRC Robert M. Burger Fellowship, 4392
SRC Simon Karecki Fellowship, 4393
SRC Technical Excellence Award, 4394
SRC Undergraduate Research Opportunities, 4395

Microprocessors
Abdus Salam ICTP Postgraduate Diploma Grants, 235
Abdus Salam ICTP Training and Research in Italian Laboratories Grants, 240

Microscopy
AOCS Young Scientist Research Award, 902
Beckman Young Investigators Program Grants, 1327
Microscopy Soc of America Student Scholarships, 2763

Microwave Theory
ARO Mobile, Wireless Communications and Networks Research Contracts, 1012
ONR Radio Frequency Sensor Processing Research Grants, 3778
ONR Radio Frequency Solid State Devices Applied Research Grants, 3779
USDD National Consortium for Measurement and Signature Intelligence Research Grants, 4759

Middle East
AIAR/Council of American Overseas Research Centers Fellowships for Advanced Multi-Country Research, 391
AIAR Annual Professorships, 393
Center for Advanced Judaic Studies Postdoctoral Fellowships, 1472
IREX MENA Media Emerging Leaders Fellows, 2421
Rumsfeld Foundation Graduate Fellowships, 4127
SSRC Our Shared Past Grants, 4437

Middle School Education
AASF Lydia Donaldson Tutt-Jones Memorial Research Grant, 203
Carnegie Corporation of New York Grants, 1432
Colonel Stanley R. McNeil Foundation Grants, 1597
DaimlerChrysler Corporation Fund Grants, 1714
Eaton Charitable Fund Grants, 1893
Helen Bader Foundation Grants, 2228
IRA Steven A. Stahl Research Grants, 2413
Joyce Foundation Education Grants, 2519
Lilly Endowment Teacher Creativity Fellowships, 2616
MAA-Tensor Foundation Women and Mathematics Grants, 2645
McCune Charitable Foundation Grants, 2719
Nellie Mae Education Foundation District-Level Change Grants, 2968
Nellie Mae Education Foundation State Level Systems Change Grants, 2971
NSF Engineering Research Centers Grants, 3561
NSF Presidential Awards for Excellence in Mathematics and Science Teaching, 3631
Reinberger Foundation Grants, 4051

Migrant Labor
Luther I. Replogle Foundation Grants, 2631

Migrants
Farmers Insurance Corporate Giving Grants, 1981
Luther I. Replogle Foundation Grants, 2631

Migration
International Studies Graduate Fellowships, 2403

Migratory Animals and Birds
Neotropical Grassland Conservancy Wijdenes Memorial Research Grant, 2978

Military History
American Historical Association Paul Birdsall Prize in European Military and Strategic History, 740
American Historical Assoc Wesley-Logan Prize, 744
Harry Frank Guggenheim Foundation Dissertation Fellowships, 2207
Marine Corps Historical Center Beeler-Raider Fellowship, 2662
Marine Corps Historical Center Dissertation Fellowship, 2663
OAH Avery O. Craven Award, 3711
OAH Richard W. Leopold Prize, 3728
Smith Richardson Foundation World Politics and Statecraft Fellowship, 4271
SSRC Our Shared Past Grants, 4437
USDD Research and Studies for the Office of Net Assessment Grants, 4782

Military Personnel
CFR Military Fellowships, 1505
Northrop Grumman Corporation Grants, 3428
USDD Care for the Critically Injured Burn Patient Grants, 4706
USDD Workpl Violence in the Military Grants, 4805

Military Sciences
AFRL/NRC Resident Research Associateship Program, 358
ARO Conference and Symposia Grants, 1007
ARO Mobile, Wireless Communications and Networks Research Contracts, 1012
ASEE Army Research Laboratory Postdoctoral Fellowships, 1059
Center for Defense Information Internships, 1473
Center for the Study of Professional Military Ethics Resident Fellowship, 1484
Cord Foundation Grants, 1645
DSO Defense Sciences Research and Technology Grants, 1863
Northrop Grumman Corporation Grants, 3428
ONR Faculty Sabbatical Leave Program Grants, 3762
ONR Marine Meteorology and Atmospheric Effects Research Grants, 3769
ORISE Faculty Research Participation at the U.S. Army Joint POW/MIA Accounting Command, Central Identification Laboratory, 3805
ORISE Postgraduate Research Participation at the US Army Central Identification Lab, Hawaii, 3822
RAND Corporation Military Fellowships, 4031
Smith Richardson Foundation Strategy and Policy Fellowship, 4270
Smith Richardson Foundation World Politics and Statecraft Fellowship, 4271
USDD High Assurance Cyber Military Systems Grants, 4731
USDD Human Performance Training and Education Grants, 4736
USDD Medical Practice Initiative Procedural Skill Decay and Maintenance Grants, 4753
USDD U.S. Army Medical Research and Materiel Command Broad Agency Announcement for Extramural Medical Research Grants, 4799

Military Training
Marine Corps Historical Center Dissertation Fellowship, 2663
Northrop Grumman Corporation Grants, 3428

Mineralogy
American Chemical Society AGRO Sterling B. Hendricks Memorial Lectureship, 635
Carnegie Postdoctoral Research Fellowships, 1436
DOE Geosciences Research Grants, 1797
DOI Geological Research and Data Acquisition Grants, 1822
Mineralogical Society of America Res Grants, 2777

Mining
American Australian Association USA to Australia Fellowships, 620
Brainerd Foundation Grants, 1368
Northwest Fund for the Environment Grants, 3431
Rocky Mountain Mineral Law Found Grants, 4094
US CRDF CGP ETF 2009 Energy Research Competition For Joint U.S.-Estonian Research Teams, 4609

Minorities
AACR-Minorities in Cancer Research Jane Cooke Wright Lectureship Awards, 119
AACR Minority-Serving Institution Faculty Scholar in Cancer Research Awards, 143
AACR Minority Scholar in Cancer Res Awards, 144
ACF Adoption Opportunities Project Grants, 270
Aetna Foundation Minority Scholars Grants, 349
AIAR Andrew W. Mellon Foundation Fellowships, 392
Akonadi Foundation Anti-Racism Grants, 477
Alex Stern Family Foundation Grants, 542
Alfred P. Sloan Foundation Research Fellowships, 560
ALSAM Foundation Grants, 579
Alston Bannerman Sabbatical Fellowships, 580
American Chemical Society Mentoring Program Grants, 685
American Psychiatric Foundation Jeanne Spurlock Congressional Fellowship, 793
American Psychiatric Foundation Minority Fellowships, 794
American Sociological Association Minority Fellowships, 805
AMS Howard Mayer Brown Fellowships, 827
Animal Behavior Society C.H. Turner Award, 849
Animal Behavior Society Diversity Grant, 851
Archer Daniels Midland Foundation Grants, 985
ASA/NSF/BLS Fellowships, 1030
ASA/NSF/Census Research Fellowships, 1031
ASME Graduate Teaching Fellowships, 1125
ASM Microbiology Undergraduate Research Fellowship, 1178
ASM William A Hinton Res Training Award, 1187
Aspen Institute William Randolph Hearst Endowed Fellowship for Minority Students, 1194
Avon Products Foundation Grants, 1275
BMW of N America Charitable Contributions, 1353
Bodenwein Public Benevolent Foundation Grants, 1355
Boston Foundation Grants, 1361
Breast Cancer Fund Grants, 1369
Carnegie Corporation of New York Grants, 1432
ConocoPhillips Grants Program, 1629
Conquer Cancer Foundation of ASCO Medical Student Rotation Grants, 1637
Conquer Cancer Foundation of ASCO Resident Travel Award for Underrepresented Populations, 1638
Dave Thomas Foundation for Adoption Grants, 1723
Edward W. and Stella C. Van Houten Memorial Fund Grants, 1899
Edwin S. Webster Foundation Grants, 1900
ETS Predoctoral Summer Program, 1954
Eugene M. Lang Foundation Grants, 1957
Eugene McDermott Foundation Grants, 1958
Farmers Insurance Corporate Giving Grants, 1981
FDHN Student Research Fellowships, 2001
Ford Motor Company Fund Grants Program, 2033
Frederick Douglass Summer Scholars Fellowship, 2053
General Service Foundation Human Rights and Economic Justice Grants, 2108
H.J. Heinz Company Foundation Grants, 2195
International Peace Research Inst Oslo Grants, 2402
Jackson Foundation Grants, 2453
MBL E.E. Just Summer Fellowship for Minority Scientists, 2697
MBL William Townsend Porter Summer Fellowships for Minority Investigators, 2716
Minority Faculty Fellowship Program, 2781
MLA Grad Scholarship for Minority Students, 2787
Motorola Foundation Innovation Generation Collaborative Grants, 2797
NAACP Willems Scholarships, 2810
NCI/NIDA Transdisciplinary Tobacco Use Research Centers Grants, 2912
Nellie Mae Education Foundation State Level Systems Change Grants, 2971
NFID Colin L. Powell Minority Postdoctoral Fellowship in Tropical Disease Research, 2987
NHLBI Biomedical Research Training Program for Individuals from Underrepresented Groups, 3019
NHLBI Lymphatics in Health & Disease in Digestive, Cardiovascular and Pulmonary Systems, 3031
NHLBI Research Supplements to Promote Diversity in Health-Related Research, 3054
NHLBI Ruth L. Kirschstein National Research Service Awards for Individual Predoctoral Fellowships to Promote Diversity in Health-Related Research, 3058

NIDDK Small Grants for Underrepresented Minority Scientists in Diabetes and Digestive and Kidney Diseases, 3183
NIEHS Small Business Innovation Research Grants, 3195
NIGMS Mentored Clinical Scientist Research Career Development Awards, 3197
NIH Enhancing Adherence to Diabetes Self-Management Behaviors, 3214
NIH Health Promotion Among Racial and Ethnic Minority Males, 3219
NINR Diabetes Self-Management in Minority Populations Grants, 3263
NMF General Electric Med Scholars Fellowships, 3322
NSF Materials World Network: Cooperative Activity in Materials Research between US Investigators and their Counterparts Abroad, 3595
NSF Postdoc Research Fellowships in Biology, 3630
NSF SBE Doctoral Dissertation Research Improvement Grants, 3641
NSF Undergraduate Research and Mentoring in the Biological Sciences, 3663
OAH Huggins-Quarles Award, 3720
Oppenstein Brothers Foundation Grants, 3792
Pfizer ASPIRE North America Rheumatology Research Awards, 3917
Playboy Foundation Grants, 3980
Prudential Foundation Education Grants, 4002
RAND Corporation Fellowships in Population Studies and the Study of Aging, 4029
Rayonier Foundation Grants, 4034
Rhode Island Foundation Grants, 4061
Richard and Susan Smith Family Found Grants, 4064
Rockefeller Foundation Humanities Fellowships, 4089
Rockefeller Foundation Scholarly Residencies, 4091
RWJF Disparities Research for Change Grants, 4137
RWJF Harold Amos Medical Faculty Development Research Grants, 4139
RWJF Harold Amos Medical Faculty Development Scholars, 4140
RWJF New Connections Jun Investigator Grants, 4146
RWJF New Conn Mid-Career Consultants Grants, 4147
Sensient Technologies Foundation Grants, 4199
Shell Oil Company Foundation Grants, 4233
Smithsonian Minority Visiting Student Awards, 4279
Sony Corporation of America Corporate Philanthropy Grants, 4329
SRC AMD/Mahboob Khan Fellowship, 4348
Teachers College Minority Postdoc Fellowships, 4498
Textron Corporate Contributions Grants, 4501
UNCFSP Summer Sabbatical Awards, 4561
UNCFSP NASA Science and Technology Institute Summer Scholars Program, 4564
United Methodist Church Asian, Black, Hispanic and Native American History Research Grants, 4570
USDA Scientific Cooperation Research Program, 4647
Virginia Foundation for the Humanities Discretionary Grants, 4813
Virginia Foundation for Humanities Open Grants, 4814
Volkswagen of America Corporate Contributions Grants, 4817
W. C. Griffith Foundation Grants, 4818
W.C. Griffith Foundation Grants, 4819
Westinghouse Charitable Giving Grants, 4856
WHOI Undergrad Fellowships for Minorities, 4866
WILD Foundation Grants, 4867

Minorities, Ethnic

Aetna Foundation Minority Scholars Grants, 349
Animal Behavior Society C.H. Turner Award, 849
ASU Graduate College Dean's Fellowships, 1236
BMW of N America Charitable Contributions, 1353
Kluge Center Kislak Fellows in American Studies, 2562
Kluge Center Kislak Short-Term Fellowships in American Studies, 2563
Rockefeller Foundation Humanities Fellowships, 4089
SRC Company-Named Scholarships, 4351
SRC Master's Scholarships, 4378
UCLA Research Grant in Ethnic Studies, 4549
United Methodist Church Asian, Black, Hispanic and Native American History Research Grants, 4570

Minority Education

3M Foundation Community Giving Grants, 5
AAA Minority Dissertation Fellowship Program, 25
AAAS Mentor Awards, 35
AACAP Jeanne Spurlock Minority Medical Student Clinical Fellowship in Child and Adolescent Psychiatry, 63
AAG Dissertation Research Grants, 171
AAUW American Postdoctoral Research Leave Fellowships, 210
AAUW Career Development Grants, 212
AAUW Selected Professions Fellowships, 217
Aetna Foundation Minority Scholars Grants, 349
Akonadi Foundation Anti-Racism Grants, 477
Alcoa Foundation Grants, 511
American Philosophical Society John Hope Franklin Dissertation Fellowships, 765
American Psychiatric Association Minority Fellowships Program, 783
American Psychiatric Foundation Jeanne Spurlock Congressional Fellowship, 793
American Psychiatric Foundation Minority Fellowships, 794
APA Planning Fellowships, 910
Appraisal Institute Education Trust Scholarships, 917
APSA Minority Fellowships, 939
ASHFoundation Graduate Student Scholarships, 1077
ASHFoundation Graduate Student Scholarships for Minority Students, 1078
ASM Robert D. Watkins Graduate Research Fellowship, 1183
Aspen Institute William Randolph Hearst Endowed Fellowship for Minority Students, 1194
Auburn University President's Graduate Opportunities Program, 1264
AUPHA Corris Boyd Scholarships, 1267
Award for Encouraging Disadvantaged Students into Careers in the Chemical Sciences, 1276
AWWA Holly A. Cornell Scholarship, 1284
BMW of N America Charitable Contributions, 1353
Chiron Foundation Community Grants, 1553
CIGNA Foundation Grants, 1568
CMS Historically Black Colleges and Universities Health Services Research Grants, 1585
Coca-Cola Foundation Grants, 1593
Consortium for Graduate Study in Management Minority Fellowships, 1642
Cruise Industry Charitable Foundation Grants, 1660
DaimlerChrysler Corporation Fund Grants, 1714
Dept of Ed Star Schools Program Grants, 1749
Dept of Ed Women's Educational Equity Program Grants, 1752
Diversifying Higher Education Faculty in Illinois Program, 1764
Edward W. and Stella C. Van Houten Memorial Fund Grants, 1899
FMC Foundation Grants, 2029
FTE Expanding Horizons Partnership Grants, 2060
FTE North American Doctoral Fellowships, 2061
Gaius Charles Bolin Fellowships for Minority Graduate Students, 2095
GEM MS Engineering Fellowships, 2104
GEM PhD Engineering Fellowships, 2105
GEM PhD Science Fellowships, 2106
Health Management Scholarships and Grants for Minorities, 2224
Helen Carr Minority Fellowships, 2229
Jeffrey Campbell Grad Fellow in Anthropology, 2482
Johnson & Johnson Corporate Contributions, 2508
Joseph L. Fisher Dissertation Awards, 2515
Josiah Macy Jr. Foundation Grants, 2516
KFF Barbara Jordan Health Policy Scholars, 2556
Martin Luther King Jr. Graduate Memorial Scholarships, 2668
McCune Charitable Foundation Grants, 2719
Miller Brewing Corporate Contributions Program Grants, 2774
Millipore Foundation Grants, 2775
MLA Grad Scholarship for Minority Students, 2787
NACC David Stevenson Fellowships, 2812
NACC William Diaz Fellowships, 2813
National Assessment of Educational Progress Visiting Scholar Program, 2846
National Black MBA Assoc Annual Scholarships, 2847
National Urban Fellows Program, 2888
NEH Awards for Faculty at Historically Black Colleges and Universities, 2949
NEH Awards for Faculty at Institutions with High Hispanic Enrollment, 2950
NEH Awards for Faculty at Tribal Colleges and Universities, 2951
Nellie Mae Education Foundation State Level Systems Change Grants, 2971
NHLBI Mentored Career Award for Faculty at Institutions that Promote Diversity, 3033
NHLBI Mentored Career Award For Faculty At Minority Institutions, 3034
NHLBI Short-Term Research Education Program to Increase Diversity in Health-Related Research, 3062
NHLBI Training Program Grants for Institutions that Promote Diversity, 3067
NIA Minority Dissertation Grants in Aging, 3110
NIH Health Promotion Among Racial and Ethnic Minority Males, 3219
NMF General Electric Med Scholars Fellowships, 3322
NMF National Medical Association Awards for Medical Journalism, 3323
NMF National Medical Association Emerging Scholar Awards, 3324
NMF National Medical Association Patti LaBelle Award, 3325
NMF Need-Based Scholarships, 3326
NMF Ralph W. Ellison Prize, 3327
NSF Louis Stokes Alliances for Minority Participation Grants, 3589
NSF Partnerships in Astronomy & Astrophysics Research and Education, 3612
NSF Research Experiences for Undergrad Grants, 3638
NSF Undergraduate Research and Mentoring in the Biological Sciences, 3663
OAH Huggins-Quarles Award, 3720
Orchid Research Fellowships, 3794
Praxair Foundation Grants, 3986
RGK Foundation Grants, 4060
RWJF New Connections Jun Investigator Grants, 4146
RWJF New Connections Mid-Career Consultants Grants, 4147
Sachs Foundation Scholarships, 4157
SRC Company-Named Scholarships, 4351
SRC Master's Scholarships, 4378
SRC Undergraduate Research Opportunities, 4395
State Farm Co Doctoral Dissertation Award, 4448
Teachers College Minority Postdoc Fellowships, 4498
U.S. Department of Education Ronald E. McNair Postbaccalaureate Achievement Grants, 4546
UNC Carolina Postdoctoral Program for Faculty Diversity, 4560
UNCFSP NASA Administrator's Fellowship, 4562
USA Funds Access to Education Scholarships, 4593
WHOI Undergrad Fellowships for Minorities, 4866
Willamette Univ Minority Grad Fellowship, 4869
Xerox Foundation Grants, 4899

Minority Employment

Akonadi Foundation Anti-Racism Grants, 477
CIC Summer Research Opportunities, 1565
M.D. Anderson Foundation Grants, 2637
Textron Corporate Contributions Grants, 4501
USDA Organic Agriculture Research and Extension Initiative, 4637

Minority Health

Aetna Foundation Integrated Health Care Grants, 348
AHRQ Independent Scientist Award, 388
Akonadi Foundation Anti-Racism Grants, 477
ASU Graduate College Martha E. Bernal Memorial Award, 1242
Avon Foundation Breast Care Fund Grants, 1274
BCBSNC Foundation Grants, 1318
BCBS of Massachusetts Foundation Grants, 1319

CDC Found Tobacco Network Lab Fellowship, 1462
Commonwealth Fund/Harvard University Fellowship in Minority Health Policy, 1604
Commonwealth Fund Harkness Fellowships in Health Care Policy and Practice, 1606
Connecticut Health Foundation Health Initiative Grants, 1628
NMF General Electric Med Scholars Fellowships, 3322
RWJF Disparities Research for Change Grants, 4137

Minority Schools
Akonadi Foundation Anti-Racism Grants, 477
Alcoa Foundation Grants, 511
Allyn Foundation Grants, 574
Coca-Cola Foundation Grants, 1593
DOT Dwight David Eisenhower Transportation Fellowship Program, 1847
Fulbright Occasional Lecturer Program Grants, 2076
NEH Awards for Faculty at Historically Black Colleges and Universities, 2949
NEH Awards for Faculty at Institutions with High Hispanic Enrollment, 2950
NEH Awards for Faculty at Tribal Colleges and Universities, 2951
Nellie Mae Education Foundation District Level Change Grants, 2968
NHLBI Mentored Career Award for Faculty at Institutions that Promote Diversity, 3033
NHLBI Training Program Grants for Institutions that Promote Diversity, 3067
NIEHS Mentored Research Scientist Development Award, 3194
NINR Mentored Research Scientist Development Award, 3265
NSF Historically Black Colleges and Universities Program Grants, 3577
NSF Partnerships in Astronomy & Astrophysics Research and Education, 3612
NSF Tribal Colleges and Universities Program, 3662
ORISE Nuclear Regulatory Commission Historically Black Colleges and Universities Faculty Research Participation, 3813
ORISE Nuclear Regulatory Commission Historically Black Colleges and Universities Student Research Participation, 3814
UNC Carolina Postdoctoral Program for Faculty Diversity, 4560
UNCFSP NASA Administrator's Fellowship, 4562

Minority/Woman-Owned Business
Akonadi Foundation Anti-Racism Grants, 477
Fitzpatrick and Francis Family Business Continuity Foundation Grants, 2021
NSF Small Business Innovation Research Grants, 3647
NSF Small Business Technology Transfer Programs Phase I Grants, 3648
USDA Small Business Innovation Research Grants - Phase I, 4648
USDA Small Business Innovation Research Grants - Phase II, 4649

Moldova
IREX Individual Advanced Research Opportunities Grants, 2419

Molecular Probes
AAAS Eppendorf Science Prize for Neurobiology, 31
NIH Solicitation of Assays for High Throughput Screening in the Molecular Libraries Probe Production Centers Network, 3246
NINDS Optimization of Small Molecule Probes for the Nervous System Grants, 3255
NINDS Optimization of Small Molecule Probes for the Nervous System Grants, 3256
NINDS Optimization of Small Molecule Probes for the Nervous System Grants, 3257

Mongolia
American Historical Association John K. Fairbank Prize in East Asian History, 733
IREX Individual Advanced Research Opportunities Grants, 2419
IREX Short-Term Travel Grants, 2424

Morphology
NSF Physiology and Ethology Research Grants, 3622
NSF Systematic Biology and Biodiversity Inventories Grants, 3657

Multiculturalism
Saint Mary's College Fellowships, 4159

Multidisciplinary Arts
NYSCA Folk Arts: Exhibitions Grants, 3703

Multiple Myeloma
ASCO/UICC International Cancer Technology Transfer Fellowships, 1044
ASCO Young Investigator Award, 1050
Conquer Cancer Foundation of ASCO Career Development Award, 1630
Robert Leet and Clara Guthrie Patterson Grants, 4076

Multiple Sclerosis
Pfizer Medical Education Track Two Grants, 3927
Pittsburgh Foundation Medical Research Grants, 3973
Seattle Foundation Medical Funds Grants, 4194

Muscular Dystrophy
MDA Development Grant, 2728
MDA Neuromuscular Disease Research Grants, 2729

Musculoskeletal Disorders
CDI Postdoctoral Fellowships, 1469
LAM Found Established Investigator Awards, 2581
LAM Foundation Pilot Project Grants, 2582
LAM Foundation Postoctoral Fellowships, 2583
NIAMS Pilot & Feasibility Clinical Grants in Arthritis & Musculoskeletal & Skin Diseases, 3111
Pfizer ASPIRE EU Dupuytren's Contracture Research Awards, 3911

Musculoskeletal System
ACSM NASA Space Physiology Research Grants, 319
CDI Postdoctoral Fellowships, 1469
NSF Physical Anthropology Research Grants, 3615

Museum Education
Colonel Stanley R. McNeil Foundation Grants, 1597
Field Museum High School Internships, 2009
IMLS Conservation Project Support Grants, 2363
McCune Charitable Foundation Grants, 2719
National Endowment for the Humanities Fellowships at Winterthur, 2867
National Gallery of Art Kress Interpretive Fellow, 2869
NYSCA Museum: Project Support Grants, 3707
SHOT Dibner Award for Excellence in Museum Exhibits, 4239
Smithsonian Minority Visiting Student Awards, 4279
The Ray Charles Foundation Grants, 4502
Winterthur Museum Dissertation Fellowships, 4887
Winterthur Research Fellowships, 4888

Museums
3M Foundation Community Giving Grants, 5
AAAS Award for Public Engagement with Science, 26
AAB Berlin Prize Fellowships, 48
ACC Ford Foundation Fellowships, 253
ACC Humanities Fellowship Program, 255
Albuquerque Community Foundation Grants, 508
American Numismatic Society Frances M. Schwartz Fellowship, 762
AMNH Graduate Student Fellowships, 815
Andy Warhol Foundation for the Visual Arts Curatorial Research Fellowships, 847
Arthur and Rochelle Belfer Foundation Grants, 1023
Auburn Foundation Grants, 1263
Ball Brothers Foundation General Grants, 1291
Barra Foundation Project Grants, 1301
Batchelor Foundation Grants, 1305
Bay and Paul Foundations, Inc Grants, 1307
Booth-Bricker Fund Grants, 1359
Burlington Northern Santa Fe Foundation Grants, 1389
Byron W. and Alice L. Lockwood Grants, 1401
CAA Professional Development Fellowships for Graduate Students, 1404
Cape Branch Foundation Grants, 1425
Carl Gellert and Celia Berta Gellert Foundation Grants, 1429
Chestnut Hill Charitable Foundation, Inc Grants, 1538
Christensen Fund Regional Grants, 1554
Cleveland-Cliffs Foundation Grants, 1580
Colonel Stanley R. McNeil Foundation Grants, 1597
Compton Foundation Grants Program, 1623
Connecticut Community Foundation Grants, 1627
Coughlin-Saunders Foundation Grants Program, 1648
Cowles Charitable Trust Grants, 1652
Elizabeth McGraw Foundation Grants, 1908
Eugene McDermott Foundation Grants, 1958
Field Museum High School Internships, 2009
Field Museum Undergraduate Internships, 2010
Francis C. Wood Scholar in Residence Program, 2047
George Foundation Grants, 2115
George W. Codrington Charitable Grants, 2122
Getty Research Grants for Institutions - Archival Grants, 2131
Getty Research Grants for Institutions - Cataloguing of Museum Collections Grants, 2133
Getty Research Institute Nonresidential Grants - Collaborative Research Grant, 2135
Getty Research Institute Nonresidential Grants - Curatorial Research Fellowships, 2136
Greenspun Family Foundation Grants, 2173
Harold R. Bechtel Charitable Remainder Uni-Trust Grants, 2201
Huie-Dellmon Trust Grants, 2295
IMLS 21st Century Museum Prof Grants, 2362
IMLS Conservation Project Support Grants, 2363
IMLS National Leadership Grants, 2366
Japan Foundation Six-Month Grants for Specialists in Cultural and Academic Fields, 2479
Joseph Alexander Foundation Grants, 2513
Kenneth T. and Eileen L. Norris Found Grants, 2554
Mabel Y. Hughes Charitable Trust Grants, 2647
Marcia and Otto Koehler Foundation Grants, 2656
McLean Contributionship Grants, 2727
Mericos Foundation Grants, 2742
Motorola Foundation Grants, 2796
NAGPRA Museum Consultation/Documentation Grants, 2822
NAGPRA Museum Repatriation Grants, 2823
National Endowment for the Humanities Fellowships at Winterthur, 2867
National Gallery of Art Kress Interpretive Fellow, 2869
NEH Fellowships at Independent Research Institutions, 2957
NHPRC Digitizing Historical Records Grants, 3072
NSF Biological Research Collections, 3504
Oppenstein Brothers Foundation Grants, 3792
Price Family Charitable Fund Grants, 3987
R.C. Baker Foundation Grants, 4006
Reinberger Foundation Grants, 4051
Richard and Susan Smith Family Found Grants, 4064
SHOT Dibner Award for Excellence in Museum Exhibits, 4239
Sid W. Richardson Foundation Grants, 4253
Sioux Falls Area Community Foundation Community Fund Grants (Unrestricted), 4262
Sioux Falls Area Comm Foundation Spot Grants, 4263
Smithsonian Minority Visiting Student Awards, 4279
Smithsonian Native American Community Scholars Awards, 4291
Smithsonian Postgraduate Fellowships in Conservation of Museum Collections, 4294
Strake Foundation Grants, 4452
Textron Corporate Contributions Grants, 4501
The Ray Charles Foundation Grants, 4502
Union Pacific Found Community & Civic Grants, 4568
Virginia Foundation for the Humanities Discretionary Grants, 4813

Virginia Foundation for Humanities Open Grants, 4814
W.C. Griffith Foundation Grants, 4819
W. C. Griffith Foundation Grants, 4818
Winterthur Museum Dissertation Fellowships, 4887
Winterthur Research Fellowships, 4888
Wolfsonian-FIU Fellowships, 4890

Music
AAB Berlin Prize Fellowships, 48
ACC Humanities Fellowship Program, 255
ACC Residency Program in Asia, 267
ACLS Charles A. Ryskamp Research Fellowships, 281
Adaptec Foundation Grants, 340
Alexander von Humboldt Foundation Hezekiah Wardwell Fellowships, 528
AMS Alvin H. Johnson 50 Dissert Fellowships, 820
AMS Paul A. Pisk Prize, 833
Anne Louise Barrett Fellowship, 861
Auburn Foundation Grants, 1263
Berea College Appalachian Music Fellowships, 1332
Berks County Community Foundation Grants, 1333
Bogliasco Foundation Liguria Study Center Residencies, 1358
Canada-U.S. Fulbright International Science and Technology Award, 1417
Canada-U.S. Fulbright Senior Specialists Program Grants, 1418
Canada-U.S. Fulbright Traditional Scholars, 1419
Clarence E. Heller Charitable Foundation Grants, 1575
Cudd Foundation Grants, 1672
DAAD Research Grants for Doctoral Candidates and Young Academics and Scientists, 1703
DAAD Research Stays for University Academics and Scientists, 1705
DAAD Scholarships for Artists: Study Visits for Academics, 1707
Diaz-Ayala Cuban and Latin American Popular Music Collection Travel Grants, 1763
Eastman School of Music Graduate Awards, 1892
Florida Division of Cultural Affairs Individual Artist Fellowships, 2026
Fulbright Alumni Initiatives Awards Grants, 2064
Fulbright Distinguished Chairs Program Grants, 2069
Fulbright European Union Scholar-in-Residence Program Grants, 2071
Fulbright mtvU Awards, 2074
Fulbright New Century Scholars Program Grants, 2075
Fulbright Occasional Lecturer Program Grants, 2076
Fulbright Specialists Program Grants, 2079
Fulbright Traditional Scholar Program in Europe and Eurasia, 2080
Fulbright Traditional Scholar Program in Sub-Saharan Africa, 2081
Fulbright Traditional Scholar Program in the East Asia/Pacific Region, 2082
Fulbright Traditional Scholar Program in the Near East and North Africa Region, 2083
Fulbright Traditional Scholar Program in the South and Central Asia Region, 2084
Fulbright Traditional Scholar Program in the Western Hemisphere, 2085
Fulbright Visiting Specialists Program: Direct Access to the Muslim World, 2087
Gladys Krieble Delmas Foundation Grants for Venetian Research, 2144
Grammy Foundation Grants, 2161
Harold R. Bechtel Charitable Remainder Grants, 2201
Harry Ransom Humanities Research Center Research Fellowships, 2209
Haverford College Mellon Postdoc Fellowships, 2220
Houghton Library Visiting Fellowships, 2284
John Ben Snow Memorial Trust Grants, 2497
Kate Neal Kinley Memorial Fellowship, 2532
M. Bastian Family Foundation Grants, 2636
Manitoba Arts Council Aboriginal Arts Creative Development Pilot Projects Grants, 2652
Margaret T. Morris Foundation Grants, 2659
NAMM Foundation Music Research Grants, 2825
National Endowment for the Arts Jazz Masters Fellowships, 2864
NL Weiss/Brown Publication Subvention Award, 3320
NYSCA Music: Services to the Field Grants, 3708
OAH James A. Rawley Prize, 3721
Ontario Arts Council Integrated Arts Grants, 3790
Surdna Foundation Arts Teachers Fellowships, 4470
T. James Kavanagh Foundation Grants, 4489
The Ray Charles Foundation Grants, 4502
W.C. Griffith Foundation Grants, 4819

Music Appreciation
Anne Louise Barrett Fellowship, 861
Fulbright mtvU Awards, 2074
National Endowment for the Arts Jazz Masters Fellowships, 2864

Music Composition
AAB Berlin Prize Fellowships, 48
AAR Rome Prize Fellowships, 202
Anne Louise Barrett Fellowship, 861
Camargo Foundation Fellowships, 1412
Cintas Foundation Fellowships, 1570
Florida Division of Cultural Affairs Individual Artist Fellowships, 2026
IES Faculty Fellowships, 2341
IIE Rockefeller Foundation Bellagio Center Residencies, 2355
NL American Society for Eighteenth-Century Studies Fellowships, 3279
NL Arthur and Lila Weinberg Fellowships for Independent Scholars and Researchers, 3280
NL Associated Colleges of the Midwest/Great Lakes College Association Faculty Fellowships, 3281
NL British Academy Fellowship for Study in the United Kingdom, 3283
NL Grad Scholar-in-Residence Opportunities, 3285
NL Herzog August Bibliothek Wolfenbuttel Fellowship, 3286
NL Lawrence Lipking Fellowship at Newberry, 3287
NL Lester J. Cappon Fellowship in Documentary Editing, 3288
NL Mellon Postdoctoral Research Fellowships, 3294
NL Midwest Modern Lang Assoc Fellowship, 3299
NL Monticello College Foundation Fellowship for Women, 3304
NL National Endowment for the Humanities Fellowships, 3310
NL Northeast Modern Language Assocation Fellowship, 3314
NL Scholar-in-Residence Opportunity, 3315
NL Short-Term Resident Fellowships for Individual Research, 3317
NL South Central Modern Languages Association Fellowship, 3318
NL Susan Kelly Power and Helen Hornbeck Tanner Fellowship, 3319
NL Weiss/Brown Publication Subvention Award, 3320
NYFA Artist Fellowships, 3698
NYSCA Music: Services to the Field Grants, 3708
Stanford Humanities Center Arts Writer/Practitioner Fellowship, 4444

Music Education
Alexander von Humboldt Foundation Hezekiah Wardwell Fellowships, 528
Berea College Appalachian Music Fellowships, 1332
Bogliasco Foundation Liguria Study Center Residencies, 1358
Clarence E. Heller Charitable Foundation Grants, 1575
Eastman School of Music Graduate Awards, 1892
Fulbright mtvU Awards, 2074
Haverford College Mellon Postdoc Fellowships, 2220
Heineman Foundation for Research, Education, Charitable and Scientific Purposes, 2227
NAMM Foundation Music Research Grants, 2825
Surdna Foundation Arts Teachers Fellowships, 4470

Music History
AAR Postdoctoral Fellowships, 201
AMS Noah Greenberg Award, 831
Clark Library Short-Term Resident Fellowships, 1577
Fulbright mtvU Awards, 2074
Haverford College Mellon Postdoc Fellowships, 2220
IES Faculty Fellowships, 2341
National Endowment for the Arts Jazz Masters Fellowships, 2864
NL American Society for Eighteenth-Century Studies Fellowships, 3279
NL Arthur and Lila Weinberg Fellowships for Independent Scholars and Researchers, 3280
NL British Academy Fellowship for Study in the United Kingdom, 3283
NL Frances C. Allen Fellowships, 3284
NL Grad Scholar-in-Residence Opportunities, 3285
NL Herzog August Bibliothek Wolfenbuttel Fellowship, 3286
NL Lawrence Lipking Fellowship at Newberry, 3287
NL Lester J. Cappon Fellowship in Documentary Editing, 3288
NL Lloyd Lewis Fellowships in American Hist, 3289
NL Mellon Postdoctoral Research Fellowships, 3294
NL Midwest Modern Lang Assoc Fellowship, 3299
NL Monticello College Foundation Fellowship for Women, 3304
NL National Endowment for the Humanities Fellowships, 3310
NL Northeast Modern Language Assocation Fellowship, 3314
NL Scholar-in-Residence Opportunity, 3315
NL Short-Term Resident Fellowships for Individual Research, 3317
NL South Central Modern Languages Association Fellowship, 3318
NL Susan Kelly Power and Helen Hornbeck Tanner Fellowship, 3319
NL Weiss/Brown Publication Subvention Award, 3320

Music Recording
NYSCA Music: Services to the Field Grants, 3708

Music Therapy
NAMM Foundation Music Research Grants, 2825

Music, Composition
NYFA Shelly Pinz Fellowship, 3699

Music, Experimental
Florida Division of Cultural Affairs Individual Artist Fellowships, 2026

Music, Vocal
NYSCA Music: Services to the Field Grants, 3708

Musicians in Residence
NYSCA Music: Services to the Field Grants, 3708
Rockefeller Foundation Scholarly Residencies, 4091

Musicology/Music Theory
AAB Berlin Prize Fellowships, 48
AAR Postdoctoral Fellowships, 201
ACC Ford Foundation Fellowships, 253
ACLS Frederick Burkhardt Residential Fellowships for Recently Tenured Scholars, 288
Alexander von Humboldt Foundation Hezekiah Wardwell Fellowships, 528
AMS Alfred Einstein Award, 819
AMS Claude V. Palisca Award, 821
AMS Eileen Southern Travel Fund Grants, 822
AMS Eugene K. Wolf Travel Fund for European Research Grants, 823
AMS H. Colin Slim Award, 825
AMS Harold Powers World Travel Fund Grants, 826
AMS Howard Mayer Brown Fellowships, 827
AMS Janet Levy Fund for Independent Scholars, 828
AMS Lewis Lockwood Award, 829
AMS M. Elizabeth C. Bartlet Fund for Research in France, 830
AMS Otto Kinkeldey Award, 832
AMS Paul A. Pisk Prize, 833
AMS Robert M. Stevenson Award, 834
AMS Ruth A. Solie Award, 835

Andrew W. Mellon/ACLS Dissertation Completion Fellowships, 844
Anne Louise Barrett Fellowship, 861
Australasian Institute of Judicial Administration Seed Funding Grants, 1268
BMI Woody Guthrie Fellowships, 1352
Haverford College Mellon Postdoc Fellowships, 2220
North Carolina Arts Council Folklife Internship, 3418
SSRC Mellon Mays Predoctoral Research Grants, 4432

Myasthenia Gravis
MDA Neuromuscular Disease Research Grants, 2729
MGFA Post-Doctoral Research Fellowships, 2751
MGFA Student Fellowships, 2752

Mycology
ASM/CDC Fellowships in Infectious Disease and Public Health Microbiology, 1100
NFID John P. Utz Medical Mycology Postdoctoral Fellowship, 2988

Myeloma
Robert Leet and Clara Guthrie Patterson Grants, 4076

Nanoscience
Kavli Foundation Research Prizes, 2534
NSF Active Nano- Structures & Systems Grants, 3459
Royal Norwegian Embassy Kavli Prizes, 4109
SRC Franco Cerrina Fellowship, 4367
SRC GRC & SEMATECH Joint Projects Grants, 4369
SRC Hans J. Coufal Fellowship, 4371
Thome Foundation Awards in Alzheimer's Disease Drug Discovery Research, 4510
USDD Nano-Bio Manufacturing Inst Grants, 4757

Nanostructures
NSF Active Nano- Structures & Systems Grants, 3459
USDD Bulk Nanostructured Materials Grants, 4704
USDD Nanoscale Electronics Grants, 4758

Nanosystems
NSF Active Nano- Structures & Systems Grants, 3459
SRC Hans J. Coufal Fellowship, 4371
USDD Nano-Bio Manufacturing Inst Grants, 4757

National Disease Organizations
Alcatel-Lucent Technologies Foundation Grants, 509
Archer Daniels Midland Foundation Grants, 985
ArvinMeritor Foundation Health Grants, 1027

National Planning/Policy
Center for National Policy Internships, 1479
Smith Richardson Foundation Domestic Public Research Fellowship, 4269
Smith Richardson Foundation Strategy and Policy Fellowship, 4270

National Security
AAAS Science and Technology Policy Fellowships: Global Health and Development, 42
Carnegie Corporation of New York Grants, 1432
Center for Defense Information Internships, 1473
Center for the Study of Professional Military Ethics Resident Fellowship, 1484
CFR International Affairs Fellowships in Nuclear Security, 1502
CIA Undergraduate Co-Op Program - Engineer/ Program Management Engineer, 1561
CIA Undergraduate Internship - Analysis, 1562
DOE Nonproliferation & Nat Security Research, 1806
DOT Transportation Statistics Research Grants, 1850
IIE David L. Boren Scholarships, 2351
IIE Rockefeller Foundation Bellagio Center Residencies, 2355
Japan Foundation Center for Global Partnership Security Studies Fellowships, 2475
JMO Strategic and International Studies Grants, 2493
National Security Education Program David L. Boren Graduate Fellowships, 2884
National Security Education Program Grants to Institutions of Higher Education, 2885
OSF Global Faculty Grants - Track 1, 3841
RAND Corporation Graduate Student Summer Associateships, 4030
RAND Corporation Military Fellowships, 4031
RAND Corporation Transatlantic Postdoctoral Fellowships for Int'l Relations & Security, 4032
Smith Richardson Foundation Strategy and Policy Fellowship, 4270
SSRC Next Generation Social Sciences in Africa: Doctoral Dissertation Research Fellowships, 4436
United States Institute of Peace - Jennings Randolph Peace Dissertation Scholarships, 4579
United States Institute of Peace - Jennings Randolph Senior Fellowships, 4580
USDD Research Interests of the US Air Force Academy Grants, 4785

Native American Education
AAA Minority Dissertation Fellowship Program, 25
American Indian College Fund STEM Leadership Fellowships, 745
APA Planning Fellowships, 910
BIA Native American Grad Student Fellowships, 1342
Burlington Northern Santa Fe Foundation Grants, 1389
Consortium for Graduate Study in Management Minority Fellowships, 1642
Diversifying Higher Education Faculty in Illinois Program, 1764
DOT Dwight David Eisenhower Transportation Fellowship Program, 1847
GEM MS Engineering Fellowships, 2104
GEM PhD Engineering Fellowships, 2105
GEM PhD Science Fellowships, 2106
General Motors Foundation Grants, 2107
John R. Oishei Foundation Grants, 2505
KFF Barbara Jordan Health Policy Scholars, 2556
MLA Grad Scholarship for Minority Students, 2787
NACC David Stevenson Fellowships, 2812
NACC William Diaz Fellowships, 2813
NHLBI Training Program Grants for Institutions that Promote Diversity, 3067
NL British Academy Fellowship for Study in the United Kingdom, 3283
NL Lawrence Lipking Fellowship at Newberry, 3287
NL Midwest Modern Lang Assoc Fellowship, 3299
NL Monticello College Foundation Fellowship for Women, 3304
NL Newberry Consortium on American Indian Studies Graduate Student Fellowships, 3312
NL South Central Modern Languages Association Fellowship, 3318
NL Susan Kelly Power and Helen Hornbeck Tanner Fellowship, 3319
NMF Need-Based Scholarships, 3326
North Dakota Indian Scholarships, 3427
NSF Tribal Colleges and Universities Program, 3662
Seneca Nation Higher Education Grants, 4198
UNC Carolina Postdoctoral Program for Faculty Diversity, 4560

Native American Languages
American Philosophical Society Phillips Fund Grants for Native American Research, 770
IES Faculty Fellowships, 2341
NL American Society for Eighteenth-Century Studies Fellowships, 3279
NL Arthur and Lila Weinberg Fellowships for Independent Scholars and Researchers, 3280
NL Associated Colleges of the Midwest/Great Lakes College Association Faculty Fellowships, 3281
NL British Academy Fellowship for Study in the United Kingdom, 3283
NL Frances C. Allen Fellowships, 3284
NL Grad Scholar-in-Residence Opportunities, 3285
NL Herzog August Bibliothek Wolfenbuttel Fellowship, 3286
NL Lawrence Lipking Fellowship at Newberry, 3287
NL Lester J. Cappon Fellowship in Documentary Editing, 3288
NL Mellon Postdoctoral Research Fellowships, 3294
NL Midwest Modern Lang Assoc Fellowship, 3299
NL Monticello College Foundation Fellowship for Women, 3304
NL National Endowment for the Humanities Fellowships, 3310
NL Newberry Consortium on American Indian Studies Graduate Student Fellowships, 3312
NL Northeast Modern Language Assocation Fellowship, 3314
NL Scholar-in-Residence Opportunity, 3315
NL Short-Term Resident Fellowships for Individual Research, 3317
NL South Central Modern Languages Association Fellowship, 3318
NL Susan Kelly Power and Helen Hornbeck Tanner Fellowship, 3319
U.S. Department of Education Foreign Languages and Area Studies Fellowships, 4539
U.S. Department of Education Native Hawaiian Education Grants, 4543

Native American Studies
American Indian College Fund STEM Leadership Fellowships, 745
American Philosophical Society Phillips Fund Grants for Native American Research, 770
Andrew W. Mellon Postdoctoral Fellowship in American Cultures, 845
Bay and Paul Foundations, Inc Grants, 1307
IES Faculty Fellowships, 2341
Institute of American Cultures Graduate and Predoctoral Fellowships, 2383
Institute of American Cultures Postdoctoral and Visiting Scholar Fellowships, 2384
NEH Awards for Faculty at Tribal Colleges and Universities, 2951
NL American Society for Eighteenth-Century Studies Fellowships, 3279
NL Arthur and Lila Weinberg Fellowships for Independent Scholars and Researchers, 3280
NL Associated Colleges of the Midwest/Great Lakes College Association Faculty Fellowships, 3281
NL British Academy Fellowship for Study in the United Kingdom, 3283
NL Frances C. Allen Fellowships, 3284
NL Grad Scholar-in-Residence Opportunities, 3285
NL Herzog August Bibliothek Wolfenbuttel Fellowship, 3286
NL Lawrence Lipking Fellowship at Newberry, 3287
NL Lester J. Cappon Fellowship in Documentary Editing, 3288
NL Lloyd Lewis Fellowships in American Hist, 3289
NL Mellon Postdoctoral Research Fellowships, 3294
NL Midwest Modern Lang Assoc Fellowship, 3299
NL Monticello College Foundation Fellowship for Women, 3304
NL National Endowment for the Humanities Fellowships, 3310
NL Newberry Consortium on American Indian Studies Faculty Fellowships, 3311
NL Newberry Consortium on American Indian Studies Graduate Student Fellowships, 3312
NL Northeast Modern Language Assocation Fellowship, 3314
NL Scholar-in-Residence Opportunity, 3315
NL Short-Term Resident Fellowships for Individual Research, 3317
NL South Central Modern Languages Association Fellowship, 3318
NL Susan Kelly Power and Helen Hornbeck Tanner Fellowship, 3319

Native Americans
American Indian College Fund STEM Leadership Fellowships, 745
American Sociological Association Minority Fellowships, 805

amfAR Fellowships, 809
amfAR Mathilde Krim Fellowships in Basic Biomedical Research, 810
amfAR Research Grants, 811
AMS Howard Mayer Brown Fellowships, 827
ASM Robert D. Watkins Graduate Research Fellowship, 1183
Dept of Ed Vocational Rehabilitation Services Demonstration and Training Programs, 1751
DHHS Emergency Medical Services for Children Program, 1755
DHHS Native American Programs, 1758
FDHN Student Research Fellowships, 2001
Honor the Earth Grants, 2279
Jacobs Research Funds Small Grants Program, 2456
Minority Faculty Fellowship Program, 2781
NAGPRA Museum Consultation/Documentation Grants, 2822
NAGPRA Museum Repatriation Grants, 2823
NEH Awards for Faculty at Tribal Colleges and Universities, 2951
NL American Society for Eighteenth-Century Studies Fellowships, 3279
NL Arthur and Lila Weinberg Fellowships for Independent Scholars and Researchers, 3280
NL Associated Colleges of the Midwest/Great Lakes College Association Faculty Fellowships, 3281
NL British Academy Fellowship for Study in the United Kingdom, 3283
NL Frances C. Allen Fellowships, 3284
NL Herzog August Bibliothek Wolfenbuttel Fellowship, 3286
NL Lawrence Lipking Fellowship at Newberry, 3287
NL Lester J. Cappon Fellowship in Documentary Editing, 3288
NL Mellon Postdoctoral Research Fellowships, 3294
NL Midwest Modern Lang Assoc Fellowship, 3299
NL Monticello College Foundation Fellowship for Women, 3304
NL National Endowment for the Humanities Fellowships, 3310
NL Northeast Modern Language Assocation Fellowship, 3314
NL Scholar-in-Residence Opportunity, 3315
NL Short-Term Resident Fellowships for Individual Research, 3317
NL South Central Modern Languages Association Fellowship, 3318
NL Susan Kelly Power and Helen Hornbeck Tanner Fellowship, 3319
Percy B. Ferebee Endowment Grants, 3897
RWJF Harold Amos Medical Faculty Development Research Grants, 4139
Smithsonian Native American Community Scholars Awards, 4291
Smithsonian Native American Visiting Student Awards, 4292
SRC AMD/Mahboob Khan Fellowship, 4348
SRC Company-Named Scholarships, 4351
SRC Master's Scholarships, 4378
UCLA Research Grant in Ethnic Studies, 4549
United Methodist Church Asian, Black, Hispanic and Native American History Research Grants, 4570
USDA Tribal College Research Grants, 4652

Natural History
American Philosophical Society Library Resident Research Fellowships, 769
Bay and Paul Foundations, Inc Grants, 1307
IMLS 21st Century Museum Prof Grants, 2362
IMLS Conservation Project Support Grants, 2363
IMLS National Leadership Grants, 2366
NIDDK Type 2 Diabetes in the Pediatric Population Research Grants, 3188
Smithsonian Senior Fellowships, 4296

Natural Products
Ernest Guenther Award in the Chemistry of Natural Products, 1950

Natural Resources
Ahn Family Foundation Grants, 379
Alces Foundation Grants, 510
ASPRS William A. Fischer Mem Scholarship, 1216
Community Found for Greater Buffalo Grants, 1615
Gilbert F. White Postdoctoral Fellowships, 2140
H.A. and Mary K. Chapman Charitable Grants, 2193
Mary K. Chapman Foundation Grants, 2673
National Wildlife Refuge System Centennial Scholarship Program, 2892
NFWF Acres for America Grants, 2996
NOAA Chesapeake Bay Office B-WET (Bay Watershed Education and Training) Grants, 3339
NOAA Chesapeake Bay Office Cooperative Research Program Grants, 3340
NOAA Chesapeake Bay Office Fisheries Science Program Grants, 3341
NOAA Chesapeake Bay Office Non-Native Oyster Research Grants, 3342
NOAA Chesapeake Bay Office Submerged Aquatic Vegetation Culture and Restoration Grants, 3343
NOAA Cooperative Unit for Fisheries Education and Research Internships, 3360
NOAA Faculty and Student Intern Research, 3371
NOAA Miami Lab Summer Internships, 3382
NOAA Monkfish Research Set-Aside Grants, 3386
NOAA Sea Scallop Research Set-Aside Program Grants, 3404
NSERC John C. Polanyi Awards, 3446
Organization for Tropical Studies Fellowships, 3796
Organization for Tropical Studies Pilot Awards for Exploratory Research, 3797
OTS Francis J. Bossuyt Fellowship, 3855
OTS Undergrad NAPIRE Summer Interns, 3858
Rhode Island Foundation Grants, 4061
Robert Schalkenbach Foundation Grants, 4079
Rocky Mountain Mineral Law Found Grants, 4094
Russell Family Foundation Environmental Sustainability Grants, 4128
SARE On-Farm Research Grants - So Region, 4173
South Florida Student Shark Program, 4330
USDA Hispanic-Serving Institutions Education Grants, 4628
USDA Norman E. Borlaug International Agricultural Science and Technology Fellowships, 4636
Vancouver Foundation Grants and Community Initiatives, 4807
WHOI Marine Policy Fellowships, 4861

Natural Sciences
AAUW American Dissertation Fellowships, 209
AAUW American Postdoctoral Research Leave Fellowships, 210
AAUW American Summer/Short-Term Research Publication Grants, 211
Abdus Salam ICTP-OEA Affil Centre Grants, 222
Abdus Salam ICTP Networks Grants, 233
ACLS Collaborative Research Fellowships, 282
AIAR/Council of American Overseas Research Centers Fellowships for Advanced Multi-Country Research, 391
Alexander von Humboldt Foundation Fellowships for Experienced Researchers, 516
Alexander von Humboldt Foundation Fellowships for Postdoctoral Researchers, 517
Alexander von Humboldt Foundation Georg Forster Research Award, 525
Alexander von Humboldt Foundation JSPS Research Fellowships, 530
Alexander von Humboldt Foundation Max Planck Research Award, 532
Alexander von Humboldt Foundation NSC Taiwan Fellowships, 533
Alexander von Humboldt Foundation Philipp Franz von Siebold Award for Japanese Researchers, 534
Alexander von Humboldt Foundation Sofja Kovalevskaja Awards, 540
American Mathematical Society AAAS Mass Media Fellowships, 751
ASU Graduate College Dissertation Fellowships, 1237
Croucher Foundation Grants, 1659
DAAD Research Grants for Doctoral Candidates and Young Academics and Scientists, 1703
DAAD Research Stays for University Academics and Scientists, 1705
Explorers Club Exploration Fund Grants, 1966
Frederick Douglass Summer Scholars Fellowship, 2053
GEM PhD Science Fellowships, 2106
Harry Frank Guggenheim Foundation Research Grants, 2208
HRF Hudson River Improvement Grants, 2286
IFUW Winifred Cullis Grants, 2347
IIE Rockefeller Foundation Bellagio Center Residencies, 2355
Inamori Foundation Research Grants, 2368
Jeffress Awards in Interdisciplinary Research, 2481
M.J. Murdock Charitable Trust College Science Research Program, 2639
M.J. Murdock Charitable Trust Scientific Research Grants, 2644
Nanyang Technological University Lee Kuan Yew Postdoctoral Fellowship, 2826
NOAA Regional Integrated Sciences and Assessments Program Grants, 3402
NSERC Gerhard Herzberg Canada Gold Medal for Science and Engineering Grants, 3440
NSF Arctic Natural Science Grants in Atmospheric Sciences, 3483
NSF Arctic Natural Science Grants in Biological Sciences, 3484
NSF Arctic Nat Science Grants in Earth Sciences, 3485
NSF Arctic Nat Science Grants in Ocean Science, 3487
NSF Arctic Natural Sciences Program Grants, 3488
NSF Arctic Research Support & Logistics Grants, 3491
NSF Physical Anthropology Research Grants, 3615
NWO Athena Grants, 3674
Organization for Tropical Studies Fellowships, 3796
Organization for Tropical Studies Pilot Awards for Exploratory Research, 3797
ORISE Higher Education Research Experiences at Oak Ridge National Laboratory for Faculty, 3808
ORISE Savannah River Site Faculty Research Participation Program, 3823
OTS Hovore-Horn Fellowships, 3856
OTS Stanley Smith Horticultural Fellowships, 3857
OTS Undergrad NAPIRE Summer Interns, 3858
Packard Foundation Fellowships for Science and Engineering, 3862
Reinberger Foundation Grants, 4051
Roy J. Carver Charitable Trust Higher Education Grants, 4113
RSC Sir John William Dawson Medal, 4125
SRC Natural and Engineering Science Grants for Postdoctoral Positions in Sweden, 4384
SRC Natural and Engineering Science Postdoctoral Fellowships, 4385
SRC Natural and Engineering Science Senior Research Positions in Sweden, 4386
SRC Natural and Engineering Sciences Grants, 4387
SSHRC-NSERC College and Community Innovation Enhancement Grants, 4400
SSHRC Graduate Scholarships-Michael Smith Foreign Study Supplements, 4408
SSRC Eurasia Program Dissertation Development Awards, 4427
Swedish Natural Science Res Council Grants, 4479
UNESCO Fellowships, 4567
W. Frank Blair Award, 4821
WHOI Undergrad Fellowships for Minorities, 4866

Nature Centers
Boston Foundation Grants, 1361
IMLS 21st Century Museum Prof Grants, 2362
IMLS National Leadership Grants, 2366

Naval History
Marine Corps Historical Center Dissertation Fellowship, 2663
USDD Research and Studies for the Office of Net Assessment Grants, 4782

Naval Research
ONR Maritime Sensing Grants, 3770
ONR Young Investigator Program Grants, 3788
USDD Acoustic Transduction Materials and Devices Grants, 4672
USDD Active Aperture Array Grants, 4673
USDD Alloys and Joining Grants, 4680
USDD Antifouling/Fouling Release Coatings, 4681
USDD Applications and Transitions Grants, 4682
USDD Applied Computational Analysis Grants, 4684
USDD Arctic and Global Prediction Grants, 4685
USDD Basic Research Challenge: Carbon Molecular Electronics Grants, 4688
USDD Basic Research Challenge: Decentralized Online Optimization Grants, 4689
USDD Bulk Nanostructured Materials Grants, 4704
USDD Capacitors for Pulsed Power Applications Grants, 4705
USDD Cellular Materials Grants, 4707
USDD Chemical and Biological Defense Innovations and Technologies Grants, 4708
USDD Command and Control Grants, 4711
USDD Computer-Aided Materials Des Grants, 4714
USDD Dielectric Films for Capacitors Grants, 4719
USDD Electrochemical Materials Grants, 4721
USDD Electromagnetic Materials Grants, 4722
USDD Entry-Level Faculty Awards in Ocean Acoustics, 4724
USDD Fires Thrust Grants, 4726
USDD Force Protection Thrust Grants, 4727
USDD Future Naval Fuels Grants, 4729
USDD Hull Grooming/Advanced Hull Cleaning Technologies Grants, 4735
USDD Image Optimization for Small Focal Plane Arrays Grants, 4739
USDD Information Integration Grants, 4741
USDD Intelligent & Autonomous Syst Grants, 4744
USDD Ionospheric Modification Grants, 4745
USDD Littoral Geosciences and Optics Grants, 4746
USDD Logistics Thrust Grants, 4747
USDD Long Range Broad Agency Announcement for Navy and Marine Corps Science and Tech, 4748
USDD Mathematical Optimization and Operations Research Grants, 4751
USDD Mixed Signal, Signal Processing Devices, Circuits and Architecture Grants, 4755
USDD Nanoscale Electronics Grants, 4758
USDD National Geospatial-Intelligency Agency Academic Research Grants, 4760
USDD Naval Research Facilities Grants, 4761
USDD Navigation and Timekeeping Grants, 4762
USDD Non-Destructive Evaluation and Prognostics: Advanced Sensors and Technologies Grants, 4763
USDD Non-Destructive Eval & Prognostics: Fatigue and Fracture Damage Protection Grants, 4764
USDD NRL Broad Agency Announcement, 4765
USDD Ocean Acoustics Research Grants, 4766
USDD Ocean Eng and Marine Systems Grants, 4767
USDD Organic Photovoltaics Grants, 4770
USDD Physical Oceanography Grants, 4771
USDD Polymeric Organic Materials: Conducting/ Semiconducting Polymers Grants, 4772
USDD Polymer Matrix Composites Grants, 4773
USDD Propulsion Materials Grants, 4778
USDD Radar and Signal Processing Grants, 4779
USDD Research and Studies for the Office of Net Assessment Grants, 4782
USDD Research Initiatives at the Naval Postgraduate School Grants, 4783
USDD RF Superconducting Technologies Grants, 4787
USDD Ship Systems and Engineering Grants, 4792
USDD Signal Processing for Networked Sensing Grants, 4793
USDD Software and Computer Systems Grants, 4795
USDD Solid and Liquid Waste Treatment Grants, 4796
USDD Solid Mechanics Grants, 4797
USDD Ultra-High Temp Materials Grants, 4800
USDD Undersea Signal Processing Grants, 4801
USDD Vacuum Devices Grants, 4802
USDD Water Desalination & Purific Grants, 4804

Naval Sciences
ONR Marine Meteorology and Atmospheric Effects Research Grants, 3769
ONR Medical Science and Technology Division Grants, 3772
ONR Navy Postdoctoral Fellowships, 3776
ONR Sea Platforms and Weapons Div Grants, 3780
USDD Air Warfare and Naval Weapons Applications Grants, 4679
USDD Chemical and Biological Defense Innovations and Technologies Grants, 4708
USDD Hull Grooming/Advanced Hull Cleaning Technologies Grants, 4735
USDD Long Range Broad Agency Announcement for Navy and Marine Corps Science and Technology Grants, 4748

Naval Technology
American Society of Naval Engineers Scholarship, 800
ASEE Naval Research Laboratory Postdoctoral Fellowships, 1060
ONR Bisson Prize for Naval Tech Achievement, 3745
ONR Electronic and Optical Ceramics Research Grants, 3756
ONR Maritime Sensing Grants, 3770
ONR Naval Research Enterprise Intern Program Grants, 3775
ONR Navy Postdoctoral Fellowships, 3776
ONR Sea Platforms and Weapons Div Grants, 3780
Society of Naval Architects and Marine Engineers Scholarships, 4328
USDD Acoustic Transduction Materials and Devices Grants, 4672
USDD Active Aperture Array Grants, 4673
USDD Air Warfare and Naval Weapons Applications Grants, 4679
USDD Alloys and Joining Grants, 4680
USDD Antifouling/Fouling Release Coatings, 4681
USDD Applications and Transitions Grants, 4682
USDD Applied & Computational Anal Grants, 4683
USDD Applied Computational Analysis Grants, 4684
USDD Arctic and Global Prediction Grants, 4685
USDD Basic Research Challenge: Decentralized Online Optimization Grants, 4689
USDD Bulk Nanostructured Materials Grants, 4704
USDD Capacitors for Pulsed Power Apps Grants, 4705
USDD Cellular Materials Grants, 4707
USDD Chemical and Biological Defense Innovations and Technologies Grants, 4708
USDD Command and Control Grants, 4711
USDD Communications and Networking Discovery and Invention Grants, 4712
USDD Computer-Aided Materials Des Grants, 4714
USDD Dielectric Films for Capacitors Grants, 4719
USDD Electrochemical Materials Grants, 4721
USDD Electromagnetic Materials Grants, 4722
USDD Entry-Level Faculty Awards in Ocean Acoustics, 4724
USDD Fires Thrust Grants, 4726
USDD Force Protection Thrust Grants, 4727
USDD Future Naval Fuels Grants, 4729
USDD Hull Grooming/Advanced Hull Cleaning Technologies Grants, 4735
USDD Information Integration Grants, 4741
USDD Intelligence, Surveillance and Reconnaissance Thrust Grants, 4743
USDD Intelligent & Autonomous Syst Grants, 4744
USDD Ionospheric Modification Grants, 4745
USDD Littoral Geosciences and Optics Grants, 4746
USDD Logistics Thrust Grants, 4747
USDD Long Range Broad Agency Announcement for Navy and Marine Corps Science and Technology Grants, 4748
USDD Mathematical Optimization and Operations Research Grants, 4751
USDD Mixed Signal, Signal Processing Devices, Circuits and Architecture Grants, 4755
USDD Nanoscale Electronics Grants, 4758
USDD Naval Research Facilities Grants, 4761
USDD Navigation and Timekeeping Grants, 4762
USDD Non-Destructive Evaluation and Prognostics: Advanced Sensors and Technologies Grants, 4763
USDD Non-Destructive Eval & Prognostics: Fatigue and Fracture Damage Protection Grants, 4764
USDD Ocean Acoustics Research Grants, 4766
USDD Ocean Engineering and Marine Systems Grants, 4767
USDD Organic Photovoltaics Grants, 4770
USDD Physical Oceanography Grants, 4771
USDD Polymeric Organic Materials: Conducting/ Semiconducting Polymers Grants, 4772
USDD Polymer Matrix Composites Grants, 4773
USDD Propulsion Materials Grants, 4778
USDD Radar and Signal Processing Grants, 4779
USDD Research and Studies for the Office of Net Assessment Grants, 4782
USDD RF Superconducting Technologies Grants, 4787
USDD Ship Systems and Engineering Grants, 4792
USDD Signal Processing for Networked Sensing Grants, 4793
USDD Software and Computer Systems Grants, 4795
USDD Solid and Liquid Waste Treatment Grants, 4796
USDD Solid Mechanics Grants, 4797
USDD Ultra-High Temp Materials Grants, 4800
USDD Undersea Signal Processing Grants, 4801
USDD Vacuum Devices Grants, 4802
USDD Water Desalination & Purific Grants, 4804

Navigation and Guidance Systems
Aviation Research Grants, 1273
ONR Computational Neuroscience & Biorobotics Grants, 3752
ONR Radio Frequency Sensor Processing Research Grants, 3778
ONR Radio Frequency Solid State Devices Applied Research Grants, 3779
USDD Fires Thrust Grants, 4726
USDD Navigation and Timekeeping Grants, 4762

Navy
ONR Young Investigator Program Grants, 3788
USDD Fires Thrust Grants, 4726
USDD Force Protection Thrust Grants, 4727
USDD Mathematical Optimization and Operations Research Grants, 4751

Near East
AIAR/Council of American Overseas Research Centers Fellowships for Advanced Multi-Country Research, 391

Neighborhood Revitalization
MGM Resorts Foundation Community Grants, 2753
Textron Corporate Contributions Grants, 4501

Neighborhoods
Allstate Foundation Safe and Vital Communities Grants, 573
Bayer Foundation Grants, 1309
Carnegie Corporation of New York Grants, 1432
Chestnut Hill Charitable Foundation, Inc Grants, 1538
Enterprise Community Partners Rose Architectural Fellowships, 1929
Farmers Insurance Corporate Giving Grants, 1981
Frances L. and Edwin L. Cummings Memorial Fund Grants, 2046
George Foundation Grants, 2115
George Gund Foundation Grants, 2117
Helen Bader Foundation Grants, 2228
Herbert & Adrian Woods Foundation Grants, 2239
Hirtzel Memorial Foundation Grants, 2269
HUD Community Outreach Partnership Centers Grants, 2292
Lynde & Harry Bradley Foundation Fellowships, 2634
Lynde and Harry Bradley Foundation Grants, 2635
MGM Resorts Foundation Community Grants, 2753
Nordson Corporation Foundation Grants, 3416
Oklahoma City Community Grants, 3737
OSF-Baltimore Community Fellowships, 3833
Russell Sage Foundation Grants, 4129

Nephrology
ACCP Pain & Palliative Care PRN Minisabbatical Program, 265
ACCP Pharmacotherapy Investigator Development Research Awards, 266
NIDDK Training in Clinical Investigation in Kidney and Urology Grants, 3185
NKF Research Fellowships, 3277
NKF Young Investigator Grants, 3278
NYAM Edward N. Gibbs Memorial Lecture and Award in Nephrology, 3682
PKD Foundation Lillian Kaplan International Prize for the Advancement in the Understanding of Polycystic Kidney Disease, 3974

Nervous System
A-T Children's Project Grants, 8
A-T Children's Project Post Doctoral Fellowships, 9
Grass Foundation Marine Biological Laboratory Advanced Imaging Fellowships, 2162
Gruber Foundation Neuroscience Prize, 2181
NINDS Optimization of Small Molecule Probes for the Nervous System Grants, 3255
NINDS Optimization of Small Molecule Probes for the Nervous System Grants, 3256
NINDS Optimization of Small Molecule Probes for the Nervous System Grants, 3257
NSF Linguistics Grants, 3587
PVA Research Foundation Grants, 4004
SfN Jacob P. Waletzky Award, 4206
SfN Ralph W. Gerard Prize in Neuroscience, 4211
SfN Swartz Prize for Theoretical and Computational Neuroscience, 4213
Simone and Cino del Duca Grand Prix Awards, 4261

Netherlands
Hague Academy of International Law Summer Scholarships, 2198

Networking
Abdus Salam ICTP Group Associates Grants, 228
Abdus Salam ICTP Junior Associates Grants, 231
Abdus Salam ICTP Regular Associates Grants, 237
Abdus Salam ICTP Senior Associates Grants, 239
NATO Collaborative Linkage Grants, 2896
NLM Exploratory/Developmental Grants, 3296
NSF CISE Community-Based Data Interoperability Networks Grants, 3528
NSF CISE Computer and Network Systems: Core Programs Grants, 3529
NSF CISE Computing and Communication Foundations: Core Programs Grants, 3530
ONR Electronics, Sensors, and Network Research Grants, 3758
RealNetworks Helix Community Grants, 4044
SRC Educational Science Network Grants, 4362
Sun Microsystems Foundation Summer Academic Enrichment Grants, 4468
USDC State Broadband Initiative Grants, 4671
USDD Applied & Computational Anal Grants, 4683
USDD Communications and Networking Grants, 4713

Neuroanatomy
Grass Foundation Marine Biological Laboratory Advanced Imaging Fellowships, 2162
Grass Foundation Marine Biological Laboratory Fellowships, 2163
NIDDK Neuroimaging in Obesity Res Grants, 3165

Neurobiology
AAAS Eppendorf Science Prize for Neurobiology, 31
AFAR Research Grants, 354
AHAF Alzheimer's Disease Research Grants, 370
Alfred P. Sloan Foundation History of Science and Technology Grants, 553
ASHA Research Mentoring-Pair Travel Award, 1073
Davis Foundation Postdoctoral Fellowships in Eating Disorders Research, 1727
Grass Foundation Marine Biological Laboratory Advanced Imaging Fellowships, 2162
Grass Foundation Marine Biological Laboratory Fellowships, 2163
Hereditary Disease Foundation John J. Wasmuth Postdoctoral Fellowships, 2244
Hereditary Disease Found Lieberman Award, 2245
Hereditary Disease Foundation Research Grants, 2246
Hunter's Hope Foundation Grants, 2298
John Merck Scholars Awards, 2502
Klarman Family Foundation Grants in Eating Disorders Research, 2558
Lumosity Human Cognition Grant, 2628
MBL Albert & Ellen Grass Summer Fellowships, 2692
MBL Associates Summer Fellowships, 2694
MBL Burr & Susie Steinbach Summer Fellow, 2696
MBL E.E. Just Summer Fellowship for Minority Scientists, 2697
MBL Erik B. Fries Summer Fellowships, 2698
MBL Evelyn & Mel Spiegal Summer Fellowship, 2700
MBL Frank R. Lillie Summer Fellowship, 2701
MBL Fred Karush Library Readership, 2703
MBL Gruss Lipper Family Foundation Summer Fellowship, 2704
MBL H. Keffer Hartline and Edward F. MacNichol, Jr. Fellowships, 2705
MBL Herbert W. Rand Summer Fellowship, 2706
MBL James E. and Faith Miller Memorial Summer Fellowship, 2707
MBL John M. Arnold Award, 2708
MBL Lucy B. Lemann Mem Summer Fellowship, 2710
MBL M.G.F. Fuortes Summer Fellowships, 2711
MBL Nikon Summer Fellowship, 2712
MBL Plum Found John E. Dowling Fellowships, 2713
MBL Stephen W. Kuffler Summer Fellowships, 2715
MBL William Townsend Porter Summer Fellowships for Minority Investigators, 2716
NARSAD Goldman-Rakic Prize for Cognitive Neuroscience, 2832
National Center for Responsible Gaming Centers of Excellence in Gambling Research Grants, 2850
National Center for Responsible Gaming Early Stage Investigator Grants, 2852
National Center for Responsible Gaming Exploration Grants, 2853
National Center for Responsible Gaming Independent Scientist Grants, 2854
National Center for Resp Gaming Large Grants, 2855
National Center for Responsible Gaming Postdoctoral Fellowships, 2856
National Center for Resp Gaming Seed Grants, 2857
National Center for Responsible Gaming Travel Grants, 2858
NHLBI Exploratory Studies in the Neurobiology of Pain in Sickle Cell Disease, 3026
NIA Role of Nuclear Receptors in Tissue and Organismal Aging Grants, 3120
NIDDK Neurobiology of Diabetic Complications Grants, 3164
NIH Research on Sleep and Sleep Disorders, 3236

Neurochemistry
Grass Foundation Marine Biological Laboratory Advanced Imaging Fellowships, 2162

Neuroendocrinology
Charles H. Hood Foundation Child Health Research Awards, 1522
Grass Foundation Marine Biological Laboratory Advanced Imaging Fellowships, 2162

Neurological Disorders
AMA Foundation Seed Grants for Research, 601
ANA Wolfe Neuropathy Research Prize, 840
Dana Foundation Science and Health Grants, 1716
Fairlawn Foundation Grants, 1974
John Merck Scholars Awards, 2502
Judith and Jean Pape Adams Charitable Foundation ALS Grants, 2528
NARSAD Goldman-Rakic Prize for Cognitive Neuroscience, 2832
NIEHS Small Business Innovation Res Grants, 3195
NIH Research on Sleep and Sleep Disorders, 3236
NINDS Support of Scientific Meetings as Cooperative Agreements, 3258
PDF Summer Fellowships, 3885
Pittsburgh Foundation Medical Research Grants, 3973

Neurology
American Philosophical Society Daland Fellowships in Clinical Investigation, 763
ANS/AAO-HNSF Herbert Silverstein Otology and Neurotology Research Award, 864
ANS Neurotology Fellows Award, 865
ANS Nicholas Torok Vestibular Award, 866
ANS Trainee Award, 867
Grass Foundation Marine Biological Laboratory Advanced Imaging Fellowships, 2162
Grass Foundation Marine Biological Laboratory Fellowships, 2163
NIA Role of Nuclear Receptors in Tissue and Organismal Aging Grants, 3120
NIDDK Neurobiology of Diabetic Complications Grants, 3164
NINDS Support of Scientific Meetings as Cooperative Agreements, 3258
NYAM Charles A. Elsberg Fellowship, 3680
PDF Postdoctoral Fellowships for Neurologists, 3884
Rett Syndrome Research Foundation Grants, 4058
USDD Basic Research Initiative Grants, 4691

Neuromuscular Disorders
ANA Wolfe Neuropathy Research Prize, 840
Judith and Jean Pape Adams Charitable Foundation ALS Grants, 2528
MDA Development Grant, 2728
MDA Neuromuscular Disease Research Grants, 2729
MGFA Post-Doctoral Research Fellowships, 2751
MGFA Student Fellowships, 2752
NAF Fellowships, 2816
NAF Kyle Bryant Translational Research Award, 2817
NAF Research Grants, 2818
NAF Young Investigator Awards, 2819

Neuropharmacology
Grass Foundation Marine Biological Laboratory Advanced Imaging Fellowships, 2162

Neurophysiology
Grass Foundation Marine Biological Laboratory Advanced Imaging Fellowships, 2162
Lumosity Human Cognition Grant, 2628
NARSAD Goldman-Rakic Prize for Cognitive Neuroscience, 2832
SfN Ralph W. Gerard Prize in Neuroscience, 4211

Neuroscience
AES-Grass Young Investigator Travel Awards, 345
AIHS Alberta/Pfizer Translational Res Grants, 435
Alfred P. Sloan Foundation Research Fellowships, 560
ANA/Grass Foundation Award in Neuroscience, 839
ANA Wolfe Neuropathy Research Prize, 840
Bristol-Myers Squibb Clinical Outcomes Grants, 1374
Dana Foundation Science and Health Grants, 1716
Dorothy Rider Pool Health Care Grants, 1841
Grass Foundation Marine Biological Laboratory Advanced Imaging Fellowships, 2162
Grass Foundation Marine Biological Laboratory Fellowships, 2163
Gruber Foundation Int'l Research Award, 2180
Gruber Foundation Neuroscience Prize, 2181
Harry Frank Guggenheim Foundation Dissertation Fellowships, 2207
IBRO/SfN International Travel Grants, 2321
IBRO Albert Aguayo Fellowship, 2322
IBRO Asia Reg APRC Exchange Fellowships, 2323
IBRO Europe Regional Grants for Western European Workshops and Symposia, 2324
IBRO InEurope Short Stay Grants, 2325
IBRO International Travel Grants, 2326
IBRO Latin America Regional Funding for Neuroscience Schools, 2327

IBRO Latin America Regional Funding for PROLAB Collaborations, 2328
IBRO Latin America Regional Travel Grants, 2329
IBRO Regional Grants for Int'l Fellowships to U.S. Laboratory Summer Neuroscience Courses, 2330
IBRO Research Fellowships, 2331
IBRO Return Home Fellowships, 2332
Kavli Foundation Research Prizes, 2534
Lumosity Human Cognition Grant, 2628
M.J. Murdock Charitable Trust College Research Program for Life Sciences, 2638
MBL Ann E. Kammer Mem Summer Fellowship, 2693
MBL James E. and Faith Miller Memorial Summer Fellowship, 2707
NARSAD Goldman-Rakic Prize for Cognitive Neuroscience, 2832
National Center for Responsible Gaming Centers of Excellence in Gambling Research Grants, 2850
National Center for Responsible Gaming Early Stage Investigator Grants, 2852
National Center for Responsible Gaming Exploration Grants, 2853
National Center for Responsible Gaming Independent Scientist Grants, 2854
National Center for Resp Gaming Large Grants, 2855
National Center for Responsible Gaming Postdoctoral Fellowships, 2856
National Center for Resp Gaming Seed Grants, 2857
National Center for Responsible Gaming Travel Grants, 2858
NCI/NIDA Transdisciplinary Tobacco Use Research Centers Grants, 2912
NIA Vulnerable Dendrites and Synapses in Aging and Alzheimer's Disease Grants, 3125
NIGMS Predoctoral Training in Bioinformatics and Computation Biology Grants, 3199
NIH Research on Sleep and Sleep Disorders, 3236
NIMH Curriculum Development Award in Neuroinformatics Research and Analysis, 3251
NIMH Jointly Sponsored Ruth L. Kirschstein National Research Service Award Institutional Predoctoral Training Program in the Neurosciences, 3253
NIMH Short Courses in Neuroinformatics, 3254
NSF CISE Collaborative Research in Computational Neuroscience Grants, 3526
ONR Computational Neuroscience & Biorobotics Grants, 3752
Pfizer Australia Neuroscience Research Grants, 3920
Rosalinde and Arthur Gilbert Foundation/AFAR New Investigator Awards in Alzheimer's Disease, 4102
Royal Norwegian Embassy Kavli Prizes, 4109
Searle Scholars Program Grants, 4193
SfN Chapters Graduate Student Travel Awards, 4202
SfN Chapters Postdoct Trainee Travel Awards, 4203
SfN Donald B. Lindsley Prize in Behavioral Neuroscience, 4204
SfN Graduate Student Travel Awards, 4205
SfN Jacob P. Waletzky Award, 4206
SfN Janett Rosenberg Trubatch Career Development Award, 4207
SfN Mika Salpeter Lifetime Achievement Award, 4209
SfN Postdoctoral Fellow Travel Awards, 4210
SfN Ralph W. Gerard Prize in Neuroscience, 4211
SfN Ricardo Miledi Neuroscience Training, 4212
SfN Swartz Prize for Theoretical and Computational Neuroscience, 4213
SfN Travel Awards for the Federation of European Neuroscience Societies Forum, 4214
SfN Travel Awards for the International Brain Research Organization World Congress, 4215
SfN Travel Awards for the Japan Neuroscience Society Meeting, 4216
SfN Young Investigator Award, 4218
Swiss National Science Foundation Fellowships for Advanced Researchers, 4483
Swiss National Science Foundation Fellowships for Prospective Researchers, 4484
Swiss National Science Foundation Marie Heim-Vogtlin Grants, 4485
USDD Basic Research Initiative Grants, 4691
USDD Defense Sciences Research and Technology Grants, 4716
W.M. Keck Foundation Science and Engineering Research Grants, 4824

Neurosurgery

AANS Neurosurgery Research and Education Foundation/Spine Section Young Clinician Investigator Award, 184
AANS Neurosurgery Research & Educ Foundation Young Clinician Investigator Award, 185
American College of Surgeons and The Triological Society Clinical Scientist Development Awards, 701

Neurotology

ANS Neurotology Fellows Award, 865
ANS Nicholas Torok Vestibular Award, 866
ANS Trainee Award, 867

Neurotransmitters

Lumosity Human Cognition Grant, 2628

New England

MHS Long Term Research Fellowships, 2755
MHS New England Regional Fellowships, 2756
MHS Short Term Research Fellowships, 2757

New Independent States

Charles Stewart Mott Foundation Grants, 1529
Eastern Europe & the Independent States of the Former Soviet Union Research & Training Grants, 1891
IREX ECA Alumni Small Grants Program, 2416
IREX Edmund S. Muskie Graduate Fellowship Program, 2417
IREX Eurasian Undergraduate Exchange Program Grants, 2418
IREX International Leadership in Education Program Grants, 2420
IREX Policy-Connect Collaborative Grants, 2422
SSRC Eurasia Program Pre-Dissertation Awards, 4428

Nitrogen Fixation

NSERC John C. Polanyi Awards, 3446

Noise Control

ASME J.P. Hartog Award, 1134
ASME Per Bruel Gold Medal for Noise Control and Acoustics, 1153

Nonfiction

AAAS Newcomb Cleveland Prize, 36
AAG John Brinkerhoff Jackson Prize, 175
Aid to Scholarly Publications, 431
American Historical Association Clarence H. Haring Prize, 716
American Historical Association George Louis Beer Prize, 720
American Historical Association Helen and Howard R. Marraro Prizes, 721
American Historical Association Herbert Baxter Adams Prize, 722
American Historical Assoc Herb Feis Award, 723
American Historical Association Joan Kelly Memorial Prize in Women's History, 730
American Historical Association John H. Dunning Prize, 732
American Historical Association John K. Fairbank Prize in East Asian History, 733
American Historical Association Littleton-Griswold Prize in American Law and Society, 735
American Historical Assoc Wesley-Logan Prize, 744
AMS Alfred Einstein Award, 819
AMS Claude V. Palisca Award, 821
AMS H. Colin Slim Award, 825
AMS Lewis Lockwood Award, 829
AMS Robert M. Stevenson Award, 834
AMS Ruth A. Solie Award, 835
ASECS James L. Clifford Prize, 1058
Bernard L. Schwartz Fellows Program, 1334
Canadian Studies Senior Fellowships, 1422
Christian Gauss Award, 1555
Forest Products Society Wood Award, 2034
International League of Antiquarian Booksellers Bibliographical Prize, 2400
MLA Murray Gottlieb Prize Essay Award, 2788
MLA Rittenhouse Award, 2790
National Endowment for the Arts Literature Fellowships: Creative Writing, 2865
NFJC Gantz-Zahler Grant, 2993
NLM Grants for Scholarly Works in Biomedicine and Health, 3298
NYFA Artist Fellowships, 3698
OAH James A. Rawley Prize, 3721
OAH Lerner-Scott Dissertation Prize, 3723
OAH Ray A. Billington Prize, 3727
Orgone Biophysical Research Lab Lou Hochberg Awards, 3798
Phi Beta Kappa Award in Science, 3935
Princeton University Lewis Center for the Arts Hodder Fellowships, 3996
Ralph Waldo Emerson Award, 4028
RSC Bancroft Award, 4118
SHOT Samuel Eleazar and Rose Tartakow Levinson Prize, 4247
SSHRC-FHSS Awards to Scholarly Publications, 4398
STC Research Grants, 4450
Trevithick Book Prize Award, 4532
W. David Smith, Jr. Grad Student Paper Award, 4820
W. Frank Blair Award, 4821
WAWH Sierra Award, 4834

Nonprofit Organizations

Alfred P. Sloan Foundation Civic Grants, 545
Bayer Foundation Grants, 1309
Callaway Golf Company Foundation Grants, 1410
Carl M. Freeman Foundation Grants, 1430
Center on Philanthropy and Civil Society's Emerging Leaders International Fellows Program, 1487
Chicago Community Trust Fellowships, 1546
Elizabeth Glaser Int'l Leadership Awards, 1906
George Foundation Grants, 2115
George Gund Foundation Grants, 2117
Harvey Randall Wickes Foundation Grants, 2218
Helen Bader Foundation Grants, 2228
Human Source Foundation Grants, 2297
IDEM Section 319(h) Nonpoint Source Program Grants, 2338
Jessie Smith Noyes Foundation Grants, 2485
Jovid Foundation Grants, 2518
M.J. Murdock Charitable Trust General Grants, 2641
Mary S. and David C. Corbin Foundation Grants, 2675
MGN Family Foundation Grants, 2754
NACC David Stevenson Fellowships, 2812
NACC William Diaz Fellowships, 2813
National Center on Philanthropy and the Law Fellowship in Nonprofit Law, 2859
Nellie Mae Education Foundation District-Level Change Grants, 2968
Nellie Mae Education Foundation Public Understanding Grants, 2969
Nellie Mae Education Foundation Research and Development Grants, 2970
Percy B. Ferebee Endowment Grants, 3897
Peter F. Drucker Award for Nonprofit Innovation, 3902
Peter F. McManus Charitable Trust Grants, 3903
Price Family Charitable Fund Grants, 3987
Rhode Island Foundation Grants, 4061
Rockefeller Archive Center Research Grants, 4086
Rockefeller Brothers Fund Fellowship in Nonprofit Law, 4088
Searle Freedom Trust Grants, 4192
Surdna Found Sustainable Environments Grants, 4471
Virginia A. Hodgkinson Research Prize, 4812

Nuclear Astrophysics

APS Hans A. Bethe Prize, 960

Nuclear Chemistry

ORISE Univ Radioactive Ion Beam Consortium, 3831

Nuclear Engineering
AINSE Honours Scholarships, 460
AINSE Postgraduate Research Awards, 463
ASME Ralph Coats Roe Medal, 1158
DOE Univ Reactor Instrumentation Program, 1818
USDD National Consortium for Measurement and Signature Intelligence Research Grants, 4759

Nuclear Materials Monitoring
ORISE Nuclear Regulatory Commission Historically Black Colleges and Universities Student Research Participation, 3814

Nuclear Medicine
AINSE Research Fellowships, 464
AINSE Research Fellowships with Institute for Science and International Security Support, 465
SNM/ Covidien Seed Grant in Molecular Imaging/ Nuclear Medicine Research, 4298
SNM Pilot Research Grants, 4300
SNM Pilot Research Grants in Nuclear Medicine/ Molecular Imaging, 4301
SNM Student Fellowship Awards, 4303
SNM Tetalman Mem Young Investigator Awards, 4304

Nuclear Physics
Abdus Salam ICTP Junior Associates Grants, 231
Abdus Salam ICTP Postgraduate Diploma Grants, 235
Abdus Salam ICTP Schools, Conferences, Workshops and Extended Research Workshops Grants, 238
Abdus Salam ICTP Senior Associates Grants, 239
Abdus Salam ICTP Training and Research in Italian Laboratories Grants, 240
APS Hans A. Bethe Prize, 960
APS Tom W. Bonner Prize in Nuclear Physics, 980
JSA JLab Sabbatical Research Leave Grants, 2522

Nuclear Power
ANL Faculty Research Participation Program, 855
ANL Graduate Thesis Parts Res Appointments, 856
ANL Laboratory-Graduate Res Appointments, 857
ASME Bernard F. Langer Nuclear Codes and Standards Award, 1109
ASME James N. Landis Medal, 1136
CFR Stanton Nuclear Security Fellowships, 1506
DOE Advanced Nuclear Energy Systems Basic Research Grants, 1769
DOE Univ Reactor Instrumentation Program, 1818
USDD National Consortium for Measurement and Signature Intelligence Research Grants, 4759

Nuclear Reactors
DOE Univ Reactor Instrumentation Program, 1818

Nuclear Receptors
NIA Role of Nuclear Receptors in Tissue and Organismal Aging Grants, 3120

Nuclear Safety
IIE Rockefeller Bellagio Center Residencies, 2355

Nuclear Sciences
AIChE Nuclear Engineering Division Outstanding Student Paper & Presentation Awards, 421
AINSE Honours Scholarships, 460
AINSE Int'l Conference Travel Scholarships, 461
AINSE John Ferris Memorial Award, 462
AINSE Research Fellowships, 464
AINSE Research Fellowships with Institute for Science and International Security Support, 465
Canada Graduate Scholarships and NSERC Postgraduate Scholarships, 1420
DOE Advanced Nuclear Energy Systems Basic Research Grants, 1769
DOE Basic Energy Sciences Grants, 1772
DOE Univ Nuclear Science & Reactor Support, 1817
NSF Particle and Nuclear Astrophysics Grants, 3608
ORISE Nuclear Regulatory Commission Historically Black Colleges and Universities Faculty Research Participation, 3813
ORISE Nuclear Regulatory Commission Historically Black Colleges and Universities Student Research Participation, 3814

Nuclear Technology
AIChE Nuclear Engineering Division Outstanding Student Paper & Presentation Awards, 421
AINSE Honours Scholarships, 460
AINSE Int'l Conference Travel Scholarships, 461
AINSE John Ferris Memorial Award, 462
AINSE Postgraduate Research Awards, 463
AINSE Research Fellowships, 464
AINSE Research Fellowships with Institute for Science and International Security Support, 465
DOE Advanced Nuclear Energy Systems Basic Research Grants, 1769

Nuclear Warfare
CFR Stanton Nuclear Security Fellowships, 1506
USDD National Consortium for Measurement and Signature Intelligence Research Grants, 4759

Nuclear Weapons
CFR International Affairs Fellowships in Nuclear Security, 1502
CFR Stanton Nuclear Security Fellowships, 1506
DOE Nonproliferation & Nat Security Research, 1806
NIEHS Hazmat Training at Doe Nuclear Weapons Complex Grants, 3191
USDD Basic Research for Combating Weapons of Mass Destruction Broad Agency Announcement Grants, 4690
USDD National Consortium for Measurement and Signature Intelligence Research Grants, 4759

Nuclear/Radioactive Waste Disposal
Honor the Earth Grants, 2279
NIEHS Hazmat Training at Doe Nuclear Weapons Complex Grants, 3191
ORISE Civilian Radioactive Waste Management Fellowship Program, 3799

Nucleic Acids
NSF Biomolecular Systems Cluster Grants, 3507

Number Theory
NSF Algebra, Number Theory and Combinatorics Grants, 3468

Numerical Analysis
Norwegian Acad of Science & Letters Abel Prize, 3433
NSF Applied Mathematics Grants, 3481

Numismatics
American Numismatic Society Doc Fellowships, 759
American Numismatic Society Donald Groves Numismatic Research Grants, 760
American Numismatic Society Eric P. Newman Graduate Seminar Grants, 761
American Numismatic Society Frances M. Schwartz Fellowship, 762
Royal Belgian Numismatic Society Grants, 4107

Nurse Practitioners
AACN-Sigma Theta Tau Critical Care Grant, 104
AACN Clinical Inquiry Fund Grants, 105
AACN End of Life/Palliative Care Small Projects Grants, 108
NAPNAP Foundation Nursing Research Grants, 2829

Nursing
AACN-Edwards Lifesciences Nurse-Driven Clinical Practice Outcomes Grants, 102
AACN-Philips Medical Systems Clinical Outcomes Grants, 103
AACN Clinical Practice Grants, 106
AACN Critical Care Grants, 107
AACN Evidence-Based Clinical Practice Grants, 109
AACN Mentorship Grant, 110
AACN Physio-Control Small Projects Grants, 111
AAHN Grant for Historical Research, 178
AAHN Pre-Doctoral Research Grant, 179
Abington Foundation Grants, 243
ACS Doctoral Scholarships in Cancer Nursing, 307
Alice Fisher Society Fellowships, 568
American Academy of Nursing Building Academic Geriatric Nursing Capacity Scholarships, 614
American Academy of Nursing Claire M. Fagin Fellowships, 615
American Academy of Nursing Mayday Grants, 616
Anthem Blue Cross and Blue Shield School Nurse Grants, 868
ASPEN C. Richard Fleming Grant, 1190
ASPEN Douglas Wilmore Grant, 1191
ASPEN Maurice Shils Grant, 1195
ASPEN Norman Yoshimura Grant, 1196
ASPEN Nutrition Research Grants, 1197
Brookdale Foundation Leadership in Aging Fellowships, 1381
CampusRN/AACN Nursing Scholarship Fund, 1416
Lillian Sholtis Brunner Summer Fellowships for Historical Research in Nursing, 2615
NAPNAP Foundation Nursing Research Grants, 2829
NCI Stages of Breast Development: Normal to Metastatic Disease Grants, 2932
NFID New Investigator Matching Grants, 2989
NHLBI Mentored Patient-Oriented Research Career Development Awards, 3037
NINR Acute and Chronic Care During Mechanical Ventilation Research Grants, 3259
NINR Mentored Research Scientist Development Award, 3265
NINR Ruth L. Kirschstein National Research Service Award for Individual Predoctoral Fellows in Nursing Research, 3267
ONF Small Grants Program, 3743
ONS Major Research Grants, 3789
PhRMA Foundation Health Outcomes Post Doctoral Fellowships, 3950
Regents Professional Opportunity Scholarships, 4049
Rehabilitation Nursing Found Research Grants, 4050
RWJF Interdisciplinary Nursing Quality Research Initiative, 4144
RWJF Partners Investing in Nursing's Future, 4149
Sid W. Richardson Foundation Grants, 4253
Sigma Theta Tau International /American Nurses Foundation Grant, 4254
Sigma Theta Tau International / Council for the Advancement of Nursing Science Grants, 4255
Sigma Theta Tau International / Oncology Nursing Society Grant, 4256
Sigma Theta Tau International Doris Bloch Research Award, 4257
Sigma Theta Tau Small Grants, 4258
Special Olympics Health Professions Stu Grants, 4332
STTI Emergency Nurses Association Foundation Grant, 4459
STTI Environment of Elder Care Nursing Research Grants, 4460
STTI Joan K. Stout RN Research Grants, 4461
STTI National League for Nursing Grants, 4462
STTI Rosemary Berkel Crisp Research Awards, 4463
STTI Virginia Henderson Clin Research Grants, 4464
U.S. Department of Education Graduate Assistance in Areas of National Need, 4540
University of Calgary Nursing Alum Scholarship, 4585

Nursing Education
AAOHN Found Experienced Researcher Grants, 195
AAOHN Foundation New Investigator Researcher Grants, 196
American Academy of Nursing Building Academic Geriatric Nursing Capacity Scholarships, 614
American Academy of Nursing Claire M. Fagin Fellowships, 615
American Academy of Nursing Mayday Grants, 616
ASU Graduate College Dean's Fellowships, 1236
CampusRN/AACN Nursing Scholarship Fund, 1416
Edward W. and Stella C. Van Houten Memorial Fund Grants, 1899

HRSA Nurse Education, Practice, Quality and Retention Grants, 2290
Josiah Macy Jr. Foundation Grants, 2516
Kosciuszko Foundation Tuition Scholarships, 2574
NAPNAP Foundation Graduate Research Grant, 2828
NAPNAP Foundation Nursing Research Grants, 2829
NHLBI Short-Term Research Education Program to Increase Diversity in Health-Related Research, 3062
NIA Mentored Clinical Scientist Research Career Development Awards, 3108
NINR Ruth L. Kirschstein National Research Service Award for Individual Predoctoral Fellows in Nursing Research, 3267
ONF Small Grants Program, 3743
Rehabilitation Nursing Found Research Grants, 4050
RWJF Interdisciplinary Nursing Quality Research Initiative, 4144
RWJF Partners Investing in Nursing's Future, 4149
San Antonio Area Foundation Grants, 4163
Stroke Association Allied Health Professional Research Bursaries, 4454
STTI/ATI Educational Assessment Nursing Research Grants, 4457
STTI Joan K. Stout RN Research Grants, 4461
STTI National League for Nursing Grants, 4462
U.S. Department of Education Graduate Assistance in Areas of National Need, 4540
Yale University Beinecke Library Research Fellowships for Yale Grad & Professional School Students, 4902

Nursing Homes
Burden Trust Grants, 1388
Jacob and Valeria Langeloth Foundation Grants, 2455
McLean Contributionship Grants, 2727
Reinberger Foundation Grants, 4051

Nutrition Education
Abbott Fund Science Education Grants, 220
Allen Foundation Educational Nutrition Grants, 570
ASPEN C. Richard Fleming Grant, 1190
ASPEN Douglas Wilmore Grant, 1191
ASPEN Dudrick Research Scholar Award, 1192
ASPEN Harry M. Vars Award, 1193
ASPEN Maurice Shils Grant, 1195
ASPEN Norman Yoshimura Grant, 1196
ASPEN Nutrition Research Grants, 1197
ASPEN Scientific Abstracts Awards for Papers or Posters, 1198
ASPEN Scientific Abstracts Promising Investigator Awards, 1199
Caesars Foundation Grants, 1407
Campbell Soup Foundation Grants, 1415
Coca-Cola Foundation Grants, 1593
Colonel Stanley R. McNeil Foundation Grants, 1597
Danone Institute Belgium Research in Human Nutrition Grants, 1719
Foundation for Seacoast Health Grants, 2041
Nestle Foundation Large Research Grants, 2979
Nestle Foundation Pilot Grants, 2980
Nestle Foundation Re-entry Grants, 2981
Nestle Foundation Small Research Grants, 2982
Nestle Foundation Training Grant, 2983
Phi Upsilon Omicron Florence Fallgatter Distinguished Service Award, 3943
USDA 1890 Institution Teaching, Research, and Extension Capacity Building Grants, 4613
USDA Distance Education Grants for Higher Education Institutions in Insular Areas, 4622
USDA Organic Agriculture Research and Extension Initiative, 4637
USDA Resident Instruction Grants for Higher Education in Insular Areas, 4645
USDA Tribal College Research Grants, 4652
Vancouver Foundation Grants and Community Initiatives, 4807

Nutrition/Dietetics
AAFCS International Graduate Fellowships, 154
AAFCS National Graduate Fellowships, 155
AAP Nutrition Award, 198
Abbott Fund Science Education Grants, 220
AFAR Research Grants, 354
Allen Foundation Educational Nutrition Grants, 570
ASPEN C. Richard Fleming Grant, 1190
ASPEN Douglas Wilmore Grant, 1191
ASPEN Dudrick Research Scholar Award, 1192
ASPEN Harry M. Vars Award, 1193
ASPEN Maurice Shils Grant, 1195
ASPEN Norman Yoshimura Grant, 1196
ASPEN Nutrition Research Grants, 1197
ASPEN Scientific Abstracts Awards for Papers or Posters, 1198
ASPEN Scientific Abstracts Promising Investigator Awards, 1199
Better Way Foundation Grants, 1338
Boeing Company Contributions Grants, 1357
Boston Foundation Grants, 1361
Bristol-Myers Squibb/Meade Johnson Award for Distinguished Achievement in Nutrition Research Grants, 1373
Bristol-Myers Squibb Foundation Health Disparities Grants, 1377
Caesars Foundation Grants, 1407
Charles H. Farnsworth Trust Grants, 1521
Chautauqua Region Community Found Grants, 1533
College of Human Sciences Scholarships, 1596
Connecticut Health Foundation Health Initiative Grants, 1628
Danone Institute Belgium Doctoral Scholarship in Nutrition, 1718
Danone Institute Belgium Research in Human Nutrition Grants, 1719
FDHN Bridging Grants, 1986
FDHN Jon I. Isenberg International Research Scholar Award, 1995
FDHN Student Research Fellowships, 2001
Gerber Foundation Grants, 2126
H.J. Heinz Company Foundation Grants, 2195
IFT Foundation Grants, 2345
NCI Cancer Educ & Career Development Grants, 2917
NCI Technologies and Software to Support Integrative Cancer Biology Research Grants, 2933
NHLBI Nutrition and Diet in the Causation, Prevention, and Management of Heart Failure Research Grants, 3044
NIA Malnutrition in Older Persons Grants, 3104
NIDDK Diet Comp & Energy Balance Grants, 3148
NIDDK Pilot and Feasibility Clinical Research Studies in Digestive Diseases and Nutrition, 3172
NIDDK Research Grants on Improving Health Care for Obese Patients, 3176
NIDDK Small Grants for Underrepresented Minority Scientists in Diabetes and Digestive and Kidney Diseases, 3183
Paul G. Allen Family Foundation Grants, 3876
Phi Upsilon Omicron Margaret Drew Alpha Fellowship, 3948
Sensient Technologies Foundation Grants, 4199
Susan G. Komen Breast Cancer Foundation Challege Grants: Investigator Initiated Research, 4475
The World Food Prize, 4503
USDA Food and Agricultural Sciences National Needs Graduate Fellowship Grants, 4624
USDA Small Business Innovation Research Grants - Phase I, 4648

Nutritional Diseases and Disorders
AACC International Graduate Fellowships, 91
Abbott Fund Science Education Grants, 220
ASPEN C. Richard Fleming Grant, 1190
ASPEN Douglas Wilmore Grant, 1191
ASPEN Dudrick Research Scholar Award, 1192
ASPEN Harry M. Vars Award, 1193
ASPEN Maurice Shils Grant, 1195
ASPEN Norman Yoshimura Grant, 1196
ASPEN Nutrition Research Grants, 1197
ASPEN Scientific Abstracts Awards for Papers or Posters, 1198
ASPEN Scientific Abstracts Promising Investigator Awards, 1199
Danone Institute Belgium Research in Human Nutrition Grants, 1719
FDHN Bridging Grants, 1986
FDHN Fellow Abstract Prizes, 1991
FDHN Graduate Student Awards, 1994
FDHN Jon I. Isenberg International Research Scholar Award, 1995
Nestle Foundation Large Research Grants, 2979
Nestle Foundation Pilot Grants, 2980
Nestle Foundation Re-entry Grants, 2981
Nestle Foundation Small Research Grants, 2982
Nestle Foundation Training Grant, 2983
NIDDK Exploratory/Developmental Clinical Research Grants in Obesity, 3154
NIDDK Pilot and Feasibility Clinical Research Studies in Digestive Diseases and Nutrition, 3172

Obesity
Anthem Blue Cross and Blue Shield School Nurse Grants, 868
Campbell Soup Foundation Grants, 1415
Charles H. Hood Foundation Child Health Research Awards, 1522
Colonel Stanley R. McNeil Foundation Grants, 1597
Gerber Foundation Grants, 2126
Institute for Agriculture and Trade Policy Food and Society Fellowships, 2381
NHLBI Bioengineering and Obesity Grants, 3016
NHLBI Bioengineering Approaches to Energy Balance and Obesity Grants for SBIR, 3017
NHLBI Bioengineering Approaches to Energy Balance and Obesity Grants for STTR, 3018
NHLBI Targeted Approaches to Weight Control for Young Adults Grants, 3065
NHLBI Translating Basic Behavioral and Social Science Discoveries into Interventions to Reduce Obesity: Centers for Behavioral Intervention Development Grants, 3068
NIDDK Adverse Metabolic Side Effects of Second Generation Psychotropic Medications Leading to Obesity and Increased Diabetes Risk Grants, 3134
NIDDK Devel of Disease Biomarkers Grants, 3145
NIDDK Education Program Grants, 3149
NIDDK Exploratory/Developmental Clinical Research Grants in Obesity, 3154
NIDDK Identifying and Reducing Diabetes and Obesity Related Health Disparities within Healthcare Systems, 3157
NIDDK Neuroimaging in Obesity Res Grants, 3165
NIDDK Planning Grants For Translational Research For The Prevention And Control Of Diabetes And Obesity, 3173
NIDDK Proteomics: Diabetes, Obesity, And Endocrine, Digestive, Kidney, Urologic, And Hematologic Diseases, 3174
NIDDK Research Grants on Improving Health Care for Obese Patients, 3176
NIDDK Role of Gastrointestinal Surgical Procedures in Amelioration of Obesity-Related Insulin Resistance and Diabetes Independent of Weight Loss, 3177
NIH School Interventions to Prevent Obesity, 3244
Northwestern Mutual Foundation Grants, 3430
Obesity Society Grants, 3732
Pfizer ASPIRE North America Broad Spectrum Antibiotics for the Treatment of Gram-Negative or Polymicrobial Infections Research Awards, 3914
Pfizer ASPIRE North America Narrow Spectrum Antibiotics for the Treatment of MRSA Research Awards, 3916
RWJF Active Living Research Grants, 4132
RWJF Childhood Obesity Grants, 4135
RWJF Healthy Eating Research Grants, 4143
USDA Agriculture and Food Research Initiative Competitive Grants - Food Security, 4614
USDA Hispanic-Serving Institutions Education Grants, 4628

Obstetrics-Gynecology
ACS Cancer Control Career Development Awards for Primary Care Physicians, 305
Charles H. Hood Foundation Child Health Research Awards, 1522
Duke Univ Obstetric Anesthesia Fellowships, 1878
Fairlawn Foundation Grants, 1974
John S. Dunn Research Foundation Grants, 2506

Occupational Health and Safety
AAOHN Found Experienced Researcher Grants, 195
AAOHN Foundation New Investigator Researcher Grants, 196
American Chemical Society Corporate Associates Seed Grants, 676
ASSE Liberty Mutual Safety Res Fellowship, 1219
ASSE Scholarships, 1220
California Wellness Foundation Work and Health Grants, 1409
CSTE CDC/CSTE Applied Epidemiology Fellowships, 1664
IIE Rockefeller Foundation Bellagio Center Residencies, 2355
ORISE Faculty Internship at the US Army Center for Health Promotion and Preventive Medicine, 3802
ORISE Faculty Research Participation at the Centers for Disease Control and Prevention, 3803
ORISE Postgraduate Internship at the US Army Center for Health Promotion and Preventive Med, 3818
ORISE Postgraduate Research Participation at the Centers for Disease Control and Prevention, 3820
ORISE Student Research Participation at the Centers for Disease Control and Prevention, 3829
Susan G. Komen Breast Cancer Foundation Challege Grants: Investigator Initiated Research, 4475
USDD Aerospace Medicine, Clinical Research, Human Performance Research, and Expeditionary Medicine Grants, 4676

Ocean Sciences
NSF CISE Accelerating Discovery in Science and Engineering Through Petascale Simulations and Analysis Grants, 3521
NSF CISE Cross-Cutting Program Grants, 3532
NSF CISE Increasing the Participation and Advancement of Women in Academic Science and Engineering Careers Grants, 3534
USDD Ship Systems and Engineering Grants, 4792
WHOI Ocean Science Journalism Fellowship, 4862

Oceanographic Facilities
NMSF Ernest F. Hollings Ocean Awareness Trust Fund Grants, 3328
NOAA Cooperative Institute: Eastern U.S. Continental Shelf Frontier Exploration, Research, and Technology Development Grants, 3356
NOAA Ocean Exploration Omnibus: Marine Archaeology Grants, 3397
NOAA Ocean Exploration Omnibus: Ocean Exploration Grants, 3398
NSF Centers for Ocean Educ Excellence Grants, 3514
NSF Oceanographic Centers & Facilities Grants, 3603
RCY - NSF Ocean Sciences Div Research Grants, 4043

Oceanography
AAAS Roger Revelle Fellowship in Global Stewardship, 38
Abdus Salam ICTP Postgraduate Diploma Grants, 235
Abdus Salam ICTP Training and Research in Italian Laboratories Grants, 240
American Australian Association USA to Australia Fellowships, 620
AMS/Industry Government Graduate Fellowships, 817
AMS/Industry Minority Scholarships, 818
AMS Grad Fellowships in History of Science, 824
AMS Undergraduate Scholarships, 837
Canada Graduate Scholarships and NSERC Postgraduate Scholarships, 1420
Cooperative Institute for Research in Environmental Sciences Visiting Fellows Program, 1643
G. Unger Vetlesen Foundation Grants, 2093
M.J. Murdock Charitable Trust College Science Research Program, 2639
National Geographic Society Expeditions Council Grants, 2872
National Geographic Society Field Research and Exploration Grants, 2873
NCAR Graduate Research Fellowships, 2901
NCAR Postdoctoral Fellowships in the Atmospheric Sciences, 2902
NMSF Ernest F. Hollings Ocean Awareness Trust Fund Grants, 3328
NOAA Coastal Hypoxia Research Grants, 3351
NOAA Coastal Ocean Program Grants, 3353
NOAA Collaborative Science, Technology, and Applied Research Program Grants, 3354
NOAA Coral Reef Ecosystem Studies Grants, 3361
NOAA Cumulative Impacts of Multiple Stressors (MultiStress) Grants, 3366
NOAA Ecological Forecasting Grants, 3368
NOAA Ecology and Oceanography of Harmful Algal Blooms Grants, 3369
NOAA Monitoring and Event Response for Harmful Algal Blooms Grants, 3385
NOAA Monkfish Research Set-Aside Grants, 3386
NOAA National Sea Grant Industry Fellowships, 3393
NOAA Northern Gulf of Mexico Ecosystems and Hypoxia Assessment Program Grants, 3396
NOAA Ocean Exploration Omnibus: Marine Archaeology Grants, 3397
NOAA Ocean Exploration Omnibus: Ocean Exploration Grants, 3398
NOAA Regional Ecosystem Prediction Program on Coral Reef Ecosystem Studies Grants, 3401
NOAA Sea Scallop Research Set-Aside Program Grants, 3404
NOAA South Florida Program Grants, 3406
NOAA Undersea Research Center Coral Reef Conservation Program Coral Research Grants, 3409
NOAA Undersea Research Center Ocean Acidification Research at Aquarius Reef Base Grants, 3410
NOAA Undersea Research Program Grants, 3411
NOAA Undersea Research Project Grants, 3412
NSF Antarctic Ocean and Atmospheric Sciences Grants, 3477
NSF Antarctic Research Grants, 3479
NSF Arctic Nat Science Grants in Ocean Science, 3487
NSF Arctic Natural Sciences Program Grants, 3488
NSF Arctic System Science Program Grants, 3493
NSF Biological Oceanography Research Grants, 3502
NSF Centers for Ocean Educ Excellence Grants, 3514
NSF Chemical Oceanography Research Grants, 3517
NSF Collaborations in Math Geosciences, 3536
NSF Hydrologic Sciences Research Grants, 3578
NSF Marine Geology & Geophysics Res Grants, 3592
NSF Ocean Drilling Program Grants, 3602
NSF Oceanographic Centers & Facilities Grants, 3603
NSF Ocean Technology and Interdisciplinary Coordination Grants, 3604
NSF OCE Education Grants, 3605
NSF Physical Oceanography Research Grants, 3617
NSF Polar Programs Division Research Grants, 3627
NSF Synthesis and Modeling Project of the US Joint Global Ocean Flux Study, 3656
ONR Electronic and Optical Ceramics Grants, 3756
ONR Optical and Biological Oceanography Research Grants, 3777
ONR Young Investigator Program Grants, 3788
ORISE National Oceanic and Atmospheric Administration Faculty Research Participation, 3811
ORISE National Oceanic and Atmospheric Admin Student Research Participation, 3812
RCY - NSF Ocean Sciences Div Research Grants, 4043
Richard and Rhoda Goldman Fund Grants, 4063
USCD/NOAA Illinois-Indiana Sea Grant Core Proposals, 4604
USDC/NOAA Nature Conservancy Community-based Matching Grants, 4661
USDD Entry-Level Faculty Awards in Ocean Acoustics, 4724
USDD Graduate Traineeship Awards in Ocean Acoustics, 4730
USDD Marine Mammals and Biological Oceanography Grants, 4750
USDD Ocean Acoustics Research Grants, 4766
USDD Ocean Engineering and Marine Systems Grants, 4767
USDD Physical Oceanography Grants, 4771
USDD Postdoctoral Fellowship Awards in Ocean Acoustics, 4774
WHOI Guest Student Internships, 4860
WHOI Ocean Science Journalism Fellowship, 4862
WHOI Postdoctoral Fellowships, 4863
WHOI Postdoctoral Scholar Awards, 4864
WHOI Summer Student Fellowships, 4865

Oceanography, Physical
Abdus Salam ICTP Joint PhD Programme in Environmental Fluid Mechanics Grants, 230
NMSF Ernest F. Hollings Ocean Awareness Trust Fund Grants, 3328
NOAA Coastal Hypoxia Research Grants, 3351
NOAA Coral Reef Ecosystem Studies Grants, 3361
NOAA Cumulative Impacts of Multiple Stressors (MultiStress) Grants, 3366
NOAA Ecological Forecasting Grants, 3368
NOAA Ecology and Oceanography of Harmful Algal Blooms Grants, 3369
NOAA Mid-Atlantic Bight National Undersea Research Center Grants, 3383
NOAA Monitoring and Event Response for Harmful Algal Blooms Grants, 3385
NOAA Northern Gulf of Mexico Ecosystems and Hypoxia Assessment Program Grants, 3396
NOAA Ocean Exploration Omnibus: Marine Archaeology Grants, 3397
NOAA Ocean Exploration Omnibus: Ocean Exploration Grants, 3398
NOAA Regional Ecosystem Prediction Program on Coral Reef Ecosystem Studies Grants, 3401
NOAA South Florida Program Grants, 3406
NOAA Undersea Research Center Coral Reef Conservation Program Coral Research Grants, 3409
NSF Antarctic Ocean and Atmospheric Sciences Grants, 3477
NSF Centers for Ocean Educ Excellence Grants, 3514
NSF Ocean Technology and Interdisciplinary Coordination Grants, 3604
NSF OCE Education Grants, 3605
RCY - NSF Ocean Sciences Div Research Grants, 4043
USDD Applied & Computational Anal Grants, 4683
USDD Entry-Level Faculty Awards in Ocean Acoustics, 4724
USDD Graduate Traineeship Awards in Ocean Acoustics, 4730
USDD Marine Mammals and Biological Oceanography Grants, 4750
USDD Physical Oceanography Grants, 4771
USDD Postdoctoral Fellowship Awards in Ocean Acoustics, 4774
WHOI Geophysical Fluid Dynamics Fellowships, 4859
WHOI Ocean Science Journalism Fellowship, 4862

Oceans and Seas
Community Foundation of the Virgin Islands Judith A. Towle Environmental Studies Grants, 1621
NMSF Ernest F. Hollings Ocean Awareness Trust Fund Grants, 3328
NOAA Assessing Atlantic Meridional Overturning Circulation Program Grants, 3331
NOAA Atlantic Salmon Conservation Grants, 3332
NOAA Coastal Hypoxia Research Grants, 3351
NOAA Cooperative Institute: Eastern U.S. Continental Shelf Frontier Exploration, Research, and Technology Development Grants, 3356
NOAA Coral Reef Ecosystem Studies Grants, 3361
NOAA Cumulative Impacts of Multiple Stressors (MultiStress) Grants, 3366
NOAA Dr. Nancy Foster Scholarships, 3367
NOAA Ecological Forecasting Grants, 3368

NOAA Ecology and Oceanography of Harmful Algal Blooms Grants, 3369
NOAA Ernest F. Hollings Scholarships, 3370
NOAA Global Carbon Cycle Program Grants, 3374
NOAA Graduate Sciences Program Grants, 3375
NOAA Mid-Atlantic Bight National Undersea Research Center Grants, 3383
NOAA Monitoring and Event Response for Harmful Algal Blooms Grants, 3385
NOAA Monkfish Research Set-Aside Grants, 3386
NOAA National Sea Grant Industry Fellowships, 3393
NOAA National Sea Grant Oyster Disease Research Program Grants, 3395
NOAA Northern Gulf of Mexico Ecosystems and Hypoxia Assessment Program Grants, 3396
NOAA Ocean Exploration Omnibus: Marine Archaeology Grants, 3397
NOAA Ocean Exploration Omnibus: Ocean Exploration Grants, 3398
NOAA Regional Ecosystem Prediction Program on Coral Reef Ecosystem Studies Grants, 3401
NOAA Sea Scallop Research Set-Aside Program Grants, 3404
NOAA South Florida Program Grants, 3406
NOAA Undersea Research Center Coral Reef Conservation Program Coral Research Grants, 3409
NOAA Undersea Research Project Grants, 3412
NSF Antarctic Ocean and Atmospheric Sciences Grants, 3477
NSF Antarctic Research Grants, 3479
NSF Centers for Ocean Educ Excellence Grants, 3514
NSF Natural, Social, and System Sciences and Research Support and Logistics Program, 3600
NSF Ocean Technology and Interdisciplinary Coordination Grants, 3604
NSF OCE Education Grants, 3605
RCY - NSF Ocean Sciences Div Research Grants, 4043
Royal Caribbean Cruises Ocean Fund, 4108
USDC/NOAA Community-Based Habitat Restoration Program Grants, 4656
USDC/NOAA Community-Based Marine Debris Removal Grants, 4657
USDD Entry-Level Faculty Awards in Ocean Acoustics, 4724
USDD Graduate Traineeship Awards in Ocean Acoustics, 4730
USDD Postdoctoral Fellowship Awards in Ocean Acoustics, 4774
USDD Ship Systems and Engineering Grants, 4792
WHOI Ocean Science Journalism Fellowship, 4862

Odontology
SRC Medical Research Grants, 4379

Oils and Fats
AOCS ACI/NBB Innovation Research Award, 872
AOCS Australasian Award for Scientific Excellence in Lipid Research, 875
AOCS Edible Applications Technology Division Outstanding Achievement Award, 877
AOCS Hans Kaunitz Award, 880
AOCS Health & Nutrition Div Student Award, 881
AOCS Health and Nutrition Poster Competition, 882
AOCS Herbert J. Dutton Award, 883
AOCS Honored Student Award, 884
AOCS Industrial Oil Products Div Student Award, 885
AOCS Lipid Oxidation and Quality Division Poster Competition, 886
AOCS Peter and Claire G. Kalustian Award, 888
AOCS Processing Distinguished Service Award, 890
AOCS Processing Division Student Award, 891
AOCS Ralph H. Potts Memorial Fellowship, 893
AOCS Ralph Holman Lifetime Award, 894
AOCS Supelco/Nicholas Pelick-AOCS Research Award, 897
AOCS Surfactants and Detergents Division Student Award, 898
AOCS Timothy L. Mounts Award, 900
AOCS Uses of Soybean Oil Award, 901
AOCS Young Scientist Research Award, 902

Olfaction
Olfactory Research Fund Grants, 3738

Oncology
AACR Clinical and Translational Research Fellowships, 135
Abbott Fund Science Education Grants, 220
ACCP Hematology/Oncology PRN Minisabbatical Program, 260
ACCP Oncology Research Fellowships, 264
ACCP Pharmacotherapy Investigator Development Research Awards, 266
ACS Doctoral Training Grants in Oncology Social Work, 308
ACS Master's Training Grants in Clinical Oncology Social Work, 311
ACS Research Scholar Grants in Basic, Preclinical, Clinical and Epidemiology Research, 328
AHNS/AAO-HNSF Surgeon Scientist Combined Award, 380
AHNS/AAO-HNSF Young Investigator Combined Award, 381
AHNS Alando J. Ballantyne Resident Research Pilot Grant, 382
AHNS Pilot Grant, 383
AMA Foundation Seed Grants for Research, 601
Ann & Robert H. Lurie Family Foundation Grants, 859
ASCO/UICC International Cancer Technology Transfer Fellowships, 1044
ASCO Advanced Clinical Research Award in Colorectal Cancer, 1045
ASCO Advanced Clinical Research Awards in Breast Cancer, 1046
ASCO Advanced Clinical Res Awards in Sarcoma, 1047
ASCO Long-Term International Fellowships, 1048
ASCO Merit Awards, 1049
ASCO Young Investigator Award, 1050
Bristol-Myers Squibb Clinical Outcomes and Research Grants, 1374
Bristol-Myers Squibb Foundation Fellowships, 1375
Conquer Cancer Foundation of ASCO Career Development Award, 1630
Conquer Cancer Foundation of ASCO Comparative Effectiveness Research Professorship in Breast Cancer, 1631
Conquer Cancer Foundation of ASCO Drug Development Research Professorship, 1632
Conquer Cancer Foundation of ASCO Improving Cancer Care Grants, 1633
Conquer Cancer Found of ASCO Int'l Development and Education Award in Palliative Care, 1634
Conquer Cancer Foundation of ASCO International Development and Education Awards, 1635
Conquer Cancer Foundation of ASCO International Innovation Grant, 1636
Conquer Cancer Foundation of ASCO Medical Student Rotation Grants, 1637
Conquer Cancer Foundation of ASCO Resident Travel Award for Underrepresented Populations, 1638
Conquer Cancer Foundation of ASCO Translational Research Professorships, 1639
M.J. Murdock Charitable Trust College Research Program for Life Sciences, 2638
NCIC Terry Fox Postdoc Research Fellowships, 2923
NCI Technologies and Software to Support Integrative Cancer Biology Research Grants, 2933
ONF Small Grants Program, 3743
ONS Major Research Grants, 3789
Pediatric Cancer Research Foundation Grants, 3889
Pfizer Australia Cancer Research Grants, 3919
Pfizer Medical Education Track Two Grants, 3927
Sigma Theta Tau International / Oncology Nursing Society Grant, 4256
STTI Rosemary Berkel Crisp Research Awards, 4463
UICC American Cancer Society International Fellowships for Beginning Investigators, 4555
UICC International Cancer Technology Transfer Fellowships, 4556
UICC Raisa Gorbachev Memorial International Cancer Fellowships, 4557

Opera/Musical Theater
Adaptec Foundation Grants, 340
Arthur F. and Alice E. Adams Charitable Foundation Grants, 1024
Boston Foundation Grants, 1361
Elizabeth McGraw Foundation Grants, 1908
Florida Division of Cultural Affairs Individual Artist Fellowships, 2026
George Frederick Jewett Foundation Grants, 2116
Mabel Y. Hughes Charitable Trust Grants, 2647
NYSCA Music: Services to the Field Grants, 3708
Seaver Institute Grants, 4195
T. James Kavanagh Foundation Grants, 4489

Operating Support
Charles H. Farnsworth Trust Grants, 1521
Expect Miracles Foundation Grants, 1965

Operations Research
AT&T Bell Labs Graduate Research Program, 1250
Aviation Research Grants, 1273
CDC Steven M. Teutsch Prevention Effectiveness Fellowships, 1467
Mayo Clinic Business Consulting Fellowship, 2691

Ophthalmology
AHAF National Glaucoma Research Grants, 372
Alcon Foundation Grants Program, 513
CNIB Baker Applied Research Fund Grants, 1588
CNIB Baker New Researcher Fund Grants, 1589
CNIB Barbara Tuck MacPhee Award, 1590
CNIB Canada Glaucoma Clinical Research Council Grants, 1591
CNIB E. & Mary Hochhausen Access Technology Research Grants, 1592
Frederick Gardner Cottrell Foundation Grants, 2054
Joseph Alexander Foundation Grants, 2513
NEI Clinical Vision Research Devel Award, 2960
NEI Vision Research Core Grants, 2966
OneSight Research Foundation Block Grants, 3742
Regents Professional Opportunity Scholarships, 4049
Thome Foundation Awards in Age-Related Macular Degeneration Research, 4509

Optical Communications
SPIE A.E. Conrady Award, 4338
USDD Electro-Optical/Infrared Sensor Processing Grants, 4720
USDD Littoral Geosciences and Optics Grants, 4746

Optics
Abdus Salam ICTP Group Associates Grants, 228
Abdus Salam ICTP Junior Associates Grants, 231
Abdus Salam ICTP Postgraduate Diploma Grants, 235
Abdus Salam ICTP Regular Associates Grants, 237
Abdus Salam ICTP Senior Associates Grants, 239
Abdus Salam ICTP Training and Research in Italian Laboratories Grants, 240
ARO Atomic, Molecular, and Optical Physics Research Contracts, 1005
ARO Theoretical Physics and Nonlinear Phenomena Research Contracts, 1017
DOE Atomic, Molecular, and Optical Sciences Research Grants, 1771
DOE Catalysis and Chemical Transformations Research Grants, 1773
DOE Chemical Energy and Chemical Engineering Research Grants, 1774
DOE Chemical Physics Research Grants, 1775
DOE Chemical Sciences Research Grants, 1776
DOE Heavy Element Chem Research Grants, 1798
DOE Photochem & Radiation Research Grants, 1809
DOE Separations and Analysis Research Grants, 1813
NSF Condensed Matter Physics Research Grants, 3540
NSF Physics Frontiers Centers Grants, 3619
NSF Physics Research Grants, 3621
NSF Upper Atmospheric Facilities Grants, 3665
ONR Electronic and Optical Ceramics Research Grants, 3756

ONR Marine Meteorology and Atmospheric Effects Research Grants, 3769
ONR Optical and Biological Oceanography Research Grants, 3777
SPIE Gold Medal of the Society Award, 4343
USDD Littoral Geosciences and Optics Grants, 4746
USDD Naval Research Facilities Grants, 4761

Optometry
Eye-Bank for Sight Restoration and Fight for Sight Summer Student Research Fellowship, 1969
Fight for Sight-Streilein Foundation for Ocular Immunology Research Award, 2012
Fight for Sight Grants-in-Aid, 2013
Fight for Sight Post-Doctoral Awards, 2014
Fight for Sight Summer Student Fellowships, 2015
OneSight Research Foundation Block Grants, 3742
Regents Professional Opportunity Scholarships, 4049
Special Olympics Health Professions Stu Grants, 4332
Streilein Foundation for Ocular Immunology Visiting Professorships, 4453
Yale University Beinecke Library Research Fellowships for Yale Grad & Professional School Students, 4902

Oral Diseases
DHHS Oral Health Promotion Research Across the Lifespan, 1759

Oral Health and Hygiene
Connecticut Health Foundation Health Initiative Grants, 1628
DHHS Oral Health Promotion Research Across the Lifespan, 1759

Oral History
American Society for Yad Vashem Grants, 798
Australasian Institute of Judicial Administration Seed Funding Grants, 1268
Institute of American Cultures Postdoctoral and Visiting Scholar Fellowships, 2384
Massachusetts Foundation for the Humanities Project Grants, 2678
NYSCA Folk Arts: Presentation Grants, 3704
Radcliffe Institute Oral History Grants, 4022
United Methodist Church World is My Parish United Methodist History Research Grant, 4576
Virginia Foundation for the Humanities Discretionary Grants, 4813
Virginia Foundation for Humanities Open Grants, 4814
Wenner-Gren Found Historical Archives Grants, 4849

Orchestras
Alcatel-Lucent Technologies Foundation Grants, 509
Clarence E. Heller Charitable Foundation Grants, 1575
George W. Codrington Charitable Grants, 2122
Kenneth T. and Eileen L. Norris Found Grants, 2554
MacDonald-Peterson Foundation Grants, 2649
San Antonio Area Foundation Grants, 4163
Sid W. Richardson Foundation Grants, 4253

Organ Transplants
Carlos and Marguerite Mason Trust Grants, 1431
MBL Eugene and Millicent Bell Fellowships, 2699
ONF Small Grants Program, 3743
Pfizer Advancing Research in Transplantation Science Research Awards, 3908
Pfizer ASPIRE North America Broad Spectrum Antibiotics for the Treatment of Gram-Negative or Polymicrobial Infections Research Awards, 3914
Robert B McMillen Foundation Grants, 4072
Roche Organ Transplant Research Found Grants, 4083
Sid W. Richardson Foundation Grants, 4253

Organic Farming
SARE Producer Research and Education Grants-Western Region, 4174
SARE Professional + Producer Grants - Western Region, 4175
SARE Research and Education Grants - Southern Region, 4179
SARE Research and Education Grants - Western Region, 4180
USDA Integrated Research Education and Extension Grants - Organic Transitions, 4630
USDA Interregional Research Project Minor Crop Pest Management Grants, 4632

Organizational Development
Center on Philanthropy and Civil Society's Emerging Leaders International Fellows Program, 1487
Mayo Clinic Administrative Fellowship, 2689
SfN Louise Hanson Marshall Specific Recognition Award, 4208

Organizational Theory and Behavior
ASTD Dissertation Award, 1225
ASTD Research Award, 1226
Brainerd Foundation Grants, 1368
Jane Beattie Memorial Scholarship, 2470
MSI Alden G. Clayton Dissertation Proposal Competition, 2800
NSF Decision, Risk, and Management Science Research Grants, 3550
NSF Innovation and Organizational Change, 3581
NSF Sociology Program Grants, 3650

Ornithology
AMNH Frank M. Chapman Fellowships, 814
International Crane Foundation Internships, 2396
Neotropical Grassland Conservancy W. J. and Virginia W. Moorhouse Research Grant, 2977

Orthopedics
John S. Dunn Research Foundation Grants, 2506
Judith Graham Pool Postdoctoral Research Fellowship in Hemophilia, 2529
Rett Syndrome Research Foundation Grants, 4058

Osteopathic Medicine
NMF Need-Based Scholarships, 3326
Scleroderma Established Investigator Grants, 4190

Osteoporosis
Fairlawn Foundation Grants, 1974
NIH Receptors and Signaling in Bone in Health and Disease, 3229

Otolaryngic
AAO-HNSF Health Services Research Grants, 187
AAO-HNSF Maureen Hannley Research Training Awards, 188
AAO-HNSF Percy Memorial Research Award, 189
AAO-HNSF Rande H. Lazar Health Services Research Grant, 190
AAO-HNSF Resident Research Awards, 191
AAOA Foundation/AAO-HNSF Combined Research Grants, 194

Otolaryngology
AAO-HNSF CORE Research Grants, 186
AAO-HNS Medical Student Research Prize, 192
AAO-HNS Resident Research Prizes, 193
ASPO Daiichi Innovative Technology Grant, 1210
ASPO Fellowships, 1211
ASPO Research Grants, 1212
Triological Society Research Career Development Awards, 4534
Xoran Technologies Resident Research Grant, 4900

Otology
AHRF Eugene L. Derlacki Research Grants, 384
AHRF Georgia Birtman Grant, 385
AHRF Wiley H. Harrison Mem Research Award, 387
ANS/AAO-HNSF Herbert Silverstein Otology and Neurotology Research Award, 864
ANS Neurotology Fellows Award, 865
ANS Nicholas Torok Vestibular Award, 866
ANS Trainee Award, 867
Triological Society Research Career Development Awards, 4534

Outpatient Care
CDC Evaluation of the Use of Rapid Testing For Influenza in Outpatient Medical Settings, 1458
USDD Medical Practice Breadth of Med Practice & Disease Frequency Exposure Grants, 4752

Ovarian Cancer
ASCO/UICC International Cancer Technology Transfer Fellowships, 1044
ASCO Young Investigator Award, 1050
Conquer Cancer Foundation of ASCO Career Development Award, 1630
Conquer Cancer Foundation of ASCO International Development and Education Awards, 1635
Mary Kay Foundation Cancer Research Grants, 2674
Penny Severns Breast, Cervical and Ovarian Cancer Research Grants, 3896
Seattle Foundation Medical Funds Grants, 4194

Pacific Islanders
American Sociological Association Minority Fellowships, 805
Asian Pacific American Institute for Congressional Studies Fellowships, 1091
ASM Robert D. Watkins Graduate Research Fellowship, 1183
DHHS Native American Programs, 1758
FDHN Student Research Fellowships, 2001
Fulbright Traditional Scholar Program in the East Asia/Pacific Region, 2082
MLA Grad Scholarship for Minority Students, 2787
NHLBI Training Program Grants for Institutions that Promote Diversity, 3067
Pacific Islanders in Communications Grants, 3861

Pain
AAHPM Hospice and Palliative Med Fellowships, 180
Duke University Pain Management Fellowships, 1879
IAFF Burn Foundation Research Grants, 2312
NIA Mechanisms, Measurement, and Management of Pain in Aging: from Molecular to Clinical, 3105
NINR Mechanisms, Models, Measurement, & Management in Pain Research Grants, 3264
NINR Quality of Life for Individuals at the End of Life Grants, 3266
Pfizer Australia Neuroscience Research Grants, 3920

Painting
ACC Residency Program in Asia, 267
ACC Starr Foundation Fellowship Program, 268
CICF Efroymson Contemporary Arts Fellows, 1563
John Simon Guggenheim Memorial Foundation Fellowships, 2507
Manitoba Arts Council Aboriginal Arts Creative Development Pilot Projects Grants, 2652
National Gallery of Art Kress Interpretive Fellow, 2869
NYFA Artist Fellowships, 3698
Pollock-Krasner Foundation Grants, 3981
Roberto Longhi Foundation Fellowships, 4078

Paleontology
American Philosophical Soc Judson Daland Prize, 766
American Philosophical Society Lewis and Clark Fund for Exploration and Field Research, 767
AMNH Graduate Student Fellowships, 815
National Geographic Society Expeditions Council Grants, 2872
National Geographic Society Field Research and Exploration Grants, 2873
NSF Arctic Nat Science Grants in Earth Sciences, 3485
NSF Arctic Natural Science Grants in Glaciology, 3486
NSF Arctic System Science Program Grants, 3493
NSF Biotic Surveys and Inventories Grants, 3510
NSF Marine Geology & Geophysics Res Grants, 3592
NSF Physical Anthropology Research Grants, 3615
NSF Sediment Geology & Paleobiology Grants, 3646
NSF Systematic Biology and Biodiversity Inventories Grants, 3657
NSF Tectonics Research Grants, 3658
Wenner-Gren Found Historical Archives Grants, 4849

Palestine
Institute for Palestine Studies, Publishing, and Research Organization Grants, 2382

Palliative Care
AACN End of Life/Palliative Care Small Projects Grants, 108
ACCP Pain & Palliative Care PRN Minisabbatical Program, 265
Aetna Foundation Integrated Health Care Grants, 348
Conquer Cancer Foundation of ASCO International Development and Education Award in Palliative Care, 1634
Piper Trust Healthcare & Med Research Grants, 3972
Sara Elizabeth O'Brien Trust Grants, 4166

Pancreas
AACR-FNAB Fellows Grant for Translational Pancreatic Cancer Research, 114
AACR-Pancreatic Cancer Action Network Career Development Award for Pancreatic Cancer Research, 122
AACR-Pancreatic Cancer Action Network Innovative Grants, 123
FDHN Designated Research Award in Research Related to Pancreatitis, 1990
FDHN Graduate Student Awards, 1994
NIAAA Mechanisms of Alcohol and Drug-Induced Pancreatitis Grants, 3079
NIDDK Pancreatic Devel & Regeneration Grants: Toward Cellular Therapies for Diabetes, 3169
NIDDK Small Business Innovation Research to Develop New Therapeutics and Monitoring Technologies for Type 1 Diabetes--Towards an Artificial Pancreas, 3181

Pancreatic Cancer
ASCO/UICC International Cancer Technology Transfer Fellowships, 1044
ASCO Young Investigator Award, 1050
Conquer Cancer Foundation of ASCO Career Development Award, 1630
Conquer Cancer Foundation of ASCO International Development and Education Awards, 1635
Mary Kay Foundation Cancer Research Grants, 2674
Pfizer Medical Education Track Two Grants, 3927
Seattle Foundation Medical Funds Grants, 4194

Paper Industry
Institute of Paper Science & Tech Fellowships, 2387

Parallel Algorithms
ARO Discrete Mathematics and Computer Science Research Contracts, 1009

Parallel Computer Architecture
ARO Discrete Mathematics and Computer Science Research Contracts, 1009

Parapsychology
Institute of Noetic Sciences Research Internships, 2386
Parapsychology Foundation Grants, 3868

Parasitology
ASM/CDC Fellowships in Infectious Disease and Public Health Microbiology, 1100
BWF Ad Hoc Grants, 1392
BWF Investigators in the Pathogenesis of Infectious Disease Awards, 1398
NFID New Investigator Matching Grants, 2989
US CRDF Leishmaniasis: Collaborative Research Opportunities in N Africa & Middle East, 4610
USDA Animal Health & Disease Grants, 4616

Parent Education
3M Foundation Community Giving Grants, 5
Allen Foundation Educational Nutrition Grants, 570
Bodman Foundation Grants, 1356
Community Foundation for S Arizona Grants, 1618
Comprehensive Health Education Found Grants, 1622
Connecticut Community Foundation Grants, 1627
Dave Thomas Foundation for Adoption Grants, 1723
FCD Child Development Grants, 1984
Gerber Foundation Grants, 2126
Helen V. Brach Foundation Grants, 2231
MGM Resorts Foundation Community Grants, 2753
Nellie Mae Education Foundation District-Level Change Grants, 2968
Plan for Social Excellence Grants, 3976

Parent Involvement
AASF Lydia Donaldson Tutt-Jones Memorial Research Grant, 203
Bernard Van Leer Foundation Grants, 1336
Bodman Foundation Grants, 1356
Vancouver Foundation Grants and Community Initiatives, 4807

Parenteral Nutrition
ASPEN C. Richard Fleming Grant, 1190
ASPEN Douglas Wilmore Grant, 1191
ASPEN Dudrick Research Scholar Award, 1192
ASPEN Harry M. Vars Award, 1193
ASPEN Maurice Shils Grant, 1195
ASPEN Norman Yoshimura Grant, 1196
ASPEN Nutrition Research Grants, 1197
ASPEN Scientific Abstracts Awards for Papers or Posters, 1198
ASPEN Scientific Abstracts Promising Investigator Awards, 1199

Parkinson's Disease
Fairlawn Foundation Grants, 1974
Michael J. Fox Found Clinical Discovery Grants, 2760
Michael J. Fox Foundation for Parkinson's Research LEAPS Program, 2761
Michael J. Fox Foundation Parkinson's Cell Line Research Grants, 2762
National Parkinson Foundation Clinical Research Fund Grants, 2881
Partnership for Cures Two Years To Cures Grants, 3872
PDF-AANF Clinician-Scientist Devel Awards, 3881
PDF International Research Grants, 3882
PDF Postdoctoral Fellowships for Basic Scientists, 3883
PDF Postdoctoral Fellowships for Neurologists, 3884
PDF Summer Fellowships, 3885
PSG Mentored Clinical Research Awards, 4003

Parks
BMW of N America Charitable Contributions, 1353
Del E. Webb Foundation Grants, 1733
HRF Hudson River Improvement Grants, 2286
Open Spaces Sacred Places National Awards, 3791
Sioux Falls Area Community Foundation Community Fund Grants (Unrestricted), 4262
Sioux Falls Area Comm Foundation Spot Grants, 4263
USDA McIntire-Stennis Cooperative Forestry Research Grants, 4634
William and Flora Hewlett Foundation Environmental Grants, 4874

Patents, Copyrights, Trademarks
ASME Heat Transfer Memorial Award, 1128
ASME Thomas A. Edison Patent Award, 1167
OSF Intellectual Property Reform Grants, 3846
Tandem Launch Technologies Grants, 4493

Pathogenesis
AHNS/AAO-HNSF Surgeon Scientist Combined Award, 380
AHNS/AAO-HNSF Young Investigator Combined Award, 381
AHNS Alando J. Ballantyne Resident Research Pilot Grant, 382
AHNS Pilot Grant, 383
ALVRE Casselberry Award, 585
ALVRE Grant, 586
ALVRE Seymour R. Cohen Award, 587
Alzheimer's Association Conference Grants, 589
Alzheimer's Association Development of New Cognitive and Functional Instruments Grants, 590
Alzheimer's Association Everyday Technologies for Alzheimer Care Grants, 591
Alzheimer's Association Mentored New Investigator Research Grants to Promote Diversity, 593
Alzheimer's Association Neuronal Hyper Excitability and Seizures in Alzheimer's Disease Grants, 594
Alzheimer's Association New Investigator Research Grants to Promote Diversity, 596
Alzheimer's Association U.S.-U.K. Young Investigator Exchange Fellowships, 597
Alzheimer's Association Zenith Fellows Awards, 598
APHL Emerging Infectious Diseases Fellowships, 914
BWF Ad Hoc Grants, 1392
BWF Investigators in the Pathogenesis of Infectious Disease Awards, 1398
DSO Controlling Pathogen Evolution Workshop Grants, 1862
Lymphatic Education & Research Network Postdoctoral Fellowships, 2633
NIA Alzheimer's Disease Core Centers Grants, 3084
Penny Severns Breast, Cervical and Ovarian Cancer Research Grants, 3896
Pfizer ASPIRE EU Emerging Mechanisms of Resistance Antibacterial Research Awards, 3912
Tourette Syndrome Association Post-Doctoral Fellowships, 4523
Tourette Syndrome Association Research Grants, 4524

Pathology
AHAF Alzheimer's Disease Research Grants, 370
NYAM Edward N. Gibbs Memorial Lecture and Award in Nephrology, 3682
PhRMA Foundation Pharmacology/Toxicology Post Doctoral Fellowships, 3962
PhRMA Foundation Pharmacology/Toxicology Sabbatical Fellowships, 3965
Simone and Cino del Duca Grand Prix Awards, 4261
UICC Raisa Gorbachev Memorial International Cancer Fellowships, 4557

Pathophysiology
AHNS/AAO-HNSF Surgeon Scientist Combined Award, 380
AHNS/AAO-HNSF Young Investigator Combined Award, 381
AHNS Alando J. Ballantyne Resident Research Pilot Grant, 382
AHNS Pilot Grant, 383
ALVRE Casselberry Award, 585
ALVRE Grant, 586
ALVRE Seymour R. Cohen Award, 587
CFF Clinical Research Grants, 1491
NARSAD Colvin Prize for Outstanding Achievement in Mood Disorders Research, 2830
NARSAD Ruane Prize for Child and Adolescent Psychiatric Research, 2835
NHLBI Lymphatics in Health and Disease in the Digestive, Urinary, Cardiovascular and Pulmonary Systems, 3032
NIDDK Type 2 Diabetes in the Pediatric Population Research Grants, 3188
Tourette Syndrome Association Post-Doctoral Fellowships, 4523
Tourette Syndrome Association Research Grants, 4524

Patient Care and Education
AAAAI Distinguished Clinician Award, 17
AAAAI Distinguished Layperson Award, 18
AACN Physio-Control Small Projects Grants, 111
AAFP Foundation Joint Grants, 156
American Psychiatric Association/Lilly Psychiatric Research Fellowship, 776
AMSSM Foundation Research Grants, 836
APA Young Investigator Grant Program, 912
Becton Dickinson and Company Grants, 1328
BWF Institutional Program Unifying Population and Laboratory Based Sciences Grants, 1397
Carlos and Marguerite Mason Trust Grants, 1431

Conquer Cancer Foundation of ASCO Improving Cancer Care Grants, 1633
Covidien Surgical Mini-Fellowships, 1651
Cystic Fibrosis Canada Clinic Incentive Grants, 1676
Fairlawn Foundation Grants, 1974
Greenwall Foundation Bioethics Grants, 2174
Healthcare Foundation of New Jersey Grants, 2222
Jacob and Valeria Langeloth Foundation Grants, 2455
Mayo Clinic Admin Fellowship Eau Claire, 2690
McCarthy Family Foundation Grants, 2717
NHLBI Mentored Patient-Oriented Research Career Development Awards, 3037
NIH Enhancing Adherence to Diabetes Self-Management Behaviors, 3214
NIH Mentored Patient-Oriented Research Career Development Award, 3226
NINR Acute and Chronic Care During Mechanical Ventilation Research Grants, 3259
NINR Quality of Life for Individuals at the End of Life Grants, 3266
Penny Severns Breast, Cervical and Ovarian Cancer Research Grants, 3896
PhRMA Foundation Health Outcomes Research Starter Grants, 3951
Rathmann Family Foundation Grants, 4033
STTI Environment of Elder Care Nursing Research Grants, 4460
USDD Care for the Critically Injured Burn Patient Grants, 4706

Patient/Physician Relationship
Conquer Cancer Foundation of ASCO Improving Cancer Care Grants, 1633
Covidien Surgical Mini-Fellowships, 1651
Jacob and Valeria Langeloth Foundation Grants, 2455
NHLBI Midcareer Investigator Award in Patient-Oriented Research, 3040
USDD Medical Practice Breadth of Med Practice & Disease Frequency Exposure Grants, 4752
USDD Medical Practice Initiative Procedural Skill Decay and Maintenance Grants, 4753

Peace/Disarmament
Cambridge Foundation For Peace Prize, 1413
Carnegie Endowment for International Peace Junior Fellowships, 1434
Center for Defense Information Internships, 1473
Compton Foundation Grants Program, 1623
DOE Nonproliferation & Nat Security Research, 1806
Donald and Sylvia Robinson Family Grants, 1830
Ford Foundation Peace and Social Justice Grants, 2031
Global Center for Dispute Resolution Res Grants, 2150
Harvard Women and Public Policy Postdoctoral Fellowship, 2217
Institute for Palestine Studies, Publishing, and Research Organization Grants, 2382
International Peace Research Inst Oslo Grants, 2402
Marine Corps Historical Center Dissertation Fellowship, 2663
Mary Ingraham Bunting Inst Peace Fellowship, 2671
NATO Science for Peace Grants, 2898
ONR Bisson Prize for Naval Tech Achievement, 3745
SRC Guest Professorship Program, 4370
SSRC African Peacebuilding Network Research Grants, 4422
SSRC Next Generation Social Sciences in Africa: Dissertation Completion Fellowships, 4434
SSRC Next Generation Social Sciences in Africa: Doctoral Dissertation Proposal Fellowships, 4435
SSRC Next Generation Social Sciences in Africa: Doctoral Dissertation Research Fellowships, 4436
United States Institute of Peace - Jennings Randolph Peace Dissertation Scholarships, 4579
United States Institute of Peace - Jennings Randolph Senior Fellowships, 4580
United States Institute of Peace - Transatlantic Post-Doc Fellowship for International Relations and Security, 4581
United States Institute of Peace Annual Grants, 4582
Ursula Thrush Peace Seed Grants, 4592

Pediatric Cancer
CDI Postdoctoral Fellowships, 1469
Conquer Cancer Foundation of ASCO International Development and Education Awards, 1635
Sara Elizabeth O'Brien Trust Grants, 4166
Seattle Foundation Medical Funds Grants, 4194
Victor E. Speas Foundation Grants, 4810

Pediatrics
1976 Foundation Grants, 7
AACAP Cancro Academic Leadership Award, 76
AACR Career Development Awards for Pediatric Cancer Research, 132
AAP Nutrition Award, 198
AAP Resident Research Grants, 199
AcademyHealth Nemours Child Health Services Research Awards, 247
ACS Cancer Control Career Development Awards for Primary Care Physicians, 305
Alexander and Margaret Stewart Trust Grants, 514
American Philosophical Society Daland Fellowships in Clinical Investigation, 763
APA Young Investigator Grant Program, 912
ASPO Daiichi Innovative Technology Grant, 1210
ASPO Fellowships, 1211
ASPO Research Grants, 1212
Blowitz-Ridgeway Foundation Grants, 1349
CDI Postdoctoral Fellowships, 1469
CFF First- and Second-Year Clinical Fellowships, 1492
CFF Leroy Matthews Physician-Scientist Awards, 1494
Charles H. Hood Foundation Child Health Research Awards, 1522
Children's Cardiomyopathy Foundation Research Grants, 1550
E.W. "Al" Thrasher Awards, 1887
Elizabeth Glaser Int'l Leadership Awards, 1906
Elizabeth Glaser Scientist Award, 1907
Gerber Foundation Grants, 2126
Lowe Foundation Grants, 2622
M. Bastian Family Foundation Grants, 2636
Mary S. and David C. Corbin Foundation Grants, 2675
MedImmune Charitable Grants, 2737
NAPNAP Foundation Graduate Student Research Grant, 2828
NAPNAP Foundation Nursing Research Grants, 2829
NHLBI Pediatric Cardiac Genomics Consortium Grants, 3046
NIDDK Planning Grants For Translational Research For The Prevention And Control Of Diabetes And Obesity, 3173
NIDDK Type 2 Diabetes in the Pediatric Population Research Grants, 3188
OneSight Research Foundation Block Grants, 3742
Peabody Foundation Grants, 3886
Pediatric Brain Tumor Found Research Grants, 3888
Pediatric Cancer Research Foundation Grants, 3889
Pfizer Australia Paediatric Endocrine Care Research Grants, 3921
Piper Trust Healthcare & Med Research Grants, 3972
Regenstrief General Internal Medicine - General Pediatrics Research Fellowship Program, 4045
STTI Rosemary Berkel Crisp Research Awards, 4463
Thrasher Research Fund Early Career Awards, 4515
Victor E. Speas Foundation Grants, 4810

Penology/Correctional Institutions and Procedures
DOJ Juvenile Justice and Delinquency Prevention Special Emphasis Grants, 1826
DOJ National Institute of Corrections Training and Staff Development Grants, 1827
OSF Soros Justice Advocacy Fellowships, 3850
OSF Soros Justice Media Fellowships, 3851

Peptides
Ralph F. Hirschmann Award in Peptide Chem, 4027

Performance Art
ACLS Frederick Burkhardt Residential Fellowships for Recently Tenured Scholars, 288
Bush Foundation Artist Fellowships, 1391
CCA Alberta Creative Development Initiative: Grants to Individuals, 1446
George S. and Dolores Dore Eccles Grants, 2120
NYSCA Folk Arts: Presentation Grants, 3704
NYSCA Folk Arts: Regional and County Folk Arts Programs Grants, 3705
Pew Fellowships in the Arts, 3905
Princess Grace Foundation Theater, Dance, and Film Grants, 3988
Stanford Humanities Center Arts Writer/Practitioner Fellowship, 4444

Performing Arts
ACC John D. Rockefeller III Award, 257
Adaptec Foundation Grants, 340
Alcatel-Lucent Technologies Foundation Grants, 509
Alces Foundation Grants, 510
Amgen Foundation Grants, 812
AMI Semiconductors Corporate Grants, 813
Andrew W. Mellon/ACLS Dissertation Completion Fellowships, 844
Andy Warhol Foundation for the Visual Arts Curatorial Research Fellowships, 847
Arizona Commission on the Arts Folklorist Residencies, 997
Arizona Public Service Corporate Giving Program Grants, 998
Arkansas Arts Council Individual Fellowships, 999
Arthur F. and Alice E. Adams Charitable Foundation Grants, 1024
Avon Products Foundation Grants, 1275
Bodenwein Public Benevolent Foundation Grants, 1355
Boeing Company Contributions Grants, 1357
Booth-Bricker Fund Grants, 1359
Burlington Northern Santa Fe Foundation Grants, 1389
Carl and Lily Pforzheimer Foundation Grants, 1427
Colonel Stanley R. McNeil Foundation Grants, 1597
Connecticut Community Foundation Grants, 1627
Cord Foundation Grants, 1645
Cowles Charitable Trust Grants, 1652
Cudd Foundation Grants, 1672
D.F. Halton Foundation Grants, 1692
Davis Family Foundation Grants, 1726
Delaware Division of the Arts Individual Artist Fellowships, 1732
DeMatteis Family Foundation Grants, 1740
Donald and Sylvia Robinson Family Foundation Grants, 1830
Elizabeth McGraw Foundation Grants, 1908
Eugene M. Lang Foundation Grants, 1957
Flinn Foundation Grants Programs, 2024
George Gund Foundation Grants, 2117
George W. Codrington Charitable Grants, 2122
Georgia Power Foundation Grants, 2125
Harold R. Bechtel Charitable Remainder Uni-Trust Grants, 2201
Lucile Horton Howe and Mitchell B. Howe Foundation Grants, 2624
Mabel Y. Hughes Charitable Trust Grants, 2647
MacDonald-Peterson Foundation Grants, 2649
Margaret T. Morris Foundation Grants, 2659
McLean Contributionship Grants, 2727
MetLife/TCG Aha! Think it Do it Grants, 2745
Meyer and Stephanie Eglin Foundation Grants, 2747
NEA/TCG Career Development Grants for Theatre Directors, 2943
Nevada Arts Council Artists Fellowships, 2984
Northwestern Mutual Foundation Grants, 3430
NYPL Short-Term Research Fellowships, 3701
NYSCA Dance: Services to the Field Grants, 3702
NYSCA Folk Arts: Presentation Grants, 3704
NYSCA Folk Arts: Regional and County Folk Arts Programs Grants, 3705
NYSCA Music: Services to the Field Grants, 3708
Oppenstein Brothers Foundation Grants, 3792
Paul G. Allen Family Foundation Grants, 3876
Rayonier Foundation Grants, 4034
Reinberger Foundation Grants, 4051
Rhode Island Foundation Grants, 4061
Rich Foundation Grants, 4068

Rockefeller Foundation MAP Fund Grants, 4090
Sensient Technologies Foundation Grants, 4199
Shastri Indo-Canadian Institute Arts Fellowships, 4219
Sid W. Richardson Foundation Grants, 4253
Stanford Humanities Center Arts Writer/Practitioner Fellowship, 4444
Surdna Foundation Arts Teachers Fellowships, 4470
Textron Corporate Contributions Grants, 4501
W.C. Griffith Foundation Grants, 4819
W. C. Griffith Foundation Grants, 4818
Xerox Foundation Grants, 4899

Perinatal Disorders
Charles H. Hood Foundation Child Health Research Awards, 1522

Periodicals
ALA Ulrich's Serials Librarianship Award, 500
JMO Law and Legal System Grants, 2491
JMO Public Policy Research Grants, 2492
JMO Strategic and International Studies Grants, 2493
SSHRC Aid to Scholarly Journals, 4402

Periodontics
Charles W. Finley Visiting Scholar Educ Grant, 1530
DHHS Oral Health Promotion Research Across the Lifespan, 1759

Personnel Studies
ASTD Research Award, 1226
SHRM Foundation Research Grants, 4249
SHRM Michael R. Losey Human Resource Research Award, 4250

Personnel Training and Development
ALA Excellence in Academic Libraries Award, 486
ASTD Dissertation Award, 1225
Piper Trust Healthcare & Med Research Grants, 3972

Peru
Eisenhower Exchange International Fellowships, 1902
Kellogg Institute for International Studies Quechua Summer Language Fellowships, 2548
OTS Francis J. Bossuyt Fellowship, 3855

Pest Control
American Chemical Society AGRO Award for Innovation in Chemistry of Agriculture, 631
American Chemical Society AGRO Educ Awards, 632
Bedding Plants Foundation Research Grants, 1330
Conservation, Food, and Health Foundation Grants for Developing Countries, 1641
SARE Producer Research and Education Grants-Western Region, 4174
SARE Research and Education Grants - North Central Region, 4177
SARE Research and Education Grants - Northeast Region, 4178
SARE Research and Education Grants - Western Region, 4180
USDA Integrated Pest Management Competitive Grants - Western Region, 4629
USDA Interregional Research Project Minor Crop Pest Management Grants, 4632
USDA Plant and Animal Disease, Pest Control, and Animal Care Grants, 4638
USDA Regional Integrated Pest Management Grants - North Central Region, 4642
USDA Regional Integrated Pest Management Grants - Northeastern Region, 4643
USDA Regional Integrated Pest Management Grants - Southern Region, 4644
USDA Special Research Grants, 4650

Pesticides
Clarence E. Heller Charitable Foundation Grants, 1575
Ekhagastiftelsen Foundation Grants, 1903
EPA Research, Development, Monitoring, Public Education, Training, Demonstrations, and Studies Grants, 1938
SARE Research and Education Grants - North Central Region, 4177
SARE Research and Educ Grants - NE Region, 4178
USDA Interregional Research Project Minor Crop Pest Management Grants, 4632
USDA Regional Integrated Pest Management Grants - Northeastern Region, 4643
USDA Regional Integrated Pest Management Grants - Southern Region, 4644

Petroleum Engineering
ASME Ralph Coats Roe Medal, 1158

Petroleum Science
American Chemical Society Green Chemistry Chapter Award for Student Affiliates, 678
American Chemical Society Green Chem Grants, 679
American Chemical Society Joseph Breen Memorial Fellowship in Green Chemistry, 682
American Chemical Society Kenneth G. Hancock Memorial Student Award in Green Chemistry, 683
American Chemical Society Overcoming Challenges Award, 686
American Chemical Society Petroleum Research Fund Doctoral New Investigator Grants, 690
American Chemical Society Petroleum Research Fund New Directions Grants, 691
American Chemical Society Petroleum Research Fund Scientific Education Grants, 692
American Chemical Society Petroleum Research Fund Undergraduate Faculty Sabbatical Grants, 693
American Chemical Society Petroleum Research Fund Undergraduate New Investigator Grants, 694
American Chemical Society Petroleum Research Fund Undergraduate Research Grants, 695
American Chemical Society Presidential Green Chemistry Challenge Awards, 696
American Chem Soc WCC Rising Star Award, 698
Archibald Waynne Dingman Memorial Graduate Scholarship, 986
NSF-NIST Interaction in Chemistry, Materials Research, Molecular Biosciences, Bioengineering, and Chemical Engineering, 3453

Petrology
Carnegie Postdoctoral Research Fellowships, 1436
NSF Sediment Geology & Paleobiology Grants, 3646
NSF Tectonics Research Grants, 3658

Pharmaceuticals
ACCP Infectious Diseases Research Fellowships, 263
ACCP Oncology Research Fellowships, 264
AIChE Food, Pharmaceutical and Bioengineering Division Distinguished Service Award in Chemical Engineering, 411
AINSE Research Fellowships, 464
AINSE Research Fellowships with Institute for Science and International Security Support, 465
American Chemical Society Award for Computers in Chemical or Pharmaceutical Research, 651
American Chemical Society GCI Pharmaceutical Roundtable Research Grants, 677
CCFF Chairman's Dist Life Sciences Award, 1449
Conquer Cancer Foundation of ASCO Drug Development Research Professorship, 1632
Hawaii Community Foundation Health Education and Research Grants, 2221
NSF Chemistry Research Experiences for Undergraduates, 3518
Pfizer Advancing Research in Transplantation Science Research Awards, 3908
Pfizer Anti Infective Research EU Grants, 3909
Pfizer ASPIRE EU Antifungal Research Awards, 3910
Pfizer ASPIRE EU Dupuytren's Contracture Research Awards, 3911
Pfizer ASPIRE EU Emerging Mechanisms of Resistance Antibacterial Research Awards, 3912
Pfizer ASPIRE EU MRSA Nosocomial Pneumonia & MRSA Complicated Skin & Soft Tissue Infections Antibacterial Research Awards, 3913
Pfizer ASPIRE North America Broad Spectrum Antibiotics for the Treatment of Gram-Negative or Polymicrobial Infections Research Awards, 3914
Pfizer ASPIRE North America Hemophilia Research Awards, 3915
Pfizer ASPIRE North America Narrow Spectrum Antibiotics for the Treatment of MRSA Research Awards, 3916
Pfizer ASPIRE North America Rheumatology Research Awards, 3917
Pfizer ASPIRE Worldwide Endocrine Young Investigator Grants, 3918
Pfizer Australia Cancer Research Grants, 3919
Pfizer Australia Neuroscience Research Grants, 3920
Pfizer Australia Paediatric Endocrine Care Research Grants, 3921
PFizer Compound Transfer Agreements, 3922
Pfizer Global Investigator-Initiated Res Grants, 3924
Pfizer Global Research Awards for Nicotine Independence, 3925
Pfizer Inflammation Competitive Res Awards, 3926
Pfizer Research Initiative Dermatology Grants, 3928
Pfizer Research Initiative Rheumatology Grants, 3929
PhRMA Foundation Health Outcomes Post Doctoral Fellowships, 3950
PhRMA Foundation Health Outcomes Research Starter Grants, 3951
PhRMA Foundation Informatics Post Doctoral Fellowships, 3953
PhRMA Foundation Pharmaceutics Postdoctoral Fellowships, 3958
PhRMA Foundation Pharmaceutics Pre Doctoral Fellowships, 3959
PhRMA Foundation Pharmaceutics Research Starter Grants, 3960
PhRMA Foundation Pharmaceutics Sabbatical Fellowships, 3961
Society of Cosmetic Chemists Allan B. Black Award Sponsored by Presperse, 4312
Society of Cosmetic Chemists Award Sponsored by McIntyre Group, Ltd., 4313
Society of Cosmetic Chemists Award Sponsored by The HallStar Company, 4314
Society of Cosmetic Chemists Chapter Best Speaker Award, 4315
Society of Cosmetic Chemists Merit Award, 4316
Society of Cosmetic Chemists Frontier of Science Award, 4317
Society of Cosmetic Chemists Schaeffer Award, 4318
Society of Cosmetic Chemists Joseph P. Ciaudelli Award, 4319
Society of Cosmetic Chemists Keynote Award Lecture Sponsored by Ruger Chemical Corporation, 4320
Society of Cosmetic Chemists Literature Award, 4321
Society of Cosmetic Chemists Maison G. de Navarre Medal, 4322
Society of Cosmetic Chemists Merit Award, 4323
Society of Cosmetic Chemists Robert A. Kramer Lifetime Service Award, 4324
Society of Cosmetic Chemists Mudge Award, 4325
Society of Cosmetic Chemists Poster Awards, 4326
Society of Cosmetic Chemists Young Scie Awards, 4327
SRC Medical Research Grants, 4379

Pharmacogenomics
NIGMS Pharmacogenomics Res Network Grants, 3198
Pfizer Advancing Research in Transplantation Science Research Awards, 3908

Pharmacology
AAA Foundation for Traffic Safety Grants, 24
ACCP Pharmacotherapy Investigator Development Research Awards, 266
AHAF Alzheimer's Disease Research Grants, 370
AHAF National Glaucoma Research Grants, 372
American Psychiatric Association/SmithKline Beecham Young Faculty Award for Research Development in Biological Psychiatry, 779
AOCS George Schroepfer Medal, 879
ASM sanofi-aventis ICAAC Award, 1184

ASPEN C. Richard Fleming Grant, 1190
ASPEN Douglas Wilmore Grant, 1191
ASPEN Maurice Shils Grant, 1195
ASPEN Norman Yoshimura Grant, 1196
ASPEN Nutrition Research Grants, 1197
ASPET-Astellas Awards in Translational Pharmacology, 1200
ASPET Brodie Award in Drug Metabolism, 1201
ASPET Division for Drug Metabolism Early Career Achievement Award, 1202
ASPET Epilepsy Research Award for Outstanding Contributions to the Pharmacology of Antiepileptic Drugs, 1203
ASPET Goodman and Gilman Award in Drug Receptor Pharmacology, 1204
ASPET John J. Abel Award in Pharmacology, 1205
ASPET P. B. Dews Lifetime Achievement Award for Research in Behavioral Pharmacology, 1206
ASPET Paul M. Vanhoutte Award, 1207
ASPET Summer Undergrad Research Fellows, 1208
ASPET T Sollmann Award in Pharmacology, 1209
Astellas Foundation Research Grants, 1227
GEM PhD Science Fellowships, 2106
GPP Charles E. Culpeper Foundation Biomedical Pilot Projects Initiative Grants, 2159
National Center for Responsible Gaming Travel Grants, 2858
NCI/NIDA Transdisciplinary Tobacco Use Research Centers Grants, 2912
NHLBI Ruth L. Kirschstein National Research Service Awards for Individual Senior Fellows, 3060
NIDA Mentored Clinical Scientist Research Career Development Awards, 3129
NIGMS Mentored Clinical Scientist Research Career Development Awards, 3197
NIGMS Pharmacogenomics Res Network Grants, 3198
NIH Ruth L. Kirschstein National Research Service Awards for Individual Postdoctoral Fellows, 3239
ORISE Faculty Research Participation at the National Center for Toxicological Research, 3804
Pfizer ASPIRE EU Antifungal Research Awards, 3910
Pfizer ASPIRE EU Emerging Mechanisms of Resistance Antibacterial Research Awards, 3912
Pfizer ASPIRE EU MRSA Nosocomial Pneumonia & MRSA Complicated Skin & Soft Tissue Infections Antibacterial Research Awards, 3913
Pfizer ASPIRE North America Hemophilia Research Awards, 3915
Pfizer ASPIRE North America Rheumatology Research Awards, 3917
Pfizer Inflammation Competitive Res Awards, 3926
Pharmacia-ASPET Award for Experimental Therapeutics, 3930
PhRMA Foundation Health Outcomes Post Doctoral Fellowships, 3950
PhRMA Foundation Health Outcomes Research Starter Grants, 3951
PhRMA Foundation Informatics Post Doctoral Fellowships, 3953
PhRMA Foundation Informatics Pre Doctoral Fellowships, 3954
PhRMA Foundation Informatics Research Starter Grants, 3955
PhRMA Foundation Informatics Sabbatical Fellowships, 3956
PhRMA Foundation Paul Calabresi Medical Student Research Fellowship, 3957
PhRMA Foundation Pharmaceutics Postdoctoral Fellowships, 3958
PhRMA Foundation Pharmaceutics Research Starter Grants, 3960
PhRMA Foundation Pharmaceutics Sabbatical Fellowships, 3961
PhRMA Foundation Pharmacology/Toxicology Post Doctoral Fellowships, 3962
PhRMA Foundation Pharmacology/Toxicology Pre Doctoral Fellowships, 3963
PhRMA Foundation Pharmacology/Toxicology Research Starter Grants, 3964
PhRMA Foundation Pharmacology/Toxicology Sabbatical Fellowships, 3965
PVA Research Foundation Grants, 4004
Rett Syndrome Research Foundation Grants, 4058
Searle Scholars Program Grants, 4193
Sigrid Juselius Foundation Grants, 4260
Simone and Cino del Duca Grand Prix Awards, 4261
USDD Developer Tools for Medical Education Public Physiology Research Platform Grant, 4717
Victor E. Speas Foundation Grants, 4810

Pharmacy
ACCP Frontiers Career Development Research Awards, 259
AIHP Sonnedecker Visiting Scholar Grants, 432
AIHP Thesis Support Grants, 433
Fischelis Grants for Research in the History of American Pharmacy, 2017
NHLBI Pathway to Independence Awards, 3045
PhRMA Foundation Informatics Pre Doctoral Fellowships, 3954
Regents Professional Opportunity Scholarships, 4049

Pharmacy Education
Yale University Beinecke Library Research Fellowships for Yale Grad & Professional School Students, 4902

Philanthropy
AFP Research Grants, 356
Alfred P. Sloan Foundation Civic Grants, 545
Ashland Corporate Contributions Grants, 1087
Carl Gellert and Celia Berta Gellert Grants, 1429
Center for the Study of Philanthropy International Fellowships, 1482
Center for the Study of Philanthropy Senior International Fellowships, 1483
Chautauqua Region Community Found Grants, 1533
Chestnut Hill Charitable Foundation, Inc Grants, 1538
Cleveland-Cliffs Foundation Grants, 1580
Commonfund Endowment Management Prize, 1601
Council on Foundations Emerging Philanthropic Leaders Fellowships, 1649
DaimlerChrysler Corporation Fund Grants, 1714
Duke Foundation Fellowships, 1868
George Foundation Grants, 2115
George W. Codrington Charitable Grants, 2122
Helen V. Brach Foundation Grants, 2231
IUCP Archives Program Fellowship, 2440
IUCP Dissertation Fellowship, 2441
IUCP Indiana Research Fund, 2442
Jackson Foundation Grants, 2453
Joel L. Fleishman Civil Society Fellowships, 2495
Josiah W. and Bessie H. Kline Foundation Grants, 2517
McCombs Foundation Grants, 2718
NACC David Stevenson Fellowships, 2812
NACC William Diaz Fellowships, 2813
Park Foundation Grants, 3870
Rockefeller Archive Center Research Grants, 4086
Sioux Falls Area Comm Foundation Spot Grants, 4263
Virginia A. Hodgkinson Research Prize, 4812

Philippines
ACC Philippines Fellowship Program, 261

Philosophy
AAB Berlin Prize Fellowships, 48
ACLS Charles A. Ryskamp Research Fellowships, 281
ACLS Frederick Burkhardt Residential Fellowships for Recently Tenured Scholars, 288
AIAR Andrew W. Mellon Foundation Fellowships, 392
Alabama Humanities Foundation Mini Grants, 480
Alabama Humanities Foundation Planning/Consultant Grants, 481
Alpha Sigma Nu Book Award Competition, 578
American Philosophical Society Franklin Research Grants, 764
Andrew W. Mellon/ACLS Dissertation Completion Fellowships, 844
APS/British Academy Fellowship for Research in London, 918
APSA Leo Strauss Award, 938
ASU Graduate College Doctoral Enrichment Fellowships, 1238
Bogliasco Foundation Liguria Study Center Residencies, 1358
Earhart Foundation Institutional Grants, 1888
Germanistic Society of America Fellowships, 2128
Gruber Foundation Cosmology Prize, 2177
Harry Frank Guggenheim Foundation Dissertation Fellowships, 2207
Houghton Library Visiting Fellowships, 2284
Inamori Foundation Kyoto Prize, 2367
ISUD Jacobsen Prizes, 2434
Mellon Post-Doctoral Fellowship in Philosophy, 2739
Metanexus Inst Templeton Research Lectures, 2744
NEH Collaborative Research Grants, 2952
NEH Digital Humanities Implementation Grants, 2953
NEH Digital Humanities Start-Up Grants, 2954
PHF Penn Faculty Research Fellowships, 3932
PHF Regional Mellon Faculty Research and Cultural Fellowships, 3933
PHF Undergraduate Humanities Forum Research Fellowships, 3934
Princeton University Center for Human Values Graduate Student Top-Up Prizes, 3990
Princeton University Center for Human Values Political Philosophy Travel Grants, 3993
Ralph Waldo Emerson Award, 4028
Robert Schalkenbach Foundation Grants, 4079
Romanell-Phi Beta Kappa Professorship, 4099
Rumsfeld Foundation Graduate Fellowships, 4127
SSHRC Insight Development Grants, 4409
SSHRC Insight Grants, 4410
SSRC Mellon Mays Predoctoral Research Grants, 4432
Stanford Humanities Center Arts Writer/Practitioner Fellowship, 4444
Stanford Humanities Center External Faculty Fellowships, 4445
University of London Visiting Fellowships, 4588
Virginia Foundation for the Humanities Residential Fellowships, 4815

Philosophy, History of
ASU Graduate College Doctoral Enrichment Fellowships, 1238

Phonology
IRA Helen M. Robinson Dissertation Res Award, 2409

Photochemistry
DOE Atomic, Molecular, and Optical Sciences Research Grants, 1771
DOE Catalysis and Chemical Transformations Research Grants, 1773
DOE Chemical Energy and Chemical Engineering Research Grants, 1774
DOE Chemical Physics Research Grants, 1775
DOE Chemical Sciences Research Grants, 1776
DOE Heavy Element Chem Research Grants, 1798
DOE Photochem & Radiation Research Grants, 1809
DOE Separations and Analysis Research Grants, 1813

Photogrammetry
ASPRS Leica Geosystems Internship, 1213
ASPRS Robert E. Altenhofen Scholarship, 1214

Photography
ACC Humanities Fellowship Program, 255
ACC Residency Program in Asia, 267
ACC Starr Foundation Fellowship Program, 268
Camargo Foundation Fellowships, 1412
CCH Documentary Project Research and Development Grants, 1451
CICF Efroymson Contemporary Arts Fellows, 1563
Clark Library Short-Term Resident Fellowships, 1577
Getty Research Grants for Institutions - Archival Grants, 2131
Harry Ransom Humanities Research Center Research Fellowships, 2209

Huntington Library Dana and David Dornsife Fellowships, 2303
Huntington Library Travel Grants and Exchange Fellowships for Study in Great Britain, 2309
John Simon Guggenheim Memorial Foundation Fellowships, 2507
NYFA Artist Fellowships, 3698
NYPL Short-Term Research Fellowships, 3701
Third Millennium Foundation Grants, 4505

Photojournalism
ASU Graduate College Herman E. Demund Memorial Fellowships, 1240

Photovoltaics
DOE Competitive Financial Assistance for Renewable Energy and Energy Efficiency Programs, 1779
DOE State Energy Program Special Projects, 1815
USDD Organic Photovoltaics Grants, 4770

Physical Activity
ACSM Carl V. Gisolfi Memorial Fund Grant, 312
ACSM Coca-Cola Company Doctoral Student Grant on Behavior Research, 313
ACSM International Student Awards, 318
ACSM Oded Bar-Or Int'l Scholars, 320
ACSM Paffenbarger-Blair Fund for Epidemiological Research on Physical Activity Grants, 321
ACSM Raymond and Rosalee Weiss Research Endowment Grant, 322

Physical Disability
Deborah Munroe Noonan Mem Research Grants, 1729
Sara Elizabeth O'Brien Trust Grants, 4166

Physical Medicine and Rehabilitation
ACSM Carl V. Gisolfi Memorial Fund Grant, 312
ACSM Coca-Cola Company Doctoral Student Grant on Behavior Research, 313
ACSM Foundation Clinical Sports Medicine Endowment Grants, 315
ACSM Found Doctoral Student Research Grants, 316
ACSM Foundation Research Grants, 317
ACSM International Student Awards, 318
ACSM Oded Bar-Or Int'l Scholars, 320
ACSM Paffenbarger-Blair Fund for Epidemiological Research on Physical Activity Grants, 321
ACSM Raymond and Rosalee Weiss Research Endowment Grant, 322
IATA Research Grants, 2319
Jacob and Valeria Langeloth Foundation Grants, 2455
NHLBI Mentored Patient-Oriented Research Career Development Awards, 3037
Regents Professional Opportunity Scholarships, 4049
Special Olympics Health Professions Stu Grants, 4332

Physical Sciences
AAAS Science and Technology Policy Fellowships: Congressional, 39
AAAS Science and Technology Policy Fellowships: Global Health and Development, 42
Abdus Salam ICTP-OEA Affil Centre Grants, 222
Abdus Salam ICTP Federation Arrangement Scheme Grants, 227
Abdus Salam ICTP Networks Grants, 233
AIAR Andrew W. Mellon Foundation Fellowships, 392
American Mathematical Society AAAS Mass Media Fellowships, 751
American Society of Naval Engineers Scholarship, 800
ANL Faculty Research Leave Program, 854
ANL Faculty Research Participation Program, 855
ANL Graduate Thesis Parts Res Appointments, 856
ANL Laboratory-Graduate Res Appointments, 857
APHL Emerging Infectious Diseases Fellowships, 914
APS Andreas Acrivos Dissertation Award in Fluid Dynamics, 940
APS Aneesur Rahman Prize for Computational Physics, 942
APS Arthur L. Schawlow Prize in Laser Science, 945
APS Davisson-Germer Prize in Atomic or Surface Physics, 952
APS Dissertation Award in Nuclear Physics, 953
APS Einstein Prize, 955
APS Fluid Dynamics Prize, 956
APS Isakson Prize for Optical Effects in Solids, 957
APS George E. Valley, Jr., Prize, 959
APS Hans A. Bethe Prize, 960
APS Herbert P. Broida Prize, 961
APS I.I. Rabi Prize in Atomic, Molecular, and Optical Physics, 962
APS Irving Langmuir Prize in Chemical Physics, 963
APS Sakurai Prize for Theoretical Particle Physics, 964
APS James C. McGroddy Prize for New Materials, 965
APS Ja Clerk Maxwell Prize for Plasma Physics, 966
APS Julius Edgar Lilienfeld Prize, 967
APS Lars Onsager Prize, 968
APS Marshall N. Rosenbluth Outstanding Doctoral Thesis Award, 969
APS Mitsuyoshi Tanaka Dissertation Award in Experimental Particle Physics, 971
APS Nicholas Metropolis Award for Outstanding Doctoral Thesis Work in Comput Physics, 972
APS Oliver E. Buckley Condensed Matter Prize, 973
APS Polymer Physics Prize, 976
APS Prize for a Faculty Member for Research in an Undergraduate Institution, 977
APS Robert R. Wilson Prize for Achievement in the Physics of Particle Accelerators, 979
APS Tom W. Bonner Prize in Nuclear Physics, 980
APS W.K.H. Panofsky Prize in Experimental Particle Physics, 981
APS Will Allis Prize for Study of Ionized Gases, 982
ASME Timoshenko Medal, 1168
AWIS Educational Foundation Predoc Awards, 1278
Bullard Fellowship in Forest Research, 1383
BWF Career Awards at the Scientific Interface, 1393
Canada-U.S. Fulbright International Science and Technology Award, 1417
Canada-U.S. Fulbright Senior Specialists Program Grants, 1418
Canada-U.S. Fulbright Traditional Scholars, 1419
Cen Cal Diving and Aquatic Studies Scholarship, 1471
Center for Defense Information Internships, 1473
Consortium for Graduate Study in Management Minority Fellowships, 1642
Fannie and John Hertz Foundation Graduate Fellowships, 1977
Fulbright Alumni Initiatives Awards Grants, 2064
Fulbright Distinguished Chairs Program Grants, 2069
Fulbright European Union Scholar-in-Residence Program Grants, 2071
Fulbright New Century Scholars Program Grants, 2075
Fulbright Occasional Lecturer Program Grants, 2076
Fulbright Specialists Program Grants, 2079
Fulbright Traditional Scholar Program in Europe and Eurasia, 2080
Fulbright Traditional Scholar Program in Sub-Saharan Africa, 2081
Fulbright Traditional Scholar Program in the East Asia/Pacific Region, 2082
Fulbright Traditional Scholar Program in the Near East and North Africa Region, 2083
Fulbright Traditional Scholar Program in the South and Central Asia Region, 2084
Fulbright Traditional Scholar Program in the Western Hemisphere, 2085
Fulbright Visiting Specialists Program: Direct Access to the Muslim World, 2087
GEM PhD Science Fellowships, 2106
HHMI-NIBIB Interfaces Initiative Grants, 2248
Horton Research Grants, 2282
John Simon Guggenheim Memorial Foundation Fellowships, 2507
L'Oreal USA Women in Science Fellowship, 2577
Metanexus Inst Templeton Research Lectures, 2744
NATO Advanced Research Workshops Program, 2894
NATO Advanced Study Institutes Program, 2895
NATO Collaborative Linkage Grants, 2896
NATO Expert Visits Program Grants, 2897
NIST National Standard Reference Data System Research Grants, 3271
NOAA Regional Integrated Sciences and Assessments Program Grants, 3402
Northrop Grumman Foundation Grants, 3429
NPSC Grad Fellowships in the Physical Sciences, 3436
NSF Alan T. Waterman Award, 3467
NSF Arctic Nat Science Grants in Ocean Science, 3487
NSF Arctic System Science Program Grants, 3493
NSF CISE Collaborative Research in Computational Neuroscience Grants, 3526
NSF Integrative Graduate Education and Research Traineeship, 3584
NSF Partnerships for International Research and Education, 3611
NSF Physical Anthropology Research Grants, 3615
NSF Sediment Geology & Paleobiology Grants, 3646
ONR Electronic and Optical Ceramics Research Grants, 3756
ORISE Fusion Energy Sciences Fellowship, 3807
ORISE Higher Education Research Experiences at Oak Ridge National Laboratory for Faculty, 3808
ORISE National Oceanic and Atmospheric Administration Faculty Research Participation, 3811
ORISE National Oceanic and Atmospheric Admin Student Research Participation, 3812
ORISE Nuclear Regulatory Commission Historically Black Colleges and Universities Student Research Participation, 3814
ORISE Postgraduate Research Participation at the National Exposure Research Laboratory, 3821
ORISE Savannah River Site Faculty Research Participation Program, 3823
Packard Foundation Fellowships for Science and Engineering, 3862
Phi Beta Kappa Award in Science, 3935
Society of Cosmetic Chemists Allan B. Black Award Sponsored by Presperse, 4312
Society of Cosmetic Chemists Award Sponsored by McIntyre Group, Ltd., 4313
Society of Cosmetic Chemists Award Sponsored by The HallStar Company, 4314
Society of Cosmetic Chemists Chapter Best Speaker Award, 4315
Society of Cosmetic Chemists Merit Award, 4316
Society of Cosmetic Chemists Frontier of Science Award, 4317
Society of Cosmetic Chemists Schaeffer Award, 4318
Society of Cosmetic Chemists Joseph P. Ciaudelli Award, 4319
Society of Cosmetic Chemists Keynote Award Lecture Sponsored by Ruger Chemical Corporation, 4320
Society of Cosmetic Chemists Literature Award, 4321
Society of Cosmetic Chemists Maison G. de Navarre Medal, 4322
Society of Cosmetic Chemists Merit Award, 4323
Society of Cosmetic Chemists Robert A. Kramer Lifetime Service Award, 4324
Society of Cosmetic Chemists Mudge Award, 4325
Society of Cosmetic Chemists Poster Awards, 4326
Society of Cosmetic Chemists Young Scie Awards, 4327
SRC Peter Verhofstadt Fellowship, 4389
Swiss National Science Foundation Fellowships for Advanced Researchers, 4483
Swiss National Science Foundation Fellowships for Prospective Researchers, 4484
Swiss National Science Foundation Marie Heim-Vogtlin Grants, 4485
UNCFSP NASA Harriett G. Jenkins Predoctoral Fellowships, 4563
USDD Defense Sciences Research & Tech Grants, 4716
W.M. Keck Foundation Science and Engineering Research Grants, 4824
Wellcome Trust Biomedical Science Grants, 4843
WHOI Undergrad Fellowships for Minorities, 4866

Physical Sciences Education
Northrop Grumman Foundation Grants, 3429
NPSC Grad Fellowships in the Physical Sciences, 3436
NSF Physics at the Information Frontier Grants, 3618

Physics

AAAS Award for Public Engagement with Science, 26
AACR Team Science Award, 148
Abdus Salam ICTP-OEA Affil Centre Grants, 222
Abdus Salam ICTP-OEA Projects Grants, 223
Abdus Salam Scientific Meetings Grants, 224
Abdus Salam ICTP-OEA Visiting Scholars/Consultants Grants, 225
Abdus Salam ICTP Federation Arrangement Scheme Grants, 227
Abdus Salam ICTP Group Associates Grants, 228
Abdus Salam ICTP Laurea Magistralis Grants, 229
Abdus Salam ICTP Joint PhD Programme in Environmental Fluid Mechanics Grants, 230
Abdus Salam ICTP Networks Grants, 233
Abdus Salam ICTP Postdoctoral Positions in Theoretical High Energy Physics, 234
Abdus Salam ICTP Schools, Conferences, Workshops and Extended Research Workshops Grants, 238
Abdus Salam ICTP Senior Associates Grants, 239
Abdus Salam Joint ICTP/SISSA PhD Programme in Physics and Mathematics Grants, 242
AIChE Excellence in Process Development Research Award, 409
Alfred P. Sloan Foundation History of Science and Technology Grants, 553
Alfred P. Sloan Foundation Research Fellowships, 560
American Philosophical Society Library Resident Research Fellowships, 769
AMS/Industry Government Graduate Fellowships, 817
AMS/Industry Minority Scholarships, 818
AMS Grad Fellowships in History of Science, 824
AMS Undergraduate Scholarships, 837
ANL Faculty Research Participation Program, 855
AOCS Samuel Rosen Memorial Award, 895
APS Abraham Pais Prize for History of Physics, 925
APS Andrei Sakharov Prize, 941
APS Dannie Heineman Prize for Math Physics, 951
APS Davisson-Germer Prize in Atomic or Surface Physics, 952
APS George E. Pake Prize, 958
APS George E. Valley, Jr., Prize, 959
APS I.I. Rabi Prize in Atomic, Molecular, and Optical Physics, 962
APS Irving Langmuir Prize in Chemical Physics, 963
APS Ja Clerk Maxwell Prize for Plasma Physics, 966
APS Julius Edgar Lilienfeld Prize, 967
APS Lars Onsager Prize, 968
APS Max Delbruck Prize in Biological Physics, 970
APS Nicholas Metropolis Award for Outstanding Doctoral Thesis Work in Comput Physics, 972
APS Oliver E. Buckley Condensed Matter Prize, 973
APS Outstanding Doctoral Thesis Research in Atomic, Molecular, or Optical Physics, 974
APS Outstanding Doctoral Thesis Research in Beam Physics Award, 975
APS Polymer Physics Prize, 976
APS Prize for a Faculty Member for Research in an Undergraduate Institution, 977
APS Prize for Industrial Applications of Physics, 978
ASEE Army Research Laboratory Postdoctoral Fellowships, 1059
ASME Van C. Mow Medal, 1169
ASME Warner T. Koiter Medal, 1170
ASME Y.C. Fung Young Investigator Award, 1173
Astronomical Society of the Pacific Priscilla and Bart Bok Awards, 1232
ASU Graduate College Science Foundation Arizona Fellowships in Science and Engineering, 1245
AT&T Bell Labs Graduate Research Program, 1250
Brookhaven National Laboratory Sambamurti Memorial Lectureships, 1382
BWF Career Awards at the Scientific Interface, 1393
Canada-U.S. Fulbright International Science and Technology Award, 1417
Canada-U.S. Fulbright Senior Specialists Program Grants, 1418
Canada-U.S. Fulbright Traditional Scholars, 1419
Canada Graduate Scholarships and NSERC Postgraduate Scholarships, 1420
Carnegie Postdoctoral Research Fellowships, 1436
Cooperative Institute for Research in Environmental Sciences Visiting Fellows Program, 1643
DOE University Coal Research Grants, 1816
Fermilab Postdoctoral Fellowship Programs, 2004
Fulbright Alumni Initiatives Awards Grants, 2064
Fulbright Distinguished Chairs Program Grants, 2069
Fulbright European Union Scholar-in-Residence Program Grants, 2071
Fulbright New Century Scholars Program Grants, 2075
Fulbright Occasional Lecturer Program Grants, 2076
Fulbright Specialists Program Grants, 2079
Fulbright Traditional Scholar Program in Europe and Eurasia, 2080
Fulbright Traditional Scholar Program in Sub-Saharan Africa, 2081
Fulbright Traditional Scholar Program in the East Asia/Pacific Region, 2082
Fulbright Traditional Scholar Program in the Near East and North Africa Region, 2083
Fulbright Traditional Scholar Program in the South and Central Asia Region, 2084
Fulbright Traditional Scholar Program in the Western Hemisphere, 2085
Fulbright Visiting Specialists Program: Direct Access to the Muslim World, 2087
GEM PhD Science Fellowships, 2106
HHMI-NIBIB Interfaces Initiative Grants, 2248
Institute of Paper Science & Tech Fellowships, 2387
ISGC Informal Education Grants, 2431
ISGC Multiple Undergraduate Research Grants, 2432
ISGC Research Seed Grants, 2433
MBL Eugene and Millicent Bell Fellowships, 2699
NCAR Graduate Research Fellowships, 2901
NCAR Postdoctoral Fellowships in the Atmospheric Sciences, 2902
NHGRI Mentored Research Scientist Development Award, 3013
NHLBI Mentored Quantitative Research Career Development Awards, 3038
NIMH Curriculum Development Award in Neuroinformatics Research and Analysis, 3251
Northwood University/Alden B. Dow Creativity Center Summer Residency Program, 3432
NPSC Grad Fellowships in the Physical Sciences, 3436
NSF Analysis Grants, 3469
NSF Applied Mathematics Grants, 3481
NSF Arctic Natural Science Grants in Atmospheric Sciences, 3483
NSF Atomic Molecular & Optical Physics Grants, 3498
NSF Biological Physics, 3503
NSF CISE Accelerating Discovery in Science and Engineering Through Petascale Simulations and Analysis Grants, 3521
NSF CISE Cross-Cutting Program Grants, 3532
NSF CISE Increasing the Participation and Advancement of Women in Academic Science and Engineering Careers Grants, 3534
NSF Condensed Matter Physics Research Grants, 3540
NSF Education & Interdisciplinary Res Grants, 3557
NSF Gravitational Physics Research Grants, 3576
NSF Magnetospheric Physics Research Grants, 3590
NSF Materials Research Grants, 3593
NSF Particle and Nuclear Astrophysics Grants, 3608
NSF Physics Frontiers Centers Grants, 3619
NSF Physics of Living Systems Grants, 3620
NSF Physics Research Grants, 3621
NSF Plasma Physics Grants, 3626
Research Corporation Cottrell Scholars Awards, 4053
Research Corporation Multi-Investigator Cottrell College Science Awards, 4054
Research Corporation Scialog Grants, 4055
Research Corporation Single-Investigator Cottrell College Science Awards, 4056
Roy J. Carver Trust Higher Education Grants, 4113
RSC Rutherford Memorial Medals, 4124
Smith Family Awards for Excellence in Biomedical Research, 4268
Smithsonian Environmental Research Center Graduate Student Fellowships, 4273
Smithsonian Environmental Research Center Internships, 4274
Smithsonian Environmental Research Center Postdoctoral/Senior Fellowships, 4275
Smithsonian Environmental Research Center Predoctoral Fellowships, 4276
SPIE C.S. Vikram Award in Optical Metrology, 4339
SPIE Dennis Gabor Award, 4340
SRC Hans J. Coufal Fellowship, 4371
SRC Natural and Engineering Sciences Grants, 4387
SRC Peter Verhofstadt Fellowship, 4389
SSRC Mellon Mays Predoctoral Research Grants, 4432
Swedish Natural Science Res Council Grants, 4479
U.S. Department of Education Graduate Assistance in Areas of National Need, 4540
UNCFSP NASA Harriett G. Jenkins Predoctoral Fellowships, 4563
US CRDF CGP-Climate Change and Energy Competition For Joint U.S.-Ukrainian Research Teams, 4607
US CRDF CGP-RFBR Climate Change and Energy Competition For Joint U.S.-Russian Research Teams, 4608
USDA Decadal and Regional Climate Prediction Using Earth System Models Grants, 4621
USDD ARO Broad Agency Announcement, 4686
USDD Ocean Acoustics Research Grants, 4766
USDD Physical Oceanography Grants, 4771
USDD Research Interest of the Air Force Office of Scientific Research Grants, 4784
Weizmann Inst of Science Postdoctoral Fellowships, 4837
WHOI Geophysical Fluid Dynamics Fellowships, 4859
WHOI Postdoctoral Scholar Awards, 4864
WHOI Summer Student Fellowships, 4865

Physics Education

APS Prize for a Faculty Member for Research in an Undergraduate Institution, 977
DAAD Research Internships in Science and Engineering, 1704
DAAD RISE Professional Internships, 1706
NAACP Willems Scholarships, 2810
NPSC Grad Fellowships in the Physical Sciences, 3436
Outstanding Junior Investigator Program in High Energy Physics, 3859
U.S. Department of Education Graduate Assistance in Areas of National Need, 4540

Physics, Applied

Abdus Salam ICTP/IAEA Sandwich Training Education Grants, 226
Abdus Salam ICTP Postgraduate Diploma Grants, 235
Abdus Salam ICTP Training and Research in Italian Laboratories Grants, 240
Abdus Salam Joint ICTP/SISSA PhD Programme in Physics and Mathematics Grants, 242
ASME Robert Henry Thurston Lecture Award, 1160
Fannie and John Hertz Foundation Graduate Fellowships, 1977

Physics, Atomic

Abdus Salam ICTP/IAEA Sandwich Training Education Grants, 226
Abdus Salam ICTP Group Associates Grants, 228
Abdus Salam ICTP Junior Associates Grants, 231
Abdus Salam ICTP Postgraduate Diploma Grants, 235
Abdus Salam ICTP Regular Associates Grants, 237
Abdus Salam ICTP Senior Associates Grants, 239
Abdus Salam ICTP Training and Research in Italian Laboratories Grants, 240
APS I.I. Rabi Prize in Atomic, Molecular, and Optical Physics, 962
APS Outstanding Doctoral Thesis Research in Atomic, Molecular, or Optical Physics, 974
APS Outstanding Doctoral Thesis Research in Beam Physics Award, 975
ARO Atomic, Molecular, and Optical Physics Research Contracts, 1005
DOE Atomic, Molecular, and Optical Sciences Research Grants, 1771

DOE Catalysis and Chemical Transformations Research Grants, 1773
DOE Chemical Energy and Chemical Engineering Research Grants, 1774
DOE Chemical Physics Research Grants, 1775
DOE Chemical Sciences Research Grants, 1776
DOE Fusion Energy Science Division Grants, 1795
DOE Heavy Element Chem Research Grants, 1798
DOE Photochem & Radiation Research Grants, 1809
DOE Separations and Analysis Research Grants, 1813
NSF Physics Frontiers Centers Grants, 3619
NSF Physics Research Grants, 3621
NSF Theoretical Atomic, Molecular, and Optical Physics Grants, 3659
USDD Atomic, Molecular and Quantum Physics Grants, 4687

Physics, Computational
Abdus Salam ICTP Training and Research in Italian Laboratories Grants, 240
APS Aneesur Rahman Prize for Computational Physics, 942
APS Nicholas Metropolis Award for Outstanding Doctoral Thesis Work in Comput Physics, 972
SfN Swartz Prize for Theoretical and Computational Neuroscience, 4213

Physics, Fluid
NSF Condensed Matter Physics Research Grants, 3540

Physics, High Energy
Abdus Salam ICTP/IAEA Sandwich Training Education Grants, 226
Abdus Salam ICTP Group Associates Grants, 228
Abdus Salam ICTP Junior Associates Grants, 231
Abdus Salam ICTP Postdoctoral Positions in Theoretical High Energy Physics, 234
Abdus Salam ICTP Regular Associates Grants, 237
Abdus Salam ICTP Schools, Conferences, Workshops and Extended Research Workshops Grants, 238
Abdus Salam ICTP Senior Associates Grants, 239
Brookhaven National Laboratory Sambamurti Memorial Lectureships, 1382
DOE Advanced Detector Research Grants, 1767
DOE High-Energy Physics Research Grants, 1799
NSF Elementary Particle Physics Grants, 3559
NSF Particle and Nuclear Astrophysics Grants, 3608
Outstanding Junior Investigator Program in High Energy Physics, 3859

Physics, Molecular
Abdus Salam ICTP Postgraduate Diploma Grants, 235
Abdus Salam ICTP Training and Research in Italian Laboratories Grants, 240
APS I.I. Rabi Prize in Atomic, Molecular, and Optical Physics, 962
APS Outstanding Doctoral Thesis Research in Atomic, Molecular, or Optical Physics, 974
APS Outstanding Doctoral Thesis Research in Beam Physics Award, 975
ARO Atomic, Molecular, and Optical Physics Research Contracts, 1005
ARO Theoretical Physics and Nonlinear Phenomena Research Contracts, 1017
NSF Physics Frontiers Centers Grants, 3619
NSF Physics of Living Systems Grants, 3620
NSF Physics Research Grants, 3621
NSF Theoretical Atomic, Molecular, and Optical Physics Grants, 3659
USDD Atomic, Molecular and Quantum Physics Grants, 4687

Physics, Nuclear
Abdus Salam ICTP/IAEA Sandwich Training Education Grants, 226
Abdus Salam ICTP Group Associates Grants, 228
Abdus Salam ICTP Regular Associates Grants, 237
AINSE Honours Scholarships, 460
AINSE Postgraduate Research Awards, 463
APS Dissertation Award in Nuclear Physics, 953
APS I.I. Rabi Prize in Atomic, Molecular, and Optical Physics, 962
Brookhaven National Laboratory Sambamurti Memorial Lectureships, 1382
DOE Nuclear Physics Research Grants, 1807
JSA JLab Graduate Fellowships, 2521
NSF Nuclear Physics Research Grants, 3601
NSF Particle and Nuclear Astrophysics Grants, 3608
NSF Physics Frontiers Centers Grants, 3619
NSF Physics Research Grants, 3621
ORISE Univ Radioactive Ion Beam Consortium, 3831

Physics, Optical
APS I.I. Rabi Prize in Atomic, Molecular, and Optical Physics, 962
APS Outstanding Doctoral Thesis Research in Atomic, Molecular, or Optical Physics, 974
NSF Atomic Molecular & Optical Physics Grants, 3498

Physics, Particle
Abdus Salam Joint ICTP/SISSA PhD Programme in Physics and Mathematics Grants, 242
APS Sakurai Prize for Theoretical Particle Physics, 964
APS Mitsuyoshi Tanaka Dissertation Award in Experimental Particle Physics, 971
APS Robert R. Wilson Prize for Achievement in the Physics of Particle Accelerators, 979
APS W.K.H. Panofsky Prize in Experimental Particle Physics, 981
ARO Theoretical Physics and Nonlinear Phenomena Research Contracts, 1017
Fermilab Postdoctoral Fellowship Programs, 2004
JSA JLab Graduate Fellowships, 2521
NSF Elementary Particle Physics Grants, 3559
NSF Nuclear Physics Research Grants, 3601
NSF Particle and Nuclear Astrophysics Grants, 3608
NSF Physics Frontiers Centers Grants, 3619
NSF Physics Research Grants, 3621
SSHRC Digging into Data Challenge Grants, 4406

Physics, Plasma
Abdus Salam ICTP Group Associates Grants, 228
Abdus Salam ICTP Junior Associates Grants, 231
Abdus Salam ICTP Postgraduate Diploma Grants, 235
Abdus Salam ICTP Regular Associates Grants, 237
Abdus Salam ICTP Senior Associates Grants, 239
Abdus Salam ICTP Training and Research in Italian Laboratories Grants, 240
APS Marshall N. Rosenbluth Outstanding Doctoral Thesis Award, 969
ARO Atomic, Molecular, and Optical Physics Research Contracts, 1005
DOE Engineering Research Grants, 1789
DOE Fusion Energy Science Division Grants, 1795
NSF/DOE Partnership in Basic Plasma Science and Engineering Grants, 3454
NSF Physics Frontiers Centers Grants, 3619
NSF Physics Research Grants, 3621
NSF Plasma Physics Grants, 3626
NSF Thermal Transport Processes Grants, 3660

Physics, Solid State
Abdus Salam ICTP Postgraduate Diploma Grants, 235
Abdus Salam ICTP Training and Research in Italian Laboratories Grants, 240
DOE Materials Sciences and Engineering Research Grants, 1803
ONR Electronics Division Research Grants, 3759

Physics, Theoretical
Abdus Salam ICTP Laurea Magistralis Grants, 229
Abdus Salam ICTP Schools, Conferences, Workshops and Extended Research Workshops Grants, 238
Abdus Salam ITCP Postgraduate Diploma Programme in General Physics, 241
APS Lars Onsager Prize, 968
ARO Theoretical Physics and Nonlinear Phenomena Research Contracts, 1017
DOE High-Energy Physics Research Grants, 1799
Fermilab Postdoctoral Fellowship Programs, 2004
NSF Physics Research Grants, 3621
NSF Theoretical Atomic, Molecular, and Optical Physics Grants, 3659
SfN Swartz Prize for Theoretical and Computational Neuroscience, 4213

Physiology
ACSM International Student Awards, 318
ACSM NASA Space Physiology Research Grants, 319
ACSM Oded Bar-Or Int'l Scholars, 320
AHAF Alzheimer's Disease Research Grants, 370
AHAF National Glaucoma Research Grants, 372
American Philosophical Society Library Resident Research Fellowships, 769
AOCS George Schroepfer Medal, 879
AOCS Health & Nutrition Div Student Award, 881
AOCS Health and Nutrition Poster Competition, 882
AOCS Ralph Holman Lifetime Award, 894
Lalor Foundation Postdoctoral Fellowships, 2580
MBL Albert & Ellen Grass Summer Fellowships, 2692
MBL Associates Summer Fellowships, 2694
MBL Burr & Susie Steinbach Summer Fellow, 2696
MBL E.E. Just Summer Fellowship for Minority Scientists, 2697
MBL Erik B. Fries Summer Fellowships, 2698
MBL Evelyn & Mel Spiegal Summer Fellowship, 2700
MBL Frank R. Lillie Summer Fellowship, 2701
MBL Fred Karush Library Readership, 2703
MBL Gruss Lipper Family Foundation Summer Fellowship, 2704
MBL H. Keffer Hartline and Edward F. MacNichol, Jr. Fellowships, 2705
MBL Herbert W. Rand Summer Fellowship, 2706
MBL James E. and Faith Miller Memorial Summer Fellowship, 2707
MBL John M. Arnold Award, 2708
MBL Lucy B. Lemann Mem Summer Fellowship, 2710
MBL M.G.F. Fuortes Summer Fellowships, 2711
MBL Nikon Summer Fellowship, 2712
MBL Stephen W. Kuffler Summer Fellowships, 2715
MBL William Townsend Porter Summer Fellowships for Minority Investigators, 2716
NIH Biobehavioral Research for Effective Sleep, 3205
NSF Ecosystem Science Grants, 3556
ONR Medical Science and Technology Division Grants, 3772
Rett Syndrome Research Foundation Grants, 4058
USDD Developer Tools for Medical Education Public Physiology Research Platform Grant, 4717

Physiology, Human
ACSM NASA Space Physiology Research Grants, 319
NHLBI Lymphatics in Health and Disease in the Digestive, Urinary, Cardiovascular and Pulmonary Systems, 3032
USDD Developer Tools for Medical Education Public Physiology Research Platform Grant, 4717
USDD Human Social, Cultural and Behavior Modeling Thrust Grants, 4737

Physiology, Vertebrate
NSF Physiology and Ethology Research Grants, 3622

Planetariums
IMLS 21st Century Museum Prof Grants, 2362
IMLS National Leadership Grants, 2366

Planetary Studies
AMNH Graduate Student Fellowships, 815
NCAR Graduate Research Fellowships, 2901
NSF Cooperative Studies Of The Earth's Deep Interior Grants, 3544
NSF Planetary Astronomy Research Grants, 3623
Smithsonian Graduate Student Fellowships, 4277
Smithsonian Minority Visiting Student Awards, 4279
Smithsonian National Air and Space Museum Postdoctoral Earth and Planetary Sciences Fellowship, 4287
Smithsonian Native American Visiting Student Awards, 4292

Smithsonian Postdoctoral Fellowships, 4293
Smithsonian Predoctoral Fellowships, 4295
WHOI Geophysical Fluid Dynamics Fellowships, 4859

Planning/Policy Studies
APA Planning Fellowships, 910
CDC Steven M. Teutsch Prevention Effectiveness Fellowships, 1467
Center for Defense Information Internships, 1473
Charles Abrams Scholarships, 1519
CTCRI Policy Research Grants, 1667
GNOF IMPACT Grants for Education, 2154
IIE Rockefeller Foundation Bellagio Center Residencies, 2355
IMLS 21st Century Museum Prof Grants, 2362
IMLS National Leadership Grants, 2366
Maine Community Foundation Keepers Preservation Education Grants, 2651
Robert Schalkenbach Foundation Grants, 4079
Searle Freedom Trust Grants, 4192
SSRC African Peacebuilding Network Research Grants, 4422

Plant Diseases/Pathology
Bedding Plants Foundation Research Grants, 1330
NIGMS/NIDCR Evolution of Infectious Diseases Research Grants, 3196
USDA Norman E. Borlaug International Agricultural Science and Technology Fellowships, 4636
USDA Plant and Animal Disease, Pest Control, and Animal Care Grants, 4638

Plant Genetics
Alfred P. Sloan Foundation Barcode of Life Grants, 543
BARD Research Grants, 1299
John D. and Catherine T. MacArthur Foundation Global Challenges Program Grants, 2498
NSF Plant Genome Research Program, 3625
Pioneer Hi-Bred Hon Plant Breeding Fellowships, 3968
Pioneer Hi-Bred International Scholarships, 3969
Pioneer Hi-Bred Society Fellowships, 3970
Pioneer Hi-Bred University Fellowships, 3971
USDA Plant Feedstock Genomics for Bioenergy: A Joint Research Grant, 4639

Plant Physiology/Morphology
American Chemical Society AGRO Sterling B. Hendricks Memorial Lectureship, 635
NSF Physiology and Ethology Research Grants, 3622
Orchid Research Fellowships, 3794

Plant Sciences
Bedding Plants Foundation Research Grants, 1330
Carnegie Plant Biology Fellowships, 1435
CPC Catherine H. Beattie Fellowship, 1653
M.J. Murdock Charitable Trust College Research Program for Life Sciences, 2638
NFWF Pulling Together Initiative Grants, 3007
NSF Animal Developmental Mechanisms Grants, 3471
NSF Cell Systems Cluster Grants, 3512
NSF Evolution of Developmental Systems Grants, 3563
NSF Hydrologic Sciences Research Grants, 3578
NSF Physiology and Ethology Research Grants, 3622
NSF Plant, Fungal and Microbial Developmental Systems Grants, 3624
Pioneer Hi-Bred Hon Plant Breeding Fellowships, 3968
Pioneer Hi-Bred International Scholarships, 3969
Pioneer Hi-Bred Society Fellowships, 3970
Pioneer Hi-Bred University Fellowships, 3971
USDA 1890 Institution Teaching, Research, and Extension Capacity Building Grants, 4613
USDA Distance Education Grants for Higher Education Institutions in Insular Areas, 4622
USDA Interregional Research Project Minor Crop Pest Management Grants, 4632
USDA Norman E. Borlaug International Agricultural Science and Technology Fellowships, 4636
USDA Plant Feedstock Genomics for Bioenergy: A Joint Research Grant, 4639
USDA Resident Instruction Grants for Higher Education in Insular Areas, 4645
USDA Small Business Innovation Res Grants, 4648
USDA Tribal College Research Grants, 4652

Plasma Physics
APS Ja Clerk Maxwell Prize for Plasma Physics, 966

Plastic Surgery
AAFPRS Ben Shuster Memorial Award, 160
AAFPRS Bernstein Grant, 161
AAFPRS Fellowships, 162
AAFPRS Investigator Development Grant, 163
AAFPRS John Orlando Roe Award, 165
AAFPRS Leslie Bernstein Resident Res Grants, 166
AAFPRS Residency Travel Awards, 167
AAFPRS Sir Harold Delf Gillies Award, 168
Plastic Surgery Educational Foundation/AAO-HNSF Combined Grant, 3977
Plastic Surgery Educational Foundation Basic Research Grants, 3978
Plastic Surgery Educational Foundation Research Fellowship Grants, 3979

Plasticity
SfN Swartz Prize for Theoretical and Computational Neuroscience, 4213

Playgrounds
William Blair and Company Foundation Grants, 4875

Podiatry
Regents Professional Opportunity Scholarships, 4049
Special Olympics Health Professions Stu Grants, 4332
Yale University Beinecke Library Research Fellowships for Yale Grad & Professional School Students, 4902

Poetry
AAB Berlin Prize Fellowships, 48
Axe-Houghton Foundation Grants, 1287
Florida Division of Cultural Affairs Individual Artist Fellowships, 2026
National Endowment for the Arts Literature Fellowships: Creative Writing, 2865
National Endowment for the Arts Literature Fellowships: Translation Projects, 2866
NHC Summer Institutes in Lit Studies Grants, 3012
North Carolina Arts Council Residency Grants, 3420
NYFA Artist Fellowships, 3698
NYFA Shelly Pinz Fellowship, 3699
Pew Fellowships in the Arts, 3905
Princeton University Lewis Center for the Arts Hodder Fellowships, 3996
Summerfield G. Roberts Award, 4465

Poison Control
USDA Animal Health & Disease Grants, 4616

Polish Americans
Kosciuszko Foundation Graduate Study and Research in Poland Scholarships, 2571
Kosciuszko Foundation Tuition Scholarships, 2574

Polish Language/Literature
Kosciuszko Found Grants for Polish Citizens, 2572
Kosciuszko Foundation Teaching Fellowships for Polish Citizens, 2573

Political Behavior
APSA Edward S. Corwin Award, 933
APSA Leonard D. White Award, 937
APSA Presidency Research Fund Fellowships, 943
APSA Special Fund for the Study of Women and Politics, 947
Carnegie Corporation of New York Grants, 1432
Century Foundation Grants, 1489
DOS Study of the United States Institutes, 1846
Harry Frank Guggenheim Foundation Research Grants, 2208
Kellogg Institute for International Studies Course Development Grants, 2537
Kellogg Institute for International Studies Dissertation Year Fellowships, 2538
Kellogg Institute for International Studies Graduate Research Grants, 2542
Kellogg Institute for International Studies Intramural Grants, 2546
Kellogg Institute for International Studies Summer Internships, 2549
Kellogg Institute for International Studies Supplemental Awards for PhD Students, 2550
Kellogg Institute for International Studies Visiting Fellowships, 2551
OAH Ellis W. Hawley Prize, 3715
Robert Sterling Clark Foundation Government Accountability Grants, 4080
RWJF Investigator Awards in Health Policy Res, 4145

Political Economics
Library Company of Philadelphia Barra Foundation International Research Fellowships, 2606
Library Company of Philadelphia Program in Early American Economy and Society Dissertation-Level Fellowships, 2609
Library Company of Philadelphia Program in Early American Economy and Society One-Month Fellowships, 2610
Library Company of Philadelphia Program in Early American Economy and Society Post-Doctoral Research Fellowships, 2611
OAH Ellis W. Hawley Prize, 3715
RAND Corporation Transatlantic Postdoctoral Fellowships for Int'l Relations & Security, 4032
Rumsfeld Foundation Graduate Fellowships, 4127
United States Institute of Peace Annual Grants, 4582

Political Science
AAB Berlin Prize Fellowships, 48
AAR Postdoctoral Fellowships, 201
ACLS Charles A. Ryskamp Research Fellowships, 281
ACLS Frederick Burkhardt Residential Fellowships for Recently Tenured Scholars, 288
ALA Marta Lange/CQ Press Award, 493
Andrew W. Mellon/ACLS Dissertation Completion Fellowships, 844
APSA Congressional Fellowships for Federal Executives, 927
APSA Congressional Fellowships for Political Scientists, 930
APSA Gabriel A. Almond Award, 934
APSA Minority Fellowships, 939
APSA Presidency Research Fund Fellowships, 943
APSA Small Research Grants, 946
APSA Special Fund for the Study of Women and Politics, 947
APSA Travel Grants, 948
APSA Warren E. Miller Fellowship in Electoral Politics, 949
ASCE Trent R. Dames and William W. Moore Engineering Fellowship, 1043
Bullard Fellowship in Forest Research, 1383
Canada-U.S. Fulbright International Science and Technology Award, 1417
Canada-U.S. Fulbright Senior Specialists Program Grants, 1418
Canada-U.S. Fulbright Traditional Scholars, 1419
Center for East Asian Studies Fellowships, 1474
Clark Library Short-Term Resident Fellowships, 1577
Douglass Institute Postdoctoral Fellowships, 1851
Earhart Foundation Institutional Grants, 1888
Fulbright Alumni Initiatives Awards Grants, 2064
Fulbright Distinguished Chairs Program Grants, 2069
Fulbright European Union Scholar-in-Residence Program Grants, 2071
Fulbright New Century Scholars Program Grants, 2075
Fulbright Occasional Lecturer Program Grants, 2076
Fulbright Specialists Program Grants, 2079
Fulbright Traditional Scholar Program in Europe and Eurasia, 2080

Fulbright Traditional Scholar Program in Sub-Saharan Africa, 2081
Fulbright Traditional Scholar Program in the East Asia/Pacific Region, 2082
Fulbright Traditional Scholar Program in the Near East and North Africa Region, 2083
Fulbright Traditional Scholar Program in the South and Central Asia Region, 2084
Fulbright Traditional Scholar Program in the Western Hemisphere, 2085
Fulbright Visiting Specialists Program: Direct Access to the Muslim World, 2087
Germanistic Society of America Fellowships, 2128
Gladys Krieble Delmas Foundation Grants for Venetian Research, 2144
Goldsmith Awards Program, 2158
Harry Frank Guggenheim Foundation Dissertation Fellowships, 2207
Herb Society of America Research and Education Grants, 2243
Horowitz Foundation for Social Policy Grants, 2280
Horowitz Found for Social Policy Special Awards, 2281
Japan Foundation Japanese Studies Fellowships, 2476
Japan Foundation New York Japanese Studies Fellowships, 2478
Marine Corps Historical Center Dissertation Fellowship, 2663
National Endowment for Democracy Discretionary Grants, 2862
NCAR Postdoctoral Fellowships in the Atmospheric Sciences, 2902
NCFR Innovation Grants, 2908
NFID Adv Vaccinology Course Travel Grant, 2986
NSF Arctic Social Sciences Grants, 3492
NSF SBE Doctoral Dissertation Research Improvement Grants, 3641
PHF Penn Faculty Research Fellowships, 3932
PHF Regional Mellon Faculty Research and Cultural Fellowships, 3933
PHF Undergraduate Humanities Forum Research Fellowships, 3934
RAND Corporation Transatlantic Postdoctoral Fellowships for Int'l Relations & Security, 4032
Rhode Island State Government Internships, 4062
Robert A. Levy Fellowship in Law and Liberty, 4070
SSHRC Insight Development Grants, 4409
SSHRC Insight Grants, 4410
SSHRC Jules and Gabrielle Leger Fellowship, 4413
SSRC Int'l Dissertation Research Fellowships, 4429

Political Science Education
Center for the Study of Democratic Politics Fellowships, 1481
Coca-Cola Foundation Grants, 1593
Hatton W. Sumners Foundation for the Study and Teaching of Self Government Grants, 2219
Kosciuszko Foundation Tuition Scholarships, 2574
United States Institute of Peace - Transatlantic Post-Doc Fellowship for International Relations and Security, 4581

Political Theory
APSA Edward S. Corwin Award, 933
Princeton University Center for Human Values Graduate Student Top-Up Prizes, 3990
Princeton University Center for Human Values Political Philosophy Travel Grants, 3993
Robert Schalkenbach Foundation Grants, 4079
SSRC Mellon Mays Predoctoral Research Grants, 4432

Politics
Alexander von Humboldt Foundation German Chancellor Fellowships for Prospective Leaders from the People's Republic of China, 526
Alexander von Humboldt Foundation German Chancellor Fellowships for Prospective Leaders from the Russian Federation, 527
APSA Leonard D. White Award, 937
APSA Leo Strauss Award, 938
APSA Presidency Research Fund Fellowships, 943
APSA Special Fund for the Study of Women and Politics, 947
APSA Warren E. Miller Fellowship in Electoral Politics, 949
Energy Foundation Climate Grants, 1923

Pollution
EPA Regional Healthy Indoor Air Projects for Community Outreach and Education, 1937
EPA Southeastern Multi-Media and Geographic Priority Project Grants, 1939
EPA Surveys, Studies, Research, Investigations, Demonstrations, and Special Purpose Activities Relating to the Clean Air Act, 1940
EPA Temporally Integrated Monitoring of Ecosystems and Long-Term Monitoring Grants, 1942
Illinois Clean Energy Community Foundation Renewable Energy Grants, 2358

Pollution Control
ARO Atomic, Molecular, and Optical Physics Research Contracts, 1005
DOE Energy from Municipal Waste Grants, 1787
Dreyfus Postdoctoral Program in Environmental Chemistry Grant, 1855
EPA Pollution Prevention Grants, 1936
EPA Regional Healthy Indoor Air Projects for Community Outreach and Education, 1937
EPA Surveys, Studies, Research, Investigations, Demonstrations, and Special Purpose Activities Relating to the Clean Air Act, 1940
EPA Temporally Integrated Monitoring of Ecosystems and Long-Term Monitoring Grants, 1942
Illinois Clean Energy Community Foundation Renewable Energy Grants, 2358
Orchard Foundation Grants, 3793
ORISE Environmental Management Participation for the US Army Environmental Center, 3801
ORISE Postgraduate Environmental Management Research Participation at the US Army Environmental Center, 3816
ORISE Student Environmental Management Participation Program for the US Army Environmental Center, 3826
Sioux Falls Area Comm Foundation Spot Grants, 4263

Pollution, Air
Bullitt Foundation Grants, 1384
Carrier Corporation Contributions Grants, 1440
DOT Transit Planning and Research Grants, 1849
EPA Regional Healthy Indoor Air Projects for Community Outreach and Education, 1937
EPA Surveys, Studies, Research, Investigations, Demonstrations, and Special Purpose Activities Relating to the Clean Air Act, 1940
EPA Temporally Integrated Monitoring of Ecosystems and Long-Term Monitoring Grants, 1942
NOAA Climate Dynamics and Experimental Prediction Program Grants, 3345
Orchard Foundation Grants, 3793
Takemi Program in International Health Research and Advanced Training Fellowships, 4491
USDA Small Business Innovation Research Grants - Phase I, 4648
Young Scientists Summer Program, 4903

Pollution, Water
Community Foundation for Greater New Haven Quinnipiac River Fund Grants, 1616
EPA National Estuary Program Grants, 1935
EPA Targeted Watersheds Grants, 1941
Freshwater Future Project Grants, 2056
Great Lakes Protection Fund Grants, 2170
Heineman Foundation for Research, Education, Charitable and Scientific Purposes, 2227
IDEM Section 319(h) Nonpoint Source Program Grants, 2338
MDEQ Beach Monitoring Grants - Inland Lakes, 2731
MDEQ Local Water Quality Monitoring Grants, 2732
Orchard Foundation Grants, 3793
Organization for Tropical Studies Fellowships, 3796
Rohm and Haas Company Grants, 4098
Takemi Program in International Health Research and Advanced Training Fellowships, 4491
USAID Palestinian Community Infrastructure Development Grants, 4600
USDA Special Research Grants, 4650

Polymer Science
American Chemical Society Award in Applied Polymer Science, 663
American Chemical Society Award in Polymer Chemistry, 669
American Chemical Society Petroleum Research Fund Undergraduate Faculty Sabbatical Grants, 693
American Chemical Society Petroleum Research Fund Undergraduate Research Grants, 695
APS Polymer Physics Prize, 976
ARO Synthesis and Processing of Materials Research Contracts, 1016
Beckman Young Investigators Program Grants, 1327
CHF Charles C. Price Fellow in Polymer History, 1539
CHF Glenn E. and Barbara Hodsdon Ullyot Scholarship, 1540
CHF Long-Term Fellowships, 1541
CHF Robert W. Gore Fellowship in Materials Innovation, 1542
CHF Societe de Chimie Industrielle (American Section) Fellowship, 1544
CHF Travel Grants, 1545
NSF Materials Research Grants, 3593
NSF Polymers Research Grants, 3628
USDD Polymeric Organic Materials: Conducting/Semiconducting Polymers Grants, 4772
USDD Polymer Matrix Composites Grants, 4773

Population Biology
NIGMS/NIDCR Evolution of Infectious Diseases Research Grants, 3196
NIGMS Predoctoral Training in Bioinformatics and Computation Biology Grants, 3199
NSF Population & Evolutionary Process Grants, 3629

Population Control
AAAS Roger Revelle Fellowship in Global Stewardship, 38
General Service Foundation Human Rights and Economic Justice Grants, 2108
Lumpkin Family Found Healthy People Grants, 2629
PC Fred H. Bixby Fellowships, 3880
William and Charlotte Parks Foundation for Animal Welfare Grants, 4872

Population Studies
Compton Foundation Grants Program, 1623
George Frederick Jewett Foundation Grants, 2116
NCI Academic-Industrial Partnerships for Devel & Validation of In Vivo Imaging Systems and Methods for Cancer Investigations, 2913
NCI Application and Use of Transformative Emerging Technologies in Cancer Research, 2914
NCI Cancer Education and Career Development Program, 2917
RAND Corporation Fellowships in Population Studies and the Study of Aging, 4029
Richard and Rhoda Goldman Fund Grants, 4063
Takemi Program in International Health Research and Advanced Training Fellowships, 4491
USDC Dissertation Fellowships, 4665
USDC Postdoctoral Research Grants, 4669
USDC Presidential Management Fellowships, 4670
Young Scientists Summer Program, 4903

Portugal
IES Faculty Fellowships, 2341
NL American Society for Eighteenth-Century Studies Fellowships, 3279
NL Arthur and Lila Weinberg Fellowships for Independent Scholars and Researchers, 3280

NL Associated Colleges of the Midwest/Great Lakes College Association Faculty Fellowships, 3281
NL Audrey Lumsden-Kouvel Fellowships, 3282
NL British Academy Fellowship for Study in the United Kingdom, 3283
NL Frances C. Allen Fellowships, 3284
NL Grad Scholar-in-Residence Opportunities, 3285
NL Herzog August Bibliothek Wolfenbuttel Fellowship, 3286
NL Lawrence Lipking Fellowship at Newberry, 3287
NL Lester J. Cappon Fellowship in Documentary Editing, 3288
NL Mellon Postdoctoral Research Fellowships, 3294
NL Midwest Modern Lang Assoc Fellowship, 3299
NL Monticello College Foundation Fellowship for Women, 3304
NL National Endowment for the Humanities Fellowships, 3310
NL Northeast Modern Language Assocation Fellowship, 3314
NL Scholar-in-Residence Opportunity, 3315
NL Short-Term Resident Fellowships for Individual Research, 3317
NL South Central Modern Languages Association Fellowship, 3318
NL Susan Kelly Power and Helen Hornbeck Tanner Fellowship, 3319

Portuguese Language/Literature
NL Audrey Lumsden-Kouvel Fellowships, 3282
NL Grad Scholar-in-Residence Opportunities, 3285
NL Herzog August Bibliothek Wolfenbuttel Fellowship, 3286
NL Lester J. Cappon Fellowship in Documentary Editing, 3288
NL National Endowment for the Humanities Fellowships, 3310
Princeton University Library Research Grants, 3997

Poultry Science
National Turkey Federation Research Award, 2887
National Wild Turkey Federation Research Grant Program, 2893

Poverty and the Poor
Achelis Foundation Grants, 274
Alice Tweed Tuohy Foundation Grants Program, 569
Allen Foundation Educational Nutrition Grants, 570
Ann Arbor Area Community Foundation Grants, 860
Avon Foundation Breast Care Fund Grants, 1274
Beatrice Laing Trust Grants, 1324
Bill and Melinda Gates Foundation Agricultural Development Grants, 1345
Bill and Melinda Gates Foundation Financial Services for the Poor Grants, 1346
Bill and Melinda Gates Foundation Water, Sanitation and Hygiene Grants, 1347
Bodman Foundation Grants, 1356
Breast Cancer Fund Grants, 1369
Burden Trust Grants, 1388
Carnegie Corporation of New York Grants, 1432
Charles Stewart Mott Foundation Anti-Poverty Program, 1528
Charles Stewart Mott Foundation Grants, 1529
Chestnut Hill Charitable Foundation, Inc Grants, 1538
DHHS Independence Demonstration Program, 1757
Elizabeth Glaser Int'l Leadership Awards, 1906
Emma B. Howe Memorial Foundation Grants, 1920
Energy Foundation Power Grants, 1924
Farmers Insurance Corporate Giving Grants, 1981
General Service Foundation Human Rights and Economic Justice Grants, 2108
George Gund Foundation Grants, 2117
Hilda and Preston Davis Foundation Grants, 2266
IIE David L. Boren Fellowships, 2350
Joint Center for Poverty Research Grants, 2511
Joint Center for Poverty Res Visiting Scholars, 2512
Jovid Foundation Grants, 2518
Luther I. Replogle Foundation Grants, 2631
Lynde & Harry Bradley Foundation Fellowships, 2634
Lynde and Harry Bradley Foundation Grants, 2635
MAP International Medical Fellowships, 2655
PC Fred H. Bixby Fellowships, 3880
Playboy Foundation Grants, 3980
Robert Schalkenbach Foundation Grants, 4079
S.E.VEN Fund Annual Grants, 4153
S.E.VEN Fund Mini-Grants, 4154
S.E.VEN Fund Open Enterprise Solutions to Poverty RFP Grants, 4155
S.E.VEN Fund WHY Prize, 4156
Shastri Indo-Canadian Institute Collaborative Research Project Grants, 4220
Sidgmore Family Foundation Grants, 4252
Sioux Falls Area Comm Foundation Spot Grants, 4263
Starr Foundation Grants, 4447
Surdna Found Sustainable Environments Grants, 4471
The World Food Prize, 4503
United Methodist Committee on Relief Global Health Grants, 4577
USDA Agriculture and Food Research Initiative Competitive Grants - Food Security, 4614
USDA Scientific Cooperation Research Program, 4647
Vanguard Public Foundation Grant Funds, 4808
W.K. Kellogg Foundation Healthy Kids Grants, 4822
William G. McGowan Charitable Fund Grants, 4877

Precipitation
DOE Program for Ecosystem Research, 1810
NOAA Collaborative Science, Technology, and Applied Research Program Grants, 3354
NSF Hydrologic Sciences Research Grants, 3578

Pregnancy
Allen Foundation Educational Nutrition Grants, 570
General Service Foundation Human Rights and Economic Justice Grants, 2108
NHLBI Prematurity and Respiratory Outcomes Program Grants, 3048
NIH Research on Sleep and Sleep Disorders, 3236
Orchard Foundation Grants, 3793
USAID Comprehensive District-Based Support for Better HIV/TB Patient Outcomes Grants, 4595
USAID Family Planning and Reproductive Health Methods Grants, 4598

Premature Births
BWF Preterm Birth Initiative Grants, 1399
NHLBI Prematurity and Respiratory Outcomes Program Grants, 3048

Presbyterian Church
Barra Foundation Project Grants, 1301
Booth-Bricker Fund Grants, 1359
MacDonald-Peterson Foundation Grants, 2649

Preschool Education
Bodenwein Public Benevolent Foundation Grants, 1355
Booth-Bricker Fund Grants, 1359
Boston Foundation Grants, 1361
Carnegie Corporation of New York Grants, 1432
Chestnut Hill Charitable Foundation, Inc Grants, 1538
College of Human Sciences Scholarships, 1596
Cowles Charitable Trust Grants, 1652
Eugene M. Lang Foundation Grants, 1957
FCD Child Development Grants, 1984
George Foundation Grants, 2115
Hutchinson Community Foundation Grants, 2310
Johnson & Johnson Corporate Contributions, 2508
Oppenstein Brothers Foundation Grants, 3792
San Antonio Area Foundation Grants, 4163
Sid W. Richardson Foundation Grants, 4253

Preventive Medicine
Alexander and Margaret Stewart Trust Grants, 514
ALVRE Casselberry Award, 585
ALVRE Grant, 586
ALVRE Seymour R. Cohen Award, 587
Alzheimer's Association Conference Grants, 589
Alzheimer's Association Development of New Cognitive and Functional Instruments Grants, 590
Alzheimer's Association Everyday Technologies for Alzheimer Care Grants, 591
Alzheimer's Association Mentored New Investigator Research Grants to Promote Diversity, 593
Alzheimer's Association Neuronal Hyper Excitability and Seizures in Alzheimer's Disease Grants, 594
Alzheimer's Association New Investigator Research Grants to Promote Diversity, 596
Alzheimer's Association U.S.-U.K. Young Investigator Exchange Fellowships, 597
Alzheimer's Association Zenith Fellows Awards, 598
APHL Emerging Infectious Diseases Fellowships, 914
Banfi Vintners Foundation Grants, 1296
BCBSM Foundation Primary and Clinical Prevention Grants, 1315
California Wellness Foundation Work and Health Grants, 1409
CDC Grants for Violence-Related Injury Prevention Research, 1463
CDC Preventive Med Residency & Fellowship, 1465
CDC Public Health Conference Support Grant Program, 1466
CMA Foundation Grants, 1582
CNIB Baker Applied Research Fund Grants, 1588
CNIB Baker New Researcher Fund Grants, 1589
CNIB Barbara Tuck MacPhee Award, 1590
CNIB Canada Glaucoma Clinical Research Council Grants, 1591
CNIB E. & Mary Hochhausen Access Technology Research Grants, 1592
Community Found AIDS Endowment Awards, 1614
Daniel Mendelsohn New Investigator Award, 1717
Emma B. Howe Memorial Foundation Grants, 1920
Fannie E. Rippel Foundation Grants, 1978
Foundation for Seacoast Health Grants, 2041
Health Foundation of Southern Florida Responsive Grants, 2223
Johnson & Johnson Corporate Contributions, 2508
Leahi Fund, 2599
NCI/NIDA Transdisciplinary Tobacco Use Research Centers Grants, 2912
NIA Health Behaviors and Aging Grants, 3098
NIAMS Pilot & Feasibility Clinical Grants in Arthritis & Musculoskeletal & Skin Diseases, 3111
NIDA Mentored Clinical Scientist Research Career Development Awards, 3129
NIDDK Type 2 Diabetes in the Pediatric Population Research Grants, 3188
NIGMS/NIDCR Evolution of Infectious Diseases Research Grants, 3196
ORISE Faculty Internship at the US Army Center for Health Promotion and Preventive Medicine, 3802
ORISE Faculty Research Participation at the Centers for Disease Control and Prevention, 3803
ORISE Postgraduate Internship at the US Army Center for Health Promotion and Preventive Med, 3818
ORISE Postgraduate Research Participation at the Centers for Disease Control and Prevention, 3820
ORISE Student Internship Program at the US Army Center for Health Promotion and Preventive Medicine, 3827
ORISE Student Research Participation at the Centers for Disease Control and Prevention, 3829
Self Foundation Grants, 4197
Sid W. Richardson Foundation Grants, 4253
Wilhelm Sander Foundation Grants, 4868

Primary Care Services
ACS Cancer Control Career Development Awards for Primary Care Physicians, 305
George Foundation Grants, 2115
Health Foundation of Southern Florida Responsive Grants, 2223
Josiah Macy Jr. Foundation Grants, 2516
NCI/NIDA Transdisciplinary Tobacco Use Research Centers Grants, 2912
RACGP Chris Silagy Research Scholarship, 4012
Union Pacific Foundation Health and Human Services Grants, 4569
W.K. Kellogg Foundation Healthy Kids Grants, 4822

Primatology
Franklin Mosher Baldwin Fellowships, 2050
NIH Nonhuman Primate Immune Tolerance Cooperative Study Group Grants, 3228
OTS Francis J. Bossuyt Fellowship, 3855

Print Media
APSA Congressional Fellowships for Communications Scholars and Journalists, 926
APSA Congressional Fellowships for International Scholars, 928
APSA Congressional Health and Aging Policy Fellowships, 931
NL Lester J. Cappon Fellowship in Documentary Editing, 3288
OSF Soros Justice Media Fellowships, 3851
WHOI Ocean Science Journalism Fellowship, 4862

Printing
Ahmanson and Getty Postdoctoral Fellowships, 377
APHA Fellowship in Printing History, 913
ASECS/Clark Fellowships, 1057
Houghton Library Visiting Fellowships, 2284
IES Faculty Fellowships, 2341
NL American Society for Eighteenth-Century Studies Fellowships, 3279
NL Arthur and Lila Weinberg Fellowships for Independent Scholars and Researchers, 3280
NL Associated Colleges of the Midwest/Great Lakes College Association Faculty Fellowships, 3281
NL British Academy Fellowship for Study in the United Kingdom, 3283
NL Frances C. Allen Fellowships, 3284
NL Herzog August Bibliothek Wolfenbuttel Fellowship, 3286
NL Lawrence Lipking Fellowship at Newberry, 3287
NL Lester J. Cappon Fellowship in Documentary Editing, 3288
NL Lloyd Lewis Fellowships in American Hist, 3289
NL Mellon Postdoctoral Research Fellowships, 3294
NL Midwest Modern Lang Assoc Fellowship, 3299
NL Monticello College Foundation Fellowship for Women, 3304
NL National Endowment for the Humanities Fellowships, 3310
NL Newberry Library/Ecole Nationale des Chartes Exchange Fellowship, 3313
NL Northeast Modern Language Assocation Fellowship, 3314
NL Short-Term Resident Fellowships for Individual Research, 3317
NL South Central Modern Languages Association Fellowship, 3318
NL Susan Kelly Power and Helen Hornbeck Tanner Fellowship, 3319

Printmaking
ACC Residency Program in Asia, 267
ACC Starr Foundation Fellowship Program, 268
CICF Efroymson Contemporary Arts Fellows, 1563
Clark Library Short-Term Resident Fellowships, 1577
Pollock-Krasner Foundation Grants, 3981

Prion Disease
CDC Cooperative Agreement for Continuing Enhanced National Surveillance for Prion Diseases in the United States, 1456

Prison Reform
OSF Human Rights and Governance Internships, 3845
OSF Soros Justice Advocacy Fellowships, 3850
OSF Soros Justice Media Fellowships, 3851

Prisoners
Achelis Foundation Grants, 274
ASA/BJS Statistical Methodological Research Program Grants, 1029
Bodman Foundation Grants, 1356
Jacob and Valeria Langeloth Foundation Grants, 2455
OSF Soros Justice Advocacy Fellowships, 3850

Prisons
OSF Soros Justice Advocacy Fellowships, 3850
OSF Soros Justice Media Fellowships, 3851

Private and Parochial Education
American Montessori Society Thesis and Dissertation Awards, 758
Coca-Cola Foundation Grants, 1593
Dept of Ed Jacob K. Javits Gifted and Talented Students Education Program-National Research and Development Center, 1746
F.W. Olin Foundation Construction Grant, 1972
Joyce Foundation Education Grants, 2519
Kenneth T. and Eileen L. Norris Found Grants, 2554
Lilly Endowment Teacher Creativity Fellowships, 2616
Lynde & Harry Bradley Foundation Fellowships, 2634
Lynde and Harry Bradley Foundation Grants, 2635
MGN Family Foundation Grants, 2754
Strake Foundation Grants, 4452
T. James Kavanagh Foundation Grants, 4489
Ursula Thrush Peace Seed Grants, 4592

Probability
Norwegian Acad of Science & Letters Abel Prize, 3433
NSF Probability Research Grants, 3634
USDD Applied & Computational Anal Grants, 4683
USDD Information Integration Grants, 4741

Problem Solving
AFRL/NRC Resident Research Associateship Program, 358
BFGoodrich Collegiate Inventors Prize, 1341
Charles Stewart Mott Foundation Grants, 1529
Durfee Foundation Stanton Fellowship, 1881
NSF Decision, Risk, and Management Science Research Grants, 3550
NSF Grant Opportunities for Academic Liaison with Industry, 3574
NSF Info & Intelligent Systems Core Grants, 3580
NSF Perception, Action & Cognition Grants, 3613
OM Association Outstanding Dissertation Awards Program, 3741
ONR Computational Neuroscience & Biorobotics Grants, 3752
Surdna Found Sustainable Environments Grants, 4471

Process Development Engineering
AIChE Excellence in Process Development Research Award, 409
AIChE Process Development Practice Award, 422
NSF Innovation and Organizational Change, 3581
NSF Interfacial Processes and Thermodynamics Research Grants, 3586
NSF Process and Reaction Engineering Grants, 3635

Process Simulation and Control
ASME Rufus T. Oldenburger Medal, 1161
NSF-NIST Interaction in Chemistry, Materials Research, Molecular Biosciences, Bioengineering, and Chemical Engineering, 3453

Production/Operations Management
NSF Grant Opportunities for Academic Liaison with Industry, 3574
USDA Norman E. Borlaug International Agricultural Science and Technology Fellowships, 4636
USDA Scientific Cooperation Exchange Program, 4646

Professional Associations
AASL Research Grant, 207
ALA Emerald Research Grant Awards, 485
ALA Gale Cengage History Research and Innovation Award, 488
ALA Jan Merrill-Oldham Professional Development Grant, 491
ALA Margaret T. Lane/Virginia F. Saunders Memorial Research Award, 492
ALA Marta Lange/CQ Press Award, 493
ALA NewsBank/Readex Cath Reynolds Award, 494
ALA Ulrich's Serials Librarianship Award, 500
American Academy of Advertising Distinguished Service Award, 607
American Automatic Control Council Control Practice Award, 621
American Automatic Control Council Donald P. Eckman Award, 622
American Automatic Control Council Hugo Schuck Award, 623
American Automatic Control Council John R. Ragazzini Education Award, 624
American Automatic Control Council Richard E. Bellman Control Heritage Award, 625
Animal Behavior Society Latin American Travel Awards, 852
AOCS A. Richard Baldwin Award, 871
AOCS ACI/NBB Innovation Research Award, 872
AOCS Analytical Division Student Award, 874
AOCS Australasian Award for Scientific Excellence in Lipid Research, 875
AOCS Biotechnology Division Student Awards, 876
AOCS Edible Applications Technology Division Outstanding Achievement Award, 877
AOCS Edible Applications Technology Division Student Award, 878
AOCS George Schroepfer Medal, 879
AOCS Hans Kaunitz Award, 880
AOCS Health & Nutrition Div Student Award, 881
AOCS Health and Nutrition Poster Competition, 882
AOCS Herbert J. Dutton Award, 883
AOCS Industrial Oil Products Division Student Award, 885
AOCS Lipid Oxidation and Quality Division Poster Competition, 886
AOCS Manuchehr Eijadi Award, 887
AOCS Peter and Claire G. Kalustian Award, 888
AOCS Processing Distinguished Service Award, 890
AOCS Processing Division Student Award, 891
AOCS Protein and Co-Products Division Student Poster Competition, 892
AOCS Ralph Holman Lifetime Award, 894
AOCS Samuel Rosen Memorial Award, 895
AOCS Stephen S. Chang Award, 896
AOCS Surfactants and Detergents Division Student Award, 898
AOCS Thomas H. Smouse Memorial Fellowship, 899
AOCS Timothy L. Mounts Award, 900
AOCS Uses of Soybean Oil Award, 901
AOCS Young Scientist Research Award, 902
ASME Allan Kraus Thermal Management Medal, 1104
ASME Arthur L. Williston Medal, 1105
ASME Barnett-Uzgiris Product Safety Award, 1106
ASME Ben C. Sparks Medal, 1107
ASME Bergles-Rohsenow Young Investigator Award in Heat Transfer, 1108
ASME Bernard F. Langer Nuclear Codes and Standards Award, 1109
ASME Blackall Machine Tool Award, 1110
ASME Burt L. Newkirk Award, 1111
ASME Calvin W. Rice Lecture Award, 1112
ASME Charles Russ Richards Memorial Award, 1113
ASME Daniel C. Drucker Medal, 1114
ASME Dixy Lee Ray Award, 1115
ASME Edward F. Obert Award, 1116
ASME Edwin F. Church Medal, 1117
ASME Engineer-Historian Award, 1118
ASME Fluids Engineering Award, 1119
ASME Frank Kreith Energy Award, 1120
ASME Frank von Flue Award, 1121
ASME Freeman Scholar Award, 1122
ASME Gas Turbine Award, 1123
ASME George Westinghouse Medals, 1124
ASME Gustus L. Larson Memorial Award, 1126
ASME H.R. Lissner Award, 1127
ASME Heat Transfer Memorial Award, 1128
ASME Henry Hess Award, 1129
ASME Henry R. Worthington Medal, 1130
ASME Internal Combustion Engine Award, 1132
ASME J. Hall Taylor Medal, 1133
ASME J.P. Hartog Award, 1134
ASME James Harry Potter Gold Medal, 1135

ASME James N. Landis Medal, 1136
ASME Kate Gleason Award, 1137
ASME M. Eugene Merchant Manuf Medal, 1139
ASME Machine Design Award, 1140
ASME Marshall B. Peterson Award, 1141
ASME Mayo D. Hersey Award, 1142
ASME McDonald Mentoring Award, 1143
ASME Medal, 1144
ASME Melville Medal, 1145
ASME Mel R. Green Codes & Standards Medal, 1146
ASME Milton C. Shaw Manufacturing Research Medal, 1147
ASME Nadai Medal, 1148
ASME Nancy DeLoye Fitzroy and Roland V. Fitzroy Medal, 1149
ASME Old Guard Early Career Award, 1150
ASME Patrick J. Higgins Award, 1152
ASME Per Bruel Gold Medal for Noise Control and Acoustics, 1153
ASME Performance Test Codes Medal, 1154
ASME Pi Tau Sigma Gold Medal, 1155
ASME Prime Movers Committee Award, 1156
ASME Ralph Coats Roe Medal, 1158
ASME Robert E. Koski Medal, 1159
ASME Robert Henry Thurston Lecture Award, 1160
ASME Rufus T. Oldenburger Medal, 1161
ASME Ruth and Joel Spira Outstanding Design Educator Award, 1162
ASME S.Y. Zamrik PVP Medal, 1163
ASME Soichiro Honda Medal, 1164
ASME Spirit of St. Louis Medal, 1165
ASME Student Section Advisor Award, 1166
ASME Thomas A. Edison Patent Award, 1167
ASME Timoshenko Medal, 1168
ASME Van C. Mow Medal, 1169
ASME Warner T. Koiter Medal, 1170
ASME William T. Ennor Manufacturing Technology Award, 1171
ASME Worcester Reed Warner Medal, 1172
ASME Y.C. Fung Young Investigator Award, 1173
Hospital Libraries Section / MLA Professional Development Grants, 2283
IRA Albert J. Harris Awards, 2405
IRA Elva Knight Research Grant, 2406
IRA Esther Zolt Academic Research Award, 2407
IRA Esther Zolt Teacher Research Award, 2408
IRA Helen M. Robinson Dissertation Res Award, 2409
IRA Outstanding Dissertation of the Year Award, 2411
IRA Pearson Foundation-IRA-Rotary Awards, 2412
IRA Steven A. Stahl Research Grants, 2413
IRA Teacher as Researcher Grants, 2414
Pfizer Medical Education Track Two Grants, 3927
Phi Upsilon Omicron Alumni Research Grant, 3940
Phi Upsilon Omicron Candle Fellowships, 3941
Phi Upsilon Omicron Diamond Anniversary Fellowships, 3942
Phi Upsilon Omicron Florence Fallgatter Distinguished Service Award, 3943
Phi Upsilon Omicron Founders Fellowship, 3944
Phi Upsilon Omicron Geraldine Clewell Doctoral Fellowship, 3945
Phi Upsilon Omicron Geraldine Clewell Masters Fellowship, 3946
Phi Upsilon Omicron Jean Dearth Dickerscheid Fellowship, 3947
Phi Upsilon Omicron Margaret Drew Alpha Fellowship, 3948
Phi Upsilon Omicron Presidents Res Fellowship, 3949
Scientific Communities Initiative Grants, 4189
SfN Chapters Postdoct Trainee Travel Awards, 4203
SfN Travel Awards for the International Brain Research Organization World Congress, 4215
SfN Travel Awards for the Japan Neuroscience Society Meeting, 4216

Professional Development

AASL Research Grant, 207
ACLS New Faculty Fellows, 290
ALA Jan Merrill-Oldham Professional Development Grant, 491
ALA NewsBank/Readex Cath Reynolds Award, 494
ANA/Grass Foundation Award in Neuroscience, 839
ANA Wolfe Neuropathy Research Prize, 840
AOCS A. Richard Baldwin Award, 871
AOCS ACI/NBB Innovation Research Award, 872
AOCS Analytical Division Student Award, 874
AOCS Australasian Award for Scientific Excellence in Lipid Research, 875
AOCS Biotechnology Division Student Awards, 876
AOCS Edible Applications Technology Division Outstanding Achievement Award, 877
AOCS Edible Applications Technology Division Student Award, 878
AOCS George Schroepfer Medal, 879
AOCS Hans Kaunitz Award, 880
AOCS Health & Nutrition Div Student Award, 881
AOCS Health and Nutrition Poster Competition, 882
AOCS Herbert J. Dutton Award, 883
AOCS Industrial Oil Products Division Student Award, 885
AOCS Lipid Oxidation and Quality Division Poster Competition, 886
AOCS Manuchehr Eijadi Award, 887
AOCS Processing Distinguished Service Award, 890
AOCS Processing Division Student Award, 891
AOCS Protein and Co-Products Division Student Poster Competition, 892
AOCS Ralph Holman Lifetime Award, 894
AOCS Samuel Rosen Memorial Award, 895
AOCS Stephen S. Chang Award, 896
AOCS Surfactants and Detergents Division Student Award, 898
AOCS Thomas H. Smouse Memorial Fellowship, 899
AOCS Timothy L. Mounts Award, 900
AOCS Uses of Soybean Oil Award, 901
AOCS Young Scientist Research Award, 902
ASME Arthur L. Williston Medal, 1105
ASME Edward F. Obert Award, 1116
ASME Edwin F. Church Medal, 1117
ASME Frank von Flue Award, 1121
ASME Medal, 1144
ASME Per Bruel Gold Medal for Noise Control and Acoustics, 1153
ASME Student Section Advisor Award, 1166
ASME Van C. Mow Medal, 1169
CCA Alberta Creative Development Initiative: Grants to Individuals, 1446
CDECD Elizabeth L. Mahaffey Arts Administration Fellowship, 1468
CFR Edward R. Murrow Press Fellowship, 1500
CFR International Affairs Fellowships, 1501
CFR International Affairs Fellowships in Nuclear Security, 1502
CFR Int'l Affairs Fellowships in South Korea, 1503
CFR International Affairs in Japan Fellowships, 1504
CFR Military Fellowships, 1505
CFR Stanton Nuclear Security Fellowships, 1506
Chicago Community Trust Fellowships, 1546
CSTE CDC/CSTE Applied Epidemiology Fellowships, 1664
DHHS Emerging Leaders Program Internships, 1756
FAMRI Young Clinical Scientist Awards, 1976
Grass Foundation Marine Biological Laboratory Advanced Imaging Fellowships, 2162
Hearst Foundations Health Grants, 2225
Hospital Libraries Section / MLA Professional Development Grants, 2283
IBRO Regional Grants for Int'l Fellowships to U.S. Laboratory Summer Neuroscience Courses, 2330
IRA Albert J. Harris Awards, 2405
IRA Elva Knight Research Grant, 2406
IRA Steven A. Stahl Research Grants, 2413
IRA Teacher as Researcher Grants, 2414
JSA Junior Scientist Travel Support, 2523
Kennedy Center Summer Fellowships for International Arts Managers, 2553
MBL Baxter Postdoctoral Summer Fellowship, 2695
MBL E.E. Just Summer Fellowship for Minority Scientists, 2697
MBL Eugene and Millicent Bell Fellowships, 2699
MBL Frederik B. and Betsy G. Bang Summer Fellowships, 2702
MBL H. Keffer Hartline and Edward F. MacNichol, Jr. Fellowships, 2705
MBL Plum Found John E. Dowling Fellowships, 2713
MBL Robert Day Allen Summer Fellowship, 2714
MBL William Townsend Porter Summer Fellowships for Minority Investigators, 2716
MLA Cunningham Memorial International Fellowship, 2783
National Endowment for the Arts Literature Fellowships: Creative Writing, 2865
NL Associated Colleges of the Midwest/Great Lakes College Association Faculty Fellowships, 3281
NL Monticello College Foundation Fellowship for Women, 3304
Pfizer Medical Education Track Two Grants, 3927
Phi Kappa Phi Found Love of Learning Awards, 3937
Phi Upsilon Omicron Florence Fallgatter Distinguished Service Award, 3943
Saint Mary's College Fellowships, 4159
SARE Professional Development Grants - Northeast Region, 4176
SfN Chapters Postdoct Trainee Travel Awards, 4203
SfN Janett Rosenberg Trubatch Career Development Award, 4207
SfN Louise Hanson Marshall Specific Recognition Award, 4208
SfN Mika Salpeter Lifetime Achievement Award, 4209
SfN Ralph W. Gerard Prize in Neuroscience, 4211
SfN Swartz Prize for Theoretical and Computational Neuroscience, 4213
SfN Travel Awards for the International Brain Research Organization World Congress, 4215
SfN Travel Awards for the Japan Neuroscience Society Meeting, 4216
SfN Young Investigator Award, 4218
SPIRE Postdoctoral Fellowships, 4345

Professorship

Alexander von Humboldt Found Professorships, 535
Cabot Corporation Foundation Grants, 1406
SPIRE Postdoctoral Fellowships, 4345
Streilein Foundation for Ocular Immunology Visiting Professorships, 4453

Program Evaluation

CDC Preventive Med Residency & Fellowship, 1465
RCPSC Royal College Fellowship for Studies in Medical Education, 4040

Project Management

USAID Devel Innovation Ventures Grants, 4597

Propulsion

ARO Atomic, Molecular, and Optical Physics Research Contracts, 1005
ASEE Army Research Laboratory Postdoctoral Fellowships, 1059
USDD Propulsion Materials Grants, 4778

Prostate Gland

JHMI Patrick C. Walsh Prostate Cancer Research Fund, 2488
NIDDK Cell-Specific Delineation of Prostate and Genitourinary Development, 3140
NIH Biology, Development, and Progression of Malignant Prostate Disease Research Grants, 3206
William G. McGowan Charitable Fund Grants, 4877

Prosthetic Devices

U.S. Department of Education Rehabilitation Engineering Research Centers Grants, 4544

Protein Pathways

AOCS Protein and Co-Products Division Student Poster Competition, 892
NCRR Networks and Pathways Collaborative Research Projects Grants, 2934

Proteins and Macromolecules
AACC International Best Student Paper and Best Student Poster Award, 83
AACC International Student Travel Grant, 97
AACC International Walter Bushuk Graduate Research Award, 99
AOCS Protein and Co-Products Division Student Poster Competition, 892
NCRR Networks and Pathways Collaborative Research Projects Grants, 2934
NHLBI Career Transition Awards, 3021
NHLBI Protein Interactions Governing Membrane Transport in Pulmonary Health Grants, 3050
NIH Structural Bio of Membrane Proteins Grant, 3247
NSF Biomolecular Systems Cluster Grants, 3507

Protestant Church
Center for the Study of Religion Fellowships, 1485
Huie-Dellmon Trust Grants, 2295
Notsew Orm Sands Foundation Grants, 3434

Psychiatry
AACAP Beatrix A. Hamburg Award for the Best New Research Poster by a Child and Adolescent Psychiatry Resident, 55
AACAP Educational Outreach Program for Child and Adolescent Psychiatry Residents, 56
AACAP Educational Outreach Program for General Psychiatry Residents, 57
AACAP Elaine Schlosser Lewis Award for Research on Attention-Deficit Disorder, 59
AACAP George Tarjan Award for Contributions in Developmental Disabilities, 60
AACAP Irving Philips Award for Prevention, 61
AACAP Jeanne Spurlock Lecture and Award on Diversity and Culture, 62
AACAP Klingenstein Third Generation Foundation Award for Research in Depression or Suicide, 66
AACAP Norbert and Charlotte Rieger Award for Scientific Achievement, 69
AACAP Pilot Research Award for Learning Disabilities, Supported by the Elaine Schlosser Lewis Fund, 71
AACAP Rieger Psychodynamic Psychothe Award, 74
AACAP Rieger Service Award for Excellence, 75
AACAP Cancro Academic Leadership Award, 76
AACAP Robinson-Cunningham Award, 77
AACAP Sidney Berman Award for the School-Based Study and Intervention for Learning Disorders and Mental Illness, 78
AACAP Simon Wile Leadership in Consultation Award, 79
AACAP Systems of Care Special Program Scholarships, 81
American Philosophical Society Daland Fellowships in Clinical Investigation, 763
American Psychiatric Association/AACDP Research Mentorship Award, 772
American Psychiatric Association/AstraZeneca Young Minds in Psychiatry International Awards, 773
American Psychiatric Association/Janssen Resident Psychiatric Research Scholars, 774
American Psychiatric Association/Kempf Fund Award for Research Development in Psychobiological Psychiatry, 775
American Psychiatric Association/Lilly Psychiatric Research Fellowship, 776
American Psychiatric Association/Lilly Resident Research Awards, 777
American Psychiatric Association/Merck Co. Early Academic Career Research Award, 778
American Psychiatric Association/SmithKline Beecham Young Faculty Award for Research Development in Biological Psychiatry, 779
American Psychiatric Association/Wyeth Pharmaceuticals M.D./Ph.D. Psychiatric Research Fellowships, 780
American Psychiatric Association Award for Research in Psychiatry, 781
American Psychiatric Assoc Early Career Award, 782
American Psychiatric Association Minority Fellowships Program, 783
American Psychiatric Association Minority Medical Student Fellowship in HIV Psychiatry, 784
American Psychiatric Assoc Minority Medical Student Summer Externship in Addiction Psychiatry, 785
American Psychiatric Association Minority Medical Student Summer Mentoring Program, 786
American Psychiatric Association Program for Minority Research Training in Psychiatry, 787
American Psychiatric Association Research Colloquium for Junior Investigators, 788
American Psychiatric Association Senior Scholar Health Services Research Award, 789
American Psychiatric Association Travel Scholarships for Minority Medical Students, 790
American Psychiatric Foundation Proposals, 791
American Psychiatric Foundation James H. Scully Jr., M.D., Educational Fund Grants, 792
American Psychiatric Foundation Jeanne Spurlock Congressional Fellowship, 793
American Psychiatric Foundation Minority Fellowships, 794
APSAA Fellowships, 920
APSAA Mini-Career Grants, 921
APSAA Research Grants, 922
APSAA Small Beginning Scholar Pre-Investigation Grants, 923
APSAA Small Beginning Scholar Visiting and Consulting Grants, 924
Baltimore Washington Center Adult Psychoanalysis Fellowships, 1293
Baltimore Washington Center Child Psychotherapy Fellowships, 1294
Blowitz-Ridgeway Foundation Grants, 1349
Boston Psychoanalytic Society and Institute Fellowship in Child Psychoanalytic Psychotherapy, 1362
Boston Psychoanalytic Society and Institute Fellowship in Psychoanalytic Psychotherapy, 1363
Chicago Institute for Psychoanalysis Fellowships, 1548
Judith Graham Pool Postdoctoral Research Fellowship in Hemophilia, 2529

Psychobiology
American Psychiatric Association/Kempf Fund Award for Research Development in Psychobiological Psychiatry, 775

Psychodynamics
AACAP Rieger Psychodynamic Psychothe Award, 74
APSAA Fellowships, 920

Psychology
ACLS Charles A. Ryskamp Research Fellowships, 281
ACLS Frederick Burkhardt Residential Fellowships for Recently Tenured Scholars, 288
APA Culture of Service in the Psychological Sciences Award, 904
APA Dissertation Research Awards, 905
APA Distinguished Scientific Award for Early Career Contribution to Psychology, 906
APA Distinguished Scientific Award for the Applications of Psychology, 907
APA Distinguished Scientific Contribution Award, 908
APA Distinguished Service to Psychological Science Award, 909
APA Scientific Conferences Grants, 911
APSAA Fellowships, 920
APSAA Mini-Career Grants, 921
APSAA Research Grants, 922
APSAA Small Beginning Scholar Pre-Investigation Grants, 923
APSAA Small Beginning Scholar Visiting and Consulting Grants, 924
Australasian Institute of Judicial Administration Seed Funding Grants, 1268
Baltimore Washington Center Adult Psychoanalysis Fellowships, 1293
Baltimore Washington Center Child Psychotherapy Fellowships, 1294
Blowitz-Ridgeway Foundation Grants, 1349
Boston Psychoanalytic Society and Institute Fellowship in Child Psychoanalytic Psychotherapy, 1362
Boston Psychoanalytic Society and Institute Fellowship in Psychoanalytic Psychotherapy, 1363
Canada-U.S. Fulbright International Science and Technology Award, 1417
Canada-U.S. Fulbright Senior Specialists Program Grants, 1418
Canada-U.S. Fulbright Traditional Scholars, 1419
Canada Graduate Scholarships and NSERC Postgraduate Scholarships, 1420
Chapin Hall International Fellowships in Children's Policy Research, 1517
Chicago Institute for Psychoanalysis Fellowships, 1548
Edward Bangs Kelley and Elza Kelley Foundation Grants, 1898
Fulbright Alumni Initiatives Awards Grants, 2064
Fulbright Distinguished Chairs Program Grants, 2069
Fulbright European Union Scholar-in-Residence Program Grants, 2071
Fulbright New Century Scholars Program Grants, 2075
Fulbright Occasional Lecturer Program Grants, 2076
Fulbright Specialists Program Grants, 2079
Fulbright Traditional Scholar Program in Europe and Eurasia, 2080
Fulbright Traditional Scholar Program in Sub-Saharan Africa, 2081
Fulbright Traditional Scholar Program in the East Asia/Pacific Region, 2082
Fulbright Traditional Scholar Program in the Near East and North Africa Region, 2083
Fulbright Traditional Scholar Program in the South and Central Asia Region, 2084
Fulbright Traditional Scholar Program in the Western Hemisphere, 2085
Fulbright Visiting Specialists Program: Direct Access to the Muslim World, 2087
Harry Frank Guggenheim Foundation Dissertation Fellowships, 2207
Horowitz Foundation for Social Policy Grants, 2280
Horowitz Found for Social Policy Special Awards, 2281
Institute of Noetic Sciences Research Internships, 2386
Jacobs Research Funds Small Grants Program, 2456
James McKeen Cattell Fund Fellowships, 2462
Martin E.P. Seligman Award for Outstanding Research in Positive Psychology, 2667
National Center for Responsible Gaming Centers of Excellence in Gambling Research Grants, 2850
National Center for Responsible Gaming Early Stage Investigator Grants, 2852
National Center for Responsible Gaming Exploration Grants, 2853
National Center for Responsible Gaming Independent Scientist Grants, 2854
National Center for Resp Gaming Large Grants, 2855
National Center for Responsible Gaming Postdoctoral Fellowships, 2856
National Center for Resp Gaming Seed Grants, 2857
National Center for Responsible Gaming Travel Grants, 2858
NCFR Innovation Grants, 2908
NCFR Ruth Hathaway Jewson Award, 2909
NCIC Canadian Cancer Society Research Grants, 2919
NIH Biobehavioral Research for Effective Sleep, 3205
NSF Arctic Social Sciences Grants, 3492
NSF CISE Collaborative Research in Computational Neuroscience Grants, 3526
NSF SBE Doctoral Dissertation Research Improvement Grants, 3641
NSF Social Psychology Research Grants, 3649
OSF Doctoral Fellows Georgia & Kyrgyzstan, 3839
Regents Professional Opportunity Scholarships, 4049
Schomburg Center Scholars-in-Residence Fellows, 4187
SSHRC Insight Development Grants, 4409
SSHRC Insight Grants, 4410
SSRC New Directions in the Study of Prayer Research Grants, 4433
UICC American Cancer Society International Fellowships for Beginning Investigators, 4555

University of Georgia Counseling and Psychological Services Postdoctoral Fellowship, 4587
USDD Basic Research Initiative Grants, 4691
USDD Workpl Violence in the Military Grants, 4805

Psychology Education
Andrew W. Mellon/ACLS Dissertation Completion Fellowships, 844
ASU Graduate College Martha E. Bernal Memorial Award, 1242
Blowitz-Ridgeway Foundation Early Childhood Development Research Award, 1348
ETS Predoctoral Summer Program, 1954
University of Georgia Counseling and Psychological Services Postdoctoral Fellowship, 4587

Psychology of Aging
Alfred P. Sloan Found Working Longer Grants, 566
NIA Health Behaviors and Aging Grants, 3098

Psychometrics
ETS Predoctoral Summer Program, 1954
USDD Basic Research Initiative Grants, 4691

Psychotherapy
AACAP Rieger Psychodynamic Psychothe Award, 74
APSAA Mini-Career Grants, 921
APSAA Research Grants, 922
APSAA Small Beginning Scholar Pre-Investigation Grants, 923
APSAA Small Beginning Scholar Visiting and Consulting Grants, 924
Baltimore Washington Center Adult Psychoanalysis Fellowships, 1293
Baltimore Washington Center Child Psychotherapy Fellowships, 1294
Boston Psychoanalytic Society and Institute Fellowship in Child Psychoanalytic Psychotherapy, 1362
Boston Psychoanalytic Society and Institute Fellowship in Psychoanalytic Psychotherapy, 1363
Chicago Institute for Psychoanalysis Fellowships, 1548
NIA Mechanisms Underlying the Links between Psychosocial Stress, Aging, the Brain and the Body Grants, 3106

Public Administration
Anderson Foundation Grants, 842
APSA Congressional Fellowships for Communications Scholars and Journalists, 926
APSA Congressional Fellowships for Federal Executives, 927
APSA Congressional Fellowships for Int'l Scholars, 928
APSA Congressional Fellowships for Journalists, 929
APSA Congressional Fellowships for Political Scientists, 930
APSA Congressional Health and Aging Policy Fellowships, 931
APSA Gabriel A. Almond Award, 934
APSA Leonard D. White Award, 937
APSA Presidency Research Fund Fellowships, 943
APSA Robert Wood Johnson Foundation Health Policy Fellowships, 944
APSA Special Fund for the Study of Women and Politics, 947
APSA Warren E. Miller Fellowship in Electoral Politics, 949
Canada-U.S. Fulbright International Science and Technology Award, 1417
Canada-U.S. Fulbright Senior Specialists Program Grants, 1418
Canada-U.S. Fulbright Traditional Scholars, 1419
Carpenter Foundation Grants, 1439
Chautauqua Region Community Found Grants, 1533
Commonwealth Fund/Harvard University Fellowship in Minority Health Policy, 1604
Fulbright/Garcia Robles Grants, 2063
Fulbright Alumni Initiatives Awards Grants, 2064
Fulbright Distinguished Chairs Program Grants, 2069
Fulbright European Union Scholar-in-Residence Program Grants, 2071
Fulbright New Century Scholars Program Grants, 2075
Fulbright Occasional Lecturer Program Grants, 2076
Fulbright Specialists Program Grants, 2079
Fulbright Traditional Scholar Program in Europe and Eurasia, 2080
Fulbright Traditional Scholar Program in Sub-Saharan Africa, 2081
Fulbright Traditional Scholar Program in the East Asia/Pacific Region, 2082
Fulbright Traditional Scholar Program in the Near East and North Africa Region, 2083
Fulbright Traditional Scholar Program in the South and Central Asia Region, 2084
Fulbright Traditional Scholar Program in the Western Hemisphere, 2085
Fulbright Visiting Specialists Program: Direct Access to the Muslim World, 2087
IREX Edmund S. Muskie Graduate Fellowship, 2417
M.D. Anderson Foundation Grants, 2637
National Urban Fellows Program, 2888
SfN Louise Hanson Marshall Specific Recognition Award, 4208
Special Olympics Health Professions Stu Grants, 4332
SSHRC Insight Development Grants, 4409
SSHRC Insight Grants, 4410

Public Affairs
AAB Berlin Prize Fellowships, 48
APSA Edward S. Corwin Award, 933
ASU Graduate College Earl and Lenore Tripke Fellowship, 1239
ASU Graduate College Lattie and Elva Coor Building Great Communities Graduate Fellowship, 1241
Axe-Houghton Foundation Grants, 1287
Banfi Vintners Foundation Grants, 1296
Burlington Northern Santa Fe Foundation Grants, 1389
Caesars Foundation Grants, 1407
Carl and Lily Pforzheimer Foundation Grants, 1427
Carpenter Foundation Grants, 1439
Eugene McDermott Foundation Grants, 1958
Fluor Foundation Grants, 2028
FMC Foundation Grants, 2029
Fulbright/Garcia Robles Grants, 2063
Fulbright Public Policy Initiative Program Grants in Mexico, 2077
Furthermore Grants in Publishing, 2091
Germanistic Society of America Fellowships, 2128
Kellogg Institute for International Studies Course Development Grants, 2537
Kellogg Institute for International Studies Dissertation Year Fellowships, 2538
Kellogg Institute for International Studies Graduate Research Grants, 2542
Kellogg Institute for International Studies Intramural Grants, 2546
Kellogg Institute for International Studies Summer Internships, 2549
Kellogg Institute for International Studies Supplemental Awards for PhD Students, 2550
Kellogg Institute for International Studies Visiting Fellowships, 2551
Kiplinger Program in Public Affairs Journalism Fellowships, 2557
Law and Public Affairs Fellowships at Princeton University, 2588
Lynde & Harry Bradley Foundation Fellowships, 2634
Lynde and Harry Bradley Foundation Grants, 2635
Meadows Foundation Grants, 2736
Meyer Memorial Trust Responsive Grants, 2748
Murphy Institute Center for Ethics and Public Affairs Graduate Fellowships, 2806
Musgrave Foundation Grants, 2807
National Rising Scholars Award to Advance Research on Higher Education for the Public Good, 2883
OAH Ellis W. Hawley Prize, 3715
Princeton University Center for Human Values Laurance S. Rockefeller Visiting Fellowships, 3992
Rajiv Gandhi Foundation Grants, 4025
Ruth and Vernon Taylor Foundation Grants, 4130
Sage Foundation Grants, 4158
Schramm Foundation Grants, 4188
Seaver Institute Grants, 4195
Shell Oil Company Foundation Grants, 4233
Starr Foundation Grants, 4447
Surdna Found Sustainable Environments Grants, 4471
Vernon K. Krieble Foundation Grants, 4809
Weyerhaeuser Company Foundation Grants, 4857
William Blair and Company Foundation Grants, 4875

Public Broadcasting
Alfred P. Sloan Foundation Public Understanding of Science & Tech-Public Television Grants, 557
ITVS Commissioned Funding, 2439
Motorola Foundation Grants, 2796
Musgrave Foundation Grants, 2807
Robert Sterling Clark Foundation Government Accountability Grants, 4080
Textron Corporate Contributions Grants, 4501

Public Education
AAAAI Distinguished Layperson Award, 18
HomeBanc Foundation Grants, 2275
HSBC Contributions Community Grants, 2291
International Human Rights Funders Grants, 2399
Joyce Foundation Education Grants, 2519
Komen Greater NYC Clinical Research Enrollment Grants, 2568
Komen Greater NYC Community Breast Health Grants, 2569
Lilly Endowment Teacher Creativity Fellowships, 2616
NEFE Grants, 2948
NFWF Acres for America Grants, 2996
Sunny and Abe Rosenberg Foundation Grants, 4469

Public Health
AAAS Award for Scientific Freedom and Responsibility, 28
Abdus Salam ICTP/IAEA Sandwich Training Education Grants, 226
AcademyHealth PHSR Research Article of the Year Awards, 248
ACSM Paffenbarger-Blair Fund for Epidemiological Research on Physical Activity Grants, 321
American Chemical Society Corporate Associates Seed Grants, 676
APHL Emerging Infectious Diseases Fellowships, 914
Appalachian Regional Commission Health Care Grants, 915
Arca Foundation Grants, 984
ASHG Public Health Genetics Fellowship, 1086
ASM/CDC Fellowships in Infectious Disease and Public Health Microbiology, 1100
ASU Graduate College Dean's Fellowships, 1236
AUPHA Corris Boyd Scholarships, 1267
BCBSM Frank J. McDevitt, DO, Excellence in Research Awards, 1317
Brinson Foundation Grants, 1372
Bupa Foundation Medical Research Grants, 1386
California Wellness Foundation Work and Health Grants, 1409
Canada-U.S. Fulbright International Science and Technology Award, 1417
Canada-U.S. Fulbright Senior Specialists Program Grants, 1418
Canada-U.S. Fulbright Traditional Scholars, 1419
CDC Cooperative Agreement for the Development, Operation, and Evaluation of an Entertainment Education Program, 1457
CDC Experience Appl Epidemiology Fellowships, 1460
CDC Found Tobacco Network Lab Fellowship, 1462
CDC National Training Information and Exchange Program, 1464
CDC Preventive Med Residency & Fellowship, 1465
CDC Public Health Conference Support Grant Program, 1466
CDC Steven M. Teutsch Prevention Effectiveness Fellowships, 1467
Charles A. King Trust Postdoctoral Research Fellowships, 1518
Colonel Stanley R. McNeil Foundation Grants, 1597

Commonwealth Fund/Harvard University Fellowship in Minority Health Policy, 1604
Commonwealth Fund Small Grants, 1611
Conservation, Food, and Health Foundation Grants for Developing Countries, 1641
Cruise Industry Charitable Foundation Grants, 1660
CSTE CDC/CSTE Applied Epidemiology Fellowships, 1664
DHHS Emerging Leaders Program Internships, 1756
Foundation for Seacoast Health Grants, 2041
Fulbright Alumni Initiatives Awards Grants, 2064
Fulbright Distinguished Chairs Program Grants, 2069
Fulbright European Union Scholar-in-Residence Program Grants, 2071
Fulbright New Century Scholars Program Grants, 2075
Fulbright Occasional Lecturer Program Grants, 2076
Fulbright Specialists Program Grants, 2079
Fulbright Traditional Scholar Program in Europe and Eurasia, 2080
Fulbright Traditional Scholar Program in Sub-Saharan Africa, 2081
Fulbright Traditional Scholar Program in the East Asia/Pacific Region, 2082
Fulbright Traditional Scholar Program in the Near East and North Africa Region, 2083
Fulbright Traditional Scholar Program in the South and Central Asia Region, 2084
Fulbright Traditional Scholar Program in the Western Hemisphere, 2085
Fulbright Visiting Specialists Program: Direct Access to the Muslim World, 2087
Greater Worcester Community Foundation Jeppson Memorial Fund for Brookfield Grants, 2165
Harry Frank Guggenheim Foundation Dissertation Fellowships, 2207
International Fellowships in Medical Education, 2398
IREX Edmund S. Muskie Graduate Fellowship Program, 2417
IWMF Public Health Fellowship, 2446
Mayo Clinic Administrative Fellowship, 2689
Mayo Clinic Admin Fellowship Eau Claire, 2690
McCarthy Family Foundation Grants, 2717
MLA Donald Lindberg Research Fellowship, 2785
NCHS National Center for Health Statistics Postdoctoral Research Appointments, 2910
NCI/NIDA Transdisciplinary Tobacco Use Research Centers Grants, 2912
Nestle Foundation Training Grant, 2983
NIA Mentored Clinical Scientist Research Career Development Awards, 3108
NOAA Climate Variability and Human Health Program Grants, 3347
Northwestern Mutual Foundation Grants, 3430
NYAM Student Essay Grants, 3693
Open Spaces Sacred Places National Awards, 3791
ORISE Student Research Participation at the Agency for Toxic Substances and Disease Registry, 3828
OSF Doctoral Fellows Georgia & Kyrgyzstan, 3839
PC Fred H. Bixby Fellowships, 3880
PhRMA Foundation Health Outcomes Post Doctoral Fellowships, 3950
Postdoctoral Research Fellowships in Gerontological Public Health, 3983
RWJF Pioneer Portfolio Grants, 4150
Samueli Institute Scientific Research Grants, 4162
Shastri Indo-Canadian Institute Collaborative Research Project Grants, 4220
Sigma Theta Tau International Doris Bloch Research Award, 4257
Siragusa Foundation Health Services & Medical Research Grants, 4264
Special Olympics Health Professions Stu Grants, 4332
Susan G. Komen Breast Cancer Foundation Challenge Grants: Career Catalyst Research, 4476
Takemi Program in International Health Research and Advanced Training Fellowships, 4491
United Methodist Committee on Relief Global Health Grants, 4577
Yale University Beinecke Library Research Fellowships for Yale Grad & Professional School Students, 4902

Public Planning/Policy

AAAS Award for Scientific Freedom and Responsibility, 28
AAAS Science and Technology Policy Fellowships: Congressional, 39
AAAS Science and Technology Policy Fellowships: Diplomacy, Security, and Development, 40
AAB Berlin Prize Fellowships, 48
Achelis Foundation Grants, 274
Alexander von Humboldt Foundation German Chancellor Fellowships for Prospective Leaders from the People's Republic of China, 526
Alexander von Humboldt Foundation German Chancellor Fellowships for Prospective Leaders from the Russian Federation, 527
Allen Lane Foundation Grants, 571
American Academy of Arts and Sciences Hellman Fellowships, 612
American Academy of Arts and Sciences Visiting Scholars, 613
American Historical Association Fellowship in Aerospace History, 717
American Sociological Association Sydney S. Spivack Program in Applied Social Research and Social Policy, 806
amfAR Fellowships, 809
amfAR Mathilde Krim Fellowships in Basic Biomedical Research, 810
amfAR Research Grants, 811
APSA Gabriel A. Almond Award, 934
APSA Harold D. Lasswell Award, 935
Arca Foundation Grants, 984
Asian Pacific American Institute for Congressional Studies Fellowships, 1091
ASU Graduate College Lattie and Elva Coor Building Great Communities Graduate Fellowship, 1241
Bernard L. Schwartz Fellows Program, 1334
Bodman Foundation Grants, 1356
Bydale Foundation Grants, 1400
Carnegie Corporation of New York Grants, 1432
CDC Steven M. Teutsch Prevention Effectiveness Fellowships, 1467
Center for Defense Information Internships, 1473
Center for the Study of Professional Military Ethics Resident Fellowship, 1484
Century Foundation Grants, 1489
Chapin Hall International Fellowships in Children's Policy Research, 1517
Charles H. Revson Foundation Grants, 1524
Charles Stewart Mott Foundation Grants, 1529
CHF Glenn E. and Barbara Hodsdon Ullyot Scholarship, 1540
CHF Long-Term Fellowships, 1541
CHF Robert W. Gore Fellowship in Materials Innovation, 1542
CHF Societe de Chimie Industrielle (American Section) Fellowship, 1544
CHF Travel Grants, 1545
CIGNA Foundation Grants, 1568
Claude Pepper Foundation Grants, 1578
Commonwealth Fund Patient-Centered Coordinated Care Program Grants, 1608
ConocoPhillips Grants Program, 1629
Conservation, Food, and Health Foundation Grants for Developing Countries, 1641
CTCRI Policy Research Grants, 1667
E.L. Wiegand Foundation Grants, 1885
Emma B. Howe Memorial Foundation Grants, 1920
Enterprise Community Partners Rose Architectural Fellowships, 1929
FCD Child Development Grants, 1984
Ford Foundation Peace and Social Justice Grants, 2031
Fulbright/Garcia Robles Grants, 2063
Fulbright Public Policy Initiative Program Grants in Mexico, 2077
General Motors Foundation Grants, 2107
George Gund Foundation Grants, 2117
Gilbert F. White Postdoctoral Fellowships, 2140
Girl's Best Friend Foundation Grants, 2143
Goldsmith Awards Program, 2158
Greenspun Family Foundation Grants, 2173
Harry Frank Guggenheim Foundation Dissertation Fellowships, 2207
Hatton W. Sumners Foundation for the Study and Teaching of Self Government Grants, 2219
Helen Kellogg Institute for International Studies Residential Fellowships, 2230
Herbert Hoover Presidential Library Association Travel Grants, 2242
Herman Kahn Fellowship, 2247
Hilton Hotels Corporate Giving Program Grants, 2268
Horowitz Foundation for Social Policy Grants, 2280
Horowitz Found for Social Policy Special Awards, 2281
HRF Tibor T. Polgar Fellowships, 2289
HUD Community Outreach Partnership Centers Grants, 2292
Ian Axford (New Zealand) Fellowships in Public Policy, 2313
IFT Foundation Grants, 2345
IREX Edmund S. Muskie Graduate Fellowship Program, 2417
IREX International Leadership in Education Program Grants, 2420
IREX Regional Policy Symposium Grants, 2423
IREX US Embassy Policy Specialist Grants, 2426
J.W. Kieckhefer Foundation Grants, 2452
Jacob and Hilda Blaustein Foundation Grants, 2454
Japan Foundation Center for Global Partnership Institutional Grants, 2473
Jewish Federation of Metropolitan Chicago Scholarships, 2486
JM Foundation Grants, 2490
JMO Public Policy Research Grants, 2492
Joel L. Fleishman Civil Society Fellowships, 2495
Joint Center for Poverty Research Grants, 2511
Kellogg Institute for International Studies Collaborative Faculty Grants, 2536
Kellogg Institute for International Studies Course Development Grants, 2537
Kellogg Institute for International Studies Dissertation Year Fellowships, 2538
Kellogg Institute for International Studies Graduate Research Grants, 2542
Kellogg Institute for International Studies Intramural Grants, 2546
Kellogg Institute for International Studies Summer Internships, 2549
Kellogg Institute for International Studies Supplemental Awards for PhD Students, 2550
Kellogg Institute for International Studies Visiting Fellowships, 2551
Library Company of Philadelphia Barra Foundation International Research Fellowships, 2606
Library Company of Philadelphia Prizes, 2608
Library Company of Philadelphia Program in Early American Economy and Society Dissertation-Level Fellowships, 2609
Library Company of Philadelphia Program in Early American Economy and Society One-Month Fellowships, 2610
Library Company of Philadelphia Program in Early American Economy and Society Post-Doctoral Research Fellowships, 2611
Lumina Foundation for Education Grants, 2627
Lynde & Harry Bradley Foundation Fellowships, 2634
Lynde and Harry Bradley Foundation Grants, 2635
M.D. Anderson Foundation Grants, 2637
Marijuana Policy Project Grants, 2661
Marine Corps Historical Center Dissertation Fellowship, 2663
Mary Ingraham Bunting Inst Peace Fellowship, 2671
Massachusetts Foundation for the Humanities Project Grants, 2678
Millipore Foundation Grants, 2775
Motorola Foundation Grants, 2796
National Wildlife Refuge System Centennial Scholarship Program, 2892
NOAA National Sea Grant John A. Knauss Marine Policy Fellowships, 3394
Nordson Corporation Foundation Grants, 3416

NSF Decision, Risk, and Management Science Research Grants, 3550
OAH Richard W. Leopold Prize, 3728
OSF Global Faculty Grants - Track 1, 3841
OSF Human Rights and Governance Internships, 3845
Playboy Foundation Grants, 3980
Praxair Foundation Grants, 3986
Princeton University Center for Human Values Postdoctoral Research Associate in Values and Public Policy, 3994
RAND Corporation Graduate Student Summer Associateships, 4030
Robert Sterling Clark Foundation Government Accountability Grants, 4080
RRF General Program Grants, 4115
Seaver Institute Grants, 4195
Sensient Technologies Foundation Grants, 4199
Shastri Indo-Canadian Institute Millennium Development Research Grants, 4227
Shastri Indo-Canadian Institute Partnership Development Seed Grants, 4228
Smith Richardson Foundation Domestic Public Research Fellowship, 4269
SRC Robert M. Burger Fellowship, 4392
Vernon K. Krieble Foundation Grants, 4809
Weyerhaeuser Company Foundation Grants, 4857
William T. Grant Foundation Scholars Program, 4882

Public Policy Systems Analysis
CDC Steven M. Teutsch Prevention Effectiveness Fellowships, 1467
Independent Institute John M. Templeton Fellowships Essay Contest, 2369
IREX Regional Policy Symposium Grants, 2423
IREX US Embassy Policy Specialist Grants, 2426
Searle Freedom Trust Grants, 4192
Shastri Indo-Canadian Institute Millennium Development Research Grants, 4227
Smith Richardson Foundation Domestic Public Research Fellowship, 4269
WHOI Marine Policy Fellowships, 4861
Yale University Beinecke Library Research Fellowships for Yale Grad & Professional School Students, 4902

Public Relations
AAAS Early Career Award for Public Engagement with Science, 29
Carrier Corporation Contributions Grants, 1440

Public Safety
Bikes Belong Foundation Research Grants, 1343
Caesars Foundation Grants, 1407
Surdna Found Sustainable Environments Grants, 4471
Westinghouse Charitable Giving Grants, 4856

Publication
AAAS Award for Public Engagement with Science, 26
AAAS Early Career Award for Public Engagement with Science, 29
AIChE W. David Smith, Jr. Graduate Student Paper Award, 428
American Academy of Advertising Doctoral Dissertation Competition Awards, 608
AOCS Phospholipid Best Paper Award, 889
APS Dannie Heineman Prize for Mathematical Physics, 951
ASCE Freeman Fellowships, 1040
ASHA Advancing Academic-Research Award, 1072
ASHA Student Research Travel Award, 1074
ASME Heat Transfer Memorial Award, 1128
Astronomical Society of the Pacific Dorothea Klumpke-Roberts Award, 1229
CAA Wyeth Publication Grant, 1405
Furthermore Grants in Publishing, 2091
George A Ohl Jr. Foundation Grants, 2112
Getty Research Institute Nonresidential Grants - Curatorial Research Fellowships, 2136
GUITS Grants for the Publication of Scholarly Books and Journals, 2186
JMO Law and Legal System Grants, 2491
JMO Public Policy Research Grants, 2492
JMO Strategic and International Studies Grants, 2493
Lynde & Harry Bradley Foundation Fellowships, 2634
Lynde and Harry Bradley Foundation Grants, 2635
NAGC Masters and Specialists Award, 2821
NEH Collaborative Research Grants, 2952
NEH Digital Humanities Implementation Grants, 2953
NEH Digital Humanities Start-Up Grants, 2954
NFJC Gantz-Zahler Grant, 2993
NLM Exploratory/Developmental Grants, 3296
NLM Grants for Scholarly Works in Biomedicine and Health, 3298
North Carolina Arts Council Folklife Public Program Grants, 3419
NYSCA Music: Services to the Field Grants, 3708
Pfizer Global Research Awards for Nicotine Independence, 3925
Potts Memorial Foundation Grants, 3985
Rocky Mountain Mineral Law Found Grants, 4094
RSC Bancroft Award, 4118
RSC Jason A. Hannah Medal, 4119
SSHRC-FHSS Awards to Scholarly Publications, 4398
SSSR Distinguished Article Award, 4440
SSSR Student Paper Award, 4441
TIAA-CREF Paul A. Samuelson Award, 4518

Publishing Industry
Center for the Study of Books and Media Visiting Research Fellowships, 1480
Furthermore Grants in Publishing, 2091
ICCS Publishing Fund, 2336
International League of Antiquarian Booksellers Bibliographical Prize, 2400
Munby Fellowship in Bibliography, 2804

Publishing, Electronic
NHPRC Publishing Historical Records Grants, 3075
NLM Grants for Scholarly Works in Biomedicine and Health, 3298
Perl 6 Microgrants, 3899
Scientific Communities Initiative Grants, 4189
SSHRC-FHSS Awards to Scholarly Publications, 4398

Puerto Rico
American Sociological Association Minority Fellowships, 805
APA Planning Fellowships, 910
Chest Found Grant in Venous Thromboembolism, 1537
Consortium for Graduate Study in Management Minority Fellowships, 1642
GEM MS Engineering Fellowships, 2104
GEM PhD Engineering Fellowships, 2105
GEM PhD Science Fellowships, 2106
Hispanic Scholarships, 2270
NMF Need-Based Scholarships, 3326
NOAA California Bay Watershed Education and Training Program Grants, 3338
RWJF Harold Amos Medical Faculty Development Research Grants, 4139

Pulmonary Diseases
AMA Foundation Seed Grants for Research, 601
CDI Postdoctoral Fellowships, 1469
CFF First- and Second-Year Clinical Fellowships, 1492
CFF Postdoctoral Research Fellowships, 1496
CFF Research Grants, 1497
CFF Third-, Fourth-, and Fifth-Year Clinical Fellowships, 1499
Chestnut Hill Charitable Foundation, Inc Grants, 1538
Duke University Adult Cardiothoracic Anesthesia and Critical Care Medicine Fellowships, 1869
Hawaii Community Foundation Health Education and Research Grants, 2221
Jacob and Valeria Langeloth Foundation Grants, 2455
LAM Found Established Investigator Awards, 2581
LAM Foundation Pilot Project Grants, 2582
LAM Foundation Postoctoral Fellowships, 2583
LAM Foundation Research Grants, 2584
Leahi Fund, 2599
M. Bastian Family Foundation Grants, 2636
NHLBI Ancillary Studies in Clinical Trials, 3015
NHLBI Circadian-Coupled Cellular Function in Heart, Lung, and Blood Tissue Grants, 3023
NHLBI Developmental Origins of Altered Lung Physiology and Immune Function Grants, 3025
NHLBI Lymphatics in Health & Disease in Digestive, Cardiovascular and Pulmonary Systems, 3031
NHLBI Lymphatics in Health and Disease in the Digestive, Urinary, Cardiovascular and Pulmonary Systems, 3032
NHLBI Mentored Career Award for Faculty at Institutions that Promote Diversity, 3033
NHLBI Mentored Career Award For Faculty At Minority Institutions, 3034
NHLBI Mentored Career Development Award to Promote Faculty Diversity/Re-Entry in Biomedical Research, 3035
NHLBI Nutrition and Diet in the Causation, Prevention, and Management of Heart Failure Research Grants, 3044
NHLBI Phase II Clinical Trials of Novel Therapies for Lung Diseases, 3047
NHLBI Protein Interactions Governing Membrane Transport in Pulmonary Health Grants, 3050
NHLBI Ruth L. Kirschstein National Research Service Awards for Individual Postdoctoral Fellows, 3057
NHLBI Ruth L. Kirschstein National Research Service Awards for Individual Predoctoral MD/PhD Fellows and Other Dual Degree Fellows, 3059
NHLBI Short-Term Research Education Program to Increase Diversity in Health-Related Research, 3062
NHLBI Training Program Grants for Institutions that Promote Diversity, 3067
NHLBI Translational Grants in Lung Diseases, 3069
NIH Research on Sleep and Sleep Disorders, 3236
Pfizer ASPIRE North America Broad Spectrum Antibiotics for the Treatment of Gram-Negative or Polymicrobial Infections Research Awards, 3914
Pfizer ASPIRE North America Narrow Spectrum Antibiotics for the Treatment of MRSA Research Awards, 3916
Pfizer Medical Education Track Two Grants, 3927

Purchasing/Procurement
ASME Patrick J. Higgins Award, 1152
NAPM Doctoral Dissertation Grant, 2827

Quality of Life
Bikes Belong Foundation Research Grants, 1343
Conquer Cancer Foundation of ASCO International Development and Education Award in Palliative Care, 1634
Donald W. Reynolds Foundation Aging and Quality of Life Grants, 1832
SARE Graduate Student Grants - North Central Region, 4168
SARE Producer Research and Education Grants-Western Region, 4174
SARE Professional + Producer Grants - Western Region, 4175
SARE Research and Education Grants - Western Region, 4180
Westinghouse Charitable Giving Grants, 4856

Quality/Product Control
NCHS National Center for Health Statistics Postdoctoral Research Appointments, 2910
ORISE Environmental Management Participation for the US Army Environmental Center, 3801
ORISE Postgraduate Environmental Management Research Participation at the US Army Environmental Center, 3816
ORISE Student Environmental Management Participation Program for the US Army Environmental Center, 3826
SSHRC-APC Automotive Partnership Canada, 4396

Quantum Cosmology
Abdus Salam ICTP Postdoctoral Positions in Theoretical High Energy Physics, 234

Alfred P. Sloan Foundation Energy and Environment Grants, 552
ISGC Graduate Fellowships, 2430

Quantum Fluids
APS Lars Onsager Prize, 968

Quantum Mechanics
Abdus Salam ICTP Postdoctoral Positions in Theoretical High Energy Physics, 234
ARO Quantum Information Science Research Contracts, 1013
ARO Theoretical Physics and Nonlinear Phenomena Research Contracts, 1017

Rabbinate
Jewish Federation of Metro Chicago Scholarships, 2486

Race Relations
CSRPC Dissertation Fellowships, 1661
CSRPC Postdoc Fellowship for Advanced Scholars, 1662
CSRPC Postdoc Fellowship for Junior Scholars, 1663

Racism
Akonadi Foundation Anti-Racism Grants, 477
Allstate Safe and Vital Communities Grants, 573
Cowles Charitable Trust Grants, 1652
Georgia Power Foundation Grants, 2125
Meadows Foundation Grants, 2736
NEI Ruth L. Kirschstein National Research Service Award Short-Term Institutional Research Training Grants, 2963
OAH James A. Rawley Prize, 3721
Rockefeller Foundation Humanities Fellowships, 4089
Rockefeller Foundation Scholarly Residencies, 4091
Russell Sage Foundation Grants, 4129
RWJF Disparities Research for Change Grants, 4137
Sioux Falls Area Comm Foundation Spot Grants, 4263
Vanguard Public Foundation Grant Funds, 4808

Radar
Arecibo Observatory Grants, 988
NSF National Astronomy and Ionosphere Center Grants, 3598
NSF Upper Atmospheric Facilities Grants, 3665
ONR Marine Meteorology and Atmospheric Effects Research Grants, 3769
USDD Radar and Signal Processing Grants, 4779

Radiation
Abdus Salam ICTP-ELETTRA Users Grants, 221
DOE Environmental Processes Research Grants, 1790
DSO Radiation Biodosimetry Grants, 1864
EPA Surveys, Studies, Research, Investigations, Demonstrations, and Special Purpose Activities Relating to the Clean Air Act, 1940
NCIC Canadian Cancer Society Research Grants, 2919
NSF Arctic Natural Science Grants in Atmospheric Sciences, 3483
NSF Physical and Dynamic Meteorology Grants, 3614
NSF Solar, Heliospheric, and INterplanetary Environment Grants, 3651
NSF Solar Terrestrial Research Grants, 3652
ONR Electronics Division Research Grants, 3759
ORISE Faculty Internship at the US Army Center for Health Promotion and Preventive Medicine, 3802
ORISE Nuclear Regulatory Commission Historically Black Colleges and Universities Student Research Participation, 3814
ORISE Postgraduate Internship at the US Army Center for Health Promotion and Preventive Med, 3818
USDD Ionospheric Modification Grants, 4745

Radiation Effects
Abdus Salam ICTP/IAEA Sandwich Training Education Grants, 226
DSO Radiation Biodosimetry Grants, 1864
EPA Surveys, Studies, Research, Investigations, Demonstrations, and Special Purpose Activities Relating to the Clean Air Act, 1940
Susan G. Komen Breast Cancer Foundation Challege Grants: Investigator Initiated Research, 4475
USDD Ionospheric Modification Grants, 4745

Radiation Instrumentation
Abdus Salam ICTP-ELETTRA Users Grants, 221
DSO Radiation Biodosimetry Grants, 1864

Radio
AAAS Kavli Science Journalism Awards, 33
AIHS Media Summer Fellowship, 453
Alfred P. Sloan Foundation Public Understanding of Science and Technology-Radio Grants, 558
ARO Mobile, Wireless Communications and Networks Research Contracts, 1012
Astronomical Society of the Pacific Dorothea Klumpke-Roberts Award, 1229
CCH Documentary Project Research and Development Grants, 1451
JMO Law and Legal System Grants, 2491
JMO Public Policy Research Grants, 2492
JMO Strategic and International Studies Grants, 2493
NEH Digital Humanities Implementation Grants, 2953
NEH Digital Humanities Start-Up Grants, 2954
North Carolina Arts Council Folklife Public Program Grants, 3419
NSF University Radio Observatories Program, 3664
OAH James A. Rawley Prize, 3721
ONR Radio Frequency Sensor Processing Research Grants, 3778
ONR Radio Frequency Solid State Devices Applied Research Grants, 3779
Union Pacific Foundation Community and Civic Grants, 4568

Radiochemistry
AINSE Research Fellowships, 464
AINSE Research Fellowships with Institute for Science and International Security Support, 465

Radiology
AINSE Research Fellowships, 464
AINSE Research Fellowships with Institute for Science and International Security Support, 465
Society for Imaging Informatics in Medicine Emeritus Mentor Grants, 4305
Society for Imaging Informatics in Medicine Micro Grants, 4306
Society for Imaging Informatics in Medicine Research Grants, 4307
Society for Imaging Informatics in Medicine Small Grant for Product Support Development, 4308
Society for Imaging Informatics in Medicine Small Training Grants, 4309

Ranching
SARE Producer Research and Education Grants-Western Region, 4174
SARE Research and Education Grants - Western Region, 4180

Range Management
USDA Rangeland Research Grants, 4641
USDA Special Research Grants, 4650

Reaction Engineering
NSF Process and Reaction Engineering Grants, 3635

Reading
BBF Fellowships, 1311
Bodenwein Public Benevolent Foundation Grants, 1355
Boston Foundation Grants, 1361
Center for the Study of Books and Media Visiting Research Fellowships, 1480
IRA Albert J. Harris Awards, 2405
IRA Elva Knight Research Grant, 2406
IRA Helen M. Robinson Dissertation Res Award, 2409
IRA Outstanding Dissertation of the Year Award, 2411
IRA Steven A. Stahl Research Grants, 2413
IRA Teacher as Researcher Grants, 2414
NSF Perception, Action & Cognition Grants, 3613
Robert Bowne Foundaion Edmund A. Stanley, Jr. Research Grants, 4073
Talbert and Leota Abrams Foundation, 4492

Reading Education
BBF Fellowships, 1311
DaimlerChrysler Corporation Fund Grants, 1714
IRA Albert J. Harris Awards, 2405
IRA Elva Knight Research Grant, 2406
IRA Esther Zolt Academic Research Award, 2407
IRA Esther Zolt Teacher Research Award, 2408
IRA Jeanne S. Chall Research Fellowship, 2410
IRA Pearson Foundation-IRA-Rotary Awards, 2412
IRA Steven A. Stahl Research Grants, 2413

Real Estate
Appraisal Institute Education Trust Scholarships, 917
Enterprise Community Partners Rose Architectural Fellowships, 1929

Reconstructive Surgery
AAFPRS Ben Shuster Memorial Award, 160
AAFPRS Bernstein Grant, 161
AAFPRS Fellowships, 162
AAFPRS Investigator Development Grant, 163
AAFPRS John Orlando Roe Award, 165
AAFPRS Leslie Bernstein Resident Res Grants, 166
AAFPRS Residency Travel Awards, 167
AAFPRS Sir Harold Delf Gillies Award, 168
Dorr Inst for Arthritis Research & Education, 1843
Plastic Surgery Educational Foundation/AAO-HNSF Combined Grant, 3977
Plastic Surgery Educational Foundation Basic Research Grants, 3978
Plastic Surgery Educational Foundation Research Fellowship Grants, 3979
Ronald McDonald House Charities Grants, 4101

Recreation and Leisure
Alice Tweed Tuohy Foundation Grants Program, 569
Boston Foundation Grants, 1361
Campbell Soup Foundation Grants, 1415
Carl C. Icahn Foundation Grants, 1428
Carrier Corporation Contributions Grants, 1440
Community Foundation for S Arizona Grants, 1618
Dept of Ed Vocational Rehabilitation Services Demonstration and Training Programs, 1751
DOT Recreational Boating Safety Grants, 1848
Emma B. Howe Memorial Foundation Grants, 1920
Frank Stanley Beveridge Foundation Grants, 2051
George A Ohl Jr. Foundation Grants, 2112
George Foundation Grants, 2115
George Kress Foundation Grants, 2119
Gil and Dody Weaver Foundation Grants, 2139
Greater Worcester Community Foundation Jeppson Memorial Fund for Brookfield Grants, 2165
Greenspun Family Foundation Grants, 2173
Harvey Randall Wickes Foundation Grants, 2218
Herbert H. and Grace A. Dow Found Grants, 2241
HRF Hudson River Improvement Grants, 2286
Illinois DNR Youth Recreation Corps Grants, 2361
McCombs Foundation Grants, 2718
Oklahoma City Community Grants, 3737
Rayonier Foundation Grants, 4034
Reinberger Foundation Grants, 4051
Sidgmore Family Foundation Grants, 4252
Sioux Falls Area Community Foundation Community Fund Grants (Unrestricted), 4262
Sioux Falls Area Comm Foundation Spot Grants, 4263
Thomas Austin Finch, Sr. Foundation Grants, 4507

Recreation and Leisure Studies
GCSAA Essay Contest, 2102
GCSAA Watson Fellowships, 2103
MDEQ Beach Monitoring Grants - Inland Lakes, 2731

Recycling
MDEQ Scrap Tire Market Development Grants, 2733

Reference Materials
ALA Marta Lange/CQ Press Award, 493
American Historical Association Waldo J. Leland Prize, 743
ATLA Bibliography Grant, 1257
Clark-Huntington Bibliographical Fellowship, 1576
Getty Research Grants for Institutions - Critical Reference Resources Grants, 2134
Gladys Krieble Delmas Foundation Grants for Venetian Research, 2144
International League of Antiquarian Booksellers Bibliographical Prize, 2400
Munby Fellowship in Bibliography, 2804
NLM Grants for Scholarly Works in Biomedicine and Health, 3298
Virginia Historical Society Research Fellowships, 4816
WAWH Barbara Penny Kanner Award, 4832

Refugees
Allen Lane Foundation Grants, 571
CCIS Visiting Research Fellowships, 1452
HIAS Scholarship Awards, 2264
Jacob and Valeria Langeloth Foundation Grants, 2455
National Book Scholarship Fund, 2849

Regional Planning/Policy
Canada-U.S. Fulbright International Science and Technology Award, 1417
Canada-U.S. Fulbright Senior Specialists Program Grants, 1418
Canada-U.S. Fulbright Traditional Scholars, 1419
Carpenter Foundation Grants, 1439
Compton Foundation Grants Program, 1623
Dorothy Rider Pool Health Care Grants, 1841
Fulbright Alumni Initiatives Awards Grants, 2064
Fulbright Distinguished Chairs Program Grants, 2069
Fulbright European Union Scholar-in-Residence Program Grants, 2071
Fulbright New Century Scholars Program Grants, 2075
Fulbright Occasional Lecturer Program Grants, 2076
Fulbright Specialists Program Grants, 2079
Fulbright Traditional Scholar Program in Europe and Eurasia, 2080
Fulbright Traditional Scholar Program in Sub-Saharan Africa, 2081
Fulbright Traditional Scholar Program in the East Asia/Pacific Region, 2082
Fulbright Traditional Scholar Program in the Near East and North Africa Region, 2083
Fulbright Traditional Scholar Program in the South and Central Asia Region, 2084
Fulbright Traditional Scholar Program in the Western Hemisphere, 2085
Fulbright Visiting Specialists Program: Direct Access to the Muslim World, 2087
Sarkeys Foundation Grants, 4181
Smith Richardson Foundation Domestic Public Research Fellowship, 4269
SSRC African Peacebuilding Network Research Grants, 4422
USDC/NOAA Community-Based Habitat Restoration Program Grants, 4656
USDC/NOAA Community-Based Marine Debris Removal Grants, 4657

Rehabilitation/Therapy
Community Foundation for S Arizona Grants, 1618
Dept of Ed Rehabilitation Fellowships, 1747
Edward W. and Stella C. Van Houten Memorial Fund Grants, 1899
IAFF Burn Foundation Research Grants, 2312
John W. Alden Trust Grants, 2510
Judith Graham Pool Postdoctoral Research Fellowship in Hemophilia, 2529
Massage Therapy Research Fund Grants, 2682
NIA Health Behaviors and Aging Grants, 3098
NIDRR Field-Initiated Projects, 3189
Peabody Foundation Grants, 3886
Potts Memorial Foundation Grants, 3985
Reuter Foundation Grants, 4059
Robert B McMillen Foundation Grants, 4072
Robert E. and Evelyn McKee Foundation Grants, 4074
U.S. Department of Education Disability and Rehabilitation Research and Related Projects, 4538
U.S. Department of Education Rehabilitation Engineering Research Centers Grants, 4544

Rehabilitation/Therapy, Emotional/Social
NCFR Family Therapy Section Best Research Paper Awards, 2907
Robert B McMillen Foundation Grants, 4072
Robert E. and Evelyn McKee Foundation Grants, 4074

Rehabilitation/Therapy, Occupational/Vocational
Achelis Foundation Grants, 274
Bodman Foundation Grants, 1356
Dept of Ed Vocational Rehabilitation Services Demonstration and Training Programs, 1751
Regents Professional Opportunity Scholarships, 4049
U.S. Department of Education Rehabilitation Research Training Centers, 4545

Rehabilitation/Therapy, Physical
PVA Research Foundation Grants, 4004
Regents Professional Opportunity Scholarships, 4049
Robert B McMillen Foundation Grants, 4072

Religion
AAB Berlin Prize Fellowships, 48
Academy of American Franciscan History Dissertation Fellowships, 249
ACLS Charles A. Ryskamp Research Fellowships, 281
Acton Institute for the Study of Religion and Liberty Calihan Academic Fellowships, 331
Acton Institute for the Study of Religion and Liberty Novak Award, 332
Ahn Family Foundation Grants, 379
AIAR Andrew W. Mellon Foundation Fellowships, 392
AIAR Ernest S. Frerichs Fellow and Program Coordinator Grant, 394
AIAR George A. Barton Fellowships, 395
Albert and Margaret Alkek Foundation Grants, 504
Alces Foundation Grants, 510
Alcoa Foundation Grants, 511
Alpha Kappa Alpha Educational Advancement Foundation Fellowships, 575
ALSAM Foundation Grants, 579
American Academy of Religion Collaborative Research Grants, 617
American Foodservice Charitable Trust Grants, 710
Anderson Foundation Grants, 842
Anthony R. Abraham Foundation Grants, 869
Archer Daniels Midland Foundation Grants, 985
Arcus Foundation Religion and Values Grants, 987
Arronson Foundation Grants, 1019
ASR Fichter Research Grant, 1217
ASR Robert J. McNamara Student Award, 1218
ATS Henry Luce III Fellowships, 1258
ATS Lilly Theological Collaborative Res Grants, 1259
ATS Lilly Theological Research Expense Grants, 1261
Banfi Vintners Foundation Grants, 1296
Barra Foundation Project Grants, 1301
Beatrice Laing Trust Grants, 1324
Bodenwein Public Benevolent Foundation Grants, 1355
Bodman Foundation Grants, 1356
Burden Trust Grants, 1388
Burton G. Bettingen Grants, 1390
Byron W. and Alice L. Lockwood Grants, 1401
BYU Religious Studies Center Research Grants, 1402
Calvin Inst of Christian Worship Renewal Grants, 1411
Carl Gellert and Celia Berta Gellert Foundation Grants, 1429
Castle Rock Foundation Grants, 1441
Catholic University Academic Scholarships, Grants, and Awards, 1443
CCCF Dora M Brown Seminary Scholarships, 1448
Center for the Study of Religion Fellowships, 1485
Charles Stewart Mott Foundation Anti-Poverty Program, 1528
Chatlos Foundation Grants Program, 1532
Chiles Foundation Grants, 1552
Cleveland-Cliffs Foundation Grants, 1580
Community Foundation of the Fox River Valley, 1620
Cord Foundation Grants, 1645
Coughlin-Saunders Foundation Grants Program, 1648
Dade Community Foundation Grants, 1712
DaimlerChrysler Corporation Fund Grants, 1714
Donald and Sylvia Robinson Family Grants, 1830
DOS Study of the United States Institutes, 1846
Evjue Foundation Grants, 1963
Frank Stanley Beveridge Foundation Grants, 2051
Fritz B. Burns Foundation Grants, 2057
FTE Expanding Horizons Partnership Grants, 2060
G. Unger Vetlesen Foundation Grants, 2093
George Frederick Jewett Foundation Grants, 2116
George Kress Foundation Grants, 2119
Gilroy and Lillian P. Roberts Charitable Foundation Grants, 2142
Gladys Krieble Delmas Foundation Grants for Venetian Research, 2144
Greater Saint Louis Community Found Grants, 2164
Harold Simmons Foundation Grants, 2203
Harry Frank Guggenheim Foundation Dissertation Fellowships, 2207
Harry Frank Guggenheim Foundation Research Grants, 2208
Harry W. Bass, Jr. Foundation Grants, 2210
Helen Bader Foundation Grants, 2228
Helen Kellogg Institute for International Studies Residential Fellowships, 2230
Helen V. Brach Foundation Grants, 2231
Henry W. Bull Foundation Grants, 2237
Houghton Library Visiting Fellowships, 2284
IES Faculty Fellowships, 2341
Institute for Advanced Studies in Culture Postdoctoral Fellowships, 2380
IREX Short-Term Travel Grants, 2424
Jacob and Hilda Blaustein Foundation Grants, 2454
James M. Collins Foundation Grants, 2461
Jenifer Altman Foundation Grants, 2484
John G. Duncan Trust Grants, 2501
Kellogg Institute for International Studies Collaborative Faculty Grants, 2536
Kellogg Institute for International Studies Course Development Grants, 2537
Kellogg Institute for International Studies Dissertation Year Fellowships, 2538
Kellogg Institute for International Studies Graduate Research Grants, 2542
Kellogg Institute for International Studies Intramural Grants, 2546
Kellogg Institute for International Studies Summer Internships, 2549
Kellogg Institute for International Studies Supplemental Awards for PhD Students, 2550
Kellogg Institute for International Studies Visiting Fellowships, 2551
Kluge Center David B. Larson Fellowship in Health and Spirituality, 2559
LI American Religion Dissertation Fellowship, 2604
LI General Grants, 2613
Lucile Horton Howe and Mitchell B. Howe Foundation Grants, 2624
Lumpkin Family Found Healthy People Grants, 2629
M. Bastian Family Foundation Grants, 2636
Maximilian E. and Marion O. Hoffman Foundation Grants, 2688
McCombs Foundation Grants, 2718
Metanexus Inst Templeton Research Lectures, 2744
MGN Family Foundation Grants, 2754
Musgrave Foundation Grants, 2807
National Foundation for Catholic Youth Grants, 2868
Nell Warren Elkin and William Simpson Elkin Foundation, 2972
NL American Society for Eighteenth-Century Studies Fellowships, 3279
NL Arthur and Lila Weinberg Fellowships for Independent Scholars and Researchers, 3280
NL Associated Colleges of the Midwest/Great Lakes College Association Faculty Fellowships, 3281

NL British Academy Fellowship for Study in the United Kingdom, 3283
NL Frances C. Allen Fellowships, 3284
NL Grad Scholar-in-Residence Opportunities, 3285
NL Herzog August Bibliothek Wolfenbuttel Fellowship, 3286
NL Lawrence Lipking Fellowship at Newberry, 3287
NL Lester J. Cappon Fellowship in Documentary Editing, 3288
NL Lloyd Lewis Fellowships in American Hist, 3289
NL Mellon Postdoctoral Research Fellowships, 3294
NL Midwest Modern Lang Assoc Fellowship, 3299
NL Monticello College Foundation Fellowship for Women, 3304
NL National Endowment for the Humanities Fellowships, 3310
NL Northeast Modern Language Assocation Fellowship, 3314
NL Scholar-in-Residence Opportunity, 3315
NL Short-Term Resident Fellowships for Individual Research, 3317
NL South Central Modern Languages Association Fellowship, 3318
NL Susan Kelly Power and Helen Hornbeck Tanner Fellowship, 3319
Norcliffe Foundation Grants, 3415
Patrick and Anna M. Cudahy Fund Grants, 3873
R.C. Baker Foundation Grants, 4006
Ralph Waldo Emerson Award, 4028
Religion and Health: Effects, Mechanisms, and Interpretation RFP, 4052
Righteous Persons Foundation Grants, 4069
Robert E. and Evelyn McKee Foundation Grants, 4074
Robert G. Cabell III and Maude Morgan Cabell Foundation Grants, 4075
Romanell-Phi Beta Kappa Professorship, 4099
Roy J. Carver Charitable Trust Medical and Science Research Grants, 4114
RSC Sir John William Dawson Medal, 4125
Salem Foundation Grants, 4160
Schomburg Center Scholars-in-Residence Fellows, 4187
Shastri Indo-Canadian Institute University of Ottawa Overseas India Summer Course Grants, 4232
Sioux Falls Area Community Foundation Community Fund Grants (Unrestricted), 4262
Sioux Falls Area Comm Foundation Spot Grants, 4263
SSRC Mellon Mays Predoctoral Research Grants, 4432
SSRC New Directions in the Study of Prayer Research Grants, 4433
SSSR Distinguished Article Award, 4440
SSSR Student Paper Award, 4441
T. James Kavanagh Foundation Grants, 4489
TD4HIM Foundation Grants, 4497
Thomas and Dorothy Leavey Foundation Grants, 4506
Thomas Austin Finch, Sr. Foundation Grants, 4507
Thompson Charitable Foundation Grants, 4511
Todd Brock Family Foundation Grants, 4522
United Methodist Church Asian, Black, Hispanic and Native American History Research Grants, 4570
United Methodist Church Center for the Evangelical UBH Audrie E. Reber Memorial Award, 4571
United Methodist Church John Harrison Ness Memorial Award, 4572
United Methodist Church Women in History Research Grant, 4574
United Methodist Church Women in History Writing Award, 4575
United Methodist Communications Stoody-West Fellowship, 4578
United Theological Seminary Merit-Based Scholarships, 4583
University of Colorado Doctoral Dissertation Fellowship in Media, Religion, and Culture, 4586
USC Pentecostal and Charismatic Research Initiative Individual Grants, 4605
USC Pentecostal and Charismatic Research Initiative Regional Center Grants, 4606
Vancouver Foundation Grants and Community Initiatives, 4807
W.C. Griffith Foundation Grants, 4819
W. C. Griffith Foundation Grants, 4818
Wayne and Gladys Valley Foundation Grants, 4835
William G. and Helen C. Hoffman Found Grants, 4876
Woodward Fund Grants, 4895

Religious Studies

AAR Postdoctoral Fellowships, 201
Academy of American Franciscan History Dissertation Fellowships, 249
ACC Asian Art and Religion Fellowships, 251
ACLS Charles A. Ryskamp Research Fellowships, 281
ACLS Collaborative Research Fellowships, 282
ACLS Fellowships, 287
ACLS Frederick Burkhardt Residential Fellowships for Recently Tenured Scholars, 288
Acton Institute for the Study of Religion and Liberty Calihan Academic Fellowships, 331
Acton Institute for the Study of Religion and Liberty Novak Award, 332
Ahmanson and Getty Postdoctoral Fellowships, 377
AIAR Andrew W. Mellon Foundation Fellowships, 392
AIAR Annual Professorships, 393
AIAR Ernest S. Frerichs Fellow and Program Coordinator Grant, 394
AIAR George A. Barton Fellowships, 395
Alabama Humanities Foundation Mini Grants, 480
Alabama Humanities Foundation Planning/Consultant Grants, 481
Alces Foundation Grants, 510
All Saints Educational Trust Grants, 572
Alpha Kappa Alpha Educational Advancement Foundation Fellowships, 575
American Academy of Religion Collaborative Research Grants, 617
American Academy of Religion Individual Research Grants, 618
Andrew W. Mellon/ACLS Dissertation Completion Fellowships, 844
Arronson Foundation Grants, 1019
ASR Fichter Research Grant, 1217
ASR Robert J. McNamara Student Award, 1218
ATLA Bibliography Grant, 1257
ATS Henry Luce III Fellowships, 1258
ATS Lilly Theological Collaborative Res Grants, 1259
ATS Lilly Theological Faculty Fellowships, 1260
ATS Lilly Theological Research Expense Grants, 1261
ATS Lilly Theological Scholars Grants, 1262
Baylor Institute for Faith and Learning Visiting Fellows Program, 1310
Booth-Bricker Fund Grants, 1359
BYU Religious Studies Center Research Grants, 1402
Catholic University Life Cycle Institute Research Fellowship, 1444
CCCF Dora M Brown Seminary Scholarships, 1448
Center for Advanced Judaic Studies Postdoctoral Fellowships, 1472
Center for Hellenic Studies Fellowships, 1476
Charlotte W. Newcombe Doctoral Dissertation Fellowships, 1531
Chatlos Foundation Grants Program, 1532
Foundation for Evangelism Harry Denman Fellowships, 2040
FTE Dissertation Fellowships, 2058
FTE Doctoral Fellowships, 2059
FTE Expanding Horizons Partnership Grants, 2060
FTE North American Doctoral Fellowships, 2061
Harvard University Women's Studies in Religion Research Associate and Visiting Faculty Grant, 2216
Institute for Advanced Studies in Culture Postdoctoral Fellowships, 2380
Martin E.P. Seligman Award for Outstanding Research in Positive Psychology, 2667
Mary Baker Eddy Library Summer Research Fellowships, 2669
Metanexus Inst Templeton Research Lectures, 2744
NCFR Innovation Grants, 2908
Nell Warren Elkin and William Simpson Elkin Foundation, 2972
NL Herzog August Bibliothek Wolfenbuttel Fellowship, 3286
NL Lawrence Lipking Fellowship at Newberry, 3287
NL Susan Kelly Power and Helen Hornbeck Tanner Fellowship, 3319
PHF Penn Faculty Research Fellowships, 3932
PHF Regional Mellon Faculty Research and Cultural Fellowships, 3933
PHF Undergraduate Humanities Forum Research Fellowships, 3934
Shastri Indo-Canadian Institute University of Ottawa Overseas India Summer Course Grants, 4232
SSHRC Insight Development Grants, 4409
SSHRC Insight Grants, 4410
SSSR Distinguished Article Award, 4440
SSSR Student Paper Award, 4441
United Methodist Church Center for the Evangelical UBH Audrie E. Reber Memorial Award, 4571
United Methodist Church John Harrison Ness Memorial Award, 4572
United Theological Seminary Merit-Based Scholarships, 4583
USC Pentecostal and Charismatic Research Initiative Individual Grants, 4605
USC Pentecostal and Charismatic Research Initiative Regional Center Grants, 4606
Wabash Center Project Grants, 4826
Wabash Center Small Project Grants, 4827
Wabash Center Waorkshop and Colloquy Participant Grants, 4828

Religious Welfare Programs

American Foodservice Charitable Trust Grants, 710
Byron W. and Alice L. Lockwood Grants, 1401
Greenspun Family Foundation Grants, 2173
Harvey Randall Wickes Foundation Grants, 2218
Herbert & Adrian Woods Foundation Grants, 2239
Margaret & James A. Elkins Jr. Found Grants, 2658
Todd Brock Family Foundation Grants, 4522
United States Institute of Peace Annual Grants, 4582
Vancouver Foundation Grants and Community Initiatives, 4807
W. C. Griffith Foundation Grants, 4818

Remedial Education

Amgen Foundation Grants, 812
IRA Jeanne S. Chall Research Fellowship, 2410

Remote Sensing

ARO Atomic, Molecular, and Optical Physics Research Contracts, 1005
ASPRS Leica Geosystems Internship, 1213
ASPRS Ta Liang Memorial Award, 1215
ASPRS William A. Fischer Mem Scholarship, 1216
Cooperative Institute for Research in Environmental Sciences Visiting Fellows Program, 1643
NSF Arctic Natural Science Grants in Glaciology, 3486
ONR Electronic and Optical Ceramics Research Grants, 3756
ONR Radio Frequency Sensor Processing Research Grants, 3778
ONR Radio Frequency Solid State Devices Applied Research Grants, 3779
USDD Signal Processing for Networked Sensing Grants, 4793

Renaissance History

AAR Rome Prize Fellowships, 202
American Historical Association George L. Mosse Prize, 719
IES Faculty Fellowships, 2341
NL American Society for Eighteenth-Century Studies Fellowships, 3279
NL Arthur and Lila Weinberg Fellowships for Independent Scholars and Researchers, 3280
NL Associated Colleges of the Midwest/Great Lakes College Association Faculty Fellowships, 3281
NL Audrey Lumsden-Kouvel Fellowships, 3282
NL British Academy Fellowship for Study in the United Kingdom, 3283
NL Frances C. Allen Fellowships, 3284
NL Grad Scholar-in-Residence Opportunities, 3285

NL Herzog August Bibliothek Wolfenbuttel Fellowship, 3286
NL Lawrence Lipking Fellowship at Newberry, 3287
NL Lester J. Cappon Fellowship in Documentary Editing, 3288
NL Mellon Postdoctoral Research Fellowships, 3294
NL Midwest Modern Lang Assoc Fellowship, 3299
NL Monticello College Foundation Fellowship for Women, 3304
NL National Endowment for the Humanities Fellowships, 3310
NL Newberry Library/Ecole Nationale des Chartes Exchange Fellowship, 3313
NL Northeast Modern Language Assocation Fellowship, 3314
NL Scholar-in-Residence Opportunity, 3315
NL Short-Term Resident Fellowships for Individual Research, 3317
NL South Central Modern Languages Association Fellowship, 3318
NL Susan Kelly Power and Helen Hornbeck Tanner Fellowship, 3319

Renewable Energy Sources
ASME Frank Kreith Energy Award, 1120
Colorado Clean Energy Solar Innovation Grants, 1599
DOE Competitive Financial Assistance for Renewable Energy and Energy Efficiency Programs, 1779
DOE Energy Biosciences Research Grants, 1784
DOE Energy from Municipal Waste Grants, 1787
DOE State Energy Program Special Projects, 1815
Energy Foundation Power Grants, 1924
US CRDF CGP ETF 2009 Energy Research Competition For Joint U.S.-Estonian Research Teams, 4609
USDA Plant Feedstock Genomics for Bioenergy: A Joint Research Grant, 4639
William and Flora Hewlett Foundation Environmental Grants, 4874

Renovation
Charles H. Farnsworth Trust Grants, 1521
Chicago Community Trust Housing Grants: Advancing Affordable Rental Housing, 1547
HRF Hudson River Improvement Grants, 2286

Reproduction
Orchard Foundation Grants, 3793
USAID Family Planning and Reproductive Health Methods Grants, 4598

Reproductive Disorders
USAID Family Planning and Reproductive Health Methods Grants, 4598

Reproductive Rights
USAID Family Planning and Reproductive Health Methods Grants, 4598

Research Participation
A-T Medical Research Foundation Grants, 10
AACR-Minorities in Cancer Research Jane Cooke Wright Lectureship Awards, 119
AACR-National Brain Tumor Foundation Fellows Grant, 120
AACR-Women in Cancer Research Charlotte Friend Memorial Lectureship Awards, 126
AACR Award for Outstanding Achievement in Cancer Research, 128
AACR Basic Cancer Research Fellowships, 130
AACR Clinical and Translational Research Fellowships, 135
AACR Minority-Serving Institution Faculty Scholar in Cancer Research Awards, 143
ACS Doctoral Training Grants in Oncology Social Work, 308
ACS Physician Training Awards in Preventive Medicine, 323
AFAR Medical Student Training in Aging Research Program, 352
AFAR Paul Beeson Career Development Awards in Aging Research for the Island of Ireland, 353
AHRF Regular Research Grants, 386
AIChE Alpha Chi Sigma Award, 396
AIChE Andrew Chase Forest Bioproducts Division Award in Chemical Engineering, 397
AIChE Clarence G. Gerhold Award, 400
AIChE Donald Q. Kern Award, 406
AIChE Environmental Division Graduate Student Paper Award, 407
AIChE Environmental Division Undergraduate Student Paper Award, 408
AIChE Food, Pharmaceutical and Bioengineering Division Awards in Chemical Engineering, 410
AIChE Institute Award for Excellence in Industrial Gases Technology, 412
AIChE Kazutoshi Fujimura Award for Lifetime Achievement, 414
AIChE Lawrence K. Cecil Award in Environmental Chemical Engineering, 415
AIChE Lectureship Award in Fluidization, 416
AIChE Nanoscale Science and Engineering Forum Young Investigator Award, 417
AIChE North American Mixing Forum Award, 418
AIChe N American Mixing Forum Startup Grant, 419
AIChE North American Mixing Forum Student Award, 420
AIChE Nuclear Engineering Division Outstanding Student Paper & Presentation Awards, 421
AIChE Process Development Practice Award, 422
AIChE Research Excellence In Sustainable Engineering Award, 424
AIChE Separations Division Graduate Student Research Award, 425
AIChE Thomas Baron Award in Fluid-Particle Systems, 426
AIChE Transport and Energy Processes Division Award, 427
AIHS Full-Time Health Research Studentships, 446
ALA ACRL Doctoral Dissertation Fellowship, 478
ALA Annual Diversity Research Grant, 479
ALA Carroll Preston Baber Research Grant, 483
ALA Ingenta Research Awards, 489
Alexander von Humboldt Foundation International Climate Protection Fellowships, 529
Alfred P. Sloan Foundation Barcode of Life Grants, 543
Alfred P. Sloan Foundation Data and Computational Research Grants, 546
Alfred P. Sloan Foundation Scholarly Communication Grants, 561
AMA-MSS Research Poster Award, 599
AMA-WPC Joan F. Giambalvo Memorial Scholarships, 600
American Philosophical Society Franklin Research Grants, 764
American Philosophical Soc Judson Daland Prize, 766
American Philosophical Society Lewis and Clark Fund for Exploration and Field Research, 767
ASHFoundation Clinical Research Grants, 1076
ASM-PAHO Infectious Diseases Epidemiology and Surveillance Fellowships, 1098
ASM/CDC Fellowships in Infectious Disease and Public Health Microbiology, 1100
ASM Abbott Laboratories Award in Clinical and Diagnostic Immunology, 1101
ASM BD Award for Res in Clinical Microbiology, 1102
ASM Eli Lilly and Company Research Award, 1138
ASM ICAAC Young Investigator Awards, 1174
ASM Merck Irving S. Sigal Memorial Awards, 1177
ASM Microbiology Undergraduate Research Fellowship, 1178
ASM Procter & Gamble Award in Applied and Environmental Microbiology, 1180
ASM Promega Biotechnology Research Award, 1181
ASM Raymond W. Sarber Awards, 1182
ASM Robert D. Watkins Graduate Research Fellowship, 1183
ASM sanofi-aventis ICAAC Award, 1184
ASM Siemens Healthcare Diagnostics Young Investigator Award, 1185
ASM Undergraduate Research Fellowship, 1186
ASM William A Hinton Res Training Award, 1187
Aspen Institute William Randolph Hearst Endowed Fellowship for Minority Students, 1194
ATA Educational Trust Grants - Curriculum Resource or Research Project, 1253
Australian Postdoctoral Research Fellowships, 1270
Australian Research Fellowships/Queen Elizabeth II Fellowships, 1272
AWM Travel Grants for Women Researchers, 1280
Ball Brothers Foundation Rapid Grants, 1292
BCBSM Foundation Excellence in Research Awards for Students, 1312
BCBSM Foundation Primary and Clinical Prevention Grants, 1315
BC Steven H. Sandell Grant Program, 1323
Carrier Corporation Contributions Grants, 1440
CCFF Chairman's Dist Life Sciences Award, 1449
CIA Undergraduate Co-Op Program - Engineer/ Program Management Engineer, 1561
CIA Undergraduate Internship - Analysis, 1562
Community Foundation for Greater New Haven Quinnipiac River Fund Grants, 1616
Community Foundation of the Virgin Islands Judith A. Towle Environmental Studies Grants, 1621
Cruise Industry Charitable Foundation Grants, 1660
Cystic Fibrosis Canada Clinical Fellowships, 1674
Cystic Fibrosis Canada Clinical Project Grants, 1675
Cystic Fibrosis Canada Clinic Incentive Grants, 1676
Cystic Fibrosis Canada Scholarships, 1679
DHHS Oral Health Promotion Research Across the Lifespan, 1759
Do Something User Experience Research Intern, 1845
Duke University Adult Cardiothoracic Anesthesia and Critical Care Medicine Fellowships, 1869
Duke University Ambulatory and Regional Anesthesia Fellowships, 1870
Duke University Clinical Cardiac Electrophysiology Fellowships, 1871
Duke University Hyperbaric Center Fellowships, 1872
Duke Univ Interventional Cardio Fellowships, 1873
Duke Univ Obstetric Anesthesia Fellowships, 1878
Duke University Pain Management Fellowships, 1879
Duke University Postdoctoral Research Fellowships in Aging, 1880
Edward W. and Stella C. Van Houten Memorial Fund Grants, 1899
Ellison Medical Foundation/AFAR Julie Martin Mid-Career Award in Aging Research, 1910
FAR Fund Grants, 1980
Foundation for the Future Grants, 2043
George A Ohl Jr. Foundation Grants, 2112
George H. Hitchings New Investigator Award in Health Research, 2118
Gertrude B. Elion Mentored Medical Student Research Awards, 2130
Getty Research Grants for Institutions - Critical Reference Resources Grants, 2134
Getty Research Institute Nonresidential Grants - Collaborative Research Grant, 2135
Getty Research Institute Nonresidential Grants - Curatorial Research Fellowships, 2136
Goldman Philanthropic Partnerships Program, 2157
GUITS Post-Doctoral Summer Travel-Research Grants, 2189
GUITS Research Grants in Comparative Studies of Modern Turkey, 2190
GUITS Sabbatical Research Grants, 2191
GUITS Summer Research Grants in Turkey for Graduate Students, 2192
Harry W. Bass, Jr. Foundation Grants, 2210
Hawaii Community Foundation Health Education and Research Grants, 2221
Henry W. Bull Foundation Grants, 2237
Herbert H. and Grace A. Dow Found Grants, 2241
Hilton Hotels Corporate Giving Program Grants, 2268
Hoglund Foundation Grants, 2271
HomeBanc Foundation Grants, 2275
IRA Albert J. Harris Awards, 2405
IRA Jeanne S. Chall Research Fellowship, 2410

IRA Outstanding Dissertation of the Year Award, 2411
IRA Steven A. Stahl Research Grants, 2413
IRA Teacher as Researcher Grants, 2414
Ireland Family Foundation Grants, 2415
John G. Duncan Trust Grants, 2501
Kellogg Institute for International Studies Global Collaboration Initiative Grants, 2541
LITA/Gaylord Award for Achievement in Library and Information Technology, 2618
Mabel H. Flory Charitable Trust Grants, 2646
Manitoba Arts Council Aboriginal Arts Creative Development Pilot Projects Grants, 2652
McCarthy Family Foundation Grants, 2717
Melville Charitable Trust Grants, 2741
MGFA Post-Doctoral Research Fellowships, 2751
MGFA Student Fellowships, 2752
Motorola Foundation Innovation Generation Collaborative Grants, 2797
NAGC Masters and Specialists Award, 2821
National Academies Postdoctoral and Senior Research Awards, 2844
NCTM Edward G. Begle PreK-6 Classroom Research Grants, 2942
NEI Mentored Clinical Scientist Development Program Award, 2961
NEI Research Grant For Secondary Data Analysis Grants, 2962
NEI Ruth L. Kirschstein National Research Service Award Short-Term Institutional Research Training Grants, 2963
Nelda C. and H.J. Lutcher Stark Found Grants, 2967
NHLBI Independent Scientist Award, 3028
NHLBI Research Supplements to Promote Diversity in Health-Related Research, 3054
NIA Independent Scientist Award, 3103
NIH Exploratory/Developmental Research Grant, 3216
NRA Foundation Grants, 3437
NSERC E.W.R. Steacie Memorial Fellowships, 3439
NSERC Gerhard Herzberg Canada Gold Medal for Science and Engineering Grants, 3440
NSERC Howard Alper Postdoctoral Prize, 3441
NSERC Industrial Research Chairs, 3443
NSERC Innovation Challenge Awards, 3445
NSERC Postdoctoral Fellowships Program, 3448
NSERC Undergrad Student Research Awards, 3451
NSF Accelerating Innovation Research, 3458
NSF Antarctic Research Grants, 3479
NSF Research Experiences for Undergrad Grants, 3638
NYAM Edward N. Gibbs Memorial Lecture and Award in Nephrology, 3682
NYCT Blood Disease Research Grants, 3696
Pancreatic Cancer Network-AACR Fellowship, 3866
Pancreatic Cancer Action Network-AACR Pathway to Leadership Grants, 3867
PDF International Research Grants, 3882
PDF Postdoctoral Fellowships for Neurologists, 3884
Peacock Foundation Grants, 3887
Percy B. Ferebee Endowment Grants, 3897
Pinkerton Foundation Grants, 3967
Pittsburgh Foundation Medical Research Grants, 3973
Pott Foundation Grants, 3984
Potts Memorial Foundation Grants, 3985
R.C. Baker Foundation Grants, 4006
R.H. Wilhelm Award in Chemical Reaction Engineering, 4007
RACGP Alan Chancellor Award, 4010
RACGP Family Medical Care, Education and Research Grants, 4013
Regenstrief General Internal Medicine - General Pediatrics Research Fellowship Program, 4045
Regenstrief Master Of Science In Clinical Research/ CITE Program, 4047
Regenstrief Medical Informatics NLM Research Fellowships, 4048
Robert B McMillen Foundation Grants, 4072
Robert M. Hearin Foundation Grants, 4077
Robert W. Deutsch Foundation Grants, 4082
Rockefeller Archive Center Research Grants, 4086
Rosalinde and Arthur Gilbert Foundation/AFAR New Investigator Awards in Alzheimer's Disease, 4102
Ross Foundation Grants, 4106
Savoy Foundation Post-Doctoral and Clinical Research Fellowships, 4184
Savoy Foundation Research Grants, 4185
SHRM Dissertation Awards, 4248
SHRM Foundation Research Grants, 4249
Sigma Theta Tau International / Council for the Advancement of Nursing Science Grants, 4255
Sigma Theta Tau Small Grants, 4258
Sigrid Juselius Foundation Grants, 4260
Skaggs Foundation Grants, 4266
Smith Richardson Foundation Domestic Public Research Fellowship, 4269
Smith Richardson Foundation Strategy and Policy Fellowship, 4270
SNM/ Covidien Seed Grant in Molecular Imaging/ Nuclear Medicine Research, 4298
SNM Molecular Imaging Research Grant For Junior Medical Faculty, 4299
SNM Pilot Research Grants, 4300
SNM Pilot Research Grants in Nuclear Medicine/ Molecular Imaging, 4301
SNM Postdoctoral Molecular Imaging Scholars, 4302
SNM Student Fellowship Awards, 4303
SNM Tetalman Mem Young Investigator Awards, 4304
SPIE A.E. Conrady Award, 4338
SPIE C.S. Vikram Award in Optical Metrology, 4339
SPIE Frits Zernike Award, 4341
SPIE George W. Goddard Award, 4342
SPIE Gold Medal of the Society Award, 4343
SPIE Technology Achievement Award, 4344
STTI/ATI Educational Assessment Nursing Research Grants, 4457
STTI Alpha Eta Collaborative Research Grants, 4458
STTI Emergency Nurses Association Foundation Grant, 4459
STTI Environment of Elder Care Nursing Research Grants, 4460
STTI Joan K. Stout RN Research Grants, 4461
STTI National League for Nursing Grants, 4462
STTI Rosemary Berkel Crisp Research Awards, 4463
STTI Virginia Henderson Clin Research Grants, 4464
Tobias Landau Fellowships in Marine Biology, 4521
UE Explore Undergraduate Research Grants, 4553
UNCFSP Summer Sabbatical Awards, 4561
UNCFSP NASA Science and Technology Institute Summer Scholars Program, 4564
USDA Policy Research Center Grants, 4640
Vancouver Foundation Grants and Community Initiatives, 4807
W. C. Griffith Foundation Grants, 4818
WILD Foundation Grants, 4867
William T. Grant Foundation Understanding the Acquisition, Interpretation, and Use of Research Evidence in Policy and Practice Grants, 4883
Woodrow Wilson MMUF Dissertation Grants, 4892
Woodrow Wilson MMUF Travel & Res Grants, 4893
Woodrow Wilson National Fellowship Foundation Travel and Research Grants, 4894
WWF Russell E. Train Education for Nature Alumni Grants, 4897

Research Resources (Health/Safety/Medical)

Cystic Fibrosis Canada Clinic Incentive Grants, 1676
Cystic Fibrosis Canada Senior Scientist Research Training Awards, 1680
IHI Quality Improvement Fellowships, 2348
NEI Research Grant For Secondary Data Analysis, 2962
NSERC Brockhouse Canada Prize for Interdisciplinary Research in Science and Engineering Grant, 3438
Pittsburgh Foundation Medical Research Grants, 3973
Regenstrief General Internal Medicine - General Pediatrics Research Fellowship Program, 4045
Regenstrief Geriatrics Fellowship, 4046
SPIRE Postdoctoral Fellowships, 4345
SRC Research Infrastructures Grants, 4391
USDA Organic Agriculture Research and Extension Initiative, 4637
USDD Clinical Functional Assessment Investigator-Initiated Grants, 4709
Woodrow Wilson MMUF Dissertation Grants, 4892
Woodrow Wilson MMUF Travel & Res Grants, 4893
Woodrow Wilson National Fellowship Foundation Travel and Research Grants, 4894

Respiratory Diseases

Fairlawn Foundation Grants, 1974
NFID New Investigator Matching Grants, 2989
NHLBI Ancillary Studies in Clinical Trials, 3015
NHLBI Developmental Origins of Altered Lung Physiology and Immune Function Grants, 3025
NHLBI Microbiome of the Lung and Respiratory Tract in HIV-Infected Individuals and HIV-Uninfected Controls, 3039
NHLBI Phase II Clinical Trials of Novel Therapies for Lung Diseases, 3047
NHLBI Prematurity and Respiratory Outcomes Program Grants, 3048
NHLBI Right Heart Function in Health and Chronic Lung Diseases Grants, 3055
NHLBI Ruth L. Kirschstein National Research Service Awards for Individual Predoctoral MD/PhD Fellows and Other Dual Degree Fellows, 3059
NHLBI Translational Grants in Lung Diseases, 3069
NIEHS Small Business Innovation Research Program Grants, 3195
Pfizer Medical Education Track Two Grants, 3927

Respiratory System

NHLBI Microbiome of the Lung and Respiratory Tract in HIV-Infected Individuals and HIV-Uninfected Controls, 3039
NHLBI Prematurity and Respiratory Outcomes Program Grants, 3048

Restoration and Preservation

ACC Residency Program in Asia, 267
ALA Jan Merrill-Oldham Professional Development Grant, 491
Albuquerque Community Foundation Grants, 508
ASID Foundation Dora Brahms Award, 1092
Auburn Foundation Grants, 1263
Boeing Company Contributions Grants, 1357
Booth-Bricker Fund Grants, 1359
Charles Edison Fund Grants, 1520
Cudd Foundation Grants, 1672
D.F. Halton Foundation Grants, 1692
Furthermore Grants in Publishing, 2091
George Foundation Grants, 2115
George Frederick Jewett Foundation Grants, 2116
Grammy Foundation Grants, 2161
HRF Hudson River Improvement Grants, 2286
HRF Mark B. Bain Graduate Fellowships, 2287
Illinois DNR Youth Recreation Corps Grants, 2361
IMLS 21st Century Museum Prof Grants, 2362
International Crane Foundation Internships, 2396
McLean Contributionship Grants, 2727
NHPRC Digitizing Historical Records Grants, 3072
NOAA Great Lakes Habitat Restoration Partnership Grants, 3376
Norcliffe Foundation Grants, 3415
NPS American Battlefield Protection Grants, 3435
Robert G. Cabell III and Maude Morgan Cabell Foundation Grants, 4075
Ross Foundation Grants, 4106
USDA Hazardous Fuels Woody Biomass Utilization Grants, 4626
USDC/NOAA Nature Conservancy Community-based Matching Grants, 4661
Winterthur Research Fellowships, 4888

Restoration and Preservation, Art Works/Artifacts

ACC Asian Art and Religion Fellowships, 251
ACC Ford Foundation Fellowships, 253
ACC Humanities Fellowship Program, 255
Andy Warhol Foundation for the Visual Arts Curatorial Research Fellowships, 847
National Endowment for the Humanities Fellowships at Winterthur, 2867
NYSCA Folk Arts: Exhibitions Grants, 3703

Winterthur Museum Dissertation Fellowships, 4887
Winterthur Research Fellowships, 4888

Restoration and Preservation, Manuscripts/Books/ Music Scores
National Endowment for the Humanities Fellowships at Winterthur, 2867
NL Newberry Library/Ecole Nationale des Chartes Exchange Fellowship, 3313
Wenner-Gren Found Historical Archives Grants, 4849
Winterthur Museum Dissertation Fellowships, 4887

Restoration and Preservation, Structural/ Architectural
AAR Rome Prize Fellowships, 202
Andy Warhol Foundation for the Visual Arts Curatorial Research Fellowships, 847
Athenaeum of Philadelphia Senior Fellowships, 1256
HRF Hudson River Improvement Grants, 2286
IMLS National Leadership Grants, 2366
Maine Community Foundation Keepers Preservation Education Grants, 2651
National Endowment for the Humanities Fellowships at Winterthur, 2867
Rhode Island Foundation Grants, 4061
Textron Corporate Contributions Grants, 4501
Winterthur Museum Dissertation Fellowships, 4887

Retirement
BC Center for Retirement Research Dissertation Fellowships, 1320
BC Steven H. Sandell Grant Program, 1323
RRF General Program Grants, 4115
TIAA-CREF Institute Research Grant Program, 4517

Rheology
AACC International Bruce Wasserman Young Investigator Award, 84
AACC Int'l Carl Wilhelm Brabender Award, 85
AACC International Excellence in Teaching Award, 89
AACC International George W. Scott Blair Lecture Award, 90
AACC International Thomas Burr Osborne Medal, 98
AACC Int'l Young Scientist Research Award, 100

Rheumatic Diseases
AMA Foundation Seed Grants for Research, 601
NIAMS Pilot & Feasibility Clinical Grants in Arthritis & Musculoskeletal & Skin Diseases, 3111
Nuffield Found Oliver Bird Rheumatism Grants, 3668
Pfizer ASPIRE North America Rheumatology Research Awards, 3917
Pfizer Inflammation Competitive Res Awards, 3926
Pfizer Research Initiative Rheumatology Grants, 3929

Rheumatology
American College of Rheumatology Fellows-in-Training Travel Scholarships, 700
Nuffield Foundation Oliver Bird Rheumatism Program Grants, 3668
Pfizer ASPIRE North America Rheumatology Research Awards, 3917
Pfizer Inflammation Competitive Res Awards, 3926
Pfizer Research Initiative Rheumatology Grants, 3929

Rhinology
AAO-HNSF Health Services Research Grants, 187
AAO-HNSF Maureen Hannley Research Training Awards, 188
AAO-HNSF Percy Memorial Research Award, 189
AAO-HNSF Rande H. Lazar Health Services Research Grant, 190
AAO-HNSF Resident Research Awards, 191
AAOA Foundation/AAO-HNSF Combined Research Grants, 194
ALVRE Casselberry Award, 585
ALVRE Grant, 586
ALVRE Seymour R. Cohen Award, 587
ARS New Investigator Award, 1020
ARS Resident Research Grants, 1021

Risk Factors/Analysis
AAAS Science and Technology Policy Fellowships: Health, Education and Human Services, 43
Abdus Salam ICTP Schools, Conferences, Workshops and Extended Research Workshops Grants, 238
American Chemical Society AGRO Educ Awards, 632
American Chemical Society Award for Creative Advances in Environmental Science and Technology, 652
DOT Transit Planning and Research Grants, 1849
Glaucoma Research Pilot Project Grants, 2146
Huebner Foundation Fellowships, 2294
NSF Decision, Risk, and Management Science Research Grants, 3550
USDA Biotechnology Risk Assessment Grants, 4618
Young Scientists Summer Program, 4903

Robotics
Microsoft Human-Robot Interaction Grants, 2768
ONR Computational Neuroscience & Biorobotics Grants, 3752

Rocket Propulsion
NASA Cooperative Agreement Notice for Technology Development Grants, 2837
NASA Graduate Student Researchers Program Fellowships, 2839
NASA Undergraduate Student Researchers Project Grants, 2843

Roman Catholic Church
MacDonald-Peterson Foundation Grants, 2649

Runaway Youth
Burton G. Bettingen Grants, 1390

Rural Areas
Greenwall Foundation Bioethics Grants, 2174
National Book Scholarship Fund, 2849
USDA Policy Research Center Grants, 4640
USDA Small Business Innovation Research Grants - Phase I, 4648
Virginia Foundation for the Humanities Discretionary Grants, 4813
Virginia Foundation for Humanities Open Grants, 4814
W.K. Kellogg Foundation Healthy Kids Grants, 4822

Rural Development
SARE Graduate Student Grants - North Central Region, 4168
USDA Agriculture and Food Research Initiative Competitive Grants - Food Security, 4614

Rural Education
USDA Policy Research Center Grants, 4640

Rural Health Care
Appalachian Regional Commission Health Care Grants, 915
Fannie E. Rippel Foundation Grants, 1978
Rajiv Gandhi Foundation Grants, 4025

Rural Planning/Policy
National Urban Fellows Program, 2888
SARE Graduate Student Grants - North Central Region, 4168
USDA Policy Research Center Grants, 4640
Virginia Foundation for the Humanities Discretionary Grants, 4813
Virginia Foundation for Humanities Open Grants, 4814

Russia
Carnegie Corporation of New York Grants, 1432
Charles Stewart Mott Foundation Grants, 1529
Eastern Europe & the Independent States of the Former Soviet Union Research & Training Grants, 1891
IREX Individual Advanced Research Opportunities Grants, 2419
IREX University Administration Support, 2425
U.S. Department of Education United States-Russia Program: Improving Research and Educational Activities in Higher Education, 4547

Russian Language/Literature
ACIE Title VIII Combined Research and Language Training Fellowships, 275
Fulbright Critical Lang Enhancement Awards, 2068
SSRC Eurasia Program Pre-Dissertation Awards, 4428

Sabbaticals
Alston Bannerman Sabbatical Fellowships, 580
American Chemical Society Petroleum Research Fund Undergraduate Faculty Sabbatical Grants, 693
JSA JLab Sabbatical Research Leave Grants, 2522
ONR Faculty Sabbatical Leave Program Grants, 3762
USDD Sabbatical Leave Fellowships, 4788

Safety
AAA Foundation for Traffic Safety Grants, 24
AAAS Award for Scientific Freedom and Responsibility, 28
Alcatel-Lucent Technologies Foundation Grants, 509
Allstate Found Safe & Vital Communities Grants, 573
ASID Foundation Irene Winifred Eno Grant, 1094
ASME Barnett-Uzgiris Product Safety Award, 1106
Aviation Research Grants, 1273
Bikes Belong Foundation Research Grants, 1343
BMW of N America Charitable Contributions, 1353
Canadian Patient Safety Inst Research Grants, 1421
Center of Excellence in General Aviation Grants, 1486
DaimlerChrysler Corporation Fund Grants, 1714
Detroit Lions Charities Grants, 1753
DOT Recreational Boating Safety Grants, 1848
DOT Transportation Statistics Research Grants, 1850
EPA Children's Health Protection Grants, 1933
Farmers Insurance Corporate Giving Grants, 1981
Ford Motor Company Fund Grants Program, 2033
Frank Stanley Beveridge Foundation Grants, 2051
Greater Worcester Community Foundation Jeppson Memorial Fund for Brookfield Grants, 2165
Helen V. Brach Foundation Grants, 2231
John Deere Foundation Grants, 2500
National Wild Turkey Federation Research Grant Program, 2893
NIST Fire Research Grants, 3268
NRA Foundation Grants, 3437
Olin Corporation Charitable Trust Grants, 3739
Rohm and Haas Company Grants, 4098
USDC/NOAA Nature Conservancy Community-based Matching Grants, 4661
William G. McGowan Charitable Fund Grants, 4877

Safety Engineering
ASSE Liberty Mutual Safety Res Fellowship, 1219
ASSE Scholarships, 1220
BMW of N America Charitable Contributions, 1353
NIST Fire Research Grants, 3268
SSHRC-APC Automotive Partnership Canada, 4396

Sanitary Engineering
Bill and Melinda Gates Foundation Water, Sanitation and Hygiene Grants, 1347
Dorr Foundation Grants, 1842
USAID Palestinian Community Infrastructure Development Grants, 4600
USDA Norman E. Borlaug International Agricultural Science and Technology Fellowships, 4636

Sarcoidosis
NIH Sarcoidosis: Research into the Cause of Multi-Organ Disease and Clinical Strategies for Therapy Grants, 3243

Sarcoma
ASCO/UICC International Cancer Technology Transfer Fellowships, 1044
ASCO Young Investigator Award, 1050
Conquer Cancer Foundation of ASCO Career Development Award, 1630

Satellite Communication
ARO Mobile, Wireless Communications and Networks Research Contracts, 1012
ONR Electronic and Optical Ceramics Research Grants, 3756
ONR Radio Frequency Sensor Processing Research Grants, 3778
ONR Radio Frequency Solid State Devices Applied Research Grants, 3779

Scandinavia
American-Scandinavian Foundation Awards for Advanced Study or Research in the U.S., 603
American-Scandinavian Foundation Fellowship for Study in Scandinavia, 604
American-Scandinavian Foundation Grant for Study in Scandinavia, 605
American-Scandinavian Foundation Short-Term Training Program Grants, 606
SASS Einar and Eva Lund Haugen Memorial Scholarship, 4183

Scandinavian Languages/Literature
SASS Birgit Baldwin Fellowship, 4182
SASS Einar and Eva Lund Haugen Memorial Scholarship, 4183

Schizophrenia
American Psychiatric Association/AstraZeneca Young Minds in Psychiatry International Awards, 773
American Psychiatric Association/Kempf Fund Award for Research Development in Psychobiological Psychiatry, 775
American Psychiatric Association/Merck Co. Early Academic Career Research Award, 778
NARSAD Distinguished Investigator Grants, 2831
NARSAD Independent Investigator Grants, 2833
NARSAD Lieber Prize for Schizophrenic Res, 2834
NARSAD Young Investigator Grants, 2836
NIMH Early Identification and Treatment of Mental Disorders in Children and Adolescents Grants, 3252

Scholarship Programs, General
Achelis Foundation Grants, 274
AHEPA Educational Foundation Scholarships, 373
AIChE Alpha Chi Sigma Award, 396
AIChE Andrew Chase Forest Bioproducts Division Award in Chemical Engineering, 397
AIChE Charles M.A. Stine Award, 399
AIChE Clarence G. Gerhold Award, 400
AIChE Computational Molecular Science and Engineering Forum Graduate Student Award, 401
AIChE Donald Q. Kern Award, 406
AIChE Excellence in Process Development Research Award, 409
AIChE Food, Pharmaceutical and Bioengineering Division Awards in Chemical Engineering, 410
AIChE Food, Pharmaceutical and Bioengineering Division Distinguished Service Award in Chemical Engineering, 411
AIChE Institute Award for Excellence in Industrial Gases Technology, 412
AIChE John A. Tallmadge Award for Contributions to Coating Technology, 413
AIChE Lectureship Award in Fluidization, 416
AIChE North American Mixing Forum Award, 418
AIChE Process Development Practice Award, 422
AIChE Research Excellence In Sustainable Engineering Award, 424
AIChE Separations Division Graduate Student Research Award, 425
AIChE Transport and Energy Processes Division Award, 427
Albuquerque Community Foundation Grants, 508
Alpha Omega Foundation Grants, 576
AMA-WPC Joan F. Giambalvo Memorial Scholarships, 600
American Family Insurance Corporate Giving Program Grants, 709
American Foodservice Charitable Trust Grants, 710
Archer Daniels Midland Foundation Grants, 985
Armenian Students' Association Scholarships, 1000
ASM BD Award for Res in Clinical Microbiology, 1102
ASM bioMerieux Sonnenwirth Award for Leadership in Clinical Microbiology, 1103
ASM Eli Lilly and Company Research Award, 1138
ASM ICAAC Young Investigator Awards, 1174
ASM Merck Irving S. Sigal Memorial Awards, 1177
ASM Procter & Gamble Award in Applied and Environmental Microbiology, 1180
ASM Promega Biotechnology Research Award, 1181
ASM Raymond W. Sarber Awards, 1182
ASM sanofi-aventis ICAAC Award, 1184
ASM Siemens Healthcare Diagnostics Young Investigator Award, 1185
ASM William A Hinton Res Training Award, 1187
ASSE Scholarships, 1220
ASTA Academic Scholarships, 1224
ATA Nadene M. Thomas Grad Research Bursary, 1255
AUPHA Corris Boyd Scholarships, 1267
Australian Postgraduate Research Fellowships, 1271
AWRA Richard A. Herbert Mem Scholarship, 1281
Bailey Family Foundation Grants, 1289
Bilateral Scholarships, 1344
Bodman Foundation Grants, 1356
Burlington Northern Santa Fe Foundation Grants, 1389
California Teachers Association Scholarships, 1408
Campbell Soup Foundation Grants, 1415
Carpenter Foundation Grants, 1439
Carrier Corporation Contributions Grants, 1440
Catholic University Academic Scholarships, Grants, and Awards, 1443
CFUW Dr. Alice E. Wilson Awards, 1510
CFUW Memorial Fellowship, 1515
Chautauqua Region Community Found Grants, 1533
Chiles Foundation Grants, 1552
CIMO Scholarships for Advanced Studies of the Finnish Language at Finnish Universities, 1569
Cleveland-Cliffs Foundation Grants, 1580
Coca-Cola Foundation Grants, 1593
Community Found for Greater Buffalo Grants, 1615
Community Found for Muskegon County Grants, 1617
Community Foundation of the Fox River Valley, 1620
Cord Foundation Grants, 1645
Corning Incorporated Found Educational Grants, 1646
Corpus Christi College Research Scholarships, 1647
Cystic Fibrosis Canada Scholarships, 1679
DAAD Bilateral Programs for Co-Operative Research Grants, 1695
DAAD Research Grants for Doctoral Candidates and Young Academics and Scientists, 1703
DAAD Research Stays for University Academics and Scientists, 1705
DAAD Scholarships for Artists: Study Visits for Academics, 1707
Danone Institute Belgium Doctoral Scholarship in Nutrition, 1718
Datatel Scholars Foundation Scholarship, 1722
DEED Student Research Grants and Internships, 1731
Dorr Foundation Grants, 1842
Edward W. and Stella C. Van Houten Memorial Fund Grants, 1899
Ensign-Bickford Foundation Grants, 1928
EPA Capacity Building Grants for Compliance Assurance and Enforcement Activities in Indian Country and Other Tribal Areas, 1932
EREF Doctoral Scholarship, 1945
EREF Fiessinger Doctoral Scholarship, 1946
Field Museum Undergraduate Internships, 2010
Flinn Foundation Grants Programs, 2024
FMC Foundation Grants, 2029
Fritz B. Burns Foundation Grants, 2057
Fulbright Scholar-in-Residence Program Grants, 2078
George Foundation Grants, 2115
Gerda Henkel Foundation Grants, 2127
Gil and Dody Weaver Foundation Grants, 2139
Greater Saint Louis Community Found Grants, 2164
HAF Riley Frazel Memorial Scholarship, 2197
Harold Simmons Foundation Grants, 2203
Helen V. Brach Foundation Grants, 2231
HIAS Scholarship Awards, 2264
Hilton Hotels Corporate Giving Program Grants, 2268
Hirtzel Memorial Foundation Grants, 2269
Hispanic Scholarships, 2270
Hoglund Foundation Grants, 2271
HomeBanc Foundation Grants, 2275
HSBC Contributions Community Grants, 2291
Huie-Dellmon Trust Grants, 2295
IIE Whitaker International Fellowships and Scholarships, 2356
ITEA Maley/FTE Scholarship, 2435
Jewish Federation of Metropolitan Chicago Scholarships, 2486
John R. Oishei Foundation Grants, 2505
Joseph Alexander Foundation Grants, 2513
Kettering Fund Grants, 2555
Kosciuszko Foundation Graduate Study and Research in Poland Scholarships, 2571
Lillian S. Wells Foundation Grants, 2614
Lumpkin Family Found Healthy People Grants, 2629
Lynde & Harry Bradley Foundation Fellowships, 2634
Lynde and Harry Bradley Foundation Grants, 2635
M. Bastian Family Foundation Grants, 2636
Marcus and Theresa Levie Educational Fund Scholarships, 2657
Marshall Scholarships, 2666
Mead Johnson Nutritionals Scholarships, 2735
Medtronic Foundation Fellowships, 2738
MGN Family Foundation Grants, 2754
Minnesota War Orphans Educational Assistance Grants, 2780
MLA Graduate Scholarship, 2786
MLA Grad Scholarship for Minority Students, 2787
NAACP Agnes Jones Jackson Grad Scholarships, 2809
NAGC Masters and Specialists Award, 2821
NASA Graduate Student Researchers Program Fellowships, 2839
NASA Undergraduate Student Researchers Project Grants, 2843
National Leadership Grant Competition, 2878
National LGBT Scholarship Fund, 2879
NCAA Postgraduate Scholarships, 2900
Nelda C. and H.J. Lutcher Stark Found Grants, 2967
New Hampshire Charitable Foundation Grants, 2985
NOAA Dr. Nancy Foster Scholarships, 3367
NOAA Graduate Sciences Program Grants, 3375
North Dakota Indian Scholarships, 3427
NSERC Industrial Undergraduate Student Research Awards, 3444
NSERC Northern Research Internships Program, 3447
NSERC Undergrad Student Research Awards, 3451
NSF Scholarships in Science, Technology, Engineering, and Mathematics, 3642
Oklahoma City Community Grants, 3737
OSF Alumni Grants, 3835
Park Foundation Grants, 3870
Patrick and Anna M. Cudahy Fund Grants, 3873
Percy B. Ferebee Endowment Grants, 3897
Pfizer Foundation Epilepsy Scholarship Award, 3923
Polytechnic University International Research Studentships, 3982
Pott Foundation Grants, 3984
Potts Memorial Foundation Grants, 3985
Price Family Charitable Fund Grants, 3987
Province of Alberta Queen Elizabeth II Graduate Scholarships, 4001
Rathmann Family Foundation Grants, 4033
Regenstrief Master Of Science In Clinical Research/CITE Program, 4047
Rhode Island Foundation Grants, 4061
Robert M. Hearin Foundation Grants, 4077
Roy J. Carver Charitable Trust Medical and Science Research Grants, 4114
Sachs Foundation Scholarships, 4157
Seneca Nation Higher Education Grants, 4198
Sir Dorabji Tata Trust Individual Education and Travel Grants, 4265
Society of Naval Architects and Marine Engineers Scholarships, 4328
SPIE A.E. Conrady Award, 4338

SPIE C.S. Vikram Award in Optical Metrology, 4339
SPIE Dennis Gabor Award, 4340
SPIE Frits Zernike Award, 4341
SPIE George W. Goddard Award, 4342
SPIE Technology Achievement Award, 4344
SSHRC Graduate Scholarships-Michael Smith Foreign Study Supplements, 4408
Starr Foundation Grants, 4447
T. James Kavanagh Foundation Grants, 4489
Thomas and Dorothy Leavey Foundation Grants, 4506
Thomas Austin Finch, Sr. Foundation Grants, 4507
Thompson Charitable Foundation Grants, 4511
UNCFSP Summer Sabbatical Awards, 4561
UNCFSP NASA Science and Technology Institute Summer Scholars Program, 4564
United States Institute of Peace - Jennings Randolph Peace Dissertation Scholarships, 4579
USA Funds Access to Education Scholarships, 4593
William G. and Helen C. Hoffman Found Grants, 4876
William G. McGowan Charitable Fund Grants, 4877
Xerox Foundation Grants, 4899

School Dental Programs
ADA Foundation Dentsply International Research Fellowships, 335
Healthcare Foundation of New Jersey Grants, 2222

School Health Programs
Anthem Blue Cross and Blue Shield School Nurse Grants, 868
Appalachian Regional Commission Health Care Grants, 915
Colonel Stanley R. McNeil Foundation Grants, 1597
DHHS Oral Health Promotion Research Across the Lifespan, 1759
Fairlawn Foundation Grants, 1974
Healthcare Foundation of New Jersey Grants, 2222

School-to-Work Transition
Rohm and Haas Company Grants, 4098
Textron Corporate Contributions Grants, 4501

Science
AAAS Award for Public Engagement with Science, 26
AAAS Award for Science Diplomacy, 27
AAAS Award for Scientific Freedom and Responsibility, 28
AAAS Early Career Award for Public Engagement with Science, 29
AAAS Entry Point Internships, 30
AAAS Kavli Science Journalism Awards, 33
AAAS Newcomb Cleveland Prize, 36
AAAS Philip Hauge Abelson Award, 37
AAAS Science and Technology Policy Fellowships: Congressional, 39
AAAS Science and Technology Policy Fellowships: Diplomacy, Security, and Development, 40
AAAS Science Prize for Inquiry-Based Instruction, 45
AFRL/NRC Resident Research Associateship Program, 358
Ahmed Zewail Award in Ultrafast Science and Technology, 378
AHRQ Independent Scientist Award, 388
AIChE Charles M.A. Stine Award, 399
AIChE Computational Molecular Science and Engineering Forum Graduate Student Award, 401
AIChE Lectureship Award in Fluidization, 416
AIChE Nanoscale Science and Engineering Forum Young Investigator Award, 417
Albert and Margaret Alkek Foundation Grants, 504
Alcoa Foundation Grants, 511
Alexander von Humboldt Feodor Lynen Research Fellowships for Experienced Researchers, 520
Alexander von Humboldt Foundation Friedrich Wilhelm Bessel Research Awards, 522
Alexander von Humboldt Foundation Research Awards, 537
Alexander von Humboldt Foundation Research Awards for Top Researchers in Germany, 538
Alfred P. Sloan Foundation Civic Grants, 545
Alfred P. Sloan Foundation Economic Analysis of Science and Technology Grants, 548
Alfred P. Sloan Foundation Energy and Environment Grants, 552
Alfred P. Sloan Foundation History of Science and Technology Grants, 553
Alfred P. Sloan Foundation Public Understanding of Science and Technology-Book Grants, 555
Alfred P. Sloan Foundation Public Understanding of Science and Technology-Other Efforts, 556
Alfred P. Sloan Foundation Research Fellowships, 560
Alfred W. Bressler Prize in Vision Science, 567
Alton Ochsner Award, 583
ALZA Corporate Contributions Grants, 588
American Indian College Fund STEM Leadership Fellowships, 745
American Schlafhorst Foundation Grants, 796
AMS Grad Fellowships in History of Science, 824
Angels Baseball Foundation Grants, 848
Applied Biosystems Grants, 916
APS Davisson-Germer Prize in Atomic or Surface Physics, 952
APS Einstein Prize, 955
APS Isakson Prize for Optical Effects in Solids, 957
APS Hans A. Bethe Prize, 960
APS Herbert P. Broida Prize, 961
APS James C. McGroddy Prize for New Materials, 965
APS Oliver E. Buckley Condensed Matter Prize, 973
APS Tom W. Bonner Prize in Nuclear Physics, 980
APS Will Allis Prize for Study of Ionized Gases, 982
Arizona Cardinals Grants, 995
ARM Faculty Fellowships, 1001
ARM Graduate Fellowships, 1002
ARM Postgraduate Fellowships, 1003
ARM Visiting Scientist Awards, 1004
ARO Conference and Symposia Grants, 1007
ARO Short-Term Innovation Research Program, 1014
ARO Young Investigator Program/Presidential Early Career Awards, 1018
ASEE Army Research Laboratory Postdoctoral Fellowships, 1059
ASEE Naval Research Laboratory Postdoctoral Fellowships, 1060
ASU Graduate College Science Foundation Arizona Bisgrove Postdoctoral Scholars, 1244
AT&T Bell Labs Graduate Research Program, 1250
Australian Academy of Science Grants, 1269
AWIS Educational Foundation Predoc Awards, 1278
Banfi Vintners Foundation Grants, 1296
BFGoodrich Collegiate Inventors Prize, 1341
Bower Award for Achievement in Science, 1364
Brinson Foundation Grants, 1372
BWF Ad Hoc Grants, 1392
BWF Career Awards for Medical Scientists, 1394
Cabot Corporation Foundation Grants, 1406
Canada-U.S. Fulbright International Science and Technology Award, 1417
Canada-U.S. Fulbright Senior Specialists Program Grants, 1418
Canada-U.S. Fulbright Traditional Scholars, 1419
Canada Graduate Scholarships and NSERC Postgraduate Scholarships, 1420
Canon Foundation Grants, 1423
Carl Gellert and Celia Berta Gellert Foundation Grants, 1429
Center for Education of Women Visiting Scholar Research Grants, 1475
Center for Inquiry Science and the Public Visiting Research Fellowships, 1477
CFUW Memorial Fellowship, 1515
Clark Library Short-Term Resident Fellowships, 1577
Cleveland-Cliffs Foundation Grants, 1580
Corning Incorporated Found Educational Grants, 1646
DAAD Research Grants for Doctoral Candidates and Young Academics and Scientists, 1703
DAAD Research Stays for University Academics and Scientists, 1705
DAAD Study & Internship Program in Germany, 1708
DaimlerChrysler Corporation Fund Grants, 1714
Dana Foundation Science and Health Grants, 1716
DOE Computational Science Grad Fellowships, 1780
DOE Experimental Program to Stimulate Competitive Research, 1791
DOE Faculty and Student Teams Grants, 1792
E.L. Wiegand Foundation Grants, 1885
Elisabeth Severance Prentiss Foundation Grants Program, 1905
EMBO / EMBL Science Writing Prize, 1913
EMBO Installation Grants, 1915
Explorers Club Scott Pearlman Field Awards, 1967
Fluor Foundation Grants, 2028
Frank Stanley Beveridge Foundation Grants, 2051
Frederick Gardner Cottrell Foundation Grants, 2054
Fulbright Alumni Initiatives Awards Grants, 2064
Fulbright Distinguished Chairs Program Grants, 2069
Fulbright European Union Scholar-in-Residence Program Grants, 2071
Fulbright German Studies Seminar Grants, 2072
Fulbright New Century Scholars Program Grants, 2075
Fulbright Occasional Lecturer Program Grants, 2076
Fulbright Specialists Program Grants, 2079
Fulbright Traditional Scholar Program in Europe and Eurasia, 2080
Fulbright Traditional Scholar Program in Sub-Saharan Africa, 2081
Fulbright Traditional Scholar Program in the East Asia/Pacific Region, 2082
Fulbright Traditional Scholar Program in the Near East and North Africa Region, 2083
Fulbright Traditional Scholar Program in the South and Central Asia Region, 2084
Fulbright Traditional Scholar Program in the Western Hemisphere, 2085
Fulbright Visiting Specialists Program: Direct Access to the Muslim World, 2087
Fund for Scientific Research--Flanders Grants, 2089
G. Harold and Leila Y. Mathers Charitable Foundation Grants, 2092
G. Unger Vetlesen Foundation Grants, 2093
Gaius Charles Bolin Fellowships for Minority Graduate Students, 2095
George A Ohl Jr. Foundation Grants, 2112
George Foundation Grants, 2115
Gruber Foundation Weizmann Institute Awards, 2184
Harry W. Bass, Jr. Foundation Grants, 2210
Heineman Foundation for Research, Education, Charitable and Scientific Purposes, 2227
Helen V. Brach Foundation Grants, 2231
Herbert H. and Grace A. Dow Found Grants, 2241
Herb Society of America Research and Education Grants, 2243
HHMI Precollege Outreach Initiative for Biomedical Research Institutions Grants, 2261
IIE Leonora Lindsley Memorial Fellowships, 2354
IIE Rockefeller Foundation Bellagio Center Residencies, 2355
IMLS 21st Century Museum Prof Grants, 2362
IMLS National Leadership Grants, 2366
Inamori Foundation Kyoto Prize, 2367
International Max Planck Research School Mathematics in the Sciences Fellowships, 2401
John Simon Guggenheim Memorial Foundation Fellowships, 2507
Josiah W. and Bessie H. Kline Foundation Grants, 2517
JSPS Bilateral Exchange Programs with Foreign Academic Institutions, 2524
JSPS Fellowships for Foreign Scientists, 2525
JSPS International Joint Research Projects and Scientific Seminars, 2526
JSPS Postdoc Fellowship for Foreign Researchers, 2527
Kavli Foundation Research Grants, 2533
Kenneth T. and Eileen L. Norris Found Grants, 2554
L'Oreal USA Women in Science Fellowship, 2577
Lady Davis Fellowships, 2578
M.J. Murdock Charitable Trust General Grants, 2641
M.J. Murdock Charitable Trust Partners in Science Grants, 2642
Marion Gardner Jackson Charitable Trust Grants, 2664
McCarthy Family Foundation Grants, 2717
McGregor Fund Grants, 2725

McKnight Doctoral Fellowships, 2726
Meadows Foundation Grants, 2736
Medtronic Foundation Fellowships, 2738
Metanexus Inst Templeton Research Lectures, 2744
MIT Knight Science Journalism Fellowships, 2782
Motorola Foundation Innovation Generation Collaborative Grants, 2797
National Geographic Society Expeditions Council Grants, 2872
National Wetlands Awards, 2889
NATO Advanced Research Workshops Program, 2894
NATO Advanced Study Institutes Program, 2895
NATO Collaborative Linkage Grants, 2896
NATO Science for Peace Grants, 2898
NEH Collaborative Research Grants, 2952
NEH Digital Humanities Implementation Grants, 2953
NEH Digital Humanities Start-Up Grants, 2954
NIGMS Support of Competitive Research Pilot Project Awards, 3202
NIGMS Support of Competitive Research Research Advancement Awards, 3203
NIH Independent Scientist Award, 3221
NIH Mentored Clinical Scientist Devel Award, 3225
NIH Mentored Research Scientist Development Awards, 3227
NIST National Standard Reference Data System Research Grants, 3271
NOAA Collaborative Science, Technology, and Applied Research Program Grants, 3354
NOAA Mid-Atlantic Research Set-Aside Program Grants, 3384
NOAA Monkfish Research Set-Aside Grants, 3386
NOAA Sea Scallop Research Set-Aside Program Grants, 3404
Northrop Grumman Foundation Grants, 3429
Northwood University/Alden B. Dow Creativity Center Summer Residency Program, 3432
NSERC Brockhouse Canada Prize for Interdisciplinary Research in Science and Engineering Grant, 3438
NSERC Industrial Research Chairs, 3443
NSERC Research Tools and Instruments Grants, 3449
NSF-CONACyT Collaborative Research Opportunities, 3452
NSF Accelerating Discovery In Science and Engineering through PETASCALE Simulations and Analysis, 3457
NSF Activities in Science, Engineering, and Mathematics for Persons with Disabilities, 3460
NSF Advanced Learning Technologies Grants, 3462
NSF Antarctic Integrated System Science Program Grants, 3476
NSF Antarctic Research Grants, 3479
NSF Antarctic Special Opportunity for Science aboard the Swedish Ice Breaker Oden Grants, 3480
NSF Applied Mathematics Grants, 3481
NSF Arctic Research and Education Grants, 3489
NSF Arctic Social Sciences Grants, 3492
NSF Centers of Research Excellence in Science and Technology and HBCU Research Infrastructure for Science and Engineering Grants, 3515
NSF CISE Accelerating Discovery in Science and Engineering Through Petascale Simulations and Analysis Grants, 3521
NSF CISE Cross-Cutting Program Grants, 3532
NSF CISE Increasing the Participation and Advancement of Women in Academic Science and Engineering Careers Grants, 3534
NSF Collaborative Research in Chemistry, 3537
NSF Communicating Res to Public Audiences, 3538
NSF Computational Math Research Grants, 3539
NSF CreativeIT Grants, 3547
NSF Deep Underground Science and Engineering Laboratory Grants, 3551
NSF East Asia and Pacific Summer Institutes for U.S. Graduate Students, 3555
NSF Focused Research Groups in the Mathematical Sciences Grants, 3566
NSF Foundations Grants, 3567
NSF Grants for the Analysis of Science and Technology Resources, 3575
NSF Industry/University Cooperative Research Centers Grants, 3579
NSF Innovation and Organizational Change, 3581
NSF Integrative Graduate Education and Research Traineeship, 3584
NSF Materials Research Science and Engineering Centers, 3594
NSF Partnerships for Innovation Grants, 3610
NSF Partnerships for International Research and Education, 3611
NSF Presidential Early Career Awards for Scientists and Engineers Grants, 3633
NSF Research on Gender in Science and Engineering Grants, 3639
NSF SBE Doctoral Dissertation Research Improvement Grants, 3641
NSF Science and Technology Studies Grants, 3644
NSF Small Business Innovation Research Grants, 3647
NSF Tribal Colleges and Universities Program, 3662
Nuffield Foundation Law and Society Grants, 3667
ONR Bisson Prize for Naval Tech Achievement, 3745
ONR Cognitive Science Basic Research Grants, 3749
ONR Multidisciplinary Research Program of the URI Grants, 3774
ONR Summer Faculty Research Program Grants, 3784
ORISE National Oceanic and Atmospheric Administration Faculty Research Participation, 3811
ORISE National Oceanic and Atmospheric Admin Student Research Participation, 3812
ORISE Nuclear Regulatory Commission Historically Black Colleges and Universities Student Research Participation, 3814
ORISE Oak Ridge National Laboratory Postdoctoral Research Associates Program, 3815
ORISE Postgraduate Research Participation at the National Exposure Research Laboratory, 3821
Phi Beta Kappa Award in Science, 3935
R.C. Baker Foundation Grants, 4006
Rajiv Gandhi Foundation Grants, 4025
Rayonier Foundation Grants, 4034
Research Corporation Cottrell Scholars Awards, 4053
Research Corporation Multi-Investigator Cottrell College Science Awards, 4054
Research Corporation Scialog Grants, 4055
Research Corporation Single-Investigator Cottrell College Science Awards, 4056
Richard and Rhoda Goldman Fund Grants, 4063
Robert Leet and Clara Guthrie Patterson Grants, 4076
Roy J. Carver Charitable Trust Medical and Science Research Grants, 4114
RSC Thomas W. Eadie Medal, 4126
San Diego Women's Foundation Grants, 4164
Schomburg Center Scholars-in-Residence Fellows, 4187
Schramm Foundation Grants, 4188
Scientific Communities Initiative Grants, 4189
Seaver Institute Grants, 4195
SFI Principal Investigator Program Grant, 4201
SHOT Abbott Payson Usher Prize, 4237
Sid W. Richardson Foundation Grants, 4253
Sigma Xi Grants in Aid of Research, 4259
SRC Educational Science Doctoral Grants, 4355
SRC Educational Science Grants, 4356
SRC Educational Science Junior Research Position Grants, 4359
SRC Educational Science Medium-Cost Scientific Equipment Grants, 4361
SRC Educational Science Network Grants, 4362
SRC Educational Science Project Res Grants, 4364
SRC Research Infrastructures Grants, 4391
Swiss National Science Foundation DORE (DO REsearch) Conference Grants, 4482
Swiss National Science Foundation Fellowships for Advanced Researchers, 4483
Swiss National Science Foundation Fellowships for Prospective Researchers, 4484
Swiss National Science Foundation Marie Heim-Vogtlin Grants, 4485
Swiss National Science Foundation Scientific Conferences Grants, 4486
UNCFSP NASA Administrator's Fellowship, 4562
UNCFSP NASA Science and Technology Institute Summer Scholars Program, 4564
USDA Cooperative Extension System Education Grants, 4620
USDD Basic Research Challenge: Decentralized Online Optimization Grants, 4689
USDD Defense Sciences Research and Technology Grants, 4716
USDD High Assurance Cyber Military Systems Grants, 4731
USDD Ocean Engineering and Marine Systems Grants, 4767
USDD Power and Thermal Management Technology Development Grants, 4775
USDD Research Initiatives at the Naval Postgraduate School Grants, 4783
USDD Sabbatical Leave Fellowships, 4788
W.M. Keck Foundation Science and Engineering Research Grants, 4824
W.M. Keck Foundation Undergraduate Education Grants, 4825
Wayne and Gladys Valley Foundation Grants, 4835
Weizmann Institute of Science Postdoctoral Fellowships, 4837
Wellcome Trust Biomedical Science Grants, 4843
Winston Churchill Foundation Scholarships, 4886
Xerox Foundation Grants, 4899

Science Education

3M Company Higher Education Grants, 4
3M Foundation Community Giving Grants, 5
AAAS Entry Point Internships, 30
AAUW Community Action Grants, 213
Abbott Fund Global AIDS Care Grants, 219
Abdus Salam ICTP-OEA Affil Centre Grants, 222
Abdus Salam ICTP Networks Grants, 233
AERA Dissertation Grants, 342
Affymetrix Corporate Contributions Grants, 355
Alcoa Foundation Grants, 511
Alfred P. Sloan Foundation Civic Grants, 545
Alfred P. Sloan Foundation History of Science and Technology Grants, 553
Alfred P. Sloan Foundation Science of Learning STEM Grants, 562
American Chemical Society Petroleum Research Fund Scientific Education Grants, 692
Amgen Foundation Grants, 812
Applied Biosystems Grants, 916
ASU Graduate College Dean's Fellowships, 1236
Bay and Paul Foundations, Inc Grants, 1307
Bayer Foundation Grants, 1309
Beckman Coulter Community Relations Grants, 1325
Bill and Melinda Gates Foundation Agricultural Development Grants, 1345
Bodman Foundation Grants, 1356
BPW Career Advancement Scholarships, 1366
Cabot Corporation Foundation Grants, 1406
Charles Edison Fund Grants, 1520
CHF Roy G. Neville Short-Term Fellowships, 1543
Chiron Foundation Community Grants, 1553
Christine Mirzayan Science and Technology Policy Graduate Fellowship Program, 1556
ConocoPhillips Grants Program, 1629
DAAD Research Internships in Science and Engineering, 1704
DAAD RISE Professional Internships, 1706
DaimlerChrysler Corporation Fund Grants, 1714
DENSO North America Foundation Grants, 1742
Dept of Ed Star Schools Program Grants, 1749
Dept of Ed Women's Educational Equity Program Grants, 1752
DOE Acad Creating Teacher Scientists Grants, 1766
DOE Albert Einstein Distinguished Educator Fellowship Program, 1770
DOE Computational Science Grad Fellowships, 1780
DOE Faculty and Student Teams Grants, 1792
Dorr Foundation Grants, 1842
Eaton Charitable Fund Grants, 1893
Education Through Astronomy and Space Science Initiative, 1896

ESF Latsis Prize, 1951
Explorers Club Scott Pearlman Field Awards, 1967
Ford Motor Company Fund Grants Program, 2033
Heineman Foundation for Research, Education, Charitable and Scientific Purposes, 2227
HHMI-NIBIB Interfaces Initiative Grants, 2248
HHMI Biomedical Research Grants for Int'l Scientists Infectious Diseases and Parasitology, 2251
HHMI Biomedical Research Grants for International Scientists in Canada and Latin America, 2252
HHMI Biomedical Research Grants for International Scientists in the Baltics, Central and Eastern Europe, Russia, and Ukraine, 2253
HHMI Grants and Fellowships Programs, 2255
HHMI Initiative for Colleges Grants, 2256
HHMI Initiative for Universities Grants, 2257
HHMI Med into Grad Initiative Grants, 2259
HHMI Precollege Outreach Initiative for Biomedical Research Institutions Grants, 2261
L'Oreal USA Women in Science Fellowship, 2577
Luther I. Replogle Foundation Grants, 2631
McCarthy Family Foundation Grants, 2717
McCune Charitable Foundation Grants, 2719
MedImmune Charitable Grants, 2737
Motorola Foundation Grants, 2796
NASA Lewis Educational Research and Collaborative Internships, 2840
NASA New Aeronautics Competition for University Students, 2842
Northrop Grumman Foundation Grants, 3429
Northwood University/Alden B. Dow Creativity Center Summer Residency Program, 3432
NSF ADVANCE: Increasing the Participation and Advancement of Women in Academic Science and Engineering Careers Grants, 3461
NSF Communicating Res to Public Audiences, 3538
NSF Course, Curriculum, and Laboratory Improvement Research Grants, 3546
NSF Developing Global Scientists & Engineers, 3552
NSF Director's Award for Distinguished Teaching Scholars, 3553
NSF Historically Black Colleges and Universities Program Grants, 3577
NSF Innovative Technology Experiences for Students and Teachers, 3582
NSF Louis Stokes Alliances for Minority Participation Grants, 3589
NSF Presidential Awards for Excellence in Mathematics and Science Teaching, 3631
NSF Presidential Awards for Excellence in Science, Math, and Engineering Mentoring, 3632
NSF Research Experiences for Undergrad Grants, 3638
NSF SBE Doctoral Dissertation Research Improvement Grants, 3641
NSF Tribal Colleges and Universities Program, 3662
Olin Corporation Charitable Trust Grants, 3739
Praxair Foundation Grants, 3986
Rathmann Family Foundation Grants, 4033
RGK Foundation Grants, 4060
Rohm and Haas Company Grants, 4098
SRC Educational Science Doctoral Grants, 4355
SRC Educational Science Grants, 4356
SRC Educational Science Junior Research Position Grants, 4359
SRC Educational Science Medium-Cost Scientific Equipment Grants, 4361
SRC Educational Science Network Grants, 4362
SRC Educational Science Project Res Grants, 4364
UNCFSP NASA Harriett G. Jenkins Predoctoral Fellowships, 4563
USDD High School Apprenticeship Program and Undergraduate Research Apprenticeship Program Grants, 4732
USDD Science, Technology, Engineering and Mathematics Grants, 4789
W.M. Keck Foundation Undergraduate Education Grants, 4825
Westinghouse Charitable Giving Grants, 4856
Xerox Foundation Grants, 4899

Science Fiction

Children's Literature Assoc Research Fellowships, 1551

Science Planning/Policy

AAAS Roger Revelle Fellowship in Global Stewardship, 38
AAAS Science and Technology Policy Fellowships: Diplomacy, Security, and Development, 40
Alfred P. Sloan Foundation Research Fellowships, 560
American Academy of Arts and Sciences Hellman Fellowships, 612
American Academy of Arts and Sciences Visiting Scholars, 613
Christine Mirzayan Science and Technology Policy Graduate Fellowship Program, 1556
NSF Science and Technology Studies Grants, 3644

Science and Technology

AAUW American Dissertation Fellowships, 209
Ahmed Zewail Award in Ultrafast Science and Technology, 378
ALA STS Innovation in Science and Technology Librarianship Award, 499
Alfred P. Sloan Foundation Economic Analysis of Science and Technology Grants, 548
Alfred P. Sloan Foundation Public Understanding of Science and Technology-Other Efforts, 556
Alfred P. Sloan Foundation Public Understanding of Science & Tech-Public Television Grants, 557
Alfred P. Sloan Foundation Public Understanding of Science and Technology-Radio Grants, 558
Alfred P. Sloan Foundation Public Understanding of Science and Technology-Theater Grants, 559
Alfred P. Sloan Foundation Science of Learning STEM Grants, 562
AOCS Surfactants and Detergents Division Student Award, 898
ASM International Fellowship for Latin America and the Caribbean, 1175
Beckman Coulter Community Relations Grants, 1325
Bill and Melinda Gates Foundation Agricultural Development Grants, 1345
Canada-U.S. Fulbright International Science and Technology Award, 1417
CIC Smithsonian Institution Fellowships, 1564
DENSO North America Foundation Grants, 1742
ESF Latsis Prize, 1951
IIE David L. Boren Fellowships, 2350
ISGC Informal Education Grants, 2431
JSA JLab Graduate Fellowships, 2521
Motorola Foundation Innovation Generation Collaborative Grants, 2797
NOAA Chesapeake Bay Office B-WET (Bay Watershed Education and Training) Grants, 3339
NOAA Chesapeake Bay Office Cooperative Research Program Grants, 3340
NOAA Chesapeake Bay Office Fisheries Science Program Grants, 3341
NOAA Chesapeake Bay Office Non-Native Oyster Research Grants, 3342
NOAA Chesapeake Bay Office Submerged Aquatic Vegetation Culture and Restoration Grants, 3343
NOAA Cooperative Unit for Fisheries Education and Research Internships, 3360
NOAA Faculty and Student Intern Research, 3371
NOAA Miami Lab Summer Internships, 3382
Northrop Grumman Foundation Grants, 3429
NSF Accelerating Discovery In Science and Engineering through PETASCALE Simulations and Analysis, 3457
NSF ADVANCE: Increasing the Participation and Advancement of Women in Academic Science and Engineering Careers Grants, 3461
NSF Biological Physics, 3503
NSF Centers of Research Excellence in Science and Technology and HBCU Research Infrastructure for Science and Engineering Grants, 3515
NSF Developing Global Scientists & Engineers, 3552
NSF Emerging Frontiers in Research and Innovation Grants, 3560
NSF Partnerships for International Research and Education, 3611
NSF Research Experiences for Undergrad Grants, 3638
NSF Science and Society, 3643
NSF Small Business Technology Transfer Programs Phase I Grants, 3648
NWO Athena Grants, 3674
ONR Multidisciplinary Research Program of the URI Grants, 3774
Parkersburg Area Community Foundation Arts in Education Special Grants, 3869
Paul G. Allen Family Foundation Grants, 3876
Rohm and Haas Company Grants, 4098
SHOT Abbott Payson Usher Prize, 4237
South Florida Student Shark Program, 4330
SPIE C.S. Vikram Award in Optical Metrology, 4339
SPIE Frits Zernike Award, 4341
SPIE Technology Achievement Award, 4344
STTI National League for Nursing Grants, 4462
UNCFSP NASA Science and Technology Institute Summer Scholars Program, 4564
USAID Development Innovation Accelerator Broad Agency Announcement Grants, 4596
US CRDF Science and Technology Entrepreneurship Program Business Partnership Grant Competition Armenia, 4611
US CRDF STEP Business Partnership Grant Competition Ukraine, 4612
USDA Agriculture and Food Research Initiative Competitive Grants - Food Security, 4614
USDA Alaska Native-Serving and Native Hawaiian-Serving Institutions Education Grants, 4615
USDA Norman E. Borlaug International Agricultural Science and Technology Fellowships, 4636
USDA Organic Agriculture Research and Extension Initiative, 4637
USDA Policy Research Center Grants, 4640
USDD Advances in Bioscience for Airmen Performance Grants, 4675
USDD ARO Broad Agency Announcement, 4686
USDD Basic Research Challenge: Carbon Molecular Electronics Grants, 4688
USDD Basic Research Challenge: Decentralized Online Optimization Grants, 4689
USDD Basic Research for Combating Weapons of Mass Destruction Broad Agency Announcement Grants, 4690
USDD Collaborative Res in Structural Sciences, 4710
USDD Crowd Sourced Formal Verif Grants, 4715
USDD Defense Sciences Research and Technology Grants, 4716
USDD Developer Tools for Medical Education Public Physiology Research Platform Grant, 4717
USDD High Assurance Cyber Military Systems Grants, 4731
USDD Image Optimization for Small Focal Plane Arrays Grants, 4739
USDD Information Innovation Office Grants, 4740
USDD Innovative Systems for Military Missions - Tactical Technology Office Grants, 4742
USDD Long Range Broad Agency Announcement for Navy and Marine Corps Science and Technology Grants, 4748
USDD Multidisciplinary Research Program of the University Research Initiative Grant, 4756
USDD National Consortium for Measurement and Signature Intelligence Research Grants, 4759
USDD National Geospatial-Intelligency Agency Academic Research Grants, 4760
USDD Office-Wide Broad Agency Announcement Microsystems Technology Office Grant, 4768
USDD Power and Thermal Management Technology Development Grants, 4775
USDD Power Efficiency Revolution for Embedded Computing Technologies Grants, 4776
USDD Radio Frequency-Field Programmable Gate Arrays Grants, 4780
USDD Research and Analytical Support for the 711th Human Performance Wing's Human Effectiveness Directorate Grants, 4781

USDD Research Initiatives at the Naval Postgraduate School Grants, 4783
USDD Research Interest of the Air Force Office of Scientific Research Grants, 4784
USDD Research Interests of the US Air Force Academy Grants, 4785
USDD Science, Technology, Engineering and Mathematics Grants, 4789
USDD Science and Tecnology for Warfighter Training and Aiding Grants, 4790
USDD Strategic Technologies Grants, 4798
USDD Warfighter Interface Technologies Advanced Research Programs Grants, 4803
W.M. Keck Foundation Science and Engineering Research Grants, 4824

Science and Technology Centers
Alfred P. Sloan Foundation Public Understanding of Science and Technology-Other Efforts, 556
Marion Gardner Jackson Charitable Trust Grants, 2664

Science and Technology, History of
Alfred P. Sloan Foundation Economic Analysis of Science and Technology Grants, 548
Alfred P. Sloan Foundation Public Understanding of Science and Technology-Other Efforts, 556
Alfred P. Sloan Foundation Public Understanding of Science & Tech-Public Television Grants, 557
American Historical Association Fellowship in Aerospace History, 717
American Philosophical Society Library Resident Research Fellowships, 769
ASME Engineer-Historian Award, 1118
CHF Glenn E. and Barbara Hodsdon Ullyot Scholarship, 1540
CHF Long-Term Fellowships, 1541
CHF Robert W. Gore Fellowship in Materials Innovation, 1542
CHF Societe de Chimie Industrielle (American Section) Fellowship, 1544
CHF Travel Grants, 1545
Gladys Krieble Delmas Foundation Grants for Venetian Research, 2144
Houghton Library Visiting Fellowships, 2284
Huntington Library Dana and David Dornsife Fellowships, 2303
Huntington Library National Endowment for the Humanities Fellowships, 2307
Huntington Library Travel Grants and Exchange Fellowships for Study in Great Britain, 2309
NSF Science and Society, 3643
SHOT Abbott Payson Usher Prize, 4237
SHOT Brooke Hindle Postdoctoral Fellowships, 4238
SHOT Eugene S. Ferguson Prize, 4240
SHOT IEEE Life Members Prize in Elec Hist, 4241
SHOT International Scholars, 4242
SHOT Joan Cahalin Robinson Prize, 4243
SHOT Leonardo da Vinci Medal, 4244
SHOT NASA Fellowship, 4246
SHOT Samuel Eleazar and Rose Tartakow Levinson Prize, 4247
Smithsonian Graduate Student Fellowships, 4277
Smithsonian Minority Visiting Student Awards, 4279
Smithsonian Native American Visiting Student Awards, 4292
Smithsonian Postdoctoral Fellowships, 4293
Smithsonian Predoctoral Fellowships, 4295
Smithsonian Senior Fellowships, 4296

Scleroderma
Scleroderma Established Investigator Grants, 4190
Scleroderma Foundation New Investigator Grants, 4191

Scotland
Royal Scottish Geographical Society Expedition Grants, 4110
Royal Scottish Geographical Society Small Research Grants, 4111
Royal Scottish Geographical Society Small Travel Grants, 4112

Scriptwriting
Alfred P. Sloan Foundation Public Understanding of Science and Technology-Theater Grants, 559
IFP Minnesota McKnight Screenwriters Fellows, 2343
Ludwig Vogelstein Foundation Grants, 2626
NYSCA Individual Artists: Film, Media and New Technology Production Grants, 3706

Sculpture
ACC Residency Program in Asia, 267
ACC Starr Foundation Fellowship Program, 268
CICF Efroymson Contemporary Arts Fellows, 1563
Florida Division of Cultural Affairs Individual Artist Fellowships, 2026
John Simon Guggenheim Memorial Foundation Fellowships, 2507
National Gallery of Art Kress Interpretive Fellow, 2869
NYFA Artist Fellowships, 3698
Pew Fellowships in the Arts, 3905
Pollock-Krasner Foundation Grants, 3981

Sea Turtles
Florida Sea Turtle Grants Program, 2027

Seafood
GNOF Gulf Coast Oil Spill Grants, 2153
USDC/NOAA Nature Conservancy Community-based Matching Grants, 4661

Secondary Education
3M Foundation Community Giving Grants, 5
A.O. Smith Community Grants, 12
AASF Lydia Donaldson Tutt-Jones Memorial Research Grant, 203
AAUW Community Action Grants, 213
Acuity Charitable Foundation Grants, 333
Affymetrix Corporate Contributions Grants, 355
AIHS Heritage Youth Researcher Summer Science Grants, 449
Alcatel-Lucent Technologies Foundation Grants, 509
Alcoa Foundation Grants, 511
Anderson Foundation Grants, 842
Applied Biosystems Grants, 916
APS/Phoenix Suns Energy and Environmental Minigrants, 919
Archer Daniels Midland Foundation Grants, 985
Astronomical Society of the Pacific Las Cumbres Amateur Outreach Award, 1230
Auburn Foundation Grants, 1263
Ball Brothers Foundation General Grants, 1291
Bay and Paul Foundations, Inc Grants, 1307
BMW of N America Charitable Contributions, 1353
Boeing Company Contributions Grants, 1357
Booth-Bricker Fund Grants, 1359
Boston Foundation Grants, 1361
Bristol-Myers Squibb Foundation Health Education Grants, 1378
CAPAA Mini Grant for Research Related to African Americans in Education, 1424
Cape Branch Foundation Grants, 1425
Carl and Lily Pforzheimer Foundation Grants, 1427
Carnegie Corporation of New York Grants, 1432
Charles H. Price II Family Foundation Grants, 1523
CIGNA Foundation Grants, 1568
CIRCLE Civic Education at the High School Level Research Grants, 1571
Clarence E. Heller Charitable Foundation Grants, 1575
CMS Historically Black Colleges and Universities Health Services Research Grants, 1585
Coca-Cola Foundation Grants, 1593
Coeta and Donald Barker Foundation Grants, 1594
Colonel Stanley R. McNeil Foundation Grants, 1597
Commonweal Foundation Community Assistance Grants, 1602
ConocoPhillips Grants Program, 1629
Cowles Charitable Trust Grants, 1652
DaimlerChrysler Corporation Fund Grants, 1714
Dept of Ed Jacob K. Javits Gifted and Talented Students Education Program-National Research and Development Center, 1746
Dept of Ed Star Schools Program Grants, 1749
DOE Acad Creating Teacher Scientists Grants, 1766
DOE Albert Einstein Distinguished Educator Fellowship Program, 1770
Dorr Foundation Grants, 1842
Earthwatch Institute Student Fellowships, 1890
Eaton Charitable Fund Grants, 1893
Elizabeth McGraw Foundation Grants, 1908
Elizabeth Nash Found Summer Research Awards, 1909
Engineering Information Foundation Grants, 1927
Eugene McDermott Foundation Grants, 1958
Field Museum High School Internships, 2009
Frances L. and Edwin L. Cummings Memorial Fund Grants, 2046
George A Ohl Jr. Foundation Grants, 2112
George Foundation Grants, 2115
George Frederick Jewett Foundation Grants, 2116
GNOF IMPACT Grants for Education, 2154
Greenspun Family Foundation Grants, 2173
Greenwall Foundation Bioethics Grants, 2174
Harold Alfond Foundation Grants, 2200
Helen Bader Foundation Grants, 2228
HHMI Grants and Fellowships Programs, 2255
Holberg Memorial Fund School Project, 2274
HSBC Contributions Community Grants, 2291
Huie-Dellmon Trust Grants, 2295
Hutchinson Community Foundation Grants, 2310
Indiana Space Grant Consortium Discovery Program Student Hands-On Research Grants, 2374
IRA Steven A. Stahl Research Grants, 2413
Joyce Foundation Education Grants, 2519
Kenneth T. and Eileen L. Norris Found Grants, 2554
Lilly Endowment Teacher Creativity Fellowships, 2616
Lumpkin Family Found Healthy People Grants, 2629
Luther I. Replogle Foundation Grants, 2631
M.J. Murdock Charitable Trust Partners in Science Grants, 2642
MAA-Tensor Foundation Women and Mathematics Grants, 2645
Margaret & James A. Elkins Jr. Found Grants, 2658
McCarthy Family Foundation Grants, 2717
McCune Charitable Foundation Grants, 2719
McGraw-Hill Companies Community Grants, 2724
Mericos Foundation Grants, 2742
Meyer Memorial Trust Responsive Grants, 2748
MGM Resorts Foundation Community Grants, 2753
Motorola Foundation Grants, 2796
Musgrave Foundation Grants, 2807
NCTM E. Glenadine Gibb 7-12 Classroom Research Grants, 2941
NEH Fellowships, 2956
Nellie Mae Education Foundation District-Level Change Grants, 2968
Nellie Mae Education Foundation Public Understanding Grants, 2969
Nellie Mae Education Foundation State Level Systems Change Grants, 2971
NHLBI Research Supplements to Promote Diversity in Health-Related Research, 3054
Norcliffe Foundation Grants, 3415
Nordson Corporation Foundation Grants, 3416
NSF Engineering Research Centers Grants, 3561
NSF Presidential Awards for Excellence in Mathematics and Science Teaching, 3631
NSF Presidential Awards for Excellence in Science, Math, and Engineering Mentoring, 3632
Oppenstein Brothers Foundation Grants, 3792
Perpetual Trust for Charitable Giving Grants, 3900
Phi Upsilon Omicron Geraldine Clewell Masters Fellowship, 3946
Price Family Charitable Fund Grants, 3987
Reinberger Foundation Grants, 4051
Richard and Susan Smith Family Found Grants, 4064
Sage Foundation Grants, 4158
Salem Foundation Grants, 4160
Sid W. Richardson Foundation Grants, 4253
Sioux Falls Area Community Foundation Community Fund Grants (Unrestricted), 4262
Sioux Falls Area Comm Foundation Spot Grants, 4263

Sony Corporation of America Corporate Philanthropy Grants, 4329
Stackpole-Hall Foundation Grants, 4442
Strake Foundation Grants, 4452
Sun Academic Excellence Grants, 4466
Sun Microsystems Foundation Summer Academic Enrichment Grants, 4468
T. James Kavanagh Foundation Grants, 4489
Tanner Humanities Center Research Fellowships, 4494
The Ray Charles Foundation Grants, 4502
Thomas and Dorothy Leavey Foundation Grants, 4506
Thornton Foundation Grants, 4514
Victor E. Speas Foundation Grants, 4810
W. C. Griffith Foundation Grants, 4818
W.C. Griffith Foundation Grants, 4819
Westinghouse Charitable Giving Grants, 4856
WHOI Guest Student Internships, 4860

Security
Harvard Women and Public Policy Postdoctoral Fellowship, 2217
IIE Rockefeller Foundation Bellagio Center Residencies, 2355
SSRC Drugs, Security & Democracy Fellows, 4426
SSRC Next Generation Social Sciences in Africa: Dissertation Completion Fellowships, 4434
SSRC Next Generation Social Sciences in Africa: Doctoral Dissertation Proposal Fellowships, 4435
SSRC Next Generation Social Sciences in Africa: Doctoral Dissertation Research Fellowships, 4436
USDD Active Authentication Grants, 4674
USDD Research Interests of the US Air Force Academy Grants, 4785

Sedimentology
DOE Geosciences Research Grants, 1797
HRF Hudson River Improvement Grants, 2286
NSF Arctic Nat Science Grants in Earth Sciences, 3485
NSF Arctic Nat Science Grants in Ocean Science, 3487
NSF Sediment Geology & Paleobiology Grants, 3646
ONR Optical and Biological Oceanography Research Grants, 3777
USDD Naval Research Facilities Grants, 4761

Seismology
Abdus Salam ICTP Postgraduate Diploma Grants, 235
Abdus Salam ICTP Training and Research in Italian Laboratories Grants, 240
DOI Earthquake Hazards Reduction Res Grants, 1821
NSF Arctic Nat Science Grants in Earth Sciences, 3485
NSF Continental Dynamics Research Grants, 3543
NSF Deep Underground Science and Engineering Laboratory Grants, 3551
NSF Geophysics Grants, 3571
NSF George E. Brown Jr. Network for Earthquake Engineering Simulation Research, 3572

Semiconductors
Abdus Salam ICTP Postgraduate Diploma Grants, 235
Abdus Salam ICTP Training and Research in Italian Laboratories Grants, 240
NSF Condensed Matter Physics Research Grants, 3540
ONR Semiconductor Devices Grants, 3781
SRC AMD/Mahboob Khan Fellowship, 4348
SRC Aristotle Award, 4349
SRC Company-Named Fellowships, 4350
SRC Company-Named Scholarships, 4351
SRC Core Research Grants, 4352
SRC Cross-Disciplinary Research Grants, 4353
SRC Franco Cerrina Fellowship, 4367
SRC Graduate Fellowships, 4368
SRC GRC & SEMATECH Joint Projects Grants, 4369
SRC Hans J. Coufal Fellowship, 4371
SRC International Fellowship, 4375
SRC Landmark Innovation Award, 4376
SRC Mahboob Khan Outstanding Industry Liaison Award, 4377
SRC Master's Scholarships, 4378
SRC Peter Verhofstadt Fellowship, 4389
SRC Research Customization Program Grants, 4390
SRC Robert M. Burger Fellowship, 4392
SRC Simon Karecki Fellowship, 4393
SRC Technical Excellence Award, 4394
SRC Undergraduate Research Opportunities, 4395
USDD RF Semiconductors, RF Solid State Amplifiers and Wide Bandgap Materials Grants, 4786

Seminars
DAAD Study Visits / Study Seminars and Practicals in Germany by Foreign Students, 1709
Greater Worcester Community Foundation Jeppson Memorial Fund for Brookfield Grants, 2165
Kennedy Center Summer Fellowships for International Arts Managers, 2553
NL Associated Colleges of the Midwest/Great Lakes College Association Faculty Fellowships, 3281
Oppenstein Brothers Foundation Grants, 3792
Potts Memorial Foundation Grants, 3985
Searle Freedom Trust Grants, 4192
Virginia Foundation for the Humanities Discretionary Grants, 4813
Virginia Foundation for Humanities Open Grants, 4814

Senile Dementia
AHAF Alzheimer's Disease Research Grants, 370
Alzheimer's Association Conference Grants, 589
Alzheimer's Association Development of New Cognitive and Functional Instruments Grants, 590
Alzheimer's Association Everyday Technologies for Alzheimer Care Grants, 591
Alzheimer's Association Investigator-Initiated Research Grants, 592
Alzheimer's Association Mentored New Investigator Research Grants to Promote Diversity, 593
Alzheimer's Association Neuronal Hyper Excitability and Seizures in Alzheimer's Disease Grants, 594
Alzheimer's Association New Investigator Research Grants, 595
Alzheimer's Association New Investigator Research Grants to Promote Diversity, 596
Alzheimer's Association U.S.-U.K. Young Investigator Exchange Fellowships, 597
Alzheimer's Association Zenith Fellows Awards, 598
Helen Bader Foundation Grants, 2228

Senior Citizen Programs and Services
Hearst Foundations Health Grants, 2225
Helen Bader Foundation Grants, 2228
M.D. Anderson Foundation Grants, 2637
Mericos Foundation Grants, 2742

Sensing Devices and Transducers
NSF Biomedical Engineering and Engineering Healthcare Grants, 3506
NSF Biophotonics, Advanced Imaging, and Sensing for Human Health Grants, 3508
ONR Marine Meteorology and Atmospheric Effects Research Grants, 3769
USDD Acoustic Transduction Materials and Devices Grants, 4672
USDD Signal Processing for Networked Sensing Grants, 4793

Sensors
ONR Electronics, Sensors, and Network Research Grants, 3758
USDD Electro-Optical/Infrared Sensor Processing Grants, 4720
USDD Non-Destructive Evaluation and Prognostics: Advanced Sensors and Technologies Grants, 4763
USDD Sensor Innovative Research Grants, 4791

Sensory System
ONR Target Tracking and Sensor Fusion Grants, 3787
SfN Swartz Prize for Theoretical and Computational Neuroscience, 4213
USDD Non-Destructive Evaluation and Prognostics: Advanced Sensors and Technologies Grants, 4763
USDD Sensor Innovative Research Grants, 4791

Separation Process
AIChE Clarence G. Gerhold Award, 400
DOE Atomic, Molecular, and Optical Sciences Research Grants, 1771
DOE Catalysis and Chemical Transformations Research Grants, 1773
DOE Chemical Energy and Chemical Engineering Research Grants, 1774
DOE Chemical Physics Research Grants, 1775
DOE Chemical Sciences Research Grants, 1776
DOE Heavy Element Chem Research Grants, 1798
DOE Photochem & Radiation Research Grants, 1809
DOE Separations and Analysis Research Grants, 1813
NSF Interfacial Processes and Thermodynamics Research Grants, 3586

Serbia
IREX Individual Advanced Research Opportunities Grants, 2419

Service Delivery Programs
CMS Research and Demonstration Grants, 1586
Eugene McDermott Foundation Grants, 1958
Health Foundation of Southern Florida Responsive Grants, 2223
ICC Scholarship of Engagement Faculty Grants, 2334
Ittleson Foundation AIDS Grants, 2436
JM Foundation Grants, 2490
Luther I. Replogle Foundation Grants, 2631
Mericos Foundation Grants, 2742
Richard and Susan Smith Family Found Grants, 4064
Robert E. and Evelyn McKee Foundation Grants, 4074
Union Pacific Foundation Health and Human Services Grants, 4569

Service Learning
AAAAI Distinguished Service Award, 20
ASME Student Section Advisor Award, 1166

Set Theory
NSF Topology Research Grants, 3661

Sex Education
Ittleson Foundation AIDS Grants, 2436

Sex Roles
Ford Foundation Peace and Social Justice Grants, 2031
International Studies Graduate Fellowships, 2403
NCSS/FASSE/CUFA Inquiry Grant, 2936
NCSS Exemplary Res in Social Studies Award, 2937
NCSS L Metcalf Exemplary Dissertation Award, 2939

Sexism
Dept of Ed Women's Educational Equity Program Grants, 1752
Sioux Falls Area Comm Foundation Spot Grants, 4263
Vanguard Public Foundation Grant Funds, 4808

Sexual Abuse
Burton G. Bettingen Grants, 1390
Community Foundation for S Arizona Grants, 1618
Ford Foundation Peace and Social Justice Grants, 2031

Sexual Behavior
amfAR Fellowships, 809
amfAR Mathilde Krim Fellowships in Basic Biomedical Research, 810
amfAR Research Grants, 811
Playboy Foundation Grants, 3980

Sexuality
Overbrook Foundation Grants, 3860

Sexually Transmitted Diseases
APHL Emerging Infectious Diseases Fellowships, 914
ASM/CDC Fellowships in Infectious Disease and Public Health Microbiology, 1100
CDC Foundation Atlanta International Health Fellowships, 1461
Ittleson Foundation AIDS Grants, 2436

NFID New Investigator Matching Grants, 2989
ORISE Faculty Research Participation at the Centers for Disease Control and Prevention, 3803
ORISE Postgraduate Research Participation at the Centers for Disease Control and Prevention, 3820
ORISE Student Research Participation at the Centers for Disease Control and Prevention, 3829
Sexually Transmitted Diseases Postdoctoral Fellowships, 4200
USAID Comprehensive District-Based Support for Better HIV/TB Patient Outcomes Grants, 4595

Shelters
Blowitz-Ridgeway Foundation Grants, 1349
Campbell Soup Foundation Grants, 1415
Chautauqua Region Community Found Grants, 1533
Helen V. Brach Foundation Grants, 2231
HomeBanc Foundation Grants, 2275
Luther I. Replogle Foundation Grants, 2631
Melville Charitable Trust Grants, 2741
Mericos Foundation Grants, 2742
MGN Family Foundation Grants, 2754
Perpetual Trust for Charitable Giving Grants, 3900
Pinkerton Foundation Grants, 3967
Reuter Foundation Grants, 4059
Textron Corporate Contributions Grants, 4501
Thompson Charitable Foundation Grants, 4511
Vancouver Foundation Grants and Community Initiatives, 4807

Sickle Cell Disease
NHLBI Exploratory Studies in the Neurobiology of Pain in Sickle Cell Disease, 3026

Signal Processing
ARO Information Processing and Fusion and Circuits Research Contracts, 1010
ARO Mobile, Wireless Communications and Networks Research Contracts, 1012
ARO Theoretical Physics and Nonlinear Phenomena Research Contracts, 1017
ASEE Army Research Laboratory Postdoctoral Fellowships, 1059
USDD Mixed Signal, Signal Processing Devices, Circuits and Architecture Grants, 4755
USDD Ocean Acoustics Research Grants, 4766
USDD Radar and Signal Processing Grants, 4779
USDD Undersea Signal Processing Grants, 4801

Single-Parent Families
Eugene M. Lang Foundation Grants, 1957

Skin Diseases
National Psoriasis Foundation Research Grants, 2882
NIAMS Pilot & Feasibility Clinical Grants in Arthritis & Musculoskeletal & Skin Diseases, 3111
Pfizer Inflammation Competitive Res Awards, 3926
Pfizer Research Initiative Dermatology Grants, 3928
Skin Cancer Foundation Research Grants, 4267

Slavic Languages/Literature
Kosciuszko Foundation Tuition Scholarships, 2574

Sleep Disorders
American Psychiatric Association/Merck Co. Early Academic Career Research Award, 778
CJ Foundation for SIDS Research Grants, 1573
National Sleep Foundation Pickwick Postdoctoral Fellowship, 2886
NHLBI Ancillary Studies in Clinical Trials, 3015
NHLBI Mentored Career Award For Faculty At Minority Institutions, 3034
NHLBI Ruth L. Kirschstein National Research Service Awards for Individual Postdoctoral Fellows, 3057
NHLBI Short-Term Research Education Program to Increase Diversity in Health-Related Research, 3062
NIH Biobehavioral Research for Effective Sleep, 3205
NIH Research on Sleep and Sleep Disorders, 3236

Small Businesses
Chautauqua Region Community Found Grants, 1533
DOE Inventions and Innovation Grant Program, 1802
DOE Small Business Innovation Research Commercialization Assistance Program, 1814
Fitzpatrick and Francis Family Business Continuity Foundation Grants, 2021
Indiana 21st Century Research and Technology Fund Awards, 2370
Indiana SBIR/STTR Commercialization Enhancement Program, 2371
McCune Charitable Foundation Grants, 2719
NCI Academic-Industrial Partnerships for Devel & Validation of In Vivo Imaging Systems and Methods for Cancer Investigations, 2913
NHLBI Bioengineering Approaches to Energy Balance and Obesity Grants for SBIR, 3017
NHLBI Bioengineering Approaches to Energy Balance and Obesity Grants for STTR, 3018
NIEHS Small Business Innovation Research Program Grants, 3195
NIH Recovery Act Limited Competition: Small Business Catalyst Awards for Accelerating Innovative Research Grants, 3233
NIST Small Business Innovation Res Grants, 3273
NLM Small Business Innovation Res Grants, 3308
NSF Small Business Innovation Research Grants, 3647
NSF Small Business Technology Transfer Programs Phase I Grants, 3648
ONR Small Business Innovation Research Grants, 3782
ONR Small Business Technology Transfer Program Grants, 3783
SSHRC-NSERC College and Community Innovation Enhancement Grants, 4400
USDA Emerging Markets Grants, 4623
USDA Foreign Market Development Grants, 4625
USDA Market Access Program Grants, 4633
USDA Small Business Innovation Research Grants - Phase I, 4648
USDA Small Business Innovation Research Grants - Phase II, 4649

Smoking Behavior
Alton Ochsner Award, 583
American Legacy Found Small Innovative Grants, 749
Bupa Foundation Multi-Country Grant, 1387
CDC Found Tobacco Network Lab Fellowship, 1462
Christine O. Gregoire Young Adult Award for Outstanding Use of Tobacco Industry Docs, 1557
CTCRI Idea Grants, 1665
CTCRI Knowledge Synthesis Grants, 1666
CTCRI Policy Research Grants, 1667
CTCRI Researcher Travel Grants, 1668
CTCRI Research Planning Grants, 1669
CTCRI Student Research Grants, 1670
CTCRI Workshop and Learning Opportunities Grants, 1671
NCI/NIDA Transdisciplinary Tobacco Use Research Centers Grants, 2912
Pfizer Global Research Awards for Nicotine Independence, 3925
Sybil G. Jacobs Award for Outstanding Use of Tobacco Industry Documents, 4487
TRDRP California Research Awards, 4525
TRDRP New Investigator Awards, 4528
TRDRP Research Grants, 4531

Social Change
Akonadi Foundation Anti-Racism Grants, 477
Alston Bannerman Sabbatical Fellowships, 580
Arcus Foundation Religion and Values Grants, 987
Ittleson Foundation Mental Health Grants, 2438
Japan Foundation Center for Global Partnership NPO Fellowships, 2474
Jenifer Altman Foundation Grants, 2484
Joel L. Fleishman Civil Society Fellowships, 2495
McCune Charitable Foundation Grants, 2719
National Endowment for Democracy Reagan-Fascell Democracy Fellowships, 2863
NL Arthur and Lila Weinberg Fellowships for Independent Scholars and Researchers, 3280
Paul G. Allen Family Foundation Grants, 3876
Playboy Foundation Grants, 3980
Radcliffe Institute Research Support Grants, 4023
S.E.VEN Fund Open Enterprise Solutions to Poverty RFP Grants, 4155
Saint Mary's College Fellowships, 4159
Shastri Indo-Canadian Institute Partnership Development Seed Grants, 4228
SSHRC Therese F. Casgrain Fellowship, 4418
Vanguard Public Foundation Grant Funds, 4808
Wolfsonian-FIU Fellowships, 4890

Social History
American Historical Association Fellowship in Aerospace History, 717
American Historical Association Michael Kraus Research Grant in American Colonial History, 738
American Society for Yad Vashem Grants, 798
ARIT/NEH Fellowships for Research in Turkey, 990
ASPRS William A. Fischer Mem Scholarship, 1216
Clark Library Short-Term Resident Fellowships, 1577
National Endowment for the Humanities Fellowships at Winterthur, 2867
OAH Merle Curti Award, 3726
Radcliffe Institute Carol K. Pforzheimer Student Fellowships, 4018
Radcliffe Institute Research Support Grants, 4023
Smithsonian Latino Studies Fellowships, 4278
Smithsonian Senior Fellowships, 4296
SSRC Our Shared Past Grants, 4437
United Methodist Church Women in History Research Grant, 4574
Winterthur Museum Dissertation Fellowships, 4887
Winterthur Research Fellowships, 4888
Wolfsonian-FIU Fellowships, 4890

Social Justice
Alston Bannerman Sabbatical Fellowships, 580
American Academy of Advertising Kim Rotzoll Award for Advertising Ethics and Social Responsibility, 610
Gruber Foundation Prize for Global Justice and Women's Rights, 2182
McCune Charitable Foundation Grants, 2719
NL Arthur and Lila Weinberg Fellowships for Independent Scholars and Researchers, 3280
OSF-Baltimore Community Fellowships, 3833
OSF Fellowships, 3840
Saint Mary's College Fellowships, 4159
SSHRC Therese F. Casgrain Fellowship, 4418

Social Measurement and Indicators
ASA/NSF/BLS Fellowships, 1030
Century Foundation Grants, 1489

Social Movements
Akonadi Foundation Anti-Racism Grants, 477
Arcus Foundation Religion and Values Grants, 987
Catholic University Life Cycle Institute Research Fellowship, 1444
Facebook Fellowships, 1973
Kellogg Institute for International Studies Course Development Grants, 2537
Kellogg Institute for International Studies Dissertation Year Fellowships, 2538
Kellogg Institute for International Studies Graduate Research Grants, 2542
Kellogg Institute for International Studies Intramural Grants, 2546
Kellogg Institute for International Studies Summer Internships, 2549
Kellogg Institute for International Studies Supplemental Awards for PhD Students, 2550
Kellogg Institute for International Studies Visiting Fellowships, 2551

Social Psychology
ASU Graduate College Martha E. Bernal Memorial Award, 1242

Judith Graham Pool Postdoctoral Research Fellowship in Hemophilia, 2529
National Center for Responsible Gaming Centers of Excellence in Gambling Research Grants, 2850
National Center for Responsible Gaming Early Stage Investigator Grants, 2852
National Center for Responsible Gaming Exploration Grants, 2853
National Center for Responsible Gaming Independent Scientist Grants, 2854
National Center for Resp Gaming Large Grants, 2855
National Center for Responsible Gaming Postdoctoral Fellowships, 2856
National Center for Resp Gaming Seed Grants, 2857
National Center for Responsible Gaming Travel Grants, 2858
NSF Social Psychology Research Grants, 3649
Olfactory Research Fund Grants, 3738
Russell Sage Foundation Grants, 4129
SSRC New Directions in the Study of Prayer Research Grants, 4433
USDD Human Social, Cultural and Behavior Modeling Thrust Grants, 4737

Social Science Education
ACLS Comparative Perspectives on Chinese Culture and Society Grants, 283
ASU Graduate College Completion Fellowships, 1235
Blowitz-Ridgeway Foundation Early Childhood Development Research Award, 1348
Fulbright-Hays Seminars Abroad Bilateral Projects Grants, 2062
Getty Scholars and Visiting Scholars Grants, 2138
Institute of Turkish Studies Research Grants in Comparative Studies of Modern Turkey, 2390
Kluge Center Fellowships, 2561
McCune Charitable Foundation Grants, 2719
NCSS/FASSE/CUFA Inquiry Grant, 2936
NCSS Exemplary Res in Social Studies Award, 2937
NCSS Jean Dresden Grambs Distinguished Career Research in Social Studies Award, 2938
NCSS L Metcalf Exemplary Dissertation Award, 2939
OSF Global Faculty Grants - Track 2, 3842
Radcliffe Inst Individual Residential Fellowships, 4021
SRC Humanities and Social Science Postdoctoral Fellowships, 4372
SRC Humanities and Social Sciences Grants for Postdoctoral Positions In Sweden, 4374
SSHRC Banting Postdoctoral Fellowships, 4404
SSHRC Digging into Data Challenge Grants, 4406

Social Sciences
AAAS Science and Technology Policy Fellowships: Congressional, 39
AAAS Science and Technology Policy Fellowships: Global Health and Development, 42
AAUW American Dissertation Fellowships, 209
AAUW American Postdoctoral Research Leave Fellowships, 210
AAUW American Summer/Short-Term Research Publication Grants, 211
ACIE Title VIII Research Scholar Fellowships, 276
ACLS/New York Public Library Fellowships, 277
ACLS African Humanities Program for Dissertation Completion, 278
ACLS American Research in the Humanities in China Fellowships, 280
ACLS Charles A. Ryskamp Research Fellowships, 281
ACLS Collaborative Research Fellowships, 282
ACLS Digital Innovation Fellowships, 284
ACLS Disse Fellowships in East European Studies, 285
ACLS Early Career Postdoctoral Fellowships in East European Studies, 286
ACLS Fellowships, 287
ACLS Frederick Burkhardt Residential Fellowships for Recently Tenured Scholars, 288
ACLS International and Area Studies Fellowships, 289
AIAR/Council of American Overseas Research Centers Fellowships for Advanced Multi-Country Research, 391
Aid to Scholarly Publications, 431
Alabama Humanities Foundation Mini Grants, 480
Alabama Humanities Foundation Planning/Consultant Grants, 481
Alexander von Humboldt Foundation Anneliese Maier Award, 515
Alexander von Humboldt Foundation Fellowships for Experienced Researchers, 516
Alexander von Humboldt Foundation Fellowships for Postdoctoral Researchers, 517
Alexander von Humboldt Foundation Georg Forster Research Award, 525
Alexander von Humboldt Foundation JSPS Research Fellowships, 530
Alexander von Humboldt Foundation NSC Taiwan Fellowships, 533
Alexander von Humboldt Foundation Reimar Lust Award, 536
Alexander von Humboldt Foundation Sofja Kovalevskaja Awards, 540
Alzheimer's Association Investigator-Initiated Research Grants, 592
American Academy of Religion Individual Research Grants, 618
American Mathematical Society AAAS Mass Media Fellowships, 751
American Numismatic Society Doc Fellowships, 759
American Philosophical Society Sabbatical Fellowship in the Humanities and Social Sciences, 771
American Sociological Association Sydney S. Spivack Program in Applied Social Research and Social Policy, 806
amfAR Fellowships, 809
amfAR Mathilde Krim Fellowships in Basic Biomedical Research, 810
amfAR Research Grants, 811
Andrew W. Mellon/ACLS Dissertation Completion Fellowships, 844
ARIT/NEH Fellowships for Research in Turkey, 990
ARIT Fellowships, 991
ARIT Mellon Foundation Postdoctoral Research Fellowships for East European Scholars, 994
ASCE Trent R. Dames and William W. Moore Engineering Fellowship, 1043
ASU Graduate College Dean's Fellowships, 1236
ASU Graduate College Dissertation Fellowships, 1237
ASU Graduate College Science Foundation Arizona Fellowships in Science and Engineering, 1245
AWIS Educational Foundation Predoc Awards, 1278
BA Larger Research Grants, 1290
Banfi Vintners Foundation Grants, 1296
BA Sir Ernest Cassel Educational Trust Grants, 1303
BA Small Research Grants, 1304
Blowitz-Ridgeway Foundation Early Childhood Development Research Award, 1348
Blowitz-Ridgeway Foundation Grants, 1349
BMI Woody Guthrie Fellowships, 1352
Brookdale Foundation Leadership in Aging Fellowships, 1381
Bullard Fellowship in Forest Research, 1383
Camargo Foundation Fellowships, 1412
Canada-U.S. Fulbright International Science and Technology Award, 1417
Canada-U.S. Fulbright Senior Specialists Program Grants, 1418
Canada-U.S. Fulbright Traditional Scholars, 1419
CCIS Visiting Research Fellowships, 1452
Center for Defense Information Internships, 1473
Center for the Study of Democratic Politics Fellowships, 1481
CFUW Margaret Dale Philp Award, 1512
Consortium for Graduate Study in Management Minority Fellowships, 1642
DAAD Research Grants for Doctoral Candidates and Young Academics and Scientists, 1703
DAAD Research Stays for University Academics and Scientists, 1705
Dartmouth College Andrew W. Mellon Postdoctoral Fellowship, 1721
Douglass Institute Predoctoral Fellowships, 1852
Duke University Postdoctoral Research Fellowships in Aging, 1880
Frederick Douglass Summer Fellowships, 2052
Frederick Douglass Summer Scholars Fellowship, 2053
Fulbright Alumni Initiatives Awards Grants, 2064
Fulbright Distinguished Chairs Program Grants, 2069
Fulbright European Union Scholar-in-Residence Program Grants, 2071
Fulbright New Century Scholars Program Grants, 2075
Fulbright Occasional Lecturer Program Grants, 2076
Fulbright Scholar-in-Residence Program Grants, 2078
Fulbright Specialists Program Grants, 2079
Fulbright Traditional Scholar Program in Europe and Eurasia, 2080
Fulbright Traditional Scholar Program in Sub-Saharan Africa, 2081
Fulbright Traditional Scholar Program in the East Asia/Pacific Region, 2082
Fulbright Traditional Scholar Program in the Near East and North Africa Region, 2083
Fulbright Traditional Scholar Program in the South and Central Asia Region, 2084
Fulbright Traditional Scholar Program in the Western Hemisphere, 2085
Fulbright Visiting Specialists Program: Direct Access to the Muslim World, 2087
Fund for Scientific Research--Flanders Grants, 2089
Gaius Charles Bolin Fellowships for Minority Graduate Students, 2095
Getty Scholars and Visiting Scholars Grants, 2138
Gilbert F. White Postdoctoral Fellowships, 2140
Harry Frank Guggenheim Foundation Research Grants, 2208
Harvard University Weatherhead Center for International Affairs Fellowships, 2215
Holberg Memorial Fund International Prize, 2272
Holberg Memorial Fund Nils Klim Prize, 2273
Holberg Memorial Fund School Project, 2274
Horowitz Foundation for Social Policy Grants, 2280
Horowitz Found for Social Policy Special Awards, 2281
HRF Hudson River Improvement Grants, 2286
IIE Rockefeller Foundation Bellagio Center Residencies, 2355
Inamori Foundation Research Grants, 2368
Indian Council of Social Science Research Grants, 2377
Institute of Turkish Studies Dissertation Writing Grants, 2389
Institute of Turkish Studies Research Grants in Comparative Studies of Modern Turkey, 2390
IREX Eurasian Undergraduate Exchange Grants, 2418
IREX Individual Advanced Research Opportunities Grants, 2419
IREX Policy-Connect Collaborative Grants, 2422
IREX Short-Term Travel Grants, 2424
Japan Foundation Japanese Studies Fellowships, 2476
Japan Foundation New York Japanese Studies Fellowships, 2478
Japan Foundation Six-Month Grants for Specialists in Cultural and Academic Fields, 2479
John B. Bene Fellowships in Social Forestry, 2496
John Ben Snow Memorial Trust Grants, 2497
Kluge Center Fellowships, 2561
Law and Social Science Doctoral Dissertation Fellowships, 2589
Library of Congress Fellowships in International Studies, 2612
Marine Corps Historical Center Dissertation Fellowship, 2663
Mary Gates Research Scholarships, 2670
McGregor Fund Grants, 2725
National Academy of Education Spencer Postdoctoral Fellowships, 2845
National Center for Responsible Gaming Centers of Excellence in Gambling Research Grants, 2850
National Center for Responsible Gaming Early Stage Investigator Grants, 2852
National Center for Responsible Gaming Exploration Grants, 2853
National Center for Responsible Gaming Independent Scientist Grants, 2854

National Center for Resp Gaming Large Grants, 2855
National Center for Responsible Gaming Postdoctoral Fellowships, 2856
National Center for Resp Gaming Seed Grants, 2857
National Center for Responsible Gaming Travel Grants, 2858
NCFR Affiliate Councils Outstanding Student Research Papers, 2906
NCFR Family Therapy Section Best Research Paper Awards, 2907
NCI Stages of Breast Development: Normal to Metastatic Disease Grants, 2932
NCSS/FASSE/CUFA Inquiry Grant, 2936
NCSS Exemplary Res in Social Studies Award, 2937
NCSS Jean Dresden Grambs Distinguished Career Research in Social Studies Award, 2938
NCSS L Metcalf Exemplary Dissertation Award, 2939
NFID Adv Vaccinology Course Travel Grant, 2986
NHLBI Translating Basic Behavioral and Social Science Discoveries into Interventions to Reduce Obesity: Centers for Behavioral Intervention Development Grants, 3068
NIA Behavioral and Social Research Grants on Disasters and Health, 3090
NIA Network Infrastructure for Emerging Behavioral and Social Research Areas in Aging, 3114
NLM Behavioral and Social Science Research on Understanding and Reducing Health Disparities Research Grants, 3293
NLM Grants for Scholarly Works in Biomedicine and Health, 3298
NLM Technological Innovations for Interdisciplinary Research Incorporating the Behavioral and Social Sciences Grants, 3309
NOAA Communicating Hurricane Information Grants, 3355
NOAA Regional Integrated Sciences and Assessments Program Grants, 3402
NSF Alan T. Waterman Award, 3467
NSF Arctic Social Sciences Grants, 3492
NSF CISE Accelerating Discovery in Science and Engineering Through Petascale Simulations and Analysis Grants, 3521
NSF CISE Cross-Cutting Program Grants, 3532
NSF CISE Increasing the Participation and Advancement of Women in Academic Science and Engineering Careers Grants, 3534
NSF Innovation and Organizational Change, 3581
NSF Integrative Graduate Education and Research Traineeship, 3584
NSF Natural, Social, and System Sciences and Research Support and Logistics Program, 3600
NSF Physical Anthropology Research Grants, 3615
NSF SBE Doctoral Dissertation Research Improvement Grants, 3641
Nuffield Foundation Law and Society Grants, 3667
OBSSR Behavioral and Social Science Research on Understanding and Reducing Health Disparities Grants, 3733
OSF Central Asia and Caucasus Research and Training Initiative - Junior Fellowships, 3836
OSF Central Asia and Caucasus Research and Training Initiative - Senior Fellowships, 3837
OSF Global Faculty Grants - Track 2, 3842
OSF Global Supplementary Grants, 3843
OSF HESP Mobility Grants, 3844
PC Fred H. Bixby Fellowships, 3880
PHF Penn Faculty Research Fellowships, 3932
PHF Regional Mellon Faculty Research and Cultural Fellowships, 3933
PHF Undergraduate Humanities Forum Research Fellowships, 3934
Radcliffe Institute Dissertation Grants, 4019
Radcliffe Inst Individual Residential Fellowships, 4021
Radcliffe Institute Oral History Grants, 4022
RAND Corporation Graduate Student Summer Associateships, 4030
RAND Corporation Transatlantic Postdoctoral Fellowships for Int'l Relations & Security, 4032
Research Program at Earthwatch Grants, 4057
SASS Einar and Eva Lund Haugen Memorial Scholarship, 4183
Savoy Foundation Studentships, 4186
Schomburg Center Scholars-in-Residence Fellows, 4187
SRC Humanities and Social Science Postdoctoral Fellowships, 4372
SRC Humanities and Social Sciences Grants, 4373
SRC Humanities and Social Sciences Grants for Postdoctoral Positions In Sweden, 4374
SSHRC-APC Automotive Partnership Canada, 4396
SSHRC-CIHR-SC Research Data Centres Joint Initiative, 4397
SSHRC-FHSS Awards to Scholarly Publications, 4398
SSHRC-IDRC International Partnerships for Sustainable Societies, 4399
SSHRC-NSERC College and Community Innovation Enhancement Grants, 4400
SSHRC-Public Safety Canada Kanishka Project Initiative, 4401
SSHRC Aid to Scholarly Journals, 4402
SSHRC Aid to Small Universities, 4403
SSHRC Banting Postdoctoral Fellowships, 4404
SSHRC Bora Laskin National Fellowship in Human Rights Research, 4405
SSHRC Digging into Data Challenge Grants, 4406
SSHRC Doctoral Fellowships, 4407
SSHRC Graduate Scholarships-Michael Smith Foreign Study Supplements, 4408
SSHRC Insight Grants, 4410
SSHRC Institutional Grants, 4411
SSHRC Joseph-Armand Bombardier CGS Doctoral Scholarships, 4412
SSHRC Jules and Gabrielle Leger Fellowship, 4413
SSHRC Postdoctoral Fellowships, 4415
SSHRC Therese F. Casgrain Fellowship, 4418
SSRC-Van Alen Fellowships, 4419
SSRC Abe Fellowship for Journalists, 4420
SSRC Dissertation Proposal Development Faculty Field Fellowships, 4424
SSRC Dissertation Proposal Development Student Fellowships, 4425
SSRC Eurasia Program Dissertation Development Awards, 4427
SSRC Eurasia Program Pre-Dissertation Awards, 4428
SSRC Int'l Dissertation Research Fellowships, 4429
SSRC Japan Society for the Promotion of Science Fellowships, 4431
SSRC Next Generation Social Sciences in Africa: Dissertation Completion Fellowships, 4434
SSRC Next Generation Social Sciences in Africa: Doctoral Dissertation Proposal Fellowships, 4435
SSRC Our Shared Past Grants, 4437
SSSP C. Wright Mills Award, 4439
Swedish Natural Science Res Council Grants, 4479
Takemi Program in International Health Research and Advanced Training Fellowships, 4491
The World Food Prize, 4503
Third Millennium Foundation Fellowships, 4504
U.S. Department of Education Jacob K. Javits Fellowships, 4541
UNESCO Fellowships, 4567
United States Institute of Peace - Transatlantic Post-Doc Fellowship for International Relations and Security, 4581
USC Pentecostal and Charismatic Research Initiative Individual Grants, 4605
USC Pentecostal and Charismatic Research Initiative Regional Center Grants, 4606
USDA Decadal and Regional Climate Prediction Using Earth System Models Grants, 4621
WHOI Marine Policy Fellowships, 4861
Woodrow Wilson Doctoral Dissertation Fellowships in Women's Studies, 4891
World Bank Graduate Scholarship Program, 4896

Social Security

BC Center for Retirement Research Dissertation Fellowships, 1320
BC Steven H. Sandell Grant Program, 1323
Young Scientists Summer Program, 4903

Social Services

A.O. Smith Community Grants, 12
AAAS Science and Technology Policy Fellowships: Health, Education and Human Services, 43
Albuquerque Community Foundation Grants, 508
Alcatel-Lucent Technologies Foundation Grants, 509
Alcoa Foundation Grants, 511
Alex Stern Family Foundation Grants, 542
Allen Lane Foundation Grants, 571
Allyn Foundation Grants, 574
ALSAM Foundation Grants, 579
ALZA Corporate Contributions Grants, 588
American Foodservice Charitable Trust Grants, 710
Amgen Foundation Grants, 812
AMI Semiconductors Corporate Grants, 813
Ann Arbor Area Community Foundation Grants, 860
Ann Peppers Foundation Grants, 863
Anthony R. Abraham Foundation Grants, 869
Archer Daniels Midland Foundation Grants, 985
Arizona Cardinals Grants, 995
Arizona Public Service Corporate Giving Grants, 998
Auburn Foundation Grants, 1263
Ball Brothers Foundation General Grants, 1291
Barra Foundation Project Grants, 1301
Batchelor Foundation Grants, 1305
Bayer Foundation Grants, 1309
Becton Dickinson and Company Grants, 1328
Beim Foundation Grants, 1331
Berks County Community Foundation Grants, 1333
Blum-Kovler Foundation Grants, 1351
Bodenwein Public Benevolent Foundation Grants, 1355
Bodman Foundation Grants, 1356
Boeing Company Contributions Grants, 1357
Booth-Bricker Fund Grants, 1359
Bosque Foundation Grants, 1360
Boston Foundation Grants, 1361
Burlington Northern Santa Fe Foundation Grants, 1389
Burton G. Bettingen Grants, 1390
Bydale Foundation Grants, 1400
Campbell Soup Foundation Grants, 1415
Carl and Eloise Pohlad Family Grants, 1426
Carpenter Foundation Grants, 1439
Chatlos Foundation Grants Program, 1532
Chautauqua Region Community Found Grants, 1533
Chestnut Hill Charitable Foundation, Inc Grants, 1538
CIGNA Foundation Grants, 1568
Claneil Foundation Grants, 1574
Cleveland-Cliffs Foundation Grants, 1580
Coeta and Donald Barker Foundation Grants, 1594
Communities Foundation of Texas Grants, 1613
Community Found for Muskegon County Grants, 1617
Community Foundation for S Arizona Grants, 1618
Community Found of E Central Illinois Grants, 1619
Compton Foundation Grants Program, 1623
ConocoPhillips Grants Program, 1629
Cord Foundation Grants, 1645
Coughlin-Saunders Foundation Grants Program, 1648
Cowles Charitable Trust Grants, 1652
D.F. Halton Foundation Grants, 1692
Dade Community Foundation Grants, 1712
DaimlerChrysler Corporation Fund Grants, 1714
Dell Foundation Open Grants, 1734
DeMatteis Family Foundation Grants, 1740
Detroit Lions Charities Grants, 1753
DOJ National Institute of Corrections Training and Staff Development Grants, 1827
Donald and Sylvia Robinson Family Foundation Grants, 1830
Duchossois Family Foundation Grants, 1867
Emerson Charitable Trust Grants, 1918
Ensign-Bickford Foundation Grants, 1928
Eugene M. Lang Foundation Grants, 1957
Evan and Susan Bayh Foundation Grants, 1960
Evjue Foundation Grants, 1963
Farmers Insurance Corporate Giving Grants, 1981
Flinn Foundation Grants Programs, 2024
Fluor Foundation Grants, 2028
FMC Foundation Grants, 2029
Ford Motor Company Fund Grants Program, 2033
Forrest C. Lattner Foundation Grants, 2035

Foundation for Enhancing Communities Grants, 2038
Frances C. and William P. Smallwood Foundation Grants, 2045
Frances L. and Edwin L. Cummings Memorial Fund Grants, 2046
Frank Stanley Beveridge Foundation Grants, 2051
Fritz B. Burns Foundation Grants, 2057
George A Ohl Jr. Foundation Grants, 2112
George Frederick Jewett Foundation Grants, 2116
George Gund Foundation Grants, 2117
Greater Saint Louis Community Found Grants, 2164
Green Foundation Human Services Grants, 2172
H.A. and Mary K. Chapman Charitable Grants, 2193
H.J. Heinz Company Foundation Grants, 2195
Harold R. Bechtel Charitable Remainder Uni-Trust Grants, 2201
Harold Simmons Foundation Grants, 2203
Harry A. and Margaret D. Towsley Found Grants, 2205
Henry W. Bull Foundation Grants, 2237
Hoglund Foundation Grants, 2271
Huie-Dellmon Trust Grants, 2295
Hutchinson Community Foundation Grants, 2310
Ireland Family Foundation Grants, 2415
J.B. Reynolds Foundation Grants, 2449
J.W. Kieckhefer Foundation Grants, 2452
James H. Cummings Foundation Grants, 2460
James M. Collins Foundation Grants, 2461
James R. Dougherty Jr. Foundation Grants, 2463
Jewish Federation of Metropolitan Chicago Scholarships, 2486
JMO Public Policy Research Grants, 2492
John Deere Foundation Grants, 2500
John G. Duncan Trust Grants, 2501
John P. McGovern Foundation Grants, 2504
John R. Oishei Foundation Grants, 2505
Joseph Alexander Foundation Grants, 2513
Josiah W. and Bessie H. Kline Foundation Grants, 2517
Lillian S. Wells Foundation Grants, 2614
Lucile Horton Howe and Mitchell B. Howe Foundation Grants, 2624
Lynde & Harry Bradley Foundation Fellowships, 2634
Lynde and Harry Bradley Foundation Grants, 2635
M. Bastian Family Foundation Grants, 2636
Mabel Y. Hughes Charitable Trust Grants, 2647
Marcia and Otto Koehler Foundation Grants, 2656
Margaret & James A. Elkins Jr. Found Grants, 2658
Margaret T. Morris Foundation Grants, 2659
Mary K. Chapman Foundation Grants, 2673
Mary S. and David C. Corbin Foundation Grants, 2675
Matilda R. Wilson Fund Grants, 2684
McCarthy Family Foundation Grants, 2717
McCune Charitable Foundation Grants, 2719
McCune Foundation Human Services Grants, 2722
Meadows Foundation Grants, 2736
Melville Charitable Trust Grants, 2741
MGN Family Foundation Grants, 2754
Millipore Foundation Grants, 2775
Millstone Foundation Grants, 2776
Motorola Foundation Grants, 2796
NCFR Affiliate Councils Outstanding Student Research Papers, 2906
Nelda C. and H.J. Lutcher Stark Found Grants, 2967
New Hampshire Charitable Foundation Grants, 2985
Norcliffe Foundation Grants, 3415
Nordson Corporation Foundation Grants, 3416
Northwestern Mutual Foundation Grants, 3430
NSF Innovation and Organizational Change, 3581
Nuffield Foundation Law and Society Grants, 3667
Nuffield Foundation Open Door Grants, 3669
Oklahoma City Community Grants, 3737
Oppenstein Brothers Foundation Grants, 3792
Patrick and Anna M. Cudahy Fund Grants, 3873
Peacock Foundation Grants, 3887
Percy B. Ferebee Endowment Grants, 3897
Perkin Fund Grants, 3898
Perpetual Trust for Charitable Giving Grants, 3900
Phi Upsilon Omicron Florence Fallgatter Distinguished Service Award, 3943
Pott Foundation Grants, 3984
R.C. Baker Foundation Grants, 4006
Rajiv Gandhi Foundation Grants, 4025
Rayonier Foundation Grants, 4034
Reinberger Foundation Grants, 4051
Rhode Island Foundation Grants, 4061
Rich Foundation Grants, 4068
Robert G. Cabell III and Maude Morgan Cabell Foundation Grants, 4075
Rockwell International Corporate Trust Grants, 4093
Roeher Institute Research Grants, 4095
Ronald McDonald House Charities Grants, 4101
RRF General Program Grants, 4115
Ruth and Vernon Taylor Foundation Grants, 4130
San Antonio Area Foundation Grants, 4163
Sarkeys Foundation Grants, 4181
Self Foundation Grants, 4197
Sensient Technologies Foundation Grants, 4199
Sid W. Richardson Foundation Grants, 4253
Sioux Falls Area Community Foundation Community Fund Grants (Unrestricted), 4262
Sioux Falls Area Comm Foundation Spot Grants, 4263
Stackpole-Hall Foundation Grants, 4442
Starr Foundation Grants, 4447
Textron Corporate Contributions Grants, 4501
Thomas and Dorothy Leavey Foundation Grants, 4506
Thomas C. Ackerman Foundation Grants, 4508
Thompson Charitable Foundation Grants, 4511
Union Pacific Foundation Community and Civic Grants, 4568
Union Pacific Foundation Health and Human Services Grants, 4569
Vancouver Foundation Grants and Community Initiatives, 4807
Wayne and Gladys Valley Foundation Grants, 4835
Weyerhaeuser Company Foundation Grants, 4857
William G. and Helen C. Hoffman Found Grants, 4876

Social Services Delivery

Albert and Margaret Alkek Foundation Grants, 504
American Schlafhorst Foundation Grants, 796
Ann Peppers Foundation Grants, 863
Bayer Foundation Grants, 1309
Byron W. and Alice L. Lockwood Grants, 1401
Catherine Holmes Wilkins Foundation Grants, 1442
Charles Lafitte Foundation Grants, 1525
Chiron Foundation Community Grants, 1553
Cudd Foundation Grants, 1672
DeMatteis Family Foundation Grants, 1740
Donald and Sylvia Robinson Family Foundation Grants, 1830
Drs. Bruce and Lee Foundation Grants, 1859
Evan and Susan Bayh Foundation Grants, 1960
Ewing Halsell Foundation Grants, 1964
George B. Storer Foundation Grants, 2113
George Foundation Grants, 2115
Gilroy and Lillian P. Roberts Charitable Foundation Grants, 2142
Green Foundation Human Services Grants, 2172
Harold Alfond Foundation Grants, 2200
Harry W. Bass, Jr. Foundation Grants, 2210
Hirtzel Memorial Foundation Grants, 2269
Ittleson Foundation Mental Health Grants, 2438
James M. Collins Foundation Grants, 2461
Ludwick Family Foundation Grants, 2625
Mary K. Chapman Foundation Grants, 2673
Max and Victoria Dreyfus Foundation Grants, 2687
McCune Charitable Foundation Grants, 2719
McCune Foundation Human Services Grants, 2722
Musgrave Foundation Grants, 2807
NCFR Family Therapy Section Best Research Paper Awards, 2907
Oscar Rennebohm Foundation Grants, 3832
Rayonier Foundation Grants, 4034
Reuter Foundation Grants, 4059
Richard and Susan Smith Family Found Grants, 4064
Ronald McDonald House Charities Grants, 4101
Sarkeys Foundation Grants, 4181
Schramm Foundation Grants, 4188
Shell Oil Company Foundation Grants, 4233
Sony Corporation of America Corporate Philanthropy Grants, 4329
Volkswagen of America Corporate Contributions Grants, 4817
W.C. Griffith Foundation Grants, 4819
Westinghouse Charitable Giving Grants, 4856
Woodward Fund Grants, 4895

Social Stratification/Mobility

RWJF Investigator Awards in Health Policy Res, 4145

Social Work

ACS Doctoral Training Grants in Oncology Social Work, 308
APSAA Fellowships, 920
ASU Graduate College Earl and Lenore Tripke Fellowship, 1239
Canada-U.S. Fulbright International Science and Technology Award, 1417
Canada-U.S. Fulbright Senior Specialists Program Grants, 1418
Canada-U.S. Fulbright Traditional Scholars, 1419
Chapin Hall International Fellowships in Children's Policy Research, 1517
DAAD Study & Internship Program in Germany, 1708
Fulbright Alumni Initiatives Awards Grants, 2064
Fulbright Distinguished Chairs Program Grants, 2069
Fulbright European Union Scholar-in-Residence Program Grants, 2071
Fulbright New Century Scholars Program Grants, 2075
Fulbright Occasional Lecturer Program Grants, 2076
Fulbright Specialists Program Grants, 2079
Fulbright Traditional Scholar Program in Europe and Eurasia, 2080
Fulbright Traditional Scholar Program in Sub-Saharan Africa, 2081
Fulbright Traditional Scholar Program in the East Asia/Pacific Region, 2082
Fulbright Traditional Scholar Program in the Near East and North Africa Region, 2083
Fulbright Traditional Scholar Program in the South and Central Asia Region, 2084
Fulbright Traditional Scholar Program in the Western Hemisphere, 2085
Fulbright Visiting Specialists Program: Direct Access to the Muslim World, 2087
NCFR Family Therapy Section Best Research Paper Awards, 2907
NCFR Ruth Hathaway Jewson Award, 2909
OSF Doctoral Fellows Georgia & Kyrgyzstan, 3839
OSF Social Work Fellowships, 3849
Regents Professional Opportunity Scholarships, 4049
SSHRC Insight Development Grants, 4409
SSHRC Insight Grants, 4410

Social Work Education

ACS Master's Training Grants in Clinical Oncology Social Work, 311
Yale University Beinecke Library Research Fellowships for Yale Grad & Professional School Students, 4902

Sociology

AAO-HNSF Rande H. Lazar Health Services Research Grant, 190
ACLS Charles A. Ryskamp Research Fellowships, 281
ACLS Frederick Burkhardt Residential Fellowships for Recently Tenured Scholars, 288
American Sociological Association Awards, 802
American Sociological Association Congressional Fellowships, 803
American Sociological Association Fund for the Advancement of the Discipline, 804
American Sociological Association Minority Fellowships, 805
Andrew W. Mellon/ACLS Dissertation Completion Fellowships, 844
ASR Robert J. McNamara Student Award, 1218
Canada-U.S. Fulbright International Science and Technology Award, 1417
Canada-U.S. Fulbright Senior Specialists Program Grants, 1418
Canada-U.S. Fulbright Traditional Scholars, 1419

Chapin Hall International Fellowships in Children's Policy Research, 1517
DOT Transit Planning and Research Grants, 1849
Douglass Institute Postdoctoral Fellowships, 1851
Edward Bangs Kelley and Elza Kelley Foundation Grants, 1898
Foundation for the Advancement of Mesoamerican Studies Grants, 2042
Fulbright Alumni Initiatives Awards Grants, 2064
Fulbright Distinguished Chairs Program Grants, 2069
Fulbright European Union Scholar-in-Residence Program Grants, 2071
Fulbright New Century Scholars Program Grants, 2075
Fulbright Occasional Lecturer Program Grants, 2076
Fulbright Specialists Program Grants, 2079
Fulbright Traditional Scholar Program in Europe and Eurasia, 2080
Fulbright Traditional Scholar Program in Sub-Saharan Africa, 2081
Fulbright Traditional Scholar Program in the East Asia/Pacific Region, 2082
Fulbright Traditional Scholar Program in the Near East and North Africa Region, 2083
Fulbright Traditional Scholar Program in the South and Central Asia Region, 2084
Fulbright Traditional Scholar Program in the Western Hemisphere, 2085
Fulbright Visiting Specialists Program: Direct Access to the Muslim World, 2087
Fund for the Advancement of the Discipline, 2090
Harry Fra Guggenheim Dissertation Fellowships, 2207
Herb Society of America Research and Education Grants, 2243
Horowitz Foundation for Social Policy Grants, 2280
Horowitz Found for Social Policy Special Awards, 2281
Horton Research Grants, 2282
Kluge Center Kislak Fellows in American Studies, 2562
Kluge Center Kislak Short-Term Fellowships in American Studies, 2563
LBI-NY Fred Grubel Fellowship, 2596
Mellon Post-Doctoral Fellowship in Sustainable Agriculture, 2740
National Council of Farmer Cooperatives Graduate Awards, 2860
NCFR Innovation Grants, 2908
NCFR Ruth Hathaway Jewson Award, 2909
NIH Biobehavioral Research for Effective Sleep, 3205
NIH Enhancing Adherence to Diabetes Self-Management Behaviors, 3214
NSF Arctic Social Sciences Grants, 3492
NSF Arctic System Science Program Grants, 3493
NSF Sociology Program Grants, 3650
OISE Muriel & Danny Fung Grad Scholarships, 3735
Orgone Biophysical Research Lab Lou Hochberg Awards, 3798
PHF Penn Faculty Research Fellowships, 3932
PHF Regional Mellon Faculty Research and Cultural Fellowships, 3933
PHF Undergraduate Humanities Forum Research Fellowships, 3934
Robert Schalkenbach Foundation Grants, 4079
SSHRC Insight Development Grants, 4409
SSHRC Insight Grants, 4410
SSRC Mellon Mays Predoctoral Research Grants, 4432
SSSR Distinguished Article Award, 4440
SSSR Student Paper Award, 4441
USC Pentecostal and Charismatic Research Initiative Individual Grants, 4605
USC Pentecostal and Charismatic Research Initiative Regional Center Grants, 4606

Sociology Education
American Sociological Association Fund for the Advancement of the Discipline, 804
OISE Muriel & Danny Fung Grad Scholarships, 3735

Sociology, Applied
American Sociological Association Sydney S. Spivack Program in Applied Social Research and Social Policy, 806

Software
SSHRC-APC Automotive Partnership Canada, 4396
USDD Crowd Sourced Formal Verif Grants, 4715
USDD High Assurance Cyber Military Systems Grants, 4731
USDD Information Innovation Office Grants, 4740

Soil Sciences
Abdus Salam ICTP Schools, Conferences, Workshops and Extended Research Workshops Grants, 238
Abdus Salam ICTP Training and Research in Italian Laboratories Grants, 240
American Chemical Society AGRO Sterling B. Hendricks Memorial Lectureship, 635
BARD Research Grants, 1299
Dreyfus Postdoctoral Program in Environmental Chemistry Grant, 1855
Mellon Post-Doctoral Fellowship in Sustainable Agriculture, 2740
NSF Hydrologic Sciences Research Grants, 3578
Organic Farming Research Foundation Grants, 3795
USDA 1890 Institution Teaching, Research, and Extension Capacity Building Grants, 4613
USDA Distance Education Grants for Higher Education Institutions in Insular Areas, 4622
USDA Norman E. Borlaug International Agricultural Science and Technology Fellowships, 4636
USDA Rangeland Research Grants, 4641
USDA Resident Instruction Grants for Higher Education in Insular Areas, 4645
USDA Small Business Innovation Research Grants - Phase I, 4648
USDA Tribal College Research Grants, 4652

Soil Sciences, Soil Genesis
Abdus Salam ICTP Postgraduate Diploma Grants, 235
American Chemical Society AGRO Sterling B. Hendricks Memorial Lectureship, 635
NSF Arctic Nat Science Grants in Earth Sciences, 3485
NSF Sediment Geology & Paleobiology Grants, 3646

Solar Studies
NASA Goddard Visiting Fellowships, 2838
NCAR Graduate Research Fellowships, 2901
NCAR Postdoctoral Fellowships in the Atmospheric Sciences, 2902
NSF Antarctic Aeronomy and Astrophysics Program Grants, 3472
NSF Arctic Natural Science Grants in Atmospheric Sciences, 3483
NSF Coupling, Energetics, and Dynamics of Atmospheric Regions Research Grants, 3545
NSF Magnetospheric Physics Research Grants, 3590
NSF National Space Weather Program, 3599
NSF Solar, Heliospheric, and INterplanetary Environment Grants, 3651
NSF Solar Terrestrial Research Grants, 3652
NSF Stellar Astronomy and Astrophysics Research Grants, 3655

Solar System
NASA Goddard Visiting Fellowships, 2838
NSF Planetary Astronomy Research Grants, 3623

Solid Mechanics
ASME Timoshenko Medal, 1168
ASME Warner T. Koiter Medal, 1170
USDD Solid Mechanics Grants, 4797

Solid Waste Disposal
EREF Solid Waste Research Grants, 1947
EREF Sustainability Research Grants, 1948
USDD Solid and Liquid Waste Treatment Grants, 4796
USDD Water Desalination & Purific Grants, 4804

Sonar Systems
ONR Maritime Sensing Grants, 3770
USDD Signal Processing for Networked Sensing Grants, 4793
USDD Undersea Signal Processing Grants, 4801

South Africa
Charles Stewart Mott Foundation Grants, 1529
USAID Comprehensive District-Based Support for Better HIV/TB Patient Outcomes Grants, 4595

South America
Fulbright Traditional Scholar Program in the Western Hemisphere, 2085

South Asia
AIAR/Council of American Overseas Research Centers Fellowships for Advanced Multi-Country Research, 391
Asia Foundation Exchange Programs, 1090

Space
ACSM NASA Space Physiology Research Grants, 319
ONR Ionosphere Interactions Research Grants, 3765
Smithsonian National Air and Space Museum Aviation Space Writers Foundation Grant, 4284
Smithsonian National Air and Space Museum Guggenheim Fellowships, 4286
USDD Innovative Systems for Military Missions - Tactical Technology Office Grants, 4742
USDD Small University Grants, 4794

Space Sciences
Abdus Salam ICTP Postgraduate Diploma Grants, 235
Abdus Salam ICTP Training and Research in Italian Laboratories Grants, 240
American Historical Association Fellowship in Aerospace History, 717
American Philosophical Soc Judson Daland Prize, 766
American Philosophical Society Lewis and Clark Fund for Exploration and Field Research, 767
Canada Graduate Scholarships and NSERC Postgraduate Scholarships, 1420
DSO Defense Sciences Research and Technology Grants, 1863
Education Through Astronomy and Space Science Initiative, 1896
Emil Buehler Perpetual Trust Grants, 1919
Indiana Space Grant Consortium Discovery Program Crosscutting Research-Team Grants, 2372
Indiana Space Grant Consortium Discovery Program Research Grants, 2373
Indiana Space Grant Consortium Discovery Program Student Hands-On Research Grants, 2374
Indiana Space Grant Consortium Fellowships, 2375
ISGC Graduate Fellowships, 2430
ISGC Informal Education Grants, 2431
ISGC Multiple Undergraduate Research Grants, 2432
ISGC Research Seed Grants, 2433
NASA Cooperative Agreement Notice for Technology Development Grants, 2837
NASA Grad Student Researchers Fellowships, 2839
NASA Lewis Educational Research and Collaborative Internships, 2840
NASA Marslink Initiative, 2841
NASA New Aeronautics Competition for University Students, 2842
NASA Undergraduate Student Researchers Project Grants, 2843
NIH Biomedical Research on the International Space Station, 3207
NSF National Space Weather Program, 3599
NSF University Radio Observatories Program, 3664
ONR Radio Frequency Sensor Processing Research Grants, 3778
ONR Radio Frequency Solid State Devices Applied Research Grants, 3779
SHOT Eugene S. Ferguson Prize, 4240
SHOT NASA Fellowship, 4246
Smithsonian National Air and Space Museum Aviation Space Writers Foundation Grant, 4284
Smithsonian National Air and Space Museum Verville Fellowships, 4288
USDD Small University Grants, 4794
Zonta International Foundation Amelia Earhart Fellowships, 4905

Space and Defense Power Systems
USDD Air Force Defense Research and Development Rapid Innovation Fund Grants, 4677
USDD Capacitors for Pulsed Power Applications Grants, 4705
USDD Dielectric Films for Capacitors Grants, 4719
USDD Innovative Systems for Military Missions - Tactical Technology Office Grants, 4742
USDD Power and Thermal Management Technology Development Grants, 4775
USDD Research and Analytical Support for the 711th Human Performance Wing's Human Effectiveness Directorate Grants, 4781
USDD Warfighter Interface Technologies Advanced Research Programs Grants, 4803

Spain
American Historical Association Premio Del Rey, 741
Fulbright Business Grants in Spain, 2067
NL Audrey Lumsden-Kouvel Fellowships, 3282

Spanish Language/Literature
NL Audrey Lumsden-Kouvel Fellowships, 3282
NL NE Modern Language Assocation Fellowship, 3314
WMU Department of Spanish Travel Study Abroad Award, 4889

Special Education
ATA Educational Trust Grants - Curriculum Resource or Research Project, 1253
ATA Nadene M. Thomas Grad Research Bursary, 1255
Becton Dickinson and Company Grants, 1328
Dorr Foundation Grants, 1842
Herbert & Adrian Woods Foundation Grants, 2239
IRA Albert J. Harris Awards, 2405
Skaggs Foundation Grants, 4266
Special Olympics Youth Fan Grants, 4333

Special Populations
Herbert & Adrian Woods Foundation Grants, 2239
NL Frances C. Allen Fellowships, 3284
NL Monticello College Foundation Fellowship for Women, 3304
USDC Presidential Management Fellowships, 4670

Specialized Museums
Japan Foundation Six-Month Grants for Specialists in Cultural and Academic Fields, 2479
National Endowment for the Humanities Fellowships at Winterthur, 2867
Smithsonian Postgraduate Fellowships in Conservation of Museum Collections, 4294
Union Pacific Foundation Community and Civic Grants, 4568
Winterthur Museum Dissertation Fellowships, 4887
Winterthur Research Fellowships, 4888
Wolfsonian-FIU Fellowships, 4890

Spectroscopy
Abdus Salam ICTP Postgraduate Diploma Grants, 235
Abdus Salam ICTP Training and Research in Italian Laboratories Grants, 240
American Chemical Society ANYL Award in Spectrochemical Analysis, 643
APS Arthur L. Schawlow Prize in Laser Science, 945
APS Plyler Prize for Molecular Spectroscopy, 954
APS Herbert P. Broida Prize, 961
ARO Degradation, Reactivity, and Protection of Materials Research Contracts, 1008
E. Bright Wilson Award in Spectroscopy, 1883
Nakanishi Prize, 2824
NSF Physical Chemistry Grants, 3616
USDD Atomic, Molecular and Quantum Physics Grants, 4687

Speech
ASME Old Guard Prizes, 1151
Axe-Houghton Foundation Grants, 1287
Claude Pepper Foundation Grants, 1578
Reinberger Foundation Grants, 4051

Speech Pathology
ASHA Student Research Travel Award, 1074
ASHFoundation Graduate Student Scholarships, 1077
ASHFoundation Graduate Student Scholarships for Minority Students, 1078
ASHFoundation Graduate Student with a Disability Scholarship, 1079
ASHFoundation New Century Scholars Program Doctoral Scholarships, 1080
ASHFoundation Res Grant for New Investigators, 1082
ASHFound Research Grant in Speech Science, 1083
Axe-Houghton Foundation Grants, 1287
Edward W. and Stella C. Van Houten Memorial Fund Grants, 1899
Grass Foundation Marine Biological Laboratory Advanced Imaging Fellowships, 2162
Mabel H. Flory Charitable Trust Grants, 2646
Regents Professional Opportunity Scholarships, 4049

Speech/Communication Education
ACLS Frederick Burkhardt Residential Fellowships for Recently Tenured Scholars, 288
Axe-Houghton Foundation Grants, 1287
Mabel H. Flory Charitable Trust Grants, 2646

Sports
ACSM Raymond and Rosalee Weiss Research Endowment Grant, 322
AMSSM Foundation Research Grants, 836
Angels Baseball Foundation Grants, 848
Callaway Golf Company Foundation Grants, 1410
Carl and Eloise Pohlad Family Grants, 1426
Carl C. Icahn Foundation Grants, 1428
Chiles Foundation Grants, 1552
Detroit Lions Charities Grants, 1753
FishAmerica Foundation Marine and Anadromous Fish Habitat Restoration Grants, 2018
FishAmerica Foundation Research Grants, 2019
GCSAA Essay Contest, 2102
George Kress Foundation Grants, 2119
Greater Worcester Community Foundation Jeppson Memorial Fund for Brookfield Grants, 2165
Helen Bader Foundation Grants, 2228
IATA Research Grants, 2319
McCombs Foundation Grants, 2718
NCAA Postgraduate Scholarships, 2900
NOCSAE Research Grants, 3414
NRA Foundation Grants, 3437
Paul Stock Foundation Grants, 3879
Sidgmore Family Foundation Grants, 4252
Special Olympics Kennedy Shriver Fellowships, 4331
SSHRC Sport Participation Research Initiative: Doctoral Award Supplements, 4416
SSHRC Sport Participation Research Initiative: Postdoctoral Fellowship Supplements, 4417

Sports Equipment
1st Touch Foundation Grants, 2
NOCSAE Research Grants, 3414

Sports Medicine
ACSM Foundation Clinical Sports Medicine Endowment Grants, 315
ACSM Found Doctoral Student Research Grants, 316
ACSM Foundation Research Grants, 317
ACSM Oded Bar-Or Int'l Scholars, 320
ACSM Paffenbarger-Blair Fund for Epidemiological Research on Physical Activity Grants, 321
ACSM Raymond and Rosalee Weiss Research Endowment Grant, 322
AMSSM Foundation Research Grants, 836
Angels Baseball Foundation Grants, 848
IATA Research Grants, 2319
NOCSAE Research Grants, 3414
SSHRC Sport Participation Research Initiative: Doctoral Award Supplements, 4416
SSHRC Sport Participation Research Initiative: Postdoctoral Fellowship Supplements, 4417

Standards and Standardization
ASME Bernard F. Langer Nuclear Codes and Standards Award, 1109
ASME J. Hall Taylor Medal, 1133
ASME Mel R. Green Codes & Standards Medal, 1146
ASME Patrick J. Higgins Award, 1152
DOE State Energy Program Special Projects, 1815
NIST Precision Measurement Grants, 3272

Starch
AACC International Alsberg-French-Schoch Memorial Lectureship Awards, 82

Statistical Mechanics
Alfred P. Sloan Foundation Energy and Environment Grants, 552

Statistics
Abdus Salam Joint ICTP/SISSA PhD Programme in Physics and Mathematics Grants, 242
Alfred P. Sloan Foundation Selected National Issues Grants, 563
ARO Theoretical Physics and Nonlinear Phenomena Research Contracts, 1017
ASA/BJS Statistical Methodological Research Program Grants, 1029
ASA/NSF/BLS Fellowships, 1030
ASA/NSF/Census Research Fellowships, 1031
ASA/NSF Federal Statistics Fellowships, 1032
ASA and USDA/NASS Research Fellow and Associate Program, 1033
ASA Deming Lecturer Award, 1034
ASA Jeanne E. Griffith Mentoring Award, 1035
ASA Outstanding Statistical Application Award, 1036
ASA Samuel S. Wilks Award, 1037
ASA W. J. Youden Award in Interlab Testing, 1038
AT&T Bell Labs Graduate Research Program, 1250
Canada Graduate Scholarships and NSERC Postgraduate Scholarships, 1420
CSTE CDC/CSTE Applied Epidemiology Fellowships, 1664
DOT Transportation Statistics Research Grants, 1850
ETS Predoctoral Summer Program, 1954
NCHS National Center for Health Statistics Postdoctoral Research Appointments, 2910
NEI Clinical Vision Research Devel Award, 2960
NHGRI Mentored Research Scientist Development Award, 3013
NHLBI Mentored Quantitative Research Career Development Awards, 3038
NIGMS Predoctoral Training in Bioinformatics and Computation Biology Grants, 3199
NIGMS Short Courses on Mathematical, Statistical, and Computational Tools for Studying Biological Systems, 3201
Norwegian Acad of Science & Letters Abel Prize, 3433
NSF Scientific Computing Research Environments for the Mathematical Sciences Grants, 3645
NSF Statistics Research Grants, 3654
ORISE Nuclear Regulatory Commission Historically Black Colleges and Universities Student Research Participation, 3814
RCPSC Royal College Fellowship for Studies in Medical Education, 4040
SSHRC-CIHR-SC Research Data Centres Joint Initiative, 4397
SSHRC Digging into Data Challenge Grants, 4406
USDD Applied & Computational Anal Grants, 4683
WHOI Marine Policy Fellowships, 4861

Steel Structures
Engineering Awards of Excellence, 1926
T.R. Higgins Lectureship Award, 4490

Stem Cell Therapy
Pediatric Cancer Research Foundation Grants, 3889

Stem Cell Transplantation
Michael J. Fox Foundation Parkinson's Cell Line Research Grants, 2762

NCRR Novel Approaches to Enhance Animal Stem Cell Research, 2935
NIDDK Intestinal Stem Cell Consortium Grants, 3160
Pediatric Cancer Research Foundation Grants, 3889
Pfizer Advancing Research in Transplantation Science Research Awards, 3908

Storytelling
Bush Foundation Artist Fellowships, 1391
Florida Division of Cultural Affairs Individual Artist Fellowships, 2026
North Carolina Arts Council Residency Grants, 3420

Strategic Planning
Covidien Surgical Mini-Fellowships, 1651

Stratigraphy
NSF Arctic Nat Science Grants in Earth Sciences, 3485
NSF Sediment Geology & Paleobiology Grants, 3646
USDD Naval Research Facilities Grants, 4761

Stress
NIA Aging, Oxidative & Cell Death Grants, 3081
NIA Mechanisms Underlying the Links between Psychosocial Stress, Aging, the Brain and the Body Grants, 3106
ONR Computational Neuroscience & Biorobotics Grants, 3752
ONR Medical Science & Technology Div Grants, 3772
Sigma Xi Grants in Aid of Research, 4259

Stroke
Goddess Scholars Grants, 2155
Jeffrey Thomas Stroke Shield Foundation Research Grants, 2483
NIH Research on Sleep and Sleep Disorders, 3236
NINDS Support of Scientific Meetings as Cooperative Agreements, 3258
NINR Clinical Interventions for Managing the Symptoms of Stroke Research Grants, 3262
Stroke Association Allied Health Professional Research Bursaries, 4454
Stroke Association Clinical Fellowships, 4455
Stroke Association Research Project Grants, 4456

Structural Design
ASME Barnett-Uzgiris Product Safety Award, 1106
USDD Ocean Engineering and Marine Systems Grants, 4767

Structural Engineering
ASCE O.H. Ammann Research Fellowship in Structural Engineering, 1042
Engineering Awards of Excellence, 1926
T.R. Higgins Lectureship Award, 4490
USDD Alloys and Joining Grants, 4680

Student Personnel Services
Higher Education Graduate Scholarships, Fellowships, and Grants, 2265

Student Support (incl. Dissertation Support)
ALA Excellence in Academic Libraries Award, 486
APSA William Anderson Award, 950
Cystic Fibrosis Canada Studentships, 1683
IRA Helen M. Robinson Dissertation Res Award, 2409
Japan Foundation Six-Month Grants for Specialists in Cultural and Academic Fields, 2479
NSF Partnerships in Astronomy & Astrophysics Research and Education, 3612
OAH Huggins-Quarles Award, 3720
Phi Kappa Phi Found Love of Learning Awards, 3937
SfN Donald B. Lindsley Prize in Behavioral Neuroscience, 4204
Trinity College Ann Plato Fellowship, 4533
UCR Humanities Grad Student Research Grants, 4551
USDC Dissertation Fellowships, 4665
Woodrow Wilson MMUF Travel & Res Grants, 4893

Substance Abuse
CSTE CDC/CSTE Applied Epidemiology Fellowships, 1664
Peter F. McManus Charitable Trust Grants, 3903
Reinberger Foundation Grants, 4051
SfN Jacob P. Waletzky Award, 4206
Sioux Falls Area Community Foundation Community Fund Grants (Unrestricted), 4262
Sioux Falls Area Comm Foundation Spot Grants, 4263

Sudden Infant Death Syndrome
CJ Foundation for SIDS Research Grants, 1573

Suicide
AACAP Klingenstein Third Generation Foundation Award for Research in Depression or Suicide, 66
CDC Grants for Violence-Related Injury Prevention Research, 1463
NARSAD Distinguished Investigator Grants, 2831
NARSAD Independent Investigator Grants, 2833
NARSAD Young Investigator Grants, 2836

Suicide Prevention
AACAP Klingenstein Third Generation Foundation Award for Research in Depression or Suicide, 66
Nick Traina Foundation Grants, 3128

Summer Camp
Gil and Dody Weaver Foundation Grants, 2139

Supercomputers
AINSE Postgraduate Research Awards, 463
NSF Scientific Computing Research Environments for the Mathematical Sciences Grants, 3645

Superconductivity
NSF Condensed Matter Physics Research Grants, 3540
ONR Superconducting Electronics Grants, 3785
USDD RF Superconducting Technologies Grants, 4787

Supportive Housing Programs
Charles H. Farnsworth Trust Grants, 1521
Chicago Community Trust Housing Grants: Advancing Affordable Rental Housing, 1547
Mericos Foundation Grants, 2742
Meyer Memorial Trust Responsive Grants, 2748
MGM Resorts Foundation Community Grants, 2753
Reinberger Foundation Grants, 4051

Surface Science
Abdus Salam ICTP Postgraduate Diploma Grants, 235
Abdus Salam ICTP Training and Research in Italian Laboratories Grants, 240
APS Davisson-Germer Prize in Atomic or Surface Physics, 952
ONR Radio Frequency Sensor Processing Research Grants, 3778
ONR Radio Frequency Solid State Devices Applied Research Grants, 3779

Surfaces and Interfaces
Abdus Salam ICTP Postgraduate Diploma Grants, 235
Abdus Salam ICTP Training and Research in Italian Laboratories Grants, 240
ARO Degradation, Reactivity, and Protection of Materials Research Contracts, 1008

Surgery
AAFPRS Bernstein Grant, 161
AAFPRS Fellowships, 162
AAFPRS Investigator Development Grant, 163
AAFPRS Ira J. Tresley Research Award, 164
AAFPRS John Orlando Roe Award, 165
AAFPRS Leslie Bernstein Resident Res Grants, 166
AAFPRS Residency Travel Awards, 167
AAFPRS Sir Harold Delf Gillies Award, 168
AANS Neurosurgery Research & Educ Foundation/ Spine Section Young Clinician Invest Award, 184
AANS Neurosurgery Research & Educ Foundation Young Clinician Investigator Award, 185
AAO-HNSF CORE Research Grants, 186
AAO-HNSF Health Services Research Grants, 187
AAO-HNSF Maureen Hannley Research Training Awards, 188
AAO-HNSF Percy Memorial Research Award, 189
AAO-HNSF Rande H. Lazar Health Services Research Grant, 190
AAO-HNSF Resident Research Awards, 191
AAO-HNS Medical Student Research Prize, 192
AAO-HNS Resident Research Prizes, 193
AAOA Foundation/AAO-HNSF Combined Research Grants, 194
AHAF Alzheimer's Disease Research Grants, 370
AHNS/AAO-HNSF Surgeon Scientist Combined Award, 380
AHNS/AAO-HNSF Young Investigator Combined Award, 381
AHNS Alando J. Ballantyne Resident Research Pilot Grant, 382
AHNS Pilot Grant, 383
American College of Surgeons and The Triological Society Clinical Scientist Development Awards, 701
American College of Surgeons Co-Sponsored K08/K23 NIH Supplement Awards, 702
American College of Surgeons Faculty Career Development Award for Neurological Surgeons, 703
American College of Surgeons Faculty Research Fellowships, 704
American College of Surgeons Resident Research Scholarships, 705
American College of Surgeons Traveling Fellowship to Germany, 706
American College of Surgeons Traveling Fellowship to Japan, 707
American College of Surgeons Wound Care Management Award, 708
American Philosophical Society Daland Fellowships in Clinical Investigation, 763
Carlos and Marguerite Mason Trust Grants, 1431
Charles H. Hood Foundation Child Health Research Awards, 1522
Clowes ACS/AAST/NIGMS Jointly Sponsored Mentored Clinical Scientist Devel Award, 1581
Covidien Advanced MIS Surgical Fellowships, 1650
Covidien Surgical Mini-Fellowships, 1651
Denton A. Cooley Foundation Grants, 1743
Dorr Inst for Arthritis Research & Education, 1843
E.L. Wiegand Foundation Grants, 1885
NYAM Charles A. Elsberg Fellowship, 3680
Pfizer ASPIRE EU Dupuytren's Contracture Research Awards, 3911
RCPSC Detweiler Traveling Fellowships, 4036
RCPSC Walt MacKenzie Fellowship in Surgery, 4042
Triological Society Research Career Development Awards, 4534
Xoran Technologies Resident Research Grant, 4900

Surveillance Systems
APHL Emerging Infectious Diseases Fellowships, 914
Aviation Research Grants, 1273
CSTE CDC/CSTE Applied Epidemiology Fellowships, 1664
ONR Surface/Aerospace Surveillance Grants, 3786
USDD Applications and Transitions Grants, 4682
USDD Chemical and Biological Defense Innovations and Technologies Grants, 4708
USDD Intelligence, Surveillance and Reconnaissance Thrust Grants, 4743
USDD RF Semiconductors, RF Solid State Amplifiers and Wide Bandgap Materials Grants, 4786
Web We Want Small Grants, 4836

Surveying and Mapping
Alfred P. Sloan Foundation Sloan Digital Sky Survey Grants, 564
ASPRS Leica Geosystems Internship, 1213

Sustainable Development
AAAS Roger Revelle Fellowship in Global Stewardship, 38

ASME Frank Kreith Energy Award, 1120
Bill and Melinda Gates Foundation Agricultural Development Grants, 1345
Boeing Company Contributions Grants, 1357
BP Conservation Programme Future Conservationist Awards, 1365
Clarence E. Heller Charitable Foundation Grants, 1575
Conservation, Food, and Health Foundation Grants for Developing Countries, 1641
DOI Rhinoceros and Tiger Conservation Grants, 1824
ESRI Conservation Program Grants, 1952
Honor the Earth Grants, 2279
HSBC Contributions Community Grants, 2291
IIE David L. Boren Fellowships, 2350
IREX International Leadership in Education Program Grants, 2420
Jenifer Altman Foundation Grants, 2484
Jessie Smith Noyes Foundation Grants, 2485
John D. and Catherine T. MacArthur Foundation Global Challenges Program Grants, 2498
MacArthur Foundation Conservation and Sustainable Development Grants, 2648
McCune Charitable Foundation Grants, 2719
McGraw-Hill Companies Community Grants, 2724
National Geographic Society/Waitt Grants, 2870
National Geographic Society Conservation Trust Grants, 2871
National Geographic Society Expeditions Council Grants, 2872
National Geographic Soc Young Explorers Grant, 2874
Nellie Mae Education Foundation District-Level Change Grants, 2968
Nellie Mae Education Foundation State Level Systems Change Grants, 2971
OTS Undergrad NAPIRE Summer Interns, 3858
SARE Graduate Student Grants - North Central Region, 4168
SARE Graduate Student Grants - So Region, 4170
SARE Graduate Student Grants in Sustainable Agriculture - Western Region, 4171
SARE Large Systems Research Grants - Southern Region, 4172
SARE On-Farm Research Grants - So Region, 4173
SARE Producer Research and Education Grants-Western Region, 4174
SARE Professional + Producer Grants - Western Region, 4175
SARE Research and Education Grants - North Central Region, 4177
SARE Research and Education Grants - Northeast Region, 4178
SARE Research and Education Grants - Southern Region, 4179
SARE Research and Education Grants - Western Region, 4180
Shastri Indo-Canadian Institute Collaborative Research Project Grants, 4220
Shastri Indo-Canadian Institute Partnership Development Seed Grants, 4228
SSHRC-APC Automotive Partnership Canada, 4396
SSHRC-IDRC International Partnerships for Sustainable Societies, 4399
USDA Agriculture and Food Research Initiative Competitive Grants - Food Security, 4614
USDA Cochran Fellowships, 4619
USDA Food and Agricultural Sciences National Needs Graduate Fellowship Grants, 4624
USDA Plant Feedstock Genomics for Bioenergy: A Joint Research Grant, 4639
USDA Regional Integrated Pest Management Grants - North Central Region, 4642
USDA Scientific Cooperation Exchange Program, 4646
USDA Scientific Cooperation Research Program, 4647
USDA Tribal College Research Grants, 4652

Sweden
NSF Antarctic Special Opportunity for Science aboard the Swedish Ice Breaker Oden Grants, 3480
Swedish Women's Educational Association International Scholarships, 4480

Swedish Language/Literature
SRC Educational Science Lang Checking Grants, 4360
Swedish Women's Educational Association International Scholarships, 4480

Symposiums
ARO Conference and Symposia Grants, 1007
FDHN Non-Career Research Awards, 1998
IBRO Europe Regional Grants for Western European Workshops and Symposia, 2324
IBRO Latin America Regional Travel Grants, 2329
SSRC Our Shared Past Grants, 4437
USDD Broad Agency Announcement Research and Development Grants, 4703

Systems Engineering
AIChE Outstanding Young Researcher Award, 398
AIChE Computing in Chem Engineering Award, 402
AIChE Computing Practice Award, 403
AIChE W. David Smith, Jr. Graduate Student Paper Award, 428
Computing in Chemical Engineering Award, 1626
DOE Energy from Municipal Waste Grants, 1787

Systems Theory
ASME Rufus T. Oldenburger Medal, 1161
Doris Duke Charitable Foundation Clinical Research Systems Grants, 1835
RSC Thomas W. Eadie Medal, 4126
Young Scientists Summer Program, 4903

Tactical Decision Making
USDD Command and Control Grants, 4711
USDD Office of Naval Research Command, Control, Computers and Communication Thrust Grants, 4769

Taiwan
ACC Taiwan Fellowship Program, 269
Alexander von Humboldt Foundation NSC Taiwan Fellowships, 533
Eisenhower Exchange International Fellowships, 1902

Tajikistan
Fulbright Traditional Scholar Program in the South and Central Asia Region, 2084
IREX Individual Advanced Research Opportunities Grants, 2419

Taxes and Taxation
JMO Public Policy Research Grants, 2492
Robert Schalkenbach Foundation Grants, 4079

Taxonomy, Animal
Alfred P. Sloan Foundation Encyclopedia of Life Grants, 551
NSF Systematic Biology and Biodiversity Inventories Grants, 3657

Taxonomy, Plant
Alfred P. Sloan Foundation Encyclopedia of Life Grants, 551
NSF Systematic Biology and Biodiversity Inventories Grants, 3657

Teacher Attitudes
NCSS/FASSE/CUFA Inquiry Grant, 2936
NCSS Exemplary Res in Social Studies Award, 2937
NCSS Jean Dresden Grambs Distinguished Career Research in Social Studies Award, 2938
NCSS L Metcalf Exemplary Dissertation Award, 2939

Teacher Certification
DaimlerChrysler Corporation Fund Grants, 1714
MfA Newton Fellowships, 2749
Nellie Mae Education Foundation State Level Systems Change Grants, 2971

Teacher Education
3M Canada National Teaching Fellowships, 3
AERA Research Grants, 344
Allen Foundation Educational Nutrition Grants, 570
All Saints Educational Trust Grants, 572
American Montessori Society Thesis and Dissertation Awards, 758
ASHA Advancing Academic-Research Award, 1072
ATA Doctoral Fellowships in Education, 1252
ATA Nadene M. Thomas Grad Research Bursary, 1255
California Teachers Association Scholarships, 1408
Carnegie Corporation of New York Grants, 1432
Clarence E. Heller Charitable Foundation Grants, 1575
Dept of Ed Jacob K. Javits Gifted and Talented Students Education Program-National Research and Development Center, 1746
Dept of Ed Star Schools Program Grants, 1749
DOE Acad Creating Teacher Scientists Grants, 1766
Earthwatch Institute Student Fellowships, 1890
Frederick Douglass Summer Scholars Fellowship, 2053
Fritz B. Burns Foundation Grants, 2057
FTE Dissertation Fellowships, 2058
FTE Doctoral Fellowships, 2059
Fulbright-Hays Seminars Abroad Bilateral Projects Grants, 2062
Fulbright English Teaching Assistantships, 2070
Harvard University Graduate School of Education Fellowships and Awards, 2213
Hatton W. Sumners Foundation for the Study and Teaching of Self Government Grants, 2219
Holberg Memorial Fund School Project, 2274
IRA Esther Zolt Academic Research Award, 2407
ISGC Informal Education Grants, 2431
ITEA Maley/FTE Scholarship, 2435
Johnson & Johnson Corporate Contributions, 2508
Kosciuszko Foundation Tuition Scholarships, 2574
Lilly Endowment Teacher Creativity Fellowships, 2616
M.J. Murdock Charitable Trust Partners in Science Grants, 2642
Martin Luther King Jr. Graduate Memorial Scholarships, 2668
Meadows Foundation Grants, 2736
MfA Newton Fellowships, 2749
MHS Teacher Fellowships, 2759
NCSS/FASSE/CUFA Inquiry Grant, 2936
NCSS Exemplary Res in Social Studies Award, 2937
NCSS Jean Dresden Grambs Distinguished Career Research in Social Studies Award, 2938
NCSS L Metcalf Exemplary Dissertation Award, 2939
NL Associated Colleges of the Midwest/Great Lakes College Association Faculty Fellowships, 3281
NSF Innovative Technology Experiences for Students and Teachers, 3582
Ohio Learning Network Grants, 3734
OISE William Waters Scholarships in Urban Education, 3736
Phi Upsilon Omicron Geraldine Clewell Doctoral Fellowship, 3945
Phi Upsilon Omicron Geraldine Clewell Masters Fellowship, 3946
Surdna Foundation Arts Teachers Fellowships, 4470
Teachers College Minority Postdoc Fellowships, 4498
TIRF Doctoral Dissertation Grants, 4520
Virginia Foundation for the Humanities Discretionary Grants, 4813
Virginia Foundation for Humanities Open Grants, 4814
Wallace Foundation Grants, 4829

Teacher Education, Inservice
Arizona Commission on the Arts Education Projects Grants, 996
Arizona Commission on the Arts Folklorist Residencies, 997
Coca-Cola Foundation Grants, 1593
Dept of Ed Star Schools Program Grants, 1749
DOE Acad Creating Teacher Scientists Grants, 1766
Flinn Foundation Grants Programs, 2024
Joyce Foundation Education Grants, 2519
NFIE Found Learning & Leadership Grants, 2992

Teacher Training
Chiron Foundation Community Grants, 1553
Joyce Foundation Education Grants, 2519

Nellie Mae Education Foundation District-Level Change Grants, 2968
NSF Innovative Technology Experiences for Students and Teachers, 3582
Phi Upsilon Omicron Geraldine Clewell Doctoral Fellowship, 3945
Phi Upsilon Omicron Geraldine Clewell Masters Fellowship, 3946
Sidgmore Family Foundation Grants, 4252
Virginia Foundation for the Humanities Discretionary Grants, 4813
Virginia Foundation for Humanities Open Grants, 4814

Technological Change
Bill and Melinda Gates Foundation Agricultural Development Grants, 1345
James S. McDonnell Foundation Complex Systems Collaborative Activity Awards, 2465

Technology
AAAS Award for Public Engagement with Science, 26
AAAS Award for Scientific Freedom and Responsibility, 28
AAAS Kavli Science Journalism Awards, 33
AACC International Bruce Wasserman Young Investigator Award, 84
AACC International Excellence in Teaching Award, 89
AACC International Thomas Burr Osborne Medal, 98
AACC Int'l Young Scientist Research Award, 100
Abdus Salam ICTP-OEA Affil Centre Grants, 222
Abdus Salam ICTP Networks Grants, 233
AIChE Outstanding Young Researcher Award, 398
AIChE Charles M.A. Stine Award, 399
AIChE Institute Award for Excellence in Industrial Gases Technology, 412
AIChE John A. Tallmadge Award for Contributions to Coating Technology, 413
AIChE Nanoscale Science and Engineering Forum Young Investigator Award, 417
Alexander von Humboldt Feodor Lynen Research Fellowships for Experienced Researchers, 520
Alfred P. Sloan Foundation Civic Grants, 545
Alfred P. Sloan Foundation Economic Analysis of Science and Technology Grants, 548
Alfred P. Sloan Foundation History of Science and Technology Grants, 553
Alfred P. Sloan Foundation Public Understanding of Science and Technology-Book Grants, 555
Alfred P. Sloan Foundation Science of Learning STEM Grants, 562
American Academy of Arts and Sciences Hellman Fellowships, 612
American Academy of Arts and Sciences Visiting Scholars, 613
American Chemical Society Award for Creative Advances in Environmental Science and Technology, 652
American Chemical Society Award in Separations Science and Technology, 671
American Indian College Fund STEM Leadership Fellowships, 745
ANL Faculty Research Participation Program, 855
Applied Biosystems Grants, 916
ARM Visiting Scientist Awards, 1004
ARO Conference and Symposia Grants, 1007
ASEE Army Research Laboratory Postdoctoral Fellowships, 1059
AT&T Industrial Ecology Faculty Fellowships, 1251
Aviation Research Grants, 1273
BARD Research Grants, 1299
BFGoodrich Collegiate Inventors Prize, 1341
Bill and Melinda Gates Foundation Financial Services for the Poor Grants, 1346
Cabot Corporation Foundation Grants, 1406
Canada-U.S. Fulbright International Science and Technology Award, 1417
CHF Charles C. Price Fellow in Polymer History, 1539
CHF Glenn E. and Barbara Hodsdon Ullyot Scholarship, 1540
CHF Long-Term Fellowships, 1541
CHF Robert W. Gore Fellowship in Materials Innovation, 1542
CHF Roy G. Neville Short-Term Fellowships, 1543
CHF Societe de Chimie Industrielle (American Section) Fellowship, 1544
CHF Travel Grants, 1545
CIA Undergraduate Co-Op Program - Engineer/ Program Management Engineer, 1561
Cleveland-Cliffs Foundation Grants, 1580
Croucher Foundation Grants, 1659
DOE Coal Liquefaction Research Grants, 1777
DOE Control Technology and Coal Preparation Research Grants, 1781
DOE Energy Conservation Research and Development Grants, 1785
DOE Energy from Municipal Waste Grants, 1787
DOE Faculty and Student Teams Grants, 1792
DOE Fusion Energy Science Technology Division Grants, 1796
DOE High-Energy Physics Research Grants, 1799
DOE Small Business Innovation Research Commercialization Assistance Program, 1814
DOE State Energy Program Special Projects, 1815
DOE University Coal Research Grants, 1816
Do Something TMI Intern, 1844
DOT Transit Planning and Research Grants, 1849
DSO Computer Science Study Group Grants, 1861
DSO Defense Sciences Research and Technology Grants, 1863
Educause Jane N. Ryland Fellowships, 1897
Engineering Information Foundation Grants, 1927
EPA Pollution Prevention Grants, 1936
ESRI Conservation Program Grants, 1952
ETS Predoctoral Summer Program, 1954
Facebook Fellowships, 1973
Ford Motor Company Fund Grants Program, 2033
Fund for Scientific Research--Flanders Grants, 2089
GNOF Gulf Coast Oil Spill Grants, 2153
Gruber Foundation Weizmann Institute Awards, 2184
IMLS 21st Century Museum Prof Grants, 2362
IMLS National Leadership Grants, 2366
Inamori Foundation Kyoto Prize, 2367
Indiana 21st Century Research and Technology Fund Awards, 2370
Indiana SBIR/STTR Commercialization Enhancement Program, 2371
Indiana Waste Tire Fund Program Grants, 2376
Institute Award for Excellence in Industrial Gases Technology, 2379
Intel Bachelor's and Master's Degree Internships, 2393
Intel Technician Internships, 2395
ITEA Maley/FTE Scholarship, 2435
John R. Oishei Foundation Grants, 2505
Kluge Center Distinguished Chairs, 2560
Library Company of Philadelphia Barra Foundation International Research Fellowships, 2606
Library Company of Philadelphia Prizes, 2608
Library Company of Philadelphia Program in Early American Economy and Society Dissertation-Level Fellowships, 2609
Library Company of Philadelphia Program in Early American Economy and Society One-Month Fellowships, 2610
Library Company of Philadelphia Program in Early American Economy and Society Post-Doctoral Research Fellowships, 2611
LITA/Gaylord Award for Achievement in Library and Information Technology, 2618
Meadows Foundation Grants, 2736
Mericos Foundation Grants, 2742
Microsoft Research A. Richard Newton Breakthrough Research Awards, 2764
Microsoft Research Biomedical Computing for Genome Wide Association Studies Grants, 2766
Microsoft Human-Robot Interaction Grants, 2768
Microsoft Research Intelligent Web 3.0 Grants, 2769
Microsoft Research Mechanisms for Safe and Scalable Multicore Computing Grants, 2770
Microsoft Res Sustainable Computing Grants, 2772
MIT Knight Science Journalism Fellowships, 2782
Monfort Family Foundation Grants, 2794
NASA Cooperative Agreement Notice for Technology Development Grants, 2837
NASA Goddard Visiting Fellowships, 2838
NASA Graduate Student Researchers Program Fellowships, 2839
NASA Undergraduate Student Researchers Project Grants, 2843
NATO Science for Peace Grants, 2898
NEH Digital Humanities Implementation Grants, 2953
NEH Digital Humanities Start-Up Grants, 2954
NFWF Marine Debris Research and Technology Grants, 3002
NFWF More Fish Partnership Fund Grants, 3003
NHGRI Mentored Research Scientist Development Award, 3013
NIH Enhancing Adherence to Diabetes Self-Management Behaviors, 3214
NIH Recovery Act Limited Competition: Academic Research Enhancement Awards, 3230
NIH Recovery Act Limited Competition: Biomedical Research, Development, and Growth to Spur the Acceleration of New Technologies Pilot Program Grants, 3231
NIH Recovery Act Limited Competition: Small Business Catalyst Awards for Accelerating Innovative Research Grants, 3233
NIH Recovery Act Limited Competition: Supporting New Faculty Recruitment to Enhance Research Resources through Biomedical Research Core Centers Grants, 3234
NIST Small Business Innovation Res Grants, 3273
NLM Technological Innovations for Interdisciplinary Research Incorporating the Behavioral and Social Sciences Grants, 3309
NOAA Mid-Atlantic Research Set-Aside Program Grants, 3384
NOAA Monkfish Research Set-Aside Grants, 3386
NOAA National Coastal and Estuarine Research and Technology Program Grants, 3388
NOAA Sea Scallop Research Set-Aside Program Grants, 3404
Northrop Grumman Foundation Grants, 3429
NSF Advanced Learning Technologies Grants, 3462
NSF Arctic Research and Education Grants, 3489
NSF Arctic Social Sciences Grants, 3492
NSF CreativeIT Grants, 3547
NSF Engineering Research Centers Grants, 3561
NSF Engineering Research Grants, 3562
NSF Grants for the Analysis of Science and Technology Resources, 3575
NSF Industry/University Cooperative Research Centers Grants, 3579
NSF Innovation and Organizational Change, 3581
NSF Ocean Technology and Interdisciplinary Coordination Grants, 3604
NSF Partnership for Advancing Technologies in Housing Grants, 3609
NSF Partnerships for Innovation Grants, 3610
NSF Research on Gender in Science and Engineering Grants, 3639
NSF Science and Technology Studies Grants, 3644
NSF Small Business Innovation Research Grants, 3647
NSF Tribal Colleges and Universities Program, 3662
NYSCA Music: Services to the Field Grants, 3708
Ohio Learning Network Grants, 3734
ONR Bisson Prize for Naval Tech Achievement, 3745
ONR Circuits/Architectures Grants, 3748
ONR Electronic Combat Technology Grants, 3757
ONR Independent Research & Devel Grants, 3764
ONR Joint Capability Technology Demonstration Program Grants, 3766
ONR Manufacturing Technology Grants, 3768
ONR Maritime Sensing Grants, 3770
ONR Multidisciplinary Research Program of the URI Grants, 3774
Plan for Social Excellence Grants, 3976
Praxair Foundation Grants, 3986
Rajiv Gandhi Foundation Grants, 4025

RAND Corporation Graduate Student Summer Associateships, 4030
Rayonier Foundation Grants, 4034
RealNetworks Helix Community Grants, 4044
Robert W. Deutsch Foundation Grants, 4082
Rockwell International Corporate Trust Grants, 4093
San Diego Women's Foundation Grants, 4164
Schomburg Center Scholars-in-Residence Fellows, 4187
Schramm Foundation Grants, 4188
Scientific Communities Initiative Grants, 4189
Seaver Institute Grants, 4195
SFI Principal Investigator Program Grant, 4201
SHOT Abbott Payson Usher Prize, 4237
SHOT Mel Kranzberg Dissertation Fellowship, 4245
SPIE A.E. Conrady Award, 4338
SPIE Dennis Gabor Award, 4340
SPIE George W. Goddard Award, 4342
SPIE Gold Medal of the Society Award, 4343
SPIE Technology Achievement Award, 4344
SRC Franco Cerrina Fellowship, 4367
STC Research Grants, 4450
STC Scholarships in Technical Communication, 4451
Sun Grid Research Grants, 4467
Tandem Launch Technologies Grants, 4493
USAID Development Innovation Accelerator Broad Agency Announcement Grants, 4596
USDA Small Business Innovation Research Grants - Phase I, 4648
USDC Low-Power Television and Translator Upgrade Grants, 4666
USDD Active Authentication Grants, 4674
USDD Air Force Defense Research and Development Rapid Innovation Fund Grants, 4677
USDD Applied & Computational Anal Grants, 4683
USDD Collaborative Res in Structural Sciences, 4710
USDD Long Range Broad Agency Announcement for Navy and Marine Corps Science and Technology Grants, 4748
USDD Maneuver Thrust Grants, 4749
USDD Multidisciplinary Research Program of the University Research Initiative Grant, 4756
USDD Ocean Engineering and Marine Systems Grants, 4767
USDD Power Efficiency Revolution for Embedded Computing Technologies Grants, 4776
USDD Signal Processing for Networked Sensing Grants, 4793
USDD Small University Grants, 4794
USDD Strategic Technologies Grants, 4798
Xerox Foundation Grants, 4899
Young Scientists Summer Program, 4903

Technology Education
AAUW Community Action Grants, 213
AERA Dissertation Grants, 342
AIHS ForeFront MBA Studentship Award, 443
Alfred P. Sloan Foundation Civic Grants, 545
Bill and Melinda Gates Foundation Agricultural Development Grants, 1345
Boeing Company Contributions Grants, 1357
Canada-U.S. Fulbright International Science and Technology Award, 1417
Charles Edison Fund Grants, 1520
Charles Lafitte Foundation Grants, 1525
DOE Acad Creating Teacher Scientists Grants, 1766
DOE Albert Einstein Distinguished Educator Fellowship Program, 1770
Eaton Charitable Fund Grants, 1893
Gerrish/FTE Technology Education Graduate Fellowship, 2129
Intel Bachelor's and Master's Degree Internships, 2393
Intel Foundation PhD Fellowships, 2394
Intel Technician Internships, 2395
ITEA Maley/FTE Scholarship, 2435
Microsoft Research Cell Phone as a Platform for Healthcare Grants, 2767
NASA Lewis Educational Research and Collaborative Internships, 2840
Northrop Grumman Foundation Grants, 3429
NSF Activities in Science, Engineering, and Mathematics for Persons with Disabilities, 3460
NSF Director's Award for Distinguished Teaching Scholars, 3553
NSF Historically Black Colleges and Universities Program Grants, 3577
NSF Innovative Technology Experiences for Students and Teachers, 3582
NSF Louis Stokes Alliances for Minority Participation Grants, 3589
NSF Tribal Colleges and Universities Program, 3662
Ohio Learning Network Grants, 3734
Olin Corporation Charitable Trust Grants, 3739
Regenstrief Medical Informatics NLM Research Fellowships, 4048
RGK Foundation Grants, 4060
Sony Corporation of America Corporate Philanthropy Grants, 4329
Sun Academic Excellence Grants, 4466
USDD High School Apprenticeship Program and Undergraduate Research Apprenticeship Program Grants, 4732
USDD Science, Technology, Engineering and Mathematics Grants, 4789
Wayne and Gladys Valley Foundation Grants, 4835
Xerox Foundation Grants, 4899

Technology Education Translation
NEH Digital Humanities Implementation Grants, 2953
NEH Digital Humanities Start-Up Grants, 2954
Northrop Grumman Foundation Grants, 3429

Technology Planning/Policy
AAAS Roger Revelle Fellowship in Global Stewardship, 38
American Academy of Arts and Sciences Hellman Fellowships, 612
American Academy of Arts and Sciences Visiting Scholars, 613
Bill and Melinda Gates Foundation Financial Services for the Poor Grants, 1346
Christine Mirzayan Science and Technology Policy Graduate Fellowship Program, 1556
Kluge Center Distinguished Chairs, 2560

Technology Transfer
AIHS ForeFront Executive-in-Residence Award, 441
ANL Faculty Research Participation Program, 855
Boeing Company Contributions Grants, 1357
Colorado Bioscience Discovery Evaluation Grants, 1598
Indiana 21st Century Research and Technology Fund Awards, 2370
Indiana SBIR/STTR Commercialization Enhancement Program, 2371
Indiana Waste Tire Fund Program Grants, 2376
NCI Academic-Industrial Partnerships for Devel & Validation of In Vivo Imaging Systems and Methods for Cancer Investigations, 2913
NHLBI Bioengineering Approaches to Energy Balance and Obesity Grants for SBIR, 3017
NHLBI Bioengineering Approaches to Energy Balance and Obesity Grants for STTR, 3018
Northrop Grumman Foundation Grants, 3429
NSF CISE Computing Res Infrastructure Grants, 3531
NSF Cyber-Enabled Discovery and Innovation, 3549
NSF Engineering Research Centers Grants, 3561
NSF Small Business Technology Transfer Programs Phase I Grants, 3648
ONR Manufacturing Technology Grants, 3768
ONR Multidisciplinary Research Program of the URI Grants, 3774
ONR Small Business Technology Transfer Program Grants, 3783
SRC Aristotle Award, 4349
SRC Landmark Innovation Award, 4376
SRC Mahboob Khan Outstanding Industry Liaison Award, 4377
SRC Technical Excellence Award, 4394
SSHRC-NSERC College and Community Innovation Enhancement Grants, 4400
U.S. Department of Education Rehabilitation Engineering Research Centers Grants, 4544
US CRDF Science and Technology Entrepreneurship Program Business Partnership Grant Competition Armenia, 4611
US CRDF STEP Business Partnership Grant Competition Ukraine, 4612
USDC/NOAA Nature Conservancy Community-based Matching Grants, 4661

Technology, Hardware and Software
Alice Tweed Tuohy Foundation Grants Program, 569
Chatlos Foundation Grants Program, 1532
Corning Incorporated Found Educational Grants, 1646
DaimlerChrysler Corporation Fund Grants, 1714
Dave Thomas Foundation for Adoption Grants, 1723
E.L. Wiegand Foundation Grants, 1885
NATO Science for Peace Grants, 2898
Northrop Grumman Foundation Grants, 3429
NSERC Research Tools and Instruments Grants, 3449
NSF CISE Broadening Participation in Computing Grants, 3524
NSF Scientific Computing Research Environments for the Mathematical Sciences Grants, 3645
NYSCA Individual Artists: Film, Media and New Technology Production Grants, 3706
Oklahoma City Community Grants, 3737
ONR Expeditionary Maneuver Warfare & Combating Terrorism Department Grants, 3761
Robert G. Cabell III and Maude Morgan Cabell Foundation Grants, 4075
Sun Microsystems Foundation Summer Academic Enrichment Grants, 4468
USDD Active Authentication Grants, 4674
USDD Applied & Computational Anal Grants, 4683
USDD Crowd Sourced Formal Verif Grants, 4715
USDD High Assurance Cyber Military Systems Grants, 4731
USDD Information Innovation Office Grants, 4740
USDD Power Efficiency Revolution for Embedded Computing Technologies Grants, 4776
USDD Strategic Technologies Grants, 4798

Tectonics
NSF Arctic Nat Science Grants in Earth Sciences, 3485
NSF Tectonics Research Grants, 3658

Teen Pregnancy
California Wellness Foundation Work and Health Grants, 1409
DHHS Welfare Reform Research, Evaluations, and National Studies Grants, 1762
Helen V. Brach Foundation Grants, 2231
Joint Center for Poverty Research Grants, 2511
Thompson Charitable Foundation Grants, 4511
USAID Family Planning and Reproductive Health Methods Grants, 4598

Telecommunications
ARO Mobile, Wireless Communications and Networks Research Contracts, 1012
CIA Undergraduate Co-Op Program - Engineer/ Program Management Engineer, 1561
Dept of Ed Star Schools Program Grants, 1749
SHOT Brooke Hindle Postdoctoral Fellowships, 4238
SHOT Eugene S. Ferguson Prize, 4240
SHOT IEEE Life Members Prize in Elec Hist, 4241
SHOT Leonardo da Vinci Medal, 4244
SHOT NASA Fellowship, 4246

Television
AAAS Kavli Science Journalism Awards, 33
Alfred P. Sloan Foundation Public Understanding of Science & Tech-Public Television Grants, 557
Astronomical Society of the Pacific Dorothea Klumpke-Roberts Award, 1229
CDC Cooperative Agreement for the Development, Operation, and Evaluation of an Entertainment Education Program, 1457
Center for Defense Information Internships, 1473

ITVS Commissioned Funding, 2439
JMO Law and Legal System Grants, 2491
JMO Public Policy Research Grants, 2492
JMO Strategic and International Studies Grants, 2493
NEH Digital Humanities Implementation Grants, 2953
NEH Digital Humanities Start-Up Grants, 2954
OAH Erik Barnouw Award, 3716
Park Foundation Grants, 3870
Righteous Persons Foundation Grants, 4069
USDC Low-Power Television and Translator Upgrade Grants, 4666

Television, Cable
OAH Erik Barnouw Award, 3716

Television, Children's
Farmers Insurance Corporate Giving Grants, 1981
Union Pacific Foundation Community and Civic Grants, 4568

Television, Public
Farmers Insurance Corporate Giving Grants, 1981
George W. Codrington Charitable Grants, 2122
John Ben Snow Memorial Trust Grants, 2497
Pacific Islanders in Communications Grants, 3861
Union Pacific Foundation Community and Civic Grants, 4568
USDC Low-Power Television and Translator Upgrade Grants, 4666

Terminally Ill
Herbert & Adrian Woods Foundation Grants, 2239

Terrestrial Physics
NASA Goddard Visiting Fellowships, 2838

Terrorism
Alfred P. Sloan Foundation Selected National Issues Grants, 563
Aviation Research Grants, 1273
SSHRC-Public Safety Canada Kanishka Project Initiative, 4401

Textile/Weaving Arts
Florida Division of Cultural Affairs Individual Artist Fellowships, 2026

Textiles
ARO Synthesis and Processing of Materials Research Contracts, 1016
Institute of Textile Technology Grad Fellowships, 2388

Texture
AACC International Bruce Wasserman Young Investigator Award, 84
AACC Int'l Carl Wilhelm Brabender Award, 85
AACC International Excellence in Teaching Award, 89
AACC International George W. Scott Blair Lecture Award, 90
AACC Int'l Young Scientist Research Award, 100
AACC Texture Techs Quality Research Award, 101

Theater
Colonel Stanley R. McNeil Foundation Grants, 1597
Houghton Library Visiting Fellowships, 2284
NL Weiss/Brown Publication Subvention Award, 3320
Ontario Arts Council Integrated Arts Grants, 3790
Sioux Falls Area Comm Foundation Spot Grants, 4263

Theater/Film Criticism
Bogliasco Foundation Liguria Study Center Residencies, 1358
Surdna Foundation Arts Teachers Fellowships, 4470

Theology
Academy of American Franciscan History Dissertation Fellowships, 249
Acton Institute for the Study of Religion and Liberty Calihan Academic Fellowships, 331
Acton Institute for the Study of Religion and Liberty Novak Award, 332
Alpha Sigma Nu Book Award Competition, 578
ATLA Bibliography Grant, 1257
ATS Henry Luce III Fellowships, 1258
ATS Lilly Theological Collaborative Res Grants, 1259
ATS Lilly Theological Faculty Fellowships, 1260
ATS Lilly Theological Research Expense Grants, 1261
ATS Lilly Theological Scholars Grants, 1262
Booth-Bricker Fund Grants, 1359
Calvin Inst of Christian Worship Renewal Grants, 1411
Clark Library Short-Term Resident Fellowships, 1577
Foundation for Evangelism Doc Scholarships, 2039
FTE Dissertation Fellowships, 2058
FTE Doctoral Fellowships, 2059
FTE Expanding Horizons Partnership Grants, 2060
FTE North American Doctoral Fellowships, 2061
Holberg Memorial Fund International Prize, 2272
Holberg Memorial Fund Nils Klim Prize, 2273
Holberg Memorial Fund School Project, 2274
Metanexus Inst Templeton Research Lectures, 2744
Wabash Center Project Grants, 4826
Wabash Center Small Project Grants, 4827
Wabash Center Waorkshop and Colloquy Participant Grants, 4828

Theology Education
Foundation for Evangelism Doc Scholarships, 2039

Therapy Evaluation
Baltimore Washington Center Adult Psychoanalysis Fellowships, 1293
Baltimore Washington Center Child Psychotherapy Fellowships, 1294
Boston Psychoanalytic Society and Institute Fellowship in Child Psychoanalytic Psychotherapy, 1362
Boston Psychoanalytic Society and Institute Fellowship in Psychoanalytic Psychotherapy, 1363
Chicago Institute for Psychoanalysis Fellowships, 1548
NHLBI Airway Smooth Muscle Function and Targeted Therapeutics in Human Asthma Grants, 3014
NICHD Innovative Therapies and Clinical Studies for Screenable Disorders Grants, 3126

Therapy/Rehabilitation
ASHFoundation New Century Scholars Program Doctoral Scholarships, 1080
ASHFoundation New Century Scholars Research Grant, 1081
ASHFoundation Res Grant for New Investigators, 1082
ASHFoundation Student Research Grants in Audiology, 1084
USDD Aerospace Medicine, Clinical Research, Human Performance Research, and Expeditionary Medicine Grants, 4676

Thermal Analysis
ASME Allan Kraus Thermal Management Medal, 1104
ASME Prime Movers Committee Award, 1156

Thermal Effects
ASME Allan Kraus Thermal Management Medal, 1104
ASME Prime Movers Committee Award, 1156
NSF Thermal Transport Processes Grants, 3660

Thermodynamics
Alfred P. Sloan Foundation Energy and Environment Grants, 552
ARO Theoretical Physics and Nonlinear Phenomena Research Contracts, 1017
ASME Allan Kraus Thermal Management Medal, 1104
ASME James Harry Potter Gold Medal, 1135
DOE Energy Conversion Tech Research Grants, 1786
DOE Fossil Advanced Research and Technology Development Grants, 1793
NSF Chemical Oceanography Research Grants, 3517
NSF Interfacial Processes and Thermodynamics Research Grants, 3586
NSF Physical Chemistry Grants, 3616
NSF Thermal Transport Processes Grants, 3660

Thermodynamics Engineering
ASME Edward F. Obert Award, 1116
ASME James Harry Potter Gold Medal, 1135

Thermosciences Engineering
ARO Theoretical Physics and Nonlinear Phenomena Research Contracts, 1017
ASME Allan Kraus Thermal Management Medal, 1104

Third World Nations
Conservation, Food, and Health Foundation Grants for Developing Countries, 1641
IIE David L. Boren Scholarships, 2351

Thoracic Medicine
Foundation of the American Thoracic Society Research Grants, 2044
Thoracic Foundation Grants, 4513

Tidal and Wave Energy
ONR Radio Frequency Sensor Processing Research Grants, 3778
ONR Radio Frequency Solid State Devices Applied Research Grants, 3779
USDD Ocean Acoustics Research Grants, 4766
USDD Physical Oceanography Grants, 4771

Tissue Culture
NIA Role of Nuclear Receptors in Tissue and Organismal Aging Grants, 3120
NSF Biomedical Engineering and Engineering Healthcare Grants, 3506

Tobacco
American Legacy Foundation National Calls for Proposals Grants, 748
American Legacy Found Small Innovative Grants, 749
CDC Found Tobacco Network Lab Fellowship, 1462
Christine O. Gregoire Young Adult Award for Outstanding Use of Tobacco Industry Docs, 1557
CTCRI Idea Grants, 1665
CTCRI Knowledge Synthesis Grants, 1666
CTCRI Policy Research Grants, 1667
CTCRI Researcher Travel Grants, 1668
CTCRI Research Planning Grants, 1669
CTCRI Student Research Grants, 1670
CTCRI Workshop and Learning Opportunities Grants, 1671
FAMRI Clinical Innovator Awards, 1975
FAMRI Young Clinical Scientist Awards, 1976
NCI/NIDA Transdisciplinary Tobacco Use Research Centers Grants, 2912
Pfizer Global Research Awards for Nicotine Independence, 3925
Sybil G. Jacobs Award for Outstanding Use of Tobacco Industry Documents, 4487
TRDRP California Research Awards, 4525
TRDRP Dissertation Research Awards, 4526
TRDRP Exploratory/Devel Research Awards, 4527
TRDRP New Investigator Awards, 4528
TRDRP Participatory Research Awards, 4529
TRDRP Postdoctoral Fellowship Awards, 4530
TRDRP Research Grants, 4531

Tolerance
Saint Mary's College Fellowships, 4159
Third Millennium Foundation Fellowships, 4504

Tolerance, Ethnic
Harry Frank Guggenheim Foundation Dissertation Fellowships, 2207

Topology
Abdus Salam ICTP Postdoctoral Positions in Theoretical High Energy Physics, 234
ARO Mobile, Wireless Communications and Networks Research Contracts, 1012
NSF Topology Research Grants, 3661

Tourette Syndrome
Tourette Syndrome Association Post-Doctoral Fellowships, 4523
Tourette Syndrome Association Research Grants, 4524

Tourism
ASTA Academic Scholarships, 1224
Bikes Belong Foundation Research Grants, 1343
National Endowment for the Humanities Fellowships at Winterthur, 2867
Winterthur Museum Dissertation Fellowships, 4887
Winterthur Research Fellowships, 4888

Toxic Substances
Bullitt Foundation Grants, 1384
Clarence E. Heller Charitable Foundation Grants, 1575
EPA Research, Development, Monitoring, Public Education, Training, Demonstrations, and Studies Grants, 1938
Orchard Foundation Grants, 3793

Toxicology
ASM sanofi-aventis ICAAC Award, 1184
Canada Graduate Scholarships and NSERC Postgraduate Scholarships, 1420
NIEHS Small Business Innovation Research Program Grants, 3195
ORISE Faculty Research Participation at the National Center for Toxicological Research, 3804
ORISE Student Research Participation at the Agency for Toxic Substances and Disease Registry, 3828
ORISE Student Research Participation at the National Center for Toxicological Research, 3830
PhRMA Foundation Health Outcomes Post Doctoral Fellowships, 3950
PhRMA Foundation Health Outcomes Research Starter Grants, 3951
PhRMA Foundation Informatics Post Doctoral Fellowships, 3953
PhRMA Foundation Pharmacology/Toxicology Post Doctoral Fellowships, 3962
PhRMA Foundation Pharmacology/Toxicology Pre Doctoral Fellowships, 3963
PhRMA Foundation Pharmacology/Toxicology Research Starter Grants, 3964
PhRMA Foundation Pharmacology/Toxicology Sabbatical Fellowships, 3965
USDD Aerospace Medicine, Clinical Research, Human Performance Research, and Expeditionary Medicine Grants, 4676

Training and Development
AALAS Foundation Grants, 182
ACC Starr Foundation Fellowship Program, 268
AES Susan S. Spencer Clinical Research Epilepsy Training Fellowship, 347
AIAR Andrew W. Mellon Foundation Fellowships, 392
AIR/NCES Fellowships for Graduate Study, 467
Allen Foundation Educational Nutrition Grants, 570
APHL Emerging Infectious Diseases Fellowships, 914
ASTD Dissertation Award, 1225
ASTD Research Award, 1226
Boeing Company Contributions Grants, 1357
Burden Trust Grants, 1388
CDC National Training Information and Exchange Program, 1464
Center of Excellence in General Aviation Grants, 1486
CIA Undergraduate Internship - Analysis, 1562
ConocoPhillips Grants Program, 1629
Conservation, Food, and Health Foundation Grants for Developing Countries, 1641
Cruise Industry Charitable Foundation Grants, 1660
Cystic Fibrosis Canada Special Travel Grants For Fellows and Students, 1682
DHHS Emerging Leaders Program Internships, 1756
DHHS Special Programs for the Aging Training, Research, and Discretionary Projects and Programs Grants, 1761
DOE Competitive Financial Assistance for Renewable Energy and Energy Efficiency Programs, 1779
DOJ National Institute of Corrections Training and Staff Development Grants, 1827
Eastern Europe & the Independent States of the Former Soviet Union Research & Training Grants, 1891
Elizabeth Glaser Int'l Leadership Awards, 1906
EPA Pollution Prevention Grants, 1936
IBRO Return Home Fellowships, 2332
IMLS 21st Century Museum Prof Grants, 2362
IMLS Conservation Project Support Grants, 2363
IMLS National Leadership Grants, 2366
IREX ECA Alumni Small Grants Program, 2416
Jacob and Hilda Blaustein Foundation Grants, 2454
John D. and Catherine T. MacArthur Foundation Global Challenges Program Grants, 2498
Jovid Foundation Grants, 2518
MAP International Medical Fellowships, 2655
MetLife/TCG Aha! Think it Do it Grants, 2745
National Housing Endowment Challenge/Build/Grow Grant, 2877
NATO Science for Peace Grants, 2898
NEA/TCG Career Development Grants for Theatre Directors, 2943
NEI Ruth L. Kirschstein National Research Service Award Short-Term Institutional Research Training Grants, 2963
New Hampshire Charitable Foundation Grants, 2985
NFIE Found Learning & Leadership Grants, 2992
NHLBI Ruth L. Kirschstein National Research Service Award Institutional Research Training Grants, 3056
NHLBI Ruth L. Kirschstein National Research Service Award Short-Term Institutional Research Training Grants, 3061
NIDCD Mentored Research Scientist Development Award, 3132
NOAA Joint Research Institutes Grants, 3379
North Carolina Biotechnology Center Collaborative Funding Grants, 3421
North Carolina Biotech Center Research Grants, 3424
NSERC Industrial Research Chairs, 3443
NSERC Research Tools and Instruments Grants, 3449
NSF Integrative Graduate Education and Research Traineeship, 3584
NSF Materials Research Science and Engineering Centers, 3594
NSF Partnerships for International Research and Education, 3611
NYAM Gladys Brooks Internship, 3686
OSF Justice Initiative Fellowships, 3847
Prudential Foundation Education Grants, 4002
Regenstrief Geriatrics Fellowship, 4046
RRF General Program Grants, 4115
SARE Professional Development Grants - Northeast Region, 4176
SfN Ricardo Miledi Neuroscience Training, 4212
Sioux Falls Area Community Foundation Community Fund Grants (Unrestricted), 4262
Sioux Falls Area Comm Foundation Spot Grants, 4263
U.S. Department of Education Disability and Rehabilitation Research and Related Projects, 4538
U.S. Department of Education Rehabilitation Research Training Centers, 4545
USDA Cochran Fellowships, 4619
USDA Organic Agriculture Research and Extension Initiative, 4637
USDC Dissertation Fellowships, 4665
WWF Russell E. Train Education for Nature Alumni Grants, 4897

Transexuals
NYPL Martin Duberman Visiting Scholars, 3700

Transfusion Medicine
AABB Dale A. Smith Memorial Award, 46
AABB Emily Cooley Memorial Award and Lectureship, 47
AABB Hemphill-Jordan Leadership Award, 49
AABB John Elliott Memorial Award, 50
AABB Karl Landsteiner Memorial Award and Lectureship, 51
AABB Sally Frank Mem Award & Lectureship, 52
AABB Tibor Greenwalt Memorial Award and Lectureship, 53
National Blood Foundation Research Grants, 2848
NHLBI Immunomodulatory, Inflammatory, and Vasoregulatory Properties of Transfused Red Blood Cell Units as a Function of Preparation and Storage Grants, 3027
NHLBI Mentored Career Award For Faculty At Minority Institutions, 3034

Transitional Students
Paul P. Fidler Research Grants, 3878

Translation
Aid to Scholarly Publications, 431
ASCE Freeman Fellowships, 1040
ICCS Publishing Fund, 2336
Japan Foundation Japanese Studies Fellowships, 2476
Japan Foundation New York Japanese Studies Fellowships, 2478
National Endowment for the Arts Literature Fellowships: Translation Projects, 2866
NCI Application of Metabolomics for Translational and Biological Research Grants, 2915
NEH Collaborative Research Grants, 2952
NL Audrey Lumsden-Kouvel Fellowships, 3282
NL Northeast Modern Language Assocation Fellowship, 3314
ONR Computational Neuroscience & Biorobotics Grants, 3752
SSHRC-FHSS Awards to Scholarly Publications, 4398
SSHRC Insight Development Grants, 4409
SSHRC Insight Grants, 4410
University of Southern Mississippi H.A. & Margret Rey Translation Fellowships, 4589

Transplantation Immunology
National Blood Foundation Research Grants, 2848
NIH Nonhuman Primate Immune Tolerance Cooperative Study Group Grants, 3228
Pfizer Advancing Research in Transplantation Science Research Awards, 3908
Roche Organ Transplant Research Found Grants, 4083

Transport Phenomena
ARO Degradation, Reactivity, and Protection of Materials Research Contracts, 1008
NSF Condensed Matter Physics Research Grants, 3540
NSF Engineering Research Grants, 3562
NSF Interfacial Processes and Thermodynamics Research Grants, 3586
NSF Thermal Transport Processes Grants, 3660

Transportation
AAA Foundation for Traffic Safety Grants, 24
AIChE CRE Travel Awards, 404
ASCE Jack E. Leisch Memorial National Graduate Fellowship, 1041
Bullitt Foundation Grants, 1384
David Bohnett Foundation Grants, 1724
DOE Competitive Financial Assistance for Renewable Energy and Energy Efficiency Programs, 1779
DOE Vehicle and Engine Research and Development Grants, 1819
DOT Dwight David Eisenhower Transportation Fellowship Program, 1847
DOT Transportation Statistics Research Grants, 1850
E.J. Sierleja Memorial Fellowship, 1884
Energy Foundation Power Grants, 1924
Energy Foundation Transportation Grants, 1925
Farmers Insurance Corporate Giving Grants, 1981
OAH Germany Residency Program Grants, 3719
OAH Japanese Residencies Program Grants, 3722
Oklahoma City Community Grants, 3737
Surdna Found Sustainable Environments Grants, 4471
USDA Animal Health & Disease Grants, 4616
USDD Ocean Engineering and Marine Systems Grants, 4767

Transportation Engineering
ASCE Jack E. Leisch Memorial National Graduate Fellowship, 1041
ASME Soichiro Honda Medal, 1164

Transportation Planning/Policy
Aviation Research Grants, 1273
Bikes Belong Foundation Research Grants, 1343
DOT Transit Planning and Research Grants, 1849
Energy Foundation Transportation Grants, 1925
William and Flora Hewlett Foundation Environmental Grants, 4874

Trauma
CDC National Training Information and Exchange Program, 1464
Dart Center Ochberg Fellowship, 1720
NIGMS Mentored Clinical Scientist Research Career Development Awards, 3197
RACGP/CONROD Research Fellowship, 4009
USDD Workpl Violence in the Military Grants, 4805

Travel
AAAAI Fellows-in-Training Travel Grants, 23
AACAP Educational Outreach Program for Child and Adolescent Psychiatry Residents, 56
AACAP Educational Outreach Program for General Psychiatry Residents, 57
AACAP Systems of Care Special Program Scholarships, 81
AACC Int'l Carl Wilhelm Brabender Award, 85
AACR-GlaxoSmithKline Clinical Cancer Research Scholar Awards, 118
AACR-NCI International Investigator Opportunity Grants, 121
AACR-WICR Scholar Awards, 125
AACR Brigid G. Leventhal Scholar in Cancer Research Awards, 131
AACR Minority Scholar in Cancer Res Awards, 144
AACR Scholar-in-Training Awards, 146
AACR Scholar-in-Training Awards - Special Conferences, 147
AACR Thomas J. Bardos Science Educ Awards, 149
AAG International Geographic Information Fund Student Travel Grants, 174
AAP International Travel Grants, 197
ACLS New Faculty Fellows, 290
AES-Grass Young Investigator Travel Awards, 345
AIChE CRE Travel Awards, 404
AIChE North American Mixing Forum Award, 418
AIHP Sonnedecker Visiting Scholar Grants, 432
ALA Frederick G. Kilgour Award for Research in Library and Information Technology, 487
ALA NewsBank/Readex Cath Reynolds Award, 494
AMA-MSS Research Poster Award, 599
AMA-WPC Joan F. Giambalvo Memorial Scholarships, 600
American Chemical Society ANYL Pfizer Graduate Travel Awards in Analytical Chemistry, 646
American College of Rheumatology Fellows-in-Training Travel Scholarships, 700
American College of Surgeons Traveling Fellowship to Germany, 706
American College of Surgeons Traveling Fellowship to Japan, 707
American Mathematical Society China Exchange Travel Grants, 753
American Math Society Simons Travel Grants, 756
American Psychiatric Association Travel Scholarships for Minority Medical Students, 790
AMS Eileen Southern Travel Fund Grants, 822
AMS Eugene K. Wolf Travel Fund for European Research Grants, 823
AMS Harold Powers World Travel Fund Grants, 826
AMS Janet Levy Fund for Independent Scholars, 828
AMS M. Elizabeth C. Bartlet Fund for Research in France, 830
APSA Travel Grants, 948
ASCO Merit Awards, 1049
ASM Undergraduate Research Fellowship, 1186
ASTA Academic Scholarships, 1224
BWF Collaborative Research Travel Grants, 1395
CFR Int'l Affairs Fellowships in South Korea, 1503
CFR International Affairs in Japan Fellowships, 1504
CHF Travel Grants, 1545
CIC Traveling Scholars, 1566
Cystic Fibrosis Canada Special Travel Grants For Fellows and Students, 1682
Cystic Fibrosis Canada Visiting Clinician Awards, 1687
Duke University Libraries Franklin Collection Travel Grants, 1874
Duke University Libraries Hartman Center Travel Grants, 1875
Duke University Libraries J. Walter Thompson Research Fellowships, 1876
Durfee Foundation Stanton Fellowship, 1881
Explorers Club Exploration Fund Grants, 1966
FDHN Moti L. & Kamla Rustgi International Travel Awards, 1997
FDHN Non-Career Research Awards, 1998
Fulbright Travel Grants, 2086
Herbert Hoover Presidential Library Association Travel Grants, 2242
IBRO Asia Reg APRC Exchange Fellowships, 2323
IBRO InEurope Short Stay Grants, 2325
IBRO Latin America Regional Travel Grants, 2329
IFUW Dr. A. Vibert Douglas Int'l Fellowship, 2346
Institute of Turkish Studies Summer Research Grants in Turkey, 2392
Jane Beattie Memorial Scholarship, 2470
Japan Foundation New York Institutional Support for Japanese Studies Grants, 2477
Kellogg Institute for International Studies Faculty Fellows' Research Grants, 2540
Kellogg Institute for International Studies Grants for Conferences and Workshops, 2543
Kellogg Institute for International Studies Grants for Events, 2545
Manitoba Arts Council Craft Grants, 2653
MAP International Medical Fellowships, 2655
MBL Gruss Lipper Family Foundation Summer Fellowship, 2704
MBL Herbert W. Rand Summer Fellowship, 2706
Mead Johnson Nutritionals Scholarships, 2735
MLA David A. Kronick Traveling Fellowship, 2784
National Center for Responsible Gaming Travel Grants, 2858
National Endowment for the Humanities Fellowships at Winterthur, 2867
NL Herzog August Wolfenbuttel Fellowship, 3286
NOAA Living Marine Resources Cooperative Science Center Mini Grants, 3381
NSERC Northern Research Internships Program, 3447
NSERC Undergrad Student Research Awards, 3451
NSF Antarctic Artists and Writers Grants, 3473
OAH/IEHS John Higham Travel Grants, 3710
OAH Germany Residency Program Grants, 3719
OAH Japanese Residencies Program Grants, 3722
Princeton University Center for Human Values Political Philosophy Travel Grants, 3993
RACGP Rex Walpole Travelling Fellowship, 4016
Rockefeller Archive Center Visiting Archivist Fellowship, 4087
Short-Term Travel to Japan Grants, 4236
Sigrid Juselius Foundation Grants, 4260
Smithsonian National Museum of History Lemelson Center Travel to Collections Awards, 4290
SRC Educational Science Grants for Invitation of Visiting Scientists, 4357
SRC Educational Science Grants for Travels to Foreign Institutions as a Guest Scientist, 4358
SRC Educational Science Travel Grants, 4365
SRC Medical Research Grants, 4379
SRC Natural and Engineering Science Conference Grants, 4383
SRC Nat & Engineering Science Travel Grants, 4388
UICC International Cancer Technology Transfer Fellowships, 4556
UICC Raisa Gorbachev Memorial International Cancer Fellowships, 4557
Winterthur Museum Dissertation Fellowships, 4887
Winterthur Research Fellowships, 4888
Woodrow Wilson Doctoral Dissertation Fellowships in Women's Studies, 4891
Woodrow Wilson MMUF Travel & Res Grants, 4893
Woodrow Wilson National Fellowship Foundation Travel and Research Grants, 4894

Tropical Medicine
Conservation, Food, and Health Foundation Grants for Developing Countries, 1641
NFID Colin L. Powell Minority Postdoctoral Fellowship in Tropical Disease Research, 2987
NFID New Investigator Matching Grants, 2989

Tropical Zones
Bernard Lowy Fund for the Study of Tropical Botany in Latin America, 1335
GCA Tropical Botany Research Awards, 2101
John D. and Catherine T. MacArthur Foundation Global Challenges Program Grants, 2498
MacArthur Foundation Conservation and Sustainable Development Grants, 2648
Neotropical Grassland Conservancy Derald G. Langham Memorial Research Grant, 2973
NOAA Cooperative Institute for Applied Meteorological Studies Grants, 3357
NOAA Cooperative Institute for Tropical Meteorology Research Grants, 3358
ONR Marine Meteorology and Atmospheric Effects Research Grants, 3769
Organization for Tropical Studies Fellowships, 3796
Organization for Tropical Studies Pilot Awards for Exploratory Research, 3797
OTS Francis J. Bossuyt Fellowship, 3855
OTS Hovore-Horn Fellowships, 3856
OTS Stanley Smith Horticultural Fellowships, 3857
OTS Undergrad NAPIRE Summer Interns, 3858
RA Kleinhans Fellowship for Research in Tropical Nontimber Forest Products, 4026
Smithsonian Senior Fellowships, 4296
Smithsonian Tropical Research Institute Short Term Fellowships, 4297

Tuberculosis
ASM/CDC Fellowships in Infectious Disease and Public Health Microbiology, 1100
NFID New Investigator Matching Grants, 2989
NYCT Biomedical Research Grants, 3694
ORISE Faculty Research Participation at the Centers for Disease Control and Prevention, 3803
ORISE Postgraduate Research Participation at the Centers for Disease Control and Prevention, 3820
ORISE Student Research Participation at the Centers for Disease Control and Prevention, 3829
Potts Memorial Foundation Grants, 3985
USAID Accelerating Progress Against Tuberculosis in Kenya Grants, 4594
USAID Comprehensive District-Based Support for Better HIV/TB Patient Outcomes Grants, 4595

Turbulence
NCAR Graduate Research Fellowships, 2901
NCAR Postdoctoral Fellowships in the Atmospheric Sciences, 2902
NSF Physical and Dynamic Meteorology Grants, 3614

Turkey
ARIT Fellowships, 991
ARIT Fellowships for Intensive Advanced Turkish Language Study, 992
ARIT Mellon Foundation Postdoctoral Research Fellowships for East European Scholars, 994
Fulbright Critical Lang Enhancement Awards, 2068
GUITS Grants for the Publication of Scholarly Books and Journals, 2186
GUITS Grants for Undergraduate Study, 2187
GUITS Matching Seed-Money Grants for New Positions in Turkish Studies, 2188

GUITS Post-Doctoral Summer Travel-Research Grants, 2189
GUITS Research Grants in Comparative Studies of Modern Turkey, 2190
GUITS Sabbatical Research Grants, 2191
GUITS Summer Research Grants in Turkey for Graduate Students, 2192
Institute of Turkish Studies Dissertation Writing Grants, 2389
Institute of Turkish Studies Research Grants in Comparative Studies of Modern Turkey, 2390
Institute of Turkish Studies Summer Language Study Grants in Turkey, 2391
Institute of Turkish Studies Summer Research Grants in Turkey, 2392
IREX Individual Advanced Research Opportunities Grants, 2419

Turkish Studies
ARIT Fellowships for Intensive Advanced Turkish Language Study, 992
Institute of Turkish Studies Dissertation Writing Grants, 2389
Institute of Turkish Studies Research Grants in Comparative Studies of Modern Turkey, 2390
Institute of Turkish Studies Summer Language Study Grants in Turkey, 2391
Institute of Turkish Studies Summer Research Grants in Turkey, 2392

Turkmenistan
IREX Individual Advanced Research Opportunities Grants, 2419

Ukraine
CIUS Neporany Res & Teaching Fellowship, 1572
IREX Individual Advanced Research Opportunities Grants, 2419

Undergraduate Education
AIHS Media Summer Fellowship, 453
AIR/NRC Summer Institute on First-Year Assessment Undergraduate Student Fellowships, 468
American Chemical Society ANYL I.M. Kolthoff Enrichment Awards for Undergrad Students, 644
American Chemical Society Award for Research at an Undergraduate Institution, 660
American Historical Association Raymond J. Cunningham Prize, 742
Animal Behavior Society C.H. Turner Award, 849
APS Prize for a Faculty Member for Research in an Undergraduate Institution, 977
BMW of N America Charitable Contributions, 1353
CIC Summer Research Opportunities, 1565
Community Foundation of the Fox River Valley, 1620
Dreyfus Senior Scientist Mentor Program Grants, 1857
Elizabeth Nash Found Summer Research Awards, 1909
Field Museum Undergraduate Internships, 2010
Flinn Foundation Grants Programs, 2024
HHMI Grants and Fellowships Programs, 2255
ISGC Multiple Undergraduate Research Grants, 2432
Jeffress Awards in Interdisciplinary Research, 2481
MBL William Townsend Porter Summer Fellowships for Minority Investigators, 2716
National Wildlife Federation Internships, 2891
NBME Stemmler Medical Education Research Fund Grants, 2899
NSF CISE Educational Innovation Grants, 3533
NSF Course, Curriculum, and Laboratory Improvement Research Grants, 3546
NSF Louis Stokes Alliances for Minority Participation Grants, 3589
NSF Presidential Awards for Excellence in Science, Math, and Engineering Mentoring, 3632
NSF Research Experiences for Undergrad Grants, 3638
NSF Undergraduate Research and Mentoring in the Biological Sciences, 3663
OTS Undergrad NAPIRE Summer Interns, 3858
SfN Undergraduate Student Travel Awards, 4217
SRC Undergraduate Research Opportunities, 4395
UE Explore Undergraduate Research Grants, 4553
WHOI Guest Student Internships, 4860
WHOI Summer Student Fellowships, 4865

Undersea Workstations/Vehicles
NOAA Mid-Atlantic Bight National Undersea Research Center Grants, 3383
NOAA Undersea Research Center Coral Reef Conservation Program Coral Research Grants, 3409
NOAA Undersea Research Center Ocean Acidification Research at Aquarius Reef Base Grants, 3410
NOAA Undersea Research Program Grants, 3411
NOAA Undersea Research Project Grants, 3412

Unitarian Church
MacDonald-Peterson Foundation Grants, 2649

United Kingdom (Great Britain and No Ireland)
Huntington Library Travel Grants and Exchange Fellowships for Study in Great Britain, 2309

United States History
AAS Kate B. and Hall J. Peterson Fellowships, 204
AAS Legacy Fellowships, 205
AAS Long-Term Visiting Academic Research Fellowships, 206
AAS Short-Term Visiting Academic Research Fellowships, 208
Ahmanson and Getty Postdoctoral Fellowships, 377
ALA Donald G. Davis Article Award, 484
American Historical Association Albert J. Beveridge Award, 712
American Historical Association Albert J. Beveridge Grants for Research in the History of the Western Hemisphere, 713
American Historical Assoc Herb Feis Award, 723
American Historical Association J. Franklin Jameson Fellowship, 725
American Historical Association James A. Rawley Prize in Atlantic History, 727
American Historical Association John H. Dunning Prize, 732
American Historical Association Littleton-Griswold Prize in American Law and Society, 735
American Historical Association Michael Kraus Research Grant in American Colonial History, 738
American Historical Association Waldo J. Leland Prize, 743
American Philosophical Society Library Resident Research Fellowships, 769
Cushwa Center Research Travel Grants, 1673
Herbert Hoover Presidential Library Association Travel Grants, 2242
Huntington Library Dana and David Dornsife Fellowships, 2303
Huntington Library National Endowment for the Humanities Fellowships, 2307
Huntington Library Travel Grants and Exchange Fellowships for Study in Great Britain, 2309
IES Faculty Fellowships, 2341
Irving and Rose Crown Scholarships and Fellowships in American History, 2427
MCEAS Dissertation Fellowships, 2723
National Endowment for the Humanities Fellowships at Winterthur, 2867
NL American Society for Eighteenth-Century Studies Fellowships, 3279
NL Arthur and Lila Weinberg Fellowships for Independent Scholars and Researchers, 3280
NL Associated Colleges of the Midwest/Great Lakes College Association Faculty Fellowships, 3281
NL British Academy Fellowship for Study in the United Kingdom, 3283
NL Frances C. Allen Fellowships, 3284
NL Grad Scholar-in-Residence Opportunities, 3285
NL Herzog August Bibliothek Wolfenbuttel Fellowship, 3286
NL Lawrence Lipking Fellowship at Newberry, 3287
NL Lester J. Cappon Fellowship in Documentary Editing, 3288
NL Lloyd Lewis Fellowships in American Hist, 3289
NL Mellon Postdoctoral Research Fellowships, 3294
NL Midwest Modern Lang Assoc Fellowship, 3299
NL Monticello College Foundation Fellowship for Women, 3304
NL National Endowment for the Humanities Fellowships, 3310
NL Northeast Modern Language Assocation Fellowship, 3314
NL Scholar-in-Residence Opportunity, 3315
NL Short-Term Resident Fellowships for Individual Research, 3317
NL South Central Modern Languages Association Fellowship, 3318
NL Susan Kelly Power and Helen Hornbeck Tanner Fellowship, 3319
OAH/IEHS John Higham Travel Grants, 3710
OAH Avery O. Craven Award, 3711
OAH Binkley-Stephenson Award, 3712
OAH Darlene Clark Hine Award, 3713
OAH David Thelen Award, 3714
OAH Ellis W. Hawley Prize, 3715
OAH Erik Barnouw Award, 3716
OAH Frederick Jackson Turner Award, 3717
OAH Friend of History Award, 3718
OAH Germany Residency Program Grants, 3719
OAH James A. Rawley Prize, 3721
OAH Japanese Residencies Program Grants, 3722
OAH Louis Pelzer Memorial Award, 3725
OAH Merle Curti Award, 3726
OAH Ray A. Billington Prize, 3727
OAH Roy Rosenzweig Dist Service Award, 3729
OAH Willi Paul Adams Award, 3730
SHGAPE Vincent P. DeSantis Prize, 4234
Smithsonian Senior Fellowships, 4296
Summerfield G. Roberts Award, 4465
Twentieth-Century Japan Research Awards, 4536
Virginia Foundation for the Humanities Discretionary Grants, 4813
Virginia Foundation for Humanities Open Grants, 4814
Winterthur Museum Dissertation Fellowships, 4887
Winterthur Research Fellowships, 4888

United States, Midwest
IES Faculty Fellowships, 2341
NL American Society for Eighteenth-Century Studies Fellowships, 3279
NL Arthur and Lila Weinberg Fellowships for Independent Scholars and Researchers, 3280
NL Associated Colleges of the Midwest/Great Lakes College Association Faculty Fellowships, 3281
NL British Academy Fellowship for Study in the United Kingdom, 3283
NL Frances C. Allen Fellowships, 3284
NL Grad Scholar-in-Residence Opportunities, 3285
NL Herzog August Bibliothek Wolfenbuttel Fellowship, 3286
NL Lawrence Lipking Fellowship at Newberry, 3287
NL Lester J. Cappon Fellowship in Documentary Editing, 3288
NL Mellon Postdoctoral Research Fellowships, 3294
NL Midwest Modern Lang Assoc Fellowship, 3299
NL Monticello College Foundation Fellowship for Women, 3304
NL National Endowment for the Humanities Fellowships, 3310
NL Northeast Modern Language Assocation Fellowship, 3314
NL Scholar-in-Residence Opportunity, 3315
NL Short-Term Resident Fellowships for Individual Research, 3317
NL South Central Modern Languages Association Fellowship, 3318
NL Susan Kelly Power and Helen Hornbeck Tanner Fellowship, 3319
OAH Ray A. Billington Prize, 3727

United States, New England
Nellie Mae Education Foundation State Level Systems Change Grants, 2971

United States, Northwest
Ahmanson and Getty Postdoctoral Fellowships, 377
ASECS/Clark Fellowships, 1057
Jacobs Research Funds Small Grants Program, 2456
OAH Ray A. Billington Prize, 3727

United States, Southwest
Ahmanson and Getty Postdoctoral Fellowships, 377
ASECS/Clark Fellowships, 1057
OAH Ray A. Billington Prize, 3727

University/Industry Cooperative Activities
AIHS ForeFront Executive-in-Residence Award, 441
American Chemical Society Chemical Technology Partnership Mini Grants, 674
American Chemical Society GCI Pharmaceutical Roundtable Research Grants, 677
Center of Excellence in General Aviation Grants, 1486
Joyce Foundation Employment Grants, 2520
Medtronic Foundation Fellowships, 2738
NSERC Industrial R&D Fellowships, 3442
NSERC Industrial Research Chairs, 3443
NSERC Synergy Awards for Innovation, 3450
NSF-CONACyT Collaborative Research Opportunities, 3452
NSF Engineering Research Centers Grants, 3561
NSF Grant Opportunities for Academic Liaison with Industry, 3574
NSF Industry/University Cooperative Research Centers Grants, 3579
NSF Louis Stokes Alliances for Minority Participation Grants, 3589
NSF Materials Research Science and Engineering Centers, 3594
SRC Aristotle Award, 4349
SRC Company-Named Fellowships, 4350
SRC Cross-Disciplinary Research Grants, 4353
SRC Graduate Fellowships, 4368
SRC International Fellowship, 4375
SRC Landmark Innovation Award, 4376
SRC Mahboob Khan Outstanding Industry Liaison Award, 4377
SRC Peter Verhofstadt Fellowship, 4389
SRC Research Customization Program Grants, 4390
SSHRC-NSERC College and Community Innovation Enhancement Grants, 4400

Urban Affairs
Charles H. Revson Foundation Grants, 1524
Chicago Community Trust Housing Grants: Advancing Affordable Rental Housing, 1547
Community Found for Muskegon County Grants, 1617
Community Found of E Central Illinois Grants, 1619
Horowitz Foundation for Social Policy Grants, 2280
Horowitz Found for Social Policy Special Awards, 2281
Jewish Federation of Metropolitan Chicago Scholarships, 2486
Joint Center for Poverty Research Grants, 2511
Nellie Mae Education Foundation District-Level Change Grants, 2968
Nordson Corporation Foundation Grants, 3416

Urban Areas
Achelis Foundation Grants, 274
Bodman Foundation Grants, 1356
Commonwealth Fund Harkness Fellowships in Health Care Policy and Practice, 1606
E.L. Wiegand Foundation Grants, 1885
HUD Community Outreach Partnership Centers Grants, 2292
ISA John Z. Duling Grants, 2429
Lynde & Harry Bradley Foundation Fellowships, 2634
Lynde and Harry Bradley Foundation Grants, 2635
National Book Scholarship Fund, 2849
Northwest Fund for the Environment Grants, 3431
Rajiv Gandhi Foundation Grants, 4025
Surdna Found Sustainable Environments Grants, 4471
William and Flora Hewlett Foundation Environmental Grants, 4874
William G. McGowan Charitable Fund Grants, 4877

Urban Education
Carnegie Corporation of New York Grants, 1432
Nellie Mae Education Foundation District-Level Change Grants, 2968
OISE William Waters Scholarships in Urban Education, 3736

Urban Geography
Open Spaces Sacred Places National Awards, 3791

Urban Planning/Policy
Andy Warhol Foundation for the Visual Arts Curatorial Research Fellowships, 847
Canada-U.S. Fulbright International Science and Technology Award, 1417
Canada-U.S. Fulbright Senior Specialists Program Grants, 1418
Canada-U.S. Fulbright Traditional Scholars, 1419
Charles Abrams Scholarships, 1519
Community Found for Muskegon County Grants, 1617
FMC Foundation Grants, 2029
Fulbright Alumni Initiatives Awards Grants, 2064
Fulbright Distinguished Chairs Program Grants, 2069
Fulbright European Union Scholar-in-Residence Program Grants, 2071
Fulbright New Century Scholars Program Grants, 2075
Fulbright Occasional Lecturer Program Grants, 2076
Fulbright Specialists Program Grants, 2079
Fulbright Traditional Scholar Program in Europe and Eurasia, 2080
Fulbright Traditional Scholar Program in Sub-Saharan Africa, 2081
Fulbright Traditional Scholar Program in the East Asia/Pacific Region, 2082
Fulbright Traditional Scholar Program in the Near East and North Africa Region, 2083
Fulbright Traditional Scholar Program in the South and Central Asia Region, 2084
Fulbright Traditional Scholar Program in the Western Hemisphere, 2085
Fulbright Visiting Specialists Program: Direct Access to the Muslim World, 2087
HUD Urban Scholars Postdoctoral Fellowships, 2293
National Urban Fellows Program, 2888
Open Spaces Sacred Places National Awards, 3791
Price Family Charitable Fund Grants, 3987
Robert Schalkenbach Foundation Grants, 4079
Sensient Technologies Foundation Grants, 4199
SSHRC Insight Development Grants, 4409
SSHRC Insight Grants, 4410

Urban Sociology
Open Spaces Sacred Places National Awards, 3791

Urinary Tract
NHLBI Lymphatics in Health and Disease in the Digestive, Urinary, Cardiovascular and Pulmonary Systems, 3032
NIDDK Basic Research Grants in the Bladder and Lower Urinary Tract, 3138

Urogenital System
NIDDK Cell-Specific Delineation of Prostate and Genitourinary Development, 3140

Urologic Diseases
NHLBI Lymphatics in Health and Disease in the Digestive, Urinary, Cardiovascular and Pulmonary Systems, 3032
NIDDK Devel of Disease Biomarkers Grants, 3145
NIDDK Education Program Grants, 3149

Urology
NIDDK Basic and Clinical Studies of Congenital Urinary Tract Obstruction Grants, 3137
NIDDK Non-Invasive Methods for Diagnosis and Progression of Diabetes, Kidney, Urological, Hematological and Digestive Diseases, 3168
NIDDK Pilot and Feasibility Clinical Research Grants in Kidney or Urologic Diseases, 3171
NIDDK Proteomics: Diabetes, Obesity, And Endocrine, Digestive, Kidney, Urologic, And Hematologic Diseases, 3174
NIDDK Small Grants for Underrepresented Minority Scientists in Diabetes and Digestive and Kidney Diseases, 3183
NIDDK Training in Clinical Investigation in Kidney and Urology Grants, 3185
NKF Research Fellowships, 3277
NKF Young Investigator Grants, 3278
NYAM Edwin Beer Research Fellowship, 3683
NYAM Ferdinand C. Valentine Fellowship, 3684
NYAM Ferdinand C. Valentine Student Research Grants, 3685
Pfizer Medical Education Track Two Grants, 3927

Utilities
Colorado Clean Energy Solar Innovation Grants, 1599
DEED Grants, 1730
DEED Student Research Grants and Internships, 1731
Energy Foundation Power Grants, 1924
MGM Resorts Foundation Community Grants, 2753

Uzbekistan
Fulbright Critical Lang Enhancement Awards, 2068
Fulbright Traditional Scholar Program in the South and Central Asia Region, 2084
IREX Individual Advanced Research Opportunities Grants, 2419

Vaccines
Alternatives Research and Development Foundation Alternatives in Education Grants, 581
Alternatives Research and Development Foundation Grants, 582
Chiron Foundation Community Grants, 1553
NFID Adv Vaccinology Course Travel Grant, 2986
Thrasher Research Fund Grants, 4516

Vacuum Electronics
USDD Vacuum Devices Grants, 4802

Vacuum Engineering
IUVSTA Prize for Science, 2443
IUVSTA Prize for Technology, 2444
IUVSTA Welch Foundation Scholarship, 2445

Vacuum Science
IUVSTA Prize for Science, 2443
IUVSTA Prize for Technology, 2444
USDD Vacuum Devices Grants, 4802

Values/Moral Education
Arcus Foundation Religion and Values Grants, 987
ConocoPhillips Grants Program, 1629
NCSS/FASSE/CUFA Inquiry Grant, 2936
NCSS Exemplary Res in Social Studies Award, 2937
NCSS Jean Dresden Grambs Distinguished Career Research in Social Studies Award, 2938
NCSS L Metcalf Exemplary Dissertation Award, 2939
Princeton University Center for Human Values Graduate Prize Fellowships, 3989
Princeton University Center for Human Values Graduate Student Top-Up Prizes, 3990
Princeton University Center for Human Values Laurance S. Rockefeller Visiting Fellowships, 3992
Princeton University Center for Human Values Postdoctoral Research Associate in Values and Public Policy, 3994
Princeton University Center for Human Values Short Movie Prize, 3995

Vehicle Design
SSHRC-APC Automotive Partnership Canada, 4396

Venereal Diseases
CDC Foundation Atlanta International Health Fellowships, 1461

Venture Capital
Barker Award, 1300
USAID Devel Innovation Ventures Grants, 4597
Washington Area Women's Foundation Strategic Opportunity and Partnership Fund Grants, 4831

Veterans
MGN Family Foundation Grants, 2754
Minnesota War Orphans Educational Assistance Grants, 2780
Northrop Grumman Corporation Grants, 3428

Veterinary Medicine
ACVIM Foundation Clinical Investigation Grants, 334
American Chemical Society AGRO Award for Innovation in Chemistry of Agriculture, 631
APHL Emerging Infectious Diseases Fellowships, 914
BARD Research Grants, 1299
Bodman Foundation Grants, 1356
International Crane Foundation Internships, 2396
NHLBI Pathway to Independence Awards, 3045
NHLBI Ruth L. Kirschstein National Research Service Awards for Individual Senior Fellows, 3060
NIGMS/NIDCR Evolution of Infectious Diseases Research Grants, 3196
Regents Professional Opportunity Scholarships, 4049
Scleroderma Foundation Established Investigator Grants, 4190
USDA Animal Health & Disease Grants, 4616
William and Charlotte Parks Foundation for Animal Welfare Grants, 4872
Winn Feline Foundation Grants, 4885

Veterinary Medicine Education
Banfield Charitable Trust Grants, 1295
Christine Mirzayan Science and Technology Policy Graduate Fellowship Program, 1556
MLA Donald Lindberg Research Fellowship, 2785
NHLBI Ruth L. Kirschstein National Research Service Award Institutional Research Training Grants, 3056
NHLBI Ruth L. Kirschstein National Research Service Award Short-Term Institutional Research Training Grants, 3061
NHLBI Short-Term Research Education Program to Increase Diversity in Health-Related Research, 3062
USDA Norman E. Borlaug International Agricultural Science and Technology Fellowships, 4636
Winn Feline Foundation Grants, 4885
Yale University Beinecke Library Research Fellowships for Yale Grad & Professional School Students, 4902

Video Production
Bogliasco Foundation Liguria Study Center Residencies, 1358
Carole Fielding Film/Video Production or Research Grants, 1438
CCH Documentary Project Research and Development Grants, 1451
Center for Defense Information Internships, 1473
Delaware Division of the Arts Individual Artist Fellowships, 1732
Florida Division of Cultural Affairs Individual Artist Fellowships, 2026
North Carolina Arts Council Folklife Public Program Grants, 3419
NYFA Artist Fellowships, 3698
NYSCA Individual Artists: Film, Media and New Technology Production Grants, 3706
OAH Erik Barnouw Award, 3716
WHOI Ocean Science Journalism Fellowship, 4862

Videos
ACC Residency Program in Asia, 267
Bogliasco Foundation Liguria Study Center Residencies, 1358
Carole Fielding Film/Video Production or Research Grants, 1438
CCH Documentary Project Research and Development Grants, 1451
OAH Erik Barnouw Award, 3716

Vietnam
American Historical Association John K. Fairbank Prize in East Asian History, 733

Violence
Allen Lane Foundation Grants, 571
AMWA Kathryn C. Bemmann New Investigator Grant on Violence against Women, 838
California Wellness Foundation Work and Health Grants, 1409
CDC Grants for Violence-Related Injury Prevention Research, 1463
Dart Center Ochberg Fellowship, 1720
David Bohnett Foundation Grants, 1724
Harry Frank Guggenheim Foundation Research Grants, 2208
Overbrook Foundation Grants, 3860
Richard and Rhoda Goldman Fund Grants, 4063
Virginia Foundation for the Humanities Discretionary Grants, 4813
Virginia Foundation for Humanities Open Grants, 4814

Violence Prevention
SSHRC-Public Safety Canada Kanishka Project Initiative, 4401
SSRC Drugs, Security & Democracy Fellows, 4426
United States Institute of Peace - Transatlantic Post-Doc Fellowship for International Relations and Security, 4581
Virginia Foundation for the Humanities Discretionary Grants, 4813
Virginia Foundation for Humanities Open Grants, 4814

Viral Studies
CDC Evaluation of the Use of Rapid Testing For Influenza in Outpatient Medical Settings, 1458
NSF Starter Grants for Postdoctoral Fellows in Microbial Biology, 3653

Virology
APHL Emerging Infectious Diseases Fellowships, 914
ASM/CDC Fellowships in Infectious Disease and Public Health Microbiology, 1100
Bristol-Myers Squibb Clinical Outcomes and Research Grants, 1374
Bristol-Myers Squibb Foundation Fellowships, 1375
Bristol-Myers Squibb Foundation Global HIV/AIDS Initiative Grants, 1376
NSF Genes and Genome Systems Cluster Grants, 3569
ORISE Faculty Research Participation at the Centers for Disease Control and Prevention, 3803
ORISE Postgraduate Research Participation at the Centers for Disease Control and Prevention, 3820

Vision
1st Touch Foundation Grants, 2
Alcon Foundation Grants Program, 513
Alfred W. Bressler Prize in Vision Science, 567
CNIB Barbara Tuck MacPhee Award, 1590
CNIB Canada Glaucoma Clinical Research Council Grants, 1591
CNIB E. & Mary Hochhausen Access Technology Research Grants, 1592
Eye-Bank for Sight Restoration and Fight for Sight Summer Student Research Fellowship, 1969
Fight for Sight-Streilein Foundation for Ocular Immunology Research Award, 2012
Fight for Sight Grants-in-Aid, 2013
Fight for Sight Post-Doctoral Awards, 2014
Fight for Sight Summer Student Fellowships, 2015
Foundation Fighting Blindness Marjorie Carr Adams Women's Career Development Awards, 2036
Hawaii Community Foundation Health Education and Research Grants, 2221
NEI Mentored Clinical Scientist Development Program Award, 2961
NEI Scholars Program, 2964
NEI Translational Research Program On Therapy For Visual Disorders, 2965
NEI Vision Research Core Grants, 2966
NYCT Blindness and Visual Disabilities Grants, 3695
OneSight Research Foundation Block Grants, 3742
Sigma Xi Grants in Aid of Research, 4259
Streilein Foundation for Ocular Immunology Visiting Professorships, 4453
Thome Foundation Awards in Age-Related Macular Degeneration Research, 4509

Visual Arts
AAR Rome Prize Fellowships, 202
ACC John D. Rockefeller III Award, 257
ACC Starr Foundation Fellowship Program, 268
Amgen Foundation Grants, 812
Andy Warhol Foundation for the Visual Arts Curatorial Research Fellowships, 847
Arkansas Arts Council Individual Fellowships, 999
ASID Foundation Joel Polsky Prize, 1096
Bogliasco Foundation Liguria Study Center Residencies, 1358
Booth-Bricker Fund Grants, 1359
Bush Foundation Artist Fellowships, 1391
CAA Professional Development Fellowships for Graduate Students, 1404
CAA Wyeth Publication Grant, 1405
Camargo Foundation Fellowships, 1412
Chestnut Hill Charitable Foundation, Inc Grants, 1538
Christensen Fund Regional Grants, 1554
Cintas Foundation Fellowships, 1570
Cord Foundation Grants, 1645
DAAD Research Grants for Doctoral Candidates and Young Academics and Scientists, 1703
DAAD Research Stays for University Academics and Scientists, 1705
DAAD Scholarships for Artists: Study Visits for Academics, 1707
Delaware Division of the Arts Individual Artist Fellowships, 1732
Fleishhacker Foundation Eureka Fellowships, 2023
Flinn Foundation Grants Programs, 2024
Florida Division of Cultural Affairs Individual Artist Fellowships, 2026
George Gund Foundation Grants, 2117
George S. and Dolores Dore Eccles Grants, 2120
Georgia Power Foundation Grants, 2125
Henry Luce Foundation/ACLS Dissertation Fellowships in American Art, 2235
NAEF Teacher Incentive Grants, 2815
Nevada Arts Council Artists Fellowships, 2984
North Carolina Arts Council Residency Grants, 3420
NYSCA Dance: Services to the Field Grants, 3702
NYSCA Folk Arts: Exhibitions Grants, 3703
NYSCA Folk Arts: Presentation Grants, 3704
NYSCA Folk Arts: Regional and County Folk Arts Programs Grants, 3705
NYSCA Individual Artists: Film, Media and New Technology Production Grants, 3706
Ontario Arts Council Integrated Arts Grants, 3790
Paul G. Allen Family Foundation Grants, 3876
Pollock-Krasner Foundation Grants, 3981
Reinberger Foundation Grants, 4051
Shastri Indo-Canadian Institute Arts Fellowships, 4219
Sid W. Richardson Foundation Grants, 4253
Stanford Humanities Center Arts Writer/Practitioner Fellowship, 4444
Surdna Foundation Arts Teachers Fellowships, 4470
Textron Corporate Contributions Grants, 4501
Winterthur Research Fellowships, 4888

Visual Impairments
AHAF National Glaucoma Research Grants, 372
Alfred W. Bressler Prize in Vision Science, 567
CNIB Baker Applied Research Fund Grants, 1588
CNIB Baker New Researcher Fund Grants, 1589
CNIB Barbara Tuck MacPhee Award, 1590
CNIB Canada Glaucoma Clinical Research Council Grants, 1591
CNIB E. & Mary Hochhausen Access Technology Research Grants, 1592
Donald and Sylvia Robinson Family Foundation Grants, 1830

Foundation Fighting Blindness Marjorie Carr Adams Women's Career Development Awards, 2036
George Gund Foundation Grants, 2117
John W. Alden Trust Grants, 2510
Josiah W. and Bessie H. Kline Foundation Grants, 2517
Mabel H. Flory Charitable Trust Grants, 2646
Marjorie C. Adams Charitable Trust Grants, 2665
Mericos Foundation Grants, 2742
NEI Clinical Vision Research Devel Award, 2960
NEI Scholars Program, 2964
NEI Translational Research Program On Therapy For Visual Disorders, 2965
NYCT Blindness and Visual Disabilities Grants, 3695
OneSight Research Foundation Block Grants, 3742
Sara Elizabeth O'Brien Trust Grants, 4166
Starr Foundation Grants, 4447
The Ray Charles Foundation Grants, 4502
William G. and Helen C. Hoffman Found Grants, 4876

Vocational Counseling
Achelis Foundation Grants, 274
Bodman Foundation Grants, 1356
GNOF Gulf Coast Oil Spill Grants, 2153

Vocational Education
BMW of N America Charitable Contributions, 1353
Catholic University Life Cycle Institute Research Fellowship, 1444
GNOF Gulf Coast Oil Spill Grants, 2153

Vocational Services
GNOF Gulf Coast Oil Spill Grants, 2153

Vocational Training
GNOF Gulf Coast Oil Spill Grants, 2153
MGM Resorts Foundation Community Grants, 2753
Sioux Falls Area Community Foundation Community Fund Grants (Unrestricted), 4262
Sioux Falls Area Comm Foundation Spot Grants, 4263
USDA Organic Agriculture Research and Extension Initiative, 4637

Vocational/Technical Education
Clay Foundation Grants, 1579
Credit Union Found of British Columbia Grants, 1654
D.F. Halton Foundation Grants, 1692
DaimlerChrysler Corporation Fund Grants, 1714
Dept of Ed Star Schools Program Grants, 1749
Dept of Ed Voc Education National Programs, 1750
ETS Predoctoral Summer Program, 1954
Frances L. and Edwin L. Cummings Memorial Fund Grants, 2046
GNOF Gulf Coast Oil Spill Grants, 2153
HAF Riley Frazel Memorial Scholarship, 2197
McCune Charitable Foundation Grants, 2719
Miller Brewing Corporate Contributions Program Grants, 2774
Minnesota War Orphans Educational Assistance Grants, 2780
National Housing Endowment Challenge/Build/Grow Grant, 2877
Norcliffe Foundation Grants, 3415
Oppenstein Brothers Foundation Grants, 3792
Stackpole-Hall Foundation Grants, 4442
U.S. Department of Education Rehabilitation Research Training Centers, 4545
WILD Foundation Grants, 4867

Volcanology
NSF Marine Geology & Geophysics Res Grants, 3592

Volunteers
AAAAI Distinguished Layperson Award, 18
ALZA Corporate Contributions Grants, 588
American Family Insurance Corporate Giving Program Grants, 709
Ann Arbor Area Community Foundation Grants, 860
Award for Volunteer Service to the American Chemical Society, 1277
Bodman Foundation Grants, 1356
Campbell Soup Foundation Grants, 1415
Catholic University Life Cycle Institute Research Fellowship, 1444
Charles Stewart Mott Foundation Grants, 1529
ConocoPhillips Grants Program, 1629
Coors Brewing Corporate Contributions Grants, 1644
Dell Foundation Open Grants, 1734
Fidelity Foundation Grants, 2007
Harry B. and Jane H. Brock Foundation Grants, 2206
Hilton Hotels Corporate Giving Program Grants, 2268
ICC Faculty Fellowships, 2333
ICC Scholarship of Engagement Faculty Grants, 2334
International Crane Foundation Internships, 2396
John Ben Snow Memorial Trust Grants, 2497
Lillian S. Wells Foundation Grants, 2614
McCombs Foundation Grants, 2718
Nordson Corporation Foundation Grants, 3416
Oklahoma City Community Grants, 3737
Olin Corporation Charitable Trust Grants, 3739
Oppenstein Brothers Foundation Grants, 3792
Phi Upsilon Omicron Florence Fallgatter Distinguished Service Award, 3943
Praxair Foundation Grants, 3986
Prudential Foundation Education Grants, 4002
Radcliffe Institute Carol K. Pforzheimer Student Fellowships, 4018
Rayonier Foundation Grants, 4034
Research Program at Earthwatch Grants, 4057
Rhode Island Foundation Grants, 4061
Rohm and Haas Company Grants, 4098
Sensient Technologies Foundation Grants, 4199
Sony Corporation of America Philanthropy Grants, 4329
Special Olympics Kennedy Shriver Fellowships, 4331
Stackpole-Hall Foundation Grants, 4442
Virginia A. Hodgkinson Research Prize, 4812
W.K. Kellogg Foundation Healthy Kids Grants, 4822
WILD Foundation Grants, 4867
Xerox Foundation Grants, 4899

Volunteers
Harry B. and Jane H. Brock Foundation Grants, 2206
ICC Faculty Fellowships, 2333
ICC Scholarship of Engagement Faculty Grants, 2334
Radcliffe Institute Carol K. Pforzheimer Student Fellowships, 4018

Voter Educational Programs
Farmers Insurance Corporate Giving Grants, 1981

Wage and Salary Administration
NYSCA Individual Artists: Film, Media and New Technology Production Grants, 3706
NYSCA Museum: Project Support Grants, 3707

Waste Management
Bill and Melinda Gates Foundation Water, Sanitation and Hygiene Grants, 1347
EREF Sustainability Research Grants, 1948
USDD Solid and Liquid Waste Treatment Grants, 4796
USDD Water Desalination & Purific Grants, 4804

Waste Management (Agriculture)
Indiana Waste Tire Fund Program Grants, 2376
SARE Research and Education Grants - North Central Region, 4177
SARE Research and Education Grants - Northeast Region, 4178
USDA 1890 Institution Teaching, Research, and Extension Capacity Building Grants, 4613
USDA Distance Education Grants for Higher Education Institutions in Insular Areas, 4622
USDA Norman E. Borlaug International Agricultural Science and Technology Fellowships, 4636
USDA Resident Instruction Grants for Higher Education in Insular Areas, 4645
USDA Tribal College Research Grants, 4652

Waste Management/Fossil Energy
DOE Energy from Municipal Waste Grants, 1787
DOE Engineering Research Grants, 1789
Honor the Earth Grants, 2279
Indiana Waste Tire Fund Program Grants, 2376
NYCT New York City Environment Grants, 3697
Rohm and Haas Company Grants, 4098
US CRDF CGP ETF Energy Research Competition For Joint U.S.-Estonian Research Teams, 4609

Wastewater Treatment
Bill and Melinda Gates Foundation Water, Sanitation and Hygiene Grants, 1347
DOE Energy from Municipal Waste Grants, 1787

Water Resources
AWRA Richard A. Herbert Mem Scholarship, 1281
Bedding Plants Foundation Research Grants, 1330
Bill and Melinda Gates Foundation Water, Sanitation and Hygiene Grants, 1347
Bullitt Foundation Grants, 1384
Bydale Foundation Grants, 1400
Cape Branch Foundation Grants, 1425
Community Found for Muskegon County Grants, 1617
DOE Energy from Municipal Waste Grants, 1787
DOI Geological Research and Data Acquisition Grants, 1822
DOI Upper Mississippi River System Long-Term Resource Monitoring Program Grants, 1825
Dreyfus Postdoctoral Program in Environmental Chemistry Grant, 1855
Edmund Niles Huyck Preserve Graduate and Postgraduate Research Grants, 1894
EPA National Estuary Program Grants, 1935
EPA Targeted Watersheds Grants, 1941
FishAmerica Foundation Marine and Anadromous Fish Habitat Restoration Grants, 2018
FishAmerica Foundation Research Grants, 2019
Freshwater Future Project Grants, 2056
Horton Research Grants, 2282
HRF Graduate Fellowships, 2285
HRF Hudson River Improvement Grants, 2286
IIE General Electric Foundation TECH Awards, 2352
MDEQ Beach Monitoring Grants - Inland Lakes, 2731
MDEQ Local Water Quality Monitoring Grants, 2732
NASA Goddard Visiting Fellowships, 2838
NFWF Marine Debris Research and Technology Grants, 3002
NFWF More Fish Partnership Fund Grants, 3003
NFWF Shell Marine Habitat Program Grants, 3008
NOAA California Bay Watershed Education and Training Program Grants, 3338
NOAA National Estuarine Research Reserve System Graduate Research Fellowships, 3390
Northwest Fund for the Environment Grants, 3431
NSF Hydrologic Sciences Research Grants, 3578
ORISE Postgrad Internship at the Office of Water, 3817
ORISE Savannah River Site Postgraduate Research Program, 3824
ORISE Savannah River Site Professional Internship Program, 3825
Padilla Bay National Estuarine Research Reserve Graduate Research Fellowships, 3864
Padilla Bay National Estuarine Research Reserve Research Assistantships, 3865
Rocky Mountain Mineral Law Found Grants, 4094
Russell Family Foundation Environmental Sustainability Grants, 4128
Thompson Charitable Foundation Grants, 4511
USAID Palestinian Community Infrastructure Development Grants, 4600
USDA McIntire-Stennis Cooperative Forestry Research Grants, 4634
USDA Small Business Innovation Research Grants - Phase I, 4648
W.K. Kellogg Foundation Healthy Kids Grants, 4822

Water Resources, Environmental Impacts
Abdus Salam ICTP/IAEA Sandwich Training Education Grants, 226
Bill and Melinda Gates Foundation Water, Sanitation and Hygiene Grants, 1347

Community Foundation for Greater New Haven Quinnipiac River Fund Grants, 1616
GNOF Gulf Coast Oil Spill Grants, 2153
IDEM Section 319(h) Nonpoint Source Program Grants, 2338
MDEQ Beach Monitoring Grants - Inland Lakes, 2731
Meadows Foundation Grants, 2736
NFWF Marine Debris Research and Technology Grants, 3002
NFWF More Fish Partnership Fund Grants, 3003
NOAA CRCP Coral Reef Ecosystem Monitoring Grants, 3362
NOAA CSCOR Regional Ecosystem Prediction Program: Cumulative Impacts of Stressors at the Land-Water Interface in the Mid-Atlantic, 3363
NOAA CSCOR Regional Ecosystem Prediction Program: Ecosystem Goal-Setting in Coastal Waters and Reefs of South Florida, 3364
NOAA CSCOR Regional Ecosystem Prediction Program: Invasive Species in the Great Lakes--A Regional Scale Approach, 3365
NOAA Mid-Atlantic Research Set-Aside Program Grants, 3384
Rocky Mountain Mineral Law Found Grants, 4094
Russell Family Foundation Environmental Sustainability Grants, 4128
Sioux Falls Area Community Foundation Community Fund Grants (Unrestricted), 4262
Sioux Falls Area Comm Foundation Spot Grants, 4263
USDA Rangeland Research Grants, 4641

Water Resources, Management/Planning
AWWA Thomas R. Camp Scholarship, 1286
BARD Research Grants, 1299
Bill and Melinda Gates Foundation Water, Sanitation and Hygiene Grants, 1347
Horton Research Grants, 2282
Illinois DNR Youth Recreation Corps Grants, 2361
Russell Family Foundation Environmental Sustainability Grants, 4128
Union Pacific Foundation Community and Civic Grants, 4568
USDA 1890 Institution Teaching, Research, and Extension Capacity Building Grants, 4613
USDA Distance Education Grants for Higher Education Institutions in Insular Areas, 4622
USDA McIntire-Stennis Cooperative Forestry Research Grants, 4634
USDA Norman E. Borlaug International Agricultural Science and Technology Fellowships, 4636
USDA Resident Instruction Grants for Higher Education in Insular Areas, 4645
USDA Tribal College Research Grants, 4652

Water Supply
AWWA Abel Wolman Doctoral Fellowships, 1282
AWWA Academic Achievement Awards, 1283
AWWA Holly A. Cornell Scholarship, 1284
Bill and Melinda Gates Foundation Water, Sanitation and Hygiene Grants, 1347
GNOF Gulf Coast Oil Spill Grants, 2153
IDEM Section 319(h) Nonpoint Source Program Grants, 2338
SARE Research and Education Grants - North Central Region, 4177
SARE Research and Education Grants - Northeast Region, 4178
USAID Palestinian Community Infrastructure Development Grants, 4600
USDA Technical Agricultural Assistance Research Grants, 4651
William and Flora Hewlett Foundation Environmental Grants, 4874

Water Treatment
AWWA Abel Wolman Doctoral Fellowships, 1282
AWWA Holly A. Cornell Scholarship, 1284
AWWA Larson Aquatic Research Support Scholarships, 1285
Bill and Melinda Gates Foundation Water, Sanitation and Hygiene Grants, 1347
USAID Palestinian Community Infrastructure Development Grants, 4600

Waterways and Harbors
BMW of N America Charitable Contributions, 1353
Freshwater Future Project Grants, 2056
Great Lakes Fishery Commission Sea Lamprey Research Grants, 2167
HRF Hudson River Improvement Grants, 2286
MDEQ Local Water Quality Monitoring Grants, 2732
USCD/NOAA Illinois-Indiana Sea Grant Core Proposals, 4604

Weapons
AAAS Science and Technology Policy Fellowships: Global Health and Development, 42
USDD Air Force Defense Research and Development Rapid Innovation Fund Grants, 4677
USDD Air Warfare and Naval Weapons Applications Grants, 4679
USDD Basic Research for Combating Weapons of Mass Destruction Broad Agency Announcement Grants, 4690
USDD Fires Thrust Grants, 4726
USDD Innovative Systems for Military Missions - Tactical Technology Office Grants, 4742
USDD National Consortium for Measurement and Signature Intelligence Research Grants, 4759

Welding
ARO Synthesis and Processing of Materials Research Contracts, 1016

Welfare Reform
DHHS Welfare Reform Research, Evaluations, and National Studies Grants, 1762
Harris Graduate School of Public Policy Studies Research Development Grants, 2204
Macquarie Bank Foundation Grants, 2650
Searle Freedom Trust Grants, 4192

Welfare-to-Work Programs
Helen Bader Foundation Grants, 2228
Prudential Foundation Education Grants, 4002
Textron Corporate Contributions Grants, 4501

Wellness
AIHS Alberta/Pfizer Translational Res Grants, 435
Bupa Foundation Medical Research Grants, 1386
Bupa Foundation Multi-Country Grant, 1387
CDC Preventive Med Residency & Fellowship, 1465
Delta Air Lines Foundation Health and Wellness Grants, 1736
Obesity Society Grants, 3732
Samueli Institute Scientific Research Grants, 4162

Wetlands
Cape Branch Foundation Grants, 1425
International Crane Foundation Internships, 2396
National Wetlands Awards, 2889
NFWF Nature Restoration Trust Grants, 3005
NFWF Puget Sound Marine Conservation Fund Grants, 3006
NFWF Shell Marine Habitat Program Grants, 3008
NOAA National Estuarine Research Reserve Land Acquisition and Construction Program Grants, 3389
Northwest Fund for the Environment Grants, 3431
NYCT New York City Environment Grants, 3697

Wildlife
A. V. Stout Fund Research Grants, 13
Banfi Vintners Foundation Grants, 1296
Batchelor Foundation Grants, 1305
Beim Foundation Grants, 1331
Bullitt Foundation Grants, 1384
Carl C. Icahn Foundation Grants, 1428
Cruise Industry Charitable Foundation Grants, 1660
Dean Witter Foundation Grants, 1728
DOI African Elephant Conservation Grants, 1820
DOI Rhinoceros and Tiger Conservation Grants, 1824
Donald and Sylvia Robinson Family Foundation Grants, 1830
FishAmerica Foundation Marine and Anadromous Fish Habitat Restoration Grants, 2018
FishAmerica Foundation Research Grants, 2019
GNOF Gulf Coast Oil Spill Grants, 2153
Great Lakes Fishery Trust Access Grants, 2168
Great Lakes Protection Fund Grants, 2170
H.A. and Mary K. Chapman Charitable Grants, 2193
International Crane Foundation Internships, 2396
M. Bastian Family Foundation Grants, 2636
Mary K. Chapman Foundation Grants, 2673
Mericos Foundation Grants, 2742
MoEF Dr. Salim Ali Nat Wildlife Fellowship, 2792
National Wildlife Refuge System Centennial Scholarship Program, 2892
National Wild Turkey Federation Research Grant, 2893
NFWF/Exxon Save the Tiger Fund Grants, 2995
NFWF Acres for America Grants, 2996
NFWF Bird Conservation Initiative Grants, 2997
NFWF Budweiser Renewable Energy and Wildlife Conservation Research Prize, 2998
NFWF International Sea Turtle Conservation Fund Grants, 3000
NFWF Marine & Coastal Conservation Initiative Grants, 3001
NFWF Marine Debris Research and Technology Grants, 3002
NFWF More Fish Partnership Fund Grants, 3003
NFWF National Whale Conservation Grants, 3004
NFWF Pulling Together Initiative Grants, 3007
NFWF Wildlife and Habitat Conservation Initiative Grants, 3009
Northwest Fund for the Environment Grants, 3431
NYZS/Wildlife Conservation Society Research Grants, 3709
Olin Corporation Charitable Trust Grants, 3739
Royal Caribbean Cruises Ocean Fund, 4108
SeaWorld and Busch Gardens Conservation Fund Grants, 4196
Sioux Falls Area Community Foundation Community Fund Grants (Unrestricted), 4262
Sioux Falls Area Comm Foundation Spot Grants, 4263
Skaggs Foundation Grants, 4266
USDA Animal Health & Disease Grants, 4616
USDA McIntire-Stennis Cooperative Forestry Research Grants, 4634
USDA Wildlife Services Grants, 4653
Whitley Awards, 4858
WILD Foundation Grants, 4867
William and Flora Hewlett Foundation Environmental Grants, 4874

Wind, Climatology
NOAA National Climatic Data Center Interns, 3387

Wine and Grape Industry
American Society for Enology and Viticulture Scholarships, 797
American Wine Society Educational Foundation Scholarship Program, 808

Wireless Communication
SSHRC-APC Automotive Partnership Canada, 4396

Women
AACR-WICR Scholar Awards, 125
AIAR Andrew W. Mellon Foundation Fellowships, 392
Alfred P. Sloan Foundation Research Fellowships, 560
AMA-WPC Joan F. Giambalvo Memorial Scholarships, 600
American Legacy Foundation National Calls for Proposals Grants, 748
AMWA Kathryn C. Bemmann New Investigator Grant on Violence against Women, 838
Animal Behavior Society C.H. Turner Award, 849
APSA Special Fund for the Study of Women and Politics, 947

Archer Daniels Midland Foundation Grants, 985
Arthur and Rochelle Belfer Foundation Grants, 1023
ASA/NSF/BLS Fellowships, 1030
ASA/NSF/Census Research Fellowships, 1031
ASME Graduate Teaching Fellowships, 1125
ASR Fichter Research Grant, 1217
Avon Products Foundation Grants, 1275
Beim Foundation Grants, 1331
Beverly Willis Architecture Foundation Grants and Fellowships, 1340
Bill and Melinda Gates Foundation Agricultural Development Grants, 1345
BMW of N America Charitable Contributions, 1353
Bodenwein Public Benevolent Foundation Grants, 1355
Boston Foundation Grants, 1361
Brico Fund Grants, 1371
Carnegie Corporation of New York Grants, 1432
Catherine Holmes Wilkins Foundation Grants, 1442
CCWH/Berkshire Conference of Women Historians Graduate Student Fellowship, 1453
CCWH Catherine Prelinger Award, 1454
CCWH Ida B. Wells Grad Student Fellowship, 1455
CDC Grants for Violence-Related Injury Prevention Research, 1463
Claneil Foundation Grants, 1574
ConocoPhillips Grants Program, 1629
Cowles Charitable Trust Grants, 1652
EFWA Accounting Research Grants, 1901
Elisabeth Severance Prentiss Foundation Grants Program, 1905
FASSE Demonstration Project Grant, 1982
Ford Foundation Peace and Social Justice Grants, 2031
Francis P. Garvan-John M. Olin Medal, 2048
General Service Foundation Human Rights and Economic Justice Grants, 2108
Georgia Power Foundation Grants, 2125
Girl's Best Friend Foundation Grants, 2143
Goddess Scholars Grants, 2155
Gruber Foundation Prize for Global Justice and Women's Rights, 2182
H.J. Heinz Company Foundation Grants, 2195
Harry B. and Jane H. Brock Foundation Grants, 2206
Heineman Foundation for Research, Education, Charitable and Scientific Purposes, 2227
Helen V. Brach Foundation Grants, 2231
Ida Smedley MacLean International Fellowship, 2337
IFUW Dr. A. Vibert Douglas Int'l Fellowship, 2346
IREX Short-Term Travel Grants, 2424
IWMF Public Health Fellowship, 2446
Jackson Foundation Grants, 2453
James R. Dougherty Jr. Foundation Grants, 2463
L'Oreal USA Women in Science Fellowship, 2577
Mary Baker Eddy Library Summer Research Fellowships, 2669
Mary Ingraham Bunting Inst Peace Fellowship, 2671
Mary Isabel Sibley Fellowship, 2672
MBL Ann E. Kammer Mem Summer Fellowship, 2693
National Book Scholarship Fund, 2849
NIGMS Mentored Clinical Scientist Research Career Development Awards, 3197
NLM Advancing Novel Science in Women's Health Research Exploratory/Developmental Research Grants, 3291
NLM Advancing Novel Science in Women's Health Research Small Project Grants, 3292
NL Monticello College Foundation Fellowship for Women, 3304
NRA Foundation Grants, 3437
NSF ADVANCE: Increasing the Participation and Advancement of Women in Academic Science and Engineering Careers Grants, 3461
NSF Research on Gender in Science and Engineering Grants, 3639
NSF SBE Doctoral Dissertation Research Improvement Grants, 3641
NWO Aspasia Grants, 3673
NWSA Abafazi-Africana Women's Studies Essay Award, 3675
NWSA Graduate Scholarship in Jewish Women's Studies, 3676
OAH Darlene Clark Hine Award, 3713
Olin Corporation Charitable Trust Grants, 3739
Overbrook Foundation Grants, 3860
Peacock Foundation Grants, 3887
Perpetual Trust for Charitable Giving Grants, 3900
Playboy Foundation Grants, 3980
Pott Foundation Grants, 3984
Radcliffe Institute Dissertation Grants, 4019
Radcliffe Institute Oral History Grants, 4022
Rajiv Gandhi Foundation Grants, 4025
Rayonier Foundation Grants, 4034
Reuter Foundation Grants, 4059
Schramm Foundation Grants, 4188
SfN Louise Hanson Marshall Specific Recognition Award, 4208
SfN Mika Salpeter Lifetime Achievement Award, 4209
SRC AMD/Mahboob Khan Fellowship, 4348
SRC Company-Named Scholarships, 4351
SRC Master's Scholarships, 4378
SRC Undergraduate Research Opportunities, 4395
SSHRC Therese F. Casgrain Fellowship, 4418
Textron Corporate Contributions Grants, 4501
United Methodist Church Women in History Research Grant, 4574
United Theological Seminary Merit-Based Scholarships, 4583
USAID Comprehensive District-Based Support for Better HIV/TB Patient Outcomes Grants, 4595
USAID Family Planning and Reproductive Health Methods Grants, 4598
USDA Norman E. Borlaug International Agricultural Science and Technology Fellowships, 4636
USDA Scientific Cooperation Research Program, 4647
Vanguard Public Foundation Grant Funds, 4808
Victor E. Speas Foundation Grants, 4810
Washington Area Women's Foundation Strategic Opportunity and Partnership Fund Grants, 4831
WAWH Founders Dissertation Fellowship, 4833
WAWH Sierra Award, 4834
Woodrow Wilson Doctoral Dissertation Fellowships in Women's Studies, 4891

Women's Education

AAAS Mentor Awards, 35
AAUW American Dissertation Fellowships, 209
AAUW American Postdoctoral Research Leave Fellowships, 210
AAUW American Summer/Short-Term Research Publication Grants, 211
AAUW Career Development Grants, 212
AAUW Community Action Grants, 213
AAUW International Doctoral Fellowships, 214
AAUW International Master's/First Professional Degree Fellowships, 215
AAUW International Postdoctoral Fellowships, 216
AAUW Selected Professions Fellowships, 217
AFUW Diamond Jubilee Bursary, 359
AFUW Doreen McCarthy, Barbara Crase, Cathy Candler & Brenda Nettle Bursaries, 360
AFUW Fellowships, 361
AFUW Georgina Sweet Fellowships, 362
AFUW Heather Latz Bursary, 363
AFUW Padnendadlu Graduate Bursary, 364
AFUW Padnendadlu Postgraduate Bursary, 365
AFUW Padnendadlu Undergraduate Bursary, 366
AFUW South Australia Jean Gilmore, Thenie Baddams & Daphne Elliott Bursaries, 367
AFUW Winifred E. Preedy Postgraduate Bursary, 368
American Chemical Society Award for Encouraging Women into Careers in the Chemical Sciences, 659
Anne Louise Barrett Fellowship, 861
Appraisal Institute Education Trust Scholarships, 917
Association for Women in Mathematics Mentoring Travel Grants, 1223
AT&T Bell Labs Graduate Research Program, 1250
Avon Products Foundation Grants, 1275
AWIS Educational Foundation Predoc Awards, 1278
AWM Mentoring Travel Grants for Women, 1279
AWM Travel Grants for Women Researchers, 1280
AWWA Holly A. Cornell Scholarship, 1284
Beryl Henderson Foundation Grants, 1337
BPW Career Advancement Scholarships, 1366
CCWH/Berkshire Conference of Women Historians Graduate Student Fellowship, 1453
CCWH Catherine Prelinger Award, 1454
CCWH Ida B. Wells Grad Student Fellowship, 1455
Center for Education of Women Visiting Scholar Research Grants, 1475
CFUW Beverley Jackson Fellowship, 1507
CFUW Bourse Georgette LeMoyne Award, 1508
CFUW Canadian Home Economics Association Fellowship, 1509
CFUW Dr. Alice E. Wilson Awards, 1510
CFUW Ecole Polytechnique Commemorative Award, 1511
CFUW Margaret Dale Philp Award, 1512
CFUW Marg McWilliams Predoc Fellowship, 1513
CFUW Marion Elder Grant Fellowship, 1514
CFUW Ruth Binnie Fellowship, 1516
Chrysalis Scholarships, 1559
Dept of Ed Women's Educational Equity Program Grants, 1752
Dorothy Leet Grants, 1840
EFWA Accounting Research Grants, 1901
Engineering Information Foundation Grants, 1927
Harvard Women and Public Policy Postdoctoral Fellowship, 2217
Heineman Foundation for Research, Education, Charitable and Scientific Purposes, 2227
IFUW Winifred Cullis Grants, 2347
International Federation of University Women Research Fellowships, 2397
Joseph L. Fisher Dissertation Awards, 2515
L'Oreal USA Women in Science Fellowship, 2577
MAA-Tensor Foundation Women and Mathematics Grants, 2645
Margaret Yardley Fellowship, 2660
National Assessment of Educational Progress Visiting Scholar Program, 2846
NCAA Postgraduate Scholarships, 2900
NL Monticello College Foundation Fellowship for Women, 3304
NSF ADVANCE: Increasing the Participation and Advancement of Women in Academic Science and Engineering Careers Grants, 3461
NSF Science and Society, 3643
NWO Aspasia Grants, 3673
NWO Athena Grants, 3674
NWSA Graduate Scholarship in Jewish Women's Studies, 3676
Orchid Research Fellowships, 3794
Radcliffe Institute Dissertation Grants, 4019
Radcliffe Inst Individual Residential Fellowships, 4021
Radcliffe Institute Research Support Grants, 4023
Radcliffe Institute Residential Group Fellowships, 4024
RGK Foundation Grants, 4060
RSC Alice Wilson Award, 4117
Saint Mary's College Fellowships, 4159
SfN Louise Hanson Marshall Specific Recognition Award, 4208
SRC Undergraduate Research Opportunities, 4395
Swedish Women's Educational Association International Scholarships, 4480
SWE Scholarships, 4481
Washington Area Women's Foundation Strategic Opportunity and Partnership Fund Grants, 4831
WAWH Founders Dissertation Fellowship, 4833
Woodrow Wilson Doctoral Dissertation Fellowships in Women's Studies, 4891
Zonta International Foundation Amelia Earhart Fellowships, 4905

Women's Employment

American Chemical Society Award for Encouraging Women into Careers in the Chemical Sciences, 659
Avon Products Foundation Grants, 1275
Bill and Melinda Gates Foundation Agricultural Development Grants, 1345
Girl's Best Friend Foundation Grants, 2143
M.D. Anderson Foundation Grants, 2637

NSF ADVANCE: Increasing the Participation and Advancement of Women in Academic Science and Engineering Careers Grants, 3461
Radcliffe Institute Dissertation Grants, 4019
Radcliffe Inst Individual Residential Fellowships, 4021
Radcliffe Institute Research Support Grants, 4023
Radcliffe Institute Residential Group Fellowships, 4024
Textron Corporate Contributions Grants, 4501
USDA Organic Agriculture Research and Extension Initiative, 4637
Woodrow Wilson Doctoral Dissertation Fellowships in Women's Studies, 4891

Women's Health
American Legacy Foundation National Calls for Proposals Grants, 748
Appalachian Regional Commission Health Care Grants, 915
Avon Foundation Breast Care Fund Grants, 1274
Avon Products Foundation Grants, 1275
Breast Cancer Fund Grants, 1369
Bristol-Myers Squibb Foundation Women's Health Grants, 1379
California Wellness Foundation Work and Health Grants, 1409
CDC Found Tobacco Network Lab Fellowship, 1462
Duke Univ Obstetric Anesthesia Fellowships, 1878
Fannie E. Rippel Foundation Grants, 1978
Foundation for Seacoast Health Grants, 2041
Goddess Scholars Grants, 2155
Harry B. and Jane H. Brock Foundation Grants, 2206
Lillian S. Wells Foundation Grants, 2614
Lowe Foundation Grants, 2622
NCIC Canadian Breast Cancer Research Grants, 2918
NCI Stages of Breast Development: Normal to Metastatic Disease Grants, 2932
NIH Building Interdisciplinary Research Careers in Women's Health Grants, 3209
NLM Advancing Novel Science in Women's Health Research Exploratory/Developmental Grants, 3291
NLM Advancing Novel Science in Women's Health Research Small Project Grants, 3292
Pfizer Medical Education Track Two Grants, 3927
Radcliffe Institute Carol K. Pforzheimer Student Fellowships, 4018
Radcliffe Institute Dissertation Grants, 4019
Radcliffe Inst Individual Residential Fellowships, 4021
Radcliffe Institute Research Support Grants, 4023
Radcliffe Institute Residential Group Fellowships, 4024
STTI Rosemary Berkel Crisp Research Awards, 4463
Susan G. Komen Breast Cancer Foundation Challege Grants: Investigator Initiated Research, 4475
Susan G. Komen Breast Cancer Foundation Challenge Grants: Career Catalyst Research, 4476
Susan G. Komen Breast Cancer Foundation Investigator Initiated Research Grants, 4477
USAID Comprehensive District-Based Support for Better HIV/TB Patient Outcomes Grants, 4595
USAID Family Planning and Reproductive Health Methods Grants, 4598
USAID Microbicides Research, Development, and Introduction Grants, 4599
Woodrow Wilson Doctoral Dissertation Fellowships in Women's Studies, 4891

Women's Rights
Bill and Melinda Gates Foundation Agricultural Development Grants, 1345
Gruber Foundation Prize for Global Justice and Women's Rights, 2182
Radcliffe Institute Dissertation Grants, 4019
Radcliffe Inst Individual Residential Fellowships, 4021
Radcliffe Institute Research Support Grants, 4023
Radcliffe Institute Residential Group Fellowships, 4024
United States Institute of Peace Annual Grants, 4582

Women's Studies
Alabama Humanities Foundation Mini Grants, 480
Alabama Humanities Foundation Planning/Consultant Grants, 481
AMA-WPC Joan F. Giambalvo Memorial Scholarships, 600
American Historical Association Joan Kelly Memorial Prize in Women's History, 730
ASU Graduate College Dean's Fellowships, 1236
CFUW Ecole Polytechnique Commemorative Award, 1511
Duke Univ Libraries Mary Lily Research Grants, 1877
Five College Women's Research Center Ford Associateships, 2022
Harvard University Women's Studies in Religion Research Associate and Visiting Faculty Grant, 2216
Indian Council of Social Science Research Grants, 2377
International Federation of University Women Research Fellowships, 2397
NCSS/FASSE/CUFA Inquiry Grant, 2936
NCSS Exemplary Res in Social Studies Award, 2937
NCSS Jean Dresden Grambs Distinguished Career Research in Social Studies Award, 2938
NCSS L Metcalf Exemplary Dissertation Award, 2939
NL Monticello College Foundation Fellowship for Women, 3304
NWSA Graduate Scholarships, 3677
NWSA Graduate Scholarships in Lesbian Studies, 3678
OAH Lerner-Scott Dissertation Prize, 3723
Radcliffe Institute Dissertation Grants, 4019
Radcliffe Inst Individual Residential Fellowships, 4021
Radcliffe Institute Oral History Grants, 4022
Radcliffe Institute Research Support Grants, 4023
Radcliffe Institute Residential Group Fellowships, 4024
United Methodist Church Women in History Writing Award, 4575
Virginia Historical Society Research Fellowships, 4816
WAWH Barbara Penny Kanner Award, 4832
Woodrow Wilson Doctoral Dissertation Fellowships in Women's Studies, 4891

Wood
Florida Division of Cultural Affairs Individual Artist Fellowships, 2026

Work Motivation
Alfred P. Sloan Found Working Longer Grants, 566

Workforce Development
Boeing Company Contributions Grants, 1357
Charles Stewart Mott Foundation Anti-Poverty Program, 1528
Helen Bader Foundation Grants, 2228
Illinois DCEO Coal Revival Grants, 2360
Nellie Mae Education Foundation State Level Systems Change Grants, 2971
NSF Innovative Technology Experiences for Students and Teachers, 3582
Reinberger Foundation Grants, 4051
SRC Mahboob Khan Outstanding Industry Liaison Award, 4377
SRC Robert M. Burger Fellowship, 4392

Workshops
ACLS Comparative Perspectives on Chinese Culture and Society Grants, 283
IBRO Europe Regional Grants for Western European Workshops and Symposia, 2324
IBRO Latin America Regional Travel Grants, 2329
Kellogg Institute for International Studies Grants for Conferences and Workshops, 2543
Kennedy Center Summer Fellowships for International Arts Managers, 2553
NAGC Masters and Specialists Award, 2821
North Carolina Arts Council Folklife Public Program Grants, 3419
SSRC Our Shared Past Grants, 4437
Virginia Foundation for the Humanities Discretionary Grants, 4813
Virginia Foundation for Humanities Open Grants, 4814
WWF Russell E. Train Education for Nature Alumni Grants, 4897

Wound Healing
IAFF Burn Foundation Research Grants, 2312
Plastic Surgery Educational Foundation Basic Research Grants, 3978

Writers in Residence
ACLS/New York Public Library Fellowships, 277
Francis C. Wood Scholar in Residence Program, 2047
IIE Rockefeller Foundation Bellagio Center Residencies, 2355
Rockefeller Foundation Scholarly Residencies, 4091

Writing
AAB Berlin Prize Fellowships, 48
AAUW American Summer/Short-Term Research Publication Grants, 211
ACLS/New York Public Library Fellowships, 277
ACLS International and Area Studies Fellowships, 289
AIHS Media Summer Fellowship, 453
American Historical Association Folger Shakespeare Library Fellowship, 718
American Historical Assoc Leo Gershoy Award, 734
IIE Rockefeller Foundation Bellagio Center Residencies, 2355
Leo Baeck Essay Prize, 2601
MLA Murray Gottlieb Prize Essay Award, 2788
National Endowment for the Arts Literature Fellowships: Creative Writing, 2865
NSF Antarctic Artists and Writers Grants, 3473
OSF Soros Justice Media Fellowships, 3851
SPJ Eugene C. Pulliam Fellowship for Editorial Writing, 4346
SPJ Pulliam/Kilgore Freedom of Information Internships, 4347
Stanford Humanities Center Arts Writer/Practitioner Fellowship, 4444
WHOI Ocean Science Journalism Fellowship, 4862

Writing/Composition Education
Bay and Paul Foundations, Inc Grants, 1307
IIE Rockefeller Foundation Bellagio Center Residencies, 2355
IRA Helen M. Robinson Dissertation Res Award, 2409
NAGC Masters and Specialists Award, 2821
Radcliffe Inst Fellowships for Creative Artists, 4020
Robert Bowne Foundaion Edmund A. Stanley, Jr. Research Grants, 4073

Young Adult Literature
Children's Literature Assoc Research Fellowships, 1551

Youth Programs
3M Foundation Community Giving Grants, 5
A.L. Mailman Family Foundation Grants, 11
AAA Foundation for Traffic Safety Grants, 24
Acuity Charitable Foundation Grants, 333
Adaptec Foundation Grants, 340
Alcatel-Lucent Technologies Foundation Grants, 509
Alice Tweed Tuohy Foundation Grants Program, 569
Allyn Foundation Grants, 574
Alvin and Lucy Owsley Foundation Grants, 584
American Foodservice Charitable Trust Grants, 710
Anderson Foundation Grants, 842
Ann & Robert H. Lurie Family Foundation Grants, 859
Ann Arbor Area Community Foundation Grants, 860
Annenberg Foundation Grants, 862
Anthony R. Abraham Foundation Grants, 869
Archer Daniels Midland Foundation Grants, 985
Arronson Foundation Grants, 1019
Batchelor Foundation Grants, 1305
Berks County Community Foundation Grants, 1333
Bodenwein Public Benevolent Foundation Grants, 1355
Bodman Foundation Grants, 1356
Boston Foundation Grants, 1361
Bullitt Foundation Grants, 1384
Callaway Golf Company Foundation Grants, 1410
Carl and Eloise Pohlad Family Grants, 1426
Carl and Lily Pforzheimer Foundation Grants, 1427
Carnegie Corporation of New York Grants, 1432
Carrier Corporation Contributions Grants, 1440

Catholic University Life Cycle Institute Research Fellowship, 1444
Chapin Hall International Fellowships in Children's Policy Research, 1517
Chestnut Hill Charitable Foundation, Inc Grants, 1538
Christine O. Gregoire Young Adult Award for Outstanding Use of Tobacco Industry Docs, 1557
Clay Foundation Grants, 1579
Coca-Cola Foundation Grants, 1593
Colonel Stanley R. McNeil Foundation Grants, 1597
Communities Foundation of Texas Grants, 1613
Community Found of E Central Illinois Grants, 1619
Connecticut Community Foundation Grants, 1627
ConocoPhillips Grants Program, 1629
Cord Foundation Grants, 1645
Cowles Charitable Trust Grants, 1652
Cruise Industry Charitable Foundation Grants, 1660
Cudd Foundation Grants, 1672
D.F. Halton Foundation Grants, 1692
Dade Community Foundation Grants, 1712
Dept of Ed Vocational Rehabilitation Services Demonstration and Training Programs, 1751
Detroit Lions Charities Grants, 1753
DOJ Juvenile Justice and Delinquency Prevention Special Emphasis Grants, 1826
Dorr Foundation Grants, 1842
Edwin S. Webster Foundation Grants, 1900
Emerson Charitable Trust Grants, 1918
Evjue Foundation Grants, 1963
Ewing Halsell Foundation Grants, 1964
Explorers Club Exploration Fund Grants, 1966
Farmers Insurance Corporate Giving Grants, 1981
Ford Motor Company Fund Grants Program, 2033
Frank Stanley Beveridge Foundation Grants, 2051
George B. Storer Foundation Grants, 2113
George Foundation Grants, 2115
Gil and Dody Weaver Foundation Grants, 2139
H.J. Heinz Company Foundation Grants, 2195
Harold Simmons Foundation Grants, 2203
Hatton W. Sumners Foundation for the Study and Teaching of Self Government Grants, 2219
Hill Crest Foundation Grants, 2267
Hilton Hotels Corporate Giving Program Grants, 2268
Hoglund Foundation Grants, 2271
Human Source Foundation Grants, 2297
J.W. Kieckhefer Foundation Grants, 2452
Josiah W. and Bessie H. Kline Foundation Grants, 2517
Kenneth T. and Eileen L. Norris Found Grants, 2554
Lillian S. Wells Foundation Grants, 2614
Lucile Horton Howe & Mitchell B. Howe Grants, 2624
Lumpkin Family Found Healthy People Grants, 2629
Luther I. Replogle Foundation Grants, 2631
M.J. Murdock Charitable Trust General Grants, 2641
Mabel Y. Hughes Charitable Trust Grants, 2647
MacDonald-Peterson Foundation Grants, 2649
Margaret & James A. Elkins Jr. Found Grants, 2658
Mary S. and David C. Corbin Foundation Grants, 2675
Matilda R. Wilson Fund Grants, 2684
Max and Victoria Dreyfus Foundation Grants, 2687
McCombs Foundation Grants, 2718
McCune Foundation Education Grants, 2721
McGregor Fund Grants, 2725
McLean Contributionship Grants, 2727
Mericos Foundation Grants, 2742
MGN Family Foundation Grants, 2754
Millipore Foundation Grants, 2775
National Foundation for Catholic Youth Grants, 2868
Norcliffe Foundation Grants, 3415
Northwestern Mutual Foundation Grants, 3430
NRA Foundation Grants, 3437
Olin Corporation Charitable Trust Grants, 3739
Oppenstein Brothers Foundation Grants, 3792
Patrick and Anna M. Cudahy Fund Grants, 3873
Paul G. Allen Family Foundation Grants, 3876
Peacock Foundation Grants, 3887
Perpetual Trust for Charitable Giving Grants, 3900
Pinkerton Foundation Grants, 3967
Pott Foundation Grants, 3984
Prudential Foundation Education Grants, 4002
R.C. Baker Foundation Grants, 4006
Rhode Island Foundation Grants, 4061
Richard and Rhoda Goldman Fund Grants, 4063
Rich Foundation Grants, 4068
Righteous Persons Foundation Grants, 4069
Robert B McMillen Foundation Grants, 4072
Robert Bowne Foundaion Edmund A. Stanley, Jr. Research Grants, 4073
Robert E. and Evelyn McKee Foundation Grants, 4074
Rockwell International Corporate Trust Grants, 4093
Rohm and Haas Company Grants, 4098
Ronald McDonald House Charities Grants, 4101
Roy J. Carver Charitable Trust Medical and Science Research Grants, 4114
SARE Professional + Producer Grants - Western Region, 4175
Self Foundation Grants, 4197
Sid W. Richardson Foundation Grants, 4253
Skaggs Foundation Grants, 4266
Stackpole-Hall Foundation Grants, 4442
Sunny and Abe Rosenberg Foundation Grants, 4469
Talbert and Leota Abrams Foundation, 4492
TD4HIM Foundation Grants, 4497
Thomas and Dorothy Leavey Foundation Grants, 4506
Thomas Austin Finch, Sr. Foundation Grants, 4507
Thomas C. Ackerman Foundation Grants, 4508
Thompson Charitable Foundation Grants, 4511
Union Pacific Foundation Community and Civic Grants, 4568
Vancouver Foundation Grants and Community Initiatives, 4807
Victor E. Speas Foundation Grants, 4810
W.K. Kellogg Foundation Healthy Kids Grants, 4822
Wayne and Gladys Valley Foundation Grants, 4835
Weyerhaeuser Company Foundation Grants, 4857
William Blair and Company Foundation Grants, 4875
William T. Grant Distinguished Fellows Awards, 4880
William T. Grant Foundation Research Grants, 4881

Youth Services
Acuity Charitable Foundation Grants, 333
Anderson Foundation Grants, 842
Caesars Foundation Grants, 1407
Gil and Dody Weaver Foundation Grants, 2139
Greenspun Family Foundation Grants, 2173
Kovler Family Foundation Grants, 2575
M.D. Anderson Foundation Grants, 2637
MacDonald-Peterson Foundation Grants, 2649
McCune Foundation Human Services Grants, 2722
Oppenstein Brothers Foundation Grants, 3792
Reinberger Foundation Grants, 4051
Richard Davoud Donchian Foundation Grants, 4066
RWJF Childhood Obesity Grants, 4135
Sunny and Abe Rosenberg Foundation Grants, 4469
Union Pacific Foundation Health and Human Services Grants, 4569

Youth Violence
Girl's Best Friend Foundation Grants, 2143
Juvenile Justice Clearing House Grants, 2530
OSF Soros Justice Advocacy Fellowships, 3850
OSF Soros Justice Media Fellowships, 3851
William G. McGowan Charitable Fund Grants, 4877
William T. Grant Foundation Research Grants, 4881

Zoology
AMNH Lerner-Gray Grants for Marine Research, 816
Field Museum Visiting Scholarships, 2011
International Crane Foundation Internships, 2396
National Geographic Society Expeditions Council Grants, 2872
National Geographic Society Field Research and Exploration Grants, 2873
Smithsonian Environmental Research Center Graduate Student Fellowships, 4273
Smithsonian Environmental Research Center Internships, 4274
Smithsonian Environmental Research Center Postdoctoral/Senior Fellowships, 4275
Smithsonian Environmental Research Center Predoctoral Fellowships, 4276

Zoos
Bay and Paul Foundations, Inc Grants, 1307
Blum-Kovler Foundation Grants, 1351
IMLS 21st Century Museum Prof Grants, 2362
IMLS National Leadership Grants, 2366
Oscar Rennebohm Foundation Grants, 3832
Reinberger Foundation Grants, 4051
Union Pacific Foundation Community and Civic Grants, 4568

Program Type Index

NOTE: Numbers refer to entry numbers

Adult Basic Education
Achelis Foundation Grants, 274
Allyn Foundation Grants, 574
Auburn Foundation Grants, 1263
Ball Brothers Foundation General Grants, 1291
Bodenwein Public Benevolent Found Grants, 1355
Bodman Foundation Grants, 1356
Booth-Bricker Fund Grants, 1359
Boston Foundation Grants, 1361
Carl and Lily Pforzheimer Foundation Grants, 1427
CIGNA Foundation Grants, 1568
Coors Brewing Corporate Contributions Grants, 1644
Cowles Charitable Trust Grants, 1652
Eaton Charitable Fund Grants, 1893
Edward W. and Stella C. Van Houten Grants, 1899
Evjue Foundation Grants, 1963
Frances L. and Edwin L. Cummings Memorial Fund Grants, 2046
George Foundation Grants, 2115
Harold R. Bechtel Testamentary Charitable Trust Grants, 2202
Harold Simmons Foundation Grants, 2203
HSBC Contributions Community Grants, 2291
Joyce Foundation Employment Grants, 2520
McCarthy Family Foundation Grants, 2717
McCune Charitable Foundation Grants, 2719
McCune Foundation Education Grants, 2721
National Book Scholarship Fund, 2849
Norcliffe Foundation Grants, 3415
Oppenstein Brothers Foundation Grants, 3792
OSF Global Faculty Grants - Track 2, 3842
Reinberger Foundation Grants, 4051
Richard and Rhoda Goldman Fund Grants, 4063
San Antonio Area Foundation Grants, 4163
SARE Graduate Student Grants in Sustainable Agriculture - Western Region, 4171
Sony Corporation of America Corporate Philanthropy Grants, 4329
Talbert and Leota Abrams Foundation, 4492
USDA Distance Education Grants for Higher Education Institutions in Insular Areas, 4622
USDA Organic Agriculture Research and Extension Initiative, 4637
USDA Regional Integrated Pest Management Grants - North Central Region, 4642
Xerox Foundation Grants, 4899

Adult/Family Literacy Training
Achelis Foundation Grants, 274
Alcatel-Lucent Technologies Foundation Grants, 509
Allyn Foundation Grants, 574
Amgen Foundation Grants, 812
Annenberg Foundation Grants, 862
Arizona Cardinals Grants, 995
Assisi Foundation of Memphis General Grants, 1221
Auburn Foundation Grants, 1263
Ball Brothers Foundation General Grants, 1291
Bayer Foundation Grants, 1309
BBF Fellowships, 1311
Bodenwein Public Benevolent Found Grants, 1355
Bodman Foundation Grants, 1356
Booth-Bricker Fund Grants, 1359
Boston Foundation Grants, 1361
Cabot Corporation Foundation Grants, 1406
Caesars Foundation Grants, 1407
Carl and Lily Pforzheimer Foundation Grants, 1427
CIGNA Foundation Grants, 1568
Cleveland-Cliffs Foundation Grants, 1580
Coca-Cola Foundation Grants, 1593
Coors Brewing Corporate Contributions Grants, 1644
Cowles Charitable Trust Grants, 1652
Cruise Industry Charitable Foundation Grants, 1660
Edward W. and Stella C. Van Houten Grants, 1899
Evjue Foundation Grants, 1963
Farmers Insurance Corporate Giving Grants, 1981
Frances L. and Edwin L. Cummings Memorial Fund Grants, 2046
Genuardi Family Foundation Grants, 2109
George Foundation Grants, 2115
Harold Simmons Foundation Grants, 2203
Helen Bader Foundation Grants, 2228
Hilda and Preston Davis Foundation Grants, 2266
Hoglund Foundation Grants, 2271
IRA Elva Knight Research Grant, 2406
John P. McGovern Foundation Grants, 2504
Joyce Foundation Employment Grants, 2520
Macquarie Bank Foundation Grants, 2650
Maximilian E. and Marion O. Hoffman Grants, 2688
McCune Charitable Foundation Grants, 2719
McCune Foundation Education Grants, 2721
McGraw-Hill Companies Community Grants, 2724
Melville Charitable Trust Grants, 2741
National Book Scholarship Fund, 2849
Norcliffe Foundation Grants, 3415
Nordson Corporation Foundation Grants, 3416
Percy B. Ferebee Endowment Grants, 3897
Pinkerton Foundation Grants, 3967
Plan for Social Excellence Grants, 3976
Reinberger Foundation Grants, 4051
RGK Foundation Grants, 4060
Richard and Rhoda Goldman Fund Grants, 4063
Richard Davoud Donchian Foundation Grants, 4066
Robert Bowne Foundaion Edmund A. Stanley, Jr. Research Grants, 4073
Russell Sage Foundation Grants, 4129
San Antonio Area Foundation Grants, 4163
San Diego Women's Foundation Grants, 4164
Seaver Institute Grants, 4195
Sony Corporation of America Corporate Philanthropy Grants, 4329
Stackpole-Hall Foundation Grants, 4442
Talbert and Leota Abrams Foundation, 4492
Textron Corporate Contributions Grants, 4501
Todd Brock Family Foundation Grants, 4522
USDA Organic Agriculture Research and Extension Initiative, 4637
Wallace Foundation Grants, 4829
Xerox Foundation Grants, 4899

Awards/Prizes
AAAAI Distinguished Layperson Award, 18
AAAAI Distinguished Scientist Award, 19
AAAAI Distinguished Service Award, 20
AAAAI Fellows-in-Training Abstract Award, 21
AAAS Award for Public Engagement with Science, 26
AAAS Award for Science Diplomacy, 27
AAAS Award for Scientific Freedom and Responsibility, 28
AAAS Early Career Award for Public Engagement with Science, 29
AAAS Eppendorf Science Prize for Neurobiology, 31
AAAS GE Science Prize for Young Life Scientists, 32
AAAS Kavli Science Journalism Awards, 33
AAAS Martin and Rose Wachtel Cancer Research Award, 34
AAAS Mentor Awards, 35
AAAS Newcomb Cleveland Prize, 36
AAAS Philip Hauge Abelson Award, 37
AAAS Science Prize for Inquiry-Based Instruction, 45
AABB Dale A. Smith Memorial Award, 46
AABB Emily Cooley Memorial Award and Lectureship, 47
AABB Hemphill-Jordan Leadership Award, 49
AABB John Elliott Memorial Award, 50
AABB Karl Landsteiner Memorial Award and Lectureship, 51
AABB Sally Frank Mem Award and Lectureship, 52
AABB Tibor Greenwalt Memorial Award and Lectureship, 53
AACAP Beatrix A. Hamburg Award for the Best New Research Poster by a Child and Adolescent Psychiatry Resident, 55
AACAP Elaine Schlosser Lewis Award for Research on Attention-Deficit Disorder, 59
AACAP George Tarjan Award for Contributions in Developmental Disabilities, 60
AACAP Irving Philips Award for Prevention, 61
AACAP Jeanne Spurlock Lecture and Award on Diversity and Culture, 62
AACAP Klingenstein Third Generation Foundation Award for Research in Depression or Suicide, 66
AACAP Norbert and Charlotte Rieger Award for Scientific Achievement, 69
AACAP Pilot Research Awards, Supported by Eli Lilly and Company, 72
AACAP Rieger PsychodynPsychotherapy Award, 74
AACAP Rieger Service Award for Excellence, 75
AACAP Robert Cancro Acad Leadership Award, 76
AACAP Robinson-Cunningham Award, 77
AACAP Sidney Berman Award for the School-Based Study and Intervention for Learning Disorders and Mental Ilness, 78
AACAP Simon Wile Leadership in Consultation Award, 79
AACC International Alsberg-French-Schoch Memorial Lectureship Awards, 82
AACC International Best Student Paper and Best Student Poster Award, 83
AACC International Bruce Wasserman Young Investigator Award, 84
AACC Int'l Carl Wilhelm Brabender Award, 85
AACC International Edith A. Christensen Award, 87
AACC International Engineering and Processing Division's Stanley Watson Award, 88
AACC International Excellence in Teaching Award, 89
AACC International George W. Scott Blair Lecture Award, 90
AACC International Phil Williams Applied Research Award, 94
AACC International Student Division's Product Development Awards, 96
AACC International Thomas Burr Osborne Medal, 98
AACC International Walter Bushuk Graduate Research Award, 99
AACC Int'l Young Scientist Research Award, 100
AACC Texture Technol Quality Research Award, 101
AACR-American Cancer Society Award for Research in Cancer Epidemiology and Prevention, 112
AACR-Minorities in Cancer Research Jane Cooke Wright Lectureship Awards, 119
AACR-Prevent Cancer Foundation Award for Excellence in Cancer Prevention Research, 124
AACR-Women in Cancer Research Charlotte Friend Memorial Lectureship Awards, 126
AACR Award for Lifetime Achievement in Cancer Research, 127
AACR Award for Outstanding Achievement in Cancer Research, 128
AACR Award for Outstanding Achievement in Chemistry in Cancer Research, 129
AACR Joseph H. Burchenal Memorial Award, 139
AACR Margaret Foti Award for Leadership and Achievements in Cancer Research, 142
AACR Outstanding Investigator Award for Breast Cancer Research, 145
AACR Scholar-in-Training Awards, 146
AACR Scholar-in-Training Awards - Special Conferences, 147
AACR Team Science Award, 148
AAES Engineering Journalism Award, 150
AAES National Engineering Award, 151
AAES Norm Augustine Award, 152
AAFP Foundation Wyeth Immunization Awards, 158
AAFPRS Ben Shuster Memorial Award, 160
AAFPRS Bernstein Grant, 161
AAFPRS Ira J. Tresley Research Award, 164
AAFPRS John Orlando Roe Award, 165
AAFPRS Leslie Bernstein Resident Res Grants, 166
AAFPRS Residency Travel Awards, 167
AAFPRS Sir Harold Delf Gillies Award, 168
AAG E. Willard and Ruby S. Miller Award in Geography, 172
AAG John Brinkerhoff Jackson Prize, 175
AAMC Award for Distinguished Research, 183

AAO-HNS Medical Student Research Prize, 192
AAO-HNS Resident Research Prizes, 193
AAP Nutrition Award, 198
Abbott-ASM Lifetime Achievement Award, 218
Abdus Salam ICTP Ramanujan Prize for Young Mathematicians, 236
AcademyHealth Alice S. Hersh New Investigator Awards, 245
AcademyHealth HSR Impact Awards, 246
AcademyHealth Nemours Child Health Services Research Awards, 247
AcademyHealth PHSR Research Article of the Year Awards, 248
ACC John D. Rockefeller III Award, 257
ACM / AAAI Allen Newell Award, 291
ACM A.M. Turing Award, 292
ACM Distinguished Service Award, 293
ACM Doctoral Dissertation Award, 294
ACM Eckert-Mauchly Award, 295
ACM Eugene L. Lawler Award, 296
ACM Gordon Bell Prize, 297
ACM Grace Murray Hopper Award, 298
ACM Karl V. Karlstrom Outstanding Ed Award, 299
ACM Outstanding Contribution Award, 300
ACM Paris Kanellakis Theory and Practice Award, 301
ACM Software System Award, 302
ACRL Academic/Research Librarian of the Year Award, 303
ACRL Instruction Section Innovation Award, 304
ACS Doctoral Training Grants in Oncology Social Work, 308
ACS Physician Training Awards in Preventive Medicine, 323
Acton Institute for the Study of Religion and Liberty Novak Award, 332
ADA Foundation George C. Paffenbarger Student Research Awards, 336
AFAR Medical Student Training in Aging Research Program, 352
Ahmed Zewail Award in Ultrafast Science and Technology, 378
AIA Latrobe Prize, 390
AIChE Alpha Chi Sigma Award, 396
AIChE Andrew Chase Forest Bioproducts Division Award in Chemical Engineering, 397
AIChE CAST Outstanding Young Res Award, 398
AIChE Charles M.A. Stine Award, 399
AIChE Clarence G. Gerhold Award, 400
AIChE Computational Molecular Science and Engineering Forum Graduate Student Award, 401
AIChE Computing in Chem Engineering Award, 402
AIChE Computing Practice Award, 403
AIChE David Himmelblau Award for Innovations in Computer-Based Chem Engineering Education, 405
AIChE Donald Q. Kern Award, 406
AIChE Environmental Division Graduate Student Paper Award, 407
AIChE Environmental Division Undergraduate Student Paper Award, 408
AIChE Excellence in Process Development Research Award, 409
AIChE Food, Pharmaceutical and Bioengineering Division Awards in Chemical Engineering, 410
AIChE Food, Pharmaceutical and Bioengineering Division Distinguished Service Award in Chemical Engineering, 411
AIChE Institute Award for Excellence in Industrial Gases Technology, 412
AIChE John A. Tallmadge Award for Contributions to Coating Technology, 413
AIChE Kazutoshi Fujimura Award for Lifetime Achievement, 414
AIChE Lawrence K. Cecil Award in Environmental Chemical Engineering, 415
AIChE Lectureship Award in Fluidization, 416
AIChE Nanoscale Science and Engineering Forum Young Investigator Award, 417
AIChE North American Mixing Forum Award, 418
AIChE North American Mixing Forum Student Award, 420
AIChE Nuclear Engineering Division Outstanding Student Paper & Presentation Awards, 421
AIChE Process Development Practice Award, 422
AIChE Professional Progress Award for Outstanding Progress in Chemical Engineering, 423
AIChE Research Excellence In Sustainable Engineering Award, 424
AIChE Separations Division Graduate Student Research Award, 425
AIChE Thomas Baron Award in Fluid-Particle Systems, 426
AIChE Transport & Energy Processes Div Award, 427
AIChE W. David Smith, Jr. Graduate Student Paper Award, 428
AIChE Warren K. Lewis Award for Contributions to Chemical Engineering Education, 429
AIChE William H. Walker Award for Excellence in Contributions to Chem Engineering Lit, 430
AIHS ForeFront Executive-in-Residence Award, 441
AIHS ForeFront MBA Studentship Award, 443
AIHS ForeFront MBT Studentship Awards, 444
AIHS Research Prize, 457
AINSE John Ferris Memorial Award, 462
ALA Annual Diversity Research Grant, 479
ALA Donald G. Davis Article Award, 484
ALA Emerald Research Grant Awards, 485
ALA Excellence in Academic Libraries Award, 486
ALA Frederick G. Kilgour Award for Research in Library and Information Technology, 487
ALA Gale Cengage History Research and Innovation Award, 488
ALA Ingenta Research Awards, 489
ALA Innovation Award, 490
ALA Marta Lange/CQ Press Award, 493
ALA NewsBank/Readex Cath J. Reynolds Award, 494
ALA Phyllis Dain Library History Dissert Award, 495
ALA ProQuest Documents to the People Award, 496
ALA STS Innovation in Science and Technology Librarianship Award, 499
ALA Ulrich's Serials Librarianship Award, 500
Albany Medical Center Prize in Medicine and Biomedical Research, 502
Alexander von Humboldt Anneliese Maier Award, 515
Alexander von Humboldt Fraunhofer-Bessel Research Awards, 521
Alexander von Humboldt Friedrich Wilhelm Bessel Research Awards, 522
Alexander von Humboldt Georg Forster Research Award, 525
Alexander von Humboldt Konrad Adenauer Research Award, 531
Alexander von Humboldt Max Planck Award, 532
Alexander von Humboldt Philipp Franz von Siebold Award for Japanese Researchers, 534
Alexander von Humboldt Reimar Lust Award, 536
Alexander von Humboldt Research Awards for Top Researchers in Germany, 538
Alfred W. Bressler Prize in Vision Science, 567
Alpha Sigma Nu Book Award Competition, 578
Alton Ochsner Award, 583
ALVRE Casselberry Award, 585
ALVRE Seymour R. Cohen Award, 587
AMA-MSS Research Poster Award, 599
American Academy of Advertising Distinguished Service Award, 607
American Academy of Advertising Ivan Preston Outstanding Contribution to Research on Advertising Award, 609
American Academy of Advertising Kim Rotzoll Award for Advertising Ethics and Social Responsibility, 610
American Automatic Control Council Control Practice Award, 621
American Automatic Control Council Donald P. Eckman Award, 622
American Automatic Control Council Hugo Schuck Award, 623
American Automatic Control Council John R. Ragazzini Education Award, 624
American Automatic Control Council Richard E. Bellman Control Heritage Award, 625
American Chemical Society AGFD Advancement of the Application of Agricultural and Food Chemistry Award, 626
American Chemical Society AGFD Donald A. Withycombe Graduate Fellowship Award in Food Chemistry, 627
American Chemical Society AGFD Teranishi Graduate Fellowship Award, 628
American Chemical Society AGFD Young Scientist Award, 630
American Chemical Society AGRO Award for Innovation in Chemistry of Agriculture, 631
American Chemical Society AGRO Ed Awards, 632
American Chemical Society AGRO International Award for Research in Agrochemicals, 634
American Chemical Society AGRO Sterling B. Hendricks Memorial Lectureship, 635
American Chemical Society AGRO Young Scientist Research Awards, 636
American Chemical Society Alfred Bader Award in BioInorganic or BioOrganic Chemistry, 637
American Chemical Society Alfred Burger Award in Medicinal Chemistry, 638
American Chemical Society ANYL Arthur F. Findeis Award for Achievements by a Young Analytical Scientist, 639
American Chemical Society ANYL Award for Distinguished Service in the Advancement of Analytical Chemistry, 640
American Chemical Society ANYL Award in Chemical Instrumentation, 641
American Chemical Society ANYL Award in Electrochemistry, 642
American Chemical Society ANYL Award in Spectrochemical Analysis, 643
American Chemical Society ANYL I.M. Kolthoff Enrichment Awards for Undergrad Students, 644
American Chemical Society ANYL J. Calvin Giddings Award for Excellence in Education, 645
American Chemical Society Art C. Cope Award, 647
American Chemical Society Arthur C. Cope Scholar Awards, 648
American Chemical Society Award for Achievement in Research for the Teaching and Learning of Chemistry, 649
American Chemical Society Award for Affordable Green Chemistry, 650
American Chemical Society Award for Computers in Chemical or Pharmaceutical Research, 651
American Chemical Society Award for Creative Advances in Environmental Science and Technology, 652
American Chemical Society Award for Creative Invention, 653
American Chemical Society Award for Creative Research and Applications of Iodine Chemistry, 654
American Chemical Society Award for Creative Work in Fluorine Chemistry, 655
American Chemical Society Award for Creative Work in Synthetic Organic Chemistry, 656
American Chemical Society Award for Distinguished Service in Advancement of Inorganic Chem, 657
American Chemical Society Award for Encouraging Disadvantaged Students into Careers in the Chemical Sciences, 658
American Chemical Society Award for Encouraging Women into Careers in the Chemical Sciences, 659
American Chemical Society Award for Research at an Undergraduate Institution, 660
American Chemical Society Award for Team Innovation, 661
American Chemical Society Award in Analytical Chemistry, 662
American Chemical Society Award in Applied Polymer Science, 663
American Chemical Soc Award in Chromatography, 664
American Chemical Society Award in Colloid and Surface Chemistry, 665
American Chemical Society Award in Industrial Chemistry, 666

American Chemical Society Award in Inorganic Chemistry, 667
American Chemical Society Award in Organometallic Chemistry, 668
American Chemical Society Award in Polymer Chemistry, 669
American Chemical Society Award in Pure Chem, 670
American Chemical Society Award in Separations Science and Technology, 671
American Chemical Society Award in the Chemistry of Materials, 672
American Chemical Society Award in Theoretical Chemistry, 673
American Chemical Society C.S. Hudson Awards, 675
American Chemical Society Green Chemistry Chapter Award for Student Affiliates, 678
American Chemical Society Irving Langmuir Award in Chemical Physics, 681
American Chemical Society Kenneth G. Hancock Memorial Student Award in Green Chemistry, 683
American Chemical Society Overcoming Challenges Award, 686
American Chemical Society PCD Award in Experimental Physical Chemistry, 687
American Chemical Society PCD Award in Theoretical Chemistry, 688
American Chemical Society Early Career Awards, 689
American Chemical Society Presidential Green Chemistry Challenge Awards, 696
American Chemical Society Rising Star Award, 698
American Hist Assoc Albert B. Corey Prize in Canadian-American Relations, 711
American Hist Assoc Albert J. Beveridge Award, 712
American Hist Assoc Award for Scholarly Dist, 714
American Hist Assoc Clarence H. Haring Prize, 716
American Hist Assoc George L. Mosse Prize, 719
American Hist Assoc George Louis Beer Prize, 720
American Hist Assoc Helen and Howard R. Marraro Prizes, 721
American Hist Assoc Herbert Baxter Adams Prize, 722
American Hist Assoc Assoc Herbert Feis Award, 723
American Hist Assoc J. Franklin Jameson Award In Outstanding Editorial Achievement, 724
American Hist Assoc Assoc J. Russell Major Prize, 726
American Hist Assoc James A. Rawley Prize in Atlantic History, 727
American Hist Assoc James H Robinson Prize, 728
American Hist Assoc James Henry Breasted Prize, 729
American Hist Assoc Joan Kelly Memorial Prize in Women's History, 730
American Hist Assoc John F. Richards Prize in South Asian History, 731
American Hist Assoc John H. Dunning Prize, 732
American Hist Assoc John K. Fairbank Prize in East Asian History, 733
American Hist Assoc Assoc Leo Gershoy Award, 734
American Hist Assoc Littleton-Griswold Prize in American Law and Society, 735
American Hist Assoc Martin A. Klein Prize in African History, 737
American Hist Assoc Morris D. Forkosch Prize, 739
American Hist Assoc Paul Birdsall Prize in European Military and Strategic History, 740
American Hist Assoc Premio Del Rey, 741
American Hist Assoc Ray J. Cunningham Prize, 742
American Hist Assoc Waldo J. Leland Prize, 743
American Hist Assoc Assoc Wesley-Logan Prize, 744
American Philosophical Soc Judson Daland Prize, 766
American Psychiatric Association/AACDP Research Mentorship Award, 772
American Psychiatric Association/AstraZeneca Young Minds in Psychiatry International Awards, 773
American Psychiatric Association/Lilly Resident Research Awards, 777
American Psychiatric Association Award for Research in Psychiatry, 781
American Psychiatric Assoc Early Career Award, 782
American Psychiatric Association Senior Scholar Health Services Research Award, 789
American Society of Criminology Gene Carte Student Paper Competition, 799
American Sociological Association Awards, 802
AMS Alfred Einstein Award, 819
AMS Claude V. Palisca Award, 821
AMS H. Colin Slim Award, 825
AMS Lewis Lockwood Award, 829
AMS Noah Greenberg Award, 831
AMS Otto Kinkeldey Award, 832
AMS Paul A. Pisk Prize, 833
AMS Robert M. Stevenson Award, 834
AMS Ruth A. Solie Award, 835
ANA/Grass Foundation Award in Neuroscience, 839
ANA Wolfe Neuropathy Research Prize, 840
ANS Neurotology Fellows Award, 865
ANS Nicholas Torok Vestibular Award, 866
ANS Trainee Award, 867
AOCS A. Richard Baldwin Award, 871
AOCS ACI/NBB Innovation Research Award, 872
AOCS Alton E. Bailey Award, 873
AOCS Analytical Division Student Award, 874
AOCS Australasian Award for Scientific Excellence in Lipid Research, 875
AOCS Biotechnology Division Student Awards, 876
AOCS Edible Applications Technology Division Outstanding Achievement Award, 877
AOCS Edible Applications Technology Division Student Award, 878
AOCS George Schroepfer Medal, 879
AOCS Hans Kaunitz Award, 880
AOCS Health & Nutrition Div Student Award, 881
AOCS Health and Nutrition Poster Competition, 882
AOCS Herbert J. Dutton Award, 883
AOCS Honored Student Award, 884
AOCS Industrial Oil Products Division Student Award, 885
AOCS Lipid Oxidation and Quality Division Poster Competition, 886
AOCS Manuchehr Eijadi Award, 887
AOCS Peter and Claire G. Kalustian Award, 888
AOCS Processing Distinguished Service Award, 890
AOCS Processing Division Student Award, 891
AOCS Protein and Co-Products Division Student Poster Competition, 892
AOCS Ralph Holman Life Achievement Award, 894
AOCS Samuel Rosen Memorial Award, 895
AOCS Stephen S. Chang Award, 896
AOCS Supelco/Nicholas Pelick-AOCS Research Award, 897
AOCS Surfactants and Detergents Division Student Award, 898
AOCS Timothy L. Mounts Award, 900
AOCS Uses of Soybean Oil Award, 901
AOCS Young Scientist Research Award, 902
APA Culture of Service in the Psychological Sciences Award, 904
APA Distinguished Scientific Award for Early Career Contribution to Psychology, 906
APA Distinguished Scientific Award for the Applications of Psychology, 907
APA Distinguished Scientific Contribution Award, 908
APA Distinguished Service to Psychological Science Award, 909
APS Abraham Pais Prize for History of Physics, 925
APSA E.E. Schattschneider Award, 932
APSA Edward S. Corwin Award, 933
APSA Gabriel A. Almond Award, 934
APSA Harold D. Lasswell Award, 935
APSA Helen Dwight Reid Award, 936
APSA Leonard D. White Award, 937
APSA Leo Strauss Award, 938
APS Andrei Sakharov Prize, 941
APS Aneesur Rahman Prize for Computational Physics, 942
APS Arthur L. Schawlow Prize in Laser Science, 945
APSA William Anderson Award, 950
APS Dannie Heineman Prize for Mathematical Physics, 951
APS Davisson-Germer Prize in Atomic or Surface Physics, 952
APS E.K. Plyler Prize for Molecular Spectroscopy, 954
APS Einstein Prize, 955
APS Fluid Dynamics Prize, 956
APS Frank Isakson Prize for Optical Eff in Solids, 957
APS George E. Pake Prize, 958
APS George E. Valley, Jr., Prize, 959
APS Hans A. Bethe Prize, 960
APS Herbert P. Broida Prize, 961
APS I.I. Rabi Prize in Atomic, Molecular, and Optical Physics, 962
APS Irving Langmuir Prize in Chemical Physics, 963
APS J.J. Sakurai Prize for Theoretical Part Physics, 964
APS James C. McGroddy Prize for New Materials, 965
APS James C Maxwell Prize for Plasma Physics, 966
APS Julius Edgar Lilienfeld Prize, 967
APS Lars Onsager Prize, 968
APS Max Delbruck Prize in Biological Physics, 970
APS Oliver E. Buckley Condensed Matter Prize, 973
APS Outstanding Doctoral Thesis Research in Beam Physics Award, 975
APS Polymer Physics Prize, 976
APS Prize for a Faculty Member for Research in an Undergraduate Institution, 977
APS Prize for Industrial Applications of Physics, 978
APS Robert R. Wilson Prize for Achievement in the Physics of Particle Accelerators, 979
APS Tom W. Bonner Prize in Nuclear Physics, 980
APS W.K.H. Panofsky Prize in Experimental Particle Physics, 981
APS Will Allis Prize for Study of Ionized Gases, 982
Arthur W. Adamson Award for Distinguished Service in the Advancement of Surface Chemistry, 1026
ASA Deming Lecturer Award, 1034
ASA Jeanne E. Griffith Mentoring Award, 1035
ASA Outstanding Statistical Application Award, 1036
ASA Samuel S. Wilks Award, 1037
ASA W. J. Youden Award in Interlab Testing, 1038
ASECS James L. Clifford Prize, 1058
ASGE / ConMed Award for Outstanding Manuscript by a Fellow/Resident, 1061
ASGE Don Wilson Award, 1062
ASGH Award for Excellence in Human Genetics Education, 1066
ASGH C.W. Cotterman Award, 1067
ASGH Charles J. Epstein Trainee Awards for Excellence in Human Genetics Research, 1068
ASGH McKusick Leadership Award, 1069
ASGH William Allan Award, 1070
ASGH William Curt Stern Award, 1071
ASHA Advancing Academic Careers Award, 1072
ASHA Research Mentoring-Pair Travel Award, 1073
ASHA Student Research Travel Award, 1074
ASHA Students Preparing for Academic & Research Careers Award, 1075
ASID Foundation Joel Polsky Academic Achievement Award, 1095
ASID Foundation Joel Polsky Prize, 1096
ASM Abbott Laboratories Award in Clinical and Diagnostic Immunology, 1101
ASM BD Award in Clinical Microbiology, 1102
ASM bioMerieux Sonnenwirth Award for Leadership in Clinical Microbiology, 1103
ASME Allan Kraus Thermal Management Medal, 1104
ASME Arthur L. Williston Medal, 1105
ASME Barnett-Uzgiris Product Safety Award, 1106
ASME Ben C. Sparks Medal, 1107
ASME Bergles-Rohsenow Young Investigator Award in Heat Transfer, 1108
ASME Bernard F. Langer Nuclear Codes and Standards Award, 1109
ASME Blackall Machine Tool Award, 1110
ASME Burt L. Newkirk Award, 1111
ASME Calvin W. Rice Lecture Award, 1112
ASME Charles Russ Richards Memorial Award, 1113
ASME Daniel C. Drucker Medal, 1114
ASME Dixy Lee Ray Award, 1115
ASME Edward F. Obert Award, 1116
ASME Edwin F. Church Medal, 1117
ASME Engineer-Historian Award, 1118
ASME Fluids Engineering Award, 1119

ASME Frank Kreith Energy Award, 1120
ASME Frank von Flue Award, 1121
ASME Freeman Scholar Award, 1122
ASME Gas Turbine Award, 1123
ASME George Westinghouse Medals, 1124
ASME Gustus L. Larson Memorial Award, 1126
ASME H.R. Lissner Award, 1127
ASME Heat Transfer Memorial Award, 1128
ASME Henry Hess Award, 1129
ASME Henry R. Worthington Medal, 1130
ASME Holley Medal, 1131
ASME Internal Combustion Engine Award, 1132
ASME J. Hall Taylor Medal, 1133
ASME J.P. Hartog Award, 1134
ASME James Harry Potter Gold Medal, 1135
ASME James N. Landis Medal, 1136
ASME Kate Gleason Award, 1137
ASM Eli Lilly and Company Research Award, 1138
ASME M. Eugene Merchant Manuf Medal, 1139
ASME Machine Design Award, 1140
ASME Marshall B. Peterson Award, 1141
ASME Mayo D. Hersey Award, 1142
ASME McDonald Mentoring Award, 1143
ASME Medal, 1144
ASME Melville Medal, 1145
ASME Melvin Green Codes & Standards Medal, 1146
ASME Milton C. Shaw Manufacturing Research Medal, 1147
ASME Nadai Medal, 1148
ASME Nancy DeLoye Fitzroy and Roland V. Fitzroy Medal, 1149
ASME Old Guard Early Career Award, 1150
ASME Old Guard Prizes, 1151
ASME Patrick J. Higgins Award, 1152
ASME Per Bruel Gold Medal for Noise Control and Acoustics, 1153
ASME Performance Test Codes Medal, 1154
ASME Pi Tau Sigma Gold Medal, 1155
ASME Prime Movers Committee Award, 1156
ASME R. Tom Sawyer Award, 1157
ASME Ralph Coats Roe Medal, 1158
ASME Robert E. Koski Medal, 1159
ASME Robert Henry Thurston Lecture Award, 1160
ASME Rufus T. Oldenburger Medal, 1161
ASME Ruth and Joel Spira Outstanding Design Educator Award, 1162
ASME S.Y. Zamrik PVP Medal, 1163
ASME Soichiro Honda Medal, 1164
ASME Spirit of St. Louis Medal, 1165
ASME Student Section Advisor Award, 1166
ASME Thomas A. Edison Patent Award, 1167
ASME Timoshenko Medal, 1168
ASME Van C. Mow Medal, 1169
ASME Warner T. Koiter Medal, 1170
ASME William T. Ennor Manufacturing Technology Award, 1171
ASME Worcester Reed Warner Medal, 1172
ASME Y.C. Fung Young Investigator Award, 1173
ASM ICAAC Young Investigator Awards, 1174
ASM Merck Irving S. Sigal Memorial Awards, 1177
ASM Procter & Gamble Award in Applied and Environmental Microbiology, 1180
ASM Promega Biotechnology Research Award, 1181
ASM Raymond W. Sarber Awards, 1182
ASM sanofi-aventis ICAAC Award, 1184
ASM Siemens Healthcare Diagnostics Young Investigator Award, 1185
ASM William A Hinton Res Training Award, 1187
ASNS Bio-Serv Award in Experimental Animal Nutrition, 1188
ASPEN Dudrick Research Scholar Award, 1192
ASPEN Harry M. Vars Award, 1193
ASPEN Scientific Abstracts Awards for Papers, 1198
ASPET Bernard B. Brodie Award in Metabolism, 1201
ASPET Division for Drug Metabolism Early Career Achievement Award, 1202
ASPET Epilepsy Research Award for Outstanding Contributions to the Pharmacology of Antiepileptic Drugs, 1203
ASPET Goodman and Gilman Award in Drug Receptor Pharmacology, 1204
ASPET John J. Abel Award in Pharmacology, 1205
ASPET P. B. Dews Lifetime Achievement Award for Research in Behavioral Pharmacology, 1206
ASPET Paul M. Vanhoutte Award, 1207
ASPET Sollmann Award in Pharmacology, 1209
ASPRS Ta Liang Memorial Award, 1215
ASR Robert J. McNamara Student Award, 1218
ASTD Research Award, 1226
Astronomical Society of the Pacific Amateur Achievement Award, 1228
Astronomical Society of the Pacific Dorothea Klumpke-Roberts Award, 1229
Astronomical Society of the Pacific Las Cumbres Amateur Outreach Award, 1230
Astronomical Society of the Pacific Maria and Eric Muhlmann Award, 1231
Astronomical Society of the Pacific Priscilla and Bart Bok Awards, 1232
Astronomical Society of the Pacific Richard H. Emmons Award for Excellence in College Astronomy Teaching, 1233
Astronomical Society of the Pacific Robert J. Trumpler Award, 1234
ASU Graduate College Martha E. Bernal Memorial Award, 1242
ATA Education Research Award, 1254
Award for Encouraging Disadvantaged Students into Careers in the Chemical Sciences, 1276
Award for Volunteer Service to the American Chemical Society, 1277
AWWA Academic Achievement Awards, 1283
Baxter International Foundation William B. Graham Prize for Health Services Research, 1306
BCBSM Foundation Excellence in Research Awards for Students, 1312
BCBSM Frank J. McDevitt, DO, Excellence in Research Awards, 1317
BFGoodrich Collegiate Inventors Prize, 1341
Board of Directors Distinguished Service Award for Senior ACS Administrators, 1354
Bower Award & Prize Achievement in Science, 1364
BP Conservation Programme Future Conservationist Awards, 1365
Bristol-Myers Squibb/Meade Johnson Award for Disting Achievement in Nutrition Research, 1373
CAA Wyeth Publication Grant, 1405
Cambridge Foundation For Peace Prize, 1413
CCA Diefenbaker Award to update, 1447
CCFF Chairman's Disting Life Sciences Award, 1449
CCFF Distinguished Ag Scientist Award, 1450
CCWH Catherine Prelinger Award, 1454
CFUW Ecole Polytechnique Commem Award, 1511
Charles Lathrop Parsons Award, 1526
Christian Gauss Award, 1555
Christine O. Gregoire Youth/Young Adult Award for Outstanding Use of Tobacco Ind Documents, 1557
Christine Wilson Award, 1558
Claude Pepper Foundation Grants, 1578
Commonfund Endowment Management Prize, 1601
Community Found AIDS Endowment Awards, 1614
Computing in Chemical Engineering Award, 1626
Conquer Cancer Foundation of ASCO International Innovation Grant, 1636
Cystic Fibrosis Canada Visiting Allied Health Professional Awards, 1686
Daniel Mendelsohn New Investigator Award, 1717
DEED Grants, 1730
Dennis Leslie Mahony Prize in Legal Theory, 1741
DOE Initial H-Prize Competition for Breakthrough Advances in Materials for Hydrogen Storage, 1801
Donald Q. Kern Award in Heat Transfer or Energy Conversion, 1831
Dreyfus Prize in the Chemical Sciences, 1856
E.B. Hershberg Award for Important Discoveries in Medicinally Active Substances, 1882
E. Bright Wilson Award in Spectroscopy, 1883
E.V. Murphree Award in Industrial and Engineering Chemistry, 1886
Earle B. Barnes Award for Leadership in Chemical Research Management, 1889
Educational Communications and Tech Awards, 1895
Elias J. Corey Award for Outstanding Original Contribution in Organic Synthesis by a Young Investigator, 1904
Elizabeth Glaser Int'l Leadership Awards, 1906
EMBO / EMBL Science Writing Prize, 1913
EMBO Award for Communication in the Life Sciences, 1914
Engineering Awards of Excellence, 1926
Ernest Guenther Award in the Chemistry of Natural Products, 1950
ESF Latsis Prize, 1951
F. Albert Cotton Award in Synthetic Inorganic Chemistry, 1971
Father James B. Macelwane Annual Awards, 1983
FDHN Designated Outcomes Award in Geriatric Gastroenterology, 1988
FDHN Designated Research Award in Research Related to Pancreatitis, 1990
FDHN Fellow Abstract Prizes, 1991
FDHN Funderburg Research Scholar Award in Gastric Biology Related to Cancer, 1993
FDHN Graduate Student Awards, 1994
FDHN Jon I. Isenberg International Research Scholar Award, 1995
FDHN Moti L. & Kamla Rustgi Travel Awards, 1997
FDHN Research Scholar Awards, 2000
Fordham Prizes, 2032
Forest Products Society Wood Award, 2034
Francis P. Garvan-John M. Olin Medal, 2048
Frank Field & J.L. Franklin Award for Outstanding Achievement in Mass Spectrometry, 2049
Frederic Stanley Kipping Award in Silicon Chem, 2055
Gabor A. Somorjai Award for Creative Research in Catalysis, 2094
GCSAA Essay Contest, 2102
George C. Pimentel Award in Chem Education, 2114
Glenn T. Seaborg Award for Nuclear Chemistry, 2149
Goldsmith Awards Program, 2158
Gruber Foundation Cosmology Prize, 2177
Gruber Foundation Gentics Prize, 2178
Gruber Foundation Int'l Research Award, 2180
Gruber Foundation Neuroscience Prize, 2181
Gruber Foundation Prize for Global Justice and Women's Rights, 2182
Gruber Foundation Rosalind Franklin Young Investigator Award, 2183
Gruber Foundation Weizmann Institute Awards, 2184
Harvard University John M. Olin Prizes, 2214
Herbert C. Brown Award for Creative Research in Synthetic Methods, 2240
Hereditary Disease Found Lieberman Award, 2245
Holberg Memorial Fund International Prize, 2272
Holberg Memorial Fund Nils Klim Prize, 2273
Holberg Memorial Fund School Project, 2274
Hydrogen Student Design Contest, 2311
IIE General Electric Foundation TECH Awards, 2352
Inamori Foundation Kyoto Prize, 2367
Institute Award for Excellence in Industrial Gases Technology, 2379
International League of Antiquarian Booksellers Bibliographical Prize, 2400
Ipatieff Prize, 2404
IRA Albert J. Harris Awards, 2405
IRA Esther Zolt Academic Research Award, 2407
IRA Esther Zolt Teacher Research Award, 2408
IRA Jeanne S. Chall Research Fellowship, 2410
IRA Outstanding Dissertation of the Year Award, 2411
IRA Pearson Foundation Rotary Lit Awards, 2412
ISUD Jacobsen Prizes, 2434
IUVSTA Prize for Science, 2443
IUVSTA Prize for Technology, 2444
James Flack Norris Award in Physical Organic Chemistry, 2458
James T. Grady-James H. Stack Award for Interpreting Chemistry for the Public, 2469
Joel Henry Hildebrand Award in the Theoretical and Experimental Chemistry of Liquids, 2494

Kavli Foundation Research Prizes, 2534
Kluge Center Kissinger Lecturer Award, 2565
Landon Foundation-AACR Innovator Award for Cancer Prevention Research, 2586
Landon Foundation-AACR Innovator Award for International Collaboration, 2587
Lawrence K. Cecil Award in Environmental Chemical Engineering, 2591
Leo Baeck Essay Prize, 2601
Library Company of Philadelphia Prizes, 2608
LITA/Gaylord Award for Achievement in Library and Information Technology, 2618
Ludwig Vogelstein Foundation Grants, 2626
Massage Therapy Foundation Practitioner Case Report Contest, 2679
Massage Therapy Foundation Student Case Report Contest, 2681
Materials Research Society Medalist Awards, 2683
MLA Murray Gottlieb Prize Essay Award, 2788
MLA Rittenhouse Award, 2790
MLA Thomas Reuters / Frank Bradway Rogers Information Advancement Award, 2791
MoEF Dr. Salim Ali Nat Wildlife Fellowship, 2792
MOEF Indira Gandhi P. Puraskar Award, 2793
MSI Alden G. Clayton Dissertation Proposal Competition, 2800
MSI Marketing Strategy Meets Wall Street Research Competition, 2801
NAGC Masters and Specialists Award, 2821
Nakanishi Prize, 2824
NARSAD Colvin Prize for Outstanding Achievement in Mood Disorders Research, 2830
NARSAD Goldman-Rakic Prize for Cognitive Neuroscience, 2832
NARSAD Lieber Prize for Schizophrenic Res, 2834
NARSAD Ruane Prize for Child and Adolescent Psychiatric Research, 2835
NASA New Aeronautics Competition for University Students, 2842
National Council of Farmer Cooperatives Graduate Awards, 2860
National Turkey Federation Research Award, 2887
National Wetlands Awards, 2889
NCFR Affiliate Councils Outstanding Student Research Papers, 2906
NCFR Family Therapy Section Best Research Paper Awards, 2907
NCFR Ruth Hathaway Jewson Award, 2909
NCSS Exemplary Research in Social St Award, 2937
NCSS Jean Dresden Grambs Distinguished Career Research in Social Studies Award, 2938
NCSS Larry Metcalf Exemplary Dissert Award, 2939
NFWF Budweiser Renewable Energy and Wildlife Conservation Research Prize, 2998
NHLBI Career Transition Awards, 3021
NHLBI National Research Service Award Programs in Cardiovascular Epidemiology and Biostatistics, 3041
NIA Mentored Clinical Scientist Research Career Development Awards, 3108
NIHCM Found Health Care Research Awards, 3210
NIHCM Foundation Health Care Television and Radio Journalism Awards, 3211
NKF Clinical Scientist Awards, 3274
NL Weiss/Brown Publication Subvention Award, 3320
NMF National Medical Association Awards for Medical Journalism, 3323
NMF National Medical Association Emerging Scholar Awards, 3324
NMF National Medical Association Patti LaBelle Award, 3325
NMF Ralph W. Ellison Prize, 3327
Nobel Laureate Signature Award for Graduate Education in Chemistry, 3413
Norwegian Acad of Science & Letters Abel Prize, 3433
NSERC Brockhouse Canada Prize for Interdisciplinary Research in Science and Engineering Grant, 3438
NSERC Howard Alper Postdoctoral Prize, 3441
NSERC Industrial Undergraduate Student Research Awards, 3444
NSERC John C. Polanyi Awards, 3446
NSERC Undergraduate Research Awards, 3451
NSF Alan T. Waterman Award, 3467
NSF Director's Award for Distinguished Teaching Scholars, 3553
NSF Presidential Awards for Excellence in Mathematics and Science Teaching, 3631
NSF Presidential Awards for Excellence in Science, Math, and Engineering Mentoring, 3632
NSF Systematic Biology and Biodiversity Inventories Grants, 3657
NWSA Abafazi-Africana Women's Studies Essay Award, 3675
NYAM Edward N. Gibbs Memorial Lecture and Award in Nephrology, 3682
NYAM Lewis Rudin Glaucoma Grants, 3689
NYAM Student Essay Grants, 3693
NYZS/Wildlife Conservation Society Research Grants, 3709
OAH Avery O. Craven Award, 3711
OAH Binkley-Stephenson Award, 3712
OAH Darlene Clark Hine Award, 3713
OAH David Thelen Award, 3714
OAH Ellis W. Hawley Prize, 3715
OAH Erik Barnouw Award, 3716
OAH Frederick Jackson Turner Award, 3717
OAH Friend of History Award, 3718
OAH Huggins-Quarles Award, 3720
OAH James A. Rawley Prize, 3721
OAH Lerner-Scott Dissertation Prize, 3723
OAH Liberty Legacy Foundation Award, 3724
OAH Louis Pelzer Memorial Award, 3725
OAH Merle Curti Award, 3726
OAH Ray A. Billington Prize, 3727
OAH Richard W. Leopold Prize, 3728
OAH Roy Rosenzweig Disting Service Award, 3729
OAH Willi Paul Adams Award, 3730
OM Association Outstanding Dissertation Awards, 3741
ONR Bisson Prize for Naval Tech Achievement, 3745
Paul Marks Prizes for Cancer Research, 3877
Peter Debye Award in Physical Chemistry, 3901
Peter F. Drucker Award for Nonprofit Innovation, 3902
Pezcoller Foundation-AACR International Award for Cancer Research, 3907
Pharmacia-ASPET Award for Experimental Therapeutics, 3930
Phi Beta Kappa Award in Science, 3935
Phi Delta Kappa Outstanding Doctoral Dissertation Awards, 3936
Phi Kappa Phi Found Love of Learning Awards, 3937
Phi Kappa Phi Scholar Award, 3939
Phi Upsilon Omicron Florence Fallgatter Distinguished Service Award, 3943
Princess Grace Foundation Theater, Dance, and Film Grants, 3988
Princeton University Center for Human Values Short Movie Prize, 3995
R.H. Wilhelm Award in Chemical Reaction Engineering, 4007
RACGP/APHCRI Indigenous Health Award, 4008
RACGP Alan Chancellor Award, 4010
RACGP Grieve Memorial Award, 4014
RACGP National Asthma Council Res Award, 4015
RACGP Rex Walpole Travelling Fellowship, 4016
Rajiv Gandhi Foundation Grants, 4025
Ralph F. Hirschmann Award in Peptide Chem, 4027
Ralph Waldo Emerson Award, 4028
Roger Adams Award in Organic Chemistry, 4096
Romanell-Phi Beta Kappa Professorship, 4099
Ronald Breslow Award for Achievement in Biomimetic Chemistry, 4100
Royal Norwegian Embassy Kavli Prizes, 4109
RSC Abbyann D. Lynch Medal in Bioethics, 4116
RSC Alice Wilson Award, 4117
RSC Bancroft Award, 4118
RSC Jason A. Hannah Medal, 4119
RSC John L. Synge Award, 4120
RSC McLaughlin Medal, 4121
RSC Miroslaw Romanowski Medal, 4122
RSC Pierre Chauveau Medal, 4123
RSC Rutherford Memorial Medals, 4124
RSC Sir John William Dawson Medal, 4125
RSC Thomas W. Eadie Medal, 4126
S.E.VEN Fund Open Enterprise Solutions to Poverty RFP Grants, 4155
S.E.VEN Fund WHY Prize, 4156
SeaWorld and Busch Gardens Conservation Fund Grants, 4196
SfN Donald B. Lindsley Prize in Behavioral Neuroscience, 4204
SfN Jacob P. Waletzky Award, 4206
SfN Louise Hanson Marshall Specific Recognition Award, 4208
SfN Mika Salpeter Lifetime Achievement Award, 4209
SfN Postdoctoral Fellow Travel Awards, 4210
SfN Ralph W. Gerard Prize in Neuroscience, 4211
SfN Swartz Prize for Theoretical and Computational Neuroscience, 4213
SfN Travel Awards for the Japan Neuroscience Society Meeting, 4216
SfN Young Investigator Award, 4218
SHGAPE Vincent P. DeSantis Prize, 4234
SHOT Abbott Payson Usher Prize, 4237
SHOT Dibner Award for Excellence in Museum Exhibits, 4239
SHOT Eugene S. Ferguson Prize, 4240
SHOT IEEE Life Members Prize in Elect History, 4241
SHOT Joan Cahalin Robinson Prize, 4243
SHOT Leonardo da Vinci Medal, 4244
SHOT Samuel Eleazar and Rose Tartakow Levinson Prize, 4247
SHRM Dissertation Awards, 4248
SIAM/ACM Award in Computational Science and Engineering, 4251
Simone and Cino del Duca Grand Prix Awards, 4261
Smith Family Awards for Excellence in Biomedical Research, 4268
SNM Tetalman Memorial Young Inv Awards, 4304
Society of Cosmetic Chemists Allan B. Black Award Sponsored by Presperse, 4312
Society of Cosmetic Chemists Award Sponsored by McIntyre Group, Ltd., 4313
Society of Cosmetic Chemists Award Sponsored by The HallStar Company, 4314
Society of Cosmetic Chemists Chapter Best Speaker Award, 4315
Society of Cosmetic Chem Chapter Merit Award, 4316
Society of Cosmetic Chem Frontier of Scie Award, 4317
Society of Cosmetic Chem Hans Schaeffer Award, 4318
Society of Cosmetic Chemists Joseph P. Ciaudelli Award, 4319
Society of Cosmetic Chemists Keynote Award Lecture Sponsored by Ruger Chemical Corporation, 4320
Society of Cosmetic Chemists Literature Award, 4321
Society of Cosmetic Chemists Maison G. de Navarre Medal, 4322
Society of Cosmetic Chemists Merit Award, 4323
Society of Cosmetic Chemists Robert A. Kramer Lifetime Service Award, 4324
Society of Cosmetic Chem Shaw Mudge Award, 4325
Society of Cosmetic Chem Student Poster Awards, 4326
Society of Cosmetic Chem Young Scientist Awards, 4327
SPIE A.E. Conrady Award, 4338
SPIE C.S. Vikram Award in Optic Metrology, 4339
SPIE Dennis Gabor Award, 4340
SPIE Frits Zernike Award, 4341
SPIE George W. Goddard Award, 4342
SPIE Gold Medal of the Society Award, 4343
SPIE Technology Achievement Award, 4344
SRC Aristotle Award, 4349
SRC Landmark Innovation Award, 4376
SRC M Khan Outstanding Industry Liaison Award, 4377
SRC Technical Excellence Award, 4394
SSSP C. Wright Mills Award, 4439
SSSR Distinguished Article Award, 4440
SSSR Student Paper Award, 4441
Summerfield G. Roberts Award, 4465
Swedish Natural Science Res Council Grants, 4479
Sybil G. Jacobs Award for Outstanding Use of Tobacco Industry Documents, 4487
T.R. Higgins Lectureship Award, 4490

Tessie Agan Award Competition, 4500
The World Food Prize, 4503
TIAA-CREF Paul A. Samuelson Award, 4518
Timoshenko Medal, 4519
TRDRP California Research Awards, 4525
TRDRP Dissertation Research Awards, 4526
TRDRP Exploratory/Devel Research Awards, 4527
TRDRP New Investigator Awards, 4528
TRDRP Participatory Research Awards, 4529
TRDRP Postdoctoral Fellowship Awards, 4530
Trevithick Book Prize Award, 4532
Twentieth-Century Japan Research Awards, 4536
UNCFSP NAI-MIRS Sabbatical Awards, 4561
United Methodist Church Asian, Black, Hispanic and Native American History Research Grants, 4570
United Methodist Church Center for the Evangelical UBH Audrie E. Reber Memorial Award, 4571
United Methodist Church John Harrison Ness Memorial Award, 4572
United Methodist Church Women in History Writing Award, 4575
United Methodist Church World is My Parish United Methodist History Research Grant, 4576
Upjohn Institute Dissertation Award, 4591
Van Alen Institute New York Prize Fellowship, 4806
Virginia A. Hodgkinson Research Prize, 4812
W. David Smith, Jr. Grad Student Paper Award, 4820
W. Frank Blair Award, 4821
WAWH Barbara Penny Kanner Award, 4832
WAWH Sierra Award, 4834
Welch Award in Chemistry, 4838
Welch Foundation Norman Hackeman Award in Chemical Research, 4841
William A. Klinger Research Award, 4871
Yves Rocher Foundation Terra Ficaria Prize for Scientific Research, 4904

Basic Research

1 in 9: Long Island Breast Cancer Action Coalition Grants, 1
1st Touch Foundation Grants, 2
3M Company Higher Education Grants, 4
25th Anniversary Foundation Grants, 6
1976 Foundation Grants, 7
A-T Children's Project Grants, 8
A-T Children's Project Post Doctoral Fellowships, 9
A-T Medical Research Foundation Grants, 10
A.L. Mailman Family Foundation Grants, 11
A.O. Smith Community Grants, 12
A. V. Stout Fund Research Grants, 13
AAAAI ARTrust Mini Grants for Allied Health, 14
AAAAI ARTrust Mini Grants for Clinical Res, 15
AAAAI ARTrust Mini Grants in Faculty Development, 16
AAAAI Distinguished Clinician Award, 17
AAAAI Distinguished Layperson Award, 18
AAAAI Distinguished Scientist Award, 19
AAAAI Distinguished Service Award, 20
AAAAI Fellows-in-Training Grants, 22
AAA Foundation for Traffic Safety Grants, 24
AAAS Eppendorf Science Prize for Neurobiology, 31
AAAS GE Science Prize for Young Life Scientists, 32
AAAS Martin and Rose Wachtel Cancer Research Award, 34
AAAS Science and Technology Policy Fellowships: Congressional, 39
AAAS Science and Technology Policy Fellowships: Diplomacy, Security, and Development, 40
AAAS Science and Technology Policy Fellowships: Health, Education and Human Services, 43
AAAS Science and Technology Policy Overseas Fellowships at USAID Missions, 44
AAAS Science Prize for Inquiry-Based Instruction, 45
AAB Berlin Prize Fellowships, 48
AACAP-NIDA K12 Career Development Awards, 54
AACAP Beatrix A. Hamburg Award for the Best New Research Poster by a Child and Adolescent Psychiatry Resident, 55
AACAP Educational Outreach Program for General Psychiatry Residents, 57
AACAP Educational Outreach Program for Residents in Alcohol Research, 58
AACAP Elaine Schlosser Lewis Award for Research on Attention-Deficit Disorder, 59
AACAP George Tarjan Award for Contributions in Developmental Disabilities, 60
AACAP Irving Philips Award for Prevention, 61
AACAP Jeanne Spurlock Lecture and Award on Diversity and Culture, 62
AACAP Jeanne Spurlock Minority Medical Student Clinical Fellowship in Child and Adolescent Psychiatry, 63
AACAP Jeanne Spurlock Research Fellowship in Substance Abuse and Addiction for Minority Medical Students, 64
AACAP Junior Investigator Awards, 65
AACAP Klingenstein Third Generation Foundation Award for Research in Depression or Suicide, 66
AACAP Life Members Mentorship Grants for Medical Students, 67
AACAP Mary Crosby Congressional Fellowships, 68
AACAP Pilot Research Award for Attention-Deficit Disorder, 70
AACAP Pilot Research Award for Learning Disabilities, Supported by the Elaine Schlosser Lewis Fund, 71
AACAP Pilot Research Awards, Supported by Eli Lilly and Company, 72
AACAP Quest for the Test Bipolar Disorder Pilot Research Award, 73
AACAP Rieger PsychodynPsychotherapy Award, 74
AACAP Rieger Service Award for Excellence, 75
AACAP Robert Cancro Acad Leadership Award, 76
AACAP Robinson-Cunningham Award, 77
AACAP Sidney Berman Award for the School-Based Study and Intervention for Learning Disorders and Mental Ilness, 78
AACAP Summer Medical Student Fellowships, 80
AACC International Alsberg-French-Schoch Memorial Lectureship Awards, 82
AACC International Best Student Paper and Best Student Poster Award, 83
AACC Int'l Carl Wilhelm Brabender Award, 85
AACC International Edith A. Christensen Award, 87
AACC International Engineering and Processing Division's Stanley Watson Award, 88
AACC International Excellence in Teaching Award, 89
AACC International George W. Scott Blair Lecture Award, 90
AACC International Graduate Fellowships, 91
AACC International Milling & Baking Division's John C. Halverson Memorial Lectureship, 92
AACC International Milling & Baking Division's M. Rella Dwyer Graduate Fellowship, 93
AACC International Phil Williams Applied Research Award, 94
AACC International Student Division's Best Research Paper Awards, 95
AACC International Student Division's Product Development Awards, 96
AACC International Thomas Burr Osborne Medal, 98
AACC International Walter Bushuk Graduate Research Award, 99
AACC Int'l Young Scientist Research Award, 100
AACC Texture Technol Quality Research Award, 101
AACN-Edwards Lifesciences Nurse-Driven Clinical Practice Outcomes Grants, 102
AACN-Philips Medical Systems Clinical Outcomes Grants, 103
AACN-Sigma Theta Tau Critical Care Grant, 104
AACN Clinical Inquiry Fund Grants, 105
AACN Clinical Practice Grants, 106
AACN Critical Care Grants, 107
AACN End of Life/Palliative Care Small Projects Grants, 108
AACN Evidence-Based Clinical Practice Grants, 109
AACN Physio-Control Small Projects Grants, 111
AACR-American Cancer Society Award for Research in Cancer Epidemiology and Prevention, 112
AACR-Colorectal Cancer Coalition Fellows Grant, 113
AACR-FNAB Fellows Grant for Translational Pancreatic Cancer Research, 114
AACR-FNAB Foundation Career Development Award for Translational Cancer Research, 115
AACR-Genentech BioOncology Career Development Award for Cancer Research on the HER Family Pathway, 116
AACR-Genentech BioOncology Fellowship for Cancer Research in Angiogenesis, 117
AACR-Minorities in Cancer Research Jane Cooke Wright Lectureship Awards, 119
AACR-National Brain Tumor Foundation Fellows Grant, 120
AACR-NCI International Investigator Opportunity Grants, 121
AACR-Pancreatic Cancer Action Network Career Development Award for Pancreatic Cancer Research, 122
AACR-Pancreatic Cancer Action Network Innovative Grants, 123
AACR-Prevent Cancer Foundation Award for Excellence in Cancer Prevention Research, 124
AACR-Women in Cancer Research Charlotte Friend Memorial Lectureship Awards, 126
AACR Award for Lifetime Achievement in Cancer Research, 127
AACR Award for Outstanding Achievement in Cancer Research, 128
AACR Award for Outstanding Achievement in Chemistry in Cancer Research, 129
AACR Basic Cancer Research Fellowships, 130
AACR Brigid G. Leventhal Scholar in Cancer Research Awards, 131
AACR Career Development Awards for Pediatric Cancer Research, 132
AACR Centennial Postdoctoral Fellowships in Cancer Research, 133
AACR Centennial Pre-doctoral Fellowships in Cancer Research, 134
AACR Fellows Grants, 136
AACR Henry Shepard Bladder Cancer Research Grants, 138
AACR Joseph H. Burchenal Memorial Award, 139
AACR Judah Folkman Career Development Award for Anti-Angiogenesis Research, 140
AACR Kirk A. Landon and Dorothy P. Landon Foundation Prizes, 141
AACR Margaret Foti Award for Leadership and Achievements in Cancer Research, 142
AACR Minority-Serving Institution Faculty Scholar in Cancer Research Awards, 143
AACR Minority Scholar in Cancer Res Awards, 144
AACR Outstanding Investigator Award for Breast Cancer Research, 145
AACR Team Science Award, 148
AACR Thomas J. Bardos Science Educ Awards, 149
AAES Engineering Journalism Award, 150
AAFA Investigator Research Grants, 153
AAFP Foundation Joint Grants, 156
AAFP Foundation Wyeth Immunization Awards, 158
AAFP Research Stimulation Grants, 159
AAFPRS Bernstein Grant, 161
AAFPRS Fellowships, 162
AAFPRS Investigator Development Grant, 163
AAFPRS Leslie Bernstein Resident Res Grants, 166
AAF Richard Morris Hunt Archit Fellowship, 169
AAG Anne U. White Fund Grants, 170
AAG Mel Marcus Fund Grants for Physical Geography, 176
AAG Research Grants, 177
AAHN Grant for Historical Research, 178
AAHN Pre-Doctoral Research Grant, 179
AAMC Award for Distinguished Research, 183
AANS Neurosurgery Research & Educ Found/Spine Section Young Clinician Investigator Award, 184
AAO-HNSF CORE Research Grants, 186
AAO-HNSF Health Services Research Grants, 187
AAO-HNSF Maureen Hannley Research Training Awards, 188
AAO-HNSF Percy Memorial Research Award, 189

AAO-HNSF Rande H. Lazar Health Services Research Grant, 190
AAO-HNSF Resident Research Awards, 191
AAO-HNS Medical Student Research Prize, 192
AAO-HNS Resident Research Prizes, 193
AAOA Foundation/AAO-HNSF Combined Research Grants, 194
AAOHN Found Experienced Researcher Grants, 195
AAOHN Foundation New Investigator Researcher Grants, 196
AAR Rome Prize Fellowships, 202
AASF Lydia Donaldson Tutt-Jones Memorial Research Grant, 203
AASL Research Grant, 207
Abbott Fund Science Education Grants, 220
Abdus Salam ICTP-ELETTRA Users Grants, 221
Abdus Salam ICTP-OEA Centre Grants, 222
Abdus Salam ICTP-OEA Projects Grants, 223
Abdus Salam ICTP-OEA Scie Meetings Grants, 224
Abdus Salam ICTP-OEA Visiting Scholars/ Consultants Grants, 225
Abdus Salam ICTP/IAEA Sandwich Training Education Grants, 226
Abdus Salam ICTP Federation Arrangement Scheme Grants, 227
Abdus Salam ICTP Group Associates Grants, 228
Abdus Salam ICTP Jnt Laurea Magistralis Grants, 229
Abdus Salam ICTP Joint PhD Programme in Environmental Fluid Mechanics Grants, 230
Abdus Salam ICTP Junior Associates Grants, 231
Abdus Salam ICTP Mathematical Research Fellowships, 232
Abdus Salam ICTP Networks Grants, 233
Abdus Salam ICTP Postdoctoral Positions in Theoretical High Energy Physics, 234
Abdus Salam ICTP Postgraduate Diploma Grants, 235
Abdus Salam ICTP Regular Associates Grants, 237
Abdus Salam ICTP Schools, Conferences, Workshops and Extended Research Workshops Grants, 238
Abdus Salam ICTP Senior Associates Grants, 239
Abdus Salam ICTP Training and Research in Italian Laboratories Grants, 240
Abdus Salam ITCP Postgraduate Diploma Programme in General Physics, 241
Abdus Salam Joint ICTP/SISSA PhD Programme in Physics and Mathematics Grants, 242
Abington Foundation Grants, 243
ACAAI Foundation Research Grants, 244
AcademyHealth HSR Impact Awards, 246
Academy of American Franciscan History Dissertation Fellowships, 249
ACC American Artists and Museum Professionals in Asia Fellowships, 250
ACC Asian Art and Religion Fellowships, 251
ACC Fellowships and Projects, 252
ACC Ford Foundation Fellowships, 253
ACC Hong Kong Arts Program Fellowships, 254
ACC Humanities Fellowship Program, 255
ACC Japan-US Arts Program Fellowships, 256
ACC John D. Rockefeller III Award, 257
ACC Mekong Region Fellowship, 258
ACCP Frontiers Career Development Research Awards, 259
ACCP Hematology/Oncology Minisabbatical, 260
ACC Philippines Fellowship Program, 261
ACCP Infectious Diseases Minisabbatical, 262
ACCP Infectious Diseases Research Fellowships, 263
ACCP Oncology Research Fellowships, 264
ACCP Pain & Palliative Care Minisabbatical, 265
ACCP Pharmacotherapy Investigator Development Research Awards, 266
ACC Residency Program in Asia, 267
ACC Starr Foundation Fellowship Program, 268
ACC Taiwan Fellowship Program, 269
ACGT Investigators Grants, 271
ACGT Young Investigator Grants, 272
A Charitable Foundation Grants, 273
Achelis Foundation Grants, 274
ACLS/New York Public Library Fellowships, 277
ACLS American Research in the Humanities in China Fellowships, 280
ACLS Collaborative Research Fellowships, 282
ACLS Early Career Postdoctoral Fellowships in East European Studies, 286
ACLS Fellowships, 287
ACLS International and Area Studies Fellowships, 289
ACS Clinical Research Professor Grants, 306
ACS Doctoral Training Grants in Oncology Social Work, 308
ACS Institutional Research Grants, 310
ACSM Carl V. Gisolfi Memorial Fund Grant, 312
ACSM Coca-Cola Company Doctoral Student Grant on Behavior Research, 313
ACS Mentored Research Scholar Grant in Applied and Clinical Research, 314
ACSM Foundation Clinical Sports Medicine Endowment Grants, 315
ACSM Foundation Doctoral Research Grants, 316
ACSM Foundation Research Grants, 317
ACSM International Student Awards, 318
ACSM NASA Space Physiology Research Grants, 319
ACSM Oded Bar-Or Int'l Scholar Awards, 320
ACSM Paffenbarger-Blair Fund for Epidemiological Research on Physical Activity Grants, 321
ACSM Raymond and Rosalee Weiss Research Endowment Grant, 322
ACS Physician Training Awards in Preventive Medicine, 323
ACS Pilot and Exploratory Projects in Palliative Care of Cancer Patients and Their Families Grants, 324
ACS Research Professor Grants, 326
ACS Research Scholar Grants for Health Services and Health Policy Research, 327
ACS Research Scholar Grants in Basic, Preclinical, Clinical and Epidemiology Research, 328
ACS Research Scholar Grants in Psychosocial and Behavioral and Cancer Control Research, 329
ACT Awards, 330
Acton Institute for the Study of Religion and Liberty Calihan Academic Fellowships, 331
Acton Institute for the Study of Religion and Liberty Novak Award, 332
ACVIM Foundation Clinical Investigation Grants, 334
Adaptec Foundation Grants, 340
AEG Foundation Legget Fund Grants, 341
AERA Dissertation Grants, 342
AERA Minority Fellowships, 343
AERA Research Grants, 344
Aetna Found Integrated Health Care Grants, 348
Aetna Foundation Minority Scholars Grants, 349
AFAR CART Fund Grants, 351
AFAR Paul Beeson Career Development Awards in Aging Research for the Island of Ireland, 353
AFAR Research Grants, 354
Affymetrix Corporate Contributions Grants, 355
AFP Research Grants, 356
African American Heritage Grants, 357
AFRL/NRC Resident Research Associateship, 358
AFUW Fellowships, 361
AFUW Georgina Sweet Fellowships, 362
AFUW Padnendadlu Graduate Bursary, 364
AFUW Padnendadlu Postgraduate Bursary, 365
AHAF Alzheimer's Disease Research Grants, 370
AHAF Macular Degeneration Research Grants, 371
AHAF National Glaucoma Research Grants, 372
AHIMA Dissertation Assistance Grants, 374
AHIMA Faculty Development Stipends, 375
AHIMA Grant-In-Aid Research Grants, 376
AHNS/AAO-HNSF Surgeon Scientist Combined Award, 380
AHNS/AAO-HNSF Young Investigator Combined Award, 381
AHNS Alando J. Ballantyne Resident Research Pilot Grant, 382
AHNS Pilot Grant, 383
AHRF Eugene L. Derlacki, M.D. Res Grants, 384
AHRF Georgia Birtman Grant, 385
AHRF Regular Research Grants, 386
AHRF Wiley H. Harrison Research Award, 387
AHRQ Independent Scientist Award, 388
AHRQ Individual Postdoc Fellows Ruth L. Kirschstein National Research Service Awards, 389
AIA Latrobe Prize, 390
AIAR Ernest S. Frerichs Fellow and Program Coordinator Grant, 394
AIChE Andrew Chase Forest Bioproducts Division Award in Chemical Engineering, 397
AIChE Clarence G. Gerhold Award, 400
AIChE Donald Q. Kern Award, 406
AIChE Excellence in Process Development Research Award, 409
AIChE Food, Pharmaceutical and Bioengineering Division Awards in Chemical Engineering, 410
AIChE Institute Award for Excellence in Industrial Gases Technology, 412
AIChE Lectureship Award in Fluidization, 416
AIChE Nanoscale Science and Engineering Forum Young Investigator Award, 417
AIChe North American Mixing Forum Grant, 419
AIChE North American Mixing Forum Student Award, 420
AIChE Nuclear Engineering Division Outstanding Student Paper & Presentation Awards, 421
AIChE Process Development Practice Award, 422
AIChE Research Excellence In Sustainable Engineering Award, 424
AIChE Separations Division Graduate Student Research Award, 425
AIChE Thomas Baron Award in Fluid-Particle Systems, 426
AIChE Transport & Energy Processes Div Award, 427
AIHP Sonnedecker Visiting Scholar Grants, 432
AIHS/Mitacs Health Pilot Partnership Internship Grants, 434
AIHS Alberta/Pfizer Translational Res Grants, 435
AIHS Clinical Fellowships, 436
AIHS Collaborative Research and Innovation Grants - Collaborative Program, 437
AIHS Collaborative Research and Innovation Grants - Collaborative Project, 438
AIHS Collaborative Research and Innovation Grants - Collaborative Team, 439
AIHS Fast-Track Fellowships, 440
AIHS ForeFront Internships, 442
AIHS ForeFront MBA Studentship Award, 443
AIHS ForeFront MBT Studentship Awards, 444
AIHS Full-Time Health Research Studentships, 446
AIHS Full-Time Studentships, 448
AIHS Interdisciplinary Team Grants, 450
AIHS Knowledge Exchange - Visiting Scientists, 452
AIHS Part-Time Fellowships, 454
AIHS Proposals for Special Initiatives, 456
AIHS Research Prize, 457
AIHS Summer Studentships, 458
AIHS Sustainability Grants, 459
AINSE Honours Scholarships, 460
AINSE Int'l Conference Travel Scholarships, 461
AINSE Postgraduate Research Awards, 463
AINSE Research Fellowships, 464
AINSE Research Fellowships with Institute for Science and International Security Support, 465
AIR/IES-NCES Postdoctoral Policy Fellowships in Washington, 466
AIR/NCES Fellowships for Graduate Study, 467
AIR/NRC Summer Institute on First-Year Assessment Undergraduate Student Fellowships, 468
AIR National Summer Data Policy Institute Fellowships, 469
AIR Research Grants/Dissertation Fellowships, 470
AJCC Cancer Staging Grants, 476
Akonadi Foundation Anti-Racism Grants, 477
ALA ACRL Doctoral Dissertation Fellowship, 478
Alabama Library Association CUS Res Grants, 482
ALA Carroll Preston Baber Research Grant, 483
ALA Emerald Research Grant Awards, 485
ALA Frederick G. Kilgour Award for Research in Library and Information Technology, 487
ALA Ingenta Research Awards, 489
ALA Innovation Award, 490

ALA Margaret T. Lane/Virginia F. Saunders Memorial Research Award, 492
ALA ProQuest Documents to the People Award, 496
Alaska Conservation Foundation Watchable Wildlife Conservation Trust Grants, 498
Albany Medical Center Prize in Medicine and Biomedical Research, 502
Alberta Law Foundation Grants, 503
Albert and Margaret Alkek Foundation Grants, 504
Albert and Mary Lasker Foundation Awards, 505
Albert and Mary Lasker Foundation Clinical Research Scholars, 506
Albert M. Greenfield Dissertation Fellowship, 507
Alcoa Foundation Grants, 511
Alcohol Misuse and Alcoholism Research Grants, 512
Alexander and Margaret Stewart Trust Grants, 514
Alexander von Humboldt Feodor-Lynen Fellowships for Experienced Researchers, 518
Alexander von Humboldt Feodor-Lynen Research Fellowships for Postdoc Researchers, 519
Alexander von Humboldt Fraunhofer-Bessel Research Awards, 521
Alexander von Humboldt Friedrich Wilhelm Bessel Research Awards, 522
Alexander von Humboldt Georg Forster Fellowships for Experienced Researchers, 523
Alexander von Humboldt Georg Forster Fellowships for Postdoctoral Researchers, 524
Alexander von Humboldt International Climate Protection Fellowships, 529
Alexander von Humboldt Max Planck Award, 532
Alexander von Humboldt Philipp Franz von Siebold Award for Japanese Researchers, 534
Alexander von Humboldt Found Professorships, 535
Alexander von Humboldt Reimar Lust Award, 536
Alexander von Humboldt Research Awards, 537
Alexander von Humboldt Research Fellowships for Postdoctoral Researchers, 539
Alexander von Humboldt Kovalevskaja Awards, 540
Alex C. Walker Educational and Charitable Foundation Grants, 541
Alex Stern Family Foundation Grants, 542
Alfred P. Sloan Barcode of Life Grants, 543
Alfred P. Sloan Behavioral Economics and Household Finance Grants, 544
Alfred P. Sloan Data and Computational Research Grants, 546
Alfred P. Sloan Deep Carbon Observatory Grants, 547
Alfred P. Sloan Economic Analysis of Science and Technology Grants, 548
Alfred P. Sloan Economic Implications of the Great Recession Grants, 549
Alfred P. Sloan Empirical Economic Research Enablers Grants, 550
Alfred P. Sloan Encyclopedia of Life Grants, 551
Alfred P. Sloan Energy and Environment Grants, 552
Alfred P. Sloan History of Science and Technology Grants, 553
Alfred P. Sloan Microbiology of the Built Environment Grants, 554
Alfred P. Sloan Public Understanding of Science and Technology-Book Grants, 555
Alfred P. Sloan Public Understanding of Science and Technology-Other Efforts, 556
Alfred P. Sloan Public Understanding of Science & Tech-Public Television Grants, 557
Alfred P. Sloan Public Understanding of Science and Technology-Theater Grants, 559
Alfred P. Sloan Research Fellowships, 560
Alfred P. Sloan Scholarly Communication Grants, 561
Alfred P. Sloan Science of Learning STEM Grants, 562
Alfred P. Sloan Selected National Issues Grants, 563
Alfred P. Sloan Sloan Digital Sky Survey Grants, 564
Alfred P. Sloan Synthetic Biology Grants, 565
Alfred P. Sloan Working Longer Grants, 566
Alfred W. Bressler Prize in Vision Science, 567
Allen Lane Foundation Grants, 571
All Saints Educational Trust Grants, 572
Allyn Foundation Grants, 574
Alpha Omega Foundation Grants, 576
Alpha Research Foundation Grants, 577
ALSAM Foundation Grants, 579
Alternatives Research and Development Foundation Alternatives in Education Grants, 581
Alternatives Research and Development Foundation Grants, 582
Alton Ochsner Award, 583
Alvin and Lucy Owsley Foundation Grants, 584
ALVRE Grant, 586
ALZA Corporate Contributions Grants, 588
Alzheimer's Association Conference Grants, 589
Alzheimer's Association Development of New Cognitive and Functional Instruments Grants, 590
Alzheimer's Association Everyday Technologies for Alzheimer Care Grants, 591
Alzheimer's Association Investigator-Initiated Research Grants, 592
Alzheimer's Association Mentored New Investigator Research Grants to Promote Diversity, 593
Alzheimer's Association Neuronal Hyper Excitability and Seizures in Alzheimer's Disease Grants, 594
Alzheimer's Association New Investigator Research Grants, 595
Alzheimer's Association New Investigator Research Grants to Promote Diversity, 596
Alzheimer's Association U.S.-U.K. Young Investigator Exchange Fellowships, 597
Alzheimer's Association Zenith Fellows Awards, 598
AMA-WPC Joan F. Giambalvo Memorial Scholarships, 600
AMA Foundation Seed Grants for Research, 601
American-Italian Cancer Foundation Fellowships, 602
American-Scandinavian Foundation Fellowship for Study in Scandinavia, 604
American Academy of Advertising Doctoral Dissertation Competition Awards, 608
American Academy of Advertising Ivan Preston Outstanding Contribution to Research on Advertising Award, 609
American Academy of Advertising Kim Rotzoll Award for Advertising Ethics and Social Responsibility, 610
American Academy of Advertising Research Fellowships, 611
American Academy of Arts and Sciences Hellman Fellowships, 612
American Academy of Arts and Sciences Visiting Scholars, 613
American Academy of Nursing Building Academic Geriatric Nursing Capacity Scholarships, 614
American Academy of Nursing Claire M. Fagin Fellowships, 615
American Academy of Nursing Mayday Grants, 616
American Academy of Religion Collaborative Research Grants, 617
American Academy of Religion Individual Research Grants, 618
American Australian Association Australia To USA Fellowships, 619
American Australian Association USA to Australia Fellowships, 620
American Chemical Society AGFD Advancement of the Application of Agricultural and Food Chemistry Award, 626
American Chemical Society AGFD Donald A. Withycombe Graduate Fellowship Award in Food Chemistry, 627
American Chemical Society AGFD Teranishi Graduate Fellowship Award, 628
American Chemical Society AGFD Undergraduate Research Symposium Travel Grants, 629
American Chemical Society AGFD Young Scientist Award, 630
American Chemical Society AGRO Award for Innovation in Chemistry of Agriculture, 631
American Chemical Society AGRO Ed Awards, 632
American Chem Society AGRO Fellow Awards, 633
American Chemical Society AGRO International Award for Research in Agrochemicals, 634
American Chemical Society AGRO Sterling B. Hendricks Memorial Lectureship, 635
American Chemical Society AGRO Young Scientist Research Awards, 636
American Chemical Society Alfred Burger Award in Medicinal Chemistry, 638
American Chemical Society ANYL Arthur F. Findeis Award for Achievements by a Young Analytical Scientist, 639
American Chemical Society ANYL Award for Distinguished Service in the Advancement of Analytical Chemistry, 640
American Chemical Society ANYL Award in Chemical Instrumentation, 641
American Chemical Society ANYL Award in Electrochemistry, 642
American Chemical Society ANYL Award in Spectrochemical Analysis, 643
American Chemical Society ANYL I.M. Kolthoff Enrichment Awards for Undergrad Students, 644
American Chemical Society ANYL J. Calvin Giddings Award for Excellence in Education, 645
American Chemical Society Art C. Cope Award, 647
American Chemical Society Arthur C. Cope Scholar Awards, 648
American Chemical Society Award for Achievement in Research for the Teaching and Learning of Chemistry, 649
American Chemical Society Award for Affordable Green Chemistry, 650
American Chemical Society Award for Computers in Chemical or Pharmaceutical Research, 651
American Chemical Society Award for Creative Advances in Environmental Science and Technology, 652
American Chemical Society Award for Creative Research and Applications of Iodine Chemistry, 654
American Chemical Society Award for Creative Work in Fluorine Chemistry, 655
American Chemical Society Award for Distinguished Service in Advancement of Inorganic Chem, 657
American Chemical Society Award for Encouraging Disadvantaged Students into Careers in the Chemical Sciences, 658
American Chemical Society Award for Research at an Undergraduate Institution, 660
American Chemical Society Award for Team Innovation, 661
American Chemical Society Award in Analytical Chemistry, 662
American Chemical Society Award in Applied Polymer Science, 663
American Chemical Society Award in Colloid and Surface Chemistry, 665
American Chemical Society Award in Industrial Chemistry, 666
American Chemical Society Award in Inorganic Chemistry, 667
American Chemical Society Award in Organometallic Chemistry, 668
American Chemical Society Award in Polymer Chemistry, 669
American Chemical Society Award in Pure Chem, 670
American Chemical Society Award in Separations Science and Technology, 671
American Chemical Society Award in the Chemistry of Materials, 672
American Chemical Society Chemical Technology Partnership Mini Grants, 674
American Chemical Society GCI Pharmaceutical Roundtable Research Grants, 677
American Chemical Society Herman Frasch Foundation Grants, 680
American Chemical Society Joseph Breen Memorial Fellowship in Green Chemistry, 682
American Chemical Society Overcoming Challenges Award, 686
American Chemical Society PCD Award in Experimental Physical Chemistry, 687
American Chemical Society PCD Award in Theoretical Chemistry, 688
American Chemical Society Early Career Awards, 689

American Chemical Society Petroleum Research Fund Doctoral New Investigator Grants, 690
American Chemical Society Petroleum Research Fund Undergraduate Faculty Sabbatical Grants, 693
American Chemical Society Petroleum Research Fund Undergraduate Research Grants, 695
American Chemical Society Rising Star Award, 698
American College of Bankruptcy Grants, 699
American College of Rheumatology Fellows-in-Training Travel Scholarships, 700
American College of Surgeons and The Triological Society Clinical Scientist Development Awards, 701
American College of Surgeons Co-Sponsored K08/K23 NIH Supplement Awards, 702
American College of Surgeons Faculty Research Fellowships, 704
American College of Surgeons Traveling Fellowship to Germany, 706
American College of Surgeons Traveling Fellowship to Japan, 707
American College of Surgeons Wound Care Management Award, 708
American Family Insurance Corporate Giving, 709
American Hist Assoc Albert B. Corey Prize in Canadian-American Relations, 711
American Hist Assoc Albert J. Beveridge Award, 712
American Hist Assoc Albert J. Beveridge Grants for Research in the History of the W Hemisphere, 713
American Hist Assoc Award for Scholarly Distin, 714
American Hist Assoc Bernadotte E. Schmitt Research Grants in European, African, or Asian History, 715
American Hist Assoc Clarence H. Haring Prize, 716
American Hist Assoc Fellow in Aerospace History, 717
American Hist Assoc Folger Shakespeare Library Fellowship, 718
American Hist Assoc George L. Mosse Prize, 719
American Hist Assoc George Louis Beer Prize, 720
American Hist Assoc Helen and Howard R. Marraro Prizes, 721
American Hist Assoc Assoc Herbert Feis Award, 723
American Hist Assoc J. Franklin Jameson Award In Outstanding Editorial Achievement, 724
American Hist Assoc J.F. Jameson Fellowship, 725
American Hist Assoc Assoc J. Russell Major Prize, 726
American Hist Assoc James A. Rawley Prize in Atlantic History, 727
American Hist Assoc James H Robinson Prize, 728
American Hist Assoc James Henry Breasted Prize, 729
American Hist Assoc Joan Kelly Memorial Prize in Women's History, 730
American Hist Assoc John F. Richards Prize in South Asian History, 731
American Hist Assoc John H. Dunning Prize, 732
American Hist Assoc John K. Fairbank Prize in East Asian History, 733
American Hist Assoc Assoc Leo Gershoy Award, 734
American Hist Assoc Littleton-Griswold Prize in American Law and Society, 735
American Hist Assoc Littleton-Griswold Research Grants in U.S. Legal History, 736
American Hist Assoc Martin A. Klein Prize in African History, 737
American Hist Assoc Michael Kraus Research Grant in American Colonial History, 738
American Hist Assoc Morris D. Forkosch Prize, 739
American Hist Assoc Paul Birdsall Prize in European Military and Strategic History, 740
American Hist Assoc Premio Del Rey, 741
American Hist Assoc Waldo J. Leland Prize, 743
American Hist Assoc Assoc Wesley-Logan Prize, 744
American Indian Studies Research Inst Fellowship, 746
American Jewish Historical Society Awards, 747
American Legacy Foundation National Calls for Proposals Grants, 748
American Legacy Found Small Innovative Grants, 749
American Mathematical Society AAAS Congressional Fellowships, 750
American Mathematical Society AAAS Mass Media Fellowships, 751
American Mathematical Society China Exchange Travel Grants, 753
American Mathematical Society Graduate Student Travel Grants, 754
American Mathematical Society International Congress of Mathematicians Travel Grants, 755
American Math Society Simons Travel Grants, 756
American Montessori Society Thesis and Dissertation Awards, 758
American Numismatic Society Donald Groves Numismatic Research Grants, 760
American Philosophical Society Franklin Grants, 764
American Philosophical Soc Judson Daland Prize, 766
American Philosophical Society Lewis and Clark Fund for Exploration and Field Research, 767
American Philosophical Society Lewis and Clark Exploration and Field Research in Astrobiology, 768
American Philosophical Society Phillips Fund Grants for Native American Research, 770
American Psychiatric Association/AstraZeneca Young Minds in Psychiatry International Awards, 773
American Psychiatric Association/Janssen Resident Psychiatric Research Scholars, 774
American Psychiatric Association/Lilly Resident Research Awards, 777
American Psychiatric Association/Merck Co. Early Academic Career Research Award, 778
American Psychiatric Association Award for Research in Psychiatry, 781
American Psychiatric Association Minority Medical Student Fellowship in HIV Psychiatry, 784
American Psychiatric Association Minority Medical Student Summer Externship in Addiction Psychiatry, 785
American Psychiatric Association Minority Medical Student Summer Mentoring Program, 786
American Psychiatric Association Program for Minority Research Training in Psychiatry, 787
American Psychiatric Association Research Colloquium for Junior Investigators, 788
American Psychiatric Foundation Grants, 791
American Psychiatric Foundation James H. Scully Jr., M.D., Educational Fund Grants, 792
American Psychiatric Foundation Jeanne Spurlock Congressional Fellowship, 793
American Psychiatric Foundation Minority Fellowships, 794
American Schlafhorst Foundation Grants, 796
American Society on Aging Graduate Student Research Award, 801
American Sociological Association Congressional Fellowships, 803
American Sociological Association Fund for the Advancement of the Discipline, 804
American Sociological Association Minority Fellowships, 805
American Sociological Association Sydney S. Spivack Applied Social Research and Social Policy, 806
American Sociological Association Teaching Enhancement Fund Grants, 807
amfAR Fellowships, 809
amfAR Mathilde Krim Fellowships in Basic Biomedical Research, 810
amfAR Research Grants, 811
Amgen Foundation Grants, 812
AMI Semiconductors Corporate Grants, 813
AMNH Graduate Student Fellowships, 815
AMNH Lerner-Gray Grants for Marine Research, 816
AMS/Industry Government Graduate Fellowships, 817
AMS Eugene K. Wolf Travel Fund for European Research Grants, 823
AMS Graduate Fellowships in History of Science, 824
AMS Harold Powers World Travel Fund Grants, 826
AMS Janet Levy Fund for Independent Scholars, 828
AMS M. Elizabeth C. Bartlet Fund for Research in France, 830
AMSSM Foundation Research Grants, 836
AMS Undergraduate Scholarships, 837
AMWA Kathryn C. Bemmann New Investigator Grant on Violence against Women, 838
ANA/Grass Foundation Award in Neuroscience, 839
Andrew Family Foundation Opportunity Knocks Grants, 843
Andrew W. Mellon/ACLS Dissertation Completion Fellowships, 844
Andy Warhol Foundation for the Visual Arts Curatorial Research Fellowships, 847
Angels Baseball Foundation Grants, 848
Animal Behavior Society Developing Nations Research Grants, 850
Animal Behavior Society Latin American Travel Awards, 852
Animal Behavior Society Student Research Grants, 853
ANL Laboratory-Grad Research Appointments, 857
Anna-Greta & Holger Crafoords Found Grants, 858
Ann & Robert H. Lurie Family Found Grants, 859
Ann Arbor Area Community Foundation Grants, 860
Ann Peppers Foundation Grants, 863
ANS/AAO-HNSF Herbert Silverstein Otology and Neurotology Research Award, 864
ANS Neurotology Fellows Award, 865
Anthem Blue Cross and Blue Shield School Nurse Grants, 868
Anthony R. Abraham Foundation Grants, 869
Anthropological Center for Training in Global and Environmental Change Grad Assistantships, 870
AOCS ACI/NBB Innovation Research Award, 872
AOCS Edible Applications Technology Division Outstanding Achievement Award, 877
AOCS Health & Nutrition Div Student Award, 881
AOCS Phospholipid Best Paper Award, 889
AOS Research Grants, 903
APA Culture of Service in the Psychological Sciences Award, 904
APA Distinguished Scientific Award for Early Career Contribution to Psychology, 906
APA Distinguished Scientific Award for the Applications of Psychology, 907
APA Distinguished Scientific Contribution Award, 908
APA Distinguished Service to Psychological Science Award, 909
APA Young Investigator Grants, 912
APHL Emerging Infectious Diseases Fellowships, 914
Applied Biosystems Grants, 916
APS/British Academy Fellowship for Research in London, 918
APSAA Fellowships, 920
APSAA Mini-Career Grants, 921
APSAA Research Grants, 922
APSAA Small Beginning Scholar Pre-Investigation Grants, 923
APSAA Small Beginning Scholar Visiting and Consulting Grants, 924
APSA Congressional Fellowships for Communications Scholars and Journalists, 926
APSA Congressional Fellowships for Federal Executives, 927
APSA Congressional Fellowships for International Scholars, 928
APSA Congressional Fellowships for Journalists, 929
APSA Congressional Fellowships for Political Scientists, 930
APSA Congressional Health and Aging Policy Fellowships, 931
APS Andreas Acrivos Dissertation Award in Fluid Dynamics, 940
APS Aneesur Rahman Prize for Computational Physics, 942
APS Arthur L. Schawlow Prize in Laser Science, 945
APSA Small Research Grants, 946
APS Dannie Heineman Prize for Mathematical Physics, 951
APS Davisson-Germer Prize in Atomic or Surface Physics, 952
APS Dissertation Award in Nuclear Physics, 953
APS E.K. Plyler Prize for Molecular Spectroscopy, 954
APS Einstein Prize, 955
APS Fluid Dynamics Prize, 956
APS Frank Isakson Prize for Optical Eff in Solids, 957
APS George E. Valley, Jr., Prize, 959

APS Hans A. Bethe Prize, 960
APS Herbert P. Broida Prize, 961
APS I.I. Rabi Prize in Atomic, Molecular, and Optical Physics, 962
APS Irving Langmuir Prize in Chemical Physics, 963
APS J.J. Sakurai Prize for Theoretical Part Physics, 964
APS James C. McGroddy Prize for New Materials, 965
APS James C Maxwell Prize for Plasma Physics, 966
APS Lars Onsager Prize, 968
APS Marshall N. Rosenbluth Outstanding Doctoral Thesis Award, 969
APS Max Delbruck Prize in Biological Physics, 970
APS Mitsuyoshi Tanaka Dissertation Award in Experimental Particle Physics, 971
APS Nicholas Metropolis Award for Outstanding Doctoral Thesis in Computational Physics, 972
APS Outstanding Doctoral Thesis Research in Atomic, Molecular, or Optical Physics, 974
APS Outstanding Doctoral Thesis Research in Beam Physics Award, 975
APS Polymer Physics Prize, 976
APS Prize for a Faculty Member for Research in an Undergraduate Institution, 977
APS Prize for Industrial Applications of Physics, 978
APS Robert R. Wilson Prize for Achievement in the Physics of Particle Accelerators, 979
APS Tom W. Bonner Prize in Nuclear Physics, 980
APS W.K.H. Panofsky Prize in Experimental Particle Physics, 981
Arcadia Foundation Grants, 983
Arecibo Observatory Grants, 988
Arie and Ida Crown Memorial Grants, 989
ARIT Kenan T. Erim Fellowship for Research at Aphrodisias, 993
Arizona Cardinals Grants, 995
Arizona Public Service Corporate Giving Grants, 998
ARO Atomic, Molecular, and Optical Physics Research Contracts, 1005
A Rocha USA Grants, 1006
ARO Degradation, Reactivity, and Protection of Materials Research Contracts, 1008
ARO Discrete Mathematics and Computer Science Research Contracts, 1009
ARO Information Processing and Fusion and Circuits Research Contracts, 1010
ARO Minimum Energy Low Power Electronics Research Contracts, 1011
ARO Mobile, Wireless Communications and Networks Research Contracts, 1012
ARO Quantum Information Science Research Contracts, 1013
ARO Short-Term Innovation Research, 1014
ARO Software and Knowledge-Based Systems Research Contracts, 1015
ARO Synthesis and Processing of Materials Research Contracts, 1016
ARO Theoretical Physics and Nonlinear Phenomena Research Contracts, 1017
Arronson Foundation Grants, 1019
ARS New Investigator Award, 1020
ARS Resident Research Grants, 1021
Arthritis Foundation of Australia Grants, 1022
Arthur F. and Alice E. Adams Charitable Foundation Grants, 1024
Arthur J. Ennis Fellowships, 1025
ArvinMeritor Foundation Health Grants, 1027
ASA/BJS Stat Methodological Research Grants, 1029
ASA/NSF/BLS Fellowships, 1030
ASA/NSF/Census Research Fellowships, 1031
ASA/NSF Federal Statistics Fellowships, 1032
ASA and USDA/NASS Research Fellow, 1033
ASCO/UICC International Cancer Technology Transfer Fellowships, 1044
ASCO Advanced Clinical Research Award in Colorectal Cancer, 1045
ASCO Advanced Clinical Research Awards in Breast Cancer, 1046
ASCO Advanced Clinical Research Awards in Sarcoma, 1047
ASCO Merit Awards, 1049
ASCO Young Investigator Award, 1050
ASEE Naval Research Laboratory Postdoctoral Fellowships, 1060
ASGE / ConMed Award for Outstanding Manuscript by a Fellow/Resident, 1061
ASGE Don Wilson Award, 1062
ASGE Endoscopic Research Awards, 1063
ASGE Endoscopic Research Career Development Awards, 1064
ASGE Given Capsule Endoscopy Award, 1065
ASHA Advancing Academic Careers Award, 1072
ASHA Research Mentoring-Pair Travel Award, 1073
ASHA Student Research Travel Award, 1074
ASHA Students Preparing for Academic & Research Careers Award, 1075
ASHFoundation Clinical Research Grants, 1076
ASHFoundation New Century Scholars Program Doctoral Scholarships, 1080
ASHFoundation New Century Scholars Research Grant, 1081
ASHFoundation Grant for New Investigators, 1082
ASHFoundation Grant in Speech Science, 1083
ASHFoundation Student Research Grants in Audiology, 1084
ASHFoundation Student Research Grants in Early Childhood Language Development, 1085
Asian Pacific American Institute for Congressional Studies Fellowships, 1091
ASID Foundation Grants, 1093
ASID Foundation Irene Winifred Eno Grant, 1094
ASID Foundation Joel Polsky Prize, 1096
ASM-PAHO Infectious Diseases Epidemiology and Surveillance Fellowships, 1098
ASM-UNESCO Leadership Grant for International Educators, 1099
ASM/CDC Fellowships in Infectious Disease and Public Health Microbiology, 1100
ASM Abbott Laboratories Award in Clinical and Diagnostic Immunology, 1101
ASME Barnett-Uzgiris Product Safety Award, 1106
ASME Daniel C. Drucker Medal, 1114
ASME Engineer-Historian Award, 1118
ASME Fluids Engineering Award, 1119
ASME Kate Gleason Award, 1137
ASME M. Eugene Merchant Manuf Medal, 1139
ASME Robert E. Koski Medal, 1159
ASME Thomas A. Edison Patent Award, 1167
ASME Warner T. Koiter Medal, 1170
ASME Y.C. Fung Young Investigator Award, 1173
ASM International Fellowship for Latin America and the Caribbean, 1175
ASM Microbiology Undergraduate Research Fellowship, 1178
ASM Millis-Colwell Postgraduate Travel Grant, 1179
ASM Robert D. Watkins Graduate Research Fellowship, 1183
ASM Undergraduate Research Fellowship, 1186
A Social Corporation Grants, 1189
ASPEN C. Richard Fleming Grant, 1190
ASPEN Douglas Wilmore Grant, 1191
Aspen Institute William Randolph Hearst Endowed Fellowship for Minority Students, 1194
ASPEN Maurice Shils Grant, 1195
ASPEN Norman Yoshimura Grant, 1196
ASPEN Nutrition Research Grants, 1197
ASPET-Astellas Awards in Translational Pharmacology, 1200
ASPET Bernard B. Brodie Award in Metabolism, 1201
ASPET Division for Drug Metabolism Early Career Achievement Award, 1202
ASPET Epilepsy Research Award for Outstanding Contributions to the Pharmacology of Antiepileptic Drugs, 1203
ASPET P. B. Dews Lifetime Achievement Award for Research in Behavioral Pharmacology, 1206
ASPET Paul M. Vanhoutte Award, 1207
ASPET Summer Undergrad Res Fellowships, 1208
ASPET Sollmann Award in Pharmacology, 1209
ASPO Daiichi Innovative Technology Grant, 1210
ASPO Fellowships, 1211
ASPO Research Grants, 1212
Assisi Foundation of Memphis General Grants, 1221
ASTA Academic Scholarships, 1224
Astellas Foundation Research Grants, 1227
ASU Grad Doctoral Enrichment Fellowships, 1238
ASU Institute for Humanities Research Grants, 1247
AT&T Industrial Ecology Faculty Fellowships, 1251
ATA Educational Trust Grants - Curriculum Resource or Research Project, 1253
ATA Education Research Award, 1254
ATA Nadene M. Thomas Grad Research Bursary, 1255
ATLA Bibliography Grant, 1257
ATS Lilly Theological Collaborative Res Grants, 1259
ATS Lilly Theological Faculty Fellowships, 1260
ATS Lilly Theological Research Expense Grants, 1261
ATS Lilly Theological Scholars Grants, 1262
Audubon Naturalist Society Research Grants, 1266
Australasian Institute of Judicial Administration Seed Funding Grants, 1268
Australian Academy of Science Grants, 1269
Australian Postgraduate Research Fellowships, 1271
Aviation Research Grants, 1273
Avon Foundation Breast Care Fund Grants, 1274
BA Elisabeth Barker Fund, 1288
Bailey Family Foundation Grants, 1289
BA Larger Research Grants, 1290
Ball Brothers Foundation General Grants, 1291
Ball Brothers Foundation Rapid Grants, 1292
Baltimore Washington Center Adult Psychoanalysis Fellowships, 1293
Baltimore Washington Center Child Psychotherapy Fellowships, 1294
Banfield Charitable Trust Grants, 1295
BARD Research Grants, 1299
Barra Foundation Project Grants, 1301
Barth Syndrome Foundation Research Grants, 1302
BA Small Research Grants, 1304
Batchelor Foundation Grants, 1305
Bayer Foundation Grants, 1309
BCBSM Foundation Investigator Initiated Research Grants, 1313
BCBSM Foundation Physician Investigator Research Awards, 1314
BCBSM Foundation Primary and Clinical Prevention Grants, 1315
BCBSM Foundation Student Award Program, 1316
BCBSM Frank J. McDevitt, DO, Excellence in Research Awards, 1317
BCBSNC Foundation Grants, 1318
BCBS of Massachusetts Foundation Grants, 1319
BCRF-AACR Grants for Translational Breast Cancer Research, 1321
BCRF Research Grants, 1322
BC Steven H. Sandell Grants, 1323
Beckman Coulter Community Relations Grants, 1325
Beckman Scholars Grants, 1326
Beckman Young Investigators Grants, 1327
Becton Dickinson and Company Grants, 1328
Bedding Plants Foundation Research Grants, 1330
Beim Foundation Grants, 1331
Berks County Community Foundation Grants, 1333
Bernard Lowy Fund for the Study of Tropical Botany in Latin America, 1335
Beryl Henderson Foundation Grants, 1337
Better Way Foundation Grants, 1338
Bikes Belong Foundation Research Grants, 1343
Bill and Melinda Gates Foundation Agricultural Development Grants, 1345
Bill and Melinda Gates Foundation Financial Services for the Poor Grants, 1346
Bill and Melinda Gates Foundation Water, Sanitation and Hygiene Grants, 1347
Blowitz-Ridgeway Foundation Early Childhood Development Research Award, 1348
Blowitz-Ridgeway Foundation Grants, 1349
Blue Shield of California Grants, 1350
Blum-Kovler Foundation Grants, 1351
BMW of North America Contributions, 1353
Bodenwein Public Benevolent Found Grants, 1355
Bodman Foundation Grants, 1356

Booth-Bricker Fund Grants, 1359
Bosque Foundation Grants, 1360
Boston Psychoanalytic Society and Institute Fellowship in Child Psychoanalytic Psychotherapy, 1362
Boston Psychoanalytic Society and Institute Fellowship in Psychoanalytic Psychotherapy, 1363
Bower Award & Prize Achievement in Science, 1364
Brainerd Foundation Grants, 1368
Breast Cancer Fund Grants, 1369
Brico Fund Grants, 1371
Bristol-Myers Squibb Clinical Outcomes and Research Grants, 1374
Bristol-Myers Squibb Foundation Fellowships, 1375
Bristol-Myers Squibb Foundation Global HIV/AIDS Initiative Grants, 1376
Bristol-Myers Squibb Foundation Health Disparities Grants, 1377
Bristol-Myers Squibb Women's Health Grants, 1379
Broad Foundation IBD Research Grants, 1380
Brookdale Foundation Leadership in Aging Fellowships, 1381
Brookhaven National Laboratory Sambamurti Memorial Lectureships, 1382
Bunting Fellowship Program, 1385
Bupa Foundation Medical Research Grants, 1386
Bupa Foundation Multi-Country Grant, 1387
Burden Trust Grants, 1388
Burton G. Bettingen Grants, 1390
BWF Ad Hoc Grants, 1392
BWF Career Awards at the Scientific Interface, 1393
BWF Career Awards for Medical Scientists, 1394
BWF Collaborative Research Travel Grants, 1395
BWF Innovation in Regulatory Science Grants, 1396
BWF Institutional Program Unifying Population and Laboratory Based Sciences Grants, 1397
BWF Investigators in the Pathogenesis of Infectious Disease Awards, 1398
BWF Preterm Birth Initiative Grants, 1399
Bydale Foundation Grants, 1400
Byron W. & Alice L. Lockwood Found Grants, 1401
BYU Religious Studies Center Research Grants, 1402
Cabot Corporation Foundation Grants, 1406
Caesars Foundation Grants, 1407
Callaway Golf Company Foundation Grants, 1410
Calvin Inst of Christian Worship Renewal Grants, 1411
Campbell Soup Foundation Grants, 1415
Canada-U.S. Fulbright Trad Scholar Awards, 1419
Canada Graduate Scholarships and NSERC Postgraduate Scholarships, 1420
Canadian Patient Safety Institute Res Grants, 1421
CAPAA Mini Grant for Research Related to African Americans in Education, 1424
Cape Branch Foundation Grants, 1425
Carl and Eloise Pohlad Family Found Grants, 1426
Carl Gellert and Celia Berta Gellert Foundation Grants, 1429
Carl M. Freeman Foundation Grants, 1430
Carnegie Corporation of New York Grants, 1432
Carole Fielding Film/Video Production or Research Grants, 1438
Carrier Corporation Contributions Grants, 1440
Castle Rock Foundation Grants, 1441
Catherine Holmes Wilkins Foundation Grants, 1442
CCFF Chairman's Disting Life Sciences Award, 1449
CCFF Distinguished Ag Scientist Award, 1450
CCWH Catherine Prelinger Award, 1454
CDC Cooperative Agreement for Continuing Enhanced National Surveillance for Prion Diseases in the United States, 1456
CDC Cooperative Agreement for the Development, Operation, and Evaluation of an Entertainment Education Program, 1457
CDC Evaluation of the Use of Rapid Testing For Influenza in Outpatient Medical Settings, 1458
CDC Evidence-Based Laboratory Medicine: Quality/ Performance Measure Evaluation, 1459
CDC Grants for Violence-Related Injury Prevention Research, 1463
CDC National Training Information and Exchange Program, 1464
CDC Preventive Med Residency Fellowship, 1465
CDC Public Health Conference Support Grants, 1466
CDC Steven M. Teutsch Prevention Effectiveness Fellowships, 1467
CDI Postdoctoral Fellowships, 1469
Cedar Tree Foundation David H. Smith Conservation Research Fellowship, 1470
Center for Education of Women Visiting Scholar Research Grants, 1475
Center of Excellence in General Aviation Grants, 1486
Center on Philanthropy and Civil Society's Emerging Leaders International Fellows Program, 1487
Century Foundation Grants, 1489
CFF-NIH Funding Grants, 1490
CFF Clinical Research Grants, 1491
CFF Harry Shwachman Clinical Invest Award, 1493
CFF Pilot and Feasibility Awards, 1495
CFF Research Grants, 1497
CFF Student Traineeships, 1498
CFR Edward R. Murrow Press Fellowship, 1500
CFR International Affairs Fellowships, 1501
CFR International Affairs Fellowships in Nuclear Security, 1502
CFR Int'l Affairs Fellowships in South Korea, 1503
CFR International Affairs in Japan Fellowships, 1504
CFR Military Fellowships, 1505
CFR Stanton Nuclear Security Fellowships, 1506
CFUW Memorial Fellowship, 1515
CFUW Ruth Binnie Fellowship, 1516
Charles A. King Postdoc Research Fellowships, 1518
Charles Edison Fund Grants, 1520
Charles H. Farnsworth Trust Grants, 1521
Charles H. Hood Foundation Child Health Research Awards, 1522
Charles H. Price II Family Foundation Grants, 1523
Charles H. Revson Foundation Grants, 1524
Charles Lafitte Foundation Grants, 1525
Charles Stewart Mott Anti-Poverty Grants, 1528
Chest Foundation/LUNGevity Foundation Clinical Research in Lung Cancer Grants, 1534
Chest Found Eli Lilly & Company Distinguished Scholar in Critical Care Medicine Award, 1535
Chest Foundation Geriatric Development Research Awards, 1536
Chest Found Grant in Venous Thromboembolism, 1537
Chestnut Hill Charitable Foundation, Inc Grants, 1538
CHF Charles C. Price Fellow in Polymer History, 1539
CHF Glenn E. and Barbara Hodsdon Ullyot Scholarship, 1540
CHF Long-Term Fellowships, 1541
CHF Robert W. Gore Fellowship in Materials Innovation, 1542
CHF Roy G. Neville Short-Term Fellowships, 1543
CHF Societe de Chimie Industrielle (American Section) Fellowship, 1544
Chicago Community Trust Housing Grants: Advancing Affordable Rental Housing, 1547
Chicago Institute for Psychoanalysis Fellowships, 1548
Children's Cardiomyopathy Foundation Research Grants, 1550
Children's Lit Association Research Fellowships, 1551
Chiles Foundation Grants, 1552
Chiron Foundation Community Grants, 1553
Christine O. Gregoire Youth/Young Adult Award for Outstanding Use of Tobacco Ind Documents, 1557
CIA Undergraduate Co-Op Program - Engineer/ Program Management Engineer, 1561
CIC Summer Research Opportunities, 1565
CIC Traveling Scholars, 1566
CIRCLE Civic Education at the High School Level Research Grants, 1571
CJ Foundation for SIDS Research Grants, 1573
Claneil Foundation Grants, 1574
Clarence E. Heller Foundation Grants, 1575
Clark-Huntington Bibliographical Fellowship, 1576
Claude Pepper Foundation Grants, 1578
Cleveland-Cliffs Foundation Grants, 1580
Clowes ACS/AAST/NIGMS Jointly Sponsored Mentored Clinical Scientist Devel Award, 1581
CMA Foundation Grants, 1582
CMS Hispanic Health Services Research Grants, 1584
CMS Historically Black Colleges and Universities Health Services Research Grants, 1585
CMS Research and Demonstration Grants, 1586
CNIB Baker Applied Research Fund Grants, 1588
CNIB Baker New Researcher Fund Grants, 1589
CNIB Barbara Tuck MacPhee Award, 1590
CNIB Canada Glaucoma Clinical Research Council Grants, 1591
CNIB E. & Mary Hochhausen Access Technology Research Grants, 1592
Coeta and Donald Barker Foundation Grants, 1594
Coleman Foundation Cancer Care Grants, 1595
Colonel Stanley R. McNeil Foundation Grants, 1597
Colorado Bioscience Discovery Eval Grants, 1598
Colorado Clean Energy Solar Innov Grants, 1599
Columbus Foundation Ann Ellis Fund Grants, 1600
Commonweal Foundation Community Assistance Grants, 1602
Commonweal Foundation Grants, 1603
Commonwealth Fund Affordable Health Insurance Grants, 1605
Commonwealth Fund Harkness Fellowships in Health Care Policy and Practice, 1606
Commonwealth Fund Health Care Quality Improvement and Efficiency Grants, 1607
Commonwealth Payment System Reform Grants, 1609
Commonwealth Fund Quality of Care for Frail Elders Grants, 1610
Commonwealth Fund Small Grants, 1611
Commonwealth Fund State High Performance Health Systems Grants, 1612
Communities Foundation of Texas Grants, 1613
Community Found AIDS Endowment Awards, 1614
Community Found for Greater Buffalo Grants, 1615
Community Fnd for Muskegon County Grants, 1617
Community Foundation for S Arizona Grants, 1618
Community Foundation of the Virgin Islands Judith A. Towle Environmental Studies Grants, 1621
Conquer Cancer Foundation of ASCO Career Development Award, 1630
Conquer Cancer Foundation of ASCO Comparative Effectiveness Research Professorship in Breast Cancer, 1631
Conquer Cancer Foundation of ASCO Drug Development Research Professorship, 1632
Conquer Cancer Foundation of ASCO Improving Cancer Care Grants, 1633
Conquer Cancer Foundation of ASCO Int'l Devel & Educ Award in Palliative Care, 1634
Conquer Cancer Foundation of ASCO International Innovation Grant, 1636
Conquer Cancer Foundation of ASCO Medical Student Rotation Grants, 1637
Conquer Cancer Foundation of ASCO Translational Research Professorships, 1639
Conservation, Food, and Health Foundation Grants for Developing Countries, 1641
Coors Brewing Corporate Contributions Grants, 1644
Cord Foundation Grants, 1645
Cowles Charitable Trust Grants, 1652
CPC Catherine H. Beattie Fellowship, 1653
CRF Project Research Grants, 1656
Crohn's Disease and Ulcerative Colitis Senior Research Grants, 1658
Croucher Foundation Grants, 1659
CSTE CDC/CSTE Applied Epidemiology Fellowships, 1664
CTCRI Idea Grants, 1665
CTCRI Knowledge Synthesis Grants, 1666
CTCRI Policy Research Grants, 1667
CTCRI Researcher Travel Grants, 1668
CTCRI Research Planning Grants, 1669
CTCRI Student Research Grants, 1670
CTCRI Workshop and Learning Opportunities Grants, 1671
Cudd Foundation Grants, 1672
Cystic Fibrosis Canada Clinical Fellowships, 1674
Cystic Fibrosis Canada Clinical Project Grants, 1675
Cystic Fibrosis Canada Clinic Incentive Grants, 1676

Cystic Fibrosis Canada Fellowships, 1677
Cystic Fibrosis Canada Research Grants, 1678
Cystic Fibrosis Canada Scholarships, 1679
Cystic Fibrosis Canada Senior Scientist Research Training Awards, 1680
Cystic Fibrosis Canada Small Conference Grants, 1681
Cystic Fibrosis Canada Studentships, 1683
Cystic Fibrosis Canada Summer Studentships, 1684
Cystic Fibrosis Canada Travel Supplement Grants, 1685
Cystic Fibrosis Canada Visiting Allied Health Professional Awards, 1686
Cystic Fibrosis Canada Visiting Clinician Awards, 1687
Cystic Fibrosis Canada Visiting Scientist Awards, 1688
Cystic Fibrosis Elizabeth Nash Fellowships, 1689
Cystic Fibrosis New Horizons Campaign Grants, 1690
Cystic Fibrosis Trust Research Grants, 1691
D.F. Halton Foundation Grants, 1692
DAAD-AICGS Research Fellowships, 1693
DAAD-Leo Baeck Institute Fellowships, 1694
DAAD Bilateral Co-Operative Research Grants, 1695
DAAD Emigre Memorial German Internship Program - Bundestag, 1696
DAAD German Studies Research Grants, 1698
DAAD Re-invitation Program for Former Scholarship Holders, 1701
DAAD Research Grants, 1702
DAAD Research Grants for Doctoral Candidates and Young Academics and Scientists, 1703
DAAD Research Internships in Science and Engineering, 1704
DAAD Research Stays for University Academics and Scientists, 1705
DAAD RISE Professional Internships, 1706
DAAD Study & Internship in Germany, 1708
Daimler and Benz Foundation Fellowships, 1713
Dana Foundation Science and Health Grants, 1716
Daniel Mendelsohn New Investigator Award, 1717
Danone Institute Belgium Research in Human Nutrition Grants, 1719
Dave Thomas Foundation for Adoption Grants, 1723
Davis Family Foundation Grants, 1726
Davis Foundation Postdoctoral Fellowships in Eating Disorders Research, 1727
Dean Witter Foundation Grants, 1728
Deborah Munroe Noonan Research Grants, 1729
DEED Grants, 1730
Del E. Webb Foundation Grants, 1733
Delta Air Lines Health and Wellness Grants, 1736
DeMatteis Family Foundation Grants, 1740
Denton A. Cooley Foundation Grants, 1743
Dept of Ed Jacob K. Javits Gifted and Talented Students Education Program-National Research and Development Center, 1746
Detroit Lions Charities Grants, 1753
DHHS Emergency Med Services for Children, 1755
DHHS Independence Demonstration Program, 1757
DHHS Native American Programs, 1758
DHHS Oral Health Promotion Research Across the Lifespan, 1759
DHHS Promoting Safe & Stable Families Grants, 1760
DHHS Special Programs for the Aging Training, Research, and Discretionary Projects Grants, 1761
DHHS Welfare Reform Research, Evaluations, and National Studies Grants, 1762
DOD HBCU/MI Partnership Training Award, 1765
DOE Advanced Detector Research Grants, 1767
DOE Advanced Energy Projects/Laboratory Technology Research Grants, 1768
DOE Advanced Nuclear Energy Systems Basic Research Grants, 1769
DOE Albert Einstein Distinguished Educator Fellowship Program, 1770
DOE Atomic, Molecular, and Optical Sciences Research Grants, 1771
DOE Basic Energy Sciences Grants, 1772
DOE Catalysis and Chemical Transformations Research Grants, 1773
DOE Chemical Energy and Chemical Engineering Research Grants, 1774
DOE Chemical Physics Research Grants, 1775
DOE Chemical Sciences Research Grants, 1776
DOE Combustion Systems Research Grants, 1778
DOE Competitive Financial Assistance for Renewable Energy and Energy Efficiency Programs, 1779
DOE Control Technology and Coal Preparation Research Grants, 1781
DOE Electric Energy Systems Research Grants, 1782
DOE Energy-Related Inventions Grants, 1783
DOE Energy Biosciences Research Grants, 1784
DOE Energy Conservation Research & Dev Grants, 1785
DOE Energy Conversion Tech Research Grants, 1786
DOE Energy from Municipal Waste Grants, 1787
DOE Energy Research Analyses Grants, 1788
DOE Engineering Research Grants, 1789
DOE Environmental Processes Research Grants, 1790
DOE Experimental Program to Stimulate Competitive Research, 1791
DOE Faculty and Student Teams Grants, 1792
DOE Fossil Advanced Research and Technology Development Grants, 1793
DOE Fossil Energy Research & Devel Grants, 1794
DOE Fusion Energy Science Division Grants, 1795
DOE Fusion Energy Science Technology Division Grants, 1796
DOE Geosciences Research Grants, 1797
DOE Heavy Element Chem Research Grants, 1798
DOE High-Energy Physics Research Grants, 1799
DOE Industrial Materials for the Future Grants, 1800
DOE Initial H-Prize Competition for Breakthrough Advances in Materials for Hydrogen Storage, 1801
DOE Inventions and Innovation Grant, 1802
DOE Materials Sciences and Engineering Research Grants, 1803
DOE Mathematical, Information, and Computational Sciences Grants, 1804
DOE Nonproliferation & National Security Research Grants, 1806
DOE Nuclear Physics Research Grants, 1807
DOE On-Line Temperature Measurement Instrumentation for Gasification Process Control Development Grants, 1808
DOE Photochem & Radiation Research Grants, 1809
DOE Program for Ecosystem Research, 1810
DOE Programs in Fusion Energy Sciences, 1811
DOE Reliable Electric Power Delivery Research Grants, 1812
DOE Separations and Analysis Research Grants, 1813
DOE Small Business Innovation Research Commercialization Assistance Program, 1814
DOE State Energy Program Special Projects, 1815
DOE University Coal Research Grants, 1816
DOE Univ Nuclear Science & Reactor Support, 1817
DOE Univ Reactor Instrumentation Grants, 1818
DOE Vehicle and Engine Research and Development Grants, 1819
DOI Earthquake Hazards Reduction Grants, 1821
DOI Geological Research and Data Acquisition Grants, 1822
DOJ Juvenile Justice and Delinquency Prevention Special Emphasis Grants, 1826
Donald and Sylvia Robinson Family Foundation Grants, 1830
Donald W. Reynolds Foundation Aging and Quality of Life Grants, 1832
Doris Duke Charitable Foundation Clinical Interfaces Award Program, 1833
Doris Duke Charitable Foundation Clinical Research Fellowships for Medical Students, 1834
Doris Duke Charitable Foundation Clinical Research Systems Grants, 1835
Doris Duke Charitable Foundation Clinical Scientist Development Award, 1836
Doris Duke Charitable Foundation Clinical Scientist Development Award Bridge Grants, 1837
Doris Duke Charitable Foundation Distinguished Clinical Scientist Award Program, 1838
Doris Duke Foundation Operations Research on AIDS Care & Treatment in Africa Grants, 1839
Dorothy Leet Grants, 1840
Dorothy Rider Pool Health Care Grants, 1841
Dorr Foundation Grants, 1842
Dorr Institute for Arthritis Research & Educ, 1843
DOT Recreational Boating Safety Grants, 1848
DOT Transit Planning and Research Grants, 1849
DOT Transportation Statistics Research Grants, 1850
Dr. Bob and Jean Smith Foundation Grants, 1853
Dreyfus Postdoctoral Program in Environmental Chemistry Grant, 1855
Dreyfus Prize in the Chemical Sciences, 1856
Dreyfus Senior Scientist Mentor Grants, 1857
Dreyfus Special Grants in the Chemical Sciences, 1858
Drs. Bruce and Lee Foundation Grants, 1859
DSO Cogn Tech Threat Warning System Grants, 1860
DSO Computer Science Study Group Grants, 1861
DSO Controlling Pathogen Evolution Workshop Grants, 1862
DSO Defense Sciences Research & Tech Grants, 1863
DSO Radiation Biodosimetry Grants, 1864
Duchossois Family Foundation Grants, 1867
Duke Foundation Fellowships, 1868
Duke University Ambulatory and Regional Anesthesia Fellowships, 1870
Duke University Clinical Cardiac Electrophysiology Fellowships, 1871
Duke University Hyperbaric Center Fellowships, 1872
Duke Univ Interventional Cardio Fellowships, 1873
Duke University Libraries Franklin Collection Travel Grants, 1874
Duke University Libraries Hartman Center Travel Grants, 1875
Duke University Libraries J. Walter Thompson Research Fellowships, 1876
Duke University Libraries Mary Lily Res Grants, 1877
Duke Univ Obstetric Anesthesia Fellowships, 1878
Duke University Pain Management Fellowships, 1879
Duke University Postdoctoral Research Fellowships in Aging, 1880
E.W. "Al" Thrasher Awards, 1887
Earle B. Barnes Award for Leadership in Chemical Research Management, 1889
Eastern Europe & the Independent States of Former Soviet Union Research & Training Grants, 1891
Edmund Niles Huyck Preserve Graduate and Postgraduate Research Grants, 1894
Education Through Astronomy and Space Science Initiative, 1896
Edward Bangs Kelley and Elza Kelley Foundation Grants, 1898
Edward W. and Stella C. Van Houten Grants, 1899
Edwin S. Webster Foundation Grants, 1900
EFWA Accounting Research Grants, 1901
Ekhagastiftelsen Foundation Grants, 1903
Elias J. Corey Award for Outstanding Original Contribution in Organic Synthesis by a Young Investigator, 1904
Elisabeth Severance Prentiss Foundation Grants, 1905
Elizabeth Glaser Int'l Leadership Awards, 1906
Elizabeth Glaser Scientist Award, 1907
Elizabeth McGraw Foundation Grants, 1908
Elizabeth Nash Found Summer Research Awards, 1909
Ellison Medical Foundation/AFAR Julie Martin Mid-Career Award in Aging Research, 1910
Ellison Medical Foundation/AFAR Postdoctoral Fellows in Aging Research Program, 1911
EMBO / EMBL Science Writing Prize, 1913
EMBO Award for Communication in the Life Sciences, 1914
EMBO Installation Grants, 1915
Emerson Charitable Trust Grants, 1918
Emil Buehler Perpetual Trust Grants, 1919
Endangered Language Fund Grants, 1921
Energy Foundation Power Grants, 1924
Engineering Information Foundation Grants, 1927
Ensign-Bickford Foundation Grants, 1928
Enterprise Community Partners Rose Architectural Fellowships, 1929
EPA Capacity Building Grants for Compliance Assurance and Enforcement Activities in Indian Country and Other Tribal Areas, 1932

EPA Compliance Assistance Support for Services to the Regulated Community and Other Assistance Providers, 1934
EPA National Estuary Grants, 1935
EPA Pollution Prevention Grants, 1936
EPA Regional Healthy Indoor Air Projects for Community Outreach and Education, 1937
EPA Surveys, Studies, Research, Investigations, Demonstrations, and Special Purpose Activities Relating to the Clean Air Act, 1940
EPA Targeted Watersheds Grants, 1941
EPA Temporally Integrated Monitoring of Ecosystems and Long-Term Monitoring Grants, 1942
Equipment Leasing and Finance Foundation Research Grants, 1943
EREF Solid Waste Research Grants, 1947
EREF Sustainability Research Grants, 1948
Eugene McDermott Foundation Grants, 1958
Ewing Halsell Foundation Grants, 1964
Explorers Club Exploration Fund Grants, 1966
ExxonMobil Found Malaria Initiative Grants, 1968
Eye-Bank for Sight Restoration and Fight for Sight Summer Student Research Fellowship, 1969
Ezra Jack Keats/Kerlan Collection Fellowship, 1970
Facebook Fellowships, 1973
Fairlawn Foundation Grants, 1974
FAMRI Clinical Innovator Awards, 1975
FAMRI Young Clinical Scientist Awards, 1976
Fannie E. Rippel Foundation Grants, 1978
FAR Fund Grants, 1980
Farmers Insurance Corporate Giving Grants, 1981
FASSE Demonstration Project Grant, 1982
FCD Child Development Grants, 1984
FDHN Bridging Grants, 1986
FDHN Centocor International Research Fellowship in Gastrointestinal Inflammation & Immunology, 1987
FDHN Designated Outcomes Award in Geriatric Gastroenterology, 1988
FDHN Designated Research Award in Geriatric Gastroenterology, 1989
FDHN Designated Research Award in Research Related to Pancreatitis, 1990
FDHN Fellow Abstract Prizes, 1991
FDHN Funderburg Research Scholar Award in Gastric Biology Related to Cancer, 1993
FDHN Graduate Student Awards, 1994
FDHN Jon I. Isenberg International Research Scholar Award, 1995
FDHN June & Donald O. Castell MD, Esophageal Clinical Research Award, 1996
FDHN Moti L. & Kamla Rustgi International Travel Awards, 1997
FDHN Non-Career Research Awards, 1998
FDHN Non-Career Research Grants, 1999
FDHN Research Scholar Awards, 2000
FDHN TAP Endowed Designated Research Award in Acid-Related Diseases, 2002
FDHN Translational Research Awards, 2003
Fetzer Institute Research Grants, 2005
Fidelity Foundation Grants, 2007
Field Museum Graduate Student Fellowships, 2008
Field Museum Visiting Scholarships, 2011
Fight for Sight-Streilein Foundation for Ocular Immunology Research Award, 2012
Fight for Sight Grants-in-Aid, 2013
Fight for Sight Post-Doctoral Awards, 2014
Fight for Sight Summer Student Fellowships, 2015
Fischelis Grants for Research in the History of American Pharmacy, 2017
FishAmerica Foundation Marine and Anadromous Fish Habitat Restoration Grants, 2018
FishAmerica Foundation Research Grants, 2019
Fishman Family Foundation Grants, 2020
Flinn Foundation Grants, 2024
Florence Tan Moeson Fellowships, 2025
Florida Sea Turtle Grants, 2027
Ford Foundation Peace and Social Justice Grants, 2031
Forrest C. Lattner Foundation Grants, 2035
Foundation Fighting Blindness Marjorie Carr Adams Women's Career Development Awards, 2036
Foundation for Endangered Languages Grants, 2037
Foundation for Enhancing Comm Grants, 2038
Foundation for Seacoast Health Grants, 2041
Foundation for the Advancement of Mesoamerican Studies Grants, 2042
Foundation for the Future Grants, 2043
Foundation of the American Thoracic Society Research Grants, 2044
Frances C. and William P. Smallwood Grants, 2045
Franklin Mosher Baldwin Fellowships, 2050
Frank Stanley Beveridge Foundation Grants, 2051
Frederick Gardner Cottrell Foundation Grants, 2054
Freshwater Future Project Grants, 2056
Fritz B. Burns Foundation Grants, 2057
Fulbright/Garcia Robles Grants, 2063
Fulbright Alumni Initiatives Awards Grants, 2064
Fulbright Beginning Prof Journalism Awards, 2065
Fulbright Binational Business Grants in Mexico, 2066
Fulbright Business Grants in Spain, 2067
Fulbright Critical Lang Enhancement Awards, 2068
Fulbright English Teaching Assistantships, 2070
Fulbright mtvU Awards, 2074
Fulbright New Century Scholars Grants, 2075
Fulbright Occasional Lecturer Grants, 2076
Fulbright Public Policy Grants in Mexico, 2077
Fulbright Travel Grants, 2086
Fulbright Visiting Specialists Program: Direct Access to the Muslim World, 2087
Fund for Folk Culture Artists Support Grants, 2088
Fund for Scientific Research--Flanders Grants, 2089
Fund for the Advancement of the Discipline, 2090
Furthermore Grants in Publishing, 2091
G. Harold and Leila Y. Mathers Charitable Foundation Grants, 2092
G. Unger Vetlesen Foundation Grants, 2093
Gabor A. Somorjai Award for Creative Research in Catalysis, 2094
Gaius Charles Bolin Fellowships for Minority Graduate Students, 2095
Gas Technology Institute Basic Research Grants, 2096
Gates Family Foundation Fellowships, 2097
General Motors Foundation Grants, 2107
George A. Olah Award in Hydrocarbon or Petroleum Chemistry, 2111
George B. Storer Foundation Grants, 2113
George Foundation Grants, 2115
George Frederick Jewett Foundation Grants, 2116
George Gund Foundation Grants, 2117
George H. Hitchings New Investigator Award in Health Research, 2118
George S. & Dolores Dore Eccles Grants, 2120
George W. Codrington Foundation Grants, 2122
Georgia-Pacific Foundation Environment Grants, 2123
Georgia Power Foundation Grants, 2125
Gerber Foundation Grants, 2126
Gertrude B. Elion Mentored Medical Student Research Awards, 2130
Getty Research Grants for Institutions - Archival Grants, 2131
Getty Research Grants for Institutions - Art History Research Grants, 2132
Getty Research Grants for Institutions - Cataloguing of Museum Collections Grants, 2133
Getty Research Grants for Institutions - Critical Reference Resources Grants, 2134
Getty Research Institute Nonresidential Grants - Collaborative Research Grant, 2135
Getty Research Institute Nonresidential Grants - Curatorial Research Fellowships, 2136
Getty Research Institute Nonresidential Grants - Postdoctoral Fellowships, 2137
Getty Scholars and Visiting Scholars Grants, 2138
Gil and Dody Weaver Foundation Grants, 2139
Girl's Best Friend Foundation Grants, 2143
Glaucoma Foundation Grants, 2145
Glaucoma Research Pilot Project Grants, 2146
Glenn/AFAR Breakthroughs in Geront Awards, 2147
Glenn/AFAR Research Grants for Postdoctoral Fellows, 2148
Global Center for Dispute Resolution Res Grants, 2150
Global Change Education Program, 2151
GNOF IMPACT Grants for Education, 2154
Goddess Scholars Grants, 2155
Goldhirsh Foundation Brain Tumor Res Grants, 2156
Goldman Philanthropic Partnerships Program, 2157
GPP Charles E. Culpeper Foundation Biomedical Pilot Projects Initiative Grants, 2159
Grammy Foundation Grants, 2161
Grass Foundation Marine Biological Laboratory Fellowships, 2163
Great Lakes Fishery Commission Pilot Research Project Grants, 2166
Great Lakes Fishery Trust Access Grants, 2168
Great Lakes Fishery Ecological & Biological Fisheries Research to Inform Management Grants, 2169
Great Lakes Protection Fund Grants, 2170
Green Diamond Charitable Contributions, 2171
Greenwall Foundation Bioethics Grants, 2174
Greenwall Foundation Faculty Scholars Program in Bioethics, 2175
Greenwall Foundation Fellowship Program in Bioethics and Health Policy, 2176
Gruber Foundation Cosmology Prize, 2177
Gruber Foundation Gentics Prize, 2178
Gruber Foundation International Astronomical Union Fellowships, 2179
Gruber Foundation Neuroscience Prize, 2181
Gruber Foundation Prize for Global Justice and Women's Rights, 2182
Gruber Foundation Rosalind Franklin Young Investigator Award, 2183
Gruber Foundation Weizmann Institute Awards, 2184
GSA Research Grants, 2185
GUITS Grants for the Publication of Scholarly Books and Journals, 2186
GUITS Post-Doctoral Summer Travel-Research Grants, 2189
GUITS Research Grants in Comparative Studies of Modern Turkey, 2190
GUITS Summer Research Grants in Turkey for Graduate Students, 2192
H.A. and Mary K. Chapman Charitable Grants, 2193
H.W. Wilson Foundation Grants, 2196
Hampton Roads Community Foundation Mental Health Research Grants, 2199
Harold R. Bechtel Charitable Remainder Uni-Trust Grants, 2201
Harold R. Bechtel Testamentary Charitable Trust Grants, 2202
Harold Simmons Foundation Grants, 2203
Harris Graduate School of Public Policy Studies Research Development Grants, 2204
Harry A. and Margaret D. Towsley Found Grants, 2205
Harry Frank Guggenheim Foundation Dissertation Fellowships, 2207
Harry Frank Guggenheim Foundation Research Grants, 2208
Harvard University Women's Studies in Religion Research Associate and Visiting Faculty, 2216
Hawaii Community Foundation Health Education and Research Grants, 2221
Healthcare Foundation of New Jersey Grants, 2222
Health Foundation of Southern Florida Responsive Grants, 2223
Hearst Foundations Health Grants, 2225
Heineman Foundation for Research, Education, Charitable and Scientific Purposes, 2227
Helen Bader Foundation Grants, 2228
Helen V. Brach Foundation Grants, 2231
Hendrick Foundation for Children Grants, 2232
Henry Dreyfus Teacher-Scholar Awards, 2233
Henry W. Bull Foundation Grants, 2237
HERA Found Ovarian Cancer OSB1 Seed Grant, 2238
Herbert A. & Adrian W. Woods Found Grants, 2239
Herbert C. Brown Award for Creative Research in Synthetic Methods, 2240
Herbert H. and Grace A. Dow Grants, 2241
Herb Society of America Research and Education Grants, 2243
Hereditary Disease Found Lieberman Award, 2245

Hereditary Disease Foundation Research Grants, 2246
HHMI-NIBIB Interfaces Initiative Grants, 2248
HHMI-NIH Cloister Research Scholars, 2249
HHMI/EMBO Start-up Grants for Cen Europe, 2250
HHMI Biomedical Research Grants for Int'l Scientists: Infectious Diseases and Parasitology, 2251
HHMI Biomedical Research Grants for International Scientists in Canada and Latin America, 2252
HHMI Biomedical Research Grants for International Scientists in the Baltics, Central and Eastern Europe, Russia, and Ukraine, 2253
HHMI Gilliam Fellowships for Advanced Study, 2254
HHMI Grants and Fellowships Programs, 2255
HHMI Initiative for Colleges Grants, 2256
HHMI Initiative for Universities Grants, 2257
HHMI International Research Scholars Program, 2258
HHMI Med into Grad Initiative Grants, 2259
HHMI Physician-Scientist Early Career Award, 2260
HHMI Precollege Outreach Initiative for Biomedical Research Institutions Grants, 2261
HHMI Undergraduate Science Education Grants for Professors, 2263
Hilda and Preston Davis Foundation Grants, 2266
Hill Crest Foundation Grants, 2267
Hilton Hotels Corporate Giving Grants, 2268
Hirtzel Memorial Foundation Grants, 2269
Hoglund Foundation Grants, 2271
Holberg Memorial Fund International Prize, 2272
Holberg Memorial Fund Nils Klim Prize, 2273
Holberg Memorial Fund School Project, 2274
HomeBanc Foundation Grants, 2275
Honeybee Health Improvement Project Grants, 2276
Honeywell Corporation Sustainable Opportunities Grants, 2278
Horowitz Foundation for Social Policy Grants, 2280
Horowitz Found for Social Policy Special Awards, 2281
Horton Research Grants, 2282
Hospital Libraries Section / MLA Professional Development Grants, 2283
HRF Graduate Fellowships, 2285
HRF New York City Environmental Grants for Newton Creek, 2288
HRSA Nurse Education, Practice, Quality and Retention Grants, 2290
HSBC Contributions Community Grants, 2291
Huie-Dellmon Trust Grants, 2295
Hunter's Hope Foundation Grants, 2298
Huntington's Disease Society of America Research Grants, 2300
Huntington Library Babara Thom Postdoctoral Fellowships, 2301
Huntington Library Clark-Huntington Joint Bibliographical Fellowship, 2302
Huntington Library Dana and David Dornsife Fellowships, 2303
Huntington Library Dibner History of Science Long-Term Fellowships, 2304
Huntington Library Mellon Fellowships, 2306
Huntington Library National Endowment for the Humanities Fellowships, 2307
Huntington Library Short-Term Fellowships, 2308
Huntington Library Travel Grants and Exchange Fellowships for Study in Great Britain, 2309
Hutchinson Community Foundation Grants, 2310
Hydrogen Student Design Contest, 2311
IAFF Burn Foundation Research Grants, 2312
IARC Expertise Transfer Fellowship, 2315
IARC Postdoctoral Fellowships for Training in Cancer Research, 2316
IARC Visiting Scientist Award for Sen Scientists, 2317
IATA Research Grants, 2319
IBM Center for Business of Government Grants, 2320
ICC Faculty Fellowships, 2333
IDEM Section 319(h) Nonpoint Source Grants, 2338
IERF Sepmeyer Research Grants, 2340
IES Faculty Fellowships, 2341
IFT Foundation Grants, 2345
IFUW Winifred Cullis Grants, 2347
IIE Whitaker International Fellowships and Scholarships, 2356
Illinois DCEO Coal Research Grants, 2359
Illinois DCEO Coal Revival Grants, 2360
IMLS Grants to State Library Administrative Agencies, 2364
IMLS Laura Bush 21st Century Librarian Grants, 2365
IMLS National Leadership Grants, 2366
Inamori Foundation Research Grants, 2368
Indiana 21st Century Research and Technology Fund Awards, 2370
Indiana SBIR/STTR Commercialization Enhancement Program, 2371
Indiana Space Grant Consortium Discovery Program Crosscutting Research-Team Grants, 2372
Indiana Space Grant Consortium Discovery Program Student Hands-On Research Grants, 2374
Indiana Space Grant Consortium Fellowships, 2375
Infinity Foundation Grants, 2378
Institute for Palestine Studies, Publishing, and Research Organization Grants, 2382
Institute of Turkish Studies Research Grants in Comparative Studies of Modern Turkey, 2390
Institute of Turkish Studies Summer Language Study Grants in Turkey, 2391
Institute of Turkish Studies Summer Research Grants in Turkey, 2392
Intel Foundation PhD Fellowships, 2394
International Human Rights Funders Grants, 2399
International Max Planck Research School Mathematics in the Sciences Fellowships, 2401
International Peace Research Inst Oslo Grants, 2402
IRA Albert J. Harris Awards, 2405
IRA Elva Knight Research Grant, 2406
IRA Esther Zolt Academic Research Award, 2407
IRA Esther Zolt Teacher Research Award, 2408
IRA Jeanne S. Chall Research Fellowship, 2410
IRA Outstanding Dissertation of the Year Award, 2411
IRA Steven A. Stahl Research Grants, 2413
IRA Teacher as Researcher Grants, 2414
Ireland Family Foundation Grants, 2415
IREX ECA Alumni Small Grants, 2416
IREX Edmund S. Muskie Graduate Fellowship, 2417
IREX Eurasian Undergraduate Exchange Grants, 2418
IREX Individual Advanced Research Opportunities Grants, 2419
IREX Policy-Connect Collaborative Res Grants, 2422
IREX Regional Policy Symposium Grants, 2423
IREX University Administration Support, 2425
IREX US Embassy Policy Specialist Grants, 2426
ISA Hyland R. Johns Grants, 2428
ISGC Multiple Undergraduate Research Grants, 2432
ISGC Research Seed Grants, 2433
ITEA Maley/FTE Scholarship, 2435
Ittleson Foundation AIDS Grants, 2436
Ittleson Foundation Environment Grants, 2437
Ittleson Foundation Mental Health Grants, 2438
ITVS Commissioned Funding, 2439
IUCP Archives Program Fellowship, 2440
IUCP Indiana Research Fund, 2442
IUVSTA Prize for Science, 2443
IUVSTA Prize for Technology, 2444
J.B. Reynolds Foundation Grants, 2449
J.W. Kieckhefer Foundation Grants, 2452
Jackson Foundation Grants, 2453
Jacob and Hilda Blaustein Foundation Grants, 2454
Jacob and Valeria Langeloth Foundation Grants, 2455
Jacobs Research Funds Small Grants, 2456
James and Sylvia Thayer Short -Term Research Fellowships, 2457
James H. Cummings Foundation Grants, 2460
James M. Collins Foundation Grants, 2461
James R. Dougherty Jr. Foundation Grants, 2463
James S. McDonnell Foundation Brain Cancer Research Collaborative Activity Awards, 2464
James S. McDonnell Foundation Complex Systems Collaborative Activity Awards, 2465
James S. McDonnell Found Research Grants, 2466
James S. McDonnell Foundation Scholar Awards, 2467
James S. McDonnell Foundation Understanding Human Cognition Awards, 2468
Jane Beattie Memorial Scholarship, 2470
Jane Bradley Pettit Foundation Health Grants, 2471
Japan Foundation Center for Global Partnership Grants, 2472
Japan Foundation Center for Global Partnership Institutional Grants, 2473
Japan Foundation Center for Global Partnership NPO Fellowships, 2474
Japan Foundation Center for Global Partnership Security Studies Fellowships, 2475
Japan Foundation Japanese Studies Fellowships, 2476
Japan Foundation New York Japanese Studies Fellowships, 2478
Japan Foundation Six-Month Grants for Specialists in Cultural and Academic Fields, 2479
Jeffress Awards in Interdisciplinary Research, 2481
Jeffrey Campbell Grad Fellow in Anthropology, 2482
Jeffrey Thomas Stroke Shield Foundation Research Grants, 2483
Jenifer Altman Foundation Grants, 2484
JHMI Patrick C. Walsh Prostate Cancer Res, 2488
JM Foundation Grants, 2490
JMO Law and Legal System Grants, 2491
JMO Public Policy Research Grants, 2492
JMO Strategic and International Studies Grants, 2493
Joel L. Fleishman Civil Society Fellowships, 2495
John D. and Catherine T. MacArthur Foundation Global Challenges Grants, 2498
John Deere Foundation Grants, 2500
John Merck Scholars Awards, 2502
John P. McGovern Foundation Grants, 2504
John S. Dunn Research Foundation Grants, 2506
Johnson & Johnson Corporate Contributions, 2508
John W. Alden Trust Grants, 2510
Joint Center for Poverty Research Grants, 2511
Joseph Alexander Foundation Grants, 2513
Josiah W. and Bessie H. Kline Foundation Grants, 2517
Joyce Foundation Education Grants, 2519
Joyce Foundation Employment Grants, 2520
JSA JLab Sabbatical Research Leave Support, 2522
JSPS Bilateral Exchange Programs with Foreign Academic Institutions, 2524
Judith and Jean Pape Adams Charitable Foundation ALS Grants, 2528
Judith Graham Pool Postdoctoral Research Fellowship in Hemophilia, 2529
Juvenile Justice Clearing House Grants, 2530
Kavli Foundation Research Grants, 2533
Kavli Foundation Research Prizes, 2534
Kellogg Institute for International Studies Andrónico Luksic Grants, 2535
Kellogg Institute for International Studies Collaborative Faculty Grants, 2536
Kellogg Institute for International Studies Dissertation Year Fellowships, 2538
Kellogg Institute for International Studies Faculty Fellows' Research Grants, 2540
Kellogg Institute for International Studies Global Collaboration Initiative Grants, 2541
Kellogg Institute for International Studies Graduate Research Grants, 2542
Kellogg Institute for International Studies Grants for Conferences and Workshops, 2543
Kellogg Institute for International Studies Grants for Distinguished Lectureships, 2544
Kellogg Inst for Int'l Studies Intramural Grants, 2546
Kellogg Institute for International Studies Kroc Undergraduate Research Grants, 2547
Kellogg Institute for International Studies Quechua Summer Language Fellowships, 2548
Kellogg Institute for International Studies Summer Internships, 2549
Kellogg Institute for International Studies Supplemental Awards for PhD Students, 2550
Kellogg Institute for International Studies Visiting Fellowships, 2551
Kenneth T. & Eileen L. Norris Found Grants, 2554
Kettering Fund Grants, 2555
Klarman Family Foundation Grants in Eating Disorders Research, 2558
Kluge Center Distinguished Chairs, 2560

Kluge Center Fellowships, 2561
Kluge Center Kissinger Chair in Foreign Policy and International Relations, 2564
Kluge Center Kissinger Lecturer Award, 2565
Knoll-MacDonald Foundation Grants, 2566
Komen Greater NYC Clinical Research Enrollment Grants, 2568
Komen Greater NYC Breast Health Grants, 2569
Kosciuszko Found Grants for Polish Citizens, 2572
Kosciuszko Foundation Teaching Fellowships for Polish Citizens, 2573
Kosciuszko Foundation Tuition Scholarships, 2574
Kovler Family Foundation Grants, 2575
Lalor Foundation Postdoctoral Fellowships, 2580
LAM Found Established Investigator Awards, 2581
LAM Foundation Pilot Project Grants, 2582
LAM Foundation Postoctoral Fellowships, 2583
LAM Foundation Research Grants, 2584
Lance Armstrong Foundation Research Grants, 2585
Law and Public Affairs Fellowships at Princeton University, 2588
LBI-L Leo Baeck/Studienstiftung Fellowships, 2592
LBI-NY/DAAD Fellowships in Germany, 2593
LBI-NY/DAAD Fellowships in New York, 2594
LBI-NY Fred Grubel Fellowship, 2596
LBI-NY Fritz Halbers Fellowships, 2597
LBI-NY Gerald Westheimer Career Development Fellowship, 2598
Leahi Fund, 2599
Leakey Foundation Paleoanthropology Award, 2600
Leo Baeck Essay Prize, 2601
Leon Lowenstein Foundation Grants, 2603
Library Company of Philadelphia and The Historical Society of Pennsylvania Visiting Research Fellowships, 2605
Library Company of Philadelphia Barra Foundation International Research Fellowships, 2606
Library Company of Philadelphia NEH Long-term Post-Doctoral Fellowships, 2607
Library Company of Philadelphia Program in Early American Economy and Society Dissertation-Level Fellowships, 2609
Library Company of Philadelphia Early American Economy & Society One-Month Fellowships, 2610
Library Company of Philadelphia Program in Early American Economy and Society Post-Doctoral Research Fellowships, 2611
LI General Grants, 2613
Lillian S. Wells Foundation Grants, 2614
LITA/Gaylord Award for Achievement in Library and Information Technology, 2618
Littlefield-AACR Grants in Metastatic Colon Cancer Research, 2619
Lloyd A. Fry Foundation Health Grants, 2620
LSAC Research Grants, 2623
Lucile Horton Howe & Mitchell B. Howe Grants, 2624
Ludwig Vogelstein Foundation Grants, 2626
Lumina Foundation for Education Grants, 2627
Lumosity Human Cognition Grant, 2628
Lumpkin Family Found Healthy People Grants, 2629
Lustgarten Foundation for Pancreatic Cancer Research Grants, 2630
Luther I. Replogle Foundation Grants, 2631
Lymphatic Education & Research Network Additional Support for NIH-funded F32 Postdoc Fellows, 2632
Lymphatic Education & Research Network Postdoctoral Fellowships, 2633
Lynde & Harry Bradley Found Fellowships, 2634
Lynde and Harry Bradley Foundation Grants, 2635
M. Bastian Family Foundation Grants, 2636
M.J. Murdock Charitable Trust College Research Program for Life Sciences, 2638
M.J. Murdock Charitable Trust College Science Research Program, 2639
M.J. Murdock Charitable Trust Exceptional Opportunity Grants, 2640
M.J. Murdock Charitable Trust General Grants, 2641
M.J. Murdock Charitable Trust Partners in Science Grants, 2642
M.J. Murdock Charitable Trust Scientific Research Grants, 2644
Mabel H. Flory Charitable Trust Grants, 2646
Mabel Y. Hughes Charitable Trust Grants, 2647
MacArthur Foundation Conservation and Sustainable Development Grants, 2648
Manitoba Arts Council Craft Grants, 2653
Manitoba Arts Council Dance Creation/Production Grants, 2654
Marcia and Otto Koehler Foundation Grants, 2656
Margaret & James A. Elkins Jr. Found Grants, 2658
Margaret T. Morris Foundation Grants, 2659
Marijuana Policy Project Grants, 2661
Marjorie C. Adams Charitable Trust Grants, 2665
Mary K. Chapman Foundation Grants, 2673
Mary Kay Foundation Cancer Research Grants, 2674
Mary S. and David C. Corbin Foundation Grants, 2675
Massachusetts Cultural Cou Artist Fellowships, 2677
Massage Therapy Foundation Practitioner Case Report Contest, 2679
Massage Therapy Foundation Research Grants, 2680
Massage Therapy Foundation Student Case Report Contest, 2681
Massage Therapy Research Fund Grants, 2682
Matilda R. Wilson Fund Grants, 2684
Mattel Children's Foundation Grants, 2685
Mattel International Grants, 2686
Max and Victoria Dreyfus Foundation Grants, 2687
Maximilian E. and Marion O. Hoffman Grants, 2688
MBL Fred Karush Library Readership, 2703
McCarthy Family Foundation Grants, 2717
McCombs Foundation Grants, 2718
McCune Charitable Foundation Grants, 2719
McCune Foundation Civic Grants, 2720
McGraw-Hill Companies Community Grants, 2724
McGregor Fund Grants, 2725
McLean Contributionship Grants, 2727
MDA Development Grant, 2728
MDA Neuromuscular Disease Research Grants, 2729
MDARD Specialty Crop Block Grants-Farm Bill, 2730
MDEQ Beach Monitoring Grants - Inland Lakes, 2731
MDEQ Local Water Quality Monitoring Grants, 2732
MDEQ Scrap Tire Market Development Grants, 2733
Mead Johnson Nutritionals Med Educ Grants, 2734
Mead Johnson Nutritionals Scholarships, 2735
Meadows Foundation Grants, 2736
MedImmune Charitable Grants, 2737
Mellon Post-Doctoral Fellowship in Philosophy, 2739
Mellon Post-Doctoral Fellowship in Sustainable Agriculture, 2740
Melville Charitable Trust Grants, 2741
Mericos Foundation Grants, 2742
Mesothelioma Applied Research Found Grants, 2743
Metanexus Institute Templeton Res Lectures, 2744
MetLife/TCG Aha! Think it Do it Grants, 2745
Meyer and Pepa Gold Family Foundation Grants, 2746
Meyer and Stephanie Eglin Foundation Grants, 2747
Meyer Memorial Trust Responsive Grants, 2748
MGFA Post-Doctoral Research Fellowships, 2751
MGFA Student Fellowships, 2752
MGM Resorts Foundation Community Grants, 2753
MGN Family Foundation Grants, 2754
MHS New England Regional Fellowships, 2756
MHS Suzanne & Caleb Loring Res Fellowship, 2758
MHS Teacher Fellowships, 2759
Michael J. Fox Found Clinical Discovery Grants, 2760
Michael J. Fox Foundation for Parkinson's Research LEAPS Program, 2761
Michael J. Fox Foundation Parkinson's Cell Line Research Grants, 2762
Microsoft Research A. Richard Newton Breakthrough Research Awards, 2764
Microsoft Research Beyond Search Semantic Computing and Internet Economics Grants, 2765
Microsoft Research Biomedical Computing for Genome Wide Association Studies Grants, 2766
Microsoft Research Cell Phone as a Platform for Healthcare Grants, 2767
Microsoft Res Human-Robot Interaction Grants, 2768
Microsoft Research Intelligent Web 3.0 Grants, 2769
Microsoft Research Mechanisms for Safe and Scalable Multicore Computing Grants, 2770
Microsoft Res Sustainable Computing Grants, 2772
Millipore Foundation Grants, 2775
Millstone Foundation Grants, 2776
Mineralogical Society of America Res Grants, 2777
Minerva Fellowships, 2778
Minerva Short-Term Research Grants, 2779
MLA Cunningham Memorial International Fellowship, 2783
MLA David A. Kronick Traveling Fellowship, 2784
MLA Donald A.B. Lindberg Res Fellowship, 2785
MLA Research, Development, and Demonstration Project Grant, 2789
MoEF Dr. Salim Ali Nat Wildlife Fellowship, 2792
Monfort Family Foundation Grants, 2794
Morgan Adams Foundation Grants, 2795
Motorola Foundation Grants, 2796
Motorola Foundation Innovation Generation Collaborative Grants, 2797
MPI for Social Anthropology Department II Postdoctoral Grants, 2798
MPI for Social Anthropology Max Planck Fellow Research Group Law, Organizations, Science, and Technology, 2799
MSI Alden G. Clayton Dissertation Competition, 2800
MSI Marketing Strategy Meets Wall Street Research Competition, 2801
MSI Research Support Standard Grants, 2802
Murphy Institute Center for Ethics and Public Affairs Graduate Fellowships, 2806
Musgrave Foundation Grants, 2807
NAAB Coordinated Research Grants, 2808
NAA Norwegian Marshall Fund Grants, 2811
NACC David Stevenson Fellowships, 2812
NACC William Diaz Fellowships, 2813
NAEA Research Grants, 2814
NAEF Teacher Incentive Grants, 2815
NAF Kyle Bryant Translational Research Award, 2817
NAF Research Grants, 2818
NAF Young Investigator Awards, 2819
NAGC Hollingworth Award, 2820
Nakanishi Prize, 2824
NAMM Foundation Music Research Grants, 2825
NAPNAP Foundation Graduate Student Research Grant, 2828
NAPNAP Foundation Nursing Research Grants, 2829
NARSAD Distinguished Investigator Grants, 2831
NARSAD Independent Investigator Grants, 2833
NARSAD Lieber Prize for Schizophrenic Res, 2834
NARSAD Ruane Prize for Child and Adolescent Psychiatric Research, 2835
NARSAD Young Investigator Grants, 2836
NASA Cooperative Agreement Notice for Technology Development Grants, 2837
NASA Goddard Visiting Fellowships, 2838
NASA Graduate Student Researchers Program Fellowships, 2839
NASA Undergraduate Student Researchers Project Grants, 2843
National Academies Postdoctoral and Senior Research Awards, 2844
National Blood Foundation Research Grants, 2848
National Book Scholarship Fund, 2849
National Center for Responsible Gaming Centers of Excellence in Gambling Research Grants, 2850
National Center for Responsible Gaming Early Stage Investigator Grants, 2852
National Center for Responsible Gaming Exploration Grants, 2853
National Center for Responsible Gaming Independent Scientist Grants, 2854
National Center for Responsible Gaming Grants, 2855
National Center for Respon Gaming Seed Grants, 2857
National Center for Responsible Gaming Travel Grants, 2858
National Center on Philanthropy and the Law Fellowship in Nonprofit Law, 2859
National Endowment for Democracy Discretionary Grants, 2862

National Foundation for Catholic Youth Grants, 2868
National Geographic Society/Waitt Grants, 2870
National Geographic Society Expeditions Council Grants, 2872
National Geographic Society Field Research and Exploration Grants, 2873
National Headache Foundation Research Grants, 2875
National Headache Foundation Seymour Diamond Clinical Fellowship in Headache Education, 2876
National MPS Society Grants, 2880
National Parkinson Foundation Clinical Research Fund Grants, 2881
National Psoriasis Foundation Research Grants, 2882
National Rising Scholars Award to Advance Research on Higher Education for the Public Good, 2883
National Sleep Foundation Pickwick Postdoctoral Fellowship, 2886
National Turkey Federation Research Award, 2887
National Wetlands Awards, 2889
National Wildlife Refuge System Centennial Scholarship Program, 2892
National Wild Turkey Federation Res Grants, 2893
NATO Science for Peace Grants, 2898
NBME Stemmler Medical Education Research Fund Grants, 2899
NCCAM Exploratory Developmental Grants for Complementary and Alternative Medicine Studies of Humans, 2903
NCCAM Translational Tools for Clinical Studies of CAM Interventions Grants, 2905
NCFR Innovation Grants, 2908
NCHS National Center for Health Statistics Postdoctoral Research Appointments, 2910
NCI/NCCAM Quick-Trials for Novel Cancer Therapies, 2911
NCI Academic-Industrial Partnerships for Devel & Validation of In Vivo Imaging Systems & Methods for Cancer Investigations, 2913
NCI Application and Use of Transformative Emerging Technologies in Cancer Research, 2914
NCI Application of Metabolomics for Translational and Biological Research Grants, 2915
NCI Basic Cancer Research in Cancer Health Disparities Grants, 2916
NCIC Canadian Cancer Society Research Grants, 2919
NCI Centers of Excellence in Cancer Communications Research, 2921
NCI Diet, Epigenetic Events, and Cancer Prevention Grants, 2925
NCI Engineered Isogenic Cell Lines with Relevant Cancer Targets Research Grants, 2926
NCI Exploratory/Developmental Grants for Diagnostic Cancer Imaging, 2927
NCI Exploratory Grants for Behavioral Research in Cancer Control, 2928
NCI Improving Diet and Physical Activity Assessment Grants, 2929
NCI Mentored Career Development Award to Promote Diversity, 2930
NCI Ruth L. Kirschstein National Research Service Award Institutional Research Training Grants, 2931
NCI Stages of Breast Development: Normal to Metastatic Disease Grants, 2932
NCRR Networks and Pathways Collaborative Research Projects Grants, 2934
NCRR Novel Approaches to Enhance Animal Stem Cell Research, 2935
NCSS/FASSE/CUFA Inquiry Grant, 2936
NCSS Exemplary Research in Social St Award, 2937
NCSS Jean Dresden Grambs Distinguished Career Research in Social Studies Award, 2938
NCTE Teacher Researcher Award, 2940
NCTM E. Glenadine Gibb 7-12 Classroom Research Grants, 2941
NCTM Edward G. Begle PreK-6 Classroom Research Grants, 2942
NEDA/AED Charron Family Research Grant, 2945
NEDA/AED Joan Wismer Research Grant, 2946
NEDA/AED Tampa Bay Eating Disorders Task Force Award, 2947
NEFE Grants, 2948
NEH Awards for Faculty at Historically Black Colleges and Universities, 2949
NEH Awards for Faculty at Institutions with High Hispanic Enrollment, 2950
NEH Awards for Faculty at Tribal Colleges and Universities, 2951
NEH Digital Humanities Implementation Grants, 2953
NEH Digital Humanities Start-Up Grants, 2954
NEH Documenting Endangered Lang Grants, 2955
NEH Fellowships at Independent Research Institutions, 2957
NEH Fellowships for Advanced Social Science Research on Japan, 2958
NEI Clinical Study Planning Grant, 2959
NEI Clinical Vision Research Devel Award, 2960
NEI Research Grant For Secondary Data Analysis Grants, 2962
NEI Ruth L. Kirschstein National Research Service Award Short-Term Institutional Research Training Grants, 2963
NEI Translational Research Program On Therapy For Visual Disorders, 2965
NEI Vision Research Core Grants, 2966
Nellie Mae Education Foundation District-Level Change Grants, 2968
Nellie Mae Education Foundation Public Understanding Grants, 2969
Nellie Mae Education Foundation Research and Development Grants, 2970
Nellie Mae Education Foundation State Level Systems Change Grants, 2971
Nell Warren Elkin and William Simpson Elkin Foundation, 2972
Neotropical Grassland Conservancy Derald G. Langham Memorial Research Grant, 2973
Neotropical Grassland Conservancy Grants, 2974
Neotropical Grassland Conservancy Remembrance Fund Grants, 2975
Neotropical Grassland Conservancy Stu Grants, 2976
Neotropical Grassland Conservancy W. J. and Virginia W. Moorhouse Research Grant, 2977
Neotropical Grassland Conservancy Wijdenes Memorial Research Grant, 2978
Nestle Foundation Large Research Grants, 2979
Nestle Foundation Pilot Grants, 2980
Nestle Foundation Re-entry Grants, 2981
Nestle Foundation Small Research Grants, 2982
Nestle Foundation Training Grant, 2983
NFID Advanced Vaccinology Travel Grant, 2986
NFL Charities Medical Grants, 2994
NFWF Acres for America Grants, 2996
NFWF Budweiser Renewable Energy and Wildlife Conservation Research Prize, 2998
NFWF International Sea Turtle Conservation Fund Grants, 3000
NFWF Marine Debris Research and Technology Grants, 3002
NFWF More Fish Partnership Fund Grants, 3003
NFWF National Whale Conservation Grants, 3004
NFWF Nature Restoration Trust Grants, 3005
NFWF Puget Sound Marine Conservation Fund Grants, 3006
NFWF Wildlife and Habitat Conservation Initiative Grants, 3009
NFX Patient-Oriented Research Grants, 3010
NHC Summer Institutes in Lit Studies Grants, 3012
NHGRI Mentored Research Scientist Development Award, 3013
NHLBI Airway Smooth Muscle Function and Targeted Therapeutics in Human Asthma Grants, 3014
NHLBI Ancillary Studies in Clinical Trials, 3015
NHLBI Bioengineering and Obesity Grants, 3016
NHLBI Bioengineering Approaches to Energy Balance and Obesity Grants for SBIR, 3017
NHLBI Bioengineering Approaches to Energy Balance and Obesity Grants for STTR, 3018
NHLBI Biomedical Research Training Program for Individuals from Underrepresented Groups, 3019
NHLBI Cardiac Devel Consortium Grants, 3020
NHLBI Career Transition Awards, 3021
NHLBI Characterizing the Blood Stem Cell Niche Grants, 3022
NHLBI Circadian-Coupled Cellular Function in Heart, Lung, and Blood Tissue Grants, 3023
NHLBI Clinical Centers for the NHLBI Asthma Network Grants, 3024
NHLBI Developmental Origins of Altered Lung Physiology and Immune Function Grants, 3025
NHLBI Exploratory Studies in the Neurobiology of Pain in Sickle Cell Disease, 3026
NHLBI Immunomodulatory, Inflammatory, and Vasoregulatory Properties of Transfused Red Blood Cell Units as a Function of Preparation and Storage Grants, 3027
NHLBI Independent Scientist Award, 3028
NHLBI Intramural Research Training Awards, 3029
NHLBI Investigator Initiated Multi-Site Clinical Trials, 3030
NHLBI Lymphatics in Health & Disease in Digestive, Cardiovascular & Pulmonary Systems, 3031
NHLBI Lymphatics in Health and Disease in the Digestive, Urinary, Cardiovascular and Pulmonary Systems, 3032
NHLBI Mentored Career Award for Faculty at Institutions that Promote Diversity (K01), 3033
NHLBI Mentored Career Award For Faculty At Minority Institutions, 3034
NHLBI Mentored Career Development Award to Promote Faculty Diversity/Re-Entry in Biomedical Research, 3035
NHLBI Mentored Clinical Scientist Research Career Development Awards, 3036
NHLBI Mentored Patient-Oriented Research Career Development Awards, 3037
NHLBI Mentored Quantitative Research Career Development Awards, 3038
NHLBI Microbiome of the Lung and Respiratory Tract in HIV-Infected Individuals and HIV-Uninfected Controls, 3039
NHLBI Midcareer Investigator Award in Patient-Oriented Research, 3040
NHLBI National Research Service Award Programs in Cardiovascular Epidemiology and Biostatistics, 3041
NHLBI New Approaches to Arrhythmia Detection and Treatment Grants for SBIR, 3042
NHLBI New Approaches to Arrhythmia Detection and Treatment Grants for STTR, 3043
NHLBI Nutrition and Diet in the Causation, Prevention, and Management of Heart Failure Research Grants, 3044
NHLBI Pathway to Independence Awards, 3045
NHLBI Pediatric Cardiac Genomics Consortium Grants, 3046
NHLBI Phase II Clinical Trials of Novel Therapies for Lung Diseases, 3047
NHLBI Prematurity & Respiratory Grants, 3048
NHLBI Progenitor Cell Biology Consortium Administrative Coordinating Center Grants, 3049
NHLBI Protein Interactions Governing Membrane Transport in Pulmonary Health Grants, 3050
NHLBI Research Demonstration and Dissemination Grants, 3051
NHLBI Research Grants on the Relationship Between Hypertension and Inflammation, 3052
NHLBI Research on the Role of Cardiomyocyte Mitochondria in Heart Disease, 3053
NHLBI Research Supplements to Promote Diversity in Health-Related Research, 3054
NHLBI Right Heart Function in Health and Chronic Lung Diseases Grants, 3055
NHLBI Ruth L. Kirschstein National Research Service Award Institutional Research Training Grants, 3056
NHLBI Ruth L. Kirschstein National Research Service Awards for Individual Postdoctoral Fellows, 3057
NHLBI Ruth L. Kirschstein National Research Service Awards for Individual Predoctoral Fellowships to Promote Diversity in Health-Related Research, 3058
NHLBI Ruth L. Kirschstein National Research Service Awards for Individual Senior Fellows, 3060

NHLBI Ruth L. Kirschstein National Research Service Award Short-Term Institutional Research Training Grants, 3061
NHLBI Short-Term Research Education Program to Increase Diversity in Health-Related Research, 3062
NHLBI Summer Inst Training in Biostatistics II, 3063
NHLBI Summer Internship Program in Biomedical Research, 3064
NHLBI Targeted Approaches to Weight Control for Young Adults Grants, 3065
NHLBI Targeting Calcium Regulatory Molecules for Arrhythmia Prevention Grants, 3066
NHLBI Translating Basic Behavioral and Social Science Discoveries into Interventions to Reduce Obesity: Centers for Behavioral Intervention Development Grants, 3068
NHLBI Translational Grants in Lung Diseases, 3069
NHPRC Archives - Basic Projects Grants, 3070
NHPRC Archives Detailed Processing Grants, 3071
NHPRC Electronic Records Projects Grants, 3073
NHPRC Historical Editing Fellowships, 3074
NHPRC Publishing Historical Records Grants, 3075
NHPRC State and National Archival Partnership Grants, 3076
NHPRC Strategies and Tools for Archives and Historical Publishing Projects Grants, 3077
NIAAA Independent Scientist Award, 3078
NIAAA Mechanisms of Alcohol and Drug-Induced Pancreatitis Grants, 3079
NIAAA Mentored Clinical Scientist Research Career Development Awards, 3080
NIA Aging, Oxidative Stress & Cell Death Grants, 3081
NIA Aging Research Dissertation Awards to Increase Diversity, 3082
NIA AIDS and Aging: Behavioral Sciences Prevention Research Grants, 3083
NIA Alzheimer's Disease Core Centers Grants, 3084
NIA Alzheimer's Disease Drug Devel Grants, 3085
NIA Alzheimer's Disease Pilot Clinical Trials, 3086
NIA Alzheimer's Disease Res Centers Grants, 3087
NIA Archiving and Development of Socialbehavioral Datasets in Aging Related Studies Grants, 3088
NIA Awards to Support Research on the Biology of Aging in Invertebrates Grants, 3089
NIA Behavioral and Social Research Grants on Disasters and Health, 3090
NIA Bioenergetics, Fatigability, and Activity Limitations in Aging, 3091
NIA Diversity in Medication Use and Outcomes in Aging Populations Grants, 3092
NIA Effects of Gene-Social Environment Interplay on Health and Behavior in Later Life Grants, 3093
NIA Factors Affecting Cognitive Function in Adults with Down Syndrome Grants, 3094
NIAF Fellowship Grants, 3095
NIA Grants for Alzheimer's Disease Drugs, 3096
NIA Harmonization of Longitudinal Cross-National Surveys of Aging Grants, 3097
NIA Health Behaviors and Aging Grants, 3098
NIA Healthy Aging through Behavioral Economic Analyses of Situations Grants, 3099
NIA Higher-Order Cognitive Functioning and Aging Grants, 3100
NIA Human Biospecimen Resources for Aging Research Grants, 3101
NIAID Independent Scientist Award, 3102
NIA Independent Scientist Award, 3103
NIA Malnutrition in Older Persons Grants, 3104
NIA Mechanisms, Measurement, and Management of Pain in Aging: from Molecular to Clinical, 3105
NIA Mechanisms Underlying the Links between Psychosocial Stress, Aging, the Brain and the Body Grants, 3106
NIA Medical Management of Older Patients with HIV/AIDS Grants, 3107
NIA Mentored Clinical Scientist Research Career Development Awards, 3108
NIA Minority Dissertation Res Grants in Aging, 3110
NIAMS Pilot and Feasibility Clinical Research Grants in Arthritis, Musculoskeletal & Skin Diseases, 3111
NIA Multicenter Study on Exceptional Survival in Families: The Long Life Family Study Grants, 3112
NIA Network Infrastructure Support for Emerging Behavioral and Social Research in Aging, 3114
NIA Postdoctoral Training in Research on Aging in Canada, 3116
NIA Promoting Careers in Aging and Health Disparities Research Grants, 3117
NIA Renal Function and Chronic Kidney Disease in Aging Grants, 3118
NIA Role of Apolipoprotein E, Lipoprotein Receptors and CNS Lipid Homeostasis in Brain Aging and Alzheimer's Disease, 3119
NIA Role of Nuclear Receptors in Tissue and Organismal Aging Grants, 3120
NIA Thyroid in Aging Grants, 3122
NIA Transdisciplinary Research on Fatigue and Fatigability in Aging Grants, 3123
NIA Translational Research at the Aging/Cancer Interface Grants, 3124
NIA Vulnerable Dendrites and Synapses in Aging and Alzheimer's Disease Grants, 3125
NICHD Innovative Therapies and Clinical Studies for Screenable Disorders Grants, 3126
NICHD Ruth L. Kirschstein National Research Service Award Institutional Predoctoral Training Program in Systems Biology of Developmental Biology & Birth Defects, 3127
NIDA Pilot and Feasibility Studies in Preparation for Drug Abuse Prevention Trials, 3130
NIDCD Mentored Clinical Scientist Research Career Development Award, 3131
NIDCD Mentored Research Scientist Development Award, 3132
NIDDK Advances in Polycystic Kidney Disease Grants, 3133
NIDDK Adverse Metabolic Side Effects of Second Generation Psychotropic Medications Leading to Obesity and Increased Diabetes Risk Grants, 3134
NIDDK Ancillary Studies of Kidney Disease Accessing Information from Clinical Trials, Epidemiological Studies, and Databases Grants, 3135
NIDDK Ancillary Studies to Major Ongoing NIDDK and NHLBI Clinical Research Studies, 3136
NIDDK Basic and Clinical Studies of Congenital Urinary Tract Obstruction Grants, 3137
NIDDK Basic Research Grants in the Bladder and Lower Urinary Tract, 3138
NIDDK Calcium Oxalate Stone Diseases Grants, 3139
NIDDK Cell-Specific Delineation of Prostate and Genitourinary Development, 3140
NIDDK Co-Activators and Co-Repressors in Gene Expression Grants, 3141
NIDDK Collaborative Interdisciplinary Team Science in Diabetes, Endocrinology and Metabolic Diseases Grants, 3142
NIDDK Developmental Biology and Regeneration of the Liver Grants, 3143
NIDDK Development of Assays for High-Throughput Drug Screening Grants, 3144
NIDDK Devel of Disease Biomarkers Grants, 3145
NIDDK Diabetes, Endocrinology, and Metabolic Diseases NRSAs--Individual, 3146
NIDDK Diabetes Research Centers, 3147
NIDDK Diet Comp & Energy Balance Grants, 3148
NIDDK Education Grants, 3149
NIDDK Endoscopic Clinical Research Grants In Pancreatic And Biliary Diseases, 3150
NIDDK Enhancing Zebrafish Research with Research Tools and Techniques, 3151
NIDDK Erythroid Line Molecular Tool Grants, 3152
NIDDK Erythropoiesis: Components and Mechanisms Grants, 3153
NIDDK Exploratory/Developmental Clinical Research Grants in Obesity, 3154
NIDDK Grants for Basic Research in Glomerular Diseases, 3155
NIDDK Health Disparities in NIDDK Diseases, 3156
NIDDK Identifying and Reducing Diabetes and Obesity Related Health Disparities within Healthcare Systems, 3157
NIDDK Insulin Signaling and Receptor Cross Talk Grants, 3158
NIDDK Intestinal Failure, Short Gut Syndrome and Small Bowel Transplantation Grants, 3159
NIDDK Intestinal Stem Cell Consortium Grants, 3160
NIDDK Mentored Research Scientist Development Award, 3161
NIDDK Multi-Center Clinical Study Cooperative Agreements, 3162
NIDDK Multi-Center Clinical Study Implementation Planning Grants, 3163
NIDDK Neurobiology of Diabetic Complications Grants, 3164
NIDDK Neuroimaging in Obesity Res Grants, 3165
NIDDK New Tech for Liver Disease SBIR, 3166
NIDDK New Tech for Liver Disease STTR, 3167
NIDDK Non-Invasive Methods for Diagnosis and Progression of Diabetes, Kidney, Urological, Hematological and Digestive Diseases, 3168
NIDDK Pancreatic Devel & Regeneration Grants: Toward Cellular Therapies for Diabetes, 3169
NIDDK Pilot Feasibility Clinical Research Grants in Diabetes, Endocrine & Metabolic Diseases, 3170
NIDDK Pilot and Feasibility Clinical Research Grants in Kidney or Urologic Diseases, 3171
NIDDK Pilot and Feasibility Clinical Research Studies in Digestive Diseases and Nutrition, 3172
NIDDK Planning Grants For Translational Research For The Prevention And Control Of Diabetes And Obesity, 3173
NIDDK Proteomics: Diabetes, Obesity, And Endocrine, Digestive, Kidney, Urologic, And Hematologic Diseases, 3174
NIDDK Research Grants for Studies of Hepatitis C in the Setting of Renal Disease, 3175
NIDDK Research Grants on Improving Health Care for Obese Patients, 3176
NIDDK Role of Gastrointestinal Surgical Procedures in Amelioration of Obesity-Related Insulin Resistance & Diabetes Ind of Weight Loss, 3177
NIDDK Secondary Analyses in Obesity, Diabetes and Digestive and Kidney Diseases Grants, 3178
NIDDK Seeding Collaborative Interdisciplinary Team Science in Diabetes, Endocrinology and Metabolic Diseases Grants, 3179
NIDDK Silvio O. Conte Digestive Diseases Research Core Centers Grants, 3180
NIDDK Small Business Innovation Research to Develop New Therapeutics and Monitoring Technologies for Type 1 Diabetes (T1D)--Towards an Artificial Pancreas, 3181
NIDDK Small Grants for K08/K23 Recipients, 3182
NIDDK Small Grants for Underrepresented Minority Scientists in Diabetes and Digestive and Kidney Diseases, 3183
NIDDK Traditional Conference Grants, 3184
NIDDK Training in Clinical Investigation in Kidney and Urology Grants, 3185
NIDDK Translational Research for the Prevention and Control of Diabetes and Obesity, 3186
NIDDK Transmission of Human Immunodeficiency Virus In Semen, 3187
NIDDK Type 2 Diabetes in the Pediatric Population Research Grants, 3188
NIDRR Field-Initiated Projects, 3189
NIEHS Hazardous Materials Worker Health and Safety Training Grants, 3190
NIEHS Hazmat Training at Doe Nuclear Weapons Complex Grants, 3191
NIEHS Independent Scientist Award, 3192
NIEHS Small Business Innovation Research Grants, 3195
NIGMS/NIDCR Evolution of Infectious Diseases Research Grants, 3196
NIGMS Mentored Clinical Scientist Research Career Development Awards, 3197
NIGMS Pharmacogenomics Res Network Grants, 3198

NIGMS Ruth L. Kirschstein National Research Service Awards for Individual Predoctoral Fellows in PharmD/PhD Grants, 3200
NIGMS Short Courses on Math, Stat, and Comp Tools for Studying Biological Systems, 3201
NIGMS Support of Competitive Research Pilot Project Awards, 3202
NIGMS Support of Competitive Research Research Advancement Awards, 3203
NIH Academic Research Enhancement Awards, 3204
NIH Biobehavioral Research for Effective Sleep, 3205
NIH Biology, Development, and Progression of Malignant Prostate Disease Research Grants, 3206
NIH Biomedical Research on the International Space Station, 3207
NIH Bone and the Hematopoietic and Immune Systems Research Grants, 3208
NIH Building Interdisciplinary Research Careers in Women's Health Grants, 3209
NIH Dietary Supplement Research Centers: Botanicals Grants, 3212
NIH Earth-Based Research Relevant to the Space Environment, 3213
NIH Enhancing Adherence to Diabetes Self-Management Behaviors, 3214
NIH Exceptional, Unconventional Research Enabling Knowledge Acceleration Grants, 3215
NIH Exploratory/Developmental Research Grant, 3216
NIH Exploratory Innovations in Biomedical Computational Science & Tech Grants, 3217
NIH Fine Mapping and Function of Genes for Type 1 Diabetes Grants, 3218
NIH Health Promotion Among Racial and Ethnic Minority Males, 3219
NIH Human Connectome Project Grants, 3220
NIH Independent Scientist Award, 3221
NIH Innovations in Biomedical Computational Science and Technology Grants, 3222
NIH Innovations in Biomedical Computational Science and Technology Initiative Grants for SBIR, 3223
NIH Innovations in Biomedical Computational Science and Technology Initiative Grants for STTR, 3224
NIH Mentored Patient-Oriented Research Career Development Award, 3226
NIH Mentored Research Scientist Development Awards, 3227
NIH Nonhuman Primate Immune Tolerance Cooperative Study Group Grants, 3228
NIH Receptors and Signaling in Bone in Health and Disease, 3229
NIH Recovery Act Limited Competition: Academic Research Enhancement Awards, 3230
NIH Recovery Act Limited Competition: Biomedical Research, Development, and Growth to Spur the Acceleration of New Technologies (BRDG-SPAN) Pilot Grants, 3231
NIH Recovery Act Limited Competition: Building Sustainable Community-Linked Infrastructure to Enable Health Science Research, 3232
NIH Recovery Act Limited Competition: Small Business Catalyst Awards for Accelerating Innovative Research Grants, 3233
NIH Recovery Act Limited Competition: Supporting New Faculty Recruitment to Enhance Research Resources through Biomedical Research Core Centers Grants, 3234
NIH Research On Ethical Issues In Human Subjects Research, 3235
NIH Research on Sleep and Sleep Disorders, 3236
NIH Research Project Grants, 3237
NIH Ruth L. Kirschstein National Research Service Awards for Individual Senior Fellows, 3238
NIH Ruth L. Kirschstein National Research Service Awards for Individual Predoctoral Fellows, 3240
NIH Ruth L. Kirschstein National Research Service Awards for Individual Predoctoral Fellowships to Promote Diversity in Health-Related Research, 3241
NIH Ruth L. Kirschstein National Research Service Awards for Individual Predoctoral MD/PhD and Other Dual Doctoral Degree Fellows, 3242
NIH Sarcoidosis: Research into the Cause of Multi-Organ Disease and Clinical Strategies for Therapy Grants, 3243
NIH School Interventions to Prevent Obesity, 3244
NIH Self-Management Strategies Across Chronic Diseases Grants, 3245
NIH Solicitation of Assays for High Throughput Screening in the Molecular Libraries Probe Production Centers Network, 3246
NIH Structural Bio of Membrane Proteins Grants, 3247
NIH Summer Internship Program in Biomedical Research, 3248
NIMH AIDS and Aging: Behavioral Sciences Prevention Research Grants, 3249
NIMH Basic and Translational Research in Emotion Grants, 3250
NIMH Curriculum Development Award in Neuroinformatics Research and Analysis, 3251
NIMH Early Identification and Treatment of Mental Disorders in Children and Adolescents Grants, 3252
NIMH Jointly Sponsored Ruth L. Kirschstein National Research Service Award Institutional Predoctoral Training Program in the Neurosciences, 3253
NIMH Short Courses in Neuroinformatics, 3254
NINDS Optimization of Small Molecule Probes for the Nervous System Grants, 3255
NINDS Optimization of Small Molecule Probes for the Nervous System Grants, 3256
NINDS Optimization of Small Molecule Probes for the Nervous System Grants, 3257
NINDS Support of Scientific Meetings as Cooperative Agreements, 3258
NINR Acute and Chronic Care During Mechanical Ventilation Research Grants, 3259
NINR AIDS and Aging: Behavioral Sciences Prevention Research Grants, 3260
NINR Chronic Illness Self-Management in Children and Adolescents Grants, 3261
NINR Clinical Interventions for Managing the Symptoms of Stroke Research Grants, 3262
NINR Diabetes Self-Management in Minority Populations Grants, 3263
NINR Mechanisms, Models, Measurement, & Management in Pain Research Grants, 3264
NINR Mentored Research Scientist Development Award, 3265
NINR Quality of Life for Individuals at the End of Life Grants, 3266
NINR Ruth L. Kirschstein National Research Service Award for Individual Predoctoral Fellows in Nursing Research, 3267
NIST Fire Research Grants, 3268
NIST Hollings Manufac Extension Partnerships, 3269
NIST Materials Science and Engineering Grants, 3270
NIST National Standard Reference Data System Research Grants, 3271
NIST Precision Measurement Grants, 3272
NIST Small Business Innovation Res Grants, 3273
NKF Clinical Scientist Awards, 3274
NKF Franklin McDonald/Fresenius Medical Care Clinical Research Grants, 3275
NKF Professional Councils Research Grants, 3276
NKF Research Fellowships, 3277
NKF Young Investigator Grants, 3278
NL American Society for Eighteenth-Century Studies Fellowships, 3279
NL Arthur and Lila Weinberg Fellowships for Independent Scholars and Researchers, 3280
NL Associated Colleges of the Midwest/Great Lakes College Association Faculty Fellowships, 3281
NL Audrey Lumsden-Kouvel Fellowships, 3282
NL British Academy Fellowship for Study in the United Kingdom, 3283
NL Frances C. Allen Fellowships, 3284
NL Grad Scholar-in-Residence Opportunities, 3285
NL Herzog August Bibliothek Wolfenbuttel Fellowship, 3286
NL Lawrence Lipking Fellowship at Newberry, 3287
NL Lester J. Cappon Fellowship in Documentary Editing, 3288
NL Lloyd Lewis Fellows in American History, 3289
NLM Academic Research Enhancement Awards, 3290
NLM Advancing Novel Science in Women's Health Research Exploratory/Developmental Grants, 3291
NLM Advancing Novel Science in Women's Health Research Small Project Grants, 3292
NLM Behavioral and Social Science Research on Understanding and Reducing Health Disparities Research Grants, 3293
NL Mellon Postdoctoral Research Fellowships, 3294
NLM Exceptional, Unconventional Research Grants Enabling Knowledge Acceleration, 3295
NLM Exploratory/Developmental Grants, 3296
NLM Express Research Grants in Biomedical Informatics and Bioinformatics, 3297
NLM Grants for Scholarly Works in Biomedicine and Health, 3298
NL Midwest Modern Lang Assoc Fellowship, 3299
NLM Limited Competition for Continuation of Biomedical Informatics/Bioinformatics Resource Grants, 3301
NLM Manufacturing Processes of Medical, Dental, and Biological Technologies Grants, 3302
NLM New Tech for Liver Disease Grants, 3303
NL Monticello College Foundation Fellowship for Women, 3304
NLM Predictive Multiscale Models of the Physiome in Health and Disease Grants, 3305
NLM Research Project Grants, 3306
NLM Research Supplements to Promote Reentry in Health-Related Research, 3307
NLM Small Business Innovation Res Grants, 3308
NLM Technological Innovations for Interdisciplinary Research Incorporating the Behavioral and Social Sciences Grants, 3309
NL National Endowment for the Humanities Fellowships, 3310
NL Newberry Consortium on American Indian Studies Faculty Fellowships, 3311
NL Newberry Library/Ecole Nationale des Chartes Exchange Fellowship, 3313
NL Northeast Modern Language Assocation Fellowship, 3314
NL Scholar-in-Residence Opportunity, 3315
NL Short-Term Fellowships in the History of Cartography, 3316
NL Short-Term Resident Fellowships for Individual Research, 3317
NL South Central Modern Languages Association Fellowship, 3318
NL Susan Kelly Power and Helen Hornbeck Tanner Fellowship, 3319
NL Weiss/Brown Publication Subvention Award, 3320
NMF General Electric Med Scholars Fellowships, 3322
NMF National Medical Association Awards for Medical Journalism, 3323
NMF Ralph W. Ellison Prize, 3327
NMSF Ernest F. Hollings Ocean Awareness Trust Fund Grants, 3328
NOAA Abrupt Climate Change Studies Grants, 3329
NOAA Arctic Research Grants, 3330
NOAA Assessing Atlantic Meridional Overturning Circulation Grants, 3331
NOAA Atlantic Salmon Conservation Grants, 3332
NOAA Atmospheric Composition and Climate Program Grants, 3333
NOAA Ballast Water Management Demonstration Program (RDTE Facility) Grants, 3334
NOAA Ballast Water Management Demonstration Program (Technologies and Practices) Grants, 3335
NOAA Ballast Water Technology Demonstration Program (Research, Development, Testing and Evaluation Facility) Grants, 3336
NOAA Ballast Water Technology Demonstration Grants, 3337
NOAA California Bay Watershed Education and Training (B-WET) Grants, 3338
NOAA Chesapeake Bay Office B-WET (Bay Watershed Education and Training) Grants, 3339

NOAA Chesapeake Bay Office Cooperative Research Grants, 3340
NOAA Chesapeake Bay Office Fisheries Science Grants, 3341
NOAA Chesapeake Bay Office Non-Native Oyster Research Grants, 3342
NOAA Chesapeake Bay Office Submerged Aquatic Vegetation Culture and Restoration Grants, 3343
NOAA Climate Dynamics and Experimental Prediction Grants, 3345
NOAA Climate Prediction Program for the Americas Grants, 3346
NOAA Cli Variability & Human Health Grants, 3347
NOAA Climate Varia & Predictability Grants, 3348
NOAA Coastal and Estuarine Land Conservation Grants, 3349
NOAA Coastal Ecosystem Research Project, 3350
NOAA Coastal Hypoxia Research Grants, 3351
NOAA Coastal Management Fellowships, 3352
NOAA Coastal Ocean Grants, 3353
NOAA Collaborative Science, Technology, and Applied Research Grants, 3354
NOAA Communicating Hurricane Info Grants, 3355
NOAA Cooperative Institute: Eastern U.S. Continental Shelf Frontier Exploration, Research, and Technology Development Grants, 3356
NOAA Cooperative Institute for Applied Meteorological Studies Grants, 3357
NOAA Cooperative Institute for Tropical Meteorology Research Grants, 3358
NOAA Cooperative Research Grants, 3359
NOAA Cooperative Unit for Fisheries Education and Research Internships, 3360
NOAA Coral Reef Ecosystem Studies Grants, 3361
NOAA CSCOR Regional Ecosystem Prediction Program: Cumulative Impacts of Stressors at the Land-Water Interface in the Mid-Atlantic, 3363
NOAA CSCOR Regional Ecosystem Prediction Program: Ecosystem Goal-Setting in Coastal Waters and Reefs of South Florida, 3364
NOAA CSCOR Regional Ecosystem Prediction Program: Invasive Species in the Great Lakes--A Regional Scale Approach, 3365
NOAA Cumulative Impacts of Multiple Stressors (MultiStress) Grants, 3366
NOAA Dr. Nancy Foster Scholarships, 3367
NOAA Ecological Forecasting Grants, 3368
NOAA Ecology and Oceanography of Harmful Algal Blooms Grants, 3369
NOAA Faculty & Student Intern Res Grants, 3371
NOAA Financial Assistance for National Centers for Central Coastal Ocean Science, 3372
NOAA General Coral Reef Conservation Grants, 3373
NOAA Global Carbon Cycle Grants, 3374
NOAA Graduate Sciences Grants, 3375
NOAA Great Lakes Habitat Restoration Partnership Grants, 3376
NOAA International Coral Reef Conservation Project Grants, 3377
NOAA Joint Hurricane Testbed Grants, 3378
NOAA Living Marine Resources Cooperative Science Center Mini Grants, 3381
NOAA Miami Lab Summer Internships, 3382
NOAA Mid-Atlantic Bight National Undersea Research Center Grants, 3383
NOAA Mid-Atlantic Research Set-Aside Grants, 3384
NOAA Monitoring and Event Response for Harmful Algal Blooms Grants, 3385
NOAA Monkfish Research Set-Aside Grants, 3386
NOAA Nat Climatic Data Center Internships, 3387
NOAA National Coastal and Estuarine Research and Technology Grants, 3388
NOAA National Estuarine Research Reserve Land Acquisition and Construction Grants, 3389
NOAA National Estuarine Research Reserve System Graduate Research Fellowships, 3390
NOAA National Height Modernization Grants, 3391
NOAA National Sea Grant Gulf of Mexico Oyster Industry Grants, 3392
NOAA National Sea Grant Oyster Disease Research Grants, 3395
NOAA Northern Gulf of Mexico Ecosystems and Hypoxia Assessment Grants, 3396
NOAA Ocean Exploration Omnibus: Marine Archaeology Grants, 3397
NOAA Ocean Exploration Omnibus: Ocean Exploration Grants, 3398
NOAA Pacific Islands Assistantships, 3399
NOAA Population Dynamics Recruiting Grants, 3400
NOAA Regional Ecosystem Prediction Program on Coral Reef Ecosystem Studies Grants, 3401
NOAA Regional Integrated Sciences and Assessments Grants, 3402
NOAA Scientific Data Stewardship Grants, 3403
NOAA Sea Scallop Research Set-Aside Grants, 3404
NOAA Sector Applications Research Grants, 3405
NOAA South Florida Grants, 3406
NOAA Spherical Display Systems for Earth System Science - Installations and Content Development Grants, 3407
NOAA Transition of Research Applications to Climate Services Grants, 3408
NOAA Undersea Research Center Coral Reef Conservation Program Coral Research Grants, 3409
NOAA Undersea Research Center Ocean Acidification Research at Aquarius Reef Base Grants, 3410
NOAA Undersea Research Grants, 3411
NOAA Undersea Research Project Grants, 3412
NOCSAE Research Grants, 3414
Norcliffe Foundation Grants, 3415
North Carolina Arts Council Artist Fellowships, 3417
North Carolina Biotechnology Center Collaborative Funding Grants, 3421
North Carolina Biotechnology Center Oliver Smithies Faculty Recruitment Grant, 3423
Northrop Grumman Foundation Grants, 3429
Northwestern Mutual Foundation Grants, 3430
Northwest Fund for the Environment Grants, 3431
Norwegian Acad of Science & Letters Abel Prize, 3433
Notsew Orm Sands Foundation Grants, 3434
NRA Foundation Grants, 3437
NSERC Gerhard Herzberg Canada Gold Medal for Science and Engineering Grants, 3440
NSERC Industrial Undergraduate Student Research Awards, 3444
NSERC Innovation Challenge Awards, 3445
NSERC Northern Research Internships Program, 3447
NSERC Postdoctoral Fellowships Program, 3448
NSERC Undergraduate Research Awards, 3451
NSF-CONACyT Collaborative Research Grants, 3452
NSF-NIST Interaction in Chemistry, Materials Research, Molecular Biosciences, Bioengineering, and Chemical Engineering, 3453
NSF/DOE Partnership in Basic Plasma Science and Engineering Grants, 3454
NSF/NEH Documenting Endangered Languages Project Grants and Fellowships: data, infrastructure and computational methods, 3455
NSF Academic Research Infrastructure Program: Recovery and Reinvestment (ARI-R^2), 3456
NSF Accelerating Discovery In Science and Engineering through PETASCALE Simulations and Analysis, 3457
NSF Accelerating Innovation Research, 3458
NSF Nanostructures & Nanosystems Grants, 3459
NSF ADVANCE: Increasing the Participation and Advancement of Women in Academic Science and Engineering Careers Grants, 3461
NSF Advanced Learning Technologies Grants, 3462
NSF Advanced Technologies and Instrumentation Grants, 3463
NSF Advances in Biological Informatics Grants, 3464
NSF Advancing Theory in Biology Grants, 3465
NSF Aeronomy Grants, 3466
NSF Alan T. Waterman Award, 3467
NSF Algebra, Number Theory & Combinatorics Grants, 3468
NSF Analysis Grants, 3469
NSF Analytical & Surface Chemistry Grants, 3470
NSF Animal Developmental Mechanisms Grants, 3471
NSF Antarctic Aeronomy & Astrophysics Grants, 3472
NSF Antarctic Artists and Writers Grants, 3473
NSF Antarctic Earth Sciences Grants, 3474
NSF Antarctic Glaciology Grants, 3475
NSF Antarctic Int System Science Grants, 3476
NSF Antarctic Ocean & Atmospheric Grants, 3477
NSF Antarctic Organisms & Ecosystems Grants, 3478
NSF Antarctic Research Grants, 3479
NSF Antarctic Special Opportunity for Science aboard the Swedish Ice Breaker Oden Grants, 3480
NSF Applied Mathematics Grants, 3481
NSF Archaeology & Archeaometry Res Grants, 3482
NSF Arctic Natural Science Grants in Atmospheric Sciences, 3483
NSF Arctic Natural Science Grants in Biology, 3484
NSF Arctic Nat Science Grants in Earth Sciences, 3485
NSF Arctic Natural Science Grants in Glaciology, 3486
NSF Arctic Nat Science Grants in Ocean Sciences, 3487
NSF Arctic Natural Sciences Grants, 3488
NSF Arctic Research and Education Grants, 3489
NSF Arctic Research Opportunities, 3490
NSF Arctic Res Support & Logistics Grants, 3491
NSF Arctic Social Sciences Grants, 3492
NSF Arctic System Science Grants, 3493
NSF Astronomy and Astrophysics Postdoctoral Fellowships, 3494
NSF Astronomy & Astrophysics Res Grants, 3495
NSF Atmospheric Chemistry Grants, 3496
NSF Atmospheric Sciences Mid-Size Infrastructure Opportunity Grants, 3497
NSF Atomic Molecular & Optical Physics Grants, 3498
NSF Bio Databases & Informatics Grants, 3499
NSF Biological Information Tech & Systems, 3501
NSF Biological Oceanography Research Grants, 3502
NSF Biological Physics, 3503
NSF Biological Research Collections, 3504
NSF Biomedical Engineering Grants, 3505
NSF Biomedical Engineering and Engineering Healthcare Grants, 3506
NSF Biomolecular Systems Cluster Grants, 3507
NSF Biophotonics, Advanced Imaging, and Sensing for Human Health Grants, 3508
NSF Biotechnology, Biochemical, and Biomass Engineering Grants, 3509
NSF Biotic Surveys and Inventories Grants, 3510
NSF Catalysis and Biocatalysis Grants, 3511
NSF Cell Systems Cluster Grants, 3512
NSF Center for Synthesis in Biological Evolution, 3513
NSF Centers for Ocean Educ Excellence Grants, 3514
NSF Centers of Research Excellence in Science and Technology and HBCU Research Infrastructure for Science and Engineering Grants, 3515
NSF Chemical & Biological Separations Grants, 3516
NSF Chemical Oceanography Research Grants, 3517
NSF Chemistry Research Experiences for Undergraduates, 3518
NSF Chemistry Research Instrumentation and Facilities: Departmental Multi-User Instrumentation, 3519
NSF Chemistry Research Instrumentation and Facilities Grants, 3520
NSF CISE Accelerating Discovery in Science and Engineering Through Petascale Simulations and Analysis Grants, 3521
NSF CISE Advanced Learning Tech Grants, 3522
NSF CISE Assembling the Tree of Life Grants, 3523
NSF CISE Broadening Participation in Computing Grants, 3524
NSF CISE Cluster Exploratory Grants, 3525
NSF CISE Collaborative Research in Computational Neuroscience Grants, 3526
NSF CISE Communicating Research to Public Audiences Grants, 3527
NSF CISE Community-Based Data Interoperability Networks Grants, 3528
NSF CISE Computer and Network Systems: Core Grants, 3529
NSF CISE Computing and Communication Foundations: Core Grants, 3530

NSF CISE Computing Res Infrastructure Grants, 3531
NSF CISE Cross-Cutting Grants, 3532
NSF CISE Increasing the Participation and Advancement of Women in Academic Science and Engineering Careers Grants, 3534
NSF Climate and Large-Scale Dynamics Grants, 3535
NSF Collaborations in Math Geosciences, 3536
NSF Collaborative Research in Chemistry, 3537
NSF Communicating Res to Public Audiences, 3538
NSF Computational Math Research Grants, 3539
NSF Condensed Matter Physics Research Grants, 3540
NSF Conferences, Workshops, and Special Meetings in the Mathematical Sciences, 3541
NSF Continental Dynamics Research Grants, 3543
NSF Cooperative Studies Of The Earth's Deep Interior Grants, 3544
NSF Coupling, Energetics, and Dynamics of Atmospheric Regions Research Grants, 3545
NSF CreativeIT Grants, 3547
NSF Cultural Anthropology Research Grants, 3548
NSF Cyber-Enabled Discovery and Innovation, 3549
NSF Decision, Risk, and Management Science Research Grants, 3550
NSF Deep Underground Science and Engineering Laboratory Grants, 3551
NSF Developing Global Scientists and Engineers, 3552
NSF Director's Award for Distinguished Teaching Scholars, 3553
NSF Doctoral Dissertation Improvement Grants in the Directorate for Biological Sciences, 3554
NSF East Asia and Pacific Summer Institutes for U.S. Graduate Students, 3555
NSF Ecosystem Science Grants, 3556
NSF Education and Interdisciplinary Res Grants, 3557
NSF Electronic and Photonic Materials Grants, 3558
NSF Elementary Particle Physics Grants, 3559
NSF Emerging Frontiers in Research and Innovation Grants, 3560
NSF Engineering Research Centers Grants, 3561
NSF Engineering Research Grants, 3562
NSF Evolution of Developmental Systems Grants, 3563
NSF Extragalactic Astronomy and Cosmology Research Grants, 3564
NSF Faculty Early Career Development Grants, 3565
NSF Focused Research Groups in the Mathematical Sciences Grants, 3566
NSF Foundations Grants, 3567
NSF Galactic Astronomy Research Grants, 3568
NSF Genes and Genome Systems Cluster Grants, 3569
NSF Geometric Analysis Research Grants, 3570
NSF Geophysics Grants, 3571
NSF George E. Brown Jr. Network for Earthquake Engineering Simulation Research, 3572
NSF Geospace Environment Modeling, 3573
NSF Grants for the Analysis of Science & Tech, 3575
NSF Gravitational Physics Research Grants, 3576
NSF Hydrologic Sciences Research Grants, 3578
NSF Industry/University Cooperative Research Centers (I/UCRC) Grants, 3579
NSF Info & Intelligent Systems Core Grants, 3580
NSF Innovation and Organizational Change, 3581
NSF Innovative Technology Experiences for Students and Teachers, 3582
NSF Interagency Opportunities in Metabolic Engineering Research Grants, 3585
NSF Interfacial Processes and Thermodynamics Research Grants, 3586
NSF Linguistics Grants, 3587
NSF Living Stock Collections Grants, 3588
NSF Magnetospheric Physics Research Grants, 3590
NSF Marine Geology & Geophysics Res Grants, 3592
NSF Materials Research Grants, 3593
NSF Materials Research Science and Engineering Centers, 3594
NSF Materials World Network: Cooperative Activity in Materials Research between US Investigators and their Counterparts Abroad, 3595
NSF Mathematical Biology Grants, 3596
NSF Microbial Observatories and Microbial Interactions and Processes, 3597
NSF National Space Weather Program, 3599
NSF Natural, Social, and System Sciences and Research Support and Logistics Program, 3600
NSF Nuclear Physics Research Grants, 3601
NSF Ocean Drilling Grants, 3602
NSF Oceanographic Centers & Facilities Grants, 3603
NSF Ocean Technology and Interdisciplinary Coordination Grants, 3604
NSF OCE Education Grants, 3605
NSF Paleoclimate Grants, 3606
NSF Paleo Perspectives on Climate Change (P2C2) Grants, 3607
NSF Partnerships for Innovation Grants, 3610
NSF Partnerships for International Research and Education, 3611
NSF Partnerships in Astronomy & Astrophysics Research and Education, 3612
NSF Perception, Action & Cognition Res Grants, 3613
NSF Physical and Dynamic Meteorology Grants, 3614
NSF Physical Anthropology Research Grants, 3615
NSF Physical Chemistry Grants, 3616
NSF Physical Oceanography Research Grants, 3617
NSF Physics at the Information Frontier Grants, 3618
NSF Physics Frontiers Centers Grants, 3619
NSF Physics of Living Systems Grants, 3620
NSF Physics Research Grants, 3621
NSF Physiology and Ethology Research Grants, 3622
NSF Planetary Astronomy Research Grants, 3623
NSF Plant, Fungal and Microbial Developmental Systems Grants, 3624
NSF Plant Genome Research Program, 3625
NSF Plasma Physics Grants, 3626
NSF Polar Programs Division Research Grants, 3627
NSF Polymers Research Grants, 3628
NSF Population & Evolutionary Processes Grants, 3629
NSF Postdoctoral Res Fellowships in Biology, 3630
NSF Presidential Early Career Awards for Scientists and Engineers Grants, 3633
NSF Probability Research Grants, 3634
NSF Process and Reaction Engineering Grants, 3635
NSF Project 2010 Grants, 3636
NSF Research Coordination Networks in Biological Sciences Grants, 3637
NSF Research Experiences for Undergrads, 3638
NSF Research on Gender in Science and Engineering Grants, 3639
NSF Research to Aid Persons with Disabilities, 3640
NSF SBE Doctoral Dissertation Research Improvement Grants, 3641
NSF Science and Society (S&S), 3643
NSF Science and Technology Studies Grants, 3644
NSF Scientific Computing Research Environments for the Mathematical Sciences Grants, 3645
NSF Sedimentary Geology & Paleobiology Grants, 3646
NSF Small Business Innovation Research Grants, 3647
NSF Small Business Technology Transfer Programs Phase I Grants, 3648
NSF Social Psychology Research Grants, 3649
NSF Sociology Grants, 3650
NSF Solar, Heliospheric, and INterplanetary Environment Grants, 3651
NSF Solar Terrestrial Research Grants, 3652
NSF Starter Grants for Postdoctoral Fellows in Microbial Biology, 3653
NSF Statistics Research Grants, 3654
NSF Stellar Astronomy and Astrophysics Grants, 3655
NSF Synthesis and Modeling Project of the US Joint Global Ocean Flux Study, 3656
NSF Systematic Biology and Biodiversity Inventories Grants, 3657
NSF Tectonics Research Grants, 3658
NSF Theoretical Atomic, Molecular, and Optical Physics Grants, 3659
NSF Thermal Transport Processes Res Grants, 3660
NSF Topology Research Grants, 3661
NSF Tribal Colleges and Universities Program, 3662
NSF Undergraduate Research and Mentoring in the Biological Sciences, 3663
NSF University Radio Observatories Grants, 3664
NSF Upper Atmospheric Facilities Grants, 3665
Nuffield Found Children & Families Grants, 3666
Nuffield Foundation Law and Society Grants, 3667
Nuffield Found Oliver Bird Rheumatism Grants, 3668
Nuffield Foundation Open Door Grants, 3669
NWHF Community-Based Participatory Research Grants, 3670
NWHF Mark O. Hatfield Research Fellowship, 3671
NWO Added Value Plus for Soc Sciences Grants, 3672
NWO Aspasia Grants, 3673
NWO Athena Grants, 3674
NYAM Audrey & William Helfand Fellowship, 3679
NYAM Charles A. Elsberg Fellowship, 3680
NYAM David E. Rogers Fellowships, 3681
NYAM Edward N. Gibbs Memorial Lecture and Award in Nephrology, 3682
NYAM Edwin Beer Research Fellowship, 3683
NYAM Ferdinand C. Valentine Fellowship, 3684
NYAM Ferdinand C. Valentine Student Research Grants, 3685
NYAM Gladys Brooks Internship, 3686
NYAM Glorney-Raisbeck Fellowships, 3687
NYAM Glorney-Raisbeck Med Student Grants, 3688
NYAM Louis L. Seaman Medical Student Research Grants, 3690
NYAM Mary and David Hoar Fellowship, 3691
NYAM Paul Klemperer Fellowship, 3692
NYCT Biomedical Research Grants, 3694
NYCT Blood Disease Research Grants, 3696
NYCT New York City Environment Grants, 3697
NYPL Martin Duberman Visiting Scholars, 3700
NYPL Short-Term Research Fellowships, 3701
NYSCA Museum: Project Support Grants, 3707
NYSCA Music: Services to the Field Grants, 3708
NYZS/Wildlife Conservation Society Research Grants, 3709
OAH/IEHS John Higham Travel Grants, 3710
OAH Avery O. Craven Award, 3711
OAH Binkley-Stephenson Award, 3712
OAH Darlene Clark Hine Award, 3713
OAH David Thelen Award, 3714
OAH Ellis W. Hawley Prize, 3715
OAH Erik Barnouw Award, 3716
OAH Frederick Jackson Turner Award, 3717
OAH Germany Residency Grants, 3719
OAH Huggins-Quarles Award, 3720
OAH James A. Rawley Prize, 3721
OAH Japanese Residencies Grants, 3722
OAH Lerner-Scott Dissertation Prize, 3723
OAH Liberty Legacy Foundation Award, 3724
OAH Louis Pelzer Memorial Award, 3725
OAH Merle Curti Award, 3726
OAH Ray A. Billington Prize, 3727
OAH Richard W. Leopold Prize, 3728
OAH Willi Paul Adams Award, 3730
Obesity Society Grants, 3732
OBSSR Behavioral and Social Science Research on Understanding Health Disparities Grants, 3733
Ohio Learning Network Grants, 3734
Oklahoma City Community Grants, 3737
Olfactory Research Fund Grants, 3738
Olin Corporation Charitable Trust Grants, 3739
OneSight Research Foundation Block Grants, 3742
ONF Small Grants, 3743
ONR Biological & Biomedical Division Grants, 3744
ONR Casualty Care and Management Research Grants, 3746
ONR Casualty Prevention Grants, 3747
ONR Circuits/Architectures Grants, 3748
ONR Cognitive Science Basic Research Grants, 3749
ONR Communications Systems Grants, 3750
ONR Computational Analysis Grants, 3751
ONR Computational Neuroscience & Biorobotics Grants, 3752
ONR Defense University Research Instrumentation Program Grants, 3753
ONR Device and Circuit Cooling Grants, 3754
ONR DOD National Defense Science & Engineering Graduate Fellowships, 3755
ONR Electronic and Optical Ceramics Research Grants, 3756

ONR Electronic Combat Technology Grants, 3757
ONR Electronics, Sensors, and Network Research Grants, 3758
ONR Electronics Division Research Grants, 3759
ONR Environmental Quality Research Grants, 3760
ONR Expeditionary Maneuver Warfare & Combating Terrorism Department Grants, 3761
ONR Faculty Sabbatical Leave Grants, 3762
ONR Historically Black Colleges and Universities Future Engineering Faculty Fellowships, 3763
ONR Independent Research & Devel Grants, 3764
ONR Ionosphere Interactions Research Grants, 3765
ONR Joint Capability Technology Demonstration Grants, 3766
ONR Magnetics Grants, 3767
ONR Manufacturing Technology Grants, 3768
ONR Marine Meteorology and Atmospheric Effects Research Grants, 3769
ONR Maritime Sensing Grants, 3770
ONR Mathematical, Computer, and Information Sciences Research Grants, 3771
ONR Medical Science and Technology Division Grants, 3772
ONR Microelectronics Applied Research Grants, 3773
ONR Multidisciplinary Research Program of the URI Grants, 3774
ONR Naval Research Enterprise Intern Program Grants, 3775
ONR Navy Postdoctoral Fellowships, 3776
ONR Optical and Biological Oceanography Research Grants, 3777
ONR Radio Frequency Sensor Processing Research Grants, 3778
ONR Radio Frequency Solid State Devices Applied Research Grants, 3779
ONR Sea Platforms & Weapons Div Grants, 3780
ONR Semiconductor Devices Grants, 3781
ONR Small Business Innovation Research (SBIR/STTR) Grants, 3782
ONR Small Business Technology Transfer Grants, 3783
ONR Summer Faculty Research Grants, 3784
ONR Superconducting Electronics Grants, 3785
ONR Surface/Aerospace Surveillance Grants, 3786
ONR Target Tracking and Sensor Fusion Grants, 3787
ONR Young Investigator Grants, 3788
ONS Major Research Grants, 3789
Ontario Arts Council Integrated Arts Grants, 3790
Open Spaces Sacred Places National Awards, 3791
Oppenstein Brothers Foundation Grants, 3792
Orchard Foundation Grants, 3793
Organic Farming Research Foundation Grants, 3795
Organization for Tropical Studies Fellowships, 3796
Organization for Tropical Studies Pilot Awards for Exploratory Research, 3797
Orgone Biophysical Research Lab Lou Hochberg Awards, 3798
ORISE Department of Commerce Internship for Postsecondary Students, 3800
ORISE Faculty Research Participation at the U.S. Army Joint POW/MIA Accounting Command, Central Identification Laboratory, 3805
ORISE National Library of Medicine, Center for Biotechnology Information Scientific Visitors Program, 3810
ORISE Postgraduate Research Participation at the US Army Central Identification Lab, Hawaii, 3822
Oscar Rennebohm Foundation Grants, 3832
OSF Fellowships, 3840
OSF Human Rights and Governance Internships, 3845
OSRI Biological Monitoring: Intertidal Resources at Risk to Oil Spills, 3854
OTS Francis J. Bossuyt Fellowship, 3855
OTS Hovore-Horn Fellowships, 3856
OTS Stanley Smith Horticultural Fellowships, 3857
OTS Undergrad NAPIRE Summer Internships, 3858
Outstanding Junior Investigator Program in High Energy Physics, 3859
Padilla Bay Foundation Grants, 3863
Padilla Bay National Estuarine Research Reserve Graduate Research Fellowships, 3864
Padilla Bay National Estuarine Research Reserve Research Assistantships, 3865
Pancreatic Cancer Action Network Fellowship, 3866
Pancreatic Cancer Action Network-AACR Pathway to Leadership Grants, 3867
Parapsychology Foundation Grants, 3868
Park Foundation Grants, 3870
Partnership for Cures Charles E. Culpeper Scholarships in Medical Science, 3871
Partnership for Cures Two Years To Cures Grants, 3872
Patron Saints Foundation Grants, 3874
Paul G. Allen Family Foundation Grants, 3876
Paul P. Fidler Research Grants, 3878
Paul Stock Foundation Grants, 3879
PDF International Research Grants, 3882
Peabody Foundation Grants, 3886
Peacock Foundation Grants, 3887
Pediatric Brain Tumor Found Research Grants, 3888
Pediatric Cancer Research Foundation Grants, 3889
Pell Grants, 3890
Penn Graduate School of Education Dean's Fellowships, 3891
Penn Graduate School of Education Foreign Language Area Studies Fellowship, 3892
Penny Severns Breast, Cervical and Ovarian Cancer Research Grants, 3896
Percy B. Ferebee Endowment Grants, 3897
Perkin Fund Grants, 3898
Perl 6 Microgrants, 3899
Perpetual Trust for Charitable Giving Grants, 3900
Peter F. McManus Charitable Trust Grants, 3903
Pew Charitable Trusts Biomedical Res Grants, 3904
Pezcoller Foundation-AACR International Award for Cancer Research, 3907
Pfizer Advancing Research in Transplantation Science Research Awards, 3908
Pfizer Anti Infective Research EU Grants, 3909
Pfizer ASPIRE EU Antifungal Research Awards, 3910
Pfizer ASPIRE EU Dupuytren's Contracture Research Awards, 3911
Pfizer ASPIRE EU Emerging Mechanisms of Resistance Antibacterial Research Awards, 3912
Pfizer ASPIRE EU MRSA Nosocomial Pneumonia & MRSA Complicated Skin & Soft Tissue Infections Antibacterial Research Awards, 3913
Pfizer ASPIRE North America Broad Spectrum Antibiotics for the Treatment of Gram-Negative or Polymicrobial Infections Research Awards, 3914
Pfizer ASPIRE North America Hemophilia Research Awards, 3915
Pfizer ASPIRE North America Narrow Spectrum Antibiotics for the Treatment of MRSA Research Awards, 3916
Pfizer ASPIRE North America Rheumatology Research Awards, 3917
Pfizer ASPIRE Worldwide Endocrine Young Investigator Grants, 3918
Pfizer Australia Cancer Research Grants, 3919
Pfizer Australia Neuroscience Research Grants, 3920
Pfizer Australia Paediatric Endocrine Care Research Grants, 3921
PFizer Compound Transfer Agreements, 3922
Pfizer Global Investigator Research Grants, 3924
Pfizer Global Research Awards for Nicotine Independence, 3925
Pfizer Inflammation Competitive Res Awards, 3926
Pfizer Medical Education Track Two Grants, 3927
Pfizer Research Initiative Dermatology Grants, 3928
Pfizer Research Initiative Rheumatology Grants, 3929
PHF Andrew W. Mellon Postdoctoral Fellowships in the Humanities, 3931
PHF Penn Faculty Research Fellowships, 3932
PHF Regional Mellon Faculty Research and Cultural Fellowships, 3933
PHF Undergraduate Humanities Forum Research Fellowships, 3934
Phi Delta Kappa Outstanding Doctoral Dissertation Awards, 3936
Phi Kappa Phi Found Love of Learning Awards, 3937
Phi Kappa Phi Scholar Award, 3939
Phi Upsilon Omicron Alumni Research Grant, 3940
Phi Upsilon Omicron Presidents Res Fellowship, 3949
PhRMA Foundation Health Outcomes Post Doctoral Fellowships, 3950
PhRMA Foundation Health Outcomes Research Starter Grants, 3951
PhRMA Foundation Health Outcomes Sabbatical Fellowships, 3952
PhRMA Foundation Informatics Research Starter Grants, 3955
PhRMA Foundation Informatics Sabbatical Fellowships, 3956
PhRMA Foundation Paul Calabresi Medical Student Research Fellowship, 3957
PhRMA Foundation Pharmaceutics Postdoctoral Fellowships, 3958
PhRMA Foundation Pharmaceutics Pre Doctoral Fellowships, 3959
PhRMA Foundation Pharmaceutics Research Starter Grants, 3960
PhRMA Foundation Pharmaceutics Sabbatical Fellowships, 3961
PhRMA Foundation Pharmacology/Toxicology Post Doctoral Fellowships, 3962
PhRMA Foundation Pharmacology/Toxicology Pre Doctoral Fellowships, 3963
PhRMA Foundation Pharmacology/Toxicology Research Starter Grants, 3964
PhRMA Foundation Pharmacology/Toxicology Sabbatical Fellowships, 3965
Pi Lambda Theta Research Grants, 3966
Pinkerton Foundation Grants, 3967
Pioneer Hi-Bred International Scholarships, 3969
Pioneer Hi-Bred Society Fellowships, 3970
Pioneer Hi-Bred University Fellowships, 3971
Piper Trust Healthcare & Med Research Grants, 3972
Pittsburgh Foundation Medical Research Grants, 3973
PKD Foundation Lillian Kaplan International Prize for the Advancement in the Understanding of Polycystic Kidney Disease, 3974
PKD Foundation Research Grants, 3975
Plan for Social Excellence Grants, 3976
Plastic Surgery Educational Foundation/AAO-HNSF Combined Grant, 3977
Plastic Surgery Educational Foundation Basic Research Grants, 3978
Plastic Surgery Educational Foundation Research Fellowship Grants, 3979
Potts Memorial Foundation Grants, 3985
Price Family Charitable Fund Grants, 3987
Princeton University Center for Human Values Graduate Prize Fellowships, 3989
Princeton University Center for Human Values Graduate Student Top-Up Prizes, 3990
Princeton University Center for Human Values Harold T. Shapiro Postdoctoral Research Associate in Bioethics, 3991
Princeton University Center for Human Values Laurance S. Rockefeller Visiting Fellowships, 3992
Princeton University Center for Human Values Political Philosophy Travel Grants, 3993
Princeton University Center for Human Values Postdoctoral Research Associate in Values and Public Policy, 3994
Princeton University Lewis Center for the Arts Hodder Fellowships, 3996
Project AWARE Foundation Grants, 3999
Prostate Cancer Foundation Research Grants, 4000
Prudential Foundation Education Grants, 4002
PVA Research Foundation Grants, 4004
R.C. Baker Foundation Grants, 4006
RACGP/APHCRI Indigenous Health Award, 4008
RACGP/CONROD Research Fellowship, 4009
RACGP Alan Chancellor Award, 4010
RACGP Cardiovascular Research Grants in General Practice, 4011
RACGP Chris Silagy Research Scholarship, 4012
RACGP Family Medical Care, Education and Research Grants, 4013
RACGP Grieve Memorial Award, 4014

RACGP Rex Walpole Travelling Fellowship, 4016
RACGP Vicki Kotsirilos Integrative Med Grants, 4017
Rajiv Gandhi Foundation Grants, 4025
RA Kleinhans Fellowship for Research in Tropical Nontimber Forest Products, 4026
RAND Corporation Fellowships in Population Studies and the Study of Aging, 4029
RAND Corporation Military Fellowships, 4031
RAND Corp Transatlantic Postdoctoral Fellowships for International Relations and Security, 4032
Rayonier Foundation Grants, 4034
RCPSC/AMS CanMEDs Research and Development Grants, 4035
RCPSC Medical Education Research Grants, 4038
RCPSC Royal College Visiting Professorships in Medical Research, 4041
RCY - NSF Ocean Sciences Div Research Grants, 4043
RealNetworks Helix Community Grants, 4044
Regenstrief General Internal Medicine - General Pediatrics Research Fellowship Program, 4045
Regenstrief Geriatrics Fellowship, 4046
Regenstrief Master Of Science In Clinical Research/ CITE Program, 4047
Rehabilitation Nursing Found Research Grants, 4050
Reinberger Foundation Grants, 4051
Religion and Health: Effects, Mechanisms, and Interpretation RFP, 4052
Research Corporation Cottrell Scholars Awards, 4053
Research Corporation Multi-Investigator Cottrell College Science Awards, 4054
Research Corporation Single-Investigator Cottrell College Science Awards, 4056
Research Program at Earthwatch Grants, 4057
Rett Syndrome Research Foundation Grants, 4058
RGK Foundation Grants, 4060
Richard & Susan Smith Family Found Grants, 4064
Richard Davoud Donchian Foundation Grants, 4066
Rich Foundation Grants, 4068
Robert B McMillen Foundation Grants, 4072
Robert E. and Evelyn McKee Foundation Grants, 4074
Robert G. Cabell III and Maude Morgan Cabell Foundation Grants, 4075
Robert Leet & Clara Guthrie Patterson Grants, 4076
Robert M. Hearin Foundation Grants, 4077
Robert Schalkenbach Foundation Grants, 4079
Robert Sterling Clark Foundation Reproductive Rights and Health Grants, 4081
Robert W. Deutsch Foundation Grants, 4082
Roche Organ Transplant Research Found Grants, 4083
Rockefeller Archive Center Ehrlich Grants, 4084
Rockefeller Archive Center Grants in Aid, 4085
Rockefeller Archive Center Research Grants, 4086
Rockefeller Foundation MAP Fund Grants, 4090
Rockford College Faculty Fellowships, 4092
Rockwell International Corporate Trust Grants, 4093
Rocky Mountain Mineral Law Found Grants, 4094
Roeher Institute Research Grants, 4095
Roger Adams Award in Organic Chemistry, 4096
Rohm and Haas Company Grants, 4098
Ronald Breslow Award for Achievement in Biomimetic Chemistry, 4100
Ronald McDonald House Charities Grants, 4101
Rosalinde and Arthur Gilbert Foundation/AFAR New Investigator Awards in Alzheimer's Disease, 4102
Rosenthal Writer-in-Residence Program, 4105
Ross Foundation Grants, 4106
Royal Belgian Numismatic Society Res Grants, 4107
Royal Caribbean Cruises Ocean Fund, 4108
Royal Scottish Geographical Society Expedition Grants, 4110
Royal Scottish Geographical Society Small Research Grants, 4111
Royal Scottish Geographical Society Small Travel Grants, 4112
Roy J. Carver Charitable Trust Higher Education Grants, 4113
Roy J. Carver Charitable Trust Medical and Science Research Grants, 4114
RRF General Grants, 4115
RSC Abbyann D. Lynch Medal in Bioethics, 4116
RSC Alice Wilson Award, 4117
RSC Bancroft Award, 4118
RSC Jason A. Hannah Medal, 4119
RSC John L. Synge Award, 4120
RSC McLaughlin Medal, 4121
RSC Pierre Chauveau Medal, 4123
RSC Rutherford Memorial Medals, 4124
RSC Sir John William Dawson Medal, 4125
Russell Family Foundation Environmental Sustainability Grants, 4128
Russell Sage Foundation Grants, 4129
Ruth Estrin Goldberg Memorial for Cancer Grants, 4131
RWJF Active Living Research Grants, 4132
RWJF Changes in Health Care Financing and Organization Grants, 4133
RWJF Changes in Health Care Financing and Organization Small Grants, 4134
RWJF Childhood Obesity Grants, 4135
RWJF Clinical Scholars Grants, 4136
RWJF Disparities Research for Change Grants, 4137
RWJF Evaluating Innovations in Nursing Education Grants, 4138
RWJF Harold Amos Medical Faculty Development Scholars, 4140
RWJF Health and Society Scholars, 4141
RWJF Health Policy Fellowships, 4142
RWJF Healthy Eating Research Grants, 4143
RWJF Interdisciplinary Nursing Quality Research Initiative, 4144
RWJF Investigator Awards in Health Policy Res, 4145
RWJF New Connections Jun Investigator Grants, 4146
RWJF New Connections Mid-Career Consultants Grants, 4147
RWJF Nurse Faculty Scholars, 4148
RWJF Partners Investing in Nursing Grants, 4149
RWJF Pioneer Portfolio Grants, 4150
Ryan Gibson Foundation Grants, 4151
S.E.VEN Fund Annual Grants, 4153
S.E.VEN Fund Mini-Grants, 4154
S.E.VEN Fund Open Enterprise Solutions to Poverty RFP Grants, 4155
S.E.VEN Fund WHY Prize, 4156
Sage Foundation Grants, 4158
Samueli Foundation Health Grants, 4161
Samueli Institute Scientific Research Grants, 4162
San Antonio Area Foundation Grants, 4163
Sandler Program for Asthma Research Grants, 4165
Sara Elizabeth O'Brien Trust Grants, 4166
Sarasota Memorial Healthcare Found Grants, 4167
SARE Graduate Student Grants - NE Region, 4169
SARE Graduate Student Grants - South Region, 4170
SARE Large Systems Research Grants - Southern Region, 4172
SARE On-Farm Res Grants - South Region, 4173
SARE Producer Research and Education Grants- Western Region, 4174
SARE Professional & Producer Grants - West, 4175
SARE Professional Devel Grants - NE Reg, 4176
SARE Research and Education Grants - Northeast Region, 4178
SARE Research and Education Grants - Southern Region, 4179
Sarkeys Foundation Grants, 4181
Savoy Foundation Post-Doctoral and Clinical Research Fellowships, 4184
Savoy Foundation Research Grants, 4185
Savoy Foundation Studentships, 4186
Schomburg Scholars-in-Residence Fellowships, 4187
Schramm Foundation Grants, 4188
Scientific Communities Initiative Grants, 4189
Scleroderma Foundation Established Investigator Grants, 4190
Scleroderma Foundation New Investigator Grants, 4191
Searle Freedom Trust Grants, 4192
Seattle Foundation Medical Funds Grants, 4194
Seaver Institute Grants, 4195
SeaWorld and Busch Gardens Conservation Fund Grants, 4196
Self Foundation Grants, 4197
Sensient Technologies Foundation Grants, 4199
SFI Principal Investigator Grants, 4201
Shastri Indo-Canadian Institute Arts Fellowships, 4219
Shastri Indo-Canadian Institute Collaborative Research Project Grants, 4220
Shastri Indo-Canadian Institute Faculty Research Fellowships, 4222
Shastri Indo-Canadian Institute Faculty Training Fellowships, 4223
Shastri Indo-Canadian Institute Millennium Development Research Grants, 4227
Shastri Indo-Canadian Institute Post-Doctoral Research Fellowships, 4229
SHGAPE Vincent P. DeSantis Prize, 4234
Shorenstein Center Fellowships, 4235
SHOT Abbott Payson Usher Prize, 4237
SHOT Brooke Hindle Postdoctoral Fellowships, 4238
SHOT International Scholars, 4242
SHOT Melvin Kranzberg Dissert Fellowship, 4245
SHOT NASA Fellowship, 4246
SHRM Dissertation Awards, 4248
SHRM Foundation Research Grants, 4249
SHRM Michael R. Losey Human Resource Research Award, 4250
Sidgmore Family Foundation Grants, 4252
Sid W. Richardson Foundation Grants, 4253
Sigma Theta Tau International /American Nurses Foundation Grant, 4254
Sigma Theta Tau International / Council for the Advancement of Nursing Science Grants, 4255
Sigma Theta Tau International / Oncology Nursing Society Grant, 4256
Sigma Theta Tau International Doris Bloch Research Award, 4257
Sigma Theta Tau Small Grants, 4258
Sigma Xi Grants in Aid of Research, 4259
Sigrid Juselius Foundation Grants, 4260
Simone and Cino del Duca Grand Prix Awards, 4261
Sioux Falls Area Community Foundation Community Fund Grants (Unrestricted), 4262
Sioux Falls Area Community Found Spot Grants, 4263
Siragusa Foundation Health Services & Medical Research Grants, 4264
Sir Dorabji Tata Trust Individual Education and Travel Grants, 4265
Skaggs Foundation Grants, 4266
Skin Cancer Foundation Research Grants, 4267
Smith Family Awards for Excellence in Biomedical Research, 4268
Smith Richardson Foundation Domestic Public Research Fellowship, 4269
Smith Richardson Foundation Strategy and Policy Fellowship, 4270
Smith Richardson Foundation World Politics and Statecraft Fellowship, 4271
Smithsonian Biodiversity Genomics and Bioinformatics Postdoctoral Fellowships, 4272
Smithsonian Environmental Research Center Graduate Student Fellowships, 4273
Smithsonian Environmental Research Center Postdoctoral/Senior Fellowships, 4275
Smithsonian Environmental Research Center Predoctoral Fellowships, 4276
Smithsonian Graduate Student Fellowships, 4277
Smithsonian Minority Visiting Student Awards, 4279
Smithsonian Molecular Evolution Fellowships, 4280
Smithsonian Museum Conservation Institute Analytical Studies Group Internships, 4281
Smithsonian Museum Conservation Institute Internships for Research Projects, 4282
Smithsonian Museum Conservation Institute Research Post-Doctorate/Post-Graduate Fellowships, 4283
Smithsonian National Air and Space Museum Aviation Space Writers Foundation Grant, 4284
Smithsonian National Air and Space Museum Charles A. Lindbergh Chair in Aerospace History, 4285
Smithsonian National Air and Space Museum Guggenheim Fellowships, 4286
Smithsonian National Air and Space Museum Postdoctoral Earth and Planetary Sciences Fellowship, 4287

Smithsonian National Air and Space Museum Verville Fellowships, 4288
Smithsonian National Museum of American History Lemelson Center Fellowships, 4289
Smithsonian National Museum of Nat Hist Lemelson Center Travel to Collections Awards, 4290
Smithsonian Native American Community Scholars Awards, 4291
Smithsonian Native American Visiting Student Awards, 4292
Smithsonian Tropical Research Institute Short Term Fellowships, 4297
SNM/ Covidien Seed Grant in Molecular Imaging/ Nuclear Medicine Research, 4298
SNM Molecular Imaging Research Grant For Junior Medical Faculty, 4299
SNM Pilot Research Grants, 4300
SNM Pilot Research Grants in Nuclear Medicine/ Molecular Imaging, 4301
SNM Postdoc Molecular Imaging Scholar Grants, 4302
SNM Tetalman Memorial Young Inv Awards, 4304
Society for Imaging Informatics in Medicine Emeritus Mentor Grants, 4305
Society for Imaging Informatics in Medicine Micro Grants, 4306
Society for Imaging Informatics in Medicine Research Grants, 4307
Society for Imaging Informatics in Medicine Small Grant for Product Support Development, 4308
Society for Imaging Informatics in Medicine Small Training Grants, 4309
Society for the Arts in Healthcare Environmental Arts Research Grant, 4311
Sony Corporation of America Corporate Philanthropy Grants, 4329
South Florida Student Shark Program, 4330
Special Olympics Youth Fan Grants, 4333
Spencer Foundation Major Research Grants, 4335
Spencer Foundation Small Research Grants, 4337
SPIE A.E. Conrady Award, 4338
SPIE C.S. Vikram Award in Optic Metrology, 4339
SPIE Dennis Gabor Award, 4340
SPIE Frits Zernike Award, 4341
SPIE George W. Goddard Award, 4342
SPIE Technology Achievement Award, 4344
SPIRE Postdoctoral Fellowships, 4345
SPJ Eugene C. Pulliam Fellowship for Editorial Writing, 4346
SRC Company-Named Scholarships, 4351
SRC Core Research Grants, 4352
SRC Cross-Disciplinary Research Grants, 4353
SRC Educational Science Conference Grants, 4354
SRC Educational Science Doctoral Grants, 4355
SRC Educational Science Grants, 4356
SRC Educational Science Grants for Invitation of Visiting Scientists, 4357
SRC Educational Science Grants for Travels to Foreign Institutions as a Guest Scientist, 4358
SRC Educational Science Junior Research Grants, 4359
SRC Educational Science Lang Checking Grants, 4360
SRC Educational Science Medium-Cost Scientific Equipment Grants, 4361
SRC Educational Science Network Grants, 4362
SRC Educational Science Postdoc Fellowships, 4363
SRC Educational Science Project Res Grants, 4364
SRC Education Science Grants for Postdoctoral Positions in Sweden, 4366
SRC GRC and SEMATECH Joint Projects, 4369
SRC Guest Professorship Program, 4370
SRC Humanities and Social Science Postdoctoral Fellowships, 4372
SRC Humanities and Social Sciences Grants, 4373
SRC Humanities and Social Sciences Grants for Postdoctoral Positions In Sweden, 4374
SRC Landmark Innovation Award, 4376
SRC Master's Scholarships, 4378
SRC Medical Research Grants, 4379
SRC Medical Science Grants for Postdoctoral Positions in Sweden, 4381
SRC Medical Science Postdoctoral Fellowships, 4382
SRC Natural and Engineering Science Conference Grants, 4383
SRC Natural and Engineering Science Grants for Postdoctoral Positions in Sweden, 4384
SRC Natural and Engineering Science Postdoctoral Fellowships, 4385
SRC Natural and Engineering Science Senior Research Positions in Sweden, 4386
SRC Natural and Engineering Sciences Grants, 4387
SRC Research Customization Program Grants, 4390
SRC Research Infrastructures Grants, 4391
SRC Simon Karecki Fellowship, 4393
SRC Undergraduate Research Opportunities, 4395
SSHRC-APC Automotive Partnership Canada, 4396
SSHRC-CIHR-SC Research Data Centres Joint Initiative, 4397
SSHRC-IDRC International Partnerships for Sustainable Societies, 4399
SSHRC-NSERC College and Community Innovation Enhancement Grants, 4400
SSHRC-Public Safety Canada Kanishka Project Initiative, 4401
SSHRC Aid to Small Universities, 4403
SSHRC Banting Postdoctoral Fellowships, 4404
SSHRC Digging into Data Challenge Grants, 4406
SSHRC Doctoral Fellowships, 4407
SSHRC Insight Development Grants, 4409
SSHRC Institutional Grants, 4411
SSHRC Joseph-Armand Bombardier CGS Doctoral Scholarships, 4412
SSHRC Metropolis Project, 4414
SSHRC Sport Participation Research Initiative: Doctoral Award Supplements, 4416
SSHRC Sport Participation Research Initiative: Postdoctoral Fellowship Supplements, 4417
SSRC-Van Alen Fellowships, 4419
SSRC African Peacebuilding Network Research Grants, 4422
SSRC Collaborative Research Grants on Environment and Health in China, 4423
SSRC Dissertation Proposal Development Faculty Field Fellowships, 4424
SSRC Dissertation Proposal Development Student Fellowships, 4425
SSRC Eurasia Program Dissertation Development Awards, 4427
SSRC Eurasia Program Pre-Dissertation Awards, 4428
SSRC Int'l Dissertation Research Fellowships, 4429
SSRC Japan Society for the Promotion of Science Fellowships, 4431
SSRC Mellon Mays Predoctoral Research Grants, 4432
SSRC New Directions in the Study of Prayer Research Grants, 4433
SSRC Next Generation Social Sciences in Africa Doc Dissertation Completion Fellowships, 4434
SSRC Next Generation Social Sciences in Africa: Doctoral Dissertation Proposal Fellowships, 4435
SSRC Next Generation Social Sciences in Africa: Doctoral Dissertation Research Fellowships, 4436
SSRC Postdoc Fellowship for Transregional Research: Inter-Asian Contexts and Connections, 4438
SSSR Distinguished Article Award, 4440
SSSR Student Paper Award, 4441
Stanford Center on Adolescence Youth Purpose Research Grants, 4443
Stanley Smith Horticultural Trust Grants, 4446
Starr Foundation Grants, 4447
STC Research Grants, 4450
Strake Foundation Grants, 4452
Streilein Foundation for Ocular Immunology Visiting Professorships, 4453
Stroke Association Allied Health Professional Research Bursaries, 4454
Stroke Association Clinical Fellowships, 4455
Stroke Association Research Project Grants, 4456
STTI/ATI Educational Assessment Nursing Research Grants, 4457
STTI Alpha Eta Collaborative Research Grants, 4458
STTI Emergency Nurses Association Foundation Grant, 4459
STTI Environment of Elder Care Nursing Research Grants, 4460
STTI Joan K. Stout RN Research Grants, 4461
STTI National League for Nursing Grants, 4462
STTI Rosemary Berkel Crisp Research Awards, 4463
STTI Virginia Henderson Clinical Res Grants, 4464
Sun Grid Research Grants, 4467
Surdna Found Sustainable Environments Grants, 4471
Susan G. Komen Breast Cancer Foundation Brinker Awards for Scientific Distinction, 4472
Susan G. Komen Breast Cancer Foundation Career Catalyst Research Grants, 4473
Susan G. Komen Breast Cancer Foundation Challege Grants: Breast Cancer and the Environment, 4474
Susan G. Komen Breast Cancer Foundation Challege Grants: Investigator Initiated Research, 4475
Susan G. Komen Breast Cancer Foundation Challenge Grants: Career Catalyst Research, 4476
Susan G. Komen Breast Cancer Foundation Investigator Initiated Research Grants, 4477
Swiss National Science Foundation Fellowships for Advanced Researchers, 4483
Swiss National Science Foundation Fellowships for Prospective Researchers, 4484
Swiss National Science Foundation Marie Heim-Vogtlin Grants, 4485
Sybil G. Jacobs Award for Outstanding Use of Tobacco Industry Documents, 4487
Syracuse University Graduate Fellowships, 4488
T. James Kavanagh Foundation Grants, 4489
Tandem Launch Technologies Grants, 4493
Taub Foundation Grants for MDS Research, 4496
TD4HIM Foundation Grants, 4497
Temple Univ George D. McDowell Fellowship, 4499
The Ray Charles Foundation Grants, 4502
Thomas and Dorothy Leavey Foundation Grants, 4506
Thomas Austin Finch, Sr. Foundation Grants, 4507
Thome Foundation Awards in Age-Related Macular Degeneration Research, 4509
Thome Foundation Awards in Alzheimer's Disease Drug Discovery Research, 4510
Thompson Charitable Foundation Grants, 4511
Thomson Reuters / MLA Doctoral Fellowships, 4512
Thoracic Foundation Grants, 4513
Thornton Foundation Grants, 4514
Thrasher Research Fund Early Career Awards, 4515
Thrasher Research Fund Grants, 4516
TIAA-CREF Institute Research Grants, 4517
TIRF Doctoral Dissertation Grants, 4520
Tobias Landau Fellowships in Marine Biology, 4521
Tourette Syndrome Association Post-Doctoral Fellowships, 4523
Tourette Syndrome Association Research Grants, 4524
TRDRP California Research Awards, 4525
TRDRP Dissertation Research Awards, 4526
TRDRP Exploratory/Devel Research Awards, 4527
TRDRP New Investigator Awards, 4528
TRDRP Participatory Research Awards, 4529
TRDRP Postdoctoral Fellowship Awards, 4530
TRDRP Research Grants, 4531
Triological Society Research Career Development Awards, 4534
Turtle Conservation Fund Grants, 4535
Twentieth-Century Japan Research Awards, 4536
U.S. Department of Education American Overseas Research Centers, 4537
U.S. Department of Education Disability and Rehabilitation Research and Related Projects, 4538
U.S. Department of Education Foreign Languages and Area Studies Fellowships, 4539
U.S. Department of Education Graduate Assistance in Areas of National Need, 4540
U.S. Department of Education National Resource Centers, 4542
U.S. Department of Education Rehabilitation Engineering Research Centers Grants, 4544
U.S. Department of Education Rehabilitation Research Training Centers, 4545
U.S. Department of Education Ronald E. McNair Postbaccalaureate Achievement Grants, 4546

U.S. Department of Education United States-Russia Program: Improving Research and Educational Activities in Higher Education, 4547
UCLA / RAND Corporation Post-Doctoral Fellowships, 4548
UCR Humanities Grad Research Grants, 4551
UE Explore Undergraduate Research Grants, 4553
UEF Advancement of Engineering Grants, 4554
UICC American Cancer Society International Fellowships for Beginning Investigators, 4555
UICC International Cancer Technology Transfer Fellowships, 4556
UICC Yamagiwa-Yoshida Memorial International Cancer Study Grants, 4558
UNC Carolina Postdoctoral Program for Faculty Diversity, 4560
UNCFSP NAI-MIRS Sabbatical Awards, 4561
UNCFSP NASA Science and Technology Institute Summer Scholars Program, 4564
Union Pacific Found Community & Civic Grants, 4568
Union Pacific Health and Human Services Grants, 4569
United Methodist Church Asian, Black, Hispanic and Native American History Research Grants, 4570
United Methodist Church Center for the Evangelical UBH Audrie E. Reber Memorial Award, 4571
United Methodist Church John Harrison Ness Memorial Award, 4572
United Methodist Church Josephine Forman Scholarship, 4573
United Methodist Church Women in History Writing Award, 4575
United Methodist Church World is My Parish United Methodist History Research Grant, 4576
United States Institute of Peace - Jennings Randolph Peace Dissertation Scholarships, 4579
United States Institute of Peace - Jennings Randolph Senior Fellowships, 4580
United States Institute of Peace Annual Grants, 4582
USAID Comprehensive District-Based Support for Better HIV/TB Patient Outcomes Grants, 4595
USAID Development Innovation Accelerator Broad Agency Announcement Grants, 4596
USAID Devel Innovation Ventures Grants, 4597
USAID Family Planning and Reproductive Health Methods Grants, 4598
USAID Microbicides Research, Development, and Introduction Grants, 4599
USAID Palestinian Community Infrastructure Development Grants, 4600
USAID Res & Innovation for Health Supply Chain Systems and Commodity Security Grants, 4601
USAID Strengthening RMNCH Through Indian Health Professional Associations and Scaling Up Uptake Grants, 4602
USCD/NOAA Illinois-Indiana Sea Grant Core Proposals, 4604
USC Pentecostal and Charismatic Research Initiative Individual Grants, 4605
USC Pentecostal and Charismatic Research Initiative Regional Center Grants, 4606
US CRDF CGP-Climate Change and Energy Competition For Joint U.S.-Ukrainian Research Teams, 4607
US CRDF CGP-RFBR Climate Change and Energy Competition For Joint U.S.-Russian Research Teams, 4608
US CRDF CGP ETF Energy Research Competition For Joint U.S.-Estonian Research Teams, 4609
US CRDF Leishmaniasis: Collaborative Opportunities in North Africa and the Middle East, 4610
US CRDF Science and Technology Entrepreneurship Program Business Partnership Grant Competition Armenia, 4611
USDA 1890 Institution Teaching, Research, and Extension Capacity Building Grants, 4613
USDA Agriculture and Food Research Initiative Competitive Grants - Food Security, 4614
USDA Alaska Native-Serving & Native Hawaiian-Serving Inst Education Competitive Grants, 4615
USDA Animal Health & Disease Res Grants, 4616
USDA Basic and Applied Research Grants, 4617
USDA Biotechnology Risk Assessment Grants, 4618
USDA Decadal and Regional Climate Prediction Using Earth System Models Grants, 4621
USDA Hazardous Fuels Woody Biomass Utilization Grants, 4626
USDA Hispanic-Serving Institutions Education Grants, 4628
USDA Integrated Research Education and Extension Grants - Organic Transitions, 4630
USDA International Science and Education Competitive Grants, 4631
USDA Interregional Research Project Minor Crop Pest Management Grants (IR-4), 4632
USDA National Integrated Food Safety Initiative Grants, 4635
USDA Plant Feedstock Genomics for Bioenergy: A Joint Research Grant, 4639
USDA Policy Research Center Grants, 4640
USDA Rangeland Research Grants, 4641
USDA Resident Instruction Grants for Higher Education in Insular Areas, 4645
USDA Scientific Cooperation Exchange Program, 4646
USDA Scientific Cooperation Research Program, 4647
USDA Small Business Innovation Research Grants - Phase II, 4649
USDA Technical Agricultural Assistance Research Grants, 4651
USDA Tribal College Research Grants, 4652
USDA Wildlife Services Grants, 4653
USDC/NOAA Anadromous Fish Conservation Act Grants, 4654
USDC/NOAA Coastal and Marine Fish Habitat Protection Grants, 4655
USDC/NOAA Community-Based Habitat Restoration Grants, 4656
USDC/NOAA Community-Based Marine Debris Removal Grants, 4657
USDC/NOAA Marine Fisheries Grants, 4658
USDC/NOAA National Marine Aquaculture Initiative Grants, 4659
USDC/NOAA National Marine Fisheries Service Sea Grant Joint Graduate Fellowships, 4660
USDC/NOAA Restoration Center National and Regional Partnership Grants, 4662
USDC/NOAA Saltonstall-Kennedy Grants, 4663
USDC Afghanistan and Iraq Investment and Reconstruction Task Forces Internships, 4664
USDC Planning and Local Technical Assistance Grants, 4668
USDC Postdoctoral Research Grants, 4669
USDC Presidential Management Fellowships, 4670
USDD Acoustic Transduction Materials and Devices Grants, 4672
USDD Active Aperture Array Grants, 4673
USDD Active Authentication Grants, 4674
USDD Advances in Bioscience for Airmen Performance Grants, 4675
USDD Aerospace Medicine, Clinical Research, Human Performance Research, and Expeditionary Medicine Grants, 4676
USDD Air Force Defense Research and Development Rapid Innovation Fund Grants, 4677
USDD Air Force Medical Support Agency/ Modernization Directorate Research/Development and Innovations Grants, 4678
USDD Air Warfare and Naval Weapons Applications Grants, 4679
USDD Alloys and Joining Grants, 4680
USDD Antifouling/Fouling Re Coatings Grants, 4681
USDD Applications and Transitions Grants, 4682
USDD Applied & Computational Analysis Grants, 4683
USDD Applied Computational Analysis Grants, 4684
USDD Arctic and Global Prediction Grants, 4685
USDD ARO Broad Agency Grants, 4686
USDD Atomic, Molecular and Quantum Physics Grants, 4687
USDD Basic Research Challenge: Carbon Molecular Electronics Grants, 4688
USDD Basic Research Challenge: Decentralized Online Optimization Grants, 4689
USDD Basic Research for Combating Weapons of Mass Destruction Broad Agency Grants, 4690
USDD Basic Research Initiative Grants, 4691
USDD Battlespace Environment Division Research Interests Grant, 4692
USDD Breast Cancer Clinical Translational Research Grants, 4693
USDD Breast Cancer Era of Hope Scholars, 4694
USDD Breast Cancer Idea Grants, 4695
USDD Breast Cancer Impact Grants, 4696
USDD Breast Cancer Innovation Grants, 4697
USDD Breast Cancer Research Postdoctoral Fellowships, 4698
USDD Breast Cancer Transform Vision Grant, 4699
USDD Broad Agency Announcement for the Naval Medical Research and Development Center-Fredrick Grants, 4700
USDD Broad Agency Announcement HIV/AIDS Prevention Grants, 4702
USDD Broad Agency Announcement Research and Development Grants, 4703
USDD Bulk Nanostructured Materials Grants, 4704
USDD Capacitors for Pulsed Power Applications Grants, 4705
USDD Care for the Critically Injured Burn Patient Grants, 4706
USDD Cellular Materials Grants, 4707
USDD Chemical and Biological Defense Innovations and Technologies Grants, 4708
USDD Clinical Functional Assessment Investigator-Initiated Grants, 4709
USDD Collaborative Res in Structural Sciences, 4710
USDD Command and Control Grants, 4711
USDD Communications and Networking Discovery and Invention Grants, 4712
USDD Communications and Networking Grants, 4713
USDD Computer Materials Design Grants, 4714
USDD Crowd Sourced Formal Verif Grants, 4715
USDD Defense Sciences Research and Technology Grants, 4716
USDD Developer Tools for Medical Education Public Physiology Research Platform Grant, 4717
USDD Development of Quantum Computing Algorithms Grants, 4718
USDD Dielectric Films for Capacitors Grants, 4719
USDD Electro-Optical/Infrared Sensor Processing Grants, 4720
USDD Electrochemical Materials Grants, 4721
USDD Electromagnetic Materials Grants, 4722
USDD Electronic Warefare Grants, 4723
USDD Entry-Level Faculty Awards in Ocean Acoustics, 4724
USDD Extended Solids Grants, 4725
USDD Fires Thrust Grants, 4726
USDD Force Protection Thrust Grants, 4727
USDD Fundamental Research to Counter Weapons of Mass Destruction Grants, 4728
USDD Future Naval Fuels Grants, 4729
USDD Graduate Traineeship Awards in Ocean Acoustics, 4730
USDD High Assurance Cyber Military Systems Grants, 4731
USDD HIV/AIDS Prevention: Military Specific HIV/ AIDS Prevention, Care, and Treatment for Non-PEPFAR Funded Countries Grants, 4733
USDD HIV/AIDS Prevention Program Information Systems Development Grants, 4734
USDD Hull Grooming/Advanced Hull Cleaning Technologies Grants, 4735
USDD Human Performance Training and Education Grants, 4736
USDD Human Social, Cultural and Behavior Modeling Thrust Grants, 4737
USDD Image Analysis & Understanding Grants, 4738
USDD Image Optimization for Small Focal Plane Arrays Grants, 4739
USDD Information Innovation Office Grants, 4740
USDD Information Integration Grants, 4741

USDD Innovative Systems for Military Missions - Tactical Technology Office Grants, 4742
USDD Intelligence, Surveillance and Reconnaissance Thrust Grants, 4743
USDD Intelligent & Autonomous Systems Grants, 4744
USDD Ionospheric Modification Grants, 4745
USDD Littoral Geosciences and Optics Grants, 4746
USDD Logistics Thrust Grants, 4747
USDD Long Range Broad Agency Announcement for Navy and Marine Corps Science and Technology Grants, 4748
USDD Maneuver Thrust Grants, 4749
USDD Marine Mammals and Biological Oceanography Grants, 4750
USDD Mathematical Optimization and Operations Research Grants, 4751
USDD Medical Practice Breadth of Medical Practice & Disease Frequency Exposure Grants, 4752
USDD Medical Practice Initiative Procedural Skill Decay and Maintenance Grants, 4753
USDD Metamaterials for Radion Frequency and Optical Applications Grants, 4754
USDD Mixed Signal, Signal Processing Devices, Circuits and Architecture Grants, 4755
USDD Multidisciplinary Research Program of the University Research Initiative Grant, 4756
USDD Nano-Bio Manufacturing Inst Grants, 4757
USDD Nanoscale Electronics Grants, 4758
USDD National Consortium for Measurement and Signature Intelligence Research Grants, 4759
USDD National Geospatial-Intelligency Agency Academic Research Grants, 4760
USDD Naval Research Facilities Grants, 4761
USDD Navigation and Timekeeping Grants, 4762
USDD Non-Destructive Evaluation and Prognostics: Advanced Sensors and Technologies Grants, 4763
USDD Non-Destructive Eval & Prognostics: Fatigue and Fracture Damage Protection Grants, 4764
USDD NRL Broad Agency Grants, 4765
USDD Ocean Acoustics Research Grants, 4766
USDD Ocean Engineering and Marine Systems Grants, 4767
USDD Office-Wide Broad Agency Announcement Microsystems Technology Office Grant, 4768
USDD Office of Naval Res Command, Control, Computers & Communication Thrust Grants, 4769
USDD Organic Photovoltaics Grants, 4770
USDD Physical Oceanography Grants, 4771
USDD Polymeric Organic Materials: Conducting/ Semiconducting Polymers Grants, 4772
USDD Polymer Matrix Composites Grants, 4773
USDD Postdoctoral Fellowship Awards in Ocean Acoustics, 4774
USDD Power and Thermal Management Technology Development Grants, 4775
USDD Power Efficiency Revolution for Embedded Computing Technologies Grants, 4776
USDD President's Emergency Plan for AIDS Relief Military Specific HIV Seroprevalence and Behavioral Epidemiology Risk Suvey Grants, 4777
USDD Propulsion Materials Grants, 4778
USDD Radar and Signal Processing Grants, 4779
USDD Radio Frequency-Field Programmable Gate Arrays Grants, 4780
USDD Research and Analytical Support for the 711th Human Performance Wing's Human Effectiveness Directorate Grants, 4781
USDD Research and Studies for the Office of Net Assessment Grants, 4782
USDD Research Initiatives at the Naval Postgraduate School Grants, 4783
USDD Research Interest of the Air Force Office of Scientific Research Grants, 4784
USDD Research Interests of the US Air Force Academy Grants, 4785
USDD RF Semiconductors, RF Solid State Amplifiers and Wide Bandgap Materials Grants, 4786
USDD RF Superconducting Technologies Grants, 4787
USDD Science, Technology, Engineering and Mathematics Grants, 4789
USDD Science and Tecnology for Warfighter Training and Aiding Grants, 4790
USDD Sensor Innovative Research Grants, 4791
USDD Ship Systems and Engineering Res Grants, 4792
USDD Signal Processing for Networked Sensing Grants, 4793
USDD Small University Grants, 4794
USDD Software and Computer Systems Grants, 4795
USDD Solid and Liquid Waste Treatment Grants, 4796
USDD Solid Mechanics Grants, 4797
USDD Strategic Technologies Grants, 4798
USDD U.S. Army Medical Research and Materiel Command Broad Agency Announcement for Extramural Medical Research Grants, 4799
USDD Ultra-High Temp Materials Grants, 4800
USDD Undersea Signal Processing Grants, 4801
USDD Vacuum Devices Grants, 4802
USDD Warfighter Interface Technologies Advanced Research Grants, 4803
USDD Water Desalination & Purification Grants, 4804
USDD Workplace Violence in Military Grants, 4805
Vancouver Foundation Grants, 4807
Victor E. Speas Foundation Grants, 4810
Viking Children's Fund Grants, 4811
Virginia A. Hodgkinson Research Prize, 4812
Virginia Foundation for the Humanities Discretionary Grants, 4813
Virginia Found for Humanities Open Grants, 4814
Virginia Foundation for the Humanities Residential Fellowships, 4815
Volkswagen of America Corporate Contributions Grants, 4817
W.M. Keck Foundation Med Research Grants, 4823
W.M. Keck Foundation Science and Engineering Research Grants, 4824
W.M. Keck Undergrad Education Grants, 4825
Walter J. Jensen Fellowship for French Language, Literature, and Culture, 4830
Washington Area Women's Foundation Strategic Opportunity and Partnership Fund Grants, 4831
Wayne and Gladys Valley Foundation Grants, 4835
Web We Want Small Grants, 4836
Welch Award in Chemistry, 4838
Welch Foundation Departmental Grants, 4839
Welch Foundation Endowed Chairs, 4840
Welch Foundation Norman Hackeman Award in Chemical Research, 4841
Welch Foundation Research Grants, 4842
Wellcome Trust Biomedical Science Grants, 4843
Wellcome Trust New Investigator Awards, 4844
Wellesley Centers for Women Postd Fellowships, 4845
Wenner-Gren Foundation African Fellowships, 4846
Wenner-Gren Foundation Hunt Postdoctoral Fellowships, 4850
Wenner-Gren Foundation International Collaborative Research Grants, 4851
Wenner-Gren Foundation Post-PhD Res Grants, 4852
Wenner-Gren Found Short-Term Fellowships, 4853
Wenner-Gren Foundation Wadsworth International Fellowships, 4854
Weyerhaeuser Company Foundation Grants, 4857
WILD Foundation Grants, 4867
Wilhelm Sander Foundation Grants, 4868
Willary Foundation Grants, 4870
William A. Klinger Research Award, 4871
William and Charlotte Parks Foundation for Animal Welfare Grants, 4872
William Blair and Company Foundation Grants, 4875
William G. & Helen C. Hoffman Found Grants, 4876
William G. McGowan Charitable Fund Grants, 4877
William H. Adams Foundation for ALS Grants, 4878
William N. and Myriam Pennington Foundation Grants, 4879
William T. Grant Foundation Distinguished Fellows Awards, 4880
William T. Grant Foundation Research Grants, 4881
William T. Grant Foundation Scholars Program, 4882
William T. Grant Foundation Understanding the Acquisition, Interpretation, and Use of Research Evidence in Policy and Practice Grants, 4883
Winn Feline Foundation/AVMF Excellence In Feline Research Award, 4884
Winn Feline Foundation Grants, 4885
Winston Churchill Foundation Scholarships, 4886
Winterthur Museum Dissertation Fellowships, 4887
Winterthur Research Fellowships, 4888
Wolfsonian-FIU Fellowships, 4890
Woodrow Wilson Doctoral Dissertation Fellowships in Women's Studies, 4891
Woodrow Wilson MMUF Dissertation Grants, 4892
Woodrow Wilson MMUF Travel & Res Grants, 4893
Woodrow Wilson National Fellowship Foundation Travel and Research Grants, 4894
WWF Russell E. Train Education for Nature Alumni Grants, 4897
WWF Russell E. Train Fellowships, 4898
Xerox Foundation Grants, 4899
Xoran Technologies Resident Research Grant, 4900
Young Scientists Summer Program, 4903
Yves Rocher Foundation Terra Ficaria Prize for Scientific Research, 4904

Building Construction and/or Renovation

3M Company Higher Education Grants, 4
A.O. Smith Community Grants, 12
Abbott Fund Global AIDS Care Grants, 219
A Charitable Foundation Grants, 273
Achelis Foundation Grants, 274
Adam Richter Charitable Trust Grants, 339
African American Heritage Grants, 357
AIHS Full-Time M.D./Ph.D. Studentships, 447
AIHS Part-Time Studentships, 455
Alcoa Foundation Grants, 511
Alex Stern Family Foundation Grants, 542
Alice Tweed Tuohy Foundation Grants, 569
Allyn Foundation Grants, 574
American Chemical Society Green Chem Grants, 679
American Chemical Society Petroleum Research Fund New Directions Grants, 691
American Chemical Society Petroleum Research Fund Undergraduate New Investigator Grants, 694
American Family Insurance Corporate Giving, 709
American Schlafhorst Foundation Grants, 796
Anderson Foundation Grants, 842
Annenberg Foundation Grants, 862
Ann Peppers Foundation Grants, 863
Anthony R. Abraham Foundation Grants, 869
Arcadia Foundation Grants, 983
Arie and Ida Crown Memorial Grants, 989
Arronson Foundation Grants, 1019
Assisi Foundation of Memphis General Grants, 1221
Auburn Foundation Grants, 1263
Bailey Family Foundation Grants, 1289
Ball Brothers Foundation General Grants, 1291
Barra Foundation Project Grants, 1301
Becton Dickinson and Company Grants, 1328
Bill and Melinda Gates Foundation Water, Sanitation and Hygiene Grants, 1347
Blowitz-Ridgeway Foundation Early Childhood Development Research Award, 1348
Bodenwein Public Benevolent Found Grants, 1355
Bodman Foundation Grants, 1356
Boeing Company Contributions Grants, 1357
Booth-Bricker Fund Grants, 1359
Bosque Foundation Grants, 1360
Brico Fund Grants, 1371
Brinson Foundation Grants, 1372
Burlington Northern Santa Fe Found Grants, 1389
Burton G. Bettingen Grants, 1390
Byron W. & Alice L. Lockwood Found Grants, 1401
Cabot Corporation Foundation Grants, 1406
Camille Dreyfus Teacher-Scholar Awards, 1414
Campbell Soup Foundation Grants, 1415
Cape Branch Foundation Grants, 1425
Carl and Eloise Pohlad Family Found Grants, 1426
Carl C. Icahn Foundation Grants, 1428
Carl Gellert and Celia Berta Gellert Foundation Grants, 1429
Carl M. Freeman Foundation Grants, 1430
Carlos and Marguerite Mason Trust Grants, 1431

CCA Diefenbaker Award to update, 1447
Charles H. Farnsworth Trust Grants, 1521
Charles M. & Mary D. Grant Found Grants, 1527
Chatlos Foundation Grants, 1532
Chautauqua Region Community Found Grants, 1533
Chestnut Hill Charitable Foundation, Inc Grants, 1538
Chiles Foundation Grants, 1552
Christensen Fund Regional Grants, 1554
Claneil Foundation Grants, 1574
Clay Foundation Grants, 1579
Cleveland-Cliffs Foundation Grants, 1580
Coeta and Donald Barker Foundation Grants, 1594
Colonel Stanley R. McNeil Foundation Grants, 1597
Communities Foundation of Texas Grants, 1613
Community Found for Greater Buffalo Grants, 1615
Community Fnd for Muskegon County Grants, 1617
Community Foundation for S Arizona Grants, 1618
Community Found of E Central Illinois Grants, 1619
Connecticut Community Foundation Grants, 1627
Cord Foundation Grants, 1645
Coughlin-Saunders Foundation Grants, 1648
Cowles Charitable Trust Grants, 1652
Cudd Foundation Grants, 1672
DaimlerChrysler Corporation Fund Grants, 1714
Dean Witter Foundation Grants, 1728
Del E. Webb Foundation Grants, 1733
DeMatteis Family Foundation Grants, 1740
DENSO North America Foundation Grants, 1742
DOE Coal Liquefaction Research Grants, 1777
DOE Million Solar Roofs Program, 1805
Donald and Sylvia Robinson Family Foundation Grants, 1830
Dr. Bob and Jean Smith Foundation Grants, 1853
Drs. Bruce and Lee Foundation Grants, 1859
E.L. Wiegand Foundation Grants, 1885
Eaton Charitable Fund Grants, 1893
Edward W. and Stella C. Van Houten Grants, 1899
Edwin S. Webster Foundation Grants, 1900
Elisabeth Severance Prentiss Foundation Grants, 1905
Energy Foundation Buildings Grants, 1922
Ensign-Bickford Foundation Grants, 1928
EPA Capacity Building Grants for Compliance Assurance and Enforcement Activities in Indian Country and Other Tribal Areas, 1932
Eugene McDermott Foundation Grants, 1958
Evan Frankel Foundation Grants, 1961
Ewing Halsell Foundation Grants, 1964
F.W. Olin Foundation Construction Grant, 1972
Fidelity Foundation Grants, 2007
Fluor Foundation Grants, 2028
FMC Foundation Grants, 2029
Ford Motor Company Fund Grants, 2033
Fritz B. Burns Foundation Grants, 2057
General Motors Foundation Grants, 2107
George B. Storer Foundation Grants, 2113
George Foundation Grants, 2115
George Frederick Jewett Foundation Grants, 2116
George Gund Foundation Grants, 2117
George Kress Foundation Grants, 2119
George S. & Dolores Dore Eccles Grants, 2120
George W. Codrington Foundation Grants, 2122
Gilroy and Lillian P. Roberts Foundation Grants, 2142
Greater St Louis Community Foundation Grants, 2164
Greater Worcester Community Foundation Jeppson Memorial Fund for Brookfield Grants, 2165
Green Diamond Charitable Contributions, 2171
Greenspun Family Foundation Grants, 2173
H.A. and Mary K. Chapman Charitable Grants, 2193
H.J. Heinz Company Foundation Grants, 2195
Harold Alfond Foundation Grants, 2200
Harold Simmons Foundation Grants, 2203
Harry A. and Margaret D. Towsley Found Grants, 2205
Harry W. Bass, Jr. Foundation Grants, 2210
Harvey Randall Wickes Foundation Grants, 2218
Helen Bader Foundation Grants, 2228
Helen V. Brach Foundation Grants, 2231
Henry W. Bull Foundation Grants, 2237
Hill Crest Foundation Grants, 2267
Hilton Hotels Corporate Giving Grants, 2268
Hirtzel Memorial Foundation Grants, 2269
Hoglund Foundation Grants, 2271
HRF Hudson River Improvement Grants, 2286
HSBC Contributions Community Grants, 2291
Huie-Dellmon Trust Grants, 2295
Huntington Library Dibner History of Science Short-Term Fellowships, 2305
IMLS 21st Century Museum Pros Grants, 2362
J.B. Reynolds Foundation Grants, 2449
J.W. Kieckhefer Foundation Grants, 2452
Jackson Foundation Grants, 2453
Jacob and Hilda Blaustein Foundation Grants, 2454
James H. Cummings Foundation Grants, 2460
James R. Dougherty Jr. Foundation Grants, 2463
John Ben Snow Memorial Trust Grants, 2497
John Deere Foundation Grants, 2500
John G. Duncan Trust Grants, 2501
John Stauffer Charitable Trust Grants, 2509
Joseph Alexander Foundation Grants, 2513
Josiah W. and Bessie H. Kline Foundation Grants, 2517
Kenneth T. & Eileen L. Norris Found Grants, 2554
Lillian S. Wells Foundation Grants, 2614
Lowe Foundation Grants, 2622
Ludwick Family Foundation Grants, 2625
Luther I. Replogle Foundation Grants, 2631
M.D. Anderson Foundation Grants, 2637
M.J. Murdock Charitable Trust General Grants, 2641
Mabel Y. Hughes Charitable Trust Grants, 2647
Macquarie Bank Foundation Grants, 2650
Marcia and Otto Koehler Foundation Grants, 2656
Margaret & James A. Elkins Jr. Found Grants, 2658
Margaret T. Morris Foundation Grants, 2659
Mary K. Chapman Foundation Grants, 2673
Mary S. and David C. Corbin Foundation Grants, 2675
Matilda R. Wilson Fund Grants, 2684
Maximilian E. & Marion O. Hoffman Grants, 2688
McGregor Fund Grants, 2725
McLean Contributionship Grants, 2727
Melville Charitable Trust Grants, 2741
Mericos Foundation Grants, 2742
Meyer Memorial Trust Responsive Grants, 2748
Mt. Sinai Health Care Foundation Academic Medicine and Bioscience Grants, 2803
Musgrave Foundation Grants, 2807
NAGPRA Museum Consultation/Documentation Grants, 2822
NAGPRA Museum Repatriation Grants, 2823
NCI Technologies and Software to Support Integrative Cancer Biology Research Grants, 2933
NEH Collaborative Research Grants, 2952
Norcliffe Foundation Grants, 3415
Nordson Corporation Foundation Grants, 3416
North Carolina Biotechnology Centers of Innovation Grants, 3425
Northwestern Mutual Foundation Grants, 3430
NPS American Battlefield Protection Grants, 3435
NSF Academic Research Infrastructure Program: Recovery and Reinvestment (ARI-R^2), 3456
NSF Biological Field Stations and Marine Laboratories Grants, 3500
Oklahoma City Community Grants, 3737
Olin Corporation Charitable Trust Grants, 3739
Oppenstein Brothers Foundation Grants, 3792
Oscar Rennebohm Foundation Grants, 3832
Overbrook Foundation Grants, 3860
Patrick and Anna M. Cudahy Fund Grants, 3873
Patron Saints Foundation Grants, 3874
Paul Stock Foundation Grants, 3879
Perkin Fund Grants, 3898
Piper Trust Healthcare & Med Research Grants, 3972
Potts Memorial Foundation Grants, 3985
Price Family Charitable Fund Grants, 3987
R.C. Baker Foundation Grants, 4006
Radcliffe Institute Indiv Residential Fellowships, 4021
Rathmann Family Foundation Grants, 4033
Rayonier Foundation Grants, 4034
Reinberger Foundation Grants, 4051
Rhode Island Foundation Grants, 4061
Rich Foundation Grants, 4068
Robert G. Cabell III and Maude Morgan Cabell Foundation Grants, 4075
Rockwell International Corporate Trust Grants, 4093
Ronald McDonald House Charities Grants, 4101
Rose Community Foundation Health Grants, 4104
Ross Foundation Grants, 4106
Roy J. Carver Charitable Trust Medical and Science Research Grants, 4114
S. D. Bechtel, Jr. Foundation / Stephen Bechtel Fund Preventive Healthcare and Selected Research Grants, 4152
Sage Foundation Grants, 4158
Salem Foundation Grants, 4160
San Antonio Area Foundation Grants, 4163
Sarkeys Foundation Grants, 4181
Schramm Foundation Grants, 4188
Sensient Technologies Foundation Grants, 4199
Shastri Indo-Canadian Ins Librarian Fellowships, 4225
Sid W. Richardson Foundation Grants, 4253
Sony Corporation of America Corporate Philanthropy Grants, 4329
Stackpole-Hall Foundation Grants, 4442
Stanley Smith Horticultural Trust Grants, 4446
Starr Foundation Grants, 4447
Strake Foundation Grants, 4452
T. James Kavanagh Foundation Grants, 4489
Textron Corporate Contributions Grants, 4501
The Ray Charles Foundation Grants, 4502
Thomas Austin Finch, Sr. Foundation Grants, 4507
Thompson Charitable Foundation Grants, 4511
Thornton Foundation Grants, 4514
Union Pacific Foundation Health and Human Services Grants, 4569
USDA McIntire-Stennis Cooperative Forestry Research Grants, 4634
USDA Special Research Grants, 4650
USDD Research Interests of the US Air Force Academy Grants, 4785
Victor E. Speas Foundation Grants, 4810
W.C. Griffith Foundation Grants, 4819
W. C. Griffith Foundation Grants, 4818
Wayne and Gladys Valley Foundation Grants, 4835
Weyerhaeuser Company Foundation Grants, 4857
William and Charlotte Parks Foundation for Animal Welfare Grants, 4872
William Blair and Company Foundation Grants, 4875
William G. McGowan Charitable Fund Grants, 4877
William N. and Myriam Pennington Foundation Grants, 4879
Woodward Fund Grants, 4895
Xerox Foundation Grants, 4899

Capital Campaigns

Abbott Fund Global AIDS Care Grants, 219
ACGT Investigators Grants, 271
Achelis Foundation Grants, 274
Alcoa Foundation Grants, 511
Alex Stern Family Foundation Grants, 542
Alice Tweed Tuohy Foundation Grants, 569
Allyn Foundation Grants, 574
American Family Insurance Corporate Giving, 709
American Schlafhorst Foundation Grants, 796
Amgen Foundation Grants, 812
Anderson Foundation Grants, 842
Angels Baseball Foundation Grants, 848
Ann Peppers Foundation Grants, 863
Arcadia Foundation Grants, 983
Arie and Ida Crown Memorial Grants, 989
Arizona Public Service Corporate Giving Grants, 998
Arronson Foundation Grants, 1019
Auburn Foundation Grants, 1263
Avon Products Foundation Grants, 1275
Ball Brothers Foundation General Grants, 1291
Barra Foundation Project Grants, 1301
Batchelor Foundation Grants, 1305
Berks County Community Foundation Grants, 1333
Bill and Melinda Gates Foundation Water, Sanitation and Hygiene Grants, 1347
Blowitz-Ridgeway Foundation Early Childhood Development Research Award, 1348
Blowitz-Ridgeway Foundation Grants, 1349
BMW of North America Contributions, 1353

Bodenwein Public Benevolent Found Grants, 1355
Bodman Foundation Grants, 1356
Boeing Company Contributions Grants, 1357
Booth-Bricker Fund Grants, 1359
Bosque Foundation Grants, 1360
Burlington Northern Santa Fe Found Grants, 1389
Burton G. Bettingen Grants, 1390
Byron W. & Alice L. Lockwood Found Grants, 1401
Cabot Corporation Foundation Grants, 1406
Carl and Eloise Pohlad Family Found Grants, 1426
Carl and Lily Pforzheimer Foundation Grants, 1427
Carl C. Icahn Foundation Grants, 1428
Carl Gellert and Celia Berta Gellert Foundation Grants, 1429
Carl M. Freeman Foundation Grants, 1430
Carpenter Foundation Grants, 1439
Carrier Corporation Contributions Grants, 1440
Charles H. Revson Foundation Grants, 1524
Charles M. & Mary D. Grant Found Grants, 1527
Charles Stewart Mott Foundation Grants, 1529
Chatlos Foundation Grants, 1532
Chestnut Hill Charitable Foundation, Inc Grants, 1538
Claneil Foundation Grants, 1574
Cleveland-Cliffs Foundation Grants, 1580
Colonel Stanley R. McNeil Foundation Grants, 1597
Communities Foundation of Texas Grants, 1613
Community Found for Greater Buffalo Grants, 1615
Community Fnd for Muskegon County Grants, 1617
Connecticut Community Foundation Grants, 1627
ConocoPhillips Grants, 1629
Coughlin-Saunders Foundation Grants, 1648
Cowles Charitable Trust Grants, 1652
Cudd Foundation Grants, 1672
D.F. Halton Foundation Grants, 1692
DeMatteis Family Foundation Grants, 1740
DENSO North America Foundation Grants, 1742
Denton A. Cooley Foundation Grants, 1743
Donald and Sylvia Robinson Family Foundation Grants, 1830
Dr. Bob and Jean Smith Foundation Grants, 1853
Drs. Bruce and Lee Foundation Grants, 1859
Duchossois Family Foundation Grants, 1867
Eaton Charitable Fund Grants, 1893
Edward W. and Stella C. Van Houten Grants, 1899
Edwin S. Webster Foundation Grants, 1900
Elisabeth Severance Prentiss Foundation Grants, 1905
Ensign-Bickford Foundation Grants, 1928
Eugene McDermott Foundation Grants, 1958
Evjue Foundation Grants, 1963
Ewing Halsell Foundation Grants, 1964
Fidelity Foundation Grants, 2007
Fluor Foundation Grants, 2028
FMC Foundation Grants, 2029
Frank Stanley Beveridge Foundation Grants, 2051
Frederick Gardner Cottrell Foundation Grants, 2054
Fritz B. Burns Foundation Grants, 2057
Genuardi Family Foundation Grants, 2109
George B. Storer Foundation Grants, 2113
George Foundation Grants, 2115
George Gund Foundation Grants, 2117
George Kress Foundation Grants, 2119
George S. & Dolores Dore Eccles Grants, 2120
George W. Codrington Foundation Grants, 2122
Georgia Power Foundation Grants, 2125
Gil and Dody Weaver Foundation Grants, 2139
Gilroy and Lillian P. Roberts Foundation Grants, 2142
Greater Worcester Community Foundation Jeppson Memorial Fund for Brookfield Grants, 2165
H.A. and Mary K. Chapman Charitable Grants, 2193
H.J. Heinz Company Foundation Grants, 2195
Harold Simmons Foundation Grants, 2203
Harry A. and Margaret D. Towsley Found Grants, 2205
Harry B. and Jane H. Brock Foundation Grants, 2206
Harry W. Bass, Jr. Foundation Grants, 2210
Hearst Foundations Health Grants, 2225
Helen Bader Foundation Grants, 2228
Henry W. Bull Foundation Grants, 2237
Herbert A. & Adrian W. Woods Found Grants, 2239
Hill Crest Foundation Grants, 2267
Hirtzel Memorial Foundation Grants, 2269
Hoglund Foundation Grants, 2271
Honor the Earth Grants, 2279
Huie-Dellmon Trust Grants, 2295
Ittleson Foundation Environment Grants, 2437
J.B. Reynolds Foundation Grants, 2449
Jackson Foundation Grants, 2453
James H. Cummings Foundation Grants, 2460
James R. Dougherty Jr. Foundation Grants, 2463
JMO Law and Legal System Grants, 2491
John R. Oishei Foundation Grants, 2505
John Stauffer Charitable Trust Grants, 2509
Joseph Alexander Foundation Grants, 2513
Josiah W. and Bessie H. Kline Foundation Grants, 2517
Kenneth T. & Eileen L. Norris Found Grants, 2554
Kettering Fund Grants, 2555
Lillian S. Wells Foundation Grants, 2614
Lowe Foundation Grants, 2622
Lumpkin Family Found Healthy People Grants, 2629
Luther I. Replogle Foundation Grants, 2631
Lynde and Harry Bradley Foundation Grants, 2635
Mabel Y. Hughes Charitable Trust Grants, 2647
Macquarie Bank Foundation Grants, 2650
Margaret & James A. Elkins Jr. Found Grants, 2658
Margaret T. Morris Foundation Grants, 2659
Marion Gardner Jackson Charitable Trust Grants, 2664
Mary K. Chapman Foundation Grants, 2673
Matilda R. Wilson Fund Grants, 2684
Maximilian E. & Marion O. Hoffman Grants, 2688
McCune Foundation Human Services Grants, 2722
McGregor Fund Grants, 2725
McLean Contributionship Grants, 2727
Musgrave Foundation Grants, 2807
New Hampshire Charitable Foundation Grants, 2985
Norcliffe Foundation Grants, 3415
Nordson Corporation Foundation Grants, 3416
Northwestern Mutual Foundation Grants, 3430
Notsew Orm Sands Foundation Grants, 3434
Oppenstein Brothers Foundation Grants, 3792
Overbrook Foundation Grants, 3860
Patrick and Anna M. Cudahy Fund Grants, 3873
Perkin Fund Grants, 3898
Piper Trust Healthcare & Med Research Grants, 3972
Prudential Foundation Education Grants, 4002
R.C. Baker Foundation Grants, 4006
Rathmann Family Foundation Grants, 4033
Rayonier Foundation Grants, 4034
Reinberger Foundation Grants, 4051
Rhode Island Foundation Grants, 4061
Richard and Rhoda Goldman Fund Grants, 4063
Richard & Susan Smith Family Found Grants, 4064
Rockwell International Corporate Trust Grants, 4093
Ruth and Vernon Taylor Foundation Grants, 4130
Sage Foundation Grants, 4158
Salem Foundation Grants, 4160
Sarkeys Foundation Grants, 4181
Self Foundation Grants, 4197
Sensient Technologies Foundation Grants, 4199
Sid W. Richardson Foundation Grants, 4253
Sony Corporation of America Corporate Philanthropy Grants, 4329
Stackpole-Hall Foundation Grants, 4442
Starr Foundation Grants, 4447
Strake Foundation Grants, 4452
T. James Kavanagh Foundation Grants, 4489
TD4HIM Foundation Grants, 4497
Textron Corporate Contributions Grants, 4501
Thompson Charitable Foundation Grants, 4511
Union Pacific Found Community & Civic Grants, 4568
Union Pacific Foundation Health and Human Services Grants, 4569
W.C. Griffith Foundation Grants, 4819
W. C. Griffith Foundation Grants, 4818
W.K. Kellogg Foundation Healthy Kids Grants, 4822
Washington Area Women's Foundation Strategic Opportunity and Partnership Fund Grants, 4831
Wayne and Gladys Valley Foundation Grants, 4835
William and Charlotte Parks Foundation for Animal Welfare Grants, 4872
William G. & Helen C. Hoffman Found Grants, 4876
Woodward Fund Grants, 4895

Centers: Research/Demonstration/Service

AAAS Science and Technology Policy Fellowships: Global Health and Development, 42
ACF Adoption Opportunities Project Grants, 270
ACLS/New York Public Library Fellowships, 277
ACSM Carl V. Gisolfi Memorial Fund Grant, 312
ACSM Coca-Cola Company Doctoral Student Grant on Behavior Research, 313
ACSM Foundation Clinical Sports Medicine Endowment Grants, 315
ACSM Foundation Research Grants, 317
ACSM International Student Awards, 318
ACSM NASA Space Physiology Research Grants, 319
ACSM Oded Bar-Or Int'l Scholar Awards, 320
ACSM Paffenbarger-Blair Fund for Epidemiological Research on Physical Activity Grants, 321
ACSM Raymond and Rosalee Weiss Research Endowment Grant, 322
AIR/IES-NCES Postdoctoral Policy Fellowships in Washington, 466
American Chemical Society Corporate Associates Seed Grants, 676
APHL Emerging Infectious Diseases Fellowships, 914
Asian Pacific American Institute for Congressional Studies Fellowships, 1091
CDC Cooperative Agreement for Continuing Enhanced National Surveillance for Prion Diseases in the United States, 1456
CDC Evaluation of the Use of Rapid Testing For Influenza in Outpatient Medical Settings, 1458
CDC Evidence-Based Laboratory Medicine: Quality/ Performance Measure Evaluation, 1459
CDC National Training Information and Exchange Program, 1464
CDI Postdoctoral Fellowships, 1469
Center of Excellence in General Aviation Grants, 1486
Chicago Community Trust Housing Grants: Advancing Affordable Rental Housing, 1547
Children's Advocacy Center Grants, 1549
CMS Hispanic Health Services Research Grants, 1584
CMS Historically Black Colleges and Universities Health Services Research Grants, 1585
CMS Research and Demonstration Grants, 1586
Colonel Stanley R. McNeil Foundation Grants, 1597
Conquer Cancer Foundation of ASCO Drug Development Research Professorship, 1632
Conquer Cancer Foundation of ASCO Improving Cancer Care Grants, 1633
Corning Foundation Educational Grants, 1646
CTCRI Research Planning Grants, 1669
CTCRI Workshop and Learning Opportunities Grants, 1671
Davis Foundation Postdoctoral Fellowships in Eating Disorders Research, 1727
Deborah Munroe Noonan Research Grants, 1729
Dept of Ed Voc Ed National Programs, 1750
EMBO Installation Grants, 1915
Evjue Foundation Grants, 1963
Expect Miracles Foundation Grants, 1965
Getty Research Grants for Institutions - Archival Grants, 2131
Getty Research Grants for Institutions - Art History Research Grants, 2132
Getty Research Grants for Institutions - Cataloguing of Museum Collections Grants, 2133
Getty Research Grants for Institutions - Critical Reference Resources Grants, 2134
Health Foundation of Southern Florida Responsive Grants, 2223
Hearst Foundations Health Grants, 2225
Henry Luce Foundation Special Projects in Higher Education Grants, 2236
Herbert A. & Adrian W. Woods Found Grants, 2239
HRF Hudson River Improvement Grants, 2286
HUD Community Outreach Partnership Centers Grants, 2292
IAFF Burn Foundation Research Grants, 2312
IIE Rockefeller Foundation Bellagio Center Residencies, 2355
Jeffress Awards in Interdisciplinary Research, 2481

JMO Law and Legal System Grants, 2491
Klarman Family Foundation Grants in Eating Disorders Research, 2558
Kluge Center Distinguished Chairs, 2560
Lymphatic Education & Research Network Additional Support for NIH-funded F32 Postdoc Fellows, 2632
Lymphatic Education & Research Network Postdoctoral Fellowships, 2633
Marion Gardner Jackson Charitable Trust Grants, 2664
McGraw-Hill Companies Community Grants, 2724
MGN Family Foundation Grants, 2754
Miles of Hope Breast Cancer Foundation Grants, 2773
Minerva Fellowships, 2778
Minerva Short-Term Research Grants, 2779
MPI for Social Anthropology Department II Postdoctoral Grants, 2798
MPI for Social Anthropology Max Planck Fellow Research Group Law, Organizations, Science, and Technology, 2799
Mt. Sinai Health Care Foundation Academic Medicine and Bioscience Grants, 2803
NASA Cooperative Agreement Notice for Technology Development Grants, 2837
National Center for Responsible Gaming Centers of Excellence in Gambling Research Grants, 2850
NCFR Innovation Grants, 2908
NCI/NIDA Transdisciplinary Tobacco Use Research Centers Grants, 2912
NCI Centers of Excellence in Cancer Communications Research, 2921
NEI Mentored Clinical Scientist Development Program Award (K12), 2961
Nellie Mae Education Foundation State Level Systems Change Grants, 2971
NFL Charities Medical Grants, 2994
NHLBI Ancillary Studies in Clinical Trials, 3015
NHLBI Career Transition Awards, 3021
NHLBI Investigator Initiated Multi-Site Clinical Trials, 3030
NHLBI Lymphatics in Health and Disease in the Digestive, Urinary, Cardiovascular and Pulmonary Systems, 3032
NHLBI Mentored Career Award for Faculty at Institutions that Promote Diversity (K01), 3033
NHLBI Mentored Career Development Award to Promote Faculty Diversity/Re-Entry in Biomedical Research, 3035
NHLBI Ruth L. Kirschstein National Research Service Awards for Individual Predoctoral MD/PhD Fellows and Other Dual Degree Fellows, 3059
NHLBI Training Grants for Institutions that Promote Diversity, 3067
NIA Alzheimer's Disease Core Centers Grants, 3084
NIA Nathan Shock Centers of Excellence in Basic Biology of Aging Grants, 3113
NIDDK Diabetes Research Centers, 3147
NIDDK Multi-Center Clinical Study Cooperative Agreements, 3162
NIDDK Multi-Center Clinical Study Implementation Planning Grants, 3163
NIDDK Pilot and Feasibility Clinical Research Grants in Kidney or Urologic Diseases, 3171
NIDDK Pilot and Feasibility Clinical Research Studies in Digestive Diseases and Nutrition, 3172
NIDDK Seeding Collaborative Interdisciplinary Team Science in Diabetes, Endocrinology and Metabolic Diseases Grants, 3179
NIH Biology, Development, and Progression of Malignant Prostate Disease Research Grants, 3206
NIH Nonhuman Primate Immune Tolerance Cooperative Study Group Grants, 3228
NOAA Collaborative Science, Technology, and Applied Research Grants, 3354
NOAA Joint Research Institutes Grants, 3379
North Carolina Biotechnology Center Institutional Development Grants, 3422
North Carolina Biotechnology Center Res Grants, 3424
North Carolina Biotechnology Centers of Innovation Grants, 3425
Northrop Grumman Foundation Grants, 3429
NSERC Brockhouse Canada Prize for Interdisciplinary Research in Science and Engineering Grant, 3438
NSF/NEH Documenting Endangered Languages Project Grants and Fellowships: data, infrastructure and computational methods, 3455
NSF Arctic Research Opportunities, 3490
NSF Biological Research Collections, 3504
NSF Center for Synthesis in Biological Evolution, 3513
NSF CISE Computing Res Infrastructure Grants, 3531
NSF Emerging Frontiers in Research and Innovation Grants, 3560
NSF Engineering Research Centers Grants, 3561
NSF Engineering Research Grants, 3562
NSF Industry/University Cooperative Research Centers (I/UCRC) Grants, 3579
NSF Living Stock Collections Grants, 3588
NSF Major Research Instrumentation Grants, 3591
NSF Materials Science and Engineering Centers, 3594
NSF Partnerships for International Research and Education, 3611
NSF Partnerships in Astronomy & Astrophysics Research and Education, 3612
NSF Physics Frontiers Centers Grants, 3619
NSF Plant Genome Research Program, 3625
NSF University Radio Observatories Grants, 3664
NYSCA Dance: Services to the Field Grants, 3702
Oklahoma City Community Grants, 3737
Piper Trust Healthcare & Med Research Grants, 3972
Radcliffe Institute Research Support Grants, 4023
RAND Corporation Fellowships in Population Studies and the Study of Aging, 4029
RAND Corporation Graduate Student Summer Associateships, 4030
RAND Corp Transatlantic Postdoctoral Fellowships for International Relations and Security, 4032
Richard Davoud Donchian Foundation Grants, 4066
Rockwell International Corporate Trust Grants, 4093
RWJF Clinical Scholars Grants, 4136
Samueli Foundation Health Grants, 4161
San Diego Women's Foundation Grants, 4164
Sidgmore Family Foundation Grants, 4252
Surdna Found Sustainable Environments Grants, 4471
Taub Foundation Grants for MDS Research, 4496
Thome Foundation Awards in Age-Related Macular Degeneration Research, 4509
Thome Foundation Awards in Alzheimer's Disease Drug Discovery Research, 4510
Turtle Conservation Fund Grants, 4535
U.S. Department of Education American Overseas Research Centers, 4537
U.S. Department of Education National Resource Centers, 4542
U.S. Department of Education Rehabilitation Engineering Research Centers Grants, 4544
U.S. Department of Education Rehabilitation Research Training Centers, 4545
UCLA/RAND Corporation Postdoc Fellowships, 4548
United States Institute of Peace - Jennings Randolph Senior Fellowships, 4580
United States Institute of Peace Annual Grants, 4582
USC Pentecostal and Charismatic Research Initiative Regional Center Grants, 4606
US CRDF STEP Business Partnership Grant Competition Ukraine, 4612
USDA Integrated Pest Management Competitive Grants - Western Region, 4629
USDA Integrated Research Education and Extension Grants - Organic Transitions, 4630
USDA Policy Research Center Grants, 4640
USDA Regional Integrated Pest Management Grants - North Central Region, 4642
USDA Regional Integrated Pest Management Grants - Northeastern Region, 4643
USDA Regional Integrated Pest Management Grants - Southern Region, 4644
USDA Resident Instruction Grants for Higher Education in Insular Areas, 4645
W.M. Keck Foundation Science and Engineering Research Grants, 4824
Web We Want Small Grants, 4836

Citizenship Instruction

Axe-Houghton Foundation Grants, 1287
Brinson Foundation Grants, 1372
CIRCLE Civic Education at the High School Level Research Grants, 1571
Fulbright Public Policy Grants in Mexico, 2077
GMFUS Black Sea Trust for Regional Corporation Grants, 2152
Lynde and Harry Bradley Foundation Grants, 2635
Miles of Hope Breast Cancer Foundation Grants, 2773
Paul & Daisy Soros Fellows for New Americans, 3875

Community Development

1st Touch Foundation Grants, 2
3M Foundation Community Giving Grants, 5
A.L. Mailman Family Foundation Grants, 11
A.O. Smith Community Grants, 12
AAAS Science and Technology Policy Fellowships: Health, Education and Human Services, 43
Aaron & Cecile Goldman Family Found Grants, 200
AAUW Community Action Grants, 213
Abbott Fund Global AIDS Care Grants, 219
Abbott Fund Science Education Grants, 220
Abington Foundation Grants, 243
ACF Adoption Opportunities Project Grants, 270
A Charitable Foundation Grants, 273
Achelis Foundation Grants, 274
Acuity Charitable Foundation Grants, 333
Adam Richter Charitable Trust Grants, 339
Aetna Found Integrated Health Care Grants, 348
African American Heritage Grants, 357
Ahn Family Foundation Grants, 379
Akonadi Foundation Anti-Racism Grants, 477
Alabama Humanities Foundation Mini Grants, 480
Alabama Humanities Foundation Planning/Consultant Grants, 481
ALA Innovation Award, 490
Alaska Conservation Foundation Watchable Wildlife Conservation Trust Grants, 498
Alberta Law Foundation Grants, 503
Albert and Margaret Alkek Foundation Grants, 504
Albuquerque Community Foundation Grants, 508
Alcatel-Lucent Technologies Foundation Grants, 509
Alces Foundation Grants, 510
Alcoa Foundation Grants, 511
Alcon Foundation Grants, 513
Alexander and Margaret Stewart Trust Grants, 514
Alex Stern Family Foundation Grants, 542
Alfred P. Sloan Civic Initiatives Grants, 545
Alfred P. Sloan Selected National Issues Grants, 563
Alice Tweed Tuohy Foundation Grants, 569
Allen Lane Foundation Grants, 571
Allstate Foundation Safe and Vital Communities Grants, 573
Allyn Foundation Grants, 574
ALSAM Foundation Grants, 579
Alvin and Lucy Owsley Foundation Grants, 584
ALZA Corporate Contributions Grants, 588
American Chemical Society Chemical Technology Partnership Mini Grants, 674
American Chemical Society Corporate Associates Seed Grants, 676
American Foodservice Charitable Trust Grants, 710
American Schlafhorst Foundation Grants, 796
American Sociological Association Sydney S. Spivack Applied Social Research and Social Policy, 806
Amgen Foundation Grants, 812
AMI Semiconductors Corporate Grants, 813
Anderson Foundation Grants, 842
Angels Baseball Foundation Grants, 848
Ann & Robert H. Lurie Family Found Grants, 859
Ann Arbor Area Community Foundation Grants, 860
Appalachian Reg Comm Health Care Grants, 915
Applied Biosystems Grants, 916
Arcadia Foundation Grants, 983
Arca Foundation Grants, 984
Archer Daniels Midland Foundation Grants, 985
Arizona Cardinals Grants, 995
Arizona Commission on the Arts Folklorist Residencies, 997

Arizona Public Service Corporate Giving Grants, 998
A Rocha USA Grants, 1006
Arthur F. and Alice E. Adams Charitable Foundation Grants, 1024
ArvinMeritor Foundation Health Grants, 1027
Ashland Corporate Contributions Grants, 1087
Assisi Foundation of Memphis General Grants, 1221
ASTA Academic Scholarships, 1224
As You Sow, 1249
Auburn Foundation Grants, 1263
Ball Brothers Foundation General Grants, 1291
Ball Brothers Foundation Rapid Grants, 1292
Banfi Vintners Foundation Grants, 1296
Bank of America Charitable Foundation Matching Gifts, 1297
Barra Foundation Project Grants, 1301
Batchelor Foundation Grants, 1305
Bay and Paul Foundations, Inc Grants, 1307
Bayer Foundation Grants, 1309
BCBSM Foundation Investigator Initiated Research Grants, 1313
BCBSNC Foundation Grants, 1318
BCBS of Massachusetts Foundation Grants, 1319
Beim Foundation Grants, 1331
Berks County Community Foundation Grants, 1333
Better Way Foundation Grants, 1338
Bikes Belong Foundation Research Grants, 1343
Bill and Melinda Gates Foundation Agricultural Development Grants, 1345
Bill and Melinda Gates Foundation Financial Services for the Poor Grants, 1346
Bill and Melinda Gates Foundation Water, Sanitation and Hygiene Grants, 1347
Blum-Kovler Foundation Grants, 1351
Bodenwein Public Benevolent Found Grants, 1355
Bodman Foundation Grants, 1356
Boeing Company Contributions Grants, 1357
Boston Foundation Grants, 1361
Brainerd Foundation Grants, 1368
Brian G. Dyson Foundation Grants, 1370
Brico Fund Grants, 1371
Bristol-Myers Squibb Foundation Health Disparities Grants, 1377
Burlington Northern Santa Fe Found Grants, 1389
Burton G. Bettingen Grants, 1390
Bydale Foundation Grants, 1400
Cabot Corporation Foundation Grants, 1406
Caesars Foundation Grants, 1407
Callaway Golf Company Foundation Grants, 1410
Campbell Soup Foundation Grants, 1415
Cape Branch Foundation Grants, 1425
Carl and Eloise Pohlad Family Found Grants, 1426
Carl and Lily Pforzheimer Foundation Grants, 1427
Carl C. Icahn Foundation Grants, 1428
Carl Gellert and Celia Berta Gellert Foundation Grants, 1429
Carl M. Freeman Foundation Grants, 1430
Carnegie Corporation of New York Grants, 1432
Carpenter Foundation Grants, 1439
Carrier Corporation Contributions Grants, 1440
CCH Documentary Project Research and Development Grants, 1451
Charles H. Farnsworth Trust Grants, 1521
Charles H. Revson Foundation Grants, 1524
Charles Lafitte Foundation Grants, 1525
Charles M. & Mary D. Grant Found Grants, 1527
Charles Stewart Mott Anti-Poverty Grants, 1528
Charles Stewart Mott Foundation Grants, 1529
Chatlos Foundation Grants, 1532
Chautauqua Region Community Found Grants, 1533
Chicago Community Trust Fellowships, 1546
Chicago Community Trust Housing Grants: Advancing Affordable Rental Housing, 1547
Chiles Foundation Grants, 1552
Churchill Fellowships, 1560
CICF Efroymson Contemporary Arts Fellows, 1563
CIRCLE Civic Education at the High School Level Research Grants, 1571
Claneil Foundation Grants, 1574
Clarence E. Heller Foundation Grants, 1575
Clay Foundation Grants, 1579
Cleveland-Cliffs Foundation Grants, 1580
CMA Foundation Grants, 1582
Coca-Cola Foundation Grants, 1593
Coeta and Donald Barker Foundation Grants, 1594
Coleman Foundation Cancer Care Grants, 1595
Colonel Stanley R. McNeil Foundation Grants, 1597
Colorado Bioscience Discovery Eval Grants, 1598
Colorado Clean Energy Solar Innov Grants, 1599
Commonweal Foundation Grants, 1603
Commonwealth Fund Patient-Centered Coordinated Care Grants, 1608
Communities Foundation of Texas Grants, 1613
Community Found AIDS Endowment Awards, 1614
Community Found for Greater Buffalo Grants, 1615
Community Foundation for Greater New Haven Quinnipiac River Fund Grants, 1616
Community Fnd for Muskegon County Grants, 1617
Community Foundation for S Arizona Grants, 1618
Community Found of E Central Illinois Grants, 1619
Compton Foundation Grants, 1623
Connecticut Community Foundation Grants, 1627
ConocoPhillips Grants, 1629
Conservation, Food, and Health Foundation Grants for Developing Countries, 1641
Coors Brewing Corporate Contributions Grants, 1644
Cord Foundation Grants, 1645
Coughlin-Saunders Foundation Grants, 1648
Cruise Industry Charitable Foundation Grants, 1660
D.F. Halton Foundation Grants, 1692
Dade Community Foundation Grants, 1712
DaimlerChrysler Corporation Fund Grants, 1714
Dave Thomas Foundation for Adoption Grants, 1723
David Bohnett Foundation Grants, 1724
Daviess County Community Health Grants, 1725
Dell Foundation Open Grants, 1734
DeMatteis Family Foundation Grants, 1740
Dept of Ed Safe and Drug-Free Schools and Communities State Grants, 1748
Detroit Lions Charities Grants, 1753
DHHS Comprehensive Community Mental Health Services Grants for Children with Serious Emotional Disturbances, 1754
DHHS Emergency Med Services for Children, 1755
DHHS Independence Demonstration Program, 1757
DOE Initial H-Prize Competition for Breakthrough Advances in Materials for Hydrogen Storage, 1801
DOI National Spatial Data Infrastructure Cooperative Agreements Grants, 1823
Dorothy Rider Pool Health Care Grants, 1841
Dorr Foundation Grants, 1842
Drs. Bruce and Lee Foundation Grants, 1859
DTE Energy Foundation Health and Human Services Grants, 1865
E.L. Wiegand Foundation Grants, 1885
Edward Bangs Kelley and Elza Kelley Grants, 1898
Edward W. and Stella C. Van Houten Grants, 1899
Edwin S. Webster Foundation Grants, 1900
Emma B. Howe Memorial Foundation Grants, 1920
Energy Foundation Buildings Grants, 1922
Energy Foundation Climate Grants, 1923
Energy Foundation Power Grants, 1924
Energy Foundation Transportation Grants, 1925
Ensign-Bickford Foundation Grants, 1928
Enterprise Community Partners Rose Architectural Fellowships, 1929
Enterprise Community Partners Terwilliger Fellowship, 1930
EPA Brownfields Area-Wide Planning Grants, 1931
EPA Children's Health Protection Grants, 1933
EPA Compliance Assistance Support for Services to the Regulated Community and Other Assistance Providers, 1934
EPA Research, Development, Monitoring, Public Education, Training, Demonstrations, and Studies Grants, 1938
EPA Surveys, Studies, Research, Investigations, Demonstrations, and Special Purpose Activities Relating to the Clean Air Act, 1940
Eugene M. Lang Foundation Grants, 1957
Evan and Susan Bayh Foundation Grants, 1960
Evan Frankel Foundation Grants, 1961
Evanston Community Foundation Grants, 1962
Evjue Foundation Grants, 1963
Expect Miracles Foundation Grants, 1965
FAR Fund Grants, 1980
Farmers Insurance Corporate Giving Grants, 1981
Fidelity Foundation Grants, 2007
Fitzpatrick and Francis Family Business Continuity Foundation Grants, 2021
Flinn Foundation Grants, 2024
Fluor Foundation Grants, 2028
FMC Foundation Grants, 2029
Ford Motor Company Fund Grants, 2033
Forrest C. Lattner Foundation Grants, 2035
Foundation for Enhancing Comm Grants, 2038
Foundation for Seacoast Health Grants, 2041
Frances C. and William P. Smallwood Foundation Grants, 2045
Frances L. and Edwin L. Cummings Memorial Fund Grants, 2046
Frank Stanley Beveridge Foundation Grants, 2051
Freshwater Future Project Grants, 2056
Fritz B. Burns Foundation Grants, 2057
Fulbright/Garcia Robles Grants, 2063
General Motors Foundation Grants, 2107
General Service Foundation Human Rights and Economic Justice Grants, 2108
Genuardi Family Foundation Grants, 2109
George A Ohl Jr. Foundation Grants, 2112
George Foundation Grants, 2115
George Frederick Jewett Foundation Grants, 2116
George Gund Foundation Grants, 2117
George S. & Dolores Dore Eccles Grants, 2120
George W. Codrington Foundation Grants, 2122
Georgia-Pacific Foundation Environment Grants, 2123
Georgia Power Foundation Grants, 2125
Gil and Dody Weaver Foundation Grants, 2139
Girl's Best Friend Foundation Grants, 2143
GMFUS Black Sea Trust for Regional Corporation Grants, 2152
GNOF Gulf Coast Oil Spill Grants, 2153
GNOF IMPACT Grants for Education, 2154
Greater St Louis Community Foundation Grants, 2164
Greater Worcester Community Foundation Jeppson Memorial Fund for Brookfield Grants, 2165
Green Diamond Charitable Contributions, 2171
Green Foundation Human Services Grants, 2172
Greenwall Foundation Bioethics Grants, 2174
H.A. and Mary K. Chapman Charitable Grants, 2193
H.J. Heinz Company Foundation Grants, 2195
Harold Alfond Foundation Grants, 2200
Harold R. Bechtel Charitable Remainder Uni-Trust Grants, 2201
Harold R. Bechtel Testamentary Charitable Trust Grants, 2202
Harold Simmons Foundation Grants, 2203
Harris Graduate School of Public Policy Studies Research Development Grants, 2204
Harry B. and Jane H. Brock Foundation Grants, 2206
Harry W. Bass, Jr. Foundation Grants, 2210
Harvey Randall Wickes Foundation Grants, 2218
Hatton W. Sumners Foundation for the Study and Teaching of Self Government Grants, 2219
Healthcare Foundation of New Jersey Grants, 2222
Health Foundation of Southern Florida Responsive Grants, 2223
Helen Bader Foundation Grants, 2228
Helen V. Brach Foundation Grants, 2231
Hendrick Foundation for Children Grants, 2232
Henry W. Bull Foundation Grants, 2237
Herbert A. & Adrian W. Woods Found Grants, 2239
Herbert H. and Grace A. Dow Grants, 2241
Hilda and Preston Davis Foundation Grants, 2266
Hill Crest Foundation Grants, 2267
Hilton Hotels Corporate Giving Grants, 2268
Hirtzel Memorial Foundation Grants, 2269
Hoglund Foundation Grants, 2271
HomeBanc Foundation Grants, 2275
Honor the Earth Grants, 2279

HRF Hudson River Improvement Grants, 2286
HSBC Contributions Community Grants, 2291
Huie-Dellmon Trust Grants, 2295
Hutchinson Community Foundation Grants, 2310
Hydrogen Student Design Contest, 2311
ICC Faculty Fellowships, 2333
ICC Scholarship of Engagement Faculty Grants, 2334
IDEM Section 319(h) Nonpoint Source Grants, 2338
IIE David L. Boren Fellowships, 2350
Illinois Clean Energy Community Foundation Renewable Energy Grants, 2358
Illinois DCEO Coal Revival Grants, 2360
Illinois DNR Youth Recreation Corps Grants, 2361
IMLS Conservation Project Support Grants, 2363
Indiana 21st Century Research and Technology Fund Awards, 2370
Indiana SBIR/STTR Commercialization Enhancement Program, 2371
Indiana Waste Tire Fund Grants, 2376
International Human Rights Funders Grants, 2399
Ireland Family Foundation Grants, 2415
ISA John Z. Duling Grants, 2429
Ittleson Foundation AIDS Grants, 2436
Ittleson Foundation Mental Health Grants, 2438
J.B. Reynolds Foundation Grants, 2449
J.W. Kieckhefer Foundation Grants, 2452
Jackson Foundation Grants, 2453
Jacob and Hilda Blaustein Foundation Grants, 2454
James M. Collins Foundation Grants, 2461
James R. Dougherty Jr. Foundation Grants, 2463
Jane Bradley Pettit Foundation Health Grants, 2471
Japan Foundation Center for Global Partnership Grants, 2472
Jenifer Altman Foundation Grants, 2484
Jessie Smith Noyes Foundation Grants, 2485
JM Foundation Grants, 2490
John Ben Snow Memorial Trust Grants, 2497
John Deere Foundation Grants, 2500
John G. Duncan Trust Grants, 2501
John P. McGovern Foundation Grants, 2504
John R. Oishei Foundation Grants, 2505
John Stauffer Charitable Trust Grants, 2509
John W. Alden Trust Grants, 2510
Josiah W. and Bessie H. Kline Foundation Grants, 2517
Jovid Foundation Grants, 2518
Joyce Foundation Employment Grants, 2520
Kettering Fund Grants, 2555
Komen Greater NYC Community Breast Health Grants, 2569
Kovler Family Foundation Grants, 2575
Leonard and Helen R. Stulman Charitable Foundation Grants, 2602
LI General Grants, 2613
Lillian S. Wells Foundation Grants, 2614
Lloyd A. Fry Foundation Health Grants, 2620
Lowe Foundation Grants, 2622
Lucile Horton Howe and Mitchell B. Howe Foundation Grants, 2624
Lumpkin Family Found Healthy People Grants, 2629
Luther I. Replogle Foundation Grants, 2631
Lynde and Harry Bradley Foundation Grants, 2635
M. Bastian Family Foundation Grants, 2636
M.J. Murdock Charitable Trust General Grants, 2641
MacDonald-Peterson Foundation Grants, 2649
Macquarie Bank Foundation Grants, 2650
Marcia and Otto Koehler Foundation Grants, 2656
Margaret & James A. Elkins Jr. Found Grants, 2658
Margaret T. Morris Foundation Grants, 2659
Marijuana Policy Project Grants, 2661
Marion Gardner Jackson Charitable Trust Grants, 2664
Mary K. Chapman Foundation Grants, 2673
Mary S. and David C. Corbin Foundation Grants, 2675
Massachusetts Foundation for the Humanities Project Grants, 2678
Maximilian E. & Marion O. Hoffman Grants, 2688
McCarthy Family Foundation Grants, 2717
McCune Charitable Foundation Grants, 2719
McCune Foundation Human Services Grants, 2722
McGraw-Hill Companies Community Grants, 2724
McGregor Fund Grants, 2725
McLean Contributionship Grants, 2727
MDARD Specialty Crop Block Grants-Farm Bill, 2730
MDEQ Beach Monitoring Grants - Inland Lakes, 2731
MDEQ Local Water Quality Monitoring Grants, 2732
Mead Johnson Nutritionals Med Educ Grants, 2734
Meadows Foundation Grants, 2736
MedImmune Charitable Grants, 2737
Melville Charitable Trust Grants, 2741
Mericos Foundation Grants, 2742
Meyer and Pepa Gold Family Foundation Grants, 2746
Meyer and Stephanie Eglin Foundation Grants, 2747
Meyer Memorial Trust Responsive Grants, 2748
MGM Resorts Foundation Community Grants, 2753
MGN Family Foundation Grants, 2754
Miller Brewing Corporate Contributions Grants, 2774
Millipore Foundation Grants, 2775
Monfort Family Foundation Grants, 2794
Motorola Foundation Grants, 2796
National Center for Responsible Gaming Conference Scholarships, 2851
National Center for Responsible Gaming Travel Grants, 2858
National Center on Philanthropy and the Law Fellowship in Nonprofit Law, 2859
National Endowment for Democracy Reagan-Fascell Democracy Fellowships, 2863
National Housing Endowment Challenge/Build/Grow Grant, 2877
NCFR Innovation Grants, 2908
NEH Digital Humanities Implementation Grants, 2953
NEH Digital Humanities Start-Up Grants, 2954
Nelda C. and H.J. Lutcher Stark Found Grants, 2967
Nellie Mae Education Foundation District-Level Change Grants, 2968
Nellie Mae Education Foundation Public Understanding Grants, 2969
Nellie Mae Education Foundation State Level Systems Change Grants, 2971
New Hampshire Charitable Foundation Grants, 2985
NFWF Pulling Together Initiative Grants, 3007
NFWF Shell Marine Habitat Grants, 3008
NHLBI Bioengineering and Obesity Grants, 3016
NHLBI Lymphatics in Health and Disease in the Digestive, Urinary, Cardiovascular and Pulmonary Systems, 3032
Nick Traina Foundation Grants, 3128
NIH Exploratory Innovations in Biomedical Computational Science & Tech Grants, 3217
NIH Innovations in Biomedical Computational Science and Technology Grants, 3222
NIH Recovery Act Limited Competition: Building Sustainable Community-Linked Infrastructure to Enable Health Science Research, 3232
Norcliffe Foundation Grants, 3415
Nordson Corporation Foundation Grants, 3416
North Carolina Arts Council Folklife Grants, 3419
North Carolina Biotechnology Center Res Grants, 3424
Northwestern Mutual Foundation Grants, 3430
Northwest Fund for the Environment Grants, 3431
NRA Foundation Grants, 3437
NSERC Brockhouse Canada Prize for Interdisciplinary Research in Science and Engineering Grant, 3438
NSF CISE Communicating Research to Public Audiences Grants, 3527
NSF CISE Community-Based Data Interoperability Networks Grants, 3528
NSF CISE Computer and Network Systems: Core Grants, 3529
NSF CISE Computing and Communication Foundations: Core Grants, 3530
NSF CISE Computing Res Infrastructure Grants, 3531
NSF Communicating Res to Public Audiences, 3538
NSF Partnership for Advancing Technologies in Housing Grants, 3609
Nuffield Found Children & Families Grants, 3666
NWHF Community-Based Participatory Research Grants, 3670
NYCT New York City Environment Grants, 3697
NYSCA Dance: Services to the Field Grants, 3702
NYSCA Museum: Project Support Grants, 3707
Oklahoma City Community Grants, 3737
Olin Corporation Charitable Trust Grants, 3739
Open Spaces Sacred Places National Awards, 3791
Oppenstein Brothers Foundation Grants, 3792
Orchard Foundation Grants, 3793
OSF-Baltimore Community Fellowships, 3833
Overbrook Foundation Grants, 3860
Park Foundation Grants, 3870
Patrick and Anna M. Cudahy Fund Grants, 3873
Patron Saints Foundation Grants, 3874
Peacock Foundation Grants, 3887
Percy B. Ferebee Endowment Grants, 3897
Perpetual Trust for Charitable Giving Grants, 3900
Pinkerton Foundation Grants, 3967
Piper Trust Healthcare & Med Research Grants, 3972
Plan for Social Excellence Grants, 3976
Playboy Foundation Grants, 3980
Pott Foundation Grants, 3984
Praxair Foundation Grants, 3986
Price Family Charitable Fund Grants, 3987
R.C. Baker Foundation Grants, 4006
Radcliffe Institute Oral History Grants, 4022
RAND Corporation Graduate Student Summer Associateships, 4030
Rathmann Family Foundation Grants, 4033
Rayonier Foundation Grants, 4034
Reinberger Foundation Grants, 4051
Rhode Island Foundation Grants, 4061
Richard and Rhoda Goldman Fund Grants, 4063
Richard & Susan Smith Family Found Grants, 4064
Richard Davoud Donchian Foundation Grants, 4066
Rich Foundation Grants, 4068
Righteous Persons Foundation Grants, 4069
Robert B McMillen Foundation Grants, 4072
Robert Bowne Foundaion Edmund A. Stanley, Jr. Research Grants, 4073
Robert G. Cabell III and Maude Morgan Cabell Foundation Grants, 4075
Robert Sterling Clark Foundation Government Accountability Grants, 4080
Robert Sterling Clark Foundation Reproductive Rights and Health Grants, 4081
Rockwell International Corporate Trust Grants, 4093
Rohm and Haas Company Grants, 4098
Ronald McDonald House Charities Grants, 4101
Rose Community Foundation Health Grants, 4104
Ross Foundation Grants, 4106
RRF General Grants, 4115
Ruth and Vernon Taylor Foundation Grants, 4130
RWJF Changes in Health Care Financing and Organization Grants, 4133
RWJF Health Policy Fellowships, 4142
RWJF Healthy Eating Research Grants, 4143
RWJF Investigator Awards in Health Policy Res, 4145
RWJF Partners Investing in Nursing Grants, 4149
RWJF Pioneer Portfolio Grants, 4150
S. D. Bechtel, Jr. Foundation / Stephen Bechtel Fund Preventive Healthcare and Selected Research Grants, 4152
S.E.VEN Fund Annual Grants, 4153
S.E.VEN Fund Mini-Grants, 4154
S.E.VEN Fund Open Enterprise Solutions to Poverty RFP Grants, 4155
S.E.VEN Fund WHY Prize, 4156
Sage Foundation Grants, 4158
San Antonio Area Foundation Grants, 4163
San Diego Women's Foundation Grants, 4164
Sarasota Memorial Healthcare Found Grants, 4167
Sarkeys Foundation Grants, 4181
Schramm Foundation Grants, 4188
Searle Freedom Trust Grants, 4192
Seattle Foundation Medical Funds Grants, 4194
Seaver Institute Grants, 4195
SeaWorld and Busch Gardens Conservation Fund Grants, 4196
Self Foundation Grants, 4197
Sensient Technologies Foundation Grants, 4199
Shastri Indo-Canadian Institute Partnership Development Seed Grants, 4228
Shell Oil Company Foundation Grants, 4233

Sidgmore Family Foundation Grants, 4252
Sid W. Richardson Foundation Grants, 4253
Sioux Falls Area Community Foundation Community Fund Grants (Unrestricted), 4262
Sioux Falls Area Community Found Spot Grants, 4263
Siragusa Foundation Health Services & Medical Research Grants, 4264
Skaggs Foundation Grants, 4266
Society for the Arts in Healthcare Environmental Arts Research Grant, 4311
Sony Corporation of America Corporate Philanthropy Grants, 4329
Special Olympics E Kennedy Shriver Fellowships, 4331
SSRC-Van Alen Fellowships, 4419
Stackpole-Hall Foundation Grants, 4442
Stanley Smith Horticultural Trust Grants, 4446
Starr Foundation Grants, 4447
Strake Foundation Grants, 4452
Sun Academic Excellence Grants, 4466
Sunny and Abe Rosenberg Foundation Grants, 4469
Surdna Foundation Arts Teachers Fellowships, 4470
Surdna Found Sustainable Environments Grants, 4471
T. James Kavanagh Foundation Grants, 4489
Talbert and Leota Abrams Foundation, 4492
Textron Corporate Contributions Grants, 4501
Third Millennium Foundation Grants, 4505
Thomas Austin Finch, Sr. Foundation Grants, 4507
Thomas C. Ackerman Foundation Grants, 4508
Thompson Charitable Foundation Grants, 4511
Thornton Foundation Grants, 4514
Todd Brock Family Foundation Grants, 4522
TRDRP Participatory Research Awards, 4529
U.S. Department of Education Disability and Rehabilitation Research and Related Projects, 4538
Union Pacific Found Community & Civic Grants, 4568
Union Pacific Foundation Health and Human Services Grants, 4569
United Methodist Committee on Relief Global Health Grants, 4577
United States Institute of Peace Annual Grants, 4582
Ursula Thrush Peace Seed Grants, 4592
USAID Accelerating Progress Against Tuberculosis in Kenya Grants, 4594
USAID Comprehensive District-Based Support for Better HIV/TB Patient Outcomes Grants, 4595
USAID Development Innovation Accelerator Broad Agency Announcement Grants, 4596
USAID Palestinian Community Infrastructure Development Grants, 4600
USAID Res & Innovation for Health Supply Chain Systems and Commodity Security Grants, 4601
USAID Strengthening RMNCH Through Indian Health Professional Associations and Scaling Up Uptake Grants, 4602
US CRDF Science and Technology Entrepreneurship Program Business Partnership Grant Competition Armenia, 4611
US CRDF STEP Business Partnership Grant Competition Ukraine, 4612
USDA Alaska Native-Serving & Native Hawaiian-Serving Inst Education Competitive Grants, 4615
USDA Cooperative Extension System Education Grants, 4620
USDC Low-Power Television and Translator Upgrade Grants, 4666
USDC Planning and Local Technical Assistance Grants, 4668
USDC State Broadband Initiative Grants, 4671
Vancouver Foundation Grants and Community Initiatives Program, 4807
Vanguard Public Foundation Grant Funds, 4808
Vernon K. Krieble Foundation Grants, 4809
Victor E. Speas Foundation Grants, 4810
Virginia Foundation for the Humanities Discretionary Grants, 4813
Virginia Found for Humanities Open Grants, 4814
Volkswagen of America Corporate Contributions Grants, 4817
W.C. Griffith Foundation Grants, 4819
W.K. Kellogg Foundation Healthy Kids Grants, 4822
Wallace Foundation Grants, 4829
Washington Area Women's Foundation Strategic Opportunity and Partnership Fund Grants, 4831
Wayne and Gladys Valley Foundation Grants, 4835
Westinghouse Charitable Giving Grants, 4856
Weyerhaeuser Company Foundation Grants, 4857
WILD Foundation Grants, 4867
Willary Foundation Grants, 4870
William Blair and Company Foundation Grants, 4875
William G. & Helen C. Hoffman Found Grants, 4876
William G. McGowan Charitable Fund Grants, 4877
Xerox Foundation Grants, 4899

Consulting/Visiting Personnel

AAAS Science and Technology Policy Fellowships: Energy, Environment, and Agriculture, 41
AAAS Science and Technology Policy Fellowships: Global Health and Development, 42
AAS Kate B. and Hall J. Peterson Fellowships, 204
AAS Legacy Fellowships, 205
AAS Long-Term Visiting Academic Research Fellowships, 206
AAS Short-Term Visiting Academic Research Fellowships, 208
Abbott Fund Global AIDS Care Grants, 219
Abdus Salam ICTP Networks Grants, 233
Abdus Salam ICTP Training and Research in Italian Laboratories Grants, 240
ACC Residency Program in Asia, 267
AIAR Ernest S. Frerichs Fellow and Program Coordinator Grant, 394
Alexander von Humboldt Konrad Adenauer Research Award, 531
American Academy of Arts and Sciences Visiting Scholars, 613
American Chemical Society Petroleum Research Fund Scientific Education Grants, 692
AMNH Graduate Student Fellowships, 815
Anderson Center Residency Program, 841
APSAA Small Beginning Scholar Visiting and Consulting Grants, 924
Arizona Comm on Arts Education Projects Grants, 996
Arizona Commission on the Arts Folklorist Residencies, 997
Asia Foundation Exchange Programs, 1090
ASU Institute for Humanities Res Fellowships, 1246
Australian Research Fellowships/Queen Elizabeth II Fellowships, 1272
Avon Products Foundation Grants, 1275
BA Elisabeth Barker Fund, 1288
Ball Brothers Foundation Rapid Grants, 1292
Baylor Inst for Faith & Learning Visiting Fellows, 1310
BMI Woody Guthrie Fellowships, 1352
Bodenwein Public Benevolent Found Grants, 1355
Bogliasco Liguria Study Center Residencies, 1358
Canon Foundation Grants, 1423
Catholic University Life Cycle Institute Research Fellowship, 1444
CBO Economic Policy Fellowships, 1445
CCIS Visiting Research Fellowships, 1452
Center for Education of Women Visiting Scholar Research Grants, 1475
Center for Inquiry Science and the Public Visiting Research Fellowships, 1477
Center for the Study of Books and Media Visiting Research Fellowships, 1480
CFR Edward R. Murrow Press Fellowship, 1500
CFR International Affairs Fellowships, 1501
CFR International Affairs Fellowships in Nuclear Security, 1502
CFR Int'l Affairs Fellowships in South Korea, 1503
CFR International Affairs in Japan Fellowships, 1504
CFR Stanton Nuclear Security Fellowships, 1506
CHF Long-Term Fellowships, 1541
Clarence E. Heller Foundation Grants, 1575
Claude Pepper Foundation Grants, 1578
Community Found for Greater Buffalo Grants, 1615
Community Fnd for Muskegon County Grants, 1617
Community Foundation for S Arizona Grants, 1618
Community Found of E Central Illinois Grants, 1619
Connecticut Community Foundation Grants, 1627
Conquer Cancer Foundation of ASCO Int'l Devel & Educ Award in Palliative Care, 1634
Conquer Cancer Foundation of ASCO International Development and Education Awards, 1635
Cooperative Institute for Research in Environmental Sciences Visiting Fellows Program, 1643
CSRPC Postdoctoral Fellowship for Advanced Scholars, 1662
CSRPC Postdoc Fellowship for Junior Scholars, 1663
Cushwa Center Research Travel Grants, 1673
Cystic Fibrosis Canada Special Travel Grants For Fellows and Students, 1682
Cystic Fibrosis Canada Visiting Allied Health Professional Awards, 1686
Cystic Fibrosis Canada Visiting Clinician Awards, 1687
DAAD Visiting Professorship Grants, 1711
Dartmouth Andrew Mellon Postdoc Fellowship, 1721
Dept of Ed Star Schools Grants, 1749
Diaz-Ayala Cuban and Latin American Popular Music Collection Travel Grants, 1763
DOS Study of the United States Institutes, 1846
Drs. Bruce and Lee Foundation Grants, 1859
Endangered Language Fund Grants, 1921
Erasmus Institute Fellowships, 1944
EU Center for Humanistic Inquiry Junior and Postdoctoral Fellowships, 1956
FAR Fund Grants, 1980
Frances L. and Edwin L. Cummings Memorial Fund Grants, 2046
Francis C. Wood Scholar in Residence Program, 2047
Fulbright European Union Scholar-in-Residence Grants, 2071
Fulbright Scholar-in-Residence Grants, 2078
Getty Scholars and Visiting Scholars Grants, 2138
GMFUS Black Sea Trust for Regional Corporation Grants, 2152
GUITS Grants for Undergraduate Study, 2187
GUITS Matching Seed-Money Grants for New Positions in Turkish Studies, 2188
GUITS Post-Doctoral Summer Travel-Research Grants, 2189
GUITS Research Grants in Comparative Studies of Modern Turkey, 2190
GUITS Sabbatical Research Grants, 2191
GUITS Summer Research Grants in Turkey for Graduate Students, 2192
Harvard Postdoctoral Fellowships on Education, 2211
Harvard University Administrative Fellowships, 2212
Harvard University Women's Studies in Religion Research Associate and Visiting Faculty, 2216
Harvard Women and Public Policy Postdoctoral Fellowship, 2217
Hawaii Community Foundation Health Education and Research Grants, 2221
HRSA Nurse Education, Practice, Quality and Retention Grants, 2290
Humanities Institute Vis Scholars Fellowships, 2296
IAS Mellon Fellowships for Assistant Professors, 2318
IES Faculty Fellowships, 2341
Institute of American Cultures Postdoctoral and Visiting Scholar Fellowships, 2384
Institute of Turkish Studies Summer Language Study Grants in Turkey, 2391
International Peace Research Inst Oslo Grants, 2402
IREX ECA Alumni Small Grants, 2416
Joel L. Fleishman Civil Society Fellowships, 2495
Joint Center for Poverty Research Visiting Scholars Grants, 2512
Juvenile Justice Clearing House Grants, 2530
Kellogg Institute for International Studies Grants for Distinguished Lectureships, 2544
Kellogg Institute for International Studies Grants for Events, 2545
KFF Barbara Jordan Health Policy Scholars, 2556
Kluge Center Kislak Fellows in American Studies, 2562
Kluge Center Kislak Short-Term Fellowships in American Studies, 2563
Law and Public Affairs Fellowships at Princeton University, 2588

Library Company of Philadelphia NEH Long-term Post-Doctoral Fellowships, 2607
Mary Baker Eddy Library Summer Research Fellowships, 2669
Massachusetts Foundation for the Humanities Project Grants, 2678
Melville Charitable Trust Grants, 2741
Metanexus Institute Templeton Res Lectures, 2744
MHS Long Term Research Fellowships, 2755
MHS New England Regional Fellowships, 2756
MHS Short Term Research Fellowships, 2757
Minerva Fellowships, 2778
National Academies Postdoctoral and Senior Research Awards, 2844
National Assessment of Educational Progress Visiting Scholar Program, 2846
National Endowment for Democracy Reagan-Fascell Democracy Fellowships, 2863
NATO Expert Visits Grants, 2897
NCI Technologies and Software to Support Integrative Cancer Biology Research Grants, 2933
Nellie Mae Education Foundation State Level Systems Change Grants, 2971
NHC Summer Institutes in Lit Studies Grants, 3012
NL American Society for Eighteenth-Century Studies Fellowships, 3279
NL Arthur and Lila Weinberg Fellowships for Independent Scholars and Researchers, 3280
NL Associated Colleges of the Midwest/Great Lakes College Association Faculty Fellowships, 3281
NL Audrey Lumsden-Kouvel Fellowships, 3282
NL British Academy Fellowship for Study in the United Kingdom, 3283
NL Frances C. Allen Fellowships, 3284
NL Herzog August Bibliothek Wolfenbuttel Fellowship, 3286
NL Lawrence Lipking Fellowship at Newberry, 3287
NL Lester J. Cappon Fellowship in Documentary Editing, 3288
NL Lloyd Lewis Fellows in American History, 3289
NL Mellon Postdoctoral Research Fellowships, 3294
NL Midwest Modern Lang Assoc Fellowship, 3299
NL Monticello College Foundation Fellowship for Women, 3304
NL National Endowment for the Humanities Fellowships, 3310
NL Newberry Consortium on American Indian Studies Faculty Fellowships, 3311
NL Newberry Consortium on American Indian Studies Graduate Student Fellowships, 3312
NL Northeast Modern Language Assocation Fellowship, 3314
NL Scholar-in-Residence Opportunity, 3315
NL Short-Term Fellowships in the History of Cartography, 3316
NL Short-Term Resident Fellowships for Individual Research, 3317
NL South Central Modern Languages Association Fellowship, 3318
NL Susan Kelly Power and Helen Hornbeck Tanner Fellowship, 3319
North Carolina Arts Council Folklife Grants, 3419
North Carolina Arts Council Res Center Grants, 3420
Northwood University/Alden B. Dow Creativity Center Summer Residency Program, 3432
NYSCA Folk Arts: Presentation Grants, 3704
NYSCA Folk Arts: Regional and County Folk Arts Grants, 3705
NYSCA Museum: Project Support Grants, 3707
OAH Germany Residency Grants, 3719
OAH Japanese Residencies Grants, 3722
Patrick and Anna M. Cudahy Fund Grants, 3873
PHF Andrew W. Mellon Postdoctoral Fellowships in the Humanities, 3931
PhRMA Foundation Health Outcomes Sabbatical Fellowships, 3952
Princeton University Lewis Center for the Arts Hodder Fellowships, 3996
Princeton University Library Research Grants, 3997
Prudential Foundation Education Grants, 4002
RCPSC Royal College Visiting Professorships in Medical Research, 4041
RCPSC Walter MacKenzie Fellow in Surgery, 4042
Rhode Island Foundation Grants, 4061
Rockefeller Archive Center Grants in Aid, 4085
Rockefeller Archive Center Visiting Archivist Fellowship, 4087
Rockefeller Foundation MAP Fund Grants, 4090
Rocky Mountain Mineral Law Found Grants, 4094
Ronald McDonald House Charities Grants, 4101
Ross Foundation Grants, 4106
Shastri Indo-Canadian Institute Partnership Development Seed Grants, 4228
Shell Oil Company Foundation Grants, 4233
SRC Guest Professorship Program, 4370
SRC Medical Science Grants for Large International Conferences in Sweden, 4380
Streilein Foundation for Ocular Immunology Visiting Professorships, 4453
Tanner Humanities Center Research Fellowships, 4494
Temple Univ George D. McDowell Fellowship, 4499
Third Millennium Foundation Fellowships, 4504
Trinity College Ann Plato Fellowship, 4533
UIC Rockefeller Foundation Resident Fellowships in the Humanities, 4559
University of London Visiting Fellowships, 4588
USC Annenberg/Getty Arts Journ Fellowships, 4603
US CRDF CGP-RFBR Climate Change and Energy Competition For Joint U.S.-Russian Research Teams, 4608
US CRDF CGP ETF 2009 Energy Research Competition For Joint U.S.-Estonian Research Teams, 4609
US CRDF Leishmaniasis: Collaborative Opportunities in North Africa and the Middle East, 4610
USDA Hazardous Fuels Woody Biomass Utilization Grants, 4626
USDA Integrated Research Education and Extension Grants - Organic Transitions, 4630
USDC Market Development Cooperator Program Grants, 4667
Virginia Foundation for the Humanities Residential Fellowships, 4815
Wolfsonian-FIU Fellowships, 4890

Cultural Outreach

3M Foundation Community Giving Grants, 5
A.O. Smith Community Grants, 12
Abington Foundation Grants, 243
Achelis Foundation Grants, 274
ACLS Comparative Perspectives on Chinese Culture and Society Grants, 283
Acuity Charitable Foundation Grants, 333
Adaptec Foundation Grants, 340
Ahn Family Foundation Grants, 379
AIAR/Council of American Overseas Research Centers Fellowships for Advanced Multi-Country Research, 391
AIAR Andrew W. Mellon Foundation Fellowships, 392
Akonadi Foundation Anti-Racism Grants, 477
Alabama Humanities Foundation Mini Grants, 480
Alabama Humanities Foundation Planning/Consultant Grants, 481
Albert and Margaret Alkek Foundation Grants, 504
Albuquerque Community Foundation Grants, 508
Alcatel-Lucent Technologies Foundation Grants, 509
Alces Foundation Grants, 510
Alex Stern Family Foundation Grants, 542
Alfred P. Sloan Civic Initiatives Grants, 545
Alfred P. Sloan Public Understanding of Science and Technology-Theater Grants, 559
ALZA Corporate Contributions Grants, 588
Amgen Foundation Grants, 812
AMS Alvin H. Johnson Dissertation Fellowships, 820
Anderson Foundation Grants, 842
Andrew W. Mellon Postdoctoral Fellowship in American Cultures, 845
Annenberg Foundation Grants, 862
Ann Peppers Foundation Grants, 863
Applied Biosystems Grants, 916
Arcadia Foundation Grants, 983
Archer Daniels Midland Foundation Grants, 985
Arie and Ida Crown Memorial Grants, 989
ARIT Kenan T. Erim Fellowship for Research at Aphrodisias, 993
Arizona Cardinals Grants, 995
Arizona Commission on the Arts Education Projects Grants, 996
Arizona Public Service Corporate Giving Grants, 998
Arthur F. and Alice E. Adams Charitable Foundation Grants, 1024
ASCSA Alison M. Frantz Fellowship, 1051
ASCSA Harry Bikakis Fellowship, 1053
Ashland Corporate Contributions Grants, 1087
Assisi Foundation of Memphis General Grants, 1221
Auburn Foundation Grants, 1263
Avon Products Foundation Grants, 1275
Axe-Houghton Foundation Grants, 1287
BA Larger Research Grants, 1290
Ball Brothers Foundation General Grants, 1291
Banfi Vintners Foundation Grants, 1296
Bank of America Charitable Foundation Matching Gifts, 1297
Barra Foundation Project Grants, 1301
BA Small Research Grants, 1304
Batchelor Foundation Grants, 1305
Bay and Paul Foundations, Inc Grants, 1307
BCBSNC Foundation Grants, 1318
Beim Foundation Grants, 1331
Berks County Community Foundation Grants, 1333
Bill and Melinda Gates Foundation Agricultural Development Grants, 1345
Bill and Melinda Gates Foundation Financial Services for the Poor Grants, 1346
Blum-Kovler Foundation Grants, 1351
BMI Woody Guthrie Fellowships, 1352
Bodenwein Public Benevolent Found Grants, 1355
Bodman Foundation Grants, 1356
Boeing Company Contributions Grants, 1357
Booth-Bricker Fund Grants, 1359
Boston Foundation Grants, 1361
Brico Fund Grants, 1371
Burlington Northern Santa Fe Found Grants, 1389
Bydale Foundation Grants, 1400
Byron W. & Alice L. Lockwood Found Grants, 1401
Caesars Foundation Grants, 1407
Campbell Soup Foundation Grants, 1415
Canada-U.S. Fulbright Senior Specialists Grants, 1418
Cape Branch Foundation Grants, 1425
Carl and Eloise Pohlad Family Found Grants, 1426
Carl and Lily Pforzheimer Foundation Grants, 1427
Carl C. Icahn Foundation Grants, 1428
Carl Gellert and Celia Berta Gellert Foundation Grants, 1429
Carl M. Freeman Foundation Grants, 1430
Carnegie Corporation of New York Grants, 1432
Carpenter Foundation Grants, 1439
CCA Alberta Creative Development Initiative: Grants to Individuals, 1446
CCH Documentary Project Research and Development Grants, 1451
Charles H. Price II Family Foundation Grants, 1523
Charles Lafitte Foundation Grants, 1525
Chatlos Foundation Grants, 1532
Chiles Foundation Grants, 1552
CIGNA Foundation Grants, 1568
Cintas Foundation Fellowships, 1570
Claneil Foundation Grants, 1574
Clarence E. Heller Foundation Grants, 1575
CMU Arts in Society Fellowship, 1587
Coca-Cola Foundation Grants, 1593
Colonel Stanley R. McNeil Foundation Grants, 1597
Community Fnd for Muskegon County Grants, 1617
Community Foundation for S Arizona Grants, 1618
Community Found of E Central Illinois Grants, 1619
Compton Foundation Grants, 1623
Connecticut Community Foundation Grants, 1627
ConocoPhillips Grants, 1629
Conseil des arts de Montreal Residency of the Americas, 1640

Cord Foundation Grants, 1645
Coughlin-Saunders Foundation Grants, 1648
Cowles Charitable Trust Grants, 1652
CRF Graduate Research Grants, 1655
Cudd Foundation Grants, 1672
D.F. Halton Foundation Grants, 1692
Dade Community Foundation Grants, 1712
David Bohnett Foundation Grants, 1724
Davis Family Foundation Grants, 1726
DeMatteis Family Foundation Grants, 1740
Dept of Ed Women's Educational Equity Grants, 1752
Diaz-Ayala Cuban and Latin American Popular Music Collection Travel Grants, 1763
Donald and Sylvia Robinson Family Foundation Grants, 1830
Dorothy Rider Pool Health Care Grants, 1841
Dorr Foundation Grants, 1842
Dr. Bob and Jean Smith Foundation Grants, 1853
Drs. Bruce and Lee Foundation Grants, 1859
E.L. Wiegand Foundation Grants, 1885
Eaton Charitable Fund Grants, 1893
Edward Bangs Kelley and Elza Kelley Foundation Grants, 1898
Edwin S. Webster Foundation Grants, 1900
Emerson Charitable Trust Grants, 1918
Emma B. Howe Memorial Foundation Grants, 1920
Endangered Language Fund Grants, 1921
Ensign-Bickford Foundation Grants, 1928
Eugene M. Lang Foundation Grants, 1957
Evan Frankel Foundation Grants, 1961
Evanston Community Foundation Grants, 1962
Evjue Foundation Grants, 1963
FAR Fund Grants, 1980
Farmers Insurance Corporate Giving Grants, 1981
FASSE Demonstration Project Grant, 1982
Fishman Family Foundation Grants, 2020
Flinn Foundation Grants, 2024
Fluor Foundation Grants, 2028
Ford Motor Company Fund Grants, 2033
Forrest C. Lattner Foundation Grants, 2035
Foundation for Enhancing Comm Grants, 2038
Foundation for the Advancement of Mesoamerican Studies Grants, 2042
Frank Stanley Beveridge Foundation Grants, 2051
Fulbright Alumni Initiatives Awards Grants, 2064
Fulbright Binational Business Grants in Mexico, 2066
Fulbright Business Grants in Spain, 2067
Fulbright Critical Lang Enhancement Awards, 2068
Fulbright Distinguished Chairs Grants, 2069
Fulbright English Teaching Assistantships, 2070
Fulbright European Union Scholar-in-Residence Grants, 2071
Fulbright German Studies Seminar Grants, 2072
Fulbright International Education Administrators Seminar Grants, 2073
Fulbright mtvU Awards, 2074
Fulbright New Century Scholars Grants, 2075
Fulbright Scholar-in-Residence Grants, 2078
Fulbright Specialists Grants, 2079
Fulbright Traditional Scholar Program in Europe and Eurasia, 2080
Fulbright Traditional Scholar Program in Sub-Saharan Africa, 2081
Fulbright Traditional Scholar Program in the East Asia/Pacific Region, 2082
Fulbright Traditional Scholar Program in the Near East and North Africa Region, 2083
Fulbright Traditional Scholar Program in the South and Central Asia Region, 2084
Fulbright Traditional Scholar Program in the Western Hemisphere, 2085
Fulbright Visiting Specialists Program: Direct Access to the Muslim World, 2087
Furthermore Grants in Publishing, 2091
G. Unger Vetlesen Foundation Grants, 2093
General Motors Foundation Grants, 2107
Genuardi Family Foundation Grants, 2109
George A Ohl Jr. Foundation Grants, 2112
George B. Storer Foundation Grants, 2113
George Foundation Grants, 2115
George Frederick Jewett Foundation Grants, 2116
George Gund Foundation Grants, 2117
George Kress Foundation Grants, 2119
George S. & Dolores Dore Eccles Grants, 2120
Georgia Power Foundation Grants, 2125
Gerda Henkel Foundation Grants, 2127
Getty Research Grants for Institutions - Archival Grants, 2131
Getty Research Grants for Institutions - Art History Research Grants, 2132
Getty Research Grants for Institutions - Cataloguing of Museum Collections Grants, 2133
Getty Research Grants for Institutions - Critical Reference Resources Grants, 2134
Getty Research Institute Nonresidential Grants - Collaborative Research Grant, 2135
Getty Research Institute Nonresidential Grants - Curatorial Research Fellowships, 2136
Getty Research Institute Nonresidential Grants - Postdoctoral Fellowships, 2137
Getty Scholars and Visiting Scholars Grants, 2138
Gil and Dody Weaver Foundation Grants, 2139
Gilroy and Lillian P. Roberts Foundation Grants, 2142
Grammy Foundation Grants, 2161
Greater St Louis Community Foundation Grants, 2164
Greater Worcester Community Foundation Jeppson Memorial Fund for Brookfield Grants, 2165
Green Diamond Charitable Contributions, 2171
GUITS Grants for Undergraduate Study, 2187
GUITS Post-Doctoral Summer Travel-Research Grants, 2189
GUITS Research Grants in Comparative Studies of Modern Turkey, 2190
GUITS Sabbatical Research Grants, 2191
GUITS Summer Research Grants in Turkey for Graduate Students, 2192
H.A. and Mary K. Chapman Charitable Grants, 2193
H.J. Heinz Company Foundation Grants, 2195
H.W. Wilson Foundation Grants, 2196
Harold Alfond Foundation Grants, 2200
Harold R. Bechtel Charitable Remainder Uni-Trust Grants, 2201
Harold Simmons Foundation Grants, 2203
Harry Frank Guggenheim Foundation Research Grants, 2208
Harry W. Bass, Jr. Foundation Grants, 2210
Haverford College Mellon Postdoc Fellowships, 2220
Heineman Foundation for Research, Education, Charitable and Scientific Purposes, 2227
Helen Bader Foundation Grants, 2228
Helen V. Brach Foundation Grants, 2231
Herbert A. & Adrian W. Woods Found Grants, 2239
Herbert H. and Grace A. Dow Grants, 2241
Hill Crest Foundation Grants, 2267
Hoglund Foundation Grants, 2271
Honor the Earth Grants, 2279
Huie-Dellmon Trust Grants, 2295
ICC Faculty Fellowships, 2333
ICCS Publishing Fund, 2336
IIE David L. Boren Fellowships, 2350
IMLS 21st Century Museum Pros Grants, 2362
IMLS Grants to State Library Administrative Agencies, 2364
IMLS National Leadership Grants, 2366
Infinity Foundation Grants, 2378
Institute of Turkish Studies Dissertation Writing Grants, 2389
Institute of Turkish Studies Summer Language Study Grants in Turkey, 2391
IREX ECA Alumni Small Grants, 2416
IREX International Leadership in Education Program Grants, 2420
IREX Short-Term Travel Grants, 2424
IREX University Administration Support, 2425
J.B. Reynolds Foundation Grants, 2449
Jackson Foundation Grants, 2453
Jacobs Research Funds Small Grants, 2456
James M. Collins Foundation Grants, 2461
Japan Foundation Center for Global Partnership Grants, 2472
Japan Foundation New York Institutional Support for Japanese Studies Grants, 2477
Japan Foundation Six-Month Grants for Specialists in Cultural and Academic Fields, 2479
Jeffrey Campbell Grad Fellow in Anthropology, 2482
John Ben Snow Memorial Trust Grants, 2497
John Deere Foundation Grants, 2500
John R. Oishei Foundation Grants, 2505
Kate Neal Kinley Memorial Fellowship, 2532
Kenneth T. & Eileen L. Norris Found Grants, 2554
Kluge Center Kislak Fellows in American Studies, 2562
Kluge Center Kislak Short-Term Fellowships in American Studies, 2563
Kovler Family Foundation Grants, 2575
Lillian S. Wells Foundation Grants, 2614
Ludwick Family Foundation Grants, 2625
Lynde and Harry Bradley Foundation Grants, 2635
M. Bastian Family Foundation Grants, 2636
M.J. Murdock Charitable Trust General Grants, 2641
Mabel Y. Hughes Charitable Trust Grants, 2647
Macquarie Bank Foundation Grants, 2650
Manitoba Arts Council Craft Grants, 2653
Marcia and Otto Koehler Foundation Grants, 2656
Margaret & James A. Elkins Jr. Found Grants, 2658
Margaret T. Morris Foundation Grants, 2659
Marion Gardner Jackson Charitable Trust Grants, 2664
Mary Baker Eddy Library Summer Research Fellowships, 2669
Mary K. Chapman Foundation Grants, 2673
Massachusetts Foundation for the Humanities Project Grants, 2678
Matilda R. Wilson Fund Grants, 2684
Mattel Children's Foundation Grants, 2685
Max and Victoria Dreyfus Foundation Grants, 2687
Maximilian E. & Marion O. Hoffman Grants, 2688
McCune Charitable Foundation Grants, 2719
McGraw-Hill Companies Community Grants, 2724
McLean Contributionship Grants, 2727
Meadows Foundation Grants, 2736
Mericos Foundation Grants, 2742
Meyer and Stephanie Eglin Foundation Grants, 2747
Meyer Memorial Trust Responsive Grants, 2748
MGN Family Foundation Grants, 2754
MHS New England Regional Fellowships, 2756
Millipore Foundation Grants, 2775
Minerva Short-Term Research Grants, 2779
Monfort Family Foundation Grants, 2794
Motorola Foundation Grants, 2796
NAGPRA Museum Consultation/Documentation Grants, 2822
NAGPRA Museum Repatriation Grants, 2823
NEH Awards for Faculty at Tribal Colleges and Universities, 2951
NFJC Gantz-Zahler Grant, 2993
NHC Summer Institutes in Lit Studies Grants, 3012
Norcliffe Foundation Grants, 3415
Nordson Corporation Foundation Grants, 3416
North Carolina Arts Council Folklife Grants, 3419
Northwestern Mutual Foundation Grants, 3430
NSF/NEH Documenting Endangered Languages Project Grants and Fellowships: data, infrastructure and computational methods, 3455
NYSCA Dance: Services to the Field Grants, 3702
NYSCA Folk Arts: Exhibitions Grants, 3703
NYSCA Folk Arts: Presentation Grants, 3704
NYSCA Folk Arts: Regional and County Folk Arts Grants, 3705
NYSCA Museum: Project Support Grants, 3707
NYSCA Music: Services to the Field Grants, 3708
Ontario Arts Council Integrated Arts Grants, 3790
Oppenstein Brothers Foundation Grants, 3792
Patrick and Anna M. Cudahy Fund Grants, 3873
Paul G. Allen Family Foundation Grants, 3876
Percy B. Ferebee Endowment Grants, 3897
PHF Andrew W. Mellon Postdoctoral Fellowships in the Humanities, 3931
Princess Grace Foundation Theater, Dance, and Film Grants, 3988
R.C. Baker Foundation Grants, 4006
Radcliffe Institute Oral History Grants, 4022

Rajiv Gandhi Foundation Grants, 4025
Rathmann Family Foundation Grants, 4033
Reinberger Foundation Grants, 4051
Rhode Island Foundation Grants, 4061
Richard and Rhoda Goldman Fund Grants, 4063
Richard & Susan Smith Family Found Grants, 4064
Rich Foundation Grants, 4068
Robert B McMillen Foundation Grants, 4072
Robert G. Cabell III and Maude Morgan Cabell Foundation Grants, 4075
Rockefeller Foundation MAP Fund Grants, 4090
Rockwell International Corporate Trust Grants, 4093
Rohm and Haas Company Grants, 4098
Ronald McDonald House Charities Grants, 4101
Ruth and Vernon Taylor Foundation Grants, 4130
Sage Foundation Grants, 4158
San Antonio Area Foundation Grants, 4163
San Diego Women's Foundation Grants, 4164
Sarkeys Foundation Grants, 4181
Schramm Foundation Grants, 4188
Sensient Technologies Foundation Grants, 4199
Shastri Indo-Canadian Institute Arts Fellowships, 4219
SHOT International Scholars, 4242
Sioux Falls Area Community Foundation Community Fund Grants (Unrestricted), 4262
Sioux Falls Area Community Found Spot Grants, 4263
Sony Corp of America Philanthropy Grants, 4329
Stackpole-Hall Foundation Grants, 4442
Starr Foundation Grants, 4447
State Scholarships Foundation Grants, 4449
Strake Foundation Grants, 4452
Surdna Foundation Arts Teachers Fellowships, 4470
Talbert and Leota Abrams Foundation, 4492
Tanner Humanities Center Research Fellowships, 4494
Textron Corporate Contributions Grants, 4501
The Ray Charles Foundation Grants, 4502
Third Millennium Foundation Grants, 4505
Thomas C. Ackerman Foundation Grants, 4508
Thompson Charitable Foundation Grants, 4511
Thornton Foundation Grants, 4514
U.S. Department of Education United States-Russia Program: Improving Research and Educational Activities in Higher Education, 4547
Union Pacific Found Community & Civic Grants, 4568
USC Annenberg/Getty Arts Journ Fellowships, 4603
USDA International Science and Education Competitive Grants, 4631
Vancouver Foundation Grants and Community Initiatives Program, 4807
Vanguard Public Foundation Grant Funds, 4808
Virginia Foundation for the Humanities Discretionary Grants, 4813
Virginia Found for Humanities Open Grants, 4814
W.C. Griffith Foundation Grants, 4819
Wallace Foundation Grants, 4829
Westinghouse Charitable Giving Grants, 4856
Weyerhaeuser Company Foundation Grants, 4857
WILD Foundation Grants, 4867
Willary Foundation Grants, 4870
William Blair and Company Foundation Grants, 4875
Wolfsonian-FIU Fellowships, 4890
Xerox Foundation Grants, 4899

Curriculum Development/Teacher Training

3M Foundation Community Giving Grants, 5
Abbott Fund Science Education Grants, 220
Abdus Salam ICTP-OEA Centre Grants, 222
Abington Foundation Grants, 243
Affymetrix Corporate Contributions Grants, 355
All Saints Educational Trust Grants, 572
ALZA Corporate Contributions Grants, 588
American Foodservice Charitable Trust Grants, 710
American Montessori Society Thesis and Dissertation Awards, 758
Amgen Foundation Grants, 812
Anderson Foundation Grants, 842
Andrew Family Foundation Opportunity Knocks Grants, 843
Angels Baseball Foundation Grants, 848
Annenberg Foundation Grants, 862
Arizona Commission on the Arts Education Projects Grants, 996
Arizona Commission on the Arts Folklorist Residencies, 997
ASHA Advancing Academic Careers Award, 1072
Ball Brothers Foundation General Grants, 1291
Bayer Foundation Grants, 1309
Boeing Company Contributions Grants, 1357
Bristol-Myers Squibb Foundation Health Education Grants, 1378
Cabot Corporation Foundation Grants, 1406
California Teachers Association Scholarships, 1408
Carl M. Freeman Foundation Grants, 1430
Carnegie Corporation of New York Grants, 1432
Charles H. Price II Family Foundation Grants, 1523
Charles Stewart Mott Foundation Grants, 1529
Chestnut Hill Charitable Foundation, Inc Grants, 1538
Chiron Foundation Community Grants, 1553
Christensen Fund Regional Grants, 1554
Clarence E. Heller Foundation Grants, 1575
Coca-Cola Foundation Grants, 1593
Comprehensive Health Education Found Grants, 1622
Corning Foundation Educational Grants, 1646
Cudd Foundation Grants, 1672
Dept of Ed Jacob K. Javits Gifted and Talented Students Education Program-National Research and Development Center, 1746
Dept of Ed Star Schools Grants, 1749
Dorr Foundation Grants, 1842
Earthwatch Institute Student Fellowships, 1890
Edward W. and Stella C. Van Houten Grants, 1899
Eugene McDermott Foundation Grants, 1958
Ford Motor Company Fund Grants, 2033
Frank Stanley Beveridge Foundation Grants, 2051
Fulbright-Hays Seminars Abroad Bilateral Projects Grants, 2062
Genuardi Family Foundation Grants, 2109
GNOF IMPACT Grants for Education, 2154
Greater Worcester Community Foundation Jeppson Memorial Fund for Brookfield Grants, 2165
Heineman Foundation for Research, Education, Charitable and Scientific Purposes, 2227
Hilton Hotels Corporate Giving Grants, 2268
Hoglund Foundation Grants, 2271
Honeywell Corp Habitat & Conservation Grants, 2277
ICC Scholarship of Engagement Faculty Grants, 2334
IRA Esther Zolt Teacher Research Award, 2408
ITEA Maley/FTE Scholarship, 2435
James R. Dougherty Jr. Foundation Grants, 2463
Japan Foundation New York Institutional Support for Japanese Studies Grants, 2477
John P. McGovern Foundation Grants, 2504
Joseph Alexander Foundation Grants, 2513
Josiah W. and Bessie H. Kline Foundation Grants, 2517
Joyce Foundation Education Grants, 2519
Kellogg Institute for International Studies Course Development Grants, 2537
M.J. Murdock Charitable Trust Partners in Science Grants, 2642
MAA-Tensor Foundation Women and Mathematics Grants, 2645
Massachusetts Foundation for the Humanities Project Grants, 2678
Mattel Children's Foundation Grants, 2685
Mead Johnson Nutritionals Med Educ Grants, 2734
Meadows Foundation Grants, 2736
Meyer Memorial Trust Responsive Grants, 2748
MfA Newton Fellowships, 2749
Motorola Foundation Innovation Generation Collaborative Grants, 2797
Mt. Sinai Health Care Foundation Academic Medicine and Bioscience Grants, 2803
NACC William Diaz Fellowships, 2813
NAEF Teacher Incentive Grants, 2815
NCTM Edward G. Begle PreK-6 Classroom Research Grants, 2942
Nellie Mae Education Foundation State Level Systems Change Grants, 2971
NFIE Found Learning & Leadership Grants, 2992
NIH School Interventions to Prevent Obesity, 3244
Northrop Grumman Foundation Grants, 3429
NRA Foundation Grants, 3437
NSF Arctic Research and Education Grants, 3489
NSF Director's Award for Distinguished Teaching Scholars, 3553
NSF Innovative Technology Experiences for Students and Teachers, 3582
Oppenstein Brothers Foundation Grants, 3792
Padilla Bay Foundation Grants, 3863
Phi Upsilon Omicron Geraldine Clewell Doctoral Fellowship, 3945
Phi Upsilon Omicron Geraldine Clewell Masters Fellowship, 3946
Rathmann Family Foundation Grants, 4033
Richard & Susan Smith Family Found Grants, 4064
Richard Davoud Donchian Foundation Grants, 4066
Robert W. Deutsch Foundation Grants, 4082
Russell Sage Foundation Grants, 4129
San Diego Women's Foundation Grants, 4164
Shell Oil Company Foundation Grants, 4233
Sidgmore Family Foundation Grants, 4252
Sony Corporation of America Corporate Philanthropy Grants, 4329
STTI/ATI Educational Assessment Nursing Research Grants, 4457
Sun Academic Excellence Grants, 4466
Sun Microsystems Foundation Summer Academic Enrichment Grants, 4468
U.S. Department of Education National Resource Centers, 4542
U.S. Department of Education Native Hawaiian Education Grants, 4543
USDA Resident Instruction Grants for Higher Education in Insular Areas, 4645
USDA Tribal College Research Grants, 4652
Vancouver Foundation Grants and Community Initiatives Program, 4807
Virginia Foundation for the Humanities Discretionary Grants, 4813
Virginia Found for Humanities Open Grants, 4814
Weyerhaeuser Company Foundation Grants, 4857
Xerox Foundation Grants, 4899

Demonstration Grants

AAA Foundation for Traffic Safety Grants, 24
ACT Awards, 330
Alfred P. Sloan Public Understanding of Science and Technology-Other Efforts, 556
Alfred P. Sloan Synthetic Biology Grants, 565
American Chemical Society Corporate Associates Seed Grants, 676
American Chemical Society Petroleum Research Fund Doctoral New Investigator Grants, 690
American Chemical Society Petroleum Research Fund New Directions Grants, 691
American Chemical Society Petroleum Research Fund Undergraduate New Investigator Grants, 694
American Legacy Found Small Innovative Grants, 749
Bank of America Charitable Foundation Matching Gifts, 1297
Barra Foundation Project Grants, 1301
BCBSM Foundation Physician Investigator Research Awards, 1314
Berks County Community Foundation Grants, 1333
Bullitt Foundation Grants, 1384
Caesars Foundation Grants, 1407
CDC Evidence-Based Laboratory Medicine: Quality/Performance Measure Evaluation, 1459
Charles Stewart Mott Anti-Poverty Grants, 1528
Charles Stewart Mott Foundation Grants, 1529
Children's Cardiomyopathy Foundation Research Grants, 1550
Colorado Bioscience Discovery Eval Grants, 1598
Commonwealth Fund Patient-Centered Coordinated Care Grants, 1608
Commonwealth Fund Quality of Care for Frail Elders Grants, 1610
CTCRI Research Planning Grants, 1669
DEED Grants, 1730

Delta Air Lines Foundation Health and Wellness Grants, 1736
Dept of Ed Star Schools Grants, 1749
Dept of Ed Vocational Rehabilitation Services Demonstration and Training Programs, 1751
DHHS Emergency Med Services for Children, 1755
DHHS Oral Health Promotion Research Across the Lifespan, 1759
DHHS Special Programs for the Aging Training, Research, and Discretionary Projects Grants, 1761
DHHS Welfare Reform Research, Evaluations, and National Studies Grants, 1762
DOI National Spatial Data Infrastructure Cooperative Agreements Grants, 1823
DOT Transit Planning and Research Grants, 1849
EPA Capacity Building Grants for Compliance Assurance and Enforcement Activities in Indian Country and Other Tribal Areas, 1932
Fight for Sight Grants-in-Aid, 2013
Great Lakes Protection Fund Grants, 2170
Harold R. Bechtel Charitable Remainder Uni-Trust Grants, 2201
Hydrogen Student Design Contest, 2311
IMLS 21st Century Museum Pros Grants, 2362
IMLS Grants to State Library Administrative Agencies, 2364
IMLS National Leadership Grants, 2366
Ittleson Foundation AIDS Grants, 2436
Ittleson Foundation Mental Health Grants, 2438
James S. McDonnell Foundation Complex Systems Collaborative Activity Awards, 2465
James S. McDonnell Foundation Scholar Awards, 2467
Jenifer Altman Foundation Grants, 2484
Juvenile Justice Clearing House Grants, 2530
Manitoba Arts Council Craft Grants, 2653
MLA Research, Development, and Demonstration Project Grant, 2789
National Psoriasis Foundation Research Grants, 2882
New Hampshire Charitable Foundation Grants, 2985
NFWF/Exxon Save the Tiger Fund Grants, 2995
NFWF Pulling Together Initiative Grants, 3007
NFWF Shell Marine Habitat Grants, 3008
NIGMS Short Courses on Mathematical, Statistical, and Computational Tools for Studying Biological Systems, 3201
NLM Exploratory/Developmental Grants, 3296
NSF CISE Cross-Cutting Grants, 3532
PhRMA Foundation Health Outcomes Research Starter Grants, 3951
Pinkerton Foundation Grants, 3967
Robert Sterling Clark Foundation Government Accountability Grants, 4080
RRF General Grants, 4115
RWJF Changes in Health Care Financing and Organization Small Grants, 4134
SARE Research and Education Grants - Northeast Region, 4178
Self Foundation Grants, 4197
Skin Cancer Foundation Research Grants, 4267
Tandem Launch Technologies Grants, 4493
USDA Plant and Animal Disease, Pest Control, and Animal Care Grants, 4638
Xerox Foundation Grants, 4899

Development (Institutional/Departmental)

Acuity Charitable Foundation Grants, 333
AIR National Summer Data Policy Institute Fellowships, 469
AIR Research Grants/Dissertation Fellowships, 470
Albuquerque Community Foundation Grants, 508
Alfred P. Sloan Public Understanding of Science and Technology-Radio Grants, 558
Alfred P. Sloan Public Understanding of Science and Technology-Theater Grants, 559
All Saints Educational Trust Grants, 572
AMA Foundation Seed Grants for Research, 601
American Chemical Society Award for Team Innovation, 661
American Chemical Society Chemical Technology Partnership Mini Grants, 674
American Chemical Society Corporate Associates Seed Grants, 676
American Chemical Society Green Chem Grants, 679
American Chemical Society Herman Frasch Foundation Grants, 680
American Chemical Society Local Section YCC Starter Grants, 684
American Chemical Society Mentoring Grants, 685
American Chemical Society Petroleum Research Fund Doctoral New Investigator Grants, 690
American Chemical Society Petroleum Research Fund Undergraduate New Investigator Grants, 694
Ann & Robert H. Lurie Family Found Grants, 859
Ann Arbor Area Community Foundation Grants, 860
Assisi Foundation of Memphis General Grants, 1221
BCBSNC Foundation Grants, 1318
Bernard Van Leer Foundation Grants, 1336
Bill and Melinda Gates Foundation Financial Services for the Poor Grants, 1346
Blowitz-Ridgeway Foundation Early Childhood Development Research Award, 1348
Brainerd Foundation Grants, 1368
California Wellness Foundation Work and Health Grants, 1409
Carl C. Icahn Foundation Grants, 1428
Carl M. Freeman Foundation Grants, 1430
Carnegie Corporation of New York Grants, 1432
CCH Documentary Project Research and Development Grants, 1451
Center on Philanthropy and Civil Society's Emerging Leaders International Fellows Program, 1487
CFF Pilot and Feasibility Awards, 1495
Chicago Community Trust Fellowships, 1546
Chicago Community Trust Housing Grants: Advancing Affordable Rental Housing, 1547
Conquer Cancer Foundation of ASCO Drug Development Research Professorship, 1632
Conquer Cancer Foundation of ASCO Improving Cancer Care Grants, 1633
Council on Foundations Emerging Philanthropic Leaders Fellowships, 1649
CTCRI Workshop and Learning Opportunities Grants, 1671
Cystic Fibrosis Canada Clinic Incentive Grants, 1676
Dave Thomas Foundation for Adoption Grants, 1723
Deborah Munroe Noonan Research Grants, 1729
DEED Grants, 1730
DOD HBCU/MI Partnership Training Award, 1765
Doris Duke Foundation Operations Research on AIDS Care & Treatment in Africa Grants, 1839
DTE Energy Found Higher Learning Grants, 1866
Edward Bangs Kelley and Elza Kelley Foundation Grants, 1898
Emil Buehler Perpetual Trust Grants, 1919
Energy Foundation Transportation Grants, 1925
EPA Capacity Building Grants for Compliance Assurance and Enforcement Activities in Indian Country and Other Tribal Areas, 1932
Evjue Foundation Grants, 1963
Fannie E. Rippel Foundation Grants, 1978
FAR Fund Grants, 1980
Farmers Insurance Corporate Giving Grants, 1981
Fidelity Foundation Grants, 2007
Greater St Louis Community Foundation Grants, 2164
Harry A. and Margaret D. Towsley Found Grants, 2205
Health Foundation of Southern Florida Responsive Grants, 2223
Honor the Earth Grants, 2279
ICC Scholarship of Engagement Faculty Grants, 2334
IMLS Laura Bush 21st Century Librarian Grants, 2365
Indiana 21st Century Research and Technology Fund Awards, 2370
IREX University Administration Support, 2425
ITVS Commissioned Funding, 2439
Jackson Foundation Grants, 2453
Jacob and Valeria Langeloth Foundation Grants, 2455
Japan Foundation New York Institutional Support for Japanese Studies Grants, 2477
JMO Law and Legal System Grants, 2491
JMO Public Policy Research Grants, 2492
JMO Strategic and International Studies Grants, 2493
John D. and Catherine T. MacArthur Foundation Global Challenges Grants, 2498
Jovid Foundation Grants, 2518
Link Foundation Energy Doctoral Fellowships, 2617
MetLife/TCG Aha! Think it Do it Grants, 2745
Miles of Hope Breast Cancer Foundation Grants, 2773
MLA David A. Kronick Traveling Fellowship, 2784
Mt. Sinai Health Care Foundation Academic Medicine and Bioscience Grants, 2803
National Housing Endowment Challenge/Build/Grow Grant, 2877
National Security Education Grants to Institutions of Higher Education, 2885
NCFR Innovation Grants, 2908
NCI Cancer Education and Career Development Program, 2917
New Hampshire Charitable Foundation Grants, 2985
NHLBI Ruth L. Kirschstein National Research Service Award Short-Term Institutional Research Training Grants, 3061
NHPRC Electronic Records Projects Grants, 3073
NIH Biology, Development, and Progression of Malignant Prostate Disease Research Grants, 3206
NOAA Collaborative Science, Technology, and Applied Research Grants, 3354
North Carolina Biotechnology Center Institutional Development Grants, 3422
North Carolina Biotechnology Center Oliver Smithies Faculty Recruitment Grant, 3423
North Carolina Biotechnology Centers of Innovation Grants, 3425
Northwest Fund for the Environment Grants, 3431
NSF Academic Research Infrastructure Program: Recovery and Reinvestment (ARI-R^2), 3456
NSF Accelerating Discovery In Science and Engineering through PETASCALE Simulations and Analysis, 3457
NSF ADVANCE: Increasing the Participation and Advancement of Women in Academic Science and Engineering Careers Grants, 3461
NSF Atmospheric Sciences Mid-Size Infrastructure Opportunity Grants, 3497
NSF Biological Research Collections, 3504
NSF Chemistry Research Experiences for Undergraduates, 3518
NSF CISE Computing Res Infrastructure Grants, 3531
NSF Developing Global Scientists and Engineers, 3552
NSF Engineering Research Centers Grants, 3561
NSF Faculty Early Career Development Grants, 3565
NSF Historically Black Colleges and Universities Grants, 3577
NSF Innovation and Organizational Change, 3581
NSF Innovative Technology Experiences for Students and Teachers, 3582
NSF Integrative Graduate Education and Research Traineeship, 3584
NSF Materials World Network: Cooperative Activity in Materials Research between US Investigators and their Counterparts Abroad, 3595
NSF Partnerships in Astronomy & Astrophysics Research and Education, 3612
NSF Research Experiences for Undergrads, 3638
NSF Research on Gender in Science and Engineering Grants, 3639
NSF Science and Society (S&S), 3643
NSF Scientific Computing Research Environments for the Mathematical Sciences Grants, 3645
NSF Sociology Grants, 3650
NSF Solar, Heliospheric, and INterplanetary Environment Grants, 3651
NSF Undergraduate Research and Mentoring in the Biological Sciences, 3663
NSF University Radio Observatories Grants, 3664
NYSCA Dance: Services to the Field Grants, 3702
NYSCA Folk Arts: Exhibitions Grants, 3703
NYSCA Individual Artists: Film, Media and New Technology Production Grants, 3706
Overbrook Foundation Grants, 3860
Perkin Fund Grants, 3898

Phi Upsilon Omicron Alumni Research Grant, 3940
Piper Trust Healthcare & Med Research Grants, 3972
Rhode Island Foundation Grants, 4061
Richard Davoud Donchian Foundation Grants, 4066
Robert W. Deutsch Foundation Grants, 4082
Rockwell International Corporate Trust Grants, 4093
Rohm and Haas Company Grants, 4098
RWJF Partners Investing in Nursing Grants, 4149
S.E.VEN Fund Annual Grants, 4153
Salem Foundation Grants, 4160
Sensient Technologies Foundation Grants, 4199
Sioux Falls Area Community Foundation Community Fund Grants (Unrestricted), 4262
Sioux Falls Area Community Found Spot Grants, 4263
SSRC Islamic Traditions and Muslim Societies in World Contexts Grants, 4430
SSRC Our Shared Past Grants, 4437
Starr Foundation Grants, 4447
Streilein Foundation for Ocular Immunology Visiting Professorships, 4453
T. James Kavanagh Foundation Grants, 4489
U.S. Department of Education American Overseas Research Centers, 4537
U.S. Department of Education Disability and Rehabilitation Research and Related Projects, 4538
U.S. Department of Education Foreign Languages and Area Studies Fellowships, 4539
U.S. Department of Education Graduate Assistance in Areas of National Need, 4540
U.S. Department of Education National Resource Centers, 4542
U.S. Department of Education Native Hawaiian Education Grants, 4543
U.S. Department of Education Rehabilitation Research Training Centers, 4545
U.S. Department of Education Ronald E. McNair Postbaccalaureate Achievement Grants, 4546
USC Pentecostal and Charismatic Research Initiative Regional Center Grants, 4606
US CRDF Science and Technology Entrepreneurship Program Business Partnership Grant Competition Armenia, 4611
US CRDF STEP Business Partnership Grant Competition Ukraine, 4612
USDA Hispanic-Serving Institutions Education Grants, 4628
USDC/NOAA Nature Conservancy Community-based Matching Grants, 4661
USDC Market Development Cooperator Program Grants, 4667
Vancouver Foundation Grants and Community Initiatives Program, 4807
Wabash Center Project Grants, 4826
Wabash Center Small Project Grants, 4827
Wabash Center Waorkshop and Colloquy Participant Grants, 4828
Wallace Foundation Grants, 4829
William T. Grant Foundation Distinguished Fellows Awards, 4880

Dissertation/Thesis Research Support

AAA Minority Dissertation Fellowship Program, 25
AAG Dissertation Research Grants, 171
AAHN Pre-Doctoral Research Grant, 179
AAS Kate B. and Hall J. Peterson Fellowships, 204
AAS Legacy Fellowships, 205
AAS Long-Term Visiting Academic Research Fellowships, 206
AAS Short-Term Visiting Academic Research Fellowships, 208
ACC Humanities Fellowship Program, 255
ACLS African Humanities Program for Dissertation Completion, 278
ACLS Dissertation Fellowships in East European Studies, 285
AERA Dissertation Grants, 342
AERA Minority Fellowships, 343
AHIMA Dissertation Assistance Grants, 374
AIHP Thesis Support Grants, 433
AIR/IES-NCES Postdoctoral Policy Fellowships in Washington, 466
AIR/NCES Fellowships for Graduate Study, 467
AIR Research Grants/Dissertation Fellowships, 470
AJA Ethel Marcus Memorial Fellowship in American Jewish Studies, 471
AJA Lowenstein-Wiener Summer Fellowship Awards in American Jewish Studies, 472
AJA Rabbi Frederic A. Doppelt Memorial Fellowship in American Jewish Studies, 474
AJA Starkoff Fellow in American Jewish Studies, 475
ALA Phyllis Dain Library History Dissert Award, 495
American Academy of Advertising Doctoral Dissertation Competition Awards, 608
American Montessori Society Thesis and Dissertation Awards, 758
American Numismatic Society Doctoral Fellows, 759
American Philosophical Society John Hope Franklin Dissertation Fellowships, 765
AMS Alvin H. Johnson Dissertation Fellowships, 820
AMS Graduate Fellowships in History of Science, 824
Andrew W. Mellon/ACLS Dissertation Completion Fellowships, 844
Animal Behavior Society Student Research Grants, 853
ANL Graduate Thesis Parts Res Appointments, 856
Anthropological Center for Training in Global and Environmental Change Grad Assistantships, 870
APA Dissertation Research Awards, 905
APSA E.E. Schattschneider Award, 932
APSA Edward S. Corwin Award, 933
APSA Gabriel A. Almond Award, 934
APSA Harold D. Lasswell Award, 935
APSA Helen Dwight Reid Award, 936
APSA Leonard D. White Award, 937
APSA Leo Strauss Award, 938
APSA William Anderson Award, 950
APS Marshall N. Rosenbluth Outstanding Doctoral Thesis Award, 969
APS Mitsuyoshi Tanaka Dissertation Award in Experimental Particle Physics, 971
APS Nicholas Metropolis Award for Outstanding Doctoral Thesis in Computational Physics, 972
APS Outstanding Doctoral Thesis Research in Atomic, Molecular, or Optical Physics, 974
ARIT Fellowships, 991
ARM Graduate Fellowships, 1002
ASA/BJS Stat Methodological Research Grants, 1029
ASCSA Fellowships, 1052
ASCSA Jacob Hirsch Fellowship, 1054
ASR Fichter Research Grant, 1217
Association for the Study of Higher Education Dissertation Fellowships, 1222
ASTD Dissertation Award, 1225
BC Center for Retirement Research Dissertation Fellowships, 1320
CFR Stanton Nuclear Security Fellowships, 1506
Charlotte W. Newcombe Doctoral Dissertation Fellowships, 1531
CIMO Scholarships for Advanced Studies of the Finnish Language at Finnish Universities, 1569
CSRPC Dissertation Fellowships, 1661
Daimler and Benz Foundation Fellowships, 1713
Delta Dental Master's Thesis Award, 1737
Delyte and Dorothy Morris Doctoral Fellowships, 1739
DOJ National Institute of Justice Graduate Research Fellowships, 1828
Dorothy Leet Grants, 1840
Educational Communications and Tech Awards, 1895
FDHN Graduate Student Awards, 1994
Field Museum Graduate Student Fellowships, 2008
Foundation for Evangelism Doctoral Scholars, 2039
FTE Dissertation Fellowships, 2058
FTE Expanding Horizons Partnership Grants, 2060
Geographic Education Dissertation Award, 2110
GSA Research Grants, 2185
GUITS Grants for Undergraduate Study, 2187
GUITS Research Grants in Comparative Studies of Modern Turkey, 2190
GUITS Sabbatical Research Grants, 2191
Harry Frank Guggenheim Foundation Dissertation Fellowships, 2207
Henry Luce Foundation/ACLS Dissertation Fellowships in American Art, 2235
Herman Kahn Fellowship, 2247
Horton Research Grants, 2282
HRF Graduate Fellowships, 2285
ICCS Graduate Students Scholarships, 2335
Ida Smedley MacLean International Fellowship, 2337
IFUW Dr. A. Vibert Douglas Int'l Fellowship, 2346
IFUW Winifred Cullis Grants, 2347
IRA Helen M. Robinson Dissertation Res Award, 2409
IUCP Dissertation Fellowship, 2441
Jacobs Research Funds Small Grants, 2456
Japan Foundation Japanese Studies Fellowships, 2476
Japan Foundation New York Japanese Studies Fellowships, 2478
Japan Foundation Six-Month Grants for Specialists in Cultural and Academic Fields, 2479
Joseph L. Fisher Dissertation Awards, 2515
Kress Foundation Art History Pre-Doctoral Fellowships, 2576
Law and Social Science Doctoral Dissertation Fellowships, 2589
LBI-L Leo Baeck/Studienstiftung Fellowships, 2592
LBI-NY/DAAD Fellowships in Germany, 2593
LBI-NY/DAAD Fellowships in New York, 2594
LBI-NY Fred Grubel Fellowship, 2596
LBI-NY Fritz Halbers Fellowships, 2597
LI American Religion Dissertation Fellowship, 2604
Library Company of Philadelphia Program in Early American Economy and Society Dissertation-Level Fellowships, 2609
Library Company of Philadelphia Early American Economy & Society One-Month Fellowships, 2610
Library Company of Philadelphia Program in Early American Economy and Society Post-Doctoral Research Fellowships, 2611
Marine Corps Historical Center Dissertation Fellowship, 2663
Martin E.P. Seligman Award for Outstanding Research in Positive Psychology, 2667
MCEAS Dissertation Fellowships, 2723
MSI Alden G. Clayton Dissertation Proposal Competition, 2800
NAPM Doctoral Dissertation Grant, 2827
NCFR Affiliate Councils Outstanding Student Research Papers, 2906
NCTE Teacher Researcher Award, 2940
Nestle Foundation Training Grant, 2983
NHLBI Ruth L. Kirschstein National Research Service Awards for Individual Predoctoral Fellowships to Promote Diversity in Health-Related Research, 3058
NIA Aging Research Dissertation Awards to Increase Diversity, 3082
NIA Minority Dissertation Res Grants in Aging, 3110
NL American Society for Eighteenth-Century Studies Fellowships, 3279
NL Frances C. Allen Fellowships, 3284
NL Grad Scholar-in-Residence Opportunities, 3285
NL Herzog August Bibliothek Wolfenbuttel Fellowship, 3286
NL Lawrence Lipking Fellowship at Newberry, 3287
NL Midwest Modern Lang Assoc Fellowship, 3299
NL Newberry Consortium on American Indian Studies Graduate Student Fellowships, 3312
NL Newberry Library/Ecole Nationale des Chartes Exchange Fellowship, 3313
NL Northeast Modern Language Assocation Fellowship, 3314
NL Short-Term Fellowships in the History of Cartography, 3316
NL South Central Modern Languages Association Fellowship, 3318
NL Susan Kelly Power and Helen Hornbeck Tanner Fellowship, 3319
NSF/NEH Documenting Endangered Languages Project Grants and Fellowships: data, infrastructure and computational methods, 3455
NSF Arctic Research Opportunities, 3490

NSF Cultural Anthropology Research Grants, 3548
NSF Developing Global Scientists and Engineers, 3552
NSF Doctoral Dissertation Improvement Grants in the Directorate for Biological Sciences, 3554
NSF Ecosystem Science Grants, 3556
NSF Linguistics Grants, 3587
NSF SBE Doctoral Dissertation Research Improvement Grants, 3641
NSF Sociology Grants, 3650
NSF Starter Grants for Postdoctoral Fellows in Microbial Biology, 3653
NWSA Graduate Scholarships, 3677
OAH Huggins-Quarles Award, 3720
ONF Small Grants, 3743
Orchid Research Fellowships, 3794
Organization for Tropical Studies Fellowships, 3796
Organization for Tropical Studies Pilot Awards for Exploratory Research, 3797
Orgone Biophysical Research Lab Lou Hochberg Awards, 3798
ORISE Savannah River Site Professional Internship Program, 3825
OTS Francis J. Bossuyt Fellowship, 3855
OTS Hovore-Horn Fellowships, 3856
OTS Stanley Smith Horticultural Fellowships, 3857
Pfizer Australia Neuroscience Research Grants, 3920
Phi Delta Kappa Outstanding Doctoral Dissertation Awards, 3936
Phi Kappa Phi Found Love of Learning Awards, 3937
Phi Upsilon Omicron Geraldine Clewell Doctoral Fellowship, 3945
Phi Upsilon Omicron Geraldine Clewell Masters Fellowship, 3946
Phi Upsilon Omicron Jean Dearth Dickerscheid Fellowship, 3947
Phi Upsilon Omicron Presidents Res Fellowship, 3949
PhRMA Foundation Health Outcomes Post Doctoral Fellowships, 3950
PhRMA Foundation Informatics Pre Doctoral Fellowships, 3954
PhRMA Foundation Pharmaceutics Pre Doctoral Fellowships, 3959
Radcliffe Institute Dissertation Grants, 4019
Radcliffe Institute Research Support Grants, 4023
SASS Einar and Eva Lund Haugen Memorial Scholarship, 4183
SfN Donald B. Lindsley Prize in Behavioral Neuroscience, 4204
Sigma Theta Tau International /American Nurses Foundation Grant, 4254
Sigma Theta Tau International Doris Bloch Research Award, 4257
Sigma Theta Tau Small Grants, 4258
Smithsonian Predoctoral Fellowships, 4295
Spencer Education Dissertation Fellowships, 4334
SRC Educational Science Doctoral Grants, 4355
SSRC Dissertation Proposal Development Faculty Field Fellowships, 4424
SSRC Dissertation Proposal Development Student Fellowships, 4425
SSRC Eurasia Program Dissertation Development Awards, 4427
SSRC Int'l Dissertation Research Fellowships, 4429
State Farm Companies Doc Dissertation Award, 4448
Swann Foundation Fellowships, 4478
Syracuse University Graduate Fellowships, 4488
Tanner Humanities Center Research Fellowships, 4494
Temple Univ George D. McDowell Fellowship, 4499
Thomson Reuters / MLA Doctoral Fellowships, 4512
TIRF Doctoral Dissertation Grants, 4520
TRDRP Dissertation Research Awards, 4526
Trinity College Ann Plato Fellowship, 4533
United States Institute of Peace - Jennings Randolph Peace Dissertation Scholarships, 4579
University of Colorado Doctoral Dissertation Fellowship in Media, Religion, and Culture, 4586
USDC Dissertation Fellowships, 4665
WAWH Founders Dissertation Fellowship, 4833
Wenner-Gren Foundation Dissertation Fieldwork Grants, 4848
Winterthur Museum Dissertation Fellowships, 4887
Winterthur Research Fellowships, 4888
Wolfsonian-FIU Fellowships, 4890
Woodrow Wilson Doctoral Dissertation Fellowships in Women's Studies, 4891
Woodrow Wilson MMUF Dissertation Grants, 4892
Woodrow Wilson MMUF Travel & Res Grants, 4893

Educational Programs

3M Canada National Teaching Fellowships, 3
3M Company Higher Education Grants, 4
3M Foundation Community Giving Grants, 5
A.L. Mailman Family Foundation Grants, 11
A.O. Smith Community Grants, 12
AAA Foundation for Traffic Safety Grants, 24
AAAS Entry Point Internships, 30
AAAS Science and Technology Policy Fellowships: Energy, Environment, and Agriculture, 41
AAAS Science and Technology Policy Fellowships: Health, Education and Human Services, 43
AAAS Science Prize for Inquiry-Based Instruction, 45
AAFA Investigator Research Grants, 153
AAFP Foundation Pfizer Teacher Development Awards, 157
AAFP Foundation Wyeth Immunization Awards, 158
AAHN Pre-Doctoral Research Grant, 179
AAHPM Hospice and Palliative Med Fellowships, 180
AALAS Foundation Grants, 182
Aaron & Cecile Goldman Family Found Grants, 200
AASF Lydia Donaldson Tutt-Jones Memorial Research Grant, 203
AAUW Community Action Grants, 213
Abbott Fund Global AIDS Care Grants, 219
Abbott Fund Science Education Grants, 220
Abington Foundation Grants, 243
ACC Ford Foundation Fellowships, 253
ACC Hong Kong Arts Program Fellowships, 254
ACC Humanities Fellowship Program, 255
ACC Japan-US Arts Program Fellowships, 256
ACC Philippines Fellowship Program, 261
ACC Starr Foundation Fellowship Program, 268
ACC Taiwan Fellowship Program, 269
A Charitable Foundation Grants, 273
Achelis Foundation Grants, 274
ACLS Charles A. Ryskamp Research Fellowships, 281
ACLS Digital Innovation Fellowships, 284
ACLS Fellowships, 287
ACLS Frederick Burkhardt Residential Fellowships for Recently Tenured Scholars, 288
ACT Awards, 330
Acuity Charitable Foundation Grants, 333
ACVIM Foundation Clinical Investigation Grants, 334
ADA Foundation Dentsply International Research Fellowships, 335
Adam Richter Charitable Trust Grants, 339
Adaptec Foundation Grants, 340
AERA Dissertation Grants, 342
Aetna Foundation Minority Scholars Grants, 349
Aetna Foundation National Medical Fellowship in Healthcare Leadership, 350
Affymetrix Corporate Contributions Grants, 355
AHEPA Educational Foundation Scholarships, 373
AHIMA Grant-In-Aid Research Grants, 376
Ahn Family Foundation Grants, 379
AIAR/Council of American Overseas Research Centers Fellowships for Advanced Multi-Country Research, 391
AIAR Annual Professorships, 393
AIAR George A. Barton Fellowships, 395
AIHS Heritage Youth Researcher Summer Science Program, 449
AIR/IES-NCES Postdoctoral Policy Fellowships in Washington, 466
AIR/NCES Fellowships for Graduate Study, 467
AIR/NRC Summer Institute on First-Year Assessment Undergraduate Student Fellowships, 468
AIR National Summer Data Policy Institute Fellowships, 469
AIR Research Grants/Dissertation Fellowships, 470
Akonadi Foundation Anti-Racism Grants, 477
ALA Ulrich's Serials Librarianship Award, 500
Alberta Law Foundation Grants, 503
Albert and Margaret Alkek Foundation Grants, 504
Albuquerque Community Foundation Grants, 508
Alcatel-Lucent Technologies Foundation Grants, 509
Alces Foundation Grants, 510
Alcoa Foundation Grants, 511
Alcohol Misuse and Alcoholism Research Grants, 512
Alcon Foundation Grants, 513
Alexander and Margaret Stewart Trust Grants, 514
Alexander von Humboldt Hezekiah Wardwell Fellowships, 528
Alex C. Walker Educational and Charitable Foundation Grants, 541
Alex Stern Family Foundation Grants, 542
Alfred P. Sloan Civic Initiatives Grants, 545
Alfred P. Sloan Energy and Environment Grants, 552
Alfred P. Sloan Public Understanding of Science and Technology-Book Grants, 555
Alfred P. Sloan Public Understanding of Science & Tech-Public Television Grants, 557
Allen Foundation Educational Nutrition Grants, 570
Allen Lane Foundation Grants, 571
Allstate Foundation Safe and Vital Communities Grants, 573
Allyn Foundation Grants, 574
Alpha Omega Foundation Grants, 576
Alvin and Lucy Owsley Foundation Grants, 584
ALZA Corporate Contributions Grants, 588
American-Italian Cancer Foundation Fellowships, 602
American Academy of Advertising Distinguished Service Award, 607
American Academy of Advertising Doctoral Dissertation Competition Awards, 608
American Academy of Advertising Kim Rotzoll Award for Advertising Ethics and Social Responsibility, 610
American Academy of Nursing Building Academic Geriatric Nursing Capacity Scholarships, 614
American Academy of Nursing Claire M. Fagin Fellowships, 615
American Chemical Society Chemical Technology Partnership Mini Grants, 674
American Chemical Society Corporate Associates Seed Grants, 676
American Chemical Society GCI Pharmaceutical Roundtable Research Grants, 677
American Chemical Society Green Chem Grants, 679
American Chemical Society Local Section YCC Starter Grants, 684
American Chemical Society Petroleum Research Fund Doctoral New Investigator Grants, 690
American Chemical Society Petroleum Research Fund Undergraduate Faculty Sabbatical Grants, 693
American Chemical Society Petroleum Research Fund Undergraduate New Investigator Grants, 694
American College of Bankruptcy Grants, 699
American College of Surgeons and The Triological Society Clinical Scientist Development Awards, 701
American Foodservice Charitable Trust Grants, 710
American Hist Assoc Award for Scholarly Distin, 714
American Legacy Foundation National Calls for Proposals Grants, 748
American Mathematical Society AAAS Mass Media Fellowships, 751
American Philosophical Society John Hope Franklin Dissertation Fellowships, 765
American Psychiatric Association/Merck Co. Early Academic Career Research Award, 778
American Psychiatric Foundation Grants, 791
American Psychiatric Foundation James H. Scully Jr., M.D., Educational Fund Grants, 792
American Psychiatric Foundation Jeanne Spurlock Congressional Fellowship, 793
American Society for Yad Vashem Grants, 798
American Society of Naval Engineers Scholarship Program, 800
American Society on Aging Graduate Student Research Award, 801
American Sociological Association Fund for the Advancement of the Discipline, 804

American Sociological Association Teaching Enhancement Fund Grants, 807
American Wine Society Educational Foundation Scholarship Program, 808
AMI Semiconductors Corporate Grants, 813
AMNH Graduate Student Fellowships, 815
AMS Alvin H. Johnson Dissertation Fellowships, 820
Anderson Foundation Grants, 842
Andrew Family Foundation Opportunity Knocks Grants, 843
Andrew W. Mellon Postdoctoral Fellowship in American Cultures, 845
Angels Baseball Foundation Grants, 848
Animal Behavior Soc Charles H. Turner Award, 849
Animal Behavior Society Diversity Grant, 851
Ann & Robert H. Lurie Family Found Grants, 859
Ann Arbor Area Community Foundation Grants, 860
Anne Louise Barrett Fellowship, 861
Annenberg Foundation Grants, 862
Anthem Blue Cross and Blue Shield School Nurse Grants, 868
Anthony R. Abraham Foundation Grants, 869
APHA Fellowship in Printing History, 913
APHL Emerging Infectious Diseases Fellowships, 914
Appalachian Regional Commission Health Care Grants, 915
Applied Biosystems Grants, 916
Appraisal Institute Education Trust Scholarships, 917
APS/Phoenix Suns Energy and Environmental Minigrants, 919
APSAA Mini-Career Grants, 921
APSAA Small Beginning Scholar Pre-Investigation Grants, 923
APSAA Small Beginning Scholar Visiting and Consulting Grants, 924
APS Mitsuyoshi Tanaka Dissertation Award in Experimental Particle Physics, 971
APS Outstanding Doctoral Thesis Research in Atomic, Molecular, or Optical Physics, 974
Arcadia Foundation Grants, 983
Archer Daniels Midland Foundation Grants, 985
Archibald Waynne Dingman Memorial Graduate Scholarship, 986
Arcus Foundation Religion and Values Grants, 987
Arie and Ida Crown Memorial Grants, 989
ARIT/NEH Fellowships for Research in Turkey, 990
ARIT Fellowships for Intensive Advanced Turkish Language Study, 992
ARIT Kenan T. Erim Fellowship for Research at Aphrodisias, 993
Arizona Cardinals Grants, 995
Arizona Commission on the Arts Education Projects Grants, 996
Arizona Public Service Corporate Giving Grants, 998
A Rocha USA Grants, 1006
Arronson Foundation Grants, 1019
Arthritis Foundation of Australia Grants, 1022
Arthur and Rochelle Belfer Foundation Grants, 1023
Arthur F. and Alice E. Adams Charitable Foundation Grants, 1024
Arthur J. Ennis Fellowships, 1025
ASCE Arthur S. Tuttle Memorial Scholarships, 1039
ASCO Long-Term International Fellowships, 1048
ASCSA Harry Bikakis Fellowship, 1053
ASCSA Jacob Hirsch Fellowship, 1054
ASECS/Clark Fellowships, 1057
ASHA Advancing Academic Careers Award, 1072
ASHA Research Mentoring-Pair Travel Award, 1073
ASHA Student Research Travel Award, 1074
ASHA Students Preparing for Academic & Research Careers Award, 1075
ASHFoundation New Century Scholars Program Doctoral Scholarships, 1080
Ashland Corporate Contributions Grants, 1087
ASID Foundation Grants, 1093
ASM Int'l Fellowships for Asia and Africa, 1176
ASM Robert D. Watkins Graduate Research Fellowship, 1183
Aspen Institute William Randolph Hearst Endowed Fellowship for Minority Students, 1194
ASPRS Robert E. Altenhofen Scholarship, 1214
ASPRS William A. Fischer Scholarship, 1216
ASR Fichter Research Grant, 1217
ASSE Scholarships, 1220
Assisi Foundation of Memphis General Grants, 1221
Association for the Study of Higher Education Dissertation Fellowships, 1222
ASU Institute for Humanities Res Fellowships, 1246
ASU Institute for Humanities Research Grants, 1247
ASU Institute for Humanities Research Visiting Fellowships, 1248
As You Sow, 1249
AT&T Industrial Ecology Faculty Fellowships, 1251
Athenaeum of Philadelphia Senior Fellowships, 1256
Auburn Foundation Grants, 1263
Auburn University President's Graduate Opportunities Program, 1264
Audio Engineering Society Grad Scholarships, 1265
Australian Academy of Science Grants, 1269
AWRA Richard A. Herbert Scholarship, 1281
AWWA Holly A. Cornell Scholarship, 1284
AWWA Larson Aquatic Research Support Scholarships, 1285
AWWA Thomas R. Camp Scholarship, 1286
Axe-Houghton Foundation Grants, 1287
BA Elisabeth Barker Fund, 1288
Bailey Family Foundation Grants, 1289
Ball Brothers Foundation General Grants, 1291
Banfield Charitable Trust Grants, 1295
Banfi Vintners Foundation Grants, 1296
Bank of America Charitable Foundation Matching Gifts, 1297
Barker Award, 1300
Barra Foundation Project Grants, 1301
Batchelor Foundation Grants, 1305
Bay and Paul Foundations, Inc Grants, 1307
Bayer Clinical Scholarship Award, 1308
Bayer Foundation Grants, 1309
BCBSM Foundation Student Award Program, 1316
BCBSNC Foundation Grants, 1318
BC Center for Retirement Research Dissertation Fellowships, 1320
BC Steven H. Sandell Grants, 1323
Beckman Coulter Community Relations Grants, 1325
Bedding Plants Found Horticulture Scholarships, 1329
Berea College Appalachian Music Fellowships, 1332
Berks County Community Foundation Grants, 1333
Beryl Henderson Foundation Grants, 1337
Better Way Foundation Grants, 1338
Bettina Bahlsen Memorial Graduate Scholarship, 1339
Beverly Willis Architecture Foundation Grants and Fellowships, 1340
Bill and Melinda Gates Foundation Agricultural Development Grants, 1345
Bill and Melinda Gates Foundation Water, Sanitation and Hygiene Grants, 1347
Blowitz-Ridgeway Foundation Early Childhood Development Research Award, 1348
Blue Shield of California Grants, 1350
Blum-Kovler Foundation Grants, 1351
BMI Woody Guthrie Fellowships, 1352
BMW of North America Contributions, 1353
Bodenwein Public Benevolent Found Grants, 1355
Bodman Foundation Grants, 1356
Boeing Company Contributions Grants, 1357
Booth-Bricker Fund Grants, 1359
Bosque Foundation Grants, 1360
Boston Foundation Grants, 1361
BPW Career Advancement Scholarships, 1366
Brian G. Dyson Foundation Grants, 1370
Brinson Foundation Grants, 1372
Bristol-Myers Squibb Foundation Global HIV/AIDS Initiative Grants, 1376
Bristol-Myers Squibb Foundation Health Disparities Grants, 1377
Bristol-Myers Squibb Foundation Health Education Grants, 1378
Bristol-Myers Squibb Women's Health Grants, 1379
Bullitt Foundation Grants, 1384
Burlington Northern Santa Fe Found Grants, 1389
Burton G. Bettingen Grants, 1390
Bush Foundation Artist Fellowships, 1391
BWF Ad Hoc Grants, 1392
BWF Career Awards for Medical Scientists, 1394
Bydale Foundation Grants, 1400
Byron W. & Alice L. Lockwood Found Grants, 1401
BYU Religious Studies Center Research Grants, 1402
C. Alma Baker Trust Grants, 1403
Caesars Foundation Grants, 1407
California Wellness Foundation Work and Health Grants, 1409
Callaway Golf Company Foundation Grants, 1410
Camargo Foundation Fellowships, 1412
Campbell Soup Foundation Grants, 1415
CampusRN/AACN Nursing Scholarship Fund, 1416
Canada-U.S. Fulbright International Science and Technology Award, 1417
Canada-U.S. Fulbright Senior Specialists Grants, 1418
Canada Graduate Scholarships and NSERC Postgraduate Scholarships, 1420
CAPAA Mini Grant for Research Related to African Americans in Education, 1424
Cape Branch Foundation Grants, 1425
Carl and Eloise Pohlad Family Found Grants, 1426
Carl and Lily Pforzheimer Foundation Grants, 1427
Carl C. Icahn Foundation Grants, 1428
Carl M. Freeman Foundation Grants, 1430
Carnegie Corporation of New York Grants, 1432
Carpenter Foundation Grants, 1439
Carrier Corporation Contributions Grants, 1440
Castle Rock Foundation Grants, 1441
Catherine Holmes Wilkins Foundation Grants, 1442
Catholic University Academic Scholarships, Grants, and Awards, 1443
Catholic University Life Cycle Institute Research Fellowship, 1444
CCCF Dora Maclellan Brown Sem Scholarships, 1448
CCIS Visiting Research Fellowships, 1452
CCWH/Berkshire Conference of Women Historians Graduate Student Fellowship, 1453
CCWH Ida B. Wells Graduate Fellowship, 1455
CDC Cooperative Agreement for the Development, Operation, and Evaluation of an Entertainment Education Program, 1457
CDC Exp Applied Epidemiology Fellowships, 1460
CDC Steven M. Teutsch Prevention Effectiveness Fellowships, 1467
Cedar Tree Foundation David H. Smith Conservation Research Fellowship, 1470
Center for Inquiry Science and the Public Visiting Research Fellowships, 1477
Center for the Study of Books and Media Visiting Research Fellowships, 1480
Center for the Study of Philanthropy International Fellowships, 1482
Center for the Study of Philanthropy Senior International Fellowships, 1483
Center of Excellence in General Aviation Grants, 1486
Center on Philanthropy and Civil Society's Emerging Leaders International Fellows Program, 1487
CFR Edward R. Murrow Press Fellowship, 1500
CFR International Affairs Fellowships, 1501
CFR International Affairs Fellowships in Nuclear Security, 1502
CFR Int'l Affairs Fellowships in South Korea, 1503
CFR Military Fellowships, 1505
CFR Stanton Nuclear Security Fellowships, 1506
CFUW Beverley Jackson Fellowship, 1507
CFUW Bourse Georgette LeMoyne Award, 1508
CFUW Canadian Home Economics Association Fellowship, 1509
CFUW Dr. Alice E. Wilson Awards, 1510
CFUW Ecole Polytechnique Commem Award, 1511
CFUW Ruth Binnie Fellowship, 1516
Charles Abrams Scholarships, 1519
Charles Edison Fund Grants, 1520
Charles H. Price II Family Foundation Grants, 1523
Charles H. Revson Foundation Grants, 1524
Charles Lafitte Foundation Grants, 1525
Charles M. & Mary D. Grant Found Grants, 1527

Charles Stewart Mott Anti-Poverty Grants, 1528
Charles Stewart Mott Foundation Grants, 1529
Chatlos Foundation Grants, 1532
Chestnut Hill Charitable Foundation, Inc Grants, 1538
CHF Roy G. Neville Short-Term Fellowships, 1543
Chiles Foundation Grants, 1552
Chiron Foundation Community Grants, 1553
Christensen Fund Regional Grants, 1554
Chrysalis Scholarships, 1559
Churchill Fellowships, 1560
CIA Undergraduate Co-Op Program - Engineer/ Program Management Engineer, 1561
CIC Summer Research Opportunities, 1565
CIC Traveling Scholars, 1566
CICU James C. Ross Research Fellowship, 1567
CIGNA Foundation Grants, 1568
CIRCLE Civic Education at the High School Level Research Grants, 1571
Clancil Foundation Grants, 1574
Clarence E. Heller Foundation Grants, 1575
Clark-Huntington Bibliographical Fellowship, 1576
Clark Library Short-Term Resident Fellowships, 1577
Claude Pepper Foundation Grants, 1578
Clay Foundation Grants, 1579
Cleveland-Cliffs Foundation Grants, 1580
CMA Foundation Grants, 1582
CMU Arts in Society Fellowship, 1587
Coca-Cola Foundation Grants, 1593
Coeta and Donald Barker Foundation Grants, 1594
Colonel Stanley R. McNeil Foundation Grants, 1597
Colorado Clean Energy Solar Innov Grants, 1599
Commonweal Foundation Community Assistance Grants, 1602
Commonweal Foundation Grants, 1603
Commonwealth Fund/Harvard University Fellowship in Minority Health Policy, 1604
Community Found AIDS Endowment Awards, 1614
Community Found for Greater Buffalo Grants, 1615
Community Foundation for Greater New Haven Quinnipiac River Fund Grants, 1616
Community Fnd for Muskegon County Grants, 1617
Community Foundation for S Arizona Grants, 1618
Community Found of E Central Illinois Grants, 1619
Community Foundation of the Fox River Valley, 1620
Community Foundation of the Virgin Islands Judith A. Towle Environmental Studies Grants, 1621
Comprehensive Health Education Found Grants, 1622
Compton Foundation International Fellowships, 1624
Compton Foundation Mentor Fellowships, 1625
Connecticut Community Foundation Grants, 1627
ConocoPhillips Grants, 1629
Conquer Cancer Foundation of ASCO Career Development Award, 1630
Conquer Cancer Foundation of ASCO Comparative Effectiveness Research Professorship in Breast Cancer, 1631
Conquer Cancer Foundation of ASCO Medical Student Rotation Grants, 1637
Conservation, Food, and Health Foundation Grants for Developing Countries, 1641
Coors Brewing Corporate Contributions Grants, 1644
Corning Foundation Educational Grants, 1646
Corpus Christi College Research Scholarships, 1647
Coughlin-Saunders Foundation Grants, 1648
Cowles Charitable Trust Grants, 1652
Credit Union Found of British Columbia Grants, 1654
CRF Graduate Research Grants, 1655
Cruise Industry Charitable Foundation Grants, 1660
CSRPC Dissertation Fellowships, 1661
CSRPC Postdoctoral Fellowship for Advanced Scholars, 1662
CSRPC Postdoc Fellowship for Junior Scholars, 1663
Cudd Foundation Grants, 1672
Cushwa Center Research Travel Grants, 1673
Cystic Fibrosis Canada Clinical Fellowships, 1674
Cystic Fibrosis Canada Clinic Incentive Grants, 1676
D.F. Halton Foundation Grants, 1692
DAAD Bilateral Programs for Co-Operative Research Grants, 1695
DAAD Emigre Memorial German Internship Program - Bundestag, 1696
DAAD German Studies Research Grants, 1698
DAAD Intensive Language Courses Grants, 1699
DAAD Learn German in Germany Grants, 1700
DAAD Re-invitation Program for Former Scholarship Holders, 1701
DAAD Research Grants, 1702
DAAD Research Internships in Science and Engineering, 1704
DAAD Research Stays for University Academics and Scientists, 1705
DAAD RISE Professional Internships, 1706
DAAD Study & Internship in Germany, 1708
DAAD Study Visits / Study Seminars and Practicals in Germany by Foreign Students, 1709
DAAD Visiting Professorship Grants, 1711
Dade Community Foundation Grants, 1712
Dana Foundation Science and Health Grants, 1716
Daniel Mendelsohn New Investigator Award, 1717
Dart Center Ochberg Fellowship, 1720
Datatel Scholars Foundation Scholarship, 1722
Davis Family Foundation Grants, 1726
DEED Student Research Grants and Internships, 1731
Del E. Webb Foundation Grants, 1733
Dell Foundation Open Grants, 1734
Deloitte Foundation Grants, 1735
Delta Dental Master's Thesis Award, 1737
DeMatteis Family Foundation Grants, 1740
DENSO North America Foundation Grants, 1742
Dept of Ed International Research and Studies Project Grants, 1745
Dept of Ed Safe and Drug-Free Schools and Communities State Grants, 1748
Dept of Ed Star Schools Grants, 1749
Dept of Ed Voc Ed National Programs, 1750
Dept of Ed Vocational Rehabilitation Services Demonstration and Training Programs, 1751
Dept of Ed Women's Educational Equity Grants, 1752
Detroit Lions Charities Grants, 1753
Diaz-Ayala Cuban and Latin American Popular Music Collection Travel Grants, 1763
DOD HBCU/MI Partnership Training Award, 1765
DOE Academies Crea Teacher Scientists Grants, 1766
DOE Albert Einstein Distinguished Educator Fellowship Program, 1770
DOE Computational Science Grad Fellowships, 1780
DOE Faculty and Student Teams Grants, 1792
Donald W. Reynolds Foundation Aging and Quality of Life Grants, 1832
Dorothy Rider Pool Health Care Grants, 1841
Dorr Foundation Grants, 1842
Dreyfus Special Grants in the Chemical Sciences, 1858
Drs. Bruce and Lee Foundation Grants, 1859
DTE Energy Foundation Health and Human Services Grants, 1865
DTE Energy Found Higher Learning Grants, 1866
Duke University Adult Cardiothoracic Anesthesia and Critical Care Medicine Fellowships, 1869
Duke University Ambulatory and Regional Anesthesia Fellowships, 1870
Duke University Clinical Cardiac Electrophysiology Fellowships, 1871
Duke University Hyperbaric Center Fellowships, 1872
Duke Univ Interventional Cardio Fellowships, 1873
Duke Univ Obstetric Anesthesia Fellowships, 1878
Duke University Pain Management Fellowships, 1879
E.L. Wiegand Foundation Grants, 1885
Earhart Foundation Institutional Grants, 1888
Eaton Charitable Fund Grants, 1893
Education Through Astronomy and Space Science Initiative, 1896
Edward Bangs Kelley and Elza Kelley Foundation Grants, 1898
Edward W. and Stella C. Van Houten Grants, 1899
Edwin S. Webster Foundation Grants, 1900
Elizabeth McGraw Foundation Grants, 1908
Elizabeth Nash Found Summer Research Awards, 1909
Emerson Charitable Trust Grants, 1918
Emil Buehler Perpetual Trust Grants, 1919
Emma B. Howe Memorial Foundation Grants, 1920
Endangered Language Fund Grants, 1921
Energy Foundation Climate Grants, 1923
Engineering Information Foundation Grants, 1927
Ensign-Bickford Foundation Grants, 1928
EPA Children's Health Protection Grants, 1933
EPA Research, Development, Monitoring, Public Education, Training, Demonstrations, and Studies Grants, 1938
EPA Southeastern Multi-Media and Geographic Priority Project Grants, 1939
Erasmus Institute Fellowships, 1944
EREF Doctoral Scholarship, 1945
EREF Fiessinger Doctoral Scholarship, 1946
EREF Solid Waste Research Grants, 1947
EREF Sustainability Research Grants, 1948
Ethics and Excellence in Journalism Grants, 1953
EU Center for Humanistic Inquiry Junior and Postdoctoral Fellowships, 1956
Eugene M. Lang Foundation Grants, 1957
Eugene McDermott Foundation Grants, 1958
Evan and Susan Bayh Foundation Grants, 1960
Evan Frankel Foundation Grants, 1961
Evanston Community Foundation Grants, 1962
Evjue Foundation Grants, 1963
Expect Miracles Foundation Grants, 1965
Explorers Club Scott Pearlman Field Awards, 1967
F.W. Olin Foundation Construction Grant, 1972
Fairlawn Foundation Grants, 1974
FAR Fund Grants, 1980
Farmers Insurance Corporate Giving Grants, 1981
FASSE Demonstration Project Grant, 1982
FCD Young Scholars Fellowships, 1985
Fetzer Institute Research Grants, 2005
Fidelity Foundation Grants, 2007
Field Museum Graduate Student Fellowships, 2008
Field Museum High School Internships, 2009
Field Museum Undergraduate Internships, 2010
Field Museum Visiting Scholarships, 2011
FINRA Investor Education Foundation General Grants, 2016
Fishman Family Foundation Grants, 2020
Fitzpatrick and Francis Family Business Continuity Foundation Grants, 2021
Florence Tan Moeson Fellowships, 2025
Florida Sea Turtle Grants, 2027
Fluor Foundation Grants, 2028
FMC Foundation Grants, 2029
Ford Foundation International Fellowships, 2030
Fordham Prizes, 2032
Ford Motor Company Fund Grants, 2033
Forrest C. Lattner Foundation Grants, 2035
Foundation for Endangered Languages Grants, 2037
Foundation for Enhancing Comm Grants, 2038
Foundation for Seacoast Health Grants, 2041
Frances C. and William P. Smallwood Foundation Grants, 2045
Frances L. and Edwin L. Cummings Memorial Fund Grants, 2046
Frank Stanley Beveridge Foundation Grants, 2051
Frederick Douglass Summer Scholars Fellowship, 2053
Frederick Gardner Cottrell Foundation Grants, 2054
Fritz B. Burns Foundation Grants, 2057
FTE Expanding Horizons Partnership Grants, 2060
Fulbright/Garcia Robles Grants, 2063
Fulbright Alumni Initiatives Awards Grants, 2064
Fulbright Beginning Prof Journalism Awards, 2065
Fulbright Binational Business Grants in Mexico, 2066
Fulbright Business Grants in Spain, 2067
Fulbright Critical Lang Enhancement Awards, 2068
Fulbright Distinguished Chairs Grants, 2069
Fulbright English Teaching Assistantships, 2070
Fulbright European Union Scholar-in-Residence Grants, 2071
Fulbright German Studies Seminar Grants, 2072
Fulbright International Education Administrators Seminar Grants, 2073
Fulbright Occasional Lecturer Grants, 2076
Fulbright Public Policy Grants in Mexico, 2077
Fulbright Specialists Grants, 2079

Fulbright Traditional Scholar Program in Europe and Eurasia, 2080
Fulbright Traditional Scholar Program in Sub-Saharan Africa, 2081
Fulbright Traditional Scholar Program in the East Asia/Pacific Region, 2082
Fulbright Traditional Scholar Program in the Near East and North Africa Region, 2083
Fulbright Traditional Scholar Program in the South and Central Asia Region, 2084
Fulbright Traditional Scholar Program in the Western Hemisphere, 2085
Fulbright Travel Grants, 2086
G. Unger Vetlesen Foundation Grants, 2093
Gates Family Foundation Fellowships, 2097
GCA Summer Environmental Scholarships, 2100
General Motors Foundation Grants, 2107
Genuardi Family Foundation Grants, 2109
George A Ohl Jr. Foundation Grants, 2112
George B. Storer Foundation Grants, 2113
George Foundation Grants, 2115
George Frederick Jewett Foundation Grants, 2116
George Gund Foundation Grants, 2117
George H. Hitchings New Investigator Award in Health Research, 2118
George Kress Foundation Grants, 2119
George S. & Dolores Dore Eccles Grants, 2120
George W. Codrington Foundation Grants, 2122
Georgia-Pacific Foundation Environment Grants, 2123
Georgia Institute of Technology Postdoctoral Fellowship, 2124
Georgia Power Foundation Grants, 2125
Getty Research Grants for Institutions - Archival Grants, 2131
Getty Research Grants for Institutions - Art History Research Grants, 2132
Getty Research Grants for Institutions - Cataloguing of Museum Collections Grants, 2133
Getty Research Institute Nonresidential Grants - Collaborative Research Grant, 2135
Getty Research Institute Nonresidential Grants - Curatorial Research Fellowships, 2136
Getty Research Institute Nonresidential Grants - Postdoctoral Fellowships, 2137
Getty Scholars and Visiting Scholars Grants, 2138
Gil and Dody Weaver Foundation Grants, 2139
Gilroy and Lillian P. Roberts Foundation Grants, 2142
GNOF Gulf Coast Oil Spill Grants, 2153
GNOF IMPACT Grants for Education, 2154
Grass Foundation Marine Biological Laboratory Advanced Imaging Fellowships, 2162
Greater St Louis Community Foundation Grants, 2164
Greater Worcester Community Foundation Jeppson Memorial Fund for Brookfield Grants, 2165
Green Diamond Charitable Contributions, 2171
Greenspun Family Foundation Grants, 2173
Greenwall Foundation Faculty Scholars Program in Bioethics, 2175
GUITS Grants for Undergraduate Study, 2187
GUITS Matching Seed-Money Grants for New Positions in Turkish Studies, 2188
GUITS Post-Doctoral Summer Travel-Research Grants, 2189
GUITS Research Grants in Comparative Studies of Modern Turkey, 2190
GUITS Sabbatical Research Grants, 2191
GUITS Summer Research Grants in Turkey for Graduate Students, 2192
H.A. and Mary K. Chapman Charitable Grants, 2193
H.J. Heinz Company Foundation Grants, 2195
H.W. Wilson Foundation Grants, 2196
Harold Alfond Foundation Grants, 2200
Harold R. Bechtel Charitable Remainder Uni-Trust Grants, 2201
Harold R. Bechtel Testamentary Charitable Trust Grants, 2202
Harry B. and Jane H. Brock Foundation Grants, 2206
Harry Frank Guggenheim Foundation Research Grants, 2208
Harry W. Bass, Jr. Foundation Grants, 2210
Harvard Postdoctoral Fellowships on Education, 2211
Harvard University Administrative Fellowships, 2212
Harvard University Graduate School of Education Fellowships and Awards, 2213
Harvard Women and Public Policy Postdoctoral Fellowship, 2217
Harvey Randall Wickes Foundation Grants, 2218
Hatton W. Sumners Foundation for the Study and Teaching of Self Government Grants, 2219
Haverford College Mellon Postdoc Fellowships, 2220
Hawaii Community Foundation Health Education and Research Grants, 2221
Healthcare Foundation of New Jersey Grants, 2222
Health Foundation of Southern Florida Responsive Grants, 2223
Hearst Foundations Health Grants, 2225
Heineman Foundation for Research, Education, Charitable and Scientific Purposes, 2227
Helen Bader Foundation Grants, 2228
Helen V. Brach Foundation Grants, 2231
Hendrick Foundation for Children Grants, 2232
Henry Dreyfus Teacher-Scholar Awards, 2233
Henry J. Kaiser Family Found Media Fellowships, 2234
Henry W. Bull Foundation Grants, 2237
Herbert A. & Adrian W. Woods Found Grants, 2239
Herbert H. and Grace A. Dow Grants, 2241
HHMI-NIBIB Interfaces Initiative Grants, 2248
HHMI Biomedical Research Grants for International Scientists in Canada and Latin America, 2252
HHMI Biomedical Research Grants for International Scientists in the Baltics, Central and Eastern Europe, Russia, and Ukraine, 2253
HHMI Grants and Fellowships Programs, 2255
HHMI Initiative for Colleges Grants, 2256
HHMI Initiative for Universities Grants, 2257
HHMI Med into Grad Initiative Grants, 2259
Hilda and Preston Davis Foundation Grants, 2266
Hill Crest Foundation Grants, 2267
Hilton Hotels Corporate Giving Grants, 2268
Hirtzel Memorial Foundation Grants, 2269
Hoglund Foundation Grants, 2271
HomeBanc Foundation Grants, 2275
Honeywell Corporation Habitat and Conservation Grants, 2277
Hospital Libraries Section / MLA Professional Development Grants, 2283
Houghton Library Visiting Fellowships, 2284
HRF Hudson River Improvement Grants, 2286
HRSA Nurse Education, Practice, Quality and Retention Grants, 2290
HSBC Contributions Community Grants, 2291
Huie-Dellmon Trust Grants, 2295
Humanities Institute Vis Scholars Fellowships, 2296
Human Source Foundation Grants, 2297
Huntington Library Babara Thom Postdoctoral Fellowships, 2301
Huntington Library Clark-Huntington Joint Bibliographical Fellowship, 2302
Huntington Library Dibner History of Science Long-Term Fellowships, 2304
Huntington Library Short-Term Fellowships, 2308
Hutchinson Community Foundation Grants, 2310
Hydrogen Student Design Contest, 2311
IAS Mellon Fellowships for Assistant Professors, 2318
IBRO Latin America Regional Funding for Neuroscience Schools, 2327
IBRO Latin America Regional Funding for PROLAB Collaborations, 2328
ICC Faculty Fellowships, 2333
ICC Scholarship of Engagement Faculty Grants, 2334
IERF Sepmeyer Research Grants, 2340
IFT Foundation Grants, 2345
IIE David L. Boren Fellowships, 2350
IIE David L. Boren Scholarships, 2351
IIPP Graduate Fellowship, 2357
Illinois DNR Youth Recreation Corps Grants, 2361
IMLS Laura Bush 21st Century Librarian Grants, 2365
IMLS National Leadership Grants, 2366
Indiana SBIR/STTR Commercialization Enhancement Program, 2371
Indiana Space Grant Consortium Discovery Program Research Grants, 2373
Indiana Space Grant Consortium Discovery Program Student Hands-On Research Grants, 2374
Infinity Foundation Grants, 2378
Institute for Advanced Studies in Culture Postdoctoral Fellowships, 2380
Institute of American Cultures Graduate and Predoctoral Fellowships, 2383
Institute of American Cultures Postdoctoral and Visiting Scholar Fellowships, 2384
Institute of Noetic Sciences Research Internships, 2386
Institute of Turkish Studies Dissertation Writing Grants, 2389
Institute of Turkish Studies Research Grants in Comparative Studies of Modern Turkey, 2390
Institute of Turkish Studies Summer Research Grants in Turkey, 2392
Intel Bachelor's and Master's Degree Internships, 2393
Intel Technician Internships, 2395
IRA Esther Zolt Teacher Research Award, 2408
IRA Pearson Foundation Rotary Lit Awards, 2412
Ireland Family Foundation Grants, 2415
IREX Eurasian Undergraduate Exchange Grants, 2418
IREX International Leadership in Education Program Grants, 2420
IREX MENA Media Emerging Leaders Fellows, 2421
IREX University Administration Support, 2425
ISA John Z. Duling Grants, 2429
ISGC Informal Education Grants, 2431
ISGC Multiple Undergraduate Research Grants, 2432
ITEA Maley/FTE Scholarship, 2435
Ittleson Foundation AIDS Grants, 2436
Ittleson Foundation Environment Grants, 2437
Ittleson Foundation Mental Health Grants, 2438
J.B. Reynolds Foundation Grants, 2449
Jackson Foundation Grants, 2453
Jacob and Hilda Blaustein Foundation Grants, 2454
Jacob and Valeria Langeloth Foundation Grants, 2455
James Graham Brown Found Brown Fellowships, 2459
James H. Cummings Foundation Grants, 2460
James M. Collins Foundation Grants, 2461
Japan Found Center for Global Partnership Grants, 2472
Japan Foundation Japanese Studies Fellowships, 2476
Japan Foundation New York Institutional Support for Japanese Studies Grants, 2477
Japan Foundation New York Japanese Studies Fellowships, 2478
Japan Foundation Six-Month Grants for Specialists in Cultural and Academic Fields, 2479
Jeffrey Campbell Grad Fellow in Anthropology, 2482
Jewish Federation of Metropolitan Chicago Scholarships, 2486
JM Foundation Grants, 2490
JMO Law and Legal System Grants, 2491
John Deere Foundation Grants, 2500
John G. Duncan Trust Grants, 2501
John P. McGovern Foundation Grants, 2504
John R. Oishei Foundation Grants, 2505
John Simon Guggenheim Memorial Foundation Fellowships, 2507
Johnson & Johnson Corporate Contributions, 2508
John W. Alden Trust Grants, 2510
Joseph Alexander Foundation Grants, 2513
Josiah Macy Jr. Foundation Grants, 2516
Josiah W. and Bessie H. Kline Foundation Grants, 2517
Jovid Foundation Grants, 2518
Joyce Foundation Education Grants, 2519
Kavli Foundation Research Grants, 2533
Kellogg Institute for International Studies Collaborative Faculty Grants, 2536
Kellogg Institute for International Studies Course Development Grants, 2537
Kellogg Inst for Int'l Studies Intramural Grants, 2546
Kellogg Institute for International Studies Kroc Undergraduate Research Grants, 2547
Kenneth T. & Eileen L. Norris Found Grants, 2554
KFF Barbara Jordan Health Policy Scholars, 2556
Kiplinger Program in Public Affairs Journalism Fellowships, 2557

Kluge Center David B. Larson Fellowship in Health and Spirituality, 2559
Kluge Center Distinguished Chairs, 2560
Kluge Center Kislak Fellows in American Studies, 2562
Kluge Center Kislak Short-Term Fellowships in American Studies, 2563
Knoll-MacDonald Foundation Grants, 2566
Komen Greater NYC Clinical Research Enrollment Grants, 2568
Komen Greater NYC Community Breast Health Grants, 2569
Kovler Family Foundation Grants, 2575
Kress Foundation Art History Pre-Doctoral Fellowships, 2576
L'Oreal USA Women in Science Fellowship, 2577
Lady Gregory Postgraduate Fellowship, 2579
Lalor Foundation Postdoctoral Fellowships, 2580
Law and Public Affairs Fellowships at Princeton University, 2588
Lawrence C. Gallen Teaching Fellowship, 2590
Leahi Fund, 2599
Leonard and Helen R. Stulman Charitable Foundation Grants, 2602
Leon Lowenstein Foundation Grants, 2603
Library Company of Philadelphia NEH Long-term Post-Doctoral Fellowships, 2607
Library of Congress Fellowships in Int'l Studies, 2612
LI General Grants, 2613
Lilly Endowment Teacher Creativity Fellowships, 2616
Lowe Foundation Grants, 2622
Lucile Horton Howe and Mitchell B. Howe Foundation Grants, 2624
Lumina Foundation for Education Grants, 2627
Luther I. Replogle Foundation Grants, 2631
Lynde and Harry Bradley Foundation Grants, 2635
M. Bastian Family Foundation Grants, 2636
M.J. Murdock Charitable Trust General Grants, 2641
M.J. Murdock Charitable Trust Partners in Science Grants, 2642
MAA-Tensor Foundation Women and Mathematics Grants, 2645
Mabel Y. Hughes Charitable Trust Grants, 2647
MacDonald-Peterson Foundation Grants, 2649
Macquarie Bank Foundation Grants, 2650
Marcia and Otto Koehler Foundation Grants, 2656
Marcus and Theresa Levie Educational Fund Scholarships, 2657
Margaret & James A. Elkins Jr. Found Grants, 2658
Margaret T. Morris Foundation Grants, 2659
Marion Gardner Jackson Charitable Trust Grants, 2664
Marshall Scholarships, 2666
Martin Luther King Jr. Graduate Memorial Scholarships, 2668
Mary Baker Eddy Library Summer Research Fellowships, 2669
Mary K. Chapman Foundation Grants, 2673
Mary S. and David C. Corbin Foundation Grants, 2675
Massachusetts Bar Foundation Legal Intern Fellowships, 2676
Massachusetts Foundation for the Humanities Project Grants, 2678
Massage Therapy Foundation Student Case Report Contest, 2681
Matilda R. Wilson Fund Grants, 2684
Mattel Children's Foundation Grants, 2685
Mattel International Grants, 2686
Max and Victoria Dreyfus Foundation Grants, 2687
Maximilian E. & Marion O. Hoffman Grants, 2688
McCarthy Family Foundation Grants, 2717
McCombs Foundation Grants, 2718
McCune Charitable Foundation Grants, 2719
McCune Foundation Civic Grants, 2720
McCune Foundation Education Grants, 2721
McGraw-Hill Companies Community Grants, 2724
McGregor Fund Grants, 2725
McLean Contributionship Grants, 2727
MDEQ Local Water Quality Monitoring Grants, 2732
Mead Johnson Nutritionals Med Educ Grants, 2734
Meadows Foundation Grants, 2736
MedImmune Charitable Grants, 2737
Melville Charitable Trust Grants, 2741
Mericos Foundation Grants, 2742
Meyer Memorial Trust Responsive Grants, 2748
MfA Newton Fellowships, 2749
MfA Newton Master Teacher Fellowships, 2750
MGFA Post-Doctoral Research Fellowships, 2751
MGM Resorts Foundation Community Grants, 2753
MGN Family Foundation Grants, 2754
MHS Long Term Research Fellowships, 2755
MHS New England Regional Fellowships, 2756
MHS Short Term Research Fellowships, 2757
MHS Teacher Fellowships, 2759
Microsoft Research A. Richard Newton Breakthrough Research Awards, 2764
Microsoft Research Mechanisms for Safe and Scalable Multicore Computing Grants, 2770
Microsoft Research New Faculty Fellowship, 2771
Miles of Hope Breast Cancer Foundation Grants, 2773
Miller Brewing Corporate Contributions Grants, 2774
Millipore Foundation Grants, 2775
Millstone Foundation Grants, 2776
MLA Cunningham Memorial International Fellowship, 2783
MLA Grad Scholarship for Minority Students, 2787
Motorola Foundation Grants, 2796
Motorola Foundation Innovation Generation Collaborative Grants, 2797
MSI Alden G. Clayton Dissertation Proposal Competition, 2800
Mt. Sinai Health Care Foundation Academic Medicine and Bioscience Grants, 2803
Musgrave Foundation Grants, 2807
NAACP Agnes Jones Jackson Grad Scholarships, 2809
NAACP Willems Scholarships, 2810
NACC David Stevenson Fellowships, 2812
NACC William Diaz Fellowships, 2813
NAEA Research Grants, 2814
NAEF Teacher Incentive Grants, 2815
NAGC Hollingworth Award, 2820
NAGC Masters and Specialists Award, 2821
Nanyang Technological University Lee Kuan Yew Postdoctoral Fellowship, 2826
NAPNAP Foundation Graduate Student Research Grant, 2828
NASA Goddard Visiting Fellowships, 2838
NASA Graduate Student Researchers Program Fellowships, 2839
NASA Marslink Initiative, 2841
NASA New Aeronautics Competition for University Students, 2842
NASA Undergrad Researchers Project Grants, 2843
National Academy of Education Spencer Postdoctoral Fellowships, 2845
National Black MBA Assoc Annual Scholarships, 2847
National Book Scholarship Fund, 2849
National Center for Responsible Gaming Conference Scholarships, 2851
National Endowment for the Humanities Fellowships at Winterthur, 2867
National Gallery of Art Kress Inter Fellowship, 2869
National Geographic Young Explorers Grants, 2874
National Housing Endowment Challenge/Build/Grow Grant, 2877
National Leadership Grant Competition, 2878
National LGBT Scholarship Fund, 2879
National Rising Scholars Award to Advance Research on Higher Education for the Public Good, 2883
National Wildlife Federation Internships, 2891
National Wildlife Refuge System Centennial Scholarship Program, 2892
NCFR Innovation Grants, 2908
NCSS/FASSE/CUFA Inquiry Grant, 2936
NCSS Exemplary Research in Social St Award, 2937
NCSS Jean Dresden Grambs Distinguished Career Research in Social Studies Award, 2938
NCTM E. Glenadine Gibb 7-12 Classroom Research Grants, 2941
NCTM Edward G. Begle PreK-6 Classroom Research Grants, 2942
NEH Awards for Faculty at Historically Black Colleges and Universities, 2949
NEH Awards for Faculty at Institutions with High Hispanic Enrollment, 2950
NEH Awards for Faculty at Tribal Colleges and Universities, 2951
NEH Digital Humanities Implementation Grants, 2953
NEH Digital Humanities Start-Up Grants, 2954
NEH Fellowships, 2956
NEH Fellowships at Independent Research Institutions, 2957
Nelda C. and H.J. Lutcher Stark Found Grants, 2967
Nellie Mae Education Foundation District-Level Change Grants, 2968
Nellie Mae Education Foundation Public Understanding Grants, 2969
Nellie Mae Education Foundation Research and Development Grants, 2970
Nellie Mae Education Foundation State Level Systems Change Grants, 2971
New Hampshire Charitable Foundation Grants, 2985
NFIE Found Learning & Leadership Grants, 2992
NFWF/Exxon Save the Tiger Fund Grants, 2995
NFWF Freshwater Fish Conservation Grants, 2999
NFWF Marine & Coastal Conservation Grants, 3001
NFWF Marine Debris Research and Technology Grants, 3002
NFWF More Fish Partnership Fund Grants, 3003
NFWF National Whale Conservation Grants, 3004
NFWF Nature Restoration Trust Grants, 3005
NFWF Puget Sound Marine Conservation Fund Grants, 3006
NFWF Pulling Together Initiative Grants, 3007
NFWF Wildlife and Habitat Conservation Initiative Grants, 3009
NFX Summer Research Fellowships, 3011
NHC Summer Institutes in Lit Studies Grants, 3012
NHLBI Investigator Initiated Multi-Site Clinical Trials, 3030
NHLBI Mentored Career Award for Faculty at Institutions that Promote Diversity (K01), 3033
NHLBI Research Supplements to Promote Diversity in Health-Related Research, 3054
NHLBI Summer Internship Program in Biomedical Research, 3064
NIEHS Hazardous Materials Worker Health and Safety Training Grants, 3190
NIEHS Hazmat Training at Doe Nuclear Weapons Complex Grants, 3191
NIH Academic Research Enhancement Awards, 3204
NIH School Interventions to Prevent Obesity, 3244
NIH Summer Internship in Biomedical Research, 3248
NIMH Jointly Sponsored Ruth L. Kirschstein National Research Service Award Institutional Predoctoral Training Program in the Neurosciences, 3253
NL Associated Colleges of the Midwest/Great Lakes College Association Faculty Fellowships, 3281
NMF General Electric Med Scholars Fellowships, 3322
NMF National Medical Association Awards for Medical Journalism, 3323
NMF National Medical Association Emerging Scholar Awards, 3324
NMF National Medical Association Patti LaBelle Award, 3325
NMSF Ernest F. Hollings Ocean Awareness Trust Fund Grants, 3328
NOAA Chesapeake Bay Office B-WET (Bay Watershed Education and Training) Grants, 3339
NOAA Chesapeake Bay Office Cooperative Research Grants, 3340
NOAA Chesapeake Bay Office Fisheries Science Grants, 3341
NOAA Ernest F. Hollings Scholarships, 3370
NOAA Faculty & Student Intern Res Grants, 3371
NOAA Living Marine Resources Cooperative Science Center Mini Grants, 3381
NOAA Miami Lab Summer Internships, 3382
NOAA Nat Climatic Data Center Internships, 3387
NOAA National Height Modernization Grants, 3391
NOAA Population Dynamics Recruiting Grants, 3400

NOAA Undersea Research Project Grants, 3412
Norcliffe Foundation Grants, 3415
Nordson Corporation Foundation Grants, 3416
North Carolina Biotechnology Center Collaborative Funding Grants, 3421
North Carolina Biotechnology Center Res Grants, 3424
Northrop Grumman Corporation Grants, 3428
Northrop Grumman Foundation Grants, 3429
Northwestern Mutual Foundation Grants, 3430
Notsew Orm Sands Foundation Grants, 3434
NPSC Grad Fellowships in the Physical Sciences, 3436
NRA Foundation Grants, 3437
NSF/NEH Documenting Endangered Languages Project Grants and Fellowships: data, infrastructure and computational methods, 3455
NSF Activities in Science, Engineering, and Mathematics for Persons with Disabilities, 3460
NSF ADVANCE: Increasing the Participation and Advancement of Women in Academic Science and Engineering Careers Grants, 3461
NSF Arctic Research and Education Grants, 3489
NSF Atmospheric Sciences Mid-Size Infrastructure Opportunity Grants, 3497
NSF Biological Research Collections, 3504
NSF Chemistry Research Experiences for Undergraduates, 3518
NSF CISE Advanced Learning Tech Grants, 3522
NSF CISE Broadening Participation in Computing Grants, 3524
NSF CISE Computing Res Infrastructure Grants, 3531
NSF CISE Educational Innovation Grants, 3533
NSF CISE Increasing the Participation and Advancement of Women in Academic Science and Engineering Careers Grants, 3534
NSF Communicating Res to Public Audiences, 3538
NSF Conferences, Workshops, and Special Meetings in the Mathematical Sciences, 3541
NSF Course, Curriculum, and Laboratory Improvement Research Grants, 3546
NSF Developing Global Scientists and Engineers, 3552
NSF Director's Award for Distinguished Teaching Scholars, 3553
NSF East Asia and Pacific Summer Institutes for U.S. Graduate Students, 3555
NSF Ecosystem Science Grants, 3556
NSF Faculty Early Career Development Grants, 3565
NSF Historically Black Colleges and Universities Grants, 3577
NSF Industry/University Cooperative Research Centers (I/UCRC) Grants, 3579
NSF Innovative Technology Experiences for Students and Teachers, 3582
NSF Instrument Devel for Biological Research, 3583
NSF Integrative Graduate Education and Research Traineeship, 3584
NSF Living Stock Collections Grants, 3588
NSF Louis Stokes Alliances for Minority Participation Grants, 3589
NSF Magnetospheric Physics Research Grants, 3590
NSF Materials World Network: Cooperative Activity in Materials Research between US Investigators and their Counterparts Abroad, 3595
NSF Particle and Nuclear Astrophysics Grants, 3608
NSF Partnerships in Astronomy & Astrophysics Research and Education, 3612
NSF Postdoctoral Res Fellowships in Biology, 3630
NSF Presidential Awards for Excellence in Mathematics and Science Teaching, 3631
NSF Presidential Awards for Excellence in Science, Math, and Engineering Mentoring, 3632
NSF Presidential Early Career Awards for Scientists and Engineers Grants, 3633
NSF Research Experiences for Undergrads, 3638
NSF Research on Gender in Science and Engineering Grants, 3639
NSF SBE Doctoral Dissertation Research Improvement Grants, 3641
NSF Science and Society (S&S), 3643
NSF Scientific Computing Research Environments for the Mathematical Sciences Grants, 3645
NSF Tribal Colleges and Universities Program, 3662
NSF Undergraduate Research and Mentoring in the Biological Sciences, 3663
NSF University Radio Observatories Grants, 3664
NWSA Graduate Scholarship in Jewish Women's Studies, 3676
NWSA Graduate Scholarships, 3677
NYAM Gladys Brooks Internship, 3686
NYCT New York City Environment Grants, 3697
NYSCA Folk Arts: Presentation Grants, 3704
NYSCA Folk Arts: Regional and County Folk Arts Grants, 3705
NYSCA Museum: Project Support Grants, 3707
OAH Germany Residency Grants, 3719
OAH Japanese Residencies Grants, 3722
OAH Louis Pelzer Memorial Award, 3725
OAH Roy Rosenzweig Disting Service Award, 3729
Oak Foundation Environmental Grants, 3731
Ohio Learning Network Grants, 3734
OISE Muriel & Danny Fung Grad Scholarships, 3735
OISE William Waters Scholarships in Urban Education, 3736
Oklahoma City Community Grants, 3737
Olin Corporation Charitable Trust Grants, 3739
OneSight Research Foundation Block Grants, 3742
Oppenstein Brothers Foundation Grants, 3792
Organization for Tropical Studies Fellowships, 3796
Organization for Tropical Studies Pilot Awards for Exploratory Research, 3797
Oscar Rennebohm Foundation Grants, 3832
OSF-Baltimore Community Fellowships, 3833
OSF Human Rights and Governance Internships, 3845
OTS Francis J. Bossuyt Fellowship, 3855
OTS Hovore-Horn Fellowships, 3856
OTS Stanley Smith Horticultural Fellowships, 3857
OTS Undergrad NAPIRE Summer Internships, 3858
Padilla Bay Foundation Grants, 3863
Parkersburg Area Community Foundation Arts in Education Special Grants, 3869
Park Foundation Grants, 3870
Partnership for Cures Charles E. Culpeper Scholarships in Medical Science, 3871
Patrick and Anna M. Cudahy Fund Grants, 3873
Paul & Daisy Soros Fellows for New Americans, 3875
Paul G. Allen Family Foundation Grants, 3876
Paul P. Fidler Research Grants, 3878
Peacock Foundation Grants, 3887
Percy B. Ferebee Endowment Grants, 3897
Perpetual Trust for Charitable Giving Grants, 3900
Pfizer Medical Education Track Two Grants, 3927
PHF Andrew W. Mellon Postdoctoral Fellowships in the Humanities, 3931
PHF Penn Faculty Research Fellowships, 3932
Phi Delta Kappa Outstanding Doctoral Dissertation Awards, 3936
Phi Upsilon Omicron Candle Fellowships, 3941
Phi Upsilon Omicron Diamond Anniversary Fellowships, 3942
Phi Upsilon Omicron Founders Fellowship, 3944
Phi Upsilon Omicron Geraldine Clewell Doctoral Fellowship, 3945
Phi Upsilon Omicron Geraldine Clewell Masters Fellowship, 3946
Phi Upsilon Omicron Jean Dearth Dickerscheid Fellowship, 3947
Phi Upsilon Omicron Margaret Drew Alpha Fellowship, 3948
Phi Upsilon Omicron Presidents Res Fellowship, 3949
PhRMA Foundation Informatics Pre Doctoral Fellowships, 3954
PhRMA Foundation Informatics Research Starter Grants, 3955
PhRMA Foundation Informatics Sabbatical Fellowships, 3956
PhRMA Foundation Paul Calabresi Medical Student Research Fellowship, 3957
PhRMA Foundation Pharmaceutics Postdoctoral Fellowships, 3958
PhRMA Foundation Pharmaceutics Pre Doctoral Fellowships, 3959
PhRMA Foundation Pharmaceutics Research Starter Grants, 3960
PhRMA Foundation Pharmaceutics Sabbatical Fellowships, 3961
PhRMA Foundation Pharmacology/Toxicology Sabbatical Fellowships, 3965
Pioneer Hi-Bred Plant Breeding Fellowships, 3968
Plan for Social Excellence Grants, 3976
Playboy Foundation Grants, 3980
Polytechnic University International Research Studentships, 3982
Postdoctoral Research Fellowships in Gerontological Public Health, 3983
Pott Foundation Grants, 3984
Potts Memorial Foundation Grants, 3985
Praxair Foundation Grants, 3986
Price Family Charitable Fund Grants, 3987
Princeton Univ Whiting Postdoc Fellowships, 3998
Project AWARE Foundation Grants, 3999
Province of Alberta Queen Elizabeth II Graduate Scholarships, 4001
Prudential Foundation Education Grants, 4002
PSG Mentored Clinical Research Awards, 4003
R.B. Terry Charitable Foundation Grants, 4005
R.C. Baker Foundation Grants, 4006
RACGP Grieve Memorial Award, 4014
RACGP Rex Walpole Travelling Fellowship, 4016
Radcliffe Institute Oral History Grants, 4022
Radcliffe Institute Research Support Grants, 4023
Rajiv Gandhi Foundation Grants, 4025
RAND Corporation Fellowships in Population Studies and the Study of Aging, 4029
Rathmann Family Foundation Grants, 4033
RCPSC Medical Education Research Grants, 4038
Regents Professional Opportunity Scholarships, 4049
Reinberger Foundation Grants, 4051
Research Corporation Cottrell Scholars Awards, 4053
Research Corporation Multi-Investigator Cottrell College Science Awards, 4054
Research Corporation Scialog Grants, 4055
Research Corporation Single-Investigator Cottrell College Science Awards, 4056
RGK Foundation Grants, 4060
Rhode Island Foundation Grants, 4061
Richard & Susan Smith Family Found Grants, 4064
Richard Davoud Donchian Foundation Grants, 4066
Rich Foundation Grants, 4068
Robert E. and Evelyn McKee Foundation Grants, 4074
Robert G. Cabell III and Maude Morgan Cabell Foundation Grants, 4075
Robert M. Hearin Foundation Grants, 4077
Robert Sterling Clark Foundation Government Accountability Grants, 4080
Robert W. Deutsch Foundation Grants, 4082
Rockford College Faculty Fellowships, 4092
Rocky Mountain Mineral Law Found Grants, 4094
Roger K. Summit Scholarship, 4097
Rohm and Haas Company Grants, 4098
Romanell-Phi Beta Kappa Professorship, 4099
Ronald McDonald House Charities Grants, 4101
Rosalynn Carter Institute John and Betty Pope Fellowships, 4103
Ross Foundation Grants, 4106
Royal Caribbean Cruises Ocean Fund, 4108
Royal Scottish Geographical Society Expedition Grants, 4110
Royal Scottish Geographical Society Small Research Grants, 4111
Roy J. Carver Charitable Trust Higher Education Grants, 4113
RRF General Grants, 4115
RSC Bancroft Award, 4118
Russell Family Foundation Environmental Sustainability Grants, 4128
Ruth and Vernon Taylor Foundation Grants, 4130
RWJF Clinical Scholars Grants, 4136
RWJF Health and Society Scholars, 4141
RWJF Partners Investing in Nursing Grants, 4149
Sachs Foundation Scholarships, 4157
Sage Foundation Grants, 4158

Samueli Foundation Health Grants, 4161
San Antonio Area Foundation Grants, 4163
San Diego Women's Foundation Grants, 4164
SARE On-Farm Res Grants - South Region, 4173
SARE Research and Education Grants - Northeast Region, 4178
Sarkeys Foundation Grants, 4181
Schomburg Scholars-in-Residence Fellowships, 4187
Schramm Foundation Grants, 4188
Scientific Communities Initiative Grants, 4189
Seaver Institute Grants, 4195
Self Foundation Grants, 4197
Sensient Technologies Foundation Grants, 4199
SfN Ricardo Miledi Neuroscience Training, 4212
Shastri Indo-Canadian Institute Graduate Degree Fellowships from an Indian Institution, 4224
Shastri Indo-Canadian Ins Librarian Fellowships, 4225
Shastri Indo-Canadian Institute Post-Doctoral Research Fellowships, 4229
Shastri Indo-Canadian Institute Research Collaboration Development Grants, 4230
Shell Oil Company Foundation Grants, 4233
Sidgmore Family Foundation Grants, 4252
Sid W. Richardson Foundation Grants, 4253
Sioux Falls Area Community Foundation Community Fund Grants (Unrestricted), 4262
Sioux Falls Area Community Found Spot Grants, 4263
Sir Dorabji Tata Trust Individual Education and Travel Grants, 4265
Skaggs Foundation Grants, 4266
Smith Richardson Foundation World Politics and Statecraft Fellowship, 4271
Smithsonian Postgraduate Fellowships in Conservation of Museum Collections, 4294
Society for Imaging Informatics in Medicine Emeritus Mentor Grants, 4305
Society for Imaging Informatics in Medicine Micro Grants, 4306
Society for Imaging Informatics in Medicine Research Grants, 4307
Society for Imaging Informatics in Medicine Small Training Grants, 4309
Society of Naval Architects and Marine Engineers Scholarships, 4328
Sony Corporation of America Corporate Philanthropy Grants, 4329
Special Olympics Health Professions Grants, 4332
Special Olympics Youth Fan Grants, 4333
Spencer Foundation Small Research Grants, 4337
SRC Educational Science Doctoral Grants, 4355
SRC Educational Science Grants, 4356
SRC Educational Science Junior Research Position Grants, 4359
SRC Educational Science Lang Checking Grants, 4360
SRC Educational Science Medium-Cost Scientific Equipment Grants, 4361
SRC Educational Science Network Grants, 4362
SRC Educational Science Project Res Grants, 4364
SRC Undergraduate Research Opportunities, 4395
SSHRC-FHSS Awards to Scholarly Publications, 4398
SSRC Abe Fellowship for Journalists, 4420
SSRC Next Generation Social Sciences in Africa: Doctoral Dissertation Proposal Fellowships, 4435
SSRC Our Shared Past Grants, 4437
Stackpole-Hall Foundation Grants, 4442
Stanford Center on Adolescence Youth Purpose Research Grants, 4443
Stanford Humanities Center Arts Writer/Practitioner Fellowship, 4444
Stanford Humanities Center External Faculty Fellowships, 4445
Stanley Smith Horticultural Trust Grants, 4446
STC Scholarships in Technical Communication, 4451
Strake Foundation Grants, 4452
Streilein Foundation for Ocular Immunology Visiting Professorships, 4453
Sun Academic Excellence Grants, 4466
Sun Grid Research Grants, 4467
Sun Microsystems Foundation Summer Academic Enrichment Grants, 4468
Sunny and Abe Rosenberg Foundation Grants, 4469
Surdna Foundation Arts Teachers Fellowships, 4470
T. James Kavanagh Foundation Grants, 4489
Talbert and Leota Abrams Foundation, 4492
Tandem Launch Technologies Grants, 4493
Tanner Humanities Center Research Fellowships, 4494
TD4HIM Foundation Grants, 4497
Teachers College Minority Postdoc Fellowships, 4498
Temple Univ George D. McDowell Fellowship, 4499
Textron Corporate Contributions Grants, 4501
The Ray Charles Foundation Grants, 4502
Third Millennium Foundation Fellowships, 4504
Third Millennium Foundation Grants, 4505
Thomas and Dorothy Leavey Foundation Grants, 4506
Thomas Austin Finch, Sr. Foundation Grants, 4507
Thomas C. Ackerman Foundation Grants, 4508
Thompson Charitable Foundation Grants, 4511
Thornton Foundation Grants, 4514
Todd Brock Family Foundation Grants, 4522
TRDRP Participatory Research Awards, 4529
Trinity College Ann Plato Fellowship, 4533
U.S. Department of Education American Overseas Research Centers, 4537
U.S. Department of Education Disability and Rehabilitation Research and Related Projects, 4538
U.S. Department of Education Graduate Assistance in Areas of National Need, 4540
U.S. Department of Education Jacob K. Javits Fellowships, 4541
U.S. Department of Education National Resource Centers, 4542
U.S. Department of Education Native Hawaiian Education Grants, 4543
U.S. Department of Education Rehabilitation Engineering Research Centers Grants, 4544
U.S. Department of Education Rehabilitation Research Training Centers, 4545
U.S. Department of Education Ronald E. McNair Postbaccalaureate Achievement Grants, 4546
U.S. Department of Education United States-Russia Program: Improving Research and Educational Activities in Higher Education, 4547
UCLA / RAND Corporation Post-Doctoral Fellowships, 4548
UCLA Research Grants in Ethnic Studies, 4549
UE Explore Undergraduate Research Grants, 4553
UIC Rockefeller Foundation Resident Fellowships in the Humanities, 4559
UNC Carolina Postdoctoral Program for Faculty Diversity, 4560
UNC School of Information and Library Science Doctoral Fellowships, 4565
United Methodist Church Center for the Evangelical UBH Audrie E. Reber Memorial Award, 4571
United Methodist Church John Harrison Ness Memorial Award, 4572
United States Institute of Peace - Transatlantic Post-Doc Fellowship for International Relations and Security, 4581
United Theological Seminary Merit-Based Scholarships, 4583
University of Colorado Doctoral Dissertation Fellowship in Media, Religion, and Culture, 4586
Ursula Thrush Peace Seed Grants, 4592
USA Funds Access to Education Scholarships, 4593
USAID Development Innovation Accelerator Broad Agency Announcement Grants, 4596
USAID Microbicides Research, Development, and Introduction Grants, 4599
USAID Strengthening RMNCH Through Indian Health Professional Associations and Scaling Up Uptake Grants, 4602
USC Annenberg/Getty Arts Journ Fellowships, 4603
USDA Agriculture and Food Research Initiative Competitive Grants - Food Security, 4614
USDA Alaska Native-Serving & Native Hawaiian-Serving Inst Education Competitive Grants, 4615
USDA Distance Education Grants for Higher Education Institutions in Insular Areas, 4622
USDA Higher Education Challenge Grants, 4627
USDA Hispanic-Serving Institutions Education Grants, 4628
USDA Integrated Research Education and Extension Grants - Organic Transitions, 4630
USDA International Science and Education Competitive Grants, 4631
USDA McIntire-Stennis Cooperative Forestry Research Grants, 4634
USDA Policy Research Center Grants, 4640
USDA Resident Instruction Grants for Higher Education in Insular Areas, 4645
USDA Small Business Innovation Research Grants - Phase II, 4649
USDA Tribal College Research Grants, 4652
USDC Afghanistan and Iraq Investment and Reconstruction Task Forces Internships, 4664
USDC Dissertation Fellowships, 4665
USDC Presidential Management Fellowships, 4670
USDC State Broadband Initiative Grants, 4671
USDD Graduate Traineeship Awards in Ocean Acoustics, 4730
USDD High School Apprenticeship Program and Undergraduate Research Apprenticeships, 4732
USDD Postdoctoral Fellowship Awards in Ocean Acoustics, 4774
Vancouver Foundation Grants and Community Initiatives Program, 4807
Victor E. Speas Foundation Grants, 4810
Viking Children's Fund Grants, 4811
Virginia Foundation for the Humanities Discretionary Grants, 4813
Virginia Found for Humanities Open Grants, 4814
Virginia Foundation for the Humanities Residential Fellowships, 4815
Virginia Historical Society Research Fellowships, 4816
Volkswagen of America Corporate Contributions Grants, 4817
W.C. Griffith Foundation Grants, 4819
W.K. Kellogg Foundation Healthy Kids Grants, 4822
W.M. Keck Undergrad Education Grants, 4825
Wabash Center Project Grants, 4826
Wabash Center Small Project Grants, 4827
Wabash Center Waorkshop and Colloquy Participant Grants, 4828
Wallace Foundation Grants, 4829
WAWH Founders Dissertation Fellowship, 4833
Wayne and Gladys Valley Foundation Grants, 4835
Weizmann Institute of Science Postdoctoral Fellowships, 4837
Welch Foundation Endowed Chairs, 4840
Welch Foundation Research Grants, 4842
Wellesley Centers for Women Postd Fellowships, 4845
Western Regional Graduate Program, 4855
Westinghouse Charitable Giving Grants, 4856
Weyerhaeuser Company Foundation Grants, 4857
WILD Foundation Grants, 4867
Willary Foundation Grants, 4870
William A. Klinger Research Award, 4871
William and Charlotte Parks Foundation for Animal Welfare Grants, 4872
William and Flora Hewlett Foundation Education Grants, 4873
William Blair and Company Foundation Grants, 4875
William G. & Helen C. Hoffman Found Grants, 4876
William G. McGowan Charitable Fund Grants, 4877
William N. and Myriam Pennington Foundation Grants, 4879
William T. Grant Foundation Distinguished Fellows Awards, 4880
William T. Grant Foundation Research Grants, 4881
William T. Grant Foundation Scholars Program, 4882
Winston Churchill Foundation Scholarships, 4886
Winterthur Museum Dissertation Fellowships, 4887
Winterthur Research Fellowships, 4888
Woodrow Wilson MMUF Dissertation Grants, 4892
Woodrow Wilson MMUF Travel & Res Grants, 4893
Woodward Fund Grants, 4895
World Bank Graduate Scholarship Program, 4896
WWF Russell E. Train Fellowships, 4898
Xerox Foundation Grants, 4899

Emergency Programs
Adaptec Foundation Grants, 340
Alcatel-Lucent Technologies Foundation Grants, 509
Alex Stern Family Foundation Grants, 542
American Foodservice Charitable Trust Grants, 710
Auburn Foundation Grants, 1263
Avon Products Foundation Grants, 1275
Banfield Charitable Trust Grants, 1295
Bank of America Charitable Foundation Matching Gifts, 1297
Becton Dickinson and Company Grants, 1328
Blowitz-Ridgeway Foundation Grants, 1349
Blum-Kovler Foundation Grants, 1351
Boeing Company Contributions Grants, 1357
Byron W. & Alice L. Lockwood Found Grants, 1401
Callaway Golf Company Foundation Grants, 1410
Campbell Soup Foundation Grants, 1415
Carpenter Foundation Grants, 1439
CFR International Affairs in Japan Fellowships, 1504
Coca-Cola Foundation Grants, 1593
Communities Foundation of Texas Grants, 1613
Community Fnd for Muskegon County Grants, 1617
Cord Foundation Grants, 1645
Coughlin-Saunders Foundation Grants, 1648
Cruise Industry Charitable Foundation Grants, 1660
Cudd Foundation Grants, 1672
DAAD Undergraduate Scholarships, 1710
Donald and Sylvia Robinson Family Foundation Grants, 1830
Doris Duke Foundation Operations Research on AIDS Care & Treatment in Africa Grants, 1839
Drs. Bruce and Lee Foundation Grants, 1859
FAR Fund Grants, 1980
Florida Sea Turtle Grants, 2027
Foundation for the Advancement of Mesoamerican Studies Grants, 2042
Frank Stanley Beveridge Foundation Grants, 2051
GNOF Gulf Coast Oil Spill Grants, 2153
Green Foundation Human Services Grants, 2172
H.J. Heinz Company Foundation Grants, 2195
Helen V. Brach Foundation Grants, 2231
Herbert A. & Adrian W. Woods Found Grants, 2239
Hoglund Foundation Grants, 2271
John Deere Foundation Grants, 2500
John G. Duncan Trust Grants, 2501
John P. McGovern Foundation Grants, 2504
Josiah W. and Bessie H. Kline Foundation Grants, 2517
Leon Lowenstein Foundation Grants, 2603
McCune Foundation Human Services Grants, 2722
Nordson Corporation Foundation Grants, 3416
Northwestern Mutual Foundation Grants, 3430
Oak Foundation Environmental Grants, 3731
Oppenstein Brothers Foundation Grants, 3792
Pinkerton Foundation Grants, 3967
Reinberger Foundation Grants, 4051
Rhode Island Foundation Grants, 4061
Richard and Rhoda Goldman Fund Grants, 4063
Ross Foundation Grants, 4106
Sioux Falls Area Community Foundation Community Fund Grants (Unrestricted), 4262
Sioux Falls Area Community Found Spot Grants, 4263
Sony Corporation of America Corporate Philanthropy Grants, 4329
Textron Corporate Contributions Grants, 4501
Todd Brock Family Foundation Grants, 4522
Union Pacific Foundation Health and Human Services Grants, 4569
Victor E. Speas Foundation Grants, 4810
William and Flora Hewlett Foundation Environmental Grants, 4874

Endowments
Achelis Foundation Grants, 274
Blum-Kovler Foundation Grants, 1351
BMW of North America Contributions, 1353
Bodman Foundation Grants, 1356
Caesars Foundation Grants, 1407
Hearst Foundations Health Grants, 2225
J.W. Kieckhefer Foundation Grants, 2452
Kovler Family Foundation Grants, 2575
Margaret T. Morris Foundation Grants, 2659
McLean Contributionship Grants, 2727
Norcliffe Foundation Grants, 3415
Perkin Fund Grants, 3898
Reinberger Foundation Grants, 4051
Sony Corporation of America Corporate Philanthropy Grants, 4329
Textron Corporate Contributions Grants, 4501
William Blair and Company Foundation Grants, 4875

Environmental Programs
3M Foundation Community Giving Grants, 5
A. V. Stout Fund Research Grants, 13
AAAS Science and Technology Policy Fellowships: Energy, Environment, and Agriculture, 41
AAAS Science and Technology Policy Fellowships: Health, Education and Human Services, 43
A Charitable Foundation Grants, 273
Achelis Foundation Grants, 274
ACS Doctoral Training Grants in Oncology Social Work, 308
Ahn Family Foundation Grants, 379
Akonadi Foundation Anti-Racism Grants, 477
Alaska Conservation Foundation Watchable Wildlife Conservation Trust Grants, 498
Albuquerque Community Foundation Grants, 508
Alces Foundation Grants, 510
Alcoa Foundation Grants, 511
Alexander von Humboldt International Climate Protection Fellowships, 529
Alfred P. Sloan Barcode of Life Grants, 543
Alfred P. Sloan Behavioral Economics and Household Finance Grants, 544
Alfred P. Sloan Civic Initiatives Grants, 545
Alfred P. Sloan Deep Carbon Observatory Grants, 547
Alfred P. Sloan Energy and Environment Grants, 552
Alfred P. Sloan Microbiology of the Built Environment Grants, 554
Alfred P. Sloan Selected National Issues Grants, 563
Allen Foundation Educational Nutrition Grants, 570
ALSAM Foundation Grants, 579
American Chemical Society GCI Pharmaceutical Roundtable Research Grants, 677
American Chemical Society Green Chemistry Chapter Award for Student Affiliates, 678
American Chemical Society Green Chem Grants, 679
American Chemical Society Joseph Breen Memorial Fellowship in Green Chemistry, 682
American Chemical Society Kenneth G. Hancock Memorial Student Award in Green Chemistry, 683
American Chemical Society Presidential Green Chemistry Challenge Awards, 696
American Psychiatric Association Minority Medical Student Summer Externship in Addiction Psychiatry, 785
Amgen Foundation Grants, 812
Anderson Foundation Grants, 842
Angels Baseball Foundation Grants, 848
Animal Behavior Soc Charles H. Turner Award, 849
Animal Behavior Society Developing Nations Research Grants, 850
Animal Behavior Society Diversity Grant, 851
Animal Behavior Society Latin American Travel Awards, 852
Animal Behavior Society Student Research Grants, 853
Ann Arbor Area Community Foundation Grants, 860
Applied Biosystems Grants, 916
APS/Phoenix Suns Energy and Environmental Minigrants, 919
Arcadia Foundation Grants, 983
Archer Daniels Midland Foundation Grants, 985
Arizona Public Service Corporate Giving Grants, 998
A Rocha USA Grants, 1006
ASCSA Alison M. Frantz Fellowship, 1051
ASCSA Weiner Laboratory Faunal Studies Research Fellowship, 1056
Ashland Corporate Contributions Grants, 1087
ASME Soichiro Honda Medal, 1164
As You Sow, 1249
AT&T Industrial Ecology Faculty Fellowships, 1251
Audubon Naturalist Society Research Grants, 1266
AWRA Richard A. Herbert Scholarship, 1281
Banfi Vintners Foundation Grants, 1296
Bank of America Charitable Foundation Matching Gifts, 1297
BARD Research Grants, 1299
Batchelor Foundation Grants, 1305
Bay and Paul Foundations, Inc Grants, 1307
Beim Foundation Grants, 1331
Berks County Community Foundation Grants, 1333
Bikes Belong Foundation Research Grants, 1343
Bill and Melinda Gates Foundation Agricultural Development Grants, 1345
Blum-Kovler Foundation Grants, 1351
Bodenwein Public Benevolent Found Grants, 1355
Bodman Foundation Grants, 1356
Boeing Company Contributions Grants, 1357
Boston Foundation Grants, 1361
BP Conservation Programme Future Conservationist Awards, 1365
Brainerd Foundation Grants, 1368
Brian G. Dyson Foundation Grants, 1370
Bullitt Foundation Grants, 1384
Burlington Northern Santa Fe Found Grants, 1389
Burton G. Bettingen Grants, 1390
Cabot Corporation Foundation Grants, 1406
California Wellness Foundation Work and Health Grants, 1409
Cape Branch Foundation Grants, 1425
Carl and Eloise Pohlad Family Found Grants, 1426
Carl C. Icahn Foundation Grants, 1428
Carl M. Freeman Foundation Grants, 1430
Carpenter Foundation Grants, 1439
Carrier Corporation Contributions Grants, 1440
Cen Cal Diving and Aquatic Studies Scholarship, 1471
Charles M. & Mary D. Grant Found Grants, 1527
Charles Stewart Mott Foundation Grants, 1529
Christensen Fund Regional Grants, 1554
Clarence E. Heller Foundation Grants, 1575
Cleveland-Cliffs Foundation Grants, 1580
Coca-Cola Foundation Grants, 1593
Coeta and Donald Barker Foundation Grants, 1594
Colorado Bioscience Discovery Eval Grants, 1598
Colorado Clean Energy Solar Innov Grants, 1599
Community Found for Greater Buffalo Grants, 1615
Community Foundation for Greater New Haven Quinnipiac River Fund Grants, 1616
Community Fnd for Muskegon County Grants, 1617
Community Found of E Central Illinois Grants, 1619
Community Foundation of the Virgin Islands Judith A. Towle Environmental Studies Grants, 1621
Compton Foundation Grants, 1623
Connecticut Community Foundation Grants, 1627
ConocoPhillips Grants, 1629
Conservation, Food, and Health Foundation Grants for Developing Countries, 1641
Coors Brewing Corporate Contributions Grants, 1644
Cowles Charitable Trust Grants, 1652
Cruise Industry Charitable Foundation Grants, 1660
Cudd Foundation Grants, 1672
Dade Community Foundation Grants, 1712
DaimlerChrysler Corporation Fund Grants, 1714
David Bohnett Foundation Grants, 1724
Dean Witter Foundation Grants, 1728
DOE Initial H-Prize Competition for Breakthrough Advances in Materials for Hydrogen Storage, 1801
DOE Program for Ecosystem Research, 1810
DOI African Elephant Conservation Grants, 1820
DOI Rhinoceros and Tiger Conservation Grants, 1824
DOI Upper Mississippi River System Long-Term Resource Monitoring Grants, 1825
Drs. Bruce and Lee Foundation Grants, 1859
Earthwatch Institute Student Fellowships, 1890
Edward Bangs Kelley and Elza Kelley Foundation Grants, 1898
Energy Foundation Buildings Grants, 1922
Energy Foundation Climate Grants, 1923
Energy Foundation Power Grants, 1924
Energy Foundation Transportation Grants, 1925
EPA Brownfields Area-Wide Planning Grants, 1931

EPA Capacity Building Grants for Compliance Assurance and Enforcement Activities in Indian Country and Other Tribal Areas, 1932
EPA Children's Health Protection Grants, 1933
EPA Compliance Assistance Support for Services to the Regulated Community and Other Assistance Providers, 1934
EPA National Estuary Grants, 1935
EPA Pollution Prevention Grants, 1936
EPA Regional Healthy Indoor Air Projects for Community Outreach and Education, 1937
EPA Research, Development, Monitoring, Public Education, Training, Demonstrations, and Studies Grants, 1938
EPA Southeastern Multi-Media and Geographic Priority Project Grants, 1939
EPA Surveys, Studies, Research, Investigations, Demonstrations, and Special Purpose Activities Relating to the Clean Air Act, 1940
EPA Targeted Watersheds Grants, 1941
EPA Temporally Integrated Monitoring of Ecosystems and Long-Term Monitoring Grants, 1942
EREF Solid Waste Research Grants, 1947
EREF Sustainability Research Grants, 1948
ESRI Conservation Grants, 1952
Evan Frankel Foundation Grants, 1961
Evjue Foundation Grants, 1963
Ewing Halsell Foundation Grants, 1964
Explorers Club Scott Pearlman Field Awards, 1967
FishAmerica Foundation Marine and Anadromous Fish Habitat Restoration Grants, 2018
FishAmerica Foundation Research Grants, 2019
Florida Sea Turtle Grants, 2027
Ford Foundation Peace and Social Justice Grants, 2031
Forrest C. Lattner Foundation Grants, 2035
Frank Stanley Beveridge Foundation Grants, 2051
Freshwater Future Project Grants, 2056
Fulbright German Studies Seminar Grants, 2072
G. Unger Vetlesen Foundation Grants, 2093
GCA Summer Environmental Scholarships, 2100
General Motors Foundation Grants, 2107
Genuardi Family Foundation Grants, 2109
George A Ohl Jr. Foundation Grants, 2112
George B. Storer Foundation Grants, 2113
George Frederick Jewett Foundation Grants, 2116
George Gund Foundation Grants, 2117
George S. & Dolores Dore Eccles Grants, 2120
Georgia-Pacific Foundation Environment Grants, 2123
Georgia Power Foundation Grants, 2125
Gerber Foundation Grants, 2126
GNOF Gulf Coast Oil Spill Grants, 2153
Greater St Louis Community Foundation Grants, 2164
Great Lakes Fishery Commission Pilot Research Project Grants, 2166
Great Lakes Fishery Commission Sea Lamprey Research Grants, 2167
Great Lakes Fishery Trust Access Grants, 2168
Great Lakes Fishery Ecological & Biological Fisheries Research to Inform Management Grants, 2169
Great Lakes Protection Fund Grants, 2170
Green Diamond Charitable Contributions, 2171
H.A. and Mary K. Chapman Charitable Grants, 2193
Harry A. and Margaret D. Towsley Found Grants, 2205
Harry Frank Guggenheim Foundation Research Grants, 2208
Heineman Foundation for Research, Education, Charitable and Scientific Purposes, 2227
Herbert A. & Adrian W. Woods Found Grants, 2239
Honeybee Health Improvement Project Grants, 2276
Honeywell Corporation Habitat and Conservation Grants, 2277
Honeywell Corporation Sustainable Opportunities Grants, 2278
Honor the Earth Grants, 2279
HRF Graduate Fellowships, 2285
HRF Hudson River Improvement Grants, 2286
HRF New York City Environmental Grants for Newton Creek, 2288
HRF Tibor T. Polgar Fellowships, 2289
HSBC Contributions Community Grants, 2291
IDEM Section 319(h) Nonpoint Source Grants, 2338
IIE David L. Boren Fellowships, 2350
Illinois Clean Energy Community Foundation Renewable Energy Grants, 2358
Illinois DCEO Coal Research Grants, 2359
Illinois DCEO Coal Revival Grants, 2360
Illinois DNR Youth Recreation Corps Grants, 2361
IMLS 21st Century Museum Pros Grants, 2362
ISA Hyland R. Johns Grants, 2428
ISA John Z. Duling Grants, 2429
Ittleson Foundation Environment Grants, 2437
J.W. Kieckhefer Foundation Grants, 2452
Jenifer Altman Foundation Grants, 2484
Jessie Smith Noyes Foundation Grants, 2485
John D. and Catherine T. MacArthur Foundation Global Challenges Grants, 2498
John Deere Foundation Grants, 2500
Kavli Foundation Research Grants, 2533
Kettering Fund Grants, 2555
Lynde and Harry Bradley Foundation Grants, 2635
MacArthur Foundation Conservation and Sustainable Development Grants, 2648
Macquarie Bank Foundation Grants, 2650
Margaret & James A. Elkins Jr. Found Grants, 2658
Margaret T. Morris Foundation Grants, 2659
Mary K. Chapman Foundation Grants, 2673
McCune Charitable Foundation Grants, 2719
McLean Contributionship Grants, 2727
MDEQ Beach Monitoring Grants - Inland Lakes, 2731
MDEQ Local Water Quality Monitoring Grants, 2732
MDEQ Scrap Tire Market Development Grants, 2733
Meyer Memorial Trust Responsive Grants, 2748
MOEF Indira Gandhi P. Puraskar Award, 2793
Motorola Foundation Grants, 2796
National Academies Postdoctoral and Senior Research Awards, 2844
National Geographic Society/Waitt Grants, 2870
National Geographic Society Conservation Trust Grants, 2871
National Geographic Society Expeditions Council Grants, 2872
National Geographic Society Field Research and Exploration Grants, 2873
National Geographic Young Explorers Grants, 2874
National Wetlands Awards, 2889
National Wildlife Federation Internships, 2891
National Wildlife Refuge System Centennial Scholarship Program, 2892
Nelda C. and H.J. Lutcher Stark Found Grants, 2967
Neotropical Grassland Conservancy Derald G. Langham Memorial Research Grant, 2973
Neotropical Grassland Conservancy Grants, 2974
Neotropical Grassland Conservancy Remembrance Fund Grants, 2975
Neotropical Grassland Conservancy Stu Grants, 2976
Neotropical Grassland Conservancy W. J. and Virginia W. Moorhouse Research Grant, 2977
Neotropical Grassland Conservancy Wijdenes Memorial Research Grant, 2978
NFWF/Exxon Save the Tiger Fund Grants, 2995
NFWF Acres for America Grants, 2996
NFWF Bird Conservation Initiative Grants, 2997
NFWF Freshwater Fish Conservation Grants, 2999
NFWF Marine & Coastal Conservation Grants, 3001
NFWF Marine Debris Research & Tech Grants, 3002
NFWF More Fish Partnership Fund Grants, 3003
NFWF National Whale Conservation Grants, 3004
NFWF Nature Restoration Trust Grants, 3005
NFWF Puget Sound Marine Conservation Fund Grants, 3006
NFWF Pulling Together Initiative Grants, 3007
NFWF Shell Marine Habitat Grants, 3008
NFWF Wildlife and Habitat Conservation Initiative Grants, 3009
NIEHS Hazardous Materials Worker Health and Safety Training Grants, 3190
NIEHS Hazmat Training at Doe Nuclear Weapons Complex Grants, 3191
NMSF Ernest F. Hollings Ocean Awareness Trust Fund Grants, 3328
NOAA Abrupt Climate Change Studies Grants, 3329
NOAA Arctic Research Grants, 3330
NOAA Assessing Atlantic Meridional Overturning Circulation Grants, 3331
NOAA Atlantic Salmon Conservation Grants, 3332
NOAA Atmospheric Composition and Climate Program Grants, 3333
NOAA Ballast Water Management Demonstration Program (RDTE Facility) Grants, 3334
NOAA Ballast Water Management Demonstration Program (Technologies and Practices) Grants, 3335
NOAA Ballast Water Technology Demonstration Program (Research, Development, Testing and Evaluation Facility) Grants, 3336
NOAA Ballast Water Technology Demonstration Grants, 3337
NOAA Chesapeake Bay Office B-WET (Bay Watershed Education and Training) Grants, 3339
NOAA Chesapeake Bay Office Cooperative Research Grants, 3340
NOAA Chesapeake Bay Office Fisheries Science Grants, 3341
NOAA Chesapeake Bay Office Non-Native Oyster Research Grants, 3342
NOAA Chesapeake Bay Office Submerged Aquatic Vegetation Culture and Restoration Grants, 3343
NOAA Climate Change & Detection Grants, 3344
NOAA Climate Dynamics and Experimental Prediction Grants, 3345
NOAA Climate Prediction Program for the Americas Grants, 3346
NOAA Climate Varia & Predictability Grants, 3348
NOAA Coastal and Estuarine Land Conservation Grants, 3349
NOAA Coastal Ecosystem Research Project, 3350
NOAA Coastal Management Fellowships, 3352
NOAA Coastal Ocean Grants, 3353
NOAA Collaborative Science, Technology, and Applied Research Grants, 3354
NOAA Communicating Hurricane Info Grants, 3355
NOAA Cooperative Institute: Eastern U.S. Continental Shelf Frontier Exploration, Research, and Technology Development Grants, 3356
NOAA Cooperative Unit for Fisheries Education and Research Internships, 3360
NOAA CRCP Coral Reef Ecosystem Monitoring Grants, 3362
NOAA CSCOR Regional Ecosystem Prediction Program: Cumulative Impacts of Stressors at the Land-Water Interface in the Mid-Atlantic, 3363
NOAA CSCOR Regional Ecosystem Prediction Program: Ecosystem Goal-Setting in Coastal Waters and Reefs of South Florida, 3364
NOAA CSCOR Regional Ecosystem Prediction Program: Invasive Species in the Great Lakes--A Regional Scale Approach, 3365
NOAA Ecological Forecasting Grants, 3368
NOAA Ernest F. Hollings Scholarships, 3370
NOAA Faculty & Student Intern Res Grants, 3371
NOAA General Coral Reef Conservation Grants, 3373
NOAA Global Carbon Cycle Grants, 3374
NOAA Graduate Sciences Grants, 3375
NOAA Great Lakes Habitat Restoration Partnership Grants, 3376
NOAA International Coral Reef Conservation Project Grants, 3377
NOAA Living Marine Resources Cooperative Science Center Internships/Fellowships, 3380
NOAA Living Marine Resources Cooperative Science Center Mini Grants, 3381
NOAA Miami Lab Summer Internships, 3382
NOAA Mid-Atlantic Research Set-Aside Grants, 3384
NOAA Monkfish Research Set-Aside Grants, 3386
NOAA National Coastal and Estuarine Research and Technology Grants, 3388
NOAA National Estuarine Research Reserve Land Acquisition and Construction Grants, 3389
NOAA National Height Modernization Grants, 3391
NOAA National Sea Grant Gulf of Mexico Oyster Industry Grants, 3392

NOAA National Sea Grant Industry Fellowships, 3393
NOAA National Sea Grant John A. Knauss Marine Policy Fellowships, 3394
NOAA Northern Gulf of Mexico Ecosystems and Hypoxia Assessment Grants, 3396
NOAA Ocean Exploration Omnibus: Marine Archaeology Grants, 3397
NOAA Ocean Exploration Omnibus: Ocean Exploration Grants, 3398
NOAA Population Dynamics Recruiting Grants, 3400
NOAA Regional Integrated Sciences and Assessments Grants, 3402
NOAA Scientific Data Stewardship Grants, 3403
NOAA Sector Applications Research Grants, 3405
NOAA South Florida Grants, 3406
NOAA Spherical Display Systems for Earth System Science - Installations and Content Development Grants, 3407
NOAA Transition of Research Applications to Climate Services Grants, 3408
NOAA Undersea Research Center Coral Reef Conservation Program Coral Research Grants, 3409
NOAA Undersea Research Center Ocean Acidification Research at Aquarius Reef Base Grants, 3410
NOAA Undersea Research Project Grants, 3412
Norcliffe Foundation Grants, 3415
Northrop Grumman Corporation Grants, 3428
Northwest Fund for the Environment Grants, 3431
NRA Foundation Grants, 3437
NSF Arctic Nat Science Grants in Earth Sciences, 3485
NSF Arctic Research Opportunities, 3490
NSF Atmospheric Sciences Mid-Size Infrastructure Opportunity Grants, 3497
NSF Climate and Large-Scale Dynamics Grants, 3535
NSF Cooperative Studies Of The Earth's Deep Interior Grants, 3544
NSF Coupling, Energetics, and Dynamics of Atmospheric Regions Research Grants, 3545
NSF Ecosystem Science Grants, 3556
NSF Natural, Social, and System Sciences and Research Support and Logistics Program, 3600
NSF Paleoclimate Grants, 3606
NSF Paleo Perspectives on Climate Change Grants, 3607
NSF Physical and Dynamic Meteorology Grants, 3614
NSF Project 2010 Grants, 3636
NSF Solar, Heliospheric, and INterplanetary Environment Grants, 3651
NSF Synthesis and Modeling Project of the US Joint Global Ocean Flux Study, 3656
NYCT New York City Environment Grants, 3697
Oak Foundation Environmental Grants, 3731
Olin Corporation Charitable Trust Grants, 3739
Open Spaces Sacred Places National Awards, 3791
Oppenstein Brothers Foundation Grants, 3792
Orchard Foundation Grants, 3793
Organization for Tropical Studies Fellowships, 3796
Organization for Tropical Studies Pilot Awards for Exploratory Research, 3797
Oscar Rennebohm Foundation Grants, 3832
OSRI Biological Monitoring: Intertidal Resources at Risk to Oil Spills, 3854
OTS Francis J. Bossuyt Fellowship, 3855
OTS Hovore-Horn Fellowships, 3856
OTS Stanley Smith Horticultural Fellowships, 3857
OTS Undergrad NAPIRE Summer Internships, 3858
Overbrook Foundation Grants, 3860
Padilla Bay Foundation Grants, 3863
Padilla Bay National Estuarine Research Reserve Graduate Research Fellowships, 3864
Padilla Bay National Estuarine Research Reserve Research Assistantships, 3865
Park Foundation Grants, 3870
Patrick and Anna M. Cudahy Fund Grants, 3873
Paul G. Allen Family Foundation Grants, 3876
Peacock Foundation Grants, 3887
Praxair Foundation Grants, 3986
Project AWARE Foundation Grants, 3999
Rajiv Gandhi Foundation Grants, 4025
RA Kleinhans Fellowship for Research in Tropical Nontimber Forest Products, 4026
Rathmann Family Foundation Grants, 4033
Rhode Island Foundation Grants, 4061
Richard and Rhoda Goldman Fund Grants, 4063
Robert & Patricia Switzer Found Fellowships, 4071
Robert Schalkenbach Foundation Grants, 4079
Rocky Mountain Mineral Law Found Grants, 4094
Rohm and Haas Company Grants, 4098
Ronald McDonald House Charities Grants, 4101
Ross Foundation Grants, 4106
Royal Caribbean Cruises Ocean Fund, 4108
RSC Miroslaw Romanowski Medal, 4122
Russell Family Foundation Environmental Sustainability Grants, 4128
Ruth and Vernon Taylor Foundation Grants, 4130
SARE Graduate Student Grants - North Central Region, 4168
SARE Graduate Student Grants - NE Region, 4169
SARE Graduate Student Grants - South Region, 4170
SARE Large Systems Research Grants - Southern Region, 4172
SARE On-Farm Res Grants - South Region, 4173
SARE Producer Research and Education Grants- Western Region, 4174
SARE Professional & Producer Grants - West, 4175
SARE Professional Devel Grants - NE Reg, 4176
SARE Research and Education Grants - North Central Region, 4177
SARE Research and Education Grants - Northeast Region, 4178
SARE Research and Education Grants - Southern Region, 4179
SARE Research and Education Grants - Western Region, 4180
SeaWorld and Busch Gardens Conservation Fund Grants, 4196
Shell Oil Company Foundation Grants, 4233
Sioux Falls Area Community Foundation Community Fund Grants (Unrestricted), 4262
Sioux Falls Area Community Found Spot Grants, 4263
Skaggs Foundation Grants, 4266
Smithsonian Museum Conservation Institute Analytical Studies Group Internships, 4281
Smithsonian Museum Conservation Institute Internships for Research Projects, 4282
Sony Corporation of America Corporate Philanthropy Grants, 4329
South Florida Student Shark Program, 4330
SRC Simon Karecki Fellowship, 4393
SSHRC-APC Automotive Partnership Canada, 4396
SSRC Collaborative Research Grants on Environment and Health in China, 4423
Stackpole-Hall Foundation Grants, 4442
Stanley Smith Horticultural Trust Grants, 4446
Surdna Found Sustainable Environments Grants, 4471
Thompson Charitable Foundation Grants, 4511
Turtle Conservation Fund Grants, 4535
U.S. Department of Education United States-Russia Program: Improving Research and Educational Activities in Higher Education, 4547
Union Pacific Found Community & Civic Grants, 4568
US CRDF CGP-Climate Change and Energy Competition For Joint U.S.-Ukrainian Research Teams, 4607
US CRDF CGP-RFBR Climate Change and Energy Competition For Joint U.S.-Russian Research Teams, 4608
US CRDF CGP ETF 2009 Energy Research Competition For Joint U.S.-Estonian Research Teams, 4609
USDA 1890 Institution Teaching, Research, and Extension Capacity Building Grants, 4613
USDA Agriculture and Food Research Initiative Competitive Grants - Food Security, 4614
USDA Alaska Native-Serving & Native Hawaiian-Serving Inst Education Competitive Grants, 4615
USDA Biotechnology Risk Assessment Grants, 4618
USDA Decadal and Regional Climate Prediction Using Earth System Models Grants, 4621
USDA Distance Education Grants for Higher Education Institutions in Insular Areas, 4622
USDA Hazardous Fuels Woody Biomass Utilization Grants, 4626
USDA Integrated Pest Management Competitive Grants - Western Region, 4629
USDA Integrated Research Education and Extension Grants - Organic Transitions, 4630
USDA International Science and Education Competitive Grants, 4631
USDA Interregional Research Project Minor Crop Pest Management Grants (IR-4), 4632
USDA McIntire-Stennis Cooperative Forestry Research Grants, 4634
USDA Organic Agriculture Research and Extension Initiative, 4637
USDA Plant and Animal Disease, Pest Control, and Animal Care Grants, 4638
USDA Plant Feedstock Genomics for Bioenergy: A Joint Research Grant, 4639
USDA Policy Research Center Grants, 4640
USDA Regional Integrated Pest Management Grants - North Central Region, 4642
USDA Regional Integrated Pest Management Grants - Northeastern Region, 4643
USDA Regional Integrated Pest Management Grants - Southern Region, 4644
USDA Resident Instruction Grants for Higher Education in Insular Areas, 4645
USDA Scientific Cooperation Research Program, 4647
USDA Small Business Innovation Research Grants - Phase I, 4648
USDA Small Business Innovation Research Grants - Phase II, 4649
USDA Tribal College Research Grants, 4652
USDA Wildlife Services Grants, 4653
USDC/NOAA Anadromous Fish Conservation Act Grants, 4654
USDC/NOAA Coastal and Marine Fish Habitat Protection Grants, 4655
USDC/NOAA Community-Based Habitat Restoration Grants, 4656
USDC/NOAA Community-Based Marine Debris Removal Grants, 4657
USDC/NOAA National Marine Aquaculture Initiative Grants, 4659
USDC/NOAA National Marine Fisheries Service Sea Grant Joint Graduate Fellowships, 4660
USDC/NOAA Restoration Center National and Regional Partnership Grants, 4662
Vancouver Foundation Grants and Community Initiatives Program, 4807
W.C. Griffith Foundation Grants, 4819
Westinghouse Charitable Giving Grants, 4856
Weyerhaeuser Company Foundation Grants, 4857
Whitley Awards, 4858
WILD Foundation Grants, 4867
William G. & Helen C. Hoffman Found Grants, 4876
WWF Russell E. Train Fellowships, 4898
Xerox Foundation Grants, 4899
Yves Rocher Foundation Terra Ficaria Prize for Scientific Research, 4904

Exchange Programs

IBRO Asia Reg APRC Exchange Fellowships, 2323
Kosciuszko Foundation Teaching Fellowships for Polish Citizens, 2573
OTS Undergrad NAPIRE Summer Internships, 3858
SSRC Collaborative Research Grants on Environment and Health in China, 4423
USDA Scientific Cooperation Exchange Program, 4646

Exhibitions, Collections, Performances, Video/Film Production

3M Foundation Community Giving Grants, 5
A.L. Mailman Family Foundation Grants, 11
Adaptec Foundation Grants, 340
Alcatel-Lucent Technologies Foundation Grants, 509
Alfred P. Sloan Public Understanding of Science & Tech-Public Television Grants, 557
Alfred P. Sloan Public Understanding of Science and Technology-Radio Grants, 558

Alfred P. Sloan Public Understanding of Science and Technology-Theater Grants, 559
Bank of America Charitable Foundation Matching Gifts, 1297
Berea College Appalachian Music Fellowships, 1332
Beverly Willis Architecture Foundation Grants and Fellowships, 1340
Boeing Company Contributions Grants, 1357
Bush Foundation Artist Fellowships, 1391
Caesars Foundation Grants, 1407
Calvin Inst of Christian Worship Renewal Grants, 1411
Carl and Lily Pforzheimer Foundation Grants, 1427
Carl M. Freeman Foundation Grants, 1430
CCA Alberta Creative Development Initiative: Grants to Individuals, 1446
Cintas Foundation Fellowships, 1570
Coca-Cola Foundation Grants, 1593
Community Fnd for Muskegon County Grants, 1617
Cowles Charitable Trust Grants, 1652
D.F. Halton Foundation Grants, 1692
DENSO North America Foundation Grants, 1742
Diaz-Ayala Cuban and Latin American Popular Music Collection Travel Grants, 1763
Donald and Sylvia Robinson Family Foundation Grants, 1830
Drs. Bruce and Lee Foundation Grants, 1859
Elizabeth McGraw Foundation Grants, 1908
Endangered Language Fund Grants, 1921
Equipment Leasing and Finance Foundation Research Grants, 1943
Flinn Foundation Grants, 2024
Fund for Folk Culture Artists Support Grants, 2088
George A Ohl Jr. Foundation Grants, 2112
George B. Storer Foundation Grants, 2113
George S. & Dolores Dore Eccles Grants, 2120
Getty Research Grants for Institutions - Archival Grants, 2131
Getty Research Grants for Institutions - Art History Research Grants, 2132
Getty Research Grants for Institutions - Cataloguing of Museum Collections Grants, 2133
Getty Research Institute Nonresidential Grants - Collaborative Research Grant, 2135
Getty Research Institute Nonresidential Grants - Curatorial Research Fellowships, 2136
Gil and Dody Weaver Foundation Grants, 2139
Gilroy and Lillian P. Roberts Foundation Grants, 2142
Greater Worcester Community Foundation Jeppson Memorial Fund for Brookfield Grants, 2165
Harold R. Bechtel Charitable Remainder Uni-Trust Grants, 2201
Haverford College Mellon Postdoc Fellowships, 2220
IFP Minnesota McKnight Film Fellowships, 2342
IFP Minnesota McKnight Screen Fellowships, 2343
IFP New York Anthony Radziwill Documentary Fund Grants, 2344
ITVS Commissioned Funding, 2439
James and Sylvia Thayer Short -Term Research Fellowships, 2457
Kate Neal Kinley Memorial Fellowship, 2532
Kiplinger Program in Public Affairs Journalism Fellowships, 2557
MacDonald-Peterson Foundation Grants, 2649
Manitoba Arts Council Aboriginal Arts Creative Development Pilot Projects Grants, 2652
Manitoba Arts Council Dance Creation/Production Grants, 2654
Mary Baker Eddy Library Summer Research Fellowships, 2669
Massachusetts Foundation for the Humanities Project Grants, 2678
McCombs Foundation Grants, 2718
Meyer and Stephanie Eglin Foundation Grants, 2747
NAMM Foundation Music Research Grants, 2825
National Endowment for the Humanities Fellowships at Winterthur, 2867
National Gallery of Art Kress Inter Fellowship, 2869
NEA/TCG Career Development Grants for Theatre Directors, 2943
NEH Digital Humanities Implementation Grants, 2953
NEH Digital Humanities Start-Up Grants, 2954
NL Grad Scholar-in-Residence Opportunities, 3285
NL Lester J. Cappon Fellowship in Documentary Editing, 3288
NL Scholar-in-Residence Opportunity, 3315
Norcliffe Foundation Grants, 3415
NSF Biological Research Collections, 3504
NYSCA Folk Arts: Exhibitions Grants, 3703
NYSCA Folk Arts: Presentation Grants, 3704
NYSCA Folk Arts: Regional and County Folk Arts Grants, 3705
NYSCA Individual Artists: Film, Media and New Technology Production Grants, 3706
NYSCA Museum: Project Support Grants, 3707
NYSCA Music: Services to the Field Grants, 3708
Ontario Arts Council Integrated Arts Grants, 3790
Oppenstein Brothers Foundation Grants, 3792
Pacific Islanders in Communications Grants, 3861
Paul G. Allen Family Foundation Grants, 3876
Pew Fellowships in the Arts, 3905
Price Family Charitable Fund Grants, 3987
Reinberger Foundation Grants, 4051
Richard and Rhoda Goldman Fund Grants, 4063
Rockefeller Foundation MAP Fund Grants, 4090
Rohm and Haas Company Grants, 4098
Royal Caribbean Cruises Ocean Fund, 4108
SHOT Dibner Award for Excellence in Museum Exhibits, 4239
Sioux Falls Area Community Foundation Community Fund Grants (Unrestricted), 4262
Sioux Falls Area Community Found Spot Grants, 4263
Smithsonian Postgraduate Fellowships in Conservation of Museum Collections, 4294
Sony Corporation of America Corporate Philanthropy Grants, 4329
Thornton Foundation Grants, 4514
Van Alen Institute New York Prize Fellowship, 4806
Virginia Foundation for the Humanities Discretionary Grants, 4813
Virginia Found for Humanities Open Grants, 4814
W.C. Griffith Foundation Grants, 4819
Winterthur Research Fellowships, 4888
Wolfsonian-FIU Fellowships, 4890

Faculty/Professional Development

AAAAI Distinguished Clinician Award, 17
AAAAI Fellows-in-Training Abstract Award, 21
AACN Physio-Control Small Projects Grants, 111
AACR Career Development Awards for Pediatric Cancer Research, 132
AACR Gertrude Elion Cancer Research Award, 137
AACR Judah Folkman Career Development Award for Anti-Angiogenesis Research, 140
AAFP Foundation Pfizer Teacher Devel Awards, 157
AALAS Foundation Grants, 182
AANS Neurosurgery Res & Educ Foundation Young Clinician Investigator Award, 185
AAS Kate B. and Hall J. Peterson Fellowships, 204
AAS Legacy Fellowships, 205
ACS Cancer Control Career Development Awards for Primary Care Physicians, 305
AFAR Paul Beeson Career Development Awards in Aging Research for the Island of Ireland, 353
AFRL/NRC Resident Research Associateship, 358
AHIMA Faculty Development Stipends, 375
AIAR Annual Professorships, 393
AIHP Sonnedecker Visiting Scholar Grants, 432
AIHS Collaborative Research and Innovation Grants - Collaborative Project, 438
AIHS Interdisciplinary Team Grants, 450
AIHS Knowledge Exchange - Visiting Scientists, 452
AIHS Research Prize, 457
AIR/NCES Fellowships for Graduate Study, 467
AIR National Summer Data Policy Institute Fellowships, 469
Alexander von Humboldt Feodor Lynen Research Fellowships for Experienced Researchers, 520
Alexander von Humboldt Research Fellowships for Postdoctoral Researchers, 539
Alfred P. Sloan Research Fellowships, 560
All Saints Educational Trust Grants, 572
American Academy of Advertising Distinguished Service Award, 607
American Academy of Advertising Kim Rotzoll Award for Advertising Ethics and Social Responsibility, 610
American Academy of Advertising Research Fellowships, 611
American Chemical Society Chemical Technology Partnership Mini Grants, 674
American Chemical Society Green Chem Grants, 679
American Chemical Society Petroleum Research Fund Doctoral New Investigator Grants, 690
American Chemical Society Petroleum Research Fund Scientific Education Grants, 692
American Chemical Society Petroleum Research Fund Undergraduate Faculty Sabbatical Grants, 693
American College of Surgeons Faculty Career Development Award for Neurological Surgeons, 703
American Psychiatric Association/AstraZeneca Young Minds in Psychiatry International Awards, 773
AMI Semiconductors Corporate Grants, 813
ANL Faculty Research Leave Program, 854
ANL Faculty Research Participation Program, 855
Annenberg Foundation Grants, 862
Arizona Commission on the Arts Education Projects Grants, 996
ARO Young Investigator Program/Presidential Early Career Awards, 1018
ASHA Research Mentoring-Pair Travel Award, 1073
ASHA Students Preparing for Academic & Research Careers Award, 1075
Asia Foundation Exchange Programs, 1090
Assisi Foundation of Memphis General Grants, 1221
ATA Educational Trust Grants - Curriculum Resource or Research Project, 1253
Ball Brothers Foundation General Grants, 1291
Ball Brothers Foundation Rapid Grants, 1292
Beckman Young Investigators Grants, 1327
Boeing Company Contributions Grants, 1357
Brookdale Leadership in Aging Fellowships, 1381
BWF Career Awards at the Scientific Interface, 1393
BWF Institutional Program Unifying Population and Laboratory Based Sciences Grants, 1397
Byron W. & Alice L. Lockwood Found Grants, 1401
Cabot Corporation Foundation Grants, 1406
Canada-U.S. Fulbright Senior Specialists Grants, 1418
Canada-U.S. Fulbright Trad Scholar Awards, 1419
Canon Foundation Grants, 1423
Carl and Lily Pforzheimer Foundation Grants, 1427
Carl M. Freeman Foundation Grants, 1430
Carpenter Foundation Grants, 1439
Carrier Corporation Contributions Grants, 1440
CBO Economic Policy Fellowships, 1445
CDECD Elizabeth L. Mahaffey Arts Administration Fellowship, 1468
CFF Harry Shwachman Clinical Invest Award, 1493
CFF Leroy Matthews Physician-Scientist Awards, 1494
CFF Research Grants, 1497
CFR Edward R. Murrow Press Fellowship, 1500
CFR International Affairs Fellowships, 1501
CFR International Affairs Fellowships in Nuclear Security, 1502
CFR Int'l Affairs Fellowships in South Korea, 1503
CFR International Affairs in Japan Fellowships, 1504
CFR Military Fellowships, 1505
CFR Stanton Nuclear Security Fellowships, 1506
Chiron Foundation Community Grants, 1553
Christensen Fund Regional Grants, 1554
CIRCLE Civic Education at the High School Level Research Grants, 1571
Clarence E. Heller Foundation Grants, 1575
Cleveland-Cliffs Foundation Grants, 1580
Clowes ACS/AAST/NIGMS Jointly Sponsored Mentored Clinical Scientist Devel Award, 1581
Community Fnd for Muskegon County Grants, 1617
Conquer Cancer Foundation of ASCO Career Development Award, 1630
Conquer Cancer Foundation of ASCO Comparative Effectiveness Research Professorship in Breast Cancer, 1631

Conquer Cancer Foundation of ASCO Drug Development Research Professorship, 1632
Conquer Cancer Foundation of ASCO Int'l Devel & Educ Award in Palliative Care, 1634
Conquer Cancer Foundation of ASCO International Development and Education Awards, 1635
Conquer Cancer Foundation of ASCO Medical Student Rotation Grants, 1637
Conquer Cancer Foundation of ASCO Resident Travel Award for Underrepresented Populations, 1638
Conquer Cancer Foundation of ASCO Translational Research Professorships, 1639
Cooperative Institute for Research in Environmental Sciences Visiting Fellows Program, 1643
Cowles Charitable Trust Grants, 1652
Cushwa Center Research Travel Grants, 1673
Cystic Fibrosis Canada Senior Scientist Research Training Awards, 1680
DAAD Faculty Research Visit Grants, 1697
Dart Center Ochberg Fellowship, 1720
Dave Thomas Foundation for Adoption Grants, 1723
Deloitte Foundation Grants, 1735
Denton A. Cooley Foundation Grants, 1743
DOE Academies Crea Teacher Scientists Grants, 1766
DOS Study of the United States Institutes, 1846
Dreyfus Special Grants in the Chemical Sciences, 1858
Drs. Bruce and Lee Foundation Grants, 1859
Eaton Charitable Fund Grants, 1893
Edward W. and Stella C. Van Houten Grants, 1899
Edwin S. Webster Foundation Grants, 1900
EFWA Accounting Research Grants, 1901
El Pomar Leadership Development and Community Engagement Postgraduate Fellowship, 1912
Eugene M. Lang Foundation Grants, 1957
Eugene McDermott Foundation Grants, 1958
Evjue Foundation Grants, 1963
FAMRI Young Clinical Scientist Awards, 1976
Fannie and John Hertz Foundation Graduate Fellowships, 1977
FAR Fund Grants, 1980
FDHN Fellowship to Faculty Transition Awards, 1992
FDHN Research Scholar Awards, 2000
Frances L. and Edwin L. Cummings Memorial Fund Grants, 2046
Fritz B. Burns Foundation Grants, 2057
Fulbright-Hays Seminars Abroad Bilateral Projects Grants, 2062
Fulbright Critical Lang Enhancement Awards, 2068
Fulbright Distinguished Chairs Grants, 2069
Fulbright English Teaching Assistantships, 2070
Fulbright International Education Administrators Seminar Grants, 2073
Fulbright New Century Scholars Grants, 2075
Fulbright Specialists Grants, 2079
Fulbright Traditional Scholar Program in Europe and Eurasia, 2080
Fulbright Traditional Scholar Program in Sub-Saharan Africa, 2081
Fulbright Traditional Scholar Program in the East Asia/Pacific Region, 2082
Fulbright Traditional Scholar Program in the Near East and North Africa Region, 2083
Fulbright Traditional Scholar Program in the South and Central Asia Region, 2084
Fulbright Traditional Scholar Program in the Western Hemisphere, 2085
Fund for Folk Culture Artists Support Grants, 2088
GCA Tropical Botany Research Awards, 2101
George A Ohl Jr. Foundation Grants, 2112
George Foundation Grants, 2115
George Kress Foundation Grants, 2119
George S. & Dolores Dore Eccles Grants, 2120
Gilroy and Lillian P. Roberts Foundation Grants, 2142
GUITS Matching Seed-Money Grants for New Positions in Turkish Studies, 2188
Harry A. and Margaret D. Towsley Found Grants, 2205
Harry Frank Guggenheim Research Grants, 2208
Harvard University Administrative Fellowships, 2212
Harvard University Graduate School of Education Fellowships and Awards, 2213
Hearst Foundations Health Grants, 2225
Henry Luce Foundation Special Projects in Higher Education Grants, 2236
HHMI/EMBO Start-up Grants for Cen Europe, 2250
Hill Crest Foundation Grants, 2267
Hospital Libraries Section / MLA Professional Development Grants, 2283
ICC Scholarship of Engagement Faculty Grants, 2334
IES Faculty Fellowships, 2341
Institute of Global Conflict and Cooperation Research, Conference, and Teaching Grants, 2385
IRA Elva Knight Research Grant, 2406
IREX ECA Alumni Small Grants, 2416
IREX Edmund S. Muskie Graduate Fellowship, 2417
IREX Individual Advanced Research Opportunities Grants, 2419
IREX International Leadership in Education Program Grants, 2420
Jacob and Hilda Blaustein Foundation Grants, 2454
James McKeen Cattell Fund Fellowships, 2462
James R. Dougherty Jr. Foundation Grants, 2463
Japan Foundation New York Institutional Support for Japanese Studies Grants, 2477
Japan Foundation Six-Month Grants for Specialists in Cultural and Academic Fields, 2479
John P. McGovern Foundation Grants, 2504
John S. Dunn Research Foundation Grants, 2506
Johnson & Johnson Corporate Contributions, 2508
Joseph Alexander Foundation Grants, 2513
Josiah Macy Jr. Foundation Grants, 2516
Joyce Foundation Education Grants, 2519
JSA JLab Sabbatical Research Leave Support, 2522
Kellogg Institute for International Studies Andrónico Luksic Grants, 2535
Kenneth T. & Eileen L. Norris Found Grants, 2554
Landon Foundation-AACR Innovator Award for International Collaboration, 2587
Lilly Endowment Teacher Creativity Fellowships, 2616
Lowe Foundation Grants, 2622
Lynde and Harry Bradley Foundation Grants, 2635
M.J. Murdock Charitable Trust Research Start-up Grants for New Science Faculty, 2643
Margaret & James A. Elkins Jr. Found Grants, 2658
Massachusetts Foundation for the Humanities Project Grants, 2678
Mead Johnson Nutritionals Med Educ Grants, 2734
Meadows Foundation Grants, 2736
Mellon Post-Doctoral Fellowship in Philosophy, 2739
Mellon Post-Doctoral Fellowship in Sustainable Agriculture, 2740
MfA Newton Fellowships, 2749
MfA Newton Master Teacher Fellowships, 2750
Minority Faculty Fellowship Program, 2781
MLA Cunningham Memorial International Fellowship, 2783
Motorola Foundation Innovation Generation Collaborative Grants, 2797
Murphy Institute Center for Ethics and Public Affairs Faculty Fellowships, 2805
NACC David Stevenson Fellowships, 2812
NACC William Diaz Fellowships, 2813
NAEF Teacher Incentive Grants, 2815
NASA Goddard Visiting Fellowships, 2838
National Academy of Education Spencer Postdoctoral Fellowships, 2845
National Endowment for the Humanities Fellowships at Winterthur, 2867
National Security Education Grants to Institutions of Higher Education, 2885
NATO Advanced Study Institutes Program, 2895
NCI Cancer Education and Career Development Program, 2917
NCIC Canadian Breast Cancer Research Initiative Grants, 2918
NCI Mentored Career Development Award to Promote Diversity, 2930
NEA/TCG Career Development Grants for Theatre Directors, 2943
NEH Fellowships, 2956
NFID New Investigator Matching Grants, 2989
NFIE Found Learning & Leadership Grants, 2992
NHGRI Mentored Research Scientist Development Award, 3013
NHLBI Mentored Career Award for Faculty at Institutions that Promote Diversity (K01), 3033
NHLBI Mentored Career Award For Faculty At Minority Institutions, 3034
NHLBI Mentored Career Development Award to Promote Faculty Diversity/Re-Entry in Biomedical Research, 3035
NHLBI Mentored Patient-Oriented Research Career Development Awards, 3037
NHLBI Mentored Quantitative Research Career Development Awards, 3038
NHLBI Pathway to Independence Awards, 3045
NHLBI Ruth L. Kirschstein National Research Service Awards for Individual Predoctoral Fellowships to Promote Diversity in Health-Related Research, 3058
NHLBI Ruth L. Kirschstein National Research Service Awards for Individual Senior Fellows, 3060
NHLBI Short-Term Research Education Program to Increase Diversity in Health-Related Research, 3062
NHLBI Training Grants for Institutions that Promote Diversity, 3067
NHPRC Electronic Records Projects Grants, 3073
NIAAA Independent Scientist Award, 3078
NIA Independent Scientist Award, 3103
NIA Mentored Research Scientist Dev Award, 3109
NIA Pilot Research Grants, 3115
NIDA Mentored Clinical Scientist Research Career Development Awards, 3129
NIDCD Mentored Clinical Scientist Research Career Development Award, 3131
NIEHS Independent Scientist Award, 3192
NIEHS Mentored Clinical Scientist Research Career Development Awards, 3193
NIEHS Mentored Research Scientist Development Award, 3194
NIGMS Mentored Clinical Scientist Research Career Development Awards, 3197
NIH Mentored Clinical Scientist Dev Award, 3225
NINR Mentored Research Scientist Dev Award, 3265
NKF Research Fellowships, 3277
NKF Young Investigator Grants, 3278
NL Associated Colleges of the Midwest/Great Lakes College Association Faculty Fellowships, 3281
NL Monticello College Foundation Fellowship for Women, 3304
NL Newberry Consortium on American Indian Studies Faculty Fellowships, 3311
North Carolina Biotechnology Center Oliver Smithies Faculty Recruitment Grant, 3423
North Carolina Biotechnology Center Res Grants, 3424
NRA Foundation Grants, 3437
NSF Course, Curriculum, and Laboratory Improvement Research Grants, 3546
NSF Director's Award for Distinguished Teaching Scholars, 3553
NSF Faculty Early Career Development Grants, 3565
NSF Historically Black Colleges and Universities Grants, 3577
NSF Industry/University Cooperative Research Centers (I/UCRC) Grants, 3579
NSF Partnerships in Astronomy & Astrophysics Research and Education, 3612
NSF Postdoctoral Res Fellowships in Biology, 3630
NSF Presidential Early Career Awards for Scientists and Engineers Grants, 3633
NSF Research Experiences for Undergrads, 3638
NSF Science and Society (S&S), 3643
NSF Undergraduate Research and Mentoring in the Biological Sciences, 3663
NYSCA Dance: Services to the Field Grants, 3702
ONR Faculty Sabbatical Leave Grants, 3762
ORISE Environmental Management Participation for the US Army Environmental Center, 3801
ORISE Faculty Research Participation at the Centers for Disease Control and Prevention, 3803
ORISE Faculty Research Participation at the National Center for Toxicological Research, 3804

ORISE Higher Education Research Experiences at Oak Ridge National Laboratory for Faculty, 3808
ORISE National Oceanic and Atmospheric Administration Faculty Research Participation, 3811
ORISE Nuclear Regulatory Commission Historically Black Colleges and Universities Faculty Research Participation, 3813
ORISE Oak Ridge National Laboratory Postdoctoral Research Associates Program, 3815
ORISE Savannah River Site Faculty Research Participation Program, 3823
Pancreatic Cancer Action Network-AACR Pathway to Leadership Grants, 3867
Parkersburg Area Community Foundation Arts in Education Special Grants, 3869
Park Foundation Grants, 3870
PDF-AANF Clinician-Scientist Devel Awards, 3881
Pfizer ASPIRE North America Hemophilia Research Awards, 3915
Pfizer Australia Cancer Research Grants, 3919
Pfizer Australia Neuroscience Research Grants, 3920
Pfizer Medical Education Track Two Grants, 3927
Phi Upsilon Omicron Florence Fallgatter Distinguished Service Award, 3943
PhRMA Foundation Health Outcomes Post Doctoral Fellowships, 3950
PhRMA Foundation Informatics Post Doctoral Fellowships, 3953
PhRMA Foundation Informatics Research Starter Grants, 3955
PhRMA Found Informatics Sabbatical Fellowships, 3956
PhRMA Foundation Pharmaceutics Postdoctoral Fellowships, 3958
PhRMA Found Pharmaceutics Starter Grants, 3960
PhRMA Foundation Pharmaceutics Sabbatical Fellowships, 3961
PhRMA Foundation Pharmacology/Toxicology Post Doctoral Fellowships, 3962
PhRMA Foundation Pharmacology/Toxicology Pre Doctoral Fellowships, 3963
PhRMA Foundation Pharmacology/Toxicology Research Starter Grants, 3964
PhRMA Foundation Pharmacology/Toxicology Sabbatical Fellowships, 3965
Pollock-Krasner Foundation Grants, 3981
Princeton University Center for Human Values Laurance S. Rockefeller Visiting Fellowships, 3992
PSG Mentored Clinical Research Awards, 4003
Research Corporation Cottrell Scholars Awards, 4053
Research Corporation Multi-Investigator Cottrell College Science Awards, 4054
Research Corporation Scialog Grants, 4055
Research Corporation Single-Investigator Cottrell College Science Awards, 4056
Rhode Island Foundation Grants, 4061
Robert M. Hearin Foundation Grants, 4077
Rockefeller Archive Center Visiting Archivist Fellowship, 4087
Rohm and Haas Company Grants, 4098
Ronald McDonald House Charities Grants, 4101
Roy J. Carver Charitable Trust Medical and Science Research Grants, 4114
RRF General Grants, 4115
RWJF Clinical Scholars Grants, 4136
RWJF Harold Amos Medical Faculty Development Scholars, 4140
RWJF New Connections Jun Investigator Grants, 4146
RWJF New Connections Mid-Career Consultants Grants, 4147
RWJF Nurse Faculty Scholars, 4148
San Antonio Area Foundation Grants, 4163
SARE Professional Devel Grants - NE Reg, 4176
Sarkeys Foundation Grants, 4181
Searle Scholars Grants, 4193
SfN Janett Rosenberg Trubatch Career Dev Award, 4207
Skaggs Foundation Grants, 4266
SNM Molecular Imaging Research Grant For Junior Medical Faculty, 4299
SNM Postdoc Molecular Imaging Scholar Grants, 4302
SPIRE Postdoctoral Fellowships, 4345
SRC Aristotle Award, 4349
SSHRC Institutional Grants, 4411
SSRC Islamic Traditions and Muslim Societies in World Contexts Grants, 4430
Stanford Humanities Center Arts Writer/Practitioner Fellowship, 4444
Stanford Humanities Center External Faculty Fellowships, 4445
Teachers College Minority Postdoc Fellowships, 4498
U.S. Department of Education Native Hawaiian Education Grants, 4543
UCLA Research Grants in Ethnic Studies, 4549
UICC Yamagiwa-Yoshida Memorial International Cancer Study Grants, 4558
UNC Carolina Postdoctoral Program for Faculty Diversity, 4560
UNCFSP NASA Administrator's Fellowship, 4562
USDA Higher Education Challenge Grants, 4627
USDA Hispanic-Serving Institutions Education Grants, 4628
USDA Integrated Research Education and Extension Grants - Organic Transitions, 4630
USDA Scientific Cooperation Research Program, 4647
USDD Sabbatical Leave Fellowships, 4788
Vancouver Foundation Grants and Community Initiatives Program, 4807
Virginia Foundation for the Humanities Residential Fellowships, 4815
Wabash Center Project Grants, 4826
Wabash Center Small Project Grants, 4827
Wabash Center Waorkshop and Colloquy Participant Grants, 4828
Wayne and Gladys Valley Foundation Grants, 4835
Welch Foundation Departmental Grants, 4839
Wellesley Centers for Women Postd Fellowships, 4845
WILD Foundation Grants, 4867
WWF Russell E. Train Education for Nature Alumni Grants, 4897
Xerox Foundation Grants, 4899

Fellowships

3M Canada National Teaching Fellowships, 3
3M Foundation Community Giving Grants, 5
A-T Children's Project Post Doctoral Fellowships, 9
AAAAI Fellows-in-Training Grants, 22
AAA Minority Dissertation Fellowship Program, 25
AAAS Roger Revelle Fellowship in Global Stewardship, 38
AAAS Science and Technology Policy Fellowships: Congressional, 39
AAAS Science and Technology Policy Fellowships: Diplomacy, Security, and Development, 40
AAAS Science and Technology Policy Fellowships: Energy, Environment, and Agriculture, 41
AAAS Science and Technology Policy Fellowships: Global Health and Development, 42
AAAS Science and Technology Policy Fellowships: Health, Education and Human Services, 43
AAAS Science and Technology Policy Overseas Fellowships at USAID Missions, 44
AAAS Science Prize for Inquiry-Based Instruction, 45
AAB Berlin Prize Fellowships, 48
AACAP-NIDA K12 Career Development Awards, 54
AACAP Educational Outreach Program for Residents in Alcohol Research, 58
AACAP Jeanne Spurlock Minority Med Student Clinical Fellowship in Child and Adolescent Psychiatry, 63
AACAP Jeanne Spurlock Research Fellowship in Substance Abuse and Addiction for Minority Medical Students, 64
AACAP Life Members Mentorship Grants for Medical Students, 67
AACAP Mary Crosby Congressional Fellowships, 68
AACAP Pilot Research Awards, Supported by Eli Lilly and Company, 72
AACAP Summer Medical Student Fellowships, 80
AACC International Graduate Fellowships, 91
AACC International Milling & Baking Division's M. Rella Dwyer Graduate Fellowship, 93
AACR-Colorectal Cancer Coalition Fellows Grant, 113
AACR-FNAB Fellows Grant for Translational Pancreatic Cancer Research, 114
AACR-Genentech BioOncology Fellowship for Cancer Research in Angiogenesis, 117
AACR-National Brain Tumor Foundation Fellows, 120
AACR Basic Cancer Research Fellowships, 130
AACR Centennial Postdoctoral Fellowships in Cancer Research, 133
AACR Centennial Pre-doctoral Fellowships in Cancer Research, 134
AACR Clinical and Translational Research Fellowships, 135
AACR Fellows Grants, 136
AAFA Investigator Research Grants, 153
AAFCS International Graduate Fellowships, 154
AAFCS National Graduate Fellowships, 155
AAFPRS Fellowships, 162
AAF Richard Morris Hunt Archit Fellowship, 169
AAHPM Hospice and Palliative Med Fellowships, 180
AAJA Fellowships, 181
AAR Postdoctoral Fellowships, 201
AAR Rome Prize Fellowships, 202
AAS Kate B. and Hall J. Peterson Fellowships, 204
AAS Legacy Fellowships, 205
AAS Long-Term Visiting Academic Research Fellowships, 206
AAS Short-Term Visiting Academic Research Fellowships, 208
AAUW American Dissertation Fellowships, 209
AAUW American Postdoctoral Research Leave Fellowships, 210
AAUW International Doctoral Fellowships, 214
AAUW International Master's/First Professional Degree Fellowships, 215
AAUW International Postdoctoral Fellowships, 216
AAUW Selected Professions Fellowships, 217
Abbott Fund Science Education Grants, 220
Abdus Salam ICTP/IAEA Sandwich Training Education Grants, 226
Abdus Salam ICTP Jnt Laurea Magistralis Grants, 229
Abdus Salam ICTP Joint PhD Programme in Environmental Fluid Mechanics Grants, 230
Abdus Salam ICTP Mathematical Research Fellowships, 232
Abdus Salam ICTP Postdoctoral Positions in Theoretical High Energy Physics, 234
Abdus Salam ITCP Postgraduate Diploma Programme in General Physics, 241
Abdus Salam Joint ICTP/SISSA PhD Programme in Physics and Mathematics Grants, 242
ACAAI Foundation Research Grants, 244
Academy of American Franciscan History Dissertation Fellowships, 249
ACC American Artists and Museum Professionals in Asia Fellowships, 250
ACC Asian Art and Religion Fellowships, 251
ACC Fellowships and Projects, 252
ACC Ford Foundation Fellowships, 253
ACC Hong Kong Arts Program Fellowships, 254
ACC Humanities Fellowship Program, 255
ACC Japan-US Arts Program Fellowships, 256
ACC Mekong Region Fellowship, 258
ACC Philippines Fellowship Program, 261
ACCP Infectious Diseases Research Fellowships, 263
ACCP Oncology Research Fellowships, 264
ACC Residency Program in Asia, 267
ACC Starr Foundation Fellowship Program, 268
ACC Taiwan Fellowship Program, 269
ACIE Title VIII Combined Research and Language Training Fellowships, 275
ACIE Title VIII Research Scholar Fellowships, 276
ACLS/New York Public Library Fellowships, 277
ACLS African Humanities Program for Dissertation Completion, 278
ACLS African Humanities Program in Ghana, Nigeria, South Africa, Tanzania, and Uganda, 279
ACLS American Research in the Humanities in China Fellowships, 280
ACLS Charles A. Ryskamp Research Fellowships, 281

ACLS Collaborative Research Fellowships, 282
ACLS Digital Innovation Fellowships, 284
ACLS Dissertation Fellowships in East European Studies, 285
ACLS Early Career Postdoctoral Fellowships in East European Studies, 286
ACLS Fellowships, 287
ACLS Frederick Burkhardt Residential Fellowships for Recently Tenured Scholars, 288
ACLS International and Area Studies Fellowships, 289
ACLS New Faculty Fellows, 290
ACSE J. Waldo Smith Hydraulic Fellowship, 309
ACS Postdoctoral Fellowships, 325
Acton Institute for the Study of Religion and Liberty Calihan Academic Fellowships, 331
ADA Foundation Dentsply International Research Fellowships, 335
ADA Foundation Scientist in Training Fellowship, 337
ADA Foundation Summer Scholars Fellowships, 338
AES-Grass Young Investigator Travel Awards, 345
AES Robert S. Morison Fellowship, 346
AES Susan S. Spencer Clinical Research Epilepsy Training Fellowship, 347
Aetna Foundation Minority Scholars Grants, 349
Aetna Foundation National Medical Fellowship in Healthcare Leadership, 350
AFAR Medical Student Training in Aging Research Program, 352
AFUW Fellowships, 361
AFUW Georgina Sweet Fellowships, 362
AGHE Graduate Scholarships and Fellowships, 369
Ahmanson and Getty Postdoctoral Fellowships, 377
AHRQ Individual Postdoc Fellows Ruth L. Kirschstein National Research Service Awards, 389
AIAR/Council of American Overseas Research Centers Fellowships for Advanced Multi-Country Research, 391
AIAR Andrew W. Mellon Foundation Fellowships, 392
AIAR Annual Professorships, 393
AIAR George A. Barton Fellowships, 395
AIHS Clinical Fellowships, 436
AIHS Fast-Track Fellowships, 440
AIHS Full-Time Fellowships, 445
AIHS Media Summer Fellowship, 453
AIHS Part-Time Fellowships, 454
AINSE Research Fellowships, 464
AINSE Research Fellowships with Institute for Science and International Security Support, 465
AIR/IES-NCES Postdoctoral Policy Fellowships in Washington, 466
AIR/NCES Fellowships for Graduate Study, 467
AIR/NRC Summer Institute on First-Year Assessment Undergraduate Student Fellowships, 468
AIR National Summer Data Policy Institute Fellowships, 469
AIR Research Grants/Dissertation Fellowships, 470
AJA Ethel Marcus Memorial Fellowship in American Jewish Studies, 471
AJA Lowenstein-Wiener Summer Fellowship Awards in American Jewish Studies, 472
AJA Marguerite R. Jacobs Memorial Postdoctoral Award in American Jewish Studies, 473
AJA Rabbi Frederic A. Doppelt Memorial Fellowship in American Jewish Studies, 474
AJA Starkoff Fellow in American Jewish Studies, 475
ALA ACRL Doctoral Dissertation Fellowship, 478
ALA Samuel Lazerow Fellowship For Research in Collections and Technical Services in Academic and Research Libraries, 497
Albert and Mary Lasker Foundation Clinical Research Scholars, 506
Albert M. Greenfield Dissertation Fellowship, 507
Alexander von Humboldt CAPES-Humboldt Fellowships for Experienced Researchers, 516
Alexander von Humboldt CAPES-Humboldt Fellowships for Postdoctoral Researchers, 517
Alexander von Humboldt Feodor-Lynen Fellowships for Experienced Researchers, 518
Alexander von Humboldt Feodor-Lynen Research Fellowships for Postdoc Researchers, 519
Alexander von Humboldt Feodor Lynen Research Fellowships for Experienced Researchers, 520
Alexander von Humboldt Georg Forster Fellowships for Experienced Researchers, 523
Alexander von Humboldt Georg Forster Fellowships for Postdoctoral Researchers, 524
Alexander von Humboldt German Chancellor Fellowships for Prospective Leaders from the People's Republic of China, 526
Alexander von Humboldt German Chancellor Fellowships for Prospective Leaders from the Russian Federation, 527
Alexander von Humboldt Wardwell Fellowships, 528
Alexander von Humboldt International Climate Protection Fellowships, 529
Alexander von Humboldt JSPS Res Fellowships, 530
Alexander von Humboldt Taiwan Fellowships, 533
Alexander von Humboldt Found Professorships, 535
Alexander von Humboldt Research Fellowships for Postdoctoral Researchers, 539
Alfred P. Sloan Economic Analysis of Science and Technology Grants, 548
Alfred P. Sloan Empirical Economic Research Enablers Grants, 550
Alfred P. Sloan History of Science and Technology Grants, 553
Alfred P. Sloan Research Fellowships, 560
Alpha Kappa Alpha Educational Advancement Foundation Fellowships, 575
Alston Bannerman Sabbatical Fellowships, 580
Alzheimer's Association U.S.-U.K. Young Investigator Exchange Fellowships, 597
Alzheimer's Association Zenith Fellows Awards, 598
American-Italian Cancer Foundation Fellowships, 602
American-Scandinavian Foundation Awards for Advanced Study or Research in the U.S., 603
American Academy of Advertising Research Fellowships, 611
American Academy of Arts and Sciences Hellman Fellowships, 612
American Academy of Arts and Sciences Visiting Scholars, 613
American Academy of Nursing Claire M. Fagin Fellowships, 615
American Australian Association Australia To USA Fellowships, 619
American Australian Association USA to Australia Fellowships, 620
American Chem Society AGRO Fellow Awards, 633
American Chemical Society Green Chem Grants, 679
American Chemical Society Petroleum Research Fund Doctoral New Investigator Grants, 690
American Chemical Society Petroleum Research Fund Undergraduate Faculty Sabbatical Grants, 693
American Chemical Society Petroleum Research Fund Undergraduate New Investigator Grants, 694
American College of Surgeons Faculty Research Fellowships, 704
American College of Surgeons Traveling Fellowship to Germany, 706
American College of Surgeons Traveling Fellowship to Japan, 707
American College of Surgeons Wound Care Management Award, 708
American Hist Assoc Fellow in Aerospace History, 717
American Hist Assoc Folger Shakespeare Library Fellowship, 718
American Hist Assoc J.F. Jameson Fellowship, 725
American Indian College Fund STEM Leadership Fellowships, 745
American Indian Studies Research Inst Fellowship, 746
American Mathematical Society Centennial Fellowships, 752
American Numismatic Society Doctoral Fellows, 759
American Numismatic Society Frances M. Schwartz Fellowship, 762
American Philosophical Society Daland Fellowships in Clinical Investigation, 763
American Philosophical Society Library Resident Research Fellowships, 769
American Philosophical Society Sabbatical Fellowship in the Humanities and Social Sciences, 771
American Psychiatric Association/Kempf Fund Award for Research Development in Psychobiological Psychiatry, 775
American Psychiatric Association/Lilly Psychiatric Research Fellowship, 776
American Psychiatric Association/SmithKline Beecham Young Faculty Award for Research Development in Biological Psychiatry, 779
American Psychiatric Association/Wyeth Pharmaceuticals M.D./Ph.D. Psychiatric Research Fellowships, 780
American Psychiatric Association Minority Fellowships Program, 783
American Psychiatric Association Minority Medical Student Fellowship in HIV Psychiatry, 784
American Psychiatric Foundation Jeanne Spurlock Congressional Fellowship, 793
American Psychiatric Minority Fellowships, 794
American Rivers Anthony A. Lapham River Conservation Fellowship, 795
American Sociological Association Minority Fellowships, 805
amfAR Fellowships, 809
amfAR Mathilde Krim Fellowships in Basic Biomedical Research, 810
amfAR Research Grants, 811
AMNH Frank M. Chapman Fellowships, 814
AMNH Graduate Student Fellowships, 815
AMS/Industry Government Graduate Fellowships, 817
AMS/Industry Minority Scholarships, 818
AMS Alvin H. Johnson Dissertation Fellowships, 820
AMS Graduate Fellowships in History of Science, 824
AMS Howard Mayer Brown Fellowships, 827
Andrew W. Mellon/ACLS Dissertation Completion Fellowships, 844
Andrew W. Mellon Postdoctoral Fellowship in American Cultures, 845
Andrew W. Mellon Postdoctoral Fellowships at UCLA Humanities Consortium, 846
Andy Warhol Foundation for the Visual Arts Curatorial Research Fellowships, 847
Anne Louise Barrett Fellowship, 861
AOCS Ralph H. Potts Memorial Fellowship, 893
AOCS Thomas H. Smouse Memorial Fellowship, 899
APA Planning Fellowships, 910
APHA Fellowship in Printing History, 913
APHL Emerging Infectious Diseases Fellowships, 914
APSAA Fellowships, 920
APSA Congressional Fellowships for Communications Scholars and Journalists, 926
APSA Congressional Fellowships for Federal Executives, 927
APSA Congressional Fellowships for International Scholars, 928
APSA Congressional Fellowships for Journalists, 929
APSA Congressional Fellowships for Political Scientists, 930
APSA Congressional Health and Aging Policy Fellowships, 931
APSA Minority Fellowships, 939
APSA Presidency Research Fund Fellowships, 943
APSA Robert Wood Johnson Foundation Health Policy Fellowships, 944
APSA Special Fund for the Study of Women and Politics, 947
APSA Warren E. Miller Fellowship in Electoral Politics, 949
Arie and Ida Crown Memorial Grants, 989
ARIT/NEH Fellowships for Research in Turkey, 990
ARIT Fellowships, 991
ARIT Fellowships for Intensive Advanced Turkish Language Study, 992
ARIT Kenan T. Erim Fellowship for Research at Aphrodisias, 993
ARIT Mellon Foundation Postdoctoral Research Fellowships for East European Scholars, 994
Arkansas Arts Council Indiv Artists Fellowships, 999
Armenian Students' Association Scholarships, 1000

ARM Faculty Fellowships, 1001
ARM Graduate Fellowships, 1002
ARM Postgraduate Fellowships, 1003
ARM Visiting Scientist Awards, 1004
Arthur and Rochelle Belfer Foundation Grants, 1023
Arthur J. Ennis Fellowships, 1025
ArvinMeritor Foundation Health Grants, 1027
ASA & NCHS Research Fellowships, 1028
ASA/NSF Federal Statistics Fellowships, 1032
ASA and USDA/NASS Research Fellowships, 1033
ASCE Freeman Fellowships, 1040
ASCE Jack E. Leisch Memorial National Graduate Fellowship, 1041
ASCE O.H. Ammann Research Fellowship in Structural Engineering, 1042
ASCE Trent R. Dames and William W. Moore Engineering Fellowship, 1043
ASCO/UICC International Cancer Technology Transfer Fellowships, 1044
ASCO Long-Term International Fellowships, 1048
ASCSA Alison M. Frantz Fellowship, 1051
ASCSA Fellowships, 1052
ASCSA Harry Bikakis Fellowship, 1053
ASCSA Jacob Hirsch Fellowship, 1054
ASCSA Weiner Laboratory Faunal Studies Research Fellowship, 1056
ASECS/Clark Fellowships, 1057
ASEE Army Research Laboratory Postdoctoral Fellowships, 1059
ASHG Public Health Genetics Fellowship, 1086
ASHRAE Grad Grant-in-Aid Fellows, 1088
ASHRAE Willis H. Carrier Graduate Research Fellowship, 1089
Asian Pacific American Institute for Congressional Studies Fellowships, 1091
ASM-PAHO Infectious Diseases Epidemiology and Surveillance Fellowships, 1098
ASM/CDC Fellowships in Infectious Disease and Public Health Microbiology, 1100
ASME Graduate Teaching Fellowships, 1125
ASM International Fellowship for Latin America and the Caribbean, 1175
ASM Int'l Fellowships for Asia and Africa, 1176
ASM Microbiology Undergraduate Research Fellowship, 1178
ASM Robert D. Watkins Graduate Research Fellowship, 1183
ASM Undergraduate Research Fellowship, 1186
Aspen Institute William Randolph Hearst Endowed Fellowship for Minority Students, 1194
ASPET Summer Undergrad Res Fellowships, 1208
ASPO Fellowships, 1211
ASPRS Leica Geosystems Internship, 1213
ASSE Liberty Mutual Safety Res Fellowship, 1219
Association for the Study of Higher Education Dissertation Fellowships, 1222
ASU Graduate College Completion Fellowships, 1235
ASU Graduate College Dean's Fellowships, 1236
ASU Graduate College Dissertation Fellowships, 1237
ASU Graduate College Doctoral Enrichment Fellowships, 1238
ASU Graduate College Earl and Lenore Tripke Fellowship, 1239
ASU Graduate College Herman E. Demund Memorial Fellowships, 1240
ASU Graduate College Lattie and Elva Coor Building Great Communities Graduate Fellowship, 1241
ASU Grad Coll Reach for the Stars Fellowship, 1243
ASU Graduate College Science Foundation Arizona Bisgrove Postdoctoral Scholars, 1244
ASU Graduate College Science Foundation Arizona Fellowships in Science and Engineering, 1245
ASU Institute for Humanities Res Fellowships, 1246
ASU Institute for Humanities Research Visiting Fellowships, 1248
AT&T Bell Labs Graduate Research Program, 1250
AT&T Industrial Ecology Faculty Fellowships, 1251
ATA Doctoral Fellowships in Education, 1252
Athenaeum of Philadelphia Senior Fellowships, 1256
ATS Henry Luce III Fellowships, 1258
ATS Lilly Theological Faculty Fellowships, 1260
Auburn University President's Grad Opportunities, 1264
Australian Academy of Science Grants, 1269
Australian Research Fellowships/Queen Elizabeth II Fellowships, 1272
AWWA Abel Wolman Doctoral Fellowships, 1282
Ball Brothers Foundation General Grants, 1291
Baltimore Washington Center Adult Psychoanalysis Fellowships, 1293
Baltimore Washington Center Child Psychotherapy Fellowships, 1294
BARD Postdoctoral Fellowships, 1298
Baylor Inst for Faith & Learning Visiting Fellows, 1310
BBF Fellowships, 1311
Berea College Appalachian Music Fellowships, 1332
Bernard L. Schwartz Fellows Program, 1334
Beverly Willis Architecture Foundation Grants and Fellowships, 1340
BIA Native American Graduate Fellowships, 1342
BMI Woody Guthrie Fellowships, 1352
Boeing Company Contributions Grants, 1357
Bogliasco Foundation Liguria Study Center Residencies, 1358
Boston Psychoanalytic Society and Institute Fellowship in Child Psychoanalytic Psychotherapy, 1362
Boston Psychoanalytic Society and Institute Fellowship in Psychoanalytic Psychotherapy, 1363
Bristol-Myers Squibb Foundation Fellowships, 1375
Brookdale Foundation Leadership in Aging Fellowships, 1381
Bullard Fellowship in Forest Research, 1383
Bunting Fellowship Program, 1385
Bush Foundation Artist Fellowships, 1391
BWF Institutional Program Unifying Population and Laboratory Based Sciences Grants, 1397
CAA Professional Development Fellowships for Graduate Students, 1404
Cabot Corporation Foundation Grants, 1406
Caesars Foundation Grants, 1407
Camargo Foundation Fellowships, 1412
Canadian Studies Senior Fellowships, 1422
Canon Foundation Grants, 1423
Carl and Lily Pforzheimer Foundation Grants, 1427
Carnegie Developmental Biology Res Fellowships, 1433
Carnegie Endowment for International Peace Junior Fellowships, 1434
Carnegie Plant Biology Fellowships, 1435
Carnegie Postdoctoral Research Fellowships, 1436
Carnegie Postdoctoral Research Fellowships in Astronomy, 1437
Carrier Corporation Contributions Grants, 1440
Catholic University Life Cycle Institute Research Fellowship, 1444
CBO Economic Policy Fellowships, 1445
CCIS Visiting Research Fellowships, 1452
CCWH/Berkshire Conference of Women Historians Graduate Student Fellowship, 1453
CCWH Ida B. Wells Graduate Fellowship, 1455
CDC Exp Applied Epidemiology Fellowships, 1460
CDC Foundation Atlanta International Health Fellowships, 1461
CDC Found Tobacco Network Lab Fellowship, 1462
CDC Preventive Med Residency Fellowship, 1465
CDC Steven M. Teutsch Prevention Effectiveness Fellowships, 1467
CDECD Elizabeth L. Mahaffey Arts Administration Fellowship, 1468
Cedar Tree Foundation David H. Smith Conservation Research Fellowship, 1470
Center for Advanced Judaic Studies Postdoctoral Fellowships, 1472
Center for East Asian Studies Fellowships, 1474
Center for Hellenic Studies Fellowships, 1476
Center for Inquiry Science and the Public Visiting Research Fellowships, 1477
Center for Judaic Studies Fellowships, 1478
Center for the Study of Books and Media Visiting Research Fellowships, 1480
Center for the Study of Democratic Politics Fellowships, 1481
Center for the Study of Philanthropy International Fellowships, 1482
Center for the Study of Philanthropy Senior International Fellowships, 1483
Center for the Study of Professional Military Ethics Resident Fellowship, 1484
Center for the Study of Religion Fellowships, 1485
Central European University Fellowships, 1488
CFF First- and Second-Year Clinical Fellowships, 1492
CFF Postdoctoral Research Fellowships, 1496
CFF Third-, Fourth-, and Fifth-Year Clinical Fellowships, 1499
CFR Edward R. Murrow Press Fellowship, 1500
CFR International Affairs Fellowships, 1501
CFR International Affairs Fellowships in Nuclear Security, 1502
CFR Int'l Affairs Fellowships in South Korea, 1503
CFR International Affairs in Japan Fellowships, 1504
CFR Military Fellowships, 1505
CFR Stanton Nuclear Security Fellowships, 1506
CFUW Beverley Jackson Fellowship, 1507
CFUW Canadian Home Economics Association Fellowship, 1509
CFUW Margaret McWilliams Predoc Fellowship, 1513
CFUW Marion Elder Grant Fellowship, 1514
CFUW Memorial Fellowship, 1515
CFUW Ruth Binnie Fellowship, 1516
Chapin Hall International Fellowships in Children's Policy Research, 1517
Charles A. King Trust Postdoctoral Research Fellowships, 1518
Charles H. Revson Foundation Grants, 1524
Charlotte W. Newcombe Doctoral Dissertation Fellowships, 1531
CHF Charles C. Price Fellow in Polymer History, 1539
CHF Glenn E. and Barbara Hodsdon Ullyot Scholarship, 1540
CHF Long-Term Fellowships, 1541
CHF Robert W. Gore Fellowship in Materials Innovation, 1542
CHF Roy G. Neville Short-Term Fellowships, 1543
CHF Societe de Chimie Industrielle (American Section) Fellowship, 1544
Chicago Community Trust Fellowships, 1546
Chicago Institute for Psychoanalysis Fellowships, 1548
Children's Lit Association Research Fellowships, 1551
Christensen Fund Regional Grants, 1554
Churchill Fellowships, 1560
CICF Efroymson Contemporary Arts Fellows, 1563
CIC Smithsonian Institution Fellowships, 1564
CICU James C. Ross Research Fellowship, 1567
CIGNA Foundation Grants, 1568
CIMO Scholarships for Advanced Studies of the Finnish Language at Finnish Universities, 1569
Cintas Foundation Fellowships, 1570
CIUS Neporany Res & Teaching Fellowship, 1572
Clark-Huntington Bibliographical Fellowship, 1576
Clark Library Short-Term Resident Fellowships, 1577
CMAP Physician Advocacy Fellowship, 1583
CMU Arts in Society Fellowship, 1587
CNIB Baker Applied Research Fund Grants, 1588
CNIB Baker New Researcher Fund Grants, 1589
CNIB E. & Mary Hochhausen Access Technology Research Grants, 1592
Coca-Cola Foundation Grants, 1593
Commonwealth Fund/Harvard University Fellowship in Minority Health Policy, 1604
Commonwealth Fund Harkness Fellowships in Health Care Policy and Practice, 1606
Community Foundation for S Arizona Grants, 1618
Compton Foundation International Fellowships, 1624
Compton Foundation Mentor Fellowships, 1625
Consortium for Graduate Study in Management Minority Fellowships, 1642
Cooperative Institute for Research in Environmental Sciences Visiting Fellows Program, 1643
Corning Foundation Educational Grants, 1646
Council on Foundations Emerging Philanthropic Leaders Fellowships, 1649
Covidien Advanced MIS Surgical Fellowships, 1650

Covidien Surgical Mini-Fellowships, 1651
CPC Catherine H. Beattie Fellowship, 1653
Crohn's Disease and Colitis Research Fellowships, 1657
CSRPC Dissertation Fellowships, 1661
CSRPC Postdoctoral Fellowship for Advanced Scholars, 1662
CSRPC Postdoc Fellowship for Junior Scholars, 1663
CSTE CDC/CSTE Applied Epidemiology Fellowships, 1664
Cystic Fibrosis Canada Clinical Fellowships, 1674
Cystic Fibrosis Canada Fellowships, 1677
Cystic Fibrosis Canada Senior Scientist Research Training Awards, 1680
DAAD-AICGS Research Fellowships, 1693
DAAD-Leo Baeck Institute Fellowships, 1694
Daimler and Benz Foundation Fellowships, 1713
Damon Runyon Cancer Foundation Fellowships, 1715
Dart Center Ochberg Fellowship, 1720
Dartmouth College Andrew W. Mellon Postdoctoral Fellowship, 1721
Delaware Division of the Arts Individual Artist Fellowships, 1732
Deloitte Foundation Grants, 1735
Delta Air Lines Foundation Health and Wellness Grants, 1736
Delta Gamma Foundation Fellowships for Graduate Study, 1738
Dept of Ed Rehabilitation Fellowships, 1747
Diversifying Higher Education Faculty in Illinois Program, 1764
DOE Albert Einstein Distinguished Educator Fellowship, 1770
DOE Computational Science Grad Fellowships, 1780
DOJ National Institute of Justice Graduate Research Fellowships, 1828
DOJ National Inst of Justice Visiting Fellowships, 1829
Doris Duke Charitable Foundation Clinical Research Fellowships for Medical Students, 1834
DOT Dwight David Eisenhower Transportation Fellowship Program, 1847
Douglass Institute Postdoctoral Fellowships, 1851
Douglass Institute Predoctoral Fellowships, 1852
Dr. Robert H. Goddard Research Fellowships, 1854
Duke Foundation Fellowships, 1868
Duke University Adult Cardiothoracic Anesthesia and Critical Care Medicine Fellowships, 1869
Duke University Ambulatory and Regional Anesthesia Fellowships, 1870
Duke University Clinical Cardiac Electrophysiology Fellowships, 1871
Duke University Hyperbaric Center Fellowships, 1872
Duke Univ Interventional Cardio Fellowships, 1873
Duke Univ Obstetric Anesthesia Fellowships, 1878
Duke University Pain Management Fellowships, 1879
Duke University Postdoctoral Research Fellowships in Aging, 1880
Durfee Foundation Stanton Fellowship, 1881
E.J. Sierleja Memorial Fellowship, 1884
Earhart Foundation Institutional Grants, 1888
Earthwatch Institute Student Fellowships, 1890
Educause Jane N. Ryland Fellowships, 1897
Edward W. and Stella C. Van Houten Grants, 1899
Edwin S. Webster Foundation Grants, 1900
Eisenhower Exchange International Fellowships, 1902
El Pomar Leadership Development and Community Engagement Postgraduate Fellowship, 1912
Enterprise Community Partners Rose Architectural Fellowships, 1929
Enterprise Community Partners Terwilliger Fellowship, 1930
EPA Capacity Building Grants for Compliance Assurance and Enforcement Activities in Indian Country and Other Tribal Areas, 1932
Erasmus Institute Fellowships, 1944
Erikson Institute Child Development Doctoral Fellowships, 1949
ETS Sylvia Taylor Johnson Minority Fellowship in Educational Measurement, 1955
EU Center for Humanistic Inquiry Junior and Postdoctoral Fellowships, 1956
Eugene M. Lang Foundation Grants, 1957
EU Robert W. Woodruff Library Res Fellowships, 1959
Evan Frankel Foundation Grants, 1961
Eye-Bank for Sight Restoration and Fight for Sight Summer Student Research Fellowship, 1969
Ezra Jack Keats/Kerlan Collection Memorial Fellowship, 1970
Facebook Fellowships, 1973
FAMRI Young Clinical Scientist Awards, 1976
Fannie and John Hertz Foundation Graduate Fellowships, 1977
Fannie Mae Foundation Kennedy School of Government Fellowships, 1979
FCD Young Scholars Fellowships, 1985
FDHN Centocor International Research Fellowship in Gastrointestinal Inflammation & Immunology, 1987
FDHN Student Research Fellowships, 2001
Fermilab Postdoctoral Fellowship Programs, 2004
Field Museum Graduate Student Fellowships, 2008
Field Museum Visiting Scholarships, 2011
Fight for Sight-Streilein Foundation for Ocular Immunology Research Award, 2012
Fight for Sight Post-Doctoral Awards, 2014
Fight for Sight Summer Student Fellowships, 2015
Fleishhacker Foundation Eureka Fellowships, 2023
Florence Tan Moeson Fellowships, 2025
Florida Division of Cultural Affairs Individual Artist Fellowships, 2026
Ford Foundation International Fellowships, 2030
Ford Foundation Peace and Social Justice Grants, 2031
Foundation for Evangelism Doctoral Scholars, 2039
Foundation for Evangelism Harry Denman Fellowships, 2040
Foundation of the American Thoracic Society Research Grants, 2044
Franklin Mosher Baldwin Fellowships, 2050
Frederick Douglass Summer Fellowships, 2052
Frederick Douglass Summer Scholars Fellowship, 2053
Fritz B. Burns Foundation Grants, 2057
FTE Dissertation Fellowships, 2058
FTE Doctoral Fellowships, 2059
FTE Expanding Horizons Partnership Grants, 2060
FTE North American Doctoral Fellowships, 2061
Fulbright Traditional Scholar Program in the East Asia/Pacific Region, 2082
Fulbright Traditional Scholar Program in the Near East and North Africa Region, 2083
Fulbright Traditional Scholar Program in the Western Hemisphere, 2085
Fund for Scientific Research--Flanders Grants, 2089
Gaius Charles Bolin Fellowships for Minority Graduate Students, 2095
Gates Family Foundation Fellowships, 2097
GCA Interchange Fellowship, 2098
GCA Katharine M. Grosscup Scholarships, 2099
GCSAA Watson Fellowships, 2103
GEM MS Engineering Fellowships, 2104
GEM PhD Engineering Fellowships, 2105
GEM PhD Science Fellowships, 2106
Georgetown University Chemistry Department Graduate Research Fellowships, 2121
Georgia Institute of Technology Postdoctoral Fellowship, 2124
Gerda Henkel Foundation Grants, 2127
Germanistic Society of America Fellowships, 2128
Gerrish/FTE Technology Education Graduate Fellowship, 2129
Getty Research Institute Nonresidential Grants - Curatorial Research Fellowships, 2136
Getty Research Institute Nonresidential Grants - Postdoctoral Fellowships, 2137
Gilbert F. White Postdoctoral Fellowships, 2140
Gilbreth Memorial Fellowships, 2141
Gilroy and Lillian P. Roberts Foundation Grants, 2142
Global Change Education Program, 2151
Graduate Institute Hodson Teacher Fellowships, 2160
Grass Foundation Marine Biological Laboratory Advanced Imaging Fellowships, 2162
Grass Foundation Marine Biological Laboratory Fellowships, 2163
Gruber Foundation International Astronomical Union Fellowships, 2179
Gruber Foundation Rosalind Franklin Young Investigator Award, 2183
GUITS Grants for Undergraduate Study, 2187
GUITS Sabbatical Research Grants, 2191
H.A. and Mary K. Chapman Charitable Grants, 2193
H.B. Earhart Fellowships, 2194
Harry Frank Guggenheim Foundation Dissertation Fellowships, 2207
Harry Ransom Humanities Research Center Research Fellowships, 2209
Harvard Postdoctoral Fellowships on Education, 2211
Harvard University Administrative Fellowships, 2212
Harvard University Graduate School of Education Fellowships and Awards, 2213
Harvard University Weatherhead Center for International Affairs Fellowships, 2215
Harvard Women and Public Policy Postdoctoral Fellowship, 2217
Haverford College Mellon Postdoc Fellowships, 2220
Hebrew University Lady Davis Trust Fellowships, 2226
Helen Carr Minority Fellowships, 2229
Helen Kellogg Institute for International Studies Residential Fellowships, 2230
Henry J. Kaiser Family Found Media Fellowships, 2234
Henry Luce Foundation/ACLS Dissertation Fellowships in American Art, 2235
Herbert Hoover Presidential Library Association Travel Grants, 2242
Hereditary Disease Foundation John J. Wasmuth Postdoctoral Fellowships, 2244
Hereditary Disease Foundation Research Grants, 2246
Herman Kahn Fellowship, 2247
HHMI Gilliam Fellowships for Advanced Study, 2254
HHMI Grants and Fellowships Programs, 2255
HHMI Research Training Fellowships, 2262
Hirtzel Memorial Foundation Grants, 2269
Houghton Library Visiting Fellowships, 2284
HRF Graduate Fellowships, 2285
HRF Mark B. Bain Graduate Fellowships, 2287
HRF Tibor T. Polgar Fellowships, 2289
HUD Urban Scholars Postdoctoral Fellowships, 2293
Huebner Foundation Fellowships, 2294
Humanities Institute Vis Scholars Fellowships, 2296
Hunter's Hope Foundation Grants, 2298
Huntington's Disease Society of America Research Fellowships, 2299
Huntington Library Babara Thom Postdoctoral Fellowships, 2301
Huntington Library Clark-Huntington Joint Bibliographical Fellowship, 2302
Huntington Library Dana and David Dornsife Fellowships, 2303
Huntington Library Dibner History of Science Long-Term Fellowships, 2304
Huntington Library Dibner History of Science Short-Term Fellowships, 2305
Huntington Library Mellon Fellowships, 2306
Huntington Library National Endowment for the Humanities Fellowships, 2307
Huntington Library Short-Term Fellowships, 2308
Huntington Library Travel Grants and Exchange Fellowships for Study in Great Britain, 2309
Ian Axford (New Zealand) Fellowships in Public Policy, 2313
Ian N. McKinnon Memorial Fellowship, 2314
IARC Expertise Transfer Fellowship, 2315
IARC Postdoctoral Fellowships for Training in Cancer Research, 2316
IAS Mellon Fellowships for Assistant Professors, 2318
IBRO Albert Aguayo Fellowship, 2322
IBRO Asia Reg APRC Exchange Fellowships, 2323
IBRO Regional Grants for Int'l Fellowships to U.S. Laboratory Summer Neuroscience Courses, 2330
IBRO Research Fellowships, 2331
IBRO Return Home Fellowships, 2332
ICC Faculty Fellowships, 2333
Ida Smedley MacLean International Fellowship, 2337
IEEE Electrical History Fellowship, 2339

IES Faculty Fellowships, 2341
IFP Minnesota McKnight Film Fellowships, 2342
IFP Minnesota McKnight Screen Fellowships, 2343
IFUW Dr. A. Vibert Douglas Int'l Fellowship, 2346
IHI Quality Improvement Fellowships, 2348
IIE Central Europe Summer Research Institute Summer Research Fellowship, 2349
IIE David L. Boren Fellowships, 2350
IIE KAUST Graduate Fellowships, 2353
IIE Leonora Lindsley Memorial Fellowships, 2354
IIE Whitaker International Fellowships and Scholarships, 2356
IIPP Graduate Fellowship, 2357
Independent Institute John M. Templeton Fellowships Essay Contest, 2369
Indiana Space Grant Consortium Fellowships, 2375
Institute for Advanced Studies in Culture Postdoctoral Fellowships, 2380
Institute for Agriculture and Trade Policy Food and Society Fellowships, 2381
Institute of American Cultures Graduate and Predoctoral Fellowships, 2383
Institute of American Cultures Postdoctoral and Visiting Scholar Fellowships, 2384
Institute of Paper Science & Tech Fellowships, 2387
Institute of Textile Tech Grad Fellowships, 2388
Institute of Turkish Studies Dissertation Writing Grants, 2389
Intel Foundation PhD Fellowships, 2394
International Federation of University Women Research Fellowships, 2397
International Fellowships in Medical Education, 2398
International Studies Graduate Fellowships, 2403
IREX Eurasian Undergraduate Exchange Grants, 2418
IREX MENA Media Emerging Leaders Fellows, 2421
IREX Policy-Connect Collaborative Res Grants, 2422
IREX University Administration Support, 2425
Irving and Rose Crown Scholarships and Fellowships in American History, 2427
ISGC Graduate Fellowships, 2430
IWMF Public Health Fellowship, 2446
Izaak Walton Killam Postdoctoral Fellowships at the University of British Columbia, 2447
Izaak Walton Killam Memorial Predoctoral Fellowships at the University of British Columbia, 2448
J.H. Stewart Reid Memorial Fellowship for Doctoral Studies, 2450
J.S. Mill Fellowships in Economics for Exceptional Students, 2451
James Graham Brown Found Brown Fellowships, 2459
James McKeen Cattell Fund Fellowships, 2462
Japan Foundation Center for Global Partnership Grants, 2472
Japan Foundation Center for Global Partnership NPO Fellowships, 2474
Japan Foundation Center for Global Partnership Security Studies Fellowships, 2475
Japan Foundation New York Japanese Studies Fellowships, 2478
Jefferson Science Fellows Program, 2480
Jeffrey Campbell Grad Fellow in Anthropology, 2482
Jewish Social Entrepreneurs Fellowships, 2487
JILA Visiting Fellowships and Postdoctoral Research Associateships, 2489
JMO Law and Legal System Grants, 2491
JMO Public Policy Research Grants, 2492
JMO Strategic and International Studies Grants, 2493
Joel L. Fleishman Civil Society Fellowships, 2495
John B. Bene Fellowships in Social Forestry, 2496
John Dana Archbold Fellowships, 2499
John Deere Foundation Grants, 2500
John Merck Scholars Awards, 2502
John Simon Guggenheim Memorial Fellowships, 2507
Johnson & Johnson Corporate Contributions, 2508
John Stauffer Charitable Trust Grants, 2509
Joseph C. Wey Fellowship in Medieval Studies, 2514
Joseph L. Fisher Dissertation Awards, 2515
JSA JLab Graduate Fellowships, 2521
JSPS Bilateral Exchange Programs with Foreign Academic Institutions, 2524
JSPS Fellowships for Foreign Scientists, 2525
JSPS Postdoc Fellowship for Foreign Researchers, 2527
Judith Graham Pool Postdoctoral Research Fellowship in Hemophilia, 2529
Juvenile Justice Clearing House Grants, 2530
Karst Research Fellowships and Grants, 2531
Kate Neal Kinley Memorial Fellowship, 2532
Kellogg Institute for International Studies Experiencing The World Fellowships, 2539
Kellogg Institute for International Studies Graduate Research Grants, 2542
Kellogg Institute for International Studies Quechua Summer Language Fellowships, 2548
Kellogg Institute for International Studies Supplemental Awards for PhD Students, 2550
Kellogg Institute for International Studies Visiting Fellowships, 2551
Kennedy Center Fellowship at the DeVos Institute of Arts Management, 2552
Kennedy Center Summer Fellowships for International Arts Managers, 2553
Kiplinger Program in Public Affairs Journalism Fellowships, 2557
Kluge Center David B. Larson Fellowship in Health and Spirituality, 2559
Kluge Center Fellowships, 2561
Kluge Center Kislak Fellows in American Studies, 2562
Kluge Center Kislak Short-Term Fellowships in American Studies, 2563
Kluge Center Kissinger Chair in Foreign Policy and International Relations, 2564
Kobe College Corporation Graduate Fellowships, 2567
Kosciuszko Found Fellowships for Polish Citizens, 2570
Kosciuszko Foundation Graduate Study and Research in Poland Scholarships, 2571
Kress Foundation Art History Pre-Doctoral Fellowships, 2576
L'Oreal USA Women in Science Fellowship, 2577
Lady Davis Fellowships, 2578
Lady Gregory Postgraduate Fellowship, 2579
Lalor Foundation Postdoctoral Fellowships, 2580
LAM Foundation Postoctoral Fellowships, 2583
Law and Public Affairs Fellowships at Princeton University, 2588
Law and Social Science Doctoral Dissertation Fellowships, 2589
Lawrence C. Gallen Teaching Fellowship, 2590
LBI-L Leo Baeck/Studienstiftung Fellowships, 2592
LBI-NY/DAAD Fellowships in Germany, 2593
LBI-NY/DAAD Fellowships in New York, 2594
LBI-NY David Baumgardt Memorial Fellowships, 2595
LBI-NY Fred Grubel Fellowship, 2596
LBI-NY Fritz Halbers Fellowships, 2597
LBI-NY Gerald Westheimer Career Development Fellowship, 2598
Library Company of Philadelphia and The Historical Society of Pennsylvania Visiting Research Fellowships, 2605
Library Company of Philadelphia Barra Foundation International Research Fellowships, 2606
Library Company of Philadelphia NEH Long-term Post-Doctoral Fellowships, 2607
Library Company of Philadelphia Program in Early American Economy and Society Dissertation-Level Fellowships, 2609
Library Company of Philadelphia Early American Economy & Society One-Month Fellowships, 2610
Library Company of Philadelphia Program in Early American Economy and Society Post-Doctoral Research Fellowships, 2611
Library of Congress Fellowships in Int'l Studies, 2612
Lillian Sholtis Brunner Summer Fellowships for Historical Research in Nursing, 2615
Lilly Endowment Teacher Creativity Fellowships, 2616
Link Foundation Energy Doctoral Fellowships, 2617
Loeb Fellowships in Advanced Environmental Design Studies, 2621
Luther I. Replogle Foundation Grants, 2631
Lymphatic Education & Research Network Postdoctoral Fellowships, 2633
Lynde & Harry Bradley Found Fellowships, 2634
Lynde and Harry Bradley Foundation Grants, 2635
MAP International Medical Fellowships, 2655
Margaret Yardley Fellowship, 2660
Marine Corps Historical Center Beeler-Raider Fellowship, 2662
Mary Baker Eddy Library Summer Research Fellowships, 2669
Mary Ingraham Bunting Inst Peace Fellowship, 2671
Mary Isabel Sibley Fellowship, 2672
Mary K. Chapman Foundation Grants, 2673
Massachusetts Bar Foundation Legal Intern Fellowships, 2676
Mayo Clinic Administrative Fellowship, 2689
Mayo Clinic Administrative Fellow Eau Claire, 2690
Mayo Clinic Business Consulting Fellowship, 2691
MBL Albert and Ellen Grass Fellowships, 2692
MBL Ann E. Kammer Summer Fellowship, 2693
MBL Associates Summer Fellowships, 2694
MBL Baxter Postdoctoral Summer Fellowship, 2695
MBL Burr and Susie Steinbach Fellowship, 2696
MBL E.E. Just Summer Fellowship for Minority Scientists, 2697
MBL Erik B. Fries Summer Fellowships, 2698
MBL Eugene and Millicent Bell Fellowships, 2699
MBL Evelyn and Melvin Spiegal Fellowship, 2700
MBL Frank R. Lillie Summer Fellowship, 2701
MBL Frederik B. and Betsy G. Bang Summer Fellowships, 2702
MBL Fred Karush Library Readership, 2703
MBL Gruss Lipper Family Foundation Summer Fellowship, 2704
MBL H. Keffer Hartline and Edward F. MacNichol, Jr. Fellowships, 2705
MBL Herbert W. Rand Summer Fellowship, 2706
MBL James E. and Faith Miller Memorial Summer Fellowship, 2707
MBL John M. Arnold Award, 2708
MBL Laura and Arthur Colwin Fellowships, 2709
MBL Lucy B. Lemann Memorial Fellowship, 2710
MBL M.G.F. Fuortes Summer Fellowships, 2711
MBL Nikon Summer Fellowship, 2712
MBL Plum Found John E. Dowling Fellowships, 2713
MBL Robert Day Allen Summer Fellowship, 2714
MBL Stephen W. Kuffler Summer Fellowships, 2715
MBL William Townsend Porter Summer Fellowships for Minority Investigators, 2716
MCEAS Dissertation Fellowships, 2723
McKnight Doctoral Fellowships, 2726
Mead Johnson Nutritionals Scholarships, 2735
MedImmune Charitable Grants, 2737
Medtronic Foundation Fellowships, 2738
Mellon Post-Doctoral Fellowship in Philosophy, 2739
Mellon Post-Doctoral Fellowship in Sustainable Agriculture, 2740
Mericos Foundation Grants, 2742
MfA Newton Fellowships, 2749
MfA Newton Master Teacher Fellowships, 2750
MGFA Post-Doctoral Research Fellowships, 2751
MGFA Student Fellowships, 2752
MHS Long Term Research Fellowships, 2755
MHS New England Regional Fellowships, 2756
MHS Short Term Research Fellowships, 2757
MHS Suzanne & Caleb Loring Res Fellowship, 2758
MHS Teacher Fellowships, 2759
Microsoft Research New Faculty Fellowship, 2771
Minerva Fellowships, 2778
Minerva Short-Term Research Grants, 2779
Minority Faculty Fellowship Program, 2781
MIT Knight Science Journalism Fellowships, 2782
MLA Cunningham Memorial International Fellowship, 2783
MLA David A. Kronick Traveling Fellowship, 2784
MLA Donald A.B. Lindberg Res Fellowship, 2785
MPI for Social Anthropology Max Planck Fellow Research Group Law, Organizations, Science, and Technology, 2799
Munby Fellowship in Bibliography, 2804
Murphy Institute Center for Ethics and Public Affairs Faculty Fellowships, 2805

Murphy Institute Center for Ethics and Public Affairs Graduate Fellowships, 2806
NACC David Stevenson Fellowships, 2812
NACC William Diaz Fellowships, 2813
NAF Fellowships, 2816
NAMM Foundation Music Research Grants, 2825
Nanyang Technological University Lee Kuan Yew Postdoctoral Fellowship, 2826
NAPM Doctoral Dissertation Grant, 2827
NASA Goddard Visiting Fellowships, 2838
NASA Graduate Student Researchers Program Fellowships, 2839
NASA Lewis Educational Research and Collaborative Internships, 2840
National Academy of Education Spencer Postdoctoral Fellowships, 2845
National Center for Responsible Gaming Postdoctoral Fellowships, 2856
National Center on Philanthropy and the Law Fellowship in Nonprofit Law, 2859
National Education Reporting Fellowships, 2861
National Endowment for Democracy Reagan-Fascell Democracy Fellowships, 2863
National Endowment for the Arts Jazz Masters Fellowships, 2864
National Endowment for the Arts Literature Fellowships: Creative Writing, 2865
National Endowment for the Humanities Fellowships at Winterthur, 2867
National Gallery of Art Kress Inter Fellowship, 2869
National Headache Foundation Seymour Diamond Clinical Fellowship in Headache Education, 2876
National MPS Society Grants, 2880
National Security Education Program David L. Boren Graduate Fellowships, 2884
National Sleep Foundation Pickwick Postdoctoral Fellowship, 2886
National Urban Fellows Program, 2888
National Wildlife Campus Ecology Fellowships, 2890
NCAR Graduate Research Fellowships, 2901
NCAR Postdoctoral Fellowships in the Atmospheric Sciences, 2902
NCCAM Ruth L. Kirschstein National Research Service Awards for Postdoctoral Training in Complementary and Alternative Medicine, 2904
NCHS National Center for Health Statistics Postdoctoral Research Appointments, 2910
NCIC Terry Fox Postdoctoral Res Fellowships, 2923
NEA National Heritage Fellowships in the Folk and Traditional Arts, 2944
NEH Documenting Endangered Lang Grants, 2955
NEH Fellowships, 2956
NEH Fellowships at Independent Research Institutions, 2957
NEH Fellowships for Advanced Social Science Research on Japan, 2958
Nevada Arts Council Artists Fellowships, 2984
New Hampshire Charitable Foundation Grants, 2985
NFID Colin L. Powell Minority Postdoctoral Fellowship in Tropical Disease Research, 2987
NFID John P. Utz Medical Mycology Postdoctoral Fellowship, 2988
NFID Nosocomial Infection Research and Training Postdoctoral Fellowship, 2990
NFID Postdoctoral Fellowships in Emerging Infectious Diseases, 2991
NFX Summer Research Fellowships, 3011
NHLBI Ruth L. Kirschstein National Research Service Awards for Individual Predoctoral Fellowships to Promote Diversity in Health-Related Research, 3058
NHLBI Ruth L. Kirschstein National Research Service Awards for Individual Predoctoral MD/PhD Fellows and Other Dual Degree Fellows, 3059
NHLBI Ruth L. Kirschstein National Research Service Awards for Individual Senior Fellows, 3060
NHPRC Historical Editing Fellowships, 3074
NIAF Fellowship Grants, 3095
NIGMS Ruth L. Kirschstein National Research Service Awards for Individual Predoctoral Fellows in PharmD/PhD Grants, 3200
NIH Ruth L. Kirschstein National Research Service Awards for Individual Senior Fellows, 3238
NIH Ruth L. Kirschstein National Research Service Awards for Individual Postdoctoral Fellows, 3239
NIH Ruth L. Kirschstein National Research Service Awards for Individual Predoctoral Fellows, 3240
NIH Ruth L. Kirschstein National Research Service Awards for Individual Predoctoral Fellowships to Promote Diversity in Health-Related Research, 3241
NIH Ruth L. Kirschstein National Research Service Awards for Individual Predoctoral MD/PhD and Other Dual Doctoral Degree Fellows, 3242
NINR Ruth L. Kirschstein National Research Service Award for Individual Predoctoral Fellows in Nursing Research, 3267
NKF Research Fellowships, 3277
NL American Society for Eighteenth-Century Studies Fellowships, 3279
NL Arthur and Lila Weinberg Fellowships for Independent Scholars and Researchers, 3280
NL Associated Colleges of the Midwest/Great Lakes College Association Faculty Fellowships, 3281
NL Audrey Lumsden-Kouvel Fellowships, 3282
NL British Academy Fellowship for Study in the United Kingdom, 3283
NL Frances C. Allen Fellowships, 3284
NL Herzog August Bibliothek Wolfenbuttel Fellowship, 3286
NL Lawrence Lipking Fellowship at Newberry, 3287
NL Lester J. Cappon Fellowship in Documentary Editing, 3288
NL Lloyd Lewis Fellows in American History, 3289
NL Mellon Postdoctoral Research Fellowships, 3294
NL Midwest Modern Lang Assoc Fellowship, 3299
NL Monticello College Foundation Fellowship for Women, 3304
NL National Endowment for the Humanities Fellowships, 3310
NL Newberry Consortium on American Indian Studies Faculty Fellowships, 3311
NL Newberry Consortium on American Indian Studies Graduate Student Fellowships, 3312
NL Newberry Library/Ecole Nationale des Chartes Exchange Fellowship, 3313
NL Northeast Modern Language Assocation Fellowship, 3314
NL Short-Term Fellowships in the History of Cartography, 3316
NL Short-Term Resident Fellowships for Individual Research, 3317
NL South Central Modern Languages Association Fellowship, 3318
NL Susan Kelly Power and Helen Hornbeck Tanner Fellowship, 3319
NMF General Electric Med Scholars Fellowships, 3322
NOAA Coastal Management Fellowships, 3352
NOAA Living Marine Resources Cooperative Science Center Internships/Fellowships, 3380
NOAA National Sea Grant Industry Fellowships, 3393
NOAA National Sea Grant John A. Knauss Marine Policy Fellowships, 3394
Norcliffe Foundation Grants, 3415
North Carolina Arts Council Artist Fellowships, 3417
North Carolina Biotechnology Center Collaborative Funding Grants, 3421
North Dakota Council on the Arts Individual Artist Fellowships, 3426
Northwood University/Alden B. Dow Creativity Center Summer Residency Program, 3432
NPSC Grad Fellowships in the Physical Sciences, 3436
NSERC E.W.R. Steacie Memorial Fellowships, 3439
NSERC Howard Alper Postdoctoral Prize, 3441
NSERC Industrial R&D Fellowships, 3442
NSERC Northern Research Internships Program, 3447
NSERC Postdoctoral Fellowships Program, 3448
NSERC Synergy Awards for Innovation, 3450
NSERC Undergraduate Research Awards, 3451
NSF/NEH Documenting Endangered Languages Project Grants and Fellowships: data, infrastructure and computational methods, 3455
NSF Astronomy and Astrophysics Postdoctoral Fellowships, 3494
NSF Grant Opportunities for Academic Liaison with Industry, 3574
NSF Linguistics Grants, 3587
NSF Postdoctoral Res Fellowships in Biology, 3630
NSF Science and Society (S&S), 3643
NSF Starter Grants for Postdoctoral Fellows in Microbial Biology, 3653
Nuffield Foundation Law and Society Grants, 3667
NWHF Mark O. Hatfield Research Fellowship, 3671
NYAM Audrey & William Helfand Fellowship, 3679
NYAM Charles A. Elsberg Fellowship, 3680
NYAM David E. Rogers Fellowships, 3681
NYAM Edwin Beer Research Fellowship, 3683
NYAM Ferdinand C. Valentine Fellowship, 3684
NYAM Glorney-Raisbeck Fellowships, 3687
NYAM Mary and David Hoar Fellowship, 3691
NYAM Paul Klemperer Fellowship, 3692
NYFA Artist Fellowships, 3698
NYFA Shelly Pinz Fellowship, 3699
NYPL Short-Term Research Fellowships, 3701
Oklahoma City Community Grants, 3737
Olin Corporation Charitable Trust Grants, 3739
Olin Fellowships, 3740
ONR DOD National Defense Science & Engineering Graduate Fellowships, 3755
ONR Historically Black Colleges and Universities Future Engineering Faculty Fellowships, 3763
ONR Navy Postdoctoral Fellowships, 3776
Orchid Research Fellowships, 3794
Organization for Tropical Studies Fellowships, 3796
Organization for Tropical Studies Pilot Awards for Exploratory Research, 3797
ORISE Civilian Radioactive Waste Management Fellowship Program, 3799
ORISE Fusion Energy Postdoctoral Research, 3806
ORISE Fusion Energy Sciences Fellowship, 3807
ORISE National Oceanic and Atmospheric Administration Student Res Participation, 3812
ORISE Nuclear Regulatory Commission Historically Black Colleges and Universities Student Research Participation, 3814
ORISE Postgraduate Environmental Management Research Participation at the US Army Environmental Center, 3816
ORISE Postgraduate Research Participation at the Center for Biologics Evaluation and Research, 3819
ORISE Postgraduate Research Participation at the Centers for Disease Control and Prevention, 3820
ORISE Postgraduate Research Participation at the National Exposure Research Laboratory, 3821
ORISE Savannah River Site Postgraduate Research Program, 3824
ORISE Student Environmental Management Participation Program for the US Army Environmental Center, 3826
ORISE Student Research Participation at the Agency for Toxic Substances and Disease Registry, 3828
ORISE Student Research Participation at the Centers for Disease Control and Prevention, 3829
ORISE Student Research Participation at the National Center for Toxicological Research, 3830
OSF-Baltimore Community Fellowships, 3833
OSF Central Asia and Caucasus Research and Training Initiative - Junior Fellowships, 3836
OSF Central Asia and Caucasus Research and Training Initiative - Senior Fellowships, 3837
OSF Civil Service Awards, 3838
OSF Doctoral Fellowships Georgia & Kyrgyzstan, 3839
OSF Fellowships, 3840
OSF Global Faculty Grants - Track 1, 3841
OSF Justice Initiative Fellowships, 3847
OSF Palestinian Rule of Law Fellowships, 3848
OSF Social Work Fellowships, 3849
OSF Soros Justice Advocacy Fellowships, 3850
OSF Soros Justice Media Fellowships, 3851
OSF Think Tank Fellowships, 3852
OSF Thomas J. Alexander Fellowships, 3853
OTS Francis J. Bossuyt Fellowship, 3855

OTS Hovore-Horn Fellowships, 3856
OTS Stanley Smith Horticultural Fellowships, 3857
Packard Foundation Fellowships for Science and Engineering, 3862
Padilla Bay National Estuarine Research Reserve Graduate Research Fellowships, 3864
Padilla Bay National Estuarine Research Reserve Research Assistantships, 3865
Pancreatic Cancer Action Network Fellowship, 3866
Pancreatic Cancer Action Network-AACR Pathway to Leadership Grants, 3867
Park Foundation Grants, 3870
Patrick and Anna M. Cudahy Fund Grants, 3873
Paul & Daisy Soros Fellows for New Americans, 3875
PC Fred H. Bixby Fellowships, 3880
PDF-AANF Clinician-Scientist Devel Awards, 3881
PDF Postdoctoral Fellowships for Basic Scientists, 3883
PDF Postdoctoral Fellowships for Neurologists, 3884
PDF Summer Fellowships, 3885
Pediatric Brain Tumor Found Research Grants, 3888
Penn Graduate School of Education Dean's Fellowships, 3891
Penn Graduate School of Education Foreign Language Area Studies Fellowship, 3892
Penn Graduate School of Education Pre-Doctoral Training in Interdisciplinary Methods for Field-Based Research in Education, 3893
Penny Severns Breast, Cervical and Ovarian Cancer Research Grants, 3896
Pew Charitable Trusts Biomedical Res Grants, 3904
Pew Fellowships in the Arts, 3905
Pew Scholars in the Biomedical Sciences, 3906
PHF Andrew W. Mellon Postdoctoral Fellowships in the Humanities, 3931
PHF Penn Faculty Research Fellowships, 3932
PHF Regional Mellon Faculty Research and Cultural Fellowships, 3933
PHF Undergraduate Humanities Forum Research Fellowships, 3934
Phi Kappa Phi Graduate Fellowships, 3938
Phi Upsilon Omicron Candle Fellowships, 3941
Phi Upsilon Omicron Diamond Anniversary Fellowships, 3942
Phi Upsilon Omicron Founders Fellowship, 3944
Phi Upsilon Omicron Geraldine Clewell Doctoral Fellowship, 3945
Phi Upsilon Omicron Geraldine Clewell Masters Fellowship, 3946
Phi Upsilon Omicron Jean Dearth Dickerscheid Fellowship, 3947
Phi Upsilon Omicron Margaret Drew Alpha Fellowship, 3948
Phi Upsilon Omicron Presidents Res Fellowship, 3949
PhRMA Foundation Health Outcomes Post Doctoral Fellowships, 3950
PhRMA Foundation Health Outcomes Sabbatical Fellowships, 3952
PhRMA Foundation Informatics Post Doctoral Fellowships, 3953
PhRMA Foundation Informatics Pre Doctoral Fellowships, 3954
PhRMA Foundation Informatics Sabbatical Fellowships, 3956
PhRMA Foundation Paul Calabresi Medical Student Research Fellowship, 3957
PhRMA Foundation Pharmaceutics Postdoctoral Fellowships, 3958
PhRMA Foundation Pharmaceutics Pre Doctoral Fellowships, 3959
PhRMA Foundation Pharmaceutics Sabbatical Fellowships, 3961
PhRMA Foundation Pharmacology/Toxicology Post Doctoral Fellowships, 3962
PhRMA Foundation Pharmacology/Toxicology Pre Doctoral Fellowships, 3963
PhRMA Foundation Pharmacology/Toxicology Sabbatical Fellowships, 3965
Pioneer Hi-Bred Plant Breeding Fellowships, 3968
Pioneer Hi-Bred International Scholarships, 3969
Pioneer Hi-Bred Society Fellowships, 3970
Pioneer Hi-Bred University Fellowships, 3971
Plastic Surgery Educational Foundation Research Fellowship Grants, 3979
Pollock-Krasner Foundation Grants, 3981
Postdoctoral Research Fellowships in Gerontological Public Health, 3983
Potts Memorial Foundation Grants, 3985
Price Family Charitable Fund Grants, 3987
Princess Grace Foundation Theater, Dance, and Film Grants, 3988
Princeton University Center for Human Values Graduate Prize Fellowships, 3989
Princeton University Center for Human Values Harold T. Shapiro Postdoctoral Research Associate in Bioethics, 3991
Princeton University Center for Human Values Laurance S. Rockefeller Visiting Fellowships, 3992
Princeton University Center for Human Values Postdoctoral Research Associate in Values and Public Policy, 3994
Princeton University Lewis Center for the Arts Hodder Fellowships, 3996
Princeton Univ Whiting Postdoc Fellowships, 3998
Province of Alberta Queen Elizabeth II Graduate Scholarships, 4001
PSG Mentored Clinical Research Awards, 4003
PVA Research Foundation Grants, 4004
R.C. Baker Foundation Grants, 4006
RACGP/CONROD Research Fellowship, 4009
RACGP Rex Walpole Travelling Fellowship, 4016
Radcliffe Institute Carol K. Pforzheimer Student Fellowships, 4018
Radcliffe Inst Fellowships for Creative Artists, 4020
Radcliffe Institute Indiv Residential Fellowships, 4021
Radcliffe Institute Residential Group Fellowships, 4024
Rajiv Gandhi Foundation Grants, 4025
RAND Corporation Fellowships in Population Studies and the Study of Aging, 4029
RAND Corporation Graduate Student Summer Associateships, 4030
RAND Corporation Military Fellowships, 4031
RAND Corp Transatlantic Postdoctoral Fellowships for International Relations and Security, 4032
Rathmann Family Foundation Grants, 4033
RCPSC International Traveling Fellowship, 4037
RCPSC Medical Educ Traveling Fellowships, 4039
RCPSC Royal College Fellowship for Studies in Medical Education, 4040
Regenstrief General Internal Medicine - General Pediatrics Research Fellowship Program, 4045
Regenstrief Geriatrics Fellowship, 4046
Regenstrief Medical Informatics NLM Research Fellowships, 4048
Rett Syndrome Research Foundation Grants, 4058
Rhode Island Foundation Grants, 4061
Richard Carley Hunt Memorial Postdoctoral Fellowships, 4065
Robert A. Levy Fellowship in Law and Liberty, 4070
Robert & Patricia Switzer Found Fellowships, 4071
Roberto Longhi Foundation Fellowships, 4078
Robert W. Deutsch Foundation Grants, 4082
Rockefeller Archive Center Visiting Archivist Fellowship, 4087
Rockefeller Brothers Fellowship in Nonprofit Law, 4088
Rockefeller Foundation Humanities Fellowships, 4089
Rockefeller Foundation Scholarly Residencies, 4091
Rockford College Faculty Fellowships, 4092
Rosalynn Carter Institute John and Betty Pope Fellowships, 4103
Rosenthal Writer-in-Residence Program, 4105
Roy J. Carver Charitable Trust Higher Education Grants, 4113
Roy J. Carver Charitable Trust Medical and Science Research Grants, 4114
Rumsfeld Foundation Graduate Fellowships, 4127
RWJF Clinical Scholars Grants, 4136
RWJF Harold Amos Medical Faculty Development Research Grants, 4139
RWJF Harold Amos Medical Faculty Development Scholars, 4140
RWJF Health and Society Scholars, 4141
RWJF Health Policy Fellowships, 4142
RWJF Investigator Awards in Health Policy Res, 4145
RWJF New Connections Jun Investigator Grants, 4146
RWJF New Connections Mid-Career Consultants Grants, 4147
RWJF Nurse Faculty Scholars, 4148
Saint Mary's College Fellowships, 4159
Sara Elizabeth O'Brien Trust Grants, 4166
SASS Birgit Baldwin Fellowship, 4182
Savoy Foundation Post-Doctoral and Clinical Research Fellowships, 4184
Schomburg Scholars-in-Residence Fellowships, 4187
Sexually Transmitted Diseases Postdoctoral Fellowships, 4200
Shastri Indo-Canadian Institute Arts Fellowships, 4219
Shastri Indo-Canadian Institute Doctoral Student Research Fellowships, 4221
Shastri Indo-Canadian Institute Faculty Research Fellowships, 4222
Shastri Indo-Canadian Institute Faculty Training Fellowships, 4223
Shastri Indo-Canadian Institute Graduate Degree Fellowships from an Indian Institution, 4224
Shastri Indo-Canadian Ins Librarian Fellowships, 4225
Shastri Indo-Canadian Institute Master's Student Research Fellowships, 4226
Shastri Indo-Canadian Institute Post-Doctoral Research Fellowships, 4229
Shastri Indo-Canadian Institute Student Language Training Fellowships, 4231
Shell Oil Company Foundation Grants, 4233
Shorenstein Center Fellowships, 4235
SHOT Brooke Hindle Postdoctoral Fellowships, 4238
SHOT Melvin Kranzberg Dissert Fellowship, 4245
SHOT NASA Fellowship, 4246
Siragusa Foundation Health Services & Medical Research Grants, 4264
Smithsonian Biodiversity Genomics and Bioinformatics Postdoctoral Fellowships, 4272
Smithsonian Environmental Research Center Graduate Student Fellowships, 4273
Smithsonian Environmental Research Center Postdoctoral/Senior Fellowships, 4275
Smithsonian Environmental Research Center Predoctoral Fellowships, 4276
Smithsonian Graduate Student Fellowships, 4277
Smithsonian Latino Studies Fellowships, 4278
Smithsonian Molecular Evolution Fellowships, 4280
Smithsonian Museum Conservation Institute Research Post-Doctorate/Post-Graduate Fellowships, 4283
Smithsonian National Air and Space Museum Charles A. Lindbergh Chair in Aerospace History, 4285
Smithsonian National Air and Space Museum Guggenheim Fellowships, 4286
Smithsonian National Air and Space Museum Postdoctoral Earth and Planetary Sciences Fellowship, 4287
Smithsonian National Air and Space Museum Verville Fellowships, 4288
Smithsonian National Museum of American History Lemelson Center Fellowships, 4289
Smithsonian Native American Community Scholars Awards, 4291
Smithsonian Postdoctoral Fellowships, 4293
Smithsonian Predoctoral Fellowships, 4295
Smithsonian Senior Fellowships, 4296
Smithsonian Tropical Research Institute Short Term Fellowships, 4297
SNM Student Fellowship Awards, 4303
Society for Research in Child Development Executive Branch Policy Fellowships, 4310
Special Olympics E Kennedy Shriver Fellowships, 4331
Spencer Foundation Education Dissertation Fellowships, 4334
Spencer Foundation Postdoctoral Fellowships, 4336
SPJ Eugene C. Pulliam Fellowship for Editorial Writing, 4346
SRC AMD/Mahboob Khan Fellowship, 4348
SRC Company-Named Fellowships, 4350

SRC Educational Science Postdoc Fellowships, 4363
SRC Education Science Grants for Postdoctoral Positions in Sweden, 4366
SRC Franco Cerrina Fellowship, 4367
SRC Graduate Fellowships, 4368
SRC Hans J. Coufal Fellowship, 4371
SRC Humanities and Social Science Postdoctoral Fellowships, 4372
SRC International Fellowship, 4375
SRC Medical Research Grants, 4379
SRC Medical Science Grants for Postdoctoral Positions in Sweden, 4381
SRC Medical Science Postdoctoral Fellowships, 4382
SRC Natural and Engineering Science Grants for Postdoctoral Positions in Sweden, 4384
SRC Natural and Engineering Science Postdoctoral Fellowships, 4385
SRC Peter Verhofstadt Fellowship, 4389
SRC Robert M. Burger Fellowship, 4392
SRC Simon Karecki Fellowship, 4393
SSHRC Bora Laskin National Fellowship in Human Rights Research, 4405
SSHRC Doctoral Fellowships, 4407
SSHRC Jules and Gabrielle Leger Fellowship, 4413
SSHRC Postdoctoral Fellowships, 4415
SSHRC Sport Participation Research Initiative: Doctoral Award Supplements, 4416
SSHRC Sport Participation Research Initiative: Postdoctoral Fellowship Supplements, 4417
SSHRC Therese F. Casgrain Fellowship, 4418
SSRC-Van Alen Fellowships, 4419
SSRC Abe Fellowship for Journalists, 4420
SSRC Abe Fellowships, 4421
SSRC Dissertation Proposal Development Faculty Field Fellowships, 4424
SSRC Drugs, Security & Democracy Fellowships, 4426
SSRC Eurasia Program Dissertation Development Awards, 4427
SSRC Eurasia Program Pre-Dissertation Awards, 4428
SSRC Int'l Dissertation Research Fellowships, 4429
SSRC Japan Society for the Promotion of Science Fellowships, 4431
SSRC Mellon Mays Predoctoral Research Grants, 4432
SSRC New Directions in the Study of Prayer Research Grants, 4433
SSRC Next Generation Social Sciences in Africa Doc Dissertation Completion Fellowships, 4434
SSRC Next Generation Social Sciences in Africa: Doctoral Dissertation Proposal Fellowships, 4435
SSRC Next Generation Social Sciences in Africa: Doctoral Dissertation Research Fellowships, 4436
SSRC Postdoc Fellowship for Transregional Research: Inter-Asian Contexts and Connections, 4438
Stanford Humanities Center Arts Writer/Practitioner Fellowship, 4444
Stanford Humanities Center External Faculty Fellowships, 4445
Stroke Association Clinical Fellowships, 4455
Surdna Foundation Arts Teachers Fellowships, 4470
Swann Foundation Fellowships, 4478
Swedish Natural Science Res Council Grants, 4479
Swedish Women's Educational Association International Scholarships, 4480
Swiss National Science Foundation Fellowships for Advanced Researchers, 4483
Swiss National Science Foundation Fellowships for Prospective Researchers, 4484
Syracuse University Graduate Fellowships, 4488
Takemi Program in International Health Research and Advanced Training Fellowships, 4491
Tanner Humanities Center Research Fellowships, 4494
Tau Beta Pi Association Fellowships, 4495
Teachers College Minority Postdoc Fellowships, 4498
Temple Univ George D. McDowell Fellowship, 4499
Third Millennium Foundation Fellowships, 4504
Thomson Reuters / MLA Doctoral Fellowships, 4512
Tourette Syndrome Association Post-Doctoral Fellowships, 4523
TRDRP Postdoctoral Fellowship Awards, 4530
U.S. Department of Education Foreign Languages and Area Studies Fellowships, 4539
U.S. Department of Education Graduate Assistance in Areas of National Need, 4540
U.S. Department of Education Jacob K. Javits Fellowships, 4541
UCLA / RAND Corporation Post-Doctoral Fellowships, 4548
UCR Grad Student in the Humanities Fellowship, 4550
UCR Resident Fellowships, 4552
UICC American Cancer Society International Fellowships for Beginning Investigators, 4555
UICC International Cancer Technology Transfer Fellowships, 4556
UIC Rockefeller Foundation Resident Fellowships in the Humanities, 4559
UNC Carolina Postdoctoral Program for Faculty Diversity, 4560
UNCFSP NASA Administrator's Fellowship, 4562
UNCFSP NASA Harriett G. Jenkins Predoctoral Fellowships, 4563
UNC School of Information and Library Science Doctoral Fellowships, 4565
UNESCO Co-Sponsored Fellowships, 4566
UNESCO Fellowships, 4567
United Methodist Church Josephine Forman Scholarship, 4573
United Methodist Communications Stoody-West Fellowship, 4578
United States Institute of Peace - Jennings Randolph Senior Fellowships, 4580
United States Institute of Peace - Transatlantic Post-Doc Fellowship for International Relations and Security, 4581
University of British Columbia Graduate Student Fellowships, 4584
University of Colorado Doctoral Dissertation Fellowship in Media, Religion, and Culture, 4586
University of Georgia Counseling and Psychological Services Postdoctoral Fellowship, 4587
University of London Visiting Fellowships, 4588
University of Southern Mississippi H.A. & Margret Rey Translation Fellowships, 4589
University of Southern Mississippi Keats/Domanska/de Grummond Children's Literature Research Fellowships, 4590
USC Annenberg/Getty Arts Journ Fellowships, 4603
USDA Cochran Fellowships, 4619
USDA Food and Agricultural Sciences National Needs Graduate Fellowship Grants, 4624
USDA Norman E. Borlaug International Agricultural Science and Technology Fellowships, 4636
USDC/NOAA National Marine Fisheries Service Sea Grant Joint Graduate Fellowships, 4660
USDD Postdoctoral Fellowship Awards in Ocean Acoustics, 4774
USDD Sabbatical Leave Fellowships, 4788
Van Alen Institute New York Prize Fellowship, 4806
Virginia Foundation for the Humanities Residential Fellowships, 4815
Virginia Historical Society Research Fellowships, 4816
W.M. Keck Foundation Med Research Grants, 4823
Walter J. Jensen Fellowship for French Language, Literature, and Culture, 4830
WAWH Founders Dissertation Fellowship, 4833
Weizmann Institute of Science Postdoctoral Fellowships, 4837
Wenner-Gren Foundation African Fellowships, 4846
Wenner-Gren Foundation Hunt Postdoctoral Fellowships, 4850
Wenner-Gren Found Short-Term Fellowships, 4853
Wenner-Gren Foundation Wadsworth International Fellowships, 4854
Weyerhaeuser Company Foundation Grants, 4857
WHOI Geophysical Fluid Dynamics Fellowships, 4859
WHOI Marine Policy Fellowships, 4861
WHOI Ocean Science Journalism Fellowship, 4862
WHOI Postdoctoral Fellowships, 4863
WHOI Postdoctoral Scholar Awards, 4864
WHOI Summer Student Fellowships, 4865
WHOI Undergrad Fellowships for Minorities, 4866
Willamette Univ Minority Grad Fellowship, 4869
William T. Grant Foundation Distinguished Fellows Awards, 4880
William T. Grant Foundation Scholars Program, 4882
Winterthur Museum Dissertation Fellowships, 4887
Winterthur Research Fellowships, 4888
Wolfsonian-FIU Fellowships, 4890
Woodrow Wilson Doctoral Dissertation Fellowships in Women's Studies, 4891
WWF Russell E. Train Education for Nature Alumni Grants, 4897
Xerox Foundation Grants, 4899
Yale University Beinecke Library Fellowships for Visiting Scholars, 4901
Yale University Beinecke Library Research Fellowships for Yale Grad & Professional School Students, 4902
Zonta International Foundation Amelia Earhart Fellowships, 4905

General Operating Support

Abbott Fund Global AIDS Care Grants, 219
Abbott Fund Science Education Grants, 220
A Charitable Foundation Grants, 273
Achelis Foundation Grants, 274
Acuity Charitable Foundation Grants, 333
Adam Richter Charitable Trust Grants, 339
AIHS Alberta/Pfizer Translational Res Grants, 435
Akonadi Foundation Anti-Racism Grants, 477
Alberta Law Foundation Grants, 503
Albuquerque Community Foundation Grants, 508
Alcatel-Lucent Technologies Foundation Grants, 509
Alcoa Foundation Grants, 511
Alcon Foundation Grants, 513
Alex Stern Family Foundation Grants, 542
Alfred P. Sloan Civic Initiatives Grants, 545
American Chemical Society Corporate Associates Seed Grants, 676
American Chemical Society Green Chem Grants, 679
American Foodservice Charitable Trust Grants, 710
American Schlafhorst Foundation Grants, 796
Amgen Foundation Grants, 812
AMI Semiconductors Corporate Grants, 813
Anderson Foundation Grants, 842
Ann Peppers Foundation Grants, 863
Anthony R. Abraham Foundation Grants, 869
Arcadia Foundation Grants, 983
Arca Foundation Grants, 984
Arie and Ida Crown Memorial Grants, 989
Arizona Public Service Corporate Giving Grants, 998
Arronson Foundation Grants, 1019
Arthur and Rochelle Belfer Foundation Grants, 1023
Arthur F. & Alice E. Adams Foundation Grants, 1024
Auburn Foundation Grants, 1263
Avon Products Foundation Grants, 1275
Ball Brothers Foundation General Grants, 1291
Ball Brothers Foundation Rapid Grants, 1292
Banfi Vintners Foundation Grants, 1296
Bank of America Charitable Foundation Matching Gifts, 1297
Batchelor Foundation Grants, 1305
Bay and Paul Foundations, Inc Grants, 1307
BCBSM Foundation Primary and Clinical Prevention Grants, 1315
Beim Foundation Grants, 1331
Berks County Community Foundation Grants, 1333
Bill and Melinda Gates Foundation Agricultural Development Grants, 1345
Bill and Melinda Gates Foundation Water, Sanitation and Hygiene Grants, 1347
Blowitz-Ridgeway Foundation Grants, 1349
Blum-Kovler Foundation Grants, 1351
BMW of North America Contributions, 1353
Bodman Foundation Grants, 1356
Boeing Company Contributions Grants, 1357
Boston Foundation Grants, 1361
Brainerd Foundation Grants, 1368
Brico Fund Grants, 1371
Bristol-Myers Squibb Foundation Health Disparities Grants, 1377

Bullitt Foundation Grants, 1384
Burlington Northern Santa Fe Found Grants, 1389
Burton G. Bettingen Grants, 1390
Bydale Foundation Grants, 1400
Byron W. & Alice L. Lockwood Found Grants, 1401
Cabot Corporation Foundation Grants, 1406
Caesars Foundation Grants, 1407
California Wellness Foundation Work and Health Grants, 1409
Callaway Golf Company Foundation Grants, 1410
Campbell Soup Foundation Grants, 1415
Cape Branch Foundation Grants, 1425
Carl and Eloise Pohlad Family Found Grants, 1426
Carl C. Icahn Foundation Grants, 1428
Carl Gellert and Celia Berta Gellert Foundation Grants, 1429
Carl M. Freeman Foundation Grants, 1430
Carnegie Corporation of New York Grants, 1432
Carpenter Foundation Grants, 1439
Carrier Corporation Contributions Grants, 1440
Castle Rock Foundation Grants, 1441
CDC Cooperative Agreement for Continuing Enhanced National Surveillance for Prion Diseases in the United States, 1456
Charles H. Farnsworth Trust Grants, 1521
Charles M. & Mary D. Grant Found Grants, 1527
Charles Stewart Mott Anti-Poverty Grants, 1528
Charles Stewart Mott Foundation Grants, 1529
Chatlos Foundation Grants, 1532
Chautauqua Region Community Found Grants, 1533
Chestnut Hill Charitable Foundation, Inc Grants, 1538
Chiles Foundation Grants, 1552
Christensen Fund Regional Grants, 1554
CIGNA Foundation Grants, 1568
Claneil Foundation Grants, 1574
Clarence E. Heller Foundation Grants, 1575
Cleveland-Cliffs Foundation Grants, 1580
CMA Foundation Grants, 1582
Coca-Cola Foundation Grants, 1593
Coeta and Donald Barker Foundation Grants, 1594
Coleman Foundation Cancer Care Grants, 1595
Colonel Stanley R. McNeil Foundation Grants, 1597
Community Found for Greater Buffalo Grants, 1615
Community Foundation for Greater New Haven Quinnipiac River Fund Grants, 1616
Community Foundation for S Arizona Grants, 1618
Comprehensive Health Education Found Grants, 1622
Compton Foundation Grants, 1623
ConocoPhillips Grants, 1629
Cord Foundation Grants, 1645
Corning Foundation Educational Grants, 1646
Coughlin-Saunders Foundation Grants, 1648
Cowles Charitable Trust Grants, 1652
Cystic Fibrosis Canada Senior Scientist Research Training Awards, 1680
D.F. Halton Foundation Grants, 1692
Dade Community Foundation Grants, 1712
DaimlerChrysler Corporation Fund Grants, 1714
David Bohnett Foundation Grants, 1724
Denton A. Cooley Foundation Grants, 1743
DHHS Oral Health Promotion Research Across the Lifespan, 1759
Donald and Sylvia Robinson Family Foundation Grants, 1830
Dorr Foundation Grants, 1842
Dr. Bob and Jean Smith Foundation Grants, 1853
DTE Energy Foundation Health and Human Services Grants, 1865
DTE Energy Found Higher Learning Grants, 1866
Duchossois Family Foundation Grants, 1867
Eaton Charitable Fund Grants, 1893
Edward W. and Stella C. Van Houten Grants, 1899
Edwin S. Webster Foundation Grants, 1900
Elisabeth Severance Prentiss Foundation Grants, 1905
EMBO Installation Grants, 1915
Emerson Charitable Trust Grants, 1918
Emma B. Howe Memorial Foundation Grants, 1920
Eugene M. Lang Foundation Grants, 1957
Eugene McDermott Foundation Grants, 1958
Evan Frankel Foundation Grants, 1961
Expect Miracles Foundation Grants, 1965
Fannie E. Rippel Foundation Grants, 1978
FAR Fund Grants, 1980
Fluor Foundation Grants, 2028
FMC Foundation Grants, 2029
Ford Motor Company Fund Grants, 2033
Frank Stanley Beveridge Foundation Grants, 2051
Fund for Folk Culture Artists Support Grants, 2088
G. Harold and Leila Y. Mathers Charitable Foundation Grants, 2092
General Motors Foundation Grants, 2107
Genuardi Family Foundation Grants, 2109
George A Ohl Jr. Foundation Grants, 2112
George B. Storer Foundation Grants, 2113
George Foundation Grants, 2115
George Frederick Jewett Foundation Grants, 2116
George Gund Foundation Grants, 2117
George S. & Dolores Dore Eccles Grants, 2120
George W. Codrington Foundation Grants, 2122
Gil and Dody Weaver Foundation Grants, 2139
Gilroy and Lillian P. Roberts Foundation Grants, 2142
Girl's Best Friend Foundation Grants, 2143
GNOF IMPACT Grants for Education, 2154
Greater St Louis Community Foundation Grants, 2164
Greater Worcester Community Foundation Jeppson Memorial Fund for Brookfield Grants, 2165
Greenspun Family Foundation Grants, 2173
H.A. and Mary K. Chapman Charitable Grants, 2193
H.J. Heinz Company Foundation Grants, 2195
Harold Simmons Foundation Grants, 2203
Harry B. and Jane H. Brock Foundation Grants, 2206
Harry Frank Guggenheim Foundation Research Grants, 2208
Harry W. Bass, Jr. Foundation Grants, 2210
Hawaii Community Foundation Health Education and Research Grants, 2221
Health Foundation of Southern Florida Responsive Grants, 2223
Helen Bader Foundation Grants, 2228
Helen V. Brach Foundation Grants, 2231
Henry W. Bull Foundation Grants, 2237
Herbert A. & Adrian W. Woods Found Grants, 2239
Herbert H. and Grace A. Dow Grants, 2241
Hilton Hotels Corporate Giving Grants, 2268
Hirtzel Memorial Foundation Grants, 2269
Hoglund Foundation Grants, 2271
Honor the Earth Grants, 2279
Huie-Dellmon Trust Grants, 2295
Human Source Foundation Grants, 2297
IDEM Section 319(h) Nonpoint Source Grants, 2338
Indiana 21st Century Research and Technology Fund Awards, 2370
Indiana SBIR/STTR Commercialization Enhancement Program, 2371
Indiana Waste Tire Fund Grants, 2376
Ittleson Foundation Environment Grants, 2437
Ittleson Foundation Mental Health Grants, 2438
J.B. Reynolds Foundation Grants, 2449
J.W. Kieckhefer Foundation Grants, 2452
Jackson Foundation Grants, 2453
Jacob and Hilda Blaustein Foundation Grants, 2454
James R. Dougherty Jr. Foundation Grants, 2463
Jessie Smith Noyes Foundation Grants, 2485
John Deere Foundation Grants, 2500
John P. McGovern Foundation Grants, 2504
Johnson & Johnson Corporate Contributions, 2508
Joseph Alexander Foundation Grants, 2513
JSPS Postdoc Fellowship for Foreign Researchers, 2527
Kenneth T. & Eileen L. Norris Found Grants, 2554
Kovler Family Foundation Grants, 2575
Leon Lowenstein Foundation Grants, 2603
Lillian S. Wells Foundation Grants, 2614
Lowe Foundation Grants, 2622
Lucile Horton Howe and Mitchell B. Howe Foundation Grants, 2624
Lumina Foundation for Education Grants, 2627
Lumpkin Family Found Healthy People Grants, 2629
Luther I. Replogle Foundation Grants, 2631
Lynde and Harry Bradley Foundation Grants, 2635
M. Bastian Family Foundation Grants, 2636
Mabel Y. Hughes Charitable Trust Grants, 2647
MacDonald-Peterson Foundation Grants, 2649
Macquarie Bank Foundation Grants, 2650
Marcia and Otto Koehler Foundation Grants, 2656
Margaret T. Morris Foundation Grants, 2659
Marion Gardner Jackson Charitable Trust Grants, 2664
Mary K. Chapman Foundation Grants, 2673
Matilda R. Wilson Fund Grants, 2684
Mattel International Grants, 2686
McCarthy Family Foundation Grants, 2717
McCune Charitable Foundation Grants, 2719
McCune Foundation Human Services Grants, 2722
McGregor Fund Grants, 2725
Melville Charitable Trust Grants, 2741
Mericos Foundation Grants, 2742
Meyer and Pepa Gold Family Foundation Grants, 2746
Meyer Memorial Trust Responsive Grants, 2748
MGN Family Foundation Grants, 2754
Miles of Hope Breast Cancer Foundation Grants, 2773
Miller Brewing Corporate Contributions Grants, 2774
Millipore Foundation Grants, 2775
Millstone Foundation Grants, 2776
Monfort Family Foundation Grants, 2794
Motorola Foundation Grants, 2796
Mt. Sinai Health Care Foundation Academic Medicine and Bioscience Grants, 2803
Musgrave Foundation Grants, 2807
NCI Technologies and Software to Support Integrative Cancer Biology Research Grants, 2933
Nellie Mae Education Foundation District-Level Change Grants, 2968
Nellie Mae Education Foundation Research and Development Grants, 2970
Nellie Mae Education Foundation State Level Systems Change Grants, 2971
NIH Biology, Development, and Progression of Malignant Prostate Disease Research Grants, 3206
Norcliffe Foundation Grants, 3415
Nordson Corporation Foundation Grants, 3416
North Carolina Arts Council Folklife Grants, 3419
Northwestern Mutual Foundation Grants, 3430
Northwest Fund for the Environment Grants, 3431
NSF Accelerating Discovery In Science and Engineering through PETASCALE Simulations and Analysis, 3457
NSF Arctic Research Opportunities, 3490
NSF Biological Research Collections, 3504
NSF Presidential Early Career Awards for Scientists and Engineers Grants, 3633
NSF Science and Society (S&S), 3643
NSF University Radio Observatories Grants, 3664
NYCT Blood Disease Research Grants, 3696
Oklahoma City Community Grants, 3737
Olin Corporation Charitable Trust Grants, 3739
Oppenstein Brothers Foundation Grants, 3792
Orchard Foundation Grants, 3793
Overbrook Foundation Grants, 3860
Park Foundation Grants, 3870
Patrick and Anna M. Cudahy Fund Grants, 3873
Percy B. Ferebee Endowment Grants, 3897
Perkin Fund Grants, 3898
Perl 6 Microgrants, 3899
Perpetual Trust for Charitable Giving Grants, 3900
Pinkerton Foundation Grants, 3967
Playboy Foundation Grants, 3980
Price Family Charitable Fund Grants, 3987
Prudential Foundation Education Grants, 4002
R.C. Baker Foundation Grants, 4006
Rathmann Family Foundation Grants, 4033
Rayonier Foundation Grants, 4034
Reinberger Foundation Grants, 4051
Rhode Island Foundation Grants, 4061
Richard and Rhoda Goldman Fund Grants, 4063
Richard & Susan Smith Family Found Grants, 4064
Rich Foundation Grants, 4068
Righteous Persons Foundation Grants, 4069
Robert B McMillen Foundation Grants, 4072
Robert G. Cabell III and Maude Morgan Cabell Foundation Grants, 4075
Robert W. Deutsch Foundation Grants, 4082

Rockwell International Corporate Trust Grants, 4093
Ross Foundation Grants, 4106
RRF General Grants, 4115
Russell Family Foundation Environmental Sustainability Grants, 4128
Ruth and Vernon Taylor Foundation Grants, 4130
RWJF Partners Investing in Nursing Grants, 4149
Sage Foundation Grants, 4158
Salem Foundation Grants, 4160
San Antonio Area Foundation Grants, 4163
San Diego Women's Foundation Grants, 4164
Schramm Foundation Grants, 4188
Seattle Foundation Medical Funds Grants, 4194
Sensient Technologies Foundation Grants, 4199
Sidgmore Family Foundation Grants, 4252
Sid W. Richardson Foundation Grants, 4253
Sioux Falls Area Community Foundation Community Fund Grants (Unrestricted), 4262
Sioux Falls Area Community Found Spot Grants, 4263
Skaggs Foundation Grants, 4266
Smith Richardson Foundation Strategy and Policy Fellowship, 4270
Sony Corporation of America Corporate Philanthropy Grants, 4329
Stanley Smith Horticultural Trust Grants, 4446
Starr Foundation Grants, 4447
Strake Foundation Grants, 4452
Surdna Foundation Arts Teachers Fellowships, 4470
Surdna Found Sustainable Environments Grants, 4471
T. James Kavanagh Foundation Grants, 4489
TD4HIM Foundation Grants, 4497
Textron Corporate Contributions Grants, 4501
The Ray Charles Foundation Grants, 4502
Thomas and Dorothy Leavey Foundation Grants, 4506
Thomas Austin Finch, Sr. Foundation Grants, 4507
Thomas C. Ackerman Foundation Grants, 4508
Thompson Charitable Foundation Grants, 4511
Thornton Foundation Grants, 4514
Todd Brock Family Foundation Grants, 4522
U.S. Department of Education American Overseas Research Centers, 4537
Union Pacific Found Community & Civic Grants, 4568
Union Pacific Foundation Health and Human Services Grants, 4569
Vanguard Public Foundation Grant Funds, 4808
Victor E. Speas Foundation Grants, 4810
W. C. Griffith Foundation Grants, 4818
Wayne and Gladys Valley Foundation Grants, 4835
Weyerhaeuser Company Foundation Grants, 4857
WILD Foundation Grants, 4867
William and Charlotte Parks Foundation for Animal Welfare Grants, 4872
William Blair and Company Foundation Grants, 4875
William G. & Helen C. Hoffman Found Grants, 4876
William G. McGowan Charitable Fund Grants, 4877
William N. and Myriam Pennington Foundation Grants, 4879
Xerox Foundation Grants, 4899

Graduate Assistantships

AAAS GE Science Prize for Young Life Scientists, 32
Abbott Fund Science Education Grants, 220
ACC Japan-US Arts Program Fellowships, 256
ACC Philippines Fellowship Program, 261
American Chemical Society Green Chem Grants, 679
Anthropological Center for Training in Global and Environmental Change Grad Assistantships, 870
Evjue Foundation Grants, 1963
Health Management Scholarships and Grants for Minorities, 2224
Higher Education Graduate Scholarships, Fellowships, and Grants, 2265
NAPNAP Foundation Graduate Student Research Grant, 2828
NCAR Graduate Research Fellowships, 2901
NCIC Terry Fox Research Studentships, 2924
NOAA Pacific Islands Assistantships, 3399
NSF Linguistics Grants, 3587
NSF Sociology Grants, 3650
Penn Graduate School of Education Research Assistantships, 3894
Pennsylvania State University Education Graduate Assistantships, 3895
Syracuse University Graduate Fellowships, 4488
UCLA Research Grants in Ethnic Studies, 4549
Vancouver Foundation Grants and Community Initiatives Program, 4807
Xerox Foundation Grants, 4899

Grants to Individuals

AAR Rome Prize Fellowships, 202
AAUW Career Development Grants, 212
A Charitable Foundation Grants, 273
AERA Minority Fellowships, 343
ALA ProQuest Documents to the People Award, 496
American Academy of Advertising Doctoral Dissertation Competition Awards, 608
Animal Behavior Soc Charles H. Turner Award, 849
Animal Behavior Society Developing Nations Research Grants, 850
Animal Behavior Society Diversity Grant, 851
Animal Behavior Society Latin American Travel Awards, 852
Animal Behavior Society Student Research Grants, 853
ASTA Academic Scholarships, 1224
CFR Edward R. Murrow Press Fellowship, 1500
CFR International Affairs Fellowships, 1501
CFR International Affairs Fellowships in Nuclear Security, 1502
CFR Int'l Affairs Fellowships in South Korea, 1503
CFR International Affairs in Japan Fellowships, 1504
CFR Military Fellowships, 1505
CFR Stanton Nuclear Security Fellowships, 1506
Do Something TMI Intern, 1844
Do Something User Experience Research Intern, 1845
Durfee Foundation Stanton Fellowship, 1881
IRA Esther Zolt Teacher Research Award, 2408
IRA Teacher as Researcher Grants, 2414
Japan Foundation Six-Month Grants for Specialists in Cultural and Academic Fields, 2479
Maine Community Foundation Keepers Preservation Education Grants, 2651
Manitoba Arts Council Dance Creation/Production Grants, 2654
MLA Cunningham Memorial International Fellowship, 2783
MLA David A. Kronick Traveling Fellowship, 2784
MLA Donald A.B. Lindberg Res Fellowship, 2785
MLA Graduate Scholarship, 2786
MLA Grad Scholarship for Minority Students, 2787
MLA Murray Gottlieb Prize Essay Award, 2788
MLA Research, Development, and Demonstration Project Grant, 2789
MLA Rittenhouse Award, 2790
MLA Thomas Reuters / Frank Bradway Rogers Information Advancement Award, 2791
NCFR Affiliate Councils Outstanding Student Research Papers, 2906
NCFR Ruth Hathaway Jewson Award, 2909
NYSCA Individual Artists: Film, Media and New Technology Production Grants, 3706
Organization for Tropical Studies Fellowships, 3796
Organization for Tropical Studies Pilot Awards for Exploratory Research, 3797
OTS Francis J. Bossuyt Fellowship, 3855
OTS Hovore-Horn Fellowships, 3856
OTS Stanley Smith Horticultural Fellowships, 3857
OTS Undergrad NAPIRE Summer Internships, 3858
Phi Kappa Phi Found Love of Learning Awards, 3937
Phi Upsilon Omicron Candle Fellowships, 3941
Phi Upsilon Omicron Diamond Anniversary Fellowships, 3942
Phi Upsilon Omicron Florence Fallgatter Distinguished Service Award, 3943
Phi Upsilon Omicron Founders Fellowship, 3944
Phi Upsilon Omicron Geraldine Clewell Doctoral Fellowship, 3945
Phi Upsilon Omicron Geraldine Clewell Masters Fellowship, 3946
Phi Upsilon Omicron Jean Dearth Dickerscheid Fellowship, 3947
Phi Upsilon Omicron Margaret Drew Alpha Fellowship, 3948
Phi Upsilon Omicron Presidents Res Fellowship, 3949
SARE Graduate Student Grants - North Central Region, 4168
SARE Graduate Student Grants - NE Region, 4169
SARE Graduate Student Grants - South Region, 4170
Tandem Launch Technologies Grants, 4493
Thomson Reuters / MLA Doctoral Fellowships, 4512
UCR Humanities Grad Research Grants, 4551
USDD Breast Cancer Era of Hope Scholars, 4694
USDD Breast Cancer Impact Grants, 4696
USDD Breast Cancer Research Postdoctoral Fellowships, 4698
Willary Foundation Grants, 4870
Winterthur Museum Dissertation Fellowships, 4887
Winterthur Research Fellowships, 4888
Wolfsonian-FIU Fellowships, 4890

International Exchange Programs

ACC Japan-US Arts Program Fellowships, 256
ACC Philippines Fellowship Program, 261
ACSM Oded Bar-Or Int'l Scholar Awards, 320
Asia Foundation Exchange Programs, 1090
ASM International Fellowship for Latin America and the Caribbean, 1175
ASM Int'l Fellowships for Asia and Africa, 1176
ASM Millis-Colwell Postgraduate Travel Grant, 1179
Australian Postdoctoral Research Fellowships, 1270
BA Elisabeth Barker Fund, 1288
BARD Postdoctoral Fellowships, 1298
Bilateral Scholarships, 1344
Canada-U.S. Fulbright International Science and Technology Award, 1417
Canada-U.S. Fulbright Senior Specialists Grants, 1418
Canon Foundation Grants, 1423
Coca-Cola Foundation Grants, 1593
DAAD Intensive Language Courses Grants, 1699
Eisenhower Exchange International Fellowships, 1902
Fulbright Alumni Initiatives Awards Grants, 2064
Fulbright Distinguished Chairs Grants, 2069
Fulbright German Studies Seminar Grants, 2072
Fulbright mtvU Awards, 2074
Fulbright Specialists Grants, 2079
Fulbright Traditional Scholar Program in Europe and Eurasia, 2080
Fulbright Traditional Scholar Program in Sub-Saharan Africa, 2081
Fulbright Traditional Scholar Program in the East Asia/Pacific Region, 2082
Fulbright Traditional Scholar Program in the Near East and North Africa Region, 2083
Fulbright Traditional Scholar Program in the South and Central Asia Region, 2084
Fulbright Traditional Scholar Program in the Western Hemisphere, 2085
Fulbright Visiting Specialists Program: Direct Access to the Muslim World, 2087
GCA Interchange Fellowship, 2098
IBRO InEurope Short Stay Grants, 2325
IES Faculty Fellowships, 2341
IREX Edmund S. Muskie Graduate Fellowship, 2417
IREX Eurasian Undergraduate Exchange Grants, 2418
IREX Individual Advanced Research Opportunities Grants, 2419
Japan Found Center for Global Partnership Grants, 2472
John Dana Archbold Fellowships, 2499
JSPS Bilateral Exchange Programs with Foreign Academic Institutions, 2524
JSPS Fellowships for Foreign Scientists, 2525
JSPS International Joint Research Projects and Scientific Seminars, 2526
JSPS Postdoc Fellowship for Foreign Researchers, 2527
Kellogg Institute for International Studies Andrónico Luksic Grants, 2535
Kosciuszko Found Fellowships for Polish Citizens, 2570
Kosciuszko Foundation Graduate Study and Research in Poland Scholarships, 2571

Kosciuszko Foundation Teaching Fellowships for Polish Citizens, 2573
LBI-NY/DAAD Fellowships in New York, 2594
NHLBI Investigator Initiated Multi-Site Clinical Trials, 3030
NL British Academy Fellowship for Study in the United Kingdom, 3283
NL Herzog August Bibliothek Wolfenbuttel Fellowship, 3286
NSF Antarctic Special Opportunity for Science aboard the Swedish Ice Breaker Oden Grants, 3480
NSF Materials World Network: Cooperative Activity in Materials Research between US Investigators and their Counterparts Abroad, 3595
Nuffield Foundation Open Door Grants, 3669
SfN Travel Awards for the Japan Neuroscience Society Meeting, 4216
SHOT International Scholars, 4242
SRC Natural and Engineering Science Conference Grants, 4383
SSRC Our Shared Past Grants, 4437
U.S. Department of Education American Overseas Research Centers, 4537
U.S. Department of Education United States-Russia Program: Improving Research and Educational Activities in Higher Education, 4547
UICC American Cancer Society International Fellowships for Beginning Investigators, 4555
US CRDF CGP-RFBR Climate Change and Energy Competition For Joint U.S.-Russian Research Teams, 4608
US CRDF CGP ETF 2009 Energy Research Competition For Joint U.S.-Estonian Research Teams, 4609
USDA Cochran Fellowships, 4619
USDA International Science and Education Competitive Grants, 4631
USDA Norman E. Borlaug International Agricultural Science and Technology Fellowships, 4636
USDA Scientific Cooperation Research Program, 4647

International Grants

3M Canada National Teaching Fellowships, 3
A-T Medical Research Foundation Grants, 10
AAAS Award for Science Diplomacy, 27
AAAS GE Science Prize for Young Life Scientists, 32
AAAS Roger Revelle Fellowship in Global Stewardship, 38
AAAS Science and Technology Policy Fellowships: Congressional, 39
AAAS Science and Technology Policy Fellowships: Global Health and Development, 42
AAB Berlin Prize Fellowships, 48
AACR Gertrude Elion Cancer Research Award, 137
AACR Kirk A. Landon and Dorothy P. Landon Foundation Prizes, 141
AAFCS International Graduate Fellowships, 154
AAG John Brinkerhoff Jackson Prize, 175
AAR Postdoctoral Fellowships, 201
AAR Rome Prize Fellowships, 202
AAS Kate B. and Hall J. Peterson Fellowships, 204
Abbott-ASM Lifetime Achievement Award, 218
Abbott Fund Global AIDS Care Grants, 219
Abdus Salam ICTP-ELETTRA Users Grants, 221
Abdus Salam ICTP-OEA Projects Grants, 223
Abdus Salam ICTP-OEA Scie Meetings Grants, 224
Abdus Salam ICTP-OEA Visiting Scholars/ Consultants Grants, 225
Abdus Salam ICTP/IAEA Sandwich Training Education Grants, 226
Abdus Salam ICTP Federation Arrangement Scheme Grants, 227
Abdus Salam ICTP Group Associates Grants, 228
Abdus Salam ICTP Jnt Laurea Magistralis Grants, 229
Abdus Salam ICTP Joint PhD Programme in Environmental Fluid Mechanics Grants, 230
Abdus Salam ICTP Junior Associates Grants, 231
Abdus Salam ICTP Mathematical Research Fellowships, 232
Abdus Salam ICTP Networks Grants, 233
Abdus Salam ICTP Postdoctoral Positions in Theoretical High Energy Physics, 234
Abdus Salam ICTP Regular Associates Grants, 237
Abdus Salam ICTP Schools, Conferences, Workshops and Extended Research Workshops Grants, 238
Abdus Salam ICTP Senior Associates Grants, 239
Abdus Salam ITCP Postgraduate Diploma Programme in General Physics, 241
Abdus Salam Joint ICTP/SISSA PhD Programme in Physics and Mathematics Grants, 242
ACC American Artists and Museum Professionals in Asia Fellowships, 250
ACC Asian Art and Religion Fellowships, 251
ACC Fellowships and Projects, 252
ACC Ford Foundation Fellowships, 253
ACC Hong Kong Arts Program Fellowships, 254
ACC Humanities Fellowship Program, 255
ACC Japan-US Arts Program Fellowships, 256
ACC John D. Rockefeller III Award, 257
ACC Mekong Region Fellowship, 258
ACC Philippines Fellowship Program, 261
ACC Residency Program in Asia, 267
ACC Starr Foundation Fellowship Program, 268
ACC Taiwan Fellowship Program, 269
ACGT Young Investigator Grants, 272
A Charitable Foundation Grants, 273
ACLS African Humanities Program in Ghana, Nigeria, South Africa, Tanzania, and Uganda, 279
ACLS Early Career Postdoctoral Fellowships in East European Studies, 286
ACM A.M. Turing Award, 292
ACM Distinguished Service Award, 293
ACM Paris Kanellakis Theory and Practice Award, 301
ACRL Academic/Research Librarian of the Year Award, 303
ACSM Carl V. Gisolfi Memorial Fund Grant, 312
ACSM Coca-Cola Company Doctoral Student Grant on Behavior Research, 313
ACSM Foundation Clinical Sports Medicine Endowment Grants, 315
ACSM Foundation Research Grants, 317
ACSM International Student Awards, 318
ACSM Oded Bar-Or Int'l Scholar Awards, 320
ACSM Raymond and Rosalee Weiss Research Endowment Grant, 322
AFAR Paul Beeson Career Development Awards in Aging Research for the Island of Ireland, 353
AFP Research Grants, 356
AFRL/NRC Resident Research Associateship, 358
AFUW Diamond Jubilee Bursary, 359
AFUW Doreen McCarthy, Barbara Crase, Cathy Candler & Brenda Nettle Bursaries, 360
AFUW Fellowships, 361
AFUW Georgina Sweet Fellowships, 362
AFUW Heather Latz Bursary, 363
AFUW Padnendadlu Graduate Bursary, 364
AFUW Padnendadlu Postgraduate Bursary, 365
AFUW Padnendadlu Undergraduate Bursary, 366
AFUW South Australia Jean Gilmore, Thenie Baddams & Daphne Elliott Bursaries, 367
AFUW Winifred E. Preedy Postgraduate Bursary, 368
AIAR/Council of American Overseas Research Centers Fellowships for Advanced Multi-Country Research, 391
AIAR Andrew W. Mellon Foundation Fellowships, 392
AIAR Annual Professorships, 393
AIAR George A. Barton Fellowships, 395
AIHS Proposals for Special Initiatives, 456
AINSE Honours Scholarships, 460
AINSE Int'l Conference Travel Scholarships, 461
AINSE John Ferris Memorial Award, 462
AINSE Postgraduate Research Awards, 463
AINSE Research Fellowships, 464
AJCC Cancer Staging Grants, 476
ALA Ingenta Research Awards, 489
ALA WESS De Gruyter European Librarianship Study Grant, 501
Albany Medical Center Prize in Medicine and Biomedical Research, 502
Alberta Law Foundation Grants, 503
Albert and Mary Lasker Foundation Awards, 505
Alcoa Foundation Grants, 511
Alcohol Misuse and Alcoholism Research Grants, 512
Alexander von Humboldt Feodor-Lynen Research Fellowships for Postdoc Researchers, 519
Alexander von Humboldt Feodor Lynen Research Fellowships for Experienced Researchers, 520
Alexander von Humboldt Georg Forster Fellowships for Experienced Researchers, 523
Alexander von Humboldt Georg Forster Fellowships for Postdoctoral Researchers, 524
Alexander von Humboldt Hezekiah Wardwell Fellowships, 528
Alexander von Humboldt International Climate Protection Fellowships, 529
Alexander von Humboldt Konrad Adenauer Research Award, 531
Alexander von Humboldt Research Fellowships for Postdoctoral Researchers, 539
Alexander von Humboldt Kovalevskaja Awards, 540
Allen Lane Foundation Grants, 571
Alternatives Research and Development Foundation Alternatives in Education Grants, 581
Alternatives Research and Development Foundation Grants, 582
Alzheimer's Association Investigator-Initiated Research Grants, 592
Alzheimer's Association New Investigator Research Grants, 595
Alzheimer's Association U.S.-U.K. Young Investigator Exchange Fellowships, 597
American-Italian Cancer Foundation Fellowships, 602
American-Scandinavian Foundation Awards for Advanced Study or Research in the U.S., 603
American-Scandinavian Foundation Fellowship for Study in Scandinavia, 604
American-Scandinavian Foundation Grant for Study in Scandinavia, 605
American Academy of Advertising Doctoral Dissertation Competition Awards, 608
American Chemical Society Petroleum Research Fund Scientific Education Grants, 692
American Chemical Society Petroleum Research Fund Undergraduate Faculty Sabbatical Grants, 693
American Hist Assoc Albert B. Corey Prize in Canadian-American Relations, 711
American Hist Assoc Paul Birdsall Prize in European Military and Strategic History, 740
American Philosophical Society Daland Fellowships in Clinical Investigation, 763
American Philosophical Society Franklin Research Grants, 764
American Philosophical Society Lewis and Clark Fund for Exploration and Field Research, 767
American Philosophical Society Lewis and Clark Exploration and Field Research in Astrobiology, 768
American Philosophical Society Library Resident Research Fellowships, 769
American Psychiatric Association/AstraZeneca Young Minds in Psychiatry International Awards, 773
American Society for Yad Vashem Grants, 798
amfAR Research Grants, 811
Amgen Foundation Grants, 812
AMNH Graduate Student Fellowships, 815
Animal Behavior Society Developing Nations Research Grants, 850
Animal Behavior Society Diversity Grant, 851
Animal Behavior Society Latin American Travel Awards, 852
Animal Behavior Society Student Research Grants, 853
Anna-Greta & Holger Crafoords Found Grants, 858
Anne Louise Barrett Fellowship, 861
AOCS Honored Student Award, 884
AOCS Phospholipid Best Paper Award, 889
APS/British Academy Fellowship for Research in London, 918
APSA Congressional Fellowships for International Scholars, 928
APS Aneesur Rahman Prize for Computational Physics, 942

APSA Travel Grants, 948
APS Davisson-Germer Prize in Atomic or Surface Physics, 952
APS E.K. Plyler Prize for Molecular Spectroscopy, 954
APS Einstein Prize, 955
APS Frank Isakson Prize for Optical Eff in Solids, 957
APS Hans A. Bethe Prize, 960
APS Herbert P. Broida Prize, 961
APS I.I. Rabi Prize in Atomic, Molecular, and Optical Physics, 962
APS J.J. Sakurai Prize for Theoretical Part Physics, 964
APS James C. McGroddy Prize for New Materials, 965
APS Lars Onsager Prize, 968
APS Nicholas Metropolis Award for Outstanding Doctoral Thesis in Computational Physics, 972
APS Oliver E. Buckley Condensed Matter Prize, 973
APS Outstanding Doctoral Thesis Research in Atomic, Molecular, or Optical Physics, 974
APS Polymer Physics Prize, 976
APS Robert R. Wilson Prize for Achievement in the Physics of Particle Accelerators, 979
APS Tom W. Bonner Prize in Nuclear Physics, 980
APS W.K.H. Panofsky Prize in Experimental Particle Physics, 981
APS Will Allis Prize for Study of Ionized Gases, 982
Arca Foundation Grants, 984
Archer Daniels Midland Foundation Grants, 985
ARIT Fellowships, 991
ARIT Fellowships for Intensive Advanced Turkish Language Study, 992
ARIT Kenan T. Erim Fellowship for Research at Aphrodisias, 993
ARIT Mellon Foundation Postdoctoral Research Fellowships for East European Scholars, 994
Arronson Foundation Grants, 1019
Arthritis Foundation of Australia Grants, 1022
ASCSA Alison M. Frantz Fellowship, 1051
ASCSA Harry Bikakis Fellowship, 1053
ASCSA Jacob Hirsch Fellowship, 1054
ASCSA Weiner Laboratory Faunal Studies Research Fellowship, 1056
ASGE Don Wilson Award, 1062
ASGE Endoscopic Research Awards, 1063
ASGE Given Capsule Endoscopy Award, 1065
Asia Foundation Exchange Programs, 1090
ASM-PAHO Infectious Diseases Epidemiology and Surveillance Fellowships, 1098
ASM-UNESCO Leadership Grant for International Educators, 1099
ASM Int'l Fellowships for Asia and Africa, 1176
ASM Millis-Colwell Postgraduate Travel Grant, 1179
ASPET Paul M. Vanhoutte Award, 1207
ASPRS Leica Geosystems Internship, 1213
ASPRS William A. Fischer Scholarship, 1216
ASTD Research Award, 1226
ASU Institute for Humanities Research Visiting Fellowships, 1248
Audio Engineering Society Grad Scholarships, 1265
Australasian Institute of Judicial Administration Seed Funding Grants, 1268
Australian Academy of Science Grants, 1269
Australian Postdoctoral Research Fellowships, 1270
Australian Postgraduate Research Fellowships, 1271
Australian Research Fellowships/Queen Elizabeth II Fellowships, 1272
Avon Foundation Breast Care Fund Grants, 1274
Avon Products Foundation Grants, 1275
AWIS Educational Found Predoctoral Awards, 1278
AWWA Abel Wolman Doctoral Fellowships, 1282
AWWA Larson Aquatic Research Support Scholarships, 1285
BA Elisabeth Barker Fund, 1288
Banfi Vintners Foundation Grants, 1296
Bank of America Charitable Foundation Matching Gifts, 1297
BARD Postdoctoral Fellowships, 1298
BARD Research Grants, 1299
Barth Syndrome Foundation Research Grants, 1302
BA Sir Ernest Cassel Educational Grants, 1303
BA Small Research Grants, 1304
Bay and Paul Foundations, Inc Grants, 1307
Beatrice Laing Trust Grants, 1324
Beckman Young Investigators Grants, 1327
Becton Dickinson and Company Grants, 1328
Bedding Plants Found Horticulture Scholarships, 1329
Bernard Lowy Fund for the Study of Tropical Botany in Latin America, 1335
Bernard Van Leer Foundation Grants, 1336
Bilateral Scholarships, 1344
Bill and Melinda Gates Foundation Agricultural Development Grants, 1345
Bill and Melinda Gates Foundation Water, Sanitation and Hygiene Grants, 1347
Bogliasco Foundation Liguria Study Center Residencies, 1358
BP Conservation Programme Future Conservationist Awards, 1365
BPW Career Advancement Scholarships, 1366
Brainerd Foundation Grants, 1368
Bristol-Myers Squibb Foundation Global HIV/AIDS Initiative Grants, 1376
Bristol-Myers Squibb Foundation Health Disparities Grants, 1377
Burden Trust Grants, 1388
BWF Career Awards at the Scientific Interface, 1393
BWF Institutional Program Unifying Population and Laboratory Based Sciences Grants, 1397
Bydale Foundation Grants, 1400
C. Alma Baker Trust Grants, 1403
Cabot Corporation Foundation Grants, 1406
Caesars Foundation Grants, 1407
Camargo Foundation Fellowships, 1412
Canada-U.S. Fulbright Senior Specialists Grants, 1418
Canada-U.S. Fulbright Trad Scholar Awards, 1419
Canada Graduate Scholarships and NSERC Postgraduate Scholarships, 1420
Canadian Studies Senior Fellowships, 1422
Canon Foundation Grants, 1423
Carnegie Corporation of New York Grants, 1432
Carnegie Endowment for International Peace Junior Fellowships, 1434
Catholic University Academic Scholarships, Grants, and Awards, 1443
CCA Diefenbaker Award to update, 1447
CDC Foundation Atlanta International Health Fellowships, 1461
Center for the Study of Philanthropy International Fellowships, 1482
Center for the Study of Philanthropy Senior International Fellowships, 1483
Center on Philanthropy and Civil Society's Emerging Leaders International Fellows Program, 1487
Central European University Fellowships, 1488
CFUW Bourse Georgette LeMoyne Award, 1508
CFUW Canadian Home Economics Association Fellowship, 1509
CFUW Ecole Polytechnique Commem Award, 1511
CFUW Margaret McWilliams Predoc Fellowship, 1513
CFUW Marion Elder Grant Fellowship, 1514
CFUW Ruth Binnie Fellowship, 1516
Chapin Hall International Fellowships in Children's Policy Research, 1517
Charles H. Revson Foundation Grants, 1524
Charles Stewart Mott Foundation Grants, 1529
Chatlos Foundation Grants, 1532
Children's Cardiomyopathy Found Research Grants, 1550
Chiles Foundation Grants, 1552
Christensen Fund Regional Grants, 1554
Churchill Fellowships, 1560
CIGNA Foundation Grants, 1568
CIMO Scholarships for Advanced Studies of the Finnish Language at Finnish Universities, 1569
CIUS Neporany Res & Teaching Fellowship, 1572
CNIB Baker Applied Research Fund Grants, 1588
CNIB Baker New Researcher Fund Grants, 1589
CNIB Barbara Tuck MacPhee Award, 1590
CNIB Canada Glaucoma Clinical Research Council Grants, 1591
CNIB E. & Mary Hochhausen Access Technology Research Grants, 1592
Coca-Cola Foundation Grants, 1593
Commonwealth Fund Harkness Fellowships in Health Care Policy and Practice, 1606
Commonwealth Fund Small Grants, 1611
Compton Foundation Grants, 1623
ConocoPhillips Grants, 1629
Conquer Cancer Foundation of ASCO Drug Development Research Professorship, 1632
Conquer Cancer Foundation of ASCO Int'l Devel & Educ Award in Palliative Care, 1634
Conquer Cancer Foundation of ASCO International Development and Education Awards, 1635
Conservation, Food, and Health Foundation Grants for Developing Countries, 1641
Corpus Christi College Research Scholarships, 1647
Croucher Foundation Grants, 1659
CTCRI Idea Grants, 1665
CTCRI Knowledge Synthesis Grants, 1666
CTCRI Policy Research Grants, 1667
CTCRI Workshop and Learning Opportunities Grants, 1671
Cystic Fibrosis Canada Clinical Project Grants, 1675
Cystic Fibrosis Canada Fellowships, 1677
Cystic Fibrosis Canada Research Grants, 1678
Cystic Fibrosis Canada Scholarships, 1679
Cystic Fibrosis Canada Studentships, 1683
Cystic Fibrosis Canada Visiting Allied Health Professional Awards, 1686
Cystic Fibrosis Canada Visiting Clinician Awards, 1687
Cystic Fibrosis Trust Research Grants, 1691
DAAD-AICGS Research Fellowships, 1693
DAAD-Leo Baeck Institute Fellowships, 1694
DAAD Faculty Research Visit Grants, 1697
DAAD German Studies Research Grants, 1698
DAAD Intensive Language Courses Grants, 1699
DAAD Learn German in Germany Grants, 1700
DAAD Research Grants, 1702
DAAD Research Grants for Doctoral Candidates and Young Academics and Scientists, 1703
DAAD Undergraduate Scholarships, 1710
Daimler and Benz Foundation Fellowships, 1713
Danone Institute Belgium Doctoral Scholarship in Nutrition, 1718
Dave Thomas Foundation for Adoption Grants, 1723
DENSO North America Foundation Grants, 1742
Diaz-Ayala Cuban and Latin American Popular Music Collection Travel Grants, 1763
DOI African Elephant Conservation Grants, 1820
Doris Duke Foundation Operations Research on AIDS Care & Treatment in Africa Grants, 1839
Dorothy Leet Grants, 1840
DOS Study of the United States Institutes, 1846
Duke University Adult Cardiothoracic Anesthesia and Critical Care Medicine Fellowships, 1869
Duke University Clinical Cardiac Electrophysiology Fellowships, 1871
Duke Univ Interventional Cardio Fellowships, 1873
E.J. Sierleja Memorial Fellowship, 1884
Eastern Europe & the Independent States of Former Soviet Union Research & Training Grants, 1891
Eisenhower Exchange International Fellowships, 1902
Ekhagastiftelsen Foundation Grants, 1903
Elizabeth Glaser Int'l Leadership Awards, 1906
Elizabeth Glaser Scientist Award, 1907
EMBO Installation Grants, 1915
EMBO Long-Term Fellowships, 1916
EMBO Short-Term Fellowships, 1917
Engineering Information Foundation Grants, 1927
ESF Latsis Prize, 1951
ESRI Conservation Grants, 1952
ExxonMobil Found Malaria Initiative Grants, 1968
Facebook Fellowships, 1973
FAMRI Clinical Innovator Awards, 1975
FDHN Moti L. & Kamla Rustgi International Travel Awards, 1997
FIC Research Collaboration Award, 2006
Fishman Family Foundation Grants, 2020
Five College Women's Research Center Ford Associateships, 2022
Fluor Foundation Grants, 2028

FMC Foundation Grants, 2029
Ford Foundation International Fellowships, 2030
Ford Foundation Peace and Social Justice Grants, 2031
Foundation for the Future Grants, 2043
Franklin Mosher Baldwin Fellowships, 2050
Fulbright-Hays Seminars Abroad Bilateral Projects Grants, 2062
Fulbright/Garcia Robles Grants, 2063
Fulbright Alumni Initiatives Awards Grants, 2064
Fulbright Beginning Prof Journalism Awards, 2065
Fulbright Binational Business Grants in Mexico, 2066
Fulbright Business Grants in Spain, 2067
Fulbright Distinguished Chairs Grants, 2069
Fulbright European Union Scholar-in-Residence Grants, 2071
Fulbright German Studies Seminar Grants, 2072
Fulbright International Education Administrators Seminar Grants, 2073
Fulbright mtvU Awards, 2074
Fulbright New Century Scholars Grants, 2075
Fulbright Occasional Lecturer Grants, 2076
Fulbright Public Policy Grants in Mexico, 2077
Fulbright Scholar-in-Residence Grants, 2078
Fulbright Specialists Grants, 2079
Fulbright Traditional Scholar Program in Europe and Eurasia, 2080
Fulbright Traditional Scholar Program in Sub-Saharan Africa, 2081
Fulbright Traditional Scholar Program in the East Asia/Pacific Region, 2082
Fulbright Traditional Scholar Program in the Near East and North Africa Region, 2083
Fulbright Traditional Scholar Program in the South and Central Asia Region, 2084
Fulbright Traditional Scholar Program in the Western Hemisphere, 2085
Fulbright Travel Grants, 2086
Fulbright Visiting Specialists Program: Direct Access to the Muslim World, 2087
Fund for Scientific Research--Flanders Grants, 2089
G. Unger Vetlesen Foundation Grants, 2093
GCA Interchange Fellowship, 2098
Gerber Foundation Grants, 2126
Gerda Henkel Foundation Grants, 2127
Germanistic Society of America Fellowships, 2128
Gerrish/FTE Technology Education Graduate Fellowship, 2129
Gilbreth Memorial Fellowships, 2141
Gladys Krieble Delmas Foundation Grants for Venetian Research, 2144
Glaucoma Foundation Grants, 2145
GMFUS Black Sea Trust for Regional Corporation Grants, 2152
Greater St Louis Community Foundation Grants, 2164
Gruber Foundation Cosmology Prize, 2177
GUITS Grants for the Publication of Scholarly Books and Journals, 2186
GUITS Grants for Undergraduate Study, 2187
GUITS Matching Seed-Money Grants for New Positions in Turkish Studies, 2188
GUITS Post-Doctoral Summer Travel-Research Grants, 2189
GUITS Research Grants in Comparative Studies of Modern Turkey, 2190
GUITS Sabbatical Research Grants, 2191
GUITS Summer Research Grants in Turkey for Graduate Students, 2192
H.J. Heinz Company Foundation Grants, 2195
Hague Academy of International Law Summer Scholarships, 2198
Harold Simmons Foundation Grants, 2203
Harry Frank Guggenheim Foundation Research Grants, 2208
Harry Ransom Humanities Research Center Research Fellowships, 2209
Harvard University Weatherhead Center for International Affairs Fellowships, 2215
Harvard Women and Public Policy Postdoctoral Fellowship, 2217
Hebrew University Lady Davis Trust Fellowships, 2226
Helen Bader Foundation Grants, 2228
Helen Kellogg Institute for International Studies Residential Fellowships, 2230
HHMI/EMBO Start-up Grants for Cen Europe, 2250
HHMI Grants and Fellowships Programs, 2255
HHMI International Research Scholars Program, 2258
HIAS Scholarship Awards, 2264
Horowitz Foundation for Social Policy Grants, 2280
Horowitz Found for Social Policy Special Awards, 2281
Hospital Libraries Section / MLA Professional Development Grants, 2283
Huebner Foundation Fellowships, 2294
Huntington Library National Endowment for the Humanities Fellowships, 2307
Hydrogen Student Design Contest, 2311
Ian Axford (New Zealand) Fellowships in Public Policy, 2313
IARC Postdoctoral Fellowships for Training in Cancer Research, 2316
IBRO InEurope Short Stay Grants, 2325
ICCS Publishing Fund, 2336
Ida Smedley MacLean International Fellowship, 2337
IERF Sepmeyer Research Grants, 2340
IES Faculty Fellowships, 2341
IFUW Dr. A. Vibert Douglas Int'l Fellowship, 2346
IFUW Winifred Cullis Grants, 2347
IIE David L. Boren Fellowships, 2350
IIE David L. Boren Scholarships, 2351
Indian Council of Social Science Research Grants, 2377
Institute for Palestine Studies, Publishing, and Research Organization Grants, 2382
Institute of Turkish Studies Summer Language Study Grants in Turkey, 2391
Institute of Turkish Studies Summer Research Grants in Turkey, 2392
International Federation of University Women Research Fellowships, 2397
International Fellowships in Medical Education, 2398
International Human Rights Funders Grants, 2399
International League of Antiquarian Booksellers Bibliographical Prize, 2400
International Max Planck Research School Mathematics in the Sciences Fellowships, 2401
International Studies Graduate Fellowships, 2403
IRA Albert J. Harris Awards, 2405
IRA Elva Knight Research Grant, 2406
IRA Esther Zolt Academic Research Award, 2407
IRA Esther Zolt Teacher Research Award, 2408
IRA Jeanne S. Chall Research Fellowship, 2410
IRA Pearson Foundation Rotary Lit Awards, 2412
IRA Steven A. Stahl Research Grants, 2413
IRA Teacher as Researcher Grants, 2414
IREX ECA Alumni Small Grants, 2416
IREX Edmund S. Muskie Graduate Fellowship, 2417
IREX Individual Advanced Research Opportunities Grants, 2419
IREX International Leadership in Education Program Grants, 2420
IREX MENA Media Emerging Leaders Fellows, 2421
IREX Policy-Connect Collaborative Res Grants, 2422
IREX Regional Policy Symposium Grants, 2423
IREX Short-Term Travel Grants, 2424
IREX University Administration Support, 2425
IREX US Embassy Policy Specialist Grants, 2426
ISUD Jacobsen Prizes, 2434
IUVSTA Prize for Science, 2443
IUVSTA Prize for Technology, 2444
IUVSTA Welch Foundation Scholarship, 2445
IWMF Public Health Fellowship, 2446
Izaak Walton Killam Postdoctoral Fellowships at the University of British Columbia, 2447
Izaak Walton Killam Memorial Predoctoral Fellowships at the University of British Columbia, 2448
Jacob and Hilda Blaustein Foundation Grants, 2454
James and Sylvia Thayer Short -Term Research Fellowships, 2457
James H. Cummings Foundation Grants, 2460
James S. McDonnell Found Research Grants, 2466
Japan Foundation Center for Global Partnership Grants, 2472
Japan Foundation Center for Global Partnership Institutional Grants, 2473
Japan Foundation Center for Global Partnership NPO Fellowships, 2474
Japan Foundation Center for Global Partnership Security Studies Fellowships, 2475
Japan Foundation Japanese Studies Fellowships, 2476
Japan Foundation New York Institutional Support for Japanese Studies Grants, 2477
Japan Foundation New York Japanese Studies Fellowships, 2478
Japan Foundation Six-Month Grants for Specialists in Cultural and Academic Fields, 2479
Jeffrey Campbell Grad Fellow in Anthropology, 2482
Jenifer Altman Foundation Grants, 2484
JMO Strategic and International Studies Grants, 2493
Joel L. Fleishman Civil Society Fellowships, 2495
John D. and Catherine T. MacArthur Foundation Global Challenges Grants, 2498
John Dana Archbold Fellowships, 2499
John Deere Foundation Grants, 2500
Johnson & Johnson Corporate Contributions, 2508
Joseph C. Wey Fellowship in Medieval Studies, 2514
JSPS Bilateral Exchange Programs with Foreign Academic Institutions, 2524
JSPS Fellowships for Foreign Scientists, 2525
JSPS International Joint Research Projects and Scientific Seminars, 2526
Kate Neal Kinley Memorial Fellowship, 2532
Kellogg Institute for International Studies Course Development Grants, 2537
Kellogg Institute for International Studies Global Collaboration Initiative Grants, 2541
Kellogg Institute for International Studies Summer Internships, 2549
Kellogg Institute for International Studies Visiting Fellowships, 2551
Kluge Center Distinguished Chairs, 2560
Kluge Center Fellowships, 2561
Kluge Center Kislak Fellows in American Studies, 2562
Kluge Center Kislak Short-Term Fellowships in American Studies, 2563
Kluge Center Kissinger Chair in Foreign Policy and International Relations, 2564
Kluge Center Kissinger Lecturer Award, 2565
Kobe College Corporation Graduate Fellowships, 2567
Kosciuszko Found Grants for Polish Citizens, 2572
Kress Foundation Art History Pre-Doctoral Fellowships, 2576
L'Oreal USA Women in Science Fellowship, 2577
Lady Davis Fellowships, 2578
LAM Found Established Investigator Awards, 2581
LAM Foundation Research Grants, 2584
Lance Armstrong Foundation Research Grants, 2585
Landon Foundation-AACR Innovator Award for International Collaboration, 2587
LBI-NY/DAAD Fellowships in New York, 2594
LBI-NY Fred Grubel Fellowship, 2596
Library of Congress Fellowships in Int'l Studies, 2612
Ludwick Family Foundation Grants, 2625
Lustgarten Foundation for Pancreatic Cancer Research Grants, 2630
MacArthur Foundation Conservation and Sustainable Development Grants, 2648
MAP International Medical Fellowships, 2655
Marshall Scholarships, 2666
Mary Isabel Sibley Fellowship, 2672
Mattel International Grants, 2686
Mead Johnson Nutritionals Med Educ Grants, 2734
Mesothelioma Applied Research Found Grants, 2743
MGFA Post-Doctoral Research Fellowships, 2751
Michael J. Fox Found Clinical Discovery Grants, 2760
Millipore Foundation Grants, 2775
Minerva Fellowships, 2778
Minerva Short-Term Research Grants, 2779
MLA Cunningham Memorial International Fellowship, 2783
MLA Murray Gottlieb Prize Essay Award, 2788
MLA Thomas Reuters / Frank Bradway Rogers Information Advancement Award, 2791

Motorola Foundation Grants, 2796
MPI for Social Anthropology Department II Postdoctoral Grants, 2798
Munby Fellowship in Bibliography, 2804
NAA Norwegian Marshall Fund Grants, 2811
NAGC Hollingworth Award, 2820
Nakanishi Prize, 2824
Nanyang Technological University Lee Kuan Yew Postdoctoral Fellowship, 2826
National Academies Postdoctoral and Senior Research Awards, 2844
National Academy of Education Spencer Postdoctoral Fellowships, 2845
National Center for Responsible Gaming Grants, 2855
National Endowment for Democracy Discretionary Grants, 2862
National Endowment for Democracy Reagan-Fascell Democracy Fellowships, 2863
National Geographic Society/Waitt Grants, 2870
National Geographic Society Conservation Trust Grants, 2871
National Geographic Society Field Research and Exploration Grants, 2873
National Geographic Young Explorers Grants, 2874
National Parkinson Foundation Clinical Research Fund Grants, 2881
National Security Education Grants to Institutions of Higher Education, 2885
NATO Advanced Research Workshops Program, 2894
NATO Advanced Study Institutes Program, 2895
NATO Collaborative Linkage Grants, 2896
NATO Expert Visits Grants, 2897
NATO Science for Peace Grants, 2898
NCI/NIDA Transdisciplinary Tobacco Use Research Centers Grants, 2912
NCIC Canadian Breast Cancer Research Initiative Grants, 2918
NCIC Canadian Cancer Society Research Grants, 2919
NCIC Canadian Cancer Society Travel Award for Senior Level PhD Students, 2920
NCIC Terry Fox Postdoctoral Res Fellowships, 2923
NCIC Terry Fox Research Studentships, 2924
NCI Exploratory/Developmental Grants for Diagnostic Cancer Imaging, 2927
NCSS Larry Metcalf Exemplary Dissert Award, 2939
Nestle Foundation Training Grant, 2983
NFWF/Exxon Save the Tiger Fund Grants, 2995
NFWF International Sea Turtle Conservation Fund Grants, 3000
NIA Malnutrition in Older Persons Grants, 3104
NIDDK Co-Activators and Co-Repressors in Gene Expression Grants, 3141
NIDDK Neurobiology of Diabetic Complications Grants, 3164
NIGMS/NIDCR Evolution of Infectious Diseases Research Grants, 3196
NIH Biology, Development, and Progression of Malignant Prostate Disease Research Grants, 3206
NIH Bone and the Hematopoietic and Immune Systems Research Grants, 3208
NIH Structural Bio of Membrane Proteins Grants, 3247
NIMH Early Identification and Treatment of Mental Disorders in Children and Adolescents Grants, 3252
NINR Acute and Chronic Care During Mechanical Ventilation Research Grants, 3259
NINR Clinical Interventions for Managing the Symptoms of Stroke Research Grants, 3262
NINR Quality of Life for Individuals at the End of Life Grants, 3266
NL American Society for Eighteenth-Century Studies Fellowships, 3279
NL Arthur and Lila Weinberg Fellowships for Independent Scholars and Researchers, 3280
NL Audrey Lumsden-Kouvel Fellowships, 3282
NL British Academy Fellowship for Study in the United Kingdom, 3283
NL Frances C. Allen Fellowships, 3284
NL Herzog August Bibliothek Wolfenbuttel Fellowship, 3286
NL Lloyd Lewis Fellows in American History, 3289
NL Mellon Postdoctoral Research Fellowships, 3294
NL National Endowment for the Humanities Fellowships, 3310
NL Short-Term Fellowships in the History of Cartography, 3316
NL Short-Term Resident Fellowships for Individual Research, 3317
NL Weiss/Brown Publication Subvention Award, 3320
NOAA International Coral Reef Conservation Project Grants, 3377
Northwood University/Alden B. Dow Creativity Center Summer Residency Program, 3432
Notsew Orm Sands Foundation Grants, 3434
NSERC Brockhouse Canada Prize for Interdisciplinary Research in Science and Engineering Grant, 3438
NSERC E.W.R. Steacie Memorial Fellowships, 3439
NSERC Industrial Research Chairs, 3443
NSERC Research Tools and Instruments Grants, 3449
NSERC Synergy Awards for Innovation, 3450
NSF-CONACyT Collaborative Research Grants, 3452
NSF Conferences, Workshops, and Special Meetings in the Mathematical Sciences, 3541
NSF Developing Global Scientists and Engineers, 3552
NSF East Asia and Pacific Summer Institutes for U.S. Graduate Students, 3555
NSF Materials World Network: Cooperative Activity in Materials Research between US Investigators and their Counterparts Abroad, 3595
NSF Ocean Drilling Grants, 3602
NSF Partnerships for International Research and Education, 3611
NSF Research Experiences for Undergrads, 3638
NSF Upper Atmospheric Facilities Grants, 3665
Nuffield Found Children & Families Grants, 3666
Nuffield Foundation Law and Society Grants, 3667
NWO Added Value Plus for Soc Sciences Grants, 3672
NWO Aspasia Grants, 3673
NWO Athena Grants, 3674
NYPL Short-Term Research Fellowships, 3701
NYZS/Wildlife Conservation Society Research Grants, 3709
Olin Corporation Charitable Trust Grants, 3739
Organization for Tropical Studies Fellowships, 3796
Organization for Tropical Studies Pilot Awards for Exploratory Research, 3797
OSF Access to Medicines Fellowships in Eastern Europe and Central Asia, 3834
OSF Alumni Grants, 3835
OSF Global Supplementary Grants, 3843
OSF HESP Mobility Grants, 3844
OSF Intellectual Property Reform Grants, 3846
OSF Palestinian Rule of Law Fellowships, 3848
OTS Francis J. Bossuyt Fellowship, 3855
OTS Hovore-Horn Fellowships, 3856
OTS Stanley Smith Horticultural Fellowships, 3857
Overbrook Foundation Grants, 3860
Patrick and Anna M. Cudahy Fund Grants, 3873
PDF International Research Grants, 3882
Peter F. Drucker Award for Nonprofit Innovation, 3902
Pfizer ASPIRE Worldwide Endocrine Young Investigator Grants, 3918
Pfizer Medical Education Track Two Grants, 3927
PHF Andrew W. Mellon Postdoctoral Fellowships in the Humanities, 3931
Phi Delta Kappa Outstanding Doctoral Dissertation Awards, 3936
Pioneer Hi-Bred Plant Breeding Fellowships, 3968
Pioneer Hi-Bred International Scholarships, 3969
PKD Foundation Lillian Kaplan International Prize for the Advancement in the Understanding of Polycystic Kidney Disease, 3974
Pollock-Krasner Foundation Grants, 3981
Polytechnic University International Research Studentships, 3982
Potts Memorial Foundation Grants, 3985
Radcliffe Institute Carol K. Pforzheimer Student Fellowships, 4018
Radcliffe Institute Indiv Residential Fellowships, 4021
Radcliffe Institute Oral History Grants, 4022
Radcliffe Institute Residential Group Fellowships, 4024
Rajiv Gandhi Foundation Grants, 4025
RA Kleinhans Fellowship for Research in Tropical Nontimber Forest Products, 4026
RAND Corp Transatlantic Postdoctoral Fellowships for International Relations and Security, 4032
RCPSC Detweiler Traveling Fellowships, 4036
RCPSC International Traveling Fellowship, 4037
RCPSC Medical Educ Traveling Fellowships, 4039
RCPSC Royal College Fellowship for Studies in Medical Education, 4040
RCPSC Walter MacKenzie Fellow in Surgery, 4042
Rett Syndrome Research Foundation Grants, 4058
Richard and Rhoda Goldman Fund Grants, 4063
Richard Carley Hunt Memorial Postdoctoral Fellowships, 4065
Roberto Longhi Foundation Fellowships, 4078
Rockefeller Archive Center Ehrlich Grants, 4084
Rockefeller Archive Center Grants in Aid, 4085
Rockefeller Archive Center Research Grants, 4086
Rockefeller Archive Center Visiting Archivist Fellowship, 4087
Roger K. Summit Scholarship, 4097
Rohm and Haas Company Grants, 4098
Ronald Breslow Award for Achievement in Biomimetic Chemistry, 4100
Ronald McDonald House Charities Grants, 4101
Rosalinde and Arthur Gilbert Foundation/AFAR New Investigator Awards in Alzheimer's Disease, 4102
Royal Belgian Numismatic Society Res Grants, 4107
Royal Scottish Geographical Society Expedition Grants, 4110
Royal Scottish Geographical Society Small Research Grants, 4111
Royal Scottish Geographical Society Travel Grants, 4112
RSC Abbyann D. Lynch Medal in Bioethics, 4116
RSC Alice Wilson Award, 4117
RSC Bancroft Award, 4118
RSC Jason A. Hannah Medal, 4119
RSC John L. Synge Award, 4120
RSC McLaughlin Medal, 4121
RSC Miroslaw Romanowski Medal, 4122
RSC Rutherford Memorial Medals, 4124
RSC Sir John William Dawson Medal, 4125
RSC Thomas W. Eadie Medal, 4126
S.E.VEN Fund Open Enterprise Solutions to Poverty RFP Grants, 4155
S.E.VEN Fund WHY Prize, 4156
Scleroderma Foundation Established Investigator Grants, 4190
SFI Principal Investigator Grants, 4201
Shastri Indo-Canadian Institute Arts Fellowships, 4219
Shastri Indo-Canadian Institute Collaborative Research Project Grants, 4220
Shastri Indo-Canadian Institute Doctoral Student Research Fellowships, 4221
Shastri Indo-Canadian Institute Faculty Research Fellowships, 4222
Shastri Indo-Canadian Institute Faculty Training Fellowships, 4223
Shastri Indo-Canadian Institute Graduate Degree Fellowships from an Indian Institution, 4224
Shastri Indo-Canadian Ins Librarian Fellowships, 4225
Shastri Indo-Canadian Institute Master's Student Research Fellowships, 4226
Shastri Indo-Canadian Institute Millennium Development Research Grants, 4227
Shastri Indo-Canadian Institute Partnership Development Seed Grants, 4228
Shastri Indo-Canadian Institute Post-Doctoral Research Fellowships, 4229
Shastri Indo-Canadian Institute Research Collaboration Development Grants, 4230
Shastri Indo-Canadian Institute Student Language Training Fellowships, 4231
Shastri Indo-Canadian Institute University of Ottawa Overseas India Summer Course Grants, 4232
SHOT International Scholars, 4242
SHRM Foundation Research Grants, 4249
Sigma Theta Tau International /American Nurses Foundation Grant, 4254

Sigma Theta Tau International / Council for the Advancement of Nursing Science Grants, 4255
Sigma Theta Tau International / Oncology Nursing Society Grant, 4256
Sigma Theta Tau International Doris Bloch Research Award, 4257
Sigma Theta Tau Small Grants, 4258
Sigrid Juselius Foundation Grants, 4260
Simone and Cino del Duca Grand Prix Awards, 4261
Smithsonian Tropical Research Institute Short Term Fellowships, 4297
SNM/ Covidien Seed Grant in Molecular Imaging/ Nuclear Medicine Research, 4298
SNM Postdoc Molecular Imaging Scholar Grants, 4302
SNM Student Fellowship Awards, 4303
Special Olympics Health Professions Grants, 4332
Spencer Foundation Small Research Grants, 4337
SRC Educational Science Conference Grants, 4354
SRC Educational Science Doctoral Grants, 4355
SRC Educational Science Grants for Invitation of Visiting Scientists, 4357
SRC Educational Science Grants for Travels to Foreign Institutions as a Guest Scientist, 4358
SRC Educational Science Junior Research Position Grants, 4359
SRC Educational Science Lang Checking Grants, 4360
SRC Educational Science Network Grants, 4362
SRC Educational Science Postdoc Fellowships, 4363
SRC Educational Science Project Res Grants, 4364
SRC Education Science Grants for Postdoctoral Positions in Sweden, 4366
SRC Humanities and Social Science Postdoctoral Fellowships, 4372
SRC Humanities and Social Sciences Grants, 4373
SRC Humanities and Social Sciences Grants for Postdoctoral Positions In Sweden, 4374
SRC Medical Research Grants, 4379
SRC Medical Science Grants for Postdoctoral Positions in Sweden, 4381
SRC Medical Science Postdoctoral Fellowships, 4382
SRC Natural and Engineering Science Grants for Postdoctoral Positions in Sweden, 4384
SRC Natural and Engineering Science Postdoctoral Fellowships, 4385
SRC Natural and Engineering Science Senior Research Positions in Sweden, 4386
SRC Natural and Engineering Sciences Grants, 4387
SRC Research Infrastructures Grants, 4391
SRC Undergraduate Research Opportunities, 4395
SSHRC-APC Automotive Partnership Canada, 4396
SSHRC-CIHR-SC Research Data Centres Joint Initiative, 4397
SSHRC-FHSS Awards to Scholarly Publications, 4398
SSHRC-IDRC International Partnerships for Sustainable Societies, 4399
SSHRC-NSERC College and Community Innovation Enhancement Grants, 4400
SSHRC-Public Safety Canada Kanishka Project Initiative, 4401
SSHRC Aid to Scholarly Journals, 4402
SSHRC Aid to Small Universities, 4403
SSHRC Banting Postdoctoral Fellowships, 4404
SSHRC Bora Laskin National Fellowship in Human Rights Research, 4405
SSHRC Digging into Data Challenge Grants, 4406
SSHRC Doctoral Fellowships, 4407
SSHRC Insight Development Grants, 4409
SSHRC Insight Grants, 4410
SSHRC Institutional Grants, 4411
SSHRC Joseph-Armand Bombardier CGS Doctoral Scholarships, 4412
SSHRC Jules and Gabrielle Leger Fellowship, 4413
SSHRC Postdoctoral Fellowships, 4415
SSHRC Therese F. Casgrain Fellowship, 4418
SSRC Abe Fellowship for Journalists, 4420
SSRC African Peacebuilding Network Research Grants, 4422
SSRC Dissertation Proposal Development Student Fellowships, 4425
SSRC Eurasia Program Pre-Dissertation Awards, 4428
SSRC Int'l Dissertation Research Fellowships, 4429
SSRC New Directions in the Study of Prayer Research Grants, 4433
SSRC Next Generation Social Sciences in Africa: Doctoral Dissertation Research Fellowships, 4436
SSRC Postdoc Fellowship for Transregional Research: Inter-Asian Contexts and Connections, 4438
Stanford Humanities Center Arts Writer/Practitioner Fellowship, 4444
Stanford Humanities Center External Faculty Fellowships, 4445
Stanley Smith Horticultural Trust Grants, 4446
Starr Foundation Grants, 4447
State Scholarships Foundation Grants, 4449
STTI Alpha Eta Collaborative Research Grants, 4458
Sun Microsystems Foundation Summer Academic Enrichment Grants, 4468
Susan G. Komen Breast Cancer Foundation Brinker Awards for Scientific Distinction, 4472
Susan G. Komen Breast Cancer Foundation Career Catalyst Research Grants, 4473
Susan G. Komen Breast Cancer Foundation Challenge Grants: Career Catalyst Research, 4476
Swann Foundation Fellowships, 4478
Swedish Natural Science Res Council Grants, 4479
Swedish Women's Educational Association International Scholarships, 4480
Swiss National Science Foundation Scientific Conferences Grants, 4486
Takemi Program in International Health Research and Advanced Training Fellowships, 4491
Third Millennium Foundation Grants, 4505
Thrasher Research Fund Grants, 4516
TRDRP Dissertation Research Awards, 4526
U.S. Department of Education American Overseas Research Centers, 4537
U.S. Department of Education Foreign Languages and Area Studies Fellowships, 4539
U.S. Department of Education National Resource Centers, 4542
U.S. Department of Education United States-Russia Program: Improving Research and Educational Activities in Higher Education, 4547
UICC American Cancer Society International Fellowships for Beginning Investigators, 4555
UICC International Cancer Technology Transfer Fellowships, 4556
UICC Yamagiwa-Yoshida Memorial International Cancer Study Grants, 4558
UNESCO Fellowships, 4567
United Methodist Church World is My Parish United Methodist History Research Grant, 4576
United Methodist Committee on Relief Global Health Grants, 4577
United States Institute of Peace - Jennings Randolph Peace Dissertation Scholarships, 4579
United States Institute of Peace - Jennings Randolph Senior Fellowships, 4580
United States Institute of Peace Annual Grants, 4582
University of London Visiting Fellowships, 4588
USAID Accelerating Progress Against Tuberculosis in Kenya Grants, 4594
USAID Comprehensive District-Based Support for Better HIV/TB Patient Outcomes Grants, 4595
USAID Development Innovation Accelerator Broad Agency Announcement Grants, 4596
USAID Devel Innovation Ventures Grants, 4597
USAID Microbicides Research, Development, and Introduction Grants, 4599
USAID Palestinian Community Infrastructure Development Grants, 4600
USAID Res & Innovation for Health Supply Chain Systems and Commodity Security Grants, 4601
USAID Strengthening RMNCH Through Indian Health Professional Associations and Scaling Up Uptake Grants, 4602
USC Annenberg/Getty Arts Journ Fellowships, 4603
US CRDF CGP-Climate Change and Energy Competition For Joint U.S.-Ukrainian Research Teams, 4607
US CRDF Science and Technology Entrepreneurship Program Business Partnership Grant Competition Armenia, 4611
USDA Emerging Markets Grants, 4623
USDA Foreign Market Development Grants, 4625
USDA Higher Education Challenge Grants, 4627
USDA Market Access Grants, 4633
USDA Plant and Animal Disease, Pest Control, and Animal Care Grants, 4638
USDC Market Development Cooperator Program Grants, 4667
USDD Broad Agency Announcement Grants, 4701
USDD Broad Agency Announcement HIV/AIDS Prevention Grants, 4702
USDD HIV/AIDS Prevention: Military Specific HIV/ AIDS Prevention, Care, and Treatment for Non-PEPFAR Funded Countries Grants, 4733
Virginia Foundation for the Humanities Residential Fellowships, 4815
W.K. Kellogg Foundation Healthy Kids Grants, 4822
Weizmann Institute of Science Postdoctoral Fellowships, 4837
Wenner-Gren Foundation African Fellowships, 4846
Wenner-Gren Foundation Conference and Workshop Grants, 4847
Wenner-Gren Foundation Dissertation Fieldwork Grants, 4848
Wenner-Gren Foundation Hunt Postdoctoral Fellowships, 4850
Wenner-Gren Foundation International Collaborative Research Grants, 4851
Wenner-Gren Foundation Post-PhD Res Grants, 4852
Whitley Awards, 4858
WILD Foundation Grants, 4867
Wilhelm Sander Foundation Grants, 4868
William and Charlotte Parks Foundation for Animal Welfare Grants, 4872
Winterthur Research Fellowships, 4888
Wolfsonian-FIU Fellowships, 4890
World Bank Graduate Scholarship Program, 4896
WWF Russell E. Train Education for Nature Alumni Grants, 4897
WWF Russell E. Train Fellowships, 4898
Xerox Foundation Grants, 4899
Young Scientists Summer Program, 4903
Zonta International Foundation Amelia Earhart Fellowships, 4905

Job Training/Adult Vocational Programs

3M Foundation Community Giving Grants, 5
Achelis Foundation Grants, 274
Adaptec Foundation Grants, 340
Alcoa Foundation Grants, 511
Allen Lane Foundation Grants, 571
Allstate Foundation Safe and Vital Communities Grants, 573
Anderson Foundation Grants, 842
ASHA Students Preparing for Academic & Research Careers Award, 1075
Ashland Corporate Contributions Grants, 1087
Assisi Foundation of Memphis General Grants, 1221
ASTA Academic Scholarships, 1224
Berks County Community Foundation Grants, 1333
Bodman Foundation Grants, 1356
Boeing Company Contributions Grants, 1357
Boston Foundation Grants, 1361
Carl M. Freeman Foundation Grants, 1430
Charles Stewart Mott Found Anti-Poverty Grant, 1528
Chiron Foundation Community Grants, 1553
Clay Foundation Grants, 1579
Coors Brewing Corporate Contributions Grants, 1644
Corning Foundation Educational Grants, 1646
Credit Union Found of British Columbia Grants, 1654
Cruise Industry Charitable Foundation Grants, 1660
DaimlerChrysler Corporation Fund Grants, 1714
Dept of Ed Vocational Rehabilitation Services Demonstration and Training Programs, 1751
Evjue Foundation Grants, 1963
Fitzpatrick and Francis Family Business Continuity Foundation Grants, 2021

Ford Motor Company Fund Grants, 2033
Frances L. and Edwin L. Cummings Memorial Fund Grants, 2046
George Gund Foundation Grants, 2117
Helen Bader Foundation Grants, 2228
Helen V. Brach Foundation Grants, 2231
Hutchinson Community Foundation Grants, 2310
JM Foundation Grants, 2490
Johnson & Johnson Corporate Contributions, 2508
Jovid Foundation Grants, 2518
Luther I. Replogle Foundation Grants, 2631
McCune Charitable Foundation Grants, 2719
McGregor Fund Grants, 2725
National Housing Endowment Challenge/Build/Grow Grant, 2877
Norcliffe Foundation Grants, 3415
Oppenstein Brothers Foundation Grants, 3792
Patrick and Anna M. Cudahy Fund Grants, 3873
Reinberger Foundation Grants, 4051
Reuter Foundation Grants, 4059
Sid W. Richardson Foundation Grants, 4253
Sioux Falls Area Community Foundation Community Fund Grants (Unrestricted), 4262
Sioux Falls Area Community Found Spot Grants, 4263
Stackpole-Hall Foundation Grants, 4442
Starr Foundation Grants, 4447
Strake Foundation Grants, 4452
Textron Corporate Contributions Grants, 4501
U.S. Department of Education Rehabilitation Research Training Centers, 4545
WILD Foundation Grants, 4867
Xerox Foundation Grants, 4899

Land Acquisition
Enterprise Community Partners Rose Architectural Fellowships, 1929
HRF Hudson River Improvement Grants, 2286
Norcliffe Foundation Grants, 3415
The Ray Charles Foundation Grants, 4502

Matching/Challenge Funds
3M Foundation Community Giving Grants, 5
A.L. Mailman Family Foundation Grants, 11
A.O. Smith Community Grants, 12
Abbott Fund Global AIDS Care Grants, 219
Adaptec Foundation Grants, 340
African American Heritage Grants, 357
Alabama Humanities Foundation Mini Grants, 480
Alabama Humanities Foundation Planning/Consultant Grants, 481
Alcoa Foundation Grants, 511
Alex Stern Family Foundation Grants, 542
Alice Tweed Tuohy Foundation Grants, 569
Allstate Found Safe & Vital Communities Grants, 573
Allyn Foundation Grants, 574
American Chemical Society Petroleum Research Fund Undergraduate Faculty Sabbatical Grants, 693
AMI Semiconductors Corporate Grants, 813
Ann Arbor Area Community Foundation Grants, 860
Ann Peppers Foundation Grants, 863
Appalachian Reg Commission Health Care Grants, 915
Arie and Ida Crown Memorial Grants, 989
Arizona Public Service Corporate Giving Grants, 998
Ashland Corporate Contributions Grants, 1087
Avon Products Foundation Grants, 1275
Ball Brothers Foundation General Grants, 1291
Bank of America Charitable Foundation Matching Gifts, 1297
Barra Foundation Project Grants, 1301
Batchelor Foundation Grants, 1305
Bay and Paul Foundations, Inc Grants, 1307
Becton Dickinson and Company Grants, 1328
Bodenwein Public Benevolent Found Grants, 1355
Boeing Company Contributions Grants, 1357
Boston Foundation Grants, 1361
Brainerd Foundation Grants, 1368
Brico Fund Grants, 1371
Bristol-Myers Squibb Foundation Health Disparities Grants, 1377
Bullitt Foundation Grants, 1384
Burlington Northern Santa Fe Found Grants, 1389
Burton G. Bettingen Grants, 1390
Bydale Foundation Grants, 1400
Cabot Corporation Foundation Grants, 1406
Callaway Golf Company Foundation Grants, 1410
Campbell Soup Foundation Grants, 1415
Carl and Eloise Pohlad Family Found Grants, 1426
Carl and Lily Pforzheimer Foundation Grants, 1427
Carl C. Icahn Foundation Grants, 1428
Carl M. Freeman Foundation Grants, 1430
Carlos and Marguerite Mason Trust Grants, 1431
Carpenter Foundation Grants, 1439
CDC Public Health Conference Support Grants, 1466
Charles H. Revson Foundation Grants, 1524
Charles Stewart Mott Anti-Poverty Grants, 1528
Chatlos Foundation Grants, 1532
Chestnut Hill Charitable Foundation, Inc Grants, 1538
Christensen Fund Regional Grants, 1554
CIGNA Foundation Grants, 1568
Claneil Foundation Grants, 1574
Clay Foundation Grants, 1579
Cleveland-Cliffs Foundation Grants, 1580
Coca-Cola Foundation Grants, 1593
Colonel Stanley R. McNeil Foundation Grants, 1597
Commonwealth Fund Small Grants, 1611
Communities Foundation of Texas Grants, 1613
Community Found for Greater Buffalo Grants, 1615
Community Fnd for Muskegon County Grants, 1617
Community Foundation for S Arizona Grants, 1618
Connecticut Community Foundation Grants, 1627
ConocoPhillips Grants, 1629
Cord Foundation Grants, 1645
Cowles Charitable Trust Grants, 1652
DaimlerChrysler Corporation Fund Grants, 1714
Dana Foundation Science and Health Grants, 1716
Dave Thomas Foundation for Adoption Grants, 1723
Deloitte Foundation Grants, 1735
DOI African Elephant Conservation Grants, 1820
DOI Rhinoceros and Tiger Conservation Grants, 1824
Dr. Bob and Jean Smith Foundation Grants, 1853
Drs. Bruce and Lee Foundation Grants, 1859
Duchossois Family Foundation Grants, 1867
Eaton Charitable Fund Grants, 1893
Edward Bangs Kelley and Elza Kelley Foundation Grants, 1898
Edward W. and Stella C. Van Houten Grants, 1899
Edwin S. Webster Foundation Grants, 1900
Emerson Charitable Trust Grants, 1918
Engineering Information Foundation Grants, 1927
Ensign-Bickford Foundation Grants, 1928
Eugene McDermott Foundation Grants, 1958
Evan Frankel Foundation Grants, 1961
Ewing Halsell Foundation Grants, 1964
Fannie E. Rippel Foundation Grants, 1978
Farmers Insurance Corporate Giving Grants, 1981
Fidelity Foundation Grants, 2007
Fluor Foundation Grants, 2028
FMC Foundation Grants, 2029
Foundation for Seacoast Health Grants, 2041
Frances L. and Edwin L. Cummings Memorial Fund Grants, 2046
George B. Storer Foundation Grants, 2113
George Foundation Grants, 2115
George Frederick Jewett Foundation Grants, 2116
George Gund Foundation Grants, 2117
George S. & Dolores Dore Eccles Grants, 2120
Georgia Power Foundation Grants, 2125
Goldman Philanthropic Partnerships Program, 2157
GUITS Post-Doctoral Summer Travel-Research Grants, 2189
GUITS Research Grants in Comparative Studies of Modern Turkey, 2190
GUITS Summer Research Grants in Turkey for Graduate Students, 2192
H.A. and Mary K. Chapman Charitable Grants, 2193
Harry A. and Margaret D. Towsley Found Grants, 2205
Harvey Randall Wickes Foundation Grants, 2218
Hatton W. Sumners Foundation for the Study and Teaching of Self Government Grants, 2219
Henry W. Bull Foundation Grants, 2237
Herbert A. & Adrian W. Woods Found Grants, 2239
Herbert H. and Grace A. Dow Grants, 2241
Hill Crest Foundation Grants, 2267
HSBC Contributions Community Grants, 2291
HUD Community Outreach Partnership Centers Grants, 2292
Huie-Dellmon Trust Grants, 2295
ICC Faculty Fellowships, 2333
ICC Scholarship of Engagement Faculty Grants, 2334
IDEM Section 319(h) Nonpoint Source Grants, 2338
Illinois DCEO Coal Research Grants, 2359
IMLS 21st Century Museum Pros Grants, 2362
IMLS Conservation Project Support Grants, 2363
IMLS National Leadership Grants, 2366
Indiana 21st Century Research and Technology Fund Awards, 2370
Indiana Space Grant Consortium Discovery Program Crosscutting Research-Team Grants, 2372
Indiana Space Grant Consortium Discovery Program Research Grants, 2373
Indiana Space Grant Consortium Discovery Program Student Hands-On Research Grants, 2374
Indiana Waste Tire Fund Grants, 2376
ISGC Informal Education Grants, 2431
ISGC Multiple Undergraduate Research Grants, 2432
ISGC Research Seed Grants, 2433
J.W. Kieckhefer Foundation Grants, 2452
Jackson Foundation Grants, 2453
James H. Cummings Foundation Grants, 2460
James McKeen Cattell Fund Fellowships, 2462
James R. Dougherty Jr. Foundation Grants, 2463
Japan Foundation New York Institutional Support for Japanese Studies Grants, 2477
JM Foundation Grants, 2490
John Ben Snow Memorial Trust Grants, 2497
John Deere Foundation Grants, 2500
John P. McGovern Foundation Grants, 2504
John R. Oishei Foundation Grants, 2505
John Stauffer Charitable Trust Grants, 2509
John W. Alden Trust Grants, 2510
Josiah W. and Bessie H. Kline Foundation Grants, 2517
JSA Junior Scientist Travel Support, 2523
Kenneth T. & Eileen L. Norris Found Grants, 2554
Lillian S. Wells Foundation Grants, 2614
Lumpkin Family Found Healthy People Grants, 2629
Lymphatic Education & Research Network Additional Support for NIH-funded F32 Postdoc Fellows, 2632
Lynde and Harry Bradley Foundation Grants, 2635
M.D. Anderson Foundation Grants, 2637
M.J. Murdock Charitable Trust Research Start-up Grants for New Science Faculty, 2643
Margaret & James A. Elkins Jr. Found Grants, 2658
Margaret T. Morris Foundation Grants, 2659
Mary K. Chapman Foundation Grants, 2673
Mary S. and David C. Corbin Foundation Grants, 2675
Matilda R. Wilson Fund Grants, 2684
McCarthy Family Foundation Grants, 2717
McCune Charitable Foundation Grants, 2719
McCune Foundation Education Grants, 2721
McGraw-Hill Companies Community Grants, 2724
McGregor Fund Grants, 2725
MDARD Specialty Crop Block Grants-Farm Bill, 2730
MDEQ Beach Monitoring Grants - Inland Lakes, 2731
MDEQ Local Water Quality Monitoring Grants, 2732
MDEQ Scrap Tire Market Development Grants, 2733
Melville Charitable Trust Grants, 2741
Mericos Foundation Grants, 2742
Meyer Memorial Trust Responsive Grants, 2748
Millipore Foundation Grants, 2775
Motorola Foundation Grants, 2796
Motorola Foundation Innovation Generation Collaborative Grants, 2797
Musgrave Foundation Grants, 2807
NASA Marslink Initiative, 2841
NEH Collaborative Research Grants, 2952
Nelda C. and H.J. Lutcher Stark Found Grants, 2967
NFID New Investigator Matching Grants, 2989
NFWF International Sea Turtle Conservation Fund Grants, 3000
NFWF Pulling Together Initiative Grants, 3007

NHLBI Midcareer Investigator Award in Patient-Oriented Research, 3040
NOAA California Bay Watershed Education and Training (B-WET) Grants, 3338
Norcliffe Foundation Grants, 3415
Nordson Corporation Foundation Grants, 3416
North Carolina Arts Council Folklife Grants, 3419
North Carolina Biotechnology Center Collaborative Funding Grants, 3421
Northwestern Mutual Foundation Grants, 3430
NSF CISE Educational Innovation Grants, 3533
NSF Grant Opportunities for Academic Liaison with Industry, 3574
Oklahoma City Community Grants, 3737
Oppenstein Brothers Foundation Grants, 3792
Padilla Bay National Estuarine Research Reserve Research Assistantships, 3865
Parkersburg Area Community Foundation Arts in Education Special Grants, 3869
Park Foundation Grants, 3870
Patron Saints Foundation Grants, 3874
Perkin Fund Grants, 3898
PhRMA Foundation Health Outcomes Sabbatical Fellowships, 3952
PhRMA Foundation Informatics Sabbatical Fellowships, 3956
PhRMA Foundation Pharmaceutics Sabbatical Fellowships, 3961
PhRMA Foundation Pharmacology/Toxicology Sabbatical Fellowships, 3965
Pinkerton Foundation Grants, 3967
Playboy Foundation Grants, 3980
Praxair Foundation Grants, 3986
Prudential Foundation Education Grants, 4002
R.C. Baker Foundation Grants, 4006
Rathmann Family Foundation Grants, 4033
Rayonier Foundation Grants, 4034
Reinberger Foundation Grants, 4051
Rhode Island Foundation Grants, 4061
Righteous Persons Foundation Grants, 4069
Robert G. Cabell III and Maude Morgan Cabell Foundation Grants, 4075
Rockwell International Corporate Trust Grants, 4093
Rohm and Haas Company Grants, 4098
Ronald McDonald House Charities Grants, 4101
Ross Foundation Grants, 4106
Roy J. Carver Charitable Trust Medical and Science Research Grants, 4114
RRF General Grants, 4115
RWJF Partners Investing in Nursing Grants, 4149
S.E.VEN Fund WHY Prize, 4156
Sage Foundation Grants, 4158
San Diego Women's Foundation Grants, 4164
Sara Elizabeth O'Brien Trust Grants, 4166
Sarkeys Foundation Grants, 4181
Schramm Foundation Grants, 4188
Self Foundation Grants, 4197
Shastri Indo-Canadian Institute University of Ottawa Overseas India Summer Course Grants, 4232
Shell Oil Company Foundation Grants, 4233
Sid W. Richardson Foundation Grants, 4253
Skaggs Foundation Grants, 4266
Sony Corporation of America Corporate Philanthropy Grants, 4329
SSHRC Aid to Scholarly Journals, 4402
Stackpole-Hall Foundation Grants, 4442
Strake Foundation Grants, 4452
Surdna Found Sustainable Environments Grants, 4471
Textron Corporate Contributions Grants, 4501
Thomas Austin Finch, Sr. Foundation Grants, 4507
Thomas C. Ackerman Foundation Grants, 4508
U.S. Department of Education Graduate Assistance in Areas of National Need, 4540
Union Pacific Found Community & Civic Grants, 4568
Union Pacific Foundation Health and Human Services Grants, 4569
USDA Higher Education Challenge Grants, 4627
USDA Scientific Cooperation Research Program, 4647
USDC/NOAA Community-Based Habitat Restoration Grants, 4656
USDC/NOAA Nature Conservancy Community-based Matching Grants, 4661
USDC Market Development Cooperator Program Grants, 4667
Victor E. Speas Foundation Grants, 4810
Virginia Foundation for the Humanities Discretionary Grants, 4813
Virginia Found for Humanities Open Grants, 4814
Wayne and Gladys Valley Foundation Grants, 4835
Welch Foundation Endowed Chairs, 4840
Weyerhaeuser Company Foundation Grants, 4857
William G. McGowan Charitable Fund Grants, 4877
Xerox Foundation Grants, 4899

Materials/Equipment Acquisition (Computers, Books, Videos, etc.)

1st Touch Foundation Grants, 2
AACN-Edwards Lifesciences Nurse-Driven Clinical Practice Outcomes Grants, 102
AACN Clinical Inquiry Fund Grants, 105
AACN End of Life/Palliative Care Small Projects Grants, 108
AACN Physio-Control Small Projects Grants, 111
Abdus Salam ICTP Federation Arrangement Scheme Grants, 227
Abdus Salam ICTP Training and Research in Italian Laboratories Grants, 240
Abington Foundation Grants, 243
ACGT Investigators Grants, 271
Achelis Foundation Grants, 274
ACSM Foundation Doctoral Research Grants, 316
AIHS Alberta/Pfizer Translational Res Grants, 435
Alexander and Margaret Stewart Trust Grants, 514
Alex Stern Family Foundation Grants, 542
Alfred P. Sloan History of Science and Technology Grants, 553
Alice Tweed Tuohy Foundation Grants, 569
Allyn Foundation Grants, 574
American Chemical Society Green Chem Grants, 679
American Family Insurance Corporate Giving, 709
American Philosophical Society Franklin Research Grants, 764
American Philosophical Society Lewis and Clark Fund for Exploration and Field Research, 767
American Philosophical Society Lewis and Clark Exploration and Field Research in Astrobiology, 768
American Schlafhorst Foundation Grants, 796
Amgen Foundation Grants, 812
Ann Arbor Area Community Foundation Grants, 860
Arcadia Foundation Grants, 983
Arie and Ida Crown Memorial Grants, 989
ASHFoundation Grant in Speech Science, 1083
Assisi Foundation of Memphis General Grants, 1221
ASTA Academic Scholarships, 1224
Auburn Foundation Grants, 1263
Ball Brothers Foundation Rapid Grants, 1292
BCBSM Foundation Primary and Clinical Prevention Grants, 1315
BCBSNC Foundation Grants, 1318
Beim Foundation Grants, 1331
Bill and Melinda Gates Foundation Agricultural Development Grants, 1345
Bill and Melinda Gates Foundation Financial Services for the Poor Grants, 1346
Blowitz-Ridgeway Foundation Early Childhood Development Research Award, 1348
Bodenwein Public Benevolent Found Grants, 1355
Bodman Foundation Grants, 1356
Boeing Company Contributions Grants, 1357
Booth-Bricker Fund Grants, 1359
Bosque Foundation Grants, 1360
Brainerd Foundation Grants, 1368
Byron W. & Alice L. Lockwood Found Grants, 1401
Cabot Corporation Foundation Grants, 1406
Caesars Foundation Grants, 1407
Callaway Golf Company Foundation Grants, 1410
Campbell Soup Foundation Grants, 1415
Carl Gellert and Celia Berta Gellert Grants, 1429
Carl M. Freeman Foundation Grants, 1430
Carlos and Marguerite Mason Trust Grants, 1431
Carpenter Foundation Grants, 1439
CCH Documentary Project Research and Development Grants, 1451
Charles H. Farnsworth Trust Grants, 1521
Chatlos Foundation Grants, 1532
Chautauqua Region Community Found Grants, 1533
Chestnut Hill Charitable Foundation, Inc Grants, 1538
Chiles Foundation Grants, 1552
Claneil Foundation Grants, 1574
Coeta and Donald Barker Foundation Grants, 1594
Communities Foundation of Texas Grants, 1613
Community Found for Greater Buffalo Grants, 1615
Community Foundation for Greater New Haven Quinnipiac River Fund Grants, 1616
Community Fnd for Muskegon County Grants, 1617
Community Foundation for S Arizona Grants, 1618
Community Found of E Central Illinois Grants, 1619
Connecticut Community Foundation Grants, 1627
Coors Brewing Corporate Contributions Grants, 1644
Cord Foundation Grants, 1645
Corning Foundation Educational Grants, 1646
Coughlin-Saunders Foundation Grants, 1648
Cowles Charitable Trust Grants, 1652
Cudd Foundation Grants, 1672
Cystic Fibrosis Canada Research Grants, 1678
Cystic Fibrosis Canada Senior Scientist Research Training Awards, 1680
Daimler and Benz Foundation Fellowships, 1713
DaimlerChrysler Corporation Fund Grants, 1714
Dean Witter Foundation Grants, 1728
DeMatteis Family Foundation Grants, 1740
DENSO North America Foundation Grants, 1742
Dept of Ed Star Schools Grants, 1749
DOE Nonproliferation & National Security Research Grants, 1806
DOE Univ Nuclear Science & Reactor Support, 1817
Drs. Bruce and Lee Foundation Grants, 1859
E.L. Wiegand Foundation Grants, 1885
Edward W. and Stella C. Van Houten Grants, 1899
Edwin S. Webster Foundation Grants, 1900
Elisabeth Severance Prentiss Foundation Grants, 1905
Ellison Medical Foundation/AFAR Postdoctoral Fellows in Aging Research Program, 1911
Emil Buehler Perpetual Trust Grants, 1919
Ensign-Bickford Foundation Grants, 1928
ESRI Conservation Grants, 1952
Eugene McDermott Foundation Grants, 1958
Evjue Foundation Grants, 1963
Ewing Halsell Foundation Grants, 1964
F.W. Olin Foundation Construction Grant, 1972
Facebook Fellowships, 1973
Fannie E. Rippel Foundation Grants, 1978
FIC Research Collaboration Award, 2006
Fight for Sight Grants-in-Aid, 2013
Florida Sea Turtle Grants, 2027
FMC Foundation Grants, 2029
Ford Motor Company Fund Grants, 2033
Foundation for Enhancing Comm Grants, 2038
Foundation for Seacoast Health Grants, 2041
Frances C. and William P. Smallwood Foundation Grants, 2045
Frank Stanley Beveridge Foundation Grants, 2051
Fritz B. Burns Foundation Grants, 2057
Fund for Folk Culture Artists Support Grants, 2088
General Motors Foundation Grants, 2107
George A Ohl Jr. Foundation Grants, 2112
George Foundation Grants, 2115
George Frederick Jewett Foundation Grants, 2116
George Kress Foundation Grants, 2119
George W. Codrington Foundation Grants, 2122
Gil and Dody Weaver Foundation Grants, 2139
Greater St Louis Community Foundation Grants, 2164
Greater Worcester Community Foundation Jeppson Memorial Fund for Brookfield Grants, 2165
H.A. and Mary K. Chapman Charitable Grants, 2193
Harris Graduate School of Public Policy Studies Research Development Grants, 2204
Harry W. Bass, Jr. Foundation Grants, 2210
Hawaii Community Foundation Health Education and Research Grants, 2221

Hearst Foundations Health Grants, 2225
Helen Bader Foundation Grants, 2228
Helen V. Brach Foundation Grants, 2231
Henry W. Bull Foundation Grants, 2237
Hill Crest Foundation Grants, 2267
Hilton Hotels Corporate Giving Grants, 2268
Hirtzel Memorial Foundation Grants, 2269
Honor the Earth Grants, 2279
HRF Hudson River Improvement Grants, 2286
Huie-Dellmon Trust Grants, 2295
IMLS Grants to State Library Administrative Agencies, 2364
IREX ECA Alumni Small Grants, 2416
J.B. Reynolds Foundation Grants, 2449
J.W. Kieckhefer Foundation Grants, 2452
Jackson Foundation Grants, 2453
Jacob and Hilda Blaustein Foundation Grants, 2454
Jeffress Awards in Interdisciplinary Research, 2481
John Ben Snow Memorial Trust Grants, 2497
John G. Duncan Trust Grants, 2501
John S. Dunn Research Foundation Grants, 2506
Joseph Alexander Foundation Grants, 2513
Josiah W. and Bessie H. Kline Foundation Grants, 2517
Kenneth T. & Eileen L. Norris Found Grants, 2554
Ludwick Family Foundation Grants, 2625
Lumpkin Family Found Healthy People Grants, 2629
Lymphatic Education & Research Network Additional Support for NIH-funded F32 Postdoc Fellows, 2632
Lynde and Harry Bradley Foundation Grants, 2635
M.D. Anderson Foundation Grants, 2637
M.J. Murdock Charitable Trust General Grants, 2641
Mabel Y. Hughes Charitable Trust Grants, 2647
Macquarie Bank Foundation Grants, 2650
Marcia and Otto Koehler Foundation Grants, 2656
Margaret & James A. Elkins Jr. Found Grants, 2658
Marion Gardner Jackson Charitable Trust Grants, 2664
Mary K. Chapman Foundation Grants, 2673
Mary S. and David C. Corbin Foundation Grants, 2675
Matilda R. Wilson Fund Grants, 2684
MBL Fred Karush Library Readership, 2703
McCune Charitable Foundation Grants, 2719
McCune Foundation Education Grants, 2721
McGregor Fund Grants, 2725
McLean Contributionship Grants, 2727
Meadows Foundation Grants, 2736
Melville Charitable Trust Grants, 2741
Mericos Foundation Grants, 2742
Meyer Memorial Trust Responsive Grants, 2748
MGN Family Foundation Grants, 2754
Mt. Sinai Health Care Foundation Academic Medicine and Bioscience Grants, 2803
Musgrave Foundation Grants, 2807
National Geographic Society Field Research and Exploration Grants, 2873
NCIC Canadian Cancer Society Research Grants, 2919
NCIC Equipment Grants, 2922
NCI Technologies and Software to Support Integrative Cancer Biology Research Grants, 2933
Nellie Mae Education Foundation State Level Systems Change Grants, 2971
NLM Internet Connection for Medical Institutions Grants, 3300
NOAA Undersea Research Grants, 3411
Norcliffe Foundation Grants, 3415
Nordson Corporation Foundation Grants, 3416
North Carolina Biotechnology Center Institutional Development Grants, 3422
NSERC Industrial Research Chairs, 3443
NSERC Research Tools and Instruments Grants, 3449
NSF-NIST Interaction in Chemistry, Materials Research, Molecular Biosciences, Bioengineering, and Chemical Engineering, 3453
NSF Atmospheric Sciences Mid-Size Infrastructure Opportunity Grants, 3497
NSF Biological Field Stations and Marine Laboratories Grants, 3500
NSF Biological Research Collections, 3504
NSF Chemistry Research Instrumentation and Facilities: Departmental Multi-User Instrumentation, 3519
NSF CISE Educational Innovation Grants, 3533
NSF Historically Black Colleges and Universities Grants, 3577
NSF Innovative Technology Experiences for Students and Teachers, 3582
NSF Instrument Devel for Biological Research, 3583
NSF Living Stock Collections Grants, 3588
NSF Materials Research Grants, 3593
NSF National Astronomy and Ionosphere Center Grants, 3598
NSF Physical and Dynamic Meteorology Grants, 3614
NSF Research Experiences for Undergrads, 3638
NSF Scientific Computing Research Environments for the Mathematical Sciences Grants, 3645
NYCT Blood Disease Research Grants, 3696
NYSCA Individual Artists: Film, Media and New Technology Production Grants, 3706
Oklahoma City Community Grants, 3737
Olin Corporation Charitable Trust Grants, 3739
OneSight Research Foundation Block Grants, 3742
Open Spaces Sacred Places National Awards, 3791
Oppenstein Brothers Foundation Grants, 3792
Oscar Rennebohm Foundation Grants, 3832
Patrick and Anna M. Cudahy Fund Grants, 3873
Patron Saints Foundation Grants, 3874
Perpetual Trust for Charitable Giving Grants, 3900
Piper Trust Healthcare & Med Research Grants, 3972
Potts Memorial Foundation Grants, 3985
Praxair Foundation Grants, 3986
Price Family Charitable Fund Grants, 3987
Prudential Foundation Education Grants, 4002
R.C. Baker Foundation Grants, 4006
Rathmann Family Foundation Grants, 4033
Rayonier Foundation Grants, 4034
RCY - NSF Ocean Sciences Div Research Grants, 4043
Reinberger Foundation Grants, 4051
Rett Syndrome Research Foundation Grants, 4058
Rhode Island Foundation Grants, 4061
Richard & Susan Smith Family Found Grants, 4064
Rich Foundation Grants, 4068
Robert G. Cabell III and Maude Morgan Cabell Foundation Grants, 4075
Robert M. Hearin Foundation Grants, 4077
Rocky Mountain Mineral Law Found Grants, 4094
Ronald McDonald House Charities Grants, 4101
Ross Foundation Grants, 4106
RWJF Partners Investing in Nursing Grants, 4149
Sage Foundation Grants, 4158
San Antonio Area Foundation Grants, 4163
Sarkeys Foundation Grants, 4181
Schramm Foundation Grants, 4188
Seattle Foundation Medical Funds Grants, 4194
Sensient Technologies Foundation Grants, 4199
Sid W. Richardson Foundation Grants, 4253
Sioux Falls Area Community Foundation Community Fund Grants (Unrestricted), 4262
Sioux Falls Area Community Found Spot Grants, 4263
Skaggs Foundation Grants, 4266
Sony Corporation of America Corporate Philanthropy Grants, 4329
SPIRE Postdoctoral Fellowships, 4345
SRC Educational Science Medium-Cost Scientific Equipment Grants, 4361
SRC Research Infrastructures Grants, 4391
Stackpole-Hall Foundation Grants, 4442
Stanley Smith Horticultural Trust Grants, 4446
Starr Foundation Grants, 4447
Strake Foundation Grants, 4452
Sun Academic Excellence Grants, 4466
Sun Microsystems Foundation Summer Academic Enrichment Grants, 4468
T. James Kavanagh Foundation Grants, 4489
Textron Corporate Contributions Grants, 4501
The Ray Charles Foundation Grants, 4502
Thomas Austin Finch, Sr. Foundation Grants, 4507
U.S. Department of Education Rehabilitation Engineering Research Centers Grants, 4544
UCLA Research Grants in Ethnic Studies, 4549
UE Explore Undergraduate Research Grants, 4553
Union Pacific Found Community & Civic Grants, 4568
United Methodist Church Women in History Research Grant, 4574
United Methodist Committee on Relief Global Health Grants, 4577
USDA Higher Education Challenge Grants, 4627
USDA Hispanic-Serving Institutions Education Grants, 4628
Vancouver Foundation Grants and Community Initiatives Program, 4807
Victor E. Speas Foundation Grants, 4810
W.M. Keck Foundation Med Research Grants, 4823
W.M. Keck Foundation Science and Engineering Research Grants, 4824
Weyerhaeuser Company Foundation Grants, 4857
William and Charlotte Parks Foundation for Animal Welfare Grants, 4872
William G. & Helen C. Hoffman Found Grants, 4876
William G. McGowan Charitable Fund Grants, 4877
William N. and Myriam Pennington Foundation Grants, 4879
Woodward Fund Grants, 4895

Preservation/Restoration

ACC Ford Foundation Fellowships, 253
African American Heritage Grants, 357
Albuquerque Community Foundation Grants, 508
Alces Foundation Grants, 510
American Society for Yad Vashem Grants, 798
Anderson Foundation Grants, 842
Annenberg Foundation Grants, 862
APA Distinguished Service to Psychological Science Award, 909
Ashland Corporate Contributions Grants, 1087
ASID Foundation Dora Brahms Award, 1092
Bank of America Charitable Foundation Matching Gifts, 1297
Batchelor Foundation Grants, 1305
Bay and Paul Foundations, Inc Grants, 1307
Booth-Bricker Fund Grants, 1359
Carl M. Freeman Foundation Grants, 1430
Charles Edison Fund Grants, 1520
Clarence E. Heller Foundation Grants, 1575
Community Found for Greater Buffalo Grants, 1615
Community Foundation for Greater New Haven Quinnipiac River Fund Grants, 1616
Cudd Foundation Grants, 1672
D.F. Halton Foundation Grants, 1692
Earhart Foundation Institutional Grants, 1888
Edwin S. Webster Foundation Grants, 1900
Endangered Language Fund Grants, 1921
Florida Sea Turtle Grants, 2027
Forrest C. Lattner Foundation Grants, 2035
George Kress Foundation Grants, 2119
Grammy Foundation Grants, 2161
Green Diamond Charitable Contributions, 2171
Harold R. Bechtel Charitable Remainder Uni-Trust Grants, 2201
Honeywell Corporation Habitat and Conservation Grants, 2277
HRF Hudson River Improvement Grants, 2286
IMLS 21st Century Museum Pros Grants, 2362
IMLS Conservation Project Support Grants, 2363
IMLS Grants to State Library Administrative Agencies, 2364
IMLS National Leadership Grants, 2366
J.W. Kieckhefer Foundation Grants, 2452
John Ben Snow Memorial Trust Grants, 2497
Luther I. Replogle Foundation Grants, 2631
Marcia and Otto Koehler Foundation Grants, 2656
Margaret & James A. Elkins Jr. Found Grants, 2658
McCombs Foundation Grants, 2718
McCune Charitable Foundation Grants, 2719
McLean Contributionship Grants, 2727
National Geographic Society/Waitt Grants, 2870
National Geographic Society Conservation Trust Grants, 2871
National Geographic Young Explorers Grants, 2874
NEH Documenting Endangered Lang Grants, 2955
NHPRC Archives - Detailed Processing Projects Grants, 3071

NHPRC Digitizing Historical Records Grants, 3072
NOAA Ocean Exploration Omnibus: Marine Archaeology Grants, 3397
Norcliffe Foundation Grants, 3415
NPS American Battlefield Protection Grants, 3435
NSF Cultural Anthropology Research Grants, 3548
NSF Physical Anthropology Research Grants, 3615
NYCT New York City Environment Grants, 3697
NYSCA Folk Arts: Exhibitions Grants, 3703
NYSCA Folk Arts: Presentation Grants, 3704
NYSCA Folk Arts: Regional and County Folk Arts Grants, 3705
NYSCA Museum: Project Support Grants, 3707
Rhode Island Foundation Grants, 4061
Robert G. Cabell III and Maude Morgan Cabell Foundation Grants, 4075
Ross Foundation Grants, 4106
Sioux Falls Area Community Foundation Community Fund Grants (Unrestricted), 4262
Sioux Falls Area Community Found Spot Grants, 4263
Smithsonian Postgraduate Fellowships in Conservation of Museum Collections, 4294
Strake Foundation Grants, 4452
Textron Corporate Contributions Grants, 4501
Thomas Austin Finch, Sr. Foundation Grants, 4507
USDA McIntire-Stennis Cooperative Forestry Research Grants, 4634
Virginia Foundation for the Humanities Discretionary Grants, 4813
Virginia Found for Humanities Open Grants, 4814
Wenner-Gren Found Historical Archives Grants, 4849
WILD Foundation Grants, 4867
Winterthur Research Fellowships, 4888

Professorships

Abdus Salam ICTP Group Associates Grants, 228
Abdus Salam ICTP Regular Associates Grants, 237
Abdus Salam ICTP Senior Associates Grants, 239
AIHS Knowledge Exchange Visiting Professors, 451
Conquer Cancer Foundation of ASCO Drug Development Research Professorship, 1632
SPIRE Postdoctoral Fellowships, 4345
Streilein Foundation for Ocular Immunology Visiting Professorships, 4453

Publishing/Editing/Translating

A.L. Mailman Family Foundation Grants, 11
AACN-Edwards Lifesciences Nurse-Driven Clinical Practice Outcomes Grants, 102
AACN Clinical Inquiry Fund Grants, 105
AACN End of Life/Palliative Care Small Projects Grants, 108
AACN Physio-Control Small Projects Grants, 111
AAUW American Summer/Short-Term Research Publication Grants, 211
AEG Foundation Legget Fund Grants, 341
Ahn Family Foundation Grants, 379
Aid to Scholarly Publications, 431
Albuquerque Community Foundation Grants, 508
Alfred P. Sloan Civic Initiatives Grants, 545
Alfred P. Sloan Data and Computational Research Grants, 546
American Academy of Advertising Doctoral Dissertation Competition Awards, 608
American Academy of Advertising Ivan Preston Outstanding Contribution to Research on Advertising Award, 609
American Academy of Advertising Kim Rotzoll Award for Advertising Ethics and Social Responsibility, 610
American Academy of Advertising Research Fellowships, 611
American Numismatic Society Donald Groves Numismatic Research Grants, 760
Anderson Foundation Grants, 842
Ann Arbor Area Community Foundation Grants, 860
AOCS Phospholipid Best Paper Award, 889
ASCE Freeman Fellowships, 1040
ASHA Advancing Academic Careers Award, 1072
ASHA Student Research Travel Award, 1074
ASME Burt L. Newkirk Award, 1111
Ball Brothers Foundation General Grants, 1291
Bay and Paul Foundations, Inc Grants, 1307
Beverly Willis Architecture Foundation Grants and Fellowships, 1340
Bodenwein Public Benevolent Found Grants, 1355
CAA Wyeth Publication Grant, 1405
Calvin Inst of Christian Worship Renewal Grants, 1411
Canadian Studies Senior Fellowships, 1422
Carl and Lily Pforzheimer Foundation Grants, 1427
Carl Gellert and Celia Berta Gellert Foundation Grants, 1429
Center for Education of Women Visiting Scholar Research Grants, 1475
Century Foundation Grants, 1489
CFR International Affairs Fellowships, 1501
Chatlos Foundation Grants, 1532
Chautauqua Region Community Found Grants, 1533
Christine O. Gregoire Youth/Young Adult Award for Outstanding Use of Tobacco Ind Documents, 1557
Christine Wilson Award, 1558
Clarence E. Heller Foundation Grants, 1575
Community Fnd for Muskegon County Grants, 1617
Community Foundation for S Arizona Grants, 1618
Community Found of E Central Illinois Grants, 1619
Community Foundation of the Virgin Islands Judith A. Towle Environmental Studies Grants, 1621
Connecticut Community Foundation Grants, 1627
Conservation, Food, and Health Foundation Grants for Developing Countries, 1641
Dean Witter Foundation Grants, 1728
Ensign-Bickford Foundation Grants, 1928
Equipment Leasing and Finance Foundation Research Grants, 1943
Ewing Halsell Foundation Grants, 1964
Ford Foundation Peace and Social Justice Grants, 2031
Foundation for Enhancing Comm Grants, 2038
Frank Stanley Beveridge Foundation Grants, 2051
Furthermore Grants in Publishing, 2091
General Motors Foundation Grants, 2107
George A Ohl Jr. Foundation Grants, 2112
George Gund Foundation Grants, 2117
Getty Research Grants for Institutions - Critical Reference Resources Grants, 2134
Getty Research Institute Nonresidential Grants - Curatorial Research Fellowships, 2136
GUITS Grants for the Publication of Scholarly Books and Journals, 2186
GUITS Grants for Undergraduate Study, 2187
GUITS Sabbatical Research Grants, 2191
Helen V. Brach Foundation Grants, 2231
Henry J. Kaiser Family Found Media Fellowships, 2234
HHMI International Research Scholars Program, 2258
Hill Crest Foundation Grants, 2267
ICCS Publishing Fund, 2336
IREX ECA Alumni Small Grants, 2416
J.B. Reynolds Foundation Grants, 2449
J.W. Kieckhefer Foundation Grants, 2452
Japan Found Center for Global Partnership Grants, 2472
Japan Foundation Japanese Studies Fellowships, 2476
Japan Foundation New York Japanese Studies Fellowships, 2478
JM Foundation Grants, 2490
JMO Law and Legal System Grants, 2491
JMO Public Policy Research Grants, 2492
JMO Strategic and International Studies Grants, 2493
John P. McGovern Foundation Grants, 2504
Link Foundation Energy Doctoral Fellowships, 2617
Ludwig Vogelstein Foundation Grants, 2626
Lynde and Harry Bradley Foundation Grants, 2635
McLean Contributionship Grants, 2727
Melville Charitable Trust Grants, 2741
MLA Murray Gottlieb Prize Essay Award, 2788
Motorola Foundation Innovation Generation Collaborative Grants, 2797
National Endowment for the Arts Literature Fellowships: Translation Projects, 2866
National Headache Foundation Seymour Diamond Clinical Fellowship in Headache Education, 2876
NCFR Affiliate Councils Outstanding Student Research Papers, 2906
NEH Documenting Endangered Lang Grants, 2955
NFJC Gantz-Zahler Grant, 2993
NHPRC Publishing Historical Records Grants, 3075
NHPRC State and National Archival Partnership Grants, 3076
NHPRC Strategies and Tools for Archives and Historical Publishing Projects Grants, 3077
NIST National Standard Reference Data System Research Grants, 3271
NL Audrey Lumsden-Kouvel Fellowships, 3282
NL Herzog August Bibliothek Wolfenbuttel Fellowship, 3286
NL Lester J. Cappon Fellowship in Documentary Editing, 3288
NLM Grants for Scholarly Works in Biomedicine and Health, 3298
NL Northeast Modern Language Assocation Fellowship, 3314
NL Weiss/Brown Publication Subvention Award, 3320
Norcliffe Foundation Grants, 3415
North Carolina Arts Council Folklife Grants, 3419
NYSCA Music: Services to the Field Grants, 3708
Olin Corporation Charitable Trust Grants, 3739
Overbrook Foundation Grants, 3860
Perl 6 Microgrants, 3899
Playboy Foundation Grants, 3980
Potts Memorial Foundation Grants, 3985
Reinberger Foundation Grants, 4051
Richard Carley Hunt Postdoc Fellowships, 4065
Rocky Mountain Mineral Law Found Grants, 4094
Ross Foundation Grants, 4106
RSC Bancroft Award, 4118
RSC Jason A. Hannah Medal, 4119
S.E.VEN Fund Mini-Grants, 4154
S.E.VEN Fund Open Enterprise Solutions to Poverty RFP Grants, 4155
S.E.VEN Fund WHY Prize, 4156
Shell Oil Company Foundation Grants, 4233
Sid W. Richardson Foundation Grants, 4253
Sigma Theta Tau International / Council for the Advancement of Nursing Science Grants, 4255
SSHRC-FHSS Awards to Scholarly Publications, 4398
SSHRC Aid to Scholarly Journals, 4402
SSRC Abe Fellowship for Journalists, 4420
SSSR Distinguished Article Award, 4440
SSSR Student Paper Award, 4441
Stanley Smith Horticultural Trust Grants, 4446
Swedish Natural Science Res Council Grants, 4479
Textron Corporate Contributions Grants, 4501
Thoracic Foundation Grants, 4513
TIAA-CREF Paul A. Samuelson Award, 4518
United Methodist Church Center for the Evangelical UBH Audrie E. Reber Memorial Award, 4571
United Methodist Church Women in History Research Grant, 4574
United Methodist Church Women in History Writing Award, 4575
United Methodist Church World is My Parish United Methodist History Research Grant, 4576
United States Institute of Peace - Jennings Randolph Peace Dissertation Scholarships, 4579
United States Institute of Peace - Jennings Randolph Senior Fellowships, 4580
Virginia Foundation for the Humanities Discretionary Grants, 4813
Virginia Found for Humanities Open Grants, 4814
Virginia Foundation for the Humanities Residential Fellowships, 4815
Web We Want Small Grants, 4836
Wenner-Gren Foundation Post-PhD Res Grants, 4852
Weyerhaeuser Company Foundation Grants, 4857
WILD Foundation Grants, 4867
Winterthur Research Fellowships, 4888

Religious Programs

AAB Berlin Prize Fellowships, 48
Aaron & Cecile Goldman Family Found Grants, 200
AAR Postdoctoral Fellowships, 201
Academy of American Franciscan History Dissertation Fellowships, 249

ACC Asian Art and Religion Fellowships, 251
Achelis Foundation Grants, 274
ACLS Charles A. Ryskamp Research Fellowships, 281
Acton Institute for the Study of Religion and Liberty Calihan Academic Fellowships, 331
Acton Institute for the Study of Religion and Liberty Novak Award, 332
Adaptec Foundation Grants, 340
AHEPA Educational Foundation Scholarships, 373
Ahmanson and Getty Postdoctoral Fellowships, 377
AIAR Annual Professorships, 393
AIAR Ernest S. Frerichs Fellow and Program Coordinator Grant, 394
AIAR George A. Barton Fellowships, 395
AJA Ethel Marcus Memorial Fellowship in American Jewish Studies, 471
AJA Lowenstein-Wiener Summer Fellowship Awards in American Jewish Studies, 472
AJA Marguerite R. Jacobs Memorial Postdoctoral Award in American Jewish Studies, 473
AJA Rabbi Frederic A. Doppelt Memorial Fellowship in American Jewish Studies, 474
AJA Starkoff Fellow in American Jewish Studies, 475
Alabama Humanities Foundation Mini Grants, 480
Alabama Humanities Foundation Planning/Consultant Grants, 481
Albert and Margaret Alkek Foundation Grants, 504
Albuquerque Community Foundation Grants, 508
Alces Foundation Grants, 510
Alcoa Foundation Grants, 511
All Saints Educational Trust Grants, 572
Allyn Foundation Grants, 574
Alpha Kappa Alpha Educational Advancement Foundation Fellowships, 575
Alpha Sigma Nu Book Award Competition, 578
ALSAM Foundation Grants, 579
Alvin and Lucy Owsley Foundation Grants, 584
American Academy of Religion Collaborative Research Grants, 617
American Academy of Religion Individual Research Grants, 618
American Foodservice Charitable Trust Grants, 710
American Jewish Historical Society Awards, 747
Anderson Foundation Grants, 842
Ann Arbor Area Community Foundation Grants, 860
Anthony R. Abraham Foundation Grants, 869
Arcadia Foundation Grants, 983
Archer Daniels Midland Foundation Grants, 985
Arcus Foundation Religion and Values Grants, 987
Arie and Ida Crown Memorial Grants, 989
A Rocha USA Grants, 1006
Arronson Foundation Grants, 1019
Arthur and Rochelle Belfer Foundation Grants, 1023
Ashland Corporate Contributions Grants, 1087
ASR Fichter Research Grant, 1217
ASR Robert J. McNamara Student Award, 1218
Assisi Foundation of Memphis General Grants, 1221
ATLA Bibliography Grant, 1257
ATS Henry Luce III Fellowships, 1258
ATS Lilly Theological Collaborative Res Grants, 1259
ATS Lilly Theological Faculty Fellowships, 1260
ATS Lilly Theological Research Expense Grants, 1261
ATS Lilly Theological Scholars Grants, 1262
Banfi Vintners Foundation Grants, 1296
Barra Foundation Project Grants, 1301
Bay and Paul Foundations, Inc Grants, 1307
Baylor Inst for Faith & Learning Visiting Fellows, 1310
BCBS of Massachusetts Foundation Grants, 1319
Beatrice Laing Trust Grants, 1324
Beim Foundation Grants, 1331
Blowitz-Ridgeway Foundation Early Childhood Development Research Award, 1348
Blowitz-Ridgeway Foundation Grants, 1349
Blue Shield of California Grants, 1350
Blum-Kovler Foundation Grants, 1351
Bodenwein Public Benevolent Found Grants, 1355
Bodman Foundation Grants, 1356
Booth-Bricker Fund Grants, 1359
Boston Foundation Grants, 1361
Brach Family Foundation Grants, 1367
Brainerd Foundation Grants, 1368
Bristol-Myers Squibb Foundation Health Disparities Grants, 1377
Burden Trust Grants, 1388
Burton G. Bettingen Grants, 1390
Byron W. & Alice L. Lockwood Found Grants, 1401
BYU Religious Studies Center Research Grants, 1402
Cabot Corporation Foundation Grants, 1406
Callaway Golf Company Foundation Grants, 1410
Calvin Inst of Christian Worship Renewal Grants, 1411
Carl C. Icahn Foundation Grants, 1428
Carl Gellert and Celia Berta Gellert Foundation Grants, 1429
Castle Rock Foundation Grants, 1441
Catholic University Academic Scholarships, Grants, and Awards, 1443
CCCF Dora Macllelan Brown Sem Scholarships, 1448
Center for Advanced Judaic Studies Postdoctoral Fellowships, 1472
Center for Hellenic Studies Fellowships, 1476
Center for Judaic Studies Fellowships, 1478
Center for the Study of Religion Fellowships, 1485
Charles H. Revson Foundation Grants, 1524
Charles Stewart Mott Foundation Grants, 1529
Charlotte W. Newcombe Doctoral Dissertation Fellowships, 1531
Chatlos Foundation Grants, 1532
Chautauqua Region Community Found Grants, 1533
Chiles Foundation Grants, 1552
Clark Library Short-Term Resident Fellowships, 1577
Cleveland-Cliffs Foundation Grants, 1580
Coeta and Donald Barker Foundation Grants, 1594
Commonwealth Fund Small Grants, 1611
Communities Foundation of Texas Grants, 1613
Community Found for Greater Buffalo Grants, 1615
Community Fnd for Muskegon County Grants, 1617
Community Found of E Central Illinois Grants, 1619
Community Foundation of the Fox River Valley, 1620
Connecticut Health Foundation Health Initiative Grants, 1628
Cord Foundation Grants, 1645
Coughlin-Saunders Foundation Grants, 1648
Cushwa Center Research Travel Grants, 1673
Dade Community Foundation Grants, 1712
DaimlerChrysler Corporation Fund Grants, 1714
Dave Thomas Foundation for Adoption Grants, 1723
Donald and Sylvia Robinson Family Grants, 1830
Energy Foundation Buildings Grants, 1922
Erasmus Institute Fellowships, 1944
Evan Frankel Foundation Grants, 1961
Evjue Foundation Grants, 1963
Fetzer Institute Research Grants, 2005
Fishman Family Foundation Grants, 2020
Foundation for Evangelism Harry Denman Fellowships, 2040
Frank Stanley Beveridge Foundation Grants, 2051
Fritz B. Burns Foundation Grants, 2057
FTE Dissertation Fellowships, 2058
FTE Doctoral Fellowships, 2059
FTE Expanding Horizons Partnership Grants, 2060
FTE North American Doctoral Fellowships, 2061
G. Unger Vetlesen Foundation Grants, 2093
George Frederick Jewett Foundation Grants, 2116
George Kress Foundation Grants, 2119
Gilroy and Lillian P. Roberts Foundation Grants, 2142
Girl's Best Friend Foundation Grants, 2143
Gladys Krieble Delmas Foundation Grants for Venetian Research, 2144
Greater St Louis Community Foundation Grants, 2164
Greenspun Family Foundation Grants, 2173
Harold Simmons Foundation Grants, 2203
Harry Frank Guggenheim Foundation Research Grants, 2208
Harry W. Bass, Jr. Foundation Grants, 2210
Harvey Randall Wickes Foundation Grants, 2218
Haverford College Mellon Postdoc Fellowships, 2220
Healthcare Foundation of New Jersey Grants, 2222
Helen Bader Foundation Grants, 2228
Helen Kellogg Institute for International Studies Residential Fellowships, 2230
Helen V. Brach Foundation Grants, 2231
Henry W. Bull Foundation Grants, 2237
Herbert A. & Adrian W. Woods Found Grants, 2239
Herbert H. and Grace A. Dow Grants, 2241
HIAS Scholarship Awards, 2264
Hilda and Preston Davis Foundation Grants, 2266
Houghton Library Visiting Fellowships, 2284
Huie-Dellmon Trust Grants, 2295
Infinity Foundation Grants, 2378
Institute for Advanced Studies in Culture Postdoctoral Fellowships, 2380
IREX Short-Term Travel Grants, 2424
Jacob and Hilda Blaustein Foundation Grants, 2454
James M. Collins Foundation Grants, 2461
Jenifer Altman Foundation Grants, 2484
Jewish Federation of Metropolitan Chicago Scholarships, 2486
Jewish Social Entrepreneurs Fellowships, 2487
John G. Duncan Trust Grants, 2501
Joseph Alexander Foundation Grants, 2513
Joseph C. Wey Fellowship in Medieval Studies, 2514
Kluge Center David B. Larson Fellowship in Health and Spirituality, 2559
Kovler Family Foundation Grants, 2575
Kress Foundation Art History Pre-Doctoral Fellowships, 2576
LI American Religion Dissertation Fellowship, 2604
LI General Grants, 2613
Lucile Horton Howe & Mitchell B. Howe Grants, 2624
Lumpkin Family Found Healthy People Grants, 2629
Lynde and Harry Bradley Foundation Grants, 2635
M. Bastian Family Foundation Grants, 2636
M.J. Murdock Charitable Trust General Grants, 2641
Margaret & James A. Elkins Jr. Found Grants, 2658
Marion Gardner Jackson Charitable Trust Grants, 2664
Martin E.P. Seligman Award for Outstanding Research in Positive Psychology, 2667
Mary Baker Eddy Library Summer Research Fellowships, 2669
McCune Foundation Human Services Grants, 2722
Metanexus Institute Templeton Res Lectures, 2744
Meyer and Stephanie Eglin Foundation Grants, 2747
MGN Family Foundation Grants, 2754
Millstone Foundation Grants, 2776
Motorola Foundation Grants, 2796
Musgrave Foundation Grants, 2807
National Foundation for Catholic Youth Grants, 2868
NCFR Innovation Grants, 2908
Nellie Mae Education Foundation State Level Systems Change Grants, 2971
Nell Warren Elkin and William Simpson Elkin Foundation, 2972
New Hampshire Charitable Foundation Grants, 2985
NFIE Found Learning & Leadership Grants, 2992
Norcliffe Foundation Grants, 3415
NWSA Graduate Scholarship in Jewish Women's Studies, 3676
Oppenstein Brothers Foundation Grants, 3792
Overbrook Foundation Grants, 3860
Patrick and Anna M. Cudahy Fund Grants, 3873
Perpetual Trust for Charitable Giving Grants, 3900
Pinkerton Foundation Grants, 3967
Princeton Univ Whiting Postdoc Fellowships, 3998
R.C. Baker Foundation Grants, 4006
Ralph Waldo Emerson Award, 4028
Rathmann Family Foundation Grants, 4033
Religion and Health: Effects, Mechanisms, and Interpretation RFP, 4052
Reuter Foundation Grants, 4059
RGK Foundation Grants, 4060
Richard and Rhoda Goldman Fund Grants, 4063
Righteous Persons Foundation Grants, 4069
Robert E. and Evelyn McKee Foundation Grants, 4074
Robert G. Cabell III and Maude Morgan Cabell Foundation Grants, 4075
Rockwell International Corporate Trust Grants, 4093
Ronald McDonald House Charities Grants, 4101
Roy J. Carver Charitable Trust Medical and Science Research Grants, 4114
RSC Sir John William Dawson Medal, 4125

Sage Foundation Grants, 4158
San Antonio Area Foundation Grants, 4163
Sioux Falls Area Community Foundation Community Fund Grants (Unrestricted), 4262
Sioux Falls Area Community Found Spot Grants, 4263
SSRC New Directions in the Study of Prayer Research Grants, 4433
SSSR Distinguished Article Award, 4440
SSSR Student Paper Award, 4441
Strake Foundation Grants, 4452
Sunny and Abe Rosenberg Foundation Grants, 4469
T. James Kavanagh Foundation Grants, 4489
Talbert and Leota Abrams Foundation, 4492
Tanner Humanities Center Research Fellowships, 4494
Thomas and Dorothy Leavey Foundation Grants, 4506
Thomas Austin Finch, Sr. Foundation Grants, 4507
Thomas C. Ackerman Foundation Grants, 4508
Thompson Charitable Foundation Grants, 4511
Todd Brock Family Foundation Grants, 4522
United Methodist Church Asian, Black, Hispanic and Native American History Research Grants, 4570
United Methodist Church Center for the Evangelical UBH Audrie E. Reber Memorial Award, 4571
United Methodist Church John Harrison Ness Memorial Award, 4572
United Methodist Church Women in History Research Grant, 4574
United Methodist Church Women in History Writing Award, 4575
United Methodist Church World is My Parish United Methodist History Research Grant, 4576
United Methodist Committee on Relief Global Health Grants, 4577
United Methodist Communications Stoody-West Fellowship, 4578
United States Institute of Peace Annual Grants, 4582
United Theological Seminary Merit-Based Scholarships, 4583
University of Colorado Doctoral Dissertation Fellowship in Media, Religion, and Culture, 4586
USC Pentecostal and Charismatic Research Initiative Individual Grants, 4605
Vancouver Foundation Grants and Community Initiatives Program, 4807
W.C. Griffith Foundation Grants, 4819
Wabash Center Project Grants, 4826
Wabash Center Small Project Grants, 4827
Wabash Center Waorkshop and Colloquy Participant Grants, 4828
Wayne and Gladys Valley Foundation Grants, 4835
Weyerhaeuser Company Foundation Grants, 4857
William G. & Helen C. Hoffman Found Grants, 4876
Woodward Fund Grants, 4895

Scholarships

A.O. Smith Community Grants, 12
AAP Resident Research Grants, 199
Abbott Fund Science Education Grants, 220
Achelis Foundation Grants, 274
ACS Doctoral Scholarships in Cancer Nursing, 307
Adaptec Foundation Grants, 340
AFUW Diamond Jubilee Bursary, 359
AFUW Doreen McCarthy, Barbara Crase, Cathy Candler & Brenda Nettle Bursaries, 360
AFUW Heather Latz Bursary, 363
AFUW Padnendadlu Graduate Bursary, 364
AFUW Padnendadlu Postgraduate Bursary, 365
AFUW Padnendadlu Undergraduate Bursary, 366
AFUW South Australia Jean Gilmore, Thenie Baddams & Daphne Elliott Bursaries, 367
AFUW Winifred E. Preedy Postgraduate Bursary, 368
AGHE Graduate Scholarships and Fellowships, 369
AHEPA Educational Foundation Scholarships, 373
AIChE Alpha Chi Sigma Award, 396
AIChE Andrew Chase Forest Bioproducts Division Award in Chemical Engineering, 397
AIChE Clarence G. Gerhold Award, 400
AIChE Donald Q. Kern Award, 406
AIChE Excellence in Process Development Research Award, 409
AIChE Food, Pharmaceutical and Bioengineering Division Awards in Chemical Engineering, 410
AIChE Institute Award for Excellence in Industrial Gases Technology, 412
AIChE John A. Tallmadge Award for Contributions to Coating Technology, 413
AIChE Lectureship Award in Fluidization, 416
AIChE Nanoscale Science and Engineering Forum Young Investigator Award, 417
AIChE North American Mixing Forum Award, 418
AIChE Nuclear Engineering Division Outstanding Student Paper & Presentation Awards, 421
AIChE Process Development Practice Award, 422
AIChE Research Excellence In Sustainable Engineering Award, 424
AIChE Thomas Baron Award in Fluid-Particle Systems, 426
AIChE Transport & Energy Processes Div Award, 427
AIChE Warren K. Lewis Award for Contributions to Chemical Engineering Education, 429
AIChE William H. Walker Award for Excellence in Contributions to Chem Engineering Lit, 430
AIHS Full-Time M.D./Ph.D. Studentships, 447
AIHS Part-Time Studentships, 455
AINSE Honours Scholarships, 460
Albert and Margaret Alkek Foundation Grants, 504
Albuquerque Community Foundation Grants, 508
Alcatel-Lucent Technologies Foundation Grants, 509
Alcoa Foundation Grants, 511
Alex Stern Family Foundation Grants, 542
Alice Fisher Society Fellowships, 568
Alice Tweed Tuohy Foundation Grants, 569
Allyn Foundation Grants, 574
Alpha Omega Foundation Grants, 576
ALSAM Foundation Grants, 579
ALZA Corporate Contributions Grants, 588
AMA-WPC Joan F. Giambalvo Scholarships, 600
American Academy of Nursing Building Academic Geriatric Nursing Capacity Scholarships, 614
American Academy of Nursing Mayday Grants, 616
American College of Surgeons and The Triological Society Clinical Scientist Development Awards, 701
American College of Surgeons Resident Research Scholarships, 705
American Family Insurance Corporate Giving, 709
American Foodservice Charitable Trust Grants, 710
American Psychiatric Association Travel Scholarships for Minority Medical Students, 790
American Schlafhorst Foundation Grants, 796
American Society for Enology and Viticulture Scholarships, 797
American Society of Naval Engineers Scholarship, 800
American Wine Society Educational Foundation Scholarship Program, 808
AMI Semiconductors Corporate Grants, 813
AMS Undergraduate Scholarships, 837
Angels Baseball Foundation Grants, 848
Ann Arbor Area Community Foundation Grants, 860
Ann Peppers Foundation Grants, 863
Appraisal Institute Education Trust Scholarships, 917
Arcadia Foundation Grants, 983
Archer Daniels Midland Foundation Grants, 985
Archibald Waynne Dingman Memorial Graduate Scholarship, 986
Arie and Ida Crown Memorial Grants, 989
Armenian Students' Association Scholarships, 1000
Arronson Foundation Grants, 1019
ASCE Arthur S. Tuttle Memorial Scholarships, 1039
ASCSA Summer Sessions Scholarships, 1055
ASHA Students Preparing for Academic & Research Careers Award, 1075
ASHFoundation Graduate Student Scholarships, 1077
ASHFoundation Graduate Student Scholarships for Minority Students, 1078
ASHFoundation Graduate Student with a Disability Scholarship, 1079
ASHFoundation New Century Scholars Program Doctoral Scholarships, 1080
ASHFoundation Student Research Grants in Early Childhood Language Development, 1085
ASID Foundation Grants, 1093
ASID Foundation Mabelle Wilhelmina Boldt Memorial Scholarship, 1097
ASPRS Robert E. Altenhofen Scholarship, 1214
ASPRS William A. Fischer Scholarship, 1216
ASSE Scholarships, 1220
ASTA Academic Scholarships, 1224
ATA Nadene M. Thomas Grad Research Bursary, 1255
Audio Engineering Society Grad Scholarships, 1265
AUPHA Corris Boyd Scholarships, 1267
Australian Academy of Science Grants, 1269
Avon Products Foundation Grants, 1275
AWIS Educational Found Predoctoral Awards, 1278
AWRA Richard A. Herbert Scholarship, 1281
AWWA Holly A. Cornell Scholarship, 1284
AWWA Larson Aquatic Research Support Scholarships, 1285
AWWA Thomas R. Camp Scholarship, 1286
Bailey Family Foundation Grants, 1289
Barker Award, 1300
Bayer Clinical Scholarship Award, 1308
Beckman Scholars Grants, 1326
Becton Dickinson and Company Grants, 1328
Bedding Plants Found Horticulture Scholarships, 1329
Berks County Community Foundation Grants, 1333
Bettina Bahlsen Memorial Graduate Scholarship, 1339
Blowitz-Ridgeway Foundation Early Childhood Development Research Award, 1348
Bodenwein Public Benevolent Found Grants, 1355
Bodman Foundation Grants, 1356
Boeing Company Contributions Grants, 1357
BPW Career Advancement Scholarships, 1366
Burlington Northern Santa Fe Found Grants, 1389
Cabot Corporation Foundation Grants, 1406
California Teachers Association Scholarships, 1408
California Wellness Foundation Work and Health Grants, 1409
Campbell Soup Foundation Grants, 1415
CampusRN/AACN Nursing Scholarship Fund, 1416
Canada-U.S. Fulbright International Science and Technology Award, 1417
Canada Graduate Scholarships and NSERC Postgraduate Scholarships, 1420
Cape Branch Foundation Grants, 1425
Carl and Eloise Pohlad Family Found Grants, 1426
Carl and Lily Pforzheimer Foundation Grants, 1427
Carl Gellert and Celia Berta Gellert Foundation Grants, 1429
Carpenter Foundation Grants, 1439
Carrier Corporation Contributions Grants, 1440
Catholic University Academic Scholarships, Grants, and Awards, 1443
CCCF Dora Macllelan Brown Sem Scholarships, 1448
Cen Cal Diving and Aquatic Studies Scholarship, 1471
CFUW Bourse Georgette LeMoyne Award, 1508
CFUW Dr. Alice E. Wilson Awards, 1510
CFUW Margaret Dale Philp Award, 1512
Charles Abrams Scholarships, 1519
Charles H. Price II Family Foundation Grants, 1523
Chatlos Foundation Grants, 1532
Chautauqua Region Community Found Grants, 1533
Children's Lit Association Research Fellowships, 1551
Chiles Foundation Grants, 1552
Christensen Fund Regional Grants, 1554
Chrysalis Scholarships, 1559
CIGNA Foundation Grants, 1568
CIMO Scholarships for Advanced Studies of the Finnish Language at Finnish Universities, 1569
Claneil Foundation Grants, 1574
Clarence E. Heller Foundation Grants, 1575
Claude Pepper Foundation Grants, 1578
Cleveland-Cliffs Foundation Grants, 1580
Coeta and Donald Barker Foundation Grants, 1594
College of Human Sciences Scholarships, 1596
Community Found for Greater Buffalo Grants, 1615
Community Fnd for Muskegon County Grants, 1617
Community Foundation for S Arizona Grants, 1618
Community Found of E Central Illinois Grants, 1619
Community Foundation of the Fox River Valley, 1620
Compton Foundation Grants, 1623

Connecticut Community Foundation Grants, 1627
Connecticut Health Foundation Health Initiative Grants, 1628
ConocoPhillips Grants, 1629
Cord Foundation Grants, 1645
Corning Foundation Educational Grants, 1646
Corpus Christi College Research Scholarships, 1647
Coughlin-Saunders Foundation Grants, 1648
Credit Union Found of British Columbia Grants, 1654
Cruise Industry Charitable Foundation Grants, 1660
Cudd Foundation Grants, 1672
Cystic Fibrosis Canada Scholarships, 1679
D.F. Halton Foundation Grants, 1692
DAAD Bilateral Programs for Co-Operative Research Grants, 1695
DAAD Learn German in Germany Grants, 1700
DAAD Research Stays for University Academics and Scientists, 1705
DAAD Scholarships for Artists: Study Visits for Academics, 1707
DAAD Undergraduate Scholarships, 1710
DaimlerChrysler Corporation Fund Grants, 1714
Danone Institute Belgium Doctoral Scholarship in Nutrition, 1718
Datatel Scholars Foundation Scholarship, 1722
DEED Student Research Grants and Internships, 1731
DENSO North America Foundation Grants, 1742
Dr. Bob and Jean Smith Foundation Grants, 1853
Earhart Foundation Institutional Grants, 1888
Eastman School of Music Graduate Awards, 1892
Edward Bangs Kelley and Elza Kelley Foundation Grants, 1898
Edward W. and Stella C. Van Houten Grants, 1899
Edwin S. Webster Foundation Grants, 1900
Emerson Charitable Trust Grants, 1918
Ensign-Bickford Foundation Grants, 1928
EREF Doctoral Scholarship, 1945
EREF Fiessinger Doctoral Scholarship, 1946
Eugene M. Lang Foundation Grants, 1957
Eugene McDermott Foundation Grants, 1958
Evan and Susan Bayh Foundation Grants, 1960
Evan Frankel Foundation Grants, 1961
Evjue Foundation Grants, 1963
FDHN Jon I. Isenberg International Research Scholar Award, 1995
Field Museum High School Internships, 2009
Field Museum Visiting Scholarships, 2011
Fishman Family Foundation Grants, 2020
Flinn Foundation Grants, 2024
Fluor Foundation Grants, 2028
FMC Foundation Grants, 2029
Foundation for Seacoast Health Grants, 2041
Fritz B. Burns Foundation Grants, 2057
GCA Summer Environmental Scholarships, 2100
Genuardi Family Foundation Grants, 2109
George Foundation Grants, 2115
George S. & Dolores Dore Eccles Grants, 2120
Gertrude B. Elion Mentored Medical Student Research Awards, 2130
Gilbreth Memorial Fellowships, 2141
Gilroy and Lillian P. Roberts Foundation Grants, 2142
Greater St Louis Community Foundation Grants, 2164
Greenwall Foundation Faculty Scholars Program in Bioethics, 2175
GSA Research Grants, 2185
H.B. Earhart Fellowships, 2194
H.W. Wilson Foundation Grants, 2196
HAF Riley Frazel Memorial Scholarship, 2197
Hague Academy of International Law Summer Scholarships, 2198
Harvard University Graduate School of Education Fellowships and Awards, 2213
Health Management Scholarships and Grants for Minorities, 2224
Helen Bader Foundation Grants, 2228
Helen V. Brach Foundation Grants, 2231
HIAS Scholarship Awards, 2264
Hilton Hotels Corporate Giving Grants, 2268
Hirtzel Memorial Foundation Grants, 2269
Hispanic Scholarships, 2270
HomeBanc Foundation Grants, 2275
HSBC Contributions Community Grants, 2291
Huie-Dellmon Trust Grants, 2295
IIE David L. Boren Scholarships, 2351
IIE General Electric Foundation TECH Awards, 2352
IIE Whitaker International Fellowships and Scholarships, 2356
International Peace Research Inst Oslo Grants, 2402
Irving and Rose Crown Scholarships and Fellowships in American History, 2427
ITEA Maley/FTE Scholarship, 2435
IUVSTA Welch Foundation Scholarship, 2445
Jacob and Hilda Blaustein Foundation Grants, 2454
James R. Dougherty Jr. Foundation Grants, 2463
Jane Beattie Memorial Scholarship, 2470
Jewish Federation of Metropolitan Chicago Scholarships, 2486
John Ben Snow Memorial Trust Grants, 2497
John Deere Foundation Grants, 2500
John O. Galloway Memorial Scholarship in Geology and Geophysics, 2503
John P. McGovern Foundation Grants, 2504
John Stauffer Charitable Trust Grants, 2509
Joseph Alexander Foundation Grants, 2513
Josiah W. and Bessie H. Kline Foundation Grants, 2517
Kenneth T. & Eileen L. Norris Found Grants, 2554
Kettering Fund Grants, 2555
Kosciuszko Foundation Graduate Study and Research in Poland Scholarships, 2571
Kosciuszko Foundation Tuition Scholarships, 2574
Lillian S. Wells Foundation Grants, 2614
Lynde and Harry Bradley Foundation Grants, 2635
M. Bastian Family Foundation Grants, 2636
Marcus and Theresa Levie Educational Fund Scholarships, 2657
Margaret T. Morris Foundation Grants, 2659
Marshall Scholarships, 2666
Martin Luther King Jr. Graduate Memorial Scholarships, 2668
Mary Gates Research Scholarships, 2670
Matilda R. Wilson Fund Grants, 2684
McCune Charitable Foundation Grants, 2719
McGraw-Hill Companies Community Grants, 2724
Mead Johnson Nutritionals Scholarships, 2735
MGN Family Foundation Grants, 2754
Microscopy Soc of America Student Scholarships, 2763
Miller Brewing Corporate Contributions Grants, 2774
Millstone Foundation Grants, 2776
Minnesota War Orphans Educational Grants, 2780
MLA Graduate Scholarship, 2786
MLA Grad Scholarship for Minority Students, 2787
Musgrave Foundation Grants, 2807
NAACP Agnes Jones Jackson Grad Scholarships, 2809
NAACP Willems Scholarships, 2810
NAGC Masters and Specialists Award, 2821
NAPNAP Foundation Nursing Research Grants, 2829
NASA Undergrad Researchers Project Grants, 2843
National Black MBA Assoc Annual Scholarships, 2847
National Book Scholarship Fund, 2849
National Leadership Grant Competition, 2878
National LGBT Scholarship Fund, 2879
National Wildlife Refuge System Centennial Scholarship, 2892
NCAA Postgraduate Scholarships, 2900
New Hampshire Charitable Foundation Grants, 2985
NMF California Comm Service Scholarships, 3321
NMF National Medical Association Emerging Scholar Awards, 3324
NMF National Medical Association Patti LaBelle Award, 3325
NMF Need-Based Scholarships, 3326
NOAA Dr. Nancy Foster Scholarships, 3367
NOAA Ernest F. Hollings Scholarships, 3370
Norcliffe Foundation Grants, 3415
Nordson Corporation Foundation Grants, 3416
North Dakota Indian Scholarships, 3427
Northwestern Mutual Foundation Grants, 3430
NSERC Industrial Undergraduate Student Research Awards, 3444
NSERC Undergraduate Research Awards, 3451
NSF Historically Black Colleges and Universities Grants, 3577
NSF Partnerships in Astronomy & Astrophysics Research and Education, 3612
NSF Scholarships in Science, Technology, Engineering, and Mathematics (S-STEM), 3642
NWSA Graduate Scholarships, 3677
NWSA Graduate Scholarships in Lesbian Studies, 3678
OISE Muriel & Danny Fung Grad Scholarships, 3735
OISE William Waters Scholarships in Urban Education, 3736
Oklahoma City Community Grants, 3737
Olin Corporation Charitable Trust Grants, 3739
OSF Alumni Grants, 3835
OSF Global Supplementary Grants, 3843
Park Foundation Grants, 3870
Partnership for Cures Charles E. Culpeper Scholarships in Medical Science, 3871
Patrick and Anna M. Cudahy Fund Grants, 3873
Paul & Daisy Soros Fellows for New Americans, 3875
Paul Stock Foundation Grants, 3879
Percy B. Ferebee Endowment Grants, 3897
Pfizer Foundation Epilepsy Scholarship Award, 3923
Pioneer Hi-Bred Plant Breeding Fellowships, 3968
Polytechnic University International Research Studentships, 3982
Pott Foundation Grants, 3984
Potts Memorial Foundation Grants, 3985
Price Family Charitable Fund Grants, 3987
Princess Grace Foundation Theater, Dance, and Film Grants, 3988
Province of Alberta Queen Elizabeth II Graduate Scholarships, 4001
Prudential Foundation Education Grants, 4002
R.C. Baker Foundation Grants, 4006
R.H. Wilhelm Award in Chemical Reaction Engineering, 4007
RACGP Chris Silagy Research Scholarship, 4012
Rajiv Gandhi Foundation Grants, 4025
Rathmann Family Foundation Grants, 4033
Rayonier Foundation Grants, 4034
Regenstrief Master Of Science In Clinical Research/ CITE Program, 4047
Regents Professional Opportunity Scholarships, 4049
Reinberger Foundation Grants, 4051
Rhode Island Foundation Grants, 4061
Richard P. Covert Scholarship, 4067
Robert E. and Evelyn McKee Foundation Grants, 4074
Robert M. Hearin Foundation Grants, 4077
Rockwell International Corporate Trust Grants, 4093
Roger K. Summit Scholarship, 4097
Ronald McDonald House Charities Grants, 4101
Roy J. Carver Charitable Trust Medical and Science Research Grants, 4114
Sachs Foundation Scholarships, 4157
Sage Foundation Grants, 4158
San Antonio Area Foundation Grants, 4163
SASS Einar and Eva Lund Haugen Memorial Scholarship, 4183
Schramm Foundation Grants, 4188
Seneca Nation Higher Education Grants, 4198
Shastri Indo-Canadian Institute University of Ottawa Overseas India Summer Course Grants, 4232
Shell Oil Company Foundation Grants, 4233
Sir Dorabji Tata Trust Individual Education and Travel Grants, 4265
Society of Naval Architects and Marine Engineers Scholarships, 4328
Sony Corporation of America Corporate Philanthropy Grants, 4329
SPIE A.E. Conrady Award, 4338
SPIE C.S. Vikram Award in Optic Metrology, 4339
SPIE Dennis Gabor Award, 4340
SPIE Frits Zernike Award, 4341
SPIE George W. Goddard Award, 4342
SPIE Technology Achievement Award, 4344
SRC Company-Named Scholarships, 4351
SRC Master's Scholarships, 4378
SSHRC Graduate Scholarships-Michael Smith Foreign Study Supplements, 4408

SSHRC Joseph-Armand Bombardier CGS Doctoral Scholarships, 4412
Stackpole-Hall Foundation Grants, 4442
Starr Foundation Grants, 4447
State Scholarships Foundation Grants, 4449
STC Scholarships in Technical Communication, 4451
Strake Foundation Grants, 4452
Swedish Natural Science Res Council Grants, 4479
SWE Scholarships, 4481
Syracuse University Graduate Fellowships, 4488
T. James Kavanagh Foundation Grants, 4489
Thomas and Dorothy Leavey Foundation Grants, 4506
Thomas Austin Finch, Sr. Foundation Grants, 4507
Thomas C. Ackerman Foundation Grants, 4508
Thompson Charitable Foundation Grants, 4511
Union Pacific Found Community & Civic Grants, 4568
United Methodist Church Center for the Evangelical UBH Audrie E. Reber Memorial Award, 4571
United States Institute of Peace - Jennings Randolph Peace Dissertation Scholarships, 4579
United Theological Seminary Merit-Based Scholarships, 4583
University of Calgary Nursing Alum Scholarship, 4585
USA Funds Access to Education Scholarships, 4593
Wayne and Gladys Valley Foundation Grants, 4835
Weizmann Institute of Science Postdoctoral Fellowships, 4837
Welch Foundation Departmental Grants, 4839
Western Regional Graduate Program, 4855
William Blair and Company Foundation Grants, 4875
William G. & Helen C. Hoffman Found Grants, 4876
William G. McGowan Charitable Fund Grants, 4877
William N. and Myriam Pennington Grants, 4879
Winston Churchill Foundation Scholarships, 4886
WMU Department of Spanish Travel Study Abroad Award, 4889
World Bank Graduate Scholarship Program, 4896
WWF Russell E. Train Fellowships, 4898
Xerox Foundation Grants, 4899

Seed Grants

A-T Medical Research Foundation Grants, 10
A.L. Mailman Family Foundation Grants, 11
AAFPRS Bernstein Grant, 161
AAUW Community Action Grants, 213
Achelis Foundation Grants, 274
ACS Institutional Research Grants, 310
AFAR CART Fund Grants, 351
Albuquerque Community Foundation Grants, 508
Alcoa Foundation Grants, 511
Alcon Foundation Grants, 513
Alexander and Margaret Stewart Trust Grants, 514
Alex C. Walker Educational and Charitable Foundation Grants, 541
Alfred P. Sloan Economic Analysis of Science and Technology Grants, 548
Alfred P. Sloan Empirical Economic Research Enablers Grants, 550
Allyn Foundation Grants, 574
Alpha Omega Foundation Grants, 576
AMA-WPC Joan F. Giambalvo Memorial Scholarships, 600
AMA Foundation Seed Grants for Research, 601
American Chemical Society Corporate Associates Seed Grants, 676
American Chemical Society Local Section YCC Starter Grants, 684
American Chemical Society Petroleum Research Fund Doctoral New Investigator Grants, 690
American Chemical Society Petroleum Research Fund New Directions Grants, 691
American Chemical Society Petroleum Research Fund Undergraduate New Investigator Grants, 694
American Family Insurance Corporate Giving, 709
American Legacy Foundation National Calls for Proposals Grants, 748
American Schlafhorst Foundation Grants, 796
amfAR Research Grants, 811
Anderson Foundation Grants, 842
Angels Baseball Foundation Grants, 848
Ann Arbor Area Community Foundation Grants, 860
Annenberg Foundation Grants, 862
Arronson Foundation Grants, 1019
ASU Institute for Humanities Research Grants, 1247
Audubon Naturalist Society Research Grants, 1266
Axe-Houghton Foundation Grants, 1287
Ball Brothers Foundation General Grants, 1291
Ball Brothers Foundation Rapid Grants, 1292
Bank of America Charitable Foundation Matching Gifts, 1297
Barra Foundation Project Grants, 1301
Barth Syndrome Foundation Research Grants, 1302
Bay and Paul Foundations, Inc Grants, 1307
BCBSM Foundation Physician Investigator Research Awards, 1314
Becton Dickinson and Company Grants, 1328
Beim Foundation Grants, 1331
Berks County Community Foundation Grants, 1333
Better Way Foundation Grants, 1338
Bill and Melinda Gates Foundation Agricultural Development Grants, 1345
Bodenwein Public Benevolent Found Grants, 1355
Bodman Foundation Grants, 1356
Boeing Company Contributions Grants, 1357
Boston Foundation Grants, 1361
Brainerd Foundation Grants, 1368
Brico Fund Grants, 1371
Bullitt Foundation Grants, 1384
Bydale Foundation Grants, 1400
Cabot Corporation Foundation Grants, 1406
Caesars Foundation Grants, 1407
Campbell Soup Foundation Grants, 1415
Carl and Lily Pforzheimer Foundation Grants, 1427
Carl M. Freeman Foundation Grants, 1430
Carlos and Marguerite Mason Trust Grants, 1431
Carpenter Foundation Grants, 1439
Charles Edison Fund Grants, 1520
Charles H. Farnsworth Trust Grants, 1521
Charles H. Hood Foundation Child Health Research Awards, 1522
Charles Stewart Mott Anti-Poverty Grants, 1528
Chautauqua Region Community Found Grants, 1533
Chestnut Hill Charitable Foundation, Inc Grants, 1538
Children's Cardiomyopathy Foundation Research Grants, 1550
Christensen Fund Regional Grants, 1554
Claneil Foundation Grants, 1574
Clarence E. Heller Foundation Grants, 1575
Clay Foundation Grants, 1579
Colonel Stanley R. McNeil Foundation Grants, 1597
Communities Foundation of Texas Grants, 1613
Community Found for Greater Buffalo Grants, 1615
Community Fnd for Muskegon County Grants, 1617
Community Foundation for S Arizona Grants, 1618
Connecticut Community Foundation Grants, 1627
Conservation, Food, and Health Foundation Grants for Developing Countries, 1641
Cowles Charitable Trust Grants, 1652
Cystic Fibrosis Canada Clinical Project Grants, 1675
Dave Thomas Foundation for Adoption Grants, 1723
David Bohnett Foundation Grants, 1724
Dorr Foundation Grants, 1842
Drs. Bruce and Lee Foundation Grants, 1859
Education Through Astronomy and Space Science Initiative, 1896
Edward Bangs Kelley and Elza Kelley Grants, 1898
Elisabeth Severance Prentiss Foundation Grants, 1905
Emma B. Howe Memorial Foundation Grants, 1920
Ensign-Bickford Foundation Grants, 1928
Eugene M. Lang Foundation Grants, 1957
Eugene McDermott Foundation Grants, 1958
Evan and Susan Bayh Foundation Grants, 1960
Evanston Community Foundation Grants, 1962
Evjue Foundation Grants, 1963
Ewing Halsell Foundation Grants, 1964
Flinn Foundation Grants, 2024
Foundation for Enhancing Comm Grants, 2038
Foundation for Seacoast Health Grants, 2041
Frances L. and Edwin L. Cummings Memorial Fund Grants, 2046
Frank Stanley Beveridge Foundation Grants, 2051
General Motors Foundation Grants, 2107
George Foundation Grants, 2115
George Frederick Jewett Foundation Grants, 2116
George Gund Foundation Grants, 2117
George H. Hitchings New Investigator Award in Health Research, 2118
Gil and Dody Weaver Foundation Grants, 2139
Girl's Best Friend Foundation Grants, 2143
Greater St Louis Community Foundation Grants, 2164
Green Foundation Human Services Grants, 2172
H.J. Heinz Company Foundation Grants, 2195
Harold R. Bechtel Testamentary Grants, 2202
Harold Simmons Foundation Grants, 2203
Harry A. and Margaret D. Towsley Found Grants, 2205
Harvey Randall Wickes Foundation Grants, 2218
Heineman Foundation for Research, Education, Charitable and Scientific Purposes, 2227
HERA Found Ovarian Cancer OSB1 Seed Grant, 2238
Herbert H. and Grace A. Dow Grants, 2241
Hereditary Disease Foundation Research Grants, 2246
Hill Crest Foundation Grants, 2267
Honeywell Corporation Habitat and Conservation Grants, 2277
Honeywell Corp Sustainable Opportunities Grants, 2278
Honor the Earth Grants, 2279
Hunter's Hope Foundation Grants, 2298
Huntington's Disease Society of America Research Grants, 2300
IFP New York Anthony Radziwill Documentary Fund Grants, 2344
Indiana 21st Century Research and Technology Fund Awards, 2370
ISA John Z. Duling Grants, 2429
ISGC Research Seed Grants, 2433
Ittleson Foundation AIDS Grants, 2436
Ittleson Foundation Environment Grants, 2437
Ittleson Foundation Mental Health Grants, 2438
James H. Cummings Foundation Grants, 2460
James R. Dougherty Jr. Foundation Grants, 2463
Jewish Social Entrepreneurs Fellowships, 2487
JM Foundation Grants, 2490
John Ben Snow Memorial Trust Grants, 2497
John Deere Foundation Grants, 2500
John G. Duncan Trust Grants, 2501
John W. Alden Trust Grants, 2510
Kellogg Institute for International Studies Dissertation Year Fellowships, 2538
Kettering Fund Grants, 2555
LAM Foundation Research Grants, 2584
Leon Lowenstein Foundation Grants, 2603
M.D. Anderson Foundation Grants, 2637
Mabel Y. Hughes Charitable Trust Grants, 2647
Macquarie Bank Foundation Grants, 2650
McCarthy Family Foundation Grants, 2717
McCune Charitable Foundation Grants, 2719
McCune Foundation Human Services Grants, 2722
McGregor Fund Grants, 2725
McLean Contributionship Grants, 2727
Meadows Foundation Grants, 2736
Melville Charitable Trust Grants, 2741
Meyer Memorial Trust Responsive Grants, 2748
NAF Research Grants, 2818
National Geographic Society Expeditions Council Grants, 2872
National Geographic Society Field Research and Exploration Grants, 2873
National Psoriasis Foundation Research Grants, 2882
Nellie Mae Education Foundation District-Level Change Grants, 2968
Nellie Mae Education Foundation State Level Systems Change Grants, 2971
New Hampshire Charitable Foundation Grants, 2985
Norcliffe Foundation Grants, 3415
Nordson Corporation Foundation Grants, 3416
Notsew Orm Sands Foundation Grants, 3434
Olin Corporation Charitable Trust Grants, 3739
Oppenstein Brothers Foundation Grants, 3792
Orchard Foundation Grants, 3793
Park Foundation Grants, 3870

Patrick and Anna M. Cudahy Fund Grants, 3873
Patron Saints Foundation Grants, 3874
Pediatric Brain Tumor Found Research Grants, 3888
PhRMA Foundation Health Outcomes Research Starter Grants, 3951
PhRMA Foundation Informatics Research Starter Grants, 3955
PhRMA Foundation Pharmaceutics Research Starter Grants, 3960
PhRMA Foundation Pharmacology/Toxicology Research Starter Grants, 3964
Pinkerton Foundation Grants, 3967
Plastic Surgery Educational Foundation Basic Research Grants, 3978
Playboy Foundation Grants, 3980
Potts Memorial Foundation Grants, 3985
Prudential Foundation Education Grants, 4002
Rathmann Family Foundation Grants, 4033
Rayonier Foundation Grants, 4034
Research Corporation Single-Investigator Cottrell College Science Awards, 4056
Rhode Island Foundation Grants, 4061
Richard and Rhoda Goldman Fund Grants, 4063
Richard & Susan Smith Family Found Grants, 4064
Righteous Persons Foundation Grants, 4069
Robert Sterling Clark Foundation Government Accountability Grants, 4080
Robert W. Deutsch Foundation Grants, 4082
Rohm and Haas Company Grants, 4098
Ronald McDonald House Charities Grants, 4101
Ross Foundation Grants, 4106
Roy J. Carver Charitable Trust Medical and Science Research Grants, 4114
S.E.VEN Fund Mini-Grants, 4154
San Antonio Area Foundation Grants, 4163
San Diego Women's Foundation Grants, 4164
Seaver Institute Grants, 4195
Sensient Technologies Foundation Grants, 4199
Sid W. Richardson Foundation Grants, 4253
Sioux Falls Area Community Foundation Community Fund Grants (Unrestricted), 4262
Sioux Falls Area Community Found Spot Grants, 4263
SNM/ Covidien Seed Grant in Molecular Imaging/ Nuclear Medicine Research, 4298
Sony Corporation of America Corporate Philanthropy Grants, 4329
Stackpole-Hall Foundation Grants, 4442
Susan G. Komen Breast Cancer Foundation Challenge Grants: Career Catalyst Research, 4476
Textron Corporate Contributions Grants, 4501
Thomas Austin Finch, Sr. Foundation Grants, 4507
Thomas C. Ackerman Foundation Grants, 4508
Todd Brock Family Foundation Grants, 4522
United Methodist Church Women in History Research Grant, 4574
USC Pentecostal and Charismatic Research Initiative Individual Grants, 4605
USC Pentecostal and Charismatic Research Initiative Regional Center Grants, 4606
USDA Basic and Applied Research Grants, 4617
Vanguard Public Foundation Grant Funds, 4808
Victor E. Speas Foundation Grants, 4810
Washington Area Women's Foundation Strategic Opportunity and Partnership Fund Grants, 4831
Wenner-Gren Foundation Post-PhD Res Grants, 4852
Weyerhaeuser Company Foundation Grants, 4857
William G. McGowan Charitable Fund Grants, 4877

Service Delivery Programs

1 in 9: Long Island Breast Cancer Action Coalition Grants, 1
3M Foundation Community Giving Grants, 5
1976 Foundation Grants, 7
A.L. Mailman Family Foundation Grants, 11
A.O. Smith Community Grants, 12
AAHPM Hospice and Palliative Med Fellowships, 180
Abbott Fund Global AIDS Care Grants, 219
Abbott Fund Science Education Grants, 220
Abington Foundation Grants, 243
ACF Adoption Opportunities Project Grants, 270
A Charitable Foundation Grants, 273
Achelis Foundation Grants, 274
Adam Richter Charitable Trust Grants, 339
Adaptec Foundation Grants, 340
Ahn Family Foundation Grants, 379
Alabama Humanities Foundation Mini Grants, 480
Alberta Law Foundation Grants, 503
Albert and Margaret Alkek Foundation Grants, 504
Albuquerque Community Foundation Grants, 508
Alcatel-Lucent Technologies Foundation Grants, 509
Alces Foundation Grants, 510
Alcoa Foundation Grants, 511
Alexander and Margaret Stewart Trust Grants, 514
Alex Stern Family Foundation Grants, 542
Alfred P. Sloan Selected National Issues Grants, 563
Allen Lane Foundation Grants, 571
Allstate Foundation Safe and Vital Communities Grants, 573
Allyn Foundation Grants, 574
ALSAM Foundation Grants, 579
Alvin and Lucy Owsley Foundation Grants, 584
ALZA Corporate Contributions Grants, 588
AMA Foundation Seed Grants for Research, 601
American Chemical Society Corporate Associates Seed Grants, 676
American Chemical Society Mentoring Grants, 685
American Family Insurance Corporate Giving, 709
American Legacy Foundation National Calls for Proposals Grants, 748
American Psychiatric Foundation Grants, 791
Amgen Foundation Grants, 812
AMI Semiconductors Corporate Grants, 813
Anderson Foundation Grants, 842
Angels Baseball Foundation Grants, 848
Ann & Robert H. Lurie Family Found Grants, 859
Ann Arbor Area Community Foundation Grants, 860
Ann Peppers Foundation Grants, 863
Anthony R. Abraham Foundation Grants, 869
APA Culture of Service in the Psychological Sciences Award, 904
Appalachian Regional Comm Health Care Grants, 915
Applied Biosystems Grants, 916
Arcadia Foundation Grants, 983
Archer Daniels Midland Foundation Grants, 985
Arie and Ida Crown Memorial Grants, 989
Arizona Cardinals Grants, 995
Arronson Foundation Grants, 1019
Arthur and Rochelle Belfer Foundation Grants, 1023
ArvinMeritor Foundation Health Grants, 1027
ASHA Advancing Academic Careers Award, 1072
Ashland Corporate Contributions Grants, 1087
Assisi Foundation of Memphis General Grants, 1221
Auburn Foundation Grants, 1263
Avon Foundation Breast Care Fund Grants, 1274
Avon Products Foundation Grants, 1275
Ball Brothers Foundation General Grants, 1291
Banfi Vintners Foundation Grants, 1296
Bank of America Charitable Foundation Matching Gifts, 1297
Barra Foundation Project Grants, 1301
Batchelor Foundation Grants, 1305
BCBSNC Foundation Grants, 1318
BCBS of Massachusetts Foundation Grants, 1319
Becton Dickinson and Company Grants, 1328
Beim Foundation Grants, 1331
Berks County Community Foundation Grants, 1333
Bernard Van Leer Foundation Grants, 1336
Better Way Foundation Grants, 1338
Bill and Melinda Gates Foundation Agricultural Development Grants, 1345
Blowitz-Ridgeway Foundation Grants, 1349
Blue Shield of California Grants, 1350
Blum-Kovler Foundation Grants, 1351
Bodenwein Public Benevolent Found Grants, 1355
Bodman Foundation Grants, 1356
Booth-Bricker Fund Grants, 1359
Boston Foundation Grants, 1361
Brian G. Dyson Foundation Grants, 1370
Bristol-Myers Squibb Foundation Global HIV/AIDS Initiative Grants, 1376
Bristol-Myers Squibb Foundation Health Disparities Grants, 1377
Bristol-Myers Squibb Women's Health Grants, 1379
Brookdale Foundation Leadership in Aging Fellowships, 1381
Burden Trust Grants, 1388
Burlington Northern Santa Fe Found Grants, 1389
Burton G. Bettingen Grants, 1390
Byron W. & Alice L. Lockwood Found Grants, 1401
Caesars Foundation Grants, 1407
California Wellness Foundation Work and Health Grants, 1409
Callaway Golf Company Foundation Grants, 1410
Carl and Eloise Pohlad Family Found Grants, 1426
Carl C. Icahn Foundation Grants, 1428
Carl M. Freeman Foundation Grants, 1430
Carlos and Marguerite Mason Trust Grants, 1431
Carpenter Foundation Grants, 1439
Catherine Holmes Wilkins Foundation Grants, 1442
CDC Evaluation of the Use of Rapid Testing For Influenza in Outpatient Medical Settings, 1458
Charles H. Farnsworth Trust Grants, 1521
Charles Lafitte Foundation Grants, 1525
Charles M. & Mary D. Grant Found Grants, 1527
Charles Stewart Mott Foundation Grants, 1529
Chatlos Foundation Grants, 1532
Chestnut Hill Charitable Foundation, Inc Grants, 1538
Chicago Community Trust Fellowships, 1546
Chicago Community Trust Housing Grants: Advancing Affordable Rental Housing, 1547
Chiron Foundation Community Grants, 1553
CIGNA Foundation Grants, 1568
Claneil Foundation Grants, 1574
Clarence E. Heller Foundation Grants, 1575
Clay Foundation Grants, 1579
Cleveland-Cliffs Foundation Grants, 1580
CMA Foundation Grants, 1582
CMS Research and Demonstration Grants, 1586
Colonel Stanley R. McNeil Foundation Grants, 1597
Colorado Clean Energy Solar Innov Grants, 1599
Commonweal Foundation Community Assistance Grants, 1602
Commonweal Foundation Grants, 1603
Commonwealth Fund Small Grants, 1611
Community Found AIDS Endowment Awards, 1614
Community Fnd for Muskegon County Grants, 1617
Community Foundation for S Arizona Grants, 1618
Community Found of E Central Illinois Grants, 1619
Comprehensive Health Education Found Grants, 1622
Compton Foundation Grants, 1623
Connecticut Community Foundation Grants, 1627
Connecticut Health Foundation Health Initiative Grants, 1628
Conquer Cancer Foundation of ASCO Improving Cancer Care Grants, 1633
Conservation, Food, and Health Foundation Grants for Developing Countries, 1641
Coors Brewing Corporate Contributions Grants, 1644
Cowles Charitable Trust Grants, 1652
Cruise Industry Charitable Foundation Grants, 1660
Cudd Foundation Grants, 1672
D.F. Halton Foundation Grants, 1692
DaimlerChrysler Corporation Fund Grants, 1714
Dave Thomas Foundation for Adoption Grants, 1723
Daviess County Community Foundation Health Grants, 1725
Davis Foundation Postdoctoral Fellowships in Eating Disorders Research, 1727
Deborah Munroe Noonan Research Grants, 1729
Dell Foundation Open Grants, 1734
DeMatteis Family Foundation Grants, 1740
DENSO North America Foundation Grants, 1742
Dept of Ed Vocational Rehabilitation Services Demonstration and Training Programs, 1751
Detroit Lions Charities Grants, 1753
DHHS Comprehensive Community Mental Health Services Grants for Children with Serious Emotional Disturbances, 1754
DHHS Oral Health Promotion Research Across the Lifespan, 1759

Donald and Sylvia Robinson Family Foundation Grants, 1830
Doris Duke Foundation Operations Research on AIDS Care & Treatment in Africa Grants, 1839
Dorothy Rider Pool Health Care Grants, 1841
Do Something User Experience Research Intern, 1845
Dr. Bob and Jean Smith Foundation Grants, 1853
Drs. Bruce and Lee Foundation Grants, 1859
DTE Energy Foundation Health and Human Services Grants, 1865
Edward Bangs Kelley and Elza Kelley Foundation Grants, 1898
Emerson Charitable Trust Grants, 1918
Emma B. Howe Memorial Foundation Grants, 1920
Ensign-Bickford Foundation Grants, 1928
EPA Children's Health Protection Grants, 1933
Eugene M. Lang Foundation Grants, 1957
Evan and Susan Bayh Foundation Grants, 1960
Evanston Community Foundation Grants, 1962
Expect Miracles Foundation Grants, 1965
ExxonMobil Found Malaria Initiative Grants, 1968
Fairlawn Foundation Grants, 1974
Fannie E. Rippel Foundation Grants, 1978
Farmers Insurance Corporate Giving Grants, 1981
FCD Child Development Grants, 1984
Fetzer Institute Research Grants, 2005
Fidelity Foundation Grants, 2007
Flinn Foundation Grants, 2024
Ford Foundation Peace and Social Justice Grants, 2031
Ford Motor Company Fund Grants, 2033
Forrest C. Lattner Foundation Grants, 2035
Foundation for Enhancing Comm Grants, 2038
Foundation for Seacoast Health Grants, 2041
Frances C. and William P. Smallwood Foundation Grants, 2045
Frances L. and Edwin L. Cummings Memorial Fund Grants, 2046
Frank Stanley Beveridge Foundation Grants, 2051
G. Unger Vetlesen Foundation Grants, 2093
Genuardi Family Foundation Grants, 2109
George B. Storer Foundation Grants, 2113
George Foundation Grants, 2115
George Frederick Jewett Foundation Grants, 2116
George Gund Foundation Grants, 2117
George Kress Foundation Grants, 2119
George W. Codrington Foundation Grants, 2122
Georgia Power Foundation Grants, 2125
Gerber Foundation Grants, 2126
Gil and Dody Weaver Foundation Grants, 2139
Gilroy and Lillian P. Roberts Foundation Grants, 2142
GNOF Gulf Coast Oil Spill Grants, 2153
Greater St Louis Community Foundation Grants, 2164
Green Diamond Charitable Contributions, 2171
Green Foundation Human Services Grants, 2172
Greenspun Family Foundation Grants, 2173
Greenwall Foundation Bioethics Grants, 2174
H.A. and Mary K. Chapman Charitable Grants, 2193
H.J. Heinz Company Foundation Grants, 2195
Harold Alfond Foundation Grants, 2200
Harold R. Bechtel Charitable Remainder Uni-Trust Grants, 2201
Harold R. Bechtel Testamentary Charitable Trust Grants, 2202
Harold Simmons Foundation Grants, 2203
Harry W. Bass, Jr. Foundation Grants, 2210
Harvey Randall Wickes Foundation Grants, 2218
Healthcare Foundation of New Jersey Grants, 2222
Health Foundation of Southern Florida Responsive Grants, 2223
Hearst Foundations Health Grants, 2225
Helen Bader Foundation Grants, 2228
Helen V. Brach Foundation Grants, 2231
Hendrick Foundation for Children Grants, 2232
Herbert A. & Adrian W. Woods Found Grants, 2239
Herbert H. and Grace A. Dow Grants, 2241
Hilda and Preston Davis Foundation Grants, 2266
Hill Crest Foundation Grants, 2267
Hirtzel Memorial Foundation Grants, 2269
Hoglund Foundation Grants, 2271
HomeBanc Foundation Grants, 2275
HSBC Contributions Community Grants, 2291
Huie-Dellmon Trust Grants, 2295
Human Source Foundation Grants, 2297
Hutchinson Community Foundation Grants, 2310
IATA Research Grants, 2319
ICC Faculty Fellowships, 2333
ICC Scholarship of Engagement Faculty Grants, 2334
IDEM Section 319(h) Nonpoint Source Grants, 2338
Indiana Waste Tire Fund Grants, 2376
International Human Rights Funders Grants, 2399
Ittleson Foundation AIDS Grants, 2436
Ittleson Foundation Mental Health Grants, 2438
J.B. Reynolds Foundation Grants, 2449
J.W. Kieckhefer Foundation Grants, 2452
Jackson Foundation Grants, 2453
Jacob and Hilda Blaustein Foundation Grants, 2454
Jacob and Valeria Langeloth Foundation Grants, 2455
James M. Collins Foundation Grants, 2461
JM Foundation Grants, 2490
John Deere Foundation Grants, 2500
John R. Oishei Foundation Grants, 2505
Johnson & Johnson Corporate Contributions, 2508
Josiah W. and Bessie H. Kline Foundation Grants, 2517
Joyce Foundation Education Grants, 2519
Juvenile Justice Clearing House Grants, 2530
Kenneth T. & Eileen L. Norris Found Grants, 2554
Kettering Fund Grants, 2555
Klarman Family Foundation Grants in Eating Disorders Research, 2558
Kovler Family Foundation Grants, 2575
Leonard and Helen R. Stulman Charitable Foundation Grants, 2602
Lloyd A. Fry Foundation Health Grants, 2620
Lowe Foundation Grants, 2622
Lucile Horton Howe and Mitchell B. Howe Foundation Grants, 2624
Ludwick Family Foundation Grants, 2625
Luther I. Replogle Foundation Grants, 2631
Lymphatic Education & Research Network Additional Support for NIH-funded F32 Postdoc Fellows, 2632
Lymphatic Education & Research Network Postdoctoral Fellowships, 2633
Lynde and Harry Bradley Foundation Grants, 2635
M. Bastian Family Foundation Grants, 2636
Mabel Y. Hughes Charitable Trust Grants, 2647
Macquarie Bank Foundation Grants, 2650
MAP International Medical Fellowships, 2655
Marcia and Otto Koehler Foundation Grants, 2656
Margaret & James A. Elkins Jr. Found Grants, 2658
Margaret T. Morris Foundation Grants, 2659
Marion Gardner Jackson Charitable Trust Grants, 2664
Mary K. Chapman Foundation Grants, 2673
Mary S. and David C. Corbin Foundation Grants, 2675
Massachusetts Bar Foundation Legal Intern Fellowships, 2676
Matilda R. Wilson Fund Grants, 2684
Mattel Children's Foundation Grants, 2685
Mattel International Grants, 2686
Max and Victoria Dreyfus Foundation Grants, 2687
Maximilian E. & Marion O. Hoffman Grants, 2688
McCarthy Family Foundation Grants, 2717
McCombs Foundation Grants, 2718
McCune Charitable Foundation Grants, 2719
McCune Foundation Human Services Grants, 2722
McGraw-Hill Companies Community Grants, 2724
McGregor Fund Grants, 2725
McLean Contributionship Grants, 2727
MDEQ Beach Monitoring Grants - Inland Lakes, 2731
MDEQ Local Water Quality Monitoring Grants, 2732
Mead Johnson Nutritionals Med Educ Grants, 2734
Meadows Foundation Grants, 2736
MedImmune Charitable Grants, 2737
Mericos Foundation Grants, 2742
Meyer Memorial Trust Responsive Grants, 2748
MGM Resorts Foundation Community Grants, 2753
MGN Family Foundation Grants, 2754
Miles of Hope Breast Cancer Foundation Grants, 2773
Miller Brewing Corporate Contributions Grants, 2774
Millipore Foundation Grants, 2775
Millstone Foundation Grants, 2776
Motorola Foundation Grants, 2796
Musgrave Foundation Grants, 2807
NAEA Research Grants, 2814
NAPNAP Foundation Nursing Research Grants, 2829
National Book Scholarship Fund, 2849
NCFR Innovation Grants, 2908
Nelda C. and H.J. Lutcher Stark Found Grants, 2967
Nell Warren Elkin and William Simpson Elkin Foundation, 2972
NHLBI Bioengineering and Obesity Grants, 3016
NHLBI Investigator Initiated Multi-Site Clinical Trials, 3030
NHLBI Lymphatics in Health and Disease in the Digestive, Urinary, Cardiovascular and Pulmonary Systems, 3032
NHLBI Midcareer Investigator Award in Patient-Oriented Research, 3040
NHLBI Research Supplements to Promote Diversity in Health-Related Research, 3054
NHLBI Short-Term Research Education Program to Increase Diversity in Health-Related Research, 3062
NMF General Electric Med Scholars Fellowships, 3322
Norcliffe Foundation Grants, 3415
Nordson Corporation Foundation Grants, 3416
Northwestern Mutual Foundation Grants, 3430
NSF Chemistry Research Experiences for Undergraduates, 3518
NSF CISE Computing Res Infrastructure Grants, 3531
NSF Communicating Res to Public Audiences, 3538
NSF Cooperative Studies Of The Earth's Deep Interior Grants, 3544
NSF Emerging Frontiers in Research and Innovation Grants, 3560
NSF Instrument Devel for Biological Research, 3583
NSF Paleoclimate Grants, 3606
NSF Solar, Heliospheric, and INterplanetary Environment Grants, 3651
NYCT Blindness and Visual Disabilities Grants, 3695
NYSCA Dance: Services to the Field Grants, 3702
Oklahoma City Community Grants, 3737
Olin Corporation Charitable Trust Grants, 3739
OneSight Research Foundation Block Grants, 3742
Oppenstein Brothers Foundation Grants, 3792
Oscar Rennebohm Foundation Grants, 3832
OSF-Baltimore Community Fellowships, 3833
OSF Access to Medicines Fellowships in Eastern Europe and Central Asia, 3834
Overbrook Foundation Grants, 3860
Park Foundation Grants, 3870
Patrick and Anna M. Cudahy Fund Grants, 3873
Patron Saints Foundation Grants, 3874
Paul G. Allen Family Foundation Grants, 3876
Peabody Foundation Grants, 3886
Percy B. Ferebee Endowment Grants, 3897
Perpetual Trust for Charitable Giving Grants, 3900
Pinkerton Foundation Grants, 3967
Piper Trust Healthcare & Med Research Grants, 3972
R.C. Baker Foundation Grants, 4006
RAND Corporation Graduate Student Summer Associateships, 4030
RAND Corporation Military Fellowships, 4031
Rathmann Family Foundation Grants, 4033
Reinberger Foundation Grants, 4051
Reuter Foundation Grants, 4059
RGK Foundation Grants, 4060
Rhode Island Foundation Grants, 4061
Richard and Rhoda Goldman Fund Grants, 4063
Richard & Susan Smith Family Found Grants, 4064
Richard Davoud Donchian Foundation Grants, 4066
Rich Foundation Grants, 4068
Robert E. and Evelyn McKee Foundation Grants, 4074
Robert G. Cabell III and Maude Morgan Cabell Foundation Grants, 4075
Robert Schalkenbach Foundation Grants, 4079
Rockwell International Corporate Trust Grants, 4093
Rohm and Haas Company Grants, 4098
Ronald McDonald House Charities Grants, 4101
Rosalynn Carter Institute John and Betty Pope Fellowships, 4103
Rose Community Foundation Health Grants, 4104

RRF General Grants, 4115
Russell Family Foundation Environmental Sustainability Grants, 4128
Ruth and Vernon Taylor Foundation Grants, 4130
RWJF Disparities Research for Change Grants, 4137
RWJF Harold Amos Medical Faculty Development Scholars, 4140
RWJF New Connections Jun Investigator Grants, 4146
RWJF Pioneer Portfolio Grants, 4150
S.E.VEN Fund Annual Grants, 4153
Sage Foundation Grants, 4158
Salem Foundation Grants, 4160
Samueli Foundation Health Grants, 4161
San Antonio Area Foundation Grants, 4163
San Diego Women's Foundation Grants, 4164
Sara Elizabeth O'Brien Trust Grants, 4166
Sarkeys Foundation Grants, 4181
Schramm Foundation Grants, 4188
Seaver Institute Grants, 4195
Self Foundation Grants, 4197
Sensient Technologies Foundation Grants, 4199
Shell Oil Company Foundation Grants, 4233
SHRM Foundation Research Grants, 4249
Sidgmore Family Foundation Grants, 4252
Sid W. Richardson Foundation Grants, 4253
Sioux Falls Area Community Foundation Community Fund Grants (Unrestricted), 4262
Sioux Falls Area Community Found Spot Grants, 4263
Sony Corporation of America Corporate Philanthropy Grants, 4329
Special Olympics E Kennedy Shriver Fellowships, 4331
Stackpole-Hall Foundation Grants, 4442
Starr Foundation Grants, 4447
Strake Foundation Grants, 4452
Sunny and Abe Rosenberg Foundation Grants, 4469
Taub Foundation Grants for MDS Research, 4496
TD4HIM Foundation Grants, 4497
Textron Corporate Contributions Grants, 4501
Thomas C. Ackerman Foundation Grants, 4508
Thompson Charitable Foundation Grants, 4511
Thoracic Foundation Grants, 4513
Todd Brock Family Foundation Grants, 4522
U.S. Department of Education American Overseas Research Centers, 4537
U.S. Department of Education Disability and Rehabilitation Research and Related Projects, 4538
U.S. Department of Education Native Hawaiian Education Grants, 4543
U.S. Department of Education Rehabilitation Engineering Research Centers Grants, 4544
U.S. Department of Education Rehabilitation Research Training Centers, 4545
UCLA / RAND Corporation Post-Doctoral Fellowships, 4548
Union Pacific Found Community & Civic Grants, 4568
Union Pacific Foundation Health and Human Services Grants, 4569
United Methodist Church World is My Parish United Methodist History Research Grant, 4576
United Methodist Committee on Relief Global Health Grants, 4577
United States Institute of Peace - Jennings Randolph Senior Fellowships, 4580
United States Institute of Peace Annual Grants, 4582
USAID Microbicides Research, Development, and Introduction Grants, 4599
USAID Strengthening RMNCH Through Indian Health Professional Associations and Scaling Up Uptake Grants, 4602
US CRDF STEP Business Partnership Grant Competition Ukraine, 4612
Vancouver Foundation Grants and Community Initiatives Program, 4807
Victor E. Speas Foundation Grants, 4810
Viking Children's Fund Grants, 4811
Volkswagen of America Corporate Contributions Grants, 4817
W.C. Griffith Foundation Grants, 4819
W.K. Kellogg Foundation Healthy Kids Grants, 4822
Wallace Foundation Grants, 4829
Wayne and Gladys Valley Foundation Grants, 4835
Westinghouse Charitable Giving Grants, 4856
Weyerhaeuser Company Foundation Grants, 4857
Willary Foundation Grants, 4870
William Blair and Company Foundation Grants, 4875
William G. McGowan Charitable Fund Grants, 4877
William N. and Myriam Pennington Foundation Grants, 4879
Woodward Fund Grants, 4895
Xerox Foundation Grants, 4899

Symposiums, Conferences, Workshops, Seminars

AACAP Systems of Care Special Program Scholarships, 81
AACR Scholar-in-Training Awards - Special Conferences, 147
Abdus Salam ICTP Networks Grants, 233
Abdus Salam ICTP Schools, Conferences, Workshops and Extended Research Workshops Grants, 238
ACLS Comparative Perspectives on Chinese Culture and Society Grants, 283
ACSM International Student Awards, 318
AERA Dissertation Grants, 342
AIR/NRC Summer Institute on First-Year Assessment Undergraduate Student Fellowships, 468
AIR National Summer Data Policy Institute Fellowships, 469
Alabama Humanities Foundation Planning/Consultant Grants, 481
Alcoa Foundation Grants, 511
Alfred P. Sloan Civic Initiatives Grants, 545
Alfred P. Sloan Economic Analysis of Science and Technology Grants, 548
Alfred P. Sloan Empirical Economic Research Enablers Grants, 550
Alfred P. Sloan Energy and Environment Grants, 552
Alfred P. Sloan History of Science and Technology Grants, 553
ALZA Corporate Contributions Grants, 588
Alzheimer's Association Conference Grants, 589
American Academy of Religion Collaborative Research Grants, 617
American Chemical Society Petroleum Research Fund Scientific Education Grants, 692
American College of Rheumatology Fellows-in-Training Travel Scholarships, 700
American Family Insurance Corporate Giving, 709
American Numismatic Society Eric P. Newman Graduate Seminar Grants, 761
AMI Semiconductors Corporate Grants, 813
Animal Behavior Soc Charles H. Turner Award, 849
Animal Behavior Society Diversity Grant, 851
Animal Behavior Society Latin American Travel Awards, 852
Ann Arbor Area Community Foundation Grants, 860
AOCS Industrial Oil Products Division Student Award, 885
APA Scientific Conferences Grants, 911
APSA Travel Grants, 948
Archer Daniels Midland Foundation Grants, 985
Arizona Commission on the Arts Folklorist Residencies, 997
Arizona Public Service Corporate Giving Grants, 998
ARO Conference and Symposia Grants, 1007
ASHA Research Mentoring-Pair Travel Award, 1073
ASHA Student Research Travel Award, 1074
ASHA Students Preparing for Academic & Research Careers Award, 1075
ASM-UNESCO Leadership Grant for International Educators, 1099
ASM Millis-Colwell Postgraduate Travel Grant, 1179
Avon Foundation Breast Care Fund Grants, 1274
BA Elisabeth Barker Fund, 1288
BA Larger Research Grants, 1290
Ball Brothers Foundation General Grants, 1291
BA Small Research Grants, 1304
Berks County Community Foundation Grants, 1333
Beverly Willis Architecture Foundation Grants and Fellowships, 1340
Bodenwein Public Benevolent Found Grants, 1355
Boeing Company Contributions Grants, 1357
Brainerd Foundation Grants, 1368
Bullitt Foundation Grants, 1384
Bydale Foundation Grants, 1400
Carl M. Freeman Foundation Grants, 1430
CDC Cooperative Agreement for the Development, Operation, and Evaluation of an Entertainment Education Program, 1457
CDC Public Health Conference Support Grants, 1466
Charles Lafitte Foundation Grants, 1525
Charles Stewart Mott Anti-Poverty Grants, 1528
Charles Stewart Mott Foundation Grants, 1529
Charles W. Finley Visiting Scholar Educ Grant, 1530
Chautauqua Region Community Found Grants, 1533
Christine Wilson Award, 1558
CIGNA Foundation Grants, 1568
Cleveland-Cliffs Foundation Grants, 1580
Community Found for Greater Buffalo Grants, 1615
Community Fnd for Muskegon County Grants, 1617
Community Foundation of the Virgin Islands Judith A. Towle Environmental Studies Grants, 1621
Connecticut Community Foundation Grants, 1627
ConocoPhillips Grants, 1629
Conquer Cancer Foundation of ASCO Int'l Devel & Educ Award in Palliative Care, 1634
Conquer Cancer Foundation of ASCO International Development and Education Awards, 1635
Conquer Cancer Foundation of ASCO Resident Travel Award for Underrepresented Populations, 1638
CTCRI Researcher Travel Grants, 1668
CTCRI Workshop and Learning Opportunities Grants, 1671
Cystic Fibrosis Canada Special Travel Grants For Fellows and Students, 1682
Cystic Fibrosis Canada Visiting Clinician Awards, 1687
DAAD Study Visits / Study Seminars and Practicals in Germany by Foreign Students, 1709
Daimler and Benz Foundation Fellowships, 1713
Dave Thomas Foundation for Adoption Grants, 1723
Dean Witter Foundation Grants, 1728
Deloitte Foundation Grants, 1735
Dept of Ed Star Schools Grants, 1749
DOJ National Institute of Corrections Training and Staff Development Grants, 1827
DSO Controlling Pathogen Evolution Workshop Grants, 1862
Engineering Information Foundation Grants, 1927
Ensign-Bickford Foundation Grants, 1928
Eugene M. Lang Foundation Grants, 1957
Evanston Community Foundation Grants, 1962
Expect Miracles Foundation Grants, 1965
Facebook Fellowships, 1973
FDHN Non-Career Research Awards, 1998
Ford Foundation Peace and Social Justice Grants, 2031
Frances C. and William P. Smallwood Foundation Grants, 2045
Fulbright-Hays Seminars Abroad Bilateral Projects Grants, 2062
Fulbright German Studies Seminar Grants, 2072
Fund for Folk Culture Artists Support Grants, 2088
George Gund Foundation Grants, 2117
GMFUS Black Sea Trust for Regional Corporation Grants, 2152
Greater St Louis Community Foundation Grants, 2164
Greater Worcester Community Foundation Jeppson Memorial Fund for Brookfield Grants, 2165
GUITS Grants for Undergraduate Study, 2187
GUITS Matching Seed-Money Grants for New Positions in Turkish Studies, 2188
GUITS Sabbatical Research Grants, 2191
Hatton W. Sumners Foundation for the Study and Teaching of Self Government Grants, 2219
Helen Bader Foundation Grants, 2228
Honor the Earth Grants, 2279
Hydrogen Student Design Contest, 2311
IBRO Europe Regional Grants for Western European Workshops and Symposia, 2324
Indian Council of Social Science Research Grants, 2377
Institute of Global Conflict and Cooperation Research, Conference, and Teaching Grants, 2385

IREX ECA Alumni Small Grants, 2416
IREX Regional Policy Symposium Grants, 2423
IREX US Embassy Policy Specialist Grants, 2426
J.W. Kieckhefer Foundation Grants, 2452
Japan Foundation Center for Global Partnership Grants, 2472
JMO Law and Legal System Grants, 2491
JMO Public Policy Research Grants, 2492
JMO Strategic and International Studies Grants, 2493
John D. and Catherine T. MacArthur Foundation Global Challenges Grants, 2498
John P. McGovern Foundation Grants, 2504
Joseph Alexander Foundation Grants, 2513
JSA Junior Scientist Travel Support, 2523
Juvenile Justice Clearing House Grants, 2530
Kellogg Institute for International Studies Faculty Fellows' Research Grants, 2540
Kellogg Institute for International Studies Grants for Conferences and Workshops, 2543
Kellogg Institute for International Studies Grants for Distinguished Lectureships, 2544
Kellogg Institute for International Studies Grants for Events, 2545
Kennedy Center Summer Fellowships for International Arts Managers, 2553
Landon Foundation-AACR Innovator Award for International Collaboration, 2587
Lillian S. Wells Foundation Grants, 2614
Link Foundation Energy Doctoral Fellowships, 2617
Lynde and Harry Bradley Foundation Grants, 2635
MacArthur Foundation Conservation and Sustainable Development Grants, 2648
Massachusetts Foundation for the Humanities Project Grants, 2678
McCune Charitable Foundation Grants, 2719
McLean Contributionship Grants, 2727
Melville Charitable Trust Grants, 2741
MGN Family Foundation Grants, 2754
NACC David Stevenson Fellowships, 2812
NACC William Diaz Fellowships, 2813
National Center for Responsible Gaming Conference Scholarships, 2851
NATO Advanced Research Workshops Program, 2894
NATO Advanced Study Institutes Program, 2895
Nellie Mae Education Foundation District-Level Change Grants, 2968
Nellie Mae Education Foundation State Level Systems Change Grants, 2971
NIA Support of Scientific Meetings as Cooperative Agreements, 3121
NIDDK Traditional Conference Grants, 3184
NINDS Support of Scientific Meetings as Cooperative Agreements, 3258
NL Associated Colleges of the Midwest/Great Lakes College Association Faculty Fellowships, 3281
Norcliffe Foundation Grants, 3415
North Carolina Arts Council Folklife Grants, 3419
Northwest Fund for the Environment Grants, 3431
NSF Biological Research Collections, 3504
NSF Communicating Res to Public Audiences, 3538
NSF Conferences, Workshops, and Special Meetings in the Mathematical Sciences Grants, 3542
NSF Linguistics Grants, 3587
NSF Partnerships for International Research and Education, 3611
NSF Science and Society (S&S), 3643
Olin Corporation Charitable Trust Grants, 3739
Oppenstein Brothers Foundation Grants, 3792
Overbrook Foundation Grants, 3860
Pfizer Medical Education Track Two Grants, 3927
Phi Kappa Phi Found Love of Learning Awards, 3937
Pioneer Hi-Bred Plant Breeding Fellowships, 3968
Potts Memorial Foundation Grants, 3985
Prudential Foundation Education Grants, 4002
PVA Research Foundation Grants, 4004
Rathmann Family Foundation Grants, 4033
RGK Foundation Grants, 4060
Rhode Island Foundation Grants, 4061
Rockwell International Corporate Trust Grants, 4093
Rocky Mountain Mineral Law Found Grants, 4094
RWJF Changes in Health Care Financing and Organization Small Grants, 4134
RWJF Partners Investing in Nursing Grants, 4149
San Antonio Area Foundation Grants, 4163
Sid W. Richardson Foundation Grants, 4253
Sigma Theta Tau International / Council for the Advancement of Nursing Science Grants, 4255
SRC Educational Science Conference Grants, 4354
SRC Medical Science Grants for Large International Conferences in Sweden, 4380
SSRC Our Shared Past Grants, 4437
Swedish Natural Science Res Council Grants, 4479
Swiss National Science Foundation DORE (DO REsearch) Conference Grants, 4482
Swiss National Science Foundation Scientific Conferences Grants, 4486
Textron Corporate Contributions Grants, 4501
Thoracic Foundation Grants, 4513
U.S. Department of Education American Overseas Research Centers, 4537
W.M. Keck Foundation Science and Engineering Research Grants, 4824
Wabash Center Project Grants, 4826
Wabash Center Small Project Grants, 4827
Wabash Center Waorkshop and Colloquy Participant Grants, 4828
Wenner-Gren Foundation Conference and Workshop Grants, 4847
Weyerhaeuser Company Foundation Grants, 4857
WWF Russell E. Train Education for Nature Alumni Grants, 4897
WWF Russell E. Train Fellowships, 4898

Technical Assistance

A.L. Mailman Family Foundation Grants, 11
AAAS Science and Technology Policy Fellowships: Global Health and Development, 42
AAAS Science and Technology Policy Fellowships: Health, Education and Human Services, 43
Abbott Fund Global AIDS Care Grants, 219
Albuquerque Community Foundation Grants, 508
Alex Stern Family Foundation Grants, 542
Anderson Foundation Grants, 842
Arie and Ida Crown Memorial Grants, 989
Assisi Foundation of Memphis General Grants, 1221
Avon Products Foundation Grants, 1275
Ball Brothers Foundation General Grants, 1291
BCBS of Massachusetts Foundation Grants, 1319
Bill and Melinda Gates Foundation Agricultural Development Grants, 1345
Bill and Melinda Gates Foundation Financial Services for the Poor Grants, 1346
Boeing Company Contributions Grants, 1357
Boston Foundation Grants, 1361
Brainerd Foundation Grants, 1368
Bullitt Foundation Grants, 1384
Caesars Foundation Grants, 1407
Carl M. Freeman Foundation Grants, 1430
Carpenter Foundation Grants, 1439
Charles H. Farnsworth Trust Grants, 1521
Charles Stewart Mott Anti-Poverty Grants, 1528
Charles Stewart Mott Foundation Grants, 1529
Children's Advocacy Center Grants, 1549
Claneil Foundation Grants, 1574
Clarence E. Heller Foundation Grants, 1575
Community Found for Greater Buffalo Grants, 1615
Community Foundation for Greater New Haven Quinnipiac River Fund Grants, 1616
Community Foundation for S Arizona Grants, 1618
Connecticut Community Foundation Grants, 1627
Conservation, Food, and Health Foundation Grants for Developing Countries, 1641
Dave Thomas Foundation for Adoption Grants, 1723
DEED Student Research Grants and Internships, 1731
Dept of Ed Jacob K. Javits Gifted and Talented Students Education Program-National Research and Development Center, 1746
Dept of Ed Star Schools Grants, 1749
DOE Inventions and Innovation Grant, 1802
DOE University Coal Research Grants, 1816
DOT Transit Planning and Research Grants, 1849
Edward Bangs Kelley and Elza Kelley Foundation Grants, 1898
EPA Brownfields Area-Wide Planning Grants, 1931
EPA Capacity Building Grants for Compliance Assurance and Enforcement Activities in Indian Country and Other Tribal Areas, 1932
EPA Surveys, Studies, Research, Investigations, Demonstrations, and Special Purpose Activities Relating to the Clean Air Act, 1940
Ewing Halsell Foundation Grants, 1964
Expect Miracles Foundation Grants, 1965
Fidelity Foundation Grants, 2007
Florence Tan Moeson Fellowships, 2025
Fluor Foundation Grants, 2028
Frances L. and Edwin L. Cummings Memorial Fund Grants, 2046
Frank Stanley Beveridge Foundation Grants, 2051
General Motors Foundation Grants, 2107
George Frederick Jewett Foundation Grants, 2116
Girl's Best Friend Foundation Grants, 2143
H.J. Heinz Company Foundation Grants, 2195
Hill Crest Foundation Grants, 2267
Honeywell Corporation Sustainable Opportunities Grants, 2278
Honor the Earth Grants, 2279
James R. Dougherty Jr. Foundation Grants, 2463
Jewish Social Entrepreneurs Fellowships, 2487
JM Foundation Grants, 2490
Manitoba Arts Council Craft Grants, 2653
McCune Charitable Foundation Grants, 2719
Melville Charitable Trust Grants, 2741
MGN Family Foundation Grants, 2754
Motorola Foundation Grants, 2796
National Book Scholarship Fund, 2849
New Hampshire Charitable Foundation Grants, 2985
NIDDK Small Business Innovation Research to Develop New Therapeutics and Monitoring Technologies for Type 1 Diabetes (T1D)--Towards an Artificial Pancreas, 3181
NOAA Cooperative Institute: Eastern U.S. Continental Shelf Frontier Exploration, Research, and Technology Development Grants, 3356
Norcliffe Foundation Grants, 3415
Nordson Corporation Foundation Grants, 3416
NSF University Radio Observatories Grants, 3664
NYCT New York City Environment Grants, 3697
NYSCA Dance: Services to the Field Grants, 3702
NYSCA Individual Artists: Film, Media and New Technology Production Grants, 3706
NYSCA Music: Services to the Field Grants, 3708
Perl 6 Microgrants, 3899
Pinkerton Foundation Grants, 3967
Prudential Foundation Education Grants, 4002
Rett Syndrome Research Foundation Grants, 4058
Rhode Island Foundation Grants, 4061
Rich Foundation Grants, 4068
Royal Scottish Geographical Society Expedition Grants, 4110
SARE Research and Education Grants - Northeast Region, 4178
Seattle Foundation Medical Funds Grants, 4194
Skaggs Foundation Grants, 4266
SSSR Student Paper Award, 4441
Streilein Foundation for Ocular Immunology Visiting Professorships, 4453
Surdna Found Sustainable Environments Grants, 4471
Textron Corporate Contributions Grants, 4501
Union Pacific Foundation Health and Human Services Grants, 4569
USAID Development Innovation Accelerator Broad Agency Announcement Grants, 4596
USDA Emerging Markets Grants, 4623
USDA Foreign Market Development Grants, 4625
USDA Market Access Grants, 4633
USDA Technical Agricultural Assistance Research Grants, 4651
USDC Low-Power Television and Translator Upgrade Grants, 4666
USDC State Broadband Initiative Grants, 4671

Vancouver Foundation Grants and Community Initiatives Program, 4807
Vanguard Public Foundation Grant Funds, 4808
Weyerhaeuser Company Foundation Grants, 4857
WILD Foundation Grants, 4867
Xerox Foundation Grants, 4899

Training Programs/Internships
AAAS Entry Point Internships, 30
AACN Mentorship Grant, 110
AAG Anne U. White Fund Grants, 170
AALAS Foundation Grants, 182
AAUW Community Action Grants, 213
ACCP Hematology/Oncology Minisabbatical, 260
ACCP Infectious Diseases Minisabbatical, 262
ACCP Pain & Palliative Care Minisabbatical, 265
Achelis Foundation Grants, 274
ACS Master's Training Grants in Clinical Oncology Social Work, 311
ACS Physician Training Awards in Preventive Medicine, 323
Aetna Foundation Minority Scholars Grants, 349
AFAR Medical Student Training in Aging Research Program, 352
AHIMA Faculty Development Stipends, 375
AHNS/AAO-HNSF Surgeon Scientist Combined Award, 380
AHNS/AAO-HNSF Young Investigator Combined Award, 381
AIHS/Mitacs Health Pilot Partnership Internship Grants, 434
AIHS ForeFront Internships, 442
Allen Foundation Educational Nutrition Grants, 570
American-Scandinavian Foundation Short-Term Training Grants, 606
American Psychiatric Association/Janssen Resident Psychiatric Research Scholars, 774
American Psychiatric Association Minority Medical Student Summer Externship in Addiction Psychiatry, 785
American Psychiatric Association Minority Medical Student Summer Mentoring Program, 786
APSA Congressional Fellowships for Federal Executives, 927
APSA Congressional Fellowships for Journalists, 929
APSA Congressional Fellowships for Political Scientists, 930
ASGE Endoscopic Research Awards, 1063
ASHA Research Mentoring-Pair Travel Award, 1073
ASHA Students Preparing for Academic & Research Careers Award, 1075
ASTA Academic Scholarships, 1224
Blowitz-Ridgeway Foundation Grants, 1349
Bodman Foundation Grants, 1356
CAA Professional Development Fellowships for Graduate Students, 1404
Carl and Lily Pforzheimer Foundation Grants, 1427
Carrier Corporation Contributions Grants, 1440
Center for Defense Information Internships, 1473
Center for National Policy Internships, 1479
CFF Student Traineeships, 1498
Charles H. Revson Foundation Grants, 1524
Charles Stewart Mott Anti-Poverty Grants, 1528
Children's Advocacy Center Grants, 1549
Christine Mirzayan Science and Technology Policy Graduate Fellowship Program, 1556
CIA Undergraduate Co-Op Program - Engineer/ Program Management Engineer, 1561
CIA Undergraduate Internship - Analysis, 1562
Community Found for Greater Buffalo Grants, 1615
Community Foundation of the Virgin Islands Judith A. Towle Environmental Studies Grants, 1621
Conservation, Food, and Health Foundation Grants for Developing Countries, 1641
Cruise Industry Charitable Foundation Grants, 1660
DAAD Emigre Memorial German Internship Program - Bundestag, 1696
DAAD Intensive Language Courses Grants, 1699
DAAD Research Internships in Science and Engineering, 1704
DAAD RISE Professional Internships, 1706
DAAD Study & Internship in Germany, 1708
DAAD Undergraduate Scholarships, 1710
DHHS Emerging Leaders Program Internships, 1756
DHHS Special Programs for the Aging Training, Research, and Discretionary Projects Grants, 1761
DOD HBCU/MI Partnership Training Award, 1765
DOJ National Institute of Corrections Training and Staff Development Grants, 1827
Donald W. Reynolds Foundation Aging and Quality of Life Grants, 1832
Do Something TMI Intern, 1844
E.L. Wiegand Foundation Grants, 1885
Edwin S. Webster Foundation Grants, 1900
Emil Buehler Perpetual Trust Grants, 1919
Engineering Information Foundation Grants, 1927
Ensign-Bickford Foundation Grants, 1928
Enterprise Community Partners Rose Architectural Fellowships, 1929
EPA Capacity Building Grants for Compliance Assurance and Enforcement Activities in Indian Country and Other Tribal Areas, 1932
ETS Predoctoral Summer Program, 1954
Eugene M. Lang Foundation Grants, 1957
Evjue Foundation Grants, 1963
FCD Child Development Grants, 1984
Field Museum High School Internships, 2009
Field Museum Undergraduate Internships, 2010
Ford Foundation Peace and Social Justice Grants, 2031
Foundation for Seacoast Health Grants, 2041
Fulbright Alumni Initiatives Awards Grants, 2064
Fulbright Critical Lang Enhancement Awards, 2068
Fulbright English Teaching Assistantships, 2070
Fulbright International Education Administrators Seminar Grants, 2073
Fulbright New Century Scholars Grants, 2075
Fulbright Traditional Scholar Program in the Near East and North Africa Region, 2083
George Gund Foundation Grants, 2117
GNOF Gulf Coast Oil Spill Grants, 2153
Hatton W. Sumners Foundation for the Study and Teaching of Self Government Grants, 2219
Hearst Foundations Health Grants, 2225
Helen Bader Foundation Grants, 2228
Henry J. Kaiser Family Found Media Fellowships, 2234
Henry Luce Foundation Special Projects in Higher Education Grants, 2236
Higher Education Graduate Scholarships, Fellowships, and Grants, 2265
IMLS Laura Bush 21st Century Librarian Grants, 2365
Institute of Noetic Sciences Research Internships, 2386
Intel Bachelor's and Master's Degree Internships, 2393
Intel Technician Internships, 2395
International Crane Foundation Internships, 2396
IREX International Leadership in Education Program Grants, 2420
Ittleson Foundation Environment Grants, 2437
JM Foundation Grants, 2490
Johnson & Johnson Corporate Contributions, 2508
Juvenile Justice Clearing House Grants, 2530
Kellogg Institute for International Studies Summer Internships, 2549
Kluge Center Kissinger Lecturer Award, 2565
Lynde and Harry Bradley Foundation Grants, 2635
Macquarie Bank Foundation Grants, 2650
Massachusetts Bar Legal Intern Fellowships, 2676
Max and Victoria Dreyfus Foundation Grants, 2687
McLean Contributionship Grants, 2727
Mead Johnson Nutritionals Scholarships, 2735
Meadows Foundation Grants, 2736
Mt. Sinai Health Care Foundation Academic Medicine and Bioscience Grants, 2803
NASA Goddard Visiting Fellowships, 2838
NASA Lewis Educational Research and Collaborative Internships, 2840
National Book Scholarship Fund, 2849
National Wildlife Federation Internships, 2891
NEI Ruth L. Kirschstein National Research Service Award Short-Term Institutional Research Training Grants, 2963
NEI Scholars Program, 2964
Nellie Mae Education Foundation State Level Systems Change Grants, 2971
New Hampshire Charitable Foundation Grants, 2985
NHLBI Biomedical Research Training Program for Individuals from Underrepresented Groups, 3019
NHLBI Mentored Career Award for Faculty at Institutions that Promote Diversity (K01), 3033
NHLBI Research Supplements to Promote Diversity in Health-Related Research, 3054
NHLBI Ruth L. Kirschstein National Research Service Award Institutional Research Training Grants, 3056
NHLBI Ruth L. Kirschstein National Research Service Award Short-Term Institutional Research Training Grants, 3061
NHLBI Short-Term Research Education Program to Increase Diversity in Health-Related Research, 3062
NHLBI Summer Internship Program in Biomedical Research, 3064
NHLBI Training Grants for Institutions that Promote Diversity, 3067
NHPRC Electronic Records Projects Grants, 3073
NIDDK Mentored Research Scientist Development Award, 3161
NIDDK Training in Clinical Investigation in Kidney and Urology Grants, 3185
NIGMS Predoctoral Training in Bioinformatics and Computation Biology Grants, 3199
NIH Summer Internship Program in Biomedical Research, 3248
NLM Exploratory/Developmental Grants, 3296
NOAA Cooperative Unit for Fisheries Education and Research Internships, 3360
NOAA Ernest F. Hollings Scholarships, 3370
NOAA Faculty & Student Intern Res Grants, 3371
NOAA Living Marine Resources Cooperative Science Center Internships/Fellowships, 3380
NOAA Miami Lab Summer Internships, 3382
NOAA Nat Climatic Data Center Internships, 3387
NOAA Pacific Islands Assistantships, 3399
NOAA Population Dynamics Recruiting Grants, 3400
North Carolina Arts Council Folklife Internship, 3418
NRA Foundation Grants, 3437
NSF Historically Black Colleges and Universities Grants, 3577
NSF Science and Society (S&S), 3643
NSF University Radio Observatories Grants, 3664
NYAM Gladys Brooks Internship, 3686
OAH Germany Residency Grants, 3719
OAH Japanese Residencies Grants, 3722
Olin Corporation Charitable Trust Grants, 3739
ONR Naval Research Enterprise Intern Program Grants, 3775
ORISE Faculty Internship at the US Army Center for Health Promotion and Preventive Medicine, 3802
ORISE Laboratory Technology Program, 3809
ORISE Postgrad Internship at the Office of Water, 3817
ORISE Postgraduate Internship at the US Army Center for Health Promotion and Preventive Medicine, 3818
ORISE Student Internship at the US Army Center for Health Promotion and Preventive Medicine, 3827
OSF Human Rights and Governance Internships, 3845
OSF Justice Initiative Fellowships, 3847
OTS Undergrad NAPIRE Summer Internships, 3858
Pinkerton Foundation Grants, 3967
Princess Grace Foundation Theater, Dance, and Film Grants, 3988
Prudential Foundation Education Grants, 4002
RAND Corporation Graduate Student Summer Associateships, 4030
Rathmann Family Foundation Grants, 4033
Regenstrief General Internal Medicine - General Pediatrics Research Fellowship Program, 4045
Regenstrief Geriatrics Fellowship, 4046
Regenstrief Master Of Science In Clinical Research/ CITE Program, 4047
Rhode Island State Government Internships, 4062
Robert W. Deutsch Foundation Grants, 4082
RWJF Clinical Scholars Grants, 4136
Savoy Foundation Studentships, 4186

SfN Ricardo Miledi Neuroscience Training, 4212
Shastri Indo-Canadian Institute University of Ottawa Overseas India Summer Course Grants, 4232
Sidgmore Family Foundation Grants, 4252
Sioux Falls Area Community Foundation Community Fund Grants (Unrestricted), 4262
Sioux Falls Area Community Found Spot Grants, 4263
Skaggs Foundation Grants, 4266
Smithsonian Environmental Research Center Internships, 4274
Smithsonian Minority Visiting Student Awards, 4279
Smithsonian Museum Conservation Institute Analytical Studies Group Internships, 4281
Smithsonian Museum Conservation Institute Internships for Research Projects, 4282
Society for Imaging Informatics in Medicine Small Training Grants, 4309
Sony Corporation of America Corporate Philanthropy Grants, 4329
South Florida Student Shark Program, 4330
SPJ Pulliam/Kilgore Freedom of Information Internships, 4347
SRC Educational Science Junior Research Position Grants, 4359
SSRC-Van Alen Fellowships, 4419
State Farm Companies Doc Dissertation Award, 4448
Textron Corporate Contributions Grants, 4501
UNCFSP NAI-MIRS Sabbatical Awards, 4561
UNCFSP NASA Science and Technology Institute Summer Scholars Program, 4564
USDA Distance Education Grants for Higher Education Institutions in Insular Areas, 4622
USDA International Science and Education Competitive Grants, 4631
USDA Organic Agriculture Research and Extension Initiative, 4637
USDA Plant and Animal Disease, Pest Control, and Animal Care Grants, 4638
USDC Afghanistan and Iraq Investment and Reconstruction Task Forces Internships, 4664
WHOI Guest Student Internships, 4860
WWF Russell E. Train Education for Nature Alumni Grants, 4897
Xerox Foundation Grants, 4899

Travel Grants

AAAAI Fellows-in-Training Travel Grants, 23
AAB Berlin Prize Fellowships, 48
AACAP Educational Outreach Program for Child and Adolescent Psychiatry Residents, 56
AACAP Educational Outreach Program for General Psychiatry Residents, 57
AACAP Educational Outreach Program for Residents in Alcohol Research, 58
AACAP Life Members Mentorship Grants for Medical Students, 67
AACAP Systems of Care Special Scholarships, 81
AACC International Cecil F. Pinney Travel Awards, 86
AACC International George W. Scott Blair Lecture Award, 90
AACC International Student Travel Grant, 97
AACR-GlaxoSmithKline Clinical Cancer Research Scholar Awards, 118
AACR-NCI Int'l Investigator Opportunity Grants, 121
AACR-WICR Scholar Awards, 125
AACR Brigid G. Leventhal Scholar in Cancer Research Awards, 131
AACR Gertrude Elion Cancer Research Award, 137
AACR Minority-Serving Institution Faculty Scholar in Cancer Research Awards, 143
AACR Minority Scholar in Cancer Res Awards, 144
AACR Scholar-in-Training Awards, 146
AACR Scholar-in-Training Awards - Special Conferences, 147
AACR Thomas J. Bardos Science Educ Awards, 149
AAFPRS Residency Travel Awards, 167
AAF Richard Morris Hunt Archit Fellowship, 169
AAG International Geographic Information Fund Student Travel Grants, 174
AAP International Travel Grants, 197
AAR Postdoctoral Fellowships, 201
AAR Rome Prize Fellowships, 202
Abbott-ASM Lifetime Achievement Award, 218
Abdus Salam ICTP-ELETTRA Users Grants, 221
ACC Asian Art and Religion Fellowships, 251
ACC Fellowships and Projects, 252
ACC Humanities Fellowship Program, 255
ACC Japan-US Arts Program Fellowships, 256
ACC John D. Rockefeller III Award, 257
ACC Mekong Region Fellowship, 258
ACC Philippines Fellowship Program, 261
ACLS Charles A. Ryskamp Research Fellowships, 281
ACLS Dissertation Fellowships in East European Studies, 285
ACLS Early Career Postdoctoral Fellowships in East European Studies, 286
ACSM International Student Awards, 318
AERA Dissertation Grants, 342
AES-Grass Young Investigator Travel Awards, 345
AIChE CRE Travel Awards, 404
AIChE Nanoscale Science and Engineering Forum Young Investigator Award, 417
AIChE North American Mixing Forum Award, 418
AIChE Nuclear Engineering Division Outstanding Student Paper & Presentation Awards, 421
AIHP Sonnedecker Visiting Scholar Grants, 432
AINSE Int'l Conference Travel Scholarships, 461
AINSE Research Fellowships with Institute for Science and International Security Support, 465
ALA Frederick G. Kilgour Award for Research in Library and Information Technology, 487
ALA Ingenta Research Awards, 489
ALA Jan Merrill-Oldham Professional Development Grant, 491
ALA NewsBank/Readex Cath J. Reynolds Award, 494
ALA Samuel Lazerow Fellowship For Research in Collections and Technical Services in Academic and Research Libraries, 497
Alexander von Humboldt International Climate Protection Fellowships, 529
Alexander von Humboldt Konrad Adenauer Research Award, 531
AMA-MSS Research Poster Award, 599
AMA-WPC Joan F. Giambalvo Memorial Scholarships, 600
American-Scandinavian Foundation Awards for Advanced Study or Research in the U.S., 603
American-Scandinavian Foundation Grant for Study in Scandinavia, 605
American Academy of Religion Individual Research Grants, 618
American Chemical Society AGFD Undergraduate Research Symposium Travel Grants, 629
American Chemical Society AGRO Sterling B. Hendricks Memorial Lectureship, 635
American Chemical Society ANYL Pfizer Graduate Travel Awards in Analytical Chemistry, 646
American Chemical Society Green Chem Grants, 679
American Chemical Society Joseph Breen Memorial Fellowship in Green Chemistry, 682
American Chemical Society Petroleum Research Fund Scientific Education Grants, 692
American Chemical Society Petroleum Research Fund Undergraduate Faculty Sabbatical Grants, 693
American Chemical Society Travel Grants for Chemists with Disabilities, 697
American College of Rheumatology Fellows-in-Training Travel Scholarships, 700
American College of Surgeons Traveling Fellowship to Germany, 706
American College of Surgeons Traveling Fellowship to Japan, 707
American Hist Assoc Albert J. Beveridge Grants for Research in the History of the W Hemisphere, 713
American Hist Assoc Bernadotte E. Schmitt Research Grants in European, African, or Asian History, 715
American Hist Assoc Littleton-Griswold Research Grants in U.S. Legal History, 736
American Hist Assoc Michael Kraus Research Grant in American Colonial History, 738
American Mathematical Society China Exchange Travel Grants, 753
American Mathematical Society Graduate Student Travel Grants, 754
American Mathematical Society International Congress of Mathematicians Travel Grants, 755
American Math Society Simons Travel Grants, 756
American Mathematical Society Travel Grant for the Mathematical Congress of the Americas, 757
American Numismatic Society Donald Groves Numismatic Research Grants, 760
American Philosophical Society Franklin Research Grants, 764
American Philosophical Society Lewis and Clark Fund for Exploration and Field Research, 767
American Philosophical Society Lewis and Clark Exploration and Field Research in Astrobiology, 768
American Psychiatric Association Travel Scholarships for Minority Medical Students, 790
amfAR Research Grants, 811
AMS Eileen Southern Travel Fund Grants, 822
AMS Eugene K. Wolf Travel Fund for European Research Grants, 823
AMS Harold Powers World Travel Fund Grants, 826
AMS Janet Levy Fund for Independent Scholars, 828
AMS M. Elizabeth C. Bartlet Fund for Research in France, 830
Andrew W. Mellon Postdoctoral Fellowship in American Cultures, 845
Animal Behavior Society Latin American Travel Awards, 852
AOCS Hans Kaunitz Award, 880
AOCS Industrial Oil Products Division Student Award, 885
AOCS Lipid Oxidation and Quality Division Poster Competition, 886
AOCS Phospholipid Best Paper Award, 889
AOCS Processing Distinguished Service Award, 890
AOCS Surfactants and Detergents Division Student Award, 898
AOCS Young Scientist Research Award, 902
APS/British Academy Fellowship for Research in London, 918
APS Andreas Acrivos Dissertation Award in Fluid Dynamics, 940
APS Aneesur Rahman Prize for Computational Physics, 942
APS Arthur L. Schawlow Prize in Laser Science, 945
APSA Travel Grants, 948
APS Dannie Heineman Prize for Mathematical Physics, 951
APS Dissertation Award in Nuclear Physics, 953
APS E.K. Plyler Prize for Molecular Spectroscopy, 954
APS Einstein Prize, 955
APS Fluid Dynamics Prize, 956
APS Frank Isakson Prize for Optical Eff in Solids, 957
APS George E. Pake Prize, 958
APS Herbert P. Broida Prize, 961
APS I.I. Rabi Prize in Atomic, Molecular, and Optical Physics, 962
APS Irving Langmuir Prize in Chemical Physics, 963
APS J.J. Sakurai Prize for Theoretical Part Physics, 964
APS James C. McGroddy Prize for New Materials, 965
APS James C Maxwell Prize for Plasma Physics, 966
APS Marshall N. Rosenbluth Outstanding Doctoral Thesis Award, 969
APS Mitsuyoshi Tanaka Dissertation Award in Experimental Particle Physics, 971
APS Nicholas Metropolis Award for Outstanding Doctoral Thesis in Computational Physics, 972
APS Outstanding Doctoral Thesis Research in Atomic, Molecular, or Optical Physics, 974
APS Prize for a Faculty Member for Research in an Undergraduate Institution, 977
APS Robert R. Wilson Prize for Achievement in the Physics of Particle Accelerators, 979
APS W.K.H. Panofsky Prize in Experimental Particle Physics, 981
ARIT Fellowships for Intensive Advanced Turkish Language Study, 992

ARM Faculty Fellowships, 1001
ARM Graduate Fellowships, 1002
ASCE Freeman Fellowships, 1040
ASCO Merit Awards, 1049
ASCSA Alison M. Frantz Fellowship, 1051
ASCSA Fellowships, 1052
ASGE Don Wilson Award, 1062
ASHA Research Mentoring-Pair Travel Award, 1073
ASHA Student Research Travel Award, 1074
ASHA Students Preparing for Academic & Research Careers Award, 1075
ASHFoundation Grant in Speech Science, 1083
Asia Foundation Exchange Programs, 1090
ASM-UNESCO Leadership Grant for International Educators, 1099
ASM Millis-Colwell Postgraduate Travel Grant, 1179
ASPEN Scientific Abstracts Awards for Papers, 1198
ASPEN Scientific Abstracts Promising Investigator Awards, 1199
ASPRS Ta Liang Memorial Award, 1215
Association for Women in Mathematics Mentoring Travel Grants, 1223
Australian Academy of Science Grants, 1269
Australian Research Fellowships/Queen Elizabeth II Fellowships, 1272
AWM Mentoring Travel Grants for Women, 1279
AWM Travel Grants for Women Researchers, 1280
BA Larger Research Grants, 1290
Ball Brothers Foundation Rapid Grants, 1292
BARD Postdoctoral Fellowships, 1298
BA Sir Ernest Cassel Educational Grants, 1303
BA Small Research Grants, 1304
Beverly Willis Architecture Foundation Grants and Fellowships, 1340
Bilateral Scholarships, 1344
BMI Woody Guthrie Fellowships, 1352
BWF Collaborative Research Travel Grants, 1395
Canada-U.S. Fulbright Senior Specialists Grants, 1418
Carl M. Freeman Foundation Grants, 1430
CDC Foundation Atlanta International Health Fellowships, 1461
CDC Public Health Conference Support Grants, 1466
CFR Edward R. Murrow Press Fellowship, 1500
CFR Int'l Affairs Fellowships in South Korea, 1503
CFR International Affairs in Japan Fellowships, 1504
Charles W. Finley Visiting Scholar Educ Grant, 1530
CHF Travel Grants, 1545
Churchill Fellowships, 1560
CIC Traveling Scholars, 1566
Conquer Cancer Foundation of ASCO Int'l Devel & Educ Award in Palliative Care, 1634
Conquer Cancer Foundation of ASCO International Development and Education Awards, 1635
Conquer Cancer Foundation of ASCO Resident Travel Award for Underrepresented Populations, 1638
CTCRI Researcher Travel Grants, 1668
CTCRI Workshop and Learning Opportunities Grants, 1671
Cushwa Center Research Travel Grants, 1673
Cystic Fibrosis Canada Senior Scientist Research Training Awards, 1680
Cystic Fibrosis Canada Travel Supplement Grants, 1685
Cystic Fibrosis Canada Visiting Allied Health Professional Awards, 1686
Cystic Fibrosis Canada Visiting Clinician Awards, 1687
Cystic Fibrosis Canada Visiting Scientist Awards, 1688
DAAD-AICGS Research Fellowships, 1693
DAAD Learn German in Germany Grants, 1700
DAAD Research Grants for Doctoral Candidates and Young Academics and Scientists, 1703
DAAD Undergraduate Scholarships, 1710
Daimler and Benz Foundation Fellowships, 1713
Diaz-Ayala Cuban and Latin American Popular Music Collection Travel Grants, 1763
DOE Albert Einstein Distinguished Educator Fellowship Program, 1770
DOE Nonproliferation & National Security Research Grants, 1806
DOE Univ Nuclear Science & Reactor Support, 1817
DOS Study of the United States Institutes, 1846
Duke University Libraries Franklin Collection Travel Grants, 1874
Duke University Libraries Hartman Center Travel Grants, 1875
Duke University Libraries J. Walter Thompson Research Fellowships, 1876
Duke University Libraries Mary Lily Res Grants, 1877
E. Bright Wilson Award in Spectroscopy, 1883
Eisenhower Exchange International Fellowships, 1902
Elizabeth Glaser Int'l Leadership Awards, 1906
Endangered Language Fund Grants, 1921
Explorers Club Exploration Fund Grants, 1966
Facebook Fellowships, 1973
FDHN Moti L. & Kamla Rustgi International Travel Awards, 1997
FDHN Non-Career Research Awards, 1998
FIC Research Collaboration Award, 2006
Florence Tan Moeson Fellowships, 2025
Fulbright-Hays Seminars Abroad Bilateral Projects Grants, 2062
Fulbright Alumni Initiatives Awards Grants, 2064
Fulbright Distinguished Chairs Grants, 2069
Fulbright German Studies Seminar Grants, 2072
Fulbright International Education Administrators Seminar Grants, 2073
Fulbright Occasional Lecturer Grants, 2076
Fulbright Specialists Grants, 2079
Fulbright Traditional Scholar Program in Europe and Eurasia, 2080
Fulbright Traditional Scholar Program in Sub-Saharan Africa, 2081
Fulbright Traditional Scholar Program in the East Asia/Pacific Region, 2082
Fulbright Traditional Scholar Program in the South and Central Asia Region, 2084
Fulbright Traditional Scholar Program in the Western Hemisphere, 2085
Fulbright Travel Grants, 2086
Fulbright Visiting Specialists Program: Direct Access to the Muslim World, 2087
Fund for Folk Culture Artists Support Grants, 2088
Germanistic Society of America Fellowships, 2128
Gladys Krieble Delmas Foundation Grants for Venetian Research, 2144
Great Lakes Protection Fund Grants, 2170
GUITS Grants for Undergraduate Study, 2187
GUITS Matching Seed-Money Grants for New Positions in Turkish Studies, 2188
GUITS Post-Doctoral Summer Travel-Research Grants, 2189
GUITS Sabbatical Research Grants, 2191
Hague Academy of International Law Summer Scholarships, 2198
Harris Graduate School of Public Policy Studies Research Development Grants, 2204
Harvard University Weatherhead Center for International Affairs Fellowships, 2215
Henry J. Kaiser Family Found Media Fellowships, 2234
Herbert Hoover Presidential Library Association Travel Grants, 2242
Hydrogen Student Design Contest, 2311
Ian Axford (New Zealand) Fellowships in Public Policy, 2313
IBRO/SfN International Travel Grants, 2321
IBRO InEurope Short Stay Grants, 2325
IBRO International Travel Grants, 2326
IBRO Latin America Regional Travel Grants, 2329
IES Faculty Fellowships, 2341
IIE Whitaker International Fellowships and Scholarships, 2356
Institute of Turkish Studies Summer Language Study Grants in Turkey, 2391
Institute of Turkish Studies Summer Research Grants in Turkey, 2392
International Fellowships in Medical Education, 2398
IREX ECA Alumni Small Grants, 2416
IREX Individual Advanced Research Opportunities Grants, 2419
IREX Short-Term Travel Grants, 2424
IWMF Public Health Fellowship, 2446
Jane Beattie Memorial Scholarship, 2470
Japan Foundation Center for Global Partnership Grants, 2472
Japan Foundation Japanese Studies Fellowships, 2476
Japan Foundation New York Institutional Support for Japanese Studies Grants, 2477
Japan Foundation New York Japanese Studies Fellowships, 2478
Japan Foundation Six-Month Grants for Specialists in Cultural and Academic Fields, 2479
John Dana Archbold Fellowships, 2499
JSA Junior Scientist Travel Support, 2523
Kellogg Institute for International Studies Faculty Fellows' Research Grants, 2540
Kellogg Institute for International Studies Grants for Conferences and Workshops, 2543
Lady Davis Fellowships, 2578
Manitoba Arts Council Craft Grants, 2653
MAP International Medical Fellowships, 2655
Marshall Scholarships, 2666
MBL M.G.F. Fuortes Summer Fellowships, 2711
Minerva Fellowships, 2778
Minerva Short-Term Research Grants, 2779
MLA Cunningham Memorial International Fellowship, 2783
MLA David A. Kronick Traveling Fellowship, 2784
NACC David Stevenson Fellowships, 2812
NACC William Diaz Fellowships, 2813
Nakanishi Prize, 2824
National Center for Responsible Gaming Conference Scholarships, 2851
National Center for Responsible Gaming Travel Grants, 2858
National Education Reporting Fellowships, 2861
National Endowment for Democracy Reagan-Fascell Democracy Fellowships, 2863
NATO Advanced Research Workshops Program, 2894
NATO Advanced Study Institutes Program, 2895
NATO Collaborative Linkage Grants, 2896
NCIC Canadian Cancer Society Research Grants, 2919
NCIC Canadian Cancer Society Travel Award for Senior Level PhD Students, 2920
NCI Technologies and Software to Support Integrative Cancer Biology Research Grants, 2933
NFID Advanced Vaccinology Travel Grant, 2986
NFID John P. Utz Medical Mycology Postdoctoral Fellowship, 2988
NFID Nosocomial Infection Research and Training Postdoctoral Fellowship, 2990
NIDDK Traditional Conference Grants, 3184
NL Herzog August Bibliothek Wolfenbuttel Fellowship, 3286
NL Newberry Consortium on American Indian Studies Graduate Student Fellowships, 3312
NL Newberry Library/Ecole Nationale des Chartes Exchange Fellowship, 3313
NOAA Living Marine Resources Cooperative Science Center Mini Grants, 3381
North Carolina Arts Council Res Center Grants, 3420
NSERC Northern Research Internships Program, 3447
NSERC Synergy Awards for Innovation, 3450
NSF-CONACyT Collaborative Research Grants, 3452
NSF-NIST Interaction in Chemistry, Materials Research, Molecular Biosciences, Bioengineering, and Chemical Engineering, 3453
NSF/NEH Documenting Endangered Languages Project Grants and Fellowships: data, infrastructure and computational methods, 3455
NSF Antarctic Artists and Writers Grants, 3473
NSF Conferences, Workshops, and Special Meetings in the Mathematical Sciences, 3541
NSF Grant Opportunities for Academic Liaison with Industry, 3574
NSF Linguistics Grants, 3587
NSF Partnerships for International Research and Education, 3611
NYAM Edward N. Gibbs Memorial Lecture and Award in Nephrology, 3682
NYPL Martin Duberman Visiting Scholars, 3700
OAH/IEHS John Higham Travel Grants, 3710

OAH Germany Residency Grants, 3719
OAH Japanese Residencies Grants, 3722
Organization for Tropical Studies Fellowships, 3796
Organization for Tropical Studies Pilot Awards for Exploratory Research, 3797
OTS Francis J. Bossuyt Fellowship, 3855
OTS Hovore-Horn Fellowships, 3856
OTS Stanley Smith Horticultural Fellowships, 3857
OTS Undergrad NAPIRE Summer Internships, 3858
Phi Kappa Phi Found Love of Learning Awards, 3937
Phi Upsilon Omicron Florence Fallgatter Distinguished Service Award, 3943
Princeton University Center for Human Values Political Philosophy Travel Grants, 3993
RCPSC Detweiler Traveling Fellowships, 4036
RCPSC Royal College Visiting Professorships in Medical Research, 4041
Rett Syndrome Research Foundation Grants, 4058
Roberto Longhi Foundation Fellowships, 4078
Rockefeller Archive Center Ehrlich Grants, 4084
Rockefeller Archive Center Grants in Aid, 4085
Rockefeller Archive Center Research Grants, 4086
Rockefeller Archive Center Visiting Archivist Fellowship, 4087
Royal Scottish Geographical Society Expedition Grants, 4110
Royal Scottish Geographical Society Small Travel Grants, 4112
SfN Chapters Graduate Student Travel Awards, 4202
SfN Chapters Postdoctoral Trainee Travel Awards, 4203
SfN Graduate Student Travel Awards, 4205
SfN Louise Hanson Marshall Specific Recognition Award, 4208
SfN Postdoctoral Fellow Travel Awards, 4210
SfN Ricardo Miledi Neuroscience Training, 4212
SfN Travel Awards for the Federation of European Neuroscience Societies Forum, 4214
SfN Travel Awards for the International Brain Research Organization World Congress, 4215
SfN Travel Awards for the Japan Neuroscience Society Meeting, 4216
SfN Undergraduate Student Travel Awards, 4217
Shastri Indo-Canadian Institute Arts Fellowships, 4219
Shastri Indo-Canadian Institute Post-Doctoral Research Fellowships, 4229
Short-Term Travel to Japan Grants, 4236
Sigrid Juselius Foundation Grants, 4260
Sir Dorabji Tata Trust Individual Education and Travel Grants, 4265
Smithsonian National Museum of Nat Hist Lemelson Center Travel to Collections Awards, 4290
Smithsonian Tropical Research Institute Short Term Fellowships, 4297
SPJ Eugene C. Pulliam Fellowship for Editorial Writing, 4346
SRC Educational Science Grants for Invitation of Visiting Scientists, 4357
SRC Educational Science Grants for Travels to Foreign Institutions as a Guest Scientist, 4358
SRC Educational Science Travel Grants, 4365
SRC Medical Research Grants, 4379
SRC Medical Science Grants for Large International Conferences in Sweden, 4380
SRC Natural and Engineering Science Conference Grants, 4383
SRC Natural & Engineering Science Travel Grants, 4388
SSRC Eurasia Program Pre-Dissertation Awards, 4428
Streilein Foundation for Ocular Immunology Visiting Professorships, 4453
Swedish Natural Science Res Council Grants, 4479
Swiss National Science Foundation DORE (DO REsearch) Conference Grants, 4482
Swiss National Science Foundation Scientific Conferences Grants, 4486
Syracuse University Graduate Fellowships, 4488
Thomson Reuters / MLA Doctoral Fellowships, 4512
TRDRP Dissertation Research Awards, 4526
U.S. Department of Education American Overseas Research Centers, 4537
UICC American Cancer Society International Fellowships for Beginning Investigators, 4555
UICC International Cancer Technology Transfer Fellowships, 4556
UICC Yamagiwa-Yoshida Memorial International Cancer Study Grants, 4558
UIC Rockefeller Foundation Resident Fellowships in the Humanities, 4559
UNESCO Fellowships, 4567
United Methodist Church Women in History Research Grant, 4574
United States Institute of Peace - Jennings Randolph Senior Fellowships, 4580
US CRDF CGP-Climate Change and Energy Competition For Joint U.S.-Ukrainian Research Teams, 4607
US CRDF CGP-RFBR Climate Change and Energy Competition For Joint U.S.-Russian Research Teams, 4608
US CRDF CGP ETF 2009 Energy Research Competition For Joint U.S.-Estonian Research Teams, 4609
US CRDF Leishmaniasis: Collaborative Opportunities in North Africa and the Middle East, 4610
Virginia Historical Society Research Fellowships, 4816
Walter J. Jensen Fellowship for French Language, Literature, and Culture, 4830
Welch Foundation Departmental Grants, 4839
Winston Churchill Foundation Scholarships, 4886
Wolfsonian-FIU Fellowships, 4890
Woodrow Wilson MMUF Travel & Res Grants, 4893
Woodrow Wilson National Fellowship Foundation Travel and Research Grants, 4894
WWF Russell E. Train Education for Nature Alumni Grants, 4897
Young Scientists Summer Program, 4903

Vocational Education

Achelis Foundation Grants, 274
ASTA Academic Scholarships, 1224
Bodman Foundation Grants, 1356
Carl M. Freeman Foundation Grants, 1430
Carrier Corporation Contributions Grants, 1440
Claude Pepper Foundation Grants, 1578
D.F. Halton Foundation Grants, 1692
Dept of Ed Adult Education--National Leadership Activities Grants, 1744
GNOF Gulf Coast Oil Spill Grants, 2153
National Housing Endowment Challenge/Build/Grow Grant, 2877
Oppenstein Brothers Foundation Grants, 3792
Reinberger Foundation Grants, 4051
Textron Corporate Contributions Grants, 4501
U.S. Department of Education Rehabilitation Research Training Centers, 4545
USDA Distance Education Grants for Higher Education Institutions in Insular Areas, 4622
USDA Organic Agriculture Research and Extension Initiative, 4637
Vancouver Foundation Grants and Community Initiatives Program, 4807
WILD Foundation Grants, 4867
Xerox Foundation Grants, 4899

Geographic Index

Note: This index lists grants for which applicants must be residents of or located in a specific geographic area. Numbers refer to entry numbers.

United States

Alabama
3M Company Higher Education Grants, 4
3M Foundation Community Giving Grants, 5
Alabama Humanities Foundation Mini Grants, 480
Alabama Humanities Foundation Planning/ Consultant Grants, 481
Alabama Library Assoc CUS Research Grants, 482
Appalachian Reg Comm Health Care Grants, 915
Boeing Company Contributions Grants, 1357
Carrier Corporation Contributions Grants, 1440
Charles M. and Mary D. Grant Found Grants, 1527
Cleveland-Cliffs Foundation Grants, 1580
DOE Experimental Program to Stimulate Competitive Research (EPSCoR), 1791
EPA Southeastern Multi-Media and Geographic Priority Project Grants, 1939
Georgia-Pacific Found Environment Grants, 2123
Harry B. and Jane H. Brock Foundation Grants, 2206
Hill Crest Foundation Grants, 2267
SARE Large Systems Grants: South Reg, 4172
SARE On-Farm Research Grants: South Reg, 4173
SARE Research & Educ Grants: South Reg, 4179
USDA 1890 Institution Teaching, Research, and Extension Capacity Building Grants, 4613
USDA Regional Integrated Pest Management Grants - Southern Region, 4644
Weyerhaeuser Company Foundation Grants, 4857
Woodward Fund Grants, 4895

Alaska
3M Company Higher Education Grants, 4
3M Foundation Community Giving Grants, 5
Alaska Conservation Foundation Watchable Wildlife Conservation Trust Grants, 498
Brainerd Foundation Grants, 1368
Bullitt Foundation Grants, 1384
ConocoPhillips Grants, 1629
Fluor Foundation Grants, 2028
M.J. Murdock Trust College Research Program for Life Sciences, 2638
M.J. Murdock Trust College Science Research, 2639
M.J. Murdock Trust Exceptional Opportunity Grants, 2640
M.J. Murdock Trust General Grants, 2641
M.J. Murdock Trust Partners in Science Grants, 2642
M.J. Murdock Trust Research Start-up Grants for New Science Faculty, 2643
M.J. Murdock Trust Scientific Research Grants, 2644
Paul G. Allen Family Foundation Grants, 3876
Robert B McMillen Foundation Grants, 4072
SARE Graduate Student Grants in Sustainable Agriculture - Western Region, 4171
SARE Producer Research and Education Grants-Western Region, 4174
SARE Pro & Producer Grants: West Reg, 4175
SARE Research & Educ Grants: West Reg, 4180
Skaggs Foundation Grants, 4266
USDA Alaska Native-Serving and Native Hawaiian-Serving Institutions Education Competitive Grants, 4615
USDA Integrated Pest Management Competitive Grants - Western Region, 4629
W.M. Keck Foundation Undergraduate Education Grants, 4825
Western Regional Graduate Program, 4855

American Samoa
Alston Bannerman Sabbatical Fellowships, 580
Bank of America Charitable Matching Gifts, 1297
EPA Pollution Prevention Grants, 1936
EPA Regional Healthy Indoor Air Projects for Community Outreach and Education, 1937
EPA Temporally Integrated Monitoring of Ecosystems and Long-Term Grants, 1942
NHLBI Ancillary Studies in Clinical Trials, 3015
NHLBI Bioengineering and Obesity Grants, 3016
NHLBI Career Transition Awards, 3021
NHLBI Mentored Career Development Award to Promote Faculty Diversity/Re-Entry in Biomedical Research, 3035
NHLBI Mentored Patient-Oriented Research Career Development Awards, 3037
NHLBI Mentored Quantitative Research Career Development Awards, 3038
NHLBI Pathway to Independence Awards, 3045
NHLBI Ruth L. Kirschstein National Research Service Awards for Individual Senior Fellows, 3060
OTS Undergrad NAPIRE Summer Internships, 3858
SARE Graduate Student Grants in Sustainable Agriculture - Western Region, 4171
SARE Producer Research and Education Grants-Western Region, 4174
SARE Pro & Producer Grants: West Reg, 4175
SARE Research & Educ Grants: West Reg, 4180
USDA Distance Education Grants for Higher Education Institutions in Insular Areas, 4622
USDA Integrated Pest Management Competitive Grants - Western Region, 4629
USDA Resident Instruction Grants for Higher Education in Insular Areas, 4645
Wellcome Trust New Investigator Awards, 4844

Arizona
Abbott Fund Global AIDS Care Grants, 219
Abbott Fund Science Education Grants, 220
APS/Phoenix Suns Energy and Environmental Minigrants, 919
Arizona Cardinals Grants, 995
Arizona Commission on the Arts Education Projects Grants, 996
Arizona Commission on the Arts Folklorist Residencies, 997
Arizona Public Service Corporate Giving Grants, 998
ASU Graduate College Dean's Fellowships, 1236
Boeing Company Contributions Grants, 1357
Caesars Foundation Grants, 1407
Carrier Corporation Contributions Grants, 1440
Community Foundation for Southern Arizona Grants, 1618
Del E. Webb Foundation Grants, 1733
E.L. Wiegand Foundation Grants, 1885
Flinn Foundation Grants, 2024
FMC Foundation Grants, 2029
Margaret T. Morris Foundation Grants, 2659
Motorola Foundation Grants, 2796
Piper Trust Healthcare & Med Research Grants, 3972
Prudential Foundation Education Grants, 4002
Rocky Mountain Mineral Law Found Grants, 4094
SARE Graduate Student Grants in Sustainable Agriculture - Western Region, 4171
SARE Producer Research and Education Grants-Western Region, 4174
SARE Pro & Producer Grants: West Reg, 4175
SARE Research & Educ Grants: West Reg, 4180
Union Pacific Foundation Community and Civic Grants, 4568
Union Pacific Foundation Health and Human Services Grants, 4569
USDA Integrated Pest Management Competitive Grants - Western Region, 4629
USDA Tribal College Research Grants, 4652
W.M. Keck Foundation Undergraduate Education Grants, 4825
Western Regional Graduate Program, 4855
Weyerhaeuser Company Foundation Grants, 4857

Arkansas
3M Company Higher Education Grants, 4
3M Foundation Community Giving Grants, 5
Archer Daniels Midland Foundation Grants, 985
Arkansas Arts Council Individual Artists Fellowships, 999
Assisi Foundation of Memphis General Grants, 1221
Eaton Charitable Fund Grants, 1893
Georgia-Pacific Found Environment Grants, 2123
Hatton W. Sumners Foundation for the Study and Teaching of Self Government Grants, 2219
Ross Foundation Grants, 4106
SARE Grad Student Grants - South Region, 4170
SARE Large Systems Grants: South Reg, 4172
SARE On-Farm Research Grants: South Reg, 4173
SARE Research & Educ Grants: South Reg, 4179
Union Pacific Foundation Community and Civic Grants, 4568
Union Pacific Foundation Health and Human Services Grants, 4569
USDA 1890 Institution Teaching, Research, and Extension Capacity Building Grants, 4613
USDA Regional Integrated Pest Management Grants - Southern Region, 4644
Weyerhaeuser Company Foundation Grants, 4857

California
3M Company Higher Education Grants, 4
3M Foundation Community Giving Grants, 5
Abbott Fund Global AIDS Care Grants, 219
Abbott Fund Science Education Grants, 220
A Charitable Foundation Grants, 273
Adam Richter Charitable Trust Grants, 339
Akonadi Foundation Anti-Racism Grants, 477
Alice Tweed Tuohy Foundation Grants, 569
ALZA Corporate Contributions Grants, 588
Angels Baseball Foundation Grants, 848
Ann Peppers Foundation Grants, 863
Archer Daniels Midland Foundation Grants, 985
Bayer Foundation Grants, 1309
Better Way Foundation Grants, 1338
Blue Shield of California Grants, 1350
Boeing Company Contributions Grants, 1357
Burton G. Bettingen Grants, 1390
Caesars Foundation Grants, 1407
California Teachers Association Scholarships, 1408
California Wellness Foundation Work and Health Grants, 1409
Callaway Golf Company Foundation Grants, 1410
Carl Gellert and Celia Berta Gellert Foundation Grants, 1429
CCH Documentary Project Research and Development Grants, 1451
Cen Cal Diving & Aquatic Studies Scholarship, 1471
Charles H. Price II Family Foundation Grants, 1523
Chiles Foundation Grants, 1552
Chiron Foundation Community Grants, 1553
Clarence E. Heller Charitable Found Grants, 1575
Coeta and Donald Barker Foundation Grants, 1594
ConocoPhillips Grants, 1629
Cystic Fibrosis Research Eliz Nash Fellowships, 1689
Dean Witter Foundation Grants, 1728
Del E. Webb Foundation Grants, 1733
Durfee Foundation Stanton Fellowship, 1881
E.L. Wiegand Foundation Grants, 1885
Expect Miracles Foundation Grants, 1965
Fleishhacker Foundation Eureka Fellowships, 2023
Fluor Foundation Grants, 2028
FMC Foundation Grants, 2029
Fritz B. Burns Foundation Grants, 2057
George Frederick Jewett Foundation Grants, 2116
Georgia-Pacific Found Environment Grants, 2123
Gil and Dody Weaver Foundation Grants, 2139
Green Diamond Charitable Contributions, 2171
HAF Riley Frazel Memorial Scholarship, 2197
Hilda and Preston Davis Foundation Grants, 2266
Hilton Hotels Corporate Giving Grants, 2268
Ireland Family Foundation Grants, 2415
John Stauffer Charitable Trust Grants, 2509
Kenneth T. and Eileen L. Norris Found Grants, 2554
Lucile Horton Howe and Mitchell B. Howe Foundation Grants, 2624
McCarthy Family Foundation Grants, 2717
Mericos Foundation Grants, 2742
Miller Brewing Corporate Contributions, 2774

Motorola Foundation Grants, 2796
NFWF Nature Restoration Trust Grants, 3005
Nick Traina Foundation Grants, 3128
NMF California Commun Service Scholarships, 3321
Nordson Corporation Foundation Grants, 3416
Patron Saints Foundation Grants, 3874
Price Family Charitable Fund Grants, 3987
Prudential Foundation Education Grants, 4002
Rathmann Family Foundation Grants, 4033
Richard and Rhoda Goldman Fund Grants, 4063
Rocky Mountain Mineral Law Found Grants, 4094
S. D. Bechtel, Jr. Foundation / Stephen Bechtel Fund Preventive Healthcare and Selected Research Grants, 4152
San Diego Women's Foundation Grants, 4164
SARE Graduate Student Grants in Sustainable Agriculture - Western Region, 4171
SARE Producer Research and Education Grants-Western Region, 4174
SARE Pro & Producer Grants: West Reg, 4175
SARE Research & Educ Grants: West Reg, 4180
Sun Microsystems Foundation Summer Academic Enrichment Grants, 4468
Thomas & Dorothy Leavey Found Grants, 4506
Thomas C. Ackerman Foundation Grants, 4508
Thornton Foundation Grants, 4514
TRDRP California Research Awards, 4525
TRDRP Dissertation Research Awards, 4526
TRDRP New Investigator Awards, 4528
TRDRP Participatory Research Awards, 4529
TRDRP Postdoctoral Fellowship Awards, 4530
TRDRP Research Grants, 4531
Union Pacific Foundation Community and Civic Grants, 4568
Union Pacific Foundation Health and Human Services Grants, 4569
USDA Integrated Pest Management Competitive Grants - Western Region, 4629
USDA Tribal College Research Grants, 4652
Vanguard Public Foundation Grant Funds, 4808
W.M. Keck Foundation Undergraduate Education Grants, 4825
Wayne and Gladys Valley Foundation Grants, 4835
Weyerhaeuser Company Foundation Grants, 4857
William G. McGowan Charitable Fund Grants, 4877

Colorado
Beim Foundation Grants, 1331
Boeing Company Contributions Grants, 1357
Colorado Bioscience Discovery Eval Grants, 1598
Colorado Clean Energy Solar Innovation Grants, 1599
Coors Brewing Corporate Contributions Grants, 1644
D.F. Halton Foundation Grants, 1692
Eaton Charitable Fund Grants, 1893
El Pomar Leadership Development and Community Engagement Postgraduate Fellowship, 1912
Gates Family Foundation Fellowships, 2097
John G. Duncan Trust Grants, 2501
Mabel Y. Hughes Charitable Trust Grants, 2647
Monfort Family Foundation Grants, 2794
NL Associated Colleges of the Midwest/Great Lakes College Association Faculty Fellowships, 3281
Rocky Mountain Mineral Law Found Grants, 4094
Rose Community Foundation Health Grants, 4104
Ruth and Vernon Taylor Foundation Grants, 4130
Sachs Foundation Scholarships, 4157
SARE Graduate Student Grants in Sustainable Agriculture - Western Region, 4171
SARE Producer Research and Education Grants-Western Region, 4174
SARE Pro & Producer Grants: West Reg, 4175
SARE Research & Educ Grants: West Reg, 4180
Schramm Foundation Grants, 4188
Sun Microsystems Foundation Summer Academic Enrichment Grants, 4468
Union Pacific Foundation Community and Civic Grants, 4568
Union Pacific Foundation Health and Human Services Grants, 4569
USDA Integrated Pest Management Competitive Grants - Western Region, 4629
W.M. Keck Foundation Undergraduate Education Grants, 4825
Western Regional Graduate Program, 4855
Weyerhaeuser Company Foundation Grants, 4857

Connecticut
3M Company Higher Education Grants, 4
3M Foundation Community Giving Grants, 5
A. V. Stout Fund Research Grants, 13
Ahn Family Foundation Grants, 379
Bay and Paul Foundations, Inc Grants, 1307
Bayer Foundation Grants, 1309
Bodenwein Public Benevolent Found Grants, 1355
Bristol-Myers Squibb Foundation Health Disparities Grants, 1377
Campbell Soup Foundation Grants, 1415
Carrier Corporation Contributions Grants, 1440
CDECD Elizabeth L. Mahaffey Arts Administration Fellowship, 1468
Charles H. Hood Foundation Child Health Research Awards, 1522
Community Foundation for Greater New Haven Quinnipiac River Fund Grants, 1616
Connecticut Community Foundation Grants, 1627
Connecticut Health Foundation Health Initiative Grants, 1628
Dorr Foundation Grants, 1842
Ensign-Bickford Foundation Grants, 1928
Expect Miracles Foundation Grants, 1965
Fannie E. Rippel Foundation Grants, 1978
Graduate Institute Hodson Trust Teacher Fellowships, 2160
Hilda and Preston Davis Foundation Grants, 2266
Maximilian E. and Marion O. Hoffman Found, 2688
Nellie Mae Education Foundation District-Level Change Grants, 2968
Orchard Foundation Grants, 3793
Perkin Fund Grants, 3898
Prudential Foundation Education Grants, 4002
Richard Davoud Donchian Foundation Grants, 4066
Robert Leet and Clara Guthrie Patterson Trust Grants, 4076
SARE Graduate Student Grants: NE Region, 4169
SARE Professional Development Grants - Northeast Region, 4176
SARE Research and Education Grants - Northeast Region, 4178
USDA Regional Integrated Pest Management Grants - Northeastern Region, 4643

Delaware
Carl M. Freeman Foundation Grants, 1430
Delaware Division of the Arts Individual Artist Fellowships, 1732
FMC Foundation Grants, 2029
Graduate Institute Hodson Trust Teacher Fellowships, 2160
Hilda and Preston Davis Foundation Grants, 2266
NOAA Chesapeake Bay Office B-WET (Bay Watershed Education and Training) Grants, 3339
NOAA Chesapeake Bay Office Cooperative Research Grants, 3340
NOAA Chesapeake Bay Office Fisheries Science Grants, 3341
NOAA Chesapeake Bay Office Submerged Aquatic Vegetation Culture and Restoration Grants, 3343
Richard Davoud Donchian Foundation Grants, 4066
SARE Graduate Student Grants: NE Region, 4169
SARE Professional Development Grants - Northeast Region, 4176
SARE Research and Education Grants - Northeast Region, 4178
USDA 1890 Institution Teaching, Research, and Extension Capacity Building Grants, 4613
USDA Regional Integrated Pest Management Grants - Northeastern Region, 4643

District of Columbia
Aaron & Cecile Goldman Family Found Grants, 200
Alexander and Margaret Stewart Trust Grants, 514
Bank of America Charitable Matching Gifts, 1297
Blum-Kovler Foundation Grants, 1351
Boeing Company Contributions Grants, 1357
Commonweal Foundation Community Assistance Grants, 1602
Commonweal Foundation Grants, 1603
E.L. Wiegand Foundation Grants, 1885
EPA Pollution Prevention Grants, 1936
EPA Regional Healthy Indoor Air Projects for Community Outreach and Education, 1937
EPA Temporally Integrated Monitoring of Ecosystems and Long-Term Grants, 1942
Expect Miracles Foundation Grants, 1965
Graduate Institute Hodson Trust Teacher Fellowships, 2160
Hilda and Preston Davis Foundation Grants, 2266
Jovid Foundation Grants, 2518
Luther I. Replogle Foundation Grants, 2631
Mabel H. Flory Charitable Trust Grants, 2646
NOAA Chesapeake Bay Office B-WET (Bay Watershed Education and Training) Grants, 3339
NOAA Chesapeake Bay Office Cooperative Research Grants, 3340
NOAA Chesapeake Bay Office Fisheries Science Grants, 3341
NOAA Chesapeake Bay Office Submerged Aquatic Vegetation Culture and Restoration Grants, 3343
Richard Davoud Donchian Foundation Grants, 4066
Robert G. Cabell III and Maude Morgan Cabell Foundation Grants, 4075
SARE Graduate Student Grants: NE Region, 4169
SARE Professional Development Grants - Northeast Region, 4176
SARE Research and Education Grants - Northeast Region, 4178
Thoracic Foundation Grants, 4513
USDA Regional Integrated Pest Management Grants - Northeastern Region, 4643
Washington Area Women's Foundation Strategic Opportunity and Partnership Fund Grants, 4831
Weyerhaeuser Company Foundation Grants, 4857
William G. McGowan Charitable Fund Grants, 4877

Florida
APSA Special Fund for the Study of Women and Politics, 947
Batchelor Foundation Grants, 1305
Boeing Company Contributions Grants, 1357
Campbell Soup Foundation Grants, 1415
Charles M. and Mary D. Grant Found Grants, 1527
Cowles Charitable Trust Grants, 1652
Dade Community Foundation Grants, 1712
Eaton Charitable Fund Grants, 1893
Elizabeth McGraw Foundation Grants, 1908
EPA Southeastern Multi-Media and Geographic Priority Project Grants, 1939
Florida Division of Cultural Affairs Individual Artist Fellowships, 2026
Florida Sea Turtle Grants, 2027
FMC Foundation Grants, 2029
Forrest C. Lattner Foundation Grants, 2035
Frances C. and William P. Smallwood Foundation Grants, 2045
Georgia-Pacific Found Environment Grants, 2123
Harold Alfond Foundation Grants, 2200
Health Foundation of Southern Florida Responsive Grants, 2223
HomeBanc Foundation Grants, 2275
Lillian S. Wells Foundation Grants, 2614
Luther I. Replogle Foundation Grants, 2631
McLean Contributionship Grants, 2727
Motorola Foundation Grants, 2796
Peacock Foundation Grants, 3887
Prudential Foundation Education Grants, 4002
Rayonier Foundation Grants, 4034
Richard Davoud Donchian Foundation Grants, 4066
RRF General Grants, 4115

Salem Foundation Grants, 4160
Sarasota Memorial Healthcare Found Grants, 4167
SARE Grad Student Grants - South Region, 4170
SARE On-Farm Research Grants: South Reg, 4173
SARE Research & Educ Grants: South Reg, 4179
USDA Regional Integrated Pest Management Grants - Southern Region, 4644
Woodward Fund Grants, 4895

Georgia
3M Company Higher Education Grants, 4
3M Foundation Community Giving Grants, 5
American Foodservice Charitable Trust Grants, 710
Appalachian Reg Comm Health Care Grants, 915
Archer Daniels Midland Foundation Grants, 985
Bayer Foundation Grants, 1309
Boeing Company Contributions Grants, 1357
Brian G. Dyson Foundation Grants, 1370
Cabot Corporation Foundation Grants, 1406
Carlos and Marguerite Mason Trust Grants, 1431
Carrier Corporation Contributions Grants, 1440
Charles M. and Mary D. Grant Found Grants, 1527
Eaton Charitable Fund Grants, 1893
EPA Southeastern Multi-Media and Geographic Priority Project Grants, 1939
Forrest C. Lattner Foundation Grants, 2035
Georgia-Pacific Found Environment Grants, 2123
Georgia Power Foundation Grants, 2125
HomeBanc Foundation Grants, 2275
John Deere Foundation Grants, 2500
Motorola Foundation Grants, 2796
NOAA Living Marine Resources Cooperative Science Center Mini Grants, 3381
Nordson Corporation Foundation Grants, 3416
OSF Civil Service Awards, 3838
Prudential Foundation Education Grants, 4002
Rayonier Foundation Grants, 4034
Rich Foundation Grants, 4068
SARE Grad Student Grants - South Region, 4170
SARE Large Systems Grants: South Reg, 4172
SARE On-Farm Research Grants: South Reg, 4173
SARE Research & Educ Grants: South Reg, 4179
Textron Corporate Contributions Grants, 4501
USDA 1890 Institution Teaching, Research, and Extension Capacity Building Grants, 4613
USDA Regional Integrated Pest Management Grants - Southern Region, 4644
Weyerhaeuser Company Foundation Grants, 4857
Woodward Fund Grants, 4895

Guam
Alston Bannerman Sabbatical Fellowships, 580
Bank of America Charitable Matching Gifts, 1297
EPA Pollution Prevention Grants, 1936
EPA Regional Healthy Indoor Air Projects for Community Outreach and Education, 1937
EPA Temporally Integrated Monitoring of Ecosystems and Long-Term Grants, 1942
NHLBI Ancillary Studies in Clinical Trials, 3015
NHLBI Bioengineering and Obesity Grants, 3016
NHLBI Career Transition Awards, 3021
NHLBI Lymphatics in Health and Disease in the Digestive, Cardio & Pulmonary Systems, 3031
NHLBI Lymphatics in Health and Disease in the Digestive, Urinary, Cardiovascular and Pulmonary Systems, 3032
NHLBI Mentored Career Development Award to Promote Faculty Diversity/Re-Entry in Biomedical Research, 3035
NHLBI Mentored Patient-Oriented Research Career Development Awards, 3037
NHLBI Mentored Quantitative Research Career Development Awards, 3038
NHLBI Pathway to Independence Awards, 3045
NHLBI Research on the Role of Cardiomyocyte Mitochondria in Heart Disease: An Integrated Approach, 3053
NHLBI Ruth L. Kirschstein National Research Service Awards for Individual Postdoctoral Fellows, 3057
NHLBI Ruth L. Kirschstein National Research Service Awards for Individual Senior Fellows, 3060
NHLBI Ruth L. Kirschstein National Research Service Award Short-Term Institutional Research Training Grants, 3061
NHLBI Short-Term Research Education to Increase Diversity in Health-Related, 3062
NHLBI Training Grants for Institutions that Promote Diversity, 3067
OTS Undergrad NAPIRE Summer Internships, 3858
SARE Graduate Student Grants in Sustainable Agriculture - Western Region, 4171
SARE Producer Research and Education Grants-Western Region, 4174
SARE Pro & Producer Grants: West Reg, 4175
SARE Research & Educ Grants: West Reg, 4180
USDA Distance Education Grants for Higher Education Institutions in Insular Areas, 4622
USDA Integrated Pest Management Competitive Grants - Western Region, 4629
USDA Resident Instruction Grants for Higher Education in Insular Areas, 4645

Hawaii
3M Company Higher Education Grants, 4
3M Foundation Community Giving Grants, 5
A Charitable Foundation Grants, 273
Boeing Company Contributions Grants, 1357
Hawaii Community Foundation Health Education and Research Grants, 2221
Leahi Fund, 2599
SARE Graduate Student Grants in Sustainable Agriculture - Western Region, 4171
SARE Producer Research and Education Grants-Western Region, 4174
SARE Pro & Producer Grants: West Reg, 4175
SARE Research & Educ Grants: West Reg, 4180
USDA Alaska Native-Serving and Native Hawaiian-Serving Institutions Education Competitive Grants, 4615
USDA Integrated Pest Management Competitive Grants - Western Region, 4629
W.M. Keck Foundation Undergraduate Education Grants, 4825
Western Regional Graduate Program, 4855

Idaho
Brainerd Foundation Grants, 1368
Bullitt Foundation Grants, 1384
Dell Foundation Open Grants, 1734
E.L. Wiegand Foundation Grants, 1885
M.J. Murdock Trust College Research Program for Life Sciences, 2638
M.J. Murdock Trust College Science Research, 2639
M.J. Murdock Trust Exceptional Opportunity Grants, 2640
M.J. Murdock Trust General Grants, 2641
M.J. Murdock Trust Partners in Science Grants, 2642
M.J. Murdock Trust Research Start-up Grants for New Science Faculty, 2643
M.J. Murdock Trust Scientific Research Grants, 2644
Paul G. Allen Family Foundation Grants, 3876
Rocky Mountain Mineral Law Found Grants, 4094
SARE Graduate Student Grants in Sustainable Agriculture - Western Region, 4171
SARE Producer Research and Education Grants-Western Region, 4174
SARE Pro & Producer Grants: West Reg, 4175
SARE Research & Educ Grants: West Reg, 4180
Union Pacific Foundation Community and Civic Grants, 4568
Union Pacific Foundation Health and Human Services Grants, 4569
USDA Integrated Pest Management Competitive Grants - Western Region, 4629
W.M. Keck Foundation Undergraduate Education Grants, 4825
Western Regional Graduate Program, 4855
Weyerhaeuser Company Foundation Grants, 4857

Illinois
3M Company Higher Education Grants, 4
3M Foundation Community Giving Grants, 5
Abbott Fund Global AIDS Care Grants, 219
Abbott Fund Science Education Grants, 220
Anderson Foundation Grants, 842
Ann & Robert H. Lurie Family Found Grants, 859
Archer Daniels Midland Foundation Grants, 985
Arie and Ida Crown Memorial Grants, 989
Blowitz-Ridgeway Foundation Grants, 1349
Blum-Kovler Foundation Grants, 1351
Boeing Company Contributions Grants, 1357
Cabot Corporation Foundation Grants, 1406
Caesars Foundation Grants, 1407
Campbell Soup Foundation Grants, 1415
Carole Fielding Film/Video Production or Research Grants, 1438
Carrier Corporation Contributions Grants, 1440
Chicago Community Trust Fellowships, 1546
Chicago Community Trust Housing Grants: Advancing Affordable Rental Housing, 1547
CICF Efroymson Contemporary Arts Fellows, 1563
Coleman Foundation Cancer Care Grants, 1595
Colonel Stanley R. McNeil Foundation Grants, 1597
Community Foundation of East Central Illinois Grants, 1619
Community Foundation of the Fox River Valley, 1620
ConocoPhillips Grants, 1629
Diversifying Higher Education Faculty in Illinois Program (DFI), 1764
DOI Upper Mississippi River System Long-Term Resource Monitoring Grants, 1825
Duchossois Family Foundation Grants, 1867
Evanston Community Foundation Grants, 1962
Expect Miracles Foundation Grants, 1965
Field Museum Graduate Student Fellowships, 2008
FMC Foundation Grants, 2029
Georgia-Pacific Found Environment Grants, 2123
Girl's Best Friend Foundation Grants, 2143
Greater St Louis Community Found Grants, 2164
Great Lakes Fishery Trust Access Grants, 2168
Great Lakes Fishery Ecological & Biological Fisheries Research to Inform Management Grants, 2169
IATA Research Grants, 2319
Illinois Clean Energy Community Foundation Renewable Energy Grants, 2358
Illinois DCEO Coal Research Grants, 2359
Illinois DCEO Coal Revival Grants, 2360
Illinois DNR Youth Recreation Corps Grants, 2361
ISGC Graduate Fellowships, 2430
ISGC Informal Education Grants, 2431
ISGC Multiple Undergraduate Research Grants, 2432
ISGC Research Seed Grants, 2433
Jewish Federation of Metropolitan Chicago Scholarships, 2486
John Deere Foundation Grants, 2500
Joyce Foundation Education Grants, 2519
Joyce Foundation Employment Grants, 2520
Kovler Family Foundation Grants, 2575
Lillian S. Wells Foundation Grants, 2614
Lloyd A. Fry Foundation Health Grants, 2620
Luther I. Replogle Foundation Grants, 2631
Marcus and Theresa Levie Educational Fund Scholarships, 2657
Marion Gardner Jackson Charitable Grants, 2664
Motorola Foundation Grants, 2796
NL Associated Colleges of the Midwest/Great Lakes College Association Faculty Fellowships, 3281
NL Grad Scholar-in-Residence Opportunities, 3285
NL Lawrence Lipking Fellowship at the Newberry, 3287
NL Scholar-in-Residence Opportunity, 3315
Patrick and Anna M. Cudahy Fund Grants, 3873
Penny Severns Breast, Cervical and Ovarian Cancer Research Grants, 3896
Peter F. McManus Charitable Trust Grants, 3903
Prudential Foundation Education Grants, 4002
Rockford College Faculty Fellowships, 4092
Roy J. Carver Charitable Trust Higher Education Grants, 4113

Roy J. Carver Charitable Trust Medical and Science Research Grants, 4114
RRF General Grants, 4115
Ruth and Vernon Taylor Foundation Grants, 4130
SARE Graduate Student Grants - North Central Region, 4168
SARE Research and Education Grants - North Central Region, 4177
Siragusa Foundation Health Services & Medical Research Grants, 4264
Textron Corporate Contributions Grants, 4501
Union Pacific Foundation Community and Civic Grants, 4568
USCD/NOAA Illinois-Indiana Sea Grant Core Proposals, 4604
USDA Regional Integrated Pest Management Grants - North Central Region, 4642
Weyerhaeuser Company Foundation Grants, 4857
William Blair and Company Foundation Grants, 4875
William G. McGowan Charitable Fund Grants, 4877

Indiana
3M Company Higher Education Grants, 4
3M Foundation Community Giving Grants, 5
African American Heritage Grants, 357
American Indian Studies Research Institute Fellowship, 746
Anderson Foundation Grants, 842
Anthem Blue Cross and Blue Shield School Nurse Grants, 868
Anthropological Center for Training in Global and Environmental Change (ACT) Grad Assistantships, 870
Archer Daniels Midland Foundation Grants, 985
Ball Brothers Foundation General Grants, 1291
Ball Brothers Foundation Rapid Grants, 1292
Bayer Foundation Grants, 1309
Better Way Foundation Grants, 1338
Bristol-Myers Squibb Foundation Health Disparities Grants, 1377
Caesars Foundation Grants, 1407
Carrier Corporation Contributions Grants, 1440
CICF Efroymson Contemporary Arts Fellows, 1563
Coleman Foundation Cancer Care Grants, 1595
Daviess County Community Foundation Health Grants, 1725
Delta Dental Master's Thesis Award, 1737
Evan and Susan Bayh Foundation Grants, 1960
Georgia-Pacific Found Environment Grants, 2123
Great Lakes Fishery Trust Access Grants, 2168
Great Lakes Fishery Ecological & Biological Fisheries Research to Inform Management Grants, 2169
ICC Faculty Fellowships, 2333
ICC Scholarship of Engagement Faculty Grants, 2334
IDEM Section 319(h) Nonpoint Source Grants, 2338
Indiana 21st Century Research and Technology Fund Awards, 2370
Indiana SBIR/STTR Commercialization Enhancement Program (ISCEP), 2371
Indiana Space Grant Consortium Discovery Program Crosscutting Research-Team Grants, 2372
Indiana Space Grant Consortium Discovery Program Research Grants, 2373
Indiana Space Grant Consortium Discovery Program Student Hands-On Research Grants, 2374
Indiana Space Grant Consortium Fellowships, 2375
Indiana Waste Tire Fund (WTF) Grants, 2376
Joyce Foundation Education Grants, 2519
Joyce Foundation Employment Grants, 2520
Lilly Endowment Teacher Creativity Fellows, 2616
NL Associated Colleges of the Midwest/Great Lakes College Association Faculty Fellowships, 3281
RRF General Grants, 4115
SARE Graduate Student Grants - North Central Region, 4168
SARE Research and Education Grants - North Central Region, 4177
Sensient Technologies Foundation Grants, 4199
USCD/NOAA Illinois-Indiana Sea Grant Core Proposals, 4604
USDA Regional Integrated Pest Management Grants - North Central Region, 4642
W. C. Griffith Foundation Grants, 4818

Iowa
3M Company Higher Education Grants, 4
3M Foundation Community Giving Grants, 5
Archer Daniels Midland Foundation Grants, 985
Caesars Foundation Grants, 1407
Coleman Foundation Cancer Care Grants, 1595
Georgia-Pacific Found Environment Grants, 2123
Great Lakes Fishery Trust Access Grants, 2168
Great Lakes Fishery Ecological & Biological Fisheries Research to Inform Management Grants, 2169
Harold R. Bechtel Charitable Remainder Uni-Trust Grants, 2201
Harold R. Bechtel Testamentary Charitable Trust Grants, 2202
John Deere Foundation Grants, 2500
NL Associated Colleges of the Midwest/Great Lakes College Association Faculty Fellowships, 3281
Prudential Foundation Education Grants, 4002
Roy J. Carver Charitable Trust Higher Education Grants, 4113
Roy J. Carver Charitable Trust Medical and Science Research Grants, 4114
RRF General Grants, 4115
SARE Graduate Student Grants - North Central Region, 4168
SARE Research and Education Grants - North Central Region, 4177
Union Pacific Foundation Community and Civic Grants, 4568
Union Pacific Foundation Health and Human Services Grants, 4569
USDA Regional Integrated Pest Management Grants - North Central Region, 4642
Viking Children's Fund Grants, 4811

Kansas
Abbott Fund Global AIDS Care Grants, 219
Abbott Fund Science Education Grants, 220
Archer Daniels Midland Foundation Grants, 985
Bayer Foundation Grants, 1309
Boeing Company Contributions Grants, 1357
Eaton Charitable Fund Grants, 1893
Forrest C. Lattner Foundation Grants, 2035
Hatton W. Sumners Foundation for the Study and Teaching of Self Government Grants, 2219
Hutchinson Community Foundation Grants, 2310
John Deere Foundation Grants, 2500
Rocky Mountain Mineral Law Found Grants, 4094
SARE Graduate Student Grants - North Central Region, 4168
SARE Research and Education Grants - North Central Region, 4177
Textron Corporate Contributions Grants, 4501
Union Pacific Foundation Community and Civic Grants, 4568
Union Pacific Foundation Health and Human Services Grants, 4569
USDA Regional Integrated Pest Management Grants - North Central Region, 4642
USDA Tribal College Research Grants, 4652
W.M. Keck Foundation Undergraduate Education Grants, 4825

Kentucky
3M Company Higher Education Grants, 4
3M Foundation Community Giving Grants, 5
Anthem Blue Cross and Blue Shield School Nurse Grants, 868
Appalachian Reg Comm Health Care Grants, 915
Archer Daniels Midland Foundation Grants, 985
Ashland Corporate Contributions Grants, 1087
Charles M. and Mary D. Grant Found Grants, 1527
CICF Efroymson Contemporary Arts Fellows, 1563
Eaton Charitable Fund Grants, 1893
EPA Southeastern Multi-Media and Geographic Priority Project Grants, 1939
Georgia-Pacific Found Environment Grants, 2123
James Graham Brown Foundation Brown Fellowships, 2459
RRF General Grants, 4115
SARE Grad Student Grants - South Region, 4170
SARE Large Systems Grants: South Reg, 4172
SARE On-Farm Research Grants: South Reg, 4173
SARE Research & Educ Grants: South Reg, 4179
Thompson Charitable Foundation Grants, 4511
USDA 1890 Institution Teaching, Research, and Extension Capacity Building Grants, 4613
USDA Regional Integrated Pest Management Grants - Southern Region, 4644
Weyerhaeuser Company Foundation Grants, 4857

Louisiana
Booth-Bricker Fund Grants, 1359
Cabot Corporation Foundation Grants, 1406
Caesars Foundation Grants, 1407
ConocoPhillips Grants, 1629
Coughlin-Saunders Foundation Grants, 1648
Cudd Foundation Grants, 1672
Eaton Charitable Fund Grants, 1893
FAR Fund Grants, 1980
Fluor Foundation Grants, 2028
FMC Foundation Grants, 2029
Georgia-Pacific Found Environment Grants, 2123
Gil and Dody Weaver Foundation Grants, 2139
GNOF Gulf Coast Oil Spill Grants, 2153
Hatton W. Sumners Foundation for the Study and Teaching of Self Government Grants, 2219
Huie-Dellmon Trust Grants, 2295
John Deere Foundation Grants, 2500
Murphy Institute Center for Ethics and Public Affairs Faculty Fellowships, 2805
Murphy Institute Center for Ethics and Public Affairs Graduate Fellowships, 2806
Prudential Foundation Education Grants, 4002
Rocky Mountain Mineral Law Found Grants, 4094
SARE Grad Student Grants - South Region, 4170
SARE Large Systems Grants: South Reg, 4172
SARE Research & Educ Grants: South Reg, 4179
Textron Corporate Contributions Grants, 4501
Union Pacific Foundation Community and Civic Grants, 4568
Union Pacific Foundation Health and Human Services Grants, 4569
USDA 1890 Institution Teaching, Research, and Extension Capacity Building Grants, 4613
USDA Regional Integrated Pest Management Grants - Southern Region, 4644
W.M. Keck Foundation Undergraduate Education Grants, 4825
Weyerhaeuser Company Foundation Grants, 4857

Maine
Bay and Paul Foundations, Inc Grants, 1307
Beim Foundation Grants, 1331
Charles H. Hood Foundation Child Health Research Awards, 1522
Davis Family Foundation Grants, 1726
Dorr Foundation Grants, 1842
Eaton Charitable Fund Grants, 1893
Expect Miracles Foundation Grants, 1965
Fannie E. Rippel Foundation Grants, 1978
FMC Foundation Grants, 2029
Foundation for Seacoast Health Grants, 2041
Frederick Gardner Cottrell Foundation Grants, 2054
Harold Alfond Foundation Grants, 2200
Orchard Foundation Grants, 3793
Richard Davoud Donchian Foundation Grants, 4066
Salem Foundation Grants, 4160
SARE Graduate Student Grants: NE Region, 4169
SARE Professional Development Grants - Northeast Region, 4176
SARE Research and Education Grants - Northeast Region, 4178
USDA Regional Integrated Pest Management Grants - Northeastern Region, 4643

Marshall Islands
Bank of America Charitable Matching Gifts, 1297
EPA Pollution Prevention Grants, 1936
EPA Regional Healthy Indoor Air Projects for Community Outreach and Education, 1937
EPA Temporally Integrated Monitoring of Ecosystems and Long-Term Grants, 1942
NHLBI Ancillary Studies in Clinical Trials, 3015
NHLBI Bioengineering and Obesity Grants, 3016
NHLBI Career Transition Awards, 3021
NHLBI Lymphatics in Health and Disease in the Digestive, Cardio & Pulmonary Systems, 3031
NHLBI Lymphatics in Health and Disease in the Digestive, Urinary, Cardiovascular and Pulmonary Systems, 3032
NHLBI Mentored Career Development Award to Promote Faculty Diversity/Re-Entry in Biomedical Research, 3035
NHLBI Mentored Patient-Oriented Research Career Development Awards, 3037
NHLBI Mentored Quantitative Research Career Development Awards, 3038
NHLBI Pathway to Independence Awards, 3045
NHLBI Research on the Role of Cardiomyocyte Mitochondria in Heart Disease: An Integrated Approach, 3053
NHLBI Ruth L. Kirschstein National Research Service Awards for Individual Postdoctoral Fellows, 3057
NHLBI Ruth L. Kirschstein National Research Service Awards for Individual Senior Fellows, 3060
NHLBI Ruth L. Kirschstein National Research Service Award Short-Term Institutional Research Training Grants, 3061
NHLBI Short-Term Research Education to Increase Diversity in Health-Related, 3062
NHLBI Training Grants for Institutions that Promote Diversity, 3067
OTS Undergrad NAPIRE Summer Internships, 3858
USDA Resident Instruction Grants for Higher Education in Insular Areas, 4645

Maryland
Alpha Research Foundation Grants, 577
Appalachian Reg Comm Health Care Grants, 915
Boeing Company Contributions Grants, 1357
Carl M. Freeman Foundation Grants, 1430
Commonweal Foundation Community Assistance Grants, 1602
Commonweal Foundation Grants, 1603
Eaton Charitable Fund Grants, 1893
Expect Miracles Foundation Grants, 1965
FMC Foundation Grants, 2029
Frederick Gardner Cottrell Foundation Grants, 2054
Graduate Institute Hodson Trust Teacher Fellowships, 2160
Hilda and Preston Davis Foundation Grants, 2266
Jacob and Hilda Blaustein Foundation Grants, 2454
Knoll-MacDonald Foundation Grants, 2566
Mabel H. Flory Charitable Trust Grants, 2646
Marjorie C. Adams Charitable Trust Grants, 2665
NOAA Chesapeake Bay Office B-WET (Bay Watershed Education and Training) Grants, 3339
NOAA Chesapeake Bay Office Cooperative Research Grants, 3340
NOAA Chesapeake Bay Office Fisheries Science Grants, 3341
NOAA Chesapeake Bay Office Non-Native Oyster Research Grants, 3342
NOAA Chesapeake Bay Office Submerged Aquatic Vegetation Culture and Restoration Grants, 3343
Rathmann Family Foundation Grants, 4033
Richard Davoud Donchian Foundation Grants, 4066
Robert G. Cabell III and Maude Morgan Cabell Foundation Grants, 4075
Robert W. Deutsch Foundation Grants, 4082
SARE Graduate Student Grants: NE Region, 4169
SARE Pro Development Grants: NE Reg, 4176
SARE Research and Education Grants - Northeast Region, 4178
Textron Corporate Contributions Grants, 4501
USDA 1890 Institution Teaching, Research, and Extension Capacity Building Grants, 4613
USDA Regional Integrated Pest Management Grants - Northeastern Region, 4643
Weyerhaeuser Company Foundation Grants, 4857

Massachusetts
3M Company Higher Education Grants, 4
3M Foundation Community Giving Grants, 5
Abbott Fund Global AIDS Care Grants, 219
Abbott Fund Science Education Grants, 220
Ahn Family Foundation Grants, 379
Alces Foundation Grants, 510
Archer Daniels Midland Foundation Grants, 985
Auburn Foundation Grants, 1263
Banfi Vintners Foundation Grants, 1296
Bay and Paul Foundations, Inc Grants, 1307
Bayer Foundation Grants, 1309
BCBS of Massachusetts Foundation Grants, 1319
Boston Foundation Grants, 1361
Bristol-Myers Squibb Foundation Health Disparities Grants, 1377
Cabot Corporation Foundation Grants, 1406
Charles A. King Trust Postdoctoral Research Fellowships, 1518
Charles H. Farnsworth Trust Grants, 1521
Charles H. Hood Foundation Child Health Research Awards, 1522
Chestnut Hill Charitable Foundation Grants, 1538
Deborah Munroe Noonan Memorial Research Grants, 1729
Dorr Foundation Grants, 1842
Edward Bangs Kelley and Elza Kelley Foundation Grants, 1898
Edwin S. Webster Foundation Grants, 1900
Elizabeth McGraw Foundation Grants, 1908
Expect Miracles Foundation Grants, 1965
Fairlawn Foundation Grants, 1974
Fannie E. Rippel Foundation Grants, 1978
Frank Stanley Beveridge Foundation Grants, 2051
Frederick Gardner Cottrell Foundation Grants, 2054
Georgia-Pacific Found Environment Grants, 2123
Greater Worcester Community Foundation Jeppson Memorial Fund for Brookfield Grants, 2165
Hilda and Preston Davis Foundation Grants, 2266
John W. Alden Trust Grants, 2510
Massachusetts Bar Foundation Legal Intern Fellowships, 2676
Massachusetts Cultural Council Artist Fellows, 2677
Massachusetts Foundation for the Humanities Project Grants, 2678
Motorola Foundation Grants, 2796
Nellie Mae Education Foundation District-Level Change Grants, 2968
Nordson Corporation Foundation Grants, 3416
Orchard Foundation Grants, 3793
Peabody Foundation Grants, 3886
Perkin Fund Grants, 3898
Perpetual Trust for Charitable Giving Grants, 3900
Richard and Susan Smith Family Found Grants, 4064
Sara Elizabeth O'Brien Trust Grants, 4166
SARE Graduate Student Grants: NE Region, 4169
SARE Professional Development Grants - Northeast Region, 4176
SARE Research and Education Grants - Northeast Region, 4178
Smith Family Awards for Excellence in Biomedical Research, 4268
Sun Microsystems Foundation Summer Academic Enrichment Grants, 4468
Textron Corporate Contributions Grants, 4501
Thoracic Foundation Grants, 4513
USDA Regional Integrated Pest Management Grants - Northeastern Region, 4643

Michigan
3M Company Higher Education Grants, 4
3M Foundation Community Giving Grants, 5
Abbott Fund Global AIDS Care Grants, 219
Abbott Fund Science Education Grants, 220
Anderson Foundation Grants, 842
Ann & Robert H. Lurie Family Found Grants, 859
Ann Arbor Area Community Foundation Grants, 860
Archer Daniels Midland Foundation Grants, 985
BCBSM Foundation Excellence in Research Awards for Students, 1312
BCBSM Foundation Investigator Initiated Research Grants, 1313
BCBSM Foundation Physician Investigator Research Awards, 1314
BCBSM Foundation Primary and Clinical Prevention Grants, 1315
BCBSM Foundation Student Award Program, 1316
BCBSM Frank J. McDevitt, DO, Excellence in Research Awards, 1317
Campbell Soup Foundation Grants, 1415
Carrier Corporation Contributions Grants, 1440
Cleveland-Cliffs Foundation Grants, 1580
Coleman Foundation Cancer Care Grants, 1595
Community Foundation for Muskegon County Grants, 1617
Delta Dental Master's Thesis Award, 1737
Detroit Lions Charities Grants, 1753
DTE Energy Foundation Health and Human Services Grants, 1865
DTE Energy Found Higher Learning Grants, 1866
Eaton Charitable Fund Grants, 1893
Georgia-Pacific Found Environment Grants, 2123
Great Lakes Fishery Trust Access Grants, 2168
Great Lakes Fishery Ecological & Biological Fisheries Research to Inform Management Grants, 2169
Harry A. and Margaret D. Towsley Foundation Grants, 2205
Harvey Randall Wickes Foundation Grants, 2218
Herbert H. & Grace A. Dow Found Grants, 2241
Knoll-MacDonald Foundation Grants, 2566
Matilda R. Wilson Fund Grants, 2684
McGregor Fund Grants, 2725
MDARD Specialty Crop Grants-Farm Bill, 2730
MDEQ Beach Monitoring Inland Lakes Grants, 2731
MDEQ Local Water Quality Grants, 2732
MDEQ Scrap Tire Market Devel Grants, 2733
MGM Resorts Foundation Community Grants, 2753
NL Associated Colleges of the Midwest/Great Lakes College Association Faculty Fellowships, 3281
SARE Graduate Student Grants - North Central Region, 4168
SARE Research and Education Grants - North Central Region, 4177
Talbert and Leota Abrams Foundation, 4492
USDA Regional Integrated Pest Management Grants - North Central Region, 4642
USDA Tribal College Research Grants, 4652
Weyerhaeuser Company Foundation Grants, 4857

Minnesota
3M Company Higher Education Grants, 4
3M Foundation Community Giving Grants, 5
Alex Stern Family Foundation Grants, 542
Archer Daniels Midland Foundation Grants, 985
Beim Foundation Grants, 1331
Better Way Foundation Grants, 1338
Bush Foundation Artist Fellowships, 1391
Carl and Eloise Pohlad Family Found Grants, 1426
Cleveland-Cliffs Foundation Grants, 1580
Eaton Charitable Fund Grants, 1893
Emma B. Howe Memorial Foundation Grants, 1920
IFP Minnesota McKnight Film Fellowships, 2342
IFP Minnesota McKnight Screen Fellowships, 2343
Joyce Foundation Education Grants, 2519
Joyce Foundation Employment Grants, 2520
Luther I. Replogle Foundation Grants, 2631
Minnesota War Orphans Educational Assistance Grants, 2780
NL Associated Colleges of the Midwest/Great Lakes College Association Faculty Fellowships, 3281
Prudential Foundation Education Grants, 4002
Rathmann Family Foundation Grants, 4033
Salem Foundation Grants, 4160

SARE Graduate Student Grants - North Central Region, 4168
SARE Research and Education Grants - North Central Region, 4177
Union Pacific Foundation Community and Civic Grants, 4568
Union Pacific Foundation Health and Human Services Grants, 4569
United Theological Seminary Merit-Based Scholarships, 4583
USDA Regional Integrated Pest Management Grants - North Central Region, 4642
USDA Tribal College Research Grants, 4652
Viking Children's Fund Grants, 4811
Weyerhaeuser Company Foundation Grants, 4857

Mississippi

Appalachian Reg Comm Health Care Grants, 915
Archer Daniels Midland Foundation Grants, 985
Assisi Foundation of Memphis General Grants, 1221
Caesars Foundation Grants, 1407
Charles M. and Mary D. Grant Found Grants, 1527
EPA Southeastern Multi-Media and Geographic Priority Project Grants, 1939
Georgia-Pacific Found Environment Grants, 2123
Gil and Dody Weaver Foundation Grants, 2139
MGM Resorts Foundation Community Grants, 2753
Robert M. Hearin Foundation Grants, 4077
SARE Grad Student Grants - South Region, 4170
SARE Large Systems Grants: South Reg, 4172
SARE On-Farm Research Grants: South Reg, 4173
SARE Research & Educ Grants: South Reg, 4179
USDA 1890 Institution Teaching, Research, and Extension Capacity Building Grants, 4613
USDA Regional Integrated Pest Management Grants - Southern Region, 4644
W.M. Keck Foundation Undergraduate Education Grants, 4825
Weyerhaeuser Company Foundation Grants, 4857

Missouri

3M Company Higher Education Grants, 4
3M Foundation Community Giving Grants, 5
Archer Daniels Midland Foundation Grants, 985
Boeing Company Contributions Grants, 1357
Caesars Foundation Grants, 1407
CDI Postdoctoral Fellowships, 1469
Charles H. Price II Family Foundation Grants, 1523
FMC Foundation Grants, 2029
Greater St Louis Community Found Grants, 2164
Hatton W. Sumners Foundation for the Study and Teaching of Self Government Grants, 2219
Herbert A. and Adrian W. Woods Foundation Grants, 2239
J.B. Reynolds Foundation Grants, 2449
John Deere Foundation Grants, 2500
Millstone Foundation Grants, 2776
Musgrave Foundation Grants, 2807
Oppenstein Brothers Foundation Grants, 3792
Pott Foundation Grants, 3984
RRF General Grants, 4115
SARE Graduate Student Grants - North Central Region, 4168
SARE Research and Education Grants - North Central Region, 4177
Sensient Technologies Foundation Grants, 4199
Union Pacific Foundation Community and Civic Grants, 4568
Union Pacific Foundation Health and Human Services Grants, 4569
USDA Regional Integrated Pest Management Grants - North Central Region, 4642
Victor E. Speas Foundation Grants, 4810
Weyerhaeuser Company Foundation Grants, 4857
William G. McGowan Charitable Fund Grants, 4877

Montana

Archer Daniels Midland Foundation Grants, 985
Brainerd Foundation Grants, 1368
Bullitt Foundation Grants, 1384
ConocoPhillips Grants, 1629
M.J. Murdock Trust College Research Program for Life Sciences, 2638
M.J. Murdock Trust College Science Research, 2639
M.J. Murdock Trust Exceptional Opportunity Grants, 2640
M.J. Murdock Trust General Grants, 2641
M.J. Murdock Trust Partners in Science Grants, 2642
M.J. Murdock Trust Research Start-up Grants for New Science Faculty, 2643
M.J. Murdock Trust Scientific Research Grants, 2644
Paul G. Allen Family Foundation Grants, 3876
Rocky Mountain Mineral Law Found Grants, 4094
Ruth and Vernon Taylor Foundation Grants, 4130
SARE Graduate Student Grants in Sustainable Agriculture - Western Region, 4171
SARE Producer Research and Education Grants- Western Region, 4174
SARE Pro & Producer Grants: West Reg, 4175
SARE Research & Educ Grants: West Reg, 4180
Union Pacific Community and Civic Grants, 4568
Union Pacific Foundation Health and Human Services Grants, 4569
USDA Integrated Pest Management Competitive Grants - Western Region, 4629
USDA Tribal College Research Grants, 4652
W.M. Keck Undergraduate Education Grants, 4825
Western Regional Graduate Program, 4855

Nebraska

3M Company Higher Education Grants, 4
3M Foundation Community Giving Grants, 5
Archer Daniels Midland Foundation Grants, 985
Hatton W. Sumners Foundation for the Study and Teaching of Self Government Grants, 2219
Rocky Mountain Mineral Law Found Grants, 4094
SARE Graduate Student Grants - North Central Region, 4168
SARE Research and Education Grants - North Central Region, 4177
Union Pacific Foundation Community and Civic Grants, 4568
Union Pacific Foundation Health and Human Services Grants, 4569
USDA 1890 Institution Teaching, Research, and Extension Capacity Building Grants, 4613
USDA Regional Integrated Pest Management Grants - North Central Region, 4642
USDA Tribal College Research Grants, 4652

Nevada

Boeing Company Contributions Grants, 1357
Caesars Foundation Grants, 1407
Carrier Corporation Contributions Grants, 1440
Cord Foundation Grants, 1645
Del E. Webb Foundation Grants, 1733
E.L. Wiegand Foundation Grants, 1885
Frances C. and William P. Smallwood Foundation Grants, 2045
Greenspun Family Foundation Grants, 2173
MGM Resorts Foundation Community Grants, 2753
Nevada Arts Council Artists Fellowships, 2984
Rocky Mountain Mineral Law Found Grants, 4094
SARE Graduate Student Grants in Sustainable Agriculture - Western Region, 4171
SARE Producer Research and Education Grants- Western Region, 4174
SARE Pro & Producer Grants: West Reg, 4175
SARE Research & Educ Grants: West Reg, 4180
Union Pacific Foundation Community and Civic Grants, 4568
Union Pacific Foundation Health and Human Services Grants, 4569
USDA Integrated Pest Management Competitive Grants - Western Region, 4629
W.M. Keck Foundation Undergraduate Education Grants, 4825
Western Regional Graduate Program, 4855
William N. and Myriam Pennington Foundation Grants, 4879

New Hampshire

Bay and Paul Foundations, Inc Grants, 1307
Charles H. Hood Foundation Child Health Research Awards, 1522
Dorr Foundation Grants, 1842
Expect Miracles Foundation Grants, 1965
Fannie E. Rippel Foundation Grants, 1978
Foundation for Seacoast Health Grants, 2041
Hilda and Preston Davis Foundation Grants, 2266
McLean Contributionship Grants, 2727
New Hampshire Charitable Foundation Grants, 2985
Orchard Foundation Grants, 3793
SARE Graduate Student Grants: NE Region, 4169
SARE Professional Development Grants - Northeast Region, 4176
SARE Research and Education Grants - Northeast Region, 4178
USDA Regional Integrated Pest Management Grants - Northeastern Region, 4643
Weyerhaeuser Company Foundation Grants, 4857

New Jersey

1 in 9: Long Island Breast Cancer Action Coalition Grants, 1
3M Company Higher Education Grants, 4
Abbott Fund Global AIDS Care Grants, 219
Abbott Fund Science Education Grants, 220
Archer Daniels Midland Foundation Grants, 985
Bay and Paul Foundations, Inc Grants, 1307
Bayer Foundation Grants, 1309
Bodman Foundation Grants, 1356
Bristol-Myers Squibb Foundation Health Disparities Grants, 1377
Caesars Foundation Grants, 1407
Campbell Soup Foundation Grants, 1415
Cape Branch Foundation Grants, 1425
Carl C. Icahn Foundation Grants, 1428
Charles Edison Fund Grants, 1520
ConocoPhillips Grants, 1629
Eaton Charitable Fund Grants, 1893
Edward W. and Stella C. Van Houten Memorial Fund Grants, 1899
Expect Miracles Foundation Grants, 1965
Fluor Foundation Grants, 2028
FMC Foundation Grants, 2029
Frances L. and Edwin L. Cummings Grants, 2046
George A Ohl Jr. Foundation Grants, 2112
Georgia-Pacific Found Environment Grants, 2123
Graduate Institute Hodson Trust Teacher Fellowships, 2160
Healthcare Foundation of New Jersey Grants, 2222
Hilda and Preston Davis Foundation Grants, 2266
Margaret Yardley Fellowship, 2660
Meyer and Pepa Gold Family Found Grants, 2746
Motorola Foundation Grants, 2796
Prudential Foundation Education Grants, 4002
Richard Davoud Donchian Foundation Grants, 4066
Robert Leet and Clara Guthrie Patterson Trust Grants, 4076
Ruth and Vernon Taylor Foundation Grants, 4130
Ruth Estrin Goldberg Memorial for Cancer Research Grants, 4131
SARE Graduate Student Grants: NE Region, 4169
SARE Professional Development Grants - Northeast Region, 4176
SARE Research and Education Grants - Northeast Region, 4178
USDA Regional Integrated Pest Management Grants - Northeastern Region, 4643
Weyerhaeuser Company Foundation Grants, 4857

New Mexico

Albuquerque Community Foundation Grants, 508
Beim Foundation Grants, 1331
Boeing Company Contributions Grants, 1357
Cabot Corporation Foundation Grants, 1406
Cudd Foundation Grants, 1672
Gil and Dody Weaver Foundation Grants, 2139
Hatton W. Sumners Foundation for the Study and Teaching of Self Government Grants, 2219

McCune Charitable Foundation Grants, 2719
Rocky Mountain Mineral Law Found Grants, 4094
SARE Graduate Student Grants in Sustainable Agriculture - Western Region, 4171
SARE Producer Research and Education Grants-Western Region, 4174
SARE Pro & Producer Grants: West Reg, 4175
SARE Research & Educ Grants: West Reg, 4180
Union Pacific Foundation Community and Civic Grants, 4568
Union Pacific Foundation Health and Human Services Grants, 4569
USDA Integrated Pest Management Competitive Grants - Western Region, 4629
USDA Tribal College Research Grants, 4652
W.M. Keck Foundation Undergraduate Education Grants, 4825
Western Regional Graduate Program, 4855
Weyerhaeuser Company Foundation Grants, 4857

New York

1 in 9: Long Island Breast Cancer Action Coalition Grants, 1
3M Company Higher Education Grants, 4
3M Foundation Community Giving Grants, 5
Abbott Fund Global AIDS Care Grants, 219
Abbott Fund Science Education Grants, 220
A Charitable Foundation Grants, 273
Achelis Foundation Grants, 274
Ahn Family Foundation Grants, 379
Alfred P. Sloan Foundation Civic Grants, 545
Allyn Foundation Grants, 574
Appalachian Reg Comm Health Care Grants, 915
Archer Daniels Midland Foundation Grants, 985
Arthur and Rochelle Belfer Foundation Grants, 1023
Axe-Houghton Foundation Grants, 1287
Banfi Vintners Foundation Grants, 1296
Bay and Paul Foundations, Inc Grants, 1307
Bayer Foundation Grants, 1309
Bodman Foundation Grants, 1356
Brach Family Foundation Grants, 1367
Bristol-Myers Squibb Foundation Health Disparities Grants, 1377
Carl C. Icahn Foundation Grants, 1428
Carrier Corporation Contributions Grants, 1440
Charles Abrams Scholarships, 1519
Charles Edison Fund Grants, 1520
Chautauqua Region Community Found Grants, 1533
Community Found for Greater Buffalo Grants, 1615
Cowles Charitable Trust Grants, 1652
DeMatteis Family Foundation Grants, 1740
Dorr Foundation Grants, 1842
E.L. Wiegand Foundation Grants, 1885
Eaton Charitable Fund Grants, 1893
Elizabeth McGraw Foundation Grants, 1908
Eugene M. Lang Foundation Grants, 1957
Expect Miracles Foundation Grants, 1965
Fluor Foundation Grants, 2028
FMC Foundation Grants, 2029
Frances L. and Edwin L. Cummings Memorial Fund Grants, 2046
Georgia-Pacific Found Environment Grants, 2123
Graduate Institute Hodson Trust Teacher Fellowships, 2160
Hilda and Preston Davis Foundation Grants, 2266
Hirtzel Memorial Foundation Grants, 2269
HRF Hudson River Improvement Grants, 2286
HRF New York City Environmental Grants for Newton Creek, 2288
Jacob and Valeria Langeloth Foundation Grants, 2455
James H. Cummings Foundation Grants, 2460
Jeffrey Campbell Grad Fellow in Anthropology, 2482
John R. Oishei Foundation Grants, 2505
Joseph Alexander Foundation Grants, 2513
Knoll-MacDonald Foundation Grants, 2566
Komen Greater NYC Clinical Research Enrollment Grants, 2568
Komen Greater NYC Community Breast Health Grants, 2569
Marjorie C. Adams Charitable Trust Grants, 2665
Meyer and Pepa Gold Family Found Grants, 2746
MfA Newton Fellowships, 2749
MfA Newton Master Teacher Fellowships, 2750
Miles of Hope Breast Cancer Found Grants, 2773
NOAA Chesapeake Bay Office B-WET (Bay Watershed Education and Training) Grants, 3339
NOAA Chesapeake Bay Office Cooperative Research Grants, 3340
NOAA Chesapeake Bay Office Fisheries Science Grants, 3341
NOAA Chesapeake Bay Office Submerged Aquatic Vegetation Culture and Restoration Grants, 3343
NYCT Biomedical Research Grants, 3694
NYCT Blindness and Visual Disabilities Grants, 3695
NYCT Blood Disease Research Grants, 3696
NYCT New York City Environment Grants, 3697
NYFA Artist Fellowships, 3698
NYFA Shelly Pinz Fellowship, 3699
NYSCA Dance: Services to the Field Grants, 3702
NYSCA Folk Arts: Exhibitions Grants, 3703
NYSCA Folk Arts: Presentation Grants, 3704
NYSCA Folk Arts: Regional and County Folk Arts Grants, 3705
NYSCA Individual Artists: Film, Media and New Technology Production Grants, 3706
NYSCA Museum: Project Support Grants, 3707
NYSCA Music: Services to the Field Grants, 3708
Orchard Foundation Grants, 3793
Perkin Fund Grants, 3898
Peter F. McManus Charitable Trust Grants, 3903
Prudential Foundation Education Grants, 4002
Regents Professional Opportunity Scholarships, 4049
Richard Davoud Donchian Foundation Grants, 4066
Robert Leet and Clara Guthrie Patterson Trust Grants, 4076
Robert Sterling Clark Foundation Government Accountability Grants, 4080
Ruth and Vernon Taylor Foundation Grants, 4130
Ruth Estrin Goldberg Memorial for Cancer Research Grants, 4131
SARE Graduate Student Grants: NE Region, 4169
SARE Professional Development Grants - Northeast Region, 4176
SARE Research and Education Grants - Northeast Region, 4178
Sunny and Abe Rosenberg Foundation Grants, 4469
Textron Corporate Contributions Grants, 4501
USDA Regional Integrated Pest Management Grants - Northeastern Region, 4643
William G. McGowan Charitable Fund Grants, 4877

North Carolina

Abbott Fund Global AIDS Care Grants, 219
Abbott Fund Science Education Grants, 220
American Schlafhorst Foundation Grants, 796
Appalachian Reg Comm Health Care Grants, 915
Archer Daniels Midland Foundation Grants, 985
Bayer Foundation Grants, 1309
BCBSNC Foundation Grants, 1318
Caesars Foundation Grants, 1407
Campbell Soup Foundation Grants, 1415
Carrier Corporation Contributions Grants, 1440
Charles M. and Mary D. Grant Found Grants, 1527
D.F. Halton Foundation Grants, 1692
Duke University Libraries Franklin Collection Travel Grants, 1874
Duke University Libraries Hartman Center Travel Grants, 1875
Duke University Libraries J. Walter Thompson Research Fellowships, 1876
Eaton Charitable Fund Grants, 1893
EPA Southeastern Multi-Media and Geographic Priority Project Grants, 1939
Fluor Foundation Grants, 2028
FMC Foundation Grants, 2029
George H. Hitchings New Investigator Award in Health Research, 2118
Georgia-Pacific Found Environment Grants, 2123
Gertrude B. Elion Mentored Medical Student Research Awards, 2130
Ireland Family Foundation Grants, 2415
James H. Cummings Foundation Grants, 2460
John Deere Foundation Grants, 2500
North Carolina Arts Council Artist Fellowships, 3417
North Carolina Arts Council Arts Folklife Internship, 3418
North Carolina Arts Council Folklife Grants, 3419
North Carolina Arts Council Residency Center Grants, 3420
North Carolina Biotechnology Center Collaborative Funding Grants, 3421
North Carolina Biotechnology Center Institutional Development Grants, 3422
North Carolina Biotechnology Center Oliver Smithies Faculty Recruitment Grant, 3423
North Carolina Biotech Center Research Grants, 3424
North Carolina Biotechnology Centers of Innovation Grants, 3425
Percy B. Ferebee Endowment Grants, 3897
Richard Davoud Donchian Foundation Grants, 4066
SARE Grad Student Grants - South Region, 4170
SARE Large Systems Grants: South Reg, 4172
SARE On-Farm Research Grants: South Reg, 4173
SARE Research & Educ Grants: South Reg, 4179
Textron Corporate Contributions Grants, 4501
Thomas Austin Finch, Sr. Foundation Grants, 4507
USDA 1890 Institution Teaching, Research, and Extension Capacity Building Grants, 4613
USDA Regional Integrated Pest Management Grants - Southern Region, 4644
Weyerhaeuser Company Foundation Grants, 4857
Woodward Fund Grants, 4895

North Dakota

Alex Stern Family Foundation Grants, 542
Archer Daniels Midland Foundation Grants, 985
Bush Foundation Artist Fellowships, 1391
North Dakota Council on the Arts Individual Artist Fellowships, 3426
North Dakota Indian Scholarships, 3427
Rocky Mountain Mineral Law Found Grants, 4094
SARE Graduate Student Grants - North Central Region, 4168
SARE Research and Education Grants - North Central Region, 4177
USDA Regional Integrated Pest Management Grants - North Central Region, 4642
USDA Tribal College Research Grants, 4652
Viking Children's Fund Grants, 4811
Western Regional Graduate Program, 4855

Northern Mariana Islands

Bank of America Charitable Matching Gifts, 1297
EPA Pollution Prevention Grants, 1936
EPA Regional Healthy Indoor Air Projects for Community Outreach and Education, 1937
EPA Temporally Integrated Monitoring of Ecosystems and Long-Term Grants, 1942
NHLBI Ancillary Studies in Clinical Trials, 3015
NHLBI Bioengineering and Obesity Grants, 3016
NHLBI Career Transition Awards, 3021
NHLBI Lymphatics in Health and Disease in the Digestive, Cardio & Pulmonary Systems, 3031
NHLBI Lymphatics in Health and Disease in the Digestive, Urinary, Cardiovascular and Pulmonary Systems, 3032
NHLBI Mentored Career Development Award to Promote Faculty Diversity/Re-Entry in Biomedical Research, 3035
NHLBI Mentored Patient-Oriented Research Career Development Awards, 3037
NHLBI Mentored Quantitative Research Career Development Awards, 3038
NHLBI Pathway to Independence Awards, 3045
NHLBI Research on the Role of Cardiomyocyte Mitochondria in Heart Disease: An Integrated Approach, 3053
NHLBI Ruth L. Kirschstein National Research Service Awards for Individual Postdoctoral Fellows, 3057

NHLBI Ruth L. Kirschstein National Research Service Awards for Individual Senior Fellows, 3060
NHLBI Ruth L. Kirschstein National Research Service Award Short-Term Institutional Research Training Grants, 3061
NHLBI Short-Term Research Education to Increase Diversity in Health-Related, 3062
NHLBI Training Grants for Institutions that Promote Diversity, 3067
OTS Undergrad NAPIRE Summer Internships, 3858
SARE Graduate Student Grants in Sustainable Agriculture - Western Region, 4171
SARE Producer Research and Education Grants-Western Region, 4174
SARE Pro & Producer Grants: West Reg, 4175
SARE Research & Educ Grants: West Reg, 4180
USDA Integrated Pest Management Competitive Grants - Western Region, 4629
USDA Resident Instruction Grants for Higher Education in Insular Areas, 4645

Ohio

3M Company Higher Education Grants, 4
3M Foundation Community Giving Grants, 5
Abbott Fund Global AIDS Care Grants, 219
Abbott Fund Science Education Grants, 220
Abington Foundation Grants, 243
Anderson Foundation Grants, 842
Anthem Blue Cross and Blue Shield School Nurse Grants, 868
Appalachian Reg Comm Health Care Grants, 915
Archer Daniels Midland Foundation Grants, 985
Ashland Corporate Contributions Grants, 1087
Bayer Foundation Grants, 1309
Boeing Company Contributions Grants, 1357
Campbell Soup Foundation Grants, 1415
CICF Efroymson Contemporary Arts Fellows, 1563
Cleveland-Cliffs Foundation Grants, 1580
CMA Foundation Grants, 1582
Coleman Foundation Cancer Care Grants, 1595
Delta Dental Master's Thesis Award, 1737
Eaton Charitable Fund Grants, 1893
Elisabeth Severance Prentiss Foundation Grants, 1905
Fordham Prizes, 2032
George W. Codrington Foundation Grants, 2122
Georgia-Pacific Found Environment Grants, 2123
Great Lakes Fishery Trust Access Grants, 2168
Great Lakes Fishery Ecological & Biological Fisheries Research to Inform Management Grants, 2169
Kettering Fund Grants, 2555
Mary S. and David C. Corbin Found Grants, 2675
Mt. Sinai Health Care Foundation Academic Medicine and Bioscience Grants, 2803
NL Associated Colleges of the Midwest/Great Lakes College Association Faculty Fellowships, 3281
Nordson Corporation Foundation Grants, 3416
Ohio Learning Network Grants, 3734
Reinberger Foundation Grants, 4051
Reuter Foundation Grants, 4059
SARE Graduate Student Grants - North Central Region, 4168
SARE Research and Education Grants - North Central Region, 4177
USDA Regional Integrated Pest Management Grants - North Central Region, 4642
Weyerhaeuser Company Foundation Grants, 4857

Oklahoma

Archer Daniels Midland Foundation Grants, 985
Boeing Company Contributions Grants, 1357
ConocoPhillips Grants, 1629
Georgia-Pacific Found Environment Grants, 2123
Gil and Dody Weaver Foundation Grants, 2139
H.A. and Mary K. Chapman Charitable Trust Grants, 2193
Hatton W. Sumners Foundation for the Study and Teaching of Self Government Grants, 2219
Mary K. Chapman Foundation Grants, 2673
Oklahoma City Community Grants, 3737
Rocky Mountain Mineral Law Found Grants, 4094
SARE Grad Student Grants - South Region, 4170
SARE Large Systems Grants: South Reg, 4172
SARE On-Farm Research Grants: South Reg, 4173
SARE Research & Educ Grants: South Reg, 4179
Sarkeys Foundation Grants, 4181
Union Pacific Foundation Community and Civic Grants, 4568
Union Pacific Foundation Health and Human Services Grants, 4569
USDA 1890 Institution Teaching, Research, and Extension Capacity Building Grants, 4613
USDA Regional Integrated Pest Management Grants - Southern Region, 4644
W.M. Keck Foundation Undergraduate Education Grants, 4825
Weyerhaeuser Company Foundation Grants, 4857

Oregon

Boeing Company Contributions Grants, 1357
Brainerd Foundation Grants, 1368
Bullitt Foundation Grants, 1384
Carpenter Foundation Grants, 1439
Chiles Foundation Grants, 1552
Coeta and Donald Barker Foundation Grants, 1594
Dell Foundation Open Grants, 1734
E.L. Wiegand Foundation Grants, 1885
Georgia-Pacific Found Environment Grants, 2123
Jackson Foundation Grants, 2453
M.J. Murdock Trust College Research Program for Life Sciences, 2638
M.J. Murdock Trust College Science Research, 2639
M.J. Murdock Trust Exceptional Opportunity Grants, 2640
M.J. Murdock Trust General Grants, 2641
M.J. Murdock Trust Partners in Science Grants, 2642
M.J. Murdock Trust Research Start-up Grants for New Science Faculty, 2643
M.J. Murdock Trust Scientific Research Grants, 2644
Meyer Memorial Trust Responsive Grants, 2748
NWHF Community-Based Participatory Research Grants, 3670
Paul G. Allen Family Foundation Grants, 3876
Rocky Mountain Mineral Law Found Grants, 4094
SARE Graduate Student Grants in Sustainable Agriculture - Western Region, 4171
SARE Producer Research and Education Grants-Western Region, 4174
SARE Pro & Producer Grants: West Reg, 4175
SARE Research & Educ Grants: West Reg, 4180
Union Pacific Foundation Community and Civic Grants, 4568
Union Pacific Foundation Health and Human Services Grants, 4569
USDA Integrated Pest Management Competitive Grants - Western Region, 4629
W.M. Keck Foundation Undergraduate Education Grants, 4825
Western Regional Graduate Program, 4855
Weyerhaeuser Company Foundation Grants, 4857

Pennsylvania

1 in 9: Long Island Breast Cancer Action Coalition Grants, 1
1976 Foundation Grants, 7
American Foodservice Charitable Trust Grants, 710
Appalachian Reg Comm Health Care Grants, 915
Arcadia Foundation Grants, 983
Archer Daniels Midland Foundation Grants, 985
Arronson Foundation Grants, 1019
Arthur J. Ennis Fellowships, 1025
Barra Foundation Project Grants, 1301
Bayer Foundation Grants, 1309
Berks County Community Foundation Grants, 1333
Boeing Company Contributions Grants, 1357
Cabot Corporation Foundation Grants, 1406
Caesars Foundation Grants, 1407
Campbell Soup Foundation Grants, 1415
Chiron Foundation Community Grants, 1553
Claneil Foundation Grants, 1574
ConocoPhillips Grants, 1629
Donald and Sylvia Robinson Family Foundation Grants, 1830
Dorothy Rider Pool Health Care Grants, 1841
Eaton Charitable Fund Grants, 1893
Eugene M. Lang Foundation Grants, 1957
Expect Miracles Foundation Grants, 1965
Fluor Foundation Grants, 2028
FMC Foundation Grants, 2029
Foundation for Enhancing Comm Grants, 2038
Genuardi Family Foundation Grants, 2109
Georgia-Pacific Found Environment Grants, 2123
Gilroy and Lillian P. Roberts Charitable Foundation Grants, 2142
Graduate Institute Hodson Trust Teacher Fellowships, 2160
Hilda and Preston Davis Foundation Grants, 2266
Hirtzel Memorial Foundation Grants, 2269
Josiah W. and Bessie H. Kline Found Grants, 2517
McLean Contributionship Grants, 2727
Meyer and Stephanie Eglin Foundation Grants, 2747
Motorola Foundation Grants, 2796
NL Associated Colleges of the Midwest/Great Lakes College Association Faculty Fellowships, 3281
NOAA Chesapeake Bay Office B-WET (Bay Watershed Education and Training) Grants, 3339
NOAA Chesapeake Bay Office Cooperative Research Grants, 3340
NOAA Chesapeake Bay Office Fisheries Science Grants, 3341
NOAA Chesapeake Bay Office Submerged Aquatic Vegetation Culture and Restoration Grants, 3343
Peter F. McManus Charitable Trust Grants, 3903
Pew Fellowships in the Arts, 3905
PHF Regional Mellon Faculty Research and Cultural Fellowships, 3933
Pittsburgh Foundation Med Research Grants, 3973
Prudential Foundation Education Grants, 4002
Rathmann Family Foundation Grants, 4033
Ruth and Vernon Taylor Foundation Grants, 4130
Ruth Estrin Goldberg Memorial for Cancer Research Grants, 4131
SARE Graduate Student Grants: NE Region, 4169
SARE Professional Development Grants - Northeast Region, 4176
SARE Research and Education Grants - Northeast Region, 4178
Stackpole-Hall Foundation Grants, 4442
Textron Corporate Contributions Grants, 4501
USDA Regional Integrated Pest Management Grants - Northeastern Region, 4643
Weyerhaeuser Company Foundation Grants, 4857
Willary Foundation Grants, 4870
William G. McGowan Charitable Fund Grants, 4877

Puerto Rico

Abbott Fund Global AIDS Care Grants, 219
Abbott Fund Science Education Grants, 220
Alston Bannerman Sabbatical Fellowships, 580
AOCS Ralph H. Potts Memorial Fellowship, 893
Archer Daniels Midland Foundation Grants, 985
Bank of America Charitable Matching Gifts, 1297
EPA Pollution Prevention Grants, 1936
EPA Regional Healthy Indoor Air Projects for Community Outreach and Education, 1937
EPA Temporally Integrated Monitoring of Ecosystems and Long-Term Grants, 1942
Gruber Foundation Int'l Research Award, 2180
IBRO Latin America Regional Funding for Neuroscience Schools, 2327
IBRO Latin America Regional Funding for PROLAB Collaborations, 2328
IBRO Latin America Regional Travel Grants, 2329
John Simon Guggenheim Memorial Foundation Fellowships, 2507
NHLBI Ancillary Studies in Clinical Trials, 3015
NHLBI Bioengineering and Obesity Grants, 3016
NHLBI Career Transition Awards, 3021
NHLBI Lymphatics in Health and Disease in the Digestive, Cardio & Pulmonary Systems, 3031

NHLBI Lymphatics in Health and Disease in the Digestive, Urinary, Cardiovascular and Pulmonary Systems, 3032
NHLBI Mentored Career Development Award to Promote Faculty Diversity/Re-Entry in Biomedical Research, 3035
NHLBI Mentored Patient-Oriented Research Career Development Awards, 3037
NHLBI Mentored Quantitative Research Career Development Awards, 3038
NHLBI Pathway to Independence Awards, 3045
NHLBI Research on the Role of Cardiomyocyte Mitochondria in Heart Disease: An Integrated Approach, 3053
NHLBI Ruth L. Kirschstein National Research Service Awards for Individual Postdoctoral Fellows, 3057
NHLBI Ruth L. Kirschstein National Research Service Awards for Individual Senior Fellows, 3060
NHLBI Ruth L. Kirschstein National Research Service Award Short-Term Institutional Research Training Grants, 3061
NHLBI Short-Term Research Education to Increase Diversity in Health-Related, 3062
NHLBI Training Grants for Institutions that Promote Diversity, 3067
SARE Grad Student Grants - South Region, 4170
SARE Large Systems Grants: South Reg, 4172
SARE On-Farm Research Grants: South Reg, 4173
SARE Research & Educ Grants: South Reg, 4179
SfN Chapters Graduate Student Travel Awards, 4202
SfN Chapters Postdoc Trainee Travel Awards, 4203
SfN Donald B. Lindsley Prize in Behavioral Neuroscience, 4204
SfN Graduate Student Travel Awards, 4205
SfN Jacob P. Waletzky Award, 4206
SfN Janett Rosenberg Trubatch Career Development Award, 4207
SfN Louise Hanson Marshall Specific Recognition Award, 4208
SfN M Salpeter Lifetime Achievement Award, 4209
SfN Postdoctoral Fellow Travel Awards, 4210
SfN Ralph W. Gerard Prize in Neuroscience, 4211
SfN Ricardo Miledi Neuroscience Training, 4212
SfN Swartz Prize for Theoretical and Computational Neuroscience, 4213
SfN Undergraduate Student Travel Awards, 4217
SfN Young Investigator Award, 4218
SSRC Drugs, Security and Democracy Fellows, 4426
USDA Distance Education Grants for Higher Education Institutions in Insular Areas, 4622
USDA Regional Integrated Pest Management Grants - Southern Region, 4644
USDA Resident Instruction Grants for Higher Education in Insular Areas, 4645

Rhode Island

Bay and Paul Foundations, Inc Grants, 1307
Charles H. Hood Foundation Child Health Research Awards, 1522
Expect Miracles Foundation Grants, 1965
Fannie E. Rippel Foundation Grants, 1978
Forrest C. Lattner Foundation Grants, 2035
Hilda and Preston Davis Foundation Grants, 2266
Knoll-MacDonald Foundation Grants, 2566
Nellie Mae Education Foundation District-Level Change Grants, 2968
Nordson Corporation Foundation Grants, 3416
Orchard Foundation Grants, 3793
Rhode Island Foundation Grants, 4061
Rhode Island State Government Internships, 4062
SARE Graduate Student Grants: NE Region, 4169
SARE Professional Development Grants - Northeast Region, 4176
SARE Research and Education Grants - Northeast Region, 4178
Textron Corporate Contributions Grants, 4501
USDA Regional Integrated Pest Management Grants - Northeastern Region, 4643

South Carolina

3M Company Higher Education Grants, 4
3M Foundation Community Giving Grants, 5
Appalachian Reg Comm Health Care Grants, 915
Archer Daniels Midland Foundation Grants, 985
Boeing Company Contributions Grants, 1357
Campbell Soup Foundation Grants, 1415
Carrier Corporation Contributions Grants, 1440
Charles M. and Mary D. Grant Found Grants, 1527
Drs. Bruce and Lee Foundation Grants, 1859
Eaton Charitable Fund Grants, 1893
EPA Southeastern Multi-Media and Geographic Priority Project Grants, 1939
Fluor Foundation Grants, 2028
Georgia-Pacific Found Environment Grants, 2123
Richard Davoud Donchian Foundation Grants, 4066
SARE Grad Student Grants - South Region, 4170
SARE Large Systems Grants: South Reg, 4172
SARE On-Farm Research Grants: South Reg, 4173
SARE Research & Educ Grants: South Reg, 4179
Self Foundation Grants, 4197
USDA 1890 Institution Teaching, Research, and Extension Capacity Building Grants, 4613
USDA Regional Integrated Pest Management Grants - Southern Region, 4644
Weyerhaeuser Company Foundation Grants, 4857
Woodward Fund Grants, 4895

South Dakota

3M Company Higher Education Grants, 4
3M Foundation Community Giving Grants, 5
Bush Foundation Artist Fellowships, 1391
Rocky Mountain Mineral Law Found Grants, 4094
SARE Graduate Student Grants - North Central Region, 4168
SARE Research and Education Grants - North Central Region, 4177
Sioux Falls Area Community Foundation Community Fund Grants (Unrestricted), 4262
Sioux Falls Area Community Foundation Spot Grants (Unrestricted), 4263
USDA Regional Integrated Pest Management Grants - North Central Region, 4642
USDA Tribal College Research Grants, 4652
Viking Children's Fund Grants, 4811
Western Regional Graduate Program, 4855

Tennessee

Appalachian Reg Comm Health Care Grants, 915
Archer Daniels Midland Foundation Grants, 985
Assisi Foundation of Memphis General Grants, 1221
Carrier Corporation Contributions Grants, 1440
CCCF Dora Maclellan Brown Seminary Scholarships, 1448
Charles M. and Mary D. Grant Found Grants, 1527
Coors Brewing Corporate Contributions Grants, 1644
Dell Foundation Open Grants, 1734
EPA Southeastern Multi-Media and Geographic Priority Project Grants, 1939
Fluor Foundation Grants, 2028
FMC Foundation Grants, 2029
Georgia-Pacific Found Environment Grants, 2123
Hilton Hotels Corporate Giving Grants, 2268
John Deere Foundation Grants, 2500
SARE Grad Student Grants - South Region, 4170
SARE Large Systems Grants: South Reg, 4172
SARE On-Farm Research Grants: South Reg, 4173
SARE Research & Educ Grants: South Reg, 4179
Thompson Charitable Foundation Grants, 4511
Union Pacific Foundation Community and Civic Grants, 4568
Union Pacific Foundation Health and Human Services Grants, 4569
USDA 1890 Institution Teaching, Research, and Extension Capacity Building Grants, 4613
USDA Regional Integrated Pest Management Grants - Southern Region, 4644
Woodward Fund Grants, 4895

Texas

3M Company Higher Education Grants, 4
3M Foundation Community Giving Grants, 5
Abbott Fund Global AIDS Care Grants, 219
Abbott Fund Science Education Grants, 220
Albert and Margaret Alkek Foundation Grants, 504
Alvin and Lucy Owsley Foundation Grants, 584
American Foodservice Charitable Trust Grants, 710
Archer Daniels Midland Foundation Grants, 985
Bayer Foundation Grants, 1309
Boeing Company Contributions Grants, 1357
Bosque Foundation Grants, 1360
Cabot Corporation Foundation Grants, 1406
Campbell Soup Foundation Grants, 1415
Carrier Corporation Contributions Grants, 1440
Communities Foundation of Texas Grants, 1613
ConocoPhillips Grants, 1629
Dell Foundation Open Grants, 1734
Denton A. Cooley Foundation Grants, 1743
Dr. Bob and Jean Smith Foundation Grants, 1853
Eaton Charitable Fund Grants, 1893
Eugene McDermott Foundation Grants, 1958
Ewing Halsell Foundation Grants, 1964
Fluor Foundation Grants, 2028
FMC Foundation Grants, 2029
Forrest C. Lattner Foundation Grants, 2035
Frances C. and William P. Smallwood Grants, 2045
Frederick Gardner Cottrell Foundation Grants, 2054
George Foundation Grants, 2115
Georgia-Pacific Found Environment Grants, 2123
Gil and Dody Weaver Foundation Grants, 2139
Harold Simmons Foundation Grants, 2203
Harry W. Bass, Jr. Foundation Grants, 2210
Hatton W. Sumners Foundation for the Study and Teaching of Self Government Grants, 2219
Hoglund Foundation Grants, 2271
Human Source Foundation Grants, 2297
James M. Collins Foundation Grants, 2461
James R. Dougherty Jr. Foundation Grants, 2463
John P. McGovern Foundation Grants, 2504
John S. Dunn Research Foundation Grants and Chairs, 2506
Lowe Foundation Grants, 2622
M.D. Anderson Foundation Grants, 2637
MacDonald-Peterson Foundation Grants, 2649
Marcia and Otto Koehler Foundation Grants, 2656
Margaret & James A. Elkins Jr. Found Grants, 2658
McCombs Foundation Grants, 2718
Meadows Foundation Grants, 2736
Motorola Foundation Grants, 2796
Peter F. McManus Charitable Trust Grants, 3903
Prudential Foundation Education Grants, 4002
Reuter Foundation Grants, 4059
Robert E. and Evelyn McKee Found Grants, 4074
Rocky Mountain Mineral Law Found Grants, 4094
Ruth and Vernon Taylor Foundation Grants, 4130
San Antonio Area Foundation Grants, 4163
SARE Grad Student Grants - South Region, 4170
SARE Large Systems Grants: South Reg, 4172
SARE On-Farm Research Grants: South Reg, 4173
SARE Research & Educ Grants: South Reg, 4179
Sid W. Richardson Foundation Grants, 4253
Textron Corporate Contributions Grants, 4501
Todd Brock Family Foundation Grants, 4522
Union Pacific Foundation Community and Civic Grants, 4568
Union Pacific Foundation Health and Human Services Grants, 4569
USDA 1890 Institution Teaching, Research, and Extension Capacity Building Grants, 4613
USDA Regional Integrated Pest Management Grants - Southern Region, 4644
W.M. Keck Foundation Undergraduate Education Grants, 4825
Welch Foundation Departmental Grants, 4839
Welch Foundation Endowed Chairs, 4840
Welch Foundation Norman Hackeman Award in Chemical Research, 4841
Welch Foundation Research Grants, 4842
Weyerhaeuser Company Foundation Grants, 4857

US Virgin Islands
Alston Bannerman Sabbatical Fellowships, 580
Bank of America Charitable Matching Gifts, 1297
Community Foundation of the Virgin Islands Judith A. Towle Environmental Studies Grants, 1621
EPA Pollution Prevention Grants, 1936
EPA Regional Healthy Indoor Air Projects for Community Outreach and Education, 1937
EPA Temporally Integrated Monitoring of Ecosystems and Long-Term Grants, 1942
NHLBI Ancillary Studies in Clinical Trials, 3015
NHLBI Bioengineering and Obesity Grants, 3016
NHLBI Career Transition Awards, 3021
NHLBI Lymphatics in Health and Disease in the Digestive, Cardio & Pulmonary Systems, 3031
NHLBI Lymphatics in Health and Disease in the Digestive, Urinary, Cardiovascular and Pulmonary Systems, 3032
NHLBI Mentored Career Development Award to Promote Faculty Diversity/Re-Entry in Biomedical Research, 3035
NHLBI Mentored Patient-Oriented Research Career Development Awards, 3037
NHLBI Mentored Quantitative Research Career Development Awards, 3038
NHLBI Pathway to Independence Awards, 3045
NHLBI Research on the Role of Cardiomyocyte Mitochondria in Heart Disease: An Integrated Approach, 3053
NHLBI Ruth L. Kirschstein National Research Service Awards for Individual Postdoctoral Fellows, 3057
NHLBI Ruth L. Kirschstein National Research Service Awards for Individual Senior Fellows, 3060
NHLBI Ruth L. Kirschstein National Research Service Award Short-Term Institutional Research Training Grants, 3061
NHLBI Short-Term Research Education to Increase Diversity in Health-Related, 3062
NHLBI Training Grants for Institutions that Promote Diversity, 3067
SARE Grad Student Grants - South Region, 4170
SARE Large Systems Grants: South Reg, 4172
SARE On-Farm Research Grants: South Reg, 4173
SfN Ricardo Miledi Neuroscience Training, 4212
USDA Distance Education Grants for Higher Education Institutions in Insular Areas, 4622
USDA Regional Integrated Pest Management Grants - Southern Region, 4644
USDA Resident Instruction Grants for Higher Education in Insular Areas, 4645

Utah
3M Company Higher Education Grants, 4
3M Foundation Community Giving Grants, 5
Abbott Fund Global AIDS Care Grants, 219
Abbott Fund Science Education Grants, 220
Boeing Company Contributions Grants, 1357
Campbell Soup Foundation Grants, 1415
E.L. Wiegand Foundation Grants, 1885
George S. and Dolores Dore Eccles Foundation Grants, 2120
Rocky Mountain Mineral Law Found Grants, 4094
SARE Graduate Student Grants in Sustainable Agriculture - Western Region, 4171
SARE Producer Research and Education Grants- Western Region, 4174
SARE Pro & Producer Grants: West Reg, 4175
SARE Research & Educ Grants: West Reg, 4180
Union Pacific Foundation Community and Civic Grants, 4568
Union Pacific Foundation Health and Human Services Grants, 4569
USDA Integrated Pest Management Competitive Grants - Western Region, 4629
W.M. Keck Foundation Undergraduate Education Grants, 4825
Western Regional Graduate Program, 4855
Weyerhaeuser Company Foundation Grants, 4857

Vermont
Bay and Paul Foundations, Inc Grants, 1307
Charles H. Hood Foundation Child Health Research Awards, 1522
Dorr Foundation Grants, 1842
Expect Miracles Foundation Grants, 1965
Fannie E. Rippel Foundation Grants, 1978
Hilda and Preston Davis Foundation Grants, 2266
Orchard Foundation Grants, 3793
Richard Davoud Donchian Foundation Grants, 4066
SARE Graduate Student Grants: NE Region, 4169
SARE Professional Development Grants - Northeast Region, 4176
SARE Research and Education Grants - Northeast Region, 4178
USDA Regional Integrated Pest Management Grants - Northeastern Region, 4643

Virginia
Abbott Fund Global AIDS Care Grants, 219
Abbott Fund Science Education Grants, 220
Appalachian Reg Comm Health Care Grants, 915
Charles M. and Mary D. Grant Found Grants, 1527
Commonweal Foundation Community Assistance Grants, 1602
Commonweal Foundation Grants, 1603
Community Found AIDS Endowment Awards, 1614
Coors Brewing Corporate Contributions Grants, 1644
Fluor Foundation Grants, 2028
Georgia-Pacific Found Environment Grants, 2123
Graduate Institute Hodson Trust Teacher Fellowships, 2160
Hampton Roads Community Foundation Mental Health Research Grants, 2199
Hilda and Preston Davis Foundation Grants, 2266
Jeffress Awards in Interdisciplinary Research, 2481
Knoll-MacDonald Foundation Grants, 2566
NOAA Chesapeake Bay Office B-WET (Bay Watershed Education and Training) Grants, 3339
NOAA Chesapeake Bay Office Cooperative Research Grants, 3340
NOAA Chesapeake Bay Office Fisheries Science Grants, 3341
NOAA Chesapeake Bay Office Non-Native Oyster Research Grants, 3342
NOAA Chesapeake Bay Office Submerged Aquatic Vegetation Culture and Restoration Grants, 3343
Richard Davoud Donchian Foundation Grants, 4066
Robert G. Cabell III and Maude Morgan Cabell Foundation Grants, 4075
SARE Grad Student Grants - South Region, 4170
SARE Large Systems Grants: South Reg, 4172
SARE On-Farm Research Grants: South Reg, 4173
SARE Research & Educ Grants: South Reg, 4179
Thompson Charitable Foundation Grants, 4511
USDA 1890 Institution Teaching, Research, and Extension Capacity Building Grants, 4613
USDA Regional Integrated Pest Management Grants - Southern Region, 4644
Virginia Historical Society Research Fellows, 4816
Weyerhaeuser Company Foundation Grants, 4857
William G. McGowan Charitable Fund Grants, 4877

Washington
Archer Daniels Midland Foundation Grants, 985
Better Way Foundation Grants, 1338
Boeing Company Contributions Grants, 1357
Brainerd Foundation Grants, 1368
Bullitt Foundation Grants, 1384
Campbell Soup Foundation Grants, 1415
Catherine Holmes Wilkins Foundation Grants, 1442
Chiron Foundation Community Grants, 1553
Comprehensive Health Education Grants, 1622
E.L. Wiegand Foundation Grants, 1885
Fluor Foundation Grants, 2028
George Frederick Jewett Foundation Grants, 2116
Georgia-Pacific Found Environment Grants, 2123
Green Diamond Charitable Contributions, 2171
M.J. Murdock Trust College Research Program for Life Sciences, 2638
M.J. Murdock Trust College Science Research, 2639
M.J. Murdock Trust Exceptional Opportunity Grants, 2640
M.J. Murdock Trust General Grants, 2641
M.J. Murdock Trust Partners in Science Grants, 2642
M.J. Murdock Trust Research Start-up Grants for New Science Faculty, 2643
M.J. Murdock Trust Scientific Research Grants, 2644
Meyer Memorial Trust Responsive Grants, 2748
Norcliffe Foundation Grants, 3415
Northwest Fund for the Environment Grants, 3431
NWHF Community-Based Participatory Research Grants, 3670
Padilla Bay Foundation Grants, 3863
Paul G. Allen Family Foundation Grants, 3876
Rathmann Family Foundation Grants, 4033
Rayonier Foundation Grants, 4034
Robert B McMillen Foundation Grants, 4072
Rocky Mountain Mineral Law Found Grants, 4094
Russell Family Foundation Environmental Sustainability Grants, 4128
SARE Graduate Student Grants in Sustainable Agriculture - Western Region, 4171
SARE Producer Research and Education Grants- Western Region, 4174
SARE Pro & Producer Grants: West Reg, 4175
SARE Research & Educ Grants: West Reg, 4180
Seattle Foundation Medical Funds Grants, 4194
Union Pacific Foundation Community and Civic Grants, 4568
Union Pacific Foundation Health and Human Services Grants, 4569
USDA Integrated Pest Management Competitive Grants - Western Region, 4629
USDA Tribal College Research Grants, 4652
W.M. Keck Foundation Undergraduate Education Grants, 4825
Western Regional Graduate Program, 4855
Weyerhaeuser Company Foundation Grants, 4857

West Virginia
Appalachian Reg Comm Health Care Grants, 915
Ashland Corporate Contributions Grants, 1087
Bayer Foundation Grants, 1309
Cabot Corporation Foundation Grants, 1406
Carl M. Freeman Foundation Grants, 1430
Charles M. and Mary D. Grant Found Grants, 1527
Clay Foundation Grants, 1579
Cleveland-Cliffs Foundation Grants, 1580
FMC Foundation Grants, 2029
Georgia-Pacific Found Environment Grants, 2123
Hilda and Preston Davis Foundation Grants, 2266
NOAA Chesapeake Bay Office B-WET (Bay Watershed Education and Training) Grants, 3339
NOAA Chesapeake Bay Office Cooperative Research Grants, 3340
NOAA Chesapeake Bay Office Fisheries Science Grants, 3341
NOAA Chesapeake Bay Office Submerged Aquatic Vegetation Culture and Restoration Grants, 3343
Parkersburg Area Community Foundation Arts in Education Special Grants, 3869
SARE Graduate Student Grants: NE Region, 4169
SARE Professional Development Grants - Northeast Region, 4176
SARE Research and Education Grants - Northeast Region, 4178
USDA 1890 Institution Teaching, Research, and Extension Capacity Building Grants, 4613
USDA Regional Integrated Pest Management Grants - Northeastern Region, 4643
Weyerhaeuser Company Foundation Grants, 4857

Wisconsin
3M Company Higher Education Grants, 4
3M Foundation Community Giving Grants, 5
Acuity Charitable Foundation Grants, 333
Archer Daniels Midland Foundation Grants, 985
A Social Corporation Grants, 1189
Brico Fund Grants, 1371

Campbell Soup Foundation Grants, 1415
Coleman Foundation Cancer Care Grants, 1595
Eaton Charitable Fund Grants, 1893
Evjue Foundation Grants, 1963
George Kress Foundation Grants, 2119
Georgia-Pacific Found Environment Grants, 2123
Great Lakes Fishery Trust Access Grants, 2168
Great Lakes Fishery Ecological & Biological Fisheries Research to Inform Management Grants, 2169
Jane Bradley Pettit Foundation Health Grants, 2471
John Deere Foundation Grants, 2500
NL Associated Colleges of the Midwest/Great Lakes College Association Faculty Fellowships, 3281
Northwestern Mutual Foundation Grants, 3430
Oscar Rennebohm Foundation Grants, 3832
Patrick and Anna M. Cudahy Fund Grants, 3873
RRF General Grants, 4115
SARE Graduate Student Grants - North Central Region, 4168
SARE Research and Education Grants - North Central Region, 4177
Sensient Technologies Foundation Grants, 4199
Union Pacific Foundation Community and Civic Grants, 4568
Union Pacific Foundation Health and Human Services Grants, 4569
USDA Regional Integrated Pest Management Grants - North Central Region, 4642
USDA Tribal College Research Grants, 4652
Viking Children's Fund Grants, 4811
Weyerhaeuser Company Foundation Grants, 4857

Wyoming
FMC Foundation Grants, 2029
Paul Stock Foundation Grants, 3879
Rocky Mountain Mineral Law Found Grants, 4094
Ruth and Vernon Taylor Foundation Grants, 4130
SARE Graduate Student Grants in Sustainable Agriculture - Western Region, 4171
SARE Producer Research and Education Grants-Western Region, 4174
SARE Pro & Producer Grants: West Reg, 4175
SARE Research & Educ Grants: West Reg, 4180
Union Pacific Foundation Community and Civic Grants, 4568
Union Pacific Foundation Health and Human Services Grants, 4569
USDA Integrated Pest Management Competitive Grants - Western Region, 4629
W.M. Keck Foundation Undergraduate Education Grants, 4825
Western Regional Graduate Program, 4855

Canada
3M Canada National Teaching Fellowships, 3
Abbott Fund Science Education Grants, 220
Academy of American Franciscan History Dissertation Fellowships, 249
ACSM Oded Bar-Or Int'l Scholar Awards, 320
Aid to Scholarly Publications (ASPP), 431
AIHS/Mitacs Health Pilot Partnership Internship Grants, 434
AIHS Alberta/Pfizer Translational Grants, 435
AIHS Clinical Fellowships, 436
AIHS Collaborative Research and Innovation Grants - Collaborative Program, 437
AIHS Collaborative Research and Innovation Grants - Collaborative Project, 438
AIHS Fast-Track Fellowships, 440
AIHS ForeFront Executive-in-Residence Award, 441
AIHS ForeFront Internships, 442
AIHS ForeFront MBA Studentship Award, 443
AIHS ForeFront MBT Studentship Awards, 444
AIHS Full-Time Fellowships, 445
AIHS Full-Time Health Research Studentships, 446
AIHS Full-Time M.D./Ph.D. Studentships, 447
AIHS Full-Time Studentships, 448
AIHS Heritage Youth Researcher Summer Science Program, 449
AIHS Interdisciplinary Team Grants, 450
AIHS Knowledge Exchange Visiting Professors, 451
AIHS Knowledge Exchange - Visiting Scientists, 452
AIHS Media Summer Fellowship, 453
AIHS Part-Time Fellowships, 454
AIHS Part-Time Studentships, 455
AIHS Proposals for Special Initiatives, 456
AIHS Research Prize, 457
AIHS Summer Studentships, 458
AIHS Sustainability Grants, 459
Alberta Law Foundation Grants, 503
Alexander von Humboldt Foundation Konrad Adenauer Research Award, 531
Alfred P. Sloan Foundation Microbiology of the Built Environment Grants, 554
Alfred P. Sloan Foundation Research Fellowships, 560
American Mathematical Society China Exchange Travel Grants, 753
American Wine Society Educational Foundation Scholarship Program, 808
AOCS Ralph H. Potts Memorial Fellowship, 893
ATA Doctoral Fellowships in Education, 1252
ATA Educational Trust Grants - Curriculum Resource or Research Project, 1253
ATA Education Research Award, 1254
ATA Nadene M. Thomas Grad Research Bursary, 1255
Bank of America Charitable Matching Gifts, 1297
Boeing Company Contributions Grants, 1357
Brainerd Foundation Grants, 1368
Bullitt Foundation Grants, 1384
BWF Ad Hoc Grants, 1392
BWF Career Awards at the Scientific Interface, 1393
BWF Career Awards for Medical Scientists, 1394
BWF Collaborative Research Travel Grants, 1395
BWF Innovation in Regulatory Science Grants, 1396
BWF Institutional Program Unifying Population and Laboratory Based Sciences Grants, 1397
BWF Investigators in the Pathogenesis of Infectious Disease Awards, 1398
BWF Preterm Birth Initiative Grants, 1399
Cabot Corporation Foundation Grants, 1406
Canada-U.S. Fulbright International Science and Technology Award, 1417
Canada-U.S. Fulbright Senior Specialists Grant, 1418
Canada-U.S. Fulbright Trad Scholar Awards, 1419
Canada Graduate Scholarships (CGS) and NSERC Postgraduate Scholarships (PGS), 1420
Canadian Patient Safety Institute (CPSI) Research Grants, 1421
Carnegie Endowment for International Peace Junior Fellowships, 1434
CCA Alberta Creative Development Initiative: Grants to Individuals, 1446
CFUW Beverley Jackson Fellowship, 1507
CFUW Bourse Georgette LeMoyne Award, 1508
CFUW Canadian Home Economics Association (CHEA) Fellowship, 1509
CFUW Dr. Alice E. Wilson Awards, 1510
CFUW Ecole Polytech Commemorative Award, 1511
CFUW Margaret Dale Philp Award, 1512
CFUW Margaret McWilliams Predoctoral Fellowship, 1513
CFUW Marion Elder Grant Fellowship, 1514
CFUW Memorial Fellowship, 1515
CFUW Ruth Binnie Fellowship, 1516
Children's Cardiomyopathy Foundation Research Grants, 1550
CNIB Baker Applied Research Fund Grants, 1588
CNIB Baker New Researcher Fund Grants, 1589
Conseil des arts de Montreal Residency of the Americas, 1640
Credit Union Found of British Columbia Grants, 1654
CTCRI Idea Grants, 1665
CTCRI Knowledge Synthesis Grants, 1666
CTCRI Policy Research Grants, 1667
CTCRI Researcher Travel Grants, 1668
CTCRI Research Planning Grants, 1669
CTCRI Student Research Grants, 1670
CTCRI Workshop and Learning Opportunities Grants, 1671
Cystic Fibrosis Canada Clinical Fellowships, 1674
Cystic Fibrosis Canada Clinical Project Grants, 1675
Cystic Fibrosis Canada Clinic Incentive Grants, 1676
Cystic Fibrosis Canada Fellowships, 1677
Cystic Fibrosis Canada Research Grants, 1678
Cystic Fibrosis Canada Scholarships, 1679
Cystic Fibrosis Canada Senior Scientist Research Training Awards, 1680
Cystic Fibrosis Canada Small Confer Grants, 1681
Cystic Fibrosis Canada Special Travel Grants For Fellows and Students, 1682
Cystic Fibrosis Canada Studentships, 1683
Cystic Fibrosis Canada Summer Studentships, 1684
Cystic Fibrosis Canada Travel Supplements, 1685
Cystic Fibrosis Canada Vis Scientist Awards, 1688
Dave Thomas Foundation for Adoption Grants, 1723
Fitzpatrick and Francis Family Business Continuity Foundation Grants, 2021
Fluor Foundation Grants, 2028
FMC Foundation Grants, 2029
Fulbright Traditional Scholar Program in the Western Hemisphere, 2085
Gruber Foundation Int'l Research Award, 2180
Honor the Earth Grants, 2279
Hospital Libraries Section / MLA Professional Development Grants, 2283
IAFF Burn Foundation Research Grants, 2312
Ian N. McKinnon Memorial Fellowship, 2314
ICCS Graduate Students Scholarships, 2335
IFUW Dr. A. Vibert Douglas International Fellowship, 2346
J.H. Stewart Reid Memorial Fellowship for Doctoral Studies, 2450
James H. Cummings Foundation Grants, 2460
John B. Bene Fellowships in Social Forestry, 2496
John Deere Foundation Grants, 2500
John Simon Guggenheim Memorial Foundation Fellowships, 2507
Klarman Family Foundation Grants in Eating Disorders Research, 2558
Manitoba Arts Council Aboriginal Arts Creative Development Pilot Projects Grants, 2652
Manitoba Arts Council Craft Grants, 2653
Manitoba Arts Council Dance Creation/Production Grants, 2654
Massage Therapy Research Fund Grants, 2682
MLA Donald Lindberg Research Fellowship, 2785
MLA Graduate Scholarship, 2786
MLA Grad Scholarship for Minority Students, 2787
MLA Research, Development, and Demonstration Project Grant, 2789
NCIC Canadian Breast Cancer Research Initiative Grants, 2918
NCIC Canadian Cancer Society Res Grants, 2919
NCIC Canadian Cancer Society Travel Award for Senior Level PhD Students, 2920
NCIC Terry Fox Postdoc Research Fellowships, 2923
NCIC Terry Fox Research Studentships, 2924
Nelda C. and H.J. Lutcher Stark Found Grants, 2967
NL Newberry Consortium on American Indian Studies Graduate Student Fellowships, 3312
NL Northeast Modern Language Assocation Fellowship, 3314
NSERC Brockhouse Canada Prize for Interdisciplinary Research in Science and Engineering Grant, 3438
NSERC E.W.R. Steacie Memorial Fellowships, 3439
NSERC Gerhard Herzberg Canada Gold Medal for Science and Engineering Grants, 3440
NSERC Howard Alper Postdoctoral Prize, 3441
NSERC Industrial R&D Fellowships, 3442
NSERC Industrial Research Chairs, 3443
NSERC Industrial Undergraduate Student Research Awards, 3444
NSERC Innovation Challenge Awards, 3445
NSERC John C. Polanyi Awards, 3446
NSERC Northern Research Internships, 3447
NSERC Postdoctoral Fellowships Program, 3448
NSERC Research Tools and Instruments (RTI) Grants, 3449

NSERC Synergy Awards for Innovation, 3450
NSERC Undergrad Student Research Awards, 3451
Oak Foundation Environmental Grants, 3731
OISE Muriel and Danny Fung Graduate Scholarships, 3735
OISE William Waters Scholarships in Urban Education, 3736
Ontario Arts Council Integrated Arts Grants, 3790
Pediatric Brain Tumor Found Research Grants, 3888
Province of Alberta Queen Elizabeth II Graduate Scholarships, 4001
PVA Research Foundation Grants, 4004
RAND Corporation Transatlantic Postdoc Fellowships for Int'l Relations & Security, 4032
RCPSC/AMS CanMEDs Research and Development Grants, 4035
RCPSC Detweiler Traveling Fellowships, 4036
RCPSC International Traveling Fellowship, 4037
RCPSC Medical Education Research Grants, 4038
RCPSC Med Education Traveling Fellowships, 4039
RCPSC Royal College Fellowship for Studies in Medical Education, 4040
RCPSC Royal College Visiting Professorships in Medical Research, 4041
RCPSC W.C. MacKenzie Fellowship in Surgery, 4042
Rocky Mountain Mineral Law Found Grants, 4094
RSC Abbyann D. Lynch Medal in Bioethics, 4116
RSC Alice Wilson Award, 4117
RSC Bancroft Award, 4118
RSC Jason A. Hannah Medal, 4119
RSC John L. Synge Award, 4120
RSC McLaughlin Medal, 4121
RSC Miroslaw Romanowski Medal, 4122
RSC Pierre Chauveau Medal, 4123
RSC Rutherford Memorial Medals, 4124
RSC Sir John William Dawson Medal, 4125
RSC Thomas W. Eadie Medal, 4126
Savoy Foundation Post-Doctoral and Clinical Research Fellowships, 4184
Savoy Foundation Research Grants, 4185
Savoy Foundation Studentships, 4186
SfN Chapters Graduate Student Travel Awards, 4202
SfN Chapters Postdoc Trainee Travel Awards, 4203
SfN Donald B. Lindsley Prize in Behavioral Neuroscience, 4204
SfN Graduate Student Travel Awards, 4205
SfN Jacob P. Waletzky Award, 4206
SfN Janett Rosenberg Trubatch Career Development Award, 4207
SfN Louise Hanson Marshall Specific Recognition Award, 4208
SfN M Salpeter Lifetime Achievement Award, 4209
SfN Postdoctoral Fellow Travel Awards, 4210
SfN Ralph W. Gerard Prize in Neuroscience, 4211
SfN Swartz Prize for Theoretical and Computational Neuroscience, 4213
SfN Travel Awards for the Federation of European Neuroscience Societies Forum, 4214
SfN Travel Awards for the International Brain Research Organization World Congress, 4215
SfN Travel Awards for the Japan Neuroscience Society Meeting, 4216
SfN Undergraduate Student Travel Awards, 4217
SfN Young Investigator Award, 4218
Shastri Indo-Canadian Inst Arts Fellowships, 4219
Shastri Indo-Canadian Institute Collaborative Research Project Grants, 4220
Shastri Indo-Canadian Institute Doctoral Student Research Fellowships, 4221
Shastri Indo-Canadian Institute Faculty Research Fellowships, 4222
Shastri Indo-Canadian Institute Faculty Training Fellowships, 4223
Shastri Indo-Canadian Institute Graduate Degree Fellowships from an Indian Institution, 4224
Shastri Indo-Canadian Institute Librarian Fellowships, 4225
Shastri Indo-Canadian Institute Master's Student Research Fellowships, 4226
Shastri Indo-Canadian Institute Millennium Development Research Grants, 4227
Shastri Indo-Canadian Institute Partnership Development Seed Grants, 4228
Shastri Indo-Canadian Institute Post-Doctoral Research Fellowships, 4229
Shastri Indo-Canadian Institute Research Collaboration Development Grants, 4230
Shastri Indo-Canadian Institute Student Language Training Fellowships, 4231
Shastri Indo-Canadian Institute University of Ottawa Overseas India Summer Course Grants, 4232
SSHRC-APC Automotive Partnership Canada, 4396
SSHRC-CIHR-SC Research Data Centres Joint Initiative, 4397
SSHRC-FHSS Awards to Scholarly Public, 4398
SSHRC-IDRC International Partnerships for Sustainable Societies, 4399
SSHRC-NSERC College and Community Innovation Enhancement Grants, 4400
SSHRC-Public Safety Canada Kanishka Project Initiative, 4401
SSHRC Aid to Scholarly Journals, 4402
SSHRC Aid to Small Universities, 4403
SSHRC Banting Postdoctoral Fellowships, 4404
SSHRC Bora Laskin National Fellowship in Human Rights Research, 4405
SSHRC Digging into Data Challenge Grants, 4406
SSHRC Doctoral Fellowships, 4407
SSHRC Graduate Scholarships-Michael Smith Foreign Study Supplements, 4408
SSHRC Insight Development Grants, 4409
SSHRC Insight Grants, 4410
SSHRC Institutional Grants, 4411
SSHRC Joseph-Armand Bombardier CGS Doctoral Scholarships, 4412
SSHRC Jules and Gabrielle Leger Fellowship, 4413
SSHRC Metropolis Project, 4414
SSHRC Postdoctoral Fellowships, 4415
SSHRC Sport Participation Research Initiative: Doctoral Award Supplements, 4416
SSHRC Sport Participation Research Initiative: Postdoctoral Fellowship Supplements, 4417
SSHRC Therese F. Casgrain Fellowship, 4418
State Scholarships Foundation Grants, 4449
Swann Foundation Fellowships, 4478
Tandem Launch Technologies Grants, 4493
Thomson Reuters / MLA Doctoral Fellowships, 4512
United States Institute of Peace - Transatlantic Post-Doc Fellowship for International Relations and Security, 4581
University of Calgary Nursing Alum Scholarship, 4585
USDD National Consortium for Measurement and Signature Intelligence Research Grants, 4759
Vancouver Foundation Grants and Community Initiatives Program, 4807
Virginia Foundation for the Humanities Open Grants, 4814
William and Flora Hewlett Foundation Environmental Grants, 4874

Other Foreign Countries

All Countries

AAAAI Fellows-in-Training Grants, 22
AAAS Award for Science Diplomacy, 27
AAAS Award for Scientific Freedom and Responsibility, 28
AAAS Mentor Awards, 35
AACC International Alsberg-French-Schoch Memorial Lectureship Awards, 82
AACC International Best Student Paper and Best Student Poster Award, 83
AACC International Bruce Wasserman Young Investigator Award, 84
AACC Int'l Carl Wilhelm Brabender Award, 85
AACC Int'l Cecil F. Pinney Travel Awards, 86
AACC International Edith A. Christensen Award, 87
AACC International Engineering and Processing Division's Stanley Watson Award, 88
AACC Int'l Excellence in Teaching Award, 89
AACC Int'l George W. Scott Blair Lecture Award, 90
AACC International Graduate Fellowships, 91
AACC International Milling & Baking Division's John C. Halverson Memorial Lectureship, 92
AACC International Milling & Baking Division's M. Rella Dwyer Graduate Fellowship, 93
AACC Int'l Phil Williams Applied Res Award, 94
AACC International Student Division's Best Research Paper Awards, 95
AACC International Student Division's Product Development Awards, 96
AACC International Student Travel Grant, 97
AACC Int'l Thomas Burr Osborne Medal, 98
AACC International Walter Bushuk Graduate Research Award, 99
AACC International Young Scientist Research Award, 100
AACC Texture Technologies Quality Research Award, 101
AAUW International Doctoral Fellowships, 214
AAUW International Master's/First Professional Degree Fellowships, 215
AAUW International Postdoctoral Fellowships, 216
Abdus Salam ICTP-ELETTRA Users Grants, 221
Abdus Salam ICTP-OEA Affiliated Grants, 222
Abdus Salam ICTP-OEA Projects Grants, 223
Abdus Salam ICTP-OEA Scientific Meetings Grants, 224
Abdus Salam ICTP-OEA Visiting Scholars/ Consultants Grants, 225
Abdus Salam ICTP/IAEA Sandwich Training Education Grants, 226
Abdus Salam ICTP Federation Arrangement Scheme Grants, 227
Abdus Salam ICTP Group Associates Grants, 228
Abdus Salam ICTP Laurea Magistralis Grants, 229
Abdus Salam ICTP Joint PhD Programme in Environmental Fluid Mechanics Grants, 230
Abdus Salam ICTP Junior Associates Grants, 231
Abdus Salam ICTP Mathematical Research Fellowships, 232
Abdus Salam ICTP Networks Grants, 233
Abdus Salam ICTP Postdoctoral Positions in Theoretical High Energy Physics, 234
Abdus Salam ICTP Postgrad Diploma Grants, 235
Abdus Salam ICTP Ramanujan Prize for Young Mathematicians, 236
Abdus Salam ICTP Regular Associates Grants, 237
Abdus Salam ICTP Schools, Conferences, Workshops and Extended Research Workshops Grants, 238
Abdus Salam ICTP Senior Associates Grants, 239
Abdus Salam ICTP Training and Research in Italian Laboratories (TRIL) Grants, 240
Abdus Salam Joint ICTP/SISSA PhD Programme in Physics and Mathematics Grants, 242
ACLS Frederick Burkhardt Residential Fellowships for Recently Tenured Scholars, 288
ACRL Academic/Research Librarian of the Year Award, 303
ACSM Carl V. Gisolfi Memorial Fund Grant, 312
ACSM Coca-Cola Company Doctoral Student Grant on Behavior Research, 313
ACSM Foundation Clinical Sports Medicine Endowment Grants, 315
ACSM Foundation Doctoral Student Research Grants, 316
ACSM Foundation Research Grants, 317
ACSM International Student Awards, 318
ACSM Paffenbarger-Blair Fund for Epidemiological Research on Physical Activity Grants, 321
ACSM Raymond and Rosalee Weiss Research Endowment Grant, 322
AES Robert S. Morison Fellowship, 346
AIHS Collaborative Research and Innovation Grants - Collaborative Team, 439
Alcatel-Lucent Technologies Foundation Grants, 509

Alexander von Humboldt Foundation Anneliese Maier Award, 515
Alexander von Humboldt Foundation Friedrich Wilhelm Bessel Research Awards, 522
Alexander von Humboldt Foundation Georg Forster Fellowships for Experienced Researchers, 523
Alexander von Humboldt Foundation Georg Forster Research Award, 525
Alexander von Humboldt Foundation International Climate Protection Fellowships, 529
Alexander von Humboldt Foundation Max Planck Research Award, 532
Alexander von Humboldt Found Professorships, 535
Alexander von Humboldt Foundation Reimar Lust Award, 536
Alexander von Humboldt Foundation Research Awards, 537
Alexander von Humboldt Foundation Research Fellowships for Postdoctoral Researchers, 539
Alexander von Humboldt Foundation Sofja Kovalevskaja Awards, 540
Alpha Omega Foundation Grants, 576
American Academy of Advertising Distinguished Service Award, 607
American Academy of Advertising Doctoral Dissertation Competition Awards, 608
American Academy of Advertising Ivan Preston Outstanding Contribution to Research on Advertising Award, 609
American Academy of Advertising Kim Rotzoll Award for Advertising Ethics and Social Responsibility, 610
American Academy of Advertising Research Fellowships, 611
Animal Behavior Society Developing Nations Research Grants, 850
Animal Behavior Society Diversity Grant, 851
Animal Behavior Society Latin American Travel Awards, 852
Animal Behavior Society Student Res Grants, 853
AOCS A. Richard Baldwin Award, 871
AOCS ACI/NBB Innovation Research Award, 872
AOCS Alton E. Bailey Award, 873
AOCS Analytical Division Student Award, 874
AOCS Biotechnology Division Student Awards, 876
AOCS Edible Applications Technology Division Outstanding Achievement Award, 877
AOCS Edible Applications Technology Division Student Award, 878
AOCS George Schroepfer Medal, 879
AOCS Health and Nutrition Div Student Award, 881
AOCS Health and Nutrition Poster Competition, 882
AOCS Herbert J. Dutton Award, 883
AOCS Honored Student Award, 884
AOCS Industrial Oil Products Division Student Award, 885
AOCS Lipid Oxidation and Quality Division Poster Competition, 886
AOCS Manuchehr Eijadi Award, 887
AOCS Peter and Claire G. Kalustian Award, 888
AOCS Phospholipid Best Paper Award, 889
AOCS Processing Distinguished Service Award, 890
AOCS Processing Division Student Award, 891
AOCS Protein and Co-Products Division Student Poster Competition, 892
AOCS R Holman Lifetime Achievement Award, 894
AOCS Samuel Rosen Memorial Award, 895
AOCS Stephen S. Chang Award, 896
AOCS Supelco/Nicholas Pelick-AOCS Research Award, 897
AOCS Surfactants and Detergents Division Student Award, 898
AOCS Thomas H. Smouse Memorial Fellowship, 899
AOCS Timothy L. Mounts Award, 900
AOCS Uses of Soybean Oil Award, 901
AOCS Young Scientist Research Award, 902
ASHG Public Health Genetics Fellowship, 1086
ASM/CDC Fellowships in Infectious Disease and Public Health Microbiology, 1100
ASME Allan Kraus Thermal Manage Medal, 1104
ASME Arthur L. Williston Medal, 1105
ASME Barnett-Uzgiris Product Safety Design Award, 1106
ASME Ben C. Sparks Medal, 1107
ASME Bergles-Rohsenow Young Investigator Award in Heat Transfer, 1108
ASME Bernard F. Langer Nuclear Codes and Standards Award, 1109
ASME Blackall Machine Tool Award, 1110
ASME Burt L. Newkirk Award, 1111
ASME Calvin W. Rice Lecture Award, 1112
ASME Charles Russ Richards Memorial Award, 1113
ASME Daniel C. Drucker Medal, 1114
ASME Dixy Lee Ray Award, 1115
ASME Edward F. Obert Award, 1116
ASME Edwin F. Church Medal, 1117
ASME Engineer-Historian Award, 1118
ASME Fluids Engineering Award, 1119
ASME Frank Kreith Energy Award, 1120
ASME Frank von Flue Award, 1121
ASME Freeman Scholar Award, 1122
ASME Gas Turbine Award, 1123
ASME George Westinghouse Medals, 1124
ASME Gustus L. Larson Memorial Award, 1126
ASME Heat Transfer Memorial Award, 1128
ASME Henry Hess Award, 1129
ASME Henry R. Worthington Medal, 1130
ASME Holley Medal, 1131
ASME Internal Combustion Engine Award, 1132
ASME J. Hall Taylor Medal, 1133
ASME J.P. Hartog Award, 1134
ASME James Harry Potter Gold Medal, 1135
ASME James N. Landis Medal, 1136
ASME Kate Gleason Award, 1137
ASME M. Eugene Merchant Manuf Medal, 1139
ASME Machine Design Award, 1140
ASME Marshall B. Peterson Award, 1141
ASME Mayo D. Hersey Award, 1142
ASME McDonald Mentoring Award, 1143
ASME Medal, 1144
ASME Melville Medal, 1145
ASME Melvin R. Green Codes and Standards Medal, 1146
ASME Milton C. Shaw Manufacturing Research Medal, 1147
ASME Nadai Medal, 1148
ASME Nancy DeLoye Fitzroy and Roland V. Fitzroy Medal, 1149
ASME Old Guard Early Career Award, 1150
ASME Old Guard Prizes, 1151
ASME Patrick J. Higgins Award, 1152
ASME Per Bruel Gold Medal for Noise Control and Acoustics, 1153
ASME Performance Test Codes Medal, 1154
ASME Pi Tau Sigma Gold Medal, 1155
ASME Prime Movers Committee Award, 1156
ASME R. Tom Sawyer Award, 1157
ASME Ralph Coats Roe Medal, 1158
ASME Robert E. Koski Medal, 1159
ASME Robert Henry Thurston Lecture Award, 1160
ASME Rufus T. Oldenburger Medal, 1161
ASME Ruth and Joel Spira Outstanding Design Educator Award, 1162
ASME S.Y. Zamrik PVP Medal, 1163
ASME Soichiro Honda Medal, 1164
ASME Spirit of St. Louis Medal, 1165
ASME Student Section Advisor Award, 1166
ASME Thomas A. Edison Patent Award, 1167
ASME Timoshenko Medal, 1168
ASME Van C. Mow Medal, 1169
ASME Warner T. Koiter Medal, 1170
ASME Worcester Reed Warner Medal, 1172
ASME Y.C. Fung Young Investigator Award, 1173
ASU Institute for Humanities Research Visiting Fellowships, 1248
Australasian Institute of Judicial Administration Seed Funding Grants, 1268
Baxter International Foundation William B. Graham Prize for Health Services Research, 1306
Bill and Melinda Gates Foundation Agricultural Development Grants, 1345
Bill and Melinda Gates Foundation Financial Services for the Poor Grants, 1346
Bill and Melinda Gates Foundation Water, Sanitation and Hygiene Grants, 1347
Bupa Foundation Medical Research Grants, 1386
CAA Wyeth Publication Grant, 1405
CDC Foundation Atlanta International Health Fellowships, 1461
CDC Foundation Tobacco Network Fellowship, 1462
CDC Steven M. Teutsch Prevention Effectiveness Fellowships, 1467
CNIB Barbara Tuck MacPhee Award, 1590
CNIB E. (Ben) & Mary Hochhausen Access Technology Research Grants, 1592
Conquer Cancer Foundation of ASCO International Development and Education Award in Palliative Care, 1634
Conquer Cancer Foundation of ASCO International Development and Education Awards, 1635
Conquer Cancer Foundation of ASCO International Innovation Grant, 1636
Conservation, Food, and Health Foundation Grants for Developing Countries, 1641
Cystic Fibrosis Canada Visiting Allied Health Professional Awards, 1686
Cystic Fibrosis Canada Vis Clinician Awards, 1687
Cystic Fibrosis Trust Research Grants, 1691
E.W. "Al" Thrasher Awards, 1887
Elizabeth Glaser Int'l Leadership Awards, 1906
Elizabeth Glaser Scientist Award, 1907
EREF Doctoral Scholarship, 1945
EREF Fiessinger Doctoral Scholarship, 1946
FAMRI Clinical Innovator Awards, 1975
FAMRI Young Clinical Scientist Awards, 1976
Grass Foundation Marine Biological Laboratory Advanced Imaging Fellowships, 2162
Grass Foundation Marine Biological Laboratory Fellowships, 2163
Gruber Foundation Cosmology Prize, 2177
Gruber Foundation International Astronomical Union Fellowships, 2179
Gruber Foundation Neuroscience Prize, 2181
Gruber Foundation Prize for Global Justice and Women's Rights, 2182
Gruber Foundation Rosalind Franklin Young Investigator Award, 2183
Gruber Foundation Weizmann Institute Awards, 2184
Harry Frank Guggenheim Foundation Dissertation Fellowships, 2207
Harry Frank Guggenheim Foundation Research Grants, 2208
Helen Bader Foundation Grants, 2228
Holberg Memorial Fund International Prize, 2272
IBRO/SfN International Travel Grants, 2321
IBRO Albert Aguayo Fellowship, 2322
IBRO International Travel Grants, 2326
IBRO Regional Grants for Int'l Fellowships to U.S. Laboratory Summer Neuroscience Courses, 2330
IBRO Research Fellowships, 2331
IBRO Return Home Fellowships, 2332
IES Faculty Fellowships, 2341
IIE KAUST Graduate Fellowships, 2353
Independent Institute John M. Templeton Fellowships Essay Contest, 2369
IRA Elva Knight Research Grant, 2406
IRA Esther Zolt Academic Research Award, 2407
IRA Esther Zolt Teacher Research Award, 2408
IRA Helen M. Robinson Dissertation Research Award, 2409
IRA Jeanne S. Chall Research Fellowship, 2410
IRA Outstanding Dissertation of Year Award, 2411
IRA Pearson Foundation-IRA-Rotary Literacy Awards, 2412
IRA Steven A. Stahl Research Grants, 2413
IRA Teacher as Researcher Grants, 2414
Japan Foundation Japanese Studies Fellowships, 2476
Japan Foundation Six-Month Grants for Specialists in Cultural and Academic Fields, 2479

JSA Junior Scientist Travel Support, 2523
Kennedy Center Summer Fellowships for International Arts Managers, 2553
LBI-L Leo Baeck/Studienstiftung Fellowships, 2592
LBI-NY David Baumgardt Mem Fellowships, 2595
LBI-NY Fred Grubel Fellowship, 2596
LBI-NY Fritz Halbers Fellowships, 2597
LBI-NY Gerald Westheimer Career Development Fellowship, 2598
Leo Baeck Essay Prize, 2601
LSAC Research Grants, 2623
Lymphatic Education & Research Network Postdoctoral Fellowships, 2633
Macquarie Bank Foundation Grants, 2650
Mary Gates Research Scholarships, 2670
MBL Albert & Ellen Grass Summer Fellowships, 2692
MBL Ann E. Kammer Memorial Summer Fellowship, 2693
MBL Associates Summer Fellowships, 2694
MBL Baxter Postdoctoral Summer Fellowship, 2695
MBL Burr & Susie Steinbach Summer Fellowship, 2696
MBL E.E. Just Summer Fellowship for Minority Scientists, 2697
MBL Erik B. Fries Summer Fellowships, 2698
MBL Eugene and Millicent Bell Summer Fellowships, 2699
MBL Evelyn and Melvin Spiegal Fellowship, 2700
MBL Frank R. Lillie Summer Fellowship, 2701
MBL Frederik B. and Betsy G. Bang Summer Fellowships, 2702
MBL Fred Karush Library Readership, 2703
MBL H. Keffer Hartline and Edward F. MacNichol, Jr. Fellowships, 2705
MBL Herbert W. Rand Summer Fellowship, 2706
MBL James E. and Faith Miller Memorial Summer Fellowship, 2707
MBL John M. Arnold Award, 2708
MBL Laura and Arthur Colwin Fellowships, 2709
MBL Lucy B. Lemann Memorial Summer Fellowship, 2710
MBL M.G.F. Fuortes Summer Fellowships, 2711
MBL Nikon Summer Fellowship, 2712
MBL Plum Found John E. Dowling Fellowships, 2713
MBL Robert Day Allen Summer Fellowship, 2714
MBL Stephen W. Kuffler Summer Fellowships, 2715
MBL William Townsend Porter Summer Fellowships for Minority Investigators, 2716
MIT Knight Science Journalism Fellowships, 2782
MLA Cunningham Memorial International Fellowship, 2783
MLA Murray Gottlieb Prize Essay Award, 2788
MLA Thomas Reuters / Frank Bradway Rogers Information Advancement Award, 2791
National Center for Responsible Gaming Centers of Excellence in Gambling Research Grants, 2850
National Center for Responsible Gaming Exploration Grants, 2853
National Center for Responsible Gaming Independent Scientist Grants, 2854
National Parkinson Foundation Clinical Research Fund Grants, 2881
Nestle Foundation Large Research Grants, 2979
Nestle Foundation Pilot Grants, 2980
Nestle Foundation Re-entry Grants, 2981
Nestle Foundation Small Research Grants, 2982
Nestle Foundation Training Grant, 2983
NHLBI Ancillary Studies in Clinical Trials, 3015
NHLBI Investigator Initiated Multi-Site Clinical Trials, 3030
NHLBI Lymphatics in Health and Disease in the Digestive, Cardio & Pulmonary Systems, 3031
NHLBI Lymphatics in Health and Disease in the Digestive, Urinary, Cardiovascular and Pulmonary Systems, 3032
NHLBI Research on the Role of Cardiomyocyte Mitochondria in Heart Disease: An Integrated Approach, 3053
NHLBI Ruth L. Kirschstein National Research Service Awards for Individual Postdoctoral Fellows, 3057
NL American Society for Eighteenth-Century Studies Fellowships, 3279
NL Arthur and Lila Weinberg Fellowships for Independent Scholars and Researchers, 3280
NL Audrey Lumsden-Kouvel Fellowships, 3282
NL Frances C. Allen Fellowships, 3284
NL Herzog August Bibliothek Wolfenbuttel Fellowship, 3286
NL Lester J. Cappon Fellowship in Documentary Editing, 3288
NL Lloyd Lewis Fellowships in American Hist, 3289
NL Mellon Postdoctoral Research Fellowships, 3294
NL Newberry Library/Ecole Nationale des Chartes Exchange Fellowship, 3313
NL Short-Term Fellowships in the History of Cartography, 3316
NL Short-Term Resident Fellowships for Individual Research, 3317
NL Weiss/Brown Publication Subvention Award, 3320
Norwegian Academy of Science and Letters Abel Prize, 3433
NSF Social Psychology Research Grants, 3649
Organization for Tropical Studies Fellowships, 3796
Organization for Tropical Studies (OTS) Pilot Awards for Exploratory Research, 3797
OSF Alumni Grants, 3835
OSF Fellowships, 3840
OSF Intellectual Property Reform Grants, 3846
OSF Justice Initiative Fellowships, 3847
OSF Thomas J. Alexander Fellowships, 3853
OTS Francis J. Bossuyt Fellowship, 3855
OTS Stanley Smith Horticultural Fellowships, 3857
PC Fred H. Bixby Fellowships, 3880
Pfizer ASPIRE Worldwide Endocrine Young Investigator Grants, 3918
Pfizer Australia Paediatric Endocrine Care Research Grants, 3921
PFizer Compound Transfer Agreements, 3922
Pfizer Global Investigator-Initiated Res Grants, 3924
Pfizer Global Research Awards for Nicotine Independence, 3925
Pfizer Medical Education Track Two Grants, 3927
Phi Kappa Phi Graduate Fellowships, 3938
Phi Kappa Phi Scholar Award, 3939
Radcliffe Institute Dissertation Grants, 4019
Radcliffe Institute Indiv Residential Fellowships, 4021
Radcliffe Institute Oral History Grants, 4022
Radcliffe Institute Research Support Grants, 4023
Radcliffe Institute Resid Group Fellowships, 4024
Roche Organ Transplant Research Foundation Grants, 4083
Royal Belgian Numismatic Society Res Grants, 4107
Royal Norwegian Embassy Kavli Prizes, 4109
Scleroderma Foundation Established Investigator Grants, 4190
SHOT International Scholars, 4242
SHOT M Kranzberg Dissertation Fellowship, 4245
SHOT NASA Fellowship, 4246
SHRM Foundation Research Grants, 4249
Sigma Theta Tau International /American Nurses Foundation Grant, 4254
Sigma Theta Tau International / Council for the Advancement of Nursing Science Grants, 4255
Sigma Theta Tau International / Oncology Nursing Society Grant, 4256
Sigma Theta Tau International Doris Bloch Research Award, 4257
Sigma Theta Tau Small Grants, 4258
SRC Cross-Disciplinary Research Grants, 4353
SRC Graduate Fellowships, 4368
SRC International Fellowship, 4375
SRC Peter Verhofstadt Fellowship, 4389
SSHRC Banting Postdoctoral Fellowships, 4404
Stanford Humanities Center Arts Writer/Practitioner Fellowship, 4444
Stanford Humanities Center External Faculty Fellowships, 4445
Stanley Smith Horticultural Trust Grants, 4446
Susan G. Komen Breast Cancer Foundation Career Catalyst Research Grants, 4473
Susan G. Komen Breast Cancer Foundation Challenge Grants: Career Catalyst Research, 4476
Susan G. Komen Breast Cancer Foundation Investigator Initiated Research Grants, 4477
Tandem Launch Technologies Grants, 4493
Thrasher Research Fund Early Career Awards, 4515
Thrasher Research Fund Grants, 4516
Tourette Syndrome Association Post-Doctoral Fellowships, 4523
Tourette Syndrome Association Res Grants, 4524
UNESCO Co-Sponsored Fellowships, 4566
UNESCO Fellowships, 4567
United Methodist Committee on Relief Global Health Grants, 4577
United States Institute of Peace - Jennings Randolph Peace Dissertation Scholarships, 4579
United States Institute of Peace - Jennings Randolph Senior Fellowships, 4580
United States Institute of Peace Annual Grants, 4582
USAID Devel Innovation Ventures Grants, 4597
USAID Microbicides Research, Development, and Introduction Grants, 4599
USDA Cochran Fellowships, 4619
USDA Higher Education Challenge Grants, 4627
USDD Basic Research for Combating Weapons of Mass Destruction Broad Agency Announcement Grants, 4690
USDD Development of Quantum Computing Algorithms Grants, 4718
USDD Fundamental Research to Counter Weapons of Mass Destruction Grants, 4728
Virginia Foundation for the Humanities Residential Fellowships, 4815
Web We Want Small Grants, 4836
Wellcome Trust Biomedical Science Grants, 4843
WHOI Ocean Science Journalism Fellowship, 4862
WHOI Summer Student Fellowships, 4865
Winterthur Museum Dissertation Fellowships, 4887
Winterthur Research Fellowships, 4888
Wolfsonian-FIU Fellowships, 4890
WWF Russell E. Train Education for Nature Alumni Grants, 4897
Zonta International Foundation Amelia Earhart Fellowships, 4905

Abkhazia

Fulbright Traditional Scholar Program in the East Asia/Pacific Region, 2082
Sir Dorabji Tata Trust Individual Education and Travel Grants, 4265

Afghanistan

Fulbright Traditional Scholar Program in the East Asia/Pacific Region, 2082
IBRO Asia Reg APRC Exchange Fellowships, 2323
OSF Central Asia and Caucasus Research and Training Initiative - Junior Fellowships, 3836
OSF Central Asia and Caucasus Research and Training Initiative - Senior Fellowships, 3837
OSF Global Faculty Grants - Track 2, 3842
OSF Global Supplementary Grants, 3843
Sir Dorabji Tata Trust Individual Education and Travel Grants, 4265
Wellcome Trust New Investigator Awards, 4844

Albania

ACLS Early Career Postdoctoral Fellowships in East European Studies, 286
Beatrice Laing Trust Grants, 1324
Canon Foundation Grants, 1423
EMBO Installation Grants, 1915
Fluor Foundation Grants, 2028
Fulbright Business Grants in Spain, 2067
Fulbright Traditional Scholar Program in Europe and Eurasia, 2080
HHMI Biomedical Research Grants for International Scientists in the Baltics, Central and Eastern Europe, Russia, and Ukraine, 2253
IBRO InEurope Short Stay Grants, 2325
Oak Foundation Environmental Grants, 3731

OSF Access to Medicines Fellowships in Eastern Europe and Central Asia, 3834
OSF Global Supplementary Grants, 3843
OSF HESP Mobility Grants, 3844
OSF Human Rights & Governance Interns, 3845
RAND Corporation Transatlantic Postdoc Fellowships for Int'l Relations & Security, 4032
Simone and Cino del Duca Grand Prix Awards, 4261
State Scholarships Foundation Grants, 4449
United States Institute of Peace - Transatlantic Post-Doc Fellowship for International Relations and Security, 4581
Wellcome Trust New Investigator Awards, 4844

Algeria
Beatrice Laing Trust Grants, 1324
Fulbright Critical Lang Enhancement Awards, 2068
Fulbright Traditional Scholar Program in Sub-Saharan Africa, 2081
Fulbright Traditional Scholar Program in the Near East and North Africa Region, 2083
State Scholarships Foundation Grants, 4449
STTI Alpha Eta Collaborative Research Grants, 4458
Wellcome Trust New Investigator Awards, 4844
Wenner-Gren Foundation African Fellowships, 4846

Andorra
Beatrice Laing Trust Grants, 1324
Canon Foundation Grants, 1423
EMBO Installation Grants, 1915
Fluor Foundation Grants, 2028
Fulbright Business Grants in Spain, 2067
Fulbright Traditional Scholar Program in Europe and Eurasia, 2080
HHMI Biomedical Research Grants for International Scientists in the Baltics, Central and Eastern Europe, Russia, and Ukraine, 2253
IBRO InEurope Short Stay Grants, 2325
Oak Foundation Environmental Grants, 3731
RAND Corporation Transatlantic Postdoc Fellowships for Int'l Relations & Security, 4032
Simone and Cino del Duca Grand Prix Awards, 4261
State Scholarships Foundation Grants, 4449
United States Institute of Peace - Transatlantic Post-Doc Fellowship for International Relations and Security, 4581

Angola
Abdus Salam ITCP Postgraduate Diploma Programme in General Physics, 241
Beatrice Laing Trust Grants, 1324
Fulbright Critical Lang Enhancement Awards, 2068
Fulbright Traditional Scholar Program in Sub-Saharan Africa, 2081
Fulbright Traditional Scholar Program in the Near East and North Africa Region, 2083
SSRC African Peacebuilding Network Research Grants, 4422
State Scholarships Foundation Grants, 4449
STTI Alpha Eta Collaborative Research Grants, 4458
Wellcome Trust New Investigator Awards, 4844
Wenner-Gren Foundation African Fellowships, 4846

Anguilla
AOCS Ralph H. Potts Memorial Fellowship, 893
John Simon Guggenheim Memorial Foundation Fellowships, 2507

Antigua & Barbuda
AOCS Ralph H. Potts Memorial Fellowship, 893
ASM International Fellowship for Latin America and the Caribbean, 1175
Fulbright Traditional Scholar Program in the Western Hemisphere, 2085
IBRO Latin America Regional Funding for Neuroscience Schools, 2327
IBRO Latin America Regional Funding for PROLAB Collaborations, 2328
IBRO Latin America Regional Travel Grants, 2329
John Simon Guggenheim Memorial Foundation Fellowships, 2507
SfN Ricardo Miledi Neuroscience Training, 4212
SSRC Drugs, Security and Democracy Fellows, 4426

Argentina
Academy of American Franciscan History Dissertation Fellowships, 249
ASM-PAHO Infectious Diseases Epidemiology and Surveillance Fellowships, 1098
ASM International Fellowship for Latin America and the Caribbean, 1175
Fulbright Traditional Scholar Program in the Western Hemisphere, 2085
Gruber Foundation Int'l Research Award, 2180
IBRO Latin America Regional Funding for Neuroscience Schools, 2327
IBRO Latin America Regional Funding for PROLAB Collaborations, 2328
IBRO Latin America Regional Travel Grants, 2329
John Simon Guggenheim Memorial Foundation Fellowships, 2507
OTS Hovore-Horn Fellowships, 3856
SfN Chapters Graduate Student Travel Awards, 4202
SfN Chapters Postdoc Trainee Travel Awards, 4203
SfN Donald B. Lindsley Prize in Behavioral Neuroscience, 4204
SfN Graduate Student Travel Awards, 4205
SfN Jacob P. Waletzky Award, 4206
SfN Janett Rosenberg Trubatch Career Development Award, 4207
SfN Louise Hanson Marshall Specific Recognition Award, 4208
SfN M Salpeter Lifetime Achievement Award, 4209
SfN Postdoctoral Fellow Travel Awards, 4210
SfN Ralph W. Gerard Prize in Neuroscience, 4211
SfN Ricardo Miledi Neuroscience Training, 4212
SfN Swartz Prize for Theoretical and Computational Neuroscience, 4213
SfN Undergraduate Student Travel Awards, 4217
SfN Young Investigator Award, 4218
SSRC Drugs, Security and Democracy Fellows, 4426
Wellcome Trust New Investigator Awards, 4844

Armenia
Beatrice Laing Trust Grants, 1324
Canon Foundation Grants, 1423
EMBO Installation Grants, 1915
Fluor Foundation Grants, 2028
Fulbright Business Grants in Spain, 2067
Fulbright Traditional Scholar Program in Europe and Eurasia, 2080
Fulbright Traditional Scholar Program in the East Asia/Pacific Region, 2082
GMFUS Black Sea Trust for Regional Corporation Grants, 2152
HHMI Biomedical Research Grants for International Scientists in the Baltics, Central and Eastern Europe, Russia, and Ukraine, 2253
IBRO InEurope Short Stay Grants, 2325
Oak Foundation Environmental Grants, 3731
OSF Central Asia and Caucasus Research and Training Initiative - Junior Fellowships, 3836
OSF Central Asia and Caucasus Research and Training Initiative - Senior Fellowships, 3837
OSF Global Faculty Grants - Track 1, 3841
OSF Global Faculty Grants - Track 2, 3842
OSF Global Supplementary Grants, 3843
OSF Human Rights & Governance Interns, 3845
OSF Think Tank Fellowships, 3852
RAND Corporation Transatlantic Postdoc Fellowships for Int'l Relations & Security, 4032
Simone and Cino del Duca Grand Prix Awards, 4261
Sir Dorabji Tata Trust Individual Education and Travel Grants, 4265
State Scholarships Foundation Grants, 4449
United States Institute of Peace - Transatlantic Post-Doc Fellowship for International Relations and Security, 4581
Wellcome Trust New Investigator Awards, 4844

Aruba
AOCS Ralph H. Potts Memorial Fellowship, 893
John Simon Guggenheim Memorial Foundation Fellowships, 2507
SfN Ricardo Miledi Neuroscience Training, 4212
SSRC Drugs, Security and Democracy Fellows, 4426

Australia
AFUW Diamond Jubilee Bursary, 359
AFUW Doreen McCarthy, Barbara Crase, Cathy Candler & Brenda Nettle Bursaries, 360
AFUW Fellowships, 361
AFUW Georgina Sweet Fellowships, 362
AFUW Heather Latz Bursary, 363
AFUW Padnendadlu Graduate Bursary, 364
AFUW Padnendadlu Postgraduate Bursary, 365
AFUW Padnendadlu Undergraduate Bursary, 366
AFUW South Australia Jean Gilmore, Thenie Baddams & Daphne Elliott Bursaries, 367
AFUW Winifred E. Preedy Postgrad Bursary, 368
AINSE Honours Scholarships, 460
AINSE Int'l Conference Travel Scholarships, 461
AINSE John Ferris Memorial Award, 462
AINSE Postgraduate Research Awards, 463
AINSE Research Fellowships, 464
American Australian Association Australia To USA Fellowships, 619
AOCS Australasian Award for Scientific Excellence in Lipid Research, 875
Australian Academy of Science Grants, 1269
Australian Postdoctoral Research Fellowships, 1270
Australian Research Fellowships/Queen Elizabeth II Fellowships, 1272
Beryl Henderson Foundation Grants, 1337
Boeing Company Contributions Grants, 1357
Bupa Foundation Multi-Country Grant, 1387
Churchill Fellowships, 1560
Fluor Foundation Grants, 2028
Fulbright Traditional Scholar Program in the East Asia/Pacific Region, 2082
Gruber Foundation Int'l Research Award, 2180
Pediatric Brain Tumor Found Research Grants, 3888
Pfizer Australia Cancer Research Grants, 3919
Pfizer Australia Neuroscience Research Grants, 3920
RACGP/APHCRI Indigenous Health Award, 4008
RACGP/CONROD Research Fellowship, 4009
RACGP Alan Chancellor Award, 4010
RACGP Cardiovascular Research Grants in General Practice, 4011
RACGP Chris Silagy Research Scholarship, 4012
RACGP Family Medical Care, Education and Research Grants, 4013
RACGP Grieve Memorial Award, 4014
RACGP National Asthma Council Res Award, 4015
RACGP Rex Walpole Travelling Fellowship, 4016
RACGP Vicki Kotsirilos Integrative Medicine Grants, 4017
SfN Chapters Graduate Student Travel Awards, 4202
SfN Chapters Postdoc Trainee Travel Awards, 4203
SfN Donald B. Lindsley Prize in Behavioral Neuroscience, 4204
SfN Graduate Student Travel Awards, 4205
SfN Jacob P. Waletzky Award, 4206
SfN Janett Rosenberg Trubatch Career Development Award, 4207
SfN Louise Hanson Marshall Specific Recognition Award, 4208
SfN M Salpeter Lifetime Achievement Award, 4209
SfN Postdoctoral Fellow Travel Awards, 4210
SfN Ralph W. Gerard Prize in Neuroscience, 4211
SfN Swartz Prize for Theoretical and Computational Neuroscience, 4213
SfN Undergraduate Student Travel Awards, 4217
SfN Young Investigator Award, 4218
State Scholarships Foundation Grants, 4449
USDD National Consortium for Measurement and Signature Intelligence Research Grants, 4759

Austria
Beatrice Laing Trust Grants, 1324
Canon Foundation Grants, 1423
Central European University Fellowships, 1488
EMBO Installation Grants, 1915
Fluor Foundation Grants, 2028
Fulbright Business Grants in Spain, 2067
Fulbright Traditional Scholar Program in Europe and Eurasia, 2080
HHMI/EMBO Start-up Grants for C Europe, 2250
HHMI Biomedical Research Grants for International Scientists in the Baltics, Central and Eastern Europe, Russia, and Ukraine, 2253
IBRO Europe Regional Grants for Western European Workshops and Symposia, 2324
IBRO InEurope Short Stay Grants, 2325
Oak Foundation Environmental Grants, 3731
OSF HESP Mobility Grants, 3844
OSF Human Rights & Governance Interns, 3845
Pfizer ASPIRE EU Antifungal Res Awards, 3910
Pfizer ASPIRE EU Dupuytren's Contracture Research Awards, 3911
Pfizer ASPIRE EU Emerging Mechanisms of Resistance Antibacterial Research Awards, 3912
Pfizer ASPIRE EU MRSA Nosocomial Pneumonia & MRSA Complicated Skin & Soft Tissue Infections Antibacterial Research Awards, 3913
RAND Corporation Transatlantic Postdoc Fellowships for Int'l Relations & Security, 4032
Simone and Cino del Duca Grand Prix Awards, 4261
State Scholarships Foundation Grants, 4449
Swiss National Science Foundation DORE (DO REsearch) Conference Grants, 4482
Swiss National Science Foundation Fellowships for Advanced Researchers, 4483
Swiss National Science Foundation Fellowships for Prospective Researchers, 4484
Swiss National Science Foundation Marie Heim-Vogtlin (MHV) Grants, 4485
Swiss National Science Foundation Scientific Conferences Grants, 4486
United States Institute of Peace - Transatlantic Post-Doc Fellowship for International Relations and Security, 4581

Azerbaijan
Beatrice Laing Trust Grants, 1324
Canon Foundation Grants, 1423
EMBO Installation Grants, 1915
Fluor Foundation Grants, 2028
Fulbright Business Grants in Spain, 2067
Fulbright Traditional Scholar Program in Europe and Eurasia, 2080
Fulbright Traditional Scholar Program in the East Asia/Pacific Region, 2082
GMFUS Black Sea Trust for Regional Corporation Grants, 2152
HHMI Biomedical Research Grants for International Scientists in the Baltics, Central and Eastern Europe, Russia, and Ukraine, 2253
IBRO InEurope Short Stay Grants, 2325
Oak Foundation Environmental Grants, 3731
OSF Central Asia and Caucasus Research and Training Initiative - Junior Fellowships, 3836
OSF Central Asia and Caucasus Research and Training Initiative - Senior Fellowships, 3837
OSF Global Faculty Grants - Track 1, 3841
OSF Global Faculty Grants - Track 2, 3842
OSF Global Supplementary Grants, 3843
OSF Think Tank Fellowships, 3852
RAND Corporation Transatlantic Postdoc Fellowships for Int'l Relations & Security, 4032
Simone and Cino del Duca Grand Prix Awards, 4261
Sir Dorabji Tata Trust Individual Education and Travel Grants, 4265
State Scholarships Foundation Grants, 4449
United States Institute of Peace - Transatlantic Post-Doc Fellowship for International Relations and Security, 4581
Wellcome Trust New Investigator Awards, 4844

Bahamas
AOCS Ralph H. Potts Memorial Fellowship, 893
ASM International Fellowship for Latin America and the Caribbean, 1175
Fulbright Traditional Scholar Program in the Western Hemisphere, 2085
IBRO Latin America Regional Funding for Neuroscience Schools, 2327
IBRO Latin America Regional Funding for PROLAB Collaborations, 2328
IBRO Latin America Regional Travel Grants, 2329
John Simon Guggenheim Memorial Foundation Fellowships, 2507
SfN Ricardo Miledi Neuroscience Training, 4212
SSRC Drugs, Security and Democracy Fellows, 4426

Bahrain
Fulbright Traditional Scholar Program in the East Asia/Pacific Region, 2082
IBRO Asia Reg APRC Exchange Fellowships, 2323
Sir Dorabji Tata Trust Individual Education and Travel Grants, 4265

Bangladesh
Fulbright Traditional Scholar Program in the East Asia/Pacific Region, 2082
IBRO Asia Reg APRC Exchange Fellowships, 2323
Sir Dorabji Tata Trust Individual Education and Travel Grants, 4265
USDA Norman E. Borlaug International Agricultural Science and Technology Fellowships, 4636
Wellcome Trust New Investigator Awards, 4844

Barbados
AOCS Ralph H. Potts Memorial Fellowship, 893
ASM International Fellowship for Latin America and the Caribbean, 1175
Fulbright Traditional Scholar Program in the Western Hemisphere, 2085
IBRO Latin America Regional Funding for Neuroscience Schools, 2327
IBRO Latin America Regional Funding for PROLAB Collaborations, 2328
IBRO Latin America Regional Travel Grants, 2329
John Simon Guggenheim Memorial Foundation Fellowships, 2507
SfN Ricardo Miledi Neuroscience Training, 4212
SSRC Drugs, Security and Democracy Fellows, 4426

Belarus
Beatrice Laing Trust Grants, 1324
Canon Foundation Grants, 1423
EMBO Installation Grants, 1915
Fluor Foundation Grants, 2028
Fulbright Business Grants in Spain, 2067
Fulbright Traditional Scholar Program in Europe and Eurasia, 2080
HHMI Biomedical Research Grants for International Scientists in the Baltics, Central and Eastern Europe, Russia, and Ukraine, 2253
IBRO InEurope Short Stay Grants, 2325
John Deere Foundation Grants, 2500
Oak Foundation Environmental Grants, 3731
OSF Access to Medicines Fellowships in Eastern Europe and Central Asia, 3834
OSF Global Supplementary Grants, 3843
OSF HESP Mobility Grants, 3844
OSF Human Rights & Governance Interns, 3845
RAND Corporation Transatlantic Postdoc Fellowships for Int'l Relations & Security, 4032
Simone and Cino del Duca Grand Prix Awards, 4261
State Scholarships Foundation Grants, 4449
United States Institute of Peace - Transatlantic Post-Doc Fellowship for International Relations and Security, 4581
Wellcome Trust New Investigator Awards, 4844

Belgium
Beatrice Laing Trust Grants, 1324
Cabot Corporation Foundation Grants, 1406
Canon Foundation Grants, 1423
Danone Institute Belgium Doctoral Scholarship in Nutrition, 1718
Danone Institute Belgium Research in Human Nutrition Grants, 1719
EMBO Installation Grants, 1915
Fluor Foundation Grants, 2028
Fulbright Business Grants in Spain, 2067
Fulbright Traditional Scholar Program in Europe and Eurasia, 2080
HHMI Biomedical Research Grants for International Scientists in the Baltics, Central and Eastern Europe, Russia, and Ukraine, 2253
IBRO Europe Regional Grants for Western European Workshops and Symposia, 2324
IBRO InEurope Short Stay Grants, 2325
Oak Foundation Environmental Grants, 3731
OSF Human Rights & Governance Interns, 3845
Pfizer ASPIRE EU Antifungal Res Awards, 3910
Pfizer ASPIRE EU Dupuytren's Contracture Research Awards, 3911
Pfizer ASPIRE EU Emerging Mechanisms of Resistance Antibacterial Research Awards, 3912
Pfizer ASPIRE EU MRSA Nosocomial Pneumonia & MRSA Complicated Skin & Soft Tissue Infections Antibacterial Research Awards, 3913
RAND Corporation Transatlantic Postdoc Fellowships for Int'l Relations & Security, 4032
Simone and Cino del Duca Grand Prix Awards, 4261
State Scholarships Foundation Grants, 4449
United States Institute of Peace - Transatlantic Post-Doc Fellowship for International Relations and Security, 4581

Belize
Academy of American Franciscan History Dissertation Fellowships, 249
AOCS Ralph H. Potts Memorial Fellowship, 893
ASM International Fellowship for Latin America and the Caribbean, 1175
Fulbright Traditional Scholar Program in the Western Hemisphere, 2085
IBRO Latin America Regional Funding for Neuroscience Schools, 2327
IBRO Latin America Regional Funding for PROLAB Collaborations, 2328
IBRO Latin America Regional Travel Grants, 2329
Oak Foundation Environmental Grants, 3731
OTS Hovore-Horn Fellowships, 3856
SfN Ricardo Miledi Neuroscience Training, 4212
Wellcome Trust New Investigator Awards, 4844

Benin
Abdus Salam ITCP Postgraduate Diploma Programme in General Physics, 241
Beatrice Laing Trust Grants, 1324
Fulbright Critical Lang Enhancement Awards, 2068
Fulbright Traditional Scholar Program in Sub-Saharan Africa, 2081
Fulbright Traditional Scholar Program in the Near East and North Africa Region, 2083
SSRC African Peacebuilding Network Research Grants, 4422
State Scholarships Foundation Grants, 4449
STTI Alpha Eta Collaborative Research Grants, 4458
Wellcome Trust New Investigator Awards, 4844
Wenner-Gren Foundation African Fellowships, 4846

Bermuda
AOCS Ralph H. Potts Memorial Fellowship, 893

Bhutan
Fulbright Traditional Scholar Program in the East Asia/Pacific Region, 2082
IBRO Asia Reg APRC Exchange Fellowships, 2323
Sir Dorabji Tata Trust Individual Education and Travel Grants, 4265
Wellcome Trust New Investigator Awards, 4844

Bolivia
Academy of American Franciscan History Dissertation Fellowships, 249
ASM-PAHO Infectious Diseases Epidemiology and Surveillance Fellowships, 1098
ASM International Fellowship for Latin America and the Caribbean, 1175
Fulbright Traditional Scholar Program in the Western Hemisphere, 2085
IBRO Latin America Regional Funding for Neuroscience Schools, 2327
IBRO Latin America Regional Funding for PROLAB Collaborations, 2328
IBRO Latin America Regional Travel Grants, 2329
John Simon Guggenheim Memorial Foundation Fellowships, 2507
OTS Hovore-Horn Fellowships, 3856
SfN Ricardo Miledi Neuroscience Training, 4212
SSRC Drugs, Security and Democracy Fellows, 4426
Wellcome Trust New Investigator Awards, 4844

Bosnia & Herzegovina
ACLS Early Career Postdoctoral Fellowships in East European Studies, 286
Beatrice Laing Trust Grants, 1324
Canon Foundation Grants, 1423
EMBO Installation Grants, 1915
Fluor Foundation Grants, 2028
Fulbright Business Grants in Spain, 2067
Fulbright Traditional Scholar Program in Europe and Eurasia, 2080
HHMI Biomedical Research Grants for International Scientists in the Baltics, Central and Eastern Europe, Russia, and Ukraine, 2253
IBRO InEurope Short Stay Grants, 2325
Oak Foundation Environmental Grants, 3731
OSF Access to Medicines Fellowships in Eastern Europe and Central Asia, 3834
OSF Global Supplementary Grants, 3843
OSF HESP Mobility Grants, 3844
RAND Corporation Transatlantic Postdoc Fellowships for Int'l Relations & Security, 4032
Simone and Cino del Duca Grand Prix Awards, 4261
State Scholarships Foundation Grants, 4449
United States Institute of Peace - Transatlantic Post-Doc Fellowship for International Relations and Security, 4581
Wellcome Trust New Investigator Awards, 4844

Botswana
Abdus Salam ITCP Postgraduate Diploma Programme in General Physics, 241
Beatrice Laing Trust Grants, 1324
Fulbright Critical Lang Enhancement Awards, 2068
Fulbright Traditional Scholar Program in Sub-Saharan Africa, 2081
Fulbright Traditional Scholar Program in the Near East and North Africa Region, 2083
SSRC African Peacebuilding Network Research Grants, 4422
State Scholarships Foundation Grants, 4449
STTI Alpha Eta Collaborative Research Grants, 4458
Wellcome Trust New Investigator Awards, 4844
Wenner-Gren Foundation African Fellowships, 4846

Brazil
Abbott Fund Science Education Grants, 220
Academy of American Franciscan History Dissertation Fellowships, 249
A Charitable Foundation Grants, 273
Alexander von Humboldt Foundation CAPES-Humboldt Fellowships for Experienced Researchers, 516
Alexander von Humboldt Foundation CAPES-Humboldt Fellowships for Postdoctoral Researchers, 517
ASM-PAHO Infectious Diseases Epidemiology and Surveillance Fellowships, 1098
ASM International Fellowship for Latin America and the Caribbean, 1175
Fulbright Traditional Scholar Program in the Western Hemisphere, 2085
Gruber Foundation Int'l Research Award, 2180
IBRO Latin America Regional Funding for Neuroscience Schools, 2327
IBRO Latin America Regional Funding for PROLAB Collaborations, 2328
John Deere Foundation Grants, 2500
John Simon Guggenheim Memorial Foundation Fellowships, 2507
OTS Hovore-Horn Fellowships, 3856
SfN Chapters Graduate Student Travel Awards, 4202
SfN Chapters Postdoc Trainee Travel Awards, 4203
SfN Donald B. Lindsley Prize in Behavioral Neuroscience, 4204
SfN Graduate Student Travel Awards, 4205
SfN Jacob P. Waletzky Award, 4206
SfN Janett Rosenberg Trubatch Career Development Award, 4207
SfN Louise Hanson Marshall Specific Recognition Award, 4208
SfN M Salpeter Lifetime Achievement Award, 4209
SfN Postdoctoral Fellow Travel Awards, 4210
SfN Ralph W. Gerard Prize in Neuroscience, 4211
SfN Ricardo Miledi Neuroscience Training, 4212
SfN Swartz Prize for Theoretical and Computational Neuroscience, 4213
SfN Undergraduate Student Travel Awards, 4217
SfN Young Investigator Award, 4218
SSRC Drugs, Security and Democracy Fellows, 4426
Wellcome Trust New Investigator Awards, 4844

British Indian Ocean Territory
Fulbright Traditional Scholar Program in the East Asia/Pacific Region, 2082
Sir Dorabji Tata Trust Individual Education and Travel Grants, 4265

British Virgin Islands
AOCS Ralph H. Potts Memorial Fellowship, 893
John Simon Guggenheim Memorial Foundation Fellowships, 2507
SSRC Drugs, Security and Democracy Fellows, 4426

Brunei
Fulbright Traditional Scholar Program in the East Asia/Pacific Region, 2082
IBRO Asia Reg APRC Exchange Fellowships, 2323
Sir Dorabji Tata Trust Individual Education and Travel Grants, 4265

Bulgaria
ACLS Early Career Postdoctoral Fellowships in East European Studies, 286
AIAR Andrew W. Mellon Found Fellowships, 392
Beatrice Laing Trust Grants, 1324
Canon Foundation Grants, 1423
EMBO Installation Grants, 1915
Fluor Foundation Grants, 2028
Fulbright Business Grants in Spain, 2067
Fulbright Traditional Scholar Program in Europe and Eurasia, 2080
GMFUS Black Sea for Regional Corp Grants, 2152
HHMI Biomedical Research Grants for International Scientists in the Baltics, Central and Eastern Europe, Russia, and Ukraine, 2253
IBRO InEurope Short Stay Grants, 2325
Oak Foundation Environmental Grants, 3731
OSF Access to Medicines Fellowships in Eastern Europe and Central Asia, 3834
OSF HESP Mobility Grants, 3844
OSF Human Rights & Governance Interns, 3845
OSF Think Tank Fellowships, 3852
RAND Corporation Transatlantic Postdoc Fellowships for Int'l Relations & Security, 4032
Simone and Cino del Duca Grand Prix Awards, 4261
State Scholarships Foundation Grants, 4449
United States Institute of Peace Transatlantic Postdoc Fellowship for Int'l Relations and Security, 4581
Wellcome Trust New Investigator Awards, 4844

Burkina Faso
Abdus Salam ITCP Postgraduate Diploma Programme in General Physics, 241
Beatrice Laing Trust Grants, 1324
Fulbright Critical Lang Enhancement Awards, 2068
Fulbright Traditional Scholar Program in Sub-Saharan Africa, 2081
Fulbright Traditional Scholar Program in the Near East and North Africa Region, 2083
SSRC African Peacebuilding Network Research Grants, 4422
State Scholarships Foundation Grants, 4449
STTI Alpha Eta Collaborative Research Grants, 4458
Wellcome Trust New Investigator Awards, 4844
Wenner-Gren Foundation African Fellowships, 4846

Burma (Myanmar)
Fulbright Traditional Scholar Program in the East Asia/Pacific Region, 2082
Sir Dorabji Tata Trust Individual Education and Travel Grants, 4265

Burundi
Abdus Salam ITCP Postgraduate Diploma Programme in General Physics, 241
Beatrice Laing Trust Grants, 1324
Fulbright Critical Lang Enhancement Awards, 2068
Fulbright Traditional Scholar Program in Sub-Saharan Africa, 2081
Fulbright Traditional Scholar Program in the Near East and North Africa Region, 2083
SSRC African Peacebuilding Network Research Grants, 4422
State Scholarships Foundation Grants, 4449
STTI Alpha Eta Collaborative Research Grants, 4458
USDA Norman E. Borlaug International Agricultural Science and Technology Fellowships, 4636
Wellcome Trust New Investigator Awards, 4844
Wenner-Gren Foundation African Fellowships, 4846

Cambodia
Fulbright Traditional Scholar Program in the East Asia/Pacific Region, 2082
IBRO Asia Reg APRC Exchange Fellowships, 2323
Sir Dorabji Tata Trust Individual Education and Travel Grants, 4265
Wellcome Trust New Investigator Awards, 4844

Cameroon
Abdus Salam ITCP Postgraduate Diploma Programme in General Physics, 241
Beatrice Laing Trust Grants, 1324
Fulbright Critical Lang Enhancement Awards, 2068
Fulbright Traditional Scholar Program in Sub-Saharan Africa, 2081
Fulbright Traditional Scholar Program in the Near East and North Africa Region, 2083
SSRC African Peacebuilding Network Research Grants, 4422
State Scholarships Foundation Grants, 4449
STTI Alpha Eta Collaborative Research Grants, 4458
Wellcome Trust New Investigator Awards, 4844
Wenner-Gren Foundation African Fellowships, 4846
WWF Russell E. Train Fellowships, 4898

Cape Verde
Abdus Salam ITCP Postgraduate Diploma Programme in General Physics, 241
Beatrice Laing Trust Grants, 1324
Fulbright Critical Lang Enhancement Awards, 2068
Fulbright Traditional Scholar Program in Sub-Saharan Africa, 2081
Fulbright Traditional Scholar Program in the Near East and North Africa Region, 2083
SSRC African Peacebuilding Network Research Grants, 4422
State Scholarships Foundation Grants, 4449
STTI Alpha Eta Collaborative Research Grants, 4458
Wellcome Trust New Investigator Awards, 4844
Wenner-Gren Foundation African Fellowships, 4846

Caribbean
Fluor Foundation Grants, 2028
OTS Hovore-Horn Fellowships, 3856

Cayman Islands
AOCS Ralph H. Potts Memorial Fellowship, 893
John Simon Guggenheim Memorial Foundation Fellowships, 2507
SfN Ricardo Miledi Neuroscience Training, 4212
SSRC Drugs, Security and Democracy Fellows, 4426

Central African Republic
Abdus Salam ITCP Postgraduate Diploma Programme in General Physics, 241
Beatrice Laing Trust Grants, 1324
Fulbright Critical Lang Enhancement Awards, 2068
Fulbright Traditional Scholar Program in Sub-Saharan Africa, 2081
Fulbright Traditional Scholar Program in the Near East and North Africa Region, 2083
SSRC African Peacebuilding Network Research Grants, 4422
State Scholarships Foundation Grants, 4449
STTI Alpha Eta Collaborative Research Grants, 4458
Wellcome Trust New Investigator Awards, 4844
Wenner-Gren Foundation African Fellowships, 4846
WWF Russell E. Train Fellowships, 4898

Chad
Abdus Salam ITCP Postgraduate Diploma Programme in General Physics, 241
Beatrice Laing Trust Grants, 1324
Fulbright Critical Lang Enhancement Awards, 2068
Fulbright Traditional Scholar Program in Sub-Saharan Africa, 2081
Fulbright Traditional Scholar Program in the Near East and North Africa Region, 2083
SSRC African Peacebuilding Network Research Grants, 4422
State Scholarships Foundation Grants, 4449
STTI Alpha Eta Collaborative Research Grants, 4458
Wellcome Trust New Investigator Awards, 4844
Wenner-Gren Foundation African Fellowships, 4846

Chile
Academy of American Franciscan History Dissertation Fellowships, 249
ASM-PAHO Infectious Diseases Epidemiology and Surveillance Fellowships, 1098
ASM International Fellowship for Latin America and the Caribbean, 1175
Fulbright Traditional Scholar Program in the Western Hemisphere, 2085
IBRO Latin America Regional Funding for Neuroscience Schools, 2327
IBRO Latin America Regional Funding for PROLAB Collaborations, 2328
IBRO Latin America Regional Travel Grants, 2329
John Simon Guggenheim Memorial Foundation Fellowships, 2507
OTS Hovore-Horn Fellowships, 3856
SfN Ricardo Miledi Neuroscience Training, 4212
SSRC Drugs, Security and Democracy Fellows, 4426

China
ACC Hong Kong Arts Program Fellowships, 254
Alexander von Humboldt Foundation German Chancellor Fellowships for Prospective Leaders from the People's Republic of China, 526
BA Small Research Grants, 1304
Cabot Corporation Foundation Grants, 1406
Fluor Foundation Grants, 2028
Fulbright Traditional Scholar Program in the East Asia/Pacific Region, 2082
Gruber Foundation Int'l Research Award, 2180
IBRO Asia Reg APRC Exchange Fellowships, 2323
IIE General Electric Found TECH Awards, 2352
SfN Chapters Graduate Student Travel Awards, 4202
SfN Chapters Postdoc Trainee Travel Awards, 4203
SfN Donald B. Lindsley Prize in Behavioral Neuroscience, 4204
SfN Graduate Student Travel Awards, 4205
SfN Janett Rosenberg Trubatch Career Development Award, 4207
SfN Louise Hanson Marshall Specific Recognition Award, 4208
SfN M Salpeter Lifetime Achievement Award, 4209
SfN Postdoctoral Fellow Travel Awards, 4210
SfN Ralph W. Gerard Prize in Neuroscience, 4211
SfN Swartz Prize for Theoretical and Computational Neuroscience, 4213
SfN Undergraduate Student Travel Awards, 4217
SfN Young Investigator Award, 4218
Sir Dorabji Tata Trust Individual Education and Travel Grants, 4265
SSRC Collaborative Research Grants on Environment and Health in China, 4423
Wellcome Trust New Investigator Awards, 4844

Christmas Island
Fulbright Traditional Scholar Program in the East Asia/Pacific Region, 2082
Sir Dorabji Tata Trust Individual Education and Travel Grants, 4265

Cocos
Fulbright Traditional Scholar Program in the East Asia/Pacific Region, 2082
Sir Dorabji Tata Trust Individual Education and Travel Grants, 4265

Colombia
Academy of American Franciscan History Dissertation Fellowships, 249
ASM-PAHO Infectious Diseases Epidemiology and Surveillance Fellowships, 1098
ASM International Fellowship for Latin America and the Caribbean, 1175
Fulbright Traditional Scholar Program in the Western Hemisphere, 2085
IBRO Latin America Regional Funding for Neuroscience Schools, 2327
IBRO Latin America Regional Funding for PROLAB Collaborations, 2328
IBRO Latin America Regional Travel Grants, 2329
John Simon Guggenheim Memorial Foundation Fellowships, 2507
OTS Hovore-Horn Fellowships, 3856
SfN Ricardo Miledi Neuroscience Training, 4212
SSRC Drugs, Security and Democracy Fellows, 4426
Wellcome Trust New Investigator Awards, 4844

Comoros
Abdus Salam ITCP Postgraduate Diploma Programme in General Physics, 241
Beatrice Laing Trust Grants, 1324
Fulbright Critical Lang Enhancement Awards, 2068
Fulbright Traditional Scholar Program in Sub-Saharan Africa, 2081
Fulbright Traditional Scholar Program in the Near East and North Africa Region, 2083
SSRC African Peacebuilding Network Research Grants, 4422
State Scholarships Foundation Grants, 4449
STTI Alpha Eta Collaborative Research Grants, 4458
USDA Norman E. Borlaug International Agricultural Science and Technology Fellowships, 4636
Wellcome Trust New Investigator Awards, 4844
Wenner-Gren Foundation African Fellowships, 4846

Congo
Abdus Salam ITCP Postgraduate Diploma Programme in General Physics, 241
Beatrice Laing Trust Grants, 1324
Fulbright Critical Lang Enhancement Awards, 2068
Fulbright Traditional Scholar Program in Sub-Saharan Africa, 2081
Fulbright Traditional Scholar Program in the Near East and North Africa Region, 2083
State Scholarships Foundation Grants, 4449
STTI Alpha Eta Collaborative Research Grants, 4458
Wellcome Trust New Investigator Awards, 4844
Wenner-Gren Foundation African Fellowships, 4846
WWF Russell E. Train Fellowships, 4898

Congo, Democratic Republic of
Abdus Salam ITCP Postgraduate Diploma Programme in General Physics, 241
Beatrice Laing Trust Grants, 1324
Fulbright Critical Lang Enhancement Awards, 2068
Fulbright Traditional Scholar Program in Sub-Saharan Africa, 2081
Fulbright Traditional Scholar Program in the Near East and North Africa Region, 2083
SSRC African Peacebuilding Network Research Grants, 4422
State Scholarships Foundation Grants, 4449
STTI Alpha Eta Collaborative Research Grants, 4458
Wellcome Trust New Investigator Awards, 4844
Wenner-Gren Foundation African Fellowships, 4846
WWF Russell E. Train Fellowships, 4898

Costa Rica
Academy of American Franciscan History Dissertation Fellowships, 249
AOCS Ralph H. Potts Memorial Fellowship, 893
ASM International Fellowship for Latin America and the Caribbean, 1175
Fulbright Traditional Scholar Program in the Western Hemisphere, 2085
IBRO Latin America Regional Funding for Neuroscience Schools, 2327
IBRO Latin America Regional Funding for PROLAB Collaborations, 2328
IBRO Latin America Regional Travel Grants, 2329
John Simon Guggenheim Memorial Foundation Fellowships, 2507
OTS Hovore-Horn Fellowships, 3856
SfN Ricardo Miledi Neuroscience Training, 4212
SSRC Drugs, Security and Democracy Fellows, 4426
Wellcome Trust New Investigator Awards, 4844

Cote d' Ivoire (Ivory Coast)
Abdus Salam ITCP Postgraduate Diploma Programme in General Physics, 241
Beatrice Laing Trust Grants, 1324
Fulbright Critical Lang Enhancement Awards, 2068
Fulbright Traditional Scholar Program in Sub-Saharan Africa, 2081
Fulbright Traditional Scholar Program in the Near East and North Africa Region, 2083
SSRC African Peacebuilding Network Research Grants, 4422
State Scholarships Foundation Grants, 4449
STTI Alpha Eta Collaborative Research Grants, 4458
Wellcome Trust New Investigator Awards, 4844
Wenner-Gren Foundation African Fellowships, 4846

Croatia
ACLS Early Career Postdoctoral Fellowships in East European Studies, 286
Beatrice Laing Trust Grants, 1324
Canon Foundation Grants, 1423
EMBO Installation Grants, 1915
Fluor Foundation Grants, 2028
Fulbright Business Grants in Spain, 2067
Fulbright Traditional Scholar Program in Europe and Eurasia, 2080
HHMI Biomedical Research Grants for International Scientists in the Baltics, Central and Eastern Europe, Russia, and Ukraine, 2253
IBRO InEurope Short Stay Grants, 2325
Oak Foundation Environmental Grants, 3731
OSF Access to Medicines Fellowships in Eastern Europe and Central Asia, 3834
OSF Global Supplementary Grants, 3843
RAND Corporation Transatlantic Postdoc Fellowships for Int'l Relations & Security, 4032
Simone and Cino del Duca Grand Prix Awards, 4261

State Scholarships Foundation Grants, 4449
United States Institute of Peace - Transatlantic Post-Doc Fellowship for International Relations and Security, 4581

Cuba
AOCS Ralph H. Potts Memorial Fellowship, 893
ASM International Fellowship for Latin America and the Caribbean, 1175
Fulbright Traditional Scholar Program in the Western Hemisphere, 2085
IBRO Latin America Regional Funding for Neuroscience Schools, 2327
IBRO Latin America Regional Funding for PROLAB Collaborations, 2328
IBRO Latin America Regional Travel Grants, 2329
John Simon Guggenheim Memorial Foundation Fellowships, 2507
OTS Hovore-Horn Fellowships, 3856
SfN Ricardo Miledi Neuroscience Training, 4212
SSRC Drugs, Security and Democracy Fellows, 4426
Wellcome Trust New Investigator Awards, 4844

Cyprus
Beatrice Laing Trust Grants, 1324
Canon Foundation Grants, 1423
EMBO Installation Grants, 1915
Fluor Foundation Grants, 2028
Fulbright Business Grants in Spain, 2067
Fulbright Traditional Scholar Program in Europe and Eurasia, 2080
Fulbright Traditional Scholar Program in the East Asia/Pacific Region, 2082
HHMI Biomedical Research Grants for International Scientists in the Baltics, Central and Eastern Europe, Russia, and Ukraine, 2253
IBRO InEurope Short Stay Grants, 2325
Oak Foundation Environmental Grants, 3731
RAND Corporation Transatlantic Postdoc Fellowships for Int'l Relations & Security, 4032
Simone and Cino del Duca Grand Prix Awards, 4261
Sir Dorabji Tata Trust Individual Education and Travel Grants, 4265
State Scholarships Foundation Grants, 4449
United States Institute of Peace - Transatlantic Post-Doc Fellowship for International Relations and Security, 4581

Czech Republic
ACLS Early Career Postdoctoral Fellowships in East European Studies, 286
AIAR Andrew W. Mellon Found Fellowships, 392
ARIT Mellon Foundation Postdoctoral Research Fellowships for East European Scholars, 994
Beatrice Laing Trust Grants, 1324
Canon Foundation Grants, 1423
Central European University Fellowships, 1488
EMBO Installation Grants, 1915
Fluor Foundation Grants, 2028
Fulbright Business Grants in Spain, 2067
Fulbright Traditional Scholar Program in Europe and Eurasia, 2080
HHMI/EMBO Start-up Grants for C Europe, 2250
HHMI Biomedical Research Grants for International Scientists in the Baltics, Central and Eastern Europe, Russia, and Ukraine, 2253
IBRO InEurope Short Stay Grants, 2325
Oak Foundation Environmental Grants, 3731
OSF Access to Medicines Fellowships in Eastern Europe and Central Asia, 3834
OSF HESP Mobility Grants, 3844
OSF Human Rights & Governance Interns, 3845
OSF Think Tank Fellowships, 3852
RAND Corporation Transatlantic Postdoc Fellowships for Int'l Relations & Security, 4032
Simone and Cino del Duca Grand Prix Awards, 4261
State Scholarships Foundation Grants, 4449
Swiss National Science Foundation DORE (DO REsearch) Conference Grants, 4482
Swiss National Science Foundation Fellowships for Advanced Researchers, 4483
Swiss National Science Foundation Fellowships for Prospective Researchers, 4484
Swiss National Science Foundation Marie Heim-Vogtlin (MHV) Grants, 4485
Swiss National Science Foundation Scientific Conferences Grants, 4486
United States Institute of Peace - Transatlantic Post-Doc Fellowship for International Relations and Security, 4581

Denmark
Beatrice Laing Trust Grants, 1324
Canon Foundation Grants, 1423
EMBO Installation Grants, 1915
Fluor Foundation Grants, 2028
Fulbright Business Grants in Spain, 2067
Fulbright Traditional Scholar Program in Europe and Eurasia, 2080
Gruber Foundation Int'l Research Award, 2180
HHMI Biomedical Research Grants for International Scientists in the Baltics, Central and Eastern Europe, Russia, and Ukraine, 2253
Holberg Memorial Fund Nils Klim Prize, 2273
IBRO Europe Regional Grants for Western European Workshops and Symposia, 2324
IBRO InEurope Short Stay Grants, 2325
Oak Foundation Environmental Grants, 3731
Pfizer ASPIRE EU Antifungal Res Awards, 3910
Pfizer ASPIRE EU Dupuytren's Contracture Research Awards, 3911
Pfizer ASPIRE EU Emerging Mechanisms of Resistance Antibacterial Research Awards, 3912
Pfizer ASPIRE EU MRSA Nosocomial Pneumonia & MRSA Complicated Skin & Soft Tissue Infections Antibacterial Research Awards, 3913
RAND Corporation Transatlantic Postdoc Fellowships for Int'l Relations & Security, 4032
SASS Einar and Eva Lund Haugen Scholarship, 4183
SfN Chapters Graduate Student Travel Awards, 4202
SfN Chapters Postdoc Trainee Travel Awards, 4203
SfN Donald B. Lindsley Prize in Behavioral Neuroscience, 4204
SfN Graduate Student Travel Awards, 4205
SfN Jacob P. Waletzky Award, 4206
SfN Janett Rosenberg Trubatch Career Development Award, 4207
SfN Louise Hanson Marshall Specific Recognition Award, 4208
SfN M Salpeter Lifetime Achievement Award, 4209
SfN Postdoctoral Fellow Travel Awards, 4210
SfN Ralph W. Gerard Prize in Neuroscience, 4211
SfN Swartz Prize for Theoretical and Computational Neuroscience, 4213
SfN Undergraduate Student Travel Awards, 4217
SfN Young Investigator Award, 4218
Simone and Cino del Duca Grand Prix Awards, 4261
State Scholarships Foundation Grants, 4449
United States Institute of Peace - Transatlantic Post-Doc Fellowship for International Relations and Security, 4581

Djibouti
Abdus Salam ITCP Postgraduate Diploma Programme in General Physics, 241
Beatrice Laing Trust Grants, 1324
Fulbright Critical Lang Enhancement Awards, 2068
Fulbright Traditional Scholar Program in Sub-Saharan Africa, 2081
Fulbright Traditional Scholar Program in the Near East and North Africa Region, 2083
SSRC African Peacebuilding Network Research Grants, 4422
State Scholarships Foundation Grants, 4449
STTI Alpha Eta Collaborative Research Grants, 4458
USDA Norman E. Borlaug International Agricultural Science and Technology Fellowships, 4636
Wellcome Trust New Investigator Awards, 4844
Wenner-Gren Foundation African Fellowships, 4846

Dominica
AOCS Ralph H. Potts Memorial Fellowship, 893
Fulbright Traditional Scholar Program in the Western Hemisphere, 2085
IBRO Latin America Regional Funding for Neuroscience Schools, 2327
IBRO Latin America Regional Funding for PROLAB Collaborations, 2328
IBRO Latin America Regional Travel Grants, 2329
John Simon Guggenheim Memorial Foundation Fellowships, 2507
SfN Ricardo Miledi Neuroscience Training, 4212
Wellcome Trust New Investigator Awards, 4844

Dominican Republic
AOCS Ralph H. Potts Memorial Fellowship, 893
Fulbright Traditional Scholar Program in the Western Hemisphere, 2085
IBRO Latin America Regional Funding for Neuroscience Schools, 2327
IBRO Latin America Regional Funding for PROLAB Collaborations, 2328
IBRO Latin America Regional Travel Grants, 2329
John Simon Guggenheim Memorial Foundation Fellowships, 2507
OTS Hovore-Horn Fellowships, 3856
SfN Ricardo Miledi Neuroscience Training, 4212
SSRC Drugs, Security and Democracy Fellows, 4426
Wellcome Trust New Investigator Awards, 4844

East Timor
IBRO Asia Reg APRC Exchange Fellowships, 2323

Ecuador
Academy of American Franciscan History Dissertation Fellowships, 249
ASM-PAHO Infectious Diseases Epidemiology and Surveillance Fellowships, 1098
ASM International Fellowship for Latin America and the Caribbean, 1175
Fulbright Traditional Scholar Program in the Western Hemisphere, 2085
IBRO Latin America Regional Funding for Neuroscience Schools, 2327
IBRO Latin America Regional Funding for PROLAB Collaborations, 2328
IBRO Latin America Regional Travel Grants, 2329
John Simon Guggenheim Memorial Foundation Fellowships, 2507
OTS Hovore-Horn Fellowships, 3856
SfN Ricardo Miledi Neuroscience Training, 4212
SSRC Drugs, Security and Democracy Fellows, 4426
Wellcome Trust New Investigator Awards, 4844

Egypt
Beatrice Laing Trust Grants, 1324
Fulbright Critical Lang Enhancement Awards, 2068
Fulbright Traditional Scholar Program in Sub-Saharan Africa, 2081
Fulbright Traditional Scholar Program in the Near East and North Africa Region, 2083
OSF Global Faculty Grants - Track 2, 3842
OSF Global Supplementary Grants, 3843
OSF Palestinian Rule of Law Fellowships, 3848
State Scholarships Foundation Grants, 4449
STTI Alpha Eta Collaborative Research Grants, 4458
Wellcome Trust New Investigator Awards, 4844
Wenner-Gren Foundation African Fellowships, 4846

El Salvador
Academy of American Franciscan History Dissertation Fellowships, 249
AOCS Ralph H. Potts Memorial Fellowship, 893
Fulbright Traditional Scholar Program in the Western Hemisphere, 2085
IBRO Latin America Regional Funding for Neuroscience Schools, 2327
IBRO Latin America Regional Funding for PROLAB Collaborations, 2328
IBRO Latin America Regional Travel Grants, 2329

John Simon Guggenheim Memorial Foundation Fellowships, 2507
OTS Hovore-Horn Fellowships, 3856
SfN Ricardo Miledi Neuroscience Training, 4212
SSRC Drugs, Security and Democracy Fellows, 4426
Wellcome Trust New Investigator Awards, 4844

Equatorial Guinea
Abdus Salam ITCP Postgraduate Diploma Programme in General Physics, 241
Beatrice Laing Trust Grants, 1324
Fulbright Critical Lang Enhancement Awards, 2068
Fulbright Traditional Scholar Program in Sub-Saharan Africa, 2081
Fulbright Traditional Scholar Program in the Near East and North Africa Region, 2083
SSRC African Peacebuilding Network Research Grants, 4422
State Scholarships Foundation Grants, 4449
STTI Alpha Eta Collaborative Research Grants, 4458
Wenner-Gren Foundation African Fellowships, 4846

Eritrea
Abdus Salam ITCP Postgraduate Diploma Programme in General Physics, 241
Beatrice Laing Trust Grants, 1324
Fulbright Critical Lang Enhancement Awards, 2068
Fulbright Traditional Scholar Program in Sub-Saharan Africa, 2081
Fulbright Traditional Scholar Program in the Near East and North Africa Region, 2083
SSRC African Peacebuilding Network Research Grants, 4422
State Scholarships Foundation Grants, 4449
STTI Alpha Eta Collaborative Research Grants, 4458
USDA Norman E. Borlaug International Agricultural Science and Technology Fellowships, 4636
Wellcome Trust New Investigator Awards, 4844
Wenner-Gren Foundation African Fellowships, 4846

Estonia
ACLS Early Career Postdoctoral Fellowships in East European Studies, 286
AIAR Andrew W. Mellon Found Fellowships, 392
Beatrice Laing Trust Grants, 1324
Canon Foundation Grants, 1423
EMBO Installation Grants, 1915
Fluor Foundation Grants, 2028
Fulbright Business Grants in Spain, 2067
Fulbright Traditional Scholar Program in Europe and Eurasia, 2080
HHMI Biomedical Research Grants for International Scientists in the Baltics, Central and Eastern Europe, Russia, and Ukraine, 2253
IBRO InEurope Short Stay Grants, 2325
John Deere Foundation Grants, 2500
Oak Foundation Environmental Grants, 3731
OSF Access to Medicines Fellowships in Eastern Europe and Central Asia, 3834
OSF Human Rights & Governance Interns, 3845
OSF Think Tank Fellowships, 3852
RAND Corporation Transatlantic Postdoc Fellowships for Int'l Relations & Security, 4032
Simone and Cino del Duca Grand Prix Awards, 4261
State Scholarships Foundation Grants, 4449
United States Institute of Peace - Transatlantic Post-Doc Fellowship for International Relations and Security, 4581
US CRDF CGP ETF 2009 Energy Research Competition For Joint U.S.-Estonian Research Teams, 4609

Ethiopia
Abdus Salam ITCP Postgraduate Diploma Programme in General Physics, 241
Beatrice Laing Trust Grants, 1324
Fulbright Critical Lang Enhancement Awards, 2068
Fulbright Traditional Scholar Program in Sub-Saharan Africa, 2081
Fulbright Traditional Scholar Program in the Near East and North Africa Region, 2083
SSRC African Peacebuilding Network Research Grants, 4422
State Scholarships Foundation Grants, 4449
STTI Alpha Eta Collaborative Research Grants, 4458
USDA Norman E. Borlaug International Agricultural Science and Technology Fellowships, 4636
Wellcome Trust New Investigator Awards, 4844
Wenner-Gren Foundation African Fellowships, 4846

Falkland Islands (Malvinas)
Academy of American Franciscan History Dissertation Fellowships, 249

Fiji
Wellcome Trust New Investigator Awards, 4844

Finland
Beatrice Laing Trust Grants, 1324
Canon Foundation Grants, 1423
EMBO Installation Grants, 1915
Fluor Foundation Grants, 2028
Fulbright Business Grants in Spain, 2067
Fulbright Traditional Scholar Program in Europe and Eurasia, 2080
HHMI Biomedical Research Grants for International Scientists in the Baltics, Central and Eastern Europe, Russia, and Ukraine, 2253
Holberg Memorial Fund Nils Klim Prize, 2273
IBRO Europe Regional Grants for Western European Workshops and Symposia, 2324
IBRO InEurope Short Stay Grants, 2325
Oak Foundation Environmental Grants, 3731
Pfizer ASPIRE EU Antifungal Res Awards, 3910
Pfizer ASPIRE EU Dupuytren's Contracture Research Awards, 3911
Pfizer ASPIRE EU Emerging Mechanisms of Resistance Antibacterial Research Awards, 3912
Pfizer ASPIRE EU MRSA Nosocomial Pneumonia & MRSA Complicated Skin & Soft Tissue Infections Antibacterial Research Awards, 3913
RAND Corporation Transatlantic Postdoc Fellowships for Int'l Relations & Security, 4032
SASS Einar and Eva Lund Haugen Memorial Scholarship, 4183
Simone and Cino del Duca Grand Prix Awards, 4261
State Scholarships Foundation Grants, 4449
United States Institute of Peace - Transatlantic Post-Doc Fellowship for International Relations and Security, 4581

France
AAF Richard Morris Hunt Architecture Fellowship, 169
Beatrice Laing Trust Grants, 1324
Canon Foundation Grants, 1423
EMBO Installation Grants, 1915
Fluor Foundation Grants, 2028
Fulbright Business Grants in Spain, 2067
Fulbright Traditional Scholar Program in Europe and Eurasia, 2080
HHMI Biomedical Research Grants for International Scientists in the Baltics, Central and Eastern Europe, Russia, and Ukraine, 2253
IBRO Europe Regional Grants for Western European Workshops and Symposia, 2324
IBRO InEurope Short Stay Grants, 2325
IIE Leonora Lindsley Memorial Fellowships, 2354
Oak Foundation Environmental Grants, 3731
Pfizer ASPIRE EU Antifungal Res Awards, 3910
Pfizer ASPIRE EU Dupuytren's Contracture Research Awards, 3911
Pfizer ASPIRE EU Emerging Mechanisms of Resistance Antibacterial Research Awards, 3912
Pfizer ASPIRE EU MRSA Nosocomial Pneumonia & MRSA Complicated Skin & Soft Tissue Infections Antibacterial Research Awards, 3913
RAND Corporation Transatlantic Postdoc Fellowships for Int'l Relations & Security, 4032
Simone and Cino del Duca Grand Prix Awards, 4261
State Scholarships Foundation Grants, 4449
United States Institute of Peace - Transatlantic Post-Doc Fellowship for International Relations and Security, 4581

French Guiana
Academy of American Franciscan History Dissertation Fellowships, 249
John Simon Guggenheim Memorial Foundation Fellowships, 2507
SfN Ricardo Miledi Neuroscience Training, 4212

Gabon
Abdus Salam ITCP Postgraduate Diploma Programme in General Physics, 241
Beatrice Laing Trust Grants, 1324
Fulbright Critical Lang Enhancement Awards, 2068
Fulbright Traditional Scholar Program in Sub-Saharan Africa, 2081
Fulbright Traditional Scholar Program in the Near East and North Africa Region, 2083
SSRC African Peacebuilding Network Research Grants, 4422
State Scholarships Foundation Grants, 4449
STTI Alpha Eta Collaborative Research Grants, 4458
Wellcome Trust New Investigator Awards, 4844
Wenner-Gren Foundation African Fellowships, 4846
WWF Russell E. Train Fellowships, 4898

Gambia
Abdus Salam ITCP Postgraduate Diploma Programme in General Physics, 241
Beatrice Laing Trust Grants, 1324
Fulbright Critical Lang Enhancement Awards, 2068
Fulbright Traditional Scholar Program in Sub-Saharan Africa, 2081
Fulbright Traditional Scholar Program in the Near East and North Africa Region, 2083
SSRC African Peacebuilding Network Research Grants, 4422
State Scholarships Foundation Grants, 4449
STTI Alpha Eta Collaborative Research Grants, 4458
Wellcome Trust New Investigator Awards, 4844
Wenner-Gren Foundation African Fellowships, 4846

Georgia
Beatrice Laing Trust Grants, 1324
Canon Foundation Grants, 1423
EMBO Installation Grants, 1915
Fluor Foundation Grants, 2028
Fulbright Business Grants in Spain, 2067
Fulbright Traditional Scholar Program in Europe and Eurasia, 2080
Fulbright Traditional Scholar Program in the East Asia/Pacific Region, 2082
HHMI Biomedical Research Grants for International Scientists in the Baltics, Central and Eastern Europe, Russia, and Ukraine, 2253
RAND Corporation Transatlantic Postdoc Fellowships for Int'l Relations & Security, 4032
Simone and Cino del Duca Grand Prix Awards, 4261
Sir Dorabji Tata Trust Individual Education and Travel Grants, 4265
State Scholarships Foundation Grants, 4449

Germany
Abbott Fund Science Education Grants, 220
Alexander von Humboldt Feodor-Lynen Research Fellowships for Experienced Researchers, 518
Alexander von Humboldt Feodor-Lynen Research Fellowships for Postdoctoral Researchers, 519
Alexander von Humboldt Feodor Lynen Research Fellowships for Experienced Researchers, 520
Alexander von Humboldt Foundation Fraunhofer-Bessel Research Awards, 521
Alexander von Humboldt Foundation Friedrich Wilhelm Bessel Research Awards, 522
Alexander von Humboldt Foundation JSPS Research Fellowships, 530

Alexander von Humboldt Foundation NSC Taiwan Fellowships, 533
Alexander von Humboldt Found Professorships, 535
Alexander von Humboldt Reimar Lust Award, 536
Alexander von Humboldt Foundation Research Awards for Top Researchers in Germany, 538
Beatrice Laing Trust Grants, 1324
Canon Foundation Grants, 1423
CCA Diefenbaker Award to update, 1447
Central European University Fellowships, 1488
Chiles Foundation Grants, 1552
Daimler and Benz Foundation Fellowships, 1713
EMBO / EMBL Science Writing Prize, 1913
EMBO Award for Communication in the Life Sciences, 1914
EMBO Installation Grants, 1915
EMBO Long-Term Fellowships, 1916
EMBO Short-Term Fellowships, 1917
Fluor Foundation Grants, 2028
Fulbright Business Grants in Spain, 2067
Fulbright German Studies Seminar Grants, 2072
Fulbright Traditional Scholar Program in Europe and Eurasia, 2080
Fulbright Travel Grants, 2086
Gerda Henkel Foundation Grants, 2127
HHMI/EMBO Start-up Grants for C Europe, 2250
HHMI Biomedical Research Grants for International Scientists in the Baltics, Central and Eastern Europe, Russia, and Ukraine, 2253
IBRO Europe Regional Grants for Western European Workshops and Symposia, 2324
IBRO InEurope Short Stay Grants, 2325
Minerva Fellowships, 2778
Minerva Short-Term Research Grants, 2779
Oak Foundation Environmental Grants, 3731
OSF HESP Mobility Grants, 3844
Pfizer ASPIRE EU Antifungal Res Awards, 3910
Pfizer ASPIRE EU Dupuytren's Contracture Research Awards, 3911
Pfizer ASPIRE EU Emerging Mechanisms of Resistance Antibacterial Research Awards, 3912
Pfizer ASPIRE EU MRSA Nosocomial Pneumonia & MRSA Complicated Skin & Soft Tissue Infections Antibacterial Research Awards, 3913
Pfizer Research Initiative Dermatology Grants (Germany), 3928
Pfizer Research Initiative Rheumatology Grants, 3929
RAND Corporation Transatlantic Postdoc Fellowships for Int'l Relations & Security, 4032
Simone and Cino del Duca Grand Prix Awards, 4261
State Scholarships Foundation Grants, 4449
Swiss National Science Foundation DORE (DO REsearch) Conference Grants, 4482
Swiss National Science Foundation Fellowships for Advanced Researchers, 4483
Swiss National Science Foundation Fellowships for Prospective Researchers, 4484
Swiss National Science Foundation Marie Heim-Vogtlin (MHV) Grants, 4485
Swiss National Science Foundation Scientific Conferences Grants, 4486
Textron Corporate Contributions Grants, 4501
United States Institute of Peace - Transatlantic Post-Doc Fellowship for International Relations and Security, 4581
Wilhelm Sander Foundation Grants, 4868

Ghana
Abdus Salam ITCP Postgraduate Diploma Programme in General Physics, 241
ACLS African Humanities Program for Dissertation Completion, 278
ACLS African Humanities Program in Ghana, Nigeria, South Africa, Tanzania, and Uganda, 279
Beatrice Laing Trust Grants, 1324
Fulbright Critical Lang Enhancement Awards, 2068
Fulbright Traditional Scholar Program in Sub-Saharan Africa, 2081
Fulbright Traditional Scholar Program in the Near East and North Africa Region, 2083
SSRC African Peacebuilding Network Research Grants, 4422
SSRC Next Generation Social Sciences in Africa: Doctoral Dissertation Compl Fellowships, 4434
SSRC Next Generation Social Sciences in Africa: Doctoral Dissertation Proposal Fellowships, 4435
SSRC Next Generation Social Sciences in Africa: Doctoral Dissertation Research Fellowships, 4436
State Scholarships Foundation Grants, 4449
STTI Alpha Eta Collaborative Research Grants, 4458
USDA Norman E. Borlaug International Agricultural Science and Technology Fellowships, 4636
Wellcome Trust New Investigator Awards, 4844
Wenner-Gren Foundation African Fellowships, 4846

Great Britain
Abbott Fund Science Education Grants, 220
Textron Corporate Contributions Grants, 4501
USDD National Consortium for Measurement and Signature Intelligence Research Grants, 4759

Greece
Beatrice Laing Trust Grants, 1324
Canon Foundation Grants, 1423
EMBO Installation Grants, 1915
Fluor Foundation Grants, 2028
Fulbright Business Grants in Spain, 2067
Fulbright Traditional Scholar Program in Europe and Eurasia, 2080
HHMI Biomedical Research Grants for International Scientists in the Baltics, Central and Eastern Europe, Russia, and Ukraine, 2253
IBRO Europe Regional Grants for Western European Workshops and Symposia, 2324
IBRO InEurope Short Stay Grants, 2325
Oak Foundation Environmental Grants, 3731
OSF HESP Mobility Grants, 3844
Pfizer ASPIRE EU Antifungal Res Awards, 3910
Pfizer ASPIRE EU Dupuytren's Contracture Research Awards, 3911
Pfizer ASPIRE EU Emerging Mechanisms of Resistance Antibacterial Research Awards, 3912
Pfizer ASPIRE EU MRSA Nosocomial Pneumonia & MRSA Complicated Skin & Soft Tissue Infections Antibacterial Research Awards, 3913
RAND Corporation Transatlantic Postdoc Fellowships for Int'l Relations & Security, 4032
Simone and Cino del Duca Grand Prix Awards, 4261
State Scholarships Foundation Grants, 4449
United States Institute of Peace - Transatlantic Post-Doc Fellowship for International Relations and Security, 4581

Grenada
AOCS Ralph H. Potts Memorial Fellowship, 893
Fulbright Traditional Scholar Program in the Western Hemisphere, 2085
IBRO Latin America Regional Funding for Neuroscience Schools, 2327
IBRO Latin America Regional Funding for PROLAB Collaborations, 2328
IBRO Latin America Regional Travel Grants, 2329
John Simon Guggenheim Memorial Foundation Fellowships, 2507
SfN Ricardo Miledi Neuroscience Training, 4212
SSRC Drugs, Security and Democracy Fellows, 4426
Wellcome Trust New Investigator Awards, 4844

Guadeloupe
John Simon Guggenheim Memorial Foundation Fellowships, 2507
SfN Ricardo Miledi Neuroscience Training, 4212
SSRC Drugs, Security and Democracy Fellows, 4426

Guatemala
Academy of American Franciscan History Dissertation Fellowships, 249
AOCS Ralph H. Potts Memorial Fellowship, 893
Fulbright Traditional Scholar Program in the Western Hemisphere, 2085
IBRO Latin America Regional Funding for Neuroscience Schools, 2327
IBRO Latin America Regional Funding for PROLAB Collaborations, 2328
IBRO Latin America Regional Travel Grants, 2329
John Simon Guggenheim Memorial Foundation Fellowships, 2507
OTS Hovore-Horn Fellowships, 3856
SfN Ricardo Miledi Neuroscience Training, 4212
SSRC Drugs, Security and Democracy Fellows, 4426
USDA Norman E. Borlaug International Agricultural Science and Technology Fellowships, 4636
Wellcome Trust New Investigator Awards, 4844

Guinea
Abdus Salam ITCP Postgraduate Diploma Programme in General Physics, 241
Beatrice Laing Trust Grants, 1324
Fulbright Critical Lang Enhancement Awards, 2068
Fulbright Traditional Scholar Program in Sub-Saharan Africa, 2081
Fulbright Traditional Scholar Program in the Near East and North Africa Region, 2083
SSRC African Peacebuilding Network Research Grants, 4422
State Scholarships Foundation Grants, 4449
STTI Alpha Eta Collaborative Research Grants, 4458
Wellcome Trust New Investigator Awards, 4844
Wenner-Gren Foundation African Fellowships, 4846

Guinea-Bissau
Abdus Salam ITCP Postgraduate Diploma Programme in General Physics, 241
Beatrice Laing Trust Grants, 1324
Fulbright Critical Lang Enhancement Awards, 2068
Fulbright Traditional Scholar Program in Sub-Saharan Africa, 2081
Fulbright Traditional Scholar Program in the Near East and North Africa Region, 2083
SSRC African Peacebuilding Network Research Grants, 4422
State Scholarships Foundation Grants, 4449
STTI Alpha Eta Collaborative Research Grants, 4458
Wellcome Trust New Investigator Awards, 4844
Wenner-Gren Foundation African Fellowships, 4846

Guyana
Academy of American Franciscan History Dissertation Fellowships, 249
ASM-PAHO Infectious Diseases Epidemiology and Surveillance Fellowships, 1098
ASM International Fellowship for Latin America and the Caribbean, 1175
Fulbright Traditional Scholar Program in the Western Hemisphere, 2085
IBRO Latin America Regional Funding for Neuroscience Schools, 2327
IBRO Latin America Regional Funding for PROLAB Collaborations, 2328
IBRO Latin America Regional Travel Grants, 2329
SfN Ricardo Miledi Neuroscience Training, 4212
Wellcome Trust New Investigator Awards, 4844

Haiti
AOCS Ralph H. Potts Memorial Fellowship, 893
Fulbright Traditional Scholar Program in the Western Hemisphere, 2085
IBRO Latin America Regional Funding for Neuroscience Schools, 2327
IBRO Latin America Regional Funding for PROLAB Collaborations, 2328
IBRO Latin America Regional Travel Grants, 2329
John Simon Guggenheim Memorial Foundation Fellowships, 2507
OTS Hovore-Horn Fellowships, 3856
SfN Ricardo Miledi Neuroscience Training, 4212
SSRC Drugs, Security and Democracy Fellows, 4426
USDA Norman E. Borlaug International Agricultural Science and Technology Fellowships, 4636
Wellcome Trust New Investigator Awards, 4844

Honduras
Academy of American Franciscan History Dissertation Fellowships, 249
AOCS Ralph H. Potts Memorial Fellowship, 893
Fulbright Traditional Scholar Program in the Western Hemisphere, 2085
IBRO Latin America Regional Funding for Neuroscience Schools, 2327
IBRO Latin America Regional Funding for PROLAB Collaborations, 2328
IBRO Latin America Regional Travel Grants, 2329
John Simon Guggenheim Memorial Foundation Fellowships, 2507
OTS Hovore-Horn Fellowships, 3856
SfN Ricardo Miledi Neuroscience Training, 4212
SSRC Drugs, Security and Democracy Fellows, 4426
USDA Norman E. Borlaug International Agricultural Science and Technology Fellowships, 4636
Wellcome Trust New Investigator Awards, 4844

Hong Kong
Fulbright Traditional Scholar Program in the East Asia/Pacific Region, 2082
Sir Dorabji Tata Trust Individual Education and Travel Grants, 4265

Hungary
ACLS Early Career Postdoctoral Fellowships in East European Studies, 286
Beatrice Laing Trust Grants, 1324
Canon Foundation Grants, 1423
Central European University Fellowships, 1488
EMBO Installation Grants, 1915
Fluor Foundation Grants, 2028
Fulbright Business Grants in Spain, 2067
Fulbright Traditional Scholar Program in Europe and Eurasia, 2080
Fulbright Travel Grants, 2086
HHMI/EMBO Start-up Grants for C Europe, 2250
HHMI Biomedical Research Grants for International Scientists in the Baltics, Central and Eastern Europe, Russia, and Ukraine, 2253
HHMI Biomedical Research Grants for International Scientists in the Baltics, Central and Eastern Europe, Russia, and Ukraine, 2253
IBRO InEurope Short Stay Grants, 2325
Oak Foundation Environmental Grants, 3731
OSF Access to Medicines Fellowships in Eastern Europe and Central Asia, 3834
OSF HESP Mobility Grants, 3844
OSF Human Rights & Governance Interns, 3845
OSF Think Tank Fellowships, 3852
RAND Corporation Transatlantic Postdoc Fellowships for Int'l Relations & Security, 4032
Simone and Cino del Duca Grand Prix Awards, 4261
State Scholarships Foundation Grants, 4449
Swiss National Science Foundation DORE (DO REsearch) Conference Grants, 4482
Swiss National Science Foundation Fellowships for Advanced Researchers, 4483
Swiss National Science Foundation Fellowships for Prospective Researchers, 4484
Swiss National Science Foundation Marie Heim-Vogtlin (MHV) Grants, 4485
Swiss National Science Foundation Scientific Conferences Grants, 4486
United States Institute of Peace - Transatlantic Post-Doc Fellowship for International Relations and Security, 4581

Iceland
Beatrice Laing Trust Grants, 1324
Canon Foundation Grants, 1423
EMBO Installation Grants, 1915
Fluor Foundation Grants, 2028
Fulbright Business Grants in Spain, 2067
Fulbright Traditional Scholar Program in Europe and Eurasia, 2080
HHMI Biomedical Research Grants for International Scientists in the Baltics, Central and Eastern Europe, Russia, and Ukraine, 2253
Holberg Memorial Fund Nils Klim Prize, 2273
IBRO Europe Regional Grants for Western European Workshops and Symposia, 2324
IBRO InEurope Short Stay Grants, 2325
Oak Foundation Environmental Grants, 3731
RAND Corporation Transatlantic Postdoc Fellowships for Int'l Relations & Security, 4032
SASS Einar and Eva Lund Haugen Memorial Scholarship, 4183
Simone and Cino del Duca Grand Prix Awards, 4261
State Scholarships Foundation Grants, 4449
United States Institute of Peace - Transatlantic Post-Doc Fellowship for International Relations and Security, 4581

India
Bupa Foundation Multi-Country Grant, 1387
Fulbright Traditional Scholar Program in the East Asia/Pacific Region, 2082
Gruber Foundation Int'l Research Award, 2180
IBRO Asia Reg APRC Exchange Fellowships, 2323
MoEF Dr. Salim Ali Nat Wildlife Fellowship, 2792
MOEF Indira Gandhi Puraskar Award, 2793
SfN Chapters Graduate Student Travel Awards, 4202
SfN Chapters Postdoc Trainee Travel Awards, 4203
SfN Donald B. Lindsley Prize in Behavioral Neuroscience, 4204
SfN Graduate Student Travel Awards, 4205
SfN Jacob P. Waletzky Award, 4206
SfN Janett Rosenberg Trubatch Career Development Award, 4207
SfN Louise Hanson Marshall Specific Recognition Award, 4208
SfN M Salpeter Lifetime Achievement Award, 4209
SfN Postdoctoral Fellow Travel Awards, 4210
SfN Ralph W. Gerard Prize in Neuroscience, 4211
SfN Swartz Prize for Theoretical and Computational Neuroscience, 4213
SfN Undergraduate Student Travel Awards, 4217
SfN Young Investigator Award, 4218
Shastri Indo-Canadian Institute Collaborative Research Project Grants, 4220
Sir Dorabji Tata Trust Individual Education and Travel Grants, 4265
USAID Strengthening RMNCH Through Indian Health Professional Associations and Scaling Up Uptake Grants, 4602
Wellcome Trust New Investigator Awards, 4844

Indonesia
Fulbright Traditional Scholar Program in the East Asia/Pacific Region, 2082
IBRO Asia Reg APRC Exchange Fellowships, 2323
Sir Dorabji Tata Trust Individual Education and Travel Grants, 4265
Wellcome Trust New Investigator Awards, 4844

Iran
Fulbright Traditional Scholar Program in the East Asia/Pacific Region, 2082
IBRO Asia Reg APRC Exchange Fellowships, 2323
Sir Dorabji Tata Trust Individual Education and Travel Grants, 4265
Wellcome Trust New Investigator Awards, 4844

Iraq
Fulbright Traditional Scholar Program in the East Asia/Pacific Region, 2082
IBRO Asia Reg APRC Exchange Fellowships, 2323
OSF Global Supplementary Grants, 3843
Sir Dorabji Tata Trust Individual Education and Travel Grants, 4265
Wellcome Trust New Investigator Awards, 4844

Ireland
Abbott Fund Science Education Grants, 220
Beatrice Laing Trust Grants, 1324
Canon Foundation Grants, 1423
Carnegie Endowment for International Peace Junior Fellowships, 1434
EMBO Installation Grants, 1915
Fluor Foundation Grants, 2028
Fulbright Business Grants in Spain, 2067
Fulbright Traditional Scholar Program in Europe and Eurasia, 2080
HHMI Biomedical Research Grants for International Scientists in the Baltics, Central and Eastern Europe, Russia, and Ukraine, 2253
IBRO Europe Regional Grants for Western European Workshops and Symposia, 2324
IBRO InEurope Short Stay Grants, 2325
Oak Foundation Environmental Grants, 3731
Pfizer ASPIRE EU Antifungal Res Awards, 3910
Pfizer ASPIRE EU Dupuytren's Contracture Research Awards, 3911
Pfizer ASPIRE EU Emerging Mechanisms of Resistance Antibacterial Research Awards, 3912
Pfizer ASPIRE EU MRSA Nosocomial Pneumonia & MRSA Complicated Skin & Soft Tissue Infections Antibacterial Research Awards, 3913
RAND Corporation Transatlantic Postdoc Fellowships for Int'l Relations & Security, 4032
Simone and Cino del Duca Grand Prix Awards, 4261
State Scholarships Foundation Grants, 4449
United States Institute of Peace - Transatlantic Post-Doc Fellowship for International Relations and Security, 4581
Wellcome Trust New Investigator Awards, 4844

Israel
American Society for Yad Vashem Grants, 798
Fulbright Traditional Scholar Program in the East Asia/Pacific Region, 2082
Fulbright Traditional Scholar Program in the Near East and North Africa Region, 2083
Gruber Foundation Int'l Research Award, 2180
IBRO Asia Reg APRC Exchange Fellowships, 2323
Klarman Family Foundation Grants in Eating Disorders Research, 2558
MBL Gruss Lipper Family Foundation Summer Fellowship, 2704
Minerva Fellowships, 2778
Minerva Short-Term Research Grants, 2779
OSF Palestinian Rule of Law Fellowships, 3848
Rosalinde & Arthur Gilbert Foundation/AFAR New Investigator Awards in Alzheimer's Disease, 4102
SfN Chapters Graduate Student Travel Awards, 4202
SfN Chapters Postdoc Trainee Travel Awards, 4203
SfN Donald B. Lindsley Prize in Behavioral Neuroscience, 4204
SfN Graduate Student Travel Awards, 4205
SfN Jacob P. Waletzky Award, 4206
SfN Janett Rosenberg Trubatch Career Development Award, 4207
SfN Louise Hanson Marshall Specific Recognition Award, 4208
SfN M Salpeter Lifetime Achievement Award, 4209
SfN Postdoctoral Fellow Travel Awards, 4210
SfN Ralph W. Gerard Prize in Neuroscience, 4211
SfN Swartz Prize for Theoretical and Computational Neuroscience, 4213
SfN Undergraduate Student Travel Awards, 4217
SfN Young Investigator Award, 4218
Sir Dorabji Tata Trust Individual Education and Travel Grants, 4265

Italy
Abbott Fund Science Education Grants, 220
Beatrice Laing Trust Grants, 1324
Canon Foundation Grants, 1423
EMBO Installation Grants, 1915
Fluor Foundation Grants, 2028
Fulbright Business Grants in Spain, 2067
Fulbright Traditional Scholar Program in Europe and Eurasia, 2080

HHMI Biomedical Research Grants for International Scientists in the Baltics, Central and Eastern Europe, Russia, and Ukraine, 2253
IBRO Europe Regional Grants for Western European Workshops and Symposia, 2324
IBRO InEurope Short Stay Grants, 2325
Oak Foundation Environmental Grants, 3731
Pfizer ASPIRE EU Antifungal Res Awards, 3910
Pfizer ASPIRE EU Dupuytren's Contracture Research Awards, 3911
Pfizer ASPIRE EU Emerging Mechanisms of Resistance Antibacterial Research Awards, 3912
Pfizer ASPIRE EU MRSA Nosocomial Pneumonia & MRSA Complicated Skin & Soft Tissue Infections Antibacterial Research Awards, 3913
RAND Corporation Transatlantic Postdoc Fellowships for Int'l Relations & Security, 4032
Simone and Cino del Duca Grand Prix Awards, 4261
State Scholarships Foundation Grants, 4449
United States Institute of Peace - Transatlantic Post-Doc Fellowship for International Relations and Security, 4581

Jamaica
AOCS Ralph H. Potts Memorial Fellowship, 893
Fulbright Traditional Scholar Program in the Western Hemisphere, 2085
IBRO Latin America Regional Funding for Neuroscience Schools, 2327
IBRO Latin America Regional Funding for PROLAB Collaborations, 2328
IBRO Latin America Regional Travel Grants, 2329
John Simon Guggenheim Memorial Foundation Fellowships, 2507
SfN Ricardo Miledi Neuroscience Training, 4212
SSRC Drugs, Security and Democracy Fellows, 4426
Wellcome Trust New Investigator Awards, 4844

Japan
Abbott Fund Science Education Grants, 220
Alexander von Humboldt Foundation Philipp Franz von Siebold Award for Japanese Researchers, 534
Fluor Foundation Grants, 2028
Fulbright Traditional Scholar Program in the East Asia/Pacific Region, 2082
IBRO Asia Reg APRC Exchange Fellowships, 2323
Inamori Foundation Kyoto Prize, 2367
Inamori Foundation Research Grants, 2368
JSPS International Joint Research Projects and Scientific Seminars, 2526
SfN Travel Awards for the Japan Neuroscience Society Meeting, 4216
Sir Dorabji Tata Trust Individual Education and Travel Grants, 4265
SSRC Abe Fellowship for Journalists, 4420
State Scholarships Foundation Grants, 4449

Jordan
Fulbright Traditional Scholar Program in the East Asia/Pacific Region, 2082
IBRO Asia Reg APRC Exchange Fellowships, 2323
OSF Global Faculty Grants - Track 2, 3842
OSF Global Supplementary Grants, 3843
OSF Palestinian Rule of Law Fellowships, 3848
OSF Social Work Fellowships, 3849
Sir Dorabji Tata Trust Individual Education and Travel Grants, 4265
Wellcome Trust New Investigator Awards, 4844

Kazakhstan
Fulbright Traditional Scholar Program in the East Asia/Pacific Region, 2082
Fulbright Traditional Scholar Program in the South and Central Asia Region, 2084
IBRO Asia Reg APRC Exchange Fellowships, 2323
OSF Access to Medicines Fellowships in Eastern Europe and Central Asia, 3834
OSF Central Asia and Caucasus Research and Training Initiative - Junior Fellowships, 3836
OSF Central Asia and Caucasus Research and Training Initiative - Senior Fellowships, 3837
OSF Global Faculty Grants - Track 1, 3841
OSF Global Faculty Grants - Track 2, 3842
OSF Global Supplementary Grants, 3843
OSF Human Rights & Governance Interns, 3845
Sir Dorabji Tata Trust Individual Education and Travel Grants, 4265
United States Institute of Peace - Transatlantic Post-Doc Fellowship for International Relations and Security, 4581
Wellcome Trust New Investigator Awards, 4844

Kenya
Abdus Salam ITCP Postgraduate Diploma Programme in General Physics, 241
Beatrice Laing Trust Grants, 1324
Fulbright Critical Lang Enhancement Awards, 2068
Fulbright Traditional Scholar Program in Sub-Saharan Africa, 2081
Fulbright Traditional Scholar Program in the Near East and North Africa Region, 2083
SSRC African Peacebuilding Network Research Grants, 4422
State Scholarships Foundation Grants, 4449
STTI Alpha Eta Collaborative Research Grants, 4458
USAID Accelerating Progress Against Tuberculosis in Kenya Grants, 4594
USDA Norman E. Borlaug International Agricultural Science and Technology Fellowships, 4636
Wellcome Trust New Investigator Awards, 4844
Wenner-Gren Foundation African Fellowships, 4846
WWF Russell E. Train Fellowships, 4898

Kiribati
Wellcome Trust New Investigator Awards, 4844

Korea
Wellcome Trust New Investigator Awards, 4844

Kosovo
ACLS Early Career Postdoctoral Fellowships in East European Studies, 286
Beatrice Laing Trust Grants, 1324
Canon Foundation Grants, 1423
EMBO Installation Grants, 1915
Fluor Foundation Grants, 2028
Fulbright Business Grants in Spain, 2067
Fulbright Traditional Scholar Program in Europe and Eurasia, 2080
HHMI Biomedical Research Grants for International Scientists in the Baltics, Central and Eastern Europe, Russia, and Ukraine, 2253
Oak Foundation Environmental Grants, 3731
OSF Access to Medicines Fellowships in Eastern Europe and Central Asia, 3834
OSF Global Supplementary Grants, 3843
OSF HESP Mobility Grants, 3844
OSF Human Rights & Governance Interns, 3845
RAND Corporation Transatlantic Postdoc Fellowships for Int'l Relations & Security, 4032
Simone and Cino del Duca Grand Prix Awards, 4261
State Scholarships Foundation Grants, 4449
Wellcome Trust New Investigator Awards, 4844

Kuwait
Fulbright Traditional Scholar Program in the East Asia/Pacific Region, 2082
IBRO Asia Reg APRC Exchange Fellowships, 2323
Sir Dorabji Tata Trust Individual Education and Travel Grants, 4265

Kyrgyzstan
Fulbright Traditional Scholar Program in the East Asia/Pacific Region, 2082
IBRO Asia Reg APRC Exchange Fellowships, 2323
OSF Access to Medicines Fellowships in Eastern Europe and Central Asia, 3834
OSF Central Asia and Caucasus Research and Training Initiative - Junior Fellowships, 3836
OSF Central Asia and Caucasus Research and Training Initiative - Senior Fellowships, 3837
OSF Doc Fellowships Georgia and Kyrgyzstan, 3839
OSF Global Faculty Grants - Track 1, 3841
OSF Global Faculty Grants - Track 2, 3842
OSF Global Supplementary Grants, 3843
Sir Dorabji Tata Trust Individual Education and Travel Grants, 4265
Wellcome Trust New Investigator Awards, 4844

Laos
Fulbright Traditional Scholar Program in the East Asia/Pacific Region, 2082
IBRO Asia Reg APRC Exchange Fellowships, 2323
Sir Dorabji Tata Trust Individual Education and Travel Grants, 4265
Wellcome Trust New Investigator Awards, 4844

Latvia
ACLS Early Career Postdoctoral Fellowships in East European Studies, 286
Beatrice Laing Trust Grants, 1324
Canon Foundation Grants, 1423
EMBO Installation Grants, 1915
Fluor Foundation Grants, 2028
Fulbright Business Grants in Spain, 2067
Fulbright Traditional Scholar Program in Europe and Eurasia, 2080
HHMI Biomedical Research Grants for International Scientists in the Baltics, Central and Eastern Europe, Russia, and Ukraine, 2253
IBRO InEurope Short Stay Grants, 2325
John Deere Foundation Grants, 2500
Oak Foundation Environmental Grants, 3731
OSF Access to Medicines Fellowships in Eastern Europe and Central Asia, 3834
OSF Human Rights & Governance Interns, 3845
OSF Think Tank Fellowships, 3852
RAND Corporation Transatlantic Postdoc Fellowships for Int'l Relations & Security, 4032
Simone and Cino del Duca Grand Prix Awards, 4261
State Scholarships Foundation Grants, 4449
United States Institute of Peace - Transatlantic Post-Doc Fellowship for International Relations and Security, 4581

Lebanon
Fulbright Traditional Scholar Program in the East Asia/Pacific Region, 2082
IBRO Asia Reg APRC Exchange Fellowships, 2323
OSF Global Supplementary Grants, 3843
Sir Dorabji Tata Trust Individual Education and Travel Grants, 4265
Wellcome Trust New Investigator Awards, 4844

Lesotho
Abdus Salam ITCP Postgraduate Diploma Programme in General Physics, 241
Beatrice Laing Trust Grants, 1324
Fulbright Critical Lang Enhancement Awards, 2068
Fulbright Traditional Scholar Program in Sub-Saharan Africa, 2081
Fulbright Traditional Scholar Program in the Near East and North Africa Region, 2083
SSRC African Peacebuilding Network Research Grants, 4422
State Scholarships Foundation Grants, 4449
STTI Alpha Eta Collaborative Research Grants, 4458
Wellcome Trust New Investigator Awards, 4844
Wenner-Gren Foundation African Fellowships, 4846

Liberia
Abdus Salam ITCP Postgraduate Diploma Programme in General Physics, 241
Beatrice Laing Trust Grants, 1324
Fulbright Critical Lang Enhancement Awards, 2068
Fulbright Traditional Scholar Program in Sub-Saharan Africa, 2081
Fulbright Traditional Scholar Program in the Near East and North Africa Region, 2083

SSRC African Peacebuilding Network Research Grants, 4422
State Scholarships Foundation Grants, 4449
STTI Alpha Eta Collaborative Research Grants, 4458
USDA Norman E. Borlaug International Agricultural Science and Technology Fellowships, 4636
Wellcome Trust New Investigator Awards, 4844
Wenner-Gren Foundation African Fellowships, 4846

Libya
Beatrice Laing Trust Grants, 1324
Fulbright Critical Lang Enhancement Awards, 2068
Fulbright Traditional Scholar Program in Sub-Saharan Africa, 2081
Fulbright Traditional Scholar Program in the Near East and North Africa Region, 2083
State Scholarships Foundation Grants, 4449
STTI Alpha Eta Collaborative Research Grants, 4458
Wellcome Trust New Investigator Awards, 4844
Wenner-Gren Foundation African Fellowships, 4846

Liechtenstein
Beatrice Laing Trust Grants, 1324
Canon Foundation Grants, 1423
Central European University Fellowships, 1488
EMBO Installation Grants, 1915
Fluor Foundation Grants, 2028
Fulbright Business Grants in Spain, 2067
Fulbright Traditional Scholar Program in Europe and Eurasia, 2080
HHMI/EMBO Start-up Grants for C Europe, 2250
HHMI Biomedical Research Grants for International Scientists in the Baltics, Central and Eastern Europe, Russia, and Ukraine, 2253
IBRO InEurope Short Stay Grants, 2325
Oak Foundation Environmental Grants, 3731
OSF HESP Mobility Grants, 3844
OSF Human Rights & Governance Interns, 3845
RAND Corporation Transatlantic Postdoc Fellowships for Int'l Relations & Security, 4032
Simone and Cino del Duca Grand Prix Awards, 4261
State Scholarships Foundation Grants, 4449
Swiss National Science Foundation DORE (DO REsearch) Conference Grants, 4482
Swiss National Science Foundation Fellowships for Advanced Researchers, 4483
Swiss National Science Foundation Fellowships for Prospective Researchers, 4484
Swiss National Science Foundation Marie Heim-Vogtlin (MHV) Grants, 4485
Swiss National Science Foundation Scientific Conferences Grants, 4486
United States Institute of Peace - Transatlantic Post-Doc Fellowship for International Relations and Security, 4581

Lithuania
ACLS Early Career Postdoctoral Fellowships in East European Studies, 286
Beatrice Laing Trust Grants, 1324
Canon Foundation Grants, 1423
EMBO Installation Grants, 1915
Fluor Foundation Grants, 2028
Fulbright Business Grants in Spain, 2067
Fulbright Traditional Scholar Program in Europe and Eurasia, 2080
HHMI Biomedical Research Grants for International Scientists in the Baltics, Central and Eastern Europe, Russia, and Ukraine, 2253
IBRO InEurope Short Stay Grants, 2325
John Deere Foundation Grants, 2500
Oak Foundation Environmental Grants, 3731
OSF Access to Medicines Fellowships in Eastern Europe and Central Asia, 3834
OSF Human Rights & Governance Interns, 3845
OSF Think Tank Fellowships, 3852
RAND Corporation Transatlantic Postdoc Fellowships for Int'l Relations & Security, 4032
Simone and Cino del Duca Grand Prix Awards, 4261
State Scholarships Foundation Grants, 4449
United States Institute of Peace - Transatlantic Post-Doc Fellowship for International Relations and Security, 4581

Luxembourg
Beatrice Laing Trust Grants, 1324
Canon Foundation Grants, 1423
EMBO Installation Grants, 1915
Fluor Foundation Grants, 2028
Fulbright Business Grants in Spain, 2067
Fulbright Traditional Scholar Program in Europe and Eurasia, 2080
HHMI Biomedical Research Grants for International Scientists in the Baltics, Central and Eastern Europe, Russia, and Ukraine, 2253
IBRO Europe Regional Grants for Western European Workshops and Symposia, 2324
IBRO InEurope Short Stay Grants, 2325
Oak Foundation Environmental Grants, 3731
RAND Corporation Transatlantic Postdoc Fellowships for Int'l Relations & Security, 4032
Simone and Cino del Duca Grand Prix Awards, 4261
State Scholarships Foundation Grants, 4449
United States Institute of Peace - Transatlantic Post-Doc Fellowship for International Relations and Security, 4581

Macau
Fulbright Traditional Scholar Program in the East Asia/Pacific Region, 2082
Sir Dorabji Tata Trust Individual Education and Travel Grants, 4265

Macedonia
ACLS Early Career Postdoctoral Fellowships in East European Studies, 286
Beatrice Laing Trust Grants, 1324
Canon Foundation Grants, 1423
EMBO Installation Grants, 1915
Fluor Foundation Grants, 2028
Fulbright Business Grants in Spain, 2067
Fulbright Traditional Scholar Program in Europe and Eurasia, 2080
HHMI Biomedical Research Grants for International Scientists in the Baltics, Central and Eastern Europe, Russia, and Ukraine, 2253
IBRO InEurope Short Stay Grants, 2325
Oak Foundation Environmental Grants, 3731
OSF Access to Medicines Fellowships in Eastern Europe and Central Asia, 3834
OSF Global Supplementary Grants, 3843
OSF HESP Mobility Grants, 3844
OSF Human Rights & Governance Interns, 3845
RAND Corporation Transatlantic Postdoc Fellowships for Int'l Relations & Security, 4032
Simone and Cino del Duca Grand Prix Awards, 4261
State Scholarships Foundation Grants, 4449
United States Institute of Peace - Transatlantic Post-Doc Fellowship for International Relations and Security, 4581
Wellcome Trust New Investigator Awards, 4844

Madagascar
Abdus Salam ITCP Postgraduate Diploma Programme in General Physics, 241
Beatrice Laing Trust Grants, 1324
Fulbright Critical Lang Enhancement Awards, 2068
Fulbright Traditional Scholar Program in Sub-Saharan Africa, 2081
Fulbright Traditional Scholar Program in the Near East and North Africa Region, 2083
SSRC African Peacebuilding Network Research Grants, 4422
State Scholarships Foundation Grants, 4449
STTI Alpha Eta Collaborative Research Grants, 4458
USDA Norman E. Borlaug International Agricultural Science and Technology Fellowships, 4636
Wellcome Trust New Investigator Awards, 4844
Wenner-Gren Foundation African Fellowships, 4846

Malawi
Abdus Salam ITCP Postgraduate Diploma Programme in General Physics, 241
Beatrice Laing Trust Grants, 1324
Fulbright Critical Lang Enhancement Awards, 2068
Fulbright Traditional Scholar Program in Sub-Saharan Africa, 2081
Fulbright Traditional Scholar Program in the Near East and North Africa Region, 2083
SSRC African Peacebuilding Network Research Grants, 4422
State Scholarships Foundation Grants, 4449
STTI Alpha Eta Collaborative Research Grants, 4458
USDA Norman E. Borlaug International Agricultural Science and Technology Fellowships, 4636
Wellcome Trust New Investigator Awards, 4844
Wenner-Gren Foundation African Fellowships, 4846
WWF Russell E. Train Fellowships, 4898

Malaysia
Fulbright Traditional Scholar Program in the East Asia/Pacific Region, 2082
Gruber Foundation Int'l Research Award, 2180
IBRO Asia Reg APRC Exchange Fellowships, 2323
SfN Chapters Graduate Student Travel Awards, 4202
SfN Chapters Postdoc Trainee Travel Awards, 4203
SfN Donald B. Lindsley Prize in Behavioral Neuroscience, 4204
SfN Graduate Student Travel Awards, 4205
SfN Jacob P. Waletzky Award, 4206
SfN Janett Rosenberg Trubatch Career Development Award, 4207
SfN Louise Hanson Marshall Specific Recognition Award, 4208
SfN M Salpeter Lifetime Achievement Award, 4209
SfN Postdoctoral Fellow Travel Awards, 4210
SfN Ralph W. Gerard Prize in Neuroscience, 4211
SfN Swartz Prize for Theoretical and Computational Neuroscience, 4213
SfN Undergraduate Student Travel Awards, 4217
SfN Young Investigator Award, 4218
Sir Dorabji Tata Trust Individual Education and Travel Grants, 4265
Wellcome Trust New Investigator Awards, 4844

Maldives
Fulbright Traditional Scholar Program in the East Asia/Pacific Region, 2082
IBRO Asia Reg APRC Exchange Fellowships, 2323
Sir Dorabji Tata Trust Individual Education and Travel Grants, 4265
Wellcome Trust New Investigator Awards, 4844

Mali
Abdus Salam ITCP Postgraduate Diploma Programme in General Physics, 241
Beatrice Laing Trust Grants, 1324
Fulbright Critical Lang Enhancement Awards, 2068
Fulbright Traditional Scholar Program in Sub-Saharan Africa, 2081
Fulbright Traditional Scholar Program in the Near East and North Africa Region, 2083
SSRC African Peacebuilding Network Research Grants, 4422
State Scholarships Foundation Grants, 4449
STTI Alpha Eta Collaborative Research Grants, 4458
USDA Norman E. Borlaug International Agricultural Science and Technology Fellowships, 4636
Wellcome Trust New Investigator Awards, 4844
Wenner-Gren Foundation African Fellowships, 4846

Malta
Beatrice Laing Trust Grants, 1324
Canon Foundation Grants, 1423
EMBO Installation Grants, 1915
Fluor Foundation Grants, 2028
Fulbright Business Grants in Spain, 2067
Fulbright Traditional Scholar Program in Europe and Eurasia, 2080

HHMI Biomedical Research Grants for International Scientists in the Baltics, Central and Eastern Europe, Russia, and Ukraine, 2253
IBRO InEurope Short Stay Grants, 2325
Oak Foundation Environmental Grants, 3731
RAND Corporation Transatlantic Postdoc Fellowships for Int'l Relations & Security, 4032
Simone and Cino del Duca Grand Prix Awards, 4261
State Scholarships Foundation Grants, 4449
United States Institute of Peace - Transatlantic Post-Doc Fellowship for International Relations and Security, 4581

Martinique
AOCS Ralph H. Potts Memorial Fellowship, 893
John Simon Guggenheim Memorial Foundation Fellowships, 2507
SfN Ricardo Miledi Neuroscience Training, 4212
SSRC Drugs, Security and Democracy Fellows, 4426

Mauritania
Abdus Salam ITCP Postgraduate Diploma Programme in General Physics, 241
Beatrice Laing Trust Grants, 1324
Fulbright Critical Lang Enhancement Awards, 2068
Fulbright Traditional Scholar Program in Sub-Saharan Africa, 2081
Fulbright Traditional Scholar Program in the Near East and North Africa Region, 2083
SSRC African Peacebuilding Network Research Grants, 4422
State Scholarships Foundation Grants, 4449
STTI Alpha Eta Collaborative Research Grants, 4458
Wellcome Trust New Investigator Awards, 4844
Wenner-Gren Foundation African Fellowships, 4846

Mauritius
Beatrice Laing Trust Grants, 1324
Fulbright Critical Lang Enhancement Awards, 2068
Fulbright Traditional Scholar Program in Sub-Saharan Africa, 2081
Fulbright Traditional Scholar Program in the Near East and North Africa Region, 2083
SSRC African Peacebuilding Network Research Grants, 4422
State Scholarships Foundation Grants, 4449
STTI Alpha Eta Collaborative Research Grants, 4458
USDA Norman E. Borlaug International Agricultural Science and Technology Fellowships, 4636
Wellcome Trust New Investigator Awards, 4844
Wenner-Gren Foundation African Fellowships, 4846

Mayotte
USDA Norman E. Borlaug International Agricultural Science and Technology Fellowships, 4636

Mexico
Academy of American Franciscan History Dissertation Fellowships, 249
AOCS Ralph H. Potts Memorial Fellowship, 893
Fluor Foundation Grants, 2028
Fulbright Binational Business Grants in Mexico, 2066
Fulbright Traditional Scholar Program in the Western Hemisphere, 2085
General Service Foundation Human Rights and Economic Justice Grants, 2108
Gruber Foundation Int'l Research Award, 2180
IBRO Latin America Regional Funding for Neuroscience Schools, 2327
IBRO Latin America Regional Funding for PROLAB Collaborations, 2328
IBRO Latin America Regional Travel Grants, 2329
John Simon Guggenheim Memorial Foundation Fellowships, 2507
NSF-CONACyT Collaborative Research Opportunities, 3452
OTS Hovore-Horn Fellowships, 3856
SfN Chapters Graduate Student Travel Awards, 4202
SfN Chapters Postdoc Trainee Travel Awards, 4203
SfN Donald B. Lindsley Prize in Behavioral Neuroscience, 4204
SfN Graduate Student Travel Awards, 4205
SfN Jacob P. Waletzky Award, 4206
SfN Janett Rosenberg Trubatch Career Development Award, 4207
SfN Louise Hanson Marshall Specific Recognition Award, 4208
SfN M Salpeter Lifetime Achievement Award, 4209
SfN Postdoctoral Fellow Travel Awards, 4210
SfN Ralph W. Gerard Prize in Neuroscience, 4211
SfN Ricardo Miledi Neuroscience Training, 4212
SfN Swartz Prize for Theoretical and Computational Neuroscience, 4213
SfN Travel Awards for the Federation of European Neuroscience Societies Forum, 4214
SfN Travel Awards for the International Brain Research Organization World Congress, 4215
SfN Travel Awards for the Japan Neuroscience Society Meeting, 4216
SfN Undergraduate Student Travel Awards, 4217
SfN Young Investigator Award, 4218
SSRC Drugs, Security and Democracy Fellows, 4426
Swann Foundation Fellowships, 4478
Wellcome Trust New Investigator Awards, 4844

Micronesia
OTS Undergrad NAPIRE Summer Internships, 3858
SARE Graduate Student Grants in Sustainable Agriculture - Western Region, 4171
SARE Producer Research and Education Grants-Western Region, 4174
SARE Pro & Producer Grants: West Reg, 4175
SARE Research & Educ Grants: West Reg, 4180
USDA Integrated Pest Management Competitive Grants - Western Region, 4629
USDA Resident Instruction Grants for Higher Education in Insular Areas, 4645
Wellcome Trust New Investigator Awards, 4844

Moldova
Beatrice Laing Trust Grants, 1324
Canon Foundation Grants, 1423
EMBO Installation Grants, 1915
Fluor Foundation Grants, 2028
Fulbright Business Grants in Spain, 2067
Fulbright Traditional Scholar Program in Europe and Eurasia, 2080
GMFUS Black Sea Trust for Regional Corporation Grants, 2152
HHMI Biomedical Research Grants for International Scientists in the Baltics, Central and Eastern Europe, Russia, and Ukraine, 2253
IBRO InEurope Short Stay Grants, 2325
John Deere Foundation Grants, 2500
Oak Foundation Environmental Grants, 3731
OSF Access to Medicines Fellowships in Eastern Europe and Central Asia, 3834
OSF Civil Service Awards, 3838
OSF Global Supplementary Grants, 3843
OSF HESP Mobility Grants, 3844
OSF Human Rights & Governance Interns, 3845
OSF Think Tank Fellowships, 3852
RAND Corporation Transatlantic Postdoc Fellowships for Int'l Relations & Security, 4032
Simone and Cino del Duca Grand Prix Awards, 4261
State Scholarships Foundation Grants, 4449
United States Institute of Peace - Transatlantic Post-Doc Fellowship for International Relations and Security, 4581
Wellcome Trust New Investigator Awards, 4844

Monaco
Beatrice Laing Trust Grants, 1324
Canon Foundation Grants, 1423
EMBO Installation Grants, 1915
Fluor Foundation Grants, 2028
Fulbright Business Grants in Spain, 2067
Fulbright Traditional Scholar Program in Europe and Eurasia, 2080
HHMI Biomedical Research Grants for International Scientists in the Baltics, Central and Eastern Europe, Russia, and Ukraine, 2253
IBRO InEurope Short Stay Grants, 2325
Oak Foundation Environmental Grants, 3731
RAND Corporation Transatlantic Postdoc Fellowships for Int'l Relations & Security, 4032
Simone and Cino del Duca Grand Prix Awards, 4261
State Scholarships Foundation Grants, 4449
United States Institute of Peace - Transatlantic Post-Doc Fellowship for International Relations and Security, 4581

Mongolia
Fulbright Traditional Scholar Program in the East Asia/Pacific Region, 2082
IBRO Asia Reg APRC Exchange Fellowships, 2323
OSF Central Asia and Caucasus Research and Training Initiative - Junior Fellowships, 3836
OSF Central Asia and Caucasus Research and Training Initiative - Senior Fellowships, 3837
OSF Global Faculty Grants - Track 1, 3841
OSF Global Faculty Grants - Track 2, 3842
OSF Global Supplementary Grants, 3843
OSF Human Rights & Governance Interns, 3845
Sir Dorabji Tata Trust Individual Education and Travel Grants, 4265
Wellcome Trust New Investigator Awards, 4844

Montenegro
ACLS Early Career Postdoctoral Fellowships in East European Studies, 286
Beatrice Laing Trust Grants, 1324
Canon Foundation Grants, 1423
EMBO Installation Grants, 1915
Fluor Foundation Grants, 2028
Fulbright Business Grants in Spain, 2067
Fulbright Traditional Scholar Program in Europe and Eurasia, 2080
HHMI Biomedical Research Grants for International Scientists in the Baltics, Central and Eastern Europe, Russia, and Ukraine, 2253
IBRO InEurope Short Stay Grants, 2325
Oak Foundation Environmental Grants, 3731
OSF Access to Medicines Fellowships in Eastern Europe and Central Asia, 3834
OSF Global Supplementary Grants, 3843
OSF HESP Mobility Grants, 3844
OSF Human Rights & Governance Interns, 3845
RAND Corporation Transatlantic Postdoc Fellowships for Int'l Relations & Security, 4032
Simone and Cino del Duca Grand Prix Awards, 4261
State Scholarships Foundation Grants, 4449
United States Institute of Peace - Transatlantic Post-Doc Fellowship for International Relations and Security, 4581
Wellcome Trust New Investigator Awards, 4844

Montserrat
AOCS Ralph H. Potts Memorial Fellowship, 893
John Simon Guggenheim Memorial Foundation Fellowships, 2507

Morocco
Beatrice Laing Trust Grants, 1324
Fulbright Critical Lang Enhancement Awards, 2068
Fulbright Traditional Scholar Program in Sub-Saharan Africa, 2081
Fulbright Traditional Scholar Program in the Near East and North Africa Region, 2083
State Scholarships Foundation Grants, 4449
STTI Alpha Eta Collaborative Research Grants, 4458
Wellcome Trust New Investigator Awards, 4844
Wenner-Gren Foundation African Fellowships, 4846

Mozambique
Abdus Salam ITCP Postgraduate Diploma Programme in General Physics, 241
Beatrice Laing Trust Grants, 1324
Fulbright Critical Lang Enhancement Awards, 2068

Fulbright Traditional Scholar Program in Sub-Saharan Africa, 2081
Fulbright Traditional Scholar Program in the Near East and North Africa Region, 2083
SSRC African Peacebuilding Network Research Grants, 4422
State Scholarships Foundation Grants, 4449
STTI Alpha Eta Collaborative Research Grants, 4458
USDA Norman E. Borlaug International Agricultural Science and Technology Fellowships, 4636
Wellcome Trust New Investigator Awards, 4844
Wenner-Gren Foundation African Fellowships, 4846
WWF Russell E. Train Fellowships, 4898

Myanmar (Burma)
IBRO Asia Reg APRC Exchange Fellowships, 2323
Wellcome Trust New Investigator Awards, 4844

Nagorno-Karabakh
Fulbright Traditional Scholar Program in the East Asia/Pacific Region, 2082
Sir Dorabji Tata Trust Individual Education and Travel Grants, 4265

Namibia
Abdus Salam ITCP Postgraduate Diploma Programme in General Physics, 241
Beatrice Laing Trust Grants, 1324
Fulbright Critical Lang Enhancement Awards, 2068
Fulbright Traditional Scholar Program in Sub-Saharan Africa, 2081
Fulbright Traditional Scholar Program in the Near East and North Africa Region, 2083
SSRC African Peacebuilding Network Research Grants, 4422
State Scholarships Foundation Grants, 4449
STTI Alpha Eta Collaborative Research Grants, 4458
Wellcome Trust New Investigator Awards, 4844
Wenner-Gren Foundation African Fellowships, 4846

Nepal
Fulbright Traditional Scholar Program in the East Asia/Pacific Region, 2082
IBRO Asia Reg APRC Exchange Fellowships, 2323
OSF Global Faculty Grants - Track 1, 3841
OSF Global Faculty Grants - Track 2, 3842
OSF Global Supplementary Grants, 3843
Sir Dorabji Tata Trust Individual Education and Travel Grants, 4265
Wellcome Trust New Investigator Awards, 4844
WWF Russell E. Train Fellowships, 4898

Netherlands
Abbott Fund Science Education Grants, 220
IBRO Europe Regional Grants for Western European Workshops and Symposia, 2324
IBRO InEurope Short Stay Grants, 2325
NWO Added Value Plus for the Social Sciences Grants, 3672
NWO Aspasia Grants, 3673
NWO Athena Grants, 3674
Oak Foundation Environmental Grants, 3731
Pfizer ASPIRE EU Antifungal Res Awards, 3910
Pfizer ASPIRE EU Dupuytren's Contracture Research Awards, 3911
Pfizer ASPIRE EU Emerging Mechanisms of Resistance Antibacterial Research Awards, 3912
Pfizer ASPIRE EU MRSA Nosocomial Pneumonia & MRSA Complicated Skin & Soft Tissue Infections Antibacterial Research Awards, 3913
SSHRC Digging into Data Challenge Grants, 4406
United States Institute of Peace - Transatlantic Post-Doc Fellowship for International Relations and Security, 4581

New Independent States
IREX Eurasian Undergraduate Exchange Program (UGRAD) Grants, 2418
IREX International Leadership in Education Program (ILEP) Grants, 2420

New Zealand
Bupa Foundation Multi-Country Grant, 1387
C. Alma Baker Trust Grants, 1403
Fluor Foundation Grants, 2028
Gruber Foundation Int'l Research Award, 2180
SfN Chapters Graduate Student Travel Awards, 4202
SfN Chapters Postdoc Trainee Travel Awards, 4203
SfN Donald B. Lindsley Prize in Behavioral Neuroscience, 4204
SfN Graduate Student Travel Awards, 4205
SfN Jacob P. Waletzky Award, 4206
SfN Janett Rosenberg Trubatch Career Development Award, 4207
SfN Louise Hanson Marshall Specific Recognition Award, 4208
SfN M Salpeter Lifetime Achievement Award, 4209
SfN Postdoctoral Fellow Travel Awards, 4210
SfN Ralph W. Gerard Prize in Neuroscience, 4211
SfN Swartz Prize for Theoretical and Computational Neuroscience, 4213
SfN Undergraduate Student Travel Awards, 4217
SfN Young Investigator Award, 4218
USDD National Consortium for Measurement and Signature Intelligence Research Grants, 4759

Nicaragua
Academy of American Franciscan History Dissertation Fellowships, 249
AOCS Ralph H. Potts Memorial Fellowship, 893
Fulbright Traditional Scholar Program in the Western Hemisphere, 2085
IBRO Latin America Regional Funding for Neuroscience Schools, 2327
IBRO Latin America Regional Funding for PROLAB Collaborations, 2328
IBRO Latin America Regional Travel Grants, 2329
John Simon Guggenheim Memorial Foundation Fellowships, 2507
OTS Hovore-Horn Fellowships, 3856
SfN Ricardo Miledi Neuroscience Training, 4212
SSRC Drugs, Security and Democracy Fellows, 4426
Wellcome Trust New Investigator Awards, 4844

Niger
Abdus Salam ITCP Postgraduate Diploma Programme in General Physics, 241
Beatrice Laing Trust Grants, 1324
Fulbright Critical Lang Enhancement Awards, 2068
Fulbright Traditional Scholar Program in Sub-Saharan Africa, 2081
Fulbright Traditional Scholar Program in the Near East and North Africa Region, 2083
SSRC African Peacebuilding Network Research Grants, 4422
State Scholarships Foundation Grants, 4449
STTI Alpha Eta Collaborative Research Grants, 4458
Wellcome Trust New Investigator Awards, 4844
Wenner-Gren Foundation African Fellowships, 4846

Nigeria
Abdus Salam ITCP Postgraduate Diploma Programme in General Physics, 241
ACLS African Humanities Program for Dissertation Completion, 278
ACLS African Humanities Program in Ghana, Nigeria, South Africa, Tanzania, and Uganda, 279
Beatrice Laing Trust Grants, 1324
Fulbright Critical Lang Enhancement Awards, 2068
Fulbright Traditional Scholar Program in Sub-Saharan Africa, 2081
Fulbright Traditional Scholar Program in the Near East and North Africa Region, 2083
Gruber Foundation Int'l Research Award, 2180
SfN Chapters Graduate Student Travel Awards, 4202
SfN Chapters Postdoc Trainee Travel Awards, 4203
SfN Donald B. Lindsley Prize in Behavioral Neuroscience, 4204
SfN Graduate Student Travel Awards, 4205
SfN Jacob P. Waletzky Award, 4206
SfN Janett Rosenberg Trubatch Career Development Award, 4207
SfN Louise Hanson Marshall Specific Recognition Award, 4208
SfN M Salpeter Lifetime Achievement Award, 4209
SfN Postdoctoral Fellow Travel Awards, 4210
SfN Ralph W. Gerard Prize in Neuroscience, 4211
SfN Swartz Prize for Theoretical and Computational Neuroscience, 4213
SfN Undergraduate Student Travel Awards, 4217
SfN Young Investigator Award, 4218
SSRC African Peacebuilding Network Research Grants, 4422
SSRC Next Generation Social Sciences in Africa: Doc Dissertation Completion Fellowships, 4434
SSRC Next Generation Social Sciences in Africa: Doctoral Dissertation Proposal Fellowships, 4435
SSRC Next Generation Social Sciences in Africa: Doctoral Dissertation Research Fellowships, 4436
State Scholarships Foundation Grants, 4449
STTI Alpha Eta Collaborative Research Grants, 4458
USDA Norman E. Borlaug International Agricultural Science and Technology Fellowships, 4636
Wellcome Trust New Investigator Awards, 4844
Wenner-Gren Foundation African Fellowships, 4846

North Korea
Fulbright Traditional Scholar Program in the East Asia/Pacific Region, 2082
IBRO Asia Reg APRC Exchange Fellowships, 2323
Sir Dorabji Tata Trust Individual Education and Travel Grants, 4265

Northern Cyprus
Fulbright Traditional Scholar Program in the East Asia/Pacific Region, 2082
Sir Dorabji Tata Trust Individual Education and Travel Grants, 4265

Norway
Beatrice Laing Trust Grants, 1324
Canon Foundation Grants, 1423
EMBO Installation Grants, 1915
Fluor Foundation Grants, 2028
Fulbright Business Grants in Spain, 2067
Fulbright Traditional Scholar Program in Europe and Eurasia, 2080
Gruber Foundation Int'l Research Award, 2180
HHMI Biomedical Research Grants for International Scientists in the Baltics, Central and Eastern Europe, Russia, and Ukraine, 2253
Holberg Memorial Fund Nils Klim Prize, 2273
Holberg Memorial Fund School Project, 2274
IBRO Europe Regional Grants for Western European Workshops and Symposia, 2324
IBRO InEurope Short Stay Grants, 2325
Oak Foundation Environmental Grants, 3731
Pfizer ASPIRE EU Antifungal Res Awards, 3910
Pfizer ASPIRE EU Dupuytren's Contracture Research Awards, 3911
Pfizer ASPIRE EU Emerging Mechanisms of Resistance Antibacterial Research Awards, 3912
Pfizer ASPIRE EU MRSA Nosocomial Pneumonia & MRSA Complicated Skin & Soft Tissue Infections Antibacterial Research Awards, 3913
RAND Corporation Transatlantic Postdoc Fellowships for Int'l Relations & Security, 4032
SASS Einar and Eva Lund Haugen Memorial Scholarship, 4183
SfN Chapters Graduate Student Travel Awards, 4202
SfN Chapters Postdoc Trainee Travel Awards, 4203
SfN Donald B. Lindsley Prize in Behavioral Neuroscience, 4204
SfN Graduate Student Travel Awards, 4205
SfN Jacob P. Waletzky Award, 4206
SfN Janett Rosenberg Trubatch Career Development Award, 4207
SfN Louise Hanson Marshall Specific Recognition Award, 4208
SfN M Salpeter Lifetime Achievement Award, 4209

SfN Postdoctoral Fellow Travel Awards, 4210
SfN Ralph W. Gerard Prize in Neuroscience, 4211
SfN Swartz Prize for Theoretical and Computational Neuroscience, 4213
SfN Undergraduate Student Travel Awards, 4217
SfN Young Investigator Award, 4218
Simone and Cino del Duca Grand Prix Awards, 4261
State Scholarships Foundation Grants, 4449
United States Institute of Peace - Transatlantic Post-Doc Fellowship for International Relations and Security, 4581

Oman
Fulbright Traditional Scholar Program in the East Asia/Pacific Region, 2082
IBRO Asia Reg APRC Exchange Fellowships, 2323
Sir Dorabji Tata Trust Individual Education and Travel Grants, 4265

Pakistan
Fulbright Traditional Scholar Program in the East Asia/Pacific Region, 2082
IBRO Asia Reg APRC Exchange Fellowships, 2323
Sir Dorabji Tata Trust Individual Education and Travel Grants, 4265
Wellcome Trust New Investigator Awards, 4844

Palau
OTS Undergrad NAPIRE Summer Internships, 3858
USDA Integrated Pest Management Competitive Grants - Western Region, 4629
USDA Resident Instruction Grants for Higher Education in Insular Areas, 4645
Wellcome Trust New Investigator Awards, 4844

Palestinian Authority
Fulbright Traditional Scholar Program in the East Asia/Pacific Region, 2082
OSF Global Faculty Grants - Track 2, 3842
Sir Dorabji Tata Trust Individual Education and Travel Grants, 4265

Palestinian Territory
OSF Global Supplementary Grants, 3843
OSF Palestinian Rule of Law Fellowships, 3848

Panama
Academy of American Franciscan History Dissertation Fellowships, 249
AOCS Ralph H. Potts Memorial Fellowship, 893
IBRO Latin America Regional Funding for Neuroscience Schools, 2327
IBRO Latin America Regional Funding for PROLAB Collaborations, 2328
IBRO Latin America Regional Travel Grants, 2329
John Simon Guggenheim Memorial Foundation Fellowships, 2507
OTS Hovore-Horn Fellowships, 3856
SfN Ricardo Miledi Neuroscience Training, 4212
SSRC Drugs, Security and Democracy Fellows, 4426
Wellcome Trust New Investigator Awards, 4844

Papua New Guinea
Wellcome Trust New Investigator Awards, 4844

Paraguay
Academy of American Franciscan History Dissertation Fellowships, 249
ASM-PAHO Infectious Diseases Epidemiology and Surveillance Fellowships, 1098
ASM International Fellowship for Latin America and the Caribbean, 1175
Fulbright Traditional Scholar Program in the Western Hemisphere, 2085
IBRO Latin America Regional Funding for Neuroscience Schools, 2327
IBRO Latin America Regional Funding for PROLAB Collaborations, 2328
IBRO Latin America Regional Travel Grants, 2329
John Simon Guggenheim Memorial Foundation Fellowships, 2507
OTS Hovore-Horn Fellowships, 3856
SfN Ricardo Miledi Neuroscience Training, 4212
SSRC Drugs, Security and Democracy Fellows, 4426
Wellcome Trust New Investigator Awards, 4844

Peru
Academy of American Franciscan History Dissertation Fellowships, 249
ASM-PAHO Infectious Diseases Epidemiology and Surveillance Fellowships, 1098
ASM International Fellowship for Latin America and the Caribbean, 1175
Fluor Foundation Grants, 2028
Fulbright Traditional Scholar Program in the Western Hemisphere, 2085
IBRO Latin America Regional Funding for Neuroscience Schools, 2327
IBRO Latin America Regional Funding for PROLAB Collaborations, 2328
IBRO Latin America Regional Travel Grants, 2329
John Simon Guggenheim Memorial Foundation Fellowships, 2507
OTS Francis J. Bossuyt Fellowship, 3855
OTS Hovore-Horn Fellowships, 3856
SfN Ricardo Miledi Neuroscience Training, 4212
SSRC Drugs, Security and Democracy Fellows, 4426
Wellcome Trust New Investigator Awards, 4844

Philippines
ACC Philippines Fellowship Program, 261
Fluor Foundation Grants, 2028
Fulbright Traditional Scholar Program in the East Asia/Pacific Region, 2082
IBRO Asia Reg APRC Exchange Fellowships, 2323
Sir Dorabji Tata Trust Individual Education and Travel Grants, 4265
Wellcome Trust New Investigator Awards, 4844

Poland
ACLS Early Career Postdoctoral Fellowships in East European Studies, 286
AIAR Andrew W. Mellon Found Fellowships, 392
Beatrice Laing Trust Grants, 1324
Canon Foundation Grants, 1423
Central European University Fellowships, 1488
EMBO Installation Grants, 1915
Fluor Foundation Grants, 2028
Fluor Foundation Grants, 2028
Fulbright Business Grants in Spain, 2067
Fulbright Traditional Scholar Program in Europe and Eurasia, 2080
HHMI/EMBO Start-up Grants for C Europe, 2250
HHMI Biomedical Research Grants for International Scientists in the Baltics, Central and Eastern Europe, Russia, and Ukraine, 2253
IBRO InEurope Short Stay Grants, 2325
Kosciuszko Foundation Fellowships for Polish Citizens, 2570
Kosciuszko Found Grants for Polish Citizens, 2572
Kosciuszko Foundation Teaching Fellowships for Polish Citizens, 2573
Oak Foundation Environmental Grants, 3731
OSF Access to Medicines Fellowships in Eastern Europe and Central Asia, 3834
OSF HESP Mobility Grants, 3844
OSF Human Rights & Governance Interns, 3845
OSF Think Tank Fellowships, 3852
RAND Corporation Transatlantic Postdoc Fellowships for Int'l Relations & Security, 4032
Simone and Cino del Duca Grand Prix Awards, 4261
State Scholarships Foundation Grants, 4449
Swiss National Science Foundation DORE (DO REsearch) Conference Grants, 4482
Swiss National Science Foundation Fellowships for Advanced Researchers, 4483
Swiss National Science Foundation Fellowships for Prospective Researchers, 4484
Swiss National Science Foundation Marie Heim-Vogtlin (MHV) Grants, 4485
Swiss National Science Foundation Scientific Conferences Grants, 4486
United States Institute of Peace - Transatlantic Post-Doc Fellowship for International Relations and Security, 4581

Portugal
Beatrice Laing Trust Grants, 1324
Canon Foundation Grants, 1423
EMBO Installation Grants, 1915
Fluor Foundation Grants, 2028
Fulbright Business Grants in Spain, 2067
Fulbright Traditional Scholar Program in Europe and Eurasia, 2080
HHMI Biomedical Research Grants for International Scientists in the Baltics, Central and Eastern Europe, Russia, and Ukraine, 2253
IBRO Europe Regional Grants for Western European Workshops and Symposia, 2324
IBRO InEurope Short Stay Grants, 2325
Oak Foundation Environmental Grants, 3731
Pfizer ASPIRE EU Antifungal Res Awards, 3910
Pfizer ASPIRE EU Dupuytren's Contracture Research Awards, 3911
Pfizer ASPIRE EU Emerging Mechanisms of Resistance Antibacterial Research Awards, 3912
Pfizer ASPIRE EU MRSA Nosocomial Pneumonia & MRSA Complicated Skin & Soft Tissue Infections Antibacterial Research Awards, 3913
RAND Corporation Transatlantic Postdoc Fellowships for Int'l Relations & Security, 4032
Simone and Cino del Duca Grand Prix Awards, 4261
State Scholarships Foundation Grants, 4449
United States Institute of Peace - Transatlantic Post-Doc Fellowship for International Relations and Security, 4581

Qatar
Fulbright Traditional Scholar Program in the East Asia/Pacific Region, 2082
IBRO Asia Reg APRC Exchange Fellowships, 2323
Sir Dorabji Tata Trust Individual Education and Travel Grants, 4265

Reunion
Abdus Salam ITCP Postgraduate Diploma Programme in General Physics, 241
USDA Norman E. Borlaug International Agricultural Science and Technology Fellowships, 4636

Romania
ACLS Early Career Postdoctoral Fellowships in East European Studies, 286
AIAR Andrew W. Mellon Found Fellowships, 392
Beatrice Laing Trust Grants, 1324
Canon Foundation Grants, 1423
EMBO Installation Grants, 1915
Fluor Foundation Grants, 2028
Fulbright Business Grants in Spain, 2067
Fulbright Traditional Scholar Program in Europe and Eurasia, 2080
GMFUS Black Sea Trust for Regional Corporation Grants, 2152
HHMI Biomedical Research Grants for International Scientists in the Baltics, Central and Eastern Europe, Russia, and Ukraine, 2253
IBRO InEurope Short Stay Grants, 2325
Oak Foundation Environmental Grants, 3731
OSF Access to Medicines Fellowships in Eastern Europe and Central Asia, 3834
OSF HESP Mobility Grants, 3844
OSF Human Rights & Governance Interns, 3845
OSF Think Tank Fellowships, 3852
RAND Corporation Transatlantic Postdoc Fellowships for Int'l Relations & Security, 4032
Simone and Cino del Duca Grand Prix Awards, 4261
State Scholarships Foundation Grants, 4449

United States Institute of Peace - Transatlantic Post-Doc Fellowship for International Relations and Security, 4581
Wellcome Trust New Investigator Awards, 4844

Russia
Beatrice Laing Trust Grants, 1324
Canon Foundation Grants, 1423
EMBO Installation Grants, 1915
Fluor Foundation Grants, 2028
Fluor Foundation Grants, 2028
Fulbright Business Grants in Spain, 2067
Fulbright Traditional Scholar Program in Europe and Eurasia, 2080
Fulbright Traditional Scholar Program in the East Asia/Pacific Region, 2082
GMFUS Black Sea Trust for Regional Corporation Grants, 2152
HHMI Biomedical Research Grants for International Scientists in the Baltics, Central and Eastern Europe, Russia, and Ukraine, 2253
HHMI Biomedical Research Grants for International Scientists in the Baltics, Central and Eastern Europe, Russia, and Ukraine, 2253
IBRO Asia Reg APRC Exchange Fellowships, 2323
IBRO InEurope Short Stay Grants, 2325
IREX University Administration Support Program (UASP), 2425
Oak Foundation Environmental Grants, 3731
OSF Access to Medicines Fellowships in Eastern Europe and Central Asia, 3834
OSF Global Supplementary Grants, 3843
OSF HESP Mobility Grants, 3844
OSF Human Rights & Governance Interns, 3845
RAND Corporation Transatlantic Postdoc Fellowships for Int'l Relations & Security, 4032
Simone and Cino del Duca Grand Prix Awards, 4261
Sir Dorabji Tata Trust Individual Education and Travel Grants, 4265
State Scholarships Foundation Grants, 4449
U.S. Department of Education United States-Russia Program: Improving Research and Educational Activities in Higher Education, 4547
United States Institute of Peace - Transatlantic Post-Doc Fellowship for International Relations and Security, 4581
US CRDF CGP-RFBR Climate Change and Energy Competition For Joint U.S.-Russian Research Teams, 4608

Russian Federation
Alexander von Humboldt Foundation German Chancellor Fellowships for Prospective Leaders from the Russian Federation, 527

Rwanda
Abdus Salam ITCP Postgraduate Diploma Programme in General Physics, 241
Beatrice Laing Trust Grants, 1324
Fulbright Critical Lang Enhancement Awards, 2068
Fulbright Traditional Scholar Program in Sub-Saharan Africa, 2081
Fulbright Traditional Scholar Program in the Near East and North Africa Region, 2083
SSRC African Peacebuilding Network Research Grants, 4422
State Scholarships Foundation Grants, 4449
STTI Alpha Eta Collaborative Research Grants, 4458
USDA Norman E. Borlaug International Agricultural Science and Technology Fellowships, 4636
Wellcome Trust New Investigator Awards, 4844
Wenner-Gren Foundation African Fellowships, 4846

Saint Kitts And Nevis
AOCS Ralph H. Potts Memorial Fellowship, 893
IBRO Latin America Regional Funding for Neuroscience Schools, 2327
IBRO Latin America Regional Funding for PROLAB Collaborations, 2328
IBRO Latin America Regional Travel Grants, 2329
John Simon Guggenheim Memorial Foundation Fellowships, 2507
SfN Ricardo Miledi Neuroscience Training, 4212
SSRC Drugs, Security and Democracy Fellows, 4426

San Marino
Beatrice Laing Trust Grants, 1324
Canon Foundation Grants, 1423
EMBO Installation Grants, 1915
Fluor Foundation Grants, 2028
Fulbright Business Grants in Spain, 2067
Fulbright Traditional Scholar Program in Europe and Eurasia, 2080
HHMI Biomedical Research Grants for International Scientists in the Baltics, Central and Eastern Europe, Russia, and Ukraine, 2253
IBRO InEurope Short Stay Grants, 2325
Oak Foundation Environmental Grants, 3731
RAND Corporation Transatlantic Postdoc Fellowships for Int'l Relations & Security, 4032
Simone and Cino del Duca Grand Prix Awards, 4261
State Scholarships Foundation Grants, 4449
United States Institute of Peace - Transatlantic Post-Doc Fellowship for International Relations and Security, 4581

Sao Tome & Principe
Abdus Salam ITCP Postgraduate Diploma Programme in General Physics, 241
Beatrice Laing Trust Grants, 1324
Fulbright Critical Lang Enhancement Awards, 2068
Fulbright Traditional Scholar Program in Sub-Saharan Africa, 2081
Fulbright Traditional Scholar Program in the Near East and North Africa Region, 2083
SSRC African Peacebuilding Network Research Grants, 4422
State Scholarships Foundation Grants, 4449
STTI Alpha Eta Collaborative Research Grants, 4458
Wellcome Trust New Investigator Awards, 4844
Wenner-Gren Foundation African Fellowships, 4846

Saudi Arabia
Fulbright Traditional Scholar Program in the East Asia/Pacific Region, 2082
IBRO Asia Reg APRC Exchange Fellowships, 2323
Sir Dorabji Tata Trust Individual Education and Travel Grants, 4265

Scandinavia
American-Scandinavian Foundation Awards for Advanced Study or Research in the U.S., 603

Senegal
Abdus Salam ITCP Postgraduate Diploma Programme in General Physics, 241
Beatrice Laing Trust Grants, 1324
Fulbright Critical Lang Enhancement Awards, 2068
Fulbright Traditional Scholar Program in Sub-Saharan Africa, 2081
Fulbright Traditional Scholar Program in the Near East and North Africa Region, 2083
SSRC African Peacebuilding Network Research Grants, 4422
State Scholarships Foundation Grants, 4449
STTI Alpha Eta Collaborative Research Grants, 4458
USDA Norman E. Borlaug International Agricultural Science and Technology Fellowships, 4636
Wellcome Trust New Investigator Awards, 4844
Wenner-Gren Foundation African Fellowships, 4846

Serbia
ACLS Early Career Postdoctoral Fellowships in East European Studies, 286
Beatrice Laing Trust Grants, 1324
Canon Foundation Grants, 1423
EMBO Installation Grants, 1915
Fluor Foundation Grants, 2028
Fulbright Business Grants in Spain, 2067
Fulbright Traditional Scholar Program in Europe and Eurasia, 2080
HHMI Biomedical Research Grants for International Scientists in the Baltics, Central and Eastern Europe, Russia, and Ukraine, 2253
IBRO InEurope Short Stay Grants, 2325
Oak Foundation Environmental Grants, 3731
OSF Access to Medicines Fellowships in Eastern Europe and Central Asia, 3834
OSF Global Supplementary Grants, 3843
OSF Human Rights & Governance Interns, 3845
RAND Corporation Transatlantic Postdoc Fellowships for Int'l Relations & Security, 4032
Simone and Cino del Duca Grand Prix Awards, 4261
State Scholarships Foundation Grants, 4449
United States Institute of Peace - Transatlantic Post-Doc Fellowship for International Relations and Security, 4581
Wellcome Trust New Investigator Awards, 4844

Seychelles
Abdus Salam ITCP Postgraduate Diploma Programme in General Physics, 241
Beatrice Laing Trust Grants, 1324
Fulbright Critical Lang Enhancement Awards, 2068
Fulbright Traditional Scholar Program in Sub-Saharan Africa, 2081
Fulbright Traditional Scholar Program in the Near East and North Africa Region, 2083
SSRC African Peacebuilding Network Research Grants, 4422
State Scholarships Foundation Grants, 4449
STTI Alpha Eta Collaborative Research Grants, 4458
USDA Norman E. Borlaug International Agricultural Science and Technology Fellowships, 4636
Wellcome Trust New Investigator Awards, 4844
Wenner-Gren Foundation African Fellowships, 4846

Sierra Leone
Abdus Salam ITCP Postgraduate Diploma Programme in General Physics, 241
Beatrice Laing Trust Grants, 1324
Fulbright Critical Lang Enhancement Awards, 2068
Fulbright Traditional Scholar Program in Sub-Saharan Africa, 2081
Fulbright Traditional Scholar Program in the Near East and North Africa Region, 2083
SSRC African Peacebuilding Network Research Grants, 4422
State Scholarships Foundation Grants, 4449
STTI Alpha Eta Collaborative Research Grants, 4458
Wellcome Trust New Investigator Awards, 4844
Wenner-Gren Foundation African Fellowships, 4846

Singapore
Fulbright Traditional Scholar Program in the East Asia/Pacific Region, 2082
Gruber Foundation Int'l Research Award, 2180
IBRO Asia Reg APRC Exchange Fellowships, 2323
SfN Chapters Graduate Student Travel Awards, 4202
SfN Chapters Postdoc Trainee Travel Awards, 4203
SfN Donald B. Lindsley Prize in Behavioral Neuroscience, 4204
SfN Graduate Student Travel Awards, 4205
SfN Jacob P. Waletzky Award, 4206
SfN Janett Rosenberg Trubatch Career Development Award, 4207
SfN Louise Hanson Marshall Specific Recognition Award, 4208
SfN M Salpeter Lifetime Achievement Award, 4209
SfN Postdoctoral Fellow Travel Awards, 4210
SfN Ralph W. Gerard Prize in Neuroscience, 4211
SfN Swartz Prize for Theoretical and Computational Neuroscience, 4213
SfN Undergraduate Student Travel Awards, 4217
SfN Young Investigator Award, 4218
Sir Dorabji Tata Trust Individual Education and Travel Grants, 4265

Slovakia
ACLS Early Career Postdoctoral Fellowships in East European Studies, 286
AIAR Andrew W. Mellon Found Fellowships, 392
Beatrice Laing Trust Grants, 1324
Canon Foundation Grants, 1423
Central European University Fellowships, 1488
EMBO Installation Grants, 1915
Fluor Foundation Grants, 2028
Fulbright Business Grants in Spain, 2067
Fulbright Traditional Scholar Program in Europe and Eurasia, 2080
HHMI/EMBO Start-up Grants for C Europe, 2250
HHMI Biomedical Research Grants for International Scientists in the Baltics, Central and Eastern Europe, Russia, and Ukraine, 2253
IBRO InEurope Short Stay Grants, 2325
Oak Foundation Environmental Grants, 3731
OSF Access to Medicines Fellowships in Eastern Europe and Central Asia, 3834
OSF HESP Mobility Grants, 3844
OSF Human Rights & Governance Interns, 3845
OSF Think Tank Fellowships, 3852
RAND Corporation Transatlantic Postdoc Fellowships for Int'l Relations & Security, 4032
Simone and Cino del Duca Grand Prix Awards, 4261
State Scholarships Foundation Grants, 4449
Swiss National Science Foundation DORE (DO REsearch) Conference Grants, 4482
Swiss National Science Foundation Fellowships for Advanced Researchers, 4483
Swiss National Science Foundation Fellowships for Prospective Researchers, 4484
Swiss National Science Foundation Marie Heim-Vogtlin (MHV) Grants, 4485
Swiss National Science Foundation Scientific Conferences Grants, 4486
United States Institute of Peace - Transatlantic Post-Doc Fellowship for International Relations and Security, 4581

Slovenia
ACLS Early Career Postdoctoral Fellowships in East European Studies, 286
Beatrice Laing Trust Grants, 1324
Canon Foundation Grants, 1423
Central European University Fellowships, 1488
EMBO Installation Grants, 1915
Fluor Foundation Grants, 2028
Fulbright Business Grants in Spain, 2067
Fulbright Traditional Scholar Program in Europe and Eurasia, 2080
HHMI/EMBO Start-up Grants for C Europe, 2250
HHMI Biomedical Research Grants for International Scientists in the Baltics, Central and Eastern Europe, Russia, and Ukraine, 2253
IBRO InEurope Short Stay Grants, 2325
Oak Foundation Environmental Grants, 3731
OSF Access to Medicines Fellowships in Eastern Europe and Central Asia, 3834
OSF HESP Mobility Grants, 3844
OSF Human Rights & Governance Interns, 3845
OSF Think Tank Fellowships, 3852
RAND Corporation Transatlantic Postdoc Fellowships for Int'l Relations & Security, 4032
Simone and Cino del Duca Grand Prix Awards, 4261
State Scholarships Foundation Grants, 4449
Swiss National Science Foundation DORE (DO REsearch) Conference Grants, 4482
Swiss National Science Foundation Fellowships for Advanced Researchers, 4483
Swiss National Science Foundation Fellowships for Prospective Researchers, 4484
Swiss National Science Foundation Marie Heim-Vogtlin (MHV) Grants, 4485
Swiss National Science Foundation Scientific Conferences Grants, 4486

Solomon Islands
Wellcome Trust New Investigator Awards, 4844

Somalia
Abdus Salam ITCP Postgraduate Diploma Programme in General Physics, 241
Beatrice Laing Trust Grants, 1324
Fulbright Critical Lang Enhancement Awards, 2068
Fulbright Traditional Scholar Program in Sub-Saharan Africa, 2081
Fulbright Traditional Scholar Program in the Near East and North Africa Region, 2083
State Scholarships Foundation Grants, 4449
STTI Alpha Eta Collaborative Research Grants, 4458
USDA Norman E. Borlaug International Agricultural Science and Technology Fellowships, 4636
Wellcome Trust New Investigator Awards, 4844
Wenner-Gren Foundation African Fellowships, 4846

South Africa
Abdus Salam ITCP Postgraduate Diploma Programme in General Physics, 241
ACLS African Humanities Program in Ghana, Nigeria, South Africa, Tanzania, and Uganda, 279
Beatrice Laing Trust Grants, 1324
Fulbright Critical Lang Enhancement Awards, 2068
Fulbright Traditional Scholar Program in Sub-Saharan Africa, 2081
Fulbright Traditional Scholar Program in the Near East and North Africa Region, 2083
SSRC African Peacebuilding Network Research Grants, 4422
SSRC Next Generation Social Sciences in Africa: Doc Dissertation Completion Fellowships, 4434
SSRC Next Generation Social Sciences in Africa: Doctoral Dissertation Proposal Fellowships, 4435
SSRC Next Generation Social Sciences in Africa: Doctoral Dissertation Research Fellowships, 4436
State Scholarships Foundation Grants, 4449
STTI Alpha Eta Collaborative Research Grants, 4458
USAID Comprehensive District-Based Support for Better HIV/TB Patient Outcomes Grants, 4595
Wellcome Trust New Investigator Awards, 4844
Wenner-Gren Foundation African Fellowships, 4846

South Korea
Fulbright Traditional Scholar Program in the East Asia/Pacific Region, 2082
Gruber Foundation Int'l Research Award, 2180
IBRO Asia Reg APRC Exchange Fellowships, 2323
SfN Chapters Graduate Student Travel Awards, 4202
SfN Chapters Postdoc Trainee Travel Awards, 4203
SfN Donald B. Lindsley Prize in Behavioral Neuroscience, 4204
SfN Graduate Student Travel Awards, 4205
SfN Jacob P. Waletzky Award, 4206
SfN Janett Rosenberg Trubatch Career Development Award, 4207
SfN Louise Hanson Marshall Specific Recognition Award, 4208
SfN M Salpeter Lifetime Achievement Award, 4209
SfN Postdoctoral Fellow Travel Awards, 4210
SfN Swartz Prize for Theoretical and Computational Neuroscience, 4213
SfN Undergraduate Student Travel Awards, 4217
SfN Young Investigator Award, 4218
Sir Dorabji Tata Trust Individual Education and Travel Grants, 4265

South Ossetia
Fulbright Traditional Scholar Program in the East Asia/Pacific Region, 2082
Sir Dorabji Tata Trust Individual Education and Travel Grants, 4265

South Sudan
Wellcome Trust New Investigator Awards, 4844

Spain
Alexander von Humboldt Foundation Hezekiah Wardwell Fellowships, 528
Beatrice Laing Trust Grants, 1324
Bupa Foundation Multi-Country Grant, 1387
Canon Foundation Grants, 1423
Carnegie Endowment for International Peace Junior Fellowships, 1434
EMBO Installation Grants, 1915
Fluor Foundation Grants, 2028
Fulbright Business Grants in Spain, 2067
Fulbright Traditional Scholar Program in Europe and Eurasia, 2080
Gruber Foundation Int'l Research Award, 2180
HHMI Biomedical Research Grants for International Scientists in the Baltics, Central and Eastern Europe, Russia, and Ukraine, 2253
IBRO Europe Regional Grants for Western European Workshops and Symposia, 2324
IBRO InEurope Short Stay Grants, 2325
Oak Foundation Environmental Grants, 3731
Pfizer ASPIRE EU Antifungal Res Awards, 3910
Pfizer ASPIRE EU Dupuytren's Contracture Research Awards, 3911
Pfizer ASPIRE EU Emerging Mechanisms of Resistance Antibacterial Research Awards, 3912
Pfizer ASPIRE EU MRSA Nosocomial Pneumonia & MRSA Complicated Skin & Soft Tissue Infections Antibacterial Research Awards, 3913
RAND Corporation Transatlantic Postdoc Fellowships for Int'l Relations & Security, 4032
SfN Chapters Graduate Student Travel Awards, 4202
SfN Chapters Postdoc Trainee Travel Awards, 4203
SfN Donald B. Lindsley Prize in Behavioral Neuroscience, 4204
SfN Graduate Student Travel Awards, 4205
SfN Jacob P. Waletzky Award, 4206
SfN Janett Rosenberg Trubatch Career Development Award, 4207
SfN Louise Hanson Marshall Specific Recognition Award, 4208
SfN M Salpeter Lifetime Achievement Award, 4209
SfN Postdoctoral Fellow Travel Awards, 4210
SfN Ralph W. Gerard Prize in Neuroscience, 4211
SfN Swartz Prize for Theoretical and Computational Neuroscience, 4213
SfN Young Investigator Award, 4218
Simone and Cino del Duca Grand Prix Awards, 4261
State Scholarships Foundation Grants, 4449
United States Institute of Peace - Transatlantic Post-Doc Fellowship for International Relations and Security, 4581

Sri Lanka
Fulbright Traditional Scholar Program in the East Asia/Pacific Region, 2082
IBRO Asia Reg APRC Exchange Fellowships, 2323
Sir Dorabji Tata Trust Individual Education and Travel Grants, 4265
Wellcome Trust New Investigator Awards, 4844

St. Lucia
AOCS Ralph H. Potts Memorial Fellowship, 893
IBRO Latin America Regional Funding for Neuroscience Schools, 2327
IBRO Latin America Regional Funding for PROLAB Collaborations, 2328
IBRO Latin America Regional Travel Grants, 2329
John Simon Guggenheim Memorial Foundation Fellowships, 2507
SfN Ricardo Miledi Neuroscience Training, 4212
SSRC Drugs, Security and Democracy Fellows, 4426
Wellcome Trust New Investigator Awards, 4844

St. Vincent and the Grenadines
AOCS Ralph H. Potts Memorial Fellowship, 893
IBRO Latin America Regional Funding for Neuroscience Schools, 2327
IBRO Latin America Regional Funding for PROLAB Collaborations, 2328
IBRO Latin America Regional Travel Grants, 2329
John Simon Guggenheim Memorial Foundation Fellowships, 2507
SfN Ricardo Miledi Neuroscience Training, 4212
Wellcome Trust New Investigator Awards, 4844

Sudan

Abdus Salam ITCP Postgraduate Diploma Programme in General Physics, 241
Beatrice Laing Trust Grants, 1324
Fulbright Critical Lang Enhancement Awards, 2068
Fulbright Traditional Scholar Program in Sub-Saharan Africa, 2081
Fulbright Traditional Scholar Program in the Near East and North Africa Region, 2083
SSRC African Peacebuilding Network Research Grants, 4422
State Scholarships Foundation Grants, 4449
STTI Alpha Eta Collaborative Research Grants, 4458
Wenner-Gren Foundation African Fellowships, 4846

Suriname

Academy of American Franciscan History Dissertation Fellowships, 249
IBRO Latin America Regional Funding for Neuroscience Schools, 2327
IBRO Latin America Regional Funding for PROLAB Collaborations, 2328
IBRO Latin America Regional Travel Grants, 2329
SfN Ricardo Miledi Neuroscience Training, 4212
Wellcome Trust New Investigator Awards, 4844

Swaziland

Abdus Salam ITCP Postgraduate Diploma Programme in General Physics, 241
Beatrice Laing Trust Grants, 1324
Fulbright Critical Lang Enhancement Awards, 2068
Fulbright Traditional Scholar Program in Sub-Saharan Africa, 2081
Fulbright Traditional Scholar Program in the Near East and North Africa Region, 2083
SSRC African Peacebuilding Network Research Grants, 4422
State Scholarships Foundation Grants, 4449
STTI Alpha Eta Collaborative Research Grants, 4458
Wellcome Trust New Investigator Awards, 4844
Wenner-Gren Foundation African Fellowships, 4846

Sweden

American-Scandinavian Foundation Awards for Advanced Study or Research in the U.S., 603
Anna-Greta and Holger Crafoords Foundation Grants, 858
Beatrice Laing Trust Grants, 1324
Canon Foundation Grants, 1423
Ekhagastiftelsen Foundation Grants, 1903
EMBO Installation Grants, 1915
Fluor Foundation Grants, 2028
Fulbright Business Grants in Spain, 2067
Fulbright Traditional Scholar Program in Europe and Eurasia, 2080
HHMI Biomedical Research Grants for International Scientists in the Baltics, Central and Eastern Europe, Russia, and Ukraine, 2253
Holberg Memorial Fund Nils Klim Prize, 2273
IBRO Europe Regional Grants for Western European Workshops and Symposia, 2324
IBRO InEurope Short Stay Grants, 2325
Oak Foundation Environmental Grants, 3731
Pfizer ASPIRE EU Antifungal Res Awards, 3910
Pfizer ASPIRE EU Dupuytren's Contracture Research Awards, 3911
Pfizer ASPIRE EU Emerging Mechanisms of Resistance Antibacterial Research Awards, 3912
Pfizer ASPIRE EU MRSA Nosocomial Pneumonia & MRSA Complicated Skin & Soft Tissue Infections Antibacterial Research Awards, 3913
RAND Corporation Transatlantic Postdoc Fellowships for Int'l Relations & Security, 4032
SASS Einar and Eva Lund Haugen Memorial Scholarship, 4183
Simone and Cino del Duca Grand Prix Awards, 4261
SRC Guest Professorship Program, 4370
State Scholarships Foundation Grants, 4449
United States Institute of Peace - Transatlantic Post-Doc Fellowship for International Relations and Security, 4581
Wilhelm Sander Foundation Grants, 4868

Switzerland

Beatrice Laing Trust Grants, 1324
Cabot Corporation Foundation Grants, 1406
Canon Foundation Grants, 1423
Central European University Fellowships, 1488
EMBO Installation Grants, 1915
Fluor Foundation Grants, 2028
Fulbright Business Grants in Spain, 2067
Fulbright Traditional Scholar Program in Europe and Eurasia, 2080
Gruber Foundation Int'l Research Award, 2180
HHMI/EMBO Start-up Grants for C Europe, 2250
HHMI Biomedical Research Grants for International Scientists in the Baltics, Central and Eastern Europe, Russia, and Ukraine, 2253
IBRO Europe Regional Grants for Western European Workshops and Symposia, 2324
IBRO InEurope Short Stay Grants, 2325
Oak Foundation Environmental Grants, 3731
OSF HESP Mobility Grants, 3844
OSF Human Rights & Governance Interns, 3845
Pfizer ASPIRE EU Antifungal Res Awards, 3910
Pfizer ASPIRE EU Dupuytren's Contracture Research Awards, 3911
Pfizer ASPIRE EU Emerging Mechanisms of Resistance Antibacterial Research Awards, 3912
Pfizer ASPIRE EU MRSA Nosocomial Pneumonia & MRSA Complicated Skin & Soft Tissue Infections Antibacterial Research Awards, 3913
RAND Corporation Transatlantic Postdoc Fellowships for Int'l Relations & Security, 4032
SfN Chapters Graduate Student Travel Awards, 4202
SfN Chapters Postdoc Trainee Travel Awards, 4203
SfN Donald B. Lindsley Prize in Behavioral Neuroscience, 4204
SfN Graduate Student Travel Awards, 4205
SfN Jacob P. Waletzky Award, 4206
SfN Janett Rosenberg Trubatch Career Development Award, 4207
SfN Louise Hanson Marshall Specific Recognition Award, 4208
SfN M Salpeter Lifetime Achievement Award, 4209
SfN Postdoctoral Fellow Travel Awards, 4210
SfN Ralph W. Gerard Prize in Neuroscience, 4211
SfN Swartz Prize for Theoretical and Computational Neuroscience, 4213
SfN Undergraduate Student Travel Awards, 4217
SfN Young Investigator Award, 4218
Simone and Cino del Duca Grand Prix Awards, 4261
State Scholarships Foundation Grants, 4449
Swiss National Science Foundation DORE (DO REsearch) Conference Grants, 4482
Swiss National Science Foundation Fellowships for Advanced Researchers, 4483
Swiss National Science Foundation Fellowships for Prospective Researchers, 4484
Swiss National Science Foundation Marie Heim-Vogtlin (MHV) Grants, 4485
Swiss National Science Foundation Scientific Conferences Grants, 4486
United States Institute of Peace - Transatlantic Post-Doc Fellowship for International Relations and Security, 4581

Syria

Fulbright Traditional Scholar Program in the East Asia/Pacific Region, 2082
IBRO Asia Reg APRC Exchange Fellowships, 2323
OSF Global Supplementary Grants, 3843
Sir Dorabji Tata Trust Individual Education and Travel Grants, 4265

Syrian Arab Republic

Wellcome Trust New Investigator Awards, 4844

Taiwan

Fulbright Traditional Scholar Program in the East Asia/Pacific Region, 2082
IBRO Asia Reg APRC Exchange Fellowships, 2323
Sir Dorabji Tata Trust Individual Education and Travel Grants, 4265

Tajikistan

Fulbright Traditional Scholar Program in the East Asia/Pacific Region, 2082
IBRO Asia Reg APRC Exchange Fellowships, 2323
OSF Access to Medicines Fellowships in Eastern Europe and Central Asia, 3834
OSF Central Asia and Caucasus Research and Training Initiative - Junior Fellowships, 3836
OSF Central Asia and Caucasus Research and Training Initiative - Senior Fellowships, 3837
OSF Global Faculty Grants - Track 1, 3841
OSF Global Faculty Grants - Track 2, 3842
OSF Global Supplementary Grants, 3843
Sir Dorabji Tata Trust Individual Education and Travel Grants, 4265
Wellcome Trust New Investigator Awards, 4844

Tanzania

Abdus Salam ITCP Postgraduate Diploma Programme in General Physics, 241
ACLS African Humanities Program for Dissertation Completion, 278
ACLS African Humanities Program in Ghana, Nigeria, South Africa, Tanzania, and Uganda, 279
Better Way Foundation Grants, 1338
SSRC African Peacebuilding Network Research Grants, 4422
SSRC Next Generation Social Sciences in Africa: Doc Dissertation Completion Fellowships, 4434
SSRC Next Generation Social Sciences in Africa: Doctoral Dissertation Proposal Fellowships, 4435
SSRC Next Generation Social Sciences in Africa: Doctoral Dissertation Research Fellowships, 4436
USDA Norman E. Borlaug International Agricultural Science and Technology Fellowships, 4636
Wellcome Trust New Investigator Awards, 4844
WWF Russell E. Train Fellowships, 4898

Thailand

Fulbright Traditional Scholar Program in the East Asia/Pacific Region, 2082
IBRO Asia Reg APRC Exchange Fellowships, 2323
Sir Dorabji Tata Trust Individual Education and Travel Grants, 4265
Wellcome Trust New Investigator Awards, 4844
WWF Russell E. Train Fellowships, 4898

The Netherlands

Beatrice Laing Trust Grants, 1324
Canon Foundation Grants, 1423
EMBO Installation Grants, 1915
Fluor Foundation Grants, 2028
Fulbright Business Grants in Spain, 2067
Fulbright Traditional Scholar Program in Europe and Eurasia, 2080
HHMI Biomedical Research Grants for International Scientists in the Baltics, Central and Eastern Europe, Russia, and Ukraine, 2253
RAND Corporation Transatlantic Postdoc Fellowships for Int'l Relations & Security, 4032
Simone and Cino del Duca Grand Prix Awards, 4261
State Scholarships Foundation Grants, 4449

Timor-Lester

Fulbright Traditional Scholar Program in the East Asia/Pacific Region, 2082
Sir Dorabji Tata Trust Individual Education and Travel Grants, 4265
Wellcome Trust New Investigator Awards, 4844

Togo

Abdus Salam ITCP Postgraduate Diploma Programme in General Physics, 241

SSRC African Peacebuilding Network Research Grants, 4422
Wellcome Trust New Investigator Awards, 4844

Tonga
Wellcome Trust New Investigator Awards, 4844

Trinidad and Tobago
AOCS Ralph H. Potts Memorial Fellowship, 893
IBRO Latin America Regional Funding for Neuroscience Schools, 2327
IBRO Latin America Regional Funding for PROLAB Collaborations, 2328
IBRO Latin America Regional Travel Grants, 2329
John Simon Guggenheim Memorial Foundation Fellowships, 2507
SfN Ricardo Miledi Neuroscience Training, 4212
SSRC Drugs, Security and Democracy Fellows, 4426

Tunisia
Wellcome Trust New Investigator Awards, 4844

Turkey
Beatrice Laing Trust Grants, 1324
Canon Foundation Grants, 1423
EMBO Installation Grants, 1915
Fluor Foundation Grants, 2028
Fulbright Business Grants in Spain, 2067
Fulbright Traditional Scholar Program in Europe and Eurasia, 2080
Fulbright Traditional Scholar Program in the East Asia/Pacific Region, 2082
GMFUS Black Sea Trust for Regional Corporation Grants, 2152
Gruber Foundation Int'l Research Award, 2180
HHMI Biomedical Research Grants for International Scientists in the Baltics, Central and Eastern Europe, Russia, and Ukraine, 2253
IBRO Asia Reg APRC Exchange Fellowships, 2323
IBRO InEurope Short Stay Grants, 2325
Oak Foundation Environmental Grants, 3731
RAND Corporation Transatlantic Postdoc Fellowships for Int'l Relations & Security, 4032
SfN Chapters Graduate Student Travel Awards, 4202
SfN Chapters Postdoc Trainee Travel Awards, 4203
SfN Donald B. Lindsley Prize in Behavioral Neuroscience, 4204
SfN Graduate Student Travel Awards, 4205
SfN Jacob P. Waletzky Award, 4206
SfN Janett Rosenberg Trubatch Career Development Award, 4207
SfN Louise Hanson Marshall Specific Recognition Award, 4208
SfN M Salpeter Lifetime Achievement Award, 4209
SfN Postdoctoral Fellow Travel Awards, 4210
SfN Ralph W. Gerard Prize in Neuroscience, 4211
SfN Swartz Prize for Theoretical and Computational Neuroscience, 4213
SfN Undergraduate Student Travel Awards, 4217
SfN Young Investigator Award, 4218
Simone and Cino del Duca Grand Prix Awards, 4261
Sir Dorabji Tata Trust Individual Education and Travel Grants, 4265
State Scholarships Foundation Grants, 4449
United States Institute of Peace - Transatlantic Post-Doc Fellowship for International Relations and Security, 4581
Wellcome Trust New Investigator Awards, 4844

Turkmenistan
Fulbright Traditional Scholar Program in the East Asia/Pacific Region, 2082
IBRO Asia Reg APRC Exchange Fellowships, 2323
OSF Access to Medicines Fellowships in Eastern Europe and Central Asia, 3834
OSF Central Asia and Caucasus Research and Training Initiative - Junior Fellowships, 3836
OSF Central Asia and Caucasus Research and Training Initiative - Senior Fellowships, 3837
OSF Global Supplementary Grants, 3843
Sir Dorabji Tata Trust Individual Education and Travel Grants, 4265
Wellcome Trust New Investigator Awards, 4844

Turks and Caicos Islands
AOCS Ralph H. Potts Memorial Fellowship, 893
John Simon Guggenheim Memorial Foundation Fellowships, 2507
SfN Ricardo Miledi Neuroscience Training, 4212
SSRC Drugs, Security and Democracy Fellows, 4426

Tuvalu
Wellcome Trust New Investigator Awards, 4844

Uganda
Abdus Salam ITCP Postgraduate Diploma Programme in General Physics, 241
ACLS African Humanities Program for Dissertation Completion, 278
ACLS African Humanities Program in Ghana, Nigeria, South Africa, Tanzania, and Uganda, 279
SSRC African Peacebuilding Network Research Grants, 4422
SSRC Next Generation Social Sciences in Africa: Doc Dissertation Completion Fellowships, 4434
SSRC Next Generation Social Sciences in Africa: Doctoral Dissertation Proposal Fellowships, 4435
SSRC Next Generation Social Sciences in Africa: Doctoral Dissertation Research Fellowships, 4436
USDA Norman E. Borlaug International Agricultural Science and Technology Fellowships, 4636
Wellcome Trust New Investigator Awards, 4844

Ukraine
Beatrice Laing Trust Grants, 1324
Canon Foundation Grants, 1423
EMBO Installation Grants, 1915
Fluor Foundation Grants, 2028
Fulbright Business Grants in Spain, 2067
Fulbright Traditional Scholar Program in Europe and Eurasia, 2080
GMFUS Black Sea Trust for Regional Corporation Grants, 2152
Gruber Foundation Int'l Research Award, 2180
HHMI Biomedical Research Grants for International Scientists in the Baltics, Central and Eastern Europe, Russia, and Ukraine, 2253
HHMI Biomedical Research Grants for International Scientists in the Baltics, Central and Eastern Europe, Russia, and Ukraine, 2253
IBRO InEurope Short Stay Grants, 2325
John Deere Foundation Grants, 2500
Oak Foundation Environmental Grants, 3731
OSF Access to Medicines Fellowships in Eastern Europe and Central Asia, 3834
OSF Global Supplementary Grants, 3843
OSF HESP Mobility Grants, 3844
OSF Human Rights & Governance Interns, 3845
OSF Think Tank Fellowships, 3852
RAND Corporation Transatlantic Postdoc Fellowships for Int'l Relations & Security, 4032
SfN Chapters Graduate Student Travel Awards, 4202
SfN Chapters Postdoc Trainee Travel Awards, 4203
SfN Donald B. Lindsley Prize in Behavioral Neuroscience, 4204
SfN Graduate Student Travel Awards, 4205
SfN Jacob P. Waletzky Award, 4206
SfN Janett Rosenberg Trubatch Career Development Award, 4207
SfN Louise Hanson Marshall Specific Recognition Award, 4208
SfN M Salpeter Lifetime Achievement Award, 4209
SfN Postdoctoral Fellow Travel Awards, 4210
SfN Ralph W. Gerard Prize in Neuroscience, 4211
SfN Swartz Prize for Theoretical and Computational Neuroscience, 4213
SfN Undergraduate Student Travel Awards, 4217
SfN Young Investigator Award, 4218
Simone and Cino del Duca Grand Prix Awards, 4261
State Scholarships Foundation Grants, 4449
United States Institute of Peace - Transatlantic Post-Doc Fellowship for International Relations and Security, 4581
US CRDF CGP-Climate Change and Energy Competition For Joint U.S.-Ukrainian Research Teams, 4607
US CRDF Science and Technology Entrepreneurship Program (STEP) Business Partnership Grant Competition Armenia, 4611
US CRDF STEP Business Partnership Grant Competition Ukraine, 4612
Wellcome Trust New Investigator Awards, 4844

United Arab Emirates
Fulbright Traditional Scholar Program in the East Asia/Pacific Region, 2082
IBRO Asia Reg APRC Exchange Fellowships, 2323
Sir Dorabji Tata Trust Individual Education and Travel Grants, 4265

United Kingdom
Allen Lane Foundation Grants, 571
All Saints Educational Trust Grants, 572
Alzheimer's Association U.S.-U.K. Young Investigator Exchange Fellowships, 597
BA Elisabeth Barker Fund, 1288
Bank of America Charitable Matching Gifts, 1297
BA Sir Ernest Cassel Educational Trust Grants, 1303
Beatrice Laing Trust Grants, 1324
Bupa Foundation Multi-Country Grant, 1387
Cabot Corporation Foundation Grants, 1406
Canon Foundation Grants, 1423
Carnegie Endowment for International Peace Junior Fellowships, 1434
Commonwealth Fund Harkness Fellowships in Health Care Policy and Practice, 1606
EMBO Installation Grants, 1915
ExxonMobil Foundation Malaria Grants, 1968
Fluor Foundation Grants, 2028
Fulbright Business Grants in Spain, 2067
Fulbright Traditional Scholar Program in Europe and Eurasia, 2080
Gruber Foundation Int'l Research Award, 2180
HHMI Biomedical Research Grants for International Scientists in the Baltics, Central and Eastern Europe, Russia, and Ukraine, 2253
IBRO Europe Regional Grants for Western European Workshops and Symposia, 2324
IBRO InEurope Short Stay Grants, 2325
Medtronic Foundation Fellowships, 2738
Nuffield Found Children & Families Grants, 3666
Nuffield Foundation Law and Society Grants, 3667
Nuffield Found Oliver Bird Rheumatism Grants, 3668
Nuffield Foundation Open Door Grants, 3669
Oak Foundation Environmental Grants, 3731
Pfizer Anti Infective Research EU Grants, 3909
Pfizer ASPIRE EU Antifungal Res Awards, 3910
Pfizer ASPIRE EU Dupuytren's Contracture Research Awards, 3911
Pfizer ASPIRE EU Emerging Mechanisms of Resistance Antibacterial Research Awards, 3912
Pfizer ASPIRE EU MRSA Nosocomial Pneumonia & MRSA Complicated Skin & Soft Tissue Infections Antibacterial Research Awards, 3913
Pfizer Inflammation Comp Research Awards, 3926
RAND Corporation Transatlantic Postdoc Fellowships for Int'l Relations & Security, 4032
SfN Chapters Graduate Student Travel Awards, 4202
SfN Chapters Postdoc Trainee Travel Awards, 4203
SfN Donald B. Lindsley Prize in Behavioral Neuroscience, 4204
SfN Graduate Student Travel Awards, 4205
SfN Jacob P. Waletzky Award, 4206
SfN Janett Rosenberg Trubatch Career Development Award, 4207
SfN Louise Hanson Marshall Specific Recognition Award, 4208
SfN M Salpeter Lifetime Achievement Award, 4209
SfN Postdoctoral Fellow Travel Awards, 4210
SfN Ralph W. Gerard Prize in Neuroscience, 4211

SfN Swartz Prize for Theoretical and Computational Neuroscience, 4213
SfN Undergraduate Student Travel Awards, 4217
SfN Young Investigator Award, 4218
Simone and Cino del Duca Grand Prix Awards, 4261
SSHRC Digging into Data Challenge Grants, 4406
State Scholarships Foundation Grants, 4449
Stroke Association Allied Health Professional Research Bursaries, 4454
Stroke Association Clinical Fellowships, 4455
Stroke Association Research Project Grants, 4456
Sun Microsystems Foundation Summer Academic Enrichment Grants, 4468
United States Institute of Peace - Transatlantic Post-Doc Fellowship for International Relations and Security, 4581
Wellcome Trust New Investigator Awards, 4844

Uruguay
Academy of American Franciscan History Dissertation Fellowships, 249
IBRO Latin America Regional Funding for Neuroscience Schools, 2327
IBRO Latin America Regional Funding for PROLAB Collaborations, 2328
IBRO Latin America Regional Travel Grants, 2329
John Simon Guggenheim Memorial Foundation Fellowships, 2507
OTS Hovore-Horn Fellowships, 3856
SfN Ricardo Miledi Neuroscience Training, 4212
SSRC Drugs, Security and Democracy Fellows, 4426

Uzbekistan
Fulbright Traditional Scholar Program in the East Asia/Pacific Region, 2082
Fulbright Traditional Scholar Program in the Near East and North Africa Region, 2083
Fulbright Traditional Scholar Program in the South and Central Asia Region, 2084
IBRO Asia Reg APRC Exchange Fellowships, 2323
OSF Access to Medicines Fellowships in Eastern Europe and Central Asia, 3834
OSF Central Asia and Caucasus Research and Training Initiative - Junior Fellowships, 3836
OSF Central Asia and Caucasus Research and Training Initiative - Senior Fellowships, 3837
OSF Global Supplementary Grants, 3843
Sir Dorabji Tata Trust Individual Education and Travel Grants, 4265
Wellcome Trust New Investigator Awards, 4844

Vanuatu
Wellcome Trust New Investigator Awards, 4844

Vatican City
Beatrice Laing Trust Grants, 1324
Canon Foundation Grants, 1423
EMBO Installation Grants, 1915
Fluor Foundation Grants, 2028
Fulbright Business Grants in Spain, 2067
Fulbright Traditional Scholar Program in Europe and Eurasia, 2080
HHMI Biomedical Research Grants for International Scientists in the Baltics, Central and Eastern Europe, Russia, and Ukraine, 2253
IBRO InEurope Short Stay Grants, 2325
Oak Foundation Environmental Grants, 3731
RAND Corporation Transatlantic Postdoc Fellowships for Int'l Relations & Security, 4032
Simone and Cino del Duca Grand Prix Awards, 4261
State Scholarships Foundation Grants, 4449
United States Institute of Peace - Transatlantic Post-Doc Fellowship for International Relations and Security, 4581

Venezuela
Academy of American Franciscan History Dissertation Fellowships, 249
Fluor Foundation Grants, 2028
IBRO Latin America Regional Funding for Neuroscience Schools, 2327
IBRO Latin America Regional Funding for PROLAB Collaborations, 2328
IBRO Latin America Regional Travel Grants, 2329
John Simon Guggenheim Memorial Foundation Fellowships, 2507
OTS Hovore-Horn Fellowships, 3856
SfN Ricardo Miledi Neuroscience Training, 4212
SSRC Drugs, Security and Democracy Fellows, 4426
Wellcome Trust New Investigator Awards, 4844

Vietnam
Fulbright Traditional Scholar Program in the East Asia/Pacific Region, 2082
IBRO Asia Reg APRC Exchange Fellowships, 2323
Sir Dorabji Tata Trust Individual Education and Travel Grants, 4265
Wellcome Trust New Investigator Awards, 4844
WWF Russell E. Train Fellowships, 4898

Virgin Islands (U.S.)
AOCS Ralph H. Potts Memorial Fellowship, 893
John Simon Guggenheim Memorial Foundation Fellowships, 2507
SARE Research & Educ Grants: South Reg, 4179
SSRC Drugs, Security and Democracy Fellows, 4426

Western Sahara
Abdus Salam ITCP Postgraduate Diploma Programme in General Physics, 241

Yemen
Fulbright Traditional Scholar Program in the East Asia/Pacific Region, 2082
IBRO Asia Reg APRC Exchange Fellowships, 2323
Sir Dorabji Tata Trust Individual Education and Travel Grants, 4265
Wellcome Trust New Investigator Awards, 4844

Zambia
Abdus Salam ITCP Postgraduate Diploma Programme in General Physics, 241
SSRC African Peacebuilding Network Research Grants, 4422
USDA Norman E. Borlaug International Agricultural Science and Technology Fellowships, 4636
Wellcome Trust New Investigator Awards, 4844

Zimbabwe
Abdus Salam ITCP Postgraduate Diploma Programme in General Physics, 241
SSRC African Peacebuilding Network Research Grants, 4422
USDA Norman E. Borlaug International Agricultural Science and Technology Fellowships, 4636
Wellcome Trust New Investigator Awards, 4844

CPSIA information can be obtained at www.ICGtesting.com
Printed in the USA
BVOW06s1450301014

372902BV00009B/52/P

9 780986 035760